JANAF Thermochemical Tables
Third Edition
Part II, Cr–Zr

Journal of

Physical and Chemical Reference Data

David R. Lide, Jr., Editor

The Journal of Physical and Chemical Reference Data (ISSN 0047-2689) is published quarterly by the American Chemical Society (1155 16th St., N. W., Washington, DC 20036-9976) and the American Institute of Physics (335 E. 45th St., New York, NY 10017-3483) for the National Bureau of Standards. Second-class postage paid at Washington, DC and additional mailing offices. POSTMASTER: Send address changes to Membership and Subscription Services, P. O. Box 3337, Columbus, Ohio 43210.

The objective of the Journal is to provide critically evaluated physical and chemical property data, fully documented as to the original sources and the criteria used for evaluation. Critical reviews of measurement techniques, whose aim is to assess the accuracy of available data in a given technical area, are also included. The Journal is not intended as a publication outlet for original experimental measurements such as are normally reported in the primary research literature, nor for review articles of a descriptive or primarily theoretical nature.

Supplements to the Journal are published at irregular intervals and are not included in subscriptions to the Journal. They contain compilations which are too lengthy for a journal format.

The Editor welcomes appropriate manuscripts for consideration by the Editorial Board. Potential contributors who are interested in preparing a compilation are invited to submit an outline of the nature and scope of the proposed compilation, with criteria for evaluation of the data and other pertinent factors, to:

David R. Lide, Jr., Editor
J. Phys. Chem. Ref. Data
National Bureau of Standards
Gaithersburg, MD 20899

One source of contributions to the Journal is The National Standard Reference Data System (NSRDS), which was established in 1963 as a means of coordinating on a national scale the production and dissemination of critically evaluated reference data in the physical sciences. Under the Standard Reference Data Act (Public Law 90-396) the National Bureau of Standards of the U.S. Department of Commerce has the primary responsibility in the Federal Government for providing reliable scientific and technical reference data. The Office of Standard Reference Data of NBS coordinates a complex of data evaluation centers, located in university, industrial, and other Government laboratories as well as within the National Bureau of Standards, which are engaged in the compilation and critical evaluation of numerical data on physical and chemical properties retrieved from the world scientific literature. The participants in this NBS-sponsored program, together with similar groups under private or other Government support which are pursuing the same ends, comprise the National Standard Reference Data System.

The primary focus of the NSRDS is on well-defined physical and chemical properties of well-characterized materials or systems. An effort is made to assess the accuracy of data reported in the primary research literature and to prepare compilations of critically evaluated data which will serve as reliable and convenient reference sources for the scientific and technical community.

Information for Contributors

Manuscripts submitted for publication must be prepared in accordance with *Instructions for Preparation of Manuscripts for the Journal of Physical and Chemical Reference Data,* available on request from the Editor.

New and renewal subscriptions should be sent with payment to the Office of the Controller at the American Chemical Society, 1155 Sixteenth Street, N.W., Washington, DC 20036-9976. **Address changes**, with at least six weeks advance notice, should be sent to Manager, Membership and Subscription Services, American Chemical Society, P.O. Box 3337, Columbus, OH 43210. Changes of address must include both old and new addresses and ZIP codes and, if possible, the address label from the mailing wrapper of a recent issue. Claims for missing numbers will not be allowed: if loss was due to failure of the change-of-address notice to be received in the time specified; if claim is dated (a) North America: more than 90 days beyond issue date, (b) all other foreign: more than one year beyond issue date.

Members of AIP member and affiliate societies requesting member subscription rates should direct subscriptions, renewals, and address changes to American Institute of Physics, Dept. S/F, 335 E. 45th St., NY 10017-3483.

Subscription Prices (1986) *(not including supplements)*	U.S.A.	Foreign (surface mail)	Optional air freight Europe Mideast N. Africa	Asia and Oceania
Members (of ACS, AIP, or affiliated society)	$ 55.00	$ 65.00	$ 80.00	$ 90.00
Regular rate	$220.00	$230.00	$245.00	$255.00

Rates above do not apply to nonmember subscribers in Japan, who must enter subscription orders with Maruzen Company Ltd., 3-10 Nihonbashi 2-chome, Chuo-ku, Tokyo 103, Japan. Tel: (03) 272-7211.

Back numbers are available at a cost of $65 per single copy and $250 per volume.

Orders for reprints, supplements, and back numbers should be addressed to the American Chemical Society, 1155 Sixteenth Street, N. W., Washington, DC 20036-9976. Prices for reprints and supplements are listed at the end of this issue.

Microfilm subscriptions of the *Journal of Physical and Chemical Reference Data* are available on 16 mm and 35 mm. This journal also appears in Sec. I of *Current Physics Microform* (CPM) along with 26 other journals published by the American Institute of Physics and its member societies. A *Microfilm Catalog* is available on request.

Journal of
Physical and Chemical Reference Data

Volume 14, 1985
Supplement No. 1

JANAF Thermochemical Tables
Third Edition
Part II, Cr–Zr

M. W. Chase, Jr., C. A. Davies, J. R. Downey, Jr., D. J. Frurip, R. A. McDonald, and A. N. Syverud

Thermal Group, 1707 Building,
Dow Chemical U.S.A.,
Midland, Michigan 48674, U.S.A.

Published by the **American Chemical Society** and the **American Institute of Physics** for the **National Bureau of Standards**

Library of Congress Catalog Card Number 85-71387

International Standard Book Number
0-88318-473-7

American Institute of Physics, Inc.
335 East 45th Street
New York, New York 10017

Printed in the United States of America

CHROMIUM (Cr)

REFERENCE STATE

0 to 2130	K	crystal
2130 to 2952.078	K	liquid
above 2952.078	K	ideal monatomic gas

Refer to the individual tables for details.

A_r = 51.996

Chromium (Cr) Cr_1(ref)

Enthalpy Reference Temperature = T_r = 298.15 K   Standard State Pressure = p° = 0.1 MPa

T/K	C_p° (J K⁻¹mol⁻¹)	S° (J K⁻¹mol⁻¹)	$-[G°-H°(T_r)]/T$ (J K⁻¹mol⁻¹)	$H°-H°(T_r)$ (kJ mol⁻¹)	$\Delta_f H°$ (kJ mol⁻¹)	$\Delta_f G°$ (kJ mol⁻¹)	Log K_f
0	0.	0.	INFINITE	-4.057	0.	0.	0.
100	9.963	4.286	41.746	-3.746	0.	0.	0.
200	19.860	14.869	25.680	-2.162	0.	0.	0.
298.15	23.434	23.618	23.618	0.	0.	0.	0.
300	23.472	23.764	23.619	0.043	0.	0.	0.
311.500	23.681	24.651	23.641	0.315	Cp LAMBDA MAXIMUM TRANSITION		
311.500	23.681	24.654	23.641	0.316			
400	25.230	30.765	24.562	2.481	0.	0.	0.
500	26.631	36.553	26.398	5.078	0.	0.	0.
600	27.719	41.508	28.513	7.797	0.	0.	0.
700	28.577	45.846	30.686	10.612	0.	0.	0.
800	29.434	49.715	32.827	13.511	0.	0.	0.
900	30.501	53.241	34.902	16.505	0.	0.	0.
1000	31.861	56.523	36.901	19.621	0.	0.	0.
1100	33.472	59.634	38.828	22.887	0.	0.	0.
1200	35.187	62.618	40.687	26.318	0.	0.	0.
1300	37.116	65.511	42.485	29.933	0.	0.	0.
1400	39.125	68.334	44.231	33.744	0.	0.	0.
1500	41.200	71.104	45.931	37.760	0.	0.	0.
1600	43.329	73.831	47.589	41.986	0.	0.	0.
1700	45.501	76.523	49.212	46.427	0.	0.	0.
1800	47.706	79.186	50.804	51.087	0.	0.	0.
1900	49.944	81.825	52.367	55.970	0.	0.	0.
2000	52.204	84.444	53.905	61.077	0.	0.	0.
2100	54.488	87.046	55.422	66.411	0.	0.	0.
2130.000	55.174	87.824	55.873	68.056	CRYSTAL <--> LIQUID TRANSITION		
2130.000	39.330	97.449	55.873	88.558			
2200	39.330	98.721	57.216	91.311	0.	0.	0.
2300	39.330	100.469	59.059	95.244	0.	0.	0.
2400	39.330	102.143	60.819	99.177	0.	0.	0.
2500	39.330	103.749	62.505	103.110	0.	0.	0.
2600	39.330	105.291	64.121	107.043	0.	0.	0.
2700	39.330	106.775	65.673	110.976	0.	0.	0.
2800	39.330	108.206	67.167	114.909	0.	0.	0.
2900	39.330	109.586	68.606	118.842	0.	0.	0.
2952.078	39.330	110.286	69.335	120.890	LIQUID <--> IDEAL GAS FUGACITY = 1 bar		
2952.078	30.533	225.302	69.335	460.427			
3000	30.788	225.796	71.830	461.897	0.	0.	0.
3100	31.302	226.814	76.813	465.001	0.	0.	0.
3200	31.796	227.815	81.517	468.156	0.	0.	0.
3300	32.275	228.801	85.965	471.360	0.	0.	0.
3400	32.745	229.772	90.180	474.611	0.	0.	0.
3500	33.212	230.728	94.182	477.909	0.	0.	0.
3600	33.680	231.670	97.988	481.253	0.	0.	0.
3700	34.156	232.599	101.614	484.645	0.	0.	0.
3800	34.646	233.516	105.073	488.085	0.	0.	0.
3900	35.157	234.423	108.378	491.575	0.	0.	0.
4000	35.692	235.320	111.540	495.117	0.	0.	0.
4100	36.260	236.208	114.570	498.715	0.	0.	0.
4200	36.864	237.089	117.477	502.370	0.	0.	0.
4300	37.511	237.964	120.269	506.089	0.	0.	0.
4400	38.205	238.834	122.954	509.874	0.	0.	0.
4500	38.954	239.701	125.538	513.732	0.	0.	0.
4600	39.760	240.566	128.030	517.667	0.	0.	0.
4700	40.632	241.430	130.433	521.686	0.	0.	0.
4800	41.572	242.295	132.755	525.795	0.	0.	0.
4900	42.587	243.163	134.999	530.003	0.	0.	0.
5000	43.681	244.034	137.171	534.315	0.	0.	0.
5200	44.964	245.697	141.308	542.822	0.	0.	0.
5400	47.221	247.435	145.206	552.036	0.	0.	0.
5600	49.697	249.197	148.889	561.724	0.	0.	0.
5800	52.370	250.986	152.378	571.925	0.	0.	0.
6000	55.230	252.809	155.696	582.682	0.	0.	0.

PREVIOUS: June 1973 (1 atm)   CURRENT: June 1973 (1 bar)

Chromium (Cr) Cr_1(ref)

CHROMIUM (Cr) CRYSTAL A_r = 51.996

$S°(298.15\ K) = 23.618 \pm 0.21\ J\ K^{-1}\ mol^{-1}$
$T_{trs} = 311.5\ K$
$T_{fus} = 2130 \pm 20\ K$

$\Delta_f H°(0\ K) = 0\ kJ\ mol^{-1}$
$\Delta_f H°(298.15\ K) = 0\ kJ\ mol^{-1}$
$\Delta_{trs} H° = 0.0008\ kJ\ mol^{-1}$
$\Delta_{fus} H° = 20.5 \pm 4.2\ kJ\ mol^{-1}$

Enthalpy of Formation

Zero by definition.

Heat Capacity and Entropy

The heat capacity values for T ≤ 10 K are chosen to be the same as those adopted by Hultgren et al. (1). A graphical integration of these Cp° data yields S°(10 K) = 0.005 kcal mol^{-1} and H°(10 K) - H°(0 K) = 0.0252 cal mol^{-1}. These data also match well with the Cp° studies of Clusius and Franzosini (2). The adopted Cp° values are based on the following studies.

Source	Points	Method	T/K
(2)	76	Cp°	14.10-274.43
(3)	23	Cp°	56.1 -291.1
(4)	Smooth	Cp°	60-300
(5)	Graph	Cp°	268-324
(6)	Smooth	Cp°	273-1073
(7)	Smooth	drop	400-1500
(8)	48	drop	1267.2108
(9)	17	drop	303-1884
(10)	Smooth	Cp°	320-1800
(11)	26	drop	673-1339

The adopted Cp° values are obtained primarily by graphical techniques to ensure smoothness of the curve and a reasonable representation of the sometimes diverse results. A ΔCp° = 0 is also adopted across T_{trs}. Below 200 K, there is excellent agreement between three independent studies (2, 3, 4). Above 200 K these three studies drift apart by at most 0.06 kcal mol^{-1}. Above 298 K, there are considerable differences in the various sets of data, differences of the order of 0.2 kcal mol^{-1} being typical. In terms of enthalpy, the data of Jaeger and Rosenbohm (11) deviates from the adopted values by +150 to +350 cal mol^{-1}; the data of Lucks and Deem (9) by 0 to -450 cal mol^{-1}; the data of Hultgren and Land (7) by -13 to +60 cal mol^{-1}; and the data of Conway and Hein (8) by -190 to +250 cal mol^{-1}. Surprisingly the enthalpy study by Kirillin et al. (12) yields enthalpies which are crudely 2 kcal mol^{-1} less than the adopted values at 2000 K.

Fusion Data

See the liquid table for details.

Transition Data

Beaumont et al. (5) measured the heat capacity of 99.998% pure Cr in the region 268-324 K. The detailed Cp° measurements gave evidence of a lambda-type anomaly, the maximum occurring at 311.5 K. Beaument et al. (5) estimated the heat associated with this transition to be 1.4 cal mol^{-1}. Garnier and Salamon (13) used an extension of the ac calorimetric method to determine latent heat and heat capacity simultaneously. They concluded that annealing the Cr sample decreases the transition temperature and sharpens the peak. They also concluded that the transition was first order with a latent heat of 0.19 ± 0.04 cal mol^{-1} at a transition temperature of 311.5 K. Garnier and Salamon (13) calculated a latent heat of 0.23 ± 0.05 cal mol^{-1} and 0.25 ± 0.08 cal mol^{-1} from expansivity data and neutron-diffraction data, respectively. Sze and Meaden (14) observed a large step in specific heat measurements (50%) and estimated a latent heat of 0.47 cal mol^{-1}.

We adopt T_{trs} = 311.5 K and $\Delta_{fus}H°$ = 0.0002 kcal mol^{-1} based on the work by Garnier and Salamon (13). In addition we adopt ΔCp° = 0 across the transition. The nature of the transition is not understood at this time.

Sublimation Data

See the ideal gas table for details.

References

1. R. Hultgren, R. L. Orr, and K. K. Kelley, Supplement to Selected Values of Thermodynamic Properties of Metals and Alloys, Cr table, October, 1966.
2. K. Clusius and P. Franzosini, Z. Naturforsch. 17a, 522 (1962).
3. C. T. Anderson, J. Amer. Chem. Soc. 59, 488 (1937).
4. G. I. Kalishevich, P. V. Gel'd, and R. P. Krentsis, Russ. J. Phys. Chem. 39, 1602 (1965).
5. R. H. Beaumont, H. Chihara, and J. A. Morrison, Phil. Mag. 5, 188 (1960).
6. L. D. Armstrong and H. Grayson-Smith, Can. J. Res. 28, 51 (1950).
7. R. Hultgren and C. Land, Trans. Met. Soc. AIME 215, 165 (1959).
8. J. B. Conway and R. A. Hein in "Advances in Thermophysical Properties at Extreme Templeratures and Pressures," ASME, 1965, pp. 131-7.
9. C. F. Lucks and H. W. Deem, WADC Tech. Rept. No. 55-496, 1956.
10. R. Kohlhaas, M. Braun, and O. Vollmer, Z. Naturforsch 20a, 1077 (1965).
11. F. M. Jaeger and R. Rosenbohm, Proc. Acad. Sci. Amsterdam 34, 489 (1934).
12. V. A. Kirillin, A. E. Sheinalin, V. Ya. Chekhovskoi, and J. A. Zhukova, Teplofiz. Vys. Temp. 5, 1124 (1967).
13. P. R. Garnier and Salamon, Phys. Rev. Lett. 27, 1523 (1971).
14. N. H. Sze and G. T. Meaden, Phys. Lett. 35A, 329 (1971).

Chromium (Cr) $Cr_1(cr)$

Enthalpy Reference Temperature = T_r = 298.15 K; Standard State Pressure = p° = 0.1 MPa

T/K	$C_p°$ (J K⁻¹mol⁻¹)	S° (J K⁻¹mol⁻¹)	-[G°-H°(T_r)]/T (J K⁻¹mol⁻¹)	H°-H°(T_r) (kJ mol⁻¹)	$\Delta_f H°$ (kJ mol⁻¹)	$\Delta_f G°$ (kJ mol⁻¹)	Log K_f
0	0.	0.	INFINITE	-4.057	0.	0.	0.
100	9.963	4.286	41.746	-3.746	0.	0.	0.
200	19.860	14.869	25.680	-2.162	0.	0.	0.
250	22.298	19.590	23.998	-1.102	0.	0.	0.
298.15	23.434	23.618	23.618	0.	0.	0.	0.
300	23.472	23.764	23.619	0.043	0.	0.	0.
311.500	23.681	24.651	23.641	0.315	Cp LAMBDA MAXIMUM TRANSITION		
311.500	23.681	24.654	23.641	0.316			
350	24.393	27.453	23.908	1.240	0.	0.	0.
400	25.230	30.765	24.562	2.481	0.	0.	0.
450	25.983	33.781	25.421	3.762	0.	0.	0.
500	26.631	36.553	26.398	5.078	0.	0.	0.
600	27.719	41.508	28.513	7.797	0.	0.	0.
700	28.577	45.846	30.686	10.612	0.	0.	0.
800	29.434	49.715	32.827	13.511	0.	0.	0.
900	30.501	53.241	34.902	16.505	0.	0.	0.
1000	31.861	56.523	36.901	19.621	0.	0.	0.
1100	33.472	59.634	38.828	22.887	0.	0.	0.
1200	35.187	62.618	40.687	26.318	0.	0.	0.
1300	37.116	65.511	42.485	29.933	0.	0.	0.
1400	39.125	68.334	44.231	33.744	0.	0.	0.
1500	41.200	71.104	45.931	37.760	0.	0.	0.
1600	43.329	73.831	47.589	41.986	0.	0.	0.
1700	45.501	76.523	49.212	46.427	0.	0.	0.
1800	47.706	79.186	50.804	51.087	0.	0.	0.
1900	49.944	81.825	52.367	55.970	0.	0.	0.
2000	52.204	84.444	53.905	61.077	0.	0.	0.
2100	54.488	87.046	55.422	66.411	0.	0.	0.
2130.000	55.174	87.824	55.873	68.056	CRYSTAL <--> LIQUID		
2200	56.785	89.634	56.918	71.975	-19.336	0.655	-0.016
2300	59.099	92.209	58.397	77.769	-17.475	1.523	-0.035
2400	61.425	94.774	59.859	83.795	-15.382	2.305	-0.050
2500	63.764	97.328	61.307	90.055	-13.055	2.995	-0.063
2600	66.111	99.875	62.741	96.548	-10.495	3.587	-0.072
2700	68.467	102.414	64.163	103.277	-7.699	4.076	-0.079
2800	70.831	104.947	65.575	110.242	-4.667	4.458	-0.083

PREVIOUS: CURRENT: June 1973

Chromium (Cr) $Cr_1(cr)$

CHROMIUM (Cr) LIQUID A_r = 51.996

$S°(298.15\ K) = [36.205]\ J\ K^{-1}\ mol^{-1}$

$T_{fus} = 2130 \pm 20\ K$

$\Delta_f H°(298.15\ K) = [26.068]\ kJ\ mol^{-1}$

$\Delta_{fus}H° = 20.5 \pm 4.2\ kJ\ mol^{-1}$

Enthalpy of Formation

The enthalpy of formation of Cr(ℓ) at 298.15 K is calculated from that of the crystal by adding $\Delta_{fus}H°$ and the difference in enthalpy, H°(2130 K) - H°(298.15 K), between the crystal and liquid.

Heat Capacity and Entropy

The heat capacity for Cr(ℓ) is estimated as 9.4 cal $K^{-1}\ mol^{-1}$ by analogy with other monatomic metals. The same value was adopted by Hutgren et al. (1). A glass transition is assumed at 1350 K. Below 1350 K, the Cp° values are those of the crystal. The entropy is calculated in a manner analogous to that used for the enthalpy of formation.

Fusion Data

The value for the enthalpy of melting is adopted as $\Delta_{fus}H° = 4.9 \pm 1.0\ kcal\ mol^{-1}$. This value is calculated from $\Delta_{fus}S°$ = 2.3 cal $K^{-1}\ mol^{-1}$. This estimated entropy of melting is consistent with the recently accepted $\Delta_{fus}S°$ values for V (2, 2.49 cal $K^{-1}\ mol^{-1}$), Nb (2, 2.33 cal $K^{-1}\ mol^{-1}$), and Ta (2, 2.68 cal $K^{-1}\ mol^{-1}$). This is in contrast to a value of $\Delta_{fus}S°$ = 1.9 cal $K^{-1}\ mol^{-1}$ used by Hultgren et al. (1).

The melting temperature is adopted as T_{fus} = 2130 K as was chosen by Hultgren et al. (1). Values reported in the literature are:

Source	Year	T, K
(3)	1949	2133
(4)	1951	2118±10
(5)	1952	2176±10
(6)	1959	2122

Vaporization Data

The value for T_{vap} calculated as the temperature for which the Gibbs energy for the reaction Cr(ℓ) = Cr(g) is zero. The difference in the enthalpy of formation of Cr(ℓ) and Cr(g) at T_{vap} is $\Delta_{vap}H°$.

References

1. R. Hultgren, R. L. Orr, and K. K. Kelley, Supplement to Selected Values of Thermodynamic Properties of Metals and Alloys, Cr table, October, 1966.
2. JANAF Thermochemical Tables: V(cr), 6-30-73; Nb(cr) and Ta(cr), 12-31-72.
3. S. J. Carlile, J. W. Christian, and W. Hume-Rothery, J. Inst. Metals 76, 169 (1949).
4. H. T. Greenaway, S. T. M. Johnstone, and M. K. McQuillan, J. Inst. Metals 80, 109 (1951).
5. D. S. Bloom, J. W. Putman, and N. J. Grant, J. Metals 4, 626 (1952).
6. C. W. Haworth and W. Hume-Rothery, J. Inst. Metals 87 265 (1959).

Chromium (Cr) $Cr_1(l)$

Enthalpy Reference Temperature = T_r = 298.15 K; Standard State Pressure = p° = 0.1 MPa

T/K	$C_p°$ (J K⁻¹mol⁻¹)	S° (J K⁻¹mol⁻¹)	$-[G°-H°(T_r)]/T$ (J K⁻¹mol⁻¹)	$H°-H°(T_r)$ (kJ mol⁻¹)	$\Delta_f H°$ (kJ mol⁻¹)	$\Delta_f G°$ (kJ mol⁻¹)	Log K_f
0							
100							
200							
250							
298.15	23.434	36.205	36.205	0.	26.068	22.316	-3.910
300	23.472	36.350	36.205	0.043	26.068	22.292	-3.881
350	24.393	40.036	36.495	1.240	26.067	21.663	-3.233
400	25.230	43.349	37.148	2.481	26.067	21.034	-2.747
450	25.983	46.365	38.007	3.761	26.067	20.405	-2.369
500	26.631	49.137	38.983	5.077	26.067	19.775	-2.066
600	27.722	54.092	41.098	7.797	26.068	18.517	-1.612
700	28.577	58.430	43.271	10.611	26.068	17.259	-1.288
800	29.434	62.299	45.412	13.510	26.068	16.000	-1.045
900	30.501	65.825	47.487	16.505	26.068	14.742	-0.856
1000	31.861	69.107	49.486	19.621	26.068	13.483	-0.704
1100	33.472	72.218	51.412	22.886	26.068	12.225	-0.581
1200	35.187	75.203	53.271	26.318	26.068	10.967	-0.477
1300	37.116	78.095	55.070	29.932	26.068	9.708	-0.390
1350.000	38.112	79.514	55.949	31.813	GLASS <--> LIQUID TRANSITION		
1350.000	39.330	79.514	55.949	31.813			
1400	39.330	80.944	56.816	33.779	26.103	8.449	-0.315
1500	39.330	83.658	58.516	37.712	26.020	7.190	-0.250
1600	39.330	86.196	60.168	41.645	25.727	5.943	-0.194
1700	39.330	88.581	61.770	45.578	25.219	4.720	-0.145
1800	39.330	90.829	63.322	49.511	24.492	3.535	-0.103
1900	39.330	92.955	64.827	53.444	23.542	2.395	-0.066
2000	39.330	94.972	66.284	57.377	22.368	1.311	-0.034
2100	39.330	96.891	67.696	61.310	20.967	0.292	-0.007
2130.000	39.330	97.449	68.111	62.490	CRYSTAL <--> LIQUID		
2200	39.330	98.721	69.065	65.243	0.	0.	0.
2300	39.330	100.469	70.393	69.176	0.	0.	0.
2400	39.330	102.143	71.681	73.109	0.	0.	0.
2500	39.330	103.749	72.932	77.042	0.	0.	0.
2600	39.330	105.291	74.147	80.975	0.	0.	0.
2700	39.330	106.775	75.328	84.908	0.	0.	0.
2800	39.330	108.206	76.477	88.841	0.	0.	0.
2900	39.330	109.586	77.595	92.774	0.	0.	0.
2952.078	39.330	110.286	78.165	94.822	FUGACITY = 1 bar		
3000	39.330	110.919	78.684	96.707	-339.122	5.508	-0.096
3100	39.330	112.209	79.744	100.640	-338.294	16.982	-0.286
3200	39.330	113.457	80.779	104.572	-337.516	28.430	-0.464
3300	39.330	114.668	81.787	108.505	-336.786	39.855	-0.631
3400	39.330	115.842	82.772	112.438	-336.104	51.258	-0.787
3500	39.330	116.982	83.733	116.371	-335.469	62.641	-0.935

PREVIOUS: CURRENT: June 1973

Chromium (Cr) $Cr_1(l)$

CHROMIUM (Cr) CRYSTAL–LIQUID A_r = 51.996 Chromium (Cr) $Cr_1(cr,l)$

0 to 2130 K crystal
above 2130 K liquid

Refer to the individual tables for details.

Enthalpy Reference Temperature = T_r = 298.15 K — Standard State Pressure = p° = 0.1 MPa

	J K⁻¹mol⁻¹			kJ mol⁻¹			
T/K	C_p°	S°	$-[G°-H°(T_r)]/T$	$H°-H°(T_r)$	$\Delta_f H°$	$\Delta_f G°$	Log K_f
0	0.	0.	INFINITE	-4.057	0.	0.	0.
100	9.963	4.286	41.746	-3.746	0.	0.	0.
200	19.860	14.869	25.680	-2.162	0.	0.	0.
250	22.298	19.590	23.998	-1.102	0.	0.	0.
298.15	23.434	23.618	23.618	0.	0.	0.	0.
300	23.472	23.764	23.619	0.043	0.	0.	0.
311.500	23.681	24.651	23.641	0.315	Cp LAMBDA MAXIMUM TRANSITION		
311.500	23.681	24.654	23.641	0.316			
350	24.393	27.453	23.908	1.240	0.	0.	0.
400	25.230	30.765	24.562	2.481	0.	0.	0.
450	25.983	33.781	25.421	3.762	0.	0.	0.
500	26.631	36.553	26.398	5.078	0.	0.	0.
600	27.719	41.508	28.513	7.797	0.	0.	0.
700	28.577	45.846	30.686	10.612	0.	0.	0.
800	29.434	49.715	32.827	13.511	0.	0.	0.
900	30.501	53.241	34.902	16.505	0.	0.	0.
1000	31.861	56.523	36.901	19.621	0.	0.	0.
1100	33.472	59.634	38.828	22.887	0.	0.	0.
1200	35.187	62.618	40.687	26.318	0.	0.	0.
1300	37.116	65.511	42.485	29.933	0.	0.	0.
1400	39.125	68.334	44.231	33.744	0.	0.	0.
1500	41.200	71.104	45.931	37.760	0.	0.	0.
1600	43.329	73.831	47.589	41.986	0.	0.	0.
1700	45.501	76.523	49.212	46.427	0.	0.	0.
1800	47.706	79.186	50.804	51.087	0.	0.	0.
1900	49.944	81.825	52.367	55.970	0.	0.	0.
2000	52.204	84.444	53.905	61.077	0.	0.	0.
2100	54.488	87.046	55.422	66.411	0.	0.	0.
2130.000	55.174	87.824	55.873	68.056	CRYSTAL <--> LIQUID TRANSITION		
2130.000	39.330	97.449	55.873	88.558			
2200	39.330	98.721	57.216	91.311	0.	0.	0.
2300	39.330	100.469	59.059	95.244	0.	0.	0.
2400	39.330	102.143	60.819	99.177	0.	0.	0.
2500	39.330	103.749	62.505	103.110	0.	0.	0.
2600	39.330	105.291	64.121	107.043	0.	0.	0.
2700	39.330	106.775	65.673	110.976	0.	0.	0.
2800	39.330	108.206	67.167	114.909	0.	0.	0.
2900	39.330	109.586	68.606	118.842	0.	0.	0.
2952.078	39.330	110.286	69.335	120.890	------- FUGACITY = 1 bar -------		
3000	39.330	110.919	69.994	122.775	-339.122	5.508	-0.096
3100	39.330	112.209	71.335	126.708	-338.294	16.982	-0.286
3200	39.330	113.457	72.632	130.641	-337.516	28.430	-0.464
3300	39.330	114.668	73.888	134.574	-336.786	39.855	-0.631
3400	39.330	115.842	75.105	138.507	-336.104	51.258	-0.787
3500	39.330	116.982	76.285	142.440	-335.469	62.641	-0.935

PREVIOUS: CURRENT: June 1973

Chromium (Cr) $Cr_1(cr,l)$

CHROMIUM (Cr) IDEAL GAS A_r = 51.996

IP(Cr, g) = 54570 ± 10 cm^{-1}
S°(298.15 K) = 174.31 ± 0.4 J K^{-1} mol^{-1}

$\Delta_f H°$(0 K) = 395.34 ± 4.2 kJ mol^{-1}
$\Delta_f H°$(298.15 K) = 397.48 ± 4.2 kJ mol^{-1}

Electronic Levels and Quantum Weights

State	ε_i, cm^{-1}	g_i
$^7S_3$	0.00	7
$^5S_2$	7593.16	5
$^5D_0$	7750.78	1
$^5D_1$	7810.82	3
$^5D_2$	7927.47	5
$^5D_3$	8095.21	7
$^5D_4$	8307.57	9

Enthalpy of Formation

The many vapor pressure studies are treated by a 2nd and 3rd law analysis and are tabulated below. In all cases the reaction of interest is Cr(cr) + Cr(g).

Source	Data Points	T/K	Method	$\Delta_f H°$(298.15 K), kcal mol^{-1} 2nd law	3rd law	Drift cal K^{-1} mol^{-1}
(1)	15*	1283-1562	Langmuir	94.05±0.84	94.82	0.54±0.60
(2)	9	1162-1282	Knudsen	98.51±2.16	95.28	-2.64±1.77
(3)	10	1292-1509	Langmuir	97.60±1.60	92.32	-3.71±1.12
	10**	1187-1550		95.15±1.00	91.82	-2.47±0.74
(4)	9	1381-1505	Knudsen	101.14±2.87	94.90	-4.29±1.97
(5)	eqn	1502-1576	Knudsen	88.08	90.06	1.28
(6)	15	1317-1558	Knudsen	93.14±5.70	93.99	0.59±3.95
(7)	eqn	1441-1672	Knudsen	92.50	94.79	1.48
(8)	9	1273-1373	Knudsen	97.54±2.93	95.97	-1.19±2.23
(9)	eqn	1376-1592	Torsion-effusion	91.79	93.85	1.40
(10)	15**	1559-1805	Knudsen	92.78±3.08	95.50	1.62±1.84

* Two points neglected due to failure of a statistical test.
** One point neglected due to failure of a statistical test.

Early studies by Greenwood (11) and Baur and Brunner (12) are not considered as their work leads to $\Delta_f H°$(298.15 K) which are 10-15 kcal mol^{-1} too low when compared to the results listed above.

We adopt $\Delta_f H°$(298.15 K) = 95.0 ± 1.0 kcal mol^{-1} for Cr(g). This value is representative of the more recent 2nd law values tabulated above (7, 8, 9, 10).

Heat Capacity and Entropy

The electronic levels and quantum weights are obtained from Moore (13); although only the lowest levels are listed above. There are predicted levels which have not been observed and/or classified. It is anticipated that these levels will not significantly alter the entropy below 3000 K. The heat capacity and entropy values are very similar to those adopted bvy Hultgren et al. (14).

References

1. R. Speiser, H. L. Johnston, and P. B. Blackburn, J. Amer. Chem. Soc. 72, 4142 (1950).
2. E. A. Gulbranson and K. F. Andrew, J. Electrochem. Soc. 99, 402 (1952).
3. V. D. Burlakov, Phys. Metals and Metallog. 5, 72 (1957).
4. C. L. McCabe, R. G. Hudson, and H. W. Paxton, Trans. Mat. Soc. AIME, 212, 102 (1958).
5. G. O. Piloyan, A. M. Evseev, and Ya. I. Gerasimov, Russ. J. Phys. Chem. 34, 842 (1960).
6. An. N. Nesmeyanov and D. Dyk Man, Dokl. Akad. Nauk SSSR 116, 230 (1959).
7. O. Kubaschewski and G. Heymer, Acta Met. 8, 416 (1960).
8. E. A. Gulbranson and K. F. Andrew, Trans. Met. Soc. AIME 221, 1247 (1961).
9. A. T. Aldred and K. M. Myles, Trans. Met. Soc. AIME 230, 736 (1964).
10. D. S. Dickson, J. R. Myers, and R. K. Saxer, J. Phys. Chem. 69, 4044 (1965).
11. H. C. Greenwood, Proc. Roy. Soc. A82, 396 (1909).
12. E. Baur and R. Brunner, Helv. Chim. Acta 17, 958 (1934).
13. C. E. Moore, U. S. Nat. Bur. Stand., NSRDS-NBS 35, Volume II, 1970 [Reprint of NBS Circular 467, Volume II, 1952].
14. R. Hultgren, R. L. Orr, and K. K. Kelley, Supplement to Selected Values of Thermodynamic Properties of Metals and Alloys, Cr table, October, 1966.

Chromium (Cr) $Cr_1(g)$

Enthalpy Reference Temperature = T_r = 298.15 K; Standard State Pressure = p° = 0.1 MPa

T/K	$C_p°$ (J $K^{-1}mol^{-1}$)	S°	$-[G°-H°(T_r)]/T$	$H°-H°(T_r)$ (kJ mol^{-1})	$\Delta_f H°$	$\Delta_f G°$	Log K_f
0	0.	0.	INFINITE	-6.197	395.340	395.340	INFINITE
100	20.786	151.604	192.791	-4.119	397.107	382.375	-199.732
200	20.786	166.012	176.212	-2.040	397.602	367.374	-95.948
250	20.786	170.650	174.653	-1.001	397.581	359.816	-75.179
298.15	20.786	174.311	174.311	0.	397.480	352.551	-61.765
300	20.786	174.440	174.312	0.038	397.475	352.272	-61.336
350	20.786	177.644	174.565	1.078	397.317	344.750	-51.451
400	20.786	180.419	175.127	2.117	397.116	337.254	-44.041
450	20.786	182.868	175.854	3.156	396.874	329.785	-38.280
500	20.786	185.058	176.666	4.196	396.598	322.346	-33.675
600	20.786	188.847	178.390	6.274	395.957	307.554	-26.775
700	20.787	192.052	180.119	8.353	395.221	292.877	-21.855
800	20.790	194.828	181.788	10.432	394.401	278.311	-18.172
900	20.803	197.277	183.375	12.511	393.486	263.854	-15.314
1000	20.835	199.470	184.877	14.593	392.452	249.504	-13.033
1100	20.901	201.459	186.296	16.679	391.272	235.265	-11.172
1200	21.015	203.282	187.636	18.775	389.937	221.140	-9.626
1300	21.191	204.970	188.905	20.884	388.432	207.134	-8.323
1400	21.440	206.549	190.110	23.015	386.751	193.250	-7.210
1500	21.768	208.039	191.256	25.175	384.895	179.492	-6.250
1600	22.173	209.457	192.350	27.372	382.865	165.864	-5.415
1700	22.653	210.815	193.396	29.612	380.665	152.368	-4.682
1800	23.198	212.125	194.400	31.904	378.297	139.006	-4.034
1900	23.798	213.395	195.367	34.254	375.764	125.780	-3.458
2000	24.439	214.632	196.299	36.665	373.068	112.692	-2.943
2100	25.109	215.840	197.201	39.142	370.211	99.743	-2.481
2200	25.794	217.024	198.075	41.688	347.856	87.589	-2.080
2300	26.483	218.186	198.925	44.301	346.537	75.788	-1.721
2400	27.165	219.328	199.751	46.984	345.287	64.044	-1.394
2500	27.831	220.450	200.557	49.734	344.104	52.350	-1.094
2600	28.476	221.554	201.343	52.549	342.986	40.702	-0.818
2700	29.095	222.641	202.112	55.428	341.932	29.096	-0.563
2800	29.687	223.710	202.864	58.367	340.939	17.528	-0.327
2900	30.250	224.761	203.601	61.364	340.003	5.994	-0.108
2952.078	0.	0.	0.	0.	------- FUGACITY = 1 bar -------		
3000	30.788	225.796	204.324	64.417	0.	0.	0.
3100	31.302	226.814	205.033	67.521	0.	0.	0.
3200	31.796	227.815	205.729	70.676	0.	0.	0.
3300	32.275	228.801	206.413	73.880	0.	0.	0.
3400	32.745	229.772	207.086	77.131	0.	0.	0.
3500	33.212	230.728	207.748	80.429	0.	0.	0.
3600	33.680	231.670	208.399	83.773	0.	0.	0.
3700	34.156	232.599	209.041	87.165	0.	0.	0.
3800	34.646	233.516	209.673	90.605	0.	0.	0.
3900	35.157	234.423	210.296	94.095	0.	0.	0.
4000	35.692	235.320	210.910	97.637	0.	0.	0.
4100	36.260	236.208	211.517	101.235	0.	0.	0.
4200	36.864	237.089	212.115	104.890	0.	0.	0.
4300	37.511	237.964	212.706	108.609	0.	0.	0.
4400	38.205	238.834	213.290	112.394	0.	0.	0.
4500	38.954	239.701	213.867	116.252	0.	0.	0.
4600	39.760	240.566	214.438	120.187	0.	0.	0.
4700	40.632	241.430	215.003	124.206	0.	0.	0.
4800	41.572	242.295	215.563	128.315	0.	0.	0.
4900	42.587	243.163	216.117	132.523	0.	0.	0.
5000	43.681	244.034	216.667	136.835	0.	0.	0.
5100	43.917	244.834	217.207	140.898	0.	0.	0.
5200	44.964	245.697	217.746	145.342	0.	0.	0.
5300	46.064	246.564	218.282	149.892	0.	0.	0.
5400	47.221	247.435	218.814	154.556	0.	0.	0.
5500	48.432	248.313	219.342	159.338	0.	0.	0.
5600	49.697	249.197	219.867	164.244	0.	0.	0.
5700	51.009	250.087	220.390	169.277	0.	0.	0.
5800	52.370	250.986	220.909	174.445	0.	0.	0.
5900	53.779	251.893	221.427	179.752	0.	0.	0.
6000	55.230	252.809	221.942	185.202	0.	0.	0.

PREVIOUS: June 1979 (1 atm) CURRENT: June 1979 (1 bar)

Chromium (Cr) $Cr_1(g)$

CHROMIUM, ION (Cr^+) IDEAL GAS M_r = 51.99545

IP(Cr^+, g) = 133060 ± 10 cm^{-1}

S°(298.15 K) = 173.029 ± 0.05 J K^{-1} mol^{-1}

$\Delta_f H°$(0 K) = 1048.14 ± 1.5 kJ mol^{-1}

$\Delta_f H°$(298.15 K) = [1056. 480] kJ mol^{-1}

Electronic Levels and Quantum Weights

State	ϵ_i cm^{-1}	g_i
$^6S_{5/2}$	0	6
$^6D_{1/2}$	1196.00	2
.	.	.
.	.	.
.	.	.
$^4G_{11/2}$	105421.9	12
IP	133060	

Enthalpy of Formation

$\Delta_f H°$(Cr^+, g, 0 K) is calculated from $\Delta_f H°$(Cr, g, 0 K) (1) using the spectroscopic value of IP(Cr) = 54570 ± 10 cm^{-1} (652.80 ± 0.12 kJ mol^{-1}) from Sugar and Corliss (2). The ionization limit is converted from cm^{-1} to kJ mol^{-1} using the factor, 1 cm^{-1} = 0.01196266 kJ cm^{-1}, which is derived from the 1973 CODATA fundamental constants (3). Rosenstock et al. (4) and Levin and Lias (5) have summarized additional ionization and appearance potential data.

$\Delta_f H°$(Cr^+, g, 298.15 K) is calculated from $\Delta_f H°$(Cr, g, 0 K) by using IP(Cr) with JANAF (1) enthalpies, H°(0 K) - H°(298.15 K), for Cr(g), Cr^+(g), and e^-(g). $\Delta_f H°$(Cr → Cr^+ + e^-, 298.15 K) differs from a room temperature threshold energy due to inclusion of these enthalpies and to threshold effects discussed by Rosenstock et al. (4). $\Delta_f H°$(298.15 K) should be changed by -6.197 kJ mol^{-1} if it is to be used in the ion convention that excludes the enthalpy of the electron.

Heat Capacity and Entropy

The information on electronic energy levels and quantum weights, given by Corliss and Sugar (2), is incomplete because many theoretically predicted levels have not been observed. Although we have listed only the ground, the first excited state, the highest observed excited state, and the ionization potential for Cr^+(g), all levels listed by Sugar and Corliss (2), as well as estimated levels, are used in the calculation. The observed levels are too numerous to list completely. The calculations indicate that for Cr^+(g), the tehrmodynamic functions are independent of the estimated missing levels (for n=4), the cut-off procedure, and the inclusion of n=6 levels up to 6000 K; the Gibbs energy function showing variations of 0.02% at this temperature. The reported uncertainty is S°(298.15 K) is due to uncertainties in the relative ionic mass, and the fundamental constants. Extension of these calculations above 6000 K may require consideration of the higher excited states (n>6), and use of different fill and cut-off procedures (6).

References

1. JANAF Thermochemical Tables: Cr(g), 6-30-73; e^-(g), 3-31-82.
2. J. Sugar and C. Corliss, J. Phys. Chem. Ref. Data 6, 317 (1977).
3. E. R. Cohen and B. N. Taylor, J. Phys. Chem. Ref. Data 2, 663 (1973).
4. H. M. Rosenstock, K. Draxl et al., J. Phys. Chem. Ref. Data 6, Supp. 1 (1977).
5. R. D. Levin and S. G. Lias, U. S. Natl. Bur. Stand. NSRDS-NBS-71, 1982.
6. J. R. Downey, Jr., The Dow Chemical Company, AFOSR-TR-78-0960, Contract No. F44620-75-1-0048, 1978.

Chromium, Ion (Cr^+) Cr_1^+(g)

Enthalpy Reference Temperature = T_r = 298.15 K — Standard State Pressure = p° = 0.1 MPa

T/K	$C_p°$ (J $K^{-1}mol^{-1}$)	S° (J $K^{-1}mol^{-1}$)	-[G°-H°(T_r)]/T (J $K^{-1}mol^{-1}$)	H°-H°(T_r) (kJ mol^{-1})	$\Delta_f H°$ (kJ mol^{-1})	$\Delta_f G°$ (kJ mol^{-1})	Log K_f
0	0.	0.	INFINITE	-6.197	1048.142		
100	20.786	150.322	191.510	-4.119			
200	20.786	164.730	174.931	-2.040			
250	20.786	169.368	173.372	-1.001			
298.15	20.786	173.029	173.029	0.	1056.480	1005.678	-176.191
300	20.786	173.158	173.030	0.038	1056.513	1005.363	-175.049
350	20.786	176.362	173.283	1.078	1057.395	996.767	-148.759
400	20.786	179.138	173.845	2.117	1058.232	988.049	-129.026
450	20.786	181.586	174.572	3.156	1059.030	979.227	-113.666
500	20.786	183.776	175.385	4.196	1059.793	970.319	-101.369
600	20.786	187.566	177.109	6.274	1061.231	952.288	-82.904
700	20.786	190.770	178.837	8.353	1062.573	934.023	-69.698
800	20.786	193.545	180.506	10.431	1063.832	915.572	-59.781
900	20.786	195.994	182.094	12.510	1064.995	896.968	-52.059
1000	20.786	198.184	183.595	14.589	1066.036	878.242	-45.875
1100	20.787	200.165	185.013	16.667	1066.927	859.418	-40.810
1200	20.790	201.974	186.352	18.746	1067.654	840.520	-36.587
1300	20.796	203.638	187.618	20.825	1068.197	821.569	-33.011
1400	20.808	205.180	188.818	22.906	1068.544	802.583	-29.945
1500	20.831	206.616	189.958	24.987	1068.689	783.580	-27.287
1600	20.868	207.961	191.041	27.072	1068.626	764.573	-24.961
1700	20.926	209.228	192.074	29.162	1068.353	745.577	-22.909
1800	21.008	210.426	193.061	31.258	1067.868	726.603	-21.085
1900	21.121	211.565	194.005	33.364	1067.170	707.662	-19.455
2000	21.269	212.652	194.910	35.484	1066.261	688.763	-17.989
2100	21.457	213.694	195.780	37.620	1065.141	669.915	-16.663
2200	21.686	214.698	196.617	39.776	1044.477	651.780	-15.475
2300	21.961	215.667	197.425	41.958	1044.805	633.923	-14.397
2400	22.283	216.609	198.204	44.170	1045.162	616.051	-13.408
2500	22.652	217.526	198.959	46.417	1045.554	598.164	-12.498
2600	23.069	218.422	199.690	48.702	1045.985	580.260	-11.658
2700	23.533	219.301	200.400	51.032	1046.461	562.338	-10.879
2800	24.044	220.166	201.091	53.410	1046.985	544.398	-10.156
2900	24.600	221.019	201.764	55.842	1047.562	526.439	-9.482
3000	25.199	221.863	202.419	58.332	709.076	513.967	-8.949
3100	25.839	222.700	203.060	60.883	710.601	507.438	-8.550
3200	26.517	223.531	203.687	63.501	712.142	500.860	-8.176
3300	27.230	224.358	204.301	66.188	713.704	494.233	-7.823
3400	27.975	225.182	204.903	68.948	715.292	487.559	-7.490
3500	28.750	226.004	205.494	71.784	716.908	480.837	-7.176
3600	29.550	226.825	206.075	74.699	718.557	474.069	-6.879
3700	30.374	227.646	206.647	77.695	720.240	467.255	-6.596
3800	31.218	228.467	207.210	80.774	721.958	460.394	-6.329
3900	32.078	229.289	207.766	83.939	723.712	453.488	-6.074
4000	32.952	230.112	208.314	87.190	725.499	446.537	-5.831
4100	33.838	230.936	208.856	90.530	727.320	439.540	-5.600
4200	34.733	231.763	209.392	93.958	729.171	432.499	-5.379
4300	35.635	232.590	209.921	97.477	731.050	425.413	-5.168
4400	36.542	233.420	210.446	101.085	732.952	418.283	-4.966
4500	37.452	234.251	210.966	104.785	734.873	411.110	-4.772
4600	38.366	235.085	211.481	108.576	736.807	403.894	-4.586
4700	39.282	235.920	211.992	112.458	738.749	396.636	-4.408
4800	40.199	236.756	212.499	116.432	740.692	389.336	-4.237
4900	41.118	237.594	213.003	120.498	742.629	381.996	-4.072
5000	42.040	238.434	213.503	124.656	744.553	374.617	-3.914
5100	42.964	239.276	214.000	128.906	746.819	367.172	-3.761
5200	43.893	240.119	214.494	133.249	748.797	359.709	-3.613
5300	44.827	240.964	214.986	137.685	750.761	352.208	-3.471
5400	45.767	241.811	215.475	142.215	752.705	344.669	-3.334
5500	46.716	242.659	215.961	146.839	754.626	337.095	-3.201
5600	47.675	243.510	216.446	151.558	756.518	329.487	-3.073
5700	48.645	244.362	216.928	156.374	758.379	321.845	-2.949
5800	49.629	245.217	217.408	161.288	760.204	314.170	-2.829
5900	50.627	246.073	217.887	166.300	761.988	306.464	-2.713
6000	51.640	246.933	218.364	171.414	763.730	298.729	-2.601

PREVIOUS: CURRENT: March 1984 (1 bar)

Chromium, Ion (Cr^+) Cr_1^+(g)

CHROMIUM, ION (Cr^-) IDEAL GAS M_r = 51.99655

EA(Cr, g) = 0.666 ± 0.012 eV

S°(298.15 K) = 173.030 ± 0.003 J K^{-1} mol^{-1}

$\Delta_f H°$(0 K) = 331.081 ± 2 kJ mol^{-1}

$\Delta_f H°$(298.15 K) = [327.024] kJ mol^{-1}

Electonic Level and Quantum Weight

State	ε_i, cm^{-1}	g_i
$^6S_{5/2}$	0.0	6

Enthalpy of Formation

$\Delta_f H°(Cr^-$, g, 0 K) is calculated from $\Delta_f H°$(Cr, g, 0 K) (1) using the adopted electron affinity of EA(Cr) = 0.666 ± 0.012 eV (64.259 ± 1.158 kJ mol^{-1}). This value, recommended by Hotop and Lineberger (2), is based on a laser photodetachment electron spectrometry study (3). Additional information on Cr^-(g) may be obtained in the critical discussions of Hotop and Lineberger (2, 4), Rosenstock et al. (5) and Massey (6).

$\Delta_f H°(Cr^-$, g, 298.15 K) is obtained from $\Delta_f H°$(Cr, g, 0 K) by using EA(Cr) with JANAF (1) enthalpies, H°(0 K) − H°(298.15 K), for Cr^-(g), Cr(g), and e^-(g). $\Delta_r H°(Cr^- \rightarrow Cr + e^-$, 298.15 K) differs from a room-temperature threshold energy due to inclusion of these enthalpies and to threshold effects discussed by Rosenstock et al. (5). $\Delta_f H°$(298.15 K) should be changed by +6.197 kJ mol^{-1} if it is to be used in the ion convention that excludes the enthalpy of the electron.

Heat Capacity of Entropy

The ground state electronic configuration for Cr^-(g) is given by Hotop and Lineberger (2, 4) and Rosenstock et al. (5). Lacking any experimental evidence as to the stability of any excited states, we assume that no stable excited states exist.

References

1. JANAF Thermochemical Tables: Cr(g), 6-30-73; e^-(g), 3-31-82.
2. H. Hotop and W. C. Lineberger, J. Phys. Chem. Ref. Data, to be published, 1985.
3. C. S. Feigerle, R. P. Corderman, S. V. Bobashev and W. C. Lineberger, J. Chem. Phys. 74, 1580 (1981).
4. H. Hotop and W. C. Lineberger, J. Phys. Chem. Ref. Data 4, 539 (1975).
5. H. M. Rosenstock, K. Draxl et al., J. Phys. Chem. Ref. Data 6, Supp. 1 (1977).
6. H. S. W. Massey, "Negative Ions", 3rd ed., Cambridge University Press, Cambridge, 1976.

Chromium, Ion (Cr^-) Cr_1^-(g)

Enthalpy Reference Temperature = T_r = 298.15 K Standard State Pressure = p° = 0.1 MPa

T/K	$C_p°$ (J $K^{-1}mol^{-1}$)	S° (J $K^{-1}mol^{-1}$)	−[G°−H°(T_r)]/T (J $K^{-1}mol^{-1}$)	H°−H°(T_r) (kJ mol^{-1})	$\Delta_f H°$ (kJ mol^{-1})	$\Delta_f G°$ (kJ mol^{-1})	Log K_f
0	0.	0.	INFINITE	-6.197	331.081		
100	20.786	150.322	191.510	-4.119			
200	20.786	164.730	174.931	-2.040			
250	20.786	169.368	173.372	-1.001			
298.15	20.786	173.030	173.030	0.	327.024	288.732	-50.585
300	20.786	173.158	173.030	0.038	326.981	288.494	-50.231
350	20.786	176.362	173.283	1.078	325.783	282.174	-42.112
400	20.786	179.138	173.845	2.117	324.543	276.028	-36.046
450	20.786	181.586	174.572	3.156	323.262	270.041	-31.345
500	20.786	183.776	175.385	4.196	321.946	264.197	-27.601
600	20.786	187.566	177.109	6.274	319.227	252.901	-22.017
700	20.786	190.770	178.837	8.353	316.412	242.068	-18.063
800	20.786	193.546	180.506	10.431	313.513	231.644	-15.125
900	20.786	195.994	182.094	12.510	310.519	221.590	-12.861
1000	20.786	198.184	183.595	14.589	307.403	211.874	-11.067
1100	20.786	200.165	185.013	16.667	304.137	202.478	-9.615
1200	20.786	201.974	186.352	18.746	300.706	193.387	-8.418
1300	20.786	203.637	187.619	20.824	297.091	184.589	-7.417
1400	20.786	205.178	188.819	22.903	293.280	176.076	-6.569
1500	20.786	206.612	189.958	24.982	289.264	167.844	-5.845
1600	20.786	207.953	191.041	27.060	285.038	159.886	-5.220
1700	20.786	209.214	192.073	29.139	280.597	152.198	-4.676
1800	20.786	210.402	193.059	31.217	275.937	144.779	-4.201
1900	20.786	211.526	194.001	33.296	271.054	137.625	-3.784
2000	20.786	212.592	194.904	35.375	265.947	130.733	-3.414
2100	20.786	213.606	195.771	37.453	260.612	124.103	-3.087
2200	20.786	214.573	196.604	39.532	235.713	118.387	-2.811
2300	20.786	215.497	197.405	41.610	231.780	113.142	-2.570
2400	20.786	216.381	198.178	43.689	227.847	108.068	-2.352
2500	20.786	217.230	198.923	45.768	223.914	103.158	-2.155
2600	20.786	218.045	199.643	47.846	219.981	98.405	-1.977
2700	20.786	218.830	200.339	49.925	216.048	93.804	-1.815
2800	20.786	219.586	201.013	52.004	212.115	89.348	-1.667
2900	20.786	220.315	201.666	54.082	208.182	85.033	-1.532
3000	20.786	221.020	202.299	56.161	-134.873	86.362	-1.504
3100	20.786	221.701	202.914	58.239	-137.977	93.788	-1.580
3200	20.786	222.361	203.512	60.318	-141.132	101.314	-1.654
3300	20.786	223.001	204.093	62.397	-144.336	108.940	-1.724
3400	20.786	223.621	204.658	64.475	-147.587	116.664	-1.792
3500	20.786	224.224	205.209	66.554	-150.885	124.484	-1.858
3600	20.786	224.809	205.745	68.632	-154.229	132.399	-1.921
3700	20.786	225.379	206.268	70.711	-157.621	140.407	-1.982
3800	20.786	225.933	206.778	72.790	-161.061	148.508	-2.041
3900	20.786	226.473	207.276	74.868	-164.551	156.700	-2.099
4000	20.786	226.999	207.763	76.947	-168.093	164.982	-2.154
4100	20.786	227.513	208.238	79.025	-171.691	173.353	-2.209
4200	20.786	228.014	208.703	81.104	-175.347	181.813	-2.261
4300	20.786	228.503	209.158	83.183	-179.065	190.361	-2.312
4400	20.786	228.981	209.603	85.261	-182.850	198.996	-2.362
4500	20.786	229.448	210.039	87.340	-186.708	207.717	-2.411
4600	20.786	229.905	210.466	89.418	-190.643	216.526	-2.459
4700	20.786	230.352	210.884	91.497	-194.662	225.420	-2.505
4800	20.786	230.789	211.294	93.576	-198.771	234.401	-2.551
4900	20.786	231.218	211.697	95.654	-202.979	243.469	-2.595
5000	20.786	231.638	212.091	97.733	-207.291	252.624	-2.639
5100	20.786	232.049	212.479	99.811	-211.354	261.839	-2.682
5200	20.786	232.453	212.859	101.890	-215.798	271.160	-2.724
5300	20.786	232.849	213.232	103.969	-220.348	280.568	-2.765
5400	20.786	233.237	213.599	106.047	-225.012	290.063	-2.806
5500	20.786	233.619	213.960	108.126	-229.794	299.645	-2.846
5600	20.786	233.993	214.314	110.204	-234.700	309.315	-2.885
5700	20.786	234.361	214.663	112.283	-239.733	319.074	-2.924
5800	20.786	234.723	215.005	114.362	-244.901	328.923	-2.962
5900	20.786	235.078	215.343	116.440	-250.208	338.862	-3.000
6000	20.786	235.428	215.674	118.519	-255.658	348.891	-3.037

PREVIOUS: CURRENT: March 1984 (1 bar)

Chromium, Ion (Cr^-) Cr_1^-(g)

CHROMIUM NITRIDE (CrN) CRYSTAL M_r = 66.0027

$S°(298.15\ K) = 37.71 \pm 2.1\ J\ K^{-1}\ mol^{-1}$

$\Delta_f H°(0\ K) = -116.46 \pm 8.4\ kJ\ mol^{-1}$

$\Delta_f H°(298.15\ K) = -117.15 \pm 8.4\ kJ\ mol^{-1}$

Enthalpy of Formation

2nd and 3rd law analyses of equilibrium nitrogen pressures for the reaction $Cr_2N(cr) + 0.5\ N_2(g) = 2\ CrN(cr)$ cited in a recent review (1) of the thermodynamic properties of the Cr-N system are made using revised thermal functions for Cr_2N [2] and CrN (see below). We have not taken into account, in our analysis, any variations in the compositions of the two nitrides (see Phase Data Section). Results of our analysis are tabulated below.

Source	Data Points	T/K	$\Delta_r H°$, kcal mol^{-1} 2nd Law	3rd Law	Drift cal K^{-1} mol^{-1}	$\Delta_f H°(298.15\ K)^a$ kcal mol^{-1}
Valensi (3)	8	1083-1288	-25.9	24.88±0.37	0.8±1.0	-27.4±1.7
Sano (4)	6	1104-1223	-28.5	26.58±0.16	1.7±0.3	-28.3±1.1
Smith (5)	10^b	1178-1351	-13.0	25.0 ±1.2	-9.5	-27.5±1.6
Mills (6)	Equation	1173-1323	-25.1	25.26±0.02	-0.2	-27.6±1.0

[a] Third law values; auxiliary data used $\Delta_f H°(Cr_2N, cr, 298.15\ K) = -30.0 \pm 3.0\ kcal\ mol^{-1}$.

[b] Data taken from reference (1); one point rejected due to failure of a statistical test.

The equilibrium pressures of Sano (4) are inconsistent with those of the other investigations, and no weight is given to these results. Reasons for these discrepancies have been presented by DeLuca and Leitnaker (1). The drifts are both positive and negative which provides support for our revised thermal functions for the two nitrides. Neumann et al. (7) measured $\Delta_f H°$ directly by combination of the elements in a bomb calorimeter. Two samples of chromium (99.2-99.7% purity) were heated to temperatures near 1200 K in 25 atmospheres of nitrogen. Under these conditions (8), it is unlikely that significant amounts of subnitride (Cr_2N) were formed. Neumann et al. (7) recommended the value $\Delta_f H°(CrN, cr, 298.15\ K) = -29.5 \pm 0.5\ kcal\ mol^{-1}$.

We adopt $\Delta_f H°(CrN, cr, 298.15\ K)$ value of $-28.0 \pm 2.0\ kcal\ mol^{-1}$ ($-117.152 \pm 8.4\ kJ\ mol^{-1}$) which is based mainly on the results from two equilibrium studies (3, 6) and to a lesser extent on the value determined by bomb calorimetry. Our adopted value agrees with that (-28.4±1.5) selected by DeLuca and Leitnaker (1) but is 1.8 kcal mol^{-1} less negative than the NBS (9) value (-29.8).

Heat Capacity and Entropy

One low temperature heat capacity study (10) has been reported for CrN which covered the temperature range 73-315 K. $C_p°$ values taken from the graph at 25° temperature intervals are used to obtain a value for S°(298.15 K) by integration. This method leads to $S°(298.15\ K) = 9.0\ cal\ K^{-1}\ mol^{-1}$ ($37.656\ J\ K^{-1}\ mol^{-1}$) with $S°(73\ K) = 0.39\ cal\ K^{-1}\ mol^{-1}$. The latter value is obtained from the Debye function D(530/T). The uncertainty in our value for S°(298.15 K) is believed to be less than ±0.5 cal $K^{-1}\ mol^{-1}$. Previously published estimates of S°(298.15 K) are 8.56 (1) and 7.85 (11). The low temperature data set (10) contains a lambda peak at 281.7 K with $C_p°$ near the maximum equal to 75.1 cal $K^{-1}\ mol^{-1}$. Other transition temperatures have been reported as 286 K (12) and 287 K (13). The anomaly arises from an antiferromagnetic-paramagnetic transformation (12, 13) which is accompanied by an orthorhombic to cubic structural change (14).

Satoh (15) has measured the heat contents of three chromium nitride samples at three temperatures each (372.6, 598.8, and 784.2 K), in an ice calorimeter. DeLuca and Leitnaker (1) have reevaluated his heat contents and presented $C_p°$ data for CrN in the temperature range 298-1400 K. Their values above 400 K are adopted and joined smoothly with the low temperature $C_p°$ data (10) near 350 K. $C_p°$ data above 1400 K are obtained by graphical extrapolation.

Phase Data

The homogeneity range of the CrN phase is not known; however, results of two recent studies (6, 12) indicate that it is probably much smaller than that for Cr_2N (2). X-ray diffraction patterns (16, 17) for CrN have been interpreted in terms of a cubic structure (NaCl type). Other information on the chromium-chromium nitride system has been reviewed by Storms (8).

Decomposition Data

No information is available on the melting point of CrN. Upon heating we assume that the compound decomposes to its elements rather than melting. T_{dcm} is the temperature at which $\Delta_r G°$ for the process $CrN(cr) = Cr(cr) + 0.5\ N_2(g)$ equals zero.

References

1. J. P. DeLuca and J. M. Leitnaker, J. Amer. Chem. Soc. 56, 126 (1973).
2. JANAF Thermochemical Table: Cr_2N(cr), 12-31-73.
3. G. Valenski, J. Chim. Phys. Physicochim. Biol. 26, 202 (1929).
4. K. Sano, Nippon Kagaku Kaishi 58, 981 (1937).
5. W. H. Smith, Ph.D. Thesis, Rensselaer Polytechnic Institute, Troy, N. Y., 1958.
6. T. Mills, J. Less-Common Metals 22, 373 (1970).
7. B. Neumann, C. Kroger, and H. Haebler, Z. Anorg. Chem. 196, 65 (1931).
8. E. K. Storms, Los Alamos Scientific Laboratory, LA-2942, Contract W-7405-Eng. 36, August, 1964.
9. U. S. Nat. Bur. Stand. Tech. Note 270-4, 1969.
10. J. Chaussy R. Georges, P. Mollard and J. Voeroir, J. Phys. (Paris), Colloq. (1) (Pt. 2), C1, 977 (1971).
11. O. Kubaschewski, E. LL. Evans, and C. B. Alcock, "Metallurgical Thermochemistry," 4th Edition, Pergamon Press, New York, 1967.
12. J. D. Browne, P. R. Liddell, R. Street, and T. Mills, Phys. Stat. Sol. 1, 715 (1970).
13. J. D. Browne, T. Mills, A. J. Dempster, and R. Street, J. Phys. (Paris), Colloq. (1) (Pt. 2), C1, 974 (1971).
14. L. M. Corliss, N. Elliott, and J. M. Hastings, Phys. Rev. 117, 929 (1960).
15. S. Satoh, Sci. Pap. Inst. Phys. Chem. Res. Tokyo 34, 1001 (1938).
16. Z. G. Pinsker and S. V. Kaverin, Kristallografiya 2, 380 (1957).
17. R. Blix, Z. Physik. Chem. 3B, 229 (1929).

Chromium Nitride (CrN) $Cr_1N_1(cr)$

Enthalpy Reference Temperature = T_r = 298.15 K; Standard State Pressure = p° = 0.1 MPa

T/K	$C_p°$ (J $K^{-1}mol^{-1}$)	S°	$-[G°-H°(T_r)]/T$	$H°-H°(T_r)$ (kJ mol^{-1})	$\Delta_f H°$	$\Delta_f G°$	Log K_f
0	0.	0.	INFINITE	-7.705	-116.465	-116.465	INFINITE
100	9.985	3.850	78.059	-7.421	-117.943	-109.909	57.410
200	29.126	16.556	43.947	-5.478	-119.039	-101.378	26.477
298.15	52.677	37.711	37.711	0.	-117.152	-92.790	16.256
300	51.393	38.033	37.712	0.096	-117.126	-92.639	16.130
400	49.078	51.780	39.599	4.873	-116.350	-84.587	11.046
500	49.748	62.804	43.176	9.814	-115.475	-76.749	8.018
600	50.417	71.933	47.229	14.822	-114.678	-69.080	6.014
700	51.087	79.756	51.330	19.898	-113.938	-61.539	4.592
800	51.714	86.618	55.321	25.037	-113.252	-54.101	3.532
900	52.384	92.748	59.145	30.242	-112.630	-46.745	2.713
1000	53.053	98.301	62.787	35.514	-112.094	-39.455	2.061
1100	53.723	103.389	66.250	40.853	-111.670	-32.212	1.530
1200	54.392	108.093	69.544	46.259	-111.369	-25.003	1.088
1300	55.020	112.471	72.679	51.729	-111.211	-17.813	0.716
1400	55.689	116.573	75.669	57.264	-111.204	-10.630	0.397
1500	56.379	120.439	78.526	62.868	-111.350	-3.442	0.120
1600	57.028	124.098	81.261	68.538	-111.655	3.761	-0.123
1700	57.697	127.575	83.884	74.275	-112.123	10.988	-0.338
1800	58.367	130.892	86.404	80.078	-112.755	18.247	-0.530
1900	59.078	134.066	88.830	85.949	-113.550	25.546	-0.702
2000	59.831	137.117	91.168	91.897	-114.504	32.891	-0.859
2100	60.375	140.049	93.427	97.906	-115.632	40.287	-1.002
2200	61.003	142.872	95.610	103.974	-136.273	48.397	-1.149
2300	61.714	145.599	97.725	110.110	-135.887	56.783	-1.290
2400	62.467	148.242	99.775	116.320	-135.433	65.150	-1.418
2500	63.074	150.805	101.765	122.598	-134.915	73.497	-1.536

PREVIOUS: CURRENT: December 1973

Chromium Nitride (CrN) $Cr_1N_1(cr)$

CHROMIUM NITRIDE (CrN) IDEAL GAS M_r = 66.0027

$\Delta_f H°(0\ K) = 504.62 \pm 20.9\ kJ\ mol^{-1}$

$S°(298.15\ K) = 230.553 \pm 8.4\ J\ K^{-1}\ mol^{-1}$ $\Delta_f H°(298.15\ K) = 505.01 \pm 20.9\ kJ\ mol^{-1}$

Electronic Levels and Quantum Weights

State	ε_i, cm^{-1}	g_i
$^4\Sigma$	0	[4]
	[15000]	[8]
	[20000]	[4]

ω_e = [1000] cm^{-1} $\omega_e x_e$ = [5.0] cm^{-1} σ = 1

B_e = [0.56115] cm^{-1} α_e = [0.00375] cm^{-1} r_e = [1.65] Å

Enthalpy of Formation

Srivastava and Farber (1) have observed mass spectrometrically CrN(g) as one of the vapor species in equilibrium with the Cr-V-N system. A substitution reaction between VN and Cr was employed in the formation of gaseous mononitride in the temperature range 1900 to 2063 K. Ion intensities for the equilibrium species CrN, VN, V, and N_2 were measured and converted to partial pressures by means of an intensity-pressure relationship for Au(g). Results of a 2nd and 3rd law analysis of their pressure data (1) are given below.

Reaction	Data Points	T/K	$\Delta_r H°$(298.15 K), kcal mol^{-1} 2nd Law	3rd Law	Drift cal K^{-1} mol^{-1}	$\Delta_f H°$(298.15 K)[a], kcal mol^{-1} 2nd Law	3rd Law
A	7	1903-2063	-35.6	-27.18±0.75	4.3±2.3	130.6	122.2
B	7	1903-2063	21.3	22.42±0.33	0.6±1.3	118.1	119.2

Reactions: (A) CrN(g) = Cr(g) + 0.5 N_2(g) (B) VN(g) + Cr(g) = CrN9g) + V(g)

[a]Auxiliary $\Delta_f H°$ data (2) used (in kcal mol^{-1}): Cr(g), 95.0±1.0; VN(g), 125±5; V(g), 123.2±2.0.

We note that several of the log Kp values listed by Srivastava and Farber (1) for the two gas phase reactions are inconsistent with their reported partial pressure data. Since the 3rd law method is relatively insensitive to errors in the equilibrium constant, we believe the two 3rd law $\Delta_f H°$ values are more reliable than those obtained by the 2nd law. Thus, we adopt the average 3rd law value of $\Delta_f H°$(CrN, g, 298.15 K) = 120.7±5.0 kcal mol^{-1} (505.009±20.9 kJ mol^{-1}). Taking into account estimated errors in the pressure determinations and thermal functions, an overall uncertainty of ±5 kcal mol^{-1} is assigned. The adopted $\Delta_f H°$ value corresponds to a dissociation energy of $\Delta_{at}H°$(CrN, g, 298.15 K) = 87.3 kcal mol^{-1}.

Although no other experimental value for $\Delta_{at}H°$ is available, two estimations have been reported. Gingerich (4), using empirical correlations of Colin and Goldfinger (3) and Pauling (6), predicted a value of $\Delta_{at}H°$(CrN, g, 298.15 K) = 96 kcal mol^{-1} for CrN(g) by both methods. This predicted value is in reasonable agreement with that determined experimentally by Srivastava and Farber (1) and reaffirms the usefulness of three empirical correlations (3-5) in predicting dissociation energies for gaseous diatomic mononitrides.

Heat Capacity and Entropy

The value of r_e(CrN) is obtained from the expression r_e(MO) - r_e(MN) = -0.023 Å with r_e(CrO) = 1.627 Å (2). Differences in the bond lengths (7) for the oxides and nitrides of As, B, P, Si, and Ti are used to establish the value of -0.023. B_e is calculated from the estimated value for r_e. The moment of inertia is calculated to be 4.9881 x 10^{-39} g cm^2. The value of α_e is obtained from the other constants assuming a Morse potential function. The values of ω_e and $\omega_e x_e$ are assumed to lie between those for VO (2) and AsN (7).

CrN is isoelectronic with VO; therefore, the ground state electronic configuration is taken to be the same as that for VO (2). Also, the electronic levels and quantum weights are estimated by analogy with those for VO (2). Futher support for the quartet ground state of CrN is provided by the fact that the atomic levels for Cr^{+3} (8) are also quartets. Our adopted upper levels and quantum weights contribute 0.13 cal K^{-1} mol^{-1} to the entropy of CrN at 4000 K.

References

1. R. D. Srivastava and M. Farber, High Temp. Sci. 5, 489 (1973).
2. JANAF Thermochemical Tables: VN(g), CrN(cr), CrO(g), and VO(g), 12-31-73; Cr(g), 6-30-73; N(g), 3-31-61.
3. R. Colin and P. Goldfinger in "Condensation and Evaporation of Solids," E. Rutner, P. Goldfinger, and J. P. Hirth, Eds., Gordon and Breach Science Publishers, Inc., New York, 1964, pp. 165.
4. K. A. Gingerich, J. Chem. Phys. 49, 19 (1968).
5. K. A. Gingerich, J. Chem. Phys. 49, 14 (1968).
6. L. Pauling, "The Nature of the Chemical Bond," 3rd Ed., Cornell University Press, Ithaca, New York, 1960.
7. B. Rosen, Ed., "Spectroscopic Data Relative to Diatomic Molecules," Pergamon Press, New York, 1970.
8. C. E. Moore, U. S. Nat. Bur. Stand. Cir. 467, Vol. II, 1952.

Chromium Nitride (CrN) $Cr_1N_1(g)$

Enthalpy Reference Temperature = T_r = 298.15 K Standard State Pressure = p° = 0.1 MPa

T/K	$C_p°$ (J $K^{-1}mol^{-1}$)	S°	-[G°-H°(T_r)]/T	H°-H°(T_r) (kJ mol^{-1})	$\Delta_f H°$	$\Delta_f G°$	Log K_f
0	0.	0.	INFINITE	-8.778	504.624	504.624	INFINITE
100	29.107	198.342	257.038	-5.870	505.769	494.354	-258.224
200	29.454	218.573	233.318	-2.949	505.651	482.908	-126.123
250	30.032	225.203	231.055	-1.463	505.349	477.256	-99.717
298.15	30.754	230.553	230.553	0.	505.009	471.876	-82.671
300	30.783	230.743	230.553	0.057	504.995	471.670	-82.125
350	31.576	235.548	230.931	1.616	504.525	466.157	-69.570
400	32.327	239.814	231.779	3.214	504.152	460.702	-60.161
450	33.000	243.661	232.889	4.847	503.772	455.293	-52.849
500	33.586	247.169	234.145	6.512	503.384	449.927	-47.004
600	34.517	253.379	236.846	9.920	502.581	439.311	-38.245
700	35.197	258.754	239.601	13.407	501.732	428.832	-32.000
800	35.698	263.488	242.297	16.953	500.824	418.479	-27.324
900	36.077	267.716	244.890	20.543	499.831	408.245	-23.694
1000	36.369	271.533	247.367	24.166	498.718	398.127	-20.796
1100	36.599	275.010	249.724	27.815	497.453	388.127	-18.431
1200	36.786	278.203	251.966	31.484	496.017	378.251	-16.465
1300	36.939	281.154	254.099	35.171	494.392	368.502	-14.807
1400	37.069	283.896	256.131	38.871	492.564	358.885	-13.390
1500	37.180	286.457	258.068	42.584	490.527	349.406	-12.167
1600	37.277	288.860	259.918	46.307	488.274	340.070	-11.102
1700	37.365	291.123	261.688	50.039	485.802	330.882	-10.167
1800	37.447	293.261	263.383	53.780	483.108	321.846	-9.340
1900	37.525	295.287	265.009	57.528	480.189	312.966	-8.604
2000	37.601	297.214	266.572	61.284	477.044	304.245	-7.946
2100	37.679	299.051	268.075	65.048	473.671	295.687	-7.355
2200	37.760	300.805	269.523	68.820	450.734	287.950	-6.837
2300	37.846	302.486	270.920	72.601	448.764	280.595	-6.373
2400	37.938	304.098	272.269	76.390	446.798	273.325	-5.949
2500	38.039	305.649	273.574	80.188	444.835	266.137	-5.561
2600	38.149	307.143	274.836	83.998	442.878	259.028	-5.204
2700	38.269	308.585	276.059	87.819	440.928	251.994	-4.875
2800	38.400	309.979	277.246	91.652	438.987	245.032	-4.571
2900	38.542	311.329	278.398	95.499	437.055	238.139	-4.289
3000	38.695	312.638	279.518	99.361	96.116	235.926	-4.108
3100	38.860	313.910	280.607	103.238	95.036	240.604	-4.054
3200	39.037	315.146	281.667	107.133	93.919	245.317	-4.004
3300	39.224	316.350	282.700	111.046	92.769	250.066	-3.958
3400	39.422	317.524	283.707	114.978	91.588	254.850	-3.915
3500	39.629	318.670	284.690	118.931	90.378	259.670	-3.875
3600	39.846	319.789	285.649	122.904	89.139	264.524	-3.838
3700	40.071	320.884	286.587	126.900	87.874	269.413	-3.803
3800	40.303	321.956	287.503	130.919	86.580	274.337	-3.771
3900	40.541	323.006	288.400	134.961	85.258	279.295	-3.741
4000	40.785	324.035	289.278	139.027	83.905	284.288	-3.712
4100	41.034	325.045	290.138	143.118	82.521	289.314	-3.686
4200	41.286	326.037	290.981	147.234	81.100	294.375	-3.661
4300	41.541	327.012	291.808	151.376	79.641	299.470	-3.638
4400	41.797	327.969	292.619	155.542	78.138	304.600	-3.616
4500	42.054	328.912	293.415	159.735	76.587	309.764	-3.596
4600	42.310	329.839	294.197	163.953	74.982	314.963	-3.577
4700	42.565	330.751	294.965	168.197	73.318	320.198	-3.559
4800	42.819	331.650	295.720	172.466	71.586	325.469	-3.542
4900	43.069	332.536	296.462	176.761	69.780	330.777	-3.526
5000	43.316	333.408	297.192	181.080	67.892	336.122	-3.511
5100	43.559	334.268	297.911	185.424	66.276	341.479	-3.497
5200	43.797	335.117	298.618	189.792	64.302	346.895	-3.485
5300	44.030	335.953	299.315	194.183	62.244	352.349	-3.473
5400	44.257	336.778	300.001	198.597	60.093	357.842	-3.461
5500	44.478	337.592	300.677	203.034	57.844	363.377	-3.451
5600	44.693	338.396	301.343	207.493	55.492	368.953	-3.441
5700	44.900	339.189	302.000	211.973	53.032	374.572	-3.433
5800	45.101	339.971	302.648	216.473	50.456	380.235	-3.424
5900	45.293	340.744	303.288	220.992	47.758	385.944	-3.417
6000	45.479	341.507	303.918	225.531	44.934	391.700	-3.410

PREVIOUS: December 1973 (1 atm) CURRENT: December 1973 (1 bar)

Chromium Nitride (CrN) $Cr_1N_1(g)$

CHROMIUM OXIDE (CrO) — IDEAL GAS — M_r = 67.9954

$S°(298.15\ K) = 239.268 \pm 6.3\ J\ K^{-1}\ mol^{-1}$

$\Delta_f H°(0\ K) = 187.85 \pm 41.8\ kJ\ mol^{-1}$
$\Delta_f H°(298.15\ K) = 188.28 \pm 41.8\ kJ\ mol^{-1}$

Electronic Levels and Quantum Weights

State	ϵ_i, cm^{-1}	g_i
$X\ ^5\pi$	0	10
$B\ ^5\pi$	16584.5	10

$\omega_e = 898.8\ cm^{-1}$ $\quad \omega_e x_e = 6.50^{-1}$ $\quad \sigma = 1$
$B_e = 0.5286\ cm^{-1}$ $\quad \alpha_e = 0.0050\ cm^{-1}$ $\quad r_e = 1.627$ Å

Enthalpy of Formation

Based on the adopted ground state vibrational constants, a linear Birge-Sponer extrapolation yields a dissociation energy, of 3.796 eV (87.54 kcal mol^{-1}). Applying a correction for the ionicity in the Cr-O bond (1), we recalculate D_0^0 = 4.017 eV (92.63 kcal mol^{-1}). This in turn corresponds to $\Delta_f H°$(298.15 K) = 60.95 kcal mol^{-1} for CrO(g), using auxiliary data (2).

Huldt and Lagerqvist (3), using a spectroscopic flame technique, reported D_0^0 = 4.4±0.5 eV for CrO(g). This value was later corrected to D_0^0 = 5.3 eV (122.9 kcal mol^{-1}) (4, 5). Grimley et al. (6) mentioned that in determining D_0^0 values by the flame technique of Huldt and Lagerqvist (3), the calculation assumes that Cr and CrO are the only gaseous species containing Cr and that the CrO_2(g) partial pressure cannot be neglected under these experimental conditions.

Grimley et al. (6) studied the vaporization of Cr_2O_3(cr) under neutral and oxidizing conditions using mass spectrometric methods. The reported partial pressure data for Cr(g), CrO(g), CrO_2(g), O(g), and O_2(g) may be analyzed in terms of many different chemical equilibria. Our analyses of four equilibria in the neutral and oxidizing conditions are tabulated below.

	Reaction	$\Delta_r H°$(298.15 K) kcal mol^{-1} 2nd Law	3rd Law	Drift cal K^{-1} mol^{-1}	$\Delta_f H°$(298.15 K) kcal mol^{-1} 2nd Law	3rd Law
Neutral conditions	A*	-39.5	-44.9	-2.8±2.1	55.5	50.1
1842-2062 K	B	-108.2	-100.4	4.0±1.4	46.3	54.2
13 points	C	174.0	178.2	2.1±1.2	38.4	42.6
	D*	206.3	205.9	-0.2±1.4	40.9	40.5
Oxidizing conditions	A	-27.2	-43.7	-8.6±13.1	67.8	51.3
1893-1945 K	B	-46.9	-99.3	-27.3±4.0	107.6	55.2
3 points	C	192.2	181.0	-5.8±4.4	56.6	45.4
	D	202.0	208.9	3.6±9.0	36.6	43.5

Reaction A: Cr(g) + 1/2 O_2(g) = CrO(g) $\quad$ C: 1/2 Cr_2O_3(cr) = CrO(g) + 1/4 O_2(g)
B: Cr(g) + O(g) = CrO(g) $\quad$ D: 1/2 Cr_2O_3(cr) = CrO(g) + 1/2 O(g)

*One point rejected due to a statistical test (1842 K).

The drifts are less severe in the netural system so that there is better agreement in the 2nd and 3rd law results for the neutral conditions than in the oxidizing system. Note that the 3rd law $\Delta_r H°$(298.15 K) values are slightly more positive for the oxidizing than the netural conditions for corresponding equilibria. The calculated $\Delta_f H°$(298.15 K) values for the gas phase (reactions A and B) are about 10 kcal mol^{-1} higher than those of the condensed phase (reactions C and D) equilibria. Grimley et al. (6) prefer the gas phase results while Brewer and Rosenblatt (7) prefer the condensed phase results. We adopt $\Delta_f H°$(CrO, cr, 298.15 K) = 45.0±10.0 kcal mol^{-1} (188.280±41.8 kJ mol^{-1}) as a value which is representative of the 2nd and 3rd law results in the neutral system. The adopted $\Delta_f H°$ value leads to D° = 4.71 eV (108.6 kcal mol^{-1}).

Chizhikov et al. (8) also studied the vaporization of Cr_2O_3 via mass spectrometry in the range 1690-2020 K. They reported the detection of gaseous Cr, CrO, CrO_2, and O during the evaporation process. The reported smoothed partial pressures of CrO(g) and CrO_2(g) lie higher than those reported by Grimley et al. (6) by a factor of 2-5, while the partial pressures of Cr(g) are less than a factor 2 different. These data were not analyzed.

Heat Capacity and Entropy

The spectroscopic work involving CrO(g) has been reviewed and referenced by Rosen (9). The adopted vibration and rotation constants as well as the electronic levels are those tabulated by Rosen (9). The ground state has not been firmly established. Weltner (10) suggested a possible $^3\Sigma^-$ ground state. The free energy functions used here are 1.02 cal K^{-1}mol^{-1} at 298 K and 1.57 cal K^{-1} mol^{-1} at 3000 K lower than those suggested by Brewer and Rosenblatt (7). The difference arises in that Brewer and Rosenblatt (7) approximated the CrO electronic levels with Cr^{++} electronic levels. The latter has a singlet ground state and many levels lying below 600 cm^{-1}.

References

1. D. L. Hildenbrand and E. Murad, J. Chem. Phys. 51, 807 (1969).
2. JANAF Thermochemical Tables: Cr(g), 6-30-73; O(g), 6-31-62.
3. L. Huldt and A. Lagerqvist, Arkiv Fysik 3, 525 (1951).
4. L. Brewer, UCRL-8356, 1957.
5. A. Lagerqvist and L. Huldt, Z. Naturforsch, 8a, 493 (1953).
6. R. T. Grimley, R. P. Burns, and M. G. Inghram, J. Chem. Phys. 34, 664 (1961).
7. L. Brewer and G. M. Rosenblatt in "Advances in High Temperature Chemistry," Vol. 2, L. Eyring, Ed., Academic Press, New York, 1967, pp. 1-83.
8. D. M. Chizhikov, Yu. V. Tsvetkov, E. K. Kazenas, and V. K. Tagirov, Russ. J. Phys. Chem. 46, 468 (1972).
9. B. Rosen, Ed., "Spectroscopic Data Relative to Diatomic Molecules," Pergamon Press, New York, 1970.
10. W. Weltner, Jr., Science 155, 155 (1967).

Chromium Oxide (CrO) $\quad Cr_1O_1(g)$

Enthalpy Reference Temperature = T_r = 298.15 K $\quad$ Standard State Pressure = p° = 0.1 MPa

T/K	$C_p°$ (J K^{-1}mol^{-1})	S°	-[G°-H°(T_r)]/T	H°-H°(T_r) (kJ mol^{-1})	$\Delta_f H°$	$\Delta_f G°$	Log K_f
0	0.	0.	INFINITE	-8.828	187.851	187.851	INFINITE
100	29.111	206.841	266.039	-5.920	188.996	177.405	-92.667
200	29.697	227.123	242.076	-2.991	188.885	165.783	-43.298
250	30.467	233.829	239.778	-1.487	188.599	160.039	-33.438
298.15	31.320	239.268	239.268	0.	188.280	154.567	-27.079
300	31.353	239.461	239.268	0.058	188.267	154.357	-26.876
350	32.214	244.360	239.653	1.647	187.818	148.746	-22.199
400	32.982	248.713	240.518	3.278	187.460	143.188	-18.698
450	33.643	252.637	241.650	4.944	187.087	137.677	-15.981
500	34.201	256.211	242.930	6.640	186.697	132.207	-13.812
600	35.062	262.527	245.684	10.106	185.863	121.387	-10.568
700	35.676	267.981	248.489	13.645	184.959	110.711	-8.261
800	36.125	272.776	251.231	17.236	183.983	100.170	-6.540
900	36.463	277.051	253.867	20.866	182.916	89.756	-5.209
1000	36.725	280.907	256.381	24.526	181.729	79.467	-4.151
1100	36.934	284.417	258.773	28.209	180.392	69.305	-3.291
1200	37.106	287.638	261.046	31.911	178.889	59.271	-2.580
1300	37.250	290.614	263.207	35.629	177.201	49.370	-1.984
1400	37.373	293.379	265.265	39.361	175.314	39.606	-1.478
1500	37.481	295.962	267.226	43.103	173.220	29.984	-1.044
1600	37.576	298.384	269.098	46.856	170.914	20.509	-0.670
1700	37.663	300.664	270.889	50.618	168.388	11.185	-0.344
1800	37.742	302.819	272.603	54.389	165.641	2.016	-0.059
1900	37.816	304.862	274.248	58.167	162.667	-6.994	0.192
2000	37.887	306.803	275.828	61.952	159.463	-15.842	0.414
2100	37.955	308.654	277.347	65.744	156.028	-24.523	0.610
2200	38.021	310.421	278.811	69.543	133.023	-32.381	0.769
2300	38.087	312.112	280.222	73.348	130.980	-39.854	0.905
2400	38.153	313.735	281.585	77.160	128.933	-47.237	1.028
2500	38.220	315.294	282.902	80.979	126.881	-54.536	1.139
2600	38.289	316.794	284.177	84.804	124.825	-61.752	1.241
2700	38.361	318.240	285.412	88.637	122.766	-68.889	1.333
2800	38.435	319.637	286.609	92.476	120.704	-75.949	1.417
2900	38.512	320.987	287.772	96.324	118.640	-82.936	1.494
3000	38.593	322.294	288.901	100.179	-222.444	-85.239	1.484
3100	38.677	323.561	289.999	104.042	-223.683	-80.645	1.359
3200	38.765	324.790	291.067	107.914	-224.973	-76.011	1.241
3300	38.856	325.984	292.107	111.795	-226.311	-71.335	1.129
3400	38.952	327.146	293.120	115.686	-227.696	-66.618	1.023
3500	39.050	328.276	294.109	119.586	-229.126	-61.860	0.923
3600	39.153	329.378	295.073	123.496	-230.600	-57.060	0.828
3700	39.258	330.452	296.015	127.416	-232.118	-52.219	0.737
3800	39.367	331.500	296.935	131.348	-233.681	-47.336	0.651
3900	39.478	332.524	297.834	135.290	-235.290	-42.411	0.568
4000	39.592	333.525	298.714	139.243	-236.946	-37.444	0.489
4100	39.709	334.504	299.575	143.208	-238.653	-32.436	0.413
4200	39.827	335.462	300.418	147.185	-240.413	-27.385	0.341
4300	39.947	336.401	301.244	151.174	-242.229	-22.291	0.271
4400	40.069	337.321	302.054	155.175	-244.107	-17.155	0.204
4500	40.191	338.223	302.848	159.188	-246.050	-11.976	0.139
4600	40.315	339.107	303.626	163.213	-248.065	-6.752	0.077
4700	40.439	339.976	304.390	167.251	-250.158	-1.483	0.016
4800	40.563	340.828	305.141	171.301	-252.335	3.830	-0.042
4900	40.687	341.666	305.878	175.363	-254.603	9.191	-0.098
5000	40.812	342.489	306.602	179.438	-256.971	14.598	-0.153
5100	40.935	343.299	307.313	183.526	-259.084	20.027	-0.205
5200	41.059	344.095	308.013	187.625	-261.572	25.523	-0.256
5300	41.181	344.878	308.701	191.737	-264.162	31.070	-0.306
5400	41.302	345.649	309.378	195.861	-266.861	36.665	-0.355
5500	41.422	346.408	310.045	199.998	-269.674	42.311	-0.402
5600	41.541	347.155	310.701	204.146	-272.608	48.010	-0.448
5700	41.658	347.892	311.347	208.306	-275.665	53.762	-0.493
5800	41.773	348.617	311.983	212.477	-278.855	59.570	-0.536
5900	41.887	349.332	312.610	216.660	-282.182	65.432	-0.579
6000	41.999	350.037	313.228	220.855	-285.652	71.353	-0.621

PREVIOUS: December 1973 (1 atm) $\quad$ CURRENT: December 1973 (1 bar)

Chromium Oxide (CrO) $\quad Cr_1O_1(g)$

CHROMIUM OXIDE (CrO_2) — IDEAL GAS — M_r = 83.9948

$S°(298.15\ K) = 269.24 \pm 12.6\ J\ K^{-1}\ mol^{-1}$

$\Delta_f H°(0\ K) = -73.78 \pm 41.8\ kJ\ mol^{-1}$
$\Delta_f H°(298.15\ K) = -75.31 \pm 41.8\ kJ\ mol^{-1}$

Vibrational Frequences and Degeneracies

ν, cm^{-1}
998 (1)
[300](1)
1008 (1)

Ground State Quantum Weight: [3] $\sigma = 2$
Point Group : [C_{2v}]
Bond Distance: Cr-O = [1.627] Å
Bond Angle: O-Cr-O = [110]°
Product of the Moments of Inertia: $I_A I_B I_C = [3.326337 \times 10^{-115}]\ g^3 cm^6$

Enthalpy of Formation

Grimley et al. (1) studied the vaporization of $Cr_2O_3(cr)$ under neutral and oxidizing conditions using mass spectrometric methods. In the range 1842-2062 K (IPTS-68) this study detected quantitatively Cr(g), CrO(g), $CrO_2(g)$, O(g), and $O_2(g)$. Of the many chemical equilibria which could be analyzed, we tabulated below our 2nd and 3rd law analyses of the equilibrium between CrO(g) and $CrO_2(g)$.

	$\Delta_r H°(298.15\ K)$ kcal mol^{-1} 2nd Law	3rd Law	Drift cal $K^{-1}mol^{-1}$	$\Delta_f H°(298.15\ K)$ kcal mol^{-1}
Neutral system, 13 points, 1843-2062 K				
Reaction A**	141.27	119.87	-10.9±3.6	-15.31
Reaction B**	71.11	64.41	-3.4±2.8	-19.41
Oxidizing system, 3 points, 1893-1945 K				
Reaction A	120.97	122.48	0.8±9.2	-17.92
Reaction B	101.21		-17.9±0.1	-21.82

** One point (1865 K) rejected due to a statistical test.

Reactions: (A) $CrO_2(g) = CrO(g) + O(g)$. (B) $CrO_2(g) = CrO(g) + 1/2\ O_2(g)$.

We adopt $\Delta_f H°(298.15\ K) = -18.0 \pm 10.0\ kcal\ mol^{-1}$ ($-75.312 \pm 41.8\ kJ\ mol^{-1}$). This value is intermediate to the four results tabulated above with extra weight given to those of the neutral conditions. Our analysis of equations analogous to those treated in the CrO(g) table (3) yields the same trends as with CrO(g). The results are consistent with our adopted value.

Using auxiliary data (3), we calculate $\Delta_{at}H° = 9.98$ eV (230.1 kcal mol^{-1}) corresponding to $CrO_2(g) = Cr(g) + 2\ O(g)$. This value is a factor of 2.12 greater than D_0^0(CrO,g) (3).

Chizhikov et al. (2) also studied the vaporization of Cr_2O_3 via mass spectrometry in the range 1690-2020 K. They reported the detection of gaseous Cr, CrO, CrO_2, and O during the evaporation process. The reported smoothed partial pressures of CrO(g) and $CrO_2(g)$ lie higher than those of Grimley et al. (1) by a factor of 2-5, while the partial pressures of Cr(g) are less than a factor of two difference. This study was not further analyzed.

Mass spectrometric studies were made by Farber and Srivastava (7) on the reactions involved with vanadium and chromium additives in potassium-seeded H_2/O_2 flames. Equilibrium values at an average flame temperature of 2250 K for the reaction for $CrO_2(g) + 2\ H_2O(g) = H_2CrO_4(g) + 2\ H(g)$ were analyzed by Farber and Srivastava (7). They calculated $\Delta_f H°(298.15\ K) = -3.0 \pm 7$ kcal mol^{-1} for $CrO_2(g)$. The experimental determination of $\Delta_f H°(298.15\ K)$ for $H_2CrO_4(g)$ also required the $\Delta_f H°(298.15\ K)$ value for CrO(g). Adjusting the calculations of Farber and Srivastava (7) to be consistent with the JANAF $\Delta_f H°(298.15\ K)$ value for CrO(g) (3), we recalculate $\Delta_f H°(298.15\ K) = -11.0$ kcal mol^{-1} for $CrO_2(g)$.

Heat Capacity and Entropy

$CrO_2(g)$ is assumed to have a molecular geometry similar to that observed for $TiO_2(g)$, $ZrO_2(g)$, and $TaO_2(g)$ (3). The O-Cr-O bond is estimated as 110° while the Cr-O bond distance is estimated to be the same as that in CrO(g) [3]. The symmetric and asymmetric stretching frequencies are obtained from the infrared spectra of gaseous tetrahedral CrO_2Cl_2 and CrO_2F_2 (4, 5). The bending frequency of 300 cm^{-1} is estimated, based on analogies with $TaO_2(g)$ and $WO_2(g)$ (3, 6). Refer also to the TiO_2 ideal gas table (3). The electronic ground state is assumed to be $^3A_1$ as in $WO_2(g)$ (6). The principal moments of inertia are: $I_A = 2.8646 \times 10^{-39}$, $I_B = 9.4383 \times 10^{-39}$, and $I_C = 12.3029 \times 10^{-39}\ g\ cm^2$.

References

1. R. F. Grimley, R. P. Burns, and M. G. Inghram, J. Chem. Phys. 34, 664 (1961).
2. D. M. Chizhikov, Yu. V. Tsvetkov, E. K. Kazenas, and V. K. Tagirov, Russ. J. Phys. Chem. 46, 468 (1972).
3. JANAF Thermochemical Tables: CrO(g), $TiO_2(g)$, and $TaO_2(g)$, 12-31-73; $ZrO_2(g)$, 12-31-65; O(g), 6-30-62; Cr(g), 6-30-73.
4. J. Hobbs, J. Chem. Phys. 28, 1220 (1958).
5. C. G. Barraclough, J. Lewis, and R. S. Nyholm, J. Chem. Soc., 1959, 3552 (1959).
6. W. Weltner, Jr., and D. McLeod, Jr., J. Mol. Spectrosc. 17, 276 (1965); J. Chem. Phys. 42, 882 (1965).
7. M. Farber and R. D. Srivastava, Combustion and Flame 20, 43 (1973).

Chromium Oxide (CrO_2) — $Cr_1O_2(g)$

Enthalpy Reference Temperature = T_r = 298.15 K — Standard State Pressure = p° = 0.1 MPa

T/K	$C_p°$ (J $K^{-1}mol^{-1}$)	S°	$-[G°-H°(T_r)]/T$	$H°-H°(T_r)$ (kJ mol^{-1})	$\Delta_f H°$	$\Delta_f G°$	Log K_f
0	0.	0.	INFINITE	-11.210	-73.781	-73.781	INFINITE
100	35.384	226.805	305.163	-7.836	-73.623	-78.544	41.027
200	39.615	252.725	273.086	-4.072	-74.354	-83.229	21.737
250	41.517	261.768	269.945	-2.044	-74.844	-85.391	17.842
298.15	43.390	269.241	269.241	0.	-75.312	-87.380	15.309
300	43.461	269.510	269.242	0.080	-75.329	-87.455	15.227
350	45.324	276.351	269.778	2.301	-75.887	-89.426	13.346
400	47.010	282.515	270.991	4.610	-76.313	-91.331	11.927
450	48.482	288.139	272.588	6.998	-76.723	-93.184	10.816
500	49.742	293.315	274.406	9.454	-77.124	-94.991	9.924
600	51.712	302.568	278.347	14.533	-77.924	-98.490	8.574
700	53.121	310.651	282.396	19.778	-78.748	-101.852	7.600
800	54.142	317.814	286.385	25.144	-79.618	-105.095	6.862
900	54.896	324.237	290.240	30.598	-80.564	-108.223	6.281
1000	55.464	330.052	293.935	36.117	-81.623	-111.241	5.811
1100	55.902	335.360	297.463	41.686	-82.829	-114.145	5.420
1200	56.245	340.239	300.827	47.294	-84.201	-116.933	5.090
1300	56.518	344.752	304.034	52.933	-85.760	-119.599	4.806
1400	56.738	348.949	307.095	58.596	-87.521	-122.137	4.557
1500	56.918	352.870	310.017	64.279	-89.495	-124.543	4.337
1600	57.068	356.548	312.811	69.979	-91.689	-126.809	4.140
1700	57.193	360.012	315.487	75.692	-94.109	-128.931	3.962
1800	57.298	363.284	318.052	81.417	-96.760	-130.904	3.799
1900	57.388	366.384	320.515	87.151	-99.647	-132.723	3.649
2000	57.466	369.330	322.883	92.894	-102.774	-134.384	3.510
2100	57.532	372.135	325.162	98.644	-106.144	-135.883	3.380
2200	57.591	374.813	327.358	104.400	-129.096	-136.559	3.242
2300	57.642	377.374	329.478	110.162	-131.098	-136.854	3.108
2400	57.686	379.828	331.525	115.928	-133.117	-137.062	2.983
2500	57.726	382.184	333.504	121.699	-135.155	-137.184	2.866
2600	57.761	384.449	335.421	127.473	-137.209	-137.225	2.757
2700	57.793	386.629	337.277	133.251	-139.282	-137.186	2.654
2800	57.821	388.732	339.077	139.032	-141.372	-137.070	2.557
2900	57.847	390.761	340.825	144.815	-143.479	-136.880	2.465
3000	57.870	392.723	342.522	150.601	-484.621	-132.003	2.298
3100	57.890	394.620	344.172	156.389	-485.933	-120.227	2.026
3200	57.909	396.459	345.778	162.179	-487.312	-108.409	1.770
3300	57.927	398.241	347.341	167.971	-488.755	-96.546	1.528
3400	57.942	399.970	348.863	173.764	-490.260	-84.638	1.300
3500	57.957	401.650	350.348	179.559	-491.827	-72.686	1.085
3600	57.970	403.283	351.795	185.355	-493.455	-60.687	0.881
3700	57.982	404.872	353.209	191.153	-495.143	-48.643	0.687
3800	57.994	406.418	354.589	196.952	-496.893	-36.552	0.502
3900	58.004	407.925	355.937	202.752	-498.705	-24.414	0.327
4000	58.014	409.393	357.255	208.553	-500.582	-12.229	0.160
4100	58.023	410.826	358.544	214.354	-502.526	0.004	-0.000
4200	58.031	412.224	359.806	220.157	-504.540	12.285	-0.153
4300	58.039	413.590	361.041	225.961	-506.629	24.615	-0.299
4400	58.046	414.924	362.250	231.765	-508.796	36.994	-0.439
4500	58.053	416.229	363.435	237.570	-511.046	49.423	-0.574
4600	58.059	417.505	364.597	243.375	-513.386	61.903	-0.703
4700	58.065	418.753	365.736	249.182	-515.821	74.436	-0.827
4800	58.071	419.976	366.853	254.988	-518.359	87.021	-0.947
4900	58.076	421.173	367.950	260.796	-521.007	99.661	-1.062
5000	58.081	422.347	369.026	266.603	-523.773	112.356	-1.174
5100	58.085	423.497	370.083	272.412	-526.302	125.080	-1.281
5200	58.090	424.625	371.121	278.221	-529.225	137.879	-1.385
5300	58.094	425.731	372.141	284.030	-532.269	150.738	-1.486
5400	58.098	426.817	373.143	289.839	-535.442	163.654	-1.583
5500	58.102	427.883	374.129	295.649	-538.748	176.630	-1.677
5600	58.105	428.930	375.098	301.460	-542.195	189.669	-1.769
5700	58.108	429.959	376.052	307.270	-545.786	202.769	-1.858
5800	58.112	430.969	376.990	313.081	-549.530	215.934	-1.945
5900	58.114	431.963	377.913	318.893	-553.433	229.164	-2.029
6000	58.117	432.940	378.822	324.704	-557.500	242.463	-2.111

PREVIOUS: December 1973 (1 atm) — CURRENT: December 1973 (1 bar)

Chromium Oxide (CrO_2) — $Cr_1O_2(g)$

CHROMIUM TRIOXIDE (CrO_3)　　IDEAL GAS　　M_r = 99.9942

$S°(298.15\ K) = [266.2 \pm 16.7]\ J\ K^{-1}\ mol^{-1}$

$\Delta_f H°(0\ K) = -288.2 \pm 41.8\ kJ\ mol^{-1}$
$\Delta_f H°(298.15\ K) = -292.88 \pm 41.8\ kJ\ mol^{-1}$

Vibrational Frequencies and Degeneracies

ν, cm^{-1}
[875](1)
[500](1)
969 (2)
[425](2)

Ground State Quantum Weight: [1]
Point Group: [D_{3h}]　　σ = [6]
Bond Distance: Cr-O = [1.627] Å
Bond Angle: O-Cr-O = [120]°
Product of the Moments of Inertia = $I_A I_B I_C$ = [2.348053 x 10^{-114}] $g^3\ cm^6$

Enthalpy of Formation

The vaporization of Cr_2O_3 under oxidizing conditions was studied by Grimley, Burns, and Inghram (1) using mass spectrometric methods. The principal gaseous species observed were Cr, CrO, CrO_2, CrO_3, O, and O_2. The vapor pressures at three different temperatures (1893, 1914, and 1942 K, IPTS-68) are analyzed for four different reactions by the 2nd and 3rd law methods.

Reaction	$\Delta_r H°$(298.15 K), kcal mol^{-1} 2nd Law	3rd Law	Drift cal K^{-1} mol^{-1}	$\Delta_f H°$(298.15 K) kcal mol^{-1}
Cr(g) + 3 O(g) = CrO_3(g)	-279.80	-332.47	-27.5±6.4	-58.79
Cr(g) + 1.5 O_2(g) = CrO_3(g)	-220.52	-165.48	28.7±21.1	-70.48
0.5 Cr_2O_3(cr) + 1.5 O(g) = CrO_3(g)	-30.81	-24.28	3.4±1.4	-70.04
0.5 Cr_2O_3(cr) + 0.75 O_2(g) = CrO_3(g)	-1.17	59.22	31.5±12.4	-75.88

The 2nd law results are considered less reliable due to the small number of data points and the small temperature range studied (52 K). We adopt $\Delta_f H°$(CrO_3, g, 298.15 K) = -70.0±10 kcal mol^{-1} (-292.880±41.8 kJ mol^{-1}).

Washburn (3), in a mass spectrometric study of the sublimation of CrO_3(cr), stated that the gaseous Cr-containing molecules are in equilibrium among themselves but not with O(g). If the same situation were to hold in the study by Grimley et al. (1), a value intermediate to -70.04 and -75.88 kcal mol^{-1} might be more appropriate. However, the adopted value with its assigned uncertainty is still consistent with this possibility.

Mass spectrometric studies were made by Farber and Srivastava (6) on the reactions involved with vanadium and chromium additives in potassium-seeded H_2/O_2 flames. Equilibrium values at an average flame temperature of 2250 K for the reaction CrO_3(g) + H_2O(g) = H_2CrO_4(g) were analyzed by Farber and Srivastava (6). They calculated $\Delta_f H°$(298.15 K) = -64.6±7 kcal mol^{-1} for CrO_3(g). The experimental determination of $\Delta_f H°$(298.15 K) for H_2CrO_4(g) also required the $\Delta_f H°$(298.15 K) value for CrO(g). Adjusting the calculations of Farber and Srivastava (6) to be consistent with the JANAF $\Delta_f H°$(298.15 K) value for CrO(g) (2), we recalculate $\Delta_f H°$(289.15 K) = -72.6 kcal mol^{-1} for CrO_3(g).

Defining the dissociation by CrO_3(g) = Cr(g) + 3O(g), we calculate $\Delta_{at}H°$ = 14.8±0.4 eV (340.3±10 kcal mol^{-1}). This compares with D_0^0 = 4.71 eV for CrO(g) and $\Delta_{at}H°$ = 9.98 eV for CrO_2(g) for similar dissociations (2).

Heat Capacity and Entropy

The molecular structure is assumed to be planar with D_{3h} symmetry. The three Cr-O bond distances are taken to be the same as in CrO(g) (2). The principal moments of inertia are $I_A = I_B$= 10.5494 x 10^{-39} and I_C = 21.0987 x 10^{-39} g cm^2.

Barraclough et al. (4) observed a vibrational frequency at 969 cm^{-1} in the infrared spectrum of CrO_3(cr) dispersed in a Nujol mull. This frequency was assigned to represent the stretching frequency of the two free Cr-O groups in crystalline CrO_3. We assume this frequency also represents the stretching mode in the vapor phase. This infrared work is supported by the infrared and Raman studies of Mattes (5). The remaining vibrational frequencies are those suggested by Grimley et al. (1). The electronic ground state is assumed to be a singlet.

References

1. R. F. Grimley, R. P. Burns, and M. G. Inghram, J. Chem. Phys. 34, 664 (1961).
2. JANAF Thermochemical Tables: CrO(g) and CrO_2(g), 12-31-73.
3. C. A. Washburn, U. S. Clearinghouse Fed. Sci. Tech. Inform., PB-183562 (1969); Chem. Abstr. 71, 755276 (1969).
4. C. G. Barraclough, J. Lewis and R. S. Nyholm, J. Chem. Soc. 1959, 3552 (1959).
5. R. Mattes, Z. Naturforsch B24, 772 (1969).
6. M. Farber and R. D. Srivastava, Combustion and Flame 20, 43 (1973).

Chromium Oxide (CrO_3)　　Cr_1O_3(g)

Enthalpy Reference Temperature = T_r = 298.15 K　　Standard State Pressure = p° = 0.1 MPa

T/K	$C_p°$ (J $K^{-1}mol^{-1}$)	S°	$-[G°-H°(T_r)]/T$	$H°-H°(T_r)$ (kJ mol^{-1})	$\Delta_f H°$	$\Delta_f G°$	Log K_f
0	0.	0.	INFINITE	-12.379	-288.176	-288.176	INFINITE
100	34.969	218.552	308.811	-9.026	-289.492	-284.922	148.828
200	45.791	245.909	270.989	-5.016	-291.432	-279.595	73.023
250	51.301	256.729	267.075	-2.586	-292.250	-276.538	57.779
298.15	56.025	266.178	266.178	0.	-292.880	-273.452	47.908
300	56.193	266.525	266.179	0.104	-292.901	-273.332	47.591
350	60.366	275.510	266.879	3.021	-293.500	-270.016	40.298
400	63.839	283.805	268.483	6.129	-293.875	-266.634	34.819
450	66.695	291.495	270.618	9.394	-294.165	-263.211	30.553
500	69.037	298.647	273.068	12.790	-294.399	-259.759	27.137
600	72.542	311.564	278.431	19.880	-294.767	-252.795	22.008
700	74.953	322.939	283.993	27.262	-295.082	-245.774	18.340
800	76.658	333.065	289.506	34.847	-295.401	-238.709	15.586
900	77.896	342.169	294.860	42.578	-295.772	-231.601	13.442
1000	78.819	350.426	300.010	50.416	-296.243	-224.447	11.724
1100	79.523	357.973	304.942	58.335	-296.854	-217.239	10.316
1200	80.071	364.917	309.654	66.315	-297.628	-209.968	9.140
1300	80.505	371.344	314.155	74.345	-298.587	-202.626	8.142
1400	80.855	377.323	318.456	82.414	-299.750	-195.202	7.283
1500	81.140	382.912	322.569	90.514	-301.128	-187.688	6.536
1600	81.375	388.156	326.506	98.640	-302.729	-180.074	5.879
1700	81.572	393.095	330.279	106.787	-304.560	-172.353	5.296
1800	81.738	397.763	333.900	114.953	-306.628	-164.518	4.774
1900	81.879	402.186	337.379	123.134	-308.939	-156.561	4.304
2000	82.000	406.389	340.725	131.328	-311.496	-148.476	3.878
2100	82.104	410.392	343.948	139.534	-314.303	-140.257	3.489
2200	82.195	414.214	347.055	147.749	-336.700	-131.242	3.116
2300	82.275	417.869	350.055	155.972	-338.155	-121.871	2.768
2400	82.345	421.373	352.955	164.203	-339.637	-112.437	2.447
2500	82.406	424.735	355.759	172.441	-341.144	-102.939	2.151
2600	82.461	427.968	358.474	180.684	-342.678	-93.381	1.876
2700	82.511	431.081	361.106	188.933	-344.238	-83.762	1.620
2800	82.555	434.083	363.659	197.186	-345.825	-74.086	1.382
2900	82.594	436.981	366.138	205.444	-347.436	-64.353	1.159
3000	82.630	439.781	368.546	213.705	-688.092	-49.951	0.870
3100	82.662	442.491	370.888	221.970	-688.925	-28.665	0.483
3200	82.692	445.116	373.167	230.237	-689.833	-7.353	0.120
3300	82.718	447.661	375.386	238.508	-690.813	13.989	-0.221
3400	82.743	450.131	377.548	246.781	-691.862	35.363	-0.543
3500	82.765	452.530	379.657	255.056	-692.981	56.767	-0.847
3600	82.786	454.862	381.713	263.334	-694.166	78.206	-1.135
3700	82.805	457.130	383.721	271.613	-695.420	99.677	-1.407
3800	82.823	459.339	385.682	279.895	-696.741	121.185	-1.666
3900	82.839	461.490	387.598	288.178	-698.131	142.728	-1.912
4000	82.854	463.588	389.472	296.463	-699.592	164.306	-2.146
4100	82.868	465.634	391.305	304.749	-701.126	185.922	-2.369
4200	82.881	467.631	393.098	313.036	-702.736	207.577	-2.582
4300	82.893	469.581	394.854	321.325	-704.427	229.271	-2.785
4400	82.904	471.487	396.575	329.615	-706.201	251.005	-2.980
4500	82.914	473.350	398.260	337.906	-708.065	272.780	-3.166
4600	82.924	475.173	399.912	346.197	-710.023	294.597	-3.345
4700	82.934	476.956	401.533	354.490	-712.083	316.459	-3.517
4800	82.942	478.702	403.122	362.784	-714.251	338.366	-3.682
4900	82.950	480.413	404.682	371.079	-716.536	360.320	-3.841
5000	82.958	482.088	406.214	379.374	-718.944	382.321	-3.994
5100	82.965	483.731	407.718	387.670	-721.123	404.344	-4.141
5200	82.972	485.342	409.195	395.967	-723.702	426.436	-4.284
5300	82.978	486.923	410.647	404.265	-726.409	448.581	-4.421
5400	82.984	488.474	412.074	412.563	-729.253	470.776	-4.554
5500	82.990	489.997	413.476	420.862	-732.237	493.025	-4.682
5600	82.996	491.492	414.856	429.161	-735.371	515.332	-4.807
5700	83.001	492.961	416.214	437.461	-738.658	537.694	-4.927
5800	83.006	494.405	417.549	445.761	-742.106	560.117	-5.044
5900	83.010	495.824	418.864	454.062	-745.723	582.598	-5.158
6000	83.015	497.219	420.158	462.363	-749.514	605.145	-5.268

PREVIOUS: December 1973 (1 atm)　　CURRENT: December 1973 (1 bar)

Chromium Oxide (CrO_3)　　Cr_1O_3(g)

CHROMIUM NITRIDE (Cr_2N) CRYSTAL M_r = 117.9987

$\Delta_f H°(0\ K)$ = Unknown

$S°(298.15\ K) = [64.85 \pm 8.4]\ J\ K^{-1}\ mol^{-1}$ $\Delta_f H°(298.15\ K) = -125.52 \pm 12.6\ kJ\ mol^{-1}$

Enthalpy of Formation

Mah (1) has measured the enthalpy of combustion of chromium subnitride at 303.15 K. From her data for the process $Cr_2N_{0.969}(cr) + 1.5O_2(g) = Cr_2O_3(cr) + 0.4845\ N_2(g)$, we calculate $\Delta_c H°(298.15\ K) = -241.29 \pm 1.5\ kcal\ mol^{-1}$ at unit fugacities of oxygen and nitrogen. Impurity corrections are based on JANAF enthalpy of formation data (2). When the value for $\Delta_c H°$ is combined with the enthalpy of formation for Cr_2O_3 (2), one obtains $\Delta_f H°(Cr_2N_{0.969}$, cr, 298.15 K) = $-29.9 \pm 3.5\ kcal\ mol^{-1}$. The large uncertainty arises from impurities in the subnitride sample, as well as from the uncertainty ($\pm 2.0\ kcal\ mol^{-1}$) in the $\Delta_f H°$ of Cr_2O_3 (2).

All available equilibrium nitrogen pressures except those of one study (5) have been recently reviewed for the reaction $2\ Cr(cr) + 0.5\ N_2(g) = Cr_2N(cr)$. These results are reanalyzed by the 2nd and 3rd law methods using revised thermal functions for Cr_2N (see below). We have not taken into account in our analysis any variations in the composition of the subnitride.

Source	Data Points	T/K	Drift cal K^{-1} mol^{-1}	$\Delta_f H°$(298.15 K),kcal mol^{-1} 2nd Law	3rd Law
Sano (4)	6	1229-1395	-3.7±0.2	-24.0	-28.8±0.5
Mozgovai and Samarin (5)	Equation	1600-1750	-14.0	-10.2	-33.8±1.7
Seybolt and Oriani (6)	5	1273-1673	-7.8±5.4	-21.3	-32.8±3.9
Tomilin and Savost'yanova (7)	Equation	1517-1724	-32.6	-20.3	-32.6±5.7
Mills (8)	Equation	1337-1653	-2.1	-29.2	-32.3±0.5
Schwerdtfeger (9)	3	1373-1583	-3.7±0.7	-27.1	-32.5±0.8
Smith (10)	2	1333-1370	-13.4	-13.6	-31.7±0.7

We adopt $\Delta_f H°(Cr_2N$, cr, 298.15 K) = $-30.0 \pm 3.0\ kcal\ mol^{-1}$ ($-125.52 \pm 12.6\ kJ\ mol^{-1}$) which is based on the 2nd and 3rd law results from the two equilibrium studies (8, 9) and the $\Delta_f H°$ value determined by combustion calorimetry (1). Our adopted value agrees with that selected by NBS (-30.5) (11) but is 1.8 kcal mol^{-1} less negative than the selected value of DeLuca and Leitnaker (3).

Heat Capacity and Entropy

No low temperature heat capacity data for Cr_2N have been reported. By analogy with entropies (2, 12, and 13) for other transition-metal subnitrides, for the reaction $2\ M(cr) + 0.5\ N_2(g) = M_2N(cr)$ where M=Fe, Mo, Nb, Ta, and V, we calculate an average $\Delta_f S°$ of $-18.7\ cal\ K^{-1}\ mol^{-1}$. This value is combined with entropies for Cr and N_2 (2) to give 15.5 cal K^{-1} mol^{-1} as S°(298.15 K) for Cr_2N(cr). Other reported estimates in cal K^{-1} mol^{-1} for S°(298.15 K) are 14.3 (3) and 17.7±1.0 (14).

The only high temperature enthalpy study reported for Cr_2N is that of Satoh (15), who measured heat contents [H°(T) - H°(273.15 K)] three nitride samples at three temperatures each (372.6, 598.8, and 784.2 K) in an ice calorimeter. DeLuca and Leitnaker (3) have shown that the subnitride sample used by Satoh (15) was substoichiometric, reevaluated his heat contents and presented $C_p°$ data for Cr_2N in the temperature range 298-1800 K. A comparison of their value (17.0 cal K^{-1} mol^{-1}) for $C_p°$(298.15 K) with similar data for other transition-metal subnitrides (2, 12, 13) indicates it is probably too large by 1-2 cal K^{-1} mol^{-1}.

We find a $\Delta_f C_p°$ of 1.1 cal K^{-1} mol^{-1} for several subnitrides (2, 12, 13) which leads to a $C_p°$ value of 15.79 cal K^{-1} mol^{-1} for Cr_2N at 298.15 K. This value is adopted and joined smoothly with the data of Deluca and Leitnaker (3) near 400 K. $C_p°$ data above 1800 K are obtained by graphical extrapolation.

Phase Data

Blix (16) has determined the structure of the subnitride phase as hexagonal by x-ray diffraction methods, and he indicated that the homogeneity limits extend from $CrN_{0.47}$ to $CrN_{0.66}$. Eriksson (17) has reported that the phase exists between $CrN_{0.38}$ and $CrN_{0.50}$. Mills (8), using a thermogravimetric technique, has examined the homogeneity limits of Cr_2N as a function of temperature. The upper limiting composition was determined to be $CrN_{0.49}$ with only a small variation with temperature. The lower limiting composition of Cr_2N varied widely with temperature in the range 1273-1373 K. Mills (8) reported the limiting composition as $CrN_{0.41}$, at 1273 K and $CrN_{0.37}$ at 1373 K. Other information on the chromium-chromium nitride system has been reviewed by Storms (18).

Decomposition Data

No information is available on the melting point of Cr_2N. We assume that the subnitride decomposes prior to melting, according to the reaction $Cr_2N(cr) \rightarrow 2\ Cr(cr) + 0.5\ N_2(g)$. T_{dcm} is the temperature at which $\Delta_r G°$ for the decomposition process equals zero.

References

1. A. D. Mah, U. S. Bur. Mines RI 5529, 1960.
2. JANAF Thermochemical Tables: N_2(g), 9-30-66; Cr(cr), 6-30-73; Cr_2O_3(cr) and $VN_{0.465}$(cr), 12-31-73.
3. J. P. DeLuca and J. M. Leitnaker, J. Amer. Ceram. Soc. 56, 126 (1973).
4. K. Sano, Nippon Kagaku Kaishi 58, 981 (1937).
5. V. S. Mozgovai and A. M. Samarin, Doklady Akad. Nauk. SSSR 74, 729 (1950).
6. A. U. Seybolt and R. A. Oriani, J. Metals 8, AIME Trans. 206, 556 (1956).
7. I. A. Tomilin and N. A. Savost'yanova, Issled. po Zharoproch. Splovam. Akad. Nauk SSSR, Inst. Met. 10, 283 (1963).
8. T. Mills, J. Less-Common Metals 22, 373 (1970).
9. K. Schwerdtfeger, Trans. AIME 239, 1432 (1967).
10. W. H. Smith, Ph.D. Thesis, Rensselaer Polytechnic Institute, Troy, N.Y., 1958.
11. U. S. Nat. Bur. Stand. Tech. Note 270-4, 1969.
12. C. E. Wicks and F. E. Block, U. S. Bur. Mines Bull. 605, 1963.
13. I. Barin and O. Knache, "Thermochemical Properties of Inorganic Substances," Springer-Verlag, Berlin, 1973.
14. O. Kubaschewski, E. LL. Evans, and C. B. Alcock," Metallurgical Thermochemistry," 4th Edition, Pergamon Press, New York, 1967.
15. S. Satoh, Sci. Pap. Inst. Phys. Chem. Res. Toyko 34, 10C1 (1938).
16. R. Blix, Z. Physik. Chem. 3B, 229 (1929).
17. S. Ericksson, Jernkont. Ann. 118, 530 (1934).
18. E. K. Storms, Los Alamos Scientific Laboratory, LA-2942, Contract W-7405-Eng. 36, August, 1964.

Chromium Nitride (Cr_2N) Cr_2N_1(cr)

Enthalpy Reference Temperature = T_r = 298.15 K; Standard State Pressure = p° = 0.1 MPa

T/K	$C_p°$ (J $K^{-1}mol^{-1}$)	S° (J $K^{-1}mol^{-1}$)	$-[G°-H°(T_r)]/T$ (J $K^{-1}mol^{-1}$)	$H°-H°(T_r)$ (kJ mol^{-1})	$\Delta_f H°$ (kJ mol^{-1})	$\Delta_f G°$ (kJ mol^{-1})	Log K_f
0							
100							
200							
298.15	66.065	64.852	64.852	0.	-125.520	-102.208	17.906
300	66.191	65.261	64.853	0.122	-125.511	-102.063	17.771
400	73.429	85.388	67.548	7.136	-125.040	-94.280	12.312
500	76.735	102.139	72.841	14.649	-124.190	-86.688	9.056
600	79.496	116.373	78.939	22.460	-123.309	-79.270	6.901
700	82.257	128.835	85.195	30.548	-122.371	-72.003	5.373
800	84.977	139.996	91.359	38.910	-121.362	-64.876	4.236
900	87.738	150.165	97.336	47.546	-120.304	-57.878	3.359
1000	90.500	159.552	103.094	56.457	-119.244	-50.999	2.664
1100	93.261	168.307	108.629	65.646	-118.236	-44.224	2.100
1200	95.981	176.538	113.949	75.108	-117.311	-37.537	1.634
1300	98.742	184.330	119.066	84.844	-116.500	-30.923	1.243
1400	101.504	191.749	123.994	94.856	-115.828	-24.366	0.909
1500	104.265	198.846	128.749	105.145	-115.305	-17.853	0.622
1600	107.027	205.663	133.345	115.709	-114.942	-11.369	0.371
1700	109.788	212.235	137.793	126.550	-114.746	-4.902	0.151
1800	112.508	218.589	142.106	137.669	-114.722	1.558	-0.045
1900	114.767	224.731	146.294	149.029	-114.912	8.022	-0.221
2000	117.487	230.688	150.366	160.644	-115.306	14.501	-0.379
2100	119.955	236.479	154.329	172.514	-115.907	21.006	-0.522
2200	122.591	242.121	158.192	184.644	-155.386	28.854	-0.685
2300	124.934	247.622	161.961	197.020	-152.693	37.169	-0.844
2400	127.277	252.989	165.643	209.630	-149.771	45.362	-0.987
2500	129.704	258.233	169.242	222.479	-146.617	53.428	-1.116

PREVIOUS: CURRENT: December 1973

Chromium Nitride (Cr_2N) Cr_2N_1(cr)

CHROMIUM OXIDE (Cr_2O_3) CRYSTAL M_r = 151.9902

$\Delta_fH°$(0 K) = -1128.84 ± 8.4 kJ mol^{-1}
S°(298.15 K) = 81.15 ± 1.3 J K^{-1} mol^{-1} $\Delta_fH°$(298.15 K) = -1134.70 ± 8.4 kJ mol^{-1}
T_{fus} = 2603 ± 15 K $\Delta_{fus}H°$ = [129.7] kJ mol^{-1}

Enthalpy of Formation

Two oxygen bomb calorimetric studies on Cr(cr) led to reported values of $\Delta_fH°(Cr_2O_3$, cr, 298.15 K) = -272.6±0.4 kcal mol^{-1} by Mah (1) and -271.0±1.8 kcal mol^{-1} by Golutvin and Chin'k'uei (2). Mah (1) reported that completions of combustion ranged from 95.30 to 98.90% and applied corrections. Of the fourteen runs by Golutvin and Chin'k'uei (2), seven had a completeness of combustion of 91.88 to 98.96%. Their reported $\Delta_fH°$(298.15 K) value is based on an average of the results for these seven runs with no correction for incomplete combustion or impurities. These latter authors (2) also stated that there was no clear dependence between the enthalpy of combustion and completeness of combustion. Early calorimetric studies led to reported values for $\Delta_fH°$ of -267.8 (298 K, 3), -288.9±1.0 (292 K, 4), and -268.9±0.6 kcal mol^{-1} (294 K, 5).

There are many equilibrium studies involving Cr_2O_3(cr). Our analysis of the more recent studies are tabulated below. The free energy of formation equation presented by Richardson and Jeffes (8) is representative of the earlier equilibrium studies (9).

Source	Reaction	T/K	Data Points	$\Delta_rH°$(298.15 K), kcal mol^{-1} 2nd Law	3rd Law	Drift cal K^{-1} mol^{-1}	$\Delta_fH°$(298.15 K), kcal mol^{-1} 2nd Law	3rd Law
(8)	A	873-1773	Eqn.	-270.62	-268.44	1.7	-270.62	-268.44
(9)	B(transpiration)	871-1427	4	-32.89±0.65	-31.61±0.65	1.1±0.5	-272.05	-268.22
(10)	B(static)	1313-1573	5(47)	-32.17±0.75	-32.36±0.19	-0.1±0.5	-269.89	-270.46
(11)	B(transpiration)	1493-1893	5	-32.85±0.94	-32.79±0.30	0.1±0.6	-271.94	-271.76
(12)	A(emf)	1000-1700	Eqn.	-264.60	-268.91	-3.3	-264.60	-268.91
(13)	A(emf)	1173-1523	Eqn.	-269.71	-270.32	-0.5	-269.71	-270.32

Reactions: (A) 2 Cr(cr) + 3/2 O_2(g) = Cr_2O_3(cr). (B) 2/3 Cr(cr) + H_2O(g) = 1/3 Cr_2O_3(cr) + H_2(g).

We adopt $\Delta_fH°(Cr_2O_3$, cr, 298.15 K) = -271.2±2.0 kcal mol^{-1} (1134.701±8.4 kJ mol^{-1}). This value is an average of two combustion studies (1, 2) and three equilibrium studies which have a small third law drift (10, 11, 13).

Kelly et al. (6) investigated experimentally the thermodynamic properties of the chromium carbides. They measured the CO(g) equilibrium pressure in four distinct systems. A combination of our analyses of these four equilibria (7) leads to $\Delta_fH°(Cr_2O_3$, cr, 298.15 K) = -269.9 kcal mol^{-1}.

Heat Capacity and Entropy

The heat capacity of Cr_2O_3(cr) was measured by Anderson [14 (56-336 K)] and by Volger (graph, 100-350 K, 15). Using the combination of Debye and Einstein functions suggested by Anderson (14), we calculate S°(50 K) = 0.386 cal $K^{-1}mol^{-1}$ and H°(50;0) = 14.3 cal mol^{-1}. Both studies (14, 15) indicated a maximum in the $C_p°$ curve at 205 K. The data of Volger (15) lies considerably higher than that of Anderson (14). Early heat capacity data by Russell (137-299 K, 16) is in good agreement with that of Anderson (14). Above 305 K the heat capacity values decrease rapidly (~5 cal $K^{-1}mol^{-1}$) until 325-340 K, at which point the values show little change with temperature. The adopted $C_p°$ values below T = 330 K are based on the study by Anderson (14).

Enthalpy data on Cr_2O_3(cr) were measured by Kelley et al. (298-1774 K, 6) and Laschschenko and Kompanskii (288-1428 K, 17). The adopted heat capacity values are those suggested by Kelley et al. (6), except that the values in the region 350-500 K are shifted slightly so as to join smoothly with the low temperature heat capacity results (14). The derived enthalpy values agree with the enthalpy data of Kelley et al. (6) to within ±0.5%, the maximum deviation of 25 cal or .5% occurring at 479 K.

Phase Data

Brewer (18) summarized the phase data and reported that Cr_2O_3 has a homogeneity range of x = 1.50-1.56 for CrO_x. Cr_2O_3 has the α-corundum structure and is antiferromagnetic with structural distortion occurring below the Curie temperature of 305 K.

Fusion Data

See the liquid table for details.

References

1. A. D. Mah, J. Amer. Chem. Soc. 76, 3363 (1954).
2. Yu. M. Golutvin and L. Chin'k'uei, Russ. J. Phys. Chem. 35, 62 (1961).
3. W. G. Mixter, Amer. J. Sci. 39, 295 (1915).
4. W. A. Roth and G. Becker, Z. Physik. Chem. A145, 461 (1929).
5. W. A. Roth and U. Wolf, Z. Elektrochem. 46, 45 (1940).
6. K. K. Kelley, F. S. Boericke, G. E. Moore, E. H. Huffman, and W. M. Bangert, U. S. Bur. Mines Tech. Paper 662 (1944).
7. JANAF Thermochemical Tables: Cr_3C_2(cr), Cr_7C_3(cr), and $Cr_{23}C_6$(cr), 12-31-73.
8. F. D. Richardson and J. H. E. Jeffes, J. Iron Steel Inst. 160, 261 (1948).
9. J. N. Ramsey, D. Caplan, and A. A. Burr, J. Electrochem. 103, 135 (1966).
10. Y. Jeannin, C. Mannerskantz, and F. D. Richardson, Trans. Met. Soc. AIME 227, 300 (1963).
11. I. A. Novokhatskii and L. M. Lenev, Russ. J. Inorg. Chem. 11, 1078 (1966).
12. K. Suzuki and K. Sambongi, Tetsu To Hagane 58, 1579 (1972).
13. F. N. Mazandarany and R. D. Pehlke, J. Electrochem. Soc., 121, 711-14(1974).
14. C. T. Anderson, J. Amer. Chem. Soc. 59, 488 (1937).
15. J. Volger, Nature 170, 1027 (1952).
16. A. S. Russell, Z. Physik. 13, 59 (1912).
17. P. N. Laschschenko and D. I. Kompanskii, J. Russ. Phys. Chem. Soc. 60, 579 (1928).
18. L. Brewer, Chem. Rev. 52, 1 (1953).

Chromium Oxide (Cr_2O_3) Cr_2O_3(cr)

Enthalpy Reference Temperature = T_r = 298.15 K Standard State Pressure = p° = 0.1 MPa

T/K	$C_p°$ (J $K^{-1}mol^{-1}$)	S°	$-[G°-H°(T_r)]/T$	$H°-H°(T_r)$ (kJ mol^{-1})	$\Delta_fH°$	$\Delta_fG°$	Log K_f
0	0.	0.	INFINITE	-15.282	-1128.844	-1128.844	INFINITE
100	24.255	10.093	155.642	-14.555	-1133.096	-1107.252	578.369
200	75.412	43.065	90.567	-9.501	-1135.576	-1080.195	282.118
298.15	120.366	81.154	81.154	0.	-1134.701	-1053.066	184.493
300	121.922	81.904	81.157	0.224	-1134.645	-1052.560	183.267
305.000	126.775	83.958	81.186	0.845	——— Cp LAMBDA MAXIMUM TRANSITION ———		
305.000	126.775	83.958	81.186	0.845			
400	112.675	113.736	85.534	11.281	-1133.129	-1025.421	133.906
500	117.721	139.503	93.829	22.837	-1131.353	-998.699	104.333
600	120.549	161.232	103.300	34.759	-1129.610	-972.333	84.649
700	122.625	179.977	112.945	46.922	-1127.958	-946.252	70.610
800	124.290	196.463	122.375	59.270	-1126.413	-920.401	60.096
900	125.725	211.187	131.440	71.773	-1125.007	-894.736	51.929
1000	127.009	224.501	140.091	84.410	-1123.795	-869.217	45.403
1100	128.202	236.663	148.325	97.171	-1122.829	-843.808	40.069
1200	129.332	247.867	156.159	110.049	-1122.137	-818.475	35.627
1300	130.411	258.261	163.618	123.036	-1121.754	-793.188	31.871
1400	131.457	267.964	170.729	136.130	-1121.703	-767.918	28.651
1500	132.482	277.069	177.518	149.327	-1121.999	-742.639	25.861
1600	133.486	285.651	184.011	162.625	-1122.654	-717.329	23.418
1700	134.478	293.774	190.230	176.024	-1123.676	-691.967	21.262
1800	135.457	301.486	196.199	189.517	-1125.077	-666.534	19.342
1900	136.880	308.848	201.935	203.135	-1126.833	-641.013	17.623
2000	138.168	315.903	207.459	216.888	-1128.937	-615.391	16.072
2100	139.056	322.665	212.785	230.749	-1131.423	-589.654	14.667
2200	139.947	329.155	217.928	244.699	-1172.985	-562.478	13.355
2300	140.839	335.395	222.900	258.739	-1172.557	-534.738	12.144
2400	141.733	341.408	227.714	272.867	-1172.074	-507.017	11.035
2500	142.628	347.212	232.378	287.085	-1171.534	-479.318	10.015
2600	143.526	352.824	236.903	301.393	-1170.937	-451.640	9.074
2603.000	143.553	352.989	237.037	301.824	------ CRYSTAL <--> LIQUID -----		
2700	144.428	358.257	241.298	315.791	-1170.281	-423.987	8.203
2800	145.333	363.526	245.570	330.279	-1169.566	-396.359	7.394
2900	146.242	368.642	249.726	344.857	-1168.789	-368.759	6.642
3000	147.155	373.615	253.773	359.527	-1845.987	-331.960	5.780

PREVIOUS: CURRENT: December 1973

Chromium Oxide (Cr_2O_3) Cr_2O_3(cr)

CHROMIUM OXIDE (Cr_2O_3) LIQUID M_r = 151.9902

S°(298.15 K) = [125.595] J K^{-1} mol^{-1}

T_{fus} = 2603 ± 15 K

$\Delta_f H$°(298.15 K) = [-1018.381] kJ mol^{-1}

$\Delta_{fus}H$° = [129.7] kJ mol^{-1}

Enthalpy of Formation

$\Delta_f H$°(Cr_2O_3, ℓ, 298.15 K) is calculated from that of the crystal by adding $\Delta_{fus}H$° and the difference in enthalpy, H°(2603 K) - H°(298.15 K), between the crystal and liquid.

Heat Capacity and Entropy

The heat capacity for Cr_2O_3(ℓ) is estimated to be 7.5 cal K^{-1} mol^{-1}. A glass transition is assumed at 1800 K. Below 1800 K, the $C°_p$ values are those of the crystal with the exception that the values below 500 K are a smooth extrapolation to $C°_p$ = 24.32 cal K^{-1} mol^{-1} at 298.15 K.

The entropy is calculated in a manner analogous to that used for the enthalpy of formation.

Fusion Data

The melting point is chosen to be T_{fus} = 2603±15 K, as suggested by Charlesworth (1). This result is based on the work by McNally et al. (2). The heat of melting is estimated by analogy with $\Delta_{fus}S$° = 12.03 for corundum (3).

References

1. J. H. Charlesworth, AFML-TR-70-137, October, 1970.
2. R. N. McNally, F. I. Peters, and P. H. Ribbs, J. Amer. Ceramic Soc. 44, 491 (1961).
3. JANAF Thermochemical Tables: Al_2O_3(cr), 6-30-72.

Chromium Oxide (Cr_2O_3) Cr_2O_3(l)

Enthalpy Reference Temperature = T_r = 298.15 K Standard State Pressure = p° = 0.1 MPa

T/K	$C°_p$ (J K⁻¹mol⁻¹)	S° (J K⁻¹mol⁻¹)	-[G°-H°(T_r)]/T (J K⁻¹mol⁻¹)	H°-H°(T_r) (kJ mol⁻¹)	$\Delta_f H$° (kJ mol⁻¹)	$\Delta_f G$° (kJ mol⁻¹)	Log K_f
0							
100							
200							
298.15	101.755	125.595	125.595	0.	-1018.381	-949.997	166.435
300	102.006	126.225	125.597	0.188	-1018.361	-949.572	165.335
400	112.717	157.188	129.745	10.977	-1017.113	-926.786	121.026
500	117.721	182.956	137.888	22.534	-1015.337	-904.409	94.483
600	120.549	204.685	147.258	34.456	-1013.593	-882.388	76.819
700	122.625	223.430	156.831	46.619	-1011.942	-860.653	64.223
800	124.290	239.916	166.207	58.967	-1010.396	-839.147	54.791
900	125.725	254.640	175.229	71.469	-1008.991	-817.827	47.465
1000	127.009	267.954	183.847	84.107	-1007.778	-796.654	41.613
1100	128.202	280.116	192.054	96.868	-1006.812	-775.590	36.830
1200	129.332	291.319	199.865	109.745	-1006.121	-754.602	32.847
1300	130.411	301.714	207.305	122.733	-1005.737	-733.660	29.479
1400	131.457	311.417	214.398	135.826	-1005.687	-712.735	26.592
1500	132.482	320.522	221.173	149.024	-1005.983	-691.802	24.091
1600	133.486	329.104	227.653	162.322	-1006.637	-670.838	21.901
1700	134.478	337.227	233.862	175.720	-1007.659	-649.821	19.967
1800	135.457	344.941	239.820	189.217	-1009.056	-628.733	18.245
1800.000	135.457	344.941	239.820	189.217	GLASS <--> LIQUID TRANSITION		
1800.000	156.900	344.941	239.820	189.217			
1900	156.900	353.424	245.578	204.907	-1008.740	-607.615	16.705
2000	156.900	361.472	251.173	220.597	-1008.908	-586.501	15.318
2100	156.900	369.127	256.609	236.287	-1009.566	-565.366	14.063
2200	156.900	376.426	261.891	251.977	-1049.387	-542.878	12.890
2300	156.900	383.401	267.024	267.667	-1047.309	-519.902	11.807
2400	156.900	390.078	272.013	283.357	-1045.265	-497.016	10.817
2500	156.900	396.483	276.864	299.047	-1043.253	-474.213	9.908
2600	156.900	402.637	281.584	314.737	-1041.273	-451.491	9.071
2603.000	156.900	402.818	281.724	315.208	------ CRYSTAL <--> LIQUID -----		
2700	156.900	408.558	286.178	330.427	-1039.325	-428.844	8.296
2800	156.900	414.265	290.651	346.117	-1037.407	-406.268	7.579
2900	156.900	419.770	295.009	361.807	-1035.520	-383.762	6.912
3000	156.900	425.090	299.257	377.497	-1711.697	-352.093	6.130
3100	156.900	430.234	303.400	393.187	-1708.210	-306.830	5.170
3200	156.900	435.216	307.441	408.877	-1704.851	-261.679	4.271
3300	156.900	440.044	311.387	424.567	-1701.615	-216.631	3.429
3400	156.900	444.728	315.240	440.257	-1698.498	-171.678	2.638
3500	156.900	449.276	319.005	455.947	-1695.500	-126.816	1.893
3600	156.900	453.696	322.685	471.637	-1692.618	-82.037	1.190
3700	156.900	457.995	326.285	487.327	-1689.852	-37.338	0.527
3800	156.900	462.179	329.806	503.017	-1687.205	7.291	-0.100
3900	156.900	466.254	333.253	518.707	-1684.678	51.850	-0.694
4000	156.900	470.227	336.628	534.397	-1682.276	96.345	-1.258
4100	156.900	474.101	339.933	550.087	-1680.003	140.782	-1.794
4200	156.900	477.882	343.173	565.777	-1677.867	185.165	-2.303
4300	156.900	481.574	346.349	581.467	-1675.874	229.500	-2.788
4400	156.900	485.181	349.463	597.157	-1674.034	273.788	-3.250
4500	156.900	488.707	352.519	612.847	-1672.356	318.037	-3.692

PREVIOUS: CURRENT: December 1973

Chromium Oxide (Cr_2O_3) Cr_2O_3(l)

CHROMIUM OXIDE (Cr_2O_3) CRYSTAL–LIQUID M_r = 151.9902

0 to 2603 K crystal
above 2603 K liquid

Refer to the individual tables for details.

Chromium Oxide (Cr_2O_3) Cr_2O_3(cr,l)

Enthalpy Reference Temperature = T_r = 298.15 K Standard State Pressure = p° = 0.1 MPa

T/K	C_p° (J K⁻¹mol⁻¹)	S° (J K⁻¹mol⁻¹)	−[G°−H°(T_r)]/T (J K⁻¹mol⁻¹)	H°−H°(T_r) (kJ mol⁻¹)	$\Delta_f H^\circ$ (kJ mol⁻¹)	$\Delta_f G^\circ$ (kJ mol⁻¹)	Log K_f
0	0.	0.	INFINITE	-15.282	-1128.844	-1128.844	INFINITE
100	24.255	10.093	155.642	-14.555	-1133.096	-1107.252	578.369
200	75.412	43.065	90.567	-9.501	-1135.576	-1080.195	282.118
298.15	120.366	81.154	81.154	0.	-1134.701	-1053.066	184.493
300	121.922	81.904	81.157	0.224	-1134.645	-1052.560	183.267
305.000	126.775	83.958	81.186	0.845	Cp LAMBDA MAXIMUM TRANSITION		
305.000	126.775	83.958	81.186	0.845			
400	112.675	113.736	85.534	11.281	-1133.129	-1025.421	133.906
500	117.721	139.503	93.829	22.837	-1131.353	-998.699	104.333
600	120.549	161.232	103.300	34.759	-1129.610	-972.333	84.649
700	122.625	179.977	112.945	46.922	-1127.958	-946.252	70.610
800	124.290	196.463	122.375	59.270	-1126.413	-920.401	60.096
900	125.725	211.187	131.440	71.773	-1125.007	-894.736	51.929
1000	127.009	224.501	140.091	84.410	-1123.795	-869.217	45.403
1100	128.202	236.663	148.325	97.171	-1122.829	-843.808	40.069
1200	129.332	247.867	156.159	110.049	-1122.137	-818.475	35.627
1300	130.411	258.261	163.618	123.036	-1121.754	-793.188	31.871
1400	131.457	267.964	170.729	136.130	-1121.703	-767.918	28.651
1500	132.482	277.069	177.518	149.327	-1121.999	-742.639	25.861
1600	133.486	285.651	184.011	162.625	-1122.654	-717.329	23.418
1700	134.478	293.774	190.230	176.024	-1123.676	-691.967	21.262
1800	135.457	301.486	196.199	189.517	-1125.077	-666.534	19.342
1900	136.880	308.848	201.935	203.135	-1126.833	-641.013	17.623
2000	138.168	315.903	207.459	216.888	-1128.937	-615.391	16.072
2100	139.056	322.665	212.785	230.749	-1131.423	-589.654	14.667
2200	139.947	329.155	217.928	244.699	-1172.985	-562.478	13.355
2300	140.839	335.395	222.900	258.739	-1172.557	-534.738	12.144
2400	141.733	341.408	227.714	272.867	-1172.074	-507.017	11.035
2500	142.628	347.212	232.378	287.085	-1171.534	-479.318	10.015
2600	143.526	352.824	236.903	301.393	-1170.937	-451.640	9.074
2603.000	143.553	352.989	237.037	301.824	CRYSTAL <--> LIQUID TRANSITION		
2603.000	156.900	402.818	237.037	431.528			
2700	156.900	408.558	243.097	446.747	-1039.325	-428.844	8.296
2800	156.900	414.265	249.109	462.437	-1037.407	-406.268	7.579
2900	156.900	419.770	254.899	478.127	-1035.520	-383.762	6.912
3000	156.900	425.090	260.484	493.817	-1711.697	-352.093	6.130
3100	156.900	430.234	265.877	509.507	-1708.210	-306.830	5.170
3200	156.900	435.216	271.092	525.197	-1704.851	-261.679	4.271
3300	156.900	440.044	276.139	540.887	-1701.615	-216.631	3.429
3400	156.900	444.728	281.029	556.577	-1698.498	-171.678	2.638
3500	156.900	449.276	285.771	572.267	-1695.500	-126.816	1.893
3600	156.900	453.696	290.374	587.957	-1692.618	-82.037	1.190
3700	156.900	457.995	294.847	603.647	-1689.852	-37.338	0.527
3800	156.900	462.179	299.196	619.337	-1687.205	7.291	-0.100
3900	156.900	466.254	303.427	635.027	-1684.678	51.850	-0.694
4000	156.900	470.227	307.548	650.717	-1682.276	96.344	-1.258
4100	156.900	474.101	311.563	666.407	-1680.003	140.782	-1.794
4200	156.900	477.882	315.478	682.097	-1677.867	185.165	-2.303
4300	156.900	481.574	319.298	697.787	-1675.874	229.500	-2.788
4400	156.900	485.181	323.027	713.477	-1674.034	273.788	-3.250
4500	156.900	488.707	326.670	729.167	-1672.356	318.037	-3.692

PREVIOUS: CURRENT: December 1973

Chromium Oxide (Cr_2O_3) Cr_2O_3(cr,l)

CESIUM (Cs) REFERENCE STATE A_r = 132.9054 Cesium (Cs) Cs_1(ref)

0 to	301.55 K	crystal
301.55 to	947.967 K	liquid
above	947.967 K	ideal monatomic gas

Refer to the individual tables for details.

Enthalpy Reference Temperature = T_r = 298.15 K   Standard State Pressure = p° = 0.1 MPa

	J K⁻¹mol⁻¹			kJ mol⁻¹			
T/K	C_p°	S°	$-[G°-H°(T_r)]/T$	$H°-H°(T_r)$	$\Delta_fH°$	$\Delta_fG°$	Log K_f
0	0.	0.	INFINITE	-7.717	0.	0.	0.
100	25.812	54.952	110.635	-5.568	0.	0.	0.
200	27.804	73.453	87.892	-2.888	0.	0.	0.
298.15	32.195	85.147	85.147	0.	0.	0.	0.
300	32.355	85.347	85.148	0.060	0.	0.	0.
301.550	32.501	85.514	85.149	0.110	CRYSTAL <--> LIQUID		
301.550	32.393	92.436	85.149	2.197	TRANSITION		
400	31.518	101.458	88.110	5.339	0.	0.	0.
500	31.154	108.448	91.506	8.471	0.	0.	0.
600	30.999	114.113	94.817	11.578	0.	0.	0.
700	30.941	118.887	97.923	14.674	0.	0.	0.
800	30.932	123.017	100.808	17.768	0.	0.	0.
900	30.957	126.662	103.482	20.862	0.	0.	0.
947.967	30.976	128.270	104.696	22.348	LIQUID <--> IDEAL GAS		
947.967	20.786	199.643	104.696	90.007	FUGACITY = 1 bar		
1000	20.786	200.754	109.665	91.089	0.	0.	0.
1100	20.788	202.735	118.037	93.167	0.	0.	0.
1200	20.791	204.544	125.172	95.246	0.	0.	0.
1300	20.799	206.208	131.342	97.326	0.	0.	0.
1400	20.814	207.750	136.746	99.406	0.	0.	0.
1500	20.841	209.187	141.528	101.489	0.	0.	0.
1600	20.886	210.533	145.799	103.575	0.	0.	0.
1700	20.952	211.802	149.645	105.667	0.	0.	0.
1800	21.048	213.002	153.131	107.767	0.	0.	0.
1900	21.178	214.143	156.313	109.878	0.	0.	0.
2000	21.349	215.233	159.232	112.004	0.	0.	0.
2100	21.564	216.280	161.924	114.149	0.	0.	0.
2200	21.830	217.289	164.417	116.318	0.	0.	0.
2300	22.150	218.266	166.737	118.517	0.	0.	0.
2400	22.527	219.217	168.904	120.750	0.	0.	0.
2500	22.960	220.145	170.935	123.023	0.	0.	0.
2600	23.456	221.055	172.846	125.344	0.	0.	0.
2700	24.016	221.950	174.648	127.717	0.	0.	0.
2800	24.639	222.835	176.353	130.149	0.	0.	0.
2900	25.326	223.711	177.971	132.646	0.	0.	0.
3000	26.082	224.582	179.510	135.216	0.	0.	0.
3100	26.905	225.451	180.978	137.864	0.	0.	0.
3200	27.796	226.319	182.381	140.599	0.	0.	0.
3300	28.599	227.175	183.725	143.384	0.	0.	0.
3400	29.572	228.043	185.016	146.293	0.	0.	0.
3500	30.600	228.915	186.258	149.301	0.	0.	0.
3600	31.680	229.792	187.455	152.414	0.	0.	0.
3700	32.770	230.671	188.610	155.625	0.	0.	0.
3800	33.902	231.557	189.729	158.949	0.	0.	0.
3900	35.036	232.447	190.812	162.375	0.	0.	0.
4000	36.250	233.349	191.864	165.939	0.	0.	0.
4100	37.488	234.259	192.887	169.626	0.	0.	0.
4200	38.744	235.178	193.883	173.438	0.	0.	0.
4300	40.010	236.104	194.854	177.375	0.	0.	0.
4400	40.717	236.965	195.795	181.147	0.	0.	0.
4500	40.901	237.756	196.706	184.728	0.	0.	0.
4600	41.940	238.667	197.608	188.871	0.	0.	0.
4700	42.950	239.580	198.491	193.115	0.	0.	0.
4800	43.924	240.494	199.357	197.459	0.	0.	0.
4900	44.859	241.409	200.206	201.899	0.	0.	0.
5000	45.748	242.325	201.039	206.430	0.	0.	0.
5100	46.588	243.239	201.857	211.047	0.	0.	0.
5200	46.722	244.034	202.647	215.214	0.	0.	0.
5300	47.407	244.931	203.436	219.921	0.	0.	0.
5400	47.729	245.763	204.205	224.415	0.	0.	0.
5500	48.117	246.611	204.964	229.060	0.	0.	0.
5600	48.599	247.483	205.716	233.896	0.	0.	0.
5700	49.022	248.347	206.456	238.778	0.	0.	0.
5800	49.385	249.203	207.186	243.699	0.	0.	0.
5900	49.689	250.049	207.905	248.653	0.	0.	0.
6000	49.935	250.887	208.614	253.634	0.	0.	0.

PREVIOUS: June 1968 (1 atm)   CURRENT: June 1968 (1 bar)

Cesium (Cs) Cs_1(ref)

CESIUM (Cs) CRYSTAL A_r = 132.9054

$S°(298.15\ K) = 85.147 \pm 0.21\ J\ K^{-1}\ mol^{-1}$

$T_{fus} = 301.55 \pm 0.01\ K$

$\Delta_f H°(0\ K) = 0\ kJ\ mol^{-1}$

$\Delta_f H°(298.15\ K) = 0\ kJ\ mol^{-1}$

$\Delta_{fus} H°(923\ K) = 2.087 \pm 0.004\ kJ\ mol^{-1}$

Enthalpy of Formation

Zero by definition.

Heat Capacity and Entropy

The low temperature heat capacities, 0.19-301.55 K, have been measured by Dauphinee (1), McCollum (2), Martin (3), Lien (4), Filby (5), and Morgan (6). The adopted values are derived based on the Cp data reported by Filby (5) and Lien (4). Lien measured the heat capacities in the temperature range 0.1874 to 1.20K in the adiabatic demagnetization cryostat and at higher temperatures, 1.2 to 4.09°K, in the liquid-helium temperature cryostat. The Cs sample was obtained from a commercial source of 99.8 per cent purity. Filby determined the Cp values in the temperature ranges 0.4-1.5 K, 3-26 K and 20-320 K, using commercial samples of 99.9 per cent purity. These two sets of measurements are joined smoothly at 4 K. S°(298.15 K) is derived from the adopted heat capacities, based on S°(0.20 K)= 0.0002 cal K^{-1} mol^{-1}. Dauphinee (1) determined the heat capacities of Cs(cr) employing a sample badly contaminated with oxygen. In the temperature range of approximately 100 to 200 K, an anomaly of Cp curve somewhat similar to those found for sodium and potassium was reported. This anomaly was not observed by Filby who used the higher purity cesium sample. The Cp values reported by Dauphines are lower than the adopted ones by about 0.5 to 1.5 per cent in the temperatures 55 to 220 K; above 270 K his reported heat capacities are unreasonably high. Therefore the data are not adopted for evaluation. McCollum (2) calculated the Debye θ values from the measured low temperature heat capacities and plotted them, since no numerical data were reported these were not used. The low temperature Cp values determined by Martin (3) and Morgan (6) are in good agreement with the adopted ones.

Fusion Data

See the liquid table for details.

Sublimation Data

$\Delta_{sub}H°(298.15\ K)$ is calculated as the difference between $\Delta_f H°(298.15\ K)$ for Cs(g) and Cs(cr).

References

1. T. M. Dauphines, D. L. Martin and H. Preston-Thomas, Proc. Roy. Soc. (London) A233, 214 (1955), 20-320K.
2. D. C. McCollum, Jr. and H. B. Silsbee, Phys. Rev. 127, 119 (1962), 1.3-12K.
3. B. D. Martin, D. A. Zych and C. V. Heer, Phys. Rev. 135, A671 (1964), 0.36-1.63K.
4. W. H. Lien and N. E. Phillips, Phys. Rev. 133, A1370 (1964), 0.19-4.09K.
5. J. D. Filby and D. L. Martin, Proc. Roy. Soc. (London) A284, 83 (1965), 0.7-320K.
6. J. A. Morgan and J. C. Thompson, J. Chem. Phys. 47, 4607 (1967), 2.1-4.2K.

Cesium (Cs) $Cs_1(cr)$

Enthalpy Reference Temperature = T_r = 298.15 K Standard State Pressure = p° = 0.1 MPa

T/K	$C_p°$	S°	$-[G°-H°(T_r)]/T$	$H°-H°(T_r)$	$\Delta_f H°$	$\Delta_f G°$	Log K_f
	J $K^{-1}mol^{-1}$			kJ mol^{-1}			
0	0.	0.	INFINITE	-7.717	0.	0.	0.
100	25.812	54.952	110.635	-5.568	0.	0.	0.
200	27.804	73.453	87.892	-2.888	0.	0.	0.
250	29.201	79.784	85.656	-1.468	0.	0.	0.
298.15	32.195	85.147	85.147	0.	0.	0.	0.
300	32.355	85.347	85.148	0.060	0.	0.	0.
301.550	32.501	85.514	85.149	0.110	------ CRYSTAL	<--> LIQUID	-----
350	34.840	90.550	85.553	1.749	-2.005	0.331	-0.049
400	36.066	95.290	86.479	3.524	-1.815	0.653	-0.085
450	36.777	99.582	87.700	5.347	-1.563	0.946	-0.110
500	37.196	103.480	89.086	7.197	-1.274	1.210	-0.126
600	37.530	110.295	92.069	10.935	-0.643	1.648	-0.144
700	37.614	116.087	95.097	14.693	0.018	1.978	-0.148
800	37.635	121.111	98.042	18.455	0.687	2.212	-0.144
900	37.656	125.545	100.856	22.220	1.358	2.363	-0.137
1000	37.656	129.513	103.527	25.986	-65.103	6.138	-0.321

PREVIOUS: CURRENT: June 1968

Cesium (Cs) $Cs_1(cr)$

CESIUM (Cs) LIQUID A_r = 132.9054

$S°(298.15\ K) = [92.069]\ J\ K^{-1}\ mol^{-1}$ $\Delta_f H°(298.15\ K) = [2.087]\ kJ\ mol^{-1}$

$T_{fus} = 301.55 \pm 0.01\ K$ $\Delta_{fus}H°(301.55\ K) = 2.087 \pm 0.004\ kJ\ mol^{-1}$

Enthalpy of Formation

$\Delta_f H°$(Cs, ℓ, 298.15 K) is calculated from that of the crystal by adding the enthalpy of fusion and the difference in enthalpy, H°(301.55 K) - H°(298.15 K), between the crystal and liquid.

Heat Capacity and Entropy

The enthalpies of Cs(ℓ) have been measured by Tepper et al. (1), Achener (2), and Lemmon et al. (3), using drop calorimetry. Lemmon et al. used a cesium sample of 99.994 per cent purity and encapsulated in Nb-1Zr alloy. They found that Cs(ℓ) does not behave as a normal liquid from about 100-300°C. The thermal conductivity measurements of Cs(ℓ) show a change of slope at about 200°C and its electrical resistivity also shows scatter in this region. The heat capacities, 301.55 - 1435.2 K, derived from the reported enthalpies (corrected for vapor condensation), are consistent with the low temperature heat capacity data on Cs(cr) at 298.15 K and are adopted.

Tepper et al. measured the enthalpies in the temperature range 351.1 - 1238 K. The reported enthalpy values are considerably higher, over 20 per cent, at the lower end of the measured temperature range than the adopted ones. However, the agreement becomes better at the higher temperature end. Achener determined the enthalpies of Cs(ℓ) at the temperatures 340.1 - 1175.7 K. Using the reported enthalpies, the average heat capacity is derived to be 9.75 kcal mol^{-1}, which is unreasonably high by comparison with those for the other alkali metals. The above two sets of data are not used for evaluation. The enthalpies of Cs(ℓ), 773 - 1423 K, were also measured by Shapiro (4 and quoted by 3). The derived constant heat capacity, 7.60 cal mol^{-1}, is in reasonable agreement with the adopted ones.

Heat capacities of Cs(ℓ), 301.55-373.2 K, were reported by Filby (5), 301.55-320 K, Dauphinee (6), 310-320 K, and Rengade (7), 301.7-373.2 K. The values obtained are consistent with the adopted ones.

S°(298.15 K) is calculated in a manner analogous to that used for the enthalpy of formation.

Fusion Data

The melting point and enthalpy of melting have been determined by many investigators. The results reported are presented in the table below. The adopted T_{fus} and $\Delta_{fus}H°$ are obtained from (5). The two determinations of fraction melted against temperature are in very good agreement with one another and lead to a linear plot of reciprocal fraction melted against temperature thus indicating that the major impurity is solid insoluble. This is confirmed by a graph of excess heat capacity against the reciprocal of $(T_0-T)^2$, where T_0 is the melting point, which indicates an impurity (mainly oxygen) concentration of 0.06 atomic per cent.

T_{fus}/K	$\Delta_{fus}H°$, kcal mol^{-1}	References
299.52	0.496	M. Eckardt and E. Graefe, Z. Anorg. Chem. 23, 378 (1900)
301.35	-	A. Guntz and W. Broniewski, J. Chim. Phys. 7, 474 (1909).
301.60±0.05	0.500	E. Rengade (7)
301.75	-	J. H. Boer, J. Broos and H. Emmens, Z. Anorg. Chem. 191, 118 (1930).
301.45	-	E. Rinck, Compt. Rend. 199, 1217 (1934).
301.95±0.1	-	L. Losana, Gazz. Chim. Ital. 65, 855 (1935).
301.75	-	J. B. Taylor and J. Langmuir, Phys. Rev. 51, 759 (1937).
301.79±0.17	-	K. Clusius and H. Stern, Z. Angew. Physik 6, 194 (1954).
-	0.510	W. D. Weatherford, Jr., et al., WADD Tech. Rept. 61-96, November 1961.
301.67	0.470	A. W. Lemmon et al. (3).
301.55±0.01	0.499±0.01	J. D. Filby and D. L. Martin (5).

Vaporization Data

T_{vap} is the temperature at which the Gibbs energy change for the reaction Cs(ℓ) = Cs(g) is zero. The differences between $\Delta_f H°$(Cs, g) and $\Delta_f H°$(Cs, ℓ) at T_{vap} is $\Delta_{vap}H°$. Due to the presence of dimer (Cs_2) in the vapor, the boiling point (1 atm) of Cs(ℓ) is calculated to be 941.3 K at which the vapor mixture contains 0.835 mol of monomer/mol of Cs(ℓ) vaporized, or 9.0 per cent of dimer. The corresponding enthalpy of vaporization is evaluated as 15.26 kcal mol^{-1} of vapor mixture.

References
1. F. Tepper, A. Murchison, J. Zelenak, and F. Roehlich, "Thermophysical Properties of Rubidium and Cesium," RTD-TDR-63-4018, Part I, MSA Research Corporation, Callery, Pennsylvania, November 1963.
2. P. Y. Achener, "The Determination of the Latent Heat of Vaporization, Vapor Pressure, Enthalpy, Specific Heat, and Density of Liquid Rubidium and Cesium up to 1800 F, "AGN-8090, Aerojet-General Nucleonics, January 1964.
3. A. W. Lemmon, Jr., H. W. Deem, E. A. Eldridge, E. H. Hall, J. Matolich, Jr., and J. F. Walling, "The Specific Heat, Thermal Conductivity, and Viscosity of Liquid Cesium," BATT-4673-T7, Battelle Memorial Institute, February 1964.
4. A. Shapiro and C. J. Meisl, General electric Company, Flight Propulsion laboratory Department Report No. R60FPD358-A, Nov. 9, 1960.
5. J. D. Filby and D. L. Martin, Proc. Roy. Soc. (London) A284, 83 (1965).
6. T. M. Dauphinee, D. L. martin, and H. Preston-Thomas, Proc. Roy. Soc. (London) A233, 214 (1955).
7. E. Rengade, Compt. Rend. 156, 1897 (1913); Bull. Soc. Chim. France 15, 130 (1914).

Cesium (Cs) $Cs_1(l)$

Enthalpy Reference Temperature = T_r = 298.15 K — Standard State Pressure = p° = 0.1 MPa

	J $K^{-1}mol^{-1}$			kJ mol^{-1}			
T/K	$C_p°$	S°	$-[G°-H°(T_r)]/T$	$H°-H°(T_r)$	$\Delta_f H°$	$\Delta_f G°$	Log K_f
0							
100							
200							
250							
298.15	32.439	92.069	92.069	0.	2.087	0.024	-0.004
300	32.413	92.269	92.069	0.060	2.087	0.011	-0.002
301.550	32.393	92.436	92.071	0.110	------ CRYSTAL <--> LIQUID -----		
350	31.898	97.225	92.462	1.667	0.	0.	0.
400	31.518	101.458	93.328	3.252	0.	0.	0.
450	31.310	105.158	94.441	4.822	0.	0.	0.
500	31.154	108.448	95.680	6.384	0.	0.	0.
600	30.999	114.113	98.295	9.491	0.	0.	0.
700	30.941	118.887	100.905	12.587	0.	0.	0.
800	30.932	123.017	103.416	15.681	0.	0.	0.
900	30.957	126.662	105.801	18.775	0.	0.	0.
947.967	30.976	128.270	106.897	20.260	------- FUGACITY = 1 bar -------		
1000	30.999	129.926	108.053	21.873	-67.129	3.699	-0.193
1100	31.054	132.883	110.178	24.975	-66.105	10.733	-0.510
1200	31.116	135.587	112.184	28.084	-65.075	17.673	-0.769
1300	31.188	138.081	114.082	31.199	-64.040	24.526	-0.985
1400	31.263	140.395	115.879	34.321	-62.998	31.300	-1.168
1500	31.342	142.554	117.587	37.452	-61.950	37.999	-1.323

PREVIOUS: CURRENT: June 1968

Cesium (Cs) $Cs_1(l)$

CESIUM (Cs) CRYSTAL-LIQUID A_r = 132.9054

0 to 301.55 K crystal
above 301.55 K liquid

Refer to the individual tables for details.

Cesium (Cs) $Cs_1(cr,l)$

Enthalpy Reference Temperature = T_r = 298.15 K Standard State Pressure = $p°$ = 0.1 MPa

	J K⁻¹mol⁻¹			kJ mol⁻¹			
T/K	C_p°	S°	$-[G°-H°(T_r)]/T$	$H°-H°(T_r)$	$\Delta_f H°$	$\Delta_f G°$	Log K_f
0	0.	0.	INFINITE	-7.717	0.	0.	0.
100	25.812	54.952	110.635	-5.568	0.	0.	0.
200	27.804	73.453	87.892	-2.888	0.	0.	0.
250	29.201	79.784	85.656	-1.468	0.	0.	0.
298.15	32.195	85.147	85.147	0.	0.	0.	0.
300	32.355	85.347	85.148	0.060	0.	0.	0.
301.550	32.501	85.514	85.149	0.110	CRYSTAL <--> LIQUID		
301.550	32.393	92.436	85.149	2.197	TRANSITION		
350	31.898	97.225	86.498	3.754	0.	0.	0.
400	31.518	101.458	88.110	5.339	0.	0.	0.
450	31.310	105.158	89.803	6.910	0.	0.	0.
500	31.154	108.448	91.506	8.471	0.	0.	0.
600	30.999	114.113	94.817	11.578	0.	0.	0.
700	30.941	118.887	97.923	14.674	0.	0.	0.
800	30.932	123.017	100.808	17.768	0.	0.	0.
900	30.957	126.662	103.482	20.862	0.	0.	0.
947.967	30.976	128.270	104.696	22.348	------- FUGACITY = 1 bar -------		
1000	30.999	129.926	105.966	23.960	-67.129	3.699	-0.193
1100	31.054	132.883	108.280	27.062	-66.105	10.733	-0.510
1200	31.116	135.587	110.445	30.171	-65.075	17.673	-0.769
1300	31.188	138.081	112.476	33.286	-64.040	24.526	-0.985
1400	31.263	140.395	114.389	36.409	-62.998	31.300	-1.168
1500	31.342	142.554	116.195	39.539	-61.950	37.999	-1.323

PREVIOUS: CURRENT: June 1968

Cesium (Cs) $Cs_1(cr,l)$

CESIUM (Cs) IDEAL GAS A_r = 132.9054

IP(Cs, g) = 31406.432 ± 0.010 cm^{-1}

S°(298.15 K) = 175.599 ± 0.025 J K^{-1} mol^{-1}

$\Delta_f H°$(298.15 K) = 78.0 ± 1.0 kJ mol^{-1}

$\Delta_f H°$(298.15 K) = 76.5 ± 1.0 kJ mol^{-1}

Electronic Levels and Quantum Weights

State	ϵ_i, cm^{-1}	g_i
$^2S_{1/2}$	0.0	2
$^2P_{1/2}$	11178.24	2
$^2P_{3/2}$	11732.35	4
$^2S_{1/2}$	18535.51	2
$^2P_{3/2}$	31383.05	4
IP	31406.432	

Enthalpy of Formation

The enthalpy of formation of cesium gas is chosen to be the value recommended by CODATA (1). This value was obtained from 2nd and 3rd law treatments of vapor pressure data. The reported vapor pressure were corrected for the effects of Cs(g), using D_0^0= 38.02 kJ mol^{-1}. At higher pressures an alternate virial treatment was employed. As stated by CODATA (1), the selected value is the weighted average of the 3rd law values of $\Delta_{sub}H°$(298.15 K) from the following:

Source	$\Delta_{sub}H°$(298.15 K), kJ mol^{-1}
Taylor and Langmuir (2)	76.71 ± 0.64
Bonilla et al. (3)	76.20 ± 1.28
Achener (4)	76.66 ± 1.08
Tepper et al. (5)	78.08 ± 1.05
Bohdansky and Schins (6)	76.10 ± 0.95
Stone et al. (7)	76.52 ± 1.08
Buck and Pauly (8)	76.77 ± 0.75
Shpil'rain and Belova (9)	76.62 ± 1.06
Volyak et al. (10)	76.48 ± 0.77
Novikov and Rosupkin (11)	76.57 ± 1.19
Ewing et al. (12)	76.44 ± 1.48
Shpil'rain et al. (13)	76.50 ± 1.12
Schins et al. (14)	76.34 ± 0.90
Shpil'rain and Nikanorov (15)	76.46 ± 1.21
Behrens et al. (16)	77.22 ± 1.15

The 2nd law calculations from the above sources led to values somewhat more positive (1)

Heat Capacity and Entropy

The thermal functions for the five alkali metal monatomic gases are calculated by the same procedure. Observed and estimated atomic energy levels are included in the partition function calculation, using an ionization potential lowering (IP-kT) technique as the cutoff procedure in the energy level summation (19).

The lowest lying levels for these metals [14904 cm^{-1} (Li), 16956 cm^{-1} (Na), 12985 cm^{-1} (K), 12578 cm^{-1} (Cs)] do not contribute to the thermal functions below ~1000 K; there is only a translational contribution below this temperature. Above this approximate temperature, the thermal functions become increasingly sensitive to the partition function cutoff procedure used, due to the combined effect of the observation of atomic energy levels of high principal quantum number and a low ionization potential. Energy levels have been observed up to n = 42 (Li), 59(Na), 79(K), 77(Rb), and 73(Cs). However, not all predicted levels have been observed for each of these principal quantum numbers. The ionization potentials vary from 43487.29 cm^{-1} for Li to 31406.1 cm^{-1} for Cs. In calculating the thermal functions with the inclusion of missing levels up to the high principal quantum numbers just mentioned, the Gibbs energy functions show significant differences (depending on the cutoff procedure) above 3000 K (17).

The atomic energy levels have been taken from the compilation of Moore (18, 19). Our calculated values for the thermal functions are similar to those recommended by CODATA (1). They do differ for two reasons, however. First, the entropy differs by 0.1094 J K^{-1} mol^{-1} because this table uses a reference pressure of 1 bar, whereas the CODATA recommendations are based on 1 atm. Second, the entropies at 298.15 K for all alkali metal gases differs by ~0.004 J K^{-1} mol^{-1}, presumably due to the use of slightly different values for auxiliary data.

References

1. J. D. Cox, Chairman, CODATA Task Group on Key Values for Thermodynamics, CODATA Special Report No. 8, 1980.
2. J. B. Taylor and I. Langmuir, Phys. Rev. 51, 753 (1937).
3. C. F. Bonilla, D. L. Sawhney and M. M. Makansi, Trans. Am. Soc. Met. Metals, 55, 877 (1962).
4. P. Y. Achener, U. S. AEC Rept. AGN-8090, 1964.
5. F. Tepper, A. Murchison, et al., Air Force Materials Lab Rept. RTD-TDR-63-4018, Part 1, 1963.
6. J. Bohdansky and H. E. J. Schins, J. Appl.Phys. 36, 3683 (1965).
7. J. P. Stone, C. P. Ewing, et al. J. Chem. Eng. Data 11, 315 (1966).
8. U. Buck and H. Pauly, Z. physik. Chem. (Frankfurt) 44, 345 (1965).
9. E. E. Sphil'rain and A. M. Belova, Teplofiz. Vys. Temp. 5, 477 (1967).
10. L. D. Volyak, Yu. K. Vonogradov, and v. M. Anisimov, Teplofiz. Vys. Temp. 6, 521 (1968).
11. I. I. Novikov and V. V. Rosupkin, Tr. Vsesoyznoi NauchnoOTekh. Konf. po Termodinam. sekts. "Teplofiz, Sv-va veshchestv", 87, 1969.
12. C. T. Ewing, J. R. Spann, et al., J. Chem. Eng. Data 15, 508 (1970).
13. E. E. Shpil'rain, E. E. Toskii and Yu. v. Karmyshin, Trudy Mosk. Energ. Inst., No. 75, 62 (1970).
14. H. E. J. Schins, R. M. W. van Wihjk and B. Dorpema, z. Metallkunde 62, 330 (1971).
15. E. E. Shpil'rain and E. v. Nikanorov, Teplofiz. Svoistva Gasdov Mater. Vses. Teplofiz. Konf. Svoistva veshchestvo Vyz. Temp. (1968), Mos. "Nauka", 171, 1970; Teplofiz, Vys. Temp. 10, 297 (1972).
16. R. G. Behrens, N. O. Woodrow and S. Aronson, J. Chem. Thermodynamics 9, 1035 (1977).
17. J. R. Downey, Jr. The Dow Chemical Compnay, Rept. AFORS-TR-78-0960,Contract No. F44620-1-0048-, 1978.
18. C. E. Moore, U. S. Nat. Bur. Stnad. NRSDS-NBS 35, 1971.
19. C. E. Moore, U. S. Nat. Bur. Stand. NRSDS-NBS 35, Vol. III, 1971. [Reprint of NBS Circular 497, vol. III, 1958].

Cesium (Cs) Cs_1(g)

Enthalpy Reference Temperature = T_r = 298.15 K — Standard State Pressure = p° = 0.1 MPa

T/K	$C_p°$ (J $K^{-1}mol^{-1}$)	S° (J $K^{-1}mol^{-1}$)	$-[G°-H°(T_r)]/T$ (J $K^{-1}mol^{-1}$)	$H°-H°(T_r)$ (kJ mol^{-1})	$\Delta_f H°$ (kJ mol^{-1})	$\Delta_f G°$ (kJ mol^{-1})	Log K_f
0	0.	0.	INFINITE	-6.197	78.019	78.019	INFINITE
100	20.786	152.892	194.080	-4.119	77.950	68.156	-35.601
200	20.786	167.300	177.501	-2.040	77.348	58.578	-15.299
250	20.786	171.938	175.942	-1.001	76.967	53.929	-11.268
298.15	20.786	175.599	175.599	0.	76.500	49.532	-8.678
300	20.786	175.728	175.600	0.038	76.479	49.364	-8.595
350	20.786	178.932	175.853	1.078	73.823	45.226	-6.750
400	20.786	181.708	176.415	2.117	73.278	41.178	-5.377
450	20.786	184.156	177.142	3.156	72.747	37.198	-4.318
500	20.786	186.346	177.955	4.196	72.225	33.276	-3.476
600	20.786	190.136	179.679	6.274	71.196	25.583	-2.227
700	20.786	193.340	181.407	8.353	70.178	18.061	-1.348
800	20.786	196.115	183.076	10.431	69.164	10.685	-0.698
900	20.786	198.564	184.664	12.510	68.148	3.436	-0.199
947.967	20.786	199.643	185.395	13.507	------- FUGACITY = 1 bar -------		
1000	20.786	200.754	186.165	14.589	0.	0.	0.
1100	20.788	202.735	187.583	16.667	0.	0.	0.
1200	20.791	204.544	188.922	18.746	0.	0.	0.
1300	20.799	206.208	190.189	20.826	0.	0.	0.
1400	20.814	207.750	191.389	22.906	0.	0.	0.
1500	20.841	209.187	192.528	24.989	0.	0.	0.
1600	20.886	210.533	193.611	27.075	0.	0.	0.
1700	20.952	211.802	194.645	29.167	0.	0.	0.
1800	21.048	213.002	195.631	31.267	0.	0.	0.
1900	21.178	214.143	196.576	33.378	0.	0.	0.
2000	21.349	215.233	197.482	35.504	0.	0.	0.
2100	21.564	216.280	198.352	37.649	0.	0.	0.
2200	21.830	217.289	199.190	39.818	0.	0.	0.
2300	22.150	218.266	199.998	42.017	0.	0.	0.
2400	22.527	219.217	200.779	44.250	0.	0.	0.
2500	22.960	220.145	201.535	46.523	0.	0.	0.
2600	23.456	221.055	202.269	48.844	0.	0.	0.
2700	24.016	221.950	202.981	51.217	0.	0.	0.
2800	24.639	222.835	203.674	53.649	0.	0.	0.
2900	25.326	223.711	204.350	56.146	0.	0.	0.
3000	26.082	224.582	205.010	58.716	0.	0.	0.
3100	26.905	225.451	205.656	61.365	0.	0.	0.
3200	27.796	226.319	206.288	64.099	0.	0.	0.
3300	28.599	227.175	206.907	66.885	0.	0.	0.
3400	29.572	228.043	207.516	69.793	0.	0.	0.
3500	30.600	228.915	208.115	72.801	0.	0.	0.
3600	31.680	229.792	208.705	75.914	0.	0.	0.
3700	32.770	230.671	209.286	79.125	0.	0.	0.
3800	33.902	231.557	209.860	82.449	0.	0.	0.
3900	35.036	232.447	210.428	85.875	0.	0.	0.
4000	36.250	233.349	210.989	89.439	0.	0.	0.
4100	37.488	234.259	211.546	93.126	0.	0.	0.
4200	38.744	235.178	212.097	96.938	0.	0.	0.
4300	40.010	236.104	212.645	100.875	0.	0.	0.
4400	40.717	236.965	213.181	104.647	0.	0.	0.
4500	40.901	237.756	213.706	108.229	0.	0.	0.
4600	41.940	238.667	214.238	112.371	0.	0.	0.
4700	42.950	239.580	214.768	116.616	0.	0.	0.
4800	43.924	240.494	215.294	120.960	0.	0.	0.
4900	44.859	241.409	215.818	125.399	0.	0.	0.
5000	45.748	242.325	216.339	129.930	0.	0.	0.
5100	46.588	243.239	216.857	134.547	0.	0.	0.
5200	46.722	244.034	217.358	138.714	0.	0.	0.
5300	47.407	244.931	217.870	143.421	0.	0.	0.
5400	47.729	245.763	218.372	147.915	0.	0.	0.
5500	48.117	246.611	218.873	152.560	0.	0.	0.
5600	48.599	247.483	219.376	157.396	0.	0.	0.
5700	49.022	248.347	219.877	162.278	0.	0.	0.
5800	49.385	249.203	220.375	167.199	0.	0.	0.
5900	49.689	250.049	220.871	172.153	0.	0.	0.
6000	49.935	250.887	221.364	177.135	0.	0.	0.

PREVIOUS: June 1968 (1 atm) CURRENT: December 1983 (1 bar)

Cesium (Cs) Cs_1(g)

CESIUM, ION (Cs^+) IDEAL GAS Mr = 132.90485

$IP(Cs^+, g) = 202263 \pm 2\ cm^{-1}$

$S°(298.15\ K) = 169.836 \pm 0.005\ J\ K^{-1}\ mol^{-1}$

$\Delta_f H°(0\ K) = 453.72 \pm 2.1\ kJ\ mol^{-1}$

$\Delta_f H°(298.15\ K) = [458.402]\ kJ\ mol^{-1}$

Electronic Level and Quantum Weight

State	ε_i, cm^{-1}	g_i
$^1S_0$	0	1

Enthalpy of Formation

$\Delta_f H°(Cs^+, g, 0\ K)$ is calculated from $\Delta_f H°(Cs, g, 0\ K)$ (1) using the spectroscopic value of IP(Cs) = 31406.432 ± 0.010 cm^{-1} (375.704 ± 0.001 kJ mol^{-1}) from Moore (2). The ionization limit is converted from cm^{-1} to kJ mol^{-1} using the factor, 1 cm^{-1} = 0.01196266 kJ cm^{-1}, which is derived from the 1973 CODATA fundamental constants (3). Rosenstock et al. (4) and Levin and Lias (5) have summarized additional ionization and appearance potential data.

$\Delta_f H°(Cs^+, g, 298.15\ K)$ is calculated from $\Delta_f H°(Cs, g, 0\ K)$ by using IP(Cs) with JANAF (1) enthalpies, H°(0 K) − H°(298.15 K), for Cs(g), $Cs^+(g)$, and $e^-(g)$. $\Delta_f H°(Cs \rightarrow Cs^+ + e^-, 298.15\ K)$ differs from a room temperature threshold energy due to inclusion of these enthalpies and to threshold effects discussed by Rosenstock et al. (4). $\Delta_f H°(298.15\ K)$ should be changed by −6.197 kJ mol^{-1} if it is to be used in the ion convention that excludes the enthalpy of the electron.

Heat Capacity and Entropy

The information on electronic energy levels and quantum weights, given by Moore (6), is incomplete because many theoretically predicted levels have not been observed. Our calculations indicate that any reasonable method of filling in these missing levels and cutting off the summation in the partition function (7) has no effect on the thermodynamic functions to 6000 K. This is a result of the high energy of all levels other than the ground state; the first excited state is approximately 107392 cm^{-1} above the ground state. Since inclusion of these excited states has no effect on the thermodynamic functions (to 6000 K), we list only the ground state. The reported uncertainty in S°(298.15 K) is due to uncertainties in the relative ionic mass and the fundamental constants. Extension of these calculations above 6000 K may require consideration of the excited states and use of different fill and cutoff procedures (7).

References

1. JANAF Thermochemical Tables: Cs(g), 6-30-68; $e^-(g)$, 3-31-82.
2. C. E. Moore, U. S. Nat. Bur. Stand., NSRDS-NBS-34, 1970.
3. E. R. Cohen and B. N. Taylor, J. Phys. Chem. Ref. Data 2, 663 (1973).
4. H. M. Rosenstock, K. Draxl et al., J. Phys. Chem. Ref. Data 6, Supp. 1 (1977).
5. R. D. Levin and S. G. Lias, U. S. Nat. Bur. Stand., NSRDS-NBS-71, 1982.
6. C. E. Moore, U. S. Nat. Bur. Stand., NSRDS-NBS-35, Volume III, 1970 [Reprint of NBS Circular 467, Volume III, 1958].
7. J. R. Downey, Jr., The Dow Chemical Company, AFOSR-TR-78-0960, Contract No. F44620-75-1-0048, 1978.

Cesium, Ion (Cs^+) $Cs_1^+(g)$

Enthalpy Reference Temperature = T_r = 298.15 K Standard State Pressure = p° = 0.1 MPa

T/K	$C_p°$ (J K⁻¹mol⁻¹)	S° (J K⁻¹mol⁻¹)	$-[G°-H°(T_r)]/T$ (J K⁻¹mol⁻¹)	$H°-H°(T_r)$ (kJ mol⁻¹)	$\Delta_f H°$ (kJ mol⁻¹)	$\Delta_f G°$ (kJ mol⁻¹)	Log K_f
0	0.	0.	INFINITE	-6.197	453.724		
100	20.786	147.129	188.317	-4.119			
200	20.786	161.537	171.738	-2.040			
250	20.786	166.175	170.178	-1.001			
298.15	20.786	169.836	169.836	0.	458.402	426.897	-74.791
300	20.786	169.965	169.837	0.038	458.419	426.701	-74.295
350	20.786	173.169	170.090	1.078	456.803	421.714	-62.937
400	20.786	175.945	170.652	2.117	457.297	416.667	-54.411
450	20.786	178.393	171.379	3.156	457.805	411.558	-47.772
500	20.786	180.583	172.192	4.196	458.322	406.392	-42.456
600	20.786	184.373	173.915	6.274	459.372	395.908	-34.467
700	20.786	187.577	175.644	8.353	460.433	385.247	-28.747
800	20.786	190.352	177.313	10.431	461.497	374.433	-24.448
900	20.786	192.801	178.900	12.510	462.560	363.486	-21.096
1000	20.786	194.991	180.402	14.589	396.491	356.121	-18.602
1100	20.786	196.972	181.820	16.667	398.569	351.983	-16.714
1200	20.786	198.780	183.159	18.746	400.647	347.656	-15.133
1300	20.786	200.444	184.425	20.824	402.725	343.156	-13.788
1400	20.786	201.985	185.625	22.903	404.802	338.496	-12.629
1500	20.786	203.419	186.764	24.982	406.876	333.687	-11.620
1600	20.786	204.760	187.847	27.060	408.947	328.741	-10.732
1700	20.786	206.020	188.880	29.139	411.013	323.664	-9.945
1800	20.786	207.208	189.865	31.217	413.070	318.467	-9.242
1900	20.786	208.332	190.808	33.296	415.116	313.155	-8.609
2000	20.786	209.398	191.711	35.375	417.148	307.736	-8.037
2100	20.786	210.413	192.578	37.453	419.160	302.216	-7.517
2200	20.786	211.379	193.410	39.532	421.148	296.601	-7.042
2300	20.786	212.303	194.212	41.610	423.106	290.896	-6.606
2400	20.786	213.188	194.984	43.689	425.030	285.106	-6.205
2500	20.786	214.037	195.730	45.768	426.914	279.237	-5.834
2600	20.786	214.852	196.449	47.846	428.751	273.293	-5.491
2700	20.786	215.636	197.146	49.925	430.535	267.280	-5.171
2800	20.786	216.392	197.820	52.004	432.260	261.202	-4.873
2900	20.786	217.122	198.473	54.082	433.920	255.063	-4.594
3000	20.786	217.826	199.106	56.161	435.508	248.868	-4.333
3100	20.786	218.508	199.721	58.239	437.016	242.622	-4.088
3200	20.786	219.168	200.319	60.318	438.439	236.328	-3.858
3300	20.786	219.807	200.899	62.397	439.810	229.988	-3.640
3400	20.786	220.428	201.465	64.475	441.060	223.611	-3.435
3500	20.786	221.031	202.015	66.554	442.209	217.198	-3.242
3600	20.786	221.616	202.552	68.632	443.252	210.755	-3.058
3700	20.786	222.186	203.075	70.711	444.198	204.282	-2.884
3800	20.786	222.740	203.585	72.790	445.032	197.785	-2.719
3900	20.786	223.280	204.083	74.868	445.763	191.267	-2.562
4000	20.786	223.806	204.569	76.947	446.356	184.734	-2.412
4100	20.786	224.319	205.045	79.025	446.826	178.187	-2.270
4200	20.786	224.820	205.510	81.104	447.172	171.631	-2.135
4300	20.786	225.309	205.965	83.183	447.392	165.067	-2.005
4400	20.786	225.787	206.410	85.261	447.777	158.467	-1.881
4500	20.786	226.254	206.846	87.340	448.353	151.826	-1.762
4600	20.786	226.711	207.272	89.418	448.368	145.236	-1.649
4700	20.786	227.158	207.691	91.497	448.280	138.647	-1.541
4800	20.786	227.596	208.101	93.576	448.093	132.061	-1.437
4900	20.786	228.024	208.503	95.654	447.811	125.480	-1.338
5000	20.786	228.444	208.898	97.733	447.438	118.905	-1.242
5100	20.786	228.856	209.285	99.811	446.978	112.339	-1.151
5200	20.786	229.260	209.665	101.890	446.968	105.705	-1.062
5300	20.786	229.656	210.039	103.969	446.418	99.147	-0.977
5400	20.786	230.044	210.406	106.047	446.081	92.558	-0.895
5500	20.786	230.426	210.766	108.126	445.593	85.992	-0.817
5600	20.786	230.800	211.121	110.204	444.914	79.460	-0.741
5700	20.786	231.168	211.469	112.283	444.190	72.940	-0.668
5800	20.786	231.529	211.812	114.362	443.426	66.433	-0.598
5900	20.786	231.885	212.149	116.440	442.629	59.940	-0.531
6000	20.786	232.234	212.481	118.519	441.805	53.461	-0.465

PREVIOUS: December 1970 (1 atm) CURRENT: December 1983 (1 bar)

Cesium, Ion (Cs^+) $Cs_1^+(g)$

CESIUM, ION (Cs^-) IDEAL GAS M_r = 132.90595

EA(Cs, g) = 0.471630 ± 0.000025 eV

S°(298.15 K) = 169.836 ± 0.005 J K^{-1} mol^{-1}

$\Delta_f H°$(0 K) = 32.51 ± 0.5 kJ mol^{-1}

$\Delta_f H°$(298.15 K) = [24.798] kJ mol^{-1}

Electonic Level and Quantum Weight

State	ε_i, cm^{-1}	g_i
$^1S_0$	0.0	1

Enthalpy of Formation

The enthalpy of formation, at 0 K, for Cs^-(g) is calculated from the adopted electron affinity, EA(Cs, g) = 0.471630 ± 0.000025 eV, (45.505 ± 0.002 kJ mol^{-1}). This value, recommended by Hotop and Lineberger (7), was measured by a tunable laser photodetachment threshold technique (4, 5). Additional discussion on Cs^-(g) may be obtained in the critical discussions of Hotop and Lineberger (1, 7), Rosenstock et al. (2), and Massey (3).

$\Delta_f H°$(Cs^-, g, 298.15 K) is obtained from $\Delta_f H°$(Cs, g, 0 K) by using EA(Cs, g) with JANAF (6) enthalpies, H°(0 K)−H°(298.15 K), for Cs^-(g), Cs(g), and e^-(g). $\Delta_r H°$(Cs^- → Cs + e^-, 298.15 K) differs from a room-temperature threshold energy due to inclusion of these enthalpies and to threshold effects discussed by Rosenstock et al. (2). $\Delta_f H°$(298.15 K) should be changed by +6.197 kJ mol^{-1} if it is to be used in the ion convention which excludes the enthalpy of the electron.

Heat Capacity and Entropy

The ground state configuration for Cs^-(g) is given by Hotop and Lineberger (1, 7) and Rosenstock et al. (2). Lacking any experimental evidence as to the stability of any excited states, we assume that no stable excited states exist.

References

1. H. Hotop and W. C. Lineberger, J. Phys. Chem. Ref. Data 4, 539 (1975).
2. H. M. Rosenstock, K. Draxl et al., J. Phys. Chem. Ref. Data 6, Supp. 1 (1977).
3. H. S. W. Massey, "Negative Ions," 3rd ed., Cambridge University Press, Cambridge, 1976.
4. J. Slater, F. H. Read, S. E. Novick, and W. C. Lineberger, Phys. Rev. A II, 201 (1978).
5. P. A. Schulz, R. D. Mead, and W. C. Lineberger, Phys. Rev. A, in preparation.
6. JANAF Thermochemical Tables: Cs(g), 3-31-82; e^-(g), 3-31-82.
7. H. Hotop and W. C. Lineberger, J. Phys. Chem. Ref. Data, to be published, 1985.

Cesium, Ion (Cs^-) Cs_1^-(g)

Enthalpy Reference Temperature = T_r = 298.15 K   Standard State Pressure = p° = 0.1 MPa

T/K	$C_p°$ (J $K^{-1}mol^{-1}$)	S°	−[G°−H°(T_r)]/T	H°−H°(T_r) (kJ mol^{-1})	$\Delta_f H°$	$\Delta_f G°$	Log K_f
0	0.	0.	INFINITE	-6.197	32.514		
100	20.786	147.129	188.317	-4.119			
200	20.786	161.537	171.738	-2.040			
250	20.786	166.175	170.179	-1.001			
298.15	20.786	169.836	169.836	0.	24.798	5.802	-1.017
300	20.786	169.965	169.837	0.038	24.738	5.685	-0.990
350	20.786	173.169	170.090	1.078	21.043	2.972	-0.444
400	20.786	175.945	170.652	2.117	19.458	0.499	-0.065
450	20.786	178.393	171.379	3.156	17.888	-1.777	0.206
500	20.786	180.583	172.192	4.196	16.327	-3.878	0.405
600	20.786	184.373	173.916	6.274	13.220	-7.627	0.664
700	20.786	187.577	175.644	8.353	10.123	-10.857	0.810
800	20.786	190.352	177.313	10.431	7.030	-13.642	0.891
900	20.786	192.801	178.901	12.510	3.935	-16.041	0.931
1000	20.786	194.991	180.402	14.589	-66.291	-14.395	0.752
1100	20.786	196.972	181.820	16.667	-68.370	-9.105	0.432
1200	20.786	198.780	183.159	18.746	-70.449	-3.625	0.158
1300	20.786	200.444	184.425	20.824	-72.528	2.028	-0.081
1400	20.786	201.985	185.625	22.903	-74.609	7.841	-0.293
1500	20.786	203.419	186.764	24.982	-76.691	13.803	-0.481
1600	20.786	204.760	187.848	27.060	-78.777	19.904	-0.650
1700	20.786	206.020	188.880	29.139	-80.869	26.135	-0.803
1800	20.786	207.208	189.865	31.217	-82.969	32.490	-0.943
1900	20.786	208.332	190.808	33.296	-85.080	38.962	-1.071
2000	20.786	209.398	191.711	35.375	-87.206	45.546	-1.190
2100	20.786	210.413	192.578	37.453	-89.351	52.236	-1.299
2200	20.786	211.380	193.411	39.532	-91.520	59.029	-1.402
2300	20.786	212.304	194.212	41.610	-93.719	65.921	-1.497
2400	20.786	213.188	194.984	43.689	-95.952	72.910	-1.587
2500	20.786	214.037	195.730	45.768	-98.226	79.992	-1.671
2600	20.786	214.852	196.450	47.846	-100.546	87.167	-1.751
2700	20.786	215.636	197.146	49.925	-102.919	94.431	-1.827
2800	20.786	216.392	197.820	52.004	-105.351	101.785	-1.899
2900	20.786	217.122	198.473	54.082	-107.848	109.226	-1.967
3000	20.786	217.826	199.106	56.161	-110.418	116.755	-2.033
3100	20.786	218.508	199.721	58.239	-113.067	124.371	-2.096
3200	20.786	219.168	200.319	60.318	-115.801	132.074	-2.156
3300	20.786	219.808	200.900	62.397	-118.587	139.860	-2.214
3400	20.786	220.428	201.465	64.475	-121.495	147.735	-2.270
3500	20.786	221.031	202.015	66.554	-124.503	155.697	-2.324
3600	20.786	221.616	202.552	68.632	-127.617	163.747	-2.376
3700	20.786	222.186	203.075	70.711	-130.828	171.883	-2.427
3800	20.786	222.740	203.585	72.790	-134.151	180.108	-2.476
3900	20.786	223.280	204.083	74.868	-137.578	188.420	-2.524
4000	20.786	223.806	204.570	76.947	-141.142	196.824	-2.570
4100	20.786	224.319	205.045	79.025	-144.828	205.319	-2.616
4200	20.786	224.820	205.510	81.104	-148.640	213.905	-2.660
4300	20.786	225.309	205.965	83.183	-152.578	222.583	-2.704
4400	20.786	225.787	206.410	85.261	-156.349	231.321	-2.746
4500	20.786	226.254	206.846	87.340	-159.931	240.113	-2.787
4600	20.786	226.711	207.273	89.418	-164.073	249.048	-2.828
4700	20.786	227.158	207.691	91.497	-168.318	258.075	-2.868
4800	20.786	227.596	208.101	93.576	-172.662	267.193	-2.908
4900	20.786	228.025	208.503	95.654	-177.101	276.402	-2.946
5000	20.786	228.445	208.898	97.733	-181.632	285.703	-2.985
5100	20.786	228.856	209.285	99.811	-186.249	295.096	-3.022
5200	20.786	229.260	209.666	101.890	-190.416	304.502	-3.059
5300	20.786	229.656	210.039	103.969	-195.123	314.064	-3.095
5400	20.786	230.044	210.406	106.047	-199.618	323.673	-3.131
5500	20.786	230.426	210.766	108.126	-204.262	333.383	-3.166
5600	20.786	230.800	211.121	110.204	-209.099	343.202	-3.201
5700	20.786	231.168	211.469	112.283	-213.980	353.107	-3.236
5800	20.786	231.530	211.812	114.362	-218.901	363.099	-3.270
5900	20.786	231.885	212.149	116.440	-223.855	373.176	-3.304
6000	20.786	232.234	212.481	118.519	-228.837	383.337	-3.337

PREVIOUS: CURRENT: December 1983 (1 bar)

Cesium, Ion (Cs^-) Cs_1^-(g)

CESIUM FLUORIDE (CsF) CRYSTAL M_r = 151.903803 Cesium Fluoride (CsF) $Cs_1F_1(cr)$

$S°(298.15\ K) = [88.3 \pm 8]\ J\ K^{-1}\ mol^{-1}$

$T_{fus} = 976\ K$

$\Delta_fH°(0\ K)$ = Unknown

$\Delta_fH°(298.15\ K) = -554.7 \pm 2\ kJ\ mol^{-1}$

$\Delta_{fus}H° = 21.7\ kJ\ mol^{-1}$

Enthalpy of Formation

The enthalpy of solution ($\Delta_{sol}H°$) of CsF(cr) in water at 298.15 K has been carefully determined with an adiabatic calorimeter by Somsen (1). Based on the reported value, $\Delta_{sol}H°(298.15\ K) = -8.62 \pm 0.04\ kcal\ mol^{-1}$ and $\Delta_fH°(298.15\ K) = -61.69$ (2) and $-79.5\ kcal\ mol^{-1}$ (3) for Cs^+(ag, ∞) and F^-(ag, ∞), respectively, the enthalpy of formation for CsF(cr) is calculated to be $-132.57\ kcal\ mol^{-1}$ which is adopted.

The value, $\Delta_{sol}H°(298.15\ K) = -8.81 \pm 0.2\ kcal\ mol^{-1}$ for CsF(cr), was reported by Parker (4) which is in agreement with the value measured by Somsen (1).

Heat Capacity and Entropy

The heat capacities of CsF(cr) are estimated according to the method of Kristov (5), based on $H°(976\ K)-H°(298.15\ K) = 9.4\ kcal\ mol^{-1}$ reported by Dworkin (6). The entropy, S°(298.15 K), is estimated such that the calculated and observed vapor pressures are in good agreement (see CsF(g) and Cs_2F_2(g) tables for details).

Fusion Data

T_{fus} and $\Delta_{fus}H°$ are obtained from Bredig (7) and Dworkin (8). Buhalova (9) reported T_{fus} = 958 K, which is not adopted.

Sublimation Data

$\Delta_{sub}H°(298.15\ K)$ is calculated as the difference between $\Delta_fH°(298.15\ K)$ for CsF(g) and CsF(cr).

References

1. G. Somsen and L. Weeda, Rec. Trav. Chim. 83, 810 (1964).
2. S. R. Gunn, J. Phys. Chem. 71, 1386 (1967).
3. U. S. Nat. Bur. Stand. Tech. Note 270-3, 1968.
4. V. B. Parker, NSRDS-NBS 2, 1965.
5. G. A. Kristov, Izv. Vycsh. Uchebn. Zavedenii, SSR, Khim. i Khim. Tech. No. 5, 751, 1964.
6. A. S. Dworkin, personal communication, Oak Ridge National Laboratory, Dec. 1, 1964.
7. M. A. Bredig, H. R. Bronstein and W. T. Smith, Jr., J. Am. Chem. Soc. 77, 1454 (1955).
8. A. S. Dworkin and M. A. Bredig, J. Phys. Chem. 64, 269 (1960).
9. G. A. Bukhalova and D. V. Sementsova, Russ. J. Inorg. Chem. 10, 1027 (1965).

Enthalpy Reference Temperature = T_r = 298.15 K; Standard State Pressure = p° = 0.1 MPa

T/K	$C_p°$ (J K⁻¹mol⁻¹)	S° (J K⁻¹mol⁻¹)	$-[G°-H°(T_r)]/T$ (J K⁻¹mol⁻¹)	$H°-H°(T_r)$ (kJ mol⁻¹)	$\Delta_fH°$ (kJ mol⁻¹)	$\Delta_fG°$ (kJ mol⁻¹)	Log K_f
0							
100							
200							
298.15	51.965	88.282	88.282	0.	-554.673	-525.377	92.044
300	52.007	88.604	88.283	0.096	-554.665	-525.195	91.445
400	53.764	103.805	90.344	5.384	-556.266	-514.758	67.221
500	55.564	115.996	94.294	10.851	-555.615	-504.454	52.700
600	57.363	126.286	98.790	16.497	-554.811	-494.295	43.032
700	59.120	135.260	103.372	22.322	-553.860	-484.282	36.138
800	60.877	143.269	107.867	28.321	-552.758	-474.415	30.976
900	62.634	150.541	112.211	34.497	-551.504	-464.696	26.970
976.000	63.952	155.672	115.397	39.308	------ CRYSTAL <--> LIQUID -----		
1000	64.350	157.231	116.383	40.848	-617.224	-451.425	23.580
1100	65.856	163.438	120.382	47.362	-614.649	-434.969	20.655
1200	66.944	169.217	124.213	54.004	-611.958	-418.752	18.228
1300	67.697	174.607	127.885	60.739	-609.188	-402.764	16.183
1400	68.157	179.642	131.404	67.533	-606.370	-386.991	14.439
1500	68.492	184.356	134.779	74.366	-603.525	-371.420	12.934
1600	68.743	188.785	138.017	81.228	-600.663	-356.039	11.623
1700	68.910	192.958	141.127	88.112	-597.794	-340.838	10.473
1800	68.994	196.899	144.117	95.007	-594.928	-325.806	9.455
1900	69.036	200.631	146.994	101.909	-592.073	-310.933	8.548
2000	69.036	204.172	149.766	108.813	-589.236	-296.209	7.736

PREVIOUS: CURRENT: June 1968

Cesium Fluoride (CsF) $Cs_1F_1(cr)$

CESIUM FLUORIDE (CsF) LIQUID M_r = 151.903803

$S°(298.15\ K)$ = [90.099] J K^{-1} mol^{-1}

T_{fus} = 976 K

$\Delta_f H°(298.15\ K)$ = [-543.849] kJ mol^{-1}

$\Delta_{fus}H°$ = 21.7 kJ mol^{-1}

Enthalpy of Formation

$\Delta_f H°$(CsF, ℓ, 298.15 K) is calculated from $\Delta_f H°$(CsF, cr, 298.15 K) by adding $\Delta_{fus}H°$ and the difference in enthalpy, H°(976 K) - H°(298.15 K), between the crystal and liquid.

Heat Capacity and Entropy

Dworkin (1) measured the heat capacity of CsF(ℓ) at the melting point to be 17.7 cal K^{-1} mol^{-1} which is adopted and assumed to be constant in the temperature range 298-3000 K.

The entropy is calculated in a manner analogous to that used for the enthalpy of formation.

Fusion Data

See the crystal table for details.

Vaporization Data

T_{vap} = 1504 K is the temperature at which the calculated vapor pressures of CsF(g) and Cs_2F_2(g) equal one atmosphere. The vapor composition at T_{vap} is calculated as CsF 80.3% and Cs_2F_2 19.7%. The enthalpy required to vaporize one mole of liquid to the above vapor mixture at T_{fus} is $\Delta_{vap}H°$.

T_{vap} has been observed as 1524 and 1526 K by Wartenberg (1) and Ruff (2), respectively, which are in fair agreement with the calculated one, 1504 K.

References

1. H. von Wartenberg and H. Schulz, Z. Elektrochem. 22, 568 F(1921).
2. O. Ruff, G. Schmidt and S. Mugdan, Z. Anorg. Allgem. Chem. 123, 83 (1922).

Cesium Fluoride (CsF) $Cs_1F_1(l)$

Enthalpy Reference Temperature = T_r = 298.15 K; Standard State Pressure = p° = 0.1 MPa

T/K	$C°_p$ (J K⁻¹mol⁻¹)	S° (J K⁻¹mol⁻¹)	-[G°-H°(T_r)]/T (J K⁻¹mol⁻¹)	H°-H°(T_r) (kJ mol⁻¹)	$\Delta_f H°$ (kJ mol⁻¹)	$\Delta_f G°$ (kJ mol⁻¹)	Log K_f
0							
100							
200							
298.15	74.057	90.099	90.099	0.	-543.849	-515.094	90.242
300	74.057	90.557	90.101	0.137	-543.801	-514.916	89.655
400	74.057	111.862	93.005	7.543	-543.284	-504.999	65.946
500	74.057	128.387	98.491	14.948	-540.693	-495.728	51.788
600	74.057	141.890	104.633	22.354	-538.131	-486.976	42.395
700	74.057	153.305	110.792	29.760	-535.598	-478.651	35.717
800	74.057	163.194	116.738	37.165	-533.090	-470.687	30.733
900	74.057	171.917	122.394	44.571	-530.606	-463.036	26.874
976.000	74.057	177.921	126.487	50.199	----- CRYSTAL <--> LIQUID -----		
1000	74.057	179.720	127.743	51.977	-595.271	-451.961	23.608
1100	74.057	186.778	132.794	59.382	-591.804	-437.798	20.789
1200	74.057	193.222	137.565	66.788	-588.351	-423.950	18.454
1300	74.057	199.150	142.077	74.194	-584.909	-410.390	16.490
1400	74.057	204.638	146.352	81.599	-581.479	-397.094	14.816
1500	74.057	209.747	150.410	89.005	-578.062	-384.043	13.374
1600	74.057	214.527	154.270	96.411	-574.656	-371.220	12.119
1700	74.057	219.016	157.948	103.817	-571.265	-358.609	11.019
1800	74.057	223.249	161.459	111.222	-567.889	-346.197	10.046
1900	74.057	227.253	164.818	118.628	-564.530	-333.973	9.182
2000	74.057	231.052	168.035	126.034	-561.191	-321.924	8.408
2100	74.057	234.665	171.123	133.439	-557.874	-310.043	7.712
2200	74.057	238.110	174.090	140.845	-554.583	-298.319	7.083
2300	74.057	241.402	176.945	148.251	-551.320	-286.743	6.512
2400	74.057	244.554	179.697	155.656	-548.089	-275.310	5.992
2500	74.057	247.577	182.352	163.062	-544.894	-264.010	5.516
2600	74.057	250.482	184.917	170.468	-541.740	-252.837	5.080
2700	74.057	253.277	187.398	177.873	-538.630	-241.784	4.678
2800	74.057	255.970	189.799	185.279	-535.570	-230.846	4.306
2900	74.057	258.569	192.126	192.685	-532.565	-220.016	3.963
3000	74.057	261.079	194.383	200.090	-529.619	-209.289	3.644

PREVIOUS: CURRENT: June 1968

Cesium Fluoride (CsF) $Cs_1F_1(l)$

CESIUM FLUORIDE (CsF) CRYSTAL-LIQUID M_r = 151.903803

0 to 976 K crystal
above 976 K liquid

Refer to the individual tables for details.

Cesium Fluoride (CsF) Cs_1F_1(cr,l)

Enthalpy Reference Temperature = T_r = 298.15 K Standard State Pressure = p° = 0.1 MPa

T/K	C_p° (J K⁻¹mol⁻¹)	S° (J K⁻¹mol⁻¹)	$-[G^\circ-H^\circ(T_r)]/T$ (J K⁻¹mol⁻¹)	$H^\circ-H^\circ(T_r)$ (kJ mol⁻¹)	Δ_fH° (kJ mol⁻¹)	Δ_fG° (kJ mol⁻¹)	Log K_f
0							
100							
200							
298.15	51.965	88.282	88.282	0.	-554.673	-525.377	92.044
300	52.007	88.604	88.283	0.096	-554.665	-525.195	91.445
400	53.764	103.805	90.344	5.384	-556.266	-514.758	67.221
500	55.564	115.996	94.294	10.851	-555.615	-504.454	52.700
600	57.363	126.286	98.790	16.497	-554.811	-494.295	43.032
700	59.120	135.260	103.372	22.322	-553.860	-484.282	36.138
800	60.877	143.269	107.867	28.321	-552.758	-474.415	30.976
900	62.634	150.541	112.211	34.497	-551.504	-464.696	26.970
976.000	63.952	155.672	115.397	39.308	CRYSTAL <--> LIQUID		
976.000	74.057	177.921	115.397	61.023	TRANSITION		
1000	74.057	179.720	116.920	62.800	-595.272	-451.962	23.608
1100	74.057	186.778	122.955	70.206	-591.805	-437.799	20.789
1200	74.057	193.222	128.546	77.611	-588.351	-423.951	18.454
1300	74.057	199.150	133.752	85.017	-584.910	-410.391	16.490
1400	74.057	204.638	138.622	92.423	-581.480	-397.095	14.816
1500	74.057	209.747	143.195	99.828	-578.062	-384.044	13.374
1600	74.057	214.527	147.505	107.234	-574.657	-371.221	12.119
1700	74.057	219.016	151.581	114.640	-571.266	-358.610	11.019
1800	74.057	223.249	155.446	122.045	-567.890	-346.198	10.046
1900	74.057	227.253	159.121	129.451	-564.531	-333.974	9.182
2000	74.057	231.052	162.624	136.857	-561.192	-321.925	8.408
2100	74.057	234.665	165.969	144.262	-557.875	-310.044	7.712
2200	74.057	238.110	169.170	151.668	-554.583	-298.320	7.083
2300	74.057	241.402	172.240	159.074	-551.320	-286.744	6.512
2400	74.057	244.554	175.188	166.479	-548.090	-275.310	5.992
2500	74.057	247.577	178.023	173.885	-544.895	-264.010	5.516
2600	74.057	250.482	180.755	181.291	-541.740	-252.837	5.080
2700	74.057	253.277	183.389	188.697	-538.631	-241.785	4.678
2800	74.057	255.970	185.933	196.102	-535.571	-230.847	4.306
2900	74.057	258.569	188.394	203.508	-532.566	-220.017	3.963
3000	74.057	261.079	190.775	210.914	-529.620	-209.290	3.644

PREVIOUS: CURRENT: June 1968

Cesium Fluoride (CsF) Cs_1F_1(cr,l)

CESIUM FLUORIDE (CsF) IDEAL GAS M_r = 151.903803

$\Delta_f H°(0\ K) = -354.0 \pm 2\ kJ\ mol^{-1}$

$S°(298.15\ K) = 243.21 \pm 0.04\ J\ K^{-1}\ mol^{-1}$ $\Delta_f H°(298.15\ K) = -356.5 \pm 2\ kJ\ mol^{-1}$

Electronic Level and Quantum Weight

State	ε_i, cm^{-1}	g_i
$^1\Sigma^+$	0	1

$\omega_e = 353 \pm 4\ cm^{-1}$ $\omega_e x_e = 1.23\ cm^{-1}$ $\sigma = 1$

$B_e = 0.18437\ cm^{-1}$ $\alpha_e = 0.001105\ cm^{-1}$ $r_e = 2.345\ Å$

Enthalpy of Formation

The vapor pressures of CsF(cr) and CsF(ℓ) were measured by the manometric method by Wartenberg (1), Russ (2) and Cantor (3), the torsion-fiber effusion method by Pugh (4) and the Knudsen effusion method by Scheer (5). Based on the reported total pressures, the corresponding partial pressures of CsF(g) and Cs_2F_2(g) were evaluated using the adjusted enthalpies of vaporization and Gibbs energy functions such that the calculated and observed total pressues are in reasonable agreement, and the derived enthalpy changes by the 2nd and 3rd law methods are consistent. The results obtained are presented in the table below. The value of $\Delta_f H°$(CsF, g, 298.15 K) is adopted as -82.5 ± 0.5 kcal mol^{-1}.

The dissociation energy (D_0^0) of CsF(g) has been reported using the spectroscopic convergence limit as 5.65 eV (130.3 kcal mol^{-1}) and 5.5 eV (126.8 kcal mol^{-1}) by Herzberg (6) and Gaydon (9), respectively. Based on the above two D_0^0, and $\Delta_f H°$(0 K) = 18.68 and 18.36 for Cs(g) and F(g), we obtain the corresponding value for $\Delta_f H°$(CsF, g, 298.15 K) as -93.86 and -90.39 kcal mol^{-1}, which are not adopted.

Source	Reaction*	Method	$\Delta_r H°$(298.15 K), kcal mol^{-1} 2nd law	3rd law	Drift cal K^{-1} mol^{-1}	$\Delta_f H°$(298.15 K), kcal mol^{-1}
(1)	B	Manometion	45.39 ±0.15	44.98	-0.28 ±0.10	-85.00
(2)	B	"	44.78 ±1.70	44.94	0.11 ±1.18	-85.04
(3)	B	"	44.83	44.89	0.05 ±0.04	-85.09
(4)	A	Torsion fibre	48.52 ±0.03	47.57	-1.18 ±0.04	-85.00
(5)	A	Kundsen	48.43 ±0.27	46.61	-2.35 ±0.35	-85.96

Reactions: (A) CsF(cr) = CsF(g) (B) CsF(ℓ) = CsF(g).

Heat Capacity and Entropy

The ground state configuration was reported by Herzberg (6). The molecular constants are obtained from Honig et al (7), who observed the microwave spectra of CsF(g) at temperatures of about 700°C. Based on the J = 1 → 2 transition, the molecular constants are derived. The vibrational frequency (ω_e) has been reported, from different kinds of spectra, in the range 270-385 cm^{-1} by many investigators, e.g. radiofrequency, infrared, ultraviolet, microwave, etc. The value of ω_e adopted was reported by Baikov and Vasilevskii (8) who also reviewed the ω_e values obtained by the earlier investigators.

References

1. H. von Wartenberg and H. Schulz, Z. Elektrochem. 27, 568 (1921), 1391-1524 K, 6 points. Two data points at 1228 and 1319 K are not adopted for evaluation due to the inconsistency with the other data.
2. O. Ruff, G. Schmidt and S. Mugdan, Z. Anorg. Allgem. Chem. 123, 83 (1922), 1306-1538 K, 7 points.
3. S. Cantor and F. Blankenship, "Vapor Pressure of the Alkali Fluorides", Oak Ridge National Laboratory, Oak Ridge, Tennessee, September 1958, 1009-1380 K, 7 points calculated using the given vapor pressure equation.
4. A. C. P. Pugh and R. F. Barrow, Trans. Faraday Soc. 54, 671 (1958), 753-856 K, 6 points.
5. M. D. Scheer and J. Fine, J. Chem. Phys. 36, 1647 (1962), 677-878 K, 32 points. Evaluation based on the reported data points.
6. G. Herzberg, "Spectra of Diatomic Molecules", D. Van Nostrand Co., Inc., New York, 1950.
7. A. Honig, M. Mandel, M. L. Stitch and C. H. Townes, Phys. Rev. 96, 629 (1954).
8. V. I. Baikov and K. P. Vasilevskii, Opt. Spectry. (USSR) (English Transl.) 22, 198 (1967).
9. A. G. Gaydon, "Dissociation Energies", Chapman and Hall Ltd., London, 1953.

Cesium Fluoride (CsF) $Cs_1F_1(g)$

Enthalpy Reference Temperature = T_r = 298.15 K Standard State Pressure = p° = 0.1 MPa

T/K	$C_p°$ (J $K^{-1}mol^{-1}$)	S°	$-[G°-H°(T_r)]/T$	$H°-H°(T_r)$ (kJ mol^{-1})	$\Delta_f H°$	$\Delta_f G°$	Log K_f
0	0.	0.	INFINITE	-9.640	-353.988	-353.988	INFINITE
100	30.506	206.799	273.829	-6.703	-354.653	-361.319	188.734
200	34.207	229.206	246.449	-3.449	-355.543	-367.628	96.015
250	35.207	236.954	243.800	-1.711	-355.977	-370.600	77.433
298.15	35.838	243.213	243.213	0.	-356.477	-373.373	65.413
300	35.857	243.435	243.214	0.066	-356.499	-373.478	65.028
350	36.298	248.998	243.652	1.871	-359.184	-375.925	56.104
400	36.611	253.866	244.631	3.694	-359.761	-378.277	49.398
450	36.842	258.192	245.902	5.531	-360.328	-380.558	44.174
500	37.019	262.083	247.329	7.377	-360.892	-382.775	39.988
600	37.276	268.857	250.369	11.093	-362.020	-387.046	33.695
700	37.456	274.617	253.432	14.830	-363.155	-391.127	29.186
800	37.594	279.628	256.400	18.583	-364.301	-395.045	25.794
900	37.707	284.063	259.232	22.348	-365.457	-398.819	23.147
1000	37.805	288.041	261.917	26.124	-433.753	-398.764	20.829
1100	37.891	291.648	264.459	29.909	-433.906	-395.257	18.769
1200	37.971	294.949	266.864	33.702	-434.065	-391.737	17.052
1300	38.046	297.991	269.143	37.503	-434.228	-388.203	15.598
1400	38.117	300.813	271.305	41.311	-434.396	-384.656	14.352
1500	38.185	303.445	273.361	45.126	-434.569	-381.098	13.271
1600	38.251	305.912	275.319	48.948	-434.747	-377.527	12.325
1700	38.316	308.233	277.188	52.776	-434.933	-373.945	11.490
1800	38.379	310.424	278.974	56.611	-435.128	-370.352	10.747
1900	38.441	312.501	280.684	60.452	-435.334	-366.748	10.083
2000	38.502	314.474	282.325	64.299	-435.554	-363.132	9.484
2100	38.563	316.355	283.901	68.152	-435.789	-359.505	8.942
2200	38.623	318.150	285.417	72.012	-436.044	-355.867	8.449
2300	38.683	319.868	286.878	75.877	-436.321	-352.216	7.999
2400	38.742	321.516	288.287	79.748	-436.625	-348.553	7.586
2500	38.801	323.098	289.648	83.625	-436.958	-344.877	7.206
2600	38.860	324.621	290.964	87.508	-437.327	-341.187	6.855
2700	38.919	326.089	292.238	91.397	-437.734	-337.481	6.529
2800	38.977	327.505	293.473	95.292	-438.185	-333.760	6.226
2900	39.035	328.874	294.670	99.193	-438.685	-330.022	5.944
3000	39.093	330.199	295.832	103.099	-439.239	-326.266	5.681
3100	39.151	331.481	296.962	107.011	-439.852	-322.490	5.434
3200	39.209	332.725	298.060	110.929	-440.532	-318.694	5.202
3300	39.266	333.933	299.129	114.853	-441.242	-314.878	4.984
3400	39.324	335.106	300.170	118.783	-442.052	-311.037	4.778
3500	39.381	336.246	301.184	122.718	-442.941	-307.171	4.584
3600	39.439	337.357	302.174	126.659	-443.913	-303.278	4.400
3700	39.496	338.438	303.139	130.606	-444.960	-299.358	4.226
3800	39.553	339.492	304.082	134.558	-446.096	-295.409	4.061
3900	39.611	340.520	305.003	138.516	-447.312	-291.430	3.903
4000	39.668	341.524	305.904	142.480	-448.644	-287.416	3.753
4100	39.725	342.504	306.785	146.450	-450.075	-283.368	3.610
4200	39.782	343.462	307.646	150.425	-451.608	-279.283	3.473
4300	39.839	344.399	308.490	154.406	-453.246	-275.161	3.343
4400	39.896	345.315	309.317	158.393	-454.695	-271.033	3.218
4500	39.953	346.213	310.127	162.385	-455.932	-266.904	3.098
4600	40.010	347.091	310.921	166.384	-457.709	-262.684	2.983
4700	40.067	347.952	311.700	170.388	-459.567	-258.424	2.872
4800	40.124	348.796	312.464	174.397	-461.503	-254.124	2.765
4900	40.181	349.624	313.214	178.412	-463.515	-249.783	2.663
5000	40.238	350.437	313.950	182.433	-465.598	-245.401	2.564
5100	40.295	351.234	314.673	186.460	-467.748	-240.975	2.468
5200	40.352	352.017	315.384	190.492	-469.428	-236.585	2.377
5300	40.408	352.786	316.082	194.530	-471.629	-232.086	2.287
5400	40.465	353.542	316.769	198.574	-473.600	-227.589	2.201
5500	40.522	354.285	317.445	202.623	-475.702	-223.038	2.118
5600	40.579	355.016	318.109	206.678	-477.979	-218.423	2.037
5700	40.636	355.735	318.763	210.739	-480.283	-213.768	1.959
5800	40.692	356.442	319.406	214.805	-482.610	-209.072	1.883
5900	40.749	357.138	320.040	218.878	-484.953	-204.335	1.809
6000	40.806	357.823	320.664	222.955	-487.308	-199.559	1.737

PREVIOUS: June 1968 (1 atm) CURRENT: June 1968 (1 bar)

CESIUM HYDROXIDE (CsOH) CRYSTAL(ALPHA-BETA-GAMMA) M_r = 149.91274

Cesium Hydroxide (CsOH) $Cs_1H_1O_1(cr)$

$S°(298.15\ K) = [98.7 \pm 4.2]\ J\ K^{-1}\ mol^{-1}$
$T_{trs1}(\alpha-\beta) = 410 \pm 1\ K$
$T_{trs2}(\beta-\gamma) = 493 \pm 1\ K$
$T_{fus} = 588 \pm 1\ K$

$\Delta_fH°(0\ K)$ = Unknown
$\Delta_fH°(298.15\ K) = -416.73 \pm 0.8\ kJ\ mol^{-1}$
$\Delta_{trs1}H° = 1.30 \pm 0.13\ kJ\ mol^{-1}$
$\Delta_{trs2}H° = 6.07 \pm 0.63\ kJ\ mol^{-1}$
$\Delta_{fus}H° = 4.56 \pm 0.42\ kJ\ mol^{-1}$

Enthalpy of Formation

The enthalpy of formation of CsOH(cr) is obtained from the enthalpy of hydrolysis of metallic cesium, the enthalpy of solution of the hydroxide in water, and appropriate auxiliary data.

Beketov (1) and de Forcrand (2) determined the enthalpy of solution of crystalline CsOH in water. Parker (3) analyzed these data and selected $\Delta_{sol}H°(CsOH, cr) = -17.1\pm0.2\ kcal\ mol^{-1}$. This value is adopted here.

Discordant values for the enthalpy of hydrolysis of metallic cesium have been reported by Vorob'yev et al. (4) and Gunn (5). For the reaction $Cs(cr) + (n+1)\ H_2O(\ell) = CsOH\cdot n\ H_2O + 0.5\ H_2(g)$, they reported enthalpies of hydrolysis of -46.5 ± 0.20 (n=1030) and -48.346 ± 0.20 (n=3000) $kcal\ mol^{-1}$, respectively. From these values $\Delta_fH°(CsOH\cdot\infty H_2O,\ 298.15\ K) = -114.82\pm0.20$ and $-116.661\pm0.020\ kcal\ mol^{-1}$ are calculated based on $\Delta_fH°(H_2O,\ \ell,\ 298.15\ K) = -68.315\ kcal\ mol^{-1}$ (6).

In both investigations the possibility of hydroxide and oxide contamination of the metal samples was reduced by repeated distillation of the metal under vacuum, and spectroscopic analysis of each sample indicated low concentrations of the lighter alkali metals. Vorob'yev et al. (4) determined the amount of reaction by both weighing the metal sample and titrating the final hydroxide solution. Gunn (5) determined the amount of reaction by both weighing the sample and recovering the hydrogen formed in the hydrolysis. The systematically high hydrogen yields (0.05-0.16%) obtained by Gunn (5) suggest the presence of lighter alkali metals in higher concentrations than indicated by the spectroscopic analysis of the cesium, and the flame-photometric analyses of the final hydroxide solutions. However, the presence of lighter alkali metals at these concentrations should not introduce any serious error in the enthalpy of hydrolysis, since their heats (5) are quite similar to cesium.

The major difference between the two measurements lies in the instrumentation used to investigate the reaction. Vorob'yev et al. (4) used a hermetically sealed static calorimeter which was not stirred. In the absence of proper agitation, temperature gradients and chemical inhomogeneities are likely to arise in the final state. Without any indication of the performance of this calorimeter, the results of Vorob'yev et al. (4) are questionable. On the other hand, Gunn [5] carried out the hydrolysis reaction in a closed bomb calorimeter, agitation being effected by rocking the assembly through an angle of approximately 150° (5, 8). Where comparisons are possible, all indications are that the results are quite reliable. Therefore, we adopt $\Delta_fH°(CsOH\cdot\infty H_2O,\ 298.15\ K) = -116.661 \pm 0.020\ kcal\ mol^{-1}$ from the work of Gunn (5). Combining this result with the enthalpy of solution of CsOH(cr) given above, we derive $\Delta_fH°(CsOH,\ cr,\ 298.15\ K) = -99.6\pm0.2\ kcal\ mol^{-1}$ ($-416.73\pm0.8\ kJ\ mol^{-1}$).

Heat Capacity and Entropy

Heat capacities for CsOH(cr) over the entire temperature range 298-1000 K are estimated by comparison with similar data for the other alkali metal hydroxides and halides (9). Likewise, $S°(298.15\ K)$ is estimated to be 23.6 cal $K^{-1}\ mol^{-1}$ (98.7 J $K^{-1}\ mol^{-1}$) for CsOH(cr) by comparison with similar data for the alkali metal halides and lighter hydroxides (9).

Transition and Fusion Data

The temperatures and enthalpies of the polymorphic transitions and melting of CsOH(cr) are from the work of Reshetnikov and Baranskaya (10). These data were determined thermographically with $\Delta_{fus}H = 1.575\ kcal\ mol^{-1}$ for NaOH(cr) as a standard.

Sublimation Data

$\Delta_{sub}H°(298.15\ K)$ for the monomer and dimer are calculated from the adopted enthalpies of formation of the crystal and the respective gaseous species.

References

1. N. Beketov, Bull. Acad. Imp. Sci. (St. Petersbourg), 34, 171 (1892).
2. R. de Forcrand, Compt. Rend., 142, 1252 (1906).
3. V. B. Parker, NSRDS-NBS 2, 1965.
4. A. F. Vorob'yev, N. A. Ibragin, and S. M. Skuratov, Vestn. Mosk. Univ., Ser. II, Khim., 20, 3 (1965).
5. S. R. Gunn, J. Phys. Chem., 71, 1386 (1967).
6. U. S. Nat. Bur. Stand. Tech. Note 270-3, 1968.
7. S. R. Gunn, J. Phys. Chem., 69, 2902 (1965).
8. S. R. Gunn and L. G. Green, J. Amer. Chem. Soc., 80, 4782 (1958).
9. JANAF Thermochemical Tables: refer to all alkali metal hydroxides and halides crystal tables.
10. N. A. Reshetnikov and E. V. Baranskaya, Izv. Vyssh. Ucheb. Zaved., Khim. Khim. Tekhnol., 10, 496 (1967).

Enthalpy Reference Temperature = T_r = 298.15 K; Standard State Pressure = p° = 0.1 MPa

T/K	$C_p°$ (J K⁻¹mol⁻¹)	S°	$-[G°-H°(T_r)]/T$	$H°-H°(T_r)$ (kJ mol⁻¹)	$\Delta_fH°$	$\Delta_fG°$	Log K_f
0							
100							
200							
298.15	67.872	98.742	98.742	0.	-416.726	-370.716	64.948
300	67.990	99.163	98.744	0.126	-416.714	-370.431	64.498
400	74.392	119.617	101.489	7.251	-417.807	-354.453	46.287
410.000	74.990	121.461	101.954	7.998	ALPHA <--> BETA	TRANSITION	
410.000	75.228	124.625	101.954	9.295			
493.000	82.174	139.114	107.010	15.827	BETA <--> GAMMA	TRANSITION	
493.000	83.680	151.420	107.010	21.894			
500	83.680	152.600	107.640	22.480	-408.701	-339.169	35.433
588.000	83.680	166.166	115.411	29.844	GAMMA <--> LIQUID		
600	83.680	167.856	116.443	30.848	-406.484	-325.471	28.335
700	83.680	180.756	124.733	39.216	-404.309	-312.142	23.292
800	83.680	191.929	132.450	47.584	-402.179	-299.121	19.531
900	83.680	201.786	139.617	55.952	-400.095	-286.364	16.620
1000	83.680	210.602	146.282	64.320	-465.187	-270.138	14.111

PREVIOUS: CURRENT June 1971

Cesium Hydroxide (CsOH) $Cs_1H_1O_1(cr)$

CESIUM HYDROXIDE (CsOH) LIQUID M_r = 149.91274

$S°(298.15\ K) = [118.513]\ J\ K^{-1}\ mol^{-1}$

$T_{fus} = 588 \pm 1\ K$

$\Delta_f H°(298.15\ K) = [-405.970]\ kJ\ mol^{-1}$

$\Delta_{fus} H° = 4.56 \pm 0.4\ kJ\ mol^{-1}$

Enthalpy of Formation

$\Delta_f H°$(CsOH, ℓ, 298.15 K) is calculated from that of the crystal by adding $\Delta_{fus}H°$ and the difference in enthalpy, H°(588 K) - H°(298.15 K), between the crystal and liquid.

Heat Capacity and Entropy

$C_p°$ for liquid CsOH is estimated to be 19.5 cal K^{-1} mol^{-1} by comparison with similar data for the other liquid alkali metal hydroxides [1]. It is assumed constant in the temperature range 298-2000 K.

S°(298.15 K) is calculated in a manner analogous to that used for the enthalpy of formation.

Fusion Data

See the crystal table for details.

Vaporization Data

T_{vap} is calculated as the temperature at which $\Delta_r G°$ for the vaporization process is zero. The difference in the enthalpy of formation of CsOH(ℓ) and CsOH(g) at the boiling point is the heat of vaporization.

Reference

1. JANAF Thermochemical Tables: refer to all alkali metal hydroxide liquid tables.

Cesium Hydroxide (CsOH) $Cs_1H_1O_1(l)$

Enthalpy Reference Temperature = T_r = 298.15 K; Standard State Pressure = p° = 0.1 MPa

T/K	$C_p°$ (J K⁻¹mol⁻¹)	S° (J K⁻¹mol⁻¹)	$-[G°-H°(T_r)]/T$ (J K⁻¹mol⁻¹)	$H°-H°(T_r)$ (kJ mol⁻¹)	$\Delta_f H°$ (kJ mol⁻¹)	$\Delta_f G°$ (kJ mol⁻¹)	Log K_f
0							
100							
200							
298.15	81.588	118.513	118.513	0.	-405.970	-365.855	64.096
300	81.588	119.017	118.514	0.151	-405.933	-365.606	63.658
400	81.588	142.489	121.715	8.310	-405.992	-351.787	45.939
500	81.588	160.695	127.758	16.469	-403.956	-338.472	35.360
588.000	81.588	173.922	133.704	23.648	------- GAMMA <--> LIQUID ------		
600	81.588	175.570	134.524	24.627	-401.949	-325.564	28.343
700	81.588	188.147	141.309	32.786	-399.983	-312.989	23.356
800	81.588	199.041	147.860	40.945	-398.062	-300.694	19.633
900	81.588	208.651	154.091	49.104	-396.187	-288.636	16.752
1000	81.588	217.247	159.985	57.263	-461.488	-273.084	14.264
1100	81.588	225.023	165.549	65.421	-458.682	-254.380	12.080
1200	81.588	232.122	170.806	73.580	-455.916	-235.930	10.270
1300	81.588	238.653	175.777	81.739	-453.188	-217.708	8.748
1400	81.588	244.699	180.487	89.898	-450.499	-199.696	7.451
1500	81.588	250.328	184.957	98.057	-447.847	-181.874	6.333
1600	81.588	255.594	189.209	106.215	-445.234	-164.228	5.361
1700	81.588	260.540	193.261	114.374	-442.659	-146.744	4.509
1800	81.588	265.204	197.130	122.533	-440.125	-129.411	3.755
1900	81.588	269.615	200.830	130.692	-437.633	-112.217	3.085
2000	81.588	273.800	204.374	138.851	-435.187	-95.153	2.485

PREVIOUS: CURRENT: June 1971

Cesium Hydroxide (CsOH) $Cs_1H_1O_1(l)$

CESIUM HYDROXIDE (CsOH)

ALPHA-BETA-GAMMA-LIQUID

0 to 410 K crystal, alpha
410 to 493 K crystal, beta
493 to 588 K crystal, gamma
above 588 K liquid

Refer to the individual tables for details.

M_r = 149.91274 Cesium Hydroxide (CsOH) $Cs_1H_1O_1$(cr,l)

Enthalpy Reference Temperature = T_r = 298.15 K Standard State Pressure = p° = 0.1 MPa

T/K	C_p° (J K⁻¹mol⁻¹)	S° (J K⁻¹mol⁻¹)	$-[G°-H°(T_r)]/T$ (J K⁻¹mol⁻¹)	$H°-H°(T_r)$ (kJ mol⁻¹)	$\Delta_fH°$ (kJ mol⁻¹)	$\Delta_fG°$ (kJ mol⁻¹)	Log K_f
0							
100							
200							
298.15	67.872	98.742	98.742	0.	-416.726	-370.716	64.948
300	67.990	99.163	98.744	0.126	-416.714	-370.431	64.498
400	74.392	119.617	101.489	7.251	-417.807	-354.453	46.287
410.000	74.990	121.461	101.954	7.998	ALPHA <--> BETA TRANSITION		
410.000	75.228	124.625	101.954	9.295			
493.000	82.174	139.114	107.010	15.827	BETA <--> GAMMA TRANSITION		
493.000	83.680	151.420	107.010	21.894			
500	83.680	152.600	107.640	22.480	-408.701	-339 169	35.433
588.000	83.680	166.166	115.411	29.844	GAMMA <--> LIQUID TRANSITION		
588.000	81.588	173.922	115.411	34.404			
600	81.588	175.570	116.598	35.383	-401.949	-325.564	28.343
700	81.588	188.147	125.944	43.542	-399.983	-312.989	23.356
800	81.588	199.041	134.415	51.701	-398.062	-300.694	19.633
900	81.588	208.651	142.140	59.860	-396.187	-288.636	16.752
1000	81.588	217.247	149.229	68.018	-461.488	-273.084	14.264
1100	81.588	225.023	155.771	76.177	-458.682	-254.380	12.080
1200	81.588	232.122	161.842	84.336	-455.916	-235.930	10.270
1300	81.588	238.653	167.503	92.495	-453.188	-217.708	8.748
1400	81.588	244.699	172.804	100.654	-450.499	-199.696	7.451
1500	81.588	250.328	177.787	108.812	-447.847	-181.874	6.333
1600	81.588	255.594	182.487	116.971	-445.234	-164.228	5.361
1700	81.588	260.540	186.934	125.130	-442.659	-146.744	4.509
1800	81.588	265.204	191.154	133.289	-440.125	-129.411	3.755
1900	81.588	269.615	195.169	141.448	-437.633	-112.217	3.085
2000	81.588	273.800	198.996	149.606	-435.187	-95.153	2.485

PREVIOUS: CURRENT: June 1971

Cesium Hydroxide (CsOH) $Cs_1H_1O_1$(cr,l)

CESIUM HYDROXIDE (CsOH) IDEAL GAS M_r = 149.91274

$S°(298.15\ K) = 254.78 \pm 0.42\ J\ K^{-1}\ mol^{-1}$

$\Delta_f H°(0\ K) = -254.9 \pm 12.6\ kJ\ mol^{-1}$

$\Delta_f H°(298.15\ K) = -259.4 \pm 12.6\ kJ\ mol^{-1}$

Vibrational Frequencies and Degeneracies

ν, cm^{-1}
335.6(1)
306 (2)
[3610](1)

Ground State Quantum Weight: 1

Point Group: $C_{\infty v}$

Bond Distances: Cs-O = 2.391 ± 0.002 Å O-H = [-0.96 ± 0.01] Å

Bond Angle: Cs-O-H = 180°

Rotational Constant: $B_0 = 0.184795\ cm^{-1}$

Enthalpy of Formation

Jensen and Padley (1) determined an equilibrium constant for the reaction $Cs(g) + H_2O(g) = CsOH(g) + H(g)$ at 2475 K by atomic absorption spectroscopy in a hydrogen-oxygen-nitrogen flame. These workers pointed out that interference from ionization of the metal had introduced some uncertainty in their equilibrium data. Using all JANAF functions (2), 3rd law analysis of the equilibrium constant gives $\Delta_r H°(298.15\ K) = 27.4\ kcal\ mol^{-1}$. This leads to an enthalpy of formation, $\Delta_f H°(CsOH, g, 298.15\ K) = -64.2\pm3.0\ kcal\ mol^{-1}$, and a corresponding bond dissociation energy of 91 $kcal\ mol^{-1}$. Cotton and Jenkins (3) investigated the same equilibrium in a fuel-rich hydrogen-oxygen-nitrogen flame and eliminated the ionization problem by working at a lower temperature (1570 K). 3rd law analysis of their equilibrium data gives $\Delta_r H°(298.15\ K) = 28.9\ kcal\ mol^{-1}$ which results in $\Delta_f H°(CsOH, g, 298.15\ K) = -62.7\pm2.0\ kcal\ mol^{-1}$ corresponding to $D_0^0(Cs\text{-}OH) = 89.6\ kcal\ mol^{-1}$. Smith and Sugden (4) had earlier reported a $D_0^0 = 91 \pm 1\ kcal\ mol^{-1}$ from flame studies. Electron impact studies (5, 6) have also led to values for the bond dissociation energy of CsOH(g). This method resulted in values of 82±3 (5) and 86±3 $kcal\ mol^{-1}$ (6).

Cotton and Jenkins (3) also reported bond dissociation energies for LiOH(g), KOH(g), and NaOH(g) which are in good agreement with JANAF enthalpy of formation data (7) for these hydroxides. Furthermore, the position of the hydroxides in the order of increasing bond dissociation energy established from their data, i.e. NaOH < KOH < RbOH < CsOH < LiOH, is that predicted by comparison with similar data on the stabilities of the alkali metal halides (8). However, trends within the bond dissociation energies for the halides suggest that the value $D_0^0 = 89.6\ kcal\ mol^{-1}$ for CsOH may be slightly high. The difference in D° for potassium and sodium compounds is from 0.1 $kcal\ mol^{-1}$ for the fluorides to 0.4 $kcal\ mol^{-1}$ for the chlorides more than the difference in D_0^0 for the cesium and potassium compounds. Assuming this same trend in D_0^0 is prevalent in the hydroxides as well, then $D_0^0(Cs\text{-}OH)\text{-}D_0^0(K\text{-}OH)$ must be less than 3.9 $kcal\ mol^{-1}$ (see NaOH(g) table). We adopt $D_0^0(CsOH) = 89.0\pm3.0\ kcal\ mol^{-1}$ which results in the difference $D_0^0(Cs\text{-}OH)\text{-}D_0^0(K\text{-}OH) = 3.6\ kcal\ mol^{-1}$ (7). This adopted bond dissociation energy corresponds to $\Delta_f H°(CsOH, g, 298.15\ K) = -62.0\pm3.0\ kcal\ mol^{-1}$.

Heat Capacity and Entropy

The microwave spectra of gaseous CsOH have been studied in a high temperature spectrometer by Kuczkowski et al. (9) and Lide et al. (10). The spectroscopic data were more consistent with patterns expected from a linear molecule, although a "quasilinear" structure could not be completely eliminated. Acquista et al. (11) observed the infrared spectra of CsOH and CsOD in an argon matrix at liquid hydrogen temperatures. The isotope shift of the bending mode, ν_2, was quite large (80 cm^{-1}), and the ratio of ν_2 for CsOH and CsOD was in excellent agreement with the value calculated for a linear molecule and a harmonic oscillator. The infrared results when used in combination with the microwave data indicate a linear equilibrium configuration for CsOH. We adopt this configuration along with the bond distances and angles determined by Lide et al. (10) from their microwave data.

The Cs-O stretching mode, ν_1, and the bending frequency, ν_2, are from the matrix isolation studies of Acquista et al. (11). The OH stretching frequency, ν_3, is an estimate from the recent work of Jensen (12).

References

1. D. E. Jensen and P. J. Padley, Trans. Faraday Soc., 62, 2132 (1966).
2. JANAF Thermochemical Tables: Cs(g), 6-30-68; H_2O(g), 3-31-61; H(g), 9-30-65.
3. D. H. Cotton and D. R. Jenkins, Trans. Faraday Soc., 63, 1537 (1969).
4. H. Smith and T. M. Sugden, Proc. Roy. Soc. A, 219, 204 (1953).
5. A. M. Emel'yanov, A. V. Gusarov, L. N. Gorokhov, and N. A. Sadovnikova, Teor. Eksp. Khim., 3, 226 (1967).
6. L. N. Gorokhov, A. V. Gusarov, and I. G. Panchenkov, Russ. J. Phys. Chem. (English Transl.), 44, 150 (1970).
7. Ref. 2; LiOH(g), 3-31-66; NaOH(g) and KOH(g), 12-31-70.
8. Ref. 2; See recent review of the fluorides by J. Chao, Thermochimica Acta, 1, 71 (1970).
9. R. L. Kuczkowski, D. R. Lide, Jr., and L. C. Krisher, J. Chem. Phys., 44, 3131 (1966).
10. D. R. Lide, Jr., and R. L. Kuczkowski, J. Chem. Phys., 46, 4768 (1967).
11. N. Acquista, S. Abramowitz, and D. R. Lide, J. Chem. Phys., 49, 780 (1968).
12. D. E. Jensen, J. Phys. Chem., 74, 207 (1970).

Cesium Hydroxide (CsOH) $Cs_1H_1O_1(g)$

Enthalpy Reference Temperature = T_r = 298.15 K Standard State Pressure = p° = 0.1 MPa

T/K	$C_p°$ ($J\ K^{-1}mol^{-1}$)	S° ($J\ K^{-1}mol^{-1}$)	$-[G°-H°(T_r)]/T$ ($J\ K^{-1}mol^{-1}$)	$H°-H°(T_r)$ ($kJ\ mol^{-1}$)	$\Delta_f H°$ ($kJ\ mol^{-1}$)	$\Delta_f G°$ ($kJ\ mol^{-1}$)	Log K_f
0	0.	0.	INFINITE	-11.834	-254.950	-254.950	INFINITE
100	34.721	207.752	295.765	-8.801	-257.018	-258.596	135.077
200	45.601	235.675	259.221	-4.709	-258.408	-259.563	67.791
250	48.173	246.150	255.588	-2.360	-258.906	-259.793	54.281
298.15	49.723	254.778	254.778	0.	-259.408	-259.920	45.537
300	49.771	255.085	254.779	0.092	-259.430	-259.923	45.257
350	50.813	262.841	255.389	2.608	-262.071	-259.625	38.747
400	51.525	269.676	256.756	5.168	-262.572	-259.241	33.853
450	52.035	275.775	258.536	7.758	-263.041	-258.797	30.040
500	52.420	281.279	260.540	10.369	-263.493	-258.300	26.984
600	52.991	290.889	264.820	15.641	-264.372	-257.179	22.389
700	53.459	299.093	269.145	20.964	-265.242	-255.912	19.096
800	53.915	306.261	273.345	26.333	-266.112	-254.519	16.618
900	54.388	312.639	277.364	31.748	-266.981	-253.018	14.685
1000	54.877	318.394	281.183	37.211	-334.977	-247.721	12.940
1100	55.371	323.648	284.809	42.723	-334.817	-239.003	11.349
1200	55.857	328.487	288.249	48.285	-334.648	-230.300	10.025
1300	56.325	332.976	291.519	53.894	-334.470	-221.611	8.904
1400	56.769	337.167	294.632	59.549	-334.285	-212.936	7.945
1500	57.184	341.098	297.600	65.247	-334.094	-204.275	7.114
1600	57.568	344.801	300.435	70.985	-333.902	-195.627	6.387
1700	57.922	348.302	303.149	76.759	-333.711	-186.991	5.746
1800	58.246	351.622	305.750	82.568	-333.527	-178.366	5.176
1900	58.543	354.779	308.248	88.408	-333.355	-169.750	4.667
2000	58.813	357.789	310.651	94.276	-333.199	-161.143	4.209
2100	59.060	360.664	312.964	100.170	-333.066	-152.544	3.794
2200	59.285	363.417	315.196	106.087	-332.961	-143.950	3.418
2300	59.490	366.057	317.350	112.026	-332.892	-135.361	3.074
2400	59.677	368.593	319.433	117.985	-332.864	-126.773	2.759
2500	59.848	371.033	321.448	123.961	-332.883	-118.186	2.469
2600	60.005	373.383	323.401	129.954	-332.958	-109.597	2.202
2700	60.148	375.650	325.294	135.961	-333.094	-101.003	1.954
2800	60.279	377.840	327.132	141.983	-333.298	-92.404	1.724
2900	60.400	379.958	328.917	148.017	-333.577	-83.796	1.509
3000	60.511	382.007	330.653	154.063	-333.938	-75.177	1.309
3100	60.614	383.993	332.342	160.119	-334.388	-66.544	1.121
3200	60.708	385.919	333.986	166.185	-334.934	-57.896	0.945
3300	60.796	387.788	335.588	172.260	-335.542	-49.232	0.779
3400	60.876	389.604	337.150	178.344	-336.282	-40.545	0.623
3500	60.951	391.370	338.674	184.435	-337.133	-31.835	0.475
3600	61.021	393.088	340.162	190.534	-338.100	-23.099	0.335
3700	61.086	394.761	341.615	196.639	-339.175	-14.335	0.202
3800	61.146	396.391	343.035	202.751	-340.373	-5.541	0.076
3900	61.203	397.980	344.424	208.869	-341.685	3.286	-0.044
4000	61.255	399.530	345.782	214.992	-343.145	12.149	-0.159
4100	61.304	401.043	347.112	221.120	-344.739	21.051	-0.268
4200	61.350	402.521	348.413	227.252	-346.468	29.994	-0.373
4300	61.394	403.965	349.689	233.390	-348.334	38.979	-0.473
4400	61.434	405.377	350.938	239.531	-350.045	47.975	-0.570
4500	61.472	406.758	352.163	245.676	-351.576	56.979	-0.661
4600	61.508	408.110	353.365	251.825	-353.680	66.081	-0.750
4700	61.542	409.433	354.544	257.978	-355.897	75.230	-0.836
4800	61.573	410.729	355.701	264.134	-358.225	84.428	-0.919
4900	61.604	411.999	356.837	270.292	-360.659	93.675	-0.999
5000	61.632	413.244	357.953	276.454	-363.196	102.973	-1.076
5100	61.659	414.464	359.049	282.619	-365.831	112.322	-1.150
5200	61.684	415.662	360.126	288.786	-368.027	121.646	-1.222
5300	61.708	416.837	361.185	294.956	-370.775	131.089	-1.292
5400	61.731	417.991	362.226	301.128	-373.323	140.541	-1.359
5500	61.753	419.124	363.251	307.302	-376.033	150.058	-1.425
5600	61.773	420.237	364.258	313.478	-378.946	159.650	-1.489
5700	61.793	421.330	365.250	319.656	-381.917	169.294	-1.551
5800	61.811	422.405	366.226	325.836	-384.939	178.991	-1.612
5900	61.829	423.462	367.187	332.018	-388.007	188.740	-1.671
6000	61.846	424.501	368.134	338.202	-391.116	198.542	-1.728

PREVIOUS: June 1971 (1 atm) CURRENT: June 1971 (1 bar)

Cesium Hydroxide (CsOH) $Cs_1H_1O_1(g)$

CESIUM HYDROXIDE, ION ($CsOH^+$) IDEAL GAS M_r = 149.91219

$S°(298.15\ K) = [268.7 \pm 8.4]\ J\ K^{-1}\ mol^{-1}$

$\Delta_f H°(0\ K) = 439.32 \pm 41.8\ kJ\ mol^{-1}$
$\Delta_f H°(298.15\ K) = [441.379]\ kJ\ mol^{-1}$

Electronic Levels and Quantum Weights

State	ε_i, cm^{-1}	g_i
$^2\Pi$	0	4
$^2\Sigma^+$	[32000]	2

Vibrational Frequencies and Degeneracies

ν, cm^{-1}
[300](1)
[275](2)
[3600](1)

Point Group: $[C_{\infty v}]$ $\sigma = 1$
Bond Distances: Cs-O = [2.40] Å O-H = [1.03] Å
Bond Angle: Cs-O-H = [180] Å
Rotational Constant: B_0 = [0.182583] cm^{-1}

Enthalpy of Formation

The appearance potential of $CsOH^+$ has been determined by Emel'yanov et al. (1) as 7.21±0.14 eV and by Gorokhov et al. (2) as 7.40±0.15 eV. These values correspond to $\Delta_r H°(0\ K)$ = 166.3±3.0 kcal mol^{-1} and 170.7±5 kcal mol^{-1}, respectively, for the process $e^- + CsOH(g) \rightarrow CsOH^+(g) + 2e^-$. Since there may be excess energy involved, we prefer the lower value. Combining this result with JANAF auxiliary data (3), we derive $\Delta_f H°(CsOH^+, g, 0\ K)$ = 105±10 kcal mol^{-1} (439.32 ± 41.8 kJ mol^{-1}).

Heat Capacity and Entropy

The correlation diagram of Walsh (4) for HAB molecules predicts that $CsOH^+$ (7 valence electrons) should be linear with the unpaired electron in a pi orbital. Therefore, we estimate the ground state of $CsOH^+$ to be $^2\Pi$. $CsOH^+$ is isoelectronic with OH (3) and SH (3). A first excited state is estimated at 32000 cm^{-1} to be $^2\Sigma$ by analogy with these isoelectronic molecules. The bonding in $CsOH^+$ is assumed to be weakened relative to that in CsOH by the loss of the bonding electron from the neutral species. The vibrational frequencies are estimated from those for CsOH with somewhat lower values due to the weaker bonding. The Cs-O and O-H bond distances are increased slightly over those for CsOH.

References

1. A. M. Emel'yanov, A. V. Gusarov, L. N. Gorokhov, and N. A. Sadovnikova, Teor. Eksp. Khim. 3, 226 (1967).
2. L. N. Gorokhov, A. V. Gusarov, and I. G. Panchenkov, Russ. J. Phys. Chem. 44, 150 (1970).
3. JANAF Thermochemical Tables: CsOH(g), 6-30-71; OH(g), 12-31-70; SH(g), 6-30-67.
4. A. D. Walsh, J. Chem. Soc., 2288 (1953).

Cesium Hydroxide, Ion ($CsOH^+$) $Cs_1H_1O_1^+(g)$

Enthalpy Reference Temperature = T_r = 298.15 K Standard State Pressure = p° = 0.1 MPa

T/K	$C_p°$ (J K^{-1}mol^{-1})	S° (J K^{-1}mol^{-1})	$-[G°-H°(T_r)]/T$ (J K^{-1}mol^{-1})	$H°-H°(T_r)$ (kJ mol^{-1})	$\Delta_f H°$ (kJ mol^{-1})	$\Delta_f G°$ (kJ mol^{-1})	Log K_f
0	0.	0.	INFINITE	-12.154	439.320		
100	36.400	220.076	310.754	-9.068			
200	46.942	249.166	273.215	-4.810			
250	49.177	259.903	269.510	-2.402			
298.15	50.492	268.686	268.686	0.	441.379	430.466	-75.416
300	50.532	268.998	268.686	0.093	441.398	430.398	-74.939
350	51.403	276.858	269.305	2.643	439.830	428.860	-64.004
400	51.993	283.762	270.690	5.229	440.394	427.255	-55.794
450	52.414	289.912	272.490	7.840	440.985	425.577	-49.400
500	52.733	295.451	274.514	10.469	441.590	423.833	-44.278
600	53.215	305.110	278.832	15.767	442.816	420.167	-36.579
700	53.628	313.345	283.188	21.110	444.044	416.295	-31.064
800	54.049	320.533	287.416	26.493	445.268	412.247	-26.917
900	54.498	326.925	291.458	31.920	446.489	408.046	-23.682
1000	54.971	332.691	295.297	37.394	380.582	407.409	-21.281
1100	55.452	337.953	298.939	42.915	382.829	409.983	-19.468
1200	55.929	342.798	302.395	48.484	385.084	412.352	-17.949
1300	56.390	347.293	305.678	54.100	387.348	414 532	-16.656
1400	56.827	351.488	308.802	59.761	389.618	416.538	-15.541
1500	57.236	355.423	311.780	65.465	391.892	418.382	-14.569
1600	57.616	359.130	314.625	71.207	394.169	420.073	-13.714
1700	57.966	362.633	317.347	76.987	396.442	421.623	-12.955
1800	58.287	365.956	319.956	82.800	398.709	423.039	-12.276
1900	58.580	369.115	322.461	88.643	400.964	424.329	-11.666
2000	58.848	372.127	324.869	94.515	403.202	425.500	-11.113
2100	59.092	375.004	327.189	100.412	405.417	426.561	-10.610
2200	59.315	377.758	329.425	106.333	407.604	427.516	-10.151
2300	59.518	380.399	331.584	112.274	409.755	428.373	-9.729
2400	59.703	382.936	333.671	118.236	411.864	429.137	-9.340
2500	59.873	385.377	335.691	124.215	413.926	429.814	-8.980
2600	60.027	387.728	337.648	130.210	415.932	430.410	-8.647
2700	60.169	389.996	339.545	136.220	417.877	430.930	-8.337
2800	60.300	392.187	341.386	142.243	419.754	431.379	-8.047
2900	60.419	394.305	343.174	148.279	421.555	431.762	-7.777
3000	60.529	396.355	344.913	154.327	423.274	432.084	-7.523
3100	60.631	398.342	346.605	160.385	424.904	432.351	-7.285
3200	60.725	400.268	348.252	166.453	426.439	432.566	-7.061
3300	60.811	402.138	349.857	172.529	427.911	432.732	-6.850
3400	60.892	403.955	351.421	178.615	429.251	432.858	-6.650
3500	60.967	405.721	352.947	184.708	430.480	432.945	-6.461
3600	61.036	407.439	354.437	190.808	431.593	433.000	-6.283
3700	61.101	409.113	355.892	196.915	432.598	433.024	-6.113
3800	61.161	410.743	357.315	203.028	433.480	433.023	-5.952
3900	61.218	412.332	358.705	209.147	434.249	432.998	-5.799
4000	61.271	413.883	360.065	215.271	434.869	432.958	-5.654
4100	61.322	415.397	361.396	221.401	435.356	432.904	-5.515
4200	61.369	416.875	362.700	227.536	435.707	432.839	-5.383
4300	61.414	418.319	363.976	233.675	435.922	432.769	-5.257
4400	61.456	419.732	365.228	239.818	436.291	432.660	-5.136
4500	61.496	421.113	366.454	245.966	436.840	432.513	-5.020
4600	61.535	422.465	367.657	252.118	436.818	432.416	-4.910
4700	61.572	423.789	368.837	258.273	436.682	432.322	-4.805
4800	61.607	425.086	369.996	264.432	436.437	432.232	-4.704
4900	61.641	426.356	371.133	270.594	436.084	432.148	-4.607
5000	61.674	427.602	372.250	276.760	435.630	432.072	-4.514
5100	61.706	428.824	373.347	282.929	435.078	432.006	-4 425
5200	61.737	430.022	374.426	289.101	434.966	431.874	-4.338
5300	61.767	431.199	375.486	295.276	434.302	431.821	-4.256
5400	61.797	432.353	376.528	301.455	433.839	431.738	-4.176
5500	61.826	433.488	377.554	307.636	433.215	431.681	-4.100
5600	61.854	434.602	378.563	313.820	432.388	431.661	-4.026
5700	61.882	435.697	379.555	320.007	431.504	431.655	-3.956
5800	61.910	436.773	380.533	326.196	430.570	431.667	-3.888
5900	61.937	437.832	381.495	332.389	429.591	431.693	-3.822
6000	61.965	438.873	382.442	338.584	428.572	431.737	-3.759

PREVIOUS: December 1971 (1 atm) CURRENT: December 1971 (1 bar)

Cesium Hydroxide, Ion ($CsOH^+$) $Cs_1H_1O_1^+(g)$

CESIUM OXIDE (CsO) IDEAL GAS M_r = 148.9048

$\Delta_f H°(0\ K)$ = [64.9 ± 41.8] kJ mol^{-1}

$S°(298.15\ K)$ = [225.5 ± 4.2] J K^{-1} mol^{-1} $\Delta_f H°(298.15\ K)$ = [62.8 ± 41.8] kJ mol^{-1}

Electronic Levels and Quantum Weights

State	ε_i, cm^{-1}	g_i
[$^2\Pi$]	0	[4]

ω_e = [280] cm^{-1} $\omega_e x_e$ = [1.0] cm^{-1} σ = 1

B_e = [0.198] cm^{-1} α_e = [0.001] cm^{-1} r_e = [2.40] Å

Enthalpy of Formation

The enthalpy of formation is estimated from three calculated values. The methods of estimation are described as follows. Brewer and Margrave (1) suggested that the $\Delta_f H°$ value for MO(g) from the gaseous ions can be taken as the mean of the corresponding $\Delta_f H°$ values for the alkali fluoride and chloride with a maximum uncertainty of 10 kcal mol^{-1}. Adopting $\Delta_r H°(298.15\ K)$ = -131.1 and -109.6 kcal mol^{-1} for the reaction $Cs^+(g) + F^-(g) = CsF(g)$ and $Cs^+(g) + Cl^-(g) = CsCl(g)$, respectively, we obtain $\Delta_r H°(298.15\ K)$ = -120.4 kcal mol^{-1} for $Cs^+(g) + O^-(g) = CsO(g)$ which leads to $\Delta_f H°$(CsO, g, 298.15 K) = 12.0 kcal mol^{-1}. The $\Delta_f H°(298.15\ K)$ values of the above species are taken from the respective JANAF Thermochemical Tables.

Based on the assumption that $D_0^0(Cs\text{-}O) = 1/2\ [D_0^0(Cs_2) + D_0^0(O_2)]$, we derive D_0^0(Cs-O, g, 298.15 K) = 65.2 kcal mol^{-1} and $\Delta_f H°(298.15\ K)$ = 12.7 kcal mol^{-1} for CsO(g).

Adopting the Cs-O bond energy as half of the atomization energy of Cs_2O, we calculate D_0^0(Cs-O) = 60±10 kcal mol^{-1} and $\Delta_f H°$(CsO, g, 298.15 K) = 17±10 kcal mol^{-1}.

The enthalpy of formation for CsO(g) is tentatively selected as 15±10 kcal mol^{-1}.

Heat Capacity and Entropy

The ground state configuration is assumed to be the same as that of the isoelectronic molecule OH(g). ω_e is estimated from those of LiO(g), CsF(g), and CsCl(g). The bond distance is taken from that in CsOH(g) reported by Lide and Kuczkowski (2). $\omega_e x_e$ is estimated by comparison with those of the other alkali oxides. The values of B_e and α_e are calculated from r_e, ω_e and $\omega_e x_e$ by the method suggested by Herzberg (3).

References

1. L. Brewer and J. L. Margrave, J. Phys. Chem. 59, 421 (1955).
2. D. R. Lide, Jr. and R. L. Kuczkowski, J. Chem. Phys. 46, 4768 (1967).
3. G. Herzberg, "Spectra of Diatomic Molecules," D. Van Nostrand Co., Inc., New York, 1950.

Cesium Oxide (CsO) $Cs_1O_1(g)$

Enthalpy Reference Temperature = T_r = 298.15 K Standard State Pressure = p° = 0.1 MPa

T/K	$C_p°$ (J K^{-1}mol^{-1})	S° (J K^{-1}mol^{-1})	-[G°-H°(T_r)]/T (J K^{-1}mol^{-1})	H°-H°(T_r) (kJ mol^{-1})	$\Delta_f H°$ (kJ mol^{-1})	$\Delta_f G°$ (kJ mol^{-1})	Log K_f
0	0.	0.	INFINITE	-9.879	64.939	64.939	INFINITE
100	31.670	217.938	287.000	-6.906	64.311	56.678	-29.606
200	35.247	241.201	258.847	-3.529	63.552	49.351	-12.889
250	36.016	249.156	256.140	-1.746	63.187	45.843	-9.578
298.15	36.481	255.542	255.542	0.	62.760	42.539	-7.453
300	36.495	255.768	255.543	0.067	62.741	42.414	-7.385
350	36.816	261.419	255.988	1.901	60.141	39.402	-5.880
400	37.044	266.351	256.982	3.748	59.656	36.473	-4.763
450	37.216	270.724	258.270	5.604	59.183	33.604	-3.901
500	37.351	274.653	259.715	7.469	58.715	30.786	-3.216
600	37.555	281.482	262.791	11.214	57.775	25.289	-2.202
700	37.709	287.283	265.886	14.978	56.814	19.950	-1.489
800	37.835	292.326	268.882	18.755	55.830	14.751	-0.963
900	37.945	296.789	271.740	22.544	54.822	9.676	-0.562
1000	38.046	300.793	274.448	26.344	-13.336	8.414	-0.440
1100	38.139	304.423	277.011	30.153	-13.360	10.590	-0.503
1200	38.228	307.745	279.436	33.972	-13.395	12.769	-0.556
1300	38.313	310.809	281.733	37.799	-13.439	14.951	-0.601
1400	38.397	313.651	283.912	41.634	-13.491	17.137	-0.639
1500	38.478	316.303	285.984	45.478	-13.550	19.327	-0.673
1600	38.558	318.789	287.958	49.330	-13.618	21.521	-0.703
1700	38.638	321.129	289.841	53.190	-13.696	23.719	-0.729
1800	38.716	323.340	291.641	57.057	-13.786	25.922	-0.752
1900	38.794	325.435	293.365	60.933	-13.891	28.131	-0.773
2000	38.871	327.427	295.019	64.816	-14.015	30.346	-0.793
2100	38.948	329.325	296.607	68.707	-14.162	32.568	-0.810
2200	39.024	331.139	298.136	72.606	-14.337	34.797	-0.826
2300	39.100	332.875	299.609	76.512	-14.545	37.035	-0.841
2400	39.176	334.541	301.030	80.426	-14.791	39.283	-0.855
2500	39.252	336.142	302.403	84.347	-15.080	41.541	-0.868
2600	39.328	337.683	303.730	88.276	-15.419	43.813	-0.880
2700	39.403	339.168	305.015	92.213	-15.814	46.098	-0.892
2800	39.478	340.603	306.261	96.157	-16.271	48.400	-0.903
2900	39.554	341.989	307.469	100.108	-16.796	50.718	-0.914
3000	39.629	343.331	308.642	104.068	-17.395	53.056	-0.924
3100	39.704	344.632	309.782	108.034	-18.075	55.416	-0.934
3200	39.779	345.894	310.891	112.008	-18.842	57.798	-0.943
3300	39.854	347.119	311.971	115.990	-19.661	60.203	-0.953
3400	39.928	348.310	313.022	119.979	-20.604	62.637	-0.962
3500	40.003	349.468	314.047	123.976	-21.648	65.100	-0.972
3600	40.078	350.596	315.046	127.980	-22.797	67.595	-0.981
3700	40.152	351.695	316.022	131.991	-24.044	70.121	-0.990
3800	40.227	352.767	316.975	136.010	-25.402	72.684	-0.999
3900	40.302	353.813	317.906	140.037	-26.863	75.282	-1.008
4000	40.376	354.834	318.817	144.070	-28.461	77.921	-1.018
4100	40.451	355.832	319.707	148.112	-30.181	80.601	-1.027
4200	40.525	356.808	320.579	152.161	-32.024	83.325	-1.036
4300	40.600	357.762	321.433	156.217	-33.992	86.095	-1.046
4400	40.674	358.697	322.269	160.281	-35.794	88.878	-1.055
4500	40.749	359.612	323.089	164.352	-37.403	91.671	-1.064
4600	40.823	360.508	323.893	168.430	-39.572	94.562	-1.074
4700	40.898	361.387	324.681	172.516	-41.842	97.503	-1.084
4800	40.972	362.249	325.455	176.610	-44.210	100.493	-1.094
4900	41.047	363.094	326.214	180.711	-46.672	103.533	-1.104
5000	41.121	363.924	326.960	184.819	-49.225	106.624	-1.114
5100	41.195	364.739	327.693	188.935	-51.864	109.767	-1.124
5200	41.270	365.540	328.413	193.058	-54.051	112.884	-1.134
5300	41.344	366.327	329.121	197.189	-56.779	116.121	-1.144
5400	41.418	367.100	329.817	201.327	-59.295	119.366	-1.155
5500	41.493	367.861	330.502	205.473	-61.961	122.675	-1.165
5600	41.567	368.609	331.176	209.626	-64.820	126.059	-1.176
5700	41.641	369.346	331.839	213.786	-67.725	129.493	-1.187
5800	41.716	370.070	332.492	217.954	-70.672	132.979	-1.198
5900	41.790	370.784	333.135	222.129	-73.655	136.515	-1.209
6000	41.864	371.487	333.768	226.312	-76.668	140.103	-1.220

PREVIOUS: December 1968 (1 atm) CURRENT: December 1968 (1 bar)

Cesium Oxide (CsO) $Cs_1O_1(g)$

CESIUM (Cs_2) IDEAL GAS M_r = 265.8108

D_0^0 = 43.389 ± 0.096 kJ mol^{-1}

S°(298.15 K) = 284.675 ± 0.084 J K^{-1} mol^{-1}

$\Delta_f H°$(0 K) = 111.77 ± 0.26 kJ mol^{-1}

$\Delta_f H°$(298.15 K) = 107.37 ± 0.30 kJ mol^{-1}

Spectroscopic Constants for Cs_2 in cm^{-1}

State	T_e	ω_e	$10^2\omega_e x_e$	$10^5\omega_e y_e$	$10^7\omega_e z_e$	$10^2 B_e$	$10^5\alpha_e$	$10^8\gamma_e$	$10^9 D_e$	$10^{10}\beta_e$	r_e, Å
$X^1\Sigma_g^+$	0	42.0194	8.191	-9.0404	-3.2912[b]	1.1743	2.215	-8.2448[c]	3.73	1.2608[d]	4.646[e]
$C^1\Pi_u$	15949[a]	29.664	4.202	--	--	1.2360	6.6503	--	[8.5833]	--	4.34 (5)

(a) The dissociation energy, D_0^0, for the $C^1\Pi_u$ state is 18228 cm^{-1}

(b) $-5.8173 \times 10^{-9}(v + 1/2)^4 - 4.5832 \times 10^{-12}(v + 1/2)^5$ for $^1\Sigma_g^+$

(c) $+5.8751 \times 10^{-10}(v + 1/2)^3 - 3.0391 \times 10^{-12}(v + 1/2)^4 + 1.8161 \times 10^{-14}(v + 1/2)^5$ for $^1\Sigma_g^+$

(d) $+9.2704 \times 10^{-13}(v + 1/2)^2 - 9.7917 \times 10^{-15}(v + 1/2)^3$ for $^1\Sigma_g^+$

(e) As determined from B_e

Enthalpy of Formation

Using the adopted value of the ground state dissociation energy for Cs_2(g), D_0^0 = 43.389 kJ mol^{-1} (1), and the recommended $\Delta_f H°$(Cs, g, 0 K) = 155.160 kJ mol^{-1} from NBS (6), we calculate the adopted value of $\Delta_f H°$(Cs_2, g, 0 K) = 111.77 kJ mol^{-1}. This value is in essential agreement with the NBS recommended value of 111.7 kJ mol^{-1}. Correcting the adopted $\Delta_f H°$(0 K) to 298.15 K using the current value of H°(298.15 K)-H°(0 K) = 11.029 kJ mol^{-1} for Cs_2(g) and the NBS value of H°(298.15 K)-H°(0 K) = 7.711 kJ mol^{-1} for Cs(cr) (6), we obtain $\Delta_f H°$(Cs_2, g, 298.15 K) = 107.38 kJ mol^{-1}.

Heat Capacity and Entropy

The heat capacity and entropy were calculated using a direct summation technique similar to the Li_2(g) calculation (7) using ground state data from Raab et al (1), as recommended by Stwalley (2). Spectroscopic data for the $C^1\Pi_u$ state of Cs_2(g) at T_e = 15949 cm^{-1} come from the work of Raab et. al. (4). The other low lying states of Cs_2(g) are the $^3\Pi_u$ state at T_e = 7850 cm^{-1} (3), the $B^1\Pi_u$ state at 13044 cm^{-1} (5), and the $^1\Sigma_u^+$ state at T_e = 16720 cm^{-1} (4). These states have not been characterized satisfactorily to date. To include the excited states in the thermodynamic calculation in an approximate manner, we proceeded as with the Na_2(g) species as follows (7). The $C^1\Pi_u$ state was included using the spectroscopic data reported by Raab (1). To account for the $^3\Pi_u$, $B^1\Pi_u$ and $^1\Sigma_u^+$ states, the rotational levels of the $C^1\Pi_u$ state were weighted x 13 to account for the sixfold degeneracy of the $^3\Pi_u$ state, the threefold degeneracy of each of the $^1\Pi_u$ states and the singly degenerate $^1\Sigma_u^+$ state. Note that the $^3\Sigma_u^+$ state is weakly bound and was neglected. Due to the minimal number of vibrational and rotational levels used by (1) to generate Dunham coefficients for the $C^1\Pi_u$ state, we artificially truncated the G(v) expression after the $\omega_e x_e$ term and the B(v) expression after the $\int_e$ term. For the same reason, we used Herzberg's method (see Li_2 table) to estimate D_e = 8.5833 x 10^{-9} cm^{-1} rather than use the value of (1).

As with the Na_2(g) calculation, the above method certainly underestimates the $^3\Sigma_u$ and $B^1\Sigma_u$ contributions but at the same time overestimates the $^1\Sigma_u^+$ contribution to the thermodynamic functions. The contribution of the excited states to the free energy function is negligible at temperatures below 2000 K and only contribute ca. 0.25 J K^{-1} mol^{-1} at 6000 K. This method of treating the excited states does result, however, in a more realistic high temperature heat capacity. The rotational levels of the ground state were weighted in accordance with the nuclear spin of the atom (I = 7/2): even J weight = .5833, odd J weight = .4167. The adopted value of S°(298.15 K) is larger than the NBS adopted value (6) by 0.74 J K^{-1} mol^{-1} but is in exact agreement with that adopted by Gurvich, et al. (8).

References

1. M. Raab, H. Weickenmeier, and W. Demtroder, Chem. Phys. Lett. 88, 377 (1982).
2. W. C. Stwalley, Director: Iowa Laser Facility and Professor of Chemistry and Physics, University of Iowa, personal communication, 1982.
3. R. P. Benedict, D. L. Drummond and L. A. Schlie, J. Chem. Phys. 66, 4600 (1977).
4. M. Raab, G. Honing, W. Demtroder, and C. R. Vidal, J. Chem. Phys. 76, 4370 (1982).
5. K. P. Huber and G. Herzberg, "Constants of Diatomic Molecules," Van Nostrand Reinhold Co., New York, 1981.
6. D. D. Wagman, et al., J. Phys. Chem. Ref. Data 11, Supp. 2 (1982).
7. JANAF Thermochemical Tables: Li_2(g), 12-31-82; Na_2(g), 12-31-82.
8. L. V. Gurvich, et al., "Thermodynamic Properties of Individual Substances," Vol. IV, 3rd ed., Nauka, Moscow, 1982.

Cesium (Cs_2) Cs_2(g)

Enthalpy Reference Temperature = T_r = 298.15 K — Standard State Pressure = p° = 0.1 MPa

T/K	$C_p°$ (J K^{-1}mol^{-1})	S° (J K^{-1}mol^{-1})	-[G°-H°(T_r)]/T (J K^{-1}mol^{-1})	H°-H°(T_r) (kJ mol^{-1})	$\Delta_f H°$ (kJ mol^{-1})	$\Delta_f G°$ (kJ mol^{-1})	Log K_f
0	0.	0.	INFINITE	-11.029	111.770	111.770	INFINITE
100	37.404	243.396	318.414	-7.502	111.000	97.651	-51.008
200	37.872	269.484	288.161	-3.735	109.406	84.890	-22.171
298.15	38.250	284.675	284.675	0.	107.366	73.263	-12.835
300	38.258	284.912	284.676	0.071	107.317	73.052	-12.719
400	38.686	295.977	286.182	3.918	100.605	63.381	-8.277
500	38.962	304.645	289.039	7.803	98.227	54.353	-5.678
600	38.757	311.739	292.249	11.694	95.904	45.796	-3.987
700	37.949	317.660	295.469	15.534	93.551	37.630	-2.808
800	36.671	322.648	298.563	19.268	91.098	29.807	-1.946
900	35.150	326.881	301.481	22.860	88.502	22.300	-1.294
1000	33.582	330.503	304.206	26.297	-48.515	22.490	-1.175
1100	32.089	333.633	306.743	29.579	-49.390	29.631	-1.407
1200	30.734	336.366	309.100	32.719	-50.408	36.859	-1.604
1300	29.537	338.777	311.292	35.731	-51.554	44.177	-1.775
1400	28.497	340.927	313.333	38.632	-52.815	51.587	-1.925
1500	27.602	342.862	315.238	41.436	-54.177	59.091	-2.058
1600	26.836	344.618	317.021	44.156	-55.628	66.690	-2.177
1700	26.183	346.225	318.692	46.807	-57.161	74.381	-2.285
1800	25.631	347.706	320.263	49.396	-58.771	82.165	-2.384
1900	25.166	349.079	321.744	51.936	-60.454	90.040	-2.475
2000	24.781	350.359	323.143	54.432	-62.209	98.006	-2.560
2100	24.468	351.561	324.468	56.894	-64.038	106.061	-2.638
2200	24.221	352.693	325.726	59.328	-65.942	114.206	-2.712
2300	24.037	353.765	326.922	61.741	-67.927	122.438	-2.781
2400	23.912	354.786	328.062	64.138	-69.997	130.759	-2.846
2500	23.842	355.760	329.150	66.525	-72.156	139.168	-2.908
2600	23.826	356.695	330.192	68.908	-74.414	147.665	-2.967
2700	23.861	357.594	331.190	71.292	-76.776	156.251	-3.023
2800	23.944	358.464	332.149	73.682	-79.250	164.927	-3.077
2900	24.073	359.306	333.071	76.082	-81.844	173.693	-3.129
3000	24.245	360.125	333.959	78.498	-84.568	182.550	-3.178
3100	24.456	360.923	334.816	80.932	-87.431	191.501	-3.227
3200	24.704	361.703	335.644	83.390	-90.442	200.546	-3.274
3300	24.986	362.468	336.445	85.874	-93.529	209.682	-3.319
3400	25.297	363.218	337.222	88.388	-96.831	218.919	-3.363
3500	25.634	363.956	337.975	90.935	-100.301	228.256	-3.407
3600	25.994	364.683	338.707	93.516	-103.947	237.694	-3.449
3700	26.371	365.401	339.419	96.134	-107.751	247.235	-3.490
3800	26.764	366.109	340.112	98.791	-111.741	256.880	-3.531
3900	27.168	366.810	340.787	101.487	-115.898	266.630	-3.571
4000	27.579	367.503	341.447	104.224	-120.289	276.494	-3.611
4100	27.994	368.189	342.090	107.003	-124.883	286.470	-3.650
4200	28.410	368.868	342.720	109.823	-129.686	296.561	-3.688
4300	28.824	369.542	343.336	112.685	-134.700	306.769	-3.726
4400	29.232	370.209	343.939	115.588	-139.340	317.028	-3.764
4500	29.632	370.870	344.530	118.531	-143.560	327.330	-3.800
4600	30.022	371.526	345.110	121.514	-148.862	337.852	-3.836
4700	30.400	372.176	345.679	124.535	-154.330	348.492	-3.873
4800	30.763	372.820	346.238	127.593	-159.960	359.249	-3.909
4900	31.110	373.458	346.787	130.687	-165.745	370.126	-3.946
5000	31.440	374.089	347.326	133.815	-171.679	381.122	-3.982
5100	31.752	374.715	347.857	136.975	-177.753	392.238	-4.017
5200	32.043	375.335	348.380	140.165	-182.897	403.319	-4.051
5300	32.315	375.948	348.894	143.383	-189.093	414.651	-4.087
5400	32.566	376.554	349.401	146.627	-194.838	426.015	-4.121
5500	32.795	377.154	349.900	149.895	-200.859	437.521	-4.155
5600	33.004	377.746	350.392	153.185	-207.241	449.185	-4.190
5700	33.192	378.332	350.877	156.495	-213.695	460.965	-4.224
5800	33.358	378.911	351.355	159.823	-220.208	472.857	-4.259
5900	33.505	379.483	351.827	163.166	-226.774	484.863	-4.293
6000	33.630	380.047	352.293	166.523	-233.380	496.980	-4.327

PREVIOUS: June 1968 (1 atm) CURRENT: December 1983 (1 bar)

Cesium (Cs_2) Cs_2(g)

CESIUM FLUORIDE $((CsF)_2)$ IDEAL GAS M_r = 303.807606

$\Delta_f H°(0\ K) = -885.12 \pm 8.4\ kJ\ mol^{-1}$

$S°(298.15\ K) = [352.3 \pm 8.4]\ J\ K^{-1}\ mol^{-1}$ $\Delta_f H°(298.15\ K) = -890.10 \pm 8.4\ kJ\ mol^{-1}$

Vibrational Frequencies and Degeneracies

ν, cm^{-1}	ν, cm^{-1}
[200](1)	[120](1)
[110](1)	[220](1)
[210](1)	[230](1)

Ground State Quantum Weight: [1]

Point Group: $[D_{2h}]$ σ = [4]

Bond Distance: Cs-F = [2.35] Å

Bond Angles: F-Cs-F = [85]° Cs-F-Cs = [95]°

Product of the Moments of Inertia: $I_A I_B I_C = [3.127422 \times 10^{-112}]\ g^3\ cm^6$

Enthalpy of Formation

Evidence from the velocity distribution in molecular beams (1) has shown that alkali halide vapors contain significant amounts of polymeric species. Using the same method Eisenstadt (2) determined the molecular composition of CsF vapor in the temperature range 838-919 K. Based on the reported equilibrium constants for the reaction $(CsF)_2(g) = 2CsF(g)$, we evaluate the enthalpy change, $\Delta_r H°(298.15\ K)$, by the 2nd and 3rd law methods to be 40.53±1.75 and 42.34±0.26 kcal mol^{-1}, respectively. Employing $\Delta_r H°(298.15\ K)$ = 42.34 and $\Delta_f H°(CsF, g, 298.15\ K)$ = -85.2 kcal mol^{-1}, we obtained $\Delta_f H°(298.15\ K)$ = -212.74 kcal mol^{-1} (-890.10±8.4 kJ mol^{-1}) for $Cs_2F_2(g)$, which is adopted.

Heat Capacity and Entropy

The Cs_2F_2 molecule is assumed to have a rhombic configuration of symmetry D_{2h} as Cs_2Cl_2 (3). The Cs-F bond distance is taken as the same as that in CsF(g). The bond angles and vibrational frequencies are estimated by comparison with those of the other dimeric alkali halides (3, 4). The principal moments of inertia are: $I_A = 15.9040 \times 10^{-39}$, $I_B = 132.5032 \times 10^{-39}$ and $I_C = 148.4071 \times 10^{-39}\ g\ cm^2$.

References

1. R. C. Miller and P. Kusch, J. Chem. Phys. 25, 860 (1953); ibid. 27, 981 (1957).
2. M. Eisenstadt, G. M. Rothberg and P. Kusch, J. Chem. Phys. 29, 797 (1958).
3. J. Berkowitz, J. Chem. Phys. 29, 1386 (1958).
4. J. Berkowitz, J. Chem. Phys. 32, 1519 (1960).

Cesium Fluoride $((CsF)_2)$ $Cs_2F_2(g)$

Enthalpy Reference Temperature = T_r = 298.15 K   Standard State Pressure = p° = 0.1 MPa

T/K	$C_p°$ (J K⁻¹mol⁻¹)	S° (J K⁻¹mol⁻¹)	$-[G°-H°(T_r)]/T$ (J K⁻¹mol⁻¹)	$H°-H°(T_r)$ (kJ mol⁻¹)	$\Delta_f H°$ (kJ mol⁻¹)	$\Delta_f G°$ (kJ mol⁻¹)	Log K_f
0	0.	0.	INFINITE	-19.273	-885.119	-885.119	INFINITE
100	62.412	272.336	420.380	-14.804	-887.854	-887.060	463.353
200	76.312	321.055	359.539	-7.697	-889.036	-885.736	231.330
250	78.605	338.355	353.628	-3.818	-889.500	-884.858	184.881
298.15	79.888	352.319	352.319	0.	-890.104	-883.913	154.858
300	79.926	352.813	352.320	0.148	-890.134	-883.875	153.896
350	80.750	365.200	353.297	4.166	-895.093	-882.098	131.646
400	81.296	376.021	355.475	8.218	-895.841	-880.190	114.941
450	81.675	385.619	358.301	12.293	-896.575	-878.190	101.938
500	81.950	394.239	361.471	16.384	-897.305	-876.108	91.526
600	82.310	409.215	368.218	24.599	-898.778	-871.730	75.891
700	82.529	421.921	375.005	32.841	-900.280	-867.104	64.704
800	82.672	432.952	381.574	41.102	-901.815	-862.260	56.300
900	82.771	442.695	387.835	49.374	-903.386	-857.222	49.752
1000	82.842	451.420	393.764	57.655	-1039.248	-844.608	44.118
1100	82.894	459.318	399.370	65.942	-1038.838	-825.164	39.184
1200	82.934	466.532	404.671	74.234	-1038.450	-805.756	35.074
1300	82.965	473.172	409.688	82.529	-1038.084	-786.380	31.597
1400	82.989	479.321	414.445	90.826	-1037.738	-767.032	28.618
1500	83.009	485.048	418.963	99.126	-1037.413	-747.707	26.037
1600	83.026	490.405	423.263	107.428	-1037.113	-728.403	23.780
1700	83.039	495.439	427.362	115.731	-1036.838	-709.117	21.789
1800	83.050	500.186	431.277	124.036	-1036.593	-689.847	20.019
1900	83.060	504.676	435.023	132.341	-1036.381	-670.589	18.436
2000	83.068	508.937	438.613	140.648	-1036.208	-651.341	17.011
2100	83.075	512.990	442.059	148.955	-1036.078	-632.101	15.723
2200	83.081	516.855	445.372	157.263	-1035.999	-612.867	14.551
2300	83.087	520.548	448.561	165.571	-1035.976	-593.634	13.482
2400	83.091	524.084	451.634	173.880	-1036.017	-574.401	12.502
2500	83.096	527.477	454.601	182.189	-1036.129	-555.165	11.600
2600	83.099	530.736	457.467	190.499	-1036.322	-535.923	10.767
2700	83.102	533.872	460.239	198.809	-1036.604	-516.672	9.996
2800	83.105	536.894	462.923	207.120	-1036.985	-497.408	9.279
2900	83.108	539.811	465.524	215.430	-1037.475	-478.130	8.612
3000	83.110	542.628	468.048	223.741	-1038.085	-458.832	7.989
3100	83.112	545.353	470.498	232.052	-1038.826	-439.512	7.406
3200	83.114	547.992	472.878	240.364	-1039.709	-420.165	6.858
3300	83.116	550.550	475.193	248.675	-1040.665	-400.795	6.344
3400	83.118	553.031	477.446	256.987	-1041.833	-381.388	5.859
3500	83.119	555.440	479.641	265.299	-1043.170	-361.944	5.402
3600	83.121	557.782	481.779	273.611	-1044.683	-342.460	4.969
3700	83.122	560.059	483.864	281.923	-1046.358	-322.933	4.559
3800	83.123	562.276	485.898	290.235	-1048.223	-303.358	4.170
3900	83.124	564.435	487.885	298.548	-1050.260	-283.734	3.800
4000	83.125	566.540	489.825	306.860	-1052.538	-264.051	3.448
4100	83.126	568.592	491.721	315.173	-1055.027	-244.309	3.113
4200	83.127	570.595	493.575	323.485	-1057.732	-224.503	2.792
4300	83.128	572.551	495.389	331.798	-1060.656	-204.630	2.486
4400	83.128	574.463	497.165	340.111	-1063.216	-184.754	2.193
4500	83.129	576.331	498.903	348.424	-1065.362	-164.881	1.914
4600	83.130	578.158	500.606	356.737	-1068.599	-144.836	1.645
4700	83.130	579.946	502.275	365.050	-1072.010	-124.717	1.386
4800	83.131	581.696	503.912	373.363	-1075.589	-104.524	1.137
4900	83.131	583.410	505.517	381.676	-1079.330	-84.256	0.898
5000	83.132	585.089	507.092	389.989	-1083.224	-63.909	0.668
5100	83.132	586.736	508.637	398.302	-1087.264	-43.483	0.445
5200	83.133	588.350	510.155	406.615	-1090.375	-23.131	0.232
5300	83.133	589.933	511.645	414.929	-1094.540	-2.568	0.025
5400	83.134	591.487	513.109	423.242	-1098.256	17.989	-0.174
5500	83.134	593.013	514.548	431.555	-1102.246	38.650	-0.367
5600	83.134	594.511	515.963	439.869	-1106.596	59.433	-0.554
5700	83.135	595.982	517.354	448.182	-1111.012	80.293	-0.736
5800	83.135	597.428	518.722	456.496	-1115.485	101.232	-0.912
5900	83.135	598.849	520.068	464.809	-1120.003	122.249	-1.082
6000	83.136	600.246	521.393	473.123	-1124.554	143.342	-1.248

PREVIOUS: June 1968 (1 atm) CURRENT: June 1968 (1 bar)

Cesium Fluoride $((CsF)_2)$ $Cs_2F_2(g)$

CESIUM HYDROXIDE ((CsOH)$_2$) IDEAL GAS M_r = 299.82548

$S°(298.15\ K) = 360.7 \pm 12.6\ J\ K^{-1}\ mol^{-1}$

$\Delta_f H°(0\ K) = -674.7 \pm 41.8\ kJ\ mol^{-1}$

$\Delta_f H°(298.15\ K) = -687.8 \pm 41.8\ kJ\ mol^{-1}$

Vibrational Frequencies and Degeneracies

ν, cm^{-1}	ν, cm^{-1}
[200](1)	[1250](1)
[110](1)	[1250](1)
[210](1)	[1250](1)
[120](1)	[1250](1)
[220](1)	[3700](1)
[230](1)	[3700](1)

Ground State Quantum Weight: 1 $\sigma = 2$

Point Group: C_{2h}

Bond Distances: Cs-O = [2.63] Å O-H = [0.96] Å

Bond Angles: Cs-O-Cs = [90]° Cs-O-H = [110]°

Product of Moments of Inertia: $I_A I_B I_C = [5.396347 \times 10^{-112}]\ g^3\ cm^6$

Enthalpy of Formation

Schoonmaker and Porter (1) analyzed the vapors in equilibrium with liquid CsOH and mixed KOH-CsOH condensed phases with a mass spectrometer and reported the presence of appreciable concentrations of dimer in the temperature range 650-700 K. By applying the method of relative equilibrium constants, these workers calculated the difference in the free energies of dimerization for KOH-CsOH. A 3rd law analysis of their free energy data for this pair leads to a difference in the enthalpies of dimerization of 4.9 kcal mol^{-1} at 298.15 K for CsOH and KOH. Based upon the adopted value for KOH(g), $\Delta_f H°$(dimerization, 298.15 K) = -45.3±3.0 kcal mol^{-1} (2), we derive $\Delta_f H°$(dimerization, 298.15 K) = -40.4±4.0 kcal mol^{-1} for 2 CsOH(g) = $Cs_2(OH)_2$(g). Combining this result with the enthalpy of formation for the gaseous monomer (3), that for the dimer is $\Delta_f H°(Cs_2(OH)_2$, g, 298.15 K) = -164.4±10.0 kcal mol^{-1} (-687.8±41.8 kJ mol^{-1}).

Heat Capacity and Entropy

Buchler et al. (4) investigated the geometries of several molecules containing cesium by the deflection of molecular beams in an inhomogeneous electric field. Their results for $Cs_2(OH)_2$(g) indicated that the molecule was nonpolar which implies a planar configuration for at least the Cs_2O_2 part of the molecule. Several structural models proposed for the dimers of the alkali metal hydroxides each incorporate this planar configuration for the alkali and oxygen atoms. Bauer et al. (5) calculated entropies of dimerization for NaOH(g) and KOH(g) with a dimer model consisting of a square planar configuration for the alkali and oxygen atoms with two hydrogen bonded bridges between the oxygens and hydrogens. Schoonmaker and Porter (1) adopted the same alkali-oxygen configuration but did not allow for hydrogen bonding between the oxygens and hydrogens. Berkowitz et al. (6) relied on the apparent similarity of the alkali halides and hydroxides in forming dimeric species to propose a structural model for $Li_2(OH)_2$(g). This model consisted of a trans-configuration with hydrogens above and below the plane of a rhombus formed by the lithium and oxygen atoms, the Li-O-Li bond angle being 100°. We adopt a structure for $Cs_2(OH)_2$(g) which is similar to the model of Berkowitz et al. (6) for $Li_2(OH)_2$(g) but with the Cs-O-Cs bond angle equal to 90°. The cesium and oxygen atoms form a square planar configuration. The Cs-O-H bond angle is assumed equal to 110°. The Cs-O bond distance is estimated as being 10% longer than that in CsOH(g). The O-H bond distance is estimated to be the same as in H_2O(g). The principal moments of inertia are: $I_A = [20.4212] \times 10^{-39}$, $I_B = [152.8696] \times 10^{-39}$, and $I_C = [172.8610] \times 10^{-39}$ g cm^2.

Acquista et al. (7) observed the infrared spectra of matrix isolated CsOH and reported bands in the region 300-200 cm^{-1} whose relative intensities decreased with increasing temperature. Since the dimer to monomer ratio decreases with increasing temperature, these bands appeared to arise from polymeric species of CsOH, but no definite assignments could be made. Similar results have been reported for RbOH and NaOH from infrared studies (8) of these molecules trapped in inert matrices. Berkowitz et al. (6) estimated vibrational frequencies for $Li_2(OH)_2$(g) by comparison with those for dimeric Li_2F_2(g). The frequencies for Li_2F_2(g) had been estimated earlier by Berkowitz (9) using an ionic model and normal coordinate analysis. It has now been established experimentally that these estimated frequencies for Li_2F_2(g) are too low (See Li_2F_2(g) table for details). Until definite assignments for the dimer hydroxides can be made from infrared studies, we tentatively adopt the six estimated frequencies for Cs_2F_2(g) (See CsF_2(g) table for details) and apply them directly to $Cs_2(OH)_2$(g). The remaining O-H stretching and bending frequencies are estimates from the work of Berkowitz et al. (6).

References

1. R. C. Schoonmaker and R. F. Porter, J. Chem. Phys., 31, 830 (1959).
2. JANAF Thermochemical Tables: KOH(g), 12-31-71; CsOH(g), 6-30-71.
3. A. Buchler, J. L. Stauffer, and W. Klemperer, J. Chem. Phys., 46, 605 (1967).
4. S. H. Bauer, R. H. Diner, and R. F. Porter, J. Chem. Phys., 29, 991 (1958).
5. J. Berkowitz, D. J. Meschi, and W. A. Chupka, J. Chem. Phys., 33, 533 (1960).
6. N. Acquista, S. Abramowitz, and D. R. Lide, J. Chem. Phys., 49, 780 (1968).
7. N. Acquista and S. Abramowitz, J. Chem. Phys., 51, 2911 (1969).
8. J. Berkowitz, J. Chem. Phys., 32, 1519 (1960).

Cesium Hydroxide ((CsOH)$_2$) $Cs_2H_2O_2$(g)

Enthalpy Reference Temperature = T_r = 298.15 K Standard State Pressure = p° = 0.1 MPa

	J $K^{-1}mol^{-1}$			kJ mol^{-1}			
T/K	$C_p°$	S°	$-[G°-H°(T_r)]/T$	$H°-H°(T_r)$	$\Delta_f H°$	$\Delta_f G°$	Log K_f
0	0.	0.	INFINITE	-19.417	-674.683	-674.683	INFINITE
100	62.412	280.202	429.685	-14.948	-680.415	-670.041	349.994
200	76.646	328.962	368.129	-7.833	-684.265	-658.097	171.877
250	79.900	346.426	362.095	-3.917	-686.043	-651.350	136.092
298.15	82.807	360.748	360.748	0.	-687.850	-644.507	112.915
300	82.919	361.260	360.749	0.153	-687.923	-644.237	112.172
350	85.963	374.271	361.770	4.375	-694.016	-636.132	94.937
400	88.968	385.947	364.074	8.749	-695.764	-627.741	81.975
450	91.824	396.592	367.104	13.270	-697.362	-619.141	71.868
500	94.461	406.406	370.550	17.928	-698.830	-610.370	63.765
600	99.017	424.045	378.029	27.609	-701.451	-592.425	51.575
700	102.729	439.597	385.735	37.703	-703.743	-574.069	42.838
800	105.812	453.521	393.353	48.134	-705.788	-555.402	36.264
900	108.443	466.140	400.751	58.850	-707.641	-536.491	31.137
1000	110.738	477.687	407.875	69.812	-843.598	-509.983	26.639
1100	112.767	488.338	414.712	80.989	-843.126	-476.643	22.634
1200	114.576	498.229	421.264	92.358	-842.542	-443.352	19.299
1300	116.193	507.465	427.544	103.898	-841.865	-410.113	16.479
1400	117.640	516.130	433.565	115.591	-841.111	-376.929	14.063
1500	118.938	524.291	439.344	127.421	-840.295	-343.801	11.972
1600	120.101	532.005	444.897	139.374	-839.433	-310.729	10.144
1700	121.144	539.318	450.238	151.437	-838.539	-277.713	8.533
1800	122.081	546.270	455.381	163.599	-837.626	-244.749	7.102
1900	122.923	552.893	460.341	175.850	-836.709	-211.837	5.824
2000	123.680	559.218	465.128	188.181	-835.803	-178.972	4.674
2100	124.364	565.269	469.753	200.583	-834.921	-146.152	3.635
2200	124.980	571.069	474.227	213.051	-834.079	-113.373	2.692
2300	125.539	576.637	478.560	225.578	-833.292	-80.632	1.831
2400	126.045	581.991	482.759	238.157	-832.573	-47.924	1.043
2500	126.504	587.146	486.832	250.785	-831.937	-15.244	0.318
2600	126.923	592.116	490.786	263.457	-831.400	17.413	-0.350
2700	127.304	596.913	494.628	276.168	-830.976	50.052	-0.968
2800	127.653	601.549	498.365	288.916	-830.679	82.677	-1.542
2900	127.972	606.034	502.000	301.698	-830.523	115.293	-2.077
3000	128.265	610.378	505.541	314.510	-830.525	147.907	-2.575
3100	128.534	614.588	508.991	327.350	-830.698	180.524	-3.042
3200	128.782	618.673	512.355	340.216	-831.057	213.149	-3.479
3300	129.010	622.639	515.637	353.106	-831.532	245.781	-3.890
3400	129.221	626.493	518.841	366.017	-832.269	278.438	-4.278
3500	129.417	630.242	521.971	378.949	-833.222	311.119	-4.643
3600	129.598	633.891	525.029	391.900	-834.402	343.831	-4.989
3700	129.767	637.444	528.020	404.869	-835.795	376.576	-5.316
3800	129.923	640.906	530.945	417.853	-837.430	409.363	-5.627
3900	130.069	644.283	533.808	430.853	-839.288	442.193	-5.923
4000	130.205	647.578	536.611	443.867	-841.441	475.078	-6.204
4100	130.333	650.795	539.357	456.894	-843.856	508.020	-6.472
4200	130.452	653.937	542.048	469.933	-846.541	541.025	-6.729
4300	130.563	657.008	544.686	482.984	-849.496	574.097	-6.974
4400	130.667	660.011	547.273	496.045	-852.140	607.173	-7.208
4500	130.766	662.948	549.811	509.117	-854.422	640.247	-7.432
4600	130.858	665.823	552.302	522.198	-857.846	673.499	-7.648
4700	130.945	668.638	554.747	535.288	-861.495	706.828	-7.856
4800	131.026	671.396	557.149	548.387	-865.363	740.238	-8.055
4900	131.104	674.099	559.508	561.494	-869.443	773.730	-8.248
5000	131.176	676.748	561.826	574.608	-873.727	807.307	-8.434
5100	131.245	679.346	564.105	587.729	-878.205	840.972	-8.613
5200	131.310	681.895	566.346	600.857	-881.803	874.570	-8.785
5300	131.372	684.397	568.550	613.991	-886.504	908.392	-8.953
5400	131.431	686.853	570.718	627.131	-890.804	942.217	-9.114
5500	131.486	689.266	572.852	640.277	-895.425	976.156	-9.271
5600	131.539	691.635	574.952	653.428	-900.453	1010.231	-9.423
5700	131.589	693.964	577.019	666.584	-905.595	1044.396	-9.571
5800	131.636	696.253	579.055	679.746	-910.839	1078.652	-9.714
5900	131.682	698.503	581.061	692.912	-916.174	1112.997	-9.854
6000	131.725	700.717	583.037	706.082	-921.588	1147.437	-9.989

PREVIOUS: June 1971 (1 atm) CURRENT June 1971 (1 bar)

Cesium Hydroxide ((CsOH)$_2$) $Cs_2H_2O_2$(g)

CESIUM OXIDE (Cs_2O) IDEAL GAS M_r = 281.8102

$S°(298.15\ K) = [-318.1 \pm 8.4]\ J\ K^{-1}\ mol^{-1}$

$\Delta_f H°(0\ K) = -86.2 \pm 41.8\ kJ\ mol^{-1}$
$\Delta_f H°(298.15\ K) = -92.0 \pm 41.8\ kJ\ mol^{-1}$

Vibrational Frequencies and Degeneracies

ν, cm^{-1}
[290](1)
[130](1)
[300](1)

Ground State Quantum Weight: [1]
Point Group: C_{2v}
Bond Distance: Cs-O = [2.4] Å
Bond Angle: Cs-O-Cs = [105]°
Product of the Moments of Inertia: $I_A I_B I_C = [1.415602 \times 10^{-112}]\ g^3\ cm^6$

σ = 2

Enthalpy of Formation

The appearance potentials of CsOH and Cs_2O ions were investigated by Emel'yanov et al. (1), using the electron-impact method with a mass spectrometer fitted with a heated ion source and a Pt effusion chamber. The enthalpy of atomization of Cs_2O was derived as 117±10 kcal mol^{-1}. Based on this value we calculate the enthalpy of formation $\Delta_f H°(298.15\ K)$ of Cs_2O to be -22±10 kcal mol^{-1} (-92.0±41.8 kJ mol^{-1}), employing $\Delta_f H°(0\ K)$ = 18.68 and 58.99 kcal mol^{-1} for Cs(g) and O(g), respectively. Brewer and Mastick (2) calculated theoretically the stability of gaseous alkali oxides according to ionic models and gave the enthalpies of formation from gaseous atoms for $Cs_2O(g)$ as -24 kcal mol^{-1}. The $\Delta_f H°(298.15\ K)$ value is evaluated to be 72.2 kcal mol^{-1} which is not adopted.

Heat Capacity and Entropy

The molecular structure of $Cs_2O(g)$ has been determined to be nonlinear by Buchler et al. (3), using electric-deflection method. The Cs-O bond distance is assumed to be the same as that in CsOH(g) reported by Lide and Kuczkowski (4). The bond angle is estimated by comparison with that in the H_2O molecule. Three vibrational frequencies are estimated from those for $H_2O(g)$. The principal moments of inertia are: $I_A = 5.3492 \times 10^{-39}$, $I_B = 160.0237 \times 10^{-39}$, and $I_C = 165.3730 \times 10^{-39}\ g\ cm^2$.

References

1. A. M. Emel'yanov, A. V. Gusarov, L. N. Gorokhov, and N. A. Sadovnikova, Teor. Eksp. Khim. 3, 226 (1967).
2. L. Brewer and D. F. Mastick, J. Amer. Chem. Soc. 73, 2045 (1951).
3. A. Buchler and J. L. Stauffer and W. Klemperer, J. Chem. Phys. 46, 605 (1967).
4. D. R. Lide, Jr. and R. L. Kuczkowski, J. Chem. Phys. 46, 4768 (1967).

Cesium Oxide (Cs_2O) $Cs_2O_1(g)$

Enthalpy Reference Temperature = T_r = 298.15 K Standard State Pressure = p° = 0.1 MPa

T/K	$C_p°$ (J K⁻¹mol⁻¹)	S° (J K⁻¹mol⁻¹)	$-[G°-H°(T_r)]/T$ (J K⁻¹mol⁻¹)	$H°-H°(T_r)$ (kJ mol⁻¹)	$\Delta_f H°$ (kJ mol⁻¹)	$\Delta_f G°$ (kJ mol⁻¹)	Log K_f
0	0.	0.	INFINITE	-13.937	-86.211	-86.211	INFINITE
100	43.945	262.828	365.083	-10.225	-88.248	-94.875	49.557
200	52.568	296.481	323.084	-5.320	-90.159	-100.726	26.307
250	54.345	308.420	318.995	-2.644	-91.051	-103.265	21.576
298.15	55.388	318.088	318.088	0.	-92.048	-105.530	18.488
300	55.420	318.431	318.089	0.102	-92.092	-105.614	18.389
350	56.109	327.029	318.767	2.892	-97.430	-107.104	15.984
400	56.574	334.554	320.280	5.710	-98.529	-108.410	14.157
450	56.902	341.237	322.244	8.547	-99.591	-109.581	12.720
500	57.141	347.246	324.449	11.398	-100.634	-110.635	11.558
600	57.457	357.694	329.145	17.130	-102.696	-112.441	9.789
700	57.651	366.567	333.873	22.886	-104.760	-113.903	8.500
800	57.778	374.274	338.452	28.658	-106.844	-115.067	7.513
900	57.866	381.085	342.818	34.440	-108.953	-115.968	6.731
1000	57.929	387.185	346.955	40.230	-245.346	-109.235	5.706
1100	57.976	392.708	350.867	46.026	-245.463	-95.618	4.541
1200	58.012	397.755	354.567	51.825	-245.596	-81.990	3.569
1300	58.040	402.399	358.070	57.628	-245.744	-68.350	2.746
1400	58.062	406.701	361.392	63.433	-245.906	-54.699	2.041
1500	58.080	410.708	364.548	69.240	-246.085	-41.035	1.429
1600	58.094	414.457	367.551	75.049	-246.283	-27.359	0.893
1700	58.106	417.979	370.415	80.859	-246.502	-13.669	0.420
1800	58.117	421.300	373.151	86.670	-246.748	0.034	-0.001
1900	58.125	424.443	375.768	92.482	-247.028	13.751	-0.378
2000	58.133	427.424	378.277	98.295	-247.348	27.485	-0.718
2100	58.139	430.261	380.686	104.108	-247.718	41.235	-1.026
2200	58.144	432.966	383.001	109.922	-248.146	55.005	-1.306
2300	58.149	435.550	385.230	115.737	-248.644	68.795	-1.562
2400	58.153	438.025	387.379	121.552	-249.222	82.609	-1.798
2500	58.157	440.399	389.452	127.368	-249.891	96.448	-2.015
2600	58.160	442.680	391.456	133.184	-250.663	110.317	-2.216
2700	58.163	444.875	393.394	139.000	-251.552	124.217	-2.403
2800	58.166	446.991	395.271	144.816	-252.568	138.153	-2.577
2900	58.168	449.032	397.089	150.633	-253.725	152.127	-2.740
3000	58.170	451.004	398.854	156.450	-255.036	166.144	-2.893
3100	58.172	452.911	400.567	162.267	-256.514	180.207	-3.036
3200	58.174	454.758	402.232	168.084	-258.173	194.321	-3.172
3300	58.176	456.548	403.851	173.902	-259.942	208.482	-3.300
3400	58.177	458.285	405.426	179.720	-261.964	222.707	-3.421
3500	58.179	459.972	406.961	185.537	-264.195	236.994	-3.537
3600	58.180	461.611	408.456	191.355	-266.644	251.348	-3.647
3700	58.181	463.205	409.915	197.173	-269.295	265.770	-3.752
3800	58.182	464.756	411.337	202.992	-272.178	280.268	-3.853
3900	58.183	466.268	412.727	208.810	-275.273	294.842	-3.949
4000	58.184	467.741	414.084	214.628	-278.651	309.503	-4.042
4100	58.185	469.177	415.410	220.446	-282.280	324.251	-4.131
4200	58.185	470.579	416.707	226.265	-286.165	339.091	-4.217
4300	58.186	471.949	417.976	232.084	-290.309	354.027	-4.301
4400	58.187	473.286	419.218	237.902	-294.127	368.994	-4.381
4500	58.187	474.594	420.434	243.721	-297.570	383.986	-4.457
4600	58.188	475.873	421.625	249.540	-302.141	399.182	-4.533
4700	58.188	477.124	422.793	255.358	-306.923	414.479	-4.606
4800	58.189	478.349	423.937	261.177	-311.910	429.881	-4.678
4900	58.189	479.549	425.060	266.996	-317.094	445.388	-4.748
5000	58.190	480.725	426.162	272.815	-322.467	461.003	-4.816
5100	58.190	481.877	427.243	278.634	-328.019	476.728	-4.883
5200	58.191	483.007	428.304	284.453	-332.678	492.407	-4.946
5300	58.191	484.115	429.347	290.272	-338.424	508.329	-5.010
5400	58.191	485.203	430.371	296.092	-343.754	524.274	-5.071
5500	58.192	486.271	431.378	301.911	-349.391	540.353	-5.132
5600	58.192	487.319	432.368	307.730	-355.420	556.585	-5.192
5700	58.192	488.349	433.341	313.549	-361.548	572.924	-5.250
5800	58.193	489.361	434.298	319.368	-367.764	589.373	-5.308
5900	58.193	490.356	435.240	325.188	-374.057	605.928	-5.364
6000	58.193	491.334	436.166	331.007	-380.415	622.592	-5.420

PREVIOUS: December 1968 (1 atm) CURRENT: December 1968 (1 bar)

Cesium Oxide (Cs_2O) $Cs_2O_1(g)$

CESIUM SULFATE, I (Cs_2SO_4, I) CRYSTAL(I) M_r = 361.8684

S°(298.15 K) = [264.834] J K^{-1} mol^{-1}
Tt(II→I) = 940 ± 5 K
T_{fus}(I→ℓ) = 1278 ± 5 K

$\Delta_fH°$(298.15 K) = [-1410.152] kJ mol^{-1}
$\Delta_{trs}H°$(II→I) = 4.310 ± 0.8 kJ mol^{-1}
$\Delta_{fus}H°$(I→ℓ) = 35.690 ± 0.8 kJ mol^{-1}

Enthalpy of Formation

$\Delta_fH°$(Cs_2SO_4, cr, I, 298.15 K) is calculated from $\Delta_fH°$(Cs_2SO_4, cr, II, 298.15 K) by adding the enthalpy of transition (II→I) and the difference in enthaply, H°(940) - H°(298.15 K), between the II and I crystalline forms (1).

Heat Capacity and Enthalpy

High-temperature enthalpies of the I-phase of Cs_2SO_4 have been determined by Denielou et al. (2, 960-1260 K) and by Polischchuk et al. (3, 950.7-1254.5 K). We have adopted heat capacities based upon the enthalpies of Denielou et al. (2). We have extrapolated the heat capacities smoothly to 298.15 K below the II→I transition and extended $C_p°$ above the melting point to 1500 K.

S°(298.15 K) is calculated in a manner similar to that used for the enthalpy of formation.

Transition Data

The high-temperature hexagonal modification of cesium sulfate is denoted Cs_2SO_4, cr, I, in accordance with the usual nomenclature. This modification has been referred to as the α-phase in contrast to our use of α to refer to the low-temperature phase for the other members of the alkali metal sulfates (1).

Fusion Data

The adopted melting point, T_{fus} = 1278±5 K, was determined from the work of Denielou et al. (2, 1274±3 K), Polishchuk et al. (3, 1281 K), Rassonskaya and Semendyaeva (4, 1283 K), and from the compilation of Stern and Weise (5, 1277 K). The enthalpy of fusion, $\Delta_{fus}H°$ = 8.530±0.2 kcal mol^{-1}, is calculated from the difference between the smoothed relative enthalpy of the liquid and the adopted enthalpy value for the I form at T_{fus}.

References

1. JANAF Thermochemical Tables: Cs_2SO_4(cr, II), 6-30-79; Na_2SO_4(cr, V), K_2SO_4(cr, α), 6-30-78.
2. L. Denielou, Y. Fournier, J. P. Petitet, and C. Tequi, C. R. Acad. Sci., Davis 270, 1854 (1970).
3. A. F. Polishchuk, T. A. Tishura, and A. N. Buderina, Ukr. Khim. Zh. 40, 120 (1974).
4. I. S. Rassonskaya and N. K. Semendyaeva, Russ. J. Inorg. Chem. 15, 27 (1970).
5. K. H. Stern and E. L. Weise, NSRDS-NBS 7, 1966.

Cesium Sulfate, I (Cs_2SO_4) $Cs_2O_4S_1$(cr)

Enthalpy Reference Temperature = T_r = 298.15 K Standard State Pressure = p° = 0.1 MPa

T/K	$C_p°$	S°	$-[G°-H°(T_r)]/T$	$H°-H°(T_r)$	$\Delta_fH°$	$\Delta_fG°$	Log K_f
	J $K^{-1}mol^{-1}$			kJ mol^{-1}			
0							
100							
200							
298.15	108.700	264.834	264.834	0.	-1410.152	-1306.452	228.885
300	108.784	265.506	264.836	0.201	-1410.221	-1305.809	227.361
400	112.131	297.150	269.129	11.209	-1420.312	-1268.991	165.713
500	120.081	322.947	277.379	22.784	-1425.046	-1230.612	128.561
600	132.256	345.866	286.911	35.373	-1428.574	-1191.370	103.718
700	147.737	367.416	296.883	49.373	-1430.624	-1151.653	85.937
800	163.218	388.154	307.003	64.921	-1431.148	-1111.747	72.590
900	178.699	408.272	317.143	82.016	-1483.308	-1070.783	62.147
940.000	184.891	416.177	321.189	89.288	-----------	II <--> I	----------
1000	194.179	427.901	327.241	100.660	-1613.897	-1017.731	53.161
1100	209.660	447.135	337.269	120.852	-1606.752	-958.448	45.513
1200	225.141	466.041	347.214	142.592	-1598.155	-899.880	39.171
1278.000	237.233	480.596	354.911	160.625	---------	I <--> LIQUID	--------
1300	240.622	484.675	357.073	165.882	-1588.093	-842.089	33.836
1400	256.103	503.073	366.846	190.718	-1576.563	-785.127	29.293
1500	271.583	521.270	376.535	217.103	-1563.560	-729.041	25.387

PREVIOUS: CURRENT: June 1979

Cesium Sulfate, I (Cs_2SO_4) $Cs_2O_4S_1$(cr)

CESIUM SULFATE, II (Cs_2SO_4, II) CRYSTAL(II) M_r = 361.8684

$S°(298.15\ K) = 211.88 \pm 0.21\ J\ K^{-1}\ mol^{-1}$

$T_{trs}(\alpha \rightarrow \beta) = 940 \pm 5\ K$

$\Delta_f H°(0\ K) = -1433.16 \pm 0.8\ kJ\ mol^{-1}$

$\Delta_f H°(298.15\ K) = -1442.64 \pm 0.8\ kJ\ mol^{-1}$

$\Delta_{trs} H°(\alpha \rightarrow \beta) = 4.310 \pm 0.8\ kJ\ mol^{-1}$

Enthalpy of Formation

The enthalpy of solution at infinite dilution of Cs_2SO_4 has been measured by Kuznetsova et al. (1, 3.980 kcal mol^{-1}) and by Borob'ev et al. (2, 4.11±0.80 kcal mol^{-1}). We adopt the mean of these values 4.05±0.1 kcal mol^{-1}. When this result is combined with enthalpies of formation of the infinitely dilute ions from CODATA (3), we obtain $\Delta_f H°(Cs_2SO_4, \infty H_2O$, 298.15 K) = -340.746±0.12 kcal mol^{-1} and $\Delta_f H°(Cs_2SO_4$, cr, 298.15 K) = -344.80±0.2 kcal mol^{-1}.

Heat Capacity and Entropy

The low temperature heat capacities of Cs_2SO_4 were measured by Paukov et al. (4) by adiabatic calorimetry over the region 12.82-308.72 K. The curve of Cp-T was graphically extrapolated to 0 K. Our adopted value of S°(298.15 K) = 50.64±0.05 cal K^{-1} mol^{-1} is based on S°(12 K) = 0.309 cal K^{-1} mol^{-1} as estimated by Paukov et al. (4). They also estimated H°(12 K)-H°(0 K) = 2.83 cal K^{-1} mol^{-1} which leads to H°(298.15 K)-H°(0 K) = 6.628±0.006 kcal mol^{-1}. These values of S°(12 K) and H°(12 K)-H°(0 K) are almost identical to estimates obtained using a Debye T^3-law extrapolation. We have smoothed the experimental data of Paukov et al. (4) by fitting to orthogonal polynomials.

High temperature heat capacities of Cs_2SO_4 have been measured by Shmidt (5) by adiabatic calorimetry over the range 297.5-774.0 K. Enthalpy measurements by drop calorimetry were carried out by Denielou et al. (6, 400-910 K) and by Polishchuk et al. (7, 803.3-933.4 K). We have joined the low temperature $C°_p$ data of Paukov (4) to the high temperature $C°_p$ data of Shmidt (5) and then merged this into $C°_p$ values derived from the measurements of Denielou et al. (6) above 700 K. We have extrapolated the data smoothly to 1200 K.

Transition Data

Cs_2SO_4(cr, II) is the low-temperature form of cesium sulfate (often denoted as Cs_2SO_4, β). The crystals are orthohombic, space group D^{16}_{2h}-Pmcn (8). The transformation to the high-temperature I-phase (hexagonal, Cs_2SO_4, I) occurs at 940±5 K (5, 6, 7). The enthalpy of transition is determined by taking the difference between the adopted enthalpy curves of II- and I-Cs_2SO_4 extended to 940 K.

References

1. G. P. Kuznetsova, G. A. Lovetskaya, V. M. Presnyakova, and B. O. Stepin, Russ. J. Phys. Chem. 48, 2141 (1974).
2. A. F. Borob'ev, N. A. Ibragim, and S. M. Skuratov, Russ. J. Inorg. Chem. 11, 13 (1966).
3. ICSU-CODATA Task Group on Key Values for Thermodynamics, J. Chem. Thermodynamics 10, 903 (1978).
4. I. E. Paukov, L. M. Khriplovich, and A. M. Korotkikh, Russ. J. Phys. Chem. 42, 661 (1968).
5. N. E. Shmidt, Russ. J. Phys. Chem. 12, 929 (1967).
6. L. Denielou, Y. Fournier, J. P. Petitet and C. Tequi, C. R. Acad. Sci., Paris, 270, 1854 (1970).
7. A. F. Polishchuk, T. A. Tishura, and A. N. Buderina, Ukr. Khim. Zh., 40, 120 (1974).
8. C. N. R. Rao and B. Prakash, NSRDS-NBS 56, 1975.

Cesium Sulfate, II (Cs_2SO_4) $Cs_2O_4S_1$(cr)

Enthalpy Reference Temperature = T_r = 298.15 K Standard State Pressure = p° = 0.1 MPa

T/K	$C°_p$ (J K⁻¹mol⁻¹)	S° (J K⁻¹mol⁻¹)	-[G°-H°(T_r)]/T (J K⁻¹mol⁻¹)	H°-H°(T_r) (kJ mol⁻¹)	$\Delta_f H°$ (kJ mol⁻¹)	$\Delta_f G°$ (kJ mol⁻¹)	Log K_f
0	0.	0.	INFINITE	-27.732	-1433.163	-1433.163	INFINITE
100	89.287	91.922	317.566	-22.564	-1438.791	-1401.080	731.849
200	114.767	162.172	223.551	-12.276	-1441.328	-1362.260	355.786
298.15	135.143	211.878	211.878	0.	-1442.643	-1323.155	231.811
300	135.478	212.715	211.880	0.250	-1442.663	-1322.413	230.253
400	151.293	253.916	217.391	14.610	-1449.401	-1280.787	167.254
500	164.557	289.120	228.298	30.411	-1449.910	-1238.562	129.392
600	176.816	320.158	241.069	47.453	-1448.986	-1196.356	104.152
700	192.255	348.512	254.416	65.867	-1446.621	-1154.417	86.144
800	212.464	375.547	267.877	86.136	-1442.424	-1112.937	72.667
900	228.697	401.504	281.293	108.190	-1489.626	-1071.010	62.160
940.000	235.337	411.592	286.624	117.470	-----------	II <--> I	----------
1000	245.266	426.458	294.570	131.888	-1615.161	-1017.551	53.151
1100	261.835	450.612	307.663	157.244	-1602.852	-958.373	45.509
1200	278.362	474.103	320.558	184.254	-1588.984	-900.384	39.193

PREVIOUS: CURRENT: June 1979

Cesium Sulfate, II (Cs_2SO_4) $Cs_2O_4S_1$(cr)

CESIUM SULFATE (Cs_2SO_4) LIQUID M_r = 361.8684

$S°(298.15\ K) = [246.449]\ J\ K^{-1}\ mol^{-1}$

$T_{fus}(I \to \ell) = 1278 \pm 5\ J\ K^{-1}\ mol^{-1}$

$\Delta_f H°(298.15\ K) = [-1400.139]\ kJ\ mol^{-1}$

$\Delta_{fus}H°(I \to \ell) = 35.690 \pm 0.8\ kJ\ mol^{-1}$

Enthalpy of Formation

$\Delta_f H°(Cs_2SO_4, \ell, 298.15\ K)$ is calculated from $\Delta_f H°(Cs_2SO_4, cr, I, 298.15\ K)$ by adding $\Delta_{fus}H°$ and the difference in enthalpy, H°(1278) - H°(298.15 K), between the I crystal and liquid (1).

Heat Capacity and Entropy

Enthalpies for $Cs_2SO_4(\ell)$ have been measured by Denielou et al. (2) and by Polishchuk et al. (3). We adopt enthalpies based on the work of Denielou et al. (2). This leads to a constant heat capacity of 49.4 cal $K^{-1}\ mol^{-1}$ at temperatures above an assumed glass transition at 780 K. Below the glass transition, the heat capacity is taken as that of the low temperature II phase.

S°(298.15 K) is calculated in a manner analogous to that used for the enthalpy of formation.

Fusion Data

See the I-crystal table for details.

References

1. JANAF Thermochemical Tables: Cs_2SO_4(cr, I), 6-30-79.
2. L. Denielou, Y. Fournier, J-P. Petitet, and C. Tequi, C. R. Acad. Sci. Paris 270, 1854 (1970).
3. A. F. Polishchuk, T. A. Tishura, and A. N. Budarina, Ukr. Khim. Zh. 40, 120 (1974).

Cesium Sulfate (Cs_2SO_4) $Cs_2O_4S_1(l)$

Enthalpy Reference Temperature = T_r = 298.15 K Standard State Pressure = p° = 0.1 MPa

T/K	$C_p°$ (J K⁻¹mol⁻¹)	S° (J K⁻¹mol⁻¹)	$-[G°-H°(T_r)]/T$ (J K⁻¹mol⁻¹)	$H°-H°(T_r)$ (kJ mol⁻¹)	$\Delta_f H°$ (kJ mol⁻¹)	$\Delta_f G°$ (kJ mol⁻¹)	Log K_f
0							
100							
200							
298.15	135.143	246.449	246.449	0.	-1400.139	-1290.958	226.171
300	135.478	247.286	246.452	0.250	-1400.159	-1290.281	224.658
400	151.293	288.488	251.962	14.610	-1406.897	-1252.112	163.509
500	164.557	323.692	262.869	30.411	-1407.406	-1213.344	126.757
600	176.816	354.729	275.640	47.453	-1406.482	-1174.595	102.258
700	192.255	383.084	288.988	65.867	-1404.117	-1136.113	84.778
800	212.464	409.874	302.434	85.952	-1400.103	-1098.078	71.697
880.000	241.626	431.400	313.174	104.039	GLASS <--> LIQUID	TRANSITION	
880.000	206.690	431.400	313.174	104.039			
900	206.690	436.045	315.853	108.173	-1447.139	-1059.609	61.498
1000	206.690	457.822	328.980	128.842	-1575.703	-1009.458	52.729
1100	206.690	477.521	341.603	149.511	-1568.081	-953.203	45.264
1200	206.690	495.506	353.689	170.180	-1560.555	-897.638	39.073
1278.000	206.690	508.522	362.746	186.301	--------- I <--> LIQUID --------		
1300	206.690	512.050	365.243	190.849	-1553.113	-842.697	33.860
1400	206.690	527.367	376.283	211.518	-1545.751	-788.326	29.413
1500	206.690	541.627	386.836	232.187	-1538.463	-734.480	25.577
1600	206.690	554.967	396.932	252.856	-1531.249	-681.116	22.236
1700	206.690	567.497	406.600	273.524	-1524.110	-628.202	19.302
1800	206.690	579.311	415.870	294.193	-1517.048	-575.706	16.707
1900	206.690	590.486	424.769	314.862	-1510.070	-523.600	14.395
2000	206.690	601.088	433.322	335.531	-1503.179	-471.859	12.324
2100	206.690	611.173	441.553	356.200	-1496.384	-420.460	10.458
2200	206.690	620.788	449.484	376.869	-1489.693	-369.382	8.770
2300	206.690	629.975	457.133	397.538	-1483.114	-318.607	7.236
2400	206.690	638.772	464.519	418.207	-1476.658	-268.116	5.835
2500	206.690	647.210	471.659	438.876	-1470.335	-217.891	4.553
2600	206.690	655.316	478.568	459.545	-1464.154	-167.915	3.373
2700	206.690	663.117	485.260	480.214	-1458.128	-118.174	2.286
2800	206.690	670.633	491.747	500.883	-1452.266	-68.654	1.281
2900	206.690	677.886	498.041	521.552	-1446.581	-19.341	0.348
3000	206.690	684.894	504.153	542.221	-1441.084	29.780	-0.519

PREVIOUS: CURRENT: June 1979

Cesium Sulfate (Cs_2SO_4) $Cs_2O_4S_1(l)$

CESIUM SULFATE (Cs_2SO_4) II-I-LIQUID

M_r = 361.8684 Cesium Sulfate (Cs_2SO_4) $Cs_2O_4S_1$(cr,l)

0 to 940 K crystal, II
940 to 1278 K crystal, I
above 1278 K liquid

Refer to the individual tables for details.

Enthalpy Reference Temperature = T_r = 298.15 K; Standard State Pressure = $p°$ = 0.1 MPa

T/K	$C_p°$ (J K⁻¹mol⁻¹)	$S°$ (J K⁻¹mol⁻¹)	$-[G°-H°(T_r)]/T$ (J K⁻¹mol⁻¹)	$H°-H°(T_r)$ (kJ mol⁻¹)	$\Delta_f H°$ (kJ mol⁻¹)	$\Delta_f G°$ (kJ mol⁻¹)	Log K_f
0	0.	0.	INFINITE	-27.732	-1433.163	-1433.163	INFINITE
100	89.287	91.922	317.566	-22.564	-1438.791	-1401.080	731.849
200	114.767	162.172	223.551	-12.276	-1441.328	-1362.260	355.786
298.15	135.143	211.878	211.878	0.	-1442.643	-1323.155	231.811
300	135.478	212.715	211.880	0.250	-1442.663	-1322.413	230.253
400	151.293	253.916	217.391	14.610	-1449.401	-1280.787	167.254
500	164.557	289.120	228.298	30.411	-1449.910	-1238.562	129.392
600	176.816	320.158	241.069	47.453	-1448.986	-1196.356	104.152
700	192.255	348.512	254.416	65.867	-1446.621	-1154.417	86.144
800	212.464	375.547	267.877	86.136	-1442.424	-1112.937	72.667
900	228.697	401.504	281.293	108.190	-1489.626	-1071.010	62.160
940.000	235.337	411.592	286.624	117.470		II <--> I	
940.000	184.891	416.177	286.624	121.780		TRANSITION	
1000	194.179	427.901	294.750	133.152	-1613.897	-1017.731	53.161
1100	209.660	447.135	307.731	153.344	-1606.752	-958.448	45.513
1200	225.141	466.041	320.138	175.084	-1598.155	-899.880	39.171
1278.000	237.233	480.596	329.488	193.116		I <--> LIQUID	
1278.000	206.690	508.522	329.488	228.806		TRANSITION	
1300	206.690	512.050	332.548	233.353	-1553.113	-842.697	33.860
1400	206.690	527.367	345.923	254.022	-1545.751	-788.326	29.413
1500	206.690	541.627	358.500	274.691	-1538.463	-734.479	25.577
1600	206.690	554.967	370.367	295.360	-1531.249	-681.116	22.236
1700	206.690	567.497	381.598	316.029	-1524.110	-628.202	19.302
1800	206.690	579.311	392.257	336.698	-1517.048	-575.706	16.707
1900	206.690	590.486	402.399	357.367	-1510.069	-523.600	14.395
2000	206.690	601.088	412.070	378.035	-1503.179	-471.859	12.324
2100	206.690	611.173	421.313	398.704	-1496.384	-420.460	10.458
2200	206.690	620.788	430.163	419.373	-1489.693	-369.382	8.770
2300	206.690	629.975	438.653	440.042	-1483.114	-318.607	7.236
2400	206.690	638.772	446.809	460.711	-1476.658	-268.116	5.835
2500	206.690	647.210	454.657	481.380	-1470.335	-217.891	4.553
2600	206.690	655.316	462.220	502.049	-1464.154	-167.915	3.373
2700	206.690	663.117	469.517	522.718	-1458.128	-118.174	2.286
2800	206.690	670.633	476.567	543.387	-1452.266	-68.654	1.281
2900	206.690	677.886	483.384	564.056	-1446.581	-19.341	0.348
3000	206.690	684.894	489.985	584.725	-1441.084	29.780	-0.519

PREVIOUS: CURRENT: June 1979

Cesium Sulfate (Cs_2SO_4) $Cs_2O_4S_1$(cr,l)

CESIUM SULFATE (Cs_2SO_4) IDEAL GAS M_r = 361.8684

$S°(298.15\ K) = 406.5 \pm 2\ J\ K^{-1}\ mol^{-1}$

$\Delta_f H°(0\ K) = -1109.3 \pm 17.0\ kJ\ mol^{-1}$

$\Delta_f H°(298.15\ K) = -1122.6 \pm 17.0\ kJ\ mol^{-1}$

Vibrational Frequencies, Symmetries, and Degeneracies

	ν, cm⁻¹		ν, cm⁻¹		ν, cm⁻¹
A_1	[970](1)	B_1	[460](1)	E	1100 (2)
	[510](1)	B_2	1125 (1)		607 (2)
	[90](1)		628 (1)		160 (2)
			190 (1)		[40](2)

Ground State Quantum Weight: 1 $\sigma = 4$

Point Group: D_{2d}

Bond Distances: S–O = 1.49±0.02 Å Ca–O = 2.60±0.07 Å

Bond Angles: O–S–O = 109.47° O–Ca–O = 55±5°

Product of the Moments of Inertia: $I_A I_B I_C = 3.287254 \times 10^{-111}\ g^3\ cm^6$

Enthalpy of Formation

Vaporization experiments on Cs_2SO_4 have been carried out by Ficalora et al. (1), Cubicciotti (2), Gorokhov and Efimova (3), and Jaganathan and Wyatt (4). Cubicciotti (2) carried out Langmuir sublimation experiments which showed that Cs_2SO_4(cr, I) vaporizes congruently and the predominant vapor species is Cs_2SO_4(g). The Knudsen effusion experiments of Cubicciotti (2) are in excellent agreement with the mass spectrometric and effusion data of Gorokhov and Efimova (3). The Knudsen weight loss results of Jagannathan and Wyatt (4) are in good agreement with the previous studies (2, 3) but the data show a substantial drift. The mass spectrometric study of Ficalora et al. (1) yields vapor pressures which are much lower than results of the other three studies. The adopted value of the enthalpy of formation, $\Delta_f H°(Cs_2SO_4$, g, 298.15 K) = $-268.3 \pm 4.0\ kcal\ mol^{-1}$, is based on JANAF analyses of the vaporization data given below. The calculated 3rd law $\Delta_r H°$(298.15 K) may have an uncertainty of 4 kcal mol⁻¹ since the JANAF Gibbs energy functions are partially based on the estimated molecular constants of Cs_2SO_4(g).

Source	Method	T/K	Data Points	δS cal K⁻¹ mol⁻¹	$\Delta_r H°$(298.15 K), kcal mol⁻¹ 2nd law	3rd law	$\Delta_f H°$(298.15 K) kcal mol⁻¹
(1)	mass spec.	950–1200	graph	-6.2 ±2.0	75.65±2.0	80.28±2.0	-264.52±4.0
(2)	Knudsen effusion	1092.6–1260.6	15	5.14±0.5	82.57±0.5	76.54±0.6	-268.26±2.0
(2)	Knudsen effusion	1148.8–1261.6	7	3.80±0.9	81.05±1.0	76.48±0.4	-268.32±2.0
(3)	mass spec.	1063–1278	eqn	1.48	78.2 ±0.5	76.48±0.5	-268.32±2.0
(4)	Knudsen effusion	1102–1269	eqn	12.84	92.28	76.95±1.8	-267.85±4.0

Reaction: $Cs_2SO_4(cr) = Cs_2SO_4(g)$

Heat Capacity and Entropy

The structure of cesium sulfate is based on the high temperature electron diffraction studies of Ugarov et al. (5) and of Spiridonov and Lutoshkin (6). These studies show a fairly rigid tetrahedral sulfate anion core but with highly mobile cesium atoms as evidenced by the very large amplitudes of vibration, ℓ(CsS) = 0.15 Å, ℓ(CsO) = 0.34 Å, (5). The average structure is of D_{2d} symmetry and can be pictured as having a central sulfur atom surrounded at the corners of a tetrahedron by four oxygen atoms with bridging cesium atoms between two pairs of oxygen atoms. The planar CsO_2S rings are mutually perpendicular. The principal moments of inertia are: $I_A = 15.7294 \times 10^{-39}$ and $I_B = I_C = 457.1516 \times 10^{-39}\ g\ cm^2$.

Belyaeva et al. (7) investigated the infrared absorption spectra (400–1500 cm⁻¹) of cesium sulfate molecules stabilized in matrices of Ne, Ar, Xe, and H_2 at 4.2 K. The bands listed above at 607, 628, 1100, and 1125 cm⁻¹ are from this work. In a later work, Belyaeva et al. (8) reported the low frequency (30–1500 cm⁻¹) infrared spectra of K_2SO_4, Rb_2SO_4, and Cs_2SO_4 molecules isolated in matrices of Ar, Kr, and Xe. The bands listed above at 160 and 140 cm⁻¹ are from this work. The band estimated at 40 cm⁻¹ is based on our upward adjustment of the calculation of Belyaeva et al. (8, 9). The band estimated at 970 cm⁻¹ is based on the Raman work of Atkins and Gingerich (10) on the similar molecules K_2SO_4 and Na_2SO_4. The remaining fundamental frequencies are estimated by analogy with the other alkali metal sulfates (11).

References

1. P. J Ficalora, O. M. Uy, D. W. Meunow and J. L. Margrave, J. Amer. Ceram. Soc. 51, 574 (1968).
2. D. Cubicciotti, High Temp. Sci. 3, 349 (1971).
3. L. N. Gorokhov and A. G. Efimova, Institute for High Temperatures, Annual Report 1975, "Nauka", 1976; as referenced in L. V. Gurich, O. V. Dorofeeva, and V. S. Yungman, Paper ANL-77-21, "Conference on High Temperature Sciences Related to Open-cycle, Coal-fired MHD Systems", Argonne National Laboratory, 1977.
4. G. V. Jagannathan and P. A. H. Wyatt, J. Chem. Research (S) 293 (1970).
5. V. V. Ugarov Y. S. Ezhov and N. G. Rambidi, J. Molec. Struct. 25, 357 (1975).
6. V. P. Spiridonov and B. I. Lutoshkin, Vest. Mosk. Univ. Khim. 11, 509 (1970).
7. A. A. Belyaeva, M. I. Dvorkin, and L. D. Shcherba, Optics and Spectroscopy 31, 309 (1971).
8. A. A. Belyaeva, M. I. Dvorkin, and L. D. Shcherba, Optics and Spectroscopy 38, 291 (1975); correction in Optics and Spectroscopy 43, 114 (1977).
9. A. A. Belyaeva, M. I. Dvorkin, and L. D. Shcherba, Optics and Spectroscopy 38, 170 (1975).
10. R. M. Atkins and K. A. Gingerich, Chem. Phys. Lett. 53, 347 (1978).
11. JANAF Thermochemical Tables: Li_2SO_4(g), 12-31-78; K_2SO_4(g), Na_2SO_4(g), 6-30-78.

Cesium Sulfate (Cs_2SO_4) $Cs_2O_4S_1$(g)

Enthalpy Reference Temperature = T_r = 298.15 K Standard State Pressure = p° = 0.1 MPa

T/K	$C_p°$ (J K⁻¹mol⁻¹)	S°	−[G°−H°(T_r)]/T	H°−H°(T_r) (kJ mol⁻¹)	$\Delta_f H°$	$\Delta_f G°$	Log K_f
0	0.	0.	INFINITE	-23.931	-1109.287	-1109.287	INFINITE
100	73.250	308.845	492.557	-18.371	-1114.522	-1098.503	573.799
200	93.381	365.960	416.040	-10.016	-1118.992	-1080.682	282.245
250	102.369	387.780	408.254	-5.118	-1120.869	-1070.882	223.749
298.15	110.070	406.484	406.484	0.	-1122.567	-1061.100	185.900
300	110.346	407.166	406.486	0.204	-1122.634	-1060.719	184.688
350	117.228	424.706	407.856	5.898	-1128.448	-1049.595	156.644
400	123.053	440.751	410.979	11.909	-1132.027	-1038.146	135.568
450	127.929	455.535	415.119	18.187	-1133.849	-1026.309	119.131
500	131.990	469.231	419.854	24.688	-1135.557	-1014.265	105.959
600	138.194	493.878	430.184	38.216	-1138.147	-989.749	86.165
700	142.552	515.527	440.861	52.266	-1140.146	-964.852	71.998
800	145.680	534.777	451.420	66.686	-1141.798	-939.695	61.356
900	147.979	552.075	461.659	81.374	-1196.366	-913.263	53.004
1000	149.708	567.760	471.497	96.263	-1330.710	-874.403	45.674
1100	151.034	582.094	480.909	111.303	-1328.717	-828.868	39.360
1200	152.072	595.282	489.898	126.460	-1326.703	-783.516	34.106
1300	152.897	607.488	498.480	141.710	-1324.680	-738.333	29.667
1400	153.563	618.844	506.676	157.034	-1322.662	-693.305	25.868
1500	154.108	629.458	514.512	172.419	-1320.659	-648.421	22.580
1600	154.559	639.418	522.011	187.853	-1318.680	-603.670	19.708
1700	154.936	648.800	529.195	203.328	-1316.734	-559.041	17.177
1800	155.254	657.665	536.089	218.838	-1314.832	-514.527	14.931
1900	155.526	666.067	542.710	234.377	-1312.983	-470.116	12.924
2000	155.759	674.050	549.080	249.942	-1311.197	-425.801	11.121
2100	155.960	681.655	555.213	265.528	-1309.485	-381.574	9.491
2200	156.136	688.914	561.127	281.133	-1307.857	-337.425	8.011
2300	156.289	695.858	566.835	296.754	-1306.326	-293.350	6.662
2400	156.424	702.513	572.350	312.390	-1304.904	-249.339	5.427
2500	156.544	708.901	577.685	328.038	-1303.600	-205.384	4.291
2600	156.650	715.043	582.851	343.698	-1302.429	-161.479	3.244
2700	156.745	720.956	587.857	359.368	-1301.402	-117.616	2.275
2800	156.830	726.658	592.713	375.047	-1300.530	-73.788	1.377
2900	156.907	732.163	597.427	390.734	-1299.827	-29.990	0.540
3000	156.976	737.484	602.008	406.428	-1299.305	13.789	-0.240
3100	157.039	742.632	606.461	422.129	-1298.976	57.553	-0.970
3200	157.096	747.619	610.795	437.836	-1298.854	101.309	-1.654
3300	157.148	752.454	615.015	453.548	-1298.867	145.057	-2.296
3400	157.195	757.146	619.127	469.265	-1299.157	188.818	-2.901
3500	157.239	761.703	623.135	484.987	-1299.680	232.588	-3.471
3600	157.279	766.133	627.046	500.713	-1300.444	276.377	-4.010
3700	157.315	770.443	630.864	516.442	-1301.433	320.188	-4.520
3800	157.349	774.639	634.592	532.176	-1302.676	364.032	-5.004
3900	157.381	778.726	638.236	547.912	-1304.154	407.908	-5.463
4000	157.410	782.711	641.798	563.652	-1305.936	451.828	-5.900
4100	157.437	786.598	645.283	579.394	-1307.990	495.797	-6.317
4200	157.462	790.393	648.693	595.139	-1310.322	539.819	-6.714
4300	157.485	794.098	652.031	610.887	-1312.933	583.901	-7.093
4400	157.507	797.719	655.302	626.636	-1315.241	627.979	-7.455
4500	157.528	801.259	658.506	642.388	-1317.195	672.046	-7.801
4600	157.547	804.721	661.647	658.142	-1320.299	716.285	-8.134
4700	157.565	808.110	664.727	673.897	-1323.637	760.595	-8.453
4800	157.582	811.427	667.749	689.655	-1327.203	804.977	-8.760
4900	157.597	814.676	670.715	705.414	-1330.990	849.438	-9.055
5000	157.612	817.860	673.626	721.174	-1334.990	893.977	-9.339
5100	157.626	820.982	676.485	736.936	-1339.196	938.598	-9.613
5200	157.639	824.043	679.293	752.699	-1342.534	983.146	-9.876
5300	157.652	827.046	682.052	768.464	-1346.987	1027.914	-10.131
5400	157.664	829.993	684.765	784.230	-1351.053	1072.680	-10.376
5500	157.675	832.886	687.432	799.996	-1355.457	1117.556	-10.614
5600	157.685	835.727	690.055	815.764	-1360.284	1162.565	-10.844
5700	157.695	838.518	692.635	831.534	-1365.243	1207.659	-11.067
5800	157.705	841.260	695.174	847.304	-1370.325	1252.843	-11.283
5900	157.714	843.956	697.673	863.074	-1375.519	1298.111	-11.493
6000	157.722	846.607	700.133	878.846	-1380.816	1343.473	-11.696

PREVIOUS: June 1979 (1 atm) CURRENT: June 1979 (1 bar)

Cesium Sulfate (Cs_2SO_4) $Cs_2O_4S_1$(g)

COPPER (Cu)

REFERENCE STATE

0 to 1358 K crystal
1358 to 2843.261 K liquid
above 2843.261 K ideal monatomic gas

Refer to the individual tables for details.

A_r = 63.546

Copper (Cu) Cu_1(ref)

Enthalpy Reference Temperature = T_r = 298.15 K — Standard State Pressure = p° = 0.1 MPa

T/K	C_p° (J K⁻¹mol⁻¹)	S° (J K⁻¹mol⁻¹)	$-[G^\circ-H^\circ(T_r)]/T$ (J K⁻¹mol⁻¹)	$H^\circ-H^\circ(T_r)$ (kJ mol⁻¹)	Δ_fH° (kJ mol⁻¹)	Δ_fG° (kJ mol⁻¹)	Log K_f
0	0.	0.	INFINITE	-5.007	0.	0.	0.
100	16.010	10.034	53.414	-4.338	0.	0.	0.
200	22.631	23.730	35.354	-2.325	0.	0.	0.
298.15	24.442	33.164	33.164	0.	0.	0.	0.
300	24.462	33.315	33.164	0.045	0.	0.	0.
400	25.318	40.484	34.136	2.539	0.	0.	0.
500	25.912	46.206	35.997	5.105	0.	0.	0.
600	26.481	50.982	38.107	7.725	0.	0.	0.
700	26.996	55.103	40.247	10.399	0.	0.	0.
800	27.494	58.739	42.336	13.123	0.	0.	0.
900	28.049	62.009	44.343	15.899	0.	0.	0.
1000	28.662	64.994	46.261	18.733	0.	0.	0.
1100	29.479	67.763	48.091	21.638	0.	0.	0.
1200	30.519	70.368	49.840	24.633	0.	0.	0.
1300	32.143	72.871	51.516	27.762	0.	0.	0.
1358.000	33.353	74.300	52.459	29.660	CRYSTAL <--> LIQUID		
1358.000	32.844	83.974	52.459	42.798	TRANSITION		
1400	32.844	84.974	53.419	44.177	0.	0.	0.
1500	32.844	87.240	55.599	47.462	0.	0.	0.
1600	32.844	89.360	57.644	50.746	0.	0.	0.
1700	32.844	91.351	59.569	54.031	0.	0.	0.
1800	32.844	93.229	61.387	57.315	0.	0.	0.
1900	32.844	95.004	63.110	60.600	0.	0.	0.
2000	32.844	96.689	64.747	63.884	0.	0.	0.
2100	32.844	98.292	66.307	67.168	0.	0.	0.
2200	32.844	99.819	67.795	70.453	0.	0.	0.
2300	32.844	101.279	69.220	73.737	0.	0.	0.
2400	32.844	102.677	70.585	77.022	0.	0.	0.
2500	32.844	104.018	71.896	80.306	0.	0.	0.
2600	32.844	105.306	73.156	83.591	0.	0.	0.
2700	32.844	106.546	74.370	86.875	0.	0.	0.
2800	32.844	107.740	75.540	90.159	0.	0.	0.
2843.261	32.844	108.244	76.034	91.580	LIQUID <--> IDEAL GAS		
2843.261	24.379	213.995	76.034	392.257	FUGACITY = 1 bar		
2900	24.653	214.479	78.738	393.648	0.	0.	0.
3000	25.147	215.323	83.277	396.138	0.	0.	0.
3100	25.652	216.156	87.550	398.678	0.	0.	0.
3200	26.162	216.978	91.582	401.268	0.	0.	0.
3300	26.673	217.791	95.394	403.910	0.	0.	0.
3400	27.180	218.595	99.006	406.603	0.	0.	0.
3500	27.680	219.390	102.434	409.346	0.	0.	0.
3600	28.168	220.177	105.694	412.138	0.	0.	0.
3700	28.641	220.955	108.799	414.979	0.	0.	0.
3800	29.097	221.725	111.760	417.866	0.	0.	0.
3900	29.533	222.486	114.590	420.798	0.	0.	0.
4000	29.948	223.239	117.296	423.772	0.	0.	0.
4100	30.339	223.984	119.889	426.786	0.	0.	0.
4200	30.707	224.719	122.377	429.839	0.	0.	0.
4300	31.051	225.446	124.765	432.927	0.	0.	0.
4400	31.370	226.163	127.062	436.048	0.	0.	0.
4500	31.665	226.872	129.272	439.200	0.	0.	0.
4600	31.936	227.571	131.401	442.380	0.	0.	0.
4700	32.185	228.260	133.455	445.587	0.	0.	0.
4800	32.413	228.940	135.437	448.817	0.	0.	0.
4900	32.621	229.611	137.352	452.069	0.	0.	0.
5000	32.809	230.272	139.204	455.340	0.	0.	0.
5100	32.981	230.923	140.996	458.630	0.	0.	0.
5200	33.137	231.565	142.731	461.936	0.	0.	0.
5300	33.279	232.198	144.413	465.257	0.	0.	0.
5400	33.409	232.821	146.045	468.591	0.	0.	0.
5500	33.530	233.435	147.628	471.938	0.	0.	0.
5600	33.641	234.040	149.166	475.297	0.	0.	0.
5700	33.746	234.637	150.660	478.666	0.	0.	0.
5800	33.845	235.224	152.113	482.046	0.	0.	0.
5900	33.941	235.804	153.527	485.435	0.	0.	0.
6000	34.035	236.375	154.903	488.834	0.	0.	0.

PREVIOUS: June 1977 (1 atm) — CURRENT: June 1977 (1 bar)

Copper (Cu) Cu_1(ref)

COPPER (Cu) CRYSTAL A_r = 63.546

$S°(298.15\ K) = 33.164 \pm 0.04\ J\ K^{-1}\ mol^{-1}$

$T_{fus} = 1358.0 \pm 0.2\ K$

$\Delta_f H°(0\ K) = 0\ kJ\ mol^{-1}$

$\Delta_f H°(298.15\ K) = 0\ kJ\ mol^{-1}$

$\Delta_{fus} H° = 13.138 \pm 0.4\ kJ\ mol^{-1}$

Enthalpy of Formation

Zero by definition.

Heat Capacity and Entropy

The low temperature heat capacity and entropy of copper have been well established by the critical review of Furukawa et al. (1). Their recommended smoothed values are adopted with minor corrections for a change in the relative atomic mass from 63.54 to 63.546 (2) and for a change to the International Practical Temperature Scale of 1968 (3). These corrections increase the entropy at 298.15 K from 7.923 to 7.926 cal K^{-1} mol^{-1} and the enthalpy difference, H°(298.15 K) - H°(0 K), from 1.1962 to 1.1967 kcal mol^{-1}. The values recommended by CODATA (4) are those of Furukawa (1).

From 300 K to the melting point, the heat capacity selected by Hultgren et al. (5) has been adopted, with correction for relative atomic mass and temperature scale, and has been joined smoothly with the adopted low temperature data. The measurements of Chekhovskoi and Gerasina (6), as given by their equation (300-900 K), are in excellent agreement with the selection of Hultgren et al. (5). Also, while the data given by Brooks et al. (313-1193, 7) were 1 to 2% higher than Hultgren's selected values, the later measurements of Yeh and Brooks (8) on the same sample did not totally confirm the earlier results and measurements on a larger sample (370-870 K) are in good agreement with Hultgren et al. (5). The same evaluation used by Hultgren et al. (5) is used by two other tabulations (9, 10), although there are small differences in the tabulated values.

Fusion Data

Righini et al. (11) reviewed the precise measurements of the freezing point of high-purity copper made since 1960. The mean value of five investigations is 1357.99±0.24 (IPTS-68). N. N. Ergardt (12) found 1358.41 K and Righini et al. (11) measured 1357.98±0.1 K. We adopt 1358.0±0.2 K for the melting point. As a secondary fixed point IPTS-68 uses 1357.6 K (13).

The adopted value for $\Delta_{fus}H°$ is based upon five measured values and an evaluation of H°(1358 K) - H°(298.15 K) for the solid and the liquid.

Source	$\Delta_{fus}H°$, kcal mol^{-1}
(14)	3.239
(15)	3.030
(16)	3.080
(17)	3.290±0.275
(18)	3.107±0.048

The average is 3.149 kcal mol^{-1}. Our least squares linear fit of the liquid enthalpy data of Wust, Meuthen, and Durrer (14) leads to $\Delta_{fus}H°$ = 3.139 kcal mol^{-1}, and with this change, the average of the five measured values is 3.129 kcal mol^{-1}.

We adopt $\Delta_{fus}H°$ = 3.140±0.1 kcal mol^{-1}.

Sublimation Data

See the ideal gas table for details.

References

1. G. T. Furukawa, W. G. Saba, and M. L. Reilly, NSRDS-NBS 18, 1968.
2. IUPAC, Pure Appl. Chem. 30, 639 (1972).
3. T. B. Douglas, J. Res. Natl. Bur. Std. 73A 451 (1969).
4. ICSU-CODATA Task Group, J. Chem. Thermodynamics 8, 603 (1976).
5. R. Hultgren, P. D. Desai, "Selected Values of the Thermodynamic Properties of the Elements", American Society for Metals, Metals Park, Ohio, 1973. Copper evaluation dated 1969.
6. V. Ya Chekhovskoi and G. Z. Gerasina, Teplofiz. Vys. Temp. 9, 938 (1971).
7. C. R. Brooks, W. E. Norem, D. E. Hendrix, J. W. Wright, and W. G. Northcutt, J. Phys. Chem. Solids 29, 565 (1968).
8. C. C. Yeh and C. R. Brooks, High Temp. Sci. 5, 403 (1973).
9. R. Hultgren and P. D. Desai, "Selected Thermodynamic Values and Phase Diagrams for Copper and Some of its Binary Alloys. INCRA Series on the Metallurgy of Copper, No. 1", INCRA, New York, 1971.
10. E. G. King, A. D. Mah, and L. B. Pankrantz, "Thermodynamic Properties of Copper and its Inorganic Compounds. INCRA Series on the Metallurgy of Copper, No. 2", INCRA, New York, 1973.
11. F. Righini, A. Russo, and G. Ruffino, High Temp.-High Pressures 4, 471 (1972).
12. N. N. ERgardt, Tr. Inst. Kom. Standartov, Mer i Izmerit. Priborov pri Sov. Min. SSSR 71, 94 (1963).
13. The International Practical Temperature Scale of 1968, Metrologia 5, 35 (1969).
14. F. Wust, A. Meuthen, and R. Durrer, Forsch. Gebiete Ingenieurw, VDI-Forschungshelft 204, (1918).
15. W. Oelsen, E. Schurmann, and D. Buchholz, Arch. Eisenhuttenw, 32, 39 (1961).
16. E. Schuermann and A. Kaune, Z. Metallk. 56, 453 (1965).
17. R. N. Dokken and J. F. Elliott, Trans. Met. Soc. AIME 233, 1351 (1965).
18. O. Vollmer and R. Kohlhaus, Z. Metallk. 59, 273 (1968).

Copper (Cu) $Cu_1(cr)$

Enthalpy Reference Temperature = T_r = 298.15 K   Standard State Pressure = p° = 0.1 MPa

T/K	$C°_p$ (J K⁻¹mol⁻¹)	S° (J K⁻¹mol⁻¹)	$-[G°-H°(T_r)]/T$ (J K⁻¹mol⁻¹)	$H°-H°(T_r)$ (kJ mol⁻¹)	$\Delta_f H°$ (kJ mol⁻¹)	$\Delta_f G°$ (kJ mol⁻¹)	Log K_f
0	0.	0.	INFINITE	-5.007	0.	0.	0.
100	16.010	10.034	53.414	-4.338	0.	0.	0.
200	22.631	23.730	35.354	-2.325	0.	0.	0.
250	23.782	28.915	33.563	-1.162	0.	0.	0.
298.15	24.442	33.164	33.164	0.	0.	0.	0.
300	24.462	33.315	33.164	0.045	0.	0.	0.
350	24.975	37.127	33.464	1.282	0.	0.	0.
400	25.318	40.484	34.136	2.539	0.	0.	0.
450	25.686	43.489	35.011	3.815	0.	0.	0.
500	25.912	46.206	35.997	5.105	0.	0.	0.
600	26.481	50.982	38.107	7.725	0.	0.	0.
700	26.996	55.103	40.247	10.399	0.	0.	0.
800	27.494	58.739	42.336	13.123	0.	0.	0.
900	28.049	62.009	44.343	15.899	0.	0.	0.
1000	28.662	64.994	46.261	18.733	0.	0.	0.
1100	29.479	67.763	48.091	21.638	0.	0.	0.
1200	30.519	70.368	49.840	24.633	0.	0.	0.
1300	32.143	72.871	51.516	27.762	0.	0.	0.
1358.000	33.353	74.300	52.459	29.660	------ CRYSTAL <--> LIQUID -----		
1400	34.258	75.330	53.129	31.081	-13.097	0.406	-0.015
1500	36.425	77.767	54.691	34.615	-12.847	1.363	-0.047
1600	38.593	80.187	56.209	38.366	-12.381	2.296	-0.075
1700	40.760	82.592	57.690	42.333	-11.697	3.194	-0.098
1800	42.928	84.983	59.140	46.518	-10.797	4.045	-0.117
1900	45.095	87.362	60.563	50.919	-9.681	4.840	-0.133
2000	47.263	89.730	61.962	55.537	-8.347	5.570	-0.145

PREVIOUS: DEcember 1965   CURRENT: June 1977

Copper (Cu) $Cu_1(cr)$

$S°(298.15\ K) = [41.620]\ J\ K^{-1}\ mol^{-1}$ $\Delta_f H°(298.15\ K) = [11.857]\ kJ\ mol^{-1}$

$T_{fus} = 1358.0 \pm 0.2\ K$ $\Delta_{fus}H° = 13.138 \pm 0.4\ kJ\ mol^{-1}$

Enthalpy of Formation

The enthalpy of formation at 298.15 is calculated from that of the crystal by adding $\Delta_{fus}H°$ and the difference in enthalpy, H°(1358 K) - H°(298.15 K), between the crystal and liquid.

Heat Capacity and Entropy

Since the critical evaluation of Hultgren et al. (1) and the selection of 7.80 cal $K^{-1}\ mol^{-1}$ for the heat capacity of liquid copper, there have been two experimental investigations. Chaudhuri et al. (2) measured enthalpies (1386-1887 K) by levitation-heating aluminum-block drop calorimetry and determined $Cp(\ell) = 7.89 \pm 0.41$ cal $K^{-1}\ mol^{-1}$. Stephens (3) found $Cp(\ell) = 8$ cal $K^{-1}\ mol^{-1}$ by levitation-heating liquid-argon calorimetry (1428-2007 K). The scatter in the new liquid argon calorimetry data is large and is considered only in that it tends to support a "chronological increase" in $Cp(\ell)$ of copper. The enthalpy differences, H°(1358 K) - H°(298.15 K), derived from the measurements of Vollmer and Kohlhaus (4) and Chaudhuri et al. (2) are in good agreement at about 10.228 kcal mol^{-1}, which we adopt. The scatter of liquid enthalpy measurements easily accommodates a wide range of values for $Cp(\ell)$; we adopt 7.85 cal $K^{-1}\ mol^{-1}$, an average of the value selected by Hultgren et al. (1) and the later measurement of Chaundhuri et al. (2). This value is assumed constant from 900 to 4000 K. A glass transition is assumed at 900 K below which the heat capacity is tha of the crystal.

The entropy is calculated in a manner similar to that used to calculate the enthalpy of formation.

Fusion Data

See the crystal table for details.

Vaporization Data

T_{vap} is the calculated temperature at which the Gibbs energy difference is zero for the reaction $Cu(\ell) = Cu(g)$. $\Delta_{vap}H°$ is the difference in $\Delta_f H°$ between gas and liquid at T_{vap}.

References

1. R. Hultgren, P. D. Desai, et al., "Selected Values of the Thermodynamic Properties of the Elements," Americal Society for Metals, Metals Park, Ohio, 1973. copper evaluation dated 1969.
2. A. K. Chaudhuri, D. W. Bonnell, L. A. Ford, and J. L. Margrave, High Temp. Sci. 2, 203 (1970).
3. H. P. Stephens, High Temp. Sci. 6, 156 (1974).
4. O. Vollmer and R. Kohlhaus, Z. Metallk. 59, 273 (1968).

Enthalpy Reference Temperature = T_r = 298.15 K Standard State Pressure = p° = 0.1 MPa

T/K	$C_p°$ (J K⁻¹mol⁻¹)	S° (J K⁻¹mol⁻¹)	$-[G°-H°(T_r)]/T$ (J K⁻¹mol⁻¹)	$H°-H°(T_r)$ (kJ mol⁻¹)	$\Delta_f H°$ (kJ mol⁻¹)	$\Delta_f G°$ (kJ mol⁻¹)	Log K_f
0							
100							
200							
298.15	24.442	41.620	41.620	0.	11.857	9.336	-1.636
300	24.460	41.772	41.621	0.045	11.857	9.320	-1.623
350	24.973	45.583	41.921	1.282	11.857	8.898	-1.328
400	25.316	48.939	42.592	2.539	11.857	8.475	-1.107
450	25.684	51.944	43.467	3.814	11.857	8.052	-0.935
500	25.910	54.661	44.453	5.104	11.857	7.629	-0.797
600	26.478	59.436	46.563	7.724	11.857	6.784	-0.591
700	26.993	63.557	48.703	10.398	11.856	5.938	-0.443
800	27.491	67.193	50.791	13.121	11.856	5.093	-0.333
900	28.047	70.463	52.798	15.898	11.856	4.248	-0.247
900.000	28.047	70.463	52.798	15.898	GLASS <--> LIQUID		
900.000	32.844	70.463	52.798	15.898	TRANSITION		
1000	32.844	73.923	54.741	19.182	12.306	3.378	-0.176
1100	32.844	77.053	56.629	22.467	12.686	2.466	-0.117
1200	32.844	79.911	58.452	25.751	12.975	1.523	-0.066
1300	32.844	82.540	60.205	29.036	13.131	0.561	-0.023
1358.000	32.844	83.974	61.190	30.941	------ CRYSTAL <--> LIQUID -----		
1400	32.844	84.974	61.889	32.320	0.	0.	0.
1500	32.844	87.240	63.504	35.604	0.	0.	0.
1600	32.844	89.360	65.055	38.889	0.	0.	0.
1700	32.844	91.351	66.543	42.173	0.	0.	0.
1800	32.844	93.229	67.974	45.458	0.	0.	0.
1900	32.844	95.004	69.351	48.742	0.	0.	0.
2000	32.844	96.689	70.676	52.027	0.	0.	0.
2100	32.844	98.292	71.953	55.311	0.	0.	0.
2200	32.844	99.819	73.185	58.596	0.	0.	0.
2300	32.844	101.279	74.375	61.880	0.	0.	0.
2400	32.844	102.677	75.525	65.164	0.	0.	0.
2500	32.844	104.018	76.639	68.449	0.	0.	0.
2600	32.844	105.306	77.717	71.733	0.	0.	0.
2700	32.844	106.546	78.761	75.018	0.	0.	0.
2800	32.844	107.740	79.775	78.302	0.	0.	0.
2843.261	32.844	108.244	80.205	79.723	------- FUGACITY = 1 bar -------		
2900	32.844	108.893	80.760	81.587	-300.204	5.996	-0.108
3000	32.844	110.006	81.716	84.871	-299.409	16.541	-0.288
3100	32.844	111.083	82.646	88.155	-298.665	27.060	-0.456
3200	32.844	112.126	83.551	91.440	-297.971	37.556	-0.613
3300	32.844	113.137	84.432	94.724	-297.328	48.031	-0.760
3400	32.844	114.117	85.291	98.009	-296.737	58.488	-0.899
3500	32.844	115.069	86.128	101.293	-296.195	68.927	-1.029
3600	32.844	115.995	86.945	104.578	-295.703	79.352	-1.151
3700	32.844	116.894	87.743	107.862	-295.259	89.764	-1.267
3800	32.844	117.770	88.521	111.147	-294.862	100.165	-1.377
3900	32.844	118.624	89.282	114.431	-294.509	110.556	-1.481
4000	32.844	119.455	90.026	117.715	-294.199	120.938	-1.579

PREVIOUS: December 1966 CURRENT: June 1977

COPPER (Cu)

CRYSTAL–LIQUID

A_r = 63.546

0 to 1358 K crystal
above 1358 K liquid

Refer to the individual tables for details.

Copper (Cu) $Cu_1(cr,l)$

Enthalpy Reference Temperature = T_r = 298.15 K   Standard State Pressure = $p°$ = 0.1 MPa

T/K	$C_p°$ (J K⁻¹mol⁻¹)	$S°$ (J K⁻¹mol⁻¹)	$-[G°-H°(T_r)]/T$ (J K⁻¹mol⁻¹)	$H°-H°(T_r)$ (kJ mol⁻¹)	$\Delta_fH°$ (kJ mol⁻¹)	$\Delta_fG°$ (kJ mol⁻¹)	Log K_f
0	0.	0.	INFINITE	-5.007	0.	0.	0.
100	16.010	10.034	53.414	-4.338	0.	0.	0.
200	22.631	23.730	35.354	-2.325	0.	0.	0.
250	23.782	28.915	33.563	-1.162	0.	0.	0.
298.15	24.442	33.164	33.164	0.	0.	0.	0.
300	24.462	33.315	33.164	0.045	0.	0.	0.
350	24.975	37.127	33.464	1.282	0.	0.	0.
400	25.318	40.484	34.136	2.539	0.	0.	0.
450	25.686	43.489	35.011	3.815	0.	0.	0.
500	25.912	46.206	35.997	5.105	0.	0.	0.
600	26.481	50.982	38.107	7.725	0.	0.	0.
700	26.996	55.103	40.247	10.399	0.	0.	0.
800	27.494	58.739	42.336	13.123	0.	0.	0.
900	28.049	62.009	44.343	15.899	0.	0.	0.
1000	28.662	64.994	46.261	18.733	0.	0.	0.
1100	29.479	67.763	48.091	21.638	0.	0.	0.
1200	30.519	70.368	49.840	24.633	0.	0.	0.
1300	32.143	72.871	51.516	27.762	0.	0.	0.
1358.000	33.353	74.300	52.459	29.660	CRYSTAL <--> LIQUID		
1358.000	32.844	83.974	52.459	42.798	TRANSITION		
1400	32.844	84.974	53.419	44.177	0.	0.	0.
1500	32.844	87.240	55.599	47.462	0.	0.	0.
1600	32.844	89.360	57.644	50.746	0.	0.	0.
1700	32.844	91.351	59.569	54.031	0.	0.	0.
1800	32.844	93.229	61.387	57.315	0.	0.	0.
1900	32.844	95.004	63.110	60.600	0.	0.	0.
2000	32.844	96.689	64.747	63.884	0.	0.	0.
2100	32.844	98.292	66.307	67.168	0.	0.	0.
2200	32.844	99.819	67.795	70.453	0.	0.	0.
2300	32.844	101.279	69.220	73.737	0.	0.	0.
2400	32.844	102.677	70.585	77.022	0.	0.	0.
2500	32.844	104.018	71.896	80.306	0.	0.	0.
2600	32.844	105.306	73.156	83.591	0.	0.	0.
2700	32.844	106.546	74.370	86.875	0.	0.	0.
2800	32.844	107.740	75.540	90.159	0.	0.	0.
2843.261	32.844	108.244	76.034	91.580	------- FUGACITY = 1 bar -------		
2900	32.844	108.893	76.671	93.444	-300.204	5.996	-0.108
3000	32.844	110.006	77.764	96.728	-299.409	16.541	-0.288
3100	32.844	111.083	78.821	100.013	-298.665	27.060	-0.456
3200	32.844	112.126	79.846	103.297	-297.971	37.556	-0.613
3300	32.844	113.137	80.839	106.582	-297.328	48.031	-0.760
3400	32.844	114.117	81.804	109.866	-296.737	58.488	-0.899
3500	32.844	115.069	82.741	113.151	-296.195	68.927	-1.029
3600	32.844	115.995	83.652	116.435	-295.703	79.352	-1.151
3700	32.844	116.894	84.538	119.719	-295.259	89.764	-1.267
3800	32.844	117.770	85.401	123.004	-294.862	100.165	-1.377
3900	32.844	118.624	86.242	126.288	-294.509	110.556	-1.481
4000	32.844	119.455	87.062	129.573	-294.199	120.938	-1.579

PREVIOUS: CURRENT: June 1977

Copper (Cu) $Cu_1(cr,l)$

COPPER (Cu) IDEAL GAS A_r = 63.546

IP(Cu, g) = 62317.2 ± 0.5 cm^{-1}
S°(298.15 K) = 166.397 ± 0.025 J K^{-1} mol^{-1}

$\Delta_f H°$(0 K) = 336.4 ± 1.2 kJ mol^{-1}
$\Delta_f H°$(298.15 K) = 337.6 ± 1.2 kJ mol^{-1}

Electronic Level and Quantum Weight

State	ε_i, cm^{-1}	g_i
$^2S_{1/2}$	0.00	2
$^2D_{5/2}$	11202.565	6
$^2D_{3/2}$	13245.432	4
$^2P_{1/2}$	30535.302	2
$^2P_{3/2}$	30783.683	4

Enthalpy of Formation

The adopted enthalpy of formation for Cu(g), $\Delta_f H°$(298.15 K) = 337.6 ± 1.2 kJ mol^{-1}, is the value recommended by CODATA (1). This value was based on the measurements of Edwards et al. (2), Hersh (3), Morris and Zellars (4), Marshall et al. (5), McCormack et al. (6), Myles and Darby (7), Grieviso et al. (8), Kirschenbaum and Cahill (9), Krupkowski and Golonka (10) and Ponslit and Bariaux (11). Additional references are given in Hultgren et al. (12).

Heat Capacity and Entropy

The information on electronic energy levels and quantum weights, given by Moore (13), is incomplete because many theoretically predicted levels have not been observed. Although we have listed only the ground state and the lowest lying levels, all levels listed by Moore (13), as well as estimated levels, are used in the calculation. The observed levels are too numerous to list completely. The calculations indicate that for Cu(g), the thermodynamic functions are independent of the estimated missing levels (for n=4-11), the cut-off procedure, and the inclusion of n<12 levels up to 6000 K; the Gibbs energy function showing variations of 0.001 J K^{-1} mol^{-1} at this temperature. The reported uncertainty in S°(298.15 K) is due to uncertainties in the relative ionic mass, and the fundamental constants. Extension of these calculations above 6000 K may require consideration of the higher excited states (n>11), and use of different fill and cut-off procedures (14).

References

1. J. D. Cox, chairman, CODATA Task Group on Key Values for Thermodynamics, J. Chem. Thermodynamics 10, 903 (1978).
2. J. W. Edwards, H. L. Johnston, and W. E. Ditmars, J. Am. Chem. Soc. 75, 2467 (1953).
3. H. N. Hersh, J. Am. Chem. Soc. 75, 1529 (1953).
4. J. P. Morris and G. R. Zellars, J. Metals 8, 1086 (1956).
5. A. L. Marshall, R. W. Dornte, and F. J. Norton, J. Am. Chem. Soc. 59, 116 (1937).
6. J. M. McCormack, J. R. Myers, and R. K. Saxer, J. Chem. Eng. Data 10, 319 (1965).
7. K. M. Myles and J. B. Darby, Acta Met. 16, 485 (1968).
8. P. Grievison, G. W. Hooper, and C. B. Alcok, Metallurg. Soc. Conf. 7, 341 (1961).
9. A. D. Kirschenbaum and J. A. Cahill, J. Inorg. Nucl. Chem. 25, 232 (1963).
10. A. Krupkowski and J. Golonka, Bull. Acad. Polon. sci Cl. [3, 12] 69 (1969).
11. A. Penslit and D. Bariaux, Bull. cl. Sci. Acad. Roy. Belgigue 52, 248 (1966).
12. R. Hultgren, R. P. Desai, "Selected Values for the Thermodynamic Properties of Metals," American Society of Metals, Met Park, Ohio, 1973.
13. C. E. Moore, U. S. Nat. Bur. Stand., NSRDS-NBS 35 Volume II, 1970 [Reprint of NBS Circular 467, Volume II, 1952].
14. J. R. Downey, The Dow Chemical Company, Rept. AFOSR-TR-78-0960, 1978.

Copper (Cu) $Cu_1(g)$

Enthalpy Reference Temperature = T_r = 298.15 K Standard State Pressure = p° = 0.1 MPa

T/K	$C_p°$ (J $K^{-1}mol^{-1}$)	S° (J $K^{-1}mol^{-1}$)	$-[G°-H°(T_r)]/T$ (J $K^{-1}mol^{-1}$)	$H°-H°(T_r)$ (kJ mol^{-1})	$\Delta_f H°$ (kJ mol^{-1})	$\Delta_f G°$ (kJ mol^{-1})	Log K_f
0	0.	0.	INFINITE	-6.197	336.409	336.409	INFINITE
100	20.786	143.690	184.877	-4.119	337.819	324.454	-169.477
200	20.786	158.097	168.298	-2.040	337.885	311.011	-81.228
250	20.786	162.736	166.739	-1.001	337.761	304.306	-63.581
298.15	20.786	166.397	166.397	0.	337.600	297.877	-52.187
300	20.786	166.525	166.397	0.038	337.593	297.630	-51.822
350	20.786	169.730	166.650	1.078	337.396	290.985	-43.427
400	20.786	172.505	167.213	2.117	337.178	284.369	-37.135
450	20.786	174.953	167.939	3.156	336.942	277.782	-32.244
500	20.786	177.144	168.752	4.196	336.691	271.222	-28.334
600	20.786	180.933	170.476	6.274	336.150	258.179	-22.476
700	20.786	184.137	172.205	8.353	335.554	245.230	-18.299
800	20.786	186.913	173.874	10.431	334.909	232.370	-15.172
900	20.786	189.361	175.461	12.510	334.211	219.594	-12.745
1000	20.787	191.551	176.963	14.589	333.456	206.898	-10.807
1100	20.788	193.533	178.380	16.667	332.629	194.282	-9.226
1200	20.793	195.342	179.719	18.746	331.713	181.745	-7.911
1300	20.803	197.006	180.986	20.826	330.665	169.289	-6.802
1400	20.823	198.549	182.186	22.907	316.330	157.326	-5.870
1500	20.856	199.986	183.325	24.991	315.130	146.011	-5.085
1600	20.909	201.334	184.409	27.079	313.933	134.775	-4.400
1700	20.985	202.604	185.443	29.174	312.743	123.614	-3.798
1800	21.091	203.806	186.430	31.277	311.562	112.523	-3.265
1900	21.230	204.950	187.375	33.393	310.394	101.497	-2.790
2000	21.407	206.043	188.281	35.525	309.241	90.533	-2.364
2100	21.622	207.093	189.152	37.676	308.107	79.625	-1.981
2200	21.878	208.104	189.990	39.850	306.998	68.771	-1.633
2300	22.173	209.083	190.799	42.053	305.915	57.967	-1.316
2400	22.507	210.034	191.581	44.286	304.865	47.209	-1.027
2500	22.878	210.960	192.338	46.555	303.849	36.495	-0.763
2600	23.282	211.865	193.071	48.863	302.873	25.820	-0.519
2700	23.715	212.752	193.784	51.213	301.938	15.182	-0.294
2800	24.174	213.622	194.477	53.607	301.048	4.578	-0.085
2843.261	24.379	213.995	194.771	54.657	------- FUGACITY = 1 bar -------		
2900	24.653	214.479	195.152	56.048	0.	0.	0.
3000	25.147	215.323	195.810	58.538	0.	0.	0.
3100	25.652	216.156	196.453	61.078	0.	0.	0.
3200	26.162	216.978	197.082	63.669	0.	0.	0.
3300	26.673	217.791	197.697	66.310	0.	0.	0.
3400	27.180	218.595	198.300	69.003	0.	0.	0.
3500	27.680	219.390	198.891	71.746	0.	0.	0.
3600	28.168	220.177	199.472	74.539	0.	0.	0.
3700	28.641	220.955	200.042	77.379	0.	0.	0.
3800	29.097	221.725	200.602	80.266	0.	0.	0.
3900	29.533	222.486	201.154	83.198	0.	0.	0.
4000	29.948	223.239	201.696	86.172	0.	0.	0.
4100	30.339	223.984	202.231	89.187	0.	0.	0.
4200	30.707	224.719	202.758	92.239	0.	0.	0.
4300	31.051	225.446	203.277	95.327	0.	0.	0.
4400	31.370	226.163	203.789	98.449	0.	0.	0.
4500	31.665	226.872	204.294	101.601	0.	0.	0.
4600	31.936	227.571	204.792	104.781	0.	0.	0.
4700	32.185	228.260	205.284	107.987	0.	0.	0.
4800	32.413	228.940	205.770	111.217	0.	0.	0.
4900	32.621	229.611	206.250	114.469	0.	0.	0.
5000	32.809	230.272	206.724	117.741	0.	0.	0.
5100	32.981	230.923	207.192	121.030	0.	0.	0.
5200	33.137	231.565	207.654	124.336	0.	0.	0.
5300	33.279	232.198	208.111	127.657	0.	0.	0.
5400	33.409	232.821	208.563	130.992	0.	0.	0.
5500	33.530	233.435	209.010	134.339	0.	0.	0.
5600	33.641	234.040	209.452	137.697	0.	0.	0.
5700	33.746	234.637	209.888	141.066	0.	0.	0.
5800	33.845	235.224	210.320	144.446	0.	0.	0.
5900	33.941	235.804	210.747	147.835	0.	0.	0.
6000	34.035	236.375	211.169	151.234	0.	0.	0.

PREVIOUS: June 1977 (1 atm) CURRENT: September 1984 (1 bar)

Copper (Cu) $Cu_1(g)$

COPPER, ION (Cu^+) IDEAL GAS M_r = 63.54545

Copper, Ion (Cu^+) $Cu_1^+(g)$

IP(Cu^+, g) = 163669.2 ± 0.5 cm^{-1}

S°(298.15 K) = 222.311 ± 0.02 J K^{-1} mol^{-1}

$\Delta_f H°$(0 K) = 1081.9 ± 2.1 kJ mol^{-1}

$\Delta_f H°$(298.15 K) = [1089.277] kJ mol^{-1}

Electronic Levels and Quantum Weights

State	ε_i, cm^{-1}	g_i
$^1S_0$	0.0	1
$^3D_0$	21928.60	7
$^3D_1$	22847.03	5
$^3D_1$	23998.31	3
$^1D_2$	26264.52	5

Enthalpy of Formation

$\Delta_f H°$(Cu^+, g, 0 K) is calculated from $\Delta_f H°$(Cu, g, 0 K) (1) using the spectroscopic value of IP(Cu) = 62317.2 ± 0.5 cm^{-1} (745.479 ± 0.006 kJ mol^{-1}) from Moore (2). The ionization limit is converted from cm^{-1} to kJ mol^{-1} using the factor, 1 cm^{-1} = 0.01196266 kJ cm^{-1}, which is derived from the 1973 CODATA fundamental constants (3). Rosenstock et al. (4) and Levin and Lias (5) have summarized additional ionization and appearance potential data.

$\Delta_f H°$(Cu^+, g, 298.15 K) is calculated from $\Delta_f H°$(Cu, g, 0 K) by using IP(Cu) with JANAF (1) enthalpies, H°(0 K) − H°(298.15 K), for Cu(g), Cu^+(g), and e^-(g). $\Delta_f H°$(Cu → Cu^+ + e^-, 298.15 K) differs from a room temperature threshold energy due to inclusion of these enthalpies and to threshold effects discussed by Rosenstock et al. (4). $\Delta_f H°$(298.15 K) should be changed by -6.197 kJ mol^{-1} if it is to be used in the ion convention that excludes the enthalpy of the electron.

Heat Capacity and Entropy

The information on electronic energy levels and quantum weights, given by Moore (2, 6) is incomplete because many theoretically predicted levels that any reasonable method of filling in these missing levels and cutting off the summation in the partition function (7) has no effect on the thermodynamic functions to 6000 K. This is a result of the high energy of all levels other than the ground state and the first four excited states. The next excited state is 66418.65 cm^{-1} above the ground state. Since inclusion of these excited states has no effect on the thermodynamic functions (to 6000 K), we list only the ground state and the first four excited states. The reported uncertainty in S°(298.15 K) is due to uncertainties in the relative ionic mass and the fundamental constants. Extension of these calculations above 6000 K may require consideration of the excited states and use of different fill and cutoff procedures (7).

References

1. JANAF Thermochemical Tables: Cu(g), 9-30-84 ; e^-(g), 3-31-82.
2. C. E. Moore, U. S. Nat. Bur. Stand., NSRDS-NBS-34, 1970.
3. E. R. Cohen and B. N. Taylor, J. Phys. Chem. Ref. Data 2, 663 (1973).
4. H. M. Rosenstock, K. Draxl et al., J. Phys. Chem. Ref. Data 6, Supp. 1 (1977).
5. R. D. Levin and S. G. Lias, U. S. Nat. Bur. Stand., NSRDS-NBS-71, 1982.
6. C. E. Moore, U. S. Nat. Bur. Stand., NSRDS-NBS-35, Volulme I, 1971 [Reprinted from NBS Circular 467, Volume I, 1949].
7. J. R. Downey, Jr., The Dow Chemical Company, AFOSR-TR-0960, Contract No. F44620-75-1-0048, 1978.

Enthalpy Reference Temperature = T_r = 298.15 K   Standard State Pressure = p° = 0.1 MPa

	J $K^{-1}mol^{-1}$			kJ mol^{-1}			
T/K	$C_p°$	S°	−[G°−H°(T_r)]/T	H°−H°(T_r)	$\Delta_f H°$	$\Delta_f G°$	Log K_f
0	0.	0.	INFINITE	-6.197	1081.889		
100	20.786	199.604	240.792	-4.119			
200	20.786	214.012	224.213	-2.040			
250	20.786	218.650	222.654	-1.001			
298.15	20.786	222.311	222.311	0.	1089.277	1026.628	-179.861
300	20.786	222.440	222.312	0.038	1089.308	1026.239	-178.684
350	20.786	225.644	222.565	1.078	1090.150	1015.660	-151.579
400	20.786	228.420	223.127	2.117	1090.972	1004.963	-131.235
450	20.786	230.868	223.854	3.156	1091.775	994.163	-115.399
500	20.786	233.058	224.667	4.196	1092.564	983.275	-102.722
600	20.786	236.848	226.391	6.274	1094.101	961.272	-83.686
700	20.786	240.052	228.119	8.353	1095.584	939.016	-70.070
800	20.786	242.828	229.788	10.431	1097.017	916.551	-59.845
900	20.786	245.276	231.376	12.510	1098.398	893.909	-51.881
1000	20.786	247.466	232.877	14.589	1099.721	871.116	-45.502
1100	20.786	249.447	234.295	16.667	1100.973	848.195	-40.277
1200	20.786	251.256	235.634	18.746	1102.135	825.163	-35.918
1300	20.786	252.919	236.901	20.824	1103.164	802.039	-32.226
1400	20.786	254.460	238.100	22.903	1090.906	779.248	-29.074
1500	20.786	255.894	239.239	24.982	1091.778	756.957	-26.360
1600	20.786	257.235	240.323	27.060	1092.651	734.607	-23.982
1700	20.786	258.495	241.355	29.139	1093.524	712.202	-21.883
1800	20.786	259.684	242.341	31.217	1094.397	689.746	-20.016
1900	20.786	260.807	243.283	33.296	1095.269	667.242	-18.344
2000	20.786	261.874	244.186	35.375	1096.142	644.692	-16.838
2100	20.786	262.888	245.053	37.453	1097.015	622.098	-15.474
2200	20.786	263.855	245.886	39.532	1097.888	599.462	-14.233
2300	20.786	264.779	246.687	41.610	1098.761	576.787	-13.099
2400	20.786	265.663	247.460	43.689	1099.633	554.074	-12.059
2500	20.786	266.512	248.205	45.768	1100.506	531.324	-11.101
2600	20.786	267.327	248.925	47.846	1101.379	508.539	-10.217
2700	20.786	268.112	249.621	49.925	1102.252	485.721	-9.397
2800	20.786	268.868	250.295	52.004	1103.124	462.871	-8.635
2900	20.786	269.597	250.948	54.082	803.793	445.985	-8.033
3000	20.786	270.302	251.581	56.161	805.461	433.618	-7.550
3100	20.786	270.983	252.196	58.239	807.078	421.197	-7.097
3200	20.786	271.643	252.794	60.318	808.645	408.724	-6.672
3300	20.786	272.283	253.375	62.397	810.160	396.203	-6.271
3400	20.787	272.903	253.940	64.475	811.625	383.637	-5.894
3500	20.787	273.506	254.490	66.554	813.039	371.028	-5.537
3600	20.787	274.091	255.027	68.633	814.404	358.380	-5.200
3700	20.787	274.661	255.550	70.711	815.720	345.694	-4.880
3800	20.787	275.215	256.060	72.790	816.990	332.974	-4.577
3900	20.787	275.755	256.558	74.869	818.216	320.220	-4.289
4000	20.788	276.282	257.045	76.948	819.399	307.436	-4.015
4100	20.788	276.795	257.520	79.026	820.542	294.623	-3.754
4200	20.788	277.296	257.985	81.105	821.647	281.782	-3.504
4300	20.789	277.785	258.440	83.184	822.716	268.915	-3.267
4400	20.789	278.263	258.885	85.263	823.753	256.024	-3.039
4500	20.789	278.730	259.321	87.342	824.758	243.110	-2.822
4600	20.790	279.187	259.748	89.421	825.735	230.174	-2.614
4700	20.790	279.634	260.166	91.500	826.687	217.216	-2.414
4800	20.791	280.072	260.576	93.579	827.614	204.239	-2.223
4900	20.791	280.501	260.979	95.658	828.520	191.243	-2.039
5000	20.792	280.921	261.373	97.737	829.406	178.228	-1.862
5100	20.792	281.332	261.761	99.816	830.274	165.196	-1.692
5200	20.793	281.736	262.141	101.895	831.126	152.147	-1.528
5300	20.793	282.132	262.514	103.975	831.963	139.081	-1.371
5400	20.794	282.521	262.881	106.054	832.787	126.000	-1.219
5500	20.794	282.902	263.242	108.133	833.598	112.904	-1.072
5600	20.795	283.277	263.596	110.213	834.397	99.793	-0.931
5700	20.796	283.645	263.945	112.292	835.186	86.668	-0.794
5800	20.796	284.007	264.288	114.372	835.965	73.530	-0.662
5900	20.797	284.362	264.625	116.452	836.734	60.378	-0.535
6000	20.798	284.712	264.957	118.531	837.493	47.213	-0.411

PREVIOUS: June 1977 (1 atm)   CURRENT: September 1984 (1 bar)

Copper, Ion (Cu^+) $Cu_1^+(g)$

COPPER, ION (Cu^-) IDEAL GAS M_r = 63.54655

EA(Cu, g) = 1.228 ± 0.010 eV

S°(298.15 K) = 160.634 ± 0.001 J K^{-1} mol^{-1}

$\Delta_f H°$(0 K) = 217.926 kJ mol^{-1}

$\Delta_f H°$(298.15 K) = [212.920] kJ mol^{-1}

Electonic Level and Quantum Weight

State	ε_i, cm^{-1}	g_i
$^1S_0$	0.0	1

Enthalpy of Formation

$\Delta_f H°(Cu^-$, g, 0 K) is calculated from $\Delta_f H°$(Cu, g, 0 K) (1) using the adopted electron affinity of EA(Cu) = 1.228 ± 0.010 eV (118.483 ± 0.965 kJ mol^{-1}). This value, recommended by Hotop and Lineberger (2), is based on a laser photodetachment election spectrometry study (3). Additional information on Cu^-(g) may be obtained in the critical discussions of Hotop and Lineberger (2, 4), Rosenstock et al. (5) and Massey (6).

$\Delta_f H°(Cu^-$, g, 298.15 K) is obtained from $\Delta_f H°$(Cu, g, 0 K) by using EA(Cu) with JANAF (1) enthalpies, H°(0 K) - H°(298.15 K), for Cu^-(g), Cu(g), and e^-(g). $\Delta_r H°(Cu^- \rightarrow Cu + e^-$, 298.15 K) differs from a room-temperature threshold energy due to inclusion of these enthalpies and to threshold effects discussed by Rosenstock et al. (5). $\Delta_f H°$(298.15 K) should be changed by +6.197 kJ mol^{-1} if it is to be used in the ion convention that excludes the enthalpy of the electron.

Heat Capacity and Entropy

The ground state electronic configuration for Cu^-(g) is given by Hotop and Lineberger (2, 4) and Rosenstock et al. (5). Lacking any experimental evidence as to the stability of any excited states, we assume that no stable excited states exist.

References

1. JANAF Thermochemical Tables: Cu(g), 9-30-84; e^-(g), 3-31-82.
2. H. Hotop and W. C. Lineberger, J. Phys. Chem. Ref. Data, to be published, 1985.
3. H. Hotop, R. A. Bennett and W. C. Lineberger, J. Chem. Phys. 58, 2373 (1973).
4. H. Hotop and W. C. Lineberger, J. Phys. Chem. Ref. Data 4, 539 (1975).
5. H. M. Rosenstock, K. Draxl et al., J. Phys. Chem. Ref. Data 6, Supp. 1 (1977).
6. H. S. W. Massey, "Negative Ions", 3rd ed., Cambridge University Press, Cambrige, 1976.

Copper, Ion (Cu^-) Cu_1^-(g)

Enthalpy Reference Temperature = T_r = 298.15 K   Standard State Pressure = p° = 0.1 MPa

	J $K^{-1}mol^{-1}$			kJ mol^{-1}			
T/K	$C_p°$	S°	-[G°-H°(T_r)]/T	H°-H°(T_r)	$\Delta_f H°$	$\Delta_f G°$	Log K_f
0	0.	0.	INFINITE	-6.197	217.926		
100	20.786	137.927	179.114	-4.119			
200	20.786	152.334	162.535	-2.040			
250	20.786	156.973	160.976	-1.001			
298.15	20.786	160.634	160.634	0.	212.920	181.169	-31.740
300	20.786	160.762	160.634	0.038	212.874	180.972	-31.510
350	20.786	163.967	160.887	1.078	211.638	175.753	-26.230
400	20.786	166.742	161.450	2.117	210.381	170.712	-22.293
450	20.786	169.190	162.176	3.156	209.105	165.830	-19.249
500	20.786	171.381	162.989	4.196	207.815	161.091	-16.829
600	20.786	175.170	164.713	6.274	205.195	151.991	-13.232
700	20.786	178.374	166.442	8.353	202.521	143.334	-10.696
800	20.786	181.150	168.111	10.431	199.797	135.064	-8.819
900	20.786	183.598	169.698	12.510	197.020	127.139	-7.379
1000	20.786	185.788	171.200	14.589	194.186	119.525	-6.243
1100	20.786	187.769	172.617	16.667	191.281	112.199	-5.328
1200	20.786	189.578	173.956	18.746	188.286	105.141	-4.577
1300	20.786	191.242	175.223	20.824	185.158	98.338	-3.951
1400	20.786	192.782	176.423	22.903	168.742	92.189	-3.440
1500	20.786	194.216	177.562	24.982	165.458	86.835	-3.024
1600	20.786	195.558	178.645	27.060	162.173	81.701	-2.667
1700	20.786	196.818	179.677	29.139	158.889	76.772	-2.359
1800	20.786	198.006	180.663	31.217	155.605	72.036	-2.090
1900	20.786	199.130	181.606	33.296	152.320	67.483	-1.855
2000	20.786	200.196	182.509	35.375	149.036	63.103	-1.648
2100	20.786	201.210	183.375	37.453	145.751	58.887	-1.465
2200	20.786	202.177	184.208	39.532	142.467	54.828	-1.302
2300	20.786	203.101	185.010	41.610	139.182	50.918	-1.156
2400	20.786	203.986	185.782	43.689	135.898	47.151	-1.026
2500	20.786	204.834	186.527	45.768	132.613	43.521	-0.909
2600	20.786	205.650	187.247	47.846	129.329	40.022	-0.804
2700	20.786	206.434	187.943	49.925	126.045	36.649	-0.709
2800	20.786	207.190	188.617	52.004	122.760	33.398	-0.623
2900	20.786	207.919	189.270	54.082	-180.728	36.261	-0.653
3000	20.786	208.624	189.904	56.161	-183.218	43.785	-0.762
3100	20.786	209.306	190.519	58.239	-185.758	51.394	-0.866
3200	20.786	209.966	191.116	60.318	-188.349	59.085	-0.964
3300	20.786	210.605	191.697	62.397	-190.990	66.858	-1.058
3400	20.786	211.226	192.262	64.475	-193.683	74.712	-1.148
3500	20.786	211.828	192.813	66.554	-196.426	82.646	-1.233
3600	20.786	212.414	193.349	68.632	-199.219	90.658	-1.315
3700	20.786	212.983	193.872	70.711	-202.059	98.750	-1.394
3800	20.786	213.538	194.382	72.790	-204.946	106.918	-1.470
3900	20.786	214.078	194.881	74.868	-207.878	115.163	-1.542
4000	20.786	214.604	195.367	76.947	-210.852	123.484	-1.613
4100	20.786	215.117	195.843	79.025	-213.867	131.880	-1.680
4200	20.786	215.618	196.308	81.104	-216.919	140.349	-1.746
4300	20.786	216.107	196.762	83.183	-220.007	148.892	-1.809
4400	20.786	216.585	197.207	85.261	-223.129	157.507	-1.870
4500	20.786	217.052	197.643	87.340	-226.281	166.194	-1.929
4600	20.786	217.509	198.070	89.418	-229.461	174.950	-1.987
4700	20.786	217.956	198.489	91.497	-232.667	183.776	-2.042
4800	20.786	218.394	198.899	93.576	-235.897	192.671	-2.097
4900	20.786	218.822	199.301	95.654	-239.149	201.633	-2.149
5000	20.786	219.242	199.696	97.733	-242.421	210.662	-2.201
5100	20.786	219.654	200.083	99.811	-245.710	219.756	-2.251
5200	20.786	220.057	200.463	101.890	-249.016	228.915	-2.299
5300	20.786	220.453	200.837	103.969	-252.337	238.138	-2.347
5400	20.786	220.842	201.203	106.047	-255.672	247.423	-2.393
5500	20.786	221.223	201.564	108.126	-259.019	256.771	-2.439
5600	20.786	221.598	201.918	110.204	-262.377	266.179	-2.483
5700	20.786	221.966	202.267	112.283	-265.747	275.647	-2.526
5800	20.786	222.327	202.610	114.362	-269.126	285.175	-2.568
5900	20.786	222.682	202.947	116.440	-272.515	294.761	-2.610
6000	20.786	223.032	203.279	118.519	-275.914	304.404	-2.650

PREVIOUS: CURRENT: September 1984 (1 bar)

Copper, Ion (Cu^-) Cu_1^-(g)

COPPER FLUORIDE (CuF) CRYSTAL M_r = 82.544403

$S°(298.15\ K) = [64.852 \pm 2.1]\ J\ K^{-1}\ mol^{-1}$

T_{fus} = Unknown

$\Delta_f H°(0\ K)$ = Unknown

$\Delta_f H°(298.15\ K) = [-280.33 \pm 41.8]\ kJ\ mol^{-1}$

$\Delta_{fus}H°$ = Unknown

Enthalpy of Formation

The enthalpy of formation is calculated by combining the adopted $\Delta_f H°$(CuF, g, 298.15 K) = -3.0±4 kcal mol^{-1} with the estimated enthalpy of sublimation of 64 kcal mol^{-1}. A comparison of enthalpies of sublimation of LiF with LiCl, NaF with NaCl, and KF with KCl (1) along with an enthalpy of sublimation of CuCl of 54.77 kcal mol^{-1} (1) indicates that $\Delta_{sub}H°$(CuF, cr, 298.15 K) might be in the range of 58 to 71 kcal mol^{-1}. The tabulations of King et al. (2) lead to $\Delta_{sub}H°$(298.15 K) = 64 kcal mol^{-1} which we adopt. We derive $\Delta_f H°$(CuF, cr, 298.15 K) = -67±10 kcal mol^{-1}.

Heat Capacity and Entropy

We adopt the heat capacities estimated by King et al. (2) and S°(298.15 K) = 15.5±0.5 kcal mol^{-1} estimated by (3).

References

1. JANAF Thermochemical Tables: LiCl(cr), 9-30-65; NaCl(cr, g), 9-30-64; KCl(cr, g), CuCl(cr, g), 3-31-66; LiF(cr, g), NaF(cr, g), 12-31-68; KF(cr, g), 6-30-69.
2. E. G. King, A. D. Mah and L. B. Pankratz, "Thermodynamic Properties of Copper and Its Inorganic Compounds. INCRA Series on a Metallurgy of Copper, No. 2," INCRA, New York, 1973.
3. K. K. Kelley and E. G. King, U. S. Bur. Mines Bull. 592, 1961.

Copper Fluoride (CuF) $Cu_1F_1(cr)$

Enthalpy Reference Temperature = T_r = 298.15 K   Standard State Pressure = p° = 0.1 MPa

T/K	$C_p°$ (J K⁻¹mol⁻¹)	S° (J K⁻¹mol⁻¹)	$-[G°-H°(T_r)]/T$ (J K⁻¹mol⁻¹)	$H°-H°(T_r)$ (kJ mol⁻¹)	$\Delta_f H°$ (kJ mol⁻¹)	$\Delta_f G°$ (kJ mol⁻¹)	Log K_f
0							
100							
200							
298.15	51.882	64.852	64.852	0.	-280.328	-259.545	45.471
300	51.965	65.173	64.853	0.096	-280.306	-259.416	45.168
400	55.145	80.574	66.931	5.457	-279.049	-252.638	32.991
500	57.656	93.160	70.956	11.102	-277.652	-246.194	25.720
600	59.580	103.850	75.570	16.968	-276.142	-240.043	20.898
700	60.961	113.143	80.288	22.998	-274.562	-234.151	17.473
800	62.007	121.354	84.918	29.149	-272.941	-228.488	14.919
900	62.802	128.706	89.382	35.391	-271.302	-223.030	12.944
1000	63.346	135.352	93.652	41.700	-269.672	-217.754	11.374
1100	63.764	141.410	97.722	48.056	-268.081	-212.639	10.097
1200	64.057	146.971	101.598	54.448	-266.557	-207.667	9.040
1300	64.266	152.107	105.288	60.865	-265.153	-202.818	8.149
1400	64.378	156.874	108.805	67.297	-277.031	-197.664	7.375
1500	64.434	161.318	112.159	73.739	-275.780	-192.038	6.687
1600	64.434	165.477	115.363	80.182	-274.536	-186.496	6.088
1700	64.434	169.383	118.427	86.625	-273.299	-181.032	5.562
1800	64.434	173.066	121.361	93.069	-272.070	-175.640	5.097
1900	64.434	176.549	124.175	99.512	-270.847	-170.316	4.682
2000	64.434	179.854	126.877	105.955	-269.629	-165.056	4.311

PREVIOUS: June 1966   CURRENT: December 1977

Copper Fluoride (CuF) $Cu_1F_1(cr)$

COPPER FLUORIDE (CuF) IDEAL GAS M_r = 82.544403

$\Delta_fH°$(0 K) = -12.22 ± 16.7 kJ mol^{-1}

$S°$(298.15 K) = 226.495 ± 0.21 J K^{-1} mol^{-1} $\Delta_fH°$(298.15 K) = -12.55 ± 16.7 kJ mol^{-1}

Electronic Levels and Quantum Weights

State	ε_i, cm^{-1}	g_i
$X(^1\Sigma)$	0	[1]
$A(^1\pi)$	17546.8	[2]
$B(^1\Sigma)$	19738.5	[1]
$C(^1\pi)$	20258.8	[2]

ω_e = 622.65 cm^{-1} ω_ex_e = 3.950 cm^{-1} σ = 1

B_e = 0.3786 cm^{-1} α_e = 0.0032 cm^{-1} r_e = 1.7449 Å

Enthalpy of Formation

Hildenbrand (1) and Ehlert and Wang (2) have reported mass spectrometric measurements of equilibrium pressures of the reactions MgF(g) and Cu(g) = Mg(g) + CuF(g) and 1/2 CuF_2(cr) + 1/2 Cu(cr) = CuF(g), respectively. Our 2nd and 3rd law analyses follow.

Source	Reaction*	T/K	Data Points	$\Delta_rH°$(298.15 K), kcal mol^{-1} 2nd law	3rd law	Drift cal K^{-1} mol^{-1}	Δ_fH(298.15 K) kcal mol^{-1}
(1)	A	1413-1590	12	11.52±2.34	7.08±0.68	-2.9±1.9	-4.15±3
(2)	B	936-948	3	75.97±3.90	62.34±0.19	-14.5±4.1	-2.06±21

*Reactions: A) MgF(g) + Cu(g) = Mg(g) + CuF(g) (B) 1/2 CuF_2 + 1/2 Cu(cr) = CuF(g)

We adopt the rounded average, $\Delta_fH°$(CuF, g, 298.15 K) = -3.0 ± 4 kcal mol^{-1} (-12.552 ± 16.7 kJ mol^{-1}), which leads to D_0^0(Cu-F, 298.15 K) = 102.5±5 kcal mol^{-1} and D_0^0(Cu-F) = 101.7±5 kcal mol^{-1} (425.513 ± 20.9 kJ mol^{-1}).

Another value extant in the literature is D_0^0(298.15 K) = 83.6±4.0 kcal mol^{-1} estimated from the heat of atomization of CuF_2, derived from sublimation studies, and the ratio $D_0^0(M\text{-}F)/\Delta_aH°(MF_2)$ = 0.46 (4). Hildenbrand (1) has discussed the valence-state concept which predicts that $D_0^0(M\text{-}F) > D_0^0(MF\text{-}F)$; therefore the ratio $D_0^0(M\text{-}F)/[D(M\text{-}F) + D_0^0(MF\text{-}F)$ must be greater than 0.5. With $\Delta_{at}H°(CuF_2$, 298.15 K) = 182.17 kcal mol^{-1} (3) and D_0^0(Cu-F, 298.15 K) = -102.5 kcal mol^{-1}, the ratio based on JANAF values is 0.56. For $\Delta_{at}H°(CuF_2$, 298.15 K) = 182.7 kcal mol^{-1}, D_0^0(Cu-F, 298.15 K) must be greater than 91.08 kcal mol^{-1} to agree with the valence-state concept.

Other calculations by Kent et al. (4) give D (298.15 K) values ranging from 84.4 to 94.5 kcal mol^{-1}. Gaydon (5) recommends D_0^0 = 81 kcal mol^{-1} from a linear Birge-Sponer extrapolation of spectroscopic data but suggests that it is probably higher. King et al. (6) used $\Delta_fH°$(298.15 K) = 14 kcal mol^{-1}, $\Delta_{at}H°$(298.15 K) = 84.7 kcal mol^{-1} in their recent tabulation.

Heat Capacity and Entropy

The values of B_e, α_e and r_e are from the microwave data of Hoeft et al. (7) adjusted for normal isotopic abundance; the electronic levels, ω_e and ω_ex_e are taken from Rosen (8).

References

1. D. L. Hildenbrand, J. Chem. Phys. 48, 2457 (1968).
2. T. C. Ehlert and J. S. Wang, J. Phys. Chem. 81, 2069 (1977).
3. JANAF Thermochemical Tables: MgF(g), 6-30-76; Mg(g), 9-30-62; Cu(cr, g), 6-30-77; CuF_2(cr, g), 12-31-77.
4. R. A. Kent, J. D. McDonald, and J. L. Margrave, J. Phys. Chem. 70, 874 (1966).
5. A. G. Gaydon, "Dissociation Energies and Spectra of Diatomic Molecules," 3rd ed., Chapman and Hall Ltd., London, 1968.
6. E. G. King, A. D. Mah and L. B. Pankratz, "Thermodynamic Properties of Copper and its Inorganic Compounds. INCRA Series on the Metallurgy of Copper, No. 2," INCRA, New York, 1973.
7. J. Hoeft, F. J. Lovas, E. Tiemann, and T. Toerring, Z. Naturforsch. A25, 35 (1970).
8. B. Rosen, Ed., "Spectroscopic Data Relative to Diatomic Molecules," Pergamon Press, New York, 1970.

Copper Fluoride (CuF) Cu_1F_1(g)

Enthalpy Reference Temperature = T_r = 298.15 K Standard State Pressure = p° = 0.1 MPa

T/K	$C_p°$ (J K^{-1}mol^{-1})	S°	-[G°-H°(T_r)]/T	H°-H°(T_r) (kJ mol^{-1})	$\Delta_fH°$	$\Delta_fG°$	Log K_f
0	0.	0.	INFINITE	-9.083	-12.216	-12.216	INFINITE
100	29.204	192.900	254.629	-6.173	-11.428	-21.196	11.072
200	31.140	213.620	229.482	-3.172	-11.905	-30.818	8.049
250	32.375	220.704	227.040	-1.584	-12.231	-35.509	7.419
298.15	33.379	226.495	226.495	0.	-12.552	-39.963	7.001
300	33.414	226.702	226.496	0.062	-12.564	-40.133	6.988
350	34.234	231.917	226.906	1.754	-12.903	-44.701	6.671
400	34.871	236.532	227.826	3.482	-13.248	-49.220	6.427
450	35.367	240.669	229.027	5.239	-13.600	-53.696	6.233
500	35.759	244.416	230.382	7.017	-13.961	-58.131	6.073
600	36.327	250.989	233.284	10.623	-14.712	-66.896	5.824
700	36.713	256.620	236.225	14.276	-15.508	-75.531	5.636
800	36.992	261.541	239.088	17.962	-16.351	-84.048	5.488
900	37.203	265.911	241.831	21.673	-17.245	-92.457	5.366
1000	37.371	269.840	244.438	25.402	-18.194	-100.764	5.263
1100	37.509	273.408	246.912	29.146	-19.215	-108.972	5.175
1200	37.627	276.677	249.258	32.903	-20.326	-117.084	5.097
1300	37.730	279.693	251.485	36.671	-21.571	-125.098	5.026
1400	37.822	282.493	253.601	40.448	-36.105	-132.603	4.947
1500	37.906	285.105	255.615	44.235	-37.508	-139.447	4.856
1600	37.984	287.554	257.536	48.029	-38.912	-146.197	4.773
1700	38.058	289.859	259.370	51.831	-40.317	-152.859	4.697
1800	38.129	292.036	261.125	55.641	-41.722	-159.439	4.627
1900	38.198	294.100	262.806	59.457	-43.126	-165.940	4.562
2000	38.267	296.061	264.421	63.280	-44.528	-172.368	4.502
2100	38.336	297.930	265.972	67.111	-45.926	-178.725	4.446
2200	38.408	299.715	267.466	70.948	-47.318	-185.016	4.393
2300	38.483	301.424	268.905	74.792	-48.702	-191.244	4.343
2400	38.564	303.063	270.294	78.645	-50.075	-197.412	4.297
2500	38.651	304.639	271.637	82.505	-51.437	-203.523	4.252
2600	38.746	306.157	272.936	86.375	-52.782	-209.580	4.211
2700	38.850	307.621	274.193	90.255	-54.110	-215.586	4.171
2800	38.965	309.036	275.413	94.145	-55.418	-221.542	4.133
2900	39.092	310.405	276.596	98.048	-356.906	-221.457	3.989
3000	39.233	311.733	277.745	101.964	-357.371	-216.778	3.774
3100	39.387	313.022	278.862	105.895	-357.857	-212.084	3.574
3200	39.556	314.275	279.949	109.842	-358.364	-207.374	3.385
3300	39.740	315.495	281.008	113.807	-358.889	-202.647	3.208
3400	39.940	316.684	282.040	117.791	-359.430	-197.904	3.040
3500	40.156	317.845	283.046	121.795	-359.984	-193.146	2.883
3600	40.387	318.980	284.029	125.822	-360.549	-188.371	2.733
3700	40.634	320.089	284.989	129.873	-361.121	-183.580	2.592
3800	40.896	321.177	285.927	133.950	-361.697	-178.774	2.457
3900	41.173	322.242	286.844	138.053	-362.273	-173.953	2.330
4000	41.464	323.288	287.742	142.185	-362.847	-169.117	2.208
4100	41.767	324.316	288.622	146.346	-363.414	-164.267	2.093
4200	42.083	325.326	289.484	150.539	-363.971	-159.402	1.982
4300	42.410	326.320	290.329	154.763	-364.516	-154.525	1.877
4400	42.746	327.299	291.158	159.021	-365.044	-149.636	1.776
4500	43.092	328.264	291.972	163.313	-365.552	-144.734	1.680
4600	43.445	329.215	292.771	167.640	-366.038	-139.822	1.588
4700	43.804	330.153	293.557	172.002	-366.499	-134.899	1.499
4800	44.168	331.079	294.329	176.400	-366.933	-129.967	1.414
4900	44.536	331.993	295.088	180.836	-367.337	-125.026	1.333
5000	44.907	332.897	295.835	185.308	-367.709	-120.077	1.254
5100	45.279	333.790	296.571	189.817	-368.049	-115.121	1.179
5200	45.651	334.673	297.295	194.364	-368.354	-110.158	1.107
5300	46.021	335.546	298.008	198.947	-368.623	-105.190	1.037
5400	46.390	336.409	298.712	203.568	-368.857	-100.218	0.969
5500	46.755	337.264	299.405	208.225	-369.054	-95.241	0.905
5600	47.115	338.110	300.088	212.919	-369.214	-90.261	0.842
5700	47.470	338.947	300.763	217.648	-369.338	-85.279	0.781
5800	47.819	339.775	301.428	222.412	-369.425	-80.295	0.723
5900	48.161	340.596	302.085	227.211	-369.477	-75.309	0.667
6000	48.496	341.408	302.734	232.044	-369.493	-70.323	0.612

PREVIOUS: December 1977 (1 atm) CURRENT: December 1977 (1 bar)

Copper Fluoride (CuF) Cu_1F_1(g)

COPPER FLUORIDE (CuF_2) CRYSTAL M_r = 101.542806

$S°(298.15\ K) = 77.452 \pm 0.08\ J\ K^{-1}\ mol^{-1}$

$T_{fus} = 1109 \pm 10\ K$

$\Delta_f H°(0\ K) = -537.2 \pm 8.4\ kJ\ mol^{-1}$

$\Delta_f H°(298.15\ K) = -538.9 \pm 8.4\ kJ\ mol^{-1}$

$\Delta_{fus} H° = 55.229 \pm 20.1\ kJ\ mol^{-1}$

Enthalpy of Formation

Pervov et al. (1) have measured a enthalpy of formation of -128.8±0.3 kcal mol^{-1} by direct fluorine combustion calorimetry. In view of the spread of values obtained from the equilibrium pressure and the emf data discussed below, we adopt $\Delta_f H°(CuF_2$, cr, 298.15 K) = -128.8±2 kcal mol^{-1} (-538.899 ± 8.4 kJ mol^{-1}).

Jellinek and Rudat (2) and Domange (3) have measured equilibrium partial pressures for reactions of CuF_2(cr) with hydrogen and with water, respectively. Koerber and De Vries (4) and Skelton and Patterson (5) have derived $\Delta_r G°$ from emf studies. Our 2nd and 3rd law analyses are as follows, using auxiliary data from the JANAF Thermochemical Tables (6).

Source	Reaction*	T/K	Data Points	$\Delta_r H°$(298.15 K), kcal mol^{-1} 2nd law	3rd law	Drift cal K^{-1} mol^{-1}	$\Delta_f H°$(298.15 K) kcal mol^{-1}
(2)	A	423	1	--	22.6	--	-152.9
(3)	B	573-723	4	23.3±1.3	20.8±0.6	-3.9±2.0	-130.6±2
(4)	C	273-283	2	-1.9	-10.6±0.4	-31.4	-126.6±4
(5)	D	661-918	Equation	-132.1	-136.4	-5.5	-136.4

*Reactions:
(A) $CuF_2(cr) + H_2(g) = Cu(cr) + 2\ HF(g)$
(B) $CuF_2(cr) + H_2O(g) = CuO(cr) + 2\ HF(g)$
(C) $Cu(cr) + Hg_2F_2 = CuF_2(cr) + 2\ Hg(\ell)$
(D) $Cu(cr) + F_2(g) = CuF_2(cr)$

The evaluation of King et al. (7), $\Delta_f H°(CuF_2$, cr, 298.15 K) = -130.2 kcal mol^{-1}, preceded the direct fluorination study of Pervov et al. (1) and is based mainly on the work of Domange (3) and Koerber and De Vries (4). We place greater reliance on the direct fluorination work (1).

Heat Capacity and Entropy

The heat capacity is based on smoothed values from the unpublished low temperature measurements of Boo and Stout (8, 10-300 K) and the differential scanning calorimetric measurements of heat capacity by Ehlert (9, 300-710 K). Our graphical joining and smoothing of the two data sets produces data which are 1 to 2% lower than Ehlert's measurements (9) in the 350-500 K range and are in agreement from 520-700 K; our extrapolation to 1000 K yields 21.60 cal K^{-1} mol^{-1} compared to 22.4 cal K^{-1} mol^{-1} by Ehlert (9).

Boo and Stout (8) show an anomaly in the heat capacity at 70.90 K. S°(298.15 K) = 18.51 cal K^{-1} mol^{-1} is based on S°(10 K) = 0.009 cal K^{-1} mol^{-1} by Boo and Stout (8).

Fusion Data

Ehlert and Wang (10) interpret their DTA measurements as indicating a solid state transition at 1065±10 K and a fusion point of 1109±10 K. They believe the observed transitions at 1028 K and 1040 K are caused by Cu_2O-CuF_2 and 10 CuO_2-20Cu-70CuF_2 weight percent eutectics, respectively. Based on the latter, and ignoring the 1065 K solid state transition, they calculate a heat of fusion of 13.2±2 kcal mol^{-1}. We adopt T_{fus} = 1109±10 K and $\Delta_{fus}H°$ = 13.2±5 kcal mol^{-1} and await confirmation of the solid state transition.

Sublimation Data

See the ideal gas table for details.

References

1. V. S. Pervov, V. Ya. Leonidov, L. I. Klyuev, and A. G. Muravina, Dokl. Akad. Nauk SSSR 214, 1088 (1974).
2. K. Jellinck and A. Rudat, Z. anorg. allgem. Chem. 175, 281 (1928).
3. L. Domange, Comp. rend. 200, 239 (1935).
4. G. G. Koerber and T. De Vries, J. Amer. Chem. Soc. 74, 5008 (1952).
5. W. H. Skelton and J. W. Patterson, J. Less-Common Metals 31, 47 (1973).
6. JANAF Thermochemical Tables: H_2(ref st), 3-31-77; Cu(cr), HF(g), 6-30-77; H_2O(g), 3-31-61; CuO(cr), 9-30-77; Hg_2F_2(cr), 3-31-62; Hg(ℓ), 12-31-61; F_2(ref st), 6-31-77.
7. E. G. King, A. D. Mah, and L. B. Pankratz, "Thermodynamic Properties of Copper and its Inorganic Compounds. INCRA Series on the Metallurgy of Copper, No. 2," INCRA, New York, 1973.
8. W. O. J. Boo and J. W. Stout, The University of Chicago, personal communication, November, 1977.
9. T. C. Ehlert, Thermochim. Acta 21, 111 (1977).
10. T. C. Ehlert and J. S. Wang, J. Phys. Chem. 81, 2069 (1977).

Copper Fluoride (CuF_2) Cu_1F_2(cr)

Enthalpy Reference Temperature = T_r = 298.15 K — Standard State Pressure = p° = 0.1 MPa

T/K	$C_p°$ (J $K^{-1}mol^{-1}$)	S°	$-[G°-H°(T_r)]/T$	$H°-H°(T_r)$ (kJ mol^{-1})	$\Delta_f H°$	$\Delta_f G°$	Log K_f
0	0.	0.	INFINITE	-12.103	-537.170	-537.170	INFINITE
100	34.229	21.852	128.307	-10.646	-539.289	-523.434	273.414
200	55.731	53.151	83.178	-6.005	-539.591	-507.345	132.505
298.15	65.552	77.452	77.452	0.	-538.899	-491.642	86.134
300	65.689	77.858	77.453	0.121	-538.881	-491.349	85.551
400	72.383	97.697	80.113	7.034	-537.682	-475.674	62.117
500	77.739	114.446	85.347	14.549	-536.097	-460.348	48.092
600	81.923	129.001	91.436	22.539	-534.201	-445.372	38.773
700	84.810	141.857	97.739	30.883	-532.083	-430.732	32.142
800	87.027	153.332	103.984	39.479	-529.821	-416.406	27.189
900	88.868	163.692	110.052	48.276	-527.454	-402.371	23.353
1000	90.374	173.136	115.895	57.241	-525.013	-388.603	20.299
1100	91.588	181.807	121.498	66.340	-522.539	-375.081	17.811
1109.000	91.713	182.554	121.991	67.164	------ CRYSTAL <--> LIQUID -----		
1200	92.550	189.820	126.862	75.550	-520.070	-361.785	15.748
1300	93.303	197.259	131.995	84.844	-517.673	-348.693	14.011
1400	93.847	204.195	136.907	94.203	-528.521	-335.374	12.513
1500	94.224	210.683	141.611	103.608	-526.211	-321.659	11.201
1600	94.433	216.771	146.121	113.041	-523.891	-308.097	10.058
1700	94.558	222.501	150.447	122.492	-521.570	-294.681	9.054
1800	94.558	227.906	154.601	131.948	-519.258	-281.402	8.166
1900	94.558	233.018	158.595	141.404	-516.959	-268.250	7.375
2000	94.558	237.868	162.439	150.859	-514.668	-255.220	6.666

PREVIOUS: June 1966 CURRENT: December 1977

Copper Fluoride (CuF_2) Cu_1F_2(cr)

COPPER FLUORIDE (CuF_2) LIQUID M_r = 101.542806

S°(298.15 K) = [122.585] J K^{-1} mol^{-1}

T_{fus} = 1109 ± 10 K

$\Delta_f H°$(298.15 K) = [-487.854] kJ mol^{-1}

$\Delta_{fus} H°$ = 55.229 ± 20.1 kJ mol^{-1}

Enthalpy of Formation

$\Delta_f H°$(CuF_2, ℓ, 298.15 K) is calculated from that of the crystal by adding $\Delta_{fus}H°$ and the difference in enthalpy, H°(1109 K)-H°(298.15 K), between the crystal and the liquid.

Heat Capacity and Entropy

The heat capacity is estimated from an average heat capacity of 8 cal K^{-1} g-atom^{-1} for several difluorides and dichlorides. A glass transition is assumed at 740 K below which the heat capacity is that of the crystal.

The entropy is calculated in a manner analogous to that used for the enthalpy of formation.

Fusion Data

See the crystal table for details.

Vaporization Data

T_{vap} = 1948.9 K is the calculated temperature at which the fugacity of CuF_2(g) is 1 atm for the reaction CuF_2(ℓ) = CuF_2(g). $\Delta_{vap}H°$ = 81.132 kJ mol^{-1} is the difference in $\Delta_f H°$ between gas and liquid at T_{vap}.

Copper Fluoride (CuF_2) Cu_1F_2(l)

Enthalpy Reference Temperature = T_r = 298.15 K Standard State Pressure = p° = 0.1 MPa

	J K⁻¹mol⁻¹			kJ mol⁻¹			
T/K	C_p°	S°	-[G°-H°(T_r)]/T	H°-H°(T_r)	$\Delta_f H°$	$\Delta_f G°$	Log K_f
0							
100							
200							
298.15	65.551	122.585	122.585	0.	-487.854	-454.053	79.548
300	65.689	122.990	122.586	0.121	-487.836	-453.844	79.021
400	72.383	142.830	125.246	7.034	-486.637	-442.682	57.808
500	77.739	159.579	130.480	14.549	-485.052	-431.870	45.117
600	81.923	174.134	136.569	22.539	-483.156	-421.407	36.687
700	84.810	186.990	142.872	30.883	-481.038	-411.280	30.690
740.000	85.772	191.730	145.386	34.295	GLASS <--> LIQUID		
740.000	100.416	191.730	145.386	34.295	TRANSITION		
800	100.416	199.558	149.159	40.320	-477.935	-401.501	26.215
900	100.416	211.386	155.429	50.361	-474.324	-392.165	22.761
1000	100.416	221.965	161.563	60.403	-470.806	-383.226	20.018
1100	100.416	231.536	167.496	70.444	-467.390	-374.634	17.790
1109.000	100.416	232.354	168.019	71.348	------ CRYSTAL <--> LIQUID -----		
1200	100.416	240.273	173.202	80.486	-464.089	-366.348	15.947
1300	100.416	248.311	178.674	90.528	-460.945	-358.332	14.398
1400	100.416	255.753	183.917	100.569	-471.110	-350.144	13.064
1500	100.416	262.681	188.940	110.611	-468.163	-341.607	11.896
1600	100.416	269.161	193.754	120.652	-465.235	-333.265	10.880
1700	100.416	275.249	198.370	130.694	-462.323	-325.106	9.989
1800	100.416	280.989	202.802	140.736	-459.426	-317.118	9.203
1900	100.416	286.418	207.061	150.777	-456.540	-309.291	8.503
2000	100.416	291.568	211.159	160.819	-453.664	-301.616	7.877
2100	100.416	296.468	215.106	170.860	-450.794	-294.084	7.315
2200	100.416	301.139	218.911	180.902	-447.926	-286.689	6.807
2300	100.416	305.603	222.584	190.944	-445.058	-279.424	6.346
2400	100.416	309.876	226.133	200.985	-442.184	-272.284	5.926
2500	100.416	313.976	229.565	211.027	-439.301	-265.264	5.542

PREVIOUS: June 1966 CURRENT: December 1977

Copper Fluoride (CuF_2) Cu_1F_2(l)

COPPER FLUORIDE (CuF_2) CRYSTAL–LIQUID M_r = 101.542806

0 to 1109 K crystal
above 1109 K liquid

Refer to the individual tables for details.

Copper Fluoride (CuF_2) Cu_1F_2(cr,l)

Enthalpy Reference Temperature = T_r = 298.15 K Standard State Pressure = $p°$ = 0.1 MPa

T/K	$C_p°$ (J K⁻¹mol⁻¹)	$S°$ (J K⁻¹mol⁻¹)	$-[G°-H°(T_r)]/T$ (J K⁻¹mol⁻¹)	$H°-H°(T_r)$ (kJ mol⁻¹)	$\Delta_fH°$ (kJ mol⁻¹)	$\Delta_fG°$ (kJ mol⁻¹)	Log K_f
0	0.	0.	INFINITE	-12.103	-537.170	-537.170	INFINITE
100	34.229	21.852	128.307	-10.646	-539.289	-523.434	273.414
200	55.731	53.151	83.178	-6.005	-539.591	-507.345	132.505
298.15	65.552	77.452	77.452	0.	-538.899	-491.642	86.134
300	65.689	77.858	77.453	0.121	-538.881	-491.349	85.551
400	72.383	97.697	80.113	7.034	-537.682	-475.674	62.117
500	77.739	114.446	85.347	14.549	-536.097	-460.348	48.092
600	81.923	129.001	91.436	22.539	-534.201	-445.372	38.773
700	84.810	141.857	97.739	30.883	-532.083	-430.732	32.142
800	87.027	153.332	103.984	39.479	-529.821	-416.406	27.189
900	88.868	163.692	110.052	48.276	-527.454	-402.371	23.353
1000	90.374	173.136	115.895	57.241	-525.013	-388.603	20.299
1100	91.588	181.807	121.498	66.340	-522.539	-375.081	17.811
1109.000	91.713	182.554	121.991	67.164	CRYSTAL <--> LIQUID TRANSITION		
1109.000	100.416	232.354	121.991	122.393			
1200	100.416	240.273	130.664	131.531	-464.089	-366.348	15.947
1300	100.416	248.311	139.409	141.573	-460.945	-358.331	14.398
1400	100.416	255.753	147.457	151.614	-471.109	-350.144	13.064
1500	100.416	262.681	154.910	161.656	-468.163	-341.607	11.896
1600	100.416	269.161	161.850	171.698	-465.234	-333.265	10.880
1700	100.416	275.249	168.344	181.739	-462.323	-325.106	9.989
1800	100.416	280.989	174.444	191.781	-459.425	-317.118	9.203
1900	100.416	286.418	180.196	201.822	-456.540	-309.291	8.503
2000	100.416	291.568	185.637	211.864	-453.664	-301.615	7.877
2100	100.416	296.468	190.798	221.906	-450.794	-294.084	7.315
2200	100.416	301.139	195.709	231.947	-447.926	-286.689	6.807
2300	100.416	305.603	200.390	241.989	-445.057	-279.424	6.346
2400	100.416	309.876	204.864	252.030	-442.183	-272.284	5.926
2500	100.416	313.976	209.147	262.072	-439.301	-265.264	5.542

PREVIOUS: CURRENT: December 1977

Copper Fluoride (CuF_2) Cu_1F_2(cr,l)

COPPER FLUORIDE (CuF_2) IDEAL GAS M_r = 101.542806

$\Delta_f H°(0\ K) = -265.16 \pm 12.6\ kJ\ mol^{-1}$

$S°(298.15\ K) = [267.09 \pm 2.1]\ J\ K^{-1}\ mol^{-1}$ $\Delta_f H°(298.15\ K) = -266.94 \pm 12.6\ kJ\ mol^{-1}$

Electronic Levels and Quantum Weights

State	ϵ_i, cm^{-1}	g_i
[$^2\Sigma^+g$]	0	[2]
[$^2\Pi g$]	[9000]	[4]
[$^2\Delta g$]	[18000]	[4]

Vibrational Frequencies and Degeneracies

ν, cm^{-1}
625 (1)
135±5(1)
782 (1)

Ground State Quantum Weight: [2] σ = 2

Point Group: C_{2v}

Bond Distance: Cu-F = [1.72] Å

Bond Angle: F-Cu-F = 165 ± 8°

Product of the Moments of Inertia: $I_A I_B I_C = [6.77288 \times 10^{-116}]\ g^3\ cm^6$

Enthalpy of Formation

Sublimation pressures, $CuF_2(cr) = CuF_2(g)$, have been measured by Kent et al. (1, 897-1026 K) and by Ehlert and Wang (2, 874-1005 K) by effusion cell-mass spectrometric techniques. Using JANAF data (6) for CuF_2(cr), we derive the following by 2nd and 3rd law analysis:

Source	T/K	Data Points	$\Delta_{sub}H°$(298.15 K), kcal mol^{-1} 2nd law	3rd law	Drift cal K^{-1} mol^{-1}	$\Delta_f H°$(cr, 298.15 K), kcal mol^{-1}
(1)	897-1026	26[a]	63.20±0.50	62.74±0.17	-0.5±0.5	-66.1
(2) via CuF_2^+	874-1005	33[b]	65.82±0.72	64.992±0.33	-0.9±0.8	-63.8
(2) via CuF^+	874-1005	33[b]	65.92±0.67	65.085±0.30	-0.9±0.7	

(a) Two data points eliminated by a statistical test.

(b) One data point eliminated by a statistical test.

We adopt $\Delta_f H°(CuF_2$, g, 298.15 K) = -63.8 ± 3 kcal mol^{-1}.

Heat Capacity and Entropy

The bond distance is taken from Brewer et al. (3). The bond distances for all of the transition metal difluorides from chromium through copper were estimated to be the same with the assessment that an error of 0.5 Å would lead to an error of less than 1 cal K^{-1} mol^{-1} in the entropy. The bond angle and vibrational frequencies are those derived by Hastie et al. (4) from infrared studies of matrix-isolated isotopic species. The electronic states and levels are assumed the same as those assigned to $CuCl_2$ (5). The principal moments of inertia are $I_A = 0.1990 \times 10^{-39}$, $I_B = 18.3484 \times 10^{-39}$, $I_C = 18.5474 \times 10^{-39}\ g\ cm^2$.

References

1. R. A. Kent, J. D. McDonald and J. L. Margrave, J. Phys. Chem. 70, 874 (1966).
2. T. C. Ehlert and J. S. Wang, J. Phys. Chem. 81, 2069 (1977).
3. L. Brewer, G. R. Somayajulu and E. Brackett, Chem. Rev. 63, 111 (1963).
4. J. W. Hastie, R. H. Hauge and J. L. Margrave, High Temp. Sci. 1, 76 (1969) and J. Chem. Soc. D 1969, 1452 (1969).
5. J. T. Hougen, G. E. Leroi and T. C. James, J. Chem. Phys. 44, 1581 (1961).
6. JANAF Thermochemical Tables: CuF_2(cr), 12-31-77.

Copper Fluoride (CuF_2) Cu_1F_2(g)

Enthalpy Reference Temperature = T_r = 298.15 K Standard State Pressure = p° = 0.1 MPa

T/K	$C_p°$ (J $K^{-1}mol^{-1}$)	S°	-[G°-H°(T_r)]/T	H°-H°(T_r) (kJ mol^{-1})	$\Delta_f H°$	$\Delta_f G°$	Log K_f
0	0.	0.	INFINITE	-12.056	-265.163	-265.163	INFINITE
100	38.109	220.861	306.489	-8.563	-265.246	-269.292	140.664
200	43.318	248.878	271.334	-4.491	-266.117	-273.016	71.304
250	45.858	258.822	267.865	-2.261	-266.552	-274.690	57.393
298.15	47.987	267.086	267.086	0.	-266.939	-276.222	48.393
300	48.061	267.384	267.087	0.089	-266.954	-276.279	48.104
350	49.859	274.932	267.679	2.539	-267.329	-277.803	41.460
400	51.289	281.687	269.016	5.069	-267.687	-279.275	36.470
450	52.419	287.796	270.768	7.662	-268.036	-280.703	32.583
500	53.315	293.367	272.754	10.307	-268.380	-282.091	29.470
600	54.609	303.210	277.031	15.707	-269.073	-284.769	24.791
700	55.466	311.697	281.391	21.214	-269.792	-287.329	21.441
800	56.057	319.144	285.654	26.792	-270.547	-289.783	18.921
900	56.481	325.773	289.751	32.420	-271.350	-292.140	16.955
1000	56.796	331.741	293.656	38.084	-272.210	-294.404	15.378
1100	57.042	337.166	297.369	43.777	-273.142	-296.579	14.083
1200	57.245	342.138	300.895	49.491	-274.169	-298.665	13.001
1300	57.426	346.727	304.246	55.225	-275.332	-300.660	12.081
1400	57.599	350.989	307.435	60.976	-289.787	-302.153	11.273
1500	57.776	354.969	310.473	66.745	-291.114	-302.990	10.551
1600	57.965	358.704	313.372	72.532	-292.440	-303.739	9.916
1700	58.170	362.224	316.143	78.339	-293.763	-304.405	9.353
1800	58.396	365.555	318.796	84.167	-295.079	-304.992	8.851
1900	58.642	368.719	321.341	90.018	-296.384	-305.508	8.399
2000	58.909	371.734	323.786	95.896	-297.672	-305.954	7.991
2100	59.195	374.615	326.138	101.801	-298.938	-306.337	7.620
2200	59.498	377.376	328.405	107.735	-300.178	-306.660	7.281
2300	59.814	380.027	330.592	113.701	-301.385	-306.928	6.971
2400	60.141	382.580	332.705	119.698	-302.555	-307.144	6.685
2500	60.476	385.042	334.750	125.729	-303.683	-307.311	6.421
2600	60.816	387.420	336.730	131.794	-304.765	-307.435	6.176
2700	61.158	389.722	338.650	137.893	-305.798	-307.518	5.949
2800	61.498	391.952	340.514	144.025	-306.777	-307.564	5.738
2900	61.836	394.116	342.326	150.192	-607.904	-301.579	5.432
3000	62.168	396.218	344.087	156.392	-607.975	-291.015	5.067
3100	62.493	398.262	345.802	162.625	-608.036	-280.449	4.726
3200	62.808	400.251	347.473	168.891	-608.088	-269.881	4.405
3300	63.114	402.188	349.101	175.187	-608.129	-259.311	4.105
3400	63.408	404.077	350.691	181.513	-608.160	-248.741	3.821
3500	63.690	405.919	352.242	187.868	-608.180	-238.169	3.554
3600	63.958	407.717	353.758	194.251	-608.188	-227.597	3.302
3700	64.214	409.473	355.241	200.659	-608.185	-217.025	3.064
3800	64.455	411.189	356.690	207.093	-608.169	-206.454	2.838
3900	64.682	412.866	358.109	213.550	-608.140	-195.882	2.624
4000	64.895	414.506	359.499	220.029	-608.098	-185.312	2.420
4100	65.094	416.111	360.860	226.528	-608.041	-174.743	2.226
4200	65.278	417.682	362.194	233.047	-607.970	-164.176	2.042
4300	65.449	419.220	363.503	239.583	-607.883	-153.610	1.866
4400	65.606	420.726	364.786	246.136	-607.780	-143.047	1.698
4500	65.749	422.202	366.046	252.704	-607.660	-132.486	1.538
4600	65.880	423.649	367.282	259.286	-607.524	-121.928	1.385
4700	65.997	425.067	368.497	265.880	-607.371	-111.373	1.238
4800	66.103	426.458	369.690	272.485	-607.200	-100.822	1.097
4900	66.197	427.822	370.862	279.100	-607.011	-90.274	0.962
5000	66.279	429.160	372.015	285.724	-606.805	-79.731	0.833
5100	66.350	430.473	373.148	292.355	-606.582	-69.191	0.709
5200	66.411	431.762	374.263	298.993	-606.341	-58.656	0.589
5300	66.462	433.027	375.360	305.637	-606.083	-48.127	0.474
5400	66.504	434.270	376.440	312.286	-605.808	-37.602	0.364
5500	66.537	435.491	377.502	318.938	-605.517	-27.082	0.257
5600	66.561	436.690	378.548	325.593	-605.211	-16.568	0.155
5700	66.577	437.868	379.579	332.249	-604.891	-6.059	0.056
5800	66.585	439.026	380.594	338.908	-604.557	4.443	-0.040
5900	66.587	440.164	381.594	345.566	-604.210	14.941	-0.132
6000	66.581	441.284	382.579	352.225	-603.852	25.432	-0.221

PREVIOUS: December 1977 (1 atm) CURRENT: December 1977 (1 bar)

Copper Fluoride (CuF_2) Cu_1F_2(g)

COPPER HYDROXIDE ($Cu(OH)_2$) CRYSTAL M_r = 97.56068

$S°(298.15\ K) = [108.366]\ J\ K^{-1}\ mol^{-1}$

$\Delta_f H°(0\ K)$ = Unknown

$\Delta_f H°(298.15\ K) = -450.37 \pm 8.4\ kJ\ mol^{-1}$

Enthalpy of Formation

The enthalpy change, $\Delta_r H°(433\ K)$, for the reaction $Cu(OH)_2(cr) = CuO(cr) + H_2O(g)$ has been determined by My (1). Based on the reported value, $\Delta_r H°(433\ K) = 12\ kcal\ mol^{-1}$. The adopted $\Delta_f H°(Cu(OH)_2,\ cr,\ 298.15\ K)$ value was derived as -107.64 kcal mol^{-1}, using $\Delta_f H°(298.15\ K)$ = -37.250 and -57.798 kcal mol^{-1} for CuO(cr) and $H_2O(g)$, respectively.

The Gibbs energy change, $\Delta_r G°(298.15\ K)$, of the reaction $Cu(OH)_2(cr) = Cu^{++}(aq) + 2\ OH^-(aq)$ was evaluated to be 26.36 and 26.26 kcal mol^{-1}, based on the solubility product for $Cu(OH)_2(cr)$ reported by Ganelina (2) and Zhuk (3), respectively. Using values of $S°(298.15\ K)$ and $\Delta_f H°(298.15\ K)$ for $Cu^{++}(aq)$ and $OH^-(aq)$ from (4) and (5), the corresponding enthalpy change $\Delta_r H°(298.15\ K)$ and enthalpy of formation, $\Delta_f H°(298.15\ K)$ for $Cu(OH)_2(cr)$ were derived. The results obtained are presented as follows:

Reaction	$\Delta_r H°(298.15\ K)$ kcal mol^{-1}	$\Delta_f H°(298.15\ K)$ kcal mol^{-1}	Source
$Cu(OH)_2(cr) = Cu^{++}(aq) + 2\ OH^-(aq)$	10.06	-104.61	(1)
$Cu(OH)_2(cr) = Cu^{++}(aq) + 2\ OH^-(aq)$	9.97	-104.52	(2)
$CuO(cr) + H_2O(\ell) = Cu(OH)_2(cr)$	-1.9	-107.5	(3)
$CuO(cr) + H_2O(\ell) = Cu(OH)_2(cr)$	0.3	-105.3	(4)

The enthalpies of solution of CuO(cr) and $Cu(OH)_2(cr)$ in $HNO_3(aq)$ have been determined by de Forcrand (6) to be -16.38 and -14.46 kcal mol^{-1}, respectively. From these data, the enthalpy change for the reaction $CuO(cr) + H_2O(\ell) = Cu(OH)_2(cr)$ was derived as -1.9 kcal mol^{-1}. For the same reaction, the enthalpy change was reported to be 0.3 kcal mol^{-1} by Sabatier (7). Based on $\Delta_f H°(298.15\ K)$ = -37.25 and -68.315 kcal mol^{-1} for CuO(cr) and $H_2O(\ell)$, respectively, the enthalpy of formation for $Cu(OH)_2$ was evaluated. The results obtained are presented in the above table.

Heat Capacity and Entropy

The heat capacities, 298.15-700 K, were estimated by comparison with those for CuO(cr), CaO(cr) and $Ca(OH)_2(cr)$. The $C_p°$ values above 700 K were obtained by graphical extrapolation. $S°(298.15\ K)$ was estimated such that the derived $\Delta_f G°(Cu(OH)_2,\ cr) = \Delta_f G°(CuO,\ cr) + \Delta_f G°(H_2O,\ g)$ at 433 K.

Decomposition Data

T_{dcm} = 433 K is the temperature at which the Gibbs energy change for the reaction $Cu(OH)_2(cr) = CuO(cr) + H_2O(g)$ is such that the fugacity of water is 1 atm.

References

1. L. V. My, Bull. soc. chim. France, No. 3, 545 (1964).
2. E. S. Ganelina, Zh. prikl. Khim. 37, 1358 (1964).
3. N. P. Zhuk, Zh. Fiz. Khim. 28, 1523 (1954).
4. U. S. Nat. Bur. Stand. Circ. 500, 1952.
5. U. S. Nat. Bur. Stand. Tech. Note 270-1, 1965.
6. de Forcrand, Compt. rend. 157, 441 (1913).
7. P. Sabatier, Compt. rend. 125, 301 (1897).

Copper Hydroxide ($Cu(OH)_2$) $Cu_1H_2O_2(cr)$

Enthalpy Reference Temperature = T_r = 298.15 K   Standard State Pressure = p° = 0.1 MPa

T/K	$C_p°$ (J $K^{-1}mol^{-1}$)	S° (J $K^{-1}mol^{-1}$)	$-[G°-H°(T_r)]/T$ (J $K^{-1}mol^{-1}$)	$H°-H°(T_r)$ (kJ mol^{-1})	$\Delta_f H°$ (kJ mol^{-1})	$\Delta_f G°$ (kJ mol^{-1})	Log K_f
0							
100							
200							
298.15	95.186	108.366	108.366	0.	-450.366	-372.660	65.289
300	95.228	108.955	108.367	0.176	-450.343	-372.178	64.802
400	99.496	136.934	112.154	9.912	-448.977	-346.323	45.225
500	103.763	159.595	119.444	20.075	-447.361	-320.841	33.518
600	108.031	178.893	127.782	30.666	-445.479	-295.709	25.744
700	112.131	195.862	136.319	41.680	-443.332	-270.913	20.216
800	115.562	211.070	144.729	53.073	-440.952	-246.442	16.091
900	117.989	224.829	152.877	64.757	-438.425	-222.279	12.901
1000	119.662	237.352	160.708	76.644	-435.838	-198.401	10.363
1100	120.750	248.811	168.204	88.668	-433.267	-174.782	8.300
1200	121.503	259.353	175.366	100.784	-430.774	-151.395	6.590
1300	121.880	269.094	182.206	112.954	-428.435	-128.209	5.152
1400	122.047	278.133	188.739	125.151	-439.431	-104.790	3.910
1500	122.173	286.558	194.983	137.363	-437.354	-80.960	2.819

PREVIOUS: CURRENT: June 1966

Copper Hydroxide ($Cu(OH)_2$) $Cu_1H_2O_2(cr)$

$S°(298.15 K) = 42.59 \pm 0.4 J K^{-1} mol^{-1}$

$\Delta_f H°(0 K) = -153.81 \pm 2.1 kJ mol^{-1}$
$\Delta_f H°(298.15 K) = -156.06 \pm 2.1 kJ mol^{-1}$

Enthalpy of Formation

The enthalpy of formation of CuO(cr) has been investigated calorimetrically (both oxidation and reduction reactions) and by extensive study of the $Cu_2O(cr) + 1/2 O_2(g) = 2 CuO(cr)$ equilibrium by pressure and by emf measurement. The more consistent data leading to the adopted value are summarized below.

Source	Method	Reaction*	T/K	Data Points	$\Delta_r H°$(298.15 K), kcal mol^{-1} 2nd law	3rd law	Drift cal K^{-1} mol^{-1}	$\Delta_f H°$(298.15 K)** kcal mol^{-1}
(1)	Aq Calorim	A	291	--	31.15		--	-37.16
(2)	Calorimetric	A	295	--	30.27±0.2		--	-38.04
(3)	Calorimetric	See Text	298.15	--	--		--	-37.23±0.15
(4)	Calorimetric	A	297.9	(Coarse grain CuO)	--		--	-38.72±0.30
		A	298	(Fine powder CuO)	--		--	-36.50±0.2
		B	297.9	(Coarse grain CuO)	--		--	-38.59±0.35
(5)	Equil Press	B	1011-1156	5	-33.84±0.31	-37.73±0.43	-3.6±0.3	-37.73
(6)	emf	B	845-1270	Equation	-37.26	-37.38	-0.1±0.0	-37.38
(7)	Equil Press	C	1223-1322	7	-32.52±0.18	-33.65±0.06	-0.9±0.1	-37.23
(8)	Equil Press	C	1189-1358	20	-33.28±0.06	-33.68±0.04	-0.3±0.0	-37.24
(9)	Equil Press	C	1193-1293	6	-33.23±0.28	-33.68±0.04	-0.4±0.2	-37.24
(10)	Equil Press	C	1047-1342	11	-32.79±0.14	-33.69±0.17	-0.7±0.1	-37.25
(11)	Equil Press	C	1026-1072	3	-35.50±0.37	-33.62±0.07	1.4±0.3	-37.21
(12)	emf	C	973-1273	Equation	-33.57	-33.64	-0.6	-37.22
(13)	emf	C	892-1320	Equation	-33.53	-33.58	-0.6	-37.19
(14)	emf	C	1023-1294	Equation	-33.39	-33.58	0.2	-37.19
(15)	emf	C	870-1170	Equation	-34.37	-35.00	0.6	-37.90
(16)	Press	C	993-1021	4	-34.18±0.06	-34.12±0.0	-0.1±0.1	-37.46
(17)	Equil Press	C	800-950	Equation	-29.92	-34.30	10.0	-37.55
(18)	Equil Press	C	873-1073	Equation	-40.06	-35.26	4.9	-38.03

*Reactions: (A) $Cu(cr) + H_2O(\ell) = CuO(cr) + H_2(g)$
(B) $Cu(cr) + 1/2 O_2(g) = CuO(cr)$
(C) $Cu_2O(cr) + 1/2 O_2(g) = 2 CuO(cr)$

**Based on third law where possible and $\Delta_f H°(Cu_2O, cr, 298.15 K)$ = -40.8 kcal mol^{-1} in Reaction C.

Thomsen's data (1) may be reduced to the enthalpy of reduction of CuO(cr) with H_2 by combining enthalpies of reaction for CuO + H_2SO_4, Fe + H_2SO_4, and Fe + $CuSO_4$(aq). Wartenberg and Werth (2) and Nunez et al. (4) have measured the direct enthalpy of reduction of CuO(cr) by H_2; Mah et al. (3) and Nunez et al. (4) have measured the direct enthalpy of oxidation of copper by O_2. Mah et al. (3) obtained a mixed Cu_2O and CuO product and used $\Delta_r H°(298.15 K) = -33.63 \pm 0.05$ kcal mol^{-1} for the equilibrium reaction $Cu_2O(cr) + 1/2 O_2(g) = 2 CuO(cr)$ to resolve their results into $\Delta_f H°$(CuO, cr, 298.15 K) and $\Delta_f H°(Cu_2O$, cr, 298.15 K). From the measurment of the reduction of coarse and fine CuO(cr), Nunez et al. (4) conclude that $\Delta_f H°$ of the copper oxides is influenced by material history, state of subdivision, and non-stoichiometry.

Eliminating three studies (5, 17, and 18) from our 3rd law analysis because of large drift and poor 2nd and 3rd law agreement, the average of the eleven remaining studies is -37.32 kcal mol^{-1}. We adopt $\Delta_f H°(298.15 K) = -37.3 \pm 0.5$ kcal mol^{-1}.

Heat Capacity and Entropy

The adopted values are based on the low temperature heat capacities by Hu and Johnston (19, 15-297 K) and the high temperature enthalpies by Mah et al. (3, 410-1400 K). Changes are made where appropriate to adjust to the 1975 atomic weights (20) and the IPTS-68 temperature scale (21). A small anomaly in the heat capacity is observed in the 210-230 K region. Magnetic measurements of O'Keefe and Stone (22) and neutron diffraction studies of Brockhouse (23) suggest that this is a Neel point associated with antiferromagnetism.

The entropy is obtained from the adopted heat capacities based on a T^3 extrapolation to obtain S°(15 K) = 0.016 kcal mol^{-1}.

Decomposition Data

T_{dcm} = 1397 K is the temperature at which fugacity (O_2) = 1 atm for the reaction $2 CuO(cr) = Cu_2O(cr) + 1/2 O_2(g)$.

References

1. J. Thomsen, "Thermochemishe Untersuchungen," Vol. I, III, Barth, Leipzig, 1883.
2. H. von Wartenberg and H. Werth, Z. Electrochem. 38, 51 (1932).
3. A. D. Mah, L. B. Pankratz, W. W. Weller, and E. G. King, U. S. Bur. Mines RI 7026, 1967.
4. L. Nunez, G. Pilcher, and H. A. Skinner, J. Chem. Thermodyn. 1, 31 (1969).
5. D. Balesdent, Compt. rend. 240, 760 (1955).
6. G. B. Barbi, Gazz. Chim. Ital. 100, 64 (1970).
7. H. W. Foote and E. K. Smith, J. Amer. Chem. Soc. 30, 1344 (1908).
8. H. S. Roberts and F. H. Smyth, J. Amer. Chem. Soc. 43, 1061 (1921); 42, 2582 (1920).
9. F. Becker, Dissertation, Darmstadt, 1927; cf. L. Wohler and N. Jochum, Z. physik. Chem. 167A, 169 (1933).
10. P. Assayag, Ann. Chim. (Paris) 12, 637 (1955).
11. A. M. Gadalla, W. F. Ford and J. White, Trans. Brit. Ceram. Soc. 62, 45 (1963).
12. L. R. Bidwell, J. Electrochem. Soc. 114, 30 (1967).
13. G. G. Charette and S. N. Flengas, J. Electrochem. Soc. 115, 796 (1968).
14. K. Hochgeschwender and T. R. Ingraham, Z. Erzbergbau Metallheuttenm. 21, 58 (1968).
15. K. Kodera, I. Kusunoki and S. Shinkichi, Bull. Chem. Soc. Jap. 41, 1039 (1968).
16. M. A. Rigdon and R. E. Grace, Trans. Met. Soc. AIME (Amer. Inst. Mining, Met., Petrol. Eng.) 242, 822 (1968).
17. E. K. Kazenas, D. M. Chizhikov and Yu. V. Tsetkov, Izv. Akad. Nauk SSSR, Metal 1969, 60 (1969).
18. V. M. Shchedrin, A. A. Telegin, V. N. Vas'kin and I. S. Kulikov, Deposited Publ. VINITI 6690-73 (1973).
19. J. Hu and H. L. Johnston, J. Amer. Chem. Soc. 75, 2471 (1953).
20. IUPAC Commission on Atomic Weights, Pure Appl. Chem. 47, 75 (1976).
21. The International Practical Temperature Scale of 1968, Metrologia 5, 35 (1969); T. B. Douglas, J. Res. Nat. Bur. Stand. 73A, 451 (1969).
22. M. O'Keefe and F. S. Stone, Phys. Chem. Solids 23, 261 (1962).
23. B. N. Brockhouse, Phys. Rev. 94, 781 (1954).

Enthalpy Reference Temperature = T_r = 298.15 K   Standard State Pressure = p° = 0.1 MPa

T/K	$C_p°$ (J $K^{-1}mol^{-1}$)	S°	$-[G°-H°(T_r)]/T$	$H°-H°(T_r)$ (kJ mol^{-1})	$\Delta_f H°$	$\Delta_f G°$	Log K_f
0	0.	0.	INFINITE	-7.092	-153.807	-153.807	INFINITE
100	16.499	9.768	74.176	-6.441	-155.277	-146.585	76.568
200	34.795	27.018	46.256	-3.848	-156.152	-137.461	35.901
298.15	42.246	42.594	42.594	0.	-156.063	-128.292	22.476
300	42.364	42.856	42.595	0.078	-156.057	-128.120	22.308
400	46.806	55.723	44.319	4.562	-155.553	-118.875	15.523
500	49.261	66.451	47.703	9.374	-154.836	-109.785	11.469
600	50.935	75.588	51.608	14.388	-154.022	-100.851	8.780
700	52.240	83.541	55.614	19.549	-153.162	-92.056	6.869
800	53.349	90.591	59.554	24.829	-152.274	-83.387	5.445
900	54.340	96.932	63.361	30.215	-151.368	-74.830	4.343
1000	55.262	102.706	67.011	35.695	-150.453	-66.375	3.467
1100	56.135	108.014	70.500	41.265	-149.542	-58.012	2.755
1200	56.978	112.935	73.834	46.921	-148.656	-49.730	2.165
1300	57.800	117.528	77.020	52.660	-147.837	-41.520	1.668
1400	58.601	121.841	80.069	58.480	-160.239	-32.963	1.230
1500	59.395	125.911	82.991	64.380	-159.444	-23.899	0.832
1600	60.181	129.769	85.795	70.359	-158.583	-14.891	0.486
1700	60.958	133.441	88.491	76.416	-157.657	-5.938	0.182
1800	61.726	136.947	91.086	82.550	-156.665	2.958	-0.086
1900	62.494	140.305	93.589	88.761	-155.608	11.798	-0.324
2000	63.259	143.530	96.006	95.049	-154.486	20.580	-0.537

PREVIOUS: June 1966   CURRENT: December 1977

COPPER OXIDE (CuO) IDEAL GAS M_r = 79.5454

$S°(298.15\ K) = 234.617\ J\ K^{-1}\ mol^{-1}$

$\Delta_f H°(0\ K) = 305.87 \pm 41.8\ kJ\ mol^{-1}$
$\Delta_f H°(298.15\ K) = 306.27 \pm 41.8\ kJ\ mol^{-1}$

Electronic Levels and Quantum Weights

State	ε_i, cm^{-1}	g_i
$X^2\Pi_{3/2}$	0	2
$X^2\Pi_{1/2}$	277.0	2
$A'^2\Sigma^+$	15531.9	2
$A^2\Sigma^+$	16492.4	2
$C^2\Pi_i$	18811.4	4
$E^2\Delta_{5/2}$	21103.7	4
$F^2\Pi_i$	21222	4
$G^2\Sigma^-$	21594.0	2
$M^2\Pi_{3/2}$	23898	4

$\omega_e = 640.14\ cm^{-1}$ $\omega_e x_e = 4.43\ cm^{-1}$ $\sigma = 1$
$B_e = 0.44454\ cm^{-1}$ $\alpha_e = 0.0046\ cm^{-1}$ $r_e = 1.7246$ Å

Enthalpy of Formation

The adopted $\Delta_f H°(298.15\ K) = 73.2 \pm 10\ kcal\ mol^{-1}$ comes from our 3rd law analysis of the mass-spectrometric Knudsen cell measurements of the equilibrium reaction CuO(g) + Ni(g) = Cu(g) + NiO(g) (1611 - 1828 K) (1). Auxiliary data (2, 3) are used in the analysis. Also considered in the evaluation are: the unpublished but quoted (4, and others] value of Burns, $D_0^0 = 62.7 \pm 3$ $kcal\ mol^{-1}$, the sublimation measurement of (5), the linear Birge-Sponer extrapolation of the ground state to obtain D_0^0, and the linear Birge-Sponer extrapolation corrected for the ionic character of CuO according to (6).

Source	D°(0 K)	$\Delta_f H°(298.15\ K)$ $kcal\ mol^{-1}$
Smoes et al. (1), Equilibrium	66.3	73.2
Burns (4), Mass spec.	62.7±3	76.8±3
Mack et al. (5), Transpiration (1273 K)	80.2	59.02
Linear Birge-Sponer	65.2	74.2
Modified Linear Birge-Sponer	61.0±12	78.5±12

Mack et al. (5) report measurements at 873, 1073, and 1173 K which lead to higher dissociation energies. Because of the scatter of the measurements and the suspect nature of the analytical measurement recognized by the authors, the D_0^0 value of 80.2 kcal mol^{-1} is discounted.

Heat Capacity and Entropy

The molecular constants and electronic levels are taken from the work of Appelblad and Lagerqvist (7, 8, 9), Lefebvre et al., (10) and Lefebvre et al. (11). The spectrum of CuO is complex and there are almost certainly excited states missing among those already analyzed (9); in fact some doubt still exists that the $X^2\Pi_i$ state is really the ground state (9) although the absence of esr signals in the matrix isolation study of Thompson et al. (12) supports this assignment. Because of the uncertain and incomplete spectral analysis, we choose to calculate the thermodynamic functions with first-order anharmonic corrections assuming the ground-state vibrational-rotational constants for all states. Use of the actual constants for each state increases the entropy at 6000 K by 0.045 cal K^{-1} mol^{-1}.

References

1. S. Smoes, F. Mandy, A. Vander Auwera-Mahieu and J. Drowart, Bull. Soc. Chim. Belg. 81, 45 (1972).
2. JANAF Thermochemical Tables: Ni(g), 12-31-76; Cu(g), 6-31-78; O(g), 3-31-77; CuO(cr), 12-31-77.
3. A. D. Mah and L. B. Pankratz, U. S. Bur. Mines Bull. 668, 1976.
4. G. J. Cheetham and R. F. Barrow in "Advances in High Temperature Chemistry," Vol. 1, L. Eyring, Ed., Academic Press, New York, pp. 7-41, 1967.
5. E. Mack, G. O. Osterhof and H. M. Kramer, J. Amer. Chem. Soc. 45, 617 (1923).
6. D. L. Hildenbrand in "Advances in High Temperature Chemistry," Vol. 1, L. Eyring, Ed., Academic Press, New York, pp. 193-217, 1967; D. L. Hildenbrand and E. Murad, J. Chem. Phys. 51, 807 (1969).
7. O. Appelblad and A. Lagerqvist, Phys. Scr. 10, 307 (1974).
8. O. Appelblad and A. Lagerqvist, Can. J. Phys. 53, 2221 (1975).
9. O. Appelblad and A. Lagerqvist, Phys. Scr. 13, 275 (1976).
10. Y. Lefebvre, B. Pinchemel and R. Bacis, Can. J. Phys. 54, 735 (1976).
11. Y. Lefebvre, B. Pinchemel and J. Schamps, J. Mol. Spectrosc. 68, 81 (1977).
12. K. R. Thompson, W. C. Easlye and L. B. Knight, J. Phys. Chem. 73, 49 (1973).

Copper Oxide (CuO) $Cu_1O_1(g)$

Enthalpy Reference Temperature = T_r = 298.15 K Standard State Pressure = p° = 0.1 MPa

T/K	$C_p°$	S°	$-[G°-H°(T_r)]/T$	$H°-H°(T_r)$	$\Delta_f H°$	$\Delta_f G°$	Log K_f
	$J\ K^{-1} mol^{-1}$			$kJ\ mol^{-1}$			
0	0.	0.	INFINITE	-9.751	305.866	305.866	INFINITE
100	31.554	197.630	265.437	-6.781	306.715	296.621	-154.939
200	34.498	220.596	237.849	-3.451	306.577	286.552	-74.840
250	35.198	228.373	235.202	-1.707	306.428	281.563	-58.829
298.15	35.693	234.617	234.617	0.	306.269	276.788	-48.492
300	35.710	234.838	234.618	0.066	306.262	276.605	-48.161
350	36.098	240.373	235.054	1.862	306.083	271.676	-40.545
400	36.399	245.214	236.028	3.674	305.891	266.774	-34.837
450	36.636	249.515	237.292	5.501	305.683	261.896	-30.400
500	36.827	253.386	238.711	7.337	305.459	257.043	-26.853
600	37.114	260.127	241.735	11.035	304.957	247.405	-21.539
700	37.320	265.864	244.782	14.757	304.378	237.858	-17.749
800	37.478	270.858	247.736	18.498	303.726	228.399	-14.913
900	37.606	275.280	250.556	22.252	303.001	219.026	-12.712
1000	37.715	279.248	253.230	26.018	302.202	209.737	-10.956
1100	37.810	282.847	255.762	29.795	301.319	200.533	-9.523
1200	37.896	286.141	258.158	33.580	300.335	191.413	-8.332
1300	37.975	289.178	260.429	37.374	299.209	182.381	-7.328
1400	38.050	291.995	262.584	41.175	284.788	173.848	-6.486
1500	38.121	294.622	264.633	44.983	283.491	165.969	-5.780
1600	38.189	297.085	266.585	48.799	282.189	158.177	-5.164
1700	38.256	299.402	268.448	52.621	280.881	150.466	-4.623
1800	38.322	301.590	270.229	56.450	279.567	142.832	-4.145
1900	38.389	303.664	271.935	60.286	278.248	135.272	-3.719
2000	38.456	305.635	273.571	64.128	276.925	127.781	-3.337
2100	38.526	307.513	275.143	67.977	275.597	120.356	-2.994
2200	38.599	309.307	276.655	71.833	274.265	112.996	-2.683
2300	38.677	311.024	278.113	75.697	272.929	105.695	-2.400
2400	38.760	312.672	279.519	79.569	271.590	98.452	-2.143
2500	38.850	314.256	280.877	83.449	270.248	91.266	-1.907
2600	38.949	315.782	282.190	87.339	268.906	84.133	-1.690
2700	39.056	317.254	283.461	91.239	267.563	77.052	-1.491
2800	39.174	318.676	284.694	95.151	266.221	70.021	-1.306
2900	39.303	320.053	285.890	99.074	-35.323	69.033	-1.243
3000	39.444	321.388	287.051	103.012	-35.864	72.640	-1.265
3100	39.598	322.684	288.179	106.964	-36.450	76.267	-1.285
3200	39.766	323.944	289.277	110.932	-37.079	79.913	-1.304
3300	39.949	325.170	290.347	114.918	-37.751	83.579	-1.323
3400	40.145	326.366	291.388	118.922	-38.463	87.266	-1.341
3500	40.357	327.532	292.404	122.947	-39.213	90.975	-1.358
3600	40.583	328.672	293.396	126.994	-39.998	94.706	-1.374
3700	40.824	329.787	294.365	131.064	-40.815	98.458	-1.390
3800	41.079	330.880	295.311	135.159	-41.662	102.234	-1.405
3900	41.348	331.950	296.237	139.280	-42.533	106.032	-1.420
4000	41.631	333.000	297.143	143.429	-43.426	109.853	-1.435
4100	41.926	334.032	298.030	147.607	-44.337	113.696	-1.449
4200	42.234	335.046	298.900	151.815	-45.263	117.562	-1.462
4300	42.552	336.043	299.752	156.054	-46.198	121.450	-1.475
4400	42.881	337.025	300.588	160.326	-47.141	125.359	-1.488
4500	43.220	337.993	301.408	164.631	-48.087	129.290	-1.501
4600	43.566	338.947	302.214	168.970	-49.033	133.242	-1.513
4700	43.920	339.887	303.006	173.344	-49.976	137.215	-1.525
4800	44.279	340.816	303.784	177.754	-50.914	141.208	-1.537
4900	44.644	341.732	304.549	182.200	-51.844	145.220	-1.548
5000	45.012	342.638	305.302	186.683	-52.763	149.251	-1.559
5100	45.383	343.533	306.042	191.203	-53.670	153.301	-1.570
5200	45.756	344.418	306.772	195.760	-54.563	157.367	-1.581
5300	46.129	345.293	307.490	200.354	-55.441	161.451	-1.591
5400	46.501	346.159	308.199	204.986	-56.303	165.552	-1.601
5500	46.872	347.015	308.897	209.654	-57.149	169.668	-1.611
5600	47.240	347.863	309.585	214.360	-57.977	173.799	-1.621
5700	47.604	348.703	310.264	219.102	-58.789	177.945	-1.631
5800	47.964	349.534	310.934	223.881	-59.584	182.106	-1.640
5900	48.319	350.357	311.595	228.695	-60.362	186.279	-1.649
6000	48.667	351.172	312.248	233.544	-61.126	190.466	-1.658

PREVIOUS: December 1977 (1 atm) CURRENT: December 1977 (1 bar)

Copper Oxide (CuO) $Cu_1O_1(g)$

COPPER SULFATE ($CuSO_4$) CRYSTAL M_r = 159.6036

$\Delta_f H°(0\ K) = -760.059 \pm 0.84\ kJ\ mol^{-1}$

$S°(298.15\ K) = 109.254 \pm 0.42\ J\ K^{-1}\ mol^{-1}$ $\Delta_f H°(298.15\ K) = -769.982 \pm 0.84\ kJ\ mol^{-1}$

Enthalpy of Formation

The enthalpy of solution, $\Delta_r H°(298.15\ K)$, of CuO(cr) in sulfuric acid to form $CuSO_4$(cr) has been measured by Adami and King (1). From the reported value, $\Delta_r H° = 5.61 \pm 0.09\ kcal\ mol^{-1}$ for the reacton CuO(cr) + $H_2SO_4 \cdot 7.068\ H_2O$(aq) = $CuSO_4$(cr) + 8.068 H_2O(ℓ), the enthalpy of formation, $\Delta_f H°(298.15\ K)$, for $CuSO_4$(cr) was evaluated as -184.03 kcal mol^{-1}, using $\Delta_f H°(298.15\ K)$ = -37.25, -209.49 and -68.315 kcal mol^{-1} for CuO(cr), $H_2SO_4 \cdot 7.068\ H_2O$(aq) and H_2O(ℓ), respectively. See $Fe_2(SO_4)_3$(cr) table for the sources of the last two $\Delta_f H°(298.15\ K)$ values.

The equilibrium pressure for the following chemical reactons: (A) 2 $CuSO_4$(cr) = CuO·$CuSO_4$(cr) + SO_3(g), (B) 3 Cu(cr) + $CuSO_4$(cr) = 2 Cu_2O(cr) + SO_2(g), and (C) 4 $CuSO_4$(cr) + Cu_2O(cr) = 3 CuO·$CuSO_4$(cr) + SO_2(g), were determined by several investigators. For reaction (A), the partial pressures of SO_3(g) at different temperatures were calculated from the measured total pressures exerted by SO_3(g), SO_2(g) and O_2(g) produced by the reaction SO_3(g) = SO_2(g) + 1/2 O_2(g). From these vapor pressure data, the corresponding enthalpy changes, $\Delta_r H°(298.15\ K)$, for the three reactions were evaluated by the 2nd and 3rd law methods. The results obtained are presented in the following table. The value of $\Delta_f H°(CuSO_4$, cr, 298.15 K) adopted was -184.03 ± 0.2 kcal mol^{-1}.

Source	Reaction	T/K	$\Delta_r H°(298.15\ K)$, kcal mol^{-1} 2nd law	3rd law	Drift cal K^{-1} mol^{-1}	$\Delta_f H°(298.15\ K)$ kcal mol^{-1}
(2)	(A)	819.15-1004.15	25.90±3.60	47.45	22±4	-181.87
(3)	(A)	953.15-1053.15	56.54±0.77	51.96	-4.55±0.79	-184.13
	(C)	825.15-921.15	42.84±1.38	42.34	-0.55±1.57	-184.42
(4)	(A)	880.15-1071.75	51.59±0.41	51.75	0.07±0.43	-184.02
(5)	(B)	525.1 - 608.1	31.71±0.76	32.27	0.84±1.30	-184.62

Heat Capacity and Entropy

The low temperature heat capacities, 52.67-296.29 K were obtained from Weller (6). The high temperature heat capacities were measured by Schottky (7, 282°C), Ewald (8, 275-373 K), Krestovnikov and Feigina (9, 288-873 K), and Chiznikov and Khirik (10, 373-1273 K). The heat capacities below 51 K were obtained from Stout (11). In that paper, the $C_p°$ values, 15-58 K, were plotted as function of temperature and the $C_p°$ curve shows a maximum at 34.8 K. In an attempt to evaluate the entropy due to the anomalous portion of the curve, a "normal" heat capacity curve was drawn. The entropy above the normal $C_p°$ curve is reported to be 0.48 cal K^{-1} mol^{-1} rather than a magnetic entropy of Rln2 = 1.377 cal K^{-1} mol^{-1}. Apparently the magnetic entropy increases gradually above 40 K so that it is not feasible to separate the heat capacity due to magnetic effects from that of the crystal lattice.

The $C_p°$ values above 298.15 K were extrapolated smoothly by comparison with those for $MnSO_4$(cr) (see $FeSO_4$(cr) table for details). The adopted $C_p°$ values are close to the average of the reported $C_p°$ values.

The value of S°(298.15 K) was derived using the low temperature $C_p°$ data reported by Weller (6), based on S°(51 K) = 2.851 cal K^{-1} mol^{-1} evaluated from $C_p°$ data determined by Stout (11).

Decomposition Data

T_{dcm} = 1078 K is the temperature at which the total pressure of the gaseous decomposition products equals one atmosphere. It was obtained by graphical extrapolation of the decomposition pressures for $CuSO_4$(cr), determined by Ingraham (4).

References

1. L. H. Adami and E. G. King, U. S. Bur. Mines RI 6617, 1965.
2. L. Wohler, W. Pluddemann and P. Wohler, Ber. Deut. Chem. Gesell. 41, 703 (1908).
3. W. Reinders and F. Goudriaan, Z. anorg. allgem. Chem. 126, 85 (1923).
4. T. R. Ingraham, Trans. Met. Soc. AIME 233, 259 (1965).
5. M. Randall, R. F. Nielsen and G. H. West, Ind. Eng. Chem. 23, 388 (1931).
6. W. W. Weller, U. S. Bur. Mines RI 6669, 1965.
7. H. Schottky, Z. phys. Chem. 64, 415 (1908).
8. R. Ewald, Ann. Physik. 44, 1213 (1914).
9. A. N. Krestovnikov and E. J. Feigina, J. Gen. Chem. USSR 6, 1481 (1936).
10. D. M. Chizhikov and A. S. Khirik, Tr. Inst. Met. im. A. A. Balkova, Akad. Nauk SSSR, No. 12 79-84 (1963).
11. J. W. Stout, J. Chem. Phys. 9, 285 (1941).

Copper Sulfate ($CuSO_4$) $Cu_1O_4S_1$(cr)

Enthalpy Reference Temperature = T_r = 298.15 K Standard State Pressure = p° = 0.1 MPa

T/K	$C_p°$ (J $K^{-1}mol^{-1}$)	S°	-[G°-H°(T_r)]/T	H°-H°(T_r) (kJ mol^{-1})	$\Delta_f H°$	$\Delta_f G°$	Log K_f
0	0.	0.	INFINITE	-16.863	-760.059	-760.059	INFINITE
100	43.718	32.670	181.424	-14.875	-765.239	-731.589	382.143
200	77.008	74.167	117.649	-8.696	-768.538	-696.504	181.908
298.15	98.868	109.254	109.254	0.	-769.982	-660.781	115.766
300	99.286	109.867	109.256	0.183	-769.994	-660.104	114.934
400	114.934	140.654	113.347	10.923	-772.288	-623.342	81.400
500	127.194	167.676	121.566	23.055	-772.767	-586.042	61.223
600	136.315	191.710	131.291	36.251	-772.095	-548.744	47.772
700	142.925	213.244	141.488	50.229	-770.647	-511.625	38.178
800	147.695	232.656	151.691	64.772	-768.713	-474.750	30.998
900	151.042	250.251	161.680	79.714	-819.616	-437.054	25.366
1000	153.762	266.309	171.352	94.958	-815.986	-394.741	20.619
1100	156.063	281.074	180.665	110.451	-812.287	-352.795	16.753
1200	158.113	294.743	189.609	126.161	-808.556	-311.188	13.546
1300	159.954	307.473	198.191	142.066	-804.849	-269.891	10.844
1400	161.628	319.388	206.427	158.146	-814.329	-228.471	8.524
1500	163.218	330.594	214.335	174.389	-810.587	-186.755	6.503
1600	164.724	341.177	221.935	190.787	-806.756	-145.290	4.743
1700	166.147	351.206	229.247	207.330	-802.843	-104.068	3.198
1800	167.569	360.743	236.289	224.017	-798.849	-63.079	1.831
1900	168.908	369.839	243.081	240.841	-794.778	-22.313	0.613
2000	170.247	378.537	249.638	257.798	-790.631	18.236	-0.476

PREVIOUS: CURRENT: June 1966

COPPER (Cu_2) IDEAL GAS M_r = 127.092

$S°(298.15\ K) = 241.72\ J\ K^{-1}\ mol^{-1}$

$\Delta_f H°(0\ K) = 485.43 \pm 12.6\ kJ\ mol^{-1}$
$\Delta_f H°(298.15\ K) = 485.34 \pm 12.6\ kJ\ mol^{-1}$

Electronic Levels and Quantum Weights

State	ϵ_i, cm^{-1}	g_i
$[^1\Sigma_g^+]$	0	[1]
	20433	[1]
	21758	[1]

$\omega_e = 265.34\ cm^{-1}$ $\omega_e x_e = 1.015\ cm^{-1}$ $\sigma = 2$
$B_e = 0.10776\ cm^{-1}$ $\alpha_e = 0.000608\ cm^{-1}$ $r_e = 2.219$ Å

Enthalpy of Formation

The enthalpy of formation is calculated from that of Cu(g) using $\Delta_r H°(298.15\ K) = 46.0\ kcal\ mol^{-1}$ for $Cu_2(g) \rightarrow 2\ Cu(g)$ obtained by 3rd law analysis of the mass spectrometric data of Ackerman et al. (1). Earlier studies of mass spectra and of the electronic band spectra yield values in reasonable agreement as shown below. The selected $\Delta_r H°$ corresponds to $D°(0\ K) = 1.97$ eV compared with the linear Birge-Sponer extrapolation of 2.1 eV.

Source	Method	T/K	$\Delta_r H°(298.15\ K)$, kcal mol^{-1} 2nd law	3rd law	Drift cal K^{-1} mol^{-1}	$\Delta_f H°(298.15\ K)$ kcal mol^{-1}
(1)	Mass Spect.	1549-1709	47.8±2.6	45.99	-1.0±1.6	116.0
(2)	Mass Spect.	1575-1720	50.3±3.7	46.76	-2.1±2.3	115.0
(3)	Mass Spect.	1440-1560	--	46.5	--	115.5
(4)	Band Spectra	--	--	49.0	--	113.0

Heat Capacity and Entropy

Molecular constants, except for $\omega_e x_e$, were taken from the rotational analysis of bands of the B-X system given by Travis and Barrow (5). The value for $\omega_e x_e$ was obtained from the earlier study of Kleman and Lindkvist (4) who showed by means of the vibrational isotope effect that the emitter was Cu_2. The spectroscopic characters of the ground and excited states are uncertain, but Travis and Barrow indicate that the transition is probably $^1\Sigma_u^+ - {}^1\Sigma_g^+$ for the B-X system. Vibrational and rotational constants were adjusted for 30.91% ^{65}Cu.

References

1. M. Ackerman, F. E. Stafford and J. Drowart, J. Chem. Phys. 33, 1784 (1960).
2. P. Schissel, J. Chem. Phys. 26, 1276 (1957).
3. J. Drowart and R. E. Honig, J. Chem. Phys. 25, 581 (1956).
4. B. Kleman and S. Lindkvist, Arkiv Fysik 8, 333 (1954).
5. D. N. Travis and R. F. Barrow, Proc. Chem. Soc. (London), 44 (1962).

Copper (Cu_2) $Cu_2(g)$

Enthalpy Reference Temperature = T_r = 298.15 K Standard State Pressure = p° = 0.1 MPa

T/K	$C_p°$ (J $K^{-1}mol^{-1}$)	S°	$-[G°-H°(T_r)]/T$	$H°-H°(T_r)$ (kJ mol^{-1})	$\Delta_f H°$	$\Delta_f G°$	Log K_f
0	0.	0.	INFINITE	-9.930	485.428	485.428	INFINITE
100	31.964	203.874	273.337	-6.946	487.074	468.693	-244.820
200	35.442	227.321	245.038	-3.543	486.450	450.478	-117.653
250	36.156	235.313	242.321	-1.752	485.916	441.545	-92.256
298.15	36.585	241.721	241.721	0.	485.344	433.051	-75.869
300	36.598	241.947	241.722	0.068	485.321	432.726	-75.344
350	36.893	247.612	242.168	1.905	484.686	424.010	-63.280
400	37.103	252.553	243.164	3.756	484.021	415.387	-54.244
450	37.261	256.933	244.455	5.615	483.329	406.849	-47.226
500	37.386	260.865	245.903	7.481	482.616	398.390	-41.619
600	37.576	267.699	248.983	11.230	481.124	381.683	-33.228
700	37.720	273.503	252.082	14.995	479.541	365.233	-27.254
800	37.840	278.548	255.082	18.773	477.872	349.016	-22.788
900	37.945	283.011	257.942	22.562	476.108	333.014	-19.328
1000	38.041	287.014	260.652	26.362	474.239	317.214	-16.570
1100	38.131	290.644	263.216	30.170	472.237	301.607	-14.322
1200	38.216	293.965	265.642	33.988	470.065	286.190	-12.458
1300	38.299	297.027	267.940	37.813	467.634	270.963	-10.887
1400	38.380	299.869	270.121	41.647	438.637	256.749	-9.579
1500	38.459	302.519	272.193	45.489	435.910	243.852	-8.492
1600	38.537	305.004	274.167	49.339	433.191	231.137	-7.546
1700	38.614	307.343	276.050	53.197	430.479	218.591	-6.716
1800	38.691	309.552	277.851	57.062	427.776	206.205	-5.984
1900	38.767	311.646	279.575	60.935	425.080	193.969	-5.333
2000	38.843	313.636	281.229	64.815	422.392	181.875	-4.750
2100	38.919	315.533	282.817	68.704	419.711	169.915	-4.226
2200	38.995	317.346	284.346	72.599	417.038	158.083	-3.753
2300	39.071	319.081	285.819	76.503	414.372	146.372	-3.324
2400	39.149	320.745	287.239	80.414	411.714	134.777	-2.933
2500	39.228	322.345	288.612	84.332	409.064	123.292	-2.576
2600	39.308	323.885	289.939	88.259	406.422	111.914	-2.248
2700	39.390	325.370	291.224	92.194	403.788	100.637	-1.947
2800	39.475	326.804	292.469	96.137	401.162	89.457	-1.669
2900	39.562	328.191	293.677	100.089	-201.863	90.362	-1.628
3000	39.653	329.533	294.850	104.050	-202.882	100.456	-1.749
3100	39.747	330.835	295.990	108.020	-203.991	110.585	-1.863
3200	39.845	332.099	297.099	111.999	-205.193	120.752	-1.971
3300	39.948	333.326	298.178	115.989	-206.487	130.958	-2.073
3400	40.055	334.520	299.230	119.989	-207.872	141.204	-2.169
3500	40.166	335.683	300.255	124.000	-209.348	151.492	-2.261
3600	40.283	336.816	301.255	128.022	-210.910	161.823	-2.348
3700	40.405	337.922	302.231	132.057	-212.557	172.200	-2.431
3800	40.533	339.001	303.184	136.104	-214.284	182.621	-2.510
3900	40.666	340.055	304.116	140.164	-216.088	193.090	-2.586
4000	40.804	341.087	305.027	144.237	-217.963	203.605	-2.659
4100	40.947	342.096	305.919	148.324	-219.904	214.168	-2.729
4200	41.096	343.085	306.793	152.427	-221.907	224.779	-2.796
4300	41.250	344.053	307.648	156.544	-223.966	235.439	-2.860
4400	41.409	345.003	308.486	160.677	-226.076	246.147	-2.922
4500	41.572	345.936	309.308	164.826	-228.231	256.904	-2.982
4600	41.740	346.851	310.114	168.991	-230.426	267.709	-3.040
4700	41.913	347.751	310.905	173.174	-232.656	278.562	-3.096
4800	42.089	348.635	311.682	177.374	-234.916	289.462	-3.150
4900	42.270	349.505	312.445	181.592	-237.201	300.411	-3.202
5000	42.454	350.361	313.195	185.828	-239.509	311.406	-3.253
5100	42.641	351.203	313.932	190.083	-241.833	322.447	-3.303
5200	42.831	352.033	314.657	194.356	-244.172	333.534	-3.350
5300	43.023	352.851	315.370	198.649	-246.521	344.666	-3.397
5400	43.218	353.657	316.071	202.961	-248.878	355.843	-3.442
5500	43.414	354.452	316.762	207.293	-251.240	367.063	-3.486
5600	43.612	355.236	317.442	211.644	-253.606	378.326	-3.529
5700	43.811	356.009	318.112	216.015	-255.973	389.632	-3.571
5800	44.012	356.773	318.772	220.406	-258.341	400.979	-3.611
5900	44.212	357.527	319.422	224.817	-260.709	412.367	-3.651
6000	44.414	358.272	320.064	229.249	-263.075	423.795	-3.689

PREVIOUS: September 1966 (1 atm) CURRENT: September 1966 (1 bar)

Copper (Cu_2) $Cu_2(g)$

$\Delta_fH°(0\ K) = -168.95 \pm 2.1\ kJ\ mol^{-1}$
$S°(298.15\ K) = 92.36 \pm 0.34\ J\ K^{-1}\ mol^{-1}$ $\Delta_fH°(298.15\ K) = -170.71 \pm 2.1\ kJ\ mol^{-1}$
$T_{fus} = 1516.7\ K$ $\Delta_{fus}H° = 64.768 \pm 0.63\ kJ\ mol^{-1}$

Enthalpy of Formation

The enthalpy of formation of $Cu_2O(cr)$ has been investigated calorimetrically (both oxidation and reduction reactions) and the oxidation reaction 2 Cu(cr) + 1/2 $O_2(g)$ = $Cu_2O(cr)$ has been studied extensively by pressure and by emf measurement. The more consistent data leading to the adopted value are summarized below.

Source	Method	Reaction*	T/K	Points	$\Delta_rH°(298.15\ K)$, kcal mol^{-1} 2nd law	3rd law	Drift cal K^{-1} mol^{-1}	$\Delta_rH°(298.15\ K)$ kcal mol^{-1}
(1)	Aq Calorim.	A	291	--	27.61		--	-40.70
(2)	Calorimetric	See Text	298.15	--	--		--	-40.83±0.30
(3)	Calorimetric	A	297.90	--	26.93		--	-41.39±0.32
(4)	Aq emf	A	273-318	5	27.47±0.04	27.58±0.02	0.4±0.1	-40.74
(5)	Aq emf	A	298-318	7	29.97±0.38	28.02±0.14	-6.4±1.3	-40.30
(6)	Aq emf	A	288-308	3	29.54±0.73	28.25±0.10	-4.3±2.5	-40.06
(7)	Equil Press	B	1011-1156	5	-35.79±0.86	-41.55±0.66	-5.3±0.8	-41.55
(8)	emf vs Fe/FexO	B	1073-1323	4	-42.29±0.59	-40.96±0.23	1.1±0.5	-40.96
(9)	Fused Salt emf	B	987	1		-41.72		-41.72
(10)	emf	B	1100-1200	2	-60.17	-39.77±2.03	14.3	-39.77
(11)	emf	B	973-1273	Eqn	-41.10	-40.75	0.31	-40.75
(12)	emf	B	892-1320	Eqn	-40.75	-40.77	-0.01	-40.77
(13)	Equil Press	B	1200-1420	Eqn	-33.37	-40.30	5.4	-40.30
(14)	Equil Press	B	1280-1450	Eqn	-36.82	-40.24	5.0	-40.24
(15)	emf	B	973-1373	Eqn	-41.83	-41.01	1.4	-41.01
(16)	emf	B	845-1270	Eqn	-40.83	-40.88	-0.1	-40.88
(17)	emf	B	976-1234	Eqn	-39.84	-40.59	-0.7	-40.59
(18)	emf	B	1073-1273	Eqn	-41.07	-39.12	1.7	-39.12
(19)	Equil Press	B	1013-1243	Eqn	-43.07	-43.86	1.4	-43.86
(20)	emf	B	973-1273	Eqn	-42.62	-41.23	1.2	-41.23
(21)	emf	B	1173-1356	Eqn	-40.71	-40.90	-0.2	-40.90

*Reactions: (A) 2 Cu(cr) + $H_2O(\ell)$ = $Co_2O(cr)$ + $H_2(g)$ (B) 2 Cu(cr) + 0.5 $O_2(g)$ = $Cu_2O(cr)$

Thomsen (1) obtained $\Delta_fH°(298.15\ K)$ values of -40180, -40.87, and -41.51 kcal mol^{-1} from aqueous calorimetry using three different paths. He favored the first path which is recalculated by combination of the heats of reaction of Cu_2O + H_2SO_4, Fe + H_2SO_4, and Fe + $CuSO_4(aq)$ to yield the heat of reduction of Cu_2O with H_2. Nunez et al (3) measured the direct heat of reduction of Cu_2O by H_2; the conclude that the heats of formation of the copper oxides are influenced by material history, state of subdivision, and non-stoichiometry. Mah et al. (2) measured the direct heat of oxidation of copper by O_2, but they obtained a mixed product of Cu_2O and CuO and used $\Delta_rH°(298.15\ K) = -33.63 \pm 0.05$ kcal mol^{-1} for the equilibrium reaction $Cu_2O(cr)$ + 1/2 $O_2(g)$ = 2 CuO(cr) to resolve their results into $\Delta_fH°(Cu_2O, cr, 298.15\ K)$ and $\Delta_fH°(CuO, cr, 298.15\ K)$. The overall average from our 3rd law analyses is -40.82 kcal mol^{-1}; without the outlier (10, 18, 19) the average is 40.81; the average of the six data sets with an absolute drift less than 1 is -40.77. We adopt $\Delta_fH°(Cu_2O, cr, 298.15\ K) = -40.8 \pm 0.5$ kcal mol^{-1}.

Heat Capacity and Entropy

The heat capacities are smoothed values adopted from a consideration of the work of Gregor (22, 2.8-21 K), Hu and Johnston (23, 15-300 K) and Mah et al. (2, 52-296 K, 406-1464 K). The data of Mah et al. (2) are favored over that of Hu and Johnston (23) above 120 K. Adjustment is made to the IPTS-68 temperature scale (24).

The entropy is based on a T^3 extrapolation to obtain $S°(2.8\ K) = 0.0015$ cal K^{-1} mol^{-1}.

Fusion Data

See the liquid table for details.

References

1. J. Thomsen, "Thermochemische Untersuchungen," Vol. III, Barth, Leipzig, 1883.
2. A. D. Mah, L. B. Pankratz, W. W. Weller and E. G. King, U. S. Bur. Mines RI 7026, 1967.
3. L. Nunez, G. Pilcher and H. A. Skinner, J. Chem. Thermodyn. 1, 31 (1969).
4. F. Ishakawa and G. Kumura, "Sexagint, Yikichi Osake," pp. 255-69, Chem. Inst. Kyoto Imp. Univ., Kyoto, Japan, 1927, cf. M. Randall, R. F. Nielsen, and G. H. West, Ind. Eng. Chem. 23, 388 (1931).
5. C. G. Maier, J. Amer. Chem. Soc. 51, 194 (1929).
6. I. A. Makolkin, J. Phys. Chem. (U.S.S.R.) 16, 13 (1942); cf. Chem. Abs. 37, 2641 (1943).
7. D. Balesdent, Compt. rend. 240, 760, 1884 (1955).
8. K. Kiukkola and C. Wagner, J. Electrochem. Soc. 104, 379 (1957).
9. D. G. Hill, B. Porter and A. S. Gillespie, Jr., J. Electrochem. Soc. 105, 408 (1958).
10. T. C. Wilder, Trans. Met. Soc. AIME 236, 1035 (1966).
11. F. E. Rizzo, L. R. Bidwell and D. F. Frank, Trans. Met. Soc. AIME 239, 593 (1967).
12. G. G. Charette and S. W. Flengas, J. Electrochem. Soc. 115, 796 (1968).
13. K. Kodera, I. Kusunoki and S. Shinkichi, Bull. Chem. Soc. Jap. 41, 1039 (1968).
14. E. K. Kazenas, D. M. Chizhikov and Yu. V. Tsetkov, Izv. Akad. Nauk SSSR, Metal 1969, 60 (1969).
15. J. Moriyama, N. Sato, H. Asao and Z. Kozuka, Mem. Fac. Eng., Kyoto Univ. 31, (Pt 2), 253 (1969).
16. G. B. Barbi, Gazz. Chim. Ital. 100, 64 (1970).
17. S. C. Shaefer, U. S. Nat. Tech. Inform. Serv. PB Rep. No. 203731 (avail. NITS), 14 pp. (1971).
18. A. A. Slobodyanyak, Yu. D. Tret'yakov and A. F. Bessonov; Russ. J. Phys. Chem. 45, 1065 (1971).
19. V. M. Shchedrin, A. A. Telegin, V. N. Vas'kin and I. S. Kalikov, Deposited Publ. VINITI 6690-73 (1973).
20. Z. Moses, K. Fitzner and W. Zakuiski, Bull. Acad. Pol. Sci., Ser. Sci. Tech. 23, 243 (1975).
21. N. Kemoir, I. Katayama and Z. Kozuka, Nippon Kinzoka Gakkaishi 41, 803 (1977).
22. L. V. Gregor, J. Phys. Chem. 66, 1645 (1962).
23. J. Hu and H. L. Johnston, J. Amer. Chem. Soc. 73, 4550 (1951).
24. The International Practical Temperature Scale of 1968, Metrologia 5, 35 (1969); T. B. Douglas, J. Res. Nat. Bur. Stand. 73A, 451 (1969).

Enthalpy Reference Temperature = T_r = 298.15 K   Standard State Pressure = p° = 0.1 MPa

T/K	$C_p°$ (J $K^{-1}mol^{-1}$)	S°	$-[G°-H°(T_r)]/T$	$H°-H°(T_r)$ (kJ mol^{-1})	$\Delta_fH°$	$\Delta_fG°$	Log K_f
0	0.	0.	INFINITE	-12.600	-168.952	-168.952	INFINITE
100	39.499	37.092	141.510	-10.442	-169.584	-162.621	84.944
200	53.619	69.143	97.826	-5.737	-170.360	-155.348	40.573
298.15	62.543	92.360	92.360	0.	-170.707	-147.886	25.909
300	62.664	92.747	92.361	0.116	-170.709	-147.745	25.725
400	67.688	111.531	94.888	6.657	-170.641	-140.092	18.294
500	70.922	127.000	99.809	13.596	-170.363	-132.484	13.840
600	73.448	140.161	105.465	20.818	-169.961	-124.944	10.877
700	75.645	151.651	111.259	28.274	-169.479	-117.478	8.766
800	77.666	161.886	116.959	35.941	-168.929	-110.087	7.188
900	79.493	171.141	122.473	43.801	-168.325	-102.767	5.964
1000	81.245	179.606	127.769	51.837	-167.688	-95.517	4.989
1100	83.385	187.443	132.842	60.061	-167.029	-88.331	4.194
1200	86.110	194.812	137.702	68.532	-166.322	-81.208	3.535
1300	89.173	201.824	142.367	77.294	-165.608	-74.144	2.979
1400	92.489	208.552	146.855	86.375	-191.165	-66.321	2.474
1500	95.987	215.052	151.186	95.798	-190.132	-57.438	2.000
1516.700	96.587	216.118	151.895	97.406	------ CRYSTAL <--> LIQUID -----		
1600	99.631	221.362	155.376	105.578	-188.755	-48.635	1.588
1700	103.380	227.514	159.439	115.728	-187.020	-39.928	1.227
1800	107.212	233.532	163.389	126.257	-184.917	-31.335	0.909
1900	111.115	239.433	167.236	137.173	-182.440	-22.868	0.629
2000	115.068	245.232	170.992	148.481	-179.581	-14.542	0.380

PREVIOUS: June 1966   CURRENT: December 1977

COPPER OXIDE (Cu_2O) LIQUID M_r = 143.0914

S°(298.15 K) = [129.957] J K^{-1} mol^{-1}

T_{fus} = 1516.7 K

$\Delta_f H°$(298.15 K) = [-111.995] kJ mol^{-1}

$\Delta_{fus}H°$ = 64.768 ± 0.63 kJ mol^{-1}

Enthalpy of Formation

$\Delta_f H°$(Cu_2O, ℓ, 298.15 K) is calculated from that of the crystal by adding $\Delta_{fus}H°$ and the difference in enthalpy, H°(1516.7 K)-H°(298.15 K), between the crystal and liquid.

Heat Capacity and Entropy

The heat capacity is taken from the enthalpy measurements of Mah et al. (1, 1521-1600 K) and is assumed constant at 23.88 cal K^{-1} mol^{-1} from 1000 to 2000 K. A glass transition is assumed at 1000 K below which the heat capacity is that of the crystal.

The entropy is calculated in a manner similar to that used for the enthalpy of formation.

Fusion Data

The melting point, 1515 K, was determined by drop calorimetry by Mah et al. (1). Correcting to the IPTS-68 temperature scale (2), we adopt T_{fus} = 1516.7 K.

$\Delta_{fus}H°$ = 15.48 ± 0.15 kcal mol^{-1} is based on our smoothing through the pre-melt region of the enthalpy measurements of Mah et al. (1).

References

1. A. D. Mah, L. B. Pankratz, W. W. Weller and E. G. King, U. S. Bur. Mines RI 7026, 1967.

2. The International Practical Temperature Scale of 1968, Methologia 5, 35 (1969); T. B. Douglas, J. Res. Nat. Bur. Stand. 73A, 451 (1969).

Copper Oxide (Cu_2O) Cu_2O_1(l)

Enthalpy Reference Temperature = T_r = 298.15 K   Standard State Pressure = p° = 0.1 MPa

T/K	$C_p°$ (J $K^{-1}mol^{-1}$)	S° (J $K^{-1}mol^{-1}$)	$-[G°-H°(T_r)]/T$ (J $K^{-1}mol^{-1}$)	$H°-H°(T_r)$ (kJ mol^{-1})	$\Delta_f H°$ (kJ mol^{-1})	$\Delta_f G°$ (kJ mol^{-1})	Log K_f
0							
100							
200							
298.15	62.542	129.957	129.957	0.	-111.995	-100.384	17.587
300	62.664	130.345	129.959	0.116	-111.997	-100.312	17.466
400	67.688	149.128	132.486	6.657	-111.929	-96.419	12.591
500	70.922	164.598	137.407	13.596	-111.651	-92.570	9.671
600	73.448	177.758	143.062	20.818	-111.249	-88.791	7.730
700	75.645	189.249	148.857	28.274	-110.767	-85.085	6.349
800	77.666	199.483	154.557	35.941	-110.217	-81.452	5.318
900	79.493	208.739	160.071	43.801	-109.613	-77.893	4.521
1000	81.245	217.203	165.367	51.837	-108.976	-74.402	3.886
1000.000	81.245	217.203	165.367	51.837	GLASS <--> LIQUID	TRANSITION	
1000.000	99.914	217.203	165.367	51.837			
1100	99.914	226.726	170.519	61.828	-106.550	-71.064	3.375
1200	99.914	235.420	175.570	71.819	-104.323	-67.938	2.957
1300	99.914	243.417	180.486	81.811	-102.380	-64.987	2.611
1400	99.914	250.822	185.249	91.802	-127.026	-61.360	2.289
1500	99.914	257.715	189.853	101.794	-125.424	-56.725	1.975
1516.700	99.914	258.821	190.606	103.462	------ CRYSTAL <-->	LIQUID -----	
1600	99.914	264.163	194.298	111.785	-123.836	-52.197	1.704
1700	99.914	270.220	198.587	121.776	-122.259	-47.768	1.468
1800	99.914	275.931	202.727	131.768	-120.694	-43.431	1.260
1900	99.914	281.333	206.723	141.759	-119.142	-39.181	1.077
2000	99.914	286.458	210.583	151.751	-117.600	-35.013	0.914

PREVIOUS: June 1966   CURRENT: December 1977

Copper Oxide (Cu_2O) Cu_2O_1(l)

COPPER OXIDE (Cu_2O) CRYSTAL-LIQUID M_r = 143.0914

0 to 1516.7 K crystal
above 1516.7 K liquid

Refer to the individual tables for details.

Copper Oxide (Cu_2O) Cu_2O_1(cr,l)

Enthalpy Reference Temperature = T_r = 298.15 K Standard State Pressure = p° = 0.1 MPa

T/K	C_p° (J K⁻¹mol⁻¹)	S° (J K⁻¹mol⁻¹)	$-[G^\circ-H^\circ(T_r)]/T$ (J K⁻¹mol⁻¹)	$H^\circ-H^\circ(T_r)$ (kJ mol⁻¹)	$\Delta_f H^\circ$ (kJ mol⁻¹)	$\Delta_f G^\circ$ (kJ mol⁻¹)	Log K_f
0	0.	0.	INFINITE	-12.600	-168.952	-168.952	INFINITE
100	39.499	37.092	141.510	-10.442	-169.584	-162.621	84.944
200	53.619	69.143	97.826	-5.737	-170.360	-155.348	40.573
298.15	62.543	92.360	92.360	0.	-170.707	-147.886	25.909
300	62.664	92.747	92.361	0.116	-170.709	-147.745	25.725
400	67.688	111.531	94.888	6.657	-170.641	-140.092	18.294
500	70.922	127.000	99.809	13.596	-170.363	-132.484	13.840
600	73.448	140.161	105.465	20.818	-169.961	-124.944	10.877
700	75.645	151.651	111.259	28.274	-169.479	-117.478	8.766
800	77.666	161.886	116.959	35.941	-168.929	-110.087	7.188
900	79.493	171.141	122.473	43.801	-168.325	-102.767	5.964
1000	81.245	179.606	127.769	51.837	-167.688	-95.517	4.989
1100	83.385	187.443	132.842	60.061	-167.029	-88.331	4.194
1200	86.110	194.812	137.702	68.532	-166.322	-81.208	3.535
1300	89.173	201.824	142.367	77.294	-165.608	-74.144	2.979
1400	92.489	208.552	146.855	86.375	-191.165	-66.321	2.474
1500	95.987	215.052	151.186	95.798	-190.132	-57.438	2.000
1516.700	96.587	216.118	151.895	97.406	CRYSTAL <--> LIQUID		
1516.700	99.914	258.821	151.895	162.174	TRANSITION		
1600	99.914	264.163	157.603	170.497	-123.836	-52.197	1.704
1700	99.914	270.220	164.051	180.488	-122.259	-47.768	1.468
1800	99.914	275.931	170.109	190.480	-120.694	-43.431	1.260
1900	99.914	281.333	175.822	200.471	-119.142	-39.181	1.077
2000	99.914	286.458	181.227	210.463	-117.600	-35.013	0.914

PREVIOUS: CURRENT: December 1977

Copper Oxide (Cu_2O) Cu_2O_1(cr,l)

COPPER OXIDE SULFATE ($CuO\cdot CuSO_4$) CRYSTAL M_r = 239.1490

$S°(298.15\ K) = [157.318 \pm 8.4]\ J\ K^{-1}\ mol^{-1}$

$\Delta_f H°(0\ K)$ = Unknown

$\Delta_f H°(298.15\ K) = -927.593 \pm 1.3\ kJ\ mol^{-1}$

Enthalpy of Formation

The equilibrium pressures for the following reactions: (A) $CuO\cdot CuSO_4(cr) = 2\ CuO(cr) + SO_3(g)$ and $SO_3(g) = SO_2(g) + 0.5\ O_2(g)$ and (B) $CuO\cdot CuSO_4(cr) + Cu_2O(cr) = 4\ CuO(cr) + SO_2(g)$, were determined by Reinders (1) and Ingraham (2). Using the reported vapor pressures, the corresponding enthalpy changes were evaluated by both the 2nd and 3rd law methods. The results obtained are presented as follows:

Reference	Reaction	T/K	$\Delta_r H°(298.15\ K)$, kcal mol^{-1} 2nd law	$\Delta_r H°(298.15\ K)$, kcal mol^{-1} 3rd law	Drift cal K^{-1} mol^{-1}	$\Delta_f H°(298.15\ K)$ kcal mol^{-1}
(1)	A*	1013-1093	53.03±2.53	52.43	-0.7±2.4	-221.5
(1)	B	917-1027	41.72±1.42	42.43	0.5±1.4	-221.7
(2)	A*	969-1138	50.92±0.16	52.57	1.5±0.2	-221.7

*Partial pressures were calculated from JANAF values for the $SO_3 - SO_2$ equilibrium.

The value of $\Delta_f H°(CuO\cdot CuSO_4$, cr, 298.15 K) adopted is $-221.7 \pm 0.3\ kcal\ mol^{-1}$.

Heat Capacity and Entropy

The heat capacities are calculated as the sum of those for CuO(cr) and $CuSO_4$(cr). The value of S°(298.15 K) is estimated by comparison with those for CuO(cr) and $CuSO_4$(cr) and adjusted so that, using the derived free energy functions, the 2nd and 3rd law values for $\Delta_r H°(298.15\ K)$ agree reasonably.

Decomposition Data

T_{dcm} = 1138 K is the temperature at which the total pressure of the gaseous decomposition products of reaction (A) equals one atmosphere. It was obtained by graphical extrapolation of the decomposition pressures for $CuO\cdot CuSO_4$(cr), determined by Ingraham (2).

References

1. W. Reinders and F. Goudriaan, Z. anorg. allgem. Chem. 126, 85 (1923).
2. T. R. Ingraham, Trans. Met. Soc. AIME 233, 359 (1965).

Copper Oxide Sulfate ($CuO\cdot CuSO_4$) $Cu_2O_5S_1(cr)$

Enthalpy Reference Temperature = T_r = 298.15 K Standard State Pressure = p° = 0.1 MPa

T/K	$C_p°$ (J $K^{-1}mol^{-1}$)	S° (J $K^{-1}mol^{-1}$)	$-[G°-H°(T_r)]/T$ (J $K^{-1}mol^{-1}$)	$H°-H°(T_r)$ (kJ mol^{-1})	$\Delta_f H°$ (kJ mol^{-1})	$\Delta_f G°$ (kJ mol^{-1})	Log K_f
0							
100							
200							
298.15	141.043	157.318	157.318	0.	-927.593	-792.253	138.799
300	141.419	158.192	157.321	0.261	-927.600	-791.413	137.797
400	162.339	201.891	163.137	15.501	-929.372	-745.953	97.411
500	176.983	239.760	174.761	32.499	-929.081	-700.122	73.141
600	187.820	273.029	188.423	50.763	-927.541	-654.457	56.976
700	195.811	302.603	202.662	69.959	-925.177	-609.121	45.453
800	202.129	329.172	216.843	89.863	-922.273	-564.163	36.836
900	207.526	353.295	230.685	110.349	-972.111	-518.512	30.094
1000	212.547	375.424	244.067	131.356	-967.283	-468.369	24.465
1100	217.150	395.897	256.951	152.841	-962.253	-418.719	19.883
1200	221.752	414.990	269.334	174.787	-957.056	-369.536	16.085
1300	226.313	432.923	281.235	197.194	-951.765	-320.790	12.889
1400	230.329	449.842	292.680	220.027	-972.716	-271.639	10.135
1500	234.304	465.869	303.696	243.259	-967.089	-221.757	7.722

PREVIOUS: CURRENT: June 1966

Copper Oxide Sulfate ($CuO\cdot CuSO_4$) $Cu_2O_5S_1(cr)$

DEUTERIUM (D) IDEAL GAS A_r = 2.014102

$S°(298.15\ K) = 123.350 \pm 0.017\ J\ K^{-1}\ mol^{-1}$

$\Delta_f H°(298.15\ K) = 219.807 \pm 0.004\ kJ\ mol^{-1}$

$\Delta_{fus} H° = 221.720 \pm 0.004\ kJ\ mol^{-1}$

Electronic Level and Quantum Weight

State	ε_1, cm^{-1}	g_1
$^2S_{1/2}$	0.00	2

Enthalpy of Formation

The adopted value for $\Delta_f H°$(D, g, 298.15 K) is derived using $D_0^0 = 36748.9 \pm 0.4\ cm^{-1}$ from Herzberg (1) and auxiliary data for D_2(2).

An earlier value for the dissociation energy was reported by Herzberg and Monfils (3) as $D_0^0 = 36743.6 \pm 0.5\ cm^{-1}$. Two theoretical studies by Kolos and Wolniewicz (4) and Bunker (5) yielded a vlue of 36748.2 cm^{-1}. More recently, LeRoy and Barwell (6), using a relationship involving the near-dissociation behavior of the rotational constants and the long range intermolecular potential, calculated $D_0^0 = 36748.88(\pm 0.3)\ cm^{-1}$.

Heat Capacity and Entropy

The electronic levels for D(g) are given in the compilation by Moore (7). Our calculations indicate that the inclusion of levels up to n = 12 has no effect on the thermodynamic functions to 6000 K. This is a result of the high energy of these levels; the first excited state lies at 82281 cm^{-1} above the ground state. Since the inclusion of these upper levels has no effect on the thermodynamic functions (to 6000 K) we list only the ground state. The reported uncertainty in S°(298.15 K) is due to uncertainties in the gram formula weight and the fundamental constants. The gram formula weight is that recommended by IUPAC (8). Extension of these calculations above 6000 K may require consideration of the higher excited states and utilization of proper fill and cut off procedures (9).

Reference

1. G. Herzberg, J. Mol. Spectrosc. 33, 147 (1970).
2. JANAF Thermochemical Tables: D_2(g), 3-31-77.
3. G. Herzberg and A. Monfils, J. Mol. Spectrosc. 5, 482 (1960).
4. W. Kolos and L. Wolniewicz, J. Mol. Spectrosc. 54, 303 (1975).
5 P. R. Bunker, private communication to G. Herzberg, Phys. Rev. Letters 23, 1081 (1969).
6. R. J. LeRoy and M. G. Barwell, Can. J. Phys. 53, 1983 (1975).
7. C. E. Moore, NSRDS-NBS 3, Section 6 (1972).
8. IUPAC Commission on Adtomic Weights, Pure Appl. Chem. 37, 589 (1974).
9. J. R. Downey, Dow Chemical Company, Thermal Research, AFOSR-TR-78-0960, Contract No. F 44620-75-0-0048, 1978.

Deuterium (D) D_1(g)

Enthalpy Reference Temperature = T_r = 298.15 K Standard State Pressure = p° = 0.1 MPa

T/K	$C_p°$ (J K⁻¹mol⁻¹)	S° (J K⁻¹mol⁻¹)	$-[G°-H°(T_r)]/T$ (J K⁻¹mol⁻¹)	$H°-H°(T_r)$ (kJ mol⁻¹)	$\Delta_f H°$ (kJ mol⁻¹)	$\Delta_f G°$ (kJ mol⁻¹)	Log K_f
0	0.	0.	INFINITE	-6.197	219.807	219.807	INFINITE
100	20.786	100.643	141.830	-4.119	220.511	216.086	-112.872
200	20.786	115.051	125.251	-2.040	221.112	211.433	-55.221
250	20.786	119.689	123.692	-1.001	221.422	208.977	-43.663
298.15	20.786	123.350	123.350	0.	221.720	206.553	-36.187
300	20.786	123.479	123.350	0.038	221.732	206.459	-35.948
350	20.786	126.683	123.603	1.078	222.041	203.889	-30.429
400	20.786	129.458	124.166	2.117	222.350	201.275	-26.284
450	20.786	131.907	124.892	3.156	222.657	198.622	-23.055
500	20.786	134.097	125.705	4.196	222.963	195.935	-20.469
600	20.786	137.886	127.429	6.274	223.568	190.473	-16.582
700	20.786	141.091	129.158	8.353	224.156	184.910	-13.798
800	20.786	143.866	130.827	10.431	224.723	179.265	-11.705
900	20.786	146.314	132.414	12.510	225.262	173.550	-10.073
1000	20.786	148.504	133.916	14.589	225.774	167.776	-8.764
1100	20.786	150.485	135.333	16.667	226.256	161.953	-7.690
1200	20.786	152.294	136.673	18.746	226.710	156 087	-6.794
1300	20.786	153.958	137.939	20.824	227.138	150.184	-6.034
1400	20.786	155.498	139.139	22.903	227.540	144.249	-5.382
1500	20.786	156.932	140.278	24.982	227.919	138.287	-4.816
1600	20.786	158.274	141.361	27.060	228.276	132.299	-4.319
1700	20.786	159.534	142.394	29.139	228.614	126.290	-3.880
1800	20.786	160.722	143.379	31.217	228.934	120.262	-3.490
1900	20.786	161.846	144.322	33.296	229.237	114.216	-3.140
2000	20.786	162.912	145.225	35.375	229.524	108.155	-2.825
2100	20.786	163.926	146.091	37.453	229.797	102.080	-2.539
2200	20.786	164.893	146.924	39.532	230.056	95.992	-2.279
2300	20.786	165.817	147.726	41.610	230.303	89.893	-2.042
2400	20.786	166.702	148.498	43.689	230.538	83.783	-1.823
2500	20.786	167.550	149.243	45.768	230.762	77.663	-1.623
2600	20.786	168.366	149.963	47.846	230.975	71.535	-1.437
2700	20.786	169.150	150.659	49.925	231.179	65.399	-1.265
2800	20.786	169.906	151.333	52.004	231.372	59.255	-1.105
2900	20.786	170.635	151.986	54.082	231.556	53.105	-0.957
3000	20.786	171.340	152.620	56.161	231.732	46.949	-0.817
3100	20.786	172.022	153.235	58.239	231.898	40.786	-0.687
3200	20.786	172.682	153.832	60.318	232.056	34.619	-0.565
3300	20.786	173.321	154.413	62.397	232.206	28.447	-0.450
3400	20.786	173.942	154.979	64.475	232.347	22.270	-0.342
3500	20.786	174.544	155.529	66.554	232.480	16.089	-0.240
3600	20.786	175.130	156.065	68.632	232.605	9.905	-0.144
3700	20.786	175.699	156.588	70.711	232.722	3.717	-0.052
3800	20.786	176.254	157.099	72.790	232.831	-2.473	0.034
3900	20.786	176.794	157.597	74.868	232.932	-8.667	0.116
4000	20.786	177.320	158.083	76.947	233.025	-14.863	0.194
4100	20.786	177.833	158.559	79.025	233.111	-21.061	0.268
4200	20.786	178.334	159.024	81.104	233.188	-27.262	0.339
4300	20.786	178.823	159.478	83.183	233.258	-33.464	0.407
4400	20.786	179.301	159.924	85.261	233.321	-39.667	0.471
4500	20.786	179.768	160.359	87.340	233.375	-45.872	0.532
4600	20.786	180.225	160.786	89.418	233.423	-52.078	0.591
4700	20.786	180.672	161.205	91.497	233.462	-58.285	0.648
4800	20.786	181.110	161.615	93.576	233.495	-64.493	0.702
4900	20.786	181.538	162.017	95.654	233.521	-70.701	0.754
5000	20.786	181.958	162.412	97.733	233.540	-76.910	0.803
5100	20.786	182.370	162.799	99.811	233.552	-83.119	0.851
5200	20.786	182.773	163.179	101.890	233.558	-89.328	0.897
5300	20.786	183.169	163.553	103.969	233.558	-95.538	0.942
5400	20.786	183.558	163.920	106.047	233.552	-101.747	0.984
5500	20.786	183.939	164.280	108.126	233.541	-107.956	1.025
5600	20.786	184.314	164.634	110.204	233.524	-114.165	1.065
5700	20.786	184.682	164.983	112.283	233.503	-120.374	1.103
5800	20.786	185.043	165.326	114.362	233.478	-126.582	1.140
5900	20.786	185.399	165.663	116.440	233.449	-132.790	1.176
6000	20.786	185.748	165.995	118.519	233.416	-138.997	1.210

PREVIOUS: March 1977 (1 atm) CURRENT: March 1982 (1 bar)

Deuterium (D) D_1(g)

DEUTERIUM, ION (D^+) IDEAL GAS M_r = 2.01355

$S°(298.15\ K) = 117.583 \pm 0.02\ JK^{-1}\ mol^{-1}$

$\Delta_f H°(0\ K) = 1532.21 \pm 0.04\ kJ\ mol^{-1}$
$\Delta_f H°(298.15\ K) = [1540.324]\ kJ\ mol^{-1}$

Enthalpy of Formation

$\Delta_f H°(D^+, g, 0\ K)$ is calculated from $\Delta_f H°(D, g, 0\ K)$ (1) using the spectroscopic value of IP(D) = 109708.608 ± 0.005 cm^{-1} (313.6727 ± 0.0001 kJ mol^{-1}) from Moore (2). The ionization limit is converted from cm^{-1} to kJ mol^{-1} using the factor, 1 cm^{-1} = 0.01196266 kJ cm^{-1}, which is derived from the 1973 CODATA fundamental constants (3). Rosenstock et al. (4) and Levin and Lias (5) have summarized additional ionization and appearance potential data.

$\Delta_f H°(D^+, g, 298.15\ K)$ is calculated from $\Delta_f H°(D, g, 0\ K)$ by using IP(D) with JANAF (1) enthalpies, H°(0 K) - H°(298.15 K), for D(g), D^+(g), and e^-(g). $\Delta_f H°(D \to D^+ + e^-, 298.15\ K)$ differs from a room temperature threshold energy due to inclusion of these enthalpies and to threshold effects discussed by Rosenstock et al. (4). $\Delta_f H°(298.15\ K)$ should be changed by -6.197 kJ mol^{-1} if it is to be used in the ion convention that excludes the enthalpy of the electron.

Heat Capacity and Entropy

The thermodynamic functions of the proton gas are calculated using the recent CODAT fundamental constants (2) and assuming that the proton is an ideal monatomic gas. Since there is no electron associated with this species, there is only a translational contribution to the thermochemical function.

References

1. JANAF Thermochemical Tables: D(g), 3-31-82 ; e^-(g), 3-31-82.
2. C. E. Moore, U. S. Nat. Bur. Stand., NSRDS-NBS-3, Section 6, 1972.
3. E. R. Cohen and B. N. Taylor, J. Phys. Chem. Ref. Data 2, 663 (1973).
4. H. M. Rosenstock, K. Draxl et al., J. Phys. Chem. Ref. Data 6, Supp. 1 (1977).
5. R. D. Levin and S. G. Lias, U. S. Nat. Bur. Stand., NSRDS-NBS-71, 1982.

Deuterium, Ion (D^+) $D_1^+(g)$

Enthalpy Reference Temperature = T_r = 298.15 K Standard State Pressure = p° = 0.1 MPa

T/K	$C_p°$ (J K⁻¹mol⁻¹)	S°	-[G°-H°(T_r)]/T	H°-H°(T_r) (kJ mol⁻¹)	$\Delta_f H°$	$\Delta_f G°$	Log K_f
0	0.	0.	INFINITE	-6.197	1532.214		
100	20.786	94.876	136.064	-4.119			
200	20.786	109.284	119.485	-2.040			
250	20.786	113.922	117.926	-1.001			
298.15	20.786	117.583	117.583	0.	1540.324	1520.622	-266.407
300	20.786	117.712	117.584	0.038	1540.374	1520.500	-264.743
350	20.786	120.916	117.837	1.078	1541.723	1517.081	-226.412
400	20.786	123.692	118.399	2.117	1543.071	1513.468	-197.639
450	20.786	126.140	119.126	3.156	1544.418	1509.687	-175.240
500	20.786	128.330	119.939	4.196	1545.763	1505.756	-157.305
600	20.786	132.120	121.663	6.274	1548.446	1497.502	-130.369
700	20.786	135.324	123.391	8.353	1551.113	1488.800	-111.096
800	20.786	138.100	125.060	10.431	1553.758	1479.718	-96.616
900	20.786	140.548	126.648	12.510	1556.377	1470.305	-85.334
1000	20.786	142.738	128.149	14.589	1558.967	1460.603	-76.294
1100	20.786	144.719	129.567	16.667	1561.528	1450.642	-68.885
1200	20.786	146.528	130.906	18.746	1564.060	1440.449	-62.701
1300	20.786	148.191	132.173	20.824	1566.566	1430.047	-57.460
1400	20.786	149.732	133.372	22.903	1569.047	1419.452	-52.960
1500	20.786	151.166	134.511	24.982	1571.505	1408.681	-49.055
1600	20.786	152.507	135.595	27.060	1573.941	1397.746	-45.632
1700	20.786	153.768	136.627	29.139	1576.357	1386.660	-42.607
1800	20.786	154.956	137.613	31.217	1578.755	1375.432	-39.914
1900	20.786	156.079	138.555	33.296	1581.137	1364.071	-37.501
2000	20.786	157.146	139.458	35.375	1583.503	1352.585	-35.326
2100	20.786	158.160	140.325	37.453	1585.854	1340.982	-33.355
2200	20.786	159.127	141.158	39.532	1588.192	1329.266	-31.561
2300	20.786	160.051	141.959	41.610	1590.518	1317.445	-29.920
2400	20.786	160.935	142.732	43.689	1592.831	1305.522	-28.414
2500	20.786	161.784	143.477	45.768	1595.134	1293.504	-27.026
2600	20.786	162.599	144.197	47.846	1597.426	1281.393	-25.744
2700	20.786	163.384	144.893	49.925	1599.708	1269.195	-24.554
2800	20.786	164.140	145.567	52.004	1601.980	1256.912	-23.448
2900	20.786	164.869	146.220	54.082	1604.243	1244.548	-22.417
3000	20.786	165.574	146.853	56.161	1606.497	1232.107	-21.453
3100	20.786	166.255	147.468	58.239	1608.742	1219.590	-20.550
3200	20.786	166.915	148.066	60.318	1610.978	1207.001	-19.702
3300	20.786	167.555	148.647	62.397	1613.206	1194.342	-18.905
3400	20.786	168.175	149.212	64.475	1615.426	1181.616	-18.153
3500	20.786	168.778	149.762	66.554	1617.638	1168.825	-17.444
3600	20.786	169.363	150.299	68.632	1619.841	1155.970	-16.773
3700	20.786	169.933	150.822	70.711	1622.037	1143.055	-16.137
3800	20.786	170.487	151.332	72.790	1624.225	1130.080	-15.534
3900	20.786	171.027	151.830	74.868	1626.404	1117.048	-14.961
4000	20.786	171.553	152.317	76.947	1628.576	1103.960	-14.416
4100	20.786	172.067	152.792	79.025	1630.740	1090.817	-13.897
4200	20.786	172.568	153.257	81.104	1632.896	1077.622	-13.402
4300	20.786	173.057	153.712	83.183	1635.045	1064.376	-12.930
4400	20.786	173.535	154.157	85.261	1637.186	1051.080	-12.478
4500	20.786	174.002	154.593	87.340	1639.319	1037.735	-12.046
4600	20.786	174.459	155.020	89.418	1641.445	1024.344	-11.632
4700	20.786	174.906	155.438	91.497	1643.564	1010.905	-11.235
4800	20.786	175.343	155.848	93.576	1645.675	997.422	-10.854
4900	20.786	175.772	156.250	95.654	1647.779	983.895	-10.488
5000	20.786	176.192	156.645	97.733	1649.877	970.325	-10.137
5100	20.786	176.603	157.032	99.811	1651.967	956.713	-9.799
5200	20.786	177.007	157.413	101.890	1654.052	943.061	-9.473
5300	20.786	177.403	157.786	103.969	1656.130	929.368	-9.159
5400	20.786	177.791	158.153	106.047	1658.203	915.636	-8.857
5500	20.786	178.173	158.514	108.126	1660.270	901.866	-8.565
5600	20.786	178.547	158.868	110.204	1662.333	888.058	-8.283
5700	20.786	178.915	159.216	112.283	1664.390	874.213	-8.011
5800	20.786	179.277	159.559	114.362	1666.444	860.332	-7.748
5900	20.786	179.632	159.896	116.440	1668.493	846.417	-7.494
6000	20.786	179.981	160.228	118.519	1670.539	832.466	-7.247

PREVIOUS: March 1977 (1 atm) CURRENT: March 1982 (1 bar)

DEUTERIUM, ION (D^-) IDEAL GAS M_r = 2.014651

EA(D, g) = 0.754209 ± 0.000003 eV

S°(298.15 K) = 117.590 ± 0.017 J K^{-1} mol^{-1}

$\Delta_f H°$(0 K) = 147.038 ± 0.04 kJ mol^{-1}

$\Delta_f H°$(298.15 K) = [142.753] kJ mol^{-1}

Electonic Level and Quantum Weight

State	ε_i, cm^{-1}	g_i
$^1S_0$	0.0	1

Enthalpy of Formation

$\Delta_f H°(D^-$, g, 0 K) is calculated from $\Delta_f H°$(D, g, 0 K) (1) using the adopted electron affinity of EA(D) = 0.754209 ± 0.000003 eV (72.7696 ± 0.0003 kJ mol^{-1}). This value is assumed to be the same as that recommended by Hotop and Lineberger (2) for H(g). Additional information on D^-(g) and H^-(g) may be obtained in the critical discussions of Hotop and Lineberger (2, 3), Rosenstock et al. (4) and Massey (5).

$\Delta_f H°(D^-$, g, 298.15 K) is obtained from $\Delta_f H°$(D, g, 0 K) by using EA(D) with JANAF (1) enthalpies, H°(0 K) − H°(298.15 K), for D^-(g), D(g), and e^-(g). $\Delta_r H°(D^- \rightarrow D + e^-$, 298.15 K) differs from a room-temperature threshold energy due to inclusion of these enthalpies and to threshold effects discussed by Rosenstock et al. (4). $\Delta_f H°$(298.15 K) should be changed by +6.197 kJ mol^{-1} if it is to be used in the ion convention that excludes the enthalpy of the electron.

Heat Capacity and Entropy

The ground state electronic configuration for D^-(g) is given by Hotop and Lineberger (2, 3) and Rosenstock et al. (4).

A comparison of the isoelectronic sequence - H^-(g), He(g), and Li^+(g) - would suggest that stable electronic states may exist at 0.8 EA(H) or roughly 6400 cm^{-1}. This would greatly affect the entropy. However, Pekeris (6) states that he was unable to find any bound states. In addition, Seman and Branscom (7) state that theoretical and semiempirical evidence suggests that atomic negative ions have very few if any excited states below the continuum. We assume that the same arguments apply for D^-(g) and that no stable excited states exist.

References

1. JANAF Thermochemical Tables: D(g), 3-31-82; e^-(g), 3-31-82.
2. H. Hotop and W. C. Lineberger, J. Phys. Chem. Ref. Data, to be published, 1985.
3. H. Hotop and W. C. Lineberger, J. Phys. Chem. Ref. Data 4, 539 (1975).
4. H. M. Rosenstock, K. Draxl et al., J. Phys. Chem. Ref. Data 6, Supp. 1 (1977).
5. H. S. W. Massey, "Negative Ions", 3rd ed., Cambridge University Press, Cambridge, 1976.
6. C. L. Perkeris, Phys. Rev. 126, 1470 (1962).
7. M. L. Seman and L. M. Branscomb, Phys. Rev. 125, 1602 (1962).

Deuterium, Ion (D^-) D_1^-(g)

Enthalpy Reference Temperature = T_r = 298.15 K Standard State Pressure = p° = 0.1 MPa

	J $K^{-1}mol^{-1}$			kJ mol^{-1}			
T/K	$C_p°$	S°	−[G° − H°(T_r)]/T	H° − H°(T_r)	$\Delta_f H°$	$\Delta_f G°$	Log K_f
0	0.	0.	INFINITE	-6.197	147.038		
100	20.786	94.883	136.071	-4.119			
200	20.786	109.291	119.492	-2.040			
250	20.786	113.929	117.933	-1.001			
298.15	20.786	117.590	117.590	0.	142.753	135.559	-23.749
300	20.786	117.719	117.591	0.038	142.726	135.514	-23.595
350	20.786	120.923	117.844	1.078	141.996	134.370	-20.054
400	20.786	123.699	118.406	2.117	141.266	133.330	-17.411
450	20.786	126.147	119.133	3.156	140.534	132.382	-15.366
500	20.786	128.337	119.946	4.196	139.801	131.515	-13.739
600	20.786	132.127	121.670	6.274	138.327	129.997	-11.317
700	20.786	135.331	123.398	8.353	136.837	128.726	-9.606
800	20.786	138.106	125.067	10.431	135.324	127.670	-8.336
900	20.786	140.555	126.655	12.510	133.785	126.805	-7.360
1000	20.786	142.745	128.156	14.589	132.218	126.113	-6.587
1100	20.786	144.726	129.574	16.667	130.622	125.580	-5.963
1200	20.786	146.534	130.913	18.746	128.998	125.193	-5.450
1300	20.786	148.198	132.179	20.824	127.346	124.943	-5.020
1400	20.786	149.739	133.379	22.903	125.670	124.821	-4.657
1500	20.786	151.173	134.518	24.982	123.970	124.819	-4.347
1600	20.786	152.514	135.601	27.060	122.249	124.932	-4.079
1700	20.786	153.774	136.634	29.139	120.508	125.153	-3.845
1800	20.786	154.962	137.619	31.217	118.749	125.477	-3.641
1900	20.786	156.086	138.562	33.296	116.974	125.899	-3.461
2000	20.786	157.152	139.465	35.375	115.182	126.415	-3.302
2100	20.786	158.167	140.332	37.453	113.376	127.021	-3.159
2200	20.786	159.134	141.165	39.532	111.557	127.713	-3.032
2300	20.786	160.058	141.966	41.610	109.726	128.488	-2.918
2400	20.786	160.942	142.738	43.689	107.882	129.343	-2.815
2500	20.786	161.791	143.484	45.768	106.027	130.276	-2.722
2600	20.786	162.606	144.204	47.846	104.162	131.282	-2.637
2700	20.786	163.390	144.900	49.925	102.287	132.361	-2.561
2800	20.786	164.146	145.574	52.004	100.402	133.509	-2.491
2900	20.786	164.876	146.227	54.082	98.507	134.725	-2.427
3000	20.786	165.580	146.860	56.161	96.604	136.006	-2.368
3100	20.786	166.262	147.475	58.239	94.692	137.351	-2.314
3200	20.786	166.922	148.073	60.318	92.771	138.758	-2.265
3300	20.786	167.562	148.654	62.397	90.842	140.225	-2.220
3400	20.786	168.182	149.219	64.475	88.905	141.751	-2.178
3500	20.786	168.785	149.769	66.554	86.959	143.333	-2.139
3600	20.786	169.370	150.306	68.632	85.006	144.972	-2.103
3700	20.786	169.940	150.829	70.711	83.044	146.664	-2.071
3800	20.786	170.494	151.339	72.790	81.075	148.410	-2.040
3900	20.786	171.034	151.837	74.868	79.097	150.208	-2.012
4000	20.786	171.560	152.324	76.947	77.112	152.056	-1.986
4100	20.786	172.073	152.799	79.025	75.119	153.955	-1.961
4200	20.786	172.574	153.264	81.104	73.118	155.902	-1.939
4300	20.786	173.063	153.719	83.183	71.109	157.896	-1.918
4400	20.786	173.541	154.164	85.261	69.093	159.938	-1.899
4500	20.786	174.008	154.600	87.340	67.069	162.025	-1.881
4600	20.786	174.465	155.027	89.418	65.037	164.158	-1.864
4700	20.786	174.912	155.445	91.497	62.999	166.335	-1.849
4800	20.786	175.350	155.855	93.576	60.953	168.555	-1.834
4900	20.786	175.779	156.257	95.654	58.900	170.818	-1.821
5000	20.786	176.198	156.652	97.733	56.840	173.123	-1.809
5100	20.786	176.610	157.039	99.811	54.774	175.469	-1.797
5200	20.786	177.014	157.419	101.890	52.701	177.856	-1.787
5300	20.786	177.410	157.793	103.969	50.622	180.283	-1.777
5400	20.786	177.798	158.160	106.047	48.538	182.749	-1.768
5500	20.786	178.180	158.520	108.126	46.448	185.253	-1.759
5600	20.786	178.554	158.875	110.204	44.353	187.796	-1.752
5700	20.786	178.922	159.223	112.283	42.253	190.376	-1.745
5800	20.786	179.284	159.566	114.362	40.149	192.993	-1.738
5900	20.786	179.639	159.903	116.440	38.042	195.646	-1.732
6000	20.786	179.988	160.235	118.519	35.931	198.335	-1.727

PREVIOUS: March 1977 (1 atm) CURRENT: March 1982 (1 bar)

Deuterium, Ion (D^-) D_1^-(g)

$S°(298.15\ K) = 179.702\ J\ K^{-1}\ mol^{-1}$

$\Delta_fH°(0\ K) = -275.458 \pm 0.8\ kJ\ mol^{-1}$
$\Delta_fH°(298.15\ K) = -275.516 \pm 0.8\ kJ\ mol^{-1}$

Electronic States and Molecular Constants σ = 1

State	ε_i, cm^{-1}	g_i	ω_e, cm^{-1}	$\omega_e x_e$, cm^{-1}	B_e, cm^{-1}	α_e, cm^{-1}	r_e, Å
$X^1\Sigma^+$	0	1	2998.19	45.76	11.000	0.2907	0.9187
$V^1\Sigma^+$	83755	1	839.4	8.90	2.121	0.00712	2.088

Enthalpy of Formation

The enthalpy of formation was calculated from $\Delta_fH°$(298.15 K) of HF(g), -65.13±0.2 kcal mol^{-1}, given in JANAF Thermochemical Tables (1) the appropriate thermal functions (see tables for H_2, D_2, DF and HF (1), and the estimated zero point energies. The energies for H_2(g) and D_2(g) are those given by Herzberg and Monfils (2). The energies for HF(g) and DF(g) include the Dunham correction and were calculated from the data given by Mann et al. (3) and Spanbauer et al. (4).

Heat Capacity and Entropy

The vibrational and rotational constants of the respective electronic levels were taken from Rosen (5). The National Bureau of Standards prepared this table (6) by critical analysis of data existing in 1972. Using molecular constants and $\Delta_fH°$ selected by NBS (6), we recalculate the table in terms of 1973 fundamental constants (7), 1975 atomic weights (8), and current JANAF reference states for the elements.

References

1. JANAF Thermochemical Tables, 2nd Edition, NSRDS-NBS 37, June, 1971.
2. G. Herzberg and A. Monfils, J. Mol. Spectrosc. 5, 482 (1960).
3. D. E. Mann, B. A. Thrush, D. R. Lide, Jr., J. J. Ball, and N. Acquista, J. Chem. Phys. 34, 420 (1961).
4. R. N. Spanbauer, K. N. Rao, and L. H. Jones, J. Mol. Spectrosc. 16, 100 (1965).
5. B. Rosen, "Spectroscopic Data Relative to Diatomic Molecules," Pergamon Press, Oxford, 1970.
6. S. Abramowitz et al., U. S. Nat. Bur. Stand. Report 10904, p. 239, July, 1972.
7. CODATA Task Group on Fundamental Constants, CODATA Bulletin 11, December, 1973.
8. IUPAC Commission on Atomic Weights, Pure Appl. Chem. 47, 75 (1976); 37, 589 (1974).

Enthalpy Reference Temperature = T_r = 298.15 K   Standard State Pressure = p° = 0.1 MPa

T/K	$C_p°$ (J K^{-1}mol^{-1})	S° (J K^{-1}mol^{-1})	$-[G°-H°(T_r)]/T$ (J K^{-1}mol^{-1})	$H°-H°(T_r)$ (kJ mol^{-1})	$\Delta_fH°$ (kJ mol^{-1})	$\Delta_fG°$ (kJ mol^{-1})	Log K_f
0	0.	0.	INFINITE	-8.638	-275.458	-275.458	INFINITE
100	29.117	147.887	205.599	-5.771	-275.419	-276.050	144.194
200	29.125	168.071	182.367	-2.859	-275.449	-276.667	72.258
250	29.130	174.571	180.182	-1.403	-275.473	-276.969	57.870
298.15	29.137	179.702	179.702	0.	-275.516	-277.254	48.574
300	29.137	179.882	179.703	0.054	-275.518	-277.265	48.276
350	29.150	184.375	180.057	1.511	-275.586	-277.551	41.422
400	29.174	188.268	180.846	2.969	-275.674	-277.825	36.280
450	29.221	191.707	181.865	4.429	-275.779	-278.088	32.280
500	29.298	194.790	183.006	5.892	-275.899	-278.339	29.078
600	29.561	200.152	185.430	8.833	-276.168	-278.802	24.272
700	29.962	204.738	187.868	11.809	-276.458	-279.218	20.836
800	30.463	208.770	190.234	14.829	-276.755	-279.592	18.255
900	31.016	212.390	192.498	17.903	-277.047	-279.929	16.247
1000	31.583	215.687	194.654	21.033	-277.329	-280.234	14.638
1100	32.137	218.724	196.706	24.219	-277.599	-280.511	13.320
1200	32.661	221.543	198.660	27.459	-277.857	-280.764	12.221
1300	33.148	224.176	200.523	30.750	-278.102	-280.997	11.291
1400	33.594	226.649	202.301	34.087	-278.336	-281.211	10.492
1500	34.001	228.981	204.003	37.468	-278.561	-281.408	9.800
1600	34.370	231.188	205.634	40.886	-278.777	-281.591	9.193
1700	34.705	233.282	207.199	44.340	-278.987	-281.760	8.657
1800	35.008	235.274	208.704	47.826	-279.190	-281.917	8.181
1900	35.284	237.174	210.153	51.341	-279.387	-282.064	7.754
2000	35.534	238.991	211.549	54.882	-279.578	-282.199	7.370
2100	35.762	240.730	212.898	58.447	-279.762	-282.326	7.022
2200	35.971	242.398	214.201	62.034	-279.939	-282.444	6.706
2300	36.162	244.002	215.462	65.641	-280.108	-282.554	6.417
2400	36.339	245.545	216.684	69.266	-280.268	-282.657	6.152
2500	36.501	247.031	217.868	72.908	-280.418	-282.753	5.908
2600	36.652	248.466	219.017	76.566	-280.556	-282.844	5.682
2700	36.792	249.852	220.134	80.238	-280.683	-282.929	5.474
2800	36.923	251.192	221.219	83.924	-280.796	-283.011	5.280
2900	37.045	252.490	222.275	87.623	-280.894	-283.088	5.099
3000	37.160	253.748	223.304	91.333	-280.978	-283.162	4.930
3100	37.269	254.968	224.306	95.054	-281.046	-283.234	4.772
3200	37.371	256.153	225.282	98.786	-281.098	-283.303	4.624
3300	37.468	257.305	226.235	102.528	-281.133	-283.372	4.485
3400	37.560	258.424	227.166	106.280	-281.151	-283.439	4.355
3500	37.647	259.514	228.074	110.040	-281.152	-283.507	4.231
3600	37.731	260.576	228.963	113.809	-281.136	-283.574	4.115
3700	37.811	261.611	229.831	117.586	-281.102	-283.642	4.004
3800	37.888	262.620	230.681	121.371	-281.052	-283.712	3.900
3900	37.961	263.606	231.512	125.164	-280.986	-283.782	3.801
4000	38.033	264.568	232.327	128.963	-280.902	-283.855	3.707
4100	38.101	265.508	233.125	132.770	-280.803	-283.930	3.617
4200	38.168	266.427	233.907	136.584	-280.688	-284.008	3.532
4300	38.232	267.325	234.673	140.404	-280.557	-284.088	3.451
4400	38.295	268.205	235.425	144.230	-280.411	-284.172	3.374
4500	38.355	269.066	236.164	148.062	-280.251	-284.259	3.300
4600	38.414	269.910	236.888	151.901	-280.077	-284.350	3.229
4700	38.472	270.737	237.599	155.745	-279.888	-284.445	3.161
4800	38.528	271.547	238.298	159.595	-279.686	-284.544	3.096
4900	38.583	272.342	238.985	163.451	-279.471	-284.648	3.034
5000	38.637	273.122	239.660	167.312	-279.242	-284.755	2.975
5100	38.690	273.888	240.324	171.178	-279.002	-284.868	2.918
5200	38.742	274.640	240.976	175.050	-278.748	-284.985	2.863
5300	38.793	275.378	241.618	178.927	-278.483	-285.108	2.810
5400	38.843	276.104	242.250	182.808	-278.205	-285.236	2.759
5500	38.892	276.817	242.872	186.695	-277.915	-285.369	2.710
5600	38.940	277.518	243.485	190.587	-277.614	-285.507	2.663
5700	38.988	278.208	244.088	194.483	-277.301	-285.651	2.618
5800	39.035	278.886	244.682	198.384	-276.976	-285.800	2.574
5900	39.081	279.554	245.267	202.290	-276.639	-285.955	2.532
6000	39.127	280.211	245.844	206.200	-276.291	-286.116	2.491

PREVIOUS: June 1977 (1 atm)   CURRENT: June 1977 (1 bar)

HYDROGEN-D_1 (HD, or $^1H^2H$) IDEAL GAS M_r = 3.022042

Hydrogen-D_1 (HD) $D_1H_1(g)$

$S°(298.15\ K) = 143.803 \pm 0.033\ J\ K^{-1}\ mol^{-1}$

$\Delta_fH°(0\ K) = 0.330 \pm 0.008\ kJ\ mol^{-1}$
$\Delta_fH°(298.15\ K) = 0.321 \pm 0.013\ kJ\ mol^{-1}$

Vibrational and Rotational Levels (cm^{-1})

Direction Summmation of Electronic Ground State:

$$E = G - G_0 + F = G - G_0 + BZ - DZ^2 + HZ^3 - LZ^4 + \ldots \simeq G - G_0 + BZ - DZ^2 + H^2Z^3/(H+LZ),$$

where $Z = J(J+1)$, $Y = v+1/2$, and we omit subscript v on G, F, B, D, H, and L

$$G = 3815.403Y - 93.61504Y^2 + 1.39630Y^3 - 0.118666Y^4 + 5.97675 \times 10^{-3}Y^5 - 2.08092 \times 10^{-4}Y^6$$

$$B = 45.66910 - 2.031845Y + 7.27360 \times 10^{-2}Y^2 - 1.82824 \times 10^{-2}Y^3 + 3.06318 \times 10^{-3}Y^4$$
$$-3.00453 \times 10^{-4}Y^5 + 1.47623 \times 10^{-5}Y^6 - 2.9887 \times 10^{-7}Y^7$$

$$D = 2.6527 \times 10^{-2} - 1.1534 \times 10^{-3}Y + 1.7075 \times 10^{-4}Y^2 - 2.1142 \times 10^{-5}Y^3 + 9.9570 \times 10^{-7}Y^4$$

$$H = 2.206 \times 10^{-5} - 2.648 \times 10^{-6}Y + 3.047 \times 10^{-7}Y^2 - 1.327 \times 10^{-8}Y^3$$

$$L = 2.122 \times 10^{-8} - 3.912 \times 10^{-9}Y + 3.298 \times 10^{-10}Y^2$$

$$v_{max} = 17, \qquad J_{max} = 46 - 42\ v/v_{max}$$

Ground State Configuration: $^1\Sigma^+$ Symmetry Number = 1

Enthalpy of Formation

$\Delta_fH°(0\ K)$ is derived from $D_0^0 = 36406.0\pm0.4\ cm^{-1}$ (104.090±0.001 kcal mol^{-1}) based on absorption limits analyzed by Herzberg (1). He derived D_0^0 values of 36406.6 and 36405.8 cm^{-1} from lower and upper absorption edges. Thorson (2), by theoretical treatment of absorption-edge doubling, concluded that the value from the upper edge is more accurate. We adopt a value closer to that from the upper edge and essentially the same as that selected by the National Bureau of Standards (3).

Heat Capacity and Entropy

These are calculated by direct summation (4) over vibrational-rotational energy levels of the electronic ground state. We performed the direct summation with an extended version of a program written by W. H. Evans and provided through cooperation of D. D. Wagman, both of the U. S. National Bureau of Standards. Contributions of excited states (T_0 > 9000 cm^{-1}, 5) are negligible at 6000 K.

Vibrational-rotational levels are represented by the polynomials given above. G, B, and D are our fits of data from Dabrowski and Herzberg (5). Data from Durie and Herzberg (6), McKellar et al. (7), Brannon et al. (8) and Stoicheff (9) were also used where appropriate. Observed HD data extend to v_{max} = 17 for G and B, v = 16 for D, but only v = 1 for H. We estimate polynomials H and L by isotopic relations from those of H_2 (4) in order to provide a similar extrapolation to high J values. Our H polynomial differs from H_0 and H_1 (8, 9) by less than the experimental uncertainty. Our combination of H and L should be consistent with our adopted approximation for the infinite series (F) of rotational levels. We assume that F has the form proposed by Khachkuruzov (10) and Woolley et al. (11). We assume a linear approximation (10) for the limiting values (J_{max}) of rotational quantum number. Values in the J_{max} equation are estimated by comparison with H_2 and D_2 (4) so that J_{max} is consistent with observation for HD (5) of a rovibrational level at J = 6, v = 16.

Thermodynamic functions are calculated using 1973 fundamental constants (12), 1975 atomic weight of H, and 1973 isotopic mass of D (13). Results apply either to $^1H^2H$ or to HD containing H of natural abundance, even though the vibrational-rotational constants are for $^1H^2H$. Maximum difference between thermodynamic functions for $^1H^2H$ and HD is 0.0002 cal $K^{-1}\ mol^{-1}$ in S°(298.15 K). Our calculations agree to within 0.002 cal $K^{-1}\ mol^{-1}$ (or kcal mol^{-1}) with those of Woolley et al. (11) up to 2000 K. The more approximate $C_p°$ values of NBS (3) differ by less than 0.04 cal $K^{-1}\ mol^{-1}$ in the range 3000 to 6000 K. We omit the nuclear-spin contribution (R ln 6) to entropy and Gibbs-energy function.

References

1. G. Herzberg, J. Mol. Spectrosc. 33, 147 (1970).
2. W. R. Thorson, J. Mol. Spectrosc. 37, 199 (1971).
3. S. Abramowitz et al., U. S. Nat. Bur. Stand. Report 10904 239, July, 1972.
4. JANAF Thermochemical Tables: H_2(g), D_2(g), 3-31-77.
5. I. Dabrowski and G. Herzberg, Can. J. Phys. 54, 525 (1976).
6. R. A. Durie and G. Herzberg, Can. J. Phys. 38, 804 (1960).
7. A. R. W. McKellar, W. Goetz, and D. A. Ramsay, Astrophys. J. 207, 663 (1976); as quoted in (5).
8. P. J. Brannon, C. H. Church, and C. W. Peters, J. Mol. Spectrosc. 27, 44 (1969).
9. B. P. Stoicheff, Can. J. Phys. 35, 730 (1957).
10. G. A. Khachkuruzov, Opt. Spectrosc. 30, 455 (1971).
11. H. W. Woolley, R. B. Scott and F. G. Brickwedde, J. Res. Nat. Bur. Stand. 41, 379 (1948).
12. CODATA Task Group on Fundamental Constants, CODATA Bulletin 11, December, 1973.
13. IUPAC Commission on Atomic Weights, Pure Appl. Chem. 47, 75 (1976); 37, 589 (1974).

Enthalpy Reference Temperature = T_r = 298.15 K   Standard State Pressure = p° = 0.1 MPa

	J K⁻¹mol⁻¹			kJ mol⁻¹			
T/K	$C_p°$	S°	$-[G°-H°(T_r)]/T$	$H°-H°(T_r)$	$\Delta_fH°$	$\Delta_fG°$	Log K_f
0	0.	0.	INFINITE	-8.509	0.330	0.330	INFINITE
100	29.288	111.898	169.759	-5.786	0.178	-0.336	0.176
200	29.188	132.147	146.473	-2.865	0.275	-0.883	0.231
298.15	29.200	143.803	143.803	0.	0.320	-1.463	0.256
300	29.202	143.983	143.803	0.054	0.321	-1.474	0.257
400	29.231	152.388	144.949	2.976	0.329	-2.074	0.271
500	29.283	158.915	147.114	5.901	0.328	-2.675	0.279
600	29.395	164.263	149.539	8.834	0.323	-3.275	0.285
700	29.594	168.808	151.975	11.783	0.312	-3.874	0.289
800	29.890	172.778	154.333	14.756	0.297	-4.471	0.292
900	30.269	176.320	156.582	17.764	0.278	-5.066	0.294
1000	30.708	179.531	158.719	20.812	0.258	-5.659	0.296
1100	31.183	182.480	160.747	23.907	0.236	-6.250	0.297
1200	31.674	185.215	162.673	27.050	0.216	-6.838	0.298
1300	32.164	187.769	164.507	30.242	0.196	-7.425	0.298
1400	32.641	190.171	166.255	33.482	0.178	-8.011	0.299
1500	33.101	192.438	167.926	36.769	0.162	-8.595	0.299
1600	33.539	194.589	169.525	40.102	0.147	-9.179	0.300
1700	33.951	196.635	171.060	43.476	0.135	-9.761	0.300
1800	34.341	198.586	172.536	46.891	0.123	-10.343	0.300
1900	34.706	200.453	173.956	50.344	0.114	-10.924	0.300
2000	35.050	202.242	175.326	53.832	0.105	-11.505	0.300
2100	35.371	203.960	176.649	57.353	0.098	-12.085	0.301
2200	35.676	205.612	177.928	60.905	0.092	-12.665	0.301
2300	35.962	207.205	179.167	64.487	0.087	-13.245	0.301
2400	36.234	208.741	180.367	68.097	0.082	-13.824	0.301
2500	36.491	210.225	181.532	71.734	0.079	-14.403	0.301
2600	36.738	211.661	182.663	75.395	0.077	-14.983	0.301
2700	36.972	213.052	183.763	79.081	0.075	-15.562	0.301
2800	37.198	214.401	184.834	82.789	0.073	-16.141	0.301
2900	37.414	215.710	185.876	86.520	0.073	-16.720	0.301
3000	37.625	216.982	186.892	90.272	0.073	-17.299	0.301
3100	37.829	218.219	187.882	94.045	0.074	-17.878	0.301
3200	38.028	219.423	188.849	97.837	0.075	-18.457	0.301
3300	38.223	220.597	189.794	101.650	0.077	-19.036	0.301
3400	38.414	221.740	190.716	105.482	0.079	-19.616	0.301
3500	38.602	222.857	191.619	109.333	0.082	-20.195	0.301
3600	38.788	223.947	192.502	113.202	0.085	-20.774	0.301
3700	38.971	225.012	193.366	117.090	0.089	-21.354	0.301
3800	39.153	226.054	194.213	120.996	0.094	-21.933	0.301
3900	39.334	227.073	195.042	124.921	0.099	-22.513	0.302
4000	39.513	228.071	195.855	128.863	0.105	-23.093	0.302
4100	39.690	229.049	196.653	132.823	0.112	-23.673	0.302
4200	39.866	230.008	197.436	136.801	0.119	-24.253	0.302
4300	40.041	230.948	198.204	140.797	0.127	-24.834	0.302
4400	40.213	231.870	198.959	144.809	0.136	-25.414	0.302
4500	40.384	232.776	199.701	148.839	0.145	-25.995	0.302
4600	40.551	233.665	200.429	152.886	0.155	-26.576	0.302
4700	40.715	234.539	201.146	156.949	0.166	-27.157	0.302
4800	40.876	235.398	201.850	161.029	0.178	-27.739	0.302
4900	41.033	236.243	202.544	165.124	0.191	-28.321	0.302
5000	41.185	237.073	203.226	169.235	0.204	-28.902	0.302
5100	41.333	237.890	203.898	173.361	0.219	-29.485	0.302
5200	41.474	238.694	204.559	177.502	0.234	-30.067	0.302
5300	41.610	239.485	205.211	181.656	0.250	-30.650	0.302
5400	41.738	240.264	205.853	185.823	0.268	-31.233	0.302
5500	41.859	241.031	206.485	190.003	0.286	-31.817	0.302
5600	41.972	241.787	207.109	194.195	0.305	-32.401	0.302
5700	42.078	242.530	207.724	198.398	0.324	-32.985	0.302
5800	42.174	243.263	208.330	202.610	0.345	-33.570	0.302
5900	42.261	243.985	208.929	206.832	0.367	-34.155	0.302
6000	42.339	244.696	209.519	211.062	0.389	-34.740	0.302

PREVIOUS: June 1977 (1 atm)   CURRENT: June 1977 (1 bar)

Hydrogen-D_1 (HD) $D_1H_1(g)$

HYDROGEN-D_1, ION (HD^+) (IDEAL GAS) M_r = 3.021493

$S°(298.15\ K) = 155.552 \pm 0.04\ J\ K^{-1}\ mol^{-1}$

$\Delta_f H°(0\ K) = 1490.499 \pm 0.021\ kJ\ mol^{-1}$

$\Delta_f H°(298.15\ K) = [1496.793]\ kJ\ mol^{-1}$

Vibrational and Rotational Levels (cm^{-1})

Direct Summation using Energy-Level Equations for the $^2\Sigma_g^+$ Electronic Ground State:

$E = G - G_o + F = G - G_o + BZ - DZ^2 + HZ^3 - LZ^4 + \ldots \approx G - G_o + BZ - DZ^2 + H^2Z^3/(H+LZ)$

where Z = N(N+1), Y = v+1/2, and we omit subscript v on G, F, B, D, H, and L

$G = 2012.19\ Y - 50.5532\ Y^2 + 0.0604244\ Y^3 - 0.0163194\ Y^4$

$B = 22.4643 - 1.03566\ Y + 0.0180076\ Y^2 - 0.000633616\ Y^3$ $\quad r_e = 1.057\ Å$

$D = 1.12041\times10^{-2} - 4.24035\times10^{-4}\ Y + 1.26643\times10^{-5}\ Y^2$

$H = 7.26086\times10^{-6} - 2.19375\times10^{-7}\ Y$ and $L = 5.06679\times10^{-9}$

$v_{max} = 21$ and $N_{max} = 47 - 38v/v_{max}$

Electronic statistical weight = 2 and rotational statistical weights = 1(even and odd N)

Enthalpy of Formation

$\Delta_f H°(0\ K) = 356.238\pm0.005\ kcal\ mol^{-1}$ is obtained from that of HD(g) (1) using IP(HD) = $124568.5\pm1.2\ cm^{-1}$($356.159\pm0.004\ kcal\ mol^{-1}$). Takezawa and Tanaka (2) determined IP = $124569.5\pm0.6\ cm^{-1}$ from Rydberg spectra of HD. Similar studies on H_2 by Herzberg and Jungen (3) revealed a bias of ~$-1.0\ cm^{-1}$ attributed to a pressure shift in the spectra. We assume an equal shift for HD, adjust IP by $-1.0\ cm^{-1}$ and increase the uncertainty due to this adjustment. $\Delta_f H°(0\ K)$ is converted to $\Delta_f H°(298.15\ K)$ by use of JANAF (1) enthalpies, H°(0 K) − H°(298.15 K), for H_2, D_2, HD^+ and e^-(g). The difference in $\Delta_f H°(298.15\ K)$ between HD^+ and HD should not be interpreted as a room-temperature ionization threshold due to inclusion of these enthalpies and to threshold effects noted by Rosenstock et al. (4).

The adopted spectroscopic IP(HD) receives strong support from a value calculated indirectly from experimental values (1) of D_0^0(HD) and IP(H) combined with the theoretical value (5) of $D_0^0(HD^+)$; this indirect value differs by only 0.3 cm^{-1}. Furthermore, IP(HD) can be predicted (6) approximately from that of H_2 (or D_2) using differences in zero-point energies (1) of the diatomic molecules and their ions. Predicted values of IP(HD) deviate from the observed value by $-16\ cm^{-1}$ (based on H_2) and $+16\ cm^{-1}$(based on D_2). Inclusion of an empirical adjustment for the difference IP(D) − IP(H) overcorrects the predictions, giving deviations of $+14\ cm^{-1}$ (based on H_2) and $-14\ cm^{-1}$ (based on D_2). The average of deviations derived from H_2 and D_2 is zero, suggesting that the slight discrepancies are due to the approximate nature of the prediction. An approximate photoionization result, reviewed by Rosenstock et al. (4), deviates from the spectroscopic result by ~$130\ cm^{-1}$.

Heat Capacity and Entropy

Thermodynamic functions for ortho-para "equilibrium" HD^+ are calculated by direct summation using the energy-level equations listed above. We use 1973 fundamental constants (7) in an extended version of a computer program written by W. H. Evans and provided by D. D. Wagman of the U. S. National Bureau of Standards. Excited electronic states are neglected because they do not contribute significantly. The electronic statistical weight is taken equal to the multiplicity since we neglect the very small rotational splitting (8).

Coefficients of the energy-level polynomials are calculated from the theoretical values of H_2^+(1) using rho = 0.866117 in isotopic equations. These polynomials are confirmed by six rotation-vibration transitions ($v\leq3$, $N\leq2$) observed between 1642 and 1869 cm^{-1} by an infrared laser-resonance method (8). Our polynomials predict these transitions within 0.1 cm^{-1}. We give the polynomial coefficients, especially higher order ones, to many more digits than are justified by their accuracy. The equations are very approximate near N_{max} but, judging by H_2^+ (1), this should have little effect on the thermodynamic functions even at 6000 K. The approximate v_{max} is derived from the G polynomial.

The thermodynamic functions include contributions from quasi-bound levels lying above $D_0^0(HD^+ \rightarrow H^+ + D) = 21516\ cm^{-1}$. At 6000 K these levels contribute 0.39 cal $K^{-1}\ mol^{-1}$ to $C_p°$ and 0.048 cal $K^{-1}\ mol^{-1}$ to the Gibbs-energy function. Their contribution is negligible below 3000 K. Limiting rotational quantum numbers N_{max} are estimated from H_2^+, H_2 and HD (1) by comparing energy increments and rotational quanta found above the respective values of D°. The crudity of these estimates makes $C_p°$ at 6000 K somewhat more uncertain than in the case of H_2^+ (1).

References

1. JANAF Thermochemical Tables: H_2(g), D_2(g), H^+(g), D^+(g), e^-(g), 3-31-77; HD(g), 6-30-77; H_2^+(g), HD^+(g), D_2^+(g), 9-30-77.
2. S. Takezawa and Y. Tanaka, J. Chem. Phys. 56, 6125 (1972).
3. G. Herzberg and Ch. Jungen, J. Mol. Spectrosc. 41, 425 (1972).
4. H. M. Rosenstock, K. Draxl, B. W. Steiner and J. T. Herron, J. Phys. Chem. Ref. Data 6, Supp. 1 (1977).
5. G. Hunter and H. O. Pritchard, J. Chem. Phys. 46, 2153 (1967).
6. D. Villarejo, J. Chem. Phys. 48, 4014 (1968).
7. E. R. Cohen and B. N. Taylor, J. Phys. Chem. Ref. Data 2, 633 (1973).
8. W. H. Wing, G. A. Ruff, W. E. Lamb and J. J. Spezeski, Phys. Rev. Lett. 36, 1488 (1976).

Hydrogen-D_1, Ion (HD^+) $D_1H_1^+$(g)

Enthalpy Reference Temperature = T_r = 298.15 K Standard State Pressure = p° = 0.1 MPa

T/K	$C_p°$ (J K⁻¹mol⁻¹)	S° (J K⁻¹mol⁻¹)	−[G°−H°(T_r)]/T (J K⁻¹mol⁻¹)	H°−H°(T_r) (kJ mol⁻¹)	$\Delta_f H°$ (kJ mol⁻¹)	$\Delta_f G°$ (kJ mol⁻¹)	Log K_f
0	0.	0.	INFINITE	-8.614	1490.499		
100	29.179	123.637	181.549	-5.791			
200	29.213	143.870	158.230	-2.872			
298.15	29.334	155.552	155.552	0.	1496.793	1485.251	-260.210
300	29.339	155.733	155.552	0.054	1496.832	1485.179	-258.593
400	29.742	164.221	156.707	3.005	1498.948	1480.977	-193.396
500	30.461	170.930	158.903	6.014	1501.108	1476.235	-154.221
600	31.354	176.561	161.389	9.104	1503.338	1471.052	-128.067
700	32.281	181.464	163.914	12.286	1505.640	1465.491	-109.356
800	33.161	185.833	166.385	15.558	1508.002	1459.594	-95.302
900	33.960	189.786	168.769	18.915	1510.412	1453.399	-84.353
1000	34.672	193.402	171.054	22.347	1512.853	1446.934	-75.580
1100	35.304	196.737	173.239	25.847	1515.316	1440.223	-68.390
1200	35.865	199.833	175.328	29.406	1517.790	1433.287	-62.389
1300	36.369	202.724	177.325	33.018	1520.269	1426.144	-57.303
1400	36.825	205.436	179.237	36.678	1522.749	1418.811	-52.936
1500	37.244	207.991	181.070	40.382	1525.228	1411.300	-49.146
1600	37.633	210.407	182.829	44.126	1527.704	1403.624	-45.824
1700	38.001	212.700	184.519	47.907	1530.177	1395.794	-42.888
1800	38.351	214.882	186.146	51.725	1532.647	1387.817	-40.273
1900	38.692	216.965	187.713	55.577	1535.115	1379.704	-37.931
2000	39.024	218.958	189.226	59.463	1537.584	1371.461	-35.819
2100	39.353	220.870	190.688	63.382	1540.053	1363.094	-33.905
2200	39.681	222.708	192.102	67.334	1542.524	1354.609	-32.163
2300	40.008	224.479	193.471	71.318	1545.000	1346.012	-30.569
2400	40.335	226.189	194.799	75.335	1547.482	1337.307	-29.106
2500	40.661	227.842	196.088	79.385	1549.970	1328.499	-27.757
2600	40.984	229.443	197.340	83.467	1552.467	1319.591	-26.511
2700	41.303	230.996	198.558	87.582	1554.973	1310.587	-25.355
2800	41.614	232.503	199.744	91.728	1557.488	1301.489	-24.280
2900	41.914	233.969	200.899	95.904	1560.012	1292.302	-23.277
3000	42.199	235.395	202.025	100.110	1562.544	1283.027	-22.339
3100	42.466	236.783	203.124	104.344	1565.084	1273.668	-21.461
3200	42.710	238.135	204.197	108.603	1567.630	1264.227	-20.636
3300	42.929	239.453	205.245	112.885	1570.180	1254.707	-19.860
3400	43.119	240.737	206.270	117.187	1572.731	1245.108	-19.129
3500	43.278	241.990	207.273	121.507	1575.282	1235.435	-18.438
3600	43.403	243.211	208.254	125.842	1577.829	1225.689	-17.784
3700	43.494	244.401	209.215	130.187	1580.369	1215.873	-17.165
3800	43.549	245.562	210.157	134.539	1582.898	1205.988	-16.577
3900	43.568	246.693	211.079	138.896	1585.414	1196.036	-16.019
4000	43.551	247.796	211.983	143.252	1587.912	1186.020	-15.488
4100	43.499	248.871	212.870	147.605	1590.390	1175.942	-14.982
4200	43.412	249.918	213.740	151.950	1592.844	1165.804	-14.499
4300	43.292	250.938	214.593	156.286	1595.271	1155.608	-14.038
4400	43.140	251.932	215.430	160.608	1597.667	1145.355	-13.597
4500	42.959	252.899	216.252	164.913	1600.030	1135.048	-13.175
4600	42.751	253.841	217.059	169.199	1602.358	1124.690	-12.771
4700	42.517	254.758	217.852	173.462	1604.648	1114.281	-12.384
4800	42.260	255.651	218.630	177.701	1606.898	1103.824	-12.012
4900	41.983	256.519	219.394	181.914	1609.106	1093.320	-11.655
5000	41.687	257.365	220.145	186.097	1611.271	1082.772	-11.312
5100	41.375	258.187	220.883	190.250	1613.391	1072.181	-10.981
5200	41.049	258.987	221.608	194.372	1615.466	1061.549	-10.663
5300	40.711	259.766	222.321	198.460	1617.495	1050.877	-10.357
5400	40.363	260.524	223.021	202.514	1619.477	1040.167	-10.062
5500	40.008	261.261	223.710	206.532	1621.412	1029.421	-9.777
5600	39.645	261.979	224.387	210.515	1623.301	1018.641	-9.501
5700	39.279	262.677	225.053	214.461	1625.143	1007.827	-9.236
5800	38.909	263.357	225.707	218.371	1626.939	996.981	-8.979
5900	38.538	264.019	226.351	222.243	1628.690	986.105	-8.730
6000	38.166	264.664	226.984	226.078	1630.396	975.199	-8.490

PREVIOUS: September 1977 (1 atm) CURRENT: September 1977 (1 bar)

Hydrogen-D_1, Ion (HD^+) $D_1H_1^+$(g)

HYDROGEN-D_1, ION (HD^-) IDEAL GAS M_r = 3.022591

$S°(298.15\ K) = [151.206 \pm 0.4]\ J\ K^{-1}\ mol^{-1}$

$\Delta_f H°(0\ K) = [241.543 \pm 41.8]\ kJ\ mol^{-1}$

$\Delta_f H°(298.15\ K) = [235.484]\ kJ\ mol^{-1}$

Electronic Level and Quantum Weight

State	ϵ_i, cm^{-1}	g_i
$X^2\Sigma_u^+$	0	2

$\omega_e = [1473]\ cm^{-1}$ $\omega_e x_e = [26.3]\ cm^{-1}$ $\sigma = 2$

$B_e = [19.0]\ cm^{-1}$ $\alpha_e = [0.259]\ cm^{-1}$ $r_e = [1.15]$ Å

Enthalpy of Formation

The enthalpy of formation at 0 K for $HD^-(g)$ is based on the electron affinity value from a Rydberg-Klein-Rees configuration-interaction calculation for H_2^- (1). This study by Sharpe (1) led to $EA(H_2) = -2.5$ eV. Assuming this value is valid for $HD^-(g)$ and using auxiliary data (2), we calculate $\Delta_f H°(0\ K) = 57.73 \pm 10$ kcal mol^{-1} and $D°(HD^-) = 29.0 \pm 10$ kcal mol^{-1}; the latter value is calculated for dissociation into ground state D(g) and $H^-(g)$ or $D^-(g)$ and H(g). The adopted EA (H_2) value is suggested to be the more reliable value of eight studies considered by Rosenstock et al. (3).

In terms of molecular orbitals, the ground state for $HD^-(g)$ is the $(1s\sigma_g)^2\ (2p\sigma_u)\ ^2\Sigma_u^+$ state involving two bonding electrons and one antibonding electron. We would expect, however, the ground state to be $(1s\sigma_g)^2(1s\sigma_u)$. As stated by Massey (4), we should expect only a weakly-bound molecule as compared with HD(g) for which the ground state has no electron in an antibonding orbital. Our adopted calculated EA (HD) value is consitent with this viewpoint. Further discussion of the $HD^-(g)$ ion may be found in Massey (4).

$\Delta_f H°(0\ K)$ is converted to $\Delta_f H°(298.15\ K)$ by use of JANAF (2) enthalpies, $H°(0\ K) - H°(298.15\ K)$, for $D_2(g)$, $H_2(g)$, $DH^-(g)$, and $e^-(g)$. $\Delta_f H°(298.15\ K)$ should not be interpreted as a room temperature electron affinity due to inclusion of these enthalpies and to threshold effects discussed by Rosenstock et al (3).

Heat Capacity and Entropy

The vibrational constants for $H_2^-(g)$ are estimated to be the same as in the isoelectronic species He_2^+ (5). The vibrational constants for $HD^-(g)$ are calculated from those for $H_2^-(g)$ using isotopic relations. The internuclear distance is estimated to be ~8% larger than that for $He_2^+(g)$. The same relationship appeared to exist for the (N_2^-, NO, O_2^+) isoelectronic series. B_e is calculated from r_e whereas α_e is calculated assuming a Morse function for the ground state.

The ground and excited electronic states are discussed in detail by Massey (4). No excited state contributions are included in this calculation.

References

1. T. E. Sharpe, Lockheed Report LMSC 5-10-69-9 (1969).
2. JANAF Thermochemical Tables: H(g), $H^-(g)$, $D_2(g)$, $H_2(g)$, and $e^-(g)$, 3-31-77; $H_2^-(g)$, 9-30-77; HD(g), 6-30-77.
3. H. M. Rosenstock, K. Draxl, B. W. Steiner, and J. T. Herron, J. Phys. Chem. Ref. Data 6, Supp. 1 (1977).
4. H. S. W. Massey, "Negative Ions," 3rd ed., Cambridge University Press, Cambridge, 1976.
5. B. Rosen, Ed., Spectroscopic Data Relative to Diatomic Molecules," Pergamon Press, New York, 1970.

Hydrogen-D_1, Ion (HD^-) $D_1H_1^-(g)$

Enthalpy Reference Temperature = T_r = 298.15 K Standard State Pressure = p° = 0.1 MPa

T/K	$C_p°$ (J K^{-1}mol^{-1})	S° (J K^{-1}mol^{-1})	$-[G°-H°(T_r)]/T$ (J K^{-1}mol^{-1})	$H°-H°(T_r)$ (kJ mol^{-1})	$\Delta_f H°$ (kJ mol^{-1})	$\Delta_f G°$ (kJ mol^{-1})	Log K_f
0	0.	0.	INFINITE	-8.656	241.543		
100	29.198	119.174	177.359	-5.818			
200	29.304	139.437	153.910	-2.895			
250	29.471	145.992	151.695	-1.426			
298.15	29.771	151.206	151.206	0.	235.484	237.747	-41.652
300	29.786	151.391	151.207	0.055	235.447	237.762	-41.398
350	30.239	156.015	151.571	1.555	234.453	238.226	-35.553
400	30.787	160.087	152.386	3.081	233.480	238.832	-31.188
450	31.383	163.748	153.449	4.635	232.533	239.558	-27.807
500	31.985	167.086	154.648	6.219	231.614	240.388	-25.113
600	33.121	173.020	157.227	9.475	229.853	242.310	-21.095
700	34.102	178.201	159.861	12.838	228.178	244.520	-18.246
800	34.921	182.810	162.447	16.290	226.563	246.965	-16.125
900	35.603	186.963	164.944	19.817	224.985	249.610	-14.487
1000	36.172	190.745	167.338	23.407	223.427	252.430	-13.186
1100	36.655	194.216	169.626	27.049	221.875	255.406	-12.128
1200	37.069	197.423	171.810	30.736	220.319	258.522	-11.253
1300	37.430	200.405	173.896	34.461	218.755	261.769	-10.518
1400	37.750	203.191	175.891	38.220	217.177	265.137	-9.892
1500	38.038	205.805	177.799	42.010	215.584	268.619	-9.354
1600	38.299	208.269	179.627	45.827	213.976	272.206	-8.887
1700	38.539	210.598	181.381	49.669	212.352	275.896	-8.477
1800	38.762	212.807	183.066	53.534	210.713	279.681	-8.116
1900	38.972	214.908	184.687	57.421	209.058	283.557	-7.796
2000	39.169	216.913	186.248	61.328	207.391	287.521	-7.509
2100	39.358	218.828	187.755	65.255	205.710	291.569	-7.252
2200	39.538	220.663	189.209	69.199	204.018	295.697	-7.021
2300	39.711	222.425	190.615	73.162	202.314	299.903	-6.811
2400	39.879	224.118	191.976	77.142	200.601	304.182	-6.620
2500	40.042	225.750	193.295	81.138	198.879	308.534	-6.446
2600	40.201	227.323	194.573	85.150	197.148	312.954	-6.287
2700	40.356	228.843	195.814	89.178	195.410	317.441	-6.141
2800	40.508	230.314	197.020	93.221	193.665	321.992	-6.007
2900	40.657	231.738	198.193	97.279	191.914	326.606	-5.883
3000	40.804	233.119	199.335	101.352	190.156	331.281	-5.768
3100	40.949	234.459	200.446	105.440	188.393	336.014	-5.662
3200	41.092	235.761	201.529	109.542	186.625	340.804	-5.563
3300	41.233	237.028	202.586	113.658	184.852	345.650	-5.471
3400	41.373	238.261	203.617	117.789	183.074	350.549	-5.386
3500	41.511	239.462	204.624	121.933	181.291	355.501	-5.306
3600	41.649	240.634	205.608	126.091	179.505	360.504	-5.231
3700	41.785	241.777	206.570	130.262	177.714	365.556	-5.161
3800	41.921	242.893	207.512	134.448	175.919	370.657	-5.095
3900	42.055	243.983	208.433	138.647	174.120	375.805	-5.033
4000	42.189	245.050	209.335	142.859	172.317	380.999	-4.975
4100	42.322	246.093	210.219	147.084	170.511	386.239	-4.921
4200	42.455	247.115	211.085	151.323	168.700	391.522	-4.869
4300	42.587	248.115	211.935	155.575	166.886	396.849	-4.821
4400	42.718	249.096	212.768	159.841	165.069	402.218	-4.775
4500	42.849	250.057	213.586	164.119	163.248	407.628	-4.732
4600	42.980	251.000	214.389	168.410	161.425	413.079	-4.691
4700	43.110	251.926	215.178	172.715	159.598	418.569	-4.652
4800	43.239	252.835	215.953	177.032	157.769	424.099	-4.615
4900	43.369	253.728	216.715	181.363	155.938	429.666	-4.580
5000	43.498	254.605	217.464	185.706	154.106	435.271	-4.547
5100	43.627	255.468	218.201	190.062	152.272	440.912	-4.516
5200	43.755	256.316	218.926	194.431	150.437	446.590	-4.486
5300	43.884	257.151	219.639	198.813	148.603	452.303	-4.458
5400	44.012	257.973	220.341	203.208	146.769	458.050	-4.431
5500	44.140	258.781	221.033	207.616	144.936	463.831	-4.405
5600	44.267	259.578	221.714	212.036	143.105	469.646	-4.381
5700	44.395	260.362	222.385	216.469	141.277	475.494	-4.357
5800	44.522	261.136	223.047	220.915	139.452	481.373	-4.335
5900	44.649	261.898	223.699	225.374	137.632	487.284	-4.314
6000	44.776	262.649	224.342	229.845	135.817	493.225	-4.294

PREVIOUS: September 1977 (1 atm) CURRENT: September 1977 (1 bar)

Hydrogen-D_1, Ion (HD^-) $D_1H_1^-(g)$

WATER-D_1 (HDO) IDEAL GAS M_r = 19.021442

S°(298.15 K) = 199.512 J K^{-1} mol^{-1}

$\Delta_fH°$(0 K) = -242.435 ± 0.063 kJ mol^{-1}
$\Delta_fH°$(298.15 K) = -245.371 ± 0.063 kJ mol^{-1}

Vibrational Frequencies and Degeneracies

ν, cm^{-1}
2723.66(1)
1402.80(1)
3707.47(1)

Ground State Quantum Weight: 1 σ = 1
Point Group: C_h
Bond Distances: O-H = 0.9584 Å O-D = 0.9854 Å
Bond Angle: H-O-D = 104.45°
Product of the Moments of Inertia: $I_AI_BI_C$ = 1.59584 x 10^{-119} g^3 cm^6

Enthalpy of Formation

3rd and 2nd law (where possible) analyses of the more recent determinations (1-5) of the experimental equilibrium constants, Keq, were made for the reacitons: (A) $H_2O(g) + HD(g) = HDO(g) + H_2(g)$, and (B) $H_2O(g) + D_2O(g)$. Spectroscopic values for the heats of reaction, $\Delta_rH°$(298.15 K), of (A) and (B) were based on the zero point energies of H_2O, D_2O, and HDO given by Hulston (6) and Wolfsberg (7) and H_2, HD given by Herzberg and Monfils (8). The earlier work on (A) cited by Kirshenbaum (9) is in poor agreement except for that of Herrick, Kirshenbaum, Brown and Herrick, Crist, Davis $\Delta_rH°$(298.15 K) = 928±30, 921 kcal mol^{-1}, respectively). $\Delta_rH°$(298.15 K) for the reaction (C) $H_2O(\ell) + D_2O(\ell) = 2\ HDO(\ell)$ has been determined to be -32±1 cal K^{-1} mol^{-1} (10-12) assuming ideal solutions (see (13)) and Keq(C) ~3.8. (We calculate Keq(C) = 3.76±0.04 from Keq(B) = 3.76±0.02, $P°(H_2O)/P°(D_2O)$ = 1.151±0.006 (14) at 298.15 K, and $P°(H_2O)/P°(HDO)$ = 1.073±0.004 (15) at 298.15 K.) $\Delta_rH°$(298.15 K) of (B) was calculated from the difference in enthalpies of vaporization at 298.15 K of $D_2O(\ell) - H_2O(\ell)$ = 331±8 cal (see D_2O table), and $HDO(\ell) - H_2O(\ell)$ = 183±20 cal K^{-1} mol^{-1} (14). Values selected for $\Delta_rH°$(298.15 K) based on non-spectroscopic and spectroscopic work for (A) and (B), underlined in the table, were used to calculate values of $\Delta_fH°$(298.15 K) of HDO(g) - H_2O(g) of -845±6, -852±12, and -851±10 cal K^{-1} mol^{-1}, respectively (see D_2O and HD tables). An average value of -850±10 cal K^{-1} mol^{-1} was added to $\Delta_fH°$(298.15 K) of H_2O(g) to obtain $\Delta_fH°$(HDO, g, 298.15 K).

Source	Reaction	T/K	$\Delta_rH°$(298.15 K), kcal mol^{-1}	Drift cal K^{-1} mol^{-1}
Cerrai et al. (1)	A	324 - 1015	-915±20	0.04±0.10 (-864±30, 0.12±0.07)[a]
Suess (2)	A	353 - 473	-921± 6	0.05±0.05 (-9 55±29, 0.09±0.09)[a]
Hulston (6)	A	Spectroscopic	-928±10	
Wolfsberg (7)	A	Spectroscopic	-937±10	
Pyper, Newbury and Barton (3)	B	273.15	56±11	(3.75±0.08)[b]
		297.95	63±11	(3.74±0.04)[b]
Friedman and Shiner (4)	B	273.15	58± 3	(3.74±0.02)[b]
		298.15	60± 3	(3.76±0.02)[b]
		358.15	62± 7	(3.80±0.04)[b]
Kresge, Chiang (5)	B	298.15	46± 5	(3.85±0.03)[b]
Reaction C, see text	B	298.15	57±22	
Hulston (6)	B	Spectroscopic	65±10	
Wolfsberg (7)	B	Spectroscopic	70±10	

[a]Second law: $\Delta_rH°$(298.15 K), S°(298.15 K) (obsv.-calc.); [b]Keq at temperature cited.

Heat Capacity and Entropy

The thermodynamic functions of this table are analogous to those in the JANAF table for H_2O(g) (March 31, 1961): both tables are tablen from Friedman and Haar (16). Friedman and Haar (16) applied their non-rigid-rotor, anharmonic-oscillator treatment (with vibrational-rotational coupling terms and low-temperature rotational corrections) to the infrared-spectra analyses of Benedict et al. (17, 18). In the present table for HDO, the values of $C_p°$ of Friedman and Haar (16) between 4000 and 5000 K were extrapolated linearly (except with a term in T^{-2}) from 5000 to 6000 K. I_A = 0.1213x 10^{-39}, I_B = 0.3071x 10^{-39}, I_C = 0.4284x 10^{-39} gm cm^2 from Friedman and Haar (16). The National Bureau of Standards prepared this table (19) by critical analysis of data existing in 1972.

References

1. I. Cerrai, C. Marchetti, R. Renzoni, L. Roseo, M. Silvestri, and S. Villani, Chem. Eng. Prog. Symp. Ser. 50, (11), 271 (1954).
2. H. Suess, Z. Naturforsch. 4a, 328 (1949).
3. J. W. Pyper, R. S. Newbury, and G. W. Barton, Jrs., J. Chem. Phys. 46, 2253 (1967).
4. L. Friedman and V. J. Shiner, Jr., J. Chem. Phys. 44, 4639 (1966).
5. A. J. Kresge and Y. Chiang, J. Chem. Phys. 49, 1439 (1968).
6. J. R. Hulston, J. Chem. Phys. 50, 1483 (1969).
7. M. Wolfsberg, J. Chem. Phys. 50, 1484 (1969).
8. G. Herzberg and A. Monfils, J. Mol. Spectrosc. 5, 482 (1960).
9. I. Kirshenbaum, "Physical Properties and Analysis of Heavy Water," McGraw-Hill, N. Y., 1951.
10. E. Doehlemann and F. Lange, Physik. Chem. 173A, 295 (1935).
11. V. P. Skripov, I. V. Povyshev, Russ. J. Phys. Chem. 36, (2), 162 (1962).
12. W. C. Duer, G. L. Bertrand, J. Chem. Phys. 53, 3020 (1970).
13. W. A. Van Hook, J. Phys. Chem. 72, 1234 (1968).
14. W. M. Jones, J. Chem. Phys. 48, 207 (1968) [see C. Liu, W. T. Lindsay, Jr., J. Chem. Eng. Data 15, 510 (1970).
15. A. Narten, J. Chem. Phys. 41, 1318 (1964).
16. A. S. Friedman and L. Haar, J. Chem. Phys. 22, 2051 (1954).
17. W. S. Benedict, N. Gailer and E. K. Plyler, J. Chem. Phys. 21, 1301 (1953).
18. W. S. Benedict, H. H. Claasen and J. H. Shaw, J. Res. Nat. Bur. Stand. 49, 91 (1952).
19. S. Abramowitz et al., U. S. Nat. Bur. Stand. Rept. 10904, 239, July, 1972.

Water-D_1 (HDO) $D_1H_1O_1$(g)

Enthalpy Reference Temperature = T_r = 298.15 K Standard State Pressure = p° = 0.1 MPa

T/K	$C_p°$ (J $K^{-1}mol^{-1}$)	S° (J $K^{-1}mol^{-1}$)	-[G°-H°(T_r)]/T (J $K^{-1}mol^{-1}$)	H°-H°(T_r) (kJ mol^{-1})	$\Delta_fH°$ (kJ mol^{-1})	$\Delta_fG°$ (kJ mol^{-1})	Log K_f
0	0.	0.	INFINITE	-9.924	-242.435	-242.435	INFINITE
100	33.300	163.034	229.267	-6.623	-243.461	-240.424	125.584
200	33.363	186.126	202.590	-3.293	-244.410	-237.015	61.902
298.15	33.786	199.511	199.511	0.	-245.371	-233.181	40.852
300	33.798	199.720	199.511	0.063	-245.389	-233.106	40.587
400	34.773	209.569	200.848	3.489	-246.362	-228.863	29.886
500	36.045	217.463	203.406	7.028	-247.278	-224.381	23.441
600	37.451	224.158	206.320	10.703	-248.122	-219.722	19.128
700	38.920	230.041	209.297	14.521	-248.890	-214.926	16.038
800	40.405	235.335	212.226	18.487	-249.581	-210.026	13.713
900	41.873	240.180	215.066	22.602	-250.195	-205.044	11.900
1000	43.283	244.665	217.805	26.860	-250.737	-199.998	10.447
1100	44.622	248.854	220.439	31.256	-251.211	-194.900	9.255
1200	45.873	252.791	222.972	35.782	-251.624	-189.763	8.260
1300	47.028	256.509	225.411	40.428	-251.981	-184.593	7.417
1400	48.087	260.033	227.759	45.184	-252.289	-179.397	6.693
1500	49.053	263.384	230.023	50.042	-252.556	-174.181	6.066
1600	49.936	266.579	232.209	54.992	-252.786	-168.949	5.516
1700	50.735	269.631	234.321	60.026	-252.985	-163.702	5.030
1800	51.463	272.551	236.364	65.137	-253.159	-158.445	4.598
1900	52.124	275.352	238.343	70.317	-253.311	-153.179	4.211
2000	52.731	278.041	240.261	75.560	-253.445	-147.906	3.863
2100	53.279	280.627	242.122	80.861	-253.566	-142.626	3.548
2200	53.785	283.118	243.929	86.214	-253.675	-137.340	3.261
2300	54.246	285.519	245.686	91.616	-253.775	-132.050	2.999
2400	54.668	287.837	247.394	97.062	-253.870	-126.756	2.759
2500	55.061	290.076	249.057	102.549	-253.961	-121.457	2.538
2600	55.421	292.243	250.676	108.073	-254.048	-116.155	2.334
2700	55.752	294.341	252.255	113.632	-254.136	-110.850	2.145
2800	56.061	296.374	253.794	119.223	-254.223	-105.541	1.969
2900	56.350	298.346	255.297	124.844	-254.313	-100.230	1.805
3000	56.618	300.261	256.764	130.492	-254.405	-94.915	1.653
3100	56.869	302.122	258.197	136.167	-254.500	-89.597	1.510
3200	57.103	303.931	259.598	141.865	-254.600	-84.276	1.376
3300	57.325	305.692	260.968	147.587	-254.705	-78.952	1.250
3400	57.530	307.406	262.309	153.330	-254.815	-73.625	1.131
3500	57.727	309.077	263.622	159.093	-254.932	-68.294	1.019
3600	57.911	310.706	264.907	164.875	-255.056	-62.960	0.914
3700	58.086	312.295	266.166	170.675	-255.187	-57.622	0.813
3800	58.250	313.846	267.401	176.491	-255.326	-52.280	0.719
3900	58.409	315.361	268.611	182.324	-255.473	-46.935	0.629
4000	58.559	316.842	269.799	188.173	-255.629	-41.586	0.543
4100	58.702	318.289	270.964	194.036	-255.794	-36.233	0.462
4200	58.835	319.706	272.107	199.913	-255.968	-30.876	0.384
4300	58.965	321.092	273.230	205.803	-256.152	-25.514	0.310
4400	59.091	322.449	274.334	211.706	-256.346	-20.148	0.239
4500	59.208	323.778	275.418	217.621	-256.551	-14.778	0.172
4600	59.321	325.080	276.483	223.547	-256.766	-9.403	0.107
4700	59.430	326.357	277.531	229.485	-256.992	-4.023	0.045
4800	59.538	327.610	278.561	235.433	-257.229	1.362	-0.015
4900	59.639	328.838	279.575	241.392	-257.477	6.752	-0.072
5000	59.739	330.044	280.572	247.361	-257.735	12.147	-0.127
5100	59.831	331.228	281.554	253.340	-258.006	17.547	-0.180
5200	59.923	332.391	282.520	259.327	-258.287	22.953	-0.231
5300	60.015	333.533	283.472	265.324	-258.580	28.364	-0.280
5400	60.099	334.656	284.410	271.330	-258.883	33.781	-0.327
5500	60.187	335.759	285.333	277.344	-259.198	39.203	-0.372
5600	60.271	336.845	286.243	283.367	-259.524	44.632	-0.416
5700	60.350	337.912	287.140	289.398	-259.860	50.066	-0.459
5800	60.430	338.962	288.025	295.437	-260.206	55.507	-0.500
5900	60.505	339.996	288.897	301.484	-260.563	60.952	-0.540
6000	60.580	341.014	289.757	307.538	-260.931	66.406	-0.578

PREVIOUS: June 1977 (1 atm) CURRENT: June 1977 (1 bar)

Water-D_1 (HDO) $D_1H_1O_1$(g)

IMIDOGEN-D_1 (ND) IDEAL GAS M_r = 16.020802

$S°(298.15\ K) = 187.234\ J\ K^{-1}\ mol^{-1}$

$\Delta_fH°(0\ K) = 375.276 \pm 20.9\ kJ\ mol^{-1}$

$\Delta_fH°(298.15\ K) = 375.305 \pm 20.9\ kJ\ mol^{-1}$

Electronic State and Molecular Constants

State	ε_i, cm^{-1}	g_i	ω_e, cm^{-1}	ω_ex_e, cm^{-1}	B_e, cm^{-1}	α_e, cm^{-1}	r_e, Å
$X^3\Sigma^-$	0	3	2422	50.6	8.993	0.252	1.040

Symmetry Number = 1

Enthalpy of Formation

$\Delta_fH°(298.15\ K)$ of ND(g) minus NH(g) was calculated from the JANAF thermal functions and the zero point energies of $H_2(g)$, $D_2(g)$ [given by Herzberg and Monfils (1)], NH(g), and ND(g) [given by Haar, et al. (2)]. $\Delta_fH°(298.15\ K)$ of ND(g) was calculated from this value and the JANAF selection for $\Delta_fH°(298.15\ K)$ of NH(g) (3).

Heat Capacity and Entropy

The molecular constants which are given for NH in the JANAF Thermochemical Tables (3) were adjusted for the isotope effect. The National Bureau of Standards prepared this table (4) by critical analysis of data existing in 1972. Using molecular constants and $\Delta_fH°$ selected by NBS (4), we recalculate the table in terms of 1973 fundamental constants (5), 1975 atomic weights (6), and current JANAF reference states for the elements.

References

1. G. Herzberg and A. Monfils, J. Mol. Spectrosc. 5, 482 (1960).
2. L. Haar, A. S. Friedman, and C. W. Beckett, NBS Monograph 20, 1961.
3. JANAF Thermochemical Tables, 2nd Edition, NSRDS-NBS 37, June, 1971.
4. S. Abramowitz et al., U. S. Nat. Bur. Stand., Rept. 10904, 239, July, 1972.
5. CODATA Task Group on Fundamental Constants, CODATA Bulletin 11, December, 1973.
6. IUPAC Commission on Atomic Weights, Pure Appl. Chem. 47, 75 (1976); 37, 589 (1974).

Imidogen-D_1 (ND) $D_1N_1(g)$

Enthalpy Reference Temperature = T_r = 298.15 K Standard State Pressure = p° = 0.1 MPa

T/K	$C_p°$	S°	$-[G°-H°(T_r)]/T$	$H°-H°(T_r)$	$\Delta_fH°$	$\Delta_fG°$	Log K_f
	J K⁻¹mol⁻¹			kJ mol⁻¹			
0	0.	0.	INFINITE	-8.648	375.276	375.276	INFINITE
100	29.118	155.412	213.138	-5.773	375.325	373.414	-195.052
200	29.131	175.599	189.900	-2.860	375.306	371.515	-97.030
250	29.140	182.100	187.714	-1.403	375.305	370.567	-77.426
298.15	29.159	187.234	187.234	0.	375.305	369.655	-64.762
300	29.160	187.414	187.234	0.054	375.305	369.620	-64.357
350	29.208	191.912	187.590	1.513	375.305	368.673	-55.021
400	29.300	195.818	188.379	2.975	375.307	367.725	-48.020
450	29.449	199.277	189.402	4.444	375.311	366.777	-42.574
500	29.655	202.390	190.547	5.921	375.318	365.828	-38.218
600	30.213	207.844	192.988	8.913	375.345	363.928	-31.683
700	30.891	212.551	195.454	11.968	375.388	362.022	-27.014
800	31.604	216.722	197.857	15.093	375.445	360.109	-23.513
900	32.298	220.485	200.165	18.288	375.514	358.188	-20.789
1000	32.943	223.922	202.372	21.551	375.589	356.259	-18.609
1100	33.528	227.090	204.477	24.875	375.668	354.322	-16.825
1200	34.050	230.030	206.485	28.254	375.749	352.378	-15.339
1300	34.514	232.774	208.403	31.683	375.829	350.427	-14.080
1400	34.925	235.347	210.236	35.155	375.908	348.470	-13.002
1500	35.290	237.770	211.992	38.666	375.986	346.508	-12.066
1600	35.615	240.058	213.675	42.212	376.061	344.540	-11.248
1700	35.905	242.226	215.292	45.788	376.134	342.568	-10.526
1800	36.166	244.286	216.846	49.392	376.204	340.591	-9.884
1900	36.401	246.247	218.342	53.020	376.271	338.611	-9.309
2000	36.614	248.120	219.784	56.671	376.337	336.627	-8.792
2100	36.809	249.911	221.177	60.343	376.400	334.640	-8.324
2200	36.987	251.628	222.522	64.033	376.461	332.650	-7.898
2300	37.152	253.276	223.823	67.740	376.519	330.657	-7.509
2400	37.304	254.860	225.084	71.463	376.576	328.662	-7.153
2500	37.446	256.386	226.306	75.200	376.631	326.664	-6.825
2600	37.579	257.857	227.491	78.952	376.684	324.665	-6.523
2700	37.703	259.278	228.642	82.716	376.735	322.663	-6.242
2800	37.821	260.651	229.761	86.492	376.784	320.659	-5.982
2900	37.932	261.980	230.849	90.280	376.831	318.654	-5.740
3000	38.038	263.268	231.908	94.078	376.877	316.647	-5.513
3100	38.139	264.517	232.940	97.887	376.920	314.639	-5.302
3200	38.235	265.729	233.946	101.706	376.962	312.629	-5.103
3300	38.328	266.907	234.927	105.534	377.001	310.618	-4.917
3400	38.417	268.053	235.885	109.371	377.039	308.606	-4.741
3500	38.503	269.167	236.820	113.217	377.075	306.593	-4.576
3600	38.586	270.253	237.733	117.072	377.108	304.578	-4.419
3700	38.667	271.312	238.627	120.934	377.140	302.563	-4.271
3800	38.745	272.344	239.500	124.805	377.169	300.547	-4.131
3900	38.822	273.351	240.356	128.683	377.195	298.531	-3.998
4000	38.896	274.335	241.193	132.569	377.219	296.513	-3.872
4100	38.969	275.296	242.013	136.463	377.241	294.495	-3.752
4200	39.040	276.236	242.817	140.363	377.260	292.477	-3.637
4300	39.110	277.156	243.605	144.271	377.276	290.458	-3.528
4400	39.178	278.056	244.377	148.185	377.290	288.439	-3.424
4500	39.246	278.937	245.136	152.106	377.302	286.419	-3.325
4600	39.312	279.800	245.880	156.034	377.310	284.400	-3.229
4700	39.377	280.646	246.611	159.969	377.316	282.380	-3.138
4800	39.441	281.476	247.328	163.910	377.320	280.360	-3.051
4900	39.505	282.290	248.034	167.857	377.322	278.340	-2.967
5000	39.567	283.089	248.727	171.810	377.321	276.320	-2.887
5100	39.629	283.873	249.408	175.770	377.318	274.300	-2.809
5200	39.690	284.643	250.078	179.736	377.313	272.280	-2.735
5300	39.751	285.400	250.738	183.708	377.306	270.260	-2.664
5400	39.811	286.143	251.386	187.686	377.298	268.240	-2.595
5500	39.870	286.874	252.025	191.670	377.289	266.221	-2.528
5600	39.929	287.593	252.654	195.660	377.279	264.202	-2.464
5700	39.987	288.300	253.273	199.656	377.269	262.182	-2.403
5800	40.045	288.996	253.883	203.658	377.258	260.163	-2.343
5900	40.103	289.681	254.484	207.665	377.247	258.144	-2.285
6000	40.160	290.356	255.076	211.678	377.236	256.126	-2.230

PREVIOUS: June 1977 (1 atm) CURRENT: June 1977 (1 bar)

Imidogen-D_1 (ND) $D_1N_1(g)$

HYDROXYL-D1 (OD) IDEAL GAS M_r = 18.013502

$S°(298.15\ K) = 189.651\ J\ K^{-1}\ mol^{-1}$

$\Delta_fH°(0\ K) = 36.021 \pm 1.3\ kJ\ mol^{-1}$
$\Delta_fH°(298.15\ K) = 36.602 \pm 1.21\ kJ\ mol^{-1}$

Electronic States and Molecular Constants

State	ε_i, cm^{-1}	g_i	ω_e, cm^{-1}	ω_ex_e, cm^{-1}	B_e, cm^{-1}	α_e, cm^{-1}	r_e, Å
$X^2\Pi_i$	0	2	2720.9	44.2	10.02	0.29	0.970
	139.7	2					
$A^2\Sigma^+$	35474	2	2322.6	55.4	9.198	0.322	1.012
$B^2\Sigma^+$	68769	2	684.3	55.6	2.91	0.25	1.80

Symmetry Number = 1

Enthalpy of Formation

The enthalpy of formation was calculated from the selected value for $\Delta_fH°(298.15\ K)$ of OH(g), $-0.27\pm3\ kcal\ mol^{-1}$ (8), the appropriate thermal functions, and zero point energies of $H_2(g)$, $D_2(g)$, OH(g), and OD(g). The zero point energies of $H_2(g)$ and $D_2(g)$ were taken from Herzberg and Monfils (1). The zero point energies of OH and OD include the Dunham correction (see Herzberg and Monfils (1)). The molecular constants are those given by Rosen (2) with the exception of the spin-coupling constants taken from Herzberg (3).

Heat Capacity and Entropy

The vibrational and rotational constants of the respective electronic levels were taken from Rosen (2). Comparison of the results with those from a more exact treatment given by Haar et al. (4) indicates errors in the table above 400 K are negligible. Below this, they may be appreciable. In particular, it is recommended that H°(0 K) − H°(298.15 K), S°(298.15 K) and $C_p°$(298.15 K) be taken as $-2.151\ kcal\ mol^{-1}$, $45.307\ cal\ K^{-1}\ mol^{-1}$, and $7.156\ cal\ K^{-1}\ mol^{-1}$, respectively. These errors result from dealing with the ground state ($X^2\Pi_i$) as two different electronic states separated by $139.7\ cm^{-1}$.

The National Bureau of Standards prepared this table (5) by critical analysis of data existing in 1972. Using molecular constants and $\Delta_fH°$ selected by NBS (5), we recalculate the table in terms of 1973 fundamental constants (6), 1975 atomic weights (7), and current JANAF reference states for the elements.

References

1. G. Herzberg and A. Monfils, J. Mol. Spectrosc. 5, 482 (1960).
2. B. Rosen, Ed., "Spectroscopic Data Relative to Diatomic Molecules," Pergamon Press, New York, 1970.
3. G. Herzberg, "Spectra of Diatomic Molecules," D. Van Nostrand Co., Inc., New York, pp. 232, 561, 1950.
4. L. Haar, A. S. Friedman, and C. W. Beckett, U. S. Nat. Bur. Stand. Monograph 20, 1961.
5. S. Abramowitz et al., U. S. Nat. Bur. Stand., Rept. 10904, 239, July, 1972.
6. CODATA Task Group on Fundamental Constants, CODATA Bulletin 11, December, 1973.
7. IUPAC Commission on Atomic Weights, Pure Appl. Chem. 47, 75 (1976); 37, 589 (1974).
8. JANAF Thermochemical Tables: OH(g), 6-30-77.

Hydroxyl-D_1 (OD) $D_1O_1(g)$

Enthalpy Reference Temperature = T_r = 298.15 K; Standard State Pressure = p° = 0.1 MPa

T/K	$C_p°$ (J K⁻¹mol⁻¹)	S° (J K⁻¹mol⁻¹)	−[G°−H°(T_r)]/T (J K⁻¹mol⁻¹)	H°−H°(T_r) (kJ mol⁻¹)	$\Delta_fH°$ (kJ mol⁻¹)	$\Delta_fG°$ (kJ mol⁻¹)	Log K_f
0	0.	0.	INFINITE	-9.207	36.021	36.021	INFINITE
100	32.617	155.537	216.921	-6.138	36.262	35.013	-18.289
200	30.775	177.535	192.414	-2.976	36.492	33.664	-8.792
250	30.282	184.345	190.147	-1.450	36.559	32.949	-6.884
298.15	29.989	189.651	189.651	0.	36.602	32.249	-5.650
300	29.980	189.837	189.652	0.055	36.603	32.222	-5.610
350	29.798	194.443	190.016	1.549	36.629	31.490	-4.700
400	29.705	198.415	190.823	3.037	36.638	30.755	-4.016
450	29.688	201.912	191.865	4.521	36.632	30.020	-3.485
500	29.740	205.042	193.029	6.007	36.614	29.286	-3.059
600	30.026	210.487	195.498	8.993	36.546	27.826	-2.422
700	30.491	215.148	197.980	12.018	36.454	26.380	-1.969
800	31.056	219.256	200.388	15.095	36.350	24.948	-1.629
900	31.657	222.949	202.693	18.230	36.244	23.529	-1.366
1000	32.252	226.315	204.889	21.426	36.141	22.122	-1.156
1100	32.815	229.416	206.980	24.680	36.044	20.725	-0.984
1200	33.335	232.293	208.971	27.987	35.953	19.336	-0.842
1300	33.809	234.981	210.869	31.345	35.868	17.955	-0.721
1400	34.237	237.502	212.682	34.748	35.787	16.580	-0.619
1500	34.622	239.878	214.417	38.191	35.710	15.211	-0.530
1600	34.968	242.123	216.079	41.671	35.635	13.847	-0.452
1700	35.279	244.253	217.674	45.183	35.561	12.487	-0.384
1800	35.559	246.277	219.208	48.726	35.486	11.132	-0.323
1900	35.813	248.207	220.683	52.294	35.410	9.781	-0.269
2000	36.043	250.050	222.106	55.887	35.330	8.434	-0.220
2100	36.252	251.813	223.479	59.502	35.246	7.092	-0.176
2200	36.444	253.504	224.806	63.137	35.158	5.753	-0.137
2300	36.619	255.128	226.089	66.790	35.064	4.419	-0.100
2400	36.782	256.690	227.332	70.461	34.965	3.088	-0.067
2500	36.932	258.195	228.536	74.146	34.858	1.762	-0.037
2600	37.071	259.646	229.705	77.846	34.745	0.440	-0.009
2700	37.201	261.047	230.840	81.560	34.625	-0.877	0.017
2800	37.323	262.403	231.943	85.286	34.497	-2.189	0.041
2900	37.438	263.714	233.016	89.025	34.362	-3.497	0.063
3000	37.546	264.985	234.061	92.774	34.219	-4.800	0.084
3100	37.648	266.218	235.078	96.534	34.069	-6.098	0.103
3200	37.745	267.415	236.070	100.303	33.911	-7.392	0.121
3300	37.838	268.578	237.038	104.082	33.746	-8.680	0.137
3400	37.926	269.709	237.982	107.871	33.573	-9.963	0.153
3500	38.011	270.809	238.904	111.668	33.392	-11.241	0.168
3600	38.092	271.881	239.806	115.473	33.204	-12.513	0.182
3700	38.171	272.926	240.687	119.286	33.009	-13.781	0.195
3800	38.246	273.945	241.549	123.107	32.806	-15.042	0.207
3900	38.320	274.940	242.392	126.935	32.596	-16.299	0.218
4000	38.391	275.911	243.218	130.771	32.378	-17.550	0.229
4100	38.460	276.859	244.027	134.613	32.153	-18.795	0.239
4200	38.528	277.787	244.820	138.463	31.921	-20.035	0.249
4300	38.594	278.694	245.597	142.319	31.682	-21.269	0.258
4400	38.659	279.582	246.359	146.181	31.435	-22.498	0.267
4500	38.722	280.452	247.107	150.050	31.181	-23.721	0.275
4600	38.784	281.304	247.842	153.926	30.920	-24.938	0.283
4700	38.846	282.138	248.562	157.807	30.652	-26.150	0.291
4800	38.907	282.957	249.270	161.695	30.376	-27.355	0.298
4900	38.967	283.760	249.966	165.589	30.093	-28.555	0.304
5000	39.026	284.548	250.650	169.488	29.802	-29.749	0.311
5100	39.085	285.321	251.322	173.394	29.504	-30.937	0.317
5200	39.144	286.081	251.983	177.305	29.199	-32.119	0.323
5300	39.202	286.827	252.634	181.223	28.886	-33.295	0.328
5400	39.261	287.560	253.274	185.146	28.565	-34.466	0.333
5500	39.319	288.281	253.904	189.075	28.237	-35.630	0.338
5600	39.378	288.990	254.524	193.010	27.902	-36.788	0.343
5700	39.437	289.687	255.135	196.950	27.558	-37.940	0.348
5800	39.496	290.374	255.736	200.897	27.207	-39.086	0.352
5900	39.555	291.049	256.329	204.849	26.849	-40.227	0.356
6000	39.615	291.715	256.913	208.808	26.482	-41.360	0.360

PREVIOUS: June 1977 (1 atm) CURRENT: June 1977 (1 bar)

Hydroxyl-D_1 (OD) $D_1O_1(g)$

MERCAPTO-D1 (SD) IDEAL GAS M_r = 34.074102

$S°(298.15\ K) = 201.484\ J\ K^{-1}\ mol^{-1}$

$\Delta_f H°(0\ K) = 137.893 \pm 5.0\ kJ\ mol^{-1}$
$\Delta_f H°(298.15\ K) = 138.490 \pm 5.0\ kJ\ mol^{-1}$

Electronic States and Molecular Constants (σ = 1)

State	ε_i, cm^{-1}	g_i	ω_e, cm^{-1}	$\omega_e x_e$, cm^{-1}	B_e, cm^{-1}	α_e, cm^{-1}	r_e, Å
$X^2\Pi_i$	0	2	1930.4	23.44	4.949	0.101	1.345
	376.8	2					
$A^2\Sigma^+$	30769	2	1417	48.85	4.392	0.172	1.423
$B^2\Sigma$	59566	2	1917.7	29.3	4.532	0.105	1.405
$C^2\Delta$	63872	4					
$D^2\Delta$	71205	4					
$E^2\Sigma$	71328	2	[1930.4]	[23.44]	[4.949]	[0.101]	
$F^2\Delta$	76717	4					
$G^2\Delta$	79320	4					
$H^2\Delta$	80858	4					

Enthalpy of Formation

The enthalpy of formation was calculated from the appropriate thermal functions (see tables for HS, DS, H_2, D_2, 1), the selected value for $\Delta_f H°(298.15\ K)$ of HS(g) (33.3±1.2 kcal mol^{-1}, see table for HS(g), 1), and the zero point energies of H_2(g) (2), D_2(g) (2), HS(g) (3), and DS(g) (3). The Dunham corrections were made in ref. (2) for H_2(g) and D_2(g). Spectroscopic constants tabulated in ref. (3) were used to calculate the zero point energies of HS and DS including Dunham corrections.

Heat Capacity and Entropy

The vibrational and rotational constants of the respective electronic levels were taken from Rosen (3). From a comparison of this approximate calculation with more exact methods (see HS(g) table, 1) it is concluded that the errors are negligible above 400 K. Below this, they may be appreciable. In particular, it is recommended that H°(0 K) - H°(298.15 K), S°(298.15 K) and $C_p°$(298.15 K) be taken as -2.171 kcal mol^{-1}, 48.138 cal K^{-1} mol^{-1}, and 7.760 cal K^{-1} mol^{-1}, respectively. These errors result from dealing with the ground state ($^2\Pi_i$) as two different electronic states separated by 376.8 cm^{-1}.

The National Bureau of Standards prepared this table (4) by critical analysis of data existing in 1972. Using molecular constants and $\Delta_f H°$ selected by NBS (4), we recalculate the table in terms of 1973 fundamental constants (5), 1975 atomic weights (6), and current JANAF reference states for the elements. Our calculated $C_p°$ values agree with the NBS table up to 4000 K; above 4000 K our values trend slightly higher and at 6000 K the difference from the original NBS table is 0.017 cal K^{-1} mol^{-1}.

References

1. JANAF Thermochemical Tables: HS(g), DS(g), 6-30-77; H_2(g), D_2(g), 3-31-77.
2. G. Herzberg and A. Monfils, J. Mol. Spectrosc. 5, 482 (1960).
3. B. Rosen, "Spectroscopic Data Relative to Diatomic Molecules," Pergamon Press, New York, 1970. For spin-coupling constants, see L. Haar, A. S. Friedman, and C. W. Beckett, U. S. Nat. Bur. Stand. Monograph 20, 1961.
4. S. Abramowitz et al., U. S. Nat. Bur. Stand., Rept. 109C4, 239, July, 1972.
5. CODATA Task Group on Fundamental Constants, CODATA Bulletin 11, December, 1973.
6. IUPAC Commission on Atomic Weights, Pure Appl. Chem. 47, 75 (1976); 37, 589 (1974).

Mercapto-D_1 (SD) D_1S_1(g)

Enthalpy Reference Temperature = T_r = 298.15 K Standard State Pressure = p° = 0.1 MPa

T/K	$C_p°$ (J $K^{-1}mol^{-1}$)	S° (J $K^{-1}mol^{-1}$)	-[G°-H°(T_r)]/T (J $K^{-1}mol^{-1}$)	H°-H°(T_r) (kJ mol^{-1})	$\Delta_f H°$ (kJ mol^{-1})	$\Delta_f G°$ (kJ mol^{-1})	Log K_f
0	0.	0.	INFINITE	-9.294	137.893	137.893	INFINITE
100	30.185	166.532	230.367	-6.383	138.738	128.977	-67.371
200	32.699	188.421	204.473	-3.210	138.793	119.167	-31.123
250	32.752	195.733	202.021	-1.572	138.682	114.271	-23.876
298.15	32.520	201.484	201.484	0.	138.490	109.585	-19.199
300	32.509	201.685	201.484	0.060	138.482	109.406	-19.049
350	32.249	206.676	201.879	1.679	138.204	104.581	-15.608
400	32.086	210.970	202.753	3.287	135.650	99.888	-13.044
450	32.049	214.746	203.880	4.890	134.596	95.472	-11.082
500	32.125	218.126	205.139	6.493	133.464	91.188	-9.526
600	32.510	224.013	207.808	9.723	131.635	82.911	-7.218
700	33.047	229.064	210.492	13.000	130.075	74.916	-5.590
800	33.613	233.514	213.097	16.334	128.685	67.133	-4.383
900	34.148	237.504	215.591	19.722	74.278	60.598	-3.517
1000	34.630	241.127	217.966	23.161	74.294	59.077	-3.086
1100	35.055	244.448	220.225	26.646	74.311	57.554	-2.733
1200	35.425	247.515	222.373	30.171	74.324	56.030	-2.439
1300	35.747	250.363	224.418	33.730	74.329	54.505	-2.190
1400	36.029	253.023	226.367	37.319	74.324	52.981	-1.977
1500	36.276	255.517	228.228	40.934	74.307	51.457	-1.792
1600	36.493	257.866	230.008	44.573	74.276	49.934	-1.630
1700	36.687	260.084	231.712	48.232	74.231	48.414	-1.488
1800	36.860	262.186	233.347	51.909	74.173	46.897	-1.361
1900	37.016	264.183	234.918	55.603	74.102	45.384	-1.248
2000	37.157	266.085	236.429	59.312	74.018	43.874	-1.146
2100	37.286	267.901	237.885	63.034	73.922	42.370	-1.054
2200	37.405	269.639	239.289	66.769	73.815	40.870	-0.970
2300	37.514	271.304	240.645	70.515	73.697	39.375	-0.894
2400	37.616	272.903	241.956	74.272	73.570	37.885	-0.825
2500	37.712	274.440	243.225	78.038	73.433	36.401	-0.761
2600	37.802	275.921	244.454	81.814	73.287	34.923	-0.702
2700	37.886	277.349	245.646	85.598	73.135	33.450	-0.647
2800	37.967	278.729	246.803	89.391	72.974	31.983	-0.597
2900	38.044	280.062	247.927	93.191	72.807	30.522	-0.550
3000	38.117	281.353	249.020	97.000	72.633	29.067	-0.506
3100	38.188	282.604	250.083	100.815	72.453	27.617	-0.465
3200	38.256	283.818	251.119	104.637	72.267	26.174	-0.427
3300	38.322	284.996	252.128	108.466	72.074	24.737	-0.392
3400	38.386	286.141	253.111	112.301	71.876	23.305	-0.358
3500	38.448	287.255	254.071	116.143	71.672	21.880	-0.327
3600	38.509	288.339	255.008	119.991	71.462	20.460	-0.297
3700	38.569	289.394	255.923	123.845	71.246	19.046	-0.269
3800	38.628	290.424	256.817	127.705	71.023	17.638	-0.242
3900	38.686	291.428	257.692	131.570	70.794	16.236	-0.217
4000	38.744	292.408	258.548	135.442	70.559	14.841	-0.194
4100	38.801	293.365	259.385	139.319	70.316	13.451	-0.171
4200	38.858	294.301	260.205	143.202	70.067	12.066	-0.150
4300	38.915	295.216	261.009	147.091	69.811	10.689	-0.130
4400	38.972	296.111	261.797	150.985	69.547	9.317	-0.111
4500	39.029	296.988	262.569	154.885	69.276	7.951	-0.092
4600	39.087	297.846	263.327	158.791	68.998	6.591	-0.075
4700	39.145	298.688	264.070	162.702	68.712	5.238	-0.058
4800	39.204	299.512	264.800	166.620	68.419	3.890	-0.042
4900	39.264	300.321	265.517	170.543	68.118	2.549	-0.027
5000	39.324	301.115	266.221	174.473	67.810	1.214	-0.013
5100	39.385	301.895	266.913	178.408	67.495	-0.115	0.001
5200	39.448	302.660	267.593	182.350	67.172	-1.438	0.014
5300	39.511	303.412	268.261	186.298	66.843	-2.754	0.027
5400	39.576	304.151	268.919	190.252	66.508	-4.064	0.039
5500	39.642	304.878	269.566	194.213	66.166	-5.368	0.051
5600	39.709	305.593	270.203	198.180	65.819	-6.665	0.062
5700	39.778	306.296	270.830	202.155	65.466	-7.956	0.073
5800	39.848	306.989	271.448	206.136	65.109	-9.242	0.083
5900	39.920	307.670	272.056	210.125	64.747	-10.520	0.093
6000	39.993	308.342	272.655	214.120	64.382	-11.793	0.103

PREVIOUS: June 1977 (1 atm) CURRENT: June 1977 (1 bar)

Mercapto-D_1 (SD) D_1S_1(g)

DEUTERIUM (D_2) REFERENCE STATE - IDEAL GAS M_r = 4.028204

0 to 6000 K Ideal Gas

D_0^0 = 439.615 ± 0.005 kJ mol^{-1}

S°(298.15 K) = 144.960 ± 0.04 J K^{-1} mol^{-1}

$\Delta_f H°$(0 K) = 0 kJ mol^{-1}

$\Delta_f H°$(298.15 K) = 0 kJ mol^{-1}

Vibrational and Rotational Levels (cm^{-1})

Direct Summation of Electronic Ground State

$E = G - G_0 + F = G - G_0 + BZ - DZ^2 + HZ^3 - LZ^4 + \ldots \simeq G - G_0 + BZ - DZ^2 + H^2Z^3/(H+LZ)$,

where $Z = J(J+1)$, $Y = v+1/2$, and we omit subscript v on G, F, B, D, H, and L

$G = 3116.334Y - 62.50516Y^2 + 0.7797565Y^3 - 5.48262\times10^{-2}Y^4 + 2.25748\times10^{-3}Y^5 - 6.30154\times10^{-5}Y^6$

$B = 30.45273 - 1.102045Y + 0.0295505Y^2 - 5.99912\times10^{-3}Y^3 + 8.37740\times10^{-4}Y^4$
$-6.87040\times10^{-5}Y^5 + 2.80134\times10^{-6}Y^6 - 4.68952\times10^{-8}Y^7$

$D = 0.011525 - 3.6876\times10^{-4}Y + 4.1199\times10^{-5}Y^2 - 4.4723\times10^{-6}Y^3 + 1.7858\times10^{-7}Y^4$

$H = 6.55\times10^{-6} - 6.42\times10^{-7}Y + 6.03\times10^{-8}Y^2 - 2.15\times10^{-9}Y^3$

$L = 4.20\times10^{-9} - 6.32\times10^{-10}Y + 4.35\times10^{-11}Y^2$

$v_{max} = 21$, $J_{max} = 54 - 47v/v_{max}$

Normalized statistical weights = 2/3 (even J) and 1/3 (odd J)

Ground State Configuration: $^1\Sigma_g^+$ r_e = 0.7415 Å

Enthalpy of Formation

Zero by definition. Refer to the monatomic deuterium gas table for a discussion of the dissociation energy.

Heat Capacity and Entropy

These are calculated by direct summation (1) over vibration-rotation energy levels of the electronic ground state. We performed the direct summation with an extended version of a program written by W. H. Evans and provided through cooperation of D. D. Wagman, both of the U. S. National Bureau of Standards. Contributions of excited states (T_0 > 90000 cm^{-1}) are negligible at 6000 K. Polynomials G, B and D are our fits of data from Stoicheff (2) and Bredohl and Herzberg (3). Polynomials H and L are estimated by isotopic relations from those of H_2 (1) in order to provide a similar extrapolation to high J values. The adopted H polynomial is not more accurate than the relatively scattered data (2-4), but it should be more consistent with our adopted approximation for the infinite-series (F) equation for rotational levels. We assume that F has the form proposed by Khachkuruzov (5) and Woolley et al. (6). We use a linear approximation for the limiting values (J_{max}) of rotational quantum number. The J_{max} equation is estimated; its second value has been revised from 52 (in the 1977 table) to 47 in order to yield $C_p°$ and S° values that agree more closely with theory (8) near 6000 K. We omit the nuclear-spin contribution (R ℓn 9) to entropy and Gibbs energy function.

We adopt ortho-para "equilibrium" D_2 as the reference state at all temperatures. Values differ from "normal" D_2 only at 100 K (6, 7). Our values can be compared with the theoretical results of Kosloff et al. (8) and Gurvich et al. (9) and with the earlier table of NBS (7). Our S° values deviate by 0.001 (8) or -0.005 (9) cal K^{-1} mol^{-1} at 6000 K. Our $C_p°$ values deviate by 0.011 (8), -0.002 (9) or 0.002 (7) at 5000 K and by 0.001 (8) or -0.005 (9) cal K^{-1} mol^{-1} at 6000 K.

References

1. JANAF Thermochemical Tables: H_2(g), 3-31-77.
2. B. P. Stoicheff, Can. J. Phys. 35, 730 (1957).
3. H. Bredohl and G. Herzberg, Can. J. Phys. 51, 867 (1973).
4. P. J. Brannon, C. H. Church, and C. W. Peters, J. Mol. Spectrosc. 27, 44 (1968).
5. G. A. Khachkuruzov, Opt. Spectrosc. 30, 455 (1971).
6. H. W. Woolley, R. B. Scott, and F. G. Brickwedde, J. Res. Natl. Bur. Stand. 41, 379 (1948).
7. S. Abramowitz et al., U. S. Natl. Bur. Stand., Rept. 10904, 239, July, 1972.
8. R. Rosloff, R. D. Levine and R. B. Bernstein, Mol. Phys. 27, 981 (1974).
9. L. V. Gurvich, I. V. Veits et al., "Thermodynamic Properties of Individual Substances", 3rd ed., Vol. I, Nauka, Moscow, 1978.

Deuterium (D_2) D_2(ref)

Enthalpy Reference Temperature = T_r = 298.15 K Standard State Pressure = p° = 0.1 MPa

T/K	$C_p°$ (J K^{-1}mol^{-1})	S°	$-[G°-H°(T_r)]/T$	$H°-H°(T_r)$ (kJ mol^{-1})	$\Delta_f H°$	$\Delta_f G°$	Log K_f
0	0.	0.	INFINITE	-8.569	0.	0.	0.
100	30.317	112.789	170.973	-5.818	0.	0.	0.
200	29.204	133.306	147.630	-2.865	0.	0.	0.
250	29.185	139.819	145.441	-1.405	0.	0.	0.
298.15	29.194	144.960	144.960	0.	0.	0.	0.
300	29.195	145.141	144.961	0.054	0.	0.	0.
350	29.213	149.642	145.316	1.514	0.	0.	0.
400	29.242	153.545	146.106	2.975	0.	0.	0.
450	29.290	156.992	147.128	4.439	0.	0.	0.
500	29.366	160.081	148.272	5.905	0.	0.	0.
600	29.619	165.456	150.701	8.853	0.	0.	0.
700	30.008	170.049	153.144	11.833	0.	0.	0.
800	30.502	174.087	155.515	14.858	0.	0.	0.
900	31.057	177.712	157.783	17.936	0.	0.	0.
1000	31.636	181.014	159.943	21.070	0.	0.	0.
1100	32.210	184.056	161.999	24.263	0.	0.	0.
1200	32.762	186.882	163.956	27.512	0.	0.	0.
1300	33.282	189.526	165.823	30.814	0.	0.	0.
1400	33.766	192.010	167.605	34.167	0.	0.	0.
1500	34.213	194.355	169.311	37.566	0.	0.	0.
1600	34.625	196.576	170.946	41.008	0.	0.	0.
1700	35.004	198.687	172.517	44.490	0.	0.	0.
1800	35.353	200.698	174.027	48.008	0.	0.	0.
1900	35.676	202.618	175.481	51.560	0.	0.	0.
2000	35.975	204.456	176.885	55.142	0.	0.	0.
2100	36.252	206.218	178.240	58.754	0.	0.	0.
2200	36.512	207.910	179.550	62.392	0.	0.	0.
2300	36.755	209.539	180.819	66.056	0.	0.	0.
2400	36.984	211.108	182.048	69.743	0.	0.	0.
2500	37.201	212.622	183.241	73.452	0.	0.	0.
2600	37.407	214.085	184.400	77.182	0.	0.	0.
2700	37.605	215.501	185.525	80.933	0.	0.	0.
2800	37.795	216.872	186.620	84.703	0.	0.	0.
2900	37.978	218.201	187.687	88.492	0.	0.	0.
3000	38.156	219.492	188.725	92.298	0.	0.	0.
3100	38.329	220.746	189.738	96.123	0.	0.	0.
3200	38.498	221.965	190.726	99.964	0.	0.	0.
3300	38.665	223.152	191.691	103.822	0.	0.	0.
3400	38.829	224.309	192.633	107.697	0.	0.	0.
3500	38.991	225.437	193.555	111.588	0.	0.	0.
3600	39.152	226.538	194.456	115.495	0.	0.	0.
3700	39.312	227.613	195.337	119.419	0.	0.	0.
3800	39.470	228.663	196.200	123.358	0.	0.	0.
3900	39.628	229.690	197.046	127.313	0.	0.	0.
4000	39.785	230.696	197.875	131.283	0.	0.	0.
4100	39.941	231.680	198.687	135.270	0.	0.	0.
4200	40.096	232.644	199.484	139.271	0.	0.	0.
4300	40.250	233.590	200.267	143.289	0.	0.	0.
4400	40.402	234.517	201.035	147.321	0.	0.	0.
4500	40.553	235.426	201.789	151.369	0.	0.	0.
4600	40.701	236.319	202.530	155.432	0.	0.	0.
4700	40.847	237.196	203.258	159.509	0.	0.	0.
4800	40.989	238.058	203.974	163.601	0.	0.	0.
4900	41.127	238.904	204.678	167.707	0.	0.	0.
5000	41.262	239.736	205.371	171.826	0.	0.	0.
5100	41.391	240.555	206.053	175.959	0.	0.	0.
5200	41.515	241.360	206.724	180.105	0.	0.	0.
5300	41.633	242.152	207.385	184.262	0.	0.	0.
5400	41.745	242.931	208.036	188.431	0.	0.	0.
5500	41.850	243.698	208.678	192.611	0.	0.	0.
5600	41.947	244.453	209.310	196.801	0.	0.	0.
5700	42.036	245.196	209.933	201.000	0.	0.	0.
5800	42.117	245.928	210.547	205.208	0.	0.	0.
5900	42.190	246.648	211.153	209.423	0.	0.	0.
6000	42.253	247.358	211.751	213.645	0.	0.	0.

PREVIOUS: CURRENT: March 1977 (1 bar)

Deuterium (D_2) D_2(ref)

DEUTERIUM, ION (D_2^+) IDEAL GAS Mr = 4.027655

$S°(298.15\ K) = 156.735 \pm 0.04\ J\ K^{-1}\ mol^{-1}$

$\Delta_fH°(0\ K) = 1492.287 \pm 0.013\ kJ\ mol^{-1}$
$\Delta_fH°(298.15\ K) = [1498.567]\ kJ\ mol^{-1}$

Vibrational and Rotational Levels (cm^{-1})

Direct Summation using Energy-Level Equations for the $^2\Sigma_g^+$ Electronic Ground State:

$$E = G - G_o + F = G - G_o + BZ - DZ^2 + HZ^3 - LZ^4 + \ldots = G - G_o + BZ - DZ^2 + H^2Z^3/(H+LZ),$$

where Z = N(N+1), Y = v+1/2, and we omit subscript v on G, F, B, D, H, and L

$$G = 1643.29\ Y - 33.7163\ Y^2 + 0.329116\ Y^3 - 0.00725916\ Y^4$$
$$B = 14.9825 - 0.564098\ Y + 0.00801011\ Y^2 - 0.000230173\ Y^3 \qquad r_3 = 1.057\ Å$$
$$D = 4.98379 \times 10^{-3} - 1.54039 \times 10^{-4}\ Y + 3.75711 \times 10^{-6}\ Y^2$$
$$H = 2.15408 \times 10^{-6} - 5.31503 \times 10^{-8}\ Y \text{ and } L = 1.00253 \times 10^{-9}$$
$$v_{max} = 26 \text{ and } N_{max} = 57 - 46v/v_{max}$$

electronic statistical weight = 2 and normalized rotational statistical weights = 2/3 (even N) or 1/3 (odd N)

Enthalpy of Formation

$\Delta_fH°(0\ K) = 356.666 \pm 0.003\ kcal\ mol^{-1}$ is the ionization potential $IP(D_2) = 124745.6 \pm 1.2\ cm^{-1}$. Takezawa and Tanaka (1) determined $IP = 124746.6 \pm 0.6\ cm^{-1}$ from Rydberg spectra of D_2. Similar studies on H_2 by Herzberg and Jungen (2) revealed a bias of $\sim 1.0\ cm^{-1}$ attributed to a pressure shift in the spectra. We assume an equal shift for D_2, adjust IP by $-1.0\ cm^{-1}$ and increase the uncertainty due to this adjustment. $\Delta_fH°(0\ K)$ is converted to $\Delta_fH°(298.15\ K)$ by use of JANAF (3) enthalpies $H°(0\ K) - H°(298.15\ K)$ for D_2, D_2^+ and $e^-(g)$. $\Delta_fH°(298.15\ K)$ should not be interpreted as a room-temperature ionization threshold due to inclusion of these enthalpies and to threshold effects noted by Rosenstock et al. (4).

The adopted spectroscopic $IP(D_2)$ receives strong support from a value calculated indirectly from experimental values (3) of $D_0^0(D_2)$ and IP(D) combined with the theoretical value (5) of $D_0^0(D_2^+)$; this indirect value differs by only $0.4\ cm^{-1}$. Rosenstock et al. (4) review several less accurate determinations of $IP(D_2)$ and recommend a value calculated from $IP(H_2)$ using differences in zero-point energies of the diatomic molecules and their ions. Repeating this calculation with JANAF values (3), we predict an IP (D_2) which differs from the observed value by $-32\ cm^{-1}$. Inclusion of an empirical adjustment for IP(D) - IP(H) overcorrects the prediction, giving a deviation of $+28\ cm^{-1}$. These slight discrepancies are due (3, see HD^+, g) to the approximate nature of the prediction. Villarejo's average photoionization result (6), $IP(D_2) = 124757 \pm 180\ cm^{-1}$, deviates by $+11\ cm^{-1}$ from the spectroscopic value.

Heat Capacity and Entropy

Thermodynamic functions for ortho-para "equilibrium" D_2^+ are calculated by direct summation using the energy-level equations listed above. We use 1973 fundamental constants (7) in an extended version of a computer program written by W. H. Evans and provided by D. D. Wagman of the U. S. National Bureau of Standards. Excited electronic states are neglected because they do not contribute significantly. The electronic statistical weight is taken equal to the multiplicity since we neglect rotational splitting which should be very small (see HD^+ and H_2^+, 3).

Coefficients of the energy-level polynomials are calculated from the theoretical values of H_2^+ (3) using rho = 0.707330 in isotopic equations. Theoretical calculations (5) on D_2^+ provide confirmation of B_e, ω_e and ω_ex_e; the latter two are confirmed by vibrational structure observed (6) in the photoionization threshold of D_2^+. We give the polynomial coefficients, especially higher order ones, to many more digits than are justified by their accuracy. The equations are very approximate near N_{max} but, judging by H_2^+ (3), this should have little effect on the thermodynamic functions even at 6000 K. The approximate v_{max} is derived from the G polynomial.

The thermodynamic functions include contributions from quasi-bound levels lying above $D_0^0 = 21712\ cm^{-1}$. At 6000 K these levels contribute $0.33\ cal\ K^{-1}\ mol^{-1}$ to Cp° and $0.041\ cal\ K^{-1}\ mol^{-1}$ to the Gibbs-energy function. Their contribution is negligible below 3000 K. Limiting rotational quantum numbers N_{max} are estimated from H_2^+, H_2 and D_2 (3) by comparing energy increments and rotational quanta found above the respective values of D_0^0. The crudity of these estimates makes Cp° at 6000 K somewhat more uncertain than in the case of H_2^+ (3).

References

1. S. Takezawa and Y. Tanaka, J. Mol. Spectrosc. 54, 379 (1975).
2. G. Herzberg and Ch. Jungen, J. Mol. Spectrosc. 41, 425 (1972).
3. JANAF Thermochemical Tables: H_2(g), D_2(g), H^+(g), D^+(g), e^-(g) 3-31-77; H_2^+(g), HD^+(g), D_2^+(g) 9-30-77.
4. H. M. Rosenstock, K. Draxl, B. W. Steiner and J. T. Herron, J. Phys. Chem. Ref. Data 6, Supp. 1 (1977).
5. G. Hunter and H. O. Pritchard, J. Chem. Phys. 46, 2153 (1967).
6. D. Villarejo, J. Chem. Phys. 48, 4014 (1968).
7. E. R. Cohen and B. N. Taylor, J. Phys. Chem. Ref. Data 2, 663 (1973).

Deuterium, Ion (D_2^+) D_2^+(g)

Enthalpy Reference Temperature = T_r = 298.15 K Standard State Pressure = p° = 0.1 MPa

T/K	$C_p°$ (J K⁻¹mol⁻¹)	S° (J K⁻¹mol⁻¹)	$-[G°-H°(T_r)]/T$ (J K⁻¹mol⁻¹)	$H°-H°(T_r)$ (kJ mol⁻¹)	$\Delta_fH°$ (kJ mol⁻¹)	$\Delta_fG°$ (kJ mol⁻¹)	Log K_f
0	0.	0.	INFINITE	-8.651	1492.287		
100	29.192	124.796	182.774	-5.798			
200	29.221	145.027	159.422	-2.879			
298.15	29.510	156.735	156.735	0.	1498.567	1488.801	-260.832
300	29.519	156.917	156.735	0.055	1498.606	1488.741	-259.213
400	30.282	165.504	157.901	3.041	1500.749	1485.131	-193.938
500	31.332	172.371	160.130	6.121	1502.978	1480.971	-154.716
600	32.420	178.181	162.666	9.309	1505.297	1476.353	-128.528
700	33.417	183.255	165.252	12.602	1507.688	1471.340	-109.793
800	34.285	187.775	167.790	15.988	1510.128	1465.982	-95.719
900	35.028	191.857	170.241	19.454	1512.595	1460.316	-84.755
1000	35.664	195.582	172.592	22.990	1515.075	1454.374	-75.969
1100	36.214	199.007	174.839	26.584	1517.556	1448.184	-68.769
1200	36.695	202.179	176.987	30.230	1520.031	1441.768	-62.759
1300	37.123	205.133	179.040	33.922	1522.499	1435.146	-57.665
1400	37.510	207.899	181.004	37.653	1524.956	1428.334	-53.292
1500	37.867	210.499	182.884	41.423	1527.405	1421.347	-49.496
1600	38.201	212.954	184.688	45.226	1529.845	1414.197	-46.169
1700	38.518	215.279	186.419	49.062	1532.278	1406.894	-43.229
1800	38.825	217.490	188.084	52.929	1534.705	1399.449	-40.611
1900	39.125	219.597	189.688	56.827	1537.130	1391.869	-38.265
2000	39.422	221.611	191.234	60.754	1539.553	1384.160	-36.151
2100	39.719	223.542	192.727	64.711	1541.977	1376.331	-34.234
2200	40.017	225.396	194.170	68.698	1544.404	1368.386	-32.490
2300	40.317	227.182	195.567	72.715	1546.836	1360.331	-30.894
2400	40.620	228.904	196.920	76.761	1549.275	1352.170	-29.429
2500	40.923	230.568	198.233	80.839	1551.721	1343.907	-28.079
2600	41.225	232.179	199.508	84.946	1554.176	1335.546	-26.831
2700	41.523	233.741	200.747	89.083	1556.642	1327.091	-25.674
2800	41.814	235.256	201.952	93.250	1559.117	1318.543	-24.598
2900	42.094	236.728	203.126	97.446	1561.603	1309.908	-23.594
3000	42.360	238.160	204.270	101.669	1564.097	1301.186	-22.656
3100	42.607	239.553	205.386	105.917	1566.600	1292.381	-21.776
3200	42.832	240.909	206.475	110.189	1569.110	1283.495	-20.951
3300	43.032	242.231	207.539	114.483	1571.624	1274.531	-20.174
3400	43.204	243.518	208.578	118.795	1574.140	1265.490	-19.442
3500	43.345	244.772	209.594	123.123	1576.655	1256.376	-18.750
3600	43.453	245.995	210.589	127.463	1579.166	1247.190	-18.096
3700	43.526	247.187	211.562	131.812	1581.671	1237.933	-17.476
3800	43.565	248.348	212.515	136.167	1584.165	1228.610	-16.888
3900	43.567	249.480	213.448	140.524	1586.646	1219.221	-16.330
4000	43.534	250.582	214.363	144.879	1589.109	1209.768	-15.798
4100	43.467	251.657	215.259	149.230	1591.552	1200.254	-15.291
4200	43.365	252.703	216.138	153.571	1593.970	1190.681	-14.808
4300	43.232	253.722	217.001	157.902	1596.362	1181.051	-14.347
4400	43.068	254.714	217.846	162.217	1598.723	1171.365	-13.906
4500	42.875	255.680	218.676	166.514	1601.051	1161.626	-13.484
4600	42.655	256.620	219.491	170.791	1603.344	1151.836	-13.080
4700	42.411	257.534	220.291	175.044	1605.598	1141.996	-12.692
4800	42.145	258.424	221.076	179.272	1607.813	1132.108	-12.320
4900	41.859	259.291	221.847	183.473	1609.986	1122.175	-11.963
5000	41.556	260.133	222.605	187.643	1612.116	1112.199	-11.619
5100	41.237	260.953	223.348	191.783	1614.202	1102.179	-11.289
5200	40.905	261.751	224.079	195.890	1616.242	1092.120	-10.970
5300	40.562	262.526	224.797	199.964	1618.237	1082.021	-10.664
5400	40.210	263.281	225.503	204.002	1620.185	1071.886	-10.368
5500	39.850	264.016	226.197	208.006	1622.087	1061.714	-10.083
5600	39.485	264.731	226.879	211.972	1623.943	1051.509	-9.808
5700	39.116	265.426	227.549	215.902	1625.752	1041.271	-9.542
5800	38.744	266.103	228.208	219.795	1627.516	1031.001	-9.285
5900	38.371	266.763	228.856	223.651	1629.235	1020.701	-9.037
6000	37.998	267.404	229.493	227.470	1630.910	1010.373	-8.796

PREVIOUS: September 1977 (1 atm) CURRENT: September 1977 (1 bar)

Deuterium, Ion (D_2^+) D_2^+(g)

DEUTERIUM, ION (D_2^-) IDEAL GAS M_r = 4.028753

$S°(298.15 K) = 158.26 \pm 0.4\ J\ K^{-1}\ mol^{-1}$

$\Delta_f H°(0 K) = [241.4 \pm 42]\ kJ\ mol^{-1}$
$\Delta_f H°(298.15 K) = [235.365]\ kJ\ mol^{-1}$

Electronic Level and Quantum Weight

State	ε_i, cm^{-1}	g_i
$X^2\Sigma_u^+$	0	2

$\omega_e = [1202]\ cm^{-1}$ $\omega_e x_e = [17.5]\ cm^{-1}$ $\sigma = 2$
$B_e = [12.7]\ cm^{-1}$ $\alpha_e = [0.141]\ cm^{-1}$ $r_e = [1.15]$ Å

Enthalpy of Formation

The enthalpy of formation at 0 K for $D_2^-(g)$ is based on the electron affinity (EA) value derived from a Rydberg-Klein-Rees configuration interaction calculation for H_2^-(1). This study by Sharpe (1) led to $EA(H_2) = -2.5$ eV. Assuming the same value for D_2 and using auxiliary data (2), we calculate $\Delta_f H°(0 K) = 57.7 \pm 10\ kcal\ mol^{-1}$ and $D_0^0(D_2^-) = 30.0 \pm 10\ kcal\ mol^{-1}$, the latter value assumes dissociation into ground state D(g) and $D^-(g)$. The adopted $EA(H_2)$ value is suggested to be the more reliable value of eight studies considered by Rosenstock et al. (3).

In terms of molecular orbitals, the ground state for $D_2^-(g)$ is the $(1s\sigma_g)^2(2p\gamma_u)^2)_u^+$ state involving two bonding electrons and one antibonding electron. As stated by Massey (4), we should expect only a weakly-bound molecule as compared with $D_2(g)$ for which the ground state has no electrons in an antibonding orbital. We would expect, however, the ground state to be $(1s\sigma_g)^2(1s\sigma_u)$. Our calculated $EA(D_2)$ value is consistent with this viewpoint. Further discussion of the $D_2^-(g)$ ion may be found in Massey (4).

$\Delta_f H(0 K)$ is converted to $\Delta_f H°(298.15 K)$ by use of JANAF (2) enthalpies, H°(0 K) - H°(298.15 K), for $D_2(g)$, $D_2^-(g)$, and e^-(g). $\Delta_f H°(298.15 K)$ should not be interpreted as a room temperature electron affinity due to inclusion of these enthalpies and to thereshold effects discussed by Rosenstock et al. (3).

Heat Capacity and Entropy

The vibrationa constants for $H_2^-(g)$ are estimated to be the same as in the isoelectronic species He_2^+(5). The vibrational constants for $D_2^-(g)$ are calculated from those of $H_2^-(g)$ using isotopic relations (2). The internuclear distance is estimated to be ~8% larger than that for $He_2^+(g)$. The same relationship appeared to exist for the (N_2^-, NO, O_2^+) isoelectronic series. B_e is calculated from r_e whereas α_e is calculated assuming a Morse function for the ground state.

The ground and excited electronic states are discussed in detail by Massey (4). No excited state contributions are included in this calculation.

References

1. T. E. Sharpe, Lockheed Report LMSC 5-10-69-9 (1969).
2. JANAF Thermochemical Tables: D(g), D^-(g), D_2(g) and e^-(g), 3-31-77; H_2^-(g), 9-30-77.
3. H. M. Rosenstock, K. Draxl, B. W. Steiner, and J. T. Herron, J. Phys. Chem. Ref. Data 6, Supp. 1 (1977).
4. H. S. W. Massey, "Negative Ions", 3rd ed., Cambridge University Press, Cambridge, 1976.
5. B. Rosen, Ed., "Spectroscopic Data Relative to Diatomic Molecules", Pergamon Press, New York, 1970.

Deuterium, Ion (D_2^-) $D_2^-(g)$

Enthalpy Reference Temperature = T_r = 298.15 K Standard State Pressure = p° = 0.1 MPa

T/K	$C_p°$ (J K⁻¹mol⁻¹)	S° (J K⁻¹mol⁻¹)	$-[G°-H°(T_r)]/T$ (J K⁻¹mol⁻¹)	$H°-H°(T_r)$ (kJ mol⁻¹)	$\Delta_f H°$ (kJ mol⁻¹)	$\Delta_f G°$ (kJ mol⁻¹)	Log K_f
0	0.	0.	INFINITE	-8.714	241.418		
100	29.190	126.100	184.605	-5.850			
200	29.403	146.375	160.997	-2.924			
250	29.771	152.971	158.755	-1.446			
298.15	30.315	158.259	158.259	0.	235.365	237.655	-41.636
300	30.338	158.447	158.260	0.056	235.329	237.669	-41.382
350	31.023	163.174	158.632	1.590	234.363	238.136	-35.540
400	31.745	167.364	159.466	3.159	233.432	238.739	-31.176
450	32.447	171.144	160.557	4.764	232.534	239.457	-27.795
500	33.101	174.597	161.791	6.403	231.667	240.272	-25.101
600	34.226	180.735	164.449	9.771	230.009	242.151	-21.081
700	35.121	186.081	167.166	13.240	228.419	244.301	-18.230
800	35.832	190.819	169.832	16.789	226.865	246.676	-16.106
900	36.404	195.073	172.404	20.402	225.322	249.245	-14.466
1000	36.875	198.934	174.867	24.067	223.773	251.986	-13.162
1100	37.270	202.468	177.218	27.775	222.210	254.883	-12.103
1200	37.610	205.725	179.459	31.519	220.627	257.923	-11.227
1300	37.907	208.748	181.598	35.295	219.022	261.095	-10.491
1400	38.173	211.567	183.639	39.100	217.395	264.393	-9.865
1500	38.413	214.209	185.590	42.929	215.746	267.807	-9.326
1600	38.634	216.695	187.457	46.782	214.078	271.332	-8.858
1700	38.840	219.044	189.246	50.655	212.392	274.962	-8.449
1800	39.033	221.269	190.964	54.549	210.689	278.692	-8.087
1900	39.216	223.384	192.615	58.462	208.971	282.517	-7.767
2000	39.390	225.400	194.204	62.392	207.240	286.432	-7.481
2100	39.558	227.326	195.736	66.339	205.497	290.435	-7.224
2200	39.721	229.170	197.214	70.303	203.744	294.520	-6.993
2300	39.878	230.940	198.642	74.283	201.982	298.686	-6.783
2400	40.032	232.640	200.024	78.279	200.212	302.928	-6.593
2500	40.182	234.277	201.361	82.290	198.435	307.245	-6.420
2600	40.329	235.856	202.658	86.315	196.652	311.632	-6.261
2700	40.474	237.381	203.916	90.355	194.862	316.089	-6.115
2800	40.616	238.855	205.138	94.410	193.068	320.611	-5.981
2900	40.757	240.283	206.325	98.479	191.270	325.198	-5.857
3000	40.896	241.667	207.480	102.561	189.467	329.847	-5.743
3100	41.034	243.010	208.605	106.658	187.661	334.556	-5.637
3200	41.170	244.315	209.700	110.768	185.851	339.324	-5.539
3300	41.305	245.584	210.769	114.892	184.038	344.147	-5.447
3400	41.439	246.819	211.811	119.029	182.222	349.027	-5.362
3500	41.573	248.022	212.828	123.180	180.403	353.959	-5.283
3600	41.705	249.195	213.822	127.344	178.581	358.944	-5.208
3700	41.837	250.340	214.794	131.521	176.756	363.979	-5.138
3800	41.968	251.457	215.744	135.711	174.929	369.063	-5.073
3900	42.099	252.549	216.674	139.914	173.099	374.196	-5.012
4000	42.229	253.617	217.584	144.131	171.266	379.376	-4.954
4100	42.359	254.661	218.476	148.360	169.430	384.601	-4.900
4200	42.488	255.683	219.349	152.602	167.592	389.871	-4.849
4300	42.617	256.685	220.206	156.858	165.751	395.185	-4.801
4400	42.745	257.666	221.046	161.126	163.908	400.542	-4.755
4500	42.873	258.628	221.871	165.407	162.063	405.941	-4.712
4600	43.001	259.572	222.680	169.700	160.215	411.381	-4.671
4700	43.129	260.498	223.475	174.007	158.366	416.861	-4.633
4800	43.256	261.407	224.256	178.326	156.514	422.381	-4.596
4900	43.383	262.300	225.023	182.658	154.662	427.939	-4.562
5000	43.510	263.178	225.778	187.003	152.809	433.535	-4.529
5100	43.637	264.041	226.519	191.360	150.955	439.167	-4.498
5200	43.763	264.889	227.249	195.730	149.101	444.837	-4.468
5300	43.890	265.724	227.967	200.113	147.247	450.542	-4.440
5400	44.016	266.546	228.674	204.508	145.395	456.282	-4.414
5500	44.142	267.355	229.370	208.916	143.544	462.056	-4.388
5600	44.268	268.151	230.055	213.336	141.696	467.864	-4.364
5700	44.394	268.936	230.731	217.769	139.852	473.704	-4.341
5800	44.519	269.709	231.396	222.215	138.011	479.577	-4.319
5900	44.645	270.471	232.052	226.673	136.175	485.482	-4.298
6000	44.770	271.222	232.698	231.144	134.345	491.419	-4.278

PREVIOUS: September 1977 (1 atm) CURRENT: September 1977 (1 bar)

Deuterium, Ion (D_2^-) $D_2^-(g)$

AMIDOGEN-D2 (ND_2) IDEAL GAS M_r = 18.034904

Amidogen-D_2 (ND_2) $D_2N_1(g)$

$\Delta_fH°(0\ K)$ = 188.276 ± 8.4 kJ mol^{-1}

$S°(298.15\ K)$ = 204.291 J K^{-1} mol^{-1} $\Delta_fH°(298.15\ K)$ = 185.351 ± 8.4 kJ mol^{-1}

Electronic Levels (Quantum Weights) and Vibrational Frequencies (Degeneracies)

State	ε, cm^{-1}	g_i	ν_1, cm^{-1}	ν_2, cm^{-1}	ν_3, cm^{-1}
$\tilde{X}^2B_1$	0	(2)	[2305](1)	1110 (1)	[2367](1)
$\tilde{A}^2A_1(\pi)$	10393	(2)	[2305](1)	[500](1)	[2367](1)

Point Group: C_{2v} σ = 2

Bond Distance: N-D = 1.024 Å

Bond Angle: D-N-D = 103.4°

Product of the Moments of Inertia: $I_AI_BI_C = 5.7948 \times 10^{-119}\ g^3\ cm^6$

Enthalpy of Formation

$\Delta_fH°(298.15\ K)$ of $ND_2(g)$ was calculated from the JANAF selection (1) for $\Delta_fH°(298.15\ K)$ of $NH_2(g)$, the JANAF thermal functions (1), and the zero point energies of $H_2(g)$, $D_2(g)$, $NH_2(g)$, and $ND_2(g)$. The zero point energies of $H_2(g)$ and $D_2(g)$ were those given by Herzberg and Monfils (2). The zero point energies of $NH_2(g)$ and $ND_2(g)$ were estimated from vibrational frequencies given by Milligan and Jacox (3) who observed the infrared and visible spectrum of matrix-isolated NH_2 and ND_2.

Heat Capacity and Entropy

The bond distance and angle are from the electronic absorption spectrum of NH_2 as summarized by Herzberg (4). The vibrational frequencies for the ground state are from Milligan and Jacox (3); those for the excited state are estimated by comparison with NH_2(1, 4).

The National Bureau of Standards prepared this table (5) by critical analysis of data existing in 1972. Using S°, $C_p°$ and $\Delta_fH°$ selected by NBS (5), we recalculate the table in terms of current JANAF reference states for the elements.

References

1. JANAF Thermochemical Tables: $NH_2(g)$, 7-31-72 (NBS)/6-3C-77.
2. G. Herzberg and A. Monfils, J. Mol. Spectrosc. 5, 482 (1960).
3. D. E. Milligan and M. E. Jacox, J. Chem. Phys. 43, 4487 (1965).
4. G. Herzberg, "Electronic Spectra of Polyatomic Molecules," D. Van Nostrand Co., Inc., New York, 1966.
5. S. Abramowitz et al., U. S. Nat. Bur. Stand. Report 10904, 239, July, 1972.

Enthalpy Reference Temperature = T_r = 298.15 K Standard State Pressure = p° = 0.1 MPa

T/K	$C_p°$ (J $K^{-1}mol^{-1}$)	S° (J $K^{-1}mol^{-1}$)	-[G°-H°(T_r)]/T (J $K^{-1}mol^{-1}$)	H°-H°(T_r) (kJ mol^{-1})	$\Delta_fH°$ (kJ mol^{-1})	$\Delta_fG°$ (kJ mol^{-1})	Log K_f
0	0.	0.	INFINITE	-9.979	188.276	188.276	INFINITE
100	33.259	167.708	234.234	-6.653	187.401	189.900	-99.193
200	33.439	190.787	207.397	-3.322	186.323	192.825	-50.361
298.15	34.422	204.291	204.291	0.	185.351	196.226	-34.378
300	34.447	204.504	204.292	0.064	185.334	196.293	-34.178
400	36.062	214.625	205.661	3.586	184.476	200.080	-26.128
500	37.949	222.871	208.302	7.285	183.775	204.065	-21.319
600	39.953	229.967	211.334	11.179	183.231	208.177	-18.123
700	41.945	236.277	214.455	15.275	182.825	212.369	-15.847
800	43.815	242.002	217.546	19.565	182.535	216.610	-14.143
900	45.509	247.262	220.559	24.033	182.337	220.882	-12.820
1000	47.007	252.136	223.476	28.660	182.210	225.173	-11.762
1100	48.308	256.679	226.290	33.427	182.136	229.473	-10.897
1200	49.442	260.932	229.002	38.316	182.1C1	233.778	-10.176
1300	50.421	264.929	231.613	43.310	182.096	238.085	-9.566
1400	51.279	268.697	234.129	48.396	182.112	242.391	-9.044
1500	52.036	272.262	236.553	53.563	182.145	246.696	-8.591
1600	52.714	275.642	238.892	58.801	182.192	250.997	-8.194
1700	53.329	278.856	241.149	64.103	182.250	255.296	-7.844
1800	53.898	281.921	243.329	69.465	182.319	259.591	-7.533
1900	54.430	284.849	245.438	74.881	182.399	263.882	-7.255
2000	54.936	287.654	247.479	80.350	182.490	268.168	-7.004
2100	55.424	290.346	249.457	85.868	182.595	272.449	-6.777
2200	55.902	292.936	251.375	91.434	182.713	276.725	-6.570
2300	56.368	295.431	253.236	97.048	182.846	280.996	-6.382
2400	56.827	297.840	255.045	102.708	182.996	285.260	-6.209
2500	57.278	300.169	256.804	108.413	183.164	289.518	-6.049
2600	57.722	302.424	258.515	114.163	183.351	293.768	-5.902
2700	58.160	304.611	260.182	119.957	183.556	298.011	-5.765
2800	58.589	306.734	261.807	125.795	183.782	302.246	-5.638
2900	59.009	308.797	263.392	131.675	184.027	306.473	-5.520
3000	59.417	310.804	264.939	137.596	184.292	310.690	-5.410
3100	59.813	312.759	266.450	143.558	184.576	314.899	-5.306
3200	60.195	314.664	267.927	149.558	184.879	319.098	-5.209
3300	60.565	316.522	269.372	155.597	185.199	323.288	-5.117
3400	60.919	318.335	270.785	161.671	185.537	327.467	-5.031
3500	61.259	320.106	272.169	167.780	185.890	331.637	-4.949
3600	61.580	321.837	273.525	173.922	186.257	335.796	-4.872
3700	61.883	323.528	274.854	180.095	186.638	339.944	-4.799
3800	62.169	325.182	276.156	186.298	187.029	344.082	-4.730
3900	62.436	326.801	277.434	192.529	187.430	348.210	-4.664
4000	62.685	328.384	278.688	198.785	187.839	352.328	-4.601
4100	62.915	329.935	279.919	205.065	188.255	356.435	-4.541
4200	63.128	331.454	281.128	211.367	188.675	360.531	-4.484
4300	63.322	332.942	282.316	217.690	189.098	364.618	-4.429
4400	63.499	334.399	283.483	224.031	189.522	368.695	-4.377
4500	63.660	335.828	284.631	230.389	189.946	372.762	-4.327
4600	63.803	337.229	285.759	236.762	190.369	376.821	-4.279
4700	63.932	338.603	286.869	243.149	190.789	380.869	-4.233
4800	64.045	339.950	287.961	249.548	191.205	384.909	-4.189
4900	64.146	341.271	289.035	255.958	191.615	388.940	-4.146
5000	64.233	342.568	290.093	262.377	192.020	392.964	-4.105
5100	64.303	343.841	291.134	268.804	192.418	396.978	-4.066
5200	64.363	345.090	292.160	275.237	192.808	400.985	-4.028
5300	64.413	346.317	293.170	281.676	193.189	404.985	-3.991
5400	64.451	347.521	294.166	288.119	193.562	408.978	-3.956
5500	64.480	348.704	295.146	294.566	193.926	412.963	-3.922
5600	64.499	349.866	296.113	301.015	194.280	416.943	-3.889
5700	64.509	351.008	297.066	307.465	194.625	420.916	-3.857
5800	64.509	352.130	298.006	313.916	194.959	424.883	-3.826
5900	64.503	353.232	298.933	320.367	195.284	428.844	-3.797
6000	64.492	354.316	299.847	326.817	195.598	432.801	-3.768

PREVIOUS: June 1977 (1 atm) CURRENT: June 1977 (1 bar)

DIAZENE-D2, CIS (N_2D_2) IDEAL GAS M_r = 32.041604

Diazene-D_2, Cis (DNND) $D_2N_2(g)$

$S°(298.15\ K) = 224.092\ J\ K^{-1}\ mol^{-1}$

$\Delta_fH°(0\ K) = 214.039 \pm 2.1\ kJ\ mol^{-1}$

$\Delta_fH°(298.15\ K) = 207.108 \pm 2.1\ kJ\ mol^{-1}$

Vibrational Frequencies and Degeneracies

ν, cm^{-1}	ν, cm^{-1}
[2300](1)	[2400](1)
[1490](1)	[1150](1)
[1058](1)	[750](1)

Ground State Quantum Weight: 1 $\sigma = 2$

Point Group: C_{2v}

Bond Distances: N-N = [1.230] Å N-D = [1.014] Å

Bond Angle: D-N-N = [100]°

Product of Moments of Inertia: $I_AI_BI_C = 3.8165 \times 10^{-117}\ g^3\ cm^6$

Enthalpy of Formation

$\Delta_fH°(N_2D_2$, g, 298.15 K) was estimated from $\Delta_fH°(N_2H_2$, g, 298.15 K) = 50.9±5 kcal mol^{-1}, given in the JANAF Thermochemical Tables (1), the appropriate thermal functions (see tables for N_2D_2, H_2, D_2, and N_2H_2, 1) and the estimated zero point energies. The zero point energies for $H_2(g)$, $D_2(g)$ are those given by Herzberg and Monfils (2). The zero point energies for $N_2D_2(g)$ and $N_2H_2(g)$ are taken to be one half the sum of the vibrational frequencies given above and for $N_2H_2(g)$ (1).

Heat Capacity and Entropy

The bond distances and angle are assumed the same as were obtained from a quantum mechanical calculation for N_2H_2 by Wheland and Chen (3). The principal moments of inertia are: $I_A = 0.5832$, $I_B = 2.1780$, $I_C = 2.7612 \times 10^{-39}\ g\ cm^2$. The infrared spectrum of $N_2H_2(g)$ has been observed using matrix isolation techniques by Rosengren and Pimentel (4); they also observed absorptions at 1481 and 1058 cm^{-1} attributed to trans-HNND and at 946 cm^{-1} attributed to trans-DNND. For cis-HNNH, they assigned $\nu_1$or ν_4 = 3074 cm^{-1} and ν_3 = 1279 cm^{-1}. Based on these observations and the vibrational frequencies adopted for $N_2H_2(g)$ (1), the vibrational frequencies for N_2D_2 are estimated.

The National Bureau of Standards prepared this table (5) by critical analysis of data existing in 1972. Using molecular constants and $\Delta_fH°$ selected by NBS (5), we recalculate the table in terms of 1973 fundamental constants (6), 1975 atomic weights (7), and current JANAF reference states for the elements.

References

1. JANAF Thermochemical Tables: $N_2H_2(g)$, 12-31-65; H_2(ref. st.), 3-31-77; $N_2D_2(g)$, 6-30-77.
2. G. Herzberg and A. Monfils, J. Mol. Spectrosc. 5, 482 (1960).
3. G. W. Wheland and P. S. K. Chen, J. Chem. Phys. 24, 482 (1960).
4. K. Rosengren and G. Pimentel, J. Chem. Phys. 24, 67 (1956).
5. S. Abramowitz et al., U. S. Nat. Bur. Stand., Report 10904, p. 239, July, 1972.
6. CODATA Task Group on Fundamental Constants, CODATA Bulletin 11, December, 1973.
7. IUPAC Commission on Atomic Weights, Pure Appl. Chem. 47, 75 (1976); 37, 589 (1974).

Enthalpy Reference Temperature = T_r = 298.15 K — Standard State Pressure = p° = 0.1 MPa

T/K	$C_p°$ (J $K^{-1}mol^{-1}$)	S° (J $K^{-1}mol^{-1}$)	$-[G°-H°(T_r)]/T$ (J $K^{-1}mol^{-1}$)	$H°-H°(T_r)$ (kJ mol^{-1})	$\Delta_fH°$ (kJ mol^{-1})	$\Delta_fG°$ (kJ mol^{-1})	Log K_f
0	0.	0.	INFINITE	-10.308	214.039	214.039	INFINITE
100	33.278	186.132	255.951	-6.982	211.712	220.359	-115.104
200	34.771	209.482	227.508	-3.605	209.225	229.987	-60.066
250	36.691	217.435	224.721	-1.821	208.094	235.310	-49.165
298.15	39.024	224.092	224.092	0.	207.108	240.643	-42.160
300	39.119	224.333	224.092	0.072	207.072	240.851	-41.936
350	41.779	230.562	224.578	2.094	206.177	246.554	-36.796
400	44.483	236.317	225.690	4.251	205.412	252.376	-32.957
450	47.125	241.710	227.174	6.541	204.774	258.286	-29.981
500	49.652	246.807	228.884	8.961	204.254	264.260	-27.607
600	54.274	256.278	232.673	14.163	203.524	276.337	-24.057
700	58.283	264.953	236.673	19.796	203.134	288.507	-21.529
800	61.703	272.965	240.715	25.800	203.004	300.715	-19.635
900	64.591	280.404	244.717	32.119	203.068	312.927	-18.162
1000	67.018	287.339	248.636	38.703	203.278	325.123	-16.983
1100	69.056	293.825	252.453	45.509	203.595	337.293	-16.017
1200	70.770	299.909	256.157	52.503	203.991	349.430	-15.210
1300	72.215	305.633	259.745	59.654	204.446	361.532	-14.527
1400	73.440	311.030	263.217	66.939	204.944	373.597	-13.939
1500	74.483	316.134	266.576	74.336	205.474	385.626	-13.429
1600	75.375	320.970	269.826	81.830	206.027	397.618	-12.981
1700	76.143	325.563	272.971	89.407	206.596	409.575	-12.585
1800	76.808	329.935	276.015	97.056	207.177	421.498	-12.232
1900	77.386	334.103	278.963	104.766	207.766	433.389	-11.915
2000	77.891	338.086	281.821	112.530	208.359	445.248	-11.629
2100	78.335	341.897	284.591	120.342	208.955	457.078	-11.369
2200	78.726	345.550	287.280	128.195	209.550	468.879	-11.133
2300	79.073	349.058	289.890	136.086	210.144	480.654	-10.916
2400	79.381	352.430	292.426	144.009	210.734	492.402	-10.717
2500	79.656	355.676	294.891	151.961	211.321	504.127	-10.533
2600	79.903	358.805	297.290	159.939	211.902	515.827	-10.363
2700	80.125	361.825	299.624	167.941	212.477	527.506	-10.205
2800	80.326	364.742	301.898	175.963	213.045	539.163	-10.058
2900	80.507	367.564	304.114	184.005	213.606	550.800	-9.921
3000	80.672	370.296	306.275	192.064	214.159	562.418	-9.793
3100	80.822	372.944	308.383	200.139	214.704	574.018	-9.672
3200	80.959	375.512	310.441	208.228	215.239	585.600	-9.559
3300	81.084	378.006	312.451	216.331	215.764	597.166	-9.452
3400	81.199	380.428	314.415	224.445	216.279	608.716	-9.352
3500	81.305	382.783	316.335	232.570	216.784	620.250	-9.257
3600	81.402	385.075	318.212	240.706	217.277	631.771	-9.167
3700	81.492	387.307	320.050	248.850	217.759	643.278	-9.081
3800	81.576	389.481	321.848	257.004	218.229	654.772	-9.000
3900	81.653	391.601	323.610	265.165	218.687	666.254	-8.923
4000	81.724	393.669	325.336	273.334	219.132	677.725	-8.850
4100	81.791	395.688	327.027	281.510	219.565	689.184	-8.780
4200	81.853	397.660	328.685	289.692	219.984	700.633	-8.714
4300	81.911	399.586	330.312	297.880	220.390	712.072	-8.650
4400	81.965	401.470	331.908	306.074	220.783	723.502	-8.589
4500	82.016	403.313	333.474	314.273	221.162	734.923	-8.531
4600	82.064	405.116	335.012	322.477	221.528	746.336	-8.475
4700	82.108	406.881	336.522	330.686	221.880	757.742	-8.421
4800	82.150	408.610	338.006	338.899	222.219	769.139	-8.370
4900	82.189	410.305	339.465	347.116	222.544	780.530	-8.321
5000	82.227	411.965	340.898	355.337	222.856	791.914	-8.273
5100	82.262	413.594	342.308	363.561	223.154	803.292	-8.227
5200	82.295	415.192	343.694	371.789	223.440	814.665	-8.183
5300	82.326	416.760	345.058	380.020	223.714	826.032	-8.141
5400	82.356	418.299	346.400	388.254	223.977	837.394	-8.100
5500	82.384	419.810	347.721	396.491	224.227	848.751	-8.061
5600	82.410	421.295	349.021	404.731	224.467	860.104	-8.023
5700	82.435	422.754	350.302	412.973	224.697	871.452	-7.986
5800	82.459	424.188	351.564	421.218	224.917	882.797	-7.950
5900	82.482	425.597	352.807	429.465	225.127	894.138	-7.916
6000	82.504	426.984	354.031	437.714	225.328	905.476	-7.883

PREVIOUS: June 1977 (1 atm) CURRENT: June 1977 (1 bar)

Diazene-D_2, Cis (DNND) $D_2N_2(g)$

WATER-D2 (D_2O) IDEAL GAS M_r = 20.020004

$S°(298.15\ K) = 198.339\ J\ K^{-1}\ mol^{-1}$

$\Delta_fH°(0\ K) = -246.251 \pm 0.067\ kJ\ mol^{-1}$
$\Delta_fH°(298.15\ K) = -249.203 \pm 0.067\ kJ\ mol^{-1}$

Vibrational Frequencies and Degeneracies

ν, cm^{-1}	ν, cm^{-1}	ν, cm^{-1}
2671.69(1)	1178.33(1)	2788.02(1)

Ground State Quantum Weight: 1 σ = 2
Point Group: C_{2v}
Bond Distance: O-D = 0.958 Å
Bond Angle: D-O-D = 104.45°
Product of the Moments of Inertia: $I_AI_BI_C = 3.9948 \times 10^{-119}\ g^3\ cm^6$

Enthalpy of Formation

Rossini et al. (1) measured the ratio of the heats of formation of $D_2O(\ell)/H_2O(\ell)$. Recalculation of their results using $\Delta_fH°(298.15\ K)$ of $H_2O(\ell) = -68.315\pm0.010\ kcal\ mol^{-1}$ (see $H_2O(g)$ table (15)), yields $\Delta_fH°(298.15\ K)$ of $D_2O(\ell) - H_2O(\ell) = -2.1098\pm0.01492 \times \Delta_fH°[HDO(\ell)-H_2O(\ell), 298.15\ K] \pm .016\ kcal\ mol^{-1}$. Using a value of 32±30 kcal mol^{-1} for the reaction $H_2O(\ell) + D_2O(\ell) = 2\ HDO(\ell)$ from (2) (some justification that the uncertainty may be smaller is given by Van Hook (3), see also HDO table (15)), one obtains $\Delta_fH°[D_2O(\ell)-H_2O(\ell), 298.15\ K] = -2.094\pm0.016\ kcal\ mol^{-1}$.

The difference between the heats of vaporization of $D_2O(\ell)$ and $H_2O(\ell)$ at 298.15 K were evaluated as follows. A. By direct calorimetry, Rossini et al. (1) determined the ratio of the heats of vaporization of $H_2O(\ell)/D_2O(\ell)$ at 298.15 K and zero pressure to be 0.969503±0.00070. B. Bartholome and Clusius (4) determined the heat of vaporization of $D_2O(\ell)$ by direct calorimetry at 0°C. Recalculation of their results, using the energy equivalent of a Bunsen ice calorimetergiven by (5) (compression correction assumed zero) and neglecting gas imperfection corrections, yields 12.637±0.026 kcal mol^{-1}. The heat of vaporization of $D_2O(\ell)$ at 25°C was calculated using a heat of fusion of $D_2O(\ell)$ selected by (6) and (7), the condensed phase of heat capacity data of $D_2O(\ell)$ given by Long and Kemp (8) and the gas phase thermal functions of Friedman and Haar (see next section). C. The difference in heats of vaporization of $D_2O(\ell)$ and $H_2O(\ell)$ was derived by differentiating the formula given by Jones (9) (see (10) for a comparison with other measurements) for the ratio of the vapor pressures of $H_2O(\ell)/D_2O(\ell)$ as a function of temperature and assuming negligible corrections for gas imperfection (see (11), however).

Source	Method	$\Delta_{vap}H°[D_2O(\ell)-H_2O(\ell), 298.15\ K]$, kcal mol^{-1}
Knowlton, Johnston, Rossini	Calorimetry	0.331±0.008*
Bartholome and Clusius	Calorimetry	0.340±0.028*
Jones	Vapor Pressure	0.307±0.020

*Calculated from $\Delta_{vap}H°(298.15\ K)$ of $H_2O(\ell) = 10.520\pm0.002\ kcal\ mol^{-1}$ (12).

Selecting 0.331±.008 kcal mol^{-1} for the difference in the heats of vaporization, one obtains −1.763±0.018 kcal mol^{-1} for $\Delta_fH°(298.15\ K)$ of $D_2O(g)-H_2O(g)$. The "spectroscopic" value for this difference was calculated to be −1.768±0.015 kcal mol^{-1} based on the zero point energies given by Hulston (13) for $H_2O(g)$ and $D_2O(g)$, and by Herzberg and Monfils (14) for $H_2(g)$ and $D_2(g)$, and the appropriate thermal functions ($H_2O(g)$, H_2(ref. st.), D_2(ref. st.), $D_2O(g)$ (15)). To close the consistency check, the "spectroscopic" value of $\Delta_fH°(298.15\ K)$ of $D_2O(g)$ minus HDO(g) is −0.916±0.0015 kcal mol^{-1} and the "nonspectroscopic" value for this difference (see HDO table, reaction A (15)) is (−1.763±0.0018) + (0.846±0.006) = −0.917±0.019 kcal mol^{-1}. The average value of −1.766±0.016 kcal mol^{-1} for $\Delta_fH°(298.15\ K)$ of $D_2O(g) - H_2O(g)$ and $\Delta_fH°(298.15\ K)$ of $H_2O(g) = -57.795\pm0.010\ kcal\ mol^{-1}$ (12) are used to obtain $\Delta_fH°(298.15\ K)$ for $D_2O(g) = -59.561\pm0.016\ kcal\ mol^{-1}$.

Heat Capacity and Entropy

The thermodynamic functions of this table are analogous to those in the JANAF table for $H_2O(g)$ (15); both tables are taken from Freidman and Haar (16). Friedman and Haar applied their non-rigid-rotor, anharmonic-oscillator treatment (with vibrational-rotational coupling terms and low-temperature rotational corrections) to the infrared-spectra analyses of Benedict et al. (17), and of Benedict et al. (18). In the present table for D_2O, the values of $C_p°$ of Friedman and Haar between 4000 and 5000 K were extrapolated linearly (except with a term in T^{-2}) from 5000 to 6000 K.

The National Bureau of Standards prepared this table (19) by critical analysis of data existing in 1972. Using S°(298.15 K) and $C_p°$ values selected by NBS (19), we recalculate the table in terms of the current JANAF reference states for the elements (15).

References

1. F. D. Rossini, J. W. Knowlton, and H. L. Johnston, J. Res. Nat. Bur. Stand. 24, 369 (1940).
2. W. C. Duer and G. L. Bertrand, J. Chem. Phys. 53, 3020 (1970).
3. W. A. Van Hook, J. Phys. Chem. 72, 1234 (1968).
4. E. Bartholome and K. Clusius, Z. Physik. Chem. B28, 167 (1935).
5. D. A. Ditmars and T. B. Douglas, J. Res. Nat. Bur. Stand. 75A, 401 (1971).
6. I. Kirshenbaum, "Physical Properties and Analysis of Heavy Water," McGraw-Hill, New York, 1951.
7. V. A. Kirillin, Editor, "Heavy Water Thermophysical Properties," National Technical Information Service, TT70-50094, 1971.
8. E. A. Long and J. D. Kemp, J. Chem. Soc. 58, 1829 (1936).
9. W. M. Jones, J. Chem. Phys. 48, 207 (1968).
10. C. Liu and W. T. Lindsay, Jr., J. Chem. Eng. Data 15, 510 (1970).
11. G. S. Kell, G. F. Laurin, and E. Whalley, J. Chem. Phys. 49, 2839 (1968).
12. CODATA Task Group on Key Values for Thermodynamics, Final Set of Key Values for Thermodynamics-part I (Bulletin No. 5) November, 1971.
13. J. R. Hulston, J. Chem. Phys. 50, 1483 (1969) see also M. Wolfsberg, J. Chem. Phys. 50, 1484 (1969).
14. G. Herzberg and A. Monfils, J. Mol. Spectrosc. 5, 482 (1960).
15. JANAF Thermochemical Tables: $H_2O(g)$, 3-31-61; HDO(g), $D_2O(g)$, 6-31-77; H_2(ref. st.), D_2(ref. st.), O_2(ref. st.), 3-31-77.
16. A. S. Friedman and L. Haar, J. Chem. Phys. 22, 2051 (1954).
17. W. S. Benedict, N. Gailar, and E. K. Plyler, J. Chem. Phys. 21, 1301 (1953).
18. W. S. Benedict, H. H. Claassen, and J. H. Shaw, J. Res. Nat. Bur. Stand., 49, 91 (1952).
19. S. Abramowitz et al., U. S. Nat. Bur. Stand., Rept. 10904, 235, July, 1972.

Water-D_2 (D_2O) $D_2O_1(g)$

Enthalpy Reference Temperature = T_r = 298.15 K; Standard State Pressure = p° = 0.1 MPa

	J $K^{-1}mol^{-1}$			kJ mol^{-1}			
T/K	$C_p°$	S°	$-[G°-H°(T_r)]/T$	$H°-H°(T_r)$	$\Delta_fH°$	$\Delta_fG°$	Log K_f
0	0.	0.	INFINITE	-9.958	-246.251	-246.251	INFINITE
100	33.299	161.756	228.282	-6.653	-247.148	-243.379	127.128
200	33.449	184.861	201.451	-3.318	-248.222	-239.185	62.469
298.15	34.255	198.339	198.339	0.	-249.203	-234.536	41.090
300	34.278	198.551	198.340	0.063	-249.221	-234.445	40.820
400	35.635	208.591	199.699	3.557	-250.135	-229.379	29.954
500	37.184	216.707	202.314	7.197	-250.954	-224.093	23.411
600	38.835	223.631	205.303	10.997	-251.681	-218.651	19.035
700	40.545	229.745	208.366	14.966	-252.320	-213.094	15.901
800	42.251	235.271	211.389	19.106	-252.873	-207.452	13.545
900	43.892	240.343	214.328	23.414	-253.346	-201.745	11.709
1000	45.424	245.048	217.168	27.881	-253.744	-195.990	10.237
1100	46.826	249.444	219.904	32.494	-254.078	-190.198	9.032
1200	48.091	253.574	222.540	37.241	-254.354	-184.378	8.026
1300	49.225	257.469	225.078	42.108	-254.581	-178.537	7.174
1400	50.236	261.155	227.525	47.082	-254.767	-172.680	6.443
1500	51.136	264.652	229.884	52.151	-254.917	-166.811	5.809
1600	51.938	267.978	232.162	57.306	-255.038	-160.934	5.254
1700	52.653	271.149	234.363	62.536	-255.136	-155.049	4.764
1800	53.292	274.177	236.491	67.834	-255.214	-149.159	4.328
1900	53.866	277.074	238.551	73.192	-255.277	-143.266	3.939
2000	54.382	279.850	240.547	78.605	-255.328	-137.369	3.588
2100	54.848	282.515	242.483	84.067	-255.371	-131.470	3.270
2200	55.270	285.076	244.361	89.573	-255.407	-125.569	2.981
2300	55.654	287.541	246.185	95.119	-255.439	-119.666	2.718
2400	56.004	289.918	247.958	100.702	-255.470	-113.763	2.476
2500	56.326	292.210	249.683	106.319	-255.500	-107.858	2.254
2600	56.621	294.425	251.361	111.967	-255.531	-101.951	2.048
2700	56.894	296.567	252.996	117.643	-255.564	-96.044	1.858
2800	57.146	298.641	254.589	123.345	-255.601	-90.135	1.681
2900	57.380	300.651	256.143	129.071	-255.642	-84.225	1.517
3000	57.598	302.600	257.660	134.820	-255.688	-78.313	1.364
3100	57.802	304.492	259.140	140.590	-255.740	-72.399	1.220
3200	57.993	306.330	260.586	146.380	-255.799	-66.484	1.085
3300	58.172	308.117	261.999	152.189	-255.864	-60.568	0.959
3400	58.342	309.856	263.381	158.014	-255.937	-54.648	0.840
3500	58.501	311.550	264.734	163.856	-256.018	-48.727	0.727
3600	58.652	313.200	266.057	169.714	-256.107	-42.803	0.621
3700	58.795	314.809	267.353	175.587	-256.205	-36.877	0.521
3800	58.931	316.379	268.623	181.473	-256.312	-30.947	0.425
3900	59.060	317.911	269.867	187.373	-256.428	-25.015	0.335
4000	59.184	319.408	271.087	193.285	-256.554	-19.080	0.249
4100	59.301	320.871	272.283	199.209	-256.690	-13.141	0.167
4200	59.414	322.301	273.457	205.145	-256.837	-7.200	0.090
4300	59.522	323.700	274.609	211.092	-256.994	-1.254	0.015
4400	59.626	325.070	275.741	217.049	-257.163	4.695	-0.056
4500	59.726	326.411	276.852	223.017	-257.342	10.648	-0.124
4600	59.822	327.725	277.944	228.994	-257.532	16.606	-0.189
4700	59.914	329.012	279.016	234.981	-257.734	22.568	-0.251
4800	60.003	330.275	280.071	240.977	-257.947	28.534	-0.311
4900	60.089	331.513	281.108	246.982	-258.173	34.505	-0.368
5000	60.172	332.728	282.129	252.995	-258.409	40.480	-0.423
5100	60.254	333.920	283.133	259.016	-258.658	46.460	-0.476
5200	60.333	335.091	284.121	265.045	-258.918	52.445	-0.527
5300	60.410	336.241	285.093	271.083	-259.190	58.436	-0.576
5400	60.484	337.371	286.051	277.127	-259.473	64.432	-0.623
5500	60.556	338.481	286.994	283.179	-259.768	70.432	-0.669
5600	60.626	339.573	287.923	289.238	-260.075	76.439	-0.713
5700	60.696	340.647	288.839	295.304	-260.392	82.450	-0.756
5800	60.764	341.703	289.741	301.377	-260.721	88.468	-0.797
5900	60.832	342.742	290.631	307.457	-261.060	94.491	-0.837
6000	60.898	343.765	291.508	313.544	-261.410	100.521	-0.875

PREVIOUS: June 1977 (1 atm) CURRENT: June 1977 (1 bar)

Water-D_2 (D_2O) $D_2O_1(g)$

HYDROGEN SULFIDE-D2 (D_2S) IDEAL GAS M_r = 36.088204

$\Delta_fH°$(0 K) = -20.997 ± 0.8 kJ mol^{-1}

S°(298.15 K) = 215.284 J K^{-1} mol^{-1} $\Delta_fH°$(298.15 K) = -23.891 ± 0.8 kJ mol^{-1}

Vibrational Frequencies and Degeneracies

ν, cm^{-1}
1896.38(1)
855.45(1)
1999 (1)

Ground State Quantum Weight: 1 σ = 2

Point Group: C_{2v}

Bond Distance: S–D = 1.328 Å

Bond Angle: D–S–D = 92.2°

Product of the Moments of Inertia: $I_AI_BI_C$ = 0.3445 x 10^{-117} g^3 cm^6

Enthalpy of Formation

$\Delta_fH°(D_2S$, g, 298.15 K) was determined by Kapustinskii and Kankovskii (1) to be -5.692±0.06 kcal mol^{-1} (recalculated) from the reaction: (a+b) $D_2S(g)$ + (1.5a+0.5b) $O_2(g)$ = a $SO_2(g)$ + (a+b) $D_2O(g)$ + b S(rhombic). The difference of $\Delta_fH°(D_2S$, g, 298.15 K) minus $H_2S(g)$ for their work was calculated using their results for $\Delta_fH°(H_2S$, g, 298.15 K), -4.92±0.08 kcal mol^{-1} (recalculated), in the hope of eliminating possible systematic error due to uncertainty in the product formed.

The "spectroscopic" value of $\Delta_fH°(D_2S$, g, 298.15 K) minus $H_2S(g)$ was calculated from the appropriate thermal functions (see H_2, D_2, H_2S, 5) and the zero point energies of $H_2S(g)$, $D_2S(g)$ (2, 3), $D_2(g)$ (4), and $H_2(g)$ (4).

The equilibrium data of Grafe et al. (5) for the exchange reaction $H_2(g) + D_2S(g) = D_2(g) + H_2S(g)$ were analyzed by third and second law methods.

Source		$\Delta_fH°$(298.15 K) of $D_2S(g)$ - $H_2S(g)$, kcal mol^{-1}
Kapustinskii and Kankovskii (1)		-0.771±0.10
Spectroscopic (2, 3, 4)		-0.810±0.01
Grafe, Clusius, and Kruis (5)	Third Law	-0.826±0.05
	Second Law	-0.887±0.02

A weighted average of -0.81±0.02 kcal mol^{-1} was selected for $\Delta_fH°$(298.15 K) of $D_2S(g)$ minus $H_2S(g)$ and was added to the selected value of $\Delta_fH°$(298.15 K) of $H_2S(g)$ (6) to determine $\Delta_fH°$(298.15 K) of $D_2S(g)$.

Heat Capacity and Entropy

The thermodynamic functions were estimated from those in the present table for $H_2S(g)$ (6) by adding those for $D_2S(g)$ and subtracting those for $H_2S(g)$, where both the added and subtracted functions were generated using the rigid-rotor harmonic oscillator approximation. In this calculation the molecular constants for D_2S were taken from reference (2).

The National Bureau of Standards prepared this table (7) by critical analysis of data existing in 1972. Using S°(298.15 K) and $C_p°$ values selected by NBS (7), we recalculate the table in terms of the current JANAF reference states for the elements.

References

1. A. F. Kapustinskii and R. T. Kankovskii, Zhur. Fiz. Khim. 32, 2810 (1958).
2. R. E. Miller, G. E. Leroi, D. F. Meggers, J. Chem. Phys. 46, 2292 (1967).
3. H. C. Allen and E. K. Plyler, J. Chem. Phys. 25, 1132 (1956).
4. G. Herzberg and A. Monfils, J. Mol. Spectrosc. 5, 482 (1960).
5. D. Grafe, K. Clusius, and A. Kruis, Z. Physik. Chem. B43, 1 (1939).
6. JANAF Thermochemical Tables: H_2(ref st), D_2(ref st), 3-31-77; $H_2S(g)$, 6-30-77.
7. S. Abramowitz et al., U. S. Nat. Bur. Stand. Report 10904, 239, July, 1972.

Hydrogen Sulfide-D_2 (D_2S) $D_2S_1(g)$

Enthalpy Reference Temperature = T_r = 298.15 K Standard State Pressure = p° = 0.1 MPa

	J K^{-1}mol^{-1}			kJ mol^{-1}			
T/K	$C_p°$	S°	-[G°-H°(T_r)]/T	H°-H°(T_r)	$\Delta_fH°$	$\Delta_fG°$	Log K_f
0	0.	0.	INFINITE	-10.088	-20.997	-20.997	INFINITE
100	33.263	177.971	245.585	-6.761	-21.111	-26.377	13.778
200	33.932	201.151	218.221	-3.414	-22.360	-31.201	8.149
298.15	35.761	215.284	215.284	0.	-23.891	-35.300	6.184
300	35.794	215.506	215.285	0.066	-23.920	-35.371	6.159
400	37.999	226.102	216.715	3.755	-27.750	-38.856	5.074
500	40.342	234.833	219.489	7.672	-30.691	-41.300	4.315
600	42.673	242.396	222.690	11.824	-33.072	-43.189	3.760
700	44.836	249.140	225.995	16.202	-35.021	-44.717	3.337
800	46.748	255.254	229.276	20.782	-36.677	-45.986	3.003
900	48.488	260.864	232.479	25.547	-91.246	-45.980	2.669
1000	49.827	266.044	235.579	30.464	-91.320	-40.946	2.139
1100	51.028	270.850	238.570	35.508	-91.339	-35.907	1.705
1200	52.049	275.335	241.449	40.664	-91.320	-30.869	1.344
1300	52.923	279.537	244.219	45.913	-91.275	-25.833	1.038
1400	53.672	283.487	246.884	51.244	-91.215	-20.801	0.776
1500	54.321	287.212	249.450	56.644	-91.147	-15.773	0.549
1600	54.886	290.737	251.921	62.105	-91.077	-10.751	0.351
1700	55.379	294.079	254.303	67.619	-91.008	-5.733	0.176
1800	55.823	297.257	256.602	73.179	-90.942	-0.718	0.021
1900	56.208	300.286	258.822	78.781	-90.881	4.292	-0.118
2000	56.568	303.178	260.968	84.420	-90.826	9.300	-0.243
2100	56.881	305.946	263.045	90.093	-90.777	14.305	-0.356
2200	57.166	308.599	265.055	95.796	-90.736	19.308	-0.458
2300	57.430	311.146	267.004	101.525	-90.701	24.309	-0.552
2400	57.676	313.595	268.895	107.281	-90.673	29.309	-0.638
2500	57.902	315.955	270.730	113.060	-90.652	34.308	-0.717
2600	58.107	318.230	272.514	118.861	-90.638	39.306	-0.790
2700	58.300	320.426	274.248	124.681	-90.630	44.304	-0.857
2800	58.480	322.550	275.935	130.520	-90.629	49.301	-0.920
2900	58.647	324.605	277.578	136.377	-90.635	54.299	-0.978
3000	58.810	326.596	279.179	142.250	-90.647	59.296	-1.032
3100	58.965	328.527	280.740	148.138	-90.666	64.295	-1.083
3200	59.112	330.401	282.263	154.042	-90.691	69.294	-1.131
3300	59.245	332.222	283.749	159.960	-90.723	74.294	-1.176
3400	59.379	333.993	285.201	165.891	-90.763	79.295	-1.218
3500	59.509	335.716	286.620	171.836	-90.810	84.297	-1.258
3600	59.630	337.394	288.007	177.793	-90.865	89.301	-1.296
3700	59.748	339.029	289.364	183.762	-90.928	94.307	-1.331
3800	59.860	340.624	290.692	189.742	-90.999	99.314	-1.365
3900	59.973	342.181	291.992	195.734	-91.079	104.323	-1.397
4000	60.082	343.700	293.266	201.737	-91.169	109.335	-1.428
4100	60.187	345.185	294.515	207.750	-91.268	114.349	-1.457
4200	60.291	346.637	295.738	213.774	-91.377	119.365	-1.485
4300	60.392	348.057	296.939	219.808	-91.497	124.384	-1.511
4400	60.488	349.446	298.116	225.852	-91.627	129.406	-1.536
4500	60.588	350.807	299.272	231.906	-91.768	134.431	-1.560
4600	60.676	352.139	300.407	237.970	-91.920	139.459	-1.584
4700	60.768	353.445	301.522	244.042	-92.084	144.491	-1.606
4800	60.865	354.726	302.617	250.123	-92.259	149.526	-1.627
4900	60.953	355.982	303.693	256.214	-92.445	154.565	-1.648
5000	61.045	357.214	304.751	262.314	-92.643	159.609	-1.667
5100	61.124	358.424	305.792	268.423	-92.851	164.655	-1.686
5200	61.216	359.611	306.815	274.540	-93.071	169.707	-1.705
5300	61.300	360.778	307.822	280.665	-93.301	174.762	-1.722
5400	61.383	361.925	308.814	286.800	-93.541	179.823	-1.739
5500	61.467	363.052	309.790	292.942	-93.791	184.887	-1.756
5600	61.551	364.160	310.751	299.093	-94.050	189.956	-1.772
5700	61.634	365.250	311.697	305.252	-94.317	195.030	-1.787
5800	61.710	366.323	312.630	311.419	-94.593	200.109	-1.802
5900	61.798	367.379	313.549	317.595	-94.875	205.192	-1.817
6000	61.873	368.418	314.455	323.779	-95.163	210.281	-1.831

PREVIOUS: June 1977 (1 atm) CURRENT: June 1977 (1 bar)

Hydrogen Sulfide-D_2 (D_2S) $D_2S_1(g)$

AMMONIA-D3 (ND_3) IDEAL GAS M_r = 20.049006

$S°(298.15\ K)$ = 203.931 J K^{-1} mol^{-1}

$\Delta_fH°(0\ K)$ = -51.622 ± 0.4 kJ mol^{-1}
$\Delta_fH°(298.15\ K)$ = -58.576 ± 0.4 kJ mol^{-1}

Vibrational Frequencies and Degeneracies

ν, cm^{-1}
2495(1)
793(1)
2652(2)
1225(2)

Ground State Quantum Weight: 1 σ(external) = 3
Point Group: C_{3v} σ(internal) = 2
Bond Length: N-D = 1.0124 Å
Bond Angle: 106.67°
Product of the Moments of Inertia; $I_AI_BI_C$ = 2.5775 x 10^{-118} g^3 cm^6

Enthalpy of Formation

A "spectroscopic" value for the difference in $\Delta_fH°(298.15\ K)$ of $ND_3(g)$ minus $NH_3(g)$ of -3.029±0.01 kcal mol^{-1} was calculated from the appropraite thermal functions (see H_2, D_2, NH_3 (6, 7)), zero point energies of $ND_3(g)$ and $NH_3(g)$ (1, 2), and the zero point energies of $H_2(g)$ and $D_2(g)$ given by Herzberg and Monfils (3).

Analysis of the equilibrium data of Schulz and Schaefer (4) for K°p (1 atm, 660-773 K) for $1/2N_2(g) + 3/2D_2(g) = ND_3(g)$ gave the following:

	$\Delta_fH°(298.15\ K)$, kcal mol^{-1}	Drift cal K^{-1} mol^{-1}	$\Delta S°$(obsv.-calc., 298.15 K)
third law	-13.80	-1.2	----
second law[a]	-13.0±0.38	--	+1.3±0.5 cal K^{-1} mol^{-1}

[a]Asuming $\Delta C°_p$=1.237-0.00608(T-700) cal K^{-1} mol^{-1}

Using the results from the same authors' data for $NH_3(g)$ (6), one obtains a difference in $\Delta_fH°(298.15\ K)$ for $ND_3(g)$ minus $NH_3(g)$ of -2.93±0.14 kcal mol^{-1} (third law) and -2.22±0.4 kcal mol^{-1} (second law). The third law value agrees with the "spectroscopic" value within combined uncertainty intervals.

A value of -3.03±0.1 kcal mol^{-1} was added to the JANAF selection for $\Delta_fH°(298.15\ K)$ of $NH_3(g)$, -10.97±0.10 kcal mol^{-1} (5, 6), to obtain $\Delta_fH°(298.15\ K)$ of $ND_3(g)$.

Heat Capacity and Entropy

The thermodynamic functions were estimated from those in the present table for $NH_3(g)$ (6) by adding those for $ND_3(g)$ and subtracting those for $NH_3(g)$, where both the added and subtracted functions were generated using the rigid-rotor harmonicoscillator approximation. In this calculation the molecular constants for ND_3 were taken from Duncan and Mills (1) and Benedict and Plyler (2).

The National Bureau of Standards prepared this table (7) by critical analysis of data existing in 1972. Using S°, $C°_p$ and $\Delta_fH°$ selected by NBS (7), we recalculate the table in terms of current JANAF reference states for the elements.

References

1. J. L. Duncan and I. M. Mills, Spectrochim. Acta 20, 523 (1964).
2. W. S. Benedict and E. K. Plyler, Can. J. Phys. 35, 1235 (1957).
3. G. Herzberg and A. Monfils, J. Mol. Spectrosc. 5, 482 (1960).
4. G. Schulz and H. Schaefer, Ber. Bunsenges, Physik. Chem. 70, 21 (1966).
5. JANAF Thermochemical Tables, Second Edition, NSRDS-NBS 37, 1971; $NH_3(g)$ 9-30-65.
6. JANAF Thermochemical Tables; $NH_3(g)$, 7-31-72 (NBS)/6-30-77.
7. S. Abramowitz et al., U. S. Nat. Bur. Stand. Report 10904, p. 239, July, 1972.

Ammonia-D_3 (ND_3) $D_3N_1(g)$

Enthalpy Reference Temperature = T_r = 298.15 K — Standard State Pressure = p° = 0.1 MPa

T/K	$C°_p$ (J $K^{-1}mol^{-1}$)	S°	$-[G°-H°(T_r)]/T$	$H°-H°(T_r)$ (kJ mol^{-1})	$\Delta_fH°$	$\Delta_fG°$	Log K_f
0	0.	0.	INFINITE	-10.234	-51.622	-51.622	INFINITE
100	33.296	166.193	235.480	-6.929	-53.893	-45.604	23.821
200	34.551	189.507	207.289	-3.556	-56.406	-36.317	9.485
298.15	38.225	203.931	203.931	0.	-58.576	-25.984	4.552
300	38.305	204.168	203.932	0.071	-58.613	-25.782	4.489
400	42.924	215.822	205.490	4.133	-60.392	-14.558	1.901
500	47.375	225.883	208.583	8.650	-61.739	-2.934	0.307
600	51.472	234.888	212.229	13.595	-62.707	8.923	-0.777
700	55.208	243.108	216.061	18.932	-63.362	20.917	-1.561
800	58.589	250.704	219.923	24.625	-63.761	32.987	-2.154
900	61.601	257.782	223.741	30.638	-63.953	45.094	-2.617
1000	64.258	264.413	227.480	36.933	-63.979	57.213	-2.989
1100	66.576	270.649	231.124	43.478	-63.873	69.328	-3.292
1200	68.592	276.530	234.665	50.238	-63.660	81.429	-3.544
1300	70.346	282.091	238.101	57.187	-63.361	93.508	-3.757
1400	71.869	287.361	241.433	64.299	-62.995	105.561	-3.939
1500	73.199	292.366	244.663	71.554	-62.573	117.587	-4.095
1600	74.362	297.128	247.795	78.933	-62.107	129.582	-4.230
1700	75.379	301.667	250.831	86.421	-61.604	141.548	-4.349
1800	76.278	306.002	253.777	94.005	-61.072	153.483	-4.454
1900	77.069	310.148	256.635	101.673	-60.516	165.387	-4.547
2000	77.768	314.119	259.411	109.416	-59.942	177.262	-4.630
2100	78.387	317.928	262.107	117.224	-59.353	189.108	-4.704
2200	78.940	321.588	264.728	125.091	-58.754	200.925	-4.771
2300	79.429	325.108	267.278	133.010	-58.147	212.715	-4.831
2400	79.868	328.498	269.758	140.975	-57.535	224.479	-4.886
2500	80.257	331.766	272.174	148.982	-56.920	236.217	-4.935
2600	80.601	334.921	274.527	157.025	-56.306	247.930	-4.981
2700	80.906	337.969	276.820	165.101	-55.694	259.620	-5.023
2800	81.182	340.916	279.057	173.205	-55.087	271.287	-5.061
2900	81.416	343.769	281.240	181.335	-54.486	282.933	-5.096
3000	81.634	346.533	283.370	189.488	-53.893	294.557	-5.129
3100	81.814	349.213	285.451	197.661	-53.310	306.163	-5.159
3200	81.973	351.813	287.484	205.850	-52.739	317.750	-5.187
3300	82.107	354.337	289.472	214.055	-52.181	329.319	-5.213
3400	82.224	356.790	291.416	222.271	-51.639	340.871	-5.237
3500	82.326	359.175	293.318	230.499	-51.112	352.408	-5.259
3600	82.408	361.496	295.180	238.736	-50.604	363.930	-5.280
3700	82.466	363.754	297.003	246.980	-50.115	375.438	-5.300
3800	82.508	365.954	298.789	255.228	-49.647	386.933	-5.319
3900	82.537	368.098	300.539	263.481	-49.201	398.416	-5.336
4000	82.550	370.188	302.254	271.735	-48.779	409.888	-5.353
4100	82.548	372.226	303.936	279.990	-48.382	421.350	-5.368
4200	82.534	374.215	305.585	288.245	-48.011	432.802	-5.383
4300	82.506	376.157	307.204	296.497	-47.667	444.246	-5.397
4400	82.467	378.053	308.793	304.745	-47.352	455.682	-5.410
4500	82.414	379.906	310.353	312.990	-47.065	467.111	-5.422
4600	82.349	381.716	311.884	321.228	-46.809	478.535	-5.434
4700	82.276	383.487	313.389	329.459	-46.583	489.953	-5.445
4800	82.186	385.218	314.868	337.682	-46.389	501.366	-5.456
4900	82.082	386.912	316.321	345.896	-46.227	512.776	-5.466
5000	81.965	388.569	317.749	354.098	-46.099	524.183	-5.476
5100	81.818	390.190	319.154	362.287	-46.006	535.587	-5.486
5200	81.780	391.779	320.535	370.467	-45.942	546.991	-5.495
5300	81.746	393.336	321.894	378.643	-45.902	558.393	-5.503
5400	81.712	394.864	323.231	386.816	-45.884	569.795	-5.512
5500	81.680	396.363	324.547	394.985	-45.887	581.196	-5.520
5600	81.649	397.834	325.843	403.152	-45.911	592.598	-5.528
5700	81.619	399.279	327.119	411.315	-45.953	603.999	-5.535
5800	81.591	400.698	328.375	419.476	-46.013	615.403	-5.542
5900	81.564	402.093	329.613	427.633	-46.089	626.807	-5.549
6000	81.538	403.464	330.832	435.789	-46.180	638.213	-5.556

PREVIOUS: June 1977 (1 atm) CURRENT: June 1977 (1 bar)

Ammonia-D_3 (ND_3) $D_3N_1(g)$

ELECTRON GAS (e^-) REFERENCE STATE M_r = 0.00054858

0 to 6000 K Ideal Gas

S°(298.15 K) = 20.979 ± 0.013 (±2.0) J K^{-1} mol^{-1}

$\Delta_f H°$(0 K) = 0 kJ mol^{-1}
$\Delta_f H°$(298.15 K) = 0 kJ mol^{-1}

Electronic Levels and Quantum Weights

ε_i, cm^{-1}	g_i
0	2

Enthalpy of Formation

Zero by definition

Heat Capacity and Entropy

As shown by Sommerfeld (1), the electron gas is a degenerate Fermi-Dirac gas and its properties will differ from the classical (Boltzmann) gas. These deviations will increase as the temperature decreases or as the density increases (2,3). Due to the low mass of the electron, these departures from classical behavior will persist to higher temperatures and lower densities than for atomic systems. Under conditions of 1 atm pressure, Gordon (3) showed that the deviation of the Fermi-Dirac gas from the Boltzmann gas is negligible above 1250 K. Below this temperature the deviation between classical and quantum statistics will be significant; Mitchell (4) calculates S°(298.15 K) = 22.72 J K^{-1} mol^{-1} for the Fermi-Dirac gas compared to the classical value of 20.87 J K^{-1} mol^{-1} for a pressure of 1 atm.

Despite these known deviations we have chosen to present the classical (Boltzmann) values here since the primary purpose of this table is to serve as a reference state for the calculation of tables of thermodynamic properties for atomic and molecular ions. When ionization equilibria are considered, the density (or partial pressure) of the electron gas will be extremely low except at very high temperatures. At an electron partial pressure of 10^{-6} bar, the deviation between classical and quantum statistics will be significant only below 5 K (3). Deviations from classical statistics are encountered at about this same temperature in some atomic and molecular systems at 1 bar pressure; these are usually ignored since the deviations in the enthalpy and entropy are quite small. Therefore, although this ideal-gas table has the formalism of 1 bar as the standard reference state, it should not be applied to real systems where the electron partial pressure exceeds 10^{-6} bar. An equivalent statement is that the electron gas will behave as an ideal gas only at electron partial pressures less than about 10^{-6} bar, with deviations from ideal gas behavior at higher pressures as a result of the departure from classical statistics.

The thermodynamic functions are calculated here via Boltzmann statistics assuming the electron gas to be an ideal monatomic gas with two equivalent spin states. The relative ionic mass is the electron rest mass as reported in the 1973 CODATA fundamental constants (5). The first uncertainty in S°(298.15 K) represents the uncertainty in the fundamental constants and electron rest mass (i.e., the uncertainty below 10^{-6} bar where classical statistics can be applied) while the value given in parenthesis represents the uncertainty due to the use of classical rather than quantum statistics at 1 bar pressure. Values below 298.15 K are omitted from the table since the classical calculation (at 1 bar pressure) results in negative values of entropy below 109 K (4). The values of the thermodynamic functions agree with those published by Gurvich et al. (6) except for two minor differences. First, the entropy differs by 0.1094 J K^{-1} mol^{-1} because this table uses a standard state pressure of 1 bar, whereas Gurvich et al. (6) used 1 atm. Second, smaller entropy differences occur due to the use of slightly different values of R.

References

1. A. Sommerfeld, Z. Physik 47, 1 (1928).
2. J. E. Mayer and M. G. Mayer, "Statistical Mechanics", Wiley, New York, 1940.
3. A. R. Gordon, J. Chem. Phys. 4, 678 (1936).
4. A. C. G. Mitchell, Z. Physik 50, 570 (1928).
5. E. R. Cohen and B. N. Taylor, J. Phys. Chem. Ref. Data 2, 663 (1973).
6. L. V. Gurvich, I. V. Veits et al., "Thermodynamic Properties of Individual Substances," 3rd ed., Vol. 1, Nauka, Moscow, 1978.

Electron Gas (e^-) e_1^-(ref)

Enthalpy Reference Temperature = T_r = 298.15 K Standard State Pressure = p° = 0.1 MPa

	J $K^{-1}mol^{-1}$			kJ mol^{-1}			
T/K	$C_p°$	S°	$-[G°-H°(T_r)]/T$	$H°-H°(T_r)$	$\Delta_f H°$	$\Delta_f G°$	Log K_f
0				-6.197			
100							
200							
250							
298.15	20.786	20.979	20.979	0.	0.	0.	0.
300	20.786	21.107	20.979	0.038	0.	0.	0.
350	20.786	24.311	21.232	1.078	0.	0.	0.
400	20.786	27.087	21.794	2.117	0.	0.	0.
450	20.786	29.535	22.521	3.156	0.	0.	0.
500	20.786	31.725	23.334	4.196	0.	0.	0.
600	20.786	35.515	25.058	6.274	0.	0.	0.
700	20.786	38.719	26.786	8.353	0.	0.	0.
800	20.786	41.495	28.455	10.431	0.	0.	0.
900	20.786	43.943	30.043	12.510	0.	0.	0.
1000	20.786	46.133	31.544	14.589	0.	0.	0.
1100	20.786	48.114	32.962	16.667	0.	0.	0.
1200	20.786	49.923	34.301	18.746	0.	0.	0.
1300	20.786	51.587	35.568	20.824	0.	0.	0.
1400	20.786	53.127	36.768	22.903	0.	0.	0.
1500	20.786	54.561	37.907	24.982	0.	0.	0.
1600	20.786	55.903	38.990	27.060	0.	0.	0.
1700	20.786	57.163	40.022	29.139	0.	0.	0.
1800	20.786	58.351	41.008	31.217	0.	0.	0.
1900	20.786	59.475	41.950	33.296	0.	0.	0.
2000	20.786	60.541	42.853	35.375	0.	0.	0.
2100	20.786	61.555	43.720	37.453	0.	0.	0.
2200	20.786	62.522	44.553	39.532	0.	0.	0.
2300	20.786	63.446	45.354	41.610	0.	0.	0.
2400	20.786	64.331	46.127	43.689	0.	0.	0.
2500	20.786	65.179	46.872	45.768	0.	0.	0.
2600	20.786	65.994	47.592	47.846	0.	0.	0.
2700	20.786	66 779	48.288	49.925	0.	0.	0.
2800	20.786	67.535	48.962	52.004	0.	0.	0.
2900	20.786	68.264	49.615	54.082	0.	0.	0.
3000	20.786	68.969	50.249	56.161	0.	0.	0.
3100	20.786	69.650	50.863	58.239	0.	0.	0.
3200	20.786	70.310	51.461	60.318	0.	0.	0.
3300	20.786	70.950	52.042	62.397	0.	0.	0.
3400	20.786	71.570	52.607	64.475	0.	0.	0.
3500	20.786	72.173	53.158	66.554	0.	0.	0.
3600	20.786	72.759	53.694	68.632	0.	0.	0.
3700	20.786	73.328	54.217	70.711	0.	0.	0.
3800	20.786	73.882	54.727	72.790	0.	0.	0.
3900	20.786	74.422	55.225	74.868	0.	0.	0.
4000	20.786	74.949	55.712	76.947	0.	0.	0.
4100	20.786	75.462	56.187	79.025	0.	0.	0.
4200	20.786	75.963	56.652	81.104	0.	0.	0.
4300	20.786	76.452	57.107	83.183	0.	0.	0.
4400	20.786	76.930	57.552	85.261	0.	0.	0.
4500	20.786	77.397	57.988	87.340	0.	0.	0.
4600	20.786	77.854	58.415	89.418	0.	0.	0.
4700	20.786	78.301	58.833	91.497	0.	0.	0.
4800	20.786	78.738	59.243	93.576	0.	0.	0.
4900	20.786	79.167	59.646	95.654	0.	0.	0.
5000	20.786	79.587	60.040	97.733	0.	0.	0.
5100	20.786	79.998	60.428	99.811	0.	0.	0.
5200	20.786	80.402	60.808	101.890	0.	0.	0.
5300	20.786	80.798	61.181	103.969	0.	0.	0.
5400	20.786	81.187	61.548	106.047	0.	0.	0.
5500	20.786	81.568	61.909	108.126	0.	0.	0.
5600	20.786	81.942	62 263	110.204	0.	0.	0.
5700	20.786	82.310	62.612	112.283	0.	0.	0.
5800	20.786	82.672	62.954	114.362	0.	0.	0.
5900	20.786	83.027	63.292	116.440	0.	0.	0.
6000	20.786	83.377	63.623	118.519	0.	0.	0.

PREVIOUS: March 1977 (1 atm) CURRENT: March 1982 (1 bar)

Electron Gas (e^-) e_1^-(ref)

FLUORINE (F) IDEAL GAS M_r = 18.998403

IP(F, g) = 140524.5 ± 0.4 cm^{-1}
S°(298.15 K) = 158.750 ± 0.003 J K^{-1} mol^{-1}

$\Delta_fH°$(0 K) = 77.28 ± 0.30 kJ mol^{-1}
$\Delta_fH°$(298.15 K) = 79.39 ± 0.30 kJ mol^{-1}

Electronic Levels and Quantum Weights

State	ϵ_i, cm^{-1}	g_i
$^2P°_{3/2}$	0.0	4
$^2P°_{1/2}$	404.1	2

Enthalpy of Formation

The adopted value for the enthalpy of formation of F(g) is the value proposed by CODATA (1); $\Delta_fH°$(F, g, 298.15 K) = 79.38±0.30 kJ mol^{-1}. This value is calculated from the dissociation energy of F_2(g), D_0^0 = 12920±50 cm^{-1} (154.56±0.60 kJ mol^{-1}), obtained from the extensive spectroscopic measurements of Colbourn et al. (2). This value is in satisfactory agreement with the results of the equilibrium measurements of Doescher (3) and the photoionization measurements of Berkowitz et al. (4). It is also in good agreement with the value calculated from the enthalpies of formation (5) and vaporization (6) of HF(ℓ) and the spectroscopic data for the dissociation energy of HF(g) (4, 7, 8). This same value is recommended by Brewer and Winn (9), Huber and Herzberg (10) and Gurvich et al. (13), whereas Wagman et al. (21) adopted a dissociation energy which is approximately 0.76 kJ mol^{-1} smaller.

Additional discussions on the enthalpy of formation may be found in the earlier reviews by the JANAF (11) and NBS (12) groups and a recent review by Gurvich et al. (13). Coxon (14) has summarized the spectroscopic information on the lower electronic states of the halogens as of 1973. The more recent spectroscopic studies are discussed by Colbourn et al. (2).

Heat Capacity and Entropy

Information on the electronic energy levels and quantum weights is taken from Moore (15, 16). Only two levels are included in the calculation. All other levels (observed and predicted) lie above 100000 cm^{-1}. Our calculations indicate that any reasonable method of filling in the missing levels and cutting off the summation in the partition function (17) has no effect on the thermodynamic properties to 6000 K. This is undoubtedly a result of the high energy of these levels. Therefore, we list only the ground state and the first excited state. Extension to higher temperatures may require consideration of the higher excited states and utilization of different fill and cutoff procedures.

The thermodynamic functions at 298.15 K are in agreement with recent CODATA recommendations (1) except for two minor differences. First, the entropy differs by 0.1094 J K^{-1} mol^{-1} because this table uses a standard-state pressure of 1 bar, whereas the CODATA recommendations are based on 1 atm. Second, entropy differences of the order of 0.001-0.004 J K^{-1} mol^{-1} for the monatomic halogens arise due to the use of slightly different values for the relative atomic mass and for R; this table uses R = 8.31441 J K^{-1} mol^{-1}. Considering these minor changes, this table agrees within the estimated uncertainty with those by Hultgren et al. (18), Gurvich et al. (13) and Wagman et al. (21). The estimated uncertainty is due to uncertainties in the relative atomic mass and fundamental constants which are based on the 1981 scale (19) and the 1973 values (20), respectively.

References

1. J. D. Cox, chairman, ICSU-CODATA Task Group on Key Values for Thermodynamics, J. Chem. Thermodynamics 10, 903 (1978).
2. E. A. Colbourn, M. Dagenais, A. E. Douglas and J. W. Raymonda, Can. J. Phys. 54, 1343 (1976).
3. R. N. Doescher, J. Chem. Phys. 20, 330 (1952).
4. J. Berkowitz, W. A. Chupka, P. M. Guyon, J. H. Holloway and R. Spohr, J. Chem. Phys. 54, 5165 (1971).
5. G. K. Johnson, P. N. Smith and W. N. Hubbard, J. Chem. Thermodynamics 5, 793, 1973.
6. C. E. Vanderzee and W. M. Rodenburg, J. Chem. Thermodynamics 3, 267 (1971).
7. G. DiLonardo and A. E. Douglas, Can. J. Phys. 51, 434 (1973).
8. J. W. C. Johns and R. F. Barrow, Proc. Roy. Soc. A251, 504 (1959).
9. L. Brewer and J. S. Winn, Faraday Symp. Chem. Soc. (GB), No. 14, 126 (1980).
10. K. P. Huber and G. Herzberg, "Constants of Diatomic Molecules," Van Nostrand Reinhold Company, New York, 1979.
11. H. Prophet and D. R. Stull, ed., "JANAF Thermochemical Tables", 2nd ed., NSRDS-NBS 37, 1971.
12. S. Abramowitz et al., U. S. Nat. Bur. Stand., Rept. 10904, 239, July, 1972.
13. L. V. Gurvich, I. V. Veits et al., "Thermodynamic Properties of Individual Substances," 3rd ed., Vol. 1, Nauka, Moscow, 1978.
14. J. A. Coxon in "Molecular Spectroscopy," Vol. 1, R. F. Barrow, D. A. Long and D. J. Miller, Senior Reporters, The Chemical Society, London, 1973, Chapter 4.
15. C. E. Moore, U. S. Nat. Bur. Stand., NSRDS-NBS-35, Volume I, 1970 [Reprint of NBS Circular 467, Volume I, 1949].
16. C. E. Moore, U. S. Nat. Bur. Stand., NSRDS-NBS-34, 1970.
17. J. R. Downey, Jr. The Dow Chemical Company, AFOSR-TR-78-0360, Contract No. F44620-75-1-0048, 1978.
18. R. Hultgren, P. D. Desai et al., "Selected Values of the Thermodynamic Properties of the Elements," American Society for Metals, Metals Park, Ohio, 1973.
19. N. E. Holden and R. L. Martin, Pure Appl. Chem. 55, 1101 (1983).
20. E. R. Cohen and B. N. Taylor, J. Phys. Chem. Ref. Data 2, 663 (1973).
21. D. D. Wagman, W. H. Evans et al., J. Phys. Chem. Ref. Data 11, Supp. 2 (1982).

Fluorine (F) F_1(g)

Enthalpy Reference Temperature = T_r = 298.15 K; Standard State Pressure = p° = 0.1 MPa

T/K	$C_p°$ (J $K^{-1}mol^{-1}$)	S° (J $K^{-1}mol^{-1}$)	$-[G°-H°(T_r)]/T$ (J $K^{-1}mol^{-1}$)	$H°-H°(T_r)$ (kJ mol^{-1})	$\Delta_fH°$ (kJ mol^{-1})	$\Delta_fG°$ (kJ mol^{-1})	Log K_f
0	0.	0.	INFINITE	-6.518	77.284	77.284	INFINITE
100	21.205	134.479	178.805	-4.433	77.916	72.987	-38.124
200	22.605	149.670	160.833	-2.233	78.652	67.783	-17.703
250	22.783	154.738	159.125	-1.097	79.037	65.021	-13.585
298.15	22.746	158.750	158.750	0.	79.390	62.289	-10.913
300	22.742	158.891	158.751	0.042	79.403	62.183	-10.827
350	22.603	162.387	159.027	1.176	79.743	59.286	-8.848
400	22.431	165.394	159.639	2.302	80.053	56.342	-7.358
450	22.259	168.026	160.428	3.419	80.337	53.361	-6.194
500	22.100	170.363	161.307	4.528	80.597	50.350	-5.260
600	21.832	174.368	163.161	6.724	81.056	44.256	-3.853
700	21.628	177.717	165.008	8.897	81.453	38.090	-2.842
800	21.474	180.594	166.780	11.051	81.803	31.871	-2.081
900	21.357	183.117	168.458	13.193	82.117	25.611	-1.486
1000	21.266	185.362	170.038	15.324	82.403	19.317	-1.009
1100	21.194	187.385	171.525	17.446	82.666	12.995	-0.617
1200	21.137	189.227	172.924	19.563	82.909	6.651	-0.290
1300	21.091	190.917	174.244	21.674	83.136	0.287	-0.012
1400	21.053	192.478	175.492	23.781	83.348	-6.095	0.227
1500	21.022	193.930	176.673	25.885	83.546	-12.490	0.435
1600	20.996	195.286	177.795	27.986	83.733	-18.899	0.617
1700	20.974	196.558	178.861	30.085	83.909	-25.319	0.778
1800	20.955	197.756	179.878	32.181	84.075	-31.749	0.921
1900	20.939	198.889	180.849	34.276	84.234	-38.188	1.050
2000	20.925	199.963	181.778	36.369	84.387	-44.635	1.166
2100	20.913	200.983	182.669	38.461	84.535	-51.090	1.271
2200	20.902	201.956	183.523	40.551	84.681	-57.551	1.366
2300	20.893	202.885	184.345	42.641	84.827	-64.020	1.454
2400	20.885	203.774	185.136	44.730	84.974	-70.494	1.534
2500	20.877	204.626	185.899	46.818	85.125	-76.975	1.608
2600	20.871	205.445	186.635	48.906	85.281	-83.462	1.677
2700	20.865	206.232	187.346	50.992	85.444	-89.956	1.740
2800	20.860	206.991	188.034	53.079	85.617	-96.455	1.799
2900	20.855	207.723	188.701	55.164	85.800	-102.961	1.855
3000	20.851	208.430	189.347	57.250	85.994	-109.473	1.906
3100	20.847	209.114	189.973	59.334	86.202	-115.992	1.954
3200	20.843	209.775	190.582	61.419	86.423	-122.518	2.000
3300	20.840	210.417	191.173	63.503	86.660	-129.051	2.043
3400	20.837	211.039	191.748	65.587	86.911	-135.592	2.083
3500	20.834	211.643	192.308	67.670	87.179	-142.140	2.121
3600	20.831	212.230	192.854	69.754	87.463	-148.695	2.158
3700	20.829	212.800	193.385	71.837	87.764	-155.259	2.192
3800	20.827	213.356	193.903	73.920	88.081	-161.832	2.225
3900	20.825	213.897	194.409	76.002	88.416	-168.413	2.256
4000	20.823	214.424	194.903	78.084	88.767	-175.003	2.285
4100	20.821	214.938	195.385	80.167	89.135	-181.602	2.314
4200	20.820	215.440	195.857	82.249	89.520	-188.209	2.341
4300	20.818	215.930	196.318	84.331	89.921	-194.827	2.367
4400	20.817	216.408	196.769	86.412	90.338	-201.454	2.392
4500	20.815	216.876	197.211	88.494	90.771	-208.090	2.415
4600	20.814	217.334	197.643	90.575	91.220	-214.736	2.438
4700	20.813	217.781	198.067	92.657	91.685	-221.392	2.461
4800	20.812	218.219	198.482	94.738	92.164	-228.059	2.482
4900	20.811	218.648	198.889	96.819	92.658	-234.735	2.502
5000	20.810	219.069	199.289	98.900	93.166	-241.422	2.522
5100	20.809	219.481	199.681	100.981	93.687	-248.119	2.541
5200	20.808	219.885	200.065	103.062	94.223	-254.826	2.560
5300	20.807	220.281	200.443	105.143	94.771	-261.544	2.578
5400	20.807	220.670	200.814	107.224	95.332	-268.272	2.595
5500	20.806	221.052	201.179	109.304	95.905	-275.011	2.612
5600	20.805	221.427	201.537	111.385	96.491	-281.760	2.628
5700	20.805	221.795	201.889	113.465	97.088	-288.520	2.644
5800	20.804	222.157	202.235	115.546	97.696	-295.290	2.659
5900	20.803	222.513	202.576	117.626	98.315	-302.071	2.674
6000	20.803	222.862	202.911	119.706	98.944	-308.862	2.689

PREVIOUS: September 1965 (1 atm) CURRENT: June 1982 (1 bar)

Fluorine (F) F_1(g)

FLUORINE, ION (F^+)　　IDEAL GAS　　M_r = 18.997854

$IP(F^+, g)$ = 282058.6 ± 1.5 cm^{-1}

S°(298.15 K) = 161.728 ± 0.005 $J\ K^{-1}\ mol^{-1}$

$\Delta_f H°$(0 K) = 1758.33 ± 0.3 $kJ\ mol^{-1}$

$\Delta_f H°$(298.15 K) = [1766.826] $kJ\ mol^{-1}$

Electronic Levels and Quantum Weights

State	ε_i, cm^{-1}	g_i
$^3P_2$	0.0	5
$^3P_1$	341.0	3
$^3P_0$	489.9	1
$^1D_2$	20873.4	5
$^1S_0$	44918.1	1

Enthalpy of Formation

$\Delta_f H°(F^+$, g, 0 K) is calculated from $\Delta_f H°$(F, g, 0 K) (1) using the spectroscopic value of IP(F) = 140524.5±0.4 cm^{-1} (1681.047±0.005 $kJ\ mol^{-1}$) from Moore (2). The ionization limit is converted from cm^{-1} to $kJ\ mol^{-1}$ using the factor, 1 cm^{-1} = 0.01196266 $kJ\ mol^{-1}$, which is derived from the 1973 CODATA fundamental constants (3). Rosenstock et al. (4) and Levin and Lias (5) have summarized additional ionization potential and appearance potential data.

$\Delta_f H°(F^+$, g, 298.15 K) is calculated from $\Delta_f H°(F^+$, g, 0 K) by using IP(F) with JANAF (1) enthalpies, H°(0 K)-H°(298.15 K), for F(g), F^+(g), and e^-(g). $\Delta_f H°(F \rightarrow F^+ + e^-$, 298.15 K) differs from a room temperature threshold energy due to inclusion of these enthalpies and to threshold effects discussed by Rosenstock et al. (4). $\Delta_f H°$(298.15 K) should be changed by -6.197 $kJ\ mol^{-1}$ if it is to be used in the ion convention that excludes the enthalpy of the electron.

Heat Capacity and Entropy

The information on electronic energy levels and quantum weights, given by Moore (2, 6), is incomplete because many theoretically predicted levels have not been observed. Our calculations indicate that any reasonable method of filling in these missing levels and cutting off the summation in the partition function (7) has no effect on the thermodynamic functions to 6000 K. This is a result of the high energy of all levels other than the ground state and the lowest four excited states; the next lowest level is approximately 164797 cm^{-1} above the ground state. Since inclusion of these upper levels has no effect on the thermodynamic functions (to 6000 K), we list only the ground state and four excited states, with the energy of these states taken from recent study by Moore (2). The reported uncertainty in S°(298.15 K) is due to uncertainties in the relative ionic mass, the fundamental constants, and the position of the four lowest excited states. Extension of these calculations above 6000 K may require consideration of the higher excited states and use of different fill and cutoff procedures (7).

References

1. JANAF Thermochemical Tables: F(g), 3-31-82; e^-(r), 3-31-82.
2. C. E. Moore, U. S. Nat. Bur. Stand., NSRDS-NBS-34, 1970.
3. E. R. Cohen and B. N. Taylor, J. Phys. Chem. Ref. Data 2, 663 (1973).
4. H. M. Rosenstock, K. Draxl et al., J. Phys. Chem. Ref. Data 6, Supp. 1 (1977).
5. R. D. Levin and S. G. Lias, U. S. Nat. Bur. Stand., NSRDS-NBS-71, 1982.
6. C. E. Moore, U. S. Nat. Bur. Stand., NSRDS-NBS-35, Volume I, 1970 [reprint of NBS Circular 467, Volume I, 1949].
7. J. R. Downey, Jr., The Dow Chemical Company, AFOSR-TR-78-0960, Contract No. F44620-75-1-0048, 1978.

Fluorine, Ion (F^+)　　$F_1^+(g)$

Enthalpy Reference Temperature = T_r = 298.15 K　　Standard State Pressure = p° = 0.1 MPa

	J K⁻¹mol⁻¹			kJ mol⁻¹			
T/K	$C_p°$	S°	$-[G°-H°(T_r)]/T$	$H°-H°(T_r)$	$\Delta_f H°$	$\Delta_f G°$	Log K_f
0	0.	0.	INFINITE	-6.711	1758.331		
100	21.738	136.478	182.608	-4.613			
200	23.642	152.281	163.889	-2.322			
250	23.694	157.569	162.116	-1.137			
298.15	23.497	161.728	161.728	0.	1766.826	1742.583	-305.293
300	23.487	161.873	161.728	0.043	1766.879	1742.433	-303.385
350	23.197	165.472	162.013	1.211	1768.291	1738.246	-259.419
400	22.904	168.550	162.643	2.363	1769.668	1733.860	-226.419
450	22.636	171.232	163.451	3.502	1771.012	1729.303	-200.732
500	22.404	173.605	164.350	4.627	1772.328	1724.597	-180.167
600	22.035	177.655	166.241	6.848	1774.891	1714.809	-149.287
700	21.770	181.031	168.120	9.038	1777.383	1704.598	-127.199
800	21.577	183.925	169.919	11.205	1779.824	1694.033	-110.609
900	21.433	186.457	171.619	13.355	1782.225	1683.164	-97.688
1000	21.324	188.710	173.217	15.492	1784.597	1672.030	-87.338
1100	21.240	190.738	174.719	17.621	1786.943	1660.659	-78.858
1200	21.174	192.583	176.132	19.741	1789.270	1649.076	-71.782
1300	21.121	194.276	177.464	21.856	1791.578	1637.300	-65.788
1400	21.078	195.840	178.721	23.966	1793.872	1625.346	-60.642
1500	21.043	197.293	179.911	26.072	1796.151	1613.229	-56.178
1600	21.014	198.650	181.041	28.174	1798.418	1600.960	-52.266
1700	20.989	199.923	182.114	30.275	1800.674	1588.550	-48.810
1800	20.968	201.122	183.137	32.372	1802.920	1576.007	-45.735
1900	20.951	202.255	184.114	34.468	1805.159	1563.340	-42.979
2000	20.935	203.329	185.048	36.563	1807.391	1550.555	-40.496
2100	20.922	204.350	185.943	38.656	1809.619	1537.658	-38.247
2200	20.911	205.324	186.802	40.747	1811.845	1524.655	-36.200
2300	20.902	206.253	187.628	42.838	1814.070	1511.551	-34.328
2400	20.894	207.142	188.422	44.928	1816.297	1498.351	-32.611
2500	20.887	207.995	189.188	47.017	1818.527	1485.057	-31.029
2600	20.882	208.814	189.928	49.105	1820.763	1471.674	-29.566
2700	20.879	209.602	190.642	51.193	1823.006	1458.205	-28.211
2800	20.876	210.361	191.333	53.281	1825.259	1444.653	-26.950
2900	20.876	211.094	192.001	55.368	1827.522	1431.020	-25.775
3000	20.876	211.802	192.650	57.456	1829.798	1417.309	-24.678
3100	20.878	212.486	193.279	59.544	1832.087	1403.521	-23.649
3200	20.882	213.149	193.889	61.632	1834.391	1389.660	-22.684
3300	20.888	213.792	194.483	63.720	1836.710	1375.726	-21.776
3400	20.895	214.416	195.060	65.809	1839.045	1361.722	-20.920
3500	20.904	215.021	195.622	67.899	1841.398	1347.649	-20.113
3600	20.914	215.610	196.169	69.990	1843.768	1333.508	-19.349
3700	20.926	216.184	196.702	72.082	1846.156	1319.302	-18.625
3800	20.940	216.742	197.222	74.175	1848.563	1305.030	-17.939
3900	20.956	217.286	197.729	76.270	1850.988	1290.695	-17.287
4000	20.974	217.817	198.225	78.367	1853.432	1276.297	-16.667
4100	20.993	218.335	198.709	80.465	1855.895	1261.839	-16.076
4200	21.014	218.841	199.183	82.565	1858.377	1247.319	-15.513
4300	21.037	219.336	199.645	84.668	1860.877	1232.741	-14.975
4400	21.061	219.820	200.099	86.773	1863.396	1218.104	-14.461
4500	21.086	220.293	200.542	88.880	1865.934	1203.410	-13.969
4600	21.113	220.757	200.976	90.990	1868.490	1188.659	-13.498
4700	21.141	221.211	201.402	93.103	1871.064	1173.852	-13.046
4800	21.171	221.657	201.820	95.218	1873.656	1158.991	-12.612
4900	21.202	222.094	202.229	97.337	1876.266	1144.075	-12.196
5000	21.234	222.522	202.630	99.459	1878.893	1129.105	-11.796
5100	21.267	222.943	203.025	101.584	1881.538	1114.083	-11.411
5200	21.300	223.356	203.412	103.712	1884.199	1099.009	-11.040
5300	21.335	223.762	203.792	105.844	1886.877	1083.884	-10.682
5400	21.371	224.162	204.165	107.979	1889.571	1068.708	-10.338
5500	21.407	224.554	204.532	110.118	1892.282	1053.482	-10.005
5600	21.443	224.940	204.893	112.261	1895.007	1038.206	-9.684
5700	21.480	225.320	205.248	114.407	1897.749	1022.882	-9.374
5800	21.518	225.694	205.598	116.557	1900.505	1007.509	-9.074
5900	21.556	226.062	205.942	118.710	1903.276	992.089	-8.783
6000	21.594	226.425	206.280	120.868	1906.061	976.622	-8.502

PREVIOUS:　　CURRENT: June 1982 (1 bar)

Fluorine, Ion (F^+)　　$F_1^+(g)$

FLUORINE, ION (F^-) IDEAL GAS M_r = 18.998952

EA(F, g) = 3.399 ± 0.003 eV
$S°$(298.15 K) = 145.576 ± 0.005 J K^{-1} mol^{-1}

$\Delta_f H°$(0 K) = -250.67 ± 0.6 kJ mol^{-1}
$\Delta_f H°$(298.15 K) = [-255.079] kJ mol^{-1}

Electronic Level and Quantum Weight

State	ε_i, cm^{-1}	g_i
$^1S_0$	0.0	1

Enthalpy of Formation

$\Delta_f H°(F^-$, g, 0 K) is calculated from $\Delta_f H°$(F, g, 0 K) (1) using the adopted election affinity of EA(F) = 3.399±0.003 eV (327.951±0.289 kJ mol^{-1}). This value, recommended by Hotop and Lineberger (10), is based on plasma absorption and plasma emission studies (5, 6). Additional information on F^-(g) may be obtained in the critical discussions of Hotop and Lineberger (2, 10) Rosenstock et al. (3) and Massey (4).

$\Delta_f H°(F^-$, g, 298.15 K) is obtained from $\Delta_f H°$(F, g, 0 K) by using EA(F) with JANAF (1) enthalpies, H°(0 K) - H°(298.15 K), for F^-(g), F(g), and e^-(g). $\Delta_r H°(F^- \rightarrow F + e^-$, 298.15 K) differs from a room-temperature threshold energy due to inclusion of these enthalpies and to threshold effects discussed by Rosenstock et al. (3). $\Delta_f H°$(298.15 K) should be changed by +6.197 kJ mol^{-1} if it is to be used in the ion convention that excludes the enthalpy of the electron.

Heat Capacity and Entropy

The ground state configuration for F^-(g) is given by Hotop and Lineberger (2, 10), Rosenstock et al. (3) and Massey (4). Lacking any experimental evidence as to the stability of any excited states, we assume that no stable excited states exist.

The entropy at all temperatures (0-6000 K) agrees within 0.001 J K^{-1} mol^{-1} with the values of Gurvich et al. (8) except for one minor difference. The entropy differs by 0.1094 J K^{-1} mol^{-1} because this table uses a standard-state pressure of 1 bar, whereas the tabulation of Gurvich et al. (8) is based on 1 atm. The estimated uncertainty is due to uncertainties in the relative ionic mass and the fundamental constants which are based on the 1981 scale (9) and the 1973 values (7), respectively.

References

1. JANAF Thermochemical Tables: F(g), 3-31-82; e^-(g), 3-31-82.
2. H. Hotop and W. C. Lineberger, J. Phys. Chem. Ref. Data 4, 539 (1975).
3. H. M. Rosenstock, K. Draxl et al., J. Phys. Chem. Ref. Data 6, Supp. 1 (1977).
4. H. S. W. Massey, "Negative Ions," 3rd ed., Cambridge University Press, Cambridge, 1976.
5. R. Milstein and R. S. Berry, J. Chem. Phys. 55, 4146 (1971).
6. H.-P. Popp, Z. Naturforsch. 22a, 254 (1967).
7. E. R. Cohen and B. N. Taylor, J. Phys. Chem. Ref. Data 2, 663 (1973).
8. L. V. Gurvich, I. V. Veits et al., "Thermodynamic Properties of Individual Substances", 3rd ed., Vol. 1, Nauka, Moscow, 1978.
9. N. E. Holden and R. L. Martin, Pure Appl. Chem. 55, 1101 (1983).
10. H. Hotop and W. C. Lineberger, J. Phys. Chem. Ref. Data, to be published, 1985.

Fluorine, Ion (F^-) F_1^-(g)

Enthalpy Reference Temperature = T_r = 298.15 K Standard State Pressure = p° = 0.1 MPa

	J $K^{-1}mol^{-1}$			kJ mol^{-1}			
T/K	$C_p°$	S°	-[G°-H°(T_r)]/T	H°-H°(T_r)	$\Delta_f H°$	$\Delta_f G°$	Log K_f
0	0.	0.	INFINITE	-6.197	-250.667		
100	20.786	122.869	164.056	-4.119			
200	20.786	137.276	147.477	-2.040			
250	20.786	141.915	145.918	-1.001			
298.15	20.786	145.576	145.576	0.	-255.079	-261.997	45.901
300	20.786	145.704	145.576	0.038	-255.108	-262.040	45.625
350	20.786	148.909	145.829	1.078	-255.903	-263.133	39.270
400	20.786	151.684	146.392	2.117	-256.718	-264.110	34.489
450	20.786	154.132	147.118	3.156	-257.552	-264.984	30.759
500	20.786	156.322	147.931	4.196	-258.401	-265.765	27.764
600	20.786	160.112	149.655	6.274	-260.138	-267.075	23.251
700	20.786	163.316	151.384	8.353	-261.913	-268.092	20.005
800	20.786	166.092	153.053	10.431	-263.718	-268.852	17.554
900	20.786	168.540	154.640	12.510	-265.545	-269.384	15.635
1000	20.786	170.730	156.142	14.589	-267.390	-269.711	14.088
1100	20.786	172.711	157.559	16.667	-269.250	-269.854	12.814
1200	20.786	174.520	158.898	18.746	-271.123	-269.826	11.745
1300	20.786	176.184	160.165	20.824	-273.008	-269.641	10.834
1400	20.786	177.724	161.365	22.903	-274.903	-269.312	10.048
1500	20.786	179.158	162.504	24.982	-276.808	-268.846	9.362
1600	20.786	180.500	163.587	27.060	-278.723	-268.253	8.758
1700	20.786	181.760	164.619	29.139	-280.645	-267.539	8.220
1800	20.786	182.948	165.605	31.217	-282.575	-266.713	7.740
1900	20.786	184.072	166.548	33.296	-284.511	-265.779	7.307
2000	20.786	185.138	167.451	35.375	-286.452	-264.743	6.914
2100	20.786	186.152	168.317	37.453	-288.395	-263.609	6.557
2200	20.786	187.119	169.150	39.532	-290.340	-262.383	6.230
2300	20.786	188.043	169.952	41.610	-292.284	-261.069	5.929
2400	20.786	188.928	170.724	43.689	-294.226	-259.670	5.652
2500	20.786	189.776	171.469	45.768	-296.163	-258.191	5.395
2600	20.786	190.591	172.189	47.846	-298.094	-256.634	5.156
2700	20.786	191.376	172.885	49.925	-300.017	-255.002	4.933
2800	20.786	192.132	173.559	52.004	-301.931	-253.300	4.725
2900	20.786	192.861	174.212	54.082	-303.834	-251.530	4.531
3000	20.786	193.566	174.846	56.161	-305.725	-249.694	4.348
3100	20.786	194.248	175.461	58.239	-307.602	-247.795	4.175
3200	20.786	194.907	176.058	60.318	-309.465	-245.836	4.013
3300	20.786	195.547	176.639	62.397	-311.313	-243.819	3.859
3400	20.786	196.168	177.204	64.475	-313.145	-241.746	3.714
3500	20.786	196.770	177.755	66.554	-314.961	-239.620	3.576
3600	20.786	197.356	178.291	68.632	-316.760	-237.442	3.445
3700	20.786	197.925	178.814	70.711	-318.542	-235.214	3.321
3800	20.786	198.480	179.324	72.790	-320.308	-232.938	3.202
3900	20.786	199.019	179.823	74.868	-322.056	-230.616	3.089
4000	20.786	199.546	180.309	76.947	-323.787	-228.249	2.981
4100	20.786	200.059	180.785	79.025	-325.501	-225.840	2.877
4200	20.786	200.560	181.249	81.104	-327.199	-223.388	2.778
4300	20.786	201.049	181.704	83.183	-328.879	-220.897	2.683
4400	20.786	201.527	182.149	85.261	-330.544	-218.367	2.592
4500	20.786	201.994	182.585	87.340	-332.192	-215.798	2.505
4600	20.786	202.451	183.012	89.418	-333.825	-213.194	2.421
4700	20.786	202.898	183.430	91.497	-335.442	-210.554	2.340
4800	20.786	203.335	183.841	93.576	-337.044	-207.880	2.262
4900	20.786	203.764	184.243	95.654	-338.631	-205.172	2.187
5000	20.786	204.184	184.637	97.733	-340.204	-202.433	2.115
5100	20.786	204.596	185.025	99.811	-341.763	-199.662	2.045
5200	20.786	204.999	185.405	101.890	-343.309	-196.860	1.977
5300	20.786	205.395	185.778	103.969	-344.841	-194.029	1.912
5400	20.786	205.784	186.145	106.047	-346.361	-191.170	1.849
5500	20.786	206.165	186.506	108.126	-347.868	-188.282	1.788
5600	20.786	206.540	186.860	110.204	-349.363	-185.367	1.729
5700	20.786	206.908	187.209	112.283	-350.847	-182.425	1.672
5800	20.786	207.269	187.552	114.362	-352.319	-179.458	1.616
5900	20.786	207.624	187.889	116.440	-353.781	-176.465	1.562
6000	20.786	207.974	188.221	118.519	-355.231	-173.447	1.510

PREVIOUS: December 1971 (1 atm) CURRENT: June 1982 (1 bar)

Fluorine, Ion (F^-) F_1^-(g)

IRON FLUORIDE (FeF) IDEAL GAS $M_r = 74.845403$

$S°(298.15\ K) = [240.337]\ J\ K^{-1}\ mol^{-1}$

$\Delta_f H°(0\ K) = [47.549 \pm 20.9]\ kJ\ mol^{-1}$
$\Delta_f H°(298.15\ K) = [47.698 \pm 20.9]\ kJ\ mol^{-1}$

Electronic Level and Quantum Weight

State	ε_i, cm^{-1}	g_i
$[^6\Sigma]$	0	[6]

$\omega_e = [630.0]\ cm^{-1}$ $\omega_e x_e = [2.7]\ cm^{-1}$ $\sigma = 1$
$B_e = [0.3670]\ cm^{-1}$ $\alpha_e = [0.0017]\ cm^{-1}$ $r_e = [1.8]$ Å

Enthalpy of Formation

The dissociation energy of FeF(g) was estimated as 4.64±0.22 eV or 107±5 kcal mol^{-1} by Margrave (1). From the value of D_0^0(FeF), the enthalpy of formation, $\Delta_f H°(298.15\ K)$, for FeF(g) is derived to be 11.4±5.0 kcal mol^{-1}.

Heat Capacity and Entropy

The ground state configuration and bond distance (r_e) were estimated by comparison with those for FeCl(g). The value of B_e was calculated using the relationship $B_e = (2.79889 \times 10^{-39})/I$ where I = moment of inertia of FeF(g), and α_e was estimated by comparison with those for AlF(g), AlCl(g) and FeCl(g). The value of ω_e was derived from the bond distance, reduced mass and number of valence electrons according to the method given by Guggenheimer (2) and the $\omega_e x_e$ was estimated from those for AlF(g), AlCl(g) and FeCl(g).

References

1. J. L. Margrave, "Optical Spectra and Molecular Parameters of Light Element Molecules", William Marsh Rice University, Houston, Texas, Progress Report No. 8, Jan. 1 to Mar. 31, 1965.
2. K. M. Guggenheimer, Proc. Phys. Soc. (London) 58, 456 (1946).

Iron Fluoride (FeF) $F_1Fe_1(g)$

Enthalpy Reference Temperature = T_r = 298.15 K Standard State Pressure = p° = 0.1 MPa

T/K	$C_p°$ (J K⁻¹mol⁻¹)	S° (J K⁻¹mol⁻¹)	$-[G°-H°(T_r)]/T$ (J K⁻¹mol⁻¹)	$H°-H°(T_r)$ (kJ mol⁻¹)	$\Delta_f H°$ (kJ mol⁻¹)	$\Delta_f G°$ (kJ mol⁻¹)	Log K_f
0	0.	0.	INFINITE	-9.067	47.549	47.549	INFINITE
100	29.193	206.817	268.389	-6.157	48.583	37.027	-19.341
200	31.049	227.505	243.314	-3.162	48.345	25.499	-6.660
250	32.266	234.566	240.880	-1.578	48.040	19.822	-4.142
298.15	33.264	240.337	240.337	0.	47.698	14.418	-2.526
300	33.298	240.543	240.338	0.062	47.684	14.211	-2.474
350	34.117	245.740	240.747	1.748	47.289	8.663	-1.293
400	34.754	250.339	241.664	3.470	46.855	3.174	-0.414
450	35.249	254.463	242.861	5.221	46.374	-2.258	0.262
500	35.639	258.197	244.211	6.993	45.842	-7.633	0.797
600	36.201	264.749	247.103	10.587	44.614	-18.217	1.586
700	36.578	270.359	250.034	14.228	43.152	-28.576	2.132
800	36.846	275.262	252.888	17.899	41.405	-38.708	2.527
900	37.045	279.614	255.620	21.594	39.244	-48.598	2.821
1000	37.200	283.525	258.218	25.307	36.364	-58.213	3.041
1100	37.324	287.077	260.683	29.033	32.588	-67.474	3.204
1200	37.428	290.329	263.020	32.771	29.363	-76.447	3.328
1300	37.516	293.328	265.237	36.518	27.782	-85.201	3.423
1400	37.594	296.111	267.344	40.274	26.115	-93.830	3.501
1500	37.663	298.708	269.350	44.037	24.362	-102.337	3.564
1600	37.725	301.140	271.261	47.806	22.523	-110.724	3.615
1700	37.783	303.429	273.087	51.582	19.647	-118.975	3.656
1800	37.837	305.590	274.833	55.363	17.302	-127.062	3.687
1900	37.887	307.637	276.506	59.149	0.774	-134.314	3.693
2000	37.935	309.582	278.112	62.940	-1.978	-141.353	3.692
2100	37.981	311.434	279.655	66.736	-4.728	-148.254	3.688
2200	38.026	313.202	281.140	70.536	-7.474	-155.024	3.681
2300	38.068	314.893	282.571	74.341	-10.216	-161.670	3.672
2400	38.110	316.514	283.952	78.150	-12.951	-168.196	3.661
2500	38.151	318.071	285.285	81.963	-15.678	-174.609	3.648
2600	38.191	319.568	286.575	85.780	-18.394	-180.912	3.635
2700	38.230	321.010	287.824	89.601	-21.099	-187.111	3.620
2800	38.268	322.401	289.034	93.426	-23.790	-193.211	3.604
2900	38.306	323.744	290.208	97.255	-26.467	-199.214	3.588
3000	38.344	325.044	291.348	101.087	-29.127	-205.126	3.572
3100	38.381	326.302	292.455	104.923	-31.771	-210.949	3.554
3200	38.417	327.521	293.532	108.763	-382.696	-209.263	3.416
3300	38.454	328.703	294.580	112.607	-383.402	-203.832	3.226
3400	38.490	329.852	295.601	116.454	-384.122	-198.380	3.048
3500	38.525	330.968	296.595	120.305	-384.855	-192.906	2.879
3600	38.561	332.054	297.565	124.159	-385.602	-187.411	2.719
3700	38.596	333.111	298.512	128.017	-386.363	-181.895	2.568
3800	38.632	334.141	299.436	131.878	-387.139	-176.359	2.424
3900	38.667	335.145	300.339	135.743	-387.931	-170.802	2.288
4000	38.702	336.124	301.221	139.612	-388.740	-165.224	2.158
4100	38.736	337.080	302.084	143.484	-389.567	-159.626	2.034
4200	38.771	338.014	302.928	147.359	-390.414	-154.007	1.915
4300	38.806	338.927	303.755	151.238	-391.280	-148.369	1.802
4400	38.840	339.819	304.565	155.120	-392.168	-142.709	1.694
4500	38.874	340.692	305.358	159.006	-393.079	-137.030	1.591
4600	38.909	341.547	306.135	162.895	-394.014	-131.329	1.491
4700	38.943	342.384	306.898	166.788	-394.976	-125.608	1.396
4800	38.977	343.205	307.645	170.684	-395.964	-119.867	1.304
4900	39.011	344.009	308.379	174.583	-396.982	-114.104	1.216
5000	39.045	344.797	309.100	178.486	-398.030	-108.321	1.132
5100	39.079	345.571	309.807	182.392	-399.110	-102.516	1.050
5200	39.113	346.330	310.503	186.301	-400.224	-96.689	0.971
5300	39.147	347.075	311.186	190.214	-401.374	-90.841	0.895
5400	39.180	347.807	311.857	194.131	-402.561	-84.971	0.822
5500	39.214	348.526	312.517	198.050	-403.789	-79.079	0.751
5600	39.248	349.233	313.167	201.974	-405.059	-73.163	0.682
5700	39.281	349.928	313.805	205.900	-406.374	-67.225	0.616
5800	39.315	350.612	314.434	209.830	-407.736	-61.264	0.552
5900	39.349	351.284	315.053	213.763	-409.148	-55.278	0.489
6000	39.382	351.946	315.662	217.700	-410.613	-49.268	0.429

PREVIOUS: September 1965 (1 atm) CURRENT: September 1965 (1 bar)

Iron Fluoride (FeF) $F_1Fe_1(g)$

HYDROGEN FLUORIDE (HF) IDEAL GAS M_r = 20.006343

$S°(298.15\ K) = 173.780 \pm 0.033\ J\ K^{-1}\ mol^{-1}$

$\Delta_f H°(0\ K) = -272.499 \pm 0.8\ kJ\ mol^{-1}$

$\Delta_f H°(298.15\ K) = -272.546 \pm 0.8\ kJ\ mol^{-1}$

Electronic States and Molecular Constants (σ = 1)

State	T_o, cm^{-1}	g_i	ω_e, cm^{-1}	$\omega_e x_e$, cm^{-1}	B_e, cm^{-1}	α_e, cm^{-1}	r_e, Å
$X^1\Sigma^+$	0	1	4138.73	90.05	20.9555	0.7958	0.9168
$B^1\Sigma^+$	83275	1	1158.46	17.718	4.0263	0.0173	2.091

Enthalpy of Formation

The enthalpy of formation was adopted from the 1968 JANAF Thermochemical Table (1). This value was obtained by simultaneous selection (2) of $\Delta_f H°$(298.15 K) for HF(g), BF_3(g), NF_3(g), CF_4(g), NaF(cr), and C_2F_4(polymer). Subsequent data have led CODATA (3) to recommend the more negative value of -65.32±0.17 kcal mol^{-1} for HF(g).

Heat Capacity and Entropy

The vibrational and rotational constants of the respective electronic levels were taken from Rosen (4). These constants do not differ appreciably from those given in the 1968 JANAF Tables (1). The values of $\Delta_f H°$, $\Delta_f G°$, and log Kp are appreciably different because of the new thermal functions for the reference state F_2 (5).

The National Bureau of Standards prepared this table (6) from data existing in 1972. Using molecular constants and $\Delta_f H°$ selected by NBS (6), we recalculate the table in terms of 1973 fundamental constants (7), 1975 atomic weights (8), and current JANAF reference states for the elements (5). A more complete analysis of electronic states and molecular constants is now available (9).

References

1. JANAF Thermochemical Tables, 2nd ed., NSRDS-NBS 37, 1971; HF(g), 6-31-68.
2. H. Prophet and A. N. Syverud, The Dow Chemical Co., AD-867352, 76 pp (1970).
3. ICSU-CODATA Task Group on Key Values for Thermodynamics, J. Chem. Thermodynamics 8, 603 (1976).
4. B. Rosen, "Spectroscopic Data Relative to Diatomic Molecules," Pergamon Press, New York, 1970.
5. JANAF Thermochemical Tables: F_2(ref), 6-30-77; H_2(ref), 3-31-77.
6. S. Abramowitz et al., U. S. Nat. Bur. Stand. Rept. 10904, 239, July, 1972.
7. CODATA Task Group on Fundamental Constants, CODATA Bulletin 11, December, 1973.
8. IUPAC Commission on Atomic Weights, Pure Appl. Chem. 47, 75 (1976).
9. G. Di Lonardo and A. E. Douglas, Can. J. Phys. 51, 434 (1973).

Enthalpy Reference Temperature = T_r = 298.15 K Standard State Pressure = p° = 0.1 MPa

T/K	$C_p°$ (J K^{-1}mol^{-1})	S° (J K^{-1}mol^{-1})	$-[G°-H°(T_r)]/T$ (J K^{-1}mol^{-1})	$H°-H°(T_r)$ (kJ mol^{-1})	$\Delta_f H°$ (kJ mol^{-1})	$\Delta_f G°$ (kJ mol^{-1})	Log K_f
0	0.	0.	INFINITE	-8.599	-272.499	-272.499	INFINITE
100	29.128	141.959	199.679	-5.772	-272.625	-273.266	142.739
200	29.128	162.148	176.445	-2.859	-272.524	-273.947	71.548
250	29.133	168.648	174.260	-1.403	-272.516	-274.304	57.313
298.15	29.138	173.780	173.780	0.	-272.546	-274.646	48.117
300	29.138	173.960	173.780	0.054	-272.548	-274.659	47.822
350	29.143	178.452	174.135	1.511	-272.609	-275.007	41.043
400	29.150	182.344	174.923	2.968	-272.696	-275.344	35.956
450	29.159	185.778	175.943	4.426	-272.802	-275.669	31.999
500	29.173	188.851	177.082	5.884	-272.924	-275.981	28.831
600	29.230	194.174	179.501	8.804	-273.205	-276.566	24.077
700	29.351	198.688	181.927	11.732	-273.522	-277.102	20.678
800	29.550	202.619	184.273	14.677	-273.858	-277.590	18.125
900	29.827	206.115	186.510	17.645	-274.205	-278.036	16.137
1000	30.169	209.275	188.631	20.644	-274.552	-278.443	14.544
1100	30.556	212.168	190.641	23.680	-274.896	-278.816	13.240
1200	30.971	214.845	192.547	26.757	-275.231	-279.157	12.151
1300	31.397	217.340	194.360	29.875	-275.558	-279.471	11.229
1400	31.822	219.683	196.086	33.036	-275.874	-279.760	10.438
1500	32.237	221.893	197.733	36.239	-276.181	-280.027	9.751
1600	32.636	223.986	199.309	39.483	-276.477	-280.273	9.150
1700	33.017	225.976	200.820	42.766	-276.763	-280.502	8.619
1800	33.376	227.873	202.270	46.085	-277.040	-280.714	8.146
1900	33.713	229.687	203.666	49.440	-277.308	-280.911	7.723
2000	34.029	231.424	205.011	52.827	-277.566	-281.093	7.341
2100	34.325	233.092	206.309	56.245	-277.815	-281.264	6.996
2200	34.601	234.695	207.563	59.692	-278.053	-281.422	6.682
2300	34.858	236.239	208.776	63.165	-278.279	-281.570	6.395
2400	35.099	237.728	209.952	66.663	-278.493	-281.709	6.131
2500	35.323	239.165	211.092	70.184	-278.694	-281.838	5.889
2600	35.533	240.555	212.198	73.727	-278.882	-281.961	5.665
2700	35.730	241.899	213.273	77.290	-279.054	-282.076	5.457
2800	35.914	243.202	214.319	80.872	-279.210	-282.185	5.264
2900	36.088	244.466	215.337	84.473	-279.349	-282.288	5.085
3000	36.251	245.692	216.329	88.090	-279.472	-282.388	4.917
3100	36.405	246.883	217.295	91.723	-279.576	-282.483	4.760
3200	36.550	248.041	218.238	95.370	-279.662	-282.575	4.613
3300	36.687	249.168	219.158	99.032	-279.730	-282.665	4.474
3400	36.818	250.265	220.057	102.708	-279.779	-282.754	4.344
3500	36.941	251.334	220.935	106.395	-279.809	-282.841	4.221
3600	37.059	252.376	221.794	110.096	-279.821	-282.927	4.105
3700	37.171	253.393	222.635	113.807	-279.814	-283.013	3.995
3800	37.279	254.386	223.457	117.530	-279.789	-283.100	3.891
3900	37.381	255.356	224.263	121.263	-279.746	-283.188	3.793
4000	37.480	256.303	225.052	125.006	-279.685	-283.277	3.699
4100	37.574	257.230	225.826	128.759	-279.607	-283.367	3.610
4200	37.665	258.137	226.584	132.521	-279.512	-283.460	3.525
4300	37.753	259.024	227.328	136.292	-279.400	-283.556	3.445
4400	37.837	259.893	228.059	140.071	-279.273	-283.654	3.367
4500	37.919	260.744	228.775	143.859	-279.130	-283.755	3.294
4600	37.998	261.578	229.479	147.655	-278.971	-283.859	3.223
4700	38.074	262.396	230.171	151.458	-278.798	-283.967	3.156
4800	38.149	263.199	230.851	155.270	-278.611	-284.079	3.091
4900	38.221	263.986	231.519	159.088	-278.410	-284.195	3.030
5000	38.291	264.759	232.176	162.914	-278.195	-284.315	2.970
5100	38.359	265.518	232.823	166.746	-277.967	-284.440	2.913
5200	38.426	266.263	233.459	170.585	-277.725	-284.569	2.859
5300	38.491	266.996	234.084	174.431	-277.471	-284.703	2.806
5400	38.555	267.716	234.701	178.284	-277.204	-284.842	2.755
5500	38.617	268.424	235.307	182.142	-276.925	-284.987	2.707
5600	38.678	269.120	235.905	186.007	-276.633	-285.135	2.660
5700	38.737	269.806	236.494	189.878	-276.329	-285.290	2.614
5800	38.796	270.480	237.074	193.754	-276.013	-285.450	2.571
5900	38.853	271.143	237.646	197.637	-275.684	-285.616	2.529
6000	38.910	271.797	238.209	201.525	-275.343	-285.787	2.488

PREVIOUS: June 1977 (1 atm) CURRENT: June 1977 (1 bar)

HYPOFLUOROUS ACID (HOF) IDEAL GAS M_r = 56.005743

$S°(298.15\ K) = 226.772 \pm 0.21\ J\ K^{-1}\ mol^{-1}$

$\Delta_f H°(0\ K) = -95.423 \pm 4.2\ kJ\ mol^{-1}$
$\Delta_f H°(298.15\ K) = -98.324 \pm 4.2\ kJ\ mol^{-1}$

Vibrational Frequencies and Degeneracies

ν, cm^{-1}
3578.5 (1)
1354.8 (1)
889.0 (1)

Ground State Quantum Weight: [1] σ =1
Point Group: C_s
Bond Distances: H-O = 0.966 Å O-F = 1.442 Å
Bond Angle: H-O-F = 96.78°
Product of the Moments of Inertia: $I_A I_B I_C = 1.4723 \times 10^{-117}\ g^3\ cm^6$

Enthalpy of Formation

$\Delta_f H°(0\ K)$ is the value derived by Berkowitz et al. (1) from photoionization data which gave 14.34 eV as the threshold of O^+ formation, $HOF + h\nu \rightarrow O^+ + HF + e^-$. The threshold of OH^+ formation, 15.07 eV for $HOF + h\nu \rightarrow OH^+ + F + e^-$, yields $\Delta_f H°(0\ K)$ = -21.7 kcal mol^{-1} assuming $\Delta_f H°(OH^+, 0\ K)$ = 307.5±0.5 kcal mol^{-1} from Dibeler et al. (2). The latter authors found 18.05 eV for $H_2O + h\nu \rightarrow OH^+ + H^+ + e^-$. Our adopted value for HOF differs by only 2.5 kcal mol^{-1} from the mean $\Delta_f H°(0\ K)$ of H_2O and F_2O, namely -25.3 kcal mol^{-1}.

Heat Capacity and Entropy

Vibrational frequencies are from gas-phase infrared spectra of Appelman and Kim (3); the values are confirmed by IR spectra of HOF isolated in a matrix of N_2 (4). Structural parameters are those derived by Pearson and Kim (5) from centrifugal distortion analysis of mm-wave rotational spectra (5, 6) of HOF and DOF. We use the atomic coordinates (5) to derive the principal moments of inertia are $I_A = 0.1431 \times 10^{-39}$, $I_B = 3.1363 \times 10^{-39}$ and $I_C = 3.2795 \times 10^{-39}\ g\ cm^2$ and their product; the corresponding bond distances are 0.96648 and 1.44215 Å, i.e., slightly longer than the rounded values (5) listed above. The earlier study (6) gave 0.964±0.01 and 1.442±0.001 Å and a bond angle of 97.2±0.6°.

References

1. J. Berkowitz, E. H. Appelman and W. A. Chupka, J. Chem. Phys., in press.
2. V. H. Dibeler, J. A. Walker and H. M. Rosenstock, J. Res. Nat. Bur. Stand. 70A, 459 (1966).
3. E. H. Appelman and H. Kim, J. Chem. Phys. 57, 3272 (1972).
4. J. A. Goleb, H. H. Claasen, M. H. Studier and E. H. Appelman, Spectrochim. Acta 28A, 65 (1972).
5. E. F. Pearson and H. Kim, J. Chem. Phys. 57, 4230 (1972).
6. H. Kim, E. F. Pearson and E. H. Appelman, J. Chem. Phys. 56, 1 (1972).

Hypofluorous Acid (HOF) $F_1H_1O_1(g)$

Enthalpy Reference Temperature = T_r = 298.15 K Standard State Pressure = p° = 0.1 MPa

T/K	$C_p°$ (J $K^{-1}mol^{-1}$)	S° (J $K^{-1}mol^{-1}$)	$-[G°-H°(T_r)]/T$ (J $K^{-1}mol^{-1}$)	$H°-H°(T_r)$ (kJ mol^{-1})	$\Delta_f H°$ (kJ mol^{-1})	$\Delta_f G°$ (kJ mol^{-1})	Log K_f
0	0.	0.	INFINITE	-10.087	-95.423	-95.423	INFINITE
100	33.261	189.739	257.352	-6.761	-96.503	-93.257	48.712
200	33.873	212.899	229.983	-3.417	-97.425	-89.650	23.414
250	34.787	220.549	227.357	-1.702	-97.889	-87.653	18.314
298.15	35.929	226.772	226.772	0.	-98.324	-85.642	15.004
300	35.976	226.994	226.772	0.067	-98.340	-85.563	14.898
350	37.277	232.637	227.215	1.898	-98.766	-83.399	12.447
400	38.571	237.700	228.215	3.794	-99.161	-81.177	10.601
450	39.790	242.314	229.528	5.753	-99.524	-78.907	9.159
500	40.907	246.565	231.022	7.771	-99.857	-76.598	8.002
600	42.830	254.199	234.263	11.962	-100.448	-71.889	6.258
700	44.409	260.924	237.601	16.326	-100.956	-67.088	5.006
800	45.744	266.943	240.899	20.835	-101.396	-62.219	4.062
900	46.906	272.400	244.101	25.469	-101.779	-57.298	3.325
1000	47.936	277.396	247.184	30.212	-102.114	-52.337	2.734
1100	48.858	282.009	250.143	35.053	-102.408	-47.345	2.248
1200	49.687	286.296	252.979	39.981	-102.666	-42.328	1.842
1300	50.432	290.303	255.698	44.987	-102.896	-37.290	1.498
1400	51.102	294.066	258.305	50.065	-103.103	-32.235	1.203
1500	51.703	297.612	260.809	55.205	-103.292	-27.167	0.946
1600	52.242	300.967	263.215	60.403	-103.468	-22.086	0.721
1700	52.726	304.148	265.530	65.652	-103.634	-16.995	0.522
1800	53.160	307.175	267.760	70.946	-103.794	-11.894	0.345
1900	53.550	310.060	269.911	76.282	-103.951	-6.784	0.186
2000	53.901	312.815	271.988	81.655	-104.105	-1.666	0.043
2100	54.217	315.453	273.995	87.061	-104.257	3.460	-0.086
2200	54.503	317.982	275.938	92.498	-104.409	8.593	-0.204
2300	54.761	320.410	277.819	97.961	-104.561	13.733	-0.312
2400	54.994	322.746	279.642	103.449	-104.712	18.879	-0.411
2500	55.206	324.995	281.412	108.959	-104.861	24.032	-0.502
2600	55.399	327.164	283.130	114.489	-105.009	29.191	-0.586
2700	55.575	329.259	284.800	120.038	-105.154	34.355	-0.665
2800	55.736	331.283	286.424	125.604	-105.296	39.525	-0.737
2900	55.883	333.241	288.005	131.185	-105.433	44.699	-0.805
3000	56.018	335.138	289.545	136.780	-105.566	49.879	-0.868
3100	56.141	336.977	291.045	142.388	-105.693	55.062	-0.928
3200	56.255	338.761	292.508	148.008	-105.814	60.250	-0.983
3300	56.360	340.494	293.936	153.639	-105.928	65.441	-1.036
3400	56.457	342.178	295.331	159.280	-106.035	70.636	-1.085
3500	56.547	343.816	296.693	164.930	-106.135	75.833	-1.132
3600	56.630	345.410	298.024	170.589	-106.228	81.034	-1.176
3700	56.708	346.962	299.326	176.256	-106.313	86.236	-1.217
3800	56.780	348.476	300.599	181.930	-106.390	91.442	-1.257
3900	56.847	349.951	301.846	187.612	-106.459	96.649	-1.294
4000	56.909	351.391	303.067	193.300	-106.522	101.858	-1.330
4100	56.967	352.797	304.262	198.994	-106.577	107.068	-1.364
4200	57.022	354.171	305.434	204.693	-106.625	112.279	-1.396
4300	57.073	355.513	306.583	210.398	-106.666	117.492	-1.427
4400	57.121	356.826	307.710	216.107	-106.702	122.705	-1.457
4500	57.166	358.110	308.816	221.822	-106.731	127.919	-1.485
4600	57.208	359.367	309.902	227.541	-106.755	133.133	-1.512
4700	57.248	360.598	310.967	233.263	-106.774	138.349	-1.538
4800	57.285	361.803	312.014	238.990	-106.789	143.565	-1.562
4900	57.320	362.985	313.042	244.720	-106.800	148.781	-1.586
5000	57.354	364.143	314.052	250.454	-106.807	153.997	-1.609
5100	57.385	365.279	315.046	256.191	-106.812	159.213	-1.631
5200	57.415	366.394	316.023	261.931	-106.814	164.429	-1.652
5300	57.443	367.488	316.983	267.674	-106.814	169.645	-1.672
5400	57.470	368.562	317.929	273.420	-106.813	174.861	-1.691
5500	57.496	369.617	318.859	279.168	-106.811	180.077	-1.710
5600	57.520	370.653	319.774	284.919	-106.809	185.294	-1.728
5700	57.543	371.671	320.676	290.672	-106.807	190.509	-1.746
5800	57.564	372.672	321.564	296.427	-106.806	195.726	-1.763
5900	57.585	373.656	322.438	302.185	-106.805	200.941	-1.779
6000	57.605	374.624	323.300	307.944	-106.807	206.158	-1.795

PREVIOUS: December 1972 (1 atm) CURRENT: December 1972 (1 bar)

Hypofluorous Acid (HOF) $F_1H_1O_1(g)$

FLUOROSULFURIC ACID (HSO_3F) IDEAL GAS M_r = 100.064543

$\Delta_fH°(0\ K) = -742.051 \pm 8.4\ kJ\ mol^{-1}$

$S°(298.15\ K) = 297.267 \pm 3.3\ J\ K^{-1}\ mol^{-1}$ $\Delta_fH°(298.15\ K) = -753.120 \pm 8.4\ kJ\ mol^{-1}$

Vibrational Frequencies and Degeneracaies

ν, cm^{-1}	ν, cm^{-1}	ν, cm^{-1}
3602(1)	825(1)	1486(1)
1243(1)	550(1)	562(1)
1150(1)	410(1)	390(1)
896(1)	330(1)	265(1)

Ground State Quantum Weight: 1 σ = 1

Point Group: [C_s]

Bond Distances: O-S = [1.405] Å S-OH = S-F = [1.53] Å O-H = [1.00] Å

Bond Angles: O-S-O = [124°] F-S-OH = S-O-H = [100°]

Product of the Moments of Inertia: $I_AI_BI_C = [4.2743 \times 10^{-114}]\ g^3\ cm^6$

Enthalpy of Formation

$\Delta_fH°(\ell$, 298.15 K) may be derived from calorimetric data of Richards and Woolf (1) which yield $\Delta_rH°(298.15\ K) = -14.94\pm0.2$ kcal mol^{-1} for $SO_3(\ell)$ + HF (real gas, 1 atm) → $HSO_3F(\ell)$. Using $\Delta_fH°(SO_3, \ell, 0\ K) = -105.41$ (2) and $\Delta_fH°$(HF, real gas, 0 K) = -70.12 (3, 4) kcal mol^{-1}, we derive $\Delta_fH° = -190.5$ kcal mol^{-1} for $HSO_3F(\ell)$ and -180.3 ± 1.5 kcal mol^{-1} for the ideal gas. The latter value agrees well with -179.2 ± 3 kcal mol^{-1} estimated for $HSO_3F(g)$ from the related molecules SO_2F_2 and H_2SO_4 (4). We adopt $\Delta_fH° = -180\pm2$ kcal mol^{-1} (-753.120 ± 8.4 kJ mol^{-1}).

Derivation of $\Delta_fH°(g)$ presumes that the heat of vaporization is 10.2 ± 1.0 kcal mol^{-1} at 25°C. Measurements of the heat of condensation (5) yield 12.8 ± 0.8 kcal mol^{-1} for the enthalpy difference between real gas at 164.4°C and the liquid at 25 C. Added to this value should be ±0.2 kcal mol^{-1} (estimated correction for gas imperfection) and -2.8 kcal mol^{-1} (calculated enthalpy difference for the ideal gas between 25°C and 164.4°C, the approximate normal boiling point).

Other calorimetric data bearing on $\Delta_fH°$ were reported by Lenskii et al. (5) and Woolf (6), but these results are much more uncertain (1) due to experimental error and to inadequate definition of the states of the reactants and products.

Heat Capacity and Entropy

Savoie and Giguere (7) reviewed possible molecular structures and preferred eclipsed (cis) forms, arbitrarily selecting one in which F-S-O-H is coplanar. We adopt this structure since the choice changes the thermodynamic functions very little. Possible structures all have σ = 1 (C_s or C_1 symmetry) and similar values of the product of the moments of inertia. Bond lengths and angles are estimated by comparison with SO_2ClF (8), SO_2CL_2 and H_2SO_4 (4). These parameters yield O=S-F and O=S-OH angles of 107.56° and principal moments of inertia are: $I_A = 15.9647 \times 10^{-39}$, $I_B = 16.1688 \times 10^{-39}$, $I_C = 16.5588 \times 10^{-39}$ g cm^2.

Vibrational fundamentals are from the assignment of Savoie and Giguere (7) as modified by Chackalackal and Stafford (9). Data include Raman spectra of the liquid (10) and infrared spectra of all three phases (7) plus the superheated vapor (9). Spectra of the superheated vapor helped to clarify the vibrational distinction between the monomer and associated molecules. Savoie and Giguere derived a barrier of about 1.3 kcal mol^{-1} for hindered interal rotation of the OH group and used this to calculate $C_p° = 17.64$ and S° = 71.38 cal K^{-1} mol^{-1} at 298.15 K. We calculated $C_p° = 17.98$ and S° = 71.02 cal K^{-1} mol^{-1} using the OH-torsional frequency (265 cm^{-1}) instead of hindered rotation.

References

1. G. W. Richards and A. A. Woolf, J. Chem. Soc. A1967, 1118 (1967).
2. U. S. Nat. Bur. Stand. Tech. Note 270-3, 1968.
3. C. E. Vanderzee and W. W. Rodenburg, J. Chem. Thermodynamics 2, 461 (1970).
4. JANAF Thermochemical Tables: HF(g), 12-31-68; H_2SO_4(g), 12-31-66; SO_2F_2(g) and SO_2Cl_2(g), 6-30-71.
5. A. S. Lenskii, A. D. Shaposhnikova and E. S. Sokolova, Russ. J. Inorg. Chem. 9, 626 (1964); 8, 1424 (1963).
6. A. A. Woolf, J. Inorg. Nucl. Chem. 14, 21 (1960).
7. R. Savoie and P. A. Giguere, Can. J. Chem. 42, 277 (1964).
8. C. W. Holt and M. C. L. Gerry, Chem. Phys. Letters 9, 621 (1971).
9. S. M. Chackalackal and F. E. Stafford, J. Amer. Chem. Soc. 88, 4815 (1966).
10. R. J. Gillespie and E. A. Robinson, Can. J. Chem. 40, 644 (1962).

Fluorosulfuric Acid (HSO_3F) $F_1H_1O_3S_1(g)$

Enthalpy Reference Temperature = T_r = 298.15 K Standard State Pressure = p° = 0.1 MPa

T/K	$C_p°$ (J $K^{-1}mol^{-1}$)	S° (J $K^{-1}mol^{-1}$)	$-[G°-H°(T_r)]/T$ (J $K^{-1}mol^{-1}$)	$H°-H°(T_r)$ (kJ mol^{-1})	$\Delta_fH°$ (kJ mol^{-1})	$\Delta_fG°$ (kJ mol^{-1})	Log K_f
0	0.	0.	INFINITE	-15.014	-742.051	-742.051	INFINITE
100	39.145	237.149	352.791	-11.564	-746.601	-729.513	381.058
200	59.278	270.467	303.667	-6.640	-750.497	-710.811	185.645
250	67.997	284.656	298.466	-3.453	-751.965	-700.713	146.406
298.15	75.243	297.267	297.267	0.	-753.120	-690.734	121.014
300	75.501	297.733	297.268	0.139	-753.160	-690.347	120.200
350	81.950	309.869	298.212	4.080	-754.119	-679.799	101.454
400	87.477	321.182	300.384	8.319	-757.096	-669.040	87.368
450	92.198	331.765	303.290	12.814	-758.366	-657.964	76.374
500	96.227	341.693	306.639	17.527	-759.549	-646.741	67.565
600	102.624	359.833	314.023	27.486	-761.115	-624.019	54.326
700	107.389	376.027	321.745	37.998	-762.077	-601.087	44.854
800	111.041	390.616	329.457	48.927	-762.646	-578.046	37.743
900	113.926	403.867	337.000	60.181	-816.071	-553.870	32.146
1000	116.264	415.996	344.302	71.694	-814.954	-524.795	27.412
1100	118.199	427.170	351.334	83.420	-813.742	-495.837	23.545
1200	119.826	437.527	358.090	95.324	-812.461	-466.992	20.328
1300	121.210	447.174	364.576	107.377	-811.130	-438.257	17.609
1400	122.397	456.201	370.802	119.559	-809.764	-409.626	15.283
1500	123.425	464.682	376.781	131.851	-808.376	-381.093	13.271
1600	124.318	472.676	382.527	144.239	-806.977	-352.653	11.513
1700	125.100	480.237	388.054	156.711	-805.575	-324.301	9.965
1800	125.786	487.407	393.376	169.256	-804.177	-296.031	8.591
1900	126.392	494.225	398.506	181.866	-802.788	-267.839	7.363
2000	126.929	500.722	403.456	194.532	-801.412	-239.719	6.261
2100	127.406	506.927	408.236	207.250	-800.052	-211.668	5.265
2200	127.832	512.864	412.858	220.012	-798.708	-183.680	4.361
2300	128.214	518.554	417.331	232.815	-797.383	-155.755	3.537
2400	128.556	524.019	421.663	245.653	-796.077	-127.887	2.783
2500	128.865	529.273	425.863	258.525	-794.789	-100.072	2.091
2600	129.144	534.333	429.938	271.425	-793.518	-72.308	1.453
2700	129.397	539.211	433.895	284.353	-792.264	-44.593	0.863
2800	129.626	543.921	437.741	297.304	-791.026	-16.924	0.316
2900	129.836	548.474	441.482	310.277	-789.802	10.700	-0.193
3000	130.027	552.879	445.122	323.271	-788.593	38.284	-0.667
3100	130.202	557.145	448.667	336.282	-787.395	65.827	-1.109
3200	130.362	561.281	452.122	349.310	-786.209	93.330	-1.523
3300	130.510	565.295	455.491	362.354	-785.034	120.798	-1.912
3400	130.645	569.193	458.778	375.412	-783.868	148.230	-2.277
3500	130.771	572.982	461.987	388.483	-782.711	175.627	-2.621
3600	130.887	576.668	465.122	401.566	-781.564	202.993	-2.945
3700	130.994	580.256	468.185	414.660	-780.424	230.325	-3.252
3800	131.094	583.750	471.181	427.764	-779.292	257.629	-3.541
3900	131.187	587.157	474.111	440.878	-778.168	284.902	-3.816
4000	131.273	590.479	476.979	454.001	-777.053	312.145	-4.076
4100	131.353	593.722	479.787	467.133	-775.945	339.361	-4.324
4200	131.429	596.888	482.537	480.272	-774.846	366.551	-4.559
4300	131.499	599.981	485.233	493.418	-773.756	393.714	-4.783
4400	131.565	603.005	487.875	506.572	-772.675	420.852	-4.996
4500	131.626	605.962	490.467	519.731	-771.604	447.964	-5.200
4600	131.684	608.856	493.009	532.897	-770.545	475.054	-5.394
4700	131.739	611.689	495.504	546.068	-769.497	502.121	-5.580
4800	131.790	614.463	497.954	559.244	-768.462	529.166	-5.759
4900	131.838	617.181	500.359	572.426	-767.440	556.190	-5.929
5000	131.884	619.845	502.722	585.612	-766.434	583.193	-6.093
5100	131.927	622.457	505.044	598.803	-765.443	610.176	-6.249
5200	131.968	625.019	507.327	611.997	-764.470	637.138	-6.400
5300	132.006	627.533	509.572	625.196	-763.516	664.084	-6.545
5400	132.043	630.001	511.779	638.399	-762.582	691.011	-6.684
5500	132.077	632.424	513.950	651.605	-761.669	717.919	-6.818
5600	132.110	634.804	516.087	664.814	-760.780	744.815	-6.947
5700	132.141	637.143	518.191	678.027	-759.915	771.692	-7.072
5800	132.171	639.441	520.261	691.242	-759.075	798.556	-7.192
5900	132.199	641.701	522.301	704.461	-758.264	825.402	-7.308
6000	132.226	643.923	524.309	717.682	-757.482	852.239	-7.419

PREVIOUS: June 1972 (1 atm) CURRENT: June 1972 (1 bar)

Fluorosulfuric Acid (HSO_3F) $F_1H_1O_3S_1(g)$

FLUOROSILANE (SiH_3F) IDEAL GAS M_r = 50.107723

$S°(298.15 K) = 238.402 \pm 0.8 J K^{-1} mol^{-1}$

$\Delta_f H°(0 K) = [-367.16 \pm 20.9] kJ mol^{-1}$
$\Delta_f H°(298.15 K) = [-376.560 \pm 20.9] kJ mol^{-1}$

Vibrational Frequencies and Degeneracies

ν, cm^{-1}	ν, cm^{-1}
2206(1)	2209(2)
991(1)	961(2)
875(1)	728(2)

Ground State Quantum Weight: [1] $\sigma = 3$
Point Group: C_{3v}
Bond Distances: Si-H = 1.468 Å Si-F = 1.593 Å
Bond Angles: H-Si-H = 110.43° H-Si-F = 108.50°
Product of the Moments of Inertia: $I_A I_B I_C = 3.42403 \times 10^{-116} g^3 cm^6$

Enthalpy of Formation

There are no reported experimental studies leading to the heat of formation of SiH_3F(g). We estimate this value via a linear interpolation between the established $\Delta_f H°$(298.15 K) values of SiH_4(g) and SiF_4(g) (1). The reasonableness of this approach has been demonstrated by Lapidus et al. (2), Hunt and Sirtl (3), and Seiter and Sirtl (4). Lapidus et al. (2) examined the trends in the thermodynamic properties of halogenated silanes and methanes. Hunt and Sirtl (3) and Seiter and Sirtl (4) studied the chlorinated silanes and proposed a linear $\Delta_f H$ relationship within the sequence SiH_4(g) to $SiCl_4$(g).

Heat Capacity and Entropy

The adopted vibrational frequencies are from the gas phase infrared spectra of SiH_3F and SiD_3F as documented by Robiette et al. (5). Except for ν_4 and ν_5, these values are within ±3 cm^{-1} of those suggested in the compilation by Shimanouchi (6). In particular, based on earlier work, Shimanouchi (6) suggested ν_4 = 2196 cm^{-1} and ν_5 = 956 cm^{-1}. The work of Robiette et al. (5) is judged to yield more accurate frequencies than the earlier work.

The adopted bond angles and bond distances are obtained from the microwave spectra study on SiH_3F and various isotopic derivatives by Kewley et al. (7). This work is deemed more accurate than earlier studies (8-11). The individual moments of inertia are calculated to be $I_A = 0.9972 \times 10^{-39} g cm^2$ and $I_B = I_C = 5.8599 \times 10^{-39} g cm^2$.

References

1. JANAF Thermochemical Tables: SiH_4(g) and SiF_4(g), 6-30-76.
2. I. I. Lapidus, L. A. Nisel'son, and A. L. Seifer in "Thermophysical Properties of Gases and Liquids," No. 1, V. A. Rabinovich, Ed., Israel Program for Scientific Translations, Jerusalem, 1970, pp. 102-135 [Translated from Russian].
3. L. P. Hunt and E. Sirtl, J. Electrochem. Soc. 119, 1741 (1972).
4. H. Seiter and E. Sirtl, Z. Naturforsch. 21a, 1956 (1966).
5. A. G. Robiette, G. J. Cartwright, A. R. Hoy, and I. M. Mills, Mol. Phys. 20, 541 (1971).
6. T. Shimanouchi, J. Phys. Chem. Ref. Data 1, 189 (1972).
7. R. Kewley, P. M. McKinney, and A. G. Robiette, J. Mol. Spectrosc. 34, 390 (1970).
8. A. H. Sharbaugh, V. G. Thomas and B. S. Pritchard, Phys. Rev. 78, 64 (1950).
9. B. Bak, J. Bruhn, and J. Rastrup-Andersen, J. Chem. Phys. 21, 752 (1953).
10. B. Bak, J. Bruhn, and J. Rastrup-Andersen, Acta Chem. Scand. 8, 459 (1954).
11. C. Newman, J. K. O'Loane, S. R. Polo, and M. K. Wilson, J. Chem. Phys. 25, 855 (1956).

Fluorosilane (SiH_3F) $F_1H_3Si_1$(g)

Enthalpy Reference Temperature = T_r = 298.15 K — Standard State Pressure = p° = 0.1 MPa

T/K	$C_p°$ (J K⁻¹mol⁻¹)	S°	$-[G°-H°(T_r)]/T$	$H°-H°(T_r)$ (kJ mol⁻¹)	$\Delta_f H°$	$\Delta_f G°$	Log K_f
0	0.	0.	INFINITE	-10.931	-367.160	-367.160	INFINITE
100	33.317	197.814	273.865	-7.605	-370.052	-365.823	191.086
200	37.439	221.708	242.386	-4.136	-373.267	-360.386	94.123
250	42.143	230.548	239.150	-2.150	-374.977	-356.968	74.584
298.15	47.196	238.402	238.402	0.	-376.560	-353.354	61.906
300	47.391	238.694	238.402	0.087	-376.619	-353.210	61.499
350	52.495	246.387	238.998	2.586	-378.123	-349.188	52.113
400	57.187	253.708	240.383	5.330	-379.466	-344.960	45.047
450	61.420	260.693	242.255	8.297	-380.647	-340.575	39.533
500	65.229	267.364	244.435	11.465	-381.676	-336.066	35.109
600	71.784	279.856	249.312	18.326	-383.321	-326.781	28.449
700	77.185	291.340	254.507	25.783	-384.494	-317.258	23.674
800	81.656	301.948	259.783	33.732	-385.288	-307.595	20.084
900	85.359	311.787	265.021	42.089	-385.782	-297.851	17.287
1000	88.429	320.944	270.161	50.783	-386.050	-288.065	15.047
1100	90.981	329.496	275.171	59.758	-386.150	-278.260	13.213
1200	93.109	337.506	280.035	68.965	-386.131	-268.452	11.685
1300	94.894	345.031	284.748	78.368	-386.034	-258.649	10.393
1400	96.398	352.120	289.310	87.935	-385.892	-248.856	9.285
1500	97.673	358.816	293.722	97.640	-385.728	-239.073	8.325
1600	98.761	365.155	297.991	107.463	-385.562	-229.301	7.486
1700	99.694	371.171	302.120	117.387	-435.586	-219.093	6.732
1800	100.500	376.893	306.116	127.398	-435.226	-206.369	5.989
1900	101.199	382.346	309.986	137.483	-434.855	-193.664	5.324
2000	101.808	387.552	313.735	147.634	-434.479	-180.980	4.727
2100	102.343	392.533	317.370	157.843	-434.102	-168.314	4.187
2200	102.814	397.305	320.895	168.101	-433.726	-155.667	3.696
2300	103.230	401.885	324.318	178.404	-433.354	-143.036	3.248
2400	103.601	406.286	327.642	188.745	-432.985	-130.421	2.839
2500	103.931	410.522	330.873	199.122	-432.620	-117.822	2.462
2600	104.227	414.604	334.016	209.530	-432.259	-105.237	2.114
2700	104.493	418.543	337.074	219.967	-431.902	-92.666	1.793
2800	104.732	422.347	340.052	230.428	-431.548	-80.108	1.494
2900	104.949	426.026	342.953	240.912	-431.197	-67.563	1.217
3000	105.146	429.588	345.782	251.417	-430.848	-55.030	0.958
3100	105.325	433.038	348.541	261.941	-430.501	-42.508	0.716
3200	105.489	436.385	351.234	272.482	-430.156	-29.998	0.490
3300	105.638	439.633	353.864	283.038	-429.812	-17.498	0.277
3400	105.776	442.789	356.433	293.609	-429.470	-5.009	0.077
3500	105.902	445.857	358.945	304.193	-429.128	7.470	-0.111
3600	106.018	448.842	361.401	314.789	-812.940	30.400	-0.441
3700	106.125	451.748	363.803	325.396	-812.188	53.816	-0.760
3800	106.224	454.580	366.155	336.014	-811.439	77.212	-1.061
3900	106.316	457.340	368.458	346.641	-810.694	100.588	-1.347
4000	106.401	460.033	370.714	357.277	-809.953	123.944	-1.619
4100	106.481	462.661	372.924	367.921	-809.215	147.282	-1.876
4200	106.554	465.228	375.092	378.573	-808.480	170.603	-2.122
4300	106.623	467.736	377.217	389.232	-807.749	193.906	-2.355
4400	106.688	470.188	379.302	399.897	-807.022	217.192	-2.578
4500	106.748	472.586	381.349	410.569	-806.300	240.461	-2.791
4600	106.805	474.933	383.358	421.247	-805.581	263.714	-2.995
4700	106.858	477.231	385.331	431.930	-804.866	286.952	-3.189
4800	106.908	479.481	387.269	442.618	-804.156	310.175	-3.375
4900	106.954	481.686	389.173	453.311	-803.449	333.383	-3.554
5000	106.999	483.847	391.045	464.009	-802.747	356.577	-3.725
5100	107.040	485.966	392.886	474.711	-802.049	379.755	-3.889
5200	107.079	488.045	394.696	485.417	-801.354	402.922	-4.047
5300	107.117	490.085	396.476	496.127	-800.663	426.074	-4.199
5400	107.152	492.088	398.229	506.840	-799.975	449.214	-4.345
5500	107.185	494.054	399.953	517.557	-799.289	472.340	-4.486
5600	107.217	495.986	401.651	528.277	-798.607	495.455	-4.621
5700	107.246	497.884	403.322	539.000	-797.926	518.557	-4.752
5800	107.275	499.749	404.969	549.726	-797.246	541.648	-4.878
5900	107.302	501.583	406.591	560.455	-796.567	564.726	-5.000
6000	107.328	503.387	408.189	571.187	-795.888	587.792	-5.117

PREVIOUS: June 1976 (1 atm) CURRENT: June 1976 (1 bar)

MERCURY FLUORIDE (HgF) IDEAL GAS M_r = 219.588403

$S°(298.15\ K) = 248.373\ J\ K^{-1}\ mol^{-1}$

$\Delta_f H°(0\ K) = 7.370\ kJ\ mol^{-1}$
$\Delta_f H°(298.15\ K) = 2.929 \pm 50.2\ kJ\ mol^{-1}$

Electronic Level and Quantum Weight

State	ε_i, cm^{-1}	g_i
$^2\Sigma$	0	2

$\omega_e = 490.8\ cm^{-1}$ $\omega_e x_e = 4.05\ cm^{-1}$ $\sigma = 1$
$B_e = [0.2732]\ cm^{-1}$ $\alpha_e = [0.0026]\ cm^{-1}$ $r_e = [1.89]$ Å

Enthalpy of Formation

Gaydon (1) gives 32±12 kcal mol^{-1} for the enthalpy of dissociation into gaseous atoms.

Heat Capacity and Entropy

Vibrational constants are from Herzberg (2). Rotational constants were estimated using relationships in Herzberg (2). Bond length estimated by comparison of the mercurous halides with thalium and cesium halides.

References

1. A. G. Gaydon, "Dissociation Energies and Spectra of Diatomic Molecules," 2nd Ed., Chapman and Hall, London 1953.
2. G. Herzberg, "Spectra of Diatomic Molecules," D. Van Nostrand Company, Inc., New York, equation III, 123 p. 108, 1950.

Mercury Fluoride (HgF) $F_1Hg_1(g)$

Enthalpy Reference Temperature = T_r = 298.15 K Standard State Pressure = p° = 0.1 MPa

T/K	$C_p°$ (J K⁻¹mol⁻¹)	S° (J K⁻¹mol⁻¹)	$-[G°-H°(T_r)]/T$ (J K⁻¹mol⁻¹)	$H°-H°(T_r)$ (kJ mol⁻¹)	$\Delta_f H°$ (kJ mol⁻¹)	$\Delta_f G°$ (kJ mol⁻¹)	Log K_f
0	0.	0.	INFINITE	-9.314	7.370	7.370	INFINITE
100	29.502	213.636	277.625	-6.399	7.126	-1.997	1.043
200	32.488	234.956	251.483	-3.305	6.170	-10.751	2.808
250	33.752	242.348	248.939	-1.648	3.380	-14.771	3.086
298.15	34.651	248.373	248.373	0.	2.929	-18.225	3.193
300	34.680	248.588	248.374	0.064	2.912	-18.356	3.196
350	35.356	253.987	248.799	1.816	2.480	-21.867	3.263
400	35.857	258.743	249.751	3.597	2.070	-25.316	3.306
450	36.236	262.989	250.990	5.400	1.672	-28.716	3.333
500	36.532	266.823	252.385	7.219	1.282	-32.071	3.350
600	36.960	273.524	255.365	10.895	0.507	-38.669	3.366
700	37.258	279.245	258.378	14.607	-59.031	-38.566	2.878
800	37.481	284.235	261.305	18.344	-59.177	-35.632	2.327
900	37.659	288.660	264.103	22.102	-59.325	-32.680	1.897
1000	37.809	292.636	266.761	25.875	-59.475	-29.711	1.552
1100	37.939	296.246	269.280	29.663	-59.626	-26.728	1.269
1200	38.055	299.552	271.667	33.462	-59.778	-23.730	1.033
1300	38.162	302.602	273.931	37.273	-59.930	-20.720	0.833
1400	38.263	305.434	276.081	41.095	-60.083	-17.698	0.660
1500	38.358	308.077	278.127	44.926	-60.236	-14.665	0.511
1600	38.449	310.556	280.077	48.766	-60.389	-11.622	0.379
1700	38.537	312.889	281.939	52.615	-60.540	-8.569	0.263
1800	38.623	315.095	283.720	56.473	-60.691	-5.508	0.160
1900	38.706	317.185	285.427	60.340	-60.839	-2.438	0.067
2000	38.788	319.173	287.065	64.215	-60.983	0.639	-0.017
2100	38.869	321.067	288.640	68.098	-61.123	3.724	-0.093
2200	38.949	322.877	290.155	71.988	-61.255	6.815	-0.162
2300	39.028	324.610	291.616	75.887	-61.379	9.912	-0.225
2400	39.106	326.273	293.025	79.794	-61.492	13.014	-0.283
2500	39.184	327.871	294.387	83.709	-61.594	16.120	-0.337
2600	39.261	329.409	295.705	87.631	-61.681	19.231	-0.386
2700	39.338	330.892	296.981	91.561	-61.753	22.344	-0.432
2800	39.414	332.324	298.218	95.498	-61.808	25.460	-0.475
2900	39.490	333.709	299.418	99.444	-61.844	28.577	-0.515
3000	39.566	335.049	300.583	103.396	-61.861	31.696	-0.552
3100	39.641	336.347	301.716	107.357	-61.856	34.814	-0.587
3200	39.716	337.607	302.818	111.325	-61.830	37.932	-0.619
3300	39.791	338.830	303.891	115.300	-61.781	41.049	-0.650
3400	39.866	340.019	304.936	119.283	-61.709	44.164	-0.678
3500	39.940	341.176	305.955	123.273	-61.613	47.277	-0.706
3600	40.015	342.302	306.949	127.271	-61.494	50.386	-0.731
3700	40.089	343.400	307.920	131.276	-61.350	53.492	-0.755
3800	40.164	344.470	308.867	135.289	-61.181	56.594	-0.778
3900	40.238	345.514	309.794	139.309	-60.988	59.691	-0.799
4000	40.312	346.534	310.700	143.336	-60.771	62.782	-0.820
4100	40.386	347.530	311.586	147.371	-60.530	65.868	-0.839
4200	40.460	348.504	312.453	151.413	-60.265	68.948	-0.857
4300	40.534	349.457	313.303	155.463	-59.976	72.021	-0.875
4400	40.607	350.390	314.135	159.520	-59.663	75.087	-0.891
4500	40.681	351.303	314.951	163.585	-59.328	78.146	-0.907
4600	40.755	352.198	315.751	167.656	-58.971	81.197	-0.922
4700	40.829	353.075	316.536	171.736	-58.591	84.240	-0.936
4800	40.902	353.936	317.306	175.822	-58.190	87.275	-0.950
4900	40.976	354.780	318.062	179.916	-57.768	90.301	-0.963
5000	41.049	355.608	318.805	184.017	-57.326	93.318	-0.975
5100	41.123	356.422	319.534	188.126	-56.864	96.327	-0.987
5200	41.196	357.221	320.252	192.242	-56.382	99.326	-0.998
5300	41.270	358.006	320.956	196.365	-55.883	102.315	-1.008
5400	41.343	358.779	321.650	200.496	-55.365	105.295	-1.019
5500	41.416	359.538	322.332	204.634	-54.831	108.266	-1.028
5600	41.490	360.285	323.003	208.779	-54.280	111.226	-1.037
5700	41.563	361.020	323.663	212.932	-53.714	114.176	-1.046
5800	41.637	361.743	324.314	217.092	-53.133	117.117	-1.055
5900	41.710	362.456	324.954	221.259	-52.538	120.047	-1.063
6000	41.783	363.157	325.585	225.434	-51.929	122.967	-1.071

PREVIOUS: December 1961 (1 atm) CURRENT: December 1961 (1 bar)

IODINE FLUORIDE (IF) IDEAL GAS M_r = 145.902903

$\Delta_f H°(0\ K) = -92.854 \pm 3.8\ kJ\ mol^{-1}$

$S°(298.15\ K) = 236.296\ J\ K^{-1}\ mol^{-1}$ $\Delta_f H°(298.15\ K) = -94.759 \pm 3.8\ kJ\ mol^{-1}$

Electronic Level and Quantum Weight

State	ε_i, cm^{-1}	g_i
$[^1\Sigma^+]$	0	1

$\omega_e = 612\ cm^{-1}$ $\omega_e x_e = 4\ cm^{-1}$ $\sigma = 1$

$B_e = 0.2799\ cm^{-1}$ $\alpha_e = 0.00763\ cm^{-1}$ $r_e = 1.906\ Å$

Enthalpy of Formation

Durie and Gaydon (1) have obtained the dissociation limit (from the spectra) of IF as 23570 cm^{-1}. The first excited states of F and I are 404 and 7598 cm^{-1} above the ground state, respectively. There are two sets of dissociation products possible: $I(^2P_{1/2}) + F(^2P_{3/2})$ or $I(^2P_{3/2}) + F(^2P_{1/2})$. When corrected to the normal atoms, the observed limit gives D_0^0(IF) 2.87±0.04 eV or 66.18±0.9 kcal mol^{-1} and 1.98±0.04 eV or 45.66±0.9 kcal mol^{-1} for dissociation energy, respectively. Durie and Gaydon (1) favored the lower value. However, Slutsky and Bauer (2) have pointed out a numerical error in their calculations which removes the chief support for the lower value and presented additional evidence favoring the higher value. More weight was given to the higher value which gives $\Delta_f H°(298.15\ K) = -22.648 \pm 0.9\ kcal\ mol^{-1}$ for the reaction $1/2\ I_2(cr) + 1/2\ F_2(g) = IF(g)$.

Heat Capacity and Entropy

All molecular constants were obtained from Evans et al. (3) except r_e = 1.906 Å which was obtained from Cole and Elverun (4) and ground state configuration which was estimated by comparison with that for IBr and ICl from Herzberg (5).

References

1. R. A. Durie and A. G. Gaydon, J. Phys. Chem. 56, 316 (1952).
2. L. Slutsky and S. H. Bauer, J. Am. Chem. Soc. 76, 270 (1954).
3. W. H. Evans, T. R. Munson and D. D. Wagman, J. Res. Nat. Bur. Stand. 55, 147 (1955).
4. L. G. Cole and G. W. Elverun Jr., J. Chem. Phys. 20, 1543 (1952).
5. G. Herzberg, "Spectra of Diatomic Molecules," D. Van Nostrand Company, Inc., New York, 1950.

Iodine Fluoride (IF) $F_1I_1(g)$

Enthalpy Reference Temperature = T_r = 298.15 K Standard State Pressure = p° = 0.1 MPa

T/K	$C_p°$ (J K^{-1}mol^{-1})	S° (J K^{-1}mol^{-1})	$-[G°-H°(T_r)]/T$ (J K^{-1}mol^{-1})	$H°-H°(T_r)$ (kJ mol^{-1})	$\Delta_f H°$ (kJ mol^{-1})	$\Delta_f G°$ (kJ mol^{-1})	Log K_f
0	0.	0.	INFINITE	-9.106	-92.854	-92.854	INFINITE
100	29.216	202.594	264.549	-6.196	-92.922	-101.613	53.077
200	31.268	223.358	239.297	-3.188	-93.845	-109.955	28.717
250	32.537	230.474	236.843	-1.592	-94.314	-113.929	23.804
298.15	33.561	236.296	236.296	0.	-94.759	-117.666	20.615
300	33.596	236.503	236.296	0.062	-94.776	-117.808	20.512
350	34.429	241.748	236.709	1.764	-95.275	-121.609	18.149
400	35.076	246.389	237.634	3.502	-103.765	-125.062	16.331
450	35.583	250.551	238.842	5.269	-104.849	-127.659	14.818
500	35.986	254.321	240.204	7.058	-125.990	-128.236	13.397
600	36.577	260.938	243.124	10.689	-125.974	-128.687	11.203
700	36.991	266.609	246.084	14.368	-125.954	-129.141	9.637
800	37.298	271.570	248.966	18.083	-125.933	-129.597	8.462
900	37.541	275.977	251.727	21.826	-125.913	-130.057	7.548
1000	37.740	279.943	254.353	25.590	-125.894	-130.518	6.818
1100	37.911	283.549	256.846	29.373	-125.879	-130.981	6.220
1200	38.062	286.854	259.211	33.172	-125.870	-131.446	5.722
1300	38.199	289.906	261.456	36.985	-125.869	-131.910	5.300
1400	38.325	292.742	263.591	40.811	-125.881	-132.375	4.939
1500	38.443	295.390	265.623	44.649	-125.912	-132.838	4.626
1600	38.555	297.874	267.562	48.499	-125.965	-133.298	4.352
1700	38.663	300.215	269.415	52.360	-126.045	-133.754	4.110
1800	38.766	302.428	271.188	56.232	-126.153	-134.204	3.895
1900	38.867	304.527	272.888	60.113	-126.293	-134.648	3.702
2000	38.965	306.523	274.520	64.005	-126.462	-135.083	3.528
2100	39.062	308.426	276.090	67.906	-126.658	-135.510	3.371
2200	39.157	310.246	277.601	71.817	-126.877	-135.926	3.227
2300	39.250	311.988	279.059	75.738	-127.114	-136.332	3.096
2400	39.342	313.661	280.466	79.667	-127.361	-136.728	2.976
2500	39.434	315.269	281.826	83.606	-127.611	-137.113	2.865
2600	39.524	316.817	283.142	87.554	-127.857	-137.488	2.762
2700	39.614	318.310	284.417	91.511	-128.092	-137.854	2.667
2800	39.704	319.753	285.654	95.477	-128.308	-138.211	2.578
2900	39.792	321.147	286.854	99.452	-128.498	-138.562	2.496
3000	39.881	322.498	288.019	103.435	-128.657	-138.906	2.419
3100	39.969	323.807	289.153	107.428	-128.778	-139.245	2.346
3200	40.056	325.077	290.256	111.429	-128.858	-139.582	2.278
3300	40.144	326.311	291.330	115.439	-128.893	-139.916	2.215
3400	40.231	327.511	292.376	119.458	-128.879	-140.250	2.155
3500	40.317	328.678	293.397	123.485	-128.813	-140.585	2.098
3600	40.404	329.815	294.393	127.521	-128.695	-140.923	2.045
3700	40.490	330.924	295.365	131.566	-128.523	-141.265	1.994
3800	40.576	332.005	296.315	135.619	-128.296	-141.612	1.947
3900	40.662	333.060	297.244	139.681	-128.013	-141.966	1.901
4000	40.748	334.090	298.152	143.752	-127.675	-142.328	1.859
4100	40.834	335.097	299.041	147.831	-127.282	-142.700	1.818
4200	40.920	336.082	299.911	151.919	-126.834	-143.081	1.779
4300	41.005	337.046	300.764	156.015	-126.333	-143.474	1.743
4400	41.091	337.990	301.599	160.120	-125.778	-143.879	1.708
4500	41.176	338.914	302.418	164.233	-125.171	-144.296	1.675
4600	41.262	339.820	303.221	168.355	-124.514	-144.729	1.643
4700	41.347	340.709	304.010	172.485	-123.806	-145.176	1.613
4800	41.432	341.580	304.783	176.624	-123.050	-145.638	1.585
4900	41.517	342.435	305.543	180.772	-122.247	-146.117	1.558
5000	41.602	343.275	306.289	184.928	-121.398	-146.613	1.532
5100	41.687	344.099	307.023	189.092	-120.504	-147.126	1.507
5200	41.772	344.910	307.743	193.265	-119.566	-147.657	1.483
5300	41.857	345.706	308.452	197.447	-118.586	-148.206	1.461
5400	41.942	346.489	309.149	201.637	-117.564	-148.775	1.439
5500	42.027	347.260	309.835	205.835	-116.503	-149.363	1.419
5600	42.112	348.018	310.510	210.042	-115.403	-149.970	1.399
5700	42.196	348.764	311.175	214.257	-114.265	-150.597	1.380
5800	42.281	349.499	311.829	218.481	-113.090	-151.245	1.362
5900	42.366	350.222	312.474	222.714	-111.880	-151.913	1.345
6000	42.450	350.935	313.109	226.954	-110.635	-152.602	1.329

PREVIOUS: December 1965 (1 atm) CURRENT: December 1965 (1 bar)

Iodine Fluoride (IF) $F_1I_1(g)$

POTASSIUM FLUORIDE (KF) CRYSTAL M_r = 58.096703

$S°(298.15\ K) = 66.547 \pm 0.21\ J\ K^{-1}\ mol^{-1}$

$T_{fus} = 1131\ K$

$\Delta_f H°(0\ K) = -567.111 \pm 0.4\ kJ\ mol^{-1}$

$\Delta_f H°(298.15\ K) = -568.606 \pm 0.4\ kJ\ mol^{-1}$

$\Delta_{fus} H° = 27.196\ kJ\ mol^{-1}$

Enthalpy of Formation

The enthalpy of solution of KF(cr) in water has been measured by several investigators. Their results have been converted to $\Delta_{sol}H°(aq, \infty)$ by Parker (5) and listed in the table below. Using $\Delta_{sol}H°(aq, \infty) = -4.24$, $\Delta_f H°(K^+, aq, \infty, 298.15\ K) = -60.32$ (6), and $\Delta_f H°(F^-, aq, \infty, 298.15\ K) = -79.82\ kcal\ mol^{-1}$ (7), we obtain $\Delta_f H°(298.15\ K) = -135.9\pm0.1\ kcal\ mol^{-1}$ ($-568.606\pm0.4\ kJ\ mol^{-1}$) which is adopted.

Source	T/K	Concentration Molality	No. of Determinations	$\Delta_{sol}H°(aq, \infty)$ kcal mol^{-1}
(1) Guntz (1884)	291	0.28	1	-3.95
(2) Forcrand (1911)	288	0.50	1	-4.54
(3) Lange and Eichler (1927)	298	3.16 - 0.21	8	-4.24±0.4
(4) Lange and Martin (1937)	298	0.31	1	-4.01

Heat Capacity and Entropy

The low temperature heat capacities, 16-323 K, were measured by Westrum and Pitzer (8). The high temperature enthalpies, 298.15-530 K and 291-1187 K, were determined by Westrum and Pitzer (8), and Lyashenko (9), respectively. However, the high temperature $C_p°$ values derived from both sets appear to be inconsistent with the low temperature $C_p°$ values. The heat capacities derived from the enthalpy data of Westrum and Pitzer are too high (3% at 530 K and less at lower temperatures) to be joined smoothly with their low temperature $C_p°$'s. The adopted enthalpy at 530 K is about 1.4% lower than the reported value. The heat capacity derived from the enthalpy data of Lyashenko is 2.5% lower than the adopted one at 900 K. The differences are smaller at lower temperatures. At 900 K, the adopted enthalpy is 1.3% higher than the reported value. The adopted heat capacities in the temperature range, 323-2000 K, are estimated by graphical comparison with other alkali fluorides and chlorides, and joined smoothly with the low temperature data at 298.15 K.

S°(298.15 K) is derived based on the adopted low temperature heat capacities.

Fusion Data

See the liquid table for details.

Sublimation Data

The value of $\Delta_{sub}H°(298.15\ K) = 241.835\ kJ\ mol^{-1}$ for the process KF(cr) = KF(g) for the process KF(cr) = KF(g) is obtained as the difference between $\Delta_f H°(298.15\ K)$ for KF(cr) and KF(g).

References

1. A. Guntz, Ann. Chim. Phys. (6) 3, 5 (1884).
2. R. de Forcrand, Compt. Rend. 152, 1073 (1911).
3. E. Lange and A. Eichler, Z. Physik. Chem. 129, 285 (1927).
4. E. Lange and W. Martin, Z. Physik. Chem. A180, 233 (1937).
5. V. B. Parker, U. S. Nat. Bur. Stand. NSRDS-NBS 2, 1965.
6. U. S. Nat. Bur. Stand. Tech. Note 270-3, 1968.
7. JANAF Thermochemical Tables: HF(g), 12-31-68.
8. E. F. Westrum, Jr. and K. S. Pitzer, J. Amer. Chem. Soc. 71, 1940 (1949).
9. V. S. Lyashenko, Metallurg (USSR) 10, 85 (1935).

Potassium Fluoride (KF) $F_1K_1(cr)$

Enthalpy Reference Temperature = T_r = 298.15 K — Standard State Pressure = p° = 0.1 MPa

T/K	$C_p°$ (J K⁻¹mol⁻¹)	S°	$-[G°-H°(T_r)]/T$	$H°-H°(T_r)$ (kJ mol⁻¹)	$\Delta_f H°$	$\Delta_f G°$	Log K_f
0	0.	0.	INFINITE	-10.000	-567.111	-567.111	INFINITE
100	31.606	20.497	107.023	-8.653	-568.949	-558.920	291.950
200	45.187	47.660	70.944	-4.657	-569.013	-548.781	143.327
298.15	48.982	66.547	66.547	0.	-568.606	-538.934	94.419
300	49.022	66.850	66.547	0.091	-568.599	-538.750	93.805
400	51.045	81.245	68.496	5.099	-570.678	-528.399	69.002
500	52.718	92.820	72.240	10.290	-570.278	-517.872	54.102
600	54.266	102.571	76.504	15.640	-569.704	-507.442	44.177
700	55.856	111.056	80.846	21.147	-568.969	-497.121	37.096
800	57.404	118.614	85.103	26.808	-568.089	-486.915	31.792
900	59.078	125.468	89.213	32.630	-567.077	-476.828	27.674
1000	61.170	131.797	93.159	38.639	-565.926	-466.860	24.386
1100	63.848	137.748	96.944	44.884	-643.560	-452.406	21.483
1131.000	64.852	139.536	98.087	46.878	------ CRYSTAL <--> LIQUID -----		
1200	66.777	143.436	100.583	51.423	-640.972	-435.140	18.941
1300	68.618	148.858	104.090	58.198	-638.160	-418.101	16.759
1400	69.789	153.989	107.472	65.123	-635.210	-401.283	14.972
1500	70.500	158.830	110.736	72.140	-632.178	-384.679	13.396
1600	70.919	163.394	113.886	79.212	-629.101	-368.280	12.023
1700	71.128	167.701	116.926	86.316	-626.002	-352.073	10.818
1800	71.128	171.766	119.861	93.429	-622.904	-336.049	9.752
1900	71.128	175.612	122.695	100.542	-619.816	-320.197	8.803
2000	71.128	179.260	125.433	107.655	-616.738	-304.507	7.953

PREVIOUS: December 1963 CURRENT: June 1969

Potassium Fluoride (KF) $F_1K_1(cr)$

POTASSIUM FLUORIDE (KF) LIQUID M_r = 58.096703

$S°(298.15\ K) = [67.634]\ J\ K^{-1}\ mol^{-1}$

$T_{fus} = 1131\ K$

$\Delta_fH°(298.15\ K) = [-554.467]\ kJ\ mol^{-1}$

$\Delta_{fus}H° = 27.196\ kJ\ mol^{-1}$

Enthalpy of Formation

$\Delta_fH°(KF,\ell, 298.15\ K)$ is calculated from $\Delta_fH°(KF, cr, 298.15\ K)$ by adding the enthalpy of fusion, $\Delta_{fus}H°$, and the difference in enthalpy, H°(1131 K) - H°(298.15 K), between the crystal and liquid.

Heat Capacity and Entropy

The heat capacity is estimated by comparison with those of the alkali fluorides and chlorides, and is assumed to be constant in the temperature range, 298-3000 K. The $C_p°$ value, 16 cal $K^{-1}\ mol^{-1}$, derived from the enthalpy data of Lyashenko (12) appears low and is not adopted.

The entropy is calculated in a manner analogous to that used for the enthalpy of formation.

Fusion Data

T_{fus} has been reported as 1119 (1, 2), 1129 (3, 4), 1131 (5, 6), and 1153 K (7). The value 1131 K is adopted. Kelley (8) evaluated the following freezing-point data of binary systems: $KF-AlF_3$ (9), $KF-K_2SO_4$ (10), KF-KCl (11), $KF-K_2CO_3$ (13), and $KF-K_3PO_4$ (14), and derived the values of $\Delta_{fus}H°$ (in kcal mol^{-1}) as 5.5, 5.55, 6.51, 7.1, and 7.1, respectively. The best value is recommended as 6.50 kcal mol^{-1}. Other reported $\Delta_{fus}H°$ values are 6.70 (7) and 6.75±0.1 kcal mol^{-1} (12). The value of heat of melting is tentatively adopted as 6.50 kcal mol^{-1}.

Vaporization Data

The boiling point (T_{vap}=1790K) is calculated to be the temperature at which the partial pressures of KF(g) and K_2F_2(g) equal one atmosphere. The composition of the gaseous mixture in equilibrium with KF(ℓ) at T_{fus} is calculated as K_2F_2 21.1% and KF 78.9%. The heat required to vaporize one mole of liquid at T_{vap} is evaluated to be 33.9 kcal mol^{-1} (141.838 kJ mol^{-1}) (141.838 kJ mol^{-1}).

The experimental values for the boiling point are 1778 K (1) and 1771 K (15).

References

1. H. von Wartenberg and H. Schulz, Z. Elektrochem. 27, 568 (1921).
2. B. S. Zakharova, L. P. Reshetnikova, and A. V. Novoselova, Vestn. Mosk. Univ., Ser. II 22, 102 (1967).
3. O. Schmitz-Dumont and E. Schmitz, Z. Anorg. Chem. 252, 329 (1944).
4. E. P. Dergunov and A. G. Bergman, Zh. Fiz. Khim. 22, 625 (1948).
5. J. W. Johnson and M. A. Bredig, J. Phys. Chem. 62, 604 (1958).
6. B. Porter and E. A. Brown, J. Amer. Ceram. Soc. 45, 49 (1962).
7. G. Petit and A. Crémieu, Compt. Rend. 243, 360 (1956).
8. K. K. Kelley, U. S. Bur. Mines Bull. 393, 1936.
9. N. A. Puschin and A. B. Baskow, J. Russ. Phys. Chem. Soc. 45, 82 (1913); Z. Anorg. Chem. 81, 347 (1913).
10. B. Karandeeff, Zentr. Min. Geol. 728, 1909.
11. W. Plato, Z. Physik, Chem. 55, 721 (1906).
12. V. S. Lyashenko, Metallurg (USSR) 10, 85 (1935).
13. M. Amadori, Atti Accad. Lincei, Ser. 5, 22, 366 (1913).
14. M. Amadori, Atti Accad. Lincei, Ser. 5, 21, 688 (1912).
15. O. Ruff, G. Schmidt, and S. Mugdan, Z. Anorg. Allg. Chem. 123, 83 (1922).

Potassium Fluoride (KF) $F_1K_1(l)$

Enthalpy Reference Temperature = T_r = 298.15 K Standard State Pressure = p° = 0.1 MPa

	J K⁻¹mol⁻¹			kJ mol⁻¹			
T/K	$C_p°$	S°	$-[G°-H°(T_r)]/T$	$H°-H°(T_r)$	$\Delta_fH°$	$\Delta_fG°$	Log K_f
0							
100							
200							
298.15	71.965	67.634	67.634	0.	-554.467	-525.120	91.999
300	71.965	68.079	67.635	0.133	-554.418	-524.938	91.400
400	71.965	88.782	70.458	7.330	-554.309	-515.045	67.258
500	71.965	104.840	75.788	14.526	-551.903	-505.507	52.810
600	71.965	117.961	81.757	21.723	-549.483	-496.455	43.220
700	71.965	129.055	87.742	28.919	-547.058	-487.809	36.401
800	71.965	138.664	93.520	36.116	-544.643	-479.510	31.309
900	71.965	147.140	99.016	43.312	-542.256	-471.512	27.366
1000	71.965	154.723	104.214	50.508	-539.918	-463.777	24.225
1100	71.965	161.582	109.123	57.705	-616.600	-451.664	21.448
1131.000	71.965	163.582	110.588	59.936	------ CRYSTAL <--> LIQUID -----		
1200	71.965	167.843	113.759	64.901	-613.355	-436.813	19.014
1300	71.965	173.604	118.144	72.098	-610.122	-422.232	16.965
1400	71.965	178.937	122.298	79.294	-606.900	-407.900	15.219
1500	71.965	183.902	126.241	86.491	-603.688	-393.798	13.713
1600	71.965	188.546	129.992	93.687	-600.487	-379.910	12.403
1700	71.965	192.909	133.566	100.884	-597.296	-366.221	11.253
1800	71.965	197.023	136.978	108.080	-594.114	-352.720	10.236
1900	71.965	200.914	140.242	115.277	-590.942	-339.396	9.331
2000	71.965	204.605	143.368	122.473	-587.781	-326.239	8.520
2100	71.965	208.116	146.368	129.670	-584.629	-313.239	7.791
2200	71.965	211.464	149.252	136.866	-581.488	-300.390	7.132
2300	71.965	214.663	152.027	144.063	-578.357	-287.683	6.533
2400	71.965	217.726	154.701	151.259	-575.238	-275.111	5.988
2500	71.965	220.663	157.281	158.456	-572.130	-262.670	5.488
2600	71.965	223.486	159.773	165.652	-569.035	-250.353	5.030
2700	71.965	226.202	162.184	172.849	-565.954	-238.154	4.607
2800	71.965	228.819	164.517	180.045	-562.889	-226.070	4.217
2900	71.965	231.344	166.778	187.242	-559.841	-214.094	3.856
3000	71.965	233.784	168.971	194.438	-556.811	-202.224	3.521

PREVIOUS: December 1963 CURRENT: June 1969

Potassium Fluoride (KF) $F_1K_1(l)$

POTASSIUM FLUORIDE (KF) CRYSTAL-LIQUID M_r = 58.096703

0 to 1131 K crystal
above 1131 K liquid

Refer to the individual tables for details.

Potassium Fluoride (KF) $F_1K_1(cr,l)$

Enthalpy Reference Temperature = T_r = 298.15 K Standard State Pressure = p° = 0.1 MPa

T/K	C_p° (J K⁻¹mol⁻¹)	S° (J K⁻¹mol⁻¹)	$-[G^\circ-H^\circ(T_r)]/T$ (J K⁻¹mol⁻¹)	$H^\circ-H^\circ(T_r)$ (kJ mol⁻¹)	Δ_fH° (kJ mol⁻¹)	Δ_fG° (kJ mol⁻¹)	Log K_f
0	0.	0.	INFINITE	-10.000	-567.111	-567.111	INFINITE
100	31.606	20.497	107.023	-8.653	-568.949	-558.920	291.950
200	45.187	47.660	70.944	-4.657	-569.013	-548.781	143.327
298.15	48.982	66.547	66.547	0.	-568.606	-538.934	94.419
300	49.022	66.850	66.547	0.091	-568.599	-538.750	93.805
400	51.045	81.245	68.496	5.099	-570.678	-528.399	69.002
500	52.718	92.820	72.240	10.290	-570.278	-517.872	54.102
600	54.266	102.571	76.504	15.640	-569.704	-507.442	44.177
700	55.856	111.056	80.846	21.147	-568.969	-497.121	37.096
800	57.404	118.614	85.103	26.808	-568.089	-486.915	31.792
900	59.078	125.468	89.213	32.630	-567.077	-476.828	27.674
1000	61.170	131.797	93.159	38.639	-565.926	-466.860	24.386
1100	63.848	137.748	96.944	44.884	-643.560	-452.406	21.483
1131.000	64.852	139.536	98.087	46.878	CRYSTAL <--> LIQUID TRANSITION		
1131.000	71.965	163.582	98.087	74.074			
1200	71.965	167.843	101.977	79.040	-613.355	-436.813	19.014
1300	71.965	173.604	107.268	86.236	-610.122	-422.232	16.96[illegible]
1400	71.965	178.937	112.199	93.433	-606.900	-407.900	15.21[illegible]
1500	71.965	183.902	116.816	100.629	-603.688	-393.798	13.713
1600	71.965	188.546	121.155	107.826	-600.487	-379.910	12.403
1700	71.965	192.909	125.249	115.022	-597.296	-366.221	11.253
1800	71.965	197.023	129.123	122.219	-594.114	-352.720	10.236
1900	71.965	200.914	132.800	129.415	-590.942	-339.396	9.331
2000	71.965	204.605	136.299	136.612	-587.781	-326.239	8.520
2100	71.965	208.116	139.636	143.808	-584.629	-313.239	7.791
2200	71.965	211.464	142.825	151.005	-581.488	-300.390	7.132
2300	71.965	214.663	145.880	158.201	-578.357	-287.683	6.533
2400	71.965	217.726	148.810	165.398	-575.238	-275.111	5.988
2500	71.965	220.663	151.626	172.594	-572.130	-262.670	5.488
2600	71.965	223.486	154.336	179.791	-569.035	-250.353	5.030
2700	71.965	226.202	156.947	186.987	-565.954	-238.154	4.607
2800	71.965	228.819	159.468	194.184	-562.889	-226.070	4.217
2900	71.965	231.344	161.903	201.380	-559.841	-214.094	3.856
3000	71.965	233.784	164.258	208.577	-556.811	-202.224	3.521

PREVIOUS: CURRENT: June 1969

Potassium Fluoride (KF) $F_1K_1(cr,l)$

POTASSIUM FLUORIDE (KF) IDEAL GAS $M_r = 58.096703$

$S°(298.15\ K) = 226.612 \pm 0.04\ J\ K^{-1}\ mol^{-1}$

$\Delta_f H°(0\ K) = -324.729 \pm 2.1\ kJ\ mol^{-1}$
$\Delta_f H°(298.15\ K) = -326.770 \pm 2.1\ kJ\ mol^{-1}$

Electronic Level and Quantum Weight

State	ε_i, cm^{-1}	g_i
$^1\Sigma$	0	1

$\omega_e = 426.04\ cm^{-1}$ $\omega_e x_e = 2.43\ cm^{-1}$ $\sigma = 1$
$B_e = 0.27994\ cm^{-1}$ $\alpha_e = 0.002335\ cm^{-1}$ $r_e = 2.17155\ Å$

Enthalpy of Formation

The total vapor pressures of KF(g) and K_2F_2(g) over KF(cr, ℓ) have been determined by many investigators using the boiling point (1, 2, 5), Knudsen-effusion (3), transpiration (4), and torsion-effusion (6) methods. A method of trial and error is employed to select the appropriate values for $\Delta_{sub}H°$(KF, cr, 298.15 K) and $\Delta_{vap}H°$(KF, ℓ, 298.15 K) such that by use of these two values and $\Delta_r H°(298.15\ K) = 50\ kcal\ mol^{-1}$ for the reaction $K_2F_2(g) = 2\ KF(g)$ and the Gibbs energy functions for KF(cr), KF(ℓ), KF(g), and K_2F_2(g), the calculated total pressures of KF(g) and K_2F_2(g) are consistent with the measured ones. Adopting $\Delta_{sub}H°(298.15\ K) = 57.8$ and $\Delta_{vap}H°(298.15\ K) = 54.42\ kcal\ mol^{-1}$, we evaluate the partial pressures of KF(g) and K_2F_2(g) from the reported total pressure measurements. Based on the derived partial pressures for KF(g), the enthalpy changes for the processes: A. KF(cr) = KF(g), and B. KF(ℓ) = KF(g) are calculated by the 2nd and 3rd law methods. Using the 3rd law $\Delta_f H°$(298.15 K) and $\Delta_f H°$(298.15 K) = -135.90 and -132.519 kcal mol^{-1} for KF(cr) and KF(ℓ), respectively, we obtain the $\Delta_f H°$(298.15 K) for KF(g). The results are presented in the table below.

The dissociation energy of KF(g) has been determined as 117±8 kcal mol^{-1} by flame photometry by Page and Sugden (7), and Bulewicz, Phillips, and Sugden (8), yielding $\Delta_f H°$(KF, g, 298.15 K) = -77.6±8 kcal mol^{-1}. Barrow and Caunt (9) measured the fluctuation bands in the ultraviolet absorption spectra of KF and reported the upper limit for the D_0^0(KF, g) value to be 121.8 kcal mol^{-1} which leads to $\Delta_f H°$(KF, g, 298.15 K) = -81.4 kcal mol^{-1} as the lower limit. Gaydon (10) reported D_0^0 = 116.9±4.6 kcal mol^{-1} for KF(g). The corresponding value of $\Delta_f H°$(298.15 K) is calculated as -77.5±4.6 kcal mol^{-1}.

The enthalpy of formation for KF(g) at 298.15 K is adopted as .78.1±0.5 kcal mol^{-1}.

Source	Method	T/K	Data Points	Process	$\Delta_r H°$, kcal mol^{-1} 2nd law	$\Delta_r H°$, kcal mol^{-1} 3rd law	Drift cal K^{-1}mol^{-1}	$\Delta_f H°$(298.15 K) kcal mol^{-1}
Wartenberg (1)	boiling point	1624-1776	5	B	54.4±0.9	54.16	-0.1±0.5	-78.36
Ruff (2)	boiling point	1551-1773	6	B	58.1±2.5	54.14	-2.4±1.5	-78.38
Niwa (3)	Knudsen-effusion	913-973	6	A	56.2±0.3	55.67	-0.6±0.3	-80.23
Naryshkin (4)	transpiration	1023-1108	2	A	--	57.23	--	-78.67
		1188-1233	2**	B	--	54.60	--	-77.92
Cantor (5)	boiling point	1200-1500	7*	B	54.4	54.38	-0.02	-78.14
Pugh (6)	torsion-effusion	910-1020	6*	A	57.7	58.12	0.47	-77.78

* Data points calculated from the given vapor pressure equation.

** Only the data point at 1233 K is adopted for evaluation.

Heat Capacity and Entropy

The rotational transition spectrum of KF(g) has been measured in the millimeter- and submillimeter-wave region with a high resolution molecular-beam absorption spectrometer by Veazey and Gordy (15). From the derived potential constants the vibrational constants were calculated. The reported values for ω_e, $\omega_e x_e$, B_e, α_e and r_e are adopted.

Other values of ω_e reported are (in cm^{-1}): 390±39 (11), 405±25 (9), 429±3 (12), and 426±4 (13). The value of $\omega_e x_e$ has been determined as 1.45 cm^{-1} by Barrow and Caunt (9), and 2.4 cm^{-1} by Ritchie and Lew (12) and Baikov and Vasilevskii (13). Using the molecular beam electric resonance method, Green and Lew (14) measured the rotational transitions between the J = 0 and J = 1 states of $K^{39}F$ and reported the following values: $B_e = 0.27994\ cm^{-1}$, $\alpha_e = 0.002335\ cm^{-1}$, and $r_e = 2.17144 \pm 0.00005\ Å$ which are in excellent agreement with the adopted values. By use of the same method, Grabner and Hughes (11) investigated the rotational Stark spectrum of $K^{39}F$ and reported the K-F bond distance for the zeroth vibrational state as 2.55±0.06 Å.

References

1. H. von Wartenberg and H. Schulz, Z. Electrochem. 27, 568 (1921).
2. O. Ruff, G. Schmidt, and S. Mugdan, Z. Anorg. Allgem. Chem. 123, 83 (1922).
3. K. Niwa, J. Fac. Sci. Hokkaido Imp. University, Ser. III 2, 201 (1938).
4. I. I. Naryshkin, Zh. Fiz. Khim. 13, 528 (1939).
5. S. Cantor and F. Blankenship, paper presented at the American Chemical Society meeting, Chicago, Sept. 7-12, 1958.
6. A. C. P. Pugh and R. F. Barrow, Trans. Faraday Soc. 54, 671 (1958).
7. F. M. Page and T. M. Sugden, Nature 183, 1672 (1959).
8. E. M. Bulewicz, L. F. Phillips, and T. M. Sugden, Trans. Faraday Soc. 57, 921 (1961).
9. R. F. Barrow and A. D. Caunt, Proc. Roy. Soc. (London) A219, 120 (1953).
10. A. G. Gaydon, "Dissociation Energies and Spectra of Diatomic Molecules," Chapman and Hall Ltd., London, 1968.
11. L. Grabner and V. Hughes, Phys. Rev. 79, 819 (1950).
12. R. K. Ritchie and H. Lew, Can. J. Phys. 42, 43 (1964).
13. V. I. Baikov and K. P. Vasilevskii, Opt. Spektrosk. 22, 198 (1967).
14. G. W. Green and H. Lew, Can. J. Phys. 38, 482 (1960).
15. S. E. Veazey and W. Gordy, Phys. Rev. 138, A1303 (1965).

Potassium Fluoride (KF) $F_1K_1(g)$

Enthalpy Reference Temperature = T_r = 298.15 K Standard State Pressure = p° = 0.1 MPa

T/K	$C_p°$ (J K^{-1}mol^{-1})	S° (J K^{-1}mol^{-1})	-[G°-H°(T_r)]/T (J K^{-1}mol^{-1})	H°-H°(T_r) (kJ mol^{-1})	$\Delta_f H°$ (kJ mol^{-1})	$\Delta_f G°$ (kJ mol^{-1})	Log K_f
0	0.	0.	INFINITE	-9.453	-324.729	-324.729	INFINITE
100	29.837	191.161	256.478	-6.532	-324.993	-332.031	173.435
200	33.260	212.918	229.781	-3.373	-325.893	-338.713	88.463
250	34.438	220.474	227.187	-1.678	-326.325	-341.868	71.429
298.15	35.228	226.612	226.612	0.	-326.770	-344.822	60.411
300	35.253	226.830	226.612	0.065	-326.789	-344.934	60.058
350	35.826	232.309	227.044	1.843	-329.697	-347.821	51.909
400	36.242	237.122	228.009	3.645	-330.297	-350.369	45.753
450	36.553	241.409	229.264	5.465	-330.874	-352.843	40.957
500	36.793	245.273	230.675	7.299	-331.433	-355.254	37.113
600	37.141	252.015	233.686	10.997	-332.511	-359.916	31.333
700	37.385	257.759	236.725	14.724	-333.556	-364.400	27.192
800	37.570	262.764	239.673	18.472	-334.590	-368.736	24.076
900	37.719	267.198	242.490	22.237	-335.635	-372.942	21.645
1000	37.846	271.179	245.163	26.015	-336.714	-377.030	19.694
1100	37.958	274.791	247.695	29.806	-416.803	-376.397	17.874
1200	38.060	278.098	250.093	33.607	-416.953	-372.717	16.224
1300	38.155	281.148	252.366	37.417	-417.106	-369.024	14.828
1400	38.244	283.979	254.524	41.237	-417.260	-365.320	13.630
1500	38.329	286.621	256.577	45.066	-417.417	-361.604	12.592
1600	38.412	289.097	258.533	48.903	-417.574	-357.878	11.684
1700	38.492	291.428	260.400	52.748	-417.734	-354.143	10.881
1800	38.570	293.631	262.185	56.601	-417.896	-350.397	10.168
1900	38.647	295.718	263.896	60.462	-418.060	-346.643	9.530
2000	38.722	297.702	265.537	64.331	-418.227	-342.880	8.955
2100	38.797	299.593	267.114	68.207	-418.396	-339.108	8.435
2200	38.871	301.400	268.632	72.090	-418.567	-335.329	7.962
2300	38.944	303.129	270.094	75.981	-418.743	-331.541	7.530
2400	39.017	304.788	271.506	79.879	-418.921	-327.746	7.133
2500	39.089	306.383	272.869	83.784	-419.105	-323.943	6.768
2600	39.160	307.917	274.188	87.697	-419.294	-320.133	6.432
2700	39.232	309.396	275.464	91.616	-419.490	-316.315	6.119
2800	39.303	310.824	276.702	95.543	-419.694	-312.490	5.830
2900	39.374	312.205	277.902	99.477	-419.909	-308.658	5.560
3000	39.444	313.541	279.068	103.418	-420.135	-304.818	5.307
3100	39.515	314.835	280.201	107.366	-420.376	-300.970	5.071
3200	39.585	316.091	281.303	111.321	-420.635	-297.114	4.850
3300	39.655	317.310	282.376	115.283	-420.905	-293.251	4.642
3400	39.725	318.495	283.421	119.252	-421.203	-289.378	4.446
3500	39.795	319.648	284.440	123.228	-421.526	-285.496	4.261
3600	39.865	320.770	285.433	127.211	-421.880	-281.605	4.086
3700	39.935	321.863	286.403	131.201	-422.269	-277.703	3.920
3800	40.004	322.929	287.350	135.198	-422.696	-273.790	3.764
3900	40.074	323.969	288.276	139.202	-423.167	-269.865	3.614
4000	40.143	324.984	289.181	143.213	-423.679	-265.929	3.473
4100	40.213	325.976	290.067	147.230	-424.245	-261.979	3.338
4200	40.282	326.946	290.933	151.255	-424.872	-258.013	3.209
4300	40.351	327.895	291.782	155.287	-425.561	-254.033	3.086
4400	40.421	328.823	292.613	159.325	-426.040	-250.062	2.969
4500	40.490	329.733	293.428	163.371	-426.800	-246.054	2.856
4600	40.559	330.623	294.227	167.423	-427.624	-242.029	2.748
4700	40.628	331.496	295.011	171.483	-428.496	-237.986	2.645
4800	40.697	332.352	295.780	175.549	-429.329	-233.937	2.546
4900	40.766	333.192	296.535	179.622	-430.335	-229.856	2.450
5000	40.835	334.016	297.276	183.702	-431.413	-225.754	2.358
5100	40.904	334.826	298.004	187.789	-432.564	-221.629	2.270
5200	40.973	335.621	298.720	191.883	-433.790	-217.481	2.185
5300	41.042	336.402	299.424	195.984	-435.090	-213.309	2.102
5400	41.111	337.170	300.116	200.091	-436.466	-209.112	2.023
5500	41.180	337.925	300.796	204.206	-437.917	-204.889	1.946
5600	41.249	338.667	301.466	208.327	-439.442	-200.638	1.871
5700	41.318	339.398	302.125	212.456	-440.697	-196.407	1.800
5800	41.387	340.117	302.774	216.591	-442.325	-192.108	1.730
5900	41.456	340.825	303.413	220.733	-444.020	-187.779	1.662
6000	41.525	341.523	304.042	224.882	-445.779	-183.421	1.597

PREVIOUS: June 1969 (1 atm) CURRENT: June 1969 (1 bar)

Potassium Fluoride (KF) $F_1K_1(g)$

$S°(298.15\ K) = 35.660 \pm 0.021\ J\ K^{-1}\ mol^{-1}$

$T_{fus} = 1121.3 \pm 1.0\ K$

$\Delta_f H°(0\ K) = -614.369 \pm 0.8\ kJ\ mol^{-1}$

$\Delta_f H°(298.15\ K) = -616.931 \pm 0.8\ kJ\ mol^{-1}$

$\Delta_{fus} H° = 27.087 \pm 0.021\ kJ\ mol^{-1}$

Enthalpy of Formation

The enthalpies of solution and reaction of LiF(cr) in water and hydrochloric acid have been measured by many investigators. The reported $\Delta_{sln}H°$ values are listed in the table below. Using $\Delta_f H°(Li^+, aq, \infty, 298.15\ K) = -66.56\ kcal\ mol^{-1}$ (8), and $\Delta_f H°(F^-, aq, \infty, 298.15\ K) = -79.82\ kcal\ mol^{-1}$ (9), we calculate the corresonding $\Delta_f H°(298.15\ K)$ values for LiF(cr). Based on the enthalpy of solution of LiOH(cr) in excess HF solution measured by Sinke (7), the value, $\Delta_f H°(LiF, cr, 298.15\ K) = -146.65 \pm 0.5\ kcal\ mol^{-1}$ is derived. The value of $\Delta_f H°(298.15\ K)$ for LiF(cr) is adopted as $-147.45 \pm 0.2\ kcal\ mol^{-1}$.

Source	Solvent	T/K	Concentration ($LiF \cdot nH_2O$)	$\Delta_r H°$, kcal mol^{-1}	$\Delta_\infty H°$, kcal mol^{-1}*	$\Delta_f H°(298.15\ K)$, kcal mol^{-1}
1. de Forcrand (1911)	H_2O	288.2	unavailable	1.04	-	-
2. Kolesov-Skuratov (1961)	H_2O	294.7	3800 H_2O	1.25 ± 0.02	1.13 ± 0.05	-147.50
3. Wulff (1962)	HCl	298.2	unavailable	unavailable	1.10 ± 0.10	-147.47
4. Stephenson et al.(1964)	H_2O	298.2	2780-5560 H_2O	unavailable	1.07 ± 0.05	-147.44
5. Cox and Harrop (1964)	H_2O	298.2	60400 H_2O	1.059 ± 0.028	1.014 ± 0.055	-147.38

*The enthalpy of dilution data of Parker (6) are used to convert $\Delta_r H°$ to $\Delta_\infty H°$.

Heat Capacity and Entropy

Clusius, et al, (10) measured the low temperature heat capacities in the temperature range 19 to 272 K and Clusius and Eichenauer (11) from 10 to 111 K. Below 20 K a number of investigators have reported values of θ_D, and the average value of 729±6 is used here. The high temperature enthalpies from 298.15 K to the melting point 1121.3 K were measured by Douglas and Dever (12), and Voskresenskaya et al. (13). All these data are taken and smoothed by computer to 400 K. Above 400 K the $C_p°$ values are obtained graphically by drawing a smooth line through the data points.

$S°(298.15\ K)$ is derived from the adopted low temperature heat capacities, based on $S°(20\ K) = 0.064\ cal\ K^{-1}\ mol^{-1}$.

Fusion Data

See the liquid table for details.

Sublimation Data

The enthalpy of sublimation, $\Delta_{sub}H°(298.15\ K) = 276.144\ kJ\ mol^{-1}$, is calculated as the difference between $\Delta_f H°(298.15\ K)$ for LiF(g) and LiF(cr).

References

1. R. de Forcrand, Compt. Rend. 152, 27 (1911).
2. V. P. Kolesov and S. M. Skuratov, Russ. J. Inorg. Chem. 6, 889 (1961).
3. C. A. Wulff, Ph. D. Thesis, Massachusetts Institute of Technology, 1962.
4. C. C. Stephenson, H. P. Hopkins and C. A. Wulff, J. Phys. Chem. 68, 1427 (1964).
5. J. D. Cox and D. Harrop, Trans. Faraday Soc. 61, 1328 (1965).
6. V. B. Parker, U. S. Nat. Bur. Stand., NSRDS-NBS 2, 1965.
7. G. C. Sinke, The Dow Chemical Company, Midland, Michigan, personal communication, February 1959.
8. S. R. Gunn and L. G. Green, J. Amer. Chem. Soc. 80, 4782 (1958).
9. Derived from JANAF value of $\Delta_f H°(HF, g, 298.15\ K)$, 12-31-68.
10. K. Clusius, J. Goldman, and A. Perlick, Z. Naturforsch. 49, 424 (1949).
11. K. Clusius and W. Eishenauer, Z. Naturforsch. 2a, 715 (1956).
12. T. B. Douglas and J. L. Dever, J. Amer. Chem. Soc. 76, 4826 (1954).
13. N. K. Voskresenskaya, V. A. Sokolov, E. I. Banashek, and N. E. Shmidt, Akad. Nauk SSSR, Izvest. Sect. Fiz. Khim. Anal. 27, 233 (1956).

Enthalpy Reference Temperature = T_r = 298.15 K; Standard State Pressure = p° = 0.1 MPa

T/K	$C_p°$ (J K^{-1}mol^{-1})	S° (J K^{-1}mol^{-1})	$-[G°-H°(T_r)]/T$ (J K^{-1}mol^{-1})	$H°-H°(T_r)$ (kJ mol^{-1})	$\Delta_f H°$ (kJ mol^{-1})	$\Delta_f G°$ (kJ mol^{-1})	Log K_f
0	0.	0.	INFINITE	-6.473	-614.369	-614.369	INFINITE
100	12.816	4.858	65.944	-6.109	-615.953	-607.186	317.162
200	32.803	20.690	39.225	-3.707	-616.858	-597.965	156.172
298.15	41.815	35.660	35.660	0.	-616.931	-588.660	103.131
300	41.936	35.919	35.661	0.077	-616.928	-588.485	102.464
400	46.543	48.706	37.373	4.533	-616.682	-579.039	75.615
500	49.329	59.401	40.738	9.331	-619.500	-569.345	59.479
600	51.589	68.597	44.633	14.378	-619.173	-559.341	48.695
700	53.723	76.711	48.647	19.645	-618.607	-549.410	40.997
800	55.731	84.016	52.619	25.118	-617.835	-539.575	35.231
900	57.697	90.695	56.483	30.790	-616.881	-529.848	30.752
1000	59.580	96.871	60.217	36.654	-615.749	-520.237	27.174
1100	61.505	102.642	63.814	42.710	-614.434	-510.748	24.253
1121.300	61.883	103.825	64.563	44.024	------ CRYSTAL <--> LIQUID -----		
1200	63.245	108.068	67.278	48.948	-612.944	-501.386	21.825
1300	64.892	113.196	70.615	55.356	-611.294	-492.155	19.775
1400	66.445	118.063	73.832	61.923	-609.487	-483.058	18.023
1500	67.906	122.697	76.936	68.642	-607.532	-474.094	16.509
1600	69.274	127.124	79.936	75.502	-605.436	-465.266	15.189
1700	70.549	131.363	82.837	82.494	-748.444	-449.397	13.808
1800	71.731	135.429	85.647	89.608	-745.339	-431.895	12.533
1900	72.821	139.337	88.370	96.837	-742.129	-414.569	11.397
2000	73.817	143.098	91.013	104.169	-738.820	-397.414	10.379
2100	74.720	146.722	93.580	111.597	-735.423	-380.427	9.463
2200	75.530	150.217	96.076	119.110	-731.944	-363.603	8.633
2300	76.248	153.590	98.503	126.700	-728.394	-346.940	7.879
2400	76.872	156.849	100.867	134.357	-724.780	-330.432	7.192
2500	77.404	159.998	103.170	142.071	-721.112	-314.076	6.562

PREVIOUS: December 1963 CURRENT: December 1968

LITHIUM FLUORIDE (LiF) LIQUID M_r = 25.939403

$S°(298.15\ K) = [42.962]\ J\ K^{-1}\ mol^{-1}$

$T_{fus} = 1121.3 \pm 1.0\ K$

$\Delta_f H°(298.15\ K) = [-598.651]\ kJ\ mol^{-1}$

$\Delta_{fus}H° = 27.087 \pm 0.021\ kJ\ mol^{-1}$

Enthalpy of Formation

$\Delta_f H°$(Lif, ℓ, 298.15 K) is calculated from $\Delta_f H°$(LiF, cr, 298.15 K) by adding the enthalpy of fusion, $\Delta_{fus}H°$, and the difference in enthalpy, H°(1121.3 K)-H°(298.15 K), between the crystal and liquid.

Heat Capacity and Entropy

Douglas and Dever (1) and Voskresenskaya et al. (2) have measured the enthalpies of LiF liquid to 1230 and 1400 K, respectively. Using the reported enthalpy data, a constant heat capacity is derived for each set. The adopted $C_p°$ for LiF(ℓ) is the mean of the two derived heat capacities, and is extended arbitrarily to the temperatures above T_{vap} and below T_{fus}. The entropy is calculated in a manner analogous to that used for the enthalpy of formation.

Fusion Data

The melting point has been determined by many investigators. See the table below for details. The adopted value, 1121.3±1.0 K, is that reported by Douglas and Dever (1). The heat of melting is the mean of the $\Delta_{fus}H°$ values reported by Douglas and Dever (1) and Voskresenskaya et al. (2).

T_{fus}/ K	Source
1115	von Wartenberg and Schulz (3)
1115	Petit and Cremieu (4)
1119	Flood et al. (5)
1120	Goryacheva et al. (6)
1121	Haendler et al. (7)
1121.2	Mateiko and Bukhalova (8)
1121.3	Douglas and Dever (9)
1123	Bukhalova and Sementsova (10)

Vaporization Data

T_{vap} = 1990 K is the temperature at which the partial pressures of LiF(g), Li_2F_2(g), and Li_3F_3(g), in equilibrium with LiF(ℓ), equal one atmosphere. The vapor composition at the boiling point is evaluated as LiF 69%, Li_2F_2 26.6%, and Li_3F_3 4.4%. The heat required to vaporize one mole of liquid to form 0.739 mole of vapor mixture of the above composition at T_{vap} is calculated to be 35.08 kcal mol^{-1} (146.775 kJ mol^{-1}).

The value of T_{vap} for LiF(ℓ) has been reported to be 1943 and 1949 K by Ruff et al. (3) and von Wartenberg and Schulz (1), respectively, which were obtained by extrapolation of their vapor pressure data, measured by the boiling point method.

References

1. T. B. Douglas and J. L. Dever, J. Am. Chem. Soc. 76, 4826 (1954).
2. N. K. Voskresenskaya, V. A. Sokolov, E. I. Banashek, and N. E. Shmidt, Akad. Nauk SSSR, Izvest. Sekt. Fiz. - Khim. Anal. 27, 233 (1956).
3. H. von Wartenberg and H. Schulz, Z. Elektrochem. 27, 568 (1921).
4. G. Petit and A. Cremieu, Compt. Rend. 243, 360 (1956).
5. H. Flood, V. Fykse, and S. Urnes, Z. Elektrochem. 59 364 (1955).
6. V. P. Goryacheva, A. G. Bergman, and A. G. Kislova, Zh. Neorg. Khim. 4, 2744 (1959).
7. H. M. Haendler, P. S. Sennett, and C. M. Wheeler, J. Electrochem. Soc. 106, 264 (1959).
8. Z. A. Mateiko and G. A. Bukhalova, Zh. Neorg. Khim. 4, 1649 (1959).
9. T. B. Douglas and J. L. Dever, J. Am. Chem. Soc. 76, 4826 (1954).
10. G. A. Bukhalova and D. V. Sementsova, Russ. J. Inorg. Chem. 10, 1027 (1965).
11. O. Ruff, G. Schmidt, and S. Mugdan, Z. anorg. allgem. Chem. 123, 83 (1922).

Lithium Fluoride (LiF) F_1Li_1(l)

Enthalpy Reference Temperature = T_r = 298.15 K Standard State Pressure = p° = 0.1 MPa

T/K	$C_p°$ (J K⁻¹mol⁻¹)	S° (J K⁻¹mol⁻¹)	-[G°-H°(T_r)]/T (J K⁻¹mol⁻¹)	H°-H°(T_r) (kJ mol⁻¹)	$\Delta_f H°$ (kJ mol⁻¹)	$\Delta_f G°$ (kJ mol⁻¹)	Log K_f
0							
100							
200							
298.15	64.183	42.962	42.962	0.	-598.651	-572.558	100.310
300	64.183	43.359	42.964	0.119	-598.607	-572.396	99.663
400	64.183	61.824	45.481	6.537	-596.399	-564.003	73.651
500	64.183	76.146	50.235	12.955	-597.597	-555.814	58.065
600	64.183	87.847	55.558	19.374	-595.898	-547.617	47.674
700	64.183	97.741	60.896	25.792	-594.181	-539.705	40.273
800	64.183	106.312	66.049	32.210	-592.464	-532.040	34.739
900	64.183	113.871	70.951	38.628	-590.764	-524.589	30.446
1000	64.183	120.634	75.587	45.047	-589.077	-517.327	27.022
1100	64.183	126.751	79.965	51.465	-587.399	-510.234	24.229
1121.300	64.183	127.982	80.865	52.832	------ CRYSTAL <--> LIQUID -----		
1200	64.183	132.335	84.100	57.883	-585.730	-503.292	21.908
1300	64.183	137.473	88.010	64.301	-584.069	-496.490	19.949
1400	64.183	142.229	91.715	70.720	-582.412	-489.815	18.275
1500	64.183	146.657	95.232	77.138	-580.757	-483.259	16.829
1600	64.183	150.800	98.577	83.556	-579.102	-476.813	15.566
1700	64.183	154.691	101.765	89.974	-722.684	-463.294	14.235
1800	64.183	158.359	104.808	96.393	-720.276	-448.106	13.004
1900	64.183	161.829	107.718	102.811	-717.875	-433.051	11.905
2000	64.183	165.122	110.507	109.229	-715.481	-418.122	10.920
2100	64.183	168.253	113.183	115.647	-713.093	-403.313	10.032
2200	64.183	171.239	115.754	122.066	-710.710	-388.617	9.227
2300	64.183	174.092	118.229	128.484	-708.331	-374.030	8.494
2400	64.183	176.823	120.614	134.902	-705.955	-359.546	7.825
2500	64.183	179.443	122.915	141.320	-703.583	-345.161	7.212
2600	64.183	181.961	125.138	147.739	-701.214	-330.871	6.647
2700	64.183	184.383	127.288	154.157	-698.848	-316.672	6.126
2800	64.183	186.717	129.369	160.575	-696.484	-302.560	5.644
2900	64.183	188.969	131.385	166.993	-694.124	-288.533	5.197
3000	64.183	191.145	133.341	173.412	-691.768	-274.588	4.781

PREVIOUS: December 1963 CURRENT: December 1968

Lithium Fluoride (LiF) F_1Li_1(l)

LITHIUM FLUORIDE (LiF) CRYSTAL-LIQUID M_r = 25.939403 Lithium Fluoride (LiF) F_1Li_1(cr,l)

0 to 1121.3 K crystal
above 1121.3 K liquid

Refer to the individual tables for details.

Enthalpy Reference Temperature = T_r = 298.15 K Standard State Pressure = $p°$ = 0.1 MPa

T/K	$C_p°$ (J K⁻¹mol⁻¹)	$S°$ (J K⁻¹mol⁻¹)	$-[G°-H°(T_r)]/T$ (J K⁻¹mol⁻¹)	$H°-H°(T_r)$ (kJ mol⁻¹)	$\Delta_fH°$ (kJ mol⁻¹)	$\Delta_fG°$ (kJ mol⁻¹)	Log K_f
0	0.	0.	INFINITE	-6.473	-614.369	-614.369	INFINITE
100	12.816	4.858	65.944	-6.109	-615.953	-607.186	317.162
200	32.803	20.690	39.225	-3.707	-616.858	-597.965	156.172
298.15	41.815	35.660	35.660	0.	-616.931	-588.660	103.131
300	41.936	35.919	35.661	0.077	-616.928	-588.485	102.464
400	46.543	48.706	37.373	4.533	-616.682	-579.039	75.615
500	49.329	59.401	40.738	9.331	-619.500	-569.345	59.479
600	51.589	68.597	44.633	14.378	-619.173	-559.341	48.695
700	53.723	76.711	48.647	19.645	-618.607	-549.410	40.997
800	55.731	84.016	52.619	25.118	-617.835	-539.575	35.231
900	57.697	90.695	56.483	30.790	-616.881	-529.848	30.752
1000	59.580	96.871	60.217	36.654	-615.749	-520.237	27.174
1100	61.505	102.642	63.814	42.710	-614.434	-510.748	24.253
1121.300	61.883	103.825	64.563	44.024	CRYSTAL <--> LIQUID		
1121.300	64.183	127.982	64.563	71.111	TRANSITION		
1200	64.183	132.335	68.867	76.162	-585.730	-503.292	21.908
1300	64.183	137.473	73.949	82.581	-584.069	-496.490	19.949
1400	64.183	142.229	78.659	88.999	-582.412	-489.815	18.275
1500	64.183	146.657	83.046	95.417	-580.757	-483.259	16.829
1600	64.183	150.800	87.152	101.835	-579.102	-476.813	15.566
1700	64.183	154.691	91.012	108.254	-722.684	-463.294	14.235
1800	64.183	158.359	94.653	114.672	-720.276	-448.106	13.004
1900	64.183	161.829	98.098	121.090	-717.875	-433.051	11.905
2000	64.183	165.122	101.367	127.508	-715.481	-418.122	10.920
2100	64.183	168.253	104.478	133.927	-713.093	-403.313	10.032
2200	64.183	171.239	107.446	140.345	-710.710	-388.617	9.227
2300	64.183	174.092	110.282	146.763	-708.331	-374.030	8.494
2400	64.183	176.823	112.998	153.181	-705.955	-359.546	7.825
2500	64.183	179.443	115.604	159.600	-703.583	-345.161	7.212
2600	64.183	181.961	118.108	166.018	-701.214	-330.871	6.647
2700	64.183	184.383	120.518	172.436	-698.848	-316.672	6.126
2800	64.183	186.717	122.841	178.854	-696.485	-302.560	5.644
2900	64.183	188.969	125.082	185.273	-694.124	-288.533	5.197
3000	64.183	191.145	127.248	191.691	-691.768	-274.588	4.781

PREVIOUS: CURRENT: December 1968

Lithium Fluoride (LiF) F_1Li_1(cr,l)

LITHIUM FLUORIDE (LiF) IDEAL GAS M_r = 25.939403

$S°(298.15\ K) = 200.278 \pm 0.008\ J\ K^{-1}\ mol^{-1}$

$\Delta_f H°(0\ K) = -340.575 \pm 8.4\ kJ\ mol^{-1}$
$\Delta_f H°(298.15\ K) = -340.787 \pm 8.4\ kJ\ mol^{-1}$

Electronic Level and Quantum Weight

State	ε_i, cm^{-1}	g_i
$^1\Sigma^+$	0	1

$\omega_e = 914.33\ cm^{-1}$ $\omega_e x_e = 8.00\ cm^{-1}$ $\sigma = 1$
$B_e = 1.3588\ cm^{-1}$ $\alpha_e = 0.0203\ cm^{-1}$ $r_e = 1.5631$ Å

Enthalpy of Formation

The molecular species produced on vaporization of lithium fluoride have been studied by many investigators during the last ten years. The occurrence of dimer and trimer molecules in addition to the monomer has been established by use of mass spectrometric (4, 7, 8), torsion effusion (3, 6), molecular beam (9, 10) and transpiration (11) methods. To evaluate $\Delta_{vap}H°$(LiF, g) we have used a trial and error variation of $\Delta_{sub}H°$ and $\Delta_{vap}H°$ for the monomer, dimer and trimer in order to obtain reasonable agreement with the observed total pressures, 997-1939 K, (1, 2, 3, 5, 6). Using, as guidelines, these total pressures and the equilibrium data for dimer and trimer (see these tables), we adopt $\Delta_{sub}H°$(298.15 K) = 66.0, 69.57 and 79.73 and $\Delta_{vap}H°$(298.15 K) = 61.63, 60.83 and 66.62 kcal mol^{-1} for LiF, Li_2F_2 and Li_3F_3, respectively. The adopted value $\Delta_{sub}H°$(1000 K) = 63.1 kcal mol^{-1} is in agreement with the reported values: 64.8±2 (12), 63.6±1 (6) and 63.0±1.5 kcal mol^{-1} (8) at 1000 K.

A 2nd and 3rd law analysis of the total pressure data is given below. Monomer pressures for this analysis are obtained by deducing dimer and trimer pressures calcualted from JANAF free energy functions and the adopted values of $\Delta_{sub}H°$(298.15 K) and $\Delta_{vap}H°$(298.15 K).

The vapor pressure data reported by Eisenstadt et al. (9, 10) are too low and so not consistent with the others and are not used for evaluation. The transpiration data of Sense and Stone (11) and Knudsen-effusion data of Evseev et al. (13) are not considered for analysis because of the complications involved in converting the report apparent pressures to the real total pressures of the gaseous mixtures containing monomer, dimer and trimer.

Herzberg (14) and Gaydon (15) reported the dissociation energies (D_0^0) of LiF(g) to be ≤ 6.6 and 5.95±0.3 eV, respectively. Using these D_0^0 values, we derive $\Delta_f H°$(LiF, g, 298.15 K) as -95.9 and -80.9±6.9 kcal mol^{-1}. By use of flame-photometric method, Page and Sugden (16) and Bulewicz et al. (17) gave D_0^0 = 141 and 136±8 kcal mol^{-1} kcal mol^{-1}, yielding $\Delta_f H°$(LiF, g, 298.15 K) = -84.7 and -79.7±8 kcal mol^{-1}.

Source	T/K	Method	Reaction†	Data Points	$\Delta_r H°$(298.15 K), kcal mol^{-1} 2nd law	3rd law	Drift cal K^{-1}mol^{-1}	$\Delta_f H°$(298.15 K) kcal mol^{-1}
Wartenberg (1921)	1626-1820	boiling point	I	5	67.6±2.4	61.4	-3.6±1.4	-81.68
Ruff (1922)	1671-1939	boiling point	I	14	65.8±0.7	60.8	-2.7±0.4	-82.28
Pugh (1958)*	1020-1120	torsion-effusion	II	6	69.1	66.8	-2.2	-80.65
Porter (1958)	1073	mass spectrometry	II	1	--	67.7	--	-79.75
Cantor (1958)*	1328-1575	boiling point	I	10	64.5	61.6	-2.0	-81.48
Hildenbrand (1964)*	997-1119	torsion-effusion	II	24	69.9	66.8	-3.0	-80.65
	1125-1159	torsion-effusion	I	19	67.1	62.3	-4.2	-80.78

*Data points are calculated from the given vapor pressure equation.
†Reaction I: LiF(ℓ) = LiF(g); II: LiF(cr) = LiF(g).

Heat Capacity and Entropy

The infrared spectrum of LiF(g) and microwave spectrum of Li^6F have been observed by Vidale (19) and Wharten and Klemperer (21), respectively. The reported vibrational and rotational constants for Li^7F and Li^6F(g) are adopted and corrected to the average isotopic species. Veazey and Gordy (18) measured the rotational transitions of Li^7F with a high resolution molecular-beam absorption spectrometer. By use of the electric resonance method of molecular beam spectroscopy, Braunstein and Trischka (22) observed the transitions between v = 0 and v = 1 vibrational states. The infrared spectrum of LiF(g) have also been measured by Vasilevskii and Baikov (20).

References

1. H. von Wartenberg and H. Schulz, Z. Electrochem. 27, 568 (1921).
2. O. Ruff, G. Schmidt, and S. Mugdan, Z. Anorg. Allg. Chem. 123, 83 (1922).
3. A. C. P. Pugh and R. F. Barrow, Trans. Faraday Soc. 54, 671 (1958).
4. R. F. Porter and R. C. Schoonmaker, J. Chem. Phys. 29, 1070, (1958); 30, 283 (1959).
5. S. Cantor and F. Blankenship, paper presented at American Chemical Society meeting at Chicago, Sept. 7-12, 1958.
6. D. L. Hildenbrand, W. F. Hall, F. Ju, and N. D. Potter, J. Chem. Phys. 40, 2882 (1964).
7. J. Berkowitz and W. A. Chupka, J. Chem. Phys. 29, 653 (1958).
8. P. A. Akishin, L. N. Gorokhov and L. N. Sidorov, Russ. J. Phys. Chem. 33, 648 (1959).
9. M. Eisenstadt, G. M. Rothberg, and P. Kusch, J. Chem. Phys. 29, 797 (1958).
10. G. M. Rothberg, M. Eisenstadt, and P. Kusch, J. Chem. Phys. 30, 517 (1959).
11. K. A. Sense and R. W. Stone, J. Phys. Chem. 62, 1411 (1958).
12. A. Buchler and J. L. Stauffer, "Thermodynamics," Vol. I, International Atomic Energy Agency, Vienna, p. 271, 1966.
13. A. M. Evseev, G. V. Pozharskaya, A. N. Nesmeyanov, and Y. I. Gerasimov, Russ. J. Inorg. Chem. 4, 997 (1959).
14. G. Herzberg, "Spectra of Diatomic Molecules," D. Van Nostrand Company, Inc., New York, 1950.
15. A. G. Gaydon, "Dissociation Energies," Chapman and Hall Ltd., London, 1968.
16. F. M. Page and T. M. Sugden, Nature, 183, 1672 (1959).
17. E. M. Bulewicz, L. F. Phillips, and T. M. Sugden, Trans. Faraday Soc. 57, 921 (1961).
18. S. E. Veazey and W. Gordy, Phys. Rev. 138, A1303 (1965).
19. G. L. Vidale, J. Phys. Chem. 64, 314 (1960).
20. K. P. Vasilevskii and V. I. Baikov, Opt. Spectry. (USSR) 11, 21 (1961).
21. L. Wharten, W. Klemperer, L. P. Gold, R. Strauch, J. J. Gallagher, and V. E. Derr, J. Chem. Phys. 38, 1203 (1963).
22. R. Braunstein and J. W. Trischka, Phys. Rev. 98, 1092 (1955).

Lithium Fluoride (LiF) $F_1Li_1(g)$

Enthalpy Reference Temperature = T_r = 298.15 K Standard State Pressure = p° = 0.1 MPa

T/K	$C_p°$ (J K^{-1}mol^{-1})	S°	-[G°-H°(T_r)]/T	H°-H°(T_r) (kJ mol^{-1})	$\Delta_f H°$	$\Delta_f G°$	Log K_f
0	0.	0.	INFINITE	-8.823	-340.575	-340.575	INFINITE
100	29.119	167.860	227.033	-5.917	-339.618	-347.151	181.333
200	29.683	188.144	203.084	-2.988	-339.995	-354.593	92.610
250	30.438	194.845	200.788	-1.486	-340.372	-358.200	74.842
298.15	31.287	200.278	200.278	0.	-340.787	-361.597	63.350
300	31.320	200.471	200.278	0.058	-340.803	-361.726	62.982
350	32.185	205.365	200.663	1.646	-341.273	-365.177	54.500
400	32.965	209.715	201.527	3.275	-341.797	-368.557	48.129
450	33.641	213.638	202.659	4.941	-342.388	-371.867	43.165
500	34.215	217.213	203.938	6.637	-346.050	-374.801	39.155
600	35.110	223.534	206.691	10.106	-347.301	-380.431	33.120
700	35.756	228.998	209.496	13.651	-348.457	-385.861	28.793
800	36.234	233.805	212.240	17.252	-349.557	-391.128	25.538
900	36.600	238.095	214.879	20.894	-350.633	-396.260	22.998
1000	36.888	241.967	217.397	24.569	-351.689	-401.273	20.960
1100	37.122	245.494	219.794	28.270	-352.729	-406.181	19.288
1200	37.318	248.733	222.072	31.992	-353.756	-410.995	17.890
1300	37.484	251.726	224.240	35.733	-354.773	-415.723	16.704
1400	37.630	254.510	226.303	39.489	-355.778	-420.374	15.684
1500	37.759	257.110	228.271	43.258	-356.771	-424.953	14.798
1600	37.876	259.551	230.151	47.040	-357.753	-429.467	14.021
1700	37.983	261.850	231.949	50.833	-503.960	-426.743	13.112
1800	38.082	264.024	233.671	54.636	-504.167	-422.195	12.252
1900	38.175	266.086	235.323	58.449	-504.372	-417.635	11.482
2000	38.263	268.046	236.911	62.271	-504.574	-413.065	10.788
2100	38.347	269.915	238.438	66.102	-504.774	-408.484	10.160
2200	38.427	271.701	239.910	69.940	-504.970	-403.894	9.590
2300	38.504	273.411	241.329	73.787	-505.163	-399.296	9.068
2400	38.579	275.051	242.701	77.641	-505.352	-394.689	8.590
2500	38.652	276.627	244.026	81.503	-505.536	-390.074	8.150
2600	38.723	278.145	245.310	85.372	-505.717	-385.452	7.744
2700	38.793	279.607	246.553	89.247	-505.893	-380.823	7.367
2800	38.861	281.020	247.759	93.130	-506.065	-376.188	7.018
2900	38.928	282.384	248.929	97.019	-506.234	-371.546	6.692
3000	38.994	283.705	250.067	100.916	-506.399	-366.899	6.388
3100	39.059	284.985	251.173	104.818	-506.562	-362.246	6.104
3200	39.124	286.226	252.249	108.727	-506.723	-357.588	5.837
3300	39.188	287.431	253.297	112.643	-506.884	-352.925	5.586
3400	39.251	288.602	254.318	116.565	-507.044	-348.257	5.350
3500	39.314	289.740	255.314	120.493	-507.206	-343.585	5.128
3600	39.376	290.849	256.286	124.428	-507.371	-338.908	4.917
3700	39.438	291.928	257.234	128.368	-507.539	-334.226	4.718
3800	39.500	292.981	258.161	132.315	-507.713	-329.539	4.530
3900	39.561	294.008	259.067	136.268	-507.893	-324.848	4.351
4000	39.622	295.010	259.953	140.227	-508.081	-320.152	4.181
4100	39.682	295.989	260.820	144.193	-508.279	-315.452	4.019
4200	39.743	296.946	261.669	148.164	-508.488	-310.746	3.865
4300	39.803	297.882	262.500	152.141	-508.711	-306.036	3.718
4400	39.863	298.798	263.315	156.124	-508.927	-301.321	3.577
4500	39.922	299.694	264.114	160.114	-509.171	-296.600	3.443
4600	39.982	300.572	264.897	164.109	-509.435	-291.873	3.314
4700	40.041	301.433	265.665	168.110	-509.718	-287.141	3.191
4800	40.101	302.277	266.419	172.117	-510.021	-282.402	3.073
4900	40.160	303.104	267.159	176.130	-510.347	-277.657	2.960
5000	40.219	303.916	267.886	180.149	-510.697	-272.904	2.851
5100	40.278	304.713	268.600	184.174	-511.069	-268.145	2.746
5200	40.336	305.496	269.303	188.205	-511.473	-263.378	2.646
5300	40.395	306.265	269.993	192.241	-511.907	-258.603	2.549
5400	40.454	307.020	270.671	196.284	-512.373	-253.819	2.455
5500	40.512	307.763	271.339	200.332	-512.874	-249.027	2.365
5600	40.570	308.494	271.996	204.386	-513.411	-244.225	2.278
5700	40.629	309.212	272.643	208.446	-513.986	-239.413	2.194
5800	40.687	309.919	273.279	212.512	-514.601	-234.591	2.113
5900	40.745	310.615	273.906	216.583	-515.258	-229.757	2.034
6000	40.803	311.301	274.524	220.661	-515.959	-224.912	1.958

PREVIOUS: December 1968 (1 atm) CURRENT: December 1968 (1 bar)

Lithium Fluoride (LiF) $F_1Li_1(g)$

LITHIUM HYPOFLUORITE (LiOF) IDEAL GAS M_r = 41.938803

$\Delta_f H°(0\ K)$ = [89.499 ± 41.8] kJ mol^{-1}

$S°(298.15\ K)$ = [246.022] J K^{-1} mol^{-1} $\Delta_f H°(298.15\ K)$ = [-92.048 ± 41.8] kJ mol^{-1}

Vibrational Frequencies and Degeneracies

ν, cm^{-1}
[800](1)
[500](1)
[950](1)

Ground State Quantum Weight: [1] σ = 1

Point Group: C_s

Bond Distance: Li-O = [1.59] Å F-O = [1.41] Å

Bond Angle: F-O-Li = [104]°

Product of the Moments of Inertia: $I_A I_B I_C$ = [4.21997 x 10^{-116}] g^3 cm^6

Enthalpy of Formation

The enthalpy of formation, $\Delta_f H°$(LiFO, g , 298.15 K), is calculated from $\Delta_r H°$(298.15 K) = 139 kcal mol^{-1} for the reaction LiFO(g) = Li(g) + F(g) + O(g) where the value of $\Delta_r H°$(298.15 K) is estimated to be the sum of the F-O and Li-O bond energies. The values of the F-O and Li-O bond energies are obtained from those in F_2O(g) and Li_2O(g) molecules, respectively.

Heat Capacity and Entropy

The F-O and Li-O bond distances are assumed to be the same as those in F_2O(g) and Li_2O(g) molecules, respectively. The F-O-Li bond angle and vibrational frequencies are estimated by comparison with the corresponding values for the same oxide molecules. The three principal moments of inertia are: I_A = 1.2732 x 10^{-39}, I_B = 5.1556 x 10^{-39} and I_C = 6.4288 x 10^{-39} g cm^2.

Lithium Hypofluorite (LiOF) $F_1Li_1O_1(g)$

Enthalpy Reference Temperature = T_r = 298.15 K Standard State Pressure = p° = 0.1 MPa

T/K	$C_p°$	S°	-[G°-H°(T_r)]/T	H°-H°(T_r)	$\Delta_f H°$	$\Delta_f G°$	Log K_f
	J K^{-1}mol^{-1}			kJ mol^{-1}			
0	0.	0.	INFINITE	-10.827	-89.499	-89.499	INFINITE
100	33.594	205.644	280.612	-7.497	-89.569	-92.216	48.168
200	37.671	229.966	249.788	-3.964	-90.799	-94.413	24.658
250	40.466	238.672	246.717	-2.011	-91.454	-95.240	19.899
298.15	43.031	246.022	246.022	0.	-92.048	-95.915	16.804
300	43.124	246.289	246.023	0.080	-92.070	-95.939	16.704
350	45.452	253.116	246.557	2.296	-92.650	-96.537	14.407
400	47.406	259.317	247.770	4.619	-93.227	-97.053	12.674
450	49.017	264.997	249.373	7.031	-93.830	-97.496	11.317
500	50.336	270.232	251.201	9.515	-97.475	-97.562	10.192
600	52.306	279.595	255.171	14.654	-98.637	-97.468	8.485
700	53.656	287.765	259.257	19.956	-99.663	-97.190	7.252
800	54.607	294.996	263.281	25.372	-100.616	-96.771	6.319
900	55.296	301.469	267.171	30.869	-101.540	-96.235	5.585
1000	55.809	307.323	270.898	36.425	-102.446	-95.597	4.993
1100	56.200	312.662	274.456	42.027	-103.340	-94.869	4.505
1200	56.504	317.565	277.846	47.662	-104.228	-94.059	4.094
1300	56.744	322.098	281.078	53.325	-105.113	-93.176	3.744
1400	56.937	326.310	284.160	59.010	-105.997	-92.225	3.441
1500	57.095	330.244	287.103	64.712	-106.879	-91.210	3.176
1600	57.225	333.933	289.916	70.428	-107.760	-90.137	2.943
1700	57.334	337.406	292.608	76.156	-253.878	-81.832	2.514
1800	57.425	340.686	295.189	81.894	-254.008	-71.709	2.081
1900	57.503	343.793	297.666	87.640	-254.149	-61.577	1.693
2000	57.570	346.744	300.047	93.394	-254.300	-51.438	1.343
2100	57.628	349.554	302.338	99.154	-254.463	-41.291	1.027
2200	57.678	352.236	304.545	104.919	-254.637	-31.135	0.739
2300	57.722	354.801	306.675	110.689	-254.822	-20.972	0.476
2400	57.760	357.258	308.732	116.464	-255.017	-10.801	0.235
2500	57.794	359.617	310.720	122.241	-255.223	-0.621	0.013
2600	57.825	361.884	312.645	128.022	-255.439	9.567	-0.192
2700	57.852	364.067	314.509	133.806	-255.666	19.764	-0.382
2800	57.876	366.172	316.317	139.593	-255.903	29.970	-0.559
2900	57.898	368.203	318.071	145.381	-256.151	40.183	-0.724
3000	57.918	370.166	319.775	151.172	-256.410	50.406	-0.878
3100	57.935	372.065	321.432	156.965	-256.681	60.638	-1.022
3200	57.952	373.905	323.043	162.759	-256.964	70.879	-1.157
3300	57.966	375.689	324.611	168.555	-257.260	81.128	-1.284
3400	57.980	377.419	326.139	174.352	-257.569	91.387	-1.404
3500	57.992	379.100	327.628	180.151	-257.893	101.655	-1.517
3600	58.004	380.734	329.081	185.951	-258.231	111.933	-1.624
3700	58.014	382.323	330.499	191.752	-258.587	122.220	-1.725
3800	58.024	383.871	331.883	197.554	-258.959	132.517	-1.822
3900	58.033	385.378	333.235	203.356	-259.351	142.824	-1.913
4000	58.041	386.847	334.557	209.160	-259.762	153.141	-2.000
4100	58.049	388.281	335.850	214.965	-260.195	163.469	-2.083
4200	58.056	389.679	337.115	220.770	-260.651	173.808	-2.162
4300	58.062	391.046	338.354	226.576	-261.131	184.158	-2.237
4400	58.069	392.381	339.566	232.382	-261.618	194.518	-2.309
4500	58.074	393.686	340.755	238.189	-262.143	204.890	-2.378
4600	58.080	394.962	341.919	243.997	-262.699	215.274	-2.445
4700	58.085	396.211	343.061	249.805	-263.286	225.671	-2.508
4800	58.090	397.434	344.181	255.614	-263.905	236.081	-2.569
4900	58.094	398.632	345.280	261.423	-264.559	246.505	-2.628
5000	58.098	399.806	346.359	267.233	-265.249	256.941	-2.684
5100	58.102	400.956	347.418	273.043	-265.972	267.392	-2.739
5200	58.106	402.084	348.459	278.853	-266.741	277.857	-2.791
5300	58.110	403.191	349.481	284.664	-267.552	288.338	-2.842
5400	58.113	404.278	350.486	290.475	-268.409	298.835	-2.891
5500	58.116	405.344	351.474	296.287	-269.314	309.347	-2.938
5600	58.119	406.391	352.445	302.098	-270.269	319.877	-2.984
5700	58.122	407.420	353.400	307.910	-271.276	330.424	-3.028
5800	58.125	408.431	354.340	313.723	-272.338	340.990	-3.071
5900	58.127	409.424	355.266	319.535	-273.458	351.573	-3.113
6000	58.130	410.401	356.177	325.348	-274.637	362.177	-3.153

PREVIOUS: September 1965 (1 atm) CURRENT: September 1965 (1 bar)

Lithium Hypofluorite (LiOF) $F_1Li_1O_1(g)$

MAGNESIUM FLUORIDE (MgF) IDEAL GAS M_r = 43.303403

$\Delta_f H°(0\ K) = -236.371 \pm 8.4$ kJ mol^{-1}

S°(298.15 K) = 221.085 ± 0.08 J K^{-1} mol^{-1} $\Delta_f H°(298.15\ K) = -236.814 \pm 8.4$ kJ mol^{-1}

Electronic Levels and Molecular Constants

State	T_0, cm^{-1}	g_i	ω_e, cm^{-1}	$\omega_e x_e$, cm^{-1}	B_e, cm^{-1}	α_e, cm^{-1}	r_e, Å
$X^2\Sigma^+$	0	2	719.56	4.91	0.51630	0.00466	1.7498
$A^2\pi$	27829.60	4	743.89	3.95	0.51711	0.00324	1.7485
$B^2\Sigma^+$	37187.45	2	759.94	5.57	0.53542	0.00507	1.7183
$C^2\Sigma^+$	42589.64	2	820.77	5.01	0.54792	0.00445	1.6986

Enthalpy of Formation

We calculate $\Delta_f H°$(MgF, g, 298.15 K) = -56.6±2.0 kcal mol^{-1} from an adopted. $\Delta_{at}H°$(MgF, 298.15 K) = 110.7±2 kcal mol^{-1} which is derived from a 3rd law analysis of three sets of mass spectral-equilibrium data reported by Murad et al. (1) and Hildenbrand (2). Also included in the analysis are independent mass spectrometer-equilibrium measurements on MgF(g) by Ehlert (3, 4).

Source	Reaction	Data Points	T/K	$\Delta_r H°$(298.15 K), kcal mol^{-1} 2nd Law	3rd Law	Drift cal K^{-1} mol^{-1}	$\Delta_f H°$(298.15 K)[c] kcal mol^{-1}	D_0^0(298.15 K) kcal mol^{-1}
Murad et al. (1)	(A)	4	1546-1585	6.2	5.6[b]±1.3	-0.4±26	-57.5±2.0	111.6
Hildenbrand (2) Set 1	(B)	4	1280-1345	22.1	25.87±0.7	2.9±8.7	-56.3±0.8	110.4
Set 2	(B)	7	1413-1493	24.3	26.58±0.2	1.6±3.7	-55.9±0.7	110.0
Ehlert (3)	(B)	5	1205-1253	-90.0	-31.5 ±2.1	47 ±18	-53.5±1.5	107.6
Ehlert et al. (4)	(C)	5	1238-1301	63.0	47.6 ±26	-12.2±26	-53.7±2.0	107.8
Ehlert et al. (4)	(D)	5	1238-1301	-91.9	-126.7 ±1.8	-28.3±17	-53.4±1.0	107.5

Reactions: (A) Tl(g) + MgF(g) = TiF(g) + Mg(g) (B) Mg(g) + MgF_2(g) = 2 MgF(g) (C) AlF_3(g) + 2 Mg(g) = AlF(g) + 2 MgF(g)
(D) 2 MgF(g) = Mg(g) + MgF_2(cr)

[a]One point rejected due to failure of a statistical test.
[b]Free-energy functions (FEF) for TiF(g) are calculated from spectral data; (5); FEF for Tl(g) are from Hultgren (6).
[c]3rd law Values; all auxillary $\Delta_f H°$(298.15 K) from JANAF Thermochemical Table (8) except for Tl and TlF (7).

We dismiss the measurements of Ehlert (3) and Ehlert et al. (4) since our analyses show that their equilibrium data most likely contain temperature dependent errors. Furthermore, their results (3, 4) yield D_0^0 values which are less than that predicted by the Rittner ionic model. Krasnov and Karaseva (9) have applied this potential function to all the alkaline earth monohalides and found that it establishes a lower limit for D_0^0. Recalculation of their results (9) sets $\Delta_{at}H°$(MgF, 298.15 K) ≥ 110.4 kcal mol^{-1}. Application of this ionic model to MgF is justified since the spin densities calculated from the ESR spectra (10) of matrix-isolated MgF show that the molecule is indeed highly ionic.

Trends in the well-established values of the ratio $\Delta_{at}H°(MX, 298.15\ K)/\Delta_{at}H°(MX_2, 298.15\ K)$ for the majority of the alkaline earth halides (8) suggests a value for MgF/MgF_2 which is less than or equal to 0.45. This sets the upper limit of $\Delta_{at}H°$(MgF, 298.15 K) at 111 kcal mol^{-1} when $\Delta_{at}H°(MgF_2$, 298.15 K) = 246.7 kcal mol^{-1} (8). If MgF_2(g) is linear then the maximum $\Delta_{at}H°$(298.15 K) value increases to 112.6 kcal mol^{-1} suggesting a possible uncertainty in the adopted D_0^0 of ± 2 kcal mol^{-1}. Our adopted results give $\Delta_{at}H°(MgF, 298.15\ K)/\Delta_{at}H(MgF_2, 298.15\ K) = 0.44_9$.

Our selected thermochemical $\Delta_{at}H°$(298.15 K) value converted to 0 K is 4.76 ev (109.8 kcal mol^{-1}). Other values for D_0^0 which were considered but believed less reliable are (all in ev 4.0 (11), 5.0 or 4.7 (11) and 4.2 (12). These spectroscopic values have been derived from linear Birge-Sponer extrapolations of the ground state ($X^2\Sigma^+$) and first excited state ($^2\pi$) vibrational levels. We obtain $D_0^0 = 3.2_2$ eV from a linear Birge-Sponer extrapolation with our adopted ground state vibrational constants. This value adjusts to 3.59 eV when corrected for the ionicity (2) of the Mg-F bond. The corrected D_0^0 value is 1.2 eV less than the thermochemical result. This discrepancy is most likely due to the use of an inaccurate value for the anharmonic constant, $\omega_e x_e$, in the Birge-Sponer extrapolation. The use of lower reported values (12, 15) for $\omega_e x_e$ brings the spectroscopic and thermochemical D_0^0 values into much better agreement but leaves the thermodynamic functions for MgF unaltered. In addition, Singh et al. (13) have employed a method to fit an electronegative potential energy function to the expermimental potential energy curves which yields D_0^0 = 4.45 eV. This value agrees with our results within probable error limits.

Heat Capacity and Entropy

The electronic levels and vibrational-rotational constants are taken from the compilation of Rosen (5). Values of these constants, tabulated for $Mg^{24}F$, are corrected for the natural isotopic abundances of magnesium. The value of α_e listed by Rosen for the $C^2\Sigma^+$ state is incorrect. The rotational constants are based on an analysis (14) of the rotational structure in the bands of the A-X, B-X, and C-X systems. Barrow and Beale (14) used the Pekeris relation (12) to calculate $\omega_e x_e$ values from their spectroscopic α_e's. These $\omega_e x_e$ values were combined with $\Delta G_{1/2}$ values to give ω_e. These estimates of ω_e and $\omega_e x_e$ differ somewhat from those obtained from a vibrational analysis (15) of band head measurements. However, thermodynamic functions calculated with the alternate set of vibrational constants are not significantly different from those adopted here. The uncertainty in the $\omega_e x_e$ values is large, probably near ±1 cm^{-1}. The partition function $Q = Q_t \sum_i Q_r^i Q_v^i g_i \exp(-c_2\varepsilon_i/T)$ is used to calculate the thermodynamic functions with first-order anharmonic corrections included for Qr and Qv.

References

1. E. Murad, D. L. Hildenbrand, and R. P. Main, J. Chem. Phys. 45, 263 (1966).
2. D. L. Hildenbrand, J. Chem. Phys. 48, 3657 (1968).
3. T. C. Ehlert, Ph.D. Thesis, University of Wisconsin, 1963.
4. T. C. Ehlert, G. D. Blue, J. W. Green and J. L. Margrave, J. Chem. Phys. 41, 2250 (1964).
5. B. Rosen, ed., "Spectroscopic Data Relative to Diatomic Molecules," Pergamon Press, New York, 1970.
6. R. Hultgren et al., "Selected Values of the Thermodynamic Properties of the Elements," Am. Soc. for Metals, Ohio, 1973.
7. U. S. Nat. Bur. Stand. Tech. Note 270-3, 1968.
8. JANAF Thermochemical Tables: Mg(g), 9-30-62; MgF_2(cr, g), 6-30-75; AlF_3(g), 6-30-70; AlF(g), 12-31-75; F^-(g), 12-31-75.
9. K. S. Krasnov and N. V. Karaseva, Optics and Spectroscopy 19, 14 (1963).
10. L. B. Knight, Jr., W. C. Easley, and W. Weltner, Jr., J. Chem. Phys. 54, 322 (1971).
11. A. G. Gaydon, "Dissociation Energies and Spectra of Diatomic Molecules," Chapman and Hall Ltd., London, 1968.
12. G. Herzberg, "Spectra of Diatomic Molecules," 2nd Edition, D. Van Nostrand Company, Inc., New York, 1950.
13. O. N. Singh, B. P. Asthana, and O. N. Singh, Spectrosc. Lett. 7, 175 (1974).
14. R. F. Barrow and J. R. Beale, Proc. Phys. Soc. 91, 483 (1967).
15. M. M. Novikov and L. V. Gurvich, Zh. Prikl. Spectrosk. 14, 820 (1971).

Magnesium Fluoride (MgF) F_1Mg_1(g)

Enthalpy Reference Temperature = T_r = 298.15 K Standard State Pressure = p° = 0.1 MPa

T/K	$C_p°$ (J K^{-1}mol^{-1})	S°	-[G°-H°(T_r)]/T	H°-H°(T_r) (kJ mol^{-1})	$\Delta_f H°$	$\Delta_f G°$	Log K_f
0	0.	0.	INFINITE	-8.968	-236.371	-236.371	INFINITE
100	29.142	188.032	248.620	-6.059	-235.558	-244.892	127.919
200	30.463	208.521	223.998	-3.095	-236.066	-254.077	66.358
250	31.569	215.437	221.616	-1.545	-236.438	-258.537	54.018
298.15	32.569	221.085	221.085	0.	-236.814	-262.759	46.034
300	32.605	221.287	221.086	0.060	-236.829	-262.920	45.778
350	33.481	226.381	221.486	1.713	-237.233	-267.237	39.883
400	34.194	230.900	222.386	3.406	-237.649	-271.495	35.454
450	34.768	234.962	223.561	5.130	-238.078	-275.700	32.002
500	35.229	238.649	224.888	6.881	-238.523	-279.857	29.236
600	35.912	245.137	227.737	10.440	-239.468	-288.036	25.076
700	36.381	250.710	230.630	14.056	-240.500	-296.050	22.092
800	36.720	255.591	233.452	17.712	-241.637	-303.910	19.843
900	36.975	259.932	236.157	21.397	-242.897	-311.620	18.086
1000	37.175	263.838	238.733	25.105	-252.890	-318.470	16.635
1100	37.338	267.389	241.179	28.831	-254.455	-324.952	15.431
1200	37.474	270.644	243.501	32.572	-256.018	-331.292	14.421
1300	37.592	273.649	245.706	36.325	-257.580	-337.501	13.561
1400	37.695	276.438	247.803	40.090	-386.551	-340.423	12.701
1500	37.787	279.042	249.800	43.864	-386.761	-337.121	11.740
1600	37.872	281.484	251.704	47.647	-386.971	-333.805	10.898
1700	37.950	283.782	253.524	51.438	-387.181	-330.475	10.154
1800	38.023	285.953	255.266	55.237	-387.391	-327.134	9.493
1900	38.092	288.011	256.936	59.042	-387.600	-323.780	8.901
2000	38.157	289.966	258.539	62.855	-387.807	-320.416	8.368
2100	38.221	291.830	260.080	66.674	-388.010	-317.042	7.886
2200	38.281	293.609	261.564	70.499	-388.209	-313.657	7.447
2300	38.341	295.312	262.995	74.330	-388.402	-310.264	7.046
2400	38.398	296.945	264.375	78.167	-388.587	-306.863	6.679
2500	38.454	298.514	265.710	82.010	-388.764	-303.454	6.340
2600	38.510	300.023	267.001	85.858	-388.930	-300.038	6.028
2700	38.564	301.477	268.251	89.712	-389.086	-296.616	5.738
2800	38.619	302.881	269.463	93.571	-389.230	-293.189	5.470
2900	38.672	304.237	270.639	97.435	-389.361	-289.757	5.219
3000	38.726	305.549	271.780	101.305	-389.480	-286.320	4.985
3100	38.779	306.820	272.890	105.180	-389.587	-282.880	4.766
3200	38.833	308.052	273.970	109.061	-389.682	-279.436	4.561
3300	38.887	309.247	275.021	112.947	-389.766	-275.989	4.369
3400	38.941	310.409	276.045	116.838	-389.840	-272.540	4.187
3500	38.997	311.539	277.043	120.735	-389.906	-269.089	4.016
3600	39.053	312.638	278.016	124.638	-389.964	-265.637	3.854
3700	39.110	313.709	278.967	128.546	-390.017	-262.182	3.701
3800	39.169	314.753	279.895	132.460	-390.066	-258.727	3.556
3900	39.230	315.771	280.802	136.380	-390.114	-255.270	3.419
4000	39.292	316.765	281.688	140.306	-390.162	-251.811	3.288
4100	39.357	317.736	282.556	144.238	-390.213	-248.352	3.164
4200	39.423	318.685	283.405	148.177	-390.270	-244.891	3.046
4300	39.492	319.613	284.236	152.123	-390.333	-241.429	2.933
4400	39.563	320.522	285.050	156.076	-390.406	-237.966	2.825
4500	39.637	321.412	285.849	160.036	-390.491	-234.500	2.722
4600	39.714	322.284	286.631	164.003	-390.590	-231.033	2.623
4700	39.794	323.139	287.399	167.979	-390.706	-227.563	2.529
4800	39.877	323.978	288.152	171.962	-390.840	-224.090	2.439
4900	39.963	324.801	288.892	175.954	-390.995	-220.615	2.352
5000	40.052	325.609	289.618	179.955	-391.172	-217.136	2.268
5100	40.145	326.403	290.332	183.965	-391.374	-213.653	2.188
5200	40.241	327.184	291.033	187.984	-391.603	-210.166	2.111
5300	40.340	327.951	291.722	192.013	-391.859	-206.675	2.037
5400	40.443	328.706	292.400	196.052	-392.146	-203.178	1.965
5500	40.549	329.449	293.067	200.101	-392.464	-199.676	1.896
5600	40.659	330.181	293.723	204.162	-392.814	-196.167	1.830
5700	40.772	330.901	294.369	208.233	-393.199	-192.653	1.765
5800	40.888	331.611	295.005	212.316	-393.620	-189.131	1.703
5900	41.008	332.311	295.632	216.411	-394.077	-185.601	1.643
6000	41.131	333.002	296.249	220.518	-394.573	-182.063	1.585

PREVIOUS: June 1976 (1 atm) CURRENT: June 1976 (1 bar)

Magnesium Fluoride (MgF) F_1Mg_1(g)

MAGNESIUM FLUORIDE, ION (MgF^+) IDEAL GAS $M_r = 43.302854$

$\Delta_f H°(0\ K) = 506.533 \pm 46.0\ kJ\ mol^{-1}$

$S°(298.15\ K) = [215.330 \pm 8.4]\ J\ K^{-1}\ mol^{-1}$ $\Delta_f H°(298.15\ K) = [512.289]\ kJ\ mol^{-1}$

Electronic Levels and Quantum Weights

State	ε_i, cm^{-1}	g_i	State	ε_i, cm^{-1}	g_i
$X^1\Sigma^+$	0	[1]	$B^1\Sigma^+$	[20000]	[1]
$^3\pi$	[2600]	[6]	$^3\pi$	[28000]	[3]
$A^1\pi$	[3500]	[2]	$^3\Delta$	[29000]	[6]
$^3\Sigma^+$	[15000]	[3]	$D^1\Delta$	[30000]	[2]

$\omega_e = [718.5]\ cm^{-1}$ $B_e = [0.5162]\ cm^{-1}$

$\omega_e x_e = [4.90]\ cm^{-1}$ $\alpha_e = [0.00463]\ cm^{-1}$

$\sigma = 1$ $r_e = [1.75]$ Å

Enthalpy of Formation

MgF^+ ionization-efficiency curves obtained from mass spectral observations on the MgF_2 (1) and MgF_2-Al (2) systems are of the two-process type. Evidence (1, 2) has been presented which shows that the low-energy portion (<13 eV) of the curve arises from the direct ionization (A) of MgF(g). These curves also show an inflection point at higher energies (~14 eV) which has been interpreted (1-3) as due to the onset of the dissociative ionization (B) of MgF_2(g). The appearance potentials, in eV, which have been observed for these two processes from measurements on various MgF_2 systems are tabulated below.

Source	Year	System	(A) Direct Ionization	(B) Dissociative Ionization
Berkowitz and Marquart (4)	1962	MgF_2		13.5±0.4
Green et al. (1)	1964	MgF_2		13.6±0.3
Ehlert et al. (2)	1964	MgF_2-Al	7.8±0.3	13.7±0.4
Murad et al. (5)	1966	MgF_2-Tl-Pd	8.0±0.5	
Hildenbrand (6)	1968	MgF_2-Cu	7.5±0.3	
Hildenbrand (3)	1968	MgF_2	7.5±0.5	

(A) $MgF(g) + e^- \rightarrow MgF^+(g) + 2\ e^-$ (B) $MgF_2(g) + e^- \rightarrow MgF^+(g) + F(g) + 2\ e^-$

Both sets of data show the normal scatter (~0.5 eV) expected for such measurements. The mean value (7.7±0.4 eV) of the four AP's for the direct ionization (A) is supported by the spectroscopic ionization potential of 7.68 eV (7). We adopt $\Delta_r H° = 177.56 \pm 9$ kcal mol^{-1} (7.7 eV) for process (A), and we obtain $\Delta_f H°(MgF^+, g) = 121.1 \pm 11.0$ kcal mol^{-1} when the $\Delta_r H°$ value is combined with $\Delta_f H°(MgF, g) = -56.5 \pm 2.0$ kcal mol^{-1} (8). The fragmentation process (B) is likely to involve unknown kinetic energy factors and would not be expected to yield as reliable $\Delta_f H°$ value as obtained from direct ionization. However, the mean value (13.58±0.5 eV) of the four AP's for process (B) leads to $\Delta_f H°(MgF^+, g, 0\ K) = 121.5 \pm 13.0$ kcal mol^{-1} which is in remarkably good with the $\Delta_f H°$ value from direct ionization (121.1 kcal mol^{-1}). This implies that the excess kinetic energy of the fragment MgF^+ is small.

Heat Capacity and Entropy

The bond dissociation energy for MgF^+ (108.6 kcal mol^{-1}) is nearly the same as that for MgF(109.8 kcal mol^{-1}, 8) which implies that the bonding in these two molecules is quite similar. Thus, it is reasonable to assume that $r_e(MgF^+) = r_e(MgF)$. We adopt the measured r_e value (1.75 Å) for MgF (8). B_e is calculated from the adopted value for r_e. The value of ω_e is obtained from Badger's rule (9) which is written in the form $\omega_e^2 = 3.159 \times 10^6/\mu(r_e{=}d_{ij})^3$. Molecular data (8) for NaF, MgF, and MgO are used to determine the constant d_{ij}. Barrow and Caunt (10) have shown that the product $x_e\mu^{1/2}$ is fairly constant within a group of similar molecules. We assume $x_e(MgF^+) = x_e(MgF)$ since $\mu^{1/2}(MgF^+) = \mu^{1/2}(MgF)$. These estimates of ω_e and $\omega_e x_e$ are only slightly different from those observed for MgF (8). α_e is obtained from the other constants assuming a Morse potential function.

Novikov and Gurvich (11) have observed a band system in the emission spectrum of MgF near 54000 cm^{-1}. An analysis of the vibrational structure showed that the system did not belong to any of those known for MgF, and they (11) assigned it to the $^1\Sigma$-$^1\Sigma$ transition of MgF^+. Band systems have appeared in the absorption spectra (12, 13) of MgF which are now known to arise from AlF. A comparison of the vibrational constants reported by Novikov and Gurvich (11) with those for AlF (8) strongly suggests that this system is the $B^1\Sigma$-$X^1\Sigma$ transition of AlF. Barrow (14) has come to the same conclusion.

MgF^+ (8 valence elecrons) is isoelectronic with the alkaline earth monoxides. We estimate the electronic states and levels for MgF^+ by analogy with those for these isoelectronic molecules (8), particularly MgO. The $^1\pi$-$^3\pi$ splitting of MgO has recently been reported as 900±200 cm^{-1} (15). We believe that our adopted electronic entropies represent minimal values. Probable upper limits for the electronic contributions can be established by decreasing the two lowest triplet levels to 1000 cm^{-1} ($^3\pi$) and 5000 cm^{-1} ($^3\Sigma$) and neglecting the other excited states. This increases the entropy to 65.1 and 78.6 cal K^{-1} mol^{-1} at 1000 K and 4000 K, respectively. The enthalpy is much more uncertain, approaching ±3.0 kcal mol^{-1} at temperatures in excess of 2000 K.

References

1. J. W. Green, G. D. Blue, T. C. Ehlert and J. L. Margrave, J. Chem. Phys. 41, 2245 (1964).
2. T. C. Ehlert, G. D. Blue, J. W. Green and J. L. Margrave, J. Chem. Phys. 41, 2250 (1964).
3. D. L. Hildenbrand, J. Chem. Phys. 48, 3657 (1968).
4. J. Berkowitz and J. R. Marquart, J. Chem. Phys. 37, 1853 (1962).
5. E. Murad, D. L. Hildenbrand and R. P. Main, J. Chem. Phys. 45, 263 (1966).
6. D. L. Hildenbrand, J. Chem. Phys. 48, 2458 (1968).
7. R. F. Barrow and J. R. Beale, Proc. Phys. Soc. 91, 483 (1967).
8. JANAF Thermochemical Tables: MgF(g), 12-31-75; NaF(g), 12-31-68; MgO(g), 12-31-74.
9. R. M. Badger, J. Chem. Phys. 2, 128 (1934).
10. R. F. Barrow and A. D. Caunt, Proc. Roy. Soc. (London) 219A, 120 (1953).
11. M. M. Novikov and L. V. Gurvich, Zh. Prikl. Spectrosk. 14, 820 (1971).
12. F. A. Jenkins and R. Grinfeld, Phys. Rev. 45, 229 (1934).
13. C. A. Fowler, Jr., Phys. Rev. 59, 645 (1941).
14. R. F. Barrow, ed., "Diatomic Molecules, A Critical Bibliography of Spectroscopic Data," Centre National de la Recherche Scientifique, Paris, 1973.
15. G. A. Capelle, H. P. Broida, and R. W. Field, J. Chem. Phys. 62, 3131 (1975).

Magnesium Fluoride, Ion (MgF^+) $F_1Mg_1^+(g)$

Enthalpy Reference Temperature = T_r = 298.15 K Standard State Pressure = p° = 0.1 MPa

T/K	$C_p°$ (J K⁻¹mol⁻¹)	S° (J K⁻¹mol⁻¹)	−[G°−H°(T_r)]/T (J K⁻¹mol⁻¹)	H°−H°(T_r) (kJ mol⁻¹)	$\Delta_f H°$ (kJ mol⁻¹)	$\Delta_f G°$ (kJ mol⁻¹)	Log K_f
0	0.	0.	INFINITE	-8.969	506.533		
100	29.142	182.270	242.875	-6.060			
200	30.469	202.761	218.245	-3.097			
250	31.580	209.679	215.862	-1.546			
298.15	32.605	215.330	215.330	0.	512.289	481.805	-84.410
300	32.643	215.532	215.331	0.060	512.313	481.616	-83.857
350	33.621	220.638	215.732	1.717	512.952	476.449	-71.106
400	34.589	225.191	216.635	3.422	513.588	471.191	-61.531
450	35.646	229.324	217.819	5.178	514.229	465.853	-54.075
500	36.868	233.142	219.162	6.990	514.885	460.443	-48.102
600	39.885	240.119	222.084	10.821	516.291	449.425	-39.126
700	43.407	246.529	225.123	14.984	517.884	438.157	-32.696
800	46.861	252.555	228.179	19.500	519.687	426.647	-27.857
900	49.709	258.246	231.207	24.336	521.655	414.900	-24.080
1000	51.662	263.593	234.181	29.412	515.109	403.642	-21.084
1100	52.688	268.572	237.084	34.637	517.121	392.398	-18.633
1200	52.931	273.172	239.902	39.923	519.183	380.969	-16.583
1300	52.600	277.399	242.626	45.204	521.226	369.368	-14.841
1400	51.901	281.273	245.251	50.431	395.797	360.779	-13.461
1500	51.001	284.824	247.772	55.577	399.038	358.164	-12.472
1600	50.020	288.084	250.191	60.629	402.175	355.336	-11.601
1700	49.035	291.087	252.510	65.581	405.204	352.315	-10.825
1800	48.095	293.863	254.731	70.437	408.131	349.119	-10.131
1900	47.225	296.439	256.859	75.203	410.960	345.763	-9.506
2000	46.436	298.841	258.899	79.885	413.701	342.261	-8.939
2100	45.731	301.090	260.855	84.493	416.365	338.623	-8.423
2200	45.107	303.202	262.732	89.034	418.961	334.860	-7.951
2300	44.559	305.195	264.536	93.517	421.498	330.980	-7.517
2400	44.081	307.081	266.269	97.948	423.986	326.991	-7.117
2500	43.665	308.872	267.938	102.335	426.432	322.899	-6.747
2600	43.306	310.577	269.545	106.683	428.844	318.710	-6.403
2700	42.997	312.206	271.095	110.998	431.229	314.428	-6.083
2800	42.732	313.765	272.592	115.284	433.590	310.059	-5.784
2900	42.506	315.260	274.037	119.545	435.935	305.606	-5.505
3000	42.315	316.698	275.436	123.786	438.265	301.072	-5.242
3100	42.154	318.082	276.789	128.009	440.585	296.461	-4.995
3200	42.021	319.419	278.101	132.218	442.896	291.775	-4.763
3300	41.913	320.710	279.372	136.414	445.201	287.017	-4.543
3400	41.825	321.960	280.607	140.601	447.501	282.189	-4.335
3500	41.757	323.171	281.806	144.780	449.796	277.293	-4.138
3600	41.706	324.347	282.971	148.953	452.087	272.332	-3.951
3700	41.670	325.489	284.105	153.122	454.373	267.307	-3.774
3800	41.648	326.600	285.208	157.288	456.654	262.221	-3.604
3900	41.638	327.682	286.284	161.452	458.929	257.075	-3.443
4000	41.639	328.736	287.332	165.616	461.197	251.870	-3.289
4100	41.651	329.764	288.354	169.780	463.457	246.609	-3.142
4200	41.671	330.768	289.352	173.946	465.707	241.293	-3.001
4300	41.700	331.749	290.327	178.114	467.944	235.923	-2.866
4400	41.736	332.708	291.279	182.286	470.169	230.501	-2.736
4500	41.780	333.646	292.210	186.462	472.379	225.029	-2.612
4600	41.829	334.565	293.121	190.642	474.571	219.508	-2.493
4700	41.884	335.465	294.013	194.828	476.744	213.940	-2.378
4800	41.944	336.348	294.885	199.019	478.896	208.326	-2.267
4900	42.009	337.213	295.740	203.217	481.026	202.667	-2.160
5000	42.079	338.063	296.578	207.421	483.130	196.965	-2.058
5100	42.152	338.897	297.400	211.633	485.209	191.221	-1.958
5200	42.229	339.716	298.206	215.852	487.258	185.436	-1.863
5300	42.309	340.521	298.997	220.079	489.278	179.613	-1.770
5400	42.393	341.313	299.773	224.314	491.266	173.751	-1.681
5500	42.479	342.091	300.535	228.557	493.222	167.853	-1.594
5600	42.567	342.858	301.284	232.810	495.142	161.920	-1.510
5700	42.658	343.612	302.020	237.071	497.025	155.953	-1.429
5800	42.751	344.354	302.744	241.341	498.870	149.953	-1.350
5900	42.845	345.086	303.455	245.621	500.676	143.922	-1.274
6000	42.941	345.807	304.155	249.910	502.441	137.860	-1.200

PREVIOUS: December 1975 (1 atm) CURRENT: December 1975 (1 bar)

Magnesium Fluoride, Ion (MgF^+) $F_1Mg_1^+(g)$

MOLYBDENUM FLUORIDE (MoF) IDEAL GAS M_r = 114.938403

D_0^0 = [451.454 ± 16.7] kJ mol^{-1} $\Delta_f H°(0\ K)$ = [282.853 ± 12.6] kJ mol^{-1}

S°(298.15 K) = [246.399 ± 8.4] J K^{-1} mol^{-1} $\Delta_f H°(298.15\ K)$ = 282.838 ± 12.6 kJ mol^{-1}

Electronic Levels and Quantum Weights

ε_i, cm^{-1}	g_i	ε_i, cm^{-1}	g_i	ε_i, cm^{-1}	g_i
0	[6]	[15199]	[6]	[15890]	[2]
[11783]	[2]	[15331]	[8]	[16796]	[2]
[12034]	[4]	[15428]	[10]	[16947]	[8]
[12417]	[6]	[15447]	[12]	[17174]	[4]
[12900]	[8]	[15691]	[6]	[17344]	[6]
[13461]	[10]	[15699]	[4]		

ω_e = [700] cm^{-1} $\omega_e x_e$ = [3.05] cm^{-1} σ = 1

B_e = [0.3209] cm^{-1} α_e = [0.0019] cm^{-1} r_e = [1.82] Å

Enthalpy of Formation

The adopted enthalpy of formation at 298.15 K is that derived by Brewer (1) from the high temperature mass spectrometric measurements of equilibria among gaseous molybdenum fluorides by Hildenbrand [2054-2242 K] (2). The value of $\Delta_f H°(0\ K)$ combined with JANAF data (3) for Mo(g) and F(g) gives D_0^0(Mo-F) = 107.9±4 kcal mol^{-1}.

Heat Capacity and Entropy

We use the ground state vibrational frequency of 700 cm^{-1} estimated by Hildenbrand (2) and adopted by Brewer (1). The Mo-F bond distance, 1.82 Å, is the Mo-F bond distance in MoF_6(g) used by Brewer (1). The electronic contributions are assumed to be the same as those for the free gaseous Mo^+ ion, as suggested by Brewer (1). All states up to 17344 cm^{-1} as listed by Moore (4) are included in the calculation. Our thermodynamic functions differ slightly from those reported by Brewer (1) because of the use of non-zero values for $\omega_e x_e$ and α_e. The estimates for $\omega_e x_e$ and α_e are based on the empirical correlations of Calder and Ruedenberg (5).

This table is based on the critical analysis of data existing in 1978 by Brewer (1).

References

1. L. Brewer, Materials and Molecular Research Division, Lawrence Berkeley Laboratory, University of California, Berkeley; personal communication, Sept. 1978, of preliminary draft of review to be submitted for publication in Atomic Energy Review, International Atomic Energy Agency, Vienna, Austria.
2. D. L. Hildenbrand, J. Chem. Phys. 65, 614 (1976).
3. JANAF Thermochemical Tables: Mo(g), 3-31-78; F(g), 9-30-65.
4. C. E. Moore, U.S. Nat. Bur. Stand., NSRDS-NBS 34 (1970); NSRDS-NBS 35, Vol. III, 1971 [Reprint of NBS Circular 467, Vol. III, 1958].
5. G. V. Calder and K. Ruedenberg, J. Chem. Phys. 49, 5399 (1968).

Molybdenum Fluoride (MoF) $F_1Mo_1(g)$

Enthalpy Reference Temperature = T_r = 298.15 K Standard State Pressure = p° = 0.1 MPa

T/K	$C_p°$ (J K^{-1}mol^{-1})	S°	$-[G°-H°(T_r)]/T$	$H°-H°(T_r)$ (kJ mol^{-1})	$\Delta_f H°$	$\Delta_f G°$	Log K_f
0	0.	0.	INFINITE	-8.982	282.853	282.853	INFINITE
100	29.145	213.279	274.009	-6.073	283.818	271.711	-141.927
200	30.549	233.795	249.321	-3.105	283.479	259.681	-67.822
250	31.674	240.733	246.931	-1.550	283.165	253.767	-53.022
298.15	32.671	246.399	246.399	0.	282.838	248.134	-43.472
300	32.706	246.601	246.399	0.060	282.826	247.919	-43.166
350	33.567	251.710	246.801	1.718	282.475	242.129	-36.136
400	34.260	256.239	247.703	3.414	282.114	236.389	-30.869
450	34.812	260.307	248.881	5.142	281.743	230.696	-26.778
500	35.253	263.999	250.211	6.894	281.362	225.044	-23.510
600	35.897	270.487	253.065	10.453	280.568	213.854	-18.618
700	36.334	276.055	255.961	14.066	279.733	202.801	-15.133
800	36.645	280.928	258.783	17.716	278.857	191.870	-12.528
900	36.875	285.258	261.489	21.393	277.940	181.051	-10.508
1000	37.053	289.153	264.064	25.089	276.979	170.337	-8.897
1100	37.195	292.692	266.508	28.802	275.969	159.721	-7.585
1200	37.313	295.933	268.827	32.527	274.903	149.200	-6.495
1300	37.416	298.924	271.029	36.264	273.774	138.770	-5.576
1400	37.511	301.700	273.121	40.010	272.575	128.430	-4.792
1500	37.605	304.291	275.114	43.766	271.300	118.177	-4.115
1600	37.707	306.722	277.014	47.532	269.944	108.013	-3.526
1700	37.825	309.011	278.830	51.308	268.503	97.936	-3.009
1800	37.967	311.177	280.567	55.097	266.971	87.946	-2.552
1900	38.144	313.234	282.233	58.903	265.346	78.044	-2.146
2000	38.363	315.196	283.832	62.728	263.624	68.231	-1.782
2100	38.634	317.074	285.371	66.577	261.803	58.505	-1.455
2200	38.965	318.879	286.853	70.457	259.882	48.869	-1.160
2300	39.362	320.619	288.284	74.372	257.859	39.323	-0.893
2400	39.830	322.304	289.666	78.331	255.733	29.866	-0.650
2500	40.373	323.941	291.005	82.341	253.502	20.501	-0.428
2600	40.993	325.536	292.302	86.408	251.153	11.226	-0.226
2700	41.688	327.096	293.562	90.542	248.656	2.045	-0.040
2800	42.457	328.626	294.787	94.749	245.958	-7.040	0.131
2900	43.296	330.130	295.980	99.036	207.102	-15.974	0.288
3000	44.198	331.613	297.143	103.410	205.820	-23.645	0.412
3100	45.157	333.078	298.279	107.877	204.644	-31.274	0.527
3200	46.163	334.527	299.389	112.443	203.581	-38.867	0.634
3300	47.206	335.964	300.476	117.111	202.636	-46.428	0.735
3400	48.277	337.389	301.540	121.885	201.812	-53.963	0.829
3500	49.363	338.804	302.585	126.767	201.112	-61.475	0.917
3600	50.453	340.210	303.610	131.758	200.538	-68.969	1.001
3700	51.537	341.607	304.618	136.857	200.090	-76.449	1.079
3800	52.603	342.995	305.610	142.064	199.766	-83.919	1.154
3900	53.640	344.375	306.586	147.377	199.565	-91.381	1.224
4000	54.640	345.746	307.548	152.791	199.482	-98.840	1.291
4100	55.593	347.107	308.497	158.303	199.515	-106.298	1.354
4200	56.491	348.458	309.432	163.908	199.656	-113.759	1.415
4300	57.330	349.797	310.355	169.599	199.902	-121.224	1.473
4400	58.102	351.124	311.267	175.372	200.244	-128.695	1.528
4500	58.805	352.438	312.167	181.218	200.676	-136.176	1.581
4600	59.436	353.737	313.057	187.130	201.190	-143.667	1.631
4700	59.992	355.021	313.936	193.102	201.780	-151.170	1.680
4800	60.474	356.290	314.805	199.126	202.436	-158.687	1.727
4900	60.882	357.541	315.664	205.195	203.151	-166.217	1.772
5000	61.217	358.774	316.514	211.300	-387.467	-168.032	1.755
5100	61.481	359.989	317.355	217.436	-388.335	-163.635	1.676
5200	61.677	361.185	318.186	223.594	-389.314	-159.219	1.599
5300	61.808	362.361	319.009	229.769	-390.406	-154.784	1.525
5400	61.879	363.517	319.822	235.954	-391.619	-150.327	1.454
5500	61.892	364.653	320.627	242.143	-392.955	-145.847	1.385
5600	61.853	365.768	321.423	248.330	-394.419	-141.341	1.318
5700	61.765	366.862	322.211	254.512	-396.011	-136.808	1.254
5800	61.632	367.935	322.990	260.682	-397.735	-132.246	1.191
5900	61.459	368.987	323.761	266.837	-399.591	-127.653	1.130
6000	61.251	370.019	324.523	272.973	-401.584	-123.028	1.071

PREVIOUS: December 1978 (1 atm) CURRENT: December 1978 (1 bar)

Molybdenum Fluoride (MoF) $F_1Mo_1(g)$

FLUOROIMIDOGEN (NF) IDEAL GAS M_r = 33.005103

$S°(298.15\ K) = 215.276\ J\ K^{-1}\ mol^{-1}$

$\Delta_f H°(0\ K) = 248.953 \pm 33\ kJ\ mol^{-1}$
$\Delta_f H°(298.15\ K) = 248.948 \pm 33\ kJ\ mol^{-1}$

Electronic Level and Quantum Weight

State	ε_i, cm^{-1}	g_i
$[^3\Sigma]$	0	3

$\omega_e = 1115\ cm^{-1}$ $\omega_e x_e = [12.175]\ cm^{-1}$ $\sigma = 1$
$B_e = [.923]\ cm^{-1}$ $\alpha_e = [.0120]\ cm^{-1}$ $r_e = [1.51\ Å]$

Enthalpy of Formation

The enthalpy of formation is obtained by taking an average of two estimates of the dissociation energy of NF. The first estimate of 3.15 eV (72.4 kcal mol^{-1}) was obtained from Price et al. (1). The second estimate of 70.4 kcal mol^{-1} is obtained by taking one-half the dissociation energy of NF_2(g) calculated from JANAF values. The $\Delta_f H°$(298.15 K) for NF(g) was then calculated with auxiliary JANAF data from a dissociation energy of 71.4 kcal mol^{-1}.

Heat Capacity and Entropy

The ω_e value was obtained from infrared studies of Milligan and Jacox (2). The r_e value was estimated from Guggenheimer's (3) relation for single bonded molecules. The anharmonicity constant x_e is estimated by assuming the separation of the vibrational levels to be a linear function of the vibrational quantum number. The α_e value is then calculated using the Morse potential function; B_e is determined from the bond length. The ground state configuration is assumed to be the same as that in NBr.

References
1. W. C. Price, T. R. Passmore and D. M. Roessler, Disc. Faraday Soc. 35, 201-11 (1963).
2. D. E. Milligan and M. E. Jacox, J. Chem. Phys. 40, 2461 (1964).
3. K. M. Guggenheimer, Proc. Phys. Soc. (London) 58, 456 (1946).

Fluoroimidogen (NF) $F_1N_1(g)$

Enthalpy Reference Temperature = T_r = 298.15 K Standard State Pressure = p° = 0.1 MPa

T/K	$C_p°$	S°	$-[G°-H°(T_r)]/T$	$H°-H°(T_r)$	$\Delta_f H°$	$\Delta_f G°$	Log K_f
	J $K^{-1}mol^{-1}$			kJ mol^{-1}			
0	0.	0.	INFINITE	-8.745	248.953	248.953	INFINITE
100	29.108	183.197	241.576	-5.838	248.953	247.142	-129.094
200	29.318	203.404	218.012	-2.922	248.949	245.332	-64.074
250	29.745	209.987	215.772	-1.446	248.946	244.429	-51.071
298.15	30.345	215.276	215.276	0.	248.948	243.558	-42.670
300	30.370	215.464	215.276	0.056	248.948	243.525	-42.401
350	31.089	220.199	215.649	1.592	248.962	242.620	-36.209
400	31.813	224.397	216.485	3.165	248.989	241.713	-31.564
450	32.492	228.184	217.578	4.773	249.030	240.801	-27.951
500	33.106	231.640	218.814	6.413	249.084	239.884	-25.060
600	34.122	237.770	221.475	9.777	249.220	238.031	-20.722
700	34.893	243.090	224.191	13.229	249.375	236.154	-17.622
800	35.480	247.790	226.853	16.749	249.536	234.255	-15.295
900	35.933	251.996	229.417	20.321	249.692	232.335	-13.484
1000	36.291	255.801	231.868	23.933	249.839	230.399	-12.035
1100	36.578	259.274	234.204	27.577	249.974	228.448	-10.848
1200	36.815	262.467	236.428	31.247	250.097	226.486	-9.859
1300	37.013	265.422	238.546	34.938	250.207	224.513	-9.021
1400	37.182	268.171	240.565	38.648	250.305	222.533	-8.303
1500	37.329	270.742	242.492	42.374	250.391	220.546	-7.680
1600	37.459	273.155	244.334	46.114	250.467	218.554	-7.135
1700	37.575	275.430	246.097	49.865	250.533	216.558	-6.654
1800	37.679	277.580	247.787	53.628	250.591	214.558	-6.226
1900	37.775	279.620	249.409	57.401	250.643	212.554	-5.844
2000	37.864	281.560	250.969	61.183	250.690	210.548	-5.499
2100	37.947	283.409	252.470	64.974	250.735	208.540	-5.187
2200	38.024	285.177	253.916	68.772	250.779	206.530	-4.904
2300	38.098	286.868	255.313	72.578	250.824	204.517	-4.645
2400	38.168	288.491	256.662	76.392	250.873	202.503	-4.407
2500	38.235	290.051	257.966	80.212	250.928	200.487	-4.189
2600	38.299	291.552	259.229	84.039	250.990	198.468	-3.987
2700	38.362	292.998	260.453	87.872	251.062	196.446	-3.800
2800	38.422	294.395	261.641	91.711	251.146	194.422	-3.627
2900	38.481	295.744	262.793	95.556	251.242	192.394	-3.465
3000	38.538	297.049	263.914	99.407	251.352	190.363	-3.315
3100	38.594	298.314	265.003	103.264	251.479	188.328	-3.173
3200	38.649	299.540	266.063	107.126	251.622	186.289	-3.041
3300	38.704	300.730	267.096	110.993	251.782	184.245	-2.916
3400	38.757	301.886	268.102	114.867	251.961	182.196	-2.799
3500	38.809	303.011	269.084	118.745	252.158	180.141	-2.688
3600	38.861	304.105	270.041	122.628	252.375	178.080	-2.584
3700	38.912	305.170	270.976	126.517	252.611	176.013	-2.485
3800	38.962	306.209	271.890	130.411	252.868	173.940	-2.391
3900	39.013	307.221	272.783	134.309	253.144	171.859	-2.302
4000	39.062	308.210	273.656	138.213	253.440	169.771	-2.217
4100	39.111	309.175	274.511	142.122	253.756	167.676	-2.136
4200	39.160	310.118	275.347	146.035	254.092	165.572	-2.059
4300	39.209	311.040	276.167	149.954	254.447	163.460	-1.986
4400	39.257	311.942	276.970	153.877	254.822	161.340	-1.915
4500	39.305	312.825	277.757	157.805	255.216	159.211	-1.848
4600	39.352	313.689	278.528	161.738	255.628	157.073	-1.784
4700	39.400	314.536	279.286	165.676	256.059	154.926	-1.722
4800	39.447	315.366	280.029	169.618	256.508	152.770	-1.662
4900	39.494	316.180	280.758	173.565	256.975	150.604	-1.605
5000	39.541	316.978	281.475	177.517	257.459	148.428	-1.551
5100	39.588	317.761	282.178	181.473	257.960	146.242	-1.498
5200	39.634	318.531	282.870	185.434	258.477	144.047	-1.447
5300	39.680	319.286	283.550	189.400	259.010	141.841	-1.398
5400	39.727	320.028	284.219	193.370	259.560	139.625	-1.351
5500	39.773	320.758	284.877	197.345	260.124	137.399	-1.305
5600	39.819	321.475	285.524	201.325	260.704	135.162	-1.261
5700	39.865	322.180	286.161	205.309	261.298	132.915	-1.218
5800	39.911	322.873	286.788	209.298	261.905	130.658	-1.177
5900	39.956	323.556	287.405	213.291	262.527	128.390	-1.137
6000	40.002	324.228	288.013	217.289	263.161	126.111	-1.098

PREVIOUS: June 1965 (1 atm) CURRENT: June 1965 (1 bar)

Fluoroimidogen (NF) $F_1N_1(g)$

NITROSYL FLUORIDE (ONF) IDEAL GAS $M_r = 49.004503$

$S°(298.15\ K) = 248.082\ J\ K^{-1} \cdot mol^{-1}$

$\Delta_f H°(0\ K) = -63.300 \pm 1.7\ kJ\ mol^{-1}$
$\Delta_f H°(298.15\ K) = -65.689 \pm 1.7\ kJ\ mol^{-1}$

Vibrational Levels and Degeneracies

ν, cm^{-1}
1844.03(1)
521 (1)
765.85(1)

Ground State Quantum Weight: 1 $\sigma = 1$
Point Group: C_s
Bond Distances: N–O = 1.13 Å N–F = 1.52 Å
Bond Angle: O–N–F = 110.2°
Product of the Moments of Inertia: $I_A I_B I_C = 49.6113 \times 10^{-117}\ g^3\ cm^6$

Enthalpy of Formation

For the reaction $2NO(g) + F_2(g) \rightarrow 2ONF(g)$, Johnston and Bertin (1) report $\Delta_r H°(298.15\ K) = -74.8 \pm 0.8\ kcal\ mol^{-1}$, determined calorimetrically. This yields $-15.7 \pm 0.4\ kcal\ mol^{-1}$ for $\Delta_f H°(298.15\ K)$.

Heat Capacity and Entropy

Stephenson and Jones (2), report the frequencies, obtained from the infrared spectrum, and moments of inertia, obtained from the microwave spectrum. The principal moments of inertia are: $I_A = 0.8778 \times 10^{-39}\ g\ cm^2$, $I_B = 7.0919 \times 10^{-39}\ g\ cm^2$, $I_C = 7.9697 \times 10^{-39}\ g\ cm^2$.

References

1. H. S. Johnston and H. J. Bertin, J. Amer. Chem. Soc. 81, 6402 (1959).
2. C. V. Stephenson and E. A. Jones, J. Chem. Phys. 20, 135 (1952).

Nitrosyl Fluoride (ONF) $F_1N_1O_1(g)$

Enthalpy Reference Temperature = T_r = 298.15 K; Standard State Pressure = $p°$ = 0.1 MPa

T/K	$C_p°$	S°	$-[G°-H°(T_r)]/T$	$H°-H°(T_r)$	$\Delta_f H°$	$\Delta_f G°$	Log K_f
	J K⁻¹mol⁻¹			kJ mol⁻¹			
0	0.	0.	INFINITE	-10.701	-63.300	-63.300	INFINITE
100	33.534	208.247	281.961	-7.371	-64.328	-59.978	31.329
200	37.177	232.429	251.737	-3.862	-65.193	-55.267	14.434
250	39.427	240.969	248.753	-1.946	-65.485	-52.750	11.021
298.15	41.348	248.082	248.082	0.	-65.689	-50.277	8.808
300	41.416	248.338	248.083	0.077	-65.695	-50.182	8.737
350	43.115	254.854	248.594	2.191	-65.842	-47.584	7.101
400	44.574	260.708	249.748	4.384	-65.942	-44.968	5.872
450	45.850	266.034	251.266	6.645	-66.005	-42.342	4.915
500	46.980	270.924	252.991	8.967	-66.041	-39.711	4.149
600	48.895	279.665	256.725	13.764	-66.052	-34.442	2.998
700	50.437	287.322	260.561	18.733	-66.007	-29.177	2.177
800	51.678	294.141	264.340	23.841	-65.927	-23.921	1.562
900	52.678	300.288	267.998	29.061	-65.826	-18.676	1.084
1000	53.486	305.881	271.511	34.370	-65.712	-13.443	0.702
1100	54.142	311.011	274.872	39.753	-65.593	-8.222	0.390
1200	54.679	315.746	278.083	45.195	-65.473	-3.012	0.131
1300	55.122	320.140	281.152	50.686	-65.355	2.188	-0.088
1400	55.491	324.239	284.085	56.217	-65.243	7.380	-0.275
1500	55.799	328.079	286.891	61.782	-65.138	12.563	-0.437
1600	56.060	331.688	289.579	67.375	-65.042	17.740	-0.579
1700	56.282	335.094	292.157	72.992	-64.956	22.911	-0.704
1800	56.472	338.316	294.633	78.630	-64.880	28.078	-0.815
1900	56.636	341.374	297.013	84.286	-64.815	33.240	-0.914
2000	56.778	344.283	299.304	89.957	-64.760	38.400	-1.003
2100	56.902	347.056	301.513	95.641	-64.715	43.556	-1.083
2200	57.011	349.706	303.644	101.337	-64.678	48.712	-1.157
2300	57.107	352.242	305.702	107.043	-64.648	53.865	-1.223
2400	57.192	354.674	307.692	112.758	-64.624	59.017	-1.284
2500	57.267	357.011	309.618	118.481	-64.604	64.168	-1.341
2600	57.335	359.258	311.485	124.211	-64.586	69.319	-1.393
2700	57.395	361.423	313.294	129.947	-64.569	74.469	-1.441
2800	57.450	363.511	315.051	135.690	-64.552	79.618	-1.485
2900	57.499	365.528	316.757	141.437	-64.532	84.766	-1.527
3000	57.544	367.478	318.415	147.189	-64.509	89.914	-1.566
3100	57.584	369.366	320.029	152.946	-64.481	95.061	-1.602
3200	57.621	371.195	321.599	158.706	-64.447	100.207	-1.636
3300	57.655	372.968	323.129	164.470	-64.406	105.352	-1.668
3400	57.686	374.690	324.620	170.237	-64.357	110.496	-1.698
3500	57.714	376.363	326.075	176.007	-64.299	115.637	-1.726
3600	57.740	377.989	327.494	181.780	-64.233	120.778	-1.752
3700	57.764	379.571	328.881	187.555	-64.157	125.915	-1.778
3800	57.786	381.112	330.235	193.332	-64.071	131.052	-1.801
3900	57.807	382.613	331.559	199.112	-63.975	136.186	-1.824
4000	57.826	384.077	332.854	204.894	-63.869	141.316	-1.845
4100	57.844	385.505	334.120	210.677	-63.752	146.444	-1.866
4200	57.860	386.899	335.360	216.462	-63.625	151.570	-1.885
4300	57.876	388.261	336.575	222.249	-63.489	156.692	-1.903
4400	57.890	389.591	337.765	228.037	-63.342	161.811	-1.921
4500	57.904	390.893	338.931	233.827	-63.185	166.926	-1.938
4600	57.916	392.165	340.074	239.618	-63.020	172.037	-1.954
4700	57.928	393.411	341.196	245.410	-62.846	177.146	-1.969
4800	57.939	394.631	342.297	251.204	-62.663	182.250	-1.983
4900	57.950	395.826	343.377	256.998	-62.473	187.351	-1.997
5000	57.959	396.996	344.438	262.794	-62.276	192.447	-2.010
5100	57.969	398.144	345.480	268.590	-62.072	197.540	-2.023
5200	57.977	399.270	346.503	274.387	-61.863	202.628	-2.035
5300	57.986	400.374	347.509	280.185	-61.648	207.712	-2.047
5400	57.993	401.458	348.498	285.984	-61.430	212.792	-2.058
5500	58.001	402.523	349.471	291.784	-61.207	217.868	-2.069
5600	58.008	403.568	350.428	297.585	-60.983	222.941	-2.080
5700	58.015	404.594	351.369	303.386	-60.756	228.008	-2.089
5800	58.021	405.603	352.295	309.187	-60.529	233.073	-2.099
5900	58.027	406.595	353.207	314.990	-60.303	238.132	-2.108
6000	58.033	407.571	354.105	320.793	-60.077	243.189	-2.117

PREVIOUS: June 1961 (1 atm) CURRENT: June 1961 (1 bar)

Nitrosyl Fluoride (ONF) $F_1N_1O_1(g)$

NITRYL FLUORIDE (NO_2F) IDEAL GAS M_r = 65.003903

$\Delta_fH°(0\ K) = 102.881 \pm 20.9\ kJ\ mol^{-1}$

$S°(298.15\ K) = 260.250\ J\ K^{-1}\ mol^{-1}$ $\Delta_fH°(298.15\ K) = -108.784 \pm 20.9\ kJ\ mol^{-1}$

Vibrational Frequencies and Degeneracies

ν, cm^{-1}	ν, cm^{-1}
1312(1)	1793(1)
822(1)	570(1)
460(1)	742(1)

Ground State Quantum Weight: 1 $\sigma = 2$

Point Group: C_{2v}

Bond Distances: N-F = 1.40 Å N-O = 1.21 Å

Bond Angles: O-N-O = 129.5° F-N-O = 115.25°

Product of the Moments of Inertia: $I_AI_BI_C = 6.400639 \times 10^{-115}\ g^3\ cm^6$

Enthalpy of Formation

The enthalpy of reaction at 25°C for direct synthesis of nitryl fluoride from nitrogen dioxide with fluorine ($NO_2(g)$ + 1/2 $F_2(g)$ = $NO_2F(g)$) has been measured calorimetrically as -33.9 kcal mol^{-1} by Breazeale and MacLaren (1). The enthalpy of formation, $\Delta_fH°(298.15\ K)$ = -26.0 kcal mol^{-1}, for $NO_2F(g)$ was then drived. The values of $\Delta_fH°(298.15\ K)$ = -20±5 and -19±2 kcal mol^{-1} were issued in their second and third quarterly tech. reports and have been quoted in the literature by Tschuikow-Roux (2).

Hetherington and Robinson (3) quoted $\Delta_fH°$ = 26 kcal mol^{-1} (presumably it has a minus sign) for $NO_2F(g)$ which was derived from the heat of solution measured by Nichols and Robinson (4).

Heat Capacity and Entropy

Smith and Magnuson (5) have measured the moments of inertia from microwave spectrum of $NO_2F(g)$. In order to calculate the bond distances, given as N-F = 1.35 Å and N-O = 1.23 Å, the bond angle O-N-O was assumed to be 125°. The selected bond distances and angle were obtained from Clayton et al. (6). Clayton et al. assumed the bond angle O-N-O in $NO_2F(g)$ should be the same as that in $NO_2Cl(g)$ and then calculated the bond distances N-F and N-O from Smith and Magnuson (5) data of the moments of inertia.

The selected vibrational frequencies were obtained from the infrared and Raman spectra measurements by Dodd et al. (7).

The three principal moments of inertia are: $I_A = 6.3641 \times 10^{-39}$, $I_B = 7.3394 \times 10^{-39}$ and $I_C = 13.7034 \times 10^{-39}\ g\ cm^2$.

References

1. J. D. Breazeale and R. O. MacLaren, "Thermochemistry of Oxygen-Fluorine Bonding", Final Tech. Summary Report, Contract No. Nonr 3433(00), ARPA Order No. 184-62, United Technology Corporation, Sunnyvale, Calif., March 1963.
2. E. Tschuikow-Roux, J. Phys. Chem. 66, 1636 (1962).
3. G. Hetherington and P. L. Robinson, Special Publication No. 10, The Chem. Soc., London, 1957.
4. Nichols and Robinson, personal communication to Hetherington and Robinson.
5. D. F. Smith and D. W. Magnuson, Phys. Rev. 87, 226 (1952).
6. L. Clayton, Q. Williams, and T. Weatherly, J. Chem. Phys. 30, 1328 (1959).
7. R. Dodd, J. Rolfe and L. Woodward, Trans. Faraday Soc. 52, 145 (1956).

Nitryl Fluoride (NO_2F) $F_1N_1O_2(g)$

Enthalpy Reference Temperature = T_r = 298.15 K Standard State Pressure = p° = 0.1 MPa

T/K	$C_p°$	S°	$-[G°-H°(T_r)]/T$	$H°-H°(T_r)$	$\Delta_fH°$	$\Delta_fG°$	Log K_f
	J $K^{-1}mol^{-1}$			kJ mol^{-1}			
0	0.	0.	INFINITE	-11.528	-102.881	-102.881	INFINITE
100	33.929	216.706	298.630	-8.192	-105.355	-93.186	48.675
200	41.239	242.134	264.542	-4.482	-107.475	-80.141	20.931
250	45.801	251.830	261.049	-2.305	-108.235	-73.215	15.298
298.15	49.865	260.250	260.250	0.	-108.784	-66.418	11.636
300	50.013	260.559	260.251	0.092	-108.802	-66.155	11.519
350	53.744	268.555	260.874	2.688	-109.205	-59.013	8.807
400	57.010	275.950	262.302	5.459	-109.475	-51.823	6.767
450	59.862	282.833	264.206	8.382	-109.635	-44.606	5.178
500	62.352	289.272	266.394	11.439	-109.706	-37.376	3.905
600	66.424	301.017	271.205	17.887	-109.646	-22.912	1.995
700	69.537	311.500	276.226	24.692	-109.393	-8.474	0.632
800	71.933	320.949	281.236	31.771	-109.011	5.918	-0.386
900	73.795	329.534	286.133	39.061	-108.541	20.256	-1.176
1000	75.256	337.387	290.871	46.516	-108.013	34.539	-1.804
1100	76.417	344.616	295.433	54.102	-107.445	48.767	-2.316
1200	77.351	351.307	299.814	61.792	-106.851	62.942	-2.740
1300	78.110	357.529	304.017	69.567	-106.241	77.067	-3.097
1400	78.734	363.342	308.049	77.410	-105.623	91.145	-3.401
1500	79.252	368.792	311.919	85.310	-105.004	105.178	-3.663
1600	79.687	373.921	315.635	93.257	-104.387	119.170	-3.890
1700	80.054	378.763	319.207	101.245	-103.777	133.123	-4.090
1800	80.366	383.348	322.645	109.266	-103.176	147.041	-4.267
1900	80.635	387.701	325.955	117.317	-102.586	160.926	-4.424
2000	80.867	391.843	329.147	125.392	-102.008	174.780	-4.565
2100	81.069	395.793	332.227	133.489	-101.442	188.605	-4.691
2200	81.245	399.569	335.203	141.605	-100.889	202.405	-4.806
2300	81.400	403.184	338.081	149.737	-100.348	216.178	-4.910
2400	81.537	406.651	340.866	157.884	-99.819	229.929	-5.004
2500	81.659	409.982	343.564	166.044	-99.299	243.658	-5.091
2600	81.767	413.187	346.181	174.216	-98.788	257.366	-5.171
2700	81.865	416.275	348.720	182.397	-98.285	271.055	-5.244
2800	81.952	419.254	351.186	190.588	-97.788	284.725	-5.312
2900	82.031	422.131	353.583	198.788	-97.295	298.377	-5.374
3000	82.102	424.913	355.915	206.994	-96.806	312.013	-5.433
3100	82.167	427.606	358.184	215.208	-96.318	325.632	-5.487
3200	82.225	430.216	360.395	223.427	-95.832	339.235	-5.537
3300	82.279	432.747	362.549	231.653	-95.345	352.823	-5.585
3400	82.328	435.204	364.650	239.883	-94.856	366.397	-5.629
3500	82.374	437.591	366.700	248.118	-94.366	379.956	-5.671
3600	82.415	439.912	368.702	256.358	-93.872	393.501	-5.710
3700	82.453	442.171	370.657	264.601	-93.375	407.032	-5.746
3800	82.489	444.370	372.568	272.848	-92.874	420.551	-5.781
3900	82.521	446.513	374.437	281.099	-92.368	434.056	-5.814
4000	82.552	448.603	376.265	289.352	-91.858	447.547	-5.844
4100	82.580	450.642	378.054	297.609	-91.342	461.026	-5.874
4200	82.606	452.632	379.806	305.868	-90.822	474.492	-5.901
4300	82.631	454.576	381.522	314.130	-90.297	487.946	-5.927
4400	82.653	456.476	383.204	322.394	-89.767	501.386	-5.952
4500	82.675	458.334	384.853	330.661	-89.233	514.816	-5.976
4600	82.695	460.151	386.471	338.929	-88.695	528.233	-5.998
4700	82.713	461.930	388.057	347.200	-88.154	541.639	-6.020
4800	82.731	463.671	389.614	355.472	-87.610	555.033	-6.040
4900	82.748	465.377	391.143	363.746	-87.064	568.416	-6.059
5000	82.763	467.049	392.645	372.021	-86.517	581.788	-6.078
5100	82.778	468.688	394.120	380.298	-85.970	595.149	-6.096
5200	82.792	470.296	395.569	388.577	-85.424	608.498	-6.112
5300	82.805	471.873	396.994	396.857	-84.879	621.838	-6.129
5400	82.817	473.421	398.395	405.138	-84.338	635.167	-6.144
5500	82.829	474.940	399.773	413.420	-83.800	648.485	-6.159
5600	82.840	476.433	401.129	421.704	-83.268	661.796	-6.173
5700	82.850	477.899	402.463	429.988	-82.743	675.095	-6.187
5800	82.860	479.340	403.776	438.274	-82.226	688.387	-6.200
5900	82.870	480.757	405.069	446.560	-81.719	701.667	-6.212
6000	82.879	482.150	406.342	454.848	-81.223	714.942	-6.224

PREVIOUS: December 1965 (1 atm) CURRENT: December 1965 (1 bar)

FLUORINE NITRATE (FNO_3) IDEAL GAS M_r = 81.003303

$\Delta_f H°(0\ K)$ = [18.049 ± 2.1] kJ mol^{-1}

$S°(298.15\ K)$ = [292.933] J K^{-1} mol^{-1} $\Delta_f H°(298.15\ K)$ = [10.460 ± 2.1] kJ mol^{-1}

Vibrational Frequencies and Degeneracies

ν, cm^{-1}	ν, cm^{-1}
[1300](1)	[760](1)
[800](1)	[980](1)
[680](1)	[370](1)
[1700](1)	[130](1)
[450](1)	

Ground State Quantum Weight: [1] σ = 1

Point Group: C_s

Bond Distances: O–N = 1.29 ± 0.05 Å N–O' = 1.39 ± 0.05 Å O'–F = 1.42 ± 0.05 Å

Bond Angles: O–N–O = 125° ± 5° F–O'–N = 105° ± 5°

Angle between NO'F plane and O_2NO' plane = 90°

Product of the Moments of Inertia: $I_A I_B I_C$ = 3.88611 x 10^{-114} g^3 cm^6

Enthalpy of Formation

The value, 2.5±0.5 kcal mol^{-1}, for the enthalpy of formation of $FNO_3(g)$ was obtained from Anderson, et al. (1). The value of $\Delta_f H°$ reported by Talakin, et al. (2) was -4.2±0.9 kcal mol^{-1} at 294 K.

Heat Capacity and Entropy

Vibrational frequencies were estimated by comparison with those of $HNO_3(g)$ reported by Palm and Kilpatrick (3). Bond distances and angles were taken from Pauling and Brockway. The three principal moments of inertia are: I_A = 8.8638 x 10^{-39}, I_B = 18.5651 x 10^{-39}, and I_C = 23.6154 10^{-39} g cm^2. The possible existence of restricted internal rotation in the molecule was not considered. Thus the calculated values of $C_p°$ and S° may be low.

References

1. R. Anderson, et al., "Thermochemistry of Oxygen-Fluorine Bonding," UTC 2002-QT4, United Technology Corporation, Sunnyvale, California, March 1, 1961 through February 28, 1962.
2. O. G. Talakin, et al., Russ. J. Phys. Chem., 36, 561 (1962).
3. A. Palm and M. Kilpatrick, J. Chem. Phys. 23, 1562 (1955).
4. L. Pauling and L. O. Brockway, J. Amer. Chem. Soc. 59, 13 (1937).

Fluorine Nitrate ($FONO_2$) $F_1N_1O_3(g)$

Enthalpy Reference Temperature = T_r = 298.15 K Standard State Pressure = p° = 0.1 MPa

T/K	$C_p°$ (J $K^{-1}mol^{-1}$)	S° (J $K^{-1}mol^{-1}$)	$-[G°-H°(T_r)]/T$ (J $K^{-1}mol^{-1}$)	$H°-H°(T_r)$ (kJ mol^{-1})	$\Delta_f H°$ (kJ mol^{-1})	$\Delta_f G°$ (kJ mol^{-1})	Log K_f
0	0.	0.	INFINITE	-14.184	18.049	18.049	INFINITE
100	41.294	237.188	342.626	-10.544	14.427	33.213	-17.349
200	53.333	269.363	298.527	-5.833	11.852	53.088	-13.865
250	59.618	281.944	293.975	-3.008	11.011	63.501	-13.268
298.15	65.223	292.933	292.933	0.	10.460	73.664	-12.906
300	65.427	293.337	292.934	0.121	10.443	74.056	-12.894
350	70.594	303.819	293.750	3.524	10.109	84.688	-12.639
400	75.105	313.547	295.624	7.169	9.967	95.354	-12.452
450	79.008	322.624	298.125	11.025	9.980	106.028	-12.307
500	82.371	331.127	301.005	15.061	10.118	116.693	-12.191
600	87.757	346.645	307.344	23.581	10.670	137.962	-12.011
700	91.766	360.489	313.965	32.567	11.477	159.116	-11.873
800	94.781	372.949	320.573	41.901	12.446	180.143	-11.762
900	97.080	384.251	327.030	51.499	13.521	201.041	-11.668
1000	98.860	394.575	333.276	61.300	14.663	221.816	-11.586
1100	100.258	404.066	339.286	71.258	15.850	242.475	-11.514
1200	101.372	412.839	345.054	81.342	17.063	263.023	-11.449
1300	102.271	420.990	350.586	91.526	18.290	283.470	-11.390
1400	103.006	428.597	355.889	101.791	19.523	303.823	-11.336
1500	103.613	435.725	360.976	112.123	20.753	324.086	-11.286
1600	104.120	442.429	365.860	122.510	21.976	344.269	-11.239
1700	104.546	448.754	370.552	132.944	23.187	364.375	-11.196
1800	104.909	454.740	375.064	143.417	24.382	384.410	-11.155
1900	105.219	460.421	379.408	153.924	25.559	404.379	-11.117
2000	105.487	465.825	383.595	164.460	26.716	424.287	-11.081
2100	105.719	470.977	387.634	175.020	27.852	444.138	-11.047
2200	105.922	475.900	391.535	185.602	28.968	463.936	-11.015
2300	106.100	480.613	395.307	196.204	30.062	483.682	-10.985
2400	106.258	485.132	398.956	206.822	31.136	503.381	-10.956
2500	106.397	489.472	402.490	217.455	32.191	523.036	-10.928
2600	106.521	493.648	405.917	228.101	33.229	542.649	-10.902
2700	106.632	497.670	409.241	238.758	34.250	562.223	-10.877
2800	106.732	501.550	412.469	249.427	35.255	581.760	-10.853
2900	106.822	505.297	415.605	260.105	36.248	601.259	-10.830
3000	106.903	508.919	418.656	270.791	37.228	620.726	-10.808
3100	106.977	512.426	421.624	281.485	38.198	640.161	-10.787
3200	107.044	515.824	424.515	292.186	39.159	659.563	-10.766
3300	107.106	519.118	427.332	302.894	40.113	678.935	-10.747
3400	107.162	522.317	430.079	313.607	41.061	698.280	-10.728
3500	107.213	525.424	432.759	324.326	42.003	717.595	-10.710
3600	107.260	528.445	435.375	335.050	42.941	736.885	-10.692
3700	107.304	531.384	437.931	345.778	43.876	756.148	-10.675
3800	107.344	534.246	440.428	356.510	44.808	775.387	-10.658
3900	107.381	537.035	442.869	367.247	45.739	794.601	-10.642
4000	107.416	539.754	445.258	377.986	46.668	813.790	-10.627
4100	107.448	542.407	447.595	388.730	47.595	832.957	-10.612
4200	107.477	544.997	449.883	399.476	48.522	852.101	-10.597
4300	107.505	547.526	452.125	410.225	49.448	871.223	-10.583
4400	107.531	549.998	454.321	420.977	50.372	890.323	-10.569
4500	107.555	552.414	456.474	431.731	51.295	909.402	-10.556
4600	107.578	554.779	458.586	442.488	52.216	928.460	-10.543
4700	107.599	557.092	460.657	453.247	53.134	947.500	-10.530
4800	107.619	559.358	462.690	464.008	54.050	966.519	-10.518
4900	107.638	561.577	464.685	474.771	54.960	985.520	-10.506
5000	107.656	563.752	466.645	485.535	55.866	1004.502	-10.494
5100	107.672	565.884	468.570	496.302	56.765	1023.466	-10.482
5200	107.688	567.975	470.462	507.070	57.657	1042.411	-10.471
5300	107.703	570.026	472.321	517.839	58.540	1061.342	-10.460
5400	107.717	572.040	474.149	528.610	59.412	1080.253	-10.449
5500	107.730	574.016	475.947	539.383	60.273	1099.149	-10.439
5600	107.743	575.958	477.715	550.156	61.119	1118.031	-10.429
5700	107.755	577.865	479.456	560.931	61.950	1136.896	-10.418
5800	107.766	579.739	481.169	571.707	62.764	1155.749	-10.409
5900	107.777	581.581	482.855	582.484	63.559	1174.585	-10.399
6000	107.787	583.393	484.516	593.263	64.331	1193.411	-10.390

PREVIOUS: March 1963 (1 atm) CURRENT: March 1963 (1 bar)

Fluorine Nitrate ($FONO_2$) $F_1N_1O_3(g)$

SODIUM FLUORIDE (NaF) CRYSTAL M_r = 41.988173

$S°$(298.15 K) = 51.212 ± 0.08 J K^{-1} mol^{-1}

T_{fus} = 1269 ± 2 K

$\Delta_f H°$(0 K) = -573.013 ± 0.8 kJ mol^{-1}

$\Delta_f H°$(298.15 K) = -575.384 ± 0.8 kJ mol^{-1}

$\Delta_{fus} H°$ = 33.346 kJ mol^{-1}

Enthalpy of Formation

The enthalpies of solution and reaction of NaF(cr) have been determined by many investigators. Incorporating their reported data with auxiliary $\Delta_f H°$(298.15 K) values (1, 2), we calculate the corresponding values of $\Delta_f H°$(NaF, cr, 298.15 K). The results are presented in the table below. The adopted $\Delta_f H°$(NaF, cr, 298.15 K) is -137.52 ± 0.2 kcal mol^{-1} (-575.384 ± 0.8 kJ mol^{-1}).

Source	Reaction*	T/K	Concentration NaF·nH_2O	No. of Determination	$\Delta_r H°$(298.15 K) kcal mol^{-1}	$\Delta_f H°$(298.15 K) kcal mol^{-1}
Guntz (1884) (1)	A	285	400	1	0.13	-137.42
Latimer - Jolly (1953) (2)	A	298	1850	2	0.23±0.01	-137.52
Hepler et al. (1953) (3)	A	298	1540	1	0.22±0.01	-137.51
Davies - Benson (1965) (4)	A	298	222	1	0.24±0.01	-137.53
Thomsen (1882) (5)	B	291	--	2	16.24±0.15	-137.38
Wartenberg - Fitzner (1926) (6)	C	293	--	1	-39.3 ±0.1	-137.62
Schmitz - Schumaker (1947) (7)	C	291	--	1	-39.5	-137.82
Coughlin (1958) (8)	D	298	--	6	-1.40±0.06	-137.58
Vorob'yev - Skuratov (1960) (9)	E	298	--	5	-325.5 ±2.2	-137.13

*Reaction A) represents NaF(cr) = NaF(n H_2O) and $\Delta_r H°$(298.15 K) = $\Delta_\infty H°$; B) NaOH(H_2O, 200 K) + HF(H_2O, 200 K) + H_2O(NaF·401 H_2O);
C) NaCl(cr) + 1/2 F_2(g) = NaF(cr) + 1/2 Cl_2(g); D) NaCl(cr) + HF(H_2O, 5.718 K) = NaF(cr) + HCl(H_2O, 12.731K);
E) 4 Na(cr) + CF_4(g) = 4 NaF(cr) + C (graphite).

Heat Capacity and Entropy

The low temperature heat capacities, 0.05 - 15 K and 54.01 - 295.86 K, were measured by Harrison et al. (10) and King (11), respectively. These two sets of data are joined smoothly with the high temperature heat capacities at 298 K which were derived from the enthalpy measurements, 406.1 - 1282.2 K, determined by O'Brien and Kelley (12). The $C_p°$ values above 1282.2 K are obtained by linear extrapolation. $S°$(298.15 K) is calculated using the adopted low temperature heat capacities and $S°$(54 K) = 0.555 cal K^{-1} mol^{-1}.

Frank (13) commented that the temperature measurements of O'Brien and Kelley (12) seemed to be about 20 K high at 1000 with a smaller error at lower temperatures. A correction for the assumed error in the temperature measurement was used in the previous JANAF table. However, Douglas and Ditmars (14) have questioned this correction, since new enthalpy data for AlF_3 yielded $C_p°$ values in better agreement with the original $C_p°$ values of O'Brien and Kelley than with the corrected values of Frank. The α → β transition of AlF_3(cr) was found in the range 723 - 6 K, while O'Brien and Kelley selected 727 K based on a point at 728.1 K in the β region and a point at 724.9 K in the pretransition region. Frank's correction yields an error of 9 at 727 K, yet the true error appears to be negligible (≤2) for AlF_3(cr). Based on the above reasoning and the comparison of the $C_p°$ curve of NaF(cr) with those of NaCl(cr), LiCl(cr) and LiF(cr) in the temperature range 300 - 1300 K, we adopt the high temperature enthalpy data of O'Brien and Kelley without any correction.

The low temperature heat capacities in the temperature ranges 81 - 190 and 198 - 273 K were also measured by Koref (15); and high temperature enthalpies at temperatures 288 - 1073 and 290 - 1287 K were reported by Krestovnikov and Karetnikov (16), and Lyashenko (17), respectively. They are not used for evaluation.

Fusion Data

See the liquid table for details.

Sublimation Data

$\Delta_{sub} H°$(298.15 K) = 284.93 kJ mol^{-1} is calculated as the difference between $\Delta_f H°$(298.15 K) for NaF(g) and NaF(cr).

References

1. A. Guntz, Ann. Chim. Phys. [6] 3, 5 (1884).
2. W. M. Latimer and W. L. Jolly, J. Amer. Chem. Soc. 75, 1548 (1953).
3. L. G. Hepler, W. L. Jolly, and W. M. Latimer, J. Amer. Chem. Soc. 75, 2809 (1953).
4. D. H. Davies and G. C. Benson, Can. J. Chem. 43, 3100 (1965).
5. J. Thomsen, Thermochemische Untersuchungen, Vol. I, J. Barth Verlag, Leipzig, p. 157, 1882.
6. H. Wartenberg and O. Fitzner, Z. Anorg. Allgem. Chem. 151, 313 (1926).
7. H. Schmitz and H. J. Schumacher, Z. Naturforsch. 2A, 362 (1947).
8. J. P. Coughlin, J. Amer. Chem. Soc. 80, 1802 (1958).
9. A. F. Vorob'yev and S. M. Skuratov, Russ. J. Inorg. Chem. 5, 679 (1960).
10. J. P. Harrison, G. Lombardo, and P. P. Peressini, J. Phys. Chem. Solids 29, 557 (1968).
11. E. G. King, J. Amer. Chem. Soc. 79, 2056 (1957).
12. C. J. O'Brien and K. K. Kelley, J. Amer. Chem. Soc. 79, 5616 (1957).
13. W. B. Frank, J. Phys. Chem. 65, 2081 (1961).
14. T. B. Douglas and D. A. Ditmars, U. S. Nat. Bur. Stand. Rept. 9389, 1966.
15. F. Koref, Ann. Physik (4) 36, 49 (1911).
16. A. N. Krestovnikov and G. A. Karetnikov, Legkie Metal. 3, 29 (1934).
17. V. S. Lyashenko, Metallurg (USSR) 10, 85 (1935).

Sodium Fluoride (NaF) F_1Na_1(cr)

Enthalpy Reference Temperature = T_r = 298.15 K; Standard State Pressure = p° = 0.1 MPa

T/K	$C_p°$ (J $K^{-1}mol^{-1}$)	S° (J $K^{-1}mol^{-1}$)	-[G°-H°(T_r)]/T (J $K^{-1}mol^{-1}$)	H°-H°(T_r) (kJ mol^{-1})	$\Delta_f H°$ (kJ mol^{-1})	$\Delta_f G°$ (kJ mol^{-1})	Log K_f
0	0.	0.	INFINITE	-8.488	-573.013	-573.013	INFINITE
100	22.836	11.067	88.086	-7.702	-575.020	-565.226	295.244
200	40.811	33.606	55.340	-4.347	-575.585	-555.101	144.978
298.15	46.852	51.212	51.212	0.	-575.384	-545.081	95.496
300	46.924	51.502	51.213	0.087	-575.378	-544.892	94.874
400	49.597	65.409	53.091	4.927	-577.769	-534.566	69.807
500	51.262	76.664	56.716	9.974	-577.506	-523.791	54.720
600	52.677	86.136	60.850	15.171	-577.062	-513.087	44.668
700	54.124	94.363	65.063	20.510	-576.451	-502.471	37.495
800	55.710	101.692	69.191	26.001	-575.674	-491.953	32.121
900	57.492	108.355	73.178	31.659	-574.730	-481.543	27.948
1000	59.505	114.514	77.007	37.507	-573.615	-471.247	24.615
1100	61.756	120.289	80.682	43.568	-572.322	-461.071	21.894
1200	64.266	125.768	84.213	49.866	-667.607	-448.581	19.526
1269.000	66.149	129.412	86.572	54.364	------ CRYSTAL <--> LIQUID -----		
1300	66.944	131.018	87.612	56.427	-665.009	-430.433	17.295
1400	69.162	136.065	90.894	63.239	-662.172	-412.494	15.390
1500	70.626	140.892	94.068	70.236	-659.158	-394.764	13.747
1600	71.128	145.469	97.139	77.328	-656.059	-377.238	12.316
1700	71.128	149.781	100.110	84.441	-652.948	-359.907	11.059
1800	71.128	153.846	102.983	91.554	-649.844	-342.759	9.947
1900	71.128	157.692	105.762	98.666	-646.747	-325.783	8.956
2000	71.128	161.340	108.451	105.779	-643.655	-308.971	8.069
2100	71.128	164.811	111.053	112.892	-640.566	-292.312	7.271
2200	71.128	168.120	113.572	120.005	-637.481	-275.801	6.548
2300	71.128	171.281	116.013	127.118	-634.396	-259.430	5.892
2400	71.128	174.309	118.379	134.230	-631.312	-243.193	5.293
2500	71.128	177.212	120.675	141.343	-628.227	-227.085	4.745
2600	71.128	180.002	122.903	148.456	-625.140	-211.101	4.241
2700	71.128	182.686	125.068	155.569	-622.051	-195.235	3.777
2800	71.128	185.273	127.172	162.682	-618.958	-179.483	3.348
2900	71.128	187.769	129.219	169.794	-615.863	-163.842	2.951
3000	71.128	190.180	131.211	176.907	-612.765	-148.308	2.582

PREVIOUS: December 1963 CURRENT: December 1968

Sodium Fluoride (NaF) F_1Na_1(cr)

SODIUM FLUORIDE (NaF) LIQUID M_r = 41.988173

$S°(298.15\ K)$ = [73.464] J K^{-1} mol^{-1} $\Delta_f H°(298.15\ K)$ = [-546.204] kJ mol^{-1}

T_{fus} = 1269 ± 2 K $\Delta_{fus}H°$ = 33.346 kJ mol^{-1}

Enthalpy of Formation

$\Delta_f H°(NaF_3, \ell, 298.15\ K)$ is calculated from $\Delta_f H°(NaF, cr, 298.15\ K)$ by adding the enthalpy of fusion, $\Delta_{fus}H°$, and the difference in enthalpy, H°(1269 K)-H°(298.15 K), between the crystal and liquid.

Heat Capacity and Entropy

The heat capacities of NaF(ℓ) are derived from the enthalpy data, 1287.6 - 1746.5 K, determined by O'Brien and Kelley (1). A glass transition temperature is assumed at 900 K. The heat capacities below 900 K are taken from those for NaF(cr). The $C_p°$ values above 1746.5 K are obtained by extrapolation. The entropy is calculated in a manner analogous to that used for the enthalpy formation.

Fusion Data

The value of T_{fus} has been reported as (K): 1313 (2), 1285 (1), 1270 (3, 4, 5), 1269 (6), 1268 (7, 8, 9), 1267 (10, 11, 12), 1265 (13), 1263 (14), and 1253 (15). We select T_{fus} = 1269 K. The value 1285 K reported by O'Brien and Kelley seems unreasonably high and is not adopted. The cause of this discrepancy is still unresolved (17). The heat of melting is calculated as the difference between the adopted enthalpies for NaF(cr) and NaF(ℓ) at 1269 K. Kelley (18) derived $\Delta_{fus}H°(1265\ K)$ to be 5.82 - 8.43 kcal mol^{-1}, based on 13 sets of molten salts equilibrium data involving NaF. The value $\Delta_{fus}H°(1285\ K)$ = 8.03 kcal mol^{-1} reported by O'Brien and Kelley is not used.

Vaporization Data

T_{vap} = 2060 K is the temperature at which the partial pressures of NaF(g) and $Na_2F_2(g)$ in equilibrium with NaF(ℓ) equal one atmosphere. The vapor composition at T_{vap} is evaluated as NaF 84.8% and Na_2F_2 15.2%. The heat required to vaporize one mole of liquid to form 0.868 mole of vapor mixture of the above composition at T_{vap} is calculated to be 42.1 kcal mol^{-1} (176.146 kJ mol^{-1})

Ruff et al. (16) measured the vapor pressure of NaF(ℓ) in the temperature range of 1699 - 1974 K and reported T_{vap} = 1978 K.

References

1. C. J. O'Brien and K. K. Kelley, J. Amer. Chem. Soc. 79, 5616 (1957).
2. N. Puschin and A. Baskow, Z. Anorg. Chem. 81, 347 (1913).
3. N. S. Kurnakow and S. F. Zemczuznyj, Z. Anorg. Chem. 52, 186 (1907).
4. G. J. Landon and A. R. Ubbelohde, Proc. Roy. Soc. (London), 240A, 160 (1957).
5. M. Rolin, Bull. Soc. Chim. France, 1960, 671 (1960).
6. K. A. Sense, C. A. Alexander, R. E. Bowman, R. W. Stone, R. B. Filbert, Jr., J. Phys. Chem. 61, 384 (1957).
7. H. S. Booth and B. A. Starrs, J. Phys. Chem. 35, 3553 (1931).
8. M. A. Bredig, J. W. Johnson and W. T. Smith, Jr., J. Amer. Chem. Soc. 77, 307 (1955).
9. S. Cantor, J. Phys. Chem. 65, 2208 (1961).
10. B. Porter and E. A. Brown, J. Amer. Ceram. Soc. 45, 49 (1962).
11. N. W. F. Phillips, R. H. Singleton, and E. A. Hollingshead, J. Electrochem. Soc. 102, 690 (1955).
12. J. Brynestad, K. Grjotheim, and S. Urnes, Metallurgia Ital. 52, No. 8, 495 (1960).
13. W. Plato, Z. Physik. Chem. 58, 350 (1907).
14. A. G. Bergman and E. P. Dergunov, Compt. Rend, Acad. Sci. U.R.S.S. 31, 753 (1941).
15. O. Ruff and W. Plato, Ber. 36, 2363 (1903).
16. O. Ruff, G. Schmidt and S. Mugdan, Z. Anorg. Allgem. Chem. 123, 83 (1922).
17. K. K. Kelley, U. S. Bur. Mines, Berkeley, California, personal communication, May 31, 1963.
18. K. K. Kelley, U. S. Bur. Mines Bull. 393, 1936.

Sodium Fluoride (NaF) $F_1Na_1(l)$

Enthalpy Reference Temperature = T_r = 298.15 K Standard State Pressure = p° = 0.1 MPa

	J $K^{-1}mol^{-1}$			kJ mol^{-1}			
T/K	$C_p°$	S°	$-[G°-H°(T_r)]/T$	$H°-H°(T_r)$	$\Delta_f H°$	$\Delta_f G°$	Log K_f
0							
100							
200							
298.15	46.852	73.464	73.464	0.000	-546.204	-522.536	91.546
300	46.924	73.754	73.465	0.087	-546.199	-522.389	90.956
400	49.597	87.661	75.343	4.927	-548.589	-514.287	67.159
500	51.262	98.915	78.968	9.974	-548.327	-505.738	52.834
600	52.677	108.387	83.102	15.171	-547.883	-497.259	43.290
700	54.124	116.615	87.314	20.510	-547.271	-488.868	36.480
800	55.710	123.944	91.443	26.001	-546.495	-480.575	31.378
900	57.492	130.606	95.430	31.659	-545.551	-472.390	27.417
900.000	57.492	130.606	95.430	31.659	GLASS <--> LIQUID TRANSITION		
900.000	76.525	130.606	95.430	31.659			
1000	74.099	138.538	99.353	39.185	-542.757	-464.413	24.258
1100	72.300	145.512	103.239	46.501	-540.209	-456.704	21.687
1200	70.961	151.744	107.026	53.662	-634.632	-446.778	19.448
1269.000	70.208	155.689	109.566	58.531	CRYSTAL <--> LIQUID		
1300	69.873	157.380	110.686	60.702	-631.555	-431.249	17.328
1400	69.036	162.526	114.207	67.646	-628.585	-415.953	15.519
1500	68.367	167.266	117.589	74.515	-625.699	-400.866	13.959
1600	67.823	171.660	120.833	81.324	-622.884	-385.969	12.601
1700	67.362	175.758	123.944	88.083	-620.127	-371.247	11.407
1800	66.986	179.597	126.931	94.800	-617.419	-356.685	10.351
1900	66.651	183.210	129.798	101.481	-614.752	-342.273	9.410
2000	66.358	186.621	132.555	108.131	-612.123	-328.000	8.566
2100	66.107	189.852	135.207	114.754	-609.525	-313.858	7.807
2200	65.898	192.923	137.761	121.354	-606.952	-299.839	7.119
2300	65.731	195.848	140.224	127.936	-604.399	-285.936	6.494
2400	65.563	198.642	142.600	134.500	-601.863	-272.144	5.923
2500	65.438	201.316	144.896	141.050	-599.341	-258.458	5.400
2600	65.312	203.880	147.115	147.588	-596.829	-244.872	4.920
2700	65.187	206.342	149.264	154.112	-594.328	-231.383	4.476
2800	65.103	208.711	151.345	160.627	-591.834	-217.986	4.067
2900	65.019	210.995	153.363	167.133	-589.345	-204.679	3.687
3000	64.936	213.197	155.321	173.631	-586.862	-191.457	3.334
3100	64.852	215.325	157.222	180.120	-584.385	-178.317	3.005
3200	64.810	217.384	159.070	186.603	-581.911	-165.258	2.698
3300	64.726	219.376	160.868	193.080	-579.442	-152.276	2.410
3400	64.685	221.308	162.617	199.550	-576.980	-139.368	2.141
3500	64.643	223.183	164.321	206.016	-574.523	-126.533	1.888

PREVIOUS: June 1964 CURRENT: December 1968

Sodium Fluoride (NaF) $F_1Na_1(l)$

SODIUM FLUORIDE (NaF) CRYSTAL-LIQUID M_r = 44.988173

0 to 1269 K crystal
above 1269 K liquid

Refer to the individual tables for details.

Sodium Fluoride (NaF) $F_1Na_1(cr,l)$

Enthalpy Reference Temperature = T_r = 298.15 K; Standard State Pressure = $p°$ = 0.1 MPa

T/K	$C_p°$ (J K⁻¹mol⁻¹)	$S°$ (J K⁻¹mol⁻¹)	$-[G°-H°(T_r)]/T$ (J K⁻¹mol⁻¹)	$H°-H°(T_r)$ (kJ mol⁻¹)	$\Delta_fH°$ (kJ mol⁻¹)	$\Delta_fG°$ (kJ mol⁻¹)	Log K_f
0	0.	0.	INFINITE	-8.488	-573.013	-573.013	INFINITE
100	22.836	11.067	88.086	-7.702	-575.020	-565.226	295.244
200	40.811	33.606	55.340	-4.347	-575.585	-555.101	144.978
298.15	46.852	51.212	51.212	0.	-575.384	-545.081	95.496
300	46.924	51.502	51.213	0.087	-575.378	-544.892	94.874
400	49.597	65.409	53.091	4.927	-577.769	-534.566	69.807
500	51.262	76.664	56.716	9.974	-577.506	-523.791	54.720
600	52.677	86.136	60.850	15.171	-577.062	-513.087	44.668
700	54.124	94.363	65.063	20.510	-576.451	-502.471	37.495
800	55.710	101.692	69.191	26.001	-575.674	-491.953	32.121
900	57.492	108.355	73.178	31.659	-574.730	-481.543	27.948
1000	59.505	114.514	77.007	37.507	-573.615	-471.247	24.615
1100	61.756	120.289	80.682	43.568	-572.322	-461.071	21.894
1200	64.266	125.768	84.213	49.866	-667.607	-448.581	19.526
1269.000	66.149	129.412	86.572	54.364	CRYSTAL <--> LIQUID		
1269.000	70.208	155.689	86.572	87.710	TRANSITION		
1300	69.873	157.380	88.240	89.882	-631.555	-431.249	17.328
1400	69.036	162.526	93.365	96.826	-628.585	-415.953	15.519
1500	68.367	167.266	98.136	103.695	-625.699	-400.866	13.959
1600	67.823	171.660	102.595	110.504	-622.884	-385.969	12.601
1700	67.362	175.758	106.780	117.262	-620.127	-371.247	11.407
1800	66.986	179.597	110.720	123.979	-617.419	-356.685	10.351
1900	66.651	183.210	114.441	130.661	-614.752	-342.273	9.410
2000	66.358	186.621	117.965	137.311	-612.123	-328.000	8.566
2100	66.107	189.852	121.312	143.934	-609.525	-313.858	7.807
2200	65.898	192.923	124.498	150.534	-606.952	-299.839	7.119
2300	65.731	195.848	127.537	157.115	-604.399	-285.936	6.494
2400	65.563	198.642	130.442	163.679	-601.863	-272.144	5.923
2500	65.438	201.316	133.224	170.229	-599.341	-258.458	5.400
2600	65.312	203.880	135.893	176.767	-596.829	-244.872	4.920
2700	65.187	206.342	138.456	183.292	-594.328	-231.383	4.476
2800	65.103	208.711	140.924	189.806	-591.834	-217.986	4.067
2900	65.019	210.995	143.301	196.312	-589.345	-204.679	3.687
3000	64.936	213.197	145.594	202.810	-586.862	-191.457	3.334
3100	64.852	215.325	147.809	209.299	-584.385	-178.317	3.005
3200	64.810	217.384	149.952	215.782	-581.911	-165.258	2.698
3300	64.726	219.376	152.025	222.259	-579.442	-152.276	2.410
3400	64.685	221.308	154.035	228.729	-576.980	-139.368	2.141
3500	64.643	223.183	155.984	235.196	-574.523	-126.533	1.888

PREVIOUS: CURRENT: December 1968

Sodium Fluoride (NaF) $F_1Na_1(cr,l)$

SODIUM FLUORIDE (NaF) IDEAL GAS M_r = 41.988173

$S°(298.15\ K) = 217.610 \pm 0.008\ J\ K^{-1}\ mol^{-1}$

$\Delta_f H°(0\ K) = -288.819 \pm 2.1\ kJ\ mol^{-1}$
$\Delta_f H°(298.15\ K) = -290.453 \pm 2.1\ kJ\ mol^{-1}$

Electronic Level and Quantum Weight

State	ε_i, cm^{-1}	g_i
$^1\Sigma$	0	1

$\omega_e = 536.10 \pm 0.35\ cm^{-1}$ $\omega_e x_e = 3.83 \pm 0.14\ cm^{-1}$ $\sigma = 1$
$B_e = 0.4369\ cm^{-1}$ $\alpha_e = 0.00456\ cm^{-1}$ $r_e = 1.99260\ Å$

Enthalpy of Formation

The total pressure of NaF(g) and Na_2F_2(g), in equilibrium with NaF(cr, ℓ) at temperatures 1020 - 1974 K, have been determined by many investigators, using manometric (1, 2, 5), torsion-effusion (6, 8), Knudsen-effusion (3, 7), transpiration (4), and molecular-beam velocity-selector (9, 10) methods. In order to evaluate $\Delta_{vap}H°$(LiF, g) we have used a trial and error variation of $\Delta_{sub}H°$ and $\Delta_{vap}H°$ for the monomer and dimer such that the $\Delta_r H°$(298.15 K) values derived by the 2nd and 3rd law methods are in reasonable agreement.

Incorporating $\Delta_{sub}H°$(298.15 K) = 68.1 and 72.7, and $\Delta_{vap}H°$(298.15 K) = 61.12 and 58.75 kcal mol^{-1} for NaF and Na_2F_2, respectively, with JANAF functions for NaF(cr), NaF(ℓ), NaF(g) and Na_2F_2(g), we derive the partial pressures of NaF monomer from the reported total pressures. Based on the calculated partial pressures of NaF(g), the corresponding enthalpy of vaporization are evaluated by the 2nd and 3rd law methods. Using the 3rd law $\Delta_r H°$(298.15 K) and $\Delta_f H°$(298.15 K) = -137.52 and -130.545 kcal mol^{-1} for NaF(cr) and NaF(ℓ), we obtain values of $\Delta_f H°$(298.15 K) for NaF(g). The results are presented in the table below. The $\Delta_f H°$(NaF, g, 298.15 K) value adopted is -69.42 kcal mol^{-1} (-290.453±2.1 kJ mol^{-1}).

Gaydon (11) and Herzberg (12) reported D_0^0 = 114.2±6.9 (4.95±0.3 eV) and ≤ 122.2 kcal mol^{-1} (5.3 eV), respectivley. Using JANAF $\Delta_f H°$(298.15 K) values for Na(g) and F(g), we obtain $\Delta_f H°$(NaF, g, 298.15 K) = -70.5±5.8 and ≥ -78.5 kcal mol^{-1}. Employing flame-photometric method, Bulewicz et al. (13) and Page and Sugden (14) derived D_0^0 = 121±7 and 119 - 120 kcal mol^{-1} for NaF(g), which lead to $\Delta_f H°$(NaF, g, 298.15 K) = -77.3±7 and -75 to -76 kcal mol^{-1}.

Source	T/K	Reaction**	Data Points	$\Delta_r H°$(298.15 K), kcal mol^{-1} 2nd law	3rd law	Drift cal K^{-1} mol^{-1}	$\Delta_f H°$(298.15 K) kcal mol^{-1}
1. Wartenberg-Schulz (1921)	1669-1830	B	5	67.76±0.11	59.90	-4.5±0.1	-70.65
2. Ruff et al. (1922)	1699-1974	B	14	64.07±0.58	59.54	-2.4±0.3	-71.01
3. Niwa (1938)	1053-1112	A	5	63.18±1.04	66.68	3.2±1.0	-70.84
4. Sense et al. (1957)	1208-1265	A	7	71.26±0.63	68.25	-2.4±0.5	-69.27
	1274-1348	B	9	65.88±0.76	61.14	-3.6±0.6	-69.41
5. Cantor-Blankenship (1958)	1381-1592	B	6*	66.13	60.84	-3.6	-69.71
6. Pugh-Barrow (1958)	1023-1166	A	7*	68.13	68.14	0.0	-69.38
7. Porter-Schoonmaker (1958)	1121	A	1	-	68.82	-	-68.70
8. Entner-Neckel (1967)	1016-1236	A	6*	68.54±0.10	67.74	-0.7±0.1	-69.78

*Data points calculated from a given total pressure equation.

**Reaction A) NaF(cr) = NaF(g) B) NaF(ℓ) = NaF(g).

Heat Capacity and Entropy

Veazey and Gordy (15) have measured the rotational transitions of NaF in the millimeter- and sub-millimeter-wave region with a high-resolution molecular-beam absorption spectrometer. The derived values of ω_e, $\omega_e x_e$, B_e, α_e and r_e are adopted here.

Using the molecular beam electric resonance method, Bauer and Lew (16) measured the transition between the J = 0 and J = 1 rotational levels of $Na^{23}F^{19}$ in the three lowest vibrational states. Radio-frequency transitions for $Na^{23}F^{19}$ have been observed by Hollowell et al. (19). Infrared spectra of NaF have been analyzed by Snelson and Pitzer (17), Ritchie and Lew (18), and Baikov and Vasilevskii (20). The molecular constants derived by the above authors are similar to the values adopted, except the value ω_e = 550 cm^{-1} (17) which is different from the value 536.1 cm^{-1} reported by Veazey and Gordy (15).

References

1. H. von Wartenberg and H. Schulz, Z. Elektrochem. 27, 568 (1921).
2. O. Ruff, G. Schmidt, and S. Mugdan, Z. Anorg. Allg. Chem. 123, 83 (1922).
3. K. Niwa, J. Fac. Sci. Hokkaido Imp. Univ. Ser. III, 2, 201 (1938).
4. K. A. Sense, C. A. Alexander, R. E. Bowman, R. W. Stone, and R. B. Filbert, Jr., J. Phys. Chem. Phys. Chem. 61, 384 (1957).
5. S. Cantor and F. Blankenship, paper presented at the American Chemical Society at Chicago, September 7-12, 1958.
6. A. C. P. Pugh and R. F. Barrow, Trans. Faraday Soc. 54, 671 (1958).
7. R. F. Porter and R. C. Schoonmaker, J. Chem. Phys. 29, 1070 (1958).
8. P. Entner and A. Neckel, Monatsh. Chem. 98, 1083 (1967).
9. R. C. Miller and P. Kusch, J. Chem. Phys. 25, 860 (1956).
10. M. Eisenstadt, G. M. Rothberg and P. Kusch, J. Chem. Phys. 29, 797 (1958).
11. A. G. Gaydon, "Dissociation Energies," Chapman and Hall Ltd., London, 1968.
12. G. Herzberg, "Spectra of Diatomic Molecules," D. van Nostrand Company, Inc., New York, 1950.
13. E. M. Bulewicz, L. F. Phillips, and T. M. Sugden, Trans. Faraday Soc. 57, 921 (1961).
14. F. M. Page and T. M. Sugden, Nature 183, 1672 (1959).
15. S. E. Veazey and W. Gordy, Phys. Rev. 138, A 1303 (1965).
16. R. K. Bauer and H. Lew, Can. J. Phys. 41, 1461 (1963).
17. A. Snelson and K. S. Pitzer, J. Phys. Chem. 67, 882 (1963).
18. R. K. Ritchie and H. Lew, Can. J. Phys. 42, 43 (1964).
19. C. D. Hollowell, A. J. Hebert and K. Street, Jr., J. Chem. Phys. 41, 3540 (1964).
20. V. I. Baikov and K. P. Vasilevskii, Opt. Spectry. 22, 198 (1967).

Sodium Fluoride (NaF) F_1Na_1(g)

Enthalpy Reference Temperature = T_r = 298.15 K Standard State Pressure = p° = 0.1 MPa

T/K	$C_p°$ (J K^{-1}mol^{-1})	S°	-[G°-H°(T_r)]/T	H°-H°(T_r) (kJ mol^{-1})	$\Delta_f H°$	$\Delta_f G°$	Log K_f
0	0.	0.	INFINITE	-9.225	-288.819	-288.819	INFINITE
100	29.357	183.316	246.446	-6.313	-288.700	-296.132	154.683
200	31.978	204.387	220.675	-3.258	-289.565	-303.238	79.198
250	33.261	211.666	218.168	-1.626	-290.009	-306.605	64.062
298.15	34.215	217.610	217.610	0.	-290.453	-309.761	54.269
300	34.247	217.821	217.610	0.063	-290.471	-309.881	53.955
350	34.984	223.159	218.030	1.795	-290.987	-313.077	46.724
400	35.538	227.868	218.971	3.559	-294.207	-315.987	41.264
450	35.962	232.079	220.198	5.347	-294.815	-318.673	36.991
500	36.293	235.886	221.579	7.153	-295.396	-321.293	33.565
600	36.773	242.548	224.535	10.808	-296.494	-326.367	28.413
700	37.103	248.243	227.525	14.503	-297.528	-331.264	24.719
800	37.348	253.214	230.432	18.226	-298.518	-336.015	21.940
900	37.541	257.625	233.213	21.971	-299.488	-340.644	19.770
1000	37.699	261.589	235.856	25.733	-300.458	-345.165	18.030
1100	37.835	265.188	238.361	29.510	-301.449	-349.588	16.601
1200	37.955	268.486	240.736	33.300	-399.243	-351.479	15.299
1300	38.064	271.528	242.989	37.101	-399.405	-347.492	13.962
1400	38.165	274.353	245.130	40.912	-399.568	-343.493	12.816
1500	38.259	276.989	247.167	44.733	-399.730	-339.482	11.822
1600	38.349	279.461	249.109	48.564	-399.893	-335.460	10.952
1700	38.436	281.789	250.963	52.403	-400.055	-331.428	10.184
1800	38.519	283.988	252.737	56.251	-400.216	-327.386	9.500
1900	38.600	286.073	254.437	60.107	-400.376	-323.336	8.889
2000	38.679	288.055	256.069	63.971	-400.532	-319.277	8.339
2100	38.757	289.944	257.638	67.843	-400.685	-315.210	7.840
2200	38.834	291.748	259.147	71.722	-400.833	-311.137	7.387
2300	38.909	293.476	260.603	75.609	-400.974	-307.056	6.973
2400	38.984	295.134	262.007	79.504	-401.108	-302.970	6.594
2500	39.058	296.727	263.364	83.406	-401.234	-298.878	6.245
2600	39.131	298.260	264.677	87.316	-401.350	-294.782	5.922
2700	39.204	299.738	265.948	91.232	-401.457	-290.681	5.624
2800	39.276	301.165	267.181	95.156	-401.553	-286.576	5.346
2900	39.348	302.545	268.377	99.088	-401.639	-282.468	5.088
3000	39.419	303.880	269.538	103.026	-401.715	-278.358	4.847
3100	39.491	305.174	270.667	106.971	-401.782	-274.245	4.621
3200	39.562	306.429	271.765	110.924	-401.839	-270.130	4.409
3300	39.632	307.647	272.834	114.884	-401.887	-266.013	4.211
3400	39.703	308.831	273.875	118.850	-401.929	-261.895	4.024
3500	39.773	309.983	274.890	122.824	-401.964	-257.776	3.847
3600	39.843	311.105	275.881	126.805	-401.996	-253.656	3.680
3700	39.913	312.197	276.848	130.793	-402.025	-249.535	3.523
3800	39.983	313.262	277.792	134.788	-402.054	-245.413	3.373
3900	40.053	314.302	278.715	138.790	-402.084	-241.291	3.232
4000	40.123	315.317	279.617	142.798	-402.119	-237.167	3.097
4100	40.192	316.308	280.500	146.814	-402.161	-233.043	2.969
4200	40.262	317.278	281.364	150.837	-402.211	-228.918	2.847
4300	40.331	318.226	282.211	154.866	-402.275	-224.791	2.731
4400	40.400	319.154	283.040	158.903	-402.313	-220.666	2.620
4500	40.470	320.063	283.852	162.946	-402.388	-216.537	2.514
4600	40.539	320.953	284.649	166.997	-402.491	-212.406	2.412
4700	40.608	321.826	285.431	171.054	-402.615	-208.273	2.315
4800	40.677	322.681	286.198	175.118	-402.765	-204.136	2.221
4900	40.746	323.521	286.951	179.190	-402.942	-199.996	2.132
5000	40.815	324.345	287.691	183.268	-403.143	-195.853	2.046
5100	40.884	325.153	288.418	187.353	-403.384	-191.705	1.963
5200	40.953	325.948	289.132	191.445	-403.661	-187.552	1.884
5300	41.022	326.729	289.834	195.543	-403.979	-183.393	1.807
5400	41.091	327.496	290.524	199.649	-404.340	-179.228	1.734
5500	41.160	328.251	291.203	203.761	-404.747	-175.056	1.663
5600	41.229	328.993	291.871	207.881	-405.202	-170.875	1.594
5700	41.297	329.723	292.529	212.007	-405.709	-166.686	1.528
5800	41.366	330.442	293.177	216.140	-406.269	-162.488	1.463
5900	41.435	331.150	293.814	220.280	-406.886	-158.280	1.401
6000	41.504	331.847	294.442	224.427	-407.404	-154.079	1.341

PREVIOUS: December 1968 (1 atm) CURRENT: December 1968 (1 bar)

Sodium Fluoride (NaF) F_1Na_1(g)

OXYGEN FLUORIDE (OF) IDEAL GAS $M_r = 34.997803$

$S°(298.15\ K) = 216.696 \pm 1.26\ J\ K^{-1}\ mol^{-1}$

$\Delta_f H°(0\ K) = [108.773 \pm 41.8]\ kJ\ mol^{-1}$
$\Delta_f H°(298.15\ K) = [108.784 \pm 41.8]\ kJ\ mol^{-1}$

Electronic Level and Quantum Weight

State	ε_i, cm^{-1}	g_i
$[^2\Pi]$	0	[4]
	[29000]	[4]

$\omega_e = 1056\ cm^{-1}$ $\omega_e x_e = [15]\ cm^{-1}$ $\sigma = 1$
$B_e = [1.148]\ cm^{-1}$ $\alpha_e = [0.020]\ cm^{-1}$ $r_e = [1.30]$ Å

Enthalpy of Formation

Several values have been proposed for the enthalpy of formation of FO(g). Dibeler et al. (1), obtained $D_0^0(FO, g) = 24\ kcal\ mol^{-1}$ from electron impact studies. However, theoretical estimates give D_0^0 values in the range 45-56 $kcal\ mol^{-1}$. Price et al. (2), estimated 52 $kcal\ mol^{-1}$ for $D_0^0(FO, g)$.

From a consideration of the central bond energies in FNNF and HNNH with that in HOOH, using all JANAF values of 101, 109 and 50 $kcal\ mol^{-1}$, we predict the bond dissocation energy in FO-OF to be 46±4 $kcal\ mol^{-1}$, which yields $D_0^0(FO, g) = 51\pm2\ kcal\ mol^{-1}$. The average F-O bond dissociation energy in $F_2O(g)$ is 49.8±0.5 $kcal\ mol^{-1}$. Also, since there is one less antibonding electron in FO than in F_2, we should expect the F-O bond to be greater than that in F_2(38 $kcal\ mol^{-1}$). Thus a median value of $D_0^0(FO, g) = 51.4\pm\ 10\ kcal\ mol^{-1}$ is adopted here which leads to $\Delta_f H°(FO, 298.15\ K) = 26\ kcal\ mol^{-1}$ (108.784±41.8 $kJ\ mol^{-1}$).

Heat Capacity and Entropy

The fundamental frequency is obtained by adding 2 $\omega_e x_e$ to the value reported by Arkell et al. (3), from matrix isolation studies. The bond length was taken equal to that in the isoelectronic molecule $O_2^-(g)$. The value of $\omega_e x_e$ was obtained from the Morse potential function and the assumed dissociation energy. The rotational constant α_e was also estimated from the above parameters and the Morse potential curve. The $^2\Pi$ ground state is that predicted by molecular orbital theory and the first excited state would also be $^2\Pi$. By analogy with the isoelectronic $O_2^-(g)$ states, the first excited level was estimated at 29000 cm^{-1}.

References

1. V. H. Dibeler, R. M. Reese and J. L. Franklin, J. Chem. Phys. 27, 1296 (1957).
2. W. C. Price, T. R. Passmore and D. M. Roessler, Disc. Faraday Soc. 35, 207 (1963).
3. A. Arkell, R. R. Reinhard and L. P. Larson, J. Am. Chem. Soc. 87, 1016 (1965).

Oxygen Fluoride (OF) $F_1O_1(g)$

Enthalpy Reference Temperature = T_r = 298.15 K Standard State Pressure = p° = 0.1 MPa

T/K	$C_p°$ (J K⁻¹mol⁻¹)	S° (J K⁻¹mol⁻¹)	$-[G°-H°(T_r)]/T$ (J K⁻¹mol⁻¹)	$H°-H°(T_r)$ (kJ mol⁻¹)	$\Delta_f H°$ (kJ mol⁻¹)	$\Delta_f G°$ (kJ mol⁻¹)	Log K_f
0	0.	0.	INFINITE	-8.765	108.773	108.773	INFINITE
100	29.111	184.528	243.120	-5.859	108.773	107.504	-56.154
200	29.408	204.752	219.452	-2.940	108.772	106.235	-27.746
250	29.935	211.367	217.196	-1.457	108.775	105.601	-22.064
298.15	30.623	216.696	216.696	0.	108.784	104.989	-18.394
300	30.651	216.886	216.697	0.057	108.785	104.966	-18.276
350	31.432	221.669	217.073	1.609	108.804	104.328	-15.570
400	32.192	225.916	217.918	3.200	108.832	103.686	-13.540
450	32.886	229.749	219.023	4.827	108.867	103.041	-11.961
500	33.501	233.246	220.273	6.487	108.907	102.392	-10.697
600	34.503	239.447	222.965	9.889	108.993	101.081	-8.800
700	35.253	244.825	225.712	13.379	109.080	99.755	-7.444
800	35.822	249.571	228.404	16.934	109.161	98.417	-6.426
900	36.262	253.817	230.996	20.539	109.237	97.070	-5.634
1000	36.612	257.656	233.473	24.183	109.305	95.714	-5.000
1100	36.897	261.160	235.833	27.859	109.367	94.352	-4.480
1200	37.134	264.381	238.079	31.561	109.421	92.984	-4.048
1300	37.336	267.361	240.219	35.285	109.469	91.613	-3.681
1400	37.511	270.134	242.258	39.028	109.509	90.238	-3.367
1500	37.666	272.728	244.203	42.786	109.542	88.860	-3.094
1600	37.805	275.163	246.063	46.560	109.568	87.480	-2.856
1700	37.931	277.459	247.843	50.347	109.586	86.099	-2.645
1800	38.047	279.630	249.549	54.146	109.597	84.717	-2.458
1900	38.156	281.690	251.187	57.956	109.602	83.334	-2.291
2000	38.257	283.650	252.762	61.777	109.601	81.952	-2.140
2100	38.354	285.519	254.277	65.607	109.595	80.570	-2.004
2200	38.445	287.305	255.738	69.447	109.586	79.188	-1.880
2300	38.534	289.016	257.148	73.296	109.576	77.806	-1.767
2400	38.619	290.658	258.511	77.154	109.565	76.425	-1.663
2500	38.701	292.236	259.828	81.020	109.557	75.044	-1.568
2600	38.781	293.756	261.104	84.894	109.551	73.664	-1.480
2700	38.859	295.221	262.341	88.776	109.552	72.284	-1.398
2800	38.935	296.635	263.540	92.666	109.559	70.904	-1.323
2900	39.011	298.003	264.705	96.563	109.574	69.523	-1.252
3000	39.084	299.327	265.837	100.468	109.600	68.141	-1.186
3100	39.158	300.610	266.939	104.380	109.637	66.759	-1.125
3200	39.230	301.854	268.010	108.299	109.686	65.375	-1.067
3300	39.302	303.062	269.054	112.226	109.750	63.989	-1.013
3400	39.373	304.236	270.072	116.160	109.827	62.601	-0.962
3500	39.444	305.379	271.064	120.101	109.920	61.211	-0.914
3600	39.515	306.491	272.033	124.048	110.029	59.818	-0.868
3700	39.586	307.575	272.979	128.003	110.155	58.421	-0.825
3800	39.657	308.631	273.903	131.966	110.298	57.021	-0.784
3900	39.728	309.662	274.807	135.935	110.458	55.617	-0.745
4000	39.800	310.669	275.691	139.911	110.635	54.209	-0.708
4100	39.872	311.653	276.556	143.895	110.830	52.796	-0.673
4200	39.945	312.614	277.403	147.886	111.043	51.378	-0.639
4300	40.018	313.555	278.233	151.884	111.274	49.955	-0.607
4400	40.093	314.476	279.047	155.889	111.522	48.525	-0.576
4500	40.168	315.378	279.844	159.902	111.788	47.091	-0.547
4600	40.244	316.261	280.626	163.923	112.071	45.650	-0.518
4700	40.321	317.128	281.393	167.951	112.370	44.203	-0.491
4800	40.400	317.977	282.147	171.987	112.687	42.749	-0.465
4900	40.479	318.811	282.887	176.031	113.020	41.289	-0.440
5000	40.560	319.630	283.613	180.083	113.368	39.821	-0.416
5100	40.642	320.434	284.327	184.143	113.732	38.347	-0.393
5200	40.725	321.224	285.029	188.212	114.110	36.865	-0.370
5300	40.810	322.000	285.720	192.288	114.503	35.376	-0.349
5400	40.896	322.764	286.399	196.374	114.910	33.879	-0.328
5500	40.983	323.515	287.067	200.468	115.329	32.374	-0.307
5600	41.072	324.255	287.724	204.570	115.761	30.863	-0.288
5700	41.162	324.982	288.371	208.682	116.205	29.342	-0.269
5800	41.253	325.699	289.009	212.803	116.659	27.815	-0.250
5900	41.345	326.405	289.637	216.933	117.124	26.278	-0.233
6000	41.439	327.101	290.255	221.072	117.599	24.735	-0.215

PREVIOUS: December 1966 (1 atm) CURRENT: December 1966 (1 bar)

Oxygen Fluoride (OF) $F_1O_1(g)$

TITANIUM FLUORIDE OXIDE (OTiF) IDEAL GAS M_r = 82.877803

$S°(298.15\ K) = [250.667]\ J\ K^{-1}\ mol^{-1}$

$\Delta_fH°(0\ K) = [-430.918]\ kJ\ mol^{-1}$
$\Delta_fH°(298.15\ K) = [-433.211]\ kJ\ mol^{-1}$

Vibrational Frequencies and Degeneracies

ν, cm^{-1}
[880](1)
[300](2)
[740](1)

Ground State Quantum Weight: [2] σ = 1
Point Group: $[C_{\infty v}]$
Bond Distances: Ti-O = [1.62] Å Ti-F = [1.74] Å
Bond Angle: O-Ti-F = [180°]
Rotational Constant: B_0 = [0.170462] cm^{-1}

Enthalpy of Formation

$\Delta_fH°(0\ K)$ was estimated as -103 kcal mol^{-1} by Gordon (1). $\Delta_fH°(298.15\ K)$ was then calculated.

Heat Capacity and Entropy

Molecular constants were estimated by Gordon (1).

References

1. J. S. Gordon, Astrosystems, Inc., Caldwell Twp., N. J., personal communication, January 10, 1963.

Titanium Fluoride Oxide (OTiF) $F_1O_1Ti_1(g)$

Enthalpy Reference Temperature = T_r = 298.15 K Standard State Pressure = p° = 0.1 MPa

T/K	$C_p°$	S°	$-[G°-H°(T_r)]/T$	$H°-H°(T_r)$	$\Delta_fH°$	$\Delta_fG°$	Log K_f
	J $K^{-1}mol^{-1}$			kJ mol^{-1}			
0	0.	0.	INFINITE	-11.291	-430.918	-430.918	INFINITE
100	33.375	206.489	289.330	-8.284	-431.378	-434.017	226.707
200	42.295	232.558	254.917	-4.472	-432.402	-436.255	113.938
250	45.706	242.373	251.450	-2.269	-432.844	-437.166	91.341
298.15	48.472	250.667	250.667	0.	-433.211	-437.964	76.729
300	48.568	250.967	250.668	0.090	-433.224	-437.993	76.261
350	50.908	258.636	251.268	2.579	-433.555	-438.761	65.482
400	52.791	265.561	252.629	5.173	-433.850	-439.484	57.391
450	54.299	271.870	254.422	7.852	-434.121	-440.172	51.094
500	55.509	277.656	256.460	10.598	-434.378	-440.831	46.053
600	57.279	287.944	260.871	16.243	-434.874	-442.075	38.486
700	58.469	296.868	265.390	22.035	-435.374	-443.235	33.075
800	59.297	304.733	269.826	27.925	-435.882	-444.324	29.011
900	59.892	311.753	274.102	33.886	-436.441	-445.347	25.847
1000	60.332	318.088	278.189	39.898	-437.126	-446.301	23.312
1100	60.666	323.854	282.082	45.949	-438.005	-447.178	21.235
1200	60.925	329.144	285.787	52.029	-443.067	-447.845	19.494
1300	61.129	334.029	289.312	58.132	-443.620	-448.222	18.010
1400	61.293	338.566	292.670	64.254	-444.260	-448.552	16.736
1500	61.427	342.799	295.873	70.390	-445.000	-448.834	15.630
1600	61.537	346.767	298.931	76.538	-445.854	-449.062	14.660
1700	61.629	350.501	301.856	82.697	-446.837	-449.233	13.803
1800	61.706	354.026	304.657	88.864	-447.962	-449.343	13.040
1900	61.772	357.364	307.344	95.038	-449.241	-449.385	12.354
2000	61.828	360.534	309.925	101.218	-465.375	-448.902	11.724
2100	61.876	363.551	312.407	107.403	-467.750	-448.020	11.144
2200	61.919	366.431	314.798	113.593	-470.133	-447.025	10.614
2300	61.956	369.184	317.103	119.786	-472.522	-445.921	10.127
2400	61.988	371.822	319.328	125.984	-474.917	-444.713	9.679
2500	62.017	374.353	321.479	132.184	-477.315	-443.406	9.264
2600	62.042	376.786	323.560	138.387	-479.715	-442.002	8.880
2700	62.065	379.127	325.575	144.592	-482.115	-440.506	8.522
2800	62.085	381.385	327.528	150.800	-484.514	-438.921	8.188
2900	62.104	383.564	329.423	157.009	-486.910	-437.250	7.876
3000	62.120	385.670	331.263	163.221	-489.301	-435.497	7.583
3100	62.135	387.707	333.051	169.433	-491.687	-433.664	7.307
3200	62.149	389.680	334.790	175.648	-494.067	-431.755	7.048
3300	62.161	391.592	336.482	181.863	-496.438	-429.771	6.803
3400	62.173	393.448	338.131	188.080	-498.801	-427.715	6.571
3500	62.183	395.251	339.737	194.298	-501.155	-425.590	6.352
3600	62.193	397.003	341.304	200.516	-503.499	-423.398	6.143
3700	62.201	398.707	342.832	206.736	-914.931	-413.352	5.835
3800	62.210	400.366	344.324	212.957	-916.019	-399.781	5.495
3900	62.217	401.982	345.782	219.178	-917.152	-386.181	5.172
4000	62.224	403.557	347.207	225.400	-918.328	-372.552	4.865
4100	62.230	405.093	348.600	231.623	-919.545	-358.893	4.572
4200	62.236	406.593	349.963	237.846	-920.802	-345.203	4.293
4300	62.242	408.058	351.297	244.070	-922.097	-331.483	4.027
4400	62.247	409.489	352.604	250.295	-923.429	-317.733	3.772
4500	62.252	410.888	353.883	256.519	-924.796	-303.951	3.528
4600	62.257	412.256	355.137	262.745	-926.197	-290.140	3.295
4700	62.261	413.595	356.367	268.971	-927.630	-276.297	3.071
4800	62.265	414.906	357.573	275.197	-929.093	-262.423	2.856
4900	62.269	416.190	358.756	281.424	-930.586	-248.519	2.649
5000	62.272	417.448	359.917	287.651	-932.105	-234.583	2.451
5100	62.275	418.681	361.058	293.878	-933.652	-220.618	2.260
5200	62.279	419.890	362.177	300.106	-935.222	-206.622	2.076
5300	62.282	421.076	363.278	306.334	-936.817	-192.594	1.898
5400	62.284	422.241	364.359	312.562	-938.434	-178.538	1.727
5500	62.287	423.383	365.422	318.791	-940.072	-164.451	1.562
5600	62.290	424.506	366.467	325.020	-941.730	-150.333	1.402
5700	62.292	425.608	367.495	331.249	-943.408	-136.187	1.248
5800	62.294	426.692	368.506	337.478	-945.103	-122.010	1.099
5900	62.296	427.757	369.501	343.707	-946.816	-107.804	0.954
6000	62.298	428.804	370.481	349.937	-948.546	-93.569	0.815

PREVIOUS: September 1963 (1 atm) CURRENT: September 1963 (1 bar)

Titanium Fluoride Oxide (OTiF) $F_1O_1Ti_1(g)$

OXYGEN FLUORIDE (O_2F) IDEAL GAS M_r = 50.997203

$\Delta_f H°(0\ K) = 14.401 \pm 2.1\ kJ\ mol^{-1}$

$S°(298.15\ K) = 259.114\ J\ K^{-1}\ mol^{-1}$ $\Delta_f H°(298.15\ K) = 12.552 \pm 2.1\ kJ\ mol^{-1}$

Vibrational Frequencies and Degeneracies

ν, cm^{-1}
1495.0(1)
584.5(1)
376.0(1)

Ground State Quantum Weight: 2 σ = 1

Point Group: C_s

Bond Distances: O–O = 1.22 Å O–F = 1.58 Å

Bond Angle: O–O–F = 109.5°

Product of the Moments of Inertia: $I_A I_B I_C = 7.75227 \times 10^{-116}\ g^3\ cm^6$

Enthalpy of Formation

The enthalpy of formation for FO_2(g) is unavailable in the literature. Based on a kinetic study of the thermal decomposition of F_2O_2(g) reported by Schumacher and Frisch (1), the enthalpy change for the reaction $F_2O_2(g) = F(g) + O_2F(g)$ was estimated to be +17.3 kcal mol^{-1} by Levy and Wesley (2), assuming the activation energy for the reverse is negligible. From this data the value of $\Delta_f H°(FO_2$, g, 298.15 K) was calculated as +3.2 kcal mol^{-1}, using $\Delta_f H°(298.15\ K) = +4.73$ kcal mol^{-1} for F_2O_2(g) determined calorimetrically by Kirshenbaum et al. (3).

Assuming the enthalpy change for the reaction $F_2O_2(g) = F(g) + O_2F(g)$ is half of the enthalpy change for the reaction $F_2O_2(g) = 2F(g) + O_2(g)$ which was calculated to be 33.0 kcal mol^{-1}, the heat of formation ($\Delta_f H°$(298.15 K) for FO_2(g) was derived as +2.4 kcal mol^{-1}.

The value of $\Delta_f H°(FO_2$, g, 298.15 K) adopted is 3±5 kcal mol^{-1} (12.552±2.1 kJ mol^{-1}).

Heat Capacity and Entropy

The vibrational frequencies for FOO(g), adopted here, were assumed to be the same as those obtained from matrix isolation by Noble and Pimentel (4). These authors have observed two more frequencies for isotopically substituted FOO and added to those observed by Spratley et al. (5), and performed a normal coordinate analysis. The results confirm the fact that the bond distances and angles in FOO are essentially the same as those reported for F_2O_2(g) by Jackson (6). The same molecular structure for FOO was also deduced from electron spin resonance and electron paramagnetic resonance data reported by Fessenden and Schuler (7) and Kasai and Kirschenbaum (8), respectively. Hence this molecular structure was adopted. Two vibrational frequencies, ν_1 and ν_3, for FOO molecule were also observed by Arkell (9).

The three principal moments of inertia are: $I_A = 1.1104 \times 10^{-39}$, $I_B = 7.8188 \times 10^{-39}$ and $I_C = 8.9292 \times 10^{-39}\ g\ cm^2$.

References

1. H. J. Schumacher and P. Frisch, Z. Physik. Chem. B24, 332 (1934).
2. J. B. Levy and B. K. Wesley Copeland, J. Phys. Chem. 69, 408 (1965).
3. A. D. Kirshenbaum, A. V. Grosse and J. G. Aston, J. Amer. Chem. Soc. 81, 6398 (1959).
4. P. N. Noble and G. C. Pimentel, J. Chem. Phys. 44, 3641 (1966).
5. R. D. Spratley, J. J. Turner and G. C. Pimentel, J. Chem. Phys. 44, 2063 (1966).
6. R. H. Jackson, J. Chem. Soc. 4585 (1962).
7. R. W. Fessenden and R. H. Schuler, J. Chem. Phys. 44, 434 (1966).
8. P. H. Kasai and A. D. Kirschenbaum, J. Amer. Chem. Soc. 87, 3069 (1965).
9. A. Arkell, J. Amer. Chem. Soc. 87, 4057 (1965).

Oxygen Fluoride (O_2F) $F_1O_2(g)$

Enthalpy Reference Temperature = T_r = 298.15 K; Standard State Pressure = p° = 0.1 MPa

T/K	$C_p°$ (J K⁻¹mol⁻¹)	S° (J K⁻¹mol⁻¹)	−[G°−H°(T_r)]/T (J K⁻¹mol⁻¹)	H°−H°(T_r) (kJ mol⁻¹)	$\Delta_f H°$ (kJ mol⁻¹)	$\Delta_f G°$ (kJ mol⁻¹)	Log K_f
0	0.	0.	INFINITE	-11.247	14.401	14.401	INFINITE
100	34.487	216.578	295.576	-7.900	13.390	17.581	-9.184
200	40.211	242.218	263.054	-4.167	12.747	22.066	-5.763
250	42.572	251.455	259.836	-2.095	12.610	24.413	-5.101
298.15	44.397	259.114	259.114	0.	12.552	26.692	-4.676
300	44.460	259.389	259.115	0.082	12.551	26.780	-4.663
350	46.022	266.364	259.663	2.345	12.543	29.153	-4.351
400	47.360	272.598	260.897	4.681	12.569	31.524	-4.117
450	48.526	278.245	262.515	7.078	12.616	33.891	-3.934
500	49.551	283.412	264.350	9.531	12.677	36.252	-3.787
600	51.244	292.603	268.312	14.574	12.824	40.954	-3.565
700	52.548	300.604	272.366	19.767	12.986	45.629	-3.405
800	53.554	307.690	276.347	25.074	13.152	50.281	-3.283
900	54.335	314.045	280.189	30.470	13.315	54.912	-3.187
1000	54.947	319.802	283.867	35.935	13.474	59.525	-3.109
1100	55.431	325.063	287.376	41.455	13.625	64.123	-3.045
1200	55.820	329.903	290.721	47.019	13.766	68.708	-2.991
1300	56.135	334.384	293.910	52.617	13.897	73.281	-2.944
1400	56.393	338.554	296.951	58.244	14.015	77.845	-2.904
1500	56.607	342.452	299.856	63.894	14.119	82.400	-2.869
1600	56.787	346.112	302.634	69.564	14.207	86.949	-2.839
1700	56.938	349.559	305.294	75.250	14.279	91.493	-2.811
1800	57.066	352.817	307.844	80.951	14.333	96.033	-2.787
1900	57.177	355.906	310.293	86.663	14.370	100.571	-2.765
2000	57.272	358.841	312.648	92.386	14.390	105.108	-2.745
2100	57.354	361.637	314.915	98.117	14.392	109.643	-2.727
2200	57.427	364.307	317.100	103.856	14.378	114.180	-2.711
2300	57.490	366.861	319.208	109.602	14.350	118.717	-2.696
2400	57.546	369.309	321.245	115.354	14.307	123.255	-2.683
2500	57.596	371.659	323.215	121.111	14.252	127.795	-2.670
2600	57.640	373.919	325.122	126.873	14.186	132.338	-2.659
2700	57.680	376.095	326.970	132.639	14.112	136.884	-2.648
2800	57.716	378.194	328.762	138.409	14.030	141.433	-2.638
2900	57.748	380.219	330.501	144.182	13.943	145.984	-2.629
3000	57.777	382.178	332.192	149.958	13.852	150.539	-2.621
3100	57.803	384.073	333.835	155.737	13.758	155.097	-2.613
3200	57.827	385.908	335.434	161.519	13.663	159.658	-2.606
3300	57.849	387.688	336.990	167.303	13.567	164.221	-2.599
3400	57.869	389.415	338.507	173.089	13.474	168.788	-2.593
3500	57.888	391.093	339.985	178.877	13.382	173.357	-2.587
3600	57.905	392.724	341.428	184.666	13.293	177.930	-2.582
3700	57.920	394.311	342.836	190.457	13.208	182.504	-2.576
3800	57.934	395.856	344.211	196.250	13.127	187.081	-2.572
3900	57.948	397.361	345.554	202.044	13.050	191.660	-2.567
4000	57.960	398.828	346.868	207.840	12.979	196.240	-2.563
4100	57.972	400.259	348.153	213.636	12.913	200.823	-2.559
4200	57.982	401.656	349.410	219.434	12.852	205.407	-2.555
4300	57.992	403.021	350.641	225.233	12.796	209.992	-2.551
4400	58.002	404.354	351.847	231.032	12.746	214.578	-2.547
4500	58.010	405.658	353.028	236.833	12.700	219.166	-2.544
4600	58.018	406.933	354.186	242.634	12.659	223.754	-2.541
4700	58.026	408.181	355.322	248.437	12.621	228.344	-2.538
4800	58.033	409.402	356.436	254.240	12.587	232.934	-2.535
4900	58.040	410.599	357.529	260.043	12.556	237.525	-2.532
5000	58.046	411.772	358.602	265.848	12.526	242.117	-2.529
5100	58.052	412.921	359.656	271.652	12.497	246.709	-2.527
5200	58.058	414.048	360.691	277.458	12.469	251.301	-2.524
5300	58.063	415.154	361.708	283.264	12.440	255.895	-2.522
5400	58.068	416.240	362.708	289.071	12.408	260.488	-2.520
5500	58.073	417.305	363.691	294.878	12.374	265.082	-2.518
5600	58.077	418.352	364.658	300.685	12.335	269.678	-2.515
5700	58.082	419.380	365.609	306.493	12.290	274.273	-2.513
5800	58.086	420.390	366.545	312.301	12.239	278.871	-2.512
5900	58.089	421.383	367.466	318.110	12.179	283.467	-2.510
6000	58.093	422.359	368.373	323.919	12.109	288.066	-2.508

PREVIOUS: September 1966 (1 atm) CURRENT: September 1966 (1 bar)

Oxygen Fluoride (O_2F) $F_1O_2(g)$

D_0^0 = 443.881 ± 19.2 kJ mol^{-1}
S°(298.15 K) = 224.968 ± 0.04 J K^{-1} mol^{-1}

$\Delta_f H°$(0 K) = -51.335 ± 20.9 kJ mol^{-1}
$\Delta_f H°$(298.15 K) = -52.250 ± 20.9 kJ mol^{-1}

Electronic and Molecular Constants (σ = 1)

State	T_0, cm^{-1}	g_i	ω_e, cm^{-1}	$\omega_e x_e$, cm^{-1}	B_e, cm^{-1}	α_e, cm^{-1}	r_e, Å
$X^3\Sigma^-$	0	3	846.75	4.489	0.5665	0.00456	1.5896
$a^1\Delta$	7090.41	2	858.79	4.438	0.5699	0.00467	1.5849
$b^1\Sigma$	13353.91	1	866.14	4.51	0.5725	0.0045	1.5812
$B^3\Pi_0$	29338.69	2	436	[1.5]	0.4632	0.004	1.752
$B^3\Pi_1$	29481.80	2	436	[1.5]	0.4663	0.0038	1.752
$B^3\Pi_2$	29623.06	2	436	[1.5]	0.4693	0.0037	1.752
$d^1\Pi$	35812.29	2	413	[1.5]	0.4848	0.0062	1.721

Enthalpy of Formation

There has been no report of a direct experimental determination of $\Delta_f H°$. Sufficient information is available, however, to allow a reasonable estimate of the P-F bond energy (D_0^0) to be made.

A linear Birge-Sponer extrapolation of the ground state vibrational constants leads to D"(LBX) = 4.90 eV. This value can be refined by taking into account the ionic character of the P-F bond. Using relationships developed by Hildenbrand (1), we adjust the value of D"(LBX) to obtain D" = 4.25 eV. A similar treatment of spectroscopic data (5) for related molecules (AlF, SiF, and ClF) which have well established bond energies indicates that the true D_0^0 value could be higher than D" by roughly 0.3 eV. Additional support for this is provided by the results of two molecular orbital studies (2, 3) which have led to estimates of D_0^0 of 4.5 eV and 4.65±0.2 eV. The latter value, obtained from Hartree-Fock energies and estimated corrections for correlation effects, is believed to be reliable within the quoted uncertainty. A CNDO/2 MO study (4), employing energy partitioning concepts, has led to a much higher estimate of D_0^0 equal to 5.17 eV.

We adopt D_0^0(PF) = 4.60±0.2 eV (106.08 kcal mol^{-1}) which gives $\Delta_f H°$(PF) = -8.546±5.0 kcal mol^{-1} when combined with JANAF auxiliary data (5) for P and F. The adopted D_0^0 value lies midway between the well-established bond energies (5) for SiF (130.3 kcal mol^{-1}) and SF (81.2 kcal mol^{-1}) and is near the mean bond energy (D = 109.1 kcal mol^{-1}) in PF_5 (5). $\Delta_f H°$ at 298.15 K is -8.345 kcal mol^{-1}.

Heat Capacity and Entropy

The thermal functions are calculated from the partition function $Q = Q_t \sum_i g_i Q_r^i Q_v^i \exp(-c_2\epsilon_i/T)$ in which Q_r^i and Q_v^i contain first order corrections for anharmonicity. The electronic and molecular constants are taken from the compilation of Suchard (6). Values of $\omega_e x_e$ placed in brackets are estimates obtained from data on isoconfigurational states of SO ($A^3\Pi$, 6) and PF($B^3\Pi$). Our entropies may contain significant errors, amounting to several tenths of a cal K^{-1} mol^{-1}, above 3500 K due to the omission of many unobserved electronic levels which have been predicted (7) to lie above the $b^1\Sigma$ state at 13353.91 cm^{-1}. In addition, Skolnik and Goodfriend (8) recently observed in emission a triple-headed band system near 5600 Å which was tentatively assigned to a $^3\Delta_{inv}$ - $^3\Pi_{inv}$ transition of PF. These inverted triplet levels are also omitted since we are unable to estimate their term values with respect to the $X^3\Sigma^-$ ground state.

Our calculations essentially update and extend those previously made on PF by O'Hare (9). The most significant difference is that our calculations are based on solid phosphorus (Red, V) as the reference state below 704 K rather than the ideal diatomic gas selected by O'Hare (9).

References

1. D. L. Hildenbrand in "Advances in High Temperature Chemistry," Leroy Eyring, ed., Vol. 1, Academic Press, New York, p. 163, 1967.
2. P. A. G. O'Hare and A. C. Wahl, J. Chem. Phys. 54, 4563 (1971).
3. P. A. G. O'Hare, J. Chem. Phys. 59, 3842 (1973).
4. A. L. Companion and Y. P. Hsia, J. Mol. Struct. 14, 117 (1972).
5. JANAF Thermochemical Tables: AlF(g), 12-31-75; SiF(g), 12-31-76; ClF(g) and F(g), 9-30-65; P(g), 6-30-62; SF(g), 6-30-76; PF_5(g), 12-31-69.
6. S. N. Suchard, The Aerospace Corporation, Aerospace Report No. TR-0074 (4641)-6, March 1974.
7. A. E. Douglas and M. Frackowiak, Can. J. Phys. 40, 832 (1962).
8. E. G. Skolnik and P. L. Goodfriend, J. Mol. Spectrosc. 50, 202 (1974).
9. P. A. G. O'Hare, Argonne National Labortory, ANL-7459, December, 1968.

Enthalpy Reference Temperature = T_r = 298.15 K Standard State Pressure = p° = 0.1 MPa

T/K	$C_p°$ (J K^{-1}mol^{-1})	S°	-[G°-H°(T_r)]/T	H°-H°(T_r) (kJ mol^{-1})	$\Delta_f H°$	$\Delta_f G°$	Log K_f
0	0.	0.	INFINITE	-8.858	-51.335	-51.335	INFINITE
100	29.115	192.409	251.908	-5.950	-50.768	-59.722	31.195
200	29.854	212.729	227.801	-3.014	-51.560	-68.616	17.921
250	30.716	219.480	225.484	-1.501	-51.890	-72.843	15.220
298.15	31.620	224.968	224.968	0.	-52.250	-76.846	13.463
300	31.654	225.164	224.969	0.059	-52.264	-76.999	13.407
350	32.528	230.110	225.357	1.663	-53.390	-81.018	12.091
400	33.289	234.505	226.231	3.309	-53.875	-84.932	11.091
450	33.928	238.464	227.374	4.990	-54.344	-88.785	10.306
500	34.461	242.067	228.666	6.700	-54.799	-92.588	9.673
600	35.270	248.426	231.443	10.189	-55.680	-100.062	8.711
700	35.838	253.908	234.270	13.746	-56.531	-107.391	8.014
800	36.250	258.721	237.032	17.352	-57.363	-114.600	7.483
900	36.563	263.010	239.684	20.993	-58.181	-121.706	7.064
1000	36.813	266.876	242.213	24.662	-58.990	-128.720	6.724
1100	37.025	270.394	244.617	28.355	-59.790	-135.655	6.442
1200	37.215	273.624	246.902	32.067	-124.156	-141.438	6.157
1300	37.395	276.610	249.074	35.797	-124.168	-142.878	5.741
1400	37.573	279.388	251.141	39.546	-124.177	-144.317	5.385
1500	37.751	281.986	253.112	43.312	-124.183	-145.755	5.076
1600	37.933	284.429	254.993	47.096	-124.183	-147.193	4.805
1700	38.120	286.734	256.793	50.899	-124.177	-148.631	4.567
1800	38.310	288.918	258.518	54.720	-124.162	-150.070	4.355
1900	38.503	290.994	260.173	58.561	-124.136	-151.510	4.165
2000	38.697	292.974	261.764	62.421	-124.098	-152.952	3.995
2100	38.892	294.867	263.296	66.300	-124.045	-154.396	3.840
2200	39.085	296.681	264.772	70.199	-123.977	-155.843	3.700
2300	39.276	298.422	266.198	74.117	-123.890	-157.293	3.572
2400	39.464	300.098	267.575	78.054	-123.784	-158.747	3.455
2500	39.647	301.713	268.909	82.010	-123.656	-160.207	3.347
2600	39.825	303.271	270.201	85.983	-123.506	-161.672	3.248
2700	39.998	304.777	271.454	89.975	-123.333	-163.143	3.156
2800	40.164	306.235	272.670	93.983	-123.134	-164.621	3.071
2900	40.324	307.647	273.852	98.007	-122.910	-166.106	2.992
3000	40.478	309.017	275.001	102.047	-122.659	-167.600	2.918
3100	40.626	310.347	276.120	106.103	-122.381	-169.103	2.849
3200	40.767	311.639	277.210	110.172	-122.076	-170.614	2.785
3300	40.902	312.895	278.272	114.256	-121.743	-172.136	2.725
3400	41.032	314.118	279.309	118.353	-121.383	-173.669	2.668
3500	41.157	315.310	280.320	122.462	-120.995	-175.212	2.615
3600	41.277	316.471	281.308	126.584	-120.579	-176.767	2.565
3700	41.393	317.603	282.274	130.717	-120.136	-178.334	2.518
3800	41.506	318.709	283.218	134.862	-119.666	-179.913	2.473
3900	41.615	319.788	284.142	139.019	-119.169	-181.505	2.431
4000	41.723	320.843	285.047	143.185	-118.645	-183.110	2.391
4100	41.828	321.875	285.932	147.363	-118.095	-184.728	2.353
4200	41.933	322.884	286.800	151.551	-117.519	-186.360	2.318
4300	42.037	323.872	287.651	155.750	-116.917	-188.006	2.284
4400	42.141	324.839	288.485	159.958	-116.290	-189.667	2.252
4500	42.245	325.788	289.304	164.178	-115.637	-191.342	2.221
4600	42.351	326.717	290.107	168.408	-114.959	-193.032	2.192
4700	42.459	327.629	290.896	172.648	-114.256	-194.736	2.164
4800	42.569	328.524	291.670	176.899	-113.529	-196.456	2.138
4900	42.682	329.403	292.431	181.162	-112.777	-198.192	2.113
5000	42.798	330.267	293.179	185.436	-112.000	-199.943	2.089
5100	42.917	331.115	293.915	189.722	-111.199	-201.709	2.066
5200	43.040	331.950	294.638	194.019	-110.374	-203.492	2.044
5300	43.167	332.771	295.350	198.330	-109.523	-205.291	2.023
5400	43.298	333.579	296.051	202.653	-108.649	-207.106	2.003
5500	43.434	334.375	296.740	206.989	-107.750	-208.938	1.984
5600	43.575	335.159	297.419	211.340	-106.826	-210.785	1.966
5700	43.720	335.931	298.088	215.705	-105.877	-212.651	1.949
5800	43.870	336.693	298.747	220.084	-104.904	-214.532	1.932
5900	44.025	337.444	299.397	224.479	-103.905	-216.431	1.916
6000	44.185	338.185	300.037	228.889	-102.881	-218.347	1.901

PREVIOUS: June 1977 (1 atm) CURRENT: June 1977 (1 bar)

PHOSPHORUS FLUORIDE, ION (PF^+) IDEAL GAS M_r = 49.971614

D_0^0 = 508.356 ± 41.8 kJ mol^{-1}

S°(298.15 K) = 224.035 ± 0.21 J K^{-1} mol^{-1}

$\Delta_f H°$(0 K) = 895.664 ± 37.7 kJ mol^{-1}

$\Delta_f H°$(298.15 K) = 901.518 ± 37.7 kJ mol^{-1}

Electronic States and Molecular Constants (σ = 1)

State	T_0, cm^{-1}	g_i	ω_e, cm^{-1}	$\omega_e x_e$, cm^{-1}	B_e, cm^{-1}	α_e, cm^{-1}	r_e, Å
$X^2\Pi_{1/2}$	0	2	1053.25	5.047	0.6360	0.0048	1.5003
$X^2\Pi_{3/2}$	323.95	2	1053.25	5.047	0.6360	0.0048	1.5003
$^2\Sigma$	35434.64	2	619.0	4.615	0.5593	0.0079	1.5998

Enthalpy of Formation

Harland et al. (1) and Torgerson and Westmore (2) reported the appearance potential of PF^+ from PF_3 as 21.0±0.3 eV and 20.5±0.2 eV, respectively. We use the mean value of AP(PF^+/PF_3) = 20.75±0.3 eV (478.50 kcal mol^{-1}) in the expression $\Delta_r H°$(0 K) = AP - E* for the process $PF_3(g) + e^- = PF^+(g) + 2\ F(g) + 2e^-$. With the assumption E* = 0, the value of $\Delta_r H°$(0 K) gives $\Delta_f H°$(PF^+, g, 0 K) = 217.823±9.0 kcal mol^{-1} with $\Delta_f H°$(PF_3, g, 0 K) = -223.97±0.9 kcal mol^{-1} and $\Delta_f H°$(F, g, 0 K) = 18.36±0.4 kcal mol^{-1} (3). We adopted this value. In the same paper, Harland et al. (1) also reported AP(PF^+/PF_2CN) = 19.1±0.2 eV which leads to $\Delta_f H°$(PF^+, g, 0 K) = 233.56±21.3 kcal mol^{-1}, in reasonable accord with the adopted result.

$\Delta_f H°$ at 298.15 K is 219.641±9.0 kcal mol^{-1}, and the adopted ionization potential (IP) is 9.82±0.61 eV. This IP compares favorably with values predicted from quantum-mechanical calculations. These have included (all in eV) 9.5±0.2 (4), 9.6 (5), and 10.35 (6). The latter value is based on IP(P) = 10.98 eV which is 0.5 eV higher than the current best value of 10.486 eV (7). Recalculation of the results of Companion and Hsia (6) gives IP(PF) = 9.85 eV.

The primary bond dissociation energy $D_0^0(PF^+)$ is calculated to be 121.5 kcal mol^{-1} which can be compared with 106.1 kcal mol^{-1} (3) in the neutral molecule. The stronger P-F bond in the ion is consistent with theoretical predictions (4-6) that the electron lost from PF occupies an antibonding orbital.

Heat Capacity and Entropy

The thermal functions are calculated from the partition function $Q = Q_t \Sigma g_i Q_r^i Q_v^i \exp(-c_2 \epsilon_i/T)$ in which Q_r^i and Q_v^i contain first order corrections for anharmonicity. The electronic and molecular constants are taken from Douglas and Frackowiak (8) who observed one band system in the emission spectrum of PF^+ formed by passing a discharge through a mixture of PF_3 and He. We treat the ground state as two distinct electronic levels with a splitting of 323.95 cm^{-1}. This procedure gives functions which are slightly biased below 400 K. The uncertainty in the value of S°(298.15 K) includes this error. By analogy with the isoelectronic molecules SiF and PO (3), other excited states are predicted to lie in the range 20000-30000 cm^{-1}. We prefer to omit these levels, but if they do exist our entropies could be slightly low above 4500 K.

References

1. P. W. Harland, D. W. H. Rankin and J. C. J. Thynne, Int. J. Mass Spectr. Ion Phys. 13, 395 (1974).
2. D. F. Torgerson and J. B. Westmore, Can. J. Chem. 53, 933 (1975).
3. JANAF Thermochemical Tables: PF_3(g), 12-31-69; F(g), 9-30-65; SiF(g), 12-31-76; PO(g), 6-30-71; PF(g), 6-30-77.
4. P. A. G. O'Hare and A. C. Wahl, J. Chem. Phys. 54, 4563 (1971).
5. P. A. G. O'Hare, J. Chem. Phys. 59, 3842 (1973).
6. A. L. Companion and Y. P. Hsia, J. Mol. Struct. 14, 117 (1972).
7. H. M. Rosenstock, K. Draxl, et al, J. Phys. Chem. Ref. Data 6, Suppl. 1 (1977).
8. A. E. Douglas and M. Frackowiak, Can. J. Phys. 40, 832 (1962).

Phosphorus Fluoride, Ion (PF^+) $F_1P_1^+(g)$

Enthalpy Reference Temperature = T_r = 298.15 K — Standard State Pressure = p° = 0.1 MPa

T/K	$C_p°$ (J K^{-1}mol^{-1})	S° (J K^{-1}mol^{-1})	-[G°-H°(T_r)]/T (J K^{-1}mol^{-1})	H°-H°(T_r) (kJ mol^{-1})	$\Delta_f H°$ (kJ mol^{-1})	$\Delta_f G°$ (kJ mol^{-1})	Log K_f
0	0.	0.	INFINITE	-9.430	895.664		
100	30.783	188.512	253.369	-6.486			
200	33.019	210.776	227.079	-3.261			
250	33.221	218.167	224.584	-1.604			
298.15	33.428	224.035	224.035	0.	901.518	870.945	-152.586
300	33.438	224.242	224.036	0.062	901.545	870.755	-151.612
350	33.734	229.418	224.444	1.741	901.533	865.638	-129.189
400	34.078	233.945	225.354	3.436	902.136	860.469	-112.366
450	34.431	237.979	226.537	5.149	902.739	855.225	-99.272
500	34.767	241.624	227.866	6.879	903.343	849.913	-88.790
600	35.352	248.017	230.707	10.386	904.559	839.113	-73.051
700	35.812	253.502	233.581	13.945	905.788	828.108	-61.794
800	36.168	258.308	236.377	17.545	907.029	816.927	-53.340
900	36.446	262.585	239.056	21.176	908.279	805.589	-46.755
1000	36.666	266.437	241.605	24.832	909.536	794.112	-41.480
1100	36.843	269.940	244.024	28.508	910.798	782.508	-37.158
1200	36.989	273.152	246.319	32.200	848.491	771.868	-33.599
1300	37.111	276.118	248.499	35.905	850.533	765.400	-30.754
1400	37.215	278.872	250.571	39.621	852.570	758.775	-28.310
1500	37.305	281.443	252.545	43.347	854.602	752.004	-26.187
1600	37.383	283.853	254.427	47.082	856.631	745.098	-24.325
1700	37.454	286.121	256.225	50.824	858.655	738.065	-22.678
1800	37.517	288.264	257.946	54.572	860.676	730.913	-21.211
1900	37.574	290.294	259.596	58.327	862.694	723.649	-19.895
2000	37.627	292.223	261.179	62.087	864.711	716.279	-18.707
2100	37.676	294.060	262.702	65.852	866.728	708.808	-17.631
2200	37.722	295.814	264.167	69.622	868.746	701.240	-16.650
2300	37.765	297.491	265.580	73.397	870.768	693.581	-15.752
2400	37.806	299.099	266.943	77.175	872.794	685.834	-14.927
2500	37.845	300.644	268.260	80.958	874.827	678.002	-14.166
2600	37.883	302.129	269.535	84.744	876.868	670.089	-13.462
2700	37.919	303.559	270.769	88.534	878.920	662.097	-12.809
2800	37.954	304.939	271.964	92.328	880.982	654.029	-12.201
2900	37.987	306.271	273.125	96.125	883.058	645.887	-11.634
3000	38.020	307.559	274.251	99.925	885.148	637.673	-11.103
3100	38.053	308.807	275.346	103.729	887.252	629.389	-10.605
3200	38.084	310.015	276.410	107.536	889.373	621.037	-10.137
3300	38.115	311.188	277.447	111.346	891.511	612.619	-9.697
3400	38.145	312.326	278.456	115.159	893.666	604.135	-9.281
3500	38.176	313.432	279.439	118.975	895.839	595.588	-8.889
3600	38.205	314.508	280.399	122.794	898.031	586.978	-8.517
3700	38.235	315.555	281.335	126.616	900.241	578.308	-8.164
3800	38.264	316.575	282.249	130.441	902.470	569.577	-7.829
3900	38.293	317.570	283.142	134.269	904.717	560.787	-7.511
4000	38.322	318.539	284.015	138.099	906.983	551.940	-7.208
4100	38.351	319.486	284.868	141.933	909.268	543.035	-6.918
4200	38.380	320.411	285.703	145.769	911.571	534.075	-6.642
4300	38.409	321.314	286.521	149.609	913.893	525.059	-6.378
4400	38.438	322.197	287.322	153.451	916.232	515.990	-6.126
4500	38.468	323.061	288.107	157.297	918.590	506.867	-5.884
4600	38.498	323.907	288.876	161.145	920.965	497.691	-5.651
4700	38.528	324.736	289.630	164.996	923.357	488.464	-5.429
4800	38.559	325.547	290.370	168.851	925.766	479.186	-5.215
4900	38.591	326.342	291.096	172.708	928.191	469.857	-5.009
5000	38.622	327.122	291.809	176.569	930.633	460.478	-4.811
5100	38.655	327.887	292.509	180.433	933.091	451.051	-4.620
5200	38.688	328.638	293.196	184.300	935.565	441.575	-4.436
5300	38.722	329.376	293.872	188.170	938.054	432.051	-4.258
5400	38.757	330.100	294.536	192.044	940.558	422.481	-4.087
5500	38.793	330.811	295.189	195.922	943.076	412.863	-3.921
5600	38.830	331.511	295.832	199.803	945.609	403.201	-3.761
5700	38.867	332.198	296.464	203.688	948.157	393.492	-3.606
5800	38.906	332.875	297.085	207.576	950.718	383.739	-3.456
5900	38.946	333.540	297.698	211.469	953.293	373.941	-3.311
6000	38.987	334.195	298.301	215.366	955.882	364.100	-3.170

PREVIOUS: June 1977 (1 atm) CURRENT: June 1977 (1 bar)

PHOSPHORUS FLUORIDE, ION (PF^-) IDEAL GAS M_r = 49.972712

D_0^0 = 222.170 ± 71.1 k J mol^{-1}
$S°(298.15\ K)$ = [225.218 ± 2.1] J K^{-1} mol^{-1}

$\Delta_fH°(0\ K)$ = -157.592 ± 69.0 kJ mol^{-1}
$\Delta_fH°(298.15\ K)$ = [-164.046 ± 69.0] kJ mol^{-1}

Electronic Levels and Quantum Weights

State	ε_i, cm^{-1}	g_i
$[^2\Pi_{3/2}]$	0	[2]
	[350]	[2]
	[25000]	[4]

ω_e = [840] cm^{-1} ω_ex_e = [4.6] cm^{-1} σ = 1
B_e = [0.56626] cm^{-1} α_e = [0.00424] cm^{-1} r_e = [1.59] Å

Enthalpy of Formation

We calculate the enthalpy of formation at 0 K as the difference between $\Delta_fH°(PF, g, 0\ K)$ = -8.55±5.0 kcal mol^{-1} (1) and the selected value of 1.1±0.5 eV (25.366 kcal mol^{-1}) for the electron affinity (EA) of PF. The value of EA refers to the vertical electron detachment process $PF^-(g) = PF(g) + e^-$ and is taken from the molecular orbital study of O'Hare (2). This value was obtained from Hartree-Fock energies and estimated corrections for correlation effects. The estimated uncertainty in EA is 0.5 eV which should be adequate to cover the possibility that the adiabatic value is lower than the vertical EA. Other theoretical predictions of EA include 2.55 eV (3) and 1.4 eV (4).

$\Delta_fH°$ at 298.15 K is -35.035 kcal mol^{-1}, and the primary bond dissociation energy (Products P(g) + F^-(g)) is 53.1 kcal mol^{-1}. The latter value can be compared with $D_0^0(PF)$ = 106.1 kcal mol^{-1} (1), indicating that the P-F bond is much weaker in the anion relative to the neutral molecule. This is consistent with MO predictions (2-4) which indicate that the electron captured by PF occupies an antibonding orbital located primarily on the phosphorus atom.

Our thermal functions indicate that PF^- is thermodynamically stable with respect to dissociation (P + F^-) below 2500 K. These predictions agree qualitatively with the results of MacNeil and Thynne (5) who observed PF^- in the negative ion mass spectrum of PF_3. Their reported appearance potential, $AP(PF^-/PF_3)$ = 11.4±0.1 eV, gives EA(PF) = -0.47 eV assuming that the dissociative electron capture process is $PF_3(g) + e^- = PF^-(g) + 2F(g)$. We believe this value is too low probably because of excess kinetic and/or excitation energies amounting to roughly 1.5 eV.

Heat Capacity and Entropy

The bond length and vibrational constants are assumed to lie between those for PF and SF (1). The value of B_e is obtained from r_e. We calculate α_e from the other constants assuming a Morse potential function.

The ground state electronic configuration ($^2\Pi_{3/2}$), ground state splitting (A = 350 cm^{-1}), and excited state ($^2\Pi$) at 25000 cm^{-1} are estimated by analogy with those for the isoelectronic molecule SF (1). The major source of error in our calculations below 500 K arises from the uncertainty in the ground state splitting. With A = 0 cm^{-1}, the value of S°(298.15 K) increases by 0.5 cal K^{-1} mol^{-1}. At higher temperatures the differences in S become smaller, indicating that the uncertainty in the adopted entropies above 1000 K is less than 0.2 cal K^{-1} mol^{-1}.

References

1. JANAF Thermochemical Tables: PF(g), 6-30-77; SF(g), 6-30-76.
2. P. A. G. O'Hare, J. Chem. Phys. 59, 3842 (1973).
3. A. L. Companion and Y. P. Hsia, J. Mol. Struct. 14, 117 (1972).
4. P. A. G. O'Hare and A. C. Wahl, J. Chem. Phys. 54, 4563 (1971).
5. K. A. G. MacNeil and J. C. J. Thynne, J. Phys. Chem. 74, 2257 (1970).

Phosphorus Fluoride, Ion (PF^-) $F_1P_1^-(g)$

Enthalpy Reference Temperature = T_r = 298.15 K Standard State Pressure = p° = 0.1 MPa

T/K	$C_p°$ (J $K^{-1}mol^{-1}$)	S° (J $K^{-1}mol^{-1}$)	$-[G°-H°(T_r)]/T$ (J $K^{-1}mol^{-1}$)	$H°-H°(T_r)$ (kJ mol^{-1})	$\Delta_fH°$ (kJ mol^{-1})	$\Delta_fG°$ (kJ mol^{-1})	Log K_f
0	0.	0.	INFINITE	-9.516	-157.592		
100	30.469	189.364	255.166	-6.580			
200	33.518	211.573	228.364	-3.358			
250	34.258	219.137	225.788	-1.663			
298.15	34.786	225.218	225.218	0.	-164.046	-182.462	31.967
300	34.804	225.433	225.218	0.064	-164.093	-182.576	31.789
350	35.242	230.832	225.644	1.816	-166.111	-185.484	27.682
400	35.603	235.562	226.594	3.587	-167.511	-188.156	24.571
450	35.903	239.774	227.829	5.375	-168.912	-190.652	22.130
500	36.152	243.570	229.216	7.177	-170.315	-192.992	20.162
600	36.535	250.197	232.177	10.812	-173.128	-197.264	17.173
700	36.810	255.850	235.165	14.480	-175.947	-201.064	15.004
800	37.016	260.780	238.065	18.172	-178.771	-204.459	13.350
900	37.175	265.149	240.836	21.882	-181.599	-207.500	12.043
1000	37.302	269.073	243.467	25.606	-184.432	-210.226	10.981
1100	37.407	272.633	245.959	29.341	-187.267	-212.669	10.099
1200	37.496	275.892	248.320	33.087	-253.678	-213.774	9.305
1300	37.574	278.896	250.558	36.840	-255.746	-210.365	8.453
1400	37.643	281.683	252.683	40.601	-257.821	-206.797	7.716
1500	37.705	284.283	254.704	44.369	-259.904	-203.079	7.072
1600	37.763	286.718	256.629	48.142	-261.994	-199.223	6.504
1700	37.816	289.009	258.467	51.921	-264.090	-195.235	5.999
1800	37.866	291.172	260.224	55.705	-266.191	-191.124	5.546
1900	37.914	293.220	261.908	59.494	-268.295	-186.897	5.138
2000	37.960	295.166	263.522	63.288	-270.402	-182.558	4.768
2100	38.004	297.019	265.074	67.086	-272.509	-178.114	4.430
2200	38.046	298.788	266.566	70.889	-274.615	-173.570	4.121
2300	38.088	300.481	268.004	74.695	-276.719	-168.930	3.837
2400	38.128	302.102	269.391	78.506	-278.817	-164.198	3.574
2500	38.168	303.660	270.731	82.321	-280.909	-159.379	3.330
2600	38.208	305.157	272.027	86.140	-282.993	-154.477	3.103
2700	38.247	306.600	273.281	89.963	-285.066	-149.495	2.892
2800	38.286	307.992	274.496	93.789	-287.128	-144.436	2.694
2900	38.325	309.336	275.674	97.620	-289.176	-139.303	2.509
3000	38.364	310.636	276.818	101.454	-291.209	-134.101	2.335
3100	38.404	311.895	277.929	105.293	-293.227	-128.831	2.171
3200	38.444	313.114	279.010	109.135	-295.228	-123.495	2.016
3300	38.484	314.298	280.061	112.981	-297.211	-118.098	1.869
3400	38.526	315.448	281.085	116.832	-299.175	-112.641	1.731
3500	38.568	316.565	282.083	120.687	-301.121	-107.126	1.599
3600	38.611	317.652	283.056	124.545	-303.046	-101.556	1.474
3700	38.656	318.710	284.005	128.409	-304.952	-95.933	1.354
3800	38.702	319.742	284.932	132.277	-306.838	-90.259	1.241
3900	38.750	320.748	285.838	136.149	-308.703	-84.535	1.132
4000	38.799	321.730	286.723	140.027	-310.547	-78.763	1.029
4100	38.849	322.688	287.588	143.909	-312.371	-72.946	0.929
4200	38.902	323.625	288.435	147.797	-314.174	-67.084	0.834
4300	38.956	324.541	289.264	151.689	-315.956	-61.180	0.743
4400	39.012	325.437	290.076	155.588	-317.718	-55.235	0.656
4500	39.070	326.315	290.872	159.492	-319.459	-49.250	0.572
4600	39.130	327.174	291.652	163.402	-321.180	-43.226	0.491
4700	39.191	328.016	292.417	167.318	-322.880	-37.165	0.413
4800	39.255	328.842	293.167	171.240	-324.560	-31.068	0.338
4900	39.320	329.652	293.903	175.169	-326.220	-24.937	0.266
5000	39.388	330.447	294.626	179.104	-327.861	-18.771	0.196
5100	39.457	331.228	295.336	183.047	-329.482	-12.574	0.129
5200	39.528	331.995	296.034	186.996	-331.084	-6.344	0.064
5300	39.601	332.748	296.719	190.952	-332.666	-0.084	0.001
5400	39.675	333.489	297.394	194.916	-334.229	6.206	-0.060
5500	39.751	334.218	298.057	198.887	-335.774	12.524	-0.119
5600	39.828	334.935	298.709	202.866	-337.300	18.871	-0.176
5700	39.907	335.640	299.350	206.853	-338.808	25.245	-0.231
5800	39.988	336.335	299.982	210.848	-340.298	31.644	-0.285
5900	40.069	337.019	300.604	214.850	-341.770	38.070	-0.337
6000	40.152	337.694	301.217	218.862	-343.224	44.520	-0.388

PREVIOUS: June 1977 (1 atm) CURRENT: June 1977 (1 bar)

Phosphorus Fluoride, Ion (PF^-) $F_1P_1^-(g)$

THIOPHOSPHORYL FLUORIDE (PSF) IDEAL GAS M_r = 82.032163

Thiophosphoryl Fluoride (PSF) $F_1P_1S_1(g)$

$S°(298.15\ K) = [277.717 \pm 12.6]\ J\ K^{-1}\ mol^{-1}$

$\Delta_f H°(0\ K) = -169.189 \pm 62.8\ kJ\ mol^{-1}$

$\Delta_f H°(298.15\ K) = [-172.268 \pm 62.8]\ kJ\ mol^{-1}$

Vibrational Frequencies and Degeneracies

ν, cm^{-1}
[940](1)
[740](1)
[400](1)

Ground State Quantum Weight: 1 σ = 1

Point Group: C_s

Product of the Moments of Inertia: $I_A I_B I_C = [7.999 \times 10^{-114}]\ g^3\ cm^6$

Enthalpy of Formation

The $\Delta_f H°(298.15\ K)$ was estimated by Henderson and Scheffee (1).

Heat Capacity and Entropy

The vibrational frequencies were estimated by comparison with the corresponding frequencies of PF(g), PS(g), and $PF_2(g)$. The vibrational levels of $PF_2(g)$ in turn were estimated by Gordon (2). The entropy was estimated by Henderson and Scheffee (1). The product of the moment of inertia was derived from the estimated entropy.

References

1. C. B. Henderson and R. S. Scheffee, "Survey of Thermochemical Data," Atlantic Research Corp., Alexandria, Virginia, January, 1960.
2. J. S. Gordon, "Thermodynamic Data for Combustion Products," Thiokol Corp., Reaction Motors Division, Denville, New Jersey, January, 1960.

Enthalpy Reference Temperature = T_r = 298.15 K; Standard State Pressure = p° = 0.1 MPa

T/K	$C_p°$ (J K⁻¹mol⁻¹)	S° (J K⁻¹mol⁻¹)	$-[G°-H°(T_r)]/T$ (J K⁻¹mol⁻¹)	$H°-H°(T_r)$ (kJ mol⁻¹)	$\Delta_f H°$ (kJ mol⁻¹)	$\Delta_f G°$ (kJ mol⁻¹)	Log K_f
0	0.	0.	INFINITE	-11.106	-169.189	-169.189	INFINITE
100	34.160	235.944	313.591	-7.765	-168.879	-180.934	94.510
200	39.208	261.056	281.619	-4.113	-170.596	-192.590	50.299
250	41.993	270.107	278.435	-2.082	-171.428	-197.993	41.368
298.15	44.439	277.717	277.717	0.	-172.268	-203.034	35.571
300	44.526	277.992	277.718	0.082	-172.301	-203.224	35.385
350	46.695	285.023	278.268	2.364	-173.915	-208.237	31.078
400	48.493	291.380	279.516	4.746	-177.096	-212.986	27.813
450	49.962	297.179	281.161	7.208	-178.709	-217.384	25.233
500	51.159	302.508	283.033	9.737	-180.348	-221.590	23.149
600	52.936	312.003	287.090	14.948	-183.091	-229.573	19.986
700	54.148	320.259	291.251	20.306	-185.489	-237.127	17.695
800	54.999	327.549	295.342	25.765	-187.677	-244.352	15.955
900	55.615	334.064	299.289	31.298	-242.862	-250.231	14.523
1000	56.073	339.949	303.065	36.883	-243.610	-251.010	13.111
1100	56.421	345.310	306.666	42.509	-244.348	-251.714	11.953
1200	56.692	350.231	310.094	48.165	-308.657	-251.273	10.938
1300	56.906	354.778	313.358	53.845	-308.621	-246.492	9.904
1400	57.078	359.002	316.470	59.545	-308.598	-241.714	9.018
1500	57.218	362.944	319.438	65.260	-308.588	-236.937	8.251
1600	57.334	366.641	322.274	70.988	-308.593	-232.160	7.579
1700	57.430	370.120	324.987	76.726	-308.614	-227.383	6.987
1800	57.512	373.405	327.586	82.473	-308.649	-222.604	6.460
1900	57.581	376.516	330.081	88.228	-308.699	-217.822	5.988
2000	57.640	379.471	332.477	93.989	-308.761	-213.037	5.564
2100	57.692	382.285	334.782	99.756	-308.834	-208.250	5.180
2200	57.736	384.970	337.003	105.527	-308.915	-203.458	4.831
2300	57.775	387.537	339.145	111.303	-309.003	-198.662	4.512
2400	57.810	389.997	341.213	117.082	-309.095	-193.863	4.219
2500	57.840	392.357	343.211	122.864	-309.189	-189.060	3.950
2600	57.867	394.626	345.146	128.650	-309.284	-184.253	3.702
2700	57.891	396.811	347.019	134.438	-309.375	-179.442	3.472
2800	57.912	398.916	348.835	140.228	-309.462	-174.628	3.258
2900	57.932	400.949	350.597	146.020	-309.544	-169.811	3.059
3000	57.949	402.913	352.309	151.814	-309.618	-164.992	2.873
3100	57.965	404.814	353.972	157.610	-309.683	-160.170	2.699
3200	57.980	406.654	355.589	163.407	-309.738	-155.346	2.536
3300	57.993	408.439	357.164	169.206	-309.782	-150.520	2.383
3400	58.005	410.170	358.698	175.006	-309.815	-145.694	2.238
3500	58.016	411.852	360.193	180.807	-309.835	-140.866	2.102
3600	58.026	413.486	361.650	186.609	-309.844	-136.038	1.974
3700	58.035	415.076	363.073	192.412	-309.840	-131.210	1.852
3800	58.044	416.624	364.462	198.216	-309.824	-126.383	1.737
3900	58.052	418.132	365.819	204.021	-309.795	-121.556	1.628
4000	58.059	419.601	367.145	209.826	-309.755	-116.730	1.524
4100	58.066	421.035	368.442	215.632	-309.702	-111.905	1.426
4200	58.072	422.435	369.711	221.439	-309.638	-107.081	1.332
4300	58.078	423.801	370.953	227.247	-309.564	-102.259	1.242
4400	58.083	425.136	372.169	233.055	-309.479	-97.439	1.157
4500	58.089	426.442	373.361	238.863	-309.384	-92.621	1.075
4600	58.093	427.718	374.529	244.672	-309.279	-87.805	0.997
4700	58.098	428.968	375.674	250.482	-309.166	-82.992	0.922
4800	58.102	430.191	376.797	256.292	-309.045	-78.181	0.851
4900	58.106	431.389	377.899	262.102	-308.916	-73.372	0.782
5000	58.110	432.563	378.980	267.913	-308.781	-68.567	0.716
5100	58.113	433.714	380.042	273.724	-308.638	-63.764	0.653
5200	58.117	434.842	381.085	279.536	-308.490	-58.964	0.592
5300	58.120	435.949	382.110	285.348	-308.337	-54.167	0.534
5400	58.123	437.036	383.117	291.160	-308.179	-49.372	0.478
5500	58.126	438.102	384.107	296.972	-308.017	-44.581	0.423
5600	58.128	439.150	385.081	302.785	-307.851	-39.793	0.371
5700	58.131	440.179	386.039	308.598	-307.681	-35.008	0.321
5800	58.133	441.190	386.981	314.411	-307.508	-30.226	0.272
5900	58.135	442.183	387.908	320.225	-307.333	-25.446	0.225
6000	58.138	443.160	388.821	326.038	-307.156	-20.670	0.180

PREVIOUS: September 1962 (1 atm) CURRENT: September 1962 (1 bar)

Thiophosphoryl Fluoride (PSF) $F_1P_1S_1(g)$

LEAD FLUORIDE (PbF) IDEAL GAS M_r = 226.198403

$S°(298.15 K) = 249.957 \pm 0.21 J K^{-1} mol^{-1}$

$\Delta_f H°(0 K) = -78.243 \pm 10.5 kJ mol^{-1}$
$\Delta_f H°(298.15 K) = -80.266 \pm 10.5 kJ mol^{-1}$

Electronic Levels and Quantum Weights

State	ε_i, cm^{-1}	g_i
$X\ ^2\Pi_{1/2}$	0	2
$^2\Pi_{3/2}$	8264.7	2
$A\ [^2\Sigma]$	22566.6	[2]
$B\ ^2\Sigma^+$	35643	2

$\omega_e = 507.2 cm^{-1}$ $\quad \omega_e x_e = 2.30 cm^{-1}$ $\quad \sigma = 1$
$B_e = 0.2288 cm^{-1}$ $\quad \alpha_e = [0.0013] cm^{-1}$ $\quad r_e = 2.058$ Å

Enthalpy of Formation

Rosen (1) lists $D° = 84.0\pm2 kcal mol^{-1}$ derived from the mass spectrometric work of Zmbov, et al. (2). The partial pressures and log K for the $PbF_2(g) + Pb(g) = 2PbF(g)$ reaction in the latter reference are inconsistent. 3rd law calculations show that the log K value corresponds to the heat of reaction given. The dissociation energy calculated by the linear Birge-Sponer extrapolation of the adopted ground state vibrational constants, as derived from emission and absorption spectra (ν, 0-13) (3, 4), is 3.44 eV. Correcting this value for the ionic character of PbF as suggested by Hildenbrand (5) gives D_0^0 = 3.64 eV ($83.9 kcal mol^{-1}$). The upper limit of the value recommended by Gaydon (6) (3±0.5 eV) falls at $80.7 kcal mol^{-1}$. An intercomparison of the dissociation energies of the lead (7), mercury (1), and potassium (7) monohalides indicates that D_0^0 = $84.0\pm2 kcal mol^{-1}$ is reasonable. Combining this with $\Delta_f H°(Pb, g, 0 K) = 46.91\pm0.13 kcal mol^{-1}$ (7) and $\Delta_f H°(F, g, 0 K) = 18.36\pm0.40 kcal mol^{-1}$ (7) gives $\Delta_f H°(PbF, g, 0 K) = -18.73\pm2.5 kcal mol^{-1}$. We adopt $\Delta_f H°(PbF, g, 0 K) = -18.70\pm2.5 kcal mol^{-1}$ which is consistent with the above. $\Delta_f H°(PbF, g, 298.15 K) = -19.18\pm2.5 kcal mol^{-1}$ is calculated from $\Delta_f H°(0 K)$.

Heat Capacity and Entropy

The vibrational constants, the electronic states and levels, and the rotational constant are those recommended in Rosen (1) with the following modifications. The first excited state, unassigned in Rosen (1), is tentatively classed as a $^2\Sigma$ state in accordance with the preference for $^2\Sigma$ over $^2\Delta$ expressed by Barrow et al. (8). If the state is $^2\Delta$ rather than $^2\Sigma$, the entropy is increased by $0.005 cal K^{-1} mol^{-1}$ at 4000 K and $0.05 cal K^{-1} mol^{-1}$ at 6000 K. The C and higher electronic states (1) have been omitted; inclusion increases the entropy only by approximately $0.005 cal K^{-1} mol^{-1}$ at 6000 K. The adopted B_e = $0.2288 cm^{-1}$ is obtained by an approximate correction to $B_0 = 0.2281 cm^{-1}$ given in Rosen (1), and r_e = 2.058 Å calculated from B_e compares with r_0 = 2.061 Å given in Rosen (1). The value for α_e is calculated assuming a Morse potential function.

References

1. B. Rosen, Ed., "Spectroscopic Data Relative to Diatomic Molecules," Pergamon Press, New York, 1970.
2. K. Zmbov, J. W. Hastie, and J. L. Margrave, Trans. Faraday Soc. 64, 861 (1968).
3. G. D. Rochester, Proc. Roy. Soc. (London) A 153, 407 (1936).
4. G. D. Rochester, proc. Roy. Soc. (London) A 167, 569 (1938).
5. D. L. Hildenbrand in "Advances in High Temperature Chemistry," Vol. 1, L. Eyring, Ed., Academic Press, New York, pp. 193-215, 1967.
6. A. G. Gaydon, "Dissociation Energies and Spectra of Diatomic Molecules," 3rd ed., Chapman and Hall, London, 1968.
7. JANAF Thermochemical Tables: Pb(g), 3-31-62; F(g), 9-30-65; PbCl(g), 6-30-73; PbBr(g), 12-31-73; PbI(g), 12-31-73; KF(g), 6-30-69; KCl(g), 3-31-66; KBr(g), 3-31-67; KI(g), 6-30-67.
8. R. F. Barrow, D. Butler, J. W. C. Johns, and J. L. Powell, Proc. Phys. Soc. (London) 73, 317 (1959).

Lead Fluoride (PbF) $F_1Pb_1(g)$

Enthalpy Reference Temperature = T_r = 298.15 K Standard State Pressure = p° = 0.1 MPa

	J K⁻¹mol⁻¹			kJ mol⁻¹			
T/K	$C_p°$	S°	$-[G°-H°(T_r)]/T$	$H°-H°(T_r)$	$\Delta_f H°$	$\Delta_f G°$	Log K_f
0	0.	0.	INFINITE	-9.268	-78.243	-78.243	INFINITE
100	29.426	215.448	278.991	-6.354	-78.549	-87.893	45.910
200	32.232	236.643	253.043	-3.280	-79.464	-96.873	25.301
250	33.491	243.976	250.519	-1.636	-79.876	-101.177	21.140
298.15	34.399	249.957	249.957	0.	-80.266	-105.244	18.438
300	34.429	250.170	249.958	0.064	-80.281	-105.399	18.352
350	35.115	255.531	250.380	1.803	-80.689	-109.553	16.350
400	35.622	260.255	251.325	3.572	-81.111	-113.648	14.841
450	36.004	264.474	252.555	5.363	-81.550	-117.689	13.661
500	36.298	268.283	253.941	7.171	-82.009	-121.680	12.712
600	36.717	274.941	256.902	10.823	-82.982	-129.524	11.276
700	36.998	280.623	259.895	14.510	-88.893	-136.401	10.178
800	37.201	285.578	262.802	18.220	-90.005	-143.112	9.344
900	37.357	289.969	265.581	21.949	-91.089	-149.685	8.687
1000	37.486	293.911	268.220	25.691	-92.146	-156.138	8.156
1100	37.601	297.490	270.721	29.445	-93.176	-162.488	7.716
1200	37.711	300.766	273.090	33.211	-94.178	-168.744	7.345
1300	37.823	303.789	275.337	36.988	-95.156	-174.918	7.028
1400	37.943	306.596	277.471	40.776	-96.126	-181.017	6.754
1500	38.075	309.219	279.501	44.577	-97.095	-187.047	6.514
1600	38.219	311.680	281.436	48.391	-98.067	-193.012	6.301
1700	38.378	314.002	283.284	52.221	-99.045	-198.917	6.112
1800	38.551	316.201	285.052	56.067	-100.030	-204.763	5.942
1900	38.737	318.290	286.747	59.932	-101.022	-210.555	5.789
2000	38.933	320.282	288.374	63.815	-102.019	-216.294	5.649
2100	39.138	322.186	289.939	67.719	-280.241	-214.868	5.345
2200	39.350	324.012	291.447	71.643	-280.854	-211.741	5.027
2300	39.567	325.766	292.901	75.589	-281.533	-208.584	4.737
2400	39.785	327.454	294.306	79.557	-282.279	-205.397	4.470
2500	40.004	329.083	295.664	83.546	-283.090	-202.177	4.224
2600	40.222	330.656	296.980	87.557	-283.963	-198.923	3.996
2700	40.436	332.178	298.256	91.590	-284.895	-195.635	3.785
2800	40.646	333.652	299.494	95.644	-285.880	-192.311	3.588
2900	40.850	335.082	300.696	99.719	-286.915	-188.951	3.403
3000	41.049	336.471	301.866	103.814	-287.991	-185.555	3.231
3100	41.240	337.820	303.004	107.929	-289.103	-182.122	3.069
3200	41.425	339.132	304.113	112.062	-290.244	-178.653	2.916
3300	41.602	340.409	305.193	116.213	-291.406	-175.148	2.772
3400	41.771	341.654	306.247	120.382	-292.584	-171.607	2.636
3500	41.933	342.867	307.276	124.567	-293.770	-168.032	2.508
3600	42.087	344.051	308.282	128.768	-294.958	-164.422	2.386
3700	42.234	345.206	309.264	132.985	-296.142	-160.780	2.270
3800	42.374	346.334	310.225	137.215	-297.316	-157.106	2.160
3900	42.508	347.436	311.165	141.459	-298.476	-153.401	2.055
4000	42.635	348.514	312.085	145.716	-299.616	-149.667	1.954
4100	42.756	349.568	312.986	149.986	-300.733	-145.904	1.859
4200	42.872	350.600	313.870	154.268	-301.823	-142.114	1.767
4300	42.983	351.610	314.736	158.560	-302.882	-138.299	1.680
4400	43.088	352.600	315.585	162.864	-303.909	-134.460	1.596
4500	43.189	353.569	316.418	167.178	-304.900	-130.597	1.516
4600	43.286	354.519	317.236	171.502	-305.855	-126.714	1.439
4700	43.379	355.451	318.040	175.835	-306.771	-122.809	1.365
4800	43.469	356.366	318.829	180.177	-307.648	-118.886	1.294
4900	43.555	357.263	319.604	184.529	-308.485	-114.944	1.225
5000	43.639	358.144	320.366	188.888	-309.281	-110.987	1.159
5100	43.720	359.009	321.115	193.256	-310.038	-107.013	1.096
5200	43.798	359.858	321.852	197.632	-310.753	-103.025	1.035
5300	43.874	360.693	322.577	202.016	-311.429	-99.024	0.976
5400	43.949	361.514	323.291	206.407	-312.066	-95.010	0.919
5500	44.021	362.321	323.993	210.805	-312.663	-90.985	0.864
5600	44.092	363.115	324.684	215.211	-313.223	-86.949	0.811
5700	44.162	363.896	325.365	219.624	-313.746	-82.904	0.760
5800	44.230	364.665	326.036	224.043	-314.234	-78.850	0.710
5900	44.297	365.421	326.698	228.470	-314.686	-74.788	0.662
6000	44.363	366.166	327.349	232.903	-315.105	-70.718	0.616

PREVIOUS: December 1973 (1 atm) CURRENT: December 1973 (1 bar)

Lead Fluoride (PbF) $F_1Pb_1(g)$

SULFUR FLUORIDE (SF) IDEAL GAS M_r = 51.058403

$S°(298.15\ K) = [225.275 \pm 0.8]\ J\ K^{-1}\ mol^{-1}$

$\Delta_fH°(0\ K) = 12.316 \pm 6.3\ kJ\ mol^{-1}$
$\Delta_fH°(298.15\ K) = 12.970 \pm 6.3\ kJ\ mol^{-1}$

Electronic Levels and Quantum Weights

State	ε_i, cm^{-1}	g_i
$X\ ^2\Pi_{3/2}$	0	2
$^2\Pi_{1/2}$	398	2
$A\ ^2\Pi_{3/2}$	24991	2
$^2\Pi_{1/2}$	25601	2

$\omega_e = [830 \pm 20]\ cm^{-1}$ $\omega_ex_e = [4.7 \pm 0.4]\ cm^{-1}$ $\sigma = 1$

$B_e = 0.55427 \pm 0.0005\ cm^{-1}$ $\alpha_e = [0.0042]\ cm^{-1}$ $r_e = 1.5967 \pm 0.001$ Å

Enthalpy of Formation

Hildenbrand (1) has investigated the isomolecular reaction $CS_2(g) + 2SF(g) = CF_2(g) + 2S_2(g)$ by mass spectrometry. The various molecular species were found to be formed as products of the reaction of gaseous SF_6 with graphite at temperatures in the range 1436-1611 K. This study employed three different effusion cell configurations which were used to optimize the reaction conditions, and ion abundances for each species were measured at 2 eV above their appearance potentials in order to eliminate fragmentation effects. We analyze the reported equilibrium data by the 2nd and 3rd law methods with the results being presented below.

Series	Cell Configuration	Data Points	T/K	$\Delta_rH°$(298.15 K), kcal mol^{-1} 2nd law	3rd law	Drift cal K^{-1} mol^{-1}	$\Delta_fH°$(298.15 K) kcal mol^{-1}
1	Mo cell/C liner/packed	2	1529-1611	-11.6	-16.2	-2.9	3.2
	with C cloth				±0.3		±1.2
2	C cell/wound Pt wire	4	1436-1564	6.8	-16.0	-15.1	3.1
					±1.8	±5.0	±2.0
3	C cell/Pt partition/packed	7	1478-1588	-37.2	-15.8	13.9	3.0
	with C cloth and Pt wire				±1.4	±5.0	±1.8

The three cell configurations used by Hildenbrand (1) yield almost identical 3rd-law $\Delta_fH°$ values. We adopt the mean value of 3.1±1.5 kcal mol^{-1}. This value corresponds to a D_0^0 of 81.2±2.0 kcal mol^{-1} which is close to the average bond energies (80.0 kcal mol^{-1}) for SF_4 and SF_3 (2). Other reported D_0^0 values include 96.9 kcal mol^{-1} (3) and $\leq$ 76.1 kcal mol^{-1} (4). The former value is based on Hartree-Foch binding energies (3) which were combined with an estimate of the extra molecular correlation energy. This estimate now appears to be about 20% too high. the upper limit value for D_0^0 was obtained from predissociation observed in the $A\ ^2\Pi$ state by Di Lonardo and Trombetti (4).

Heat Capacity and Entropy

The electronic states and levels (T_0) are taken from the recent compilation of Barrow (5). His results were derived from the spectral measurements of Di Lonardo and Trombetti (4). Further confirmation that the ground state is inverted $^2\Pi$ is provided by the results of an analysis of the gas-phase electron resonance spectra by Carrington et al. (6). The rotational constant (B_e) is calculated from the microwave data (B_{eff}) of Amano and Hirota (7). Other values for B_e determined from EPR (6) measurements and from a rotational analysis (4) of the $A\ ^2\Pi - X\ ^2\Pi$ band system agree with the microwave results but are less precise. The value of r_e is calculated from B_e.

The ground state vibrational constants have not been determined experimentally. O'Hare (8) computed Hartree-Fock total energies for the ground states of NF, SiF, PF, and SF and derived spectroscopic constants for each monofluoride by a Dunham analysis. He adjusted the computed values for SF for differences found between calculated and experimental data for the other three monofluorides. We adopt his predicted vibrational constants along with his computed value for α_e. We do not include the rotational and vibrational constants (4, 5) for the $A\ ^2\Pi$ state since they have a negligible effect.

Thermal functions have been previously reported for SF by Wilkins (9) and O'Hare (10). These two sets of functions agree reasonably well, and presumably they are both based on a ground state quantum weight of 2. Our entropies are consistently higher than the literature data (9, 10) at all temperatures by roughly 1.0-1.5 cal K^{-1} mol^{-1}. This increase can be attributed primarily to the entropy contribution from the $X\ ^2\Pi_{1/2}$ state which lies at 398 cm^{-1}.

References

1. D. L. Hildenbrand, J. Phys. Chem. 77, 897 (1973).
2. JANAF Thermochemical Tables: $CS_2(g)$, 6-30-61; $CF_2(g)$, 6-30-70; $S_2(g)$, 12-31-65; $SF_4(g)$ and $SF_3(g)$, 6-30-76.
3. P. A. G. O'Hare and A. C. Wahl, J. Chem. Phys. 53, 2834 (1970).
4. G. Di Lonardo and A. Trombetti, Trans. Faraday Soc. 66, 2694 (1970).
5. R. F. Barrow, "Diatomic Molecules: A Critical Bibliography of Spectroscopic Data," Centre National De La Recherche Scientifique, Paris, 1973.
6. A Carrington, G. N. Currie, T. A. Miller, and D. H. Levy, J. Chem. Phys. 50, 2726 (1969).
7. T. Amano and E. Hirota, J. Mol. Spectrosc. 45, 417 (1973).
8. P. A. G. O'Hare, J. Chem. Phys. 59, 3842 (1973).
9. R. L. Wilkins, Aerospace Corporation, Report No. TR-0158 (3240-20)-19, July, 1968.
10. P. A. G. O'Hare, Argonne National Laboratory, ANL-7315, July, 1968.

Sulfur Fluoride (SF) $F_1S_1(g)$

Enthalpy Reference Temperature = T_r = 298.15 K Standard State Pressure = p° = 0.1 MPa

T/K	$C_p°$ (J $K^{-1}mol^{-1}$)	S°	$-[G°-H°(T_r)]/T$	$H°-H°(T_r)$ (kJ mol^{-1})	$\Delta_fH°$	$\Delta_fG°$	Log K_f
0	0.	0.	INFINITE	-9.478	12.316	12.316	INFINITE
100	29.999	189.668	255.214	-6.555	13.097	3.901	-2.038
200	33.400	211.569	228.443	-3.375	13.170	-5.351	1.398
250	34.454	219.143	225.850	-1.677	13.098	-9.974	2.084
298.15	35.152	225.275	225.275	0.	12.970	-14.407	2.524
300	35.174	225.493	225.276	0.065	12.964	-14.577	2.538
350	35.683	230.955	225.706	1.837	12.776	-19.153	2.858
400	36.055	235.745	226.668	3.631	10.324	-23.611	3.083
450	36.337	240.009	227.917	5.441	9.375	-27.805	3.227
500	36.556	243.849	229.322	7.264	8.346	-31.878	3.330
600	36.872	250.544	232.317	10.936	6.697	-39.763	3.462
700	37.088	256.245	235.338	14.635	5.273	-47.391	3.536
800	37.246	261.209	238.269	18.352	3.974	-54.825	3.580
900	37.368	265.603	241.066	22.083	-50.379	-61.019	3.541
1000	37.466	269.545	243.720	25.825	-50.338	-62.204	3.249
1100	37.549	273.120	246.233	29.576	-50.319	-63.392	3.010
1200	37.620	276.390	248.612	33.334	-50.320	-64.580	2.811
1300	37.684	279.404	250.866	37.099	-50.342	-65.768	2.643
1400	37.741	282.199	253.006	40.871	-50.384	-66.953	2.498
1500	37.795	284.805	255.040	44.648	-50.446	-68.135	2.373
1600	37.844	287.245	256.977	48.430	-50.527	-69.311	2.263
1700	37.892	289.541	258.826	52.216	-50.625	-70.482	2.166
1800	37.937	291.708	260.593	56.008	-50.740	-71.647	2.079
1900	37.980	293.761	262.285	59.804	-50.870	-72.805	2.002
2000	38.023	295.710	263.908	63.604	-51.011	-73.956	1.932
2100	38.064	297.566	265.467	67.408	-51.163	-75.100	1.868
2200	38.104	299.338	266.966	71.217	-51.322	-76.236	1.810
2300	38.143	301.032	268.411	75.029	-51.486	-77.365	1.757
2400	38.182	302.657	269.804	78.845	-51.652	-78.486	1.708
2500	38.221	304.216	271.150	82.665	-51.818	-79.601	1.663
2600	38.259	305.716	272.451	86.489	-51.981	-80.709	1.621
2700	38.298	307.160	273.710	90.317	-52.138	-81.811	1.583
2800	38.336	308.554	274.929	94.149	-52.288	-82.907	1.547
2900	38.374	309.900	276.112	97.984	-52.429	-83.998	1.513
3000	38.413	311.201	277.260	101.824	-52.559	-85.085	1.481
3100	38.452	312.462	278.375	105.667	-52.676	-86.167	1.452
3200	38.491	313.683	279.460	109.514	-52.780	-87.246	1.424
3300	38.531	314.868	280.515	113.365	-52.868	-88.321	1.398
3400	38.573	316.019	281.542	117.220	-52.941	-89.394	1.373
3500	38.615	317.138	282.543	121.080	-52.998	-90.466	1.350
3600	38.658	318.226	283.520	124.943	-53.039	-91.535	1.328
3700	38.702	319.286	284.472	128.811	-53.061	-92.604	1.307
3800	38.748	320.319	285.402	132.684	-53.067	-93.673	1.288
3900	38.795	321.326	286.310	136.561	-53.055	-94.742	1.269
4000	38.843	322.308	287.198	140.443	-53.026	-95.811	1.251
4100	38.893	323.268	288.066	144.330	-52.980	-96.881	1.234
4200	38.945	324.206	288.915	148.222	-52.917	-97.953	1.218
4300	38.999	325.123	289.747	152.119	-52.836	-99.026	1.203
4400	39.055	326.020	290.561	156.021	-52.740	-100.101	1.188
4500	39.112	326.899	291.359	159.930	-52.627	-101.179	1.174
4600	39.171	327.759	292.141	163.844	-52.498	-102.259	1.161
4700	39.232	328.602	292.907	167.764	-52.354	-103.342	1.149
4800	39.295	329.429	293.660	171.690	-52.194	-104.429	1.136
4900	39.360	330.239	294.398	175.623	-52.020	-105.519	1.125
5000	39.427	331.035	295.123	179.562	-51.832	-106.613	1.114
5100	39.495	331.817	295.835	183.508	-51.629	-107.710	1.103
5200	39.565	332.584	296.534	187.461	-51.413	-108.812	1.093
5300	39.638	333.339	297.221	191.422	-51.184	-109.918	1.083
5400	39.711	334.080	297.897	195.389	-50.941	-111.028	1.074
5500	39.787	334.810	298.562	199.364	-50.686	-112.143	1.065
5600	39.863	335.527	299.215	203.346	-50.419	-113.263	1.056
5700	39.942	336.233	299.859	207.337	-50.140	-114.388	1.048
5800	40.022	336.929	300.492	211.335	-49.848	-115.518	1.040
5900	40.103	337.614	301.115	215.341	-49.546	-116.652	1.033
6000	40.185	338.288	301.729	219.355	-49.232	-117.793	1.025

PREVIOUS: June 1976 (1 atm) CURRENT: June 1976 (1 bar)

Sulfur Fluoride (SF) $F_1S_1(g)$

SULFUR FLUORIDE, ION (SF^+) IDEAL GAS M_r = 51.057854

$\Delta_f H°(0\ K)$ = 985.664 ± 16.7 kJ mol^{-1}

$S°(298.15\ K)$ = [225.405 ± 9.2] J K^{-1} mol^{-1} $\Delta_f H°(298.15\ K)$ = [991.901] mol^{-1}

Electronic Levels and Quantum Weights

State	ε_i, cm^{-1}	g_i
[X $^3\Sigma$]	0	[3]
[a $^1\Delta$]	[8000]	[2]
[b $^1\Sigma$]	[14500]	[1]
[B $^3\Pi$]	[30000]	[6]

ω_e = [838] cm^{-1} $\omega_e x_e$ = [4.6] cm^{-1} σ = 1

B_e = [0.55686] cm^{-1} α_e = [0.0044] cm^{-1} r_e = [1.593] Å

Enthalpy of Formation

The electron-impact appearance potential of SF^+ from SF has been measured by Hildenbrand (1) as 10.09±0.1 eV. Normally, this measured AP would be identical to the vertical ionization potential of SF; however, formation of SF^+ involves the loss of an antibonding electron which appears to be situated primarily on the sulfur atom (IP(S) = 10.36 eV, 2). Thus, the geometrical changes produced upon ionization should be minimal, and therefore, the measured AP is probably very close to the true (adiabatic) ionization potential, as suggested by Hildenbrand (1). We employ the experimental IP value as the heat of reaction, $\Delta_r H°$ = 232.68±2.3 kcal mol^{-1}, for the process $SF(g) + e^- = SF^+(g) + 2e^-$ at 0 K, and we calculate $\Delta_f H°(SF^+, g, 0\ K)$ = 235.6±4.0 kcal mol^{-1} by combining the value of $\Delta_r H°(0\ K)$ with $\Delta_f H°(SF, g, 0\ K)$ = 2.9±1.5 kcal mol^{-1} (3). Independent values of IP include 10.0 eV (4) and 9.9 eV (5). Both of these values were obtained from semiempirical molecular orbital calculations and are in excellent agreement with the results adopted here.

Earlier electron-impact studies which can also be used to derive $\Delta_f H°$ of SF^+ have been previously analyzed by O'Hare and Wahl (6). We do not reanalyze these results since their analysis (6) has shown that the data are seriously in error.

Heat Capacity and Entropy

As discussed in the enthalpy of formation section, one would expect that the bonding in SF^+ is probably not very different from that in SF, since the ionizing orbital is antibonding. Additional support for this is provided by the fact that the bond dissociation energies for SF^+ (87.4 kcal mol^{-1}) and SF (81.2 kcal mol^{-1}) are very similar. In addition, SF^+ is isoelectronic with PF, and one might therefore expect that the molecular characteristics of these two isoelectronic molecules would be quite similar. We estimate the vibrational and rotational constants to be intermediate between those for SF (3) and PF (7). By analogy with other twelve valence electron molecules (7), we predict that the ground state is $^3\Sigma$ and expect several excited states to exist below 35000 cm^{-1}. The two singlet levels are estimated from those observed for NF and O_2 (7). We also include a triplet level at 30000 cm^{-1} based on that observed for PF (7). We assume that the vibrational and rotational constants for the excited states are identical with those for the ground state since a comparison of the constants (7) that are available for the various states of NF and PF shows that they are not significantly different. The uncertainty in our value of S°(298.15 K) is estimated as ±2.2 cal K^{-1} mol^{-1} and arises primarily from the effect of the ground state quantum weight. If the SF^+ ground state is singlet, then our entropy value at 298.15 K should be decreased by 2.18 cal K^{-1} mol^{-1}. The electronic contribution to the entropy from the estimated excited states is negligible below 2000 K but increases to 0.4 cal K^{-1} mol^{-1} at 4500 K.

References

1. D. L. Hildenbrand, J. Phys. Chem. 77, 897 (1973).
2. J. L. Franklin, J. G. Dillard et al., U.S. Nat. Bur. Stand., NSRDS-NBS 26, 1969.
3. JANAF Thermochemical Tables: SF(g), 6-30-76.
4. A. L. Companion, Theoret. Chim. Acta, 25, 268 (1972).
5. P. A. G. O'Hare, J. Chem. Phys. 59, 3842 (1973).
6. P. A. G. O'Hare and A. C. Wahl, J. Chem. Phys. 53, 2834 (1970).
7. B. Rosen, ed., "Spectroscopic Data Relative to Diatomic Molecules", Pergamon Press, New York, 1970.

Sulfur Fluoride, Ion (SF^+) $F_1S_1^+(g)$

Enthalpy Reference Temperature = T_r = 298.15 K Standard State Pressure = p° = 0.1 MPa

T/K	$C_p°$ (J $K^{-1}mol^{-1}$)	S° (J $K^{-1}mol^{-1}$)	$-[G°-H°(T_r)]/T$ (J $K^{-1}mol^{-1}$)	$H°-H°(T_r)$ (kJ mol^{-1})	$\Delta_f H°$ (kJ mol^{-1})	$\Delta_f G°$ (kJ mol^{-1})	Log K_f
0	0.	0.	INFINITE	-8.864	985.664		
100	29.116	192.820	252.379	-5.956			
200	29.886	213.148	228.243	-3.019			
250	30.766	219.908	225.922	-1.503			
298.15	31.679	225.405	225.405	0.	991.901	958.230	-167.878
300	31.713	225.601	225.406	0.059	991.927	958.021	-166.806
350	32.590	230.557	225.795	1.667	992.614	952.315	-142.125
400	33.348	234.960	226.671	3.316	991.056	946.601	-123.613
450	33.985	238.926	227.816	4.999	991.020	941.037	-109.233
500	34.513	242.535	229.110	6.712	990.920	935.491	-97.730
600	35.313	248.902	231.892	10.206	991.171	924.388	-80.475
700	35.874	254.390	234.723	13.767	991.688	913.219	-68.145
800	36.280	259.208	237.489	17.376	992.359	901.965	-58.892
900	36.585	263.500	240.145	21.020	939.998	891.702	-51.753
1000	36.824	267.367	242.677	24.691	942.047	886.225	-46.292
1100	37.021	270.887	245.084	28.383	944.086	880.544	-41.814
1200	37.191	274.115	247.371	32.094	946.116	874.678	-38.074
1300	37.346	277.098	249.544	35.821	948.134	868.643	-34.903
1400	37.494	279.871	251.612	39.563	950.141	862.453	-32.179
1500	37.640	282.463	253.584	43.319	952.138	856.120	-29.813
1600	37.789	284.897	255.465	47.091	954.125	849.654	-27.738
1700	37.942	287.193	257.265	50.877	956.105	843.064	-25.904
1800	38.100	289.366	258.988	54.679	958.079	836.357	-24.270
1900	38.263	291.430	260.642	58.498	960.051	829.541	-22.806
2000	38.431	293.397	262.231	62.332	962.022	822.621	-21.485
2100	38.602	295.276	263.760	66.184	963.996	815.603	-20.287
2200	38.776	297.076	265.234	70.053	965.977	808.490	-19.196
2300	38.952	298.804	266.656	73.939	967.965	801.287	-18.198
2400	39.129	300.465	268.030	77.843	969.966	793.997	-17.281
2500	39.305	302.066	269.360	81.765	971.980	786.624	-16.436
2600	39.480	303.611	270.648	85.704	974.011	779.170	-15.654
2700	39.653	305.104	271.897	89.661	976.061	771.637	-14.928
2800	39.823	306.549	273.108	93.635	978.132	764.028	-14.253
2900	39.989	307.950	274.286	97.625	980.225	756.344	-13.623
3000	40.152	309.308	275.431	101.632	982.341	748.588	-13.034
3100	40.310	310.627	276.545	105.656	984.482	740.761	-12.482
3200	40.463	311.910	277.630	109.694	986.649	732.865	-11.963
3300	40.611	313.157	278.688	113.748	988.841	724.900	-11.474
3400	40.755	314.371	279.720	117.816	991.060	716.869	-11.013
3500	40.893	315.555	280.727	121.899	993.305	708.772	-10.578
3600	41.026	316.709	281.710	125.995	995.575	700.610	-10.166
3700	41.155	317.835	282.671	130.104	997.872	692.385	-9.775
3800	41.278	318.934	283.611	134.225	1000.195	684.098	-9.404
3900	41.397	320.008	284.531	138.359	1002.542	675.749	-9.051
4000	41.512	321.057	285.431	142.505	1004.913	667.340	-8.715
4100	41.623	322.083	286.312	146.662	1007.308	658.871	-8.394
4200	41.729	323.088	287.176	150.829	1009.725	650.343	-8.088
4300	41.832	324.071	288.023	155.007	1012.165	641.757	-7.796
4400	41.932	325.034	288.853	159.196	1014.626	633.115	-7.516
4500	42.029	325.977	289.667	163.394	1017.107	624.416	-7.248
4600	42.122	326.902	290.467	167.601	1019.608	615.662	-6.991
4700	42.213	327.809	291.252	171.818	1022.128	606.854	-6.744
4800	42.302	328.699	292.023	176.044	1024.665	597.991	-6.507
4900	42.388	329.572	292.780	180.278	1027.220	589.076	-6.280
5000	42.473	330.429	293.525	184.521	1029.791	580.108	-6.060
5100	42.556	331.271	294.256	188.773	1032.377	571.089	-5.849
5200	42.638	332.098	294.976	193.033	1034.979	562.019	-5.646
5300	42.718	332.911	295.684	197.300	1037.594	552.898	-5.449
5400	42.797	333.710	296.381	201.576	1040.223	543.728	-5.260
5500	42.875	334.496	297.067	205.860	1042.866	534.510	-5.076
5600	42.952	335.269	297.742	210.151	1045.521	525.243	-4.899
5700	43.029	336.030	298.407	214.450	1048.187	515.928	-4.728
5800	43.105	336.779	299.063	218.757	1050.866	506.567	-4.562
5900	43.180	337.517	299.708	223.071	1053.555	497.160	-4.402
6000	43.255	338.243	300.344	227.393	1056.255	487.707	-4.246

PREVIOUS: June 1976 (1 atm) CURRENT: June 1976 (1 bar)

Sulfur Fluoride, Ion (SF^+) $F_1S_1^+(g)$

SULFUR FLUORIDE, ION (SF^-) IDEAL GAS M_r = 51.058952

$\Delta_fH°(0\ K)$ = [-180.649 ± 54.4] kJ mol^{-1}

S°(298.15 K) = [218.956 ± 0.8] J K^{-1} mol^{-1} $\Delta_fH°(298.15\ K)$ = [-186.590] kJ mol^{-1}

Electronic Levels and Quantum Weights

State	ϵ_i, cm^{-1}	g_i
[$X\ ^1\Sigma^+$]	[0]	[1]
[$A\ ^3\Pi$]	[20000]	[6]

ω_e = [627] cm^{-1} ω_ex_e = [5.06] cm^{-1} σ = 1

B_e = [0.45620] cm^{-1} α_e = [0.0046] cm^{-1} r_e = [1.76] Å

Enthalpy of Formation

The adopted value of $\Delta_fH°(0\ K)$ = -43.2±13.0 kcal mol^{-1} is calculated as the difference between the $\Delta_fH°$(SF, g, 0 K) = 2.94±1.5 kcal mol^{-1} (1) and an estimated EA (SF) = 2.0±0.5 eV (46.121 kcal mol^{-1}). We estimate EA based on a consideration of trends in the electron affinities for other S-F species (1). These data follow a regular pattern among the even- and odd-electron molecules, with the odd-electron molecules having, as expected, the higher EA's. Within each even- and odd-electron series, the EA values become progressively larger as the number of fluorine atoms in the ion is increased. Our $\Delta_fH°(0\ K)$ value corresponds to a fluoride-ion affinity IA (S) and bond dissociation energy (SF^- = S + F^-) of 48.9±13.5 kcal mol^{-1}. Correcting $\Delta_fH°(0\ K)$ to a temperature of 298.15 K, we obtain the value of -44.6 kcal mol^{-1}.

Several pieces of information are available which tend to provide support for our estimated EA value. Formation of SF^- involves the addition of an electron to an antibonding orbital of SF (2) which is located primarily on the sulfur atom. It is reasonable, therefore, to assume that EA(SF)≃EA(S) = 2.0772±0.005 eV (3). O'Hare (2) has pointed out that Koopmans' theorem (4) gives reliable estimates of electron affinities and ionization potentials provided that the parent molecule or ion have a closed-shell electronic configuration. Application of this theorem to SF^- which has the required closed-shell structure leads to EA = 1.7 eV (2). Other estimates of EA based on MO calculations have included (all in eV) 2.8 (5), 2.5±0.5 (6), and <3.2 (2).

JANAF data (1) suggest that SF^- should be reasonably stable at moderate temperatures, with dissociative ionization becoming thermodynamically favorable above 2000 K. These predictions are supported qualitatively by the observations of Harland and Thynne (7, 8) who have identified SF^- in the negative ion mass spectra of SF_4, SF_6, and SF_5Cl. The temperature of the ion source was reported as around 325 K (8).

Heat Capacity and Entropy

The additional antibonding electron in SF^- should weaken the bonding in the ion relative to the parent molecule, as evidenced by a reduction (~32 kcal mol^{-1}) in the S-F bond strength for SF^- when compared with that for SF (1). In addition, MO calculations (9) on OF and OF^- show that the O-F bond length in OF increases by 0.159 Å upon forming the negative ion. We apply this difference to r_e(SF) = 1.5967 Å (1), and we obtain $r_e(SF^-)$ = 1.76 Å. The value of B_e is calculated from our estimate of r_e. We use Badger's rule (10) to estimate a value for ω_e. We write the equation in the form $\omega_e^2 = 3.159 \times 10^6/\mu(r_e-d_{ij})^3$, and we set d_{ij} equal to 0.8833 based on molecular data for ClF (11). Barrow and Caunt (12) have shown that the product $x_e\mu^{1/2}$ is reasonably constant for a series of similar molecules. Using $x_e\mu^{1/2}$ = 0.0279 from ClF (11), we calculate x_e = 0.008068 and ω_ex_e = 5.06 cm^{-1} for SF^-. The value of α_e is obtained from the other constants, assuming a Morse potential function (13).

The electronic states and levels are estimated by analogy with those for ClF (11). The effect of the triplet level at 20000 cm^{-1} on our entropies is insignificant below 3000 K, but this level does contribute 0.15 cal K^{-1} mol^{-1} to S at 4500 K. We assign an uncertainty of ±0.2 cal K^{-1} mol^{-1} to the value of S°(298.15 K) based on conservative estimates of the uncertainties in the vibrational and rotational data.

References

1. JANAF Thermochemical Tables: SF(g), 6-30-76; SF_2^-(g), SF_3^-(g), SF_4^-(g), and SF_5^-(g), 12-31-76; F^-(g), 12-31-71; S(g), 6-30-71; SF_6^-(g), 6-30-77.
2. P. A. G. O'Hare, J. Chem. Phys. 59, 3842 (1973).
3. J. L. Franklin and P. W. Harland, Ann. Rev. Phys. Chem. 25, 485 (1974).
4. T. Koopmans, Physica (Utr.) 1, 104 (1934).
5. A. L. Companion, Theoret. Chim. Acta (Berl.) 25, 268 (1972).
6. P. A. G. O'Hare and A. C. Wahl, J. Chem. Phys. 53, 2834 (1970).
7. P. W. Harland and J. C. J. Thynne, J. Phys. Chem. 73, 4031 (1969).
8. P. W. Harland and J. C. J. Thynne, J. Phys. Chem. 75, 3517 (1971).
9. P. A. G. O'Hare and A. C. Wahl, J. Chem. Phys. 53, 2469 (1970).
10. R. M. Badger, J. Chem. Phys. 2, 128 (1934).
11. B. Rosen, ed., "Spectroscopic Data Relative to Diatomic Molecules," Pergamon Press, New York, 1970.
12. R. F. Barrow and A. D. Caunt, Proc. Roy. Soc. (London) 219A, 120 (1953).
13. G. Herzberg, "Spectra of Diatomic Molecules," 2nd. ed., D. Van Nostrand Co., Inc., New York, 1963.

Sulfur Fluoride, Ion (SF^-) $F_1S_1^-(g)$

Enthalpy Reference Temperature = T_r = 298.15 K Standard State Pressure = p° = 0.1 MPa

T/K	$C_p°$ (J $K^{-1}mol^{-1}$)	S°	$-[G°-H°(T_r)]/T$	$H°-H°(T_r)$ (kJ mol^{-1})	$\Delta_fH°$	$\Delta_fG°$	Log K_f
0	0.	0.	INFINITE	-9.082	-180.649		
100	29.204	185.367	247.084	-6.172			
200	31.130	206.084	221.942	-3.172			
250	32.368	213.166	219.500	-1.584			
298.15	33.380	218.956	218.956	0.	-186.590	-205.828	36.060
300	33.414	219.163	218.957	0.062	-186.637	-205.948	35.859
350	34.243	224.379	219.367	1.754	-187.945	-209.063	31.201
400	34.889	228.995	220.287	3.483	-191.501	-211.901	27.671
450	35.395	233.135	221.489	5.241	-193.542	-214.337	24.880
500	35.795	236.886	222.844	7.021	-195.653	-216.533	22.621
600	36.378	243.467	225.748	10.631	-199.443	-220.346	19.183
700	36.779	249.106	228.692	14.290	-202.985	-223.548	16.681
800	37.071	254.038	231.558	17.984	-206.386	-226.252	14.773
900	37.295	258.417	234.304	21.702	-262.830	-227.455	13.201
1000	37.476	262.356	236.915	25.441	-264.871	-223.415	11.670
1100	37.626	265.936	239.393	29.196	-266.925	-219.170	10.408
1200	37.756	269.215	241.744	32.966	-268.995	-214.737	9.347
1300	37.871	272.242	243.975	36.747	-271.079	-210.131	8.443
1400	37.974	275.052	246.095	40.539	-273.179	-205.364	7.662
1500	38.070	277.675	248.114	44.342	-275.294	-200.447	6.980
1600	38.159	280.135	250.040	48.153	-277.423	-195.388	6.379
1700	38.244	282.451	251.879	51.973	-279.567	-190.195	5.844
1800	38.326	284.640	253.638	55.802	-281.724	-184.875	5.365
1900	38.404	286.714	255.325	59.638	-283.891	-179.436	4.933
2000	38.482	288.686	256.944	63.483	-286.067	-173.882	4.541
2100	38.560	290.565	258.501	67.335	-288.250	-168.219	4.184
2200	38.638	292.361	259.999	71.195	-290.436	-162.452	3.857
2300	38.719	294.080	261.444	75.063	-292.623	-156.586	3.556
2400	38.803	295.730	262.839	78.939	-294.807	-150.624	3.278
2500	38.893	297.315	264.186	82.823	-296.987	-144.572	3.021
2600	38.988	298.843	265.490	86.717	-299.159	-138.432	2.781
2700	39.092	300.316	266.753	90.621	-301.319	-132.209	2.558
2800	39.206	301.740	267.977	94.536	-303.464	-125.907	2.349
2900	39.330	303.118	269.165	98.463	-305.592	-119.528	2.153
3000	39.467	304.453	270.319	102.403	-307.701	-113.076	1.969
3100	39.617	305.750	271.441	106.357	-309.786	-106.554	1.795
3200	39.782	307.010	272.533	110.327	-311.845	-99.965	1.632
3300	39.963	308.237	273.597	114.314	-313.877	-93.313	1.477
3400	40.160	309.433	274.633	118.320	-315.877	-86.599	1.330
3500	40.374	310.600	275.644	122.346	-317.846	-79.826	1.191
3600	40.605	311.741	276.631	126.395	-319.779	-72.998	1.059
3700	40.854	312.857	277.595	130.468	-321.676	-66.117	0.933
3800	41.120	313.950	278.537	134.566	-323.534	-59.185	0.814
3900	41.403	315.021	279.459	138.692	-325.352	-52.205	0.699
4000	41.703	316.073	280.361	142.847	-327.128	-45.178	0.590
4100	42.019	317.107	281.245	147.033	-328.862	-38.108	0.486
4200	42.351	318.123	282.111	151.252	-330.550	-30.996	0.385
4300	42.697	319.124	282.960	155.504	-332.194	-23.844	0.290
4400	43.057	320.110	283.793	159.792	-333.791	-16.655	0.198
4500	43.429	321.081	284.611	164.116	-335.340	-9.429	0.109
4600	43.813	322.040	285.415	168.478	-336.842	-2.171	0.025
4700	44.206	322.987	286.204	172.879	-338.296	5.121	-0.057
4800	44.608	323.921	286.980	177.319	-339.701	12.442	-0.135
4900	45.018	324.845	287.743	181.801	-341.057	19.793	-0.211
5000	45.433	325.759	288.494	186.323	-342.364	27.171	-0.284
5100	45.853	326.663	289.234	190.887	-343.622	34.574	-0.354
5200	46.275	327.557	289.962	195.494	-344.831	42.001	-0.422
5300	46.700	328.443	290.680	200.143	-345.991	49.452	-0.487
5400	47.125	329.320	291.388	204.834	-347.104	56.923	-0.551
5500	47.548	330.188	292.085	209.567	-348.169	64.415	-0.612
5600	47.970	331.049	292.773	214.343	-349.186	71.926	-0.671
5700	48.388	331.902	293.452	219.161	-350.158	79.454	-0.728
5800	48.801	332.747	294.123	224.021	-351.084	86.999	-0.784
5900	49.208	333.584	294.784	228.921	-351.966	94.560	-0.837
6000	49.608	334.415	295.438	233.862	-352.804	102.136	-0.889

PREVIOUS: December 1976 (1 atm) CURRENT: December 1976 (1 bar)

Sulfur Fluoride, Ion (SF^-) $F_1S_1^-(g)$

FLUOROSILYLIDYNE (SiF) IDEAL GAS M_r = 47.083903

S°(298.15 K) = 225.786 ± 0.17 J K^{-1} mol^{-1}

$\Delta_f H°$(0 K) = -21.912 ± 12.6 kJ mol^{-1}
$\Delta_f H°$(298.15 K) = -20.083 ± 12.6 kJ mol^{-1}

Electronic Levels and Molecular Constants (σ = 1)

State	ε_i, cm^{-1}	g_i	r_e, Å	B_e, cm^{-1}	α_e, cm^{-1}	ω_e, cm^{-1}	$\omega_e x_e$, cm^{-1}
X $^2\pi_{1/2}$	0	2	1.6008	0.58138	0.00494	857.20	4.735
$^2\pi_{3/2}$	161.93	2	1.6008	0.58138	0.00494	857.20	4.735
A $^2\Sigma$	22858.84	2	1.6049	0.57839	0.00941	718.5	10.167
a $^4\Sigma^-$	29805.06	4	1.604	0.5786	0.00502	863.16	5.730
B $^2\Sigma^+$	34561.5	2	1.5714	0.62707	0.00462	1011.2	4.825
C $^2\Delta$	39438.4	4	1.5714	0.60338	0.00539	829.41	7.01
C' $^2\pi$	41964.9	4	-	0.6376	0.0039	1031.9	4.45
D' $^2\pi$	46606.7	4	-	0.6329	0.0044	1032.9	5.28
D $^2\Sigma^+$	47418.6	2	1.54	0.625	0.005	1003.2	5.64

Enthalpy of Formation

Ehlert and Margrave (1) have studied mass-spectrometrically the reaction Si(g) + SiF_2(g) + 2 SiF(g) by monitoring the intensities of the Si^+, SiF_2^+ and SiF^+ ions in an equilibrium system containing CaF_2 and Si. The ionization cross-sections are assumed to cancel so that equilibrium constants can be obtained directly from the ion intensities. Eleven data points were reported in the 1395 to 1543 K range. We have analyzed their data using present JANAF functions (2) and obtain $\Delta_r H°$(298.15 K) = 23.5±3 kcal mol^{-1} by the 3rd law technique with a drift of -0.8±11 cal K^{-1} mol^{-1}; the 2nd law value is 23.9±7 kcal mol^{-1}. Farber (6) also has determined equilibrium constants mass-spectrometrically for the Si(g) + SiF_2(g) = 2 SiF(g) reaction; ten data points over a 1590 to 1782 K range were obtained. Our analysis yields a 3rd law $\Delta_r H°$(298.15 K) = 21.08±0.59 kcal mol^{-1} with a drift of -1.9±1.5 cal K^{-1} mol^{-1}; the 2nd law value is 24.18±2.5 kcal mol^{-1}. The two groups of data are in good agreement. We adopt an average value and calculate $\Delta_f H°$(SiF, g, 298.15 K) = -4.8±3 kcal mol^{-1}.

D_0^0 = 130±4 kcal mol^{-1} is calculated from the resulting $\Delta_f H°$(0 K) and auxiliary data (2). Johns and Barrow (3) recommended 125±10 kcal mol^{-1} from Birge-Sponer extrapolations of several states. Gaydon (4) preferred 115±12 kcal mol^{-1} from a Birge-Sponer extrapolation of the ground state.

Heat Capacity and Entropy

The molecular constants and electronic levels are those given by Suchard (5). Five higher levels are given but are not used since they do not significantly affect the calculations. The ground state is treated as two distinct levels due to the splitting of this state as expressed by the spin coupling constant (A=161.93 cm^{-1}). This approximation gives slightly biased results at lower temperatures. An error of ±0.04 cal K^{-1} mol^{-1} has been assigned to S°(298.15 K) to allow for this.

References

1. T. C. Ehlert and J. L. Margrave, J. Chem. Phys. 41, 1066 (1964).
2. JANAF Thermochemical Tables: Si(g), 3-31-67; SiF_2(g), 6-31-76; F(g), 9-30-65.
3. J. W. C. Johns and R. F. Barrow, Proc. Phys. Soc. (London) 71, 476 (1958).
4. A. G. Gaydon, "Dissociation Energies and Spectra of Diatomic Molecules," 3rd ed., Chapman and Hall Ltd., London, 1968.
5. S. N. Suchard, The Aerospace Corporation, Air Force Space and Missile Systems Organization Contract F04701-73-C-0074, March 29, 1974.
6. M. Farber, Space Sciences, Inc., personal communication, October 18, 1976.

Fluorosilylidyne (SiF) F_1Si_1(g)

Enthalpy Reference Temperature = T_r = 298.15 K   Standard State Pressure = p° = 0.1 MPa

T/K	$C_p°$	S°	-[G°-H°(T_r)]/T	H°-H°(T_r)	$\Delta_f H°$	$\Delta_f G°$	Log K_f
	J $K^{-1}$$mol^{-1}$			kJ mol^{-1}			
0	0.	0.	INFINITE	-9.460	-21.912	-21.912	INFINITE
100	32.762	190.572	254.370	-6.380	-20.552	-30.708	16.040
200	31.864	212.954	228.738	-3.157	-19.972	-41.165	10.751
250	32.126	220.085	226.320	-1.559	-19.976	-46.466	9.709
298.15	32.650	225.786	225.786	0.	-20.083	-51.559	9.033
300	32.672	225.988	225.786	0.060	-20.089	-51.755	9.011
350	33.289	231.070	226.186	1.709	-20.269	-57.019	8.510
400	33.874	235.554	227.083	3.389	-20.492	-62.254	8.130
450	34.392	239.575	228.251	5.096	-20.742	-67.459	7.830
500	34.836	243.222	229.569	6.827	-21.014	-72.636	7.588
600	35.530	249.638	232.393	10.347	-21.607	-82.905	7.218
700	36.029	255.155	235.260	13.926	-22.251	-93.071	6.945
800	36.396	259.991	238.055	17.548	-22.943	-103.142	6.734
900	36.674	264.295	240.736	21.203	-23.678	-113.123	6.565
1000	36.891	268.170	243.289	24.881	-24.455	-123.020	6.426
1100	37.065	271.695	245.714	28.579	-25.273	-132.837	6.308
1200	37.208	274.926	248.015	32.293	-26.131	-142.578	6.206
1300	37.329	277.909	250.201	36.020	-27.029	-152.246	6.117
1400	37.432	280.679	252.281	39.758	-27.968	-161.844	6.038
1500	37.523	283.265	254.261	43.506	-28.950	-171.372	5.968
1600	37.604	285.690	256.150	47.263	-29.974	-180.834	5.904
1700	37.678	287.972	257.956	51.027	-81.217	-189.783	5.831
1800	37.745	290.127	259.684	54.798	-82.096	-196.144	5.692
1900	37.807	292.170	261.340	58.576	-82.973	-202.456	5.566
2000	37.866	294.110	262.931	62.359	-83.850	-208.722	5.451
2100	37.921	295.959	264.460	66.149	-84.724	-214.944	5.346
2200	37.973	297.724	265.932	69.943	-85.593	-221.125	5.250
2300	38.024	299.414	267.351	73.743	-86.457	-227.266	5.161
2400	38.072	301.033	268.721	77.548	-87.314	-233.370	5.079
2500	38.120	302.588	270.045	81.358	-88.161	-239.438	5.003
2600	38.166	304.084	271.325	85.172	-88.997	-245.472	4.932
2700	38.211	305.525	272.566	88.991	-89.821	-251.475	4.865
2800	38.256	306.916	273.768	92.814	-90.631	-257.447	4.803
2900	38.301	308.259	274.934	96.642	-91.426	-263.391	4.744
3000	38.346	309.558	276.067	100.474	-92.204	-269.307	4.689
3100	38.391	310.816	277.167	104.311	-92.964	-275.198	4.637
3200	38.436	312.036	278.238	108.153	-93.705	-281.065	4.588
3300	38.483	313.219	279.280	111.998	-94.427	-286.909	4.541
3400	38.530	314.369	280.295	115.849	-95.128	-292.731	4.497
3500	38.579	315.486	281.285	119.705	-95.808	-298.533	4.455
3600	38.629	316.574	282.250	123.565	-480.618	-293.855	4.264
3700	38.681	317.633	283.192	127.430	-480.841	-288.664	4.075
3800	38.735	318.665	284.112	131.301	-481.042	-283.467	3.897
3900	38.791	319.672	285.011	135.177	-481.223	-278.265	3.727
4000	38.850	320.655	285.890	139.060	-481.382	-273.059	3.566
4100	38.911	321.615	286.750	142.947	-481.519	-267.849	3.412
4200	38.974	322.553	287.591	146.842	-481.633	-262.637	3.266
4300	39.041	323.471	288.415	150.742	-481.724	-257.421	3.127
4400	39.111	324.370	289.222	154.650	-481.792	-252.204	2.994
4500	39.184	325.249	290.013	158.565	-481.837	-246.986	2.867
4600	39.260	326.111	290.788	162.487	-481.858	-241.766	2.745
4700	39.340	326.957	291.549	166.417	-481.856	-236.547	2.629
4800	39.423	327.786	292.295	170.355	-481.830	-231.328	2.517
4900	39.510	328.599	293.028	174.302	-481.781	-226.110	2.410
5000	39.601	329.399	293.747	178.257	-481.708	-220.893	2.308
5100	39.696	330.184	294.454	182.222	-481.611	-215.677	2.209
5200	39.794	330.955	295.148	186.196	-481.491	-210.464	2.114
5300	39.897	331.714	295.831	190.181	-481.347	-205.253	2.023
5400	40.004	332.461	296.503	194.176	-481.180	-200.045	1.935
5500	40.114	333.196	297.163	198.182	-480.990	-194.841	1.850
5600	40.229	333.920	297.813	202.199	-480.777	-189.640	1.769
5700	40.348	334.633	298.453	206.228	-480.541	-184.443	1.690
5800	40.471	335.336	299.083	210.269	-480.282	-179.251	1.614
5900	40.597	336.029	299.703	214.322	-480.002	-174.063	1.541
6000	40.728	336.712	300.314	218.388	-479.698	-168.880	1.470

PREVIOUS: December 1976 (1 atm)   CURRENT: December 1976 (1 bar)

Fluorosilylidyne (SiF) F_1Si_1(g)

STRONTIUM FLUORIDE (SrF) IDEAL GAS M_r = 106.618403

$S°(298.15\ K) = 239.919 \pm 0.21\ J\ K^{-1}\ mol^{-1}$

$\Delta_f H°(0\ K) = -292.859 \pm 8.4\ kJ\ mol^{-1}$

$\Delta_f H°(298.15\ K) = -294.554 \pm 8.4\ kJ\ mol^{-1}$

Electronic Levels and Quantum Weights

State	ε_i, cm^{-1}	g_i	State	ε_i, cm^{-1}	g_i
$X\ ^2\Sigma^+$	0	2	$D\ ^2\Sigma$	28296.6	2
$A\ ^2\Pi$	15067.8	2	$E\ ^2\Pi$	31528.7	2
	15348.5	2		31614.8	2
$B\ ^2\Sigma$	17303.4	2	$F\ ^2\Sigma$	32822.0	2
$C\ ^2\Pi$	27383.7	4	$G\ ^2\Pi$	34759.2	4

$\omega_e = 498.0\ cm^{-1}$ $\omega_e x_e = 2.15\ cm^{-1}$ $\sigma = 1$

$B_e = 0.2506\ cm^{-1}$ $\alpha_e = [0.00148]\ cm^{-1}$ $r_e = 2.0757 \pm 0.0005$ Å

Enthalpy of Formation

Equilibrium constants for four dissociation equilibria have been determined from observations made on the condensed phase systems SrF_2-B (1), SrF_2-Al (2), and SrF_2-BaF_2-Al (2) by the mass spectrometric-Knudsen effusion method. Enthalpies of reaction have been calculated from these Kp data by both the 2nd and 3rd law methods. Results of our analysis are given below, using JANAF auxiliary data (4). Also included below is a value for the enthalpy of dissociation of SrF which was determined from spectrophotometric studies (3) of reaction equilibria in H_2-O_2-N_2 flames.

Source	Reaction	T/K	Data Points	$\Delta_r H°(298.15)$, kcal mol^{-1} 2nd law	3rd law	Drift cal K^{-1} mol^{-1}	$\Delta_f H°(298.15\ K)$ kcal mol^{-1}
Hildenbrand (1)	A	1476-1585	10	23.4	25.1±0.9	1.2±4.7	-70.1±2.7
Hildenbrand (1)	B	1436-1525	9	4.3	2.0±0.7	-1.5±4.0	-71.0±1.0
Ehlert et al. (2)	C	1174-1290	9[a]	8.1	10.6±0.3	2.0±1.0	-70.0±4.0
Ehlert et al. (2)	D	1125-1292	10	-127.6	-116.5±1.5	9.1±2.7	-67.6±1.3
Ryabova and Gurvich (3)	E			132 ± 7			-73.9±8.0

Reactions:
(A) $2Sr(g) + BF_3(g) = 2SrF(g) + BF(g)$
(B) $Sr(g) + SrF_2(g) = 2SrF(g)$
(C) $Sr(g) + BaF(g) = Ba(g) + SrF(g)$
(D) $2SrF(g) = Sr(g) + SrF_2(cr)$
(E) $SrF(g) = Sr(g) + F(g)$

[a]One point rejected due to failure of a statistical test.

It is believed that the Kp data for the three homogeneous reactions (A, B, and C) are more reliable than those for the heterogeneous reaction (D) due to the need for a machine calibration constant which appears in the equilibrium expression for the latter. Ehlert et al. (2) determined this constant from vaporization experiments performed with silver contained in their Knudsen cell. This belief is further substantiated by the appearance of the rather large positive drift in the third law analysis of their Kp data for reaction (D). Also, it is felt that the $\Delta_f H°$ value obtained from the flame-spectrophotometric studies (3) is perhaps somewhat less reliable than the three values derived from the Kp data for the homogeneous reactions. Thus, a median value $\Delta_f H°(SrF, g, 298.15\ K) = -70.4 \pm 2.0\ kcal\ mol^{-1}$ of the latter results is adopted. The adopted value corresponds to a dissociation energy of $D_0^0(Sr\text{-}F) = 127.7\ kcal\ mol^{-1}$ (5.54 cal K^{-1} mol^{-1}).

Spectroscopic values for D_0^0 based on a linear Birge-Sponer extrapolation of the ground state ($X\ ^2\Sigma$) vibrational levels (v, 0-4) have been reported as 3.2 (5) and 3.5 eV (6). However, it is well established now that the linear extrapolation (5, 6) generally underestimates D_0^0 if the binding of the molecule is partially ionic. Based on the ionic character correction suggested by Hildenbrand (6), these values adjust to 4.9 and 5.4 eV. The latter value is now in reasonable agreement with the adopted thermochemical value. Other spectroscopic values (7) include 4.6 (from $X\ ^2\Sigma$), 4.7 (from $A\ ^2\Pi$), and 5.2 eV (from $C\ ^2\Pi$). The products of dissociation from the two excited states of SrF were assumed to be $Sr(^3P) + F(^2P)$.

Heat Capacity and Entropy

The spectroscopic data with the exception of the value for α_e are those given by Rosen (8). α_e is calculated from the other constants assuming a Morse potential function. The electronic states and levels are also taken from the compilation of Rosen (8).

References

1. D. L. Hildenbrand, J. Chem. Phys. 48, 3657 (1968).
2. T. C. Ehlert, G. D. Blue, J. W. Green, and J. L. Margrave, J. Chem. Phys. 41, 2250 (1964).
3. V. G. Ryabova and L. V. Gurvich, Teplofiz. Vysokikh Temperature, Akad. Nauk SSSR 2, 834 (1964).
4. JANAF Thermochemical Tables: Sr(g) and Ba(g), 12-31-70; BF_3(g), 6-30-69; BF(g), 12-31-64; SrF_2(cr, g) and BaF(g), 12-31-72; F(g), 9-30-65.
5. A. G. Gaydon, "Dissociation Energies and Spectra of Diatomic Molecules," Third Ed., Chapman and Hill Ltd., London, 1968.
6. D. L. Hildenbrand in "Advances in High Temperature Chemistry," Vol. 1, Ed. by L. Eyring, Academic Press, New York, 1967.
7. M. M. Novikov and L. V. Gurvich, Optics and Spectroscopy 22, 395 (1967).
8. B. Rosen, "Spectroscopic Data Relative to Diatomic Molecules," Pergamon Press, New York, 1970.

Strontium Fluoride (SrF) $F_1Sr_1(g)$

Enthalpy Reference Temperature = T_r = 298.15 K; Standard State Pressure = p° = 0.1 MPa

T/K	$C_p°$ (J K^{-1}mol^{-1})	S°	$-[G°-H°(T_r)]/T$	$H°-H°(T_r)$ (kJ mol^{-1})	$\Delta_f H°$	$\Delta_f G°$	Log K_f
0	0.	0.	INFINITE	-9.286	-292.859	-292.859	INFINITE
100	29.458	205.319	269.038	-6.372	-292.898	-302.106	157.804
200	32.336	226.565	243.014	-3.290	-293.768	-310.974	81.218
250	33.590	233.922	240.482	-1.640	-294.172	-315.234	65.865
298.15	34.487	239.919	239.919	0.	-294.554	-319.250	55.931
300	34.516	240.133	239.920	0.064	-294.568	-319.403	55.613
350	35.190	245.507	240.343	1.807	-294.972	-323.510	48.281
400	35.687	250.240	241.290	3.580	-295.391	-327.559	42.775
450	36.060	254.466	242.523	5.374	-295.832	-331.554	38.486
500	36.347	258.280	243.911	7.184	-296.297	-335.498	35.049
600	36.756	264.946	246.878	10.841	-297.307	-343.245	29.882
700	37.031	270.634	249.875	14.531	-298.426	-350.814	26.178
800	37.229	275.592	252.786	18.245	-299.657	-358.216	23.389
900	37.381	279.987	255.569	21.976	-301.628	-365.384	21.206
1000	37.503	283.932	258.212	25.720	-302.762	-372.407	19.453
1100	37.606	287.511	260.715	29.476	-311.803	-378.951	17.995
1200	37.694	290.787	263.086	33.241	-313.857	-384.965	16.757
1300	37.773	293.807	265.335	37.014	-315.914	-390.807	15.703
1400	37.845	296.609	267.470	40.795	-317.975	-396.491	14.793
1500	37.912	299.223	269.501	44.583	-320.038	-402.027	14.000
1600	37.977	301.671	271.436	48.377	-322.104	-407.426	13.301
1700	38.040	303.976	273.283	52.178	-461.087	-411.516	12.644
1800	38.103	306.152	275.049	55.985	-461.296	-408.594	11.857
1900	38.168	308.214	276.740	59.799	-461.509	-405.660	11.152
2000	38.236	310.173	278.364	63.619	-461.729	-402.715	10.518
2100	38.310	312.040	279.923	67.446	-461.957	-399.759	9.943
2200	38.390	313.824	281.424	71.281	-462.195	-396.791	9.421
2300	38.479	315.533	282.870	75.125	-462.447	-393.813	8.944
2400	38.577	317.172	284.265	78.977	-462.715	-390.823	8.506
2500	38.686	318.749	285.613	82.840	-463.004	-387.822	8.103
2600	38.808	320.269	286.917	86.715	-463.321	-384.808	7.731
2700	38.942	321.736	288.180	90.602	-463.670	-381.782	7.386
2800	39.091	323.155	289.404	94.504	-464.058	-378.742	7.066
2900	39.253	324.530	290.591	98.421	-464.493	-375.687	6.767
3000	39.431	325.863	291.745	102.355	-464.981	-372.617	6.488
3100	39.623	327.159	292.867	106.308	-465.532	-369.529	6.227
3200	39.829	328.421	293.958	110.280	-466.151	-366.423	5.981
3300	40.051	329.649	295.021	114.274	-466.846	-363.296	5.750
3400	40.286	330.849	296.057	118.291	-467.624	-360.147	5.533
3500	40.536	332.020	297.068	122.332	-468.492	-356.973	5.328
3600	40.798	333.166	298.055	126.398	-469.456	-353.773	5.133
3700	41.073	334.287	299.019	130.492	-470.520	-350.545	4.949
3800	41.360	335.386	299.962	134.613	-471.689	-347.288	4.774
3900	41.658	336.464	300.884	138.764	-472.965	-343.997	4.607
4000	41.966	337.523	301.787	142.945	-474.350	-340.673	4.449
4100	42.284	338.563	302.671	147.158	-475.845	-337.313	4.297
4200	42.611	339.586	303.538	151.402	-477.447	-333.915	4.153
4300	42.946	340.592	304.388	155.680	-479.159	-330.478	4.015
4400	43.288	341.584	305.222	159.992	-480.976	-326.999	3.882
4500	43.636	342.560	306.041	164.338	-482.896	-323.478	3.755
4600	43.990	343.523	306.845	168.719	-484.848	-319.919	3.633
4700	44.348	344.473	307.636	173.136	-486.939	-316.311	3.515
4800	44.711	345.411	308.413	177.589	-489.111	-312.658	3.402
4900	45.076	346.336	309.177	182.078	-491.359	-308.959	3.294
5000	45.444	347.251	309.930	186.604	-493.670	-305.213	3.189
5100	45.813	348.154	310.670	191.167	-496.023	-301.423	3.087
5200	46.183	349.047	311.400	195.767	-498.445	-297.583	2.989
5300	46.553	349.931	312.119	200.404	-500.911	-293.697	2.895
5400	46.923	350.804	312.827	205.078	-503.412	-289.764	2.803
5500	47.291	351.669	313.525	209.788	-505.940	-285.785	2.714
5600	47.658	352.524	314.214	214.536	-508.486	-281.759	2.628
5700	48.021	353.371	314.894	219.320	-511.043	-277.687	2.545
5800	48.382	354.209	315.564	224.140	-513.604	-273.571	2.464
5900	48.739	355.039	316.226	228.996	-515.951	-269.435	2.385
6000	49.092	355.861	316.880	233.888	-518.467	-265.235	2.309

PREVIOUS: December 1972 (1 atm) CURRENT: December 1972 (1 bar)

Strontium Fluoride (SrF) $F_1Sr_1(g)$

STRONTIUM FLUORIDE, ION (SrF^+) IDEAL GAS M_r = 106.617854

$\Delta_f H°(0\ K) = 192.679 \pm 41.8\ kJ\ mol^{-1}$

$S°(298.15\ K) = [232.846 \pm 8.4]\ J\ K^{-1}\ mol^{-1}$ $\Delta_f H°(298.15\ K) = [197.066]\ kJ\ mol^{-1}$

Electronic Levels and Quantum Weights

State	ε_i, cm^{-1}	g_i
$X\ ^1\Sigma^+$	0	1
$a\ ^3\pi$	[10000]	6
$A'\ ^1\pi$	[12000]	2
$B'\ ^1\Delta$	[15000]	2
$A^1\ \Sigma^+$	[18000]	1
$b\ ^3\Sigma^-$	[20000]	3
$B\ ^1\Sigma^+$	[35000]	1

$\omega_e = [561]\ cm^{-1}$ $\omega_e x_e = [2.42]\ cm^{-1}$ $\sigma = 1$

$B_e = [0.27265]\ cm^{-1}$ $\alpha_e = [0.00157]\ cm^{-1}$ $r_e = [1.99]$ Å

Enthalpy of Formation

The appearance potential of SrF^+ from SrF(g) has been measured mass spectrometrically by Hildenbrand (1), Green et al. (2) and Ehlert et al. (3) as 5.0±0.3 eV, 4.9±0.3 eV, and 5.2±0.3 eV, respectively. For the process $e^- + SrF(g) \rightarrow SrF^+(g) + 2e^-$, the corresponding $\Delta_f H°(SrF^+, g, 0\ K)$ are derived to be 45.3±9.0, and 49.9±9.0 kcal mol^{-1} with $\Delta_f H°(SrF, g, 0\ K) = -70.0\pm2.0$ kcal mol^{-1} (4).

Hildenbrand (1) has also reported AP = 13.0±1.0 eV for SrF^+ from $SrF_2(g)$. Green et al. (2) obtained an ionization efficiency curve for SrF^+ which extrapolates (high-energy portion) to 10.5 eV at zero ion current. Assuming the process to be $e^- + SrF_2(g) \rightarrow SrF^+(g) + F(g) + 2e^-$, we obtain $\Delta_f H°(SrF^+, g, 0\ K)$ equal to 98.7±25.0 and 41.1 kcal mol^{-1} by combining the above results with $\Delta_f H°(SrF_2, g, 0\ K) = -182.7\pm1.0$ kcal mol^{-1} and $\Delta_f H°(F, g, 0\ K) = 18.36\pm0.4$ kcal mol^{-1} (4).

The adopted $\Delta_f H°(0\ K)$ is based on the three mass spectral results (1, 2, 3) for the direct ionization process of SrF(g). The value of $\Delta_f H°$ at 298.15 K is 47.1±10.0 kcal mol^{-1}, and the corresponding ionization potential for SrF(g) is 5.03 eV.

Heat Capacity and Entropy

The internuclear distance (r_e) is the value estimated by Barrow and Beale (5) from spectroscopic work on SrF. The value of ω_e is calculated from Badger's rule (6) which can be written in the form $\omega_e^2 = 3.159 \times 10^6/\mu(r_e - d_{ij})^3$. Molecular data for the ground states of SrF (4), SrO (7), and RbF (8) were used to determine the constant d_{ij}. The product $x_e\mu^{1/2}$ has been shown to be constant within a group of similar molecules by Barrow and Caunt (9). Since $\mu^{1/2}(SrF^+) \equiv \mu^{1/2}(SrF)$, we assume $x_e(SrF^+) = x_e(SrF)$ and obtain $\omega_e x_e$ equal to 2.42 cm^{-1} with our estimated value $\omega_e = 561\ cm^{-1}$. The value of α_e is obtained from the other constants assuming a Morse potential function.

SrF^+ has eight valence electrons; thus, the ground state configuration ($^1\Sigma^+$) and electronic level for the B state are estimated from those for the isoelectronic molecule SrS (7). By analogy with the gaseous alkaline earth oxides (10), C_2 (4), BeS (11), and BaS (12), low lying triplet states are to be expected. The two estimated triplet levels are based on those which have been observed directly in C_2 (4), BeS (11), and indirectly (through perturbative effects) in BaS (12). Also included are additional states at 12,000 cm^{-1} ($^1\pi$), 15,000 cm^{-1} ($^1\Delta$), and 18,000 cm^{-1} ($^1\Sigma$). These three levels are also estimated from those for C_2 (4), Bes (11), and BaS (12). The adopted electronic entropies are believed to represent minimal values. Probable upper limits for the electronic contributions can be established by decreasing the two triplet levels to 1000 cm^{-1} ($^3\pi$) and 5000 cm^{-1} ($^3\Sigma$) and neglecting the other excited states. This increases the entropy by 3.9 cal K^{-1} mol^{-1} and 2.6 cal K^{-1} mol^{-1} at 2000 K and 4000 K, respectively.

References

1. D. L. Hildenbrand, J. Chem. Phys. 48, 3657 (1968).
2. J. W. Green, G. D. Blue, T. C. Ehlert, and J. L. Margrave, J. Chem. Phys. 41, 2245 (1964).
3. T. C. Ehlert, G. D. Blue, J. W. Green, and J. L. Margrave, J. Chem. Phys. 41, 2250 (1964).
4. JANAF Thermochemical Tables: SrF(g) and SrF_2(g), 12-31-72; F(g), 9-30-65; C_2(g), 12-31-69.
5. R. F. Barrow and J. R. Beale, Chem. Comm. 1967, 606 (1967).
6. R. M. Badger, J. Chem. Phys. 2, 128 (1934).
7. M. Marcano and R. F. Barrow, Trans. Faraday Soc. 66, 1917 (1970).
8. B. Rosen, "Spectroscopic Data Relative to Diatomic Molecules," Pergamon Press, New York, 1970.
9. R. F. Barrow and A. D. Caunt, Proc. Roy. Soc. (London) 219A, 120 (1953).
10. L. Brewer and G. M. Rosenblatt in "Advances in High Temperature Chemistry," Vol. 2, Ed. by L. Eyring, Academic Press, New York, 1969.
11. C. J. Cheetham, W. J. M. Gissane, and R. F. Barrow, Trans. Faraday Soc. 61, 1308 (1965).
12. R. F. Barrow, W. G. Burton, and P. A. Jones, Trans. Faraday Soc. 67, 902 (1971).

Strontium Fluoride, Ion (SrF^+) $F_1Sr_1^+(g)$

Enthalpy Reference Temperature = T_r = 298.15 K Standard State Pressure = p° = 0.1 MPa

T/K	$C_p°$	S°	$-[G°-H°(T_r)]/T$	$H°-H°(T_r)$	$\Delta_f H°$	$\Delta_f G°$	Log K_f
	J $K^{-1}mol^{-1}$			kJ mol^{-1}			
0	0.	0.	INFINITE	-9.171	192.679		
100	29.288	198.822	261.416	-6.259			
200	31.664	219.755	235.882	-3.225			
250	32.928	226.961	233.399	-1.610			
298.15	33.891	232.846	232.846	0.	197.066	168.224	-29.472
300	33.923	233.056	232.847	0.063	197.089	168.045	-29.259
350	34.677	238.345	233.263	1.779	197.697	163.157	-24.350
400	35.245	243.014	234.196	3.527	198.294	158.181	-20.656
450	35.679	247.191	235.412	5.301	198.871	153.132	-17.775
500	36.016	250.969	236.782	7.094	199.428	148.020	-15.464
600	36.497	257.581	239.713	10.721	200.467	137.639	-11.983
700	36.819	263.233	242.679	14.388	201.403	127.093	-9.484
800	37.049	268.165	245.563	18.082	202.231	116.419	-7.601
900	37.223	272.539	248.322	21.796	202.322	105.720	-6.136
1000	37.364	276.469	250.943	25.526	203.252	94.936	-4.959
1100	37.488	280.036	253.428	29.268	196.277	84.426	-4.009
1200	37.611	283.303	255.784	33.023	196.292	74.257	-3.232
1300	37.748	286.319	258.018	36.791	196.307	64.086	-2.575
1400	37.913	289.122	260.141	40.574	196.327	53.915	-2.012
1500	38.124	291.745	262.161	44.375	196.355	43.742	-1.523
1600	38.394	294.213	264.088	48.200	196.399	33.566	-1.096
1700	38.734	296.551	265.930	52.056	59.550	24.567	-0.755
1800	39.153	298.776	267.693	55.950	61.506	22.453	-0.652
1900	39.657	300.906	269.385	59.890	63.497	20.229	-0.556
2000	40.247	302.955	271.013	63.884	65.530	17.899	-0.467
2100	40.923	304.935	272.581	67.942	67.611	15.467	-0.385
2200	41.678	306.856	274.096	72.071	69.746	12.934	-0.307
2300	42.506	308.726	275.561	76.280	71.939	10.303	-0.234
2400	43.398	310.554	276.981	80.575	74.192	7.575	-0.165
2500	44.340	312.344	278.360	84.961	76.504	4.752	-0.099
2600	45.322	314.103	279.701	89.444	78.875	1.835	-0.037
2700	46.328	315.832	281.007	94.026	81.299	-1.174	0.023
2800	47.345	317.535	282.281	98.710	83.771	-4.273	0.080
2900	48.361	319.214	283.526	103.495	86.284	-7.462	0.134
3000	49.361	320.871	284.743	108.382	88.826	-10.738	0.187
3100	50.335	322.505	285.935	113.367	91.387	-14.099	0.238
3200	51.271	324.118	287.103	118.447	93.955	-17.543	0.286
3300	52.160	325.709	288.249	123.619	96.516	-21.067	0.333
3400	52.994	327.279	289.374	128.877	99.058	-24.668	0.379
3500	53.768	328.827	290.479	134.216	101.566	-28.344	0.423
3600	54.475	330.351	291.566	139.629	104.027	-32.090	0.466
3700	55.114	331.853	292.634	145.109	106.428	-35.905	0.507
3800	55.682	333.330	293.686	150.649	108.757	-39.783	0.547
3900	56.179	334.783	294.721	156.243	111.002	-43.721	0.586
4000	56.606	336.211	295.740	161.883	113.154	-47.716	0.623
4100	56.963	337.613	296.745	167.562	115.204	-51.763	0.659
4200	57.253	338.990	297.734	173.273	117.148	-55.860	0.695
4300	57.481	340.340	298.709	179.010	118.974	-60.001	0.729
4400	57.647	341.663	299.671	184.767	120.680	-64.183	0.762
4500	57.758	342.960	300.618	190.538	122.264	-68.402	0.794
4600	57.817	344.230	301.553	196.317	123.788	-72.661	0.825
4700	57.828	345.474	302.474	202.100	125.142	-76.946	0.855
4800	57.796	346.691	303.382	207.881	126.377	-81.259	0.884
4900	57.724	347.882	304.278	213.658	127.494	-85.597	0.912
5000	57.618	349.047	305.162	219.425	128.503	-89.957	0.940
5100	57.480	350.187	306.034	225.180	129.422	-94.337	0.966
5200	57.315	351.302	306.894	230.920	130.218	-98.732	0.992
5300	57.127	352.392	307.742	236.642	130.916	-103.142	1.017
5400	56.919	353.457	308.579	242.345	131.522	-107.564	1.040
5500	56.693	354.500	309.404	248.026	132.043	-111.997	1.064
5600	56.454	355.519	310.219	253.683	132.486	-116.438	1.086
5700	56.203	356.516	311.022	259.316	132.856	-120.886	1.108
5800	55.942	357.491	311.815	264.923	133.161	-125.341	1.129
5900	55.675	358.445	312.597	270.504	133.617	-129.824	1.149
6000	55.403	359.379	313.369	276.058	133.842	-134.291	1.169

PREVIOUS: December 1972 (1 atm) CURRENT: December 1972 (1 bar)

Strontium Fluoride, Ion (SrF^+) $F_1Sr_1^+(g)$

TITANIUM FLUORIDE (TiF) IDEAL GAS M_r = 66.878403

Titanium Fluoride (TiF) $F_1Ti_1(g)$

$S°(298.15\ K) = 237.31 \pm 8.4\ J\ K^{-1}\ mol^{-1}$

$\Delta_fH°(0\ K) = [-66.941 \pm 41.8]\ kJ\ mol^{-1}$

$\Delta_fH°(298.15\ K) = [-66.944 \pm 41.8]\ kJ\ mol^{-1}$

Electronic Levels and Quantum Weights

ε_i, cm^{-1}	g_i	ε_i, cm^{-1}	g_i
0	[4]	[6000]	[4]
[1000]	[4]	[8000]	[4]
[2000]	[4]	[10000]	[4]
[4000]	[4]		

$\omega_e = [590]\ cm^{-1}$ $\omega_ex_e = [6.6]\ cm^{-1}$ $\sigma = 1$

$B_e = [0.328]\ cm^{-1}$ $\alpha_e = [0.0036]\ cm^{-1}$ $r_e = [1.95]$ Å

Enthalpy of Formation

The enthalpy of formation, $\Delta_fH°(TiF, g, 298.15\ K)$, is calculated from the dissociation energy. The dissociation energy is estimated as 148±6 kcal mol^{-1} using the corresponding quantity for $TiF_2(g)$ and the relation $D_0^0(MF)/\Delta_{at}H°(MF_2) = 0.46 \pm 0.02$ which Zmbov and Margrave (1) found held for other mono- and di-fluorides.

Heat Capacity and Entropy

The vibrational frequency, ω_e, and anharmonic vibrational term, ω_ex_e, are estimated from those of the alkali and mercury monohalides. The interatomic distance, r_e, is estimated from those of TiCl(g), $TiF_2(g)$ and $TiF_3(g)$. B_e is calculated from r_e. The ground state term and electronic levels are estimated from the ground state multiplet of Ti^+ (2).

References

1. K. F. Zmbov and J. L. Margrave, J. Phys. Chem. 71, 2893 (1967).
2. C. E. Moore, U. S. Nat. Bur. Stand. Circ. 467, 1949.

Enthalpy Reference Temperature = T_r = 298.15 K; Standard State Pressure = p° = 0.1 MPa

T/K	$C_p°$ (J K⁻¹mol⁻¹)	S° (J K⁻¹mol⁻¹)	$-[G°-H°(T_r)]/T$ (J K⁻¹mol⁻¹)	$H°-H°(T_r)$ (kJ mol⁻¹)	$\Delta_fH°$ (kJ mol⁻¹)	$\Delta_fG°$ (kJ mol⁻¹)	Log K_f
0	0.	0.	INFINITE	-9.239	-66.941	-66.941	INFINITE
100	29.254	203.037	266.316	-6.328	-66.044	-77.003	40.222
200	31.809	223.943	240.429	-3.297	-66.394	-87.872	22.950
250	33.638	231.236	237.881	-1.661	-66.673	-93.210	19.475
298.15	35.339	237.309	237.309	0.	-66.944	-98.296	17.221
300	35.401	237.528	237.309	0.065	-66.954	-98.490	17.149
350	36.971	243.105	237.746	1.876	-67.226	-103.725	15.480
400	38.309	248.132	238.736	3.759	-67.484	-108.921	14.224
450	39.417	252.711	240.038	5.703	-67.731	-114.086	13.243
500	40.316	256.912	241.518	7.697	-67.969	-119.224	12.455
600	41.599	264.385	244.722	11.798	-68.430	-129.431	11.268
700	42.376	270.861	248.004	16.000	-68.892	-139.561	10.414
800	42.830	276.552	251.224	20.262	-69.359	-149.625	9.770
900	43.089	281.613	254.325	24.559	-69.880	-159.628	9.265
1000	43.240	286.161	257.285	28.876	-70.529	-169.567	8.857
1100	43.333	290.287	260.100	33.205	-71.375	-179.432	8.521
1200	43.402	294.061	262.775	37.542	-76.406	-189.090	8.231
1300	43.461	297.537	265.318	41.885	-76.928	-198.460	7.974
1400	43.520	300.760	267.735	46.235	-77.533	-207.786	7.753
1500	43.584	303.765	270.038	50.590	-78.233	-217.066	7.559
1600	43.652	306.580	272.235	54.951	-79.041	-226.296	7.388
1700	43.725	309.228	274.334	59.320	-79.967	-235.472	7.235
1800	43.801	311.730	276.343	63.696	-81.025	-244.589	7.098
1900	43.879	314.100	278.268	68.080	-82.224	-253.644	6.973
2000	43.958	316.353	280.117	72.472	-98.265	-262.178	6.847
2100	44.037	318.499	281.894	76.872	-100.533	-270.318	6.724
2200	44.113	320.550	283.604	81.280	-102.794	-278.350	6.609
2300	44.186	322.512	285.254	85.695	-105.047	-286.280	6.502
2400	44.255	324.394	286.846	90.117	-107.290	-294.111	6.401
2500	44.320	326.202	288.384	94.545	-109.522	-301.850	6.307
2600	44.381	327.942	289.872	98.981	-111.742	-309.499	6.218
2700	44.437	329.618	291.313	103.422	-113.948	-317.063	6.134
2800	44.488	331.235	292.710	107.868	-116.139	-324.546	6.054
2900	44.534	332.797	294.066	112.319	-118.315	-331.950	5.979
3000	44.575	334.307	295.382	116.774	-120.474	-339.280	5.907
3100	44.612	335.769	296.662	121.234	-122.615	-346.539	5.839
3200	44.645	337.186	297.906	125.697	-124.739	-353.728	5.774
3300	44.674	338.560	299.117	130.163	-126.845	-360.851	5.712
3400	44.700	339.894	300.297	134.631	-128.932	-367.911	5.652
3500	44.723	341.191	301.447	139.103	-131.000	-374.910	5.595
3600	44.742	342.451	302.569	143.576	-133.050	-381.849	5.540
3700	44.760	343.677	303.663	148.051	-544.179	-380.944	5.378
3800	44.775	344.871	304.732	152.528	-544.957	-376.522	5.176
3900	44.789	346.034	305.776	157.006	-545.772	-372.079	4.983
4000	44.800	347.168	306.797	161.485	-546.623	-367.614	4.801
4100	44.811	348.274	307.795	165.966	-547.508	-363.128	4.626
4200	44.821	349.354	308.772	170.448	-548.426	-358.620	4.460
4300	44.830	350.409	309.728	174.930	-549.376	-354.090	4.301
4400	44.838	351.440	310.664	179.414	-550.356	-349.537	4.150
4500	44.847	352.448	311.581	183.898	-551.364	-344.962	4.004
4600	44.855	353.433	312.481	188.383	-552.400	-340.364	3.865
4700	44.863	354.398	313.362	192.869	-553.462	-335.743	3.731
4800	44.871	355.343	314.227	197.355	-554.547	-331.099	3.603
4900	44.880	356.268	315.076	201.843	-555.655	-326.432	3.480
5000	44.889	357.175	315.909	206.331	-556.783	-321.743	3.361
5100	44.899	358.064	316.726	210.821	-557.930	-317.031	3.247
5200	44.909	358.936	317.530	215.311	-559.094	-312.296	3.137
5300	44.920	359.791	318.319	219.803	-560.274	-307.539	3.031
5400	44.932	360.631	319.095	224.295	-561.467	-302.759	2.929
5500	44.944	361.456	319.858	228.789	-562.673	-297.957	2.830
5600	44.958	362.266	320.608	233.284	-563.889	-293.133	2.734
5700	44.972	363.061	321.346	237.781	-565.115	-288.287	2.642
5800	44.988	363.844	322.072	242.279	-566.348	-283.420	2.552
5900	45.004	364.613	322.786	246.778	-567.587	-278.531	2.466
6000	45.021	365.369	323.490	251.279	-568.832	-273.621	2.382

PREVIOUS: December 1967 (1 atm) CURRENT: December 1967 (1 bar)

Titanium Fluoride (TiF) $F_1Ti_1(g)$

TUNGSTEN FLUORIDE (WF) IDEAL GAS M_r = 202.848403

$S°(298.15\ K) = [251.141]\ J\ K^{-1}\ mol^{-1}$

$\Delta_f H°(0\ K) = [386.518 \pm 62.8]\ kJ\ mol^{-1}$
$\Delta_f H°(298.15\ K) = [386.183 \pm 62.8]\ kJ\ mol^{-1}$

Electronic Levels and Quantum Weights

State	ε_i, cm^{-1}	g_i
$[^2\Delta]$	0	[4]
	[5000]	[2]
	[15000]	[20]

$\omega_e = [726.5]\ cm^{-1}$ $\omega_e x_e = [2.9]\ cm^{-1}$ $\sigma = 1$
$B_e = [0.2923]\ cm^{-1}$ $\alpha_e = [0.0015]\ cm^{-1}$ $r_e = [1.83]$ Å

Enthalpy of Formation

The enthalpy of formation, $\Delta_f H°(WF,\ g,\ 298.15\ K) = 92.3\ kcal\ mol^{-1}$, is calculated from the bond dissociation energy $\Delta_{at}H°(WF,\ 298.15\ K) = 130 \pm 15\ kcal\ mol^{-1}$. This value of $\Delta_{at}H°(298.15\ K)$ is estimated to be slightly higher than the average bond dissociation energy of $WF_6(g)$ by analogy with the WCl_x system.

Heat Capacity and Entropy

The bond distance is estimated to be the same as that in $WF_6(g)$. This distance is then used with Guggenheimer's (1) relation to calculate the fundamental vibrational frequency, ω_e (1). The anharmonicity correction x_e is estimated roughly by assuming $x_e = \omega_e/4(D_0^0 + 0.5\ \omega_e) = 0.0042$. The rotational constant, B_e, is calculated from the estimated bond distance. The value of α_e is calculated form the Morse potential function.

The ground state configuration, low lying electronic levels and their quantum weights are estimated based on the tentative molecular orbital diagram for TaO given by Weltner and McLeod (2). Using this scheme, we obtain the ground state configuration as $\pi^4\sigma\delta^4$ which yields a $^2\Sigma$ state. Higher levels are considered to be formed by transition to π^* antibonding orbitals, and by analogy with TaO these are assumed to lie in the 15000 cm^{-1} region. The degeneracy of these levels is estimated roughly from the large number of possible configurations.

References

1. K. M. Guggenheimer, Proc. Phys. Soc. 58, 456 (1946).
2. W. Weltner, Jr., and D. McLeod, Jr., J. Chem. Phys. 42, 882 (1965).

Tungsten Fluoride (WF) $F_1W_1(g)$

Enthalpy Reference Temperature = T_r = 298.15 K — Standard State Pressure = p° = 0.1 MPa

T/K	$C_p°$ (J K⁻¹mol⁻¹)	S° (J K⁻¹mol⁻¹)	$-[G°-H°(T_r)]/T$ (J K⁻¹mol⁻¹)	$H°-H°(T_r)$ (kJ mol⁻¹)	$\Delta_f H°$ (kJ mol⁻¹)	$\Delta_f G°$ (kJ mol⁻¹)	Log K_f
0	0.	0.	INFINITE	-9.051	386.518	386.518	INFINITE
100	29.135	217.762	279.175	-6.141	387.322	375.025	-195.893
200	30.708	238.278	254.142	-3.173	386.818	362.882	-94.775
250	32.340	245.300	251.691	-1.598	386.485	356.937	-74.578
298.15	34.026	251.141	251.141	0.	386.183	351.274	-61.542
300	34.090	251.351	251.141	0.063	386.172	351.057	-61.125
350	35.728	256.732	251.563	1.809	385.900	345.227	-51.522
400	37.157	261.598	252.518	3.632	385.668	339.433	-44.325
450	38.353	266.046	253.777	5.521	385.472	333.665	-38.731
500	39.326	270.139	255.212	7.464	385.303	327.918	-34.257
600	40.693	277.441	258.323	11.471	385.015	316.469	-27.551
700	41.464	283.778	261.517	15.583	384.751	305.066	-22.764
800	41.824	289.342	264.655	19.750	384.468	293.701	-19.177
900	41.920	294.275	267.677	23.939	384.141	282.375	-16.389
1000	41.854	298.690	270.561	28.128	383.752	271.088	-14.160
1100	41.697	302.672	273.303	32.306	383.291	259.843	-12.339
1200	41.493	306.292	275.903	36.466	382.753	248.643	-10.823
1300	41.271	309.604	278.370	40.604	382.137	237.492	-9.543
1400	41.049	312.654	280.711	44.720	381.441	226.391	-8.447
1500	40.835	315.479	282.936	48.815	380.668	215.343	-7.499
1600	40.635	318.108	285.053	52.888	379.817	204.348	-6.671
1700	40.451	320.566	287.071	56.942	378.891	193.409	-5.943
1800	40.284	322.873	288.996	60.979	377.893	182.527	-5.297
1900	40.133	325.047	290.837	64.999	376.826	171.702	-4.720
2000	39.998	327.102	292.599	69.006	375.691	160.935	-4.203
2100	39.877	329.051	294.289	72.999	374.491	150.227	-3.737
2200	39.769	330.903	295.912	76.982	373.229	139.577	-3.314
2300	39.674	332.669	297.472	80.954	371.909	128.986	-2.929
2400	39.589	334.356	298.974	84.917	370.532	118.453	-2.578
2500	39.514	335.970	300.422	88.872	369.101	107.979	-2.256
2600	39.449	337.519	301.819	92.820	367.623	97.563	-1.960
2700	39.391	339.006	303.169	96.762	366.072	87.206	-1.687
2800	39.340	340.438	304.474	100.698	364.416	76.907	-1.435
2900	39.295	341.818	305.738	104.630	362.618	66.670	-1.201
3000	39.257	343.149	306.963	108.558	360.654	56.498	-0.984
3100	39.223	344.436	308.152	112.481	358.480	46.395	-0.782
3200	39.195	345.681	309.305	116.402	356.068	36.365	-0.594
3300	39.170	346.886	310.426	120.321	353.393	26.415	-0.418
3400	39.149	348.055	311.515	124.236	350.421	16.550	-0.254
3500	39.132	349.190	312.576	128.151	347.089	6.778	-0.101
3600	39.118	350.292	313.608	132.063	343.301	-2.892	0.042
3700	39.107	351.364	314.614	135.974	304.192	-12.260	0.173
3800	39.098	352.407	315.595	139.885	302.780	-20.794	0.286
3900	39.092	353.422	316.552	143.794	301.385	-29.290	0.392
4000	39.088	354.412	317.486	147.703	300.007	-37.752	0.493
4100	39.086	355.377	318.399	151.612	298.645	-46.179	0.588
4200	39.086	356.319	319.290	155.520	297.300	-54.573	0.679
4300	39.088	357.239	320.162	159.429	295.971	-62.935	0.765
4400	39.091	358.137	321.015	163.338	294.659	-71.267	0.846
4500	39.095	359.016	321.850	167.247	293.364	-79.568	0.924
4600	39.101	359.875	322.667	171.157	292.085	-87.842	0.997
4700	39.108	360.716	323.468	175.067	290.822	-96.087	1.068
4800	39.116	361.540	324.252	178.979	289.574	-104.306	1.135
4900	39.125	362.346	325.022	182.891	288.343	-112.499	1.199
5000	39.135	363.137	325.776	186.804	287.126	-120.667	1.261
5100	39.146	363.912	326.516	190.718	285.925	-128.811	1.319
5200	39.158	364.672	327.243	194.633	284.738	-136.932	1.375
5300	39.170	365.418	327.956	198.549	283.565	-145.030	1.429
5400	39.183	366.150	328.656	202.467	282.407	-153.105	1.481
5500	39.197	366.869	329.345	206.386	281.262	-161.160	1.531
5600	39.212	367.576	330.021	210.306	280.131	-169.194	1.578
5700	39.227	368.270	330.686	214.228	279.013	-177.208	1.624
5800	39.242	368.952	331.340	218.152	277.908	-185.202	1.668
5900	39.259	369.623	331.983	222.077	276.815	-193.177	1.710
6000	39.275	370.283	332.616	226.003	275.735	-201.134	1.751

PREVIOUS: March 1967 (1 atm) CURRENT: March 1967 (1 bar)

Tungsten Fluoride (WF) $F_1W_1(g)$

ZIRCONIUM FLUORIDE (ZrF) IDEAL GAS M_r =110.218403

$S°(298.15\ K) = [243.736 \pm 8.4]\ J\ K^{-1}\ mol^{-1}$

$\Delta_f H°[83.669 \pm 20.9]\ kJ\ mol^{-1}$
$\Delta_f H°(298.15\ K) = [82.843 \pm 20.9]\ kJ\ mol^{-1}$

Electronic Levels and Quantum Weights

ε_i, cm^{-1}	g_i	ε_i, cm^{-1}	g_i
0	[4]	[18000]	[4]
[3000]	[4]	[24000]	[4]
[6000]	[4]	[30000]	[4]
[12000]	[4]		

$\omega_e = [828.0]\ cm^{-1}$ $\omega_e x_e = [4.0]\ cm^{-1}$ $\sigma = 1$
$B_e = [0.29699]\ cm^{-1}$ $\alpha_e = [0.009]\ cm^{-1}$ $r_e = [1.90]$ Å

Enthalpy of Formation

Based on $[D_0^0(ZrF)/\Delta_{at}H°(ZrF_2)] = 0.46$, as indicated by recent work on the fluorides of Groups II and IV elements (1, 2, 3, 4, 5), and $\Delta_{at}H°(ZrF_2) = 318.1\ kcal\ mol^{-1}$, the $D_0^0(ZrF)$ is calculated as 146.3 kcal mol^{-1}. Using $\Delta_f H°(ZrF, g, 0\ K) = 147.99$, $\Delta_f H°(F, g, 0\ K) = 18.36$, and $D_0^0(ZrF, g) = 146.3\ kcal\ mol^{-1}$, we derive $\Delta_f H°(ZrF, g, 0\ K) = 20.0$ or $\Delta_f H°(ZrF, g, 298.15\ K) = 19.8 \pm 5\ kcal\ mol^{-1}$ (82.843 ± 20.9 kJ mol^{-1}) which is adopted.

The $\Delta_f H°(ZrF, g, 298.15\ K)$ was evaluated as approximately 16±5 kcal mol^{-1} by Murad and Hildenbrand (8), employing a different $\Delta_{at}H°(ZrF_2)$ value. Their value is in agreement with the adopted one, within the assigned uncertainty.

Heat Capacity and Entropy

The ground state configuration is taken from the ground state multiplet of Zr^+ (6). The electronic levels are estimated as three times higher than those of TiF(g) and the quantum weights are assumed to be the same. ω_e is taken to be approximately the same as the ZrF_2 stretching frequency ν_1 for ZrF_4(g). B_e is calculated from the formula $B_e = 2.799076 \times 10^{-39}/I_e$ where $I_e = \mu r_e$, μ = reduced mass of ZrF(g), r_e = 1.90 Å which is estimated from those of ZrF_4, CF_4, CF and other related gaseous molecules, α_e is derived from B_e, ω_e and $\omega_e x_e$ (estimated), according to the method suggested by Herzberg (7).

References

1. G. D. Blue, J. W. Green, R. G. Bautista, and J. L. Margrave, J. Phys. Chem. 67, 877 (1963).
2. H. L. Hildenbrand and E. Murad, J. Chem. Phys. 44, 1524 (1966).
3. G. D. Blue, J. W. Green, T. C. Ehlert, and J. L. Margrave, Nature 119, 804 (1963).
4. T. C. Ehlert and J. L. Margrave, J. Chem. Phys. 41, 1066 (1964).
5. R. A. Kent, T. C. Ehlert, and J. L. Margrave, J. Amer. Chem. Soc. 86, 5080 (1964).
6. C. E. Moore, U. S. Nat. Bur. Stand. Circ. 467, Vol. II, 1952.
7. G. Herzberg, "Spectra of Diatomic Molecules," D. Van Nostrand Co., Inc., New York, 1950.
8. E. Murad and D. L. Hildenbrand, J. Chem. Phys. 45, 4751 (1966).

Zirconium Fluoride (ZrF) $F_1Zr_1(g)$

Enthalpy Reference Temperature = T_r = 298.15 K Standard State Pressure = p° = 0.1 MPa

T/K	$C_p°$ (J K⁻¹mol⁻¹)	S° (J K⁻¹mol⁻¹)	$-[G°-H°(T_r)]/T$ (J K⁻¹mol⁻¹)	$H°-H°(T_r)$ (kJ mol⁻¹)	$\Delta_f H°$ (kJ mol⁻¹)	$\Delta_f G°$ (kJ mol⁻¹)	Log K_f
0	0.	0.	INFINITE	-9.084	83.669	83.669	INFINITE
100	29.199	210.139	271.878	-6.174	84.239	73.149	-38.209
200	31.136	230.853	246.724	-3.174	83.581	62.286	-16.267
250	32.394	237.939	244.280	-1.585	83.203	57.006	-11.911
298.15	33.427	243.736	243.736	0.	82.843	51.993	-9.109
300	33.462	243.942	243.736	0.062	82.829	51.802	-9.019
350	34.316	249.167	244.147	1.757	82.459	46.660	-6.964
400	34.996	253.796	245.069	3.491	82.089	41.571	-5.429
450	35.552	257.951	246.274	5.255	81.715	36.528	-4.240
500	36.027	261.722	247.633	7.044	81.335	31.528	-3.294
600	36.856	268.365	250.549	10.689	80.552	21.639	-1.884
700	37.629	274.105	253.513	14.414	79.734	11.885	-0.887
800	38.392	279.179	256.410	18.215	78.881	2.250	-0.147
900	39.142	283.744	259.198	22.092	77.983	-7.276	0.422
1000	39.866	287.906	261.864	26.042	77.033	-16.698	0.872
1100	40.548	291.738	264.408	30.064	76.023	-26.023	1.236
1200	41.178	295.294	266.835	34.150	71.235	-35.031	1.525
1300	41.752	298.613	269.153	38.297	70.629	-43.862	1.762
1400	42.268	301.726	271.370	42.499	70.024	-52.647	1.964
1500	42.727	304.659	273.492	46.749	69.406	-61.387	2.138
1600	43.132	307.429	275.528	51.042	68.758	-70.086	2.288
1700	43.486	310.055	277.482	55.374	68.066	-78.743	2.419
1800	43.794	312.549	279.362	59.738	67.314	-87.357	2.535
1900	44.063	314.925	281.171	64.131	66.491	-95.928	2.637
2000	44.295	317.191	282.916	68.549	65.582	-104.453	2.728
2100	44.496	319.357	284.600	72.989	64.575	-112.931	2.809
2200	44.671	321.431	286.228	77.448	42.135	-120.612	2.864
2300	44.824	323.420	287.802	81.923	40.482	-127.973	2.906
2400	44.957	325.331	289.326	86.412	38.846	-135.262	2.944
2500	45.076	327.168	290.803	90.914	37.226	-142.483	2.977
2600	45.181	328.938	292.236	95.427	35.624	-149.639	3.006
2700	45.276	330.645	293.627	99.950	34.040	-156.735	3.032
2800	45.364	332.294	294.979	104.482	32.474	-163.772	3.055
2900	45.445	333.887	296.293	109.022	30.927	-170.753	3.076
3000	45.521	335.429	297.572	113.570	29.401	-177.682	3.094
3100	45.593	336.923	298.817	118.126	27.896	-184.560	3.110
3200	45.663	338.371	300.031	122.689	26.411	-191.389	3.124
3300	45.731	339.777	301.214	127.259	24.949	-198.173	3.137
3400	45.798	341.144	302.369	131.835	23.510	-204.912	3.148
3500	45.865	342.472	303.496	136.418	22.093	-211.610	3.158
3600	45.932	343.765	304.596	141.008	20.699	-218.267	3.167
3700	45.999	345.025	305.672	145.605	19.330	-224.886	3.175
3800	46.067	346.252	306.724	150.208	17.984	-231.469	3.182
3900	46.136	347.450	307.753	154.818	16.662	-238.016	3.188
4000	46.205	348.619	308.760	159.435	15.363	-244.530	3.193
4100	46.276	349.760	309.746	164.059	14.089	-251.011	3.198
4200	46.348	350.876	310.712	168.690	12.839	-257.462	3.202
4300	46.421	351.968	311.659	173.329	11.613	-263.883	3.206
4400	46.495	353.036	312.587	177.975	10.410	-270.276	3.209
4500	46.570	354.082	313.498	182.628	9.231	-276.642	3.211
4600	46.646	355.106	314.391	187.289	8.075	-282.982	3.213
4700	46.724	356.110	315.268	191.957	6.943	-289.297	3.215
4800	46.802	357.095	316.129	196.633	-555.103	-283.971	3.090
4900	46.881	358.060	316.975	201.317	-555.842	-278.315	2.967
5000	46.962	359.008	317.806	206.010	-556.576	-272.643	2.848
5100	47.043	359.939	318.623	210.710	-557.303	-266.957	2.734
5200	47.124	360.853	319.427	215.418	-558.023	-261.257	2.624
5300	47.207	361.752	320.217	220.135	-558.735	-255.543	2.519
5400	47.290	362.635	320.994	224.859	-559.437	-249.816	2.416
5500	47.373	363.503	321.759	229.593	-560.129	-244.076	2.318
5600	47.457	364.358	322.512	234.334	-560.809	-238.324	2.223
5700	47.541	365.198	323.254	239.084	-561.477	-232.559	2.131
5800	47.625	366.026	323.984	243.842	-562.132	-226.783	2.042
5900	47.710	366.841	324.704	248.609	-562.773	-220.995	1.957
6000	47.795	367.643	325.413	253.384	-563.398	-215.197	1.873

PREVIOUS: June 1969 (1 atm) CURRENT: June 1969 (1 bar)

Zirconium Fluoride (ZrF) $F_1Zr_1(g)$

0 to 6000 K Ideal Diatomic Gas

D_0^0 = 154.56 ± 0.60 kJ mol^{-1}
S°(298.15 K) = 202.789 ± 0.040 J K^{-1} mol^{-1}

$\Delta_f H°$(0 K) = 0 kJ mol^{-1}
$\Delta_f H°$(298.15 K) = 0 kJ mol^{-1}

Spectroscopic Data for F_2 in cm^{-1}

v =	0	1	2	3	4	5	6	7	8	9	10	11
G(v)-G(0)	0.00	893.90	1764.15	2610.22	3431.53	4227.43	4997.19	5740.05	6455.17	7141.63	7798.48	8424.67
B_v	0.8833	0.8696	0.8560	0.8423	0.8284	0.8142	0.7996	0.7484	0.7685	0.7518	0.7343	0.7156
D_v x 10^6	3.3	3.1	2.7	4.4	3.6	2.2	4.6	4.5	5.2	5.4	6.5	6.7

v =	12	13	14	15	16	17	18	19	20	21	22
G(v)-G(0)	9019.11	9580.63	10108.02	10599.62	11053.90	11468.96	11842.62	12172.25	12452.98	12678.00	12830.38
B_v	0.6958	0.6747	0.6522	0.6282	0.6025	0.5750	0.5449	0.5094	0.4711	0.4185	0.3865
D_v x 10^6	3.7	9.2	8.5	5.5	8.6	7.0	9.8	15.0	19.6	32.7	86.4

r_e = 1.41193 Å (10)

Enthalpy of Formation

Zero by definition. Refer to the monatomic fluorine gas table for a discussion of the dissociation energy.

Heat Capacity and Entropy

The thermal functions are calculated using a direct summation technique. Due to the absence of known low-lying bound excited states (i.e. <93000 cm^{-1}), only the ground state is included in the calculation.

The G(v), B_v and D_v spectroscopic constants for the ground state and the dissociation energy were determined by Colbourn et al. (1). These are used in the computation as given (and not as the usual fitting coefficients ω_e, $\omega_e x_e$, etc.). The rotational levels are weighted in accordance with the nuclear spin-rotation interaction (5) as follows: even J weight = 0.25, odd J weight = 0.75. Also, the rotational levels are extrapolated to high J values according to the method of Khachkuruzov (6) who proposed a simpler form of Woolley's method (7). Note that there is only one stable isotope of fluorine, ^{19}F.

The so-called quasibound rotational levels above the dissociation limit (2,3) are also included. For a given vibrational quantum number, v, the summation over rotational states is carried out up to a rotational quantum number, J, given by $J = J_{lim}(1-v/v_{max})$ where J_{lim} and v_{max} were determined by Gurvich et al. (4) to be 178 and 23, respectively. Our adopted value for S°(298.15 K) differs from values recommended by CODATA (9), Gurvich et al. (3) and Wagman et al. (11) by less than 0.01 J K^{-1} mol^{-1}. For this comparison, the entropy values reported by CODATA (9) and Gurvich et al. (3) are increased by 0.1094 J K^{-1} mol^{-1} to account for a change in the standard-state pressure from 1 atm to 1 bar.

Phase Data

Hultgren et al. (8) recommended the $\alpha \to \beta$ transition temperature of 45.55 K, the triple point of 53.48 K (0.000249 atm) and the boiling point of 84.95 K (1 atm). These values should prove reasonably accurate, although they have not been evaluated by the present authors, and are furnished for the convenience of the reader. As a result of the low values, the reference state for fluorine is chosen to be the ideal diatomic gas at all temperatures.

References

1. E. A. Colbourn, M. Dagenais, A. E. Douglas and J. W. Raymonda, Can. J. Phys. 54, 1343 (1976).
2. F. M. Mies and P. S. Julienne, J. Chem. Phys. 77, 6162 (1982).
3. L. V. Gurvich, I. V. Veits et al., "Thermodynamic Properties of Individual Substances," 3rd ed., Vol. I, Part 1, pp. 38-40, Nauka, Moscow, 1978.
4. Reference (3), Part 1, pp. 146-9.
5. G. Herzberg, "Spectra of Diatomic Molecules," 2nd ed., Chapter III, D. Van Nostrand Co., Inc., New York, 1950.
6. G. A. Khachkuruzov, Opt. Spectrosc. 30, 455 (1971).
7. H. W. Woolley, R. B. Scott and F. G. Brickwedde, J. Res. Nat. Bur. Stand. 41, 379 (1948).
8. R. Hultgren, P. D. Desai et al., "Selected Values of the Thermodynamic Properties of the Elements," American Society for Metals, Metals Park, Ohio, 1973.
9. J. D. Cox, chairman, ICSU-CODATA Task Group on Key Values for Thermodynamics, J. Chem. Thermodynamics 10, 903 (1978).
10. K. P. Huber and G. Herzberg, "Constants of Diatomic Molecules," Van Nostrand-Reinhold Co., New York, 1979.
11. D. D. Wagman, W. H. Evans et al., J. Phys. Chem. Ref. Data 11, Supp. 2 (1982).

Enthalpy Reference Temperature = T_r = 298.15 K; Standard State Pressure = p° = 0.1 MPa

T/K	$C_p°$ (J K^{-1}mol^{-1})	S°	$-[G°-H°(T_r)]/T$	$H°-H°(T_r)$ (kJ mol^{-1})	$\Delta_f H°$	$\Delta_f G°$	Log K_f
0	0.	0.	INFINITE	-8.825	0.	0.	0.
100	29.114	170.370	229.549	-5.918	0.	0.	0.
200	29.685	190.652	205.596	-2.989	0.	0.	0.
250	30.447	197.354	203.300	-1.486	0.	0.	0.
298.15	31.302	202.789	202.789	0.	0.	0.	0.
300	31.336	202.983	202.790	0.058	0.	0.	0.
350	32.207	207.880	203.175	1.647	0.	0.	0.
400	32.992	212.233	204.040	3.277	0.	0.	0.
450	33.674	216.159	205.172	4.944	0.	0.	0.
500	34.255	219.738	206.452	6.643	0.	0.	0.
600	35.166	226.068	209.208	10.116	0.	0.	0.
700	35.832	231.542	212.016	13.668	0.	0.	0.
800	36.336	236.361	214.764	17.277	0.	0.	0.
900	36.732	240.664	217.407	20.932	0.	0.	0.
1000	37.057	244.552	219.930	24.622	0.	0.	0.
1100	37.334	248.097	222.332	28.341	0.	0.	0.
1200	37.579	251.356	224.617	32.087	0.	0.	0.
1300	37.802	254.373	226.791	35.857	0.	0.	0.
1400	38.008	257.182	228.863	39.647	0.	0.	0.
1500	38.199	259.811	230.839	43.458	0.	0.	0.
1600	38.374	262.282	232.728	47.287	0.	0.	0.
1700	38.530	264.613	234.536	51.132	0.	0.	0.
1800	38.663	266.819	236.268	54.992	0.	0.	0.
1900	38.770	268.913	237.932	58.864	0.	0.	0.
2000	38.846	270.904	239.531	62.745	0.	0.	0.
2100	38.889	272.800	241.071	66.632	0.	0.	0.
2200	38.895	274.609	242.554	70.521	0.	0.	0.
2300	38.864	276.338	243.986	74.409	0.	0.	0.
2400	38.795	277.991	245.369	78.293	0.	0.	0.
2500	38.690	279.572	246.705	82.167	0.	0.	0.
2600	38.549	281.087	247.999	86.030	0.	0.	0.
2700	38.375	282.539	249.251	89.876	0.	0.	0.
2800	38.170	283.931	250.465	93.704	0.	0.	0.
2900	37.939	285.266	251.642	97.509	0.	0.	0.
3000	37.683	286.548	252.785	101.290	0.	0.	0.
3100	37.406	287.779	253.894	105.045	0.	0.	0.
3200	37.111	288.962	254.971	108.771	0.	0.	0.
3300	36.802	290.100	256.019	112.467	0.	0.	0.
3400	36.482	291.193	257.037	116.131	0.	0.	0.
3500	36.152	292.246	258.028	119.763	0.	0.	0.
3600	35.817	293.260	258.993	123.361	0.	0.	0.
3700	35.477	294.237	259.932	126.926	0.	0.	0.
3800	35.135	295.178	260.848	130.457	0.	0.	0.
3900	34.793	296.087	261.740	133.953	0.	0.	0.
4000	34.452	296.963	262.609	137.415	0.	0.	0.
4100	34.114	297.810	263.458	140.844	0.	0.	0.
4200	33.780	298.628	264.285	144.238	0.	0.	0.
4300	33.450	299.419	265.093	147.600	0.	0.	0.
4400	33.125	300.184	265.882	150.929	0.	0.	0.
4500	32.807	300.925	266.653	154.225	0.	0.	0.
4600	32.495	301.642	267.405	157.490	0.	0.	0.
4700	32.191	302.338	268.141	160.724	0.	0.	0.
4800	31.893	303.013	268.861	163.929	0.	0.	0.
4900	31.603	303.667	269.565	167.103	0.	0.	0.
5000	31.321	304.303	270.253	170.249	0.	0.	0.
5100	31.046	304.920	270.927	173.368	0.	0.	0.
5200	30.779	305.521	271.586	176.459	0.	0.	0.
5300	30.519	306.104	272.232	179.524	0.	0.	0.
5400	30.268	306.673	272.865	182.563	0.	0.	0.
5500	30.023	307.226	273.484	185.577	0.	0.	0.
5600	29.787	307.765	274.092	188.568	0.	0.	0.
5700	29.557	308.290	274.687	191.535	0.	0.	0.
5800	29.335	308.802	275.271	194.480	0.	0.	0.
5900	29.119	309.301	275.843	197.402	0.	0.	0.
6000	28.911	309.789	276.405	200.304	0.	0.	0.

PREVIOUS: June 1977 (1 atm) CURRENT: June 1982 (1 bar)

IRON FLUORIDE (FeF_2) CRYSTAL M_r = 93.843806

$S°(298.15\ K) = 86.985 \pm 0.17\ J\ K^{-1}\ mol^{-1}$

T_{fus} = [1373] K

$\Delta_f H°(0\ K) = -705.266 \pm 41.8\ kJ\ mol^{-1}$

$\Delta_f H°(298.15\ K) = -705.841 \pm 41.8\ kJ\ mol^{-1}$

$\Delta_{fus} H°$ = [51.882] kJ mol^{-1}

Enthalpy of Formation

The equilibrium pressures for the reaction $FeF_2(cr) + H_2(g) = Fe(cr) + 2HF(g)$ have been measured by Jellinek and Rudat (1) at 773, 873 and 973 K. The enthalpy change of the reaction, $\Delta_r H°(298.15\ K)$, was evaluated by the 2nd and 3rd law methods to be 27.19 and 39.08 kcal mol^{-1}, respectively. Based on the 3rd law value of $\Delta_r H°(298.15\ K)$, the $\Delta_f H°(FeF_2$, cr, 298.15 K) was calculated to be -168.7±10 kcal mol^{-1}. The drift in the 3rd law values of $\Delta_r H°(298.15\ K)$ was evaluated to be 13.3±1.8 cal K^{-1} mol^{-1}.

Heat Capacity and Entropy

The low temperature (11.33 - 307.30 K) heat capacities were determined by Catalano and Stout (2). The heat capacities above 307 K were estimated by comparison with those of $FeCl_2(cr)$, $MgCl_2(cr)$ and $MgF_2(cr)$. The value of S°(298.15 K) was obtained from Catalano and Stout (2), in which a sharp maximum $C_p°$ of 17.8 cal K^{-1} mol^{-1} at 78.35 K was reported. The S°(298.15 K) value was based on S°(15 K) = 0.039 cal K^{-1} mol^{-1}.

Fusion Data

T_{fus} was estimated as 1100°C based on the data reported by Biltz and Rahlfs (3). The value of $\Delta_{fus}H°$ was calculated using $\Delta_m S°$ = 9 cal K^{-1} mol^{-1} which was assumed to be the same as the corresponding value for $MgF_2(cr)$.

Sublimation Data

The value of $\Delta_{sub}H°(298.15\ K)$ = 316.31±8.4 kJ mol^{-1} was obtained from Margrave (4). See $FeF_2(g)$ table for details.

References

1. K. Jellinek and A. Rudat, Z. Anorg. Allg. Chem. 175, 281 (1928).
2. E. Catalano and J. W. Stout, J. Chem. Phys. 23, 1803 (1955).
3. W. Biltz and E. Rahlfs, Z. Anorg. Allg. Chem. 166, 363 (1927).
4. J. L. Margrave, "Research in Fluorine Chemistry", William Rice University, Houston, Texas, Progress Report No. 7, January 1 to March 31, 1965.

Iron Fluoride (FeF_2) $F_2Fe_1(cr)$

Enthalpy Reference Temperature = T_r = 298.15 K Standard State Pressure = p° = 0.1 MPa

T/K	$C_p°$ (J K⁻¹mol⁻¹)	S° (J K⁻¹mol⁻¹)	$-[G°-H°(T_r)]/T$ (J K⁻¹mol⁻¹)	$H°-H°(T_r)$ (kJ mol⁻¹)	$\Delta_f H°$ (kJ mol⁻¹)	$\Delta_f G°$ (kJ mol⁻¹)	Log K_f
0	0.	0.	INFINITE	-12.757	-705.266	-705.266	INFINITE
100	33.447	30.853	138.926	-10.807	-706.646	-692.088	361.510
200	56.610	62.024	92.902	-6.176	-706.712	-677.397	176.918
298.15	68.116	86.985	86.985	0.	-705.841	-663.168	116.184
300	68.199	87.407	86.987	0.126	-705.819	-662.903	115.422
400	71.965	107.562	89.709	7.141	-704.651	-648.774	84.721
500	74.894	123.948	94.968	14.490	-703.521	-634.937	66.331
600	77.153	137.811	100.983	22.097	-702.473	-621.321	54.091
700	78.868	149.837	107.122	29.900	-701.547	-607.871	45.360
800	80.249	160.462	113.138	37.859	-700.813	-594.541	38.820
900	81.337	169.979	118.934	45.941	-700.414	-581.286	33.737
1000	82.132	178.592	124.476	54.115	-700.676	-568.045	29.672
1100	82.717	186.448	129.758	62.359	-701.795	-554.713	26.341
1200	83.178	193.666	134.787	70.655	-702.335	-541.337	23.564
1300	83.516	200.338	139.576	78.990	-701.212	-527.966	21.214
1373.000	83.685	204.906	142.929	85.094	------ CRYSTAL <--> LIQUID -----		
1400	83.764	206.536	144.140	87.354	-700.166	-514.679	19.203
1500	83.983	212.323	148.495	95.742	-699.200	-501.464	17.463
1600	84.182	217.750	152.656	104.150	-698.315	-488.311	15.942
1700	84.365	222.859	156.636	112.578	-698.461	-475.192	14.601
1800	84.517	227.685	160.451	121.022	-698.073	-462.071	13.409
1900	84.637	232.258	164.111	129.480	-711.865	-448.266	12.324
2000	84.726	236.602	167.628	137.949	-711.880	-434.392	11.345

PREVIOUS: CURRENT: September 1965

Iron Fluoride (FeF_2) $F_2Fe_1(cr)$

IRON FLUORIDE (FeF_2) LIQUID M_r = 93.843806

$S°(298.15\ K) = [92.537]\ J\ K^{-1}\ mol^{-1}$ $\Delta_f H°(298.15\ K) = [-674.549]\ kJ\ mol^{-1}$

$T_{fus} = [1373]\ K$ $\Delta_{fus}H° = [51.882]\ kJ\ mol^{-1}$

Enthalpy of Formation

$\Delta_f H°(FeF_2, \ell, 298.15\ K)$ is calculated from $\Delta_f H°(FeF_2, cr, 298.15\ K)$ by adding the enthalpy of fusion, $\Delta_{fus}H°$, and the difference in enthalpy, H°(1373 K) - H°(298.15 K), between the crystal and liquid.

Heat Capacity and Entropy

The heat capacity for $FeF_2(\ell)$ is estimated by comparison with those for $FeCl_2(cr)$, $MgCl_2(cr)$, $MgF_2(cr)$, $CaCl_2(cr)$ and $CaF_2(cr)$. The $C_p°$ (FeF_2, ℓ) value is assumed to be constant in the temperature range, 298.15 - 3000 K. S°(298.15 K) is calculated in a manner analogous to that used for the enthalpy of formation.

Fusion Data

See the crystal table for details.

Vaporization Data

T_{vap} = 2110 K is calculated as the temperature at which the fugacity of the reaction $FeF_2(\ell) = FeF_2(g)$ is 1 atm. The difference between $\Delta_f H°$ for FeF_2 (ℓ) and $FeF_2(g)$ at T_{vap} is $\Delta_{vap}H° = 224.43\ kJ\ mol^{-1}$.

Iron Fluoride (FeF_2) $F_2Fe_1(l)$

Enthalpy Reference Temperature = T_r = 298.15 K Standard State Pressure = p° = 0.1 MPa

	J K⁻¹mol⁻¹			kJ mol⁻¹			
T/K	$C_p°$	S°	-[G°-H°(T_r)]/T	H°-H°(T_r)	$\Delta_f H°$	$\Delta_f G°$	Log K_f
0							
100							
200							
298.15	98.324	92.537	92.537	0.	-674.549	-633.532	110.992
300	98.324	93.145	92.539	0.182	-674.472	-633.277	110.263
400	98.324	121.431	96.395	10.014	-670.486	-620.157	80.984
500	98.324	143.371	103.678	19.847	-666.872	-608.000	63.517
600	98.324	161.298	111.833	29.679	-663.599	-596.539	51.933
700	98.324	176.455	120.010	39.511	-660.644	-585.600	43.698
800	98.324	189.584	127.904	49.344	-658.036	-575.062	37.548
900	98.324	201.165	135.413	59.176	-655.886	-564.825	32.782
1000	98.324	211.524	142.516	69.009	-654.491	-554.792	28.979
1100	98.324	220.896	149.222	78.841	-654.021	-544.831	25.872
1200	98.324	229.451	155.556	88.673	-653.025	-534.968	23.287
1300	98.324	237.321	161.547	98.506	-650.405	-525.237	21.104
1373.000	98.324	242.693	165.720	105.684	------ CRYSTAL <--> LIQUID -----		
1400	98.324	244.608	167.223	108.338	-647.891	-515.703	19.241
1500	98.324	251.391	172.611	118.171	-645.480	-506.346	17.633
1600	98.324	257.737	177.735	128.003	-643.170	-497.147	16.230
1700	98.324	263.698	182.618	137.835	-641.911	-488.069	14.997
1800	98.324	269.318	187.280	147.668	-640.135	-479.072	13.902
1900	98.324	274.634	191.739	157.500	-652.553	-469.468	12.907
2000	98.324	279.677	196.011	167.333	-651.204	-459.867	12.010
2100	98.324	284.475	200.110	177.165	-649.861	-450.333	11.201
2200	98.324	289.049	204.050	186.997	-648.521	-440.863	10.467
2300	98.324	293.419	207.841	196.830	-647.179	-431.454	9.799
2400	98.324	297.604	211.495	206.662	-645.832	-422.104	9.187
2500	98.324	301.618	215.020	216.495	-644.477	-412.809	8.625
2600	98.324	305.474	218.425	226.327	-643.109	-403.570	8.108
2700	98.324	309.185	221.718	236.159	-641.725	-394.383	7.630
2800	98.324	312.761	224.906	245.992	-640.323	-385.248	7.187
2900	98.324	316.211	227.996	255.824	-638.899	-376.163	6.775
3000	98.324	319.544	230.992	265.657	-637.450	-367.128	6.392

PREVIOUS: CURRENT: September 1965

Iron Fluoride (FeF_2) $F_2Fe_1(l)$

IRON FLUORIDE (FeF_2) CRYSTAL-LIQUID M_r = 93.843806

0 to 1373 K crystal
above 1373 K liquid

Refer to the individual tables for details.

Iron Fluoride (FeF_2) F_2Fe_1(cr,l)

Enthalpy Reference Temperature = T_r = 298.15 K Standard State Pressure = p° = 0.1 MPa

T/K	C_p° (J K⁻¹mol⁻¹)	S° (J K⁻¹mol⁻¹)	$-[G°-H°(T_r)]/T$ (J K⁻¹mol⁻¹)	$H°-H°(T_r)$ (kJ mol⁻¹)	$\Delta_fH°$ (kJ mol⁻¹)	$\Delta_fG°$ (kJ mol⁻¹)	Log K_f
0	0.	0.	INFINITE	-12.757	-705.266	-705.266	INFINITE
100	33.447	30.853	138.926	-10.807	-706.646	-692.088	361.510
200	56.610	62.024	92.902	-6.176	-706.712	-677.397	176.918
298.15	68.116	86.985	86.985	0.	-705.841	-663.168	116.184
300	68.199	87.407	86.987	0.126	-705.819	-662.903	115.422
400	71.965	107.562	89.709	7.141	-704.651	-648.774	84.721
500	74.894	123.948	94.968	14.490	-703.521	-634.937	66.331
600	77.153	137.811	100.983	22.097	-702.473	-621.321	54.091
700	78.868	149.837	107.122	29.900	-701.547	-607.871	45.360
800	80.249	160.462	113.138	37.859	-700.813	-594.541	38.820
900	81.337	169.979	118.934	45.941	-700.414	-581.286	33.737
1000	82.132	178.592	124.476	54.115	-700.676	-568.045	29.672
1100	82.717	186.448	129.758	62.359	-701.795	-554.713	26.341
1200	83.178	193.666	134.787	70.655	-702.335	-541.337	23.564
1300	83.516	200.338	139.576	78.990	-701.212	-527.966	21.214
1373.000	83.685	204.906	142.929	85.094	CRYSTAL <--> LIQUID TRANSITION		
1373.000	98.324	242.693	142.929	136.975			
1400	98.324	244.608	144.872	139.630	-647.891	-515.703	19.241
1500	98.324	251.391	151.750	149.462	-645.480	-506.346	17.633
1600	98.324	257.737	158.178	159.295	-643.170	-497.147	16.230
1700	98.324	263.698	164.211	169.127	-641.911	-488.069	14.997
1800	98.324	269.318	169.896	178.960	-640.135	-479.072	13.902
1900	98.324	274.634	175.270	188.792	-652.553	-469.468	12.907
2000	98.324	279.677	180.365	198.624	-651.204	-459.867	12.010
2100	98.324	284.475	185.209	208.457	-649.861	-450.333	11.201
2200	98.324	289.049	189.826	218.289	-648.521	-440.863	10.467
2300	98.324	293.419	194.236	228.122	-647.179	-431.454	9.799
2400	98.324	297.604	198.456	237.954	-645.832	-422.104	9.187
2500	98.324	301.618	202.503	247.786	-644.477	-412.809	8.625
2600	98.324	305.474	206.390	257.619	-643.109	-403.570	8.108
2700	98.324	309.185	210.129	267.451	-641.725	-394.383	7.630
2800	98.324	312.761	213.731	277.284	-640.323	-385.248	7.187
2900	98.324	316.211	217.205	287.116	-638.899	-376.163	6.775
3000	98.324	319.544	220.561	296.948	-637.450	-367.128	6.392

PREVIOUS: CURRENT: September 1965

Iron Fluoride (FeF_2) F_2Fe_1(cr,l)

IRON FLUORIDE (FeF_2) IDEAL GAS M_r = 93.843806

S°(298.15 K) = [265.205] $J K^{-1} mol^{-1}$

$\Delta_f H°$(0 K) = -388.870 ± 14.2 $kJ mol^{-1}$
$\Delta_f H°$(298.15 K) = -389.530 ± 14.2 $kJ mol^{-1}$

Electronic Levels and Quantum Weights

ε_i, cm^{-1}	g_i
0	[5]
[1000]	[10]
[4000]	[5]
[6000]	[5]

Vibrational Frequencies and Degeneracies

ν, cm^{-1}
[550](1)
[190](2)
[714](1)

Point Group: [$D_{\infty h}$] σ = 2
Bond Distance: Fe-F = [1.8] Å
Bond Angle: F-Fe-F = [180]°
Rotational constant: B_0 = 0.136932 cm^{-1}

Enthalpy of Formation

The rate of sublimation of FeF_2(cr) was measured between 965 and 1149 K by the Knudsen technique employing the high temperature magnetic mass spectrometer by Margrave (1). The only vapor species of importance is FeF_2(g). From the vapor pressure data, the 3rd law enthalpy of sublimation, $\Delta_{sub}H°$(298.15 K), was reported as 75.6±2.0 kcal mol^{-1}. The sum of the values $\Delta_{sub}H°$(298.15 K) and $\Delta_f H°$(298.15 K) for FeF_2(cr) is $\Delta_f H°$(FeF_2, g, 298.15 K).

Heat Capacity and Entropy

The molecular structure was assumed to be linear. The Fe-F bond distance was estimated from that in FeF_2(cr) reported by Baur (2). The vibrational frequencies, ν_1 and ν_3, were estimated by Brewer et al. (3). The values of ν_2, electronic levels and quantum weights were estimated by comparison with those for $FeCl_2$(g). See the ideal gas table for details.

References

1. J. L. Margrave, "Research in Fluorine Chemistry", William Marsh Rice University, Houston, Texas, Progress Report No. 7, Jan. 1 to March 31, 1965.
2. W. H. Baur, Acta Cryst. 11, 488 (1958).
3. L. Brewer, G. R. Somayajulu and E. Brackett, Chem. Rev. 63, 111 (1963).

Iron Fluoride (FeF_2) F_2Fe_1(g)

Enthalpy Reference Temperature = T_r = 298.15 K Standard State Pressure = p° = 0.1 MPa

T/K	$C_p°$ (J K⁻¹mol⁻¹)	S° (J K⁻¹mol⁻¹)	-[G°-H°(T_r)]/T (J K⁻¹mol⁻¹)	H°-H°(T_r) (kJ mol⁻¹)	$\Delta_f H°$ (kJ mol⁻¹)	$\Delta_f G°$ (kJ mol⁻¹)	Log K_f
0	0.	0.	INFINITE	-12.671	-388.870	-388.870	INFINITE
100	38.559	214.824	309.258	-9.443	-388.972	-392.811	205.183
200	47.902	244.563	270.063	-5.100	-389.326	-396.519	103.560
250	52.104	255.709	266.103	-2.599	-389.465	-398.300	83.220
298.15	55.747	265.205	265.205	0.	-389.530	-399.994	70.077
300	55.876	265.551	265.206	0.103	-389.532	-400.059	69.656
350	58.995	274.407	265.899	2.978	-389.532	-401.813	59.967
400	61.390	282.449	267.473	5.991	-389.491	-403.570	52.701
450	63.115	289.786	269.551	9.106	-389.441	-405.332	47.050
500	64.284	296.500	271.915	12.293	-389.408	-407.100	42.529
600	65.466	308.344	277.026	18.791	-389.469	-410.636	35.749
700	65.785	318.466	282.241	25.358	-389.780	-414.144	30.904
800	65.747	327.251	287.331	31.936	-390.425	-417.585	27.265
900	65.601	334.986	292.205	38.504	-391.541	-420.919	24.429
1000	65.452	341.890	296.834	45.056	-393.425	-424.092	22.152
1100	65.339	348.123	301.218	51.595	-396.248	-427.008	20.277
1200	65.266	353.805	305.367	58.125	-398.555	-429.722	18.705
1300	65.229	359.027	309.296	64.650	-399.243	-432.292	17.370
1400	65.219	363.861	313.023	71.172	-400.038	-434.805	16.223
1500	65.226	368.360	316.564	77.694	-400.937	-437.258	15.227
1600	65.244	372.570	319.935	84.218	-401.937	-439.647	14.353
1700	65.265	376.526	323.148	90.743	-403.985	-441.952	13.580
1800	65.286	380.257	326.218	97.270	-405.514	-444.142	12.889
1900	65.304	383.788	329.156	103.800	-421.235	-445.542	12.249
2000	65.316	387.138	331.972	110.331	-423.187	-446.771	11.668
2100	65.323	390.325	334.676	116.863	-425.144	-447.902	11.141
2200	65.323	393.364	337.275	123.395	-427.104	-448.940	10.659
2300	65.316	396.267	339.777	129.927	-429.063	-449.888	10.217
2400	65.302	399.047	342.189	136.458	-431.017	-450.752	9.810
2500	65.283	401.712	344.517	142.988	-432.965	-451.534	9.434
2600	65.258	404.272	346.766	149.515	-434.903	-452.239	9.086
2700	65.228	406.734	348.942	156.039	-436.827	-452.869	8.761
2800	65.193	409.106	351.049	162.560	-438.736	-453.428	8.459
2900	65.155	411.393	353.090	169.078	-440.627	-453.919	8.176
3000	65.114	413.601	355.071	175.591	-442.497	-454.346	7.911
3100	65.070	415.736	356.994	182.100	-444.345	-454.710	7.662
3200	65.024	417.801	358.862	188.605	-794.468	-447.591	7.306
3300	64.976	419.801	360.678	195.105	-794.366	-436.753	6.913
3400	64.927	421.740	362.446	201.600	-794.269	-425.918	6.543
3500	64.877	423.621	364.167	208.090	-794.179	-415.085	6.195
3600	64.827	425.448	365.844	214.576	-794.094	-404.255	5.866
3700	64.776	427.224	367.479	221.056	-794.015	-393.427	5.554
3800	64.726	428.951	369.074	227.531	-793.943	-382.602	5.259
3900	64.675	430.631	370.631	234.001	-793.879	-371.778	4.979
4000	64.625	432.268	372.151	240.466	-793.822	-360.955	4.714
4100	64.575	433.863	373.637	246.926	-793.775	-350.134	4.461
4200	64.525	435.419	375.090	253.381	-793.739	-339.314	4.220
4300	64.477	436.936	376.511	259.831	-793.715	-328.495	3.990
4400	64.429	438.418	377.901	266.276	-793.704	-317.676	3.771
4500	64.382	439.865	379.262	272.717	-793.709	-306.857	3.562
4600	64.336	441.280	380.595	279.152	-793.730	-296.038	3.362
4700	64.290	442.663	381.901	285.584	-793.770	-285.218	3.170
4800	64.246	444.016	383.181	292.011	-793.829	-274.397	2.986
4900	64.203	445.340	384.436	298.433	-793.911	-263.575	2.810
5000	64.161	446.637	385.667	304.851	-794.017	-252.751	2.640
5100	64.119	447.907	386.875	311.265	-794.148	-241.924	2.478
5200	64.079	449.152	388.061	317.675	-794.308	-231.095	2.321
5300	64.040	450.372	389.225	324.081	-794.497	-220.262	2.171
5400	64.002	451.569	390.368	330.483	-794.719	-209.425	2.026
5500	63.965	452.743	391.492	336.881	-794.975	-198.584	1.886
5600	63.928	453.895	392.596	343.276	-795.269	-187.738	1.751
5700	63.893	455.026	393.681	349.667	-795.602	-176.887	1.621
5800	63.859	456.137	394.748	356.055	-795.979	-166.029	1.495
5900	63.825	457.228	395.798	362.439	-796.401	-155.164	1.374
6000	63.793	458.301	396.831	368.820	-796.873	-144.292	1.256

PREVIOUS: September 1965 (1 atm) CURRENT: September 1965 (1 bar)

Iron Fluoride (FeF_2) F_2Fe_1(g)

POTASSIUM FLUORIDE (KHF_2)   CRYSTAL M_r = 78.103046

$S°(298.15\ K) = 104.264 \pm 0.4\ J\ K^{-1}\ mol^{-1}$

$T_{trs} = 469.85\ K$

$T_{fus} = 511.95\ K$

$\Delta_f H°(0\ K) = -926.386 \pm 1.26\ kJ\ mol^{-1}$

$\Delta_f H°(298.15\ K) = -931.233 \pm 1.26\ kJ\ mol^{-1}$

$\Delta_{trs} H° = 11.221 \pm 0.04\ kJ\ mol^{-1}$

$\Delta_{fus} H° = 6.619 \pm 0.013\ kJ\ mol^{-1}$

Enthalpy of Formation

Davis and Westrum (1) measured directly the enthalpy of the decomposition reaction $KHF_2(cr) \rightarrow KF(cr) + HF(g)$ in an adiabatic calorimeter as $\Delta_{dcm}H° = 18.52\pm0.05$ kcal mol^{-1} at 500 K. Reducing to 298.15 K, we obtain $\Delta_{dcm}H°(298.15\ K) = 21.533\pm0.05$ kcal mol^{-1} which leads to the enthalpy of formation, $\Delta_f H°(KHF_2,\ cr,\ 298.15\ K) = -222.57\pm0.3$ kcal mol^{-1}, using JANAF auxiliary data. This value is adopted in the tabulation.

Westrum and Pitzer (3) measured the dissociation vapor pressures of KHF_2(cr, ℓ) by a static method. Their pressure data have been analyzed by both 2nd law and 3rd law methods as given below. The enthalpy of formation derived from the 3rd law analysis differs by about 0.3 kcal mol^{-1} from the value adopted, which is greater than the combined experimental errors.

Reaction	T/K	Data Points	$\Delta_{dcm}H°(298.15\ K)$, kcal mol^{-1} 2nd law	3rd law	Drift cal K^{-1} mol^{-1}	$\Delta_f H°(298.15\ K)$ kcal mol^{-1}
$KHF_2(cr) \rightarrow KF(cr) + HF(g)$	478-500	14	21.21±0.12	21.31±0.01	0.2±0.2	-222.35
$KHF_2(\ell) \rightarrow KF(cr) + HF(g)$	504-577	5	19.82±0.08	17.73±0.24	-3.9±0.2	-222.28**

**This value is derived from the adopted heat of melting, $\Delta_{fus}H°(298.15\ K) = 3.504$ kcal mol^{-1} and $\Delta_f H°(KHF_2,\ \ell,\ 298.15\ K) = -218.773$ kcal mol^{-1}. The latter is calculated from the 3rd law $\Delta_{dcm}H°(298.15\ K)$.

Heat Capacity and Entropy

Westrum and Pitzer (3) measured both low temperature $C°_p$ data (16-315 K) in an adiabatic calorimeter and high temperature enthalpy data (321.6 - 465 K for α phase and 471.3 - 505.9 K for β phase) in a drop calorimeter. We use their experimental $C°_p$ data to derive $S°(298.15\ K) = 24.92\pm0.1$ cal K^{-1} mol^{-1}, based on $S° = 0.0643$ cal K^{-1} mol^{-1} at 15 K. The low temperature $C°_p$ and high temperature enthalpy data for the α phase are smoothly joined at 298.15 K by a polynomial curve fitting method. The deviations of the observed enthalpy data from the adopted values are 0.68% at 321.6 K and 0.09% at 463.7 K. The constant heat capacity, 23.96 cal K^{-1} mol^{-1}, in the β phase is derived from the measured enthalpy data (3), and is assumed to be the same when the temperature is above the melting point.

Transition and Fusion Data

Davis and Westrum (1) measured the heat of the α-β transition, $\Delta_{trs}H° = 2.682\pm0.01$ kcal mol^{-1} at 196.7°C, and the heat of melting, $\Delta_{fus}H° = 1.582\pm0.003$ kcal mol^{-1} at 238.8°C in an adiabatic calorimeter. Their data are adopted in the tabulation. Based on the drop calorimetric measurements (3), we have derived $\Delta_{fus}H° = 1.554$ kcal mol^{-1} which is in good agreement with the value adopted.

References

1. M. L. Davis and E. F. Westrum, Jr., J. Phys. Chem. 65, 338 (1961).
2. JANAF Thermochemical Tables: HF(g), 12-31-68; KF(cr), 6-30-69.
3. E. F. Westrum, Jr. and K. S. Pitzer, J. Amer. Chem. Soc. 71, 1940 (1949).

Potassium Fluoride ($K(HF_2)$) $F_2H_1K_1(cr)$

Enthalpy Reference Temperature = T_r = 298.15 K   Standard State Pressure = p° = 0.1 MPa

T/K	$C°_p$ (J $K^{-1}mol^{-1}$)	S°	$-[G°-H°(T_r)]/T$	$H°-H°(T_r)$ (kJ mol^{-1})	$\Delta_f H°$	$\Delta_f G°$	Log K_f
0	0.	0.	INFINITE	-15.294	-926.386	-926.386	INFINITE
100	48.961	34.988	165.420	-13.043	-930.274	-908.140	474.363
200	67.070	75.565	110.999	-7.087	-931.188	-885.531	231.277
298.15	76.940	104.264	104.264	0.	-931.233	-863.095	151.211
300	77.111	104.741	104.266	0.142	-931.230	-862.672	150.205
400	86.065	128.152	107.398	8.302	-933.221	-839.415	109.616
469.850	92.450	142.501	111.566	14.535	ALPHA <--> BETA		
469.850	100.249	166.384	111.566	25.757	TRANSITION		
500	100.249	172.619	115.061	28.779	-920.678	-816.803	85.331
511.950	100.249	174.987	116.432	29.977	------- BETA <--> LIQUID -------		
600	100.249	190.897	126.224	38.804	-918.631	-796.221	69.317
700	100.249	206.350	136.595	48.829	-916.622	-775.979	57.904
800	100.249	219.737	146.169	58.854	-914.660	-756.021	49.363
900	100.249	231.544	155.012	68.879	-912.760	-736.306	42.734
1000	100.249	242.106	163.203	78.903	-910.939	-716.799	37.442
1100	100.249	251.661	170.817	88.928	-988.173	-692.859	32.901
1200	100.249	260.384	177.923	98.953	-985.512	-666.130	28.996
1300	100.249	268.408	184.579	108.978	-982.895	-639.621	25.700
1400	100.249	275.837	190.835	119.003	-980.322	-613.312	22.883
1500	100.249	282.754	196.735	129.028	-977.791	-587.186	20.448
1600	100.249	289.224	202.316	139.053	-975.301	-561.227	18.322
1700	100.249	295.301	207.609	149.077	-972.851	-535.423	16.452
1800	100.249	301.031	212.641	159.102	-970.438	-509.761	14.793
1900	100.249	306.451	217.437	169.127	-968.060	-484.233	13.313
2000	100.249	311.594	222.017	179.152	-965.716	-458.829	11.983

PREVIOUS: December 1963   CURRENT: June 1971

Potassium Fluoride ($K(HF_2)$) $F_2H_1K_1(cr)$

POTASSIUM FLUORIDE (KHF_2) LIQUID M_r = 78.103046

$S°(298.15\ K)$ = [132.764] J K^{-1} mol^{-1} $\Delta_f H°(298.15\ K)$ = [-916.573] kJ mol^{-1}

T_{fus} = 511.95 K $\Delta_{fus}H°$ = 6.619 ± 0.013 kJ mol^{-1}

Enthalpy of Formation

The enthalpy of formation is calculated from that of the crystal by adding the enthalpy of fusion, $\Delta_{fus}H°$, and the difference in enthalpy, H°(511.95 K) - H°(298.15 K), between the crystal and liquid.

Heat Capacity and Entropy

Westrum and Pitzer (1) measured high temperature enthalpy data by drop calorimetry in a narrow range of 510.6-523.2 K and derived a constant heat capacity of approximately 25 cal K^{-1} mol^{-1} which is adopted in the tabulation.

A glass transition is assumed at 314 K. Below 314 K, the heat capacities are assumed to be the same as those of the crystal. Between 314 K and the melting point, the $C_p°$ is assumed to be the same as that of the liquid.

The entropy is calculated in a manner similar to that used for the enthalpy of formation.

Fusion Data

See the crystal table for details.

Decomposition Data

The decomposition temperature, T_{dcm} = 750 K, is calculated as the temperature at which $\Delta_f G$ = 0 for $KHF_2(\ell) \rightarrow KF(cr) + HF(g)$. At T_{dcm}, the vapor pressure of HF(g) reaches one atmosphere.

Westrum and Pitzer (1) determined the decomposition vapor pressures of KHF_2(cr, ℓ) by a static method. Their data lead to the decomposition temperature, 700 K when the pressure of HF reaches one atmosphere. JANAF analyses of their data are given in KHF_2(cr) table. Their data over the crystal phase are suspected to be high by approximately 11.5%; and the data over the liquid phase are probably in error as indicated by the 3rd law drift (3.9 cal K^{-1} mol^{-1}) in the JANAF analyses.

Reference

1. E. F. Westrum, Jr. and K. S. Pitzer, J. Amer. Chem. Soc. 71, 1940 (1949).

Potassium Fluoride ($K(HF_2)$) $F_2H_1K_1(l)$

Enthalpy Reference Temperature = T_r = 298.15 K Standard State Pressure = p° = 0.1 MPa

	J $K^{-1}mol^{-1}$			kJ mol^{-1}			
T/K	$C_p°$	S°	-[G°-H°(T_r)]/T	H°-H°(T_r)	$\Delta_f H°$	$\Delta_f G°$	Log K_f
0							
100							
200							
298.15	76.944	132.764	132.764	0.	-916.573	-856.932	150.131
300	76.944	133.240	132.766	0.142	-916.569	-856.562	149.141
314.000	78.408	136.784	132.866	1.230	GLASS <--> LIQUID TRANSITION		
314.000	104.600	136.784	132.866	1.230			
400	104.600	162.105	136.540	10.226	-916.636	-836.412	109.224
500	104.600	185.446	144.074	20.686	-914.111	-816.649	85.315
511.950	104.600	187.916	145.069	21.936	BETA <--> LIQUID		
600	104.600	204.517	152.607	31.146	-911.629	-797.391	69.419
700	104.600	220.641	161.204	41.606	-909.185	-778.545	58.096
800	104.600	234.608	169.526	52.066	-906.788	-760.046	49.626
900	104.600	246.928	177.455	62.526	-904.452	-741.844	43.056
1000	104.600	257.949	184.963	72.986	-902.197	-723.898	37.813
1100	104.600	267.918	192.059	83.446	-978.995	-701.564	33.314
1200	104.600	277.020	198.765	93.906	-975.898	-676.480	29.446
1300	104.600	285.392	205.111	104.366	-972.847	-651.652	26.184
1400	104.600	293.144	211.125	114.826	-969.839	-627.058	23.396
1500	104.600	300.361	216.837	125.286	-966.873	-602.678	20.987
1600	104.600	307.111	222.270	135.746	-963.948	-578.494	18.886
1700	104.600	313.453	227.449	146.206	-961.062	-554.491	17.037
1800	104.600	319.431	232.395	156.666	-958.214	-530.658	15.399
1900	104.600	325.087	237.126	167.126	-955.401	-506.982	13.938
2000	104.600	330.452	241.659	177.586	-952.622	-483.452	12.626

PREVIOUS: December 1963 CURRENT: June 1971

Potassium Fluoride ($K(HF_2)$) $F_2H_1K_1(l)$

POTASSIUM FLUORIDE (KHF_2) ALPHA-BETA-LIQUID M_r = 78.103046

Potassium Fluoride ($K(HF_2)$) $F_2H_1K_1$(cr,l)

0 to 469.85 K	crystal, alpha	
467.85 to 511.95 K	crystal, beta	
above 511.95 K	liquid	

Refer to the individual tables for details.

Enthalpy Reference Temperature = T_r = 298.15 K; Standard State Pressure = p° = 0.1 MPa

	J K⁻¹mol⁻¹			kJ mol⁻¹			
T/K	C_p°	S°	$-[G°-H°(T_r)]/T$	$H°-H°(T_r)$	$\Delta_f H°$	$\Delta_f G°$	Log K_f
0	0.	0.	INFINITE	-15.294	-926.386	-926.386	INFINITE
100	48.961	34.988	165.420	-13.043	-930.274	-908.140	474.363
200	67.070	75.565	110.999	-7.087	-931.188	-885.531	231.277
298.15	76.940	104.264	104.264	0.	-931.233	-863.095	151.211
300	77.111	104.741	104.266	0.142	-931.230	-862.672	150.205
400	86.065	128.152	107.398	8.302	-933.221	-839.415	109.616
469.850	92.450	142.501	111.566	14.535	ALPHA <--> BETA		
469.850	100.249	166.384	111.566	25.757	TRANSITION		
500	100.249	172.619	115.061	28.779	-920.678	-816.803	85.331
511.950	100.249	174.987	116.432	29.977	BETA <--> LIQUID		
511.950	104.600	187.916	116.432	36.596	TRANSITION		
600	104.600	204.517	128.173	45.806	-911.629	-797.391	69.419
700	104.600	220.641	140.261	56.266	-909.185	-778.545	58.096
800	104.600	234.608	151.200	66.726	-906.788	-760.046	49.626
900	104.600	246.928	161.166	77.186	-904.452	-741.844	43.056
1000	104.600	257.949	170.303	87.646	-902.197	-723.898	37.813
1100	104.600	267.918	178.731	98.106	-978.995	-701.564	33.314
1200	104.600	277.020	186.548	108.566	-975.898	-676.480	29.446
1300	104.600	285.392	193.834	119.026	-972.847	-651.652	26.184
1400	104.600	293.144	200.654	129.486	-969.839	-627.058	23.396
1500	104.600	300.361	207.063	139.946	-966.873	-602.678	20.987
1600	104.600	307.111	213.107	150.406	-963.948	-578.494	18.886
1700	104.600	313.453	218.826	160.866	-961.062	-554.491	17.037
1800	104.600	319.431	224.250	171.326	-958.214	-530.658	15.399
1900	104.600	325.087	229.410	181.786	-955.401	-506.982	13.938
2000	104.600	330.452	234.329	192.246	-952.622	-483.452	12.626

PREVIOUS: CURRENT: June 1971

Potassium Fluoride ($K(HF_2)$) $F_2H_1K_1$(cr,l)

HYDROGEN FLUORIDE, DIMER $((HF)_2)$ IDEAL GAS M_r = 40.012686

S°(298.15 K) = 238.849 ± 8.4 J K^{-1} mol^{-1}

$\Delta_f H°$(0 K) = -566.600 ± 3.3 kJ mol^{-1}
$\Delta_f H°$(298.15 K) = -572.664 ± 3.3 kJ mol^{-1}

Vibrational Frequencies and Degeneracies

ν, cm^{-1}
[4000](1)
[3400](1)
[1200](2)
[720](1)
[350](1)

Ground State Quantum Weight: 1 σ = 2
Point Group: C_s
Bond Distances: F-H = 0.92 Å H ... F = 1.87 Å
Bond Angles: H...F-H = 108° F-H ... F = 180°
Product of the Moments of Inertia: $I_A I_B I_C = 1.98556 \times 10^{-116}$ g^3 cm^6

Enthalpy of Formation

The enthalpy of 2 HF(g) → H_2F_2(g) was taken as the mean of the 3rd law values which were obtained from the following equilibrium data. Briegleb and Strohmeier (1) measured the vapor density of associated HF(g) between 26 and 56 C and between 30 and 700 torr. Franck and Meyer (2) measured the heat capacity between -20 and 100°C and between 100 and 700 torr. Their 2nd law values differed by 1.2 kcal mol^{-1}; and using the molecular constants discussed below gave respectively mean 3rd law values which differed by 0.34 kcal mol^{-1}, a 0.14 kcal mol^{-1} temperature trend for Briegleb and Strohmeier, and a 0.02 kcal mol^{-1} temperature trend for Franck and Meyer. $\Delta_f H°(H_2F_2$, g, 298.15 K) was calculated using auxiliary JANAF data (3).

Heat Capacity and Entropy

The structure of the semi-rigid hydrogen bonded dimer is taken from Dyke et al. (4). Using bond distances and angles given by these authors we obtain $I_A = 0.1203 \times 10^{-39}$, $I_B = 12.7869 \times 10^{-39}$ and $I_C = 12.9022 \times 10^{-39}$ g cm^2. Dyke et al (4) observed a doubling of the rotational energy levels of the dimer and attributed this to a tunneling motion involving breaking of the hydrogen bond, interchange of the two HF units, and formation of a new hydrogen bond. Thus there are two identical configurations for the dimer separated by a low energy barrier so that quantum mechanical tunneling occurs at a significant rate. Therefore the overall symmetry number is 2 even though the external symmetry number is 1.

Reliable experimental values of the vibrational frequencies of H_2F_2(g) are not available. The estimated frequencies are similar to those used for the higher polymers which were obtained from data on HF solid. The infrared absorption bands observed in the vapor (350 to 400 cm^{-1}, 700 to 800 cm^{-1}, and 1000-1200 cm^{-1} regions) are largely due to the higher polymers, such as the tetramer and hexamer. Calculated values (610, 443, and 144 cm^{-1}) were obtained from force constants given in a paper on theory of molecular interactions of the HF polymers by Del Bene and Pople (5). The potential energy surface computed for the dimer suggests very anharmonic low frequency motions for the external hydrogen (estimated at 600±200 cm^{-1}) and for the hydrogen bond stretching mode.

The National Bureau of Standards prepared this table (6) by critical analysis of data existing in 1972. Using our $I_A I_B I_C$ with other molecular constants and $\Delta_f H°$ selected by NBS (6), we recalculate the table in terms of 1973 fundamental constants (7), 1975 atomic weights (8) and current JANAF reference states for the elements.

References
1. G. Briegleb and W. Strohmeier, Z. Elektrochem. 57, 668 (1953).
2. E. U. Franck and F. Meyer, Z. Elektrochem. 63, 571 (1959).
3. JANAF Thermochemical Tables: HF(g), 6-30-77.
4. T. R. Dyke, B. J. Howard and W. Klemperer, J. Chem. Phys. 56, 2442 (1972).
5. J. E. Del Bene and J. A. Pople, J. Chem. Phys. 55, 2296 (1971).
6. S. Abramowitz et al., U.S. Nat. Bur. Stand. Report 10904, 239, 1972.
7. CODATA Task Group on Fundamental Constants, CODATA Bulletin 11, December, 1973.
8. IUPAC Commission on Atomic Weights, Pure Appl. Chem. 47, 75 (1976).

Hydrogen Fluoride $((HF)_2)$ F_2H_2(g)

Enthalpy Reference Temperature = T_r = 298.15 K Standard State Pressure = p° = 0.1 MPa

T/K	$C_p°$ (J $K^{-1}mol^{-1}$)	S° (J $K^{-1}mol^{-1}$)	$-[G°-H°(T_r)]/T$ (J $K^{-1}mol^{-1}$)	$H°-H°(T_r)$ (kJ mol^{-1})	$\Delta_f H°$ (kJ mol^{-1})	$\Delta_f G°$ (kJ mol^{-1})	Log K_f
0	0.	0.	INFINITE	-11.228	-566.600	-566.600	INFINITE
100	34.675	196.439	275.181	-7.874	-569.152	-561.686	293.395
200	39.777	222.014	242.789	-4.155	-571.056	-553.446	144.545
250	42.385	231.170	239.574	-2.101	-571.901	-548.945	114.696
298.15	44.874	238.849	238.849	0.	-572.664	-544.453	95.386
300	44.968	239.127	238.850	0.083	-572.692	-544.278	94.767
350	47.433	246.246	239.407	2.394	-573.419	-539.484	80.514
400	49.702	252.731	240.673	4.823	-574.077	-534.590	69.810
450	51.736	258.705	242.349	7.360	-574.668	-529.618	61.477
500	53.536	264.251	244.265	9.993	-575.196	-524.584	54.803
600	56.536	274.287	248.450	15.502	-576.089	-514.374	44.780
700	58.952	283.190	252.789	21.281	-576.800	-504.030	37.611
800	61.000	291.199	257.098	27.281	-577.363	-493.594	32.228
900	62.809	298.490	261.298	33.473	-577.799	-483.096	28.038
1000	64.449	305.194	265.357	39.837	-578.129	-472.555	24.684
1100	65.953	311.408	269.264	46.358	-578.366	-461.985	21.938
1200	67.335	317.207	273.020	53.023	-578.525	-451.397	19.649
1300	68.602	322.647	276.631	59.821	-578.617	-440.799	17.712
1400	69.761	327.774	280.103	66.740	-578.653	-430.196	16.051
1500	70.818	332.624	283.444	73.770	-578.642	-419.592	14.611
1600	71.778	337.225	286.663	80.900	-578.591	-408.990	13.352
1700	72.651	341.603	289.767	88.123	-578.508	-398.392	12.241
1800	73.442	345.779	292.763	95.428	-578.396	-387.801	11.254
1900	74.159	349.769	295.659	102.808	-578.261	-377.216	10.370
2000	74.809	353.590	298.461	110.257	-578.103	-366.638	9.576
2100	75.399	357.254	301.174	117.768	-577.924	-356.069	8.857
2200	75.934	360.774	303.804	125.335	-577.726	-345.509	8.203
2300	76.421	364.161	306.355	132.953	-577.507	-334.959	7.607
2400	76.864	367.423	308.832	140.618	-577.267	-324.418	7.061
2500	77.268	370.569	311.239	148.325	-577.004	-313.888	6.558
2600	77.637	373.606	313.579	156.071	-576.719	-303.369	6.095
2700	77.975	376.543	315.857	163.852	-576.409	-292.862	5.666
2800	78.284	379.384	318.076	171.665	-576.072	-282.366	5.268
2900	78.568	382.137	320.237	179.507	-575.709	-271.883	4.897
3000	78.829	384.805	322.345	187.377	-575.317	-261.413	4.552
3100	79.069	387.393	324.402	195.273	-574.897	-250.956	4.229
3200	79.291	389.907	326.410	203.191	-574.447	-240.514	3.926
3300	79.495	392.350	328.371	211.130	-573.967	-230.085	3.642
3400	79.685	394.726	330.288	219.089	-573.456	-219.672	3.375
3500	79.860	397.039	332.163	227.067	-572.915	-209.275	3.123
3600	80.023	399.291	333.996	235.061	-572.344	-198.893	2.886
3700	80.175	401.485	335.791	243.071	-571.744	-188.527	2.662
3800	80.316	403.625	337.548	251.095	-571.114	-178.179	2.449
3900	80.448	405.713	339.269	259.134	-570.455	-167.847	2.248
4000	80.571	407.752	340.956	267.185	-569.769	-157.532	2.057
4100	80.686	409.743	342.609	275.248	-569.055	-147.235	1.876
4200	80.793	411.688	344.231	283.322	-568.315	-136.955	1.703
4300	80.894	413.591	345.822	291.406	-567.550	-126.694	1.539
4400	80.989	415.451	347.383	299.500	-566.760	-116.451	1.382
4500	81.077	417.272	348.916	307.604	-565.946	-106.226	1.233
4600	81.161	419.055	350.422	315.716	-565.109	-96.019	1.090
4700	81.240	420.802	351.900	323.836	-564.251	-85.830	0.954
4800	81.314	422.513	353.354	331.963	-563.371	-75.660	0.823
4900	81.384	424.190	354.782	340.098	-562.470	-65.509	0.698
5000	81.450	425.835	356.187	348.240	-561.550	-55.376	0.579
5100	81.513	427.449	357.569	356.388	-560.610	-45.262	0.464
5200	81.572	429.032	358.928	364.542	-559.652	-35.166	0.353
5300	81.628	430.586	360.265	372.703	-558.675	-25.089	0.247
5400	81.681	432.113	361.582	380.868	-557.681	-15.031	0.145
5500	81.732	433.612	362.878	389.039	-556.668	-4.992	0.047
5600	81.780	435.085	364.154	397.214	-555.639	5.030	-0.047
5700	81.825	436.533	365.411	405.395	-554.592	15.032	-0.138
5800	81.869	437.956	366.650	413.579	-553.527	25.016	-0.225
5900	81.910	439.356	367.870	421.768	-552.446	34.982	-0.310
6000	81.949	440.733	369.073	429.961	-551.347	44.929	-0.391

PREVIOUS: June 1977 (1 atm) CURRENT: June 1977 (1 bar)

Hydrogen Fluoride $((HF)_2)$ F_2H_2(g)

DIFLUOROSILANE (SiH_2F_2) IDEAL GAS M_r = 68.098186

$\Delta_fH°(0\ K) = [-782.254 \pm 20.9]\ kJ\ mol^{-1}$

$S°(298.15\ K) = 262.130 \pm 2.1\ J\ K^{-1}\ mol^{-1}$ $\Delta_fH°(298.15\ K) = [-790.776 \pm 20.9]\ kJ\ mol^{-1}$

Vibrational Frequencies and Degeneracies

ν, cm^{-1}	ν, cm^{-1}	ν, cm^{-1}
2246(1)	322(1)	730(1)
980(1)	730(1)	981(1)
869(1)	2251(1)	903(1)

Ground State Quantum Weight: [1] $\sigma = 2$

Point Group: C_{2v}

Bond Distances: Si-H = 1.471 Å Si-F = 1.5767 Å

Bond Angles: H-Si-H = 112.02° F-Si-F = 107.93°

Product of the Moments of Inertia: $I_AI_BI_C = 4.815754 \times 10^{-115}\ g^3\ cm^6$

Enthalpy of Formation

There are no reported experimental studies leading to the enthalpy of formation of SiH_2F_2(g). We estimate this value via a linear interpolation between the established $\Delta_fH°$(298.15 K) values of SiH_4(g) and SiF_4(g) (1). The reasonableness of this approach has been demonstrated by Lapidus et al. (2), Hunt and Sirtl (3), and Seiter and Sirtl (4).

Heat Capacity and Entropy

The adopted vibrational frequencies are from the gas phase infrared study of Cradock et al. (6). Two frequencies (ν_5 and ν_6) were quite uncertain and were discussed in terms of a strong Coriolis interaction.

These frequencies appear quite reasonable when a comparison of the dihalosilanes and dihalomethanes (halogen is fluorine, chlorine, and bromine) is made (1, 5).

The adopted bond distances and bond angles are obtained from the microwave spectrum study by Laurie (7). The individual moments of inertia are: $I_A = 3.4021 \times 10^{-39}\ g\ cm^2$, $I_B = 10.7552 \times 10^{-39}\ g\ cm^2$, and $I_C = 13.1614 \times 10^{-39}\ g\ cm^2$.

References

1. JANAF Thermochemical Tables: SiH_4(g) and SiF_4(g), 6-30-76; CH_2F_2, 12-31-69.
2. I. I. Lapidus, L. A. Nisel'son, and A. L. Seifer in "Thermophysical Properties of Gases and Liquids," No. 1, V. A. Rabinovich, Ed., Israel Program for Scientific Translations, Jerusalem, pp. 102-135, 1970.
3. L. P. Hunt and E. Sirtl, J. Electrochem. Soc. 119, 1741 (1972).
4. H. Seiter and E. Sirtl, Z. Naturforsch. 21a, 1956 (1966).
5. T. Shimanouchi, NSRDS-NBS-39, 1972; J. Phys. Chem. Ref. Data 2, 121 (1973).
6. S. Cradock, E. A. V. Ebsworth, and A. G. Robiette, Trans. Faraday Soc. 60, 1502 (1964).
7. V. W. Laurie, J. Chem. Phys. 26, 1359 (1957).

Difluorosilane (SiH_2F_2) $F_2H_2Si_1$(g)

Enthalpy Reference Temperature = T_r = 298.15 K Standard State Pressure = p° = 0.1 MPa

	J K⁻¹mol⁻¹			kJ mol⁻¹			
T/K	$C_p°$	S°	$-[G°-H°(T_r)]/T$	$H°-H°(T_r)$	$\Delta_fH°$	$\Delta_fG°$	Log K_f
0	0.	0.	INFINITE	-11.988	-782.254	-782.254	INFINITE
100	35.089	216.460	302.698	-8.624	-785.062	-779.215	407.020
200	42.974	242.890	266.723	-4.766	-788.006	-772.238	201.688
250	48.646	253.078	262.992	-2.478	-789.467	-768.126	160.491
298.15	54.268	262.130	262.130	0.	-790.776	-763.895	133.831
300	54.479	262.466	262.131	0.101	-790.824	-763.728	132.977
350	59.903	271.278	262.814	2.962	-792.035	-759.114	113.292
400	64.714	279.598	264.397	6.081	-793.090	-754.338	98.506
450	68.914	287.469	266.527	9.424	-793.999	-749.437	86.992
500	72.575	294.923	268.998	12.963	-794.774	-744.443	77.771
600	78.609	308.712	274.487	20.535	-795.981	-734.257	63.923
700	83.350	321.200	280.283	28.642	-796.811	-723.899	54.018
800	87.139	332.586	286.120	37.173	-797.351	-713.443	46.583
900	90.201	343.033	291.871	46.045	-797.670	-702.934	40.797
1000	92.697	352.670	297.475	55.195	-797.825	-692.398	36.167
1100	94.746	361.604	302.904	64.570	-797.865	-681.853	32.378
1200	96.440	369.923	308.146	74.132	-797.826	-671.307	29.221
1300	97.850	377.699	313.201	83.848	-797.739	-660.767	26.550
1400	99.033	384.995	318.071	93.694	-797.631	-650.235	24.261
1500	100.033	391.863	322.764	103.649	-797.519	-639.711	22.277
1600	100.882	398.347	327.287	113.696	-797.418	-629.194	20.541
1700	101.609	404.485	331.649	123.821	-847.516	-618.235	18.996
1800	102.236	410.311	335.859	134.014	-847.236	-604.756	17.550
1900	102.778	415.854	339.924	144.266	-846.949	-591.293	16.256
2000	103.251	421.138	343.854	154.568	-846.658	-577.845	15.092
2100	103.665	426.186	347.655	164.914	-846.364	-564.411	14.039
2200	104.029	431.017	351.335	175.299	-846.067	-550.992	13.082
2300	104.351	435.648	354.901	185.718	-845.766	-537.586	12.209
2400	104.637	440.096	358.359	196.168	-845.461	-524.194	11.409
2500	104.892	444.372	361.714	206.645	-845.148	-510.814	10.673
2600	105.120	448.491	364.973	217.145	-844.827	-497.447	9.994
2700	105.325	452.462	368.140	227.668	-844.495	-484.092	9.365
2800	105.510	456.296	371.221	238.210	-844.150	-470.750	8.782
2900	105.677	460.001	374.219	248.769	-843.789	-457.421	8.239
3000	105.828	463.586	377.138	259.345	-843.412	-444.104	7.733
3100	105.966	467.059	379.983	269.934	-843.016	-430.801	7.259
3200	106.092	470.425	382.757	280.537	-842.601	-417.510	6.815
3300	106.207	473.691	385.463	291.152	-842.165	-404.232	6.398
3400	106.313	476.864	388.105	301.779	-841.707	-390.969	6.007
3500	106.410	479.947	390.685	312.415	-841.227	-377.719	5.637
3600	106.499	482.946	393.207	323.060	-1224.876	-354.022	5.137
3700	106.581	485.865	395.672	333.714	-1223.937	-329.844	4.657
3800	106.657	488.708	398.083	344.376	-1222.977	-305.692	4.202
3900	106.728	491.480	400.442	355.046	-1221.996	-281.566	3.771
4000	106.794	494.183	402.752	365.722	-1220.994	-257.465	3.362
4100	106.855	496.820	405.014	376.404	-1219.972	-233.389	2.973
4200	106.911	499.396	407.231	387.092	-1218.928	-209.339	2.604
4300	106.964	501.912	409.404	397.786	-1217.865	-185.314	2.251
4400	107.014	504.372	411.534	408.485	-1216.781	-161.313	1.915
4500	107.060	506.777	413.624	419.189	-1215.678	-137.338	1.594
4600	107.103	509.131	415.675	429.897	-1214.556	-113.387	1.288
4700	107.144	511.435	417.688	440.609	-1213.416	-89.461	0.994
4800	107.182	513.691	419.665	451.326	-1212.258	-65.559	0.713
4900	107.218	515.901	421.606	462.046	-1211.082	-41.682	0.444
5000	107.252	518.068	423.514	472.769	-1209.889	-17.828	0.186
5100	107.284	520.192	425.389	483.496	-1208.680	6.001	-0.061
5200	107.314	522.275	427.232	494.226	-1207.455	29.806	-0.299
5300	107.343	524.320	429.045	504.959	-1206.214	53.588	-0.528
5400	107.370	526.327	430.828	515.695	-1204.957	77.346	-0.748
5500	107.395	528.297	432.582	526.433	-1203.686	101.080	-0.960
5600	107.420	530.232	434.308	537.174	-1202.400	124.792	-1.164
5700	107.443	532.134	436.008	547.917	-1201.099	148.480	-1.361
5800	107.464	534.003	437.681	558.662	-1199.785	172.146	-1.550
5900	107.485	535.840	439.330	569.410	-1198.456	195.788	-1.733
6000	107.505	537.646	440.953	580.159	-1197.113	219.408	-1.910

PREVIOUS: June 1976 (1 atm) CURRENT: June 1976 (1 bar)

MERCURY FLUORIDE (HgF_2) CRYSTAL M_r = 238.586806

S°(298.15 K) = [116.315 ± 8.4] J K^{-1} mol^{-1} $\Delta_f H°(298.15\ K)$ = [-422.584 ± 41.8] kJ mol^{-1}

T_{fus} = 918 K $\Delta_{fus}H°$ = [23.012 ± 4.2] kJ mol^{-1}

Enthalpy of Formation

Estimated by comparison of the mercurous and mercuric halides.

Heat Capacity and Entropy

The heat capacity is assumed to be a linear function of temperature and was estimated to be 7.0 cal K^{-1} g-atom^{-1} at the melting point. The entropy is estimated by comparison with the other mercuric and mercurous halides and use of additive entropy constants for the halogens (1).

Fusion Data

T_{fus} was given by Ruff and Bahlau (2). The heat of melting was from Brewer (3).

References

1. K. K. Kelley, U.S. Bur. Mines, personal communication, 1960.
2. O. Ruff and G. Bahlau, Ber. 51, 1752 (1918).
3. L. Brewer, in "Thermodynamics", McGraw-Hill, National Nuclear Energy Series 4-19B paper 7 (1950).

Mercury Fluoride (HgF_2) $F_2Hg_1(cr)$

Enthalpy Reference Temperature = T_r = 298.15 K Standard State Pressure = p° = 0.1 MPa

T/K	$C_p°$ (J $K^{-1}mol^{-1}$)	S° (J $K^{-1}mol^{-1}$)	$-[G°-H°(T_r)]/T$ (J $K^{-1}mol^{-1}$)	$H°-H°(T_r)$ (kJ mol^{-1})	$\Delta_f H°$ (kJ mol^{-1})	$\Delta_f G°$ (kJ mol^{-1})	Log K_f
0							
100							
200							
298.15	74.856	116.315	116.315	0.	-422.584	-374.134	65.547
300	74.894	116.778	116.317	0.139	-422.555	-373.833	65.090
400	76.986	138.610	119.279	7.732	-420.946	-357.833	46.728
500	79.078	156.014	124.943	15.536	-419.236	-342.250	35.755
600	81.170	170.616	131.370	23.548	-417.411	-327.022	28.470
700	83.262	183.286	137.901	31.770	-474.215	-305.539	22.800
800	85.354	194.541	144.290	40.200	-471.472	-281.627	18.388
900	87.446	204.714	150.448	48.840	-468.565	-258.070	14.978
918.000	87.822	206.450	151.529	50.418	------ CRYSTAL <--> LIQUID -----		
1000	89.538	214.036	156.347	57.689	-465.485	-234.845	12.267
1100	91.630	222.668	161.988	66.748	-462.225	-211.937	10.064
1200	93.722	230.731	167.384	76.015	-458.782	-189.334	8.241
1300	95.814	238.315	172.552	85.492	-455.153	-167.026	6.711
1400	97.906	245.492	177.508	95.178	-451.336	-145.004	5.410
1500	99.998	252.318	182.269	105.073	-447.330	-123.262	4.292
1600	102.090	258.839	186.853	115.178	-443.133	-101.793	3.323
1700	104.182	265.091	191.272	125.491	-438.743	-80.593	2.476
1800	106.274	271.105	195.541	136.014	-434.159	-59.657	1.731
1900	108.366	276.907	199.672	146.746	-429.378	-38.981	1.072
2000	110.458	282.518	203.675	157.687	-424.396	-18.561	0.485

PREVIOUS: CURRENT: March 1962

Mercury Fluoride (HgF_2) $F_2Hg_1(cr)$

MERCURY FLUORIDE (HgF_2) LIQUID M_r = 238.586806

$S°(298.15\ K) = [134.246]\ J\ K^{-1}\ mol^{-1}$

$T_{fus} = 918\ K$

$\Delta_f H°(298.15\ K) = [-404.897]\ kJ\ mol^{-1}$

$\Delta_{fus} H° = [23.012 \pm 4.2]\ kJ\ mol^{-1}$

Enthalpy of Formation

$\Delta_f H°(HgF_2,\ \ell,\ 298.15\ K)$ is calculated from $\Delta_f H°(HgF_2,\ cr,\ 298.15\ K)$ by adding the enthalpy of fusion, $\Delta_{fus} H°$, and the difference in enthalpy, H°(918 K)-H°(298.15 K), between the crystal and liquid.

Heat Capacity and Entropy

$C_p°$ is estimated by analogy with mercuric iodide and bromide. A glass type transitiion is assumed at 613 K below which the heat capacity is that of HgF_2(cr). The entropy is calculated in a manner analogous to that used for the enthalpy of formation.

Fusion and Vaporization Data

T_{fus} was given by Ruff and Bahlan (1). T_{vap} = 920 K, $\Delta_{f\ell s} H° = 23.0 \pm 4.2\ kJ\ mol^{-1}$, and $\Delta_{vap} H° = 92.0 \pm 16.7\ kJ\ mol^{-1}$ were all taken from Brewer (2).

References

1. O. Ruff and G. Bahlan, Ber. 51, 1752 (1918).
2. L. Brewer, in "Thermodynamics" McGraw-Hill, New York, National Nuclear Energy Series 4-19B paper 7, 1950.

Mercury Fluoride (HgF_2) $F_2Hg_1(l)$

Enthalpy Reference Temperature = T_r = 298.15 K — Standard State Pressure = p° = 0.1 MPa

T/K	$C_p°$ (J K⁻¹mol⁻¹)	S° (J K⁻¹mol⁻¹)	$-[G°-H°(T_r)]/T$ (J K⁻¹mol⁻¹)	$H°-H°(T_r)$ (kJ mol⁻¹)	$\Delta_f H°$ (kJ mol⁻¹)	$\Delta_f G°$ (kJ mol⁻¹)	Log K_f
0							
100							
200							
298.15	74.856	134.246	134.246	0.	-404.897	-361.793	63.385
300	74.894	134.709	134.248	0.139	-404.868	-361.525	62.947
400	76.986	156.541	137.210	7.732	-403.259	-347.318	45.355
500	79.078	173.945	142.874	15.536	-401.549	-333.528	34.843
600	81.170	188.548	149.301	23.548	-399.724	-320.093	27.867
613.000	81.442	190.290	150.152	24.605	GLASS <--> LIQUID	TRANSITION	
613.000	102.090	190.290	150.152	24.605			
700	102.090	203.839	156.001	33.487	-454.810	-300.521	22.425
800	102.090	217.471	162.852	43.696	-450.290	-278.789	18.203
900	102.090	229.496	169.602	53.905	-445.813	-257.621	14.952
918.000	102.090	231.517	170.796	55.742	------ CRYSTAL	<--> LIQUID -----	
1000	102.090	240.252	176.138	64.114	-441.373	-236.949	12.377
1100	102.090	249.982	182.416	74.323	-436.963	-216.720	10.291
1200	102.090	258.865	188.422	84.532	-432.578	-196.892	8.570
1300	102.090	267.037	194.159	94.741	-428.217	-177.428	7.129
1400	102.090	274.602	199.638	104.949	-423.877	-158.300	5.906
1500	102.090	281.646	204.873	115.158	-419.557	-139.481	4.857
1600	102.090	288.234	209.880	125.367	-415.256	-120.950	3.949
1700	102.090	294.424	214.673	135.576	-410.971	-102.687	3.155
1800	102.090	300.259	219.267	145.785	-406.700	-84.676	2.457
1900	102.090	305.779	223.676	155.994	-402.442	-66.902	1.839
2000	102.090	311.015	227.913	166.203	-398.193	-49.351	1.289

PREVIOUS: CURRENT: March 1962

Mercury Fluoride (HgF_2) $F_2Hg_1(l)$

MERCURY FLUORIDE (HgF_2)

CRYSTAL-LIQUID

0 to 918 K crystal
above 918 K liquid

Refer to the individual tables for details.

M_r = 238.586806 Mercury Fluoride (HgF_2) $F_2Hg_1(cr,l)$

Enthalpy Reference Temperature = T_r = 298.15 K Standard State Pressure = $p°$ = 0.1 MPa

	J K⁻¹mol⁻¹			kJ mol⁻¹			
T/K	$C_p°$	S°	$-[G°-H°(T_r)]/T$	$H°-H°(T_r)$	$\Delta_fH°$	$\Delta_fG°$	Log K_f
0							
100							
200							
298.15	74.856	116.315	116.315	0.	-422.584	-374.134	65.547
300	74.894	116.778	116.317	0.139	-422.555	-373.833	65.090
400	76.986	138.610	119.279	7.732	-420.946	-357.833	46.728
500	79.078	156.014	124.943	15.536	-419.236	-342.250	35.755
600	81.170	170.616	131.370	23.548	-417.411	-327.022	28.470
700	83.262	183.286	137.901	31.770	-474.215	-305.539	22.800
800	85.354	194.541	144.290	40.200	-471.472	-281.627	18.388
900	87.446	204.714	150.448	48.840	-468.565	-258.070	14.978
918.000	87.822	206.450	151.529	50.418	CRYSTAL <--> LIQUID		
918.000	102.090	231.517	151.529	73.430	TRANSITION		
1000	102.090	240.252	158.451	81.801	-441.373	-236.949	12.377
1100	102.090	249.982	166.337	92.010	-436.963	-216.720	10.291
1200	102.090	258.865	173.683	102.219	-432.578	-196.892	8.570
1300	102.090	267.037	180.554	112.428	-428.217	-177.428	7.129
1400	102.090	274.602	187.005	122.637	-423.877	-158.300	5.906
1500	102.090	281.646	193.082	132.846	-419.557	-139.481	4.857
1600	102.090	288.234	198.825	143.055	-415.256	-120.950	3.949
1700	102.090	294.424	204.268	153.264	-410.971	-102.687	3.155
1800	102.090	300.259	209.441	163.473	-406.700	-84.676	2.457
1900	102.090	305.779	214.367	173.682	-402.442	-66.902	1.839
2000	102.090	311.015	219.070	183.891	-398.193	-49.351	1.289

PREVIOUS: CURRENT: March 1962

Mercury Fluoride (HgF_2) $F_2Hg_1(cr,l)$

MERCURY FLUORIDE (HgF_2) IDEAL GAS M_r = 238.586806

$\Delta_f H°(0\ K)$ = -288.672 kJ mol^{-1}

S°(298.15 K) = 271.737 J K^{-1} mol^{-1} $\Delta_f H°(298.15\ K)$ = [-293.654 ± 41.8] kJ mol^{-1}

Vibrational Frequencies and Degeneracies

ν, cm^{-1}
[600](1)
[120](2)
[593](1)

Ground State Quantum Weight: 1 σ = 2

Point Group: $D_{\infty h}$

Bond Distance: Hg-F = [1.96] Å

Bond Angle: F-Hg-F = [180°]

Rotational Constant: B_0 = [0.115488] cm^{-1}

Enthalpy of Formation

Estimated by comparison with the other gaseous mercuric halides and with the crystalline mercurous mercuric fluorides.

Heat Capacity and Entropy

Vibrational constants were estimated by analogy with the other mercuric halides and the mercurous halides. The bond length was estimated by assuming the mercurous bond to be 0.965 of the mercuric bond by analogy with the mercury chlorides.

Mercury Fluoride (HgF_2) $F_2Hg_1(g)$

Enthalpy Reference Temperature = T_r = 298.15 K Standard State Pressure = p° = 0.1 MPa

T/K	$C_p°$ (J K^{-1}mol^{-1})	S° (J K^{-1}mol^{-1})	-[G°-H°(T_r)]/T (J K^{-1}mol^{-1})	H°-H°(T_r) (kJ mol^{-1})	$\Delta_f H°$ (kJ mol^{-1})	$\Delta_f G°$ (kJ mol^{-1})	Log K_f
0	0.	0.	INFINITE	-13.186	-288.672	-288.672	INFINITE
100	42.378	219.689	316.215	-9.653	-289.751	-290.961	151.983
200	49.043	251.161	276.531	-5.074	-290.687	-291.784	76.206
250	51.842	262.416	272.614	-2.550	-293.362	-291.860	60.981
298.15	53.968	271.737	271.737	0.	-293.654	-291.543	51.077
300	54.040	272.071	271.738	0.100	-293.664	-291.530	50.760
350	55.705	280.532	272.402	2.845	-293.897	-291.155	43.453
400	56.961	288.056	273.897	5.664	-294.085	-290.750	37.968
450	57.915	294.823	275.853	8.537	-294.246	-290.323	33.700
500	58.649	300.965	278.062	11.451	-294.390	-289.880	30.284
600	59.677	311.756	282.803	17.372	-294.658	-288.952	25.156
700	60.339	321.008	287.616	23.375	-353.680	-281.409	20.999
800	60.787	329.096	292.306	29.432	-353.311	-271.110	17.702
900	61.102	336.275	296.800	35.528	-352.948	-260.857	15.140
1000	61.333	342.725	301.075	41.650	-352.594	-250.643	13.092
1100	61.505	348.579	305.132	47.792	-352.250	-240.465	11.419
1200	61.638	353.937	308.979	53.950	-351.917	-230.318	10.025
1300	61.743	358.875	312.630	60.119	-351.596	-220.197	8.848
1400	61.826	363.454	316.098	66.298	-351.287	-210.102	7.839
1500	61.893	367.722	319.399	72.484	-350.990	-200.027	6.966
1600	61.949	371.718	322.546	78.676	-350.705	-189.972	6.202
1700	61.995	375.475	325.550	84.873	-350.432	-179.935	5.529
1800	62.034	379.020	328.423	91.075	-350.168	-169.913	4.931
1900	62.067	382.375	331.175	97.280	-349.914	-159.906	4.396
2000	62.095	385.559	333.815	103.488	-349.665	-149.912	3.915
2100	62.119	388.589	336.352	109.699	-349.420	-139.931	3.481
2200	62.140	391.480	338.793	115.912	-349.175	-129.961	3.086
2300	62.159	394.242	341.144	122.126	-348.927	-120.002	2.725
2400	62.175	396.888	343.412	128.343	-348.673	-110.054	2.395
2500	62.189	399.426	345.602	134.561	-348.408	-100.117	2.092
2600	62.202	401.866	347.719	140.781	-348.129	-90.191	1.812
2700	62.213	404.214	349.768	147.002	-347.833	-80.276	1.553
2800	62.223	406.476	351.754	153.224	-347.517	-70.372	1.313
2900	62.232	408.660	353.678	159.446	-347.179	-60.480	1.089
3000	62.241	410.770	355.546	165.670	-346.815	-50.600	0.881
3100	62.248	412.811	357.361	171.894	-346.424	-40.733	0.686
3200	62.255	414.787	359.125	178.120	-346.003	-30.878	0.504
3300	62.261	416.703	360.841	184.345	-345.552	-21.037	0.333
3400	62.266	418.562	362.511	190.572	-345.069	-11.211	0.172
3500	62.272	420.367	364.139	196.799	-344.552	-1.399	0.021
3600	62.276	422.121	365.725	203.026	-344.002	8.398	-0.122
3700	62.281	423.827	367.272	209.254	-343.418	18.179	-0.257
3800	62.285	425.488	368.783	215.482	-342.799	27.943	-0.384
3900	62.288	427.106	370.257	221.711	-342.146	37.692	-0.505
4000	62.292	428.683	371.698	227.940	-341.458	47.422	-0.619
4100	62.295	430.222	373.107	234.169	-340.737	57.135	-0.728
4200	62.298	431.723	374.485	240.399	-339.981	66.831	-0.831
4300	62.301	433.189	375.833	246.629	-339.193	76.507	-0.929
4400	62.303	434.621	377.153	252.859	-338.372	86.165	-1.023
4500	62.306	436.021	378.446	259.089	-337.519	95.805	-1.112
4600	62.308	437.391	379.712	265.320	-336.635	105.424	-1.197
4700	62.310	438.731	380.954	271.551	-335.721	115.024	-1.278
4800	62.312	440.042	382.171	277.782	-334.777	124.605	-1.356
4900	62.314	441.327	383.365	284.013	-333.805	134.165	-1.430
5000	62.316	442.586	384.537	290.245	-332.806	143.705	-1.501
5100	62.317	443.820	385.688	296.477	-331.780	153.225	-1.569
5200	62.319	445.030	386.817	302.708	-330.728	162.725	-1.635
5300	62.320	446.217	387.927	308.940	-329.652	172.205	-1.697
5400	62.322	447.382	389.017	315.172	-328.553	181.663	-1.757
5500	62.323	448.526	390.089	321.405	-327.432	191.101	-1.815
5600	62.324	449.649	391.142	327.637	-326.289	200.519	-1.870
5700	62.325	450.752	392.178	333.869	-325.126	209.916	-1.924
5800	62.327	451.836	393.198	340.102	-323.945	219.292	-1.975
5900	62.328	452.901	394.201	346.335	-322.746	228.648	-2.024
6000	62.329	453.949	395.188	352.568	-321.530	237.984	-2.072

PREVIOUS: December 1961 (1 atm) CURRENT: December 1961 (1 bar)

Mercury Fluoride (HgF_2) $F_2Hg_1(g)$

MERCURY FLUORIDE (Hg_2F_2) CRYSTAL M_r = 439.176806

$S°$(298.15 K) = [160.666 ± 8.4] J K^{-1} mol^{-1}

$\Delta_f H°$(0 K) = unknown

$\Delta_f H°$(298.15 K) = [-485.344 ± 12.6] kJ mol^{-1}

Enthalpy of Formation

A value of $\Delta_f G°$(298.15 K) was obtained from the cell measurements of Koerber and de Vries (1). This was combined with $\Delta_f S°$ from the estimated entropy.

Heat Capacity and Entropy

The heat capacity is estimated by comparison with mercurous chloride. The entropy is estimated by comparison with the other mercurous and mercuric halides and use of additive entropy constants for the halogens from Kelley (2).

Sublimation Data

Sublimation with decomposition to Hg(g) and HgF_2(g) is assumed at 949 K from the Gibbs energy change of the reaction.

References

1. G. G. Koerber and T. de Vries, J. Amer. Chem. Soc. 74, 5008 (1952).
2. K. K. Kelley, U.S. Bur. Mines, personal communication, 1960.

Mercury Fluoride (Hg_2F_2) F_2Hg_2(cr)

Enthalpy Reference Temperature = T_r = 298.15 K; Standard State Pressure = p° = 0.1 MPa

T/K	$C_p°$ (J $K^{-1}mol^{-1}$)	S° (J $K^{-1}mol^{-1}$)	$-[G°-H°(T_r)]/T$ (J $K^{-1}mol^{-1}$)	$H°-H°(T_r)$ (kJ mol^{-1})	$\Delta_f H°$ (kJ mol^{-1})	$\Delta_f G°$ (kJ mol^{-1})	Log K_f
0							
100							
200							
298.15	100.416	160.666	160.666	0.	-485.344	-468.629	82.102
300	100.500	161.287	160.668	0.186	-485.316	-468.526	81.578
400	104.684	190.783	164.661	10.449	-486.063	-462.717	60.425
500	108.407	214.551	172.337	21.107	-483.967	-457.113	47.754
600	111.671	234.611	181.086	32.115	-481.427	-451.974	39.348
700	114.516	252.044	190.005	43.427	-596.058	-434.146	32.396
800	116.943	267.499	198.744	55.004	-591.616	-411.316	26.856
900	118.951	281.391	207.168	66.800	-586.959	-389.057	22.580
1000	120.748	294.018	215.231	78.787	-582.142	-367.325	19.187
1100	122.334	305.603	222.927	90.943	-656.162	-341.471	16.215
1200	123.709	316.308	230.269	103.247	-650.094	-313.129	13.630
1300	124.872	326.257	237.274	115.678	-643.900	-285.299	11.463
1400	125.825	335.547	243.966	128.214	-637.599	-257.950	9.624
1500	126.566	344.255	250.364	140.836	-631.215	-231.055	8.046

PREVIOUS: CURRENT: March 1962

Mercury Fluoride (Hg_2F_2) F_2Hg_2(cr)

POTASSIUM FLUORIDE, ION (KF_2^-) IDEAL GAS M_r = 77.095655

$S°(298.15\ K) = [261.868 \pm 8.4]\ J\ K^{-1}\ mol^{-1}$

$\Delta_fH°(0\ K) = -686.217 \pm 41.8\ kJ\ mol^{-1}$
$\Delta_fH°(298.15\ K) = -694.544 \pm 41.8\ kJ\ mol^{-1}$

Vibrational Frequencies and Degeneracies

ν, cm^{-1}
[35C](1)
[13C](2)
[50C](1)

Ground State Quantum Weight: [1] σ = 2
Point Group: $[D_{\infty h}]$
Bond Distance: K-F = [2.17] Å
Bond Angle: F-K-F = [180°]
Rotational Constant: B_0 = [0.094216] cm^{-1}

Enthalpy of Formation

The enthalpy of formation is estimated from three calculated values. The methods of calculation are described as follows. In a manner analogous to $LiF_2^-(g)$, the first reaction considered is $KF(g) + F^-(g) = KF_2^-(g)$ and the $\Delta_rH°(298.15\ K)$ value is evaluated to be between 0 and 1/2(-50) = 25 kcal mol^{-1} (see $LiF_2^-(g)$ table for details), yielding $\Delta_fH°(KF_2^-, g, 298.15\ K)$ between -140 and -165 kcal mol^{-1}. The $\Delta_fH°(298.15\ K)$ values for the other species involved are taken from JANAF tables.

The second reaction employed for estimation is $K_2F_2(g) = KF_2^-(g) + K^+(g)$. The enthalpy change of this reaction is calculated as $\Delta_rH°(298.15\ K)$ = (1/2)(50.0) + 138.8 = 163.8 kcal mol^{-1} where the KF_2^- ion is considered to be converted from $K_2F_2(g)$ by breaking one of the two dimer association bonds, $\Delta_rH°(298.15\ K)$ = 50.0 kcal mol^{-1}, and one K^+-F^- ionic bond, $\Delta_rH°(298.15\ K)$ = 138.8 kcal mol^{-1}. Using $\Delta_fH°(298.15\ K)$ = -206.2 and 122.9 kcal mol^{-1} for $K_2F_2(g)$ and $K^+(g)$, we obtain the enthalpy of formation for $KF_2^-(g)$ as -165 kcal mol^{-1}.

Based on an assumption that the electron affinity of $KF_2(g)$ is similar to that of F (see $LiF_2^-(g)$ table for details), i.e., $\Delta_rH°(298.15\ K)$ = 83.5 kcal mol^{-1} (1) for the reaction $KF_2^-(g) = KF_2(g) + e^-(g)$, we derive $\Delta_fH°(KF_2^-, g, 298.15\ K)$ = -167.7 kcal mol^{-1}. The $\Delta_fH°(KF_2, g, 298.15\ K)$ is calculated as -84.2 kcal mol^{-1} from the value $\Delta_rH°(298.15\ K)$ = 143.3 kcal mol^{-1} for the reaction $K_2F_2(g) = KF_2(g) + K(g)$ where the $\Delta_rH°(298.15\ K)$ value is calculated as the sum of two enthalpy changes, i.e. half of the heat of dimer dissociation into two monomers or 1/2(50.0) = 25.0 kcal mol^{-1}, and the heat of dissociation of KF(g) or 118.3 kcal mol^{-1}.

The value of $\Delta_fH°(KF_2^-, 298.15\ K)$ is tentatively adopted as -166±10 kcal mol^{-1}.

Heat Capacity and Entropy

The molecular structure is assumed to be linear, according to the suggestion given by Walsh (2) that molecules with not more than 16 valency electrons are linear in their ground states. The K-F bond distance is taken to be the same as that of KF(g) reported by Veazey and Gordy (3). The vibrational frequencies are estimated from the vibrational frequency of KF(g) using the method suggested by Herzberg (4).

References

1. H. O. Pritchard, Chem. Rev. 52, 529 (1953).
2. A. D. Walsh, J. Chem. Soc. 1953, 2266 (1953).
3. S. E. Veazey and W. Gordy, Phys. Rev. 138, A1303 (1965).
4. G. Herzberg, "Infrared and Raman Spectra," p. 172, D. Van Nostrand Co., Inc., New York, 1945.

Potassium Fluoride, Ion (KF_2^-) $F_2K_1^-(g)$

Enthalpy Reference Temperature = T_r = 298.15 K Standard State Pressure = p° = 0.1 MPa

	J $K^{-1}mol^{-1}$			kJ mol^{-1}			
T/K	$C_p°$	S°	$-[G°-H°(T_r)]/T$	$H°-H°(T_r)$	$\Delta_fH°$	$\Delta_fG°$	Log K_f
0	0.	0.	INFINITE	-13.777	-686.217		
100	43.337	206.619	309.310	-10.269			
200	52.713	239.908	266.969	-5.412			
250	55.318	251.970	262.798	-2.707			
298.15	57.015	261.868	261.868	0.	-694.544	-686.622	120.293
300	57.069	262.221	261.869	0.106	-694.589	-686.573	119.543
350	58.272	271.114	262.569	2.991	-698.224	-685.041	102.237
400	59.122	278.954	264.137	5.927	-699.544	-683.068	89.200
450	59.739	285.955	266.179	8.899	-700.842	-680.930	79.040
500	60.200	292.274	268.478	11.898	-702.125	-678.649	70.898
600	60.824	303.310	273.390	17.952	-704.663	-673.715	58.652
700	61.215	312.717	278.352	24.055	-707.185	-668.357	49.873
800	61.475	320.909	283.171	30.191	-709.715	-662.638	43.266
900	61.656	328.161	287.775	36.348	-712.273	-656.600	38.108
1000	61.787	334.664	292.144	42.520	-714.882	-650.275	33.967
1100	61.884	340.558	296.282	48.704	-796.516	-639.075	30.347
1200	61.959	345.946	300.199	54.896	-798.227	-624.687	27.192
1300	62.017	350.908	303.912	61.095	-799.954	-610.155	24.516
1400	62.064	355.506	307.435	67.299	-801.699	-595.490	22.218
1500	62.101	359.789	310.784	73.508	-803.459	-580.699	20.222
1600	62.132	363.798	313.973	79.719	-805.235	-565.791	18.471
1700	62.158	367.565	317.016	85.934	-807.027	-550.771	16.923
1800	62.179	371.119	319.924	92.151	-808.834	-535.645	15.544
1900	62.198	374.481	322.708	98.370	-810.654	-520.418	14.307
2000	62.213	377.672	325.377	104.590	-812.488	-505.095	13.192
2100	62.227	380.708	327.940	110.812	-814.333	-489.680	12.180
2200	62.238	383.603	330.405	117.036	-816.188	-474.177	11.258
2300	62.248	386.370	332.778	123.260	-818.052	-458.589	10.415
2400	62.257	389.019	335.067	129.485	-819.924	-442.920	9.640
2500	62.265	391.561	337.276	135.711	-821.803	-427.173	8.925
2600	62.272	394.003	339.411	141.938	-823.687	-411.351	8.264
2700	62.278	396.353	341.477	148.166	-825.577	-395.456	7.651
2800	62.284	398.618	343.477	154.394	-827.473	-379.491	7.079
2900	62.289	400.804	345.417	160.622	-829.373	-363.458	6.547
3000	62.293	402.916	347.298	166.852	-831.281	-347.359	6.048
3100	62.298	404.958	349.126	173.081	-833.196	-331.197	5.581
3200	62.301	406.936	350.902	179.311	-835.122	-314.973	5.141
3300	62.305	408.853	352.629	185.541	-837.050	-298.689	4.728
3400	62.308	410.713	354.310	191.772	-838.997	-282.346	4.338
3500	62.311	412.520	355.947	198.003	-840.959	-265.945	3.969
3600	62.313	414.275	357.543	204.234	-842.944	-249.488	3.620
3700	62.316	415.982	359.100	210.466	-844.952	-232.976	3.289
3800	62.318	417.644	360.619	216.697	-846.989	-216.409	2.975
3900	62.320	419.263	362.102	222.929	-849.058	-199.788	2.676
4000	62.322	420.841	363.550	229.161	-851.159	-183.113	2.391
4100	62.323	422.380	364.967	235.393	-853.303	-166.386	2.120
4200	62.325	423.882	366.352	241.626	-855.498	-149.606	1.861
4300	62.327	425.348	367.707	247.858	-857.745	-132.773	1.613
4400	62.328	426.781	369.033	254.091	-859.773	-115.913	1.376
4500	62.329	428.182	370.332	260.324	-862.073	-98.981	1.149
4600	62.331	429.552	371.604	266.557	-864.428	-81.998	0.931
4700	62.332	430.892	372.852	272.790	-866.822	-64.965	0.722
4800	62.333	432.204	374.075	279.023	-869.168	-47.892	0.521
4900	62.334	433.490	375.274	285.257	-871.680	-30.756	0.328
5000	62.335	434.749	376.451	291.490	-874.256	-13.569	0.142
5100	62.336	435.983	377.606	297.724	-876.899	3.671	-0.038
5200	62.337	437.194	378.741	303.957	-879.609	20.964	-0.211
5300	62.337	438.381	379.855	310.191	-882.387	38.309	-0.378
5400	62.338	439.547	380.949	316.425	-885.235	55.707	-0.539
5500	62.339	440.690	382.025	322.659	-888.153	73.158	-0.695
5600	62.339	441.814	383.083	328.892	-891.139	90.664	-0.846
5700	62.340	442.917	384.123	335.126	-893.850	108.175	-0.991
5800	62.341	444.001	385.146	341.361	-896.930	125.781	-1.133
5900	62.341	445.067	386.153	347.595	-900.073	143.442	-1.270
6000	62.342	446.115	387.143	353.829	-903.277	161.155	-1.403

PREVIOUS: December 1968 (1 atm) CURRENT: December 1968 (1 bar)

Potassium Fluoride, Ion (KF_2^-) $F_2K_1^-(g)$

POTASSIUM FLUORIDE, DIMER $((KF)_2)$ IDEAL GAS M_r = 116.193406

$\Delta_f H°(0\ K)$ = -858.080 ± 8.4 kJ mol^{-1}

S°(298.15 K) = [319.955 ± 8.4] J K^{-1} mol^{-1} $\Delta_f H°(298.15\ K)$ = -862.741 ± 8.4 kJ mol^{-1}

Vibrational Frequencies and Degeneracies

ν, cm^{-1}	ν, cm^{-1}
[273](1)	[150](1)
[150](1)	[255](1)
[235](1)	[258](1)

Ground State Quantum Weight: [1] σ = 4

Point Group: D_{2h}

Bond Distance: K-F = 2.2 Å

Bond Angles: F-K-F = [90°] K-F-K = [90°]

Product of the Moments of Inertia: $I_A I_B I_C$ = [2.240426 x 10^{-113}] g^3 cm^6

Enthalpy of Formation

The equilibrium constants (K_c) for the reaction $K_2F_2(g) = 2\ KF(g)$, in the temperature range 974-1052 K, were determined by Eisenstadt et al. (1). Although the reported absolute values of the vapor pressures of KF(cr) are not adopted for evaluation, the reported K_c values are used to calculate the enthalpy change, $\Delta_r H°(298.15\ K)$, for the above reaction. The 2nd and 3rd law values of $\Delta_r H°(298.15\ K)$ are derived as 51.0±1.6 and 50.0 kcal mol^{-1}, respectively. Based on the 3rd law $\Delta_r H°(298.15\ K)$ and $\Delta_f H°(KF,\ g,\ 298.15\ K)$ = -78.1 kcal mol^{-1}, we obtain $\Delta_f H°(298.15\ K)$ = -206.2 kcal mol^{-1} (-862.74 ± 8.4 kJ mol^{-1}) for $K_2F_2(g)$ which is tentatively adopted.

Heat Capacity and Enthalpy

The molecular structure and K-F bond distance were determined by Akishin and Rambidi (2), using electron diffraction with sector microphotometry. The bond angles are estimated by comparison with other alkali halide dimers. The vibrational frequencies are taken from Berkowitz (3) which were calculated based on an ionic model. The three principal moments of inertia are: I_A = 15.2693 x 10^{-39}, I_B = 31.4238 x 10^{-39}, and I_C = 46.6931 x 10^{-39} g cm^2.

References

1. M. Eisenstadt, G. M. Rothberg, and P. Kusch, J. Chem. Phys. 29, 797 (1958).
2. P. A. Akishin and G. Rambidi, Z. Physik. Chem. 213, 111 (1960).
3. J. Berkowitz, J. Chem. Phys. 32, 1519 (1960).

Potassium Fluoride $((KF)_2)$ $F_2K_2(g)$

Enthalpy Reference Temperature = T_r = 298.15 K Standard State Pressure = p° = 0.1 MPa

T/K	$C_p°$ (J $K^{-1}mol^{-1}$)	S° (J $K^{-1}mol^{-1}$)	$-[G°-H°(T_r)]/T$ (J $K^{-1}mol^{-1}$)	$H°-H°(T_r)$ (kJ mol^{-1})	$\Delta_f H°$ (kJ mol^{-1})	$\Delta_f G°$ (kJ mol^{-1})	Log K_f
0	0.	0.	INFINITE	-18.329	-858.080	-858.080	INFINITE
100	56.735	243.679	385.805	-14.213	-860.336	-860.546	449.503
200	73.655	289.471	327.017	-7.509	-861.750	-860.117	224.640
250	76.754	306.272	321.238	-3.742	-862.234	-859.652	179.615
298.15	78.525	319.955	319.955	0.	-862.741	-859.111	150.513
300	78.578	320.440	319.956	0.145	-862.763	-859.088	149.581
350	79.731	332.646	320.917	4.105	-868.174	-858.232	128.084
400	80.500	343.346	323.066	8.112	-868.972	-856.757	111.881
450	81.039	352.861	325.858	12.151	-869.727	-855.184	99.267
500	81.429	361.420	328.993	16.213	-870.451	-853.529	89.168
600	81.944	376.316	335.675	24.384	-871.833	-850.014	74.000
700	82.259	388.973	342.408	32.596	-873.165	-846.272	63.150
800	82.464	399.972	348.931	40.833	-874.492	-842.339	54.999
900	82.606	409.693	355.153	49.086	-875.857	-838.239	48.650
1000	82.707	418.402	361.050	57.352	-877.307	-833.982	43.563
1100	82.783	426.289	366.628	65.627	-1036.789	-820.355	38.955
1200	82.840	433.494	371.904	73.908	-1036.411	-800.696	34.853
1300	82.885	440.127	376.900	82.195	-1036.052	-781.068	31.384
1400	82.921	446.271	381.639	90.485	-1035.710	-761.467	28.411
1500	82.949	451.993	386.140	98.779	-1035.386	-741.889	25.835
1600	82.973	457.347	390.425	107.075	-1035.080	-722.333	23.582
1700	82.992	462.378	394.511	115.373	-1034.792	-702.795	21.594
1800	83.009	467.122	398.415	123.673	-1034.522	-683.273	19.828
1900	83.023	471.610	402.150	131.975	-1034.271	-663.767	18.248
2000	83.034	475.869	405.730	140.278	-1034.037	-644.272	16.827
2100	83.045	479.921	409.168	148.582	-1033.823	-624.790	15.541
2200	83.053	483.784	412.472	156.886	-1033.629	-605.316	14.372
2300	83.061	487.476	415.653	165.192	-1033.455	-585.852	13.305
2400	83.068	491.011	418.720	173.499	-1033.302	-566.394	12.327
2500	83.074	494.402	421.680	181.806	-1033.173	-546.942	11.428
2600	83.079	497.661	424.540	190.113	-1033.068	-527.495	10.598
2700	83.084	500.796	427.307	198.422	-1032.991	-508.052	9.829
2800	83.088	503.818	429.986	206.730	-1032.945	-488.610	9.115
2900	83.092	506.734	432.582	215.039	-1032.932	-469.170	8.451
3000	83.095	509.551	435.101	223.349	-1032.957	-449.730	7.830
3100	83.098	512.275	437.547	231.658	-1033.026	-430.288	7.250
3200	83.101	514.914	439.924	239.968	-1033.143	-410.843	6.706
3300	83.104	517.471	442.235	248.278	-1033.296	-391.395	6.195
3400	83.106	519.952	444.485	256.589	-1033.521	-371.940	5.714
3500	83.108	522.361	446.675	264.900	-1033.807	-352.478	5.260
3600	83.110	524.702	448.810	273.211	-1034.172	-333.007	4.832
3700	83.112	526.979	450.892	281.522	-1034.618	-313.524	4.426
3800	83.114	529.196	452.924	289.833	-1035.156	-294.028	4.042
3900	83.115	531.355	454.907	298.144	-1035.794	-274.516	3.677
4000	83.117	533.459	456.845	306.456	-1036.528	-254.988	3.330
4100	83.118	535.511	458.739	314.768	-1037.383	-235.441	3.000
4200	83.119	537.514	460.591	323.080	-1038.374	-215.869	2.685
4300	83.120	539.470	462.402	331.392	-1039.503	-196.273	2.384
4400	83.121	541.381	464.176	339.704	-1040.227	-176.703	2.098
4500	83.122	543.249	465.912	348.016	-1041.525	-157.062	1.823
4600	83.123	545.076	467.613	356.328	-1042.967	-137.392	1.560
4700	83.124	546.864	469.281	364.641	-1044.518	-117.693	1.308
4800	83.125	548.614	470.915	372.953	-1046.002	-97.983	1.066
4900	83.126	550.328	472.519	381.266	-1047.849	-78.214	0.834
5000	83.127	552.007	474.092	389.578	-1049.853	-58.406	0.610
5100	83.127	553.653	475.636	397.891	-1052.016	-38.555	0.395
5200	83.128	555.268	477.151	406.204	-1054.342	-18.661	0.187
5300	83.128	556.851	478.640	414.516	-1056.832	1.279	-0.013
5400	83.129	558.405	480.103	422.829	-1059.486	21.268	-0.206
5500	83.130	559.930	481.541	431.142	-1062.304	41.308	-0.392
5600	83.130	561.428	482.954	439.455	-1065.285	61.400	-0.573
5700	83.131	562.899	484.344	447.768	-1067.737	81.452	-0.746
5800	83.131	564.345	485.711	456.081	-1070.950	101.641	-0.915
5900	83.131	565.766	487.055	464.394	-1074.311	121.887	-1.079
6000	83.132	567.164	488.379	472.708	-1077.815	142.190	-1.238

PREVIOUS: June 1969 (1 atm) CURRENT: June 1969 (1 bar)

LITHIUM FLUORIDE, ION (LiF_2^-) IDEAL GAS M_r = 44.938355

$\Delta_fH°$(0 K) = [-702.735 ± 62.8] kJ mol^{-1}

S°(298.15 K) = [229.618 ± 8.4] J K^{-1} mol^{-1} $\Delta_fH°$(298.15 K) = [-711.280 ± 62.8] kJ mol^{-1}

Vibrational Frequencies and Degeneracies

ν, cm^{-1}
[1300](1)
[350](2)
[550](1)

Ground State Quantum Weight: [1] σ = [2]
Point Group [$D_{\infty h}$]
Bond Distance: Li-F = [1.57] Å
Bond Angle: F-Li-F = [180]°
Rotational Constant: B_0 = [0.179988] cm^{-1}

Enthalpy of Formation

Due to lack of thermal data, the enthalpy of formation for LiF_2^-(g) is arbitrarily selected from the following three estimated values. For the reaction (1) LiF(g) + F^-(g) = LiF_2(g), we expect the enthalpy change would be between 0 to -31 kcal mol^{-1}. Comparison with the other molecules like CO_2, BOF, BeF_2, etc. having 16 valence electrons leads us to assume that the isoelectronic molecule LiF_2^- should also be quite stable. This indicates that reaction (1) is favorable or $\Delta_rH°$(298.15 K) is negative. However, the LiF-F^- bond strength is expected to be weaker than half of the two ionic bonds between Li^+ and F^- in Li_2F_2 molecule (see Li_2F_2(g) table for the molecular structure). In other words, the $\Delta_rH°$(298.15 K) value of reaction (1) is less negative than half of the enthalpy of dimerization (-62.4 kcal mol^{-1}). Incorporating $\Delta_rH°$(298.15 K) = 0 and -31 kcal mol^{-1}, respectively, for reaction (1), with $\Delta_fH°$(298.15 K) = -81.45 and -62.2 kcal mol^{-1} for LiF(g) and F^-(g), we derived the corresponding values for $\Delta_fH°$(LiF_2^-, g, 298.15 K) as -144 and -175 kcal mol^{-1}, which are the upper and lower limits.

The electron affinities of Li and F atoms are reported as 0.7 and 3.448 eV, respectively, by Beckett and Cassidy (1). From this data we estimate that the electron affinity of LiF_2(g) should be closer to that of F atom than to that of Li. Assuming $\Delta_rH°$(298.15 K) = electron affinity of F atom = 83.5 kcal mol^{-1} for the reaction LiF_2^-(g) = LiF_2(g) + e^-(g), we obtain the enthalpy of formation for LiF_2^-(g) as -177.3 kcal mol^{-1}. The $\Delta_fH°$(LiF_2, g, 298.15 K) used for calculation is derived from an estimated $\Delta_rH°$(298.15 K) = 169.9 kcal mol^{-1} for the decomposition of LiF dimer of Li_2F_2(g) = LiF_2(g) + Li(g) as the sum of two enthalpy changes, namely, half of the enthalpy of dissociation of the dimer and the enthalpy of dissociation of the monomer.

The third estimated value for the enthalpy of formation of LiF_2^-(g) is calculated as follows. Based on the two enthalpies of reaction, i.e. $\Delta_rH°$(298.15 K) = 62.4 kcal mol^{-1} for Li_2F_2(g) = LiF(g), and $\Delta_rH°$(298.15 K) = 183.5 kcal mol^{-1} for LiF(g) = Li^+(g) + F^-(g), we assume the enthalpy change of the reaction Li_2F_2(g) = LiF_2^-(g) + Li^+(g) to be 1/2(62.4) + 183.5 = 214.7 kcal mol^{-1}. In other words, we consider the LiF_2^-(g) molecule is formed by breaking half of the dimer dissociation bonds and one monomer dissociation bond to produce two ions. Using $\Delta_fH°$(298.15 K) = -225.33 and 164.24 kcal mol^{-1} for Li_2F_2(g) and Li^+(g), we evaluate $\Delta_fH°$(LiF_2^-, g, 298.15 K) = -174.9 kcal mol^{-1}.

From the above three estimated values, we tentatively choose $\Delta_fH°$(LiF_2^-, g, 298.15 K) as -170±15 kcal mol^{-1}.

Heat Capacity and Entropy

According to Walsh (3), molecules with not more than 16 valence electrons are linear in their ground states. Since LiF_2^-(g) has 16 valence electrons, we assume its molecular structure is linear. The Li-F bond distance is estimated by comparison with that of LiF(g) reported by Wharton et al. (2). The vibrational frequencies are estimated from those of the isoelectronic gaseous molecules, CO_2 and BeF_2.

References

1. C. W. Beckett and E. C. Cassidy, U. S. Nat. Bur. Stand. Report 8628, 1965.
2. L. Wharton, W. Klemperer, L. P. Gold, R. Strauch, J. J. Gallagher, and V. E. Derr, J. Chem. Phys. 38, 1203 (1963).
3. A. D. Walsh, J. Chem. Soc. 1953, 2266 (1953).

Lithium Fluoride, Ion (LiF_2^-) $F_2Li_1^-$(g)

Enthalpy Reference Temperature = T_r = 298.15 K Standard State Pressure = p° = 0.1 MPa

T/K	$C_p°$	S°	-[G°-H°(T_r)]/T	H°-H°(T_r)	$\Delta_fH°$	$\Delta_fG°$	Log K_f
	J $K^{-1}mol^{-1}$			kJ mol^{-1}			
0	0.	0.	INFINITE	-11.100	-702.735		
100	32.069	186.362	267.697	-8.133			
200	41.808	211.729	233.812	-4.417			
250	45.181	221.439	230.390	-2.238			
298.15	47.670	229.618	229.618	0.	-711.280	-704.353	123.400
300	47.754	229.914	229.619	0.088	-711.334	-704.309	122.631
350	49.802	237.434	230.209	2.529	-712.784	-703.024	104.921
400	51.483	244.198	231.542	5.062	-714.258	-701.530	91.610
450	52.886	250.345	233.295	7.673	-715.777	-699.848	81.236
500	54.066	255.980	235.285	10.347	-720.350	-697.687	72.887
600	55.910	266.009	239.591	15.851	-723.382	-692.868	60.319
700	57.248	274.734	244.002	21.513	-726.276	-687.551	51.306
800	58.233	282.446	248.334	27.289	-729.084	-681.827	44.519
900	58.973	289.349	252.515	33.151	-731.845	-675.754	39.220
1000	59.538	295.593	256.516	39.078	-734.573	-669.375	34.965
1100	59.977	301.289	260.331	45.054	-737.276	-662.724	31.470
1200	60.324	306.523	263.965	51.070	-739.961	-655.828	28.547
1300	60.602	311.363	267.427	57.117	-742.634	-648.708	26.065
1400	60.828	315.863	270.728	63.189	-745.298	-641.383	23.930
1500	61.014	320.066	273.879	69.281	-747.952	-633.868	22.073
1600	61.169	324.009	276.890	75.391	-750.599	-626.176	20.443
1700	61.298	327.721	279.772	81.514	-898.477	-611.142	18.778
1800	61.408	331.228	282.534	87.650	-900.361	-594.186	17.243
1900	61.502	334.551	285.185	93.795	-902.247	-577.125	15.866
2000	61.582	337.708	287.733	99.950	-904.136	-559.964	14.625
2100	61.652	340.714	290.185	106.111	-906.027	-542.710	13.499
2200	61.713	343.584	292.547	112.280	-907.916	-525.364	12.474
2300	61.767	346.328	294.827	118.454	-909.804	-507.933	11.536
2400	61.814	348.958	297.028	124.633	-911.689	-490.420	10.674
2500	61.856	351.482	299.156	130.816	-913.567	-472.829	9.879
2600	61.893	353.909	301.215	137.004	-915.438	-455.162	9.144
2700	61.926	356.245	303.210	143.195	-917.301	-437.424	8.462
2800	61.956	358.498	305.145	149.389	-919.154	-419.617	7.828
2900	61.983	360.673	307.022	155.586	-920.997	-401.744	7.236
3000	62.007	362.774	308.846	161.785	-922.828	-383.807	6.683
3100	62.029	364.808	310.619	167.987	-924.648	-365.810	6.164
3200	62.049	366.778	312.343	174.191	-926.456	-347.754	5.676
3300	62.067	368.687	314.022	180.397	-928.253	-329.641	5.218
3400	62.084	370.540	315.657	186.605	-930.039	-311.475	4.785
3500	62.099	372.340	317.251	192.814	-931.814	-293.256	4.377
3600	62.113	374.090	318.805	199.024	-933.581	-274.986	3.990
3700	62.126	375.792	320.323	205.236	-935.339	-256.668	3.624
3800	62.138	377.449	321.804	211.449	-937.090	-238.302	3.276
3900	62.149	379.063	323.252	217.664	-938.835	-219.890	2.945
4000	62.159	380.637	324.667	223.879	-940.577	-201.433	2.630
4100	62.169	382.172	326.051	230.096	-942.316	-182.933	2.331
4200	62.178	383.670	327.405	236.313	-944.055	-164.391	2.044
4300	62.186	385.133	328.730	242.531	-945.796	-145.807	1.771
4400	62.193	386.563	330.029	248.750	-947.520	-127.184	1.510
4500	62.201	387.960	331.301	254.970	-949.261	-108.520	1.260
4600	62.207	389.328	332.547	261.190	-951.010	-89.818	1.020
4700	62.214	390.666	333.770	267.411	-952.769	-71.077	0.790
4800	62.220	391.975	334.969	273.633	-954.538	-52.299	0.569
4900	62.225	393.258	336.145	279.855	-956.321	-33.484	0.357
5000	62.230	394.516	337.300	286.078	-958.119	-14.633	0.153
5100	62.235	395.748	338.434	292.301	-959.930	4.254	-0.044
5200	62.240	396.956	339.548	298.525	-961.765	23.178	-0.233
5300	62.244	398.142	340.642	304.749	-963.622	42.137	-0.415
5400	62.249	399.306	341.718	310.974	-965.505	61.131	-0.591
5500	62.252	400.448	342.775	317.199	-967.414	80.160	-0.761
5600	62.256	401.570	343.815	323.424	-969.354	99.225	-0.926
5700	62.260	402.672	344.838	329.650	-971.325	118.324	-1.084
5800	62.263	403.754	345.845	335.876	-973.331	137.458	-1.238
5900	62.266	404.819	346.835	342.103	-975.373	156.627	-1.387
6000	62.269	405.865	347.810	348.330	-977.454	175.831	-1.531

PREVIOUS: December 1968 (1 atm) CURRENT: December 1968 (1 bar)

Lithium Fluoride, Ion (LiF_2^-) $F_2Li_1^-$(g)

$S°(298.15 K) = 258.627 \pm 8.4 J K^{-1} mol^{-1}$

$\Delta_f H°(0 K) = -938.070 \pm 16.7 kJ mol^{-1}$
$\Delta_f H°(298.15 K) = -942.781 \pm 16.7 kJ mol^{-1}$

Vibrational Frequencies and Degeneracies

ν, cm^{-1}	ν, cm^{-1}
[570](1)	301.5(1)
[570](1)	571.5(1)
[490](1)	658.0(1)

Ground State Quantum Weight: [1] $\sigma = 4$
Point Group: D_{2h}
Bond Distances: Li-F = 1.68 Å F-F = 2.67 Å
Bond Angles: Li-F-Li = 74.76° F-Li-F = 105.24°
Product of the Moments of Inertia: $I_A I_B I_C = 3.678682 \times 10^{-115} g^3 cm^6$

Enthalpy of Formation

$\Delta_f H°(Li_2F_2$, g, 298.15 K) is calculated based on the adopted value $\Delta_r H°(1000) = 61.4 kcal mol^{-1}$ for the reaction $Li_2F_2(g)$ = 2LiF(g) which is consistent with the literature values listed in the table below. Due to lack of knowledge of the fragmentation patterns, the relative ionization cross sections of the ionic species formed in the mass spectrometer, and the molecular constants of the LiF polymers, the evaluation of equilibrium constants used to calculate $\Delta_r H°$ involves considerable uncertainty. In order to solve this problem more research work seems necessary.

Source	$\Delta_r H°(1000 K)$, kcal mol^{-1}	Method
1. Berkowitz et al. (1962)	61.6 ± 2	mass spectrometry with a double-oven apparatus
2. Buchler and Stauffer (1966)	61.3 ± 3	mass spectrometry with a two-piece Ni Knudsen cell
3. Eisenstadt et al. (1958)	61.2	molecular-beam velocity-selector
4. Hildenbrand et al. (1964)	60.4 ± 3	torsion effusion
5. Akishin et al. (1959)	59.5 ± 3	mass spectrometry with a double effusion chamber
6. Porter and Schoonmaker (1958)	65.0 *	mass spectrometry with a Mo Knudsen cell

*derived from reported vapor pressures of LiF(g) and $Li_2F_2(g)$ at 1073 K.

Heat Capacity and Entropy

The molecular structure of $Li_2F_2(g)$ was determined as D_{2h} by Bushler et al. (12), using electric deflection and mass spectrometric detection. Akishin and Rambidi (7) studied the structure of $Li_2F_2(g)$ by electron deflection with sector microphotometry and assigned the bond distances of Li-F and F-F which are adopted. Evidence for a matrix stabilized linear structure of Li_2F_2 has been obtained by extending the spectral measurements for matrix isolated lithium fluoride into the far infrared region by Redington (8) and Abramowitz et al. (9).

Klemperer and Norris (10) observed two infrared-active vibrational frequencies, 640 and 460 cm^{-1}, and assigned them as $\nu_5(B_{2u})$ and $\nu_6(B_{3u})$. The infrared absorption spectra of lithium fluoride monomer and dimer isolated in solid inert gas matrices were investigated by Linevsky (11), Snelson and Pitzer (13), and Schlick and Schnepp (14). Redington (8) obtained infrared spectra of LiF species suspended in solid Ne, O_2, CO, and CH_4 at liquid-helium temperature. The observed solvent shifts for LiF and Li_2F_2 spectra exhibit several regularities. The vibrational frequencies used for calculation are those reported by Snelson (15) who re-examined the spectrum of LiF species over the range 200-4000 cm^{-1}, using sample purity of 99.3%. The saturated vapor was isolated at 900°C and the superheated unsaturated vapor at 1200°C. A variety of deposition rates and matrix dilutions was tried. Based on the frequencies assigned to the isotopic species, the vibrational frequencies of the natural isotopic $Li_2F_2(g)$ are calculated. It appears that the vibrational frequencies calculated on the basis of the potential function for an ionic model by Berkowitz (16) are too low.

The three principal moments of inertia are: $I_A = 2.3980 \times 10^{-39}$, $I_B = 11.2447 \times 10^{-39}$ and $I_C = 13.6427 \times 10^{-39} g cm^2$.

References

1. J. Berkowitz, H. A. Tasman, and W. A. Chupka, J. Chem. Phys. 36, 2170 (1962).
2. A. Buchler and J. L. Stauffer, "Thermodynamics," Vol. I, International Atomic Energy Agency, Vienna, 1966.
3. M. Eisenstadt, G. M. Rothberg, and P. Kusch, J. Chem. Phys. 29, 797 (1958).
4. D. L. Hildenbrand, W. F. Hall, F. Ju, and N. D. Potter, J. Chem. Phys. 40, 2882 (1964).
5. P. A. Akishin, L. N. Gorokhov, and L. N. Sidorov, Russ. J. Phys. Chem. 33, 648 (1959).
6. R. F. Porter and R. C. Schoonmaker, J. Chem. Phys. 29, 1070 (1958).
7. P. A. Akishin and N. G. Rambidi, Z. Physik. Chem. 213, 111 (1960).
8. R. L. Redington, J. Chem. Phys. 44, 1238 (1966).
9. S. Abramowitz, N. Acquista, and I. W. Levin, U.S. Nat. Bur. Stand. Report 9601, July 1, 1967.
10. W. Klemperer and W. G. Norris, J. Chem. Phys. 34, 1071 (1961).
11. M. J. Linevsky, J. Chem. Phys. 34, 587 (1961); 38, 658 (1963).
12. A. Buchler, J. L. Stauffer, and W. Klemperer, J. Amer. Chem. Soc. 86, 4544 (1964).
13. A. Snelson and K. S. Pitzer, J. Phys. Chem. 67, 882 (1963).
14. S. Schlick and O. Schnepp, J. Chem. Phys. 41, 463 (1964).
15. A. Snelson, Illinois Institute of Technology Research Institute, IITRI-U6001-13, 14 May 1963 - 12 August 1966.
16. J. Berkowitz, J. Chem. Phys. 32, 1519 (1960).

Enthalpy Reference Temperature = T_r = 298.15 K Standard State Pressure = p° = 0.1 MPa

T/K	$C_p°$ (J $K^{-1}mol^{-1}$)	S° (J $K^{-1}mol^{-1}$)	$-[G°-H°(T_r)]/T$ (J $K^{-1}mol^{-1}$)	$H°-H°(T_r)$ (kJ mol^{-1})	$\Delta_f H°$ (kJ mol^{-1})	$\Delta_f G°$ (kJ mol^{-1})	Log K_f
0	0.	0.	INFINITE	-13.359	-938.070	-938.070	INFINITE
100	36.231	206.434	306.175	-9.974	-938.582	-940.720	491.382
200	51.023	235.802	264.055	-5.651	-940.871	-941.971	246.017
250	57.938	247.958	259.639	-2.920	-941.900	-942.124	196.846
298.15	63.131	258.627	258.627	0.	-942.781	-942.086	165.049
300	63.304	259.018	258.629	0.117	-942.813	-942.081	164.031
350	67.331	269.094	259.415	3.388	-943.657	-941.892	140.570
400	70.346	278.291	261.208	6.833	-944.517	-941.582	122.958
450	72.626	286.715	263.581	10.410	-945.454	-941.160	109.247
500	74.375	294.461	266.287	14.087	-952.495	-940.014	98.203
600	76.816	308.254	272.161	21.656	-954.366	-937.337	81.602
700	78.383	320.222	278.191	29.421	-956.002	-934.367	69.723
800	79.442	330.762	284.117	37.316	-957.510	-931.173	60.799
900	80.187	340.164	289.831	45.299	-958.962	-927.793	53.848
1000	80.730	348.642	295.296	53.347	-960.377	-924.254	48.278
1100	81.138	356.357	300.501	61.441	-961.765	-920.574	43.714
1200	81.451	363.430	305.455	69.571	-963.133	-916.769	39.906
1300	81.696	369.960	310.169	77.729	-964.489	-912.850	36.679
1400	81.892	376.022	314.659	85.909	-965.832	-908.827	33.909
1500	82.051	381.677	318.940	94.106	-967.160	-904.709	31.505
1600	82.182	386.977	323.028	102.318	-968.476	-900.503	29.398
1700	82.290	391.963	326.938	110.542	-1260.252	-881.862	27.096
1800	82.382	396.669	330.683	118.775	-1260.039	-859.610	24.945
1900	82.459	401.125	334.274	127.018	-1259.832	-837.370	23.021
2000	82.525	405.357	337.723	135.267	-1259.631	-815.140	21.289
2100	82.582	409.385	341.041	143.522	-1259.436	-792.921	19.723
2200	82.632	413.227	344.235	151.783	-1259.245	-770.710	18.299
2300	82.675	416.902	347.315	160.049	-1259.058	-748.508	16.999
2400	82.713	420.421	350.288	168.318	-1258.875	-726.314	15.808
2500	82.747	423.798	353.162	176.591	-1258.694	-704.128	14.712
2600	82.777	427.044	355.941	184.867	-1258.516	-681.949	13.701
2700	82.803	430.169	358.633	193.146	-1258.341	-659.776	12.764
2800	82.827	433.180	361.242	201.428	-1258.169	-637.610	11.895
2900	82.848	436.087	363.773	209.712	-1258.002	-615.450	11.085
3000	82.868	438.896	366.231	217.997	-1257.839	-593.296	10.330
3100	82.885	441.614	368.619	226.285	-1257.683	-571.148	9.624
3200	82.901	444.246	370.941	234.574	-1257.534	-549.004	8.962
3300	82.916	446.797	373.201	242.865	-1257.395	-526.864	8.340
3400	82.929	449.272	375.403	251.158	-1257.268	-504.729	7.754
3500	82.941	451.676	377.548	259.451	-1257.155	-482.597	7.202
3600	82.952	454.013	379.639	267.746	-1257.059	-460.468	6.681
3700	82.962	456.286	381.680	276.041	-1256.981	-438.342	6.188
3800	82.972	458.499	383.673	284.338	-1256.925	-416.218	5.721
3900	82.980	460.654	385.619	292.636	-1256.893	-394.094	5.278
4000	82.988	462.755	387.521	300.934	-1256.890	-371.971	4.857
4100	82.996	464.804	389.382	309.233	-1256.917	-349.848	4.457
4200	83.003	466.804	391.201	317.533	-1256.978	-327.723	4.076
4300	83.009	468.758	392.982	325.834	-1257.077	-305.597	3.712
4400	83.015	470.666	394.726	334.135	-1257.175	-283.472	3.365
4500	83.021	472.532	396.434	342.437	-1257.340	-261.340	3.034
4600	83.026	474.356	398.109	350.739	-1257.556	-239.205	2.716
4700	83.031	476.142	399.750	359.042	-1257.820	-217.064	2.412
4800	83.036	477.890	401.360	367.346	-1258.138	-194.917	2.121
4900	83.040	479.602	402.939	375.649	-1258.512	-172.763	1.842
5000	83.044	481.280	404.489	383.954	-1258.946	-150.600	1.573
5100	83.048	482.925	406.011	392.258	-1259.434	-128.430	1.315
5200	83.052	484.537	407.506	400.563	-1259.998	-106.248	1.067
5300	83.055	486.119	408.974	408.869	-1260.634	-84.054	0.828
5400	83.059	487.672	410.417	417.174	-1261.346	-61.848	0.598
5500	83.062	489.196	411.836	425.480	-1262.138	-39.628	0.376
5600	83.065	490.693	413.231	433.787	-1263.014	-17.393	0.162
5700	83.067	492.163	414.603	442.093	-1263.977	4.859	-0.045
5800	83.070	493.607	415.952	450.400	-1265.032	27.128	-0.244
5900	83.072	495.028	417.281	458.707	-1266.182	49.417	-0.438
6000	83.075	496.424	418.588	467.015	-1267.431	71.725	-0.624

PREVIOUS: December 1968 (1 atm) CURRENT: December 1968 (1 bar)

MAGNESIUM FLUORIDE (MgF_2) CRYSTAL M_r = 62.301806

$\Delta_fH°(0\ K) = -1120.332 \pm 1.3$ kJ mol^{-1}

$S°(298.15\ K) = 57.255 \pm 0.21$ J K^{-1} mol^{-1} $\Delta_fH°(298.15\ K) = -1124.241 \pm 1.3$ kJ mol^{-1}

$T_{fus} = 1536 \pm 5$ K $\Delta_{fus}H° = 58.702 \pm 0.4$ kJ mol^{-1}

Enthalpy of Formation

Rudzitis et al. (1) used fluorine bomb calorimetry to study the heat of combustion of a highly pure (99.91 mole %) sample of magnesium. Six combustion experiments were performed, and the sole combustion product was identified as MgF_2 by x-ray diffraction. Values for the completeness of combustion ranged from 98.4 to 99.99%. Corrections for the unburned magnesium were based on analysis performed by hydrogen evolution. This study gave $\Delta_fH°(MgF_2$, cr, 298.15 K) = -268.7±0.3 kcal mol^{-1} (-1124.241±1.3 kJ mol^{-1}) which is adopted. The quoted uncertainty is twice the combined standard deviation arising from the scatter in the six results and from the analytical and calibration data.

More uncertain values for $\Delta_fH°$ derived indirectly from reaction calorimetry (2, 3), high-temperature equilibria (4), and emf's of solid electrolyte galvanic cells (5, 6) are tabulated below. The original data are reanalyzed with the use of the most recent auxiliary data. These sources are included in the table.

Source	Method	Reaction	T/K	$\Delta_rH°(298.15\ K)$ kcal mol^{-1}	Drift cal K^{-1} mol^{-1}	$\Delta_fH°(298.15\ K)$ kcal mol^{-1}
(2)	Reaction Calorimetry	$Mg(OH)_2(cr)+2(HF\cdot4.5H_2O)=MgF_2(cr)+H_2O(\ell)$	346.9	-30.85[b]		-268.4±1.0
(3)	Reaction Calorimetry	$Mg(cr)+2(HF\cdot80H_2O)=MgF_2(cr)+H_2(g)$	293	-109.46±0.7[b]		-263.1±1.0
(4)	Equilibria	$MgF_2(cr)+H_2O(g)=MgO(cr)+2HF(g)$	1173-1373	51.58±0.9[a]	4.6±0.3	-268.0±1.5
(5)	emf	$AlF_3(cr)+1.5Mg(cr)=Al(cr)+1.5MgF_2(cr)$	720-860	-41.5±0.2[a]	-1.5	-268.3±1.6
(6)	emf	$AlF_3(cr)+1.5Mg(cr)=Al(cr)+1.5MgF_2(cr)$	750-900	-40.6±0.2[a]	1.8	-267.7±1.8

[a] 3rd Law Values

[b] $\Delta_fH°(HF, aq)$ consistent with JANAF value for HF(g); also, $\phi_c(HF, aq)$ taken from reference 9.

Auxiliary $\Delta_fH°(298.15\ K)$ values (kcal mol^{-1}): $Mg(OH)_2$(cr), -221.0±0.5 (7); HF·4.5 H_2O, -76.6±0.1[c]; $H_2O(\ell)$, -68.315 (8); HF·80H_2O, -76.8±0.1[c]; H_2O(g), -57.7979 (7); MgO(cr), -143.7+0.15 (7); HF(g), -65.14±0.2 (7); AlF_3(cr), -361.0±0.3 (7). With the exception of the older calorimetric measurements of Wartenberg (3), these results are in reasonable agreement with the directly measured $\Delta_fH°$ value (1).

Heat Capacity and Entropy

Todd (10) measured the low-temperature heat capacities (54.2-296.5 K) by adiabatic calorimetry. These $C_p°$ data extrapolate to 14.71 cal K^{-1} mol^{-1} at 298.15 K. Integration of these $C_p°$'s gives S°(298.15 K) = 13.68 cal K^{-1} mol^{-1} when combined with S°(51 K) = 0.539 cal K^{-1} mol^{-1}. The latter value is calculated from a combination of Debye and Einstein functions which fits all the experimental $C_p°$ data with an average deviation of ±1.0%.

Relative enthalpies H°(T;298.15) have been measured (411.3-1516 K) by the "drop" method (11) on a portion of the same sample used for the $C_p°$ study (10). Gravimetric analysis for Mg as $MgSO_4$ indicated a MgF_2 purity of 99.87 wt%. However, premelting effects which appear in the experimental enthalpies about 80° below the melting point suggest that the sample was probably somewhat less pure. Insufficient information is reported to attempt to correct for the impurities. High-temperature heat capacities (300-2000 K) are derived from the measured enthalpies (11) by curve fitting with orthogonal polynomials. The curve is constrained to join smoothly with the low-temperature $C_p°$ data near 298.15 K. Three enthalpy points which appear to involve premelting are not included in the fit. Also, we omit the enthalpy point at 411.3 K which shows an usually large positive deviation (+2.9%) from the adopted curve. The maximum deviation of the seven enthalpy points used in the fit is -0.5% and occurs at 1032.7 K. The older heat capacity measurements (288-1278 K) reported by Krestovnikov and Karentnikov (12) are considered less reliable and are not included in our evaluation.

Fusion Data

T_{fus} is the value measured by drop calorimetry (11). Melting point determinations (1529 and 1534 K) on two commercially available materials which had been further purified by hydrofluorination (13) suggest an uncertainty in the adopted T_{fus} value (1536 K) of ±5 K. Other literature values include 1525 K (14) and 1528 K (15).

$\Delta_{fus}H°$ is calculated as the difference between JANAF enthalpies for the liquid and crystal at T_{fus}. Other reported values for $\Delta_{fus}H°$ are 13.9 (11) and 13.15 kcal mol^{-1} (14).

Sublimation Data

See the ideal gas table for details.

References

See the liquid table for details.

Magnesium Fluoride (MgF_2) F_2Mg_1(cr)

Enthalpy Reference Temperature = T_r = 298.15 K — Standard State Pressure = p° = 0.1 MPa

T/K	$C_p°$ (J K^{-1}mol^{-1})	S° (J K^{-1}mol^{-1})	$-[G°-H°(T_r)]/T$ (J K^{-1}mol^{-1})	$H°-H°(T_r)$ (kJ mol^{-1})	$\Delta_fH°$ (kJ mol^{-1})	$\Delta_fG°$ (kJ mol^{-1})	Log K_f
0	0.	0.	INFINITE	-9.914	-1120.332	-1120.332	INFINITE
100	21.722	10.686	102.413	-9.173	-1123.140	-1106.221	577.830
200	48.817	35.056	62.549	-5.499	-1124.402	-1088.654	284.327
298.15	61.547	57.255	57.255	0.	-1124.241	-1071.109	187.654
300	61.689	57.636	57.256	0.114	-1124.231	-1070.779	186.439
400	68.484	76.377	59.765	6.645	-1123.475	-1053.065	137.516
500	72.634	92.132	64.706	13.713	-1122.439	-1035.579	108.186
600	75.320	105.624	70.429	21.117	-1121.275	-1018.315	88.652
700	77.178	117.381	76.315	28.746	-1120.070	-1001.250	74.714
800	78.555	127.780	82.110	36.535	-1118.878	-984.358	64.272
900	79.626	137.096	87.711	44.446	-1117.740	-967.612	56.159
1000	80.496	145.532	93.078	52.453	-1125.279	-950.276	49.637
1100	81.237	153.239	98.202	60.541	-1124.342	-932.821	44.296
1200	81.877	160.336	103.088	68.697	-1123.363	-915.452	39.849
1300	82.454	166.912	107.748	76.914	-1122.346	-898.167	36.089
1400	82.977	173.042	112.195	85.186	-1248.705	-877.795	32.751
1500	83.467	178.784	116.445	93.508	-1246.272	-851.386	29.648
1536.000	83.635	180.766	117.929	96.516	------ CRYSTAL <--> LIQUID -----		
1600	83.923	184.186	120.512	101.878	-1243.809	-825.141	26.938
1700	84.358	189.286	124.409	110.292	-1241.319	-799.050	24.552
1800	84.776	194.120	128.148	118.749	-1238.801	-773.107	22.435
1900	85.178	198.715	131.743	127.247	-1236.254	-747.304	20.545
2000	85.567	203.094	135.201	135.784	-1233.676	-721.636	18.847

PREVIOUS: March 1966 CURRENT: June 1975

Magnesium Fluoride (MgF_2) F_2Mg_1(cr)

MAGNESIUM FLUORIDE (MgF_2) LIQUID M_r = 62.301806

$S°(298.15\ K) = [89.967]\ J\ K^{-1}\ mol^{-1}$ $\Delta_f H°(298.15\ K) = [-1072.356]\ kJ\ mol^{-1}$

$T_{fus} = 1536 \pm 5\ K$ $\Delta_{fus}H° = 8.702 \pm 0.4\ kJ\ mol^{-1}$

Enthalpy of Formation

$\Delta_f H°(MgF_2,\ \ell,\ 298.15\ K)$ is calculated from $\Delta_f H°(MgF_2,\ cr,\ 298.15\ K)$ by adding the enthalpy of fusion, $\Delta_{fus}H°$, and the difference in enthalpy, H°(1536 K)−H°(298.15 K), between the crystal and liquid.

Heat Capacity and Entropy

Naylor (1) has measured relative enthalpies (1539-1760 K) for the liquid by drop calorimetry. These results give a $C_p°(\ell)$ equal to 22.687 cal K^{-1} mol^{-1}. This value is used in the temperature range 1000-3000 K. Below 1000 K, the assumed glass transition temperature, $C_p°$ is assumed to be that of the crystal. S°(298.15 K) is calculated in a manner analogous to that used for the enthalpy of formation.

Fusion Data

See the crystal table for details.

Vaporization Data

T_{vap} = 2536 K is the temperature at which $\Delta_r G°$ for the process $MgF_2(\ell) = MgF_2(g)$ is zero. $\Delta_{vap}H° = 274.052\ kJ\ mol^{-1}$ is the difference between the $\Delta_f H°$ values for the gas and liquid at T_{vap}. Several methods (2-6) predict an insignificant amount of dimer (~1%) in the saturated vapor below 2000 K.

References - $MgF_2(\ell)$

1. B. F. Naylor, J. Amer. Chem. Soc. 67, 150 (1945).
2. J. Berkowitz and J. R. Marquart, J. Chem. Phys. 37, 1853 (1962).
3. R. R. Hammer and J. A. Pask, J. Amer. Ceram. Soc. 47, 264 (1964).
4. M. A. Greenbaum, H. C. Ko, M. Wong, and M. Farber, J. Phys. Chem. 68, 965 (1964).
5. D. L. Hildenbrand, W. F. Hall, F. Ju, and N. D. Potter, J. Chem. Phys. 40, 2882 (1964).
6. J. W. Green, G. D. Blue, T. C. Ehlert and J. L. Margrave, J. Chem. Phys. 41, 2245 (1964).

References - $MgF_2(cr)$

1. E. Rudzitis, H. M. Feder, and W. N. Hubbard, J. Phys. Chem. 68, 2978 (1964).
2. D. R. Torgeson and Th. G. Sahama, J. Amer. Chem. Soc. 70, 2156 (1948).
3. H. v. Wartenberg, Z. Anorg. Allg. Chem. 249, 100 (1942).
4. L. Domange, Ann. Chim. Phys. 7, 225 (1937).
5. T. N. Rezukhina, T. F. Sisoeva, L. I. Halokhonova, and E. G. Ippolitor, J. Chem. Thermodynamics 6, 883 (1974).
6. W. H. Skelton and J. W. Patterson, J. Less-Common Metals 31, 47 (1973).
7. JANAF Thermochemical Tables: $Mg(OH)_2(cr)$, 3-31-67; $H_2O(g)$, 3-31-61; HF(g), 12-31-68; $AlF_3(cr)$, 6-30-70; MgO(cr), 12-31-74.
8. U. S. Nat. Bur. Stand. Tech. Note 270-3, 1968.
9. V. B. Parker, NSRDS-NBS 2, April, 1965.
10. S. S. Todd, J. Amer. Chem. Soc. 71, 4115 (1949).
11. B. F. Taylor, J. Amer. Chem. Soc. 67, 150 (1945).
12. A. N. Krestovnikov and G. A. Karetnikov, Legkie Metal. 3, 29 (1934).
13. H. Kojima, S. G. Whiteway, and C. R. Masson, Can. J. Chem. 46, 2968 (1968).
14. G. Petit and F. Delbove, Compt. rend. 254, 1388 (1962).
15. A. Duncanson and R. W. H. Stevenson, Proc. Phys. Soc. 72, 1001 (1958).

References - $MgF_2(g)$

1. O. Ruff and L. LeBoucher, Z. Anorg. Allg. Chem. 219, 376 (1934).
2. K. G. Gunther, Glastech. Ber. 31, 9 (1958).
3. J. Berkowitz and J. R. Marquart, J. Chem. Phys. 37, 1853 (1962).
4. G. Layne, Dow Chemical Co., personal communication, May, 1975.
5. R. R. Hammer, Ph.D. Thesis, University of California, 1961.
6. R. R. Hammer and J. A. Pask, J. Amer. Ceram. Soc. 47, 264 (1964).
7. M. A. Greenbaum, H. C. Ko, M. Wong, and M. Farber, J. Phys. Chem. 68, 965 (1964).
8. D. L. Hildenbrand, Standard Research Institute, personal communication, September, 1975.
9. J. W. Green, G. D. Blue, T. C. Ehlert and J. L. Margrave, J. Chem. Phys. 41, 2245 (1964).
10. JANAF Thermochemical Tables: $BeF_2(g)$, 6-30-70.
11. P. A. Akishin, V. P. Spiridonov, G. A. Sobolev, and V. A. Naumov, Zhur. Fiz. Khim. 31, 461 (1957).
12. A. Buchler, J. L. Stauffer, and W. Klemperer, J. Amer. Chem. Soc. 86, 4544 (1964).
13. A. Snelson, J. Phys. Chem. 70, 3208 (1966).
14. M. Astier and P. Millie, J. Organometal. Chem. 31, 139 (1971).
15. M. Astier, G. Berthier, and P. Millie, J. Chem. Phys. 57, 5008 (1972).
16. J. L. Gole, A. K. Q. Siu and E. F. Hayes, J. Chem. Phys. 58, 857 (1973).
17. D. E. Mann, G. V. Calder, K. S. Seshadri, D. White and M. J. Linevsky, J. Chem. Phys. 46, 1138 (1967).
18. V. Calder, D. E. Mann, K. S. Seshadri, M. Allaveua and D. White, J. Chem. Phys. 51, 2093 (1969).
19. J. F. Labarre and C. Leibovici, J. Chim. Phys. Physiochim. Biol. 69, 541 (1972).
20. V. I. Baikov, Optics and Spectroscopy 25, 194 (1968).
21. L. Brewer, G. R. Somayajulu and E. Brackett, Chem. Rev. 63, 111 (1963).
22. R. H. Hauge, J. L. Margrave, and A. S. Kana'an, J. Chem. Soc., Faraday Transaction II 71, 1082 (1975).
23. M. L. Lesiecki and J. W. Nibler, to be published.
24. A. Allavena and S. Besnainov, J. Mol. Struct. 11, 439 (1975).

Magnesium Fluoride (MgF_2) $F_2Mg_1(l)$

Enthalpy Reference Temperature = T_r = 298.15 K Standard State Pressure = p° = 0.1 MPa

T/K	$C_p°$	S°	$-[G°-H°(T_r)]/T$	$H°-H°(T_r)$	$\Delta_f H°$	$\Delta_f G°$	Log K_f
	J K⁻¹mol⁻¹			kJ mol⁻¹			
0							
100							
200							
298.15	61.547	89.967	89.967	0.	-1072.356	-1028.977	180.272
300	61.689	90.348	89.968	0.114	-1072.345	-1028.707	179.114
400	68.484	109.088	92.477	6.645	-1071.589	-1014.265	132.449
500	72.634	124.843	97.418	13.713	-1070.553	-1000.049	104.474
600	75.320	138.336	103.140	21.117	-1069.390	-986.057	85.844
700	77.178	150.092	109.026	28.746	-1068.184	-972.263	72.551
800	78.555	160.491	114.822	36.535	-1066.993	-958.642	62.593
900	79.626	169.807	120.423	44.446	-1065.855	-945.167	54.856
1000	80.496	178.244	125.790	52.454	-1073.393	-931.102	48.636
1000.000	80.496	178.244	125.790	52.454	GLASS <--> LIQUID		
1000.000	94.922	178.244	125.790	52.454	TRANSITION		
1100	94.922	187.291	130.976	61.946	-1071.052	-916.987	43.544
1200	94.922	195.550	136.018	71.438	-1068.736	-903.083	39.310
1300	94.922	203.148	140.894	80.931	-1066.444	-889.372	35.735
1400	94.922	210.182	145.595	90.423	-1191.583	-872.669	32.560
1500	94.922	216.731	150.121	99.915	-1187.980	-850.015	29.600
1536.000	94.922	218.983	151.709	103.332	CRYSTAL <--> LIQUID		
1600	94.922	222.858	154.478	109.407	-1184.395	-827.601	27.018
1700	94.922	228.612	158.671	118.900	-1180.826	-805.411	24.747
1800	94.922	234.038	162.709	128.392	-1177.273	-783.430	22.735
1900	94.922	239.170	166.599	137.884	-1173.731	-761.647	20.939
2000	94.922	244.039	170.351	147.376	-1170.199	-740.049	19.328
2100	94.922	248.670	173.971	156.869	-1166.672	-718.629	17.875
2200	94.922	253.086	177.467	166.361	-1163.149	-697.376	16.558
2300	94.922	257.305	180.848	175.853	-1159.625	-676.283	15.359
2400	94.922	261.345	184.118	185.345	-1156.097	-655.344	14.263
2500	94.922	265.220	187.285	194.838	-1152.561	-634.552	13.258
2600	94.922	268.943	190.355	204.330	-1149.014	-613.902	12.333
2700	94.922	272.526	193.332	213.822	-1145.455	-593.388	11.480
2800	94.922	275.978	196.223	223.314	-1141.879	-573.006	10.690
2900	94.922	279.309	199.030	232.807	-1138.285	-552.753	9.956
3000	94.922	282.527	201.760	242.299	-1134.673	-532.624	9.274

PREVIOUS: March 1966 CURRENT: June 1975

Magnesium Fluoride (MgF_2) $F_2Mg_1(l)$

MAGNESIUM FLUORIDE (MgF_2) CRYSTAL-LIQUID M_r = 62.301806

0 to 1536 K crystal
above 1536 K liquid

Refer to the individual tables for details.

Magnesium Fluoride (MgF_2) F_2Mg_1(cr,l)

Enthalpy Reference Temperature = T_r = 298.15 K Standard State Pressure = p° = 0.1 MPa

T/K	C_p° (J K⁻¹mol⁻¹)	S° (J K⁻¹mol⁻¹)	$-[G^\circ-H^\circ(T_r)]/T$ (J K⁻¹mol⁻¹)	$H^\circ-H^\circ(T_r)$ (kJ mol⁻¹)	Δ_fH° (kJ mol⁻¹)	Δ_fG° (kJ mol⁻¹)	Log K_f
0	0.	0.	INFINITE	-9.914	-1120.332	-1120.332	INFINITE
100	21.722	10.686	102.413	-9.173	-1123.140	-1106.221	577.830
200	48.817	35.056	62.549	-5.499	-1124.402	-1088.654	284.327
298.15	61.547	57.255	57.255	0.	-1124.241	-1071.109	187.654
300	61.689	57.636	57.256	0.114	-1124.231	-1070.779	186.439
400	68.484	76.377	59.765	6.645	-1123.475	-1053.065	137.516
500	72.634	92.132	64.706	13.713	-1122.439	-1035.579	108.186
600	75.320	105.624	70.429	21.117	-1121.275	-1018.315	88.652
700	77.178	117.381	76.315	28.746	-1120.070	-1001.250	74.714
800	78.555	127.780	82.110	36.535	-1118.878	-984.358	64.272
900	79.626	137.096	87.711	44.446	-1117.740	-967.612	56.159
1000	80.496	145.532	93.078	52.453	-1125.279	-950.276	49.637
1100	81.237	153.239	98.202	60.541	-1124.342	-932.821	44.296
1200	81.877	160.336	103.088	68.697	-1123.363	-915.452	39.849
1300	82.454	166.912	107.748	76.914	-1122.346	-898.167	36.089
1400	82.977	173.042	112.195	85.186	-1248.705	-877.795	32.751
1500	83.467	178.784	116.445	93.508	-1246.272	-851.386	29.648
1536.000	83.635	180.766	117.929	96.516	CRYSTAL <--> LIQUID	TRANSITION	
1536.000	94.922	218.983	117.929	155.218			
1600	94.922	222.858	122.050	161.293	-1184.395	-827.601	27.018
1700	94.922	228.612	128.150	170.785	-1180.826	-805.411	24.747
1800	94.922	234.038	133.884	180.277	-1177.273	-783.430	22.735
1900	94.922	239.170	139.291	189.769	-1173.731	-761.647	20.939
2000	94.922	244.039	144.408	199.262	-1170.199	-740.049	19.328
2100	94.922	248.670	149.264	208.754	-1166.672	-718.629	17.875
2200	94.922	253.086	153.883	218.246	-1163.149	-697.376	16.558
2300	94.922	257.305	158.289	227.738	-1159.625	-676.283	15.359
2400	94.922	261.345	162.499	237.231	-1156.097	-655.344	14.263
2500	94.922	265.220	166.531	246.723	-1152.561	-634.552	13.258
2600	94.922	268.943	170.399	256.215	-1149.014	-613.902	12.333
2700	94.922	272.526	174.115	265.707	-1145.454	-593.388	11.480
2800	94.922	275.978	177.692	275.200	-1141.879	-573.006	10.690
2900	94.922	279.309	181.139	284.692	-1138.285	-552.753	9.956
3000	94.922	282.527	184.465	294.184	-1134.673	-532.624	9.274

PREVIOUS: CURRENT: June 1975

Magnesium Fluoride (MgF_2) F_2Mg_1(cr,l)

MAGNESIUM FLUORIDE (MgF_2) IDEAL GAS M_r = 62.301806

$S°(298.15\ K) = 256.508 \pm 2.1$ (or ±12.6) $J\ K^{-1}\ mol^{-1}$

$\Delta_f H°(0\ K) = -724.994 \pm 3.3$ (or ±16.7) $kJ\ mol^{-1}$
$\Delta_f H°(298.15\ K) = 726.761 \pm 3.3$ (or 16.7) $kJ\ mol^{-1}$

Vibrational Frequencies and Degeneracies

ν, cm^{-1}
508 (1)
215 (1)
875 (1)

Ground State Quantum Weight = [1] $\sigma = 2$
Point Group = C_{2v}
Bond Distance: Mg-F = 1.77 ± 0.02 Å Bond Angle: F-Mg-F = 158°
Product of the Principal Moments of Inertia: $I_A I_B I_C = 1.033673 \times 10^{-115}\ g^3\ cm^6$

Enthalpy of Formation

$\Delta_{sub}H°(298.15\ K)$ is added to $\Delta_f H°(MgF_2$, cr, 298.15 K) to give $\Delta_f H°(g, 298.15\ K) = -173.7 \pm 0.8\ kcal\ mol^{-1}$. $\Delta_{sub}H°(298.15\ K)$ is selected from the results of a 3rd law analysis of twenty sets (1-9) of vapor pressure data for MgF_2(cr, ℓ). These measurements cover an extended temperature range (~1200°) and include the use of manometric (1, 4) Knudsen effusion (2, 5, 6), torsion-effusion (5-8) and mass spectrometric (3, 9) methods. The amount of dimer present in the saturated vapor is predicted not to exceed 1% below 2000 K by several methods (3, 6, 7-9).

Source	Reaction	Method	T/K	Data Points	$\Delta_r H°$ 2nd law	kcal mol^{-1} 3rd law	Drift cal K^{-1} mol^{-1}	$-\Delta_f H°(298.15\ K)$ kcal mol^{-1}
(1)	A	Manometric	1934-2129	6	89.2	82.5±1.0	-3.3±3.4	173.8±1.5
(2) Set 1	B	Knudsen Effusion	1337-1480	8	93.7	94.6±0.7	0.6±2.6	174.1±1.0
2	B	"	1440-1530	4	91.3	95.0±0.4	2.5±2.9	173.7±0.7
3	B	"	1284-1357	6	100.3	94.8±0.3	-4.1±1.9	173.9±0.6
(3)	B	Mass Spec	1450	1	91.4[a]	93.9		174.8
(4) Set 1	A	Manometric	2155-2327	Equation	82.9	82.40±0.03	-0.2	173.9±0.5
2	A	Manometric	2087-2357	Equation	84.8	82.53±0.2	-1.0	173.8±0.7
(5, 6) Set 1	B	Knudsen Effusion	1451-1533	6	98.7	95.3 ±0.5	-2.3±3.5	173.4±0.8
2	A	"	1568-1613	4	81.9	82.9 ±0.2	0.6±4.2	
3	B	Torsion Effusion	1413-1518	11[b]	92.9	95.3 ±0.6	1.7±2.3	173.4±0.9
4	A	"	1539-1614	11[c]	80.6	82.8 ±0.3	1.4±1.6	173.5±0.8
(7) Set 1	B	"	1273-1433	21	72.7	90.6 ±1.4	13.2±1.0	178.1±1.7
2	B	"	1373-1513	18[b]	76.6	91.0 ±1.0	10.0±0.8	177.7±1.3
(8) Set 1	B	"	1340-1500	Equation	97.7	95.8 ±0.2	-1.4	172.9±0.5
2	B	"	1348-1482	"	98.3	96.0 ±0.2	-1.6	172.7±0.5
3	B	"	1324-1465	"	96.7	95.7 ±0.1	-0.7	173.0±0.4
4	B	"	1376-1517	"	98.9	95.6 ±0.3	-2.2	173.1±0.7
5	B	"	1345-1515	"	97.9	95.7 ±0.2	-1.5	173.0±0.5
6	B	"	1341-1520	"	96.9	95.6 ±0.1	-1.0	173.1±0.4
(9)	B	Mass Spec	1241-1492	Equation	93.9	95.4 ±0.2	1.1	173.3±0.5

Reactions: (A) $MgF_2(\ell) = MgF_2(g)$; (B) $MgF_2(cr) = MgF_2(g)$.
[a] 2nd law value by slope method.
[b] One point rejected due to failure of a statistical test.
[c] Two points rejected due to failure of a statistical test.

These data are in agreement on $\Delta_{sub}H°(298.15\ K) = 95.0 \pm 0.5\ kcal\ mol^{-1}$. We adopt this value but assign an alternate uncertainty to $\Delta_f H°$ to include the possibility that MgF_2 is linear (see below).

Heat Capacity and Entropy

Considerable confusion exists as to whether MgF_2 is bent or linear. Electron diffraction (11), electric deflection (12), matrix-isolation Raman (23) and infrared (13, 22, 23) spectral studies, and MO calculatons (14-16, 24) predict a linear configuration. Other matrix-isolation IR studies (17, 18) and MO calculations (19, 24) favor the nonlinear structure. Isotopic shifts (18) in the matrix frequencies suggest that the most probable value of the apex angle is 158°. An energy minimum has been reported at 140° with a CNDO approximation (19). Unfortunately, the vapor pressure data are not sufficiently precise to be of use in the selection of a structure. The entropies of these two forms differ by only about 1.5 cal K^{-1} mol^{-1} at temperatures in the range of the vapor pressure data. We tentatively adopt a bond angle of 158° but assign an alternate uncertainty (±3 cal K^{-1} mol^{-1}) to S°(298.15 K) to include the possibility that the angle is 180°. The Mg-F bond length is taken from the electron-diffraction studies of Akishin et al. (11). The individual moments of inertia are: $I_A = 0.2808 \times 10^{-39}$, $I_B = 19.0477 \times 10^{-39}$, and $I_C = 19.3285 \times 10^{-39}\ g\ cm^2$.

The symmetric stretching frequency is a compromise between the krypton matrix infrared ($\nu_1 = 478\ cm^{-1}$, 17) and Raman ($\nu_1 = 545\ cm^{-1}$, 23) values. Also, an intermediate value ($\nu_2 = 215\ cm^{-1}$) is selected for the bending frequency from the estimated gas-phase value (270 cm^{-1}) of Snelson (13) and that (165 cm^{-1}) which was observed in the high temperature (~2000 K) infrared spectrum (20) of the vapor. ν_3 is the gas-phase value that was estimated by Snelson (13) from frequencies measured in three matrices. The krypton frequencies for ν_2 and ν_3 reported by Mann et al. (17) and Lesiecki and Nibler (23) agree quite well with those of Snelson (13). Also, the results of Hauge et al. (22) provide further support for the adopted ν_3 value. If MgF_2 is assumed to be linear and ν_2 is reduced to 165 cm^{-1}, then our free energy functions are decreased by 2.1 cal K^{-1} mol^{-1} at 298.15 K, 1.6 cal K^{-1} mol^{-1} at 1000 K and 1.0 cal K^{-1} mol^{-1} at 2000 K. These alternate functions are also reasonably consistent with the reported vapor pressure data. Previously published free energy functions (21) for MgF_2(g) which were based on a linear structure are several units (2-5 cal K^{-1} mol^{-1}) too low due to the use of an incorrect bending frequency (470 cm^{-1}).

References

References are listed on the liquid table.

Magnesium Fluoride (MgF_2) F_2Mg_1(g)

Enthalpy Reference Temperature = T_r = 298.15 K Standard State Pressure = p° = 0.1 MPa

T/K	$C_p°$ (J $K^{-1}mol^{-1}$)	S° (J $K^{-1}mol^{-1}$)	$-[G°-H°(T_r)]/T$ (J $K^{-1}mol^{-1}$)	$H°-H°(T_r)$ (kJ mol^{-1})	$\Delta_f H°$ (kJ mol^{-1})	$\Delta_f G°$ (kJ mol^{-1})	Log K_f
0	0.	0.	INFINITE	-12.056	-724.994	-724.994	INFINITE
100	37.519	210.143	296.182	-8.604	-725.091	-728.117	380.330
200	43.734	238.149	260.788	-4.528	-725.951	-730.821	190.871
250	46.242	248.185	257.291	-2.277	-726.373	-731.990	152.941
298.15	48.259	256.508	256.508	0.	-726.761	-733.036	128.425
300	48.329	256.806	256.509	0.089	-726.775	-733.075	127.640
350	50.021	264.388	257.103	2.550	-727.166	-734.094	109.558
400	51.375	271.159	258.445	5.086	-727.553	-735.057	95.989
450	52.455	277.275	260.203	7.683	-727.944	-735.972	85.429
500	53.320	282.848	262.193	10.328	-728.344	-736.842	76.977
600	54.583	292.689	266.477	15.727	-729.185	-738.464	64.289
700	55.431	301.171	270.842	21.231	-730.106	-739.939	55.215
800	56.020	308.613	275.107	26.805	-731.129	-741.275	48.400
900	56.444	315.237	279.205	32.429	-732.277	-742.476	43.092
1000	56.758	321.201	283.111	38.090	-742.162	-742.829	38.801
1100	56.996	326.623	286.824	43.778	-743.625	-742.825	35.274
1200	57.180	331.590	290.351	49.487	-745.093	-742.687	32.328
1300	57.326	336.173	293.702	55.213	-746.567	-742.427	29.831
1400	57.442	340.426	296.889	60.952	-875.459	-738.886	27.568
1500	57.538	344.392	299.925	66.701	-875.599	-729.126	25.390
1600	57.616	348.108	302.822	72.459	-875.749	-719.356	23.485
1700	57.681	351.603	305.589	78.224	-875.908	-709.577	21.803
1800	57.736	354.902	308.238	83.994	-876.075	-699.788	20.307
1900	57.783	358.025	310.777	89.771	-876.250	-689.990	18.969
2000	57.823	360.990	313.214	95.551	-876.429	-680.182	17.765
2100	57.858	363.812	315.557	101.335	-876.611	-670.365	16.674
2200	57.888	366.504	317.812	107.122	-876.793	-660.539	15.683
2300	57.914	369.078	319.985	112.912	-876.971	-650.706	14.778
2400	57.937	371.543	322.083	118.705	-877.142	-640.864	13.948
2500	57.958	373.909	324.109	124.500	-877.304	-631.016	13.184
2600	57.976	376.182	326.068	130.297	-877.453	-621.162	12.479
2700	57.992	378.370	327.965	136.095	-877.587	-611.302	11.826
2800	58.007	380.480	329.803	141.895	-877.704	-601.437	11.220
2900	58.020	382.516	331.586	147.696	-877.801	-591.569	10.655
3000	58.032	384.483	333.316	153.499	-877.878	-581.697	10.128
3100	58.042	386.386	334.998	159.302	-877.934	-571.824	9.635
3200	58.052	388.229	336.633	165.107	-877.968	-561.948	9.173
3300	58.061	390.015	338.223	170.913	-877.980	-552.072	8.739
3400	58.069	391.749	339.772	176.719	-877.971	-542.197	8.330
3500	58.076	393.432	341.281	182.527	-877.942	-532.321	7.944
3600	58.083	395.068	342.753	188.335	-877.894	-522.447	7.581
3700	58.089	396.660	344.188	194.143	-877.829	-512.574	7.236
3800	58.095	398.209	345.590	199.952	-877.748	-502.704	6.910
3900	58.100	399.718	346.958	205.762	-877.655	-492.835	6.601
4000	58.105	401.189	348.296	211.572	-877.550	-482.969	6.307
4100	58.110	402.624	349.603	217.383	-877.437	-473.106	6.027
4200	58.114	404.024	350.883	223.194	-877.318	-463.246	5.761
4300	58.118	405.392	352.134	229.006	-877.196	-453.389	5.508
4400	58.122	406.728	353.360	234.818	-877.075	-443.534	5.265
4500	58.125	408.034	354.561	240.630	-876.955	-433.682	5.034
4600	58.129	409.312	355.737	246.443	-876.842	-423.833	4.813
4700	58.132	410.562	356.890	252.256	-876.737	-413.986	4.601
4800	58.134	411.786	358.021	258.069	-876.643	-404.141	4.398
4900	58.137	412.984	359.131	263.883	-876.564	-394.298	4.203
5000	58.140	414.159	360.219	269.697	-876.501	-384.457	4.016
5100	58.142	415.310	361.288	275.511	-876.458	-374.616	3.837
5200	58.144	416.439	362.338	281.325	-876.438	-364.776	3.664
5300	58.146	417.547	363.369	287.140	-876.441	-354.937	3.498
5400	58.148	418.634	364.383	292.955	-876.471	-345.097	3.338
5500	58.150	419.701	365.379	298.769	-876.531	-335.256	3.184
5600	58.152	420.748	366.358	304.585	-876.622	-325.414	3.035
5700	58.154	421.778	367.322	310.400	-876.747	-315.570	2.892
5800	58.155	422.789	368.269	316.215	-876.907	-305.724	2.753
5900	58.157	423.783	369.202	322.031	-877.105	-295.874	2.619
6000	58.158	424.761	370.120	327.847	-877.343	-286.021	2.490

PREVIOUS: June 1975 (1 atm) CURRENT: June 1975 (1 bar)

Magnesium Fluoride (MgF_2) F_2Mg_1(g)

MAGNESIUM FLUORIDE, ION (MgF_2^+) IDEAL GAS M_r = 62.301257

$\Delta_fH°(0\ K)$ = 587.193 ± 20.9 (or ±37.7) kJ mol^{-1}

S°(298.15 K) = [258.145 ± 12.6] J K^{-1} mol^{-1} $\Delta_fH°(298.15\ K)$ = 591.982 ± 20.9 (or ±37.7) kJ mol^{-1}

Electronic Levels and Quantum Weights

State	ε_i, cm^{-1}	g_i
$^2\Pi$	0	4
$^2\Pi$	[20000]	4
$^2\Sigma$	[25000]	2

Vibrational Frequencies and Degeneracies

ν, cm^{-1}
[500](1)
[200](2)
[800](1)

Point Group: [$D_{\infty h}$] σ = 2

Bond Distance: Mg-F = [1.95] Å

Bond Angle: F-Mg-F = [180]°

Rotational Constant: B_0 = [0.116677] cm^{-1}

Enthalpy of Formation

The MgF_2^+ ion has been identified as an important species in the vapor mass spectra of the MgF_2 (1-3), MgF_2-Tℓ-Pd (4), and MgF_2-Cu (5) systems. The observed appearance potentials (AP) for this ion are tabulated below.

Source	Year	AP, eV
Berkowitz and Marquart (1)	1962	13.5±0.4
Green et al. (2)	1964	13.6
Murad et al. (4)	1966	14.0±0.5
Hildenbrand (5)	1968	13.5
Hildenbrand (3)	1968	13.3±0.3

These results show the normal scatter expected for such measurements; the largest deviation is 0.7 eV. We adopt an average value of 13.6±0.2 eV (313.62±4.6 kcal mol^{-1}), and we assume that this value refers to the direct ionization process $MgF_2(g)$ + e- = $MgF_2^+(g)$ + $2e^-$ occurring at 0 K. Combining the selected AP value with $\Delta_fH°(MgF_2,g)$ = -173.3 ± 0.8 kcal mol^{-1} (6), we obtain $\Delta_fH°(MgF_2^+, g, 0\ K)$ = 140.3±5 kcal mol^{-1}. The $\Delta_fH°$ value at 298.15 K is 141.5 kcal mol^{-1}. An alternate uncertainty of ±9 kcal mol^{-1} is included to cover the possibility that the MgF_2 molecule is linear (6).

Heat Capacity and Entropy

A comparison of the atomization energies, $\Delta_{at}H°(0\ K)$ in kcal mol^{-1}, for MgF_2(245, 6) and MgF_2^+ (107.7) suggests weaker bonding in the ion relative to the neutral molecule. One might therefore expect that the fundamental frequencies of MgF_2^+ would be somewhat less than those for MgF_2. We assume a slight decrease in the three vibrational frequencies of MgF_2 (6). MgF_2^+ (fifteen valence electrons) is isoelectronic with the molecules BO_2, N_3, NCO, and N_2O^+. By analogy with the ground states (6) for these isoelectronic molecules, we assume that the ground state for MgF_2^+ is $^2\Pi$. We also include two excited states ($^2\Pi$ and $^2\Sigma$) which would be expected to exist based on those observed for BO_2 (6).

According to the diagram of Walsh (7), MgF_2^+ is predicted to be linear. This prediction is supported by the fact that several other fifteen valence electron molecules (BO_2, NCO, N_2O^+, and N_3) are now known to be linear in their ground states. We adopt a linear configuration. Due to the weaker bonding in MgF_2^+ relative to MgF_2, it is reasonable to assume that $r_e(MgF_2^+) > r_e(MgF_2)$. We assume a 10% increase in the bond length for MgF_2 (6).

References

1. J. Berkowitz and J. R. Marquart, J. Chem. Phys. 37, 1853 (1962).
2. J. W. Green, G. D. Blue, T. C. Ehlert, and J. L. Margrave, J. Chem. Phys. 41, 2245 (1964).
3. D. L. Hildenbrand, J. Chem. Phys. 48, 3657 (1968).
4. E. Murad, D. L. Hildenbrand, and R. P. Main, J. Chem. Phys. 45, 263 (1966).
5. D. L. Hildenbrand, J. Chem. Phys. 48, 2457 (1968).
6. JANAF Thermochemical Tables: $MgF_2(g)$, 6-30-75; $BO_2(g)$, 6-30-68; CNO(g), $N_2O^+(g)$ and $N_3(g)$, 12-31-70.
7. A. D. Walsh, J. Chem. Soc. 1953, 2266 (1953).

Magnesium Fluoride, Ion (MgF_2^+) $F_2Mg_1^+(g)$

Enthalpy Reference Temperature = T_r = 298.15 K Standard State Pressure = p° = 0.1 MPa

T/K	$C_p°$	S°	$-[G°-H°(T_r)]/T$	$H°-H°(T_r)$	$\Delta_fH°$	$\Delta_fG°$	Log K_f
	J $K^{-1}mol^{-1}$			kJ mol^{-1}			
0	0.	0.	INFINITE	-12.414	587.193		
100	38.135	208.753	300.902	-9.215			
200	47.129	238.250	262.789	-4.908			
250	50.159	249.106	258.996	-2.473			
298.15	52.449	258.145	258.145	0.	591.982	578.963	-101.432
300	52.526	258.469	258.146	0.097	592.013	578.883	-100.792
350	54.358	266.710	258.792	2.771	592.876	576.626	-86.057
400	55.774	274.065	260.250	5.526	593.746	574.245	-74.989
450	56.875	280.700	262.160	8.343	594.615	571.756	-66.368
500	57.738	286.739	264.320	11.210	595.476	569.169	-59.461
600	58.973	297.384	268.967	17.050	597.154	563.749	-49.079
700	59.786	306.539	273.696	22.990	598.749	558.054	-41.643
800	60.344	314.561	278.313	28.998	600.239	552.138	-36.051
900	60.742	321.693	282.745	35.054	601.600	546.042	-31.691
1000	61.034	328.109	286.966	41.143	594.222	540.515	-28.234
1100	61.255	333.937	290.975	47.258	595.265	535.094	-25.409
1200	61.425	339.274	294.780	53.393	596.301	529.578	-23.052
1300	61.559	344.196	298.395	59.542	597.329	523.976	-21.054
1400	61.666	348.763	301.831	65.703	470.938	521.462	-19.456
1500	61.753	353.020	305.104	71.875	473.299	524.989	-18.282
1600	61.825	357.008	308.224	78.054	475.649	528.358	-17.249
1700	61.885	360.758	311.205	84.239	477.989	531.580	-16.333
1800	61.936	364.297	314.058	90.430	480.320	534.666	-15.516
1900	61.979	367.647	316.791	96.626	482.644	537.621	-14.780
2000	62.016	370.827	319.414	102.826	484.963	540.455	-14.115
2100	62.048	373.853	321.935	109.029	487.278	543.172	-13.511
2200	62.077	376.740	324.361	115.235	489.595	545.780	-12.958
2300	62.103	379.500	326.698	121.444	491.914	548.282	-12.452
2400	62.126	382.144	328.954	127.656	494.240	550.683	-11.985
2500	62.149	384.680	331.133	133.870	496.576	552.987	-11.554
2600	62.170	387.118	333.239	140.086	498.925	555.197	-11.154
2700	62.192	389.465	335.279	146.304	501.289	557.316	-10.782
2800	62.214	391.727	337.254	152.524	503.671	559.348	-10.435
2900	62.236	393.911	339.171	158.746	506.074	561.294	-10.110
3000	62.260	396.021	341.031	164.971	508.497	563.157	-9.805
3100	62.286	398.063	342.838	171.199	510.944	564.938	-9.519
3200	62.313	400.041	344.595	177.428	513.414	566.641	-9.249
3300	62.343	401.959	346.304	183.661	515.907	568.266	-8.995
3400	62.376	403.820	347.968	189.897	518.424	569.815	-8.754
3500	62.411	405.629	349.590	196.137	520.964	571.289	-8.526
3600	62.449	407.388	351.171	202.380	523.525	572.691	-8.310
3700	62.491	409.099	352.714	208.627	526.108	574.021	-8.104
3800	62.536	410.767	354.220	214.878	528.709	575.281	-7.908
3900	62.584	412.392	355.691	221.134	531.328	576.473	-7.721
4000	62.636	413.977	357.128	227.395	533.962	577.597	-7.543
4100	62.691	415.524	358.534	233.661	536.609	578.655	-7.372
4200	62.749	417.035	359.908	239.933	539.267	579.648	-7.209
4300	62.811	418.513	361.254	246.211	541.934	580.578	-7.053
4400	62.876	419.957	362.572	252.495	544.606	581.446	-6.903
4500	62.945	421.371	363.863	258.786	547.283	582.253	-6.759
4600	63.016	422.755	365.128	265.084	549.960	583.001	-6.620
4700	63.090	424.111	366.369	271.390	552.636	583.690	-6.487
4800	63.167	425.440	367.586	277.703	555.308	584.323	-6.359
4900	63.247	426.744	368.780	284.023	557.973	584.900	-6.235
5000	63.329	428.022	369.952	290.352	560.629	585.422	-6.116
5100	63.413	429.277	371.103	296.689	563.274	585.892	-6.001
5200	63.499	430.509	372.234	303.035	565.904	586.310	-5.890
5300	63.587	431.720	373.345	309.389	568.520	586.677	-5.782
5400	63.677	432.909	374.437	315.752	571.116	586.995	-5.678
5500	63.767	434.078	375.510	322.124	573.692	587.265	-5.577
5600	63.859	435.228	376.567	328.506	576.246	587.489	-5.480
5700	63.953	436.359	377.606	334.896	578.775	587.667	-5.385
5800	64.046	437.472	378.628	341.296	581.278	587.801	-5.294
5900	64.141	438.568	379.635	347.706	583.752	587.893	-5.205
6000	64.235	439.647	380.626	354.124	586.196	587.942	-5.118

PREVIOUS: December 1975 (1 atm) CURRENT: December 1975 (1 bar)

Magnesium Fluoride, Ion (MgF_2^+) $F_2Mg_1^+(g)$

MOLYBDENUM FLUORIDE (MoF_2) IDEAL GAS M_r = 133.93680

$\Delta_{at}H°(0\ K)$ = 974.035 ± 25.1 kJ mol^{-1}
$S°(298.15\ K)$ = [271.932 ± 8.4] J K^{-1} mol^{-1}

$\Delta_fH°(0\ K)$ = [-162.853 ± 16.7] kJ mol^{-1}
$\Delta_fH°(298.15\ K)$ = -162.758 ± 16.7] kJ mol^{-1}

Electronic Levels and Quantum Weights

ε_i, cm^{-1}	g_i
0	[3]
[450]	[5]

Vibrational Frequencies and Degeneracies

ν, cm^{-1}
[700](1)
[150](2)
[700](1)

Point Group: $[D_{\infty h}]$ σ = [2]
Bond Distance: Mo-F = [1.82] Å
Bond Angle: F-Mo-F = [180]°
Rotational Constant: B_0 = [0.133939] cm^{-1}

Enthalpy of Formation

The adopted enthalpy of formation at 298.15 K is that derived by Brewer (1) from the high temperature mass spectrometric measurements of equilibria among gaseous fluorides by Hildenbrand [1818-2242 K] (2). The value of $\Delta_fH°(0\ K)$ combined with JANAF data (3) for Mo(g) and F(g) gives $\Delta_{at}H°(0\ K)$ = 232.8±6 kcal mol^{-1} and an average Mo-F bond energy of $\Delta_{at}H°(0\ K)/2$ = 116.4±3 kcal mol^{-1}.

Heat Capacity and Entropy

We use the vibrational frequencies of 700(2) and 150(2) cm^{-1} estimated by Hildenbrand (2) and adopted by Brewer (1). Brewer (1) assumed a linear molecule with a Mo-F distance of 1.82 Å as in MoF_6(g). No electronic level data are available for the dihalides of molybdenum. Brewer (1) uses g = 3 for the ground state and one excited level at 450 cm^{-1} with g = 5. Additional electronic contributions are expected, but the electronic contribution was kept small (1) to be as consistent as possible with the smaller entropy values derived for MoF(g) and MoF_2(g) from the 2nd law treatment of Hildenbrand's equilibria data (2).

This table is based on the critical analysis of data existing in 1978 by Brewer (1).

References

1. L. Brewer, Materials and Molecular Research Division, Lawrence Berkeley Laboratory, University of California, Berkeley; personal communication, Sept. 1978, of preliminary draft of review to be submitted for publication in Atomic Energy Review, International Atomic Energy Agency, Vienna, Austria.
2. D. L. Hildenbrand, J. Chem. Phys. 65, 614 (1976).
3. JANAF Thermochemical Tables: Mo(g), 3-31-78; F(g), 9-30-65.

Molybdenum Fluoride (MoF_2) $F_2Mo_1(g)$

Enthalpy Reference Temperature = T_r = 298.15 K Standard State Pressure = p° = 0.1 MPa

T/K	$C_p°$ (J $K^{-1}mol^{-1}$)	S° (J $K^{-1}mol^{-1}$)	$-[G°-H°(T_r)]/T$ (J $K^{-1}mol^{-1}$)	$H°-H°(T_r)$ (kJ mol^{-1})	$\Delta_fH°$ (kJ mol^{-1})	$\Delta_fG°$ (kJ mol^{-1})	Log K_f
0	0.	0.	INFINITE	-13.505	-162.853	-162.853	INFINITE
100	41.502	217.787	318.908	-10.112	-162.858	-166.897	87.178
200	52.007	250.046	277.031	-5.397	-162.914	-170.897	44.634
250	55.223	262.022	272.864	-2.711	-162.848	-172.900	36.125
298.15	57.221	271.932	271.932	0.	-162.758	-174.844	30.632
300	57.282	272.286	271.933	0.106	-162.754	-174.919	30.456
350	58.594	281.222	272.636	3.005	-162.657	-176.954	26.409
400	59.454	289.106	274.212	5.958	-162.577	-179.002	23.375
450	60.042	296.145	276.265	8.946	-162.521	-181.059	21.017
500	60.460	302.494	278.575	11.959	-162.490	-183.121	19.131
600	61.005	313.569	283.511	18.035	-162.504	-187.247	16.301
700	61.337	323.000	288.496	24.153	-162.610	-191.364	14.280
800	61.557	331.206	293.333	30.298	-162.796	-195.460	12.762
900	61.711	338.466	297.952	36.462	-163.052	-199.528	11.580
1000	61.824	344.974	302.334	42.639	-163.378	-203.564	10.633
1100	61.909	350.870	306.483	48.826	-163.774	-207.564	9.856
1200	61.976	356.260	310.410	55.021	-164.244	-211.525	9.207
1300	62.028	361.223	314.130	61.221	-164.794	-215.444	8.657
1400	62.071	365.821	317.660	67.426	-165.429	-219.316	8.183
1500	62.105	370.105	321.015	73.635	-166.156	-223.141	7.770
1600	62.134	374.114	324.210	79.847	-166.980	-226.913	7.408
1700	62.158	377.882	327.257	86.061	-167.906	-230.631	7.086
1800	62.178	381.435	330.169	92.278	-168.940	-234.292	6.799
1900	62.196	384.797	332.957	98.497	-170.088	-237.892	6.540
2000	62.211	387.988	335.629	104.717	-171.355	-241.428	6.305
2100	62.224	391.024	338.195	110.939	-172.747	-244.898	6.092
2200	62.235	393.918	340.663	117.162	-174.269	-248.299	5.895
2300	62.245	396.685	343.039	123.386	-175.928	-251.627	5.715
2400	62.254	399.335	345.330	129.611	-177.729	-254.880	5.547
2500	62.262	401.876	347.541	135.837	-179.682	-258.055	5.392
2600	62.269	404.318	349.678	142.063	-181.803	-261.149	5.247
2700	62.275	406.668	351.746	148.291	-184.129	-264.157	5.110
2800	62.281	408.933	353.748	154.518	-186.720	-267.075	4.982
2900	62.286	411.119	355.689	160.747	-225.538	-269.845	4.860
3000	62.290	413.230	357.572	166.975	-226.856	-271.350	4.725
3100	62.294	415.273	359.401	173.205	-228.147	-272.812	4.597
3200	62.298	417.251	361.178	179.434	-229.409	-274.232	4.476
3300	62.302	419.168	362.906	185.664	-230.640	-275.614	4.363
3400	62.305	421.028	364.588	191.895	-231.840	-276.959	4.255
3500	62.308	422.834	366.227	198.125	-233.007	-278.269	4.153
3600	62.310	424.589	367.824	204.356	-234.140	-279.546	4.056
3700	62.313	426.297	369.381	210.587	-235.239	-280.792	3.964
3800	62.315	427.958	370.901	216.819	-236.304	-282.009	3.876
3900	62.317	429.577	372.385	223.050	-237.334	-283.198	3.793
4000	62.319	431.155	373.834	229.282	-238.330	-284.361	3.713
4100	62.321	432.694	375.251	235.514	-239.292	-285.500	3.637
4200	62.323	434.195	376.637	241.747	-240.220	-286.615	3.565
4300	62.324	435.662	377.992	247.979	-241.115	-287.710	3.495
4400	62.326	437.095	379.319	254.211	-241.977	-288.783	3.428
4500	62.327	438.495	380.619	260.444	-242.806	-289.837	3.364
4600	62.329	439.865	381.892	266.677	-243.604	-290.874	3.303
4700	62.330	441.206	383.140	272.910	-244.371	-291.893	3.244
4800	62.331	442.518	384.363	279.143	-245.108	-292.896	3.187
4900	62.332	443.803	385.563	285.376	-245.815	-293.885	3.133
5000	62.333	445.063	386.741	291.609	-837.878	-289.127	3.020
5100	62.334	446.297	387.896	297.843	-840.208	-278.130	2.849
5200	62.335	447.507	389.031	304.076	-842.657	-267.084	2.683
5300	62.336	448.695	390.146	310.310	-845.224	-255.991	2.523
5400	62.337	449.860	391.241	316.543	-847.907	-244.849	2.368
5500	62.337	451.004	392.317	322.777	-850.706	-233.656	2.219
5600	62.338	452.127	393.375	329.011	-853.618	-222.410	2.075
5700	62.339	453.230	394.416	335.244	-856.641	-211.113	1.935
5800	62.339	454.315	395.439	341.478	-859.774	-199.761	1.799
5900	62.340	455.380	396.446	347.712	-863.012	-188.353	1.668
6000	62.341	456.428	397.437	353.946	-866.358	-176.891	1.540

PREVIOUS: December 1978 (1 atm) CURRENT: December 1978 (1 bar)

Molybdenum Fluoride (MoF_2) $F_2Mo_1(g)$

DIFLUOROAMIDOGEN (NF_2) IDEAL GAS M_r = 52.003506

$\Delta_f H°(0\ K)$ = 44.8 ± 8 kJ mol^{-1}

$S°(298.15\ K)$ = 249.950 J K^{-1} mol^{-1} $\Delta_f H°(298.15\ K)$ = 42.3 ± 8 kJ mol^{-1}

Vibrational Frequencies and Degeneracies

ν, cm^{-1}
1074(1)
573(1)
931(1)

Ground State Quantum Weight: 2 σ = 2

Point Group: C_{2v}

Bond Distance: N-F = 1.37 Å

Bond Angle: F-N-F = 104.2°

Product of the Moments of Inertia: $I_A I_B I_C = 7.61262 \times 10^{-116}\ g^3\ cm^6$

Enthalpy of Formation

Colburn and Johnson (1) reported the study of the dissociation reaction $N_2F_4(g) = 2\ NF_2(\ell)$ by two independent methods: (a) measurement of the pressure variation with temperature at constant volume and (b) a spectrophotometric method based on the temperature dependence of the ultraviolet absorption of NF_2 radical. The authors gave a more detailed account of their data in a later publication (2). The contents of these articles indicate that $2\ NF_2(\ell)$ should read $2\ NF_2(g)$ in the above reaction. The first method gave a value of 19.85 kcal mol^{-1} for the enthalpy of dissociation in the temperature range 373 to 423 K which corresponds to 19.87 kcal mol^{-1} at 298.15 K. From their ΔH and ΔS data, ΔG was calculated and used in a 3rd law evaluation of the enthalpy of dissociation. The $\Delta_r H°(298.15\ K)$ so calculated showed an increasing trend with increasing temperature; the average value was 22.87 kcal mol^{-1} which is not in agreement with their value of 19.87 kcal mol^{-1}. The spectrophotometric value of 21.7 kcal mol^{-1} at 298.15 K was obtained from the integrated Van't Hoff equation.

Piette, Johnson, Booman and Colburn (3) studied the temperature dependence of the EPR spectrum of the NF_2 radical. They obtained a 2nd law value of 19.3 kcal mol^{-1} for the enthalpy of dissociation. This value, however, depends upon the degree of dissociation of N_2F_4 which they calculated from the (dP/dT) measurements of Colburn and Johnson (1). Any error in the (dP/dT) measurements would manifest itself in like manner here.

Doorenbos and Loy (4) investigated the ESR spectrum of the N_2F_4-NF_2 equilibrium from which they calculated a 2nd law value of 19.8 kcal mol^{-1} for the enthalpy of dissociation. They report that at 25°C and 13 atm, N_2F_4 is 0.022% dissociated in the vapor phase. A 3rd law calculation using this datum and the relationship $K_p = (4\alpha^2/1-\alpha^2)P_T$, where α is the degree of dissociation of N_2F_4 and P_T is the total pressure, gave a heat of dissociation of 21.79 kcal mol^{-1} at 298.15 K; this is not in agreement with their 2nd law value.

Herron and Diebler (5) report thermal dissociation data for N_2F_4. From their data they calculated a 2nd law value of 21.5 kcal mol^{-1} for the heat of dissociation of N_2F_4. A 2nd and 3rd law evaluation of their data was performed. It is found that if the k factor, relating their measured ion abundances to partial pressure, is taken as 4.72 x 10^{-8}, a 3rd law is obtained that shows slight scatter but no temperature dependence. Excluding those points that showed scatter, the 3rd law heat of dissociation at 298.15 K is 22.27 kcal mol^{-1}. A 2nd law plot gives a heat value of 22.23 kcal mol^{-1} at an average temperature of 400 K or 22.25 kcal mol^{-1} at 298.15 K. Those points that deviated from the chosen straight line in the 2nd law plot shows corresponding deviations in the 3rd law calculation. Those points lying on the 2nd law slope give a constant 3rd law value.

The enthalpy of dissociation for $N_2F_4(g) = 2\ NF_2(g)$ is taken to be 22.26 kcal mol^{-1}. Utilizing the enthalpy of formation of $N_2F_4(g)$, the enthalpy of formation of the difluoroamino radical becomes 10.1 ± 2.0 kcal mol^{-1}. All values for the enthalpy of dissociation fall within the limits of error assigned to the enthalpy of formation of NF_2.

Heat Capacity and Entropy

See table on tetrafluorohydrazine, $N_2F_4(g)$, for discussion.

References

1. C. B. Colburn and F. A. Johnson, J. Chem. Phys. 33, 1869 (1960).
2. C. B. Colburn and F. A. Johnson, J. Am. Chem. Soc. 83, 3043 (1961).
3. L. H. Piette, F. A. Johnson, K. A. Booman and C. B. Colburn, J. Chem. Phys. 35, 1481 (1961).
4. H. E. Doorenbos and B. R. Loy, J. Chem. Phys. 39, 2393 (1963).
5. J. T. Herron and V. H. Dibeler, J. Res. Nat. Bur. Stand. 65A, 405 (1961).

Difluoroamidogen (NF_2) $F_2N_1(g)$

Enthalpy Reference Temperature = T_r = 298.15 K Standard State Pressure = p° = 0.1 MPa

T/K	$C_p°$ (J K^{-1}mol^{-1})	S° (J K^{-1}mol^{-1})	$-[G°-H°(T_r)]/T$ (J K^{-1}mol^{-1})	$H°-H°(T_r)$ (kJ mol^{-1})	$\Delta_f H°$ (kJ mol^{-1})	$\Delta_f G°$ (kJ mol^{-1})	Log K_f
0	0.	0.	INFINITE	-10.575	44.844	44.844	INFINITE
100	33.409	210.747	283.220	-7.247	43.813	47.766	-24.950
200	36.304	234.591	253.546	-3.791	42.885	52.096	-13.606
250	38.656	242.940	250.612	-1.918	42.528	54.441	-11.375
298.15	41.005	249.950	249.950	0.	42.258	56.762	-9.944
300	41.093	250.204	249.951	0.076	42.249	56.852	-9.899
350	43.377	256.713	250.460	2.189	42.045	59.302	-8.850
400	45.399	262.640	251.617	4.409	41.905	61.778	-8.067
450	47.136	268.091	253.149	6.724	41.820	64.268	-7.460
500	48.604	273.135	254.898	9.118	41.779	66.765	-6.975
600	50.874	282.209	258.711	14.099	41.794	71.762	-6.247
700	52.481	290.179	262.649	19.271	41.893	76.750	-5.727
800	53.639	297.267	266.542	24.580	42.038	81.720	-5.336
900	54.491	303.636	270.315	29.989	42.204	86.670	-5.030
1000	55.132	309.412	273.941	35.471	42.377	91.602	-4.785
1100	55.625	314.691	277.409	41.010	42.547	96.516	-4.583
1200	56.010	319.548	280.721	46.593	42.709	101.415	-4.414
1300	56.316	324.044	283.883	52.210	42.860	106.301	-4.271
1400	56.563	328.227	286.902	57.854	42.997	111.176	-4.148
1500	56.766	332.136	289.789	63.521	43.119	116.041	-4.041
1600	56.933	335.805	292.552	69.206	43.226	120.899	-3.947
1700	57.073	339.261	295.199	74.907	43.319	125.751	-3.864
1800	57.191	342.527	297.738	80.620	43.398	130.598	-3.790
1900	57.292	345.622	300.177	86.344	43.465	135.440	-3.724
2000	57.379	348.563	302.524	92.078	43.523	140.279	-3.664
2100	57.453	351.364	304.783	97.820	43.575	145.116	-3.610
2200	57.519	354.038	306.962	103.568	43.625	149.950	-3.560
2300	57.576	356.596	309.065	109.323	43.675	154.782	-3.515
2400	57.626	359.048	311.097	115.083	43.729	159.611	-3.474
2500	57.670	361.401	313.062	120.848	43.791	164.439	-3.436
2600	57.710	363.664	314.965	126.617	43.864	169.263	-3.401
2700	57.745	365.843	316.809	132.390	43.953	174.085	-3.368
2800	57.776	367.943	318.598	138.166	44.059	178.902	-3.337
2900	57.805	369.971	320.335	143.945	44.187	183.716	-3.309
3000	57.831	371.931	322.022	149.727	44.337	188.525	-3.283
3100	57.854	373.828	323.663	155.511	44.514	193.328	-3.258
3200	57.875	375.665	325.260	161.297	44.718	198.125	-3.234
3300	57.894	377.446	326.814	167.086	44.951	202.916	-3.212
3400	57.912	379.175	328.329	172.876	45.215	207.699	-3.191
3500	57.928	380.854	329.806	178.668	45.510	212.474	-3.171
3600	57.943	382.486	331.246	184.462	45.838	217.240	-3.152
3700	57.957	384.074	332.653	190.257	46.199	221.996	-3.134
3800	57.969	385.619	334.027	196.053	46.592	226.742	-3.117
3900	57.981	387.125	335.369	201.851	47.019	231.477	-3.100
4000	57.992	388.593	336.681	207.649	47.479	236.201	-3.084
4100	58.002	390.026	337.965	213.449	47.972	240.913	-3.069
4200	58.011	391.423	339.221	219.249	48.497	245.613	-3.055
4300	58.020	392.788	340.451	225.051	49.055	250.299	-3.041
4400	58.028	394.122	341.656	230.853	49.644	254.973	-3.027
4500	58.035	395.427	342.836	236.656	50.265	259.632	-3.014
4600	58.042	396.702	343.993	242.460	50.916	264.278	-3.001
4700	58.049	397.950	345.128	248.265	51.597	268.909	-2.989
4800	58.055	399.173	346.241	254.070	52.306	273.525	-2.977
4900	58.061	400.370	347.334	259.876	53.045	278.126	-2.965
5000	58.067	401.543	348.406	265.682	53.810	282.712	-2.953
5100	58.072	402.693	349.460	271.489	54.602	287.282	-2.942
5200	58.077	403.820	350.494	277.297	55.420	291.837	-2.932
5300	58.081	404.927	351.511	283.105	56.264	296.375	-2.921
5400	58.086	406.012	352.510	288.913	57.131	300.897	-2.911
5500	58.090	407.078	353.493	294.722	58.022	305.403	-2.900
5600	58.094	408.125	354.459	300.531	58.936	309.893	-2.891
5700	58.097	409.153	355.409	306.341	59.872	314.366	-2.881
5800	58.101	410.164	356.345	312.150	60.829	318.822	-2.871
5900	58.104	411.157	357.265	317.961	61.805	323.263	-2.862
6000	58.108	412.134	358.172	323.771	62.802	327.685	-2.853

PREVIOUS: March 1964 (1 atm) CURRENT: March 1964 (1 bar)

DIFLUORODIAZINE, Cis-(N_2F_2) IDEAL GAS M_r = 66.010206

$\Delta_fH°(0\ K)$ = 74.711 ± 5.0 kJ mol^{-1}

S°(298.15 K) = 259.810 J K^{-1} mol^{-1} $\Delta_fH°(298.15\ K)$ = 68.618 ± 5.0 kJ mol^{-1}

Vibrational Frequencies and Degeneracies

ν, cm^{-1}	ν, cm^{-1}
896(1)	952(1)
1524(1)	573(1)
552(1)	737(1)

Ground State Quantum Weight: = 1 σ = 2

Point Group: C_{2v}

Bond Distances: N-F = 1.384 Å N-N = 1.214 Å

Bond Angle: F-N-N = 114.5 ± 0.5°

Product of the Moments of Inertia: $I_AI_BI_C$ = 6.590443 x 10^{-115} g^3 cm^6

Enthalpy of Formation

The enthalpy of formation was derived from the heat of reaction $N_2F_2(g) + 8/3\ NH_3(g) = 2\ NH_4F(cr) + 4/3\ N_2(g)$ which was measured calorimetrically by Armstrong and Marantz (1). There are two isomers, the trans and the active form of difluoridiazine. The enthalpy of formation were determined to be 19.4 and 16.4 kcal mol^{-1} for the trans and the active isomers, respectively. The active form of N_2F_2 has been interpreted here as the cis isomer which will be discussed later. The enthalpy of isomerization has also been reported as 3.0 kcal mol^{-1} by Armstrong and Marantz. The value, $\Delta_fH°$(cis N_2F_2, g, 298.15 K) = 16.4 kcal mol^{-1}, is adopted in the tabulation.

Pankratov et al. (2) have measured the enthalpy of reaction $N_2F_2(g) + 3\ KI = 2\ KF + N_2 + KI_3$ in acidic solution (pH = 3), and reported $\Delta_fH°$(active N_2F_2, g, 298.15 K) = 25.3 ± 2.0 kcal mol^{-1}.

The "heat of isomerization" has been reported as 27.5 ± 5.0 kcal mol^{-1} by Colburn et al. (9). The derivation of this value is unknown. The authors observed three values of the equilibrium constant for the isomerization reaction. Third law analysis of their data gives $\Delta_rH°$(298.15 K) = 2.52 ± 0.05 kcal mol^{-1} for the reaction cis $N_2F_2(g)$ → trans $N_2F_2(g)$ which is in good agreement with Armstrong and Sidney's data.

Heat Capacity and Entropy

Although the infrared spectrum (7, 8) of the active form of $N_2F_2(g)$ has been interpreted as favoring the 1,1-di-flurodiazine over the cis-configuration (3, 4), n.m.r. (5) and microwave (6) studies definitely indicate the latter. The vibrational frequencies (7) previously assigned to the 1,1-difluorodiazine have been adopted here for the cis isomer. The selected bond distances and angle were taken from the isotopic nitrogen microwave studies (6). The principal moments of inertia were: I_A = 4.2469 x 10^{-39}, I_B = 10.5134 x 10^{-39} and I_C = 14.7603 x 10^{-39} g cm^2.

See the trans-N_2F_2(g) table for the comparison of the different physical properties between trans and cis isomers.

References

1. G. T. Armstrong and S. Marantz, J. Chem. Phys. 38, 169 (1963).
2. A. V. Pankratov, A. N. Zercheninov, O. G. Talakin, O. M. Sokolov, and N. A. Knyazeva, Zh. Fiz. Khim. 37, 1399 (1963).
3. R. Ettinger, F. A. Johnson and C. B. Colburn, J. Chem. Phys. 34, 2187 (1961).
4. J. T. Herron and V. H. Dibeler, J. Res. Nat. Bur. Stand. 65A, 405 (1961).
5. J. H. Noggle, J. D. Baldeschwieler, and C. B. Colburn, J. Chem. Phsy. 37, 182 (1962).
6. R. L. Kuczkowski and E. B. Wilson, Jr., J. Chem. Phys. 39, 1030 (1963).
7. R. H. Sanborn, J. Chem. Phys. 33, 1855 (1960).
8. R. H. Sanborn, J. Chem. Phys. 34, 2188 (1961).
9. C. B. Colburn, F. A. Johnson, A. Kennedy, K. McCallum, L. C. Metzger and C. O. Parker, J. Amer. Chem. Soc. 81, 6397 (1959).

Nitrogen Fluoride, Cis (FNNF) $F_2N_2(g)$

Enthalpy Reference Temperature = T_r = 298.15 K Standard State Pressure = p° = 0.1 MPa

	J $K^{-1}mol^{-1}$			kJ mol^{-1}			
T/K	$C_p°$	S°	$-[G°-H°(T_r)]/T$	$H°-H°(T_r)$	$\Delta_fH°$	$\Delta_fG°$	Log K_f
0	0.	0.	INFINITE	-11.402	74.711	74.711	INFINITE
100	33.621	216.960	297.678	-8.072	72.232	83.554	-43.644
200	40.350	241.888	264.073	-4.437	70.027	95.777	-25.014
250	45.327	251.424	260.607	-2.296	69.210	102.313	-21.377
298.15	49.961	259.810	259.810	0.	68.618	108.745	-19.052
300	50.131	260.119	259.811	0.093	68.598	108.994	-18.978
350	54.432	268.177	260.437	2.709	68.169	115.763	-17.277
400	58.158	275.695	261.880	5.526	67.895	122.583	-16.008
450	61.339	282.734	263.810	8.516	67.752	129.429	-15.024
500	64.039	289.340	266.036	11.652	67.716	136.284	-14.238
600	68.275	301.411	270.946	18.279	67.887	149.987	-13.057
700	71.349	312.179	276.081	25.268	68.281	163.641	-12.211
800	73.616	321.862	281.209	32.522	68.816	177.229	-11.572
900	75.317	330.635	286.221	39.973	69.436	190.744	-11.070
1000	76.619	338.641	291.069	47.572	70.106	204.187	-10.666
1100	77.631	345.993	295.732	55.287	70.803	217.561	-10.331
1200	78.432	352.784	300.207	63.092	71.513	230.872	-10.050
1300	79.074	359.088	304.497	70.968	72.226	244.123	-9.809
1400	79.596	364.968	308.609	78.902	72.936	257.319	-9.601
1500	80.026	370.474	312.551	86.884	73.639	270.464	-9.418
1600	80.383	375.651	316.335	94.905	74.333	283.564	-9.257
1700	80.683	380.533	319.969	102.959	75.016	296.620	-9.114
1800	80.938	385.152	323.463	111.040	75.688	309.636	-8.985
1900	81.155	389.534	326.826	119.145	76.351	322.614	-8.869
2000	81.342	393.702	330.067	127.270	77.007	335.559	-8.764
2100	81.505	397.675	333.192	135.413	77.657	348.470	-8.668
2200	81.646	401.470	336.210	143.571	78.306	361.351	-8.580
2300	81.770	405.102	339.127	151.742	78.955	374.202	-8.498
2400	81.879	408.584	341.949	159.924	79.609	387.025	-8.423
2500	81.976	411.929	344.682	168.117	80.271	399.820	-8.354
2600	82.062	415.146	347.331	176.319	80.944	412.589	-8.289
2700	82.140	418.244	349.900	184.529	81.632	425.332	-8.229
2800	82.209	421.233	352.394	192.747	82.338	438.049	-8.172
2900	82.271	424.119	354.818	200.971	83.064	450.740	-8.119
3000	82.327	426.909	357.175	209.201	83.813	463.405	-8.069
3100	82.378	429.609	359.468	217.436	84.588	476.046	-8.021
3200	82.425	432.225	361.701	225.676	85.389	488.661	-7.977
3300	82.467	434.762	363.877	233.921	86.219	501.250	-7.934
3400	82.506	437.225	365.998	242.170	87.079	513.814	-7.894
3500	82.542	439.617	368.068	250.422	87.970	526.352	-7.855
3600	82.574	441.942	370.088	258.678	88.893	538.865	-7.819
3700	82.604	444.205	372.060	266.937	89.847	551.351	-7.784
3800	82.632	446.409	373.988	275.199	90.834	563.810	-7.750
3900	82.658	448.555	375.873	283.463	91.854	576.244	-7.718
4000	82.681	450.648	377.716	291.730	92.905	588.651	-7.687
4100	82.704	452.690	379.520	299.999	93.989	601.031	-7.657
4200	82.724	454.683	381.286	308.271	95.105	613.385	-7.629
4300	82.743	456.630	383.015	316.544	96.252	625.711	-7.601
4400	82.761	458.533	384.710	324.819	97.430	638.011	-7.574
4500	82.778	460.393	386.371	333.096	98.639	650.283	-7.548
4600	82.794	462.212	388.000	341.375	99.877	662.528	-7.523
4700	82.808	463.993	389.598	349.655	101.144	674.746	-7.499
4800	82.822	465.737	391.166	357.936	102.438	686.937	-7.475
4900	82.835	467.444	392.706	366.219	103.760	699.100	-7.453
5000	82.847	469.118	394.217	374.503	105.109	711.236	-7.430
5100	82.859	470.759	395.702	382.789	106.483	723.345	-7.409
5200	82.869	472.368	397.161	391.075	107.882	735.427	-7.387
5300	82.880	473.946	398.595	399.362	109.305	747.482	-7.367
5400	82.889	475.496	400.005	407.651	110.751	759.509	-7.347
5500	82.898	477.017	401.391	415.940	112.220	771.510	-7.327
5600	82.907	478.511	402.755	424.231	113.710	783.484	-7.308
5700	82.915	479.978	404.097	432.522	115.220	795.430	-7.289
5800	82.923	481.420	405.418	440.814	116.750	807.350	-7.271
5900	82.930	482.838	406.718	449.106	118.299	819.244	-7.253
6000	82.937	484.232	407.998	457.400	119.865	831.111	-7.235

PREVIOUS: December 1965 (1 atm) CURRENT: December 1965 (1 bar)

Nitrogen Fluoride, Cis (FNNF) $F_2N_2(g)$

DIFLUORODIAZINE, Trans-(N_2F_2) IDEAL GAS M_r = 66.010206

$S°(298.15\ K) = 262.554\ J\ K^{-1}\ mol^{-1}$

$\Delta_fH°(0\ K) = 86.437 \pm 5.0\ kJ\ mol^{-1}$
$\Delta_fH°(298.15\ K) = 81.170 \pm 5.0\ kJ\ mol^{-1}$

Vibrational Frequencies and Degeneracies

ν, cm^{-1}	ν, cm^{-1}
1010(1)	989(1)
1636(1)	421(1)
592(1)	360(1)

Ground State Quantum Weight = 1 $\sigma = 2$

Point Group: C_{2h}

Bond Distances: N-F = 1.44 Å N-N = 1.25 Å

Bond Angle: F-N-N = 115°

Product of the Moments of Inertia: $I_AI_BI_C = 432.6784 \times 10^{-117}\ g^3\ cm^6$

Heat of Formation

Armstrong and Marantz (1) have measured calorimetrically the enthalpy of reaction $N_2F_2(g) + 8/3\ NH_3(g) = 2\ NH_4F(cr) + 4/3\ N_2(g)$, and reported $\Delta_fH°$(trans N_2F_2, g, 298.15 K) = 19.4±1.2 kcal mol^{-1} which is adopted in this tabulation.

Heat Capacity and Entropy

The bond distances and bond angle were obtained from Bauer (2) by electron diffraction. The fundamental vibrational frequences were obtained from Sanborn (3) by infrared spectroscopy. The three principal moments of inertia are $I_A = 0.919 \times 10^{-39}$, $I_B = 21.2433 \times 10^{-39}$ and $I_C = 22.1623 \times 10^{-39}\ g\ cm^2$.

As pointed out by Ettinger et al. (4), the trans N_2F_2 is the less active isomer which is different from the active isomer by its (1) vapor pressure (2) boiling point (3) heat of vaporization (4) critical temperature (5) melting point (6) mass spectral cracking pattern (7) NMR spectrum (8) infrared spectrum (9) heat of isomerization (10) reactivity with mercury and glass.

References

1. G. T. Armstrong and S. Marantz, J. Chem. Phys. 38, 169 (1963).
2. S. H. Bauer, J. Am. Chem. Soc. 69, 3104 (1947).
3. R. H. Sanborn, J. Chem. Phys. 33, 1855.
4. R. Ettinger, F. A. Johnson and C. B. Colburn, J. Chem. Phys. 34, 2187 (1961).

Nitrogen Fluoride, Trans (FNNF) $F_2N_2(g)$

Enthalpy Reference Temperature = T_r = 298.15 K Standard State Pressure = p° = 0.1 MPa

	J K⁻¹mol⁻¹			kJ mol⁻¹			
T/K	$C_p°$	S°	$-[G°-H°(T_r)]/T$	$H°-H°(T_r)$	$\Delta_fH°$	$\Delta_fG°$	Log K_f
0	0.	0.	INFINITE	-12.229	86.436	86.436	INFINITE
100	35.368	215.607	304.259	-8.865	83.990	95.448	-49.857
200	45.076	243.033	267.293	-4.852	82.164	107.685	-28.124
250	49.593	253.586	263.519	-2.483	81.575	114.137	-23.848
298.15	53.466	262.660	262.660	0.	81.170	120.448	-21.102
300	53.606	262.991	262.661	0.099	81.157	120.691	-21.014
350	57.134	271.525	263.326	2.870	80.881	127.304	-18.999
400	60.214	279.360	264.848	5.805	80.726	133.948	-17.492
450	62.886	286.610	266.868	8.884	80.673	140.605	-16.321
500	65.199	293.359	269.183	12.088	80.704	147.263	-15.384
600	68.922	305.591	274.254	18.803	80.962	160.554	-13.977
700	71.707	316.435	279.520	25.841	81.406	173.786	-12.968
800	73.810	326.154	284.752	33.121	81.967	186.946	-12.206
900	75.417	334.944	289.849	40.586	82.601	200.031	-11.610
1000	76.663	342.958	294.765	48.193	83.278	213.043	-11.128
1100	77.643	350.312	299.485	55.910	83.978	225.986	-10.731
1200	78.424	357.103	304.007	63.715	84.688	238.864	-10.397
1300	79.055	363.406	308.337	71.590	85.400	251.683	-10.113
1400	79.571	369.284	312.482	79.522	86.108	264.448	-9.867
1500	79.998	374.789	316.455	87.501	86.808	277.162	-9.652
1600	80.354	379.963	320.264	95.519	87.499	289.829	-9.462
1700	80.654	384.844	323.920	103.570	88.179	302.454	-9.293
1800	80.908	389.462	327.435	111.649	88.848	315.039	-9.142
1900	81.127	393.842	330.815	119.751	89.508	327.587	-9.006
2000	81.315	398.008	334.072	127.873	90.161	340.101	-8.883
2100	81.478	401.980	337.212	136.013	90.809	352.582	-8.770
2200	81.621	405.773	340.242	144.168	91.455	365.032	-8.667
2300	81.746	409.404	343.171	152.336	92.102	377.453	-8.572
2400	81.857	412.886	346.004	160.517	92.753	389.846	-8.485
2500	81.955	416.229	348.746	168.707	93.413	402.211	-8.404
2600	82.042	419.445	351.404	176.907	94.084	414.550	-8.328
2700	82.120	422.543	353.982	185.115	94.770	426.862	-8.258
2800	82.191	425.531	356.484	193.331	95.474	439.149	-8.192
2900	82.254	428.416	358.915	201.553	96.199	451.411	-8.131
3000	82.311	431.206	361.279	209.782	96.946	463.647	-8.073
3100	82.363	433.906	363.578	218.015	97.719	475.857	-8.018
3200	82.410	436.521	365.817	226.254	98.519	488.042	-7.966
3300	82.453	439.058	367.998	234.497	99.348	500.202	-7.918
3400	82.493	441.520	370.125	242.745	100.206	512.337	-7.871
3500	82.529	443.912	372.199	250.996	101.096	524.445	-7.827
3600	82.562	446.237	374.223	259.250	102.017	536.528	-7.785
3700	82.593	448.500	376.200	267.508	102.971	548.585	-7.745
3800	82.621	450.703	378.132	275.769	103.956	560.615	-7.706
3900	82.647	452.849	380.020	284.032	104.975	572.619	-7.669
4000	82.671	454.942	381.867	292.298	106.026	584.597	-7.634
4100	82.694	456.984	383.675	300.566	107.108	596.548	-7.600
4200	82.715	458.977	385.444	308.837	108.223	608.472	-7.567
4300	82.734	460.923	387.177	317.109	109.370	620.369	-7.536
4400	82.753	462.825	388.875	325.384	110.547	632.239	-7.506
4500	82.770	464.685	390.539	333.660	111.754	644.083	-7.476
4600	82.786	466.505	392.170	341.938	112.992	655.898	-7.448
4700	82.801	468.285	393.771	350.217	114.258	667.687	-7.421
4800	82.815	470.029	395.342	358.498	115.552	679.449	-7.394
4900	82.828	471.736	396.883	366.780	116.873	691.183	-7.368
5000	82.840	473.410	398.397	375.063	118.221	702.890	-7.343
5100	82.852	475.050	399.884	383.348	119.594	714.570	-7.319
5200	82.863	476.659	401.345	391.634	120.992	726.222	-7.295
5300	82.874	478.238	402.781	399.921	122.415	737.848	-7.272
5400	82.883	479.787	404.193	408.208	123.860	749.446	-7.249
5500	82.893	481.308	405.581	416.497	125.328	761.017	-7.228
5600	82.902	482.802	406.947	424.787	126.818	772.562	-7.206
5700	82.910	484.269	408.290	433.078	128.328	784.080	-7.185
5800	82.918	485.711	409.613	441.369	129.857	795.571	-7.165
5900	82.926	487.128	410.915	449.661	131.406	807.036	-7.145
6000	82.933	488.522	412.197	457.954	132.972	818.474	-7.125

PREVIOUS: December 1965 (1 atm) CURRENT: December 1965 (1 bar)

Nitrogen Fluoride, Trans (FNNF) $F_2N_2(g)$

SODIUM FLUORIDE, ION (NaF_2^-) IDEAL GAS M_r = 60.987125

$S°(298.15\ K) = [250.672 \pm 8.4]\ J\ K^{-1}\ mol^{-1}$

$\Delta_f H°(0\ K) = -661.011 \pm 41.8\ kJ\ mol^{-1}$
$\Delta_f H°(298.15\ K) = -669.440 \pm 41.8\ kJ\ mol^{-1}$

Vibrational Frequencies and Degeneracies

ν, cm^{-1}
[400](1)
[170](2)
[650](1)

Ground State Quantum Weight: [1] σ = [2]
Point Group: $[D_{\infty h}]$
Bond Distance: Na-F = [1.93] Å
Bond Angle: F-Na-F = [180]°
Rotational Constant: B_0 = [0.119105] cm^{-1}

Enthalpy of Formation

The enthalpy of formation for $NaF_2^-(g)$ is estimated from three calculated values. The methods of calculation are described as follows. The first reaction we take into consideration is $NaF(g) + F^-(g) = NaF_2^-(g)$. In a manner analogous to $LiF_2^-(g)$ molecule (see $LiF_2^-(g)$ table for details), we assume $NaF_2^-(g)$ is stable and $\Delta_r H°(298.15\ K)$ for the reaction $NaF(g) + F^-(g) = NaF_2^-(g)$ is somewhere between 0 to 1/2(-64) = -32 kcal mol^{-1} where the value -56 kcal mol^{-1} is the enthalpy of dimerization. Based on $\Delta_r H°(298.15\ K)$ = 0 and -28 kcal mol^{-1}, respectively, with JANAF $\Delta_f H°(298.15\ K)$ values for the species involved, we obtain $\Delta_f H°(NaF_2^-$, g, 298.15 K) = -132 and -164 kcal mol^{-1} as the upper and lower limits.

The enthalpy of formation, based on the reaction $NaF_2^-(g) = NaF_2(g) + e^-(g)$, is -165.8 kcal mol^{-1}, using $\Delta_r H°(298.15\ K)$ and $\Delta_f H°(NaF_2$, g, 298.15 K) estimated as 83.5 and -82.3 kcal mol^{-1}, respectively. The $\Delta_r H°(298.15\ K)$ value is obtained from an assumption that the enthalpy change for removing an electron from $NaF_2^-(g)$ is comparable to that from $F^-(g)$. Due to the fact that the Na atom has smaller electronegativity (0.9) than that of the F atom (4.0), we expect that there is less possibility that the electron will associate with Na atom in the NaF_2^- molecule. Therefore, we adopt EA(F,g), reported by Pritchard (1), or $\Delta_r H°(298.15\ K)$ = 83.5 kcal mol^{-1} for the reaction $F^-(g) = F(g) + e^-(g)$ as the $\Delta_r H°(298.15\ K)$ for $NaF_2^-(g) = NaF_2(g) + e^-(g)$. The value of $\Delta_f H°(298.15\ K)$ for $NaF_2(g)$ is calculated using the enthalpy change for the reaction $Na_2F_2(g) = NaF_2(g) + Na(g)$ as 145.8 kcal mol^{-1} which is derived from the sum of two enthalpy changes, namely, half of the $\Delta_r H°(298.15\ K)$ value for the dimer dissociation reaction $Na_2F_2(g)$ = 2 NaF(g), and the $\Delta_r H°(298.15\ K)$ for the monomer dissociation reaction NaF(g) = Na(g) + F(g).

Employing $\Delta_r H°(298.15\ K)$ = 63.5 kcal mol^{-1} for $Na_2F_2(g)$ = 2 NaF(g) and 153.0 kcal mol^{-1} for $NaF(g) = Na^+(g) + F^-(g)$, we estimate $\Delta_r H°(298.15\ K)$ = (1/2)(63.5) + 153.0 = 184.7 kcal mol^{-1} for the reaction $Na_2F_2(g) = NaF_2^-(g) + Na^+(g)$, which leads to $\Delta_f H°(298.15\ K)$ = 163.4 kcal mol^{-1} for $NaF_2^-(g)$. For all calculations the $\Delta_f H°(298.15\ K)$ values of the chemical species involved are taken from JANAF Thermochemical Tables.

$\Delta_f H°(NaF_2^-$, g, 298.15 K) is tentatively adopted as - 160±10 kcal mol^{-1} (-669.440±41.8 kJ mol^{-1}).

Heat Capacity and Entropy

The molecular structure is assumed to be linear, according to the suggestion by Walsh (2) that molecules with not more than 16 valence electrons are linear in their ground state. The Na-F bond distance is estimated by comparison with that of NaF(g) reported by Veazey and Gordy (3). The vibrational frequencies are estimated from those for the isoelectronic molecule $SiO_2(g)$.

References

1. H. O. Pritchard, Chem. Rev. 52, 529 (1953).
2. A. D. Walsh, J. Chem. Soc. 1953, 2266 (1953).
3. S. E. Veazey and W. Gordy, Phys. Rev. 138, A1303 (1965).

Sodium Fluoride, Ion (NaF_2^-) $F_2Na_1^-(g)$

Enthalpy Reference Temperature = T_r = 298.15 K Standard State Pressure = p° = 0.1 MPa

T/K	$C_p°$	S°	$-[G°-H°(T_r)]/T$	$H°-H°(T_r)$	$\Delta_f H°$	$\Delta_f G°$	Log K_f
	J $K^{-1}mol^{-1}$			kJ mol^{-1}			
0	0.	0.	INFINITE	-13.040	-661.011		
100	40.374	198.472	295.762	-9.729			
200	49.878	229.724	255.552	-5.166			
250	52.805	241.186	251.564	-2.595			
298.15	54.860	250.672	250.672	0.	-669.440	-662.120	116.001
300	54.927	251.011	250.673	0.102	-669.487	-662.074	115.277
350	56.473	259.600	251.347	2.889	-670.781	-660.738	98.610
400	57.612	267.219	252.864	5.742	-674.766	-659.006	86.057
450	58.466	274.057	254.846	8.645	-676.132	-656.953	76.257
500	59.118	280.252	257.081	11.585	-677.468	-654.750	68.401
600	60.024	291.117	261.874	17.546	-680.076	-649.960	56.584
700	60.603	300.416	266.732	23.579	-682.625	-644.739	48.111
800	60.994	308.536	271.460	29.660	-685.141	-639.155	41.733
900	61.269	315.737	275.987	35.774	-687.647	-633.256	36.753
1000	61.469	322.203	280.291	41.912	-690.166	-627.078	32.755
1100	61.619	328.069	284.372	48.066	-692.718	-620.646	29.472
1200	61.735	333.436	288.240	54.234	-792.085	-611.540	26.620
1300	61.825	338.381	291.910	60.413	-793.833	-596.423	23.965
1400	61.897	342.965	295.395	66.599	-795.595	-581.172	21.684
1500	61.956	347.238	298.710	72.792	-797.370	-565.794	19.703
1600	62.004	351.238	301.869	78.990	-799.158	-550.298	17.965
1700	62.044	354.998	304.885	85.192	-800.958	-534.689	16.429
1800	62.077	358.545	307.769	91.398	-802.769	-518.974	15.060
1900	62.106	361.903	310.530	97.607	-804.590	-503.158	13.833
2000	62.130	365.089	313.179	103.819	-806.418	-487.246	12.726
2100	62.151	368.121	315.724	110.033	-808.250	-471.242	11.722
2200	62.170	371.012	318.172	116.250	-810.085	-455.151	10.807
2300	62.186	373.776	320.530	122.467	-811.918	-438.976	9.969
2400	62.200	376.423	322.804	128.687	-813.748	-422.722	9.200
2500	62.212	378.963	325.000	134.907	-815.571	-406.392	8.491
2600	62.223	381.403	327.122	141.129	-817.385	-389.989	7.835
2700	62.233	383.751	329.177	147.352	-819.187	-373.517	7.226
2800	62.241	386.015	331.166	153.575	-820.976	-356.978	6.659
2900	62.249	388.199	333.096	159.800	-822.751	-340.375	6.131
3000	62.256	390.309	334.968	166.025	-824.509	-323.711	5.636
3100	62.263	392.351	336.786	172.251	-826.251	-306.989	5.173
3200	62.269	394.328	338.554	178.478	-827.975	-290.211	4.737
3300	62.274	396.244	340.273	184.705	-829.683	-273.380	4.327
3400	62.279	398.103	341.947	190.933	-831.374	-256.497	3.941
3500	62.283	399.909	343.577	197.161	-833.050	-239.564	3.575
3600	62.287	401.663	345.166	203.389	-834.711	-222.584	3.230
3700	62.291	403.370	346.716	209.618	-836.360	-205.557	2.902
3800	62.295	405.031	348.229	215.847	-837.998	-188.487	2.591
3900	62.298	406.649	349.706	222.077	-839.628	-171.373	2.295
4000	62.301	408.227	351.150	228.307	-841.252	-154.218	2.014
4100	62.304	409.765	352.561	234.537	-842.872	-137.022	1.746
4200	62.306	411.266	353.941	240.768	-844.490	-119.786	1.490
4300	62.308	412.732	355.291	246.998	-846.112	-102.513	1.245
4400	62.311	414.165	356.613	253.229	-847.699	-85.204	1.012
4500	62.313	415.565	357.907	259.461	-849.313	-67.857	0.788
4600	62.315	416.935	359.176	265.692	-850.946	-50.474	0.573
4700	62.317	418.275	360.419	271.923	-852.592	-33.054	0.367
4800	62.318	419.587	361.638	278.155	-854.255	-15.600	0.170
4900	62.320	420.872	362.834	284.387	-855.937	1.890	-0.020
5000	62.321	422.131	364.007	290.619	-857.635	19.413	-0.203
5100	62.323	423.365	365.159	296.851	-859.367	36.971	-0.379
5200	62.324	424.575	366.290	303.084	-861.128	54.564	-0.548
5300	62.325	425.763	367.401	309.316	-862.923	72.190	-0.711
5400	62.327	426.928	368.493	315.549	-864.755	89.851	-0.869
5500	62.328	428.071	369.565	321.782	-866.628	107.546	-1.021
5600	62.329	429.194	370.620	328.014	-868.543	125.275	-1.169
5700	62.330	430.297	371.658	334.247	-870.506	143.039	-1.311
5800	62.331	431.381	372.678	340.480	-872.517	160.838	-1.449
5900	62.332	432.447	373.682	346.713	-874.581	178.672	-1.582
6000	62.333	433.495	374.670	352.947	-876.542	196.523	-1.711

PREVIOUS: December 1968 (1 atm) CURRENT: December 1968 (1 bar)

Sodium Fluoride, Ion (NaF_2^-) $F_2Na_1^-(g)$

SODIUM FLUORIDE ((NaF)$_2$) IDEAL GAS M_r = 83.976346

$\Delta_f H°(0\ K)$ = [-839.955 ± 12.6] kJ mol^{-1}

S°(298.15 K) = [287.366 ± 16.7] J K^{-1} mol^{-1} $\Delta_f H°(298.15\ K)$ = [-846.423 ± 12.6] kJ mol^{-1}

Vibrational Frequencies and Degeneracies

ν, cm^{-1}	ν, cm^{-1}
[420](1)	[200](1)
[400](1)	[425](1)
[350](1)	[500](1)

Ground State Quantum Weight: [1] σ = [4]

Point Group: [D_{2h}]

Bond Distance: Na-F = 2.02 Å

Bond Angle: Na-F-Na = [85]°

Product of the Moments of Inertia: $I_A I_B I_C$ = [5.614743 x 10^{-114}] g^3 cm^6

Enthalpy of Formation

Although the absolute values of the vapor pressure measurements are not used for evaluation, the ratios of the numbers of dimeric and monomeric NaF molecules that effuse from the oven at the temperatures 1115 - 1191 K determined by Eisenstadt et al. (1) by use of the electron-beam velocity-selector method are used. Based on the reported equilibrium data, we evaluate the enthalpy change of the reaction (A) Na_2F_2(g) = 2 NaF(g) by the 2nd and 3rd law methods. The results are presented in the table below.

Porter and Schoonmaker (2) analyzed the vapor effusing from a Knudsen cell at 1121 K mass spectrometrically and derived the partial pressures of NaF(g) and Na_2F_2(g) at that temperature. Using these reported partial pressures we calculate, by the 3rd law method, the value $\Delta_r H°$(298.15 K) = 64.1 kcal mol^{-1} for reaction (A).

Sidorov et al. (3) studied the NaF - ZrF_4 system by the mass spectrometric method and derived the partial pressures of NaF(g) and Na_2F_2(g) as a function of composition and temperature. Employing the data points reported at 1074 and 1167 K we obtain the 3rd law value $\Delta_r H°$(298.15 K) = 62.0±1.6 kcal mol^{-1} for the same reaction.

The $\Delta_f H°$(Na_2F_2, g, 298.15 K) is calculated based on the adopted 3rd law value $\Delta_r H°$(298.15 K) = 63.5 kcal mol^{-1} for the reaction Na_2F_2(g) = 2 NaF(g), using $\Delta_f H°$(NaF, g, 298.15 K) = -69.5 kcal mol^{-1}. Brewer and Brackett (9) reported $\Delta_r H°$(1000 K) = 56.9 kcal mol^{-1} for the above reaction. According to our tables the $\Delta_r H°$(1000 K) value is 62.5 kcal mol^{-1}.

Source	T/K	Data Points	$\Delta_r H°$(298.15 K), kcal mol^{-1} 2nd law	3rd law	Drift cal K^{-1} mol^{-1}	$\Delta_f H°$(298.15 K) kcal mol^{-1}
Eisenstadt et al. (1958)	1115 - 1191	6	58.1 ± 2.9	63.5	4.7 ± 2.5	-202.4
Porter and Schoonmaker (1958)	1121	1	--	64.1	--	-203.0
Sidorov et al. (1964)	1074 - 1167	8	--	62.0 ± 1.6	--	-200.9

Heat Capacity and Entropy

The Na_2F_2(g) molecule is assumed to have a rhombic structure with D_{2h} (v_h) symmetry (4, 5). The Na-F bond distance is reported by Akishin and Rambidi (5), who studied the molecular structure of alkali halide vapors by electron diffraction with sector microphotometry. The bond angle is estimated by comparison with those of the other alkali halides. Berkowitz (4, 6) has calculated the vibration frequencies of Na_2F_2 based on an ionic model. By comparison of the calculated frequencies of Li_2F_2 (6) with those of the observed ones (see Li_2F_2(g) table for details), we re-estimate the vibrational frequencies for Na_2F_2(g) to be higher than the ones reported by Berkowitz (6). Bauer et al. (7) used a pseudo-electrostatic model, similar to the one suggested by O'Konski and Higuchi (8), to calculate the entropies of dimerization for a number of gaseous alkali halides and estimated the six vibrational frequencies as 477, 318, 291, 477, 291 and 291 cm^{-1}. Bauer et al. (7) calculated $\Delta_{sub}H°$(1146 K) = -32.5 cal K^{-1} mol^{-1} for the reaction 2 NaF(g) = Na_2F_2(g), in reasonable agreement with the value -33.5 cal K^{-1} mol^{-1} which is derived using JANAF S°(1146 K) values for NaF(g) and Na_2F_2(g).

The three principal moments of inertia are: I_A = 13.9943 x 10^{-39}, I_B = 14.2197 x 10^{-39}, and I_C = 28.2145 x 10^{-39} g cm^2.

References

1. M. Eisenstadt, G. M. Rothberg, and P. Kusch, J. Chem. Phys. 29, 797 (1958).
2. R. F. Porter and R. C. Schoonmaker, J. Chem. Phys. 29, 1070 (1958).
3. L. N. Sidorov, P. A. Akishin, V. I. Belousov, and V. V. Shol'ts, Russ. J. Phys. Chem. 38, 641 (1964).
4. J. Berkowitz, J. Chem. Phys. 29, 1386 (1958).
5. P. A. Akishin and N. G. Rambidi, Z. Physik. Chem. 213, 111 (1960).
6. J. Berkowitz, J. Chem. Phys. 32, 1519 (1960).
7. S. H. Bauer, R. M. Diner and R. F. Porter, J. Chem. Phys. 29, 991 (1958).
8. C. T. O'Konski and W. I. Higuchi, J. Chem. Phys. 23, 1175 (1955).
9. L. Brewer and E. Brackett, Chem. Rev. 61, 425 (1961).

Sodium Fluoride ((NaF)$_2$) F_2Na_2(g)

Enthalpy Reference Temperature = T_r = 298.15 K; Standard State Pressure = p° = 0.1 MPa

T/K	$C_p°$ (J K^{-1}mol^{-1})	S°	-[G°-H°(T_r)]/T	H°-H°(T_r) (kJ mol^{-1})	$\Delta_f H°$	$\Delta_f G°$	Log K_f
0	0.	0.	INFINITE	-15.250	-839.955	-839.955	INFINITE
100	41.612	225.792	342.904	-11.711	-842.002	-842.782	440.224
200	60.880	260.954	293.579	-6.525	-844.657	-842.438	220.022
250	66.886	275.228	288.512	-3.321	-845.604	-841.771	175.878
298.15	70.821	287.366	287.366	0.	-846.423	-840.957	147.332
300	70.945	287.805	287.368	0.131	-846.454	-840.923	146.418
350	73.732	298.963	288.243	3.752	-847.327	-839.934	125.353
400	75.698	308.945	290.218	7.491	-853.557	-838.401	109.484
450	77.125	317.947	292.807	11.313	-854.527	-836.448	97.092
500	78.187	326.131	295.736	15.197	-855.419	-834.391	87.168
600	79.625	340.524	302.034	23.094	-857.028	-830.030	72.261
700	80.524	352.871	308.436	31.105	-858.473	-825.415	61.593
800	81.120	363.665	314.679	39.189	-859.817	-820.599	53.580
900	81.536	373.245	320.664	47.323	-861.111	-815.619	47.337
1000	81.835	381.852	326.360	55.492	-862.407	-810.495	42.336
1100	82.059	389.662	331.765	63.687	-863.748	-805.240	38.238
1200	82.230	396.810	336.892	71.902	-1058.700	-794.979	34.605
1300	82.364	403.398	341.758	80.132	-1058.397	-773.014	31.060
1400	82.470	409.505	346.381	88.374	-1058.103	-751.073	28.023
1500	82.556	415.198	350.781	96.625	-1057.819	-729.152	25.391
1600	82.627	420.529	354.976	104.885	-1057.546	-707.250	23.089
1700	82.686	425.540	358.981	113.150	-1057.283	-685.365	21.059
1800	82.735	430.267	362.811	121.422	-1057.030	-663.494	19.254
1900	82.777	434.742	366.480	129.697	-1056.785	-641.638	17.640
2000	82.812	438.989	370.000	137.977	-1056.547	-619.794	16.187
2100	82.843	443.030	373.382	146.260	-1056.313	-597.963	14.874
2200	82.870	446.884	376.637	154.545	-1056.081	-576.142	13.679
2300	82.893	450.569	379.771	162.833	-1055.850	-554.331	12.589
2400	82.913	454.097	382.795	171.124	-1055.617	-532.531	11.590
2500	82.931	457.482	385.716	179.416	-1055.380	-510.741	10.671
2600	82.947	460.735	388.539	187.710	-1055.138	-488.960	9.823
2700	82.962	463.866	391.271	196.005	-1054.890	-467.189	9.038
2800	82.974	466.883	393.918	204.302	-1054.634	-445.427	8.310
2900	82.986	469.795	396.484	212.600	-1054.370	-423.674	7.631
3000	82.996	472.608	398.975	220.899	-1054.100	-401.930	6.998
3100	83.006	475.330	401.395	229.199	-1053.823	-380.196	6.406
3200	83.014	477.965	403.747	237.500	-1053.542	-358.471	5.851
3300	83.022	480.520	406.035	245.802	-1053.256	-336.754	5.330
3400	83.029	482.999	408.262	254.105	-1052.970	-315.046	4.840
3500	83.035	485.406	410.432	262.408	-1052.686	-293.347	4.378
3600	83.041	487.745	412.547	270.712	-1052.406	-271.655	3.942
3700	83.047	490.020	414.610	279.016	-1052.136	-249.971	3.529
3800	83.052	492.235	416.624	287.321	-1051.878	-228.295	3.138
3900	83.057	494.392	418.591	295.627	-1051.637	-206.625	2.767
4000	83.061	496.495	420.512	303.932	-1051.419	-184.961	2.415
4100	83.065	498.546	422.390	312.239	-1051.229	-163.302	2.080
4200	83.069	500.548	424.228	320.545	-1051.067	-141.647	1.762
4300	83.072	502.503	426.025	328.852	-1050.948	-119.996	1.458
4400	83.075	504.412	427.785	337.160	-1050.788	-98.354	1.168
4500	83.078	506.279	429.509	345.468	-1050.719	-76.710	0.890
4600	83.081	508.105	431.198	353.776	-1050.717	-55.065	0.625
4700	83.084	509.892	432.853	362.084	-1050.772	-33.420	0.371
4800	83.086	511.641	434.476	370.392	-1050.891	-11.773	0.128
4900	83.089	513.355	436.069	378.701	-1051.079	9.877	-0.105
5000	83.091	515.033	437.631	387.010	-1051.328	31.530	-0.329
5100	83.093	516.679	439.165	395.319	-1051.670	53.191	-0.545
5200	83.095	518.292	440.671	403.629	-1052.100	74.859	-0.752
5300	83.097	519.875	442.151	411.938	-1052.623	96.536	-0.951
5400	83.098	521.428	443.605	420.248	-1053.246	118.224	-1.144
5500	83.100	522.953	445.033	428.558	-1053.975	139.924	-1.329
5600	83.102	524.450	446.438	436.868	-1054.814	161.639	-1.508
5700	83.103	525.921	447.820	445.178	-1055.770	183.369	-1.680
5800	83.104	527.367	449.179	453.489	-1056.847	205.118	-1.847
5900	83.106	528.787	450.516	461.799	-1058.051	226.886	-2.009
6000	83.107	530.184	451.832	470.110	-1059.069	248.639	-2.165

PREVIOUS: December 1968 (1 atm) CURRENT: December 1968 (1 bar)

Sodium Fluoride ((NaF)$_2$) F_2Na_2(g)

OXYGEN FLUORIDE (OF_2) IDEAL GAS M_r = 53.996206

$S°(298.15\ K) = 247.457 \pm 0.4\ J\ K^{-1}\ mol^{-1}$

$\Delta_f H°(0\ K) = 26.790 \pm 1.59\ kJ\ mol^{-1}$
$\Delta_f H°(298.15\ K) = 24.518 \pm 1.59\ kJ\ mol^{-1}$

Vibrational Frequencies and Degeneracies

ν, cm^{-1}
928(1)
461(1)
831(1)

Ground State Quantum Weight: [1] $\sigma = 2$
Point Group: C_{2v}
Bond Distance: F–O = 1.412 Å
Bond Angle: F–O–F = 103°10'
Product of the Moments of Inertia: $I_A I_B I_C = 1.018282 \times 10^{-115}\ g^3\ cm^6$

Enthalpy of Formation

King and Armstrong (1) have established the enthalpy of formation with a series of reactions in a flame calorimeter. They burned F_2O in hydrogen to give HF aqueous, in addition they burned F_2 in hydrogen and O_2 in hydrogen so that the heat of formation, 5.86 kcal mol^{-1}, is directly obtainable. This value is in good agreement with the recalculated values (1) from Wartenberg and Klinkott (2), 5.7±2 kcal mol^{-1}, and Ruff and Menzel (3), 4.7±2 kcal mol^{-1} but differs from that of Bisbee et al. (4), −4.06±2.2 kcal mol^{-1}. There were several factors in the experiments of Bisbee that might cause a too negative enthalpy of formation and we choose to adopt the definitive value of 5.86 kcal mol^{-1} reported by King and Armstrong.

Heat Capacity and Entropy

The structural parameters are those reported by Pierce et al. (5) for the average ground state molecule from the microwave spectrum. Earlier measurements of Hilton et al. (6) disagree but according to Pierce et al. (7) the line assignments used by Hilton are incorrect. The vibrational frequencies were reported by Jones et al. (8), several other investigators report similar values (9, 10, 11). Nebgen et al. (12) have obtained the anharmonic constants for the three vibrations. Individual moments of inertia are $I_A = 1.4392 \times 10^{-39}$, $I_B = 7.7226 \times 10^{-39}$ and $I_C = 9.1618 \times 10^{-39}\ g\ cm^2$.

References

1. R. C. King and G. T. Armstrong, J. Res. Nat. Bur. Stand. 72A, 113 (1968).
2. H. Wartenberg and G. Z. Klinkott, Z. Anorg. Chem. 193, 409 (1930).
3. O. Ruff and W. Menzel, Z. Anorg. Chem. 190, 257 (1930).
4. W. R. Bisbee, J. V. Hamilton, R. Rushworth, T. J. Houser, and J. M. Gerhauser, Advan. Chem. Ser. 54, 215 (1965).
5. L. Pierce, N. DiCianni and R. H. Jackson, J. Chem. Phys. 38, 730 (1963).
6. A. R. Hilton, A. W. Jache, J. B. Beal, W. D. Henderson and R. J. Robinson, J. Chem. Phys. 34, 1137 (1961).
7. L. Pierce, R. H. Jackson and N. DiCianni, J. Chem. Phys. 35, 2240 (1961).
8. E. A. Jones, J. S. Kirby-Smith, P. J. H. Woltz and A. H. Nielsen, J. Chem. Phys. 19, 337 (1951).
9. H. J. Bernstein and J. Powling, J. Chem. Phys. 18, 685 (1950).
10. G. Hettner, R. Pohlman, and H. J. Schumacher, Z. Physik. 96, 203 (1935).
11. G. B. B. M. Sutherland and W. G. Penney, Proc. Roy. Soc. (London) 156A, 678 (1936).
12. J. W. Nebgen, F. I. Metz and W. B. Rose, J. Mol. Spectrosc. 21, 99 (1966).

Oxygen Fluoride (OF_2) $F_2O_1(g)$

Enthalpy Reference Temperature = T_r = 298.15 K Standard State Pressure = p° = 0.1 MPa

T/K	$C_p°$ (J K⁻¹mol⁻¹)	S° (J K⁻¹mol⁻¹)	−[G°−H°(T_r)]/T (J K⁻¹mol⁻¹)	H°−H°(T_r) (kJ mol⁻¹)	$\Delta_f H°$ (kJ mol⁻¹)	$\Delta_f G°$ (kJ mol⁻¹)	Log K_f
0	0.	0.	INFINITE	-10.895	26.790	26.790	INFINITE
100	33.750	206.726	282.348	-7.562	25.763	30.793	-16.085
200	38.055	231.267	251.252	-3.997	24.944	36.170	-9.447
250	40.795	240.054	248.156	-2.026	24.684	39.008	-8.150
298.15	43.301	247.457	247.457	0.	24.518	41.783	-7.320
300	43.392	247.725	247.458	0.080	24.513	41.890	-7.294
350	45.669	254.589	247.994	2.308	24.414	44.795	-6.685
400	47.585	260.817	249.214	4.641	24.369	47.710	-6.230
450	49.166	266.516	250.824	7.061	24.364	50.628	-5.877
500	50.462	271.765	252.659	9.553	24.386	53.546	-5.594
600	52.399	281.148	256.644	14.702	24.482	59.369	-5.169
700	53.727	289.332	260.742	20.013	24.614	65.174	-4.863
800	54.663	296.570	264.777	25.435	24.758	70.958	-4.633
900	55.342	303.050	268.676	30.937	24.903	76.725	-4.453
1000	55.847	308.908	272.411	36.497	25.043	82.475	-4.308
1100	56.231	314.250	275.975	42.102	25.173	88.212	-4.189
1200	56.530	319.156	279.372	47.741	25.291	93.938	-4.089
1300	56.767	323.690	282.609	53.406	25.396	99.654	-4.004
1400	56.957	327.905	285.696	59.093	25.485	105.363	-3.931
1500	57.112	331.840	288.642	64.796	25.557	111.065	-3.868
1600	57.240	335.530	291.458	70.514	25.613	116.764	-3.812
1700	57.347	339.003	294.154	76.244	25.651	122.460	-3.763
1800	57.437	342.284	296.738	81.983	25.673	128.154	-3.719
1900	57.514	345.391	299.217	87.731	25.679	133.847	-3.680
2000	57.580	348.343	301.600	93.485	25.671	139.540	-3.644
2100	57.637	351.154	303.894	99.246	25.652	145.234	-3.612
2200	57.686	353.836	306.103	105.013	25.625	150.929	-3.584
2300	57.729	356.402	308.235	110.783	25.592	156.625	-3.557
2400	57.767	358.859	310.293	116.558	25.557	162.323	-3.533
2500	57.801	361.218	312.284	122.337	25.524	168.022	-3.511
2600	57.831	363.486	314.209	128.118	25.495	173.723	-3.490
2700	57.857	365.669	316.075	133.903	25.474	179.424	-3.471
2800	57.881	367.773	317.884	139.690	25.465	185.127	-3.454
2900	57.903	369.805	319.640	145.479	25.470	190.828	-3.437
3000	57.922	371.768	321.345	151.270	25.491	196.530	-3.422
3100	57.940	373.668	323.002	157.063	25.532	202.231	-3.408
3200	57.956	375.507	324.614	162.858	25.594	207.930	-3.394
3300	57.970	377.291	326.184	168.654	25.679	213.626	-3.381
3400	57.983	379.022	327.713	174.452	25.788	219.320	-3.369
3500	57.996	380.703	329.203	180.251	25.923	225.011	-3.358
3600	58.007	382.337	330.656	186.051	26.085	230.697	-3.347
3700	58.017	383.926	332.074	191.852	26.275	236.378	-3.337
3800	58.027	385.474	333.459	197.654	26.492	242.053	-3.327
3900	58.035	386.981	334.812	203.457	26.738	247.723	-3.318
4000	58.043	388.450	336.135	209.261	27.012	253.386	-3.309
4100	58.051	389.884	337.429	215.066	27.314	259.041	-3.300
4200	58.058	391.283	338.694	220.872	27.644	264.690	-3.292
4300	58.065	392.649	339.933	226.678	28.002	270.329	-3.284
4400	58.071	393.984	341.147	232.484	28.387	275.960	-3.276
4500	58.076	395.289	342.335	238.292	28.799	281.582	-3.269
4600	58.082	396.565	343.500	244.100	29.236	287.195	-3.261
4700	58.087	397.815	344.643	249.908	29.699	292.798	-3.254
4800	58.091	399.038	345.763	255.717	30.187	298.391	-3.247
4900	58.096	400.235	346.863	261.526	30.697	303.973	-3.240
5000	58.100	401.409	347.942	267.336	31.231	309.545	-3.234
5100	58.104	402.560	349.002	273.146	31.785	315.106	-3.227
5200	58.108	403.688	350.043	278.957	32.361	320.655	-3.221
5300	58.111	404.795	351.065	284.768	32.955	326.194	-3.215
5400	58.114	405.881	352.070	290.579	33.568	331.721	-3.209
5500	58.117	406.948	353.058	296.391	34.198	337.236	-3.203
5600	58.120	407.995	354.030	302.203	34.844	342.741	-3.197
5700	58.123	409.024	354.986	308.015	35.504	348.232	-3.191
5800	58.126	410.034	355.926	313.827	36.178	353.713	-3.186
5900	58.128	411.028	356.852	319.640	36.865	359.182	-3.180
6000	58.131	412.005	357.763	325.453	37.562	364.639	-3.174

PREVIOUS: December 1969 (1 atm) CURRENT: December 1969 (1 bar)

THIONYL FLUORIDE (SOF_2) IDEAL GAS M_r = 86.056206

$S°(298.15\ K) = 279.132\ J\ K^{-1}\ mol^{-1}$

$\Delta_f H°(0\ K) = [-538.967 \pm 105]\ kJ\ mol^{-1}$
$\Delta_f H°(298.15\ K) = -543.920 \pm 105]\ kJ\ mol^{-1}$

Vibrational Frequencies and Degeneracies

ν, cm^{-1}	ν, cm^{-1}
1330 (1)	377.8(1)
808.2(1)	747.0(1)
530.4(1)	392.5(1)

Ground State Quantum Weight: 1 $\sigma = 1$

Point Group: C_s

Bond Distances: O-S = 1.412 Å S-F = 1.585 Å

Bond Angles: O-S-F = 106.82° F-S-F = 92.82°

Product of the Moments of Inertia: $I_A I_B I_C = 1.655344 \times 10^{-114}\ g^3\ cm^6$

Enthalpy of Formation

The increment in $\Delta_f H°$ at 298.15 K between gaseous SOF_2 and SO_2F_2 is estimated from increments for the four pairs a) S and SO, b) SO and SO_2, c) SO_2 and SO_3, and d) $SOCl_2$ and SO_2Cl_2. Based on $\Delta_f H° = -181.3$ kcal mol^{-1} for SO_2F_2 (1), we derive a) -116, b) -109, c) -158, and d) -148 for $\Delta_f H°$ of SOF_2(g). The derivation is based on $\Delta_f H°(SOCl_2$, g, 298.15 K) from (2) and other values from (1). Giving lesser weight to c) which involves SO_3, we adopt an average of -130±25 kcal mol^{-1}.

Heat Capacity and Entropy

The molecular structure is from the microwave study of Ferguson (3). Principal moments of inertia are $I_A = 9.7369 \times 10^{-39}$, $I_B = 10.0399 \times 10^{-39}$, and $I_C = 16.9332 \times 10^{-39}$ g cm^2. Vibrational frequencies are those assigned by Pace and Samuelson (4) from their gas-phase infrared spectra. These authors also measured Raman spectra of liquid and gas and summarized earlier spectral studies. We adopt ν_1 as the lower wave number of the Fermi resonance doublet at 1330 and 1340 cm^{-1}. Uncertainty is greatest for the overlapping bands near 378 and 392 cm^{-1}.

Pace and Turnbull (5) derived S° = 63.56±0.15 cal K^{-1} mol^{-1} for the ideal gas at T_{vap} = 228.84 K from calorimetric measurements of heat capacity (12 to 230 K), $\Delta_{fus}H°$, and $\Delta_{vap}H°$. Spectroscopic data of this table yield the corresponding value of S° = 63.29 which is 0.27 cal K^{-1} mol^{-1} smaller than observed. Pace and Turnbull attributed the minor discrepancy to bias in the measurements of $\Delta_{vap}H°$.

References

1. JANAF Thermochemical Tables: F_2O_2S(g), Cl_2O_2S(g), OS(g), and S(g) 6-30-71; O_2S(g) 6-30-61; O_3S(g) 9-30-65.
2. U. S. Nat. Bur. Stand. Tech. Note 270-3, 1968.
3. R. C. Ferguson, J. Am. Chem. Soc. 76, 850 (1954).
4. E. L. Pace and H. V. Samuelson, J. Chem. Phys. 44, 3682 (1966).
5. E. L. Pace and B. F. Turnbull, J. Chem. Phys. 43, 1953 (1965).

Thionyl Fluoride (OSF_2) $F_2O_1S_1(g)$

Enthalpy Reference Temperature = T_r = 298.15 K Standard State Pressure = p° = 0.1 MPa

T/K	$C_p°$ (J $K^{-1}mol^{-1}$)	S° (J $K^{-1}mol^{-1}$)	$-[G°-H°(T_r)]/T$ (J $K^{-1}mol^{-1}$)	$H°-H°(T_r)$ (kJ mol^{-1})	$\Delta_f H°$ (kJ mol^{-1})	$\Delta_f G°$ (kJ mol^{-1})	Log K_f
0	0.	0.	INFINITE	-12.625	-538.967	-538.967	INFINITE
100	35.545	230.278	322.875	-9.260	-540.650	-536.723	280.356
200	47.164	258.341	284.057	-5.143	-542.560	-532.022	138.950
250	52.631	269.468	280.048	-2.645	-543.313	-529.298	110.591
298.15	57.105	279.132	279.132	0.	-543.920	-526.542	92.248
300	57.262	279.485	279.133	0.106	-543.941	-526.434	91.660
350	61.105	288.610	279.844	3.068	-544.472	-523.473	78.124
400	64.276	296.983	281.471	6.205	-547.144	-520.353	67.951
450	66.894	304.710	283.629	9.486	-548.214	-516.948	60.006
500	69.060	311.874	286.100	12.887	-549.285	-513.414	53.636
600	72.365	324.775	291.495	19.968	-550.842	-506.084	44.059
700	74.697	336.115	297.076	27.328	-552.008	-498.528	37.201
800	76.381	346.206	302.598	34.886	-552.939	-490.823	32.047
900	77.626	355.277	307.956	42.589	-606.849	-481.929	27.970
1000	78.568	363.507	313.106	50.401	-606.315	-468.077	24.450
1100	79.293	371.031	318.035	58.296	-605.766	-454.280	21.572
1200	79.863	377.956	322.743	66.255	-605.214	-440.533	19.176
1300	80.317	384.367	327.240	74.265	-604.668	-426.832	17.150
1400	80.685	390.333	331.536	82.315	-604.132	-413.172	15.416
1500	80.987	395.910	335.644	90.399	-603.613	-399.551	13.914
1600	81.237	401.145	339.576	98.511	-603.112	-385.963	12.600
1700	81.446	406.077	343.344	106.645	-602.631	-372.406	11.443
1800	81.623	410.737	346.960	114.799	-602.172	-358.877	10.414
1900	81.774	415.154	350.434	122.969	-601.733	-345.372	9.495
2000	81.904	419.352	353.776	131.153	-601.312	-331.891	8.668
2100	82.016	423.351	356.994	139.349	-600.909	-318.430	7.921
2200	82.114	427.169	360.098	147.556	-600.518	-304.987	7.241
2300	82.200	430.821	363.094	155.772	-600.138	-291.562	6.622
2400	82.276	434.321	365.989	163.996	-599.764	-278.154	6.054
2500	82.342	437.681	368.790	172.227	-599.394	-264.762	5.532
2600	82.402	440.912	371.502	180.464	-599.023	-251.384	5.050
2700	82.455	444.023	374.131	188.707	-598.647	-238.020	4.605
2800	82.503	447.022	376.681	196.955	-598.264	-224.670	4.191
2900	82.546	449.918	379.157	205.207	-597.869	-211.335	3.807
3000	82.584	452.717	381.563	213.464	-597.461	-198.013	3.448
3100	82.619	455.426	383.902	221.724	-597.037	-184.705	3.112
3200	82.651	458.049	386.178	229.988	-596.594	-171.411	2.798
3300	82.680	460.593	388.395	238.254	-596.130	-158.131	2.503
3400	82.707	463.062	390.555	246.524	-595.645	-144.865	2.226
3500	82.731	465.459	392.661	254.795	-595.137	-131.615	1.964
3600	82.754	467.790	394.715	263.070	-594.606	-118.378	1.718
3700	82.774	470.058	396.721	271.346	-594.050	-105.158	1.485
3800	82.793	472.266	398.680	279.625	-593.469	-91.952	1.264
3900	82.811	474.417	400.595	287.905	-592.863	-78.762	1.055
4000	82.827	476.513	402.467	296.187	-592.233	-65.588	0.856
4100	82.843	478.559	404.298	304.470	-591.578	-52.430	0.668
4200	82.857	480.555	406.090	312.755	-590.900	-39.288	0.489
4300	82.870	482.505	407.844	321.042	-590.198	-26.163	0.318
4400	82.882	484.410	409.563	329.329	-589.474	-13.054	0.155
4500	82.894	486.273	411.247	337.618	-588.728	0.038	-0.000
4600	82.904	488.095	412.898	345.908	-587.961	13.113	-0.149
4700	82.914	489.878	414.517	354.199	-587.174	26.172	-0.291
4800	82.924	491.624	416.105	362.491	-586.369	39.213	-0.427
4900	82.933	493.334	417.664	370.783	-585.546	52.238	-0.557
5000	82.941	495.009	419.194	379.077	-584.706	65.245	-0.682
5100	82.949	496.652	420.697	387.372	-583.852	78.236	-0.801
5200	82.956	498.263	422.173	395.667	-582.983	91.210	-0.916
5300	82.963	499.843	423.624	403.963	-582.102	104.167	-1.027
5400	82.970	501.394	425.049	412.259	-581.209	117.107	-1.133
5500	82.976	502.916	426.451	420.557	-580.306	130.030	-1.235
5600	82.982	504.411	427.830	428.855	-579.394	142.937	-1.333
5700	82.988	505.880	429.187	437.153	-578.475	155.827	-1.428
5800	82.993	507.324	430.521	445.452	-577.549	168.702	-1.519
5900	82.998	508.742	431.835	453.752	-576.618	181.560	-1.607
6000	83.003	510.137	433.129	462.052	-575.683	194.403	-1.692

PREVIOUS: June 1972 (1 atm) CURRENT: June 1972 (1 bar)

Thionyl Fluoride (OSF_2) $F_2O_1S_1(g)$

DIFLUOROOXOSILANE ($SiOF_2$) IDEAL GAS M_r = 82.081706

$S°(298.15\ K) = [271.247]\ J\ K^{-1}\ mol^{-1}$

$\Delta_f H°(0\ K) = [-962.125 \pm 25.1]\ kJ\ mol^{-1}$
$\Delta_f H°(298.15\ K) = [-966.504 \pm 25.1]\ kJ\ mol^{-1}$

Vibrational Frequencies and Degeneracies

ν, cm	ν, cm
[440](1)	[760](1)
[490](1)	[980](1)
[610](1)	[1420](1)

Ground State Quantum Weight: [1] σ = 2
Point Group: [C_{2v}]
Bond Distances: Si-O = [1.60] Å Si-F = [1.54] Å
Bond Angles: F-Si-O = [120]° F-Si-F = [120]°
Product of the Moments of Inertia: $I_A I_B I_C = 2.565383 \times 10^{-114}\ g^3\ cm^6$

Enthalpy of Formation

$\Delta_f H°(SiOF_2$, g, 298.15 K) is estimated by comparison with those for SiF_4(g) and SiO_2(g).

Heat Capacity and Entropy

The vibrational frequencies were estimated by comparison with those for COF_2(g) and TiOF(g). The vibrational frequencies for COF_2(g) and $TiOF_2$(g) were obtained from Lovell et al. (1) and Gordon (2), respectively. The bond distances of Si-O and Si-F were assumed to be the same as those for SiO(g) and SiF_4(g), respectively. Point group, ground state multiplicity, and bond angles were estimated.

References

1. R. J. Lovell, C. V. Stephenson, and E. A. Jones, J. Chem. Phys. 22, 1953 (1954).
2. J. S. Gordon, personal communication, May 12, 1961.

Difluorooxosilane ($OSiF_2$) $F_2O_1Si_1$(g)

Enthalpy Reference Temperature = T_r = 298.15 K Standard State Pressure = p° = 0.1 MPa

T/K	$C_p°$ (J K⁻¹mol⁻¹)	S° (J K⁻¹mol⁻¹)	-[G°-H°(T_r)]/T (J K⁻¹mol⁻¹)	H°-H°(T_r) (kJ mol⁻¹)	$\Delta_f H°$ (kJ mol⁻¹)	$\Delta_f G°$ (kJ mol⁻¹)	Log K_f
0	0.	0.	INFINITE	-12.006	-962.125	-962.125	INFINITE
100	34.330	225.460	312.100	-8.664	-963.409	-959.869	501.384
200	43.808	251.839	275.855	-4.803	-965.111	-955.667	249.595
250	49.127	262.195	272.106	-2.478	-965.867	-953.217	199.164
298.15	53.689	271.247	271.247	0.	-966.504	-950.721	166.562
300	53.851	271.579	271.248	0.099	-966.527	-950.623	165.518
350	57.899	280.193	271.919	2.896	-967.092	-947.927	141.470
400	61.320	288.154	273.457	5.879	-967.574	-945.155	123.425
450	64.198	295.548	275.506	9.019	-967.983	-942.327	109.383
500	66.615	302.441	277.858	12.291	-968.334	-939.457	98.144
600	70.366	314.937	283.019	19.151	-968.904	-933.626	81.279
700	73.058	325.997	288.384	26.329	-969.352	-927.709	69.227
800	75.027	335.888	293.715	33.738	-969.730	-921.734	60.183
900	76.497	344.814	298.905	41.318	-970.069	-915.714	53.147
1000	77.615	352.934	303.908	49.026	-970.393	-909.657	47.516
1100	78.481	360.374	308.708	56.833	-970.717	-903.568	42.907
1200	79.164	367.233	313.303	64.716	-971.052	-897.448	39.065
1300	79.710	373.592	317.699	72.661	-971.409	-891.301	35.813
1400	80.154	379.516	321.905	80.655	-971.795	-885.124	33.024
1500	80.518	385.058	325.932	88.689	-972.216	-878.919	30.607
1600	80.821	390.265	329.792	96.756	-972.677	-872.684	28.490
1700	81.074	395.172	333.495	104.851	-1023.358	-865.973	26.608
1800	81.289	399.813	337.052	112.970	-1023.677	-856.706	24.861
1900	81.473	404.213	340.472	121.108	-1024.000	-847.421	23.297
2000	81.630	408.396	343.764	129.264	-1024.326	-838.119	21.889
2100	81.767	412.382	346.938	137.434	-1024.656	-828.800	20.615
2200	81.886	416.189	349.999	145.616	-1024.986	-819.466	19.457
2300	81.991	419.831	352.957	153.810	-1025.315	-810.117	18.398
2400	82.083	423.323	355.817	162.014	-1025.641	-800.753	17.428
2500	82.164	426.675	358.584	170.227	-1025.960	-791.376	16.535
2600	82.237	429.899	361.266	178.447	-1026.270	-781.987	15.710
2700	82.301	433.004	363.866	186.674	-1026.568	-772.586	14.947
2800	82.360	435.998	366.389	194.907	-1026.851	-763.174	14.237
2900	82.412	438.889	368.839	203.145	-1027.116	-753.752	13.577
3000	82.459	441.684	371.221	211.389	-1027.362	-744.321	12.960
3100	82.502	444.388	373.538	219.637	-1027.586	-734.883	12.383
3200	82.541	447.008	375.793	227.889	-1027.786	-725.438	11.842
3300	82.577	449.549	377.990	236.145	-1027.961	-715.987	11.333
3400	82.609	452.014	380.131	244.405	-1028.110	-706.530	10.855
3500	82.639	454.410	382.219	252.667	-1028.230	-697.071	10.403
3600	82.666	456.738	384.257	260.932	-1412.475	-677.146	9.825
3700	82.692	459.003	386.246	269.200	-1412.124	-656.726	9.271
3800	82.715	461.209	388.190	277.471	-1411.746	-636.314	8.747
3900	82.736	463.358	390.090	285.743	-1411.340	-615.913	8.249
4000	82.756	465.453	391.948	294.018	-1410.905	-595.523	7.777
4100	82.775	467.496	393.766	302.294	-1410.441	-575.145	7.327
4200	82.792	469.491	395.545	310.573	-1409.949	-554.777	6.900
4300	82.808	471.440	397.288	318.853	-1409.429	-534.422	6.492
4400	82.823	473.343	398.995	327.134	-1408.880	-514.080	6.103
4500	82.837	475.205	400.668	335.417	-1408.304	-493.750	5.731
4600	82.850	477.026	402.308	343.702	-1407.700	-473.433	5.376
4700	82.863	478.808	403.917	351.987	-1407.071	-453.129	5.036
4800	82.874	480.552	405.495	360.274	-1406.416	-432.840	4.710
4900	82.885	482.261	407.044	368.562	-1405.736	-412.564	4.398
5000	82.895	483.936	408.566	376.851	-1405.033	-392.302	4.098
5100	82.905	485.577	410.060	385.141	-1404.308	-372.054	3.811
5200	82.914	487.187	411.527	393.432	-1403.561	-351.821	3.534
5300	82.922	488.767	412.970	401.724	-1402.794	-331.603	3.268
5400	82.931	490.317	414.388	410.017	-1402.008	-311.399	3.012
5500	82.938	491.839	415.782	418.310	-1401.205	-291.211	2.766
5600	82.946	493.333	417.154	426.604	-1400.385	-271.035	2.528
5700	82.952	494.801	418.503	434.899	-1399.551	-250.877	2.299
5800	82.959	496.244	419.831	443.195	-1398.704	-230.731	2.078
5900	82.965	497.662	421.138	451.491	-1397.846	-210.602	1.865
6000	82.971	499.057	422.425	459.788	-1396.976	-190.486	1.658

PREVIOUS: September 1963 (1 atm) CURRENT: September 1963 (1 bar)

Difluorooxosilane ($OSiF_2$) $F_2O_1Si_1$(g)

TITANIUM FLUORIDE OXIDE ($TiOF_2$) IDEAL GAS M_r = 101.896205

$\Delta_f H°(0\ K)$ = [-920.454] kJ mol^{-1}

S°(298.15 K) = [284.678] J K^{-1} mol^{-1} $\Delta_f H°(298.15\ K)$ = [-924.873] kJ mol^{-1}

Vibrational Frequencies and Degeneracies

ν, cm^{-1}	ν, cm^{-1}
[553](1)	[715](1)
[888](1)	[300](1)
[350](1)	[452](1)

Ground State Quantum Weight: [1] σ = 2

Point Group: [C_{2v}]

Bond Distances: Ti-O = [1.62] Å Ti-F = [1.74] Å

Bond Angle: O-Ti-F = [120°]

Product of the Moments of Inertia: $I_A I_B I_C$ = [4.344131 x 10^{-114}] g^3 cm^6

Enthalpy of Formation

$\Delta_f H°(0\ K)$ was estimated as -220 kcal mol^{-1} by Gordon (1). $\Delta_f H°(298.15\ K)$ was then calculated.

Heat Capacity and Entropy

Molecular constants were estimated by Gordon (1). The principal moments of inertia are: I_A = 11.6652 x 10^{-39}, I_B = 14.3272 x 10^{-39}, and I_C = 25.9925 x 10^{-39} g cm^2.

Reference

1. J. S. Gordon, Astrosystems, Inc., Caldwell Twp., N. J., personal communication, January 10, 1963.

Titanium Fluoride Oxide ($OTiF_2$) $F_2O_1Ti_1(g)$

Enthalpy Reference Temperature = T_r = 298.15 K Standard State Pressure = p° = 0.1 MPa

	J K^{-1}mol^{-1}			kJ mol^{-1}			
T/K	$C_p°$	S°	-[G°-H°(T_r)]/T	H°-H°(T_r)	$\Delta_f H°$	$\Delta_f G°$	Log K_f
0	0.	0.	INFINITE	-13.577	-920.454	-920.454	INFINITE
100	37.517	231.211	332.858	-10.165	-921.961	-918.554	479.803
200	52.117	261.723	290.111	-5.678	-923.776	-914.396	238.816
250	58.147	274.024	285.687	-2.916	-924.409	-911.974	190.547
298.15	62.779	284.678	284.678	0.	-924.873	-909.535	159.347
300	62.936	285.067	284.679	0.116	-924.889	-909.440	158.348
350	66.665	295.060	285.460	3.360	-925.259	-906.835	135.338
400	69.558	304.159	287.237	6.769	-925.555	-904.182	118.074
450	71.812	312.488	289.587	10.305	-925.801	-901.495	104.643
500	73.585	320.149	292.265	13.942	-926.017	-898.782	93.895
600	76.128	333.807	298.079	21.437	-926.401	-893.299	77.769
700	77.806	345.676	304.049	29.139	-926.765	-887.753	66.245
800	78.961	356.146	309.920	36.981	-927.126	-882.155	57.599
900	79.784	365.496	315.585	44.920	-927.535	-876.510	50.871
1000	80.390	373.935	321.005	52.930	-928.067	-870.813	45.487
1100	80.848	381.620	326.171	60.993	-928.793	-865.055	41.078
1200	81.202	388.670	331.090	69.097	-933.705	-859.101	37.396
1300	81.480	395.181	335.773	77.231	-934.112	-852.868	34.269
1400	81.703	401.228	340.235	85.391	-934.608	-846.600	31.587
1500	81.885	406.871	344.491	93.571	-935.210	-840.294	29.262
1600	82.034	412.161	348.557	101.767	-935.931	-833.943	27.225
1700	82.159	417.138	352.446	109.977	-936.785	-827.544	25.427
1800	82.263	421.837	356.172	118.198	-937.785	-821.090	23.827
1900	82.352	426.288	359.746	126.429	-938.944	-814.576	22.394
2000	82.428	430.514	363.180	134.668	-954.959	-807.543	21.091
2100	82.494	434.537	366.483	142.914	-957.216	-800.117	19.902
2200	82.551	438.376	369.664	151.166	-959.481	-792.582	18.818
2300	82.601	442.047	372.732	159.424	-961.751	-784.945	17.827
2400	82.645	445.563	375.694	167.686	-964.022	-777.210	16.916
2500	82.684	448.938	378.556	175.953	-966.291	-769.379	16.075
2600	82.718	452.181	381.326	184.223	-968.555	-761.458	15.298
2700	82.749	455.304	384.009	192.496	-970.811	-753.449	14.576
2800	82.777	458.313	386.609	200.773	-973.055	-745.358	13.905
2900	82.801	461.219	389.132	209.052	-975.284	-737.187	13.278
3000	82.824	464.026	391.582	217.333	-977.496	-728.939	12.692
3100	82.844	466.742	393.963	225.616	-979.689	-720.617	12.142
3200	82.862	469.373	396.278	233.902	-981.860	-712.226	11.626
3300	82.879	471.923	398.532	242.189	-984.008	-703.766	11.140
3400	82.894	474.397	400.727	250.477	-986.131	-695.242	10.681
3500	82.908	476.800	402.867	258.768	-988.229	-686.656	10.248
3600	82.921	479.136	404.953	267.059	-990.299	-678.010	9.838
3700	82.933	481.408	406.989	275.352	-1401.440	-661.519	9.339
3800	82.944	483.620	408.976	283.646	-1402.221	-641.511	8.818
3900	82.954	485.775	410.918	291.941	-1403.028	-621.481	8.324
4000	82.963	487.875	412.816	300.236	-1403.861	-601.431	7.854
4100	82.972	489.924	414.672	308.533	-1404.718	-581.360	7.407
4200	82.980	491.923	416.487	316.831	-1405.598	-561.267	6.980
4300	82.988	493.876	418.264	325.129	-1406.500	-541.154	6.574
4400	82.995	495.784	420.005	333.428	-1407.422	-521.019	6.185
4500	83.001	497.649	421.709	341.728	-1408.362	-500.863	5.814
4600	83.007	499.473	423.380	350.029	-1409.320	-480.686	5.458
4700	83.013	501.259	425.018	358.330	-1410.295	-460.487	5.118
4800	83.018	503.006	426.625	366.631	-1411.285	-440.268	4.791
4900	83.024	504.718	428.201	374.933	-1412.290	-420.028	4.478
5000	83.028	506.396	429.748	383.236	-1413.307	-399.768	4.176
5100	83.033	508.040	431.267	391.539	-1414.336	-379.486	3.887
5200	83.037	509.652	432.759	399.842	-1415.377	-359.185	3.608
5300	83.041	511.234	434.225	408.146	-1416.428	-338.864	3.340
5400	83.045	512.786	435.666	416.451	-1417.489	-318.523	3.081
5500	83.048	514.310	437.082	424.755	-1418.558	-298.162	2.832
5600	83.052	515.806	438.474	433.060	-1419.635	-277.781	2.591
5700	83.055	517.276	439.844	441.366	-1420.720	-257.382	2.359
5800	83.058	518.721	441.191	449.671	-1421.812	-236.962	2.134
5900	83.061	520.141	442.518	457.977	-1422.910	-216.525	1.917
6000	83.064	521.537	443.823	466.283	-1424.014	-196.068	1.707

PREVIOUS: September 1963 (1 atm) CURRENT: September 1963 (1 bar)

Titanium Fluoride Oxide ($OTiF_2$) $F_2O_1Ti_1(g)$

SULFURYL FLUORIDE (SO_2F_2) IDEAL GAS M_r = 102.055606

$S°(298.15\ K) = 283.623 \pm 0.4\ J\ K^{-1}\ mol^{-1}$

$\Delta_fH°(0\ K) = -750.138 \pm 8.4\ kJ\ mol^{-1}$
$\Delta_fH°(298.15\ K) = -758.559 \pm 8.4\ kJ\ mol^{-1}$

Vibrational Frequencies and Degeneracies

ν, cm^{-1}	ν, cm^{-1}
1269(1)	1502(1)
848(1)	553(1)
544(1)	885(1)
384(1)	539(1)
388(1)	

Ground State Quantum Weight: [1] $\sigma = 2$
Point Group: C_{2v}
Bond Distances: S–O = 1.405 ± 0.003 Å S–F = 1.530 ± 0.003 Å
Bond Angles: O–S–O = 123.97 ± 0.20° F–S–F = 96.12 ± 0.17°
Product of the Moments of Inertia: $I_AI_BI_C = 4.490464 \times 10^{-114}\ g^3\ cm^6$

Enthalpy of Formation

Ruh et al. (1) reported percent conversions for the reaction $SO_2(g) + Cl_2(g) + 2\ HF(g) \rightarrow SO_2F_2(g) + 2\ HCl(g)$ at 110 to 700°C and at pressures slightly above atmospheric. The data allow calculation of apparent values of K_p from the assumption of exact stoichiometry. Two temperatures may be excluded, namely 110 °C for failure to approach equilibrium and 597 °C due to power failure. The remaining values (503 to 973 K) yield $\Delta_rH°$(3rd law) = −24.15±2 and $\Delta_rH°$(2nd law) = −20.7±1.6 kcal mol^{-1} at 298.15 K. The 3rd law values show a minor trend with temperature, corresponding to $\Delta_rS°$(2nd law) −$\Delta_rS°$(3rd law) = 5.4±2.5 cal K^{-1} mol^{-1}, but this inconsistency is small enough to suggest that there was reasonable approach to equilibrium. An example given in the text of the patent (1) reveals that the actual recovery of SO_2 was somewhat less than that predicted on the assumption of stoichiometry. Use of this actual recovery of SO_2 would make $\Delta_rH°$(3rd law) more negative by less than 0.4 kcal mol^{-1}. We adopt $\Delta_rH°$ = −24.15 kcal mol^{-1} and $\Delta_fH°$ = −181.3±2 kcal mol^{-1} at 298.15 K.

Reese et al. (2) derived $\Delta_fH°$ = −205 kcal mol^{-1} from electron-impact studies. They measured potentials of 12.44±0.1 and 19.9±0.3 eV for the appearance of SO_2^+ from SO_2 and SO_2F_2, respectively. Combination of these values implies that $\Delta_fH°$ of SO_2F_2 should be about 1.0 eV more negative than the adopted value. This discrepancy could reasonably arise frm neglect of the unknown excess energies of the products of the two reactions. Bond-energy comparisons tend to confirm this view. Average values of D(S–F) may be calculated from $\Delta_rH°/2$ for $SF_6 \rightarrow SF_4 + 2\ F$ and $SO_2F_2 \rightarrow SO_2 + 2F$. D(S–F) = 72 kcal mol^{-1} from SF_6 is much more consistent with D_0^0(S–F) = 74 kcal mol^{-1} from the adopted $\Delta_fH°$ than with D_0^0(S–F) = 86 kcal mol^{-1} from the electron-impact value.

Heat Capacity and Entropy

Bond lengths and bond angles are from the microwave study of Lide et al. (3). The resulting O–S–F angle is 108.3° and the principal moments of inertia are $I_A = 16.3467 \times 10^{-39}$, $I_B = 16.5727 \times 10^{-39}$ and $I_C = 16.5756 \times 10^{-39}\ g\ cm^2$. An early electron-diffraction study (4) is in reasonable agreement.

Vibrational frequencies are those assigned by Lide et al. (5) based on additional microwave and infrared data for the gaseous and matrix-isolated phases. Assignment of the three fundamentals near 544 cm^{-1} and the two near 388 cm^{-1} is supported by gas-phase Raman spectra (6), by combination and overtones in the infrared (6) and by force-field calculations (7).

Calculated values of $C_p°$ agree within 0.1% with those derived by Gehri (8) from measurements of $C_p°$ at temperatures of 273 to 323 K and pressures up to 1 atm. The calculated entropy, S° = 63.25 cal K^{-1} mol^{-1} at the normal boiling point of 217.78 K, may be compared with 62.66±0.15 cal K^{-1} mol^{-1} derived from low-temperature calorimetric data by Bockhoff et al. (9). The discrepancy is 0.59 cal K^{-1} mol^{-1}; it was attributed (9) to randomness in the solid at absolute zero. Bockhoff and Pace (10) later reported an entropy discrepancy of 1.48 cal K^{-1} mol^{-1}, but this value resulted from an incorrect vibrational assignment (5–7).

References

1. R. P. Ruh, R. A. Davis and K. A. Allswede, U. S. Patent 3,092,458, June 4, 1963.
2. R. M. Reese, V. H. Dibeler and J. L. Franklin, J. Chem. Phys. 29, 880 (1958).
3. D. R. Lide, D. E. Mann and R. M. Fristrom, J. Chem. Phys. 26, 734 (1957).
4. D. P. Stevenson and H. Russell, J. Amer. Chem. Soc. 61, 3264 (1939).
5. D. R. Lide, D. E. Mann and J. J. Comeford, Spectrochim. Acta 21, 497 (1965).
6. H. V. Samuelson, Dissertation Abstr. B27, 443 (1966).
7. H. Toyuki and K. Shimizu, Bull. Chem. Soc. Japan 39, 2364 (1966).
8. D. C. Gehri, Dissertation Abstr. B29, 133 (1968).
9. F. J. Bockhoff, R. V. Petrella and E. L. Pace, J. Chem. Phys. 32, 799 (1960).
10. F. J. Bockhoff and E. L. Pace, J. Chem. Phys. 36, 3502 (1962).

Sulfuryl Fluoride (SO_2F_2) $F_2O_2S_1(g)$

Enthalpy Reference Temperature = T_r = 298.15 K Standard State Pressure = p° = 0.1 MPa

T/K	$C_p°$ (J K⁻¹mol⁻¹)	S° (J K⁻¹mol⁻¹)	−[G°−H°(T_r)]/T (J K⁻¹mol⁻¹)	H°−H°(T_r) (kJ mol⁻¹)	$\Delta_fH°$ (kJ mol⁻¹)	$\Delta_fG°$ (kJ mol⁻¹)	Log K_f
0	0.	0.	INFINITE	-13.499	-750.138	-750.138	INFINITE
100	35.872	230.833	332.132	-10.130	-753.270	-740.734	386.920
200	51.609	260.231	289.215	-5.797	-756.419	-726.911	189.850
250	59.368	272.600	284.672	-3.018	-757.620	-719.389	150.308
298.15	65.825	283.623	283.623	0.	-758.559	-711.938	124.729
300	66.054	284.031	283.625	0.122	-758.592	-711.648	123.909
350	71.723	294.651	284.450	3.570	-759.375	-703.760	105.030
400	76.513	304.550	286.351	7.280	-762.221	-695.683	90.847
450	80.558	313.803	288.893	11.209	-763.401	-687.302	79.780
500	83.975	322.472	291.822	15.325	-764.529	-678.783	70.912
600	89.324	338.282	298.274	24.005	-766.067	-661.478	57.587
700	93.205	352.358	305.014	33.141	-767.083	-643.960	48.053
800	96.066	365.000	311.736	42.612	-767.770	-626.321	40.895
900	98.215	376.445	318.300	52.331	-821.368	-607.529	35.260
1000	99.858	386.882	324.644	62.238	-820.469	-583.817	30.495
1100	101.136	396.462	330.744	72.290	-819.517	-560.198	26.602
1200	102.146	405.307	336.593	82.456	-818.532	-536.666	23.360
1300	102.956	413.516	342.198	92.713	-817.531	-513.218	20.621
1400	103.615	421.171	347.569	103.042	-816.523	-489.847	18.276
1500	104.157	428.339	352.717	113.432	-815.519	-466.549	16.247
1600	104.608	435.075	357.656	123.871	-814.524	-443.316	14.473
1700	104.986	441.429	362.399	134.351	-813.544	-420.146	12.910
1800	105.307	447.439	366.958	144.866	-812.581	-397.033	11.522
1900	105.581	453.140	371.345	155.411	-811.637	-373.973	10.281
2000	105.817	458.562	375.572	165.981	-810.712	-350.962	9.166
2100	106.021	463.730	379.648	176.573	-809.804	-327.997	8.158
2200	106.200	468.666	383.582	187.184	-808.913	-305.074	7.243
2300	106.356	473.391	387.385	197.812	-808.036	-282.192	6.409
2400	106.494	477.920	391.064	208.455	-807.171	-259.348	5.645
2500	106.616	482.270	394.626	219.111	-806.313	-236.540	4.942
2600	106.725	486.454	398.078	229.778	-805.460	-213.766	4.295
2700	106.822	490.483	401.426	240.455	-804.609	-191.024	3.696
2800	106.909	494.370	404.676	251.142	-803.755	-168.314	3.140
2900	106.988	498.123	407.834	261.837	-802.897	-145.635	2.623
3000	107.058	501.751	410.905	272.539	-802.032	-122.986	2.141
3100	107.123	505.262	413.892	283.248	-801.156	-100.365	1.691
3200	107.181	508.664	416.801	293.963	-800.268	-77.773	1.270
3300	107.235	511.963	419.635	304.684	-799.366	-55.210	0.874
3400	107.284	515.165	422.398	315.410	-798.448	-32.673	0.502
3500	107.328	518.276	425.093	326.141	-797.514	-10.164	0.152
3600	107.369	521.300	427.723	336.876	-796.561	12.319	-0.179
3700	107.407	524.242	430.292	347.615	-795.590	34.774	-0.491
3800	107.442	527.107	432.803	358.357	-794.599	57.204	-0.786
3900	107.475	529.898	435.257	369.103	-793.589	79.607	-1.066
4000	107.505	532.620	437.657	379.852	-792.559	101.983	-1.332
4100	107.532	535.275	440.006	390.604	-791.511	124.333	-1.584
4200	107.558	537.866	442.305	401.358	-790.443	146.658	-1.824
4300	107.582	540.398	444.557	412.115	-789.358	168.958	-2.052
4400	107.605	542.871	446.763	422.875	-788.255	191.231	-2.270
4500	107.626	545.290	448.926	433.636	-787.135	213.479	-2.478
4600	107.646	547.655	451.047	444.400	-785.999	235.702	-2.676
4700	107.664	549.971	453.127	455.165	-784.849	257.901	-2.866
4800	107.682	552.237	455.168	465.933	-783.686	280.075	-3.048
4900	107.698	554.458	457.172	476.702	-782.511	302.225	-3.222
5000	107.713	556.634	459.139	487.472	-781.325	324.351	-3.388
5100	107.728	558.767	461.072	498.244	-780.130	346.452	-3.548
5200	107.741	560.859	462.971	509.018	-778.927	368.530	-3.702
5300	107.754	562.911	464.837	519.793	-777.718	390.586	-3.849
5400	107.766	564.926	466.672	530.569	-776.506	412.617	-3.991
5500	107.778	566.903	468.477	541.346	-775.290	434.626	-4.128
5600	107.789	568.845	470.252	552.124	-774.073	456.615	-4.259
5700	107.799	570.753	471.998	562.904	-772.857	478.580	-4.386
5800	107.809	572.628	473.717	573.684	-771.644	500.525	-4.508
5900	107.818	574.471	475.409	584.465	-770.434	522.447	-4.625
6000	107.827	576.283	477.075	595.248	-769.232	544.351	-4.739

PREVIOUS: June 1971 (1 atm) CURRENT: June 1971 (1 bar)

Sulfuryl Fluoride (SO_2F_2) $F_2O_2S_1(g)$

PHOSPHORUS FLUORIDE (PF_2) IDEAL GAS M_r = 68.970566

$\Delta_{at}H°(0\ K)$ = [954.370 ± 25.1] kJ mol^{-1}
$S°(298.15\ K)$ = [262.957 ± 0.8] J K^{-1} mol^{-1}

$\Delta_f H°(0\ K)$ = [-485.280 ± 20.9] kJ mol^{-1}
$\Delta_f H°(298.15\ K)$ = [-488.256 ± 20.9] kJ mol^{-1}

Electronic Levels and Quantum Weights

ϵ_i, cm^{-1}	g_i
0	2
[25000]	[2]
[28000]	[2]
[30000]	[2]

Vibrational Frequencies and Degeneracies

ν, cm^{-1}
831 (1)
[353](1)
852 (1)

Point Group: C_{2v} σ = 2
Bond Distance: P-F = [1.58] Å
Bond Angle: F-P-F = [99°]
Product of the Moments of Inertia: $I_A I_B I_C$ = [3.285613 x 10^{-115}] g^3 cm^6

Enthalpy of Formation

The adopted results are $\Delta_f H°(0\ K)$ = -112.2±5.0 kcal mol^{-1}, D_0^0(PF-F) = 122.0±10.0 kcal mol^{-1}, and $\Delta_{at}H°$ = 228.1±6.0 kcal mol^{-1}. We estimate $\Delta_f H°$ such that the value gives D_0^0(PF-F)/$\Delta_{at}H°(0\ K)$ = 0.535. The value of this ratio is selected from those for SiF_2 (0.54±0.03) and SF_2(0.53±0.05) which are calculated from the most recent JANAF tables (1). Two previous estimates of $\Delta_f H°$ are available for comparison. O'Hare (2) used the relationship $D_0^0(PF_2-F)$ = 1.1 $\Delta_{at}H°(PF_3)$ to obtain a $\Delta_f H°$ of -115.7±5 kcal mol^{-1}. This estimate agrees quite well with the results adopted here. An earlier JANAF table (3) gives $\Delta_f H°(PF_2)$ = -104.7±15.0 kcal mol^{-1} which was based on an estimate value for the $\Delta_{at}H°$ of PF_3. We estimate that the uncertainty in the value of the ratio $D_0^0/\Delta_{at}H°$ should not exceed ±0.06 which corresponds to an uncertainty in $\Delta_f H°$ of ±5.0 kcal mol^{-1}.

Heat Capacity and Entropy

The results of electron spin resonance experiments (4, 5) and molecular orbital calculations (4, 6, and 7) are consistent with PF_2 having C_{2v} symmetry; however, no experimental determination of the structural parameters has been reported. Numerous values for the bond length and angle have been proposed based on energy-optimization (6, 7) and other estimation techniques (2, 4, 8, and 9). These estimates are widely scattered with the bond lengths ranging from 1.52-1.73 Å and the bond angles ranging from 98.6-112°. We prefer to use molecular parameters similar to those measured (10) for the related molecule PF_2H. The adopted values appear quite reasonable by comparison with structural data (1) for SiF_2 and SF_2. The principal moments of inertia are: I_A = 2.9836 x 10^{-39}, I_B = 9.1077 x 10^{-39}, and I_C = 12.0913 x 10^{-39} g cm^2.

The two stretching frequencies (ν_1 and ν_3) have been observed (11) in the infrared spectrum of PF_2 formed in an argon matrix by photolysis of PF_2H. The adopted values are rounded to the nearest cm^{-1}. The bending frequency (ν_2) is calculated from the force constant f_α/r^2 = 0.495 mdyn/Å. We obtain the bending force constant from the ratio $f_r/(f_\alpha/r^2)$ equal to 9.59 which is the value calculated for SiF_2 (1).

The ground state is expected to be $^2B_1$, as shown by Wei et al. (4). We assume that the upper electronic levels and their corrresponding quantum weights are the same as those previously estimated for the isoelectronic SF_2^+ ion. The procedure used to estimate these levels is thoroughly discussed on the SF_2^+ table (1).

This table essentially updates the functions previously tabulated by O'Hare (2).

References

1. JANAF Thermochemical Tables: SF_2(g) and SiF_2(g), 6-30-76; SF_2^+(g), 12-31-76; PF_3(g) and PF_5(g), 12-31-69.
2. P. A. G. O'Hare, Argonne National Laboratory, ANL-7459, December, 1968.
3. JANAF Thermochemical Tables, 2nd ed., NSRDS-NBS 37, 1971.
4. M. S. Wei, J. H. Current and J. Gendell, J. Chem. Phys. 52, 1592 (1970).
5. W. Nelson, G. Jackel, and W. Gordy, J. Chem. Phys. 52, 4572 (1970).
6. A. Hudson and J. T. Wiffen, Chem. Phys. Lett. 29, 113 (1974).
7. D. Kilcast and C. Thomson, J. Chem. Soc., Faraday Trans. II, 68, 535 (1972).
8. J. Kendrick, I. H. Hillier and M. F. Guest, Chem. Phys. Lett. 33, 173 (1975).
9. J. C. Cobb and A. Hinchliffe, Chem. Phys. Lett. 24, 75 (1974).
10. R. L. Kuczkowski, J. Amer. Chem. Soc. 90, 1705 (1968).
11. J. K. Burdett, L. Hodges, V. Dunning and J. H. Current, J. Phys. Chem. 74, 4053 (1970).

Phosphorus Fluoride (PF_2) F_2P_1(g)

Enthalpy Reference Temperature = T_r = 298.15 K Standard State Pressure = p° = 0.1 MPa

	J K^{-1}mol^{-1}			kJ mol^{-1}			
T/K	$C_p°$	S°	-[G°-H°(T_r)]/T	H°-H°(T_r)	$\Delta_f H°$	$\Delta_f G°$	Log K_f
0	0.	0.	INFINITE	-11.209	-485.280	-485.280	INFINITE
100	34.623	220.645	299.211	-7.857	-485.722	-488.981	255.417
200	39.685	246.149	266.890	-4.148	-487.206	-491.880	128.466
250	42.344	255.291	263.680	-2.097	-487.749	-492.985	103.004
298.15	44.717	262.957	262.957	0.	-488.256	-493.948	86.538
300	44.803	263.233	262.957	0.083	-488.275	-493.983	86.010
350	46.927	270.304	263.511	2.378	-489.505	-494.823	73.848
400	48.696	276.690	264.766	4.770	-490.060	-495.544	64.711
450	50.142	282.511	266.419	7.242	-490.571	-496.198	57.597
500	51.319	287.858	268.299	9.779	-491.048	-496.797	51.900
600	53.065	297.379	272.372	15.004	-491.929	-497.863	43.343
700	54.253	305.654	276.548	20.374	-492.744	-498.787	37.220
800	55.086	312.956	280.652	25.843	-493.517	-499.597	32.620
900	55.687	319.481	284.610	31.383	-494.263	-500.312	29.037
1000	56.134	325.372	288.397	36.976	-494.994	-500.945	26.167
1100	56.473	330.739	292.006	42.607	-495.715	-501.505	23.814
1200	56.736	335.664	295.441	48.268	-560.005	-500.921	21.805
1300	56.944	340.214	298.713	53.952	-559.947	-496.000	19.930
1400	57.111	344.441	301.830	59.655	-559.898	-491.083	18.323
1500	57.248	348.386	304.804	65.373	-559.857	-486.169	16.930
1600	57.360	352.084	307.644	71.104	-559.825	-481.258	15.711
1700	57.454	355.564	310.362	76.845	-559.803	-476.348	14.636
1800	57.533	358.851	312.965	82.594	-559.790	-471.439	13.681
1900	57.600	361.963	315.463	88.351	-559.784	-466.531	12.826
2000	57.658	364.919	317.862	94.114	-559.783	-461.623	12.056
2100	57.707	367.733	320.171	99.882	-559.786	-456.715	11.360
2200	57.751	370.419	322.394	105.655	-559.787	-451.807	10.727
2300	57.789	372.987	324.538	111.432	-559.786	-446.899	10.149
2400	57.822	375.447	326.609	117.213	-559.778	-441.991	9.620
2500	57.852	377.808	328.610	122.996	-559.760	-437.083	9.132
2600	57.879	380.078	330.546	128.783	-559.728	-432.177	8.683
2700	57.904	382.263	332.421	134.572	-559.679	-427.272	8.266
2800	57.926	384.369	334.239	140.364	-559.611	-422.369	7.879
2900	57.948	386.402	336.003	146.157	-559.520	-417.469	7.519
3000	57.968	388.367	337.716	151.953	-559.405	-412.573	7.184
3100	57.987	390.268	339.381	157.751	-559.262	-407.681	6.869
3200	58.007	392.109	341.000	163.551	-559.089	-402.794	6.575
3300	58.027	393.895	342.576	169.352	-558.886	-397.913	6.298
3400	58.047	395.627	344.111	175.156	-558.651	-393.038	6.038
3500	58.068	397.310	345.607	180.962	-558.383	-388.171	5.793
3600	58.091	398.946	347.066	186.770	-558.080	-383.312	5.562
3700	58.115	400.538	348.490	192.580	-557.743	-378.462	5.343
3800	58.142	402.088	349.880	198.393	-557.370	-373.621	5.136
3900	58.171	403.599	351.238	204.208	-556.962	-368.791	4.939
4000	58.202	405.072	352.565	210.027	-556.518	-363.972	4.753
4100	58.237	406.510	353.864	215.849	-556.037	-359.164	4.576
4200	58.275	407.914	355.134	221.675	-555.521	-354.368	4.407
4300	58.316	409.285	356.377	227.504	-554.969	-349.586	4.247
4400	58.361	410.626	357.595	233.338	-554.381	-344.816	4.093
4500	58.410	411.938	358.788	239.176	-553.757	-340.060	3.947
4600	58.462	413.223	359.958	245.020	-553.098	-335.318	3.808
4700	58.519	414.481	361.104	250.869	-552.404	-330.591	3.674
4800	58.580	415.713	362.229	256.724	-551.675	-325.879	3.546
4900	58.646	416.922	363.333	262.585	-550.911	-321.183	3.424
5000	58.716	418.107	364.417	268.453	-550.114	-316.503	3.306
5100	58.790	419.271	365.481	274.329	-549.282	-311.839	3.194
5200	58.869	420.413	366.526	280.211	-548.417	-307.191	3.086
5300	58.952	421.535	367.554	286.102	-547.519	-302.561	2.982
5400	59.039	422.638	368.564	292.002	-546.588	-297.948	2.882
5500	59.131	423.722	369.557	297.910	-545.624	-293.353	2.786
5600	59.226	424.789	370.534	303.828	-544.628	-288.774	2.694
5700	59.326	425.838	371.495	309.756	-543.600	-284.215	2.605
5800	59.430	426.870	372.440	315.694	-542.540	-279.674	2.519
5900	59.537	427.887	373.372	321.642	-541.449	-275.150	2.436
6000	59.649	428.889	374.289	327.601	-540.327	-270.647	2.356

PREVIOUS: June 1977 (1 atm) CURRENT: June 1977 (1 bar)

Phosphorus Fluoride (PF_2) F_2P_1(g)

PHOSPHORUS FLUORIDE, ION (PF_2^+) IDEAL GAS M_r = 68.970017

S°(298.15 K) = [257.358 ± 8.4] J K^{-1} mol^{-1}

$\Delta_f H°$(0 K) = 469.054 ± 33.5 kJ mol^{-1}
$\Delta_f H°$(298.15 K) = 472.273 ± 33.5 kJ mol^{-1}

Electronic Levels and Quantum Weights

ϵ_i, cm^{-1}	g_i
0	[1]
[25000]	[3]
[40000]	[1]

Vibrational Frequencies and Degeneracies

ν, cm^{-1}
[840](1)
[350](1)
[860](1)

Point Group: C_{2v} σ = [2]
Bond Distance: P-F = [1.59] Å
Bond Angle: F-P-F = [101]°
Product of the Moments of Inertia: $I_A I_B I_C$ = [3.412079 x 10^{-115}] g^3 cm^6

Enthalpy of Formation

A summary of the information available on the appearance potential of PF_2^+ from PF_3 is presented below. We analyze these electron-impact results using the expression $\Delta_r H°$(0 K) = AP-E* by assuming that the dissociative ionization process is PF_3(g) + e^- = PF_2^+(g) + F(g) + 2 e^-. Values of $\Delta_f H°$ and the ionization potential are derived by setting E* in the above expression equal to zero. As a result, the values are listed in the table as inequalities to emphasize that they are limiting (upper) values.

Source	AP(PF_2^+/PF_3) eV	$\Delta_f H°$(PF_2^+, g, 0 K) kcal mol^{-1}	IP(PF_2) eV
(1)	15.4±0.2	<112.78	<9.76
(2)	15.7±0.5	<119.70	<10.06
(3)	15.5±0.2	<115.09	<9.86
(4)	13.5	<68.97	<7.86

The results derived from the first three studies (1-3) are in excellent accord, and we adopt the mean values of $\Delta_f H°$ = 115.861±8.0 kcal mol^{-1} and IP = 9.89 eV. Dean et al. (4) were unable to offer an explanation as to why their results were significantly lower than the other literature data. This study (4) indicates that IP(PF_2)<IP(PCl_2) (3) which we believe is very unlikely. Harland et al. (1) have reported the AP for PF_2^+ from PF_2CN as 13.4±0.2 eV, suggesting the possibility that the precursor responsible for the PF_2^+ ion in the measurements of Dean et al. (4) was PF_2CN rather than PF_3. A MO study (5) gave IP(PF_2) = 9.88 eV based on IP(P) - 10.98 eV. Recalculation of their results using a more recent value of IP(P) = 10.486 eV (6) gives IP(PF_2) = 9.88 eV. Our adopted value of IP (9.89 eV) is close to that for atomic phosphorus (10.49 eV). This is consistent with the results of electron spin resonance experiments (7, 8) and MO calculations (5, 7, 9, and 10) which show that the ionizing orbital is slightly antibonding and located primarily on the phosphorus atom.

The mean bond energy (products P^+(g)+2F(g)) is 120.9 kcal mol^{-1} which when compared with that for PF_2(114.1 kcal mol^{-1}, 11) suggests that the bonding is somewhat stronger in the positive ion. This predicted increase in bond strength upon ionization is also consistent with the antibonding nature of the ionizing orbital. $\Delta_f H°$ (298.15 K) is calculated to be 117.049 kcal mol^{-1}.

Heat Capacity and Entropy

The molecular parameters and vibrational frequencies are assumed to lie between those for PF_2 and SiF_2 (11). The electronic levels and quantum weights are estimated by analogy with those for the isoelectronic molecule SiF_2 (11). The principal moments of inertia are: I_A = 2.8983 x 10^{-39}, I_B = 9.4974 x 10^{-39}, and I_C = 12.3957 x 10^{-39} g cm^2.

References

1. P. W. Harland, D. W. H. Rankin and J. C. J. Thynne, Int. J. Mas Spectr. Ion Phys. 13, 395 (1974).
2. F. E. Stafford, personal communication quoted in Harland et al. (1).
3. D. F. Torgerson and J. B. Westmore, Can. J. Chem. 53, 933 (1975).
4. C. R. S. Dean, A. Finch, P. J. Gardner, and D. W. Payling, J. Chem. Soc., Faraday Trans. I, 70, 1921 (1974).
5. A. L. Companion and Y. P. Hsia, J. Mol. Struct. 14, 117 (1972).
6. H. M. Rosenstock, K. Draxl, B. W. Steiner, and J. T. Herron, J. Phys. Chem. Ref. Data, 6, Suppl. 1 (1977).
7. M. S. Wei, J. H. Current and J. Gendell, J. Chem. Phys. 52, 1592 (1970).
8. W. Nelson, G. Jackel and W. Gordy, J. Chem. Phys. 52, 1592 (1970).
9. D. Kilcast and C . Thomson, J. Chem. Soc., Faraday Trans. II, 68, 435 (1972).
10. A. Hudson and J. T. Wiffen, Chem. Phys. Lett. 29, 113 (1974).
11. JANAF Thermochemical Tables: PF_2(g), 6-30-77; SiF_2(g), 6-30-76.

Phosphorus Fluoride, Ion (PF_2^+) $F_2P_1^+$(g)

Enthalpy Reference Temperature = T_r = 298.15 K Standard State Pressure = p° = 0.1 MPa

T/K	$C_p°$	S°	-[G°-H°(T_r)]/T	H°-H°(T_r)	$\Delta_f H°$	$\Delta_f G°$	Log K_f
	J $K^{-1}$$mol^{-1}$			kJ mol^{-1}			
0	0.	0.	INFINITE	-11.207	469.054		
100	34.659	215.050	293.584	-7.853			
200	39.667	240.571	261.285	-4.143			
250	42.285	249.705	258.080	-2.094			
298.15	44.635	257.358	257.358	0.	472.273	461.996	-80.940
300	44.720	257.634	257.359	0.083	472.292	461.932	-80.430
350	46.835	264.691	257.911	2.373	472.097	460.235	-68.686
400	48.602	271.064	259.163	4.760	472.577	458.509	-59.875
450	50.053	276.875	260.813	7.228	473.101	456.719	-53.015
500	51.236	282.212	262.690	9.761	473.658	454.869	-47.520
600	52.996	291.720	266.755	14.979	474.848	451.001	-39.263
700	54.196	299.985	270.925	20.342	476.106	446.928	-33.350
800	55.039	307.280	275.022	25.806	477.407	442.671	-28.903
900	55.648	313.800	278.975	31.342	478.735	438.250	-25.435
1000	56.101	319.687	282.757	36.930	480.079	433.680	-22.653
1100	56.445	325.051	286.362	42.558	481.433	428.974	-20.370
1200	56.712	329.974	289.794	48.217	419.220	425.224	-18.510
1300	56.924	334.523	293.062	53.899	421.353	425.637	-17.102
1400	57.093	338.747	296.176	59.600	423.480	425.887	-15.890
1500	57.232	342.691	299.147	65.317	425.598	425.985	-14.834
1600	57.346	346.389	301.985	71.046	427.706	425.942	-13.906
1700	57.441	349.868	304.700	76.785	429.806	425.767	-13.082
1800	57.522	353.154	307.302	82.534	431.896	425.470	-12.347
1900	57.590	356.266	309.798	88.289	433.980	425.056	-11.686
2000	57.649	359.221	312.196	94.051	436.058	424.532	-11.088
2100	57.699	362.035	314.503	99.819	438.134	423.905	-10.544
2200	57.743	364.720	316.725	105.591	440.209	423.178	-10.048
2300	57.782	367.288	318.868	111.367	442.289	422.358	-9.592
2400	57.817	369.748	320.937	117.147	444.375	421.447	-9.173
2500	57.848	372.109	322.937	122.930	446.471	420.448	-8.785
2600	57.876	374.378	324.872	128.717	448.581	419.366	-8.425
2700	57.903	376.563	326.746	134.506	450.708	418.202	-8.091
2800	57.927	378.669	328.563	140.297	452.855	416.959	-7.778
2900	57.951	380.703	330.326	146.091	455.025	415.638	-7.486
3000	57.975	382.668	332.038	151.887	457.220	414.243	-7.213
3100	57.999	384.569	333.702	157.686	459.442	412.774	-6.955
3200	58.024	386.411	335.321	163.487	461.695	411.232	-6.713
3300	58.050	388.197	336.896	169.291	463.978	409.620	-6.484
3400	58.079	389.930	338.431	175.097	466.295	407.938	-6.267
3500	58.109	391.614	339.926	180.907	468.645	406.188	-6.062
3600	58.143	393.251	341.385	186.719	471.031	404.370	-5.867
3700	58.179	394.845	342.808	192.535	473.453	402.485	-5.682
3800	58.219	396.397	344.198	198.355	475.911	400.534	-5.506
3900	58.264	397.910	345.556	204.179	478.406	398.518	-5.338
4000	58.313	399.386	346.884	210.008	480.939	396.437	-5.177
4100	58.366	400.826	348.182	215.842	483.510	394.293	-5.023
4200	58.424	402.233	349.452	221.682	486.119	392.086	-4.876
4300	58.488	403.609	350.696	227.527	488.766	389.815	-4.735
4400	58.557	404.954	351.913	233.379	491.451	387.483	-4.600
4500	58.631	406.271	353.107	239.239	494.174	385.090	-4.470
4600	58.711	407.560	354.277	245.106	496.935	382.635	-4.345
4700	58.797	408.824	355.424	250.981	499.734	380.120	-4.225
4800	58.888	410.063	356.549	256.865	502.571	377.546	-4.109
4900	58.985	411.278	357.654	262.759	505.446	374.911	-3.997
5000	59.087	412.471	358.738	268.663	508.357	372.218	-3.889
5100	59.195	413.642	359.803	274.577	511.306	369.466	-3.784
5200	59.309	414.792	360.850	280.502	514.292	366.656	-3.683
5300	59.428	415.923	361.878	286.439	517.315	363.788	-3.585
5400	59.551	417.035	362.889	292.387	520.374	360.862	-3.491
5500	59.680	418.129	363.884	298.349	523.470	357.880	-3.399
5600	59.814	419.206	364.862	304.324	526.601	354.841	-3.310
5700	59.952	420.266	365.825	310.312	529.769	351.745	-3.223
5800	60.094	421.309	366.773	316.314	532.971	348.594	-3.139
5900	60.240	422.338	367.706	322.331	536.209	345.388	-3.058
6000	60.390	423.352	368.625	328.362	539.482	342.126	-2.978

PREVIOUS: June 1977 (1 atm) CURRENT: June 1977 (1 bar)

Phosphorus Fluoride, Ion (PF_2^+) $F_2P_1^+$(g)

PHOSPHORUS FLUORIDE, ION (PF_2^-) IDEAL GAS M_r = 68.971115

$S°(298.15\ K)$ = [257.364 ± 0.8] J K^{-1} mol^{-1}

$\Delta_f H°(0\ K)$ = [630.010 ± 69.0] kJ mol^{-1}
$\Delta_f H°(298.15\ K)$ = [-639.177 ± 69.0] kJ mol^{-1}

Vibrational Frequencies and Degeneracies

ν, cm^{-1}
[820](1)
[355](1)
[845](1)

Ground State Quantum Weight: [1] σ = [2]
Point Group: C_{2v}
Bond Distance: P-F = [1.59] Å
Bond Angle: F-P-F = [98]°
Product of the Moments of Inertia: $I_A I_B I_C$ = [3.409086 x 10^{-115}] g^3 cm^6

Enthalpy of Formation

The PF_2^- anion has been observed in the negative ion mass spectra of PF_3 (1) and PF_2CN (2). Both studies included thermochemical calculations on plausible ion formation processes, showing that the measured appearance potentials contained large excess energy contributions. At present, definite values cannot be assigned to these contributions; thus, their studies are of little value in defining $\Delta_f H°$.

Two estimates of the electron affinity of PF_2 have been reported (2, 3). Harland et al. (2) considered differences in bond energies for diatomic and triatomic species to obtain a value of EA(PF_2) = 1.4 eV. Rescaled CNDO calculations (3) have led to an EA = 1.71 eV. JANAF data (4) on the S-F system indicate that the EA values alternate between the even- and odd- electron molecules, with the odd-electron molecules having the higher EA values. Assuming the same trend is followed in the P-F system, we predict that EA(PF_2) > EA(PF) which is equal to 1.1±0.5 eV (4). We adopt an EA of 1.5±0.5 eV for the reaction PF_2^-(g) = PF_2(g) + e^- at 0 K which leads to $\Delta_f H°$(PF_2^-, g, 0 K) = -146.822±16.5 kcal mol^{-1} with $\Delta_f H°$(PF_2, g, 0 K) = -112.23±5.0 kcal mol^{-1} (4). The adopted value of EA is close to that for atomic phosphorus (0.77±0.05 eV, 5). This is in agreement with results of MO calculations (3, 6) and electron spin resonance experiments (6, 7) which show that the electron captured by PF_2 goes into an antibonding orbital which is located primarily on the phosphorus atom.

$\Delta_f H°$ at 298.15 K and the fluoride-ion affinity (IA) are calculated to be -148.594 kcal mol^{-1} and 78.2 kcal mol^{-1}, respectively. The mean bond energy $\Delta_{at}H°$/2(products P(g)+F(g)+F^-(g)) is 92.2 kcal mol^{-1} which is 21.9 kcal mol^{-1} less than that for PF_2 (114.1 kcal mol^{-1}, 4). The predicted decrease in the bond strength for the anion is also consistent with the antibonding nature of the electron orbital.

Heat Capacity and Entropy

The molecular parameters and vibrational frequencies are assumed to lie between those for PF_2 and SF_2 (4). The ground state is assumed to be a singlet since PF_2^- has no unpaired electrons. Also, we neglect excited states since these probably lie at energies which would not contribute to the partition function (see the table for the isoelectronic species SF_2). The principal moments of inertia are: I_A = 3.0833 x 10^{-39}, I_B = 9.0858 x 10^{-39}, and I_C = 12.1691 x 10^{-39} g cm^2.

References

1. K. A. G. MacNeil and J. C. J. Thynne, J. Phys. Chem. 74, 2256 (1970).
2. P. W. Harland, D. W. H. Rankin, and J. C. J. Thynne, Int. J. Mass Spectr. Ion Phys. 13, 395 (1974).
3. A. L. Companion and Y. P. Hsia, J. Mol. Struct. 14, 117 (1972).
4. JANAF Thermochemical Tables: PF^-(g) and PF_2(g), 6-30-77; SF_2(g), 6-30-76.
5. H. S. W. Massey, "Negative Ions," 3rd ed., Cambridge University Press, Cambridge, 1976.
6. M. S. Wei, J. H. Current, and J. Gendell, J. Chem. Phys. 52, 1592 (1970).
7. W. Nelson, G. Jackel, and W. Gordy, J. Chem. Phys. 52, 4572 (1970).

Phosphorus Fluoride, Ion (PF_2^-) $F_2P_1^-(g)$

Enthalpy Reference Temperature = T_r = 298.15 K Standard State Pressure = p° = 0.1 MPa

T/K	$C_p°$ (J $K^{-1}mol^{-1}$)	S°	$-[G°-H°(T_r)]/T$	$H°-H°(T_r)$ (kJ mol^{-1})	$\Delta_f H°$	$\Delta_f G°$	Log K_f
0	0.	0.	INFINITE	-11.215	-630.010		
100	34.601	215.028	293.664	-7.864			
200	39.722	240.527	261.304	-4.155			
250	42.421	249.683	258.088	-2.101			
298.15	44.817	257.364	257.364	0.	-639.177	-636.946	111.590
300	44.903	257.641	257.365	0.083	-639.235	-636.932	110.900
350	47.035	264.728	257.919	2.383	-641.499	-636.355	94.971
400	48.802	271.128	259.177	4.780	-643.087	-635.512	82.989
450	50.242	276.962	260.834	7.258	-644.632	-634.471	73.648
500	51.412	282.318	262.718	9.800	-646.144	-633.261	66.156
600	53.142	291.855	266.799	15.034	-649.096	-630.406	54.882
700	54.315	300.141	270.983	20.410	-651.982	-627.062	46.792
800	55.137	307.450	275.094	25.885	-654.828	-623.308	40.698
900	55.729	313.980	279.058	31.430	-657.648	-619.198	35.937
1000	56.169	319.876	282.850	37.026	-660.453	-614.775	32.113
1100	56.503	325.246	286.464	42.660	-663.250	-610.072	28.970
1200	56.762	330.174	289.904	48.324	-729.615	-604.036	26.293
1300	56.966	334.725	293.179	54.011	-731.634	-593.489	23.847
1400	57.131	338.953	296.299	59.716	-733.661	-582.787	21.744
1500	57.265	342.900	299.276	65.436	-735.697	-571.939	19.917
1600	57.375	346.599	302.119	71.168	-737.742	-560.955	18.313
1700	57.467	350.080	304.839	76.910	-739.798	-549.843	16.895
1800	57.545	353.367	307.444	82.661	-741.862	-538.610	15.630
1900	57.611	356.480	309.944	88.419	-743.933	-527.262	14.495
2000	57.667	359.437	312.345	94.183	-746.010	-515.804	13.471
2100	57.716	362.252	314.655	99.952	-748.090	-504.243	12.542
2200	57.759	364.938	316.880	105.726	-750.170	-492.582	11.695
2300	57.796	367.506	319.026	111.504	-752.246	-480.827	10.920
2400	57.829	369.966	321.098	117.285	-754.316	-468.982	10.207
2500	57.857	372.328	323.100	123.069	-756.376	-457.050	9.550
2600	57.883	374.597	325.037	128.856	-758.422	-445.037	8.941
2700	57.906	376.782	326.914	134.646	-760.452	-432.945	8.376
2800	57.927	378.889	328.732	140.437	-762.462	-420.778	7.850
2900	57.945	380.922	330.497	146.231	-764.450	-408.540	7.359
3000	57.962	382.886	332.211	152.026	-766.413	-396.234	6.899
3100	57.977	384.787	333.877	157.823	-768.350	-383.863	6.468
3200	57.990	386.628	335.496	163.622	-770.257	-371.429	6.063
3300	58.003	388.413	337.073	169.421	-772.135	-358.937	5.681
3400	58.014	390.145	338.609	175.222	-773.981	-346.388	5.322
3500	58.025	391.826	340.105	181.024	-775.795	-333.785	4.981
3600	58.034	393.461	341.565	186.827	-777.576	-321.131	4.659
3700	58.043	395.051	342.989	192.631	-779.324	-308.428	4.354
3800	58.051	396.599	344.379	198.436	-781.038	-295.678	4.064
3900	58.059	398.107	345.738	204.241	-782.718	-282.883	3.789
4000	58.066	399.577	347.066	210.047	-784.365	-270.046	3.526
4100	58.072	401.011	348.364	215.854	-785.978	-257.168	3.276
4200	58.078	402.411	349.634	221.662	-787.559	-244.251	3.038
4300	58.084	403.777	350.877	227.470	-789.107	-231.297	2.810
4400	58.089	405.113	352.095	233.279	-790.622	-218.307	2.592
4500	58.094	406.418	353.288	239.088	-792.106	-205.283	2.383
4600	58.099	407.695	354.457	244.897	-793.560	-192.226	2.183
4700	58.103	408.945	355.603	250.708	-794.983	-179.138	1.991
4800	58.107	410.168	356.727	256.518	-796.377	-166.020	1.807
4900	58.111	411.366	357.830	262.329	-797.743	-152.874	1.630
5000	58.114	412.540	358.912	268.140	-799.081	-139.700	1.459
5100	58.118	413.691	359.975	273.952	-800.391	-126.499	1.296
5200	58.121	414.820	361.019	279.764	-801.676	-113.272	1.138
5300	58.124	415.927	362.044	285.576	-802.935	-100.022	0.986
5400	58.127	417.013	363.052	291.389	-804.169	-86.748	0.839
5500	58.129	418.080	364.043	297.201	-805.380	-73.451	0.698
5600	58.132	419.127	365.018	303.014	-806.567	-60.133	0.561
5700	58.134	420.156	365.976	308.828	-807.732	-46.793	0.429
5800	58.137	421.167	366.919	314.641	-808.875	-33.434	0.301
5900	58.139	422.161	367.847	320.455	-809.998	-20.054	0.178
6000	58.141	423.138	368.760	326.269	-811.100	-6.656	0.058

PREVIOUS: June 1977 (1 atm) CURRENT: June 1977 (1 bar)

Phosphorus Fluoride, Ion (PF_2^-) $F_2P_1^-(g)$

LEAD FLUORIDE, ALPHA (α-PbF_2) CRYSTAL M_r = 245.196805

Lead Fluoride, Alpha (PbF_2) $F_2Pb_1(cr)$

$S°(298.15\ K) = [-112.968 \pm 8.4]\ J\ K^{-1}\ mol^{-1}$

$T_{trs} = 583 \pm 1\ K$

$\Delta_f H°(0\ K)$ = Unknown

$\Delta_f H°(298.15\ K) = [-676.971 \pm 4.2]\ kJ\ mol^{-1}$

$\Delta_{trs} H° = 1.463\ kJ\ mol^{-1}$

Enthalpy of Formation

Enthalpies of formation derived from the enthalpy changes of different chemical reactions are tabulated below. Due to a lack of sample characterization we can only assume from sample preparation that the following reactions involve α-$PbF_2(cr)$.

Source	Reaction	Method	$\Delta_r H°(298.15\ K)$ kcal mol^{-1}	$\Delta_f H°(298.15\ K)$ kcal mol^{-1}
(1)	$Pb(cr) + F_2(g) = PbF_2(\alpha)$	emf	-159.7±0.5	-159.7±0.5
(2)	$PbO(cr) + 2\ HF(40\%\ aq) = PbF_2(\alpha) + H_2O(\ell)$	calor.	- 22.9±0.3	-159.4±0.5
(3)	$PbF_2(\alpha) + 2\ Na(cr) = Pb(cr) + 2\ NaF(aq)$	emf	-112.2±0.2	-162.2±0.4
(4)	$PbF_2(\alpha) + Mg(cr) = Pb(cr) + MgF_2(cr)$	calor.	-109.5±1.5	-159.2±2.0
(5)	$3/2\ PbF_2(cr) + Al(cr) = 3/2\ Pb(cr) + AlF_3(cr)$	calor.	-118.5±0.1	-161.6±1.0
(6)	$3/2\ PbF_2(cr) + Al(cr) = 3/2\ Pb(cr) + AlF_3(cr)$	calor.	-117.7±0.6	-162.2±1.0

The samples used by Gross et al. (4, 5) could easily be β-PbF_2 rather than α-PbF_2. Other reactions (7, 8, 9) have led to values near $\Delta_f H°(298.15\ K) = -161.8\ kcal\ mol^{-1}$ (10); this value is adopted and is assigned an uncertainty of ±2 kcal mol^{-1}.

The adopted enthalpy of formation is supported by measurements (11) of the reaction $PbF_2(cr) + Be(cr) = BeF_2(cr) + Pb(cr)$ from which the enthalpy of formation of $BeF_2(cr)$ agrees with that from the direct fluorination of beryllium metal if the adopted value for $\Delta_f H°(PbF_2, \alpha, 298.15\ K)$ is used.

Heat Capacity and Entropy

The enthalpy of α-$PbF_2(cr)$ was measured by Linsey (12). Twenty data points (320-728 K) were obtained using a sample encapsulated in gold while fifteen were obtained (370-595 K) using a sample encapsulated in Nichrome V. We adopt the heat capacity values calculated by Linsey (12) and extrapolate linearly to 1000 K. It should be noted, however, that Linsey (12) adjusted his experimental enthalpy for α-$PbF_2(cr)$ so that when the orthorhombic sample was heated above T_{trs}, the resulting enthalpy was consistent with the cubic PbF_2 enthalpy.

There is no low temperature heat capacity data for α-$PbF_2(cr)$. We estimate $S°(298.15\ K) = 27.0 \pm 2.0\ cal\ K^{-1}\ mol^{-1}$.

Transition Data

The transition temperature is that obtained by Linsey (12) via enthalpy measurements on orthorhombic PbF_2. When heated above T_{trs}, orthorhombic PbF_2 changes quickly to the cubic form. Upon cooling the transformation is very sluggish.

The heat of transition at room temperature, 240±9 kcal mol^{-1}, was determined by Linsey (12) by the differences in the heat of solution of the orthorhombic and cubic modifications in 1.0 M $Fe(NO_3)_3$. $\Delta_{trs}H°$ is calculated from this value using the enthalpy differences H°(583 K)-H°(298.15 K) for the cubic and orthorhombic modifications.

References

1. W. Jahn-Held and K. Jellinek, Z. Elektrochem. 42, 401 (1936).
2. H. V. Wartenberg, Z. Anorg. Allg. Chem. 244, 337 (1940).
3. R. W. Ivett and T. D. Vries, J. Amer. Chem. Soc. 63, 2821 (1941).
4. P. Gross, C. Hayman, and D. L. Levi, Trans. Faraday Soc. 50, 477 (1954).
5. P. Gross, C. Hayman, and D. L. Levi, "Physical Chemistry of Process Metallurgy," Part 2, Interscience Publishers, Inc., New York, 1961.
6. V. P. Kolesov, A. M. Martynov, and S. M. Skuratov, Russ. J. Inorg. Chem. 6, 1326 (1961).
7. H. H. Broene and T. DeVries, J. Amer. Chem. Soc. 69, 1644 (1947).
8. G. G. Koerber and T. DeVries, J. Amer. Chem. Soc. 74, 5008 (1952).
9. P. Schindler and M. Reinert, Chimia (Switz) 15, 534 (1961).
10. E. Rudzitis, H. M. Feder, and W. N. Hubbard, Inorg. Chem. 6, 1716 (1967).
11. P. Gross, C. Hayman, and J. T. Bingham, Fulmer Research Institute, AD 721476, 1971.
12. C. W. Linsey, Ph.D. Dissertation, North Texas State University, 1970.

Enthalpy Reference Temperature = T_r = 298.15 K; Standard State Pressure = p° = 0.1 MPa

T/K	$C_p°$ (J $K^{-1}mol^{-1}$)	S° (J $K^{-1}mol^{-1}$)	$-[G°-H°(T_r)]/T$ (J $K^{-1}mol^{-1}$)	$H°-H°(T_r)$ (kJ mol^{-1})	$\Delta_f H°$ (kJ mol^{-1})	$\Delta_f G°$ (kJ mol^{-1})	Log K_f
0							
100							
200							
298.15	72.258	112.968	112.968	0.	-676.971	-630.875	110.527
300	72.383	113.415	112.969	0.134	-676.945	-630.589	109.795
400	76.065	134.642	115.838	7.522	-675.505	-615.351	80.357
500	79.705	152.035	121.389	15.323	-673.883	-600.497	62.733
583.000	81.965	164.425	126.653	22.021	--------	ALPHA <--> BETA	-------
600	82.488	166.788	127.757	23.419	-672.150	-585.980	51.014
700	85.772	179.751	134.276	31.832	-675.110	-570.968	42.606
800	89.056	191.418	140.701	40.574	-672.996	-556.231	36.318
900	92.341	202.097	146.938	49.643	-670.565	-541.778	31.444
1000	95.625	211.996	152.955	59.042	-667.812	-527.613	27.560

PREVIOUS: June 1969 CURRENT: December 1973

Lead Fluoride, Alpha (PbF_2) $F_2Pb_1(cr)$

LEAD FLUORIDE, BETA (β-PbF_2) CRYSTAL M_r = 245.196805

$S°$(298.15 K) = [114.398] J K^{-1} mol^{-1}

T_{trs} = 583 ± 1 K

T_{fus} = 1103 ± 2 K

$\Delta_f H°$(298.15 K) = [-675.990] kJ mol^{-1}

$\Delta_{trs} H°$ = 1.463 kJ mol^{-1}

$\Delta_{fus} H°$ = 14.728 ± 1.3 kJ mol^{-1}

Enthalpy of Formation

Linsey (1) determined the heat of transition at room temperature to be 240±9 kcal mol^{-1} from heat of solution measurements of the orthorhombic and cubic modifications in 1 M $Fe(NO_3)_3$. We calculate and adopt $\Delta_f H°$(PbF_2, β, 298.15 K) by adding this heat of transition value to the $\Delta_f H°$(298.15 K) value for PbF_2(cr, α).

Other enthalpy of formation values derived from the enthalpy changes of different chemical reactions are tabulated below. Due to a lack of sample characterization we can only assume from the sample preparation and experimental conditions that the following reactions involve β-PbF_2(cr).

Source	Reaction	Method	$\Delta_r H°$(298.15 K) kcal mol^{-1}	$\Delta_f H°$(298.15 K) kcal mol^{-1}
(2)	$PbF_2(\beta) + H_2(g) = Pb(\ell) + 2\ HF(g)$	Press	32.0	-161.3
(3)	$PbF_2(\beta) + H_2O(g) = PbO(cr) + 2\ HF(g)$	Press	30.06	-154.6
(4)	$1/2\ PbF_2(\beta) + 1/3\ Al(cr) = 1/2\ Pb(\ell) + 1/3\ AlF_3(cr)$	emf	-116.39	-162.0

The two studies by Gross et al. (5, 6) may well involve β-PbF_2. Kolesov et al (7) used two PbF_2 samples which had been fused. These samples should then be in the β-modification. The results are considerably different from those given in the α-PbF_2 table (8) for α-PbF_2.

Heat Capacity and Entropy

Linsey (1) measured the enthalpy of cubic PbF_2. The enthalpy data (61 data points, 321-1164 K) revealed an anomaly in the enthalpy and heat capacity in the range 573-823 K. Linsey (1) interpreted the results in terms of a lambda-type (cooperative) transition in the heat capacity with a maximum at 716 K. We adopt the heat capacity values calculated and reported by Linsey (1).

Banashek et al. (8) measured the high temperature enthalpies in the range 670-1165 K. Their data is consistent with the interpretation of Linsey (1).

The entropy is calculated in a manner analogous to that used for the enthalpy of formation.

Fusion Data

See the liquid table for details.

References

1. C. W. Linsey, Ph.D. Dissertation, North Texas State University, 1970.
2. K. Jellinek and A. Rudat, Z. Anorg. Chem. 175, 281 (1928).
3. M. L. Domange, Compt. Rend. 200, 2180 (1935).
4. R. J. Heus and J. J. Egan, Z. Phys. Chem. 49, S. 38 (1966).
5. P. Gross, C. Hayman, and D. L. Levi, Trans. Faraday Soc. 50, 477 (1954).
6. P. Gross, C. Hayman, and D. L. Levi, "Physical Chemistry of Process Metallurgy," Part 2, Interscience Publishers, Inc., New York, 1961.
7. V. P. Kolesov, A. M. Martynov, and S. M. Skuratov, Russ. J. Inorg. Chem. 6, 1326 (1961).
8. E. I. Banashek, N. N. Patsukova, and I. S. Rassonskaya, Izvest. Sektora Fiz.-Khim. Anal., Inst. Obshchei i Neorg. Khim., Akad. Nauk SSSR 27, 223 (1956).

Lead Fluoride, Beta (PbF_2) F_2Pb_1(cr)

Enthalpy Reference Temperature = T_r = 298.15 K Standard State Pressure = p° = 0.1 MPa

T/K	$C_p°$ (J $K^{-1}mol^{-1}$)	S° (J $K^{-1}mol^{-1}$)	$-[G°-H°(T_r)]/T$ (J $K^{-1}mol^{-1}$)	$H°-H°(T_r)$ (kJ mol^{-1})	$\Delta_f H°$ (kJ mol^{-1})	$\Delta_f G°$ (kJ mol^{-1})	Log K_f
0							
100							
200							
298.15	74.266	114.398	114.398	0.	-675.990	-630.320	110.429
300	74.392	114.857	114.399	0.138	-675.960	-630.037	109.699
400	76.776	136.571	117.344	7.691	-674.354	-614.971	80.307
500	80.584	154.051	122.988	15.531	-672.694	-600.315	62.714
583.000	88.868	166.934	128.336	22.502	-------- ALPHA <--> BETA -------		
600	91.128	169.518	129.466	24.031	-670.557	-586.025	51.018
700	147.486	186.826	136.333	35.345	-670.616	-571.426	42.640
800	105.939	203.976	143.801	48.140	-664.448	-557.729	36.416
900	94.433	215.393	151.145	57.823	-661.404	-544.583	31.607
1000	94.433	225.342	158.076	67.266	-658.606	-531.753	27.776
1100	94.433	234.342	164.607	76.709	-655.807	-519.203	24.655
1103.000	94.433	234.600	164.797	76.993	------- BETA <--> LIQUID -------		
1200	94.433	242.559	170.765	86.153	-653.003	-506.908	22.065
1300	94.433	250.118	176.583	95.596	-650.200	-494.847	19.883
1400	94.433	257.116	182.088	105.039	-647.410	-483.002	18.021
1500	94.433	263.631	187.310	114.482	-644.642	-471.355	16.414

PREVIOUS: CURRENT: December 1973

Lead Fluoride, Beta (PbF_2) F_2Pb_1(cr)

LEAD FLUORIDE (PbF_2) LIQUID M_r = 245.196806

$S°(298.15\ K) = [130.927]\ J\ K^{-1}\ mol^{-1}$

$T_{fus} = 1103 \pm 4\ K$

$\Delta_f H°(298.15\ K) = [-660.320]\ kJ\ mol^{-1}$

$\Delta_{fus} H° = 14.728 \pm 2.1\ kJ\ mol^{-1}$

Enthalpy of Formation

$\Delta_f H°(PbF_2, \ell, 298.15\ K)$ is calculated from $\Delta_f H°(PbF_2, cr, \beta, 298.15\ K)$ by adding the enthalpy of fusion, $\Delta_{fus} H°$, and the difference in enthalpy, H°(1103 K)-H°(298.25 K), between the β - crystal and liquid.

Heat Capacity and Entropy

The enthalpies of $PbF_2(\ell)$ in the temperature range 1104-1165 K were determined by Banashek et al (1). Based on their reported results, the heat capacity of $PbF_2(\ell)$ is evaluated to be 24 ± 2 cal $K^{-1}\ mol^{-1}$ by the least square method. Linsey (13) measured the enthalpy of $PbF_2(\ell)$ in the range 1105-1164 K (8 data points) and reported a constant heat capacity value of 26.10 cal $K^{-1}\ mol^{-1}$. We adopt this value and extrapolate to 700 K and 1500 K. A glass transition is assumed at 700 K below which the heat capacity values of cubic PbF_2 are used.

S°(298.15 K) is calculated in a manner analogous to that used for the enthalpy of formation.

Fusion Data

T_{fus} and $\Delta_{fus}H°$ are taken from the enthalpy data of Linsey (13). Our analysis of the enthalpy data of Banashek et al. (1) yields a heat of melting value of 2.98 kcal mol^{-1}.

Other values of T_{fus}(K) reported are 1091 (2, 11), 1097 (3, 4, 5) and 1128 (6). Using the freezing point data of binary systems PbF_2-$PbBr_2$ (5), PbF_2-$PbCl_2$ (5, 7), PbF_2-PbI_2 (5, 8), PbF_2-PbO (9), and PbF_2-NaF (10), Kelley (3) evaluated the corresponding values of $\Delta_{fus}H°$ kcal mol^{-1} as 1.70, 1.70, 1.96, 3.87, and 2.47. Other reported values of $\Delta_{fus}H°$ are 3.00 (1) and 4.16 (2) kcal mol^{-1}.

Vaporization Data

From mass spectroscopic study on the vapor species in PbF_2, Zmbov, et al. (4) have shown that molten PbF_2 vaporizes partly as molecular PbF_2. There is considerable disproportionation, and at 988 K the vapor composition is approximately 66% PbF_4, 25% PbF_2, 2% PbF, and 7% Pb (4, 12). T_{vap} has been reported as 1565 K (11) where the vapor is expected to be a mixture of the above species.

References

1. E. I. Banashek, N. N. Patsukova, and I. S. Rassonskaya, Izvest. Sektora Fiz. Khim. Anal., Inst. Obshchei i Neorg. Khim., Akad. Nauk SSSR 27, 223 (1956).
2. M. Blanc, Diplome and d'Etudes Superieures, Paris, 1956, quoted by M. Blanc and G. Petit, Compt. Rend. 248, 1305 (1959).
3. K. K. Kelley, U. S. Bur. Mines Bull. 393, 1936.
4. K. Zmbov, J. W. Hastie, and J. L. Margrave, Trans. Faraday Soc. 64, 861 (1968).
5. C. Sandonini, Gazz. Chim. Ital. 41, II 144 (1911).
6. Amadori, Gazz. Chim. Ital. 49, I 38 (1919).
7. C. Sandonini, Atti Acad. Lincei, Ser. 5, 20, I 172 (1911).
8. C. Sanconini, Atti Acad. Lincei, Ser. 5, 20, I 253 (1911).
9. C. Sandonini, Atti Acad. Lincei, Ser. 5, 23, I 959 (1914).
10. N. A. Puschin and A. B. Baskow, J. Russ. Phys. Chem. Soc. 45, 82 (1913); Z. Anorg. Chem. 81, 347 (1913).
11. H. V. Wartenberg and O. Bosse, Z. Elektrochem. 28, 384 (1922).
12. R. H. Hauge, J. W. Hastie, and J. L. Margrave, J. Phys. Chem. 72, 3510 (1968).
13. C. W. Linsey, Ph.D. Dissertation, North Texas State University, 1970.

Lead Fluoride (PbF_2) $F_2Pb_1(l)$

Enthalpy Reference Temperature = T_r = 298.15 K — Standard State Pressure = p° = 0.1 MPa

T/K	$C_p°$	S°	$-[G°-H°(T_r)]/T$	$H°-H°(T_r)$	$\Delta_f H°$	$\Delta_f G°$	Log K_f
	J $K^{-1}mol^{-1}$			kJ mol^{-1}			
0							
100							
200							
298.15	74.266	130.927	130.927	0.	-660.320	-619.579	108.548
300	74.392	131.386	130.928	0.138	-660.290	-619.326	107.834
400	76.776	153.101	133.873	7.691	-658.685	-605.914	79.124
500	79.705	170.548	139.517	15.515	-657.040	-592.909	61.941
600	82.634	185.339	145.951	23.632	-655.286	-580.246	50.515
700	85.563	198.297	152.522	32.042	-658.249	-567.089	42.317
700.000	85.563	198.297	152.522	32.042	GLASS <--> LIQUID TRANSITION		
700.000	109.202	198.297	152.522	32.042			
800	109.202	212.879	159.176	42.962	-653.956	-554.359	36.196
900	109.202	225.741	165.871	53.883	-649.675	-542.167	31.467
1000	109.202	237.247	172.444	64.803	-645.399	-530.451	27.708
1100	109.202	247.655	178.815	75.723	-641.123	-519.163	24.653
1103.000	109.202	247.952	179.003	76.051	BETA <--> LIQUID		
1200	109.202	257.157	184.954	86.643	-636.843	-508.265	22.124
1300	109.202	265.897	190.848	97.564	-632.563	-497.723	19.999
1400	109.202	273.990	196.502	108.484	-628.296	-487.511	18.189
1500	109.202	281.524	201.922	119.404	-624.050	-477.604	16.632

PREVIOUS: June 1969 CURRENT: December 1973

Lead Fluoride (PbF_2) $F_2Pb_1(l)$

LEAD FLUORIDE (PbF_2) ALPHA-BETA-LIQUID M_r = 245.196806 Lead Fluoride (PbF_2) F_2Pb_1(cr,l)

0 to 583 K crystal, alpha
583 to 1103 K crystal, beta
above 1103 K liquid

Refer to the individual tables for details.

Enthalpy Reference Temperature = T_r = 298.15 K; Standard State Pressure = p° = 0.1 MPa

T/K	C_p° (J K⁻¹mol⁻¹)	S° (J K⁻¹mol⁻¹)	−[G°−H°(T_r)]/T (J K⁻¹mol⁻¹)	H°−H°(T_r) (kJ mol⁻¹)	$\Delta_f H^\circ$ (kJ mol⁻¹)	$\Delta_f G^\circ$ (kJ mol⁻¹)	Log K_f
0							
100							
200							
298.15	72.258	112.968	112.968	0.	-676.971	-630.875	110.527
300	72.383	113.415	112.969	0.134	-676.945	-630.589	109.795
400	76.065	134.642	115.838	7.522	-675.505	-615.351	80.357
500	79.705	152.035	121.389	15.323	-673.883	-600.497	62.733
583.000	81.965	164.425	126.653	22.021	ALPHA <--> BETA	TRANSITION	
583.000	88.868	166.934	126.653	23.484			
600	91.128	169.518	127.831	25.012	-670.557	-586.025	51.018
700	147.486	186.826	134.931	36.327	-670.616	-571.426	42.640
800	105.939	203.976	142.574	49.121	-664.448	-557.729	36.416
900	94.433	215.393	150.055	58.804	-661.404	-544.583	31.607
1000	94.433	225.342	157.095	68.248	-658.606	-531.753	27.776
1100	94.433	234.343	163.714	77.691	-655.807	-519.203	24.655
1103.000	94.433	234.600	163.907	77.974	BETA <--> LIQUID	TRANSITION	
1103.000	109.202	247.952	163.907	92.702			
1200	109.202	257.157	171.078	103.294	-636.843	-508.265	22.124
1300	109.202	265.897	178.040	114.215	-632.563	-497.723	19.999
1400	109.202	273.990	184.608	125.135	-628.296	-487.511	18.189
1500	109.202	281.524	190.821	136.055	-624.050	-477.604	16.632

PREVIOUS: CURRENT: December 1973

Lead Fluoride (PbF_2) F_2Pb_1(cr,l)

LEAD FLUORIDE (PbF_2) IDEAL GAS M_r = 245.196806

$S°$(298.15 K) = 292.692 J K^{-1} mol^{-1}

$\Delta_f H°$(0 K) = -431.999 ± 8.4 kJ mol^{-1}

$\Delta_f H°$(298.15 K) = -435.136 ± 8.4 kJ mol^{-1}

Vibrational Frequencies and Degeneracies

ν, cm^{-1}
545.7(1)
170 (1)
522.5(1)

Ground State Quantum Weight: [1] $\sigma = 2$

Point Group: C_{2v}

Bond Distance: Pb-F = 2.13 ± 0.02 Å

Bond Angle: F-Pb-F = [90°]

Product of the Moments of Inertia: $I_A I_B I_C$ = [4.571688 x 10^{-114}] g^3 cm^6

Enthalpy of Formation

Vapor pressure data for PbF_2 have previously been interpreted on the assumption that PbF_2 was the only species present in the gas phase (1, 2). Zmbov, et al. (3) have studied the vapor species over PbF_2(cr, ℓ) mass spectrometrically and shown that PbF_2 vaporizes only partly as molecular PbF_2 and there is considerable disproportionation. The vapor composition at 988 K is approximately 66% PbF_4, 25% PbF_2, 2% PbF, and 7% Pb (3, 4). From extrapolation of the ion intensity data from 1043 to 988 K to the temperature region in which Nesmeyanov and Iofa (2) measured the saturated vapor pressures of PbF_2(cr), 792-988 K, by Knudsen effusion method, the partial pressure of PbF_2(g), at 988 K was calculated to be 0.0435±0.005 torr. Based on this calculated vapor pressure, the enthalpy of sublimation $\Delta_{sub}H°$(298.15 K) of PbF_2(cr, β) is derived as 57.53 kcal mol^{-1} by the 3rd law method, yielding $\Delta_f H°$(PbF_2, g, 298.15 K) = -104.0±2 kcal mol^{-1}.

Due to lack of data on vapor compositions over PbF_2 at various temperatures, the total pressure measurements reported by Wartenberg and Bosse (1), using boiling point method, and those of Nesmeyanov and Iofa (2) are not used for evaluation.

The adopted $\Delta_f H°$(298.15 K) leads to an atomization energy $\Delta_{at}H°$(0 K) = 8.10 eV for the process PbF_2(g) = Pb(g) + 2F(g), using auxiliary data (5).

Heat Capacity and Entropy

The bond distance was determined by Akishin, et al. (6). The bond angle was estimated by Hauge et al. (7) from a Group IV A difluoride comparison. Hauge et al. (7) determined the vibrational frequencies from the infrared spectra of neon matrix isolated PbF_2(g).

References

1. H. V. Wartenberg and O. Bosse, Z. Elektrochem. 28, 384 (1922).
2. A. N. Nesmeyanov and B. Z. Iofa, Zhur. Neorg. Khim. 4, 486 (1959).
3. K. Zmbov, J. W. Hastie, and J. L. Margrave, Trans. Faraday Soc. 64, 861 (1968).
4. R. H. Hauge, J. W. Hastie, and J. L. Margrave, J. Phys. Chem. 72, 3510 (1968).
5. JANAF Thermochemical Tables: Pb(g), 3-31-62; F(g), 9-30-65.
6. P. A. Akishin, V. P. Spiridonov, and A. N. Khodchenkov, Zhur. Fiz. Khim. 32, 1679 (1958).
7. R. H. Hauge, J. W. Hastie, and J. L. Margrave, J. Mol. Spectrosc. 45, 420 (1973).

Lead Fluoride (PbF_2) F_2Pb_1(g)

Enthalpy Reference Temperature = T_r = 298.15 K Standard State Pressure = p° = 0.1 MPa

T/K	$C_p°$ (J K⁻¹mol⁻¹)	S° (J K⁻¹mol⁻¹)	-[G°-H°(T_r)]/T (J K⁻¹mol⁻¹)	H°-H°(T_r) (kJ mol⁻¹)	$\Delta_f H°$ (kJ mol⁻¹)	$\Delta_f G°$ (kJ mol⁻¹)	Log K_f
0	0.	0.	INFINITE	-12.567	-431.999	-431.999	INFINITE
100	38.879	244.078	334.499	-9.042	-433.148	-436.836	228.180
200	46.110	273.286	297.217	-4.786	-434.346	-440.018	114.921
250	48.928	283.892	293.521	-2.407	-434.774	-441.385	92.222
298.15	50.950	292.692	292.692	0.	-435.136	-442.625	77.546
300	51.016	293.008	292.693	0.094	-435.149	-442.671	77.076
350	52.537	300.992	293.320	2.685	-435.501	-443.897	66.248
400	53.653	308.084	294.731	5.341	-435.850	-445.073	58.121
450	54.485	314.454	296.575	8.046	-436.210	-446.204	51.794
500	55.116	320.229	298.656	10.786	-436.585	-447.295	46.729
600	55.988	330.361	303.119	16.345	-437.389	-449.362	39.120
700	56.542	339.036	307.645	21.974	-443.134	-450.491	33.616
800	56.914	346.612	312.053	27.648	-444.086	-451.477	29.478
900	57.175	353.332	316.273	33.353	-445.021	-452.344	26.253
1000	57.365	359.366	320.286	39.080	-445.938	-453.109	23.668
1100	57.506	364.840	324.091	44.824	-446.838	-453.782	21.548
1200	57.615	369.849	327.699	50.580	-447.722	-454.374	19.778
1300	57.700	374.464	331.121	56.346	-448.596	-454.893	18.278
1400	57.768	378.743	334.371	62.120	-449.476	-455.345	16.989
1500	57.823	382.730	337.464	67.900	-450.371	-455.733	15.870
1600	57.869	386.464	340.411	73.684	-451.287	-456.060	14.889
1700	57.906	389.973	343.224	79.473	-452.229	-456.330	14.021
1800	57.938	393.284	345.914	85.265	-453.199	-456.543	13.249
1900	57.964	396.417	348.490	91.060	-454.195	-456.702	12.556
2000	57.987	399.391	350.962	96.858	-455.219	-456.808	11.931
2100	58.007	402.220	353.336	102.658	-633.488	-449.747	11.187
2200	58.024	404.919	355.620	108.459	-634.169	-440.981	10.470
2300	58.039	407.499	357.820	114.263	-634.935	-432.183	9.815
2400	58.052	409.969	359.941	120.067	-635.785	-423.350	9.214
2500	58.064	412.339	361.990	125.873	-636.717	-414.480	8.660
2600	58.074	414.617	363.971	131.680	-637.725	-405.570	8.148
2700	58.083	416.809	365.887	137.488	-638.805	-396.621	7.673
2800	58.092	418.921	367.744	143.297	-639.950	-387.631	7.231
2900	58.099	420.960	369.544	149.106	-641.153	-378.598	6.819
3000	58.106	422.930	371.291	154.916	-642.404	-369.523	6.434
3100	58.112	424.835	372.988	160.727	-643.697	-360.406	6.073
3200	58.117	426.680	374.637	166.539	-645.023	-351.247	5.734
3300	58.122	428.469	376.241	172.351	-646.373	-342.045	5.414
3400	58.127	430.204	377.803	178.163	-647.739	-332.803	5.113
3500	58.131	431.889	379.324	183.976	-649.113	-323.520	4.828
3600	58.135	433.526	380.807	189.789	-650.488	-314.198	4.559
3700	58.138	435.119	382.254	195.603	-651.857	-304.837	4.304
3800	58.141	436.670	383.665	201.417	-653.213	-295.441	4.061
3900	58.144	438.180	385.044	207.231	-654.551	-286.008	3.831
4000	58.147	439.652	386.391	213.046	-655.865	-276.541	3.611
4100	58.150	441.088	387.707	218.860	-657.151	-267.042	3.402
4200	58.152	442.489	388.995	224.676	-658.404	-257.512	3.203
4300	58.154	443.858	390.255	230.491	-659.622	-247.953	3.012
4400	58.157	445.195	391.489	236.306	-660.800	-238.365	2.830
4500	58.158	446.502	392.697	242.122	-661.938	-228.751	2.655
4600	58.160	447.780	393.880	247.938	-663.033	-219.113	2.488
4700	58.162	449.031	395.041	253.754	-664.084	-209.451	2.328
4800	58.164	450.255	396.178	259.571	-665.089	-199.767	2.174
4900	58.165	451.455	397.294	265.387	-666.048	-190.063	2.026
5000	58.166	452.630	398.389	271.204	-666.961	-180.340	1.884
5100	58.168	453.782	399.464	277.020	-667.827	-170.599	1.747
5200	58.169	454.911	400.519	282.837	-668.648	-160.841	1.616
5300	58.170	456.019	401.556	288.654	-669.423	-151.068	1.489
5400	58.171	457.106	402.575	294.471	-670.153	-141.281	1.367
5500	58.172	458.174	403.576	300.288	-670.839	-131.481	1.249
5600	58.173	459.222	404.560	306.106	-671.483	-121.668	1.135
5700	58.174	460.252	405.528	311.923	-672.085	-111.845	1.025
5800	58.175	461.263	406.481	317.741	-672.647	-102.011	0.919
5900	58.176	462.258	407.418	323.558	-673.169	-92.168	0.816
6000	58.177	463.236	408.340	329.376	-673.654	-82.317	0.717

PREVIOUS: December 1973 (1 atm) CURRENT: December 1973 (1 bar)

Lead Fluoride (PbF_2) F_2Pb_1(g)

SULFUR FLUORIDE (SF_2) IDEAL GAS M_r = 70.056806

$\Delta_f H°(0\ K) = -294.629 \pm 16.7\ kJ\ mol^{-1}$
$S°(298.15\ K) = [257.703 \pm 0.084]\ J\ K^{-1}\ mol^{-1}$ $\Delta_f H°(298.15\ K) = -296.646 \pm 16.7\ kJ\ mol^{-1}$

Vibrational Frequencies and Degeneracies

ν, cm^{-1}
[840 ± 20] (1)
[357 ± 2] (1)
[809 ± 10] (1)

Ground State Quantum Weight = [1] $\sigma = 2$
Point Group: C_{2v}
Bond Distance: S–F = 1.59208 ± 0.00008 Å
Bond Angle: F–S–F = 98.197 ± 0.011°
Product of the Moments of Inertia: $I_A I_B I_C = 3.518837 \times 10^{-115}\ g^3\ cm^6$

Enthalpy of Formation

We calculate $\Delta_f H°$ from a 3rd law analysis of two sets of equilibrium data for the gaseous isomolecular reaction $S + SF_2 = 2\ SF$. The equilibrium data are obtained from ion currents reported in a mass spectrometric study (1) of the molecular species formed from the reaction of gaseous SF_6 with graphite. We do not include in the equilibrium constants a proportionality constant involving ion multiplier gain and ionization cross sections since Hildenbrand (1) has recently demonstrated that these factors tend to cancel for isomolecular reactions. Results of our analyses are presented below.

Series	Cell Configuration	Data Points	T/K	$\Delta_r H°$(298.15 K), kcal mol^{-1} 2nd law	3rd law	Drift cal K^{-1} mol^{-1}	$\Delta_f H°$(298.15 K) kcal mol^{-1}
2	C cell/wound Pt wire	4	1436–1564	3.1	10.68±0.7	5.0±2.3	70.77±4.0
3	C cell/Pt partition/packed with cloth and Pt wire	7	1478–1588	7.7	10.87±0.9	2.1±5.0	70.96±4.0

The two cell configurations used by Hildenbrand (1) yield nearly identical 3rd law $\Delta_f H°$ values; thus, we adopt the rounded average of −70.9±4.0 kcal mol^{-1}. Our adopted $\Delta_f H°$ value corresponds to a enthalpy of atomization and average bond dissociation energy of 172.9 kcal mol^{-1} and 86.4 kcal mol^{-1}, respectively. In addition, the primary bond dissociation energy of SF_2 is calculated to be $D_0 0$(SF-F) = 91.7 kcal mol^{-1}, suggesting that the individual S-F bonds are somewhat stronger than that in the SF (81.2 kcal mol^{-1}, 2) radical. This increase in bond strength is predicted by MO calculations (3) and has been qualitatively accounted for in terms of the valence-state model (1) of covalent bonding.

Two additional pieces of information exist which provide some support for our selected $\Delta_f H°$ value. Kinetic studies (4) on the dissociation of SF_3 indicate that $D_0^0(SF_2\text{-}F) \simeq 72$ kcal mol^{-1}. We calculate $D_0^0(SF_2\text{-}F) = 67.3$ kcal mol^{-1} from JANAF data (2), indicating that our heats of formation for SF_3 and SF_2 are consistent with these kinetic results. Di Lonardo and Trombetti (5) obtained D_0^0(SF-F) = 92.2±12 kcal mol^{-1} from electron-impact threshold measurements on the COS/F and SF_6 systems. This value is to be compared with our calculated value of 91.7 kcal mol^{-1}.

Heat Capacity and Entropy

We adopt molecular data which refer to an average ground state structure for SF_2. These results are based on microwave measurements and force field calculations performed by Kirchhoff et al. (6). Earlier microwave spectroscopic observations (7) on the isotopic $^{34}SF_2$ species support the assignment of the spectrum to the difluoride and confirm its angular structure. The individual moments of inertia are: $I_A = 3.1377 \times 10^{-39}$, $I_B = 9.1367 \times 10^{-39}$, and $I_C = 12.2744 \times 10^{-39}\ g\ cm^2$.

A recent review on the chemistry of the lower sulfur fluorides includes a discussion on the vibrational spectrum of SF_2 (8). The only observed vibrational frequency (830 cm^{-1}) which has so far been attributed to the difluoride appeared as a shoulder to a band at 810 cm^{-1} in the infrared spectra of a mixture of S-F compounds. The band at 810 cm^{-1} has been assigned (8) to the S-F equatorial stretching frequency of F_3SSF. We prefer not to adopt the 830 cm^{-1} frequency as a fundamental since the band cannot be positively identified as due to SF_2. Four sets (6, 9-11) of estimated frequencies have been reported. We believe that the best estimates are those of Kirchhoff et al. (6) which we adopt. These frequencies were obtained from force field calculations and include corrections which were made to account for the differences found between predicted and observed fundamentals for the related molecules of OF_2, SO_2, SiF_2, and CF_2. By analogy with other even electron S-F species (2), we predict that the ground state is singlet and that there are no low-lying electronic levels which would contribute to the partition function.

Our thermal functions essentially update those reported by Wilkins (10). Another set of functions which cover a rather short (0–2000 K) temperature interval has been reported by O'Hare (11). We believe, however, that his estimate (ν_2 = 523 cm^{-1}, 11) of the bending frequency is too high based on that observed for SO_2(ν_2 = 517 cm^{-1}, 2), indicating that his entropies are probably too low by as much as 0.6 – 0.8 cal K^{-1} mol^{-1}.

References

1. D. L. Hildenbrand, J. Phys. Chem. 77, 897 (1973).
2. JANAF Thermochemical Tables: S(g), 6-30-71, SF_6(g), SF_4(g), SF_3(g), and SF(g), 6-30-76, F(g), 9-30-65, SO_2(g), 6-30-6
3. A. L. Companion, Theoret. Chim. Acta, 25, 268 (1972).
4. J. F. Bott, J. Chem. Phys. 54, 181 (1971).
5. G. Di Lonardo and A. Trombetti, Trans. Faraday Soc. 66, 2694 (1970).
6. W. H. Kirchhoff, D. R. Johnson, and F. X. Powell, J. Mol. Spectro. 48, 157 (1973).
7. D. R. Johnson and F. X. Powell, Science, 164, 950 (1969).
8. F. Seel, in "Advances in Inorganic Chemistry and Radiochemistry", H. J. Emeleus and A. G. Sharpe, eds., Vol. 16, Academic Press, New York, 1974.
9. C. Bliefert and K. P. Wanczek, Z. Naturforsch. 259, 1770 (1970).
10. R. L. Wilkins, Aerospace Corporation, Report No. TR-0158 (3240-20)-19, July, 1968.
11. P. A. G. O'Hare, Argonne National Laboratory, ANL-7315, July, 1968.

Sulfur Fluoride (SF_2) $F_2S_1(g)$

Enthalpy Reference Temperature = T_r = 298.15 K Standard State Pressure = p° = 0.1 MPa

T/K	$C_p°$ (J $K^{-1}mol^{-1}$)	S°	−[G°−H°(T_r)]/T	H°−H°(T_r) (kJ mol^{-1})	$\Delta_f H°$	$\Delta_f G°$	Log K_f
0	0.	0.	INFINITE	-11.221	-294.629	-294.629	INFINITE
100	34.579	215.347	294.044	-7.870	-294.875	-298.121	155.722
200	39.755	240.841	261.651	-4.162	-295.738	-301.049	78.626
250	42.491	250.008	258.429	-2.105	-296.204	-302.322	63.167
298.15	44.906	257.703	257.703	0.	-296.646	-303.461	53.165
300	44.992	257.981	257.704	0.083	-296.662	-303.503	52.845
350	47.131	265.082	258.260	2.388	-297.113	-304.607	45.460
400	48.896	271.495	259.520	4.790	-299.772	-305.560	39.902
450	50.331	277.340	261.180	7.272	-300.882	-306.225	35.546
500	51.493	282.705	263.068	9.818	-302.037	-306.755	32.046
600	53.208	292.256	267.157	15.059	-303.855	-307.520	26.772
700	54.370	300.550	271.348	20.442	-305.370	-308.008	22.984
800	55.181	307.867	275.464	25.922	-306.711	-308.292	20.129
900	55.766	314.402	279.434	31.471	-361.073	-307.334	17.837
1000	56.200	320.301	283.231	37.070	-361.020	-301.365	15.742
1100	56.529	325.673	286.849	42.707	-360.974	-295.402	14.027
1200	56.784	330.603	290.292	48.373	-360.941	-289.443	12.599
1300	56.986	335.157	293.570	54.062	-360.924	-283.485	11.391
1400	57.148	339.386	296.694	59.769	-360.925	-277.529	10.355
1500	57.280	343.333	299.673	65.491	-360.948	-271.571	9.457
1600	57.388	347.034	302.519	71.224	-360.991	-265.611	8.671
1700	57.479	350.516	305.240	76.968	-361.056	-259.648	7.978
1800	57.555	353.803	307.848	82.720	-361.140	-253.681	7.362
1900	57.620	356.917	310.349	88.478	-361.243	-247.708	6.810
2000	57.676	359.874	312.752	94.243	-361.360	-241.729	6.313
2100	57.724	362.689	315.064	100.013	-361.490	-235.745	5.864
2200	57.766	365.375	317.290	105.788	-361.627	-229.754	5.455
2300	57.802	367.944	319.437	111.566	-361.769	-223.756	5.082
2400	57.835	370.405	321.510	117.348	-361.911	-217.753	4.739
2500	57.863	372.766	323.513	123.133	-362.049	-211.743	4.424
2600	57.888	375.036	325.451	128.921	-362.180	-205.728	4.133
2700	57.911	377.221	327.329	134.711	-362.298	-199.709	3.864
2800	57.931	379.328	329.148	140.503	-362.402	-193.685	3.613
2900	57.949	381.361	330.914	146.297	-362.487	-187.658	3.380
3000	57.966	383.326	332.628	152.093	-362.551	-181.628	3.162
3100	57.980	385.227	334.295	157.890	-362.592	-175.597	2.959
3200	57.994	387.068	335.915	163.689	-362.607	-169.564	2.768
3300	58.006	388.853	337.492	169.489	-362.594	-163.532	2.588
3400	58.017	390.584	339.029	175.290	-362.554	-157.500	2.420
3500	58.028	392.266	340.526	181.092	-362.483	-151.470	2.261
3600	58.037	393.901	341.986	186.895	-362.383	-145.442	2.110
3700	58.046	395.491	343.410	192.700	-362.252	-139.418	1.968
3800	58.054	397.040	344.802	198.505	-362.091	-133.398	1.834
3900	58.061	398.548	346.160	204.310	-361.899	-127.382	1.706
4000	58.068	400.018	347.488	210.117	-361.676	-121.371	1.585
4100	58.074	401.452	348.787	215.924	-361.424	-115.367	1.470
4200	58.080	402.851	350.058	221.732	-361.142	-109.369	1.360
4300	58.086	404.218	351.302	227.540	-360.831	-103.378	1.256
4400	58.091	405.553	352.519	233.349	-360.493	-97.394	1.156
4500	58.096	406.859	353.713	239.158	-360.127	-91.419	1.061
4600	58.100	408.136	354.882	244.968	-359.735	-85.452	0.970
4700	58.105	409.385	356.028	250.778	-359.318	-79.494	0.883
4800	58.109	410.609	357.153	256.589	-358.876	-73.544	0.800
4900	58.112	411.807	358.256	262.400	-358.411	-67.605	0.721
5000	58.116	412.981	359.339	268.211	-357.923	-61.675	0.644
5100	58.119	414.132	360.402	274.023	-357.414	-55.755	0.571
5200	58.122	415.260	361.446	279.835	-356.885	-49.845	0.501
5300	58.125	416.368	362.472	285.647	-356.336	-43.946	0.433
5400	58.128	417.454	363.480	291.460	-355.768	-38.057	0.368
5500	58.130	418.521	364.471	297.273	-355.182	-32.179	0.306
5600	58.133	419.568	365.446	303.086	-354.579	-26.311	0.245
5700	58.135	420.597	366.404	308.900	-353.960	-20.455	0.187
5800	58.138	421.608	367.347	314.713	-353.326	-14.610	0.132
5900	58.140	422.602	368.275	320.527	-352.677	-8.775	0.078
6000	58.142	423.579	369.189	326.341	-352.014	-2.952	0.026

PREVIOUS: June 1976 (1 atm) CURRENT: June 1976 (1 bar)

Sulfur Fluoride (SF_2) $F_2S_1(g)$

SULFUR FLUORIDE, ION (SF_2^+) IDEAL GAS M_r = 70.056257

$\Delta_fH°(0\ K) = 698.195 \pm 46.0\ kJ\ mol^{-1}$

$S°(298.15\ K) = [263.523 \pm 0.42]\ J\ K^{-1}\ mol^{-1}$ $\Delta_fH°(298.15\ K) = 702.389 \pm 46.0\ kJ\ mol^{-1}$

Electronic Levels and Quantum Weights

State	ε_i, cm^{-1}	g_i
$[^2B_1]$	[0]	[2]
$[^2A_1]$	[25000]	[2]
$[^2A_1]$	[28000]	[2]
$[^2A_2]$	[30000]	[2]

Vibrational Frequencies and Degeneracies

ν, cm^{-1}
[807] (1)
[352] (1)
[844] (1)

Point Group: $[C_{2v}]$ σ = [2]
Bond Distance: S-F = [1.59] Å
Bond Angle: F-S-F = [100]°
Product of the Moments of Inertia: $I_AI_BI_C = [3.4931 \times 10^{-115}]\ g^3\ cm^6$

Enthalpy of Formation

$\Delta_fH°(0\ K) = 166.9\pm11.0\ kcal\ mol^{-1}$ is calculated as the sum of $\Delta_fH°(SF_3, g, 0\ K) = -70.42\pm4\ kcal\ mol^{-1}$ (1) and $IP(SF_2) = 10.29\pm0.3$ eV (237.29 kcal mol^{-1}). IP is the electron-impact appearance potential of SF_2^+ which was measured by Hildenbrand (2) in a mass spectral study of the products formed from the reaction of SF_6 with C at about 1500 K. We increase the uncertainty in IP to ±0.3 eV from the value (±0.1 eV) assigned by Hildenbrand (2) since recent photoelectron (PE) spectrometric studies on the related dihalides SCl_2 (3) and SiF_2 (4) show that their adiabatic ionization potentials lie 0.22 and 0.40 eV, respectively, below the corresponding vertical values. The adopted value of IP is bracketed by the accurately known adiabatic values for SCl_2(9.45±0.05 eV, 3) and SiF_2(10.78±0.05 eV, 4) and is close to the IP for atomic sulfur (10.36 eV, 5). MO calculations (6) predict $IP(SF_2)$ = 9.6 eV.

Several fragmentation appearnce potentials for the ion-molecule systems SF_2^+/SF_6 (2, 7), SF_2^+/SF_4 (2), and SF_2^+/NSF_3 (8) have been reported and provide alternate paths leading to $\Delta_fH°$ once the ion formation processes are assigned. We use plausible reactions for the formation of SF_2^+ to calculate $\Delta_fH°$ values which cover the range 146-342 kcal mol^{-1}. The scatter in these values indicate that either the dissociative ionization processes have been incorrectly assigned, or perhaps more likely, the processes involve excess kinetic and/or excitation energies.

We find that the primary bond dissociation energy $D_0^0(SF^+-F)$ is 87.1 kcal mol^{-1} which when compared with D_0^0 values (1) for SF_2(91.7 kcal mol^{-1}) and SF_2^-(68.7 kcal mol^{-1}) indicates that the trend is $D_0^0(SF_2^+) \simeq D_0^0(SF_2) > D_0^0(SF_2^-)$. Similar trends in bond dissociation energies have been reported (9) for the ions and neutrals of phosphorus and nitrogen difluorides.

Heat Capacity and Entropy

According to the Walsh correlation diagram (10) for non-hydride AB_2 molecules, the most loosely bound pair of electrons in SF_2 is in a b_1'' orbital. The Walsh prediction is supported by the results of theoretical studies (3, 11, and 12) and assignments of the bands in the PE spectra (3, 12) for the isoelectronic molecules SCl_2, OF_2, and OCl_2. The b_1'' orbital is S-F antibonding and F-F bonding, but its binding energy (10) shows only a small positive angular dependence. We would expect then that removal of an electron from this orbital should be accompanied by a small increase in the bond angle. We adopt a bond angle of 100° which corresponds to roughly a 2° increase from that for SF_2 (1). The antibonding nature of the b_1'' orbital is also expected to produce upon ionization a decrease in the S-F bond length. However, a comparison of the S-F bond strengths in SF_2^+(D_0^0 = 87.1 kcal mol^{-1}) and SF_2(D_0^0 = 91.7 kcal mol^{-1}) would seem to indicate that any changes in molecular parameters are relatively small. We adopt $r_e(SF_2^+)$ = 1.59 Å which amounts to only a 0.002 Å reduction from that for SF_2 (1). Our estimated molecular parameters appear reasonable in comparison with those for SiF_2 (1). The individual moments of inertia are: $I_A = 1.2375 \times 10^{-38}$, $I_B = 9.3597 \times 10^{-39}$, and $I_C = 3.0157 \times 10^{-39}\ g\ cm^2$.

We calculate the vibrational frequencies from the force constants f_r = 4.7 and f_ω/r^2 = 0.49 mdyn/Å. The stretching force constant (f_r) is estimated from that for SiF_2 (1) and PF_2 (13). We obtain the bending force constant from the ratio $f_r/(f_\omega/r^2)$ = 9.59 which is the value calculated for SiF_2 (1). The changes in our frequencies on going from SF_2 to SF_2^+ are insignficant, in agreement with the view that the bonding in the two molecules is very similar.

The electronic configurations of the ground and excited states are predicted from the Walsh correlation diagram (10). We also predict from this diagram that the two $^2A_1 - ^2B_1$ spacings will strongly depend on the bond angle. This is supported at least for the first spacing by PE studies (3, 12) on SCl_2, OCl_2 and OF_2. Based on the similarity in the bond angles for OF_2 (3) and SF_2 (1), we estimate the first $^2A_1 \leftarrow ^2B_1$ spacing to be somewhat greater than that for OF_2^+ (23471 cm^{-1}, 12). The PE spectra for SCl_2 (3) and OF_2 (12) do not provide support for the existence of a second $^2A_1$ state. This state has been observed, however, in the PE spectrum of OCl_2 (12). These results show for OCl_2^+ that the second $^2A_1$ state lies 2258 cm^{-1} above the first $^2A_1$ state. We include a second $^2A_1 - ^2A_1$ spacing of 3000 cm^{-1}. We estimate a somewhat larger spacing for SF_2^+ based on the fact that the bond angle in SF_2^+ is predicted to be smaller than that for OCl_2^+. The relative term value for the upper most state ($^2A_2$) is estimated from that for OF_2^+ (12).

References

1. JANAF Thermochemical Tables: SF_2(g) and SiF_2(g), 6-30-76; SF_2(g), 12-31-76.
2. D. L. Hildenbrand, J. Phys. Chem. 77, 897 (1973).
3. B. Solouki, P. Rosmus, and H. Bock, Chem. Phys. Letters, 26, 20 (1974).
4. N. P. C. Westwood, Chem. Phys. Letters, 25, 558 (1974).
5. U.S. Nat. Bur. Stand. NSRDS-NBS 26, June, 1969.
6. A. L. Companion, Theoret. Chim. Acta (Berl.) 25, 268 (1972).
7. V. H. Dibeler and F. L. Mohler, J. Res. Nat. Bur. Stand. 40, 25 (1948).
8. O. Glemser, A. Muller, D. Bohler, and B. Krebs, Z. Anorg. Allg. Chem. 357, 184 (1968).
9. P. W. Harland, D. W. H. Rankin, and J. C. J. Thynne, Int. J. Spec. Ion Phys. 13, 395 (1974).
10. A. D. Walsh, J. Chem. Soc. 1953, 2266 (1953).
11. R. J. Buenker and S. D. Peyerimhoff, J. Chem. Phys. 45, 3682 (1966).
12. A. B. Cornford, D. C. Frost, F. G. Herring, and C. A. McDowell, J. Chem. Phys. 55, 2820 (1971).
13. J. K. Burdett, L. Hodges, V. Dunning, and J. H. Current, J. Phys. Chem. 74, 4053 (1970).

Sulfur Fluoride, Ion (SF_2^+) $F_2S_1^+(g)$

Enthalpy Reference Temperature = T_r = 298.15 K Standard State Pressure = p° = 0.1 MPa

T/K	$C_p°$ (J K⁻¹mol⁻¹)	S° (J K⁻¹mol⁻¹)	−[G°−H°(T_r)]/T (J K⁻¹mol⁻¹)	H°−H°(T_r) (kJ mol⁻¹)	$\Delta_fH°$ (kJ mol⁻¹)	$\Delta_fG°$ (kJ mol⁻¹)	Log K_f
0	0.	0.	INFINITE	-11.234	698.195		
100	34.639	221.099	299.912	-7.881			
200	39.818	246.642	267.474	-4.166			
250	42.536	255.822	264.250	-2.107			
298.15	44.937	263.523	263.523	0.	702.389	687.584	-120.462
300	45.023	263.801	263.524	0.083	702.411	687.492	-119.703
350	47.151	270.906	264.080	2.389	703.001	684.959	-102.225
400	48.910	277.321	265.341	4.792	701.382	682.429	-89.116
450	50.341	283.168	267.002	7.275	701.311	680.055	-78.939
500	51.500	288.534	268.890	9.822	701.196	677.702	-70.799
600	53.212	298.085	272.980	15.063	701.458	672.986	-58.589
700	54.372	306.381	277.172	20.446	702.021	668.199	-49.862
800	55.183	313.697	281.290	25.926	702.759	663.318	-43.310
900	55.767	320.232	285.260	31.475	650.476	659.419	-38.272
1000	56.200	326.131	289.057	37.074	652.608	660.299	-34.490
1100	56.529	331.504	292.675	42.712	654.732	660.965	-31.387
1200	56.785	336.434	296.119	48.378	656.844	661.438	-28.792
1300	56.986	340.987	299.398	54.067	658.940	661.736	-26.589
1400	57.148	345.217	302.521	59.774	661.017	661.873	-24.695
1500	57.280	349.164	305.501	65.495	663.073	661.862	-23.048
1600	57.388	352.864	308.346	71.229	665.108	661.715	-21.603
1700	57.479	356.346	311.069	76.972	667.122	661.441	-20.324
1800	57.556	359.634	313.676	82.724	669.117	661.049	-19.183
1900	57.621	362.748	316.178	88.483	671.093	660.547	-18.160
2000	57.676	365.705	318.581	94.248	673.054	659.941	-17.236
2100	57.724	368.520	320.892	100.018	675.003	659.238	-16.398
2200	57.766	371.206	323.119	105.793	676.944	658.441	-15.633
2300	57.803	373.775	325.266	111.571	678.881	657.557	-14.934
2400	57.835	376.236	327.339	117.353	680.818	656.588	-14.290
2500	57.864	378.597	329.342	123.138	682.758	655.539	-13.697
2600	57.890	380.867	331.280	128.926	684.706	654.412	-13.147
2700	57.914	383.052	333.158	134.716	686.667	653.210	-12.637
2800	57.936	385.159	334.977	140.509	688.642	651.934	-12.162
2900	57.957	387.193	336.743	146.303	690.636	650.588	-11.718
3000	57.976	389.158	338.458	152.100	692.652	649.173	-11.303
3100	57.995	391.059	340.124	157.899	694.691	647.690	-10.913
3200	58.014	392.901	341.745	163.699	696.756	646.141	-10.547
3300	58.034	394.686	343.322	169.501	698.849	644.527	-10.202
3400	58.053	396.419	344.858	175.306	700.972	642.849	-9.876
3500	58.074	398.102	346.356	181.112	703.125	641.108	-9.568
3600	58.097	399.738	347.816	186.921	705.309	639.306	-9.276
3700	58.121	401.330	349.241	192.732	707.525	637.442	-8.999
3800	58.147	402.881	350.632	198.545	709.774	635.517	-8.736
3900	58.176	404.391	351.991	204.361	712.055	633.534	-8.485
4000	58.207	405.865	353.320	210.180	714.369	631.491	-8.246
4100	58.242	407.302	354.619	216.003	716.715	629.390	-8.019
4200	58.279	408.706	355.890	221.829	719.094	627.232	-7.801
4300	58.320	410.078	357.134	227.659	721.505	625.016	-7.592
4400	58.365	411.419	358.353	233.493	723.947	622.744	-7.393
4500	58.413	412.732	359.547	239.332	726.421	620.416	-7.202
4600	58.466	414.016	360.717	245.176	728.926	618.033	-7.018
4700	58.523	415.274	361.864	251.025	731.461	615.595	-6.842
4800	58.584	416.507	362.990	256.880	734.026	613.102	-6.672
4900	58.649	417.715	364.095	262.742	736.620	610.556	-6.509
5000	58.719	418.901	365.179	268.610	739.243	607.957	-6.351
5100	58.793	420.064	366.244	274.486	741.894	605.305	-6.200
5200	58.872	421.207	367.290	280.369	744.574	602.601	-6.053
5300	58.954	422.329	368.318	286.260	747.280	599.845	-5.912
5400	59.042	423.432	369.328	292.160	750.014	597.037	-5.775
5500	59.133	424.516	370.322	298.069	752.774	594.179	-5.643
5600	59.229	425.582	371.299	303.987	755.560	591.271	-5.515
5700	59.329	426.632	372.261	309.915	758.373	588.311	-5.391
5800	59.432	427.664	373.207	315.853	761.210	585.303	-5.271
5900	59.540	428.681	374.139	321.801	764.072	582.246	-5.155
6000	59.651	429.683	375.056	327.761	766.959	579.140	-5.042

PREVIOUS: December 1976 (1 atm) CURRENT: December 1976 (1 bar)

Sulfur Fluoride, Ion (SF_2^+) $F_2S_1^+(g)$

SULFUR FLUORIDE, ION (SF_2^-) IDEAL GAS M_r = 70.057355

$\Delta_f H°(0\ K)$ = [-391.116 ± 64.9] kJ mol^{-1}

S°(298.15 K) = [267.445 ± 0.8] J K^{-1} mol^{-1} $\Delta_f H°(298.15\ K)$ = [-398.426 ± 64.9] kJ mol^{-1}

Vibrational Frequencies and Degeneracies

ν cm^{-1}

[479](1)

[286](1)

[594](1)

Ground State Quantum Weight: [2] σ = [2]

Point Group: [C_{2v}]

Bond Distance: S-F = [1.63] Å

Bond Angle: F-S-F = [130]°

Product of the Moments of Inertia: $I_A I_B I_C$ = [2.856674 x 10^{-115}] g^3 cm^6

Enthalpy of Formation

We calculate the adopted value of $\Delta_f H°(0\ K)$ = -93.5±15.5 kcal mol^{-1} as the difference between $\Delta_f H°(SF_2, g, 0\ K)$ = -70.42±4.0 kcal mol^{-1} (1) and an estimate EA(SF_2) = 1.0±0.5 eV (23.06 kcal mol^{-1}). The value of EA is estimated from a consideration of trends in the electron affinities for other S-F species (1). These data follow a regular pattern among the even-and odd-electron molecules, with the odd-electron molecules having, as expected, the higher electron affinities. Within each even- and odd-electron series, the EA values progressively increase as the number of fluorine atoms are increased in the ions. The only other estimate available to compare with our result is EA = 1.2±0.5 eV. This estimate was made by Hildenbrand (2) who used a method similar to ours. Our $\Delta_f H°$ value corresponds to a fluoride-ion affinity for SF of IA = 36.4±17.0 kcal mol^{-1}, and $\Delta_f H°$ at 298.15 K is -95.2 kcal mol^{-1}. In addition, we find that the primary bond dissociation energy $D_0^0(SF^- - F)$ is 68.7 kcal mol^{-1} which when compared with the values for SF_2(91.7 kcal mol^{-1}) and SF_2^+(87.1 kcal mol^{-1}) indicates that the trend is $D_0^0(SF_2^+) \simeq D_0^0(SF_2) > D_0^0(SF_2^-)$. Similar trends in bond dissociation energies have been reported (3) for the ions and neutrals of phosphorus and nitrogen difluorides.

JANAF data (1) predict that SF_2^- should be reasonably stable at moderate temperatures, with dissociative ionization (SF + F^-) becoming thermodynamically favorable above 1500 K. These predictions are qualitatively supported by the observations of Harland and Thynne (4, 5) who have observed SF_2^- in the negative ion mass spectra of SF_6, SF_4, and SF_5Cl. The temperature of the ion source was reported as around 325 K.

Heat Capacity and Entropy

According to the Walsh correlation diagram (6) for non-hydride AB_2 molecules, the most loosely bound electron in SF_2^- is in an $\bar{a}_1$' orbital whose binding energy falls markedly with an increase in bond angle. We predict then that electron capture by SF_2 should be accompanied by a large increase in bond angle. Based on the bond angle reported for the isoelectronic valence (21 electrons) molecule ClF_2(136±15°, 7), we estimate an angle for SF_2^- of 130° which represents a 32° increase from that for SF_2 (1). The addition of an antibonding electron to SF_2 should also produce an increase in the S-F bond length. We estimate $r_e(SF_2^-) = r_e(Cl-F)$ = 1.63 Å (1) which corresponds to $r_e(SF_2^-) - r_e(SF_2)$ = 0.038 Å. The individual moments of inertia are: I_A = 1.3702 x 10^{-39}, I_B = 13.7701 x 10^{-39}, and I_C = 15.1403 x 10^{-39} g cm^2.

We calculate vibrational frequencies from the force constants f_r = 2.0 and f_α/r^2 = 0.27 mdyn/Å. These force constants are estimated from those for ClF_2(f_r = 1.94 and f_α/r^2 = 0.26 mdyn/Å) which are obtained from the bond angle and vibrational frequencies reported by Mamantov et al. (7). Our frequencies on going from SF_2 to SF_2^- show a large decrease, in agreement with the view that the additional antibonding electron weakens the bonding in the radical anion relative to the neutral molecule.

The Walsh correlation diagram (6) predicts a ground state electronic configuration of $^2A_1$ for a 21 valence electron molecule. In addition, the existence of two doublet excited states is highly probable. We omit these excited states since no spectroscopic information is available on these states for isoelectronic molecules which would allow us to estimate the relative term values. We speculate that these levels lie above 20000 cm^{-1}, and thus, their contributions to the thermal functions are probably small below 3000 K.

References

1. JANAF Thermochemical Tables: SF_2(g), 6-30-76; SF^-(g), SF_3^-(g), SF_4^-(g), SF_5^-(g), and SF_2^+(g), 12-31-76; ClF(g), 9-30-65.
2. D. L. Hildenbrand, J. Phys. Chem. 77, 897 (1973).
3. P. W. Harland, D. W. H. Rankin, and J. C. J. Thynne, Int. J. Mass Spec. and Ion Phys. 13, 395 (1974).
4. P. Harland and J. C. J. Thynne, J. Phys. Chem. 73, 4031 (1969).
5. P. W. Harland and J. C. J. Thynne, J. Phys. Chem. 75, 3517 (1971).
6. A. D. Walsh, J. Chem. Soc. 1953, 2266 (1953).
7. G. Mamantov, E. J. Vasini, M. C. Moulton, D. G. Vickroy, and T. Maekawa, J. Chem. Phys. 54, 3419 (1971).

Sulfur Fluoride, Ion (SF_2^-) $F_2S_1^-$(g)

Enthalpy Reference Temperature = T_r = 298.15 K Standard State Pressure = p° = 0.1 MPa

T/K	$C_p°$ (J K^{-1}mol^{-1})	S° (J K^{-1}mol^{-1})	-[G°-H°(T_r)]/T (J K^{-1}mol^{-1})	H°-H°(T_r) (kJ mol^{-1})	$\Delta_f H°$ (kJ mol^{-1})	$\Delta_f G°$ (kJ mol^{-1})	Log K_f
0	0.	0.	INFINITE	-12.125	-391.116		
100	36.153	220.728	308.077	-8.735			
200	44.703	248.468	271.882	-4.683			
250	47.901	258.804	268.259	-2.364			
298.15	50.164	267.445	267.445	0.	-398.426	-401.890	70.410
300	50.238	267.756	267.446	0.093	-398.471	-401.912	69.979
350	51.930	275.634	268.064	2.649	-399.709	-402.388	60.053
400	53.167	282.653	269.457	5.278	-403.181	-402.597	52.574
450	54.088	288.971	271.280	7.961	-405.130	-402.416	46.711
500	54.787	294.707	273.341	10.683	-407.148	-402.004	41.997
600	55.752	304.789	277.765	16.214	-410.754	-400.630	34.878
700	56.365	313.432	282.258	21.822	-414.123	-398.674	29.749
800	56.777	320.987	286.636	27.481	-417.364	-396.245	25.872
900	57.065	327.692	290.833	33.174	-473.660	-392.334	22.770
1000	57.275	333.716	294.825	38.891	-475.568	-383.195	20.016
1100	57.432	339.183	298.613	44.627	-477.502	-373.865	17.753
1200	57.552	344.185	302.205	50.376	-479.464	-364.357	15.860
1300	57.647	348.796	305.614	56.136	-481.454	-354.684	14.251
1400	57.722	353.071	308.853	61.905	-483.473	-344.857	12.867
1500	57.783	357.055	311.935	67.680	-485.520	-334.885	11.662
1600	57.833	360.786	314.873	73.461	-487.595	-324.775	10.603
1700	57.875	364.294	317.678	79.247	-489.696	-314.534	9.664
1800	57.909	367.603	320.360	85.036	-491.821	-304.169	8.827
1900	57.939	370.734	322.930	90.828	-493.969	-293.686	8.074
2000	57.964	373.707	325.395	96.624	-496.135	-283.088	7.394
2100	57.986	376.536	327.764	102.421	-498.315	-272.383	6.775
2200	58.005	379.234	330.042	108.221	-500.506	-261.573	6.211
2300	58.022	381.812	332.238	114.022	-502.704	-250.663	5.693
2400	58.036	384.282	334.355	119.825	-504.903	-239.657	5.216
2500	58.049	386.651	336.400	125.629	-507.101	-228.560	4.776
2600	58.061	388.928	338.377	131.435	-509.292	-217.375	4.367
2700	58.071	391.120	340.290	137.241	-511.473	-206.106	3.987
2800	58.080	393.232	342.143	143.049	-513.639	-194.757	3.633
2900	58.088	395.270	343.940	148.857	-515.789	-183.330	3.302
3000	58.095	397.240	345.684	154.666	-517.918	-171.830	2.992
3100	58.102	399.145	347.378	160.476	-520.025	-160.259	2.700
3200	58.108	400.989	349.025	166.287	-522.106	-148.620	2.426
3300	58.114	402.778	350.627	172.098	-524.162	-136.917	2.167
3400	58.119	404.512	352.186	177.910	-526.189	-125.152	1.923
3500	58.123	406.197	353.705	183.722	-528.188	-113.327	1.691
3600	58.128	407.835	355.186	189.534	-530.157	-101.446	1.472
3700	58.131	409.427	356.631	195.347	-532.096	-89.510	1.264
3800	58.135	410.978	358.041	201.160	-534.004	-77.523	1.066
3900	58.138	412.488	359.418	206.974	-535.883	-65.486	0.877
4000	58.141	413.960	360.763	212.788	-537.731	-53.401	0.697
4100	58.144	415.396	362.078	218.602	-539.550	-41.270	0.526
4200	58.147	416.797	363.364	224.417	-541.340	-29.095	0.362
4300	58.149	418.165	364.623	230.232	-543.102	-16.878	0.205
4400	58.152	419.502	365.855	236.047	-544.836	-4.620	0.055
4500	58.154	420.809	367.061	241.862	-546.543	7.677	-0.089
4600	58.156	422.087	368.244	247.678	-548.224	20.011	-0.227
4700	58.158	423.338	369.403	253.493	-549.879	32.382	-0.360
4800	58.160	424.562	370.539	259.309	-551.511	44.788	-0.487
4900	58.161	425.761	371.654	265.125	-553.120	57.228	-0.610
5000	58.163	426.936	372.748	270.941	-554.706	69.700	-0.728
5100	58.164	428.088	373.822	276.758	-556.271	82.203	-0.842
5200	58.166	429.217	374.876	282.574	-557.815	94.738	-0.952
5300	58.167	430.325	375.912	288.391	-559.341	107.302	-1.058
5400	58.168	431.413	376.930	294.208	-560.847	119.894	-1.160
5500	58.169	432.480	377.930	300.025	-562.336	132.514	-1.259
5600	58.171	433.528	378.914	305.842	-563.808	145.161	-1.354
5700	58.172	434.558	379.881	311.659	-565.264	157.834	-1.446
5800	58.173	435.570	380.832	317.476	-566.705	170.532	-1.536
5900	58.174	436.564	381.769	323.293	-568.131	183.256	-1.622
6000	58.174	437.542	382.690	329.111	-569.543	196.003	-1.706

PREVIOUS: December 1976 (1 atm) CURRENT: December 1976 (1 bar)

Sulfur Fluoride, Ion (SF_2^-) $F_2S_1^-$(g)

FLUORODISULFANE (FS_2F) IDEAL GAS M_r = 102.116806

$\Delta_f H°(0\ K) = [-333.382 \pm 41.8]\ kJ\ mol^{-1}$

$S°(298.15\ K) = [294.082 \pm 0.08]\ J\ K^{-1}\ mol^{-1}$ $\Delta_f H°(298.15\ K) = [336.435 \pm 41.8]\ kJ\ mol^{-1}$

Vibrational Frequencies and Degeneracies

ν, cm^{-1}	ν, cm^{-1}
717.0(1)	182.5(1)
614.6(1)	680.8(1)
319.8(1)	301.0(1)

Ground State Quantum Weight: [1] σ = 2

Point Group: C_2

Bond Distances: S-F = 1.635 0.01 A S-S = 1.888 0.01 A

Bond Angles: F-S-S = 108.3 ± 0.5° Dihedral = 87.9 ± 1.5°

Product of Moments of Inertia: $I_AI_BI_C = 7.332228 \times 10^{-114}\ g^3\ cm^6$

Enthalpy of Formation

No direct experimental measurement of $\Delta_f H°$ is available. We estimate a value from bond energy considerations. From a comparison of bond lengths and force constants for FSSF (1) and SF_4(2), we expect that the S-F bond energies in FSSF would be quite similar to those for the axial S-F bonds in SF_4. We assume $D_0^0(FSS\text{-}F) = D_0^0(SF_3\text{-}F)_{axial}$ = 73.6 kcal mol^{-1} which leads to the adopted value of $\Delta_f H°(S_2F_2, g, 0\ K) = -79.7$ kcal mol^{-1}. We use auxiliary heat of formation data from JANAF (3) and estimate the uncertainty in $\Delta_f H°$ as ±10.0 kcal mol^{-1}. A previous estimate (-83.5 kcal mol^{-1}) has been reported by O'Hare (4) who derived this value by a procedure identical with that used here. $\Delta_f H°$(298.15 K) corresponds to -80.4 kcal mol^{-1}, and the atomization energy is calculated from $\Delta_f H°$ to be 247.9 kcal mol^{-1}.

FSSF is isomeric with SSF_2 (3). Qualitative evidence has been presented (1) which indicates that SSF_2 is the more stable isomer. From our estimated free energy data, we predict a Gibbs energy change for the isomerization reaction FSSF(g) = SSF_2(g) of -15.4 kcal mol^{-1} at 298.15 K, indicating that SSF_2 is substantially more stable than FSSF. This conflicts with relative stability predictions based on results from two independent MO studies (5, 6). These semiempirical (CNDO/2) calculations indicate that the instability of FSSF relative to SSF_2 amounts to only about 2.0 kcal mol^{-1}. If we assume that these MO calculations are correct, we calculate that $D_0^0(FS\text{-}SF) > D_0^0(S{=}SF_2)$ by roughly 7.0 kcal mol^{-1}. This seems very unlikely in view of what is known (1) about the S-S linkages in the two isomers. Our adopted results give D_0^0(FS-SF) = 85.6 kcal mol^{-1} which lies between that for SSF_2(90.4 kcal mol^{-1}) and S_2O(79.9 kcal mol^{-1}) (3).

Heat Capacity and Entropy

The structural data and vibrational frequencies are taken from the recent review of Seel (1) and are based on reliable experimental measurements. We note that the molecular characteristics of FSSF are unusual in that the S-S bond length (1.888 Å) is much shorter than that in HSSH(2.05 Å, 1) and nearly equal to those in S_2(1.889 Å), SSF_2(1.860 Å), and S_2O(1.884 Å) (3). In contrast, the S-F bonds are longer than those in most S-F species (~1.58 Å, 3) and nearly as long as the axial S-F bonds (1.646 Å, 3) in SF_4. Rationale (1) has been presented to explain these unique molecular features in terms of double bond ($F^-S{=}S^+\text{-}F$) formation. The individual moments of inertia are: $I_A = 7.3579 \times 10^{-39}$, $I_B = 30.4921 \times 10^{-39}$, and $I_C = 32.6808 \times 10^{-39}\ g\ cm^2$.

We assume that there is no free internal rotation in the molecule, and the contribution from the torsional oscillation (ν_4 = 182.5 cm^{-1}) is included in the vibrational partition function. Extended Huckel calculations (7) show that the potential barrier to internal rotation is high (~40 kcal mol^{-1}), providing some justification for our treatment. We estimate that the uncertainty in our entropies above 2000 K should not exceed 1.0 cal K^{-1} mol^{-1}. The UV spectra (1) of FSSF reveal no excited states which would be significant in our calculations. Since S_2F_2 has no unpaired electrons, we predict that the ground state is singlet.

Our thermal functions essentially extend those reported by Brown and Pez (8). Other published functions (4, 9) are based on estimated vibrational frequencies and are considered less reliable.

References

1. F. Seel, in "Advances in Inorganic Chemistry and Radiochemistry," Vol. 16, H. J. Emeleus and A. G. Sharpe, eds., Academic Press, New York, 1974.
2. I. W. Levin, J. Chem. Phys. 55, 5393 (1971).
3. JANAF Thermochemical Tables: S_2(g), 12-31-65, F(g), 9-30-65, SSF_2(g) and SF_4(g), 6-30-76, S_2O(g), 9-30-65.
4. P. A. G. O'Hare, Argonne National Laboratory, ANL-7315, July, 1968.
5. A. L. Companion, Theoret. Chim. Acta, 25, 268 (1972).
6. G. Wagner, H. Bock, R. Budenz, and F. Seel, Chem. Ber. 106, 1285 (1973).
7. R. M. Minyaev, V. I. Minkin, I. I. Zakharov, and I. D. Sadekov, Teor. Eksp. Khim. 9, 816 (1973).
8. R. D. Brown and G. P. Pez, Spectrochimica Acta, 26A, 1375 (1970).
9. B. J. McBride, S. Heimel, J. G. Ehlers, and S. Gordon, NASA Rep. SP-3001, July, 1963.

Difluorodisulfane (FSSF) $F_2S_2(g)$

Enthalpy Reference Temperature = T_r = 298.15 K Standard State Pressure = p° = 0.1 MPa

T/K	$C_p°$	S°	$-[G°-H°(T_r)]/T$	$H°-H°(T_r)$	$\Delta_f H°$	$\Delta_f G°$	Log K_f
	J $K^{-1}mol^{-1}$			kJ mol^{-1}			
0	0.	0.	INFINITE	-14.595	-333.382	-333.382	INFINITE
100	42.158	235.778	345.882	-11.010	-334.084	-338.120	176.616
200	56.425	269.610	299.845	-6.047	-335.333	-341.670	89.235
250	61.892	282.811	295.147	-3.084	-335.912	-343.186	71.705
298.15	66.041	294.082	294.082	0.	-336.435	-344.539	60.362
300	66.180	294.491	294.083	0.122	-336.455	-344.590	59.998
350	69.466	304.951	294.902	3.517	-336.981	-345.904	51.623
400	71.974	314.398	296.758	7.056	-341.935	-346.967	45.309
450	73.899	322.991	299.203	10.705	-343.804	-347.501	40.337
500	75.395	330.858	301.980	14.439	-345.774	-347.801	36.335
600	77.509	344.805	307.985	22.092	-348.764	-347.912	30.288
700	78.883	356.864	314.126	29.916	-351.185	-347.573	25.936
800	79.819	367.462	320.144	37.854	-353.279	-346.912	22.651
900	80.482	376.904	325.936	45.871	-461.429	-343.838	19.956
1000	80.968	385.410	331.465	53.945	-460.758	-330.809	17.280
1100	81.333	393.145	336.726	62.061	-460.105	-317.846	15.093
1200	81.614	400.234	341.727	70.208	-459.477	-304.941	13.274
1300	81.835	406.776	346.483	78.381	-458.878	-292.088	11.736
1400	82.011	412.847	351.009	86.574	-458.312	-279.279	10.420
1500	82.154	418.510	355.322	94.782	-457.781	-266.510	9.281
1600	82.272	423.816	359.439	103.004	-457.285	-253.775	8.285
1700	82.370	428.807	363.374	111.236	-456.823	-241.070	7.407
1800	82.453	433.518	367.141	119.477	-456.395	-228.391	6.628
1900	82.523	437.978	370.753	127.726	-455.996	-215.735	5.931
2000	82.583	442.212	374.221	135.982	-455.625	-203.099	5.304
2100	82.635	446.243	377.556	144.243	-455.276	-190.482	4.738
2200	82.679	450.088	380.766	152.508	-454.945	-177.880	4.223
2300	82.719	453.764	383.860	160.778	-454.627	-165.293	3.754
2400	82.753	457.285	386.847	169.052	-454.318	-152.720	3.324
2500	82.784	460.664	389.732	177.329	-454.013	-140.160	2.928
2600	82.811	463.911	392.523	185.609	-453.708	-127.612	2.564
2700	82.835	467.037	395.226	193.891	-453.396	-115.076	2.226
2800	82.856	470.050	397.844	202.175	-453.074	-102.551	1.913
2900	82.876	472.958	400.385	210.462	-452.740	-90.038	1.622
3000	82.893	475.768	402.851	218.751	-452.391	-77.537	1.350
3100	82.909	478.486	405.247	227.041	-452.022	-65.048	1.096
3200	82.923	481.119	407.577	235.332	-451.632	-52.571	0.858
3300	82.937	483.670	409.845	243.625	-451.218	-40.107	0.635
3400	82.949	486.147	412.053	251.920	-450.780	-27.656	0.425
3500	82.960	488.551	414.204	260.215	-450.318	-15.218	0.227
3600	82.970	490.888	416.302	268.511	-449.829	-2.793	0.041
3700	82.979	493.162	418.349	276.809	-449.313	9.618	-0.136
3800	82.987	495.375	420.347	285.107	-448.771	22.014	-0.303
3900	82.995	497.531	422.298	293.406	-448.202	34.396	-0.461
4000	83.003	499.632	424.205	301.706	-447.608	46.763	-0.611
4100	83.009	501.682	426.070	310.007	-446.989	59.114	-0.753
4200	83.016	503.682	427.894	318.308	-446.344	71.451	-0.889
4300	83.022	505.635	429.680	326.610	-445.677	83.771	-1.018
4400	83.027	507.544	431.428	334.913	-444.986	96.076	-1.141
4500	83.032	509.410	433.140	343.215	-444.274	108.365	-1.258
4600	83.037	511.235	434.818	351.519	-443.541	120.637	-1.370
4700	83.042	513.021	436.463	359.823	-442.789	132.894	-1.477
4800	83.046	514.769	438.076	368.127	-442.018	145.135	-1.579
4900	83.050	516.482	439.659	376.432	-441.231	157.359	-1.677
5000	83.054	518.159	441.212	384.737	-440.427	169.567	-1.771
5100	83.057	519.804	442.737	393.043	-439.609	181.759	-1.862
5200	83.060	521.417	444.235	401.349	-438.776	193.935	-1.948
5300	83.063	522.999	445.706	409.655	-437.932	206.094	-2.031
5400	83.066	524.552	447.152	417.961	-437.076	218.237	-2.111
5500	83.069	526.076	448.573	426.268	-436.209	230.364	-2.188
5600	83.072	527.573	449.970	434.575	-435.332	242.476	-2.262
5700	83.074	529.043	451.345	442.882	-434.446	254.572	-2.333
5800	83.077	530.488	452.697	451.190	-433.553	266.652	-2.401
5900	83.079	531.908	454.027	459.498	-432.652	278.717	-2.468
6000	83.081	533.305	455.337	467.806	-431.745	290.766	-2.531

PREVIOUS: June 1976 (1 atm) CURRENT: June 1976 (1 bar)

Difluorodisulfane (FSSF) $F_2S_2(g)$

THIOTHIONYL FLUORIDE (SSF_2) IDEAL GAS M_r = 102.116806

$S°(298.15 K) = 292.827 \pm 0.42 J K^{-1} mol^{-1}$

$\Delta_f H°(0 K) = [-397.482 \pm 41.8] kJ mol^{-1}$
$\Delta_f H°(298.15 K) = [-401.413 \pm 41.8] kJ mol^{-1}$

Electronic Levels and Quantum Weights

ϵ_i, cm^{-1}	g_i
0	[1]
34000	[1]

Vibrational Frequencies and Degeneracies

ν, cm^{-1}	ν, cm^{-1}
760.5(1)	330.0(1)
718.5(1)	692.3(1)
411.2(1)	274.0(1)

Point Group C_s σ = 1
Bond Distances: S-S = 1.860 ± 0.015 Å S-F = 1.598 ± 0.012 Å
Bond Angles: S-S-F = 107.5 ± 1.0° F-S-F = 92.5 ± 1.0°
Product of the Moments of Inertia: $I_AI_BI_C = 5.957541 \times 10^{-114}$ g^3 cm^6

Heat of Formation

No direct experimental measurement of $\Delta_f H°$ is available. We estimate two values from bond energy considerations. Seel (1) has presented a detailed discussion on the bonding in S_2, S_2O, and SSF_2 and has shown from bond length and force constant correlations that the S=S linkages in these molecules are very similar. We assume that $D_0^0(S{=}SF_2) = (D_0^0(S_2) + D_0^0(S_2O))/2$ = 90±10 kcal mol^{-1} which gives $\Delta_f H°(SSF_2, g, 0 K)$ = -94.7±14 kcal mol^{-1} when the value of D_0^0 is combined with $\Delta_f H°(SF_2, g, 0 K)$ = -70.4±4 kcal mol^{-1} and $\Delta_f H°(S, g, 0 K)$ = 65.75±.01 kcal mol^{-1} (2). From a comparison of bond lengths and force constants (1, 3) for the S-F bonds in SSF_2, and SF_4, we would expect that the mean S-F bond energies for these molecules are probably not too different. Values (2) for OSF_2 and SF_4 are 83.4 kcal mol^{-1} and 80.0 kcal mol^{-1}, respectively. Using $D_0^0(S_2F{-}F)$ = 81.5±2.5 kcal mol^{-1}, we calculate $\Delta_r H°(0 K)$ = 163.0 kcal mol^{-1} for the dissociation process $S_2F_2 = S_2{+}2F$ which leads to $\Delta_f H°(SSF_2, g, 0 K)$ = -95.5±5.0 kcal mol^{-1}. These two estimates are in reasonable agreement and we choose to adopt the value of -95.0±10.0 kcal mol^{-1}.

The heat of atomization $\Delta_{at}H°(0 K)$ is calculated from our adopted $\Delta_f H°$ value to be 263.2 kcal mol^{-1} which is close to the value of 262.0 kcal mol^{-1} predicted by CNDO/2 MO calculations and energy partitioning methods (4).

Published estimates (1, 5, and 6) of $\Delta_f H°$ based on bond energy schemes similar to those used here include in kcal mol^{-1} -92±7, -80.2, and -86.4. Wilkins (6) assumed that the mean S-F bond energy for SSF_2 was 80.0 kcal mol^{-1} and obtained $\Delta_f H°$ = -86.4 kcal mol^{-1}. We believe that this estimate is in error. Recalculation of his data gives $\Delta_f H°$ = -91.6 kcal mol^{-1} which is in much better agreement with our results.

Heat Capacity and Entropy

The structural data and vibrational frequencies are taken from the recent review by Seel (1) and are based on reliable experimental measurements. Since SSF_2 has no unpaired electrons, we predict that the ground state is singlet. The ultra-violet spectra (1) show absorption peaks at roughly 34000 cm^{-1} and 42000 cm^{-1}, suggesting the existence of at least two excited states. We do not include the level near 42000 cm^{-1} since its effect is negligible. The individual moments of inertia are: $I_A = 10.2965 \times 10^{-39}$, $I_B = 21.0146 \times 10^{-39}$, and $I_C = 27.5332 \times 10^{-39}$ g cm^2.

Our thermal functions essentially extend those reported by Brown and Pez (7). Other published functions (5, 6) are based on an older set of vibrational frequencies (8) and include a bias due to incorrect analysis of the SSF_2 structure. O'Hare's analysis (5) of the structural data resulted in a value for the product of the moments of inertia which is about 7% lower than our results. In addition, he assigned SSF_2 to point group C_{2v} and used a symmetry number of two rather than the correct value of one. The moments of inertia used by Wilkins (6) are in units of amu $Å^2$, not g cm^2 as assumed.

References

1. F. Seel, in "Advances in Inorganic Chemistry and Radio Chemistry", Vol. 16, H. J. Emeleus and A. G. Sharpe, eds., Academic Press, New York, 1974.
2. JANAF Thermochemical Tables: S_2(g), 12-31-65, S_2O(g), 6-30-65, SF_2(g) and SF_4(g), 6-30-76, S(g), 6-30-71, OSF_2(g), 6-30-72.
3. I. W. Levin, J. Chem. Phys. 55, 5393 (1971).
4. A. L. Companion, Theoret. Chim. Acta, 25, 268 (1972).
5. P. A. G. O'Hare, Argonne National Laboratory, ANL-7315, July, 1968.
6. R. L. Wilkins, Aerospace Corporation, Report No. TR-0158(3240-20)-19, July, 1968.
7. R. D. Brown and G. P. Pez, Spectrochimica Acta, 26A, 1375 (1970).
8. R. D. Brown, G. P. Pez, and M. F. O'Dwyer, Aust. J. Chem. 18, 627 (1965).

Thiothionyl Fluoride (SSF_2) F_2S_2(g)

Enthalpy Reference Temperature = T_r = 298.15 K Standard State Pressure = p° = 0.1 MPa

T/K	$C_p°$	S°	$-[G°-H°(T_r)]/T$	$H°-H°(T_r)$	$\Delta_f H°$	$\Delta_f G°$	Log K_f
	J $K^{-1}mol^{-1}$			kJ mol^{-1}			
0	0.	0.	INFINITE	-13.718	-397.482	-397.482	INFINITE
100	38.393	238.662	341.446	-10.278	-398.329	-402.654	210.325
200	52.610	269.714	298.294	-5.716	-399.980	-406.337	106.124
250	58.524	282.110	293.842	-2.933	-400.738	-407.837	85.213
298.15	63.127	292.827	292.827	0.	-401.413	-409.143	71.680
300	63.284	293.218	292.828	0.117	-401.438	-409.191	71.246
350	67.002	303.264	293.613	3.378	-402.098	-410.430	61.253
400	69.882	312.408	295.400	6.803	-407.165	-411.401	53.723
450	72.118	320.773	297.761	10.356	-409.131	-411.830	47.804
500	73.869	328.466	300.452	14.007	-411.183	-412.015	43.043
600	76.365	342.171	306.291	21.528	-414.305	-411.873	35.857
700	78.003	354.074	312.286	29.252	-416.827	-411.263	30.689
800	79.124	364.568	318.178	37.111	-418.999	-410.316	26.791
900	79.921	373.936	323.863	45.066	-527.212	-406.950	23.619
1000	80.505	382.388	329.299	53.088	-526.592	-393.621	20.561
1100	80.946	390.082	334.481	61.162	-525.981	-380.353	18.061
1200	81.286	397.141	339.412	69.274	-525.389	-367.141	15.981
1300	81.553	403.658	344.107	77.417	-524.820	-353.977	14.223
1400	81.767	409.710	348.579	85.583	-524.281	-340.855	12.717
1500	81.941	415.357	352.845	93.769	-523.772	-327.771	11.414
1600	82.084	420.650	356.919	101.970	-523.296	-314.720	10.275
1700	82.203	425.630	360.816	110.185	-522.852	-301.698	9.270
1800	82.303	430.332	364.548	118.410	-522.439	-288.701	8.378
1900	82.388	434.784	368.129	126.645	-522.055	-275.726	7.580
2000	82.461	439.012	371.568	134.888	-521.697	-262.771	6.863
2100	82.524	443.037	374.877	143.137	-521.359	-249.833	6.214
2200	82.579	446.877	378.063	151.392	-521.039	-236.911	5.625
2300	82.626	450.549	381.135	159.652	-520.731	-224.003	5.087
2400	82.668	454.067	384.101	167.917	-520.431	-211.108	4.595
2500	82.705	457.442	386.968	176.186	-520.134	-198.226	4.142
2600	82.738	460.686	389.741	184.458	-519.836	-185.355	3.724
2700	82.767	463.810	392.427	192.733	-519.531	-172.496	3.337
2800	82.794	466.820	395.030	201.011	-519.216	-159.649	2.978
2900	82.817	469.726	397.556	209.292	-518.888	-146.813	2.644
3000	82.839	472.534	400.009	217.575	-518.544	-133.989	2.333
3100	82.858	475.250	402.392	225.860	-518.180	-121.176	2.042
3200	82.876	477.881	404.711	234.146	-517.795	-108.376	1.769
3300	82.892	480.432	406.967	242.435	-517.386	-95.588	1.513
3400	82.907	482.907	409.164	250.725	-516.953	-82.812	1.272
3500	82.921	485.310	411.305	259.016	-516.494	-70.050	1.045
3600	82.934	487.646	413.394	267.309	-516.009	-57.301	0.831
3700	82.946	489.919	415.431	275.603	-515.497	-44.566	0.629
3800	82.957	492.131	417.421	283.898	-514.958	-31.846	0.438
3900	82.968	494.286	419.364	292.194	-514.392	-19.139	0.256
4000	82.978	496.387	421.264	300.491	-513.800	-6.448	0.084
4100	82.988	498.436	423.121	308.790	-513.183	6.228	-0.079
4200	82.998	500.436	424.938	317.089	-512.541	18.889	-0.235
4300	83.007	502.389	426.717	325.389	-511.875	31.534	-0.383
4400	83.017	504.297	428.458	333.691	-511.185	44.164	-0.524
4500	83.027	506.163	430.164	341.993	-510.474	56.777	-0.659
4600	83.036	507.988	431.836	350.296	-509.742	69.375	-0.788
4700	83.046	509.774	433.476	358.600	-508.989	81.956	-0.911
4800	83.057	511.522	435.084	366.905	-508.218	94.521	-1.029
4900	83.067	513.235	436.661	375.211	-507.429	107.070	-1.141
5000	83.079	514.913	438.209	383.519	-506.623	119.603	-1.249
5100	83.090	516.558	439.730	391.827	-505.802	132.119	-1.353
5200	83.102	518.172	441.223	400.137	-504.966	144.620	-1.453
5300	83.115	519.755	442.689	408.448	-504.117	157.104	-1.548
5400	83.129	521.309	444.131	416.760	-503.255	169.571	-1.640
5500	83.143	522.834	445.548	425.073	-502.381	182.023	-1.729
5600	83.158	524.332	446.942	433.388	-501.496	194.459	-1.814
5700	83.174	525.804	448.312	441.705	-500.601	206.878	-1.896
5800	83.191	527.251	449.661	450.023	-499.697	219.282	-1.975
5900	83.208	528.673	450.988	458.343	-498.784	231.671	-2.051
6000	83.226	530.072	452.295	466.665	-497.864	244.043	-2.125

PREVIOUS: June 1976 (1 atm) CURRENT: June 1976 (1 bar)

Thiothionyl Fluoride (SSF_2) F_2S_2(g)

$S°(298.15\ K) = 256.579 \pm 0.42\ J\ K^{-1}\ mol^{-1}$

$\Delta_f H°(0\ K) = -587.013 \pm 12.6\ kJ\ mol^{-1}$

$\Delta_f H°(298.15\ K) = -587.852 \pm 12.6\ kJ\ mol^{-1}$

Electronic Levels and Quantum Weights

State	ε_i, cm^{-1}	g_i
$^1A_1$	0	1
$^3B_1$	26310	3
$^1B_1$	44109	1
$^1B_2$	62280	1

Vibrational Frequencies and Degeneracies

ν, cm^{-1}
855 (1)
345 (1)
872 (1)

Point Group: C_{2v} $\sigma = 2$

Bond Distance: Si-F = 1.591 Å

Bond Angle: F-Si-F = 100° 59'

Product of the Moments of Inertia: $I_A I_B I_C = 3.200830 \times 10^{-115}\ g^3\ cm^6$

Enthalpy of Formation

Margrave et al. (1) have reported some approximate equilibrium constants for the reaction $Si(cr) + SiF_4(g) = 2\ SiF_2(g)$. These were calculated from yields of polymerized products and a knowledge of the total system pressure given in U.S. Patent No. 2, 840,588 (1958). The data are subjected to a 3rd law analysis using the present JANAF functions (2) and yield $\Delta_r H°(298.15\ K) = 92\pm11$ kcal mol^{-1}, which gives $\Delta_f H°(SiF_2,\ g,\ 298.15\ K) = -147\pm5$ kcal mol^{-1}.

Ehlert and Margrave (3) and Farber (13) have determined equilibrium constants mass spectrometrically. A 2nd and 3rd law analysis of the data is shown below. We adopt $\Delta_f H°(SiF_2,\ g,\ 298.15\ K) = -140.5\pm3$ kcal mol^{-1} (-587.852 ± 12.6 kJ mol^{-1}).

Source	Reaction	T/K	Data Points	$\Delta_r H°(298.15\ K)$, kcal mol^{-1} 2nd law	3rd law	Drift cal K^{-1} mol^{-1}	$\Delta_f H°(298.15\ K)$ kcal mol^{-1}
(3)	$CaF_2(cr) + Si(g) = Ca(g) + SiF_2(g)$	1395-1543	12	80.2±14.1	88.0±3.0	4.6±9.7	-140.1
(3)	$CaF_2(g) + Si(g) = Ca(g) + SiF_2(g)$	1395-1543	12	-8.3±14.1	-18.0±3.0	-6.6±9.6	-140.7
(3)	$2\ CaF(g) + Si(g) = 2Ca(g) + SiF_2(g)$	1395-1543	12	-18.6± 7.5	-33.4±1.8	-10.0±5.1	-141.0
(13)	$SiF_4(g) + Si(g) = 2SiF_2(g)$	1590-1782	10	-2.0± 0.3	-3.2±0.3	0.7±0.8	-140.7

Using JANAF values (2), we calculate $\Delta_{at} H°(0\ K) = 283.7\pm5$ kcal mol^{-1} for the process $SiF_2(g) = Si(g) + 2\ F(g)$. Within 2 kcal mol^{-1} the average bond energies in $SiF_2(g)$, $SiF_3(g)$, and $SiF_4(g)$ are the same.

Heat Capacity and Entropy

The electronic levels are adopted from the ultraviolet spectral work of Khanna et al. (4), Rao (5) and Gole et al. (6). The vibrational assignments of $\nu_1 = 855\ cm^{-1}$ (symmetric stretch) and $\nu_3 = 872\ cm^{-1}$ (asymmetric stretch) of Khanna et al. (7) are adopted. These assignments are supported by the argon matrix studies of Milligan and Jacox (8), the neon matrix and the argon matrix measurements of Hastie et al. (9) and the excited state microwave work of Shoji et al. (10). The bending frequency assignment, $\nu_2 = 345\ cm^{-1}$, is adopted from the ultraviolet and microwave study of Rao and Curl (11). This assignment is also confirmed by Milligan and Jacox (8).

The gas phase geometry was established by Rao et al. (12) from microwave spectra. The principal moments of inertia are: $I_A = 2.7474 \times 10^{-39}$, $I_B = 9.5072 \times 10^{-39}$, and $I_C = 12.2545 \times 10^{-39}\ g\ cm^2$.

References

1. J. L. Margrave, A. S. Kannan, and D. C. Pease, J. Phys. Chem. 66, 1200 (1962).
2. JANAF Thermochemical Tables: Si(cr), 12-13-66; Si(g), 3-31-67; Ca(g), CaF(g), CaF_2(cr, g), 12-31-68; SiF_4(g), 6-30-76; F(g), 9-30-65; SiF_3(g) 12-31-77.
3. T. C. Ehlert and J. L. Margrave, J. Chem. Phys. 4, 1066 (1964).
4. V. M. Khanna, G. Besenbruch, and J. L. Margrave, J. Chem. Phys. 46, 2310 (1967).
5. D. R. Rao, J. Mol. Spectrosc. 34, 284 (1970).
6. J. L. Gole, R. H. Hauge, J. L. Margrave, and J. W. Hastie, J. Mol. Spectrosc. 43, 441 (1972).
7. V. M. Khanna, R. H. Hauge, R. F. Curle, Jr., and J. L. Margrave, J. Chem. Phys. 47, 5031 (1967).
8. D. E. Milligan and M. E. Jacox, J. Chem. Phys. 49, 4269 (1968).
9. J. W. Hastie, R. H. Hauge, and J. L. Margrave, J. Amer. Chem. Soc. 91, 2536 (1969).
10. H. Shoji, T. Tanaka, and E. Hirota, J. Mol. Spectrosc. 47, 268 (1973).
11. V. M. Rao and R. F. Curl, Jr., J. Chem. Phys. 45, 2032 (1966).
12. V. M. Rao, R. F. Curl, Jr., P. L. Timms, and J. L. Margrave, J. Chem. Phys. 43, 2557 (1965).
13. M. Farber, Space Sciences, Inc., personal communication, October 18, 1976.

Enthalpy Reference Temperature = T_r = 298.15 K Standard State Pressure = p° = 0.1 MPa

T/K	$C°_p$ (J $K^{-1}mol^{-1}$)	S° (J $K^{-1}mol^{-1}$)	$-[G°-H°(T_r)]/T$ (J $K^{-1}mol^{-1}$)	$H°-H°(T_r)$ (kJ mol^{-1})	$\Delta_f H°$ (kJ mol^{-1})	$\Delta_f G°$ (kJ mol^{-1})	Log K_f
0	0.	0.	INFINITE	-11.204	-587.013	-587.013	INFINITE
100	34.719	214.272	292.766	-7.849	-586.832	-590.839	308.622
200	39.648	239.824	260.498	-4.135	-587.225	-594.726	155.327
250	42.199	248.946	257.299	-2.088	-587.531	-596.567	124.646
298.15	44.509	256.579	256.579	0.	-587.852	-598.278	104.816
300	44.593	256.855	256.580	0.082	-587.865	-598.343	104.181
350	46.691	263.891	257.131	2.366	-588.205	-600.063	89.554
400	48.456	270.244	258.379	4.746	-588.542	-601.733	78.578
450	49.912	276.038	260.024	7.206	-588.873	-603.362	70.036
500	51.105	281.361	261.896	9.733	-589.199	-604.955	63.199
600	52.886	290.847	265.950	14.938	-589.843	-608.046	52.935
700	54.106	299.096	270.109	20.291	-590.489	-611.029	45.596
800	54.964	306.380	274.196	25.747	-591.151	-613.918	40.085
900	55.586	312.892	278.140	31.276	-591.839	-616.723	35.794
1000	56.049	318.774	281.914	36.859	-592.556	-619.449	32.357
1100	56.401	324.133	285.513	42.483	-593.309	-622.103	29.541
1200	56.674	329.053	288.939	48.137	-594.099	-624.685	27.192
1300	56.891	333.598	292.201	53.815	-594.930	-627.201	25.201
1400	57.065	337.821	295.311	59.514	-595.805	-629.651	23.493
1500	57.206	341.763	298.278	65.227	-596.726	-632.036	22.009
1600	57.324	345.459	301.112	70.954	-597.694	-634.359	20.710
1700	57.421	348.937	303.824	76.691	-648.887	-636.173	19.547
1800	57.504	352.221	306.423	82.438	-649.720	-635.401	18.439
1900	57.574	355.332	308.916	88.192	-650.558	-634.583	17.446
2000	57.634	358.287	311.311	93.952	-651.398	-633.720	16.551
2100	57.686	361.100	313.615	99.718	-652.239	-632.816	15.740
2200	57.731	363.785	315.835	105.489	-653.077	-631.871	15.003
2300	57.771	366.352	317.976	111.264	-653.910	-630.888	14.328
2400	57.806	368.811	320.043	117.043	-654.734	-629.870	13.709
2500	57.837	371.172	322.042	122.825	-655.546	-628.817	13.138
2600	57.865	373.441	323.975	128.610	-656.343	-627.732	12.611
2700	57.891	375.625	325.848	134.398	-657.121	-626.617	12.123
2800	57.915	377.731	327.664	140.188	-657.878	-625.473	11.668
2900	57.938	379.764	329.425	145.981	-658.610	-624.303	11.245
3000	57.959	381.728	331.136	151.776	-659.316	-623.108	10.849
3100	57.981	383.629	332.799	157.573	-659.993	-621.890	10.479
3200	58.002	385.470	334.416	163.372	-660.640	-620.650	10.131
3300	58.024	387.255	335.991	169.174	-661.254	-619.391	9.804
3400	58.047	388.988	337.524	174.977	-661.834	-618.113	9.496
3500	58.072	390.671	339.019	180.783	-662.380	-616.819	9.206
3600	58.098	392.307	340.476	186.592	-1047.041	-605.049	8.779
3700	58.127	393.899	341.899	192.403	-1047.100	-592.771	8.368
3800	58.159	395.450	343.288	198.217	-1047.124	-580.492	7.979
3900	58.193	396.961	344.645	204.035	-1047.111	-568.212	7.610
4000	58.231	398.435	345.971	209.856	-1047.062	-555.933	7.260
4100	58.272	399.873	347.268	215.681	-1046.976	-543.656	6.926
4200	58.318	401.278	348.537	221.510	-1046.852	-531.381	6.609
4300	58.367	402.651	349.780	227.345	-1046.691	-519.110	6.306
4400	58.421	403.993	350.997	233.184	-1046.491	-506.843	6.017
4500	58.480	405.307	352.189	239.029	-1046.254	-494.580	5.741
4600	58.544	406.593	353.358	244.880	-1045.979	-482.324	5.477
4700	58.612	407.853	354.504	250.738	-1045.666	-470.074	5.224
4800	58.685	409.087	355.628	256.603	-1045.315	-457.831	4.982
4900	58.764	410.298	356.732	262.475	-1044.927	-445.596	4.750
5000	58.847	411.486	357.815	268.356	-1044.502	-433.369	4.527
5100	58.936	412.652	358.879	274.245	-1044.041	-421.151	4.313
5200	59.029	413.798	359.924	280.143	-1043.542	-408.942	4.108
5300	59.128	414.923	360.951	286.051	-1043.008	-396.743	3.910
5400	59.232	416.029	361.961	291.969	-1042.438	-384.555	3.720
5500	59.340	417.117	362.954	297.897	-1041.832	-372.377	3.537
5600	59.453	418.187	363.931	303.837	-1041.192	-360.211	3.360
5700	59.570	419.241	364.892	309.788	-1040.517	-348.057	3.190
5800	59.692	420.278	365.838	315.751	-1039.809	-335.915	3.025
5900	59.818	421.299	366.769	321.726	-1039.067	-323.784	2.867
6000	59.948	422.306	367.686	327.715	-1038.292	-311.668	2.713

PREVIOUS: December 1977 (1 atm) CURRENT: December 1977 (1 bar)

STRONTIUM FLUORIDE (SrF_2) CRYSTAL M_r = 125.616806

$S°(298.15\ K) = 82.123 \pm 0.21\ J\ K^{-1}\ mol^{-1}$
$T_{trs}H° = [1421]\ K$
$T_{trs}H° = [1484]\ K$
$T_{fus} = 1750 \pm 2\ K$

$\Delta_f H°(0\ K) = -1214.807 \pm 2.9\ kJ\ mol^{-1}$
$\Delta_f H°(298.15\ K) = -1217.126 \pm 2.9\ kcal\ mol^{-1}$
$\Delta_{trs}H° = [0.0]\ kJ\ mol^{-1}$
$\Delta_{trs}H° = [0.0]\ kJ\ mol^{-1}$
$\Delta_{fus}H° = 29.673 \pm 0.628\ kJ\ mol^{-1}$

Enthalpy of Formation

Guntz (1) measured the enthalpy of neutralization of $Sr(OH)_2$ (aq) with HF(aq) as ΔH = -35.8 kcal mol^{-1} at 284 K. We estimate a $\Delta C°_p$ = 105 cal K^{-1} mol^{-1} for $Sr(OH)_2 \cdot 555\ H_2O + 2(HF \cdot 110\ H_2O) \rightarrow SrF_2(cr) + 2\ H_2O(\ell)$ which gives ΔH(298.15 K) = -34.1 kcal mol^{-1}. This value leads to $\Delta_f H°(SrF_2$, cr, 298.15 K) = -291.3 kcal mol^{-1} when combined with the following enthalpies of formation: $\Delta_f H°(H_2O$, ℓ, 298.15 K) = -68.315 (2), $\Delta_f H°(Sr(OH)_2 \cdot 555\ H_2O$, 298.15 K) = -240.1 (3), and $\Delta_f H°(HF \cdot 110\ H_2O$, 298.15 K) = -76.78 kcal mol^{-1} (4).

Petersen (5) measured the enthalpy of precipitation of SrF_2 from an aqueous $SrCl_2$ solution with AgF(aq) as ΔH -34.5 kcal mol^{-1} at 293 K. We estimate a $\Delta C°_p$ = 108 cal K^{-1} mol^{-1} for $2(AgF \cdot 200\ H_2O) + SrCl_2 \cdot 400\ H_2O \rightarrow SrF_2(cr) + 2\ AgCl(cr)$ which gives ΔH(298.15 K) = -34.0 kcal mol^{-1}. This value leads to $\Delta_f H°(SrF_2$, cr, 298.15 K) = -290.9 kcal mol^{-1} when combined with the following enthalpies of formation: $\Delta_f H°$(AgCl, cr, 298.15 K) = -30.37 (6), $\Delta_f H°(AgF \cdot 200\ H_2O$, 298.15 K) = -53.398 (6), and $\Delta_f H°(SrCl_2 \cdot 400\ H_2O$, 298.15 K) = -209.871 kcal mol^{-1} (3).

Stephen and Stephen (7) have selected as a "best" value for the solubility of SrF_2 0.119 g/ℓ at 298.15 K which corresponds to $K_{sp} = 3.397 \times 10^{-9}$. Using $\Delta_f G°(Sr^{+2}$, aq ∞, 298.15 K) = -133.71 (3) and $\Delta_f G°(F^-$, aq ∞, 298.15 K) = -66.96 kcal mol^{-1} (4), we calculate $\Delta_f G°(SrF_2$, cr, 298.15 K) = -279.18 kcal mol^{-1}. Based on the adopted functions for SrF_2, this value gives $\Delta_f H°(SrF_2$, cr, 298.15 K) = -291.5 kcal mol^{-1}.

Very recently, Finch et al. (9) measured the enthalpy of precipitation of SrF_2 from an aqueous $SrCl_2$ solution by addition of a slight excess of NaF. X-ray diffraction patterns showed the precipitate to be crystalline SrF_2. Reduction of their data gives $\Delta_{sln}H°$ = 0.42 kcal mol^{-1} for $SrF_2(cr) + Sr^{+2}(aq\ \infty) + 2\ F^-(aq\ \infty)$ from which we obtain $\Delta_f H°(SrF_2$, cr, 298.15 K) = -290.5±0.5 kcal mol^{-1} with $\Delta_f H°(Sr^{+2}$, aq ∞, 298.15 K) = -130.45 (3) and $\Delta_f H°(F^-$, aq ∞, 298.15 K) = -79.82 kcal mol^{-1}.

The adopted enthalpy of formation, $\Delta_f H°(SrF_2$, cr, 298.15 K) = -290.9±0.7 kcal mol^{-1}, is a weighted value of these four results. Barany et al. (8) determined the enthalpy of solution of SrO(cr) in 20.1 wt % aqueous HF, which contained some dissolved silica, as -65.41±0.30 kcal mol^{-1}. We assume this value refers to the process $SrO(cr) + 2\ HF \cdot 4.41\ H_2O \rightarrow SrF_2(cr) + 2\ H_2O(\ell)$ at 298.15 K, and we calculate $\Delta_f H°(SrF_2$, cr, 298.15 K) = -291.8±2.0 kcal mol^{-1} with $\Delta_f H°$(SrO, cr, 298.15 K) = -141.5 kcal mol^{-1} (4). These results substantiate our adopted $\Delta_f H°$ value.

Heat Capacity and Entropy

$C°_p$ below 300 K is based on the low-temperature (11-300 K) heat capacities for SrF_2 reported by Smith et al. (10). The sample purity was 99.9% by petrographic examination. We have fitted a smooth polynomial curve through their data and obtained S°(298.15 K) = 19.63±0.05 cal K^{-1} mol^{-1} based on S(10.9) = 0.021 cal K^{-1} mol^{-1}.

The only high temperature enthalpy data is that recently reported by Efremova and Matizen (11) in the temperature range 813-2126 K. Spectroscopic analysis showed the SrF_2 sample contained less than 0.04% of foreign metals. A second order transition was observed between the temperatures of 1421 and 1484 K. We adopt their smoothed $C°_p$ data in the temperature range 900-1750 K; $C°_p$'s above the melting point are obtained by graphical extrapolation. $C°_p$ is assumed constant at 39.5 cal K^{-1} mol^{-1} over the temperature interval (1421-1484 K) of the transition (see further discussion in Transition Data). The observed enthalpies of Efremova and Matizen (11) deviate from our adopted ones by 1.4% at the lowest temperature (813 K) of their measurements and by only a few tenths of one percent at the higher temperatures.

Transition Data

Efremova and Matizen (11) reported that the phase transition for SrF_2 occurs over a short temperature interval (1421-1484 K). There are no discontinuities in their enthalpy data in this temperature range. Other alkaline earth dihalides (12), which have the calcium fluoride structure, are known to exhibit similar behavior, and this fact would seem to rule out the possibility that the observed transition resulted wholly from impurities. We speculate that the two crystalline forms, α and β, of SrF_2 are practically of the same energy content and can therefore, coexist over this temperature interval.

We choose to tentatively adopt two transitions with zero enthalpies, pending resolution of the phases present at these temperatures by X-ray diffraction measurements. The adopted values are placed in brackets above to emphasize that there is no confirmatory evidence for their existence.

Fusion Data

See the liquid table for details.

Sublimation Data

See the ideal gas table for details.

References

1. M. Guntz, Ann. Chim. Phys. 3, 5 (1884).
2. U. S. Nat. Bur. Stand. Tech. Note 270-3, 1968.
3. U. S. Nat. Bur. Stand. Tech. Note 270-6, 1971.
4. JANAF Thermochemical Tables: Consistent with HF(g), 12-31-68; SrO(cr) and $BaCl_2$(cr), 12-31-72; CaF_2, 12-31-68.
5. E. Petersen, Z. Physik. Chem. 4, 384 (1889).
6. U. S. Nat. Bur. Stand. Tech. Note 270-4, 1969.
7. H. Stephen and T. Stephen, "Solubilities of Inorganic and Organic Compounds," Vol. I, The Pergamon Press, Ltd., London, 1963.
8. R. Barany, E. G. King, and S. S. Todd, J. Amer. Chem. Soc. 79, 3639 (1957).
9. A. Finch, P. J. Gardner, and C. J. Steadman, Can. J. Chem. 46, 3447 (1968).
10. D. F. Smith, T. E. Gardner, B. B. Letson, and A. R. Taylor, U. S. Bur. Mines RI-6316, 1963.
11. R. I. Efremova and E. V. Matizen, Izv. Sib. Akad. Nauk SSSR, Ser. Khim. Nauk, (1), 3 (1970).
12. A. S. Dworkina and M. A. Bredig, J. Chem. Eng. Data, 8, 416 (1963).

Strontium Fluoride (SrF_2) F_2Sr_1(cr)

Enthalpy Reference Temperature = T_r = 298.15 K   Standard State Pressure = p° = 0.1 MPa

T/K	$C°_p$ (J K⁻¹mol⁻¹)	S° (J K⁻¹mol⁻¹)	-[G°-H°(T_r)]/T (J K⁻¹mol⁻¹)	H°-H°(T_r) (kJ mol⁻¹)	$\Delta_f H°$ (kJ mol⁻¹)	$\Delta_f G°$ (kJ mol⁻¹)	Log K_f
0	0.	0.	INFINITE	-13.075	-1214.807	-1214.807	INFINITE
100	36.671	21.017	137.682	-11.666	-1217.805	-1200.065	626.850
200	61.636	55.736	88.303	-6.513	-1218.070	-1182.045	308.718
298.15	69.985	82.123	82.123	0.	-1217.126	-1164.544	204.023
300	70.103	82.556	82.125	0.130	-1217.104	-1164.218	202.708
400	74.726	103.406	84.936	7.388	-1215.793	-1146.781	149.754
500	77.320	120.379	90.380	14.999	-1214.376	-1129.691	118.018
600	78.827	134.616	96.599	22.810	-1212.968	-1112.887	96.885
700	79.914	146.849	102.924	30.748	-1211.616	-1096.315	81.808
800	80.960	157.585	109.099	38.789	-1210.324	-1079.932	70.512
900	82.341	167.180	115.028	46.937	-1209.705	-1063.636	61.732
1000	85.772	176.024	120.691	55.334	-1208.031	-1047.493	54.715
1100	90.207	184.388	126.104	64.113	-1213.908	-1031.169	48.966
1200	97.069	192.513	131.300	73.455	-1212.258	-1014.624	44.165
1300	108.784	200.659	136.321	83.640	-1209.788	-998.246	40.110
1400	143.930	209.735	141.230	95.907	-1205.258	-982.123	36.643
1421.000	165.687	212.027	142.259	99.141	ALPHA <--> BETA	TRANSITION	
1421.000	165.268	212.027	142.259	99.141			
1484.000	165.268	219.197	145.374	109.553	UNDEFINED	TRANSITION	
1484.000	165.268	219.197	145.374	109.553			
1500	154.808	220.911	146.171	112.110	-1196.812	-966.475	33.656
1600	125.938	229.799	151.132	125.867	-1190.831	-951.330	31.058
1700	113.805	237.012	155.978	137.758	-1323.645	-935.314	28.739
1750.000	109.441	240.249	158.340	143.342	------- BETA <--> LIQUID -------		
1800	107.320	243.303	160.658	148.762	-1318.587	-912.620	26.484
1900	102.717	248.976	165.159	159.251	-1314.061	-890.192	24.473
2000	99.579	254.159	169.481	169.356	-1309.937	-867.991	22.670
2100	97.696	258.969	173.629	179.213	-1306.079	-845.990	21.043
2200	96.650	263.487	177.612	188.925	-1302.384	-824.168	19.568
2300	96.232	267.772	181.440	198.565	-1298.783	-802.511	18.226
2400	96.232	271.868	185.123	208.189	-1295.222	-781.011	16.998
2500	96.232	275.796	188.672	217.812	-1291.689	-759.658	15.872

PREVIOUS: CURRENT: December 1972

Strontium Fluoride (SrF_2) F_2Sr_1(cr)

STRONTIUM FLUORIDE (SrF_2) LIQUID M_r = 125.616806

$S°(298.15\ K) = [109.445]\ J\ k^{-1}\ mol^{-1}$

$T_{fus} = 1750 \pm 2\ K$

$\Delta_f H°(298.15\ K) = [-1172.043]\ kJ\ mol^{-1}$

$\Delta_{fus} H° = 29.673 \pm 0.628\ kJ\ mol^{-1}$

Enthalpy of Formation

$\Delta_f H°(SrF_2, \ell, 298.15\ K)$ is calculated from $\Delta_f H°(SrF_2, cr, 298.15\ K)$ by adding the enthalpy of fusion, $\Delta_{fus}H°$, and the difference in enthalpy, H°(1750 K)-H°(298.15 K), between the crystal and liquid.

Heat Capacity and Entropy

The heat capacity for liquid SrF_2 is obtained from the enthalpy measurements of Efremova and Matizen (1). $C_p°$ is assumed to be constant in the temperature range 1200-3500 K. At 1200 K a glass transition is assumed below which $C_p°$ follows that of the crystal. The entropy is calculated in a manner analogous to that used for the enthalpy of formation.

Fusion Data

Kojima et al. (2) determined the melting point of a SrF_2 sample, which had been purified by passing anhydrous HF through the melt, as 1746±1 K. Porter and Brown (3) found a melting point of 1736±5 K for a sample of 99.8+ % purity, while Petit and Delbove (4) reported 1735 K. The adopted melting point, 1750±2 K, is that determined by Efremova and Matizen (1) in their calorimetric studies of a high purity SrF_2 sample. The enthalpy of melting, 7.092±0.150 kcal mol^{-1}, is calculated as the difference between the adopted enthalpies of the liquid and crystal at T_{fus}.

Vaporization Data

The boiling point is calculated as the temperature at which the fugacity is one atmosphere for the vaporization process. The heat of vaporization is the difference in $\Delta_f H°(T_{vap})$ for the liquid and gas.

Reference

1. R. I. Efremova and E. V. Matizen, Izv. Sib. Otd. Akad. Nauk SSSR, Ser. Khim. Nauk, (1), 3 (1970).
2. H. Kojima, S. G. Whiteway, and C. R. Masson, Can. J. Chem. 46, 2968 (1968).
3. B. Porter and E. A. Brown, J. Amer. Ceram. Soc. 45, 49 (1962).
4. G. Petit and F. Delbove, Compt. rend. 254, 1388 (1962).

Strontium Fluoride (SrF_2) $F_2Sr_1(l)$

Enthalpy Reference Temperature = T_r = 298.15 K   Standard State Pressure = p° = 0.1 MPa

T/K	$C_p°$ (J K⁻¹mol⁻¹)	S°	$-[G°-H°(T_r)]/T$	$H°-H°(T_r)$ (kJ mol⁻¹)	$\Delta_f H°$	$\Delta_f G°$	Log K_f
0							
100							
200							
298.15	69.977	109.445	109.445	0.	-1172.043	-1127.607	197.552
300	70.103	109.878	109.446	0.130	-1172.021	-1127.331	196.286
400	74.726	130.727	112.257	7.388	-1170.711	-1112.627	145.294
500	77.320	147.701	117.702	14.999	-1169.293	-1098.269	114.735
600	78.827	161.937	123.920	22.810	-1167.885	-1084.197	94.388
700	79.914	174.171	130.246	30.748	-1166.533	-1070.357	79.871
800	80.960	184.906	136.421	38.789	-1165.242	-1056.706	68.996
900	82.341	194.502	142.350	46.937	-1164.622	-1043.142	60.542
1000	85.772	203.346	148.012	55.334	-1162.948	-1029.732	53.788
1100	90.207	211.710	153.425	64.113	-1168.825	-1016.139	48.252
1200	97.069	219.835	158.622	73.456	-1167.174	-1002.327	43.630
1200.000	97.069	219.835	158.622	73.456	GLASS <--> LIQUID	TRANSITION	
1200.000	99.048	219.835	158.622	73.456			
1300	99.048	227.763	163.640	83.361	-1164.985	-988.678	39.726
1400	99.048	235.104	168.485	93.266	-1162.817	-975.198	36.385
1500	99.048	241.937	173.157	103.170	-1160.669	-961.872	33.495
1600	99.048	248.330	177.658	113.075	-1158.539	-948.688	30.971
1700	99.048	254.334	181.993	122.980	-1293.340	-934.457	28.712
1750.000	99.048	257.205	184.101	127.932	------- BETA <-->	LIQUID -------	
1800	99.048	259.996	186.171	132.885	-1289.381	-913.461	26.508
1900	99.048	265.351	190.199	142.790	-1285.440	-892.684	24.542
2000	99.048	270.431	194.084	152.694	-1281.516	-872.114	22.777
2100	99.048	275.264	197.836	162.599	-1277.610	-851.740	21.186
2200	99.048	279.872	201.461	172.504	-1273.722	-831.552	19.744
2300	99.048	284.275	204.966	182.409	-1269.856	-811.540	18.431
2400	99.048	288.490	208.359	192.313	-1266.014	-791.696	17.231
2500	99.048	292.533	211.646	202.218	-1262.199	-772.011	16.130
2600	99.048	296.418	214.832	212.123	-1258.416	-752.478	15.117
2700	99.048	300.156	217.924	222.028	-1254.671	-733.090	14.182
2800	99.048	303.758	220.925	231.933	-1250.970	-713.840	13.317
2900	99.048	307.234	223.842	241.837	-1247.320	-694.721	12.513
3000	99.048	310.592	226.678	251.742	-1243.729	-675.728	11.765
3100	99.048	313.840	229.437	261.647	-1240.204	-656.853	11.068
3200	99.048	316.984	232.124	271.552	-1236.754	-638.090	10.416
3300	99.048	320.032	234.742	281.457	-1233.386	-619.434	9.805
3400	99.048	322.989	237.294	291.361	-1230.108	-600.879	9.231
3500	99.048	325.860	239.784	301.266	-1226.929	-582.419	8.692

PREVIOUS: CURRENT: December 1972

Strontium Fluoride (SrF_2) $F_2Sr_1(l)$

STRONTIUM FLUORIDE (SrF_2) CRYSTAL-CRYSTAL-CRYSTAL-LIQUID

M_r = 125.616806 Strontium Fluoride (SrF_2) F_2Sr_1(cr,l)

0 to 1421 K crystal
1421 to 1484 K crystal
1484 to 1750 K crystal
above 1750 K liquid

Refer to the individual tables for details.

Enthalpy Reference Temperature = T_r = 298.15 K; Standard State Pressure = $p°$ = 0.1 MPa

T/K	$C_p°$ (J K⁻¹mol⁻¹)	S° (J K⁻¹mol⁻¹)	$-[G°-H°(T_r)]/T$ (J K⁻¹mol⁻¹)	$H°-H°(T_r)$ (kJ mol⁻¹)	$\Delta_fH°$ (kJ mol⁻¹)	$\Delta_fG°$ (kJ mol⁻¹)	Log K_f
0	0.	0.	INFINITE	-13.075	-1214.807	-1214.807	INFINITE
100	36.671	21.017	137.682	-11.666	-1217.805	-1200.065	626.850
200	61.636	55.736	88.303	-6.513	-1218.070	-1182.045	308.718
298.15	69.985	82.123	82.123	0.	-1217.126	-1164.544	204.023
300	70.103	82.556	82.125	0.130	-1217.104	-1164.218	202.708
400	74.726	103.406	84.936	7.388	-1215.793	-1146.781	149.754
500	77.320	120.379	90.380	14.999	-1214.376	-1129.691	118.018
600	78.827	134.616	96.599	22.810	-1212.968	-1112.887	96.885
700	79.914	146.849	102.924	30.748	-1211.616	-1096.315	81.808
800	80.960	157.585	109.099	38.789	-1210.324	-1079.932	70.512
900	82.341	167.180	115.028	46.937	-1209.705	-1063.636	61.732
1000	85.772	176.024	120.691	55.334	-1208.031	-1047.493	54.715
1100	90.207	184.388	126.104	64.113	-1213.908	-1031.169	48.966
1200	97.069	192.513	131.300	73.455	-1212.258	-1014.624	44.165
1300	108.784	200.659	136.321	83.640	-1209.788	-998.246	40.110
1400	143.930	209.735	141.230	95.907	-1205.258	-982.123	36.643
1421.000	165.687	212.027	142.259	99.141	ALPHA <--> BETA		
1421.000	165.268	212.027	142.259	99.141	TRANSITION		
1484.000	165.268	219.197	145.374	109.553	UNDEFINED		
1484.000	165.268	219.197	145.374	109.553	TRANSITION		
1500	154.808	220.911	146.171	112.110	-1196.812	-966.475	33.656
1600	125.938	229.799	151.132	125.867	-1190.831	-951.330	31.058
1700	113.805	237.012	155.978	137.758	-1323.645	-935.314	28.739
1750.000	109.441	240.249	158.340	143.342	BETA <--> LIQUID		
1750.000	99.048	257.205	158.340	173.015	TRANSITION		
1800	99.048	259.996	161.125	177.968	-1289.381	-913.461	26.508
1900	99.048	265.351	166.471	187.872	-1285.440	-892.684	24.542
2000	99.048	270.431	171.543	197.777	-1281.516	-872.114	22.777
2100	99.048	275.264	176.368	207.682	-1277.610	-851.740	21.186
2200	99.048	279.872	180.969	217.587	-1273.722	-831.552	19.744
2300	99.048	284.275	185.365	227.492	-1269.856	-811.540	18.431
2400	99.048	288.490	189.575	237.396	-1266.014	-791.695	17.231
2500	99.048	292.533	193.613	247.301	-1262.199	-772.011	16.130
2600	99.048	296.418	197.493	257.206	-1258.416	-752.478	15.117
2700	99.048	300.156	201.226	267.111	-1254.671	-733.090	14.182
2800	99.048	303.758	204.824	277.015	-1250.970	-713.840	13.317
2900	99.048	307.234	208.296	286.920	-1247.320	-694.721	12.513
3000	99.048	310.592	211.650	296.825	-1243.729	-675.728	11.765
3100	99.048	313.840	214.895	306.730	-1240.204	-656.853	11.068
3200	99.048	316.984	218.036	316.635	-1236.754	-638.090	10.416
3300	99.048	320.032	221.081	326.539	-1233.386	-619.434	9.805
3400	99.048	322.989	224.035	336.444	-1230.108	-600.879	9.231
3500	99.048	325.860	226.903	346.349	-1226.929	-582.419	8.692

PREVIOUS: CURRENT: December 1972

Strontium Fluoride (SrF_2) F_2Sr_1(cr,l)

STRONTIUM FLUORIDE (SrF_2) IDEAL GAS M_r = 125.616806

$S°(298.15\ K) = 291.715 \pm 2.1\ J\ K^{-1}\ mol^{-1}$

$\Delta_f H°(0\ K) = -764.053 \pm 4.2\ kJ\ mol^{-1}$

$\Delta_f H°(298.15\ K) = -766.090 \pm 4.2\ kJ\ mol^{-1}$

Vibrational Frequencies and Degeneracies

ν, cm^{-1}
442 (1)
82 (1)
443 (1)

Ground State Quantum Weight: 1 $\sigma = 2$

Point Group: C_{2v}

Bond Distance: Sr-F = 2.20 ± 0.03 Å

Bond Angle: F-Sr-F = 108°

Product of the Moments of Inertia: $I_A I_B I_C = 4.022689 \times 10^{-114}\ g^3\ cm^6$

Enthalpy of Formation

$\Delta_f H°$ is obtained from that of the crystal by addition of the adopted $\Delta_{sub}H°$ = 107.8 ± 0.3 kcal mol^{-1}. 2nd and 3rd law analyses of four sets of vapor pressure data for SrF_2(cr, ℓ) are listed below.

Source	Method	T/K	$\Delta_{sub}H°$(298.15 K), kcal mol^{-1} 2nd law	3rd law	Drift cal K^{-1} mol^{-1}	$\Delta_f H°$(298.15 K) kcal mol^{-1}
Ruff and Le Boucher (1)	Dynamic B.P.	2095-2232	106.4[a]	96.1±0.8	-4.8 ±3.1	-184.0±1.5
Bautista and Margrave (2)	Langmuir	1235-1319	104.5	108.0±0.5	2.8 ±2.9	-182.9±1.2
Green et al. (3)	Knudsen	1207-1563[b]	107.9	107.8±0.3	-0.07±0.38	-183.1±1.0
Hildenbrand (4)	Torsion-Effusion	1420-1710	103.8	106.5±0.4	1.7	-184.4±1.1

[a] $\Delta_{vap}H°$(298.15 K)

[b] Six of the 30 experimental points were rejected due to failure of a statistical test.

Note that the drifts are both positive and negative indicating the correctness of the functions used in the analysis. A weighted value of these four results is adopted.

Heat Capacity and Entropy

The bond length is that measured by Akishin and Spiridonov (5) in their high-temperature electron diffraction studies of the alkaline earth dihalide vapors. Although it was inferred from these measurements that all the dihalides are linear, recent results of electric-deflection experiments (6) and matrix isolation infrared spectra (1, 8) for SrF_2 can only be satisfactorily interpreted in terms of a molecule with a bent configuration. We adopt the vibrational frequencies and bond angle determined by Calder et al. (8) in their thorough investigation of the infrared spectra of isotopically enriched samples of SrF_2 trapped in krypton matrices. The uncertainty assigned to the entropy reflects possible errors due to matrix-shift effects in these frequencies. The adopted symmetric (ν_1) and antisymmetric (ν_3) stretching frequencies are in reasonable agreement with those determined by Snelson (7) for SrF_2 trapped in neon, argon, and krypton matrices. Also, the antisymmetric frequency is quite consistent with the value 445 ± 7 cm^{-1} observed by Baikov (9) in the infrared spectra of SrF_2 vapor at 2050 K. In an earlier paper, Baikov (10) reported observing under similar conditions a broad absorption band at 105 cm^{-1} which he assigned to the symmetric bending (ν_2) mode of SrF_2. We prefer the matrix value (82 cm^{-1}), since Baikov (10) recorded only the high-frequency (>90 cm^{-1}) portion of this band. Furthermore, Hastie et al. (11) have pointed out that, in general, bending frequencies are less influenced by matrix effects and in some cases may be shifted slightly upwards from the gas phase values.

The individual moments of inertia are: $I_A = 7.3594 \times 10^{-39}$, $I_B = 19.9877 \times 10^{-39}$, and $I_C = 27.3471 \times 10^{-39}\ g\ cm^2$.

References

1. O. Ruff and L. Le Boucher, Z. Anorg. Chem. 219, 376 (1934).
2. R. G. Bautista and J. L. Margrave, J. Phys. Chem. 69, 1770 (1965).
3. J. W. Green, G. D. Blue, T. C. Ehlert and J. L. Margrave, J. Chem. Phys. 41, 2245 (1964).
4. D. L. Hildenbrand, E. Murad, N. D. Potter, L. P. Theard, and W. F. Hall, Aeronutronic Report No. U-3183, Ford Motor Company, June 30, 1965.
5. P. A. Akishin and V. P. Spiridonov, Kristallografiya 2, 475 (1957).
6. L. Wharton, R. A. Berg and W. Klemperer, J. Chem. Phys. 39, 2023 (1963).
7. A. Snelson, J. Phys. Chem. 70, 3208 (1966).
8. V. Calder, D. E. Mann, K. S. Seshadri, M. Allavena, and D. White, J. Chem. Phys. 51, 2093 (1969).
9. V. I. Baikov, Opt. Spektrosk. 27, 923 (1969).
10. V. I. Baikov, Opt. Spektrosk. 25, 356 (1968).
11. J. W. Hastie, R. H. Hange and J. L. Margrave in "Spectroscopy in Inorganic Chemistry," Ed. by C. N. R. Rao and J. R. Ferraro, Academic Press, New York, 1970.

Strontium Fluoride (SrF_2) F_2Sr_1(g)

Enthalpy Reference Temperature = T_r = 298.15 K Standard State Pressure = p° = 0.1 MPa

T/K	$C_p°$ (J $K^{-1}mol^{-1}$)	S°	$-[G°-H°(T_r)]/T$	$H°-H°(T_r)$ (kJ mol^{-1})	$\Delta_f H°$	$\Delta_f G°$	Log K_f
0	0.	0.	INFINITE	-13.356	-764.053	-764.053	INFINITE
100	41.833	240.072	335.836	-9.576	-764.680	-768.845	401.604
200	48.936	271.323	296.453	-5.026	-765.547	-772.639	201.793
250	51.366	282.520	292.579	-2.515	-765.841	-774.382	161.798
298.15	52.991	291.715	291.715	0.	-766.090	-775.998	135.952
300	53.042	292.043	291.716	0.098	-766.100	-776.060	135.124
350	54.207	300.312	292.366	2.781	-766.358	-777.700	116.065
400	55.034	307.607	293.825	5.513	-766.633	-779.301	101.766
450	55.637	314.126	295.725	8.281	-766.934	-780.867	90.641
500	56.087	320.012	297.864	11.074	-767.266	-782.398	81.737
600	56.698	330.297	302.437	16.716	-768.027	-785.355	68.371
700	57.081	339.068	307.059	22.406	-768.922	-788.174	58.814
800	57.335	346.708	311.548	28.128	-769.950	-790.856	51.638
900	57.513	353.472	315.837	33.871	-771.735	-793.329	46.044
1000	57.641	359.538	319.909	39.629	-772.701	-795.677	41.562
1100	57.736	365.037	323.766	45.398	-781.588	-797.561	37.873
1200	57.810	370.064	327.417	51.175	-783.502	-798.929	34.776
1300	57.867	374.693	330.878	56.959	-785.434	-800.136	32.150
1400	57.912	378.983	334.163	62.748	-787.382	-801.194	29.893
1500	57.949	382.980	337.286	68.542	-789.345	-802.112	27.932
1600	57.979	386.721	340.260	74.338	-791.324	-802.899	26.212
1700	58.005	390.237	343.097	80.137	-930.231	-802.382	24.654
1800	58.026	393.553	345.809	85.939	-930.375	-794.857	23.066
1900	58.043	396.691	348.405	91.742	-930.535	-787.324	21.645
2000	58.059	399.668	350.895	97.547	-930.710	-779.782	20.366
2100	58.072	402.501	353.285	103.354	-930.902	-772.231	19.208
2200	58.083	405.203	355.584	109.162	-931.112	-764.671	18.156
2300	58.093	407.785	357.798	114.971	-931.342	-757.100	17.194
2400	58.102	410.258	359.933	120.780	-931.595	-749.519	16.313
2500	58.110	412.630	361.993	126.591	-931.874	-741.927	15.502
2600	58.117	414.909	363.985	132.402	-932.185	-734.323	14.753
2700	58.123	417.102	365.912	138.214	-932.533	-726.706	14.059
2800	58.128	419.216	367.778	144.027	-932.924	-719.076	13.415
2900	58.133	421.256	369.587	149.840	-933.365	-711.431	12.814
3000	58.138	423.227	371.343	155.653	-933.865	-703.770	12.254
3100	58.142	425.134	373.047	161.467	-934.431	-696.091	11.729
3200	58.145	426.979	374.704	167.282	-935.071	-688.393	11.237
3300	58.148	428.769	376.315	173.096	-935.794	-680.673	10.774
3400	58.152	430.505	377.884	178.911	-936.606	-672.930	10.338
3500	58.154	432.190	379.411	184.727	-937.516	-665.162	9.927
3600	58.157	433.829	380.900	190.542	-938.530	-657.366	9.538
3700	58.159	435.422	382.352	196.358	-939.654	-649.541	9.170
3800	58.161	436.973	383.770	202.174	-940.893	-641.684	8.821
3900	58.163	438.484	385.153	207.990	-942.252	-633.792	8.489
4000	58.165	439.957	386.505	213.807	-943.733	-625.864	8.173
4100	58.167	441.393	387.826	219.623	-945.338	-617.898	7.872
4200	58.169	442.795	389.118	225.440	-947.065	-609.891	7.585
4300	58.170	444.163	390.383	231.257	-948.918	-601.842	7.311
4400	58.171	445.501	391.620	237.074	-950.895	-593.748	7.049
4500	58.173	446.808	392.832	242.891	-952.992	-585.607	6.798
4600	58.174	448.087	394.019	248.709	-955.140	-577.424	6.557
4700	58.175	449.338	395.183	254.526	-957.448	-569.188	6.326
4800	58.176	450.562	396.324	260.344	-959.857	-560.902	6.104
4900	58.177	451.762	397.443	266.161	-962.365	-552.565	5.890
5000	58.178	452.937	398.542	271.979	-964.957	-544.176	5.685
5100	58.179	454.089	399.619	277.797	-967.614	-535.736	5.487
5200	58.180	455.219	400.678	283.615	-970.364	-527.241	5.296
5300	58.181	456.327	401.717	289.433	-973.181	-518.693	5.112
5400	58.181	457.415	402.739	295.251	-976.057	-510.091	4.934
5500	58.182	458.482	403.743	301.069	-978.984	-501.435	4.762
5600	58.183	459.531	404.730	306.887	-981.955	-492.725	4.596
5700	58.183	460.561	405.700	312.706	-984.962	-483.963	4.435
5800	58.184	461.573	406.655	318.524	-987.997	-475.147	4.279
5900	58.184	462.567	407.594	324.342	-990.843	-466.302	4.128
6000	58.185	463.545	408.518	330.161	-993.883	-457.386	3.982

PREVIOUS: December 1972 (1 atm) CURRENT: December 1972 (1 bar)

Strontium Fluoride (SrF_2) F_2Sr_1(g)

TITANIUM FLUORIDE (TiF_2) IDEAL GAS M_r = 85.876806

$S°(298.15\ K) = 255.690 \pm 6.3\ J\ K^{-1}\ mol^{-1}$

$\Delta_f H°(0\ K) = -686.627 \pm 41.8\ kJ\ mol^{-1}$
$\Delta_f H°(298.15\ K) = -688.268 \pm 41.8\ kJ\ mol^{-1}$

Electronic Levels and Quantum Weights

ϵ_i, cm^{-1}	g_i
0	[3]
[7000]	[6]
[17000]	[6]
[22000]	[15]

Vibrational Frequencies and Degeneracies

ν, cm^{-1}
[473](1)
[283](2)
[633](1)

Ground State Quantum Weight: 3 σ = 2
Point Group: [$D_{\infty h}$]
Bond Distance: Ti-F = [1.95] Å
Bond Angle: F-Ti-F = [180]°
Rotational Constant: B_0 = [0.116676] cm^{-1}

Enthalpy of Formation

Zmbov and Margrave (1) have reported ion intensities and the corresponding equilibrium constants for the reaction $Ca(g) + TiF_3(g) = CaF(g) + TiF_2(g)$. 2nd and 3rd law analyses of their data give a 2nd law $\Delta_r H°(298.15\ K)$ of 7.0 kcal mol^{-1}, a 3rd law $\Delta_r H°(298.15\ K)$ of 11.76 kcal mol^{-1}, and a 3rd law drift of 3.1 ± 3.8 cal K^{-1} mol^{-1}. The enthalpy of formation, $\Delta_f H°(298.15\ K)$, of $TiF_2(g)$ is calculated from the 3rd law $\Delta_r H°(298.15\ K)$, above, and the JANAF enthalpies of formation of $TiF_3(g)$, CaF(g) and Ca(g).

Heat Capacity and Entropy

The interatomic distances are estimated from those of $TiCl_2$, $TiCl_4$, and TiF_4. The vibrational frequencies are estimated from a valence force field model.

The electronic levels are assumed to be the same as $TiCl_2(g)$. The levels of $TiCl_2(g)$ are estimated by assuming they correspond to the inverted states of $NiCl_2(g)$ (2). The linear configuration is assumed because experimental evidence (3) indicates that other transition metal difluorides, viz those of Mn, Co, Ni, Cu, and Zn, are linear.

References

1. K. F. Zmbov and J. L. Margrave, J. Phys. Chem. 71, 2893 (1967).
2. C. W. DeKock and D. M. Gruen, J. Chem. Phys. 44, 4387 (1966).
3. A. Buchler, J. L. Stauffer, and W. Klemperer, J. Chem. Phys. 40, 3471 (1964).

Titanium Fluoride (TiF_2) $F_2Ti_1(g)$

Enthalpy Reference Temperature = T_r = 298.15 K   Standard State Pressure = p° = 0.1 MPa

	J $K^{-1}mol^{-1}$			kJ mol^{-1}			
T/K	$C_p°$	S°	$-[G°-H°(T_r)]/T$	$H°-H°(T_r)$	$\Delta_f H°$	$\Delta_f G°$	Log K_f
0	0.	0.	INFINITE	-12.013	-686.627	-686.627	INFINITE
100	34.469	208.051	297.843	-8.979	-687.060	-690.002	360.420
200	46.288	235.837	260.345	-4.902	-687.828	-692.620	180.894
250	50.179	246.608	256.546	-2.485	-688.077	-693.787	144.959
298.15	52.878	255.690	255.690	0.	-688.268	-694.869	121.738
300	52.965	256.017	255.691	0.098	-688.275	-694.910	120.995
350	54.966	264.340	256.343	2.799	-688.450	-696.002	103.873
400	56.425	271.780	257.816	5.586	-688.620	-697.070	91.028
450	57.510	278.492	259.747	8.435	-688.795	-698.115	81.035
500	58.333	284.595	261.931	11.332	-688.979	-699.141	73.039
600	59.470	295.340	266.628	17.227	-689.383	-701.136	61.039
700	60.194	304.565	271.404	23.213	-689.837	-703.060	52.463
800	60.686	312.637	276.064	29.258	-690.326	-704.916	46.026
900	61.046	319.807	280.534	35.345	-690.884	-706.707	41.016
1000	61.336	326.254	284.789	41.465	-691.575	-708.430	37.005
1100	61.597	332.112	288.829	47.612	-692.464	-710.074	33.719
1200	61.857	337.483	292.663	53.784	-697.532	-711.509	30.971
1300	62.134	342.445	296.303	59.984	-698.082	-712.651	28.635
1400	62.437	347.060	299.766	66.212	-698.703	-713.749	26.630
1500	62.770	351.379	303.065	72.472	-699.404	-714.800	24.892
1600	63.132	355.442	306.212	78.767	-700.192	-715.801	23.369
1700	63.518	359.281	309.222	85.099	-701.078	-716.750	22.023
1800	63.924	362.923	312.105	91.471	-702.069	-717.644	20.825
1900	64.341	366.390	314.872	97.885	-703.176	-718.480	19.752
2000	64.764	369.701	317.531	104.340	-719.094	-718.800	18.773
2100	65.185	372.871	320.091	110.837	-721.207	-718.734	17.878
2200	65.599	375.913	322.560	117.377	-723.281	-718.567	17.061
2300	66.002	378.838	324.944	123.957	-725.313	-718.307	16.313
2400	66.391	381.655	327.248	130.577	-727.300	-717.960	15.626
2500	66.762	384.373	329.479	137.234	-729.241	-717.530	14.992
2600	67.113	386.998	331.641	143.928	-731.133	-717.025	14.405
2700	67.445	389.538	333.739	150.656	-732.975	-716.447	13.860
2800	67.758	391.996	335.776	157.417	-734.766	-715.802	13.353
2900	68.051	394.379	337.756	164.207	-736.505	-715.093	12.880
3000	68.325	396.691	339.682	171.026	-738.191	-714.326	12.438
3100	68.582	398.935	341.557	177.872	-739.824	-713.504	12.022
3200	68.823	401.117	343.385	184.742	-741.403	-712.629	11.632
3300	69.050	403.238	345.166	191.636	-742.929	-711.707	11.265
3400	69.263	405.302	346.905	198.552	-744.401	-710.738	10.919
3500	69.465	407.313	348.602	205.488	-745.820	-709.727	10.592
3600	69.656	409.273	350.260	212.444	-747.186	-708.676	10.283
3700	69.838	411.184	351.881	219.419	-1157.598	-699.800	9.879
3800	70.012	413.049	353.466	226.412	-1157.626	-687.427	9.449
3900	70.179	414.869	355.018	233.421	-1157.657	-675.053	9.041
4000	70.339	416.648	356.536	240.447	-1157.692	-662.678	8.654
4100	70.494	418.387	358.024	247.489	-1157.730	-650.302	8.285
4200	70.643	420.087	359.481	254.546	-1157.771	-637.925	7.934
4300	70.788	421.751	360.910	261.618	-1157.812	-625.548	7.599
4400	70.928	423.380	362.311	268.703	-1157.854	-613.169	7.279
4500	71.063	424.976	363.686	275.803	-1157.896	-600.789	6.974
4600	71.195	426.539	365.036	282.916	-1157.937	-588.409	6.682
4700	71.322	428.072	366.361	290.042	-1157.975	-576.027	6.402
4800	71.445	429.575	367.662	297.180	-1158.011	-563.645	6.134
4900	71.564	431.049	368.941	304.331	-1158.043	-551.262	5.877
5000	71.678	432.496	370.197	311.493	-1158.071	-538.878	5.630
5100	71.788	433.916	371.433	318.666	-1158.093	-526.494	5.392
5200	71.893	435.311	372.648	325.850	-1158.109	-514.110	5.164
5300	71.994	436.682	373.843	333.044	-1158.118	-501.725	4.945
5400	72.089	438.028	375.019	340.249	-1158.119	-489.340	4.733
5500	72.180	439.352	376.177	347.462	-1158.112	-476.956	4.530
5600	72.266	440.653	377.317	354.685	-1158.097	-464.571	4.333
5700	72.346	441.933	378.439	361.915	-1158.072	-452.187	4.144
5800	72.421	443.192	379.545	369.154	-1158.037	-439.804	3.961
5900	72.490	444.431	380.634	376.399	-1157.992	-427.420	3.784
6000	72.554	445.649	381.708	383.651	-1157.936	-415.038	3.613

PREVIOUS: December 1968 (1 atm)   CURRENT: December 1968 (1 bar)

Titanium Fluoride (TiF_2) $F_2Ti_1(g)$

ZIRCONIUM FLUORIDE (ZrF_2) CRYSTAL M_r = 129.216806

$S°(298.15\ K) = [75.312 \pm 8.4]\ J\ K^{-1}\ mol^{-1}$

$T_{fus} = [1175 \pm 150]\ K$

$\Delta_f H°(298.15\ K) = [-962.320 \pm 63]\ kJ\ mol^{-1}$

$\Delta_{fus}H° = [38 \pm 8]\ kJ\ mol^{-1}$

$\Delta_{sub}H°(298.15\ K) = [404.174]\ kJ\ mol^{-1}$

Enthalpy of Formation

The adopted $\Delta_f H°(ZrF_2$, cr, 298.15 K) value is taken from NBS (1) which was estimated by Brewer (2). More recent thermal data related to the enthalpy of formation of ZrF_2(cr) is unavailable.

Heat Capacity and Entropy

The heat capacities in the temperature range 298-2000 K are estimated by comparison with those of ZrF_4(cr) and Zr(cr). The S°(298.15 K) is calculated as 17.2 and 17.0 cal $K^{-1}\ mol^{-1}$ based on the assumptions that $S(ZrF_2) = 1/2[S(ZrF_4) + S(Zr)]$ and $S(ZrF_2) = S(ZnF_2) - S(Zn) + S(Zr)$. Using the additive entropy constants S°(298.15 K) = 12.1 and 4.7 cal $K^{-1}\ mol^{-1}$ for Zr and F suggested by Kubaschewski and Evans (3) we obtain $S°(ZrF_2$, cr, 298.15 K) = 12.1 + 2 (4.7) = 21.5 cal $K^{-1}\ mol^{-1}$. Based on S°(298.15 K) = 11.5 and 3.9 cal $K^{-1}\ mol^{-1}$ for Zr^{4+} and F^- recommended by Kelley (4), the value $S°(ZrF_2$, cr, 298.15 K) is derived as 19.3 cal $K^{-1}\ mol^{-1}$. The entropy at 298 K for ZrF_2(cr) is tentatively adopted as 18 ± 2 cal $K^{-1}\ mol^{-1}$.

Fusion Data

Both T_{fus} and $\Delta_{fus}H°$ are estimated. The derived entropy of melting, 2.55 cal K^{-1} g-atom^{-1}, is the same as that of ZrF_4(cr) which was obtained from experimental measurements (5). Therefore these estimated values are tentatively adopted.

Sublimation Data

The difference between $\Delta_f H°(298.15\ K)$ for ZrF(g) and ZrF_2(cr) is $\Delta_{sub}H°(298.15\ K)$.

References

1. U. S. Nat. Bur. Stand. Circ. 500, 1952.
2. L. Brewer, University of California, Berkeley, California, unpublished data.
3. O. Kubaschewski and E. L. Evans, "Metallurgical Thermochemistry," Pergamon Press, New York, 1958.
4. K. K. Kelley, U.S. Bur. Mines, personal communication, June, 1960.
5. R. A. McDonald, G. C. Sinke, and D. R. Stull, J. Chem. Eng. Data 7, 83 (1962).

Zirconium Fluoride (ZrF_2) F_2Zr_1(cr)

Enthalpy Reference Temperature = T_r = 298.15 K Standard State Pressure = p° = 0.1 MPa

T/K	$C_p°$	S°	$-[G°-H°(T_r)]/T$	$H°-H°(T_r)$	$\Delta_f H°$	$\Delta_f G°$	Log K_f
	J K⁻¹mol⁻¹			kJ mol⁻¹			
0							
100							
200							
298.15	65.940	75.312	75.312	0.	-962.320	-912.724	159.905
300	66.024	75.720	75.313	0.122	-962.303	-912.416	158.866
400	69.873	95.253	77.950	6.921	-961.282	-895.937	116.997
500	73.220	111.212	83.054	14.079	-960.115	-879.733	91.905
600	76.149	124.827	88.908	21.551	-958.807	-863.777	75.198
700	78.659	136.761	94.909	29.296	-957.380	-848.050	63.282
800	80.626	147.395	100.817	37.262	-955.874	-832.533	54.359
900	82.425	156.997	106.535	45.416	-954.322	-817.209	47.430
1000	84.015	165.766	112.026	53.741	-952.743	-802.058	41.895
1100	85.270	173.834	117.283	62.206	-951.168	-787.066	37.375
1175.000	86.107	179.486	121.074	68.633	------ CRYSTAL <--> LIQUID -----		
1200	86.400	181.302	122.310	70.790	-953.333	-771.995	33.604
1300	87.404	188.258	127.119	79.481	-951.279	-756.966	30.415
1400	88.282	194.770	131.721	88.268	-949.193	-742.097	27.688
1500	88.826	200.879	136.130	97.124	-947.112	-727.377	25.330
1600	89.328	206.629	140.358	106.033	-945.058	-712.796	23.270
1700	89.621	212.053	144.418	114.981	-943.056	-698.341	21.457
1800	89.914	217.185	148.319	123.960	-941.123	-684.002	19.849
1900	89.956	222.048	152.073	132.954	-939.282	-669.768	18.413
2000	89.956	226.663	155.688	141.950	-937.554	-655.628	17.123

PREVIOUS: June 1961 CURRENT: June 1969

Zirconium Fluoride (ZrF_2) F_2Zr_1(cr)

ZIRCONIUM FLUORIDE (ZrF_2) LIQUID M_r = 129.216806

$S°(298.15 K)$ = [100.846] J K^{-1} mol^{-1} $\Delta_f H°(298.15 K)$ = [-936.951] kJ mol^{-1}

T_{fus} = [1175 ± 150] K $\Delta_{fus}H°$ = [38 ± 8] kJ mol^{-1}

Enthalpy of Formation

$\Delta_f H°(ZrF_2, \ell, 298.15 K)$ is calculated from $\Delta_f H°(ZrF_2$, cr, 298.15 K) by adding the enthalpy of fusion, $\Delta_{fus}H°$, and the difference in enthalpy, H°(1175 K)-H°(298.15 K), between the crystal and liquid.

Heat Capacity and Entropy

The heat capacity is estimated from those of $ZrF_4(\ell)$, $CaF_2(\ell)$, $Zr(\ell)$ and $Ca(\ell)$ and is assumed to be constant in the temperature range 800-3000 K. A glass transition temperature is assumed at 800 K. The $C_p°$ values below 800 K are adopted from the ZrF_2(cr) table. The entropy is calculated in a manner analogous to that used for the enthalpy of formation.

Fusion Data

Both T_{fus} and $\Delta_{fus}H°$ are estimated. The derived entorpy of melting, 2.55 cal K^{-1} mol^{-1}/atom, is the same as that of ZrF_4(cr) which was obtained from experimental measurements (1). Therefore these estimated values are tentatively adopted.

Vaporization Data

The boiling point, T_{vap}, is calculated as the temperature at which the fugacity is one atmosphere for the process $ZrF_2(\ell)$ = $ZrF_2(g)$.

Reference

1. R. A. McDonald, G. C. Sinke and D. R. Stull, J. Chem. Eng. Data 7, 83 (1962).

Zirconium Fluoride (ZrF_2) $F_2Zr_1(l)$

Enthalpy Reference Temperature = T_r = 298.15 K   Standard State Pressure = p° = 0.1 MPa

T/K	$C_p°$ (J $K^{-1}mol^{-1}$)	S° (J $K^{-1}mol^{-1}$)	$-[G°-H°(T_r)]/T$ (J $K^{-1}mol^{-1}$)	$H°-H°(T_r)$ (kJ mol^{-1})	$\Delta_f H°$ (kJ mol^{-1})	$\Delta_f G°$ (kJ mol^{-1})	Log K_f
0							
100							
200							
298.15	65.940	100.846	100.846	0.	-936.951	-894.968	156.795
300	66.024	101.254	100.847	0.122	-936.934	-894.708	155.782
400	69.873	120.787	103.484	6.921	-935.914	-880.782	115.018
500	73.220	136.746	108.588	14.079	-934.746	-867.131	90.589
600	76.149	150.361	114.442	21.551	-933.439	-853.729	74.324
700	78.659	162.295	120.443	29.296	-932.012	-840.555	62.723
800	80.626	172.933	126.351	37.265	-930.502	-827.592	54.036
800.000	80.626	172.933	126.351	37.265	GLASS <--> LIQUID		
800.000	100.416	172.933	126.351	37.265	TRANSITION		
900	100.416	184.760	132.197	47.306	-927.063	-814.936	47.298
1000	100.416	195.340	137.992	57.348	-923.767	-802.656	41.926
1100	100.416	204.910	143.647	67.390	-920.616	-790.698	37.547
1175.000	100.416	211.534	147.771	74.921	------ CRYSTAL <--> LIQUID -----		
1200	100.416	213.648	149.122	77.431	-921.322	-778.800	33.900
1300	100.416	221.685	154.398	87.473	-917.919	-767.061	30.821
1400	100.416	229.127	159.474	97.514	-914.578	-755.583	28.191
1500	100.416	236.055	164.351	107.556	-911.311	-744.340	25.920
1600	100.416	242.536	169.037	117.598	-908.125	-733.313	23.940
1700	100.416	248.623	173.541	127.639	-905.029	-722.483	22.199
1800	100.416	254.363	177.874	137.681	-902.033	-711.832	20.657
1900	100.416	259.792	182.043	147.722	-899.144	-701.344	19.281
2000	100.416	264.943	186.061	157.764	-896.371	-691.006	18.047
2100	100.416	269.842	189.935	167.806	-893.719	-680.804	16.934
2200	100.416	274.513	193.674	177.847	-912.521	-669.979	15.907
2300	100.416	278.977	197.286	187.889	-910.551	-658.998	14.966
2400	100.416	283.251	200.780	197.930	-908.577	-648.103	14.106
2500	100.416	287.350	204.161	207.972	-906.594	-637.291	13.315
2600	100.416	291.288	207.437	218.014	-904.599	-626.558	12.588
2700	100.416	295.078	210.613	228.055	-902.587	-615.902	11.915
2800	100.416	298.730	213.695	238.097	-900.557	-605.321	11.292
2900	100.416	302.254	216.689	248.138	-898.505	-594.813	10.714
3000	100.416	305.658	219.598	258.180	-896.429	-584.377	10.175

PREVIOUS: June 1961 CURRENT: June 1969

Zirconium Fluoride (ZrF_2) $F_2Zr_1(l)$

ZIRCONIUM FLUORIDE (ZrF_2) CRYSTAL–LIQUID M_r = 129.216806 Zirconium Fluoride (ZrF_2) F_2Zr_1(cr,l)

0 to 1175 K crystal
above 1175 K liquid

Refer to the individual tables for details.

Enthalpy Reference Temperature = T_r = 298.15 K Standard State Pressure = p° = 0.1 MPa

T/K	C_p° (J K⁻¹mol⁻¹)	S° (J K⁻¹mol⁻¹)	$-[G^\circ-H^\circ(T_r)]/T$ (J K⁻¹mol⁻¹)	$H^\circ-H^\circ(T_r)$ (kJ mol⁻¹)	Δ_fH° (kJ mol⁻¹)	Δ_fG° (kJ mol⁻¹)	Log K_f
0							
100							
200							
298.15	65.940	75.312	75.312	0.	-962.320	-912.724	159.905
300	66.024	75.720	75.313	0.122	-962.303	-912.416	158.866
400	69.873	95.253	77.950	6.921	-961.282	-895.937	116.997
500	73.220	111.212	83.054	14.079	-960.115	-879.733	91.905
600	76.149	124.827	88.908	21.551	-958.807	-863.777	75.198
700	78.659	136.761	94.909	29.296	-957.380	-848.050	63.282
800	80.626	147.395	100.817	37.262	-955.874	-832.533	54.359
900	82.425	156.997	106.535	45.416	-954.322	-817.209	47.430
1000	84.015	165.766	112.026	53.741	-952.743	-802.058	41.895
1100	85.270	173.834	117.283	62.206	-951.168	-787.066	37.375
1175.000	86.107	179.486	121.074	68.633	CRYSTAL <--> LIQUID	TRANSITION	
1175.000	100.416	211.534	121.074	106.289			
1200	100.416	213.648	122.981	108.800	-915.322	-772.800	33.639
1300	100.416	221.685	130.269	118.841	-911.919	-761.061	30.580
1400	100.416	229.127	137.068	128.883	-908.578	-749.583	27.967
1500	100.416	236.055	143.438	138.925	-905.311	-738.340	25.711
1600	100.416	242.536	149.432	148.966	-902.125	-727.313	23.744
1700	100.416	248.623	155.089	159.008	-899.029	-716.483	22.015
1800	100.416	254.363	160.447	169.049	-896.033	-705.832	20.483
1900	100.416	259.792	165.534	179.091	-893.144	-695.344	19.116
2000	100.416	264.943	170.376	189.133	-890.371	-685.006	17.891
2100	100.416	269.842	174.997	199.174	-887.719	-674.803	16.785
2200	100.416	274.513	179.415	209.216	-906.521	-663.979	15.765
2300	100.416	278.977	183.648	219.257	-904.551	-652.998	14.830
2400	100.416	283.251	187.709	229.299	-902.577	-642.103	13.975
2500	100.416	287.350	191.614	239.341	-900.594	-631.291	13.190
2600	100.416	291.288	195.372	249.382	-898.599	-620.558	12.467
2700	100.416	295.078	198.995	259.424	-896.587	-609.902	11.799
2800	100.416	298.730	202.492	269.465	-894.557	-599.321	11.180
2900	100.416	302.254	205.872	279.507	-892.505	-588.813	10.606
3000	100.416	305.658	209.142	289.549	-890.429	-578.377	10.070

PREVIOUS: CURRENT: June 1969

Zirconium Fluoride (ZrF_2) F_2Zr_1(cr,l)

ZIRCONIUM FLUORIDE (ZrF_2) IDEAL GAS M_r = 129.216806

$S°(298.15\ K) = [283.434 \pm 8.4]\ J\ K^{-1}\ mol^{-1}$

$\Delta_f H°(0\ K) = -555.653 \pm 20.9\ kJ\ mol^{-1}$
$\Delta_f H°(298.15\ K) = -558.146 \pm 20.9\ kJ\ mol^{-1}$

Electronic Levels and Quantum Weights

ε_i, cm^{-1}	g_i
0	[3]
[7000]	[6]
[17000]	[6]
[22000]	[15]

Vibrational Frequencies and Degeneracies

ν, cm^{-1}
[560](1)
[300](1)
[670](1)

Ground State Quantum Weight: [3] σ = 2
Point Group: C_{2v}
Bond Distance: Zr-F = [1.92] Å
Bond Angle: F-Zr-F = [125]°
Product of the Moments of Inertia: $I_A I_B I_C = [1.396810 \times 10^{-114}]\ g^3\ cm^6$

Enthalpy of Formation

Murad and Hildenbrand (1) studied the gaseous equilibria involving ZrF_4, ZrF_3, ZrF_2, Ca and CaF, mass-spectrometrically. Ion intensities were measured 3 cal K^{-1} mol^{-1} above threshold over the temperature range 1665-1747 K, and the equilibrium constants for the reaction 2 Ca(g) + ZrF_4(g) = 2 CaF(g) + ZrF_2(g) were calculated. Using the reported equilibrium constants, the enthalpy changes $\Delta_r H°(298.15\ K)$ of this reaction are evaluated by the 2nd and 3rd law methods to be 15.4 and 50.9 kcal mol^{-1}, respectively. The drift in 3rd law values is 20.4 ± 17.6 cal K^{-1} mol^{-1}. Based on the 3rd law $\Delta_r H°(298.15\ K)$ value and $\Delta_f H°(298.15\ K)$ = 42.85±0.3, -65.0±2, and -400.0±0.5 kcal mol^{-1} for Ca(g), CaF(g), and ZrF_4, respectively, we obtain $\Delta_f H°(ZrF_2$, g, 298.15 K) = -133.4±5 kcal mol^{-1} (-558.146±20.9 kJ mol^{-1}) which is adopted.

Using Gibbs energy functions derived from different molecular constants for the reactants and products, Murad and Hildenbrand (1) derived a 3rd law $\Delta_r H°(298.15\ K)$ = 42.6±5 and $\Delta_f H°(ZrF_2$, g, 298.15 K) = -141±5 kcal mol^{-1}, which are in fair agreement with the adopted values.

Heat Capacity and Entropy

According to the correlations of Charkin and Dyatkina (2), the ZrF_2(g) molecule has an angular configuration. The F-Zr-F angle is estimated at 125°, which is similar to the value reported by Hastie et al. (4) for TiF_2(g). The F-Zr bond distance is estimated from that of ZrF_4(g). The vibrational frequencies are calculated based on the method suggested by Herzberg (3). The value ν_1 is calculated from the vibrational frequency (ν) of ZrF(g) by use of the relation $\nu_1 = \nu[Zr/(Zr+F)]^{1/2}$ where Zr and F are gram atomic weights of zirconium and fluorine. ν_3 is derived from ν_1 using the formula $\nu_3 = \nu_1[(Zr+2F)/Zr]^{1/2}$, and ν_2 is calculated as 0.44 ν_3 where the value 0.44 is estimated by comparison with the ratios of ν_2/ν_3 for other related compounds. Ground state quantum weight is taken to be the same as that of TiF_2(g). The three principal moments of inertia are: $I_A = 3.5010 \times 10^{-39}$, $I_B = 18.3005 \times 10^{-39}$, and $I_C = 21.8015 \times 10^{-39}\ g\ cm^2$. The electronic levels and quantum weights are estimated to be the same as those of TiF_2(g).

References

1. E. Murad and D. L. Hildenbrand, J. Chem. Phys. 45, 4751 (1966).
2. O. P. Charkin and M. E. Dyatkina, Russ. J. Struct. Chem. 6, 550 (1965).
3. G. Herzberg, "Infrared and Raman Spectra," D. Van Nostrand Co., Inc., New York, 1945.
4. J. W. Hastie, R. H. Hauge and J. L. Margrave, J. Chem. Phys. 51, 2648 (1969).

Zirconium Fluoride (ZrF_2) F_2Zr_1(g)

Enthalpy Reference Temperature = T_r = 298.15 K Standard State Pressure = p° = 0.1 MPa

T/K	$C_p°$	S°	$-[G°-H°(T_r)]/T$	$H°-H°(T_r)$	$\Delta_f H°$	$\Delta_f G°$	Log K_f
	J $K^{-1}mol^{-1}$			kJ mol^{-1}			
0	0.	0.	INFINITE	-11.829	-555.653	-555.653	INFINITE
100	35.603	238.172	322.695	-8.452	-556.069	-561.444	293.268
200	43.051	265.112	287.723	-4.522	-557.261	-566.343	147.914
250	46.243	275.074	284.222	-2.287	-557.744	-568.556	118.793
298.15	48.652	283.434	283.434	0.	-558.146	-570.601	99.967
300	48.732	283.735	283.435	0.090	-558.160	-570.678	99.364
350	50.617	291.396	284.036	2.576	-558.534	-572.735	85.476
400	52.039	298.252	285.392	5.144	-558.885	-574.739	75.053
450	53.121	304.447	287.170	7.774	-559.226	-576.700	66.942
500	53.956	310.089	289.184	10.452	-559.567	-578.624	60.448
600	55.128	320.038	293.520	15.911	-560.273	-582.369	50.700
700	55.886	328.597	297.934	21.464	-561.038	-585.992	43.727
800	56.405	336.095	302.245	27.080	-561.881	-589.501	38.490
900	56.787	342.762	306.383	32.741	-562.823	-592.898	34.411
1000	57.094	348.761	310.326	38.435	-563.874	-596.184	31.141
1100	57.368	354.216	314.072	44.158	-565.042	-599.360	28.461
1200	57.638	359.219	317.628	49.908	-570.039	-602.202	26.213
1300	57.923	363.844	321.008	55.686	-570.899	-604.848	24.303
1400	58.233	368.147	324.223	61.494	-571.793	-607.426	22.663
1500	58.572	372.176	327.287	67.334	-572.727	-609.939	21.240
1600	58.938	375.968	330.212	73.209	-573.707	-612.388	19.992
1700	59.329	379.553	333.010	79.123	-574.740	-614.774	18.890
1800	59.738	382.955	335.691	85.076	-575.833	-617.097	17.908
1900	60.158	386.197	338.265	91.070	-576.991	-619.359	17.027
2000	60.583	389.293	340.739	97.107	-578.221	-621.557	16.233
2100	61.006	392.259	343.122	103.187	-579.532	-623.692	15.513
2200	61.423	395.107	345.421	109.308	-602.253	-625.017	14.840
2300	61.827	397.846	347.641	115.471	-604.163	-626.009	14.217
2400	62.217	400.486	349.788	121.673	-606.028	-626.918	13.645
2500	62.589	403.033	351.868	127.914	-607.846	-627.751	13.116
2600	62.942	405.495	353.883	134.191	-609.616	-628.512	12.627
2700	63.275	407.877	355.839	140.502	-611.335	-629.206	12.173
2800	63.588	410.184	357.739	146.845	-613.003	-629.838	11.750
2900	63.882	412.420	359.586	153.219	-614.619	-630.410	11.355
3000	64.157	414.591	361.384	159.621	-616.183	-630.928	10.985
3100	64.415	416.698	363.134	166.050	-617.692	-631.394	10.639
3200	64.657	418.747	364.840	172.503	-619.149	-631.813	10.313
3300	64.884	420.741	366.504	178.980	-620.551	-632.187	10.007
3400	65.098	422.681	368.128	185.480	-621.900	-632.519	9.717
3500	65.300	424.571	369.714	192.000	-623.196	-632.812	9.444
3600	65.492	426.413	371.263	198.539	-624.439	-633.069	9.186
3700	65.674	428.210	372.778	205.098	-625.629	-633.293	8.940
3800	65.848	429.964	374.260	211.674	-626.768	-633.485	8.708
3900	66.016	431.676	375.710	218.267	-627.855	-633.647	8.487
4000	66.176	433.350	377.130	224.877	-628.891	-633.782	8.276
4100	66.331	434.986	378.522	231.502	-629.878	-633.892	8.076
4200	66.481	436.586	379.885	238.143	-630.816	-633.978	7.885
4300	66.625	438.152	381.222	244.798	-631.707	-634.043	7.702
4400	66.765	439.685	382.533	251.468	-632.550	-634.088	7.528
4500	66.901	441.187	383.820	258.151	-633.347	-634.113	7.361
4600	67.033	442.659	385.083	264.848	-634.099	-634.122	7.201
4700	67.160	444.102	386.324	271.557	-634.808	-634.115	7.047
4800	67.283	445.517	387.542	278.280	-1196.410	-622.476	6.774
4900	67.402	446.906	388.740	285.014	-1196.686	-610.516	6.508
5000	67.517	448.269	389.917	291.760	-1196.939	-598.551	6.253
5100	67.627	449.607	391.074	298.517	-1197.169	-586.581	6.008
5200	67.732	450.921	392.212	305.285	-1197.375	-574.606	5.772
5300	67.833	452.212	393.332	312.064	-1197.557	-562.628	5.545
5400	67.929	453.481	394.434	318.852	-1197.715	-550.647	5.326
5500	68.020	454.728	395.519	325.649	-1197.850	-538.663	5.116
5600	68.105	455.955	396.588	332.455	-1197.960	-526.677	4.913
5700	68.186	457.161	397.640	339.270	-1198.047	-514.689	4.717
5800	68.261	458.347	398.676	346.092	-1198.110	-502.700	4.527
5900	68.330	459.515	399.698	352.922	-1198.150	-490.709	4.344
6000	68.394	460.664	400.704	359.758	-1198.165	-478.719	4.168

PREVIOUS: December 1969 (1 atm) CURRENT: December 1969 (1 bar)

Zirconium Fluoride (ZrF_2) F_2Zr_1(g)

IRON FLUORIDE (FeF_3) CRYSTAL M_r = 112.842209

$S°(298.15\ K) = [98.324 \pm 8.4]\ J\ K^{-1}\ mol^{-1}$

$\Delta_fH°(0\ K)$ = Unknown

$\Delta_fH°(298.15\ K) = -1041.816 \pm 12.6\ kJ\ mol^{-1}$

Enthalpy of Formation

The equilibrium pressures for the reaction 2 FeF_3(cr) + H_2(g) = 2 FeF_2(cr) + 2 HF(g) have been measured at temperatures 623, 723 and 823 K by Jellinek and Rudat (1). Using the reported data, the enthalpy change of the reaction was evaluated as 37.9 and 31.3 kcal mol^{-1} by the 3rd and 2nd law, respectively. The corresponding $\Delta_fH°(FeF_3$, cr, 298.15 K) values are calculated to be -252.4 and -249.2 kcal mol^{-1}. The equilibrium constants for the reaction 2 FeF_3(cr) + 3 H_2O(g) = Fe_2O_3(cr) + 6 HF(g) were determined by Domange (2). Based on the 3rd law value of $\Delta_rH°$(298.15 K), 62.0 kcal mol^{-1}, the $\Delta_fH°(FeF_3$, cr, 298.15 K) was calculated as -237.4 kcal mol^{-1}. The value of $\Delta_fH°$(298.15 K) for FeF_3(cr) is selected to be -249±3 kcal mol^{-1}.

Heat Capacity and Entropy

The heat capacities of FeF_3(cr) at temperatures 100-450 K were determined by Bizette et al. (3). The results were presented graphically by plotting $C_p°$ vs T. The $C_p°$ curve shows a peak at 367 K. An antiferromagnetic transition at about 394 K was reported by Wollan et al. (4). The heat capacities above 450 K and S°(298.15 K) were estimated by comparison with those for $FeCl_3$(cr), $AlCl_3$(cr) and AlF_3(cr).

Sublimation Data

The phenomenon that FeF_3(cr) sublimes without fusion at temperatures near 1000°C has been observed and reported by Poulenc (5). Based on this data, the value of T_{sub} was estimated. The value of heat of sublimation, $\Delta_{sub}H°$(298.15 K), is calculated using an estimated entropy of sublimation, $\Delta_{sub}S°$ = 40 cal K^{-1} mol^{-1} in order to have $\Delta_rG°$(1200 K) = 0 for the reaction FeF_3(cr) = FeF_3(g).

References

1. K. Jellinek and A. Rudat, Z. anorg. allg. Chem. 175, 281 (1928).
2. L. Domange, Ann. Chim. 7, 225 (1937).
3. H. Bizette, R. Mainard and J. Picard, Comp. Rend. 5508 (1965).
4. E. O. Wollan, H. R. Child, W. C. Koehler and M. K. Wilkinson, Phys. Rev. 112, 1132 (1958).
5. Poulenc, Ann. Chim. Phys. (7), 2, 1 (1894).

Iron Fluoride (FeF_3) F_3Fe_1(cr)

Enthalpy Reference Temperature = T_r = 298.15 K Standard State Pressure = p° = 0.1 MPa

T/K	$C_p°$	S°	$-[G°-H°(T_r)]/T$	$H°-H°(T_r)$	$\Delta_fH°$	$\Delta_fG°$	Log K_f
	J $K^{-1}mol^{-1}$			kJ mol^{-1}			
0							
100							
200							
298.15	91.002	98.324	98.324	0.	-1041.816	-972.293	170.342
300	91.630	98.889	98.326	0.169	-1041.780	-971.862	169.216
367.000	113.047	119.010	100.268	6.878	Cp LAMBDA MAXIMUM	TRANSITION	
367.000	113.408	119.021	100.268	6.882			
400	96.371	127.516	102.176	10.136	-1039.270	-948.929	123.917
500	95.667	148.882	109.462	19.710	-1037.598	-926.547	96.796
600	96.818	166.429	117.536	29.336	-1036.268	-904.466	78.741
700	98.073	181.448	125.619	39.080	-1035.177	-882.588	65.860
800	99.328	194.626	133.438	48.950	-1034.335	-860.851	56.208
900	100.583	206.397	140.902	58.946	-1033.850	-839.199	48.706
1000	101.839	217.060	147.993	69.067	-1034.011	-817.571	42.706
1100	103.094	226.825	154.722	79.314	-1034.986	-795.865	37.792
1200	104.349	235.849	161.111	89.686	-1035.323	-774.130	33.697
1300	105.604	244.251	167.187	100.183	-1033.923	-752.421	30.233
1400	106.859	252.123	172.976	110.807	-1032.513	-730.820	27.267
1500	108.115	259.539	178.502	121.555	-1031.091	-709.320	24.701
1600	109.370	266.556	183.788	132.429	-1029.654	-687.915	22.458
1700	110.625	273.224	188.854	143.429	-1029.151	-666.582	20.482
1800	111.880	279.583	193.719	154.554	-1028.012	-645.288	18.726
1900	113.135	285.666	198.400	165.805	-1040.947	-623.355	17.137
2000	114.391	291.501	202.910	177.182	-1039.994	-601.400	15.707

PREVIOUS: CURRENT: September 1965

Iron Fluoride (FeF_3) F_3Fe_1(cr)

IRON FLUORIDE (FeF_3) IDEAL GAS M_r = 112.842209

$\Delta_f H°(0\ K)$ = [-817.965 ± 20.9] kJ mol^{-1}

S°(298.15 K) = [304.236] J K^{-1} mol^{-1} $\Delta_f H°(298.15\ K)$ = [-820.901 ± 20.9] kJ mol^{-1}

Vibrational Frequencies and Degeneracies

ν, cm^{-1}	ν, cm^{-1}
[600](1)	[800](2)
[250](1)	[220](2)

Ground State Quantum Weight: [1] σ = [6]

Point Group: [D_{3h}]

Bond Distance: Fe-F = [1.9] Å

Bond Angle: F-Fe-F = [120]°

Product of the Moments of Inertia: $I_A I_B I_C$ = [9.971132 x 10^{-114}] g^3 cm^6

Enthalpy of Formation

The value of $\Delta_f H°(FeF_3$, g, 298.15 K) is calculated from $\Delta_{sub}H°$(298.15 K) and $\Delta_f H°$(298.15 K) for FeF_3(cr). The value $\Delta_{sub}H°$(298.15 K) = 65.5 kcal mol^{-1} is derived from $\Delta_{sub}H°$(1230 K). See the crystal table for details.

Heat Capacity and Entropy

The molecular structure, bond distance and angle are estimated by comparison with those for AlF_3(g). The vibrational frequencies are estimated by comparison with those for AlF_3(g), TiF_3(g) and ZrF_3(g). The three principal moments of inertia are: $I_A = I_B$ = 17.0833 x 10^{-39} and I_C = 34.1666 x 10^{-39} g cm^2.

Iron Fluoride (FeF_3) F_3Fe_1(g)

Enthalpy Reference Temperature = T_r = 298.15 K Standard State Pressure = p° = 0.1 MPa

T/K	$C_p°$ (J $K^{-1}mol^{-1}$)	S° (J $K^{-1}mol^{-1}$)	-[G°-H°(T_r)]/T (J $K^{-1}mol^{-1}$)	H°-H°(T_r) (kJ mol^{-1})	$\Delta_f H°$ (kJ mol^{-1})	$\Delta_f G°$ (kJ mol^{-1})	Log K_f
0	0.	0.	INFINITE	-14.808	-817.965	-817.965	INFINITE
100	44.171	244.781	356.426	-11.164	-819.104	-817.420	426.977
200	57.084	279.760	309.982	-6.044	-820.147	-815.313	212.938
250	61.796	293.020	305.295	-3.069	-820.563	-814.055	170.088
298.15	65.546	304.236	304.236	0.	-820.901	-812.771	142.394
300	65.676	304.642	304.238	0.121	-820.913	-812.720	141.507
350	68.783	315.009	305.049	3.486	-821.218	-811.330	121.084
400	71.241	324.360	306.888	6.989	-821.502	-809.898	105.762
450	73.180	332.868	309.309	10.601	-821.788	-808.431	93.840
500	74.717	340.661	312.060	14.300	-822.092	-806.930	84.299
600	76.939	354.494	318.009	21.891	-822.797	-803.834	69.980
700	78.415	366.472	324.096	29.663	-823.678	-800.607	59.742
800	79.434	377.014	330.065	37.559	-824.812	-797.237	52.054
900	80.162	386.414	335.813	45.541	-826.340	-793.704	46.065
1000	80.699	394.889	341.304	53.585	-828.578	-789.968	41.264
1100	81.105	402.600	346.531	61.676	-831.709	-785.941	37.321
1200	81.418	409.672	351.503	69.803	-834.291	-781.685	34.026
1300	81.665	416.199	356.231	77.957	-835.234	-777.264	31.231
1400	81.863	422.258	360.734	86.134	-836.270	-772.766	28.832
1500	82.025	427.912	365.026	94.329	-837.402	-768.191	26.751
1600	82.157	433.210	369.124	102.538	-838.630	-763.538	24.927
1700	82.268	438.194	373.041	110.760	-840.905	-758.785	23.315
1800	82.361	442.899	376.793	118.991	-842.660	-753.905	21.878
1900	82.440	447.354	380.390	127.231	-858.605	-748.222	20.570
2000	82.508	451.585	383.845	135.479	-860.782	-742.356	19.388
2100	82.566	455.612	387.168	143.733	-862.961	-736.381	18.316
2200	82.617	459.454	390.367	151.992	-865.139	-730.303	17.340
2300	82.661	463.127	393.451	160.256	-867.309	-724.125	16.445
2400	82.700	466.646	396.428	168.524	-869.469	-717.853	15.624
2500	82.735	470.023	399.305	176.796	-871.611	-711.492	14.866
2600	82.765	473.268	402.087	185.071	-873.732	-705.045	14.165
2700	82.793	476.393	404.782	193.349	-875.826	-698.517	13.514
2800	82.817	479.404	407.394	201.629	-877.889	-691.912	12.908
2900	82.839	482.311	409.927	209.912	-879.917	-685.234	12.342
3000	82.859	485.119	412.387	218.197	-881.907	-678.487	11.814
3100	82.877	487.837	414.777	226.484	-883.854	-671.675	11.318
3200	82.893	490.468	417.102	234.772	-1234.057	-657.376	10.731
3300	82.908	493.019	419.364	243.062	-1234.012	-639.355	10.120
3400	82.922	495.494	421.567	251.354	-1233.952	-621.336	9.546
3500	82.934	497.898	423.713	259.647	-1233.874	-603.319	9.004
3600	82.946	500.235	425.807	267.941	-1233.780	-585.304	8.493
3700	82.956	502.507	427.849	276.236	-1233.669	-567.292	8.009
3800	82.966	504.720	429.843	284.532	-1233.541	-549.284	7.550
3900	82.975	506.875	431.791	292.829	-1233.397	-531.279	7.116
4000	82.983	508.976	433.694	301.127	-1233.239	-513.278	6.703
4100	82.991	511.025	435.555	309.426	-1233.068	-495.281	6.310
4200	82.998	513.025	437.376	317.725	-1232.884	-477.288	5.936
4300	83.005	514.978	439.158	326.025	-1232.691	-459.301	5.579
4400	83.011	516.886	440.903	334.326	-1232.489	-441.317	5.239
4500	83.017	518.752	442.613	342.627	-1232.281	-423.338	4.914
4600	83.022	520.577	444.288	350.929	-1232.069	-405.364	4.603
4700	83.028	522.362	445.930	359.232	-1231.854	-387.394	4.305
4800	83.032	524.110	447.541	367.535	-1231.640	-369.429	4.020
4900	83.037	525.822	449.121	375.838	-1231.428	-351.469	3.747
5000	83.041	527.500	450.672	384.142	-1231.221	-333.512	3.484
5100	83.045	529.145	452.194	392.447	-1231.021	-315.560	3.232
5200	83.049	530.757	453.690	400.751	-1230.831	-297.612	2.990
5300	83.052	532.339	455.159	409.056	-1230.654	-279.667	2.756
5400	83.056	533.892	456.602	417.362	-1230.492	-261.725	2.532
5500	83.059	535.416	458.021	425.668	-1230.348	-243.787	2.315
5600	83.062	536.912	459.417	433.974	-1230.225	-225.850	2.107
5700	83.065	538.382	460.789	442.280	-1230.127	-207.916	1.905
5800	83.068	539.827	462.140	450.587	-1230.057	-189.984	1.711
5900	83.070	541.247	463.469	458.893	-1230.018	-172.051	1.523
6000	83.073	542.643	464.776	467.201	-1230.014	-154.120	1.342

PREVIOUS: September 1965 (1 atm) CURRENT: September 1965 (1 bar)

Iron Fluoride (FeF_3) F_3Fe_1(g)

TRIFLUOROSILANE ($SiHF_3$) IDEAL GAS M_r = 86.088649

$S°(298.15\ K) = 277.268 \pm 0.8\ J\ K^{-1}\ mol^{-1}$

$\Delta_f H°(0\ K) = -1193.663 \pm 20.9]\ kJ\ mol^{-1}$
$\Delta_f H°(298.15\ K) = [-1200.808 \pm 20.9]\ kJ\ mol^{-1}$

Vibrational Frequencies and Degeneracies

ν, cm^{-1}	ν, cm^{-1}
2316(1)	998(2)
858(1)	844(2)
425(1)	306(2)

Ground State Quantum Weight: [1] $\sigma = 3$
Point Group: C_{3v}
Bond Distances: Si-H = 1.4468 Å Si-F = 1.5624 Å
Bond Angles: H-Si-F = 110.64° F-Si-F = 108.28°
Product of the Moments of Inertia: $I_A I_B I_C = 2.726959 \times 10^{-114}\ g^3\ cm^6$

Enthalpy of Formation

There are no reported experimental studies leading to the enthalpy of formation of $SiHF_3$(g). We estimate this value via a linear interpolation between the established $\Delta_f H°$(298.15 K) values of SiH_4(g) and SiF_4(g) (1). The reasonableness of this approach has been demonstrated by Lapidus et al. (2), Hunt and Sirtl (3), and Seiter and Sirtl (4). Lapidus et al (2) comparatively examined the thermodynamic properties of halogenated silanes and methanes. Hunt and Sirtl (3) and Seiter and Sirtl (4) studied the chlorinated silanes and proposed a linear $\Delta_f H°$ relationship within the sequence SiH_4(g) to $SiCl_4$(g).

Heat Capacity and Entropy

The adopted vibrational frequencies are from the infrared gas phase studies of Buerger et al. (5, 6). These values are all within ±1 cm^{-1} of the suggested values in the compilation by Shimanouchi (7), as derived from Newman et al. (8).

The adopted bond angles and bond distances are obtained from the microwave spectrum study of six isotopic species of $HSiF_3$ (9). Earlier work by Sheridan and Gordy (10) and Heath et al. (11) is in good agreement with these values. The moments of inertia are calculated to be $I_A = I_B = 11.6094 \times 10^{-39}\ g\ cm^2$ and $I_C = 20.2329 \times 10^{-39}\ g\ cm^2$.

References

1. JANAF Thermochemical Tables: SiH_4(g) and SiF_4(g), 6-30-76.
2. I. I. Lapidus, L. A. Nisel'son, and A. L. Seifer in "Thermophysical Properties of Gases and Liquids," No. 1, V. A. Rabinovich, Ed., Israel Program for Scientific Translations, Jerusalem, pp 102-135, 1970.
3. L. P. Hunt and E. Sirtl, J. Electrochem. Soc. 119, 1741 (1972).
4. H. Seiter and E. Sirtl, Z. Naturforsch. 21a, 1956 (1966).
5. H. Buerger, S. Biedermann, and A. Ruoff, Specrochim. Acta, Part A 27, 1687 (1971).
6. H. Buerger, S. Biedermann, and A. Ruoff, Spectrochim. Acta, Part A 30, 1655 (1974).
7. T. Shimanouchi, J. Phys. Chem. Ref. Data 1, 189 (1972).
8. C. Newman, S. R. Polo, and M. K. Wilson, Spectrochim. Acta 15, 793 (1959).
9. A. R. Hoy, M. Bertram, and I. M. Mills, J. Mol. Spectrosc. 46, 429 (1973).
10. J. Sheridan and W. Gordy, J. Chem. Phys. 19, 965 (1951).
11. G. A. Heath, L. F. Thomas, and J. Sheridan, Trans. Faraday Soc. 50, 779 (1954).

Trifluorosilane ($SiHF_3$) $F_3H_1Si_1$(g)

Enthalpy Reference Temperature = T_r = 298.15 K Standard State Pressure = p° = 0.1 MPa

T/K	$C_p°$	S°	$-[G°-H°(T_r)]/T$	$H°-H°(T_r)$	$\Delta_f H°$	$\Delta_f G°$	Log K_f
	J K⁻¹mol⁻¹			kJ mol⁻¹			
0	0.	0.	INFINITE	-13.544	-1193.663	-1193.663	INFINITE
100	38.013	224.001	325.165	-10.116	-1196.362	-1187.787	620.436
200	51.293	254.467	282.691	-5.645	-1198.809	-1178.233	307.723
250	57.690	266.604	278.281	-2.919	-1199.886	-1172.962	245.077
298.15	63.485	277.268	277.268	0.	-1200.808	-1167.691	204.575
300	63.696	277.662	277.269	0.118	-1200.841	-1167.485	203.277
350	69.048	287.892	278.065	3.439	-1201.662	-1161.859	173.398
400	73.663	297.421	279.896	7.010	-1202.352	-1156.125	150.974
450	77.588	306.330	282.343	10.794	-1202.923	-1150.311	133.525
500	80.918	314.682	285.164	14.759	-1203.391	-1144.439	119.559
600	86.189	329.926	291.379	23.128	-1204.073	-1132.579	98.600
700	90.118	343.521	297.875	31.952	-1204.492	-1120.628	83.622
800	93.123	355.759	304.358	41.121	-1204.724	-1108.629	72.386
900	95.469	366.868	310.696	50.555	-1204.820	-1096.610	63.646
1000	97.332	377.027	316.829	60.198	-1204.824	-1084.586	56.653
1100	98.831	386.377	322.732	70.009	-1204.769	-1072.565	50.932
1200	100.052	395.030	328.401	79.955	-1204.679	-1060.550	46.165
1300	101.057	403.080	333.839	90.012	-1204.577	-1048.543	42.131
1400	101.891	410.600	339.056	100.161	-1204.478	-1036.545	38.674
1500	102.591	417.655	344.064	110.386	-1204.397	-1024.552	35.678
1600	103.183	424.295	348.873	120.676	-1204.343	-1012.565	33.057
1700	103.687	430.566	353.495	131.020	-1254.498	-1000.133	30.730
1800	104.119	436.505	357.944	141.411	-1254.284	-985.176	28.589
1900	104.492	442.145	362.228	151.842	-1254.066	-970.232	26.674
2000	104.816	447.513	366.359	162.307	-1253.847	-955.298	24.950
2100	105.099	452.634	370.346	172.804	-1253.624	-940.377	23.391
2200	105.347	457.529	374.199	183.326	-1253.394	-925.466	21.973
2300	105.567	462.217	377.925	193.872	-1253.156	-910.565	20.680
2400	105.761	466.714	381.531	204.439	-1252.905	-895.675	19.494
2500	105.934	471.035	385.025	215.024	-1252.636	-880.796	18.403
2600	106.089	475.193	388.414	225.625	-1252.347	-865.928	17.397
2700	106.228	479.199	391.702	236.241	-1252.032	-851.072	16.465
2800	106.353	483.065	394.897	246.870	-1251.688	-836.227	15.600
2900	106.466	486.799	398.002	257.511	-1251.312	-821.396	14.795
3000	106.569	490.410	401.022	268.163	-1250.901	-806.578	14.044
3100	106.662	493.906	403.962	278.825	-1250.450	-791.775	13.341
3200	106.747	497.294	406.826	289.495	-1249.960	-776.987	12.683
3300	106.825	500.580	409.618	300.174	-1249.426	-762.214	12.065
3400	106.896	503.770	412.340	310.860	-1248.848	-747.459	11.483
3500	106.961	506.869	414.997	321.553	-1248.225	-732.721	10.935
3600	107.021	509.883	417.591	332.252	-1631.707	-707.541	10.266
3700	107.077	512.816	420.125	342.957	-1630.577	-681.885	9.626
3800	107.128	515.673	422.602	353.667	-1629.402	-656.261	9.021
3900	107.175	518.456	425.025	364.382	-1628.182	-630.667	8.447
4000	107.220	521.170	427.394	375.102	-1626.917	-605.106	7.902
4100	107.261	523.818	429.714	385.826	-1625.606	-579.577	7.384
4200	107.299	526.403	431.986	396.554	-1624.251	-554.080	6.891
4300	107.334	528.928	434.211	407.286	-1622.851	-528.617	6.421
4400	107.368	531.396	436.392	418.021	-1621.408	-503.186	5.974
4500	107.399	533.810	438.530	428.759	-1619.922	-477.788	5.546
4600	107.428	536.170	440.627	439.501	-1618.395	-452.424	5.137
4700	107.455	538.481	442.684	450.245	-1616.826	-427.094	4.747
4800	107.481	540.744	444.704	460.992	-1615.217	-401.797	4.372
4900	107.505	542.960	446.686	471.741	-1613.570	-376.535	4.014
5000	107.528	545.132	448.634	482.492	-1611.885	-351.306	3.670
5100	107.549	547.262	450.547	493.246	-1610.162	-326.112	3.340
5200	107.569	549.350	452.427	504.002	-1608.404	-300.951	3.023
5300	107.589	551.399	454.275	514.760	-1606.611	-275.824	2.718
5400	107.607	553.411	456.092	525.520	-1604.785	-250.732	2.425
5500	107.624	555.385	457.880	536.282	-1602.925	-225.675	2.143
5600	107.640	557.325	459.638	547.045	-1601.034	-200.651	1.872
5700	107.655	559.230	461.369	557.809	-1599.113	-175.661	1.610
5800	107.670	561.102	463.072	568.576	-1597.161	-150.705	1.357
5900	107.684	562.943	464.749	579.343	-1595.181	-125.783	1.114
6000	107.697	564.753	466.401	590.113	-1593.173	-100.896	0.878

PREVIOUS: June 1976 (1 atm) CURRENT: June 1976 (1 bar)

Trifluorosilane ($SiHF_3$) $F_3H_1Si_1$(g)

HYDROGEN FLUORIDE (H_3F_3) IDEAL GAS M_r = 60.019029

$\Delta_fH°(0\ K) = -869.289 \pm 12.6\ kJ\ mol^{-1}$

$S°(298.15\ K) = [288.458 \pm 12.6]\ J\ K^{-1}\ mol^{-1}$ $\Delta_fH°(298.15\ K) = -879.100 \pm 12.6\ kJ\ mol^{-1}$

Vibrational Frequencies and Degeneracies

ν, cm^{-1}
[202](3)
[962](3)
[552](3)
[3060](3)

Ground State Quantum Weight: 1 σ = [3]

Point Group: C_3

Interatomic Distance: F-F = [2.5] Å

Product of the Moments of Inertia: $I_AI_BI_C = [2.237956 \times 10^{-114}]\ g^3\ cm^6$

Enthalpy of Formation

The enthalpy of 3 HF(g) = H_3F_3(g) was taken as the mean of 3rd law values which were determined from the equilibrium data of two investigations. Briegleb and Strohmeier (1) measured the vapor density of HF between 20 and 60°C and between 50 and 650 torr. Franck and Meyer (2) measured $C_p°$ between -20 and 100°C and between 100 and 700 torr. Each investigation evaluted K_p at n = 2, 3, 4, ... for the reactions n(HF) → $(HF)_n$ and reported 2nd law values of $\Delta_rH°$ and $\Delta_rS°$. At n = 3, their 2nd law values differed by 2.7 kcal mol^{-1} which was taken as an estimate of error, while the 3rd law values differed by 0.20 kcal mol^{-1}. Using $\Delta_rH°(298.15\ K)$ = -14.69 kcal mol^{-1} and auxiliary JANAF data (3) gives the enthalpy of formation of H_3F_3(g).

Heat Capacity and Entropy

The molecular structure of H_3F_3 was assumed as planar with the F atoms forming the vertices of a regular triangle and with the H atoms also lying on the circumscribed circle. That rings are more stable than open chains was confirmed by Del Bene and Pople's (4) theoretical molecular-orbital studies on HF polymers. The length of side (F-F axis) was taken from Atoji and Lipscomb's (5) X-studies of solid HF (F-F = 2.49 Å) and agrees with 2.52 Å which Janzen and Bartell (6) determined for HF gaseous polymers by electron diffraction. Vibrational frequencies were taken from Kittelberger and Hornig's (7) work on crystalline HF. Huong and Couzi (8) and Smith (9) have made spectral studies of the gas phase in the range from 350 to 4000 cm^{-1}.

The National Bureau of Standards prepared this table (10) by critical analysis of data existing in 1972. Using molecular constants and $\Delta_fH°$ selected by NBS (10), we recalculate the table in terms of 1973 fundamental constants (11), 1975 atomic weights (12), and current JANAF reference states for the elements.

References

1. G. Briegleb and W. Strohmeier, Z. Elektrochem. 57, 668 (1953).
2. E. U. Franck and F. Meyer, Z. Elektrochem. 63, 571 (1959).
3. JANAF Thermochemical Tables: HF(g), 6-30-77.
4. J. E. Del Bene and J. A. Pople, J. Chem. Phys. 55, 2296 (1971).
5. M. Atoji and W. N. Lipscomb, Acta Cryst. 7, 173 (1954).
6. J. Janzen and L. S. Bartell, J. Chem. Phys. 50, 3611 (1969).
7. J. S. Kittelberger and D. F. Hornig, J. Chem. Phys. 46, 3099 (1967).
8. P. V. Huong and M. Couzi, J. Chim. Phys. 66, 1309 (1969).
9. D. F. Smith, J. Chem. Phys. 28, 1040 (1958); 48, 1429 (1968).
10. S. Abramowitz et al., U.S. Nat. Bur. Stand. Report 10904, 239, 1972.
11. CODATA Task Group on Fundamental Constants, CODATA Bulletin 11, December, 1973.
12. IUPAC Commission on Atomic Weights, Pure Appl. Chem. 47, 75 (1976).

Hydrogen Fluoride (H_3F_3) F_3H_3(g)

Enthalpy Reference Temperature = T_r = 298.15 K Standard State Pressure = p° = 0.1 MPa

T/K	$C_p°$ (J $K^{-1}mol^{-1}$)	S°	$-[G°-H°(T_r)]/T$	$H°-H°(T_r)$ (kJ mol^{-1})	$\Delta_fH°$	$\Delta_fG°$	Log K_f
0	0.	0.	INFINITE	-16.127	-869.289	-869.289	INFINITE
100	46.713	223.110	346.863	-12.375	-874.397	-856.043	447.151
200	63.124	260.767	295.010	-6.849	-877.304	-836.438	218.455
250	70.045	275.612	289.675	-3.516	-878.320	-826.100	172.604
298.15	75.849	288.458	288.458	0.	-879.100	-815.968	142.954
300	76.055	288.928	288.460	0.141	-879.127	-815.577	142.005
350	81.120	301.044	289.405	4.074	-879.750	-804.933	120.130
400	85.327	312.160	291.564	8.238	-880.217	-794.212	103.713
450	88.821	322.418	294.430	12.595	-880.552	-783.440	90.939
500	91.760	331.932	297.710	17.111	-880.776	-772.636	80.717
600	96.495	349.099	304.876	26.534	-880.957	-750.986	65.379
700	100.303	364.269	312.298	36.380	-880.846	-729.329	54.423
800	103.575	377.881	319.659	46.577	-880.492	-707.705	46.208
900	106.478	390.251	326.826	57.082	-879.930	-686.139	39.822
1000	109.082	401.607	333.744	67.863	-879.190	-664.645	34.717
1100	111.416	412.115	340.397	78.890	-878.301	-643.232	30.545
1200	113.502	421.900	346.786	90.138	-877.290	-621.906	27.071
1300	115.359	431.060	352.920	101.582	-876.180	-600.668	24.135
1400	117.008	439.671	358.812	113.202	-874.992	-579.519	21.622
1500	118.470	447.794	364.476	124.978	-873.744	-558.457	19.447
1600	119.767	455.482	369.925	136.891	-872.451	-537.480	17.547
1700	120.917	462.778	375.175	148.926	-871.124	-516.585	15.873
1800	121.938	469.719	380.236	161.070	-869.771	-495.769	14.387
1900	122.847	476.337	385.121	173.310	-868.398	-475.028	13.059
2000	123.657	482.659	389.841	185.636	-867.008	-454.360	11.867
2100	124.380	488.710	394.406	198.039	-865.604	-433.763	10.789
2200	125.029	494.512	398.825	210.510	-864.186	-413.232	9.811
2300	125.611	500.082	403.108	223.042	-862.752	-392.766	8.920
2400	126.135	505.440	407.260	235.630	-861.301	-372.363	8.104
2500	126.608	510.598	411.291	248.267	-859.831	-352.020	7.355
2600	127.037	515.573	415.207	260.950	-858.338	-331.737	6.665
2700	127.425	520.374	419.014	273.673	-856.821	-311.512	6.027
2800	127.779	525.015	422.717	286.434	-855.276	-291.344	5.435
2900	128.101	529.505	426.323	299.228	-853.701	-271.231	4.885
3000	128.395	533.853	429.835	312.053	-852.093	-251.173	4.373
3100	128.665	538.067	433.259	324.906	-850.452	-231.170	3.895
3200	128.912	542.156	436.598	337.785	-848.775	-211.220	3.448
3300	129.139	546.126	439.857	350.688	-847.061	-191.322	3.028
3400	129.349	549.985	443.040	363.613	-845.309	-171.478	2.634
3500	129.543	553.737	446.149	376.557	-843.520	-151.685	2.264
3600	129.722	557.389	449.189	389.521	-841.691	-131.944	1.914
3700	129.887	560.945	452.161	402.501	-839.825	-112.254	1.585
3800	130.041	564.411	455.070	415.498	-837.920	-92.616	1.273
3900	130.184	567.791	457.917	428.509	-835.979	-73.027	0.978
4000	130.318	571.089	460.705	441.534	-834.000	-53.490	0.699
4100	130.442	574.308	463.437	454.572	-831.986	-34.002	0.433
4200	130.558	577.453	466.114	467.622	-829.937	-14.563	0.181
4300	130.667	580.526	468.739	480.684	-827.854	4.825	-0.059
4400	130.768	583.531	471.314	493.756	-825.738	24.166	-0.287
4500	130.864	586.471	473.841	506.837	-823.591	43.458	-0.504
4600	130.953	589.349	476.321	519.928	-821.413	62.701	-0.712
4700	131.037	592.166	478.756	533.028	-819.206	81.897	-0.910
4800	131.116	594.925	481.147	546.135	-816.970	101.047	-1.100
4900	131.191	597.630	483.497	559.251	-814.706	120.148	-1.281
5000	131.261	600.281	485.806	572.373	-812.415	139.204	-1.454
5100	131.328	602.881	488.076	585.503	-810.099	158.213	-1.620
5200	131.391	605.432	490.309	598.639	-807.757	177.177	-1.780
5300	131.450	607.935	492.504	611.781	-805.390	196.096	-1.933
5400	131.507	610.392	494.665	624.929	-802.998	214.969	-2.079
5500	131.560	612.806	496.791	638.082	-800.582	233.797	-2.220
5600	131.611	615.177	498.884	651.241	-798.143	252.583	-2.356
5700	131.659	617.507	500.945	664.404	-795.679	271.324	-2.486
5800	131.704	619.797	502.974	677.572	-793.192	290.022	-2.612
5900	131.748	622.049	504.973	690.745	-790.681	308.676	-2.733
6000	131.789	624.263	506.943	703.922	-788.145	327.287	-2.849

PREVIOUS: June 1977 (1 atm) CURRENT: June 1977 (1 bar)

Hydrogen Fluoride (H_3F_3) F_3H_3(g)

LITHIUM FLUORIDE $((LiF)_3)$ IDEAL GAS $M_r = 77.818209$

$\Delta_f H°(0\ K) = 1510.529 \pm 25.1\ kJ\ mol^{-1}$

$S°(298.15\ K) = [318.087 \pm 8.4]\ J\ K^{-1}\ mol^{-1}$ $\Delta_f H°(298.15\ K) = -1517.202 \pm 25.1\ kJ\ mol^{-1}$

Vibrational Frequencies and Degeneracies

ν, cm^{-1}	ν, cm^{-1}
280 (1)	530 (2)
[430](1)	755 (2)
[640](1)	245 (2)
[480](1)	[140](2)

Ground State Quantum Weight: [1] $\sigma = 6$

Point Group: $[D_{3h}]$

Bond Distance: Li-F = [1.68] Å

Bond Angles: F-Li-F = [120]° Li-F-Li = [120]°

Product of the Moments of Inertia: $I_A I_B I_C = [1.212852 \times 10^{-113}]\ g^3\ cm^6$

Enthalpy of Formation

$\Delta_f H°(Li_3F_3$, g, 298.15 K) is derived based on the adopted value $\Delta_r H°(298.15\ K) = 79.73\ kcal\ mol^{-1}$ for the reaction 3LiF(cr) = Li_3F_3(g), using $\Delta_f H$(LiF, cr, 298.15 K) = $-147.45\ kcal\ mol^{-1}$. The $\Delta_r H°(298.15\ K)$ value is calculated to be 73.9 kcal mol^{-1} at 1000 K, which is consistent with the values $\Delta_s H°(1000\ K) = 73.9 \pm 3$ and $74.9 \pm 1\ kcal\ mol^{-1}$ reported by Buchler and Stauffer (1) and Akishin et al. (2), respectively. Using the adopted $\Delta_f H°(298.15\ K)$ values for monomer, dimer, and trimer, we also evaluate the value $\Delta_r H°(1073\ K) = 53.8\ kcal\ mol^{-1}$ for the reaction $Li_3F_3(g) = Li_2F_2(g) + LiF(g)$ which is in agreement with the value $50 < \Delta_r H°(1073\ K) < 65$ kcal mol^{-1} reported by Porter and Schoonmaker (3). Rothberg et al. (4) have determined the partial pressures of LiF monomer, dimer, and trimer in the temperature range 980 - 1070 K, using the molecular-beam velocity-selector method (5). However, the total pressures derived from these data are too low in comparison with the other reported vapor pressures. Thus their data are not adopted for evaluation.

Heat Capacity and Entropy

The molecular structure is assumed to be the same as that of the BeO trimer. The Li-F bond distance is taken from that of the Li_2F_2 molecule determined by Akishin and Rambidi (6). Four vibrational frequencies, A_2'' and three E', are reported by Snelson (7) from analysis of the infrared spectra of LiF Using the matrix isolation technique. The reported frequencies are corrected to the average isotopic species. The other four frequencies, E_2'' and three A', are estimated by comparison with those for the isoelectronic molecule Be_3O_3. The three principal moments of inertia are: $I_A = I_B = 18.2358 \times 10^{-39}$ and $I_C = 36.4717 \times 10^{-39}$ g cm^2.

References

1. A. Buchler and J. C. Stauffer, "Thermodynamics," Vol. I, International Atomic Agency, Vienna, p. 271, 1966.
2. P. A. Akishin, L. N. Gorokhov, and L. N. Sidorov, Russ. J. Phys. Chem. 33, 648 (1959).
3. R. F. Porter and R. C. Schoonmaker, J. Chem. Phys. 29, 1070 (1958).
4. G. M. Rothberg, M. Eisenstadt, and P. Kusch, J. Chem. Phys. 30, 517 (1959).
5. R. C. Miller and P. Kusch, J. Chem. Phys. 25, 860 (1956); 27, 981 (1957).
6. P. A. Akinshin and N. G. Rambid, Z. Physik. Chem. 213, 111 (1960).
7. A. Snelson, IITRI-U6001-13, 14 May 1963 - 12 August 1966, Illinois Institute of Technology Research Institute, Chicago, Illinois.

Lithium Fluoride $((LiF)_3)$ $F_3Li_3(g)$

Enthalpy Reference Temperature = T_r = 298.15 K Standard State Pressure = p° = 0.1 MPa

T/K	$C_p°$	S°	$-[G°-H°(T_r)]/T$	$H°-H°(T_r)$	$\Delta_f H°$	$\Delta_f G°$	Log K_f
	J $K^{-1}mol^{-1}$			kJ mol^{-1}			
0	0.	0.	INFINITE	-20.431	-1510.529	-1510.529	INFINITE
100	55.829	232.778	396.049	-16.327	-1512.271	-1507.790	787.589
200	84.315	280.714	326.946	-9.247	-1515.108	-1502.160	392.324
250	94.780	300.702	319.734	-4.758	-1516.258	-1498.785	313.154
298.15	102.501	318.087	318.087	0.	-1517.202	-1495.332	261.976
300	102.757	318.722	318.089	0.190	-1517.236	-1495.197	260.337
350	108.755	335.035	319.364	5.485	-1518.113	-1491.453	222.587
400	113.278	349.866	322.264	11.041	-1519.015	-1487.584	194.259
450	116.724	363.416	326.095	16.795	-1520.032	-1483.596	172.211
500	119.386	375.858	330.457	22.700	-1530.203	-1478.565	154.465
600	123.131	397.982	339.914	34.840	-1532.223	-1468.041	127.804
700	125.556	417.157	349.610	47.283	-1533.883	-1457.208	108.738
800	127.203	434.037	359.130	59.926	-1535.343	-1446.154	94.424
900	128.368	449.091	368.304	72.708	-1536.716	-1434.922	83.281
1000	129.220	462.662	377.073	85.589	-1538.028	-1423.541	74.358
1100	129.860	475.009	385.423	98.545	-1539.296	-1412.030	67.052
1200	130.352	486.330	393.367	111.556	-1540.531	-1400.406	60.958
1300	130.739	496.780	400.925	124.612	-1541.746	-1388.680	55.798
1400	131.048	506.481	408.122	137.702	-1542.940	-1376.860	51.371
1500	131.299	515.531	414.985	150.819	-1544.111	-1364.957	47.532
1600	131.506	524.012	421.537	163.960	-1545.262	-1352.975	44.170
1700	131.677	531.989	427.801	177.119	-1982.103	-1319.394	40.540
1800	131.822	539.520	433.801	190.294	-1980.958	-1280.444	37.158
1900	131.944	546.651	439.554	203.483	-1979.823	-1241.558	34.133
2000	132.049	553.421	445.080	216.683	-1978.696	-1202.731	31.412
2100	132.140	559.866	450.394	229.892	-1977.576	-1163.961	28.952
2200	132.218	566.015	455.510	243.110	-1976.463	-1125.243	26.717
2300	132.287	571.894	460.444	256.336	-1975.356	-1086.576	24.677
2400	132.347	577.525	465.206	269.567	-1974.253	-1047.957	22.808
2500	132.400	582.929	469.807	282.805	-1973.154	-1009.384	21.090
2600	132.447	588.123	474.259	296.047	-1972.059	-970.855	19.505
2700	132.490	593.122	478.569	309.294	-1970.968	-932.368	18.038
2800	132.527	597.941	482.747	322.545	-1969.882	-893.921	16.676
2900	132.561	602.593	486.800	335.799	-1968.801	-855.513	15.409
3000	132.592	607.087	490.735	349.057	-1967.728	-817.142	14.228
3100	132.620	611.435	494.559	362.318	-1966.665	-778.807	13.123
3200	132.645	615.646	498.277	375.581	-1965.613	-740.506	12.088
3300	132.668	619.728	501.896	388.847	-1964.575	-702.237	11.115
3400	132.689	623.689	505.420	402.114	-1963.555	-664.000	10.201
3500	132.708	627.536	508.854	415.384	-1962.556	-625.792	9.339
3600	132.726	631.274	512.203	428.656	-1961.582	-587.612	8.526
3700	132.742	634.911	515.471	441.929	-1960.635	-549.460	7.757
3800	132.757	638.451	518.661	455.204	-1959.721	-511.333	7.029
3900	132.771	641.900	521.777	468.481	-1958.844	-473.228	6.338
4000	132.783	645.262	524.822	481.758	-1958.008	-435.146	5.682
4100	132.795	648.540	527.800	495.037	-1957.218	-397.085	5.059
4200	132.806	651.741	530.713	508.317	-1956.480	-359.042	4.465
4300	132.817	654.866	533.564	521.599	-1955.798	-321.016	3.900
4400	132.826	657.919	536.355	534.881	-1955.115	-283.010	3.360
4500	132.835	660.904	539.090	548.164	-1954.533	-245.015	2.844
4600	132.844	663.824	541.770	561.448	-1954.026	-207.031	2.351
4700	132.851	666.681	544.398	574.733	-1953.593	-169.058	1.879
4800	132.859	669.478	546.974	588.018	-1953.238	-131.093	1.427
4900	132.866	672.218	549.503	601.304	-1952.969	-93.134	0.993
5000	132.872	674.902	551.984	614.591	-1952.790	-55.181	0.576
5100	132.878	677.533	554.420	627.879	-1952.690	-17.231	0.176
5200	132.884	680.114	556.812	641.167	-1952.706	20.720	-0.208
5300	132.890	682.645	559.163	654.455	-1952.830	58.671	-0.578
5400	132.895	685.129	561.472	667.745	-1953.067	96.626	-0.935
5500	132.900	687.567	563.743	681.034	-1953.424	134.586	-1.278
5600	132.904	689.962	565.976	694.325	-1953.907	172.554	-1.610
5700	132.909	692.314	568.171	707.615	-1954.521	210.531	-1.929
5800	132.913	694.626	570.332	720.906	-1955.273	248.521	-2.238
5900	132.917	696.898	572.458	734.198	-1956.167	286.526	-2.537
6000	132.921	699.132	574.550	747.490	-1957.210	324.546	-2.825

PREVIOUS: December 1968 (1 atm) CURRENT: December 1968 (1 bar)

Lithium Fluoride $((LiF)_3)$ $F_3Li_3(g)$

MOLYBDENUM FLUORIDE (MoF_3) IDEAL GAS M_r = 152.935209

$\Delta_f H°(0\ K)$ = [-603.975 ± 16.7] kJ mol^{-1}

$S°(298.15\ K)$ = [295.968 ± 8.4] J K^{-1} mol^{-1} $\Delta_f H°(298.15\ K)$ = -607.098 ± 16.7 kJ mol^{-1}

Electronic Levels and Quantum Weights

ε_i, cm^{-1}	g_i
0	[2]
[3505]	[2]
[10861]	[4]
[11062]	[4]

Vibrational Frequencies and Degeneracies

ν, cm^{-1}	ν, cm^{-1}
[600](1)	[650](2)
[300](1)	[250](2)

Ground State Quantum Weight: [2] σ = [6]

Point Group: D_{3h}

Bond Distance: Mo-F = [1.82] Å

Bond Angle: F-Mo-F = [120°]

Product of the Moments of Inertia: $I_A I_B I_C$ = [7.702846 x 10^{-114}] g^3 cm^6

Enthalpy of Formation

The adopted enthalpy of formation at 298.15 K is that derived by Brewer (1) from the high temperature mass spectrometric measurements of equilibria among gaseous molybdenum fluorides by Hildenbrand [1664 and 2146 K] (2) and among gaseous and hetrogeneous phases by Alikhanyan et al. [911 and 964 K] (3). The value of $\Delta_f H°(0\ K)$ combined with JANAF data (4) for Mo(g) and F(g) gives $\Delta_{at}H°(0\ K)$ = 356.5±6 kcal mol^{-1} and an average Mo-F bond energy of $\Delta_{at}H°(0\ K)/3$ = 118.8±2 kcal mol^{-1}.

Heat Capacity and Entropy

We use the vibrational frequencies and degeneracies estimated by Hildenbrand (2) and adopted by Brewer (1). The correlation of Drake and Rosenblatt (5) leads to a planar configuration for MoF_3(g). The Mo-F distances of 1.82 Å from MoF_6(g) is used. The observed electronic levels (4) of the isoelectronic TaO(g) are assumed, as suggested by Brewer (1), up through 11062 cm^{-1}. The principal moments of inertia are: $I_A = I_B$ = 15.6750 x 10^{-39} g cm^2, I_C = 31.3500 x 10^{-39}.

This table is based on the critical analysis of data existing in 1978 by Brewer (1).

References

1. L. Brewer, Materials and Molecular Research Division, Lawrence Berkeley Laboratory, University of California, Berkeley; personal communication, Sept. 1978, of preliminary draft of review to be submitted for publication in Atomic Energy Review International Atomic Energy Agency, Vienna, Austria.
2. D. L. Hildenbrand, J. Chem. Phys. 65, 614 (1976).
3. A. S. Ailikhanyan, V. S. Pervov, I. P. Malkerova, V. D. Butskii, and V. I. Gorgoraki, Russ. J. Inorg. Chem. 23, 817 (1978).
4. JANAF Thermochemical Tables: Mo(g), 3-31-78; F(g), 9-30-65; TaO(g), 9-31-73.
5. M. C. Drake and G. M. Rosenblatt in Proc. Symp. on High Temperature Metal Halide Chemistry, D. L. Hildenbrand and D. D. Cubicciotti, Eds., The Electrochem. Soc., Vol. 78-1, pp. 199-209, 1978.

Molybdenum Fluoride (MoF_3) F_3Mo_1(g)

Enthalpy Reference Temperature = T_r = 298.15 K Standard State Pressure = p° = 0.1 MPa

T/K	$C_p°$ (J K^{-1}mol^{-1})	S°	-[G°-H°(T_r)]/T	H°-H°(T_r) (kJ mol^{-1})	$\Delta_f H°$	$\Delta_f G°$	Log K_f
0	0.	0.	INFINITE	-14.699	-603.975	-603.975	INFINITE
100	41.852	237.013	348.551	-11.154	-605.282	-602.725	314.831
200	57.340	271.083	301.828	-6.149	-606.513	-599.638	156.609
250	62.957	284.508	297.050	-3.135	-606.871	-597.875	124.919
298.15	67.093	295.968	295.968	0.	-607.098	-596.120	104.438
300	67.230	296.383	295.969	0.124	-607.105	-596.052	103.782
350	70.430	306.999	296.800	3.570	-607.257	-594.197	88.679
400	72.834	316.568	298.683	7.154	-607.360	-592.324	77.350
450	74.665	325.257	301.160	10.844	-607.436	-590.439	68.536
500	76.087	333.200	303.973	14.614	-607.498	-588.547	61.485
600	78.145	347.267	310.047	22.332	-607.606	-584.747	50.907
700	79.599	359.428	316.252	30.223	-607.715	-580.929	43.349
800	80.737	370.134	322.332	38.242	-607.832	-577.094	37.680
900	81.694	379.700	328.184	46.364	-607.957	-573.244	33.270
1000	82.528	388.351	333.775	54.576	-608.092	-569.380	29.741
1100	83.262	396.252	339.101	62.867	-608.245	-565.502	26.853
1200	83.908	403.525	344.170	71.226	-608.423	-561.609	24.446
1300	84.472	410.264	348.998	79.645	-608.638	-557.699	22.409
1400	84.963	416.543	353.601	88.118	-608.902	-553.771	20.661
1500	85.391	422.419	357.995	96.636	-609.224	-549.822	19.147
1600	85.766	427.943	362.196	105.194	-609.616	-545.850	17.820
1700	86.101	433.152	366.218	113.788	-610.086	-541.850	16.649
1800	86.405	438.082	370.075	122.413	-610.642	-537.821	15.607
1900	86.689	442.762	373.779	131.068	-611.289	-533.758	14.674
2000	86.958	447.215	377.340	139.751	-612.035	-529.659	13.833
2100	87.220	451.464	380.769	148.460	-612.883	-525.520	13.072
2200	87.479	455.528	384.076	157.195	-613.838	-521.338	12.378
2300	87.736	459.422	387.268	165.955	-614.904	-517.109	11.744
2400	87.993	463.161	390.352	174.742	-616.086	-512.833	11.162
2500	88.250	466.759	393.337	183.554	-617.389	-508.504	10.625
2600	88.506	470.225	396.228	192.392	-618.830	-504.121	10.128
2700	88.760	473.570	399.031	201.255	-620.444	-499.679	9.667
2800	89.011	476.803	401.751	210.144	-622.287	-495.173	9.238
2900	89.256	479.930	404.393	219.057	-660.323	-490.547	8.836
3000	89.493	482.960	406.962	227.995	-660.823	-484.684	8.439
3100	89.721	485.898	409.461	236.955	-661.260	-478.806	8.068
3200	89.938	488.750	411.895	245.938	-661.631	-472.914	7.720
3300	90.144	491.521	414.266	254.943	-661.936	-467.011	7.392
3400	90.335	494.215	416.578	263.967	-662.174	-461.101	7.084
3500	90.512	496.836	418.834	273.009	-662.345	-455.184	6.793
3600	90.674	499.388	421.036	282.069	-662.449	-449.263	6.519
3700	90.820	501.875	423.187	291.144	-662.487	-443.341	6.259
3800	90.950	504.299	425.290	300.232	-662.460	-437.419	6.013
3900	91.064	506.663	427.346	309.333	-662.369	-431.497	5.779
4000	91.162	508.969	429.358	318.445	-662.216	-425.579	5.557
4100	91.245	511.221	431.328	327.565	-662.004	-419.666	5.347
4200	91.312	513.421	433.256	336.693	-661.734	-413.758	5.146
4300	91.364	515.570	435.145	345.827	-661.408	-407.858	4.954
4400	91.402	517.671	436.997	354.965	-661.028	-401.966	4.772
4500	91.426	519.726	438.813	364.107	-660.597	-396.083	4.598
4600	91.438	521.735	440.594	373.250	-660.117	-390.210	4.431
4700	91.437	523.702	442.341	382.394	-659.590	-384.348	4.272
4800	91.425	525.627	444.056	391.537	-659.018	-378.498	4.119
4900	91.403	527.512	445.740	400.679	-658.405	-372.660	3.973
5000	91.371	529.358	447.394	409.817	-1249.136	-361.104	3.772
5100	91.329	531.167	449.019	418.953	-1250.123	-343.333	3.516
5200	91.280	532.940	450.616	428.083	-1251.221	-325.542	3.270
5300	91.223	534.678	452.186	437.208	-1252.428	-307.730	3.033
5400	91.159	536.383	453.729	446.327	-1253.745	-289.893	2.804
5500	91.089	538.055	455.247	455.440	-1255.172	-272.031	2.584
5600	91.013	539.695	456.741	464.545	-1256.709	-254.142	2.371
5700	90.933	541.305	458.210	473.642	-1258.352	-236.225	2.165
5800	90.848	542.886	459.657	482.731	-1260.102	-218.278	1.966
5900	90.759	544.438	461.080	491.812	-1261.955	-200.300	1.773
6000	90.667	545.963	462.483	500.883	-1263.914	-182.290	1.587

PREVIOUS: December 1978 (1 atm) CURRENT: December 1978 (1 bar)

Molybdenum Fluoride (MoF_3) F_3Mo_1(g)

NITROGEN FLUORIDE (NF_3) IDEAL GAS M_r = 71.001909

$\Delta_f H°(0\ K) = -126.366 \pm 1.13\ kJ\ mol^{-1}$

$S°(298.15\ K) = 260.773 \pm 0.21\ J\ K^{-1}\ mol^{-1}$ $\Delta_f H°(298.15\ K) = -132.089 \pm 1.13\ kJ\ mol^{-1}$

Vibrational Frequencies and Degeneracies

ν, cm^{-1}
1032(1)
642(1)
906(2)
492(2)

Ground State Quantum Weight: 1 $\sigma = 3$

Point Group: C_{3v}

Bond Distance: N-F = 1.371 Å

Bond Angle: F-N-F = 102°9'

Product of the Moments of Inertia: $I_A I_B I_C = 8.858632 \times 10^{-115}\ g^3\ cm^6$

Enthalpy of Formation

The adopted enthalpy of formation is the mean of values calculated from the reaction processes listed below and weighted inversely as the squares of the standard deviations. Auxiliary enthalpies of formation were taken from NBS (1) or from simultaneous adjustment of several interconnecting pieces of data (2). Also given below are the references, individual and averaged measured values of enthalpies of reaction and their uncertainties, the value of enthalpy of formation calculated from each process and its uncertainty, and the weighted mean.

Reaction	$\Delta_r H°(298.15\ K)$, kcal mol^{-1}	$\Delta_f H°(298.15\ K)$, kcal mol^{-1}
$NF_3(g)+3/2\ H_2(g)=3\ HF(aq,\ 50\ H_2O)+1/2\ N_2(g)$	-199.42±0.22	-30.92±1.72
(3) =196.3±64; (4) -199.49±0.22; (5) we derive -196.23±0.77		
$S(cr,\ rh)+2\ NF_3(g)=SF_6(g)+N_2(g)$ (6)	-228.26±0.2	-31.77±0.33
$NF_3(g)+1/2\ N_2(g)+3/2\ F_2(g)$ (7)	31.44±0.3	-31.44±0.3
$8\ NF_3(g)+3\ C_2N_2(g)=6\ CF_4(g)+7\ N_2(g)$ (8)	-1308.8 ±1.3	-31.4 ±4.4
$B(cr)+NF_3(g)=BF_3(g)+1/2\ N_2(g)$ (9)	-239.7 ±1.2	-31.7 ±1.6
$NF_3(g)+4\ NH_3(g)=3\ NH_4F(cr)+N_2(g)$ (3)	-259.5 ±1.0	-29.0 ±3.0
	Weighted Mean	-31.57±0.27

Heat Capacity and Entropy

The vibrational frequencies are from the infrared studies of Schatz and Levin (10) and are in good agreement with those of Pace and Pierce (11), and Wilson and Polo (12). The adopted values are also in agreement with the Raman studies of Kotov and Tatevskii (13). The bond length and angle are from the microwave studies of Sheridan and Gordy (14), whose values are in substantial agreement with the electron diffraction results of Schomaker and Lu (15).

The principal moments of inertia are $I_A = I_B = 7.8554 \times 10^{-39}\ g\ cm^2$ and $I_C = 14.3560 \times 10^{-39}\ g\ cm^2$.

Pierce and Pace (16) have obtained the entropy of the gas at 144.15 K from solid heat capacity measurements and the heats of melting and vaporization. Their reported value of 54.50 cal $K^{-1}\ mol^{-1}$ is in good agreement with our calculated value of 54.56 cal $K^{-1}\ mol^{-1}$.

The National Bureau of Standards prepared this table (17) by critical analysis of data existing in 1972. Using molecular constants and $\Delta_f H°$ selected by NBS (17), we recalculate the table in terms of 1973 fundamental constants (18), 1975 atomic weights (19), and current JANAF reference states for the elements.

References

1. U. S. Nat. Bur. Stand. Tech. Note 270-3, 1968.
2. A. N. Syverud, The Dow Chemical Company, AD867952, 76 pp. (1970).
3. G. T. Armstrong, S. Marantz, and C. F. Coyle, J. Amer. Chem. Soc. 81, 3798 (1959).
4. G. C. Sinke, J. Chem. Eng. Data 10, 295 (1965).
5. A. N. Zercheninov, V. I. Chesnokov, and A. V. Pankratov, Zh. Fiz. Khim, 43, 390 (1969).
6. L. C. Walker, J. Phys. Chem. 71, 361 (1967).
7. G. C. Sinke, J. Phys. Chem. 71, 359 (1967).
8. L. C. Walker, Paper presented at 22nd Annual Calorimetry Conference, Thousand Oaks, California, June, 1967.
9. J. R. Ludwig and W. J. Cooper, J. Chem. Eng. Data 8, 76 (1963).
10. P. N. Schatz and I. W. Levin, J. Chem. Phys. 29, 475 (1958).
11. E. L. Pace and L. Pierce, J. Chem. Phys. 23, 1248 (1955).
12. M. K. Wilson and S. R. Polo, J. Chem. Phys. 20, 1716 (1952).
13. Yu. I. Kotov and W. N. Tatevskii, Opt. Spectry 14, 160, Supplement 2 Mol. Spectrosc. (1963).
14. J. Sheridan and W. Gordy, Phys. Rev. 79, 513 (1950).
15. V. Schomaker and C. Lu, J. Amer. Chem. Soc. 72, 1182 (1950).
16. L. Pierce and E. L. Pace, J. Chem. Phys. 23, 551 (1955).
17. S. Abramowitz et al., U. S. Nat. Bur. Stand. Report 10904, 239, 1972.
18. CODATA Task Group on Fundamental Constants, CODATA Bulletin 11, December, 1973.
19. IUPAC Commission on Atomic Weights, Pure Appl. Chem. 47, 75 (1976).

Nitrogen Fluoride (NF_3) $F_3N_1(g)$

Enthalpy Reference Temperature = T_r = 298.15 K Standard State Pressure = p° = 0.1 MPa

T/K	$C_p°$ (J $K^{-1}mol^{-1}$)	S° (J $K^{-1}mol^{-1}$)	$-[G°-H°(T_r)]/T$ (J $K^{-1}mol^{-1}$)	$H°-H°(T_r)$ (kJ mol^{-1})	$\Delta_f H°$ (kJ mol^{-1})	$\Delta_f G°$ (kJ mol^{-1})	Log K_f
0	0.	0.	INFINITE	-11.850	-126.366	-126.366	INFINITE
100	34.037	215.800	300.931	-8.513	-128.841	-116.875	61.049
200	42.777	241.672	265.323	-4.730	-130.907	-104.047	27.174
250	48.357	251.819	261.624	-2.451	-131.609	-97.246	20.319
298.15	53.370	260.773	260.773	0.	-132.089	-90.582	15.870
300	53.551	261.104	260.774	0.099	-132.104	-90.324	15.727
350	58.068	269.707	261.443	2.892	-132.422	-83.334	12.437
400	61.860	277.717	262.983	5.894	-132.597	-76.307	9.965
450	64.994	285.190	265.040	9.067	-132.656	-69.267	8.040
500	67.568	292.175	267.408	12.384	-132.625	-62.224	6.501
600	71.430	304.858	272.615	19.346	-132.365	-48.165	4.193
700	74.090	316.081	278.039	26.630	-131.929	-34.164	2.549
800	75.971	326.104	283.432	34.138	-131.390	-20.234	1.321
900	77.340	335.135	288.683	41.807	-130.791	-6.375	0.370
1000	78.360	343.339	293.745	49.594	-130.158	7.416	-0.387
1100	79.138	350.846	298.599	57.471	-129.510	21.142	-1.004
1200	79.744	357.759	303.245	65.416	-128.858	34.809	-1.515
1300	80.224	364.161	307.688	73.416	-128.210	48.421	-1.946
1400	80.611	370.121	311.937	81.458	-127.570	61.984	-2.313
1500	80.926	375.694	316.004	89.535	-126.942	75.501	-2.629
1600	81.186	380.925	319.899	97.641	-126.329	88.978	-2.905
1700	81.404	385.854	323.636	105.771	-125.730	102.416	-3.147
1800	81.587	390.512	327.223	113.921	-125.144	115.819	-3.361
1900	81.744	394.928	330.671	122.088	-124.570	129.190	-3.552
2000	81.877	399.124	333.990	130.269	-124.005	142.532	-3.723
2100	81.993	403.122	337.187	138.463	-123.444	155.845	-3.876
2200	82.094	406.939	340.272	146.667	-122.884	169.131	-4.016
2300	82.182	410.590	343.250	154.881	-122.319	182.392	-4.142
2400	82.259	414.089	346.129	163.103	-121.745	195.628	-4.258
2500	82.328	417.448	348.915	171.333	-121.156	208.840	-4.363
2600	82.389	420.679	351.614	179.568	-120.546	222.028	-4.461
2700	82.443	423.789	354.230	187.810	-119.912	235.192	-4.550
2800	82.492	426.788	356.768	196.057	-119.249	248.332	-4.633
2900	82.535	429.684	359.233	204.308	-118.552	261.447	-4.709
3000	82.575	432.482	361.628	212.564	-117.818	274.538	-4.780
3100	82.611	435.191	363.957	220.823	-117.044	287.604	-4.846
3200	82.643	437.814	366.225	229.086	-116.226	300.644	-4.908
3300	82.673	440.357	368.433	237.352	-115.364	313.658	-4.965
3400	82.700	442.826	370.585	245.620	-114.454	326.645	-5.018
3500	82.725	445.224	372.683	253.892	-113.495	339.605	-5.068
3600	82.748	447.554	374.731	262.165	-112.486	352.537	-5.115
3700	82.769	449.822	376.730	270.441	-111.427	365.440	-5.159
3800	82.788	452.029	378.682	278.719	-110.318	378.313	-5.200
3900	82.806	454.180	380.591	286.999	-109.157	391.157	-5.239
4000	82.823	456.277	382.457	295.280	-107.945	403.970	-5.275
4100	82.838	458.322	384.282	303.563	-106.683	416.752	-5.309
4200	82.852	460.318	386.069	311.848	-105.371	429.504	-5.342
4300	82.866	462.268	387.818	320.134	-104.010	442.222	-5.372
4400	82.878	464.173	389.532	328.421	-102.600	454.909	-5.400
4500	82.890	466.036	391.212	336.709	-101.142	467.564	-5.427
4600	82.901	467.858	392.858	344.999	-99.638	480.185	-5.453
4700	82.911	469.641	394.473	353.289	-98.088	492.774	-5.477
4800	82.921	471.387	396.057	361.581	-96.494	505.328	-5.499
4900	82.930	473.096	397.612	369.873	-94.857	517.849	-5.520
5000	82.938	474.772	399.139	378.167	-93.177	530.337	-5.540
5100	82.946	476.414	400.638	386.461	-91.457	542.790	-5.559
5200	82.954	478.025	402.110	394.756	-89.697	555.209	-5.577
5300	82.961	479.605	403.558	403.052	-87.898	567.594	-5.594
5400	82.967	481.156	404.980	411.348	-86.062	579.945	-5.610
5500	82.974	482.678	406.379	419.645	-84.190	592.261	-5.625
5600	82.980	484.174	407.755	427.943	-82.283	604.543	-5.639
5700	82.985	485.642	409.109	436.241	-80.342	616.790	-5.652
5800	82.991	487.086	410.441	444.540	-78.369	629.003	-5.665
5900	82.996	488.504	411.752	452.839	-76.364	641.183	-5.677
6000	83.001	489.899	413.043	461.139	-74.329	653.327	-5.688

PREVIOUS: June 1977 (1 atm) CURRENT: June 1977 (1 bar)

Nitrogen Fluoride (NF_3) $F_3N_1(g)$

NITROGEN FLUORIDE OXIDE (NF_3O) IDEAL GAS M_r = 87.001309

Nitrogen Fluoride Oxide (NF_3O) $F_3N_1O_1(g)$

$\Delta_fH°$(0 K) = -154.926 ± 20.9 kJ mol^{-1}

S°(298.15 K) = 278.490 ± 0.21 J K^{-1} mol^{-1} $\Delta_fH°$(298.15 K) = -163.176 ± 20.9 kJ mol^{-1}

Vibrational Frequencies and Degeneracies

ν, cm^{-1}	ν, cm^{-1}
1687 (1)	887 (2)
743 (1)	528 (2)
558 (1)	398 (2)

Ground State Quantum Weight: 1 σ = 3

Point Group: C_{3v}

Bond Distances: N-O = [1.15] Å N-F = 1.48 ± 0.1 Å

Bond Angles: O-N-F = [109° 28'] F-N-F = [109° 28']

Product of the Moments of Inertia: $I_AI_BI_C$ = [4.064027 x 10^{-114}] g^3 cm^6

Enthalpy of Formation

Bougon et al. (1) determined equilibrium constants for the reaction NOF(g) + F_2(g) ⇄ NOF_3(g) as a function of temperature and total pressure. 2nd and 3rd law analyses of these equilibrium data are given below:

	Data Points	T/K	$\Delta_rH°$(298.15 K), kcal mol^{-1} 2nd law	3rd law	Drift cal K^{-1} mol^{-1}
Set I	7	533-643	-17.1	-18.7	-2.4±2.1
Set II	1	533.2		-18.2	

Set I contains their equilibrium data as a function of temperature, while Set II is their adopted equilibrium constant, $K_p°$ (553.2 K) = 0.015 ± 0.005 atm^{-1}, corresponding to the lowest pressure (103 atm.) that was studied. Estimated corrections for nonideality are small and neglected. Adopting $\Delta_rH°$(298.15 K) = -18.7 ± 1.0 kcal mol^{-1}, we derive $\Delta_fH°$(NOF_3, g, 298.15 K) = -34.4 ± 1.4 kcal mol^{-1} with $\Delta_fH°$(NOF, g, 298.15 K) = -15.7 ± 0.4 kcal mol^{-1} (2).

Armstrong and King (3) determined $\Delta_rH°$(304 K) = -253.7 ± 0.6 kcal mol^{-1} for the reaction NOF_3(g) + 5/2 H_2(g) + 122 H_2O(ℓ) + 1/2 N_2(g) + [3 HF·123 H_2O](ℓ) in a constant pressure flame calorimeter. Combining this result with their earlier enthalpy data for reduction reactions involving F_2(g) and O_2(g) with H_2(g), they obtained $\Delta_fH°$(NOF_3, g, 304 K) = -44.7 ± 1.5 kcal mol^{-1}.

Very recently, Dibeler and Walker (4) investigated the photoionization of NOF_3 by mass spectrometry. From their ionization threshold data, they determined $\Delta_fH°$(NOF_3, g, 0 K) = -69.6 ± 1.2 kcal mol^{-1}.

The value for the enthalpy of formation of gaseous NOF_3 is tentatively selected as -39.0 ± 5.0 kcal mol^{-1}.

Heat Capacity and Entropy

Several recent infrared (5, 6, 7) and Raman (8) studies of NOF_3 have been reported in the literature. The fundamental vibrational frequencies observed in the infrared spectrum of the molecule by the various investigators are in good agreement, although the assignment of the ν_3 and ν_5 bending modes has varied. Fox et al. (6) assigned ν_3 = 558 cm^{-1} and ν_5 = 528 cm^{-1}, while Curtis et al. (7) reversed the assignments. Very recently, Abramowitz and Levin (8) investigated the Raman spectrum of polycrystalline film of NOF_3 at 78 K and assigned ν_3 = 535 cm^{-1} and ν_5 = 522 cm^{-1} consistent with the infrared work of Fox et al. (6). The fundamental vibrational frequencies and assignments given by Fox et al. (6) are adopted here.

Fox et al. (9) also investigated the nuclear magnetic resonance spectrum of NOF_3 which indicated that the molecule contained three equivalent fluorine atoms. This result, along with the general appearance of the infrared spectrum, clearly establishes that the molecule has C_{3v} symmetry. The N-O bond distance and bond angles are estimated quantities from the work of Curtis et al. (7). These estimated data have been used in calculating the N-F bond distance from the rotational constants given by Curtis et al. (7). Individual moments of inertia calculated from the structural data given above are: $I_A = I_B$ = 14.8502 x 10^{-39} g cm^2, and I_C = 18.4285 x 10^{-39} g cm^2.

References

1. R. Bougon, J. Chatelet, J. P. Desmoulin, and P. Plurien, Compt. Rend. 226c, 760 (1968).
2. JANAF Thermochemical Tables: NOF(g), 6-30-61.
3. G. T. Armstrong and R. C. King, U. S. Nat. Bur. Stand., AFOSR 69-2154 TR, July 1, 1969.
4. V. H. Dibeler and J. A. Walker, Inorg. Chem. 8, 1728 (1969).
5. N. Bartlett, J. Passmore, and E. J. Wells, Chem. Commun., 213, 1966.
6. W. B. Fox, J. S. MacKenzie, N. Vanderkooi, B. S. Sukornick, C. A. Wamser, J. R. Holmes, R. E. Eibeck, and B. B. Stewart, J. Amer. Chem. Soc. 88, 2604 (1966).
7. E. C. Curtis, D. Pilipovich, and W. H. Moberly, J. Chem. Phys. 46, 2904 (1967).
8. S. Abramowitz and I. W. Levin, J. Chem. Phys. 51, 463 (1969).
9. W. B. Fox, J. S. MacKenzie, E. R. McCarthy, J. R. Holmes, R. F. Stahl, and R. Juurik, Inorg. Chem. 7, 2064 (1968).

Enthalpy Reference Temperature = T_r = 298.15 K Standard State Pressure = p° = 0.1 MPa

T/K	$C_p°$ (J $K^{-1}mol^{-1}$)	S°	-[G°-H°(T_r)]/T	H°-H°(T_r) (kJ mol^{-1})	$\Delta_fH°$	$\Delta_fG°$	Log K_f
0	0.	0.	INFINITE	-13.664	-154.926	-154.926	INFINITE
100	35.733	225.011	327.998	-10.299	-158.825	-139.114	72.666
200	52.446	254.510	284.237	-5.945	-161.776	-118.135	30.854
250	60.935	267.147	279.570	-3.106	-162.646	-107.116	22.381
298.15	67.862	278.490	278.490	0.	-163.176	-96.369	16.883
300	68.103	278.911	278.492	0.126	-163.191	-95.954	16.707
350	73.975	289.865	279.343	3.683	-163.484	-84.722	12.644
400	78.762	300.067	281.304	7.505	-163.585	-73.461	9.593
450	82.682	309.577	283.923	11.545	-163.538	-62.197	7.220
500	85.920	318.462	286.937	15.762	-163.376	-50.945	5.322
600	90.882	334.592	293.564	24.617	-162.803	-28.508	2.482
700	94.429	348.882	300.465	33.892	-162.004	-6.186	0.462
800	97.033	361.670	307.331	43.471	-161.062	16.010	-1.045
900	98.988	373.217	314.021	53.277	-160.029	38.083	-2.210
1000	100.486	383.727	320.474	63.254	-158.937	60.037	-3.136
1100	101.654	393.362	326.668	73.363	-157.811	81.880	-3.888
1200	102.579	402.248	332.601	83.576	-156.666	103.620	-4.510
1300	103.324	410.489	338.279	93.873	-155.511	125.264	-5.033
1400	103.930	418.169	343.714	104.236	-154.357	146.819	-5.478
1500	104.430	425.357	348.920	114.655	-153.209	168.291	-5.860
1600	104.846	432.110	353.911	125.120	-152.071	189.687	-6.193
1700	105.197	438.478	358.700	135.622	-150.945	211.013	-6.484
1800	105.494	444.499	363.301	146.157	-149.832	232.272	-6.740
1900	105.748	450.210	367.726	156.720	-148.732	253.470	-6.968
2000	105.967	455.640	371.987	167.306	-147.643	274.610	-7.172
2100	106.157	460.815	376.095	177.912	-146.563	295.696	-7.355
2200	106.323	465.757	380.059	188.536	-145.487	316.731	-7.520
2300	106.469	470.486	383.888	199.176	-144.411	337.717	-7.670
2400	106.597	475.021	387.592	209.830	-143.332	358.656	-7.806
2500	106.711	479.374	391.176	220.495	-142.244	379.550	-7.930
2600	106.812	483.562	394.650	231.171	-141.142	400.400	-8.044
2700	106.903	487.595	398.018	241.857	-140.023	421.207	-8.149
2800	106.984	491.484	401.287	252.552	-138.881	441.973	-8.245
2900	107.058	495.239	404.462	263.254	-137.712	462.697	-8.334
3000	107.124	498.870	407.549	273.963	-136.513	483.380	-8.416
3100	107.184	502.383	410.552	284.678	-135.280	504.023	-8.493
3200	107.239	505.787	413.475	295.399	-134.011	524.625	-8.564
3300	107.289	509.088	416.323	306.126	-132.703	545.188	-8.630
3400	107.334	512.292	419.098	316.857	-131.355	565.710	-8.691
3500	107.376	515.404	421.806	327.593	-129.964	586.191	-8.748
3600	107.415	518.429	424.448	338.332	-128.529	606.633	-8.802
3700	107.450	521.372	427.028	349.075	-127.050	627.034	-8.852
3800	107.483	524.238	429.548	359.822	-125.525	647.394	-8.899
3900	107.513	527.031	432.012	370.572	-123.955	667.714	-8.943
4000	107.541	529.753	434.422	381.325	-122.340	687.993	-8.984
4100	107.567	532.409	436.780	392.080	-120.680	708.230	-9.023
4200	107.591	535.001	439.087	402.838	-118.975	728.427	-9.059
4300	107.614	537.533	441.348	413.598	-117.226	748.583	-9.093
4400	107.635	540.007	443.562	424.361	-115.434	768.697	-9.126
4500	107.655	542.427	445.732	435.125	-113.600	788.771	-9.156
4600	107.673	544.793	447.860	445.892	-111.724	808.802	-9.184
4700	107.691	547.109	449.947	456.660	-109.808	828.794	-9.211
4800	107.707	549.376	451.995	467.430	-107.853	848.743	-9.236
4900	107.722	551.597	454.005	478.201	-105.860	868.652	-9.260
5000	107.736	553.774	455.979	488.974	-103.832	888.520	-9.282
5100	107.750	555.907	457.917	499.748	-101.769	908.346	-9.303
5200	107.763	558.000	459.822	510.524	-99.672	928.132	-9.323
5300	107.775	560.052	461.694	521.301	-97.544	947.877	-9.342
5400	107.786	562.067	463.534	532.079	-95.385	967.582	-9.359
5500	107.797	564.045	465.343	542.858	-93.198	987.246	-9.376
5600	107.807	565.987	467.123	553.638	-90.984	1006.871	-9.392
5700	107.817	567.896	468.875	564.420	-88.745	1026.455	-9.406
5800	107.826	569.771	470.598	575.202	-86.482	1046.000	-9.420
5900	107.835	571.614	472.295	585.985	-84.197	1065.506	-9.433
6000	107.843	573.427	473.965	596.769	-81.892	1084.973	-9.446

PREVIOUS: June 1970 (1 atm) CURRENT: June 1970 (1 bar)

Nitrogen Fluoride Oxide (NF_3O) $F_3N_1O_1(g)$

PHOSPHORYL FLUORIDE (POF_3) IDEAL GAS M_r = 103.968369

$S°(298.15\ K) = 285.413 \pm 0.30\ J\ K^{-1}\ mol^{-1}$

$\Delta_f H°(0\ K) = -1245.468 \pm 8.0\ kJ\ mol^{-1}$
$\Delta_f H°(298.15\ K) = -1254.250 \pm 8.0\ kJ\ mol^{-1}$

Vibrational Frequencies and Degeneracies

ν, cm⁻¹	ν, cm⁻¹
1417.7(1)	994 (2)
872.8(1)	488.2(2)
473.2(1)	335.6(2)

Ground State Quantum Weight: [1] σ = 3

Point Group: C_{3v}

Bond Distances: P–O = 1.45 ± 0.03 Å P–F = 1.52 ± 0.02 Å

Bond Angles: F–P–F = 102.5° ± 2° F–P–O = 115.8° ± 2°

Product of the Moments of Inertia: $I_A I_B I_C = 5.907318 \times 10^{-114}\ g^3\ cm^6$

Enthalpy of Formation

Ebel and Bretscher (1) measured the enthalpy change for the following reaction: $PF_3(g) + 1/2\ O_2(g) \rightarrow POF_3(g)$. They reported a heat of reaction of $-70.6\pm1\ kcal\ mol^{-1}$. Combining this result with the enthalpy of formation of PF_3, $\Delta_f H°(298.15\ K) = -224.9\pm0.9\ kcal\ mol^{-1}$ (2), we derive, $\Delta_f H°(POF_3,\ g,\ 298.15\ K) = -295.6\pm1.9\ kcal\ mol^{-1}$.

Heat Capacity and Entropy

The vibrational frequencies are from the infrared studies of Selig and Claasen (3) and are in good agreement with the Raman and infrared studies of Gutowsky and Liehr (4). Several microwave studies (5, 6, 7, 8) of POF_3 have been reported. As pointed out by Williams et al. (7), the rotational constant calculated from their microwave studies leads to a moment of inertia, I_B, which is in disagreement with the moment of inertia calculated from the molecular constants obtained from electron-diffraction studies by Brockway and Beach (9). Since insufficient microwave data were available for an independent determination, Williams et al. (7) sought to find parameters which were consistent with both measurements. Their molecular constants are adopted for this tabulation. Individual moments of inertia are $I_A = 17.7329 \times 10^{-39}$, $I_B = I_C = 18.2518 \times 10^{-39}\ g\ cm^2$.

References

1. F. Ebel and E. Bretscher, Helv. Chim. Acta. 12, 450 (1929).
2. JANAF Thermochemical Tables: PF_3(g), 12-31-69.
3. H. Selig and H. H. Claassen, J. Chem. Phys. 44, 1404 (1966).
4. H. S. Gutowsky and A. D. Liehr, J. Chem. Phys. 20, 1652 (1952).
5. S. J. Senatore, Phys. Rev. 78, 293 (1950).
6. N. J. Hawkins, V. W. Cohen, and W. S. Koski, J. Chem. Phys. 20, 528 (1952).
7. Q. Williams, J. Sheridan, and W. Gordy, J. Chem. Phys. 20, 164 (1952).
8. C. A. Burrus and W. Gordy, J. Chem. Phys. 26, 391 (1957).
9. L. O. Brockway and J. Y. Beach, J. Amer. Chem. Soc. 60, 1836 (1938).

Phosphoryl Fluoride (POF_3) $F_3O_1P_1(g)$

Enthalpy Reference Temperature = T_r = 298.15 K Standard State Pressure = p° = 0.1 MPa

T/K	$C_p°$ (J K⁻¹mol⁻¹)	S° (J K⁻¹mol⁻¹)	−[G°−H°(T_r)]/T (J K⁻¹mol⁻¹)	H°−H°(T_r) (kJ mol⁻¹)	$\Delta_f H°$ (kJ mol⁻¹)	$\Delta_f G°$ (kJ mol⁻¹)	Log K_f
0	0.	0.	INFINITE	-14.157	-1245.468	-1245.468	INFINITE
100	37.612	229.302	336.796	-10.749	-1248.761	-1235.701	645.464
200	55.073	260.721	291.294	-6.115	-1252.238	-1221.413	319.000
250	62.605	273.841	286.512	-3.168	-1253.366	-1213.571	253.562
298.15	68.815	285.413	285.413	0.	-1254.250	-1205.824	211.255
300	69.034	285.839	285.414	0.128	-1254.281	-1205.523	209.900
350	74.490	296.902	286.275	3.719	-1255.746	-1197.265	178.682
400	79.094	307.158	288.252	7.562	-1256.413	-1188.863	155.250
450	82.962	316.704	290.890	11.617	-1256.934	-1180.387	137.016
500	86.209	325.618	293.922	15.848	-1257.337	-1171.859	122.423
600	91.231	341.806	300.582	24.735	-1257.873	-1154.708	100.526
700	94.821	356.154	307.516	34.047	-1258.148	-1137.489	84.880
800	97.435	368.995	314.412	43.666	-1258.244	-1120.243	73.144
900	99.378	380.589	321.132	53.512	-1258.215	-1102.994	64.016
1000	100.853	391.139	327.613	63.527	-1258.099	-1085.753	56.714
1100	101.993	400.807	333.834	73.671	-1257.921	-1068.526	50.740
1200	102.890	409.722	339.791	83.917	-1321.273	-1050.239	45.716
1300	103.607	417.987	345.492	94.243	-1320.251	-1027.695	41.293
1400	104.188	425.687	350.948	104.634	-1319.215	-1005.229	37.506
1500	104.664	432.892	356.174	115.077	-1318.175	-982.838	34.225
1600	105.060	439.660	361.182	125.564	-1317.135	-960.516	31.358
1700	105.391	446.039	365.988	136.087	-1316.100	-938.259	28.829
1800	105.672	452.071	370.604	146.641	-1315.070	-916.063	26.583
1900	105.911	457.791	375.044	157.220	-1314.047	-893.924	24.576
2000	106.117	463.229	379.318	167.822	-1313.030	-871.838	22.770
2100	106.295	468.411	383.438	178.443	-1312.015	-849.804	21.138
2200	106.451	473.360	387.414	189.080	-1311.002	-827.817	19.655
2300	106.587	478.095	391.255	199.732	-1309.985	-805.877	18.302
2400	106.707	482.633	394.968	210.397	-1308.961	-783.981	17.063
2500	106.813	486.992	398.562	221.073	-1307.925	-762.128	15.924
2600	106.907	491.183	402.045	231.759	-1306.873	-740.317	14.873
2700	106.992	495.219	405.421	242.454	-1305.800	-718.547	13.901
2800	107.067	499.112	408.698	253.157	-1304.703	-696.817	12.999
2900	107.135	502.870	411.881	263.867	-1303.578	-675.126	12.160
3000	107.197	506.503	414.975	274.584	-1302.420	-653.475	11.378
3100	107.253	510.019	417.985	285.306	-1301.227	-631.863	10.647
3200	107.303	513.425	420.914	296.034	-1299.997	-610.291	9.962
3300	107.350	516.727	423.768	306.767	-1298.726	-588.757	9.319
3400	107.392	519.933	426.549	317.504	-1297.414	-567.262	8.715
3500	107.431	523.046	429.262	328.245	-1296.058	-545.807	8.146
3600	107.466	526.073	431.909	338.990	-1294.657	-524.390	7.609
3700	107.499	529.018	434.494	349.738	-1293.211	-503.014	7.101
3800	107.530	531.885	437.020	360.490	-1291.720	-481.677	6.621
3900	107.558	534.679	439.488	371.244	-1290.182	-460.380	6.166
4000	107.583	537.402	441.902	382.001	-1288.598	-439.123	5.734
4100	107.608	540.059	444.264	392.761	-1286.968	-417.907	5.324
4200	107.630	542.653	446.576	403.523	-1285.294	-396.730	4.934
4300	107.651	545.185	448.840	414.287	-1283.574	-375.594	4.563
4400	107.670	547.660	451.058	425.053	-1281.811	-354.499	4.208
4500	107.689	550.080	453.231	435.821	-1280.006	-333.444	3.871
4600	107.706	552.447	455.362	446.591	-1278.158	-312.430	3.548
4700	107.722	554.764	457.453	457.362	-1276.270	-291.455	3.239
4800	107.737	557.032	459.504	468.135	-1274.343	-270.522	2.944
4900	107.751	559.254	461.517	478.909	-1272.377	-249.629	2.661
5000	107.764	561.431	463.494	489.685	-1270.375	-228.777	2.390
5100	107.776	563.565	465.435	500.462	-1268.338	-207.965	2.130
5200	107.788	565.658	467.342	511.240	-1266.268	-187.194	1.880
5300	107.799	567.711	469.217	522.020	-1264.165	-166.462	1.641
5400	107.810	569.726	471.059	532.800	-1262.032	-145.771	1.410
5500	107.820	571.704	472.871	543.582	-1259.869	-125.120	1.188
5600	107.829	573.647	474.654	554.364	-1257.679	-104.507	0.975
5700	107.838	575.556	476.407	565.147	-1255.464	-83.935	0.769
5800	107.847	577.431	478.133	575.932	-1253.223	-63.401	0.571
5900	107.855	579.275	479.831	586.717	-1250.961	-42.907	0.380
6000	107.862	581.088	481.504	597.503	-1248.677	-22.450	0.195

PREVIOUS: December 1969 (1 atm) CURRENT: December 1969 (1 bar)

Phosphoryl Fluoride (POF_3) $F_3O_1P_1(g)$

PHOSPHORUS FLUORIDE (PF_3) IDEAL GAS M_r = 87.968969

$S°(298.15 K) = 273.064 \pm 0.13 J K^{-1} mol^{-1}$

$\Delta_f H°(0 K) = -952.782 \pm 3.8 kJ mol^{-1}$
$\Delta_f H°(298.15 K) = -958.441 \pm 3.8 kJ mol^{-1}$

Vibrational Frequencies and Degeneracies

ν, cm^{-1}	ν, cm^{-1}
892 (1)	860 (2)
487 (1)	344 (2)

Ground State Quantum Weight: [1] σ = 3
Point Group: C_{3v}
Bond Distances: P-F = 1.5700 ± 0.0012 Å F....F = 2.3643 ± 0.0028 Å
Bond Angle: F-P-F = 97.8 ± 0.2°
Product of the Moments of Inertia: $I_A I_B I_C = 2.070369 \times 10^{-114} g^3 cm^6$

Enthalpy of Formation

Berthelot (1) measured the enthalpy of hydrolysis of $PF_3(g)$ and $PCl_3(\ell)$. These results can be combined to give, $\Delta_r H°$ = 24.5 kcal mol^{-1} for the reaction

$$PF_3(g) + 3 KCl\cdot157 H_2O + PCl_3(\ell) + 3 KF\cdot112 H_2O.$$

Combining this result with the following enthalpy of formation data,

$\Delta_f H°(PCl_3, \ell, 298.15 K) = -72.2 kcal mol^{-1}$ (2)
$\Delta_f H°(KCl\cdot 157 H_2O, 298.15 K) = -100.0 kcal mol^{-1}$ (3)
$\Delta_f H°(KF\cdot112 H_2O, 298.15 K) = -138.5 kcal mol^{-1}$ (3)

we derive, $\Delta_f H°(PF_3, g, 298.15 K) = -212.2 kcal mol^{-1}$.

Duus and Mykytink (4) reported an enthalpy of reaction of -2.361 kcal mol^{-1} (corrected to 25°C) for the following reaction

$$PCl_3(g) + 3/2 CaF_2(cr) + PF_3(g) + 3/2 CaCl_2(cr).$$

Combining this result with the following enthalpy of formation data,

$\Delta_f H°(PCl_3, g, 298.15 K) = -64.4 kcal mol^{-1}$ (2)
$\Delta_f H°(CaF_2, cr, 298.15 K) = 293.0 \pm 1.5 kcal mol^{-1}$ (5)
$\Delta_f H°(CaCl_2, cr, 298.15 K) = -190.0 kcal mol^{-1}$ (3)

we derive, $\Delta_f H°(PF_3, g, 298.15 K) = -221.9 kcal mol^{-1}$.

Recently, Rudzitis et al. (6) measured the enthalpy of combustion of PF_3 in fluorine in a two-chambered bomb calorimeter at 25°C. They reported an enthalpy of combustion of -152.0 ± 0.2 kcal mol^{-1} for the reaction

$$PF_3(g) - F_2(g) + PF_5(g).$$

Combining this result with $\Delta_f H°(PF_5, g, 298.15 K) = -376.9 \pm 0.7 kcal mol^{-1}$ (5), we derive $\Delta_f H°(PF_3, g, 298.15 K) = -224.9 \pm 0.9$ kcal mol^{-1}. This result is adopted for this tabulation.

Heat Capacity and Entropy

The vibrational frequencies are those reported by Wilson and Palo (8). Several different values for the P-F bond length and F-P-F bond angle have been reported in the literature. This problem was recently discussed by Morino et al. (9) who also reported new data obtained from electron-diffraction studies. Their results are adopted for this tabulation. Individual moments of inertia calculated according to their structure are: $I_A = I_B = 10.8265 \times 10^{-39} g cm^2$ and $I_C = 17.6633 \times 10^{-39} g cm^2$.

References

1. M. Berthelot, Compt. Rend. 100, 81 (1885). M. Berthelot and W. Lounginine, Ann. Chim. Phys. [5], 6, 305 (1875).
2. U. S. Nat. Bur. Stand. Tech. Note 270-3, 1968.
3. U. S. Nat. Bur. Stand. Circ. 500, 1952.
4. H. C. Duus and D. P. Mykytink, J. Chem. Eng. Data 9, 585 (1964).
5. JANAF Thermochemical Tables: CaF_2(cr), 12-31-68.
6. E. Rudzitis, Argonne National Laboratory, personal communication, 1969.
7. M. Kent Wilson and S. R. Palo, J. Chem. Phys. 20, 1716 (1952).
8. Y. Morino, K. Kuchitsu, and T. Moritani, Inorg. Chem. 8, 867 (1969).

Phosphorus Fluoride (PF_3) $F_3P_1(g)$

Enthalpy Reference Temperature = T_r = 298.15 K Standard State Pressure = p° = 0.1 MPa

T/K	$C_p°$ (J $K^{-1}mol^{-1}$)	S° (J $K^{-1}mol^{-1}$)	$-[G°-H°(T_r)]/T$ (J $K^{-1}mol^{-1}$)	$H°-H°(T_r)$ (kJ mol^{-1})	$\Delta_f H°$ (kJ mol^{-1})	$\Delta_f G°$ (kJ mol^{-1})	Log K_f
0	0.	0.	INFINITE	-12.938	-952.782	-952.782	INFINITE
100	36.571	222.647	318.125	-9.548	-954.640	-949.580	496.010
200	48.620	251.683	278.127	-5.289	-957.038	-943.753	246.483
250	54.098	263.132	274.006	-2.718	-957.812	-940.340	196.473
298.15	58.691	273.064	273.064	0.	-958.441	-936.916	164.144
300	58.853	273.428	273.065	0.109	-958.464	-936.782	163.108
350	62.825	282.808	273.797	3.154	-959.738	-933.052	139.250
400	66.071	291.417	275.469	6.379	-960.274	-929.202	121.341
450	68.700	299.356	277.688	9.751	-960.720	-925.291	107.405
500	70.827	306.708	280.227	13.241	-961.094	-921.334	96.251
600	73.966	319.918	285.767	20.491	-961.686	-913.323	79.512
700	76.094	331.490	291.489	28.001	-962.137	-905.225	67.549
800	77.584	341.754	297.143	35.689	-962.495	-897.070	58.573
900	78.658	350.957	302.620	43.504	-962.794	-888.873	51.589
1000	79.455	359.288	307.877	51.411	-963.054	-880.646	46.000
1100	80.061	366.891	312.901	59.388	-963.290	-872.393	41.427
1200	80.531	373.878	317.695	67.419	-1027.083	-863.042	37.567
1300	80.902	380.339	322.269	75.491	-1026.522	-849.394	34.129
1400	81.201	386.346	326.634	83.597	-1025.965	-835.790	31.184
1500	81.444	391.957	330.804	91.729	-1025.415	-822.225	28.632
1600	81.644	397.219	334.792	99.884	-1024.874	-808.697	26.401
1700	81.811	402.174	338.611	108.057	-1024.342	-795.202	24.434
1800	81.952	406.855	342.274	116.245	-1023.820	-781.739	22.685
1900	82.072	411.289	345.790	124.447	-1023.305	-768.304	21.122
2000	82.175	415.501	349.171	132.659	-1022.795	-754.896	19.716
2100	82.264	419.513	352.426	140.881	-1022.287	-741.514	18.444
2200	82.341	423.341	355.563	149.112	-1021.777	-728.155	17.289
2300	82.408	427.003	358.590	157.349	-1021.259	-714.820	16.234
2400	82.468	430.512	361.515	165.593	-1020.729	-701.508	15.268
2500	82.520	433.879	364.342	173.843	-1020.183	-688.218	14.380
2600	82.567	437.117	367.079	182.097	-1019.614	-674.951	13.560
2700	82.608	440.234	369.731	190.356	-1019.019	-661.706	12.801
2800	82.646	443.239	372.303	198.618	-1018.394	-648.483	12.098
2900	82.679	446.139	374.800	206.885	-1017.733	-635.284	11.443
3000	82.709	448.943	377.225	215.154	-1017.034	-622.108	10.832
3100	82.737	451.655	379.582	223.426	-1016.294	-608.956	10.261
3200	82.762	454.282	381.876	231.701	-1015.509	-595.828	9.726
3300	82.784	456.829	384.109	239.979	-1014.679	-582.726	9.224
3400	82.805	459.301	386.284	248.258	-1013.800	-569.650	8.752
3500	82.824	461.702	388.405	256.540	-1012.872	-556.600	8.307
3600	82.841	464.035	390.473	264.823	-1011.893	-543.577	7.887
3700	82.858	466.305	392.492	273.108	-1010.864	-530.582	7.490
3800	82.872	468.515	394.464	281.394	-1009.782	-517.616	7.115
3900	82.886	470.668	396.390	289.682	-1008.650	-504.679	6.759
4000	82.899	472.767	398.274	297.972	-1007.466	-491.771	6.422
4100	82.911	474.814	400.116	306.262	-1006.232	-478.895	6.101
4200	82.921	476.812	401.918	314.554	-1004.947	-466.048	5.796
4300	82.932	478.763	403.683	322.846	-1003.612	-453.233	5.506
4400	82.941	480.670	405.411	331.140	-1002.228	-440.449	5.229
4500	82.950	482.534	407.104	339.435	-1000.797	-427.697	4.965
4600	82.958	484.357	408.764	347.730	-999.318	-414.978	4.712
4700	82.966	486.141	410.391	356.026	-997.794	-402.291	4.471
4800	82.973	487.888	411.987	364.323	-996.225	-389.637	4.240
4900	82.980	489.599	413.554	372.621	-994.613	-377.017	4.019
5000	82.987	491.275	415.092	380.919	-992.958	-364.430	3.807
5100	82.993	492.919	416.602	389.218	-991.262	-351.876	3.604
5200	82.999	494.530	418.085	397.518	-989.526	-339.356	3.409
5300	83.004	496.112	419.542	405.818	-987.750	-326.869	3.221
5400	83.009	497.663	420.974	414.119	-985.938	-314.417	3.041
5500	83.014	499.186	422.383	422.420	-984.088	-301.999	2.868
5600	83.019	500.682	423.768	430.721	-982.204	-289.613	2.701
5700	83.023	502.152	425.130	439.024	-980.285	-277.263	2.541
5800	83.027	503.595	426.470	447.326	-978.333	-264.947	2.386
5900	83.031	505.015	427.790	455.629	-976.349	-252.663	2.237
6000	83.035	506.410	429.088	463.932	-974.334	-240.415	2.093

PREVIOUS: December 1969 (1 atm) CURRENT: December 1969 (1 bar)

Phosphorus Fluoride (PF_3) $F_3P_1(g)$

THIOPHOSPHORYL FLUORIDE (PSF_3) IDEAL GAS M_r = 120.028969

$S°(298.15\ K) = 298.139 \pm 0.21\ J\ K^{-1}\ mol^{-1}$

$\Delta_f H°(0\ K) = [-1001.318 \pm 62.8]\ kJ\ mol^{-1}$
$\Delta_f H°(298.15\ K) = [-1009.068 \pm 62.8]\ kJ\ mol^{-1}$

Vibrational Frequencies and Degeneracies

ν, cm^{-1}	ν, cm^{-1}
981 (1)	945 (2)
442 (1)	275 (2)
695 (1)	402 (2)

Ground State Quantum Weight: [1] $\sigma = 3$
Point Group: C_{3v}
Bond Distances: P-S = 1.87 ± 0.03 Å P-F = 1.53 ± 0.02 Å
Bond Angles: F-P-F = 100.3° ± 2° F-P-S = 117.6° ± 2°
Product of the Moments of Inertia: $I_A I_B I_C = 1.731454 \times 10^{-113}\ g^3\ cm^6$

Enthalpy of Formation

The enthalpy of formation of PSF_3 is calculated from the enthalpy of formation of POF_3 and estimated P=O and P=S bond strengths. Mortimer (1) calculated several P=O and P=S bond dissociation energies from enthalpy of reaction data reported in the literature. Some of these data are given below.

Reaction	D(P=O)	D(P=S)
	kcal mol^{-1}	
ΔH oxidation PF_3 to POF_3	130.3	
ΔH oxidation $P(OEt)_3$ to $PS(OEt)_3$		81.6
ΔH oxidation PPr_3 to $PSPr_3$		82.6

Mole and McCoubrey (2) reported the following enthalpy of reaction:

$$PCl_3(\ell) + S(cr) \rightarrow PSCl_3(\ell). \quad \Delta_r H = -5.6\ kcal\ mol^{-1}.$$

Combining this result with enthalpy of vaporization data for PCl_3 (3) and $PSCl_3$ (4), along with the enthalpy of sublimation of S (3), we derive D(P=S) = 71 kcal mol^{-1}. Averaging this result with the two values given above, we obtain $\bar{D}$(P=S) = 78.4 kcal mol^{-1}. We assume that this result represents the enthalpy change for reaction $PSF_3(g) \rightarrow PF_3(g) + S(g)$. Combining this result with similar data for POF_3, namely D(P=O) = 130.3 kcal mol^{-1} and $\Delta_f H°(POF_3$, g, 298.15 K) = -295.6 ± 1.9 kcal mol^{-1} (3), we derive $\Delta_f H°(PSF_3$, g, 298.15 K) = -237 ± 15 kcal mol^{-1}, from the relationship $\Delta_f H°(PSF_3$, g, 298.15 K) = $\Delta_f H°(POF_3$, g, 298.15 K) + $\Delta_f H°$(S, g, 298.15 K) $- \Delta_f H°$(O, g, 298.15 K) - D(P=O) + D(P=S).

Heat Capacity and Entropy

Several recent infrared and Raman studies (7, 8, 9, 10) of PSF_3 have been reported. These data are in good agreement, except for the τ_1 fundamental. Dewaulle and Francois (11) assigned this fundamental to the 869 cm^{-1} band observed in their spectra. The recent work indicates that this fundamental should be assigned to the 981 cm^{-1} band. Average values of the vibrational frequencies reported by (8, 9, 10) are adopted for this tabulation. The molecular structure of this molecule was determined by Williams et al. (12) and Hawkins and Cohen (13) from microwave studies. The individual moments of inertia calculated from the constants given by Williams et al. (12) are $I_A = 1.74111 \times 10^{-39}\ g\ cm^2$, $I_B = I_C = 31.5349 \times 10^{-39}\ g\ cm^2$.

References

1. C. T. Mortimer, Pure Appl. Chem. 2, 71 (1961).
2. M. F. Mole and J. C. McCoubrey, Nature 202, 450 (1964).
3. JANAF Thermochemical Tables: $PCl_3(\ell)$, 9-30-62; S(cr), 12-31-65; POF_3(g), 12-31-69.
4. D. E. Rogers and G. Nickless in "Inorganic Sulphur Chemistry," G. Nickless, Ed., Elsevier Publishing Co., New York, N. Y., chapter 9, 1968.
5. H. H. Horn, A. Muller and O. Glemser, Z. Naturforsch 20a, 746 (1965).
6. R. G. Cavell, Spectrochim. Acta 23a, 249 (1967).
7. A. Muller, B. Krebs, E. Niecke, and A. Ruoff, Ber. Bunsenges Phys. Chem. 71 (6), 571 (1967).
8. J. R. Durig and J. W. Clark, J. Chem. Phys. 46, 3057 (1967).
9. M. L. Dewaulle and F. Francois, Compt. Rend. 226, 894 (1948).
10. Q. Williams, J. Sheridan and W. Gordy, J. Chem. Phys. 20, 164 (1952).
11. N. J. Hawkins, V. W. Cohen and W. S. Koski, J. Chem. Phys. 20, 528 (1952).

Thiophosphoryl Fluoride (PSF_3) $F_3P_1S_1(g)$

Enthalpy Reference Temperature = T_r = 298.15 K Standard State Pressure = p° = 0.1 MPa

T/K	$C_p°$	S°	$-[G°-H°(T_r)]/T$	$H°-H°(T_r)$	$\Delta_f H°$	$\Delta_f G°$	Log K_f
	J $K^{-1}mol^{-1}$			kJ mol^{-1}			
0	0.	0.	INFINITE	-15.260	-1001.318	-1001.318	INFINITE
100	40.782	236.647	354.315	-11.767	-1003.763	-998.851	521.746
200	60.470	271.207	304.539	-6.666	-1006.961	-992.854	259.307
250	68.259	285.562	299.333	-3.443	-1008.103	-989.192	206.680
298.15	74.551	298.139	298.139	0.	-1009.068	-985.461	172.649
300	74.771	298.601	298.140	0.138	-1009.103	-985.314	171.558
350	80.142	310.544	299.071	4.015	-1010.711	-981.207	146.437
400	84.525	321.542	301.202	8.136	-1013.783	-976.844	127.563
450	88.085	331.710	304.034	12.454	-1015.207	-972.149	112.844
500	90.978	341.146	307.279	16.933	-1016.595	-967.287	101.052
600	95.275	358.139	314.372	26.260	-1018.696	-957.218	83.333
700	98.210	373.059	321.712	35.943	-1020.319	-946.838	70.654
800	100.276	386.316	328.975	45.873	-1021.647	-936.247	61.131
900	101.774	398.218	336.018	55.980	-1075.911	-924.422	53.652
1000	102.888	409.001	342.786	66.215	-1075.699	-907.600	47.408
1100	103.737	418.849	349.260	76.548	-1075.450	-890.802	42.301
1200	104.396	427.905	355.441	86.956	-1138.753	-872.950	37.999
1300	104.919	436.282	361.342	97.423	-1137.700	-850.842	34.187
1400	105.339	444.074	366.976	107.937	-1136.653	-828.815	30.923
1500	105.682	451.353	372.361	118.488	-1135.617	-806.863	28.097
1600	105.965	458.183	377.514	129.071	-1134.596	-784.980	25.627
1700	106.201	464.615	382.450	139.680	-1133.592	-763.159	23.449
1800	106.400	470.691	387.185	150.310	-1132.604	-741.398	21.515
1900	106.569	476.448	391.733	160.959	-1131.632	-719.691	19.786
2000	106.714	481.918	396.107	171.623	-1130.672	-698.034	18.231
2100	106.840	487.128	400.318	182.301	-1129.720	-676.426	16.825
2200	106.949	492.101	404.378	192.990	-1128.773	-654.863	15.548
2300	107.045	496.857	408.296	203.690	-1127.825	-633.343	14.384
2400	107.128	501.414	412.082	214.399	-1126.871	-611.864	13.317
2500	107.203	505.789	415.743	225.116	-1125.906	-590.425	12.336
2600	107.269	509.995	419.288	235.839	-1124.924	-569.025	11.432
2700	107.328	514.045	422.723	246.569	-1123.920	-547.663	10.595
2800	107.380	517.949	426.054	257.305	-1122.889	-526.339	9.819
2900	107.428	521.718	429.288	268.045	-1121.828	-505.053	9.097
3000	107.471	525.360	432.430	278.790	-1120.732	-483.803	8.424
3100	107.510	528.885	435.485	289.539	-1119.599	-462.591	7.795
3200	107.545	532.299	438.458	300.292	-1118.424	-441.416	7.205
3300	107.577	535.609	441.352	311.048	-1117.207	-420.278	6.652
3400	107.606	538.821	444.172	321.807	-1115.944	-399.178	6.133
3500	107.633	541.940	446.921	332.569	-1114.636	-378.115	5.643
3600	107.658	544.973	449.602	343.334	-1113.281	-357.091	5.181
3700	107.681	547.923	452.220	354.101	-1111.877	-336.105	4.745
3800	107.702	550.795	454.776	364.870	-1110.426	-315.157	4.332
3900	107.721	553.593	457.274	375.641	-1108.928	-294.248	3.941
4000	107.739	556.320	459.717	386.414	-1107.382	-273.379	3.570
4100	107.756	558.981	462.105	397.189	-1105.789	-252.549	3.218
4200	107.771	561.578	464.443	407.965	-1104.151	-231.757	2.882
4300	107.786	564.114	466.732	418.743	-1102.467	-211.006	2.563
4400	107.799	566.592	468.973	429.522	-1100.740	-190.294	2.259
4500	107.812	569.014	471.169	440.303	-1098.969	-169.622	1.969
4600	107.824	571.384	473.322	451.085	-1097.157	-148.990	1.692
4700	107.835	573.703	475.433	461.868	-1095.305	-128.397	1.427
4800	107.845	575.974	477.504	472.652	-1093.414	-107.845	1.174
4900	107.855	578.197	479.537	483.437	-1091.485	-87.332	0.931
5000	107.864	580.376	481.532	494.223	-1089.521	-66.859	0.698
5100	107.873	582.512	483.491	505.010	-1087.521	-46.425	0.475
5200	107.881	584.607	485.415	515.797	-1085.488	-26.031	0.261
5300	107.889	586.662	487.306	526.586	-1083.423	-5.677	0.056
5400	107.896	588.679	489.165	537.375	-1081.327	14.638	-0.142
5500	107.903	590.659	490.993	548.165	-1079.202	34.914	-0.332
5600	107.909	592.603	492.790	558.955	-1077.048	55.152	-0.514
5700	107.916	594.513	494.558	569.747	-1074.867	75.350	-0.691
5800	107.921	596.390	496.297	580.539	-1072.661	95.510	-0.860
5900	107.927	598.235	498.009	591.331	-1070.429	115.632	-1.024
6000	107.932	600.049	499.695	602.124	-1068.174	135.715	-1.182

PREVIOUS: December 1969 (1 atm) CURRENT: December 1969 (1 bar)

Thiophosphoryl Fluoride (PSF_3) $F_3P_1S_1(g)$

SULFUR FLUORIDE (SF_3) IDEAL GAS M_r = 89.055209

$S°(298.15\ K) = [286.174 \pm 4.2]\ J\ K^{-1}\ mol^{-1}$

$\Delta_f H°(0\ K) = -498.980 \pm 33.5\ kJ\ mol^{-1}$
$\Delta_f H°(298.15\ K) = -503.030 \pm 33.5\ kJ\ mol^{-1}$

Electronic Levels and Quantum Weights

ε_i, cm^{-1}	g_i
0	[2]
[25000]	[2]

Vibrational Frequencies and Degeneracies

ν, cm^{-1}	ν, cm^{-1}
[850](1)	[725](1)
[550](1)	[450](1)
[300](1)	[350](1)

Point Group: C_{2v} σ = 2

Bond Distances: S-F* = [1.545] Å S-F** = [1.646] Å

Bond Angles: F*-S-F** = [89°] F**-S-F** = [178°]

(* - equatorial ** - axial)

Product of the Moments of Inertia: $I_A I_B I_C = [2.272514 \times 10^{-114}]\ g^3\ cm^6$

Enthalpy of Formation

$\Delta_f H°(0\ K) = -119.26 \pm 8.0\ kcal\ mol^{-1}$ is calculated from the primary bond dissociation energy for SF_4 of $D_0^0(SF_3-F) = 80.0 \pm 3.0\ kcal\ mol^{-1}$ with JANAF (1) $\Delta_f H°$ data for F(g) and SF_4(g). The value of D_0^0 is taken to be equal to one-fourth the enthalpy of atomization of SF_4 (1). We calculate the heat of atomization, $\Delta_{at}H°(0\ K)$, and mean bond dissocation energy (D_0^0) for SF_3 to be 240.1 kcal mol^{-1} and 80.0 kcal mol^{-1}, respectively. $\Delta_f H°$ is -120.227 kcal mol^{-1} at 298.15 K.

Several pieces of information are available which are relatively imprecise but do support our assumption that $D_0^0(SF_3-F) = \Delta_{at}H°(SF_4)/4$. Bott (2) has reported the results of shock-tube experiments on the dissociation of SF_4 over the temperature range 1650-1950 K. Rate constants based on spectroscopic measurements were correlated with the Rice-Ramsperger-Kassel (RRK) theory. This study strongly suggests that $D_0^0(SF_3-F) = 79.0 \pm 3.0\ kcal\ mol^{-1}$.

Electron-impact threshold measurements for SF_3^+ from SF_4 have been reported by Hildenbrand (3) and Glemser et al. (4). The measured appearance potentials (AP) are in good agreement and indicate $AP(SF_3^+/SF_4) = 12.67 \pm 0.1$ eV. Combining this value with an upper limit for the ionization potential of SF_3(IP≤9.25 eV, 5), we obtain a lower limit for $D_0^0(SF_3-F)$ of 78.9 kcal mol^{-1}.

Two upper limit values for $D_0^0(SF_3-F)$ can be obtained from threshold measurements by Harland and Thynne (6) on dissociative attachment in SF_6 and SF_4. We combine their $AP(F^-/SF_4) = 0.20 \pm 0.05$ eV with the electron affinity of F(EA = 3.399 eV, 1) to obtain $D_0^0(SF_3-F) \leq 83.0\ kcal\ mol^{-1}$. In addition, their appearance potentials (6) for the formation of F^- from SF_6 by two distinct processes are combined directly to give $D_0^0(SF_3-F) \leq 80.7\ kcal\ mol^{-1}$.

The AP of SF_3^+ from SF_6 has been measured by electron-impact (3, 7) and photoelectron spectroscopy (8). These three threshold values are in agreement with $AP(SF_3^+/SF_6) = 19.97 \pm 0.1$ eV. Assuming the ionization process to be $SF_6(g) + e^- = SF_3^+(g) + 3\ F(g) + 2\ e^-$, we combine the mean AP value with $IP(SF_3) \leq 9.25$ eV (5) to obtain $\Delta_f H°(SF_3, g, 0\ K) \leq -96.3\ kcal\ mol^{-1}$. This result conflicts with all of the other available information; thus, we are led to conclude that the formation of SF_3^+ from SF_6 probably involves significant kinetic and/or excitation energies.

Heat Capacity and Entropy

We assume a molecular structure in which the four atoms form a slightly distorted T-shaped molecule having bond angles similar to those for ClF_3 (1). We eliminate a nonplanar pyramid (C_{3v} symmetry) as a plausible structure since recent ESR studies (9) show that the radical has only two equivalent fluorine nuclei. The equatorial and axial S-F bond lengths are assumed to be equal to those for SF_4 (1). The principal moments of inertia are: $I_A = 5.8076 \times 10^{-39}$, $I_B = 17.0895 \times 10^{-39}$, and $I_C = 22.8971 \times 10^{-39}$ g cm^2.

The vibrational representation for the adopted structure is $3A_1 + 2B_1 + B_2$. We estimate the six fundamentals according to symmetry species by analogy with the frequencies and assignments for ClF_3 and SF_4 (1, 10). Since several of the assignments for SF_4 (11) are uncertain, more weight is given to the results for ClF_3. Bott (2) detected IR radiation in emission at 11.0-12.6 μ in SF_4/Ar mixtures which were shock-heated to temperatures above 2050 K. He attributed this radiation to SF_3, indicating that at least one fundamental lies in the region 794-909 cm^{-1}. We assume that the unpaired electron will form all doublet states, and we include an excited level at 25000 cm^{-1} by analogy with that which has been observed for SF_4^+ (8).

Wilkins (12) has previously reported thermal functions for SF_3 which cover an extended temperature range (0-6000 K). These functions are based on a pyramidal structure and vibrational frequencies obtained from estimated force constants. Since this structure now appears to be incorrect, our functions are believed to be more reliable.

References

1. JANAF Thermochemical Tables: F(g), 9-30-65; SF_4(g) and SF_6(g), 6-30-76; F^-(g), 12-31-71; ClF_3(g), 9-30-65.
2. J. F. Bott, J. Chem. Phys. 54, 181 (1971).
3. D. L. Hildenbrand, J. Phys. Chem. 77, 897 (1973).
4. O. Glemser, A. Muller, D. Bohler and B. Krebs, Z. Anorg. Allg. Chem. 357, 184 (1968).
5. F. C. Fehsenfeld, J. Chem. Phys. 54, 438 (1971).
6. P. W. Harland and J. C. J. Thynne, J. Phys. Chem. 75, 3517 (1971).
7. V. H. Dibeler and F. L. Mohler, J. Res. Nat. Bur. Stand. 40, 25 (1948).
8. J. Delwiche, Dynamic Mass Spectrom. 1, 71 (1971).
9. A. J. Colussi, J. R. Morton, K. F. Preston, and R. W. Fessenden, J. Chem. Phys. 61, 1247 (1974).
10. R. A. Frey, R. L. Redington, and A. L. K. Aljbury, J. Chem. Phys. 54, 344 (1971).
11. I. W. Levin, J. Chem. Phys. 55, 5393 (1971).
12. R. L. Wilkins, Aerospace Corporation, Report No. TR-0158 (3240-20)-19, July, 1968.

Sulfur Fluoride (SF_3) F_3S_1(g)

Enthalpy Reference Temperature = T_r = 298.15 K Standard State Pressure = p° = 0.1 MPa

T/K	$C_p°$ (J $K^{-1}mol^{-1}$)	S° (J $K^{-1}mol^{-1}$)	-[G°-H°(T_r)]/T (J $K^{-1}mol^{-1}$)	H°-H°(T_r) (kJ mol^{-1})	$\Delta_f H°$ (kJ mol^{-1})	$\Delta_f G°$ (kJ mol^{-1})	Log K_f
0	0.	0.	INFINITE	-13.599	-498.980	-498.980	INFINITE
100	37.533	232.606	334.474	-10.187	-500.617	-497.070	259.643
200	52.232	263.155	291.624	-5.694	-502.160	-492.868	128.724
250	58.316	275.488	287.187	-2.925	-502.664	-490.483	102.481
298.15	62.978	286.174	286.174	0.	-503.030	-488.103	85.514
300	63.136	286.564	286.176	0.117	-503.042	-488.010	84.970
350	66.875	296.590	286.959	3.371	-503.337	-485.480	72.454
400	69.763	305.717	288.742	6.790	-505.795	-482.825	63.050
450	72.006	314.069	291.099	10.336	-506.674	-479.909	55.706
500	73.764	321.750	293.785	13.982	-507.579	-476.884	49.820
600	76.276	335.437	299.615	21.494	-508.863	-470.617	40.971
700	77.927	347.328	305.600	29.209	-509.821	-464.163	34.636
800	79.060	357.812	311.484	37.062	-510.594	-457.586	29.877
900	79.867	367.173	317.161	45.010	-564.383	-449.839	26.108
1000	80.460	375.620	322.592	53.028	-563.757	-437.145	22.834
1100	80.907	383.311	327.767	61.098	-563.139	-424.514	20.159
1200	81.252	390.366	332.694	69.206	-562.536	-411.939	17.931
1300	81.524	396.881	337.384	77.346	-561.953	-399.413	16.049
1400	81.742	402.930	341.852	85.509	-561.393	-386.931	14.437
1500	81.918	408.576	346.115	93.693	-560.859	-374.488	13.041
1600	82.064	413.868	350.185	101.892	-560.351	-362.080	11.821
1700	82.185	418.847	354.079	110.105	-559.869	-349.703	10.745
1800	82.287	423.547	357.809	118.328	-559.411	-337.354	9.790
1900	82.374	427.999	361.387	126.562	-558.975	-325.029	8.936
2000	82.448	432.226	364.825	134.803	-558.557	-312.727	8.168
2100	82.512	436.250	368.131	143.051	-558.152	-300.446	7.473
2200	82.568	440.090	371.315	151.305	-557.755	-288.183	6.842
2300	82.616	443.761	374.386	159.564	-557.360	-275.939	6.267
2400	82.659	447.278	377.350	167.828	-556.962	-263.712	5.740
2500	82.698	450.654	380.215	176.096	-556.555	-251.501	5.255
2600	82.732	453.898	382.987	184.367	-556.132	-239.307	4.808
2700	82.762	457.021	385.672	192.642	-555.689	-227.130	4.394
2800	82.790	460.031	388.274	200.920	-555.221	-214.970	4.010
2900	82.816	462.937	390.799	209.200	-554.722	-202.826	3.653
3000	82.840	465.745	393.250	217.483	-554.190	-190.701	3.320
3100	82.862	468.461	395.633	225.768	-553.620	-178.594	3.009
3200	82.884	471.092	397.950	234.055	-553.010	-166.506	2.718
3300	82.905	473.643	400.205	242.345	-552.356	-154.438	2.445
3400	82.925	476.118	402.402	250.636	-551.657	-142.390	2.188
3500	82.945	478.523	404.543	258.930	-550.911	-130.364	1.946
3600	82.965	480.859	406.630	267.225	-550.118	-118.359	1.717
3700	82.986	483.133	408.667	275.523	-549.276	-106.377	1.502
3800	83.007	485.346	410.656	283.823	-548.385	-94.419	1.298
3900	83.029	487.503	412.599	292.124	-547.445	-82.484	1.105
4000	83.052	489.605	414.498	300.428	-546.456	-70.575	0.922
4100	83.076	491.656	416.355	308.735	-545.419	-58.691	0.748
4200	83.101	493.658	418.172	317.044	-544.333	-46.832	0.582
4300	83.127	495.614	419.950	325.355	-543.200	-35.001	0.425
4400	83.155	497.526	421.692	333.669	-542.021	-23.195	0.275
4500	83.185	499.395	423.398	341.986	-540.796	-11.417	0.133
4600	83.215	501.223	425.070	350.306	-539.526	0.332	-0.004
4700	83.248	503.013	426.709	358.629	-538.213	12.055	-0.134
4800	83.282	504.766	428.317	366.956	-536.858	23.748	-0.258
4900	83.317	506.484	429.895	375.286	-535.461	35.413	-0.378
5000	83.354	508.167	431.444	383.619	-534.024	47.049	-0.492
5100	83.393	509.818	432.964	391.957	-532.549	58.656	-0.601
5200	83.433	511.438	434.458	400.298	-531.036	70.234	-0.706
5300	83.475	513.028	435.925	408.643	-529.486	81.782	-0.806
5400	83.518	514.588	437.368	416.993	-527.901	93.300	-0.903
5500	83.563	516.121	438.786	425.347	-526.281	104.789	-0.995
5600	83.609	517.627	440.180	433.706	-524.628	116.248	-1.084
5700	83.657	519.108	441.552	442.069	-522.942	127.677	-1.170
5800	83.706	520.563	442.902	450.437	-521.226	139.076	-1.253
5900	83.756	521.994	444.230	458.810	-519.479	150.447	-1.332
6000	83.807	523.403	445.538	467.188	-517.703	161.786	-1.408

PREVIOUS: June 1977 (1 atm) CURRENT: June 1977 (1 bar)

Sulfur Fluoride (SF_3) F_3S_1(g)

SULFUR FLUORIDE, ION (SF_3^+) IDEAL GAS M_r = 89.054660

S°(298.15 K) = 268.972 ± 2.1 J K^{-1} mol^{-1}

$\Delta_f H°$(0 K) = 392.632 ± 33.5 kJ mol^{-1}
$\Delta_f H°$(298.15 K) = 393.589 ± 33.5 kJ mol^{-1}

Electronic Levels and Quantum Weights

ϵ_i, cm^{-1}	g_i
0	1
21777	1

Vibrational Frequencies and Degeneracies

ν, cm^{-1}	ν, cm^{-1}
938 (1)	914 (2)
530 (1)	411 (2)

Point Group: C_{3v} σ = 3
Bond Distance: S-F = 1.497 ± 0.03 Å
Bond Angle: F-S-F = 97.51 ± 0.5°
Product of the Moments of Inertia: $I_A I_B I_C$ = 1.5584 x 10^{-114} g^3 cm^6

Enthalpy of Formation

Glemser et al. (1) determined AP(SF_3^+) = 12.70±0.03 eV from electron-impact measurements on SF_4. Their results (1) are confirmed by the subsequent measurements of Hildenbrand (2) who reported AP values of 12.63±0.1 eV (from SF_4) and 12.8 eV (from SF_4 formed as a reaction product in SF_6 + C). From these results we arrive at a $\Delta_r H°$(0 K) = 293.1±2.3 kcal mol^{-1} (mean AP value) for the process SF_4(g) + e^- = SF_3^+(g) + F(g) + 2 e^-. Assuming that the products are formed with no excess energies, we combine the value of $\Delta_r H°$(0 K) with JANAF enthalpy of formation (3) for SF_4 and F to obtain the result, $\Delta_f H°$(SF_3^+, g, 0 K) = 93.84±8.0 kcal mol^{-1} which is adopted. Our $\Delta_f H°$ value corresponds to IP(SF_3) = 213.1±16.0 kcal mol^{-1} (9.24±0.7 eV). This value is consistent with the observations of Fehsenfeld (4) who found no charge transfer reaction between NO and SF_3^+, suggesting that IP(SF_3) < 9.25 eV. In addition, we predict that IP(SF_3) << IP(SF_2) due to a stabilization of the positive charge in the even-electron SF_3^+ cation by resonance structures such as $F_2S{=}F^+$. JANAF data (3) gives IP(SF_2) - IP(SF_3) = 1.05 eV which compares favorably with the 1-2 eV lowerings (5) observed in other di- and tri-fluoride molecules and radicals.

Paulett and Lustig (6) reported appearance potentials for SF_3^+ and SF_4^+ from SF_5NF_2. We combine their results to arrive at $\Delta_r H°$(0 K) = 2:.31±4.6 kcal mol^{-1} for the process SF_4^+(g) + NF_3(g) = SF_3^+(g) + NF_2(g) + F_2(g) which leads to $\Delta_f H°$(SF_3^+, g, 0 K) = 61.5 kcal mol^{-1}. This value is in poor agreement with the adopted result, suggesting that the fragmentation processes involving SF_5NF_2 probably include unknown excess energy terms. Other reported fragment-ion appearance potentials for SF^+ have been previously analyzed (see SF_3(g) and SF_4(g) tables) and were shown to be seriously in error.

$\Delta_f H°$(298.15 K) is calculated to be 94.07 kcal mol^{-1}. Using JANAF data (3), we find that the primary bond dissociation energy D_0^0(SF_2^+-F) is 91.4 kcal mol^{-1} which when compared with D_0^0 values for SF_3 (67.3 kcal mol^{-1}) and SF_3^- (106.4 kcal mol^{-1}) indicates that the trend is D_0^0(SF_3^-) > D_0^0(SF_3^+) >> D_0^0(SF_3). The observed increased bond strengths in both ions relative to the SF_3 radical provide additional support to the view that the positive and negative charges are delocalized in the even-electron ions. Furthermore, available structural information shows that the bond length in SF_3^+ is the shortest S-F bond that has so far been reported. The heat of atomization, $\Delta_{at}H°$(0 K), and average bond dissociation energy are 265.9 kcal mol^{-1} and 88.6 kcal mol^{-1}, respectively.

Heat Capacity and Entropy

The molecular structure and vibrational frequencies are those measured by Gibler et al. (7) for the SF_3^+ ion in the solid $SF_4 \cdot BF_3$ adduct. We adopt average values of the bond lengths and angles reported by Gibler et al. (7) since spectroscopic evidence (8) indicates that in crystal adducts the SF_3^+ symmetry is lowered somewhat as a result of flourine bridging with the anion. The adopted bond length is estimated to be within ±2.0% of the true gas-phase value, based on a comparison of structural data for gaseous and crystalline ClF_3 (9). The adopted bond angle (97.51°) is close to the gas-phase value (97.8±0.2°) found for the isoelectronic molecule PF_3 (3), suggesting that the gas-phase bond angle for SF_3^+ is probably within ±0.5° of the adopted solid-state value. Our frequencies show good agreement with those determined by Azeem et al. (8) in their infrared and Raman studies of solid adducts of SF_4 with several fluoride acceptors. Their results (8) indicate that the SF_3^+ frequencies are relatively independent of the anion and are not significantly different when measured in the liquid state. Thus, we believe that the error introduced in the SF_3^+ thermal functions by use of solid-state frequencies is probably small. The individual moments of inertia are: I_A = 1.5987 x 10^{-38}, I_B = 9.8733 x 10^{-39}, and I_C = 9.8733 x 10^{-39} g cm^2.

Delwiche (10) has reported an ionization efficiency curve for SF_3^+ which was obtained by retarded potential difference (RPD) measurements. This curve reveals the existence of an excited state, lying 2177 cm^{-1} above the ground state. We include this level and assume that the ground and excited states are both singlets since SF_3^+ has no unpaired electrons.

References

1. O. Glemser, A. Muller, D. Bohler, and B. Krebs, Z. Anorg. Allg. Chem. 357, 184 (1968).
2. D. L. Hildenbrand, J. Phys. Chem. 77, 897 (1973).
3. JANAF Thermochemical Tables: SF_4(g) and SF_3(g), 6-30-76, F(g), 9-30-65; SF_2^+(g), 12-31-76; PF_3(g), 12-31-69.
4. F. C. Fehsenfeld, J. Chem. Phys. 54, 438 (1971).
5. D. F. Torgerson and J. B. Westmore, Can. J. Chem. 53, 933 (1975).
6. G. S. Paulett and M. Lustig, J. Amer. Chem. Soc. 87, 1020 (1965).
7. D. D. Gibler, C. J. Adams, M. Fischer, A. Zalkin, and N. Bartlett, Inorg. Chem. 11, 2325 (1972).
8. M. Azeem, M. Brownstein, and R. J. Gillespie, Can. J. Chem. 47, 4159 (1969).
9. R. A. Frey, R. L. Redington, and A. L. K. Aljibury, J. Chem. Phys. 54, 344 (1971).
10. J. Delwiche, Dynamic Mass Spectrom. 1, 71 (1970).

Sulfur Fluoride, Ion (SF_3^+) $F_3S_1^+$(g)

Enthalpy Reference Temperature = T_r = 298.15 K Standard State Pressure = p° = 0.1 MPa

T/K	$C_p°$ (J $K^{-1}mol^{-1}$)	S° (J $K^{-1}mol^{-1}$)	-[G°-H°(T_r)]/T (J $K^{-1}mol^{-1}$)	H°-H°(T_r) (kJ mol^{-1})	$\Delta_f H°$ (kJ mol^{-1})	$\Delta_f G°$ (kJ mol^{-1})	Log K_f
0	0.	0.	INFINITE	-12.409	392.632		
100	35.082	221.200	311.734	-9.053			
200	45.909	248.643	273.799	-5.031			
250	51.457	259.492	269.872	-2.595			
298.15	56.223	268.972	268.972	0.	393.589	407.390	-71.373
300	56.393	269.321	268.974	0.104	393.603	407.476	-70.948
350	60.593	278.339	269.676	3.032	394.020	409.756	-61.153
400	64.076	286.665	271.286	6.151	392.303	412.059	-53.809
450	66.931	294.382	273.429	9.429	392.193	414.526	-48.117
500	69.262	301.559	275.887	12.836	392.089	417.016	-43.565
600	72.739	314.515	281.270	19.947	392.484	421.974	-36.736
700	75.120	325.917	286.850	27.347	393.288	426.830	-31.850
800	76.797	336.064	292.379	34.948	394.342	431.552	-28.177
900	78.014	345.183	297.748	42.691	342.426	437.213	-25.375
1000	78.919	353.452	302.912	50.540	344.963	447.609	-23.381
1100	79.609	361.008	307.855	58.468	347.518	457.750	-21.737
1200	80.145	367.958	312.578	66.457	350.079	467.658	-20.357
1300	80.570	374.391	317.088	74.493	352.638	477.352	-19.180
1400	80.911	380.375	321.397	82.568	355.187	486.850	-18.165
1500	81.189	385.967	325.518	90.673	357.722	496.166	-17.278
1600	81.419	391.214	329.461	98.804	360.240	505.313	-16.497
1700	81.611	396.156	333.241	106.956	362.740	514.304	-15.803
1800	81.773	400.825	336.867	115.125	365.222	523.148	-15.181
1900	81.911	405.250	340.351	123.310	367.688	531.854	-14.622
2000	82.029	409.455	343.702	131.507	370.140	540.431	-14.115
2100	82.131	413.460	346.929	139.715	372.584	548.885	-13.653
2200	82.220	417.283	350.040	147.933	375.023	557.223	-13.230
2300	82.299	420.939	353.044	156.159	377.464	565.451	-12.842
2400	82.368	424.443	355.947	164.392	379.910	573.571	-12.483
2500	82.430	427.807	358.754	172.632	382.368	581.590	-12.152
2600	82.486	431.041	361.473	180.878	384.843	589.510	-11.843
2700	82.537	434.155	364.107	189.129	387.341	597.335	-11.556
2800	82.584	437.158	366.663	197.385	389.867	605.066	-11.288
2900	82.628	440.056	369.144	205.646	392.424	612.707	-11.036
3000	82.670	442.858	371.555	213.911	395.017	620.259	-10.800
3100	82.709	445.570	373.899	222.180	397.649	627.723	-10.577
3200	82.748	448.196	376.180	230.453	400.324	635.102	-10.367
3300	82.786	450.743	378.401	238.729	403.044	642.397	-10.168
3400	82.824	453.215	380.565	247.010	405.810	649.609	-9.980
3500	82.862	455.616	382.675	255.294	408.625	656.739	-9.801
3600	82.900	457.951	384.734	263.582	411.490	663.788	-9.631
3700	82.939	460.223	386.744	271.874	414.405	670.756	-9.469
3800	82.979	462.436	388.707	280.170	417.370	677.644	-9.315
3900	83.021	464.592	390.625	288.470	420.387	684.455	-9.167
4000	83.063	466.694	392.500	296.774	423.455	691.187	-9.026
4100	83.108	468.746	394.335	305.083	426.573	697.841	-8.891
4200	83.154	470.749	396.131	313.396	429.742	704.420	-8.761
4300	83.201	472.706	397.889	321.714	432.960	710.921	-8.636
4400	83.250	474.619	399.611	330.036	436.226	717.348	-8.516
4500	83.301	476.491	401.299	338.364	439.540	723.700	-8.400
4600	83.354	478.322	402.953	346.696	442.901	729.978	-8.289
4700	83.408	480.115	404.576	355.034	446.308	736.182	-8.182
4800	83.464	481.872	406.168	363.378	449.759	742.313	-8.078
4900	83.522	483.594	407.731	371.727	453.253	748.372	-7.978
5000	83.580	485.282	409.265	380.082	456.790	754.359	-7.881
5100	83.641	486.937	410.772	388.443	460.368	760.275	-7.787
5200	83.702	488.562	412.252	396.811	463.986	766.120	-7.696
5300	83.765	490.157	413.707	405.184	467.642	771.896	-7.607
5400	83.829	491.723	415.137	413.564	471.336	777.602	-7.522
5500	83.894	493.262	416.544	421.950	475.066	783.239	-7.439
5600	83.960	494.774	417.927	430.343	478.832	788.808	-7.358
5700	84.026	496.261	419.289	438.742	482.632	794.309	-7.279
5800	84.093	497.723	420.628	447.148	486.465	799.744	-7.202
5900	84.161	499.161	421.947	455.560	490.330	805.113	-7.128
6000	84.228	500.576	423.246	463.980	494.226	810.415	-7.055

PREVIOUS: December 1976 (1 atm) CURRENT: December 1976 (1 bar)

Sulfur Fluoride, Ion (SF_3^+) $F_3S_1^+$(g)

SULFUR FLUORIDE, ION (SF_3^-) IDEAL GAS M_r = 89.055758

$S°(298.15 K) = [280.198 \pm 4.2]\ J\ K^{-1}\ mol^{-1}$

$\Delta_fH°(0\ K) = -766.148 \pm 29.3\ kJ\ mol^{-1}$
$\Delta_fH°(298.15\ K) = -776.396 \pm 29.3\ kJ\ mol^{-1}$

Vibrational Frequencies and Degeneracies

ν, cm^{-1}	ν, cm^{-1}	ν, cm^{-1}
[850](1)	[300](1)	[450](1)
[550](1)	[725](1)	[350](1)

Ground State Quantum Weight: [1] σ = [2]
Point Group: [C_{2v}]
Bond Distances: S-F* = [1.545] Å S-F** = [1.646] Å
Bond Angles: F*-S-F** = [87°] F**-S-F** = [174°]
(* - equatorial ** - axial)
Product of the Moments of Inertia: $I_AI_BI_C = [2.158616 \times 10^{-114}]\ g^3\ cm^6$

Enthalpy of Formation

Using time-of-flight measurements, Harland and Thynne (1, 2) have identified SF_3^- in the negative ion mass spectra of SF_4, SF_6, and SF_5Cl. Ionization efficiency curves obtained for a dissociative resonance capture process involving SF_4 were analyzed by a deconvolution procedure to give an appearance potential (AP) for SF_3^- of 0.70±0.05 eV. Harland and Thynne (1) suggested that the process responsible for ion formation was $SF_4(g) + e^- = SF_3^-(g) + F(g)$. Assuming that the products are formed with no excess energies, we assign the measured value of AP to $\Delta_rH°$ of the proposed ion-formation process, and we obtain the adopted value, $\Delta_fH°(SF_3^-, g, 0\ K) = -183.1 \pm 7.0\ kcal\ mol^{-1}$ by combining $\Delta_rH°(0\ K)$ with JANAF enthalpies of formation (3) for SF_4 and F. Using this $\Delta_fH°$ value, we calculate $EA(SF_3)$ = 2.72±0.65 eV and a fluoride-ion affinity (IA) of $IA(SF_2)$ = 52.7 kcal mol^{-1}. This value of EA is in excellent agreement with the result (EA = 2.71 eV) obtained by a magnetron technique (4), suggesting that the electron-impact data are reliable and that the interpretation is correct.

The electron-impact AP for SF_3^- from SF_5Cl has been reported (2) as 7.0±0.2 eV. Harland and Thynne (1, 2) have suggested two ion formation processes; namely, (A) $SF_5Cl(g) + e^- = SF_3^-(g) + F(g) + FCl(g)$ and (B) $SF_5Cl(g) + e^- = SF_3^-(g) + 2\ F(g) + Cl(g)$. We eliminate (A) as a plausible process since it leads to $EA(SF_3)$ = -2.1 eV. We are unable to reproduce the result (EA = 2.9±0.1 eV) obtained for process (A) by Harland and Thynne (1) and believe their calculations to be in error. Process (B) leads to $EA(SF_3)$ = 0.5 eV which suggests that $EA(SF_3) < EA(SF_4)$. We believe that this is very unlikely based on trends in the electron affinities for S-F species (3). We conclude that process (B) very likely involves excess energy terms, amounting to about 2.2 eV.

$\Delta_fH°$(298.15 K) is calculated to be -185.6 kcal mol^{-1} at 298.15 K. In addition, we find that the primary bond dissociation energy $D_0^0(SF_2^- -F)$ is 108.0 kcal mol^{-1} which when compared with JANAF D_0^0 values for SF_3^+(91.4 kcal mol^{-1}) and SF_3(67.3 kcal mol^{-1}) indicates that the trend is $D_0^0(SF_3^-) > D_0^0(SF_3^+) \gg D_0^0(SF_3)$. The observed increased bond strengths in both ions relative to the SF_3 radical suggest that the positive and negative charges are delocalized in the even-electron ions. Further evidence supporting this view is presented on the SF_3^+ (g) table (3). The enthalpy of atomization and average bond dissociation energy are 225.6 kcal mol^{-1} and 75.2 kcal mol^{-1}, respectively.

Heat Capacity and Entropy

SF_3^- has 28 valence electrons, and according to the Walsh correlation diagram (5), the ion should have a planar structure. We predict that this planar structure will have a slightly distorted T-shape (C_{2v} symmetry) by analogy with that for the isoelectronic molecule ClF_3 (3). As pointed out by Frey et al. (6), the slightly distorted T-shaped structure of ClF_3 is closely related to the trigonal bipyramidal structure of SF_4. We estimate structural data for SF_3^- from that for SF_4 (3) by considering the effects of removing one of its equatorial fluorine atoms. We assume that this removal process would produce no change in the S-F bond lengths and only minor changes in the bond angles. The structural data obtained in this way appear reasonable in comparison with that for ClF_3 (3). The individual moments of inertia are: $I_A = 5.5926 \times 10^{-39}$, $I_B = 17.0479 \times 10^{-39}$, and $I_C = 22.6406 \times 10^{-39}\ g\ cm^2$.

The vibrational representation for the adopted structure of SF_3^- is $3A_1 + 2B_1 + B_2$. We estimate the fundamentals according to symmetry species by analogy with the frequencies and assignments for ClF_3 and SF_4 (3, 6). Since several of the assignments for SF_4 (7) are uncertain, more weight is given to the results for ClF_3. We assume a singlet ground state (no unpaired electrons) and neglect excited states.

References

1. P. W. Harland and J. C. J. Thynne, J. Phys. Chem. 75, 3517 (1971).
2. P. Harland and J. C. J. Thynne, J. Phys. Chem. 73, 4031 (1969).
3. JANAF Thermochemical Tables: SF_4(g) and SF_3(g), 6-30-76; F(g) and ClF_3(g), 9-30-65; SF_3^+(g), SF^-(g), SF_2^-, SF_4^-(g), SF_5^-(g), 12-31-76; SF_6^-(g), 6-30-77.
4. F. M. Page and G. C. Goode, "Negative Ions and the Magnetron," Wiley, New York, N.Y., 1970.
5. A. D. Walsh, J. Chem. Soc. 1953, 2301 (1953).
6. R. A. Frey, R. L. Redington, and A. L. K. Aljibury, J. Chem. Phys. 54, 344 (1971).
7. I. W. Levin, J. Chem. Phys. 55, 5393 (1971).

Sulfur Fluoride, Ion (SF_3^-) $F_3S_1^-(g)$

Enthalpy Reference Temperature = T_r = 298.15 K Standard State Pressure = p° = 0.1 MPa

T/K	$C_p°$ (J $K^{-1}mol^{-1}$)	S° (J $K^{-1}mol^{-1}$)	$-[G°-H°(T_r)]/T$ (J $K^{-1}mol^{-1}$)	$H°-H°(T_r)$ (kJ mol^{-1})	$\Delta_fH°$ (kJ mol^{-1})	$\Delta_fG°$ (kJ mol^{-1})	Log K_f
0	0.	0.	INFINITE	-13.599	-766.148		
100	37.533	226.630	328.497	-10.187			
200	52.232	257.178	285.647	-5.694			
250	58.316	269.511	281.210	-2.925			
298.15	62.978	280.198	280.198	0.	-776.396	-753.432	131.998
300	63.136	280.588	280.199	0.117	-776.446	-753.289	131.159
350	66.875	290.613	280.982	3.371	-777.781	-749.323	111.830
400	69.763	299.740	282.765	6.790	-781.277	-745.082	97.298
450	72.006	308.092	285.122	10.336	-783.196	-740.451	85.949
500	73.764	315.773	287.808	13.982	-785.140	-735.594	76.847
600	76.276	329.461	293.638	21.494	-788.503	-725.362	63.148
700	77.927	341.351	299.623	29.209	-791.540	-714.595	53.324
800	79.060	351.835	305.508	37.062	-794.391	-703.406	45.928
900	79.867	361.196	311.184	45.010	-850.259	-690.787	40.092
1000	80.460	369.643	316.615	53.028	-851.711	-672.990	35.153
1100	80.907	377.334	321.790	61.098	-853.172	-655.047	31.106
1200	81.252	384.389	326.717	69.206	-854.647	-636.971	27.727
1300	81.524	390.904	331.407	77.346	-856.143	-618.771	24.863
1400	81.742	396.954	335.876	85.509	-857.662	-600.455	22.403
1500	81.918	402.599	340.138	93.693	-859.206	-582.029	20.268
1600	82.064	407.891	344.209	101.892	-860.777	-563.499	18.396
1700	82.185	412.870	348.103	110.105	-862.374	-544.871	16.742
1800	82.287	417.571	351.833	118.328	-863.995	-526.147	15.268
1900	82.374	422.022	355.411	126.562	-865.637	-507.334	13.948
2000	82.448	426.249	358.848	134.803	-867.298	-488.432	12.757
2100	82.512	430.273	362.154	143.051	-868.971	-469.448	11.677
2200	82.568	434.113	365.338	151.305	-870.653	-450.384	10.693
2300	82.616	437.785	368.409	159.564	-872.336	-431.243	9.794
2400	82.659	441.302	371.373	167.828	-874.017	-412.029	8.968
2500	82.697	444.677	374.238	176.096	-875.688	-392.745	8.206
2600	82.730	447.921	377.010	184.367	-877.345	-373.395	7.502
2700	82.760	451.044	379.695	192.642	-878.980	-353.981	6.848
2800	82.787	454.054	382.297	200.919	-880.591	-334.507	6.240
2900	82.811	456.959	384.822	209.199	-882.172	-314.976	5.673
3000	82.833	459.767	387.273	217.481	-883.719	-295.391	5.143
3100	82.852	462.484	389.656	225.765	-885.228	-275.755	4.646
3200	82.870	465.114	391.973	234.051	-886.697	-256.071	4.180
3300	82.886	467.665	394.228	242.339	-888.124	-236.342	3.741
3400	82.901	470.139	396.425	250.629	-889.505	-216.570	3.327
3500	82.915	472.543	398.566	258.920	-890.841	-196.758	2.936
3600	82.927	474.878	400.653	267.212	-892.130	-176.909	2.567
3700	82.939	477.151	402.690	275.505	-893.371	-157.024	2.217
3800	82.950	479.363	404.679	283.799	-894.564	-137.107	1.885
3900	82.959	481.518	406.621	292.095	-895.709	-117.159	1.569
4000	82.968	483.618	408.520	300.391	-896.806	-97.182	1.269
4100	82.977	485.667	410.377	308.689	-897.856	-77.178	0.983
4200	82.985	487.666	412.193	316.987	-898.860	-57.149	0.711
4300	82.992	489.619	413.971	325.285	-899.818	-37.098	0.451
4400	82.999	491.527	415.712	333.585	-900.732	-17.024	0.202
4500	83.005	493.393	417.418	341.885	-901.602	3.071	-0.036
4600	83.011	495.217	419.090	350.186	-902.430	23.184	-0.263
4700	83.017	497.002	420.728	358.487	-903.218	43.315	-0.481
4800	83.022	498.750	422.336	366.789	-903.965	63.462	-0.691
4900	83.027	500.462	423.913	375.092	-904.675	83.624	-0.891
5000	83.032	502.139	425.460	383.395	-905.347	103.800	-1.084
5100	83.036	503.784	426.980	391.698	-905.984	123.989	-1.270
5200	83.040	505.396	428.473	400.002	-906.587	144.191	-1.448
5300	83.044	506.978	429.939	408.306	-907.157	164.404	-1.620
5400	83.048	508.530	431.380	416.611	-907.696	184.627	-1.786
5500	83.051	510.054	432.797	424.916	-908.204	204.859	-1.946
5600	83.054	511.551	434.190	433.221	-908.683	225.102	-2.100
5700	83.057	513.021	435.560	441.526	-909.134	245.351	-2.248
5800	83.060	514.465	436.908	449.832	-909.558	265.609	-2.392
5900	83.063	515.885	438.235	458.139	-909.957	285.875	-2.531
6000	83.066	517.281	439.540	466.445	-910.331	306.146	-2.665

PREVIOUS: December 1976 (1 atm) CURRENT: December 1976 (1 bar)

Sulfur Fluoride, Ion (SF_3^-) $F_3S_1^-(g)$

TRIFLUOROSILYL (SiF_3) IDEAL GAS M_r = 85.080709

$\Delta_{at}H°(0\ K) = 1758.954 \pm 16.7\ kJ\ mol^{-1}$
$S°(298.15\ K) = [282.381 \pm 0.4]\ J\ K^{-1}\ mol^{-1}$

$\Delta_fH°(0\ K) = -1082.273 \pm 16.7\ kJ\ mol^{-1}$
$\Delta_fH°(298.15\ K) = -1085.330 \pm 16.7\ kJ\ mol^{-1}$

Electronic Levels and Quantum Weights

State	ε_i, cm^{-1}	g_i
$[X^2A_1]$	0	[2]
$[^2\Sigma]$	44120	[2]

Vibratonal Frequencies and Degeneracies

ν, cm^{-1}
832 (1)
406 (1)
954 (2)
290 (2)

Point Group: C_{3v} $\sigma = 3$
Bond Distance: Si-F = [1.5624] Å
Bond Angle: F-Si-F = 109.4712°
Product of the Moments of Inertia: $I_AI_BI_C = [2.537436 \times 10^{-114}]\ g^3\ cm^6$

Enthalpy of Formation

$\Delta_fH°(SiF_3,\ g,\ 298.15\ K) = -259.4\pm2\ kcal\ mol^{-1}$ is derived from the mass spectrometric equilibrium constants determined by Farber (9) for the reaction $SiF_4(g) + SiF_2(g) = 2\ SiF_3(g)$. Our 2nd and 3rd law analysis of the ten points in the 1590 to 1782 K range, using auxiliary data (1), gives $\Delta_rH°(298.15\ K) = 7.47\pm0.19\ kcal\ mol^{-1}$ (2nd law) and $\Delta_rH°(298.15\ K) = 7.61\pm0.04\ kcal\ mol^{-1}$ (3rd law) with a drift of $0.09\pm0.11\ cal\ K^{-1}\ mol^{-1}$ (3rd law). The 3rd law heat of reaction leads to the adopted enthalpy of formation. The adopted value is consistent with the total pressure data above the silicon-magnesium fluoride system observed by Layne (2). The value of $\Delta_fH° = -235\pm20\ kcal\ mol^{-1}$, derived by McDonald et al. (3) from appearance and ionization potentials and by Wang et al. (4) from appearance potentials and translational energies, appears to be too positive. A later paper by Franklin (10) suggests that the assumption of the fraction of active vibrations as a constant may have caused a bias of 15 kcal mol^{-1} or more in the derived enthalpy of formation.

$\Delta_fH°(0\ K) = -258.7\ kcal\ mol^{-1}$ corresponds to $\Delta_{at}H°(0\ K) = 420.4\ kcal\ mol^{-1}$, using auxiliary data (1). The average bond energies in $SiF_2(g)$, $SiF_3(g)$, and $SiF_4(g)$ are all within ~ 2 kcal mol^{-1}.

Heat Capacity and Entropy

The electronic levels are those tentatively assigned by Wang et al. (5) from emission spectral observations. The first excited state is designated as ^2E, rather than ^2B, as stated by So (11) based on ab initio Hartree-Fock calculations and group theory arguments. The vibrational frequencies are derived from matrix-isolation UV and IR spectra of Milligan et al. (6). They also determined the angle between the Si-F bonds and the three-fold axis to be 71±2° which is consistent with the adopted tetrahedral F-Si-F bond angle. Theoretical calculations (7, 8, 11) indicate that this angle should be somewhat less than the tetrahedral angle. Calculations based on a bond angle of 108° decrease S°(298.15 K) by 0.029 cal K^{-1} mol^{-1}. As suggested by Milligan et al. (6), the Si-F bond length is assumed to be that in $SiHF_3$ (1). The C_{3v} structure is well established by ESR studies (12, 13). The principal moments of inertia are $I_A = I_B = 11.1156 \times 10^{-39}\ g\ cm^2$ and $I_C = 20.5365 \times 10^{-39}\ g\ cm^2$.

References

1. JANAF Thermochemical Tables: $SiHF_3(g)$ and $SiF_4(g)$, 6-30-76; F(g), 9-30-65; Si(g), 3-31-67; $SiF_2(g)$, 12-31-77.
2. G. S. Layne, The Dow Chemical Company, personal communication, Aug., 1969.
3. J. D. McDonald, C. H. Williams, J. C. Thompson, and J. L. Margrave, Advan. Chem. Ser. No. 72, 261 (1968).
4. J. Ling-Fai Wang, J. L. Margrave, and J. L. Franklin, J. Chem. Phys. 58, 5417 (1973).
5. J. Ling-Fai Wang, C. N. Krishnan, and J. L. Margrave, J. Mol. Spectrosc. 48, 346 (1973).
6. D. E. Milligan, M. E. Jacox, and W. A. Guillory, J. Chem. Phys. 49, 5330 (1968).
7. L. J. Aarons, I. H. Hillier, and M. F. Guest, J. Chem. Soc. Faraday Trans. II 70, 167 (1974).
8. Y. I. Gorlov, I. I. Ukrainskii and V. V. Pen'Kovskii, Theor. Chim. Acta 34, 31 (1974).
9. M. Farber, Space Sciences, Inc., personal communication, October 18, 1976.
10. J. L. Franklin, Science 193, 725 (1976).
11. S. P. So, J. Mol. Struct. 39, 133 (1977).
12. A. Hasegawa, K. Sogabe, and M. Miura, Mol. Phys. 30, 1889 (1975).
13. M. V. Merritt and R. W. Fessenden, J. Chem. Phys. 56, 2353 (1972).

Trifluorosilyl (SiF_3) $F_3Si_1(g)$

Enthalpy Reference Temperature = T_r = 298.15 K Standard State Pressure = $p°$ = 0.1 MPa

T/K	$C_p°$ (J K⁻¹mol⁻¹)	S° (J K⁻¹mol⁻¹)	$-[G°-H°(T_r)]/T$ (J K⁻¹mol⁻¹)	$H°-H°(T_r)$ (kJ mol⁻¹)	$\Delta_fH°$ (kJ mol⁻¹)	$\Delta_fG°$ (kJ mol⁻¹)	Log K_f
0	0.	0.	INFINITE	-13.399	-1082.273	-1082.273	INFINITE
100	38.701	229.583	329.089	-9.951	-1083.451	-1080.471	564.380
200	50.816	260.374	287.566	-5.438	-1084.512	-1077.058	281.299
250	55.586	272.237	283.340	-2.776	-1084.953	-1075.143	224.639
298.15	59.625	282.381	282.381	0.	-1085.330	-1073.218	188.023
300	59.769	282.750	282.382	0.110	-1085.343	-1073.143	186.851
350	63.348	292.240	283.123	3.191	-1085.681	-1071.082	159.850
400	66.341	300.901	284.812	6.435	-1085.968	-1068.976	139.594
450	68.812	308.862	287.048	9.816	-1086.212	-1066.837	123.835
500	70.843	316.221	289.602	13.309	-1086.421	-1064.673	111.226
600	73.891	329.424	295.164	20.556	-1086.761	-1060.290	92.306
700	75.992	340.982	300.901	28.056	-1087.036	-1055.855	78.789
800	77.478	351.231	306.564	35.734	-1087.281	-1051.384	68.648
900	78.559	360.423	312.047	43.538	-1087.520	-1046.883	60.759
1000	79.365	368.743	317.307	51.436	-1087.768	-1042.354	54.447
1100	79.980	376.338	322.333	59.405	-1088.035	-1037.800	49.281
1200	80.458	383.318	327.128	67.428	-1088.330	-1033.221	44.975
1300	80.838	389.774	331.702	75.493	-1088.659	-1028.615	41.330
1400	81.143	395.776	336.067	83.593	-1089.027	-1023.983	38.205
1500	81.392	401.383	340.237	91.720	-1089.440	-1019.323	35.496
1600	81.598	406.643	344.224	99.870	-1089.899	-1014.634	33.124
1700	81.770	411.595	348.043	108.039	-1140.584	-1009.467	31.017
1800	81.915	416.273	351.705	116.223	-1140.909	-1001.745	29.070
1900	82.038	420.705	355.221	124.421	-1141.238	-994.005	27.327
2000	82.144	424.916	358.601	132.630	-1141.570	-986.247	25.758
2100	82.235	428.926	361.855	140.849	-1141.901	-978.473	24.338
2200	82.314	432.754	364.992	149.077	-1142.228	-970.683	23.047
2300	82.384	436.414	368.018	157.312	-1142.545	-962.878	21.868
2400	82.445	439.922	370.941	165.553	-1142.848	-955.060	20.786
2500	82.499	443.289	373.768	173.800	-1143.132	-947.230	19.791
2600	82.547	446.525	376.505	182.053	-1143.393	-939.388	18.873
2700	82.590	449.641	379.156	190.310	-1143.625	-931.538	18.022
2800	82.629	452.646	381.728	198.571	-1143.825	-923.679	17.231
2900	82.663	455.546	384.223	206.835	-1143.988	-915.813	16.496
3000	82.695	458.349	386.648	215.103	-1144.112	-907.943	15.809
3100	82.723	461.061	389.005	223.374	-1144.193	-900.070	15.166
3200	82.749	463.688	391.298	231.648	-1144.227	-892.194	14.564
3300	82.772	466.234	393.530	239.924	-1144.215	-884.318	13.998
3400	82.793	468.706	395.705	248.202	-1144.152	-876.443	13.465
3500	82.813	471.106	397.825	256.482	-1144.039	-868.571	12.963
3600	82.831	473.439	399.893	264.765	-1528.026	-850.241	12.337
3700	82.848	475.709	401.912	273.049	-1527.395	-831.423	11.738
3800	82.863	477.918	403.883	281.334	-1526.713	-812.622	11.170
3900	82.877	480.071	405.809	289.621	-1525.979	-793.839	10.632
4000	82.891	482.169	407.692	297.910	-1525.194	-775.076	10.121
4100	82.903	484.216	409.534	306.199	-1524.357	-756.334	9.636
4200	82.914	486.214	411.336	314.490	-1523.469	-737.612	9.174
4300	82.925	488.165	413.100	322.782	-1522.531	-718.912	8.733
4400	82.935	490.072	414.828	331.075	-1521.542	-700.235	8.313
4500	82.945	491.936	416.520	339.369	-1520.504	-681.580	7.912
4600	82.954	493.759	418.180	347.664	-1519.418	-662.949	7.528
4700	82.962	495.543	419.807	355.960	-1518.284	-644.342	7.161
4800	82.970	497.290	421.403	364.256	-1517.104	-625.760	6.810
4900	82.978	499.001	422.969	372.554	-1515.878	-607.203	6.473
5000	82.985	500.677	424.507	380.852	-1514.608	-588.672	6.150
5100	82.993	502.320	426.016	389.151	-1513.296	-570.166	5.840
5200	83.000	503.932	427.499	397.450	-1511.942	-551.686	5.542
5300	83.007	505.513	428.956	405.751	-1510.547	-533.233	5.255
5400	83.014	507.065	430.389	414.052	-1509.114	-514.807	4.980
5500	83.020	508.588	431.797	422.353	-1507.642	-496.407	4.714
5600	83.027	510.084	433.181	430.656	-1506.134	-478.034	4.459
5700	83.034	511.554	434.543	438.959	-1504.591	-459.690	4.213
5800	83.041	512.998	435.884	447.263	-1503.015	-441.372	3.975
5900	83.048	514.417	437.203	455.567	-1501.405	-423.081	3.746
6000	83.055	515.813	438.501	463.872	-1499.764	-404.818	3.524

PREVIOUS: December 1977 (1 atm) CURRENT: December 1977 (1 bar)

Trifluorosilyl (SiF_3) $F_3Si_1(g)$

TITANIUM FLUORIDE (TiF_3) CRYSTAL M_r = 104.875209

$\Delta_fH°(0 K)$ = Unknown

S°(298.15 K) = [87.864 ± 12.6] J K^{-1} mol^{-1} $\Delta_fH°(298.15 K)$ = -1435.530 ± 41.8 kJ mol^{-1}

Enthalpy of Formation

Blocher and Hall (1) have reported vapor pressure data for the process 1/2 Hg_2F_2 + $TiF_3(cr)$ = $TiF_4(g)$ + Hg(ℓ). 2nd and 3rd law analyses of their data give a 2nd law $\Delta_rH°(298.15 K)$ of 23.7 kcal mol^{-1}, a 3rd law $\Delta_rH°(298.15 K)$ of 29.864, and a 3rd law drift of 11.9±1.0 cal K^{-1} mol^{-1}. The 3rd law value is combined with the JANAF values for the enthalpy of formation of $Hg_2F_2(cr)$ and $TiF_4(g)$ to give a $\Delta_fH°(298.15 K)$ of -343.1±10.0 kcal mol^{-1}. The large uncertainty on $\Delta_fH°(298.15 K)$ is assigned because Blocher and Hall were not able to identify $TiF_3(cr)$ as the reaction product in the above process. They reported that the X-ray pattern of the products contained lines corresponding to $Hg_2F_2(cr)$ plus others that did not correspond to $TiF_3(cr)$.

Heat Capacity and Entropy

The heat capacity of $TiF_3(cr)$ is estimated from the heat capacities of ZrF_4 and TiF_4. The entropy, S°(298.15 K), is estimated from CrF_3 and from ionic entropy contributions. Both methods give values which are within ±0.5 cal K^{-1} mol^{-1} of the chosen value of 21±3 cal K^{-1} mol^{-1}.

Sublimation Data

The heat of sublimation is calculated from the vapor pressure data of Zmbov and Margrave (2). See $TiF_3(g)$ table for details. The sublimation temperature is taken as temperature at which the fugacity is one atmosphere for the process $TiF_3(cr)$ = $TiF_3(g)$.

References

1. J. N. Blocher, Jr. and E. N. Hall, J. Phys. Chem. 63, 127 (1959).
2. K. F. Zmbov and J. L. Margrave, J. Phys. Chem. 71, 2893 (1967).

Titanium Fluoride (TiF_3) $F_3Ti_1(cr)$

Enthalpy Reference Temperature = T_r = 298.15 K Standard State Pressure = p° = 0.1 MPa

T/K	$C_p°$	S°	$-[G°-H°(T_r)]/T$	$H°-H°(T_r)$	$\Delta_fH°$	$\Delta_fG°$	Log K_f
	J $K^{-1}mol^{-1}$			kJ mol^{-1}			
0							
100							
200							
298.15	92.023	87.864	87.864	0.	-1435.530	-1361.864	238.593
300	92.048	88.433	87.866	0.170	-1435.494	-1361.407	237.042
400	93.303	115.049	91.486	9.425	-1433.682	-1336.992	174.593
500	96.106	136.170	98.379	18.895	-1432.000	-1313.014	137.170
600	98.475	153.900	106.194	28.624	-1430.307	-1289.376	112.250
700	100.952	169.264	114.130	38.594	-1428.552	-1266.025	94.472
800	103.537	182.912	121.890	48.817	-1426.668	-1242.933	81.155
900	106.232	195.262	129.367	59.305	-1424.653	-1220.086	70.812
1000	109.035	206.598	136.531	70.067	-1422.546	-1197.469	62.549
1100	111.947	217.126	143.385	81.116	-1420.393	-1175.066	55.799
1200	114.968	226.996	149.945	92.461	-1422.162	-1152.740	50.177
1300	118.098	236.321	156.234	104.113	-1419.144	-1130.409	45.420
1400	121.336	245.191	162.274	116.084	-1415.918	-1108.319	41.352
1500	124.683	253.691	168.087	128.406	-1412.462	-1086.467	37.834
1600	125.520	261.764	173.692	140.916	-1408.949	-1064.849	34.764
1700	126.357	269.399	179.099	153.510	-1405.496	-1043.448	32.061
1800	127.194	276.645	184.319	166.187	-1402.112	-1022.249	29.665
1900	128.030	283.544	189.361	178.948	-1398.807	-1001.237	27.526
2000	128.867	290.133	194.236	191.793	-1410.275	-979.942	25.593

PREVIOUS: December 1967 CURRENT: June 1968

Titanium Fluoride (TiF_3) $F_3Ti_1(cr)$

TITANIUM FLUORIDE (TiF_3) IDEAL GAS M_r = 104.895209

$S°(298.15\ K) = [291.440 \pm 4.2]\ J\ K^{-1}\ mol^{-1}$

$\Delta_f H°(0\ K) = -1184.510 \pm 41.8\ kJ\ mol^{-1}$

$\Delta_f H°(298.15\ K) = -1188.674 \pm 41.8\ kJ\ mol^{-1}$

Electronic Levels and Quantum Weights

ε_i, cm^{-1}	g_i
0	[2]
[600]	[2]
[1500]	[2]
[4000]	[2]
[10000]	[2]

Vibrational Frequencies and Degeneracies

ν, cm^{-1}
[800](1)
[450](1)
[750](2)
[300](2)

Ground State Quantum Weight: [2] $\sigma = 3$

Point Group: C_{3v}

Bond Distance: Ti-F = [1.9] Å

Bond Angle: F-Ti-F = [100]°

Product of the Moments of Inertia: $I_A I_B I_C = [7.509460 \times 10^{-114}]\ g^3\ cm^6$

Enthalpy of Formation

The enthalpy of formation of TiF_3(g) is calculated from that of TiF_3(cr) and the enthalpy of sublimation determined from the mass spectrometric vapor pressure data of Zmbov and Margrave (1). Their data give nine points over the temperature range 759 - 865 K.

2nd and 3rd law analyses of their data give a 2nd law $\Delta_{sub}H°(298.15\ K)$ of 61.2±0.4 kcal mol^{-1} and a 3rd law $\Delta_{sub}H°(298.15\ K)$ of 59.0 kcal mol^{-1} with an entropy drift of -2.7±0.5 units. The 3rd law value is used to calculate $\Delta_f H°(298.15\ K)$.

Heat Capacity and Entropy

The interatomic distance is estimated from those of TiF_4(g), and $TiCl_4$(g), and $TiCl_3$(g). The pyramidal bond angle is estimated assuming TiF_3(g) similar to the group V trihalides. The principal moments of inertia are $I_A = I_B = 16.7602 \times 10^{-39}\ g\ cm^2$ and $I_C = 26.7330 \times 10^{-39}\ g\ cm^2$. The vibrational frequencies are estimated from those of PF_3 and AsF_3. The electronic levels are estimated from the levels of Ti^{+3} (2).

References

1. K. F. Zmbov and J. L. Margrave, J. Phys. Chem. 71, 2893 (1967).
2. C. E. Moore, U. S. Nat. Bur. Stand. Circ. 467, 1949.

Titanium Fluoride (TiF_3) F_3Ti_1(g)

Enthalpy Reference Temperature = T_r = 298.15 K Standard State Pressure = p° = 0.1 MPa

T/K	$C_p°$ (J $K^{-1}mol^{-1}$)	S° (J $K^{-1}mol^{-1}$)	$-[G°-H°(T_r)]/T$ (J $K^{-1}mol^{-1}$)	$H°-H°(T_r)$ (kJ mol^{-1})	$\Delta_f H°$ (kJ mol^{-1})	$\Delta_f G°$ (kJ mol^{-1})	Log K_f
0	0.	0.	INFINITE	-13.903	-1184.510	-1184.510	INFINITE
100	38.206	236.501	341.197	-10.470	-1185.998	-1183.266	618.075
200	53.531	267.655	297.086	-5.886	-1187.725	-1179.815	308.136
250	60.296	280.349	292.491	-3.035	-1188.291	-1177.768	246.081
298.15	65.585	291.440	291.440	0.	-1188.674	-1175.704	205.978
300	65.764	291.846	291.441	0.121	-1188.687	-1175.623	204.694
350	70.016	302.317	292.258	3.521	-1188.958	-1173.423	175.124
400	73.278	311.889	294.123	7.107	-1189.144	-1171.191	152.942
450	75.772	320.671	296.591	10.836	-1189.273	-1168.938	135.687
500	77.684	328.757	299.409	14.674	-1189.365	-1166.673	121.882
600	80.293	343.172	305.531	22.584	-1189.491	-1162.122	101.172
700	81.878	355.678	311.822	30.699	-1189.591	-1157.553	86.378
800	82.877	366.681	318.006	38.940	-1189.689	-1152.970	75.281
900	83.536	376.483	323.968	47.263	-1189.839	-1148.372	66.650
1000	83.989	385.309	329.669	55.641	-1190.117	-1143.751	59.743
1100	84.314	393.330	335.097	64.057	-1190.596	-1139.093	54.091
1200	84.554	400.678	340.260	72.501	-1195.265	-1134.262	49.373
1300	84.733	407.453	345.172	80.966	-1195.435	-1129.172	45.371
1400	84.869	413.738	349.847	89.446	-1195.699	-1124.066	41.939
1500	84.973	419.597	354.304	97.938	-1196.073	-1118.937	38.965
1600	85.052	425.083	358.558	106.440	-1196.569	-1113.779	36.361
1700	85.112	430.241	362.625	114.948	-1197.202	-1108.586	34.063
1800	85.157	435.108	366.518	123.462	-1197.981	-1103.351	32.018
1900	85.192	439.713	370.250	131.979	-1198.920	-1098.070	30.188
2000	85.218	444.083	373.833	140.500	-1214.713	-1092.280	28.527
2100	85.238	448.241	377.278	149.023	-1216.744	-1086.109	27.015
2200	85.253	452.207	380.595	157.547	-1218.778	-1079.840	25.639
2300	85.265	455.997	383.791	166.073	-1220.808	-1073.479	24.379
2400	85.273	459.626	386.876	174.600	-1222.830	-1067.030	23.223
2500	85.280	463.107	389.856	183.128	-1224.838	-1060.497	22.158
2600	85.285	466.452	392.738	191.656	-1226.827	-1053.884	21.173
2700	85.288	469.671	395.528	200.184	-1228.791	-1047.195	20.259
2800	85.290	472.773	398.232	208.713	-1230.728	-1040.434	19.410
2900	85.291	475.765	400.854	217.242	-1232.631	-1033.604	18.617
3000	85.290	478.657	403.400	225.771	-1234.497	-1026.709	17.877
3100	85.289	481.454	405.873	234.300	-1236.324	-1019.753	17.183
3200	85.287	484.161	408.277	242.829	-1238.108	-1012.738	16.531
3300	85.283	486.786	410.617	251.358	-1239.847	-1005.668	15.918
3400	85.279	489.332	412.895	259.886	-1241.539	-998.546	15.341
3500	85.274	491.804	415.114	268.413	-1243.183	-991.376	14.795
3600	85.268	494.206	417.278	276.941	-1244.777	-984.158	14.280
3700	85.260	496.542	419.389	285.467	-1655.420	-969.109	13.681
3800	85.252	498.816	421.449	293.993	-1655.680	-950.557	13.066
3900	85.243	501.030	423.461	302.517	-1655.944	-931.997	12.483
4000	85.234	503.188	425.428	311.041	-1656.213	-913.431	11.928
4100	85.223	505.292	427.350	319.564	-1656.484	-894.859	11.401
4200	85.212	507.346	429.230	328.086	-1656.756	-876.279	10.898
4300	85.199	509.351	431.070	336.606	-1657.030	-857.693	10.419
4400	85.187	511.309	432.872	345.126	-1657.302	-839.101	9.961
4500	85.173	513.224	434.636	353.644	-1657.574	-820.502	9.524
4600	85.159	515.096	436.365	362.160	-1657.844	-801.897	9.106
4700	85.144	516.927	438.060	370.675	-1658.110	-783.287	8.705
4800	85.129	518.719	439.722	379.189	-1658.373	-764.671	8.321
4900	85.113	520.474	441.352	387.701	-1658.630	-746.049	7.953
5000	85.097	522.194	442.951	396.212	-1658.883	-727.423	7.599
5100	85.080	523.879	444.522	404.721	-1659.128	-708.791	7.260
5200	85.063	525.531	446.064	413.228	-1659.367	-690.154	6.933
5300	85.046	527.151	447.579	421.733	-1659.597	-671.513	6.618
5400	85.028	528.740	449.067	430.237	-1659.819	-652.869	6.315
5500	85.010	530.300	450.530	438.739	-1660.031	-634.220	6.023
5600	84.992	531.832	451.968	447.239	-1660.233	-615.566	5.742
5700	84.974	533.336	453.382	455.737	-1660.424	-596.910	5.470
5800	84.955	534.814	454.774	464.234	-1660.603	-578.251	5.208
5900	84.937	536.266	456.142	472.728	-1660.770	-559.587	4.954
6000	84.918	537.693	457.490	481.221	-1660.924	-540.922	4.709

PREVIOUS: June 1968 (1 atm) CURRENT: June 1968 (1 bar)

Titanium Fluoride (TiF_3) F_3Ti_1(g)

ZIRCONIUM FLUORIDE (ZrF_3) CRYSTAL M_r = 148.215209

$\Delta_f H°(0\ K)$ = Unknown

$S°(298.15\ K) = [87.864 \pm 8.4]\ J\ K^{-1}\ mol^{-1}$ $\Delta_f H°(298.15\ K) = [-1401.640 \pm 83.7]\ kJ\ mol^{-1}$

Enthalpy of Formation

The enthalpy of formation at 298.15 K for ZrF_3(cr) was reported as -350 kcal mol^{-1} (1) which was estimated by Brewer (2). Based on this $\Delta_f H°$(298.15 K) value, the crystal lattice energy (U) of ZrF_3(cr) is calculated to be 1371 kcal mol^{-1} by the Born-Haber cycle method (3). For the evaluation of U from the relation: $\Delta_f H°(ZrF_3$, cr, 298.15 K) = $\Sigma I_i + L - 3(E - D/2) - U$, the following values (kcal mol^{-1}) are used: ΣI_i (the sum of the successive three ionization potentials of Zr metal) = 1055 (4), L (the enthalpy of sublimation of Zr metal) = 148.3 (5), E (the electron affinity of fluorine) = 79.5 (6), and D (the enthalpy of dissociation of fluorine) = 37.7 (5). The derived value of U(ZrF_3, cr is of the same order of magnitude of the value, U(TiF_3, cr) = 1376 kcal mol^{-1}, calculated by Cavell and Clark (3) using the same method.

From $\Delta_f H°(ZrF_3$, cr, 298.15 K) = -350 kcal mol^{-1}, the derived $\Delta_f G(ZrF_3$, cr) values indicate that ZrF_3(cr) is thermodynamically more stable than the system ZrF_4(cr) + Zr(cr). For example, in the temperature range 298-1200 K, the Gibbs energy changes for the reaction 3 ZrF_4(cr) + Zr(cr) = 4 ZrF_3(cr) are about -29 ± 1 kcal mol^{-1}. However, this product ZrF_3(cr) has never been prepared successfully from this reaction in many laboratories (7). In order to make the $\Delta_f G°(ZrF_3$, cr, 298.15 K) values consistent with the experimental observations, the enthalpy of formation of ZrF_3(cr) is arbitrarily adjusted to be 15 kcal mol^{-1} less negative, or $\Delta_f H°(ZrF_3$, cr, 298.15 K) = -335 kcal mol^{-1} with uncertainty as high as ±20 kcal mol^{-1}. This value is tentatively adopted.

The Gibbs energy changes for the reaction 3 ZrF_4(g) + Zr(cr) = 4 ZrF_3(cr) are evaluated to be very favorable at temperatures below 900 K, e.g., $\Delta_r G°$ = -93.9 and -6.7 kcal mol^{-1} at 298 and 900 K, respectively. But according to Larsen and Leddy (8), this reaction does not occur in the temperature range 473-973 K, and pressure range 5-15 atm. One possible explanation is that these heterogeneous reactants may require very high activation energy to initiate the reaction. In other words, this reaction is unfavorable at these temperatures probably due to kinetic rather than thermodynamic reasons.

Heat Capacity and Entropy

The heat capacities of ZrF_3(cr) are not available in the literature. They are estimated from the $C_p°$ values of ZrF_4(cr) by deduction of one F atom $C_p°$ contributions which are calculated as 1/4 [$C_p°(ZrF_4$, cr)- $C_p°$(Zr, cr). The S°(298.15 K) is taken to be 1/2 [S°(ZrF_4, cr, 298.15 K) + S°(ZrF_2, cr, 298.15 K)] of which the value S°(ZrF_2, cr, 298.15 K) is estimated; see ZrF_2(cr) table for details.

Sublimation Data

The difference between $\Delta_f H°$(298.15 K) for ZrF_3(g) and ZrF_3(cr) is $\Delta_{sub}H°$(298.15 K). Actually ZrF_3(cr) is thermodynamically unstable, e.g., the Gibbs energy changes for the reaction 4 ZrF_3(cr) = 3 ZrF_4(cr) + Zr(cr) are about -30 kcal mol^{-1} in the temperature range 298-1400 K. In other words, ZrF_3(cr) decomposes into ZrF_4(cr) + Zr(cr) before sublimation occurs.

References

1. U. S. Nat. Bur. Stand. Circ. 500, 1952.
2. L. Brewer, University of California, Berkeley, California, personal communication; cited in reference 1.
3. R. G. Cavell and H. C. Clark, J. Chem. Soc. 1965, 444 (1965).
4. C. E. Moore, U. S. Nat. Bur. Stand. Circ. 467, Vol. III, 1958.
5. JANAF Thermochemical Tables: Zr(cr), 12-31-67; F(g), 9-30-65.
6. U. S. Nat. Bur. Stand. Report 8628, 1965.
7. T. B. Douglas, U. S. Nat. Bur. Stand., personal communication, July 10, 1961.
8. E. M. Larsen and J. J. Leddy, J. Amer. Chem. Soc. 78, 5983 (1956).

Zirconium Fluoride (ZrF_3) F_3Zr_1(cr)

Enthalpy Reference Temperature = T_r = 298.15 K Standard State Pressure = p° = 0.1 MPa

T/K	$C_p°$ (J K⁻¹mol⁻¹)	S°	-[G°-H°(T_r)]/T	H°-H°(T_r) (kJ mol⁻¹)	$\Delta_f H°$	$\Delta_f G°$	Log K_f
0							
100							
200							
298.15	83.680	87.864	87.864	0.	-1401.640	-1325.555	232.232
300	84.308	88.384	87.866	0.155	-1401.618	-1325.083	230.718
400	91.839	113.729	91.270	8.984	-1400.178	-1299.777	169.733
500	96.860	134.789	97.927	18.431	-1398.404	-1274.876	133.185
600	100.416	152.778	105.606	28.303	-1396.433	-1250.353	108.853
700	102.926	168.451	113.489	38.474	-1394.357	-1226.170	91.498
800	105.018	182.339	121.244	48.876	-1392.218	-1202.289	78.501
900	106.483	194.795	128.736	59.453	-1390.070	-1178.677	68.409
1000	107.738	206.083	135.915	70.168	-1387.946	-1155.303	60.347
1100	108.491	216.390	142.769	80.983	-1385.882	-1132.139	53.761
1200	108.868	225.847	149.304	91.851	-1387.635	-1108.937	48.271
1300	109.202	234.574	155.532	102.755	-1385.253	-1085.809	43.628
1400	109.495	242.678	161.471	113.690	-1382.915	-1062.863	39.656
1500	109.746	250.241	167.139	124.653	-1380.632	-1040.082	36.219

PREVIOUS: June 1964 CURRENT: June 1969

Zirconium Fluoride (ZrF_3) F_3Zr_1(cr)

ZIRCONIUM FLUORIDE (ZrF_3) IDEAL GAS M_r = 148.215209 Zirconium Fluoride (ZrF_3) $F_3Zr_1(g)$

S°(298.15 K) = [305.742 ± 8.4] J K^{-1} mol^{-1}

$\Delta_fH°$(0 K) = -1101.784 ± 20.9 kJ mol^{-1}
$\Delta_fH°$(298.15 K) = -1105.413 ± 20.9 kJ mol^{-1}

Electronic Levels and Quantum Weights

ε_i, cm^{-1}	g_i
0	[2]
[1800]	[2]
[4500]	[2]
[12000]	[2]
[30000]	[2]

Vibrational Frequencies and Degeneracies

ν, cm^{-1}	ν, cm^{-1}
[611](1)	[655](2)
[203](1)	[240](2)

Ground State Quantum Weight: [2] σ = [3]
Point Group: C_{3v}
Bond Distance: Zr-F = [1.93] Å
Bond Angle: F-Zr-F = [105]°
Product of the Moments of Inertia: $I_AI_BI_C$ = [9.87893 x 10^{-114}] g^3 cm^6

Enthalpy of Formation

Murad and Hildenbrand (1) studied the gaseous equilibria involving ZrF_4, ZrF_3, ZrF_2, Ca and CaF. These species were generated at high temperatures in a graphite Knudsen cell containing powdered calcium fluoride and elemental zirconium. Ion intensities were measured 3 eV above threshold over the temperature range 1665-1747 K and the equilibrium constants were calculated for the following isomolecular reaction: Ca(g) + ZrF_4(g) = CaF(g) + ZrF_3(g). Based on the reported equilibrum constants the enthalpy change of this reaction is evaluated by the 2nd and 3rd law methods to be -8.0±25 and 28.0±2 kcal mol^{-1}, respectively. The 3rd law entropy drift is 20±15 cal K^{-1} mol^{-1}. Using the 3rd law value and $\Delta_fH°$(298.15 K) = 42.85±0.3, -400.0±0.5, and -65.0±2 kcal mol^{-1} for Ca(g), ZrF_4(g), and CaF(g) we obtain $\Delta_fH°$(ZrF_3, g, 298.15 K) = -264.2±5 kcal mol^{-1} which is adopted. Employing slightly different Gibbs energy functions and $\Delta_fH°$(298.15 K) values for the products and reactants, Murad and Hildenbrand (1) evaluated the 3rd law enthalpy change of the above reaction as 29.4±5 and $\Delta_fH°$(ZrF_3, g, 298.15 K) = -262±5 kcal mol^{-1} which are in reasonable agreement with the values adopted here.

Heat Capacity and Entropy

The molecular structure is assumed to be the same as that of TiF_3(g). The Zr-F bond distance is estimated by comparison with those for ZrF_4(g) and ZrF_2(g). The F-Zr-F bond angle is taken to be slightly bigger than the F-Ti-F bond angle in TiF_3(g) molecule. The vibrational frequencies are calculated by the valence force method (2) using force constants transferred from those for the ZrF_2(g) molecule. The three principal moments of inertia are: $I_A = I_B$ = 18.2815x 10^{-39}, and I_C = 29.5856x 10^{-39} g cm^2. The electronic levels are estimated to be three times higher than those of TiF_3(g), based on the ratio of the first splitting interval between Ti(IV) and Zr(IV), given by Moore (3). The quantum weights are assumed to be the same as those of TiF_3(g).

References

1. E. Murad and D. L. Hildenbrand, J. Chem. Phys. 45, 4751 (1966).
2. G. Herzberg, "Infrared and Raman Spectra," D. Van Nostrand Co., Inc., New York, 1945.
3. C. E. Moore, U. S. Nat. Bur. Stand. Circ. 467, Vol. III, 1958.

Enthalpy Reference Temperature = T_r = 298.15 K Standard State Pressure = p° = 0.1 MPa

	J $K^{-1}mol^{-1}$			kJ mol^{-1}			
T/K	$C_p°$	S°	-[G°-H°(T_r)]/T	H°-H°(T_r)	$\Delta_fH°$	$\Delta_fG°$	Log K_f
0	0.	0.	INFINITE	-15.106	-1101.784	-1101.784	INFINITE
100	44.437	244.931	359.453	-11.452	-1103.377	-1100.909	575.056
200	58.700	280.460	311.683	-6.245	-1104.756	-1097.842	286.727
250	63.889	294.138	306.838	-3.175	-1105.156	-1096.064	229.010
298.15	67.832	305.742	305.742	0.	-1105.413	-1094.288	191.715
300	67.965	306.162	305.743	0.126	-1105.421	-1094.219	190.521
350	71.158	316.888	306.583	3.607	-1105.594	-1092.338	163.022
400	73.704	326.562	308.486	7.231	-1105.704	-1090.436	142.396
450	75.781	335.367	310.991	10.969	-1105.770	-1088.523	126.352
500	77.506	343.444	313.838	14.803	-1105.805	-1086.604	113.517
600	80.188	357.825	320.000	22.695	-1105.814	-1082.762	94.263
700	82.126	370.340	326.317	30.816	-1105.787	-1078.922	80.510
800	83.534	381.403	332.525	39.103	-1105.765	-1075.086	70.196
900	84.561	391.304	338.515	47.510	-1105.786	-1071.251	62.174
1000	85.314	400.255	344.249	56.006	-1105.881	-1067.409	55.756
1100	85.869	408.413	349.717	64.566	-1106.072	-1063.554	50.504
1200	86.277	415.903	354.924	73.175	-1110.084	-1059.454	46.117
1300	86.576	422.822	359.885	81.818	-1109.963	-1055.240	42.400
1400	86.791	429.246	364.612	90.487	-1109.891	-1051.034	39.215
1500	86.941	435.239	369.123	99.174	-1109.883	-1046.831	36.454
1600	87.040	440.854	373.433	107.873	-1109.954	-1042.625	34.038
1700	87.100	446.132	377.556	116.581	-1110.115	-1038.413	31.907
1800	87.129	451.112	381.505	125.292	-1110.379	-1034.188	30.011
1900	87.134	455.823	385.294	134.006	-1110.754	-1029.946	28.315
2000	87.122	460.292	388.933	142.719	-1111.250	-1025.680	26.788
2100	87.097	464.542	392.433	151.430	-1111.873	-1021.387	25.406
2200	87.062	468.593	395.803	160.138	-1133.952	-1016.315	24.130
2300	87.021	472.462	399.053	168.842	-1135.264	-1010.938	22.959
2400	86.976	476.165	402.189	177.542	-1136.573	-1005.505	21.834
2500	86.928	479.715	405.220	186.237	-1137.874	-1000.017	20.894
2600	86.880	483.123	408.151	194.927	-1139.161	-994.477	19.979
2700	86.832	486.401	410.989	203.613	-1140.429	-988.888	19.131
2800	86.784	489.558	413.739	212.294	-1141.674	-983.253	18.343
2900	86.738	492.602	416.406	220.970	-1142.890	-977.573	17.608
3000	86.693	495.542	418.995	229.641	-1144.074	-971.853	16.921
3100	86.651	498.384	421.511	238.309	-1145.223	-966.093	16.279
3200	86.610	501.135	423.956	246.972	-1146.333	-960.297	15.675
3300	86.570	503.799	426.335	255.631	-1147.402	-954.467	15.108
3400	86.533	506.383	428.652	264.286	-1148.427	-948.605	14.574
3500	86.497	508.891	430.909	272.937	-1149.407	-942.713	14.069
3600	86.463	511.327	433.109	281.585	-1150.341	-936.794	13.593
3700	86.431	513.696	435.255	290.230	-1151.227	-930.850	13.141
3800	86.400	516.000	437.350	298.871	-1152.066	-924.883	12.713
3900	86.370	518.244	439.395	307.510	-1152.856	-918.893	12.307
4000	86.341	520.430	441.394	316.145	-1153.598	-912.885	11.921
4100	86.313	522.562	443.348	324.778	-1154.291	-906.859	11.554
4200	86.287	524.642	445.259	333.408	-1154.937	-900.815	11.203
4300	86.261	526.672	447.129	342.036	-1155.536	-894.758	10.869
4400	86.236	528.655	448.959	350.660	-1156.088	-888.687	10.550
4500	86.212	530.592	450.752	359.283	-1156.595	-882.603	10.245
4600	86.189	532.487	452.508	367.903	-1157.056	-876.510	9.953
4700	86.166	534.340	454.229	376.521	-1157.474	-870.406	9.573
4800	86.144	536.154	455.917	385.136	-1718.785	-852.677	9.279
4900	86.123	537.930	457.573	393.749	-1718.770	-834.633	8.897
5000	86.102	539.670	459.198	402.361	-1718.730	-816.590	8.531
5100	86.082	541.375	460.792	410.970	-1718.667	-798.548	8.179
5200	86.062	543.046	462.358	419.577	-1718.579	-780.507	7.840
5300	86.043	544.685	463.896	428.182	-1718.467	-762.468	7.515
5400	86.024	546.293	465.407	436.786	-1718.330	-744.432	7.201
5500	86.006	547.872	466.892	445.387	-1718.168	-726.399	6.899
5600	85.989	549.421	468.352	453.987	-1717.980	-708.368	6.607
5700	85.971	550.943	469.788	462.585	-1717.767	-690.341	6.326
5800	85.955	552.438	471.200	471.181	-1717.528	-672.318	6.055
5900	85.938	553.907	472.589	479.776	-1717.264	-654.299	5.793
6000	85.923	555.351	473.957	488.369	-1716.974	-636.286	5.539

PREVIOUS: December 1969 (1 atm) CURRENT: December 1969 (1 bar)

Zirconium Fluoride (ZrF_3) $F_3Zr_1(g)$

HYDROGEN FLUORIDE (H_4F_4) IDEAL GAS M_r = 80.025372

$\Delta_fH°(0\ K) = -1171.786 \pm -20.9\ kJ\ mol^{-1}$

$S°(298.15\ K) = [348.892 \pm 16.7]\ J\ K^{-1}\ mol^{-1}$ $\Delta_fH°(298.15\ K) = -1183.821 \pm 20.9\ kJ\ mol^{-1}$

Vibrational Frequencies and Degeneracies

ν, cm^{-1}
[202](4)
[53](2)
[962](4)
[552](4)
[3060](4)

Ground State Quantum Weight: 1 σ = [4]

Point Group: C_4

Bond Distance: F-F = [2.5] Å

Product of the Moments of Inertia: $I_AI_BI_C = [1.658443 \times 10^{-113}]\ g^3\ cm^6$

Enthalpy of Formation

The enthalpy of 4 HF(g) → H_4F_4(g) was taken as the mean of 3rd law values which were determined from the equilibrium data of two investigations. Briegleb and Strohmeier (1) measured the vapor density of associated HF between 20 and 60°C and between 50 and 650 torr. Franck and Meyer (2) measured $C_p°$ between -20 and 100°C and between 100 and 700 torr. Each investigation evaluated K_p at n = 2, 3,4, ... for the reactions n(HF) → $(HF)_n$ and reported 2nd law values of ΔH° and ΔS°. At n = 4 their 2nd law values differed by 4.2 kcal mol^{-1}, which was taken as an estimate of error, while the calculated 3rd law values differed by 1.5 kcal mol^{-1}. Using $\Delta_rH°(298.15\ K) = -22.38\ kcal\ mol^{-1}$ and auxiliary JANAF data (3) gives the enthalpy of formation of H_4F_4(g).

Heat Capacity and Entropy

The molecular structure of H_4F_4 was assumed as planar with the F atoms forming the vertices of a square and with the H atoms also lying in the circumscribed circle. That rings are more stable than open chains was confirmed by Del Bene and Pople's (4) theoretical molecular-orbital studies on HF polymers. The length of side (F-F axis) was taken from Atoji and Lipscomb's (5) X-ray studies of solid HF (F-F = 2.49 Å) and agrees with the 2.52 Å which Janzen and Bartell (6) determined for the gaseous polymers by electron diffraction. The low F-bending frequency (53 cm^{-1}) was taken from Boutin et al. (7). The other vibrational frequencies were taken from Kittelberger and Hornig's (8) work on crystalline HF. Huong and Couzi (9) and Smith (10) have made spectral studies of the gas phase in the range from 350 to 4000 cm^{-1}.

The National Bureau of Standards prepared this table (11) by critical analysis of data existing in 1972. Using molecular constants and $\Delta_fH°$ selected by NBS (11), we recalculate the table in terms of 1973 fundamental constants (12), 1975 atomic weights (13), and current JANAF reference states for the elements.

References

1. G. Briegleb and W. Strohmeier, Z. Elektrochem. 57, 668 (1953).
2. E. U. Franck and F. Meyer, Z. Elektrochem. 63, 571 (1959).
3. JANAF Thermochemical Tables: HF(g), 6-30-77.
4. J. E. Del Bene and J. A. Pople, J. Chem. Phys. 55, 2296 (1971).
5. M. Atoji and W. N. Lipscomb, Acta Cryst. 7, 173 (1954).
6. J. Janzen and L. S. Bartell, J. Chem. Phys. 50, 3611 (1969).
7. H. Boutin, G. J. Stafford, and V. Brajovic, J. :Chem. Phys. 39, 3135 (1963).
8. J. S. Kittelberger and D. F. Hornig, J. Chem. Phys. 46, 3099 (1967).
9. P. V. Huong and M. Couzi, J. Chim. Phys. 66, 1309 (1969).
10. D. F. Smith, J. Chem. Phys. 28, 1040 (1958); 48, 1429 (1968).
11. S. Abramowitz et al., U. S. Nat. Bur. Stand, Report 10904, 239, 1972.
12. CODATA Task Group on Fundamental Constants, CODATA Bulletin 11, December, 1973.
13. IUPAC Commission on Atomic Weights, Pure Appl. Chem. 47, 75 (1976).

Hydrogen Fluoride (H_4F_4) F_4H_4(g)

Enthalpy Reference Temperature = T_r = 298.15 K Standard State Pressure = p° = 0.1 MPa

T/K	$C_p°$ (J $K^{-1}mol^{-1}$)	S° (J $K^{-1}mol^{-1}$)	-[G°-H°(T_r)]/T (J $K^{-1}mol^{-1}$)	H°-H°(T_r) (kJ mol^{-1})	$\Delta_fH°$ (kJ mol^{-1})	$\Delta_fG°$ (kJ mol^{-1})	Log K_f
0	0.	0.	INFINITE	-22.549	-1171.786	-1171.786	INFINITE
100	67.044	256.058	431.517	-17.546	-1178.595	-1149.981	600.689
200	89.508	309.812	358.124	-9.662	-1181.957	-1119.894	292.486
250	98.807	330.806	350.605	-4.950	-1183.042	-1104.246	230.720
298.15	106.585	348.892	348.892	0.	-1183.821	-1088.996	190.788
300	106.860	349.552	348.894	0.197	-1183.846	-1088.407	189.509
350	113.638	366.549	350.221	5.715	-1184.404	-1072.453	160.055
400	119.262	382.103	353.248	11.542	-1184.752	-1056.434	137.956
450	123.931	396.428	357.260	17.625	-1184.923	-1040.383	120.764
500	127.857	409.694	361.848	23.923	-1184.948	-1024.320	107.010
600	134.181	433.589	371.858	37.038	-1184.637	-992.216	86.380
700	139.264	454.666	382.212	50.718	-1183.937	-960.196	71.651
800	143.630	473.554	392.469	64.868	-1182.911	-928.300	60.612
900	147.504	490.699	402.446	79.428	-1181.609	-896.549	52.034
1000	150.977	506.423	412.068	94.355	-1180.069	-864.956	45.181
1100	154.091	520.961	421.315	109.611	-1178.330	-833.528	39.581
1200	156.873	534.491	430.189	125.162	-1176.428	-802.265	34.922
1300	159.350	547.147	438.704	140.976	-1174.394	-771.166	30.986
1400	161.549	559.038	446.879	157.023	-1172.257	-740.229	27.618
1500	163.500	570.252	454.734	173.277	-1170.039	-709.446	24.705
1600	165.229	580.860	462.288	189.715	-1167.761	-678.814	22.161
1700	166.762	590.924	469.561	206.316	-1165.437	-648.326	19.921
1800	168.124	600.495	476.572	223.062	-1163.079	-617.975	17.933
1900	169.336	609.618	483.336	239.936	-1160.695	-587.757	16.159
2000	170.416	618.332	489.870	256.925	-1158.288	-557.664	14.565
2100	171.382	626.671	496.187	274.016	-1155.862	-527.693	13.126
2200	172.246	634.664	502.301	291.198	-1153.417	-497.837	11.820
2300	173.023	642.338	508.224	308.462	-1150.951	-468.093	10.631
2400	173.722	649.716	513.966	325.800	-1148.462	-438.457	9.543
2500	174.353	656.821	519.539	343.204	-1145.947	-408.925	8.544
2600	174.924	663.671	524.952	360.669	-1143.403	-379.494	7.624
2700	175.442	670.282	530.213	378.187	-1140.826	-350.162	6.774
2800	175.913	676.671	535.330	395.755	-1138.211	-320.926	5.987
2900	176.343	682.852	540.311	413.369	-1135.557	-291.784	5.256
3000	176.736	688.837	545.163	431.023	-1132.860	-262.734	4.575
3100	177.095	694.638	549.891	448.715	-1130.117	-233.775	3.939
3200	177.425	700.266	554.503	466.441	-1127.327	-204.905	3.345
3300	177.728	705.730	559.003	484.199	-1124.487	-176.123	2.788
3400	178.008	711.040	563.397	501.986	-1121.598	-147.428	2.265
3500	178.266	716.204	567.690	519.799	-1118.657	-118.819	1.773
3600	178.504	721.229	571.885	537.638	-1115.665	-90.294	1.310
3700	178.726	726.123	575.988	555.500	-1112.622	-61.853	0.873
3800	178.931	730.892	580.002	573.383	-1109.529	-33.496	0.460
3900	179.122	735.543	583.931	591.285	-1106.386	-5.219	0.070
4000	179.299	740.080	587.778	609.207	-1103.193	22.975	-0.300
4100	179.465	744.509	591.547	627.145	-1099.953	51.089	-0.651
4200	179.620	748.836	595.241	645.099	-1096.667	79.124	-0.984
4300	179.765	753.064	598.862	663.069	-1093.336	107.079	-1.301
4400	179.900	757.198	602.414	681.052	-1089.961	134.957	-1.602
4500	180.027	761.243	605.898	699.048	-1086.543	162.758	-1.889
4600	180.147	765.201	609.319	717.057	-1083.085	190.481	-2.163
4700	180.259	769.076	612.677	735.077	-1079.588	218.130	-2.424
4800	180.364	772.872	615.975	753.109	-1076.052	245.705	-2.674
4900	180.464	776.592	619.215	771.150	-1072.480	273.204	-2.912
5000	180.558	780.239	622.399	789.201	-1068.871	300.631	-3.141
5100	180.647	783.816	625.529	807.262	-1065.228	327.984	-3.359
5200	180.730	787.324	628.607	825.330	-1061.551	355.267	-3.569
5300	180.810	790.768	631.634	843.407	-1057.841	382.478	-3.770
5400	180.885	794.148	634.612	861.492	-1054.098	409.619	-3.962
5500	180.956	797.468	637.543	879.584	-1050.323	436.689	-4.147
5600	181.023	800.729	640.428	897.683	-1046.515	463.692	-4.325
5700	181.088	803.934	643.269	915.789	-1042.676	490.625	-4.496
5800	181.149	807.083	646.066	933.901	-1038.806	517.492	-4.661
5900	181.207	810.181	648.822	952.018	-1034.903	544.291	-4.819
6000	181.262	813.227	651.536	970.142	-1030.968	571.023	-4.971

PREVIOUS: June 1977 (1 atm) CURRENT: June 1977 (1 bar)

Hydrogen Fluoride (H_4F_4) F_4H_4(g)

MAGNESIUM FLUORIDE $(MgF_2)_2$ IDEAL GAS M_r = 124.603612

$\Delta_fH°$(0 K) = -1711.866 ± 20.9 (or ± 37.7) kJ mol^{-1}

S°(298.15 K) = [337.014 ± 20.9] J K^{-1} mol^{-1} $\Delta_fH°$(298.15 K) = -1718.369 ± 20.9 (or ± 37.7) kJ mol^{-1}

Vibrational Frequencies and Degeneracies

ν, cm^{-1}	ν, cm^{-1}	ν, cm^{-1}
585 (1)	[235](1)	487 (1)
353 (1)	[220](1)	[220](1)
[400](1)	[240](1)	747 (1)
[508](1)	[200](1)	450 (1)

Ground State Quantum Weight: [1] σ = 4

Point Group: $[D_{2h}]$

Bond Distances: Mg-F = [1.77] Å Mg-F' = [1.96] Å

Bond Angles: Mg-F'-Mg = [90]° F'-Mg-F' = [90]° F-Mg-F' = [135]°

(F' = Ring Fluorine)

Product of Moments of Inertia: $I_AI_BI_C$ = [8.5873 x 10^{-113}] g^3 cm^6

Enthalpy of Formation

Berkowitz and Marquart (1) have found approximately 1% dimer in the equilibrium vapor over MgF_2(cr) at about 1400 K by mass spectrometry. Independent measurements which are discussed elsewhere (See MgF_2(g) table) support their results. Berkowitz and Marquart (1) reported a 2nd law $\Delta_rH°$(1440 K) of 30 kcal mol^{-1} for the reaction MgF_2(cr) + MgF_2(g) = Mg_2F_2(g). This value corrected to 298.15 K gives $\Delta_rH°$ = 31.7 kcal mol^{-1} which leads to $\Delta_fH°(Mg_2F_4$, g, 298.15 K) = -410.7 kcal mol^{-1}. The corresponding heat of dimerization at 298.15 K is -63.3 kcal mol^{-1}. Berkowitz and Marquart (1) also reported approximate absolute pressures for MgF_2(g) and Mg_2F_4(g) at 1429 K. A 3rd law analysis of these reported pressures gives $\Delta_rH°$(298.15 K) of -69.5 kcal mol^{-1} for the process 2 MgF_2(g) = Mg_2F_4(g), or $\Delta_fH°(Mg_2F_4$, g, 298.15 K) = -416.9 kcal mol^{-1}. We prefer the 2nd law result $\Delta_fH°$ = 410.7 kcal mol^{-1} since the else negative (-63.3 kcal mol^{-1}) heat of dimerization is more consistent with the dimerization energies for the alkali metal fluorides (2). The uncertainty in $\Delta_fH°$ is estimated as ±5.0 kcal mol^{-1}, however, we also include an alternate uncertainty (±9 kcal mol^{-1}) to cover the possibility that M_gF_2(g) is linear (2).

Heat Capacity and Entropy

The dimer is assumed to have a bridge-bond structure of D_{2h} symmetry similar to that suggested by Thompson and Carlson (3) for the dimers of several transition metal dichlorides. The two outer Mg-F bond lengths are assumed to be the same as that for MgF_2 (2). The four ring Mg-F' bond lengths are assumed to be 10% longer. The four atoms which lie in the ring form a square. The F'-Mg'F bond angle is estimated as 135°. The individual moments of inertia are: I_A = 9.0458 x 10^{-38}, I_B = 7.8340 x 10^{-38} and I_C = 1.2118 x 10^{-38} g cm^2.

This planar bridge-bond structure will have six Raman ($3A_g + 2B_{1g} + 2B_{2g}$) and six infrared ($2B_{1u} + 2B_{2u} + 2B_{3u}$) active fundamentals. The Raman spectra of MgF_2 isolated in an argon matrix has been recorded by Lesiecki and Nibler (4). They assigned two bands observed at 585 and 353 cm^{-1} to ν_1 and ν_2, respectively, of Mg_2F_4. Lesiecki and Nibler (4), using a standard high-low frequency separation method and a diagonal force field, calculated ν_4 as 508 cm^{-1}. The other three Raman active frequencies (ν_3, ν_5, ν_6) are estimated by analogy with those for MgF_2 and Na_2F_2 (2). Four of the six infrared active fundamentals have been osbserved and assigned by Lesiecki and Nibler (4). These frequencies were recorded in an argon matrix. Results of three independent infrared studies (5-7) conducted with various matrices support these assignments. The other two infrared active fundamentals (ν_8 and ν_{10}) are estimated from the bending frequency (ν_2 = 215 cm^{-1}) for MgF_2 (2). Following the observations made by Thompson and Carlson (3), all the estimated frequencies are taken to lie above the bending frequency of the monomer.

References

1. J. Berkowitz and J. R. Marquart, J. Chem. Phys. 37, 1853 (1962).
2. JANAF Thermochemical Tables: MgF_2(g), 6-30-75; Li_2F_2(g) and Na_2F_2(g), 12-31-68; K_2F_2(g), 6-30-69.
3. K. R. Thompson and K. D. Carlson, J. Chem. Phys. 49, 4379 (1968).
4. M. L. Lesiecki and J. W. Nibler, to be published.
5. R. H. Hauge, J. L. Margrave, and A. S. Kana'an, J. Chem. Soc. Faraday Trans. II 71, 1082 (1975).
6. A. Snelson, J. Phys. Chem. 70, 3208 (1966).
7. D. E. Mann, G. V. Calder, K. S. Seshadri, D. White, and M. J. Linevsky, J. Chem. Phys. 46, 1138 (1967).

Magnesium Fluoride $((MgF_2)_2)$ F_4Mg_2(g)

Enthalpy Reference Temperature = T_r = 298.15 K Standard State Pressure = p° = 0.1 MPa

T/K	$C_p°$ (J $K^{-1}mol^{-1}$)	S° (J $K^{-1}mol^{-1}$)	$-[G°-H°(T_r)]/T$ (J $K^{-1}mol^{-1}$)	$H°-H°(T_r)$ (kJ mol^{-1})	$\Delta_fH°$ (kJ mol^{-1})	$\Delta_fG°$ (kJ mol^{-1})	Log K_f
0	0.	0.	INFINITE	-21.143	-1711.866	-1711.866	INFINITE
100	55.670	247.462	419.349	-17.189	-1715.010	-1703.781	889.964
200	89.656	297.494	346.359	-9.773	-1717.466	-1691.447	441.761
250	100.233	318.704	338.746	-5.011	-1718.051	-1684.868	352.034
298.15	107.501	337.014	337.014	0.	-1718.369	-1678.445	294.057
300	107.735	337.680	337.016	0.199	-1718.378	-1678.197	292.200
350	113.108	354.713	338.349	5.727	-1718.551	-1671.485	249.455
400	117.026	370.086	341.372	11.486	-1718.640	-1664.754	217.395
450	119.941	384.046	345.350	17.413	-1718.687	-1658.015	192.457
500	122.154	396.803	349.866	23.468	-1718.722	-1651.272	172.507
600	125.214	419.367	359.619	35.849	-1718.823	-1637.774	142.581
700	127.164	438.825	369.576	48.474	-1719.045	-1624.251	121.205
800	128.475	455.897	379.321	61.260	-1719.454	-1610.683	105.167
900	129.395	471.085	388.689	74.157	-1720.103	-1597.051	92.690
1000	130.065	484.755	397.624	87.131	-1738.221	-1581.906	82.630
1100	130.567	497.176	406.118	100.164	-1739.490	-1566.213	74.373
1200	130.952	508.554	414.187	113.241	-1740.767	-1550.404	67.487
1300	131.254	519.048	421.855	126.352	-1742.056	-1534.489	61.657
1400	131.495	528.784	429.149	139.489	-1998.180	-1512.139	56.419
1500	131.690	537.863	436.097	152.649	-1996.798	-1477.470	51.450
1600	131.850	546.368	442.726	165.826	-1995.436	-1442.893	47.106
1700	131.984	554.365	449.060	179.018	-1994.092	-1408.400	43.275
1800	132.096	561.912	455.122	192.222	-1992.765	-1373.986	39.872
1900	132.191	569.057	460.932	205.437	-1991.451	-1339.645	36.829
2000	132.272	575.840	466.510	218.660	-1990.148	-1305.373	34.093
2100	132.342	582.295	471.871	231.891	-1988.849	-1271.167	31.619
2200	132.403	588.453	477.031	245.128	-1987.549	-1237.022	29.371
2300	132.456	594.340	482.004	258.371	-1986.243	-1202.936	27.320
2400	132.502	599.978	486.803	271.619	-1984.923	-1168.907	25.441
2500	132.544	605.388	491.439	284.871	-1983.583	-1134.934	23.713
2600	132.580	610.587	495.923	298.128	-1982.218	-1101.015	22.120
2700	132.613	615.591	500.263	311.387	-1980.824	-1067.149	20.645
2800	132.642	620.415	504.468	324.650	-1979.394	-1033.335	19.277
2900	132.668	625.070	508.547	337.916	-1977.926	-999.573	18.004
3000	132.692	629.568	512.507	351.184	-1976.417	-965.863	16.817
3100	132.713	633.919	516.353	364.454	-1974.866	-932.203	15.708
3200	132.733	638.133	520.093	377.726	-1973.271	-898.594	14.668
3300	132.751	642.218	523.733	391.000	-1971.633	-865.035	13.692
3400	132.767	646.181	527.276	404.276	-1969.952	-831.527	12.775
3500	132.782	650.030	530.728	417.554	-1968.231	-798.069	11.911
3600	132.795	653.770	534.095	430.833	-1966.473	-764.661	11.095
3700	132.808	657.409	537.378	444.113	-1964.679	-731.302	10.324
3800	132.819	660.951	540.584	457.394	-1962.855	-697.992	9.595
3900	132.830	664.401	543.715	470.677	-1961.004	-664.730	8.903
4000	132.840	667.764	546.774	483.960	-1959.132	-631.516	8.247
4100	132.849	671.044	549.765	497.245	-1957.242	-598.349	7.623
4200	132.857	674.246	552.691	510.530	-1955.342	-565.228	7.030
4300	132.865	677.372	555.554	523.816	-1953.436	-532.153	6.464
4400	132.873	680.427	558.358	537.103	-1951.529	-499.122	5.925
4500	132.880	683.413	561.104	550.391	-1949.628	-466.133	5.411
4600	132.886	686.333	563.795	563.679	-1947.738	-433.188	4.919
4700	132.892	689.191	566.432	576.968	-1945.865	-400.283	4.449
4800	132.898	691.989	569.019	590.257	-1944.015	-367.418	3.998
4900	132.903	694.730	571.557	603.547	-1942.194	-334.591	3.567
5000	132.908	697.415	574.047	616.838	-1940.406	-301.801	3.153
5100	132.913	700.047	576.492	630.129	-1938.657	-269.047	2.756
5200	132.918	702.628	578.893	643.421	-1936.952	-236.325	2.374
5300	132.922	705.160	581.251	656.713	-1935.296	-203.637	2.007
5400	132.926	707.644	583.569	670.005	-1933.694	-170.980	1.654
5500	132.930	710.083	585.847	683.298	-1932.150	-138.352	1.314
5600	132.933	712.478	588.087	696.591	-1930.669	-105.750	0.986
5700	132.937	714.831	590.290	709.884	-1929.256	-73.176	0.671
5800	132.940	717.143	592.457	723.178	-1927.914	-40.625	0.366
5900	132.943	719.416	594.590	736.472	-1926.647	-8.095	0.072
6000	132.946	721.650	596.689	749.767	-1925.459	24.412	-0.213

PREVIOUS: December 1975 (1 atm) CURRENT: December 1975 (1 bar)

Magnesium Fluoride $((MgF_2)_2)$ F_4Mg_2(g)

MOLYBDENUM FLUORIDE (MoF_4) IDEAL GAS M_r = 171.933612

$\Delta_fH°(0\ K)$ = -944.351 ± 16.7 kJ mol^{-1}

$S°(298.15\ K)$ = [328.940 ± 8.4] J K^{-1} mol^{-1} $\Delta_fH°(298.15\ K)$ = -947.676 ± 16.7 kJ mol^{-1}

Electronic Levels and Quantum Weights

ε_i, cm^{-1}	g_i
0	[3]
[15000]	[11]
[20000]	[9]

Vibrational Frequencies and Degeneracies

ν, cm^{-1}	ν, cm^{-1}
[700](1)	[750](3)
[160](2)	[180](2)

Point Group: [T_d] σ = [12]

Bond Distance: Mo-F = [1.82] Å

Bond Angle: F-Mo-F = [109.47°]

Product of the Moments of Inertia: $I_AI_BI_C$ = [2.163977 x 10^{-113}] g^3 cm^6

Enthalpy of Formation

We adopt the weighted average value derived by Brewer (1) from several divergent equilibria measurements. Weaver et al. (2) reported a partial pressure of 4.6 x 10^{-6} atm MoF_4(g) in equilibrium with Mo and MoF_3 solids at 973 K. Hildenbrand (3) has measured several gaseous equilibria for reactions involving MoF_4(g) [1143-2146 K]; and Alikhanyan (4) has measured partial pressures for the reaction $MoF_5(g) + MoF_3(g) = 2\ MoF_4(g)$ [911 and 964 K]. Our 3rd law analyses of the equilibrium reactions, using auxiliary data from Brewer (1) for the MoF_x compounds other than MoF_4(g) and other auxiliary data from the JANAF Thermochemical Tables (5), leads to $\Delta_fH°(MoF_4$, g, 298.15 K) values ranging from -216.0 to -226.6 kcal mol^{-1}. The value of $\Delta_fH°$(0 K) combined with JANAF data (5) for Mo(g) and F(g) gives $\Delta_{at}H°$(0 K) = 456.357 kcal mol^{-1} and an average Mo-F bond energy of $\Delta_{at}H°$(0 K)/4 = 114.1±2 kcal mol^{-1}.

Heat Capacity and Entropy

We use the vibrational frequencies and degeneracies estimated by Hildenbrand (3) and adopted by Brewer (1). These are in reasonable agreement with the estimates of Tumanov and Galkin (6). The structure is assumed to be tetrahedral and a Mo-F distance of 1.82 Å from MoF_6(g) is used. The electronic contributions are those estimated by Brewer (1) based on crystal field theory. The principal moments of inertia are: $I_A = I_B = I_C$ = 27.8666 x 10^{39} g cm^2.

This table is based on the critical analysis of data existing in 1978 by Brewer (1).

References

1. L. Brewer, Materials and Molecular Research Division, Lawrence Berkeley Laboratory, University of California, Berkeley; personal communication, Sept. 1978, of preliminary draft of review to be submitted for publication in Atomic Energy Review, International Atomic Eneergy Agency, Vienna, Austria.
2. C. F. Weaver, H. A. Friedman, J. D. Redman, R. A. Strehlcw, J. W. Gooch, Jr., and J. S. Gill, Oak Ridge National Laboratory Reports ONRL-4191, pp 142-146 (1967); ORNL-4254, pp 129-136 (1968); ORNL-4396, pp 157-162 (1969); ORNL-4400, pp 33-39 1970; ORNL-4449, pp 113-121 (1970); ORNL-4586, p 17 (1970); ORNL-4801, p 21 (1972).
3. D. L. Hildenbrand, J. Chem. Phys. 65, 614 (1976).
4. A. S. Alikhanyan, V. S. Pervov, I. P. Malkervoa, V. D. Butskii, and V. I. Gorgoraki, Russ. J. Inorg. Chem. 23, 817 (1978).
5. JANAF Thermochemical Tables: S(g), 9-30-77; SF(g), 6-30-76; Mo(cr) and Mo(g), 3-31-78; F(g) 9-30-65.
6. Yu. N. Tumanov and N. P. Galkin, Russ. J. Phys. Chem. 43, 464 (1969).

Molybdenum Fluoride (MoF_4) F_4Mo_1(g)

Enthalpy Reference Temperature = T_r = 298.15 K Standard State Pressure = p° = 0.1 MPa

T/K	$C_p°$ (J K^{-1}mol^{-1})	S° (J K^{-1}mol^{-1})	-[G°-H°(T_r)]/T (J K^{-1}mol^{-1})	H°-H°(T_r) (kJ mol^{-1})	$\Delta_fH°$ (kJ mol^{-1})	$\Delta_fG°$ (kJ mol^{-1})	Log K_f
0	0.	0.	INFINITE	-18.909	-944.351	-944.351	INFINITE
100	58.912	250.806	397.136	-14.633	-946.380	-936.684	489.273
200	74.611	297.057	336.418	-7.872	-947.319	-926.574	241.996
250	80.449	314.348	330.317	-3.992	-947.562	-921.357	192.507
298.15	85.232	328.940	328.940	0.	-947.676	-916.298	160.532
300	85.398	329.468	328.942	0.158	-947.678	-916.103	159.508
350	89.417	342.945	329.997	4.532	-947.696	-910.838	135.935
400	92.611	355.102	332.388	9.086	-947.645	-905.575	118.256
450	95.137	366.162	335.535	13.782	-947.547	-900.322	104.507
500	97.141	376.293	339.111	18.591	-947.420	-895.081	93.508
600	100.035	394.278	346.845	28.460	-947.114	-884.641	77.015
700	101.955	409.853	354.759	38.565	-946.784	-874.255	65.238
800	103.279	423.558	362.520	48.831	-946.459	-863.916	56.408
900	104.225	435.780	369.993	59.209	-946.156	-853.617	49.543
1000	104.921	446.799	377.132	69.668	-945.889	-843.350	44.052
1100	105.447	456.825	383.928	80.187	-945.672	-833.107	39.561
1200	105.854	466.018	390.391	90.753	-945.517	-822.881	35.819
1300	106.174	474.504	396.539	101.355	-945.434	-812.665	32.653
1400	106.432	482.383	402.393	111.986	-945.435	-802.453	29.940
1500	106.643	489.733	407.973	122.640	-945.527	-792.237	27.588
1600	106.819	496.621	413.301	133.313	-945.718	-782.012	25.530
1700	106.969	503.102	418.394	144.003	-946.015	-771.772	23.714
1800	107.102	509.220	423.272	154.707	-946.422	-761.511	22.098
1900	107.224	515.014	427.949	165.423	-946.944	-751.225	20.653
2000	107.340	520.517	432.441	176.151	-947.584	-740.907	19.351
2100	107.456	525.757	436.761	186.891	-948.345	-730.556	18.172
2200	107.575	530.758	440.921	197.642	-949.228	-720.165	17.099
2300	107.703	535.543	444.931	208.406	-950.235	-709.730	16.118
2400	107.842	540.130	448.803	219.183	-951.368	-699.250	15.219
2500	107.995	544.535	452.545	229.975	-952.629	-688.719	14.390
2600	108.166	548.774	456.165	240.783	-954.031	-678.136	13.624
2700	108.356	552.860	459.671	251.609	-955.605	-667.495	12.913
2800	108.566	556.804	463.070	262.455	-957.406	-656.792	12.253
2900	108.798	560.618	466.368	273.323	-995.389	-645.971	11.635
3000	109.053	564.310	469.572	284.215	-995.825	-633.914	11.037
3100	109.330	567.891	472.686	295.134	-996.181	-621.845	10.478
3200	109.629	571.366	475.716	306.082	-996.451	-609.765	9.953
3300	109.951	574.745	478.666	317.061	-996.629	-597.678	9.460
3400	110.293	578.032	481.540	328.073	-996.711	-585.587	8.996
3500	110.654	581.234	484.343	339.120	-996.693	-573.495	8.559
3600	111.034	584.357	487.078	350.204	-996.572	-561.405	8.146
3700	111.429	587.404	489.748	361.327	-996.344	-549.320	7.755
3800	111.839	590.382	492.358	372.491	-996.007	-537.243	7.385
3900	112.261	593.292	494.909	383.696	-995.561	-525.175	7.034
4000	112.693	596.140	497.404	394.943	-995.003	-513.121	6.701
4100	113.133	598.928	499.846	406.235	-994.334	-501.082	6.384
4200	113.578	601.659	502.238	417.570	-993.553	-489.060	6.082
4300	114.026	604.337	504.581	428.950	-992.662	-477.059	5.795
4400	114.474	606.964	506.878	440.375	-991.660	-465.080	5.521
4500	114.921	609.541	509.131	451.845	-990.549	-453.124	5.260
4600	115.365	612.072	511.342	463.359	-989.330	-441.195	5.010
4700	115.803	614.558	513.511	474.918	-988.006	-429.293	4.771
4800	116.233	617.000	515.642	486.520	-986.578	-417.420	4.542
4900	116.654	619.401	517.735	498.164	-985.048	-405.578	4.324
5000	117.064	621.762	519.792	509.850	-1574.805	-388.038	4.054
5100	117.461	624.084	521.814	521.576	-1574.760	-364.303	3.731
5200	117.845	626.369	523.803	533.342	-1574.769	-340.568	3.421
5300	118.213	628.617	525.760	545.145	-1574.830	-316.833	3.123
5400	118.565	630.830	527.685	556.984	-1574.948	-293.096	2.835
5500	118.900	633.009	529.580	568.857	-1575.121	-269.357	2.558
5600	119.217	635.154	531.446	580.763	-1575.352	-245.613	2.291
5700	119.516	637.267	533.284	592.700	-1575.639	-221.867	2.033
5800	119.796	639.348	535.095	604.666	-1575.985	-198.113	1.784
5900	120.057	641.398	536.879	616.659	-1576.387	-174.353	1.544
6000	120.299	643.418	538.638	628.677	-1576.849	-150.587	1.311

PREVIOUS: December 1978 (1 atm) CURRENT: December 1978 (1 bar)

Molybdenum Fluoride (MoF_4) F_4Mo_1(g)

MOLYBDENUM FLUORIDE OXIDE (MoF_4O)   IDEAL GAS M_r = 187.933012

$\Delta_fH°(0\ K)$ = [-1248.030 ± 126] kJ mol^{-1}

$S°(298.15\ K)$ = [330.617 ± 8.4] J K^{-1} mol^{-1} $\Delta_fH°(298.15\ K)$ = [1255.200 ± 126] kJ mol^{-1}

Vibrational Frequencies and Degeneracies

ν, cm^{-1}	ν, cm^{-1}	ν, cm^{-1}
1045 (1)	[600](1)	720 (2)
[300](1)	[300](1)	530 (2)
680 (1)	[200](1)	[150](2)

Ground State Quantum Weight: [1]   σ = [4]

Point Group: [C_{4v}]

Bond Distance: Mo-O = [1.64] Å   Mo-F = [1.82] Å

Bond Angles: O-Mo-F = [90]°   F-Mo-F = [90]°

Product of the Moments of Inertia: $I_AI_BI_C$ = [3.1468 x 10^{-113}] g^3 cm^6

Enthalpy of Formation

Zmbov et al. (1) investigated the stabilities of some molybdenum and tungsten oxyfluorides by means of mass spectrometry. These studies led to enthalpies of formation for $MoO_2F_2(g)$, $WO_2F_2(g)$, and $WOF_4(g)$ which were used by the authors to develop a correlation among the enthalpies of formation of the oxides, oxyfluorides, and fluorides of Mo and W. $\Delta_fH°(298.15\ K)$ varies almost linearly with the number of oxygen atoms (1) for the series WF_6, WOF_4 and WO_2F_2. Assuming a similar variation for MoF_6, $MoOF_4$ and MoO_2F_2, we estimate $\Delta_fH°(MoOF_4,\ g)$ = -320 kcal mol^{-1}.

Enthalpy of formation data (2) for $MoF_6(g)$, MoO(g), $MoO_2(g)$, and $MoO_3(g)$ have been used to derive the following average bond dissociation energies:

$$\Delta_{at}H°(MoF,\ g)/6 = 107 \text{ kcal mol}^{-1}$$

$$\Delta_{at}H°(MoOn,\ g)/n = 135 \text{ kcal mol}^{-1}$$

From these data we derive $\Delta_fH°(MoOF_4,\ g)$ = -271 kcal mol^{-1}.

The enthalpy of formation for $MoOF_4$ is tentatively selected as 300±30 kcal mol^{-1}.

Heat Capacity and Entropy

Blanchard (3) suggested that $MoOF_4(g)$ possessed C_{4v} symmetry by reason of the similarity of its infrared spectra with that of $XeOF_4(g)$. X-ray diffraction studies of crystalline $MoOF_4$ by Edwards et al. (4, 5) support this view. The structure of the molecule is assumed to be square-pyramidal. The bond lengths are from the crystallographic studies of $MoOF_4$ by Edwards et al. (4, 5). Individual moments of inertia are $I_A = I_B$ = 27.437 x 10^{-39}, I_C = 41.8010 x 10^{-39} g cm^2.

Blanchard (3) reported the following vibrational frequencies and assignments:

$$\nu_1 = 1045,\ \nu_7 = 720,\ \nu_3 = 680, \text{ and } \nu_8 = 530 \text{ cm}^{-1}.$$

Similar frequencies have been reported by Edwards et al. (4) from their study of the infrared spectra of gaseous $MoOF_4$. The rest of the frequencies and assignments are estimated by comparison with data for $XeOF_4$ (6).

References

1. K. F. Zmbov, O. M. Uy and J. L. Margrave, J. Phys. Chem. 73, 3008 (1969).
2. JANAF Thermochemical Tables.
3. S. Blanchard, Commissariat a l'Energie Atomique Report CEA-R 3194, 1967.
4. A. J. Edwards, G. R. Jones and B. R. Steventon, Chem. Commun. 462 (1967).
5. A. J. Edwards and B. R. Steventon, J. Chem. Soc. 2503 (1968).
6. G. M. Begun, W. H. Fletcher and D. F. Smith, J. Chem. Phys. 42, 2236 (1965).

Molybdenum Fluoride Oxide (MoF_4O) $F_4Mo_1O_1(g)$

Enthalpy Reference Temperature = T_r = 298.15 K   Standard State Pressure = p° = 0.1 MPa

T/K	$C_p°$ (J $K^{-1}mol^{-1}$)	S° (J $K^{-1}mol^{-1}$)	$-[G°-H°(T_r)]/T$ (J $K^{-1}mol^{-1}$)	$H°-H°(T_r)$ (kJ mol^{-1})	$\Delta_fH°$ (kJ mol^{-1})	$\Delta_fG°$ (kJ mol^{-1})	Log K_f
0	0.	0.	INFINITE	-19.407	-1248.030	-1248.030	INFINITE
100	53.979	250.195	403.835	-15.364	-1251.745	-1233.323	644.222
200	78.838	295.560	338.941	-8.676	-1254.213	-1213.820	317.017
250	88.874	314.265	332.167	-4.476	-1254.864	-1203.639	251.487
298.15	96.740	330.617	330.617	0.	-1255.200	-1193.739	209.138
300	97.009	331.216	330.619	0.179	-1255.208	-1193.358	207.782
350	103.440	346.673	331.824	5.197	-1255.320	-1183.038	176.559
400	108.488	360.828	334.578	10.500	-1255.267	-1172.714	153.141
450	112.460	373.845	338.228	16.028	-1255.097	-1162.404	134.928
500	115.608	385.863	342.398	21.732	-1254.844	-1152.117	120.361
600	120.165	407.373	351.478	33.537	-1254.183	-1131.632	98.517
700	123.201	426.140	360.832	45.715	-1253.408	-1111.267	82.924
800	125.304	442.736	370.053	58.147	-1252.585	-1091.017	71.236
900	126.812	457.587	378.968	70.756	-1251.753	-1070.871	62.152
1000	127.925	471.008	387.512	83.496	-1250.937	-1050.817	54.889
1100	128.769	483.242	395.667	96.333	-1250.157	-1030.843	48.951
1200	129.422	494.476	403.439	109.243	-1249.431	-1010.937	44.005
1300	129.937	504.856	410.846	122.212	-1248.773	-991.090	39.822
1400	130.350	514.501	417.910	135.227	-1248.196	-971.291	36.239
1500	130.686	523.506	424.653	148.280	-1247.710	-951.529	33.135
1600	130.964	531.949	431.098	161.363	-1247.326	-931.797	30.420
1700	131.195	539.896	437.266	174.471	-1247.049	-912.085	28.025
1800	131.389	547.401	443.178	187.600	-1246.889	-892.387	25.896
1900	131.555	554.509	448.852	200.748	-1246.849	-872.694	23.992
2000	131.696	561.261	454.305	213.911	-1246.936	-853.000	22.278
2100	131.819	567.689	459.553	227.087	-1247.154	-833.299	20.727
2200	131.925	573.824	464.609	240.274	-1247.505	-813.583	19.317
2300	132.018	579.690	469.486	253.471	-1247.995	-793.849	18.029
2400	132.100	585.311	474.195	266.677	-1248.625	-774.091	16.848
2500	132.172	590.705	478.749	279.891	-1249.401	-754.303	15.760
2600	132.236	595.890	483.155	293.111	-1250.339	-734.481	14.756
2700	132.293	600.882	487.423	306.338	-1251.471	-714.620	13.825
2800	132.345	605.694	491.562	319.570	-1252.854	-694.712	12.960
2900	132.391	610.339	495.578	332.806	-1290.448	-674.700	12.153
3000	132.433	614.828	499.479	346.048	-1290.524	-653.466	11.378
3100	132.470	619.171	503.270	359.293	-1290.551	-632.230	10.653
3200	132.505	623.377	506.958	372.542	-1290.527	-610.994	9.973
3300	132.536	627.455	510.548	385.794	-1290.447	-589.760	9.335
3400	132.564	631.412	514.045	399.049	-1290.310	-568.529	8.734
3500	132.590	635.255	517.453	412.306	-1290.114	-547.303	8.168
3600	132.614	638.991	520.778	425.567	-1289.856	-526.083	7.633
3700	132.637	642.625	524.022	438.829	-1289.535	-504.871	7.128
3800	132.657	646.162	527.190	452.094	-1289.151	-483.669	6.648
3900	132.676	649.608	530.285	465.361	-1288.704	-462.477	6.194
4000	132.693	652.967	533.310	478.629	-1288.194	-441.299	5.763
4100	132.709	656.244	536.269	491.899	-1287.620	-420.134	5.353
4200	132.724	659.442	539.163	505.171	-1286.984	-398.983	4.962
4300	132.738	662.565	541.997	518.444	-1286.286	-377.848	4.590
4400	132.752	665.617	544.772	531.719	-1285.528	-356.731	4.235
4500	132.764	668.601	547.491	544.994	-1284.710	-335.630	3.896
4600	132.775	671.519	550.155	558.271	-1283.833	-314.550	3.572
4700	132.786	674.374	552.768	571.549	-1282.901	-293.488	3.262
4800	132.796	677.170	555.331	584.828	-1281.913	-272.447	2.965
4900	132.805	679.908	557.845	598.109	-1280.872	-251.427	2.680
5000	132.814	682.591	560.314	611.390	-1871.164	-224.698	2.347
5100	132.823	685.222	562.737	624.671	-1871.701	-191.763	1.964
5200	132.831	687.801	565.117	637.954	-1872.336	-158.817	1.595
5300	132.838	690.331	567.456	651.237	-1873.069	-125.857	1.240
5400	132.845	692.814	569.755	664.522	-1873.901	-92.884	0.898
5500	132.852	695.252	572.014	677.807	-1874.829	-59.895	0.569
5600	132.858	697.646	574.236	691.092	-1875.856	-26.886	0.251
5700	132.864	699.997	576.422	704.378	-1876.979	6.140	-0.056
5800	132.870	702.308	578.573	717.665	-1878.197	39.188	-0.353
5900	132.875	704.579	580.689	730.952	-1879.508	72.257	-0.640
6000	132.880	706.813	582.773	744.240	-1880.915	105.350	-0.917

PREVIOUS: June 1970 (1 atm)   CURRENT: June 1970 (1 bar)

Molybdenum Fluoride Oxide (MoF_4O) $F_4Mo_1O_1(g)$

TETRAFLUOROHYDRAZINE (N_2F_4) IDEAL GAS M_r = 104.007012

$S°(298.15 K) = 301.191 J K^{-1} mol^{-1}$

$\Delta_f H°(0 K) = 2.430 \pm 10.5 kJ mol^{-1}$

$\Delta_f H°(298.15 K) = -8.368 \pm 10.5 kJ mol^{-1}$

Vibrational Frequencies and Degeneracies

ν, cm^{-1}	ν, cm^{-1}	ν, cm^{-1}
1010 (1)	549 (1)	946 (1)
998 (1)	390 (1)	737 (1)
933 (1)	122 (1)	519 (1)
591 (1)	959 (1)	467 (1)

Ground State Quantum Weight: 1 $\sigma = 2$

Point Group: C_2

Bond Distances: N-N = [1.47] Å N-F = [1.37] Å

Bond Angles: F-N-F = [108]° N-N-F = [104]° Dihedral Angle = [65]°

Product of the Moments of Inertia: $I_A I_B I_C = [1.1817 \times 10^{-113}] g^3 cm^6$

Enthalpy of Formation

Armstrong et al. (1) studied the reaction $N_2F_4(g) + 16/3\ NH_3(g) = 4\ NH_4F(s) + 5/3\ N_2(g)$. From this reaction, they calculate the enthalpy of formation of N_2F_4 to be -2.0 ± 2.5 kcal mol^{-1}. The recalculated enthalpy of formation of $NH_4F(s)$ (see NF_2 table) differs from the value quoted in Circular 500 by 0.06 kcal mol^{-1}. This correction would make the enthalpy of formation of tetrafluorohydrazine more negative by 0.24 kcal mol^{-1}. In view of the large uncertainty in the enthalpy of formation, however, this correction was not applied.

Heat Capacity and Entropy - N_2F_4(g)

The fundamental frequencies were taken from the vapor phase infrared work of Durig and Lord (2). The principal moments of inertia were calculated from the rotational constants reported by Lide and Mann (3). Lide and Mann calculated the bond angles for assumed values of the NN and NF distance.

The principal moments of inertia are: $I_A = 1.5051 \times 10^{-38}$ g cm^2, $I_B = 2.6315 \times 10^{-38}$ g cm^2, $I_C = 2.9836 \times 10^{-38}$ g cm^2.

Heat Capacity and Entropy - NF_2(g)

Harmony et al. (4) investigated the infrared spectrum of NF_2. They were able to assign the band center at 1074 cm^{-1} to the symmetric stretching mode ν_1. The absorption in the 930-940 cm^{-1} region was assigned to ν_3.

Harmony and Myers (5) performed matrix isolation experiments on NF_2. They arrived at the assignment $\nu_1 = 1070$ cm^{-1}, $\nu_2 = 573$ cm^{-1} and $\nu_3 = 931$ cm^{-1}. The stretching frequencies for similar molecules in an N_2 matrix are about 5 cm^{-1} below the gas phase values. A comparison of the gas phase and matrix values of ν_1 show excellent agreement.

Johnson and Colburn (6) report infrared studies on NF_2 but they draw no definite conclusions.

The gas phase value for ν_1 and the matrix values of ν_2 and ν_3 were taken to represent the fundamental vibrational frequencies of NF_2.

Harmony et al. (4) estimated the N-F distance as 1.37 Å and used this with their data to calculate an F-N-F angle of 104.2°. Using these data the moments of inertia are: $I_A = 7.3728 \times 10^{-39}$ g cm^2, $I_B = 1.2035 \times 10^{-39}$ g cm^2, $I_C = 8.5763 \times 10^{-39}$ g cm^2.

References

1. G. T. Armstrong, S. Marantz, and C. F. Coyle, U. S. Nat. Bur. Stand. Report 6584, 1959.
2. J. R. Durig and R. C. Lord, Spectrochim. Acta 19, 1877 (1963).
3. D. R. Lide, Jr., and D. E. Mann, J. Chem. Phys. 31, 1129 (1959).
4. M. D. Harmony, R. J. Myers, L. J. Schoen, D. R. Lide, Jr., and D. E. Mann, J. Chem. Phys. 35, 1129 (1961).
5. M. D. Harmony and R. J. Myers, J. Chem. Phys. 37, 636 (1962).
6. F. A. Johnson and C. B. Colburn, Inorg. Chem. 1, 431 (1962).

Tetrafluorohydrazine (N_2F_4) $F_4N_2(g)$

Enthalpy Reference Temperature = T_r = 298.15 K Standard State Pressure = p° = 0.1 MPa

T/K	$C_p°$ (J $K^{-1}mol^{-1}$)	S° (J $K^{-1}mol^{-1}$)	$-[G°-H°(T_r)]/T$ (J $K^{-1}mol^{-1}$)	$H°-H°(T_r)$ (kJ mol^{-1})	$\Delta_f H°$ (kJ mol^{-1})	$\Delta_f G°$ (kJ mol^{-1})	Log K_f
0	0.	0.	INFINITE	-15.522	2.430	2.430	INFINITE
100	41.759	239.576	358.172	-11.860	-2.624	23.474	-12.261
200	59.845	273.592	307.831	-6.848	-6.381	51.159	-13.361
250	70.089	288.055	302.445	-3.598	-7.591	65.693	-13.726
298.15	79.169	301.191	301.191	0.	-8.368	79.883	-13.995
300	79.496	301.682	301.193	0.147	-8.391	80.431	-14.004
350	87.647	314.566	302.192	4.331	-8.842	95.275	-14.219
400	94.490	326.730	304.506	8.890	-9.004	110.163	-14.386
450	100.146	338.197	307.619	13.760	-8.933	125.056	-14.516
500	104.796	348.997	311.222	18.887	-8.677	139.932	-14.619
600	111.781	368.762	319.197	29.739	-7.756	169.575	-14.763
700	116.598	386.377	327.560	41.172	-6.469	199.032	-14.852
800	120.009	402.181	335.917	53.012	-4.957	228.288	-14.906
900	122.490	416.467	344.086	65.143	-3.311	257.346	-14.936
1000	124.341	429.473	351.984	77.489	-1.585	286.216	-14.950
1100	125.754	441.393	359.578	89.996	0.186	314.911	-14.954
1200	126.854	452.384	366.860	102.629	1.978	343.443	-14.950
1300	127.726	462.574	373.835	115.360	3.776	371.826	-14.940
1400	128.427	472.066	380.517	128.169	5.570	400.070	-14.927
1500	129.000	480.946	386.919	141.041	7.353	428.186	-14.911
1600	129.473	489.287	393.059	153.965	9.121	456.184	-14.893
1700	129.868	497.149	398.953	166.933	10.872	484.072	-14.874
1800	130.202	504.582	404.617	179.937	12.607	511.857	-14.854
1900	130.485	511.629	410.065	192.972	14.328	539.546	-14.833
2000	130.728	518.328	415.312	206.033	16.039	567.145	-14.812
2100	130.939	524.712	420.371	219.116	17.743	594.658	-14.791
2200	131.121	530.807	425.253	232.219	19.448	622.090	-14.770
2300	131.281	536.640	429.970	245.340	21.158	649.445	-14.749
2400	131.422	542.230	434.532	258.475	22.881	676.725	-14.729
2500	131.547	547.597	438.948	271.624	24.625	703.933	-14.708
2600	131.657	552.759	443.227	284.784	26.394	731.070	-14.687
2700	131.756	557.730	447.376	297.955	28.196	758.139	-14.667
2800	131.845	562.523	451.403	311.135	30.037	785.141	-14.647
2900	131.924	567.151	455.315	324.323	31.922	812.075	-14.627
3000	131.996	571.625	459.118	337.519	33.856	838.944	-14.607
3100	132.061	575.954	462.818	350.722	35.843	865.748	-14.588
3200	132.120	580.148	466.419	363.931	37.888	892.487	-14.568
3300	132.174	584.214	469.927	377.146	39.992	919.160	-14.549
3400	132.224	588.161	473.347	390.366	42.159	945.769	-14.530
3500	132.269	591.994	476.682	403.591	44.390	972.314	-14.511
3600	132.310	595.721	479.938	416.820	46.688	998.794	-14.492
3700	132.348	599.346	483.116	430.053	49.051	1025.209	-14.473
3800	132.384	602.876	486.221	443.289	51.483	1051.559	-14.455
3900	132.416	606.316	489.257	456.529	53.981	1077.845	-14.436
4000	132.446	609.668	492.225	469.772	56.547	1104.065	-14.418
4100	132.474	612.939	495.130	483.018	59.179	1130.220	-14.399
4200	132.500	616.132	497.973	496.267	61.878	1156.311	-14.381
4300	132.525	619.250	500.757	509.518	64.641	1182.336	-14.363
4400	132.547	622.297	503.485	522.772	67.469	1208.296	-14.344
4500	132.568	625.276	506.159	536.028	70.360	1234.192	-14.326
4600	132.588	628.190	508.780	549.286	73.312	1260.022	-14.308
4700	132.607	631.041	511.351	562.545	76.324	1285.788	-14.290
4800	132.624	633.833	513.874	575.807	79.395	1311.489	-14.272
4900	132.641	636.568	516.350	589.070	82.523	1337.126	-14.254
5000	132.656	639.248	518.781	602.335	85.706	1362.697	-14.236
5100	132.671	641.875	521.169	615.601	88.943	1388.205	-14.218
5200	132.684	644.452	523.515	628.869	92.232	1413.649	-14.200
5300	132.697	646.979	525.821	642.138	95.571	1439.029	-14.182
5400	132.709	649.460	528.088	655.409	98.960	1464.346	-14.165
5500	132.721	651.895	530.317	668.680	102.396	1489.599	-14.147
5600	132.732	654.286	532.509	681.953	105.878	1514.790	-14.129
5700	132.742	656.636	534.666	695.226	109.404	1539.917	-14.112
5800	132.752	658.944	536.789	708.501	112.972	1564.982	-14.094
5900	132.761	661.214	538.879	721.777	116.581	1589.987	-14.077
6000	132.770	663.445	540.936	735.053	120.230	1614.929	-14.059

PREVIOUS: March 1964 (1 atm) CURRENT: March 1964 (1 bar)

Tetrafluorohydrazine (N_2F_4) $F_4N_2(g)$

TUNGSTEN FLUORIDE OXIDE (WF_4O) CRYSTAL M_r = 275.843012

$\Delta_f H°(0\ K)$ = Unknown

$S°(298.15\ K) = 175.728 \pm 4.2\ J\ K^{-1}\ mol^{-1}$ $\Delta_f H°(298.15\ K) = -1406.908 \pm 62.8\ kJ\ mol^{-1}$

T_{fus} = 379 K $\Delta_{fus}H° = 5.021 \pm 4.2\ kJ\ mol^{-1}$

Enthalpy of Formation

The enthalpy of formation, $\Delta_f H°(WOF_4$, cr, 298.15 K) = -333.26 kcal mol^{-1}, is calculated from the enthalpy of formation of WOF_4(g) less the enthalpy of sublimation, $\Delta_{sub}H°$(298.15 K) = 16.80 kcal mol^{-1}. This latter quantity is calculated by the 3rd law method from the vapor pressure equation determined by Cady and Hargreaves (1). The 2nd law value is $\Delta_{sub}H°$(298.15 K) = 16.94 kcal mol^{-1}.

Heat Capacity and Entropy

The heat capacities are estimated from those of $WOCl_4$(cr), WO_3(cr), WO_2(cr), WCl_6(cr), and WF_6(cr, ℓ) (3).

The entropy, S°(298.15 K) = 42.0 cal K^{-1} mol^{-1} is calculated from the vapor pressure equation of Cady and Hargreaves (1) using a 2nd law analysis.

This entropy is larger than the corresponding value for $WOCl_4$(cr) (2). In both cases the entropies of the condensed phases are derived from gas phase values using 2nd law analysis of sublimation and vaporization data. The entropy of melting for $WOCl_4$ is an order of magnitude larger than that of WOF_4. This suggests some inconsistency in the data for $WOCl_4$ since all values for WOF_4(cr, ℓ, g) appear internally reasonable.

Fusion Data

T_{fus} is calculated as the temperature at which the Gibbs energy of reaction $WOF_4(cr) \rightarrow WOF_4(ℓ)$ approaches zero. The difference between the enthalpies of formation for the crystal and liquid at the melting point is $\Delta_{fus}H°$.

Ruff et al. (3) reported the melting point as 383 K. Cady and Hargreaves (1) calculated the melting point as 378 K, and the heat of fusion as 2.26 kcal mol^{-1} from vapor pressure equations for the crystal and liquid. This 2nd law value of $\Delta_{fus}H°$, when corrected for $\Delta C°_p$ of vaporization and sublimation, is in good agreement with the value adopted in this table.

References

1. G. H. Cady and G. B. Hargreaves, J. Chem. Soc. (London) 1961, 1563 (1961).
2. JANAF Thermochemical Tables: $WOCl_4$(cr), 3-31-67, WCl_6(cr), 12-31-66, WF_6(cr), 3-31-67; WO_2(cr), 9-30-66; WO_3(cr), 9-30-66.
3. O. Ruff, F. Eisner, and W. Heller, Z. Anorg. Chem. 52, 256 (1907).

Tungsten Fluoride Oxide (WF_4O) $F_4O_1W_1$(cr)

Enthalpy Reference Temperature = T_r = 298.15 K Standard State Pressure = p° = 0.1 MPa

T/K	$C°_p$ (J K⁻¹mol⁻¹)	S°	$-[G°-H°(T_r)]/T$	$H°-H°(T_r)$ (kJ mol⁻¹)	$\Delta_f H°$	$\Delta_f G°$	Log K_f
0							
100							
200							
298.15	133.578	175.728	175.728	0.	-1406.908	-1298.058	227.414
300	133.888	176.555	175.731	0.247	-1406.848	-1297.383	225.894
379.000	147.109	209.337	179.398	11.347	------ CRYSTAL <--> LIQUID -----		
400	150.624	217.357	181.182	14.470	-1403.013	-1261.438	164.727
500	166.523	252.746	192.028	30.359	-1397.900	-1226.610	128.143
600	178.657	284.242	204.821	47.653	-1391.690	-1192.920	103.853
700	186.606	312.412	218.216	65.937	-1384.737	-1160.334	86.585
800	192.046	337.703	231.599	84.883	-1377.323	-1128.778	73.702
900	195.811	360.557	244.679	104.290	-1369.616	-1098.170	63.736
1000	197.903	381.310	257.321	123.990	-1361.761	-1068.430	55.809

PREVIOUS: March 1967 CURRENT: December 1971

Tungsten Fluoride Oxide (WF_4O) $F_4O_1W_1$(cr)

TUNGSTEN FLUORIDE OXIDE (WF_4O) LIQUID M_r = 275.843012

$S°(298.15\ K) = 178.915 \pm 4.2\ J\ K^{-1}\ mol^{-1}$
$\Delta_f H°(298.15\ K) = -1405.255 \pm 62.8\ kJ\ mol^{-1}$

$T_{fus} = 379\ K$
$\Delta_{fus}H° = 5.021 \pm 4.2\ kJ\ mol^{-1}$

Enthalpy of Formation

The enthalpy of formation, $\Delta_f H°(WOF_4, \ell, 298.15\ K) = -335.84\ kcal\ mol^{-1}$ is calculated from that of the gas less the heat of vaporization, $\Delta_{vap}H°(298.15\ K) = 16.38\ kcal\ mol^{-1}$. This latter value is calculated by the 3rd law method from the vapor pressure equation determined by Cady and Hargreaves (1). The 2nd law value is $\Delta_{vap}H°(298.15\ K) = 16.50\ kcal\ mol^{-1}$.

Heat Capacity and Entropy

The heat capacity is assumed to be constant at 7.25 cal K^{-1} g-atom^{-1} as suggested by Kubaschewski and Evans (2). The entropy, $S°(298.15\ K) = 42.8\ cal\ K^{-1}\ mol^{-1}$ is calculated by a 2nd law analysis of the vapor pressure equation given by Cady and Hargreaves (1) for the system $WOF_4(\ell) \rightarrow WOF_4(g)$.

Fusion Data

See the crystal table for details.

Vaporization Data

T_{vap} is calculated as the temperature at which the fugacity is one atmosphere for the process $WOF_4(\ell) \rightarrow WOF_4(g)$. The difference between the enthalpies of formation for the liquid and gas at T_{vap} is $\Delta_{vap}H°$.

Ruff et al. (3) reported the boiling point of $WOF_4(\ell)$ as 459 K. Cady and Hargreaves (1) calculated the boiling point as 459 K, and the heat of vaporization as 14.23 kcal mol^{-1} from the vapor pressure equation for the liquid. This 2nd law value of $\Delta_{vap}H°$, when corrected for $\Delta C°_p$ of vaporization, is in good agreement with the value adopted in the tabulation.

References

1. G. H. Cady and G. B. Hargreaves, J. Chem. Soc. (London) 1961, 1563 (1961).
2. O. Kubaschewski and E. L. Evans, "Metallurgical Thermochemistry," Pergamon Press, New York, 1958.
3. O. Ruff, F. Eisner, and W. Zeller, Z. Anorg. Chem. 52, 256 (1907).

Tungsten Fluoride Oxide (WF_4O) $F_4O_1W_1(l)$

Enthalpy Reference Temperature = T_r = 298.15 K — Standard State Pressure = p° = 0.1 MPa

T/K	$C°_p$ (J K⁻¹mol⁻¹)	S° (J K⁻¹mol⁻¹)	$-[G°-H°(T_r)]/T$ (J K⁻¹mol⁻¹)	$H°-H°(T_r)$ (kJ mol⁻¹)	$\Delta_f H°$ (kJ mol⁻¹)	$\Delta_f G°$ (kJ mol⁻¹)	Log K_f
0							
100							
200							
298.15	182.004	178.915	178.915	0.	-1405.255	-1297.355	227.291
300	182.004	180.041	178.918	0.337	-1405.106	-1296.686	225.773
379.000	182.004	222.585	183.759	14.715	------ CRYSTAL <--> LIQUID -----		
400	182.004	232.400	186.057	18.537	-1397.294	-1261.736	164.766
500	182.004	273.013	199.538	36.738	-1389.868	-1228.713	128.363
600	182.004	306.196	214.633	54.938	-1382.752	-1197.155	104.222
700	182.004	334.252	229.769	73.138	-1375.883	-1166.768	87.065
800	182.004	358.555	244.382	91.339	-1369.215	-1137.351	74.261
900	182.004	379.992	258.282	109.539	-1362.715	-1108.760	64.351
1000	182.004	399.168	271.429	127.740	-1356.359	-1080.885	56.460

PREVIOUS: March 1967 CURRENT: December 1971

Tungsten Fluoride Oxide (WF_4O) $F_4O_1W_1(l)$

TUNGSTEN FLUORIDE OXIDE (WF_4O) CRYSTAL-LIQUID M_r = 275.843012

0 to 379 K crystal
above 379 K liquid

Refer to the individual tables for details.

Tungsten Fluoride Oxide (WF_4O) $F_4O_1W_1$(cr,l)

Enthalpy Reference Temperature = T_r = 298.15 K Standard State Pressure = p° = 0.1 MPa

T/K	C_p° (J K⁻¹mol⁻¹)	S° (J K⁻¹mol⁻¹)	−[G°−H°(T_r)]/T (J K⁻¹mol⁻¹)	H°−H°(T_r) (kJ mol⁻¹)	$\Delta_f H^\circ$ (kJ mol⁻¹)	$\Delta_f G^\circ$ (kJ mol⁻¹)	Log K_f
0							
100							
200							
298.15	133.578	175.728	175.728	0.	-1406.908	-1298.058	227.414
300	133.888	176.555	175.731	0.247	-1406.848	-1297.383	225.894
379.000	147.109	209.337	179.398	11.347	CRYSTAL <--> LIQUID		
379.000	182.004	222.585	179.398	16.368	TRANSITION		
400	182.004	232.400	181.926	20.190	-1397.294	-1261.736	164.766
500	182.004	273.013	196.233	38.390	-1389.868	-1228.713	128.363
600	182.004	306.196	211.879	56.590	-1382.752	-1197.155	104.222
700	182.004	334.252	227.408	74.791	-1375.883	-1166.768	87.065
800	182.004	358.555	242.316	92.991	-1369.215	-1137.351	74.261
900	182.004	379.992	256.446	111.192	-1362.715	-1108.760	64.351
1000	182.004	399.168	269.776	129.392	-1356.359	-1080.885	56.460

PREVIOUS: CURRENT: December 1971

Tungsten Fluoride Oxide (WF_4O) $F_4O_1W_1$(cr,l)

TUNGSTEN FLUORIDE OXIDE (WF_4O) IDEAL GAS M_r = 275.843012

$\Delta_f H°$(0 K) = -1328.914 ± 62.8 kJ mol^{-1}

S°(298.15 K) = 334.704 ± 4.2 J K^{-1} mol^{-1} $\Delta_f H°$(298.15 K) = -1336.621 ± 62.8 kJ mol^{-1}

Vibrational Frequencies and Degeneracies

ν, cm^{-1}	ν, cm^{-1}	ν, cm^{-1}
1055 (1)	620 (1)	700 (2)
[300](1)	[300](1)	560 (2)
735 (1)	[200](1)	[150](2)

Ground State Quantum Weight: [1] σ = [4]

Point Group: [C_{4v}]

Bond Distances: W-O = [1.64] Å W-F = [1.82] Å

Bond Angles: O-W-F = [90]° F-W-F = [90]°

Product of the Moments of Inertia: $I_A I_B I_C$ = [3.1909 x 10^{-113}] g^3 cm^6

Enthalpy of Formation

The enthalpy of formation, $\Delta_f H°(WOF_4$, g, 298.15 K) = -319.46 kcal mol^{-1}, is calculated from the enthalpy of reaction data of Zmbov et al. (1). This data came from a mass spectrometric study of the equilibrium in the reaction:

$$2\ TAF_5(g) + WO_3(cr) = 2\ TAOF_3(g) + WOF_4(g).$$

Heat Capacity and Entropy

The infrared spectra of WOF_4 and $MoOF_4$ in the vapor state are very similar (2, 3, 4) and have been interpreted on the basis of a monomeric unit with C_{4v} symmetry (5) by comparison with the spectrum of X_eOF_4 (6). Thus, as with X_eOF_4 and $MoOF_4$, gaseous WOF_4 is assumed to be C_{4v} symmetry with a square-pyamidal structure.

Blanchard (5) and Reynes (10) reported the following vibrational assignments and frequencies, the Q frequency being adopted in this tabulation:

ν_1 = 1064(R), 1055(Q), 1047(P); ν_7 = 705(R), 700(Q), 690(P); ν_3 = 745(R), 735(Q), 720(P);
ν_8 = 560 cm^{-1}

Similar frequencies, as well as additional frequencies, have been reported by Ward and Stafford (4), Edwards et al. (3), and Edwards and Jones (2). The remaining frequencies and assignments are estimated with regard to the preceding references, the data for X_3OF_4 (6), and results for $MoOF_4$ (7).

There exists some controversy as the crystallographic structure of WOF_4. Beattie et al. (8) suggest fluorine-bridged tetramers as in $MoOF_4$ (9) rather than an oxygen-bridged structure as suggested by Edwards and Jones (2). This leads to significantly different bond lengths for W-F and W-O. Due to the similarities in the melt, solid, and vapor phase infrared spectra of $MoOF_4$ and WOF_4, one would suspect similar structures (4, 8). The assumption is made that the structures are similar and thus the bond lengths are taken to be the same as those of $MoOF_4$. Further support for the bond length assumption is the similarity of bond lengths in MoO_3 and WO_3 and in MoF_6 and WF_6 (7).

References

1. K. F. Zmbov, O. M. Uy, and J. L. Margrave, J. Phys. Chem. 73, 3008 (1969).
2. A. J. Edwards and G. R. Jones, J. Chem. Soc. (London) A1968, 2074 (1968).
3. A. J. Edwards, G. R. Jones, and B. R. Stevenson, Chem. Comm. 462 (1967).
4. B. G. Ward and F. E. Stafford, J. Inorg. Chem. 7, 2569 (1968).
5. S. Blanchard, Commissariat a l'Energie Atomique Rep. CEA-R 3194 (1967); CA 67 48663d.
6. G. M. Begun, W. A. Fletcher, and D. F. Smith, J. Chem. Phys. 42, 2236 (1965).
7. JANAF Thermochemical Tables: $MoOF_4$(g), 6-30-70; MoO_3(g), 6-30-68; WO_3(g), 9-30-66; MoF_6(g), 6-30-70; WF_6(g), 3-31-67.
8. I. R. Beattie, R. M. S. Livingston, D. J. Reynolds, and G. A. Ozin, J. Chem. Soc. (London) A1970, 1210 (1970).
9. A. J. Edwards and B. R. Stevenson, J. Chem. Soc. (London) A1968, 2503 (1968).
10. J. Reynes, Commissariat a l'Energie Atomique, Bibliographic CEA-BIB-143 (1969).

Tungsten Fluoride Oxide (WF_4O) $F_4O_1W_1$(g)

Enthalpy Reference Temperature = T_r = 298.15 K Standard State Pressure = p° = 0.1 MPa

T/K	$C_p°$ (J K^{-1}mol^{-1})	S° (J K^{-1}mol^{-1})	-[G°-H°(T_r)]/T (J K^{-1}mol^{-1})	H°-H°(T_r) (kJ mol^{-1})	$\Delta_f H°$ (kJ mol^{-1})	$\Delta_f G°$ (kJ mol^{-1})	Log K_f
0	0.	0.	INFINITE	-19.257	-1328.914	-1328.914	INFINITE
100	53.821	255.013	407.183	-15.217	-1332.792	-1314.592	686.673
200	77.962	300.013	342.944	-8.586	-1335.482	-1295.221	338.277
250	87.938	318.514	336.240	-4.431	-1336.218	-1285.063	268.499
298.15	95.853	334.704	334.704	0.	-1336.621	-1275.170	223.404
300	96.125	335.298	334.706	0.178	-1336.631	-1274.788	221.960
350	102.646	350.625	335.901	5.153	-1336.795	-1264.465	188.711
400	107.789	364.681	338.633	10.419	-1336.777	-1254.132	163.773
450	111.850	377.620	342.255	15.914	-1336.627	-1243.809	144.378
500	115.076	389.578	346.397	21.590	-1336.381	-1233.508	128.864
600	119.756	411.003	355.422	33.348	-1335.707	-1212.994	105.600
700	122.882	429.713	364.727	45.490	-1334.897	-1192.604	88.993
800	125.049	446.272	373.905	57.893	-1334.026	-1172.336	76.546
900	126.604	461.095	382.784	70.480	-1333.140	-1152.177	66.871
1000	127.753	474.496	391.296	83.200	-1332.264	-1132.118	59.136
1100	128.624	486.715	399.423	96.021	-1331.416	-1112.144	52.811
1200	129.299	497.937	407.171	108.919	-1330.609	-1092.246	47.544
1300	129.831	508.308	414.557	121.876	-1329.852	-1072.414	43.090
1400	130.258	517.946	421.602	134.882	-1329.151	-1052.638	39.274
1500	130.606	526.945	428.328	147.925	-1328.511	-1032.910	35.969
1600	130.892	535.384	434.758	161.001	-1327.937	-1013.222	33.078
1700	131.131	543.326	440.913	174.102	-1327.429	-993.568	30.529
1800	131.333	550.828	446.813	187.226	-1326.988	-973.942	28.263
1900	131.504	557.933	452.476	200.368	-1326.611	-954.340	26.237
2000	131.650	564.682	457.919	213.526	-1326.298	-934.754	24.413
2100	131.777	571.109	463.158	226.697	-1326.043	-915.184	22.764
2200	131.887	577.241	468.205	239.880	-1325.842	-895.624	21.265
2300	131.983	583.106	473.074	253.074	-1325.689	-876.072	19.896
2400	132.067	588.725	477.777	266.277	-1325.578	-856.526	18.642
2500	132.142	594.118	482.323	279.487	-1325.502	-836.984	17.488
2600	132.209	599.302	486.723	292.705	-1325.453	-817.444	16.423
2700	132.268	604.293	490.986	305.929	-1325.450	-797.906	15.436
2800	132.321	609.104	495.119	319.158	-1325.523	-778.366	14.521
2900	132.369	613.748	499.130	332.393	-1325.705	-758.822	13.668
3000	132.412	618.236	503.026	345.632	-1326.018	-739.269	12.872
3100	132.451	622.579	506.813	358.875	-1326.502	-719.704	12.127
3200	132.486	626.784	510.496	372.122	-1327.184	-700.119	11.428
3300	132.519	630.862	514.082	385.372	-1328.087	-680.510	10.772
3400	132.548	634.818	517.576	398.625	-1329.241	-660.870	10.153
3500	132.575	638.661	520.980	411.882	-1330.711	-641.192	9.569
3600	132.600	642.396	524.301	425.140	-1332.590	-621.465	9.017
3700	132.623	646.029	527.543	438.401	-1369.743	-601.490	8.492
3800	132.644	649.567	530.707	451.665	-1369.152	-580.734	7.983
3900	132.663	653.012	533.799	464.930	-1368.497	-559.994	7.500
4000	132.681	656.371	536.822	478.197	-1367.778	-539.272	7.042
4100	132.698	659.648	539.778	491.466	-1366.997	-518.569	6.607
4200	132.714	662.846	542.670	504.737	-1366.152	-497.886	6.192
4300	132.728	665.969	545.501	518.009	-1365.246	-477.224	5.797
4400	132.742	669.020	548.274	531.283	-1364.280	-456.583	5.420
4500	132.754	672.003	550.991	544.558	-1363.253	-435.965	5.061
4600	132.766	674.921	553.653	557.834	-1362.169	-415.371	4.717
4700	132.777	677.777	556.264	571.111	-1361.028	-394.800	4.388
4800	132.788	680.572	558.825	584.389	-1359.832	-374.254	4.073
4900	132.798	683.310	561.337	597.668	-1358.582	-353.734	3.771
5000	132.807	685.993	563.804	610.949	-1357.281	-333.240	3.481
5100	132.815	688.623	566.225	624.230	-1355.930	-312.773	3.203
5200	132.824	691.202	568.604	637.512	-1354.531	-292.332	2.937
5300	132.831	693.733	570.941	650.794	-1353.086	-271.919	2.680
5400	132.839	696.216	573.238	664.078	-1351.597	-251.534	2.433
5500	132.846	698.653	575.496	677.362	-1350.065	-231.177	2.196
5600	132.852	701.047	577.717	690.647	-1348.493	-210.846	1.967
5700	132.858	703.398	579.901	703.932	-1346.883	-190.547	1.746
5800	132.864	705.709	582.051	717.219	-1345.236	-170.274	1.533
5900	132.870	707.980	584.166	730.505	-1343.554	-150.031	1.328
6000	132.875	710.213	586.248	743.793	-1341.841	-129.816	1.130

PREVIOUS: December 1971 (1 atm) CURRENT: December 1971 (1 bar)

Tungsten Fluoride Oxide (WF_4O) $F_4O_1W_1$(g)

LEAD FLUORIDE (PbF_4) IDEAL GAS M_r = 283.193612

$S°(298.15\ K) = [333.602 \pm 12.6]\ J\ K^{-1}\ mol^{-1}$

$\Delta_f H°(0\ K) = -1128.678 \pm 20.9\ kJ\ mol^{-1}$
$\Delta_f H°(298.15\ K) = -1133.446 \pm 20.9\ kJ\ mol^{-1}$

Vibrational Frequencies and Degeneracies

ν, cm^{-1}
[564](1)
[158](2)
[570](3)
[180](3)

Ground State Quantum Weight: [1] σ = [12]
Point Group: [T_d]
Bond Distance: Pb-F = [2.08] Å
Bond Angle: F-Pb-F = [109° 28']
Product of the Moments of Inertia: $I_A I_B I_C = [4.821770 \times 10^{-113}]\ g^3\ cm^6$

Enthalpy of Formation

A 3rd law analysis of the partial pressure data at 988 K reported by Zmbov et al. (1) for the equilibrium 2 $PbF_2(g) = PbF_4(g)$ + Pb(g) gives $\Delta_r H°(298.15\ K) = -16.15\ kcal\ mol^{-1}$. There are inconsistencies in the published results but the partial pressure data are correctly given (2). The 2nd law enthalpy of reaction derived by Zmbov et al. (1) from a graphical plot of six ion current quotient values vs. reciprocal temperature (approximately 1020 to 1110 K), when corrected to 298.15 K, is in agreement with the 3rd law value of $-16.15\ kcal\ mol^{-1}$. Combining this enthalpy of reaction with $\Delta_f H°(PbF_2, g, 298.15\ K) = -104.0\pm1\ kcal\ mol^{-1}$ (3) and $\Delta_f H°(Pb, g, 298.15\ K) = 46.75\pm0.13\ kcal\ mol^{-1}$ (3) gives the adopted $\Delta_f H°(PbF_4, g, 298.15\ K) = -270.9\pm5\ kcal\ mol^{-1}$. The adopted enthalpy of formation leads to a dissociation energy $\Delta_{at}H°$= 16.96 eV for the process $PbF_4(g) = Pb(g) + 4\ F(g)$. This is consistent with $\Delta_{at}H°$= 8.10 eV for $PbF_2(g)$ (3) and D_0^0 = 3.64 eV for PbF(g) (3).

Heat Capacity and Entropy

The molecular constants and vibrational frequencies are those calculated by Aleksandrovskaya et al. (4). Later works (5, 6, 7) mistakenly cite this work as observed data. A band at 663 cm^{-1} was recently observed in an infrared study by Hauge et al. (8) and was assigned to matrix isolated PbF_4. Calculation with infrared active ν_3 = 663 cm^{-1} decreases S°(298.15 K) by 0.45 cal $K^{-1}\ mol^{-1}$. Only the calculated frequencies are used for the present table pending more complete observation.

References

1. K. Zmbov, J. W. Hastie, and J. L. Margrave, Trans. Faraday Soc. 64, 861 (1968).
2. J. W. Hastie, U. S. Nat. Bur. Stand., personal communication, Dec. 26, 1973.
3. JANAF Thermochemical Tables: Pb(g), 3-31-62; $PbF_2(g)$, 12-31-73; PbF(g), 12-31-73; F(g), 9-30-65.
4. A. M. Aleksandrovskaya, I. V. Rigina, and I. N. Godnev, Optics and Spectrosc. 7, 495 (1959).
5. G. Nagarajan, Bull. Soc. Chim. Belges 71, 119 (1962).
6. M. Radhakrishnan, Indian J. Pure Appl. Phys. 1, 402 (1963).
7. S. P. Singh, Labdev, Part A 7, 185 (1969).
8. R. H. Hauge, J. W. Hastie, and J. L. Margrave, J. Mol. Spectrosc. 45, 420 (1973).

Lead Fluoride (PbF_4) $F_4Pb_1(g)$

Enthalpy Reference Temperature = T_r = 298.15 K Standard State Pressure = p° = 0.1 MPa

T/K	$C_p°$ (J K⁻¹mol⁻¹)	S° (J K⁻¹mol⁻¹)	$-[G°-H°(T_r)]/T$ (J K⁻¹mol⁻¹)	$H°-H°(T_r)$ (kJ mol⁻¹)	$\Delta_f H°$ (kJ mol⁻¹)	$\Delta_f G°$ (kJ mol⁻¹)	Log K_f
0	0.	0.	INFINITE	-19.761	-1128.678	-1128.678	INFINITE
100	59.551	251.440	406.137	-15.470	-1131.968	-1119.354	584.691
200	79.572	299.488	341.603	-8.423	-1133.303	-1106.086	288.880
250	86.207	317.993	335.075	-4.270	-1133.461	-1099.258	229.677
298.15	90.935	333.602	333.602	0.	-1133.446	-1092.670	191.431
300	91.090	334.165	333.604	0.168	-1133.443	-1092.417	190.207
350	94.654	348.489	334.727	4.817	-1133.326	-1085.588	162.015
400	97.277	361.308	337.263	9.618	-1133.160	-1078.779	140.874
450	99.239	372.884	340.588	14.533	-1132.976	-1071.992	124.434
500	100.733	383.421	344.352	19.534	-1132.789	-1065.226	111.283
600	102.801	401.984	352.453	29.719	-1132.441	-1051.747	91.563
700	104.120	417.937	360.695	40.070	-1137.015	-1037.524	77.421
800	105.007	431.902	368.741	50.529	-1136.792	-1023.326	66.816
900	105.630	444.308	376.461	61.063	-1136.552	-1009.157	58.570
1000	106.083	455.462	383.813	71.649	-1136.300	-995.015	51.974
1100	106.422	465.590	390.794	82.275	-1136.038	-980.900	46.579
1200	106.683	474.861	397.419	92.931	-1135.768	-966.808	42.084
1300	106.887	483.409	403.709	103.610	-1135.498	-952.739	38.282
1400	107.050	491.336	409.688	114.307	-1135.245	-938.690	35.023
1500	107.182	498.727	415.381	125.019	-1135.019	-924.659	32.199
1600	107.290	505.648	420.808	135.743	-1134.825	-910.641	29.729
1700	107.380	512.155	425.992	146.476	-1134.668	-896.634	27.550
1800	107.456	518.295	430.951	157.218	-1134.547	-882.636	25.613
1900	107.520	524.106	435.702	167.967	-1134.462	-868.644	23.881
2000	107.575	529.623	440.262	178.722	-1134.409	-854.655	22.321
2100	107.622	534.872	444.643	189.482	-1311.605	-833.553	20.733
2200	107.663	539.880	448.859	200.246	-1311.212	-810.798	19.251
2300	107.699	544.667	452.921	211.014	-1310.902	-788.059	17.897
2400	107.731	549.251	456.840	221.786	-1310.669	-765.332	16.657
2500	107.758	553.649	460.625	232.560	-1310.506	-742.613	15.516
2600	107.783	557.876	464.285	243.338	-1310.407	-719.899	14.463
2700	107.805	561.944	467.827	254.117	-1310.362	-697.189	13.488
2800	107.825	565.865	471.259	264.898	-1310.362	-674.479	12.583
2900	107.843	569.649	474.587	275.682	-1310.396	-651.768	11.740
3000	107.859	573.306	477.817	286.467	-1310.454	-629.056	10.953
3100	107.873	576.843	480.954	297.254	-1310.525	-606.342	10.217
3200	107.886	580.268	484.005	308.042	-1310.601	-583.625	9.527
3300	107.898	583.588	486.972	318.831	-1310.669	-560.906	8.878
3400	107.909	586.809	489.862	329.621	-1310.721	-538.185	8.268
3500	107.919	589.937	492.676	340.413	-1310.749	-515.463	7.693
3600	107.928	592.977	495.420	351.205	-1310.743	-492.740	7.149
3700	107.937	595.935	498.097	361.998	-1310.697	-470.018	6.635
3800	107.945	598.813	500.710	372.792	-1310.604	-447.299	6.149
3900	107.952	601.617	503.262	383.587	-1310.458	-424.581	5.687
4000	107.958	604.350	505.755	394.383	-1310.253	-401.869	5.248
4100	107.965	607.016	508.192	405.179	-1309.986	-379.163	4.831
4200	107.970	609.618	510.576	415.976	-1309.652	-356.463	4.433
4300	107.976	612.159	512.909	426.773	-1309.249	-333.774	4.055
4400	107.981	614.641	515.193	437.571	-1308.774	-311.093	3.693
4500	107.985	617.068	517.430	448.369	-1308.226	-288.424	3.348
4600	107.990	619.441	519.622	459.168	-1307.603	-265.769	3.018
4700	107.994	621.764	521.771	469.967	-1306.905	-243.128	2.702
4800	107.998	624.037	523.878	480.767	-1306.131	-220.502	2.400
4900	108.001	626.264	525.945	491.567	-1305.281	-197.894	2.110
5000	108.005	628.446	527.973	502.367	-1304.357	-175.303	1.831
5100	108.008	630.585	529.964	513.167	-1303.357	-152.732	1.564
5200	108.011	632.682	531.919	523.968	-1302.285	-130.181	1.308
5300	108.014	634.740	533.840	534.770	-1301.141	-107.651	1.061
5400	108.017	636.759	535.727	545.571	-1299.926	-85.144	0.824
5500	108.019	638.741	537.582	556.373	-1298.642	-62.660	0.595
5600	108.022	640.687	539.406	567.175	-1297.291	-40.199	0.375
5700	108.024	642.599	541.200	577.977	-1295.875	-17.764	0.163
5800	108.026	644.478	542.964	588.780	-1294.397	4.646	-0.042
5900	108.028	646.325	544.700	599.582	-1292.856	27.031	-0.239
6000	108.030	648.140	546.409	610.385	-1291.258	49.387	-0.430

PREVIOUS: December 1973 (1 atm) CURRENT: December 1973 (1 bar)

Lead Fluoride (PbF_4) $F_4Pb_1(g)$

SULFUR FLUORIDE (SF_4) IDEAL GAS M_r = 108.053612

$S°(298.15\ K) = 299.644 \pm 0.4\ J\ K^{-1}\ mol^{-1}$

$\Delta_f H°(0\ K) = -756.882 \pm 20.9\ kJ\ mol^{-1}$

$\Delta_f H°(298.15\ K) = -763.162 \pm 20.9\ kJ\ mol^{-1}$

Vibrational Frequencies and Degeneracies

ν, cm^{-1}	ν, cm^{-1}	ν, cm^{-1}
891.5(1)	228.0(1)	532.5(1)
558.4(1)	475.0(1)	867.0(1)
353.0(1)	728.0(1)	233.0(1)

Ground State Quantum Weight: 1 Point Group: C_{2v} σ = 2

Bond Distances: S-F* = 1.545±0.003 Å S-F** = 1.646±0.003 Å

Bond Angles: F*-S-F* = 101.55±0.5° F*-S-F** = 87.81±0.5°

F**-S-F** = 173.07±0.5° (* - equatorial ** - axial)

Product of Moments of Inertia: $I_A I_B I_C = 6.723859 \times 10^{-114}\ g^3\ cm^6$

Enthalpy of Formation

Data on $\Delta_f H°$ published prior to 1970 have been reviewed by O'Hare et al. (1). We reanalyze this earlier work together with more recent studies. Unless otherwise indicated, all auxiliary data are taken from JANAF tables (2). We also include, for those reactions involving the formation of HF, $\Delta_f H°$ values which have been derived in previous analyses (11, 12, 13).

Source	Method	Reaction	$\Delta_r H°(T/K)$ kcal mol^{-1}	$\Delta_f H°(SF_4$, g, 298.15 K), kcal mol^{-1} From (Ref.)	This Work
(3)	Calorimetric	$SF_4(g) + 2\ H_2(g) = 4\ HF(\ell) + S(cr)$	-104.8(298)[b] ±4.0	185.2(11, 12) 188.5(13)	184.7[a] ±6.0
(4)	Calorimetric	$SF_4(g) + 3\ H_2O(\ell) = 4\ HF(aq) + H_2SO_3(aq)$	c	179.0(11)	180.8
(5-8)	Appearance Potentials[d]	$SF_6(g) = SF_4(g) + 2\ F(g)$	145.0±4.6(0)		181.6±5.0
(8)	Appearance Potentials[e]	$NSF_3(g) + F(g) = SF_4(g) + N(g)$	17.5(0)		161.8
(9)	Appearance Potentials[f]	$SF_6(g) = SF_4(g) + 2\ F(g)$	116.2(0)		210.4

[a]Auxiliary $\Delta_f H°(298.15\ K)$ data in kcal mol^{-1}: $NSF_3(g)$, -85.2±0.5(1); HF(ℓ), -72.37(2, 10).
[b]Results reduced with HF PVT and calorimetric data from (10).
[c]Hydrolysis data unavailable; $\Delta_f H°$ value from (11) adjusted for changes in $\Delta_f H°$ of HF(aq).
[d]Ion Processes: $SF_6(g) + e^- = SF_4^+(g) + 2\ F(g) + 2\ e^-$ $SF_4(g) + e^- = SF_4^+(g) + 2\ e^-$
[e]Ion Processes: $SF_4(g) + e^- = SF_3^+(g) + F(g) + 2\ e^-$ $NSF_3(g) + e^- = SF_3^+(g) + N(g) + 2\ e^-$
[f]Ion Processes: $SF_5(g) + e^- = SF_4(g) + F^-(g)$ $SF_5(g) + e^- = SF_5^-(g)$ $SF_6(g) + e^- = SF_5^-(g) + F(g)$

All of these studies are relatively imprecise, suggesting the need for an unequivocal redetermination of $\Delta_f H°$. Both calorimetric results (3, 4) are based on measurements made with impure samples (~ 98% SF_4) and suffer from the additional uncertainties introduced by the formation of HF. The ionization processes involve unknown kinetic and excitation energies, as evidenced by the widely scattered results. The first three $\Delta_f H°$ values listed above are reasonably consistent; thus, we are led to adopt their mean value of $\Delta_f H°(SF_4$, g, 298.15 K) = -182.4 kcal mol^{-1}. The estimated uncertainty is ±5.0 kcal mol^{-1}. Our $\Delta_f H°$ value corresponds to a enthalpy of atomization and average bond dissociation energy of 320.1 and 80.0 kcal mol^{-1}, respectively.

Heat Capacity and Entropy

Detailed structural data determined by electron diffraction (14) and microwave spectroscopy (15) show good agreement. We arbitrarily adopt the microwave data which indicate a trigonal bipyramid structure with two inequivalent sets of fluorine atoms. The inequivalence of the fluorines is also supported by nuclear resonance (16, 17) and x-ray photoelectron spectroscopic (18, 19) studies. The principal moments of inertia are: $I_A = 12.5525 \times 10^{-39}$, $I_B = 20.5464 \times 10^{-39}$ and $I_C = 26.0707 \times 10^{-39}\ g\ cm^2$.

Various spectroscopic methods have been used to study the vibrational spectrum of SF_4. These studies include crystal (20), liquid (21) and gas (22) phase Raman and gas (21-23) phase and argon matrix-isolation (24) infrared spectra. The measured frequencies show good agreement but various interpretations of the observed spectra have been presented. We choose to adopt the results of the latest interpretation by Levin (20). It should be noted that the vibrational assignments are immaterial for the purpose of evaluating the thermal functions since none of the fundamentals are degenerate. MO calculations (25) predict a singlet ground state with no excited states lying at energies which would be significant in the statistical calculations.

Our thermal functions essentially extend and update those reported by O'Hare (11) but are in disagreement with other literature data (26, 27). The functions reported by Radhakrishnan (26) are based on an incorrect structure and older vibrational frequencies (21). The set of frequencies used by Wilkins (27) are very similar to ours but his moments of inertia appear unrealistic. We conclude that his thermal functions (27) are erroneous.

References

1. P. A. O'Hare, W. N. Hubbard, O. Glemser, and J. Wegener, J. Chem. Thermodynamics 2, 71 (1970).
2. JANAF Thermochemical Tables: HF(g), 12-31-68; F(g), 9-30-65; SF_6(g), 6-30-76; F^-(g), 12-31-71; N(g), 3-31-61.
3. J. D. Vaugh and E. L. Muetterties, J. Phys. Chem. 64, 1787 (1960).
4. M. J. Nichols, Ph.D. Thesis, University of Durham, England, 1958.
5. D. L. Hildenbrand, J. Phys. Chem. 77, 897 (1973).
6. V. H. Dibeler and F. L. Mohler, J. Res. Nat. Bur. Stand. 40, 25 (1948).
7. J. Delwiche, Dynamic Mass Spectrom. 1, 71 (1970).
8. O. Glemser, A. Muller, D. Bohler, and B. Krebs, Z. Anorg. Allgem. Chem. 357, 184 (1968).
9. J. Kay and F. M. Page, Trans. Faraday Soc. 60, 1042 (1964).
10. C. E. Vanderzee and W. Wm. Rodenburg, J. Chem. Thermodynamics 2, 461 (1970).
11. P. A. G. O'Hare, Argonne National Laboratory, ANL-7315, July, 1968.
12. U. S. Nat. Bur. Stand. Tech. Note 270-3, 1968.
13. V. B. Parker, D. D. Wagman, and D. Garvin, NBSIR 75-968, January, 1976.
14. K. Kimura and S. H. Bauer, J. Chem. Phys. 39, 3172 (1963).
15. W. M. Tolles and W. D. Gwinn, J. Chem. Phys. 36, 1119 (1962).
16. F. A. Cotton, J. W. George, and J. S. Waugh, J. Chem. Phys. 28, 944 (1958).
17. J. Bacon, R. J. Gillespie, and J. W. Quail, Can. J. Chem. 41, 1016 (1963).
18. R. W. Shaw, Jr., T. X. Carroll, and T. D. Thomas, J. Amer. Chem. Soc. 95, 2033 (1973).
19. R. W. Shaw, Jr., T. X. Carroll, and T. D. Thomas, J. Amer. Chem. Soc. 95, 5870 (1973).
20. I. W. Levin, J. Chem. Phys. 55, 5393 (1971).
21. R. E. Dodd, L. A. Woodward, and H. L. Roberts, Trans. Faraday Soc. 52, 1052 (1956).
22. K. O. Christe and W. Sawodny, J. Chem. Phys. 52, 6320 (1970).
23. I. W. Levin and C. V. Berney, J. Chem. Phys. 44, 2557 (1966).
24. R. L. Redington and C. V. Berney, J. Chem. Phys. 43, 2020 (1965).
25. R. D. Brown and J. B. Peel, Aust. J. Chem. 21, 2617 (1968).
26. M. Radhakrishnan, Z. Naturforsch. 18a, 103 (1963).
27. R. L. Wilkins, Aerospace Corporation, Report No. TR-0158 (3240-20)-19, July, 1968.

Sulfur Fluoride (SF_4) $F_4S_1(g)$

Enthalpy Reference Temperature = T_r = 298.15 K Standard State Pressure = p° = 0.1 MPa

T/K	$C_p°$ (J $K^{-1}mol^{-1}$)	S° (J $K^{-1}mol^{-1}$)	$-[G°-H°(T_r)]/T$ (J $K^{-1}mol^{-1}$)	$H°-H°(T_r)$ (kJ mol^{-1})	$\Delta_f H°$ (kJ mol^{-1})	$\Delta_f G°$ (kJ mol^{-1})	Log K_f
0	0.	0.	INFINITE	-15.782	-756.882	-756.882	INFINITE
100	42.613	235.853	357.961	-12.211	-759.814	-748.073	390.754
200	62.633	271.650	306.302	-6.931	-762.034	-735.376	192.060
250	71.003	286.554	300.887	-3.583	-762.711	-728.628	152.238
298.15	77.598	299.644	299.644	0.	-763.162	-722.020	126.495
300	77.824	300.125	299.646	0.144	-763.176	-721.764	125.670
350	83.238	312.545	300.614	4.176	-763.488	-714.836	106.683
400	87.497	323.949	302.828	8.448	-765.907	-707.783	92.427
450	90.849	334.456	305.766	12.910	-766.704	-700.478	81.309
500	93.504	344.171	309.127	17.522	-767.493	-693.074	72.405
600	97.341	361.583	316.453	27.078	-768.469	-678.089	59.033
700	99.891	376.792	324.010	36.947	-769.049	-662.976	49.472
800	101.653	390.252	331.465	47.030	-769.397	-647.797	42.297
900	102.914	402.302	338.678	57.261	-822.730	-631.503	36.651
1000	103.844	413.196	345.594	67.602	-821.626	-610.315	31.880
1100	104.547	423.128	352.198	78.023	-820.516	-589.237	27.981
1200	105.091	432.249	358.494	88.506	-819.412	-568.260	24.736
1300	105.519	440.678	364.496	99.037	-818.322	-547.376	21.994
1400	105.863	448.511	370.220	109.607	-817.251	-526.574	19.647
1500	106.143	455.824	375.686	120.207	-816.205	-505.848	17.615
1600	106.373	462.682	380.911	130.834	-815.184	-485.191	15.840
1700	106.565	469.137	385.913	141.481	-814.191	-464.597	14.275
1800	106.726	475.233	390.708	152.146	-813.222	-444.060	12.886
1900	106.864	481.007	395.310	162.825	-812.275	-423.578	11.645
2000	106.981	486.491	399.733	173.518	-811.347	-403.144	10.529
2100	107.083	491.714	403.989	184.221	-810.430	-382.757	9.521
2200	107.171	496.697	408.091	194.934	-809.519	-362.412	8.605
2300	107.248	501.463	412.048	205.655	-808.606	-342.109	7.770
2400	107.316	506.029	415.869	216.383	-807.685	-321.847	7.005
2500	107.376	510.411	419.564	227.118	-806.748	-301.623	6.302
2600	107.429	514.623	423.139	237.858	-805.788	-281.437	5.654
2700	107.477	518.679	426.603	248.603	-804.798	-261.288	5.055
2800	107.519	522.588	429.962	259.353	-803.771	-241.177	4.499
2900	107.557	526.362	433.221	270.107	-802.702	-221.103	3.982
3000	107.592	530.009	436.387	280.865	-801.586	-201.067	3.501
3100	107.623	533.537	439.464	291.625	-800.417	-181.069	3.051
3200	107.652	536.954	442.458	302.389	-799.193	-161.109	2.630
3300	107.677	540.267	445.372	313.156	-797.910	-141.188	2.235
3400	107.701	543.482	448.210	323.925	-796.566	-121.308	1.864
3500	107.723	546.605	450.977	334.696	-795.159	-101.467	1.514
3600	107.743	549.640	453.676	345.469	-793.687	-81.668	1.185
3700	107.761	552.592	456.310	356.244	-792.150	-61.911	0.874
3800	107.778	555.466	458.881	367.021	-790.547	-42.196	0.580
3900	107.793	558.266	461.394	377.800	-788.878	-22.524	0.302
4000	107.808	560.995	463.850	388.580	-787.144	-2.896	0.038
4100	107.821	563.657	466.252	399.361	-785.346	16.688	-0.213
4200	107.834	566.256	468.602	410.144	-783.484	36.228	-0.451
4300	107.845	568.793	470.903	420.928	-781.559	55.721	-0.677
4400	107.856	571.273	473.156	431.713	-779.573	75.170	-0.892
4500	107.866	573.696	475.363	442.499	-777.527	94.574	-1.098
4600	107.876	576.067	477.527	453.286	-775.423	113.930	-1.294
4700	107.885	578.387	479.648	464.074	-773.262	133.241	-1.481
4800	107.893	580.659	481.729	474.863	-771.046	152.505	-1.660
4900	107.901	582.884	483.771	485.653	-768.777	171.722	-1.831
5000	107.908	585.064	485.775	496.443	-766.457	190.893	-1.994
5100	107.915	587.201	487.743	507.235	-764.086	210.016	-2.151
5200	107.922	589.296	489.676	518.026	-761.668	229.093	-2.301
5300	107.928	591.352	491.575	528.819	-759.204	248.123	-2.445
5400	107.934	593.369	493.441	539.612	-756.695	267.105	-2.584
5500	107.939	595.350	495.276	550.406	-754.143	286.041	-2.717
5600	107.945	597.295	497.081	561.200	-751.549	304.930	-2.844
5700	107.949	599.205	498.856	571.995	-748.916	323.772	-2.967
5800	107.954	601.083	500.602	582.790	-746.245	342.568	-3.085
5900	107.959	602.928	502.321	593.585	-743.537	361.318	-3.199
6000	107.963	604.743	504.013	604.382	-740.794	380.021	-3.308

PREVIOUS: June 1976 (1 atm) CURRENT: June 1976 (1 bar)

Sulfur Fluoride (SF_4) $F_4S_1(g)$

SULFUR FLUORIDE, ION (SF_4^+) IDEAL GAS M_r=108.053063

$\Delta_f H°(0\ K) = 415.4 \pm 50\ kJ\ mol^{-1}$

$S°(298.15\ K) = [311.7 \pm 8]\ J\ K^{-1}mol^{-1}$ $\Delta_f H°(298.15\ K) = 416.0 \pm 50\ kJ\ mol^{-1}$

Electronic Levels and Quantum Weights

ε_i, cm^{-1}	g_i
0	[2]
26213	[2]

Vibrational Frequencies and Degeneracies

ν, cm^{-1}	ν, cm^{-1}	ν, cm^{-1}
[850](1)	[200](1)	[500](1)
[500](1)	[450](1)	[800](1)
[300](1)	[700](1)	[200](1)

Point Group: $[C_{2v}]$ σ = [2]

Bond Distances: S-F* = [1.62] Å S-F** = [1.73] Å

Bond Angles: F*-S-F* = [109]° F*-S-F** = [88.3]° F**-S-F** = [174]°

(* - equatorial ** - axial)

Product of Moments of Inertia: $I_A I_B I_C = [9.391199 \times 10^{-114}]\ g^3\ cm^6$

Enthalpy of Formation

The electron-impact AP of SF_4^* from SF_4 has been reported as 12.28±0.03 eV (1), 12.08±0.1 eV (2), and 12.1±0.5 eV (2). Using the mean value (12.15 eV) of these three measured AP as $\Delta_r H°(0\ K)$ for the direct ionization process $SF_4(g) + e^- \rightarrow SF_4^+ + 2\ e^-$, we obtain the adopted value $\Delta_f H°(SF_4^+, g, 0\ K) = 99.3\ kcal\ mol^{-1}$, by combining $\Delta_r H°(0\ K)$ with $\Delta_f H°(SF_4, g, 0\ K) = -180.9 \pm 5.0\ kcal\ mol^{-1}$(3). We estimate that the adiabatic AP (or IP) could be as much as 0.3 eV lower than the measured vertical values (1, 2), suggesting uncertainties in $\Delta_r H°(0\ K)$ and $\Delta_f H°(0\ K)$ of ±7.0 kcal mol^{-1} and ±12.0 kcal mol^{-1}, respectively. Our predictions are based on the fact that photoelectron studies (4) on the species ClF_3 show that the first adiabatic ionization potential is 0.23 eV lower than the vertical value and 0.35 eV lower than the results obtained by the electron-impact method. The similarity in the structures of SF_4 and ClF_3 has been noted (5), and we predict that the geometry changes produced by ionization should also be very similar for the two fluorides.

Additional information is available which tends to provide some support for our adopted value of $IP(SF_4)$ = 12.15±0.3 eV. Fehsenfeld (6) has observed a charge-transfer reaction between SF_4^+ and NO which sets the lower limit of IP at 9.25 eV. Binding energies (7) for SF_4 and SF_4^+, computed by CNDO/2 calculations and rescaled according to energy partitioning concepts, have been combined to give IP values of 11.1 eV and 10.6 eV. Three independent measurements (2, 8, and 9) of the fragment-ion AP for SF_4^+ from SF_6 are in agreement with the result AP = 18.61 eV. We combine this $AP(SF_4^+/SF_6)$ with our adopted IP value and obtain $\Delta_r H°(0\ K) = 149.0\ kcal\ mol^{-1}$ for the process $SF_6(g) = SF_4(g) + 2F(g)$. This leads to a $\Delta_f H°$ of $-176.2\ kcal\ mol^{-1}$ for SF_4 which compares favorably with JANAF data (3).

$\Delta_f H°$ is calculated to be 99.4 kcal mol^{-1} at 298.15 K. Also, we find that the primary bond dissociation energy $D_0^0(SF_3^+ -F)$ is 12.9 kcal mol^{-1} which when compared with D_0^0 values (3) for SF_4 (80.0 kcal mol^{-1}) and SF_4^- (46.6 kcal mol^{-1}) indicates that the trend is $D_0^0(SF_4) >> D_0^0(SF_4^-) >> D_0^0(SF_4^+)$. The observed instability of the radical cation towards dissociation to SF_3^+ and F is consistent with ion abundances (9, 10) obtained for SF_4^+ and SF_3^+ in the mass spectra of SF_5Cl and SF_6. Values of the ratio $I(SF_3^+)/I(SF_4^+)$ lie in the range 3.1-7.0. (See the $SF_3^+(g)$ table for a discussion on its stability.) The enthalpy of atomization and mean bond dissociation energy are 278.8 kcal mol^{-1} and 69.7 kcal mol^{-1}, respectively.

Heat Capacity and Entropy

Delwiche (8) has reported an ionization efficiency curve for SF_4^+ which was obtained from retarded potential difference (RPD) measurements on SF_6. This curve shows the existence of two excited states, lying 26213 cm^{-1} and 45167 cm^{-1} above the ground state. The level at 45167 cm^{-1} is not included here since it has a negligible effect on the thermal functions. We assume that the one unpaired electron forms two doublet states which are non-degenerate.

The ESR spectrum (11) of SF_4^+, recorded during the radiolysis of SF_6 with 2.8 MeV electrons, suggests that the four fluorine nuclei are equivalent. However, the assignment of this spectrum to SF_4^+ has been questioned by Morton and Preston (12) who presented evidence which indicates that the radical responsible is more likely be be SF_5. Their re-assignment (12) is substantiated by the results of SCF MO calculations (13) on SF_5. By analogy with the isoelectronic radical PF_4 (14, 15), we predict that the SF_4^+ structure is trigonal bipyramid with the unpaired electron occupying one of the equatorial positions. We use bond angles from PF_4 (14) and assume that the loss of an electron from SF_4 increases both the equatorial and axial bond lengths by 5%. Our view that the bonding in SF_4^+ is weaker than that in SF_4 follows the predictions (14) made for PF_4 and is further confirmed by the fact that the mean bond dissociation energy in SF_4 (80.0 kcal mol^{-1}) is roughly 10 kcal mol^{-1} greater than in SF_4^+ (69.7 kcal mol^{-1}). The individual moments of inertia are: $I_A = 14.2509 \times 10^{-39}$, $I_B = 22.1083 \times 10^{-39}$, and $I_C = 29.8073 \times 10^{-39}\ g\ cm^2$.

We estimate vibrational frequencies from those for SF_4 (3) by considering the effect of the structural changes brought about by ionization. On going from SF_4 to SF_4^+, we expect a decrease in all frequencies to reflect the weaker bonding in the cation.

References

1. O. Glemser, A. Muller, D. Bohler, and B. Krebs, Z. Anorg. Allgem. Chem. 357, 184 (1968).
2. D. L. Hildenbrand, J. Phys. Chem. 77, 897 (1973).
3. JANAF Thermochemical Tables: SF_4(g), 6-30-76; SF_4^-(g) and SF_3^+(g), 12-31-76.
4. R. L. Dekock, B. R. Higginson, D. R. Lloyd, A. Breeze, D. W. J. Cruickshank, and D. R. Armstrong, Mol. Phys. 24, 1059 (1972).
5. R. A. Frey, R. L. Redington, and A. L. K. Aljibury, J. Chem. Phys. 54, 344 (1971).
6. F. C. Fehsenfeld, J. Chem. Phys. 54, 438 (1971).
7. A. L. Companion, Theoret. Chim. Acta (Berl.) 25, 268 (1972).
8. J. Delwiche, dynamic Mass Spectron. 1, 71 (1970).
9. V. H. Dibeler and F. L. Molhler, J. Res. Nat. Bur. Stand. 40, 25 (1948).
10. P. Harland and J. C. J. Thynne, J. Phys. Chem. 73, 4031 (1969).
11. R. W. Fessenden and R. H. Schuler, J. Chem. Phys. 45, 1845 (1966).
12. J. R. Morton and K. F. Preston, Chem. Phys. Letters, 18, 98 (1973).
13. A. R. Gregory, Chem. Phys. Letters, 28, 552 (1974).
14. J. Higuchi, J. Chem. Phys. 50, 1001 (1969).
15. A. J. Colussi, J. R. Morton, and K. F. Preston, J. Phys. Chem 79, 651 (1975).

Sulfur Fluoride, Ion (SF_4^+) $F_4S_1^+$(g)

Enthalpy Reference Temperature = T_r = 298.15 K Standard State Pressure = p° = 0.1 MPa

T/K	$C_p°$ (J $K^{-1}mol^{-1}$)	S° (J $K^{-1}mol^{-1}$)	$-[G°-H°(T_r)]/T$ (J $K^{-1}mol^{-1}$)	$H°-H°(T_r)$ (kJ mol^{-1})	$\Delta_f H°$ (kJ mol^{-1})	$\Delta_f G°$ (kJ mol^{-1})	Log K_f
0	0.	0.	INFINITE	-16.505	415.404		
100	45.321	244.482	372.760	-12.828			
200	65.909	282.444	318.604	-7.232			
250	74.087	298.061	312.961	-3.725			
298.15	80.395	311.671	311.671	0.	416.044	447.346	-78.373
300	80.609	312.169	311.672	0.149	416.074	447.540	-77.924
350	85.696	324.993	312.674	4.312	416.932	452.718	-67.564
400	89.643	336.705	314.956	8.700	415.668	457.854	-59.790
450	92.718	347.449	317.978	13.262	416.010	463.099	-53.755
500	95.133	357.347	321.427	17.960	416.348	468.315	-48.925
600	98.591	375.021	328.922	27.659	417.593	478.601	-41.666
700	100.869	390.401	336.630	37.640	419.202	488.645	-36.463
800	102.435	403.979	344.217	47.809	421.021	498.443	-32.545
900	103.550	416.112	351.543	58.112	369.836	509.086	-29.547
1000	104.370	427.066	358.557	68.510	373.077	524.384	-27.391
1100	104.989	437.044	365.245	78.979	376.314	539.358	-25.612
1200	105.467	446.201	371.615	89.503	379.537	554.038	-24.117
1300	105.843	454.658	377.682	100.069	382.741	568.450	-22.841
1400	106.144	462.513	383.464	110.669	385.920	582.616	-21.738
1500	106.389	469.845	388.981	121.296	389.071	596.555	-20.774
1600	106.591	476.718	394.252	131.945	392.193	610.286	-19.924
1700	106.758	483.185	399.295	142.613	395.286	623.822	-19.168
1800	106.900	489.291	404.127	153.296	398.352	637.177	-18.490
1900	107.020	495.075	408.763	163.992	401.394	650.361	-17.880
2000	107.123	500.567	413.217	174.699	404.416	663.387	-17.326
2100	107.211	505.795	417.502	185.416	407.425	676.261	-16.821
2200	107.288	510.785	421.630	196.141	410.427	688.992	-16.359
2300	107.356	515.555	425.610	206.874	413.429	701.588	-15.934
2400	107.415	520.126	429.454	217.612	416.439	714.052	-15.541
2500	107.468	524.512	433.169	228.356	419.464	726.390	-15.177
2600	107.514	528.727	436.764	239.105	422.511	738.607	-14.839
2700	107.556	532.786	440.246	249.859	425.589	750.706	-14.523
2800	107.594	536.698	443.621	260.617	428.702	762.691	-14.228
2900	107.629	540.474	446.896	271.378	431.857	774.563	-13.951
3000	107.660	544.124	450.076	282.142	435.059	786.326	-13.691
3100	107.689	547.654	453.167	292.910	438.312	797.982	-13.446
3200	107.716	551.074	456.174	303.680	441.621	809.531	-13.214
3300	107.742	554.389	459.100	314.453	444.989	820.977	-12.995
3400	107.766	557.606	461.950	325.228	448.419	832.319	-12.787
3500	107.790	560.730	464.728	336.006	451.911	843.559	-12.589
3600	107.813	563.767	467.437	346.786	455.469	854.700	-12.401
3700	107.835	566.721	470.081	357.569	459.092	865.739	-12.222
3800	107.858	569.597	472.662	368.353	462.781	876.680	-12.051
3900	107.880	572.399	475.183	379.140	466.536	887.524	-11.887
4000	107.903	575.131	477.648	389.929	470.358	898.270	-11.730
4100	107.927	577.795	480.058	400.721	474.245	908.919	-11.580
4200	107.951	580.396	482.417	411.515	478.197	919.474	-11.435
4300	107.976	582.937	484.725	422.311	482.213	929.933	-11.296
4400	108.002	585.419	486.985	433.110	486.291	940.298	-11.163
4500	108.029	587.847	489.200	443.911	490.431	950.570	-11.034
4600	108.057	590.221	491.370	454.716	494.631	960.749	-10.910
4700	108.086	592.546	493.498	465.523	498.889	970.836	-10.790
4800	108.117	594.821	495.585	476.333	503.205	980.832	-10.674
4900	108.148	597.051	497.633	487.146	507.576	990.738	-10.561
5000	108.182	599.236	499.644	497.963	512.001	1000.553	-10.453
5100	108.216	601.379	501.618	508.783	516.479	1010.280	-10.347
5200	108.252	603.481	503.556	519.606	521.007	1019.919	-10.245
5300	108.290	605.543	505.461	530.433	525.585	1029.469	-10.146
5400	108.329	607.567	507.333	541.264	530.210	1038.933	-10.050
5500	108.369	609.556	509.174	552.099	534.882	1048.311	-9.956
5600	108.410	611.509	510.984	562.938	539.599	1057.604	-9.865
5700	108.453	613.428	512.764	573.781	544.359	1066.811	-9.776
5800	108.497	615.314	514.516	584.628	549.161	1075.935	-9.690
5900	108.542	617.169	516.241	595.480	554.004	1084.976	-9.606
6000	108.589	618.994	517.938	606.337	558.887	1093.934	-9.524

PREVIOUS: December 1976 (1 atm) CURRENT: December 1976 (1 bar)

Sulfur Fluoride, Ion (SF_4^+) $F_4S_1^+$(g)

SULFUR FLUORIDE, ION (SF_4^-) IDEAL GAS M_r = 108.054161

$\Delta_fH°(0\ K) = -877.685 \pm 33.5\ kJ\ mol^{-1}$

$S°(298.15\ K) = [312.962 \pm 8.4]\ J\ K^{-1}\ mol^{-1}$ $\Delta_fH°(298.15\ K) = -887.464 \pm 33.5\ kJ\ mol^{-1}$

Electronic Levels and Quantum Weights

State	ε_i, cm^{-1}	g_i
$[^2A_1]$	0	[2]
$[^2B_1]$	[15000]	[2]

Vibrational Frequencies and Degeneracies

ν, cm^{-1}	ν, cm^{-1}	ν, cm^{-1}
[500](1)	[375](1)	[150](2)
[430](1)	[325](1)	
[300](1)	[595](2)	

Point Group: $[D_{4h}]$ σ = [8]

Bond Distance: S-F = [1.63 ± 0.1] Å

Bond Angle: F-S-F = [90]°

Product of the Moments of Inertia: $I_AI_BI_C = [9.4210 \times 10^{-114}]\ g^3\ cm^6$

Enthalpy of Formation

We adopt $\Delta_fH°(SF_4^-, g, 0\ K) = -209.8\pm8.0\ kcal\ mol^{-1}$ and $EA(SF_4) = 1.25\pm0.56$ eV. The value of $\Delta_fH°(0\ K)$ is obtained from the electron-impact appearance potential for SF_4^- from SF_6, reported by Harland and Thynne (1) as AP = 5.0±0.1 eV. This value has been verified in an independent study (9). We assume that the negative ion is formed by the process $SF_6(g) + e^- = SF_4^-(g) + 2\ F(g)$ with no excess energies. We estimate the error in the electron-impact experiments to be ±0.3 eV (7.0 kcal mol^{-1}) which when combined with the uncertainties in the auxiliary data (2) gives an overall uncertainty in $\Delta_fH°(0\ K)$ of ±8.0 kcal mol^{-1}. Our adopted value of EA = 1.25 eV is supported by EA = 1.5 eV (3) and EA = 1.26 eV (10) which were predicted by theoretical calculations.

In the same paper, Harland and Thynne (1) also reported the AP(SF_4^-/SF_5Cl) as 4.1±0.1 eV. This leads to $\Delta_fH°(SF_4^-, g, 0\ K) = -199.9\pm5.0\ kcal\ mol^{-1}$ and $EA(SF_4) = 0.82\pm0.4$ eV. They (1) also reported appearance potentials for SF_3^- and SF_5^- from SF_5Cl which give EA values that are several tenths of an electron volt below JANAF data (2). Thus, we are led to conclude that dissociative ionization of SF_5Cl probably involves excess energies and that the derived electron affinities are lower limit values.

Our adopted results give $\Delta_fH°(298.15\ K) = -212.1\ kcal\ mol^{-1}$ and a fluoride-ion affinity (IA) of $IA(SF_3) = 30.5\ kcal\ mol^{-1}$. Also, we find that the primary bond dissociation energy $D_0^0(SF_3^- - F)$ is 46.6 kcal mol^{-1} which when compared with D_0^0 values (2) for SF_4 (80.0 kcal mol^{-1}) and SF_4^+ (12.9 kcal mol^{-1}) indicates that the trend is $D_0^0(SF_4) >> D_0^0(SF_4^-) >> D_0^0(SF_4^+)$. These thermochemical predictions of the SF_4^- stability are consistent with an autodetachment lifetime of 16.3 μsec, as measured by Harland and Thynne (4) using time-of-flight techniques. The heat of atomization ($\Delta_{at}H°$) for the products $S(g) + F^-(g) + 3\ F(g)$ and mean bond dissociation energy are 270.5 kcal mol^{-1} and 67.7 kcal mol^{-1}, respectively.

Heat Capacity and Entropy

We assume a square-planar configuration (D_{4h} symmetry) similar to that found for the isoelectronic radical ClF_4 (5-7). We estimate the bond length to be slightly greater (0.03 Å) than the mean (1.60 Å) of the equatorial and axial S-F bonds in SF_4 (2). An upper limit of 1.70 Å can be established from results of MO calculations on ClF_4 (7). The individual moments of inertia are: $I_A = 3.3526 \times 10^{-38}$ and $I_B = I_C = 1.6763 \times 10^{-38}\ g\ cm^2$.

The square-planar structure follows the vibrational representation $A_1 + A_2 + 2B_1 + B_2 + 2E$. We estimate the seven fundamentals by comparison with the observed frequencies for the approximately square-planar SF_4 part of SF_5^- (2). Comparison of the experimentally measured frequencies for the related species ClF_5 (2) and ClF_4^- (8) suggests that this procedure should give reliable estimates.

The electronic states and levels are based on those for ClF_4 which were predicted by Ungemach and Schaefer (7) from MO calculations. An independent MO study of ClF_4 by Gregory (6) indicates that the upper level is an $^2A_2$ state. This, however, would not alter our functions since both configurations have the same degeneracy.

References

1. P. Harland and J. C. J. Thynne, J. Phys. Chem. 73, 4031 (1969).
2. JANAF Thermochemical Tables: SF_4(g) and SF_6(g), 6-30-76; SF_3^-(g), SF_5^-(g) and SF_4^+(g), 12-31-76; ClF_5(g), 6-30-66; F(g), 9-30-65.
3. A. L. Companion, Theoret. Chim. Acta (Berl.) 25, 268 (1972).
4. P. W. Harland and J. C. J. Thynne, J. Phys. Chem. 75, 3517 (1971).
5. J. R. Morton and K. F. Preston, J. Chem. Phys. 58, 3112 (1973).
6. A. R. Gregory, J. Chem. Phys. 60, 3713 (1974).
7. S. R. Ungemach and H. F. Schaefer, J. Amer. Chem. Soc. 98, 1958 (1976).
8. K. O. Christe and W. Sawodny, Z. Anorg. Allg. Chem. 374, 306 (1970).
9. D. L. Hildenbrand, Stanford Research Institute, personal communication, March, 1977.
10. R. J. Donovan, P. W. Harland, J. H. Knox, J. A. Makowski, and J. C. J. Thynne, Int. J. Mass. Spectrom. Ion Phys. 13, 464 (1974).

Sulfur Fluoride, Ion (SF_4^-) $F_4S_1^-(g)$

Enthalpy Reference Temperature = T_r = 298.15 K Standard State Pressure = p° = 0.1 MPa

T/K	$C_p°$	S°	$-[G°-H°(T_r)]/T$	$H°-H°(T_r)$	$\Delta_fH°$	$\Delta_fG°$	Log K_f
	J K^{-1}mol^{-1}			kJ mol^{-1}			
0	0.	0.	INFINITE	-18.480	-877.685		
100	50.869	236.570	382.190	-14.562			
200	75.553	279.949	320.747	-8.160			
250	83.644	297.728	314.402	-4.168			
298.15	89.187	312.962	312.962	0.	-887.464	-844.038	147.872
300	89.365	313.514	312.964	0.165	-887.495	-843.769	146.913
350	93.431	327.613	314.068	4.741	-888.303	-836.416	124.828
400	96.371	340.291	316.567	9.490	-891.285	-828.863	108.238
450	98.542	351.773	319.851	14.365	-892.708	-820.984	95.297
500	100.180	362.244	323.574	19.335	-894.178	-812.933	84.927
600	102.430	380.725	331.601	29.475	-896.648	-796.445	69.337
700	103.854	396.629	339.781	39.794	-898.858	-779.568	58.172
800	104.807	410.563	347.776	50.230	-900.931	-762.384	49.779
900	105.473	422.948	355.453	60.746	-956.058	-743.864	43.173
1000	105.957	434.087	362.769	71.318	-956.800	-720.248	37.622
1100	106.319	444.204	369.719	81.933	-957.575	-696.555	33.077
1200	106.596	453.467	376.318	92.579	-958.386	-672.790	29.286
1300	106.814	462.009	382.585	103.250	-959.235	-648.956	26.075
1400	106.987	469.931	388.545	113.941	-960.123	-625.056	23.321
1500	107.128	477.317	394.219	124.647	-961.050	-601.090	20.932
1600	107.244	484.235	399.631	135.366	-962.015	-577.062	18.839
1700	107.342	490.740	404.801	146.095	-963.018	-552.972	16.991
1800	107.425	496.877	409.748	156.833	-964.054	-528.821	15.346
1900	107.499	502.688	414.488	167.580	-965.119	-504.613	13.873
2000	107.564	508.203	419.037	178.333	-966.208	-480.348	12.545
2100	107.625	513.453	423.409	189.092	-967.314	-456.028	11.343
2200	107.682	518.461	427.616	199.858	-968.429	-431.655	10.249
2300	107.738	523.249	431.671	210.629	-969.545	-407.231	9.248
2400	107.793	527.835	435.583	221.405	-970.654	-382.758	8.331
2500	107.849	532.237	439.362	232.187	-971.749	-358.240	7.485
2600	107.907	536.468	443.016	242.975	-972.820	-333.679	6.704
2700	107.968	540.541	446.553	253.769	-973.860	-309.076	5.979
2800	108.031	544.469	449.980	264.569	-974.861	-284.436	5.306
2900	108.098	548.261	453.304	275.375	-975.818	-259.761	4.679
3000	108.168	551.927	456.531	286.189	-976.725	-235.054	4.093
3100	108.241	555.475	459.666	297.009	-977.576	-210.318	3.544
3200	108.319	558.913	462.714	307.837	-978.366	-185.555	3.029
3300	108.399	562.247	465.680	318.673	-979.092	-160.768	2.545
3400	108.483	565.484	468.568	329.517	-979.751	-135.960	2.089
3500	108.570	568.630	471.382	340.370	-980.341	-111.134	1.659
3600	108.660	571.690	474.126	351.231	-980.860	-86.292	1.252
3700	108.752	574.668	476.803	362.102	-981.306	-61.436	0.867
3800	108.846	577.570	479.417	372.981	-981.679	-36.570	0.503
3900	108.942	580.398	481.970	383.871	-981.978	-11.694	0.157
4000	109.039	583.158	484.465	394.770	-982.203	13.188	-0.172
4100	109.138	585.852	486.905	405.679	-982.356	38.074	-0.485
4200	109.236	588.483	489.293	416.597	-982.437	62.964	-0.783
4300	109.336	591.054	491.630	427.526	-982.446	87.854	-1.067
4400	109.435	593.569	493.918	438.465	-982.385	112.745	-1.338
4500	109.534	596.029	496.160	449.413	-982.255	137.633	-1.598
4600	109.632	598.438	498.357	460.371	-982.059	162.517	-1.845
4700	109.730	600.797	500.512	471.339	-981.796	187.397	-2.083
4800	109.826	603.108	502.625	482.317	-981.470	212.270	-2.310
4900	109.921	605.373	504.699	493.305	-981.082	237.135	-2.528
5000	110.014	607.595	506.735	504.301	-980.634	261.992	-2.737
5100	110.105	609.774	508.734	515.307	-980.128	286.840	-2.938
5200	110.194	611.913	510.698	526.322	-979.565	311.678	-3.131
5300	110.281	614.013	512.627	537.346	-978.948	336.503	-3.316
5400	110.366	616.075	514.524	548.378	-978.278	361.317	-3.495
5500	110.448	618.101	516.389	559.419	-977.558	386.117	-3.667
5600	110.528	620.092	518.223	570.468	-976.788	410.905	-3.833
5700	110.605	622.049	520.027	581.524	-975.972	435.677	-3.993
5800	110.679	623.973	521.803	592.589	-975.110	460.435	-4.147
5900	110.751	625.866	523.551	603.660	-974.205	485.179	-4.295
6000	110.820	627.728	525.271	614.739	-973.257	509.906	-4.439

PREVIOUS: December 1976 (1 atm) CURRENT: December 1976 (1 bar)

Sulfur Fluoride, Ion (SF_4^-) $F_4S_1^-(g)$

TETRAFLUOROSILANE (SiF_4) IDEAL GAS M_r = 104.079112

$S°(298.15\ K) = 282.760 \pm 0.42\ J\ K^{-1}\ mol^{-1}$

$\Delta_f H°(0\ K) = -1609.428 \pm 0.84\ kJ\ mol^{-1}$
$\Delta_f H°(298.15\ K) = -1614.940 \pm 0.84\ kJ\ mol^{-1}$

Vibrational Frequencies and Degeneracies

ν, cm^{-1}	ν, cm^{-1}
800.8(1)	1029.6(3)
264.2(2)	388.7(3)

Ground State Quantum Weight: [1] σ = 12
Point Group: T_d
Bond Distance: Si-F = 1.552 ± 0.002 Å
Bond Angle: F-Si-F = 109° 28'
Product of the Moments of Inertia: $I_A I_B I_C = 8.321027 \times 10^{-114}\ g^3\ cm^6$

Enthalpy of Formation

The adopted $\Delta_f H°(SiF_4$, g, 298.15 K) = -385.98±0.20 kcal mol^{-1} is the result of measurement of the direct combination of the elements in a bomb calorimeter by Wise et al. (1) and is the value recommended by CODATA (2). Bousquet et al. (3) determined $\Delta_f H°$(298.15 K) = -385±4 kcal mol^{-1}, also by fluorine combustion calorimetry, which is less precise but is in agreement with the adopted value. An average result, $\Delta_f H°$(298.15 K) = -372.4±0.4 kcal mol^{-1} by Vorob'ev et al. (4) from the reaction of SiF_4(g) and Na(cr) and from two rather involved enthalpy of solution schemes, is not used.

Using JANAF auxiliary values (5), $\Delta_{at}H°$(0 K) = 564.76±3 kcal mol^{-1} is calculated for the process $SiF_4(g) = Si(g) + 4\ F(g)$.

Heat Capacity and Entropy

The four fundamental vibrational frequencies are adopted from the vapor-phase Raman spectral investigation of Clark and Rippon (6). Jones et al. (7) photographed the Raman spectrum of the liquid and assigned values of 800, 268, 1010, and 390 cm^{-1} to the four fundamental vibrational frequencies. Yost et al. (8) determined only ν_1 at 800 cm^{-1}. Shimanouchi (9) references four articles including that of Jones et al. (7) and has selected 800, 268, 1032, and 389 cm^{-1}. All of these assignments are in good agreement.

The interatomic distances from electron diffraction studies were found to be 1.555±0.002 and 2.534±0.003 Å for Si-F and F...F, respectively, by Beagley et al. (10) and 1.552±0.002 and 2.534±0.003 Å by Hagen and Hedberg (11). The value of Si-F = 1.552±0.002 Å is adopted because it is in agreement with the F...F distance and the tetrahedral bond angle. The three principal moments of inertia are: $I_A = I_B = I_C = 20.2640 \times 10^{-39}\ g\ cm^2$.

Our calculation of S°(298.15 K) = 67.55±0.10 cal K^{-1} mol^{-1} agrees with the value recommended by CODATA (2).

References

1. S. S. Wise, J. L. Margrave, H. M. Feder, and W. N. Hubbard, J. Phys. Chem. 67, 815 (1963).
2. ICSU-CODATA Task Group on Key Values for Thermodynamics, CODATA Bulletin No. 17, 1976.
3. J. Bousquet, J. Carre, P. Claudy, J. Etienne, P. Provencal, J. Thourey, and P. Barberi, J. Chem. Phys. Physicochim. Biol. 69, 1065 (1972).
4. A. F. Vorob'ev, V. P. Kolesov, and S. M. Skuratov, Zh. Neorgan. Khim. 5, 1402 (1960).
5. JANAF Thermochemical Tables: F(g), 9-30-65; Si(g), 3-31-67.
6. R. J. H. Clark and D. M. Rippon, J. Mol. Spectrosc. 44, 479 (1972).
7. E. A. Jones, J. S. Kirby-Smith, P. J. H. Woltz, and A. H. Nielsen, J. Chem. Phys. 19, 242 (1951).
8. D. M. Yost, E. N. Lassettre, and S. T. Gross, J. Chem. Phys. 4, 325 (1936).
9. T. Shimanouchi, NSRDS-NBS 39, 1972.
10. B. Beagley, D. P. Brown, and J. M. Freeman, J. Mol. Struct. 18, 337 (1973).
11. K. Hagen and K. Hedberg, J. Chem. Phys. 59, 1549 (1973).

Tetrafluorosilane (SiF_4) F_4Si_1(g)

Enthalpy Reference Temperature = T_r = 298.15 K Standard State Pressure = p° = 0.1 MPa

T/K	$C_p°$ (J K^{-1}mol^{-1})	S° (J K^{-1}mol^{-1})	$-[G°-H°(T_r)]/T$ (J K^{-1}mol^{-1})	$H°-H°(T_r)$ (kJ mol^{-1})	$\Delta_f H°$ (kJ mol^{-1})	$\Delta_f G°$ (kJ mol^{-1})	Log K_f
0	0.	0.	INFINITE	-15.356	-1609.428	-1609.428	INFINITE
100	41.815	220.670	339.001	-11.833	-1611.986	-1599.596	835.543
200	60.878	255.943	289.111	-6.634	-1613.823	-1586.418	414.330
250	67.873	270.301	283.941	-3.410	-1614.454	-1579.491	330.017
298.15	73.621	282.760	282.760	0.	-1614.940	-1572.711	275.532
300	73.825	283.216	282.761	0.136	-1614.957	-1572.449	273.788
350	78.887	294.987	283.679	3.958	-1615.348	-1565.332	233.613
400	83.143	305.807	285.778	8.012	-1615.642	-1558.165	203.476
450	86.687	315.811	288.566	12.260	-1615.851	-1550.967	180.032
500	89.626	325.102	291.761	16.670	-1615.992	-1543.750	161.274
600	94.090	341.862	298.746	25.870	-1616.116	-1529.287	133.136
700	97.207	356.614	305.980	35.444	-1616.093	-1514.816	113.037
800	99.433	369.747	313.145	45.282	-1615.983	-1500.354	97.963
900	101.063	381.558	320.102	55.311	-1615.824	-1485.910	86.240
1000	102.285	392.272	326.791	65.481	-1615.645	-1471.484	76.862
1100	103.221	402.067	333.196	75.758	-1615.463	-1457.077	69.191
1200	103.952	411.081	339.316	86.118	-1615.294	-1442.686	62.798
1300	104.532	419.425	345.161	96.543	-1615.147	-1428.308	57.390
1400	105.000	427.190	350.746	107.021	-1615.034	-1413.941	52.755
1500	105.383	434.447	356.087	117.541	-1614.959	-1399.580	48.738
1600	105.699	441.259	361.199	128.095	-1614.928	-1385.222	45.223
1700	105.963	447.675	366.099	138.679	-1665.120	-1370.418	42.108
1800	106.186	453.738	370.801	149.286	-1664.952	-1353.088	39.266
1900	106.376	459.485	375.319	159.915	-1664.787	-1335.767	36.723
2000	106.540	464.945	379.665	170.561	-1664.622	-1318.454	34.434
2100	106.681	470.147	383.851	181.222	-1664.455	-1301.150	32.364
2200	106.803	475.113	387.887	191.896	-1664.279	-1283.854	30.483
2300	106.911	479.863	391.783	202.582	-1664.090	-1266.565	28.765
2400	107.005	484.415	395.549	213.278	-1663.880	-1249.286	27.190
2500	107.089	488.785	399.191	223.983	-1663.644	-1232.016	25.742
2600	107.163	492.986	402.719	234.696	-1663.375	-1214.756	24.405
2700	107.229	497.032	406.137	245.415	-1663.068	-1197.508	23.167
2800	107.289	500.933	409.454	256.141	-1662.717	-1180.271	22.018
2900	107.342	504.698	412.673	266.873	-1662.316	-1163.047	20.949
3000	107.391	508.338	415.802	277.610	-1661.862	-1145.839	19.951
3100	107.435	511.860	418.844	288.351	-1661.349	-1128.647	19.018
3200	107.474	515.272	421.804	299.096	-1660.775	-1111.472	18.143
3300	107.511	518.580	424.687	309.846	-1660.137	-1094.316	17.322
3400	107.544	521.790	427.496	320.598	-1659.432	-1077.180	16.549
3500	107.574	524.908	430.235	331.354	-1658.660	-1060.066	15.821
3600	107.602	527.938	432.907	342.113	-2041.969	-1032.514	14.981
3700	107.628	530.887	435.515	352.875	-2040.643	-1004.492	14.181
3800	107.652	533.758	438.063	363.639	-2039.247	-976.507	13.423
3900	107.674	536.554	440.553	374.405	-2037.783	-948.558	12.705
4000	107.694	539.280	442.987	385.173	-2036.248	-920.649	12.022
4100	107.713	541.940	445.368	395.944	-2034.645	-892.779	11.374
4200	107.730	544.536	447.699	406.716	-2032.973	-864.948	10.757
4300	107.747	547.071	449.980	417.490	-2031.234	-837.159	10.169
4400	107.762	549.548	452.215	428.265	-2029.427	-809.410	9.609
4500	107.776	551.970	454.405	439.042	-2027.554	-781.703	9.074
4600	107.789	554.339	456.552	449.820	-2025.617	-754.039	8.562
4700	107.802	556.657	458.657	460.600	-2023.617	-726.417	8.073
4800	107.814	558.927	460.723	471.381	-2021.554	-698.839	7.605
4900	107.825	561.150	462.750	482.162	-2019.432	-671.305	7.156
5000	107.835	563.329	464.739	492.945	-2017.250	-643.814	6.726
5100	107.845	565.464	466.694	503.729	-2015.012	-616.367	6.313
5200	107.854	567.558	468.613	514.514	-2012.718	-588.965	5.916
5300	107.863	569.613	470.500	525.300	-2010.371	-561.607	5.535
5400	107.871	571.629	472.354	536.087	-2007.971	-534.295	5.168
5500	107.879	573.608	474.177	546.874	-2005.521	-507.027	4.815
5600	107.886	575.552	475.970	557.663	-2003.022	-479.804	4.475
5700	107.893	577.462	477.734	568.452	-2000.477	-452.627	4.148
5800	107.900	579.338	479.469	579.241	-1997.886	-425.494	3.832
5900	107.906	581.183	481.178	590.031	-1995.253	-398.406	3.527
6000	107.912	582.997	482.860	600.822	-1992.577	-371.364	3.233

PREVIOUS: June 1976 (1 atm) CURRENT: June 1976 (1 bar)

Tetrafluorosilane (SiF_4) F_4Si_1(g)

TITANIUM FLUORIDE (TiF_4) CRYSTAL M_r = 123.873612

$\Delta_f H°(0\ K)$ = -1647.108 ± 3.8 kJ mol^{-1}
$\Delta_f H°(298.15\ K)$ = -1649.333 ± 3.8 kJ mol^{-1}

S°(298.15 K) = 133.955 J K^{-1} mol^{-1}

T_{fus} = 558.6 K

Enthalpy of Formation

The enthalpy of formation of TiF_4(cr) was measured using a fluorine bomb calorimeter by Greenberg et al. (1). Gross et al. (2) reported a value of -393.4 kcal mol^{-1} for $\Delta_f H°(298.15\ K)$. The latter measurements were not corrected for impurities in the titanium sample, thus the value reported by Greenberg et al. is adopted.

Heat Capacity and Entropy

The heat capacity of TiF_4(cr) has been measured over the temperature range 6 to 304 K by Euler and Westrum (3). They obtained the value of S°(298.15 K) based on S°(10 K) = 0.11 cal K^{-1} mol^{-1}. Heat capacities above 300 K are estimated from those of ZrF_4(cr) (4).

Sublimation Data

The sublimation point is calculated as the temperature at which the fugacity is one atmosphere for the process TiF_4(cr) = TiF_4(g). The heat of sublimation is obtained from the data of Hall et al. (5).

References

1. E. Greenberg, J. L. Settle, and W. N. Hubbard, J. Phys. Chem. 66, 1345 (1962).
2. P. Gross, C. Hayman, and D. L. Levi, XVIIth Interim. Cong. Pure Appl. Chem. Abstr. 1, 90 (1959).
3. R. D. Euler and E. F. Westrum, Jr., J. Phys. Chem. 65, 132 (1961).
4. R. A. McDonald, G. C. Sinke, and D. R. Stull, J. Chem. Eng. Data 7, 83 (1962).
5. E. H. Hall, J. M. Blocher, Jr., and I. E. Campbell, J. Electrochem. Soc. 105, 275 (1958).

Titanium Fluoride (TiF_4) F_4Ti_1(cr)

Enthalpy Reference Temperature = T_r = 298.15 K Standard State Pressure = p° = 0.1 MPa

T/K	$C_p°$ (J $K^{-1}mol^{-1}$)	S°	$-[G°-H°(T_r)]/T$	$H°-H°(T_r)$ (kJ mol^{-1})	$\Delta_f H°$	$\Delta_f G°$	Log K_f
0	0.	0.	INFINITE	-20.255	-1647.108	-1647.108	INFINITE
100	54.518	41.970	219.476	-17.751	-1650.978	-1620.275	846.345
200	92.383	92.772	143.789	-10.204	-1651.206	-1589.254	415.071
298.15	114.265	133.955	133.955	0.	-1649.333	-1559.177	273.161
300	114.600	134.663	133.957	0.212	-1649.284	-1558.618	271.380
400	126.650	169.380	138.609	12.309	-1646.239	-1528.835	199.645
500	134.557	198.535	147.757	25.389	-1642.630	-1499.893	156.693
600	140.080	223.578	158.357	39.133	-1638.659	-1471.714	128.124
700	144.139	245.489	169.272	53.352	-1634.431	-1444.221	107.769
800	147.235	264.946	180.038	67.926	-1630.000	-1417.348	92.543
900	149.662	282.432	190.460	82.775	-1625.451	-1391.039	80.734
1000	151.670	298.307	200.463	97.844	-1620.883	-1365.239	71.313
1100	153.302	312.842	210.028	113.095	-1616.386	-1339.893	63.626
1200	154.641	326.239	219.161	128.494	-1615.974	-1314.832	57.233
1300	155.812	338.665	227.882	144.018	-1610.969	-1289.939	51.830
1400	156.816	350.249	236.213	159.650	-1605.977	-1265.432	47.214
1500	157.695	361.099	244.181	175.377	-1601.022	-1241.281	43.225

PREVIOUS: June 1964 CURRENT: September 1967

Titanium Fluoride (TiF_4) F_4Ti_1(cr)

TITANIUM FLUORIDE (TiF_4) IDEAL GAS M_r = 123.873612

S°(298.15 K) = 314.928 ± 2.1 J K^{-1} mol^{-1}

Δ_fH°(0 K) = -1547.811 ± 4.2 kJ mol^{-1}
Δ_fH°(298.15 K) = -1551.427 ± 4.2 kJ mol^{-1}

Vibrational Frequencies and Degeneracies

ν, cm^{-1}	ν, cm^{-1}
[701](1)	[750](3)
[168](2)	[178](3)

Ground State Quantum Weight: [1] σ = [12]
Point Group: [T_d]
Bond Distance: Ti-F = [1.80] Å
Bond Angle: F-Ti-F = [109°28']
Product of the Moments of Inertia: $I_AI_BI_C$ = [2.025160 x 10^{-113}] g^3 cm^6

Enthalpy of Formation

The enthalpy of formation of TiF_4(g) is calculated from that of the crystal and the heat of sublimation, Δ_{sub}H°(298.15 K), obtained from vapor pressure measurements. Hall et al. (1) have reported extensive vapor pressure data for the reaction TiF_4(cr) = TiF_4(g). 2nd and 3rd law analyses of their data yielded the following results.

	Method	Data Points	T/K	Δ_rH°(298.15 K), kcal mol^{-1} 2nd law	3rd law	Drift cal K^{-1} mol^{-1}	Δ_fH°(298.15 K) kcal mol^{-1}
1.	Spoon Gauge	25**	441-552	23.5 ± 0.1	23.37	-0.3 ± 0.1	-370.8
2.	Ruff-Fischer	6	435-498	23.0 ± 0.4	23.40	+0.9 ± 0.9	-370.8
3.	Transpiration	3	426-512	25.8 ± 2.0	23.72	-4.5 ± 4.3	-370.5
4.	Rodebush	6	473-537	23.7 ± 0.6	23.39	-0.5 ± 1.2	-370.8
5.	1, 2, and 4	36**	435-552	23.4 ± 0.1	23.37	-0.1 ± 0.1	-370.8

** Four points rejected due to failure of a statistical test.
*** Eight points rejected due to failure of a statistical test.

Heat Capacity and Entropy

The interatomic distance is obtained from a correlation of the measured values for $TiCl_4$(g), $TiBr_4$(g), $TiCl_6^{-2}$, and TiF_6^{-2}. The principal moments of inertia are $I_A = I_B = I_C$ = 27.2575 x 10^{-39} g cm^2.

The vibrational frequencies are estimated by a valence force field treatment of estimated force constants. The force constants are estimated from a correlation with the constants of CF_4, SiF_4, CCl_4, $SiCl_4$, CBr_4, $SiBr_4$, $TiCl_4$, and $TiBr_4$. The resulting frequencies are adjusted so that calculations of Δ_{sub}H°(298.15 K) by both 2nd and 3rd law methods are in agreement.

References

1. E. H. Hall, J. M. Blocher, Jr., and I. E. Campbell, J. Electrochem. Soc. 105, 275 (1958).

Titanium Fluoride (TiF_4) F_4Ti_1(g)

Enthalpy Reference Temperature = T_r = 298.15 K; Standard State Pressure = p° = 0.1 MPa

T/K	$C_p°$ (J $K^{-1}mol^{-1}$)	S°	-[G°-H°(T_r)]/T	H°-H°(T_r) (kJ mol^{-1})	Δ_fH°	Δ_fG°	Log K_f
0	0.	0.	INFINITE	-18.863	-1547.811	-1547.811	INFINITE
100	58.633	236.952	383.035	-14.608	-1549.930	-1538.725	803.748
200	74.506	283.076	322.400	-7.865	-1550.962	-1527.071	398.830
250	80.375	300.348	316.305	-3.989	-1551.255	-1521.061	317.808
298.15	85.177	314.928	314.928	0.	-1551.427	-1515.229	265.462
300	85.344	315.455	314.930	0.158	-1551.432	-1515.004	263.786
350	89.374	328.925	315.984	4.529	-1551.525	-1508.925	225.195
400	92.577	341.077	318.374	9.081	-1551.561	-1502.836	196.250
450	95.109	352.133	321.520	13.776	-1551.558	-1496.745	173.738
500	97.118	362.262	325.095	18.584	-1551.530	-1490.656	155.728
600	100.018	380.244	332.826	28.451	-1551.435	-1478.490	128.714
700	101.942	395.816	340.737	38.555	-1551.322	-1466.341	109.420
800	103.269	409.520	348.496	48.819	-1551.202	-1454.209	94.950
900	104.217	421.741	355.968	59.196	-1551.124	-1442.090	83.697
1000	104.915	432.759	363.105	69.655	-1551.167	-1429.975	74.694
1100	105.442	442.785	369.900	80.174	-1551.403	-1417.846	67.328
1200	105.849	451.977	376.362	90.739	-1555.824	-1405.567	61.183
1300	106.170	460.463	382.509	101.340	-1555.741	-1393.049	55.973
1400	106.427	468.341	388.362	111.971	-1555.751	-1380.535	51.508
1500	106.636	475.691	393.941	122.624	-1555.869	-1368.016	47.639
1600	106.808	482.579	399.268	133.297	-1556.108	-1355.485	44.252
1700	106.952	489.058	404.361	143.985	-1556.483	-1342.935	41.263
1800	107.072	495.175	409.238	154.686	-1557.005	-1330.359	38.606
1900	107.175	500.967	413.915	165.399	-1557.684	-1317.751	36.227
2000	107.263	506.467	418.406	176.121	-1573.217	-1304.647	34.074
2100	107.338	511.702	422.725	186.851	-1574.985	-1291.176	32.116
2200	107.404	516.697	426.884	197.588	-1576.750	-1277.620	30.335
2300	107.462	521.472	430.894	208.332	-1578.507	-1263.983	28.706
2400	107.512	526.047	434.764	219.080	-1580.249	-1250.271	27.211
2500	107.557	530.437	438.503	229.834	-1581.968	-1236.486	25.835
2600	107.597	534.656	442.121	240.592	-1583.658	-1222.634	24.563
2700	107.632	538.718	445.624	251.353	-1585.314	-1208.716	23.384
2800	107.664	542.632	449.019	262.118	-1586.928	-1194.739	22.288
2900	107.692	546.411	452.313	272.886	-1588.495	-1180.704	21.267
3000	107.718	550.062	455.510	283.656	-1590.011	-1166.617	20.313
3100	107.741	553.595	458.618	294.429	-1591.470	-1152.479	19.419
3200	107.763	557.016	461.639	305.204	-1592.871	-1138.296	18.581
3300	107.782	560.332	464.580	315.982	-1594.209	-1124.069	17.793
3400	107.800	563.550	467.444	326.761	-1595.482	-1109.804	17.050
3500	107.816	566.675	470.235	337.542	-1596.689	-1095.501	16.349
3600	107.830	569.713	472.956	348.324	-1597.827	-1081.165	15.687
3700	107.844	572.667	475.611	359.108	-2007.995	-1059.010	14.951
3800	107.857	575.543	478.203	369.893	-2007.761	-1033.365	14.205
3900	107.868	578.345	480.735	380.679	-2007.512	-1007.726	13.497
4000	107.879	581.076	483.210	391.466	-2007.248	-982.094	12.825
4100	107.889	583.740	485.629	402.255	-2006.968	-956.469	12.186
4200	107.898	586.340	487.996	413.044	-2006.670	-930.850	11.577
4300	107.907	588.879	490.313	423.834	-2006.354	-905.240	10.996
4400	107.915	591.360	492.582	434.625	-2006.020	-879.636	10.443
4500	107.923	593.785	494.804	445.417	-2005.666	-854.040	9.913
4600	107.930	596.157	496.981	456.210	-2005.292	-828.452	9.407
4700	107.936	598.479	499.116	467.003	-2004.897	-802.873	8.923
4800	107.943	600.751	501.210	477.797	-2004.482	-777.302	8.459
4900	107.948	602.977	503.264	488.592	-2004.044	-751.741	8.014
5000	107.954	605.158	505.280	499.387	-2003.585	-726.188	7.586
5100	107.959	607.296	507.260	510.183	-2003.103	-700.645	7.176
5200	107.964	609.392	509.204	520.979	-2002.598	-675.111	6.782
5300	107.969	611.449	511.114	531.775	-2002.070	-649.587	6.402
5400	107.973	613.467	512.990	542.572	-2001.517	-624.074	6.037
5500	107.977	615.448	514.835	553.370	-2000.941	-598.572	5.685
5600	107.981	617.394	516.649	564.168	-2000.340	-573.079	5.345
5700	107.985	619.305	518.434	574.966	-1999.715	-547.598	5.018
5800	107.988	621.183	520.189	585.765	-1999.064	-522.128	4.702
5900	107.991	623.029	521.917	596.564	-1998.388	-496.669	4.397
6000	107.995	624.844	523.617	607.363	-1997.687	-471.223	4.102

PREVIOUS: September 1967 (1 atm) CURRENT: September 1967 (1 bar)

Titanium Fluoride (TiF_4) F_4Ti_1(g)

ZIRCONIUM FLUORIDE (ZrF_4) CRYSTAL M_r = 167.213612

Zirconium Fluoride (ZrF_4) F_4Zr_1(cr)

$S°(298.15\ K) = 104.700 \pm 0.21\ J\ K^{-1}\ mol^{-1}$

$T_{trs}\ (\alpha \rightarrow \beta) = 723\ K$

$T_{fus} = 1205 \pm 2\ K$

$\Delta_f H°(0\ K) = -1905.601 \pm 0.11\ kJ\ mol^{-1}$

$\Delta_f H°(298.15\ K) = -1911.251 \pm 0.11\ kJ\ mol^{-1}$

$\Delta_{trs} H°$ = Unknown

$\Delta_{fus} H° = 64.224 \pm 0.42\ kJ\ mol^{-1}$

Enthalpy of Formation

The enthalpy of formation of zirconium tetrafluoride was measured by direct combination of the elements in a bomb calorimeter by Greenberg et al. (1). The reported value, $\Delta_f H°(ZrF_4$, cr, 298.15 K) = -456.80 ± 0.25 kcal mol^{-1}, is adopted.

Heat Capacity and Entropy

The low temperature heat capacities in the temperature range 5 to 307 K were determined by adiabatic calorimetry by Westrum (2). Using these low temperature $C_p°$ data the value of S°(298.15 K) is derived as 25.024 ± 0.05 cal K^{-1} mol^{-1} based on S°(10 K) = 0.0074 cal K^{-1} mol^{-1}.

McDonald et al. (3) measured the high temperature enthalpies of ZrF_4(cr) at temperatures 283.9-1225.8 K in a copper block drop calorimeter. Smith et al. (4) used a Bunsen ice calorimeter for the enthalpy measurements in the temperature range 273-1150 K. These two sets of enthalpy data are not in good agreement. It is possible that the discrepancies are due to the difference in crystal structure of the samples used (see "Transition Data" for more information). In order to join smoothly with the low temperature heat capacities at 298.15 K, the high temperature heat capacities derived from the enthalpy data of McDonald et al. (3) are adopted. The $C_p°$ values above 1200 K are obtained by smooth extrapolation.

Fontana and Winand (8) also measured the enthalpy of ZrF_4 in the range 773-1350 K. Their reported average heat capacity in this range and their reported enthalpies are considerably lower than those adopted.

Transition Data

Gaudreau (5) presented evidence that ZrF_4 has three crystal structures (α, β, and τ) and one amorphous form. Chretien and Gaudreau (6) found that ZrF_4(cr) has an α (tetragonal) and β monoclinic) form with a transition temperature of T_{trs} = 723 K. The crystal data compilation of Donnay and Ondik (7) tabulated two monoclinic structures for ZrF_4; one of which was specified as the β form while no mention was made of a tetragonal form.

The ZrF_4 sample employed by McDonald et al. (3) for enthalpy measurement was prepared by dissolving hafnium-free zirconium metal in 48% aqueous HF, and the resulting solution was evaporated to dryness. The crystalline product was heated slowly to 773 K in a platinum boat in a slow current of anhydrous HF. X-ray diffraction showed only crysalline ZrF_4. Wet analysis indicated 54.6% Zr (theory 54.55) and 44.9% F (theory 45.45). Due to the above facts we are uncertain whether the sample prepared is a mixture of α and β forms of a pure ZrF_4(β). Smith et al. (4) obtained their ZrF_4 sample from the Oak Ridge National Laboratories, Oak Ridge, Tenn. Since the method of preparation of the compound is unavailable from the report, we do not know what kind of sample they used for measurement.

Because of the above unresolved situation, we emphasize that this ZrF_4(cr) table is not strictly a α, β-combined phase table. However, the differences in the calculated functions are probably not significant.

Fusion Data

T_{fus} and $\Delta_{fus}H°$ are taken from McDonald et al. (3). The values were obtained under conditions greater than one atmosphere. Fontana and Winand (8) reported a heat of melting of 11.81 ± 2.40 kcal mol^{-1} (under pressure).

Sublimation Data

The sublimation temperature (T_{sub} = 1179) is calculated as the temperature at which the fugacity is one atmosphere for the process $ZrF_4(cr) = ZrF_4(g)$. Since the sublimation temperature is lower than the melting point, the ZrF_4 sublimes before it melts.

References

1. E. Greenberg, J. L. Settle, H. M. Feder and W. N. Hubbard, J. Phys. Chem. 65, 1168 (1961).
2. E. F. Westrum, Jr., J. Chem. Eng. Data 10, 140 (1965).
3. R. A. McDonald, G. C. Sinke and D. R. Stull, J. Chem. Eng. Data 7, 83 (1962).
4. D. F. Smith, W. C. Miller and A. R. Taylor, Jr., U. S. Bur. Mines RI 5964, 1962.
5. B. Gaudreau, Rev. Chim. Minerale 2, 1 (1965); CA 63 17245d.
6. A. Chretien and B. Gaudreau, Compt. Rend. 246, 2266 (1958); Compt. Rend. 248, 2878 (1959).
7. J. D. H. Donnay and H. M. Ondik, "Crystal Data, Determinative Tables," 3rd edition, Vol. 2, Nat. Bur. Stand. and Joint Committee on Powder Diffraction Standards, 1973.
8. A. Fontana and R. Winand J. Nucl. Mat. 3, 295 (1972).

Enthalpy Reference Temperature = T_r = 298.15 K Standard State Pressure = p° = 0.1 MPa

	J K⁻¹mol⁻¹			kJ mol⁻¹			
T/K	$C_p°$	S°	$-[G°-H°(T_r)]/T$	$H°-H°(T_r)$	$\Delta_f H°$	$\Delta_f G°$	Log K_f
0	0.	0.	INFINITE	-17.497	-1905.601	-1905.601	INFINITE
100	42.091	22.932	182.385	-15.945	-1910.750	-1877.563	980.738
200	84.559	66.906	113.663	-9.351	-1912.207	-1843.517	481.477
298.15	103.596	104.700	104.700	0.	-1911.251	-1809.955	317.097
300	103.805	105.342	104.702	0.192	-1911.222	-1809.327	315.032
400	113.470	136.661	108.907	11.101	-1909.310	-1775.635	231.874
500	119.704	162.685	117.134	22.776	-1906.992	-1742.478	182.035
600	123.972	184.911	126.623	34.973	-1904.433	-1709.812	148.852
700	126.984	204.264	136.362	47.531	-1901.745	-1677.587	125.183
800	129.411	221.379	145.940	60.351	-1898.994	-1645.751	107.457
900	131.838	236.761	155.191	73.413	-1896.187	-1614.264	93.689
1000	134.097	250.772	164.059	86.713	-1893.323	-1583.092	82.692
1100	136.231	263.652	172.535	100.228	-1890.419	-1552.210	73.708
1200	138.281	275.592	180.632	113.953	-1891.188	-1521.371	66.224
1300	140.373	286.743	188.370	127.885	-1887.663	-1490.695	59.897
1400	142.507	297.225	195.774	142.031	-1884.009	-1460.296	54.484

PREVIOUS: June 1969 CURRENT: June 1975

Zirconium Fluoride (ZrF_4) F_4Zr_1(cr)

ZIRCONIUM FLUORIDE (ZrF_4) IDEAL GAS M_r = 167.213612

$S°(298.15 K) = [319.516 \pm 4.2] J K^{-1} mol^{-1}$

$\Delta_f H°(0 K) = -1669.395 \pm 3.3 kJ mol^{-1}$
$\Delta_f H°(298.15 K) = [1673.600 \pm 3.3] kJ mol^{-1}$

Vibrational Frequencies and Degeneracies

ν, cm^{-1}	ν, cm^{-1}
[630](1)	668(3)
[180](2)	190(3)

Ground State Quantum Weight: [1] σ = 12
Point Group: T_d
Bond Distance: Zr-F = 1.94 ± 0.02 Å
Bond Angle: F-Zr-F = 109° 28'
Products of the Moments of Inertia: $I_A I_B I_C = 3.174220 \times 10^{-113} g^3 cm^6$

Enthalpy of Formation

The vapor pressures of ZrF_4(cr) in the temperature range 617-1150 K were determined by many investigators using various methods. Based on the reported vapor pressure data, the corresponding heats of sublimation are evaluated by the 2nd and 3rd law methods. Using the 3rd law $\Delta_{sub}H$(298.15 K) and $\Delta_f H°(ZrF_4$, cr, 298.15 K) = -456.8 kcal mol^{-1}, the enthalpies of formation at 298.15 K for ZrF_4(g) are evaluated. The results are presented in the table below. The values of $\Delta_f H°(ZrF_4$, g, 298.15 K), listed in the last column of this table are in excellent agreement which indicate that the estimated missing vibrational frequencies ν_1 and ν_2 are reasonable. The enthalpy of formation at 298.15 K for ZrF_4(g) is adopted as -400.0 kcal mol^{-1}.

Source	Method	T/K	Points	$\Delta_{sub}H°$(298.15 K), kcal mol^{-1} 2nd law	3rd law	Drift J K^{-1} mol^{-1}	$\Delta_f H°$(298.15 K) kcal mol^{-1}
(1)	unavailable	950-1150	eqn	56.49±0.12	56.83	0.3	-399.97
(2)	transpiration	617-881	14	60.67±0.09	57.13	-3.4±0.1	-399.67
(3)	transpiration	900-1150	6	60.64±0.13	57.24	-3.3±0.1	-399.56
(4)	quasistatic	710-808	29	56.11±0.62	56.65	0.5±0.6	-400.15
(5)	torsion/effusion	748-849	15	55.81±1.44	57.05	1.5±1.8	-399.75
(6)	torsion/effusion	735-825	10	52.59±0.39	57.09	5.7±0.5	-399.71
(7)	Knudsen-effusion	720-825	5	51.53±0.05	54.79	4.1±0.1	-402.01
(8)	mass spectrometry	700-900	5	59.25±2.34	57.30	-2.4±2.9	-399.50
(9)	"bell method"	1000-1150	eqn	57.38±0.12	56.89	-0.5	-399.91
(10)	mass spectrometry	769	1	---	57.58	--	-399.22

Heat Capacity and Entropy

The molecular structure of ZrF_4(g) has been studied by electron diffraction by Spiridonov (11). The configuration was found to be a regular tetrahedron with interatomic distance Zr-F = 1.94±0.02 Å and F-Zr-F bond angle = 109.47° which are adopted.

Employing the estimated interaction coefficients and interatomic distance Zr-F = 1.85 Å, Godnev et al. (12) calculated three vibrational frequencies for ZrF_4(g) as ν_2 = 150-200, and ν_4 = 180±230 cm^{-1}. Bucher et al. (13) observed the infrared spectra of some group IV halides and assigned ν_3 = 668 and ν_4 = 190±20 cm^{-1} for ZrF_4(g). The frequency ν_3 has also been reported as 670 cm^{-1} (14). In studying matrix isolated ZrF_4, Hauge et al. (15) found a strong infrared band at 668.0 cm^{-1} in an argon matrix and 677.0 cm^{-1} in a neon matrix; both of which were attributed to the ν_3 stretch frequency. The values of ν_3 and ν_4 adopted here are those reported by Buchler et al. (13). The adopted ν_1 is obtained by correlating the corresponding frequencies of the tetrafluorides of C, Pb, Si, Ti and Ge with their respective interatomic distances. ν_2 is calculated from ν_1, ν_3 and ν_4 using the relationship (15) $\nu_3\nu_4/\nu_1\nu_2 = 2/3 + 4$ F/Zr where F and Zr are gram atomic weights of fluorine and zirconium, respectively. Vibrational frequencies (635, 178, 668, and 190 cm^{-1}) were suggested by Godnev et al. (17) who also critically examined the sublimation data in order to achieve internal consistency. These values are in good agreement with our adopted values. The three principal moments of inertia are $I_A = I_B = I_C = 31.6625 \times 10^{-39} g cm^2$.

References

1. S. Lauter, Dissertation Technische Hohschule Hannover, 1948. (Cited in reference 9).
2. K. A. Sense, M. Snyder and R. B. Filbert, Jr., J. Phys. Chem. 58, 995 (1954).
3. K. A. Sense, C. A. Alexander, R. E. Bowman and R. B. Filbert, Jr., J. Phys. Chem. 61, 337 (1957).
4. S. Cantor, R. F. Newton, W. R. Grimes and F. F. Blankenship, J. Phys. Chem. 62, 96 (1958).
5. D. L. Hildenbrand and L. P. Theard, Aeronutronic, U-1187, Ford Motor Company, 15 March 1961.
6. D. L. Hildenbrand and L. P. Theard, Aeronutronic, U-1274, Ford Motor Company, 15 June 1961.
7. N. P. Galkin, Y. N. Tumanov, V. I. Tarasov and Y. D. Shishhov, Russ. J. Inorg. Chem. 8, 1054 (1963).
8. P. A. Akishin, V. I. Belousov and L. N. Sidorov, Russ. J. Inorg. Chem. 8, 1520 (1963).
9. W. Fischer, T. Petzel and S. Lauter, Z. Anorg. Allg. Chem. 333, 226 (1964).
10. L. N. Sirorov, P. A. Akishim, V. B. Shil'ts and Y. M. Korenev, Russ. J. Phys. Chem. 39, 1146 (1965).
11. V. P. Spiridonov, Vestn. Mosk. Univ. Ser. II, Khim. 23, 113 (1968).
12. I. N. Godnev, A. M. Aleksandrovskaia and I. V. Rigina, Opt. Spectry, USSR 7, 172 (1959).
13. A. Buchler, J. B. Berkowitz-Mattuck and D. H. Dugre, J. Chem. Phys. 34, 2202 (1961).
14. M. Blander and H. W. Morgan, presented at the 138th Meeting, American Chemical Society, New York, September 1960.
15. R. H. Hauge, J. L. Margrave and J. W. Hastie, High Temp. Sci. 5, 89 (1973).
16. G. Herzberg, "Infrared and Raman Spectra," D. Van Nostrand Co., Inc., New York, p. 182, 1945.
17. I. N. Godnev, A. M. Aleksandrooskaya and A. S. Sverdlin, Opt. Spektrosk. 29, 676 (1970).

Zirconium Fluoride (ZrF_4) F_4Zr_1(g)

Enthalpy Reference Temperature = T_r = 298.15 K Standard State Pressure = p° = 0.1 MPa

T/K	$C_p°$	S°	$-[G°-H°(T_r)]/T$	$H°-H°(T_r)$	$\Delta_f H°$	$\Delta_f G°$	Log K_f
	J $K^{-1}mol^{-1}$			kJ mol^{-1}			
0	0.	0.	INFINITE	-18.942	-1669.395	-1669.395	INFINITE
100	57.127	240.945	388.852	-14.791	-1671.944	-1660.559	867.387
200	75.813	286.909	327.175	-8.053	-1673.258	-1648.568	430.562
250	82.385	304.559	320.928	-4.092	-1673.517	-1642.362	343.153
298.15	87.393	319.516	319.516	0.	-1673.600	-1636.352	286.682
300	87.561	320.057	319.518	0.162	-1673.601	-1636.120	284.874
350	91.536	333.867	320.599	4.644	-1673.567	-1629.875	243.246
400	94.573	346.297	323.048	9.300	-1673.461	-1623.640	212.076
450	96.905	357.577	326.267	14.089	-1673.310	-1617.421	187.745
500	98.716	367.885	329.921	18.982	-1673.135	-1611.220	168.323
600	101.275	386.127	337.808	28.991	-1672.764	-1598.872	139.194
700	102.939	401.872	345.861	39.207	-1672.417	-1586.585	118.392
800	104.071	415.696	353.744	49.561	-1672.132	-1574.344	102.794
900	104.872	428.002	361.324	60.011	-1671.939	-1562.133	90.664
1000	105.459	439.084	368.555	70.529	-1671.856	-1549.938	80.960
1100	105.900	449.157	375.432	81.097	-1671.899	-1537.745	73.021
1200	106.239	458.386	381.965	91.705	-1675.785	-1525.320	66.395
1300	106.506	466.901	388.176	102.343	-1675.554	-1512.791	60.785
1400	106.719	474.802	394.084	113.005	-1675.384	-1500.278	55.976
1500	106.893	482.171	399.714	123.685	-1675.288	-1487.775	51.809
1600	107.035	489.074	405.086	134.382	-1675.276	-1475.275	48.163
1700	107.153	495.567	410.219	145.092	-1675.357	-1462.772	44.946
1800	107.253	501.695	415.132	155.812	-1675.542	-1450.262	42.086
1900	107.338	507.496	419.842	166.542	-1675.837	-1437.740	39.526
2000	107.410	513.003	424.364	177.279	-1676.249	-1425.198	37.222
2100	107.472	518.245	428.710	188.023	-1676.782	-1412.633	35.137
2200	107.526	523.246	432.895	198.773	-1698.764	-1399.294	33.223
2300	107.574	528.027	436.928	209.528	-1699.970	-1385.654	31.469
2400	107.615	532.606	440.820	220.288	-1701.161	-1371.963	29.860
2500	107.652	537.000	444.580	231.051	-1702.330	-1358.222	28.379
2600	107.685	541.223	448.216	241.818	-1703.472	-1344.435	27.010
2700	107.714	545.288	451.736	252.588	-1704.579	-1330.605	25.742
2800	107.740	549.205	455.148	263.361	-1705.645	-1316.735	24.564
2900	107.763	552.987	458.457	274.136	-1706.666	-1302.827	23.466
3000	107.785	556.640	461.669	284.914	-1707.635	-1288.885	22.441
3100	107.804	560.175	464.790	295.693	-1708.548	-1274.911	21.482
3200	107.821	563.598	467.825	306.474	-1709.403	-1260.909	20.582
3300	107.837	566.916	470.777	317.257	-1710.196	-1246.881	19.736
3400	107.851	570.135	473.652	328.042	-1710.924	-1232.830	18.940
3500	107.865	573.262	476.454	338.827	-1711.586	-1218.759	18.189
3600	107.877	576.301	479.186	349.614	-1712.180	-1204.669	17.479
3700	107.888	579.257	481.850	360.403	-1712.705	-1190.564	16.808
3800	107.898	582.134	484.452	371.192	-1713.161	-1176.447	16.171
3900	107.908	584.937	486.992	381.982	-1713.547	-1162.317	15.567
4000	107.917	587.669	489.475	392.774	-1713.864	-1148.179	14.994
4100	107.925	590.334	491.903	403.566	-1714.113	-1134.034	14.448
4200	107.932	592.934	494.278	414.359	-1714.293	-1119.883	13.928
4300	107.940	595.474	496.602	425.152	-1714.407	-1105.729	13.432
4400	107.946	597.956	498.877	435.946	-1714.454	-1091.573	12.959
4500	107.952	600.382	501.106	446.741	-1714.436	-1077.416	12.506
4600	107.958	602.754	503.290	457.537	-1714.355	-1063.261	12.074
4700	107.964	605.076	505.431	468.333	-1714.211	-1049.108	11.660
4800	107.969	607.349	507.531	479.130	-2274.943	-1023.342	11.136
4900	107.974	609.576	509.591	489.927	-2274.331	-997.273	10.631
5000	107.978	611.757	511.612	500.724	-2273.679	-971.218	10.146
5100	107.982	613.895	513.597	511.522	-2272.986	-945.175	9.681
5200	107.986	615.992	515.546	522.321	-2272.252	-919.146	9.233
5300	107.990	618.049	517.461	533.120	-2271.479	-893.132	8.802
5400	107.994	620.068	519.342	543.919	-2270.665	-867.134	8.388
5500	107.997	622.049	521.191	554.718	-2269.812	-841.151	7.989
5600	108.000	623.995	523.010	565.518	-2268.920	-815.182	7.604
5700	108.003	625.907	524.798	576.318	-2267.988	-789.232	7.232
5800	108.006	627.785	526.558	587.119	-2267.018	-763.297	6.874
5900	108.009	629.632	528.289	597.920	-2266.009	-737.379	6.528
6000	108.011	631.447	529.994	608.721	-2264.961	-711.479	6.194

PREVIOUS: June 1975 (1 atm) CURRENT: June 1975 (1 bar)

Zirconium Fluoride (ZrF_4) F_4Zr_1(g)

HYDROGEN FLUORIDE (H_5F_5) IDEAL GAS M_r = 100.031715

$S°(298.15\ K) = [408.442 \pm 20.9]\ J\ K^{-1}\ mol^{-1}$

$\Delta_fH°(0\ K) = -1475.453 \pm 25.1\ kJ\ mol^{-1}$

$\Delta_fH°(298.15\ K) = [1489.713 \pm 25.1]\ kJ\ mol^{-1}$

Vibrational Frequencies and Degeneracies

ν, cm^{-1}
[202](5)
[53](4)
[962](5)
[552](5)
[3060](5)

Ground State Quantum Weight: 1 σ = [5]

Point Group: $[C_5]$

Bond Distance: F-F = [2.5] Å

Product of the Moments of Inertia: $I_AI_BI_C = [1.059909 \times 10^{-112}]\ g^3\ cm^6$

Enthalpy of Formation

The enthalpy of 5 HF(g) → H_5F_5(g) was taken as the mean of the 3rd law values which were determined from the equilibrium data of two investigations. Briegleb and Strohmeier (1) measured the vapor density of associated HF between 20 and 60 C and between 50 and 650 torr. Franck and Meyer (2) measured $C_p°$ between -20 and 100 C and between 100 and 700 torr. Each investigation evaluated K_p at n = 2, 3, 4, ... for the reactions n(HF) → $(HF)_n$ and reported 2nd law values of Δ_fH and Δ_fS. At n = 5 their 2nd law values differed by 5.2 kcal mol^{-1}, which was taken as an estimate of error, while the calculated 3rd law values differed by 3.1 kcal mol^{-1}. Using $\Delta_fH°(298.15\ K) = -30.35\ kcal\ mol^{-1}$ and auxiliary JANAF data (3) gives the enthalpy of formation of H_5F_5(g).

Heat Capacity and Entropy

The molecular structure of H_5F_5 was assumed as planar with the F atoms forming the vertices of a regular pentagon and with the H atoms also lying on the circumscribed circle. That rings are more stable than open chains was confirmed by Del Bene and Pople's (4) theoretical molecular-orbital studies on HF polymers. The length of side (F-F axis) was taken from Atoji and Lipscomb's (5) X-ray studies of solid HF (F-F - 2.49 Å) and agrees with the 2.52 Å which Janzen and Bartell (6) determined for HF gaseous polymers by electron diffraction. The low F-bending frequency (53 cm^{-1}) was taken from Boutin et al. (7). The other vibrational frequencies were taken from Kittelberger and Hornig's (8) work on crystalline HF. Huong and Couzi (9) and Smith (10) have made spectral studies of the gas phase in the range from 350 to 4000 cm^{-1}.

The National Bureau of Standards prepared this table (11) by critical analysis of data existing in 1972. Using molecular constants and $\Delta_fH°$ selected by NBS (11), we recalculate the table in terms of 1973 fundamental constants (12), 1975 atomic weights (13), and current JANAF reference states for the elements.

References

1. G. Briegleb and W. Strohmeier, Z. Elektrochem. 57, 668 (1953).
2. E. U. Franck and F. Meyer, Z. Elektrochem. 63, 571 (1959).
3. JANAF Thermochemical Tables: HF(g), 6-30-77.
4. J. E. Del Bene and J. A. Pople, J. Chem. Phys. 55, 2296 (1971).
5. M. Atoji and W. N. Lipscomb, Acta Cryst. 7, 173 (1954).
6. J. Janze and L. S. Bartell, J. Chem. Phys. 50, 3611 (1969).
7. H. Boutin, G. J. Safford and V. Brajovic, J. Chem. Phys. 39, 3135 (1963).
8. J. S. Kittelberger and D. F. Hornig, J. Chem. Phys. 46, 3099 (1967).
9. P. V. Huong and M. Couzi, J. Chim. Phys. 66, 1309 (1969).
10. D. F. Smith, J. Chem. Phys. 28, 1040 (1958); 48, 1429 (1968).
11. S. Abramowitz et al., U. S. Nat. Bur. Stand. Report 10904, 239, 1972.
12. CODATA Task Group on Fundamental Constants, CODATA Bulletin 11, December, 1973.
13. IUPAC Commission on Atomic Weights, Pure Appl. Chem. 47, 75 (1976).

Hydrogen Fluoride (H_5F_5) F_5H_5(g)

Enthalpy Reference Temperature = T_r = 298.15 K; Standard State Pressure = p° = 0.1 MPa

	J K⁻¹mol⁻¹			kJ mol⁻¹			
T/K	$C_p°$	S°	$-[G°-H°(T_r)]/T$	$H°-H°(T_r)$	$\Delta_fH°$	$\Delta_fG°$	Log K_f
0	0.	0.	INFINITE	-28.970	-1475.453	-1475.453	INFINITE
100	87.374	288.122	515.288	-22.717	-1483.965	-1445.003	754.792
200	115.893	357.974	420.354	-12.476	-1487.781	-1404.344	366.777
250	127.570	385.117	410.650	-6.383	-1488.935	-1383.343	289.034
298.15	137.321	408.442	408.442	0.	-1489.713	-1362.931	238.780
300	137.665	409.293	408.445	0.254	-1489.737	-1362.144	237.170
350	146.155	431.171	410.153	7.356	-1490.230	-1340.836	200.109
400	153.197	451.162	414.048	14.846	-1490.459	-1319.475	172.306
450	159.042	469.554	419.207	22.656	-1490.467	-1298.100	150.679
500	163.954	486.572	425.103	30.734	-1490.291	-1276.733	133.379
600	171.866	517.195	437.958	47.543	-1489.489	-1234.088	107.437
700	178.225	544.180	451.243	65.057	-1488.199	-1191.617	88.920
800	183.685	568.343	464.396	83.158	-1486.503	-1149.359	75.045
900	188.530	590.263	477.182	101.773	-1484.459	-1107.336	64.268
1000	192.872	610.356	489.509	120.847	-1482.120	-1065.556	55.659
1100	196.766	628.925	501.349	140.333	-1479.531	-1024.023	48.627
1200	200.244	646.198	512.709	160.187	-1476.738	-982.735	42.777
1300	203.341	662.350	523.605	180.369	-1473.781	-941.687	37.837
1400	206.091	677.522	534.063	200.843	-1470.693	-900.872	33.612
1500	208.529	691.826	544.108	221.577	-1467.506	-860.281	29.958
1600	210.691	705.354	553.767	242.540	-1464.243	-819.906	26.767
1700	212.608	718.186	563.065	263.707	-1460.923	-779.736	23.958
1800	214.311	730.388	572.024	285.054	-1457.560	-739.763	21.467
1900	215.826	742.016	580.667	306.563	-1454.163	-699.978	19.244
2000	217.176	753.122	589.015	328.214	-1450.740	-660.372	17.247
2100	218.383	763.747	597.084	349.993	-1447.291	-620.939	15.445
2200	219.464	773.932	604.893	371.886	-1443.819	-581.670	13.811
2300	220.435	783.709	612.456	393.882	-1440.321	-542.560	12.322
2400	221.309	793.110	619.789	415.970	-1436.795	-503.602	10.961
2500	222.097	802.160	626.904	438.141	-1433.236	-464.792	9.711
2600	222.811	810.885	633.813	460.387	-1429.640	-426.125	8.561
2700	223.459	819.307	640.528	482.701	-1426.002	-387.598	7.499
2800	224.048	827.444	647.059	505.077	-1422.319	-349.205	6.515
2900	224.585	835.316	653.416	527.509	-1418.585	-310.945	5.601
3000	225.076	842.938	659.607	549.992	-1414.798	-272.815	4.750
3100	225.525	850.325	665.641	572.523	-1410.954	-234.813	3.957
3200	225.938	857.492	671.525	595.096	-1407.050	-196.935	3.215
3300	226.317	864.451	677.266	617.709	-1403.085	-159.180	2.520
3400	226.666	871.212	682.871	640.359	-1399.058	-121.546	1.867
3500	226.989	877.787	688.347	663.042	-1394.966	-84.032	1.254
3600	227.287	884.186	693.698	685.756	-1390.811	-46.635	0.677
3700	227.564	890.417	698.931	708.498	-1386.591	-9.354	0.132
3800	227.820	896.489	704.051	731.268	-1382.309	27.811	-0.382
3900	228.059	902.410	709.061	754.062	-1377.964	64.863	-0.869
4000	228.281	908.187	713.967	776.879	-1373.558	101.803	-1.329
4100	228.488	913.827	718.773	799.717	-1369.092	138.631	-1.766
4200	228.682	919.335	723.483	822.576	-1364.569	175.351	-2.181
4300	228.863	924.718	728.101	845.453	-1359.989	211.961	-2.575
4400	229.032	929.981	732.630	868.348	-1355.354	248.465	-2.950
4500	229.191	935.130	737.073	891.259	-1350.667	284.863	-3.307
4600	229.340	940.169	741.433	914.186	-1345.929	321.155	-3.647
4700	229.481	945.103	745.714	937.127	-1341.141	357.345	-3.971
4800	229.613	949.936	749.919	960.082	-1336.306	393.433	-4.281
4900	229.737	954.671	754.049	983.050	-1331.425	429.417	-4.578
5000	229.854	959.314	758.108	1006.029	-1326.499	465.304	-4.861
5100	229.965	963.867	762.098	1029.020	-1321.529	501.090	-5.132
5200	230.070	968.333	766.021	1052.022	-1316.517	536.780	-5.392
5300	230.169	972.717	769.880	1075.034	-1311.463	572.372	-5.641
5400	230.263	977.020	773.676	1098.056	-1306.369	607.868	-5.880
5500	230.352	981.246	777.412	1121.086	-1301.234	643.268	-6.109
5600	230.436	985.397	781.089	1144.126	-1296.059	678.578	-6.330
5700	230.516	989.477	784.709	1167.173	-1290.845	713.792	-6.541
5800	230.593	993.486	788.274	1190.229	-1285.591	748.915	-6.745
5900	230.665	997.429	791.786	1213.292	-1280.297	783.948	-6.941
6000	230.734	1001.306	795.246	1236.362	-1274.963	818.889	-7.129

PREVIOUS: June 1977 (1 atm) CURRENT: June 1977 (1 bar)

Hydrogen Fluoride (H_5F_5) F_5H_5(g)

IODINE FLUORIDE (IF_5) IDEAL GAS M_r = 221.896515

$S°(298.15\ K) = 334.663 \pm 2.1\ J\ K^{-1}\ mol^{-1}$

$\Delta_f H°(0\ K) = -831.702 \pm 1.7\ kJ\ mol^{-1}$

$\Delta_f H°(298.15\ K) = -840.315 \pm 1.5\ kJ\ mol^{-1}$

Vibrational Frequencies and Degeneracies

ν, cm^{-1}	ν, cm^{-1}	ν, cm^{-1}
710(1)	604 (1)	631(2)
616(1)	[257](1)	372(2)
318(1)	276 (1)	190(2)

Ground State Quantum Weight: [1] $\sigma = 4$

Point Group: C_{4v}

Bond Distances: I-F(equatorial) = [1.86] Å I-F(axial) = [1.75] Å

Bond Angle: F-I-F = [90]°

Product of the Moments of Inertia: $I_A I_B I_C = [4.104805 \times 10^{-113}]\ g^3\ cm^6$

Enthalpy of Formation

The adopted value is derived from calorimetric data of Settle et al. (1) for the combustion of iodine in fluorine. Although the combustions gave various mixtures of IF_5 and IF_7, the proportions ranged up to almost pure pentafluoride. Reduction of the data gave $\Delta_f H° = -210.8_0 \pm 0.3_2$ kcal mol^{-1} for $IF_5(\ell)$ at 298.15 K. The result is essentially independent of the heptafluoride.

Comparison values of -208.9±[≥1] and -208.8±[≥1.7] kcal mol^{-1} may be obtained from calorimetric data of Woolf (2) for hydrolysis in water and in aqueous KOH. Uncertainties for these values include large contributions due to auxiliary data (3). In particular, we estimate lower limits of ±0.8 kcal mol^{-1} for $\Delta_f H°$ of $HIO_3(1250\ H_2O)$ and $KIO_3(1250\ H_2O)$ by comparison with recent data (4).

$\Delta_f H°$ for the ideal gas is derived from that of the liquid by use of $\Delta_{vap}H° = 9.96$ kcal mol^{-1} at 298.15 K. $\Delta_{vap}H°$ was derived by Osborne et al. (5) from measurements of vapor pressure and the second virial coefficient.

Heat Capacity and Entropy

There is no published structural determination for gaseous IF_5, but spectral data are consistent with C_{4v} symmetry and a configuration which is approximately a square pyramid. It is uncertain whether the iodine is located above, in, or below the base of the pyramid and whether or not the axial and equatorial I-F distances are equal. We adopt the structure used by Begun et al. (7) in correlation of infrared and Raman spectra of BrF_5, IF_5, ClF_5, and $XeOF_4$ by normal coordinate analysis. In this structure the axial bond length is shorter than the equatorial length and the values are close to those observed experimentally for IF_7. The iodine is presumed to be in the base of the pyramid. Simplified Huckel-molecular-orbital calculations (8) predict that C_{4v} is the most stable symmetry, that the axial bond is shorter, and that the equatorial flourines are distorted toward the axial fluorine. It is not clear to what extent these predictions depend on the simplified HMO model used in the calculations. Begun et al. (7) note that the normal coordinate calculations are insensitive to changes in the bond angles. Principal moments are $I_A = 43.6575 \times 10^{-39}$, $I_B = I_C = 30.6631 \times 10^{-39}\ g\ cm^2$.

Vibrational frequencies are based on the assignments of Begun et al. (7) as modified by the vapor-phase Raman spectra observed by Selig (6). Location of the band center of ν_9 (~200 cm^{-1}) is difficult, so we substitute the liquid value of 190 cm^{-1}. The Raman-active fundamental ν_5 is not observed for IF_5, or any of the similar molecules, so its value is taken from the calculations (7).

Osborne et al. (5) used calorimetric data at low temperatures to derive S° = 57.96±0.06 cal K^{-1} mol^{-1} for the liquid at 330 K. Combining this with $\Delta_{vap}S°$ based on their vapor pressure equation, the authors obtained S° = 83.06±0.20 cal K^{-1} mol^{-1} for the ideal gas at 330 K. The corresponding value from this table (82.50) is lower by 0.56 cal K^{-1} mol^{-1}. Part of this difference arises from uncertainties in ν_5, ν_9 and the product of the moments of inertia. We estimate this uncertainty at about 0.5 cal K^{-1} mol^{-1}. Another source of minor uncertainty is the neglect of anharmonicity in the calculations.

References

1. J. L. Settle, J. H. E. Jeffes, P. A. G. O'Hare and W. N. Hubbard, presented at the 23rd Annual Calorimetry Conference, Midland, Michigan, August, 1968.
2. A. A. Woolf, J. Chem. Soc. 231 (1951).
3. $\Delta_f H°$ values for $H_2O(\ell)$ and $KIO_3(1250\ H_2O)$ are derived from U. S. Nat. Bur. Stand. Tech. Note 270-3, 1968, in combination with V. B. Parker, NSRDS-NBS 2, 1965.
4. P. B. Howard and H. A. Skinner, J. Chem. Soc. A1967, 269 (1967); C. Wu, M. M. Birky and L. G. Hepler, J. Phys. Chem. 67, 1202 (1963).
5. D. W. Osborne, F. Schreiner, H. Selig, to be published.
6. H. Selig and H. Holzman, Isreal J. Chem. 7, 417 (1969).
7. G. M. Begun, W. H. Fletcher and D. F. Smith, J. Chem. Phys. 42, 2236 (1965).
8. R. M. Gavin, J. Chem. Educ. 46, 413 (1969).

Iodine Fluoride (IF_5) $F_5I_1(g)$

Enthalpy Reference Temperature = T_r = 298.15 K Standard State Pressure = p° = 0.1 MPa

T/K	$C_p°$ (J $K^{-1}mol^{-1}$)	S° (J $K^{-1}mol^{-1}$)	$-[G°-H°(T_r)]/T$ (J $K^{-1}mol^{-1}$)	$H°-H°(T_r)$ (kJ mol^{-1})	$\Delta_f H°$ (kJ mol^{-1})	$\Delta_f G°$ (kJ mol^{-1})	Log K_f
0	0.	0.	INFINITE	-20.049	-831.702	-831.702	INFINITE
100	52.519	250.536	412.327	-16.179	-836.625	-816.036	426.254
200	83.943	297.260	343.540	-9.256	-839.491	-794.121	207.403
250	94.929	317.229	336.315	-4.771	-840.076	-782.702	163.537
298.15	102.878	334.663	334.663	0.	-840.315	-771.626	135.186
300	103.139	335.300	334.664	0.191	-840.319	-771.200	134.278
350	109.222	351.679	335.944	5.507	-840.381	-759.674	113.375
400	113.756	366.574	338.856	11.087	-848.290	-747.874	97.662
450	117.181	380.179	342.703	16.864	-848.697	-735.297	85.351
500	119.808	392.667	347.084	22.792	-869.098	-720.779	75.299
600	123.478	414.861	356.577	34.970	-867.480	-691.265	60.180
700	125.838	434.085	366.308	47.444	-865.769	-662.030	49.401
800	127.434	450.999	375.858	60.113	-864.013	-633.044	41.334
900	128.559	466.078	385.060	72.916	-862.241	-604.279	35.071
1000	129.379	479.667	393.853	85.815	-860.468	-575.712	30.072
1100	129.994	492.029	402.224	98.785	-858.705	-547.322	25.990
1200	130.468	503.361	410.187	111.809	-856.962	-519.091	22.595
1300	130.839	513.819	417.761	124.875	-855.247	-491.005	19.729
1400	131.135	523.526	424.974	137.974	-853.568	-463.050	17.277
1500	131.376	532.582	431.849	151.100	-851.932	-435.213	15.155
1600	131.573	541.068	438.413	164.248	-850.345	-407.484	13.303
1700	131.738	549.049	444.688	177.414	-848.810	-379.852	11.671
1800	131.876	556.583	450.697	190.594	-847.329	-352.309	10.224
1900	131.993	563.717	456.460	203.788	-845.901	-324.848	8.931
2000	132.093	570.490	461.993	216.993	-844.519	-297.460	7.769
2100	132.180	576.937	467.315	230.206	-843.177	-270.140	6.719
2200	132.255	583.087	472.438	243.428	-841.864	-242.884	5.767
2300	132.320	588.968	477.378	256.657	-840.569	-215.686	4.898
2400	132.378	594.601	482.146	269.892	-839.277	-188.545	4.104
2500	132.429	600.006	486.753	283.132	-837.975	-161.458	3.373
2600	132.474	605.200	491.209	296.378	-836.649	-134.423	2.701
2700	132.514	610.201	495.524	309.627	-835.284	-107.440	2.079
2800	132.550	615.021	499.706	322.880	-833.867	-80.509	1.502
2900	132.583	619.673	503.763	336.137	-832.387	-53.629	0.966
3000	132.612	624.168	507.702	349.397	-830.832	-26.802	0.467
3100	132.638	628.517	511.530	362.659	-829.193	-0.028	0.
3200	132.662	632.728	515.252	375.924	-827.461	26.692	-0.436
3300	132.684	636.811	518.874	389.192	-825.629	53.357	-0.845
3400	132.704	640.772	522.401	402.461	-823.693	79.964	-1.228
3500	132.723	644.619	525.838	415.732	-821.647	106.512	-1.590
3600	132.740	648.358	529.190	429.006	-819.489	133.001	-1.930
3700	132.755	651.995	532.460	442.280	-817.216	159.428	-2.251
3800	132.769	655.536	535.653	455.557	-814.827	185.792	-2.554
3900	132.782	658.985	538.771	468.834	-812.322	212.092	-2.841
4000	132.795	662.347	541.818	482.113	-809.700	238.326	-3.112
4100	132.806	665.626	544.798	495.393	-806.963	264.492	-3.370
4200	132.817	668.826	547.713	508.674	-804.111	290.592	-3.614
4300	132.826	671.952	550.567	521.956	-801.146	316.621	-3.846
4400	132.836	675.005	553.360	535.239	-798.071	342.580	-4.067
4500	132.844	677.991	556.097	548.523	-794.887	368.469	-4.277
4600	132.852	680.911	558.778	561.808	-791.596	394.285	-4.477
4700	132.860	683.768	561.407	575.094	-788.202	420.029	-4.668
4800	132.867	686.565	563.986	588.380	-784.707	445.699	-4.850
4900	132.873	689.305	566.515	601.667	-781.114	471.295	-5.024
5000	132.879	691.989	568.998	614.955	-777.425	496.817	-5.190
5100	132.885	694.621	571.436	628.243	-773.644	522.265	-5.349
5200	132.891	697.201	573.830	641.532	-769.772	547.638	-5.501
5300	132.896	699.732	576.181	654.821	-765.814	572.935	-5.647
5400	132.901	702.217	578.492	668.111	-761.771	598.156	-5.786
5500	132.906	704.655	580.764	681.401	-757.647	623.301	-5.920
5600	132.910	707.050	582.998	694.692	-753.444	648.373	-6.048
5700	132.914	709.403	585.195	707.983	-749.164	673.366	-6.171
5800	132.918	711.714	587.356	721.275	-744.811	698.285	-6.289
5900	132.922	713.986	589.483	734.567	-740.386	723.129	-6.402
6000	132.926	716.220	591.577	747.859	-735.893	747.896	-6.511

PREVIOUS: December 1969 (1 atm) CURRENT: December 1969 (1 bar)

Iodine Fluoride (IF_5) $F_5I_1(g)$

MOLYBDENUM FLUORIDE (MoF_5) IDEAL GAS M_r = 190.932015

$\Delta_f H°(0\ K) = -1235.878 \pm 4.2\ kJ\ mol^{-1}$

$S°(298.15\ K) = 347.655 \pm 8.4\ J\ K^{-1}\ mol^{-1}$ $\Delta_f H°(298.15\ K) = -1241.393 \pm 4.2\ kJ\ mol^{-1}$

Electronic Levels and Quantum Weights

ε_i, cm^{-1}	g_i
0	[2]
[1000]	[2]
[7500]	[2]
[25000]	[4]

Vibrational Frequencies and Degeneracies

ν, cm^{-1}	ν, cm^{-1}
759 (1)	713 (2)
738 (1)	261 (2)
683 (1)	112 (2)
[500](1)	[200](2)

Point Group: $[D_{3h}]$ σ = [6]

Bond Distance: Mo-F = [1.78] Å

Bond Angle: F-Mo-F = [90]°

Product of the Moments of Inertia: $I_A I_B I_C = [3.670258 \times 10^{-113}]\ g^3\ cm^6$

Enthalpy of Formation

We adopt the enthalpy of formation at 298.15 K derived by Brewer (1) from the high temperature mass spectrometric measurments of equilibria among gaseous molybdenum fluorides by Hildenbrand [1143-1664 K] (2) and by Kleinschmidt et al. [460-525 K] (8). The value of $\Delta_f H°(0\ K)$ combined with JANAF data (3) for Mo(g) and F(g) gives $\Delta_{at}H°(0\ K) = 544.3 \pm 4\ kcal\ mol^{-1}$ and an average Mo-F bond energy of $\Delta_{at}H°(0\ K)/5 = 108.9 \pm 1\ kcal\ mol^{-1}$.

Heat Capacity and Entropy

Hildenbrand (4), Douglas (5) and Krause (6) have calculated somewhat different thermodynamic properties of MoF_5(g) monomer. All have used the vibrational frequencies, cm^{-1}, and degeneracies determined by Acquista and Abramowitz (7)--759(1), 738(1), 683(1), 713(2), 261(2) and 112(2)--and an estimated frequency of 500(1). For the remaining doubly degenerate frequency, Hildenbrand (4) estimated 200 cm^{-1} and both Douglas (5) and Krause (6) estimated 400 cm^{-1}. We follow Brewer (1) who has used the determined vibrational frequencies and the estimated frequencies [500](1) and [200](2), as had Hildenbrand (4).

From the following analysis, Brewer (1) has attempted to define the electronic contribution more accurately then just a ground state degeneracy of 2 (4, 5) or 4 (6). Mo^{5+} has only the low-lying electronic state ^2D with a spin orbit splitting of 2578 cm^{-1} and a total degeneracy of 10. From the liquid field treatment of ReF_6 by Moffitt et al. (9), one might expect an intermediate coupling case with a $t_{3/2}$ quartet ground state and a low-lying $t_{1/2}$ state, with the next quartet state well above 20000 cm^{-1}. Peacock and Sleight (10) observed an absorption in liquid MoF_5 at 7500 cm^{-1}. For MoF_5 monomer, the actual levels would be split more. On this basis, the following electronic energy levels, cm^{-1}, and quantum weights were assigned (1): 0 (2), [1000] (2), [7500] (2), and [25000] (4).

The trigonal bipyramid symmetry tentatively assigned by Acquista and Abramowitz (7) is used with the Mo-F distance of 1.78 Å adopted by Brewer (1). The principal moments of inertia are: $I_A = 29.9871 \times 10^{-39}\ g\ cm^2$, $I_B = I_C = 34.9849 \times 10^{-39}\ g\ cm^2$. This table is based on the critical analysis of data existing in 1978 by Brewer (1).

References

1. L. Brewer, Materials and Molecular Research Division, Lawrence Berkeley Laboratory, University of California, berkeley; personal communication, Sept. 1978, of preliminary draft of review to be submitted for publication in Atomic Energy Review, International Atomic Energy Agency, Vienna, Austria.
2. D. L. Hildenbrand, J. Chem. Phys. 65, 614 (1976).
3. JANAF Thermochemical Tables: Mo(g), 3-31-78; F(g), 9-30-65.
4. D. L. Hildenbrand, Stanford Research Institute, personal communication to L. Brewer.
5. T. B. Douglas, J. Chem. Thermodyn. 9, 1165 (1977).
6. R. F. Krause, Jr., in Proc. Symp. on High Temperature Metal Halide Chemistry, D. L. Hildenbrand and D. D. Cubicciotti, Eds., The Electrochem. Soc., Vol. 78-1, pp. 199-209, 1978, .
7. N. Acquista and S. Abramowitz, J. Chem. Phys. 58, 5484 (1973).
8. P. D. Kleinschmidt, K. H. Lau, and D. L. Hildenbrand, J. Chem. Thermodyn. 11, 765 (1979).
9. W. Moffitt, G. L. Goodman, M. Fred, and B. Weinstock, Molecular Physics 2, 109 (1959) and G. L. Goodman, Ph. D. Thesis, Harvard, 1959.
10. R. D. Peacock and J. P. Sleight, J. Fluorine Chem. 1, 253 (1971).

Molybdenum Fluoride (MoF_5) F_5Mo_1(g)

Enthalpy Reference Temperature = T_r = 298.15 K Standard State Pressure = p° = 0.1 MPa

T/K	$C_p°$ (J $K^{-1}mol^{-1}$)	S° (J $K^{-1}mol^{-1}$)	$-[G°-H°(T_r)]/T$ (J $K^{-1}mol^{-1}$)	$H°-H°(T_r)$ (kJ mol^{-1})	$\Delta_f H°$ (kJ mol^{-1})	$\Delta_f G°$ (kJ mol^{-1})	Log K_f
0	0.	0.	INFINITE	-21.132	-1235.878	-1235.878	INFINITE
100	61.633	259.654	426.629	-16.697	-1239.202	-1221.872	638.241
200	85.339	310.140	356.534	-9.279	-1240.948	-1203.754	314.388
250	94.948	330.241	349.304	-4.766	-1241.309	-1194.408	249.558
298.15	102.763	347.655	347.655	0.	-1241.393	-1185.363	207.671
300	103.034	348.291	347.657	0.190	-1241.392	-1185.016	206.330
350	109.545	364.682	348.936	5.511	-1241.257	-1175.628	175.453
400	114.655	379.658	351.854	11.121	-1240.964	-1166.271	152.299
450	118.614	393.401	355.718	16.957	-1240.561	-1156.957	134.296
500	121.664	406.063	360.127	22.968	-1240.081	-1147.693	119.899
600	125.825	428.645	369.712	35.360	-1238.989	-1129.316	98.316
700	128.324	448.245	379.562	48.078	-1237.823	-1111.129	82.913
800	129.862	465.489	389.247	60.993	-1236.652	-1093.110	71.373
900	130.837	480.845	398.587	74.032	-1235.516	-1075.236	62.405
1000	131.478	494.665	407.516	87.150	-1234.435	-1057.485	55.237
1100	131.918	507.218	416.018	100.321	-1233.426	-1039.840	49.378
1200	132.235	518.711	424.103	113.529	-1232.501	-1022.282	44.499
1300	132.478	529.305	431.794	126.765	-1231.669	-1004.799	40.373
1400	132.676	539.130	439.114	140.023	-1230.938	-987.375	36.839
1500	132.847	548.290	446.090	153.300	-1230.313	-970.001	33.778
1600	133.001	556.869	452.749	166.592	-1229.799	-952.664	31.101
1700	133.147	564.936	459.113	179.899	-1229.401	-935.355	28.740
1800	133.285	572.551	465.206	193.221	-1229.120	-918.067	26.642
1900	133.420	579.761	471.047	206.556	-1228.959	-900.792	24.764
2000	133.550	586.608	476.655	219.905	-1228.919	-883.521	23.075
2100	133.676	593.127	482.047	233.266	-1229.002	-866.250	21.547
2200	133.799	599.348	487.239	246.640	-1229.208	-848.972	20.157
2300	133.916	605.298	492.244	260.026	-1229.537	-831.681	18.888
2400	134.028	611.000	497.074	273.423	-1229.992	-814.374	17.724
2500	134.135	616.474	501.741	286.831	-1230.573	-797.045	16.653
2600	134.236	621.736	506.256	300.250	-1231.296	-779.690	15.664
2700	134.331	626.804	510.627	313.678	-1232.191	-762.304	14.748
2800	134.419	631.691	514.864	327.116	-1233.313	-744.881	13.896
2900	134.502	636.410	518.974	340.562	-1270.622	-727.363	13.101
3000	134.579	640.971	522.965	354.016	-1270.386	-708.634	12.338
3100	134.649	645.385	526.844	367.478	-1270.077	-689.915	11.625
3200	134.714	649.661	530.615	380.946	-1269.689	-671.205	10.956
3300	134.774	653.807	534.286	394.420	-1269.220	-652.510	10.328
3400	134.829	657.831	537.861	407.901	-1268.666	-633.830	9.738
3500	134.880	661.740	541.344	421.386	-1268.026	-615.167	9.181
3600	134.926	665.541	544.742	434.876	-1267.297	-596.524	8.655
3700	134.969	669.238	548.057	448.371	-1266.480	-577.902	8.159
3800	135.008	672.838	551.293	461.870	-1265.573	-559.305	7.688
3900	135.044	676.345	554.455	475.373	-1264.577	-540.731	7.242
4000	135.078	679.765	557.545	488.879	-1263.492	-522.184	6.819
4100	135.109	683.101	560.567	502.388	-1262.319	-503.666	6.417
4200	135.138	686.357	563.523	515.900	-1261.059	-485.177	6.034
4300	135.166	689.537	566.417	529.416	-1259.713	-466.720	5.670
4400	135.192	692.645	569.251	542.934	-1258.282	-448.295	5.322
4500	135.217	695.683	572.027	556.454	-1256.769	-429.902	4.990
4600	135.241	698.655	574.747	569.977	-1255.174	-411.545	4.673
4700	135.265	701.564	577.415	583.502	-1253.500	-393.223	4.370
4800	135.288	704.412	580.031	597.030	-1251.748	-374.938	4.080
4900	135.310	707.202	582.598	610.560	-1249.921	-356.690	3.802
5000	135.332	709.936	585.117	624.092	-1839.405	-332.748	3.476
5100	135.354	712.616	587.591	637.626	-1839.111	-302.618	3.099
5200	135.376	715.245	590.021	651.163	-1838.894	-272.493	2.737
5300	135.398	717.823	592.408	664.702	-1838.752	-242.371	2.389
5400	135.420	720.355	594.754	678.243	-1838.688	-212.252	2.053
5500	135.443	722.840	597.060	691.786	-1838.698	-182.133	1.730
5600	135.465	725.280	599.328	705.331	-1838.785	-152.012	1.418
5700	135.488	727.678	601.559	718.879	-1838.945	-121.891	1.117
5800	135.510	730.035	603.754	732.429	-1839.179	-91.766	0.826
5900	135.533	732.351	605.914	745.981	-1839.483	-61.635	0.546
6000	135.556	734.629	608.040	759.535	-1839.859	-31.500	0.274

PREVIOUS: December 1978 (1 atm) CURRENT: December 1978 (1 bar)

Molybdenum Fluoride (MoF_5) F_5Mo_1(g)

PHOSPHORUS FLUORIDE (PF_5) IDEAL GAS M_r = 125.965775

$S°(298.15 K) = 300.810 \pm 2.1 J K^{-1} mol^{-1}$

$\Delta_f H°(0 K) = -1583.514 \pm 2.9 kJ mol^{-1}$
$\Delta_f H°(298.15 K) = -1594.409 \pm 2.9 kJ mol^{-1}$

Vibrational Frequencies and Degeneracies

ν, cm^{-1}	ν, cm^{-1}
817 (1)	1025 (2)
640 (1)	532.5(2)
946.6(1)	179 (2)
575.1(1)	514 (2)

Ground State Quantum Weight: [1] $\sigma = 6$
Point Group: D_{3h}
Bond Distance: $P\text{-}F_{eq} = 1.534 \pm 0.004$ Å $P\text{-}F_{ax} = 1.577 \pm 0.005$ Å
Bond Angles: $F_{eq}\text{-}P\text{-}F_{eq} = 120° \pm 0.4°$ $F_{eq}\text{-}P\text{-}F_{ax} = 74.3° \pm 0.4°$
Product of the Moments of Inertia: $I_A I_B I_C = 1.614 \times 10^{-113} g^3 cm^6$

Enthalpy of Formation

Gross et al. (1) measured the heat of combustion of α-white phosphorus in fluorine at 25°C. They reported an enthalpy of formation of $-381.4\pm0.8 kcal mol^{-1}$ (based on α-white phosphorus standard state) calculated from their combustion studies. Similar combustion studies were performed by O'Hare and Hubbard (2). They reported a enthalpy of formation of $-380.8\pm0.3 kcal mol^{-1}$ (α-white phosphorus standard state), which is in excellent agreement with the work of Gross et al. (1). An average value of $-381.1\pm0.5 kcal mol^{-1}$ is adopted for this tabulation. Converting to the P(red,V) standard state with a $\Delta_f H$ trans = 4.2 ± 0.2 kcal mol^{-1} (3), we derive, $\Delta_f H°(PF_5, g, 298.15 K) = -376.9\pm0.7 kcal mol^{-1}$ ($-1576.950 \pm 2.9 kJ mol^{-1}$).

Heat Capacity and Entropy

The molecular structure and bond lengths given above for PF_5 were recently determined by Hansen and Bartell (4) from electron-diffraction studies. These results indicate that PF_5 has a trigonal bipyramid structure with nonequivalent axial and equatorial bonds. Wyatt et al. (5) determined rotational constants for PF_5 from a study of the infrared vibration-rotation band contour of the ν_4 (575 cm^{-1}) fundamental. Their analysis led to the following P-F bond lengths which are in excellent agreement with the electron-diffraction results: $r_{eq.} = 1.534\pm0.024$ Å and $r_{ax.} = 1.581\pm0.025$ Å. Individual moments of inertia, calculated from the rotational constants given by Wyatt et al. (5) and adopted for this table, are $I_A = 2.230 \times 10^{-38} g cm^2$ and $I_B = I_C = 2.690 \times 10^{-38} g cm^2$. The electron diffraction data lead to a product of moments of inertia of $1.604 \times 10^{-113} g^3 cm^6$.

Several recent infrared and Raman studies (6, 7, and 8) have been reported for PF_5. These data are in good agreement except for the ν_7 fundamental. Griffiths et al. (6) originally assigned this fundamental to the 126 cm^{-1} band observed in their infrared spectra. Hoskins (9) later attributed this band to one of the pure rotational lines of hydrogen fluoride, which he felt was present in the sample as an impurity. Further work by Griffiths (10) on the pressure dependence of the infrared spectra of PF_5 indicated that the missing fundamental should be assigned to the weak band at 301 cm^{-1}. Deiters and Holmes (11) recently showed that this band is not a fundamental of PF_5 but a difference band (ν_3-ν_2). Very recently, Levin (8) investigated the Raman spectra of solid phosphorus pentafluoride and assigned the ν_7 fundamental to the 179 cm^{-1} band. Also, Beattie et al. (12) investigated the Raman spectra of PF_5 in the gas phase at 25°C and 125°C and assigned the ν_7 fundamental to 175 cm^{-1} band which is in excellent agreement with Levin's assignment (8). The fundamental vibrational frequencies of Holmes and Lord (7) are adopted along with Levin's assignment (8) of the ν_7 fundamental (179 cm^{-1}).

References

1. P. Gross, C. Hayman and M. C. Stuart, Trans. Faraday Soc. 62, 2716 (1966).
2. P. A. G. O'Hare and W. N. Hubbard, Trans. Faraday Soc. 62, 2709 (1966).
3. JANAF Thermochemical Tables: P(red, v), 6-30-61.
4. K. W. Hansen and L. S. Bartell, Inorg. Chem. 4, 1775 (1965).
5. R. Wyatt, J. T. Roberts, R. E. Wentz and R. M. Wilt, J. Chem. Phys. 50, 2552 (1969).
6. J. E. Griffiths, R. P. Carter and R. R. Holmes, J. Chem. Phys. 41, 63 (1964).
7. L. C. Hoskins and R. C. Lord, J. Chem. Phys. 46, 2402 (1967).
8. I. W. Levin, J. Chem. Phys. 50, 1031 (1969).
9. L. C. Hoskins, J. Chem. Phys. 42, 2631 (1965).
10. J. E. Griffiths, J. Chem. Phys. 42, 2631 (1965).
11. R. M. Deiters and R. R. Holmes, J. Chem. Phys. 48, 4796 (1968).
12. I. R. Beattie, K. M. S. Livingston and D. J. Reynolds, J. Chem. Phys. 51, 4269 (1969).

Phosphorus Fluoride (PF_5) $F_5P_1(g)$

Enthalpy Reference Temperature = T_r = 298.15 K Standard State Pressure = p° = 0.1 MPa

T/K	$C_p°$ (J K⁻¹mol⁻¹)	S° (J K⁻¹mol⁻¹)	$-[G°-H°(T_r)]/T$ (J K⁻¹mol⁻¹)	$H°-H°(T_r)$ (kJ mol⁻¹)	$\Delta_f H°$ (kJ mol⁻¹)	$\Delta_f G°$ (kJ mol⁻¹)	Log K_f
0	0.	0.	INFINITE	-16.527	-1583.514	-1583.514	INFINITE
100	44.341	234.160	362.483	-12.832	-1587.974	-1567.029	818.532
200	65.173	270.951	307.973	-7.404	-1592.132	-1544.571	403.401
250	75.849	286.661	302.158	-3.874	-1593.450	-1532.521	320.203
298.15	84.836	300.810	300.810	0.	-1594.409	-1520.695	266.419
300	85.152	301.336	300.811	0.157	-1594.441	-1520.237	264.697
350	92.905	315.064	301.878	4.615	-1595.891	-1507.738	225.018
400	99.238	327.897	304.337	9.424	-1596.474	-1495.102	195.240
450	104.373	339.893	307.629	14.519	-1596.864	-1482.405	172.073
500	108.537	351.113	311.423	19.845	-1597.100	-1469.673	153.536
600	114.710	371.484	319.771	31.028	-1597.234	-1444.169	125.726
700	118.917	389.503	328.471	42.722	-1597.051	-1418.669	105.862
800	121.874	405.586	337.124	54.770	-1596.660	-1393.211	90.967
900	124.015	420.071	345.549	67.070	-1596.127	-1367.811	79.386
1000	125.608	433.224	353.669	79.555	-1595.500	-1342.475	70.124
1100	126.821	445.255	361.456	92.179	-1594.808	-1317.206	62.549
1200	127.763	456.332	368.906	104.910	-1657.647	-1290.923	56.192
1300	128.509	466.589	376.031	117.725	-1656.113	-1260.424	50.644
1400	129.109	476.135	382.844	130.607	-1654.570	-1230.045	45.894
1500	129.598	485.060	389.364	143.543	-1653.027	-1199.776	41.780
1600	130.002	493.437	395.610	156.524	-1651.488	-1169.609	38.184
1700	130.339	501.329	401.599	169.541	-1649.958	-1139.538	35.014
1800	130.623	508.787	407.348	182.590	-1648.435	-1109.557	32.199
1900	130.865	515.856	412.875	195.664	-1646.919	-1079.662	29.682
2000	131.072	522.574	418.193	208.762	-1645.406	-1049.845	27.419
2100	131.251	528.973	423.317	221.878	-1643.891	-1020.104	25.374
2200	131.407	535.083	428.260	235.011	-1642.367	-990.436	23.516
2300	131.543	540.927	433.032	248.159	-1640.827	-960.836	21.821
2400	131.663	546.528	437.645	261.319	-1639.264	-931.305	20.269
2500	131.769	551.905	442.109	274.491	-1637.670	-901.840	18.843
2600	131.863	557.075	446.432	287.672	-1636.036	-872.439	17.528
2700	131.947	562.053	450.623	300.863	-1634.356	-843.101	16.311
2800	132.022	566.853	454.688	314.061	-1632.622	-813.827	15.182
2900	132.090	571.487	458.637	327.267	-1630.828	-784.616	14.132
3000	132.151	575.966	462.473	340.479	-1628.968	-755.468	13.154
3100	132.206	580.300	466.205	353.697	-1627.036	-726.383	12.239
3200	132.257	584.499	469.836	366.920	-1625.030	-697.361	11.383
3300	132.303	588.569	473.373	380.148	-1622.944	-668.403	10.580
3400	132.345	592.519	476.819	393.381	-1620.777	-639.510	9.825
3500	132.383	596.356	480.180	406.617	-1618.525	-610.682	9.114
3600	132.418	600.086	483.459	419.857	-1616.188	-581.919	8.443
3700	132.451	603.715	486.660	433.101	-1613.765	-553.223	7.810
3800	132.481	607.247	489.788	446.347	-1611.254	-524.593	7.211
3900	132.508	610.689	492.844	459.597	-1608.657	-496.030	6.644
4000	132.534	614.044	495.832	472.849	-1605.972	-467.535	6.105
4100	132.558	617.317	498.755	486.103	-1603.202	-439.109	5.594
4200	132.580	620.512	501.616	499.360	-1600.346	-410.750	5.108
4300	132.601	623.632	504.418	512.619	-1597.407	-382.462	4.646
4400	132.620	626.680	507.162	525.880	-1594.384	-354.242	4.205
4500	132.638	629.661	509.851	539.143	-1591.281	-326.091	3.785
4600	132.655	632.576	512.487	552.408	-1588.099	-298.011	3.384
4700	132.670	635.429	515.073	565.674	-1584.839	-270.000	3.001
4800	132.685	638.223	517.610	578.942	-1581.503	-242.060	2.634
4900	132.699	640.959	520.099	592.211	-1578.094	-214.190	2.283
5000	132.712	643.640	522.543	605.482	-1574.613	-186.391	1.947
5100	132.725	646.268	524.943	618.754	-1571.062	-158.661	1.625
5200	132.736	648.845	527.301	632.027	-1567.443	-131.002	1.316
5300	132.747	651.374	529.619	645.301	-1563.759	-103.413	1.019
5400	132.758	653.855	531.896	658.576	-1560.011	-75.895	0.734
5500	132.767	656.291	534.136	671.852	-1556.201	-48.447	0.460
5600	132.777	658.683	536.339	685.130	-1552.332	-21.067	0.197
5700	132.785	661.034	538.506	698.408	-1548.404	6.241	-0.057
5800	132.794	663.343	540.638	711.687	-1544.420	33.481	-0.302
5900	132.802	665.613	542.737	724.966	-1540.382	60.653	-0.537
6000	132.809	667.845	544.804	738.247	-1536.291	87.754	-0.764

PREVIOUS: December 1969 (1 atm) CURRENT: December 1969 (1 bar)

Phosphorus Fluoride (PF_5) $F_5P_1(g)$

SULFUR FLUORIDE (SF_5) IDEAL GAS $M_r = 127.052015$

$S°(298.15\ K) = [304.752 \pm 8.4]\ J\ K^{-1}\ mol^{-1}$

$\Delta_fH°(0\ K) = -898.295 \pm 15.1\ kJ\ mol^{-1}$
$\Delta_fH°(298.15\ K) = -908.447 \pm 15.1\ kJ\ mol^{-1}$

Electronic Levels and Quantum Weights

ε_i, cm^{-1}	g_i	ε_i, cm^{-1}	g_i
0	[2]	[25000]	[2]
[10000]	[2]	[30000]	[2]
[20000]	[2]		

Vibrational Frequencies and Degeneracies

ν, cm^{-1}	ν, cm^{-1}	ν, cm^{-1}
[800](1)	[550](1)	812 (2)
[600](1)	[400](1)	552 (2)
[550](1)	[450](1)	[350)(2)

Point Group: C_{4v} σ = 4
Bond Distances: S-F* = [1.58] Å S-F** = [1.51] Å
Bond Angles: F*-S-F* = 90° F**-S-F* = [90°]
(* = equatorial, ** = axial)
Product of the Moments of Inertia: $I_AI_BI_C = [1.415983 \times 10^{-113}]\ g^3\ cm^6$

Enthalpy of Formation

$\Delta_fH°(0\ K)$ is calculated from the bond dissociation process $SF_6(g) = SF_5(g) + F(g)$ with $D_0^0 = 92.0\pm3.0\ kcal\ mol^{-1}$ and JANAF auxiliary enthapies of formation (1). The information which was considered in arriving at the selected D_0^0 value is reviewed below. Another recent review (2) of the thermodynamics of SF_5 led to selected values of $D_0^0(SF_5\text{-}F) = 92.6\pm3.0\ kcal\ mol^{-1}$ and $\Delta_fH°(298.15\ K) = -218\pm\ 3\ kcal\ mol^{-1}$ which agree with the results adopted here.

The electron-impact appearance potential of F^- from SF_6 has been reported as approximately zero by Harland and Thynne (3) and Curran (4). In addition, Curran (4) measured the kinetic energy of the negative ion and found K.E. (Total) = 0.23±0.05 eV. Combining these results with EA(F) = 3.399 eV (1), we obtain an upper limit value for D_0^0 of <3.2 eV (73.1 kcal mol^{-1}).

Modica (5) performed shock tube measurements on mixtures of SF_6/Ar, SF_5Cl/Ar, and SF_6/H_2 and analyzed the equilibrium data with a thermochemical equilibrium computer program. With $\Delta_fH°(SF_4,\ g,\ 0\ K) = -172.2\ kcal\ mol^{-1}$, Modica (5) adjusted the value of $\Delta_fH°$ for SF_5 to bring the calculated and experimental SF_4/HF concentrations into good agreement. These results suggest $\Delta_fH°(SF_5,\ g,\ 0\ K) = -241.7\ kcal\ mol^{-1}$ or $D_0^0(SF_5\text{-}F) = 65.2\ kcal\ mol^{-1}$. We are unable to reanalyze these shock tube equilibrium data but presumably the $\Delta_fH°$ and D_0^0 values should be adjusted for the changes in $\Delta_fH°$ of SF_4 (1).

Bott and Jacobs (6) used infrared and ultraviolet absorption techniques to study the thermal (1650-2050 K) dissociation of SF_6 in a shock tube. They interpreted their rate data in terms of the RRK theory and showed that the results were consistent with $D_0^0(SF_5\text{-}F) = 75.9\pm4.0\ kcal\ mol^{-1}$. Lyman (7) has re-evaluated the SF_6 dissociation data in terms of the more complete RRKM unimolecular reaction rate theory and presented evidence which indicates that the $D_0^0(SF_5\text{-}F)$ is considerably higher than previously reported. His analysis of the rate data yielded a D_0^0 value of 92 kcal mol^{-1} which we adopt.

The enthalpy of atomization and mean bond dissociation energy are calculated to be 372.2 kcal mol^{-1} and 74.4 kcal mol^{-1}, respectively. $\Delta_fH°(0\ K)$ combined with that for SF_5^+ (1) gives IP = 11.14 eV which appears reasonable in comparison with IP values for the other sulfur fluorides (1).

Heat Capacity and Entropy

We assume that the five fluorine atoms form a tetragonal pyramid around the central sulfur atom based on evidence from an EPR study (8) which shows that four of the fluorine atoms are equivalent. This structure is further supported by recent SCF MO calculations (9) which predict that the F**-S-F* bond angle is near 90°. The geometry optimized bond lengths (9) appear to be about 0.2 Å too large in comparison with experimentally measured S-F bond lengths for other sulfur-fluorine species (1). We assume that the S-F* bond length is the same as that reported for SF_6 and S_2F_{10} (1). The axial bond length is taken to be slightly shorter than the equatorial bonds, in accordance with the pattern followed in the related molecules ClF_5 (1) and TeF_5^- (10). The principal moments of inertia are: $I_A = 30.7102 \times 10^{-39}$ and $I_B = I_C = 21.4728 \times 10^{-39}\ g\ cm^2$.

Smardzewski and Fox (11) reported the matrix infrared spectra of SF_5 formed during the vacuum ultraviolet photolysis of SF_6 and its derivatives, SF_5X (X=F, Cl, Br, SF_5), in argon matrices at 8 K. Values of ν_7 and ν_8 are taken from this study. The other seven fundamentals are estimated by analogy with those for SF_5^- (1) and ClF_5 (12). Double (ν_1 and ν_7) and triple (ν_3, ν_4, and ν_8) coincidences are predicted at about 800 cm^{-1} and 550 cm^{-1}, respectively, based on the infrared spectra of ClF_5 (12). The four upper electronic levels are estimated from those observed for SF_6^+ (13) and are assummed to be doublets.

Our thermal functions essentially update and extend those previously reported by O'Hare (14). Wilkins (15) has also calculated functions covering the temperature range 50 - 6000 K. He assumed that the structure of the radical was a trigonal bipyramid (D_{3h} symmetry). We cannot completely rule out this structure but consider it to be unlikely.

References

1. JANAF Thermochemical Tables: SF_6(g), SF_4(g), SF_3(g), SF_2(g) and SF(g), 6-30-76; F(g), 9-30-65; F^-(g), 12-31-71; S_2F_{10}(g), SF_5^+(g) and SF_5^-(g), 12-31-77; ClF_5(g), 6-30-66.
2. S. W. Benson, Chem. Rev. (to be published).
3. P. Harland and J. C. J. Thynne, J. Phys. Chem. 73, 4031 (1969).
4. R. K. Curran, J. Chem. Phys. 34, 1069 (1961).
5. A. P. Modica, J. Phys. Chem. 77, 2713 (1973).
6. J. F. Bott and T. A. Jacobs, J. Chem. Phys. 50, 3850 (1969).
7. J. L. Lyman, J. Chem. Phys. 67, 1868 (1977).
8. J. R. Morton and K. F. Preston, Chem. Phys. Lett. 18, 98 (1973).
9. A. R. Gregory, Chem. Phys. Lett. 28, 552 (1974).
10. H. A. Edwards and M. A. Mouty, J. Chem. Soc. 1969, 703 (1969).
11. R. R. Smardzewski and W. B. Fox, J. Chem. Phys. 67, 2309 (1977).
12. K. O. Christe, Spectrochim. Acta. 27A, 631 (1971).
13. J. Delwiche, Dynamic Mass Spectrom. 1, 71 (1971).
14. P. A. O'Hare, Argonne National Laboratory, ANL-7315, July, 1968.
15. R. L. Wilkins, Aerospace Corporation, Report No. TR-0158 (3240-20)-19, July, 1968.

Sulfur Fluoride (SF_5) F_5S_1(g)

Enthalpy Reference Temperature = T_r = 298.15 K Standard State Pressure = p° = 0.1 MPa

	J $K^{-1}mol^{-1}$			kJ mol^{-1}			
T/K	$C_p°$	S°	$-[G°-H°(T_r)]/T$	$H°-H°(T_r)$	$\Delta_fH°$	$\Delta_fG°$	Log K_f
0	0.	0.	INFINITE	-16.323	-898.295	-898.295	INFINITE
100	38.347	238.884	367.958	-12.907	-902.837	-882.881	461.170
200	66.342	273.589	312.273	-7.737	-906.631	-861.295	224.947
250	79.469	289.850	306.174	-4.081	-907.751	-849.822	177.561
298.15	89.642	304.752	304.752	0.	-908.447	-838.597	146.919
300	89.986	305.308	304.754	0.166	-908.468	-838.163	145.937
350	98.138	319.818	305.881	4.878	-908.894	-826.409	123.335
400	104.412	333.351	308.479	9.949	-911.330	-814.521	106.365
450	109.265	345.941	311.950	15.296	-912.076	-802.382	93.138
500	113.054	357.657	315.942	20.857	-912.764	-790.154	82.547
600	118.445	378.782	324.695	32.452	-913.437	-765.557	66.648
700	121.973	397.324	333.774	44.485	-913.630	-740.890	55.286
800	124.386	413.778	342.765	56.810	-913.540	-716.217	46.764
900	126.101	428.533	351.490	69.339	-966.403	-690.485	40.075
1000	127.359	441.888	359.872	82.015	-964.808	-659.913	34.470
1100	128.309	454.073	367.890	94.801	-963.194	-629.501	29.893
1200	129.044	465.270	375.545	107.670	-961.577	-599.237	26.084
1300	129.627	475.623	382.850	120.604	-959.968	-569.107	22.867
1400	130.102	485.247	389.824	133.592	-958.375	-539.101	20.114
1500	130.497	494.237	396.489	146.622	-956.804	-509.208	17.732
1600	130.836	502.670	402.864	159.689	-955.257	-479.419	15.651
1700	131.134	510.611	408.971	172.788	-953.735	-449.725	13.818
1800	131.403	518.114	414.828	185.915	-952.233	-420.121	12.192
1900	131.652	525.226	420.453	199.068	-950.750	-390.600	10.738
2000	131.887	531.985	425.862	212.245	-949.277	-361.156	9.432
2100	132.112	538.425	431.070	225.445	-947.807	-331.787	8.253
2200	132.331	544.576	436.091	238.667	-946.331	-302.487	7.182
2300	132.545	550.463	440.936	251.911	-944.840	-273.255	6.206
2400	132.757	556.108	445.618	265.176	-943.324	-244.088	5.312
2500	132.966	561.532	450.147	278.462	-941.773	-214.985	4.492
2600	133.174	566.751	454.532	291.769	-940.177	-185.945	3.736
2700	133.380	571.781	458.782	305.097	-938.528	-156.967	3.037
2800	133.584	576.635	462.905	318.445	-936.816	-128.051	2.389
2900	133.787	581.327	466.908	331.814	-935.035	-99.198	1.787
3000	133.988	585.866	470.798	345.202	-933.178	-70.407	1.226
3100	134.187	590.262	474.581	358.611	-931.239	-41.680	0.702
3200	134.384	594.526	478.263	372.040	-929.213	-13.017	0.212
3300	134.578	598.664	481.849	385.488	-927.097	15.582	-0.247
3400	134.769	602.684	485.344	398.955	-924.886	44.115	-0.678
3500	134.958	606.594	488.753	412.442	-922.580	72.582	-1.083
3600	135.143	610.398	492.079	425.947	-920.175	100.982	-1.465
3700	135.326	614.103	495.328	439.470	-917.672	129.313	-1.826
3800	135.505	617.715	498.501	453.012	-915.070	157.574	-2.166
3900	135.682	621.237	501.603	466.571	-912.369	185.767	-2.488
4000	135.855	624.674	504.637	480.148	-909.569	213.889	-2.793
4100	136.026	628.031	507.606	493.742	-906.672	241.940	-3.082
4200	136.193	631.311	510.512	507.353	-903.679	269.919	-3.357
4300	136.357	634.517	513.359	520.981	-900.592	297.825	-3.618
4400	136.518	637.654	516.148	534.624	-897.411	325.659	-3.866
4500	136.675	640.724	518.883	548.284	-894.140	353.420	-4.102
4600	136.830	643.729	521.564	561.959	-890.780	381.106	-4.328
4700	136.982	646.674	524.195	575.650	-887.334	408.719	-4.542
4800	137.131	649.559	526.777	589.356	-883.803	436.257	-4.747
4900	137.277	652.388	529.311	603.076	-880.191	463.721	-4.943
5000	137.419	655.163	531.801	616.811	-876.499	491.111	-5.131
5100	137.559	657.886	534.246	630.560	-872.730	518.425	-5.310
5200	137.696	660.558	536.650	644.323	-868.887	545.666	-5.481
5300	137.830	663.182	539.013	658.099	-864.971	572.832	-5.646
5400	137.961	665.760	541.336	671.888	-860.985	599.923	-5.803
5500	138.088	668.292	543.621	685.691	-856.931	626.939	-5.954
5600	138.213	670.782	545.870	699.506	-852.812	653.882	-6.099
5700	138.335	673.229	548.083	713.333	-848.630	680.749	-6.238
5800	138.454	675.636	550.261	727.173	-844.387	707.543	-6.372
5900	138.569	678.004	552.406	741.024	-840.085	734.265	-6.501
6000	138.681	680.334	554.519	754.887	-835.726	760.911	-6.624

PREVIOUS: December 1977 (1 atm) CURRENT: December 1977 (1 bar)

Sulfur Fluoride (SF_5) F_5S_1(g)

SULFUR FLUORIDE, ION (SF_5^+) IDEAL GAS M_r = 127.051466

$S°(298.15\ K)$ = [298.151 ± 8.4] $J\ K^{-1}\ mol^{-1}$

$\Delta_f H°(0\ K)$ = 176.574 ± 20.9 $kJ\ mol^{-1}$
$\Delta_f H°(298.15\ K)$ = [172.644] $kJ\ mol^{-1}$

Electronic Levels and Quantum Weights

ϵ_i, cm^{-1}	g_i	ϵ_i, cm^{-1}	g_i
0	[1]	29036	[1]
9275	[1]	37101	[1]
17341	[1]		

Vibrational Frequencies and Degeneracies

ν, cm^{-1}	ν, cm^{-1}	ν, cm^{-1}
[800](1)	[550](1)	[800](2)
[600](1)	[400](1)	[550](2)
[550](1)	[450](1)	[350](2)

Point Group: $[C_{4v}]$ σ = [4]
Bond Distances: S-F* = [1.50] Å S-F** = [1.46] Å
Bond Angles: F**-S-F* = [90°] F*-S-F* = [90°]
(* - equatorial ** - axial)
Product of the Moments of Inertia: $I_A I_B I_C$ = [1.126194 x 10^{-113}] $g^3\ cm^6$

Enthalpy of Formation

The threshold energy for the formation of SF_5^+ from SF_6 has been measured by photoionization (1), electron-impact (2-5), and photoelectron spectroscopy (4, 6-8). Presented below is a summary of the reported appearance potentials (AP) along with results arrived at from an analysis of the data. Also included in the summary are results derived from an electron-impact study (9) on the formation of SF_5^+ from SF_5Cl. All auxiliary data are from the most recent JANAF tables (10) with the exception of the value for $\Delta_f H°(0\ K)$ of SF_5Cl. We use $\Delta_f H°(SF_5Cl, g, 0\ K)$ = -247.4 kcal mol^{-1}.

Source	Method	AP[a], eV	$\Delta_f H°(0\ K)$ kcal mol^{-1}	IP, eV
(1)	Photoionization	<15.29	<45.88	<11.3
(2)	Electron-Impact	15.9 ±0.2	<59.94	<11.9
(3)	Electron-Impact	15.85±0.15	<59.79	<11.9
(4)	RPD	15.75±0.05	<56.48	<11.8
	Photoelectron Spectroscopy	15.30	46.11	11.31
(5)	Electron-Impact	15.50±0.1	<50.72	<11.5
(6)	Photoelectron Spectroscopy	15.35±0.2	47.26	11.36
(7)	Photoelectron Spectroscopy	15.15	42.65	11.16
(8)	Photoelectron Spectroscopy	14.72	32.73	10.73
(9)	Electron-Impact	13.2 ±0.2[b]	<28.33	<10.5

[a] $SF_6(g) + e^- = SF_5^+(g) + F(g) + 2e^-$ [b] $SF_5Cl(g) + e^- = SF_5^+(g) + Cl(g) + 2e^-$

No study has been made of the kinetic energies of the fragments produced by photoionization or electron-impact of SF_6 and SF_5Cl. As a result, the ionization potentials (IP) derived from these studies (1-5, 9) represent upper limits to the true value. SF_5^+ is produced in the photoionization of SF_6 by rapid dissociation of SF_6^+ whose ground state is expected to exhibit Jahn-Teller instability (1). As a result of this, an accurate determination of the threshold energy for the dissociation process is difficult. We choose to adopt a mean $\Delta_f H°(0\ K)$ value of 42.2±5.0 kcal mol^{-1} based on the results measured in four photoelectron spectroscopic studies (4, 6-8) of the SF_6/SF_5^+ dissociation. Our adopted enthalpies of formation correspond to $IF(SF_5)$ = 11.14±0.37 eV which is in agreement with the limiting values of 9.25<IP<12.05 eV (13).

$\Delta_f H°$ is calculated to be 41.263 kcal mol^{-1} at 298.15 K. Using JANAF data (10), we find that the primary bond dissociation energy, $D_0^0(SF_4^+\text{-}F)$, is 75.4 kcal mol^{-1}. This value appears reasonable in comparison with D_0^0 values for other even-electron positive and negative S-F species (10), providing some additional support for our adopted results.

Heat Capacity and Entropy

By analog with SF_5 (14, 15) and SF_5^- (10), we adopt a tetragonal pyramidal structure (C_{4v} symmetry). The SF_4 part of this configuration is predicted to be planar as a result of a reduction in the repulsive forces (16) upon ionization. The four equatorial S-F bond lengths are assumed equal to that for SF_3^+ (10). The axial bond length is assumed to be slightly shorter than the equatorial bonds, in accordance with the pattern followed in the related molecules ClF_5 (10) and TeF_5^- (17). The individual moments of inertia are: I_A = 28.3933 x 10^{-39} and $I_B = I_C$ = 19.9158 x 10^{-39} g cm^2. We estimate vibrational frequencies from those for SF_5 (10), SF_5^- (10) and ClF_5 (18). Double (ν_1 and ν_7) and triple (ν_3, ν_4, and ν_8) coincidences are predicted at 800 cm^{-1} and 550 cm^{-1}, respectively, based on the infrared spectra of ClF_5 (18). The electronic levels are calculated from the RPD measurements of Delwiche (4). We do not include levels predicted to lie above 40,000 cm^{-1} since these have a negligible effect on the thermal functions. The existence of the first two excited states is verified by an independent RPD study (3).

References

1. V. H. Dibeler and J. A. Walker, J. Chem. Phys. 44, 4405 (1966).
2. V. H. Dibeler and F. L. Mohler, J. Res. Nat. Bur. Stand. 40, 25 (1948).
3. R. E. Fox and R. K. Curran, J. Chem. Phys. 34, 1595 (1961).
4. J. Delwiche, Dynamic Mass Spectrom. 1, 71 (1970).
5. D. L. Hildenbrand, J. Phys. Chem. 77, 897 (1973).
6. D. C. Frost, C. A. McDowell, J. S. Sandhu, and D. A. Vroom, Advan. Mass Spectrom. 4, 781 (1968).
7. A. W. Potts, H. J. Lempka, D. G. Streets, and W. C. Price, Phil. Trans. Roy. Soc. Lond. A 268, 59 (1970).
8. B. P. Pullen, Ph.D. Thesis, University of Tennessee, 1970.
9. P. Harland and J. C. J. Thynne, J. Phys. Chem. 73, 4031 (1969).
10. JANAF Thermochemical Tables: SF_6(g), 6-30-76; F(g), 9-30-65; Cl(g), 6-30-72; SF_3^-(g), SF_3^+(g) and SF_4^+(g), 12-31-76; SF_5(g) and SF_5^-(g), 12-31-77; Cl_5(g), 6-30-66.
11. U. S. Nat. Bur. Stand. Tech. Note 270-3, 1968.
12. D. F. Torgerson and J. B. Westmore, Can. J. Chem. 53, 933 (1975).
13. F. C. Fehsenfeld, J. Chem. Phys. 54, 438 (1971).
14. J. R. Morton and K. F. Preston, Chem. Phys. Lett. 18, 98 (1973).
15. A. R. Gregory, Chem. Phys. Lett. 28, 552 (1974).
16. L. F. Drullinger and J. E. Griffiths, Spectrochim. Acta. 27, 1793 (1971).
17. H. A. Edwards and M. A. Mouty, J. Chem. Soc. 1969, 703 (1969).
18. K. O. Christe, Spectrochim. Acta, 27A, 631 (1971).

Sulfur Fluoride, Ion (SF_5^+) $F_5S_1^+$(g)

Enthalpy Reference Temperature = T_r = 298.15 K Standard State Pressure = p° = 0.1 MPa

T/K	$C_p°$	S°	$-[G°-H°(T_r)]/T$	$H°-H°(T_r)$	$\Delta_f H°$	$\Delta_f G°$	Log K_f
	J $K^{-1}mol^{-1}$			kJ mol^{-1}			
0	0.	0.	INFINITE	-16.347	176.574		
100	38.358	232.170	361.488	-12.932			
200	66.479	266.918	305.689	-7.754			
250	79.653	283.215	299.576	-4.090			
298.15	89.842	298.151	298.151	0.	172.644	238.207	-41.733
300	90.187	298.708	298.153	0.167	172.662	238.614	-41.546
350	98.336	313.250	299.283	4.889	173.286	249.560	-37.245
400	104.598	326.808	301.886	9.969	171.898	260.490	-34.017
450	109.434	339.419	305.364	15.325	172.200	271.538	-31.519
500	113.207	351.152	309.363	20.894	172.560	282.560	-29.519
600	118.567	372.303	318.131	32.503	173.979	304.438	-26.504
700	122.071	390.861	327.223	44.547	175.875	326.036	-24.329
800	124.466	407.327	336.226	56.881	178.053	347.341	-22.679
900	126.167	422.091	344.961	69.417	127.276	369.443	-21.442
1000	127.415	435.451	353.352	82.099	130.955	396.154	-20.693
1100	128.358	447.642	361.378	94.890	134.653	422.495	-20.063
1200	129.091	458.843	369.040	107.764	138.354	448.499	-19.523
1300	129.676	469.200	376.351	120.703	142.047	474.195	-19.053
1400	130.156	478.828	383.331	133.695	145.723	499.606	-18.641
1500	130.561	487.822	390.001	146.732	149.378	524.756	-18.274
1600	130.914	496.259	396.381	159.806	153.011	549.662	-17.945
1700	131.229	504.206	402.492	172.913	156.621	574.343	-17.647
1800	131.519	511.715	408.353	186.051	160.211	598.811	-17.377
1900	131.791	518.833	413.982	199.216	163.786	623.080	-17.130
2000	132.050	525.600	419.395	212.408	167.353	647.161	-16.902
2100	132.301	532.049	424.608	225.626	170.919	671.063	-16.692
2200	132.547	538.209	429.632	238.869	174.494	694.796	-16.497
2300	132.789	544.106	434.482	252.135	178.086	718.366	-16.315
2400	133.027	549.763	439.168	265.426	181.707	741.779	-16.144
2500	133.263	555.198	443.702	278.741	185.365	765.041	-15.985
2600	133.496	560.429	448.091	292.079	189.070	788.154	-15.834
2700	133.727	565.472	452.346	305.440	192.831	811.125	-15.692
2800	133.954	570.339	456.473	318.824	196.657	833.955	-15.558
2900	134.179	575.044	460.481	332.231	200.555	856.647	-15.430
3000	134.399	579.596	464.376	345.660	204.531	879.203	-15.308
3100	134.616	584.007	468.165	359.110	208.590	901.625	-15.192
3200	134.829	588.284	471.852	372.583	212.739	923.916	-15.081
3300	135.037	592.436	475.443	386.076	216.979	946.075	-14.975
3400	135.240	596.470	478.944	399.590	221.315	968.103	-14.873
3500	135.437	600.393	482.358	413.124	225.747	990.004	-14.775
3600	135.630	604.212	485.690	426.677	230.279	1011.776	-14.680
3700	135.816	607.930	488.944	440.249	234.909	1033.421	-14.589
3800	135.997	611.555	492.123	453.840	239.639	1054.938	-14.501
3900	136.172	615.089	495.231	467.449	244.468	1076.331	-14.416
4000	136.341	618.539	498.271	481.074	249.395	1097.599	-14.333
4100	136.504	621.908	501.245	494.717	254.419	1118.741	-14.253
4200	136.662	625.199	504.157	508.375	259.538	1139.761	-14.175
4300	136.813	628.417	507.010	522.049	264.751	1160.657	-14.099
4400	136.959	631.564	509.805	535.738	270.054	1181.431	-14.025
4500	137.099	634.643	512.545	549.440	275.447	1202.084	-13.953
4600	137.233	637.658	515.232	563.157	280.927	1222.615	-13.883
4700	137.361	640.611	517.869	576.887	286.491	1243.027	-13.815
4800	137.485	643.504	520.456	590.629	292.137	1263.320	-13.748
4900	137.603	646.340	522.996	604.384	297.862	1283.493	-13.682
5000	137.715	649.121	525.491	618.150	303.663	1303.549	-13.618
5100	137.823	651.849	527.942	631.926	309.539	1323.489	-13.555
5200	137.926	654.526	530.351	645.714	315.486	1343.313	-13.494
5300	138.024	657.155	532.718	659.512	321.502	1363.021	-13.433
5400	138.118	659.735	535.047	673.319	327.584	1382.616	-13.374
5500	138.208	662.271	537.337	687.135	333.730	1402.096	-13.316
5600	138.293	664.762	539.590	700.960	339.937	1421.466	-13.259
5700	138.374	667.210	541.808	714.793	346.204	1440.722	-13.203
5800	138.451	669.617	543.991	728.635	352.528	1459.869	-13.148
5900	138.524	671.985	546.140	742.484	358.906	1478.908	-13.093
6000	138.594	674.313	548.257	756.339	365.337	1497.836	-13.040

PREVIOUS: December 1977 (1 atm) CURRENT: December 1977 (1 bar)

Sulfur Fluoride, Ion (SF_5^+) $F_5S_1^+$(g)

SULFUR FLUORIDE, ION (SF_5^-) IDEAL GAS M_r = 127.052564

$\Delta_a H°(0\ K)$ = 1586.154 ± 41.8 kJ mol^{-1}
S°(298.15 K) = [315.571 ± 2.1] J K^{-1} mol^{-1}

$\Delta_f H°(0\ K)$ = 1255.192 ± 33.5 kJ mol^{-1}
$\Delta_f H°(298.15\ K)$ = [-1268.706] kJ mol^{-1}

Vibrational Frecuencies and Degeneracies

ν, cm^{-1}	ν, cm^{-1}	ν, cm^{-1}
796(1)	435(1)	595 (2)
522(1)	269(1)	[435](2)
470(1)	342(1)	241 (2)

Ground State Quantum Weight: [1] σ = 4
Point Group: C_{4v}
Bond Distances: S-F* = [1.56] Å S-F** = [1.46] Å
Bond Angles: F**-S-F* = [85°] F*-S-F* = 90°
(* - equatorial ** - axial)
Product of the Moments of Inertia: $I_A I_B I_C$ = [1.256203 x 10^{-113}] g^3 cm^6

Enthalpy of Formation

The threshold energy for the dissociative electron attachment process $e^- + SF_6(g) = SF_5^-(g) + F(g)$ has been measured directly by means of mass spectrometry (1-4). In addition, mass spectrometric studies (5) have been reported on various electron transfer reactions which provide independent estimates of the enthalpy for the above process. Presented below is a summary of the results derived from these measurements. Also included in the summary are results obtained from an electron-impact study (1) of the process $e^- + SF_5Cl(g) = SF_5^-(g) + Cl(g)$. We assume that the fragment-ions are formed in each process with no excess or kinetic energies. As a result, the derived electron affinities represent lower limits to the true value. Other reported values of EA(SF_5) include >1.465 eV (9), 3.2 eV (5) and 3.66±0.04 eV (10). The first two results are based on charge-transfer studies (5-9) while the latter value represents a direct measurement of EA using the magnetron technique (10).

Source	Reaction	Method	AP or $\Delta_r H°(0\ K)$ kcal mol^{-1}	$\Delta_f H°(SF_5^-, g, 0\ K)^a$ kcal mol^{-1}	EA(SF_5)^a eV
(1)	A	Electron-impact	2.31	-304.41	>3.89
(2)	A	RPD	1.15	-305.56	>3.94
(3)	A	RPD	5.77	-300.95	>3.74
(4)	A	Electron-impact	~0	-308.72	>3.99
(5)	A	Electron transfer	11.53	-295.19	>3.49
(6)	A	Collisional ionization	11.45	-295.07	>3.49
(1)	B	Electron-impact	4.61	-271.45	>2.46

Reactions: (A) $e^- + SF_6(g) = SF_5^-(g) + F(g)$
(B) $e^- + SF_5(g) = SF_5^-(g) + Cl(g)$

^{a}Auxiliary $\Delta_f H°(0\ K)$ data (in kcal mol^{-1}): SF_6 (7), -288.36±0.2; F (7), 18.36±0.4; Cl (7), 28.587; SF_5Cl (8), -247.48; SF_5 (7), -214.7±3.6.

We reject the electron-impact results on the dissociative electron attachment process for SF_5Cl since Harland and Thynne (1) also reported results for SF_3^- and SF_4^- which lead to EA that are too low (7). The other results are reasonably consistent with the mean EA value of 3.7±0.2 eV. This value corresponds to $\Delta_f H°(SF_5^-, g, 0\ K)$ = -300.0±8.0 kcal mol^{-1} which we adopt.

$\Delta_f H°$ (298.15 K) is -303.228 kcal mol^{-1} and the fluoride-ion affinity is IA(SF_4) = 59.1 kcal mol^{-1}.

Heat Capacity and Entropy

The infrared and Raman spectra (11, 12) of the solid adduct $CsF·SF_4$ are consistent with a slightly distorted tetragonal pyramidal structure for the SF_5^- anion. We adopt this configuration assuming the length of the four equivalent bonds is the same as that in SF_6 (7). The axial bond length is taken to be slightly shorter (0.1 Å) than the equatorial bonds, in accordance with the pattern followed in the related molecules ClF_5 (7) and TeF_5^- (13). The F**-S-F* bond angle has been estimated by Drullinger and Griffiths (11). The principal moments of inertia are: I_A = 30.4769 x 10^{-39} and $I_B = I_C$ = 20.3023 x 10^{-39} g cm^2.

Smardzewski and Fox (14) reported the matrix infrared spectra of SF_5^- formed during the vacuum ultraviolet photolysis of SF_6 and S_2F_{10} in argon matrices at 8 K. Values of ν_1, ν_3 and ν_4 are taken from this recent study. The other six fundamentals are those measured by Christe et al. (12) for the SF_5^- anion observed in the Raman spectra of the solid adduct $CsF·SF_4$. An independent set (11) of vibrational frequencies for the SF_5^- anion show good agreement with those adopted here except for ν_8. Drullinger and Griffiths (11) observed a weak band at 388 cm^{-1} in the infrared spectra of solid $Cs^+SF_5^-$ which they assigned to ν_8. This band does not appear in the spectrum reported by Christe et al. (12) who assumed a double coincidence between ν_4 and ν_8 as observed for the related molecule ClF_5 (12). We adopt this latter assignment.

Drullinger and Griffiths (11) have previously evaluated thermal functions for SF_5^- at selected temperatures in the range 200-2000 K. Using the bond lengths estimated by Drullinger and Griffiths and their vibrational frequencies, we calculate entropies which are about 0.8 cal K^{-1} mol^{-1} lower than those reported. The reason for this discrepancy is not apparent.

References

1. P. Harland and J. C. J. Thynne, J. Phys. Chem. 73, 4031 (1969).
2. W. M. Hickman and D. Berg, Advan. Mass. Spec. 1958, 458 (1958).
3. C. Lifshitz, A. M. Peers, R. Grajower, and M. Weiss, J. Chem. Phys. 53, 4605 (1970).
4. C. L. Chen and P. J. Chantry, Bull. Amer. Phys. Soc. 15, 418 (1970).
5. C. Lifshitz, T. O. Tiernan, and B. M. Hughes, J. Chem. Phys. 59, 3182 (1973).
6. R. N. Compton and C. D. Cooper, J. Chem. Phys. 59, 4140 (1973).
7. JANAF Thermochemical Tables: SF_6(g), 6-30-76; F(g), 9-30-65; Cl(g), 6-30-72; SF_3^-(g) and SF_4^-(g), 12-31-76; ClF_5(g), 6-30-66; SF_5(g) and SF_5^+(g), 12-31-77.
8. U. S. Nat. Bur. Stand. Tech. Note 270-3, 1968.
9. F. C. Fehsenfeld, J. Chem. Phys. 54, 438 (1971).
10. J. Kay and F. M. Page, Trans. Faraday Soc. 60, 1042 (1964).
11. L. F. Drullinger and J. E. Griffiths, Spectrochim. Acta. 27A, 1793 (1971).
12. K. O. Christe, E. C. Curtis, C. J. Schack, and D. Pilipovich, Inorg. Chem. 11, 1679 (1972).
13. H. A. Edwards and M. A. Mouty, J. Chem. Soc. 1969, 703 (1969).
14. R. R. Smardzewski and W. B. Fox, J. Chem. Phys. 67, 2309 (1977).

Sulfur Fluoride, Ion (SF_5^-) $F_5S_1^-(g)$

Enthalpy Reference Temperature = T_r = 298.15 K — Standard State Pressure = p° = 0.1 MPa

T/K	$C_p°$ (J K^{-1}mol^{-1})	S° (J K^{-1}mol^{-1})	-[G°-H°(T_r)]/T (J K^{-1}mol^{-1})	H°-H°(T_r) (kJ mol^{-1})	$\Delta_f H°$ (kJ mol^{-1})	$\Delta_f G°$ (kJ mol^{-1})	Log K_f
0	0.	0.	INFINITE	-19.158	-1255.192		
100	46.921	235.666	390.718	-15.505			
200	80.815	279.023	324.282	-9.052			
250	92.943	298.427	317.198	-4.693			
298.15	101.535	315.571	315.571	0.	-1268.706	-1195.827	209.504
300	101.815	316.200	315.573	0.188	-1268.743	-1195.375	208.133
350	108.286	332.407	316.838	5.449	-1269.660	-1183.072	176.564
400	113.061	347.194	319.722	10.989	-1272.666	-1170.559	152.859
450	116.644	360.727	323.537	16.735	-1274.052	-1157.721	134.385
500	119.381	373.165	327.887	22.639	-1275.437	-1144.718	119.588
600	123.187	395.294	337.324	34.782	-1277.641	-1118.359	97.362
700	125.627	414.480	347.006	47.231	-1279.496	-1091.662	81.461
800	127.273	431.369	356.517	59.881	-1281.160	-1064.713	69.519
900	128.432	446.430	365.686	72.670	-1335.842	-1036.482	60.156
1000	129.276	460.008	374.450	85.557	-1336.114	-1003.206	52.402
1100	129.910	472.360	382.798	98.518	-1336.403	-969.901	46.057
1200	130.396	483.686	390.740	111.534	-1336.717	-936.569	40.768
1300	130.778	494.139	398.297	124.594	-1337.062	-903.209	36.291
1400	131.083	503.842	405.494	137.688	-1337.441	-869.822	32.453
1500	131.330	512.894	412.355	150.809	-1337.858	-836.407	29.126
1600	131.533	521.377	418.907	163.952	-1338.314	-802.962	26.214
1700	131.702	529.356	425.171	177.114	-1338.806	-769.487	23.643
1800	131.844	536.888	431.171	190.292	-1339.333	-735.983	21.358
1900	131.965	544.020	436.924	203.482	-1339.890	-702.449	19.312
2000	132.068	550.792	442.450	216.684	-1340.471	-668.884	17.469
2100	132.157	557.238	447.763	229.895	-1341.069	-635.290	15.802
2200	132.234	563.387	452.880	243.115	-1341.674	-601.668	14.285
2300	132.301	569.267	457.814	256.342	-1342.279	-568.017	12.900
2400	132.360	574.899	462.576	269.575	-1342.873	-534.341	11.630
2500	132.412	580.303	467.178	282.814	-1343.448	-500.640	10.460
2600	132.459	585.497	471.629	296.057	-1343.994	-466.917	9.380
2700	132.500	590.497	475.940	309.305	-1344.503	-433.173	8.380
2800	132.537	595.317	480.118	322.557	-1344.967	-399.412	7.451
2900	132.571	599.968	484.171	335.813	-1345.377	-365.634	6.586
3000	132.601	604.463	488.106	349.071	-1345.729	-331.844	5.778
3100	132.628	608.811	491.930	362.333	-1346.016	-298.043	5.022
3200	132.652	613.022	495.648	375.597	-1346.234	-264.234	4.313
3300	132.675	617.105	499.267	388.863	-1346.377	-230.419	3.647
3400	132.696	621.066	502.792	402.132	-1346.444	-196.601	3.020
3500	132.714	624.913	506.226	415.402	-1346.432	-162.782	2.429
3600	132.732	628.651	509.575	428.674	-1346.339	-128.964	1.871
3700	132.748	632.288	512.843	441.948	-1346.164	-95.150	1.343
3800	132.762	635.829	516.033	455.224	-1345.906	-61.343	0.843
3900	132.776	639.278	519.149	468.501	-1345.566	-27.542	0.369
4000	132.788	642.639	522.194	481.779	-1345.144	6.248	-0.082
4100	132.800	645.918	525.172	495.058	-1344.640	40.026	-0.510
4200	132.811	649.119	528.085	508.339	-1344.056	73.792	-0.918
4300	132.821	652.244	530.937	521.621	-1343.393	107.542	-1.306
4400	132.830	655.297	533.728	534.903	-1342.653	141.277	-1.677
4500	132.839	658.283	536.463	548.187	-1341.836	174.994	-2.031
4600	132.847	661.202	539.143	561.471	-1340.946	208.691	-2.370
4700	132.855	664.059	541.771	574.756	-1339.984	242.369	-2.694
4800	132.862	666.857	544.348	588.042	-1338.952	276.025	-3.004
4900	132.869	669.596	546.876	601.329	-1337.852	309.659	-3.301
5000	132.875	672.281	549.357	614.616	-1336.686	343.270	-3.586
5100	132.881	674.912	551.794	627.904	-1335.457	376.857	-3.860
5200	132.887	677.492	554.186	641.192	-1334.166	410.420	-4.123
5300	132.892	680.024	556.537	654.481	-1332.816	443.957	-4.375
5400	132.898	682.508	558.846	667.770	-1331.409	477.467	-4.619
5500	132.902	684.946	561.117	681.060	-1329.947	510.951	-4.853
5600	132.907	687.341	563.350	694.351	-1328.431	544.409	-5.078
5700	132.911	689.693	565.546	707.642	-1326.864	577.838	-5.295
5800	132.915	692.005	567.706	720.933	-1325.247	611.239	-5.505
5900	132.919	694.277	569.832	734.225	-1323.583	644.614	-5.707
6000	132.923	696.511	571.925	747.517	-1321.873	677.958	-5.902

PREVIOUS: December 1977 (1 atm) CURRENT: December 1977 (1 bar)

Sulfur Fluoride, Ion (SF_5^-) $F_5S_1^-(g)$

HYDROGEN FLUORIDE (H_6F_6) IDEAL GAS M_r = 120.038058

$S°(298.15\ K)$ = [466.420 ± 25.1] $J\ K^{-1}\ mol^{-1}$

$\Delta_fH°(0\ K) = -1791.171 \pm 8.4\ kJ\ mol^{-1}$

$\Delta_fH°(298.15\ K) = -1807.655 \pm 8.4\ kJ\ mol^{-1}$

Vibrational Frequencies and Degeneracies

ν, cm^{-1}
[202](6)
[53](6)
[962](6)
[552](6)
[3060](6)

Ground State Quantum Weight: 1 σ = [6]

Point Group: [C_6]

Bond Distance: F-F = [2.5] Å

Product of the Moments of Inertia: $I_AI_BI_C = [4.833984 \times 10^{-112}]\ g^3\ cm^6$

Enthalpy of Formation

The enthalpy of 6 HF(g) → H_6F_6(g) was taken as the mean of 3rd law values which were determined from the equilibrium data of two investigations. Briegleb and Storhmeier (1) measured the vapor density of associated HF between 20 and 60°C and between 50 and 650 torr. Franck and Meyer (2) measured $C_p°$ between -20 and 100°C and between 100 and 700 torr. Each investigation evaluated K_p at n = 2, 3, 4, ... for the reactions n(HF) → $(HF)_n$ and reported 2nd law values of Δ_fH and Δ_fS. At n = 6 their 2nd and 3rd law values differed by 0.8 kcal mol^{-1}, which was taken as an estimate of error, while the calculated 3rd law values differed by 0.6 kcal mol^{-1}. Using $\Delta_rH°(298.15\ K) = -41.20$ kcal mol^{-1} and auxiliary JANAF data (3) gives the enthalpy of formation of H_6F_6(g).

Heat Capacity and Entropy

The molecular structure of H_6F_6 was assumed as planar with the F atoms forming the vertices of a regular hexagon and with the H atoms also lying on the circumscribed circle. That rings are more stable than open chains was confirmed by Del Bene and Pople's (4) theoretical molecular-orbital studies on HF polymers. The length of side (F-F axis) was taken from Atoji and Lipscomb's (5) X-ray studies of solid HF(F-F = 2.49 Å) and agrees with the 2.52 Å which Janzen and Bartell (6) determined for HF gaseous polymers by electron diffraction. The low F-bending frequency (53 cm^{-1}) was taken from Boutin et al. (7). The other vibrational frequencies were taken from Kittelberger and Hornig's (8) work on crystalline HF. Huong and Couzi (9) and Smith (10) have made spectral studies of the gas phase in the range from 350 to 4000 cm^{-1}.

The National Bureau of Standards prepared this table (11) by critical analysis of data existing in 1972. Using molecular constants and $\Delta_fH°$ selected by NBS (11), we recalculate the table in terms of 1973 fundamental constants (12), 1975 atomic weights (13), and current JANAF reference states for the elements.

References

1. G. Briegleb and W. Strohmeier, Z. Elektrochem. 57, 668 (1953).
2. E. U. Franck and F. Meyer, Z. Electrochem. 63, 571 (1959).
3. JANAF Thermochemical Tables: HF(g), 6-30-77.
4. J. E. Del Bene and J. A. Pole, J. Chem. Phys. 55, 2296 (1971).
5. M. Atoji and W. N. Lipscomb, Acta Cryst. 7, 173 (1954).
6. J. Janzen and L. S. Bartell, J. Chem. Phys. 50 3611 (1969).
7. H. Boutin, G. J. Safford, and V. Brajovic, J. Chem. Phys. 39, 3135 (1963).
8. J. S. Kittelberger and D. F. Hornig, J. Chem. Phys. 46, 3099 (1967).
9. P. V. Huong and M. Couzi, J. Chim. Phys. 66, 1309 (1969).
10. D. F. Smith, J. Chem. Phys. 28, 1040 (1958), 48, 1429 (1968).
11. S. Abramowitz et al., U. S. Nat. Bur. Stand. Report 10904, 239, 1972.
12. CODATA Task Group on Fundamental Constants, CODATA Bulletin 11, December, 1973.
13. IUPAC Commission on Atomic Weights, Pure Appl. Chem. 47, 75 (1976).

Hydrogen Fluoride (H_6F_6) F_6H_6(g)

Enthalpy Reference Temperature = T_r = 298.15 K — Standard State Pressure = p° = 0.1 MPa

T/K	$C_p°$ (J K⁻¹mol⁻¹)	S° (J K⁻¹mol⁻¹)	-[G°-H°(T_r)]/T (J K⁻¹mol⁻¹)	H°-H°(T_r) (kJ mol⁻¹)	$\Delta_fH°$ (kJ mol⁻¹)	$\Delta_fG°$ (kJ mol⁻¹)	Log K_f
0	0.	0.	INFINITE	-35.392	-1791.171	-1791.171	INFINITE
100	107.705	318.614	597.486	-27.887	-1801.385	-1751.917	915.107
200	142.277	404.563	481.012	-15.290	-1805.656	-1700.530	444.133
250	156.333	437.855	469.123	-7.817	-1806.879	-1674.097	349.783
298.15	168.056	466.420	466.420	0.	-1807.655	-1648.447	288.801
300	168.470	467.461	466.423	0.311	-1807.678	-1647.459	286.848
350	178.673	494.221	468.513	8.998	-1808.105	-1620.717	241.879
400	187.132	518.649	473.276	18.149	-1808.215	-1593.937	208.147
450	194.152	541.108	479.582	27.687	-1808.060	-1567.159	181.911
500	200.051	561.878	486.786	37.546	-1807.684	-1540.410	160.926
600	209.552	599.229	502.484	58.047	-1806.391	-1487.066	129.461
700	217.186	632.122	518.701	79.395	-1804.511	-1433.987	107.005
800	223.740	661.561	534.750	101.448	-1802.144	-1381.211	90.184
900	229.555	688.256	550.346	124.119	-1799.360	-1328.757	77.119
1000	234.768	712.716	565.377	147.340	-1796.220	-1276.633	66.684
1100	239.440	735.316	579.812	171.054	-1792.782	-1224.839	58.163
1200	243.615	756.332	593.656	195.211	-1789.098	-1173.368	51.075
1300	247.332	775.981	606.934	219.762	-1785.217	-1122.214	45.091
1400	250.632	794.434	619.674	244.664	-1781.180	-1071.365	39.973
1500	253.559	811.827	631.910	269.876	-1777.023	-1020.808	35.548
1600	256.153	828.276	643.674	295.364	-1772.775	-970.532	31.685
1700	258.454	843.876	654.995	321.097	-1768.458	-920.523	28.284
1800	260.497	858.708	665.904	347.046	-1764.090	-870.770	25.269
1900	262.315	872.842	676.427	373.189	-1759.682	-821.262	22.578
2000	263.936	886.339	686.587	399.503	-1755.241	-771.986	20.162
2100	265.384	899.252	696.409	425.970	-1750.771	-722.933	17.982
2200	266.681	911.628	705.912	452.575	-1746.271	-674.094	16.005
2300	267.846	923.509	715.116	479.302	-1741.741	-625.460	14.205
2400	268.895	934.931	724.039	506.140	-1737.177	-577.023	12.559
2500	269.842	945.927	732.696	533.078	-1732.574	-528.778	11.048
2600	270.699	956.527	741.102	560.105	-1727.926	-480.718	9.658
2700	271.476	966.758	749.271	587.215	-1723.228	-432.838	8.374
2800	272.183	976.644	757.216	614.398	-1718.476	-385.132	7.185
2900	272.828	986.207	764.949	641.649	-1713.663	-337.597	6.081
3000	273.416	995.466	772.479	668.962	-1708.786	-290.230	5.053
3100	273.956	1004.441	779.818	696.331	-1703.841	-243.027	4.095
3200	274.451	1013.146	786.974	723.752	-1698.824	-195.983	3.199
3300	274.906	1021.599	793.956	751.220	-1693.733	-149.097	2.360
3400	275.325	1029.812	800.773	778.732	-1688.567	-102.368	1.573
3500	275.712	1037.798	807.432	806.284	-1683.325	-55.791	0.833
3600	276.070	1045.570	813.939	833.873	-1678.006	-9.365	0.136
3700	276.402	1053.139	820.302	861.497	-1672.610	36.912	-0.521
3800	276.710	1060.514	826.527	889.153	-1667.139	83.042	-1.141
3900	276.996	1067.706	832.619	916.838	-1661.592	129.028	-1.728
4000	277.263	1074.722	838.584	944.551	-1655.973	174.870	-2.284
4100	277.511	1081.572	844.428	972.290	-1650.281	220.570	-2.810
4200	277.744	1088.262	850.154	1000.053	-1644.521	266.132	-3.310
4300	277.961	1094.800	855.768	1027.838	-1638.692	311.554	-3.785
4400	278.164	1101.192	861.273	1055.645	-1632.798	356.841	-4.236
4500	278.355	1107.446	866.674	1083.471	-1626.841	401.993	-4.666
4600	278.534	1113.565	871.975	1111.315	-1620.822	447.011	-5.076
4700	278.702	1119.557	877.179	1139.177	-1614.745	491.899	-5.467
4800	278.861	1125.427	882.290	1167.055	-1608.610	536.658	-5.840
4900	279.010	1131.178	887.311	1194.949	-1602.420	581.286	-6.197
5000	279.151	1136.816	892.245	1222.857	-1596.176	625.789	-6.538
5100	279.284	1142.346	897.095	1250.779	-1589.880	670.164	-6.864
5200	279.409	1147.770	901.864	1278.713	-1583.533	714.419	-7.176
5300	279.528	1153.093	906.554	1306.660	-1577.136	758.549	-7.476
5400	279.641	1158.319	911.168	1334.619	-1570.690	802.558	-7.763
5500	279.748	1163.452	915.708	1362.588	-1564.196	846.446	-8.039
5600	279.849	1168.493	920.177	1390.568	-1557.654	890.218	-8.304
5700	279.945	1173.447	924.577	1418.558	-1551.064	933.871	-8.558
5800	280.037	1178.317	928.910	1446.557	-1544.426	977.409	-8.803
5900	280.124	1183.105	933.178	1474.565	-1537.741	1020.831	-9.038
6000	280.207	1187.813	937.383	1502.582	-1531.007	1064.139	-9.264

PREVIOUS: June 1977 (1 atm) CURRENT: June 1977 (1 bar)

Hydrogen Fluoride (H_6F_6) F_6H_6(g)

MOLYBDENUM FLUORIDE (MoF_6) LIQUID M_r = 209.930418

$S°(298.15\ K)$ = 259.693 ± 0.25 J K^{-1} mol^{-1}
T_{trs} = 263.5 ± 0.02 K
T_{fus} = 290.76 ± 0.02 K

$\Delta_f H°(298.15\ K)$ = -1585.656 ± 0.11 kJ mol^{-1}
$\Delta_{trs} H°$ = 8.171 ± 0.008 kJ mol^{-1}
$\Delta_{fus} H°$ = 4.326 ± 0.004 kJ mol^{-1}

Enthalpy of Formation

The adopted $\Delta_f H°(MoF_6$, ℓ, 298.15 K) = -378.98±0.25 kcal mol^{-1}(-1585.657±0.11 kJ mol^{-1}) is calculated from $\Delta_f H°(MoF_6$, g, 298.15 K) = - 372.29±0.22 kcal mol^{-1} (1) minus $\Delta_{vap}H°(298.15\ K)$ = 6.69±0.03 kcal mol^{-1} from our 3rd law analysis of vapor pressure data.

The enthalpy of reaction of MoF_6(ℓ) with excess of OH^-(aq) has been measured by Myers and Brady (2) and by Nuttall et al. (3); the derived $\Delta_f H°(MoF_6$, ℓ, 298.15 K) is -393.2 and -380.81±1.3-1.0 kcal mol^{-1}, respectively. The agreement of the latter value with the adopted value is not quite within the assigned uncertainties.

Heat Capacity and Entropy

The heat capacity of MoF_6 has been measured by Brady et al. (50-300 K, 4) and by Osborne et al. (4-350 K, 5). We adopt the more extensive data of Osborne et al. (5), extrapolating $C_p°$ linearly with temperature outside the observed range. $S°(298.15\ K)$ = 62.07±0.06 cal K^{-1} mol^{-1} is derived from heat capacity, heat of transition, and heat of melting data and $S°(5\ K)$ = 0.024 cal K^{-1} mol^{-1} from the T^3 extrapolation of Osborne et al. (5).

Transition Data

Osborne et al. (5) confirmed the solid state transition (263.5±0.02 K, $\Delta_{trs}H°$ = 1.953±0.002 kcal mol^{-1} found by Brady et al. (263.6 K, $\Delta_{trs}H°$ = 1.957±0.010 kcal mol^{-1}, 4).

Fusion Data

Melting point and enthalpy of fusion data have been reported by several investigators. These results are summarized below:

Source	Method	T_{fus}/K	$\Delta_{fus}H°$, kcal mol^{-1}
(4)	Calorimetric	290.7	1.059±0.010
(5)	Calorimetric	290.76±0.02	1.034±0.001
(6)	Vapor Pressure	290.65	0.858*
(7)	Vapor Pressure	290.55	0.902*
(8)	Vapor Pressure	290.95	0.987*

*Calculated as the difference between the 2nd law heat of sublimation at 291 K and the adopted 3rd law $\Delta_{vap}H°(298.15\ K)$ corrected to 291 K.

We adopt the calorimetric values of Osborne et al. (5).

Vaporization Data

Our 2nd and 3rd law analysis of four sets of vapor pressure data produces the following results. The second virial coefficient determined by Osborne et al. (5) is used to convert all four vapor pressure data sets to ideal gas fugacity.

Source	T/K	Data Points	$\Delta_{vap}H°(298.15\ K)$, kcal mol^{-1} 2nd law	3rd law	Drift cal K^{-1} mol^{-1}
(6)	291-295	4	5.981±0.467	6.691±0.012	2.42±1.59
(7)	291-307	Eqn	6.737	6.692±0.002	-0.15
(5)	291-320	21	6.691±0.002	6.691±0.000	0.00±0.01
(8)	291-335	Eqn	6.696	6.691±0.001	-0.02

The agreement in 3rd law values is excellent and we adopt $\Delta_{vap}H°(298.15\ K)$ = 6.69±0.03 kcal mol^{-1}. T_{vap} = 308.0 K and $\Delta_{vap}H°(308\ K)$ = 6.57 kcal mol^{-1} are calculated for a fugacity of one atmosphere for the process MoF_6(ℓ) = MoF_6(g). For vaporization to the real gas, T_{vap} = 307.04 K and $\Delta_{vap}H°(307\ K)$ = 6.51 kcal mol^{-1}.

This table is based on the critical analysis of data existing in 1978 by Brewer (9).

References

1. JANAF Thermochemical Tables: MoF_6(g), 9-30-78.
2. O. E. Myers and A. P. Brady, J. Phys. Chem. 64, 591 (1960).
3. R. L. Nuttall, K. L. Churney and M. V. Kilday, J. Res. Nat. Bur. Stand. 83, 335 (1978).
4. A. P. Brady, O. E. Myers, and J. K. Clauss, J. Phys. Chem. 64, 558 (1960).
5. D. W. Osborne, F. Schreiner, J. G. Malm, H. Selig, and L. Rochester, J. Chem. Phys. 44, 2802 (1966).
6. O. Ruff and E. Ascher, Z. Anorg. Allg. Chem. 196, 413 (1931).
7. G. H. Cady and G. B. Hargreaves, J. Chem. Soc. 1563 (1961).
8. D. Meixner, A. Heintz, and R. N. Lichtenthaler, Ber. Bunsenges, Phys. Chem. 82, 220 (1978).
9. L. Brewer, Materials and Molecular Research Division, Lawrence Berkeley Laboratory, University of California, Berkeley; personal communication, Sept. 1978, of preliminary draft of review to be submitted for publication in Atomic Energy Review, International Atomic Energy Agency, Vienna, Austria.

Molybdenum Fluoride (MoF_6) F_6Mo_1(l)

Enthalpy Reference Temperature = T_r = 298.15 K — Standard State Pressure = p° = 0.1 MPa

T/K	$C_p°$ (J $K^{-1}mol^{-1}$)	S°	$-[G°-H°(T_r)]/T$	$H°-H°(T_r)$ (kJ mol^{-1})	$\Delta_f H°$	$\Delta_f G°$	Log K_f
0							
100							
200	152.720	195.532	274.669	-15.827	-1590.266	-1511.085	394.655
298.15	169.795	259.693	259.693	0.	-1585.656	-1473.170	258.093
300	170.117	260.744	259.696	0.314	-1585.560	-1472.473	256.380
400	187.512	312.066	266.576	18.196	-1579.792	-1435.615	187.472
500	204.907	355.777	280.143	37.817	-1572.817	-1400.351	146.294
600	222.302	394.673	296.044	59.177	-1564.493	-1366.617	118.975
700	239.697	430.247	312.709	82.277	-1554.721	-1334.389	99.573
800	257.092	463.390	329.494	107.117	-1543.431	-1303.665	85.121
900	274.487	494.675	346.125	133.696	-1530.581	-1274.450	73.967
1000	291.882	524.496	362.482	162.014	-1516.145	-1246.750	65.124

PREVIOUS: June 1970 CURRENT: September 1978

MOLYBDENUM FLUORIDE (MoF_6) IDEAL GAS M_r = 209.930418

$S°(298.15 K) = 350.710 \pm 0.42 J K^{-1} mol^{-1}$

$\Delta_f H°(0 K) = -1550.595 \pm 0.92 kJ mol^{-1}$
$\Delta_f H°(298.15 K) = -1557.661 \pm 0.92 kJ mol^{-1}$

Vibrational Frequencies and Degeneracies

ν, cm^{-1}	ν, cm^{-1}
741.8(1)	264(3)
652.0(2)	317(3)
742 (3)	117(3)

Ground State Quantum Weight: 1 σ= 24
Point Group: O_h
Bond Distance: Mo-F = 1.82 ± 0.01 Å
Bond Angle: F-Mo-F = 90°
Product of the Moments of Inertia: $I_A I_B I_C = 7.303442 \times 10^{-113} g^3 cm^6$

Enthalpy of Formation

The adopted $\Delta_f H°(MoF_6, g, 298.15 K) = -372.29 \pm 0.22 kcal mol^{-1}$ is from the measurements of the direct combination of the elements by Settle et al. (1) with the minor corrections (2) for a change in atomic weights and new values for the enthalpy of formation of MoF_5(cr) and NiF_2(cr).

Burgess et al. (3) reacted MoF_6(g) with excess NaOH solution and derived $\Delta_f H°(MoF_6, g, 298.15 K) = -373.8 kcal mol^{-1}$. The result of the direct measurement of Settle et al. (1) is preferred. The value of $\Delta_f H°(0 K)$ combined with JANAF data (15) for Mo(g) and F(g) gives $\Delta_{at} H°(0 K) = 637.9 \pm 1.5 kcal mol^{-1}$ and an average Mo-F bond energy of $\Delta_{at} H°(0 K)/6 = 106.3 \pm 0.3 kcal mol^{-1}$.

Heat Capacity and Entropy

From the Raman spectra of liquid MoF_6, Tanner and Duncan (4) concluded that the structure of the molecule was a regular octahedron with O_h symmetry rather than the unsymmetrical structure deduced by Braune and Pinnow (5) from electron diffraction studies. This discrepancy was explained by Glauber and Schomaker (6).

Values of the Mo-F bond distance vary from 1.84 Å from electron diffraction (7) to 1.80 Å from x-ray diffraction of the cubic solid (8). Seip and Seip (9) calculated a bond distance of 1.82 Å from electron diffraction measurements. We adopt this value with an uncertainty of ±0.01 Å (10). The individual moments of inertia are $I_A = I_B = I_C = 41.8000 \times 10^{-39} g cm^2$.

The vibrational frequencies are those adopted by Brewer (10) from the assignment by Weinstock and Goodman (11) and the more recent studies of Claasen et al. (12) and McDowell et al. (13).

$S°(MoF_6, g, 298.15 K) = 83.796 cal K^{-1} mol^{-1}$ calculated from the molecular data is in excellent agreement with $S°(MoF_6, \ell, 298.15 K) = 62.068 cal K^{-1} mol^{-1}$ plus $\Delta_{vap} S°(298.15 K) = 21.727 cal K^{-1} mol^{-1}$ to give $S°(MoF_6, g, 298.15 K) = 83.795 \pm 0.10 cal K^{-1} mol^{-1}$ based on our analysis of the data of Osborne et al. (14).

This table is based on the critical analysis of data existing in 1978 by Brewer (10).

References

1. J. L. Settle, H. M. Feder and W. N. Hubbard, J. Phys. Chem. 65, 1337 (1961).
2. R. L. Nuttall, K. L. Churney and M. V. Kilday, J. Res. Nat. Bur. Stand. 83, 335 (1978).
3. J. Burgess, I. Haigh and R. D. Peacock, J. Chem. Soc., Dalton Trans. 1062 (1974).
4. K. N. Tanner and A. B. F. Duncan, J. Amer. Chem. Soc. 73, 1164 (1951).
5. H. Braune and P. Pinnow, Z. Phys. 35B, 239 (1937).
6. R. Glauber and V. Schomaker, Phys. Rev. 89, 667 (1953).
7. G. M. Nazarian, Ph. D. Thesis, California Institute of Technology, 1957.
8. J. H. Levy, P. L. Sanger, J. C. Taylor and P. W. Wilson, Acta Cryst. B31, 1065 (1965).
9. H. M. Seip and R. Seip, Acta Chem. Scand. 20, 2698 (1966); Select. Top. Struct. Chem., 26 (1967).
10. L. Brewer, Materials and Molecular Research Division, Lawrence Berkeley Laboratory, University of California, Berkeley; personal communication, Sept. 1978, of preliminary draft of review to be submitted for publication in Atomic Energy Review, International Atomic Energy Agency, Vienna, Austria.
11. B. Weinstock and G. L. Goodman, Advan. Chem. Phys. 9, 169 (1965).
12. H. H. Claasen, G. L. Goodman, J. H. Halloway and H. J. Selig, J. Chem. Phys. 53, 341 (1970).
13. R. S. McDowell, R. J. Sherman, B. Asprey and R. J. Kennedy, J. Chem. Phys. 62, 3974 (1975).
14. D. W. Osborne, F. Schreiner, J. G. Malm, H. Selig and L. Rochester, J. Chem. Phys. 44, 2802 (1966).
15. JANAF Thermochemical Tables: Mo(g), 3-31-78; F(g), 9-30-65.

Molybdenum Fluoride (MoF_6) $F_6Mo_1(g)$

Enthalpy Reference Temperature = T_r = 298.15 K Standard State Pressure = p° = 0.1 MPa

T/K	$C_p°$ (J $K^{-1}mol^{-1}$)	S° (J $K^{-1}mol^{-1}$)	$-[G°-H°(T_r)]/T$ (J $K^{-1}mol^{-1}$)	$H°-H°(T_r)$ (kJ mol^{-1})	$\Delta_f H°$ (kJ mol^{-1})	$\Delta_f G°$ (kJ mol^{-1})	Log K_f
0	0.	0.	INFINITE	-23.993	-1550.595	-1550.595	INFINITE
100	67.247	249.179	442.877	-19.370	-1555.184	-1528.288	798.296
200	99.881	306.715	361.123	-10.882	-1557.325	-1500.381	391.859
250	111.452	330.294	352.644	-5.587	-1557.656	-1486.099	310.503
298.15	120.276	350.710	350.710	0.	-1557.661	-1472.312	257.943
300	120.574	351.455	350.713	0.223	-1557.657	-1471.782	256.260
350	127.635	370.596	352.208	6.436	-1557.424	-1457.486	217.518
400	133.077	388.010	355.611	12.959	-1557.033	-1443.234	188.467
450	137.291	403.938	360.109	19.723	-1556.536	-1429.038	165.878
500	140.585	418.580	365.234	26.673	-1555.965	-1414.902	147.814
600	145.275	444.658	376.352	40.984	-1554.692	-1386.806	120.732
700	148.347	467.299	387.763	55.675	-1553.328	-1358.932	101.405
800	150.448	487.253	398.977	70.621	-1551.932	-1331.257	86.922
900	151.941	505.064	409.793	85.744	-1550.537	-1303.756	75.668
1000	153.036	521.132	420.137	100.996	-1549.168	-1276.410	66.673
1100	153.861	535.759	429.993	116.342	-1547.844	-1249.199	59.320
1200	154.497	549.175	439.373	131.762	-1546.581	-1222.105	53.197
1300	154.997	561.562	448.302	147.237	-1545.394	-1195.114	48.020
1400	155.398	573.063	456.808	162.758	-1544.295	-1168.212	43.587
1500	155.723	583.796	464.920	178.314	-1543.295	-1141.384	39.747
1600	155.991	593.855	472.667	193.901	-1542.403	-1114.620	36.389
1700	156.214	603.319	480.077	209.511	-1541.624	-1087.907	33.427
1800	156.401	612.253	487.174	225.142	-1540.963	-1061.238	30.796
1900	156.560	620.714	493.982	240.790	-1540.425	-1034.602	28.443
2000	156.696	628.748	500.521	256.453	-1540.012	-1007.991	26.326
2100	156.814	636.396	506.811	272.129	-1539.724	-981.398	24.411
2200	156.916	643.693	512.868	287.816	-1539.562	-954.815	22.670
2300	157.005	650.671	518.709	303.512	-1539.525	-928.236	21.081
2400	157.084	657.354	524.348	319.216	-1539.613	-901.657	19.624
2500	157.153	663.768	529.797	334.928	-1539.829	-875.072	18.284
2600	157.214	669.933	535.069	350.647	-1540.182	-848.475	17.046
2700	157.269	675.868	540.175	366.371	-1540.705	-821.862	15.900
2800	157.318	681.588	545.124	382.100	-1541.449	-795.225	14.835
2900	157.363	687.109	549.925	397.834	-1578.373	-768.507	13.842
3000	157.402	692.445	554.587	413.573	-1577.744	-740.592	12.895
3100	157.439	697.607	559.118	429.315	-1577.031	-712.698	12.009
3200	157.471	702.606	563.524	445.060	-1576.229	-684.829	11.179
3300	157.501	707.452	567.813	460.809	-1575.333	-656.986	10.399
3400	157.528	712.154	571.989	476.560	-1574.340	-629.172	9.666
3500	157.553	716.721	576.059	492.314	-1573.247	-601.389	8.975
3600	157.576	721.159	580.029	508.071	-1572.052	-573.638	8.323
3700	157.597	725.477	583.902	523.830	-1570.753	-545.922	7.707
3800	157.617	729.680	587.683	539.590	-1569.349	-518.243	7.124
3900	157.635	733.775	591.377	555.353	-1567.842	-490.601	6.571
4000	157.652	737.766	594.987	571.117	-1566.230	-462.999	6.046
4100	157.667	741.659	598.517	586.883	-1564.514	-435.440	5.548
4200	157.682	745.459	601.970	602.651	-1562.696	-407.923	5.073
4300	157.695	749.169	605.350	618.420	-1560.778	-380.452	4.622
4400	157.707	752.794	608.660	634.190	-1558.759	-353.025	4.191
4500	157.719	756.339	611.903	649.961	-1556.643	-325.645	3.780
4600	157.730	759.805	615.081	665.734	-1554.431	-298.314	3.387
4700	157.740	763.198	618.196	681.507	-1552.126	-271.032	3.012
4800	157.750	766.519	621.252	697.282	-1549.730	-243.799	2.653
4900	157.759	769.771	624.250	713.057	-1547.244	-216.618	2.309
5000	157.767	772.959	627.192	728.833	-2136.057	-183.758	1.920
5100	157.775	776.083	630.081	744.610	-2135.079	-144.722	1.482
5200	157.783	779.147	632.918	760.388	-2134.166	-105.703	1.062
5300	157.790	782.152	635.706	776.167	-2133.317	-66.703	0.657
5400	157.797	785.102	638.445	791.946	-2132.534	-27.718	0.268
5500	157.803	787.997	641.138	807.726	-2131.815	11.253	-0.107
5600	157.809	790.841	643.786	823.507	-2131.161	50.213	-0.468
5700	157.815	793.634	646.390	839.288	-2130.571	89.159	-0.817
5800	157.820	796.379	648.953	855.070	-2130.046	128.097	-1.154
5900	157.825	799.077	651.475	870.852	-2129.581	167.027	-1.479
6000	157.830	801.729	653.957	886.635	-2129.180	205.948	-1.793

PREVIOUS: September 1978 (1 atm) CURRENT: September 1978 (1 bar)

Molybdenum Fluoride (MoF_6) $F_6Mo_1(g)$

SULFUR FLUORIDE (SF_6) IDEAL GAS M_r = 146.050418

$S°(298.15 K) = 291.535 ± 0.4 J K^{-1} mol^{-1}$

$\Delta_f H°(0 K) = -1206.501 ± 0.8 kJ mol^{-1}$
$\Delta_f H°(298.15 K) = -1220.473 ± 0.8 kJ mol^{-1}$

Vibrational Frequencies and Degeneracies

ν, cm^{-1}	ν, cm^{-1}	ν, cm^{-1}
773.5(1)	947.5(3)	525.0(3)
641.7(2)	615.5(3)	347.0(3)

Ground State Quantum Weight: 1 σ = 24
Point Group: O_h
Bond Distance: S-F = 1.564 ± 0.01 A
Bond Angle: F-S-F = 90°
Product of the Moments of Inertia: $I_A I_B I_C = 2.941168 \times 10^{-113} g^3 cm^6$

Enthalpy of Formation

The calorimetric measurement of $\Delta_f H°$ by direct combination of the elements has been the subject of numerous investigations (1-6). O'Hare et al. (1) performed six combustion experiments on a sample of rhombic sulfur which contained 60 ppm oxygen as the major impurity. The sulfur (~0.5 gram) was burned in 5-6 atmospheres of fluorine which had a purity of 99.97%. Spectrometric and gas chromatographic analyses of the product gases indicated that the sole combustion product was SF_6. We adjust their result $\Delta_f H°(298.15 k) = -271.7±0.24 kcal mol^{-1}$ to correspond to an atomic weight of sulfur equal to 32.06, and we obtain $\Delta_f H°(SF_6, g, 298.15 k) = -291.7_3±0.2 kcal mol^{-1}$. The rounded value of $-291.7±0.2 kcal mol^{-1}$ is adopted in this tabulation. Less extensive measurements (3 determinations) by Schroder and Sieben (2) on a portion of the same sample used by O'Hare et al. (1) gave $\Delta_f H°(SF_6, g, 298.15 K) = -291.4 kcal mol^{-1}$ which provides confirmation for our selected value.

Other reported $\Delta_f H°$ values which are consistently less negative than the value selected here include (in kcal mol^{-1}) -291.0±0.3 (3), -288.9±0.7 (4), -288.5±0.7 (5) and -262.0 (6). The fluorine used by Leonidov et al. (3) contained 0.13 wt. % oxygen, and this impurity may be responsible for their low results. Gross et al. (4) studied the sulfur fluorination reaction in a glass apparatus and also applied their technique to titanium. Their reported $\Delta_f H°$ value for TiF_4 is in satisfactory agreement (±0.8 kcal mol^{-1}) with JANAF data (7). Insufficient information has been reported (4) to resolve the discrepancy for SF_6.

Our adopted $\Delta_f H°$ value is essentially the same as that recently selected by NBS (8). The earlier NBS selection (9) of -289.0 kcal mol^{-1} based on the work of Gross et al. (4, 5) is unreliable.

Heat Capacity and Entropy

The product of the moments of inertia is calculated for a structure in which the fluorine atoms lie at the corners of a regular octahedran. This structure belongs to the highest symmetry point group O_h (σ = 24) and is consistent with observations made from electron diffraction (10-12), Raman (13-15) and infrared (16-18) studies. The bond length and angle are from the electron diffraction work of Ewing and Sutton (10). Earlier and less precise work (11-12) led to r_e(S-F) values which are only slightly different from the adopted value. The individual moments of inertia are: $I_A = I_B = I_C = 30.8679 \times 10^{-39} g cm^2$.

The SF_6 molecule has been extensively characterized by vibrational spectroscopy. Weinstock and Goodman (19) reviewed work prior to 1965. Recent studies include gas-phase Raman (13-14), matrix-isolation Raman (15), and gas-phase infrared (16-18). These spectrometric results are in good agreement; the largest deviation arising in the Raman and infrared inactive fundamental ν_6. Weinstock and Goodman (19) obtained $\nu_6 = 349 cm^{-1}$ from the combination (2 + 6) band. Claassen et al. (13) reported $\nu_6 = 347 cm^{-1}$ from the overtone $2\nu_6$ band, and Holzer and Ouillon (14) have directly observed this forbidden fundamental in the collision-induced Raman spectra at 336 cm^{-1}. Rosenberg and Birnbaum (20) recorded the far-infrared absorption spectrum of gaseous SF_6 and assigned a band centered at 173 cm^{-1} to ν_5-ν_6. The difference band gives $\nu_6 = 352 cm^{-1}$ with $\nu_5 = 525 cm^{-1}$ (13) and supports the higher value for ν_6. All frequencies used in this tabulation except for the two infrared active (ν_3 and ν_4) fundamentals are taken from the work of Claassen et al. (13). Values for the infrared active frequencies are from Abramowitz and Levin (16). Recent MO calculations (21) predict a singlet ground state with no low-lying electronic levels, and are supported by the x-ray emission (22) and absorption (22, 23) spectra of gaseous SF_6. Gianturco (28) has presented a description of the electronic properties.

Kelley and King (24) analyzed $C_p°$ data (13-230 K) reported by Eucken and Schroder (25) and obtained S°(298.15 K) = 70.3±0.7 cal K^{-1} mol^{-1} for the gas phase. This value is in reasonable agreement with the spectroscopic result (69.7±0.1 cal K^{-1} mol^{-1}). Our thermal functions essentially extend and update those reported in the literature (1, 26, 27).

References

1. P. A. G. O'Hare, J. L. Settle and W. N. Hubbard, Trans. Faraday Soc. 62, 558 (1966).
2. J. Schroder and F. J. Sieben, Chem. Ber. 103, 76 (1970).
3. V. Ya. Leonidov, V. S. Pervov, O. M. Gaisinskaya and L. I. Klyuev, Dokl. Akad. Nauk SSSR, 211, 901 (1973).
4. P. Gross, Bull. Chem. Thermodyanmics, No. 3, 14 (1960).
5. P. Gross, Bull. Chem. Thermodynamics, No. 3, 14 (1960).
6. D. M. Yost and W. H. Claassen, J. Amer. Chem. Soc. 55, 885 (1933).
7. JANAF Thermochemical Tables: TiF_4(cr), 9-30-67.
8. V. B. Parker, D. D. Wagman and D. Garvin, NBSIR 75-968, 1976.
9. U. S. Nat. Bur. Stand. Tech. Note 270-3, 1968.
10. V. C. Ewing and L. E. Sutton, Trans. Faraday Soc. 59, 1241 (19630.
11. L. O. Brockway and L. Pauling, Proc. Natl. Acad. Sci. 19, 68 (1933).
12. H. Braune and S. Knobe, Z. Physik. Chem. 21, 297 (1933).
13. H. H. Claassen, G. L. Goodman, J. H. Holloway and H. Selig, J. Chem. Phys. 53, 341 (1970).
14. W. Holzer and R. Ouillon, Chem. Phys. Letters, 24, 589 (1974).
15. J. S. Shirk and H. H. Claassen, J. Chem. Phys. 54, 3237 (1971).
16. S. Abramowitz and I. W. Levin, J. Chem. Phys. 44, 3353 (1966).
17. H. Brunet and M. Perez, J. Mol. Spectrosc. 29, 472 (1969).
18. V. D. Klimov and E. A. Lobikov, Opt. Spektrosk. 30, 25 (1971).
19. B. Weinstock and G. L. Goodman, in "Advances in Chemical Physics," Vol. 9, Ed. by I. Prigogine, Interscience, 1965.
20. A. Rosenberg and G. Birnbaum, J. Chem. Phys. 52, 683 (1970).
21. E. L. Rozenberg and M. E. Dyatkina, Zhur. Strukt. Khim. 11, 299 (1970).
22. R. E. LaVilla, J. Chem. Phys. 57, 899 (1972).
23. J. L. Dehmer, J. Chem. Phys. 56, 4496 (1972).
24. K. K. Kelley and E. G. King, U. S. Bur. Mines Bull. 592, 1961.
25. A. Eucken and E. Schroder, Z. Physik. Chem. 41B, 307 (1938).
26. G. Nagarajin and D. C. Brinkley, Z. Naturforsch. 26a, 1658 (1971).
27. R. L. Wilkins, Aerospace Corporation, Report No. TR-0158(3240-20)-19, July, 1968.
28. F. A. Gianturco, Chem. Phys. Letters 17, 127 (1972).

Sulfur Fluoride (SF_6) F_6S_1(g)

Enthalpy Reference Temperature = T_r = 298.15 K Standard State Pressure = p° = 0.1 MPa

T/K	$C_p°$ (J $K^{-1}mol^{-1}$)	S°	$-[G°-H°(T_r)]/T$	$H°-H°(T_r)$ (kJ mol^{-1})	$\Delta_f H°$	$\Delta_f G°$	Log K_f
0	0.	0.	INFINITE	-16.915	-1206.501	-1206.501	INFINITE
100	38.721	223.132	358.027	-13.490	-1212.486	-1182.436	617.641
200	68.822	258.508	299.562	-8.211	-1217.637	-1150.220	300.407
250	84.315	275.568	293.063	-4.374	-1219.327	-1133.158	236.760
298.15	96.960	291.535	291.535	0.	-1220.473	-1116.451	195.598
300	97.398	292.136	291.537	0.180	-1220.509	-1115.806	194.279
350	107.968	307.974	292.763	5.324	-1221.297	-1098.288	163.911
400	116.373	322.962	295.610	10.940	-1224.003	-1080.591	141.111
450	123.040	337.068	299.441	16.932	-1224.938	-1062.615	123.345
500	128.351	350.317	303.873	23.222	-1225.747	-1044.532	109.122
600	136.069	374.450	313.667	36.470	-1226.504	-1008.204	87.772
700	141.228	395.838	323.907	50.351	-1226.624	-971.804	72.517
800	144.808	414.944	334.114	64.663	-1226.352	-935.416	61.076
900	147.375	432.156	344.068	79.279	-1278.955	-897.998	52.118
1000	149.272	447.786	353.670	94.116	-1277.044	-855.771	44.701
1100	150.709	462.084	362.885	109.118	-1275.073	-813.739	38.641
1200	151.822	475.247	371.707	124.247	-1273.069	-771.888	33.599
1300	152.700	487.435	380.147	139.475	-1271.051	-730.204	29.340
1400	153.404	498.778	388.220	154.782	-1269.035	-688.677	25.695
1500	153.978	509.382	395.948	170.152	-1267.030	-647.293	22.541
1600	154.450	519.335	403.352	185.574	-1265.042	-606.042	19.785
1700	154.844	528.711	410.453	201.039	-1263.075	-564.914	17.358
1800	155.176	537.571	417.271	216.540	-1261.130	-523.902	15.203
1900	155.458	545.969	423.825	232.073	-1259.203	-482.998	13.279
2000	155.700	553.949	430.134	247.631	-1257.289	-442.195	11.549
2100	155.908	561.551	436.212	263.211	-1255.382	-401.487	9.986
2200	156.089	568.808	442.076	278.811	-1253.473	-360.870	8.568
2300	156.248	575.750	447.738	294.428	-1251.553	-320.340	7.275
2400	156.387	582.403	453.211	310.060	-1249.612	-279.894	6.092
2500	156.510	588.789	458.507	325.705	-1247.639	-239.529	5.005
2600	156.620	594.930	463.637	341.362	-1245.625	-199.245	4.003
2700	156.717	600.843	468.610	357.029	-1243.560	-159.038	3.077
2800	156.805	606.544	473.435	372.705	-1241.434	-118.910	2.218
2900	156.883	612.048	478.120	388.390	-1239.240	-78.858	1.420
3000	156.954	617.368	482.674	404.082	-1236.970	-38.883	0.677
3100	157.019	622.515	487.102	419.780	-1234.619	1.014	-0.017
3200	157.077	627.501	491.412	435.485	-1232.180	40.835	-0.667
3300	157.130	632.336	495.610	451.195	-1229.649	80.578	-1.275
3400	157.179	637.027	499.700	466.911	-1227.022	120.242	-1.847
3500	157.224	641.584	503.689	482.631	-1224.298	159.828	-2.385
3600	157.264	646.014	507.582	498.356	-1221.473	199.336	-2.892
3700	157.302	650.323	511.381	514.084	-1218.547	238.762	-3.371
3800	157.337	654.519	515.093	529.816	-1215.520	278.108	-3.823
3900	157.369	658.606	518.721	545.551	-1212.391	317.374	-4.251
4000	157.399	662.590	522.268	561.290	-1209.161	356.558	-4.656
4100	157.426	666.477	525.738	577.031	-1205.831	395.659	-5.041
4200	157.452	670.271	529.134	592.775	-1202.403	434.680	-5.406
4300	157.476	673.976	532.460	608.521	-1198.877	473.615	-5.753
4400	157.498	677.597	535.718	624.270	-1195.256	512.469	-6.084
4500	157.519	681.137	538.910	640.021	-1191.542	551.240	-6.399
4600	157.539	684.599	542.040	655.774	-1187.737	589.926	-6.699
4700	157.557	687.987	545.109	671.528	-1183.843	628.530	-6.985
4800	157.574	691.305	548.120	687.285	-1179.864	667.049	-7.259
4900	157.590	694.554	551.076	703.043	-1175.801	705.484	-7.521
5000	157.605	697.738	553.977	718.803	-1171.658	743.836	-7.771
5100	157.620	700.859	556.827	734.564	-1167.436	782.104	-8.010
5200	157.633	703.920	559.626	750.327	-1163.138	820.289	-8.240
5300	157.646	706.922	562.377	766.091	-1158.767	858.390	-8.460
5400	157.658	709.869	565.081	781.856	-1154.325	896.408	-8.671
5500	157.669	712.762	567.740	797.622	-1149.815	934.342	-8.874
5600	157.680	715.603	570.355	813.390	-1145.239	972.196	-9.068
5700	157.690	718.394	572.928	829.158	-1140.599	1009.965	-9.255
5800	157.700	721.137	575.460	844.928	-1135.898	1047.653	-9.435
5900	157.709	723.833	577.952	860.698	-1131.137	1085.261	-9.608
6000	157.718	726.483	580.405	876.470	-1126.320	1122.785	-9.775

PREVIOUS: June 1976 (1 atm) CURRENT: June 1976 (1 bar)

Sulfur Fluoride (SF_6) F_6S_1(g)

SULFUR FLUORIDE, ION (SF_6^-) IDEAL GAS M_r = 146.050967

$S°(298.15\ K) = [302.860 \pm 2.1]\ J\ K^{-1}\ mol^{-1}$

$\Delta_f H°(0\ K) = -1322.282 \pm 29.3\ kJ\ mol^{-1}$

$\Delta_f H°(298.15\ K) = -1341.876 \pm 29.3\ kJ\ mol^{-1}$

Vibrational Frequencies and Degeneracies

ν, cm^{-1}	ν, cm^{-1}	ν, cm^{-1}
[700](1)	[925](3)	[500](3)
[625](2)	594 (3)	[325](3)

Ground State Quantum Weight: 2 σ = 24

Point Group: O_h

Bond Distance: S-F = [1.72] Å

Bond Angle: F-S-F = 90°

Product of the Moments of Inertia: $I_A I_B I_C = [5.203193 \times 10^{-113}]\ g^3\ cm^6$

Enthalpy of Formation

We calculate the enthalpy of formation at 0 K as the difference between $\Delta_f H°(SF_6, g) = -288.36\pm0.2\ kcal\ mol^{-1}$ (1) and the selected value of the electron affinity (EA) equal to 1.2±0.3 eV (27.67 kcal mol^{-1}). There have been numerous values proposed for EA(SF_6). Franklin and Harland (2) have recently reviewed several of the published values; other values of EA not included in the review are tabulated below. The only direct measurement of EA = 1.49 eV by the magnetron technique (3) is suspect since mass spectral studies (4) have shown that SF_5^- rather than SF_6^- is the predominant ion formed under conditions used in the magnetron experiments. An EA = 1.2 eV is consistent with most of the available information and is adopted here as the best current value.

Source	Method	EA(SF_6), eV
(5)	Electron-Transfer	<1.18
(6)	Collisional Ionization Cs/SF_6	>0.54
(7)	Rate Measurements	>0.7
(8)	Electron-Transfer	>0.43
		<1.47
(9)	Solution Electron-Transfer	0.6
(10)	MO Calculation	1.7

From the adopted results, we calculate the primary bond dissociation energy $D_0^0(SF_6^- - F)$ to be 34.4 kcal mol^{-1} which agrees with the value of 24.2 kcal mol^{-1} obtained from threshold energies for SF_6^- and SF_5^- measured in collisional ionization studies (6) on the Cs/SF_6 system. $\Delta_f H°(298.15\ K)$ is -320.7 kcal mol^{-1}, and the fluoride ion affinity (IA) is IA = 41.3 kcal mol^{-1}.

Heat Capacity and Entropy

The six fluorine atoms are assumed to form a regular octahedron around the central sulfur atom, in accordance with observations (11) on the ESR spectrum of SF_6^-. The value of the bond length is taken from theoretical calculations (11). This value (1.72 Å) is intermediate between the bond lengths for SF_6 (1) and XeF_6 (12) and therefore appears to be a reasonable estimate. The principal moments of inertia are: $I_A = I_B = I_C = 37.3328 \times 10^{-39}\ g\ cm^2$.

MO calculations (11) in the INDO approximation indicate that the unpaired electron occupies a totally symmetric antibonding orbital (A_{1g}). Thus, we predict that the bonding in SF_6^- is somewhat weaker than that in the neutral molecule. Additional support for this is provided by the fact that the mean S-F bond energy (products $S(g)+F^-(g) +5F(g)$) in SF_6^- is 8.5 kcal mol^{-1} less than that in SF_6 (1). Due to the weaker bonding in SF_6^-, all vibrational frequencies are expected to show a lowering relative to those for SF_6 (1). Very recently, Barefield and Guillory (13) have assigned an absorption band at 594 cm^{-1} observed in the matrix-isolated infrared spectra of NO, Na, and K photoionized in the presence of SF_6 as the ν_4 fundamental of SF_6^-. This represents a lowering of 21.3 cm^{-1} from the gas-phase frequency for SF_6 (1). The other infrared active mode (ν_3) for SF_6^- expected to lie near 925 cm^{-1} was not observed but was believed to be hidden beneath the matrix-isolated ν_3 band (937 cm^{-1}) of SF_6. We estimate the eight unobserved fundamental from the gas-phase values for SF_6 (1).

MO calculations (11) indicate that the ground state electronic configuration is A_{1g}. There is, however, little information available on the excited states for SF_6^- or for isoelectronic molecules such as ClF_6 and BrF_6. Indirect experimental evidence for the existence of an excited state at 0.5 eV (4033 cm^{-1}) has been reported by Lifshitz et al. (14) from mass spectral studies of charge-transfer reactions. As shown by Fehsenfeld (15), stable excited states are expected to exist only if $D°(SF_5^- - F) > EA(SF_6)$. Our adopted results give a difference in $D_0^0(SF_5^- - F)$ and EA(SF_6) of only 0.053 eV, suggesting that perhaps the SF_6^- ion observed by Lifshitz et al. (14) was vibrationally rather than electronically excited. We prefer not to include this level, but if it does exist, our entropies should be increased by 0.41 and 1.0 cal K^{-1} mol^{-1} at 2000 K and 4500 K.

References

1. JANAF Thermochemical Tables: SF_6(g), 6-30-76.
2. J. L. Franklin and P. W. Marland, Ann. Rev. Phys. Chem. 25, 485 (1974).
3. J. Kay and F. M. Page, Trans. Faraday Soc. 60, 1042 (1964).
4. J. T. Herron, H. M. Rosenstock and W. R. Shields, Nature, 206, 611 (1965).
5. C. Lifshitz, B. M. Hughes and T. O. Tiernan, Chem. Phys. Letters, 7, 469 (1970).
6. R. N. Compton and C. D. Cooper, J. Chem. Phys. 59, 4140 (1973).
7. E. Chen, R. D. George and W. E. Wentworth, J. Chem. Phys. 49, 1973 (1968).
8. F. C. Fehsenfeld, J. Chem. Phys. 54, 438 (1971).
9. P. R. Hammond, J. Chem. Phys. 55, 3468 (1971).
10. A. L. Companion, Theoret. Chem. Acta (Berl.) 25, 268 (1972).
11. J. R. Morton, K. F. Preston and J. C. Tait, J. Chem. Phys. 62, 2029 (1975).
12. L. S. Bartell and R. M. Gavin, J. Chem. Phys. 48, 2460 (1968).
13. J. E. Barefield and W. A. Guillory, J. Phys. Chem. 81, 634 (1977).
14. C. Lifshitz, T. O. Tiernan and B. M. Hughes, J. Chem. Phys. 59, 3182 (1973).
15. F. C. Fehsenfeld, J. Chem. Phys. 53, 2000 (1970).

Sulfur Fluoride, Ion (SF_6^-) $F_6S_1^-$(g)

Enthalpy Reference Temperature = T_r = 298.15 K Standard State Pressure = p° = 0.1 MPa

T/K	$C_p°$	S°	$-[G°-H°(T_r)]/T$	$H°-H°(T_r)$	$\Delta_f H°$	$\Delta_f G°$	Log K_f
	J $K^{-1}mol^{-1}$			kJ mol^{-1}			
0	0.	0.	INFINITE	-17.490	-1322.282		
100	39.970	231.635	371.985	-14.035			
200	71.908	268.576	311.174	-8.520			
250	87.504	286.343	304.439	-4.524			
298.15	99.984	302.860	302.860	0.	-1341.876	-1234.976	216.363
300	100.413	303.480	302.862	0.185	-1341.945	-1234.313	214.913
350	110.698	319.762	304.124	5.473	-1343.629	-1216.236	181.513
400	118.794	335.094	307.047	11.219	-1347.245	-1197.851	156.423
450	125.168	349.469	310.971	17.324	-1349.105	-1179.072	136.863
500	130.218	362.928	315.501	23.713	-1350.854	-1160.082	121.193
600	137.516	387.362	325.485	37.126	-1353.525	-1121.663	97.649
700	142.367	408.948	335.897	51.136	-1355.595	-1082.850	80.803
800	145.720	428.191	346.253	65.550	-1357.300	-1043.766	68.151
900	148.119	445.500	356.335	80.248	-1411.899	-1003.403	58.236
1000	149.889	461.202	366.049	95.153	-1411.999	-958.009	50.041
1100	151.228	475.554	375.361	110.212	-1412.050	-912.607	43.336
1200	152.264	488.758	384.268	125.388	-1412.076	-867.202	37.748
1300	153.080	500.980	392.782	140.657	-1412.097	-821.795	33.020
1400	153.735	512.349	400.921	155.999	-1412.123	-776.387	28.967
1500	154.268	522.974	408.707	171.400	-1412.166	-730.975	25.455
1600	154.707	532.945	416.164	186.850	-1412.229	-685.561	22.381
1700	155.072	542.335	423.312	202.339	-1412.317	-640.141	19.669
1800	155.380	551.208	430.173	217.862	-1412.429	-594.716	17.258
1900	155.642	559.616	436.767	233.414	-1412.561	-549.284	15.101
2000	155.866	567.605	443.111	248.989	-1412.708	-503.844	13.159
2100	156.059	575.215	449.222	264.586	-1412.864	-458.398	11.402
2200	156.227	582.479	455.115	280.200	-1413.019	-412.943	9.805
2300	156.374	589.427	460.805	295.830	-1413.164	-367.482	8.346
2400	156.504	596.085	466.303	311.475	-1413.290	-322.015	7.008
2500	156.618	602.476	471.623	327.131	-1413.385	-276.543	5.778
2600	156.719	608.620	476.775	342.798	-1413.439	-231.068	4.642
2700	156.810	614.537	481.768	358.474	-1413.442	-185.592	3.590
2800	156.891	620.241	486.613	374.159	-1413.386	-140.117	2.614
2900	156.964	625.748	491.316	389.852	-1413.262	-94.645	1.705
3000	157.029	631.070	495.886	405.552	-1413.064	-49.178	0.856
3100	157.089	636.220	500.331	421.258	-1412.784	-3.720	0.063
3200	157.143	641.208	504.656	436.969	-1412.416	41.728	-0.681
3300	157.192	646.045	508.867	452.686	-1411.957	87.163	-1.380
3400	157.238	650.738	512.971	468.408	-1411.403	132.583	-2.037
3500	157.279	655.297	516.973	484.134	-1410.752	177.985	-2.656
3600	157.317	659.728	520.877	499.863	-1410.001	223.368	-3.241
3700	157.352	664.039	524.688	515.597	-1409.148	268.727	-3.794
3800	157.384	668.235	528.411	531.334	-1408.195	314.062	-4.317
3900	157.414	672.324	532.049	547.074	-1407.140	359.372	-4.813
4000	157.441	676.310	535.606	562.816	-1405.984	404.653	-5.284
4100	157.467	680.198	539.085	578.562	-1404.729	449.903	-5.732
4200	157.490	683.992	542.490	594.310	-1403.375	495.122	-6.158
4300	157.513	687.699	545.824	610.060	-1401.924	540.306	-6.563
4400	157.533	691.320	549.090	625.812	-1400.378	585.457	-6.950
4500	157.553	694.860	552.290	641.566	-1398.739	630.572	-7.319
4600	157.571	698.323	555.427	657.323	-1397.009	675.649	-7.672
4700	157.588	701.712	558.504	673.080	-1395.191	720.687	-8.010
4800	157.603	705.030	561.522	688.840	-1393.288	765.686	-8.332
4900	157.618	708.280	564.484	704.601	-1391.301	810.644	-8.642
5000	157.632	711.465	567.392	720.364	-1389.233	855.561	-8.938
5100	157.646	714.586	570.247	736.128	-1387.087	900.435	-9.222
5200	157.658	717.647	573.053	751.893	-1384.865	945.268	-9.495
5300	157.670	720.651	575.809	767.659	-1382.570	990.056	-9.758
5400	157.681	723.598	578.519	783.427	-1380.204	1034.800	-10.010
5500	157.692	726.491	581.183	799.195	-1377.770	1079.500	-10.252
5600	157.702	729.333	583.803	814.965	-1375.271	1124.156	-10.486
5700	157.711	732.124	586.381	830.736	-1372.707	1168.765	-10.711
5800	157.720	734.867	588.918	846.507	-1370.083	1213.329	-10.927
5900	157.728	737.563	591.414	862.280	-1367.399	1257.849	-11.136
6000	157.737	740.214	593.872	878.053	-1364.659	1302.321	-11.338

PREVIOUS: June 1977 (1 atm) CURRENT: June 1977 (1 bar)

Sulfur Fluoride, Ion (SF_6^-) $F_6S_1^-$(g)

TUNGSTEN FLUORIDE (WF_6) LIQUID M_r = 297.840418

$S°(298.15\ K) = 249.216 \pm 0.8\ J\ K^{-1}\ mol^{-1}$

$T_{fus} = 275.2\ K$

$\Delta_fH°(298.15\ K) = -1748.368 \pm 2.1\ kJ\ mol^{-1}$

$\Delta_{fus}H° = 4.100\ kJ\ mol^{-1}$

Enthalpy of Formation

The adopted enthalpy of formation, $-417.87\ kcal\ mol^{-1}$, is calculated from the chosen $\Delta_{vap}H°(290.3\ K) = 6.465\pm0.08\ kcal\ mol^{-1}$ with $\Delta_fH°(WF_6,\ g,\ 298.15\ K) = -411.5\pm0.4\ kcal\ mol^{-1}$. The value of $\Delta_{vap}H°(290.3\ K)$ is obtained by corrected 2nd law analysis of vapor pressure data (6, 7, 8) which are given in the vaporization section of this table.

Myers and Brady (1) measured calorimetrically $\Delta_rH°(298.15\ K) = -150.1\ kcal\ mol^{-1}$ for the reaction $WF_6(\ell)$ + 8 NaOH(aq, 500 H_2O) + Na_2WO_4(aq, 4900 H_2O) + 4 $H_2O(\ell)$ + 6 NaF(aq, 800 H_2O). Spitsyn and Patsukova (2) measurd calorimetrically $\Delta_rH°(298.15\ K) = -13.8\ kcal\ mol^{-1}$ for the reaction WO_3(cr) + 2 $NaOH_4$(aq, 66 H_2O) + Na_2WO_4(aq, 4200 H_2O) + $H_2O(\ell)$. Based on these data and auxiliary data for NaOH(aq) (3, 9), NaF(aq) (3, 10) and Na_2WO_4(aq) (4), we obtain $\Delta_rH°(298.15\ K) = -136.2\ kcal\ mol^{-1}$ for $WF_6(\ell)$ + 6 NaOH(aq, ∞) + WO_3(cr) + 3 $H_2O(\ell)$ + 6 NaF(aq, ∞). This leads to $\Delta_fH°(WF_6,\ \ell,\ 298.15\ K) = -416.9\pm1\ kcal\ mol^{-1}$ which is in good agreement with the value adopted.

Heat Capacity and Entropy

The heat capacity, $40.5\ cal\ K^{-1}\ mol^{-1}$, is obtained from a graph of the $C_p°$ versus temperature curve determined calorimetrically by Westrum (5).

The entropy, $S°(298.15\ K) = 59.564\ cal\ K^{-1}\ mol^{-1}$, is calculated from the chosen $\Delta_{vap}H°(290.3) = 6.465\ kcal\ mol^{-1}$ with $S°(WF_6,\ g,\ 298.15\ K) = 81.50\ cal\ K^{-1}\ mol^{-1}$.

Fusion Data

The melting point and heat of melting have been reported by the following investigators, and their results are summarized below. Ruff and Ascher's value, $2.40\ kcal\ mol^{-1}$, is too high because they did not know of the existence of a solid-solid transition (5) at 264.7 K with $\Delta_{trs}H° = 2.067\ kcal\ mol^{-1}$. Westrum's data are adopted in the tabulation.

Source	Method	T_{fus}/K	$\Delta_{fus}H°$ kcal mol^{-1}
(5)	Calorimetric	275.2	0.980
(7)	Vapor pressure	275.4	2.400*
(8)	Warming curve	275.1	0.500
(6)	Vapor pressure	275.2	0.420*

*This value is the difference between the 2nd law heats of sublimation and vaporization without correction for $\Delta C_p°$ for both processes.

Vaporization Data

2nd law analyses of the data (6, 7, 8), neglecting corrections for nonideality of the vapor, are summarized below. The value of $\Delta_{vap}H°(298.15\ K) = 6.37\ kcal\ mol^{-1}$ ($\Delta_{vap}H°(290.3\ K) = 6.465\ kcal\ mol^{-1}$ is selected in the tabulation.

Source	T/K	Data Points	T_{vap}(1 atm)/K	$\Delta_{vap}H°(298.15\ K)$ kcal mol^{-1}	$\Delta_{vap}S°(298.15\ K)$* cal K^{-1} mol^{-1}
(7)	273-387.5	5	290.7±0.2	6.30±0.12	21.70
(8)	285-324.5	16	290.2±0.05	6.37±0.01	21.94
(6)	275-290	Equation	290.3±0.05	6.12	21.09

*From 2nd law analysis using $\Delta C_p° = -12.4\ cal\ K^{-1}\ mol^{-1}$

References

1. O. E. Myers and A. P. Brady, J. Phys. Chem. 64, 591 (1960).
2. V. I. Spitsyn and N. N. Patsukova, Russ. J. Inorg. Chem. 10, 1304 (1965).
3. V. B. Parker, U.S. Nat. Bur. Stand., NSRDS-NBS 2, 1965.
4. We have assumed $\Delta_{dil}H° = 0$ for Na_2WO_4(aq, 4200 H_2O) + Na_2WO_4(aq, 4900 H_2O) based on the dilution data of R. L. Graham and L. G. Hepler, J. Amer. Chem. Soc. 80, 3538 (1958).
5. E. F. Westrum, Jr., Pure & Appl. Chem. 8, 187 (1964); J. Chem. Ed. 39, 443 (1962).
6. G. H. Cady and G. B. Hargreaves, J. Chem. Soc., 1563 (1961).
7. O. Ruff and E. Ascher, Z. Anorg. Allg. Chem. 196, 413 (1931).
8. E. J. Barber and G. H. Cady, J. Phys. Chem. 60, 505 (1956).
9. JANAF Thermochemical Table: NaOH(cr), 3-31-66.
10. U. S. Nat. Bur. Stand. Tech. Note 270-1, 1965.

Tungsten Fluoride (WF_6) F_6W_1(l)

Enthalpy Reference Temperature = T_r = 298.15 K; Standard State Pressure = p° = 0.1 MPa

T/K	$C_p°$ (J $K^{-1}mol^{-1}$)	S° (J $K^{-1}mol^{-1}$)	$-[G°-H°(T_r)]/T$ (J $K^{-1}mol^{-1}$)	$H°-H°(T_r)$ (kJ mol^{-1})	$\Delta_fH°$ (kJ mol^{-1})	$\Delta_fG°$ (kJ mol^{-1})	Log K_f
0							
100							
200							
298.15	169.452	249.216	249.216	0.	-1748.368	-1631.549	285.841
300	169.452	250.264	249.219	0.313	-1748.273	-1630.825	283.952
400	169.452	299.012	255.866	17.259	-1743.449	-1592.418	207.948
500	169.452	336.824	268.417	34.204	-1739.116	-1555.170	162.468
600	169.452	367.719	282.471	51.149	-1735.148	-1518.759	132.220
700	169.452	393.840	296.563	68.094	-1731.459	-1482.989	110.662
800	169.452	416.467	310.168	85.039	-1727.987	-1447.733	94.527
900	169.452	436.426	323.110	101.985	-1724.693	-1412.900	82.003
1000	169.452	454.280	335.350	118.930	-1721.552	-1378.426	72.002

PREVIOUS: December 1963 CURRENT: March 1967

TUNGSTEN FLUORIDE (WF_6) IDEAL GAS M_r = 297.840418

$\Delta_f H°(0\ K) = -1713.009 \pm 1.7\ kJ\ mol^{-1}$

$S°(298.15\ K) = 341.127\ J\ K^{-1}\ mol^{-1}$ $\Delta_f H°(298.15\ K) = -1721.716 \pm 1.7\ kJ\ mol^{-1}$

Vibrational Frequencies and Degeneracies

ν, cm^{-1}	ν, cm^{-1}
769 (1)	256 (3)
670 (2)	322 (3)
712 (3)	216 (3)

Ground State Quantum Weight: 1 $\sigma = 24$
Point Group: O_h
Bond Distance: W-F = 1.83 Å
Bond Angle: F-W-F = 90°
Product of the Moments of Inertia: $I_A I_B I_C = 7.547547 \times 10^{-113}\ g^3\ cm^6$

Enthalpy of Formation

O'Hare and Hubbard (7) determined the enthalpy of formation, $\Delta_f H°(WF_6, g, 298.15\ K) = -411.5 \pm 0.4\ kcal\ mol^{-1}$, by combustion of tungsten in fluorine in a bomb calorimeter. This value is adopted in the tabulation.

Heat Capacity and Entropy

The molecular configuration and vibrational frequencies were obtained from the infrared and Raman spectroscopic studies by Gaunt (1), Burke et al. (2) and Tanner and Duncan (3).

The bond distance was reported by Claasen (4), based on the preliminary result of Schomaker et al. (5) who determined the bond distance by electron diffraction. Gaunt (6) derived W-F = 1.89 Å from correlations with other hexafluorides and a valence force constant calculation. The three principal moments of inertia are $I_A = I_B = I_C = 42.2606 \times 10^{-39}\ g\ cm^2$.

References

1. J. Gaunt, Trans. Faraday Soc. 49, 1122 (1953).
2. T. G. Burke, D. F. Smith and A. H. Nielsen, J. Chem. Phys. 20, 447 (1952).
3. K. N. Tanner and A. B. F. Duncan, J. Am. Chem. Soc. 73, 1164 (1951).
4. H. H. Classen, J. Chem. Phys. 30, 968 (1959).
5. V. Schomaker, M. Kimura and B. Weinstock, Proceedings of 2nd U.N. International Conference on Peaceful Uses of Atomic Energy, Geneva, 28, 125 (1958).
6. J. Gaunt, Trans. Faraday Soc. 50, 546 (1954).
7. P. A. G. O'Hare and W. N. Hubbard, J. Phys. Chem. 70, 3353 (1966).

Tungsten Fluoride (WF_6) $F_6W_1(g)$

Enthalpy Reference Temperature = T_r = 298.15 K Standard State Pressure = p° = 0.1 MPa

T/K	$C°_p$ (J K⁻¹mol⁻¹)	S° (J K⁻¹mol⁻¹)	$-[G°-H°(T_r)]/T$ (J K⁻¹mol⁻¹)	$H°-H°(T_r)$ (kJ mol⁻¹)	$\Delta_f H°$ (kJ mol⁻¹)	$\Delta_f G°$ (kJ mol⁻¹)	Log K_f
0	0.	0.	INFINITE	-22.741	-1713.009	-1713.009	INFINITE
100	59.529	243.924	430.923	-18.700	-1718.341	-1690.661	883.111
200	97.058	297.911	351.387	-10.695	-1721.131	-1661.668	433.983
250	109.635	320.979	343.037	-5.515	-1721.615	-1646.736	344.067
298.15	119.029	341.127	341.127	0.	-1721.716	-1632.301	285.972
300	119.343	341.865	341.130	0.220	-1721.714	-1631.746	284.112
350	126.763	360.844	342.612	6.382	-1721.544	-1616.762	241.288
400	132.434	378.158	345.989	12.868	-1721.188	-1601.815	209.176
450	136.800	394.020	350.457	19.603	-1720.706	-1586.921	184.205
500	140.200	408.616	355.553	26.532	-1720.136	-1572.086	164.235
600	145.020	434.637	366.619	40.811	-1718.835	-1542.596	134.295
700	148.166	457.244	377.986	55.481	-1717.421	-1513.333	112.926
800	150.313	477.178	389.164	70.411	-1715.964	-1484.277	96.913
900	151.836	494.975	399.950	85.523	-1714.503	-1455.404	84.469
1000	152.952	511.033	410.268	100.765	-1713.065	-1426.693	74.523
1100	153.793	525.652	420.103	116.104	-1711.664	-1398.124	66.391
1200	154.440	539.063	429.466	131.517	-1710.314	-1369.680	59.621
1300	154.950	551.446	438.378	146.987	-1709.021	-1341.347	53.896
1400	155.357	562.944	446.870	162.503	-1707.793	-1313.111	48.993
1500	155.688	573.674	454.970	178.056	-1706.634	-1284.960	44.746
1600	155.960	583.731	462.707	193.639	-1705.548	-1256.885	41.033
1700	156.186	593.193	470.107	209.247	-1704.533	-1228.874	37.759
1800	156.377	602.126	477.195	224.875	-1703.589	-1200.922	34.850
1900	156.538	610.585	483.995	240.521	-1702.711	-1173.021	32.249
2000	156.677	618.618	490.527	256.182	-1701.894	-1145.163	29.909
2100	156.796	626.265	496.810	271.856	-1701.131	-1117.346	27.792
2200	156.900	633.562	502.862	287.541	-1700.414	-1089.564	25.870
2300	156.990	640.539	508.697	303.235	-1699.733	-1061.813	24.115
2400	157.070	647.222	514.331	318.938	-1699.078	-1034.091	22.506
2500	157.140	653.635	519.776	334.649	-1698.439	-1006.397	21.028
2600	157.203	659.800	525.043	350.366	-1697.804	-978.728	19.663
2700	157.259	665.733	530.145	366.089	-1697.190	-951.083	18.400
2800	157.308	671.454	535.090	381.818	-1696.623	-923.460	17.227
2900	157.353	676.974	539.888	397.551	-1696.133	-895.855	16.136
3000	157.394	682.310	544.547	413.288	-1695.741	-868.267	15.118
3100	157.430	687.471	549.075	429.029	-1695.484	-840.689	14.166
3200	157.464	692.470	553.478	444.774	-1695.387	-813.117	13.273
3300	157.494	697.316	557.764	460.522	-1695.472	-785.545	12.434
3400	157.522	702.018	561.938	476.273	-1695.770	-757.968	11.645
3500	157.547	706.584	566.006	492.026	-1696.342	-730.378	10.900
3600	157.570	711.023	569.972	507.782	-1697.283	-702.766	10.197
3700	157.592	715.341	573.843	523.540	-1733.457	-674.932	9.528
3800	157.612	719.544	577.622	539.300	-1731.845	-646.345	8.885
3900	157.630	723.638	581.314	555.063	-1730.128	-617.801	8.275
4000	157.647	727.629	584.922	570.826	-1728.308	-589.302	7.696
4100	157.663	731.522	588.451	586.592	-1726.384	-560.852	7.145
4200	157.677	735.321	591.903	602.359	-1724.357	-532.448	6.622
4300	157.691	739.032	595.281	618.127	-1722.229	-504.095	6.124
4400	157.703	742.657	598.590	633.897	-1720.002	-475.791	5.648
4500	157.715	746.201	601.831	649.668	-1717.677	-447.539	5.195
4600	157.726	749.668	605.007	665.440	-1715.257	-419.341	4.762
4700	157.737	753.060	608.121	681.213	-1712.743	-391.195	4.348
4800	157.747	756.381	611.175	696.987	-1710.137	-363.104	3.951
4900	157.756	759.634	614.172	712.762	-1707.443	-335.069	3.572
5000	157.764	762.821	617.113	728.538	-1704.662	-307.090	3.208
5100	157.772	765.945	620.001	744.315	-1701.796	-279.167	2.859
5200	157.780	769.009	622.837	760.093	-1698.848	-251.300	2.524
5300	157.787	772.014	625.623	775.871	-1695.821	-223.491	2.203
5400	157.794	774.964	628.362	791.650	-1692.716	-195.741	1.893
5500	157.801	777.859	631.054	807.430	-1689.536	-168.049	1.596
5600	157.807	780.703	633.701	823.210	-1686.284	-140.414	1.310
5700	157.813	783.496	636.304	838.991	-1682.960	-112.840	1.034
5800	157.818	786.240	638.866	854.773	-1679.569	-85.324	0.768
5900	157.823	788.938	641.387	870.555	-1676.111	-57.865	0.512
6000	157.828	791.591	643.868	886.338	-1672.589	-30.467	0.265

PREVIOUS: March 1967 (1 atm) CURRENT: March 1967 (1 bar)

Tungsten Fluoride (WF_6) $F_6W_1(g)$

HYDROGEN FLUORIDE (H_7F_7) IDEAL GAS M_r = 140.044401

$\Delta_fH°(0\ K) = -2083.834 \pm 37.7\ kJ\ mol^{-1}$

$S°(298.15\ K) = [523.432 \pm 29.3]\ J\ K^{-1}\ mol^{-1}$ $\Delta_fH°(298.15\ K) = -2102.544 \pm 37.7\ kJ\ mol^{-1}$

Vibrational Frequencies and Degeneracies

ν, cm^{-1}
[202](7)
[53](8)
[962](7)
[552](7)
[3060](7)

Ground State Quantum Weight: 1 σ = [7]

Point Group: $[C_7]$

Interatomic Distance: F-F = [2.5] Å

Product of the Moments of Inertia: $I_AI_BI_C = [1.797733 \times 10^{-111}]\ g^3\ cm^6$

Enthalpy of Formation

The enthalpy of 7 HF(g) → H_7F_7(g) was taken as the mean of 3rd law values which were determined from the equilibrium data of two investigations. Briegleb and Strohmeier (1) measured the vapor density of associated HF between 20 and 60°C and between 50 and 650 torr. Franck and Meyer (2) measured $C_p°$ between -20 and 100°C and between 100 and 700 torr. Each investigation evaluated K_p at n = 2, 3, 4, ... for the reactions n(HF) → $(HF)_n$ and reported 2nd law values of ΔH and ΔS. At n = 7 their 2nd law values differed by 7.3 kcal mol^{-1}, which was taken as an estimate of error, while the calculated 3rd law values differed by 6.4 kcal mol^{-1}. Using $\Delta_rH°(298.15\ K)$ = 046.54 kcal mol^{-1} and auxiliary JANAF data (3) gives the enthalpy of formation of H_7F_7(g).

Heat Capacity and Entropy

The molecular structure of H_7F_7 was assumed as planar with the F atoms forming the vertices of a regular heptagon and the H atoms also lying on the circumscribed circle. That rings are more stable than open chains was confirmed by Del Bene and Pople's (4) theoretical molecular-orbital studies on HF polymers. The length of side (F-F axis) was taken from Atoji and Lipscomb's (5) X-ray studies of solid HF (F-F = 2.49 Å) and agrees with the 2.52 Å which Janzen and Bartell (6) determined for HF gaseous polymers by electron diffraction. The low F-bending frequency (53 cm^{-1}) was taken from Boutin et al. (7). The other vibrational frequencies were taken from Kittelberger and Hornig's (8) work on crystalline HF. Huong and Couzi (9) and Smith (10) have made spectral studies of the gas phase in the range from 350 to 4000 cm^{-1}.

The National Bureau of Standards prepared this table (11) by critical analysis of data existing in 1972. Using molecular constants and $\Delta_fH°$ selected by NBS (11), we recalculate the table in terms of 1973 fundamental constants (12), 1975 atomic weights (13) and current JANAF reference states for the elements.

References

1. G. Briegleb and W. Strohmeier, Z. Elektrochem. 57, 668 (1953).
2. E. U. Franck and F. Meyer, Z. Elektrochem. 63, 571 (1959).
3. JANAF Thermochemical Tables: HF(g), 6-30-77.
4. J. E. Del Bene and J. A. Pople, J. Chem. Phys. 55, 2296 (1971).
5. M. Atoji and W. N. Lipscomb, Acta Cryst. 7, 173 (1954).
6. J. Janzen and L. S. Bartell, J. Chem. Phys. 50, 3611 (1969).
7. H. Boutin, G. J. Safford, and V. Brajovic, J. Chem. Phys. 39, 3135 (1963).
8. J. S. Kittelberger and D. F. Hornig, J. Chem. Phys. 46, 3099 (1967).
9. P. V. Huong and M. Couzi, J. Chim. Phys. 66, 1309 (1969).
10. D. F. Smith, J. Chem. Phys. 28, 1040 (1958); 48, 1429 (1968).
11. S. Abramowitz et al., U. S. Nat. Bur. Stand. Report 10904, 239, 1972.
12. CODATA Task Group on Fundamental Constants, CODATA Bulletin 11, December, 1973.
13. IUPAC Commission on Atomic Weights, Pure Appl. Chem. 47, 75 (1976).

Hydrogen Fluoride (H_7F_7) F_7H_7(g)

Enthalpy Reference Temperature = T_r = 298.15 K; Standard State Pressure = p° = 0.1 MPa

T/K	$C_p°$ (J $K^{-1}mol^{-1}$)	S° (J $K^{-1}mol^{-1}$)	$-[G°-H°(T_r)]/T$ (J $K^{-1}mol^{-1}$)	$H°-H°(T_r)$ (kJ mol^{-1})	$\Delta_fH°$ (kJ mol^{-1})	$\Delta_fG°$ (kJ mol^{-1})	Log K_f
0	0.	0.	INFINITE	-41.813	-2083.834	-2083.834	INFINITE
100	128.036	348.140	678.718	-33.058	-2095.751	-2035.680	1063.330
200	168.661	450.187	540.705	-18.104	-2100.476	-1973.468	515.417
250	185.096	489.628	526.631	-9.251	-2101.769	-1941.556	405.666
298.15	198.792	523.432	523.432	0.	-2102.544	-1910.622	334.733
300	199.275	524.663	523.436	0.368	-2102.565	-1909.431	332.461
350	211.190	556.305	525.908	10.639	-2102.927	-1877.208	280.158
400	221.067	585.171	531.538	21.453	-2102.918	-1844.959	240.927
450	229.262	611.697	538.991	32.718	-2102.600	-1812.730	210.416
500	236.149	636.218	547.503	44.358	-2102.023	-1780.551	186.013
600	247.237	680.297	566.046	68.551	-2100.239	-1716.411	149.427
700	256.147	719.099	585.194	93.733	-2097.769	-1652.627	123.321
800	263.796	753.813	604.139	119.739	-2094.731	-1589.236	103.766
900	270.581	785.283	622.545	146.464	-2091.207	-1526.256	88.582
1000	276.663	814.112	640.280	173.832	-2087.267	-1463.692	76.455
1100	282.115	840.741	657.308	201.776	-2082.979	-1401.539	66.554
1200	286.986	865.501	673.638	230.236	-2078.404	-1339.789	58.320
1300	291.323	888.647	689.297	259.155	-2073.599	-1278.432	51.368
1400	295.174	910.380	704.320	288.484	-2068.612	-1217.451	45.424
1500	298.588	930.864	718.747	318.175	-2063.486	-1156.832	40.284
1600	301.615	950.233	732.615	348.189	-2058.252	-1096.559	35.799
1700	304.300	968.600	745.961	378.487	-2052.939	-1036.615	31.851
1800	306.684	986.062	758.819	409.038	-2047.566	-976.986	28.351
1900	308.805	1002.702	771.220	439.815	-2042.146	-917.658	25.228
2000	310.696	1018.590	783.194	470.792	-2036.688	-858.614	22.425
2100	312.385	1033.791	794.768	501.947	-2031.196	-799.846	19.895
2200	313.899	1048.359	805.966	533.263	-2025.670	-741.340	17.602
2300	315.258	1062.343	816.811	564.722	-2020.108	-683.085	15.513
2400	316.482	1075.786	827.324	596.310	-2014.506	-625.074	13.604
2500	317.586	1088.728	837.523	628.014	-2008.858	-567.297	11.853
2600	318.586	1101.204	847.426	659.824	-2003.158	-509.747	10.241
2700	319.493	1113.245	857.049	691.729	-1997.401	-452.417	8.753
2800	320.318	1124.879	866.408	723.720	-1991.579	-395.302	7.374
2900	321.070	1136.133	875.516	755.790	-1985.687	-338.395	6.095
3000	321.757	1147.029	884.386	787.932	-1979.720	-281.694	4.905
3100	322.386	1157.590	893.029	820.139	-1973.673	-225.193	3.794
3200	322.963	1167.835	901.458	852.407	-1967.543	-168.888	2.757
3300	323.494	1177.781	909.681	884.730	-1961.327	-112.775	1.785
3400	323.984	1187.446	917.709	917.105	-1955.023	-56.853	0.873
3500	324.435	1196.844	925.551	949.526	-1948.630	-1.118	0.017
3600	324.853	1205.989	933.214	981.990	-1942.147	54.435	-0.790
3700	325.240	1214.895	940.707	1014.495	-1935.575	109.805	-1.550
3800	325.599	1223.574	948.038	1047.038	-1928.915	164.996	-2.268
3900	325.933	1232.036	955.212	1079.614	-1922.167	220.012	-2.947
4000	326.244	1240.292	962.236	1112.223	-1915.334	274.852	-3.589
4100	326.534	1248.351	969.116	1144.863	-1908.417	329.521	-4.198
4200	326.805	1256.223	975.859	1177.530	-1901.418	384.022	-4.776
4300	327.059	1263.916	982.469	1210.223	-1894.341	438.352	-5.325
4400	327.296	1271.438	988.951	1242.941	-1887.188	492.519	-5.847
4500	327.519	1278.795	995.311	1275.682	-1879.961	546.522	-6.344
4600	327.727	1285.996	1001.552	1308.444	-1872.662	600.362	-6.817
4700	327.924	1293.047	1007.679	1341.227	-1865.294	654.045	-7.269
4800	328.109	1299.952	1013.696	1374.029	-1857.860	707.571	-7.700
4900	328.283	1306.720	1019.608	1406.848	-1850.361	760.938	-8.112
5000	328.447	1313.353	1025.417	1439.685	-1842.799	814.155	-8.505
5100	328.602	1319.859	1031.126	1472.537	-1835.177	867.217	-8.882
5200	328.749	1326.241	1036.740	1505.405	-1827.495	920.132	-9.243
5300	328.888	1332.505	1042.262	1538.287	-1819.754	972.897	-9.588
5400	329.019	1338.654	1047.694	1571.182	-1811.957	1025.515	-9.920
5500	329.144	1344.692	1053.039	1604.091	-1804.103	1077.987	-10.238
5600	329.262	1350.624	1058.300	1637.011	-1796.194	1130.319	-10.543
5700	329.374	1356.453	1063.480	1669.943	-1788.228	1182.506	-10.836
5800	329.481	1362.182	1068.581	1702.886	-1780.207	1234.555	-11.118
5900	329.583	1367.815	1073.605	1735.839	-1772.131	1286.464	-11.389
6000	329.679	1373.355	1078.555	1768.802	-1763.998	1338.235	-11.650

PREVIOUS: June 1977 (1 atm) CURRENT: June 1977 (1 bar)

IODINE FLUORIDE (IF_7) IDEAL GAS M_r = 259.893321

$S°(298.15\ K) = 347.740 \pm 1.3\ J\ K^{-1}\ mol^{-1}$

$\Delta_f H°(0\ K) = -947.050 \pm 2.5\ kJ\ mol^{-1}$

$\Delta_f H°(298.15\ K) = -961.065 \pm 2.5\ kJ\ mol^{-1}$

Vibrational Frequencies and Degeneracies

ν, cm⁻¹	ν, cm⁻¹	ν, cm⁻¹
676 (1)	746 (2)	510 (2)
635 (1)	425 (2)	352 (2)
670 (1)	257 (2)	310 (2)
365 (1)		[340] (2)

Ground State Quantum Weight: [1] $\sigma = 10$

Point Group: D_{5h}

Bond Distances: I-F(equatorial) = 1.86 Å I-F(axial) = 1.76 Å

Bond Angles: F(eq)-I-F(ax) = 90° F(eq)-I-F(eq) = 72°

Product of the Moments of Inertia: $I_A I_B I_C = 1.196821 \times 10^{-112}\ g^3\ cm^6$

Enthalpy of Formation

Settle et al. (1) derived $\Delta_f H° = -229.7 \pm 0.6$ kcal mol⁻¹ for IF_7(g) at 298.15 K from treatment of their calorimetric data on combustion of iodine in fluorine. The combustion products, consisting of mixtures of IF_5 and IF_7, were expanded, treated to remove excess fluorine, and then hydrolyzed. The solutions were analyzed for iodate and the sum of iodate plus periodate. Since the recovery of iodine was not complete (about 70 to 99%), it was necessary to try various assumptions about the form of the unrecovered iodine. Consistent results were obtained only by assigning the unrecovered iodine to IF_7, the less volatile product. This treatment indicated that IF_7 was formed in yields ranging from about 1 to 70%.

$\Delta_r H°$ for the reaction $IF_7(g) \rightarrow IF_5(g) + F_2(g)$ may be checked via equilibrium data (2). Using six points read from Figure 2 of Bernstein and Katz, we obtain $\Delta_r H°$(3rd law) = 29.36 ± 0.2 kcal mol⁻¹ and $\Delta_r H°$(2nd law) = 29.53 ± 0.7 kcal mol⁻¹, both at 298.15 K. The entropy test, $\Delta_r S°$(2nd law) - $\Delta_r S°$(JANAF Functions) = 0.3 ± 1.4 cal K⁻¹ mol⁻¹, indicates consistency with the JANAF functions. Combination of $\Delta_r H°$(3rd law) with $\Delta_f H°$ for IF_5(g) yields $\Delta_f H°(IF_7, g) = -230.2 \pm 0.6$ kcal mol⁻¹. In deriving the uncertainty, we assume that $\Delta_r H°$(3rd law) is uncertain by ±0.5 kcal mol⁻¹ due to an estimated uncertainty of about 1.0 cal K⁻¹ mol⁻¹ in the JANAF value of $\Delta_r S°$ at 500 K.

Heat Capacity and Entropy

Recent evidence favors D_{5h} symmetry and a pentagonal bipyramidal structure for the heptafluoride. Claassen et al. (3) review the earlier debate about structure of lower symmetry. They provide convincing evidence of D_{5h} symmetry from the first observation of Raman spectra of the vapor state and re-examination of the infrared spectra. Their data--including five fundamentals in Raman (two polarized), five fundamentals in infrared, no coincidences between Raman and infrared, and one fundamental inactive--are consistent only with D_{5h} symmetry. This is confirmed by Falconer et al. (4) who interpret their electric-deflection experiments as indicating a symmetry-forbidden dipole moment. Extended Huckel-molecular-orbital calculations (5) also predict D_{5h} symmetry. The adopted structural parameters are from our approximate analysis of the electron-diffraction data of Thompson and Bartell (6). The authors gave a radial distribution curve and suggested only a gross (unrefined) structure because of the probable presence of a fluorocarbon impurity in their sample. Principal moments of inertia are $I_A = I_B = 46.8306 \times 10^{-39}$ and $I_C = 54.5719 \times 10^{-39}$ g cm².

Observed vibrational fundamentals are from Claassen et al. (3). The inactive fundamental is estimated from the calculations of Khanna (7).

References

1. J. L. Settle, J. H. E. Jeffes, P. A. G. O'Hare and W. N. Hubbard, presented at the 23rd Annual Calorimetry Conference, Midland, Mich., August, 1968.
2. R. B. Bernstein and J. J. Katz, J. Phys. Chem. 56, 885 (1952).
3. H. H. Claassen, E. L. Gasner and H. Selig, J. Chem. Phys. 49, 1803 (1968).
4. W. E. Falconer, A. Buchler, J. L. Stauffer and W. Klemperer, J. Chem. Phys. 48, 312 (1968).
5. R. L. Oakland and G. H. Duffey, J. Chem. Phys. 46, 19 (1967).
6. H. B. Thompson and L. S. Bartell, Trans. Amer. Crystallogr. Ass. 2, 190 (1966).
7. R. K. Khanna, J. Mol. Spectrosc. 8, 134 (1962).

Iodine Fluoride (IF_7) $F_7I_1(g)$

Enthalpy Reference Temperature = T_r = 298.15 K Standard State Pressure = p° = 0.1 MPa

T/K	$C_p°$ (J K⁻¹mol⁻¹)	S° (J K⁻¹mol⁻¹)	$-[G°-H°(T_r)]/T$ (J K⁻¹mol⁻¹)	$H°-H°(T_r)$ (kJ mol⁻¹)	$\Delta_f H°$ (kJ mol⁻¹)	$\Delta_f G°$ (kJ mol⁻¹)	Log K_f
0	0.	0.	INFINITE	-23.471	-947.050	-947.050	INFINITE
100	52.248	247.318	444.637	-19.732	-955.010	-917.063	479.024
200	103.461	300.045	359.173	-11.826	-959.821	-876.878	229.017
250	121.537	325.173	349.886	-6.178	-960.746	-856.020	178.856
298.15	134.451	347.740	347.740	0.	-961.065	-835.814	146.431
300	134.875	348.573	347.743	0.249	-961.069	-835.037	145.393
350	144.688	370.140	349.424	7.251	-961.034	-814.031	121.487
400	151.979	389.959	353.269	14.676	-968.728	-792.774	103.526
450	157.478	408.192	358.373	22.419	-968.837	-770.771	89.469
500	161.694	425.012	364.207	30.403	-988.880	-746.865	78.024
600	167.580	455.054	376.906	46.889	-986.428	-698.688	60.826
700	171.367	481.191	389.978	63.850	-983.782	-650.938	48.574
800	173.928	504.252	402.849	81.122	-981.032	-603.576	39.409
900	175.733	524.848	415.281	98.610	-978.229	-556.562	32.302
1000	177.050	543.434	427.182	116.252	-975.402	-509.861	26.632
1100	178.038	560.358	438.531	134.009	-972.573	-463.444	22.007
1200	178.798	575.883	449.339	151.853	-969.756	-417.284	18.164
1300	179.395	590.219	459.632	169.763	-966.966	-371.358	14.921
1400	179.871	603.531	469.440	187.727	-964.212	-325.646	12.150
1500	180.257	615.955	478.799	205.734	-961.506	-280.129	9.755
1600	180.575	627.599	487.739	223.777	-958.853	-234.791	7.665
1700	180.839	638.554	496.291	241.848	-956.258	-189.616	5.826
1800	181.061	648.897	504.485	259.943	-953.723	-144.593	4.196
1900	181.249	658.692	512.345	278.059	-951.244	-99.710	2.741
2000	181.410	667.993	519.897	296.192	-948.815	-54.955	1.435
2100	181.549	676.847	527.162	314.340	-946.425	-10.321	0.257
2200	181.670	685.296	534.159	332.501	-944.062	34.200	-0.812
2300	181.775	693.374	540.907	350.674	-941.712	78.614	-1.785
2400	181.868	701.112	547.422	368.856	-939.356	122.925	-2.675
2500	181.950	708.538	553.719	387.047	-936.978	167.138	-3.492
2600	182.022	715.676	559.812	405.245	-934.560	211.256	-4.244
2700	182.087	722.547	565.713	423.451	-932.086	255.279	-4.939
2800	182.145	729.170	571.433	441.663	-929.539	299.209	-5.582
2900	182.197	735.562	576.983	459.880	-926.904	343.046	-6.179
3000	182.244	741.740	582.373	478.102	-924.168	386.791	-6.735
3100	182.287	747.716	587.610	496.328	-921.319	430.443	-7.253
3200	182.325	753.504	592.705	514.559	-918.347	474.001	-7.737
3300	182.360	759.115	597.663	532.793	-915.245	517.465	-8.191
3400	182.393	764.560	602.492	551.031	-912.005	560.832	-8.616
3500	182.422	769.847	607.198	569.272	-908.621	604.102	-9.016
3600	182.449	774.987	611.788	587.515	-905.091	647.273	-9.392
3700	182.474	779.986	616.267	605.761	-901.412	690.344	-9.746
3800	182.497	784.853	620.639	624.010	-897.581	733.313	-10.080
3900	182.518	789.593	624.911	642.261	-893.599	776.180	-10.396
4000	182.538	794.214	629.086	660.514	-889.465	818.943	-10.694
4100	182.556	798.722	633.169	678.768	-885.181	861.599	-10.977
4200	182.573	803.121	637.163	697.025	-880.749	904.151	-11.245
4300	182.589	807.418	641.073	715.283	-876.170	946.594	-11.499
4400	182.603	811.615	644.901	733.542	-871.446	988.930	-11.740
4500	182.617	815.719	648.652	751.804	-866.582	1031.158	-11.969
4600	182.630	819.733	652.327	770.066	-861.579	1073.274	-12.187
4700	182.642	823.661	655.931	788.330	-856.441	1115.281	-12.395
4800	182.653	827.506	659.466	806.594	-851.172	1157.178	-12.593
4900	182.664	831.273	662.934	824.860	-845.774	1198.962	-12.781
5000	182.674	834.963	666.338	843.127	-840.252	1240.635	-12.961
5100	182.684	838.580	669.679	861.395	-834.609	1282.197	-13.132
5200	182.692	842.128	672.962	879.664	-828.849	1323.648	-13.296
5300	182.701	845.608	676.187	897.933	-822.976	1364.986	-13.453
5400	182.709	849.023	679.356	916.204	-816.992	1406.212	-13.602
5500	182.716	852.376	682.471	934.475	-810.901	1447.326	-13.746
5600	182.723	855.668	685.535	952.747	-804.707	1488.330	-13.883
5700	182.730	858.902	688.548	971.020	-798.413	1529.220	-14.014
5800	182.736	862.080	691.513	989.293	-792.023	1570.000	-14.139
5900	182.742	865.204	694.430	1007.567	-785.539	1610.671	-14.260
6000	182.748	868.276	697.302	1025.842	-778.964	1651.228	-14.375

PREVIOUS: December 1969 (1 atm) CURRENT: December 1969 (1 bar)

Iodine Fluoride (IF_7) $F_7I_1(g)$

MOLYBDENUM FLUORIDE $((MoF_5)_2)$ IDEAL GAS M_r = 381.864030

$\Delta_fH°(0\ K) = -2686.418 \pm 8.4\ kJ\ mol^{-1}$

$S°(298.15\ K) = [531.554 \pm 16.7]\ J\ K^{-1}\ mol^{-1}$ $\Delta_fH°(298.15\ K) = 2697.843 \pm 8.4\ kJ\ mol^{-1}$

Vibrational Frequencies and Degeneracies

ν, cm^{-1}	ν, cm^{-1}
759 (2)	261 (4)
738 (2)	112 (4)
683 (2)	[400](4)
[500](2)	[90](3)
713 (4)	[60](3)

Ground State Quantum Weight: [3] σ = [4]

Point Group: $[D_{2h}]$

Bond Distances: Mo-F = [1.78] Å $Mo\text{-}F_{bridge}$ = [2.06] Å

Bond Angles: F-Mo-F = [90° or 180°] $F_{bridge}\text{-}Mo\text{-}F_{bridge}$ = [90]°

Product of the Moments of Inertia: $I_AI_BI_C = [3.683573 \times 10^{-111}]\ g^3\ cm^6$

Enthalpy of Formation

We adopt $\Delta_fH°(298.15\ K) = -644.8\pm2\ kcal\ mol^{-1}$ in agreement with Brewer (1). This is derived by proceeding from $\Delta_fH°(MoF_5$, cr, 298.15 K) to $\Delta_fH°(MoF_5$, ℓ, 298.15 K) to $\Delta_fH°$(dimer, g, 298.15 K). Brewer (1) adopted $\Delta_fH°(MoF_5$, cr, 298.15 K) = $-331.9\pm1\ kcal\ mol^{-1}$ from the measurements of Burgess et al. (2) and Nuttall et al. (3). Krause and Douglas (4) measured $\Delta_{fus}H°(318.82\ K) = 1.46\pm0.07\ kcal\ mol^{-1}$ and the $\Delta C_p°$ correction to 298.15 K is only -4 cal $K^{-1}\ mol^{-1}$. The heat of vaporization to the dimer $\Delta_{vap}H°(298.15\ K) = 15.99\ kcal\ mol^{-1}$ was derived by Krause (5) from vapor density data.

The value of $\Delta_fH°(0\ K)$ combined with JANAF data (11) for Mo(g) and F(g) gives $\Delta_{at}H°(0\ K) = 1139.9\pm8\ kcal\ mol^{-1}$ and an average bond energy of $\Delta_{at}H°(0\ K)/10 = 114.0\pm1\ kcal\ mol^{-1}$.

Heat Capacity and Entropy

For Mo_2F_{10}(g), the dimer of MoF_5(g), Douglas (6), Krause (5), and Brewer (1) have used the MoF_5(g) frequencies with doubled degeneracy along with estimates of two triply degenerate frequencies. The frequencies, cm^{-1}, and degeneracies for MoF_5(g) were assigned from the spectroscopic data of Acquista and Abramowitz (7) and two fundamentals were estimated (8) -- 759(1), 738(1), 683(1), [500](1), 713(2), 261(2), 112(2), and [200](2). Douglas (6) and Krause (5) elected to use an estimate of [400](2) rather than [200](2) while, for MoF_5, Brewer (1) and Hildenbrand (9) retained the [200](2) estimate. The derived functions for Mo_2F_{10}(g) presented here and by Brewer (1) use the [400](4) estimate. For the six remaining frequencies, Douglas (6) has chosen the estimates [121](3) and [90](3) and Krause (5) has chosen [60](6); Brewer (1) has used [90](3) and [60](3), one of the alternative estimates of Douglas (6), in order to match the selected entropy of the liquid and the entropy of vaporization to the dimer (5). We adopt Brewer's (1) evaluation of the vibrational assignment and the estimate of an electronic ground state quantum weight of [3] (1, 6).

By analogy with the MoF_5(cr) structure (10), the dimer(g) molecule was assumed to have the structure of two octahedra sharing a common edge (5, 6). The two Mo atoms at the centers of the octahedra and the two bridging F atoms at the ends of the common edge form Mo-F-Mo angles of 90°. The bridging Mo-F bond distance of 2.06 Å and the non-bridging Mo-F distance of 1.78 Å from MoF_3(cr) (10) were assumed for the gaseous dimer. All adjacent F-Mo-F angles are 90°.

The principal moments of inertia are: $I_A = 73.3619 \times 10^{-39}$, $I_B = 220.8005 \times 10^{-39}$, $I_C = 227.4042 \times 10^{-39}\ g\ cm^2$.

This table is in good agreement with the critical evaluation by Brewer (1) on which it is based.

References

1. L. Brewer, Materials and Molecular Research Division, Lawrence Berkeley Laboratory, University of California, Berkeley; personal communication, Sept. 1978; a preliminary draft of review to be submitted for publication in Atomic Energy Review, International Atomic Energy Agency, Vienna, Austria.
2. J. Burgess, I. Haigh, and R. D. Peacock, J. Chem. Soc., Dalton Trans. 1062 (1974).
3. R. L. Nuttall, M. V. Kilday, and K. L. Churney in Report NBSIR 73-281, 1 July 1973 (AD Report 782028, 1974), Chapter 5.
4. R. F. Krause, Jr., and T. B. Douglas, J. Chem. Thermodyn. 9, 1149 (1977).
5. R. F. Krause, Jr., in "Proc. Symp. on High Temperature Metal Halide Chemistry," D. L. Hildenbrand and D. D. Cubicciotti, Eds., The Electrochem. Soc. Vol. 78-1, pp. 199-209, 1978.
6. T. B. Douglas, J. Chem. Thermodyn. 9, 1165 (1977).
7. N. Acquista and S. Abramowitz, J. Chem. Phys. 58, 5484 (1973).
8. T. B. Douglas, R. F. Krause, Jr., N. Acquista, and S. Abramowitz, in Report NBSIR 73-280, 1 January 1973 (AD Report 782094, 1973), Chapter 3.
9. D. L. Hildenbrand, Stanford Research Institute, personal communication to L. Brewer. Also see P. D. Kleinschmidt, K. H. Lau, and D. L. Hildenbrand, J. Chem. Thermodyn. 11, 765 (1979).
10. A. J. Edwards, R. D. Peacock, and R. W. H. Small, J. Chem. Soc. 4486 (1962).
11. JANAF Thermochemical Tables: Mo(g), 3-31-78; F(g), 9-30-65.

Molybdenum Fluoride $((MoF_5)_2)$ $F_{10}Mo_2$(g)

Enthalpy Reference Temperature = T_r = 298.15 K Standard State Pressure = p° = 0.1 MPa

T/K	$C_p°$ (J K⁻¹mol⁻¹)	S° (J K⁻¹mol⁻¹)	$-[G°-H°(T_r)]/T$ (J K⁻¹mol⁻¹)	$H°-H°(T_r)$ (kJ mol⁻¹)	$\Delta_fH°$ (kJ mol⁻¹)	$\Delta_fG°$ (kJ mol⁻¹)	Log K_f
0	0.	0.	INFINITE	-41.869	-2686.418	-2686.418	INFINITE
100	121.299	352.980	693.437	-34.046	-2694.112	-2642.820	1380.468
200	174.754	454.231	549.884	-19.131	-2697.527	-2589.929	676.420
250	195.952	495.583	534.963	-9.845	-2697.989	-2562.962	535.502
298.15	212.349	531.554	531.554	0.	-2697.843	-2536.961	444.465
300	212.904	532.870	531.558	0.393	-2697.828	-2535.962	441.550
350	226.083	566.722	534.202	11.382	-2697.211	-2509.028	374.452
400	236.249	597.604	540.226	22.951	-2696.277	-2482.206	324.143
450	244.122	625.905	548.196	34.969	-2695.125	-2455.514	285.029
500	250.274	651.957	557.287	47.335	-2693.820	-2428.959	253.752
600	259.028	698.420	577.032	72.833	-2690.923	-2376.255	206.871
700	264.757	738.809	597.322	99.041	-2687.818	-2324.053	173.423
800	268.674	774.434	617.279	125.724	-2684.624	-2272.304	148.366
900	271.456	806.248	636.539	152.738	-2681.414	-2220.957	128.901
1000	273.496	834.960	654.969	179.991	-2678.236	-2169.967	113.347
1100	275.033	861.102	672.538	207.421	-2675.131	-2119.291	100.637
1200	276.218	885.086	689.265	234.986	-2672.133	-2068.893	90.057
1300	277.150	907.234	705.191	262.656	-2669.271	-2018.740	81.114
1400	277.895	927.801	720.366	290.409	-2666.570	-1968.802	73.457
1500	278.501	946.995	734.842	318.230	-2664.052	-1919.050	66.827
1600	278.999	964.986	748.669	346.106	-2661.735	-1869.460	61.032
1700	279.414	981.913	761.897	374.027	-2659.631	-1820.008	55.922
1800	279.763	997.894	774.568	401.987	-2657.753	-1770.673	51.384
1900	280.059	1013.028	786.724	429.978	-2656.110	-1721.438	47.326
2000	280.313	1027.400	798.401	457.997	-2654.709	-1672.281	43.676
2100	280.531	1041.082	809.634	486.040	-2653.555	-1623.190	40.375
2200	280.721	1054.136	820.453	514.102	-2652.651	-1574.147	37.375
2300	280.887	1066.619	830.887	542.183	-2652.001	-1525.139	34.637
2400	281.033	1078.576	840.960	570.279	-2651.608	-1476.155	32.128
2500	281.162	1090.051	850.696	598.389	-2651.478	-1427.181	29.819
2600	281.277	1101.081	860.115	626.511	-2651.638	-1378.207	27.689
2700	281.379	1111.698	869.238	654.644	-2652.152	-1329.220	25.715
2800	281.470	1121.933	878.081	682.787	-2653.129	-1280.207	23.883
2900	281.553	1131.812	886.661	710.938	-2726.487	-1231.049	22.174
3000	281.627	1141.358	894.992	739.097	-2724.766	-1179.511	20.537
3100	281.694	1150.594	903.090	767.263	-2722.904	-1128.034	19.007
3200	281.755	1159.538	910.965	795.435	-2720.893	-1076.618	17.574
3300	281.810	1168.209	918.629	823.614	-2718.724	-1025.267	16.229
3400	281.861	1176.623	926.094	851.797	-2716.393	-973.985	14.963
3500	281.908	1184.794	933.369	879.986	-2713.895	-922.773	13.772
3600	281.950	1192.736	940.464	908.179	-2711.226	-871.635	12.647
3700	281.990	1200.462	947.387	936.376	-2708.383	-820.574	11.584
3800	282.026	1207.982	954.146	964.577	-2705.367	-769.594	10.579
3900	282.059	1215.309	960.749	992.781	-2702.176	-718.692	9.626
4000	282.091	1222.450	967.203	1020.988	-2698.811	-667.876	8.722
4100	282.119	1229.416	973.514	1049.199	-2695.273	-617.147	7.863
4200	282.146	1236.215	979.688	1077.412	-2691.564	-566.505	7.046
4300	282.171	1242.854	985.731	1105.628	-2687.687	-515.955	6.268
4400	282.194	1249.341	991.649	1133.846	-2683.643	-465.495	5.526
4500	282.216	1255.683	997.446	1162.067	-2679.437	-415.129	4.819
4600	282.236	1261.886	1003.128	1190.290	-2675.071	-364.860	4.143
4700	282.255	1267.956	1008.698	1218.514	-2670.549	-314.686	3.497
4800	282.273	1273.899	1014.161	1246.741	-2665.874	-264.610	2.880
4900	282.290	1279.719	1019.522	1274.969	-2661.051	-214.633	2.288
5000	282.306	1285.423	1024.783	1303.199	-3838.853	-153.294	1.601
5100	282.321	1291.013	1029.948	1331.430	-3837.103	-79.601	0.815
5200	282.335	1296.495	1035.022	1359.663	-3835.509	-5.939	0.060
5300	282.348	1301.873	1040.006	1387.897	-3834.068	67.692	-0.667
5400	282.361	1307.151	1044.905	1416.132	-3832.786	141.298	-1.367
5500	282.372	1312.332	1049.720	1444.369	-3831.657	214.880	-2.041
5600	282.384	1317.421	1054.455	1472.607	-3830.683	288.446	-2.691
5700	282.394	1322.419	1059.112	1500.846	-3829.859	361.992	-3.317
5800	282.404	1327.330	1063.695	1529.086	-3829.186	435.526	-3.922
5900	282.414	1332.158	1068.204	1557.327	-3828.658	509.053	-4.507
6000	282.423	1336.904	1072.643	1585.568	-3828.279	582.568	-5.072

PREVIOUS: March 1979 (1 atm) CURRENT: March 1979 (1 bar)

Molybdenum Fluoride $((MoF_5)_2)$ $F_{10}Mo_2$(g)

SULFUR FLUORIDE (S_2F_{10}) IDEAL GAS M_r = 254.104030

$S°(298.15 K) = [396.943 \pm 8.4] kJ mol^{-1}$

$\Delta_f H°(0 K) = -2040.744 \pm 29.3 kJ mol^{-1}$

$\Delta_f H°(298.15 K) = -2064.386 \pm 29.3 kJ mol^{-1}$

Vibrational Frequencies and Degeneracies

ν, cm^{-1}	ν, cm^{-1}	ν, cm^{-1}	ν, cm^{-1}
913 (1)	938 (1)	410 (2)	860 (2)
690 (1)	684 (1)	[150](2)	634 (2)
[642](1)	571 (1)	728 (2)	425 (2)
247 (1)	826 (2)	624 (2)	188 (2)
Torsion	544 (2)	509 (2)	

Ground State Quantum Weight: [1] σ = 8

Point Group: [D_{4h}]

Bond Distances: S-F = 1.56 ± 0.02 Å S-S = 2.21 ± 0.03 Å

Bond Angles: F*-S-F* = 90° F*-S-S = 90°

F**-S-S* = 180° (* - equatorial ** - axial)

Product of the Moments of Inertia: $I_A I_B I_C = 8.747547 \times 10^{-112} g^3 cm^6$

Reduced Moment of Inertia: $I_{red} = 1.5354 \times 10^{-38} g cm^2$

Barrier to Internal Rotation: V = [8.0] kcal mol^{-1}

Enthalpy of Formation

The kinetics of the decomposition of gaseous S_2F_{10} have been studied by Trost and McIntosh (1) from 434-455 K. Benson and Bott (2) have reevaluated these kinetic data and have reported two values for the equilibrium constant of the dissociation process $S_2F_{10}(g) = SF_4(g) + SF_6(g)$. Both values refer to a temperature of 444 K. We analyze these two Kp values by the 3rd law method, and we obtain the result $\Delta_r H°(298.15 K) = 19.26 \pm 1.4$ kcal mol^{-1}. The adopted value of $\Delta_f H°(S_2F_{10}$, g, 298.15 K) = -493.4±7 kcal mol^{-1} is obtained by combining the value of $\Delta_r H°$ with JANAF enthalpies of formation (3) for SF_4 and SF_6.

The heat of atomization, $\Delta_{at}H°(0 K)$, and bond dissocation energy $D_0^0(F_5S\text{-}SF_5)$ are calculated to be 802.8±11.0 kcal mol^{-1} and 58.4±14.0 kcal mol^{-1}, respectively. The latter value agrees with an upper limit value of 62.6 kcal mol^{-1} for the bond energy of a covalent S-S linkage established from thermochemical data (3) for the S_8 ring.

Heat Capacity and Entropy

The results of electron diffraction experiments (4) show that the S_2F_{10} molecule consists of two SF_5 octahedral groups sharing a common S-S bond. The adopted bond lengths and angles are taken from this study (4). The relative orientation of the two SF_5 groups has not been experimentally established (5-7), but the staggered configuration of D_{4d} symmetry has been predicted to be more stable (8) than the eclipsed form (D_{4h} symmetry). We assume that the point group is D_{4d}. The principal moments of inertia are: $I_A = 61.4204 \times 10^{-39}$ and $I_B = I_C = 119.3402 \times 10^{-39}$ g cm^2.

The vibrational spectra of S_2F_{10} have been studied in detail in three independent investigations (5-7). The reported fundamental frequencies show reasonable agreement but the assignments of the observed fundamentals made by Wilmshurst and Bernstein (5) differ somewhat from those made by Dodd et al. (6). The adopted frequencies and assignments except for the torsion (ν_5) and SF_5 rocking frequency (ν_{12}) are taken from Dodd et al. (6). We estimate the unobserved ν_{12} fundamental at 150 cm^{-1}. Other estimated values that have been reported include 200 cm^{-1} (2) and 37 cm^{-1} (5). The inactive torsional frequency is treated as a hindered internal rotation. We use an estimated potential barrier of 8.0 kcal mol^{-1} (5) to calculate heat capacity contributions for hindered rotation from the table of Pitzer and Brewer (9). Contributions below 201 K could not be obtained by this procedure since the values of the ratio V/RT were outside the range of the table. $C_p°$ value between 150-200 K are estimated by graphical interpolation of the hindered rotor $C_p°$ data above 210 K and the corresponding data for a harmonic oscillator (ν_5 = 89 cm^{-1}, 5) below 150 K. This treatment assumes that the internal rotation approaches harmonic oscillation as V/RT >> 1. The uncertainty (±2 cal K^{-1} mol^{-1}) in the value of S°(298.15 K) includes the possibility that the potential barrier is as low as 4.0 kcal mol^{-1} (8) and contains a contribution from the uncertainty due to the two estimated vibrational frequencies.

Previously published thermodynamic functions (10), 11) for S_2F_{10} are based on the assumption that free internal rotation exists in the molecule and on different estimates (5) for the SF_5 rocking fundamentals. Benson and Bott (2) have shown that these earlier frequencies (5) give values for $S°(S_2F_{10}$, g) which are too high by several units. Our treatment of internal rotation is consistent with the interpretation of electron diffraction data (4) which indicates that the rotation is not completely restricted. It should also be noted that the $C_p°$ data reported by Pass (11) are erroneous above 400 K.

References

1. W. R. Trost and R. L. McIntosh, Can. J. Chem. 29, 508 (1951).
2. S. W. Benson and J. Bott, Int. J. Chem. Kinet. 1, 451 (1969).
3. JANAF Thermochemical Tables: SF_4(g) and SF_6(g), 6-30-76; S_8(g), 6-30-64.
4. R. B. Harvey and S. H. Bauer, J. Amer. Chem. Soc. 75, 2840 (1953).
5. J. K. Wilmshurst and H. J. Berstein, Can. J. Chem. 35, 191 (1957).
6. R. E. Dodd, L. A. Woodward, and H. L. Roberts, Trans. Faraday Soc. 53, 1545 (1957).
7. R. R. Smardzewski, R. E. Noftle and W. B. Fox, J. Mol. Spectrosc. 62, 449 (1976).
8. F. Crasnier, J. F. Labarre, and C. Leibovici, J. Fluorine Chem. 3, 307 (1973).
9. K. S. Pitzer and L. Brewer, "Thermodynamics," 2nd ed., McGraw-Hill Book Company, New York, 1961.
10. P. A. G. O'Hare, Argonne National Laboratory, ANL-7315, July, 1968.
11. G. Pass, J. Appl. Chem. (London), 19, 77 (1969).

Sulfur Fluoride (S_2F_{10}) $F_{10}S_2$(g)

Enthalpy Reference Temperature = T_r = 298.15 K Standard State Pressure = p° = 0.1 MPa

T/K	$C_p°$ (J $K^{-1}mol^{-1}$)	S° (J $K^{-1}mol^{-1}$)	$-[G°-H°(T_r)]/T$ (J $K^{-1}mol^{-1}$)	$H°-H°(T_r)$ (kJ mol^{-1})	$\Delta_f H°$ (kJ mol^{-1})	$\Delta_f G°$ (kJ mol^{-1})	Log K_f
0	0.	0.	INFINITE	-29.307	-2040.744	-2040.744	INFINITE
100	69.170	271.813	518.037	-24.622	-2051.974	-1991.466	1040.235
200	124.892	336.657	411.591	-14.987	-2060.267	-1927.492	503.409
298.15	176.682	396.943	396.943	0.	-2064.386	-1861.311	326.094
300	177.473	398.038	396.946	0.328	-2064.432	-1860.051	323.864
400	211.418	454.110	404.360	19.900	-2070.149	-1791.494	233.945
500	232.735	503.752	419.380	42.186	-2072.548	-1721.547	179.849
600	246.438	547.482	437.162	66.192	-2073.079	-1651.270	143.756
700	255.525	586.198	455.742	91.319	-2072.404	-1581.009	117.976
800	261.768	620.750	474.248	117.202	-2070.991	-1510.899	98.652
900	266.123	651.847	492.282	143.608	-2175.369	-1438.835	83.508
1000	269.228	680.054	509.671	170.383	-2170.756	-1357.244	70.895
1100	271.377	705.819	526.349	197.417	-2166.065	-1276.120	60.598
1200	273.094	729.508	542.305	224.644	-2161.341	-1195.424	52.035
1300	274.275	751.416	557.558	252.014	-2156.622	-1115.122	44.806
1400	275.232	771.778	572.141	279.492	-2151.934	-1035.184	38.623
1500	275.907	790.791	586.091	307.049	-2147.295	-955.578	33.276
1600	276.470	808.616	599.448	334.669	-2142.716	-876.279	28.608
1700	276.813	825.387	612.250	362.334	-2138.203	-797.265	24.497
1800	277.117	841.218	624.535	390.030	-2133.759	-718.515	20.851
1900	277.384	856.208	636.337	417.756	-2129.371	-640.011	17.595
2000	277.613	870.442	647.689	445.506	-2125.029	-561.735	14.671
2100	277.754	883.991	658.622	473.274	-2120.721	-483.678	12.031
2200	277.883	896.915	669.162	501.056	-2116.432	-405.824	9.635
2300	277.997	909.270	679.335	528.851	-2112.143	-328.164	7.453
2400	278.098	921.103	689.164	556.655	-2107.836	-250.692	5.456
2500	278.186	932.458	698.670	584.470	-2103.492	-173.400	3.623
2600	278.245	943.370	707.873	612.291	-2099.093	-96.283	1.934
2700	278.298	953.872	716.791	640.119	-2094.622	-19.337	0.374
2800	278.345	963.994	725.440	667.951	-2090.064	57.443	-1.072
2900	278.386	973.762	733.835	695.787	-2085.402	134.057	-2.415
3000	278.420	983.200	741.991	723.628	-2080.626	210.509	-3.665
3100	278.441	992.330	749.920	751.471	-2075.722	286.799	-4.833
3200	278.459	1001.170	757.634	779.316	-2070.683	362.930	-5.924
3300	278.475	1009.739	765.144	807.162	-2065.499	438.901	-6.947
3400	278.488	1018.053	772.461	835.011	-2060.164	514.712	-7.908
3500	278.500	1026.125	779.594	862.860	-2054.674	590.364	-8.811
3600	278.504	1033.971	786.552	890.710	-2049.026	665.858	-9.661
3700	278.508	1041.602	793.342	918.561	-2043.216	741.191	-10.464
3800	278.512	1049.029	799.973	946.412	-2037.243	816.365	-11.222
3900	278.516	1056.264	806.453	974.263	-2031.108	891.381	-11.939
4000	278.521	1063.315	812.786	1002.115	-2024.811	966.238	-12.618
4100	278.526	1070.193	818.981	1029.967	-2018.353	1040.933	-13.262
4200	278.531	1076.905	825.043	1057.820	-2011.736	1115.471	-13.873
4300	278.535	1083.459	830.976	1085.674	-2004.963	1189.847	-14.454
4400	278.539	1089.862	836.788	1113.527	-1998.036	1264.066	-15.006
4500	278.541	1096.122	842.481	1141.381	-1990.959	1338.126	-15.533
4600	278.542	1102.244	848.062	1169.235	-1983.735	1412.025	-16.034
4700	278.543	1108.234	853.534	1197.090	-1976.370	1485.768	-16.512
4800	278.543	1114.098	858.902	1224.944	-1968.866	1559.351	-16.969
4900	278.544	1119.842	864.169	1252.798	-1961.228	1632.777	-17.406
5000	278.546	1125.469	869.338	1280.653	-1953.459	1706.045	-17.823
5100	278.548	1130.985	874.415	1308.507	-1945.565	1779.157	-18.222
5200	278.549	1136.394	879.401	1336.362	-1937.548	1852.114	-18.605
5300	278.551	1141.700	884.300	1364.217	-1929.414	1924.914	-18.971
5400	278.553	1146.906	889.115	1392.073	-1921.166	1997.560	-19.323
5500	278.554	1152.018	893.849	1419.928	-1912.809	2070.051	-19.660
5600	278.555	1157.037	898.504	1447.783	-1904.345	2142.392	-19.983
5700	278.556	1161.967	903.083	1475.639	-1895.780	2214.577	-20.294
5800	278.557	1166.812	907.589	1503.495	-1887.117	2286.613	-20.593
5900	278.558	1171.574	912.023	1531.350	-1878.359	2358.501	-20.881
6000	278.558	1176.255	916.388	1559.206	-1869.510	2430.236	-21.157

PREVIOUS: CURRENT: December 1977

MOLYBDENUM FLUORIDE $((MoF_5)_3)$ IDEAL GAS M_r = 572.796045

$S°(298.15 K) = [706.844 \pm 29.3] J K^{-1} mol^{-1}$

$\Delta_f H°(0 K) = -4049.417 \pm 20.9 kJ mol^{-1}$
$\Delta_f H°(298.15 K) = -4065.593 \pm 20.9 kJ mol^{-1}$

Vibrational Frequencies and Degeneracies

ν, cm^{-1}	ν, cm^{-1}
759 (3)	261 (6)
738 (3)	112 (6)
683 (3)	[400](6)
[500](3)	[106](6)
713 (6)	[53](6)

Ground State Quantum Weight: [4] σ = [6]
Point Group: $[D_{3h}]$
Bond Distances: Mo-F = [1.78] Å $Mo\text{-}F_{bridge}$ = [2.06] Å
Bond Angles: F-Mo-F = [90° or 180°] $F_{bridge}\text{-}Mo\text{-}F_{bridge}$ = [90°]
Product of the Moments of Inertia: $I_A I_B I_C = [9.5696 \times 10^{-110}] g^3 cm^6$

Enthalpy of Formation

We adopt $\Delta_f H°(Mo_3F_{15}$, g, 298.15 K) = -971.7 ± 5 kcal mol^{-1} to agree with Brewer (1). This is based on 1) the $\Delta_f H°(MoF_5$, cr, 298.15 K) = -331.9 ± 1 kcal mol^{-1} adopted by Brewer from the measurements of Burgess et al. (2) and Nuttall et al. (3), 2) the heat of fusion $\Delta_{fus}H°(318.82 K) = 1.46 \pm 0.07$ kcal mol^{-1} measured by Krause and Douglas (4), and 3) a heat of vaporization to the trimer $\Delta_{vap}H°(298.15 K)$ = 18.9 kcal mol^{-1}. This heat of vaporization was derived by Brewer (1) from a re-evaluation of the vapor density data of Douglas (5)--the proportion of $Mo_3F_{15}(g)$ in the saturated vapor was reduced to be consistent with no appreciable monomer and a vapor density corresponding to $Mo_2F_{10}(g)$. Douglas (5) had chosen $\Delta_{vap}H°(298.15 K)$ of $MoF_5(\ell)$ to $Mo_3F_{15}(g)$ equal to 17.7 kcal mol^{-1} from a possible range of 16 to 21 kcal mol^{-1}. The value of $\Delta_f H°(0 K)$ combined with JANAF data (6) for Mo(g) and F(g) gives $\Delta_{at}H°(0 K) = 1714.5 \pm 14$ kcal mol^{-1} and an average Mo-F bond energy of $\Delta_{at}H°(0 K)/15 = 114.3 \pm 1$ kcal mol^{-1}.

Heat Capacity and Entropy

Brewer (1) has followed Douglas's (5) choice of [4] for the electronic ground state quantum weight, 2) of vibrational frequencies and 3) of molecular configuration.

The $MoF_5(g)$ monomer vibrational frequencies were used, with triple the degeneracy, for the trimer, Mo_3F_{15}, and the 12 remaining frequencies were estimated as [106](6) and [53](6) (5). The vibrational frequencies for $MoF_5(g)$ had been assigned from the spectroscopic data of Aquista and Abramowitz (7) and two fundamentals had been estimated (8); these frequencies, cm^{-1}, and degeneracies are 759(1), 738(1), 683(1), [500](1), 713(2), 261(2), and [200](2). Douglas (5) used [400](2) rather than [200](2).

The structure assumed by Douglas (5) consists of three tetragonal bipyramids joined by three bridging F atoms such that the three Mo atoms and the three bridging F atoms form a planar, irregular hexagon. The $F_{bridge}\text{-}Mo\text{-}F_{bridge}$ angles are 90°, the $Mo\text{-}F_{bridge}\text{-}Mo$ angles are 150°; the remaining F-Mo-F angles are either 90° or 180°. The $Mo\text{-}F_{bridge}$ bond distance, 2.06 Å, and the Mo-F bond distance, 1.78 Å, are taken from the x-ray study of $MoF_5(cr)$ by Edwards et al. (9).

The principal moments of inertia are $I_A = I_B = 3.8417 \times 10^{-37}$, $I_C = 6.4840 \times 10^{-37}$ g cm^2.

This table is in good agreement with the critical evaluation of Brewer (1) on which it is based.

References

1. L. Brewer, Materials and Molecular Research Division, Lawrence Berkeley Laboratory, University of California, Berkeley; personal communication, Sept. 1978; a preliminary draft of review to be submitted for publication in Atomic Energy Review, International Atomic Energy Agency, Vienna, Austria.
2. J. Burgess, I. Haigh, and R. D. Peacock, J. Chem. Soc., Dalton Trans. 1062 (1974).
3. R. E. Nuttall, M. V. Kilday, and K. L. Churney in NBSIR 73-281, (AD Report 782028, 1974), Chap. 5., 1 July 1973
4. R. F. Krause, Jr., and T. B. Douglas, J. Chem. Thermodyn. 9, 1149 (1977).
5. T. B. Douglas, J. Chem. Thermodyn. 9, 1165 (1977).
6. JANAF Thermochemical Tables: Mo(g), 3-31-78; F(g), 9-30-65.
7. N. Aquista and S. Abramowitz, J. Chem. Phys. 58, 5484 (1973).
8. T. B. Douglas, R. F. Krause, Jr., N. Aquista, and S. Abramowitz in NBSIR 73-380 (AD Report 782094, Chap. 3) 1 January 1973.
9. A. J. Edwards, R. D. Peacock, and R. W. H. Small, J. Chem. Soc. 4486 (1962).

Molybdenum Fluoride $((MoF_5)_3)$ $F_{15}Mo_3(g)$

Enthalpy Reference Temperature = T_r = 298.15 K Standard State Pressure = p° = 0.1 MPa

T/K	$C_p°$ (J $K^{-1}mol^{-1}$)	S° (J $K^{-1}mol^{-1}$)	$-[G°-H°(T_r)]/T$ (J $K^{-1}mol^{-1}$)	$H°-H°(T_r)$ (kJ mol^{-1})	$\Delta_f H°$ (kJ mol^{-1})	$\Delta_f G°$ (kJ mol^{-1})	Log K_f
0	0.	0.	INFINITE	-63.765	-4049.417	-4049.417	INFINITE
100	186.331	431.758	956.105	-52.435	-4061.363	-3974.653	2076.146
200	269.353	587.843	735.038	-29.439	-4065.861	-3885.765	1014.858
250	301.534	651.527	712.085	-15.139	-4066.184	-3840.681	802.466
298.15	326.335	706.844	706.844	0.	-4065.593	-3797.290	665.269
300	327.175	708.865	706.850	0.604	-4065.556	-3795.625	660.877
350	347.072	760.860	710.911	17.482	-4064.236	-3750.733	559.766
400	362.408	808.250	720.161	35.236	-4062.436	-3706.066	483.963
450	374.275	851.650	732.394	53.665	-4060.304	-3661.644	425.033
500	383.545	891.583	746.343	72.620	-4057.940	-3617.474	377.915
600	396.731	962.767	776.629	111.683	-4052.778	-3529.858	307.301
700	405.357	1024.615	807.735	151.816	-4047.300	-3443.135	256.930
800	411.255	1079.152	838.321	192.664	-4041.687	-3357.208	219.203
900	415.443	1127.846	867.834	234.010	-4036.046	-3271.987	189.901
1000	418.513	1171.784	896.068	275.716	-4030.453	-3187.392	166.492
1100	420.826	1211.786	922.979	317.688	-4024.968	-3103.353	147.366
1200	422.610	1248.482	948.596	359.863	-4019.642	-3019.805	131.449
1300	424.012	1282.366	972.984	402.197	-4014.521	-2936.695	117.998
1400	425.134	1313.832	996.220	444.656	-4009.641	-2853.971	106.483
1500	426.046	1343.195	1018.384	487.217	-4005.035	-2771.585	96.515
1600	426.795	1370.716	1039.553	529.860	-4000.729	-2689.497	87.803
1700	427.420	1396.610	1059.803	572.572	-3996.743	-2607.668	80.124
1800	427.945	1421.056	1079.200	615.341	-3993.097	-2526.064	73.304
1900	428.391	1444.206	1097.807	658.158	-3989.802	-2444.656	67.208
2000	428.772	1466.189	1115.681	701.017	-3986.871	-2363.408	61.726
2100	429.101	1487.117	1132.874	743.911	-3984.309	-2282.301	56.769
2200	429.387	1507.086	1149.433	786.836	-3982.123	-2201.306	52.266
2300	429.637	1526.178	1165.401	829.787	-3980.317	-2120.400	48.156
2400	429.857	1544.468	1180.817	872.762	-3978.897	-2039.566	44.390
2500	430.051	1562.020	1195.717	915.758	-3977.871	-1958.783	40.927
2600	430.223	1578.890	1210.132	958.771	-3977.280	-1878.033	37.730
2700	430.376	1595.130	1224.092	1001.802	-3977.221	-1797.296	34.771
2800	430.514	1610.784	1237.625	1044.846	-3977.856	-1716.549	32.023
2900	430.638	1625.894	1250.754	1087.904	-4087.062	-1635.615	29.461
3000	430.749	1640.495	1263.504	1130.973	-4083.649	-1551.140	27.008
3100	430.850	1654.621	1275.894	1174.053	-4080.025	-1466.784	24.715
3200	430.942	1668.301	1287.944	1217.143	-4076.177	-1382.546	22.568
3300	431.026	1681.563	1299.672	1260.242	-4072.094	-1298.432	20.552
3400	431.102	1694.432	1311.094	1303.348	-4067.766	-1214.446	18.658
3500	431.172	1706.929	1322.226	1346.462	-4063.187	-1130.591	16.873
3600	431.237	1719.077	1333.082	1389.582	-4058.352	-1046.869	15.190
3700	431.296	1730.893	1343.674	1432.709	-4053.258	-963.286	13.599
3800	431.350	1742.396	1354.016	1475.841	-4047.902	-879.847	12.094
3900	431.401	1753.601	1364.119	1518.979	-4042.284	-796.548	10.669
4000	431.447	1764.524	1373.993	1562.121	-4036.406	-713.397	9.316
4100	431.491	1775.178	1383.649	1605.268	-4030.268	-630.399	8.031
4200	431.531	1785.576	1393.095	1648.419	-4023.873	-547.551	6.810
4300	431.569	1795.731	1402.341	1691.574	-4017.226	-464.862	5.647
4400	431.604	1805.653	1411.395	1734.733	-4010.330	-382.327	4.539
4500	431.636	1815.352	1420.265	1777.895	-4003.189	-299.951	3.482
4600	431.667	1824.840	1428.957	1821.060	-3995.808	-217.739	2.473
4700	431.696	1834.123	1437.479	1864.228	-3988.194	-135.688	1.508
4800	431.722	1843.212	1445.838	1907.399	-3980.351	-53.801	0.585
4900	431.748	1852.114	1454.038	1950.573	-3972.285	27.918	-0.298
5000	431.771	1860.837	1462.087	1993.749	-5738.157	126.664	-1.323
5100	431.794	1869.388	1469.990	2036.927	-5734.700	243.926	-2.498
5200	431.815	1877.772	1477.752	2080.108	-5731.478	361.124	-3.628
5300	431.835	1885.998	1485.377	2123.290	-5728.486	478.260	-4.714
5400	431.854	1894.070	1492.871	2166.474	-5725.731	595.341	-5.759
5500	431.872	1901.994	1500.238	2209.661	-5723.206	712.373	-6.766
5600	431.888	1909.776	1507.482	2252.849	-5720.914	829.365	-7.736
5700	431.905	1917.421	1514.607	2296.038	-5718.847	946.312	-8.672
5800	431.920	1924.932	1521.617	2339.230	-5717.007	1063.227	-9.575
5900	431.934	1932.316	1528.515	2382.422	-5715.382	1180.116	-10.448
6000	431.948	1939.576	1535.306	2425.616	-5713.982	1296.974	-11.291

PREVIOUS: March 1979 (1 atm) CURRENT: March 1979 (1 bar)

IRON OXIDE, WUSTITE ($Fe_{0.947}O$) CRYSTAL M_r = 68.8865

$S°(298.15\ K) = 57.588 \pm 0.42\ J\ K^{-1}\ mol^{-1}$

$T_{fus} = 1650\ K$

$\Delta_fH°(0\ K) = -267.123 \pm 0.84\ kJ\ mol^{-1}$

$\Delta_fH°(298.15\ K) = -266.270 \pm 0.84\ kJ\ mol^{-1}$

$\Delta_{fus}H° = 31.338\ kJ\ mol^{-1}$

Enthalpy of Formation

The chemical equilibria in the Fe-C-O and Fe-H-O system have been studied by many investigators. Using the reported equilibrium constants for the following two reactions (A) $Fe_{0.947}O(cr) + CO(g) = 0.947Fe(cr) + CO_2(g)$ and (B) $Fe_{0.947}O(cr, \ell) + H_2(g) = 0.947Fe(cr) + H_2O(g)$, the corresponding enthalpy changes $\Delta_rH°$ were evaluated by both the 2nd and 3rd law methods. Based on the 3rd law values for $\Delta_rH°$, the $\Delta_fH°$(Wustite, cr, 298.15 K) values were also calculated. The results obtained are presented as follows.

Source	Reaction	T/K	$\Delta_rH°(298.15\ K)$, kcal mol^{-1} 2nd law	3rd law	$\Delta_fH°(298.15\ K)$ kcal mol^{-1}
Eastman (1)	A	873.2-1273.2	-3.80±0.01	-4.01	-63.63
Eastman - Evans (2)	A	973.2-1273.2	-4.02±0.01	-4.03	-63.61
Eastman (1)	B	873.2-1273.2	6.50±0.12	5.90	-63.70
Eastman - Evans (2)	B	973.2-1273.2	5.81±0.01	5.07	-62.86
Emmett - Schultz (3)	B	873.2-1273.2	5.95±0.07	5.82	-63.62
Britzke et al. (4)	B	1123.2-1498.2	-2.25±0.03	4.26	-62.06
Jominy - Murphy (5)	B	1360.0-1646.0	5.01±0.61	6.01	-63.81
Chipman - Marshall (6)	B	1438.2-1633.2	5.03±0.34	5.85	-62.83
Chipman - Marshall (6)	B	1653.2-1787.2	-2.61±0.77	-0.02	-61.87
Jominy - Murphy (5)	B	1698.0-1700.0	---	0.87	-62.72
Britzke et al. (4)	C	1123.0-1473.0	-129.08±0.18	-126.28	-63.14

Britzke et al. (4) also determined the equilibrium pressures for the reaction (C) $1.894\ Fe(cr) + O_2(g) = 2Fe_{0.947}O(cr)$. The corresponding $\Delta_fH°(298.15\ K)$ for Wustite was evaluated based on the 3rd law value for $\Delta_rH°(298.15\ K)$. The adopted value of $\Delta_fH°$(Wustite, cr, 298.15 K) is the weighted average of the $\Delta_fH°(298.15\ K)$ listed in the above table.

Heat Capacity and Entropy

The low temperature (54.37-298.16 K) heat capacities were measured by Todd and Bonnickson (7). The high temperture (298-1650 K) heat capacities were determined by Coughlin et al. (8). The two sets of $C_p°$ data were plotted and joined smoothly at 298.15 K. The $C_p°$ values above 1650 K were obtained by graphical extrapolation. The low temperature (70.7-279.8 K) $C_p°$'s were also reported by Millar (9). S°(298.15 K) was derived from the low temperature heat capacities, based on $S°(52\ K) = 0.696\ cal\ K^{-1}\ mol^{-1}$. Todd and Bonnickson (7) added 0.41 cal $K^{-1}\ mol^{-1}$, the amount calculated for completely random distribution of the vacant Fe spaces in the Wustite lattice. However, the Gibbs energy functions derived from $S°(52\ K) = 0.696\ cal\ K^{-1}\ mol^{-1}$ give better agreement between 2nd and 3rd law values for $\Delta_rH°$ than those derived from $S°(52\ K) = 1.11\ cal\ K^{-1}\ mol^{-1}$. Therefore, the value 0.41 cal $K^{-1}\ mol^{-1}$ was not added here.

Fusion Data

T_{fus} and $\Delta_{fus}H°$ were taken from Coughlin et al. (8). T_{fus} was reported as 1372°C (1645 K) by Darken and Gurry (9) and as 1369°C (1642 K) by Chipman and Marshall (6).

References

1. E. D. Eastman, J. Amer. Chem. Soc. 44, 975 (1922).
2. E. D. Eastman and R. M. Evans, J. Amer. Chem. Soc. 46, 888 (1924). The K_p value at 1273.2 K for reaction (1) is rejected by a statistical criterion.
3. P. H. Emmett and J. F. Schultz, J. Amer. Chem. Soc. 52, 4268 (1930).
4. E. V. Britzke, A. F. Kapustinsky and T. I. Schaschkina, Z. Anorg. Allg. Chem. 219, 287 (1934). The K_p value at 1123 K for reaction (3) is rejected by a statistical criterion.
5. W. E. Jominy and D. W. Murphy, Ind. Eng. Chem. 23, 384 (1931). The K_p values at 1590 and 1645 K are rejected by a statistical criterion.
6. J. Chipman and S. Marshall, J. Amer. Chem. Soc. 62, 299 (1940).
7. S. S. Todd and K. R. Bonnickson, J. Amer. Chem. Soc. 73, 3894 (1951).
8. J. P. Coughlin, E. G. King and K. R. Bonnickson, J. Amer. Chem. Soc. 73, 3891 (1951).
9. R. W. Millar, J. Amer. Chem. Soc. 51, 215 (1929).
10. L. S. Darken and R. W. Gurry, J. Amer. Chem. Soc. 68, 798 (1946).

Iron Oxide, Wustite ($Fe_{0.947}O$) $Fe_{0.947}O_1(cr)$

Enthalpy Reference Temperature = T_r = 298.15 K Standard State Pressure = p° = 0.1 MPa

T/K	$C_p°$ (J $K^{-1}mol^{-1}$)	S°	$-[G°-H°(T_r)]/T$	$H°-H°(T_r)$ (kJ mol^{-1})	$\Delta_fH°$	$\Delta_fG°$	Log K_f
0	0.	0.	INFINITE	-9.462	-267.123	-267.123	INFINITE
100	23.336	12.294	98.526	-8.623	-268.136	-260.126	135.876
187.670	58.206	35.752	63.498	-5.207	Cp LAMBDA MAXIMUM TRANSITION		
187.670	58.196	35.754	63.498	-5.207			
200	45.966	38.989	61.893	-4.581	-267.224	-252.274	65.887
298.15	48.116	57.588	57.588	0.	-266.270	-245.143	42.948
300	48.183	57.886	57.589	0.089	-266.252	-245.012	42.660
400	50.380	72.065	59.507	5.023	-265.292	-238.078	31.090
500	51.848	83.470	63.196	10.137	-264.409	-231.379	24.172
600	53.028	93.030	67.393	15.382	-263.666	-224.845	19.575
700	54.070	101.283	71.658	20.737	-263.088	-218.423	16.299
800	55.041	108.567	75.825	26.193	-262.723	-212.071	13.847
900	55.969	115.104	79.833	31.744	-262.690	-205.745	11.941
1000	56.873	121.048	83.661	37.386	-263.275	-199.395	10.415
1100	57.760	126.510	87.312	43.118	-264.641	-192.927	9.161
1200	58.630	131.573	90.792	48.938	-265.416	-186.391	8.113
1300	59.496	136.300	94.112	54.844	-264.562	-179.840	7.226
1400	60.358	140.741	97.286	60.837	-263.716	-173.355	6.468
1500	61.212	144.934	100.324	66.915	-262.878	-166.930	5.813
1600	62.065	148.912	103.238	73.079	-262.046	-160.561	5.242
1700	62.787	152.699	106.037	79.325	-262.124	-154.225	4.739
1800	63.304	156.303	108.730	85.631	-261.650	-147.892	4.292
1900	63.696	159.736	111.325	91.982	-274.573	-140.919	3.874
2000	64.015	163.012	113.828	98.368	-274.427	-133.889	3.497
2100	64.238	166.141	116.245	104.781	-274.265	-126.866	3.156
2200	64.434	169.134	118.582	111.214	-274.094	-119.850	2.846
2300	64.601	172.002	120.842	117.666	-273.916	-112.844	2.563
2400	64.740	174.754	123.032	124.134	-273.733	-105.844	2.304
2500	64.852	177.399	125.154	130.614	-273.549	-98.853	2.065

PREVIOUS: CURRENT: June 1965

Iron Oxide, Wustite ($Fe_{0.947}O$) $Fe_{0.947}O_1(cr)$

IRON SULFIDE, PYRRHOTITE ($Fe_{0.877}S$) CRYSTAL M_r = 81.038

$S°(298.15\ K) = 60.799 \pm 0.4\ J\ K^{-1}\ mol^{-1}$

$T_{trs2} = 598 \pm 3\ K$

$\Delta_f H°(0\ K) = -106.31 \pm 1.7\ kJ\ mol^{-1}$

$\Delta_f H°(298.15\ K) = -105.44 \pm 1.7\ kJ\ mol^{-1}$

$\Delta_{trs2}H° = 0.398 \pm 0.84\ kJ\ mol^{-1}$

Enthalpy of Formation

The adopted value of $\Delta_f H°(298.15\ K)$ is based on the calorimetric determination of Stolyarova and Bezmen (1) assuming that their reported value refers to 298.15 K. This value came from an average of eight determinations of $\Delta_r H°$ for $Fe_{0.877}S(cr)$ + Zn(cr) → ZnS(cr) + 0.877 Fe(cr).

In the composition range y = 0.095 to 0.145, Bugli et al. (2) determined $\Delta_r H°(433\ K) = -14.4(y-0.025)\ kcal\ mol^{-1}$ for FeS(cr) + yS(ℓ) → (1+y)$Fe_{1/(1+y)}S(cr)$. Using y = x/(1-x) we calculate that $\Delta_r H°(433\ K) = -14.76x + 0.36\ kcal\ mol^{-1}$ for (1-x)FeS(cr) + xS(ℓ) → $Fe_{1-x}S(cr)$ in this same composition range. Using x = 0.123 and auxiliary JANAF data (3), we can then calculate $\Delta_f H°(Fe_{0.877}S,\ cr) = -22.1\ kcal\ mol^{-1}$ based on Bugli's data. The difference between this value of $\Delta_f H°(298.15\ K)$ and that of Stolyarova and Bezmen is probably due to different morphology of the products; Bugli et al. state they are uncertain as to the exact nature of their product. Due to the complicated and kinetically slow phase behavior of $Fe_{0.877}S$, it appears they may not have obtained the pure monoclinic product.

Heat Capacity and Entropy

The adopted heat capacities from 6-350 K are based on our analysis of the $C_p°$ data of Gronvold et al. (4). Two small anomalies in the $C_p°$ curve, for which there is no obvious explanation, were observed near 8 and 30 K (4). These lead to entropy increments of 0.003 and 0.03 cal $K^{-1}\ mol^{-1}$, respectively. Since there are no measured $C_p°$ or enthalpy data above 350 K, the $C_p°$ is estimated via the following procedure. Above Tβ, $C_p°$ is estimated as $C_p°(Fe_{0.877}S,\ cr) - C_p°(FeS,\ cr) = 0.123\ C_p°(Fe,\ cr)$ using auxiliary JANAF data (3). Since the Tα transition present in FeS(cr, troilite) is not present in pyrrhotite of composition $Fe_{0.877}S$ (3), the heat capacity is assumed to be linear between 350 and 598 K where $C_p°(598\ K)$ is estimated via the procedure outlined above. Hirone et al. (5) observed non-linear behavior for $C_p°$ of $Fe_{0.90}S$ from 513 to 593 K.

The value of S°(298.15 K) is derived from the appropriate integration of $C_p°$ data assuming $S°(6\ K) = 0.0007\ cal\ K^{-1}\ mol^{-1}$.

Phase Data

The mineral pyrrhotite has a composition range extending from FeS to $Fe_{0.877}S$ (~Fe_7S_8); the name troilite is reserved for the stoichiometric FeS composition. The data presented in this table refer to the iron poor composition limit which is found to be $Fe_{0.877}S$ below T_{trs2} (6). For stoichiometric FeS there are 3 crystalline polymorphs designated α, β, γ from low to high temperature, where the α-β transition is designated T_{trs1} and the β-γ transition is designated T_{trs2}. We retain this nomenclature for $Fe_{0.877}S$ while recognizing that the α phase does not exist for this composition. Hirone et al. (5) found Tβ and ΔHβ° to be independent of composition so we adopt the same values as for FeS(cr, troilite) (3). The transition at 411 K (T_{trs1}) in FeS was found to be non-existent i compositions containing less iron than $Fe_{0.95}S$ (2), so it will not be present in $Fe_{0.877}S$. The crystal structure of $Fe_{0.877}S$ is hexagonal above T_{trs2} while the stable form below T_{trs2} is monoclinic (6, 7). The conversion from hexagonal to monoclinic when cooling through T_{trs2} is kinetically slow and even annealing for several months may not be sufficient for complete conversion (6, 7).

References

1. T. A. Stolyarova and N. I. Bezmen, Russ. J. Phys. Chem. 50, 330 (1976).
2. G. Bugli, L. Abello and G. Pannetier, Bull. Soc. Chim. Fr. 497 (1972).
3. JANAF Thermochemical Tables: FeS(cr) 9-30-77; Fe(cr) 3-31-65; S(ℓ) 12-30-65.
4. F. Gronvold, E. F. Westrum and C. Chou, J. Chem. Phys. 23, 528 (1959).
5. T. Hirone, S. Maeda, S. Chiba and N. Tsuya, J. Phys. Soc. Japan 9, 500 (1954).
6. F. Gronvold and H. Haraldsen, Acta Chem. Scand. 6, 1452 (1952).
7. H. Nakazawa and N. Morimoto, Mat. Res. Bull. 6, 345 (1971).

Iron Sulfide, Pyrrhotite ($Fe_{0.877}S$) $Fe_{0.877}S_1(cr)$

Enthalpy Reference Temperature = T_r = 298.15 K; Standard State Pressure = p° = 0.1 MPa

T/K	$C_p°$	S°	$-[G°-H°(T_r)]/T$	$H°-H°(T_r)$	$\Delta_f H°$	$\Delta_f G°$	Log K_f
	J K⁻¹mol⁻¹			kJ mol⁻¹			
0	0.	0.	INFINITE	-9.241	-106.313	-106.313	INFINITE
100	25.899	18.565	99.374	-8.081	-106.214	-106.286	55.518
200	42.257	42.394	65.139	-4.549	-105.875	-106.478	27.809
298.15	49.883	60.799	60.799	0.	-105.437	-106.863	18.722
300	50.000	61.108	60.800	0.092	-105.427	-106.872	18.608
400	56.149	76.334	62.835	5.399	-107.022	-107.353	14.019
500	62.383	89.531	66.880	11.325	-107.526	-107.381	11.218
598.000	68.563	101.232	71.565	17.741	BETA <--> GAMMA TRANSITION		
598.000	58.175	101.897	71.565	18.139			
600	58.116	102.091	71.666	18.255	-106.887	-107.365	9.347
700	55.564	110.847	76.659	23.932	-107.474	-107.403	8.015
800	53.848	118.146	81.401	29.396	-108.391	-107.335	7.008
900	52.886	124.432	85.841	34.732	-162.845	-106.047	6.155
1000	52.007	129.922	89.980	39.941	-163.655	-99.702	5.208
1100	55.396	135.035	93.845	45.309	-165.108	-93.221	4.427
1200	59.120	140.016	97.486	51.036	-165.731	-86.673	3.773
1300	62.676	144.882	100.945	57.118	-164.572	-80.130	3.220
1400	67.195	149.686	104.255	63.603	-163.098	-73.687	2.749
1500	72.676	154.503	107.444	70.589	-161.213	-67.364	2.346
1600	79.161	159.398	110.538	78.177	-158.813	-61.182	1.997
1700	86.107	164.398	113.558	86.428	-156.671	-55.152	1.695
1800	94.517	169.551	116.525	95.447	-153.309	-49.274	1.430

PREVIOUS: CURRENT: September 1977

Iron Sulfide, Pyrrhotite ($Fe_{0.877}S$) $Fe_{0.877}S_1(cr)$

IRON (Fe) REFERENCE STATE A_r = 55.847

0 to 1184	K	crystal, alpha
1184 to 1665	K	crystal, gamma
1665 to 1809	K	crystal, delta
1809 to 3133.345	K	liquid
above 3133.345	K	ideal monatomic gas

Refer to the individual tables for details.

Iron (Fe) Fe_1(ref)

Enthalpy Reference Temperature = T_r = 298.15 K Standard State Pressure = $p°$ = 0.1 MPa

	J K⁻¹mol⁻¹			kJ mol⁻¹			
T/K	C_p°	S°	$-[G°-H°(T_r)]/T$	$H°-H°(T_r)$	$\Delta_fH°$	$\Delta_fG°$	Log K_f
0	0.	0.	INFINITE	-4.507	0.	0.	0.
100	12.101	6.065	46.905	-4.084	0.	0.	0.
200	21.588	17.949	29.524	-2.315	0.	0.	0.
298.15	25.094	27.321	27.321	0.	0.	0.	0.
300	25.140	27.476	27.321	0.046	0.	0.	0.
400	27.386	35.021	28.335	2.674	0.	0.	0.
500	29.702	41.377	30.323	5.527	0.	0.	0.
600	32.049	46.997	32.642	8.613	0.	0.	0.
700	34.602	52.119	35.063	11.939	0.	0.	0.
800	37.949	56.940	37.498	15.553	0.	0.	0.
900	43.095	61.679	39.922	19.582	0.	0.	0.
1000	54.434	66.672	42.342	24.329	0.	0.	0.
1042.000	83.680	69.245	43.372	26.960	Cp LAMBDA MAXIMUM		
1042.000	83.681	69.245	43.372	26.960	TRANSITION		
1100	46.401	72.062	44.815	29.972	0.	0.	0.
1184.000	41.422	75.260	46.865	33.619	ALPHA <--> GAMMA		
1184.000	33.882	76.020	46.865	34.519	TRANSITION		
1200	34.016	76.476	47.257	35.062	0.	0.	0.
1300	34.853	79.231	49.612	38.505	0.	0.	0.
1400	35.690	81.845	51.822	42.033	0.	0.	0.
1500	36.526	84.336	53.907	45.643	0.	0.	0.
1600	37.363	86.720	55.884	49.338	0.	0.	0.
1665.000	37.907	88.218	57.117	51.784	GAMMA <--> DELTA		
1665.000	41.112	88.721	57.117	52.621	TRANSITION		
1700	41.463	89.580	57.776	54.066	0.	0.	0.
1800	42.468	91.978	59.610	58.263	0.	0.	0.
1809.000	42.558	92.190	59.772	58.645	DELTA <--> LIQUID		
1809.000	46.024	99.823	59.772	72.452	TRANSITION		
1900	46.024	102.082	61.744	76.641	0.	0.	0.
2000	46.024	104.442	63.821	81.243	0.	0.	0.
2100	46.024	106.688	65.809	85.845	0.	0.	0.
2200	46.024	108.829	67.716	90.448	0.	0.	0.
2300	46.024	110.875	69.549	95.050	0.	0.	0.
2400	46.024	112.833	71.312	99.653	0.	0.	0.
2500	46.024	114.712	73.010	104.255	0.	0.	0.
2600	46.024	116.517	74.649	108.857	0.	0.	0.
2700	46.024	118.254	76.232	113.460	0.	0.	0.
2800	46.024	119.928	77.763	118.062	0.	0.	0.
2900	46.024	121.543	79.245	122.665	0.	0.	0.
3000	46.024	123.103	80.681	127.267	0.	0.	0.
3133.345	46.024	125.105	82.529	133.404	LIQUID <--> IDEAL GAS		
3133.345	26.642	236.674	82.530	482.989	FUGACITY = 1 bar		
3200	26.856	237.237	85.746	484.772	0.	0.	0.
3400	27.511	238.885	94.706	490.208	0.	0.	0.
3600	28.192	240.477	102.761	495.778	0.	0.	0.
3800	28.905	242.020	110.050	501.487	0.	0.	0.
4000	29.654	243.521	116.686	507.342	0.	0.	0.
4200	30.442	244.987	122.761	513.351	0.	0.	0.
4400	31.270	246.422	128.349	519.521	0.	0.	0.
4600	32.142	247.831	133.514	525.862	0.	0.	0.
4800	33.058	249.219	138.306	532.381	0.	0.	0.
5000	34.022	250.587	142.770	539.088	0.	0.	0.
5200	35.040	251.941	146.943	545.993	0.	0.	0.
5400	36.120	253.284	150.856	553.108	0.	0.	0.
5600	37.274	254.618	154.538	560.446	0.	0.	0.
5800	38.513	255.947	158.012	568.024	0.	0.	0.
6000	39.853	257.275	161.299	575.858	0.	0.	0.

PREVIOUS: March 1978 (1 atm) CURRENT: March 1978 (1 (bar)

Iron (Fe) Fe_1(ref)

IRON, ALPHA - DELTA (Fe) CRYSTAL A_r = 55.847 Iron, Alpha-Delta (Fe) $Fe_1(cr)$

$S°(298.15\ K) = 27.32 \pm 0.13\ J\ K^{-1}\ mol^{-1}$
$T_{trs1} = 1042 \pm 1\ K$ (Curie temp)
$T_{trs2} = 1184 \pm 3\ K\ (\alpha - \gamma)$
$T_{trs3} = 1665 \pm 3\ K\ (\gamma - \delta)$
$T_{fus} = 1809 \pm 5\ K\ (\delta - \ell)$

$\Delta_f H°(0\ K) = 0\ kJ\ mol^{-1}$
$\Delta_f H°(298.15\ K) = 0\ kJ\ mol^{-1}$
$\Delta_{trs1} H° = 0\ kJ\ mol^{-1}$
$\Delta_{trs2} H° = 0.900 \pm 0.042\ kJ\ mol^{-1}$
$\Delta_{trs3} H° = 0.837 \pm 0.084\ kJ\ mol^{-1}$
$\Delta_{fus} H° = 13.807 \pm 0.837\ kJ\ mol^{-1}$

Enthalpy of Formation

Zero by definition.

Heat Capacity and Entropy

The current literature was surveyed by Hultgren (1). Smoothed low temperature heat capacity values are obtained by fitting the data (2 - 7) graphically and with orthogonal polynomials over selected overlapping temperature intervals. Using $S°(20\ K) = 0.0364\ cal\ mol^{-1}\ K^{-1}$, we integrate over the adopted smoothed heat capacity values to obtain $S°(298.15\ K) = 6.529\ cal\ mol^{-1}\ K^{-1}$. An uncertainty of $0.015\ cal\ mol^{-1}\ K^{-1}$ arises due to the choice of heat capacity values in the region 20 - 55 K. In this region there is considerable disagreement between the reported heat capacity values of Eucken and Werth (2), Simon and Swain (3), and those of Duyckaerts (4) and Keesom and Kurrelmeyer (5). An additional uncertainty of $0.01\ cal\ mol^{-1}\ K^{-1}$ results from the differences in the range 50-298 K in the heat capacity data of Kelley (6) and Stepakoff and Kaufman(8).

The heat capacity and enthalpy studies above room temperature and numerous. A bibliography of all known studies (including low temperature studies) is available on request. The adopted heat capacity values above 400 K are those selected by Hultgren (8) and Orr and Chipman (9). The method of orthogonal polynomials is used to join smoothly the low and high temperature heat capacity values. The smooth heat capacity values for metastable α-Fe in the range 1184-1665 K are adjusted from those of Hultgren (8) so as to yield the proper entropy values for γ-Fe. The available heat capacity/enthalpy data show differences up to 10% in Cp° for δ-Fe with the major discrepancies being different slopes in dCp°/dT.

Transition Data

The enthalpies of the α - γ transition at 1184 K and the γ - δ transition at 1665 K are those selected by Hultgren (8) and Orr and Chipman (9). The α and δ phase has a bcc structure whereas the γ phase has a fcc structure.

Fusion Data

Refer to the liquid table for details.

Sublimation Data

The value adopted for the enthalpy of sublimation is that selected by Hultgren (8) and is based on a 2nd and 3rd law analysis of sublimation data over γ- Fe. Refer to the ideal gas table for details.

References

1. R. Hultgren, Materials and Molecular Research Division, Lawrence Berkeley Laboratory, University of Califormia, private communication.
2. A. Eucken and H. Werth, Z. Anorg. Chem. 188, 153 (1930).
3. F. Simon and R. C. Swain , Z. Physik. Chem. 28b, 189 (1935).
4. G. Duyckaerts, Physica 6, 401 (1939); Compt. Rend. 208, 979 (1939).
5. H. W. Keesom and B. Kurrelmeyer, Physica 6, 633 (1939).
6. K. K. Kelley, J. Chem. Phys. 11, 16 (1943).
7. G. L. Stepakoff and L. Kaufman, Acta Met. 16, 13 (1968).
8. R. Hultgren et al., "Selected Values of the Thermodynamic Properties of the Elements," American Society for Metals, Metals Park, Ohio, 1973.
9. R. L. Orr and J. Chipman, Trans. Met. Soc. AIME 239, 630 (1967).

Enthalpy Reference Temperature = T_r = 298.15 K Standard State Pressure = p° = 0.1 MPa

	$J\ K^{-1} mol^{-1}$			$kJ\ mol^{-1}$			
T/K	$C_p°$	S°	$-[G°-H°(T_r)]/T$	$H°-H°(T_r)$	$\Delta_f H°$	$\Delta_f G°$	Log K_f
0	0.	0.	INFINITE	-4.507	0.	0.	0.
100	12.101	6.065	46.905	-4.084	0.	0.	0.
200	21.588	17.949	29.524	-2.315	0.	0.	0.
250	23.742	23.018	27.727	-1.177	0.	0.	0.
298.15	25.094	27.321	27.321	0.	0.	0.	0.
300	25.140	27.476	27.321	0.046	0.	0.	0.
350	26.287	31.439	27.632	1.333	0.	0.	0.
400	27.386	35.021	28.335	2.674	0.	0.	0.
450	28.518	38.312	29.263	4.072	0.	0.	0.
500	29.702	41.377	30.323	5.527	0.	0.	0.
600	32.049	46.997	32.642	8.613	0.	0.	0.
700	34.602	52.119	35.063	11.939	0.	0.	0.
800	37.949	56.940	37.498	15.553	0.	0.	0.
900	43.095	61.679	39.922	19.582	0.	0.	0.
1000	54.434	66.672	42.342	24.329	0.	0.	0.
1042.000	83.680	69.245	43.372	26.960	Cp LAMBDA MAXIMUM TRANSITION		
1042.000	83.681	69.245	43.372	26.960			
1100	46.401	72.062	44.815	29.972	0.	0.	0.
1184.000	41.422	75.260	46.865	33.619	------- ALPHA <--> GAMMA -------		
1200	40.794	75.830	47.249	34.298	-0.764	0.010	0.001
1300	38.493	78.985	49.571	38.238	-0.267	0.052	-0.002
1400	38.535	81.830	51.775	42.077	0.044	0.065	-0.002
1500	39.455	84.518	53.869	45.974	0.330	0.056	-0.002
1600	40.459	87.097	55.866	49.969	0.631	0.029	-0.001
1665.000	41.112	88.721	57.117	52.620	------- GAMMA <--> DELTA -------		
1700	41.463	89.580	57.776	54.065	-0.001	0.	0.
1800	42.468	91.978	59.610	58.262	-0.001	0.	0.
1809.000	42.558	92.190	59.772	58.645	------- DELTA <--> LIQUID ------		
1900	43.472	94.301	61.375	62.559	-14.082	0.702	-0.019
2000	44.476	96.556	63.078	66.956	-14.287	1.486	-0.039
2100	45.480	98.751	64.725	71.454	-14.391	2.277	-0.057
2200	46.484	100.889	66.320	76.052	-14.395	3.071	-0.073

PREVIOUS: March 1965 CURRENT: March 1978

Iron, Alpha-Delta (Fe) $Fe_1(cr)$

IRON, GAMMA (Fe) CRYSTAL A_r = 55.847

$S°(298.15\ K) = 35.545\ J\ K^{-1}\ mol^{-1}$

$T_{trs1} = 1184 \pm 3\ K\ (\alpha - \gamma)$

$T_{trs2} = 1665 \pm 3\ K\ (\gamma - \delta)$

$\Delta_f H°(298.15\ K) = 7.788\ kJ\ mol^{-1}$

$\Delta_{trs1}H° = 0.900 \pm 0.042\ kJ\ mol^{-1}$

$\Delta_{trs2}H° = 0.837 \pm 0.084\ kJ\ mol^{-1}$

Enthalpy of Formation

The enthalpy of formation at 298.15 K is obtained from that of Fe(α) by adding $\Delta_{trs1}H°$ and the difference between H°(1184) - H°(298.15 K) of Fe(α) and Fe(γ).

Heat Capacity and Entropy

A current literature survey by Hultgren (1) yielded approximately twenty heat capacity/enthalpy studies which gave information as to the heat capacity of γ-Fe in the stable and metastable region. A listing of these studies is available on request. We adopt the heat capacity values chosen by Hultgren (2) and Orr and Chipman (3). These values are based primarily on the studies by Dench and Kubaschewski (4), Braun and Kohlhaas (5), and Wallace et al. (6). However, all the available heat capacity/enthalpy data show discrepancies up to 5% (in Cp°); with the major discrepancies appearing as different slopes in Cp° vs T plots.

Values are not extended below 298.15 K due to the lack of substantial data in this metastable region. The entropy at 298.15 K is calculated in a manner analogous to that used for the enthalpy of formation.

Transition Data

Refer to the Fe(α, δ) table (7).

References

1. R. Hultgren, Materials and Molecular Research Division, Lawrence Berkeley Laboratory, University of California, personal communication.
2. R. Hultgren, et al., "Selected Values of the Thermodynamic Properties of the Elements," American Society for Metals, Metals Park, Ohio, 1973.
3. R. L. Orr and J. Chipman, Trans. Met. Soc. AIME 239, 630 (1967).
4. W. A. Dench and O. Kubaschewski, J. Iron steel Inst. 201, 140 (1963).
5. M. Braun and R. Kohlhaas, Phys. Stat. Sol. 12, 429 (1965).
6. D. C. Wallace, P. H. Sidles, and A. C. Danielson, J. Appl. Phys. 31, 168 (1960).
7. JANAF Thermochemical Tables: Fe(c, α) 3-31-78.

Iron, Gamma (Fe) $Fe_1(cr)$

Enthalpy Reference Temperature = T_r = 298.15 K — Standard State Pressure = p° = 0.1 MPa

T/K	$C_p°$ (J K⁻¹mol⁻¹)	S° (J K⁻¹mol⁻¹)	$-[G°-H°(T_r)]/T$ (J K⁻¹mol⁻¹)	$H°-H°(T_r)$ (kJ mol⁻¹)	$\Delta_f H°$ (kJ mol⁻¹)	$\Delta_f G°$ (kJ mol⁻¹)	Log K_f
0							
100							
200							
250							
298.15	26.469	35.545	35.545	0.	7.788	5.336	-0.935
300	26.485	35.709	35.546	0.049	7.790	5.321	-0.926
350	26.903	39.823	35.870	1.384	7.839	4.905	-0.732
400	27.322	43.443	36.594	2.739	7.853	4.484	-0.586
450	27.740	46.685	37.539	4.116	7.832	4.064	-0.472
500	28.158	49.629	38.603	5.513	7.774	3.648	-0.381
600	28.995	54.837	40.885	8.371	7.546	2.842	-0.247
700	29.832	59.369	43.209	11.312	7.161	2.086	-0.156
800	30.669	63.408	45.486	14.337	6.572	1.398	-0.091
900	31.506	67.068	47.684	17.446	5.652	0.802	-0.047
1000	32.342	70.431	49.792	20.638	4.097	0.338	-0.018
1100	33.179	73.553	51.812	23.915	1.730	0.091	-0.004
1184.000	33.882	76.020	53.443	26.731	------- ALPHA <--> GAMMA -------		
1200	34.016	76.476	53.747	27.274	0.	0.	0.
1300	34.853	79.231	55.602	30.718	0.	0.	0.
1400	35.690	81.845	57.384	34.245	0.	0.	0.
1500	36.526	84.336	59.099	37.856	0.	0.	0.
1600	37.363	86.720	60.751	41.550	0.	0.	0.
1665.000	37.907	88.218	61.794	43.996	------- GAMMA <--> DELTA -------		
1700	38.200	89.010	62.346	45.328	-0.950	0.019	-0.001
1800	39.037	91.217	63.889	49.190	-1.285	0.085	-0.002
1900	39.874	93.350	65.384	53.136	-15.717	0.873	-0.024
2000	40.710	95.417	66.834	57.165	-16.290	1.761	-0.046
2100	41.547	97.423	68.243	61.278	-16.780	2.676	-0.067
2200	42.384	99.375	69.614	65.474	-17.186	3.612	-0.086

PREVIOUS: CURRENT: March 1978

Iron, Gamma (Fe) $Fe_1(cr)$

$S°(298.15 K) = [34.763] J K^{-1} mol^{-1}$ $\Delta_f H°(298.15 K) = [12.395] kJ mol^{-1}$

$T_{fus} = 1809 \pm 5 K (\delta - \ell)$ $\Delta_{fus}H° = 13.807 \pm 0.4 kJ mol^{-1}$

Enthalpy of Formation

The enthalpy of formation at 298.15 K is calculated from that of the combined phase crystal (α-γ-δ) by adding $\Delta_{fus}H°$ and the difference in enthalpy, H°(1809 K) - H°(298.15 K), between the crystal and liquid.

Heat Capaci and Entropy

The heat capacity of the liquid is taken to be 11.0 cal $mol^{-1} K^{-1}$ as in Hultgren (1) and Orr and Chipman (2). We adopt this value in the real liquid region as well as in the extrapolated region to 1200 K and to 3500 K. Below 1200 K, heat capacity values are chosen so as to form a curve which joins smoothly the α-Fe heat capacity values in the range 298 - 500 K. The entropy is calculated in a manner analogous to that used for the enthalpy of formation. A current literature survey (3) revealed numerous heat capacity and enthalpy studies in the liquid region which yielded heat capacity values in the range 8.0 to 12.4 cal $mol^{-1} K^{-1}$ as given in the table below. The more recent studies by Braun et al. (10), using a quasi-adiabatic calorimeter suggest Cp° = 10.938 cal $mol^{-1} K^{-1}$ whereas the levitation calorimetric enthalpy measurements of Margrave et al. (12, 14) suggest 10.29 ± 0.29 and 11.138 ± 0.43 cal $mol^{-1} K^{-1}$ respectively.

Fusion Data

The adopted T_{fus} and $\Delta_{fus}H°$ values are those selected by Hultgren (1) and Orr and Chipman (2); 1809 ± 5 K and 3300 ± 100 cal mol^{-1}, respectively. Reported values for $\Delta_{fus}H°$ vary from a low of 2760 to a high of 3865 cal mol^{-1}, as given in the table below.

Vaporization Data

The boiling point is calculated from the adopted thermodynamic functions and the chosen enthalpy of sublimation so that the Gibbs energy functions calculated by integration of the crystal-liquid data and by statistical methods for the gas phase are equal at T_{vap}.

Reference	Year	Calorimetric Method	T_{fus}, K	$\Delta_{fus}H°$ kcal mol^{-1}	$Cp°(\ell)$ cal $mol^{-1} K^{-1}$	T/K
(4)	1918	drop	1801	2760	11.96	373 - 1873
(5)	1926, 1929	mixture	1800	3865	12.398	373 - 1906
(6)	1927	drop	1801	3595	11.2	370 - 1860
(7)	1956	mixture	1806	3658	8.0	1273 - 1923
(8)	1962	drop	1809±1	3292±80	11.226	1665 - 2210
(9)	1966	drop	1808	3298±100	9.766	1720 - 1875
(10)	1965 - 1968	quasi adiab.	1809	3442±50	10.038	300 - 1900
(11)	1971	exploding wire		3256		
(12)	1971	levitation		3307±70	10.29±0.29	1804 - 2142
(13)	1974	pulse	1808±5			
(14)	1975	levitation	1811	3370±90	11.138±0.43	

References

1. R. Hultgren et al., "Selected Values of the Thermodynamic Properties of the Elements," American Society for Metals", Metals Park, Ohio, 1973.
2. R. L. Orr and J. Chipman, Trans. Met. soc. AIME 239, 630 (1967).
3. R. Hultgren, Materials and Molecular Research Division, Lawrence Berkeley laboratory, University of california, Personal communication.
4. F. Wust, A. Meuthern, and P. Durrer, Forsch. Gebiete Ingenieurw., VDI-Forschungsh. 204 (1918).
5. S. Umino, Sci. repts. Tohoku Imp. Univ. 18, 91 (1929); 15, 597 (1926).
6. P. Oberhoffer and W. Grosse, Stahl u. Eisen 47, 576 (1927).
7. J. R. Pattison and P. W. Willows, J. Iron Steel Inst. 183, 390 (1956).
8. A. Ferrier and M. Olette, Compt. Rend. 254, 2322 (1962).
9. J. P. Mooris, P. F. Foerster, C. W. Schultz, and G. R. Zellars, U. S. Bur. Mines RI 6823, 1966.
10. M. Braun and R. Kohlhaas, Phys. Stat. Sol 12, 429 (1965); O. Vollmer, R. Kohlhaas, and M. Braun. Z. Nturforsch 21, 181 (1966); M. Braun, R. Kohlhaas, and O. Vollmer, Z. Angew. Phys. 25, 365 (1968).
11. S. V. Lebedev, A. I. Savvatimskii, and Yu. B. Smirnov, High Temp. 9, 578 (1971).
12. J. A. Treverton and J. L. Margrave, J. Chem. Thermodynamics 3, 473 (1971).
13. A. Cezairliyan and J. L. McClure, J. Res. Nat. Bur. Std. 78A, 1 (1974).
14. J. L. Margrave, Report ORO-2907-92, 1972.

Enthalpy Reference Temperature = T_r = 298.15 K; Standard State Pressure = p° = 0.1 MPa

T/K	$C_p°$ (J $K^{-1}mol^{-1}$)	S° (J $K^{-1}mol^{-1}$)	$-[G°-H°(T_r)]/T$ (J $K^{-1}mol^{-1}$)	$H°-H°(T_r)$ (kJ mol^{-1})	$\Delta_f H°$ (kJ mol^{-1})	$\Delta_f G°$ (kJ mol^{-1})	Log K_f
0							
100							
200							
250							
298.15	25.096	34.763	34.763	0.	12.395	10.177	-1.783
300	25.142	34.918	34.764	0.046	12.395	10.163	-1.769
350	26.254	38.877	35.074	1.331	12.394	9.791	-1.461
400	27.384	42.457	35.776	2.672	12.393	9.419	-1.230
450	28.542	45.749	36.704	4.070	12.394	9.047	-1.050
500	29.702	48.816	37.763	5.526	12.395	8.675	-0.906
600	32.033	54.437	40.082	8.613	12.396	7.931	-0.690
700	34.363	59.550	42.503	11.933	12.389	7.187	-0.536
800	36.694	64.291	44.934	15.486	12.328	6.447	-0.421
900	39.024	68.747	47.334	19.272	12.085	5.724	-0.332
1000	41.355	72.980	49.689	23.291	11.356	5.049	-0.264
1100	43.689	77.031	51.992	27.543	9.966	4.501	-0.214
1200	46.024	80.932	54.242	32.028	9.362	4.014	-0.175
1200.000	46.024	80.932	54.242	32.028	GLASS <--> LIQUID TRANSITION		
1200.000	46.024	80.932	54.242	32.028			
1300	46.024	84.616	56.438	36.631	10.521	3.521	-0.141
1400	46.024	88.027	58.574	41.233	11.596	2.941	-0.110
1500	46.024	91.202	60.645	45.836	12.588	2.288	-0.080
1600	46.024	94.172	62.649	50.438	13.495	1.571	-0.051
1700	46.024	96.963	64.586	55.040	13.370	0.819	-0.025
1800	46.024	99.593	66.458	59.643	13.776	0.069	-0.002
1809.000	46.024	99.823	66.624	60.057	------- DELTA <--> LIQUID ------		
1900	46.024	102.082	68.268	64.245	0.	0.	0.
2000	46.024	104.442	70.019	68.848	0.	0.	0.
2100	46.024	106.688	71.712	73.450	0.	0.	0.
2200	46.024	108.829	73.351	78.052	0.	0.	0.
2300	46.024	110.875	74.938	82.655	0.	0.	0.
2400	46.024	112.834	76.476	87.257	0.	0.	0.
2500	46.024	114.712	77.968	91.860	0.	0.	0.
2600	46.024	116.517	79.417	96.462	0.	0.	0.
2700	46.024	118.254	80.823	101.064	0.	0.	0.
2800	46.024	119.928	82.190	105.667	0.	0.	0.
2900	46.024	121.543	83.519	110.269	0.	0.	0.
3000	46.024	123.103	84.813	114.872	0.	0.	0.
3100	46.024	124.613	86.073	119.474	0.	0.	0.
3133.345	46.024	125.105	86.485	121.009	------- FUGACITY = 1 bar -------		
3200	46.024	126.074	87.300	124.076	-348.300	7.424	-0.121
3300	46.024	127.490	88.496	128.679	-346.399	18.511	-0.293
3400	46.024	128.864	89.664	133.281	-344.531	29.541	-0.454
3500	46.024	130.198	90.803	137.884	-342.697	40.516	-0.605
3600	46.024	131.495	91.915	142.486	-340.896	51.439	-0.746
3700	46.024	132.756	93.002	147.088	-339.131	62.313	-0.880
3800	46.024	133.983	94.064	151.691	-337.401	73.140	-1.005
3900	46.024	135.179	95.103	156.293	-335.707	83.921	-1.124
4000	46.024	136.344	96.120	160.896	-334.051	94.660	-1.236

PREVIOUS: March 1965 CURRENT: March 1978

IRON (Fe) ALPHA-GAMMA-DELTA-LIQUID A_r = 55.847

0 to 1184 K crystal, alpha
1184 to 1665 K crystal, gamma
1665 to 1809 K crystal, delta
above 1809 K liqiud

Refer to the individual tables for details.

Iron (Fe) $Fe_1(cr,l)$

Enthalpy Reference Temperature = T_r = 298.15 K   Standard State Pressure = p° = 0.1 MPa

T/K	C_p°	S°	$-[G^\circ - H^\circ(T_r)]/T$	$H^\circ - H^\circ(T_r)$	$\Delta_f H^\circ$	$\Delta_f G^\circ$	Log K_f
	J K⁻¹mol⁻¹			kJ mol⁻¹			
0	0.	0.	INFINITE	-4.507	0.	0.	0.
100	12.101	6.065	46.905	-4.084	0.	0.	0.
200	21.588	17.949	29.524	-2.315	0.	0.	0.
298.15	25.094	27.321	27.321	0.	0.	0.	0.
300	25.140	27.476	27.321	0.046	0.	0.	0.
400	27.386	35.021	28.335	2.674	0.	0.	0.
500	29.702	41.377	30.323	5.527	0.	0.	0.
600	32.049	46.997	32.642	8.613	0.	0.	0.
700	34.602	52.119	35.063	11.939	0.	0.	0.
800	37.949	56.940	37.498	15.553	0.	0.	0.
900	43.095	61.679	39.922	19.582	0.	0.	0.
1000	54.434	66.672	42.342	24.329	0.	0.	0.
1042.000	83.680	69.245	43.372	26.960	Cp LAMBDA MAXIMUM TRANSITION		
1042.000	83.681	69.245	43.372	26.960			
1100	46.401	72.062	44.815	29.972	0.	0.	0.
1184.000	41.422	75.260	46.865	33.619	ALPHA <--> GAMMA TRANSITION		
1184.000	33.882	76.020	46.865	34.519			
1200	34.016	76.476	47.257	35.062	0.	0.	0.
1300	34.853	79.231	49.612	38.505	0.	0.	0.
1400	35.690	81.845	51.822	42.033	0.	0.	0.
1500	36.526	84.336	53.907	45.643	0.	0.	0.
1600	37.363	86.720	55.884	49.338	0.	0.	0.
1665.000	37.907	88.218	57.117	51.784	GAMMA <--> DELTA TRANSITION		
1665.000	41.112	88.721	57.117	52.621			
1700	41.463	89.580	57.776	54.066	0.	0.	0.
1800	42.468	91.978	59.610	58.263	0.	0.	0.
1809.000	42.558	92.190	59.772	58.645	DELTA <--> LIQUID TRANSITION		
1809.000	46.024	99.823	59.772	72.452			
1900	46.024	102.082	61.744	76.641	0.	0.	0.
2000	46.024	104.442	63.821	81.243	0.	0.	0.
2100	46.024	106.688	65.809	85.845	0.	0.	0.
2200	46.024	108.829	67.716	90.448	0.	0.	0.
2300	46.024	110.875	69.549	95.050	0.	0.	0.
2400	46.024	112.833	71.312	99.653	0.	0.	0.
2500	46.024	114.712	73.010	104.255	0.	0.	0.
2600	46.024	116.517	74.649	108.857	0.	0.	0.
2700	46.024	118.254	76.232	113.460	0.	0.	0.
2800	46.024	119.928	77.763	118.062	0.	0.	0.
2900	46.024	121.543	79.245	122.665	0.	0.	0.
3000	46.024	123.103	80.681	127.267	0.	0.	0.
3100	46.024	124.613	82.074	131.869	0.	0.	0.
3133.345	46.024	125.105	82.529	133.404	------- FUGACITY = 1 bar -------		
3200	46.024	126.074	83.426	136.472	-348.300	7.424	-0.121
3300	46.024	127.490	84.740	141.074	-346.399	18.511	-0.293
3400	46.024	128.864	86.018	145.677	-344.531	29.541	-0.454
3500	46.024	130.198	87.261	150.279	-342.697	40.516	-0.605
3600	46.024	131.495	88.472	154.881	-340.896	51.439	-0.746
3700	46.024	132.756	89.652	159.484	-339.131	62.313	-0.880
3800	46.024	133.983	90.802	164.086	-337.401	73.140	-1.005
3900	46.024	135.179	91.925	168.689	-335.707	83.921	-1.124
4000	46.024	136.344	93.021	173.291	-334.051	94.660	-1.236

PREVIOUS: CURRENT: March 1978

Iron (Fe) $Fe_1(cr,l)$

IRON, (Fe) IDEAL GAS A_r = 55.847

IP(Fe, g) = 63480 ± 500 cm^{-1} $\Delta_f H°$(0 K) = 413.1 ± 1.3 kJ mol^{-1}

S°(298.15 K) = 180.49 ± 0.04 J K^{-1} mol^{-1} $\Delta_f H°$(298.15 K) = 415.5 ± 1.3 kJ mol^{-1}

Electronic Levels and Quantum Weights

State	ϵ_i, cm^{-1}	g_i
$^5D_4$	0.000	9
$^5D_3$	415.932	7
$^5D_2$	704.004	5
$^5D_1$	888.129	3
$^5D_0$	978.072	1
.	.	.
.	.	.
.	.	.
$^3G_2°$	62081.27	5
IP	63480	

Enthalpy of Formation

The sublimation and vaporization studies on iron are too numerous to tabulate. A current literature survey by Hultgren yielded approximately thirty investigations (1). As in Hultgren (2), we adopt $\Delta_f H°$(Fe, g, 298.15 K) = 99.3±0.3 kcal mol^{-1} based on the studies of Morris et al. (3) and Myles and Aldred (4). A complete listing of all vapor pressure studies is available on request. Many of the more recent studies do not provide sufficient information to enable calculation of enthalpy of formation values via the 2nd and 3rd law approach.

Our 2nd and 3rd law analyses of the vaporization data of Morris et al. (3) and the sublimation data of Myles and Aldred (4) are given below.

Source	Data Points	T/K	$\Delta_r H°$(298.15 K), kcal, mol^{-1} 2nd law	3rd law	drift cal $K^{-1}mol^{-1}$	$\Delta_f H°$(298.15 K) kcal mol^{-1}
Fe(ℓ) = Fe(g)	21	1812-1891	96.8±1.1	96.5±0.1	0.1±6	99.5
Fe(γ) = Fe(g)	eqn	1450-1650	98.1	97.1	-0.6	99.0

Heat Capacity and Entropy

The electronic energy levels are given in the compilation by Reader and Sugar (5). Although we have only listed the ground state, the lowest four excited states, the highest observed excited state, and the ionization limit for Fe(g), all levels listed by Reader and Sugar (5), as well as estimated missing levels, are used in the calculations. The observed levels are too numerous to list completely. Our calculations indicate that for Fe(g) the thermochemical functions are independent of the estimated missing levels (for n = 4, 5, 6, 7) and the cutoff procedure up to 3500 K. The Gibbs energy function is essentially unaffected up to 5200 K. The reported uncertainty in S°(298.15 K) is due to uncertainties in the relative atomic mass and the fundamental constants. Extension of these calculations above 6000 K may require consideration of the higher excited states and utilization of proper fill and cutoff procedures (6). In fact, the inclusion of some higher states (n = 4, 5, 6, 7) and consideration of various cut-off procedures leads to calculational differences in the Gibbs energy function of roughly 0.1 cal K^{-1} mol^{-1} or greater at temperatures in excess of 8000 K.

References

1. R. Hultgren, Materials and Molecular Research Division, Lawrence Berkeley Laboratory, University of California, personal communication.
2. R. Hultgren et al., "Selected Values of the Thermodynamic Properties of the Elements", American Society for Metals, Metals Park, Ohio, 1973.
3. J. P. Morris, G. R. Zellars, S. L. Payne, and R. L. Kipp, U.S. Bur. Mines RI 5364, 1957.
4. K. M. Myles and A. T. Aldred, J. Phys. Chem. 68, 64 (1964).
5. J. Reader and J. Sugar, J. Phys. Chem. Ref. Data 4, 353 (1975).
6. J. R. Downey, The Dow Chemical Company, Report AFOSR-TR-78-0960, Contract No. F44670-75-1-0048, 1978.

Iron (Fe) $Fe_1(g)$

Enthalpy Reference Temperature = T_r = 298.15 K — Standard State Pressure = p° = 0.1 MPa

T/K	$C_p°$ (J $K^{-1}mol^{-1}$)	S°	-[G°-H°(T_r)]/T	H°-H°(T_r) (kJ mol^{-1})	$\Delta_f H°$	$\Delta_f G°$	Log K_f
0	0.	0.	INFINITE	-6.850	413.127	413.127	INFINITE
100	21.387	154.700	202.317	-4.762	414.794	399.930	-208.902
200	24.362	170.450	182.809	-2.472	415.314	384.814	-100.503
250	25.279	175.995	180.909	-1.228	415.420	377.176	-78.807
298.15	25.675	180.488	180.488	0.	415.471	369.804	-64.788
300	25.682	180.647	180.488	0.048	415.472	369.521	-64.339
350	25.718	184.612	180.801	1.334	415.472	361.862	-54.005
400	25.533	188.036	181.497	2.616	415.412	354.207	-46.255
450	25.233	191.027	182.393	3.885	415.284	346.563	-40.228
500	24.889	193.668	183.391	5.138	415.082	338.937	-35.409
600	24.202	198.144	185.490	7.592	414.451	323.762	-28.186
700	23.610	201.829	187.569	9.982	413.514	308.717	-23.037
800	23.133	204.949	189.551	12.318	412.236	293.829	-19.185
900	22.763	207.651	191.416	14.612	410.501	279.127	-16.200
1000	22.488	210.035	193.161	16.874	408.016	264.653	-13.824
1100	22.297	212.168	194.793	19.113	404.612	250.495	-11.895
1200	22.183	214.103	196.323	21.336	401.745	236.592	-10.299
1300	22.137	215.876	197.760	23.551	400.517	222.879	-8.955
1400	22.156	217.517	199.113	25.766	399.204	209.263	-7.808
1500	22.231	219.048	200.392	27.984	397.812	195.744	-6.816
1600	22.356	220.487	201.603	30.213	396.347	182.320	-5.952
1700	22.523	221.847	202.754	32.457	393.862	169.009	-5.193
1800	22.726	223.140	203.851	34.719	391.928	155.837	-4.522
1900	22.958	224.374	204.899	37.003	375.834	143.477	-3.944
2000	23.213	225.558	205.903	39.312	373.540	131.308	-3.429
2100	23.485	226.698	206.866	41.646	371.272	119.252	-2.966
2200	23.771	227.797	207.793	44.009	369.033	107.303	-2.548
2300	24.065	228.860	208.685	46.401	366.822	95.456	-2.168
2400	24.365	229.890	209.548	48.822	364.641	83.705	-1.822
2500	24.669	230.891	210.381	51.274	362.490	72.043	-1.505
2600	24.976	231.865	211.189	53.756	360.370	60.467	-1.215
2700	25.284	232.813	211.973	56.269	358.280	48.972	-0.947
2800	25.594	233.738	212.733	58.813	356.222	37.554	-0.701
2900	25.905	234.642	213.473	61.388	354.195	26.209	-0.472
3000	26.219	235.525	214.194	63.994	352.198	14.933	-0.260
3100	26.536	236.390	214.896	66.632	350.234	3.724	-0.063
3133.345	26.642	236.674	215.126	67.518	------- FUGACITY = 1 bar -------		
3200	26.856	237.237	215.581	69.301	0.	0.	0.
3300	27.181	238.069	216.250	72.003	0.	0.	0.
3400	27.511	238.885	216.903	74.738	0.	0.	0.
3500	27.848	239.687	217.543	77.506	0.	0.	0.
3600	28.192	240.477	218.169	80.308	0.	0.	0.
3700	28.544	241.254	218.783	83.144	0.	0.	0.
3800	28.905	242.020	219.384	86.017	0.	0.	0.
3900	29.274	242.776	219.974	88.925	0.	0.	0.
4000	29.654	243.521	220.554	91.872	0.	0.	0.
4100	30.043	244.258	221.123	94.856	0.	0.	0.
4200	30.442	244.987	221.682	97.881	0.	0.	0.
4300	30.851	245.708	222.233	100.945	0.	0.	0.
4400	31.270	246.422	222.774	104.051	0.	0.	0.
4500	31.701	247.130	223.308	107.200	0.	0.	0.
4600	32.142	247.831	223.833	110.392	0.	0.	0.
4700	32.594	248.527	224.351	113.628	0.	0.	0.
4800	33.058	249.219	224.862	116.911	0.	0.	0.
4900	33.533	249.905	225.366	120.240	0.	0.	0.
5000	34.022	250.587	225.864	123.618	0.	0.	0.
5100	34.524	251.266	226.355	127.045	0.	0.	0.
5200	35.040	251.941	226.841	130.523	0.	0.	0.
5300	35.572	252.614	227.321	134.054	0.	0.	0.
5400	36.120	253.284	227.795	137.638	0.	0.	0.
5500	36.687	253.952	228.265	141.278	0.	0.	0.
5600	37.274	254.618	228.729	144.976	0.	0.	0.
5700	37.882	255.283	229.189	148.734	0.	0.	0.
5800	38.513	255.947	229.645	152.553	0.	0.	0.
5900	39.170	256.611	230.097	156.437	0.	0.	0.
6000	39.853	257.275	230.544	160.388	0.	0.	0.

PREVIOUS: March 1978 (1 atm) CURRENT: March 1978 (1 bar)

Iron (Fe) $Fe_1(g)$

IRON, ION (Fe^+) IDEAL GAS M_r = 55.84645

$IP(Fe^+, g) = 130563 \pm 10\ cm^{-1}$
$S°(298.15\ K) = 181.86 \pm 0.05\ J\ K^{-1}\ mol^{-1}$

$\Delta_f H°(0\ K) = 1172.52 \pm 7.5\ kJ\ mol^{-1}$
$\Delta_f H°(298.15\ K) = [1181.144]\ kJ\ mol^{-1}$

Electronic Levels and Quantum Weights

State	ε_i, cm^{-1}	g_i
$^6D_{9/2}$	0.000	10
$^6D_{7/2}$	384.790	8
$^6D_{5/2}$	667.683	6
$^6D_{3/2}$	862.613	4
$^6D_{1/2}$	977.053	2
⋮	⋮	⋮

Enthalpy of Formation

$\Delta_f H°(Fe^+, g, 0\ K)$ is calculated from $\Delta_f H°(Fe, g, 0\ K)$ (1) using the spectroscopic value of IP(Fe) = 63480 ± 500 cm^{-1} (759.390 ± 5.98 $kJ\ mol^{-1}$) from Corliss and Sugar (2). The ionization limit is converted from cm^{-1} to $kJ\ mol^{-1}$ using the factor, 1 cm^{-1}= 0.01196266 $kJ\ cm^{-1}$, which is derived from the 1973 CODATA fundamental constants (3). Rosenstock et al. (4) and Levin and Lias (5) have summarized additional ionization and appearance potential data.

$\Delta_f H°(Fe^+, g, 298.15\ K)$ is calculated from $\Delta_f H°(Fe, g, 0\ K)$ by using IP(Fe) with JANAF (1) enthalpies, H°(0 K) - H°(298.15 K), for Fe(g), $Fe^+(g)$, and $e^-(g)$. $\Delta_f H°(Fe \rightarrow Fe^+ + e^-, 298.15\ K)$ differs from a room temperature threshold energy due to inclusion of these enthalpies and to threshold effects discussed by Rosenstock et al. (4). $\Delta_f H°(298.15\ K)$ should be changed by -6.197 $kJ\ mol^{-1}$ if it is to be used in the ion convention that excludes the enthalpy of the electron.

Heat Capacity and Entropy

The information on electronic energy levels and quantum weights, given by Corliss and Sugar (2), is incomplete because many theoretically predicted levels have not been observed. Although we have listed only five levels for $Fe^+(g)$, all levels listed by Corliss and Sugar (2), as well as estimated levels, are used in the calculation. The observed levels are too numerous to list completely. The calculations indicate that for $Fe^+(g)$, the thermodynamic functions are independent of the estimated missing levels (for n=4), the cut-off procedure, and the inclusion of n=5 levels up to 6000 K; the Gibbs energy function showing no significant variations at this temperature. The reported uncertainty in S°(298.15 K) is due to uncertainties in the relative ionic mass, and the fundamental constants. Extension of these calculations above 6000 K may require consideration of the higher excited states (n>5), and use of different fill and cut-off procedures (6).

References

1. JANAF Thermochemical Tables: Fe(g), 3-31-78; $e^-(g)$, 3-31-82.
2. C. Corliss and J. Sugar, J. Phys. Chem. Ref. Data 11, 135 (1982).
3. E. R. Cohen and B. N. Taylor, J. Phys. Chem. Ref. Data 2, 663 (1973).
4. H. M. Rosenstock, K. Draxl et al., J. Phys. Chem. Ref. Data 6, Supp. 1 (1977).
5. R. D. Levin and S. G. Lias, U. S. Natl. Bur. Stand., NSRDS-NBS-71, 1982.
6. J. R. Downey, Jr., The Dow Chemical Company, AFOSR-TR-78-0960, Contract No. F44620-75-1-0048, 1978.

Iron, Ion (Fe^+) $Fe_1^+(g)$

Enthalpy Reference Temperature = T_r = 298.15 K Standard State Pressure = p° = 0.1 MPa

	J K⁻¹mol⁻¹			kJ mol⁻¹			
T/K	$C_p°$	S°	$-[G°-H°(T_r)]/T$	$H°-H°(T_r)$	$\Delta_f H°$	$\Delta_f G°$	Log K_f
0	0.	0.	INFINITE	-6.936	1172.517		
100	21.618	155.635	204.058	-4.842			
200	24.813	171.652	184.215	-2.512			
250	25.689	177.294	182.284	-1.248			
298.15	26.068	181.857	181.857	0.	1181.144	1128.814	-197.764
300	26.075	182.018	181.857	0.048	1181.184	1128.489	-196.487
350	26.163	186.047	182.175	1.355	1182.244	1119.622	-167.094
400	26.112	189.538	182.882	2.662	1183.249	1110.607	-145.030
450	26.018	192.608	183.796	3.966	1184.194	1101.470	-127.855
500	25.931	195.345	184.816	5.264	1185.076	1092.230	-114.104
600	25.830	200.062	186.976	7.851	1186.656	1073.508	-93.457
700	25.802	204.041	189.137	10.433	1187.990	1054.541	-78.691
800	25.786	207.485	191.220	13.012	1189.034	1035.402	-67.605
900	25.744	210.521	193.199	15.589	1189.661	1016.155	-58.976
1000	25.661	213.229	195.069	18.160	1189.562	996.872	-52.071
1100	25.539	215.669	196.833	20.720	1188.559	977.666	-46.425
1200	25.390	217.885	198.497	23.266	1188.094	958.495	-41.722
1300	25.225	219.911	200.067	25.797	1189.260	939.314	-37.742
1400	25.054	221.774	201.552	28.311	1190.325	920.047	-34.327
1500	24.885	223.497	202.958	30.808	1191.290	900.707	-31.365
1600	24.726	225.098	204.293	33.289	1192.155	881.306	-28.772
1700	24.579	226.592	205.561	35.754	1191.970	861.872	-26.482
1800	24.447	227.993	206.769	38.205	1192.303	842.445	-24.447
1900	24.330	229.312	207.921	40.644	1178.443	823.703	-22.645
2000	24.229	230.557	209.022	43.071	1178.347	805.035	-21.025
2100	24.144	231.737	210.076	45.490	1178.242	786.372	-19.560
2200	24.074	232.859	211.086	47.901	1178.129	767.714	-18.228
2300	24.018	233.928	212.056	50.305	1178.009	749.062	-17.012
2400	23.975	234.949	212.989	52.705	1177.885	730.415	-15.897
2500	23.945	235.927	213.887	55.101	1177.757	711.773	-14.872
2600	23.927	236.866	214.753	57.494	1177.627	693.136	-13.925
2700	23.920	237.769	215.588	59.886	1177.495	674.504	-13.049
2800	23.925	238.639	216.396	62.278	1177.363	655.877	-12.236
2900	23.940	239.478	217.178	64.672	1177.233	637.255	-11.478
3000	23.966	240.290	217.935	67.067	1177.104	618.637	-10.771
3100	24.002	241.077	218.669	69.465	1176.979	600.024	-10.110
3200	24.049	241.840	219.381	71.868	828.558	588.838	-9.612
3300	24.104	242.580	220.073	74.275	830.342	581.319	-9.202
3400	24.170	243.301	220.745	76.689	832.100	573.747	-8.815
3500	24.245	244.003	221.400	79.110	833.831	566.123	-8.449
3600	24.329	244.687	222.037	81.538	835.536	558.450	-8.103
3700	24.423	245.355	222.658	83.976	837.216	550.730	-7.775
3800	24.525	246.007	223.264	86.423	838.869	542.965	-7.464
3900	24.636	246.646	223.856	88.881	840.497	535.156	-7.168
4000	24.756	247.271	224.433	91.351	842.099	527.307	-6.886
4100	24.883	247.884	224.998	93.833	843.675	519.417	-6.617
4200	25.018	248.485	225.550	96.328	845.224	511.490	-6.361
4300	25.161	249.075	226.090	98.836	846.747	503.526	-6.117
4400	25.310	249.656	226.619	101.360	848.243	495.526	-5.883
4500	25.467	250.226	227.137	103.899	849.712	487.493	-5.659
4600	25.629	250.788	227.646	106.453	851.154	479.428	-5.444
4700	25.797	251.341	228.144	109.025	852.567	471.332	-5.238
4800	25.970	251.885	228.633	111.613	853.951	463.206	-5.041
4900	26.149	252.423	229.113	114.219	855.306	455.051	-4.851
5000	26.331	252.953	229.584	116.843	856.631	446.869	-4.668
5100	26.518	253.476	230.048	119.485	857.925	438.661	-4.493
5200	26.708	253.993	230.503	122.147	859.187	430.428	-4.324
5300	26.901	254.503	230.951	124.827	860.415	422.171	-4.161
5400	27.096	255.008	231.392	127.527	861.609	413.891	-4.004
5500	27.293	255.507	231.826	130.246	862.767	405.589	-3.852
5600	27.493	256.001	232.253	132.985	863.887	397.266	-3.706
5700	27.693	256.489	232.674	135.745	864.967	388.924	-3.564
5800	27.894	256.972	233.089	138.524	866.006	380.564	-3.427
5900	28.095	257.451	233.498	141.324	867.000	372.185	-3.295
6000	28.297	257.925	233.901	144.143	867.947	363.791	-3.167

PREVIOUS: March 1978 (1 atm) CURRENT: June 1984 (1 bar)

Iron, Ion (Fe^+) $Fe_1^+(g)$

IRON, ION (Fe^-) IDEAL GAS M_r = 55.84755

EA(Fe, g) = 0.163 ± 0.035 eV
S°(298.15 K) = 180.198 ± 0.005 J K^{-1} mol^{-1}

$\Delta_f H°$(0 K) = 397.400 kJ mol^{-1}
$\Delta_f H°$(298.15 K) = [393.338] kJ mol^{-1}

Electonic Levels and Quantum Weights

State	ε_i, cm^{-1}	g_i
$^4F_{9/2}$	0.0	10
$^4F_{7/2}$	540	8
$^4F_{5/2}$	930	6
$^4F_{3/2}$	1200	4

Enthalpy of Formation

$\Delta_f H°(Fe^-$, g, 0 K) is calculated from $\Delta_f H°$(Fe, g, 0 K) (1) using the adopted electron affinity of EA(Fe) = 0.163 ± 0.035 eV (15.727 ± 3.377 kJ mol^{-1}). This value, recommended by Hotop and Lineberger (2), is based on a laser photodetachment electron spectrometry study (3). Additional information on Fe^-(g) may be obtained in the critical discussions of Hotop and Lineberger (2, 4), Rosenstock et al. (5) and Massey (6).

$\Delta_f H°(Fe^-$, g, 298.15 K) is obtained from $\Delta_f H°$(Fe, g, 0 K) by using EA(Fe) with JANAF (1) enthalpies, H°(0 K) - H°(298.15 K), for Fe^-(g), Fe(g), and e^-(g). $\Delta_r H°(Fe^- \rightarrow Fe + e^-$, 298.15 K) differs from a room-temperature threshold energy due to inclusion of these enthalpies and to threshold effects discussed by Rosenstock et al. (5). $\Delta_f H°$(298.15 K) should be changed by +6.197 kJ mol^{-1} if it is to be used in the ion convention that excludes the enthalpy of the electron.

Heat Capacity and Entropy

The ground state electronic configuration for Fe^-(g) is given by Hotop and Lineberger (2, 4) and Rosenstock et al. (5). The fine structure separations in the ground state have been calculated by an isoelectronic extropolation of ratios of fine structure separations (7).

References

1. JANAF Thermochemical Tables: Fe(g), 3-31-78; e^-(g), 3-31-82.
2. H. Hotop and W. C. Lineberger, J. Phys. Chem. Ref. Data, 14, 731 (1985).
3. P. C. Engelking and W. C. Lineberger, Phys. Rev. A 19, 149 (1979).
4. H. Hotop and W. C. Lineberger, J. Phys. Chem. Ref. Data 4, 539 (1975).
5. H. M. Rosenstock, K. Draxl et al., J. Phys. Chem. Ref. Data 6, Supp. 1 (1977).
6. H. S. W. Massey, "Negative Ions", 3rd ed., Cambridge University Press, Cambridge, 1976.
7. C. S. Feigerle, R. R. Corderman, S. V. Babashev, and W. C. Linberger, J. Chem. Phys. 74, 1580 (1981).

Iron, Ion (Fe^-) Fe_1^-(g)

Enthalpy Reference Temperature = T_r = 298.15 K Standard State Pressure = p° = 0.1 MPa

T/K	$C_p°$	S°	$-[G°-H°(T_r)]/T$	$H°-H°(T_r)$	$\Delta_f H°$	$\Delta_f G°$	Log K_f
	J $K^{-1}mol^{-1}$			kJ mol^{-1}			
0	0.	0.	INFINITE	-6.642	397.400		
100	20.957	155.485	201.095	-4.561			
200	23.092	170.579	182.433	-2.371			
250	24.226	175.858	180.606	-1.187			
298.15	25.023	180.198	180.198	0.	393.338	354.013	-62.022
300	25.047	180.353	180.198	0.046	393.300	353.769	-61.597
350	25.536	184.254	180.505	1.312	392.240	347.264	-51.826
400	25.748	187.681	181.193	2.595	391.142	340.913	-44.519
450	25.758	190.716	182.085	3.884	389.994	334.703	-38.851
500	25.632	193.424	183.086	5.169	388.784	328.624	-34.331
600	25.172	198.059	185.208	7.710	386.162	316.833	-27.583
700	24.627	201.899	187.327	10.201	383.247	305.505	-22.797
800	24.108	205.153	189.357	12.637	379.990	294.616	-19.236
900	23.652	207.965	191.272	15.024	376.271	284.162	-16.492
1000	23.264	210.437	193.067	17.370	371.790	274.158	-14.321
1100	22.938	212.638	194.748	19.679	366.378	264.671	-12.568
1200	22.666	214.622	196.323	21.959	361.489	255.621	-11.127
1300	22.438	216.427	197.801	24.214	358.222	246.930	-9.922
1400	22.246	218.083	199.191	26.448	354.851	238.495	-8.898
1500	22.084	219.612	200.502	28.664	351.377	230.305	-8.020
1600	21.946	221.033	201.742	30.865	347.806	222.349	-7.259
1700	21.829	222.360	202.916	33.054	343.188	214.639	-6.595
1800	21.727	223.604	204.031	35.232	339.090	207.195	-6.013
1900	21.640	224.777	205.092	37.400	320.802	200.683	-5.517
2000	21.563	225.885	206.105	39.560	316.281	194.478	-5.079
2100	21.496	226.935	207.072	41.713	311.753	188.499	-4.689
2200	21.438	227.934	207.997	43.860	307.218	182.736	-4.339
2300	21.386	228.885	208.885	46.001	302.678	177.179	-4.024
2400	21.340	229.795	209.738	48.137	298.134	171.820	-3.740
2500	21.299	230.665	210.557	50.269	293.585	166.651	-3.482
2600	21.262	231.500	211.347	52.397	289.032	161.663	-3.248
2700	21.229	232.301	212.108	54.521	284.475	156.851	-3.034
2800	21.199	233.073	212.843	56.643	279.916	152.208	-2.839
2900	21.173	233.816	213.554	58.761	275.353	147.727	-2.661
3000	21.148	234.534	214.241	60.877	270.788	143.404	-2.497
3100	21.126	235.227	214.907	62.991	266.221	139.233	-2.346
3200	21.106	235.897	215.553	65.103	-86.649	142.634	-2.328
3300	21.088	236.546	216.179	67.212	-89.319	149.840	-2.372
3400	21.071	237.176	216.787	69.320	-92.024	157.128	-2.414
3500	21.055	237.786	217.379	71.427	-94.765	164.495	-2.455
3600	21.041	238.379	217.954	73.531	-97.540	171.942	-2.495
3700	21.028	238.955	218.514	75.635	-100.352	179.466	-2.534
3800	21.015	239.516	219.059	77.737	-103.201	187.067	-2.571
3900	21.004	240.062	219.590	79.838	-106.088	194.743	-2.608
4000	20.993	240.593	220.109	81.938	-109.013	202.494	-2.644
4100	20.984	241.112	220.615	84.037	-111.977	210.318	-2.679
4200	20.975	241.617	221.109	86.134	-114.982	218.215	-2.714
4300	20.966	242.111	221.592	88.232	-118.028	226.184	-2.748
4400	20.958	242.593	222.063	90.328	-121.116	234.225	-2.781
4500	20.951	243.063	222.525	92.423	-124.248	242.336	-2.813
4600	20.944	243.524	222.976	94.518	-127.424	250.517	-2.845
4700	20.937	243.974	223.418	96.612	-130.645	258.768	-2.876
4800	20.931	244.415	223.851	98.705	-133.913	267.088	-2.907
4900	20.925	244.846	224.275	100.798	-137.228	275.476	-2.937
5000	20.920	245.269	224.691	102.890	-140.592	283.933	-2.966
5100	20.915	245.683	225.099	104.982	-144.006	292.457	-2.995
5200	20.910	246.089	225.498	107.073	-147.472	301.049	-3.024
5300	20.905	246.488	225.891	109.164	-150.990	309.708	-3.052
5400	20.901	246.878	226.276	111.255	-154.563	318.434	-3.080
5500	20.897	247.262	226.654	113.344	-158.191	327.226	-3.108
5600	20.893	247.638	227.025	115.434	-161.878	336.085	-3.135
5700	20.890	248.008	227.390	117.523	-165.626	345.011	-3.162
5800	20.886	248.371	227.749	119.612	-169.435	354.003	-3.188
5900	20.883	248.728	228.101	121.700	-173.309	363.060	-3.214
6000	20.880	249.079	228.448	123.788	-177.250	372.184	-3.240

PREVIOUS: March 1978 (1 atm) CURRENT: June 1984 (1 bar)

Iron, Ion (Fe^-) Fe_1^-(g)

IRON HYDROXIDE ($Fe(OH)_2$) CRYSTAL M_r = 89.86168

S°(298.15 K) = [87.864 ± 8.4] J K^{-1} mol^{-1}

T_{fus} = Unknown

$\Delta_f H°$(0 K) = Unknown

$\Delta_f H°$(298.15 K) = -574.04 ± 2.9 kJ mol^{-1}

$\Delta_{fus} H°$ = Unknown

Enthalpy of Formation

The enthalpy of combustion of $Fe(OH)_2$(cr), according to the chemical reaction $Fe(OH)_2$(cr) + 1/4 O_2(g) = 1/2 Fe_2O_3(cr) + H_2O (1), was determined to be -29.8 ± 0.65 kcal mol^{-1} by Fricke and Rihl (1). From these data the adopted enthalpy of formation, $\Delta_f H°$(298.15 K), was evaluated to be -137.2 ± 0.7 kcal mol^{-1}.

The enthalpy changes for the following three reactions were determined by Thomsen (2).

Reaction	$\Delta_r H°$(291 K), kcal mol^{-1}
$FeCl_2$(cr) = $FeCl_2$ (400 H_2O)	-17.9
$FeCl_2$(200 H_2O) + H_2SO_4(200 H_2O) = $FeSO_4$(200 H_2O) + 2 HCl(100 H_2O)	-3.6
$FeSO_4$(aq) + 2 KOH(aq) = $Fe(OH)_2$(cr) + K_2SO_4(aq)	-6.34

Using $\Delta_f H°$(298.15 K) = -81.7 kcal mol^{-1} for $FeCl_2$(cr) and those for the other compounds in aqueous solution (assuming aq. = 200 H_2O in Reaction 3) obtained from Wagman et al. (3), the enthalpy of formation at 298.15 K for $Fe(OH)_2$(cr) was derived as -135.8 kcal mol^{-1} which agrees reasonably with the adopted value.

Heat Capacity and Entropy

The heat capacities, 298.15 - 700 K and S°(298.15 K) were estimated by comparison with those for $FeCl_2$(cr), $CaCl_2$(cr) and $Ca(OH)_2$(cr). The $C_p°$ values above 700 K were obtained by graphical extrapolation of the $C_p°$ curve plotted using the above $C_p°$ values.

Sublimation Data

The value of $\Delta_{sub} H°$(298.15 K) was calculated as the difference between $\Delta_f H°$(298.15 K) for $Fe(OH)_2$(g) and $Fe(OH)_2$(cr).

References

1. R. Fricke and S. Rihl, Z. Anorg. Allg. Chem. 251, 414 (1943).
2. J. Thomsen, "Thermochemische Untersuchungen," Barth, Leipzig, 1882-1886.
3. U. S. Nat. Bur. Stand. Tech. Note 270-1, 1965.

Iron Hydroxide ($Fe(OH)_2$) $Fe_1H_2O_2$(cr)

Enthalpy Reference Temperature = T_r = 298.15 K Standard State Pressure = p° = 0.1 MPa

T/K	$C_p°$ (J $K^{-1}mol^{-1}$)	S° (J $K^{-1}mol^{-1}$)	$-[G°-H°(T_r)]/T$ (J $K^{-1}mol^{-1}$)	$H°-H°(T_r)$ (kJ mol^{-1})	$\Delta_f H°$ (kJ mol^{-1})	$\Delta_f G°$ (kJ mol^{-1})	Log K_f
0							
100							
200							
298.15	97.069	87.864	87.864	0.	-574.045	-491.969	86.191
300	97.152	88.465	87.866	0.180	-574.019	-491.460	85.571
400	102.090	117.096	91.736	10.144	-572.560	-464.155	60.612
500	106.776	140.383	99.206	20.589	-570.950	-437.238	45.678
600	111.294	160.255	107.764	31.495	-569.218	-410.656	35.751
700	115.478	177.734	116.535	42.839	-567.392	-384.373	28.682
800	118.951	193.391	125.181	54.568	-565.567	-358.353	23.398
900	121.545	207.559	133.559	66.599	-563.945	-332.552	19.301
1000	123.386	220.466	141.614	78.852	-562.905	-306.906	16.031
1100	124.516	232.283	149.328	91.251	-562.696	-281.301	13.358
1200	125.102	243.145	156.699	103.735	-561.930	-255.773	11.134
1300	125.353	253.170	163.740	116.259	-559.553	-230.357	9.256
1400	125.436	262.462	170.463	128.799	-557.318	-205.119	7.653
1500	125.520	271.119	176.889	141.346	-555.230	-180.036	6.269

PREVIOUS: CURRENT: June 1966

Iron Hydroxide ($Fe(OH)_2$) $Fe_1H_2O_2$(cr)

IRON HYDROXIDE ($Fe(OH)_2$) — IDEAL GAS — M_r = 89.86168

$S°(298.15\ K)$ = [283.088] $J\ K^{-1}\ mol^{-1}$

$\Delta_f H°(0\ K)$ = [-323.09 ± 2.1] $kJ\ mol^{-1}$
$\Delta_f H°(298.15\ K)$ = [-330.54 ± 2.1 kJ] mol^{-1}

Electronic Levels and Quantum Weights

ϵ_i, cm^{-1}	g_i
0	[5]
[700]	[10]
[2500]	[5]
[4000]	[5]

Vibrational Frequencies and Degeneracies

ν, cm^{-1}	ν, cm^{-1}	ν, cm^{-1}
[2300](1)	[450](1)	[800](1)
[750](1)	[320](1)	[700](1)
[400](1)	[2600](1)	[570](1)

Point Group: [C_{2h}] σ = 2
Bond Distances: Fe-O = [1.8] Å O-H = [0.96] Å
Bond Angles: O-Fe-O = [180]° Fe-O-H = [105]°
Product of the Moments of Inertia: $I_A I_B I_C$ = [9.90391 x 10^{-116}] $g^3\ cm^6$

Enthalpy of Formation

The equilibrium constants, at temperatures between 1300 and 1460°C, for the chemical reaction $Fe(cr) + 2\ H_2O(g) = Fe(OH)_2(g) + H_2(g)$ were determined by Belton and Richardson. (1) Using these data, the enthalpy change, $\Delta_r H°$ (298.15 K), of the reaction is evaluated to be 35.53 ± 3.0 and 36.57 kcal mol^{-1} by the 2nd and 3rd law method, respectively. Based on the 3rd law value for $\Delta_r H°$(298.15 K), the enthalpy of formation, $\Delta_f H°(Fe[OH]_2$, g, 298.15 K), is calculated to be -79 ± 0.5 kcal mol^{-1}.

Heat Capacity and Entropy

The molecular structure is assumed to be the same as that for $B(OH)_2(g)$. The Fe-O bond distance is taken from FeO(g). The O-H bond distance and Fe-O-H bond angle are estimated from those for $H_2O(g)$. The vibrational frequencies, and electronic levels and quantum weights are obtained by comparison with those for $B(OH)_2(g)$ and $FeCl_2(g)$, respectively. These values are adjusted to give reasonable 2nd and 3rd law agreements. The principal moments of inertia are: I_A = 0.2814 x 10^{-39}, I_B = 18.6208 10^{-39} and I_C = 18.9022 x 10^{-39} g cm^2.

Reference

1. G. R. Belton and F. D. Richardson, Trans. Faraday Soc. 58, 1562 (1962).

Iron Hydroxide ($Fe(OH)_2$) — $Fe_1H_2O_2(g)$

Enthalpy Reference Temperature = T_r = 298.15 K — Standard State Pressure = p° = 0.1 MPa

T/K	$C_p°$ (J $K^{-1}mol^{-1}$)	S° (J $K^{-1}mol^{-1}$)	$-[G°-H°(T_r)]/T$ (J $K^{-1}mol^{-1}$)	$H°-H°(T_r)$ (kJ mol^{-1})	$\Delta_f H°$ (kJ mol^{-1})	$\Delta_f G°$ (kJ mol^{-1})	Log K_f
0	0.	0.	INFINITE	-14.209	-323.088	-323.088	INFINITE
100	36.766	227.041	335.221	-10.818	-326.023	-320.717	167.526
200	55.199	257.765	289.157	-6.278	-328.857	-314.241	82.071
250	64.437	271.106	284.228	-3.280	-329.851	-310.466	64.868
298.15	71.504	283.088	283.088	0.	-330.536	-306.666	53.727
300	71.738	283.531	283.089	0.132	-330.558	-306.518	53.370
350	77.115	295.015	283.984	3.861	-331.041	-302.471	45.141
400	81.011	305.581	286.032	7.819	-331.376	-298.365	38.962
450	83.894	315.297	288.752	11.945	-331.625	-294.223	34.153
500	86.129	324.256	291.860	16.198	-331.831	-290.056	30.302
600	89.528	340.275	298.626	24.989	-332.215	-281.665	24.521
700	92.212	354.284	305.598	34.080	-332.642	-273.207	20.387
800	94.509	366.751	312.477	43.419	-333.207	-264.681	17.282
900	96.515	378.001	319.142	52.973	-334.062	-256.067	14.862
1000	98.263	388.262	325.549	62.714	-335.535	-247.331	12.919
1100	99.773	397.700	331.685	72.617	-337.821	-238.385	11.320
1200	101.068	406.439	337.554	82.661	-339.495	-229.290	9.981
1300	102.172	414.573	343.170	92.825	-339.479	-220.108	8.844
1400	103.108	422.180	348.545	103.090	-339.518	-210.924	7.870
1500	103.901	429.322	353.694	113.441	-339.627	-201.736	7.025
1600	104.571	436.049	358.633	123.866	-339.815	-192.537	6.286
1700	105.137	442.406	363.376	134.352	-341.042	-183.306	5.632
1800	105.614	448.430	367.935	144.890	-340.078	-174.031	5.050
1900	106.018	454.151	372.323	155.473	-353.286	-164.056	4.510
2000	106.360	459.598	376.552	166.092	-354.404	-153.881	4.019
2100	106.650	464.795	380.631	176.743	-355.551	-143.647	3.573
2200	106.895	469.762	384.571	187.420	-356.729	-133.357	3.166
2300	107.104	474.518	388.379	198.121	-357.941	-123.011	2.794
2400	107.281	479.080	392.064	208.840	-357.428	-112.708	2.453
2500	107.432	483.463	395.633	219.576	-356.929	-102.384	2.139
2600	107.561	487.679	399.092	230.326	-358.219	-91.869	1.846
2700	107.671	491.741	402.449	241.088	-359.547	-81.302	1.573
2800	107.765	495.658	405.708	251.860	-360.913	-70.685	1.319
2900	107.845	499.441	408.876	262.640	-362.317	-60.017	1.081
3000	107.914	503.098	411.956	273.428	-361.930	-49.446	0.861
3100	107.973	506.638	414.953	284.223	-361.561	-38.852	0.655
3200	108.023	510.067	417.872	295.022	-711.351	-20.602	0.336
3300	108.066	513.391	420.717	305.827	-710.977	1.360	-0.022
3400	108.103	516.618	423.490	316.635	-710.670	23.313	-0.358
3500	108.134	519.752	426.196	327.447	-710.432	45.258	-0.675
3600	108.161	522.799	428.837	338.262	-710.262	67.199	-0.975
3700	108.184	525.763	431.417	349.079	-710.160	89.137	-1.258
3800	108.203	528.648	433.938	359.899	-710.126	111.075	-1.527
3900	108.220	531.459	436.402	370.720	-710.161	133.014	-1.782
4000	108.234	534.199	438.813	381.543	-710.265	154.954	-2.023
4100	108.245	536.872	441.172	392.367	-710.438	176.900	-2.254
4200	108.255	539.480	443.482	403.192	-710.682	198.851	-2.473
4300	108.263	542.028	445.744	414.018	-710.996	220.811	-2.682
4400	108.270	544.517	447.961	424.844	-711.383	242.778	-2.882
4500	108.275	546.950	450.134	435.671	-711.841	264.757	-3.073
4600	108.279	549.330	452.264	446.499	-712.373	286.746	-3.256
4700	108.283	551.658	454.355	457.327	-712.980	308.750	-3.431
4800	108.285	553.938	456.406	468.156	-713.662	330.768	-3.599
4900	108.287	556.171	458.419	478.984	-714.421	352.803	-3.761
5000	108.288	558.358	460.396	489.813	-715.258	374.854	-3.916
5100	108.289	560.503	462.338	500.642	-716.174	396.924	-4.065
5200	108.289	562.606	464.246	511.471	-717.171	419.011	-4.209
5300	108.288	564.668	466.121	522.300	-718.251	441.121	-4.348
5400	108.288	566.692	467.965	533.128	-719.416	463.251	-4.481
5500	108.287	568.679	469.778	543.957	-720.666	485.405	-4.610
5600	108.286	570.631	471.562	554.786	-722.005	507.584	-4.735
5700	108.284	572.547	473.317	565.614	-723.435	529.787	-4.855
5800	108.283	574.430	475.044	576.443	-724.957	552.017	-4.971
5900	108.281	576.281	476.744	587.271	-726.576	574.272	-5.084
6000	108.279	578.101	478.418	598.099	-728.293	596.559	-5.194

PREVIOUS: December 1966 (1 atm) — CURRENT: December 1966 (1 bar)

IRON HYDROXIDE ($Fe(OH)_3$) CRYSTAL M_r = 106.86902

$S°(298.15\ K) = [104.600 \pm 8.4]\ J\ K^{-1}\ mol^{-1}$

T_{fus} = Unknown

$\Delta_f H°(0\ K)$ = Unknown

$\Delta_f H°(298.15\ K) = -832.62 \pm 12.6\ kJ\ mol^{-1}$

$\Delta_{fus} H°$ = Unknown

Enthalpy of Formation

Schindler et al. (1) investigated the solubility of aged $Fe(OH)_3$ precipitates by determining the Fe^{+++} and H^+ ion concentration of solution in contact with the solid phase. The ion concentration was measured by the EMF - method at the constant ionic strength 3M $NaClO_4$(solution). The precipitates were investigated by X-ray and electron microscope methods. The equilibrium constant was derived as $\log [Fe^{+++}][H^+]^{-3} = \log K = 3.55 \pm 0.1$ for amorphous inactive hydroxide at 25°C. From these data the quantity $\log Ks = -39.1 \pm 0.2$ was evaluated where Ks is the solubility product of $Fe(OH)_3$. The Gibbs energy change, $\Delta_r G°(298.15\ K)$, of the reaction $Fe(OH)_3(cr) = Fe^{+++}(aq) + 3\ OH^-(aq)$ was calculated to be $53.34 \pm 0.27\ kcal\ mol^{-1}$. Using $S°(298.15\ K) = -70.1$ and $2.57\ cal\ K^{-1}\ mol^{-1}$ for $Fe^{+++}(aq)$ and $OH^-(aq)$, respectively, and an estimated $S°(Fe(OH)_3$, cr, 298.15 K) = 25 cal $K^{-1}\ mol^{-1}$, the enthalpy change, $\Delta_r H°(298.15\ K)$ of the reaction is calculated to be 22.69 kcal mol^{-1}, according to the relationship $\Delta H = \Delta F + T\Delta S$. The entropy values for Fe^{+++} and OH^- ions, were obtained from (2) and (3), respectively. Based on $\Delta_f H°(Fe^{+++}$, aq, 298.15 K) = -11.4 and $\Delta_f H°(OH^-$, aq, 298.15 K) = -54.97 kcal mol^{-1}, taken from the same sources, the enthalpy of formation, $\Delta_f H°(298.15\ K)$, for $Fe(OH)_3(cr)$ was evaluated to be -199 ± 3 kcal mol^{-1}, which was adopted here.

The divergent values for solubility product (SP) of $Fe(OH)_3$ found in the literature may be due, not only to the various forms of precipitate but also to the tendency to form colloidal solutions. The SP values for $Fe(OH)_3$ were also reported by Evans and Pryor (4), Kriukor and Awsejewitsch (5), Ruff and Hirsch (6), Britton (7), and Jellinek and Gorden (8).

The enthalpy changes, $\Delta_r H°(291\ K)$, for the reactions: (1) $FeCl_3(aq) + 3\ NaOH(aq) = Fe(OH)_3(cr) + 3\ NaCl(aq)$ and (2) $FeCl_3(cr) = FeCl_3(aq)$ were determined to be -24.50 and -31.68 kcal mol^{-1}, respectively by Thomsen (9). Assuming the aqueous solutions all contain 200 mol of H_2O, the enthalpy of formation, $\Delta_f H°(298.15\ K)$, for $Fe(OH)_3$ was calculated as -196.9 kcal mol^{-1} from Reaction (1) and (2). The $\Delta_f H°(298.15\ K)$ values for NaOH(aq) and NaCl(aq) were obtained from Parker (10), and JANAF $\Delta_f H°(298.15\ K)$ values for NaOH(cr) and NaCl(cr).

Heat Capacity and Entropy

The heat capacities, 298.15 - 1000 K, were estimated by comparison with those for $B_2O_3(cr)$, $Fe_2O_3(cr)$ and $B(OH)_3(cr)$. The $C_p°$ values above 1000 K were obtained by graphical extrapolation. The $S°(Fe(OH)_3$, cr, 298.15 K) value was estimated by comparison with that for $B(OH)_3(cr)$.

References

1. P. Schindler, W. Michaelis and W. Feitknect, Helv. Chim. Acta 46, 444 (1963).
2. U. S. Nat. Bur. Stand. Circ. 500, 1952.
3. U. S. Nat. Bur. Stand. Tech. Note 270-1, 1965.
4. Evans and Pryor, J. Chem. Soc. S157 (1949).
5. Kriukor and Awsejewitsch, Z. Elektrochem. 39, 884 (1933).
6. Ruff and Hirsch, Z. Anorg. Chem. 146, 338 (1925).
7. Britton, J. Chem. Soc. 2148 (1925).
8. Jellinek and Gorden, Z. Phys. Chem. 112, 207 (1924).
9. J. Thomsen, "Thermochemische Untersuchungen," Barth, Leipzig, 1882 - 1886.
10. V. B. Parker, NSRDS-NBS-2, 1965.

Iron Hydroxide ($Fe(OH)_3$) $Fe_1H_3O_3(cr)$

Enthalpy Reference Temperature = T_r = 298.15 K Standard State Pressure = $p°$ = 0.1 MPa

	J K⁻¹mol⁻¹			kJ mol⁻¹			
T/K	$C_p°$	S°	$-[G°-H°(T_r)]/T$	$H°-H°(T_r)$	$\Delta_f H°$	$\Delta_f G°$	Log K_f
0							
100							
200							
298.15	101.671	104.600	104.600	0.	-832.616	-705.467	123.595
300	102.090	105.230	104.602	0.188	-832.636	-704.678	122.695
400	117.989	136.877	108.809	11.227	-833.040	-661.930	86.439
500	129.704	164.480	117.242	23.619	-832.474	-619.202	64.688
600	140.582	189.140	127.205	37.161	-831.150	-576.661	50.203
700	148.114	211.387	137.669	51.603	-829.323	-534.385	39.876
800	154.808	231.616	148.166	66.759	-827.216	-492.393	32.150
900	160.247	250.172	158.484	82.519	-825.055	-450.671	26.156
1000	164.850	267.303	168.520	98.783	-823.237	-409.177	21.373
1100	168.406	283.188	178.231	115.452	-822.032	-367.813	17.466
1200	171.126	297.962	187.601	132.434	-820.081	-326.624	14.218
1300	173.218	311.747	196.626	149.657	-816.357	-285.654	11.478
1400	174.682	324.640	205.315	167.055	-812.653	-244.970	9.140
1500	175.728	336.729	213.677	184.579	-809.013	-204.549	7.123

PREVIOUS: CURRENT: June 1966

Iron Hydroxide ($Fe(OH)_3$) $Fe_1H_3O_3(cr)$

IRON IODIDE (FeI_2) CRYSTAL M_r = 309.6560

$S°(298.15\ K) = [167.360 \pm 8.4]\ J\ K^{-1}\ mol^{-1}$
$T_{trs}H° = 650\ K$
$T_{fus} = 860 \pm 2\ K$

$\Delta_f H°(0\ K)$ = Unknown
$\Delta_f H°(298.15\ K) = -104.60 \pm 8.4\ kJ\ mol^{-1}$
$\Delta_{trs}H° = 0.8\ kJ\ mol^{-1}$
$\Delta_{fus}H° = [44.769]\ kJ\ mol^{-1}$

Enthalpy of Formation

The chemical equilibrium of the decomposition of FeI_2(cr), 771.15 - 858.15 K, has been studied by Zaugg and Gregory (1). Using the reported partial pressures for I_2(g), the enthalpy change, $\Delta_r H°$(298.15 K), for the reaction FeI_2(cr) = Fe(cr) + I_2(g) was evaluated by both the 2nd and 3rd law methods to be 39.79±1.0 and 39.69±0.3 kcal mol^{-1}, respectively. Based on the 3rd law $\Delta_r H°$(298.15 K) value, the enthalpy of formation, $\Delta_f H°$(298.15 K) for FeI_2(cr) was calculated as -25 kcal mol^{-1} which was adopted here.

The enthalpy change for the reaction FeI_2(cr) = Fe^{++}(aq) = 2 I^-(aq) was determined to be -19.46±0.03 kcal mol^{-1} by Paoletti et al. (2), using a solution calorimetric method. Adopting $\Delta_f H°$(298.15 K) = -21.3 and -13.19 kcal mol^{-1} for Fe^{++}(aq) and I^-(aq), respectively, the value of $\Delta_f H$(FeI_2, cr, 298.15 K) was derived as -28.22±1 kcal mol^{-1}. The value of $\Delta_f H°$(I^-, aq, 298.15 K) was taken from Wagman et al. (3). The value of $\Delta_f H°$(Fe^{++}, aq, 298.15 K) was derived from enthalpy of solution and formation for $FeCl_2$(cr).

The enthalpies of solution of Fe(cr), I_2(cr) and FeI_2(cr) in aqueous Br_2-KBr solution were measured by Hieber and Woerner (4), using an ice calorimeter. From the results obtained the enthalpy of formation for FeI_2(cr) was reported as -30.1 kcal mol^{-1}.

Heat Capacity and Entropy

The heat capacities, 343.15 - 773.15 K, were measured by Oetting and Gregory (5). The $C_p°$ value at temperatures below 343.15 and above 773.15 K were estimated by graphic extrapolation. The low temperature heat capacities, 11-130 K were determined by Miljutin and Parfenowa (6). These data appear to be inadequate for the derivation of the entropy at 298.15 K. The value of S°(298.15 K) adopted was calculated from the entropy change, $\Delta_r S°$(298.15 K) = 28.7 cal K^{-1} mol^{-1}, for the decomposition reaction FeI_2(cr) = Fe(cr) + I_2(g), obtained by the second law analysis on the equilibrium pressure data reported by Zaugg and Gregory (1).

Transition Data

A minor transition in the vicinity of 370°C (360-385°C) has been reported by Oetting and Gregory (5). Subsequent studies by Wydeven (7) have shown the magnitude of the transition to be sensitive to trace amounts of impurities, but the nature of the transition has not been clearly established. A magnetic transition at 10 K was reported by Bizette et al. (8). The $\Delta_{trs}H°$ value was evaluated from the heat capacity - temperature plot given by Oetting and Gregory (5).

Fusion Data

T_{fus} was reported by Fischer and Gewehr (9). $\Delta_{fus}H°$ was estimated by comparison with that for $FeBr_2$(cr).

Sublimation Data

$\Delta_{sub}H°$(cr → monomer, 298.15 K) is calculated as the difference between $\Delta_f H°$(298.15 K) for FeI_2(g) and FeI_2(cr). $\Delta_{sub}H°$(cr → dimer, 298.15 K) was calculated as the difference between those for Fe_2I_4(g) and 2 FeI_2(cr).

References

1. W. E. Zaugg and N. W. Gregory, J. Phys. Chem. 70, 486 (1966).
2. P. Paoletti, A. Sabatini and A. Vacca, Trans. Faraday Soc. 61, 2417 (1965).
3. U. S. Nat. Bur. Stand. Tech. Note 270-1, 1965.
4. W. Hieber and A. Woerner, Z. Elektrochem. 40, 287 (1934).
5. F. L. Oetting and N. W. Gregory, J. Phys. Chem. 65, 173 (1961).
6. G. Miljutin and E. A. Parfenowa, Phys. Trans. Ukrain, Acad. Sci. 9, 81 (1940).
7. T. J. Wydeven, Ph.D. Thesis, University of Washington, Seattle, Wash., 1964.
8. H. Bizette, C. Terrier and B. Tasi, Compt. rend. 245, 507 (1957).
9. W. Fischer and R. Gewehr, Z. Anorg. Allg. Chem. 222, 303 (1935).

Iron Iodide (FeI_2) Fe_1I_2(cr)

Enthalpy Reference Temperature = T_r = 298.15 K; Standard State Pressure = p° = 0.1 MPa

T/K	$C_p°$ (J K⁻¹mol⁻¹)	S° (J K⁻¹mol⁻¹)	-[G°-H°(T_r)]/T (J K⁻¹mol⁻¹)	H°-H°(T_r) (kJ mol⁻¹)	$\Delta_f H°$ (kJ mol⁻¹)	$\Delta_f G°$ (kJ mol⁻¹)	Log K_f
0							
100							
200							
298.15	83.680	167.360	167.360	0.	-104.600	-111.725	19.574
300	83.680	167.878	167.362	0.155	-104.592	-111.769	19.461
400	83.931	191.985	170.647	8.535	-120.478	-113.639	14.840
500	84.182	210.741	176.859	16.941	-163.121	-107.843	11.266
600	84.425	226.111	183.825	25.371	-161.531	-96.941	8.439
650.000	84.546	232.873	187.341	29.596	I <--> II TRANSITION		
650.000	87.864	234.160	187.341	30.432			
700	95.395	240.947	190.927	35.013	-158.983	-86.362	6.444
800	110.876	254.699	198.037	45.330	-156.060	-76.173	4.974
860.000	120.005	263.045	202.281	52.257	II <--> LIQUID		
900	126.224	268.644	205.106	57.184	-152.024	-66.419	3.855
1000	136.817	282.555	212.158	70.397	-147.361	-57.157	2.986
1100	141.188	295.834	219.168	84.333	-142.883	-48.336	2.295
1200	142.256	308.176	226.078	98.517	-137.621	-39.991	1.741
1300	142.256	319.562	232.837	112.743	-130.695	-32.137	1.291
1400	142.256	330.105	239.413	126.969	-123.883	-24.812	0.926
1500	142.256	339.919	245.790	141.194	-117.196	-17.970	0.626

PREVIOUS: CURRENT: September 1966

Iron Iodide (FeI_2) Fe_1I_2(cr)

IRON IODIDE (FeI_2) LIQUID M_r = 309.6560

$S°(298.15\ K)$ = [195.430] J K^{-1} mol^{-1} $\Delta_f H°(298.15\ K)$ = [-71.046] kJ mol^{-1}

T_{fus} = 860 ± 2 K $\Delta_{fus}H°$ = 44.769 kJ mol^{-1}

Enthalpy of Formation

$\Delta_f H°(FeI_2$, ℓ, 298.15 K) is calculated from $\Delta_f H°(FeI_2$, cr, 298.15 K) by adding the enthalpy of fusion, $\Delta_{fus}H°$, and the difference in enthalpy, H°(860 K) - H°(298.15 K), between the crystal and liquid.

Heat Capacity and Entropy

The heat capacity is assumed to be constant in the temperature range from 298.15 to 2000 K and is calculated on the assumption that $C_p°$ = 9.0 cal K^{-1} g-$atom^{-1}$ for FeI_2(ℓ). S°(FeI_2, ℓ, 298.15 K) is calculated in a manner analogous to that used for the enthalpy of formation.

Vaporization Data

T_{vap} is the temperature at which the vapor pressure of FeI_2(g) and Fe_2I_4(g) over FeI_2(ℓ) equals one bar. The value of $\Delta_{vap}H°$ is derived from the vapor composition and the enthalpies of vaporization of FeI_2(ℓ) to monomer and dimer at T_{vap}.

Iron Iodide (FeI_2) Fe_1I_2(l)

Enthalpy Reference Temperature = T_r = 298.15 K Standard State Pressure = p° = 0.1 MPa

T/K	$C_p°$	S°	$-[G°-H°(T_r)]/T$	$H°-H°(T_r)$	$\Delta_f H°$	$\Delta_f G°$	Log K_f
	J $K^{-1}mol^{-1}$			kJ mol^{-1}			
0							
100							
200							
298.15	112.968	195.430	195.430	0.	-71.046	-86.540	15.161
300	112.968	196.129	195.433	0.209	-70.984	-86.636	15.085
400	112.968	228.628	199.864	11.506	-83.953	-91.771	11.984
500	112.968	253.836	208.231	22.803	-123.705	-89.975	9.400
600	112.968	274.433	217.600	34.099	-119.249	-83.652	7.283
700	112.968	291.847	226.995	45.396	-115.045	-78.055	5.825
800	112.968	306.931	236.065	56.693	-111.142	-73.041	4.769
860.000	112.968	315.101	241.298	63.471	-------- II <--> LIQUID --------		
900	112.968	320.237	244.693	67.990	-107.664	-68.493	3.975
1000	112.968	332.140	252.853	79.287	-104.917	-64.297	3.359
1100	112.968	342.907	260.558	90.583	-103.078	-60.310	2.864
1200	112.968	352.736	267.836	101.880	-100.704	-56.546	2.461
1300	112.968	361.778	274.719	113.177	-96.706	-53.029	2.131
1400	112.968	370.150	281.240	124.474	-92.823	-49.816	1.859
1500	112.968	377.944	287.430	135.771	-89.065	-46.876	1.632
1600	112.968	385.235	293.318	147.067	-85.440	-44.183	1.442
1700	112.968	392.084	298.928	158.364	-82.907	-41.693	1.281
1800	112.968	398.541	304.285	169.661	-79.907	-39.356	1.142
1900	112.968	404.649	309.408	180.958	-91.159	-36.479	1.003
2000	112.968	410.443	314.316	192.255	-88.705	-33.665	0.879

PREVIOUS: CURRENT: September 1966

Iron Iodide (FeI_2) Fe_1I_2(l)

IRON IODIDE (FeI_2) CRYSTAL–CRYSTAL–LIQUID M_r = 309.6560

0 to 650 K crystal, I
650 to 860 K crystal, II
above 860 K liquid

Refer to the individual tables for details.

Iron Iodide (FeI_2) Fe_1I_2(cr,l)

Enthalpy Reference Temperature = T_r = 298.15 K   Standard State Pressure = p° = 0.1 MPa

T/K	C_p° (J K⁻¹mol⁻¹)	S° (J K⁻¹mol⁻¹)	−[G°−H°(T_r)]/T (J K⁻¹mol⁻¹)	H°−H°(T_r) (kJ mol⁻¹)	$\Delta_f H^\circ$ (kJ mol⁻¹)	$\Delta_f G^\circ$ (kJ mol⁻¹)	Log K_f
0							
100							
200							
298.15	83.680	167.360	167.360	0.	-104.600	-111.725	19.574
300	83.680	167.878	167.362	0.155	-104.592	-111.769	19.461
400	83.931	191.985	170.647	8.535	-120.478	-113.639	14.840
500	84.182	210.741	176.859	16.941	-163.121	-107.843	11.266
600	84.425	226.111	183.825	25.371	-161.531	-96.941	8.439
650.000	84.546	232.873	187.341	29.596	I <--> II		
650.000	87.864	234.160	187.341	30.432	TRANSITION		
700	95.395	240.947	190.927	35.013	-158.983	-86.362	6.444
800	110.876	254.699	198.037	45.330	-156.060	-76.173	4.974
860.000	120.005	263.045	202.281	52.257	II <--> LIQUID		
860.000	112.968	315.101	202.281	97.026	TRANSITION		
900	112.968	320.237	207.410	101.544	-107.664	-68.493	3.975
1000	112.968	332.140	219.298	112.841	-104.917	-64.297	3.359
1100	112.968	342.907	230.054	124.138	-103.078	-60.310	2.864
1200	112.968	352.736	239.874	135.435	-100.704	-56.546	2.461
1300	112.968	361.778	248.908	146.731	-96.706	-53.029	2.131
1400	112.968	370.150	257.273	158.028	-92.823	-49.816	1.859
1500	112.968	377.944	265.061	169.325	-89.065	-46.876	1.632
1600	112.968	385.235	272.346	180.622	-85.440	-44.183	1.442
1700	112.968	392.084	279.190	191.919	-82.907	-41.693	1.281
1800	112.968	398.541	285.643	203.215	-79.907	-39.356	1.142
1900	112.968	404.648	291.747	214.512	-91.159	-36.479	1.003
2000	112.968	410.443	297.538	225.809	-88.705	-33.665	0.879

PREVIOUS: CURRENT: September 1966

Iron Iodide (FeI_2) Fe_1I_2(cr,l)

IRON IODIDE (FeI_2) IDEAL GAS M_r = 309.6560

$\Delta_f H°(0\ K)$ = [89.40 ± 12.6] kJ mol^{-1}

S°(298.15 K) = [349.597] J K^{-1} mol^{-1} $\Delta_f H°(298.15\ K)$ = [87.86 ± 12.6] kJ mol^{-1}

Electronic Levels and Quantum Weights

ε_i, cm^{-1}	g_i
0	[10]
[4200]	[10]
[6800]	[5]

Vibrational Frequencies and Degeneracies

ν, cm^{-1}
[140] (1)
[32] (2)
[280] (1)

Point Group: [$D_{\infty h}$] σ = 2
Bond Distance: Fe-I = [2.43] Å
Bond Angle: I-Fe-I = [180]°
Rotational Constant: B_0 = [0.011248] cm^{-1}

Enthalpy of Formation

The chemical equilibria for the reactions: (A) $FeI_2(cr) = FeI_2(g)$, (B) $FeI_2(\ell) = FeI_2(g)$, and (C) Fe(cr) + 2 I(g) = $FeI_2(g)$, have been studied by several investigators. Based on the equilibrium pressures reported, the corresponding enthalpy changes were evaluated by both the 2nd and 3rd law method. The results obtained are presented in the table below. Using the third law $\Delta_r H°(298.15\ K)$ values, the enthalpy of formation for $FeI_2(g)$ were calculated. The adopted value for $\Delta_f H°(FeI_2, g, 298.15\ K)$ is 21.0±3 kcal mol^{-1}. The disagreement between the sets of 2nd and 3rd law $\Delta_r H°(298.15\ K)$ values listed in the table may be due to the presence of dimer, $Fe_2I_4(g)$ which was not accounted for in deriving the partial pressure of $FeI_2(g)$. The data obtained from Schoonmaker et al. was corrected for the presence of dimer. However, only one point value was given.

Source	Reaction	T/K	Method	$\Delta_r H°(298.15\ K)$, kcal mol^{-1} 2nd law	3rd law	Drift cal K^{-1}mol^{-1}	$\Delta_f H°(298.15\ K)$, kcal mol^{-1}
Schafer and Honest* (1)	A	790.15-850.15	Transpiration	41.81	45.58	+5.78	21.58
	B	874.15-959.15	Transpiration	43.56	35.58	-8.75	18.6
Schoonmaker et al. (2)	A	714	Mass Spectrometric	-	46.78	-	21.78
Sime and Gregory* (3)	A	670.0 -740.0	Torsion-Effusion	45.13	45.70	0.85	20.7
Zaugg and Gregory (4)	C	865.15-1023.15	Transpiration	-32.53	-28.99	3.75	22.1

*The vapor pressure data used for evaluation were those corrected by (4).

Heat Capacity and Entropy

The molecular structure was assumed to be linear. The Fe-I bond distance was estimated by Brewer et al. (5). The vibrational frequencies were estimated so that the derived Gibbs energy fucntions yielded 2nd and 3rd law $\Delta_r H°(298.15\ K)$ values in reasonable agreements, see the above table. The electronic levels and quantum weights were estimated from those for $FeCl_2(g)$ reported by DeKock and Gruen (6).

References

1. H. Schafer and W. J. Hones, Z. Anorg. Allgem. Chem. 288, 62 (1956).
2. R. C. Schoonmaker, A. H. Friedman and F. R. Porter, J. Chem. Phys. 31, 1586 (1959).
3. R. J. Sime and N. W. Gregory, J. Phys. Chem. 64, 86 (1960). The vapor pressure equation used for calculation is log P(mmHg) = -(9760)/T + 11.82; In the original paper, the term A = 960 instead of 9760 is in error.
4. W. E. Zaugg and N. W. Gregory, J. Phys. Chem. 70, 490 (1966).
5. L. Brewer, G. R. Somayajulu and E. Brackett, Chem. Rev. 63, 111 (1963).
6. C. W. DeKock and D. M. Gruen, J. Chem. Phys. 44, 4387 (1966).

Iron Iodide (FeI_2) $Fe_1I_2(g)$

Enthalpy Reference Temperature = T_r = 298.15 K Standard State Pressure = p° = 0.1 MPa

T/K	$C_p°$ (J K^{-1}mol^{-1})	S°	$-[G°-H°(T_r)]/T$	$H°-H°(T_r)$ (kJ mol^{-1})	$\Delta_f H°$	$\Delta_f G°$	Log K_f
0	0.	0.	INFINITE	-16.167	89.402	89.402	INFINITE
100	53.921	286.148	402.435	-11.629	90.467	68.558	-35.811
200	59.295	325.577	355.125	-5.910	89.485	46.951	-12.262
250	60.288	338.925	350.596	-2.918	88.712	36.404	-7.606
298.15	60.859	349.597	349.597	0.	87.864	26.405	-4.626
300	60.876	349.973	349.598	0.113	87.829	26.024	-4.531
350	61.248	359.387	350.341	3.166	86.785	15.802	-2.358
400	61.498	367.583	351.995	6.235	69.686	6.286	-0.821
450	61.675	374.838	354.138	9.315	67.335	-1.499	0.174
500	61.806	381.343	356.538	12.402	24.804	-5.219	0.545
600	62.003	392.630	361.641	18.593	24.154	-11.167	0.972
700	62.180	402.200	366.769	24.802	23.270	-16.987	1.268
800	62.388	410.517	371.729	31.030	22.104	-22.663	1.480
900	62.645	417.879	376.455	37.281	20.537	-28.169	1.635
1000	62.955	424.495	380.934	43.561	18.267	-33.469	1.748
1100	63.308	430.512	385.172	49.874	15.122	-38.476	1.827
1200	63.690	436.036	389.183	56.224	12.549	-43.254	1.883
1300	64.085	441.150	392.987	62.612	11.639	-47.868	1.923
1400	64.479	445.914	396.599	69.041	10.653	-52.409	1.955
1500	64.861	450.375	400.037	75.508	9.582	-56.876	1.981
1600	65.220	454.573	403.315	82.012	8.414	-61.269	2.000
1700	65.551	458.537	406.448	88.551	6.189	-65.567	2.015
1800	65.850	462.292	409.447	95.121	4.462	-69.739	2.024
1900	66.113	465.860	412.323	101.720	-11.487	-73.109	2.010
2000	66.340	469.257	415.086	108.343	-13.707	-76.295	1.993
2100	66.533	472.499	417.743	114.987	-15.974	-79.369	1.974
2200	66.691	475.597	420.303	121.648	-18.285	-82.334	1.955
2300	66.818	478.565	422.772	128.324	-20.637	-85.193	1.935
2400	66.916	481.411	425.156	135.011	-23.023	-87.949	1.914
2500	66.987	484.144	427.462	141.706	-25.434	-90.605	1.893
2600	67.034	486.772	429.693	148.407	-27.861	-93.163	1.872
2700	67.059	489.303	431.854	155.112	-30.296	-95.629	1.850
2800	67.065	491.742	433.949	161.819	-32.728	-98.004	1.828
2900	67.055	494.095	435.983	168.525	-35.148	-100.293	1.806
3000	67.031	496.368	437.958	175.229	-37.549	-102.498	1.785
3100	66.994	498.565	439.878	181.930	-39.924	-104.624	1.763
3200	66.947	500.691	441.745	188.628	-390.566	-99.249	1.620
3300	66.892	502.751	443.563	195.320	-390.968	-90.140	1.427
3400	66.829	504.747	445.333	202.006	-391.362	-81.018	1.245
3500	66.760	506.683	447.059	208.685	-391.743	-71.884	1.073
3600	66.686	508.563	448.741	215.358	-392.110	-62.740	0.910
3700	66.609	510.389	450.383	222.022	-392.462	-53.586	0.756
3800	66.529	512.164	451.985	228.679	-392.799	-44.423	0.611
3900	66.447	513.891	453.551	235.328	-393.121	-35.251	0.472
4000	66.363	515.572	455.080	241.969	-393.430	-26.071	0.340
4100	66.278	517.210	456.576	248.601	-393.726	-16.883	0.215
4200	66.193	518.806	458.038	255.224	-394.012	-7.688	0.096
4300	66.108	520.363	459.470	261.839	-394.289	1.513	-0.018
4400	66.024	521.881	460.871	268.446	-394.560	10.721	-0.127
4500	65.940	523.364	462.243	275.044	-394.827	19.935	-0.231
4600	65.857	524.813	463.588	281.634	-395.093	29.155	-0.331
4700	65.776	526.228	464.906	288.216	-395.360	38.380	-0.427
4800	65.695	527.612	466.198	294.789	-395.631	47.612	-0.518
4900	65.616	528.966	467.465	301.355	-395.908	56.849	-0.606
5000	65.539	530.291	468.708	307.912	-396.195	66.092	-0.690
5100	65.463	531.588	469.928	314.463	-396.495	75.341	-0.772
5200	65.389	532.858	471.126	321.005	-396.809	84.596	-0.850
5300	65.316	534.103	472.303	327.540	-397.142	93.856	-0.925
5400	65.246	535.323	473.459	334.068	-397.497	103.124	-0.998
5500	65.177	536.520	474.594	340.589	-397.875	112.398	-1.067
5600	65.109	537.694	475.711	347.104	-398.282	121.679	-1.135
5700	65.044	538.845	476.808	353.611	-398.719	130.968	-1.200
5800	64.980	539.976	477.888	360.113	-399.192	140.265	-1.263
5900	64.918	541.086	478.950	366.607	-399.702	149.570	-1.324
6000	64.858	542.177	479.994	373.096	-400.254	158.884	-1.383

PREVIOUS: September 1966 (1 atm) CURRENT: September 1966 (1 bar)

Iron Iodide (FeI_2) $Fe_1I_2(g)$

IRON OXIDE (FeO) CRYSTAL M_r = 71.8464

$\Delta_f H°$(0 K) = Unknown

$S°$(298.15 K) = [60.752] J K^{-1} mol^{-1} $\Delta_f H°$(298.15 K) = [-272.044] kJ mol^{-1}

T_{fus} = [1650] K $\Delta_{fus} H°$ = [24.058] kJ mol^{-1}

Enthalpy of Formation

The value of $\Delta_f H°$(298.15 K) for FeO(cr) was derived based on an assumption that at 1650 K, $\Delta_f F°$(FeO, cr) = $\Delta_f F°$(FeO, ℓ). From the value $\Delta_f F°$(FeO, cr, 1650 K) = -39.77 kcal mol^{-1}, $\Delta_f H°$(FeO, cr, 1650 K) was calculated to be -63.97 kcal mol^{-1}, yielding $\Delta_f H°$(FeO, cr, 298.15 K) = -65.02 kcal mol^{-1} (-272.044 k J mol^{-1}).

Heat Capacity and Entropy

The $C_p°$ values were estimated assuming $C_p°$ (FeO, cr) = $C_p°$(Wustite, cr) + 0.053 $C_p°$(Fe, cr). The $C_p°$ values obtained were plotted. The adopted $C_p°$ points were taken from the smoothed $C_p°$ curve. S°(298.15 K) for FeO(cr) was calculated as S°(FeO, cr) = S° (Wustite, cr) + 0.053 S°(Fe, cr) + ΔS°(mixing) where ΔS° is the entropy of mixing (0.41 cal k^{-1} mol^{-1}).

Fusion Data

T_{fus} is assumed to be the same as that for Wustite. The difference between $\Delta_f H°$(1650 K) for FeO(ℓ) and FeO(cr) is $\Delta_{fus} H°$.

Sublimation Data

$\Delta_{sub} H°$(298.15 K) is calculated as the difference between $\Delta_f H°$(298.15 K) for FeO(g) and FeO(cr).

Iron Oxide (FeO) Fe_1O_1(cr)

Enthalpy Reference Temperature = T_r = 298.15 K — Standard State Pressure = p° = 0.1 MPa

	J $K^{-1}mol^{-1}$			kJ mol^{-1}			
T/K	$C_p°$	S°	-[G°-H°(T_r)]/T	H°-H°(T_r)	$\Delta_f H°$	$\Delta_f G°$	Log K_f
0							
100							
200							
298.15	49.915	60.752	60.752	0.	-272.044	-251.429	44.049
300	49.999	61.061	60.753	0.092	-272.025	-251.301	43.755
400	51.840	75.704	62.737	5.187	-271.044	-244.543	31.934
500	53.388	87.438	66.541	10.449	-270.164	-238.022	24.866
600	54.894	97.309	70.868	15.865	-269.414	-231.666	20.168
700	56.149	105.866	75.270	21.418	-268.814	-225.425	16.821
800	57.321	113.443	79.577	27.093	-268.422	-219.256	14.316
900	58.283	120.249	83.724	32.872	-268.374	-213.118	12.369
1000	59.371	126.447	87.691	38.756	-268.969	-206.955	10.810
1100	60.234	132.146	91.477	44.736	-270.385	-200.670	9.529
1200	61.086	137.424	95.089	50.802	-271.184	-194.316	8.458
1300	61.975	142.349	98.537	56.956	-270.265	-187.947	7.552
1400	62.760	146.971	101.833	63.194	-269.361	-181.649	6.777
1500	63.396	151.323	104.989	69.502	-268.484	-175.415	6.108
1600	64.015	155.435	108.014	75.873	-267.642	-169.238	5.525
1650.000	64.319	157.409	109.481	79.081	------ CRYSTAL <--> LIQUID -----		
1700	64.631	159.334	110.919	82.305	-267.784	-163.094	5.011
1800	65.229	163.045	113.713	88.798	-267.345	-156.949	4.555
1900	65.790	166.587	116.403	95.349	-281.041	-150.125	4.127
2000	66.316	169.975	118.998	101.955	-280.919	-143.237	3.741

PREVIOUS: CURRENT: June 1965

Iron Oxide (FeO) Fe_1O_1(cr)

IRON OXIDE (FeO) LIQUID M_r = 71.8464

$S°(298.15\ K) = 75.415\ J\ K^{-1}\ mol^{-1}$

$T_{fus} = [1650]\ K$

$\Delta_f H°(298.15\ K) = [-249.532]\ kJ\ mol^{-1}$

$\Delta_{fus} H° = [24.058]\ kJ\ mol^{-1}$

Enthalpy of Formation

The enthalpy of formation, $\Delta_f H°$(FeO, ℓ, 298.15 K), is evaluated based on an assumption that the enthalpy of fusion of Wustite, $\Delta_{fus} H°$(1650 K) = 7.49 kcal mol^{-1}, represents the enthalpy change of the reaction $Fe_{0.947}O$(cr) = 0.947 FeO(ℓ) + (0.053/2) O_2(g). In other words, during melting the Wustite releases O_2(g) and converts to FeO(ℓ).

Heat Capacity and Entropy

The enthalpy changes, H°(T)-H°(298.15 K), for 1650-1800 K of liquid iron oxide were determined by Coughlin et al. (1). The $C_p°$ (FeO, ℓ) was derived as 16.3 cal K^{-1} mol^{-1}. This value was adopted as the heat capacity for FeO(ℓ) up to 4000 K. A glass transition temperature was assumed at 1100 K. S°(FeO, ℓ, 298.15 K) is calculated based on an assumption that the entropy of fusion of Wustite represents also the entropy change of the reaction. $Fe_{0.947}O$(cr) = 0.947 FeO(ℓ) + (0.053/2) O_2(g). From the entropy of fusion, $\Delta_{fus} H°$(1650 K) = 4.539 cal K^{-1} mol^{-1}, the value S°(FeO, ℓ, 298.15 K) = 18.029 cal K^{-1} mol^{-1} was derived.

Fusion Data

The melting point (T_{fus}) for Wustite ($Fe_{0.947}O$, cr) has been reported by many investigators. However, the composition of the liquid state of Wustite was not identified. An assumption was made that during melting the reaction $Fe_{0.947}O$(cr) = 0.947 FeO(ℓ) + (0.053/2) O_2(g) occurs. Hence the related thermodynamic properties were derived.

Decomposition Data

T_{dcm} is the temperature at which $\Delta_f G°$ equals zero.

Reference

1. J. P. Coughlin, E. G. King and K. R. Bonnickson, J. Amer. Chem. Soc. 73, 3891 (1951).

Iron Oxide (FeO) $Fe_1O_1(l)$

Enthalpy Reference Temperature = T_r = 298.15 K Standard State Pressure = p° = 0.1 MPa

T/K	$C_p°$ (J K⁻¹mol⁻¹)	S° (J K⁻¹mol⁻¹)	−[G°−H°(T_r)]/T (J K⁻¹mol⁻¹)	H°−H°(T_r) (kJ mol⁻¹)	$\Delta_f H°$ (kJ mol⁻¹)	$\Delta_f G°$ (kJ mol⁻¹)	Log K_f
0							
100							
200							
298.15	48.116	75.415	75.415	0.	-249.532	-233.289	40.871
300	48.183	75.713	75.416	0.089	-249.517	-233.189	40.602
400	50.380	89.893	77.334	5.023	-248.696	-227.871	29.757
500	51.848	101.298	81.023	10.137	-247.965	-222.752	23.271
600	53.028	110.857	85.220	15.382	-247.385	-217.766	18.958
700	54.070	119.110	89.485	20.737	-246.983	-212.864	15.884
800	55.041	126.394	93.653	26.193	-246.810	-208.005	13.581
900	55.969	132.931	97.660	31.744	-246.991	-203.148	11.790
1000	56.873	138.875	101.489	37.386	-247.827	-198.241	10.355
1100	57.761	144.337	105.139	43.118	-249.492	-193.187	9.174
1100.000	57.761	144.337	105.139	43.118	GLASS <--> LIQUID TRANSITION		
1100.000	68.199	144.337	105.139	43.118			
1200	68.199	150.271	108.656	49.938	-249.537	-188.086	8.187
1300	68.199	155.730	112.070	56.758	-247.952	-183.030	7.354
1400	68.199	160.784	115.371	63.578	-246.466	-178.092	6.645
1500	68.199	165.490	118.558	70.398	-245.077	-173.257	6.033
1600	68.199	169.891	121.630	77.218	-243.785	-168.512	5.501
1650.000	68.199	171.990	123.124	80.628	------ CRYSTAL <--> LIQUID -----		
1700	68.199	174.026	124.592	84.038	-243.540	-163.826	5.034
1800	68.199	177.924	127.447	90.858	-242.774	-159.160	4.619
1900	68.199	181.611	130.202	97.678	-256.202	-153.831	4.229
2000	68.199	185.109	132.861	104.497	-255.866	-148.452	3.877
2100	68.199	188.437	135.428	111.317	-255.541	-143.089	3.559
2200	68.199	191.609	137.911	118.137	-255.227	-137.741	3.270
2300	68.199	194.641	140.312	124.957	-254.925	-132.408	3.007
2400	68.199	197.543	142.636	131.777	-254.634	-127.088	2.766
2500	68.199	200.327	144.889	138.597	-254.354	-121.779	2.544
2600	68.199	203.002	147.073	145.417	-254.085	-116.481	2.340
2700	68.199	205.576	149.192	152.237	-253.826	-111.194	2.151
2800	68.199	208.056	151.250	159.057	-253.577	-105.915	1.976
2900	68.199	210.450	153.251	165.877	-253.338	-100.646	1.813
3000	68.199	212.762	155.196	172.697	-253.109	-95.385	1.661
3100	68.199	214.998	157.089	179.517	-252.890	-90.131	1.519
3200	68.199	217.163	158.933	186.336	-600.979	-77.460	1.264
3300	68.199	219.262	160.729	193.156	-598.876	-61.133	0.968
3400	68.199	221.298	162.481	199.976	-596.815	-44.869	0.689
3500	68.199	223.275	164.190	206.796	-594.795	-28.665	0.428
3600	68.199	225.196	165.858	213.616	-592.816	-12.518	0.182
3700	68.199	227.064	167.487	220.436	-590.880	3.575	-0.050
3800	68.199	228.883	169.079	227.256	-588.987	19.616	-0.270
3900	68.199	230.655	170.635	234.076	-587.137	35.608	-0.477
4000	68.199	232.381	172.157	240.896	-585.331	51.553	-0.673
4100	68.199	234.065	173.647	247.716	-583.570	67.453	-0.859
4200	68.199	235.709	175.105	254.536	-581.855	83.311	-1.036
4300	68.199	237.314	176.533	261.356	-580.186	99.128	-1.204
4400	68.199	238.881	177.932	268.176	-578.566	114.907	-1.364
4500	68.199	240.414	179.304	274.995	-576.993	130.650	-1.517
4600	68.199	241.913	180.649	281.815	-575.470	146.358	-1.662
4700	68.199	243.380	181.968	288.635	-573.998	162.034	-1.801
4800	68.199	244.816	183.262	295.455	-572.578	177.680	-1.934
4900	68.199	246.222	184.533	302.275	-571.212	193.296	-2.061
5000	68.199	247.600	185.781	309.095	-569.900	208.885	-2.182

PREVIOUS: CURRENT: June 1965

Iron Oxide (FeO) $Fe_1O_1(l)$

IRON OXIDE (FeO) CRYSTAL-LIQUID M_r = 71.8464

0 to 1650 K crystal
above 1650 K liquid

Refer to the individual tables for details.

Iron Oxide (FeO) $Fe_1O_1(cr,l)$

Enthalpy Reference Temperature = T_r = 298.15 K — Standard State Pressure = p° = 0.1 MPa

T/K	C_p° (J K⁻¹mol⁻¹)	S° (J K⁻¹mol⁻¹)	$-[G^\circ-H^\circ(T_r)]/T$ (J K⁻¹mol⁻¹)	$H^\circ-H^\circ(T_r)$ (kJ mol⁻¹)	Δ_fH° (kJ mol⁻¹)	Δ_fG° (kJ mol⁻¹)	Log K_f
0							
100							
200							
298.15	49.915	60.752	60.752	0.	-272.044	-251'.429	44.049
300	49.999	61.061	60.753	0.092	-272.025	-251.301	43.755
400	51.840	75.704	62.737	5.187	-271.044	-244.543	31.934
500	53.388	87.438	66.541	10.449	-270.164	-238.022	24.866
600	54.894	97.309	70.868	15.865	-269.414	-231.666	20.168
700	56.149	105.866	75.270	21.418	-268.814	-225.425	16.821
800	57.321	113.443	79.577	27.093	-268.422	-219.256	14.316
900	58.283	120.249	83.724	32.872	-268.374	-213.118	12.369
1000	59.371	126.447	87.691	38.756	-268.969	-206.955	10.810
1100	60.234	132.146	91.477	44.736	-270.385	-200.670	9.529
1200	61.086	137.424	95.089	50.802	-271.184	-194.316	8.458
1300	61.975	142.349	98.537	56.956	-270.265	-187.947	7.552
1400	62.760	146.971	101.833	63.194	-269.361	-181.649	6.777
1500	63.396	151.323	104.989	69.502	-268.484	-175.415	6.108
1600	64.015	155.435	108.014	75.873	-267.642	-169.238	5.525
1650.000	64.319	157.409	109.481	79.081	CRYSTAL <--> LIQUID TRANSITION		
1650.000	68.199	171.990	109.481	103.139			
1700	68.199	174.026	111.350	106.549	-243.540	-163.826	5.034
1800	68.199	177.924	114.941	113.369	-242.774	-159.160	4.619
1900	68.199	181.611	118.354	120.189	-256.202	-153.831	4.229
2000	68.199	185.109	121.605	127.009	-255.866	-148.452	3.877
2100	68.199	188.437	124.709	133.829	-255.541	-143.089	3.559
2200	68.199	191.609	127.678	140.649	-255.227	-137.741	3.270
2300	68.199	194.641	130.524	147.468	-254.925	-132.408	3.007
2400	68.199	197.543	133.257	154.288	-254.634	-127.088	2.766
2500	68.199	200.327	135.884	161.108	-254.354	-121.779	2.544
2600	68.199	203.002	138.415	167.928	-254.085	-116.481	2.340
2700	68.199	205.576	140.855	174.748	-253.826	-111.194	2.151
2800	68.199	208.056	143.211	181.568	-253.577	-105.915	1.976
2900	68.199	210.450	145.488	188.388	-253.338	-100.646	1.813
3000	68.199	212.762	147.692	195.208	-253.109	-95.385	1.661
3100	68.199	214.998	149.828	202.028	-252.890	-90.131	1.519
3200	68.199	217.163	151.898	208.848	-600.979	-77.460	1.264
3300	68.199	219.262	153.908	215.668	-598.876	-61.133	0.968
3400	68.199	221.298	155.860	222.488	-596.815	-44.869	0.689
3500	68.199	223.275	157.758	229.308	-594.795	-28.665	0.428
3600	68.199	225.196	159.605	236.127	-592.816	-12.518	0.182
3700	68.199	227.064	161.403	242.947	-590.880	3.575	-0.050
3800	68.199	228.883	163.155	249.767	-588.987	19.616	-0.270
3900	68.199	230.655	164.863	256.587	-587.137	35.608	-0.477
4000	68.199	232.381	166.530	263.407	-585.331	51.552	-0.673
4100	68.199	234.065	168.156	270.227	-583.570	67.453	-0.859
4200	68.199	235.709	169.745	277.047	-581.855	83.310	-1.036
4300	68.199	237.314	171.298	283.867	-580.186	99.128	-1.204
4400	68.199	238.881	172.816	290.687	-578.566	114.907	-1.364
4500	68.199	240.414	174.301	297.507	-576.993	130.650	-1.517
4600	68.199	241.913	175.755	304.327	-575.470	146.358	-1.662
4700	68.199	243.380	177.178	311.147	-573.998	162.034	-1.801
4800	68.199	244.816	178.573	317.966	-572.578	177.679	-1.934
4900	68.199	246.222	179.939	324.786	-571.212	193.296	-2.061
5000	68.199	247.600	181.278	331.606	-569.900	208.885	-2.182

PREVIOUS: CURRENT: June 1965

Iron Oxide (FeO) $Fe_1O_1(cr,l)$

IRON OXIDE (FeO) IDEAL GAS M_r = 71.8464

$\Delta_f H°(0\ K)$ = 251.05 ± 20.9 kJ mol^{-1}

$S°(298.15\ K)$ = [241.924] J K^{-1} mol^{-1} $\Delta_f H°(298.15\ K)$ = 251.04 ± 20.9 kJ mol^{-1}

Electronic Levels and Quantum Weights

State	ε_i, cm^{-1}	g_i
[$^5\Delta$]	0	[10]
	[10000]	[10]
	[16000]	[5]

ω_e = 880.0 cm^{-1} $\omega_e x_e$ = 5.0 cm^{-1} σ = 1

B_e = [0.4184] cm^{-1} α_e = [0.00293] cm^{-1} r_e = [1.8] Å

Enthalpy of Formation

The enthalpy of formation for FeO(g) is not well established at the present time. The values of $\Delta_f H°$(FeO, g, 298.15 K) derived from the following reactions: (A) FeO(g) = Fe(g) + O(g) , (B) FeO(g) + 1/2 O_2(g), and (C) FeO(ℓ) = FeO(g), are not in agreement. There are three D_0^0(Fe-O) values reported for reaction (1). The enthalpy change for reaction (2) was evaluated based on the partial pressure data (P_{FeO}/P_{Fe}) = 0.2 at 1600°C, measured by Washburn (4), using a mass spectrometric method (6), and P_{O_2} = 1.66 x 10^{-6} atm reported by Darken and Gurry (7). The enthalpy value for reaction (3) was calculated based on the reported value, $\Delta_{vap}H°$(2785 K) = 96.096 kcal mol^{-1} determined by Burtsev et al (5). The results obtained are presented as follows. The value of $\Delta_f H°$(298.15 K) for FeO(g) adopted is 60±5 kcal mol^{-1}.

Source	Reaction	Method	$\Delta_r H°$(298.15 K) kcal mol^{-1}	$\Delta_f H°$(298.15 K) kcal mol^{-1}
Herzberg (1)	A	Spectroscopic	111.9	47.2
Gaydon (2)	A	Spectroscopic	93.4±23.1	65.7±23.1
Lagerqvist and Huldt (3)	A	Spectroscopic	99.0±11.6	60.1±11.6
Washburn et al. (4)	B	Spectroscopic	99.3±5	59.8±5
Burtsev et al. (5)	C	Transpiration	111.8	52.2

Heat Capacity and Entropy

Since the ground state configuration is mainly that due to Fe^{++} in a ligand field, it was assumed to be the same as that for $FeCl_2$(g) reported by DeKock and Gruen (8). The electronic levels and quantum weights were estimated by comparison with those for $FeCl_2$(g). The values of ω_e and $\omega_e x_e$ were taken from Herzberg (1). The bond distance was calculated according to the method suggested by Guggenheimer (9), assuming FeO(g) as a polar molecule. The value of B_e was calculated by use of the relationship: B_e = (2.799076 x 10^{-39})/I where I is the moment of inertia of FeO(g). The value of α_e was derived from ω_e, $\omega_e x_e$ and B_e by the method suggested by Herzberg (1).

References

1. G. Herzberg, "Spectra of Diatomic Molecules," D. Van Nostrand Company, Inc., New York, 1950.
2. A. G. Gaydon, "Dissociation Energies," Chapman and Hall, Ltd., London, 1953.
3. A. Lagerqvist and L. Huldt, Z. Naturforsch. 8A, 493 (1953). D_0^0 was derived from atomic spectra of Fe produced by acetylene-air flame method.
4. Washburn et al., UCRL 10991, August 1963, obtained from L. Brewer, personal communication, October 18, 1966.
5. V. T. Burtsev, R. A. Karasev and A. M. Samarin, Fiz.-Khim. Osnovy Proizv. Stali, Akad. Nauk SSSR, Inst. Met., Tr. 6-01 [Shestoi] Konf., Moscow, 366 (1961); published in 1964.
6. U. S. Atomic Energy Commission, UCRL-10991, August 1963.
7. L. S. Darken and R. W. Gurry, J. Amer. Chem. Soc. 68, 798 (1946).
8. C. W. DeKock and D. M. Gruen, J. Chem. Phsy. 44, 4387 (1966).
9. K. M. Guggenheimer, Proc. Phys. Soc. (London) 58, 456 (1946).

Iron Oxide (FeO) Fe_1O_1(g)

Enthalpy Reference Temperature = T_r = 298.15 K Standard State Pressure = p° = 0.1 MPa

T/K	$C_p°$ (J K^{-1}mol^{-1})	S° (J K^{-1}mol^{-1})	$-[G°-H°(T_r)]/T$ (J K^{-1}mol^{-1})	$H°-H°(T_r)$ (kJ mol^{-1})	$\Delta_f H°$ (kJ mol^{-1})	$\Delta_f G°$ (kJ mol^{-1})	Log K_f
0	0.	0.	INFINITE	-8.837	251.051	251.051	INFINITE
100	29.110	209.460	268.743	-5.928	252.085	240.411	-125.578
200	29.741	229.752	244.739	-2.997	251.791	228.779	-59.751
250	30.538	236.471	242.436	-1.491	251.431	223.067	-46.607
298.15	31.406	241.924	241.924	0.	251.040	217.639	-38.129
300	31.439	242.118	241.924	0.058	251.024	217.431	-37.858
350	32.302	247.030	242.310	1.652	250.594	211.866	-31.619
400	33.065	251.394	243.178	3.287	250.139	206.364	-26.948
450	33.715	255.328	244.313	4.957	249.653	200.921	-23.322
500	34.261	258.909	245.596	6.656	249.127	195.534	-20.427
600	35.098	265.234	248.356	10.127	247.932	184.925	-16.099
700	35.689	270.691	251.166	13.668	246.520	174.532	-13.024
800	36.118	275.486	253.912	17.259	244.828	164.360	-10.732
900	36.439	279.760	256.551	20.888	242.725	154.422	-8.962
1000	36.687	283.612	259.068	24.545	239.904	144.752	-7.561
1100	36.884	287.119	261.461	28.223	236.186	135.431	-6.431
1200	37.048	290.335	263.735	31.920	233.018	126.392	-5.502
1300	37.188	293.306	265.897	35.632	231.495	117.568	-4.724
1400	37.315	296.067	267.954	39.357	229.886	108.864	-4.062
1500	37.434	298.645	269.915	43.095	228.192	100.278	-3.492
1600	37.550	301.065	271.787	46.844	226.413	91.808	-2.997
1700	37.669	303.345	273.577	50.605	223.600	83.471	-2.565
1800	37.792	305.502	275.291	54.378	221.319	75.293	-2.185
1900	37.922	307.548	276.936	58.164	204.857	67.947	-1.868
2000	38.060	309.497	278.516	61.963	202.172	60.811	-1.588
2100	38.207	311.357	280.035	65.776	199.490	53.809	-1.338
2200	38.363	313.138	281.500	69.605	196.812	46.934	-1.114
2300	38.527	314.847	282.913	73.449	194.139	40.182	-0.913
2400	38.699	316.491	284.278	77.310	191.471	33.545	-0.730
2500	38.877	318.074	285.598	81.189	188.810	27.019	-0.565
2600	39.063	319.602	286.877	85.086	186.157	20.600	-0.414
2700	39.253	321.080	288.117	89.002	183.511	14.283	-0.276
2800	39.447	322.511	289.319	92.936	180.875	8.063	-0.150
2900	39.644	323.899	290.488	96.891	178.248	1.938	-0.035
3000	39.842	325.246	291.624	100.865	175.632	-4.097	0.071
3100	40.041	326.556	292.730	104.859	173.025	-10.045	0.169
3200	40.240	327.830	293.807	108.873	-177.870	-8.485	0.139
3300	40.437	329.071	294.857	112.907	-178.553	-3.182	0.050
3400	40.633	330.281	295.881	116.961	-179.258	2.143	-0.033
3500	40.825	331.462	296.881	121.034	-179.985	7.489	-0.112
3600	41.014	332.615	297.858	125.126	-180.734	12.856	-0.187
3700	41.198	333.741	298.812	129.236	-181.508	18.244	-0.258
3800	41.378	334.842	299.746	133.365	-182.305	23.654	-0.325
3900	41.553	335.919	300.660	137.512	-183.129	29.084	-0.390
4000	41.722	336.973	301.555	141.676	-183.979	34.536	-0.451
4100	41.885	338.006	302.431	145.856	-184.858	40.010	-0.510
4200	42.042	339.017	303.290	150.052	-185.766	45.505	-0.566
4300	42.193	340.008	304.133	154.264	-186.706	51.023	-0.620
4400	42.338	340.980	304.959	158.491	-187.678	56.563	-0.671
4500	42.476	341.933	305.770	162.732	-188.685	62.125	-0.721
4600	42.608	342.868	306.566	166.986	-189.727	67.709	-0.769
4700	42.734	343.785	307.349	171.253	-190.808	73.318	-0.815
4800	42.853	344.686	308.117	175.532	-191.929	78.949	-0.859
4900	42.966	345.571	308.872	179.823	-193.091	84.605	-0.902
5000	43.072	346.440	309.615	184.125	-194.297	90.284	-0.943
5100	43.173	347.294	310.346	188.438	-195.549	95.988	-0.983
5200	43.267	348.133	311.064	192.760	-196.849	101.717	-1.022
5300	43.356	348.958	311.771	197.091	-198.200	107.472	-1.059
5400	43.440	349.770	312.468	201.431	-199.604	113.252	-1.095
5500	43.518	350.567	313.153	205.779	-201.063	119.059	-1.131
5600	43.590	351.352	313.828	210.134	-202.581	124.893	-1.165
5700	43.658	352.124	314.493	214.497	-204.161	130.755	-1.198
5800	43.721	352.884	315.149	218.866	-205.805	136.645	-1.231
5900	43.779	353.632	315.795	223.241	-207.518	142.563	-1.262
6000	43.833	354.368	316.431	227.621	-209.302	148.512	-1.293

PREVIOUS: September 1966 (1 atm) CURRENT: September 1966 (1 bar)

Iron Oxide (FeO) Fe_1O_1(g)

IRON SULFATE ($FeSO_4$) CRYSTAL M_r = 151.9046

$\Delta_fH°(0\ K) = -919.33 \pm 8.4$ kJ mol^{-1}

$S°(298.15\ K) = 120.957 \pm 1.3$ J K^{-1} mol^{-1} $\Delta_fH°(298.15\ K) = -928.85 \pm 8.4$ kJ mol^{-1}

Enthalpy of Formation

The enthalpy change for the reaction $FeSO_4(cr) = FeSO_4(110 \infty H_2O)$ was determined to be -14.9 kcal mol^{-1} by Forcrand (1). Using $\Delta_fH°(298.15\ K) = -345.85$ kcal mol^{-1} for $FeSO_4(110 \infty H_2O)$, the enthalpy of formation (298.15 K) for $FeSO_4(cr)$ was calculated as -221 kcal mol^{-1}. The value of $\Delta_fH°(FeSO_4$, cr, 298.15 K) was estimated based on $\Delta_fH°(298.15\ K) = 235.9$ kcal mol^{-1} for $FeSO_4(200 \infty H_2O) + H_2SO_4(200\ H_2O) = FeSO_4(200\ H_2O) + 2HCl(100 \infty H_2O)$ measured by Thomsen (2). See the $Fe(OH)_2(cr)$ table for details.

The decomposition pressures of $FeSO_4(cr)$ at different temperatures were investigated by D'Ans (3) and Greulich (4). Using their reported partial pressures for $SO_3(g)$ and $SO_2(g)$, the enthalpy change for the reaction $2FeSO_4(cr) = Fe_2O_3(cr) + SO_3(g) + SO_2(g)$ was evaluated by both the 2nd and 3rd law methods. The results obtained are presented in the following table.

Source	T/K	$\Delta_rH°(298.15\ K)$, kcal mol^{-1} 2nd law	3rd law	$\Delta_fH°(298.15\ K)$ kcal mol^{-1}
D'Ans (3, 1905)	753.15-908.15	83.2±4.6	82.7±1.1	-222.7
Greulich (4, 1927)	887.15-971.15	64.8±8.8	80.1±1.4	-221.5
Neumann and Heintke (5, 1937)	633.15-862.15	72.4±5.4	75.3±0.5	-219.1

The decomposition pressures asumed for the same reaction have been determined by Neumann and Heintke (5). From their data the partial pressures for $SO_3(g)$ and $SO_2(g)$ were evaluated. Using the derived $SO_3(g)$ and $SO_2(g)$ partial pressures, the corresponding enthalpy change for the reaction were calculated by both the 2nd and 3rd law methods. The results obtained are also listed in the same table.

The value of $\Delta_fH°(298.15\ K)$ for $FeSO_4(cr)$ is selected as -222±2 kcal mol^{-1}.

Heat Capacity and Entropy

The low temperature heat capacities, 50.12-294.9 K, were determined by Moore and Kelley (6). Based on S°(50.12 K) = 2.10 cal K^{-1} mol^{-1}, the value of $S°(FeSO_4$, cr, 298.15 K) was reported to be 25.71±0.2 cal K^{-1} mol^{-1}. Using the Gibbs energy functions for $FeSO_4(cr)$ based on this S°(298.15 K) value, to evaluate the vapor pressure data for Reaction (1), the 2nd and 3rd law values of $\Delta_fH°(298.15\ K)$ were derived as 72.36±7.60 and 86.51±1.26 kcal mol^{-1}, respectively. Since the report by Moore and Kelley (6) did not mention the magnetic entropy contribution, an attempt was made to add 3.20 (=Rln5) cal K^{-1} mol^{-1} to $S°(FeSO_4$, cr, 298.15 K) and reevaluate the decomposition pressure data. The results obtained were better than before, see the paragraph on "Enthalpy of Formation," for details. Therefore the value, S°(298.15 K) = 25.71 + 3.2 = 28.91 cal K^{-1} mol^{-1} for $FeSO_4(cr)$, was adopted. The heat capacities above 294.9 K were estimated by comparison with those for $MnSO_4(cr)$. The high temperature heat capacities, 870.3-1082.3 K, were determined by Southard and Shomate (7). The two sets of data were joined smoothly at 298.15 K by use of Shomate-function plot.

Decomposition Data

T_{dcm} = 944 K is the temperature at which the vapor pressure of the gaseous decomposition products equals one atmosphere, which was obtained by graphical interpolation of the decomposition pressure data on $FeSO_4(cr)$, reported by Greulich (4).

References

1. R. de Forcrand, Comp. Rend. 158, 20 (1914).
2. J. Thomsen, "Thermochemische Untersuchunger," Barth, Leipzig, 1882-1886.
3. J. D'Ans, Dissertation, Darmstadt, 1905. Data quoted by B. Neumann and G. Heintke, loc. cit.
4. E. Greulich, Z. Anorg. Chem. 168, 197 (1927). Only the last seven high temperature points were adopted for evaluation.
5. B. Neumann and G. Heintke, Z. Elektrochem. 43, 246 (1937), based on the last 18 high temperature points adopted.
6. G. E. Moore and K. K. Kelley, J. Amer. Chem. Soc., 64, 2949 (1942).
7. J. C. Southard and C. H. Shomate, J. Amer. Chem. Soc. 64, 1770 (1942).

Iron Sulfate ($FeSO_4$) $Fe_1O_4S_1(cr)$

Enthalpy Reference Temperature = T_r = 298.15 K Standard State Pressure = p° = 0.1 MPa

T/K	$C_p°$ (J $K^{-1}mol^{-1}$)	S°	$-[G°-H°(T_r)]/T$	$H°-H°(T_r)$ (kJ mol^{-1})	$\Delta_fH°$	$\Delta_fG°$	Log K_f
0	0.	0.	INFINITE	-16.769	-919.331	-919.331	INFINITE
100	44.074	42.771	194.810	-15.204	-924.688	-892.445	466.166
200	78.848	85.016	129.562	-8.909	-927.627	-858.919	224.327
298.15	100.583	120.957	120.957	0.	-928.848	-824.879	144.515
300	101.002	121.580	120.959	0.186	-928.859	-824.234	143.512
400	116.650	152.861	125.118	11.097	-931.115	-789.237	103.064
500	128.909	180.266	133.463	23.401	-931.710	-753.695	78.738
600	138.030	204.613	143.331	36.769	-931.332	-718.113	62.517
700	144.641	226.410	153.670	50.918	-930.365	-682.648	50.940
800	149.411	246.052	164.011	65.633	-929.150	-647.342	42.267
900	152.758	263.849	174.131	80.746	-981.132	-611.106	35.468
1000	155.477	280.088	183.927	96.161	-979.245	-570.101	29.779
1100	157.779	295.017	193.356	111.826	-978.112	-529.226	25.131
1200	159.829	308.835	202.411	127.709	-976.304	-488.516	21.265
1300	161.670	321.702	211.098	143.785	-972.740	-448.012	18.001
1400	163.343	333.744	219.433	160.036	-969.161	-407.782	15.215
1500	164.933	345.069	227.435	176.451	-965.573	-367.809	12.808
1600	166.440	355.762	235.124	193.020	-961.981	-328.076	10.711
1700	167.862	365.895	242.521	209.735	-959.340	-288.548	8.866
1800	169.285	375.530	249.645	226.593	-956.087	-249.184	7.231
1900	170.624	384.719	256.514	243.589	-966.938	-209.298	5.754
2000	171.962	393.505	263.146	260.718	-963.937	-169.499	4.427

PREVIOUS: CURRENT: June 1966

Iron Sulfate ($FeSO_4$) $Fe_1O_4S_1(cr)$

IRON SULFIDE, TROILITE (FeS) CRYSTAL M_r = 87.907

$S°(298.15\ K) = 60.321 \pm 0.04\ J\ K^{-1}\ mol^{-1}$
$T_{trs1} = 411 \pm 3\ K$
$T_{trs2} = 598 \pm 3\ K$
$T_{fus} = 1463 \pm 3\ K$

$\Delta_f H°(0\ K) = -102.17 \pm 0.8\ kJ\ mol^{-1}$
$\Delta_f H°(298.15\ K) = -101.67 \pm 0.8\ kJ\ mol^{-1}$
$\Delta_{trs1}H° = 1.665 \pm 0.8\ kJ\ mol^{-1}$
$\Delta_{trs2}H° = 0.397 \pm 0.08\ kJ\ mol^{-1}$
$\Delta_{fus}H° = 31.464 \pm 2.1\ kJ\ mol^{-1}$

Enthalpy of Formation

The adopted $\Delta_f H°(298.15\ K)$ is -24.30±0.2 kcal mol^{-1} and is based on our 3rd law analysis of gas phase H_2S/H_2 equilibrium data of (1-7). Additional studies of a similar nature are reported by Mills (8). The gas phase equilibrium results are favored over calorimetric (8) and emf (9-11) results because of greater certainty in the stoichiometry of the iron sulfide. Of the calorimetric studies (8), only those of Adami and King ($\Delta_f H°(298.15\ K)$ = -23.89±0.41 kcal mol^{-1}) and the annealed sample of Ariya et al. ($\Delta_f H°(298.15\ K)$ = -24.4 kcal mol^{-1}) are in reasonable agreement with the adopted value. Of the emf results, all refer to less than stoichiometric composition $Fe_{0.95}S$ (9), $Fe_{0.98}S$ (10) or to impure sample, 97.58% (11). $\Delta_f H°(298.15\ K)$ has been shown to vary significantly as a function of stoichiometry (12) which is a result of the variation of $\Delta_a H°$ with composition (12).

Source	Reaction†	Method	Data Points	T/K	$\Delta_r H°(298.15\ K)$, kcal mol^{-1} 2nd law	3rd law	Drift cal K^{-1} mol^{-1}	$\Delta_f H°(298.15\ K)$ kcal mol^{-1}
Rosenqvist (1)	C	H_2S/H_2	11	773 - 1257	-17.89±0.41	-19.40±0.53	-1.45±0.40	-24.30±0.6
Blaise (2)	C	"	equation	923 - 1043	-17.87	-19.41±0.14	-1.57	-24.31±0.2
Turkdogan (3)	C	"	3	943 - 1173	-17.24±0.39	-19.40±0.48	-2.03±0.36	-24.30±0.5
Alcock (4)	B	"	equation	720 - 1261	-38.30	-39.68±0.56	-1.38	-24.26±0.6
McCabe (5)	B	"	10	1224 - 1255	-33.76±2.84	-39.70±0.15	4.81±2.3	-24.28±0.2
Sudo (6)	C	"	4	1048 - 1145	-17.39±2.19	-19.51±0.29	-1.95±2.01	-24.41±0.4
Rau (7)	C	"	9	820 - 1257	-19.40±0.23	-19.31±0.20	0.09±0.22	-24.15±0.3
Berner (9)	A	emf	1	298.15		-24.08		-24.08
DeRanter (10)	A	H titration	1	298.15				-20.55
DeRanter (10)	A	emf	1	298.15		-20.56		-20.56
Guldin (11)	B	emf	1	973		36.445		-21.02

†Reactions: (A) Fe(cr) + S(cr) → FeS(cr), (B) Fe(cr) + 1/2 S_2(g) → FeS(cr), (C) Fe(cr) + H_2S(g) → FeS(cr) + H_2(g)

Heat Capacity and Entropy

The adopted heat capacities are obtained by merging the $C°_p$ data (7-345 K) of Gronvold et al. (13) and the enthalpy data (356-1488 K) of Coughlin (14). The data are difficult to merge smoothly because of the proximity of the α-β transition at 411 K. Coughlin gives the composition of his sample as $FeS_{1.02}$ and this introduces some uncertainty at the higher temperatures.

The value of S°(298.15 K) is obtained from the appropriate integration of $C°_p$ values assuming S°(7 K) = 0.0028 cal K^{-1} mol^{-1}.

Phase Data

The mineral pyrrhotite has a composition range extending from FeS to $Fe_{0.875}S$ (Fe_7S_8); the name troilite is reserved for the stoichiometric FeS composition. The phase diagram (15) of this system shows that the iron-rich limit of this phase corresponds to stoichiometric FeS below ~1370 K; above this temperature only an iron-deficient material exists so that stoichiometric FeS will melt incongruently.

There are two solid state phase transitions, both associated with electronic-magnetic phenomena. The adopted temperature of the α-β transition is T_{trs1} = 411±3 K (15). Our analysis of the adopted $C°_p$ data yields $\Delta_{trs1}H°$ = 0.398 kcal mol^{-1}; in view of the uncertainties in these data we adopt an uncertainty of ±0.2 kcal mol^{-1} in $\Delta_{trs1}H°$. The adopted temperature of the β-γ transition is T_{trs2} = 598±3 K (15) and our analysis of Coughlin's high temperature enthalpy data (14) yields $\Delta_{trs2}H°$ = 0.095±0.02 kcal mol^{-1}.

Fusion Data

See the liquid table for details.

References

1. T. Rosenqvist, J. Iron Steel Inst. 176, 37 (1954).
2. B. Blaise and A. Gentry, C. R. Acad. Sci., Ser. C 273, 1125 (1971).
3. E. T. Turkdogan, Trans. AIME 242, 1665 (1968).
4. C. B. Alcock and F. D. Richardson, Nature 168, 661 (1951).
5. C. L. McCabe, C. B. Alcock and R. G. Hudson, J. Metals 8, 693 (1956).
6. K. Sudo, Sci. Rept. Res. Inst. Tohoku Univ., Ser. A 2, 312 (1950).
7. H. Rau, J. Phys. Chem. Solids 37, 425 (1976).
8. K. C. Mills, "Thermodynamic Data for Inorganic Sulphides, Selenides and Tellurides," Butterworths, London, 1974.
9. R. A. Berner, Amer. J. Sci. 265, 773 (1967).
10. C. DeRanter and R. Breckpot, Bull. Soc. Chim. Belg. 78, 503 (1969).
11. I. T. Guldin and A. V. Buzhinskaya, J. Appl. Chem. USSR 37, 2221 (1964).
12. G. Bugli, L. Abello and G. Pannetier, Bull. Soc. Chim. Fr., 497 (1972).
13. F. Gronvold, E. F. Westrum, Jr., and C. Chou, J. Chem. Phys. 30, 528 (1959).
14. J. P. Coughlin, J. Amer. Chem. Soc. 72, 5445 (1950).
15. M. Hansen and K. Anderko, "Constitution of Binary Alloys," 2nd ed., McGraw-Hill, New York, 1958.

Iron Sulfide, Troilite (FeS) $Fe_1S_1(cr)$

Enthalpy Reference Temperature = T_r = 298.15 K — Standard State Pressure = p° = 0.1 MPa

T/K	$C°_p$ (J $K^{-1}mol^{-1}$)	S°	$-[G°-H°(T_r)]/T$	$H°-H°(T_r)$ (kJ mol^{-1})	$\Delta_f H°$	$\Delta_f G°$	Log K_f
0	0.	0.	INFINITE	-9.413	-102.166	-102.166	INFINITE
100	26.670	17.216	99.528	-8.231	-102.096	-101.959	53.258
200	43.052	41.665	64.714	-4.610	-101.886	-101.901	26.614
298.15	50.517	60.321	60.321	0.	-101.671	-101.953	17.862
300	50.671	60.634	60.322	0.094	-101.666	-101.954	17.752
400	89.203	79.060	62.567	6.597	-102.388	-102.086	13.331
411.000	95.051	81.559	63.042	7.610	ALPHA <--> BETA TRANSITION		
411.000	72.358	85.611	63.042	9.276			
500	72.358	99.794	68.363	15.715	-100.050	-102.492	10.707
598.000	72.358	112.745	74.607	22.807	BETA <--> GAMMA TRANSITION		
598.000	62.085	113.409	74.607	23.204			
600	62.030	113.617	74.736	23.328	-99.108	-103.033	8.970
700	59.808	122.999	80.982	29.412	-99.697	-103.645	7.734
800	58.560	130.893	86.740	35.322	-100.612	-104.151	6.800
900	58.286	137.765	92.036	41.156	-155.064	-103.437	6.003
1000	58.986	143.934	96.922	47.012	-155.812	-97.670	5.102
1100	60.662	149.627	101.457	52.986	-157.351	-91.766	4.358
1200	63.311	155.011	105.697	59.177	-158.138	-85.786	3.734
1300	66.935	160.216	109.692	65.681	-156.980	-79.802	3.206
1400	71.533	165.338	113.484	72.596	-155.509	-73.918	2.758
1463.000	74.930	168.560	115.786	77.208	GAMMA <--> LIQUID		
1500	77.106	170.458	117.111	80.020	-153.630	-68.153	2.373
1600	83.653	175.638	120.607	88.050	-151.243	-62.529	2.041
1700	91.174	180.930	123.999	96.783	-149.200	-57.054	1.753
1800	99.670	186.377	127.312	106.317	-145.839	-51.727	1.501
1900	109.140	192.016	130.568	116.749	-155.774	-45.906	1.262

PREVIOUS: CURRENT: September 1977

Iron Sulfide, Troilite (FeS) $Fe_1S_1(cr)$

IRON SULFIDE (FeS) LIQUID M_r = 87.907

$S°(298.15\ K) = [82.811]\ J\ K^{-1}\ mol^{-1}$

$T_{fus} = 1463 \pm 3\ K$

$\Delta_f H°(298.15\ K) = [-68.811]\ kJ\ mol^{-1}$

$\Delta_{fus} H° = 31.464 \pm 2.1\ kJ\ mol^{-1}$

Enthalpy of Formation

$\Delta_f H°$(FeS, ℓ, 298.15 K) is calculated from that of the crystal by adding the enthalpy of fusion, $\Delta_{fus}H°$, and the difference in enthalpy, H°(1463 K) - H°(298.15 K), between the crystal and liquid.

Heat Capacity and Entropy

The heat capacity of liquid FeS above an assumed glass transition temperature of 1050 K is taken from Vaisburd and Zedina (2). Their constant value of $C_p° = 14.95\ cal\ K^{-1}\ mol^{-1}$ is lower than $C_p° = 17.8\ cal\ K^{-1}\ mol^{-1}$ which we derive from the enthalpy data of Coughlin (3). The latter value is based on two points separated by only 9 K while the results of Vaisburd and Zedina are quoted as covering the range from the mp to 1723 K. The heat capacity below the glass transition is that of the crystal (1).

S°(FeS, ℓ, 298.15 K) is calculated in a manner analogous to that used for the enthalpy of formation.

Fusion Data

The adopted melting point of 1463±3 K is taken from the phase diagram of Hansen and Anderko (4). The adopted enthalpy of fusion, $\Delta_{fus}H° = 7.52 \pm 0.5\ kcal\ mol^{-1}$, is from our analysis of the high temperature enthalpy data of Coughlin (3).

References

1. JANAF Thermochemical Tables: FeS(cr) 9-30-77.
2. S. E. Vaisburd and I. N. Zedina, Russ. J. Phys. Chem. 45, 1156 (1971).
3. J. P. Coughlin, J. Amer. Chem. Soc. 72, 5445 (1950).
4. M. Hansen and K. Anderko, "Constitution of Binary Alloys," 2nd ed., McGraw-Hill, New York, 1958.

Iron Sulfide (FeS) $Fe_1S_1(l)$

Enthalpy Reference Temperature = T_r = 298.15 K Standard State Pressure = p° = 0.1 MPa

T/K	$C_p°$ (J K⁻¹mol⁻¹)	S° (J K⁻¹mol⁻¹)	-[G°-H°(T_r)]/T (J K⁻¹mol⁻¹)	H°-H°(T_r) (kJ mol⁻¹)	$\Delta_f H°$ (kJ mol⁻¹)	$\Delta_f G°$ (kJ mol⁻¹)	Log K_f
0							
100							
200							
298.15	50.517	82.811	82.811	0.	-68.811	-75.798	13.280
300	50.671	83.124	82.812	0.094	-68.806	-75.842	13.205
400	89.203	101.550	85.057	6.597	-69.528	-78.222	10.215
411.000	95.051	104.049	85.532	7.610	ALPHA <--> BETA TRANSITION		
411.000	72.358	108.100	85.532	9.276			
500	72.358	122.284	90.853	15.715	-67.190	-80.877	8.449
598.000	72.358	135.235	97.096	22.807	BETA <--> GAMMA TRANSITION		
598.000	62.085	135.899	97.096	23.204			
600	62.030	136.106	97.226	23.328	-66.248	-83.667	7.284
700	59.808	145.489	103.472	29.412	-66.837	-86.528	6.457
800	58.560	153.383	109.230	35.322	-67.752	-89.283	5.830
900	58.286	160.255	114.526	41.156	-122.204	-90.818	5.271
1000	58.986	166.424	119.412	47.012	-122.952	-87.300	4.560
1050.000	59.702	169.318	121.720	49.978	GLASS <--> LIQUID TRANSITION		
1050.000	62.551	169.318	121.720	49.978			
1100	62.551	172.228	123.950	53.106	-124.372	-83.648	3.972
1200	62.551	177.670	128.203	59.361	-125.094	-79.933	3.479
1300	62.551	182.677	132.203	65.616	-124.185	-76.207	3.062
1400	62.551	187.313	135.976	71.871	-123.375	-72.548	2.707
1463.000	62.551	190.066	138.247	75.812	GAMMA <--> LIQUID		
1500	62.551	191.628	139.544	78.126	-122.664	-68.943	2.401
1600	62.551	195.665	142.927	84.381	-122.052	-65.382	2.134
1700	62.551	199.457	146.142	90.636	-122.487	-61.837	1.900
1800	62.551	203.033	149.204	96.891	-122.405	-58.273	1.691
1900	62.551	206.415	152.127	103.146	-136.518	-54.008	1.485
2000	62.551	209.623	154.922	109.401	-136.866	-49.656	1.297
2100	62.551	212.675	157.600	115.656	-137.226	-45.287	1.126
2200	62.551	215.585	160.170	121.911	-137.596	-40.900	0.971
2300	62.551	218.365	162.641	128.167	-137.975	-36.497	0.829
2400	62.551	221.027	165.018	134.422	-138.363	-32.076	0.698
2500	62.551	223.581	167.310	140.677	-138.759	-27.639	0.577
2600	62.551	226.034	169.522	146.932	-139.163	-23.187	0.466
2700	62.551	228.395	171.659	153.187	-139.572	-18.718	0.362
2800	62.551	230.670	173.726	159.442	-139.987	-14.235	0.266
2900	62.551	232.865	175.728	165.697	-140.408	-9.736	0.175
3000	62.551	234.985	177.668	171.952	-140.834	-5.223	0.091
3100	62.551	237.036	179.550	178.207	-141.265	-0.695	0.012
3200	62.551	239.022	181.378	184.462	-490.000	11.270	-0.184
3300	62.551	240.947	183.154	190.717	-488.538	26.912	-0.426
3400	62.551	242.814	184.881	196.972	-487.114	42.510	-0.653
3500	62.551	244.627	186.562	203.228	-485.727	58.067	-0.867
3600	62.551	246.389	188.200	209.483	-484.378	73.585	-1.068
3700	62.551	248.103	189.796	215.738	-483.069	89.066	-1.257
3800	62.551	249.771	191.352	221.993	-481.799	104.512	-1.437

PREVIOUS: CURRENT: September 1977

Iron Sulfide (FeS) $Fe_1S_1(l)$

IRON SULFIDE, TROILITE (FeS) ALPHA-BETA-GAMMA-LIQUID M_r = 87.907

0 to 41 K	crystal, alpha
411 to 598 K	crystal, beta
598 to 1463 K	crystal, gamma
above 1463 K	liquid

Refer to the individual tables for details.

Iron Sulfide, Troilite (FeS) $Fe_1S_1(cr,l)$

Enthalpy Reference Temperature = T_r = 298.15 K Standard State Pressure = p° = 0.1 MPa

T/K	C_p° (J K⁻¹mol⁻¹)	S° (J K⁻¹mol⁻¹)	$-[G^\circ-H^\circ(T_r)]/T$ (J K⁻¹mol⁻¹)	$H^\circ-H^\circ(T_r)$ (kJ mol⁻¹)	Δ_fH° (kJ mol⁻¹)	Δ_fG° (kJ mol⁻¹)	Log K_f
0	0.	0.	INFINITE	-9.413	-102.166	-102.166	INFINITE
100	26.670	17.216	99.528	-8.231	-102.096	-101.959	53.258
200	43.052	41.665	64.714	-4.610	-101.886	-101.901	26.614
298.15	50.517	60.321	60.321	0.	-101.671	-101.953	17.862
300	50.671	60.634	60.322	0.094	-101.666	-101.954	17.752
400	89.203	79.060	62.567	6.597	-102.388	-102.086	13.331
411.000	95.051	81.559	63.042	7.610	ALPHA <--> BETA		
411.000	72.358	85.611	63.042	9.276	TRANSITION		
500	72.358	99.794	68.363	15.715	-100.050	-102.492	10.707
598.000	72.358	112.745	74.607	22.807	BETA <--> GAMMA		
598.000	62.085	113.409	74.607	23.204	TRANSITION		
600	62.030	113.617	74.736	23.328	-99.108	-103.033	8.970
700	59.808	122.999	80.982	29.412	-99.697	-103.645	7.734
800	58.560	130.893	86.740	35.322	-100.612	-104.151	6.800
900	58.286	137.765	92.036	41.156	-155.064	-103.437	6.003
1000	58.986	143.934	96.922	47.012	-155.812	-97.670	5.102
1100	60.662	149.627	101.457	52.986	-157.351	-91.766	4.358
1200	63.311	155.011	105.697	59.177	-158.138	-85.786	3.734
1300	66.935	160.216	109.692	65.681	-156.980	-79.802	3.206
1400	71.533	165.338	113.484	72.596	-155.509	-73.918	2.758
1463.000	74.930	168.560	115.786	77.208	GAMMA <--> LIQUID		
1463.000	62.551	190.066	115.786	108.671	TRANSITION		
1500	62.551	191.628	117.638	110.986	-122.664	-68.943	2.401
1600	62.551	195.665	122.390	117.241	-122.052	-65.382	2.134
1700	62.551	199.457	126.813	123.496	-122.487	-61.837	1.900
1800	62.551	203.033	130.949	129.751	-122.405	-58.273	1.691
1900	62.551	206.415	134.832	136.006	-136.518	-54.008	1.485
2000	62.551	209.623	138.492	142.261	-136.866	-49.656	1.297
2100	62.551	212.675	141.953	148.516	-137.226	-45.287	1.126
2200	62.551	215.585	145.234	154.771	-137.596	-40.900	0.971
2300	62.551	218.365	148.354	161.026	-137.975	-36.496	0.829
2400	62.551	221.027	151.327	167.281	-138.363	-32.076	0.698
2500	62.551	223.581	154.166	173.537	-138.759	-27.639	0.577
2600	62.551	226.034	156.883	179.792	-139.163	-23.187	0.466
2700	62.551	228.395	159.489	186.047	-139.572	-18.718	0.362
2800	62.551	230.670	161.990	192.302	-139.987	-14.235	0.266
2900	62.551	232.865	164.397	198.557	-140.408	-9.736	0.175
3000	62.551	234.985	166.714	204.812	-140.834	-5.223	0.091
3100	62.551	237.036	168.950	211.067	-141.265	-0.695	0.012
3200	62.551	239.022	171.109	217.322	-490.000	11.270	-0.184
3300	62.551	240.947	173.196	223.577	-488.538	26.912	-0.426
3400	62.551	242.814	175.216	229.832	-487.114	42.510	-0.653
3500	62.551	244.627	177.174	236.087	-485.727	58.067	-0.867
3600	62.551	246.389	179.072	242.342	-484.378	73.585	-1.068
3700	62.551	248.103	180.915	248.597	-483.069	89.066	-1.257
3800	62.551	249.771	182.705	254.853	-481.799	104.512	-1.437

PREVIOUS: CURRENT: September 1977

Iron Sulfide, Troilite (FeS) $Fe_1S_1(cr,l)$

IRON SULFIDE (FeS) IDEAL GAS M_r = 87.907

$\Delta_f H°(0\ K)$ = 370.49 ± 16.3 kJ mol^{-1}

$S°(298.15\ K)$ = 252.344 ± 12.6 J K^{-1} mol^{-1} $\Delta_f H°(298.15\ K)$ = 370.77 ± 16.3 kJ mol^{-1}

Electronic Levels and Quantum Weights

ε_i, cm^{-1}	g_i	ε_i, cm^{-1}	g_i	ε_i, cm^{-1}	g_i
[0]	[9]	[20481.9]	[9]	[30716.2]	[5]
[436.2]	[7]	[21462.2]	[9]	[30857.8]	[7]
[738.9]	[5]	[21699.9]	[7]	[30886.4]	[9]
[932.4]	[3]	[21857.2]	[5]	[34812.4]	[1]
[1027.3]	[1]	[24558.8]	[11]	[35803.7]	[5]
[19404.8]	[5]	[24940.9]	[9]	[40999.9]	[5]
[20688.4]	[3]	[25142.4]	[7]	[42896.9]	[7]
[21208.5]	[1]	[30088.8]	[7]	[49148.0]	[1]
[20051.1]	[13]	[30356.2]	[13]	[49576.9]	[3]
[20300.8]	[11]	[30725.8]	[3]	[50412.3]	[5]

ω_e = 550 cm^{-1} $\omega_e x_e$ = [2.8] cm^{-1} σ = 1

B_e = [0.198] cm^{-1} α_e = [0.0012] cm^{-1} r_e = [2.05] Å

Enthalpy of Formation

Drowart et al. (1) used a modified Knudsen cell and a mass spectrometric technique to study the equilibrium FeS(g) + Mn(g) = MnS(g) + Fe(g). We adopt a value of D_0^0 = 76.3±3.5 kcal mol^{-1} based on their 3rd law analysis of this equilibrium. With auxiliary JANAF data (2) this yields $\Delta_f H°(298.15\ K)$ = 88.6±3.9 kcal mol^{-1}. Trevedi (3) studied the absorption spectrum of FeS and determined D_0^0 = 93 kcal mol^{-1} from the onset of continuous absorption due to photodissociation. This value is likely to be too high in view of the experimental difficulties in determining the exact position of the onset of continuous absorption. Complications arise because of background emission from the furnace. Marquardt and Berkowitz (4) determined an upper limit D_0^0 < 77 kcal mol^{-1} by mass-spectrometric meaurements. These results may be biased since they used ν = 412 cm^{-1} for the vibrational frequency of FeS in determining their Gibbs energy functions. A Birge-Sponer extrapolation of the vibrational data to determine D_0^0 is not meaningful since DeVore and Franzen (5) used the reverse procedure to arrive at the reported value of $\omega_e x_e$.

Heat Capacity and Entropy

With the exception of the observance of one excited state at 17992 cm^{-1} with a vibrational spacing of ~497 cm^{-1} by DeVore and Franzen (5), there is no information on the electronic spectrum of FeS(g). The identity and quantum weight of this state, as well as the ground state, are unknown. As a result, the electronic states are estimated to be identical to those of the Fe^{2+} ion (6) following the procedure of Brewer and Rosenblatt (7) for the transition metal oxides. As stated by Brewer and Rosenblatt (7), this procedure usually overestimates the contribution of electronic states. Nevertheless; it appears to be the best method available. The uncertainty assigned to S°(298.15 K) is based mainly on the uncertainties in this approximation since the presence of the sulfide ion will undoubtedly alter the electronic structure of the iron ion. Levels above 50,000 cm^{-1} are not included since they have a negligible effect on the thermodynamic properties. The vibrational constants are taken from a study of the matrix isolation spectrum by DeVore and Franzen (5). The rotational constants are estimated based on an oxide-sulfide correlation due to Barrow and Cousins (8), r(FeS) = 0.237 ± 1.116 r(FeO). The value r(FeO) = 1.626 Å is taken from Suchard (9). The value of α_e is estimated assuming a Morse potential function using the expression given by Herzberg (10).

References

1. J. Drowart, A. Pattoret, and S. Smoes, Proc. Brit. Ceram. Soc. 8, 67 (1967).
2. JANAF Thermochemical Tables: Fe(g), 3-31-65; S(cr), 12-31-65; S(g), 6-30-71.
3. H. Trevedi, Proc. Acad. Sci. United Provinces Agra Oudh, India 5, 34 (1935).
4. J. R. Marquardt and J. Berkowitz, J. Chem. Phys. 39, 283 (1963).
5. T. C. DeVore and H. F. Franzen, High Temp. Sci. 7, 220 (1975).
6. C. E. Moore, U. S. Nat. Bur. Stand. Circ. 467, Vol. 2, 1952.
7. L. Brewer and G. M. Rosenblatt, "Advances in High Temperature Chemistry," L. Eyring, ed., Vol. 2, Academic Press, New York, 1969.
8. R. F. Barrow and C. Cousins, "Advances in High Temperature Chemistry," L. Eyring, ed., Vol. 4, Academic Press, New York, 1971.
9. S. N. Suchard, "Spectroscopic Data," Vol. 1, Plenum Press, New York, 1975.
10. G. Herzberg, "Spectra of Diatomic Molecules," Second Edition, Van Nostrand, New York, p. 108, 1950.

Iron Sulfide (FeS) $Fe_1S_1(g)$

Enthalpy Reference Temperature = T_r = 298.15 K Standard State Pressure = p° = 0.1 MPa

T/K	$C_p°$	S°	$-[G°-H°(T_r)]/T$	$H°-H°(T_r)$	$\Delta_f H°$	$\Delta_f G°$	Log K_f
	J $K^{-1}mol^{-1}$			kJ mol^{-1}			
0	0.	0.	INFINITE	-9.191	370.494	370.494	INFINITE
100	29.310	218.223	281.010	-6.279	372.294	352.331	-184.039
200	31.780	239.206	255.391	-3.237	371.925	332.401	-86.814
250	33.047	246.438	252.899	-1.615	371.390	322.579	-67.399
298.15	34.002	252.344	252.344	0.	370.767	313.234	-54.877
300	34.033	252.555	252.345	0.063	370.741	312.877	-54.477
350	34.775	257.860	252.762	1.784	370.010	303.289	-45.263
400	35.332	262.541	253.698	3.537	366.991	293.900	-38.379
450	35.756	266.728	254.917	5.315	365.446	284.847	-33.064
500	36.085	270.513	256.291	7.111	363.784	275.982	-28.832
600	36.554	277.137	259.229	10.745	360.747	258.710	-22.523
700	36.870	282.797	262.201	14.417	357.746	241.940	-18.054
800	37.095	287.736	265.091	18.116	354.619	225.607	-14.731
900	37.266	292.115	267.855	21.834	298.052	210.764	-12.232
1000	37.402	296.049	270.481	25.568	295.183	201.209	-10.510
1100	37.516	299.619	272.970	29.314	291.415	192.008	-9.118
1200	37.615	302.888	275.329	33.071	288.194	183.094	-7.970
1300	37.707	305.902	277.566	36.837	286.614	174.399	-7.007
1400	37.796	308.700	279.691	40.612	284.944	165.830	-6.187
1500	37.886	311.310	281.713	44.396	283.184	157.382	-5.481
1600	37.981	313.759	283.640	48.189	281.335	149.055	-4.866
1700	38.083	316.064	285.480	51.992	278.447	140.865	-4.328
1800	38.194	318.244	287.241	55.806	276.088	132.840	-3.855
1900	38.315	320.312	288.927	59.632	259.546	125.650	-3.454
2000	38.448	322.281	290.546	63.470	256.780	118.674	-3.099
2100	38.590	324.160	292.102	67.322	254.017	111.837	-2.782
2200	38.744	325.959	293.601	71.188	251.259	105.131	-2.496
2300	38.908	327.685	295.045	75.071	248.507	98.551	-2.238
2400	39.081	329.344	296.440	78.970	245.763	92.090	-2.004
2500	39.262	330.943	297.788	82.887	243.029	85.743	-1.791
2600	39.451	332.487	299.093	86.823	240.307	79.505	-1.597
2700	39.645	333.979	300.358	90.778	237.597	73.372	-1.419
2800	39.844	335.425	301.585	94.752	234.901	67.339	-1.256
2900	40.046	336.826	302.776	98.746	232.220	61.402	-1.106
3000	40.251	338.188	303.934	102.761	229.553	55.557	-0.967
3100	40.456	339.511	305.060	106.797	226.903	49.801	-0.839
3200	40.662	340.798	306.157	110.852	-124.031	51.554	-0.842
3300	40.866	342.053	307.226	114.929	-124.748	57.052	-0.903
3400	41.069	343.276	308.268	119.026	-125.482	62.572	-0.961
3500	41.270	344.469	309.286	123.143	-126.234	68.114	-1.017
3600	41.467	345.634	310.279	127.279	-127.003	73.678	-1.069
3700	41.659	346.773	311.250	131.436	-127.792	79.263	-1.119
3800	41.848	347.887	312.200	135.611	-128.602	84.870	-1.167
3900	42.031	348.976	313.129	139.805	-129.434	90.499	-1.212
4000	42.209	350.043	314.038	144.017	-130.290	96.149	-1.256
4100	42.382	351.087	314.929	148.247	-131.171	101.820	-1.297
4200	42.548	352.110	315.802	152.494	-132.080	107.514	-1.337
4300	42.709	353.113	316.658	156.756	-133.018	113.230	-1.375
4400	42.863	354.097	317.498	161.035	-133.987	118.968	-1.412
4500	43.011	355.062	318.322	165.329	-134.989	124.728	-1.448
4600	43.153	356.009	319.131	169.637	-136.025	130.511	-1.482
4700	43.288	356.938	319.926	173.959	-137.099	136.316	-1.515
4800	43.417	357.851	320.706	178.294	-138.211	142.145	-1.547
4900	43.540	358.748	321.474	182.642	-139.363	147.998	-1.578
5000	43.656	359.628	322.228	187.002	-140.559	153.875	-1.608
5100	43.767	360.494	322.970	191.373	-141.799	159.776	-1.636
5200	43.872	361.345	323.700	195.755	-143.087	165.701	-1.664
5300	43.971	362.182	324.418	200.148	-144.423	171.652	-1.692
5400	44.065	363.004	325.125	204.549	-145.811	177.629	-1.718
5500	44.153	363.814	325.821	208.960	-147.253	183.631	-1.744
5600	44.236	364.610	326.506	213.380	-148.751	189.661	-1.769
5700	44.314	365.394	327.182	217.807	-150.309	195.718	-1.794
5800	44.387	366.165	327.847	222.243	-151.928	201.803	-1.817
5900	44.456	366.924	328.503	226.685	-153.612	207.916	-1.841
6000	44.521	367.672	329.150	231.134	-155.364	214.058	-1.864

PREVIOUS: September 1977 (1 atm) CURRENT: September 1977 (1 bar)

Iron Sulfide (FeS) $Fe_1S_1(g)$

IRON SULFIDE, MARCASITE (FeS_2) CRYSTAL M_r = 119.967

$S°(298.15 K) = 53.865 \pm 0.13\ J\ K^{-1}\ mol^{-1}$

$\Delta_f H°(0 K) = -163.78 \pm 2.1\ kJ\ mol^{-1}$

$\Delta_f H°(298.15 K) = -167.36 \pm 2.1\ kJ\ mol^{-1}$

Enthalpy of Formation

Based on highly accurate adiabatic-shield calorimetry measurements, Gronvold and Westrum (1) reported that the enthalpy of transformation of marcasite to pyrite is -1.05±0.05 kcal mol^{-1} at 700 K. The adopted value of $\Delta_f H°$(298.15 K) is selected to reproduce this enthalpy of reaction within the reported uncertainty. Lipin et al. (2), based on combustion calorimetry, reported a value of -5.6 kcal mol^{-1} for the marcasite-pyrite transformation at 298.15 K. Due to the state of the art in combustion calorimetry at the time of this measurement and uncertainty in the products (oxides of sulfur), this value must have a high uncertainty and is given no weight in our selection process.

Heat Capacity and Entropy

The adopted heat capacities of marcasite are based on our analysis of the $C_p°$ measurements (6 - 700 K) of Gronvold and Westrum (1). Values above 700 K are extrapolated assuming $C_p°$(marcasite) + 0.075 cal $K^{-1}\ mol^{-1}$.

The adopted value of S°(298.15 K) is obtained from the appropriate integration of the adopted $C_p°$ data assuming S°(6.06 K) = 0.0008 cal $K^{-1}\ mol^{-1}$.

Phase Data

Marcasite has an orthorhombic structure, V_h^{12} = Pnnm, and exists as stoichiometric FeS_2 within the limits of experimental error (3). Marcasite is known to be metastable with respect to the other known FeS_2 polymorph, cubic pyrite, above 673 K (3); the JANAF Thermochemical Tables (4) indicate it will also be metastable below 673 K. The marcasite to pyrite transformation is kinetically slow at low temperatures and this accounts for the persistence of marcasite in nature. Kjekshus and Rakke (5) found no conversion to pyrite using heat treatments at 573 K for periods up to 14 months. At 673 K the conversion to pyrite is complete in 4 months and is even more rapid (< 1 hour) at 973 K (5). Indications are that the conversion to pyrite will be complete before the pyrite decomposition temperature of 1016 K is reached (3).

References

1. F. Gronvold and E. F. Westrum, J. Chem. Thermodyn. 8, 1039 (1976).
2. S. V. Lipin, V. S. Uskov and V. R. Klokman, Zhur. Prikl. Khim. 15, 411 (1942).
3. G. Kullerud and H. S. Yoder, Econ. Geol. 54, 533 (1959).
4. JANAF Thermochemical Tables: FeS_2(cr, pyrite) 9-30-77.
5. A. Kjekshus and T. Rakke, Acta Chem. Scand. Ser. A A29, 443 (1975).

Iron Sulfide, Marcasite (FeS_2) Fe_1S_2(cr)

Enthalpy Reference Temperature = T_r = 298.15 K Standard State Pressure = p° = 0.1 MPa

	J $K^{-1}mol^{-1}$			kJ mol^{-1}			
T/K	$C_p°$	S°	$-[G°-H°(T_r)]/T$	$H°-H°(T_r)$	$\Delta_f H°$	$\Delta_f G°$	Log K_f
0	0.	0.	INFINITE	-9.753	-163.782	-163.782	INFINITE
100	19.269	7.576	99.375	-9.180	-165.012	-162.658	84.964
200	49.399	31.380	59.227	-5.569	-166.454	-159.685	41.705
298.15	62.390	53.865	53.865	0.	-167.360	-156.159	27.358
300	62.554	54.251	53.866	0.116	-167.375	-156.090	27.178
400	69.169	73.244	56.411	6.733	-172.579	-152.034	19.854
500	72.447	89.075	61.407	13.834	-176.187	-146.504	15.305
600	74.568	102.472	67.163	21.185	-179.092	-140.283	12.213
700	76.959	114.139	73.058	28.757	-181.539	-133.617	9.971
800	78.668	124.569	78.857	36.569	-183.764	-126.619	8.267
900	80.718	133.953	84.466	44.538	-292.337	-117.177	6.801
1000	82.764	142.563	89.851	52.712	-292.623	-97.707	5.104
1100	84.810	150.547	95.010	61.091	-293.630	-78.151	3.711
1200	86.860	158.014	99.952	69.674	-293.910	-58.568	2.549
1300	88.910	165.048	104.692	78.463	-292.370	-39.018	1.568
1400	90.960	171.712	109.243	87.456	-290.740	-19.589	0.731

PREVIOUS: CURRENT: September 1977

Iron Sulfide, Marcasite (FeS_2) Fe_1S_2(cr)

$S°(298.15 K) = 52.916 \pm 0.13 J K^{-1} mol^{-1}$

$\Delta_fH°(0 K) = -167.86 \pm 2.1 kJ mol^{-1}$
$\Delta_fH°(298.15 K) = -171.54 \pm 2.1 kJ mol^{-1}$

Enthalpy of Formation

The adopted value of $\Delta_fH°(298.15 K) = -41.0\pm0.5 kcal mol^{-1}$ is based on our 3rd law analysis of the equilibrium data of Toulmin and Barton (1). These authors made corrections for the changing stoichiometry of pyrrhotite as a function of temperature in the pyrrhotite-pyrite equilibrium, and also accounted for the changing activity of pyrrhotite as a function of temperature and stoichiometry. These corrections have not been applied to our analysis of the rest of the equilibrium data shown below.

Source	Method	Reaction*	T/K	Data Points	$\Delta_rH°(298.15 K)$, kcal mol^{-1} 2nd law	3rd law	Drift cal $K^{-1}mol^{-1}$	$-\Delta_fH°(298.15 K)$ kcal mol^{-1}
Toulmin (1)	Electrum-tarnish	A	598 - 1016	17†	-71.77±0.12	-71.89±0.16	-0.15±0.15	41.0±0.5
DeRudder (2)	Manometric	B	873 - 962	7	-32.47±1.27	-27.74±0.43	5.17±1.38	43.2±0.6
Ugai (3)	Gravimetric	B	923 - 1016	equation	-16.70	-26.81±0.93	-10.28	36.0
DeRanter (4)	emf	C	298.15	1		-36.00		42.3±1.0
Schneeberg (5)	emf	B	597 - 711	4	-28.98±1.10	-28.16±0.26	1.26±1.68	43.6±0.5
Rosenqvist (6)	Dew point	B	869 - 1030	22	-34.03±1.22	-28.04±0.63	6.35±1.29	43.5±0.7

†Point at 1009 K rejected by statistical test.

*Reaction: (A) $Fe(cr) + S_2(g) \rightarrow FeS_2(cr, pyrite)$
(B) $1.14\ Fe_{0.877}S(cr) + 0.43\ S_2(g) \rightarrow FeS_2(cr, pyrite)$
(C) $Fe(cr) + 2\ S(cr) \rightarrow FeS_2(cr, pyrite)$

The value reported by DeRanter (4) is biased since FeS_2 is a semiconductor and DeRanter acknowledged that semiconductors lead to incorrect results in his emf technique. The other values are believed to be biased because the equilibrium is treated as if the composition ($Fe_{0.877}S$) and activity (a=1) of pyrrhotite remain unchanged.

Heat Capacity and Entropy

The adopted values of $C_p°$ are based on our analysis of the calorimetrically measured values of Gronvold and Westrum, 4 - 346 K (7) and 314 - 688 K (8). At higher temperatures the $C_p°$ values are based on our analysis of the enthalpy data of Coughlin (9), 405 - 980 K, which merges smoothly with the $C_p°$ data of Gronvold and Westrum. Values above 980 K are extrapolated.

The adopted value of S°(298.15 K) is obtained from the appropriate integration of the adopted $C_p°$ data assuming $S°(4.6 K) = 0.0007 cal K^{-1} mol^{-1}$.

Phase Data

Pyrite has a cubic structure, T_h^6 = Pa3, and exits as stoichiometric FeS_2 within the limits of experimental error (10). The other known FeS_2 polymorph is orthorhombic marcasite which is known to be metastable with respect to pyrite above 673 K (10); the JANAF Thermochemical Tables (11) indicate that pyrite will also be the stable form below 673 K.

Decomposition Data

Pyrite melts incongruently to pyrrhotite, a liquid whose composition is $\sim FeS_4$ (3), and vapor at 1016Å2 K at the vapor pressure of the system, ~10 atm (3, 10). Assuming the vapor to be composed only of $S_2(g)$, we calculate (11) a decomposition temperature of 906 K at 1 atm for the process $2.3263\ FeS_2(pyrite) \rightarrow S_2(g) + 2.6525\ Fe_{0.87}S(cr)$.

References

1. P. Toulmin and P. B. Barton, Geochim. Cosmochim. Acta 28, 641 (1964).
2. F. DeRudder, Bull. Soc. Chim. Fr. 47, 1225 (1930).
3. Y. A. Ugai, V. Z. Anokhin, A. N. Prokin and V. R. Pshestanchik, Izv. Akad. Nauk SSSR, Neorg. Mater. 10, 250 (1974).
4. C. DeRanter and R. Breckpot, Bull. Soc. Chim. Belg. 78, 503 (1969).
5. E. P. Schneeberg, Econ. Geol. 68, 507 (1973).
6. T. Rosenqvist, J. Iron Steel Inst. (London) 176, 37 (1957).
7. F. Gronvold and E. F. Westrum, Inorg. Chem. 1, 36 (1962).
8. F. Gronvold and E. F. Westrum, J. Chem. Thermodyn. 8, 1039 (1976).
9. J. P. Coughlin, J. Amer. Chem. Soc. 72, 5445 (1950).
10. G. Kullerud and H. S. Yoder, Econ. Geol. 54, 533 (1959).
11. JANAF Thermochemical Tables: FeS_2(cr, marcasite), 9-30-77; $Fe_{0.877}S$(cr), 9-30-77.

Enthalpy Reference Temperature = T_r = 298.15 K Standard State Pressure = p° = 0.1 MPa

T/K	$C_p°$ (J $K^{-1}mol^{-1}$)	S°	$-[G°-H°(T_r)]/T$	$H°-H°(T_r)$ (kJ mol^{-1})	$\Delta_fH°$	$\Delta_fG°$	Log K_f
0	0.	0.	INFINITE	-9.641	-167.855	-167.855	INFINITE
100	18.712	7.091	98.120	-9.103	-169.119	-166.717	87.084
200	49.056	30.556	58.251	-5.539	-170.607	-163.674	42.747
298.15	62.124	52.916	52.916	0.	-171.544	-160.060	28.042
300	62.289	53.301	52.918	0.115	-171.559	-159.989	27.857
400	68.848	72.213	55.451	6.705	-176.792	-155.835	20.350
500	72.040	87.953	60.425	13.764	-180.441	-150.197	15.691
600	74.307	101.295	66.152	21.086	-183.375	-143.860	12.524
700	76.308	112.899	72.019	28.616	-185.865	-137.074	10.229
800	78.354	123.222	77.786	36.349	-188.169	-129.946	8.485
900	80.404	132.569	83.362	44.287	-296.773	-120.367	6.986
1000	82.450	141.146	88.717	52.429	-297.090	-100.757	5.263
1100	84.496	149.101	93.849	60.777	-298.128	-81.058	3.849
1200	86.546	156.541	98.767	69.329	-298.440	-61.329	2.670
1300	88.592	163.549	103.483	78.086	-296.931	-41.630	1.673
1400	90.642	170.189	108.012	87.047	-295.333	-22.051	0.823

PREVIOUS: CURRENT: September 1977

IRON IODIDE $[(FeI_2)_2]$ IDEAL GAS M_r = 619.3120

$\Delta_fH°(0\ K) = 9.75 \pm 20.9\ kJ\ mol^{-1}$

$S°(298.15\ K) = [543.471]\ J\ K^{-1}\ mol^{-1}$ $\Delta_fH°(298.15\ K) = 8.37 \pm 20.9\ kJ\ mol^{-1}$

Electronic Levels and Quantum Weights

ε_i, cm^{-1}	g_i
0	[5]
[4200]	[5]
[6800]	[5]

Vibrational Frequencies and Degeneracies

ν, cm^{-1}	ν, cm^{-1}	ν, cm^{-1}
[100](1)	[70](1)	[80](2)
[50](1)	[90](1)	[20](2)
[95](1)	[105](1)	[160](2)

Point Group: $[D_{2h}]$ σ = 4

Bond Distance: Fe-I = [2.43] Å

Bond Angles: $I\text{-}Fe\text{-}I_{bridge}$ = [135]° $Fe\text{-}I_{bridge}\text{-}Fe$ = [90]° $I_{bridge}\text{-}Fe\text{-}I_{bridge}$ = [90]°

Product of the Moments of Inertia: $I_AI_BI_C = [8.778988 \times 10^{-110}]\ g^3\ cm^6$

Enthalpy of Formation

The chemical equilibrium of the interaction of Fe(cr) and I(g) has been studied by a transpiration method by Zaugg and Gregory (1). Using the reported equilibrium constants in the temperature range from 591 to 751 K, the enthalpy change $\Delta_rH°$(298.15 K) of the reaction $2Fe(cr) + 4I(g) = Fe_2I_4(g)$ was evaluated by both the 2nd and 3rd law method to be -85.3 and -103.75 kcal mol^{-1}, respectively. The corresponding values of the enthalpy of formation, $\Delta_fH°(Fe_2I_4$, g, 298.15 K), are calculated as +16.9 and -1.6 kcal mol^{-1}. Mass spectrometric and Knudsen effusion techniques have been used to study the vaporization of $FeI_2(cr)$ by Schoonmaker et al. (2). Using the reported vapor pressure, $p = 1.21 \times 10^{-6}$ atm at 714 K for $Fe_2I_4(g)$, the enthalpy change $\Delta_rH°$(298.15 K) for the reaction $2FeI_2(cr) = Fe_2I_4(g)$ was evaluated by the 3rd law method as 52.8 kcal mol^{-1}, yielding $\Delta_fH°$(298.15 K) = 2.8 kcal mol^{-1} for $Fe_2I_4(g)$.

The adopted value of $\Delta_fH°$(298.15 K) for $Fe_2I_4(g)$ is 2.0±5 kcal mol^{-1} (8.368±20.9 kJ mol^{-1}).

Heat Capacity and Entropy

The molecular structure was assumed to be the same as that for $Fe_2Br_4(g)$. The Fe-I bond distance was estimated from that in $FeI_2(g)$ molecule. Four vibrational frequencies (ν_i), i.e. i = 2, 4, 5 and 6, were estimated by comparison with those for $K_2I_2(g)$, calculated by Berkowitz (3). The other frequency values were estimated from those for $FeI_2(g)$. The electronic levels and quantum weights were estimated from those for $FeI_2(g)$. The principal moments of inertia are: $I_A = 124.4358 \times 10^{-39}$, $I_B = 780.0258 \times 10^{-39}$ and $I_C = 904.4616 \times 10^{-39}\ g\ cm^2$.

References

1. W. E. Zaugg and N. W. Gregory, J. Phys. Chem. 70, 490 (1966).
2. R. C. Schoonmaker, A. H. Friedman and R. F. Porter, J. Chem. Phys. 31, 1586 (1959).
3. J. Berkowitz, J. Chem. Phys. 32, 1519 (1960).

Iron Iodide $((FeI_2)_2)$ $Fe_2I_4(g)$

Enthalpy Reference Temperature = T_r = 298.15 K Standard State Pressure = p° = 0.1 MPa

T/K	$C°_p$ (J $K^{-1}mol^{-1}$)	S° (J $K^{-1}mol^{-1}$)	$-[G°-H°(T_r)]/T$ (J $K^{-1}mol^{-1}$)	$H°-H°(T_r)$ (kJ mol^{-1})	$\Delta_fH°$ (kJ mol^{-1})	$\Delta_fG°$ (kJ mol^{-1})	Log K_f
0	0.	0.	INFINITE	-34.030	9.747	9.747	INFINITE
100	119.470	404.580	658.433	-25.385	11.447	-15.600	8.149
200	129.233	491.411	555.431	-12.804	10.625	-42.494	11.098
250	130.564	520.408	545.629	-6.305	9.593	-55.661	11.630
298.15	131.283	543.471	543.471	0.	8.368	-68.121	11.935
300	131.304	544.284	543.474	0.243	8.316	-68.596	11.944
350	131.756	564.561	545.074	6.820	6.698	-81.293	12.132
400	132.052	582.175	548.635	13.416	-27.043	-92.647	12.098
450	132.258	597.741	553.243	20.024	-31.296	-100.593	11.676
500	132.409	611.684	558.402	26.641	-115.916	-100.461	10.495
600	132.629	635.845	569.357	39.893	-116.344	-97.338	8.474
700	132.820	656.305	580.354	53.166	-117.258	-94.105	7.022
800	133.038	674.054	590.981	66.458	-118.753	-90.704	5.922
900	133.308	689.739	601.100	79.775	-121.074	-87.069	5.053
1000	133.635	703.800	610.679	93.122	-124.826	-83.108	4.341
1100	134.013	716.555	619.733	106.504	-130.359	-78.640	3.734
1200	134.430	728.233	628.295	119.926	-134.783	-73.781	3.212
1300	134.872	739.010	636.402	133.391	-135.917	-68.653	2.759
1400	135.325	749.022	644.093	146.900	-137.235	-63.431	2.367
1500	135.774	758.374	651.404	160.455	-138.756	-58.107	2.023
1600	136.209	767.150	658.366	174.055	-140.500	-52.675	1.720
1700	136.620	775.421	665.011	187.696	-144.386	-47.089	1.447
1800	137.001	783.240	671.364	201.378	-147.300	-41.283	1.198
1900	137.347	790.657	677.449	215.095	-178.679	-33.904	0.932
2000	137.655	797.710	683.287	228.846	-182.614	-26.182	0.684
2100	137.924	804.433	688.897	242.625	-186.656	-18.262	0.454
2200	138.153	810.855	694.296	256.429	-190.797	-10.147	0.241
2300	138.345	817.000	699.498	270.254	-195.027	-1.841	0.042
2400	138.500	822.891	704.518	284.097	-199.330	6.651	-0.145
2500	138.621	828.548	709.367	297.953	-203.686	15.322	-0.320
2600	138.710	833.987	714.056	311.820	-208.077	24.170	-0.486
2700	138.771	839.223	718.595	325.694	-212.481	33.186	-0.642
2800	138.806	844.270	722.994	339.573	-216.879	42.365	-0.790
2900	138.817	849.141	727.260	353.455	-221.251	51.701	-0.931
3000	138.808	853.847	731.402	367.336	-225.581	61.187	-1.065
3100	138.782	858.398	735.426	381.216	-229.853	70.817	-1.193
3200	138.740	862.804	739.338	395.092	-930.654	95.431	-1.558
3300	138.685	867.072	743.144	408.963	-930.973	127.501	-2.018
3400	138.618	871.212	746.850	422.829	-931.267	159.580	-2.452
3500	138.543	875.229	750.461	436.687	-931.529	191.668	-2.860
3600	138.459	879.130	753.981	450.537	-931.758	223.762	-3.247
3700	138.369	882.923	757.415	464.378	-931.950	255.863	-3.612
3800	138.275	886.612	760.767	478.211	-932.106	287.969	-3.958
3900	138.176	890.202	764.040	492.033	-932.225	320.078	-4.287
4000	138.074	893.699	767.238	505.846	-932.311	352.189	-4.599
4100	137.970	897.107	770.364	519.648	-932.365	384.302	-4.896
4200	137.865	900.431	773.421	533.440	-932.392	416.416	-5.179
4300	137.760	903.674	776.413	547.221	-932.396	448.531	-5.449
4400	137.654	906.839	779.341	560.992	-932.380	480.645	-5.706
4500	137.549	909.932	782.209	574.752	-932.351	512.759	-5.952
4600	137.444	912.954	785.019	588.501	-932.312	544.872	-6.187
4700	137.340	915.909	787.772	602.241	-932.270	576.984	-6.412
4800	137.238	918.799	790.472	615.969	-932.230	609.096	-6.628
4900	137.137	921.628	793.120	629.688	-932.197	641.207	-6.835
5000	137.038	924.397	795.718	643.397	-932.178	673.316	-7.034
5100	136.941	927.110	798.268	657.096	-932.178	705.427	-7.225
5200	136.845	929.768	800.771	670.785	-932.204	737.537	-7.409
5300	136.752	932.374	803.230	684.465	-932.260	769.647	-7.585
5400	136.661	934.929	805.645	698.136	-932.354	801.760	-7.755
5500	136.572	937.436	808.018	711.797	-932.492	833.874	-7.919
5600	136.486	939.896	810.351	725.450	-932.681	865.992	-8.078
5700	136.401	942.311	812.645	739.094	-932.927	898.113	-8.230
5800	136.319	944.683	814.902	752.730	-933.238	930.239	-8.378
5900	136.239	947.012	817.121	766.358	-933.621	962.371	-8.520
6000	136.161	949.301	819.305	779.978	-934.083	994.510	-8.658

PREVIOUS: September 1966 (1 atm) CURRENT: September 1966 (1 bar)

IRON OXIDE, HEMATITE (Fe_2O_3) CRYSTAL M_r = 159.6922

S°(298.15 K) = 87.400 J K^{-1} mol^{-1}

$\Delta_f H°$(0 K) = -819.03 ± 1.3 kJ mol^{-1}
$\Delta_f H°$(298.15 K) = -825.50 ± 1.3 kJ mol^{-1}

Enthalpy of Formation

The chemical equilibria for the reaction 4 Fe_3O_4(cr) + O_2(g) = 6 Fe_2O_3(cr) have been studied by several investigators. From the reported data the corresponding values of $\Delta_r H°$(298.15 K) and $\Delta_f H°$(Fe_2O_3, cr, 298.15 K) were evaluated. The results obtained are given as follows:

Source	T/K	$\Delta_r H°$(298.15 K), kcal mol^{-1} 2nd law	3rd law	$\Delta_f H°$(298.15 K) kcal mol^{-1}
Tretyakov and Khomyakov (1)	1373-1728	-106.32±0.19	-111.05	-197.11
Schmahl (2)	1583-1683	-123.95±1.37	-112.94	-197.42
Smiltens (3)	1373-1673	-117.88±0.38	-114.25	-197.64
Darken and Gurry (4)	1723	-110.29	---	-196.98

The value of $\Delta_f H°$(298.15 K) adopted for Fe_2O_3(cr) is the weighted average of the $\Delta_f H°$(298.15 K) values listed in the above table.

Heat Capacity and Entropy

The low temperature (5.58-345.42 K) heat capacities were determined by Gronvold and Westrum (5). The high temperature (298-1750 K) $C_p°$ values were obtained from Coughlin et al. (6). The two sets of data were plotted and joined smoothly at 298 K. The $C_p°$ values above 1750 K were obtained by graphical extrapolation. The heat capacities of Fe_2O_3(cr) have also been reported by Brown and Furnas (0-650°C, 7) and Parks and Kelley (88-289 K, 8). S°(298.15 K) for Fe_2O_3(cr) was derived from the smoothed $C_p°$ data, based on S°(5 K) = 0.0003 cal K^{-1} mol^{-1}. Coughlin et al. (6) measured the enthalpy for hematite and found two anomalies, presumably of megnetic nature, at about 950 and 1050 K, respectively. Both appear to involve maxima in the heat-capacity curves rather than isothermal heats of transformation. The complex magnetic properties at low temperatures were discussed by Gronvold and Westrum (5). No heat capacity anomaly was observed at the magnetic transition about 250 K. The Curie points of the antiferromagnetic α-Fe_2O_3 were reported by Aharoni et al. (9) and Lielmezs and Chaklader (10).

Decomposition Data

T_{dcm} = 1735 K is the temperature at which the pressure is 1 atm for the reaction 6 Fe_2O_3(cr) = 4 Fe_3O_4(cr) + O_2(g). In other words, Fe_2O_3(cr) decomposes into Fe_3O_4(cr) and O_2(g) at 1735 K.

References

1. Y. D. Tretyakov and K. G. Khomyakov, Russ. J. Inorg. Chem. 7, 628 (1962).
2. N. G. Schmahl, Z. Elektrochem. 47, 821 (1941).
3. J. Smilten, J. Amer. Chem. Soc. 79, 4877 (1957).
4. L. S. Darken and R. W. Gurry, J. Amer. Chem. Soc. 68, 798 (1946).
5. F. Gronvold and E. F. Westrum, Jr., J. Amer. Chem. Soc. 81, 1780 (1959).
6. J. P. Coughlin, E. G. King and K. R. Bonnickson, J. Amer. Chem. Soc. 73, 3891 (1951).
7. G. G. Brown and C. C. Furnas, Trans. Amer. Inst. Chem. Eng. 18, 309 (1926).
8. G. S. Parks and K. K. Kelley, J. Phys. Chem. 30, 47 (1926).
9. A. Aharoni, E. H. Frei and M. Schieber, Phys. Rev. 127, 439 (1962).
10. J. Lielmezs and A. C. D. Chaklader, Phys. Rev. Letters 13, 866 (1965).

Iron Oxide, Hematite (Fe_2O_3) Fe_2O_3(cr)

Enthalpy Reference Temperature = T_r = 298.15 K Standard State Pressure = p° = 0.1 MPa

T/K	$C_p°$ (J $K^{-1}mol^{-1}$)	S°	-[G°-H°(T_r)]/T	H°-H°(T_r) (kJ mol^{-1})	$\Delta_f H°$	$\Delta_f G°$	Log K_f
0	0.	0.	INFINITE	-15.560	-819.025	-819.025	INFINITE
100	31.497	14.611	159.795	-14.518	-823.186	-797.438	416.539
200	76.567	51.279	96.132	-8.970	-825.542	-770.573	201.253
298.15	103.763	87.400	87.400	0.	-825.503	-743.523	130.262
300	104.182	88.043	87.402	0.192	-825.485	-743.014	129.370
400	120.123	120.299	91.691	11.443	-823.947	-715.727	93.464
500	131.796	148.400	100.284	24.058	-821.626	-688.929	71.972
600	141.168	173.270	110.415	37.713	-818.882	-662.642	57.688
700	149.729	195.675	121.020	52.259	-815.870	-636.837	47.521
800	158.218	216.225	131.653	67.658	-812.705	-611.477	39.925
900	166.490	235.340	142.124	83.894	-809.634	-586.510	34.040
950.000	170.570	244.451	147.271	92.321		I <--> II	
950.000	150.624	245.156	147.271	92.990		TRANSITION	
1000	150.624	252.882	152.360	100.522	-807.695	-561.866	29.349
1050.000	150.624	260.231	157.323	108.053		II <--> III	
1050.000	140.407	260.231	157.323	108.053		TRANSITION	
1100	140.775	266.771	162.150	115.083	-809.682	-537.171	25.508
1200	141.511	279.052	171.388	129.197	-811.071	-512.374	22.303
1300	142.248	290.408	180.112	143.385	-809.145	-487.562	19.590
1400	142.984	300.976	188.372	157.647	-807.358	-462.893	17.271
1500	143.720	310.866	196.212	171.982	-805.706	-438.347	15.265
1600	144.566	320.170	203.671	186.397	-804.180	-413.906	13.513
1700	145.266	328.956	210.785	200.890	-804.681	-389.520	11.968
1800	145.780	337.275	217.583	215.445	-804.094	-365.117	10.595
1900	146.231	345.169	224.092	230.045	-831.859	-339.338	9.329
2000	146.789	352.684	230.335	244.696	-832.056	-313.410	8.185
2100	147.277	359.858	236.334	259.400	-832.235	-287.474	7.151
2200	147.695	366.719	242.105	274.149	-832.403	-261.528	6.209
2300	148.044	373.292	247.667	288.937	-832.566	-235.576	5.350
2400	148.323	379.599	253.034	303.756	-832.731	-209.617	4.562
2500	148.532	385.658	258.219	318.599	-832.905	-183.650	3.837

PREVIOUS: CURRENT: June 1965

Iron Oxide, Hematite (Fe_2O_3) Fe_2O_3(cr)

IRON SULFATE ($Fe_2(SO_4)_3$) CRYSTAL M_r = 399.8668

$\Delta_f H°(0\ K)$ = Unknown

S°(298.15 K) = [307.524 ± 8.4] $J\ K^{-1}\ mol^{-1}$ $\Delta_f H°(298.15\ K) = -2582.99 \pm 1.67\ kJ\ mol^{-1}$

Enthalpy of Formation

The enthalpy change, $\Delta_r H°(298.15\ K)$, for the chemical reaction $Fe_2O_3(cr) + 3(H_2SO_4 \cdot 14.855\ H_2O)(sol.) = Fe_2(SO_4)_3(cr) + 47.565\ H_2O(\ell)$ was measured to be 8.54 + 0.18 kcal mol^{-1} by solution calorimetry by Barany and Adami (1). Using $\Delta_f H°(298.15\ K) = -211.178$ kcal mol^{-1} for $H_2SO_4 \cdot 14.855\ H_2O$(sol.) and -68.315 kcal mol^{-1} for $H_2O(\ell)$, obtained from (2) and -197.3 kcal mol^{-1} for Fe_2O_3(cr), the enthalpy of formation, $\Delta_f H°(298.15\ K)$, for $Fe_2(SO_4)_3$(cr) was evaluated as -617.35±0.40 kcal mol^{-1} which was adopted here.

The equilibrium pressures for the decomposition of $Fe_2(SO_4)_3$(cr) have been determined by several investigators at different temperatures. The chemical reactions are: (1) $Fe_2(SO_4)_3(cr) = Fe_2O_3(cr) + 3\ SO_3(g)$ and (2) $SO_3(g) = SO_2(g) + 0.5\ O_2(g)$. However, the measured pressure of the chemical equilibrium is the total pressure of three partial pressures, namely, $P(SO_3)$, $P(SO_2)$, and $P(O_2)$. In order to calculate the enthalpy change of Reaction (1), the partial pressures of $SO_3(g)$ were evaluated from the total vapor pressure data at each temperature. Based on the derived values for $P(SO_3)$, the $\Delta_r H°(298.15\ K)$ value for Reaction (1) was calculated by both the 2nd and 3rd law methods. The results obtained are presented in the following table. The $\Delta_f H°(Fe_2(SO_4)_3$, cr, 298.15 K) values were derived, using the 3rd law $\Delta_r H°(298.15\ K)$. These determinations were not given any weight.

Source	T/K	$\Delta_r H°$(298.15 K), kcal mol^{-1} 2nd law	3rd law	Drift cal K^{-1} mol^{-1}	$\Delta_f H°$(298.15 K) kcal mol^{-1}
Keppeler and D'Ans (3)	753.15-1013.15	136.8±0.5	139.9±0.5	3.4±0.6	-621.0
Wohler, et al. (4)	826.15- 980.15	114.2±6.2	133.1±2.3	19.2±6.9	-614.2
Bodenstein and Suzuki (5)	903.65- 997.15	140.7±6.6	139.1±1.3	-1.2±7.1	-620.2
Wohler and Grunzweig (6)	909.15- 994.15	155.1±1.3	135.5±1.3	-20.4±1.4	-616.6
Neumann and Heintke (7)	793.15- 973.15	136.4±2.3	136.2±1.2	-1.2±2.6	-617.3
Warner and Ingraham (8)	903.15- 997.15	133.9±2.6	135.4±0.5	1.7±2.7	-616.5

The decomposition pressures of $Fe_2(SO_4)_3$(cr) have also been investigated by: Grunzweig (793-994 K, 9), Blanks (845-976 K, 10) and Ingraham (856-995 K, 11). These data together with some of the data mentioned previously were critically reviewed by Kellogg (12).

Heat Capacity and Entropy

The heat capacities were estimated by comparison with those for $FeSO_4$(cr), assuming their average specific heats, cal K^{-1} g^{-1}, to be the same. The value of S°(298.15 K) was estimated so that the 2nd and 3rd law $\Delta_r H°(298.15\ K)$ values, derived from decomposition pressure data, are in reasonable agreement.

Decomposition Data

T_{dcm} is the temperature (1451 K) at which the total pressure of the gaseous products equals one atmosphere, which was obtained by graphical extrapolation of the decomposition pressure data measured by Warner and Ingraham (8).

References

1. R. Barany and L. H. Adami, U. S. Bur. Mines RI 6687, 1965.
2. U. S. Nat. Bur. Stand. Tech. Note 270-1, 1965.
3. G. Keppeler and J. D'Ans, Z. Physik. Chem. 62, 89 (1908).
4. L. Wohler, W. Pluddemann and P. Wohler, Ber. Deut. Chem. Gesell. 41, 703 (1908).
5. M. Bodenstein and T. Suzuki, Z. Elektrochem. 16, 912 (1910).
6. L. Wohler and M. Grunzweig, Ber. Deut. Chem. Gesell. 43, 1587 (1913).
7. B. Neumann and G. Heintke, Z. Elektrochem. 43, 246 (1937).
8. N. A. Warner and T. R. Ingraham, Can. J. Chem. 38, 2196 (1960).
9. M. Grunzweig, Ph.D. dissertation, Darmstadt, Germany, 1913.
10. R. F. Blanks, Ph.D. dissertation, University of Melbourne, Australia, 1961.
11. T. R. Ingraham, Internal Rept. EMT-63-17, Department of Mines and Technical Surveys, Ottawa, Canada, 1963.
12. H. H. Kellogg, Trans. AIME 230, 1622 (1964).

Iron Sulfate ($Fe_2(SO_4)_3$) $Fe_2O_{12}S_3$(cr)

Enthalpy Reference Temperature = T_r = 298.15 K Standard State Pressure = p° = 0.1 MPa

T/K	$C_p°$ (J $K^{-1}mol^{-1}$)	S°	$-[G°-H°(T_r)]/T$	$H°-H°(T_r)$ (kJ mol^{-1})	$\Delta_f H°$	$\Delta_f G°$	Log K_f
0							
100							
200							
298.15	264.722	307.524	307.524	0.	-2582.992	-2262.729	396.421
300	265.851	309.165	307.529	0.491	-2583.047	-2260.742	393.630
400	307.022	391.497	318.475	29.209	-2591.203	-2152.742	281.119
500	339.281	463.626	340.441	61.593	-2594.662	-2042.719	213.401
600	363.297	527.707	366.413	96.776	-2595.362	-1932.225	168.215
700	380.702	585.079	393.626	134.017	-2594.341	-1821.767	135.942
800	393.254	636.776	420.843	172.746	-2592.496	-1711.521	111.751
900	402.041	683.617	447.479	212.524	-2749.978	-1598.273	92.761
1000	409.237	726.359	473.261	253.098	-2745.239	-1470.576	76.815
1100	415.262	765.651	498.080	294.328	-2741.963	-1343.236	63.785
1200	420.659	802.019	521.912	336.128	-2737.297	-1216.322	52.945
1300	425.513	835.884	544.776	378.441	-2729.077	-1089.907	43.793
1400	429.906	867.580	566.714	421.214	-2720.794	-964.127	35.972
1500	434.090	897.385	587.775	464.415	-2712.461	-838.943	29.215
1600	438.065	925.528	608.013	508.025	-2704.089	-714.314	23.320
1700	441.830	952.199	627.482	552.020	-2697.589	-590.172	18.134
1800	445.554	977.560	646.232	596.391	-2689.835	-466.432	13.536
1900	449.069	1001.744	664.311	641.121	-2710.265	-341.717	9.394
2000	452.625	1024.868	681.765	686.206	-2702.965	-217.244	5.674

PREVIOUS: CURRENT: June 1966

Iron Sulfate ($Fe_2(SO_4)_3$) $Fe_2O_{12}S_3$(cr)

IRON OXIDE, MAGNETITE (Fe_3O_4) CRYSTAL M_r = 231.5386

$S°$(298.15 K) = 145.266 J K^{-1} mol^{-1}

T_{fus} = 1870 ± 2 K

$\Delta_f H°$(0 K) = -1114.58 ± 0.8 kJ mol^{-1}

$\Delta_f H°$(298.15 K) = -1120.89 ± 0.8 kJ mol^{-1}

$\Delta_{fus} H°$ = 138.070 ± 8.4 kJ mol^{-1}

Enthalpy of Formation

The equilibrium constants for the following four reactions, i.e. (A) 3Fe(cr) + $4H_2O$(g) = Fe_3O_4(cr) + $4H_2$(g); (B) Fe_3O_4(cr) + H_2(g) = 3FeO(cr) + H_2O(g); (C) Fe_3O_4(cr) + CO(g) = 3FeO(cr) + CO_2(g); and (D) $2Fe_3O_4$(cr) = 6FeO(cr) + O_2(g), have been studied by several investigators. Using their reported equilibrium data, the corresponding enthalpy changes, $\Delta_r H°$(298.15 K), were evaluated by both 2nd and 3rd law methods. Based on the 3rd law value for $\Delta_r H°$(298.15 K), the respective $\Delta_f H°$(298.15 K) values for Fe_3O_4(cr) were also derived. The results obtained are presented as follows.

Source	Reaction	T/K	$\Delta_r H°$(298.15 K), kcal mol^{-1} 2nd law	3rd law	$\Delta_f H°$(298.15 K) kcal mol^{-1}
Fricke, et al. (1)	A	633.2-823.2	-29.68±1.30	-36.91	-268.10
Eastman (2)	B	873.2-1273.2	22.41±0.09	19.29	-268.01
Eastman and Evans (3)	B	973.2-1273.2	21.56±0.07	18.54	-267.26
Emmett and Shultz (4)	B	973.2-1073.2	---	19.18	-267.90
Eastman (2)	C	873.2-1273.2	11.87±0.03	9.38	-267.93
Eastman and Evans (3)	C	973.2-1273.2	11.63±0.03	9.38	-267.94
Salmon (5)	D	1473.2-1673.2	---	153.73	-267.79

The enthalpy change of the reaction $Fe_{0.947}O$(cr) + 0.8215 Fe(cr) + 0.6787 O_2(g) = 0.5895 Fe_3O_4(cr) has been determined by Humphrey et al. (6). Based on $\Delta_r H°$(298.15 K) = -93.46 kcal mol^{-1} (average of two separate values derived from two experimental determinations) and $\Delta_f H°$(wustite, cr, 298.15 K) = -63.64 kcal mol^{-1}, the value $\Delta_f H°$(Fe_3O_4, cr, 298.15 K) was calculated as -266.5 kcal mol^{-1}. The enthalpy change, $\Delta_r H°$(298.15 K) = 4.33 kcal mol^{-1}, for the reaction 3Fe(cr) + $4CO_2$(g) = Fe_3O_4(cr) + 4CO(g) was reported by Darken and Gurry (7). The corresponding $\Delta_f H°$(Fe_3O_4, cr, 298.15 K) was evaluated as -266.22 kcal mol^{-1}. The adopted value of $\Delta_f H°$(298.15 K) for Fe_3O_4(cr) is selected as -267.9±0.2 kcal mol^{-1}.

Heat Capacity and Entropy

The low temperature heat capacities (60.5-299.7 K) were reported by Millar (8). The high temperature heat capacities (298-1800 K) were obtained from Coughlin et al. (9). These two sets of data were plotted and joined smoothly at 298.15 K. The $C_p°$ values above 1800 K were obtained by graphical extrapolation. Heat capacities (90-295 K) for Fe_3O_4(cr) were also reported by Parks and Kelley (10). $S°$(Fe_3O_4, cr, 298.15 K) was derived from the smoothed $C_p°$ data mentioned previously, using $S°$(44.7 K) = 0.969 cal K^{-1} mol^{-1}. Coughlin et al. (9) measured the heat-content for magnetite and found a magnetic anomaly around 880 K, corresponding to a maximum in heat capacity. A maximum heat capacity of 37.1 cal K^{-1} mol^{-1} at 114.15 K was also reported by Millar (8).

Fusion Data

T_{fus} and $\Delta_{fus}H°$ were obtained from Darken and Gurry (7).

References

1. R. Fricke, K. Walter and W. Lohrer, Z. Elektrochem. 47, 487 (1941). The two enthalpies of reaction values reported are the average values of three separate results derived from three sets of experimental data.
2. E. D. Eastman, J. Amer. Chem. Soc. 44, 975 (1922).
3. E. D. Eastman and R. M. Evans, J. Amer. Chem. Soc. 46, 888 (1924).
4. P. H. Emmett and J. F. Schultz, J. Amer. Chem. Soc. 52, 4268 (1930).
5. O. N. Salmon, J. Phys. Chem. 65, 550 (1961).
6. G. L. Humphrey, E. G. King and K. K. Kelley, U. S. Bur. Mines RI 4870, 1952.
7. L. S. Darken and R. W. Gurry, J. Amer. Chem. Soc. 68, 798 (1946).
8. R. W. Millar, J. Amer. Chem. Soc. 51, 215 (1929).
9. J. P. Coughlin, E. G. King and K. R. Bonnickson, J. Amer. Chem. Soc. 73, 3891 (1951).
10. G. S. Parks and K. K. Kelley, J. Phys. Chem. 30, 47 (1926).

Iron Oxide, Magnetite (Fe_3O_4) Fe_3O_4(cr)

Enthalpy Reference Temperature = T_r = 298.15 K; Standard State Pressure = p° = 0.1 MPa

T/K	$C_p°$ (J K^{-1}mol^{-1})	S°	-[G°-H°(T_r)]/T	H°-H°(T_r) (kJ mol^{-1})	$\Delta_f H°$	$\Delta_f G°$	Log K_f
0	0.	0.	INFINITE	-24.571	-1114.579	-1114.579	INFINITE
100	56.335	27.611	254.203	-22.659	-1119.743	-1086.024	567.281
114.150	155.535	37.694	226.660	-21.570	Cp LAMBDA MAXIMUM		
114.150	155.279	37.698	226.660	-21.570	TRANSITION		
200	116.945	92.367	157.872	-13.101	-1121.314	-1051.624	274.656
298.15	147.235	145.266	145.266	0.	-1120.894	-1017.438	178.251
300	147.695	146.179	145.269	0.273	-1120.869	-1016.797	177.040
400	171.126	191.952	151.352	16.240	-1118.728	-982.386	128.286
500	192.380	232.441	163.593	34.424	-1115.219	-948.681	99.108
600	212.547	269.299	178.181	54.671	-1110.550	-915.794	79.727
700	232.714	303.578	193.672	76.934	-1104.774	-883.776	65.948
800	252.881	335.969	209.452	101.214	-1098.011	-852.657	55.673
900	273.050	366.919	225.241	127.510	-1090.611	-822.428	47.732
900.000	273.050	366.919	225.241	127.511	Cp LAMBDA MAXIMUM		
900.000	200.832	366.919	225.241	127.511	TRANSITION		
1000	200.832	388.079	240.485	147.594	-1091.694	-792.602	41.401
1100	200.832	407.220	254.786	167.677	-1095.556	-762.463	36.206
1200	200.832	424.695	268.228	187.760	-1097.841	-732.139	31.869
1300	200.832	440.770	280.890	207.843	-1095.254	-701.771	28.197
1400	200.832	455.653	292.848	227.927	-1092.980	-671.590	25.057
1500	200.832	469.509	304.169	248.010	-1091.011	-641.562	22.341
1600	200.832	482.471	314.912	268.093	-1089.346	-611.654	19.968
1700	200.832	494.646	325.131	288.176	-1090.831	-581.785	17.876
1800	200.832	506.125	334.870	308.259	-1090.769	-551.846	16.014
1900	200.832	516.984	344.172	328.343	-1133.299	-519.794	14.290
2000	200.832	527.285	353.072	348.426	-1134.548	-487.472	12.731
2100	200.832	537.084	361.603	368.509	-1135.843	-455.087	11.320
2200	200.832	546.426	369.793	388.592	-1137.183	-422.637	10.035
2300	200.832	555.354	377.669	408.675	-1138.568	-390.128	8.860
2400	200.832	563.901	385.252	428.759	-1139.998	-357.557	7.782
2500	200.832	572.099	392.563	448.842	-1141.472	-324.925	6.789
2600	200.832	579.976	399.620	468.925	-1142.988	-292.234	5.871
2700	200.832	587.556	406.441	489.008	-1144.547	-259.482	5.020
2800	200.832	594.859	413.041	509.091	-1146.146	-226.673	4.229
2900	200.832	601.907	419.433	529.175	-1147.786	-193.807	3.491
3000	200.832	608.715	425.629	549.258	-1149.464	-160.882	2.801

PREVIOUS: CURRENT: June 1965

Iron Oxide, Magnetite (Fe_3O_4) Fe_3O_4(cr)

GALLIUM (Ga) REFERENCE STATE A_r = 69.72

0 to 302.920 K	crystal
302.92 to 2476.570 K	liquid
above 2476.570 K	ideal monatomic gas

Refer to the individual tables for details.

Gallium (Ga) Ga_1(ref)

Enthalpy Reference Temperature = T_r = 298.15 K   Standard State Pressure = p° = 0.1 MPa

T/K	C_p° (J K⁻¹mol⁻¹)	S° (J K⁻¹mol⁻¹)	$-[G^\circ-H^\circ(T_r)]/T$ (J K⁻¹mol⁻¹)	$H^\circ-H^\circ(T_r)$ (kJ mol⁻¹)	Δ_fH° (kJ mol⁻¹)	Δ_fG° (kJ mol⁻¹)	Log K_f
0	0.	0.	INFINITE	-5.561	0.	0.	0.
100	18.484	16.085	62.261	-4.618	0.	0.	0.
200	23.815	30.912	43.151	-2.448	0.	0.	0.
298.15	26.064	40.838	40.838	0.	0.	0.	0.
300	26.091	41.000	40.839	0.048	0.	0.	0.
302.920	26.133	41.253	40.842	0.124	CRYSTAL <--> LIQUID TRANSITION		
302.920	28.493	59.706	40.842	5.714			
400	27.154	67.426	46.408	8.407	0.	0.	0.
500	26.819	73.445	51.237	11.104	0.	0.	0.
600	26.694	78.324	55.358	13.780	0.	0.	0.
700	26.568	82.428	58.939	16.442	0.	0.	0.
800	26.568	85.976	62.102	19.099	0.	0.	0.
900	26.568	89.105	64.932	21.756	0.	0.	0.
1000	26.568	91.904	67.492	24.413	0.	0.	0.
1100	26.568	94.436	69.828	27.069	0.	0.	0.
1200	26.568	96.748	71.976	29.726	0.	0.	0.
1300	26.568	98.875	73.965	32.383	0.	0.	0.
1400	26.568	100.844	75.815	35.040	0.	0.	0.
1500	26.568	102.677	77.546	37.697	0.	0.	0.
1600	26.568	104.391	79.171	40.354	0.	0.	0.
1700	26.568	106.002	80.702	43.010	0.	0.	0.
1800	26.568	107.521	82.150	45.667	0.	0.	0.
1900	26.568	108.957	83.524	48.324	0.	0.	0.
2000	26.568	110.320	84.830	50.981	0.	0.	0.
2100	26.568	111.616	86.075	53.638	0.	0.	0.
2200	26.568	112.852	87.264	56.295	0.	0.	0.
2300	26.568	114.033	88.402	58.951	0.	0.	0.
2400	26.568	115.164	89.494	61.608	0.	0.	0.
2476.570	26.568	115.998	90.301	63.643	LIQUID <--> IDEAL GAS FUGACITY = 1 bar		
2476.570	21.260	220.462	90.301	322.353			
2500	21.251	220.662	91.522	322.851	0.	0.	0.
2600	21.215	221.495	96.505	324.974	0.	0.	0.
2700	21.184	222.295	101.149	327.094	0.	0.	0.
2800	21.156	223.065	105.489	329.211	0.	0.	0.
2900	21.131	223.807	109.556	331.325	0.	0.	0.
3000	21.109	224.523	113.377	333.437	0.	0.	0.
3100	21.090	225.214	116.973	335.547	0.	0.	0.
3200	21.073	225.884	120.366	337.655	0.	0.	0.
3300	21.059	226.532	123.574	339.762	0.	0.	0.
3400	21.047	227.160	126.611	341.867	0.	0.	0.
3500	21.036	227.770	129.493	343.971	0.	0.	0.
3600	21.029	228.363	132.231	346.075	0.	0.	0.
3700	21.023	228.939	134.837	348.177	0.	0.	0.
3800	21.020	229.500	137.321	350.279	0.	0.	0.
3900	21.018	230.045	139.691	352.381	0.	0.	0.
4000	21.020	230.578	141.957	354.483	0.	0.	0.
4100	21.024	231.097	144.125	356.585	0.	0.	0.
4200	21.030	231.603	146.202	358.688	0.	0.	0.
4300	21.040	232.098	148.193	360.791	0.	0.	0.
4400	21.052	232.582	150.106	362.896	0.	0.	0.
4500	21.068	233.055	151.944	365.002	0.	0.	0.
4600	21.087	233.519	153.712	367.110	0.	0.	0.
4700	21.109	233.972	155.415	369.219	0.	0.	0.
4800	21.135	234.417	157.056	371.332	0.	0.	0.
4900	21.165	234.853	158.640	373.446	0.	0.	0.
5000	21.198	235.281	160.168	375.565	0.	0.	0.
5100	21.236	235.701	161.645	377.686	0.	0.	0.
5200	21.278	236.114	163.073	379.812	0.	0.	0.
5300	21.324	236.520	164.455	381.942	0.	0.	0.
5400	21.375	236.919	165.793	384.077	0.	0.	0.
5500	21.431	237.312	167.090	386.217	0.	0.	0.
5600	21.491	237.698	168.348	388.363	0.	0.	0.
5700	21.556	238.079	169.568	390.515	0.	0.	0.
5800	21.625	238.455	170.752	392.674	0.	0.	0.
5900	21.700	238.825	171.903	394.841	0.	0.	0.
6000	21.780	239.190	173.021	397.015	0.	0.	0.

PREVIOUS: CURRENT: June 1970 (1 bar)

Gallium (Ga) Ga_1(ref)

$S°(298.15\ K) = 40.838 \pm 0.20\ J\ K^{-1}\ mol^{-1}$

$T_{fus} = 302.92 \pm 0.02\ K$

$\Delta_f H°(0\ K) = 0\ kJ\ mol^{-1}$

$\Delta_f H°(298.15\ K) = 0\ kJ\ mol^{-1}$

$\Delta_{fus} H° = 5.59 \pm 0.04\ kJ\ mol^{-1}$

Enthalpy of Formation

Zero by definition.

Heat Capacity and Entropy

The adopted thermal functions are those recommended by Hultgren et al. (1). The heat capacity values were based on the studies of Phillips (0.1 - 4.2 K, 2), Seidel and Kiesom (0.35 - 4.38 K, 3), Eichenaur and Schulze (5-22 K, 4), Adams, et al. (15 - 323 K, 5) and Clusius and Hartick (15 - 200 K, 6). Refer to Hultgren et al. (1) for details.

Enthalpy of Fusion

See the liquid table for details.

Enthalpy of Sublimation

The enthalpy of sublimation is the enthalpy of formation of the gas, $\Delta_f H°(298.15\ K) = \Delta_f H°(g,\ 298.15\ K)$.

References

1. R. Hultgren, P. D. Desai, et al., "Selected Values of the Thermodynamics Properties of the Elements," american Society for Metals, Metals Parks, Ohio 1973.
2. N. E. Phillips, Phys. Rev. 134, A385 (1964).
3. G. Seidel and P. H. Keesom, Phys. Rev. 112, 1083 (1958).
4. W. Eichenauer and M. Schulze, Z. Naturforsch. 14a, 962 (1959).
5. G. B. Adams, H. L. Johnston and E. C. Kerr, J. Am. Chem. Soc. 74, 4784 (1952).
6. K. Clusius and P. Hartick, Z. Physik. Chem. 134, 243 91928).

Enthalpy Reference Temperature = T_r = 298.15 K   Standard State Pressure = p° = 0.1 MPa

T/K	$C_p°$ (J K⁻¹mol⁻¹)	S° (J K⁻¹mol⁻¹)	$-[G°-H°(T_r)]/T$ (J K⁻¹mol⁻¹)	$H°-H°(T_r)$ (kJ mol⁻¹)	$\Delta_f H°$ (kJ mol⁻¹)	$\Delta_f G°$ (kJ mol⁻¹)	Log K_f
0	0.	0.	INFINITE	-5.561	0.	0.	0.
100	18.484	16.085	62.261	-4.618	0.	0.	0.
200	23.815	30.912	43.151	-2.448	0.	0.	0.
250	24.907	36.338	41.262	-1.231	0.	0.	0.
298.15	26.064	40.838	40.838	0.	0.	0.	0.
300	26.091	41.000	40.839	0.048	0.	0.	0.
302.920	26.133	41.253	40.842	0.124	------ CRYSTAL <--> LIQUID -----		
350	27.004	45.091	41.160	1.376	-5.661	0.875	-0.131
400	27.460	48.726	41.883	2.737	-5.670	1.810	-0.236
450	27.916	51.987	42.827	4.122	-5.638	2.743	-0.318
500	28.372	54.952	43.894	5.529	-5.575	3.672	-0.384

PREVIOUS: CURRENT: June 1970

GALLIUM (Ga) LIQUID A_r = 69.72

$S°(298.15\ K)$ = [59.253] J K^{-1} mol^{-1} $\Delta_f H°(298.15\ K)$ = [5.578] kJ mol^{-1}

T_{fus} = 302.92 ± 0.02 K $\Delta_{fus}H°$ = 5.59 ± 0.04 kJ mol^{-1}

Enthalpy of Formation

The enthalpy of formation, $\Delta_f H°(\ell,\ 298.15\ K)$, is calculated from that of the crystal by adding the enthalpy of fusion and the difference in enthalpy, H°(302.92 K) - H°(298.15 K), between the crystal and liquid.

Heat Capacity and Entropy

The heat capacity of the liquid is based on the recommendations of Hultgren, et al. (1). These values are based on the enthalpy data of Kochethova and Rezukhina (315-712 K, 2) and the heat capacity data of Chen and Turnbull (245-611 K, 3).

Enthalpy of Fusion

The adopted value for the enthalpy of fusion is derived from the direct calorimetric measurement of Roth, Meyer, and Zeumir (4).

The temperature of fusion, T_{fus} = 302.92 ± 0.2 K (corrected to IPTS-68), was derived from the study of Boedtker et al. (5).

Enthalpy of Vaporization

The temperature at which $\Delta_r G°$ = 0 (1 bar) for the process Ga(ℓ) → Ga(g) is the adopted T_{vap}(1 bar). The difference in the enthalpies of formation of the ideal gas and liquid at T_{vap} is $\Delta_{vap}H°$. Refer to the ideal gas table for additional information.

References

1. R. Hultgren, P. D. Desai, et al., "Selected Values of the Thermodynamics Properties of the Elements," American Society for Metals, Metals Parks, Ohio 1973.
2. H. M. Kochetkova and T. N. Rezukhina, Proc. 4th All-Union Conf. on Semiconductor Materials (Eng. transl.), 26 (1961).
3. H. S. Chem. and S. Turnbell, Acta Met. 16, 369 (1968).
4. W. A. Roth, I. Meyer and H. Zeumar, Z. Anorg. Chem. 214, 309 (1933); Z. Anorg. Chem. 216, 303 (1934).
5. O. A. Boedtker, R. C. LaForce, et al., Trans. Faraday Soc. 61, 665 (1965).

Gallium (Ga) $Ga_1(l)$

Enthalpy Reference Temperature = T_r = 298.15 K Standard State Pressure = p° = 0.1 MPa

T/K	$C_p°$ (J K⁻¹mol⁻¹)	S°	$-[G°-H°(T_r)]/T$	$H°-H°(T_r)$ (kJ mol⁻¹)	$\Delta_f H°$	$\Delta_f G°$	Log K_f
0							
100							
200							
250							
298.15	28.559	59.253	59.253	0.	5.578	0.088	-0.015
300	28.533	59.429	59.253	0.053	5.583	0.054	-0.009
302.920	28.493	59.706	59.256	0.136	------ CRYSTAL <--> LIQUID -----		
350	27.719	63.764	59.598	1.458	0.	0.	0.
400	27.154	67.426	60.353	2.829	0.	0.	0.
450	26.961	70.613	61.320	4.182	0.	0.	0.
500	26.819	73.445	62.393	5.526	0.	0.	0.
600	26.694	78.324	64.655	8.202	0.	0.	0.
700	26.568	82.428	66.908	10.864	0.	0.	0.
800	26.568	85.976	69.075	13.521	0.	0.	0.
900	26.568	89.105	71.130	16.177	0.	0.	0.
1000	26.568	91.904	73.070	18.834	0.	0.	0.
1100	26.568	94.436	74.899	21.491	0.	0.	0.
1200	26.568	96.748	76.625	24.148	0.	0.	0.
1300	26.568	98.875	78.256	26.805	0.	0.	0.
1400	26.568	100.844	79.800	29.462	0.	0.	0.
1500	26.568	102.677	81.264	32.118	0.	0.	0.
1600	26.568	104.391	82.657	34.775	0.	0.	0.
1700	26.568	106.002	83.983	37.432	0.	0.	0.
1800	26.568	107.521	85.249	40.089	0.	0.	0.
1900	26.568	108.957	86.459	42.746	0.	0.	0.
2000	26.568	110.320	87.619	45.403	0.	0.	0.
2100	26.568	111.616	88.731	48.060	0.	0.	0.
2200	26.568	112.852	89.799	50.716	0.	0.	0.
2300	26.568	114.033	90.828	53.373	0.	0.	0.
2400	26.568	115.164	91.818	56.030	0.	0.	0.
2476.570	26.568	115.998	92.553	58.064	------- FUGACITY = 1 bar -------		
2500	26.568	116.249	92.774	58.687	-258.586	2.447	-0.051
2600	26.568	117.291	93.697	61.344	-258.052	12.878	-0.259
2700	26.568	118.293	94.589	64.001	-257.515	23.288	-0.451
2800	26.568	119.260	95.453	66.657	-256.975	33.679	-0.628
2900	26.568	120.192	96.290	69.314	-256.433	44.050	-0.793
3000	26.568	121.093	97.102	71.971	-255.888	54.402	-0.947

PREVIOUS: CURRENT: June 1970

Gallium (Ga) $Ga_1(l)$

GALLIUM (Ga) CRYSTAL-LIQUID $A_r = 69.72$ Gallium (Ga) $Ga_1(cr,l)$

0 to 302.92 K crystal
above 302.92 K liquid

Refer to the individual tables for details.

Enthalpy Reference Temperature = T_r = 298.15 K Standard State Pressure = $p°$ = 0.1 MPa

T/K	$C_p°$ (J K⁻¹mol⁻¹)	$S°$ (J K⁻¹mol⁻¹)	$-[G°-H°(T_r)]/T$ (J K⁻¹mol⁻¹)	$H°-H°(T_r)$ (kJ mol⁻¹)	$\Delta_f H°$ (kJ mol⁻¹)	$\Delta_f G°$ (kJ mol⁻¹)	Log K_f
0	0.	0.	INFINITE	-5.561	0.	0.	0.
100	18.484	16.085	62.261	-4.618	0.	0.	0.
200	23.815	30.912	43.151	-2.448	0.	0.	0.
250	24.907	36.338	41.262	-1.231	0.	0.	0.
298.15	26.064	40.838	40.838	0.	0.	0.	0.
300	26.091	41.000	40.839	0.048	0.	0.	0.
302.920	26.133	41.253	40.842	0.124	CRYSTAL <--> LIQUID		
302.920	28.493	59.706	40.842	5.714	TRANSITION		
350	27.719	63.764	43.660	7.037	0.	0.	0.
400	27.154	67.426	46.408	8.407	0.	0.	0.
450	26.961	70.613	48.924	9.760	0.	0.	0.
500	26.819	73.445	51.237	11.104	0.	0.	0.
600	26.694	78.324	55.358	13.780	0.	0.	0.
700	26.568	82.428	58.939	16.442	0.	0.	0.
800	26.568	85.976	62.102	19.099	0.	0.	0.
900	26.568	89.105	64.932	21.756	0.	0.	0.
1000	26.568	91.904	67.492	24.413	0.	0.	0.
1100	26.568	94.436	69.828	27.069	0.	0.	0.
1200	26.568	96.748	71.976	29.726	0.	0.	0.
1300	26.568	98.875	73.965	32.383	0.	0.	0.
1400	26.568	100.844	75.815	35.040	0.	0.	0.
1500	26.568	102.677	77.546	37.697	0.	0.	0.
1600	26.568	104.391	79.171	40.354	0.	0.	0.
1700	26.568	106.002	80.702	43.010	0.	0.	0.
1800	26.568	107.521	82.150	45.667	0.	0.	0.
1900	26.568	108.957	83.524	48.324	0.	0.	0.
2000	26.568	110.320	84.830	50.981	0.	0.	0.
2100	26.568	111.616	86.075	53.638	0.	0.	0.
2200	26.568	112.852	87.264	56.295	0.	0.	0.
2300	26.568	114.033	88.402	58.951	0.	0.	0.
2400	26.568	115.164	89.494	61.608	0.	0.	0.
2476.570	26.568	115.998	90.301	63.643	------- FUGACITY = 1 bar -------		
2500	26.568	116.249	90.543	64.265	-258.586	2.447	-0.051
2600	26.568	117.291	91.551	66.922	-258.052	12.878	-0.259
2700	26.568	118.293	92.523	69.579	-257.515	23.288	-0.451
2800	26.568	119.260	93.461	72.236	-256.975	33.679	-0.628
2900	26.568	120.192	94.367	74.892	-256.433	44.050	-0.793
3000	26.568	121.093	95.243	77.549	-255.888	54.402	-0.947

PREVIOUS: CURRENT: June 1970

Gallium (Ga) $Ga_1(cr,l)$

GALLIUM (Ga) IDEAL GAS A_r = 51.996

IP(Ga, g) = 48380 ± 5 cm^{-1}

S°(298.15 K) = 169.042 ± 0.01 J K^{-1} mol^{-1}

$\Delta_f H°$(0 K) = 270.97 ± 2.1 kJ mol^{-1}

$\Delta_f H°$(298.15 K) = 271.96 ± 2.1 kJ mol^{-1}

Electronic Levels and Quantum Weights

State	ε_i, cm^{-1}	g_i
$^3P_{1/2}$	0	2
$^3P_{3/2}$	826.24	4
$^2S_{1/2}$	24788.58	2
$^2P_{1/2}$	33044.06	2
$^2P_{3/2}$	33155.03	4

Enthalpy of Formation

The enthalpy of formation of Ga(g), as recommended by Hultgren et al. (1), was derived from the third law analysis of seven vaporization studies and a spectroscopic calculation. The studies are: Matern et al. (1081-1213 K, 2), Macur et al. (1196-1473 K, 3), Alcock et al. (1173-1437 K, 4), Munir and Searcy (1174-1603 K, 5), Cochran and Foster (1179-1383 K, 6), Speiser and Johnston (1230-1518 K, 7), Hartick (1198-1391 K, 8), and the spectroscopic calculaion by Gurvich (9).

Heat Capacity and Entropy

The thermal functions were calculated using the atomic levels as given by Moore (10). The resulting thermal functions are those recommended by Hultgren et al. (1).

References

1. R. Hultgren, P. D. Desai, et al., "Selected Values of the Thermodynamic Properties of the Elements," American Society for Metals, Metals Park, Ohio, 1973.
2. G. Matern, Yu. A. Sapozhnikov, et al, Zzvest. Akad. Naukl SSSR, Metally (3), 210 (1969).
3. G. J. Macur, R. K. Edwards, and P. G. Wahlbick, J. Phys. Chem. 70, 2956 (1966).
4. C. B. Alcock, J. B. Cormish and P. Grievson, "Thermodynamics," Proc. Symp. Vienna, Vol. 1, 211-30, 367-8 (1965).
5. Z. A. Munir and A. W. Searcy, J. Electrochem. Soc. 111, 1170 (1964).
6. C. N. Cochran and L. M. Foster, J. Electrochem. Soc. 109, 144 (1962).
7. R. Speiser and H. L. Johnston, J. Am. Chem. Soc. 75, 1469 (1953).
8. P. Hartick, Z. Physik. Chem. 134, 1 (1960).
9. L. V. Gurvich, Zh. Fiz. Khim. 34, 1690 (1960).
10. C. E. Moore, U. S. Nat. Bur. Stnad., NSRDS-NBS-35, vol II, 1971 [Reprint of NBS Circ. 497, 1952.

Gallium (Ga) Ga_1(g)

Enthalpy Reference Temperature = T_r = 298.15 K   Standard State Pressure = p° = 0.1 MPa

T/K	$C_p°$ (J $K^{-1}mol^{-1}$)	S°	$-[G°-H°(T_r)]/T$	$H°-H°(T_r)$ (kJ mol^{-1})	$\Delta_f H°$	$\Delta_f G°$	Log K_f
0	0.	0.	INFINITE	-6.551	270.970	270.970	INFINITE
100	20.802	144.848	189.570	-4.472	272.105	259.229	-135.407
200	22.311	159.555	171.266	-2.342	272.066	246.337	-64.337
250	23.914	164.703	169.452	-1.187	272.004	239.913	-50.127
298.15	25.346	169.042	169.042	0.	271.960	233.736	-40.950
300	25.394	169.199	169.043	0.047	271.959	233.499	-40.656
350	26.430	173.198	169.356	1.345	266.268	227.966	-34.022
400	26.974	176.768	170.064	2.682	266.234	222.497	-29.055
450	27.120	179.957	170.989	4.035	266.235	217.030	-25.192
500	26.991	182.809	172.031	5.389	266.245	211.563	-22.102
600	26.316	187.676	174.247	8.057	266.237	200.626	-17.466
700	25.489	191.671	176.460	10.648	266.166	189.696	-14.155
800	24.724	195.023	178.576	13.157	266.018	178.781	-11.673
900	24.077	197.896	180.567	15.596	265.801	167.888	-9.744
1000	23.550	200.405	182.428	17.977	265.524	157.023	-8.202
1100	23.125	202.629	184.166	20.310	265.200	146.189	-6.942
1200	22.781	204.626	185.789	22.604	264.838	135.385	-5.893
1300	22.503	206.438	187.309	24.868	264.445	124.613	-5.007
1400	22.275	208.097	188.735	27.107	264.027	113.872	-4.249
1500	22.088	209.627	190.077	29.324	263.588	103.162	-3.592
1600	21.933	211.048	191.344	31.525	263.132	92.482	-3.019
1700	21.802	212.373	192.543	33.712	262.661	81.831	-2.514
1800	21.692	213.616	193.679	35.886	262.179	71.208	-2.066
1900	21.599	214.786	194.760	38.051	261.687	60.611	-1.666
2000	21.519	215.892	195.789	40.207	261.186	50.041	-1.307
2100	21.449	216.940	196.771	42.355	260.677	39.497	-0.982
2200	21.389	217.937	197.711	44.497	260.162	28.976	-0.688
2300	21.337	218.886	198.611	46.633	259.642	18.479	-0.420
2400	21.291	219.794	199.475	48.764	259.116	8.005	-0.174
2476.570	21.260	220.462	200.114	50.394	------- FUGACITY = 1 bar -------		
2500	21.251	220.662	200.305	50.892	0.	0.	0.
2600	21.215	221.495	201.104	53.015	0.	0.	0.
2700	21.184	222.295	201.874	55.135	0.	0.	0.
2800	21.156	223.065	202.618	57.252	0.	0.	0.
2900	21.131	223.807	203.335	59.366	0.	0.	0.
3000	21.109	224.523	204.030	61.478	0.	0.	0.
3100	21.090	225.214	204.702	63.588	0.	0.	0.
3200	21.073	225.884	205.354	65.696	0.	0.	0.
3300	21.059	226.532	205.986	67.803	0.	0.	0.
3400	21.047	227.160	206.599	69.908	0.	0.	0.
3500	21.036	227.770	207.195	72.012	0.	0.	0.
3600	21.029	228.363	207.775	74.115	0.	0.	0.
3700	21.023	228.939	208.339	76.218	0.	0.	0.
3800	21.020	229.500	208.889	78.320	0.	0.	0.
3900	21.018	230.045	209.424	80.422	0.	0.	0.
4000	21.020	230.578	209.947	82.524	0.	0.	0.
4100	21.024	231.097	210.456	84.626	0.	0.	0.
4200	21.030	231.603	210.954	86.729	0.	0.	0.
4300	21.040	232.098	211.440	88.832	0.	0.	0.
4400	21.052	232.582	211.915	90.937	0.	0.	0.
4500	21.068	233.055	212.379	93.043	0.	0.	0.
4600	21.087	233.519	212.834	95.150	0.	0.	0.
4700	21.109	233.972	213.279	97.260	0.	0.	0.
4800	21.135	234.417	213.715	99.372	0.	0.	0.
4900	21.165	234.853	214.142	101.487	0.	0.	0.
5000	21.198	235.281	214.560	103.605	0.	0.	0.
5100	21.236	235.701	214.970	105.727	0.	0.	0.
5200	21.278	236.114	215.373	107.853	0.	0.	0.
5300	21.324	236.520	215.768	109.983	0.	0.	0.
5400	21.375	236.919	216.156	112.118	0.	0.	0.
5500	21.431	237.312	216.537	114.258	0.	0.	0.
5600	21.491	237.698	216.912	116.404	0.	0.	0.
5700	21.556	238.079	217.280	118.556	0.	0.	0.
5800	21.625	238.455	217.642	120.715	0.	0.	0.
5900	21.700	238.825	217.998	122.881	0.	0.	0.
6000	21.780	239.190	218.348	125.055	0.	0.	0.

PREVIOUS: CURRENT: June 1970 (1 bar)

Gallium (Ga) Ga_1(g)

GALLIUM, ION (Ga^+) IDEAL GAS M_r = 69.71945

$IP(Ga^+, g)$ = 165458 ± 5 cm^{-1}

$S°(298.15 K)$ = 161.790 ± 0.04 $J K^{-1} mol^{-1}$

$\Delta_f H°(0 K)$ = 849.8 ± 4.0 $kJ mol^{-1}$

$\Delta_f H°(298.15 K)$ = [856.648] $kJ mol^{-1}$

Electronic Levels and Quantum Weights

State	ε_i, cm^{-1}	g_i
$^1S_0$	0	1
$^3P_0$	47370	1
$^3P_1$	47816	3
$^3P_2$	48750	5

Enthalpy of Formation

$\Delta_f H°(Ga^+, g, 0 K)$ is calculated from $\Delta_f H°(Ga, g, 0 K)$ (1) using the spectroscopic value of IP(Ga) = 48387.63 ± 0.1 cm^{-1} (578.837 ± 0.001 $kJ mol^{-1}$) from Moore (2). The ionization limit is converted from cm^{-1} to $kJ mol^{-1}$ using the factor, 1 cm^{-1} = 0.01196266 $kJ cm^{-1}$, which is derived from the 1973 CODATA fundamental constants (3). Rosenstock et al. (4) and Levin and Lias (5) have summarized additional ionization and appearance potential data.

$\Delta_f H°(Ga^+, g, 298.15 K)$ is calculated from $\Delta_f H°(Ga, g, 0 K)$ by using IP(Ga) with JANAF (1) enthalpies, H°(0 K) - H°(298.15 K), for Ga(g), $Ga^+(g)$, and $e^-(g)$. $\Delta_r H°(Ga \rightarrow Ga^+ + e^-, 298.15 K)$ differs from a room temperature threshold energy due to inclusion of these enthalpies and to threshold effects discussed by Rosenstock et al. (4). $\Delta_f H°(298.15 K)$ should be changed by -6.197 $kJ mol^{-1}$ if it is to be used in the ion convention that excludes the enthalpy of the electron.

Heat Capacity and Entropy

The information on electronic energy levels and quantum weights, given by Moore (2, 6) is incomplete because many theoretically predicted levels have not been observed. Our calcualtions indicate that any reasonable method of filling in these missing levels and cutting off the summation in the partition function (7) has no effect on the thermodynamic functions to 6000 K. This is a result of the high energy of all levels other than the ground state and the first three excited states; the next excited state is 70700 cm^{-1} above the ground state. Since inclusion of these excited states has no effect on the thermodynamic functions (to 6000 K), we list only the ground state and the first three excited states. The position of the $^3P_0$ levels are uncertain by 249 cm^{-1}. The reported uncertainty in S°(298.15 K) is due to uncertainties in the relative ionic mass and the fundamental constants. Extension of these calculations above 6000 K may require consideration of the excited states and use of different fill and cutoff procedures (7).

References

1. JANAF Thermochemical Tables: Ga(g); $e^-(g)$, 3-31-82.
2. C. E. Moore, U. S. Nat. Bur. Stand., NSRDS-NBS-34, 1970.
3. E. R. Cohen and B. N. Taylor, J. Phys. Chem. Ref. Data 2, 663 (1973).
4. H. M. Rosenstock, K. Draxl et al., J. Phys. Chem. Ref. Data 6, Supp. 1 (1977).
5. R. D. Levin and S. G. Lias, U. S. Nat. Bur. Stand., NSRDS-NBS-71, 1982.
6. C. E. Moore, U. S. Nat. Bur. Stand., NSRDS-NBS-35, Volume II, 1970 [Reprint of NBS Circular 467, Volume II, 1952].
7. J. R. Downey, Jr., The Dow Chemical Company, AFOSR-TR-0960, Contract No. F44620-75-1-0048, 1978.

Gallium, Ion (Ga^+) $Ga_1^+(g)$

Enthalpy Reference Temperature = T_r = 298.15 K   Standard State Pressure = p° = 0.1 MPa

T/K	$C_p°$ (J K⁻¹mol⁻¹)	S° (J K⁻¹mol⁻¹)	$-[G°-H°(T_r)]/T$ (J K⁻¹mol⁻¹)	$H°-H°(T_r)$ (kJ mol⁻¹)	$\Delta_f H°$ (kJ mol⁻¹)	$\Delta_f G°$ (kJ mol⁻¹)	Log K_f
0	0.	0.	INFINITE	-6.197	849.815		
100	20.786	139.083	180.270	-4.119			
200	20.786	153.491	163.691	-2.040			
250	20.786	158.129	162.132	-1.001			
298.15	20.786	161.790	161.790	0.	856.648	814.332	-142.668
300	20.786	161.919	161.790	0.038	856.677	814.069	-141.742
350	20.786	165.123	162.044	1.078	851.767	807.783	-120.555
400	20.786	167.898	162.606	2.117	852.475	801.451	-104.659
450	20.786	170.347	163.333	3.156	853.201	795.030	-92.285
500	20.786	172.537	164.145	4.196	853.935	788.527	-82.377
600	20.786	176.326	165.869	6.274	855.417	775.307	-67.496
700	20.786	179.531	167.598	8.353	856.912	761.837	-56.849
800	20.786	182.306	169.267	10.431	858.413	748.152	-48.849
900	20.786	184.754	170.854	12.510	859.913	734.280	-42.616
1000	20.786	186.944	172.356	14.589	861.413	720.240	-37.621
1100	20.786	188.926	173.774	16.667	862.914	706.050	-33.528
1200	20.786	190.734	175.113	18.746	864.414	691.723	-30.110
1300	20.786	192.398	176.379	20.824	865.914	677.272	-27.213
1400	20.786	193.938	177.579	22.903	867.415	662.705	-24.726
1500	20.786	195.372	178.718	24.982	868.915	648.030	-22.566
1600	20.786	196.714	179.801	27.060	870.415	633.255	-20.674
1700	20.786	197.974	180.834	29.139	871.916	618.387	-19.001
1800	20.786	199.162	181.819	31.217	873.416	603.430	-17.511
1900	20.786	200.286	182.762	33.296	874.917	588.390	-16.176
2000	20.786	201.352	183.665	35.375	876.417	573.271	-14.972
2100	20.786	202.366	184.532	37.453	877.917	558.077	-13.881
2200	20.786	203.333	185.364	39.532	879.418	542.811	-12.888
2300	20.786	204.257	186.166	41.610	880.918	527.477	-11.979
2400	20.786	205.142	186.938	43.689	882.418	512.078	-11.145
2500	20.786	205.991	187.683	45.768	625.333	499.064	-10.427
2600	20.786	206.806	188.403	47.846	627.367	493.973	-9.924
2700	20.786	207.590	189.100	49.925	629.404	488.804	-9.456
2800	20.786	208.346	189.773	52.004	631.444	483.559	-9.021
2900	20.786	209.076	190.427	54.082	633.487	478.241	-8.614
3000	20.786	209.780	191.060	56.161	635.533	472.853	-8.233
3100	20.786	210.462	191.675	58.239	637.580	467.397	-7.876
3200	20.786	211.122	192.272	60.318	639.629	461.874	-7.539
3300	20.786	211.761	192.853	62.397	641.679	456.288	-7.222
3400	20.786	212.382	193.419	64.475	643.731	450.639	-6.923
3500	20.786	212.984	193.969	66.554	645.785	444.930	-6.640
3600	20.786	213.570	194.505	68.632	647.838	439.162	-6.372
3700	20.786	214.140	195.028	70.711	649.893	433.337	-6.118
3800	20.786	214.694	195.539	72.790	651.948	427.457	-5.876
3900	20.786	215.234	196.037	74.868	654.004	421.522	-5.646
4000	20.787	215.760	196.523	76.947	656.059	415.535	-5.426
4100	20.787	216.273	196.999	79.026	658.114	409.497	-5.217
4200	20.787	216.774	197.464	81.104	660.169	403.408	-5.017
4300	20.788	217.263	197.919	83.183	662.223	397.270	-4.826
4400	20.789	217.741	198.364	85.262	664.275	391.085	-4.643
4500	20.790	218.209	198.799	87.341	666.327	384.853	-4.467
4600	20.791	218.665	199.226	89.420	668.377	378.575	-4.299
4700	20.792	219.113	199.645	91.499	670.425	372.253	-4.137
4800	20.794	219.550	200.055	93.578	672.471	365.887	-3.982
4900	20.797	219.979	200.457	95.658	674.514	359.479	-3.832
5000	20.799	220.399	200.852	97.738	676.554	353.029	-3.688
5100	20.803	220.811	201.239	99.818	678.591	346.538	-3.549
5200	20.807	221.215	201.619	101.898	680.625	340.008	-3.415
5300	20.812	221.612	201.993	103.979	682.654	333.438	-3.286
5400	20.818	222.001	202.360	106.061	684.679	326.830	-3.161
5500	20.825	222.383	202.720	108.143	686.700	320.184	-3.041
5600	20.833	222.758	203.075	110.226	688.715	313.502	-2.924
5700	20.843	223.127	203.423	112.310	690.726	306.784	-2.811
5800	20.854	223.489	203.766	114.394	692.730	300.031	-2.702
5900	20.866	223.846	204.104	116.480	694.728	293.243	-2.596
6000	20.880	224.197	204.436	118.568	696.720	286.422	-2.494

PREVIOUS: CURRENT: June 1983 (1 bar)

Gallium, Ion (Ga^+) $Ga_1^+(g)$

GALLIUM, ION (Ga^-) IDEAL GAS M_r = 69.72055

EA(Ga, g) = 0.30 ± 0.15 eV

$S°(298.15\ K)$ = [175.801] J K^{-1} mol^{-1}

$\Delta_f H°(0\ K)$ = 242.025 ± kJ mol^{-1}

$\Delta_f H°(298.15\ K)$ = [238.531] kJ mol^{-1}

Electonic Levels and Quantum Weights

State	ϵ_i, cm^{-1}	g_i
$^3P_0$	0.0	1
$^3P_1$	220	3
$^3P_2$	580	5

Enthalpy of Formation

$\Delta_f H°(Ga^-, g, 0\ K)$ is calculated from $\Delta_f H°(Ga, g, 0\ K)$ (1) using the adopted electron affinity of EA(Ga) = 0.30 ± 0.15 eV (28.945 ± 14.473 kJ mol^{-1}). This value, recommended by Hotop and Lineberger (2), is based on a photodetachment threshold (using conventional light sources) and semiempirical extrapolation (3, 4, 5). Additional information on $Ga^-(g)$ may be obtained in the critical discussions of Hotop and Lineberger (2, 5), Rosenstock et al. (6) and Massey (7).

$\Delta_f H°(Ga^-, g, 298.15\ K)$ is obtained from $\Delta_f H°(Ga, g, 0\ K)$ by using EA(Ga) with JANAF (1) enthalpies, H°(0 K) - H°(298.15 K), for $Ga^-(g)$, Ga(g), and $e^-(g)$. $\Delta_r H°(Ga^- \rightarrow Ga + e^-, 298.15\ K)$ differs from a room-temperature threshold energy due to inclusion of these enthalpies and to threshold effects discussed by Rosenstock et al. (6). $\Delta_f H°(298.15\ K)$ should be changed by +6.197 kJ mol^{-1} if it is to be used in the ion convention that excludes the enthalpy of the electron.

Heat Capacity and Entropy

The ground state electronic configuration for $Ga^-(g)$ is given by Hotop and Lineberger (2, 5), Rosenstocsk et al. (6), and Massey (7). The fine-structure separation has been calculated via an isoelectronic extrapolation of ratios of fine structure separations and a quadratic isoelectronic extrapolation (4, 5) and is that recommended by Hotop and Lineberger (2).

A comparison of the isoelectronic sequences - $Ga^-(g)$, Ge(g), $As^+(g)$, $Se^{++}(g)$ - would suggest that a state, $^1D_2$, may exist at low wave numbers. However, this is most likely a metasable state lying at an energy greater than the electron affinity, as in the case of $Al^-(g)$. As discussed by Hotop and Lineberger (2, 5), Rosenstock et al., (6) and Massey (7), no stable excited states have been observed. Thus, we assume no stable excited states exist.

References

1. JANAF Thermochemical Tables: Ga(g); $e^-(g)$, 3-31-82.
2. H. Hotop and W. C. Lineberger, J. Phys. Chem. Ref. Data, to be published, 1985.
3. D. Feldmann, R. Rackwitz, E. Heinicke and H. J. Kaisu, Z. Naturf. 32a, 302 (1977).
4. R. J. Zollweg, J. Chem. Phys. 50, 4251 (1969).
5. H. Hotop and W. C. Lineberger, J. Phys. Chem. Ref. Data 4, 539 (1975).
6. H. M. Rosenstock, K. Draxl et al., J. Phys. Chem. Ref. Data 6, Supp. 1 (1977).
7. H. S. W. Massey, "Negative Ions", 3rd ed., Cambridge University Press, Cambridge, 1976.

Gallium, Ion (Ga^-) $Ga_1^-(g)$

Enthalpy Reference Temperature = T_r = 298.15 K Standard State Pressure = p° = 0.1 MPa

T/K	$C_p°$ (J $K^{-1}mol^{-1}$)	S°	-[G°-H°(T_r)]/T	H°-H°(T_r) (kJ mol^{-1})	$\Delta_f H°$	$\Delta_f G°$	Log K_f
0	0.	0.	INFINITE	-8.265	242.025		
100	29.645	143.110	201.948	-5.884			
200	30.076	164.238	178.410	-2.834			
250	28.867	170.824	176.263	-1.360			
298.15	27.618	175.801	175.801	0.	238.531	204.547	-35.836
300	27.572	175.972	175.802	0.051	238.496	204.336	-35.578
350	26.407	180.132	176.133	1.400	231.817	199.597	-29.788
400	25.437	183.593	176.855	2.695	230.702	195.070	-25.474
450	24.656	186.542	177.772	3.947	229.562	190.684	-22.134
500	24.034	189.107	178.780	5.163	228.395	186.427	-19.476
600	23.155	193.405	180.872	7.519	225.996	178.257	-15.519
700	22.612	196.930	182.922	9.805	223.542	170.494	-12.722
800	22.287	199.926	184.865	12.049	221.050	163.086	-10.648
900	22.106	202.539	186.687	14.268	218.533	155.991	-9.053
1000	22.021	204.863	188.390	16.473	216.003	149.177	-7.792
1100	21.998	206.961	189.985	18.674	213.469	142.617	-6.772
1200	22.012	208.875	191.480	20.874	210.933	136.288	-5.932
1300	22.049	210.638	192.887	23.077	208.401	130.171	-5.230
1400	22.097	212.274	194.214	25.284	205.873	124.248	-4.636
1500	22.149	213.801	195.469	27.497	203.350	118.505	-4.127
1600	22.199	215.232	196.660	29.714	200.832	112.931	-3.687
1700	22.246	216.579	197.793	31.936	198.318	107.515	-3.304
1800	22.287	217.852	198.872	34.163	195.810	102.246	-2.967
1900	22.323	219.058	199.903	36.394	193.305	97.116	-2.670
2000	22.353	220.203	200.890	38.627	190.803	92.118	-2.406
2100	22.377	221.295	201.836	40.864	188.304	87.245	-2.170
2200	22.395	222.336	202.744	43.103	185.807	82.491	-1.959
2300	22.408	223.332	203.618	45.343	183.312	77.851	-1.768
2400	22.417	224.286	204.459	47.584	180.818	73.319	-1.596
2500	22.422	225.201	205.270	49.826	-80.261	71.339	-1.491
2600	22.423	226.080	206.054	52.068	-82.221	77.442	-1.556
2700	22.422	226.927	206.812	54.311	-84.177	83.620	-1.618
2800	22.417	227.742	207.545	56.553	-86.131	89.870	-1.677
2900	22.410	228.528	208.255	58.794	-88.082	96.190	-1.733
3000	22.402	229.288	208.943	61.035	-90.032	102.578	-1.786
3100	22.391	230.022	209.611	63.274	-91.981	109.030	-1.837
3200	22.379	230.733	210.260	65.513	-93.929	115.545	-1.886
3300	22.366	231.422	210.891	67.750	-95.877	122.122	-1.933
3400	22.351	232.089	211.505	69.986	-97.825	128.757	-1.978
3500	22.336	232.737	212.102	72.220	-99.774	135.449	-2.021
3600	22.320	233.366	212.684	74.453	-101.723	142.198	-2.063
3700	22.303	233.977	213.252	76.684	-103.673	149.000	-2.104
3800	22.286	234.572	213.805	78.914	-105.624	155.855	-2.142
3900	22.268	235.150	214.345	81.141	-107.577	162.762	-2.180
4000	22.250	235.714	214.872	83.367	-109.531	169.718	-2.216
4100	22.232	236.263	215.387	85.591	-111.488	176.724	-2.251
4200	22.214	236.799	215.891	87.814	-113.447	183.777	-2.286
4300	22.195	237.321	216.383	90.034	-115.408	190.877	-2.319
4400	22.176	237.831	216.865	92.253	-117.373	198.022	-2.351
4500	22.158	238.329	217.336	94.469	-119.341	205.213	-2.382
4600	22.139	238.816	217.798	96.684	-121.312	212.447	-2.412
4700	22.120	239.292	218.250	98.897	-123.288	219.724	-2.442
4800	22.102	239.757	218.693	101.108	-125.267	227.043	-2.471
4900	22.083	240.213	219.128	103.318	-127.252	234.403	-2.499
5000	22.065	240.659	219.554	105.525	-129.241	241.804	-2.526
5100	22.047	241.096	219.972	107.731	-131.236	249.245	-2.553
5200	22.029	241.524	220.382	109.934	-133.236	256.724	-2.579
5300	22.011	241.943	220.785	112.136	-135.243	264.243	-2.604
5400	21.993	242.354	221.181	114.336	-137.256	271.799	-2.629
5500	21.976	242.758	221.570	116.535	-139.277	279.393	-2.653
5600	21.958	243.154	221.951	118.732	-141.305	287.023	-2.677
5700	21.941	243.542	222.327	120.927	-143.341	294.690	-2.701
5800	21.924	243.924	222.696	123.120	-145.385	302.393	-2.723
5900	21.908	244.298	223.059	125.311	-147.438	310.131	-2.746
6000	21.891	244.666	223.416	127.501	-149.501	317.903	-2.768

PREVIOUS: CURRENT: June 1983 (1 bar)

Gallium, Ion (Ga^-) $Ga_1^-(g)$

HYDROGEN (H) IDEAL GAS $A_r = 1.00794$ Hydrogen (H) $H_1(g)$

$S°(298.15\ K) = 114.716 \pm 0.017\ J\ K^{-1}\ mol^{-1}$

$\Delta_f H°(0\ K) = 216.035 \pm 0.006\ kJ\ mol^{-1}$

$\Delta_f H°(298.15\ K) = 217.999 \pm 0.006\ kJ\ mol^{-1}$

Electronic Level and Quantum Weight

State	ε_i, cm^{-1}	g_i
$^2S_{1/2}$	0.00	2

Enthalpy of Formation

The enthalpy of formation is calculated from the dissociation energy, $D_0^0(H_2)$ = 36118.3 ± 1 cm^{-1} (432.071 ± 0.012 kJ mol^{-1}), from Herzberg (2) and auxiliary data from H_2 (3). The adopted value for the dissociation energy of hydrogen is the value recommended by CODATA (1).

Earlier experimental values for $D_0^0(H_2)$ were obtained by Herzberg and Monfils (36113.0±0.3 cm^{-1}, 4) and Beutler (36116±6 cm^{-1}, 5). Kolos and Wolniewicz (6) calculated the adiabatic dissociation energy of H_2, corrected for relativistic and radiative effects, to be 36117.4 cm^{-1}.

Heat Capacity and Entropy

The electronic levels for H(g) are given in the compilation by Moore (7). Our calculations indicate that the inclusion of levels through n = 12 has no effect on the thermodynamic functions to 6000 K. This is a result of the high energy of these levels; the first excited state lies at 82258 cm^{-1} above the ground state. Since the inclusion of these upper levels has no effect on the thermodynamic functions (to 6000 K) we list only the ground state. The reported uncertainty in S°(298.15 K) is due to uncertainties in the relative atomic mass and the fundamental constants. Extension of these calculations above 6000 K may require consideration of the higher excited states and utilization of proper fill and cut off procedures (8).

The thermal functions at 298.15 K differ from the CODATA recommendations for two reasons: a difference of 0.001 J K^{-1} mol^{-1} in the entropy due to the use of more current fundamental constants and a difference of 0.1094 J K^{-1} mol^{-1} in the entropy due to the use of a different standard state pressure.

References

1. J. D. Cox, chairman, ICSU-CODATA Task Group on Key Values for Thermodynamics, J. Chem. Thermodynamics 10, 903 (1978).
2. G. Herzberg, J. Mol. Spectrosc. 33, 147 (1970).
3. JANAF Thermochemical Tables: $H_2(g)$, 3-31-77.
4. G. Herzberg and A. Monfils, J. Mol. Spectrosc. 5, 482 (1960).
5. H. Beutler, Z. Phys. Chem. B39, 315 (1935).
6. W. Kolos and L. Wolniewicz, J. Chem. Phys. 49, 404 (1968).
7. C. E. Moore, NSRDS-NBS 3, Section 6 (1972).
8. J. R. Downey, Jr., Dow Chemical Company, Thermal Research, to be published, 1977.
9. E. R. Cohen and B. Taylor, J. Phys. Chem. Ref. Data 2, 663 (1973).

Enthalpy Reference Temperature = T_r = 298.15 K — Standard State Pressure = p° = 0.1 MPa

T/K	$C_p°$ (J $K^{-1}mol^{-1}$)	S° (J $K^{-1}mol^{-1}$)	$-[G°-H°(T_r)]/T$ (J $K^{-1}mol^{-1}$)	$H°-H°(T_r)$ (kJ mol^{-1})	$\Delta_f H°$ (kJ mol^{-1})	$\Delta_f G°$ (kJ mol^{-1})	Log K_f
0	0.	0.	INFINITE	-6.197	216.035	216.035	INFINITE
100	20.786	92.009	133.197	-4.119	216.614	212.450	-110.972
200	20.786	106.417	116.618	-2.040	217.346	208.004	-54.325
250	20.786	111.055	115.059	-1.001	217.687	205.629	-42.964
298.15	20.786	114.716	114.716	0.	217.999	203.278	-35.613
300	20.786	114.845	114.717	0.038	218.011	203.186	-35.378
350	20.786	118.049	114.970	1.078	218.326	200.690	-29.951
400	20.786	120.825	115.532	2.117	218.637	198.150	-25.876
450	20.786	123.273	116.259	3.156	218.946	195.570	-22.701
500	20.786	125.463	117.072	4.196	219.254	192.957	-20.158
600	20.786	129.253	118.796	6.274	219.868	187.640	-16.335
700	20.786	132.457	120.524	8.353	220.478	182.220	-13.597
800	20.786	135.232	122.193	10.431	221.080	176.713	-11.538
900	20.786	137.681	123.781	12.510	221.671	171.132	-9.932
1000	20.786	139.871	125.282	14.589	222.248	165.485	-8.644
1100	20.786	141.852	126.700	16.667	222.807	159.782	-7.587
1200	20.786	143.660	128.039	18.746	223.346	154.028	-6.705
1300	20.786	145.324	129.305	20.824	223.865	148.230	-5.956
1400	20.786	146.865	130.505	22.903	224.361	142.394	-5.313
1500	20.786	148.299	131.644	24.982	224.836	136.522	-4.754
1600	20.786	149.640	132.728	27.060	225.289	130.620	-4.264
1700	20.786	150.900	133.760	29.139	225.721	124.689	-3.831
1800	20.786	152.088	134.745	31.217	226.132	118.734	-3.446
1900	20.786	153.212	135.688	33.296	226.525	112.757	-3.100
2000	20.786	154.278	136.591	35.375	226.898	106.760	-2.788
2100	20.786	155.293	137.458	37.453	227.254	100.744	-2.506
2200	20.786	156.260	138.291	39.532	227.593	94.712	-2.249
2300	20.786	157.184	139.092	41.610	227.916	88.664	-2.014
2400	20.786	158.068	139.864	43.689	228.224	82.603	-1.798
2500	20.786	158.917	140.610	45.768	228.518	76.530	-1.599
2600	20.786	159.732	141.330	47.846	228.798	70.444	-1.415
2700	20.786	160.516	142.026	49.925	229.064	64.349	-1.245
2800	20.786	161.272	142.700	52.004	229.318	58.243	-1.087
2900	20.786	162.002	143.353	54.082	229.560	52.129	-0.939
3000	20.786	162.706	143.986	56.161	229.790	46.007	-0.801
3100	20.786	163.388	144.601	58.239	230.008	39.877	-0.672
3200	20.786	164.048	145.199	60.318	230.216	33.741	-0.551
3300	20.786	164.688	145.780	62.397	230.413	27.598	-0.437
3400	20.786	165.308	146.345	64.475	230.599	21.449	-0.330
3500	20.786	165.911	146.895	66.554	230.776	15.295	-0.228
3600	20.786	166.496	147.432	68.632	230.942	9.136	-0.133
3700	20.786	167.066	147.955	70.711	231.098	2.973	-0.042
3800	20.786	167.620	148.465	72.790	231.244	-3.195	0.044
3900	20.786	168.160	148.963	74.868	231.381	-9.366	0.125
4000	20.786	168.686	149.450	76.947	231.509	-15.541	0.203
4100	20.786	169.200	149.925	79.025	231.627	-21.718	0.277
4200	20.786	169.700	150.390	81.104	231.736	-27.899	0.347
4300	20.786	170.190	150.845	83.183	231.836	-34.082	0.414
4400	20.786	170.667	151.290	85.261	231.927	-40.267	0.478
4500	20.786	171.135	151.726	87.340	232.009	-46.454	0.539
4600	20.786	171.591	152.153	89.418	232.082	-52.643	0.598
4700	20.786	172.038	152.571	91.497	232.147	-58.834	0.654
4800	20.786	172.476	152.981	93.576	232.204	-65.025	0.708
4900	20.786	172.905	153.383	95.654	232.253	-71.218	0.759
5000	20.786	173.325	153.778	97.733	232.294	-77.412	0.809
5100	20.786	173.736	154.165	99.811	232.327	-83.606	0.856
5200	20.786	174.140	154.546	101.890	232.353	-89.801	0.902
5300	20.786	174.536	154.919	103.969	232.373	-95.997	0.946
5400	20.786	174.924	155.286	106.047	232.386	-102.192	0.989
5500	20.786	175.306	155.646	108.126	232.392	-108.389	1.029
5600	20.786	175.680	156.001	110.204	232.393	-114.584	1.069
5700	20.786	176.048	156.349	112.283	232.389	-120.780	1.107
5800	20.786	176.410	156.692	114.362	232.379	-126.976	1.144
5900	20.786	176.765	157.029	116.440	232.365	-133.172	1.179
6000	20.786	177.114	157.361	118.519	232.348	-139.368	1.213

PREVIOUS: March 1977 (1 atm) CURRENT: March 1982 (1 bar)

Hydrogen (H) $H_1(g)$

HYDROGEN, ION (H^+) IDEAL GAS Mr = 1.00739

$S°(298.15\ K) = 108.946 \pm 0.02\ JK^{-1}\ mol^{-1}$

$\Delta_f H°(0\ K) = 1528.085 \pm 0.04\ kJ\ mol^{-1}$
$\Delta_f H°(298.15\ K) = [1536.246]\ kJ\ mol^{-1}$

Enthalpy of Formation

$\Delta_f H°(H^+, g, 0\ K)$ is calculated from $\Delta_f H°(H, g, 0\ K)$ (1) using the spectroscopic value of IP(H) = 109678.764 ± 0.005 cm^{-1} (1312.0498 ± 0.0001 kJ mol^{-1}) from Moore (2). The ionization limit is converted from cm^{-1} to kJ mol^{-1} using the factor, 1 cm^{-1} = 0.01196266 kJ cm^{-1}, which is derived from the 1973 CODATA fundamental constants (3). Rosenstock et al. (4) and Levin and Lias (5) have summarized additional ionization and appearance potential data.

$\Delta_f H°(H^+, g, 298.15\ K)$ is calculated from $\Delta_f H°(H, g, 0\ K)$ by using IP(H) with JANAF (1) enthalpies, H°(0 K) - H°(298.15 K), for H(g), $H^+(g)$, and $e^-(g)$. $\Delta_f H°(H \rightarrow H^+ + e^-, 298.15\ K)$ differs from a room temperature threshold energy due to inclusion of these enthalpies and to threshold effects discussed by Rosenstock et al. (4). $\Delta_f H°(298.15\ K)$ should be changed by -6.197 kJ mol^{-1} if it is to be used in the ion convention that excludes the enthalpy of the electron.

Heat Capacity and Entropy

The thermodynamic functions of the proton gas are calculated using the recent CODATA fundamental constants (2) and assuming that the proton is an ideal monatomic gas. Since there is no electron associated with this species, there is only a translational contribution to the thermochemical function.

References

1. JANAF Thermochemical Tables: H(g), 3-31-82 ; $e^-(g)$, 3-31-82.
2. C. E. Moore, U. S. Nat. Bur. Stand., NSRDS-NBS-3, Section 6, 1972.
3. E. R. Cohen and B. N. Taylor, J. Phys. Chem. Ref. Data 2, 663 (1973).
4. H. M. Rosenstock, K. Draxl et al., J. Phys. Chem. Ref. Data 6, Supp. 1 (1977).
5. R. D. Levin and S. G. Lias, U. S. Nat. Bur. Stand., NSRDS-NBS-71, 1982.

Hydrogen, Ion (H^+) $H_1^+(g)$

Enthalpy Reference Temperature = T_r = 298.15 K — Standard State Pressure = p° = 0.1 MPa

	J K⁻¹mol⁻¹			kJ mol⁻¹			
T/K	$C_p°$	S°	$-[G°-H°(T_r)]/T$	$H°-H°(T_r)$	$\Delta_f H°$	$\Delta_f G°$	Log K_f
0	0.	0.	INFINITE	-6.197	1528.085		
100	20.786	86.239	127.427	-4.119			
200	20.786	100.647	110.848	-2.040			
250	20.786	105.285	109.289	-1.001			
298.15	20.786	108.946	108.946	0.	1536.246	1516.990	-265.770
300	20.786	109.075	108.947	0.038	1536.297	1516.871	-264.111
350	20.786	112.279	109.200	1.078	1537.651	1513.526	-225.881
400	20.786	115.055	109.762	2.117	1539.001	1509.987	-197.184
450	20.786	117.503	110.489	3.156	1540.349	1506.280	-174.844
500	20.786	119.693	111.302	4.196	1541.697	1502.422	-156.957
600	20.786	123.483	113.026	6.274	1544.389	1494.314	-130.092
700	20.786	126.687	114.754	8.353	1547.078	1485.755	-110.868
800	20.786	129.463	116.423	10.431	1549.758	1476.812	-96.426
900	20.786	131.911	118.011	12.510	1552.428	1467.533	-85.173
1000	20.786	134.101	119.512	14.589	1555.084	1457.958	-76.156
1100	20.786	136.082	120.930	16.667	1557.721	1448.117	-68.765
1200	20.786	137.891	122.269	18.746	1560.339	1438.038	-62.596
1300	20.786	139.554	123.536	20.824	1562.936	1427.740	-57.367
1400	20.786	141.095	124.735	22.903	1565.511	1417.244	-52.878
1500	20.786	142.529	125.874	24.982	1568.065	1406.564	-48.981
1600	20.786	143.870	126.958	27.060	1570.596	1395.715	-45.565
1700	20.786	145.130	127.990	29.139	1573.107	1384.708	-42.547
1800	20.786	146.319	128.976	31.217	1575.597	1373.554	-39.859
1900	20.786	147.442	129.918	33.296	1578.068	1362.261	-37.451
2000	20.786	148.509	130.821	35.375	1580.520	1350.840	-35.280
2100	20.786	149.523	131.688	37.453	1582.954	1339.296	-33.313
2200	20.786	150.490	132.521	39.532	1585.372	1327.636	-31.522
2300	20.786	151.414	133.322	41.610	1587.774	1315.867	-29.884
2400	20.786	152.298	134.095	43.689	1590.161	1303.994	-28.381
2500	20.786	153.147	134.840	45.768	1592.533	1292.021	-26.995
2600	20.786	153.962	135.560	47.846	1594.891	1279.954	-25.715
2700	20.786	154.747	136.256	49.925	1597.236	1267.797	-24.527
2800	20.786	155.503	136.930	52.004	1599.569	1255.553	-23.423
2900	20.786	156.232	137.583	54.082	1601.889	1243.225	-22.393
3000	20.786	156.937	138.216	56.161	1604.198	1230.818	-21.430
3100	20.786	157.618	138.831	58.239	1606.495	1218.334	-20.529
3200	20.786	158.278	139.429	60.318	1608.781	1205.777	-19.682
3300	20.786	158.918	140.010	62.397	1611.057	1193.148	-18.886
3400	20.786	159.538	140.575	64.475	1613.322	1180.450	-18.135
3500	20.786	160.141	141.125	66.554	1615.576	1167.685	-17.427
3600	20.786	160.726	141.662	68.632	1617.821	1154.857	-16.757
3700	20.786	161.296	142.185	70.711	1620.056	1141.966	-16.122
3800	20.786	161.850	142.695	72.790	1622.281	1129.014	-15.519
3900	20.786	162.390	143.193	74.868	1624.497	1116.005	-14.947
4000	20.786	162.916	143.680	76.947	1626.703	1102.938	-14.403
4100	20.786	163.430	144.155	79.025	1628.899	1089.817	-13.884
4200	20.786	163.931	144.620	81.104	1631.087	1076.642	-13.390
4300	20.786	164.420	145.075	83.183	1633.265	1063.416	-12.918
4400	20.786	164.897	145.520	85.261	1635.435	1050.138	-12.467
4500	20.786	165.365	145.956	87.340	1637.596	1036.812	-12.035
4600	20.786	165.821	146.383	89.418	1639.748	1023.437	-11.621
4700	20.786	166.268	146.801	91.497	1641.891	1010.016	-11.225
4800	20.786	166.706	147.211	93.576	1644.027	996.549	-10.845
4900	20.786	167.135	147.613	95.654	1646.154	983.038	-10.479
5000	20.786	167.555	148.008	97.733	1648.274	969.484	-10.128
5100	20.786	167.966	148.395	99.811	1650.386	955.887	-9.790
5200	20.786	168.370	148.776	101.890	1652.491	942.249	-9.465
5300	20.786	168.766	149.149	103.969	1654.588	928.570	-9.152
5400	20.786	169.154	149.516	106.047	1656.680	914.852	-8.849
5500	20.786	169.536	149.877	108.126	1658.765	901.095	-8.558
5600	20.786	169.910	150.231	110.204	1660.845	887.301	-8.276
5700	20.786	170.278	150.579	112.283	1662.919	873.469	-8.004
5800	20.786	170.640	150.922	114.362	1664.988	859.601	-7.742
5900	20.786	170.995	151.259	116.440	1667.053	845.697	-7.487
6000	20.786	171.344	151.591	118.519	1669.113	831.758	-7.241

PREVIOUS: March 1977 (1 atm) CURRENT: March 1982 (1 bar)

Hydrogen, Ion (H^+) $H_1^+(g)$

HYDROGEN, ION (H^-) IDEAL GAS M_r = 1.00849

EA(H, g) = 0.754209 ± 0.000003 eV
S°(298.15 K) = 108.960 ± 0.017 J K^{-1} mol^{-1}

$\Delta_f H°$(0 K) = 143.27 ± 0.02 kJ mol^{-1}
$\Delta_f H°$(298.15 K) = [139.032] kJ mol^{-1}

Electronic Level and Quantum Weight

State	ε_i, cm^{-1}	g_i
$^1S_0$	0.0	1

Enthalpy of Formation

$\Delta_f H°(H^-$, g, 0 K) is calculated from $\Delta_f H°$(H, g, 0 K) (1) using the adopted electron affinity of EA(H) = 0.754209 ± 0.000003 eV (72.7695 ± 0.0003 kJ mol^{-1}). This value, recommended by Hotop and Lineberger (2), is based on extensive Hylleras-type variational calculations on two election systems (3, 4, 5). Additional information on H^-(g) may be obtained in the critical discussions of Hotop and Lineberger (2, 6), Rosenstock et al. (7) and Massey (8). Experimentally, Dehmer and Chupka (9) have reported EA(H) ≥ 0.7540 ± 0.0003 eV.

$\Delta_f H°(H^-$, g, 298.15 K) is obtained from $\Delta_f H°$(H, g, 0 K) by using EA(H) with JANAF (1) enthalpies, H°(0 K) - H°(298.15 K), for H^-(g), H(g), and e^-(g). $\Delta_r H°(H^- \rightarrow H + e^-$, 298.15 K) differs from a room-temperature threshold energy due to inclusion of these enthalpies and to threshold effects discussed by Rosenstock et al (7). $\Delta_f H°$(298.15 K) should be changed by +6.197 kJ mol^{-1} if it is to be used in the ion convention that excludes the enthalpy of the electron.

Heat Capacity and Entropy

The ground state electronic configuration for H^-(g) is given by Hotop and Lineberger (2, 6) and Rosenstock et al. (7). A comparison of the isoelectronic sequence - H^-(g), He(g), Li^+(g) - would suggest that stable electronic states may exist at 0.8 EA(H) or roughly 6400 cm^{-1}. This would greatly affect the entropy. However, Pekeris (4) states that he was unable to find any bound states. In addition, Seman and Branscomb (10) state that theoretical and semiempirical evidence suggests that atomic negative ions have very few if any excited states below the continuum. We assume no stable excited states exist.

Reference

1. JANAF Thermochemical Tables: H(g) and e^-(g), 3-31-82.
2. H. Hotop and W. C. Lineberger, J. Phys. Chem. Ref. Data, to be published, 1985.
3. C. L. Pekeris, Phys. Rev. 112, 1649 (1958).
4. C. L. Pekeris, Phys. Rev. 126, 1470 (1962).
5. K. Aaskamar, Nucl. Instr. Meth. 90, 263 (1970).
6. H. Hotop and W. C. Lineberger, J. Phys. Chem. Ref. Data 4, 539 (1975).
7. H. M. Rosenstock, K. Draxel et al., J. Phys. Chem. Ref. Data 6, Supp. 1 (1977).
8. H. S. W. Massey, "Negative Ions", 3rd ed., Cambridge University Press, Cambridge, 1976.
9. P. M. Dehmer and W. A. Chupka, Bull. Am. Phys. Soc. 20, 729 (1975).
10. M. L. Seman and L. M. Branscomb, Phys. Rev. 125, 1602 (1962).

Hydrogen, Ion (H^-) H_1^-(g)

Enthalpy Reference Temperature = T_r = 298.15 K Standard State Pressure = p° = 0.1 MPa

T/K	$C_p°$ (J $K^{-1}mol^{-1}$)	S°	$-[G°-H°(T_r)]/T$	$H°-H°(T_r)$ (kJ mol^{-1})	$\Delta_f H°$	$\Delta_f G°$	Log K_f
0	0.	0.	INFINITE	-6.197	143.266		
100	20.786	86.253	127.440	-4.119			
200	20.786	100.661	110.861	-2.040			
250	20.786	105.299	109.302	-1.001			
298.15	20.786	108.960	108.960	0.	139.032	132.282	-23.175
300	20.786	109.089	108.960	0.038	139.006	132.240	-23.025
350	20.786	112.293	109.213	1.078	138.281	131.170	-19.576
400	20.786	115.068	109.776	2.117	137.553	130.203	-17.003
450	20.786	117.517	110.502	3.156	136.823	129.328	-15.012
500	20.786	119.707	111.315	4.196	136.091	128.535	-13.428
600	20.786	123.496	113.039	6.274	134.627	127.161	-11.070
700	20.786	126.701	114.768	8.353	133.158	126.033	-9.405
800	20.786	129.476	116.437	10.431	131.682	125.116	-8.169
900	20.786	131.924	118.024	12.510	130.194	124.384	-7.219
1000	20.786	134.114	119.526	14.589	128.692	123.819	-6.468
1100	20.786	136.096	120.943	16.667	127.173	123.405	-5.860
1200	20.786	137.904	122.283	18.746	125.634	123.130	-5.360
1300	20.786	139.568	123.549	20.824	124.073	122.985	-4.942
1400	20.786	141.108	124.749	22.903	122.491	122.960	-4.588
1500	20.786	142.542	125.888	24.982	120.887	123.050	-4.285
1600	20.786	143.884	126.971	27.060	119.262	123.247	-4.024
1700	20.786	145.144	128.004	29.139	117.615	123.546	-3.796
1800	20.786	146.332	128.989	31.217	115.948	123.943	-3.597
1900	20.786	147.456	129.932	33.296	114.262	124.433	-3.421
2000	20.786	148.522	130.835	35.375	112.557	125.012	-3.265
2100	20.786	149.536	131.701	37.453	110.834	125.677	-3.126
2200	20.786	150.503	132.534	39.532	109.094	126.425	-3.002
2300	20.786	151.427	133.336	41.610	107.339	127.252	-2.890
2400	20.786	152.312	134.108	43.689	105.568	128.156	-2.789
2500	20.786	153.160	134.853	45.768	103.783	129.133	-2.698
2600	20.786	153.976	135.573	47.846	101.984	130.183	-2.615
2700	20.786	154.760	136.269	49.925	100.172	131.302	-2.540
2800	20.786	155.516	136.943	52.004	98.348	132.488	-2.472
2900	20.786	156.245	137.596	54.082	96.511	133.740	-2.409
3000	20.786	156.950	138.230	56.161	94.662	135.055	-2.352
3100	20.786	157.632	138.845	58.239	92.802	136.432	-2.299
3200	20.786	158.292	139.442	60.318	90.931	137.869	-2.250
3300	20.786	158.931	140.023	62.397	89.049	139.365	-2.206
3400	20.786	159.552	140.589	64.475	87.157	140.918	-2.165
3500	20.786	160.154	141.139	66.554	85.255	142.527	-2.127
3600	20.786	160.740	141.675	68.632	83.342	144.190	-2.092
3700	20.786	161.309	142.198	70.711	81.420	145.907	-2.060
3800	20.786	161.864	142.709	72.790	79.488	147.676	-2.030
3900	20.786	162.404	143.207	74.868	77.546	149.496	-2.002
4000	20.786	162.930	143.693	76.947	75.595	151.365	-1.977
4100	20.786	163.443	144.169	79.025	73.635	153.284	-1.953
4200	20.786	163.944	144.634	81.104	71.665	155.250	-1.931
4300	20.786	164.433	145.088	83.183	69.686	157.264	-1.910
4400	20.786	164.911	145.534	85.261	67.699	159.323	-1.891
4500	20.786	165.378	145.969	87.340	65.702	161.428	-1.874
4600	20.786	165.835	146.396	89.418	63.697	163.577	-1.857
4700	20.786	166.282	146.815	91.497	61.684	165.770	-1.842
4800	20.786	166.720	147.225	93.576	59.662	168.007	-1.828
4900	20.786	167.148	147.627	95.654	57.632	170.285	-1.815
5000	20.786	167.568	148.022	97.733	55.594	172.604	-1.803
5100	20.786	167.980	148.409	99.811	53.549	174.965	-1.792
5200	20.786	168.383	148.789	101.890	51.497	177.366	-1.782
5300	20.786	168.779	149.163	103.969	49.437	179.806	-1.772
5400	20.786	169.168	149.530	106.047	47.372	182.285	-1.763
5500	20.786	169.549	149.890	108.126	45.300	184.802	-1.755
5600	20.786	169.924	150.245	110.204	43.222	187.358	-1.748
5700	20.786	170.292	150.593	112.283	41.139	189.950	-1.741
5800	20.786	170.653	150.936	114.362	39.051	192.579	-1.734
5900	20.786	171.009	151.273	116.440	36.958	195.244	-1.729
6000	20.786	171.358	151.605	118.519	34.862	197.944	-1.723

PREVIOUS: March 1977 (1 atm) CURRENT: March 1982 (1 bar)

Hydrogen, Ion (H^-) H_1^-(g)

MERCURY HYDRIDE (HgH) IDEAL GAS M_r = 201.59794

$S°(298.15\ K) = [219.705]\ J\ K^{-1}\ mol^{-1}$

$\Delta_f H°(0\ K) = 243.35 \pm 16.7\ kJ\ mol^{-1}$
$\Delta_f H°(298.15\ K) = 238.49 \pm 16.7\ kJ\ mol^{-1}$

Electronic Level and Quantum Weight

State	ε_i, cm^{-1}	g_i
$[^2\Sigma^+]$	0	[2]

$\omega_e = 1387\ cm^{-1}$ $\omega_e x_e = 83.01\ cm^{-1}$ $\sigma = 1$
$B_e = 5.549\ cm^{-1}$ $\alpha_e = 0.312\ cm^{-1}$ $r_e = 1.74\ Å$

Enthalpy of Formation

The enthalpy of formation is calculated from D^0_0 = 8.671 kcal mol^{-1} = taken from Herzberg (1).

Heat Capacity and Entropy

Molecular constants are taken from Herzberg (1).

Reference

1. G. Herzberg, "Spectra of Diatomic Molecules," 2nd Edition, D. Van Nostrand Company, Inc., New York (1950).

Mercury Hydride (HgH) $H_1Hg_1(g)$

Enthalpy Reference Temperature = T_r = 298.15 K Standard State Pressure = p° = 0.1 MPa

T/K	$C_p°$ (J K⁻¹mol⁻¹)	S° (J K⁻¹mol⁻¹)	$-[G°-H°(T_r)]/T$ (J K⁻¹mol⁻¹)	$H°-H°(T_r)$ (kJ mol⁻¹)	$\Delta_f H°$ (kJ mol⁻¹)	$\Delta_f G°$ (kJ mol⁻¹)	Log K_f
0	0.	0.	INFINITE	-8.711	243.353	243.353	INFINITE
100	29.132	187.686	245.898	-5.821	243.038	233.028	-121.721
200	29.265	207.901	222.422	-2.904	242.023	223.389	-58.343
250	29.563	214.459	220.197	-1.434	239.098	218.955	-45.748
298.15	30.043	219.705	219.705	0.	238.488	215.132	-37.690
300	30.064	219.891	219.706	0.056	238.465	214.987	-37.433
350	30.708	224.572	220.074	1.574	237.870	211.121	-31.508
400	31.417	228.719	220.900	3.127	237.318	207.338	-27.076
450	32.133	232.461	221.980	4.716	236.810	203.622	-23.636
500	32.822	235.882	223.202	6.340	236.342	199.960	-20.890
600	34.057	241.979	225.835	9.686	235.510	192.763	-16.782
700	35.084	247.308	228.530	13.145	176.026	192.269	-14.347
800	35.931	252.050	231.179	16.697	176.023	194.591	-12.705
900	36.636	256.324	233.739	20.326	176.086	196.908	-11.428
1000	37.233	260.216	236.195	24.020	176.200	199.216	-10.406
1100	37.747	263.789	238 544	27.770	176.352	201.511	-9.569
1200	38.199	267.093	240.787	31.568	176.531	203.790	-8.871
1300	38.603	270.167	242.930	35.408	176.733	206.054	-8.279
1400	38.970	273.042	244.979	39.287	176.951	208.301	-7.772
1500	39.308	275.742	246.941	43.201	177.183	210.533	-7.331
1600	39.623	278.289	248.822	47.148	177.425	212.748	-6.946
1700	39.919	280.700	250.626	51.125	177.677	214.948	-6.605
1800	40.201	282.990	252.361	55.131	177.938	217.133	-6.301
1900	40.471	285.171	254.031	59.165	178.206	219.303	-6.029
2000	40.731	287.253	255.641	63.225	178.483	221.459	-5.784
2100	40.983	289.247	257.194	67.311	178.767	223.601	-5.562
2200	41.227	291.159	258.695	71.422	179.060	225.729	-5.359
2300	41.466	292.997	260.146	75.556	179.361	227.844	-5.174
2400	41.700	294.767	261.552	79.715	179.670	229.945	-5.005
2500	41.930	296.473	262.915	83.896	179.988	232.033	-4.848
2600	42.157	298.122	264.238	88.101	180.315	234.109	-4.703
2700	42.380	299.718	265.522	92.327	180.651	236.171	-4.569
2800	42.601	301.263	266.771	96.576	180.996	238.221	-4.444
2900	42.819	302.762	267.987	100.847	181.352	240.259	-4.328
3000	43.035	304.217	269.170	105.140	181.717	242.284	-4.219
3100	43.249	305.631	270.324	109.454	182.093	244.297	-4.116
3200	43 462	307.008	271.449	113.790	182.479	246.297	-4.020
3300	43.674	308.349	272.547	118.147	182.875	248.285	-3.930
3400	43.884	309.656	273.619	122.525	183.283	250.261	-3.845
3500	44.093	310.931	274.667	126.924	183.700	252.225	-3.764
3600	44.301	312.176	275.691	131.343	184.129	254.177	-3.688
3700	44.508	313.392	276.694	135.784	184.568	256.117	-3.616
3800	44.715	314.582	277.675	140.245	185.018	258.044	-3.547
3900	44.921	315.746	278.637	144.727	185.479	259.960	-3.482
4000	45.126	316.886	279.579	149.229	185.951	261.864	-3.420
4100	45.330	318.003	280.502	153.752	186.434	263.756	-3.360
4200	45.534	319.098	281.408	158.295	186.928	265.636	-3.304
4300	45.738	320.171	282.297	162.859	187.433	267.504	-3.250
4400	45.941	321.225	283.170	167.443	187.949	269.360	-3.198
4500	46.143	322.260	284.027	172.047	188.476	271.204	-3.148
4600	46.346	323.276	284.870	176.671	189.013	273.037	-3.100
4700	46.548	324.275	285.697	181.316	189.562	274.858	-3.055
4800	46.749	325.257	286.511	185.981	190.121	276.667	-3.011
4900	46.950	326.223	287.312	190.666	190.692	278.464	-2.968
5000	47.152	327.174	288.100	195.371	191.274	280.249	-2.928
5100	47.352	328.110	288.875	200.096	191.866	282.022	-2.888
5200	47.553	329.031	289.638	204.841	192.470	283.785	-2.851
5300	47.753	329.939	290.390	209.607	193.085	285.535	-2.814
5400	47.954	330.833	291.131	214.392	193.711	287.273	-2.779
5500	48.154	331.715	291.861	219.197	194.348	289.000	-2.745
5600	48.353	332.584	292.580	224.023	194.996	290.715	-2.712
5700	48.553	333.442	293.290	228.868	195.656	292.419	-2.680
5800	48.753	334.288	293.989	233.733	196.326	294.110	-2.649
5900	48.952	335.123	294.679	238.618	197.008	295.790	-2.619
6000	49.151	335.948	295.360	243.524	197.701	297.459	-2.590

PREVIOUS: June 1963 (1 atm) CURRENT: June 1963 (1 bar)

Mercury Hydride (HgH) $H_1Hg_1(g)$

HYDROGEN IODIDE (HI) IDEAL GAS M_r = 127.91244

$S°(298.15\ K) = 206.589\ J\ K^{-1}\ mol^{-1}$

$\Delta_fH°(0\ K) = 28.535 \pm 0.21\ kJ\ mol^{-1}$
$\Delta_fH°(298.15\ K) = 26.359 \pm 0.21\ kJ\ mol^{-1}$

Electronic Level and Quantum Weight

State	ε_i, cm^{-1}	g_i
$^1\Sigma$	0	1

$\omega_e = 2309.06\ cm^{-1}$ $\omega_ex_e = 39.73\ cm^{-1}$ $\sigma = 2$
$B_e = 6.512\ cm^{-1}$ $\alpha_e = 0.1715\ cm^{-1}$

Enthalpy of Formation

Using the calculated thermal functions and functions for $H_2(g)$ and $I_2(g)$ given in the JANAF tables, equilibrium data from the literature are used to calculate $\Delta_fH°(298.15\ K)$ for the reaction $H_2(g) + I_2(g) \rightarrow 2\ HI(g)$.

Source	$\Delta_fH°(298.15\ K)$, kcal mol^{-1}
Taylor and Crist (4) (667 to 764 K)	-2.375 ± .005
Bodenstin (5) (556 to 781 K)	-2.053 ± .133
Bright and Hagerty (6) (696 and 779 K)	-2.219 ± .070
Rittenberg and Urey (7) (671 and 741 K)	-2.156 ± .072

The data of Taylor and Crist are the most concordant and are given the most weight. Converting to the solid state of iodine, which has been adopted as a JANAF reference state gives the standard enthalpy of formation of hydrogen iodide.

Heat Capacity and Entropy

Molecular constants are from Cowan and Gordy (1) and Boyd and Thompson (2). The calculated functions are in reasonable agreement with the compilation of Kelley (3).

References

1. M. Cowan and W. Gordy, Phys. Rev. 104, 551 (1956).
2. D. R. J. Boyd and H. W. Thompson, Spectrochim. Acta 5, 308 (1952).
3. K. K. Kelley, U. S. Bur. Mines Bull. 584, 1960.
4. A. H. Taylor and R. H. Crist, J. Amer. Chem. Soc. 65, 1377 (1941).
5. M. Bodenstein, Z. physikal. Chem. 29, 295 (1899).
6. N. F. H. Bright and R. P. Hagerty, Trans. Faraday Soc. 43, 697 (1947).
7. D. Rittenberg and H. C. Urey, J. Amer. Chem. Soc. 56, 1885 (1934).

Hydrogen Iodide (HI) $H_1I_1(g)$

Enthalpy Reference Temperature = T_r = 298.15 K Standard State Pressure = p° = 0.1 MPa

T/K	$C_p°$ (J $K^{-1}mol^{-1}$)	S°	$-[G°-H°(T_r)]/T$	$H°-H°(T_r)$ (kJ mol^{-1})	$\Delta_fH°$	$\Delta_fG°$	Log K_f
0	0.	0.	INFINITE	-8.656	28.535	28.535	INFINITE
100	29.114	174.773	232.489	-5.772	28.396	19.004	-9.927
200	29.124	194.956	209.255	-2.860	27.494	9.940	-2.596
250	29.134	201.456	207.069	-1.403	26.939	5.615	-1.173
298.15	29 156	206.589	206.589	0.	26.359	1.560	-0.273
300	29.158	206.770	206.590	0.054	26.336	1.406	-0.245
350	29.216	211.268	206.945	1.513	25,665	-2.698	0.403
400	29.328	215.176	207.735	2.976	16.987	-6.428	0.839
450	29.502	218.640	208.758	4.447	15.710	-9.279	1.077
500	29.736	221.760	209.905	5.928	-5.622	-10.088	1.054
600	30.348	227.233	212.349	8.931	-5.961	-10.948	0.953
700	31.064	231.964	214.820	12.001	-6.243	-11.756	0.877
800	31.798	236.160	217.230	15.144	-6.466	-12.528	0.818
900	32.496	239.946	219.547	18.359	-6.633	-13.275	0.770
1000	33.135	243.404	221.763	21.641	-6.754	-14.006	0.732
1100	33.706	246.589	223.877	24.984	-6.838	-14.727	0.699
1200	34.211	249.544	225.894	28.380	-6.897	-15.441	0.672
1300	34.655	252.300	227.821	31.824	-6.942	-16.151	0.649
1400	35.046	254.883	229.662	35.309	-6.982	-16.858	0.629
1500	35.391	257.313	231.425	38.832	-7.027	-17.562	0.612
1600	35.696	259.607	233.116	42.386	-7.087	-18.263	0.596
1700	35.967	261.780	234.739	45.970	-7.168	-18.959	0.583
1800	36.209	263.842	236.299	49.579	-7.277	-19.649	0.570
1900	36.427	265.806	237.800	53.211	-7.416	-20.333	0.559
2000	36.623	267.680	239.248	56.863	-7.589	-21.009	0 549
2100	36.802	269.471	240.645	60.535	-7.794	-21.675	0.539
2200	36.966	271.187	241.994	64.223	-8.030	-22.330	0.530
2300	37.116	272.833	243.300	67.927	-8.294	-22.975	0.522
2400	37.255	274.416	244.563	71.646	-8.581	-23.607	0.514
2500	37.384	275.939	245.788	75.378	-8.886	-24.227	0.506
2600	37.504	277.408	246.976	79.123	-9.204	-24.834	0.499
2700	37.617	278.825	248.130	82.879	-9.528	-25.429	0.492
2800	37.723	280.195	249.251	86.646	-9.854	-26.012	0.485
2900	37.823	281.521	250.341	90.423	-10.176	-26.583	0.479
3000	37.918	282.805	251.401	94.210	-10.489	-27.144	0.473
3100	38.009	284.050	252.435	98.006	-10.789	-27.694	0.467
3200	38.095	285.258	253.442	101.812	-11.073	-28.235	0.461
3300	38.178	286.431	254.424	105.625	-11.338	-28.767	0.455
3400	38.258	287.572	255.382	109.447	-11.580	-29.291	0.450
3500	38.335	288.682	256.317	113.277	-11.799	-29.809	0.445
3600	38.409	289.763	257.232	117.114	-11.993	-30 321	0.440
3700	38.481	290.817	258.125	120.959	-12.161	-30 828	0.435
3800	38.550	291.844	258.999	124.810	-12.303	-31.330	0.431
3900	38.618	292.846	259.854	128.669	-12.417	-31.829	0.426
4000	38.684	293 825	260.691	132.534	-12.504	-32.326	0.422
4100	38.749	294.781	261.511	136.405	-12.565	-32.821	0.418
4200	38.812	295.715	262.314	140.283	-12.599	-33.314	0.414
4300	38.874	296.629	263.102	144.168	-12.607	-33.807	0.411
4400	38.934	297.523	263.874	148.058	-12.590	-34.300	0.407
4500	38.994	298.399	264.631	151.955	-12.549	-34.794	0.404
4600	39.052	299.257	265.375	155.857	-12.484	-35.289	0.401
4700	39.110	300.097	266.105	159.765	-12.395	-35.786	0.398
4800	39.166	300.921	266.822	163.679	-12.284	-36.284	0.395
4900	39.222	301.729	267.526	167.598	-12.151	-36.786	0.392
5000	39.277	302.522	268.218	171.523	-11.998	-37.290	0.390
5100	39.332	303.301	268.898	175.454	-11.823	-37.798	0.387
5200	39.385	304.065	269.567	179.389	-11.630	-38.309	0.385
5300	39.439	304.816	270.225	183.331	-11.416	-38.824	0.383
5400	39.491	305.553	270.872	187.277	-11.185	-39.343	0.381
5500	39.544	306.278	271.510	191.229	-10.935	-39.867	0.379
5600	39.595	306.991	272.137	195.186	-10.667	-40.395	0.377
5700	39.647	307.693	272.754	199.148	-10.382	-40.929	0.375
5800	39.698	308.383	273.363	203.115	-10.080	-41.467	0.373
5900	39.748	309.062	273.962	207.087	-9.760	-42.011	0.372
6000	39.798	309.730	274.553	211.065	-9.425	-42.561	0.371

PREVIOUS: September 1961 (1 atm) CURRENT: September 1961 (1 bar)

Hydrogen Iodide (HI) $H_1I_1(g)$

TRIIODOSILANE ($SiHI_3$) IDEAL GAS M_r = 409.80694

$S°(298.15\ K) = [375.009 \pm 2.1]\ J\ K^{-1}\ mol^{-1}$

$\Delta_fH°(0\ K) = [-66.36 \pm 20.9]\ kJ\ mol^{-1}$

$\Delta_fH°(298.15\ K) = [-74.48 \pm 20.9]\ kJ\ mol^{-1}$

Vibrational Frequencies and Degeneracies

ν, cm^{-1}	ν, cm^{-1}
2190(1)	717(2)
286(1)	399(2)
122(1)	80(2)

Ground State Quantum Weight: [1] $\sigma = 3$

Point Group: C_{3v}

Bond Distances: SiH = [1.50] Å Si-I = [2.435] Å

Bond Angles: H-Si-I = [106.81]° I-Si-I = [112]°

Product of the Moments of Inertia: $I_AI_BI_C = [1.04795 \times 10^{-110}]\ g^3\ cm^6$

Enthalpy of Formation

$\Delta_fH°$(298.15 K) is estimated by linear interpolation between the values (1) of SiI_4(g) and SiH_4(g). There are no experimental $\Delta_fH°$ data for SiH_3I, SiH_2I_2, and $SiHI_3$. Data for the iodomethanes (2) have a surprising progression (1, SiH_3I, g); they are of doubtful use in predicting $\Delta_fH°$ of the iodosilanes. We conclude, as did Hunt and Sirtl (3), that the available data justify only linear interpolation.

Heat Capacity and Entropy

The molecular structure is estimated by comparison with SiH_3I, SiI_4 and the various bormo-, chloro-, and fluorosilanes (1). The principal moments of inertia are: $I_A = I_B = 174.6 \times 10^{-39}$ and $I_C = 343.48 \times 10^{-39}\ g\ cm^2$.

Vibrational frequencies are from liquid-phase Raman spectra of Hengge and Hoefler (4). Infrared data, also presumably for the liquid phase, were reported (4) for ν_1 and ν_4.

We neglect excited states and assume the electronic ground state to be a singlet. We expect that contributions from excited states should be unimportant, as discussed on the tables for SiH_3Br and SiH_2Cl_2 (1).

References

1. JANAF Thermochemical Tables: $H_3ISi(g)$, $HI_3Si(g)$, $BrH_3Si(g)$, $Br_2H_2Si(g)$, $Br_3HSi(g)$, $ClH_3Si(g)$, $Cl_2H_2Si(g)$, $Cl_3HSi(g)$ 12-31-76; $FH_3Si(g)$, $F_2H_2Si(g)$, $F_3HSi(g)$, $I_4Si(g)$, $H_4Si(g)$ 6-30-76.
2. S. A. Kudchadker and A. P. Kudchadker, J. Phys. Chem. Ref. Data 4, 457 (1975).
3. L. P. Hunt and E. Sirtl, J. Electrochem. Soc. 120, 806 (1973).
4. E. Hengge and F. Hoefler, Z. Naturforsch. A26, 768 (1971).

Triiodosilane ($SiHI_3$) $H_1I_3Si_1(g)$

Enthalpy Reference Temperature = T_r = 298.15 K Standard State Pressure = p° = 0.1 MPa

T/K	$C_p°$ (J K⁻¹mol⁻¹)	S° (J K⁻¹mol⁻¹)	$-[G°-H°(T_r)]/T$ (J K⁻¹mol⁻¹)	$H°-H°(T_r)$ (kJ mol⁻¹)	$\Delta_fH°$ (kJ mol⁻¹)	$\Delta_fG°$ (kJ mol⁻¹)	Log K_f
0	0.	0.	INFINITE	-19.135	-66.361	-66.361	INFINITE
100	58.844	297.407	442.714	-14.531	-68.098	-83.270	43.496
200	74.448	343.359	382.414	-7.811	-71.302	-97.213	25.389
250	79.878	360.582	376.369	-3.947	-72.928	-103.503	21.626
298.15	83.884	375.009	375.009	0.	-74.475	-109.250	19.140
300	84.018	375.528	375.011	0.155	-74.535	-109.466	19.060
350	87.179	388.728	376.046	4.439	-76.229	-115.157	17.186
400	89.637	400.536	378.382	8.862	-101.860	-119.782	15.642
450	91.605	411.211	381.446	13.394	-105.232	-121.819	14.140
500	93.228	420.949	384.917	18.016	-168.739	-117.785	12.305
600	95.791	438.185	392.396	27.474	-168.754	-107.590	9.367
700	97.757	453.105	400.026	37.155	-168.641	-97.404	7.268
800	99.322	466.264	407.500	47.012	-168.438	-87.240	5.696
900	100.590	478.039	414.695	57.009	-168.175	-77.105	4.475
1000	101.625	488.692	421.570	67.122	-167.878	-67.002	3.500
1100	102.479	498.419	428.121	77.328	-167.570	-56.930	2.703
1200	103.186	507.367	434.357	87.613	-167.272	-46.885	2.041
1300	103.776	515.651	440.296	97.962	-167.008	-36.863	1.481
1400	104.272	523.360	445.957	108.365	-166.800	-26.860	1.002
1500	104.691	530.569	451.360	118.814	-166.670	-16.870	0.587
1600	105.048	537.337	456.524	129.301	-166.640	-6.885	0.225
1700	105.354	543.715	461.467	139.821	-216.905	3.549	-0.109
1800	105.617	549.744	466.205	150.370	-216.894	16.516	-0.479
1900	105.846	555.461	470.754	160.944	-216.982	29.485	-0.811
2000	106.045	560.896	475.126	171.538	-217.172	42.462	-1.109
2100	106.219	566.074	479.335	182.152	-217.463	55.450	-1.379
2200	106.373	571.019	483.391	192.782	-217.849	68.454	-1.625
2300	106.509	575.750	487.304	203.426	-218.317	81.478	-1.850
2400	106.629	580.286	491.085	214.083	-218.856	94.524	-2.057
2500	106.737	584.641	494.740	224.751	-219.448	107.594	-2.248
2600	106.833	588.829	498.279	235.430	-220.077	120.688	-2.425
2700	106.920	592.863	501.708	246.118	-220.726	133.807	-2.589
2800	106.998	596.752	505.033	256.814	-221.379	146.949	-2.741
2900	107.068	600.508	508.261	267.517	-222.018	160.115	-2.884
3000	107.132	604.139	511.397	278.227	-222.631	173.303	-3.017
3100	107.191	607.653	514.446	288.943	-223.205	186.510	-3.143
3200	107.244	611.057	517.412	299.665	-223.729	199.735	-3.260
3300	107.292	614.358	520.300	310.392	-224.193	212.976	-3.371
3400	107.337	617.562	523.114	321.123	-224.590	226.230	-3.476
3500	107.378	620.674	525.857	331.859	-224.916	239.494	-3.574
3600	107.416	623.699	528.533	342.599	-609.316	263.227	-3.819
3700	107.450	626.643	531.145	353.342	-609.071	287.462	-4.058
3800	107.483	629.509	533.696	364.089	-608.746	311.688	-4.284
3900	107.513	632.301	536.188	374.839	-608.340	335.905	-4.499
4000	107.540	635.023	538.625	385.591	-607.852	360.110	-4.703
4100	107.566	637.679	541.009	396.347	-607.283	384.302	-4.896
4200	107.590	640.271	543.342	407.104	-606.634	408.480	-5.080
4300	107.612	642.803	545.626	417.865	-605.905	432.640	-5.256
4400	107.633	645.278	547.862	428.627	-605.099	456.783	-5.423
4500	107.653	647.697	550.054	439.391	-604.218	480.907	-5.582
4600	107.671	650.063	552.203	450.157	-603.263	505.011	-5.735
4700	107.688	652.379	554.309	460.925	-602.237	529.093	-5.880
4800	107.705	654.646	556.376	471.695	-601.142	553.153	-6.020
4900	107.720	656.867	558.405	482.466	-599.981	577.188	-6.153
5000	107.734	659.043	560.396	493.239	-598.756	601.200	-6.281
5100	107.748	661.177	562.351	504.013	-597.470	625.186	-6.403
5200	107.760	663.269	564.272	514.788	-596.124	649.147	-6.521
5300	107.772	665.322	566.159	525.565	-594.722	673.081	-6.634
5400	107.784	667.337	568.014	536.343	-593.266	696.988	-6.742
5500	107.795	669.315	569.838	547.122	-591.756	720.868	-6.846
5600	107.805	671.257	571 632	557.902	-590.197	744.720	-6.946
5700	107.815	673.165	573.396	568.683	-588.590	768.544	-7.043
5800	107.824	675.040	575.133	579.465	-586.937	792.339	-7.136
5900	107.833	676.884	576.842	590.248	-585.240	816.105	-7.225
6000	107.841	678.696	578.524	601.031	-583.500	839.842	-7.311

PREVIOUS: December 1976 (1 atm) CURRENT: December 1976 (1 bar)

Triiodosilane ($SiHI_3$) $H_1I_3Si_1(g)$

POTASSIUM HYDRIDE (KH) CRYSTAL M_r = 40.10624

$\Delta_f H°$(0 K) = Unknown

S°(298.15 K) = [50.208] J K^{-1} mol^{-1} $\Delta_f H°$(298.15 K) = -57.820 ± 0.046 kJ mol^{-1}

Enthalpy of Formation

The value of $\Delta_f H°$(298.15 K) is taken from Gunn and Green (1). The corresponding value reported by Messer et al. (2) was -15.16±0.6 kcal mol^{-1}. In this latter paper, the results of the previous investigations were also reviewed.

Heat Capacity and Entropy

$C_p°$ and S°(298.15 K) are estimated by comparison with those for NaH(cr) and LiH(cr).

Decomposition Data

T_{dcm} = 690 K was assumed to be the temperature at which the value of $\Delta_f G°$ changes from negative to positive.

References

1. S. R. Gunn and L. G. Green, J. Amer. Chem. Soc. 80, 4782 (1958).
2. C. E. Messler, L. G. Fasolino and C. E. Thalmayer, J. Amer. Chem. Soc. 77, 4524 (1955).

Potassium Hydride (KH) H_1K_1(cr)

Enthalpy Reference Temperature = T_r = 298.15 K Standard State Pressure = p° = 0.1 MPa

T/K	$C_p°$ (J K⁻¹mol⁻¹)	S° (J K⁻¹mol⁻¹)	$-[G°-H°(T_r)]/T$ (J K⁻¹mol⁻¹)	$H°-H°(T_r)$ (kJ mol⁻¹)	$\Delta_f H°$ (kJ mol⁻¹)	$\Delta_f G°$ (kJ mol⁻¹)	Log K_f
0							
100							
200							
298.15	37.907	50.208	50.208	0.	-57.820	-53.010	9.287
300	38.074	50.443	50.209	0.070	-57.776	-52.980	9.225
400	44.141	62.267	51.780	4.195	-55.104	-51.768	6.760
500	48.534	72.608	54.933	8.837	-51.924	-51.293	5.359
600	51.923	81.766	58.657	13.865	-48.360	-51.496	4.483
700	54.606	89.977	62.555	19.196	-44.498	-52.320	3.904
800	56.819	97.418	66.455	24.771	-40.400	-53.715	3.507
900	58.440	104.207	70.278	30.536	-36.122	-55.635	3.229
1000	59.831	110.438	73.987	36.451	-31.708	-58.038	3.032
1100	60.966	116.195	77.565	42.493	-27.187	-60.889	2.891
1200	61.923	121.542	81.010	48.639	-22.580	-64.156	2.793
1300	62.701	126.531	84.322	54.871	-17.907	-67.810	2.725
1400	63.301	131.200	87.505	61.173	-13.188	-71.825	2.680
1500	63.722	135.583	90.566	67.526	-8.439	-76.179	2.653

PREVIOUS: CURRENT: March 1963

Potassium Hydride (KH) H_1K_1(cr)

POTASSIUM HYDRIDE (KH) IDEAL GAS M_r = 40.10624

S°(298.15 K) = 198.024 J K^{-1} mol^{-1}

$\Delta_f H°$(0 K) = 125.53 ± 14.6 kJ mol^{-1}
$\Delta_f H°$(298.15 K) = 123.01 ± 14.6 kJ mol^{-1}

Electronic Level and Quantum Weight

State	ε_i, cm^{-1}	g_i
$^1\Sigma^+$	0	1

ω_e = 985.0 cm^{-1} $\omega_e x_e$ = 14.65 cm^{-1} σ = 1
B_e = 3407 cm^{-1} α_e = 0.0673 cm^{-1}

Enthalpy of Formation

$\Delta_f H°$(298.15 K) is calculated from D_0^0 = 1.86±0.15 eV as reported by Gaydon (1).

Heat Capacity and Entropy

All molecular and spectroscopic constants are obtained from Herzberg (2).

References

1. A. G. Gaydon, "Dissociation Energies," Chapman and Hall, Ltd., 1953.
2. G. Herzberg, "Spectra of Diatomic Molecules," D. Van Nostrand Company, Inc., 1950.

Potassium Hydride (KH) $H_1K_1(g)$

Enthalpy Reference Temperature = T_r = 298.15 K   Standard State Pressure = p° = 0.1 MPa

T/K	$C_p°$ (J $K^{-1}mol^{-1}$)	S° (J $K^{-1}mol^{-1}$)	$-[G°-H°(T_r)]/T$ (J $K^{-1}mol^{-1}$)	$H°-H°(T_r)$ (kJ mol^{-1})	$\Delta_f H°$ (kJ mol^{-1})	$\Delta_f G°$ (kJ mol^{-1})	Log K_f
0	0.	0.	INFINITE	-8.795	125.530	125.530	INFINITE
100	29.136	165.689	224.657	-5.897	125.197	117.224	-61.232
200	29.583	185.963	200.811	-2.970	124.183	109.630	-28.632
250	30.243	192.630	198.530	-1.475	123.605	106.058	-22.160
298.15	31.036	198.024	198.024	0.	123.010	102.731	-17.998
300	31.068	198.216	198.025	0.057	122.986	102.606	-17.865
350	31.917	203.069	198.406	1.632	119.945	99.357	-14.828
400	32.711	207.384	199.263	3.248	119.245	96.465	-12.597
450	33.419	211.279	200.386	4.902	118.605	93.656	-10.871
500	34.034	214.832	201.655	6.589	118.017	90.916	-9.498
600	35.019	221.129	204.389	10.044	116.968	85.597	-7.452
700	35.752	226.585	207.179	13.584	116.043	80.443	-6.003
800	36.309	231.397	209.912	17.188	115.194	75.416	-4.924
900	36.746	235.700	212.542	20.842	114.378	70.493	-4.091
1000	37.098	239.590	215.055	24.535	113.556	65.661	-3.430
1100	37.391	243.140	217.450	28.259	33.742	65.523	-3.111
1200	37.640	246.404	219.728	32.011	33.876	68.406	-2.978
1300	37.858	249.426	221.898	35.786	34.012	71.278	-2.864
1400	38.052	252.239	223.966	39.582	34.147	74.140	-2.766
1500	38.228	254.870	225.939	43.396	34.277	76.992	-2.681
1600	38.389	257.342	227.826	47.227	34.402	79.835	-2.606
1700	38.540	259.674	229.631	51.074	34.520	82.671	-2.540
1800	38.681	261.881	231.362	54.935	34.629	85.500	-2.481
1900	38.816	263.976	233.024	58.810	34.728	88.324	-2.428
2000	38.944	265.971	234.622	62.698	34 817	91.142	-2.380
2100	39.068	267.874	236.160	66.598	34.893	93 956	-2.337
2200	39.188	269.694	237.643	70.511	34.956	96.767	-2.298
2300	39.305	271.438	239.075	74.436	35.004	99.576	-2.261
2400	39.419	273.114	240.459	78.372	35.034	102.383	-2.228
2500	39.530	274.725	241.797	82.319	35.045	105.189	-2.198
2600	39.640	276.278	243.094	86.278	35.034	107.994	-2.170
2700	39.747	277.776	244.351	90.247	34.999	110.801	-2.144
2800	39.854	279.223	245.571	94.227	34.937	113.610	-2.119
2900	39.959	280.623	246.755	98.218	34.845	116.421	-2.097
3000	40.062	281.980	247.907	102.219	34.721	119.236	-2.076
3100	40.165	283.295	249.027	106.230	34.561	122.056	-2.057
3200	40.267	284.572	250.118	110.252	34.361	124.881	-2.038
3300	40.369	285.813	251.181	114.284	34.127	127.713	-2.022
3400	40.469	287.019	252.218	118.326	33.841	130.553	-2.006
3500	40.569	288.194	253.229	122.378	33.508	133 402	-1.991
3600	40.669	289.338	254.216	126.440	33.119	136.261	-1.977
3700	40.767	290.454	255.180	130.511	32.672	139.132	-1.964
3800	40.866	291.542	256.123	134.593	32.163	142.016	-1.952
3900	40.964	292.605	257.045	138.685	31.586	144.915	-1.941
4000	41.062	293.643	257.947	142.786	30.944	147.828	-1.930
4100	41.159	294.659	258.830	146.897	30.226	150.759	-1.921
4200	41.256	295.652	259.695	151.018	29.423	153.708	-1.912
4300	41.353	296.623	260.542	155.148	28.535	156.677	-1.903
4400	41.450	297.575	261.373	159 288	27.834	159.641	-1.895
4500	41.546	298.508	262.188	163.438	26.830	162.648	-1.888
4600	41.643	299.422	262.988	167.598	25.740	165.678	-1.881
4700	41.739	300.319	263.772	171.767	24.581	168.731	-1.875
4800	41.835	301.198	264.543	175.945	23.441	171.797	-1.870
4900	41.930	302.062	265.300	180.134	22.107	174.901	-1.864
5000	42.026	302.910	266.044	184.331	20.683	178.034	-1.860
5100	42.121	303.743	266.775	188.539	19.166	181.196	-1.856
5200	42.217	304.562	267.494	192.756	17.557	184.389	-1.852
5300	42 312	305.367	268.201	196.982	15.855	187.613	-1.849
5400	42.407	306.159	268.896	201.218	14.061	190.870	-1.846
5500	42.502	306.938	269.581	205.464	12.177	194.162	-1.844
5600	42.597	307.705	270.255	209.718	10.202	197.488	-1.842
5700	42.692	308.459	270.918	213.983	8.484	200.803	-1.840
5800	42.787	309.203	271.572	218.257	6.379	204.196	-1.839
5900	42.881	309.935	272.216	222.540	4 195	207.625	-1.838
6000	42.976	310.656	272.851	226.833	1.933	211.092	-1.838

PREVIOUS: March 1963 (1 atm)   CURRENT: March 1963 (1 bar)

Potassium Hydride (KH) $H_1K_1(g)$

$S°(298.15\ K) = 78.907 \pm 0.84\ J\ K^{-1}\ mol^{-1}$

$T_{trs} = 516\ K$

$T_{fus} = 679\ K$

$\Delta_f H°(0\ K) = -421.22 \pm 0.4\ kJ\ mol^{-1}$

$\Delta_f H°(298.15\ K) = -424.72 \pm 0.4\ kJ\ mol^{-1}$

$\Delta_{trs} H° = 6.443 \pm 0.63\ kJ\ mol^{-1}$

$\Delta_{fus} H° = 8.619 \pm 0.63\ kJ\ mol^{-1}$

Enthalpy of Formation

The enthalpy of formation of KOH(cr) is obtained from its enthalpy of solution in water, the enthalpy of hydrolysis of metallic potassium, and appropriate auxiliary data.

Reshetnikov (1) determined calorimetrically the enthalpy of solution of KOH(cr) in 650 H_2O as -13.665 ± 0.009 kcal mol^{-1}. Combining this result with enthalpy of dilution data for KOH reported by Parker (2) we derive $\Delta_{sln}H°(KOH, cr) = -13.77 \pm 0.01$ kcal mol^{-1}.

The enthalpy of hydrolysis of metallic potassium has been determined by various investigators (3-5). The data are summarized below for the reaction $K(cr) + (n + 1)H_2O(\ell) \rightarrow KOH \cdot nH_2O + 0.5\ H_2(g)$.

Source	n moles of H_2O	$\Delta_r H°(298.15\ K)$ kcal mol^{-1}	$\Delta_f H°(KOH, aq, \infty, 298.15\ K)$* kcal mol^{-1}	$\Delta_f H°(KOH, cr, 298.15\ K)$ kcal mol^{-1}
Gunn and Green (3a, 3b)	∞	-46.958±0.22	-115.273±0.22	-101.50±0.02
Messer et al. (4)	∞	-47.02 ±0.15	-115.34 ±0.15	-101.57±0.16
Ketchen and Wallace (5)	∞	-46.89 ±0.24	-115.21 ±0.24	-101.44±0.25

*Based on $\Delta_f H°(H_2O, \ell, 298.15\ K) = -68.315$ kcal mol^{-1} (6).

The values of $\Delta_f H°(KOH, cr, 298.15\ K)$ given in the last column are calculated from $\Delta_f H°(KOH, aq, \infty, 298.15\ K)$ and the enthalpy of solution of KOH(cr) at infinite dilution given above. The adopted value for $\Delta_f H°(KOH, cr, 298.15\ K)$ is from the work of Gunn and Green (3a, 3b).

Earlier investigations pertaining to the determination of $\Delta_f H°(KOH, cr, 298.15\ K)$ have been reviewed by Bichowsky and Rossini (7) and Messer et al. (4) and are only of historical interest.

Heat Capacity and Entropy

The low temperature heat capacities for KOH(cr) in the temperature range 20-298 K are from the recent work of Stull et al. (8). These data contain a broad lambda type transition with a heat capacity peak at 227.5 K. Powers and Blalock (9) measured high temperature enthalpy data for KOH(cr) in both the α and β phases in a Bunsen ice calorimeter. Their enthalpy data are scattered and not precise enough to accurately define the heat capacities for the α phase. Therefore, the selected heat capacities between 298 and 516 K are estimated by graphical extrapolation of the low temperature heat capacity data. Heat capacities for the β phase are from Powers and Blalock (9).

$S°(298.15\ K)$ is calculated from the smoothed $C_p°$ data and is based on an extrapolation of $S°(20\ K) = 0.199$ cal K^{-1} mol^{-1}.

Transition and Fusion Data

Several values for the enthalpy of the α-β transition and the enthalpy of melting of KOH have been reported in the literature.

Source	Method	T_{trs}/K	$\Delta_{trs}H°$ kcal mol^{-1}	T_{fus}/K	$\Delta_{fus}H°$ kcal mol^{-1}
Powers and Blalock (9)	Calorimetric	522	1.346	673	2.244
Seward and Martin (10)	Freezing Point			683	1.830
Kelley (11)	Freezing Point			633	1.980
Reshetnikov and Baranskaya (12)	Thermographic	515	1.540±0.5	679	1.980±0.15
Maurice (13)	Freezing Point	517		678	2.13 ±0.07

The selected values are from the work of Reshetnikov and Baranskaya (12) and Maurice (13).

Sublimation Data

$\Delta_{sub}H°(298.15\ K)$ for the monomer and dimer are calculated from the selected enthalpies of vaporization and the enthalpy of melting at 298.15 K. See the ideal gas table for details.

References

1. N. A. Reshetnikov, Zhur. Neorg. Khim. 6, 682 (1961).
2. V. B. Parker, NSRDS-NBS 2, 1965.
3. a) S. R. Gunn and L. G. Green, J. Amer. Chem. Soc. 80, 4782 (1958). b) S. R. Gunn, J. Phys. Chem. 71, 1386 (1967).
4. C. E. Messer, L. C. Fasolino and C. O. Thalmayer, J. Amer. Chem. Soc. 77, 4524 (1955).
5. E. E. Ketchen and W. E. Wallace, J. Amer. Chem. Soc. 76, 4736 (1954).
6. U. S. Nat. Bur. Stand. Tech. Note 270-3, 1968.
7. F. R. Bichowsky and F. D. Rossini, "The Thermochemistry of the Chemical Substances," Reinhold Publishing Corp., New York, 1936.
8. D. R. Stull, D. L. Hildenbrand, F. L. Oetting and G. C. Sinke, J. Chem. Eng. Data 15, 52 (1970).
9. W. D. Powers and G. C. Blalock, Oak Ridge Natl. Lab., ONRL 1653, January, 1954.
10. P. S. Seward and K. E. Martin, J. Amer. Chem. Soc. 71, 3564 (1949).
11. K. K. Kelley, U. S. Bur. Mines Bull. 393, 1936.
12. N. A. Reshetnikov and E. V. Baranskaya, Izv. Vyssh. Ucheb. Zaved., Khim. Khim. Tekhnol. 10, 496 (1967); CA 67 94551.
13. M. Maurice, Rev. Chim. Miner. 5, 89 (1968); CA 69 99905.

Enthalpy Reference Temperature = T_r = 298.15 K — Standard State Pressure = p° = 0.1 MPa

T/K	$C_p°$ (J $K^{-1}mol^{-1}$)	S° (J $K^{-1}mol^{-1}$)	$-[G°-H°(T_r)]/T$ (J $K^{-1}mol^{-1}$)	$H°-H°(T_r)$ (kJ mol^{-1})	$\Delta_f H°$ (kJ mol^{-1})	$\Delta_f G°$ (kJ mol^{-1})	Log K_f
0	0.	0.	INFINITE	-12.163	-421.223	-421.223	INFINITE
100	34.100	23.857	130.178	-10.632	-424.377	-409.500	213.901
200	53.932	53.620	84.784	-6.233	-425.375	-394.110	102.931
298.15	64.894	78.907	78.907	0.	-424.718	-378.899	66.382
300	65.011	79.309	78.909	0.120	-424.706	-378.615	65.923
400	71.672	98.894	81.534	6.944	-426.299	-362.909	47.391
500	79.580	115.714	86.721	14.496	-424.845	-347.213	36.273
516.000	80.961	118.242	87.660	15.780	ALPHA <--> BETA	TRANSITION	
516.000	78.659	130.729	87.660	22.224			
600	78.659	142.593	94.541	28.831	-416.594	-332.908	28.982
679.000	78.659	152.322	100.709	35.045	BETA <--> LIQUID		
700	78.659	154.718	102.294	36.697	-414.821	-319.100	23.812
800	78.659	165.221	109.518	44.563	-413.076	-305.545	19.950
900	78.659	174.486	116.232	52.429	-411.383	-292.206	16.959
1000	78.659	182.774	122.479	60.295	-409.763	-279.052	14.576
1100	78.659	190.271	128.306	68.161	-487.190	-261.446	12.415
1200	78.659	197.115	133.759	76.027	-484.716	-241.033	10.492
1300	78.659	203.411	138.878	83.893	-482.281	-220.825	8.873
1400	78.659	209.240	143.699	91.758	-479.883	-200.803	7.492
1500	78.659	214.667	148.251	99.624	-477.521	-180.952	6.301

PREVIOUS: March 1966 CURRENT: December 1970

POTASSIUM HYDROXIDE (KOH) LIQUID M_r = 56.10564

Potassium Hydroxide (KOH) $H_1K_1O_1(l)$

$S°(298.15\ K) = [96.617]\ J\ K^{-1}\ mol^{-1}$

$T_{fus} = 679\ K$

$\Delta_f H°(298.15\ K) = [-412.705]\ kJ\ mol^{-1}$

$\Delta_{fus} H° = 8.619 \pm 0.63\ kJ\ mol^{-1}$

Enthalpy of Formation

$\Delta_f H°(KOH, \ell, 298.15\ K)$ is calculated from that of the crystal by adding $\Delta_{fus}H°$ and the difference in enthalpy, $H°(679\ K) - H°(298.15\ K)$, between the crystal and liquid.

Heat Capacity and Entropy

The heat capacity of the liquid phase is obtained from the high temperature enthalpy measurements of Powers and Blalock (1). The constant $C_p°$ is extrapolated below the melting point and up to 2000 K.

S°(298.15 K) is calculated in a manner similar to that used for the enthalpy of formation.

Fusion Data

See the crystal table for details.

Vaporization Data

T_{vap} = 1596 K for the process $KOH(\ell) = KOH(g)$ is calculated as the temperature at which the fugacity of KOH(g) is 1 atm. The difference in the enthalpies of formation of $KOH(\ell)$ and KOH(g) at T_{vap} is the enthalpy of vaporization. If $[KOH]_2(g)$ is also considered a minor component in the vapor mixture (approximately 2% of the total vapor pressure at the boiling point) the calculated boiling point is unchanged.

Wartenberg and Albrecht (2) reported a boiling point of 1597 K for KOH from static vapor pressure measurements, while very recently Dubois and Millet (3) determined a boiling point of 1540 K from transpiration vapor pressure measurements.

References

1. W. D. Powers and G. C. Blalock, Oak Ridge Natl. Lab., ORNL 1653, January, 1954.
2. H. von Wartenberg and P. Albrecht, Z. Elektrochem. 27, 162 (1921).
3. J. Dubois and J. Millet, Compt. Rend. 269, 1336 (1969).

Enthalpy Reference Temperature = T_r = 298.15 K   Standard State Pressure = p° = 0.1 MPa

T/K	$C_p°$ (J K⁻¹mol⁻¹)	S° (J K⁻¹mol⁻¹)	$-[G°-H°(T_r)]/T$ (J K⁻¹mol⁻¹)	$H°-H°(T_r)$ (kJ mol⁻¹)	$\Delta_f H°$ (kJ mol⁻¹)	$\Delta_f G°$ (kJ mol⁻¹)	Log K_f
0							
100							
200							
298.15	83.107	96.617	96.617	0.	-412.705	-372.166	65.202
300	83.107	97.131	96.618	0.154	-412.660	-371.915	64.756
400	83.107	121.039	99.878	8.464	-412.766	-358.234	46.781
500	83.107	139.584	106.034	16.775	-410.553	-344.856	36.027
600	83.107	154.736	112.926	25.086	-408.327	-331.926	28.897
679.000	83.107	165.016	118.401	31.651	------- BETA	<--> LIQUID	-------
700	83.107	167.547	119.838	33.396	-406.108	-319.368	23.832
800	83.107	178.644	126.511	41.707	-403.919	-307.127	20.053
900	83.107	188.433	132.858	50.018	-401.781	-295.157	17.130
1000	83.107	197.189	138.861	58.328	-399.716	-283.421	14.804
1100	83.107	205.110	144.529	66.639	-476.698	-267.278	12.692
1200	83.107	212.341	149.883	74.950	-473.780	-248.368	10.811
1300	83.107	218.993	154.947	83.261	-470.900	-229.701	9.230
1400	83.107	225.152	159.744	91.571	-468.057	-211.254	7.882
1500	83.107	230.886	164.298	99.882	-465.250	-193.010	6.721
1600	83.107	236.250	168.629	108.193	-462.480	-174.951	5.712
1700	83.107	241.288	172.757	116.503	-459.744	-157.064	4.826
1800	83.107	246.038	176.697	124.814	-457.043	-139.338	4.043
1900	83.107	250.532	180.466	133.125	-454.378	-121.760	3.347
2000	83.107	254.794	184.077	141.435	-451.748	-104.322	2.725

PREVIOUS: March 1966   CURRENT: December 1970

Potassium Hydroxide (KOH) $H_1K_1O_1(l)$

POTASSIUM HYDROXIDE (KOH) ALPHA-BETA-LIQUID M_r = 56.10564

0 to 516 K crystal, alpha
516 to 679 K crystal, beta
above 679 K liquid

Refer to the individual tables for details.

Potassium Hydroxide (KOH) $H_1K_1O_1(cr,l)$

Enthalpy Reference Temperature = T_r = 298.15 K — Standard State Pressure = p° = 0.1 MPa

T/K	C_p° (J K⁻¹mol⁻¹)	S° (J K⁻¹mol⁻¹)	$-[G^\circ - H^\circ(T_r)]/T$ (J K⁻¹mol⁻¹)	$H^\circ - H^\circ(T_r)$ (kJ mol⁻¹)	$\Delta_f H^\circ$ (kJ mol⁻¹)	$\Delta_f G^\circ$ (kJ mol⁻¹)	Log K_f
0	0.	0.	INFINITE	-12.163	-421.223	-421.223	INFINITE
100	34.100	23.857	130.178	-10.632	-424.377	-409.500	213.901
200	53.932	53.620	84.784	-6.233	-425.375	-394.110	102.931
298.15	64.894	78.907	78.907	0.	-424.718	-378.899	66.382
300	65.011	79.309	78.909	0.120	-424.706	-378.615	65.923
400	71.672	98.894	81.534	6.944	-426.299	-362.909	47.391
500	79.580	115.714	86.721	14.496	-424.845	-347.213	36.273
516.000	80.961	118.242	87.660	15.780		ALPHA <--> BETA	
516.000	78.659	130.729	87.660	22.224		TRANSITION	
600	78.659	142.593	94.541	28.831	-416.594	-332.908	28.982
679.000	78.659	152.322	100.709	35.045		BETA <--> LIQUID	
679.000	83.107	165.016	100.709	43.664		TRANSITION	
700	83.107	167.547	102.676	45.409	-406.108	-319.368	23.832
800	83.107	178.644	111.494	53.720	-403.919	-307.127	20.053
900	83.107	188.433	119.510	62.031	-401.781	-295.157	17.130
1000	83.107	197.189	126.848	70.342	-399.716	-283.421	14.804
1100	83.107	205.110	133.608	78.652	-476.698	-267.278	12.692
1200	83.107	212.341	139.872	86.963	-473.780	-248.369	10.811
1300	83.107	218.993	145.706	95.274	-470.900	-229.701	9.230
1400	83.107	225.152	151.164	103.584	-468.057	-211.254	7.882
1500	83.107	230.886	156.289	111.895	-465.251	-193.010	6.721
1600	83.107	236.250	161.121	120.206	-462.480	-174.951	5.712
1700	83.107	241.288	165.690	128.516	-459.744	-157.064	4.826
1800	83.107	246.038	170.023	136.827	-457.043	-139.338	4.043
1900	83.107	250.532	174.143	145.138	-454.378	-121.760	3.347
2000	83.107	254.794	178.070	153.448	-451.748	-104.322	2.725

PREVIOUS: CURRENT: December 1970

Potassium Hydroxide (KOH) $H_1K_1O_1(cr,l)$

POTASSIUM HYDROXIDE (KOH) IDEAL GAS M_r = 56.10564

$S°(298.15\ K) = [236.379 \pm 1.3]\ J\ K^{-1}\ mol^{-1}$

$\Delta_f H°(0\ K) = -228.40 \pm 12.6\ kJ\ mol^{-1}$
$\Delta_f H°(298.15\ K) = -232.63 \pm 12.6\ kJ\ mol^{-1}$

Vibrational Frequencies and Degeneracies

ν, cm^{-1}
408 (1)
[340](2)
[3610](1)

Ground State Quantum Weight: 1 σ = 1
Point Group: $C_{\infty v}$
Bond Distances: K-O = 2.18 ± 0.01 Å O-H = [0.97 ± 0.05] Å
Bond Angle: K-O-H = 180°
Rotational Constant: B_0 = 0.279902 cm^{-1}

Enthalpy of Formation

Mass spectrometric studies (1-5) of the equilibrium gases over pure KOH(cr, ℓ) and mixed KOH-NaOH condensed phases have unequivocally identified the vapor species as monomer and dimer in the temperature range 600-700 K. Absolute partial pressures for KOH(g) and $K_2(OH)_2$ (g) have been determined from peak intensity data by Porter and Schoonmaker (3) and Gusarov and Gorokhov (5). These data are analyzed by the 3rd law method with JANAF Gibbs energy functions (6) in order to evaluate an enthalpy of dimerization at 298 K. The adopted value is $\Delta_f H°(298.15\ K) = -45.3 \pm 3.0\ kcal\ mol^{-1}$ for the reaction 2 KOH(g) = $K_2(OH)_2$ (g).

Vapor pressures for liquid KOH have been determined by static (1443-1600 K) (7) and transpiration (873-1323 K) (8) methods. In order to evaluate $\Delta_{vap}H°$(KOH, ℓ, 298.15 K) we have used a trial and error variation of $\Delta_{vap}H°$(298.15 K) for the monomer and dimer, such that these values are in accordance with the adopted enthalpy of dimerization given above, and the sum of the calculated partial pressures for KOH(g) and $[KOH]_2$(g) is in good agreement with the experimental vapor pressure data. Since the dimer to monomer ratio decreases with increasing temperature, more weight is given to the static data of Wartenberg and Albrecht (7). The selected values are: $\Delta_{vap}H°$(to monomer, 298.15 K) = 43.0 ± 3.0 kcal mol^{-1} and $\Delta_{vap}H°$(to dimer, 298.15 K) = 40.8 ± 3.0 kcal mol^{-1}. From these results we derive $\Delta_f H°$(KOH, g, 298.15 K) = -55.6 ± 3.0 kcal mol^{-1} with $\Delta_f H°$(KOH, ℓ, 298.15 K) = -98.64 ± 0.50 kcal mol^{-1} (6). This value is adopted here and when used in combination with the enthalpy of dimerization results in $\Delta_f H°[KOH]_2$, g, 298.15 K) = -156.5 ± 3.0 kcal mol^{-1}.

The bond dissociation energies of the alkali metal hydroxides have been determined by several workers from flame studies (9-11) and by mass spectrometry (12). The data for KOH are summarized below.

Source	Method	$\Delta_r H°$(298.15 K), kcal mol^{-1} 3rd law	D_0^0(K-OH) kcal mol^{-1}
Smith and Sugden (9)	Flame		86 ± 1
Jensen and Padley (10)	Flame	36.4	82 ± 2
Cotton and Jenkins (11)	Flame	32.5	86 ± 2
Gorokhov et al. (12)	Mass Spec.		80 ± 3

Reaction: $K(g) + H_2O(g) = KOH(g) + H(g)$

Our adopted enthalpy of formation for KOH(g) leads to D_0^0 = 85.4 ± 3.0 kcal mol^{-1}, which is in reasonable agreement with the results of the flame studies (9-11), particularly when one considers the difficulties inherent in this method.

Heat Capacity and Entropy

Kuczkowski et al. (13) observed the microwave spectra of gaseous KOH and interpreted the results in terms of a linear geometry. Very recently, microwave studies of gaseous CsOH (14a) and RbOH (14b), along with infrared spectra of matrix-isolated CsOH (15a), RbOH (15b) and NaOH (15b) have been reported and interpreted in terms of linear structures for all these molecules. We adopt a linear configuration for KOH(g), along with the bond distances determined by Kuczkowski et al. (13) from their microwave studies.

Acquista and Abramowitz (15b) attempted to measure the infrared spectrum of matrix-isolated monomeric KOH but observed bands which could only be ascribable to polymeric species. Spinar and Margrave (16) investigated the infrared spectrum of the equilibrium gases over liquid KOH at temperatures up to 1000 K. An absorption band at 408 ± 10 cm^{-1} was assigned to the K-OH stretching by these workers, and this assignment is adopted here. The doubly degenerate bending and O-H stretching frequencies are estimates taken from the recent work of Jensen (17).

References

1. R. F. Porter and R. C. Schoonmaker, J. Chem. Phys. 28, 168 (1958).
2. R. F. Porter and R. C. Schoonmaker, J. Phys. Chem. 62, 234 (1958).
3. R. F. Porter and R. C. Schoonmaker, J. Phys. Chem. 62, 486 (1958).
4. R. C. Schoonmaker and R. F. Porter, J. Chem. Phys. 31, 830 (1959).
5. A. V. Gusarov and L. N. Gorokhov, Russ. J. Phys. Chem. 42, 449 (1968).
6. JANAF Thermochemical Tables: KOH(g), $K_2(OH)_2$(g), and KOH(ℓ), 12-31-70.
7. H. von Wartenberg and P. Albrecht, Z. Elektrochem. 27, 162 (1921).
8. J. Dubois and J. Millet, Compt. Rend. 269, 1336 (1969).
9. H. Smith and T. M. Sugden, Proc. Roy. Soc. A 219, 304 (1953).
10. D. E. Jensen and P. J. Padley, Trans. Faraday Soc. 62, 2132 (1966).
11. D. H. Cotton and D. R. Jenkins, Trans. Faraday Soc. 65, 1537 (1969).
12. L. N. Gorokhov, A. V. Gusarov and I. G. Panchenkov, Russ. J. Phys. Chem. 44, 150 (1970).
13. R. L. Kuczkowski, D. R. Lide, Jr., and L. C. Krisher, J. Chem. Phys. 44, 3131 (1966).
14. a) D. R. Lide, Jr., and R. L. Kuczkowski, J. Chem. Phys. 46, 4768 (1967); b) C. Matsumura and D. R. Lide, Jr., J. Chem. Phys. 50, 71 (1969).
15. a) N. Acquista, S. Abramowitz and D. R. Lide, J. Chem. Phys. 49, 780 (1968); b) N. Acquista and S. Abramowtiz, J. Chem. Phys. 51, 2911 (1969).
16. L. H. Spinar and J. L. Margrave, Spectrochimia Acta 12, 244 (1958).
17. D. E. Jensen, J. Phys. Chem. 74, 207 (1970).

Potassium Hydroxide (KOH) $H_1K_1O_1(g)$

Enthalpy Reference Temperature = T_r = 298.15 K Standard State Pressure = p° = 0.1 MPa

T/K	$C_p°$ (J K^{-1}mol^{-1})	S°	-[G°-H°(T_r)]/T	H°-H°(T_r) (kJ mol^{-1})	$\Delta_f H°$	$\Delta_f G°$	Log K_f
0	0.	0.	INFINITE	-11.423	-228.396	-228.396	INFINITE
100	32.946	191.442	275.830	-8.439	-230.096	-231.978	121.173
200	43.677	217.899	240.692	-4.559	-231.613	-233.204	60.907
250	46.658	227.990	237.169	-2.295	-232.150	-233.538	48.795
298.15	48.531	236.379	236.379	0.	-232.630	-233.762	40.954
300	48.589	236.679	236.380	0.090	-232.649	-233.769	40.703
350	49.880	244.273	236.977	2.554	-235.539	-233.819	34.896
400	50.776	250.995	238.317	5.071	-236.084	-233.535	30.497
450	51.424	257.015	240.066	7.627	-236.581	-233.186	27.068
500	51.913	262.460	242.037	10.211	-237.043	-232.784	24.319
600	52.628	271.992	246.258	15.440	-237.898	-231.850	20.184
700	53.187	280.147	250.531	20.732	-238.699	-230.779	17.221
800	53.704	287.283	254.688	26.076	-239.476	-229.594	14.991
900	54.220	293.638	258.669	31.473	-240.252	-228.312	13.251
1000	54.740	299.378	262.457	36.920	-241.050	-226.943	11.854
1100	55.257	304.620	266.056	42.420	-320.843	-220.882	10.489
1200	55.761	309.449	269.473	47.971	-320.684	-211.802	9.220
1300	56.244	313.932	272.723	53.572	-320.514	-202.735	8.146
1400	56.698	318.117	275.817	59.219	-320.335	-193.682	7.226
1500	57.122	322.043	278.769	64.911	-320.147	-184.642	6.430
1600	57.514	325.742	281.591	70.643	-319.955	-175.615	5.733
1700	57.874	329.240	284.292	76.412	-319.761	-166.599	5.119
1800	58.204	332.558	286.882	82.216	-319.567	-157.596	4.573
1900	58.504	335.713	289.369	88.052	-319.376	-148.603	4.085
2000	58.779	338.721	291.762	93.916	-319.192	-139.619	3.646
2100	59.028	341.595	294.067	99.807	-319.018	-130.645	3.250
2200	59.256	344.346	296.291	105.721	-318.858	-121.678	2.889
2300	59.464	346.985	298.438	111.657	-318.714	-112.719	2.560
2400	59.653	349.519	300.514	117.613	-318.591	-103.766	2.258
2500	59.826	351.958	302.523	123.588	-318.491	-94.817	1.981
2600	59.984	354.308	304.470	129.578	-318.418	-85.871	1.725
2700	60.129	356.574	306.358	135.584	-318.375	-76.928	1.488
2800	60.262	358.763	308.191	141.603	-318.366	-67.986	1.268
2900	60.383	360.880	309.971	147.636	-318.395	-59.043	1.063
3000	60.496	362.929	311.703	153.680	-318.465	-50.099	0.872
3100	60.599	364.915	313.387	159.735	-318.579	-41.151	0.693
3200	60.695	366.840	315.028	165.799	-318.743	-32.200	0.526
3300	60.783	368.709	316.626	171.873	-318.950	-23.243	0.368
3400	60.864	370.525	318.185	177.956	-319.220	-14.278	0.219
3500	60.940	372.290	319.706	184.046	-319.546	-5.305	0.079
3600	61.010	374.008	321.190	190.144	-319.939	3.679	-0.053
3700	61.076	375.681	322.641	196.248	-320.401	12.674	-0.179
3800	61.137	377.310	324.058	202.359	-320.935	21.684	-0.298
3900	61.193	378.899	325.444	208.475	-321.548	30.708	-0.411
4000	61.246	380.449	326.800	214.597	-322.237	39.748	-0.519
4100	61.296	381.962	328.127	220.724	-323.014	48.807	-0.622
4200	61.342	383.440	329.426	226.856	-323.886	57.886	-0.720
4300	61.386	384.883	330.699	232.993	-324.855	66.987	-0.814
4400	61.427	386.295	331.947	239.133	-325.648	76.084	-0.903
4500	61.465	387.676	333.170	245.278	-326.756	85.227	-0.989
4600	61.501	389.027	334.369	251.426	-327.963	94.395	-1.072
4700	61.535	390.350	335.547	257.578	-329.250	103.588	-1.151
4800	61.567	391.646	336.702	263.733	-330.531	112.799	-1.227
4900	61.598	392.916	337.836	269.892	-332.019	122.050	-1.301
5000	61.626	394.161	338.950	276.053	-333.611	131.333	-1.372
5100	61.653	395.381	340.045	282.217	-335.308	140.648	-1.441
5200	61.679	396.579	341.121	288.383	-337.111	149.998	-1.507
5300	61.703	397.754	342.178	294.553	-339.022	159.384	-1.571
5400	61.726	398.908	343.218	300.724	-341.039	168.806	-1.633
5500	61.748	400.040	344.241	306.898	-343.163	178.267	-1.693
5600	61.769	401.153	345.247	313.074	-345.392	187.768	-1.751
5700	61.788	402.247	346.238	319.251	-347.381	197.261	-1.808
5800	61.807	403.321	347.213	325.431	-349.774	206.838	-1.863
5900	61.825	404.378	348.173	331.613	-352.264	216.455	-1.916
6000	61.842	405.417	349.118	337.796	-354.849	226.116	-1.969

PREVIOUS: December 1970 (1 atm) CURRENT December 1970 (1 bar)

Potassium Hydroxide (KOH) $H_1K_1O_1(g)$

POTASSIUM HYDROXIDE, ION (KOH^+) IDEAL GAS M_r = 56.105091

$S°(298.15 K) = [250.331 \pm 8.4] J K^{-1} mol^{-1}$

$\Delta_f H°(0 K) = 495.60 \pm 41.8 kJ mol^{-1}$
$\Delta_f H°(298.15 K) = 497.90 \pm 41.8 kJ mol^{-1}$

Electronic Levels and Quantum Weights

State	ε_i, cm^{-1}	g_i
$^2\Pi$	0	4
$^2\Sigma^+$	[32000]	2

Vibrational Frequencies and Degeneracies

ν, cm^{-1}
[375](1)
[300](2)
[3600](1)

Point Group: $[C_{\infty v}]$ σ = 1
Bond Distances: K-O = [2.20] Å O-H = [1.03] Å
Bond Angle: Na-O-H = [180]°
Rotational Constant: B_0 = [0.273684] cm^{-1}

Enthalpy of Formation

Porter and Schoonmaker (1) obtained an appearance potential curve near threshold for KOH which extrapolates to 7.8 eV at zero ion current. They presented evidence which indicated that a large fraction of KOH^+ was formed by dissociative ionization of the dimer. in a later study, these same workers (2) rejected their earlier results, since it was found that the presence of magnesia in their cell had reduced the activity of the hydroxide. Qusarov and Gorokhov (3) in a similar mass spectrometric study of the evaporation products of KOH have shown that KOH^+ is formed primarily from direct ionization of the hydroxide. Very recently, Gorokhov et al. (4) reported the appearance potential of KOH^+ from KOH as 7.5±0.2 eV. We adopt this result as $\Delta_r H°$ for the process $e^- + KOH(g) \rightarrow KOH^+(g) + 2e^-$, since it is consistent with the expected mode of ionization. This value leads to $\Delta_f H°(KOH^+, g, 0 K) = 118 \pm 10 kcal mol^{-1}$ when used in conjunction with $\Delta_f H°(KOH, g, 0 K) = -54.6 \pm 3.0 kcal mol^{-1}$ (5).

Heat Capacity and Entropy

The configuration of KOH^+ is predicted to be linear by Walsh (6), since it has seven valence electrons. The same correlation predicts that the unpaired electron is in a pi orbital, thus, the ground state of KOH^+ is $^2\Pi$. A first excited state is estimated at 32000 cm^{-1} to be $^2\Sigma$ by analogy with the isoelectronic molecules OH (5) and SH (5). The vibrational frequencies are estimated from those for KOH with somewhat lower values to account for the effect of weaker bonding predicted for the positive ion. Similarly, the K-O and O-H bond distances are increased slightly over those for KOH.

References

1. R. F. Porter and R. C. Schoonmaker, J. Phys. Chem. 62, 234 (1958).
2. R. C. Schoonmaker and R. F. Porter, J. Chem. Phys. 31, 830 (1959).
3. A. V. Gusarov and L. N. Gorokhov, Russ. J. Phys. Chem. 42, 449 (1968).
4. L. N. Gorokhov, A. V. Gusarov, and I. G. Panchenkov, Russ. J. Phys. Chem. 44, 150 (1970).
5. JANAF Thermochemical Tables: KOH(g) and OH(g) 12-31-70; SH(g) 6-30-67.
6. A. D. Walsh, J. Chem. Soc. 2288 (1953).

Potassium Hydroxide, Ion (KOH^+) $H_1K_1O_1^+(g)$

Enthalpy Reference Temperature = T_r = 298.15 K Standard State Pressure = p° = 0.1 MPa

T/K	$C_p°$ (J K⁻¹mol⁻¹)	S° (J K⁻¹mol⁻¹)	$-[G°-H°(T_r)]/T$ (J K⁻¹mol⁻¹)	$H°-H°(T_r)$ (kJ mol⁻¹)	$\Delta_f H°$ (kJ mol⁻¹)	$\Delta_f G°$ (kJ mol⁻¹)	Log K_f
0	0.	0.	INFINITE	-11.760	495.596		
100	34.457	203.689	291.020	-8.733			
200	45.225	231.352	254.747	-4.679			
250	47.868	241.752	251.137	-2.346			
298.15	49.479	250.331	250.331	0.	497.896	486.350	-85.207
300	49.529	250.637	250.332	0.092	497.918	486.278	-84.669
350	50.620	258.359	250.939	2.597	496.108	484.390	-72.291
400	51.369	265.171	252.301	5.148	496.636	482.680	-63.032
450	51.908	271.254	254.075	7.731	497.205	480.902	-55.822
500	52.315	276.745	256.072	10.337	497.804	479.059	-50.047
600	52.917	286.339	260.339	15.600	499.062	475.192	-41.369
700	53.406	294.533	264.653	20.916	500.365	471.111	-35.155
800	53.877	301.696	268.845	26.280	501.686	466.842	-30.482
900	54.362	308.069	272.856	31.692	503.004	462.407	-26.837
1000	54.859	313.822	276.669	37.153	504.298	457.827	-23.914
1100	55.360	319.075	280.289	42.664	426.595	457.729	-21.736
1200	55.851	323.913	283.725	48.225	428.842	460.460	-20.043
1300	56.323	328.402	286.991	53.834	431.099	463.004	-18.604
1400	56.769	332.592	290.101	59.489	433.364	465.373	-17.363
1500	57.186	336.523	293.066	65.187	435.637	467.580	-16.283
1600	57.572	340.227	295.899	70.925	437.914	469.635	-15.332
1700	57.927	343.728	298.610	76.700	440.192	471.548	-14.489
1800	58.252	347.048	301.210	82.509	442.470	473.327	-13.736
1900	58.549	350.206	303.706	88.349	444.744	474.979	-13.058
2000	58.820	353.216	306.107	94.218	447.011	476.512	-12.445
2100	59.067	356.092	308.419	100.113	449.267	477.931	-11.888
2200	59.291	358.845	310.649	106.031	451.510	479.244	-11.379
2300	59.496	361.485	312.802	111.970	453.735	480.454	-10.911
2400	59.684	364.021	314.884	117.929	455.940	481.568	-10.481
2500	59.854	366.461	316 899	123.906	458.122	482.591	-10.083
2600	60.011	368.812	318.850	129.900	460.277	483.528	-9.714
2700	60.154	371.079	320.743	135.908	462.400	484.382	-9.371
2800	60.285	373.269	322.580	141.930	464.490	485.157	-9.051
2900	60.406	375.387	324.365	147.965	466.543	485.859	-8.751
3000	60.517	377.437	326.100	154.011	468.553	486.491	-8.471
3100	60.619	379.423	327.788	160.068	470.519	487.056	-8.207
3200	60.714	381.349	329.432	166.134	472.436	487.559	-7.959
3300	60.801	383.218	331.033	172.210	474.309	488.002	-7.724
3400	60.882	385.035	332.595	178.294	476.121	488.389	-7.503
3500	60.957	386.801	334.119	184.386	477.874	488.724	-7.294
3600	61.027	388.519	335.606	190.486	479.562	489.010	-7.095
3700	61.093	390.192	337.059	196.592	481.181	489.250	-6.907
3800	61.154	391.822	338.479	202.704	482.726	489.448	-6.728
3900	61.211	393.411	339.867	208.822	484.194	489.605	-6.558
4000	61.264	394.962	341.225	214.946	485.585	489.725	-6.395
4100	61.315	396.475	342.554	221.075	486.889	489.812	-6.240
4200	61.362	397.953	343.856	227.209	488.097	489.869	-6.092
4300	61.407	399.397	345.131	233.348	489.209	489.898	-5.951
4400	61.450	400.810	346.380	239.490	490.496	489.874	-5.816
4500	61.491	402.191	347.605	245.638	491.469	489.849	-5.686
4600	61.530	403.543	348.806	251.789	492.344	489.803	-5.562
4700	61.567	404.867	349.985	257.943	493.138	489.737	-5.443
4800	61.602	406.163	351.142	264.102	493.940	489.644	-5.328
4900	61.637	407.434	352.278	270.264	494.534	489.548	-5.219
5000	61.670	408.679	353.394	276.429	495.025	489.441	-5.113
5100	61.702	409.901	354.490	282.598	495.411	489.326	-5.012
5200	61.733	411.099	355.567	288.769	495.691	489.203	-4.914
5300	61.763	412.276	356.626	294.944	495.865	489.077	-4.820
5400	61.793	413.430	357.667	301.122	495.932	488.948	-4.730
5500	61.822	414.564	358.691	307.303	495.895	488.818	-4.642
5600	61.851	415.679	359.699	313.487	495.752	488.691	-4.558
5700	61.879	416.774	360.691	319.673	495.850	488.519	-4.477
5800	61.907	417.850	361.667	325.862	495.545	488.394	-4.398
5900	61.934	418.909	362.628	332.054	495.144	488.273	-4.323
6000	61.961	419.950	363.575	338.249	494.649	488.161	-4.250

PREVIOUS: December 1971 (1 atm) CURRENT: December 1971 (1 bar)

Potassium Hydroxide, Ion (KOH^+) $H_1K_1O_1^+(g)$

LITHIUM HYDRIDE (LiH) CRYSTAL M_r = 7.94894

$S°(298.15\ K) = 20.037 \pm 0.13\ J\ K^{-1}\ mol^{-1}$
$T_{fus} = 961.8\ K$

$\Delta_f H°(0\ K) = -85.548 \pm 0.13\ kJ\ mol^{-1}$
$\Delta_f H°(298.15\ K) = -90.625 \pm 0.13\ kJ\ mol^{-1}$
$\Delta_{fus} H° = 22.594 \pm 0.42\ kJ\ mol^{-1}$

Enthalpy of Formation

The enthalpies of solution of Li(cr) and LiH(cr) in water have been measured by Guntz (1) and Moers (2) using an open Dewar flask calorimeter and by Messer (3) and Gunn (4) using closed bomb calorimetry. Based on the reported values of the enthalpy changes ($\Delta_r H°$) for the reaction (A) Li(cr) + H_2O(ℓ) = LiOH(aq) + 1/2 H_2(g), and (B) LiH(cr) + H_2O(ℓ) = LiOH(aq) + H_2(g), the corresponding $\Delta_f H°$(298.15 K) for LiH(cr) are calculated. The results are presented in the table below.

The decomposition pressures of LiH(cr) have been determined by Hurd (5), at temperatures 782-926 K. By use of the 2nd and 3rd law methods, the enthalpy changes for the decomposition reaction (C) LiH(cr) = Li(ℓ) + 1/2 H_2(g) are evaluated as 21.87±0.3 and 22.34 kcal mol^{-1}, respectively. Based on the 3rd law $\Delta_r H°$(298.15 K) and $\Delta_f H°$(Li, ℓ, 298.15 K) = 0.569 kcal mol^{-1}, we derive the value $\Delta_f H°$(LiH, cr, 298.15 K) = -21.77±0.2 kcal mol^{-1}, which is in good agreement with the values obtained by calorimetric method. Gunts (1) measured the decomposition pressure at 953 K for the same reaction. The 3rd law $\Delta_r H°$(298.15 K) and $\Delta_f H°$(LiH, cr, 298.15 K) are evaluated as 21.86 and -21.29 kcal mol^{-1}, respectively.

Johnson (6) has determined the standard Gibbs energy of formation for LiH(cr) from emf measurements over the temperature range 675 - 885 K. The cell utilized an Armco iron flag over which H_2 was passed as the cathode and a molten Li anode. LiCl, LiBr and LiI, each saturated with LiH were used as electrolytes. Using the reported emf values at different temperatures (47 data points) for reaction (C) we derive $\Delta_f H°$(298.15 K) = 21.81 kcal mol^{-1} for LiH(cr).

The $\Delta_f H°$(LiH, cr, 298.15 K) value reported by Gunn (4) is adopted.

Source	$\Delta_r H°$(298.15 K), kcal mol^{-1} Reaction A	Reaction B	Reaction C	Drift cal K^{-1} mol^{-1}	$\Delta_f H°$(298.15 K) kcal mol^{-1}
Guntz (1)	-53.2	-31.6	-	-	-21.6
	-	-	21.86	-	-21.29
Moers (2)	-52.7 ±0.2	-31.1 ±0.1	-	-	-21.6 ±0.25
Messer (3)	-53.10±0.11	-31.76±0.10	-	-	-21.34±0.15
Gunn (4)	-53.14±0.02	-31.48±0.02	-	-	-21.66±0.03
Hurd (5)	-	-	22.34	0.5±0.3	-21.77±0.2
Johnson (6)	-	-	22.38	1.0±0.2	-21.81±0.15

Heat Capacity and Entropy

The low temperature heat capacities are obtained from the data of Kostyukov (7), 3.72 - 295.5 K. The $C_p°$ data reported by Gunther (8) are consistently lower than the adopted ones in the temperature range 74.0 - 90.5 K but considerably higher at 292.7 K. The high temperature enthalpies have been determined by Fieldhouse (9), 413.2 - 914.3 K and Vogt (14), 878.15 - 953.15 K. The derived high temperature heat capacities are joined smoothly with the low temperature values and extrapolated smoothly above 953.15 K.

S°(298.15 K) is calculated based on the adopted low temperature heat capacities, using S°(7 K) = 0.0002 cal K^{-1} mol^{-1}.

Fusion Data

The value of T_{fus} has been reported as 953 (1), 961±1 (10, 12), 959.6 (11) and 961.8±0.3 (13). The value reported by Messer (13) is adopted. The heat of melting, $\Delta_{fus}H°$, has been determined as 5.095±0.46 (13) and 5.237±0.040 kcal mol (14). The first $\Delta_{fus}H°$ value was evaluated cryogenically from the lowering of melting point in the systems LiH-LiO_2O and LiH-CaH_2. The second value was derived from enthalpy data measured with copper block calorimeter. The adopted $\Delta_{fus}H°$ value is derived from the enthalpy data of Vogt (14) using the adopted heat capacities for LiH(cr) and LiH(ℓ).

Sublimation Data

The value of $\Delta_{sub}H°$(298.15 K) is calculated as the difference between $\Delta_f H°$(298.15 K) for LiH(g) and LIH(cr).

References

1. A. Guntz, Compt. Rend. 123, 694 (1896).
2. K. Moers, Z. Anorg. Chem. 113, 179 (1920).
3. C. E. Messer, L. G. Fasolino and C. E. Thalmayer, J. Amer. Chem. Soc. 77, 4524 (1955).
4. S. R. Gunn and L. G. Green, J. Amer. Chem. Soc. 80, 4782 (1958); UCRL-7992, Univ. of California, July 29, 1964.
5. C. B. Hurd and G. A. Moore, Jr., J. Amer. Chem. Soc. 57, 332 (1935).
6. C. E. Johnson, R. R. Heinrich and C. E. Crouthamel, J. Phys. Chem. 70, 242 (1966).
7. V. N. Kostryukov, Russ. J. Phys. Chem. 35, 865 (1961).
8. P. Gunther, Ann. Physik 63, 476 (1920).
9. I. B. Fieldhouse, J. C. Hedge and J. I. Lang, Armour Research Foundation, PB 151583, November, 1958.
10. C. E. Messer, E. B. Damon, P. C. Maybury, J. Mellor and R. A. Seales, J. Phys. Chem. 62, 220 (1958).
11. C. E. Messer and J. Mellor, J. Phys. Chem. 64, 503 (1960).
12. C. E. Messer, Tufts University, Mass., NYO-8027, May 5, 1960.
13. C. E. Messer, J. Mellor, J. A. Krol and I. S. Levy, J. Chem. Eng. Data 6, 328 (1961).
14. J. W. Vogt, Thompson Ramo Wooldridge Inc., Cleveland, Ohio, NASA N63-22167, June 1962.

Lithium Hydride (LiH) H_1Li_1(cr)

Enthalpy Reference Temperature = T_r = 298.15 K; Standard State Pressure = p° = 0.1 MPa

T/K	$C_p°$ (J $K^{-1}mol^{-1}$)	S°	$-[G°-H°(T_r)]/T$	$H°-H°(T_r)$ (kJ mol^{-1})	$\Delta_f H°$	$\Delta_f G°$	Log K_f
0	0.	0.	INFINITE	-3.778	-85.548	-85.548	INFINITE
100	6.381	2.322	38.346	-3.602	-87.367	-81.828	42.743
200	18.778	10.715	22.326	-2.322	-89.275	-75.511	19.722
298.15	27.957	20.037	20.037	0.	-90.625	-68.447	11.992
300	28.083	20.210	20.038	0.052	-90.646	-68.309	11.894
400	34.794	29.236	21.223	3.205	-91.546	-60.718	7.929
500	40.735	37.646	23.676	6.985	-95.160	-52.628	5.498
600	46.380	45.575	26.671	11.342	-95.251	-44.103	3.839
700	51.877	53.138	29.915	16.256	-94.731	-35.611	2.657
800	57.300	60.420	33.276	21.715	-93.644	-27.233	1.778
900	62.676	67.480	36.686	27.715	-92.024	-19.023	1.104
961.800	65.985	71.751	38.802	31.690	------ CRYSTAL <--> LIQUID -----		
1000	68.028	74.361	40.111	34.250	-89.876	-11.022	0.576
1100	73.362	81.095	43.531	41.320	-87.207	-3.261	0.155
1200	78.680	87.706	46.938	48.922	-84.020	4.233	-0.184
1300	83.994	94.214	50.325	57.056	-80.319	11.441	-0.460
1400	89.299	100.633	53.690	65.720	-76.103	18.345	-0.684
1500	94.600	106.974	57.031	74.915	-71.369	24.929	-0.868

PREVIOUS: September 1962 CURRENT: September 1967

Lithium Hydride (LiH) H_1Li_1(cr)

LITHIUM HYDRIDE (LiH) LIQUID M_r = 7.94894

$S°(298.15\ K) = [22.227]\ J\ K^{-1}\ mol^{-1}$

$T_{fus} = 961.8 \pm 0.3\ K$

$\Delta_f H°(298.15\ K) = [-77.715]\ kJ\ mol^{-1}$

$\Delta_{fus} H° = 22.594 \pm 0.42\ kJ\ mol^{-1}$

Enthalpy of Formation

$\Delta_f H°$(LiH, ℓ, 298.15 K) is calculated from that of the crystal by adding $\Delta_{fus}H°$ and the difference in enthalpy, H°(961.8 K) - H°(298.15 K), between the crystal and liquid.

The decomposition pressures of LiH(ℓ) at different temperatures have been investigated by Hill (1), Perlow (2), Johnson, and Gibb (4). Based on their reported equilibrium pressures, the enthalpy changes for the reaction LiH(ℓ) = Li(ℓ) + 1/2H_2(g) are evaluated by the 2nd and 3rd law methods. The results are presented in the table below. It should be noted that the measured decomposition pressures are highly sensitive to composition and change extremely rapidly between 99 and 100 percent purity. Therefore we give no weight to the equilibrium data.

Source	T/K	Data Points	$\Delta_r H°$(298.15 K), kcal mol^{-1} 2nd law	3rd law	Drift cal K^{-1} mol^{-1}	$\Delta_f H°$(298.15 K) kcal mol^{-1}
Hill (1938)	973.2	1	-	16.93	-	-16.36
Perlow (1941)	1043-1098	2	20.10	17.69	-2.2	-17.12
Johnson*	1050-1090	-	24.69	18.36	-5.9	-17.79
Gibb (1951)	1073-1173	2	5.66	15.2	+8.5	-14.63

*The vapor pressure values are calculated from an equation quoted by Gibb (3), but derived from the vapor pressures measured by W. C. Johnson, University of Chicago.

Heat Capacity and Entropy

The enthalpies of LiH(ℓ) in the temperature range 975.15-1075.15 K have been measured by Vogt (5). Using the reported enthalpy values, the heat capacity of LiH(ℓ) is derived as 14.9 cal K^{-1} mol^{-1}. This $C_p°$ value is adopted and extrapolated above and below the measured range.

S°(298.15 K) is calculated in a manner similar to that used for the enthalpy of formation.

Fusion Data

See the crystal table for details.

Decomposition Data

T_{dcm} = 1223 K is calculated as the temperature at which LiH(ℓ) decomposes into Li(ℓ) and H_2(g) where the fugucity of H_2(g) equals one atmosphere.

References

1. L. L. Hill, Ph. D. Thesis, University of Chicago, 1938.
2. M. R. J. Perlow, Ph. D. Thesis, University of Chicago, 1941.
3. T. R. P. Gibb, J. Electrochem. Soc. 93, 198 (1948).
4. T. R. P. Gibb, NEPA-1841, Apr. 30, 1951.
5. J. W. Vogt, Thompson Ramo Wooldridge Inc., Cleveland, Ohio, NASA N63-22167, June 1962.

Lithium Hydride (LiH) $H_1Li_1(l)$

Enthalpy Reference Temperature = T_r = 298.15 K   Standard State Pressure = p° = 0.1 MPa

T/K	$C_p°$ (J $K^{-1}mol^{-1}$)	S°	$-[G°-H°(T_r)]/T$	$H°-H°(T_r)$ (kJ mol^{-1})	$\Delta_f H°$	$\Delta_f G°$	Log K_f
0							
100							
200							
298.15	62.342	22.227	22.227	0.	-77.714	-56.189	9.844
300	62.342	22.613	22.228	0.115	-77.671	-56.055	9.760
400	62.342	40.547	24.673	6.349	-75.491	-49.187	6.423
500	62.342	54.458	29.291	12.584	-76.651	-42.525	4.443
600	62.342	65.825	34.462	18.818	-74.865	-35.866	3.122
700	62.342	75.435	39.646	25.052	-73.024	-29.511	2.202
800	62.342	83.759	44.651	31.286	-71.163	-23.422	1.529
900	62.342	91.102	49.413	37.520	-69.307	-17.566	1.020
961.800	62.342	95.242	52.226	41.373	------ CRYSTAL <--> LIQUID -----		
1000	62.342	97.670	53.916	43.754	-67.461	-11.916	0.622
1100	62.342	103.612	58.168	49.989	-65.627	-6.451	0.306
1200	62.342	109.036	62.184	56.223	-63.808	-1.151	0.050
1300	62.342	114.026	65.983	62.457	-62.007	3.997	-0.161
1400	62.342	118.646	69.581	68.691	-60.221	9.007	-0.336
1500	62.342	122.948	72.997	74.925	-58.448	13.890	-0.484
1600	62.342	126.971	76.246	81.159	-56.689	18.655	-0.609
1700	62.342	130.750	79.342	87.394	-200.179	30.487	-0.937
1800	62.342	134.314	82.298	93.628	-197.692	43.983	-1.276
1900	62.342	137.684	85.126	99.862	-195.226	57.342	-1.576
2000	62.342	140.882	87.834	106.096	-192.781	70.572	-1.843

PREVIOUS: September 1962   CURRENT: September 1967

Lithium Hydride (LiH) $H_1Li_1(l)$

LITHIUM HYDRIDE (LiH) CRYSTAL-LIQUID M_r = 7.94894

0 to 961.8 K crystal
above 961.8 K liquid

Refer to the individual tables for details.

Lithium Hydride (LiH) $H_1Li_1(cr,l)$

Enthalpy Reference Temperature = T_r = 298.15 K   Standard State Pressure = p° = 0.1 MPa

T/K	C_p° (J K⁻¹mol⁻¹)	S° (J K⁻¹mol⁻¹)	$-[G^\circ-H^\circ(T_r)]/T$ (J K⁻¹mol⁻¹)	$H^\circ-H^\circ(T_r)$ (kJ mol⁻¹)	Δ_fH° (kJ mol⁻¹)	Δ_fG° (kJ mol⁻¹)	Log K_f
0	0.	0.	INFINITE	-3.778	-85.548	-85.548	INFINITE
100	6.381	2.322	38.346	-3.602	-87.367	-81.828	42.743
200	18.778	10.715	22.326	-2.322	-89.275	-75.511	19.722
298.15	27.957	20.037	20.037	0.	-90.625	-68.447	11.992
300	28.083	20.210	20.038	0.052	-90.646	-68.309	11.894
400	34.794	29.236	21.223	3.205	-91.546	-60.718	7.929
500	40.735	37.646	23.676	6.985	-95.160	-52.628	5.498
600	46.380	45.575	26.671	11.342	-95.251	-44.103	3.839
700	51.877	53.138	29.915	16.256	-94.731	-35.611	2.657
800	57.300	60.420	33.276	21.715	-93.644	-27.233	1.779
900	62.676	67.480	36.686	27.715	-92.024	-19.023	1.104
961.800	65.985	71.751	38.802	31.690	CRYSTAL <--> LIQUID TRANSITION		
961.800	62.342	95.242	38.802	54.284			
1000	62.342	97.670	41.005	56.665	-67.461	-11.916	0.622
1100	62.342	103.612	46.431	62.900	-65.627	-6.451	0.306
1200	62.342	109.036	51.425	69.134	-63.808	-1.151	0.050
1300	62.342	114.026	56.051	75.368	-62.007	3.997	-0.161
1400	62.342	118.646	60.359	81.602	-60.221	9.007	-0.336
1500	62.342	122.948	64.390	87.836	-58.448	13.890	-0.484
1600	62.342	126.971	68.177	94.070	-56.689	18.655	-0.609
1700	62.342	130.750	71.748	100.305	-200.179	30.487	-0.937
1800	62.342	134.314	75.126	106.539	-197.692	43.983	-1.276
1900	62.342	137.684	78.330	112.773	-195.226	57.342	-1.576
2000	62.342	140.882	81.379	119.007	-192.781	70.572	-1.843

PREVIOUS: CURRENT: September 1967

Lithium Hydride (LiH) $H_1Li_1(cr,l)$

LITHIUM HYDRIDE (LiH) IDEAL GAS M_r = 7.94894

$S°(298.15\ K) = 170.909\ J\ K^{-1}\ mol^{-1}$

$\Delta_f H°(0\ K) = 140.794 \pm 0.04\ kJ\ mol^{-1}$
$\Delta_f H°(298.15\ K) = 140.624 \pm 0.04\ kJ\ mol^{-1}$

Electronic Levels and Quantum Weights

State	ε_i, cm^{-1}	g_i
$^1\Sigma^+$	0	1
	26513.7	1
	34912.0	2

$\omega_e = 1406.73\ cm^{-1}$ $\omega_e x_e = 23.24\ cm^{-1}$ $\sigma = 1$
$B_e = 7.5247\ cm^{-1}$ $\alpha_e = 0.2137\ cm^{-1}$ $r_e = 1.595$ Å

Enthalpy of Formation

Velasco (1), observed the absorption spectra of LiH(g) in the near ultraviolet with high dispersion and absorbing path lengths up to 16 meters. He found a new band system involving the ground state and a $^1\Pi$ excited state. Rotational and vibrational analyses of this system were carried out and the observed breaking off of the rotational structure of the bands was interpreted as due to predissociation by rotation. With this assumption very accurate dissociation limits of the $B^1\Pi$ state were obtained. From these dissociation limits the dissociation energy of the ground state of LiH was found to be 2.4288±0.0002 eV. Based on this value, we adopt $\Delta_f H°(298.15\ K) = 33.61\ kcal\ mol^{-1}$ for LiH(g)

Kemperer (2), observed the infrared spectrum of LiH(g) in emission in the frequency region 1500 to 970 cm^{-1}, using a Perkin-Elmer double-pass monochromator with NaCl optics. Estimating the minimum LiH pressure as 10 mm and assuming the maximum possible Li pressure, he reported K = 0.005 (atm) at 1400 K for the reaction $2Li(g) + H_2(g) = 2LiH(g)$. We calculate the corresponding enthalpy change, $\Delta_r H°(1400\ K)$, to be $-9.63\ kcal\ mol^{-1}$, employing $\Delta_r S°(1400\ K) = -17.407\ cal\ K^{-1}\ mol^{-1}$. Incorporating this $\Delta_r H°(1400\ K)$ value with $\Delta_f H°(1400\ K) = 35.601\ kcal\ mol^{-1}$ for Li(g), we obtain $\Delta_f H°(298.15\ K) \leq 33.92\ kcal\ mol^{-1}$ for LiH(g), which is in good agreement with the adopted value.

Gaydon (3), derived the value $D_0^0(LiH) = 2.5 \pm 0.2$ eV from earlier spectroscopic data.

Heat Capacity and Entropy

The third electronic level and quantum weight are obtained from Velasco (1). All the other constants are taken from Herzberg (4), and corrected to the average isotopic species.

References

1. R. Velasco, Can. J. Phys. 35, 1204 (1957).
2. W. Kemperer, J. Chem. Phys. 23, 2452 (1955).
3. A. G. Gaydon, "Dissociation Energies," Chapman and Hall Ltd., London, 1953.
4. G. Herzberg, "Spectra of Diatomic Molecules," D. Van Nostrand Co., Inc., New York, 1950.

Lithium Hydride (LiH) $H_1Li_1(g)$

Enthalpy Reference Temperature = T_r = 298.15 K Standard State Pressure = p° = 0.1 MPa

T/K	$C_p°$ (J K⁻¹mol⁻¹)	S° (J K⁻¹mol⁻¹)	$-[G°-H°(T_r)]/T$ (J K⁻¹mol⁻¹)	$H°-H°(T_r)$ (kJ mol⁻¹)	$\Delta_f H°$ (kJ mol⁻¹)	$\Delta_f G°$ (kJ mol⁻¹)	Log K_f
0	0.	0.	INFINITE	-8.686	140.794	140.794	INFINITE
100	29.139	138.955	196.998	-5.804	141.681	133.556	-69.763
200	29.220	159.168	173.607	-2.888	141.409	125.482	-32.773
250	29.401	165.705	171.396	-1.423	141.048	121.540	-25.394
298.15	29.731	170.909	170.909	0.	140.624	117.820	-20.642
300	29.746	171.093	170.910	0.055	140.607	117.679	-20.490
350	30.235	175.714	171.273	1.554	140.119	113.896	-16.998
400	30.811	179.788	172.088	3.080	139.578	110.186	-14.389
450	31.422	183.452	173.150	4.636	138.981	106.547	-12.368
500	32.029	186.794	174.350	6.222	135.326	103.285	-10.790
600	33.144	192.735	176.931	9.482	134.138	96.991	-8.444
700	34.081	197.917	179.567	12.845	133.107	90.883	-6.782
800	34.848	202.519	182.154	16.293	132.182	84.915	-5.544
900	35.472	206.661	184.650	19.810	131.321	79.059	-4.588
1000	35.986	210.426	187.043	23.383	130.507	73.296	-3.829
1100	36.414	213.876	189.327	27.004	129.727	67.613	-3.211
1200	36.776	217.061	191.507	30.664	128.972	62.000	-2.699
1300	37.088	220.017	193.588	34.358	128.233	56.449	-2.268
1400	37.359	222.776	195.576	38.080	127.507	50.954	-1.901
1500	37.600	225.362	197.476	41.828	126.794	45.511	-1.585
1600	37.816	227.795	199.296	45.599	126.090	40.115	-1.310
1700	38.013	230.094	201.040	49.391	-19.843	41.939	-1.289
1800	38.193	232.272	202.715	53.201	-19.780	45.571	-1.322
1900	38.361	234.341	204.326	57.029	-19.720	49.200	-1.353
2000	38.517	236.313	205.876	60.873	-19.665	52.826	-1.380
2100	38.665	238.196	207 371	64.732	-19.615	56.450	-1.404
2200	38.806	239.998	208.813	68.606	-19.571	60.071	-1.426
2300	38.940	241.726	210.207	72.493	-19.534	63.690	-1.446
2400	39.069	243.386	211.555	76.394	-19.506	67.308	-1.465
2500	39.193	244.983	212.860	80.307	-19.486	70.925	-1.482
2600	39.314	246.523	214 126	84.232	-19.478	74.541	-1.498
2700	39.432	248.009	215.353	88.170	-19.481	78.157	-1.512
2800	39.546	249.445	216.545	92.119	-19.498	81.773	-1.526
2900	39.659	250.835	217.704	96.079	-19.530	85.391	-1.538
3000	39.770	252.181	218.831	100.050	-19 578	89.010	-1.550
3100	39.879	253.487	219.928	104.033	-19.644	92.630	-1.561
3200	39.988	254.755	220.996	108.026	-19.729	96.253	-1.571
3300	40.095	255.987	222.038	112.030	-19.835	99.879	-1.581
3400	40.202	257.185	223.054	116.045	-19.963	103.509	-1.590
3500	40.308	258.352	224.046	120.071	-20.114	107.142	-1.599
3600	40.415	259.489	225.015	124.107	-20.290	110.781	-1.607
3700	40.522	260.598	225.962	128.154	-20.492	114.424	-1.615
3800	40.630	261.680	226.888	132.211	-20.722	118.074	-1.623
3900	40.738	262.737	227.793	136.280	-20.980	121.730	-1.630
4000	40.847	263.770	228.680	140.359	-21.268	125.392	-1.637
4100	40.958	264.780	229.548	144.449	-21.587	129.063	-1.644
4200	41.070	265.768	230.399	148.550	-21.939	132.741	-1.651
4300	41.183	266.736	231.233	152.663	-22.324	136.429	-1.657
4400	41.299	267.684	232.050	156.787	-22.723	140.124	-1.663
4500	41.416	268.613	232.852	160.923	-23.169	143.830	-1.670
4600	41.536	269 525	233.640	165.070	-23.653	147.546	-1.675
4700	41.658	270.419	234.413	169.230	-24.173	151.273	-1.681
4800	41.782	271.298	235.172	173.402	-24.732	155.012	-1.687
4900	41.909	272.160	235.918	177.586	-25.329	158.763	-1.692
5000	42.039	273.008	236.652	181.784	-25.965	162.526	-1.698
5100	42.171	273.842	237.373	185.994	-26.637	166.302	-1.703
5200	42.306	274.662	238.082	190.218	-27.354	170.092	-1.709
5300	42.445	275.469	238.780	194.456	-28.114	173.896	-1.714
5400	42.586	276.264	239.467	198.707	-28.918	177.715	-1.719
5500	42.731	277.047	240.143	202.973	-29.766	181.549	-1.724
5600	42.878	277.818	240.809	207.253	-30.659	185.400	-1.729
5700	43.029	278.578	241.465	211.549	-31.598	189.266	-1.734
5800	43.182	279.328	242.111	215.859	-32.584	193.150	-1.740
5900	43.339	280.068	242.748	220.185	-33.618	197.050	-1.745
6000	43.499	280.797	243.376	224.527	-34.700	200.969	-1.750

PREVIOUS: September 1967 (1 atm) CURRENT: September 1967 (1 bar)

Lithium Hydride (LiH) $H_1Li_1(g)$

LITHIUM HYDROXIDE (LiOH) CRYSTAL M_r = 23.94834

$S°(298.15 K) = 42.821 \pm 0.21 J K^{-1} mol^{-1}$

$\Delta_f H°(0 K) = -479.14 \pm 0.4 kJ mol^{-1}$

$\Delta_f H°(298.15 K) = -484.93 \pm 0.4 kJ mol^{-1}$

Enthalpy of Formation

The enthalpy of formation of crystalline LiOH is determined from calorimetric studies on the enthalpy of hydrolysis of metallic lithium and the enthalpy of solution of the hydroxide in water.

Reshetnikov (1) determined the enthalpy of solution of LiOH(cr) in (aq, 400 H_2O) as -5.479 ± 0.007 kcal mol^{-1}. Combining this result with enthalpy of dilution data for LiOH tabulated by Parker (2), we derive $\Delta_{sol}H°$(LiOH, cr) = -5.632 ± 0.020 kcal mol^{-1}, which is essentially the value recommended by Parker (2). Lower values by some 600-650 cal mol^{-1} for the enthalpy of solution of LiOH(cr) are summarized by Parker (2). The discrepancy in these values probably arises from carbonate and water impurities in the LiOH samples. Since their presence would tend to reduce the enthalpy of solution, we adopt the higher results of Reshetnikov (1).

Gunn and Green (3a, 3b) and Messer et al. (4) measured the enthalpy of hydrolysis of metallic lithium. Their results for the reaction Li(cr) + (n+1) H_2O → LiOH·n H_2O + 0.5 H_2(g) are summarized below.

Source	Moles of H_2O	$\Delta_{hyd}H°$(298.15 K)* kcal mol^{-1}	$\Delta_f H°$(LiOH, aq, ∞, 298.15 K)** kcal mol^{-1}	$\Delta_f H°$(298.15 K) kcal mol^{-1}
Gunn and Green (3a, 3b)	1000	-53.210±0.040	-121.525±0.040	-115.893±0.060
Messer et al. (4)	70	-53.05 ±0.20	-121.37 ±0.20	-115.74 ±0.20

*Dilution data from Parker (2)

**Based on $\Delta_f H°(H_2O$, ℓ, 298.15 K) = -68.315 kcal mol^{-1} (5)

Combination of the $\Delta_f H°$(LiOH, aq, ∞, 298.15 K) values with the enthalpy of solution of LiOH(cr) given above results in the values for $\Delta_f H°$(LiOH, cr, 298.15 K) given in the last column. The adopted value is from the work of Gunn and Green (3a, 3b) with minor adjustments in the uncertainty interval to include possible errors in some of the auxiliary data used in the calculations.

Heat Capacity and Entropy

Heat capacities for LiOH(cr) in the temperature range 15-300 K are those of Bauer et al. (6). Heat capacities above 300 K are calculated from the high temperature (418-879 K) enthalpy data of Shomate and Cohen (7). Both sets of data were smoothed by computer and joined at 298.15 K. The rapid rise in the enthalpy data of Shomate and Cohen (7) near the melting point is attributed to premelting and data in the temperature range 725-732 K were not used in the fit. Powers and Blalock (8) also measured high temperature (397-1213 K) enthalpy data for LiOH(cr) in a Bunsen ice calorimeter. Although their results are less precise than those of Shomate and Cohen (7), the two sets of measurements are in reasonable agreement.

S°(298.15 K) is determined from the smoothed $C_p°$ data of Bauer et al. (6) and is based on an extrapolation of S°(16 K) = 0.027 cal K^{-1} mol^{-1}.

Transition and Fusion Data

The adopted enthalpy of melting and melting point for LiOH are from the enthalpy measurements of Shomate and Cohen (7). Powers and Blalock (8) reported $\Delta_{fus}H°$ = 5.029 kcal mol^{-1} at 746 K from their enthalpy measurements on LiOH. Very recently, Reshetnikov and Baranskaya (9) reported a slightly higher value of 5.29 kcal mol^{-1} at 747 K which was determined by a thermographic method.

Sublimation Data

$\Delta_{sub}H°$(298.15 K) values for the monomer and dimer are calculated from the adopted enthalpies of formation for the gaseous species and the crystal.

References

1. N. A. Reshetnikov, Zhur. Neorg. Khim. 6, 682 (1961).
2. V. B. Parker, NSRDS-NBS 2, 1965.
3. a. S. R. Gunn and L. G. Green, J. Amer. Chem. Soc. 80, 4782 (1958).
   b. S. R. Gunn, J. Phys. Chem. 71, 1386 (1967).
4. C. E. Messer, L. C. Fasolino, and C. E. Thalmayer, J. Amer. Chem. Soc. 77, 4524 (1955).
5. U. S. Nat. Bur. Stand. Tech. Note 270-3, 1968.
6. T. W. Bauer, H. L. Johnston, and E. C. Kerr, J. Amer. Chem. Soc. 72, 5174 (1950).
7. C. H. Shomate and A. J. Cohen, J. Amer. Chem. Soc. 77, 285 (1955).
8. W. D. Powers and G. C. Blalock, Oak Ridge Natl. Lab., ORNL 1653, January, 1954.
9. N. A. Reshetnikov and E. V. Baranskaya, Izv. Vyssh. Ucheb. Zaved., Khim. Khim. Tekhnol. 10, 496 (1967).

Lithium Hydroxide (LiOH) $H_1Li_1O_1(cr)$

Enthalpy Reference Temperature = T_r = 298.15 K   Standard State Pressure = p° = 0.1 MPa

T/K	$C_p°$ (J $K^{-1}mol^{-1}$)	S° (J $K^{-1}mol^{-1}$)	$-[G°-H°(T_r)]/T$ (J $K^{-1}mol^{-1}$)	$H°-H°(T_r)$ (kJ mol^{-1})	$\Delta_f H°$ (kJ mol^{-1})	$\Delta_f G°$ (kJ mol^{-1})	Log K_f
0	0.	0.	INFINITE	-7.416	-479.144	-479.144	INFINITE
100	14.644	8.561	77.138	-6.858	-482.033	-468.453	244.695
200	35.977	25.632	46.982	-4.270	-484.089	-453.960	118.562
298.15	49.591	42.821	42.821	0.	-484.926	-438.958	76.904
300	49.761	43.128	42.822	0.092	-484.933	-438.672	76.380
400	58.053	58.678	44.886	5.517	-485.047	-423.222	55.267
500	63.647	72.260	49.032	11.614	-487.874	-407.475	42.569
600	68.208	84.277	53.924	18.212	-487.303	-391.441	34.078
700	72.279	95.101	59.045	25.239	-486.298	-375.538	28.023
744.300	73.970	99.588	61.326	28.479	------ CRYSTAL <--> LIQUID -----		
800	76.086	105.002	64.180	32.658	-484.920	-359.805	23.493
900	79.747	114.176	69.231	40.450	-483.208	-344.265	19.981
1000	83.308	122.763	74.159	48.604	-481.174	-328.933	17.182
1100	86.814	130.868	78.950	57.110	-478.823	-313.820	14.902
1200	90.274	138.570	83.600	65.965	-476.158	-298.935	13.012
1300	93.705	145.932	88.113	75.164	-473.183	-284.285	11.423
1400	97.115	153.001	92.497	84.705	-469.897	-269.876	10.069
1500	100.508	159.817	96.759	94.586	-466.298	-255.713	8.905

PREVIOUS: March 1966   CURRENT: June 1971

Lithium Hydroxide (LiOH) $H_1Li_1O_1(cr)$

LITHIUM HYDROXIDE (LiOH) LIQUID M_r = 94834

$S°(298.15\ K) = [47.969]\ J\ K^{-1}\ mol^{-1}$

$T_{fus} = 744.3\ K$

$\Delta_f H°(298.15\ K) = [-474.422]\ kJ\ mol^{-1}$

$\Delta_{fus}H° = 20.878 \pm 0.21\ kJ\ mol^{-1}$

Enthalpy of Formation

$\Delta_f H°$(LiOH, ℓ, 298.15 K) is calculated from that of the crystal by adding $\Delta_{fus}H°$ and the difference in enthalpy, H°(744.3 K) - H°(298.15 K), between the crystal and liquid.

Heat Capacity and ENtropy

The heat capacity of the liquid phase is obtained from the high temperature enthalpy measurements (750-875 K) of Shomate and Cohen (1). We derive a constant $C_p°$ of 20.814 cal $K^{-1}\ mol^{-1}$ for LiOH(ℓ) from these enthalpy data. The average percent deviation of the linear fit of the data was 0.2%. Powers and Blalock (2) also determined a constant $C_p°$ = 22.03 cal $K^{-1}\ mol^{-1}$ for LiOH(ℓ) from enthalpy data (746-1200 K) determined by drop calorimetry.

S°(298.15 K) is calculated in a manner analogous to that of the enthalpy of formation.

Fusion Data

See the crystal table for details.

Vaporization Data

Dissociation pressures for LiOH(cr, ℓ) have been measured by static (4), effusion (5, 6), and transpiration (6) methods. These data are combined with the equilibrium studies of Berkowitz et al. (3) to give either $\Delta_{sub}H°$(298.15 K) or $\Delta_{vap}H°$(298.15 K). An analysis of the data is summarized below, and further discussion of these results is presented on the ideal gas table under Enthalpy of Formation.

Source	Reaction[a]	Method	T/K	Data Points	$\Delta_r H°$(298.15 K),kcal mol^{-1} 2nd law	3rd law	Drift cal $K^{-1}mol^{-1}$	$\Delta_{sub}H°$(298.15 K) kcal mol^{-1}	$\Delta_{vap}H°$(298.15 K) kcal mol^{-1}
Berkowitz et al.	A	Mass. Spec.	1120-1450	18	82.3	82.9±1.9	0.5±2.6		
Johnston (4)	B	Static	793-1176	9[b]	34.7	25.9±2.1	-8.5±0.4		54.4±5.0
Ditmars and Johnston (5)	C	Effusion	648-736	11	37.8	39.5±1.0	2.6±5.4	61.2±3.0	
	B	Effusion	752-795	7	35.5	34.3±0.9	-1.9±12.9		58.6±3.0
Gregory and Mohr (6)	C	Transpiration and Effusion	525-675	Equation	32.1	32.7±0.1	0.9±0.1	57.8±2.0	

[a]A) $Li_2O(cr) + H_2O(g) = 2LiOH(g)$ B) $2LiOH(ℓ) = Li_2O(cr) + H_2O(g)$ C) $2LiOH(cr) = Li_2O(cr) + H_2O(g)$

[b]One point rejected due to failure of a statistical test.

The boiling point, T_{vap} = 1899 K, is determined as the temperature at which the fugacity of LiOH(g) becomes 1 atm for the process LiOH(ℓ) = LiOH(g). The difference in their enthalpies of formation at the boiling point is the enthalpy of vaporization, $\Delta_{vap}H°$ = 44.9 kcal mol^{-1}.

References

1. C. H. Shomate and A. J. Cohen, J. Amer. Chem. Soc. 77, 285 (1955).
2. W. D. Powers and G. C. Blalock, Oak Ridge Nat. Lab., ORNL 1653, January, 1954.
3. J. Berkowitz, D. J. Meschi, and W. A. Chupka, J. Chem. Phys. 33, 533 (1960).
4. J. Johnston, Z. Phys. Chem. 62, 339 (1908).
5. W. E. Ditmars and H. L. Johnston, J. Amer. Chem. Soc. 75, 1830 (1953).
6. N. W. Gregory and R. H. Mohr, J. Amer. Chem. Soc. 77, 2142 (1955).

Lithium Hydroxide (LiOH) $H_1Li_1O_1(l)$

Enthalpy Reference Temperature = T_r = 298.15 K   Standard State Pressure = p° = 0.1 MPa

T/K	$C_p°$ (J $K^{-1}mol^{-1}$)	S°	$-[G°-H°(T_r)]/T$	$H°-H°(T_r)$ (kJ mol^{-1})	$\Delta_f H°$	$\Delta_f G°$	Log K_f
0							
100							
200							
298.15	87.086	47.969	47.969	0.	-474.422	-429.989	75.332
300	87.086	48.508	47.971	0.161	-474.360	-429.713	74.820
400	87.086	73.561	51.386	8.870	-471.191	-415.318	54.235
500	87.086	92.993	57.837	17.578	-471.406	-401.374	41.931
600	87.086	108.871	65.060	26.287	-468.725	-387.619	33.745
700	87.086	122.295	72.302	34.995	-466.038	-374.314	27.932
744.300	87.086	127.639	75.438	38.853	------ CRYSTAL <--> LIQUID -----		
800	87.086	133.924	79.294	43.704	-463.370	-361.393	23.597
900	87.086	144.181	85.945	52.413	-460.743	-348.804	20.244
1000	87.086	153.357	92.235	61.121	-458.153	-336.506	17.577
1100	87.086	161.657	98.175	69.830	-455.600	-324.465	15.408
1200	87.086	169.234	103.786	78.538	-453.081	-312.655	13.610
1300	87.086	176.205	109.092	87.247	-450.596	-301.054	12.096
1400	87.086	182.659	114.119	95.955	-448.142	-289.642	10.807
1500	87.086	188.667	118.891	104.664	-445.716	-278.406	9.695
1600	87.086	194.287	123.429	113.373	-443.316	-267.330	8.727
1700	87.086	199.567	127.754	122.081	-586.177	-249.228	7.658
1800	87.086	204.544	131.883	130.790	-583.074	-229.497	6.660
1900	87.086	209.253	135.833	139.498	-580.004	-209.938	5.772
2000	87.086	213.720	139.616	148.207	-576.965	-190.540	4.976
2100	87.086	217.969	143.247	156.916	-573.958	-171.292	4.261
2200	87.086	222.020	146.736	165.624	-570.983	-152.187	3.613
2300	87.086	225.891	150.094	174.333	-568.041	-133.217	3.025
2400	87.086	229.597	153.330	183.041	-565.130	-114.375	2.489
2500	87.086	233.152	156.453	191.750	-562.253	-95.652	1.999

PREVIOUS: March 1966   CURRENT: June 1971

Lithium Hydroxide (LiOH) $H_1Li_1O_1(l)$

LITHIUM HYDROXIDE (LiOH) CRYSTAL-LIQUID M_r = 23.94834

0 to 744.3 K crystal
above 744.3 K liquid

Refer to the individual tables for details.

Lithium Hydroxide (LiOH) $H_1Li_1O_1$(cr,l)

Enthalpy Reference Temperature = T_r = 298.15 K   Standard State Pressure = $p°$ = 0.1 MPa

T/K	$C_p°$ (J K⁻¹mol⁻¹)	$S°$ (J K⁻¹mol⁻¹)	$-[G°-H°(T_r)]/T$ (J K⁻¹mol⁻¹)	$H°-H°(T_r)$ (kJ mol⁻¹)	$\Delta_f H°$ (kJ mol⁻¹)	$\Delta_f G°$ (kJ mol⁻¹)	Log K_f
0	0.	0.	INFINITE	-7.416	-479.144	-479.144	INFINITE
100	14.644	8.561	77.138	-6.858	-482.033	-468.453	244.695
200	35.977	25.632	46.982	-4.270	-484.089	-453.960	118.562
298.15	49.591	42.821	42.821	0.	-484.926	-438.958	76.904
300	49.761	43.128	42.822	0.092	-484.933	-438.672	76.380
400	58.053	58.678	44.886	5.517	-485.047	-423.222	55.267
500	63.647	72.260	49.032	11.614	-487.874	-407.475	42.569
600	68.208	84.277	53.924	18.212	-487.303	-391.441	34.078
700	72.279	95.101	59.045	25.239	-486.298	-375.538	28.023
744.300	73.970	99.588	61.326	28.479	CRYSTAL <--> LIQUID	TRANSITION	
744.300	87.086	127.639	61.326	49.357			
800	87.086	133.924	66.164	54.208	-463.370	-361.393	23.597
900	87.086	144.181	74.274	62.916	-460.743	-348.804	20.244
1000	87.086	153.357	81.732	71.625	-458.153	-336.506	17.577
1100	87.086	161.657	88.626	80.333	-455.600	-324.465	15.408
1200	87.086	169.234	95.032	89.042	-453.081	-312.655	13.610
1300	87.086	176.205	101.012	97.751	-450.596	-301.054	12.096
1400	87.086	182.659	106.616	106.459	-448.142	-289.642	10.807
1500	87.086	188.667	111.888	115.168	-445.716	-278.406	9.695
1600	87.086	194.287	116.864	123.876	-443.316	-267.330	8.727
1700	87.086	199.567	121.576	132.585	-586.177	-249.228	7.658
1800	87.086	204.544	126.048	141.294	-583.074	-229.497	6.660
1900	87.086	209.253	130.304	150.002	-580.003	-209.938	5.772
2000	87.086	213.720	134.365	158.711	-576.965	-190.540	4.976
2100	87.086	217.969	138.245	167.419	-573.958	-171.292	4.261
2200	87.086	222.020	141.962	176.128	-570.983	-152.187	3.613
2300	87.086	225.891	145.527	184.836	-568.041	-133.217	3.025
2400	87.086	229.597	148.954	193.545	-565.130	-114.375	2.489
2500	87.086	233.152	152.251	202.254	-562.253	-95.652	1.999

PREVIOUS: CURRENT: June 1971

Lithium Hydroxide (LiOH) $H_1Li_1O_1$(cr,l)

LITHIUM HYDROXIDE (LiOH) IDEAL GAS M_r = 23.94834

$\Delta_f H°(0\ K)$ = -231.99 ± 6.3 kJ mol^{-1}

S°(298.15 K) = [210.659 ± 2.1] J K^{-1} mol^{-1} $\Delta_f H°(298.15\ K)$ = -234.30 ± 6.3 kJ mol^{-1}

Vibrational Frequencies and Degeneracies

ν, cm^{-1}
[630](1)
[362](2)
[3666](1)

Ground State Quantum Weight: 1 σ = 1
Point Group: $C_{\infty v}$
Bond Distances: Li-O = 1.582 Å O-H = [0.97] Å
Bond Angle: Li-O-H = 180°
Rotational Constant: B_0 = 1.192055 cm^{-1}

Enthalpy of Formation

Smith and Sugden (1) provided the first information on the stability of gaseous LiOH by studying its formation in hydrogen-air flames. Very recently, other flame photometric studies (3-5) have been reported on LiOH(g), primarily in connection with establishing its bond dissociation energy. Mass spectrometry (6-7) has also provided additional information on the stability of this species. These investigations have resulted in a wealth of equilibrium data involving this compound. 2nd and 3rd law analyses of these data are presented in tabular form below. Vapor pressure data for LiOH(cr, ℓ) are analyzed under Vaporization Data (see LiOH(ℓ) table); these data lead to $\Delta_f H°$(LiOH, g, 298.15 K) values of -59.0, -54.7, -54.8 and -58.1 kcal mol^{-1}.

Source	Reaction*	Method	T/K	Data Points	$\Delta_r H°$(298.15 K), kcal mol^{-1} 2nd law	3rd law	Drift cal K^{-1} mol^{-1}	$\Delta_f H°$(298.15 K) kcal mol^{-1}
Smith and Sugden (1)	A	Flame	2200	1		99.6		-51.7
James and Sugden (2)	A	Flame	2200	1		100.0		-53.2
Jensen and Padley (3)	B	Flame	2475	1		15.9		-55.6 ± 2.0
Cotton and Jenkins (4)	B	Flame	2370	1		15.1		-56.4 ± 1.5
McEwan and Phillips (5)	B	Flame	1400-2600	4	15.8	16.7±0.6	0.5 ± 0.2	-54.8 ± 1.6

Reactions: (A) LiOH(g) = Li(g) + OH(g) (B) Li(g) + H_2O(g) + H(g)

The equilibrium data of Sugden and co-workers (1, 2) are likely to contain significant errors, since the dissociation constants were calculated with OH concentrations determined from measured flame temperatures and known gas compositions. It is now well established (8, 9) that flame radical concentrations vary greatly with the distance from the reaction zone of the flame. The dissociation pressure data of Johnston and Ditmars and Johnston (see LiOH(ℓ) table) show excessive drifts which usually are indications of nonequilibrium measurements. However, even the remaining flame work (3-5) and vapor pressure data still show a scatter of over 3 kcal mol^{-1} in $\Delta_f H°$(298.15 K) of LiOH(g). Cotton and Jenkins (4) investigated the other alkali metal hydroxides, and their data lead to $\Delta_f H°$(298.15 K) values for these compounds which are quite consistent with JANAF data (10). On the other hand, the flame studies of McEwan and Phillips (5) as a function of temperature do not show significant drift; yet, the discrepancy between their $\Delta_f H°$(298.15 K) value and that of Cotton and Jenkins (4) is 1.6 kcal mol^{-1}. We choose to adopt for $\Delta_f H°$(LiOH, g, 298.15 K) a weighted average of the results from the flame work (3-5) and the vapor pressure data. The adopted value $\Delta_f H°$(LiOH, g, 298.15 K) = -56.0±1.5 kcal mol^{-1} corresponds to a bond dissociation energy of D_0^0(Li-OH) = 102.8±1.5 kcal mol^{-1}.

Heat Capacity and Entropy

Freund et al. (11) measured the molecular beam electric resonance spectra of LiOH and calculated molecular constants on the basis of a linear equilibrium structure. Linear structures for the other alkali metal hydroxides have been proposed from microwave and infrared studies on these molecules. (See the respective tables for details.) We adopt a linear configuration for LiOH and calculate a Li-O bond length from the rotational constant given by Freund et al. (11). The O-H bond length is estimated to be the same as for H_2O.

Experimental vibrational frequencies for LiOH have not been reported in the literature at the present time. Acquista and Abramowitz (12) have made matrix-isolation studies of monomeric CsOH, RbOH, and NaOH and provided assignments for the metal-oxygen stretch (ν_1) and the bending modes (ν_2) of these molecules. A Valence Force Treatment of the frequencies by these workers led to metal-oxygen (F_{11}), O-H (F_{33}), and bending (F_{22}) force constants. We extrapolate their data to LiOH with the result; F_{11} = 1.15 mdyn/Å, F_{33} assumed equal to 7.5 mdyn/Å with the interaction constant F_{13} = 0, and F_{22} = 0.060 mdyn Å. These estimated force constants are used to calculate the adopted frequencies by the Valence Force Treatment. Our adopted frequencies when used in an analysis of the equilibrium data from McEwan and Phillips (5) and Berkowitz et al. (7) (see LiOH(ℓ) table) lead to reasonable agreement between the 2nd and 3rd law $\Delta_r H°$(298.15 K) values.

References

1. H. Smith and T. M. Sugden, Proc. Roy. Soc. A219, 204 (1953).
2. C. G. James and T. M. Sugden, Proc. Roy. Soc. A227, 312 (1955).
3. D. E. Jensen and P. J. Padley, Trans. Faraday Soc. 62, 2132 (1966).
4. D. H. Cotton and D. R. Jenkins, Trans. Faraday Soc. 65, 1537 (1969).
5. M. J. McEwan and L. F. Phillips, Combustion and Flame 9, 420 (1965).
6. R. C. Schoonmaker and R. F. Porter, J. Phys. Chem. 64, 457 (1960).
7. J. Berkowitz, D. J. Meschi and W. A. Chupka, J. Chem. Phys. 33, 533 (1960).
8. P. J. Th. Zeegers and C. Th. J. Alkemade, Combustion and Flame 9, 247 (1965).
9. M. J. McEwan and L. F. Phillips, Combustion and Flame 11, 63 (1967).
10. JANAF Thermochemical Tables: NaOH(g), KOH(g), 12-31-70; CsOH(g), 6-30-71.
11. S. M. Freund, P. D. Godfrey and W. Klemperer, Twenty-Fifth Symposium on Molecular Structure and Spectroscoy, The Ohio State University, Columbus, Ohio, September, 1970.
12. N. Acquista and S. Abramowitz, J. Chem. Phys. 51, 2911 (1969).

Lithium Hydroxide (LiOH) $H_1Li_1O_1(g)$

Enthalpy Reference Temperature = T_r = 298.15 K Standard State Pressure = p° = 0.1 MPa

T/K	$C_p°$ (J K^{-1}mol^{-1})	S°	-[G°-H°(T_r)]/T	H°-H°(T_r) (kJ mol^{-1})	$\Delta_f H°$	$\Delta_f G°$	Log K_f
0	0.	0.	INFINITE	-10.879	-231.985	-231.985	INFINITE
100	31.674	168.451	247.699	-7.925	-232.478	-234.887	122.693
200	40.704	193.281	214.729	-4.289	-233.487	-236.888	61.869
250	43.897	202.726	211.407	-2.170	-233.923	-237.687	49.662
298.15	46.129	210.659	210.659	0.	-234.304	-238.377	41.763
300	46.201	210.945	210.660	0.085	-234.318	-238.402	41.510
350	47.864	218.198	211.229	2.439	-234.690	-239.053	35.677
400	49.081	224.673	212.512	4.864	-235.078	-239.651	31.295
450	49.993	230.509	214.193	7.342	-235.512	-240.197	27.881
500	50.696	235.814	216.094	9.860	-239.006	-240.384	25.113
600	51.719	245.154	220.180	14.984	-239.910	-240.573	20.944
700	52.479	253.185	224.335	20.196	-240.720	-240.619	17.955
800	53.130	260.236	228.391	25.477	-241.480	-240.552	15.706
900	53.737	266.529	232.285	30.820	-242.217	-240.392	13.952
1000	54.323	272.222	235.998	36.223	-242.933	-240.151	12.544
1100	54.889	277.426	239.531	41.684	-243.627	-239.838	11.389
1200	55.432	282.225	242.892	47.200	-244.301	-239.464	10.424
1300	55.945	286.683	246.091	52.769	-244.956	-239.034	9.605
1400	56.426	290.846	249.141	58.388	-245.592	-238.555	8.901
1500	56.873	294.755	252.053	64.054	-246.209	-238.031	8.289
1600	57.285	298.439	254.838	69.762	-246.809	-237.466	7.752
1700	57.662	301.923	257.506	75.509	-392.631	-229.688	7.057
1800	58.008	305.229	260.066	81.293	-392.453	-220.108	6.387
1900	58.322	308.374	262.527	87.110	-392.274	-210.538	5.788
2000	58.609	311.373	264.894	92.957	-392.097	-200.978	5.249
2100	58.871	314.239	267.176	98.831	-391.925	-191.426	4.761
2200	59.109	316.983	269.378	104.730	-391.760	-181.882	4.318
2300	59.326	319.615	271.506	110.652	-391.604	-172.346	3.914
2400	59.524	322.145	273.563	116.595	-391.459	-162.816	3.544
2500	59.705	324.578	275.556	122.556	-391.329	-153.292	3.203
2600	59.870	326.923	277.487	128.535	-391.215	-143.773	2.888
2700	60.022	329.186	279.360	134.530	-391.120	-134.258	2.597
2800	60.161	331.371	281.178	140.539	-391.045	-124.746	2.327
2900	60.289	333.484	282.946	146.562	-390.994	-115.236	2.076
3000	60.406	335.530	284.665	152.596	-390.967	-105.728	1.841
3100	60.514	337.513	286.338	158.643	-390.967	-96.220	1.621
3200	60.614	339.436	287.967	164.699	-390.996	-86.712	1.415
3300	60.706	341.302	289.555	170.765	-391.055	-77.202	1.222
3400	60.792	343.116	291.104	176.840	-391.147	-67.690	1.040
3500	60.871	344.879	292.615	182.923	-391.272	-58.175	0.868
3600	60.945	346.595	294.091	189.014	-391.433	-48.656	0.706
3700	61.013	348.266	295.533	195.112	-391.631	-39.132	0.552
3800	61.077	349.894	296.942	201.217	-391.868	-29.601	0.407
3900	61.137	351.481	298.320	207.327	-392.145	-20.064	0.269
4000	61.192	353.030	299.669	213.444	-392.463	-10.520	0.137
4100	61.244	354.541	300.989	219.566	-392.825	-0.967	0.012
4200	61.293	356.018	302.281	225.693	-393.232	8.596	-0.107
4300	61.338	357.460	303.548	231.824	-393.685	18.168	-0.221
4400	61.381	358.871	304.789	237.960	-394.165	27.750	-0.329
4500	61.422	360.251	306.006	244.100	-394.705	37.344	-0.433
4600	61.459	361.601	307.200	250.244	-395.298	46.952	-0.533
4700	61.495	362.923	308.372	256.392	-395.942	56.573	-0.629
4800	61.529	364.218	309.522	262.543	-396.639	66.209	-0.720
4900	61.560	365.487	310.651	268.698	-397.389	75.859	-0.809
5000	61.590	366.731	311.760	274.855	-398.196	85.525	-0.893
5100	61.619	367.951	312.850	281.016	-399.055	95.208	-0.975
5200	61.645	369.148	313.921	287.179	-399.977	104.908	-1.054
5300	61.671	370.323	314.975	293.345	-400.960	114.627	-1.130
5400	61.695	371.476	316.010	299.513	-402.006	124.364	-1.203
5500	61.718	372.608	317.029	305.684	-403.116	134.122	-1.274
5600	61.740	373.720	318.031	311.857	-404.293	143.901	-1.342
5700	61.760	374.813	319.018	318.032	-405.537	153.701	-1.409
5800	61.780	375.887	319.989	324.209	-406.850	163.524	-1.473
5900	61.799	376.944	320.946	330.388	-408.234	173.368	-1.535
6000	61.816	377.982	321.888	336.569	-409.692	183.239	-1.595

PREVIOUS: June 1971 (1 atm) CURRENT June 1971 (1 bar)

Lithium Hydroxide (LiOH) $H_1Li_1O_1(g)$

LITHIUM HYDROXIDE, ION ($LiOH^+$) IDEAL GAS M_r = 23.947791

$S°(298.15\ K) = [221.685 \pm 8.4]\ J\ K^{-1}\ mol^{-1}$

$\Delta_fH°(0\ K) = [774.1 \pm 105]\ kJ\ mol^{-1}$
$\Delta_fH°(298.15\ K) = [778.2 \pm 105]\ kJ\ mol^{-1}$

Electronic Levels and Quantum Weights

State	ε_i, cm^{-1}	g_i
$^2\Pi$	0	4
$^2\Sigma^+$	[32000]	2

Vibrational Frequencies and Degeneracies

ν, cm^{-1}
[600](1)
[325](2)
[3600](1)

Point Group: $[C_{\infty v}]$ $\sigma = 1$
Bond Distances: Li-O = [1.60] Å O-H = [1.03] Å
Bond Angle: Li-O-H = [180]°
Rotational Constant: $B_0 = [1.15435]\ cm^{-1}$

Enthalpy of Formation

The enthalpy of formation, $\Delta_fH°(LiOH^+, g, 298.15\ K) = 186 \pm 25\ kcal\ mol^{-1}$, is calculated from the ionization potential for LiOH and $\Delta_fH°(LiOH, g, 298.15\ K) = -56.0 \pm 1.5\ kcal\ mol^{-1}$ (1). The ionization potential is estimated to be 10.5 ± 1.0 eV for LiOH by comparison with similar data (2) for the isoelectronic molecules LiF and LiCl. We note that this estimated value is in accord with the ionization potential of LiO (IP = 9.0 eV) (3), which should be less than LiOH since the electron in the monoxide is unpaired.

Heat Capacity and Entropy

The correlation diagram of Walsh (4) for HAB molecules predicts a linear configuration for $LiOH^+$ (seven valence electrons) with the unpaired electron in a pi orbital. Therefore, the electronic ground state of $LiOH^+$ should be $^2\pi$. A first excited state ($^2\Sigma$) is estimated at 32000 cm^{-1} by analogy with the isoelectronic molecules OH (1) and SH (1). The ionization of LiOH is assumed to lead to weaker bonding in $LiOH^+$, since the electron lost is a bonding one. The Li-O and O-H bond distances are increased slightly over those for LiOH to account for this weaker bonding. Similarly, the vibrational frequencies are estimated from those for LiOH with somewhat lower values.

References

1. JANAF Thermochemical Tables: LiOH(g), 6-30-71; OH(g), 12-31-70; SH(g), 6-30-67.
2. J. L. Franklin, J. G. Dillard, H. M. Rosenstock, J. T. Herron, K. Draxl, and F. H. Field, NSRDS-NBS 26, 1969.
3. C. W. Beckett and E. C. Cassidy, U. S. Nat. Bur. Stand. Report 8628, 1965.
4. A. D. Walsh, J. Chem. Soc. 2288 (1953).

Lithium Hydroxide, Ion ($LiOH^+$) $H_1Li_1O_1^+(g)$

Enthalpy Reference Temperature = T_r = 298.15 K Standard State Pressure = p° = 0.1 MPa

T/K	$C_p°$ (J K⁻¹mol⁻¹)	S° (J K⁻¹mol⁻¹)	$-[G°-H°(T_r)]/T$ (J K⁻¹mol⁻¹)	$H°-H°(T_r)$ (kJ mol⁻¹)	$\Delta_fH°$ (kJ mol⁻¹)	$\Delta_fG°$ (kJ mol⁻¹)	Log K_f
0	0.	0.	INFINITE	-11.137	774.087		
100	32.662	178.163	259.751	-8.159			
200	41.973	203.872	225.848	-4.395			
250	44.971	213.579	222.449	-2.217			
298.15	47.019	221.685	221.685	0.	778.224	764.609	-133.956
300	47.085	221.976	221.686	0.087	778.250	764.524	-133.116
350	48.589	229.354	222.265	2.481	778.958	762.181	-113.749
400	49.681	235.917	223.569	4.939	779.642	759.737	-99.211
450	50.495	241.818	225.274	7.444	780.274	757.210	-87.895
500	51.121	247.172	227.201	9.985	777.843	754.923	-78.866
600	52.040	256.578	231.334	15.146	779.055	750.227	-65.313
700	52.739	264.655	235.531	20.386	780.352	745.321	-55.617
800	53 355	271.738	239.623	25.691	781.695	740.225	-48.332
900	53.943	278.056	243.549	31.057	783.057	734.960	-42.656
1000	54.516	283.769	247.290	36.480	784.440	729.542	-38.107
1100	55.074	288 992	250.847	41.959	785.843	723.984	-34.379
1200	55.609	293.807	254.229	47.494	787.267	718.298	-31.267
1300	56.116	298.278	257.447	53.080	788.708	712.492	-28.628
1400	56.590	302.454	260.514	58.716	790.167	706.575	-26.363
1500	57.030	306.374	263.442	64.397	791.645	700.553	-24.395
1600	57.434	310.068	266.242	70.121	793.138	694 431	-22.671
1700	57.805	313.561	268.924	75.883	649.409	695.392	-21.367
1800	58.143	316.875	271.497	81.681	651.680	698.031	-20.256
1900	58.451	320.027	273.968	87 510	653.951	700.544	-19.259
2000	58.731	323.032	276.347	93.370	656.218	702.938	-18.359
2100	58.986	325.904	278.639	99.256	658.481	705.218	-17.541
2200	59.218	328.653	280.850	105.166	660.736	707.391	-16.796
2300	59.429	331.290	282.986	111.099	662.982	709.461	-16.112
2400	59.622	333.824	285.052	117.051	665.215	711.434	-15.484
2500	59.797	336.261	287.052	123.022	667.433	713.314	-14.904
2600	59.958	338.610	288.990	129.010	669.634	715.106	-14.367
2700	60.105	340.875	290.870	135.014	671.817	716.814	-13.868
2800	60.240	343.064	292.695	141.031	673.978	718.441	-13.403
2900	60.363	345.180	294.469	147.061	676.116	719.991	-12.968
3000	60.477	347.228	296.194	153.103	678.228	721.468	-12.562
3100	60.582	349.213	297.872	159.156	680.314	722.875	-12 180
3200	60.679	351.138	299.507	165.219	682.370	724.214	-11.822
3300	60.768	353.006	301.100	171.292	684.396	725.491	-11.484
3400	60.851	354.822	302.653	177.373	686.389	726.706	-11,164
3500	60.928	356.587	304.169	183.462	688.348	727.863	-10.863
3600	61.000	358.304	305.649	189.558	690.271	728.965	-10.577
3700	61.066	359.976	307.095	195.661	692.157	730.013	-10.306
3800	61.129	361.606	308.508	201.771	694.004	731.012	-10.048
3900	61.187	363.194	309.890	207.887	695.811	731.963	-9.804
4000	61.242	364.744	311.242	214.009	697.576	732.867	-9.570
4100	61.294	366.257	312.565	220.135	699.298	733.728	-9.348
4200	61.342	367.735	313.861	226.267	700.975	734.547	-9.135
4300	61.388	369.179	315.131	232.404	702.605	735.327	-8.932
4400	61.432	370.590	316.376	238.545	704.209	736.068	-8.738
4500	61.473	371.971	317.596	244.690	705.752	736.774	-8.552
4600	61.513	373.323	318.793	250.839	707.243	737.447	-8.374
4700	61.550	374.646	319.967	256.992	708.683	738.088	-8.203
4800	61.587	375.942	321.120	263.149	710.071	738.699	-8.039
4900	61.622	377.213	322.252	269.310	711.405	739.281	-7.881
5000	61.655	378.458	323.363	275.474	712.683	739.838	-7.729
5100	61.688	379.679	324 456	281.641	713.909	740.368	-7.583
5200	61.720	380.877	325.529	287.811	715.073	740.875	-7.442
5300	61.751	382.053	326.585	293.985	716.176	741.361	-7.307
5400	61.781	383.208	327.622	300.161	717.217	741.826	-7.176
5500	61.810	384.342	328.643	306.341	718.194	742.272	-7.050
5600	61.839	385.456	329.648	312.523	719.106	742.702	-6.928
5700	61.868	386.551	330.637	318.709	719.951	743.116	-6.810
5800	61.896	387.627	331.610	324.897	720.728	743.516	-6.696
5900	61.924	388.685	332.569	331.088	721.434	743.901	-6.586
6000	61.951	389.726	333.512	337.282	722.068	744.277	-6.480

PREVIOUS: December 1971 (1 atm) CURRENT: December 1971 (1 bar)

Lithium Hydroxide, Ion ($LiOH^+$) $H_1Li_1O_1^+(g)$

MAGNESIUM HYDRIDE (MgH) IDEAL GAS M_r = 25.31294

$\Delta_f H°(0\ K)$ = 169.58 ± 48.1 kJ mol^{-1}

$S°(298.15\ K)$ = 193.200 J K^{-1} mol^{-1} $\Delta_f H°(298.15\ K)$ = 169.03 ± 48.1 kJ mol^{-1}

Electronic Levels and Quantum Weights

ε_i, cm^{-1}	g_i	ε_i, cm^{-1}	g_i	ε_i, cm^{-1}	g_i
0	2	35400	2	42070	2
19221.2	2	[38730]	2	48000	4
19224.7	2	41120	4		

ω_e = 1495.7 cm^{-1} $\omega_e x_e$ = 31.5 cm^{-1} σ = 1

B_e = 5.8181 cm^{-1} α_e = 0.1668 cm^{-1} r_e = 1.73 Å

Enthalpy of Formation

The enthalpy of formation is calculated using D_0^0 = 2.0±0.5 eV derived from a graphical extrapolation of the $\nu(0-3)$ energy levels of the $X^2\Sigma^+$ state reported by Gaydon (1). Herzberg (2) reported $D_0^0 \leq$ 2.49 eV assuming predissociation of the $C^2\pi$ state of MgH into Mg(^{3}P) and H(^{2}S).

Heat Capacity and Entropy

The molecular constants and electronic levels are obtained from Herzberg (2) except for the electronic levels 35400 and 48000 obtained from Khan (3).

References

1. A. G. Gaydon, "Dissociation Energies and Spectra of Diatomic Molecules," 2nd Edition, Chapman and Hall Ltd., London, 1953.
2. G. Herzberg, "Spectra of Diatomic Molecules", 2nd Edition, D. Van Nostrand Co., Inc., New York, 1950.
3. M. A. Khan, Proc. Phys. Soc. (London) 80, 209-21 (1962).

Magnesium Hydride (MgH) $H_1Mg_1(g)$

Enthalpy Reference Temperature = T_r = 298.15 K Standard State Pressure = p° = 0.1 MPa

T/K	$C_p°$ (J K⁻¹mol⁻¹)	S° (J K⁻¹mol⁻¹)	$-[G°-H°(T_r)]/T$ (J K⁻¹mol⁻¹)	$H°-H°(T_r)$ (kJ mol⁻¹)	$\Delta_f H°$ (kJ mol⁻¹)	$\Delta_f G°$ (kJ mol⁻¹)	Log K_f
0	0.	0.	INFINITE	-8.682	169.583	169.583	INFINITE
100	29.127	161.292	219.230	-5.794	170.330	160.188	-83.673
200	29.181	181.492	195.889	-2.879	169.890	150.162	-39.218
250	29.316	188.016	193.685	-1.417	169.483	145.275	-30.354
298.15	29.583	193.200	193.200	0.	169.034	140.653	-24.642
300	29.596	193.383	193.201	0.055	169.016	140.477	-24.459
350	30.016	197.975	193.562	1.545	168.519	135 759	-20.261
400	30.535	202.016	194.371	3.058	168.011	131.114	-17.122
450	31.103	205.645	195.426	4.599	167.501	126.533	-14.688
500	31.683	208.952	196.615	6.169	166.993	122.008	-12.746
600	32.778	214.828	199.173	9.393	165.986	113.106	-9.847
700	33.724	219.954	201.783	12.719	164.971	104.372	-7.788
800	34.510	224.510	204.345	16.132	163.919	95.786	-6.254
900	35.158	228.613	206.817	19.616	162.798	87.336	-5.069
1000	35.694	232.346	209.186	23.160	152.983	79.728	-4.165
1100	36.141	235.770	211.449	26.752	151.625	72.469	-3.441
1200	36.519	238.931	213.609	30.386	150.289	65.332	-2.844
1300	36.844	241.867	215.671	34.054	148.966	58.306	-2.343
1400	37.127	244.608	217.642	37.753	20.243	54.549	-2.035
1500	37.376	247.178	219.526	41.479	20.286	56.998	-1.985
1600	37.599	249.598	221.330	45.228	20.330	59.443	-1.941
1700	37.800	251.883	223.061	48.998	20.375	61.887	-1.902
1800	37.985	254.049	224.723	52.787	20.419	64.327	-1.867
1900	38.155	256.108	226.321	56.594	20.461	66.765	-1.836
2000	38.314	258.069	227.860	60.418	20.501	69.202	-1.807
2100	38.465	259.942	229.343	64.257	20.538	71.636	-1.782
2200	38.608	261.734	230.775	68.110	20.573	74.068	-1.759
2300	38.746	263.454	232.159	71.978	20.605	76.499	-1.737
2400	38.881	265.106	233.497	75.860	20.636	78.928	-1.718
2500	39.013	266.695	234.794	79.754	20.663	81.357	-1.700
2600	39.144	268.228	236.050	83.662	20.689	83.784	-1.683
2700	39.275	269.708	237.270	87.583	20.712	86.210	-1.668
2800	39.407	271.139	238.454	91.517	20.732	88.636	-1.654
2900	39.541	272.524	239.605	95.464	20.749	91.060	-1.640
3000	39.678	273.867	240.725	99.425	20.763	93.485	-1.628
3100	39.818	275.170	241.815	103.400	20.773	95.909	-1.616
3200	39.962	276.436	242.877	107.389	20.778	98 332	-1.605
3300	40.110	277.668	243.913	111.393	20.778	100.756	-1.595
3400	40.264	278.868	244.923	115.411	20.771	103.179	-1.585
3500	40.423	280.037	245.910	119.445	20.756	105.603	-1.576
3600	40.587	281.178	246.874	123.496	20.733	108.028	-1.567
3700	40.757	282.293	247.816	127.563	20.699	110.453	-1.559
3800	40.933	283.382	248.738	131.648	20.653	112.880	-1.552
3900	41.114	284.448	249.640	135.750	20.595	115.307	-1.544
4000	41.301	285.491	250.523	139.871	20.521	117.737	-1.537
4100	41.494	286.513	251.389	144.010	20.431	120.168	-1.531
4200	41.693	287.515	252.237	148.170	20.323	122.602	-1.525
4300	41.897	288.499	253.069	152.349	20.195	125.039	-1.519
4400	42.106	289.464	253.885	156.549	20.046	127.479	-1.513
4500	42.320	290.413	254.686	160.770	19.874	129.922	-1.508
4600	42.538	291.345	255.473	165.013	19.678	132.370	-1.503
4700	42.761	292.263	256.246	169.278	19.455	134.822	-1.498
4800	42.989	293.165	257.006	173.566	19.205	137.279	-1.494
4900	43.220	294.054	257.753	177.876	18.926	139.742	-1.490
5000	43.455	294.930	258.488	182.210	18.617	142.211	-1.486
5100	43.694	295.792	259.211	186.567	18.277	144.686	-1.482
5200	43.936	296.643	259.922	190.949	17.903	147.168	-1.478
5300	44.181	297.482	260.623	195.355	17.497	149.658	-1 475
5400	44.428	298.311	261.313	199.785	17.056	152.155	-1.472
5500	44.678	299.128	261.994	204.240	16.579	154.661	-1.469
5600	44.931	299.935	262.664	208.721	16.066	157.177	-1.466
5700	45.185	300.733	263.325	213.226	15.516	159.701	-1.463
5800	45.441	301.521	263.977	217.758	14.928	162.236	-1.461
5900	45.699	302.300	264.620	222.315	14.301	164.781	-1.459
6000	45.958	303.070	265.254	226.898	13.636	167.337	-1.457

PREVIOUS: December 1966 (1 atm) CURRENT: December 1966 (1 bar)

Magnesium Hydride (MgH) $H_1Mg_1(g)$

MAGNESIUM HYDROXIDE (MgOH) IDEAL GAS $M_r = 41.31234$

$S°(298.15\ K) = [226.458 \pm 8.4]\ J\ K^{-1}\ mol^{-1}$

$\Delta_f H°(0\ K) = [-161.34 \pm 37.7]\ kJ\ mol^{-1}$
$\Delta_f H°(298.15\ K) = [-164.75 \pm 37.7]\ kJ\ mol^{-1}$

Electronic Levels and Quantum Weights

ϵ_i, cm^{-1}	g_i
0	2
[27000]	[4]

Vibrational Frequencies and Degeneracies

ν, cm^{-1}
[718](1)
[492](2)
[3650](1)

Point Group: $[C_{\infty v}]$ $\sigma = 1$
Bond Distances: Mg-O = [1.77] Å O-H = [0.96] Å
Bond Angle: Mg-O-H = [180]°
Rotational Constant: B_0 = [0.491925] cm^{-1}

Enthalpy of Formation

From observations made during a flame spectra study of the bond dissociation energy of magnesium oxide, Cotton and Jenkins (1) concluded that D_0^0(Mg-OH) is less than 90 kcal mol^{-1}. The analogy between gaseous monohydroxides and monohalides has been recognized (2-5). Hildenbrand (6) has found that the ratio of $D_0^0(Mg\text{-}F)/D_0^0(F\text{-}Mg\text{-}F)$ is 0.45; current JANAF values (7) give 0.448. From the latter ratio and current JANAF data (7), D_0^0(Mg-OH) = 84.49 kcal mol^{-1} is derived. Adding the difference, -16.1 kcal mol^{-1}, between D_0^0(Mg-F) and D_0^0(Ca-F) (7) to D_0^0(Ca-OH) = 97.5 kcal mol^{-1} (7) gives D_0^0(Mg-OH) = 81.4 kcal mol^{-1}. The average D_0^0(Mg-OH) = 82.9±5 kcal mol^{-1} is adopted from which $\Delta_f H°$(MgOH, g, 0 K) = -38.56±9 kcal mol^{-1} is derived.

The value of D_0^0(Mg-OH) = 56±5 kcal mol^{-1} derived by Bulewicz and Sugden (8) from flame spectra studies appears to be too low.

Heat Capacity and Entropy

The molecular configuration is assumed to be linear in accordance with the prediction of Walsh (9), the evidence that the alkali metal monohydroxides are linear (10-13), and the ESR spectrum study of MgOH in argon and neon matrices from which Brom and Weltner (13) concluded that MgOH is probably linear with a $^2\Sigma$ ground state. The analogy with MgF and MgCl (7) also indicates a $^2\Sigma^+$ ground state, which is assumed.

Ultraviolet emission spectra near 3800 Å were observed from MgOH and MgOD by Pesic and Gaydon (14) and by Brewer and Trajmar (15). Brom and Weltner (13) observed absorption spectra of MgOH in an argon matrix in the 3100-3700 Å range. From these observations and from the comparison with MgF and MgCl, the first excited state at 27000 cm^{-1} is estimated.

The Mg-O bond distance is estimated to be slightly larger, 0.02 Å, than the MgF bond distances (7) after noting the close similarity in bond distance of the alkali metal fluorides and hydroxides. The O-H bond distance is that in water (7).

The Mg-O stretching frequency, 718 cm^{-1}, is estimated to be the same as in MgF (5, 7). The O-H stretching frequency, 3650 cm^{-1}, is estimated from the alkali hydroxide series. The bending frequency, 492 cm^{-1}, is estimated by assuming that the ratio of the bending force constant to the stretching force constant is 0.022, which is the average ratio found in the alkali metal monohydroxides by Acquista and Abramowitz (5, 12).

The entropy in the present table is lower by 0.20 cal K^{-1} mol^{-1} at 298 K and 0.25 cal K^{-1} mol^{-1} at 1000 K than that proposed by Jackson (5); the data relevant to the calculation are nearly the same.

References

1. D. H. Cotton and D. R. Jenkins, Trans. Faraday Soc. 65, 376 (1969).
2. R. C. Schoonmaker and R. F. Porter, J. Chem. Phys. 31, 830 (1959).
3. J. Berkowitz, D. J. Meschi and W. A. Chupka, J. Chem. Phys. 33, 533 (1960).
4. F. E. Stafford and J. Berkowitz, J. Chem. Phys. 40, 2963 (1964).
5. D. D. Jackson, Lawrence Livermore Laboratory, Univ. of Calif., UCRL-51137, 1971.
6. D. L. Hildenbrand, personal communication, Stanford Research Institute, July 2, 1974.
7. JANAF Thermochemical Tables: $Mg(OH)_2$(g), CaOH(g), 12-31-75; MgF_2(g), 6-30-75; Mg(g), 9-30-62; OH(g), 12-31-70; CaF(g), 12-31-68; H_2O(g), 3-31-61.
8. E. M. Bulewicz and T. M. Sugden, Trans. Faraday Soc. 55, 720 (1959).
9. A. D. Walsh, J. Chem. Soc. 1953, 2288 (1953).
10. N. Acquista, S. Abramowitz, and R. Lide, J. Chem. Phys. 49, 780 (1968).
11. R. L. Kuczkowski and R. Lide, J. Chem. Phys. 44, 3131 (1966).
12. N. Acquista and S. Abramowitz, U. S. Nat. Bur. Stand. Report 9905, July, 1968.
13. J. M. Brom, Jr., and W. Weltner, Jr., J. Chem. Phys. 58, 5322 (1973).
14. D. Pesic and A. G. Gaydon, Proc. Phys. Soc. (London) 73, 244 (1959).
15. L. Brewer and S. Trajmar, J. Chem. Phys. 36, 1585 (1962).

Magnesium Hydroxide (MgOH) $H_1Mg_1O_1(g)$

Enthalpy Reference Temperature = T_r = 298.15 K Standard State Pressure = p° = 0.1 MPa

T/K	$C_p°$ (J $K^{-1}mol^{-1}$)	S° (J $K^{-1}mol^{-1}$)	$-[G°-H°(T_r)]/T$ (J $K^{-1}mol^{-1}$)	$H°-H°(T_r)$ (kJ mol^{-1})	$\Delta_f H°$ (kJ mol^{-1})	$\Delta_f G°$ (kJ mol^{-1})	Log K_f
0	0.	0.	INFINITE	-10.160	-161.341	-161.341	INFINITE
100	29.833	187.913	260.329	-7.242	-162.017	-166.156	86.791
200	36.797	210.512	230.214	-3.940	-163.525	-169.709	44.323
250	40.324	219.114	227.152	-2.010	-164.192	-171.176	35.765
298.15	43.036	226.458	226.458	0.	-164.755	-172.469	30.216
300	43.127	226.725	226.459	0.080	-164.775	-172.517	30.038
350	45.270	233.541	226.992	2.292	-165.287	-173.766	25.933
400	46.900	239.698	228.202	4.598	-165.750	-174.945	22.845
450	48.151	245.297	229.795	6.976	-166.182	-176.068	20.437
500	49.130	250.423	231.605	9.409	-166.597	-177.144	18.506
600	50.561	259.515	235.519	14.398	-167.419	-179.177	15.599
700	51.597	267.390	239.522	19.508	-168.278	-181.069	13.512
800	52.441	274.337	243.448	24.711	-169.208	-182.834	11.938
900	53.189	280.557	247.232	29.993	-170.234	-184.477	10.707
1000	53.879	286.197	250.851	35.347	-179.969	-185.287	9.678
1100	54.524	291.363	254.302	40.767	-181.254	-185.756	8.821
1200	55.128	296.134	257.592	46.250	-182.515	-186.109	8.101
1300	55.690	300.569	260.729	51.792	-183.757	-186.358	7.488
1400	56.209	304.715	263.724	57.387	-312.390	-183.345	6.841
1500	56.686	308.610	266.588	63.032	-312.249	-174.133	6.064
1600	57.123	312.282	269.330	68.723	-312.096	-164.930	5.384
1700	57.521	315.757	271.960	74.455	-311.934	-155.737	4.785
1800	57.883	319.055	274.486	80.226	-311.767	-146.554	4.253
1900	58.212	322.194	276.915	86.031	-311.597	-137.380	3.777
2000	58.511	325.188	279.254	91.867	-311.426	-128.215	3.349
2100	58.783	328.049	281.510	97.732	-311.255	-119.059	2.961
2200	59.030	330.789	283.688	103.623	-311.088	-109.910	2.610
2300	59.254	333.418	285.794	109.537	-310.924	-100.770	2.289
2400	59.459	335.945	287.831	115.473	-310.766	-91.636	1.994
2500	59.646	338.376	289.804	121.428	-310.615	-82.509	1.724
2600	59.817	340.718	291.718	127.401	-310.472	-73.388	1.474
2700	59.973	342.979	293.575	133.391	-310.339	-64.271	1.243
2800	60.117	345.163	295.378	139.396	-310.217	-55.160	1.029
2900	60.250	347.275	297.132	145.414	-310.107	-46.053	0.830
3000	60.373	349.319	298.837	151.445	-310.012	-36.949	0.643
3100	60.486	351.301	300.498	157.488	-309.931	-27.848	0.469
3200	60.592	353.223	302.116	163.542	-309.868	-18.750	0.306
3300	60.692	355.089	303.693	169.607	-309.823	-9.654	0.153
3400	60.785	356.902	305.231	175.681	-309.799	-0.558	0.009
3500	60.874	358.665	306.733	181.764	-309.797	8.537	-0.127
3600	60.958	360.381	308.199	187.855	-309.819	17.633	-0.256
3700	61.038	362.053	309.632	193.955	-309.867	26.729	-0.377
3800	61.116	363.682	311.033	200.063	-309.943	35.828	-0.492
3900	61.192	365.270	312.404	206.178	-310.050	44.929	-0.602
4000	61.266	366.820	313.745	212.301	-310.189	54.032	-0.706
4100	61.338	368.334	315.058	218.431	-310.363	63.140	-0.804
4200	61.410	369.813	316.344	224.569	-310.574	72.252	-0.899
4300	61.482	371.259	317.605	230.713	-310.823	81.370	-0.988
4400	61.553	372.673	318.840	236.865	-311.114	90.494	-1.074
4500	61.625	374.057	320.052	243.024	-311.447	99.624	-1.156
4600	61.698	375.412	321.241	249.190	-311.825	108.763	-1.235
4700	61.772	376.740	322.407	255.364	-312.250	117.911	-1.310
4800	61.846	378.041	323.553	261.544	-312.724	127.069	-1.383
4900	61.923	379.317	324.678	267.733	-313.249	136.236	-1.452
5000	62.000	380.569	325.783	273.929	-313.826	145.416	-1.519
5100	62.079	381.798	326.870	280.133	-314.457	154.607	-1.583
5200	62.160	383.004	327.938	286.345	-315.144	163.810	-1.645
5300	62.243	384.189	328.988	292.565	-315.888	173.028	-1.705
5400	62.328	385.353	330.021	298.794	-316.690	182.260	-1.763
5500	62.415	386.497	331.037	305.031	-317.552	191.507	-1.819
5600	62.503	387.623	332.038	311.277	-318.475	200.772	-1.873
5700	62.593	388.730	333.023	317.531	-319.461	210.053	-1.925
5800	62.686	389.819	333.993	323.795	-320.510	219.352	-1.975
5900	62.780	390.892	334.948	330.069	-321.624	228.668	-2.024
6000	62.876	391.948	335.889	336.351	-322.804	238.006	-2.072

PREVIOUS: December 1975 (1 atm) CURRENT: December 1975 (1 bar)

Magnesium Hydroxide (MgOH) $H_1Mg_1O_1(g)$

MAGNESIUM HYDROXIDE, ION ($MgOH^+$) IDEAL GAS M_r = 41.311791

$\Delta_f H°(0\ K)$ = [581.61 ± 62.8] kJ mol^{-1}

S°(298.15 K) = [220.824 ± 8.4] J K^{-1} mol^{-1} $\Delta_f H°(298.15\ K)$ = [584.42 ± 62.8] kJ mol^{-1}

Vibrational Frequencies and Degeneracies

ν, cm^{-1}
[700](1)
[490](2)
[3650](1)

Ground State Quantum Weight: [1] σ = 1

Point Group: [$C_{\infty v}$]

Bond Distances: Mg-O = [1.77] Å O-H = [0.96] Å

Bond Angle: Mg-O-H = [180]°

Rotational Constant: B_0 = [0.491933] cm^{-1}

Enthalpy of Formation

The enthalpy of formation of $MgOH^+(g)$ is estimated based on trends suggested by the alkaline earth monohydroxide unipositive ions. The estimate is based on the close agreement of the appearance potentials of the alkaline earth monohydroxides and the corresponding monofluoride (1). In addition, the ionization potentials of the alkaline earth elements are fairly close to the appearance potential of the monohydroxide. We adopt, as the ionization potential for MgOH(g), a value of 7.7 eV (177.57 kcal mol^{-1}). This value is identical to the appearance potential of MgF(g), (1). The ionization potential of Mg(g) is 7.65 eV (1), which is very close to our adopted value.

The adopted ionization potential of 7.7 eV refers to the process $MgOH(g) + e^- = MgOH^+(g) + 2\ e^-$. Using auxiliary data (1), we calculate $\Delta_f H°(0\ K)$ = 139.01 ± 15 kcal mol^{-1} for $MgOH^+(g)$. This leads to $\Delta_f H°(298.15\ K)$ = 139.68 ± 15.0 kcal mol^{-1}.

Heat Capacity and Entropy

The molecular configuration is assumed to be linear, since experimental evidence indicates that the gaseous alkali metal hydroxides are linear (1, 2, 3, 4). In addition, Walsh (5) had predicted that BAH molecules (H = hydrogen atom) with ten or less valence electrons ($MgOH^+$ has 8 valence electrons) will be linear in their ground state. The molecule $MgOH^+$ is isoelectronic with NaOH.

The bond dissociation energy for $MgOH^+$ (81.7 kcal mol^{-1}) (1) for the process $MgOH^+(g) = Mg^+(g) + OH(g)$ is essentially identical to that for MgOH (82.0 kcal mol^{-1}) (1). This suggests a similar bonding in these two molecules. Thus, the bond distances are assumed to be the same as those adopted for MgOH(g) (1). The vibrational frequencies are assumed to be similar to those adopted for MgOH(g) (1). The ground state quantum weight is assumed to be the same as that of NaOH(g) (1).

References

1. JANAF Thermochemical Tables: e^-, 3-31-65; NaOH(g) and $Mg^+(g)$, 12-31-70; MgOH(g) and $MgF^+(g)$, 12-31-75.
2. N. Acquista, S. Abramowitz, and D. R. Lide, J. Chem. Phys. 49, 780 (1968).
3. R. L. Kuczkowski and D. R. Like, J. Chem. Phys. 44, 313 (1966).
4. N. Acquista and S. Abramowitz, U. S. Nat. Bur. Stand. Report 9905, July 1968.
5. A. D. Walsh, J. Chem. Soc. 1953, 2288 (1953).

Magnesium Hydroxide, Ion ($MgOH^+$) $H_1Mg_1O_1{}^+(g)$

Enthalpy Reference Temperature = T_r = 298.15 K Standard State Pressure = p° = 0.1 MPa

T/K	$C_p°$	S°	$-[G°-H°(T_r)]/T$	$H°-H°(T_r)$	$\Delta_f H°$	$\Delta_f G°$	Log K_f
	J $K^{-1}mol^{-1}$			kJ mol^{-1}			
0	0.	0.	INFINITE	-10.186	581.611		
100	29.854	182.153	254.830	-7.268			
200	36.949	204.808	224.596	-3.958			
250	40.503	213.448	221.522	-2.018			
298.15	43.216	220.824	220.824	0.	584.421	572.132	-100.235
300	43.306	221.092	220.825	0.080	584.440	572.055	-99.604
350	45.437	227.935	221.361	2.301	584.975	569.950	-85.060
400	47.050	234.113	222.575	4.615	585.560	567.764	-74.142
450	48.284	239.729	224.174	7.000	586.174	565.503	-65.642
500	49.246	244.868	225.990	9.439	586.805	563.172	-58.834
600	50.651	253.979	229.915	14.438	588.071	558.327	-48.607
700	51.668	261.867	233.929	19.557	589.299	553.271	-41.286
800	52.498	268.822	237.864	24.766	590.455	548.044	-35.784
900	53.235	275.048	241.656	30.053	591.512	542.679	-31.496
1000	53.917	280.693	245.282	35.411	583.859	537.914	-28.098
1100	54.556	285.862	248.739	40.835	584.657	533.281	-25.323
1200	55.155	290.635	252.034	46.321	585.477	528.574	-23.008
1300	55.713	295.072	255.176	51.865	586.317	523.798	-21.046
1400	56.229	299.220	258.175	57.462	459.764	522.124	-19.481
1500	56.704	303.116	261.043	63.109	461.986	526.501	-18.334
1600	57.138	306.789	263.788	68.802	464.219	530.729	-17.327
1700	57.535	310.266	266.421	74.536	466.461	534.818	-16.433
1800	57.895	313.564	268.949	80.308	468.708	538.774	-15.635
1900	58.223	316.704	271.381	86.114	470.958	542.605	-14.917
2000	58.521	319.698	273.722	91.951	473.209	546.318	-14.268
2100	58.792	322.560	275.980	97.817	475.459	549.918	-13.678
2200	59.038	325.301	278.160	103.709	477.706	553.411	-13.140
2300	59.262	327.930	280.267	109.624	479.949	556.802	-12.645
2400	59.466	330.456	282.306	115.560	482.187	560.095	-12.190
2500	59.652	332.888	284.281	121.516	484.417	563.295	-11.769
2600	59.822	335.231	286.196	127.490	486.639	566.406	-11.379
2700	59.977	337.491	288.054	133.480	488.851	569.432	-11.016
2800	60.120	339.675	289.859	139.485	491.052	572.377	-10.678
2900	60.250	341.787	291.613	145.504	493.240	575.242	-10.361
3000	60.371	343.832	293.320	151.535	495.414	578.033	-10.064
3100	60.481	345.813	294.982	157.578	497.573	580.752	-9.786
3200	60.583	347.735	296.600	163.631	499.714	583.400	-9.523
3300	60.678	349.601	298.178	169.694	501.837	585.982	-9.275
3400	60.765	351.413	299.717	175.766	503.938	588.501	-9.041
3500	60.846	353.176	301.220	181.847	506.016	590.957	-8.820
3600	60.921	354.891	302.687	187.935	508.070	593.355	-8.609
3700	60.991	356.561	304.121	194.031	510.096	595.696	-8.410
3800	61.056	358.189	305.522	200.133	512.093	597.983	-8.220
3900	61.117	359.775	306.893	206.242	514.058	600.218	-8.039
4000	61.173	361.324	308.234	212.357	515.989	602.402	-7.867
4100	61.226	362.835	309.548	218.477	517.883	604.539	-7.702
4200	61.276	364.311	310.834	224.602	519.739	606.630	-7.545
4300	61.322	365.753	312.095	230.732	521.553	608.678	-7.394
4400	61.366	367.163	313.330	236.866	523.324	610.683	-7.250
4500	61 407	368.543	314.542	243.005	525.050	612.649	-7.111
4600	61.445	369.893	315.731	249.147	526.726	614.577	-6.979
4700	61.482	371.215	316.897	255.294	528.353	616.469	-6.851
4800	61.516	372.510	318.042	261.444	529.926	618.327	-6.729
4900	61.548	373.778	319.167	267.597	531.445	620.153	-6.611
5000	61.579	375.022	320.272	273.753	532.906	621.949	-6.497
5100	61.607	376.242	321.357	279.912	534.309	623.716	-6.388
5200	61.635	377.438	322.424	286.075	535.651	625.455	-6.283
5300	61.660	378.613	323.473	292.239	536.931	627.170	-6.181
5400	61.685	379.765	324.505	298.407	538.146	628.861	-6.083
5500	61.708	380.898	325.520	304.576	539.295	630.530	-5.988
5600	61.730	382.010	326.519	310.748	540.376	632.180	-5.897
5700	61.751	383.102	327.502	316.922	541.389	633.810	-5.808
5800	61.771	384.177	328.470	323.098	542.330	635.423	-5.723
5900	61.790	385.233	329.423	329.276	543.200	637.020	-5.640
6000	61.808	386.271	330.362	335.456	543.996	638.603	-5.560

PREVIOUS: December 1975 (1 atm) CURRENT: December 1975 (1 bar)

Magnesium Hydroxide, Ion ($MgOH^+$) $H_1Mg_1O_1{}^+(g)$

IMIDOGEN (NH) IDEAL GAS M_r = 15.01464

$S°(298.15\ K) = 181.253 \pm 0.04\ J\ K^{-1}\ mol^{-1}$

$\Delta_f H°(0\ K) = 376.51 \pm 16.7\ kJ\ mol^{-1}$
$\Delta_f H°(298.15\ K) = 376.56 \pm 16.7\ kJ\ mol^{-1}$

Electronic States and Molecular Constants (σ = 1)

Source	State	ε_i, cm^{-1}	g_i	r_e, A	B_e, cm^{-1}	α_e, cm^{-1}	ω_e, cm^{-1}	$\omega_e x_e$, cm^{-1}
(13, 15)	$X^3\Sigma^-$	0.0	3	1.038	16.666	0.648	3282.09	78.3
(13, 14)	$a^1\Sigma$	12590	2	1.041	16.439	[0.7]	3303	55
(13, 16)	$b^1\Sigma^+$	21240	1	1.035	16.7326	0.6049	3347	70.7
(13, 15)	$A^3\Pi$	29777	6	1.036	16.6901	0.7440	3231.0	98.5
(16, 17)	$c^1\Pi$	43345	2	1.1005	14.7985	1.267	2503	194
(16, 17)	$d^1\Sigma^+$	82856	1	1.1165	14.3809	0.6119	2665	71.0

Enthalpy of Formation

The electron impact appearance potential of N_2^+ from HN_3 determined by Franklin et al. (1), leads to a value $\Delta_f H°$(NH, g, 298.15 K) = 81.7±0.5 kcal mol^{-1}. Reed and Snedden (2) determined the electron impact appearance potential of NH^+ from NH_3. This, combined with the directly measured electron impact ionization potential of NH, 13.1 eV, gives $\Delta_f H°$ = 82.9 kcal mol^{-1}. This value of the ionization potential of NH has also been obtained by Foner and Hudson (3). However, all of these electron impact experiments are subject to errors of several tenths of an electron volt or more.

More recently, Seal and Gaydon (4) measured the concentration of NH in reflected shock waves in nitrogen-hydrogen-krypton and ammonia-krypton mixtures. These led to D_0^0(NH) = 3.21±0.16 eV, corresponding to $\Delta_f H°$(NH, g) = 90.13±3 kcal mol^{-1}. Kaskan and Nadler (5) determined NH, NH_3, and OH concentrations in a flat NH_3-D_2-N_2 flame and concluded $\Delta_f H°$(NH, g) = 90±4 kcal mol^{-1}. Stedman (6) studied the NH emission spectrum obtained by collision of metastable rare gas atoms with HN_3. Taking the highest level of NH as observed in emission as a limit, he concluded that $\Delta_f H°$(NH, g) > 80 kcal mol^{-1}. From analogous consideration on N_2 emission observed by collision of metastable argon atoms with HN_3 he concluded that $\Delta_f H°$(NH, g) $\leq$ 94 kcal mol^{-1}.

Quantum chemical calculation of the dissociation energy of small molecules is possible with estimated accuracies of 0.1 to 0.2 eV. A theoretical calculation of the dissociation energy of NH($X^3\Sigma^-$), applying the techniques of Wahl and Das (7) has been determined by Stevens (8) to yield D_e = 3.4 eV. This result corroborates the results of Seal and Gaydon and would support $\Delta_f H°$(NH, g) = 90±4 kcal mol^{-1}. Theoretical calculations of this type can be used to distinquish between disparate experimental results that differ by more than 0.2 eV.

Heat Capacity and Entropy

The vibrational and rotational constants for the ground and excited states are taken from the sources indicated. The splitting between the $X^3\Sigma^-$ and $a^1\Delta$ states is based on the work of Gilles et al. (14) who observed the $b^1\Sigma^+ - X^3\Sigma^-$ band in emission.

The thermodynamic functions are calculated using first-order anharmonic corrections to Q_r^i and Q_v^i in the partition function $Q = Q_t \sum_i Q_r^i Q_v^i g_i \exp(-c_2\varepsilon_i/T)$. The National Bureau of Standards prepared this table (9) by critical analysis of data existing in 1972 using only the ground electronic state. Using the molecular constants selected here and $\Delta_f H°$ selected by NBS (9), we recalculate the table in terms of 1973 fundamental constants (10), 1975 atomic weights (11) and current JANAF reference states for the elements.

References

1. J. L. Franklin, V. H. Dibeler, R. M. Reese, and M. Krauss, J. Amer. Chem. Soc. 80, 298 (1958).
2. R. I. Reed and W. Snedden, J. Chem. Soc. 4132 (1959).
3. S. N. Foner and R. L. Hudson, J. Chem. Phys. 45, 40 (1966).
4. K. E. Seal and A. G. Gaydon, Proc. Phys. Soc. 89, 459 (1966).
5. W. E. Kaskan and M. P. Nadler, J. Chem. Phys. 56, 2220 (1972).
6. D. H. Stedman, J. Chem. Phys. 52, 3966 (1970).
7. A. C. Wahl and G. Das, Advan. Quantum Chem. 5, 261 (1970).
8. W. J. Stevens, J. Chem. Phys. 58, 1264 (1973).
9. S. Abramowitz et al., U. S. Nat. Bur. Stand. Report 10904, 239, July, 1972.
10. CODATA Task Group on Fundamental Constants, CODATA Bulletin 11, December, 1973.
11. IUPAC Commission on Atomic Weights, Pure Appl. Chem. 47, 75 (1976); 37, 589 (1974).
12. JANAF Thermochemical Tables, 2nd ed., NSRDS-NBS 37, 1971.
13. B. Rosen, "Spectroscopic Data Relative to Diatomic Molecules," Pergamon Press, Oxford, 1970.
14. A. Gilles, C. Vermeil, and J. Masanet, J. Photochem. 3, 417 (1974/75).
15. J. Malicet, J. Brain, and H. Guenebaut, J. Chim. Phys. 67, 25 (1970).
16. F. L. Whittaker, J. Phys. B1, 977 (1968); Can. J. Phys. 47, 1291 (1969).
17. N. A. Narasimham and G. Krishnamurty, Proc. Indian Acad. Sci. 64A, 97 (1966).

Imidogen (NH) $H_1N_1(g)$

Enthalpy Reference Temperature = T_r = 298.15 K Standard State Pressure = p° = 0.1 MPa

T/K	$C_p°$ (J K^{-1}mol^{-1})	S°	$-[G°-H°(T_r)]/T$	$H°-H°(T_r)$ (kJ mol^{-1})	$\Delta_f H°$	$\Delta_f G°$	Log K_f
0	0.	0.	INFINITE	-8.617	376.511	376.511	INFINITE
100	29.126	149.428	207.157	-5,773	376.405	374.489	-195.613
200	29.133	169.618	183.918	-2.860	376.516	372.532	-97.295
250	29.140	176.119	181.732	-1.403	376.547	371.532	-77.627
298.15	29.147	181.253	181.253	0.	376.560	370.565	-64.921
300	29.147	181.433	181.253	0.054	376.560	370.528	-64 515
350	29.158	185.927	181.608	1.512	376.565	369.522	-55.148
400	29.175	189.821	182.397	2.970	376.565	368.515	-48.123
450	29.208	193.259	183.416	4.429	376.561	367.509	-42.659
500	29.263	196.339	184.557	5.891	376 555	366.504	-38.288
600	29.461	201.690	186.980	8.826	376.534	364.496	-31.732
700	29.787	206.254	189.415	11.787	376.505	362.492	-27.049
800	30.219	210.259	191.775	14.787	376.473	360.492	-23.538
900	30.722	213.847	194.032	17.834	376.444	358.496	-20.807
1000	31.258	217.111	196.179	20.933	376.421	356.503	-18.622
1100	31.798	220.116	198.220	24.085	376.406	354.512	-16.834
1200	32.325	222.906	200.162	27.292	376.399	352.522	-15.345
1300	32.825	225.513	202.013	30.549	376.399	350.532	-14.085
1400	33.294	227.963	203.780	33.856	376.407	348.542	-13.004
1500	33.730	230.275	205.470	37.207	376.420	346.552	-12.068
1600	34.134	232.465	207.089	40.601	376.438	344.560	-11 249
1700	34.507	234.546	208.644	44.033	376.461	342.567	-10.526
1800	34.852	236.528	210.138	47.501	376.488	340.572	-9.883
1900	35.173	238.421	211.577	51.003	376.518	338.576	-9.308
2000	35.473	240.233	212.965	54.535	376.551	336.578	-8.791
2100	35.755	241.970	214.305	58.097	376.587	334.579	-8.322
2200	36.021	243.640	215.601	61.685	376.627	332.578	-7.896
2300	36.274	245.247	216.855	65.300	376.670	330.574	-7.508
2400	36.516	246.796	218.071	68.940	376.716	328.569	-7.151
2500	36.749	248.291	219.250	72.603	376.766	326.562	-6.823
2600	36.975	249.737	220.395	76.290	376.820	324.553	-6.520
2700	37.194	251.136	221.508	79.998	376.879	322.541	-6.240
2800	37.407	252.493	222.590	83.728	376.942	320.528	-5.980
2900	37.616	253.809	223,644	87.479	377.010	318.512	-5.737
3000	37.821	255.088	224.671	91.251	377.084	316.493	-5.511
3100	38.022	256.331	225.672	95.043	377.163	314.472	-5.299
3200	38.220	257.542	226.649	98.855	377 247	312.449	-5.100
3300	38.416	258.721	227.603	102.687	377.338	310.423	-4.914
3400	38.609	259.870	228.536	106.538	377.435	308.393	-4.738
3500	38.799	260.992	229.447	110.409	377.538	306.361	-4.572
3600	38.988	262.088	230.339	114.298	377.648	304.326	-4.416
3700	39.175	263.159	231.211	118.206	377.764	302.288	-4.268
3800	39.360	264.206	232.066	122.133	377.886	300.246	-4.127
3900	39.543	265.231	232.903	126.078	378.015	298.201	-3.994
4000	39.725	266.234	233.724	130.042	378.151	296.153	-3.867
4100	39.905	267.217	234.529	134.023	378.294	294.102	-3.747
4200	40.084	268.181	235.319	138.023	378.443	292.046	-3.632
4300	40.262	269.126	236.094	142.040	378.599	289.987	-3.523
4400	40.439	270.054	236.855	146.075	378.763	287.925	-3.418
4500	40.615	270.965	237.603	150.128	378.933	285.858	-3.318
4600	40.789	271.859	238.338	154.198	379.110	283.788	-3.223
4700	40.963	272.738	239.061	158.286	379.295	281.714	-3.131
4800	41.136	273.603	239.771	162.391	379.486	279.636	-3.043
4900	41.309	274.453	240.470	166.513	379.686	277.553	-2.959
5000	41.480	275.289	241.158	170.652	379.893	275.467	-2.878
5100	41.652	276.112	241.836	174.809	380.108	273.376	-2 800
5200	41.823	276.922	242.503	178.983	380.331	271.281	-2.725
5300	41.993	277.721	243.160	183.173	380.563	269.182	-2.653
5400	42.164	278.507	243.807	187.381	380.803	267.078	-2.583
5500	42.334	279.283	244.445	191.606	381.053	264.970	-2.516
5600	42.504	280.047	245.074	195.848	381.312	262.857	-2.452
5700	42.674	280.801	245.694	200.107	381.581	260.740	-2.389
5800	42.844	281.544	246.306	204.383	381.861	258.617	-2.329
5900	43.015	282.278	246.909	208.676	382.151	256.490	-2.271
6000	43.185	283.003	247.505	212.986	382.451	254 357	-2.214

PREVIOUS: June 1977 (1 atm) CURRENT: June 1977 (1 bar)

Imidogen (NH) $H_1N_1(g)$

NITROXYL (HNO) IDEAL GAS M_r = 31.01404

Nitrosyl Hydride (HNO) $H_1N_1O_1(g)$

$\Delta_fH°(0\ K)$ = 102.501 kJ mol^{-1}

S°(298.15 K) = 220.722 J K^{-1} mol^{-1} $\Delta_fH°(298.15\ K)$ = 99.579 kJ mol^{-1}

Vibrational Frequencies and Degeneracies

ν, cm^{-1}
2854.17(1)
1570 (1)
1110 (1)

Ground State Quantum Weight: [1] σ = 1

Point Group: C_s

Bond Distances: H-N = 1.020 ± 0.02 Å N=O = 1.239 ± 0.005 Å

Bond Angle: H-N=O = 114° 25' ± 2°

Product of the Moments of Inertia: $I_AI_BI_C = 5.939 \times 10^{-118}\ g^3\ cm^6$

Enthalpy of Formation

$\Delta_fH°$(HNO, g, 298.15 K) is calculated from D_0^0(H-NO) = 48.6 kcal mol^{-1} reported by Clement and Ramsay (1).

Heat Capacity and Entropy

Vibrational frequencies were assigned by Bancroft et al. (2) and Brown and Pimentel (3). Bond distances and bond angles were taken from Bancroft et al. (2). The molecular constants given by Dalby (4) were: r(N=O) = 1.211_6 Å; r(N-H) = 1.0628 Å, and a H-N=O bond angle of 108.5_8°. The principal moments of inertia are: 0.1262×10^{-39}, 2.1070×10^{-39}, and 2.2332×10^{-39} g cm^2.

References

1. M. J. Y. Clement and D. A. Ramsay, Can. J. Phys. 39, 205 (1961).
2. J. L. Bancroft, J. M. Hollas and D. A. Ramsay, Can. J. Phys. 40, 322 (1962).
3. H. Brown and G. Pimentel, J. Chem. Phys. 29, 883 (1958).
4. F. W. Dalby, Can. J. Phys. 36, 1336 (1958).

Nitrosyl Hydride (HNO) $H_1N_1O_1(g)$

Enthalpy Reference Temperature = T_r = 298.15 K Standard State Pressure = p° = 0.1 MPa

T/K	$C_p°$ (J $K^{-1}mol^{-1}$)	S° (J $K^{-1}mol^{-1}$)	$-[G°-H°(T_r)]/T$ (J $K^{-1}mol^{-1}$)	$H°-H°(T_r)$ (kJ mol^{-1})	$\Delta_fH°$ (kJ mol^{-1})	$\Delta_fG°$ (kJ mol^{-1})	Log K_f
0	0.	0.	INFINITE	-9.988	102.501	102.501	INFINITE
100	33.258	184.104	250.730	-6.663	101.424	104.706	-54.693
200	33.451	207.183	223.843	-3.332	100.497	108.348	-28.298
250	33.911	214.691	221.288	-1.649	100.025	110.366	-23.060
298.15	34.640	220.722	220.722	0.	99.579	112.398	-19.692
300	34.673	220.937	220.723	0.064	99.562	112.478	-19.584
350	35.658	226.353	221.148	1.822	99.129	114.665	-17.113
400	36.768	231.186	222.106	3.632	98.733	116.912	-15.267
450	37.928	235.584	223.363	5.499	98.379	119.206	-13.837
500	39.085	239.640	224.790	7.425	98.065	121.538	-12.697
600	41.289	246.964	227.889	11.445	97.550	126.283	-10.994
700	43.281	253.481	231.087	15.675	97.162	131.104	-9.783
800	45.042	259.378	234.261	20.093	96.881	135.973	-8.878
900	46.583	264.774	237.356	24.676	96.686	140.872	-8.176
1000	47.922	269.753	240.350	29.403	96.560	145.789	-7.615
1100	49.081	274.376	243.235	34.255	96.489	150.716	-7.157
1200	50.083	278.691	246.012	39.214	96.460	155.647	-6.775
1300	50.948	282.734	248.683	44.267	96.464	160.580	-6.452
1400	51.697	286.538	251.252	49.400	96.491	165.511	-6.175
1500	52.346	290.128	253.726	54.603	96.535	170.439	-5.935
1600	52.911	293.524	256.108	59.866	96.590	175.364	-5.725
1700	53.403	296.747	258.404	65.183	96.651	180.286	-5.540
1800	53.834	299.812	260.620	70.545	96.714	185.204	-5.374
1900	54.213	302.733	262.761	75.948	96.776	190.118	-5.227
2000	54.548	305.523	264.829	81.386	96.834	195.029	-5.094
2100	54.843	308.191	266.831	86.856	96.886	199.938	-4.973
2200	55.106	310.749	268.770	92.354	96.930	204.844	-4.864
2300	55.341	313.204	270.649	97.876	96.965	209.748	-4.764
2400	55.550	315.563	272.471	103.421	96.990	214.651	-4.672
2500	55.739	317.835	274.241	108.986	97.004	219.553	-4.587
2600	55.908	320.024	275.960	114.568	97.006	224.455	-4.509
2700	56.061	322.137	277.631	120.167	96.996	229.358	-4.437
2800	56.200	324.179	279.257	125.780	96.974	234.260	-4.370
2900	56.326	326.153	280.840	131.406	96.938	239.164	-4.308
3000	56.440	328.064	282.383	137.045	96.890	244.069	-4.250
3100	56.545	329.917	283.887	142.694	96.828	248.976	-4.195
3200	56.641	331.714	285.353	148.353	96.753	253.885	-4.144
3300	56.729	333.458	286.785	154.022	96.665	258.797	-4.096
3400	56.809	335.153	288.182	159.699	96.564	263.711	-4.051
3500	56.884	336.801	289.548	165.384	96.449	268.629	-4.009
3600	56.953	338.404	290.883	171.075	96.322	273.550	-3.969
3700	57.016	339.965	292.189	176.774	96.181	278.475	-3.931
3800	57.075	341.487	293.466	182.478	96.027	283.404	-3.896
3900	57.130	342.970	294.716	188.189	95.860	288.337	-3.862
4000	57.180	344.417	295.941	193.904	95.681	293.275	-3.830
4100	57.228	345.829	297.140	199.625	95.488	298.217	-3.799
4200	57.272	347.209	298.316	205.350	95.282	303.164	-3.770
4300	57.313	348.557	299.469	211.079	95.063	308.116	-3.743
4400	57.352	349.875	300.600	216.812	94.832	313.073	-3.717
4500	57.388	351.164	301.709	222.549	94.587	318.036	-3.692
4600	57.422	352.426	302.798	228.290	94.330	323.004	-3.668
4700	57.454	353.661	303.867	234.034	94.059	327.978	-3.645
4800	57.484	354.871	304.917	239.780	93.775	332.959	-3.623
4900	57.512	356.057	305.949	245.530	93.478	337.945	-3.603
5000	57.539	357.219	306.963	251.283	93.168	342.937	-3.583
5100	57.564	358.359	307.959	257.038	92.844	347.936	-3.564
5200	57.587	359.477	308.939	262.796	92.507	352.940	-3.545
5300	57.610	360.574	309.903	268.555	92.157	357.953	-3.528
5400	57.631	361.651	310.851	274.317	91.792	362.971	-3.511
5500	57.651	362.709	311.785	280.082	91.414	367.996	-3.495
5600	57.670	363.748	312.703	285.848	91.022	373.028	-3.479
5700	57.688	364.768	313.608	291.616	90.615	378.068	-3.465
5800	57.705	365.772	314.499	297.385	90.195	383.115	-3.450
5900	57.722	366.759	315.376	303.157	89.759	388.168	-3.437
6000	57.737	367.729	316.241	308.930	89.309	393.230	-3.423

PREVIOUS: March 1963 (1 atm) CURRENT: March 1963 (1 bar)

Nitrosyl Hydride (HNO) $H_1N_1O_1(g)$

NITROUS ACID, CIS-(HNO_2) IDEAL GAS M_r = 47.01344

$\Delta_f H°(0\ K) = -70.509 \pm 1.34$ kJ mol^{-1}

$S°(298.15\ K) = 249.416$ J K^{-1} mol^{-1} $\Delta_f H°(298.15\ K) = -76.735 \pm 1.34$ kJ mol^{-1}

Vibrational Frequencies and Degeneracies

ν, cm^{-1}	ν, cm^{-1}
3462(1)	855(1)
1639(1)	525(1)
1330(1)	638(1)

Ground State Quantum Weight: [1] σ = 1

Point Group: C_s

Bond Distances: O–N = 1.20 Å N–O' = 1.46 Å O–H = 0.98 Å

Bond Angles: O–N–O' = 114° N–O'–H = 103°

Product of the Moments of Inertia: $I_A I_B I_C = 5.51348 \times 10^{-116}$ g^3 cm^6

Enthalpy of Formation

Ashmore and Tyler (1) reported $\Delta_r H°(298.15\ K) = -9.06\pm0.32$ kcal mol^{-1} for the reaction: $NO(g) + NO_2(g) + H_2O(g) = 2\ HNO_2(g)$, yielding $\Delta_f H°(298.15\ K) = -18.57\pm0.20$ kcal mol^{-1} for $HNO_2(g)$. Jones et al. estimated the difference in energy of cis- and trans-$HNO_2(g)$ to be 506±250 cal K^{-1} mol^{-1}. Hence the value of $\Delta_f H°(298.15\ K)$ for cis-$HNO_2(g)$ was calculated.

The heat of reaction and the equilibrium constants for the formation of $HNO_2(g)$ have been calculated by Karavaev and Skvortsov (3). The value of $\Delta_f H°(298.15\ K)$ reported is -18.8 kcal mol^{-1}. However, based on the enthalpy of formation of liquid HNO_2, Ashmore and Levitt (4) give $\Delta_f H° = -13.7$ kcal mol^{-1}. The corresponding value reported by Rosser and Wise (5) is -20.0 kcal mol^{-1}, based on the experimental data of Wayne and Yost (6), and the entropy of the equilibrium mixture of the trans- and cis-$HNO_2(g)$ calculated by Jones, et al. (2).

Heat Capacity and Entropy

Vibrational frequencies, bond distances, and bond angles were obtained from Dor and Tarte (7) and Jones et al. (2). The principal moments of inertia are: 1.0282×10^{-39}, 6.8267×10^{-39}, and 7.8549×10^{-39} g cm^2.

References

1. P. G. Ashmore and B. J. Tyler, J. Chem. Soc. 1017 (1961).
2. L. H. Jones, R. M. Badger and G. E. Moore, J. Chem. Phys. 19, 1599 (1951).
3. M. M. Karavaev and G. A. Skvortsov, Russ. J. Phys. Chem. 36, 566 (1962).
4. P. G. Ashmore and B. P. Levitt, Trans. Faraday Soc. 53, 945 (1957).
5. W. A. Rosser and H. Wise, J. Chem. Phys. 26, 571 (1957).
6. L. G. Wayne and D. M. Yost, J. Chem. Phys. 19, 41 (1951).
7. L. Dor and P. Tarte, Bull. Soc. Roy. Sci. Lg. 478 (1951).

Nitrous Acid, Cis (HONO) $H_1N_1O_2(g)$

Enthalpy Reference Temperature = T_r = 298.15 K Standard State Pressure = p° = 0.1 MPa

T/K	$C_p°$ (J K^{-1}mol^{-1})	S°	$-[G°-H°(T_r)]/T$	$H°-H°(T_r)$ (kJ mol^{-1})	$\Delta_f H°$	$\Delta_f G°$	Log K_f
0	0.	0.	INFINITE	-11.026	-70.509	-70.509	INFINITE
100	33.585	208.174	285.140	-7.697	-73.034	-63.494	33.166
200	38.653	232.718	253.350	-4.126	-75.178	-53.084	13.864
250	42.127	241.717	250.144	-2.107	-76.042	-47.458	9.916
298.15	45.354	249.416	249.416	0.	-76.735	-41.888	7.339
300	45.473	249.696	249.416	0.084	-76.759	-41.672	7.256
350	48.560	256.942	249.982	2.436	-77.336	-35.776	5.339
400	51 357	263.612	251.274	4.935	-77.790	-29.807	3.892
450	53.870	269.809	252.993	7.567	-78.138	-23.787	2.761
500	56.115	275.603	254.967	10.318	-78.398	-17.733	1.853
600	59.904	286.182	259.305	16.126	-78.705	-5.568	0.485
700	62.943	295.652	263.833	22.273	-78.802	6.632	-0.495
800	65.424	304.225	268.355	28.696	-78.748	18.835	-1.230
900	67.490	312.053	272.782	35.344	-78.580	31.024	-1.801
1000	69.236	319.257	277.074	42.183	-78.326	43.189	-2.256
1100	70.730	325.928	281.215	49.183	-78.003	55.325	-2.627
1200	72.017	332.138	285.203	56.322	-77.626	67.430	-2.935
1300	73.133	337.948	289.039	63.581	-77.208	79.501	-3.194
1400	74.105	343.404	292.730	70.944	-76.757	91.539	-3.415
1500	74.955	348.546	296.281	78.398	-76.283	103.543	-3.606
1600	75.700	353.408	299.701	85.931	-75.791	115.516	-3.771
1700	76.356	358.018	302.997	93.535	-75.289	127.457	-3.916
1800	76.935	362.399	306.176	101.200	-74.781	139.369	-4.044
1900	77.448	366.572	309.246	108.920	-74.272	151.252	-4.158
2000	77.904	370.557	312.213	116.688	-73.766	163.109	-4.260
2100	78.311	374.368	315.083	124.499	-73.266	174.940	-4.351
2200	78.675	378.019	317.861	132.349	-72.774	186.749	-4.434
2300	79.001	381.524	320.553	140.233	-72.292	198.534	-4.509
2400	79.294	384.892	323.164	148.148	-71.823	210.298	-4.577
2500	79.558	388.135	325.699	156.091	-71.369	222.044	-4.639
2600	79.797	391.260	328.160	164.059	-70.929	233.772	-4 697
2700	80.014	394.276	330.554	172.049	-70.506	245.483	-4.749
2800	80.212	397.189	332.882	180.061	-70.099	257.179	-4.798
2900	80.391	400.007	335.148	188.091	-69.709	268.860	-4.843
3000	80.556	402.735	337.356	196.139	-69.337	280.529	-4.884
3100	80.706	405.379	339.508	204.202	-68.983	292.186	-4.923
3200	80.844	407.944	341.606	212.279	-68.646	303.830	-4.960
3300	80.971	410.433	343.655	220.370	-68.327	315.465	-4.993
3400	81.088	412 852	345.654	228.473	-68.026	327.091	-5.025
3500	81.196	415.204	347.608	236.587	-67.743	338.707	-5.055
3600	81.296	417.493	349.518	244.712	-67.478	350.317	-5.083
3700	81.388	419.722	351.385	252.846	-67.230	361.918	-5.109
3800	81.474	421.894	353.212	260.990	-66.999	373.515	-5.134
3900	81.554	424.011	355.000	269.141	-66.786	385.105	-5.158
4000	81.628	426.077	356.752	277.300	-66.590	396.688	-5.180
4100	81.698	428.093	358.467	285.466	-66.411	408.268	-5.201
4200	81.762	430.063	360.149	293.640	-66.249	419.844	-5.222
4300	81.823	431.987	361.797	301.819	-66.105	431.416	-5.241
4400	81.880	433.869	363.414	310.004	-65.977	442.984	-5.259
4500	81.933	435.710	365.000	318.195	-65.867	454.550	-5.276
4600	81.983	437.511	366.557	326.391	-65.775	466.113	-5.293
4700	82.030	439.275	368.085	334.591	-65.700	477.676	-5.309
4800	82.074	441.002	369.586	342.796	-65.643	489.236	-5.324
4900	82.116	442.695	371.061	351.006	-65.604	500.796	-5.339
5000	82.155	444.354	372.510	359.220	-65.583	512.355	-5.353
5100	82.192	445.982	373.935	367.437	-65.582	523.914	-5.366
5200	82.227	447.578	375.336	375.658	-65.600	535.472	-5.379
5300	82.261	449.145	376.714	383.882	-65.637	547.033	-5.391
5400	82.292	450.682	378.069	392.110	-65.696	558.592	-5.403
5500	82.322	452.193	379.403	400.341	-65.774	570.153	-5.415
5600	82.350	453.676	380.717	408.574	-65.874	581.718	-5.426
5700	82.377	455.134	382.009	416.811	-65.997	593.282	-5.437
5800	82.403	456.567	383.283	425.050	-66.142	604.851	-5.447
5900	82.427	457 976	384.537	433.291	-66.311	616.419	-5.457
6000	82.450	459.361	385.772	441.535	-66.504	627.993	-5.467

PREVIOUS: June 1963 (1 atm) CURRENT June 1963 (1 bar)

Nitrous Acid, Cis (HONO) $H_1N_1O_2(g)$

NITROUS ACID, TRANS-(HNO_2) IDEAL GAS M_r = 47.01344

$S°(298.15\ K) = 249.261\ J\ K^{-1}\ mol^{-1}$

$\Delta_f H°(0\ K) = -72.674 \pm 1.34\ kJ\ mol^{-1}$
$\Delta_f H°(298.15\ K) = -78.827 \pm 1.34\ kJ\ mol^{-1}$

Vibrational Frequencies and Degeneracies

ν, cm^{-1}	ν, cm^{-1}
3590(1)	794(1)
1696(1)	598(1)
1260(1)	543(1)

Ground State Quantum Weight: [1] σ = 1
Point Group: C_s
Bond Distances: O-N = 1.20 Å N-O' = 1.46 Å O'-H = 0.98 Å
Bond Angles: O-N-O' = 118° N-O'-H = 105°
Product of the Moments of Inertia: $I_A I_B I_C = 4.91660 \times 10^{-116}\ g^3\ cm^6$

Enthalpy of Formation

Ashmore and Tyler (1) reported $\Delta_r H°(298.15\ K) = -9.06 \pm 0.32\ kcal\ mol^{-1}$ for the reaction: $NO(g) + NO_2(g) + H_2O(g) \rightleftarrows 2HNO_2(g)$, yielding $\Delta_f H°(298.15\ K) = -18.57 \pm 0.20\ kcal\ mol^{-1}$ for $HNO_2(g)$. Jones et al. (2) estimated the difference in energy of cis- and trans- $HNO_2(g)$ to be 506±250 cal $K^{-1}\ mol^{-1}$. Hence the value of $\Delta_f H°(298.15\ K)$ for trans-$HNO_2(g)$ was calculated.

The heat of reaction and the equilibrium constants for the formation of $HNO_2(g)$ have been calculated by Karavaev and Skvortsov (3). The value of $\Delta_f H°(298.15\ K)$ reported is -18.8 kcal mol^{-1}. However, based on the enthalply of formation of liquid HNO_2, Ashmore and Levitt (4) give $\Delta_f H° = -13.7\ kcal\ mol^{-1}$. The corresponding value reported by Rosser and Wise (5) is -20.0 kcal mol^{-1}, based on the experimental data of Wayne and Yost (6)and the entropy of the equilibrium mixture of the trans- and cis- $HNO_2(g)$ calculated by Jones et al. (2).

Heat Capacity and Entropy

Vibrational frequencies, bond distances, and bond angles were taken from Jones et al. (2). The values of vibrational frequencies were in good agreement to those reported by Dor and Tarte (7). The principal moments of inertia are: 0.7855×10^{-39}, 7.5286×10^{-39}, and $8.3141 \times 10^{-39}\ g\ cm^2$.

References

1. P. G. Ashmore and B. J. Tyler, J. Chem. Soc. 1017 (1961).
2. L. H. Jones, R. M. Badger and G. E. Moore, J. Chem. Phys. 19, 1599 (1951).
3. M. M. Karavaev and G. A. Skvortsov, Russ. J. Phys. Chem. 36, 566 (1962).
4. P. G. Ashmore and B. P. Levitt, Trans. Faraday Soc. 53, 945 (1957).
5. W. A. Rosser and H. Wise, J. Chem. Phys. 26, 571 (1957).
6. L. G. Wayne and D. M. Yost, J. Chem. Phys. 19, 41 (1951).
7. L. Dor and P. Tarte, Bull. Soc. Roy. Sci. Lg. 478 (1951).

Nitrous Acid, Trans (HONO) $H_1N_1O_2(g)$

Enthalpy Reference Temperature = T_r = 298.15 K Standard State Pressure = p° = 0.1 MPa

T/K	$C_p°$	S°	$-[G°-H°(T_r)]/T$	$H°-H°(T_r)$	$\Delta_f H°$	$\Delta_f G°$	Log K_f
	J $K^{-1}mol^{-1}$			kJ mol^{-1}			
0	0.	0.	INFINITE	-11.099	-72.674	-72.674	INFINITE
100	33.588	207.696	285.389	-7.769	-75.199	-65.611	34.272
200	39.045	232.339	253.252	-4.183	-77.326	-55.157	14.405
250	42.718	241.447	250.001	-2.138	-78.165	-49.514	10.345
298.15	46.051	249.261	249.261	0.	-78.827	-43.934	7.697
300	46.173	249.546	249.262	0.085	-78.849	-43.717	7.612
350	49.290	256.902	249.835	2.473	-79.391	-37.817	5.644
400	52.066	263.669	251.147	5.009	-79.809	-31.848	4.159
450	54 528	269.946	252.891	7.675	-80.123	-25.833	2.999
500	56.707	275.807	254.893	10.457	-80.351	-19 788	2.067
600	60.354	286.481	259.286	16.317	-80.606	-7.648	0.666
700	63.260	296.011	263.864	22.503	-80.665	4.519	-0.337
800	65.628	304.618	268.429	28.951	-80.585	16.683	-1.089
900	67.601	312.465	272.892	35.615	-80.402	28.832	-1.673
1000	69.275	319.676	277.215	42.461	-80.139	40.956	-2.139
1100	70.713	326.348	281.382	49.462	-79.816	53.050	-2.519
1200	71.959	332.555	285.391	56.597	-79.443	65.113	-2.834
1300	73.046	338.359	289.245	63.849	-79.032	77.143	-3.100
1400	73.998	343.808	292.949	71.202	-78.591	89.140	-3.326
1500	74.835	348.943	296.513	78 645	-78.128	101.104	-3.521
1600	75.572	353.796	299.943	86.166	-77.649	113.037	-3.690
1700	76.224	358.398	303.247	93.756	-77.160	124.940	-3.839
1800	76.802	362.771	306.433	101.408	-76.665	136.814	-3.970
1900	77.316	366.938	309.509	109.115	-76.170	148.661	-4.087
2000	77.774	370.916	312.481	116.870	-75.676	160.481	-4.191
2100	78.184	374.720	315.355	124.668	-75.189	172.277	-4.285
2200	78.551	378.366	318.136	132.505	-74.709	184.050	-4.370
2300	78.881	381.865	320.832	140.377	-74.240	195.801	-4.447
2400	79.179	385.229	323.445	148.280	-73.783	207.532	-4.517
2500	79.448	388.466	325.982	156.212	-73 340	219.244	-4.581
2600	79.692	391.587	328.445	164.169	-72.911	230.939	-4.640
2700	79.913	394.599	330.840	172.149	-72.498	242.618	-4.694
2800	80.115	397.509	333.169	180.151	-72.101	254.282	-4.744
2900	80.298	400.324	335.437	188.172	-71.720	265.931	-4.790
3000	80.467	403.049	337.645	196.210	-71.358	277.568	-4.833
3100	80.621	405.690	339.798	204.264	-71.012	289.194	-4.873
3200	80.763	408.252	341.897	212.334	-70.684	300.807	-4.910
3300	80.893	410.739	343.946	220.417	-70.373	312.411	-4.945
3400	81.013	413.156	345.946	228.512	-70.080	324.007	-4.978
3500	81.124	415.506	347.900	236.619	-69.804	335.593	-5.008
3600	81.227	417.792	349.810	244.737	-69.545	347.172	-5.037
3700	81.323	420.019	351.678	252.864	-69.304	358.744	-5.065
3800	81.411	422.189	353.505	261.001	-69.080	370.311	-5.090
3900	81.494	424.305	355.293	269.146	-68.873	381.871	-5.115
4000	81.570	426.369	357.044	277.299	-68.683	393.426	-5.138
4100	81.642	428.384	358.760	285.460	-68.509	404.976	-5.159
4200	81.709	430.352	360.441	293.628	-68.353	416.523	-5.180
4300	81.772	432.276	362.089	301.802	-68.214	428.066	-5.200
4400	81.830	434.156	363.706	309.982	-68.092	439.606	-5.219
4500	81.885	435.996	365.292	318.168	-67.986	451.143	-5.237
4600	81.937	437.796	366.849	326.359	-67.898	462.677	-5.254
4700	81.986	439.559	368.377	334.555	-67.828	474.211	-5.270
4800	82.032	441.286	369.878	342.756	-67.775	485.744	-5.286
4900	82.075	442.977	371.353	350.961	-67.740	497.275	-5.301
5000	82.115	444.636	372.802	359.171	-67.724	508.806	-5.315
5100	82.154	446.262	374.226	367.384	-67.727	520.337	-5.329
5200	82.190	447.858	375.627	375.602	-67.748	531.867	-5.343
5300	82.225	449.424	377.005	383.822	-67.789	543.399	-5.356
5400	82.257	450.961	378.360	392.046	-67.851	554.931	-5.368
5500	82.288	452.471	379 694	400.274	-67.933	566.464	-5.380
5600	82.318	453.954	381.007	408.504	-68.037	578.001	-5.391
5700	82.346	455.411	382.299	416.737	-68.162	589.538	-5.403
5800	82.372	456.843	383.572	424.973	-68.311	601.078	-5.413
5900	82.397	458.252	384.826	433.212	-68.482	612.619	-5.424
6000	82.421	459.637	386.061	441.453	-68.679	624.166	-5.434

PREVIOUS: June 1963 (1 atm) CURRENT: June 1963 (1 bar)

Nitrous Acid, Trans (HONO) $H_1N_1O_2(g)$

NITRIC ACID (HNO_3) IDEAL GAS M_r = 63.01284

$S°(298.15\ K) = 266.400\ J\ K^{-1}\ mol^{-1}$

$\Delta_f H°(0\ K) = -124.492 \pm 0.42\ kJ\ mol^{-1}$
$\Delta_f H°(298.15\ K) = -134.306 \pm 0.42\ kJ\ mol^{-1}$

Vibrational Frequencies and Degeneracies

ν, cm^{-1}	ν, cm^{-1}
1320(1)	765(1)
886(1)	3560(1)
680(1)	1335(1)
1710(1)	465(1)
583(1)	

Ground State Quantum Weight: [1] $\sigma = 1$
Point Group: C_s
Bond Distances: O-N = 1.206 ± 0.005 Å N-O' = 1.405 ± 0.005 Å O-H = [0.96] Å
Bond Angles: O-N-O = 130° ± 20' N-O'-H = 102° ± 30'
Product of the Moments of Inertia: $I_A I_B I_C = 6.08612 \times 10^{-115}\ g^3\ cm^6$

Enthalpy of Formation

$\Delta_f H°(298.15\ K)$ was taken from (1). Equilibria involving HNO_3 were measured by Jones (2) Feick (3) and Abel et al. (4), yielding $\Delta_f H°$ = -32.06, -32.02, and -32.09 kcal mol^{-1}, respectively. Becker and Roth (5) and Thomsen (6) measured the enthalpy of neutralization of HNO_3(aq) with NH_3, combining these values with the data for NH_4NO_3 yields $\Delta_f H°$ = -32.27 and -32.07 kcal mol^{-1}. Berthelot (7) measured the enthalpy of reaction of N_2O_4(g) and Cl_2(g), obtaining $\Delta_f H°$ = -32.10 kcal mol^{-1}. The value of $\Delta_f H°(298.15\ K)$ reported by Forsythe and Giauque (8) was -31.994 kcal mol^{-1}. The value adopted is the weighted average of these six values.

Heat Capacity and Entropy

Vibrational frequencies were obtained from Palm and Kilpatrick (9). Bond distances and bond angles were taken from Millen and Morton (10). The basic parameters of HNO_3(g) reported by Akishin et al. (1) were: r(N=O) = 1.22±0.01 Å; r(N-O) = 1.40 ± 0.01 A; and the bond angle (O-N-O) = 135°±2.5°. Forsythe and Giauque (8) have calculated the absolute entropy of pure nitric acid (g) from calorimetric data. The value obtained was S°(298.15 K) = 63.62 cal K^{-1} mol^{-1}. The entropy of nitric acid vapor at 298.15 K has also been computed by Cohn et al. (9) to be 63.70 cal K^{-1} mol^{-1}. The prinicpal moments of inertia are: 6.3868 x 10^{-39}, 7.0025 x 10^{-39}, and 13.3893 x 10^{-39} g cm^2.

References

1. D. D. Wagman, U. S. Nat. Bur. Stand. Report 7437, Chapter B4, 83, January 1, 1962.
2. E. J. Jones, J. Amer. Chem. Soc. 65, 2274 (1943).
3. G. Feick, J. Amer. Chem. Soc. 76, 5858 (1954).
4. H. E. Abel, H. Schmid and M. Stein-Wien, Z. Elektrochem. 36, 692 (1930).
5. G. Becker and W. A. Roth, Z. Elektrochem. 40, 836 (1934).
6. J. Thomsen, "Thermochemische Untersuchungen," Barth, Leipsiz, 1882-1886.
7. M. Berthelot, "Sur la Force des Matieres Explosives," 3rd ed., Paris, 1883.
8. W. R. Forsythe and W. F. Giauque, J. Amer. Chem. Soc. 64, 48 (1942).
9. A. Palm and M. Kilpatrick, J. Chem. Phys. 23, 1562 (1955).
10. D. J. Millen and J. R. Morton, J. Chem. Soc. 1523 (1960).
11. P. A. Akishin, L. V. Vilkov and V. Y. Rosolovskii, Zhur. Strukt. Khim. 1, 1 (1960).
12. H. Cohn, C. K. Ingold, and H. G. Poole, J. Chem. Soc. 4272 (1952).

Nitric Acid ($HONO_2$) $H_1N_1O_3$(g)

Enthalpy Reference Temperature = T_r = 298.15 K Standard State Pressure = p° = 0.1 MPa

T/K	$C_p°$ (J $K^{-1}mol^{-1}$)	S° (J $K^{-1}mol^{-1}$)	$-[G°-H°(T_r)]/T$ (J $K^{-1}mol^{-1}$)	$H°-H°(T_r)$ (kJ mol^{-1})	$\Delta_f H°$ (kJ mol^{-1})	$\Delta_f G°$ (kJ mol^{-1})	Log K_f
0	0.	0.	INFINITE	-11.780	-124.492	-124.492	INFINITE
100	33.920	221.800	306.248	-8.445	-128.465	-111.622	58.305
200	42.298	247.433	270.926	-4.699	-131.888	-93.389	24.391
250	48.008	257.484	267.248	-2.441	-133.243	-83.602	17.468
298.15	53.336	266.400	266.400	0.	-134.306	-73.941	12.954
300	53.534	266.730	266.401	0.099	-134.343	-73.567	12.809
350	58.618	275.371	267.072	2.905	-135.205	-63.366	9.457
400	63.190	283.502	268.622	5.952	-135.858	-53.057	6.928
450	67.252	291.185	270.706	9.215	-136.333	-42.676	4.954
500	70.836	298.460	273.121	12.669	-136.660	-32.251	3.369
600	76.765	311.922	278.485	20.062	-136.963	-11.334	0.987
700	81.382	324.116	284.147	27.979	-136.918	9.605	-0.717
800	85.039	335.231	289.848	36.306	-136.627	30.519	-1.993
900	87.994	345.424	295.465	44.963	-136.154	51.386	-2.982
1000	90.425	354.824	300.937	53.888	-135.544	72.192	-3.771
1100	92.455	363.541	306.237	63.035	-134.829	92.932	-4.413
1200	94.172	371.661	311.354	72.368	-134.032	113.603	-4.945
1300	95.636	379.258	316.289	81.861	-133.172	134.204	-5.392
1400	96.895	386.393	321.044	91.489	-132.263	154.738	-5.773
1500	97.984	393.116	325.627	101.234	-131.318	175.205	-6.101
1600	98.931	399.471	330.045	111.081	-130.347	195.608	-6.386
1700	99.758	405.494	334.308	121.016	-129.358	215.950	-6.635
1800	100.485	411.217	338.423	131.029	-128.361	236.233	-6.855
1900	101.125	416.667	342.399	141.110	-127.360	256.461	-7.051
2000	101.692	421.869	346.243	151.252	-126.362	276.637	-7.225
2100	102.196	426.843	349.964	161.447	-125.370	296.762	-7.382
2200	102.645	431.608	353.567	171.689	-124.389	316.842	-7.523
2300	103.047	436.179	357.060	181.974	-123.423	336.875	-7.651
2400	103.407	440.573	360.449	192.297	-122.472	356.867	-7.767
2500	103.732	444.801	363.739	202.654	-121.541	376.821	-7.873
2600	104.025	448.875	366.936	213.042	-120.629	396.737	-7.971
2700	104.291	452.806	370.044	223.458	-119.739	416.619	-8.060
2800	104.532	456.603	373.068	233.900	-118.871	436.469	-8.142
2900	104.751	460.275	376.012	244.364	-118.026	456.286	-8.219
3000	104.952	463.830	378.880	254.849	-117.205	476.076	-8.289
3100	105.135	467.274	381.676	265.354	-116.407	495.840	-8.355
3200	105.303	470.615	384.404	275.876	-115.633	515.576	-8.416
3300	105.457	473.858	387.066	286.414	-114.882	535.290	-8.473
3400	105.599	477.008	389.665	296.967	-114.155	554.982	-8.526
3500	105.731	480.071	392.204	307.534	-113.452	574.651	-8.576
3600	105.852	483.051	394.687	318.113	-112.771	594.302	-8.623
3700	105.964	485.953	397.114	328.704	-112.114	613.933	-8.667
3800	106.069	488.780	399.489	339.305	-111.479	633.549	-8.709
3900	106.165	491.537	401.814	349.917	-110.866	653.147	-8.748
4000	106.256	494.226	404.091	360.538	-110.276	672.729	-8.785
4100	106.340	496.851	406.322	371.168	-109.708	692.297	-8.820
4200	106.418	499.414	408.508	381.806	-109.162	711.851	-8.853
4300	106.492	501.919	410.651	392.451	-108.638	731.394	-8.885
4400	106.560	504.368	412.753	403.104	-108.136	750.923	-8.915
4500	106.625	506.763	414.816	413.763	-107.657	770.441	-8.943
4600	106.685	509.108	416.840	424.429	-107.199	789.948	-8.970
4700	106.742	511.403	418.828	435.100	-106.765	809.447	-8.996
4800	106.796	513.650	420.780	445.777	-106.354	828.937	-9.021
4900	106.846	515.853	422.698	456.459	-105.966	848.419	-9.044
5000	106.894	518.012	424.583	467.146	-105.603	867.893	-9.067
5100	106.938	520.129	426.436	477.838	-105.264	887.360	-9.088
5200	106.981	522.206	428.257	488.534	-104.951	906.818	-9.109
5300	107.021	524.244	430.049	499.234	-104.665	926.274	-9.129
5400	107.059	526.245	431.812	509.938	-104.406	945.722	-9.148
5500	107.095	528.210	433.547	520.646	-104.174	965.166	-9.166
5600	107.129	530.140	435.255	531.357	-103.973	984.608	-9.184
5700	107.162	532.036	436.936	542.072	-103.801	1004.045	-9.201
5800	107.193	533.900	438.592	552.789	-103.662	1023.481	-9.217
5900	107.222	535.733	440.223	563.510	-103.555	1042.911	-9.233
6000	107.250	537.535	441.830	574.234	-103.482	1062.343	-9.249

PREVIOUS: June 1963 (1 atm) CURRENT: June 1963 (1 bar)

Nitric Acid ($HONO_2$) $H_1N_1O_3$(g)

$\Delta_f H°(0\ K) = -52.021 \pm 0.08\ kJ\ mol^{-1}$

$S°(298.15\ K) = 40.016\ J\ K^{-1}\ mol^{-1}$ $\Delta_f H°(298.15\ K) = -56.442 \pm 0.08\ kJ\ mol^{-1}$

Enthalpy of Formation

$\Delta_f H°$(298.15 K) was obtained from Gunn and Green (1). The value of $\Delta_f H°$(298.15 K) reported by Messer et al. was -13.60±0.27 kcal mol^{-1} which is in excellent agreement with the value given by Gunn and Green (1). The results of the previous investigations were also reviewed by Messer, et al. (2).

Heat Capacity and Entropy

$C_p°$ below 350 K were measured by Westrum, et al. (3). $C_p°$ above 350 K were estimated by comparison with those for LiH(cr). $C_p°$ (59.94 ~ 91.84 K) were also measured by Sayre and Beaver (4). S°(T) at 100, 200, and 298.15 K were taken from Westrum et al. (3) using the extrapolated value S°(10 K) = 0.008 cal K^{-1} mol^{-1}.

Decomposition Data

T_{dcm} was estimated from the value of $\Delta_f G°$ in the table. The table of T_{dcm} given by Banus et al. (5) was 940 K. It was assumed that NaH would behave similarly to LiH and that decomposition pressures would be purity dependent, and this would produce apparently high decomposition temperatures.

References

1. S. R. Gunn and L. G. Green, J. Amer. Chem. Soc. 80, 4782 (1958).
2. C. E. Messer, L. G. Fasolino and C. E. Thalmayer, J. Amer. Chem. Soc. 77, 4524 (1955).
3. E. F. Westrum, et al., University of Michigan, Ann Arbor, Michigan, personal communication, May 19, 1960.
4. E. V. Sayre and J. J. Beaver, J. Chem. Phys. 18, 584 (1950).
5. D. M. Banus, J. J. McSharry and E. A. Sullivan, J. Amer. Chem. Soc. 77, 2007 (1955).

Enthalpy Reference Temperature = T_r = 298.15 K Standard State Pressure = p° = 0.1 MPa

T/K	$C_p°$ (J K⁻¹mol⁻¹)	S° (J K⁻¹mol⁻¹)	$-[G°-H°(T_r)]/T$ (J K⁻¹mol⁻¹)	$H°-H°(T_r)$ (kJ mol⁻¹)	$\Delta_f H°$ (kJ mol⁻¹)	$\Delta_f G°$ (kJ mol⁻¹)	Log K_f
0	0.	0.	INFINITE	-6.259	-52.021	-52.021	INFINITE
100	17.497	11.560	66.580	-5.502	-54.103	-47.841	24.990
200	28.234	27.125	43.108	-3.197	-55.600	-40.945	10.694
298.15	36.392	40.016	40.016	0.	-56.442	-33.551	5.878
300	36.535	40.241	40.016	0.067	-56.454	-33.408	5.817
400	42.468	51.636	41.530	4.043	-59.553	-25.444	3.323
500	47.154	61.637	44.571	8.533	-59.625	-16.897	1.765
600	50.710	70.557	48.172	13.431	-59.208	-8.384	0.730
700	53.723	78.609	51.955	18.658	-58.402	0.028	-0.002
800	56.066	85.940	55.751	24.151	-57.294	8.303	-0.542
900	57.948	92.657	59.484	29.855	-55.965	16.425	-0.953
1000	59.413	98.840	63.115	35.725	-54.484	24.390	-1.274
1100	60.668	104.563	66.626	41.731	-52.906	32.201	-1.529
1200	61.641	109.885	70.012	47.848	-148.038	42.307	-1.842
1300	62.425	114.851	73.272	54.053	-145.472	58.065	-2.333
1400	63.021	119.501	76.410	60.327	-142.859	73.624	-2.747
1500	63.429	123.864	79.430	66.651	-140.218	88.995	-3.099

PREVIOUS: CURRENT: March 1963

SODIUM HYDRIDE (NaH) IDEAL GAS M_r = 23.99771

Sodium Hydride (NaH) $H_1Na_1(g)$

$S°(298.15 K) = 188.387 J K^{-1} mol^{-1}$

$\Delta_f H°(0 K) = 126.21 \pm 19.2 kJ mol^{-1}$
$\Delta_f H°(298.15 K) = 124.26 \pm 19.2 kJ mol^{-1}$

Electronic Level and Quantum Weight

State	ε_i, cm^{-1}	g_i
$^1\Sigma^+$	0	1

$\omega_e = 1172.2 cm^{-1}$ $\omega_e x_e = 19.72 cm^{-1}$ $\sigma = 1$
$B_e = 4.9012 cm^{-1}$ $\alpha_e = 0.1353 cm^{-1}$

Enthalpy of Formation

$\Delta_f H°(298.15 K)$ is calculated from $D_0^0 = 2.05 \pm 0.2$ eV reported by Gaydon (1).

Heat Capacity and Entropy

All molecular and spectroscopic constants were obtained from Herzberg (2).

References

1. A. G. Gaydon, "Dissociation Energies and Spectra of Diatomic Molecules," Chapman and Hall Ltd., London, 1953.
2. G. Herzberg, "Spectra of Diatomic Molecules," D. Van Nostrand, Inc., New York, 1950.

Enthalpy Reference Temperature = T_r = 298.15 K Standard State Pressure = p° = 0.1 MPa

T/K	$C_p°$ (J K⁻¹mol⁻¹)	S° (J K⁻¹mol⁻¹)	$-[G°-H°(T_r)]/T$ (J K⁻¹mol⁻¹)	$H°-H°(T_r)$ (kJ mol⁻¹)	$\Delta_f H°$ (kJ mol⁻¹)	$\Delta_f G°$ (kJ mol⁻¹)	Log K_f
0	0.	0.	INFINITE	-8.731	126.214	126.214	INFINITE
100	29.136	156.294	214.684	-5.839	126.267	118.055	-61.666
200	29.333	176.522	191.121	-2.920	125.384	110.160	-28.771
250	29.724	183.105	188.882	-1.444	124.836	106.416	-22.234
298.15	30.293	188.387	188.387	0.	124.265	102.919	-18.031
300	30.318	188.575	188.388	0.056	124.242	102.787	-17.897
350	31.023	193.300	188.759	1.589	123.598	99.261	-14.814
400	31.753	197.490	189.594	3.159	120.270	96.037	-12.541
450	32.453	201.271	190.684	4.764	119.583	93.050	-10.801
500	33.098	204.725	191.918	6.403	118.952	90.136	-9.416
600	34.191	210.860	194.577	9.770	117.838	84.481	-7.355
700	35.047	216.197	197.292	13.233	116.880	78.999	-5.895
800	35.716	220.923	199.956	16.773	116.035	73.646	-4.809
900	36.249	225.161	202.525	20.372	115.259	68.395	-3.970
1000	36.681	229.003	204.984	24.019	114.517	63.227	-3.303
1100	37.039	232.517	207.330	27.706	113.776	58.134	-2.761
1200	37.344	235.753	209.565	31.425	16.246	55.549	-2.418
1300	37.608	238.753	211.696	35.173	16.355	58.820	-2.363
1400	37.840	241.549	213.730	38.946	16.467	62.083	-2.316
1500	38.049	244.166	215.673	42.741	16.579	65.337	-2.275
1600	38.239	246.628	217.531	46.555	16.689	68.584	-2.239
1700	38.414	248.952	219.312	50.388	16.796	71.824	-2.207
1800	38.577	251.152	221.020	54.237	16.900	75.058	-2.178
1900	38.730	253.242	222.662	58.103	16.999	78.286	-2.152
2000	38.876	255.232	224.241	61.983	17.094	81.510	-2.129
2100	39.015	257.132	225.762	65.878	17.185	84.728	-2.107
2200	39.148	258.950	227.230	69.786	17.272	87.942	-2.088
2300	39.277	260.694	228.647	73.707	17.353	91.153	-2.070
2400	39.402	262.368	230.017	77.641	17.430	94.360	-2.054
2500	39.524	263.979	231.344	81.588	17.500	97.564	-2.038
2600	39.644	265.531	232.629	85.546	17.565	100.765	-2.024
2700	39.760	267.030	233.876	89.516	17.623	103.964	-2.011
2800	39.875	268.478	235.086	93.498	17.674	107.161	-1.999
2900	39.988	269.879	236.261	97.491	17.715	110.356	-1.988
3000	40.099	271.237	237.405	101.496	17.747	113.550	-1.977
3100	40.209	272.553	238.517	105.511	17.768	116.743	-1.967
3200	40.318	273 831	239.601	109.537	17.777	119.935	-1 958
3300	40.426	275.074	240.657	113.575	17.772	123.128	-1.949
3400	40.533	276.282	241.687	117.623	17.752	126.321	-1.941
3500	40.639	277.459	242.693	121.681	17.715	129.514	-1 933
3600	40.744	278.605	243.674	125.750	17.658	132.709	-1.926
3700	40.848	279.723	244.634	129.830	17.581	135.906	-1.919
3800	40.952	280.814	245.571	133.920	17.481	139.106	-1.912
3900	41.056	281.879	246.489	138.020	17.355	142.308	-1.906
4000	41.158	282.919	247.387	142.131	17.202	145.513	-1.900
4100	41.261	283.937	248.266	146.252	17.019	148.724	-1.895
4200	41.363	284.932	249.127	150.383	16.805	151.938	-1.890
4300	41.464	285.907	249.971	154.525	16.555	155.159	-1.885
4400	41.566	286.861	250.799	158.676	16.309	158.382	-1.880
4500	41 667	287.797	251.610	162.838	16.003	161.613	-1.876
4600	41.768	288.713	252.407	167.009	15.649	164.853	-1.872
4700	41.868	289.613	253.189	171.191	15.253	168.101	-1.868
4800	41.968	290.495	253.957	175.383	14.811	171.357	-1.865
4900	42.068	291.362	254.712	179.585	14.322	174.624	-1.862
5000	42.168	292.213	255.453	183.797	13.791	177.900	-1.859
5100	42.268	293.049	256.182	188.018	13.201	181.188	-1.856
5200	42.367	293.870	256.899	192.250	12.556	184.488	-1.853
5300	42.466	294.678	257.604	196.492	11.854	187.801	-1.851
5400	42.566	295.473	258.298	200.743	11.093	191.128	-1.849
5500	42.665	296.255	258.981	205.005	10.271	194.469	-1.847
5600	42.764	297.025	259.654	209.276	9.385	197.826	-1.845
5700	42.862	297.782	260.316	213.558	8.434	201.199	-1.844
5800	42.961	298.529	260.968	217.849	7.416	204.590	-1.843
5900	43.060	299.264	261.611	222.150	6.328	207.999	-1.841
6000	43.158	299.988	262.245	226.461	5.329	211.408	-1.840

PREVIOUS: March 1963 (1 atm) CURRENT: March 1963 (1 bar)

Sodium Hydride (NaH) $H_1Na_1(g)$

SODIUM HYDROXIDE (NaOH) CRYSTAL(ALPHA-BETA) M_r = 39.99711

$\Delta_f H°(0 K) = -421.40 \pm 0.4$ kJ mol^{-1}
$S°(298.15 K) = 64.445 \pm 0.84$ J K^{-1} mol^{-1} $\Delta_f H°(298.15 K) = -425.93 \pm 0.4$ kJ mol^{-1}
T_{trs} = 572 K $\Delta_{trs}H° = 7.196 \pm 0.84$ kJ mol^{-1}
T_{fus} = 596 K $\Delta_{fus}H° = 6.611 \pm 0.84$ kJ mol^{-1}

Enthalpy of Formation

The enthalpy of formation of NaOH(cr) is obtained from its enthalpy of solution in water, the enthalpy of hydrolysis of metallic sodium, and appropriate auxiliary data.

Reshetnikov (1) determined calorimetrically the enthalpy of solution of NaOH in aq, 400 H_2O as -10.445 ± 0.015 kcal mol^{-1}. Combining this result with enthalpy of dilution data for aqueous sodium hydroxide solutions tabulated by Parker (2), we derive $\Delta_{sln}H°$(NaOH, cr) = -10.56 ± 0.02 kcal mol^{-1}. Murch and Giauque (3) measured the heats of solution in water of a series of solids which contained various amounts of water in the range NaOH·(0.1 to 1) H_2O. Their data indicated a linear relationship between the heat of solution and the ratio of the moles of water to the moles of sodium hydroxide. Upon extrapolation their data gave $\Delta_{sln}H°$(NaOH, cr) = -10.637 ± 0.010 kcal mol^{-1} (2). This latter value is adopted here, since the extrapolation to zero moles of water tends to eliminate its effect on the heat of solution.

The enthalpy of hydrolysis of metallic sodium has been determined by various investigators (4-6). For the reaction Na(cr) + (n+1)H_2O → NaOH·n H_2O + 0.5 H_2(g) data are summarized below.

Source	n moles of H_2O	$\Delta_h H°$(298.15 K) kcal mol^{-1}	$\Delta_f H°$(NaOH·aq, ∞)* kcal mol^{-1}	$\Delta_f H°$(NaOH, cr, 298.15 K) kcal mol^{-1}
Gunn and Green (4a, 4b)	∞	-44.124±0.015	-112.439	-101.80
Messer et al. (5)	∞	-44.22 ±0.20	-112.53	-101.88
Ketchen and Wallace (6)	∞	-44.05 ±0.20	-112.37	-101.72

* Based on $\Delta_f H°(H_2O$, ℓ, 298.15 K) = -68.315 kcal mol^{-1} (7).

Combination of the $\Delta_f H°$(NaOH·aq, ∞, 298.15 K) values with the enthalpy of solution of NaOH(cr) at infinite dilution given above, results in the $\Delta_f H°$(NaOH, cr, 298.15 K) values given in the last column. The value determined from the work of Gunn and Green (4) is adopted.

Earlier determinations of $\Delta_f H°$(NaOH, cr, 298.15 K) have been summarized by Bichowsky and Rossini (8) and are not listed here.

Heat Capacity and Entropy

Murch and Giauque (3) determined low temperature heat capacities for NaOH·0.04014 H_2O and NaOH·0.97776 H_2O from 15 to 320 K. From these data heat capacities for anhydrous NaOH(cr) were calculated and joined smoothly at 298.15 K with the high temperature (273-973 K) enthalpy data of Douglas and Dever (9). These smoothed heat capacities given by Murch and Giauque are adopted. Low temperature heat capacity data for NaOH(cr) in the temperature range 60 to 300 K have also been reported by Kelley and Snyder (10). The deviations of their data from the adopted values range from a few tenths of a percent (150-300 K) to up to about 30% at the lowest temperatures investigated (60 K). Popov and Ginzburg (11) also determined high temperture enthalpy data (293-1016 K) for NaOH by drop calorimetry. Their reported value H°(500 K)-H°(298.15 K) = 3.112 kcal mol^{-1} (11) is in very good agreement with the tabulated value of 3.176 kcal mol^{-1}.

S°(298.15 K) is based upon the extrapolated value of 0.020 cal K^{-1} mol^{-1} at 15 K.

Transition and Fusion Data

The adopted enthalpies of the polymorphic transformation and fusion of NaOH are mean values from the work of Douglas and Dever (9), Popov and Ginzburg (11) and Reshetnikov and Baranskaya (12). T_{trs} and T_{fus} are from Reshetnikov and Baranskaya (12).

Sublimation Data

$\Delta_{sub}H°$(298.15 K) for the monomer and dimer are calculated from the adopted enthalpies of formation of the crystal and the respective gaseous species.

References

1. N. A. Reshetnikov, Zhur. Neorg. Khim. 6, 682 (1961).
2. V. B. Parker, NSRDS-NBS 2, 1965.
3. L. E. Murch and W. F. Giauque, J. Phys. Chem. 66, 2052 (1962).
4. a) S. R. Gunn and L. G. Green, J. Amer. Chem. Soc. 80, 4782 (1958); b) S. R. Gunn, J. Phys. Chem. 71, 1386 (1967).
5. C. E. Messer, L. G. Fasolino, and C. E. Thalmayer, J. Amer. Chem. Soc. 77, 4524 (1965).
6. E. E. Ketchen and W. E. Wallace, J. Amer. Chem. Soc. 76, 4736 (1954).
7. U. S. Nat. Bur. Stand. Tech. Note 270-3, 1968.
8. F. R. Bichowsky and F. D. Rossini, "The Thermochemistry of the Chemical Substances," Reinhold Publishing Corp., New York, 1936.
9. T. B. Douglas and J. L. Dever, J. Res. Nat. Bur. Stand. 53, 81 (1954).
10. a) J. C. R. Kelley and P. E. Snyder, J. Amer. Chem. Soc. 73, 4114 (1951); b) J. C. R. Kelley and P. E. Snyder, J. Amer. Chem. Soc. 73, 5933 (1951).
11. a) M. M. Popov and D. M. Ginzburg, Zhur. Obshchei Khim. 26, 971 (1956); b) D. M. Ginzburg, Zhur. Obshchei Khim. 26, 968 (1956).
12. N. A. Reshetnikov and E. V. Baranskaya, Izv. Vyssh. Ucheb. Zaved., Khim. Tekhnol. 10, 496 (1967); C.A. 67:94551.

Sodium Hydroxide (NaOH) $H_1Na_1O_1$(cr)

Enthalpy Reference Temperature = T_r = 298.15 K Standard State Pressure = p° = 0.1 MPa

T/K	$C_p°$ (J K^{-1}mol^{-1})	S°	-[G°-H°(T_r)]/T	H°-H°(T_r) (kJ mol^{-1})	$\Delta_f H°$	$\Delta_f G°$	Log K_f
0	0.	0.	INFINITE	-10.487	-421.396	-421.396	INFINITE
100	27.744	15.513	109.879	-9.437	-424.637	-410.105	214.217
200	49.580	42.588	69.615	-5.406	-425.864	-394.953	103.151
298.15	59.530	64.445	64.445	0.	-425.931	-379.741	66.529
300	59.664	64.814	64.447	0.110	-425.927	-379.454	66.069
400	64.936	82.739	66.856	6.353	-428.244	-363.802	47.508
500	75.157	98.172	71.595	13.288	-427.401	-347.767	36.331
572.000	85.887	108.971	75.620	19.077	I <--> II	TRANSITION	
572.000	86.023	121.574	75.620	26.286			
596.000	86.023	125.110	77.542	28.350	II <--> LIQUID		
600	86.023	125.685	77.861	28.694	-418.055	-332.373	28.936
700	86.023	138.946	85.665	37.297	-415.502	-318.294	23.751
800	86.023	150.433	93.059	45.899	-412.953	-304.581	19.887
900	86.023	160.565	100.008	54.501	-410.428	-291.187	16.900
1000	86.023	169.628	106.525	63.104	-407.946	-278.071	14.525
1100	86.023	177.827	112.640	71.706	-405.526	-265.202	12.593
1200	86.023	185.312	118.388	80.308	-499.948	-250.109	10.887
1300	86.023	192.198	123.805	88.911	-496.776	-229.418	9.218
1400	86.023	198.573	128.921	97.513	-493.641	-208.970	7.797
1500	86.023	204.508	133.764	106.115	-490.542	-188.745	6.573

PREVIOUS: March 1966 CURRENT: December 1970

Sodium Hydroxide (NaOH) $H_1Na_1O_1$(cr)

SODIUM HYDROXIDE (NaOH) LIQUID M_r = 39.99711

S°(298.15 K) = [75.891] J K^{-1} mol^{-1} $\Delta_f H°$(298.15 K) = [-416.878] kJ mol^{-1}

T_{fus} = 596 K $\Delta_{fus}H°$ = 6.611 ± 0.84 kJ mol^{-1}

Enthalpy of Formation

$\Delta_f H°$(NaOH, ℓ, 298.15 K) is calculated from that of the crystal by adding $\Delta_{fus}H°$ and the difference in enthalpy, H°(596 K) - H°(298.15 K), between the crystal and liquid.

Heat Capacity and Entropy

The adopted heat capacities for NaOH(ℓ) in the temperature range 596 to 1000 K are from the enthalpy measurements of Douglas and Dever (1). The heat capacities below the melting point and above 1000 K are extrapolated from the experimental heat capacity curve. Powers and Blalock (2) reported $C_p°$(ℓ) = 19.6 cal K^{-1} mol^{-1} from enthalpy measurements, in a short temperature range, by drop calorimetry. The smoothed enthalpy data for NaOH(ℓ) reported by Popov and Ginzburg (3a, 3b) are from 0.2 to 0.5 kcal mol^{-1} less than the adopted values in the temperature range 700 to 1000 K.

S°(298.15 K) is calculated in a manner analogous to that used for the enthalpy of formation.

Fusion Data

See the crystal table for details.

Vaporization Data

T_{vap} = 1830 K (for monomer) is calculated as the temperature at which the fugacity of NaOH(g) is 1 atm for the process NaOH(ℓ) = NaOH(g). The difference in the enthalpies of formation of NaOH(ℓ) and NaOH(g) at the T_{vap} is the enthalpy of vaporization, $\Delta_{vap}H°$ = 175.310 kJ mol^{-1}.

Wartenberg and Albrecht (4) reported a boiling point of 1661 K from their static vapor pressure data. However, these workers observed the presence of water after each experiment which may indicate some decomposition of the sample. See the ideal gas table for further discussion.

References

1. T. B. Douglas and J. L. Dever, J. Res. Nat. Bur. Stand. 53, 81 (1954).
2. W. D. Powers and G. C. Blalock, Oak Ridge Nat. Lab., ORNL 1653, 1954.
3. a. M. M. Popov and D. M. Ginzburg, Zhur. Obshchei Khim. 26, 971 (1956).
   b. D. M. Ginzburg, Zhur. Obshchei Khim. 26, 968 (1956).
4. H. von Wartenberg and P. Albrecht, Z. Elektrochem. 27, 162 (1921).

Sodium Hydroxide (NaOH) $H_1Na_1O_1(l)$

Enthalpy Reference Temperature = T_r = 298.15 K Standard State Pressure = p° = 0.1 MPa

T/K	$C_p°$ (J $K^{-1}mol^{-1}$)	S° (J $K^{-1}mol^{-1}$)	$-[G°-H°(T_r)]/T$ (J $K^{-1}mol^{-1}$)	$H°-H°(T_r)$ (kJ mol^{-1})	$\Delta_f H°$ (kJ mol^{-1})	$\Delta_f G°$ (kJ mol^{-1})	Log K_f
0							
100							
200							
298.15	87.822	75.891	75.891	0.	-416.878	-374.100	65.541
300	87.814	76.434	75.892	0.162	-416.821	-373.835	65.090
400	87.211	101.613	79.329	8.913	-416.631	-359.738	46.977
500	86.651	121.013	85.800	17.606	-414.030	-345.816	36.127
596.000	86.111	136.202	92.791	25.873	-------- II	<--> LIQUID	--------
600	86.065	136.778	93.024	26.252	-411.445	-332.417	28.939
700	85.479	150.001	100.245	34.829	-408.916	-319.446	23.837
800	84.893	161.377	107.192	43.348	-406.451	-306.834	20.034
900	84.308	171.342	113.777	51.808	-404.068	-294.526	17.094
1000	83.722	180.194	119.985	60.209	-401.787	-282.478	14.755
1100	83.136	188.146	125.826	68.552	-399.626	-270.653	12.852
1200	82.550	195.355	131.324	76.837	-494.366	-256.579	11.169
1300	81.965	201.939	136.507	85.062	-491.571	-236.877	9.518
1400	81.379	207.992	141.400	93.230	-488.871	-217.387	8.111
1500	80.793	213.587	146.028	101.338	-486.266	-198.087	6.898
1600	80.207	218.782	150.415	109.388	-483.754	-178.957	5.842
1700	79.622	223.627	154.580	117.379	-481.334	-159.982	4.916
1800	79.119	228.164	158.544	125.316	-479.001	-141.147	4.096
1900	78.638	232.428	162.321	133.203	-476.749	-122.439	3.366
2000	78.199	236.450	165.928	141.045	-474.574	-103.848	2.712
2100	77.801	240.256	169.378	148.844	-472.471	-85.363	2.123
2200	77.467	243.867	172.682	156.607	-470.434	-66.977	1.590
2300	77.182	247.304	175.853	164.339	-468.457	-48.683	1.106
2400	76.956	250.584	178.899	172.046	-466.535	-30.474	0.663
2500	76.760	253.722	181.829	179.731	-464.662	-12.343	0.258

PREVIOUS: March 1966 CURRENT: December 1970

Sodium Hydroxide (NaOH) $H_1Na_1O_1(l)$

SODIUM HYDROXIDE (NaOH) ALPHA-BETA-LIQUID M_r = 39.99711 Sodium Hydroxide (NaOH) $H_1Na_1O_1(cr,l)$

0 to 572 K crystal, alpha
572 to 596 K crystal, beta
above 596 K liquid

Refer to the individual tables for details.

Enthalpy Reference Temperature = T_r = 298.15 K Standard State Pressure = $p°$ = 0.1 MPa

	J K⁻¹mol⁻¹			kJ mol⁻¹			
T/K	$C_p^°$	S°	$-[G°-H°(T_r)]/T$	$H°-H°(T_r)$	$\Delta_f H°$	$\Delta_f G°$	Log K_f
0	0.	0.	INFINITE	-10.487	-421.396	-421.396	INFINITE
100	27.744	15.513	109.879	-9.437	-424.637	-410.105	214.217
200	49.580	42.588	69.615	-5.406	-425.864	-394.953	103.151
298.15	59.530	64.445	64.445	0.	-425.931	-379.741	66.529
300	59.664	64.814	64.447	0.110	-425.927	-379.454	66.069
400	64.936	82.739	66.856	6.353	-428.244	-363.802	47.508
500	75.157	98.172	71.595	13.288	-427.401	-347.767	36.331
572.000	85.887	108.971	75.620	19.077		I <--> II	
572.000	86.023	121.574	75.620	26.286		TRANSITION	
596.000	86.023	125.110	77.542	28.350		II <--> LIQUID	
596.000	86.111	136.202	77.542	34.961		TRANSITION	
600	86.065	136.778	77.935	35.305	-411.445	-332.417	28.939
700	85.479	150.001	87.311	43.883	-408.916	-319.446	23.837
800	84.893	161.377	95.875	52.401	-406.451	-306.834	20.034
900	84.308	171.342	103.718	60.861	-404.068	-294.526	17.094
1000	83.722	180.194	110.931	69.263	-401.787	-282.478	14.755
1100	83.136	188.146	117.596	77.606	-399.626	-270.653	12.852
1200	82.550	195.355	123.780	85.890	-494.366	-256.579	11.169
1300	81.965	201.939	129.543	94.116	-491.571	-236.877	9.518
1400	81.379	207.992	134.933	102.283	-488.871	-217.387	8.111
1500	80.793	213.587	139.992	110.392	-486.266	-198.087	6.898
1600	80.207	218.782	144.756	118.442	-483.754	-178.957	5.842
1700	79.622	223.627	149.255	126.432	-481.334	-159.982	4.916
1800	79.119	228.164	153.514	134.369	-479.001	-141.147	4.096
1900	78.638	232.428	157.556	142.257	-476.749	-122.439	3.366
2000	78.199	236.450	161.401	150.098	-474.574	-103.848	2.712
2100	77.801	240.256	165.067	157.898	-472.471	-85.363	2.123
2200	77.467	243.867	168.567	165.661	-470.434	-66.977	1.590
2300	77.182	247.304	171.916	173.393	-468.457	-48.683	1.106
2400	76.956	250.584	175.126	181.099	-466.535	-30.474	0.663
2500	76.760	253.722	178.208	188.785	-464.662	-12.343	0.258

PREVIOUS: CURRENT: December 1970

Sodium Hydroxide (NaOH) $H_1Na_1O_1(cr,l)$

SODIUM HYDROXIDE (NaOH) IDEAL GAS M_r = 39.99711

$\Delta_f H°(0\ K) = -194.12 \pm 12.6\ kJ\ mol^{-1}$

$S°(298.15\ K) = 228.443 \pm 0.42\ J\ K^{-1}\ mol^{-1}$ $\Delta_f H°(298.15\ K) = -197.76 \pm 12.6\ kJ\ mol^{-1}$

Vibrational Frequencies and Degeneracies

ν, cm^{-1}
431 (1)
337 (2)
[3650](1)

Ground State Quantum Weight: 1 $\sigma = 1$

Point Group: $C_{\infty v}$

Bond Distances: Na-O = [1.93] Å O-H = [0.97] Å

Bond Angle: Na-O-H = 180°

Rotational Constant: $B_0 = 0.426703\ cm^{-1}$

Enthalpy of Formation

The bond dissociation energies of the alkali metal hydroxides have been the subject of a number of investigations (1-4) in recent years. The majority of this work has involved equilibrium studies on the reaction $A(g) + H_2O(g) = AOH(g) + H(g)$, where A is an alkali metal, in hydrogen-oxygen-nitrogen flames containing water. In the case of NaOH, data are summarized below:

Source	Method	Reaction*	Data Points	T/K	$\Delta_r H°(298.15\ K)$, kcal mol^{-1} 3rd Law	D°(Na-OH, 0 K) kcal mol^{-1}
Jensen and Padley (2a)	Flames	A	1	2475	40.5	77.9
Cotton and Jenkins (4)	Flames	A	1	2030	36.6	81.8

*(A) $Na(g) + H_2O(g) = NaOH(g) + H(g)$

We note that Cotton and Jenkins (4) in the same paper reported bond dissociation energies for LiOH(g) and KOH(g) which are in reasonable agreement with JANAF values (5) for the enthalpy of formation of these two compounds. Furthermore, bond dissociation data for the alkali metal halides (6) clearly establish the sodium compound as the least stable within each halide series. The potassium and sodium compounds differ by from 4.3 kcal mol^{-1} for the fluorides and bromides to 3.6 kcal mol^{-1} for the chlorides. It seems most likely that this same trend in the bond dissociation energies would apply to the hydroxides as well. Based upon these correlations, we adopt $D_0^0(Na\text{-}OH) = 81.5 \pm 3.0$ kcal mol^{-1} with $D_0^0(K\text{-}OH) - D_0^0(Na\text{-}OH) = 3.9$ kcal mol^{-1}. This value corresponds to $\Delta_f H°(NaOH, g, 298.15\ K) = -47.3 \pm 3.0$ kcal mol^{-1}.

Wartenberg and Albrecht (7) determined vapor pressures for NaOH(ℓ) in the temperature range 1283-1681 K by a static method. In order to evaluate $\Delta_{vap}H°(NaOH, ℓ, 298.15\ K)$, we have used a trial and error variation of $\Delta_{vap}H°(298.15\ K)$ for the monomer and dimer, such that these values are in accordance with the adopted enthalpy of dimerization (see $Na_2(OH)_2(g)$ table), and the sum of the calculated partial pressures for NaOH(g) and $Na_2(OH)_2(g)$ in good agreement with the experimental vapor pressure data. The resulting values are: $\Delta_{vap}H°$(to monomer) = 48.25 kcal mol^{-1} and $\Delta_{vap}H°$(to dimer) = 48.8 kcal mol^{-1}. From the enthalpy of vaporization of the monomer, we derive $\Delta_f H°(NaOH, g, 298.15\ K) = -51.4$ kcal mol^{-1}. This value corresponds to a bond dissociation energy for NaOH(g) that is slightly greater than the adopted value for KOH(g) (5), and thus not in line with the established trends in D_0^0 discussed above. Buchler and Berkowitz-Mattuck (8) in their review on gaseous ternary compounds of the alkali metals point to the decomposition of NaOH in the condensed phase with a resulting change in composition as the major problem in the interpretation of thermodynamic measurements on this system. Until the nature of the condensed phase and the relationship between the vapor and condensed phase composition for NaOH(ℓ) are established, an accurate enthalpy of vaporization will remain undetermined.

Heat Capacity and Entropy

The infrared spectra of matrix-isolated NaOH and NaOD have been recently observed by Acquista and Abramowitz (9). The alkali-metal-oxygen stretching frequency, ν_1, and the bending mode, ν_2, for NaOH were assigned as 431 and 337 cm^{-1}, respectively, by these workers. The isotope shift of ν_2 was consistent with a linear structure for NaOH having bond lengths of 1.93 and 0.97 Å for the Na-O and O-H bonds. We have adopted a linear configuration for NaOH with the bond lengths and frequencies given above. Spinar and Margrave (10) assigned ν_1 as 437 ± 10 cm^{-1} for NaOH which is in good agreement with the adopted value. The O-H stretching frequency, ν_3, was estimated by Jensen (2b).

References

1. H. Smith and T. M. Sugden, Proc. Roy. Soc. A 219, 204 (1953).
2. a) D. E. Jensen and P. J. Padley, Trans. Faraday Soc. 62, 2132 (1966); b) D. E. Jensen, J. Phys. Chem. 74, 207 (1970).
3. M. J. McEwan and L. F. Phillips, Combustion and Flame 11, 63 (1967).
4. D. H. Cotton and D. R. Jenkins, Trans. Faraday Soc. 65, 1537 (1969).
5. JANAF Thermochemical Tables: LiOH(g), 3-31-66; KOH(g), 12-31-70.
6. JANAF DATA; See recent review of the fluorides by J. Chao, Thermochimica Acta 1, 71 (1970).
7. H. von Wartenberg and P. Albrecht, Z. Elektrochem. 27, 162 (1921).
8. A. Buchler and J. B. Berkowitz-Mattuck in "Advances in High Temperature Chemistry," Volume I, Ed. by L. Eyring, Academic Press, New York, 1967.
9. N. Acquista and S. Abramowitz, J. Chem. Phys. 51, 2911 (1969).
10. L. H. Spinar and J. L. Margrave, Spectrochimia Acta 12, 244 (1958).

Sodium Hydroxide (NaOH) $H_1Na_1O_1(g)$

Enthalpy Reference Temperature = T_r = 298.15 K Standard State Pressure = p° = 0.1 MPa

T/K	$C_p°$ (J $K^{-1}mol^{-1}$)	S° (J $K^{-1}mol^{-1}$)	$-[G°-H°(T_r)]/T$ (J $K^{-1}mol^{-1}$)	$H°-H°(T_r)$ (kJ mol^{-1})	$\Delta_f H°$ (kJ mol^{-1})	$\Delta_f G°$ (kJ mol^{-1})	Log K_f
0	0.	0.	INFINITE	-11.385	-194.119	-194.119	INFINITE
100	32.864	183.702	267.723	-8.402	-195.428	-197.715	103.276
200	43.462	210.040	232.739	-4.540	-196.824	-199.403	52.079
250	46.468	220.085	229.230	-2.286	-197.322	-199.989	41.785
298.15	48.371	228.443	228.443	0.	-197.757	-200.463	35.120
300	48.431	228.743	228.444	0.090	-197.773	-200.479	34.907
350	49.750	236.314	229.039	2.546	-198.232	-200.895	29.982
400	50.669	243.021	230.376	5.058	-201.365	-201.036	26.253
450	51.334	249.029	232.120	7.609	-201.866	-200.964	23.327
500	51.836	254.465	234.087	10.189	-202.326	-200.839	20.981
600	52.566	263.984	238.299	15.411	-203.165	-200.461	17.452
700	53.130	272.131	242.564	20.696	-203.928	-199.949	14.920
800	53.646	279.259	246.715	26.035	-204.642	-199.332	13.015
900	54.158	285.607	250.690	31.426	-205.329	-198.626	11.528
1000	54.674	291.340	254.473	36.867	-206.008	-197.845	10.334
1100	55.188	296.575	258.066	42.360	-206.697	-196.996	9.355
1200	55.690	301.399	261.478	47.904	-304.177	-193.642	8.429
1300	56.171	305.875	264.723	53.498	-304.014	-184.438	7.411
1400	56.626	310.055	267.814	59.138	-303.842	-175.246	6.539
1500	57.051	313.976	270.762	64.822	-303.661	-166.067	5.783
1600	57.444	317.671	273.579	70.547	-303.474	-156.900	5.122
1700	57.806	321.165	276.277	76.310	-303.282	-147.745	4.540
1800	58.138	324.478	278.863	82.107	-303.089	-138.601	4.022
1900	58.441	327.630	281.348	87.936	-302.895	-129.468	3.559
2000	58.718	330.635	283.738	93.795	-302.703	-120.346	3.143
2100	58.971	333.506	286.040	99.679	-302.515	-111.232	2.767
2200	59.201	336.255	288.260	105.588	-302.332	-102.128	2.425
2300	59.411	338.891	290.405	111.519	-302.157	-93.032	2.113
2400	59.603	341.424	292.478	117.470	-301.990	-83.943	1.827
2500	59.779	343.860	294.485	123.439	-301.834	-74.861	1.564
2600	59.939	346.208	296.429	129.425	-301.689	-65.785	1.322
2700	60.086	348.473	298.315	135.426	-301.559	-56.714	1.097
2800	60.221	350.661	300.146	141.442	-301.444	-47.648	0.889
2900	60.345	352.776	301.924	147.470	-301.346	-38.586	0.695
3000	60.459	354.824	303.654	153.510	-301.266	-29.526	0.514
3100	60.564	356.808	305.336	159.561	-301.207	-20.469	0.345
3200	60.661	358.732	306.975	165.623	-301.171	-11.414	0.186
3300	60.751	360.600	308.572	171.693	-301.157	-2.360	0.037
3400	60.834	362.415	310.129	177.773	-301.170	6.695	-0.103
3500	60.911	364.180	311.648	183.860	-301.211	15.750	-0.235
3600	60.983	365.897	313.131	189.955	-301.281	24.807	-0.360
3700	61.049	367.568	314.580	196.056	-301.383	33.867	-0.478
3800	61.111	369.197	315.996	202.164	-301.520	42.930	-0.590
3900	61.169	370.785	317.381	208.278	-301.693	51.996	-0.696
4000	61.223	372.335	318.735	214.398	-301.905	61.068	-0.797
4100	61.274	373.847	320.061	220.523	-302.159	70.145	-0.894
4200	61.321	375.324	321.359	226.653	-302.454	79.229	-0.985
4300	61.365	376.768	322.631	232.787	-302.798	88.321	-1.073
4400	61.407	378.179	323.878	238.926	-303.150	97.417	-1.156
4500	61.446	379.559	325.100	245.068	-303.574	106.525	-1.237
4600	61.483	380.910	326.298	251.215	-304.058	115.643	-1.313
4700	61.518	382.233	327.475	257.365	-304.597	124.773	-1.387
4800	61.550	383.529	328.629	263.518	-305.195	133.915	-1.457
4900	61.581	384.798	329.762	269.675	-305.853	143.070	-1.525
5000	61.610	386.042	330.875	275.834	-306.567	152.238	-1.590
5100	61.638	387.263	331.969	281.997	-307.354	161.422	-1.653
5200	61.664	388.460	333.044	288.162	-308.209	170.622	-1.714
5300	61.689	389.635	334.101	294.330	-309.137	179.839	-1.772
5400	61.712	390.788	335.140	300.500	-310.138	189.074	-1.829
5500	61.735	391.921	336.162	306.672	-311.217	198.328	-1.884
5600	61.756	393.033	337.168	312.847	-312.375	207.604	-1.936
5700	61.776	394.126	338.157	319.023	-313.616	216.900	-1.988
5800	61.795	395.201	339.132	325.202	-314.940	226.219	-2.037
5900	61.813	396.257	340.091	331.382	-316.352	235.560	-2.085
6000	61.831	397.296	341.036	337.564	-317.694	244.909	-2.132

PREVIOUS: December 1970 (1 atm) CURRENT: December 1970 (1 bar)

Sodium Hydroxide (NaOH) $H_1Na_1O_1(g)$

$S°(298.15\ K) = [242.626 \pm 8.4]\ J\ K^{-1}\ mol^{-1}$

$\Delta_fH°(0\ K) = [674.9 \pm 105]\ kJ\ mol^{-1}$

$\Delta_fH°(298.15\ K) = [677.8 \pm 105]\ kJ\ mol^{-1}$

Electronic Levels and Quantum Weights

State	ε_i, cm^{-1}	g_i
$^2\Pi$	0	4
$^2\Sigma^+$	[32000]	2

Vibrational Frequencies and Degeneracies

ν, cm^{-1}
[400](1)
[300](2)
[3600](1)

Point Group: $[C_{\infty v}]$ $\sigma = 1$

Bond Distance: Na-O = [2.00] Å O-H = [1.03] Å

Bond Angle: Na-O-H = [180]°

Rotational Constant: $B_0 = [0.396362]\ cm^{-1}$

Enthalpy of Formation

Beckett and Cassidy (1) selected the ionization potential of NaOH as 9.0 ± 1.0 eV (207.6 ± 23 kcal mol^{-1}). A measured value of roughly 9 eV has been reported by Schoonmaker and Porter (2). Combining this result with JANAF auxiliary data (3), we derive $\Delta_fH°(NaOH^+, g, 0\ K) = 161 \pm 25\ kcal\ mol^{-1}$, which is tentatively adopted.

Heat Capacity and Entropy

$NaOH^+$ has seven valence electrons. According to the correlation diagram of Walsh (4) for HAB type molecules, $NaOH^+$ should be linear with the unpaired electron in a pi orbital. Therefore, we assume the ground state to be $^2\Pi$. A first excited state is estimated at 32000 cm^{-1} to be $^2\Sigma$ by analogy with the isoelectronic molecules OH (3) and SH (3). The removal of a bonding electron from NaOH to form $NaOH^+$ should lead to weaker bonding in the positive ion. Consequently, the vibrational frequencies are estimated to be somewhat lower than those for NaOH. Similarly, the Na-O and O-H bond distances are increased slightly over those for NaOH.

References

1. C. W. Beckett and E. C. Cassidy, U. S. Nat. Bur. Stand. Report 8628, 1965.
2. R. C. Schoonmaker and R. F. Porter, J. Chem. Phys. 28, 454 (1958).
3. JANAF Thermochemical Tables: NaOH(g), 12-31-70; OH(g), 12-31-70; SH(g), 6-30-67.
4. A. D. Walsh, J. Chem. Soc. 2288 (1953).

Enthalpy Reference Temperature = T_r = 298.15 K Standard State Pressure = p° = 0.1 MPa

T/K	$C_p°$ (J $K^{-1}mol^{-1}$)	S° (J $K^{-1}mol^{-1}$)	$-[G°-H°(T_r)]/T$ (J $K^{-1}mol^{-1}$)	$H°-H°(T_r)$ (kJ mol^{-1})	$\Delta_fH°$ (kJ mol^{-1})	$\Delta_fG°$ (kJ mol^{-1})	Log K_f
0	0.	0.	INFINITE	-11.693	674.940		
100	34.226	196.325	283.048	-8.672			
200	44.889	223.761	247.019	-4.652			
250	47.591	234.091	243.429	-2.334			
298.15	49.257	242.626	242.626	0.	677.808	664.619	-116.438
300	49.308	242.931	242.627	0.091	677.832	664.537	-115.706
350	50.443	250.623	243.232	2.587	678.451	662.271	-98.839
400	51.226	257.413	244.589	5.130	676.389	660.126	-86.204
450	51.790	263.481	246.357	7.706	676.953	658.060	-76.386
500	52.216	268.960	248.348	10.306	677 552	655.929	-68.524
600	52.846	278.539	252.604	15.561	678.825	651.486	-56.717
700	53.352	286.724	256.908	20.872	680.165	646.825	-48.267
800	53.836	293.880	261.091	26.231	681.550	641.968	-41.916
900	54.328	300.249	265.095	31.639	682.959	636.935	-36.967
1000	54.832	305.999	268.902	37.097	684.375	631.746	-32.999
1100	55.337	311.249	272.517	42.606	685.780	626.415	-29.746
1200	55.832	316.085	275.949	48.164	590.393	623.397	-27.136
1300	56.307	320.573	279.211	53.771	592.648	626.056	-25.155
1400	56.755	324.763	282.316	59.425	594.913	628.541	-23.451
1500	57.174	328.693	285.279	65.121	597.185	630.863	-21.969
1600	57.561	332.395	288.109	70 858	599.463	633.034	-20.666
1700	57.918	335.896	290.818	76.633	601.744	635.062	-19.513
1800	58.244	339.216	293.415	82.441	604.027	636.956	-18.484
1900	58.541	342.373	295.909	88.280	606.310	638.723	-17.560
2000	58.813	345.383	298.308	94.148	608.590	640.371	-16.725
2100	59.060	348.258	300.619	100.042	610.866	641.904	-15.966
2200	59.286	351.011	302.847	105.960	613.136	643.329	-15.275
2300	59.491	353.651	304.999	111.899	615.399	644.650	-14.640
2400	59.679	356.187	307.080	117.857	617.652	645.873	-14.057
2500	59.850	358.627	309.093	123.834	619.894	647.003	-13.518
2600	60.007	360.977	311.044	129.827	622.124	648.043	-13.019
2700	60.150	363.244	312.935	135.835	624.339	648.998	-12.556
2800	60.282	365.434	314.771	141.856	626.539	649.872	-12.124
2900	60.402	367.552	316.555	147.891	628.722	650.666	-11.720
3000	60.514	369.602	318.289	153.937	630.886	651.386	-11.342
3100	60.616	371.587	319.977	159.993	633.029	652.034	-10.987
3200	60.711	373.513	321.620	166.060	635.149	652.613	-10.653
3300	60.798	375.383	323.221	172.135	637.246	653.126	-10.338
3400	60.880	377.199	324.782	178.219	639.316	653.576	-10.041
3500	60.955	378.965	326.305	184.311	641.359	653.966	-9.760
3600	61.025	380.683	327.792	190.410	643.371	654.297	-9.494
3700	61.091	382.356	329.244	196.516	645.352	654.573	-9.241
3800	61.152	383.986	330.663	202.628	647.298	654.797	-9.001
3900	61.209	385.575	332.051	208.746	649.208	654.970	-8.772
4000	61.263	387.126	333.408	214.870	651.078	655.093	-8.555
4100	61.313	388 639	334.737	220.998	652.907	655.171	-8.347
4200	61.361	390.117	336.038	227.132	654.694	655.204	-8.149
4300	61.406	391.562	337.313	233.270	656.432	655.196	-7.959
4400	61.449	392.974	338.562	239.413	658.163	655.144	-7.778
4500	61.489	394.355	339.786	245.560	659.823	655.055	-7.604
4600	61.528	395.707	340.987	251.711	661.421	654.931	-7.437
4700	61.565	397.031	342.166	257.866	662.965	654.774	-7.277
4800	61.601	398.327	343.322	264.024	664.451	654.584	-7.123
4900	61.635	399.598	344.458	270.186	665.877	654.364	-6.976
5000	61.669	400.843	345.573	276.351	667.247	654.114	-6.833
5100	61.701	402.065	346.669	282.520	668.545	653.838	-6.697
5200	61.732	403.263	347.746	288.691	669.775	653.538	-6.565
5300	61.762	404.439	348.804	294.866	670.933	653.215	-6.438
5400	61.792	405.594	349.845	301.044	672.018	652.870	-6.315
5500	61.821	406.728	350.869	307.224	673.026	652.505	-6.197
5600	61.850	407 842	351.877	313.408	673.955	652.124	-6.083
5700	61.878	408.937	352.868	319.594	674.803	651.727	-5.972
5800	61.906	410.014	353.844	325.784	675.568	651.316	-5.866
5900	61.933	411.072	354.805	331.975	676.246	650.890	-5.763
6000	61.961	412.113	355.752	338.170	676.996	650.438	-5.663

PREVIOUS: December 1971 (1 atm) CURRENT: December 1971 (1 bar)

HYDROXYL (OH) IDEAL GAS $M_r = 17.00734$

$S°(298.15\ K) = 183.708 \pm 0.04\ J\ K^{-1}\ mol^{-1}$

$\Delta_f H°(0\ K) = 38.390 \pm 1.21\ kJ\ mol^{-1}$
$\Delta_f H°(298.15\ K) = 38.987 \pm 1.21\ kJ\ mol^{-1}$

Electronic States and Molecular Constants (σ = 1)

State	ε_i, cm^{-1}	g_i	ω_e, cm^{-1}	$\omega_e x_e$, cm^{-1}	B_e, cm^{-1}	α_e, cm^{-1}	r_e, Å
$X^2\Pi_i$	0	2	3735.21	82.81	18.871	0.714	0.9706
	139.7	2					
$A^2\Sigma^+$	32403	2	3184.28	97.84	17.355	0.807	1.0121
$B^2\Sigma^+$	68372	2	940	105	5.54	0.65	1.80
$C^2\Sigma^+$	89420	2	2339	32	4.20	0.16	2.16

Enthalpy of Formation

$\Delta_f H°(OH, g, 0\ K)$ is calculated from the relation $\Delta_f H°(OH, g, 0\ K) = 1/2\ D_0^0(O_2) + 1/2\ D_0^0(H_2) = D_0^0(OH)$. The values employed are as follows: $D_0^0(O_2) = 117.967 \pm 0.042$ kcal mol^{-1} and $D_0^0(H_2) = 103.267 \pm 0.003$ kcal mol^{-1} were taken from the CODATA (1) selection. Barrow (2) in a refinement of the work of Barrow and Downie (3) obtains a value of D_0^0 for $OH(X^2\Pi_{3/2}) \rightarrow O(^3P_2) + H(^2S_{1/2})$ of 35427 cm^{-1} from an extrapolation of $\Delta_{vap}G$ versus v; it was increased to 35450±100 cm^{-1} to account for the fact that $\Delta_{vap}G$ yields slightly low values at high v. Fehlenbok (4) obtains a value of D_e for $OH(B^2\Sigma^+)$ of 1315 cm^{-1} and G(O) for this state of 441 cm^{-1}. Using $T_e(B^2\Sigma^+)$ given by Rosen (5) and the zero point energy of HO (including the Dunham correction, see Herzberg (6)) of 847.0 cm^{-1}, this yields $D_0^0(OH) = 35451$ cm^{-1} with an estimated uncertainty of 100 cm^{-1}. A value of $D_0^0(OH) = 35450 \pm 100$ cm^{-1} = 101.356±0.29 kcal mol^{-1} was adopted. Combining the above values, one obtains $\Delta_f H°(OH, g, 0\ K) = 9.261 \pm 0.29$ kcal mol^{-1} which is in good agreement with the last JANAF (7) selection.

A review of earlier work is given in references (7-9)

Heat Capacity and Entropy

The vibrational and rotational constants of the respective electronic levels were taken from Rosen (5). Comparison of the results of these calculations with those of the more exact treatment given by Haar et al. (10) suggests that errors in the tables due to approximations in our calculations may be neglected above 400 K. Below this, they may be appreciable. It is recommended that H°(0 K) - H°(298.15 K), S°(298.15 K) and $C_p°$(298.15 K) be taken as -2.107 kcal mol^{-1}, 43.890 kJ mol^{-1}, and 7.144 kJ mol^{-1}, respectively. These errors result from dealing with the ground state ($X^2\Pi_i$) as two separate electronic states separated by 139.7 cm^{-1}

The thermodynamic functions are calculated using first-order anharmonic corrections to Q_r^i and Q_v^i in the partition function $Q = Q_t \sum_i Q_r^i Q_v^i g_i \exp(-c_2 \varepsilon_i/T)$. The National Bureau of Standards prepared this table (11) by critical analysis of data existing in 1972. Using molecular constants and $\Delta_f H°$ selected by NBS (11), we recalculate the table in terms of 1973 fundamental constants (12), 1975 atomic weights (13) and current JANAF reference states for the elements.

References

1. CODATA Task Group on Key Values for Thermodynamics, Final Set of Key Values for Thermodyanmics - Part I, November, 1971 (Bulletin No. 5).
2. R. F. Barrow, Arkiv Fysik 11, 281 (1956).
3. R. F. Barrow and A. R. Downie, Proc. Phys. Soc. (London) A69, 178 (1956).
4. P. Felenbok, Ann. d'Astr. 26, 393 (1963).
5. B. Rosen, "Spectroscopic Data Relative to Diatomic Molecules," Pergamon Press, Oxford, 1970.
6. G. Herzberg and A. Monfils, J. Mol. Spectrosc. 5, 482 (1960).
7. JANAF Thermochemical Tables, 2nd ed., NSRDS-NBS 37, 1971.
8. P. Gray, Trans. Faraday Soc. 55, 408 (1959).
9. R. Edse, "Third Combustion Symposium," Williams and Wilkins Co., Baltimore, p. 611, 1949.
10. L. Haar, A. S. Friedman, and C. W. Beckett, U.S. Nat. Bur. Stand. Monograph 20, 1961.
11. S. Abramowitz et al., U. S. Nat. Bur. Stand. Report 10904, 239, 1972.
12. CODATA Task Group on Fundamental Constants, CODATA Bulletin 11, December, 1973.
13. IUPAC Commission on Atomic Weights, Pure Appl. Chem. 47, 75 (1976); 37, 589 (1974).

Hydroxyl (OH) $H_1O_1(g)$

Enthalpy Reference Temperature = T_r = 298.15 K Standard State Pressure = p° = 0.1 MPa

T/K	$C_p°$ (J K^{-1}mol^{-1})	S° (J K^{-1}mol^{-1})	$-[G°-H°(T_r)]/T$ (J K^{-1}mol^{-1})	$H°-H°(T_r)$ (kJ mol^{-1})	$\Delta_f H°$ (kJ mol^{-1})	$\Delta_f G°$ (kJ mol^{-1})	Log K_f
0	0.	0.	INFINITE	-9.172	38.390	38.390	INFINITE
100	32.627	149.590	210.980	-6.139	38.471	37.214	-19.438
200	30.777	171.592	186.471	-2.976	38.832	35.803	-9.351
250	30.283	178.402	184.204	-1.450	38.930	35.033	-7.320
298.15	29.986	183.708	183.708	0.	38.987	34.277	-6.005
300	29.977	183.894	183.709	0.055	38.988	34.248	-5.963
350	29.780	188.499	184.073	1.549	39.019	33.455	-4.993
400	29.650	192.466	184.880	3.035	39.029	32.660	-4.265
450	29.567	195.954	185.921	4.515	39.020	31.864	-3.699
500	29.521	199.066	187.082	5.992	38.995	31.070	-3.246
600	29.527	204.447	189.542	8.943	38.902	29.493	-2.568
700	29.663	209.007	192.005	11.902	38.764	27.935	-2.085
800	29.917	212.983	194.384	14.880	38.598	26.399	-1.724
900	30.264	216.526	196.651	17.888	38.416	24.884	-1.444
1000	30.676	219.736	198.801	20.935	38.230	23.391	-1.222
1100	31.124	222.680	200.840	24.024	38.046	21.916	-1.041
1200	31.586	225.408	202.775	27.160	37.867	20.458	-0.891
1300	32.046	227.955	204.615	30.342	37.697	19.014	-0.764
1400	32.492	230.346	206.368	33.569	37.535	17.583	-0.656
1500	32.917	232.602	208.043	36.839	37.381	16.163	-0.563
1600	33.319	234.740	209.645	40.151	37.234	14.753	-0.482
1700	33.694	236.771	211.182	43.502	37.093	13.352	-0.410
1800	34.044	238.707	212.657	46.889	36.955	11.960	-0.347
1900	34.369	240 557	214.078	50.310	36.819	10.575	-0.291
2000	34.670	242.327	215.446	53.762	36.685	9.197	-0.240
2100	34.950	244.026	216.767	57.243	36.551	7.826	-0.195
2200	35.209	245.658	218.043	60.752	36.416	6.462	-0.153
2300	35.449	247.228	219.278	64.285	36.278	5.103	-0.116
2400	35.673	248.741	220.474	67.841	36.137	3.750	-0.082
2500	35.881	250.202	221.635	71.419	35.992	2.404	-0.050
2600	36.075	251.613	222.761	75.017	35.843	1.063	-0.021
2700	36.256	252.978	223.855	78.633	35.689	-0.271	0.005
2800	36.426	254.300	224.918	82.267	35.530	-1.600	0.030
2900	36.586	255.581	225.954	85.918	35.365	-2.924	0.053
3000	36.736	256.824	226.962	89.584	35.194	-4.241	0.074
3100	36.878	258.031	227.945	93.265	35.017	-5.552	0.094
3200	37.013	259.203	228.904	96.960	34.834	-6.858	0.112
3300	37.140	260.344	229.839	100.667	34.644	-8.158	0.129
3400	37.261	261.455	230.753	104.387	34.448	-9.452	0.145
3500	37.376	262.537	231.645	108.119	34.246	-10.741	0.160
3600	37.486	263.591	232.518	111.863	34.037	-12.023	0.174
3700	37.592	264.620	233.372	115.617	33.821	-13.300	0.188
3800	37.693	265.624	234.208	119.381	33.599	-14.570	0.200
3900	37.791	266.604	235.026	123.155	33.371	-15.834	0.212
4000	37.885	267.562	235.827	126.939	33.136	-17.093	0.223
4100	37.976	268.499	236.613	130.732	32.894	-18.346	0.234
4200	38.064	269.415	237.383	134.534	32.646	-19.593	0.244
4300	38.150	270.311	238.138	138.345	32.391	-20.833	0.253
4400	38.233	271.189	238.879	142.164	32.130	-22.068	0.262
4500	38.315	272.050	239.607	145.991	31.862	-23.297	0.270
4600	38.394	272.893	240.322	149.827	31.587	-24.520	0.278
4700	38.472	273.719	241.023	153.670	31.305	-25.737	0.286
4800	38.549	274.530	241.713	157.521	31.017	-26.947	0.293
4900	38.625	275.326	242.391	161,380	30.722	-28.152	0.300
5000	38.699	276.107	243.057	165.246	30.420	-29.350	0.307
5100	38.773	276.874	243.713	169.120	30.111	-30.542	0.313
5200	38.846	277.627	244.358	173.001	29.796	-31.729	0.319
5300	38.919	278.368	244.993	176.889	29.473	-32.909	0.324
5400	38.991	279.096	245.617	180.784	29.144	-34.083	0.330
5500	39.062	279.812	246.233	184.687	28.807	-35.251	0.335
5600	39.134	280.517	246.839	188.597	28.464	-36.412	0.340
5700	39.206	281.210	247.436	192.514	28.113	-37.568	0.344
5800	39.278	281.892	248.024	196.438	27.756	-38.716	0.349
5900	39.350	282.564	248.604	200.369	27.391	-39.860	0.353
6000	39.423	283.226	249.175	204.308	27.019	-40.997	0.357

PREVIOUS: June 1977 (1 atm) CURRENT: June 1977 (1 bar)

Hydroxyl (OH) $H_1O_1(g)$

HYDROXYL, ION (OH^+) IDEAL GAS M_r = 17.006791

$\Delta_f H°(0\ K)$ = 1310.88 ± 0.5 kJ mol^{-1}

$S°(298.15\ K)$ = 182.776 ± 0.21 J K^{-1} mol^{-1} $\Delta_f H°(298.15\ K)$ = 1317.12 ± 10.5 kJ mol^{-1}

Electronic Levels and Quantum Weights

State	ε_i, cm^{-1}	g_i
$X^3\Sigma^-$	0	3
$a^1\Delta$	[17700]	2
$A^3\pi$	22949	6
$b^1\Sigma^+$	29150	1

ω_e = [3135 ± 100] cm^{-1} $\omega_e x_e$ = [89.3 ± 50] cm^{-1} σ = 1

B_e = 16.776 cm^{-1} α_e = 0.728 cm^{-1} r_e = 1.0289 Å

Enthalpy of Formation

The ionization potential of hydroxyl has been determined by Mann et al. (1) as 13.6 eV and by Foner and Hudson (2) as 13.18 ± 0.1 eV, both by electron impact methods. These values correspond to $\Delta_f H°(OH^+, g, 0\ K)$ = 322.97 kcal mol^{-1} and 313.28 ± 2.5 kcal mol^{-1}, respectively. Metzger and Cook (3) have reported the photoionization onset of the process $H_2O(g) + h\nu \rightarrow OH^+ + H + e^-$ as 18.3 eV, with JANAF auxiliary data (4) this gives $\Delta_f H°(OH^+, g, 0\ K)$ = 313.3 kcal mol^{-1}, which is adopted.

Heat Capacity and Entropy

The measured electronic levels are those reported by Merer et al. (5), the rotational constants are derived from this analysis also. The estimated electronic level is from the Hartree-Fock calculations of Cade (6).

The vibrational constants are estimated so that $\Delta G_{1/2}$ is equal to that reported by Merer et al. (5). The anharmonicity, $\omega_e x_e$, is estimated by comparison with OH and NH. Calculations of $\omega_e x_e$ based on the Morse potential curve and the relation $\omega_e^2 = 4B_e^3/D_e$, or $\omega_o^2 = 4\omega_o x_o D_0^0$ (7) do not appear to yield values which are preferable to our estimates; this is due to the inexactness of the potential for ionic molecules.

References

1. M. M. Mann, A. Hustrulid and J. T. Tate, Phys. Rev. 58, 340 (1940).
2. S. F. Foner and R. L. Hudson, J. Chem. Phys. 25, 602 (1956).
3. P. H. Metzger and G. R. Cook, J. Chem. Phys. 41, 642 (1964).
4. JANAF Thermochemical Tables: $H_2O(g)$, 3-31-61; H(g), 9-30-65.
5. A. J. Merer, E. K. Achter, M. Horani and J. Rostas, Can. J. Phys. 47, 1723 (1969).
6. P. E. Cade, Can. J. Phys. 46, 1989 (1968).
7. V. S. Yungman, Russ. J. Phys. Chem. 8, 147 (1960).

Hydroxyl, Ion (OH^+) $H_1O_1^+(g)$

Enthalpy Reference Temperature = T_r = 298.15 K Standard State Pressure = p° = 0.1 MPa

T/K	$C_p°$ (J K^{-1}mol^{-1})	S°	$-[G°-H°(T_r)]/T$	$H°-H°(T_r)$ (kJ mol^{-1})	$\Delta_f H°$	$\Delta_f G°$	Log K_f
0	0.	0.	INFINITE	-8.618	1310.883		
100	29.127	150.948	208.684	-5.774			
200	29.136	171.140	185.442	-2.860			
250	29.144	177.642	183.256	-1.403			
298.15	29.153	182.776	182.776	0.	1317.123	1306 437	-228.882
300	29.153	182.956	182.777	0.054	1317.162	1306.371	-227.459
350	29.167	187.451	183.132	1.512	1318.196	1304.490	-194.685
400	29.192	191.347	183.920	2.971	1319.219	1302.462	-170.084
450	29.237	194.788	184.940	4.431	1320.230	1300.307	-150.936
500	29.311	197.872	186.082	5.895	1321.231	1298.040	-135.605
600	29.566	203.236	188.507	8.838	1323.208	1293.215	-112.584
700	29.959	207.822	190.946	11.813	1325.165	1288.062	-96.116
800	30.456	211.854	193.313	14.833	1327.119	1282.628	-83.747
900	31.013	215.473	195.577	17.906	1329.081	1276.949	-74.112
1000	31.590	218.770	197.734	21.036	1331.057	1271.050	-66.393
1100	32.160	221.808	199.786	24.224	1333.049	1264.954	-60.068
1200	32.706	224.630	201.740	27.467	1335.057	1258.675	-54.789
1300	33.217	227.268	203.604	30.764	1337.081	1252.227	-50.315
1400	33.691	229.747	205.383	34.110	1339.116	1245.624	-46.475
1500	34.127	232.087	207.086	37.501	1341.161	1238.874	-43.141
1600	34.526	234.302	208.719	40.934	1343.214	1231.988	-40.220
1700	34.891	236.407	210.286	44.405	1345.271	1224.974	-37.639
1800	35.226	238.411	211.793	47.911	1347.331	1217.838	-35.341
1900	35.532	240.324	213.245	51.449	1349.391	1210.588	-33.281
2000	35.814	242.153	214.645	55.017	1351.451	1203.229	-31.425
2100	36.074	243.907	215.997	58.611	1353.509	1195.767	-29.743
2200	36.315	245.591	217.304	62.231	1355.563	1188.208	-28.212
2300	36.539	247.210	218.569	65.874	1357.614	1180.555	-26.811
2400	36.749	248.770	219.795	69.538	1359.660	1172.813	-25.526
2500	36.946	250.274	220.985	73.223	1361.701	1164.985	-24.341
2600	37.133	251.727	222.139	76.927	1363.737	1157.076	-23.246
2700	37.311	253.131	223.261	80.649	1365.767	1149.090	-22.230
2800	37.482	254.491	224.352	84.389	1367.792	1141.028	-21.286
2900	37.647	255.810	225.415	88.146	1369.811	1132.893	-20.406
3000	37.806	257.089	226.449	91.918	1371.825	1124.689	-19.583
3100	37.962	258.331	227.458	95.707	1373.834	1116.418	-18.812
3200	38.114	259.538	228.441	99.510	1375.839	1108.082	-18.088
3300	38.265	260.714	229.402	103.329	1377.839	1099.684	-17.407
3400	38.413	261.858	230.340	107.163	1379.836	1091.225	-16.765
3500	38.562	262.974	231.256	111.012	1381.829	1082.707	-16.159
3600	38.709	264.062	232.152	114.876	1383.819	1074.133	-15.585
3700	38.857	265.125	233.029	118.754	1385.806	1065.503	-15.042
3800	39.006	266.163	233.887	122.647	1387.792	1056.820	-14.527
3900	39.156	267.178	234.728	126.555	1389.776	1048.084	-14.038
4000	39.307	268.171	235.552	130.478	1391.759	1039.298	-13.572
4100	39.460	269.144	236.359	134.417	1393.741	1030.462	-13.128
4200	39.615	270.097	237.151	138.371	1395.723	1021.577	-12.705
4300	39.773	271.031	237.928	142.340	1397.706	1012.646	-12.301
4400	39.932	271.947	238.691	146.325	1399.689	1003.668	-11.915
4500	40.094	272.846	239.440	150.326	1401.673	994.644	-11.546
4600	40.258	273.729	240.176	154.344	1403 659	985.577	-11.192
4700	40.425	274.597	240.899	158.378	1405.647	976.467	-10.852
4800	40.595	275.449	241.610	162.429	1407.637	967.315	-10.527
4900	40.767	276.288	242.309	166.497	1409.630	958.121	-10.214
5000	40.942	277.114	242.997	170.583	1411.626	948.887	-9.913
5100	41.120	277.926	243.674	174.686	1413.625	939.612	-9.624
5200	41.299	278.726	244.340	178.807	1415.628	930.298	-9.345
5300	41.482	279.515	244.997	182.946	1417.635	920.946	-9.076
5400	41.666	280.292	245.643	187.103	1419.646	911.555	-8.818
5500	41.853	281.058	246.280	191.279	1421.662	902.127	-8.568
5600	42.042	281.814	246.908	195.474	1423.682	892.663	-8.326
5700	42.233	282.560	247.527	199.688	1425.706	883.163	-8.093
5800	42.425	283.296	248.137	203.920	1427.736	873.627	-7.868
5900	42.619	284.023	248.739	208.173	1429.771	864.055	-7.650
6000	42.815	284.741	249.333	212.444	1431.811	854.450	-7.439

PREVIOUS: December 1970 (1 atm) CURRENT: December 1970 (1 bar)

Hydroxyl, Ion (OH^+) $H_1O_1^+(g)$

HYDROXYL, ION (OH^-) IDEAL GAS M_r = 17.007889

$S°(298.15\ K) = [172.469 \pm 0.8]\ J\ K^{-1}\ mol^{-1}$

$\Delta_f H°(0\ K) = -137.43 \pm 3.8\ kJ\ mol^{-1}$
$\Delta_f H°(298.15\ K) = -143.59 \pm 3.8\ kJ\ mol^{-1}$

Electronic Level and Quantum Weight

state	ε_i, cm^{-1}	g_i
$^1\Sigma^+$	0	1

$\omega_e = [3770]\ cm^{-1}$ $\omega_e x_e = [74.7]\ cm^{-1}$ $\sigma = 1$
$B_e = [19.2]\ cm^{-1}$ $\alpha_e = [0.65]\ cm^{-1}$ $r_e = [0.9628]$ Å

Enthalpy of Formation

Branscomb (1) has obtained an electron affinity of OH of 1.83±0.04 eV (42.2±0.9 kcal mol^{-1}) from the photodetachment spectrum of OH^-. The vibrational state was identified as the ground state since no other absorption sequence could be detected by a thorough search. Kay and Page (2) have also determined the electron affinity by the magnetron method and report 49.1 at 1820 K, which reduces to 40±2.7 kcal mol^{-1} at 0 K. They also disclaim prior papers, Page (3) and Page and Sugden (4), which supported higher values. Feugier and Queraud (5) have reported an electron affinity at 2250 K of 2.13 eV (49.1 kcal mol^{-1}) which reduces to 37.7 kcal mol^{-1} at 0 K. We adopt the value reported by Branscomb (1) which yields $\Delta_f H°(OH^-, g)$ 0 K = -32.85 kcal mol^{-1}.

Heat Capacity and Entropy

Branscomb (1) deduced from his photodetachment spectra that ω_e, B_e and r_e are very close to the values for OH(g). Our adopted values are from the calculations of Cade (6) corrected for the mean deviations in OH and HF. The values of $\omega_e x_e$ and α_e were estimated by comparison with OH(g) and other related hydrides. The ground state configuration was assumed to be that of HF(g) with which OH^- is isoelectronic.

References

1. L. M. Branscomb, Phys. Rev. 148, 11 (1966).
2. J. Kay and F. M. Page, Trans. Faraday Soc. 62, 3081 (1966).
3. F. M. Page, Discussions Faraday Soc. 19, 87 (1955).
4. F. M. Page and T. M. Sugden, Trans. Faraday Soc. 53, 1092 (1957).
5. A. Feugier and A. Queraud, Elec. MHD, Proc. Symp. 4, 2129 (1968).
6. P. E. Cade, unpublished calculations, quoted in ref. 1.

Hydroxyl, Ion (OH^-) $H_1O_1^-(g)$

Enthalpy Reference Temperature = T_r = 298.15 K Standard State Pressure = p° = 0.1 MPa

T/K	$C_p°$ (J K⁻¹mol⁻¹)	S° (J K⁻¹mol⁻¹)	$-[G°-H°(T_r)]/T$ (J K⁻¹mol⁻¹)	$H°-H°(T_r)$ (kJ mol⁻¹)	$\Delta_f H°$ (kJ mol⁻¹)	$\Delta_f G°$ (kJ mol⁻¹)	Log K_f
0	0.	0.	INFINITE	-8 606	-137.429		
100	29.127	140.647	198.369	-5.772			
200	29.130	160.836	175.134	-2.860			
250	29.135	167.336	172.948	-1.403			
298.15	29.141	172.469	172.469	0.	-143.595	-138.698	24.299
300	29.141	172.649	172.469	0.054	-143.633	-138.668	24.144
350	29.148	177.141	172.824	1.511	-144.678	-137.758	20.559
400	29.157	181.034	173.612	2.969	-145.736	-136.697	17.851
450	29.170	184.469	174.631	4.427	-146.806	-135.503	15.729
500	29.194	187.543	175.772	5.886	-147.888	-134.189	14.019
600	29.289	192.873	178.191	8.809	-150.087	-131.244	11 426
700	29.471	197.401	180.620	11.747	-152.325	-127.927	9.546
800	29.749	201.353	182.970	14.707	-154.588	-124.287	8.115
900	30.108	204.877	185.211	17.699	-156.864	-120.363	6.986
1000	30.525	208.070	187.340	20.730	-159.145	-116.185	6.069
1100	30.975	211.001	189.360	23.805	-161.422	-111.779	5.308
1200	31.439	213.716	191.278	26.926	-163.694	-107.165	4.665
1300	31.900	216.250	193.102	30.093	-165.957	-102.363	4.113
1400	32.347	218.631	194.841	33.305	-168.212	-97.386	3.634
1500	32.773	220.877	196.503	36.562	-170.459	-92.249	3.212
1600	33.176	223.005	198.093	39.859	-172.700	-86.962	2.839
1700	33.553	225.028	199.619	43.196	-174.934	-81.535	2.505
1800	33.904	226.956	201.084	46.569	-177.165	-75.976	2.205
1900	34.230	228.798	202.495	49.976	-179.392	-70.294	1.933
2000	34.532	230.561	203.855	53.414	-181.619	-64.494	1.684
2100	34.812	232.253	205.167	56.881	-183.846	-58.583	1.457
2200	35.072	233.879	206.435	60.376	-186.074	-52.566	1.248
2300	35.312	235.443	207.663	63.895	-188.304	-46.448	1.055
2400	35.536	236.951	208.852	67.438	-190.537	-40.232	0.876
2500	35.744	238.406	210.005	71.002	-192.774	-33.924	0.709
2600	35.938	239.811	211.124	74.586	-195.015	-27.525	0.553
2700	36.119	241.171	212.212	78.189	-197.261	-21.041	0.407
2800	36.288	242.488	213.270	81.809	-199.513	-14.473	0.270
2900	36.447	243.764	214.300	85.446	-201.771	-7.825	0.141
3000	36.596	245.002	215.303	89.098	-204.034	-1.098	0.019
3100	36.737	246.204	216.280	92.765	-206.304	5.704	-0.096
3200	36.870	247.373	217.234	96.446	-208.580	12.579	-0.205
3300	36.996	248.509	218.164	100.139	-210.862	19.525	-0.309
3400	37.115	249.616	219.073	103.844	-213.151	26.541	-0.408
3500	37.228	250.693	219.961	107.562	-215.447	33.624	-0.502
3600	37 336	251.743	220.830	111.290	-217.750	40.773	-0.592
3700	37.439	252.768	221.679	115.029	-220.059	47.986	-0.677
3800	37.538	253.768	222.510	118.778	-222.375	55.262	-0.760
3900	37.632	254.744	223.324	122.536	-224.698	62.599	-0.838
4000	37.723	255.698	224.122	126.304	-227.027	69.995	-0.914
4100	37.810	256.630	224.903	130.081	-229.364	77.449	-0.987
4200	37.894	257.542	225.670	133.866	-231.708	84.961	-1.057
4300	37.975	258.435	226.421	137.659	-234.058	92.528	-1.124
4400	38.053	259.309	227.159	141.461	-236.416	100.150	-1.189
4500	38.129	260.165	227.883	145.270	-238.781	107.826	-1.252
4600	38 203	261.004	228.594	149.086	-241.153	115.554	-1.312
4700	38.274	261.826	229.292	152.910	-243.533	123.335	-1.371
4800	38.344	262.633	229.978	156.741	-245.920	131.166	-1.427
4900	38.411	263.424	230.653	160.579	-248.315	139.047	-1.482
5000	38.477	264.201	231.316	164.423	-250.717	146.977	-1.535
5100	38.541	264.963	231.968	168.274	-253.127	154.954	-1.587
5200	38.604	265.712	232.610	172.132	-255.545	162.979	-1.637
5300	38.666	266.448	233.242	175.995	-257.970	171.051	-1.686
5400	38.726	267.172	233.863	179.865	-260.405	179.168	-1.733
5500	38.785	267.883	234.475	183.740	-262.847	187.331	-1.779
5600	38.843	268.582	235.078	187.622	-265.297	195.539	-1.824
5700	38 900	269.270	235.672	191.509	-267.756	203.789	-1.868
5800	38.955	269.947	236.257	195.402	-270.224	212.084	-1.910
5900	39.010	270.613	236.834	199.300	-272.700	220.420	-1.951
6000	39.064	271.270	237.402	203.204	-275.186	228.799	-1.992

PREVIOUS: December 1970 (1 atm) CURRENT: December 1970 (1 bar)

Hydroxyl, Ion (OH^-) $H_1O_1^-(g)$

STRONTIUM HYDROXIDE (SrOH) IDEAL GAS M_r = 104.62734

$\Delta_f H°(0\ K) = -201.08 \pm 20.9\ kJ\ mol^{-1}$

$S°(298.15\ K) = [246.511 \pm 8.4]\ J\ K^{-1}\ mol^{-1}$ $\Delta_f H°(298.15\ K) = -205.52 \pm 20.9\ kJ\ mol^{-1}$

Electronic Levels and Quantum Weights		Vibrational Frequencies and Degeneracies
ϵ_i, cm^{-1}	g_i	ν, cm^{-1}
0	[2]	[498](1)
[14700]	[2]	[438](2)
[15100]	[2]	[3650](1)
[16000]	[2]	

Point Group: $[C_{\infty v}]$ $\sigma = 1$

Bond Distances: Sr-O = [2.10] Å O-H = [0.96] Å

Bond Angle: Sr-O-H = [180]°

Rotational Constant: $B_0 = [0.251104]\ cm^{-1}$

Enthalpy of Formation

The adopted enthalpy of formation is based on assessment of D_0^0 values derived from flame spectra of CaOH, SrOH, and BaOH. Cotton and Jenkins (1) found both the alkaline earth monohydroxides and dihydroxides to be present in significant amounts in fuel-rich hydrogen-oxygen-nitrogen flames. They determined equilibrium constants for the reactions $M(g) + H_2O(g) = MOH(g) + H(g)$ and $M(g) + 2\ H_2O(g) = M(OH)_2(g) + 2\ H(g)$ and derived D_0^0 values. For D_0^0(Sr-OH), Cotton and Jenkins (1) gave 103±5 kcal mol^{-1}. Ryabova and Gurvich (2) considered SrOH to be the dominant compound and reported D_0^0 = 95±8 kcal mol^{-1}. Sugden and Schofield (3) interpreted $Sr(OH)_2$ as dominant. Cotton and Jenkins (1) have recalculated the work of Ryabova and Gurvich (2) and of Sugden and Schofield (3) considering both SrOH and $Sr(OH)_2$ to be present and have obtained the recalculated D_0^0(Sr-OH) values of 96 and 101 kcal mol^{-1}, respectively. Gurvich et al. (4) made further measurements, interpreted SrOH as dominant, and reported D_0^0(Sr-OH) = 93±3 kcal mol^{-1}. Kalff and Alkemade (5) chose flame conditions to minimize the dihydroxide formation and determined a 3rd law analysis of the experimental equilibrium constants tabulated by Cotton and Jenkins (1) using current JANAF auxiliary data (7) leads to D_0^0(Sr-OH) = 103.7 kcal mol^{-1} which is 0.7 kcal mol^{-1} higher than the 103 kcal mol^{-1} given by Cotton and Jenkins (1). Applying the correction to the value of Ryabova and Gurvich (2) and Sugden and Schofield (3) as recalculated by Cotton and Jenkins (1) gives D_0^0(Sr-OH) = 96.7 and 101.7 kcal mol^{-1}, respectively.

Because the corrected dissociation energy of Ryabova and Gurvich was near the adopted value for $Ba(OH)_2(g)$ (7), their corrected value was adopted for $Sr(OH)_2(g)$ (7). While their corrected value for BaOH(g) (7) is not as close to the adopted value as in the dihydroxide case, it is nevertheless the nearest one, and therefore D_0^0(Sr-OH) = 96.7 kcal mol^{-1} is adopted.

The ratio of the dissociation energies of the alkaline earth monohalides to those of the corresponding dihalides range from 0.40 to 0.51 with the ratio for the strontium fluorides being 0.49 (7). The similarity between the halides and hydroxides has been established (8-11). The ratio of the adopted values for the dissociation energies of SrOH(g) and $Sr(OH)_2$ is 0.49 where D_0^0 of the dihydroxide is defined by the reaction $Sr(OH)_2(g) = Sr(g) + 2\ OH(g)$ and is 198.6 kcal mol^{-1} (7).

$\Delta_f H°$(SrOH, g, 0 K) = -48.11±5 kcal mol^{-1} is calculated from the adopted dissociation energy.

Heat Capacity and Entropy

The analogy between gaseous monohydroxides and monohalides, particularly the monofluorides, has been recognized (8-11). The molecular configuration is assumed to be linear in accordance with the prediction of Walsh (12) and the evidence that the gaseous alkali metal hydroxide are linear (13-15). The ground state is assumed to be $^2\Sigma^+$ by analogy with SrF and SrCl (7). The electronic levels are estimated from the band spectra observed by James and Sugden (16), Lagerqvist and Huldt (17), Charton and Gaydon (18), Zhitkevich et al. (19), and Van der Hurk et al. (20), and the comparison with SrF and SrCl (7).

The Sr-O bond distance is estimated to be slightly larger, 0.02 Å, than the Sr-F distance (7) after noting the close similarity in the bond distance of the alkali metal fluorides and hydroxides. The O-H bond distance is that in water (7).

The Sr-O stretching frequency, 498 cm^{-1}, is estimated to be the same as the Sr-F stretching frequency (7, 11). The O-H stretching frequency, 3650 cm^{-1}, is estimated from the alkali hydroxide series. The bending frequency, 438 cm^{-1}, is estimated by assuming that the ratio of the bending force constant to the stretching force constant is 0.022, which is the average ratio found in the alkali metal monohydroxides by Acquista and Abramowitz (11, 15).

The entropy in the present table is lower by 0.06 cal K^{-1} mol^{-1} at 298.15 K and 0.10 cal K^{-1} mol^{-1} at 1000 K than that proposed by Jackson (11); the data relevant to the calculation are nearly the same.

References

1. D. H. Cotton and D. R. Jenkins, Trans. Faraday Soc. 64, 2988 (1968).
2. V. G. Ryabova and L. V. Gurvich, Teplofiz. Vys. Temp. 3, 318 (1965).
3. T. M. Sugden and K. Schofield, Trans. Faraday Soc. 62, 566 (1966).
4. L. V. Gurvich, V. G. Ryabova, A. N. Khitrov, and E. M. Starovoitov, Teplofiz. Vys. Temp. 9, 250 (1971).
5. P. J. Kalff and C. Th. J. Alkemade, J. Chem. Phys. 59, 2572 (1973).
6. V. G. Ryabova, A. N. Khitrov and L. V. Gurvich, Teplofiz. Vys. Temp. 10, 744 (1972).
7. JANAF Thermochemical Tables: Sr(g), 12-31-70; H_2O(g), 3-31-61; H(g), 6-30-74; SrCl(g), SrF(g), 12-31-72; $Sr(OH)_2$, BaOH, $Ba(OH)_2$, 12-31-75.
8. R. C. Schoonmaker and R. F. Porter, J. Chem. Phys. 31, 830 (1959).
9. J. Berkowitz, D. J. Meschi, and W. A. Chupka, J. Chem. Phys. 33, 533 (1960).
10. F. E. Stafford and J. Berkowitz, J. Chem. Phys. 40, 2963 (1964).
11. D. D. Jackson, Lawrence Livermore Laboratory, Univ. of Calif., UCRL-51137, Dec. 8, 1971.
12. A. D. Walsh, J. Chem. Soc. 1953, 2288 (1953).
13. N. Acquista, S. Abramowitz and R. Lide, J. Chem. Phys. 49, 780 (1968).
14. R. L. Kuczkowski and R. Lide, J. Chem. Phys. 44, 3131 (1966).
15. N. Acquista and S. Abramowitz, U. S. Nat. Bur. Stand. Report 9905, 1968.
16. C. G. James and T. M. Sugden, Nature 175, 333 (1955).
17. A. Lagerqvist and L. Huldt, Naturwissenschaften 42, 365 (1955).
18. M. Charton and A. G. Gaydon, Proc. Phys. Soc. (London) 69A, 520 (1952).
19. V. F. Zhitkevich, A. I. Lyutyi, N. A. Nesterko, V. S. Rossikhin, and I. L. Tsikora, Optika i Spektroskopiya 14, 35 (1963).
20. J. Van der Hurk, Tj. Hollander and C. Th. J. Alkemade, J. Quant. Spectrosc. Radiat. Transfer 13, 273 (1973).

Strontium Hydroxide (SrOH) $H_1O_1Sr_1(g)$

Enthalpy Reference Temperature = T_r = 298.15 K Standard State Pressure = p° = 0.1 MPa

T/K	$C_p°$	S°	$-[G°-H°(T_r)]/T$	$H°-H°(T_r)$	$\Delta_f H°$	$\Delta_f G°$	Log K_f
	J $K^{-1}mol^{-1}$			kJ mol^{-1}			
0	0.	0.	INFINITE	-10.707	-201.082	-201.082	INFINITE
100	30.646	205.253	282.997	-7.774	-202.600	-206.619	107.927
200	39.954	229.305	250.551	-4.249	-204.366	-209.895	54.819
250	43.517	238.626	247.255	-2.157	-205.004	-211.206	44.129
298.15	45.958	246.511	246.511	0.	-205.518	-212.347	37.202
300	46.036	246.796	246.512	0.085	-205.536	-212.389	36.980
350	47.812	254.034	247.080	2.434	-206.003	-213.494	31 862
400	49.088	260.506	248.361	4.858	-206.431	-214.534	28.015
450	50.028	266.345	250.040	7.337	-206.842	-215.523	25 017
500	50.744	271.654	251.940	9.857	-207.251	-216.465	22.614
600	51.775	281.003	256.026	14.986	-208.096	-218.230	18.999
700	52.534	289.043	260.182	20.203	-209.009	-219.848	16.405
800	53.182	296.101	264.239	25 490	-210.007	-221.329	14.451
900	53.787	302.401	268.136	30.838	-211.722	-222.608	12.920
1000	54.371	308.098	271.852	36.246	-212.581	-223.772	11.689
1100	54.936	313.307	275.387	41.712	-221.325	-224.484	10 660
1200	55.477	318.110	278.750	47.233	-223.064	-224.694	9.781
1300	55.990	322.571	281.951	52.806	-224.789	-224.760	9 031
1400	56.470	326.738	285.003	58.430	-226.501	-224.694	8.383
1500	56.916	330.650	287.917	64.099	-228.202	-224.505	7.818
1600	57.329	334.336	290.704	69.812	-229.895	-224.203	7.319
1700	57.710	337.824	293.374	75.564	-368.496	-222.616	6.840
1800	58.062	341.132	295.936	81.353	-368.318	-214.040	6.211
1900	58.389	344.280	298.398	87.176	-368.143	-205.474	5.649
2000	58.694	347.283	300.768	93 030	-367.974	-196.917	5.143
2100	58.982	350.154	303.052	98.914	-367.817	-188.368	4.685
2200	59.256	352.904	305.256	104.826	-367.677	-179.826	4.270
2300	59.520	355.544	307.385	110.765	-367.560	-171.290	3.890
2400	59.777	358.083	309.445	116.730	-367.471	-162.759	3.542
2500	60.031	360.528	311.440	122.720	-367.418	-154.231	3.222
2600	60.283	362.887	313.374	128.736	-367.409	-145.703	2.927
2700	60.537	365.167	315.250	134.777	-367.452	-137.175	2.654
2800	60.793	367.373	317.072	140.843	-367.556	-128.645	2.400
2900	61.053	369.511	318.844	146.935	-367.728	-120.110	2.163
3000	61.319	371.585	320.567	153.054	-367.979	-111.567	1.943
3100	61.590	373.600	322.246	159.199	-368.317	-103.014	1.736
3200	61.867	375.560	323.881	165.372	-368.750	-94.450	1.542
3300	62.149	377.468	325.477	171.573	-369.288	-85.870	1.359
3400	62.438	379.328	327.033	177.802	-369.938	-77.272	1.187
3500	62.731	381.142	328.553	184.060	-370.707	-68.654	1.025
3600	63.029	382.913	330.039	190.348	-371.602	-60.011	0.871
3700	63 330	384.644	331.491	196.666	-372.629	-51.342	0.725
3800	63.634	386.337	332.912	203.014	-373.792	-42.643	0.586
3900	63 940	387.994	334.304	209.393	-375.094	-33.912	0.454
4000	64.247	389.617	335.666	215.803	-376.539	-25.146	0.328
4100	64.553	391.207	337.002	222.243	-378.127	-16.341	0.208
4200	64.858	392.766	338.311	228.713	-379.856	-7.497	0.093
4300	65.161	394.296	339.595	235.214	-381.729	1.391	-0.017
4400	65.461	395.797	340.855	241.745	-383.744	10.323	-0.123
4500	65.756	397.272	342.093	248.306	-385.896	19.303	-0.224
4600	66.046	398.720	343.308	254.896	-388.117	28.327	-0.322
4700	66.330	400.144	344.502	261.515	-390.513	37.406	-0.416
4800	66.608	401.543	345.676	268.162	-393.029	46.537	-0.506
4900	66.878	402.919	346.830	274.837	-395.659	55.722	-0.594
5000	67.140	404.273	347.966	281.537	-398.390	64.961	-0.679
5100	67.393	405.605	349.083	288.264	-401.201	74.254	-0.761
5200	67.637	406.916	350.182	295.016	-404.123	83.604	-0.840
5300	67.871	408.207	351.265	301.791	-407.129	93.013	-0.917
5400	68.096	409.478	352.331	308.590	-410.210	102.478	-0.991
5500	68.311	410.729	353.382	315.410	-413.360	112.001	-1.064
5600	68.515	411.962	354.417	322.251	-416.571	121.582	-1.134
5700	68.708	413.176	355.437	329.113	-419.835	131.221	-1.203
5800	68.891	414.373	356.443	335.993	-423.145	140.918	-1.269
5900	69.063	415.552	357.435	342.891	-426.285	150.647	-1.334
6000	69.225	416.714	358.413	349.805	-429 638	160.454	-1.397

PREVIOUS: December 1975 (1 atm) CURRENT: December 1975 (1 bar)

Strontium Hydroxide (SrOH) $H_1O_1Sr_1(g)$

STRONTIUM HYDROXIDE, ION ($SrOH^+$) IDEAL GAS M_r = 104.626791

$S°(298.15\ K) = [241.049 \pm 8.4]\ J\ K^{-1}\ mol^{-1}$

$\Delta_f H°(0\ K) = 318.99 \pm 62.8\ kJ\ mol^{-1}$
$\Delta_f H°(298.15\ K) = 320.80 \pm 62.8\ kJ\ mol^{-1}$

Vibrational Frequencies and Degeneracies

ν, cm^{-1}
[470](1)
[430](2)
[3650](1)

Ground State Quantum Weight: [1] $\sigma = 1$
Point Group: [$C_{\infty v}$]
Bond Distances: Sr–O = [2.1] Å O–H = [0.96] Å
Bond Angle: Sr–O–H = [180]°
Rotational Constant: B_0 = [0.251111] cm^{-1}

Enthalpy of Formation

The ionization potential of SrOH(g) was deduced by Kelly an Padley (1) to be 5.55±0.1 eV. Using a rotating single probe, these authors quantitatively examined the total positive ion concentrations produced from Sr aqueous salt solutions in fuel rich, premixed $H_2 + O_2 + N_2$ flames. Using current JANAF auxiliary data (2), we recalculate the ionization potential to be 5.75 eV.

Jensen (3) determined $\Delta_r H°(0\ K) = 25\pm8\ kcal\ mol^{-1}$ for $Sr(g) + OH(g) = SrOH^+(g) + e^-$ in atmospheric pressure $H_2 + O_2 + N_2$ flames using the microwave cavity resonance method. This value was calculated assuming a bent molecule for $SrOH^+$; the value is not significantly changed, within the uncertainty, by the change in the configuration. Using auxiliary data (2), we derive an ionization potential of 5.28 eV, which is in fair agreement with the value derived from the data of Kelly and Padly (1).

We adopt IP = 5.39 eV (124.30 kcal mol^{-1}) which is an average of the above two studies (1, 3). This leads to $\Delta_f H°(0\ K)$ = 76.19 kcal mol^{-1} and $\Delta_f H°(298.15\ K)$ = 76.67 kcal mol^{-1} for $SrOH^+(g)$. We assign an uncertainty of ±15 kcal mol^{-1}.

In comparison, the ionization potential of SrF(g) is 5.03 eV (2) while that of Sr(g) is 5.692 eV (4). Note that the average of these two values is very close to our adopted ionization potential.

Heat Capacity and Entropy

The molecular configuration is assumed to be linear, since experimental evidence indicates that the gaseous alkali metal hydroxides are linear (2, 5, 6, 7). In addition, Walsh (8) had predicted that BAH molecules (H = hydrogen atom) with ten or less valence electrons ($SrOH^+$ has eight valence electrons) will be linear in their ground state. The molecule $SrOH^+$ is isoelectronic with RbOH.

The bond dissociation energy for $SrOH^+$ (103.7 kcal mol^{-1}, 2, 4) for the process $SrOH^+(g) = Sr^+(g) + OH(g)$ is fairly close to that for SrOH (96.7 kcal mol^{-1}, 2). This suggests that the bonding in these two molecules may be quite similar. Thus bond distances are assumed to be the same as those adopted for SrOH(g, 2). The vibrational frequencies are assumed to be similar to those adopted for SrOH(g, 2). The ground state quantum weight is assumed to be the same as that of KOH (g, 2).

References

1. R. Kelly and P. J. Padley, Trans. Faraday Soc. 67, 1384 (1971).
2. JANAF Thermochemical Tables: e^-, 3-31-65; H_2(g) and H_2O(g), 3-31-61; Sr(g), 12-31-70; OH(g), 12-31-70; H(g), 6-30-74; SrOH(g) and $Sr(OH)_2$(g), 12-31-75.
3. D. E. Jensen, Combustion and Flame 12, 261 (1968).
4. C. E. Moore, U. S. Nat. Bur. Stand. Circ. 467, Vol. II, 1958.
5. N. Acquista, S. Abramowitz, and D. R. Lide, J. Chem. Phys. 49, 780 (1968).
6. R. L. Kuczkowski and D. R. Lide, J. Chem. Phys. 44, 3131 (1966).
7. N. Acquista and S. Abramowitz, U. S. Nat. Bur. Stand. Report 9905, 1968.
8. A. D. Walsh, J. Chem. Soc. 1953, 2288 (1953).

Strontium Hydroxide, Ion ($SrOH^+$) $H_1O_1Sr_1{}^+(g)$

Enthalpy Reference Temperature = T_r = 298.15 K Standard State Pressure = p° = 0.1 MPa

T/K	$C_p°$ (J K⁻¹mol⁻¹)	S° (J K⁻¹mol⁻¹)	$-[G°-H°(T_r)]/T$ (J K⁻¹mol⁻¹)	$H°-H°(T_r)$ (kJ mol⁻¹)	$\Delta_f H°$ (kJ mol⁻¹)	$\Delta_f G°$ (kJ mol⁻¹)	Log K_f
0	0.	0.	INFINITE	-10.760	318.986		
100	30.774	199.518	277.769	-7.825			
200	40.248	223.738	245.112	-4.275			
250	43.779	233.120	241.797	-2.169			
298.15	46.180	241.049	241.049	0.	320.800	309.345	-54.196
300	46.256	241.335	241.050	0.085	320.820	309.274	-53.849
350	47.994	248.603	241.620	2.444	321.403	307.304	-45.863
400	49.238	255.098	242.906	4.877	322.023	305.248	-39.861
450	50.154	260.953	244.591	7.363	322.658	303.113	-35.184
500	50.850	266.275	246.498	9.888	323.294	300.907	-31.436
600	51.852	275.640	250.595	15.027	324.537	296.312	-25.796
700	52.592	283.691	254.762	20.250	325.709	291.514	-21.753
800	53.228	290.756	258.828	25.542	326.795	286.553	-18 710
900	53.824	297.060	262.732	30.895	327.162	281.535	-16.340
1000	54.401	302.760	266.454	36.306	328.386	276.399	-14.438
1100	54.961	307.972	269.995	41.774	321.722	271.506	-12 893
1200	55.498	312.777	273.363	47.297	322.064	266.926	-11.619
1300	56.007	317.240	276.568	52.873	322.420	262.317	-10.540
1400	56.484	321.408	279.624	58.498	322.788	257.680	-9.614
1500	56.927	325.320	282.541	64.169	323.167	253.016	-8 811
1600	57.335	329.008	285.331	69.882	323.554	248.327	-8.107
1700	57.710	332.495	288.004	75.635	187.031	244.793	-7.522
1800	58.052	335.803	290.568	81.423	189.288	248.126	-7,200
1900	58.364	338.951	293.033	87.244	191.540	251.333	-6.910
2000	58.648	341.952	295.404	93.095	193.783	254.422	-6.645
2100	58.907	344.819	297.690	98.973	196.013	257.399	-6.402
2200	59.143	347.565	299.895	104.876	198.222	260.271	-6.180
2300	59.358	350.199	302.025	110.801	200.405	263.042	-5.974
2400	59.554	352.730	304.085	116.747	202.553	265.719	-5.783
2500	59.734	355.164	306.080	122.711	204.658	268.307	-5.606
2600	59.897	357.511	308.013	128.693	206.712	270.812	-5.441
2700	60.047	359.774	309.889	134.690	208.704	273.240	-5.286
2800	60.185	361.960	311.710	140.702	210.625	275.595	-5.141
2900	60.311	364.074	313.479	146.727	212.463	277.882	-5.005
3000	60.428	366.121	315.200	152.764	214.210	280.108	-4.877
3100	60.535	368.104	316 875	158.812	215.853	282.278	-4.756
3200	60.634	370.028	318.506	164.870	217.384	284.395	-4.642
3300	60.725	371.895	320.095	170.938	218.792	286.467	-4.534
3400	60.810	373.709	321.646	177.015	220.069	288.499	-4.432
3500	60.888	375.473	323.159	183.100	221.204	290.494	-4.335
3600	60.961	377.189	324.636	189.193	222.192	292.460	-4.243
3700	61.029	378.860	326.079	195.292	223.026	294.399	-4.156
3800	61.092	380.489	327.489	201.398	223.699	296.320	-4.073
3900	61.151	382.076	328.869	207.510	224.209	298.224	-3.994
4000	61.205	383.625	330.218	213.628	224.551	300.116	-3.919
4100	61.257	385.137	331.539	219.751	224.725	302.003	-3.848
4200	61.305	386.614	332.833	225.879	224.732	303.887	-3.779
4300	61.350	388.057	334.101	232.012	224.569	305.774	-3.714
4400	61.392	389.468	335.343	238.149	224.239	307.666	-3.652
4500	61.432	390.848	336.561	244.291	223.746	309.567	-3.593
4600	61.470	392.199	337.756	250.436	223.159	311.475	-3.537
4700	61.505	393.521	338.929	256.584	222.371	313.404	-3.483
4800	61.538	394.816	340.080	262.737	221.439	315.350	-3.432
4900	61.570	396.086	341.210	268.892	220.369	317.318	-3.383
5000	61.599	397.330	342.320	275.050	219.174	319.308	-3.336
5100	61.627	398.550	343.410	281.212	217.876	321.321	-3.291
5200	61.654	399.747	344.482	287.376	216.445	323.363	-3.248
5300	61.679	400.921	345.536	293.542	214.909	325.434	-3.207
5400	61.703	402.074	346.572	299.712	213.277	327.534	-3.168
5500	61.725	403.207	347.592	305.883	211.557	329.666	-3.131
5600	61.747	404.319	348.595	312.057	209.757	331 830	-3.095
5700	61.767	405.412	349.582	318.232	207.886	334.026	-3.061
5800	61.787	406.487	350.554	324.410	205.952	336.256	-3.028
5900	61.805	407.543	351.511	330.590	204.172	338.495	-2.997
6000	61.823	408.582	352.454	336.771	202.164	340.789	-2.967

PREVIOUS: June 1976 (1 atm) CURRENT June 1976 (1 bar)

Strontium Hydroxide, Ion ($SrOH^+$) $H_1O_1Sr_1{}^+(g)$

$S°(298.15\ K) = 229.106 \pm 0.08\ J\ K^{-1}\ mol^{-1}$

$\Delta_f H°(0\ K) = 5.01 \pm 8.4\ kJ\ mol^{-1}$
$\Delta_f H°(298.15\ K) = 2.09 \pm 8.4\ kJ\ mol^{-1}$

Electronic Levels and Quantum Weights

State	ε_i, cm^{-1}	g_i
$[\tilde{X}^2A'']$	0	2
$[^2A']$	7029.764	2

Vibrational Frequencies and Degeneracies

ν, cm^{-1}
3410(1)
1390(1)

Point Group: C_1 $\sigma = 1$
Bond Distances: H–O = 0.977 Å O–O = 1.335 Å
Bond Angle: H–O–O = 104.1°
Prodcut of the Moments of Inertia: $I_AI_BI_C = 0.9128 \times 10^{-117}\ g^3\ cm^6$

Enthalpy of Formation

Wagman (1) and Cohen and Westburg (2) have reviewed the data which leads to the enthalpy of formation for HO_2(g). We adopt the value suggested by Wagman (1), $\Delta_f H°(HO_2$, g, 298.15 K) = 0.5±2.0 kcal mol^{-1}. Gurvich et al. (3) have also critically reviewed many of the same sources and have adopted the same value for the enthaply of formation. The analysis by Wagman (1) follows.

Kochubei and Moin (4) made a kinetic determination of the enthaply of formation of the HO_2 radical by studying the bimolecular radical formation reactions (forward only) HX(g) + O_2(g) = HO_2(g) + X(g), where X = Cl, Br, and I. The basic rate data are not given for this study. However, Wagman corrected their reported results by a small amount due to the improperly used term designated as $\overline{T}/d\Delta H/dT$). In the recalculation, the reported E_t-E_r values were used as the $\Delta_r H°$ at the stated mean temperature. Hence the $\Delta_f H°(298.15\ K)$ for HO_2(g) from this study is about 0.7 kcal mol^{-1}.

Calculations for $\Delta_f H°(HO_2$, g) were also made using the reaction H_3^+(g) + O_2(g) = HO_2^+(g) + H_2 (g). For this reaction, Pennelly, et al. (5) obtained $\Delta_r H°(0\ K)$ = -0.05 + 0.07 kcal mol^{-1}. Kim et al. (6), from rate constants for the above, obtain $\Delta_r G°(298.15\ K)$ = -0.41 kcal mol^{-1}. Estimating $\Delta S°(298.15\ K)$ to be +0.40 cal K^{-1} mol^{-1}, and hence $\Delta_r H°(298.15\ K)$ = -0.3±0.10 kcal mol^{-1}. The best value is $\Delta_r H°(298.15\ K)$ = -0.2±0.2 kcal mol^{-1}. Quantum mechanical calculations of the energy of H_3^+ have been made by a number of people (7, 8). Using the results of Duben and Lowe (8) and ionization potential of HO_2 from Foner and Hudson (9), Wagman obtained $\Delta_f H°(HO_2^+$, g) = 1.1 kcal mol^{-1}. This yields $\Delta_f H°(HO_2$, g, 298.15 K) = 0.4 kcal mol^{-1}.

An additional calculation can be obtained from the ionization and appearance potentials for HO_2 from H_2O_2. Recent measurements by photoionization at NBS yield $\Delta_r H°$ = 348.7±0.2 kcal mol^{-1} for H_2O_2(g), H_2O_2 + e$^-$ = HO_2^+(g) + H^+ + 2 e$^-$. Combining with the ionization potential of HO_2(g) leads to $\Delta_r H°$ for H_2O_2(g) = HO_2(g) + H(g) and $\Delta_f H°(HO_2$, g, 298.15 K) = 0.6 kcal mol^{-1}.

In these last two calculations, the principal uncertainty lies in the value for the ionization potential of HO_2 reported by Foner and Hudson (9). However the general consistency of the various paths indicates that a reasonable value for $\Delta_f H°(HO_2$, g, 298.15 K) is 0.5±2.0 kcal mol^{-1}.

Heat Capacity and Entropy

The vibrational frequencies adopted are those determined by Paukert and Johnston (11) from infrared absorption measurements on gas phase HO_2. Matrix studies of HO_2 yielded similar results (12, 13, 14). The bond lengths and bond angles are taken from the microwave spectroscopic study of Beers and Howard (15). Other measurements are in agreement with these values (11, 14, 16, 17).

The energy separation between the ground and lowest excited electronic state was determined by Freedman and Jones (18). Their measurements also yield $I_AI_BI_C$ to be $1.090 \times 10^{-117}\ g^3\ cm^6$ and ν_3 = 983 cm^{-1} for the $^2A^1$ state. This indicates that the structure of this excited state is quite similar to the ground state; thus, for convenience in calculating the entropy and heat capacity, the vibrational frequencies and moments of inertia of the excited state are taken to be the same as the ground state. For the ground electronic state, the principal moments of inertia are: $I_A = 0.1375 \times 10^{-39}$, $I_B = 2.5088 \times 10^{-39}$ and $I_C = 2.6463 \times 10^{-39}\ g\ cm^2$.

References

1. D. D. Wagman, U. S. Nat. Bur. Stand., personal communication to Dr. Norman Cohen, March 21, 1978.
2. N. Cohen and K. Westburg, Aerospace Corporation, personal communication to Dr. Joseph Masi, February 22, 1978.
3. L. V. Gurvich et al., "Thermodynamic Properties of Individual Substances, Vol. I, Books 1 and 2," Nauka, Moscow, USSR, 1978.
4. V. F. Kochubie and J. B. Moin, Dokl. Akad. Nauk SSSR 219, 141 (1974).
5. P. F. Fennelly, R. S. Hensworth, H. I. Schiff, and D. K. Bohme, J. Chem. Phys. 59, 6405 (1973).
6. J. K. Kim, L. P. Theard, and W. T. Huntress, Chem. Phys. Letters 32, 610 (1975).
7. M. E. Schwartz and L. J. Schaad, J. Chem. Phys. 47, 5325 (1967).
8. A. J. Duben and J. P. Lowe, J. Chem. Phys. 55, 4270 (1971).
9. S. N. Foner and R. L. Hudson, J. Chem. Phys. 36, 2681 (1962).
10. K. McCulloh, to be published, see reference (1).
11. T. T. Paukert and H. S. Johnston, J. Chem. Phys. 56, 2824 (1972).
12. D. W. Smith and L. Andrews, J. Chem. Phys. 60, 81 (1974).
13. D. E. Milligan and M. E. Jacox, J. Chem. Phys. 38, 2677 (1963).
14. M. E. Jacox and D. E. Milligan, J. Mol. Spectrosc. 42, 495 (1972).
15. Y. Beers and C. J. Howard, J. Chem. Phys. 64, 1541 (1976).
16. S. Saito, J. Mol. Spectrosc. 65, 229 (1977).
17. C. E. Barnes, J. M. Brown, A. Carrington, J. Pinkstone, and T. J. Sears, J. Mol. Spectrosc. 72, 86 (1978).
18. P. A. Freedman and W. J. Jones, J. Chem. Soc., Faraday II 72, 2071 (1976).

Enthalpy Reference Temperature = T_r = 298.15 K — Standard State Pressure = p° = 0.1 MPa

T/K	$C_p°$ (J K^{-1}mol^{-1})	S° (J K^{-1}mol^{-1})	$-[G°-H°(T_r)]/T$ (J K^{-1}mol^{-1})	$H°-H°(T_r)$ (kJ mol^{-1})	$\Delta_f H°$ (kJ mol^{-1})	$\Delta_f G°$ (kJ mol^{-1})	Log K_f
0	0.	0.	INFINITE	-10.003	5.006	5.006	INFINITE
100	33.258	192.430	259.201	-6.677	3.928	7.052	-3.683
200	33.491	215.515	232.243	-3.346	3.001	10.536	-2.752
250	34.044	223.040	229.676	-1.659	2.532	12.475	-2.606
298.15	34.905	229.106	229.106	0.	2.092	14.430	-2.528
300	34.943	229.322	229.107	0.065	2.076	14.506	-2.526
350	36.072	234.791	229.536	1.839	1.649	16.612	-2.479
400	37.296	239.688	230.504	3.673	1.260	18.777	-2.452
450	38.519	244.151	231.776	5.569	0.908	20.988	-2.436
500	39.687	248.271	233.222	7.524	0.591	23.236	-2.427
600	41.781	255.697	236.363	11.601	0.043	27.819	-2.422
700	43.558	262.275	239.603	15.870	-0.411	32.485	-2.424
800	45.084	268.193	242.813	20.304	-0.790	37.211	-2.430
900	46.418	273.582	245.937	24.880	-1.107	41.981	-2.436
1000	47.604	278.535	248.952	29.583	-1.368	46.783	-2.444
1100	48.672	283.123	251.853	34.397	-1.582	51.609	-2.451
1200	49.643	287.400	254.639	39.314	-1.754	56.452	-2.457
1300	50.535	291.410	257.315	44.323	-1.887	61.308	-2.463
1400	51.360	295.185	259.886	49.419	-1.988	66.173	-2.469
1500	52.128	298.755	262.360	54.593	-2.058	71.045	-2.474
1600	52.845	302.143	264.741	59.842	-2.102	75.920	-2.479
1700	53.518	305.367	267.037	65.161	-2.122	80.797	-2.483
1800	54.149	308.444	269.252	70.545	-2.121	85.674	-2.486
1900	54.742	311.388	271.393	75.989	-2.102	90.551	-2.489
2000	55.299	314.210	273.464	81.492	-2.067	95.427	-2.492
2100	55.820	316.921	275.469	87.048	-2.019	100.301	-2.495
2200	56.308	319.529	277.413	92.655	-1.960	105.172	-2.497
2300	56.763	322.042	279.299	98.309	-1.893	110.040	-2.499
2400	57.186	324.467	281.131	104.006	-1.818	114.905	-2.501
2500	57.578	326.809	282.911	109.745	-1.740	119.767	-2.502
2600	57.940	329.075	284.644	115.521	-1.659	124.625	-2.504
2700	58.274	331.268	286.330	121.332	-1.577	129.481	-2.505
2800	58.580	333.393	287.973	127.175	-1.497	134.334	-2.506
2900	58.860	335.453	289.575	133.047	-1.419	139.183	-2.507
3000	59.115	337.453	291.138	138.946	-1.346	144.030	-2.508
3100	59.347	339.395	292.663	144.869	-1.278	148.876	-2.509
3200	59.557	341.283	294.153	150.814	-1.218	153.718	-2.509
3300	59.745	343.118	295.609	156.780	-1.165	158.558	-2.510
3400	59.915	344.904	297.033	162.763	-1.122	163.398	-2.510
3500	60.066	346.643	298.426	168.762	-1.089	168.236	-2.511
3600	60.200	348.337	299.789	174.775	-1.067	173.074	-2.511
3700	60.318	349.989	301.123	180.801	-1.058	177.910	-2.512
3800	60.423	351.599	302.431	186.839	-1.061	182.748	-2.512
3900	60.513	353.169	303.711	192.886	-1.078	187.586	-2.512
4000	60.592	354.702	304.967	198.941	-1.109	192.423	-2.513
4100	60.659	356.199	306.199	205.004	-1.156	197.262	-2.513
4200	60.716	357.662	307.406	211.072	-1.217	202.102	-2.514
4300	60.764	359.091	308.592	217.146	-1.296	206.945	-2.514
4400	60.803	360.488	309.756	223.225	-1.391	211.788	-2.514
4500	60.834	361.855	310.898	229.307	-1.504	216.634	-2.515
4600	60.858	363.193	312.021	235.391	-1.634	221.482	-2.515
4700	60.876	364.502	313.123	241.478	-1.784	226.335	-2.515
4800	60.888	365.783	314.207	247.566	-1.952	231.190	-2.516
4900	60.895	367.039	315.272	253.656	-2.141	236.050	-2.516
5000	60.897	368.269	316.320	259.745	-2.350	240.913	-2.517
5100	60.894	369.475	317.351	265.835	-2.580	245.780	-2.517
5200	60.888	370.658	318.364	271.924	-2.831	250.652	-2.518
5300	60.879	371.817	319.362	278.012	-3.105	255.530	-2.518
5400	60.866	372.955	320.344	284.100	-3.402	260.412	-2.519
5500	60.851	374.072	321.311	290.185	-3.722	265.300	-2.520
5600	60.834	375.168	322.263	296.270	-4.067	270.195	-2.520
5700	60.814	376.245	323.200	302.352	-4.437	275.096	-2.521
5800	60.792	377.302	324.124	308.432	-4.832	280.004	-2.522
5900	60.769	378.341	325.034	314.511	-5.254	284.917	-2.522
6000	60.745	379.362	325.931	320.586	-5.702	289.839	-2.523

PREVIOUS: September 1978 (1 atm) CURRENT: September 1978 (1 bar)

PHOSPHINIDENE (PH) IDEAL GAS $M_r = 31.98170$ Phosphinidene (PH) $H_1P_1(g)$

$S°(298.15\ K) = 196.344\ J\ K^{-1}\ mol^{-1}$

$\Delta_f H°(0\ K) = 254.49 \pm 33.5\ kJ\ mol^{-1}$
$\Delta_f H°(298.15\ K) = 253.55 \pm 33.5\ kJ\ mol^{-1}$

Electronic Levels and Quantum Weights

State	ε_i, cm^{-1}	g_i
$^3\Sigma^-$	0	3
	[7650]	2
	[15150]	1
	29560	6
	[38110]	2
	[57490]	1

$\omega_e = 2380\ cm^{-1}$ $\omega_e x_e = [55]\ cm^{-1}$ $\sigma = 1$
$B_e = 8.412\ cm^{-1}$ $\alpha_e = [0.28]\ cm^{-1}$ $r_e = 1.43\ Å$

Enthalpy of Formation

Jordan (1) estimated the dissociation energy of PH(g) as 70.44 kcal mol^{-1} using a semiempirical theoretical method. The corresponding value for the enthalpy of formation $\Delta_f H°(298.15\ K)$ is 60.6±8 kcal mol^{-1}.

Heat Capacity and Entropy

Jordan (1) predicted five electronic levels ($X^3\Sigma$, $A^1\Delta$, $B^1\Sigma$, $D^1\Pi$, $E^1\Sigma$) based on the reported value (2) of 29560 cm^{-1} for the $C^3\Pi$ level (this level is designated $A^3\Pi$ by Herzberg (2)). Several qualitative spectroscopic investigations of PH(g) have been reported and are in general accord with the predictions of Jordan (1). Ishaq and Pearse (3) reported the values of the rotational constant B_e and the fundamental vibrational frequency ω_e. The value of α_e is calculated from the Morse potential function. The bond distance is calculated from B_e. The value of the anharmonic vibrational term $\omega_e x_e$ is estimated from the relation $\omega_e x_e = \omega_e^2 /(4D° + 2\omega_e)$.

References

1. P. Jordan, J. Chem. Phys. 41, 1442 (1964).
2. G. Herzberg, "Spectra of Diatomic Molecules," D. Van Nostrand Co., Inc., New York, 1950.
3. M. Ishaq and R. W. B. Pearse, Proc. Roy. Soc. (London) A173, 265 (1939), and R. W. B. Pearse, ibid, A129, 328 (1930).

Enthalpy Reference Temperature = T_r = 298.15 K Standard State Pressure = p° = 0.1 MPa

T/K	$C_p°$ (J $K^{-1}mol^{-1}$)	S°	$-[G°-H°(T_r)]/T$	$H°-H°(T_r)$ (kJ mol^{-1})	$\Delta_f H°$	$\Delta_f G°$	Log K_f
0	0.	0.	INFINITE	-8.650	254 494	254.494	INFINITE
100	29.117	164.523	222.248	-5.773	254.985	245.337	-128.151
200	29.130	184.709	199.010	-2.860	254.287	235.711	-61.561
250	29.140	191.210	196.824	-1.404	253 953	231.104	-48.287
298.15	29.162	196.344	196.344	0.	253.550	226.739	-39.724
300	29.163	196.525	196.345	0.054	253.534	226.572	-39.450
350	29.217	201.024	196.700	1.513	252.333	222.187	-33.160
400	29.321	204.931	197.490	2.977	251.751	217.921	-28.458
450	29.485	208.394	198.513	4.446	251.175	213.726	-24.809
500	29.709	211.511	199.659	5.926	250.607	209.596	-21 896
600	30.302	216.978	202.102	8.925	249.509	201.498	-17.542
700	31.008	221.701	204.572	11.990	248.472	193.579	-14.445
800	31.742	225.889	206.980	15.128	247.501	185.804	-12.132
900	32.452	229.669	209.294	18.338	246.591	178.147	-10.339
1000	33.111	233.123	211.507	21.616	245.735	170.589	-8.911
1100	33.713	236.308	213.619	24.958	244.925	163.114	-7.746
1200	34.260	239.265	215.634	28.357	180.580	156.789	-6.825
1300	34.759	242.027	217.559	31.808	180.613	154.806	-6.220
1400	35.218	244.620	219.400	35.307	180.668	152.818	-5.702
1500	35.645	247.065	221.164	38.851	180.740	150.827	-5.252
1600	36.046	249.378	222.856	42.436	180.829	148.830	-4.859
1700	36.424	251.575	224.481	46.059	180.932	146.827	-4.511
1800	36.785	253.667	226.045	49.720	181.050	144.817	-4.202
1900	37.131	255.665	227.552	53.416	181.180	142.801	-3.926
2000	37.463	257.578	229.005	57.146	181.324	140.777	-3.677
2100	37.783	259.414	230.410	60.908	181.480	138.746	-3.451
2200	38.090	261.179	231.769	64.702	181.649	136.707	-3.246
2300	38.387	262.878	233.085	68.526	181.830	134.660	-3.058
2400	38.672	264.518	234.360	72.379	182.023	132.605	-2.886
2500	38.947	266.103	235.599	76.260	182.229	130.542	-2.728
2600	39.212	267.635	236.801	80.168	182.445	128.470	-2.581
2700	39.466	269.120	237.971	84.102	182.673	126.390	-2.445
2800	39.709	270.560	239.109	88.061	182.911	124.301	-2.319
2900	39.943	271.957	240.218	92.043	183.160	122.204	-2.201
3000	40.167	273.315	241.299	96.049	183.418	120.097	-2.091
3100	40.382	274.636	242.353	100.077	183.685	117.982	-1.988
3200	40.588	275.921	243.382	104.125	183.961	115.858	-1.891
3300	40.785	277.173	244.387	108.194	184.246	113.726	-1.800
3400	40.973	278.393	245.369	112.282	184.537	111.584	-1.714
3500	41.154	279.584	246.330	116.388	184.835	109.434	-1.633
3600	41.328	280.746	247.270	120.512	185.140	107.276	-1.557
3700	41.494	281.880	248.190	124.654	185.451	105.109	-1.484
3800	41.654	282.989	249.091	128.811	185.767	102.933	-1.415
3900	41.808	284.073	249.974	132.984	186.087	100.749	-1.349
4000	41.956	285.133	250.840	137.172	186.412	98.557	-1.287
4100	42.098	286.171	251.689	141.375	186.741	96.356	-1.228
4200	42.237	287.187	252.522	145.592	187.074	94.148	-1.171
4300	42.370	288.183	253.340	149.822	187.410	91.931	-1.117
4400	42.500	289.158	254.143	154.066	187.748	89.707	-1.065
4500	42 627	290.115	254.932	158.322	188.090	87.475	-1.015
4600	42.750	291.053	255.707	162.591	188.434	85.235	-0.968
4700	42.870	291.974	256.469	166.872	188.781	82.988	-0.922
4800	42.988	292.877	257.218	171.165	189.130	80.734	-0.879
4900	43.104	293.765	257.955	175.470	189.482	78.471	-0.837
5000	43.218	294.637	258.680	179.786	189.836	76.203	-0.796
5100	43.331	295.494	259.393	184.113	190.193	73.926	-0.757
5200	43.442	296.336	260.096	188.452	190.553	71.643	-0.720
5300	43.552	297.165	260.787	192.802	190.915	69.353	-0.684
5400	43.661	297.980	261.468	197.162	191.281	67.056	-0.649
5500	43.770	298.782	262.140	201.534	191.651	64.752	-0.615
5600	43.878	299.572	262.801	205.916	192.024	62.441	-0.582
5700	43.985	300.349	263.453	210.309	192.402	60.124	-0.551
5800	44.092	301.115	264.096	214.713	192.784	57.800	-0.521
5900	44.199	301.870	264.730	219.128	193.171	55.469	-0.491
6000	44.307	302.614	265.355	223.553	193.564	53.132	-0.463

PREVIOUS: June 1967 (1 atm) CURRENT: June 1967 (1 bar)

Phosphinidene (PH) $H_1P_1(g)$

LEAD HYDRIDE (PbH) IDEAL GAS $M_r = 208.20794$

$S°(298.15\ K) = 220.756\ J\ K^{-1}\ mol^{-1}$

$\Delta_f H°(0\ K) = 238.62 \pm 19.2\ kJ\ mol^{-1}$
$\Delta_f H°(298.15\ K) = 236.19 \pm 19.2\ kJ\ mol^{-1}$

Electronic Levels and Quantum Weights

State	ε_i, cm^{-1}	g_i
$^2\Pi_{1/2}$	0	2
$^2\Pi_{3/2}$	[8000]	2

$\omega_e = 1564.1\ cm^{-1}$ $\omega_e x_e = 29.75\ cm^{-1}$ $\sigma = 1$
$B_e = 4.971\ cm^{-1}$ $\alpha_e = 0.144\ cm^{-1}$ $r_e = 1.8388\ Å$

Enthalpy of Formation

$\Delta_f H°(298.15\ K)$ is calculated from the dissociation energy of PbH(g) reported by Gaydon (1).

Heat Capacity and Entropy

Ground state configuration and spectroscopic data were obtained from Herzberg (2). Electronic levels and multiplicity were estimated by comparison with those for PbF(g).

References

1. A. G. Gaydon, "Dissociation Energies and Spectra of Diatomic Molecules," Chapman and Hall, Ltd., London, 1953.
2. G. Herzberg, "Spectra of Diatomic Molecules," Van Nostrand, New York (1950).

Lead Hydride (PbH) $H_1Pb_1(g)$

Enthalpy Reference Temperature = T_r = 298.15 K   Standard State Pressure = p° = 0.1 MPa

T/K	$C_p°$	S°	$-[G°-H°(T_r)]/T$	$H°-H°(T_r)$	$\Delta_f H°$	$\Delta_f G°$	Log K_f
	J K⁻¹mol⁻¹			kJ mol⁻¹			
0	0.	0.	INFINITE	-8.678	238.620	238.620	INFINITE
100	29.121	188.877	246.747	-5.787	238.246	228.077	-119.135
200	29.160	209.070	223.438	-2.874	237.288	218.270	-57.006
250	29.261	215.585	221.240	-1.414	236.744	213.577	-44.624
298.15	29.476	220.756	220.756	0.	236.187	209.165	-36.645
300	29.487	220.938	220.756	0.055	236.165	208.998	-36.390
350	29.847	225.509	221.116	1.537	235.570	204.517	-30.522
400	30.311	229.524	221.921	3.041	234.970	200.121	-26.133
450	30.836	233.124	222.969	4.569	234.371	195.801	-22.728
500	31.384	236.401	224.151	6.125	233.778	191.548	-20.011
600	32.448	242.218	226.690	9.317	232.617	183.211	-15.950
700	33.390	247.293	229.278	12.610	226.619	175.865	-13.123
800	34.184	251.805	231.817	15.990	225.506	168.692	-11.014
900	34.846	255.871	234.267	19.443	224.486	161.652	-9.382
1000	35.400	259.571	236.616	22.956	223.542	154.722	-8.082
1100	35.872	262.968	238.859	26.520	222.662	147.883	-7.022
1200	36.285	266.107	241.001	30.128	221.837	141.122	-6.143
1300	36.656	269.027	243.045	33.776	221.054	134.427	-5.401
1400	36.999	271.756	245.000	37.459	220.292	127.792	-4.768
1500	37.326	274.320	246.870	41.175	219.540	121.212	-4.221
1600	37.644	276.739	248.662	44.924	218.791	114.681	-3.744
1700	37.957	279.031	250.381	48.704	218.038	108.197	-3.324
1800	38.267	281.209	252.034	52.515	217.281	101.757	-2.953
1900	38.577	283.286	253.625	56.357	216.517	95.360	-2.622
2000	38.886	285.273	255.158	60.230	215.745	89.003	-2.325
2100	39.194	287.178	256.637	64.134	37.745	89.801	-2.234
2200	39.499	289.008	258.067	68.069	37.347	92.288	-2.191
2300	39.800	290.770	259.451	72.034	36.876	94.796	-2.153
2400	40.096	292.471	260.792	76.029	36.328	97.326	-2.118
2500	40.386	294.113	262.092	80.053	35.704	99.880	-2.087
2600	40.667	295.703	263.354	84.106	35.005	102.461	-2.058
2700	40.940	297.243	264.581	88.186	34.232	105.069	-2.033
2800	41.203	298.736	265.774	92.293	33.388	107.708	-2.009
2900	41.456	300.187	266.936	96.426	32.478	110.378	-1.988
3000	41.698	301.596	268.068	100.584	31.507	113.081	-1.969
3100	41.929	302.967	269.172	104.766	30.479	115.817	-1.952
3200	42.149	304.302	270.249	108.970	29.401	118.587	-1.936
3300	42.357	305.602	271.301	113.195	28.279	121.391	-1.921
3400	42.554	306.870	272.328	117.441	27.118	124.230	-1.909
3500	42.739	308.106	273.333	121.705	25.925	127.104	-1.897
3600	42.914	309.312	274.316	125.988	24.705	130.012	-1.886
3700	43.079	310.490	275.277	130.288	23.465	132.954	-1.877
3800	43.233	311.641	276.219	134.604	22.209	135.930	-1.868
3900	43.378	312.766	277.142	138.934	20.942	138.939	-1.861
4000	43.513	313.866	278.046	143.279	19.670	141.981	-1.854
4100	43.640	314.942	278.933	147.637	18.395	145.054	-1.848
4200	43.759	315.995	279.803	152.007	17.121	148.159	-1.843
4300	43.870	317.026	280.657	156.388	15.852	151.294	-1.838
4400	43.973	318.036	281.495	160.780	14.591	154.458	-1.834
4500	44.070	319.025	282.318	165.183	13.340	157.651	-1.830
4600	44.161	319.995	283.127	169.594	12.100	160.872	-1.827
4700	44.246	320.945	283.921	174.014	10.875	164.119	-1.824
4800	44.325	321.878	284.702	178.443	9.664	167.393	-1.822
4900	44.400	322.793	285.470	182.879	8.470	170.691	-1.820
5000	44.470	323.690	286.226	187.323	7.293	174.014	-1.818
5100	44.536	324.572	286.969	191.773	6.133	177.359	-1.817
5200	44.598	325.437	287.700	196.230	4.991	180.728	-1.815
5300	44.657	326.287	288.420	200.693	3.867	184.118	-1.815
5400	44.712	327.122	289.129	205.161	2.762	187.530	-1.814
5500	44.765	327.943	289.828	209.635	1.675	190.961	-1.814
5600	44.815	328.750	290.516	214.114	0.606	194.413	-1.813
5700	44.863	329.544	291.193	218.598	-0.445	197.883	-1.813
5800	44.909	330.325	291.861	223.087	-1.480	201.371	-1.814
5900	44.953	331.093	292.520	227.580	-2.496	204.877	-1.814
6000	44.995	331.849	293.169	232.077	-3.496	208 400	-1.814

PREVIOUS: June 1962 (1 atm)   CURRENT June 1962 (1 bar)

Lead Hydride (PbH) $H_1Pb_1(g)$

MERCAPTO (SH) IDEAL GAS M_r = 33.06794

$\Delta_f H°(0\ K) = 136.49 \pm 5.0\ kJ\ mol^{-1}$

$S°(298.15\ K) = 195.628 \pm 0.04\ J\ K^{-1}\ mol^{-1}$(*)

$\Delta_f H°(298.15\ K) = 139.33 \pm 5.0\ kJ\ mol^{-1}$

Electronic States and Molecular Constants (σ = 1)

State	ε_i, cm^{-1}	g_i	ω_e, cm^{-1}	$\omega_e x_e$, cm^{-1}	B_e, cm^{-1}	α_e, cm^{-1}	r_e, Å
	0	2					
$X^2\Pi_i$			2689.6	45.5	9.601	0.285	1.345
	377.01	2					
$A^2\Sigma^+$	30663	2	1979.8	97.65	8.521	0.464	1.423
$B^2\Sigma$	59622	2	2670.6	56.8	8.785	0.259	1.428
$C^2\Delta$	[63900]	4					
$D^2\Delta$	71195	4					
$E^2\Sigma$	71318	2					
$F^2\Delta$	76708	4	[2689.6]	[45.5]	[9.601]	[0.285]	
$G^2\Delta$	79343	4					
$H^2\Delta$	80848	4					

Enthalpy of Formation

The previous JANAF (1) selection for $\Delta_f H°$(HS, g, 298.15 K) was Mackle's (2) estimate of 34.6±4 kcal mol^{-1} which he derived from the average of three independent determinations. Of these, the uncertainty can be reduced in the determination involving the calculation of $\Delta H_f°$(HS, g, 298.15 K) from $\Delta_f H°$(H_2S, g, 298.15 K), the ionization potential of HS(g), and the appearance potential of HS^+(g) from H_2S(g). A summary of previous work on the ionization potential is given below. From the spectroscopic and photoionization work we arrive at a value of $\Delta_f H°$(0 K) for the reaction $H_2S(g) = HS(g) + H(g)$ of 89.02±1.5 kcal mol^{-1}. Using the appropriate thermal functions for each of the species (see tables for H_2S(g), HS(g), and H(g) (1)) and the selected enthalpies of formation of H_2S(g) of -4.90±2 kcal mol^{-1} (see H_2S(g) table (1)) and H(g) of 52.103±0.001 kcal mol^{-1} (8), one obtains 33.3±1.2 kcal mol^{-1} for $\Delta_f H°$(HS, g, 298.15 K) versus Mackle's (2) values of 33.7±3 kcal mol^{-1} for this determination. The new value is preferred because it is based on photoionization rather than electron impact studies.

Source	Method	Potential(eV)
$HS(g) \rightarrow HS^+(g) + e^-(g)$:		
Morrow (3)	Rydberg Extrapolation	10.40±0.03
Palmer and Loosing (4)	Electron Impact	10.5 ±0.1
$H_2S(g) \rightarrow HS^+(g) + H(g) + e^-(g)$		
Dibeler and Liston (5)	Photoionization	14.27±0.04†
Dibeler and Rosenstock (6)	Electron Impact	14.43±0.1
Palmer and Lossing (4)	Electron Impact	14.43±0.1
Neuert and Clasen (7)	Electron Impact	15.2 ±0.5

†The uncertainty given by the authors was doubled because the threshold value was not corrected to absolute zero.

Heat Capacity and Entropy*

Calculations were made using the vibrational and rotational constants for the respective electronic levels from Rosen (12). Comparison of the results of the more exact treatment of Haar et al. (13) with those given in the previous JANAF (1) calculations (both used the same molecular constants for the ground state) suggests errors in our calculations are negligible above 400 K. Below this, they can be appreciable. In particular, it is recommended that 0.045 kcal mol^{-1}, 0.012 cal K^{-1} mol^{-1} and -0.028 cal K^{-1} mol^{-1} be added to our values of H°(0 K)-H°(298.5 K), S°(298.15 K), and $C_p°$(298.15 K). These errors result from dealing with the ground state ($X^2\pi_i$) as two separate electronic states separated by 377.01 cm^{-1}.

The thermodynamic functions are calculated using first-order anharmonic corrections to Q_r^i and Q_v^i in the partition function $Q = Q_t \sum_i Q_r^i Q_v^i g_i \exp(-c_2\varepsilon_i/T)$. The National Bureau of Standards prepared this table (9) by critical analysis of data existing in 1972. Using molecular constants and $\Delta_f H°$ selected by NBS (9), we recalculate the table in terms of 1973 fundamental constants (10), 1975 atomic weights (11) and current JANAF reference states for the elements.

References

1. JANAF Thermochemical Tables, 2nd ed., NSRDS-NBS 37, 1971.
2. H. Mackle, Tetrahedron 19, 1159 (1963).
3. B. A. Morrow, Can. J. Phys. 44, 2447 (1966).
4. T. F. Palmer and F. P. Lossing, J. Am. Chem. Soc. 84, 4661 (1962).
5. V. H. Dibeler and S. K. Liston, J. Chem. Phys. 49, 482 (1968).
6. V. H. Dibeler and H. M. Rosenstock, J. Chem. Phys. 39, 3106 (1963).
7. H. Neuert and H. Clasen, Z. Naturforschung 79, 410 (1952).
8. ISCU CODATA Task Group on Key Values for Thermodynamics Final Set of Key Values for Thermodynamics - Part I, November, 1971.
9. S. Abramowitz et al., U. S. Nat. Bur. Stand., Report 10904, 239, 1972.
10. CODATA Task Group on Fundamental Constants, CODATA Bulletin 11, December, 1973.
11. IUPAC Commission on Atomic Weights, Pure Appl. Chem. 47, 75 (1976).
12. B. Rosen, "Spectroscopic Data Relative to Diatomic Molecules", Pergamon Press, Oxford, 1970.
13. L. Haar, A. S. Friedman, and C. N. Beckett, U. S. Nat. Bur. Stand. Monograph 20, 1961.

Mercapto (HS) $H_1S_1(g)$

Enthalpy Reference Temperature = T_r = 298.15 K — Standard State Pressure = p° = 0.1 MPa

T/K	$C_p°$ (J K^{-1}mol^{-1})	S° (J K^{-1}mol^{-1})	-[G°-H°(T_r)]/T (J K^{-1}mol^{-1})	H°-H°(T_r) (kJ mol^{-1})	$\Delta_f H°$ (kJ mol^{-1})	$\Delta_f G°$ (kJ mol^{-1})	Log K_f
0	0.	0.	INFINITE	-9.273	136.493	136.493	INFINITE
100	30.185	160.686	224.496	-6.381	137.541	127.135	-66.409
200	32.697	182.575	198.614	-3.208	138.547	116.337	-30.384
250	32.734	189.885	196.164	-1.570	138.977	110.733	-23.136
298.15	32.447	195.628	195.628	0.	139.327	105 260	-18.441
300	32.434	195.829	195.629	0.060	139.340	105.049	-18.291
350	32.052	200.800	196.022	1.672	139.644	99.309	-14.821
400	31.707	205.057	196.892	3.266	138.794	93.573	-12.219
450	31.446	208.775	198.010	4.844	138.679	87.922	-10.206
500	31.283	212.079	199.255	6.412	138.515	82.293	-8.597
600	31.228	217.773	201.881	9 535	138.381	71.064	-6.187
700	31.452	222.601	204.505	12.667	138.370	59.846	-4.466
800	31.844	226.825	207.036	15.831	138.452	48.624	-3.175
900	32.318	230.602	209.448	19.039	112.044	37.927	-2.201
1000	32.814	234.033	211.738	22.295	112.871	29.648	-1.549
1100	33.301	237.184	213.910	25.601	113.722	21.284	-1.011
1200	33.761	240.101	215.972	28.955	114.592	12.843	-0.559
1300	34.185	242.820	217.934	32.352	115.478	4.328	-0.174
1400	34.571	245.368	219.804	35.790	116.376	-4.256	0.159
1500	34.922	247.765	221.589	39.265	117.280	-12.904	0.449
1600	35.239	250.030	223.296	42.774	118.188	-21.613	0.706
1700	35.526	252.175	224.932	46.312	119.099	-30.378	0.933
1800	35.786	254.213	226.503	49.878	120.009	-39.197	1.137
1900	36.021	256.154	228.013	53.468	120.919	-48.067	1.321
2000	36.236	258.007	229.467	57.081	121.826	-56.984	1.488
2100	36.432	259.780	230.868	60.715	122.731	-65.947	1.640
2200	36.611	261.479	232.221	64.367	123.632	-74.953	1.780
2300	36.777	263.110	233.529	68.037	124.530	-83.999	1.908
2400	36.930	264.679	234.794	71.722	125.425	-93.085	2.026
2500	37.072	266.189	236.020	75.422	126.316	-102.208	2.136
2600	37.205	267.646	237.209	79.136	127.203	-111.366	2.237
2700	37.329	269.052	238.362	82.863	128.086	-120.559	2.332
2800	37.445	270.412	239.483	86.602	128.967	-129.784	2.421
2900	37.555	271.728	240.572	90.352	129.843	-139.041	2.504
3000	37.659	273.003	241.632	94.113	130.716	-148.327	2.583
3100	37.758	274.239	242.664	97.883	131.585	-157.643	2.656
3200	37.852	275.439	243.669	101.664	132.451	-166.987	2.726
3300	37.942	276.606	244.650	105.454	133.313	-176.358	2.792
3400	38.028	277.740	245.607	109.252	134.171	-185.755	2.854
3500	38.112	278.843	246.540	113.059	135.025	-195.177	2.913
3600	38.192	279.918	247.453	116.874	135.876	-204.623	2.969
3700	38.270	280.965	248.344	120.698	136.722	-214.093	3.022
3800	38.346	281.987	249.216	124.528	137.565	-223.586	3.073
3900	38.420	282.984	250.069	128.367	138.403	-233.101	3.122
4000	38.492	283.958	250.905	132.212	139.237	-242.637	3.169
4100	38.564	284.909	251.722	136.065	140.066	-252.194	3.213
4200	38.634	285.839	252.524	139.925	140.890	-261.771	3.256
4300	38.703	286.749	253.309	143.792	141.710	-271.368	3.296
4400	38.772	287.640	254.079	147.666	142.526	-280.984	3.336
4500	38.841	288.512	254.835	151.546	143.336	-290.619	3.373
4600	38.909	289.366	255.576	155.434	144.142	-300.271	3.410
4700	38.977	290.204	256.304	159.328	144.943	-309.941	3.445
4800	39.045	291.025	257.019	163.229	145.740	-319.627	3.478
4900	39.114	291.831	257.721	167.137	146.533	-329.331	3.511
5000	39.183	292.622	258.411	171.052	147.321	-339.050	3.542
5100	39.253	293.398	259.090	174.974	148.106	-348.786	3.572
5200	39.323	294.161	259.757	178.902	148.886	-358.536	3.602
5300	39.394	294.911	260.413	182.838	149.664	-368.302	3.630
5400	39.466	295.648	261.059	186.781	150.438	-378.082	3.657
5500	39.539	296.373	261.694	190.732	151.210	-387.877	3.684
5600	39.614	297.086	262.320	194.689	151.980	-397.685	3.709
5700	39.689	297.788	262.936	198.654	152.748	-407.507	3.734
5800	39.766	298.478	263.543	202.627	153.516	-417.343	3.759
5900	39.844	299.159	264.141	206.608	154.283	-427.192	3.782
6000	39.924	299.829	264.730	210.596	155.050	-437.054	3.805

PREVIOUS: June 1977 (1 atm) CURRENT: June 1977 (1 bar)

Mercapto (HS) $H_1S_1(g)$

SILYLIDYNE (SiH) IDEAL GAS M_r = 29.09344

Silylidyne (SiH) $H_1Si_1(g)$

$S°(298.15\ K) = 198.040 \pm 0.21\ J\ K^{-1}\ mol^{-1}$

$\Delta_f H°(0\ K) = 374.89 \pm 8.4\ kJ\ mol^{-1}$
$\Delta_f H°(298.15\ K) = 376.66 \pm 8.4\ kJ\ mol^{-1}$

Electronic Levels and Quantum Weights

State	ε_i, cm^{-1}	g_i	Ref
A $^2\Pi$	0	2	(1)
	142.83	2	(1)
$^4\Sigma^-$	[21510]	4	(2)
A $^2\Delta$	24193.04	4	(1)
$^2\Sigma^-$	[24730]	2	(2)
C $^2\Sigma^+$	31899.8	2	(3)
B $^2\Sigma^+$	31909.7	2	(3)
D $^2\Delta$	48510.11	4	(1)
E $^2\Sigma^+$	52399.19	2	(1)

$\omega_e = 2041.80\ cm^{-1}$ $\omega_e x_e = 35.51\ cm^{-1}$ $\sigma = 1$
$B_e = 7.4996\ cm^{-1}$ $\alpha_e = 0.2190\ cm^{-1}$ $r_e = 1.5201\ Å$

Enthalpy of Formation

Verma (4) has derived an upper limit for the dissociation energy of 24680 cm^{-1} (70.6 kcal mol^{-1}) from predissociation in the B $^2\Sigma^+$ state. Hildenbrand (5) has applied an ionic-covalent correction to the linear Birge-Sponer extrapolation and obtained D_0^0 = 70 kcal mol^{-1}. Rao and Lakshman (6) have estimated the ground state dissociation energy of 24040 cm^{-1} (68.7 kcal mol^{-1}) from an evaluation of the potential energy curves and Franck-Condon factors. This last number, D_0^0 = 68.7 kcal mol^{-1}, is adopted and may be compared with the average per-bond atomization energy of SiH_4(g) of 75.7 kcal mol^{-1}. The $D_0^0/\Delta_{at}H°(0\ K)$ ratio is 0.227. For SiF/SiF the ratio is 0.230, for $SiCl/SiCl_4$ it is 0.233 and for SiI/SiI_4 it is 0.245 (7). The adopted D_0^0 and JANAF auxiliary data (7) lead to $\Delta_f H°(0\ K)$ = 89.6 kcal mol^{-1}. The allowed error of ±2 kcal mol^{-1} includes the upper limit value of D_0^0.

Heat Capacity and Entropy

The observed and estimated electronic levels are taken from Herzberg et al. (1), Jordan (2), and Bollmark et al. (3) as indicated. Wirsam's (8) ab initio self-consistent-field and configuration-interaction calculations predicted the electronic level for the $^4\Sigma^-$ state at 6430 and 8275 cm^{-1} respectively. Substitution of 6430 cm^{-1} for the adopted Jordan estimate of 21510 cm^{-1} increases the derived entropy starting near 700 K; at 6000 K, the entropy is 0.85 cal K^{-1} mol^{-1} higher.

The rotational and vibrational constants are those selected by Rosen (9). The ground state is treated as two distinct levels because of the splitting of the ground state as expressed by the spin coupling constant (A = 142.83 cm^{-1}). This approximation gives slightly biased results at lower temperatures which are allowed for in the ±0.05 cal K^{-1} mol^{-1} error assigned to S°(298.15 K).

References

1. G. Herzberg, A. Lagerqvist, and B. J. McKenzie, Can. J. Phys. 47, 1891 (1969).
2. P. C. Jordan, J. Chem. Phys. 44, 3400 (1966).
3. P. Bollmark, L. Klynning, and P. Pages, Phys. Scr. 3, 219 (1971).
4. R. D. Verma, Can. J. Phys. 43, 2136 (1965).
5. D. L. Hildenbrand, Douglas Advanced Research Laboratories, personal communication, Dec. 30, 1969.
6. T. V. Ramakrishna Rao and S. V. J. Lakshman, Physica (Utrecht) 56, 322 (1971).
7. JANAF Thermochemical Tables: Si(g), 3-31-67; H(g), 6-30-74; SiF(g), SiCl(g), SII(g), 12-31-76; SiF_4(g), SiI_4(g), 6-30-76; $SiCl_4$(g), 12-31-70.
8. B. Wirsam, Chem. Phys. Lett. 10, 180 (1971).
9. B. Rosen, Ed., "Spectroscopic Data Relative to Diatomic Molecules," Pergamon Press, New York, 1970.

Enthalpy Reference Temperature = T_r = 298.15 K Standard State Pressure = p° = 0.1 MPa

T/K	$C_p°$ (J $K^{-1}mol^{-1}$)	S° (J $K^{-1}mol^{-1}$)	$-[G°-H°(T_r)]/T$ (J $K^{-1}mol^{-1}$)	$H°-H°(T_r)$ (kJ mol^{-1})	$\Delta_f H°$ (kJ mol^{-1})	$\Delta_f G°$ (kJ mol^{-1})	Log K_f
0	0.	0.	INFINITE	-9.227	374.886	374.886	INFINITE
100	32.654	163.854	225.373	-6.152	376.196	365.230	-190.776
200	30.840	185.896	200.810	-2.983	376.839	353.934	-92.438
250	30.349	192.720	198.537	-1.454	376.819	348.206	-72.754
298.15	30.088	198.040	198.040	0.	376.662	342.708	-60 041
300	30.082	198.226	198.041	0.056	376.654	342.497	-59.634
350	29.989	202.855	198.407	1.557	376.395	336.824	-50.268
400	30.043	206.861	199.219	3.057	376.080	331.192	-43.249
450	30.216	210.409	200.269	4.563	375.732	325.602	-37.795
500	30.480	213.605	201.445	6.080	375 364	320.051	-33.436
600	31.174	219.221	203.952	9.162	374.605	309.060	-26 906
700	31.950	224.085	206.488	12.318	373.844	298.196	-22.252
800	32.707	228.401	208.963	15.551	373 092	287.440	-18.768
900	33.400	232.294	211.342	18.857	372.349	276.778	-16 064
1000	34.013	235.846	213.618	22.228	371.608	266.199	-13.905
1100	34.549	239.113	215 789	25.657	370.861	255.694	-12.142
1200	35.015	242.140	217.860	29.136	370.102	245.258	-10.676
1300	35.420	244.959	219.838	32.658	369.323	234.886	-9.438
1400	35.774	247.597	221.727	36.218	368.519	224.574	-8.379
1500	36.084	250.076	223.535	39.811	367.684	214.321	-7.463
1600	36.359	252.414	225.268	43.434	366.815	204.125	-6.664
1700	36.603	254.626	226.930	47.082	315.731	194.432	-5.974
1800	36.822	256.724	228.528	50.753	315.016	187.317	-5.436
1900	37.020	258.720	230.065	54.446	314.303	180.242	-4.955
2000	37.201	260.624	231.545	58.157	313.589	173 205	-4.524
2100	37.366	262.443	232.974	61.885	312.875	166.203	-4.134
2200	37.519	264.185	234.353	65.630	312.161	159.235	-3.781
2300	37.661	265.856	235.687	69.389	311.445	152.300	-3.459
2400	37.794	267.461	236.977	73.162	310.727	145.396	-3.164
2500	37.920	269.007	238.228	76.947	310.008	138.522	-2.894
2600	38.040	270.496	239.441	80.745	309.288	131.677	-2.645
2700	38.156	271.934	240.618	84.555	308.566	124.860	-2.416
2800	38.268	273.324	241.761	88.376	307.843	118.069	-2.203
2900	38.377	274.669	242.873	92.209	307.119	111.304	-2.005
3000	38.485	275.972	243.954	96.052	306.394	104.564	-1.821
3100	38.592	277.235	245.008	99.906	305.668	97.848	-1.649
3200	38.699	278.462	246.034	103.770	304.942	91.156	-1.488
3300	38.806	279.655	247.035	107.645	304.216	84.487	-1.337
3400	38.916	280.815	248.011	111.531	303.490	77.839	-1.196
3500	39.027	281.944	248.965	115.429	302.765	71.213	-1.063
3600	39.142	283.045	249.896	119.337	-82.110	75.068	-1.089
3700	39.259	284.119	250.807	123.257	-82.418	79.438	-1.121
3800	39.380	285.168	251.697	127.189	-82.726	83.817	-1.152
3900	39 505	286.193	252.569	131.133	-83.032	88.204	-1.181
4000	39.635	287.194	253.422	135.090	-83.336	92.598	-1.209
4100	39.769	288.175	254.258	139 060	-83.637	97.000	-1.236
4200	39.908	289.135	255.077	143.044	-83.934	101.410	-1.261
4300	40.053	290.075	255.880	147.042	-84.226	105.826	-1.286
4400	40.202	290.998	256.667	151.055	-84.512	110.249	-1.309
4500	40.357	291.903	257.440	155.083	-84.792	114.679	-1.331
4600	40.518	292.792	258.199	159.126	-85.064	119.114	-1.353
4700	40.683	293.665	258.944	163.186	-85.328	123.556	-1.373
4800	40.854	294.523	259.677	167.263	-85.583	128.003	-1.393
4900	41.030	295.367	260 397	171.357	-85.829	132.455	-1.412
5000	41.212	296.198	261.104	175.469	-86.064	136.913	-1.430
5100	41.398	297.016	261.800	179 600	-86.288	141.374	-1.448
5200	41.589	297.822	262.485	183.749	-86.499	145.840	-1.465
5300	41.785	298.616	263.160	187.918	-86.699	150.310	-1.481
5400	41.985	299.399	263.824	192.106	-86.885	154.784	-1.497
5500	42.189	300.171	264.477	196.315	-87.056	159.261	-1.513
5600	42.397	300.933	265.122	200.544	-87.213	163.741	-1.527
5700	42.608	301.685	265.757	204.794	-87.355	168.223	-1.542
5800	42.823	302.428	266.382	209.066	-87.482	172.709	-1.555
5900	43.041	303.162	267.000	213.359	-87.592	177.195	-1.569
6000	43.262	303.887	267.608	217.674	-87.686	181.684	-1.582

PREVIOUS: December 1976 (1 atm) CURRENT: December 1976 (1 bar)

Silylidyne (SiH) $H_1Si_1(g)$

SILYLYDYNE, ION (SiH^+) IDEAL GAS M_r = 29.09289

Silylidyne, Ion (SiH^+) $H_1Si_1{}^+$(g)

$S°(298.15\ K) = 186.789 \pm 0.004\ J\ K^{-1}\ mol^{-1}$

$\Delta_f H°(0\ K) = 1140.27 \pm 11.3\ kJ\ mol^{-1}$

$\Delta_f H°(298.15\ K) = 1147.67 \pm 11.3\ kJ\ mol^{-1}$

Electronic Levels and Quantum Weights

State	ε_i, cm^{-1}	g_i
$X^1\Sigma^+$	0	1
$A^1\Pi$	25025.2	2

$\omega_e = 2157.1\ cm^{-1}$ $\omega_e x_e = 34.21\ cm^{-1}$ $\sigma = 1$

$B_e = 7.6603\ cm^{-1}$ $\alpha_e = 0.2096\ cm^{-1}$ $r_e = 1.499\ Å$

Enthalpy of Formation

According to Douglas and Lutz (1) the A state must have an extremely shallow potential curve since the observed first vibrational quantum is only $\Delta_{1/2}G' = 390.2\ cm^{-1}$. Since no other levels were observed they estimated ω_e' and $\omega_e x_e'$ for the A state from $\Delta_{1/2}G'$, α_e' and B_e'. From the estimated ω_e' and $\omega_e x_e'$ they obtained a dissociation limit of 1180 cm^{-1}. Thus, the dissociation limit of the A state is likely to lie between 390 and 1180 cm^{-1}, on this basis Douglas and Lutz select 780±600 cm^{-1}. Since SiH^+ can correlate with normal products $Si^+(^2P)$ and $H(^2S)$ in both the X and A states these are assumed to be the products. Thus, the dissociation energy of the ground state to these 294.626±1.0 kcal mol^{-1} (2) we obtain $\Delta_f H°(SiH^+, g)$ = 272.5±2.7 kcal mol^{-1}, which is adopted. This gives an ionization potential of 182.9 kcal mol^{-1} (7.93 eV) with the JANAF value for SiH (2).

Heat Capacity and Entropy

The electronic levels and vibrational and rotational constants are all from Douglas and Lutz (1).

References

1. A. E. Douglas and B. L. Lutz, Can. J. Phys. 48, 247 (1970).
2. JANAF Thermochemical Tables; H(g) 9-30-65; Si(g) 3-31-67; SiH(g) 12-31-69; Si^+(g) 12-31-71.

Enthalpy Reference Temperature = T_r = 298.15 K Standard State Pressure = p° = 0.1 MPa

T/K	$C_p°$ (J K⁻¹mol⁻¹)	S° (J K⁻¹mol⁻¹)	$-[G°-H°(T_r)]/T$ (J K⁻¹mol⁻¹)	$H°-H°(T_r)$ (kJ mol⁻¹)	$\Delta_f H°$ (kJ mol⁻¹)	$\Delta_f G°$ (kJ mol⁻¹)	Log K_f
0	0.	0.	INFINITE	-8.654	1140.271		
100	29.118	154.962	212.699	-5.774			
200	29.133	175.150	189.456	-2 861			
250	29.148	181.652	187.269	-1.404			
298.15	29.184	186.789	186.789	0.	1147.671	1110.818	-194.611
300	29.186	186.969	186.789	0.054	1147.700	1110.589	-193.371
350	29.273	191.474	187.145	1.515	1148.441	1104.344	-164.814
400	29.427	195.392	187.936	2.982	1149.132	1097.997	-143.384
450	29.652	198.870	188.961	4.459	1149.794	1091.565	-126.706
500	29.941	202.009	190.112	5.949	1150.438	1085.060	-113.355
600	30.651	207.529	192.567	8.977	1151.705	1071.866	-93.314
700	31.436	212.312	195.053	12.081	1152.971	1058.459	-78.983
800	32.208	216.561	197.481	15.264	1154.247	1044.871	-68.223
900	32.922	220.396	199.817	18.521	1155.533	1031.122	-59.845
1000	33.560	223.899	202.053	21.846	1156.823	1017.229	-53.135
1100	34 121	227.124	204.187	25.230	1158.111	1003.207	-47.638
1200	34.611	230.114	206.225	28.668	1159.389	989.068	-43.053
1300	35.039	232.902	208.171	32.151	1160.650	974.824	-39.169
1400	35.413	235.513	210.032	35.674	1161.887	960.483	-35.836
1500	35.741	237.967	211.813	39.232	1163.095	946.054	-32.945
1600	36.031	240.284	213.521	42.820	1164.271	931.546	-30.412
1700	36.288	242.476	215.160	46.437	1115.234	917.413	-28.189
1800	36.518	244.557	216.736	50.077	1116.567	905.738	-26.284
1900	36.725	246.537	218.253	53.739	1117.902	893.989	-24.577
2000	36.912	248.425	219.714	57.421	1119.238	882.169	-23.040
2100	37.083	250.230	221.125	61.121	1120.574	870.283	-21.647
2200	37.240	251.959	222.487	64.838	1121.910	858.333	-20.379
2300	37.385	253.618	223.805	68.569	1123.245	846.322	-19.221
2400	37.520	255.212	225.081	72.314	1124 579	834.254	-18.157
2500	37.646	256.746	226.317	76.073	1125.911	822.130	-17.177
2600	37.764	258.225	227.516	79.843	1127.242	809.952	-16.272
2700	37.876	259.652	228.680	83.625	1128.571	797.723	-15.433
2800	37.983	261.032	229.811	87.418	1129.898	785.445	-14.653
2900	38.085	262.366	230.910	91.222	1131.223	773.120	-13.925
3000	38.184	263.659	231.981	95.035	1132.547	760.749	-13.246
3100	38.280	264.913	233.023	98.859	1133.870	748.334	-12 609
3200	38.374	266.130	234.038	102.691	1135.190	735.876	-12.012
3300	38.467	267.312	235.029	106.533	1136.510	723.377	-11.450
3400	38.558	268.461	235.995	110.385	1137.828	710.838	-10.921
3500	38.649	269.581	236.939	114.245	1139.145	698.261	-10.421
3600	38.740	270.671	237.861	118.114	756.309	696.106	-10.100
3700	38.832	271.733	238.762	121.993	758.038	694.410	-9.803
3800	38.925	272.770	239.644	125.881	759.765	692.667	-9.521
3900	39.019	273.782	240.506	129.778	761.491	690.879	-9.253
4000	39.115	274.771	241.350	133.685	763.215	689.046	-8.998
4100	39.213	275.739	242.177	137.601	764.939	687.171	-8.755
4200	39.314	276.685	242.988	141.528	766.663	685.253	-8.522
4300	39.416	277.611	243.782	145.464	768.389	683.295	-8.300
4400	39.522	278.518	244.561	149.411	770.115	681.296	-8.088
4500	39.630	279.408	245.326	153 369	771.844	679.258	-7.885
4600	39.741	280.280	246.076	157.337	773.575	677.181	-7.690
4700	39.855	281.136	246.813	161.317	775.309	675.067	-7.503
4800	39.973	281.976	247.537	165.308	777.047	672.916	-7.323
4900	40.093	282.802	248.248	169.311	778.789	670.728	-7.150
5000	40.216	283.613	248.947	173.327	780.536	668.505	-6.984
5100	40.343	284.410	249.635	177.355	782.288	666.247	-6.824
5200	40.472	285.195	250.311	181.396	784.047	663.955	-6.670
5300	40.605	285.967	250.977	185.449	785.811	661.629	-6.521
5400	40.740	286.727	251.632	189.517	787.583	659.269	-6.377
5500	40.879	287.476	252.277	193.598	789.362	656.876	-6.238
5600	41.019	288.214	252.912	197.692	791.149	654.451	-6.104
5700	41.163	288.941	253.538	201.802	792.944	651.994	-5.975
5800	41.308	289.659	254.154	205.925	794.748	649.506	-5.849
5900	41.456	290.366	254.762	210.063	796.561	646.986	-5.728
6000	41.606	291.064	255.361	214.216	798.385	644.435	-5.610

PREVIOUS: December 1971 (1 atm) CURRENT: December 1971 (1 bar)

Silylidyne, Ion (SiH^+) $H_1Si_1{}^+$(g)

ZIRCONIUM HYDRIDE (ZrH) IDEAL GAS $M_r = 92.22794$

$S°(298.15\ K) = [216.158]\ J\ K^{-1}\ mol^{-1}$

$\Delta_f H°(0\ K) = [517.346]\ kJ\ mol^{-1}$
$\Delta_f H°(298.15\ K) = [516.306]\ kJ\ mol^{-1}$

Electronic Level and Quantum Weight

State	ε_i, cm^{-1}	g_i
$[^2\Pi]$	0	[4]

$\omega_e = [1400]\ cm^{-1}$ $\omega_e x_e = [19]\ cm^{-1}$ $\sigma = 1$
$B_e = [5.1]\ cm^{-1}$ $\alpha_e = [0.09]\ cm^{-1}$ $r_e = [1.821]$ Å

Enthalpy of Formation

$\Delta_f H°(0\ K)$ was estimated as 123.7 kcal mol^{-1} by Gordon (1). $\Delta_f H°(298.15\ K)$ is then calculated.

Heat Capacity and Entropy

Molecular constants were estimated by Gordon (1). r_e is calculated from the relationship $r_e = (16.858778/\mu B_e)^{1/2}$ Å.

Reference

1. J. S. Gordon, Astrosystems International, personal communication, January 10, 1963.

Zirconium Hydride (ZrH) $H_1Zr_1(g)$

Enthalpy Reference Temperature = T_r = 298.15 K Standard State Pressure = p° = 0.1 MPa

T/K	$C_p°$ (J K⁻¹mol⁻¹)	S° (J K⁻¹mol⁻¹)	$-[G°-H°(T_r)]/T$ (J K⁻¹mol⁻¹)	$H°-H°(T_r)$ (kJ mol⁻¹)	$\Delta_f H°$ (kJ mol⁻¹)	$\Delta_f G°$ (kJ mol⁻¹)	Log K_f
0	0.	0.	INFINITE	-8.690	517.346	517.346	INFINITE
100	29.126	184.229	242.222	-5.799	517.851	505.870	-264.240
200	29.195	204.430	218.853	-2.885	517.226	494.091	-129.043
250	29.368	210.961	216.645	-1.421	516.775	488.359	-102.037
298.15	29.687	216.158	216.158	0.	516.306	482.928	-84.607
300	29.702	216.342	216.159	0.055	516.287	482.721	-84.049
350	30.178	220.955	216.522	1.551	515.788	477.166	-71.213
400	30.741	225.021	217.335	3.074	515.294	471.682	-61.595
450	31.338	228.676	218.396	4.626	514.811	466.260	-54.122
500	31.930	232.008	219.593	6.208	514.342	460.891	-48.149
600	33.015	237.928	222.168	9.457	513.435	450.286	-39.201
700	33.923	243.088	224.795	12.805	512.548	439.832	-32.821
800	34.661	247.668	227.374	16.236	511.652	429.505	-28.044
900	35.258	251.786	229.861	19.732	510.714	419.292	-24.335
1000	35.743	255.527	232.244	23.283	509.707	409.187	-21.374
1100	36.143	258.953	234.518	26.878	508.611	399.187	-18.956
1200	36.478	262.113	236.688	30.510	503.702	389.513	-16.955
1300	36.762	265.044	238.758	34.172	502.935	380.029	-15.270
1400	37.007	267.778	240.734	37.861	502.131	370.604	-13.827
1500	37.221	270.338	242.623	41.573	501.276	361.239	-12.579
1600	37.411	272.747	244.431	45.304	500.355	351.933	-11.489
1700	37.581	275.020	246.164	49.054	499.357	342.686	-10.529
1800	37.735	277.172	247.828	52.820	498.270	333.501	-9.678
1900	37.876	279.216	249.426	56.601	497.084	324.379	-8.918
2000	38.006	281.162	250.965	60.395	495.786	315.323	-8.235
2100	38.127	283.020	252.448	64.202	494.367	306.334	-7.620
2200	38.242	284.796	253.878	68.020	471.493	298.161	-7.079
2300	38.350	286.498	255.259	71.850	469.383	290.330	-6.594
2400	38.452	288.133	256.595	75.690	467.268	282.590	-6.150
2500	38.551	289.704	257.888	79.540	465.149	274.939	-5.745
2600	38.645	291.218	259.141	83.400	463.026	267.373	-5.372
2700	38.736	292.678	260.357	87.269	460.899	259.888	-5.028
2800	38.824	294.089	261.536	91.147	458.769	252.482	-4.710
2900	38.910	295.453	262.682	95.034	456.634	245.152	-4.416
3000	38.993	296.773	263.797	98.929	454.497	237.896	-4.142
3100	39.075	298.053	264.881	102.832	452.356	230.711	-3.887
3200	39.155	299.295	265.938	106.744	450.213	223.596	-3.650
3300	39.233	300.501	266.967	110.663	448.067	216.547	-3.428
3400	39.310	301.673	267.970	114.590	445.918	209.564	-3.220
3500	39.386	302.814	268.950	118.525	443.766	202.643	-3.024
3600	39.461	303.925	269.906	122.467	441.612	195.785	-2.841
3700	39.535	305.007	270.840	126.417	439.455	188.986	-2.668
3800	39.608	306.062	271.753	130.374	437.296	182.245	-2.505
3900	39.680	307.092	272.646	134.339	435.135	175.561	-2.351
4000	39.752	308.097	273.520	138.310	432.972	168.933	-2.206
4100	39.823	309.080	274.375	142.289	430.806	162.359	-2.068
4200	39.893	310.040	275.213	146.275	428.638	155.838	-1.938
4300	39.963	310.980	276.034	150.268	426.468	149.368	-1.814
4400	40.032	311.899	276.839	154.267	424.296	142.949	-1.697
4500	40.101	312.800	277.628	158.274	422.122	136.579	-1.585
4600	40.170	313.682	278.402	162.288	419.947	130.258	-1.479
4700	40.238	314.546	279.162	166.308	417.770	123.984	-1.378
4800	40.306	315.394	279.908	170.335	-145.346	129.373	-1.408
4900	40.373	316.226	280.641	174.369	-147.177	135.115	-1.440
5000	40.441	317.042	281.360	178.410	-149.027	140.895	-1.472
5100	40.508	317.844	282.068	182.457	-150.893	146.712	-1.503
5200	40.574	318.631	282.764	186.511	-152.774	152.566	-1.533
5300	40.641	319.405	283.448	190.572	-154.668	158.456	-1.562
5400	40.707	320.165	284.121	194.640	-156.574	164.382	-1.590
5500	40.773	320.912	284.783	198.714	-158.489	170.343	-1.618
5600	40.839	321.648	285.434	202.794	-160.413	176.339	-1.645
5700	40.905	322.371	286.076	206.882	-162.343	182.370	-1.671
5800	40.971	323.083	286.708	210.975	-164.278	188.434	-1.697
5900	41.036	323.784	287.331	215.076	-166.216	194.532	-1.722
6000	41.102	324.474	287.944	219.183	-168.156	200.663	-1.747

PREVIOUS: June 1963 (1 atm) CURRENT: June 1963 (1 bar)

Zirconium Hydride (ZrH) $H_1Zr_1(g)$

HYDROGEN (H_2) REFERENCE STATE - IDEAL GAS M_r = 2.01588

0 to 6000 K Ideal Gas

D_0^0 = 432.071 ± 0.012 kJ mol^{-1} $\Delta_f H°(0\ K)$ = 0 kJ mol^{-1}

S°(298.15 K) = 130.680 ± 0.033 J K^{-1} mol^{-1} $\Delta_f H°(298.15\ K)$ = 0 kJ mol^{-1}

Vibrational and Rotational Levels (cm^{-1})

Direct Summation of Electronic Ground State

$$E = G - G_0 + F = G - G_0 + BZ - DZ^2 + HZ^3 - LZ^4 + \ldots \approx G - G_0 + BZ - DZ^2 + H^2Z^3/(H+LZ),$$

where Z = J(J+1), Y = v+1/2, and we omit subscript v on G, F, B, D, H, and L

$$G = 4403.566Y - 123.8573Y^2 + 1.87269Y^3 - 0.173514Y^4 + 9.93128\times10^{-3}Y^5 - 4.38015\times10^{-4}Y^6$$

$$B = 60.8904 - 3.16997Y + 0.155932Y^2 - 4.60094\times10^{-2}Y^3 + 8.72205\times10^{-3}Y^4 - 9.59207\times10^{-4}Y^5 + 5.31722\times10^{-5}Y^6 - 1.21393\times10^{-6}Y^7$$

$$D = 4.6573\times10^{-2} - 1.5085\times10^{-3}Y - 2.7385\times10^{-4}Y^2 + 1.0242\times10^{-4}Y^3 - 1.172\times10^{-5}Y^4 + 4.684\times10^{-7}Y^5$$

$$H = 5.224\times10^{-5} - 7.240\times10^{-6}Y + 9.619\times10^{-7}Y^2 - 4.838\times10^{-8}Y^3$$

$$L = 6.70\times10^{-8} - 1.426\times10^{-8}Y + 1.388\times10^{-9}Y^2$$

v_{max} = 14, J_{max} = 38-32v/v_{max}

Normalized statistical weights = 1/4 (even J) and 3/4 (odd J)

Ground State Configuration: $^1\Sigma_g^+$ r_e = 0.7414 Å

Enthalpy of Formation

Zero by definition. Refer to the monatomic hydrogen gas table for a discussion of the dissociation energy.

Heat Capacity and Entropy

These are calculated by direct summation over vibration-rotation energy levels of the electronic ground state. We performed the direct summation with an extended version of a program written by W. H. Evans and provided through cooperation of D. D. Wagman, both of the U. S. National Bureau of Standards. Contributions of excited states (T_0>90000 cm^{-1}) are negligible at 6000 K. Polynomials G, B, D, and H are our fits of data from Stoicheff (1), Herzberg and Howe (2) and Rank et al. (3, 4). We estimate polynomial L such that our approximation (5, 6) for the infinite series F yields high-J rotational levels in reasonable agreement with the theoretical values of Waech and Bernstein (7). Maximum deviations in our F values are about ±400 cm^{-1}; these occur near 1≤v≤4 and 29≤J≤36 which is far into the extrapolated region of F. Only about one-third of the vibration-rotation levels have been observed spectroscopically and the theoretical calculations (7, 6) provide the best available extrapolation to high values of J. Accuracy of the thermodynamic functions near 6000 K depends on this extrapolation and on the rotational cutoff procedure. We assume a linear approximation (5) for the limiting values (J_{max}) of rotational quantum number. Values in the J_{max} equation are estimated from theoretical calculations (7, 6). We omit the nuclear-spin contribution (R ℓn 4) to entroy and Gibbs energy function.

We adopt ortho-para "equilibrium" H_2 as the reference state at all temperatures. The previous JANAF reference state (8) referred to "normal" H_2 (75% ortho and 25% para). Our new reference state has signficiant changes in $C_p°$, S° and H°(T)-H°(298.15 K) at 100 K and slight changes at 200 K. Use of "equilibrium" H_2 as a reference state was proposed on the NBS H_2 table (9) which discusses three alternatives. Preferred alternatives are either "equilibrium" H_2 or "normal" H_2. "Normal" H_2 is the form always encountered except in low-temperature generation or catalytic ortho-para equilibrium. Use of "normal" H_2 involves a possible complication, depending on the chice of zero energy for ortho-H_2. If we chose the lowest allowed level (v=0, J=1 instead of v=0, J=0), then H°(298.15 K)-H°(0 K) would be 0.254 kcal mol^{-1} less for "normal" than for "equilibrium" H_2. This would change the difference between $\Delta_f H°(0\ K)$ and $\Delta_f H°(298.15\ K)$ for all species involving hydrogen (8, 10). No such change would occur if we chose the lowest level (v=0, J=0) as the energy zero for ortho-H_2. "Equilibrium" H_2 is the form which parallels most substances, i.e., those maintaining equilibrium among all rotational levels (9).

JANAF values and uncertainties at 298.15 K are the same as those selected by CODATA (10). Previous thermochemical tables based on direct-summation calculations of Woolley et al. (6) include that of Gurvich et al. (11) and that of NBS-JANAF (9, 8). Differences of the new JANAF values from the NBS table are greatest near 5300 K, reaching maximum of 0.005 cal K^{-1} mol^{-1} in S° and 0.018 kcal mol^{-1} in H°(T)-H°(298.15 K). Differences from Gurvich et al. at 6000 K are 0.006 cal K^{-1} mol^{-1} in S° and 0.030 kcal mol^{-1} in H°(T)-H°(298.15 K). Errors larger than these differences arise from uncertainty in extrapolation of the rotational levels.

References

1. B. P. Stoicheff, Can. J. Phys. 35, 730 (1957).
2. G. Herzberg and L. L. Howe, J. Phys. 37, 636 (1959).
3. J. V. Foltz, D. H. Rank, and T. A. Wiggins, J. Mol. Spectrosc. 21, 203 (1966).
4. U. Fink, T. A. Wiggins and D. H. Rank, J. Mol. Spectrosc. 18, 384 (1965).
5. G. A. Khachkuruzov, Opt. Spectrosc. 30, 455 (1971).
6. H. W. Woolley, R. B. Scott, and F. G. Brickwedde, J. Res. Natl. Bur. Stand. 41, 379 (1948).
7. T. G. Waech and R. B. Bernstein, J. Chem. Phys. 46, 4905 (1967).
8. H. Prophet and D. R. Stull, ed., JANAF Thermochemical Tables, 2 ed., NSRDS-NBS 37, 1971.
9. S. Abramowitz et al., U. S. Natl. Bur. Stand. Rept. 10904, 239, July, 1972.
10. J. D. Cox, chairman, ICSU-CODATA Task Group on Key Values for Thermodynamics, J. Chem. Thermodynamics 4, 331 (1972); 8, 603 (1976).
11. L. V. Gurvich, G. A. Khachkuruzov et al., "Thermodynamic Properties of Individual Substances", 2nd ed., Vol. II, Nauka, Moscow, 1962.

Hydrogen (H_2) H_2(ref)

Enthalpy Reference Temperature = T_r = 298.15 K Standard State Pressure = p° = 0.1 MPa

T/K	$C_p°$ (J K^{-1}mol^{-1})	S°	-[G°-H°(T_r)]/T	H°-H°(T_r) (kJ mol^{-1})	$\Delta_f H°$	$\Delta_f G°$	Log K_f
0	0.	0.	INFINITE	-8.467	0.	0.	0.
100	28.154	100.727	155.408	-5.468	0.	0.	0.
200	27.447	119.412	133.284	-2.774	0.	0.	0.
250	28.344	125.640	131.152	-1.378	0.	0.	0.
298.15	28.836	130.680	130.680	0.	0.	0.	0.
300	28.849	130.858	130.680	0.053	0.	0.	0.
350	29.081	135.325	131.032	1.502	0.	0.	0.
400	29.181	139.216	131.817	2.959	0.	0.	0.
450	29.229	142.656	132.834	4.420	0.	0.	0.
500	29.260	145.737	133.973	5.882	0.	0.	0.
600	29.327	151.077	136.392	8.811	0.	0.	0.
700	29.441	155.606	138.822	11.749	0.	0.	0.
800	29.624	159.548	141.171	14.702	0.	0.	0.
900	29.881	163.051	143.411	17.676	0.	0.	0.
1000	30.205	166.216	145.536	20.680	0.	0.	0.
1100	30.581	169.112	147.549	23.719	0.	0.	0.
1200	30.992	171.790	149.459	26.797	0.	0.	0.
1300	31.423	174.288	151.274	29.918	0.	0.	0.
1400	31.861	176.633	153.003	33.082	0.	0.	0.
1500	32.298	178.846	154.652	36.290	0.	0.	0.
1600	32.725	180.944	156.231	39.541	0.	0.	0.
1700	33.139	182.940	157.743	42.835	0.	0.	0.
1800	33.537	184.846	159.197	46.169	0.	0.	0.
1900	33.917	186.669	160.595	49.541	0.	0.	0.
2000	34.280	188.418	161.943	52.951	0.	0.	0.
2100	34.624	190.099	163.244	56.397	0.	0.	0.
2200	34.952	191.718	164.501	59.876	0.	0.	0.
2300	35.263	193.278	165.719	63.387	0.	0.	0.
2400	35.559	194.785	166.899	66.928	0.	0.	0.
2500	35.842	196.243	168.044	70.498	0.	0.	0.
2600	36.111	197.654	169.155	74.096	0.	0.	0.
2700	36.370	199.021	170.236	77.720	0.	0.	0.
2800	36.618	200.349	171.288	81.369	0.	0.	0.
2900	36.856	201.638	172.313	85.043	0.	0.	0.
3000	37.087	202.891	173.311	88.740	0.	0.	0.
3100	37.311	204.111	174.285	92.460	0.	0.	0.
3200	37.528	205.299	175.236	96.202	0.	0.	0.
3300	37.740	206.457	176.164	99.966	0.	0.	0.
3400	37.946	207.587	177.072	103.750	0.	0.	0.
3500	38.149	208.690	177.960	107.555	0.	0.	0.
3600	38.348	209.767	178.828	111.380	0.	0.	0.
3700	38.544	210.821	179.679	115.224	0.	0.	0.
3800	38.738	211.851	180.512	119.089	0.	0.	0.
3900	38.928	212.860	181.328	122.972	0.	0.	0.
4000	39.116	213.848	182.129	126.874	0.	0.	0.
4100	39.301	214.816	182.915	130.795	0.	0.	0.
4200	39.484	215.765	183.686	134.734	0.	0.	0.
4300	39.665	216.696	184.442	138.692	0.	0.	0.
4400	39.842	217.610	185.186	142.667	0.	0.	0.
4500	40.017	218.508	185.916	146.660	0.	0.	0.
4600	40.188	219.389	186.635	150.670	0.	0.	0.
4700	40.355	220.255	187.341	154.698	0.	0.	0.
4800	40.518	221.106	188.035	158.741	0.	0.	0.
4900	40.676	221.943	188.719	162.801	0.	0.	0.
5000	40.829	222.767	189.392	166.876	0.	0.	0.
5100	40.976	223.577	190.054	170.967	0.	0.	0.
5200	41.117	224.374	190.706	175.071	0.	0.	0.
5300	41.252	225.158	191.349	179.190	0.	0.	0.
5400	41.379	225.931	191.982	183.322	0.	0.	0.
5500	41.498	226.691	192.606	187.465	0.	0.	0.
5600	41.609	227.440	193.222	191.621	0.	0.	0.
5700	41.712	228.177	193.829	195.787	0.	0.	0.
5800	41.806	228.903	194.427	199.963	0.	0.	0.
5900	41.890	229.619	195.017	204.148	0.	0.	0.
6000	41.965	230.323	195.600	208.341	0.	0.	0.

PREVIOUS: March 1961 (1 atm) CURRENT: March 1977 (1 bar)

Hydrogen (H_2) H_2(ref)

HYDROGEN, ION (H_2^+) IDEAL GAS M_r = 2.01525

D_0^0 = 255.760 ± 0.004 kJ mol^{-1} $\Delta_f H°(0\ K)$ = 1488.358 ± 0.004 kJ mol^{-1}

S°(298.15 K) = 142.370 ± 0.04 J K^{-1} mol^{-1} $\Delta_f H°(298.15\ K)$ = [1494.671] kJ mol^{-1}

Vibrational and Rotational Levels (cm^{-1})

Direct Summation using Energy-Level Equations for the $^2\Sigma_g^+$ Electronic Ground State:

$$E = G - G_0 + F = G - G_0 + BZ - DZ^2 + HZ^3 - LZ^4 + \ldots \cong G - G_0 + BZ - DZ^2 + H^2Z^3/(H+LZ),$$

where Z = N(N+1), Y = v+1/2, and we omit subscript v on G, F, B, D, H, and L

$$G = 2323.23\ Y = 67.39\ Y^2 + 0.93\ Y^3 - 0.029\ Y^4$$

$$B = 29.946 - 1.594\ Y + 0.032\ Y^2 - 0.013\ Y^3 \qquad r_e = 1.057\ Å\ (1,\ 13)$$

$$D = 0.01991 - 0.00087\ Y + 0.00003\ Y^2; \quad H = 1.75 \times 10^{-5} - 6\times10^{-7}\ Y; \quad \text{and } L = 1.6\times10^{-8}$$

$$v_{max} = 18 \text{ and } N_{max} = 41.33v/v_{max}$$

Electronic statistical weight = 2 and normalized rotational statistical weights = 1/4 (even N) or 3/4 (odd N)

Enthalpy of Formation

The adopted value for the enthalpy of formation of H_2^+(g), $\Delta_f H°(0\ K)$ = 355.727±0.001 kcal mol^{-1}, is based on the ionization potential of H_2(g) determined by Herzberg and Jungen (1). These authors studied in high resolution absorption two Rydberg series of ortho- and para-hydrogen which converge to the $X^2\Sigma_g^+$ ground state of H_2^+(g). This ionization potential, IP = 124417.2±0.4 cm^{-1} is recommended by Rosenstock (2) as the definitive value. The IP or $\Delta_f H°(0\ K)$ value is converted to a $\Delta_f H°(298.15\ K)$ value for H_2^+(g) by including JANAF (3) enthalpy differences, H°(0 K) - H°(298.15 K), for H_2^+(g), H_2(g) and e^-(g). $\Delta_f H°(298.15\ K)$ should not be interpreted as a room-temperature ionization threshold due to inclusion of these enthalpies and to threshold effects noted by Rosenstock et al. (4).

Rosenstock et al. (4) had earlier recommened IP = 16.4256 eV (124418.4 cm^{-1}) based on the spectroscopic studies of Takezawa (5) and Herzberg (6); in both studies the results were derived from extended Rydberg series. The more recent work of Herzberg and Jungen (1) provides a more accurate experimental value for the ionization potential due to the recognition of a pressure shift which was felt to be negligible in the earlier study (6). The theroetical value, IP = 124417.3 cm^{-1}, is in excellent agreement (0.1 cm^{-1}) with our adopted value. This theoretical value, as discussed by Herzberg and Jungen (1), is essentially the value of Hunter and Pritchard (7) with the relativistic and Lamb shift corrections of Jeziorski and Kolos (8) and the nonadiabatic correction of Bunker (9).

Heat Capacity and Entropy

Thermodynamic functions for ortho-para "equilibrim" H_2^+ are calculated, as they were for H_2 (3), by direct summation over energy levels of the electronic ground state. We use 1973 fundamental constants (10) in an extended version of a computer program written by W. H. Evans and provided by D. D. Wagman of the U. S. National Bureau of Standards. Excited electronic states are neglected because they are either repulsive or too high lying (11) to contribute significantly. The electronic statistical weight is taken equal to the multiplicity since we neglect rotational splitting. Assumption of negligible splitting is common practice in theoretical calculation of H_2^+ energy levels (11 - 13) and it is justified thermodynamically by the small splitting constant (14).

Coefficients of the energy-level polynomials are those derived from theoretical levels by Beckel et al. (13). New calculations (15) yield slightly different levels. The polynomials are confirmed by vibrational (v≤3) and rotational (N≤4) levels obtained from Rydberg spectra (1). The value of L was derived (13) from a fit of theoretical levels with v = 0 to 14. Our adopted v_{max}, derived from the G polynomial, is one unit too low (13, 11). Limiting rotaitonal quantum numbers N_{max} are an approximation of values from theoretical calculations (11, 13). The resulting energy-level equations are very approximate near N_{max}, but they do closely reproduce thermodynamic functions (12, 11) calculated directly from energy levels. More accurate equations do not signficantly improve the thermodynamic functions up to 6000 K.

We adopt functions closely reproducing those of Iorish and Yungman (12). The effect of slightly different fundamental constants (12) is barely detectable. Our Gibbs-energy function deviates by less than 0.001 cal K^{-1} mol^{-1} at all temperatures. Cp°, S° and H° begin to deviate near 2000 K; differences range from +0.005 to -0.012 cal K^{-1} mol^{-1} in Cp°, ±0.001 cal K^{-1} mol^{-1} in S° and ±5 cal K^{-1} mol^{-1} in H°. Cp° of Patch and McBride (11) is slightly lower at 4000 to 6000 K but well within the probable uncertainty. Due to the high thermal energy (~20% of D_0^0 at 6000 K), Cp° depends on the number of levels included in the summation. We follow the published references (11, 12) which include contributions from quasibound levels lying above D_0^0. At 6000 K these levels contribute 0.42 cal K^{-1} mol^{-1} to Cp° and 0.05 cal K^{-1} mol^{-1} to the Gibbs-energy function.

References

1. G. Herzberg and Ch. Jungen, J. Mol. Spectrosc. 41, 425 (1972).
2. H. M. Rosenstock, U. S. Nat. Bur. Stand., personal communication, November 29, 1977.
3. JANAF Thermochemical Tables: H_2(g), e^-(g) 3-31-77.
4. H. M. Rosenstock, K. Draxl, B. W. Steiner, and J. T. Herron, J. Phys. Chem. Ref. Data 6, Supp. 1 (1977).
5. S. Takezawa, J. Chem. Phys. 52, 5793 (1970).
6. G. Herzberg, Phys. Rev. Lett. 23, 1081 (1969).
7. G. Hunter and H. O. Pritchard, J. Chem. Phys. 46, 2153 (1967).
8. B. Jeziorski and W. Kolos, Chem. Phys. Lett. 3, 677 (1969).
9. P. R. Bunker, J. Mol. Spectrosc. 46, 504 (1973).
10. E. R. Cohen and B. N. Taylor, J. Phys. Chem. Ref. Data 2, 663 (1973).
11. R. W. Patch and B. J. McBride, NASA TN D-4523 (avail. NTIS), 73 pp (1968); J. Chem. Phys. 49, 961 (1968).
12. V. S. Iorish and V. S. Yungman, Russ. J. Phys. Chem. 50, 1168 (1976).
13. C. L. Beckel, B. D. Hansen III, and J. M. Peek, J. Chem. Phys. 53, 3681 (1970).
14. G. Herzberg, "The Spectra and Structure of Simple Free Radicals", Cornell University Press, Ithaca, 1971, pp. 59-60.
15. T. Murai, J. Phys. Soc. Jap. 38, 514 (1975); M. Cohen et al., Chem. Phys. Lett. 49, 374 (1977).

Hydrogen, Ion (H_2^+) H_2^+(g)

Enthalpy Reference Temperature = T_r = 298.15 K Standard State Pressure = p° = 0.1 MPa

	J K^{-1}mol^{-1}			kJ mol^{-1}			
T/K	$C_p°$	S°	-[G°-H°(T_r)]/T	H°-H°(T_r)	$\Delta_f H°$	$\Delta_f G°$	Log K_f
0	0.	0.	INFINITE	-8.583	1488.358	1488.358	INFINITE
100	27.510	110.873	168.264	-5.739	1490.282	1489.440	-778.004
200	29.175	130.698	145.046	-2.870	1492.536	1487.743	-388.559
298.15	29.288	142.370	142.370	0.	1494.671	1484.931	-260.154
300	29.292	142.551	142.371	0.054	1494.711	1484.870	-258.539
400	29.525	151.004	143.522	2.993	1496.822	1481.272	-193.434
500	30.016	157.641	145.705	5.968	1498.953	1477.139	-154.316
600	30.719	163.173	148.167	9.003	1501.138	1472.571	-128.199
700	31.529	167.968	150.661	12.115	1503.391	1467.633	-109.516
800	32.357	172.233	153.095	15.310	1505.711	1462.367	-95.483
900	33.152	176.090	155.439	18.586	1508.091	1456.807	-84.551
1000	33.887	179.622	157.684	21.938	1510.518	1450.979	-75.791
1100	34.559	182.884	159.828	25.361	1512.981	1444.906	-68.613
1200	35.168	185.917	161.877	28.848	1515.468	1438.608	-62.621
1300	35.720	188.754	163.837	32.393	1517.971	1432.102	-57.543
1400	36.226	191.420	165.713	35.990	1520.483	1425.402	-53.182
1500	36.691	193.935	167.511	39.636	1522.999	1418.523	-49.397
1600	37.123	196.317	169.238	43.327	1525.518	1411.476	-46.080
1700	37.530	198.580	170.898	47.060	1528.036	1404.271	-43.148
1800	37.918	200.737	172.496	50.833	1530.553	1396.918	-40.538
1900	38.291	202.797	174.037	54.643	1533.069	1389.426	-38.198
2000	38.654	204.770	175.525	58.491	1535.585	1381.800	-36.089
2100	39.011	206.665	176.963	62.374	1538.102	1374.049	-34.178
2200	39.365	208.488	178.355	66.293	1540.620	1366.178	-32.437
2300	39.716	210.245	179.703	70.247	1543.142	1358.193	-30.846
2400	40.065	211.943	181.011	74.236	1545.668	1350.097	-29.384
2500	40.413	213.585	182.282	78.260	1548.201	1341.896	-28.037
2600	40.757	215.177	183.516	82.318	1550.740	1333.594	-26.792
2700	41.096	216.722	184.718	86.411	1553.287	1325.194	-25.637
2800	41.427	218.222	185.888	90.537	1555.843	1316.699	-24.563
2900	41.746	219.682	187.028	94.696	1558.406	1308.113	-23.562
3000	42.050	221.102	188.140	98.886	1560.978	1299.439	-22.625
3100	42.335	222.486	189.226	103.105	1563.556	1290.678	-21.748
3200	42.599	223.834	190.286	107.352	1566.139	1281.834	-20.924
3300	42.836	225.148	191.323	111.624	1568.726	1272.910	-20.148
3400	43.045	226.430	192.337	115.919	1571.315	1263.907	-19.418
3500	43.222	227.681	193.329	120.232	1573.902	1254.828	-18.727
3600	43.366	228.901	194.300	124.562	1576.486	1245.675	-18.074
3700	43.475	230.090	195.251	128.904	1579.062	1236.450	-17.456
3800	43.547	231.251	196.183	133.256	1581.628	1227.156	-16.868
3900	43.583	232.382	197.097	137.612	1584.180	1217.795	-16.311
4000	43.582	233.486	197.993	141.971	1586.715	1208.368	-15.780
4100	43.545	234.562	198.872	146.328	1589.229	1198.878	-15.274
4200	43.473	235.610	199.734	150.679	1591.720	1189.327	-14.791
4300	43.367	236.632	200.581	155.021	1594.183	1179.717	-14.331
4400	43.229	237.627	201.411	159.351	1596.616	1170.050	-13.890
4500	43.060	238.597	202.227	163.666	1599.017	1160.328	-13.469
4600	42.862	239.541	203.028	167.962	1601.381	1150.554	-13.065
4700	42.639	240.461	203.815	172.237	1603.708	1140.728	-12.678
4800	42.391	241.356	204.587	176.489	1605.994	1130.853	-12.306
4900	42.122	242.227	205.347	180.715	1608.239	1120.931	-11.949
5000	41.834	243.075	206.093	184.913	1610.440	1110.964	-11.606
5100	41.528	243.901	206.826	189.081	1612.597	1100.952	-11.276
5200	41.208	244.704	207.547	193.218	1614.708	1090.900	-10.958
5300	40.875	245.486	208.255	197.322	1616.772	1080.807	-10.652
5400	40.532	246.247	208.952	201.393	1618.790	1070.675	-10.357
5500	40.179	246.987	209.637	205.428	1620.760	1060.506	-10.072
5600	39.820	247.708	210.310	209.428	1622.683	1050.303	-9.797
5700	39.456	248.410	210.972	213.392	1624.559	1040.065	-9.531
5800	39.087	249.093	211.624	217.319	1626.389	1029.795	-9.274
5900	38.717	249.758	212.265	221.209	1628.173	1019.493	-9.026
6000	38.346	250.405	212.895	225.063	1629.912	1009.161	-8.786

PREVIOUS: September 1977 (1 atm) CURRENT: September 1977 (1 bar)

Hydrogen, Ion (H_2^+) H_2^+(g)

HYDROGEN, ION (H_2^-) IDEAL GAS M_r = 2.01635

D_0^0 = 118.0 ± 42 kJ mol^{-1}
$S°$(298.15 K) = 143.75 ± 0.4 J K^{-1} mol^{-1}

$\Delta_f H°$(0 K) = 241.4 ± 42 kJ mol^{-1}
$\Delta_f H°$(298.15 K) = [235.373] kJ mol^{-1}

Electonic Levels and Quantum Weights

State	ε_i, cm^{-1}	g_i
$^2\Sigma_u^+$	0	2

ω_e = [1700] cm^{-1} $\omega_e x_e$ = [35] cm^{-1} σ = 2
B_e = [25.3] cm^{-1} α_e = [0.398] cm^{-1} r_e = [1.15] Å

Enthalpy of Formation

The heat of formation at 0 K for H_2^-(g) is based on the electron affinity value derived from a Rydberg-Klein-Rees configuration-interaction calculation (1). This study by Sharpe (1) led to EA (H_2) = -2.5 eV. Using auxiliary data (2), we calculate $\Delta_f H°$(0 K) kcal mol^{-1} and $D_0^0(H_2^-)$ = 28.2 ± 10 kcal mol^{-1}; the latter value assumes dissociation into ground state H(g) and H^-(g). This adopted EA(H_2) value is suggested to be the more reliable value of eight studies considered by Rosenstock et al. (3).

In terms of molecular orbitals, the ground state for H_2^-(g) is the $(1s\sigma_g)^2(2p\sigma_u)$ $^2\Sigma_u^+$ state involving two bonding electrons and one antibonding electron (4). We would expect, however, the ground state to be $(1s\sigma_g)^2(2s\sigma_u)$. As stated by Massey (4), we should expect only a weakly-bound molecule as compared with H_2(g) for which the ground state has no electron in an antibonding orbital. Our calculated EA(H_2) value is consistent with this viewpoint. Further discussion of the H_2^-(g) ion may be found in Massey (4).

$\Delta_f H°$(0 K) is converted to $\Delta_f H°$(298.15 K) by use of JANAF (2) enthalpies, H°(0 K) - H°(298.15 K), for H_2(g), H_2^-(g), and e^-(g). $\Delta_f H°$(298.15 K) should not be interpreted as a room temperature electron affintiy due to inclusion of these enthalpies and to threshold effects discussed by Rosenstock et al. (3)

Heat Capacity and Entropy

The vibrational constants for H_2^-(g) are estimated to be the same as in the isoelectronic species He_2^+(5). The internuclear distance is estimated to be ~8% larger than that for He_2^+(g). The same relationship appeared to exist for the (N_2^-, NO, O_2^+) isoelectronic series. B_e is calculated from r_e whereas α_e is calculated assuming a Morse function for the ground state.

The ground and excited electronic states are discussed in detail by Massey (4). No excited state contributions are included in this calculation. By assuming the same force constant to exist for H_2^- as in He_2^+, we would calculate ω_e = 3384 cm^{-1}. This leads to an entropy at 298.15 K which is 0.06 cal K^{-1} mol^{-1} smaller than our adopted value. The vibrational constants derived by Comer and Read (6) for some resonant state of H_2^- suggests ω_e values of 1500, 2300, and 2800 cm^{-1}. Thus, the ω_e value is very uncertain.

References

1. T. E. Sharpe, Lockheed Report LMSC 5-10-69-9 (1969).
2. JANAF Thermochemical Tables: H(g) 3-31-82, H^-(g) 3-31-82, H_2(g) 3-31-77, e^-(g) 3-31-82.
3. H. M. Rosenstock, K. Draxl, B. W. Steiner, and J. T. Herron, J. Phys. Chem. Ref. Data 6, Supp. 1 (1977).
4. H. S. W. Massey, "Negative Ions," 3rd ed., Cambridge University Press, Cambridge, 1976.
5. B. Rosen, Ed., "Spectroscopic Data Relative to Diatomic Molecules," Pergamon Press, New York, 1970.
6. J. Comer and F. H. Read, J. Phys. B. 4, 368 (1971).

Hydrogen, Ion (H_2^-) H_2^-(g)

Enthalpy Reference Temperature = T_r = 298.15 K Standard State Pressure = p° = 0.1 MPa

T/K	$C_p°$ (J $K^{-1}mol^{-1}$)	S° (J $K^{-1}mol^{-1}$)	$-[G°-H°(T_r)]/T$ (J $K^{-1}mol^{-1}$)	$H°-H°(T_r)$ (kJ mol^{-1})	$\Delta_f H°$ (kJ mol^{-1})	$\Delta_f G°$ (kJ mol^{-1})	Log K_f
0	0.	0.	INFINITE	-8.621	241.417		
100	29.209	111.749	169.836	-5.809			
200	29.285	132.014	146.437	-2.885			
250	29.378	138.557	144.231	-1.418			
298.15	29.556	143.746	143.746	0.	235.373	237.732	-41.650
300	29.564	143.928	143.746	0.055	235.335	237.747	-41.395
350	29.866	148.507	144.107	1.540	234.332	238.228	-35.554
400	30.271	152.520	144.913	3.043	233.339	238.852	-31.191
450	30.751	156.113	145.961	4.568	232.365	239.600	-27.812
500	31.272	159.380	147.142	6 119	231.414	240.455	-25.120
600	32.333	165.175	149.677	9.299	229.587	242.437	-21.106
700	33.320	170.235	152.260	12.583	227.854	244.717	-18.261
800	34.187	174.742	154.793	15.959	226.199	247.239	-16.143
900	34.932	178.813	157.240	19.416	224.602	249 966	-14.508
1000	35.568	182 527	159.585	22.942	223.046	252.868	-13.208
1100	36.114	185.944	161.828	26.527	221.513	255.924	-12 153
1200	36.586	189.107	163.971	30.162	219.992	259.119	-11.279
1300	36.999	192.052	166.019	33.842	218.472	262.442	-10.545
1400	37.365	194.807	167.978	37.560	216.948	265.881	-9.920
1500	37.692	197.397	169 854	41.314	215.414	269.430	-9.382
1600	37.988	199.839	171.653	45.098	213.869	273.081	-8.915
1700	38.259	202.150	173.379	48.910	212.309	276.830	-8 506
1800	38.509	204.344	175.039	52.749	210.735	280.670	-8.145
1900	38.742	206.432	176.637	56.612	209.147	284.599	-7.824
2000	38.961	208.425	178.177	60.497	207.543	288.611	-7 538
2100	39.168	210.331	179.663	64.403	205.926	292.705	-7.281
2200	39.365	212.158	181.099	68 330	204 295	296.875	-7.049
2300	39.553	213.912	182.487	72 276	202.652	301.120	-6.839
2400	39.735	215.599	183.832	76.241	200.996	305.437	-6.648
2500	39.910	217.225	185.136	80.223	199.330	309.822	-6.473
2600	40.079	218.793	186.400	84.222	197.653	314.275	-6.314
2700	40.244	220.309	187.628	88.239	195 966	318.793	-6.167
2800	40.406	221.776	188.822	92.271	194 271	323 373	-6 033
2900	40.563	223.196	189.983	96 320	192.567	328.014	-5 908
3000	40.718	224.574	191.113	100.384	190.855	332 713	-5.793
3100	40.869	225.912	192.214	104.463	189 136	337 470	-5.686
3200	41.019	227.212	193.287	108.557	187.410	342.283	-5.587
3300	41.166	228.476	194.335	112.667	185.677	347.149	-5.495
3400	41.311	229.707	195.357	116.791	183.938	352.068	-5.409
3500	41.455	230.907	196.356	120.929	182 193	357.039	-5.329
3600	41.597	232.077	197.332	125.082	180.442	362.059	-5.253
3700	41 738	233.218	198.286	129.248	178.686	367 128	-5.183
3800	41.878	234.333	199.220	133.429	176.924	372.245	-5.117
3900	42.016	235.423	200.135	137.624	175.157	377.408	-5.055
4000	42.154	236.488	201.030	141.832	173.384	382.616	-4.996
4100	42.291	237.531	201.908	146.055	171.607	387.869	-4.942
4200	42.426	238.552	202.768	150.291	169.825	393.166	-4.890
4300	42.562	239.551	203.612	154.540	168.038	398.504	-4.841
4400	42.696	240.531	204.440	158.803	166 247	403.885	-4.795
4500	42.830	241.492	205.253	163.079	164.452	409 306	-4.751
4600	42.963	242.435	206.051	167.369	162.653	414 767	-4.710
4700	43.096	243.361	206.835	171 672	160 850	420.267	-4.671
4800	43.228	244.269	207.605	175.988	159.044	425.806	-4.634
4900	43 360	245.162	208.363	180.317	157.235	431.382	-4.599
5000	43 491	246.039	209.107	184.660	155.423	436 995	-4.565
5100	43.622	246.902	209.840	189.015	153.610	442 644	-4.534
5200	43.753	247.750	210.561	193.384	151.795	448.329	-4.504
5300	43.883	248.585	211.270	197.766	149.980	454.049	-4 475
5400	44.013	249.406	211.969	202.161	148.165	459.804	-4.448
5500	44.143	250.215	212.657	206.569	146.350	465.591	-4 422
5600	44.272	251.012	213.335	210.989	144.537	471.412	-4.397
5700	44.401	251.796	214 003	215.423	142 726	477 265	-4.374
5800	44.530	252.570	214 661	219.870	140.918	483.150	-4.351
5900	44.659	253.332	215.310	224.329	139.114	489.066	-4.330
6000	44.788	254.084	215.950	228.801	137.314	495 013	-4.309

PREVIOUS: September 1977 (1 atm) CURRENT: September 1977 (1 bar)

Hydrogen, Ion (H_2^-) H_2^-(g)

DIIODOSILANE (SiH_2I_2) IDEAL GAS M_r = 283.91038

$\Delta_fH°$(0 K) = [-28.11 ± 20.9] kJ mol^{-1}

S°(298.15 K) = [326.797 ± 3.3] J K^{-1} mol^{-1} $\Delta_fH°$(298.15 K) = [-38.07 ± 20.9] kJ mol^{-1}

Vibrational Frequencies and Degeneracies

ν, cm^{-1}	ν, cm^{-1}
2205 (1)	[2230](1)
925 (1)	497 (1)
[350](1)	796 (1)
[90](1)	[410](1)
[650](1)	

Ground State Quantum Weight: [1] σ = [2]

Point Group: [C_{2v}]

Bond Distances: Si-H = [1.49] Å Si-I = [2.44] Å

Bond Angles: H-Si-H = [111]° I-Si-I = [110]°

Product of the Moments of Inertia: $I_AI_BI_C$ = [3.067590 x 10^{-112}] g^3 cm^6

Enthalpy of Formation

$\Delta_fH°$(298.15 K) is estimated by linear interpolation between the values (1) of SiI_4(g) and SiH_4(g). There are no experimental $\Delta_fH°$ data for SiH_3I, SiH_2I_2 and $SiHI_3$. Data for the iodomethanes (2) have a surprising progression (1, SiH_3I, g); they are of doubtful use in predicting $\Delta_fH°$ of the iodosilanes. We conclude, as did Hunt and Sirtl (3), that the available data justify only linear interpolation of $\Delta_fH°$.

Heat Capacity and Entropy

The molecular structure is estimated by comparison with SiH_3I, SiI_4 and the various bromo-, chloro- and flurosilanes (1). The principal moments of inertia are: I_A = 10.2009 x 10^{-39}, I_B = 168.8775 x 10^{-39} and I_C = 178.0688 x 10^{-39} g cm^2.

Five vibrational frequencies are estimated by comparison with the dihalo-, monohalo- and trihalosilanes (1) plus $SiHBrI_2$ and $SiHClI_2$ (4). Frequencies ν_1, ν_2 and ν_8 were observed in gas-phase infrared spectra (5), while ν_7 was observed in liquid-phase infrared spectra (6).

We neglect excited states and assume the electronic ground state to be a singlet. We expect that contributions from excited states should be unimportant, as discussed on the tables for SiH_3Br and SiH_2Cl_2 (1).

References

1. JANAF Thermochemical Tables: H_3ISi(g), HI_3Si(g), BrH_3Si(g), Br_2H_2Si(g), Br_3HSi(g), ClH_3Si(g), Cl_2H_2Si(g), Cl_3HSi(g) 12-31-76; FH_3Si(g), F_2H_2Si(g), F_3HSi(g), I_4Si(g) 6-30-76.
2. S. A. Kudchadker and A. P. Kudchadker, J. Phys. Chem. Ref. Data 4, 457 (1975).
3. L. P. Hunt and E. Sirtl, J. Electrochem. Soc. 120, 806 (1973).
4. E. Hengge and F. Hoefler, Z. Naturforsch. A26, 768 (1971).
5. H. J. Emeleus, A. G. MacDiarmid and A. G. Maddock, J. Inorg. Nucl. Chem. 1, 194 (1955).
6. E. A. V. Ebsworth, M. Onyszchuk and N. Sheppard, J. Chem. Soc. 1958, 1453 (1958).

Diiodosilane (SiH_2I_2) $H_2I_2Si_1$(g)

Enthalpy Reference Temperature = T_r = 298.15 K Standard State Pressure = p° = 0.1 MPa

T/K	$C_p°$	S°	-[G°-H°(T_r)]/T	H°-H°(T_r)	$\Delta_fH°$	$\Delta_fG°$	Log K_f
	J $K^{-1}mol^{-1}$			kJ mol^{-1}			
0	0.	0.	INFINITE	-14.917	-28.108	-28.108	INFINITE
100	43.099	267.914	379.301	-11.139	-30.645	-40.881	21.354
200	56.780	301.906	332.682	-6.155	-34.466	-49.641	12.965
250	62.986	315.261	327.889	-3.157	-36.342	-53.217	11.119
298.15	67.980	326.797	326.797	0.	-38.074	-56.308	9.865
300	68.153	327.218	326.798	0.126	-38.140	-56.421	9.824
350	72.371	338.051	327.644	3.643	-39.919	-59.328	8.854
400	75.838	347.949	329.572	7.351	-57.581	-61.436	8.023
450	78.748	357.054	332.127	11.217	-60.332	-61.752	7.168
500	81.249	365.483	335.046	15.219	-103.110	-57.968	6.056
600	85.398	380.678	341.413	23.559	-103.829	-48.867	4.254
700	88.749	394.102	348.000	32.272	-104.269	-39.668	2.960
800	91.512	406.139	354.528	41.289	-104.492	-30.422	1.986
900	93.803	417.054	360.878	50.558	-104.550	-21.158	1.228
1000	95.709	427.039	367.002	60.037	-104.488	-11.895	0.621
1100	97.298	436.238	372.884	69.690	-104.345	-2.642	0.125
1200	98.628	444.763	378.523	79.488	-104.157	6.596	-0.287
1300	99.745	452.702	383.927	89.408	-103.954	15.817	-0.636
1400	100.689	460.130	389.108	99.431	-103.764	25.023	-0.934
1500	101.491	467.105	394.077	109.541	-103.614	34.217	-1.192
1600	102.176	473.677	398.849	119.725	-103.524	43 402	-1.417
1700	102.764	479.890	403.435	129.973	-153.690	53.031	-1.629
1800	103.273	485.778	407.847	140.276	-153.542	65.186	-1.892
1900	103.714	491.374	412.097	150.625	-153.455	77.335	-2.126
2000	104.100	496.704	416.196	161.017	-153.434	89.481	-2.337
2100	104.438	501.791	420.151	171.444	-153.480	101.628	-2.528
2200	104.736	506.657	423.974	181.903	-153.590	113.778	-2.701
2300	105.000	511.318	427.671	192.390	-153.758	125.935	-2.860
2400	105.235	515.792	431.250	202.902	-153.977	138.100	-3.006
2500	105.444	520.092	434.718	213.436	-154.237	150.275	-3.140
2600	105.632	524.232	438.082	223.990	-154.527	162.461	-3.264
2700	105.801	528.222	441.347	234.562	-154.835	174.659	-3.379
2800	105.953	532.072	444.519	245.150	-155.153	186.868	-3.486
2900	106.091	535.793	447.602	255.752	-155.468	199.089	-3.586
3000	106.216	539.391	450.602	266.367	-155.773	211.320	-3.679
3100	106.330	542.876	453.523	276.995	-156.059	223.561	-3.767
3200	106.433	546.253	456.368	287.633	-156.318	235.811	-3.849
3300	106.528	549.530	459.142	298.281	-156.546	248.069	-3.927
3400	106.616	552.712	461.847	308.938	-156.738	260.333	-4.000
3500	106.696	555.803	464.488	319.604	-156.889	272.602	-4.068
3600	106.770	558.810	467.066	330.277	-541.150	295.336	-4 285
3700	106.838	561.736	469.586	340.958	-540.800	318.567	-4.497
3800	106.901	564.586	472.048	351.645	-540.406	341.788	-4.698
3900	106.959	567.364	474.457	362.338	-539.967	364.998	-4.889
4000	107.014	570.073	476.814	373.036	-539.483	388.196	-5.069
4100	107.064	572.716	479.121	383.740	-538.954	411.381	-5.241
4200	107 111	575.296	481.380	394.449	-538 381	434.554	-5.404
4300	107.155	577.817	483.593	405.162	-537.764	457.711	-5.560
4400	107.196	580.281	485.763	415.880	-537.105	480.854	-5.708
4500	107.234	582.691	487.890	426.601	-536.404	503.982	-5.850
4600	107.270	585 048	489.977	437.327	-535 664	527.093	-5.985
4700	107.304	587.355	492.024	448.055	-534.885	550.189	-6.115
4800	107.336	589.615	494.034	458.787	-534.068	573.267	-6.238
4900	107.366	591.828	496.007	469.523	-533.216	596.328	-6.357
5000	107.394	593.997	497.945	480.261	-532.330	619.371	-6.471
5100	107.420	596.124	499.850	491.001	-531.411	642.396	-6.579
5200	107.445	598.211	501.721	501.745	-530.461	665.402	-6.684
5300	107.469	600.257	503.561	512.490	-529.480	688.390	-6.784
5400	107.491	602.266	505.370	523.238	-528.470	711.360	-6.881
5500	107.512	604.239	507.150	533.988	-527.431	734.309	-6.974
5600	107.533	606.176	508.901	544.741	-526.366	757.241	-7.063
5700	107.552	608.080	510.625	555.495	-525.275	780.153	-7.149
5800	107.570	609.951	512.321	566.251	-524.159	803.046	-7.232
5900	107.587	611.790	513.991	577.009	-523.019	825.918	-7.312
6000	107.603	613.598	515.636	587.768	-521.854	848.772	-7.389

PREVIOUS: December 1976 (1 atm) CURRENT: December 1976 (1 bar)

Diiodosilane (SiH_2I_2) $H_2I_2Si_1$(g)

POTASSIUM HYDROXIDE ((KOH)$_2$) IDEAL GAS M_r = 112.21128

$S°(298.15\ K) = [327.996 \pm 8.4]\ J\ K^{-1}\ mol^{-1}$

$\Delta_f H°(0\ K) = -641.95 \pm 12.6\ kJ\ mol^{-1}$
$\Delta_f H°(298.15\ K) = -654.80 \pm 12.6\ kJ\ mol^{-1}$

Vibrational Frequencies and Degeneracies

ν, cm^{-1}	ν, cm^{-1}
[273](1)	[3700](1)
[150](1)	[3700](1)
[235](1)	[1250](1)
[150](1)	[1250](1)
[255](1)	[1250](1)
[258](1)	[1250](1)

Ground State Quantum Weight: 1 $\sigma = 2$
Point Group: C_{2h}
Bond Distances: K-O = [2.43] Å O-H = [0.96] Å
Bond Angles: K-O-K = [90]° K-O-H = [110]°
Product of the Moments of Inertia: $I_A I_B I_C = [3.758094] \times 10^{-113}\ g^3\ cm^6$

Enthalpy of Formation

$\Delta_f H°(K_2(OH)_2$, g, 298.15 K) is calculated from the enthalpy of dimerization, $\Delta_{dim}H°(298.15\ K) = -45.3 \pm 3.0\ kcal\ mol^{-1}$, and the enthalpy of formation of the monomer, $\Delta_f H°$(KOH, g, 298.15 K) = $-55.6 \pm 3.0\ kcal\ mol^{-1}$. See KOH(g) table for details.

Heat Capacity and Entropy

A key to the molecular structure of $K_2(OH)_2(g)$ is provided by the work of Buchler et al. (1) which involves the electric deflection of molecular beams of $Cs_2(OH)_2(g)$ and $Na_2(OH)_2(g)$. Their results indicate that these dimers are nonpolar which implies a planar structure for at least the M_2O_2 part of the molecule. The three models proposed thus far for the dimeric alkali metal hydroxides each incorporate this planar configuration for the alkali and oxygen atoms. Bauer et al (2) proposed a model consisting of a square planar configuration for the alkali and oxygen atoms with two hydrogen bonded bridges between the oxygens and hydrogens. Schoonmaker and Porter (3) adopted the square planar model for the M_2O_2 configuration but did not allow for hydrogen bonding. Berkowitz et al. (4) proposed a model for $Li_2(OH)_2(g)$ which consisted of a trans configuration with hydrogens above and below the plane of a rhombus formed by the lighium and oxygen atoms, the O-Li-O bond angle being 100°. The adopted molecular structure for $K_2(OH)_2(g)$ is similar to the model of Berkowitz et al. 94) but with the O-K-O bond angle equal to 90°. The K-O-H bond angle is assumed equal to 110°. The K-O bond distance is estimated as being 10% longer than that in KOH(g). The O-H bond distance is estimated to be the same as in $H_2O(g)$. The principal moments of inertia are: $I_A = [17.5192] \times 10^{-39}$, $I_B = [38.5520] \times 10^{-39}$, and $I_C = [55.6426] \times 10^{-39}\ g\ cm^2$.

The first six vibrational frequencies are taken equal to those for dimeric $K_2F_2(g)$ (See $K_2F_2(g)$ table), and the remaining O-H stretching and bending modes are estimates taken from the work of Berkowitz et al. (4).

References

1. A. Buchler, J. L. Stauffer, and W. Klemperer, J. Chem. Phys. 46, 605 (1967).
2. S. H. Bauer, R. H. Diner, and R. F. Porter, J. Chem. Phys. 29, 991 (1958).
3. R. C. Schoonmaker and R. F. Porter, J. Chem. Phys. 31, 830 (1959).
4. J. Berkowitz, D. J. Meschi, and W. A. chupka, J. Chem. Phys. 33, 533 (1960).

Potassium Hydroxide ((KOH)$_2$) $H_2K_2O_2(g)$

Enthalpy Reference Temperature = T_r = 298.15 K Standard State Pressure = p° = 0.1 MPa

T/K	$C_p°$ (J K^{-1}mol^{-1})	S° (J K^{-1}mol^{-1})	$-[G°-H°(T_r)]/T$ (J K^{-1}mol^{-1})	$H°-H°(T_r)$ (kJ mol^{-1})	$\Delta_f H°$ (kJ mol^{-1})	$\Delta_f G°$ (kJ mol^{-1})	Log K_f
0	0.	0.	INFINITE	-18.473	-641.954	-641.954	INFINITE
100	56.735	251.157	394.722	-14.357	-647.206	-637.798	333.151
200	73.990	296.990	335.219	-7.646	-651.289	-626.711	163.680
250	78.049	313.955	329.318	-3.841	-653.087	-620.357	129.616
298.15	81.444	327.996	327.996	0.	-654.796	-613.898	107.552
300	81.571	328.500	327.997	0.151	-654.862	-613.644	106.845
350	84.944	341.329	329.003	4.314	-661.407	-606.440	90.506
400	88.173	352.884	331.277	8.643	-663.204	-598.462	78.151
450	91.187	363.446	334.273	13.128	-664.824	-590.271	68.517
500	93.941	373.198	337.684	17.757	-666.286	-581.907	60.791
600	98.651	390.758	345.099	27.395	-668.816	-564.786	49.169
700	102.458	406.261	352.750	37.457	-670.938	-547.275	40.838
800	105.603	420.153	360.322	47.865	-672.774	-529.481	34.572
900	108.278	432.750	367.681	58.562	-674.422	-511.469	29.685
1000	110.604	444.281	374.773	69.509	-675.967	-493.279	25.766
1100	112.656	454.921	381.581	80.674	-835.387	-465.717	22.115
1200	114.482	464.803	388.110	92.033	-834.813	-432.135	18.810
1300	116.113	474.032	394.368	103.564	-834.143	-398.606	16.016
1400	117.572	482.692	400.371	115.249	-833.393	-365.130	13.623
1500	118.878	490.849	406.133	127.073	-832.578	-331.711	11.551
1600	120.048	498.559	411.671	139.020	-831.711	-298.348	9.740
1700	121.097	505.869	416.999	151.079	-830.802	-265.041	8.144
1800	122.039	512.818	422.131	163.236	-829.865	-231.788	6.726
1900	122.885	519.439	427.080	175.483	-828.908	-198.587	5.460
2000	123.647	525.762	431.857	187.810	-827.942	-165.437	4.321
2100	124.333	531.812	436.474	200.210	-826.976	-132.336	3.292
2200	124.953	537.610	440.940	212.675	-826.019	-99.280	2.357
2300	125.513	543.177	445.265	225.199	-825.080	-66.268	1.505
2400	126.021	548.530	449.457	237.776	-824.168	-33.296	0.725
2500	126.483	553.684	453.523	250.401	-823.291	-0.361	0.008
2600	126.903	558.653	457.472	263.071	-822.456	32.540	-0.654
2700	127.286	563.449	461.308	275.781	-821.673	65.409	-1.265
2800	127.635	568.085	465.040	288.527	-820.948	98.251	-1.833
2900	127.956	572.569	468.671	301.307	-820.290	131.067	-2.361
3000	128.250	576.912	472.207	314.117	-819.707	163.863	-2.853
3100	128.520	581.122	475.653	326.956	-819.208	196.640	-3.313
3200	128.768	585.206	479.013	339.820	-818.800	229.402	-3.745
3300	128.998	589.172	482.291	352.709	-818.474	262.152	-4.150
3400	129.210	593.027	485.491	365.619	-818.266	294.895	-4.531
3500	129.406	596.775	488.618	378.550	-818.170	327.633	-4.890
3600	129.588	600.423	491.673	391.500	-818.200	360.370	-5.229
3700	129.757	603.976	494.660	404.467	-818.365	393.110	-5.550
3800	129.914	607.438	497.583	417.451	-818.672	425.857	-5.854
3900	130.060	610.815	500.443	430.450	-819.132	458.614	-6.142
4000	130.197	614.109	503.244	443.463	-819.740	491.382	-6.417
4100	130.325	617.326	505.987	456.489	-820.522	524.169	-6.678
4200	130.444	620.468	508.676	469.527	-821.492	556.978	-6.927
4300	130.556	623.539	511.311	482.577	-822.653	589.812	-7.165
4400	130.660	626.541	513.896	495.638	-823.460	622.621	-7.391
4500	130.759	629.479	516.432	508.709	-824.895	655.501	-7.609
4600	130.851	632.354	518.921	521.790	-826.524	688.416	-7.817
4700	130.939	635.169	521.365	534.879	-828.312	721.365	-8.017
4800	131.021	637.926	523.764	547.977	-830.086	754.331	-8.209
4900	131.098	640.629	526.122	561.083	-832.272	787.362	-8.393
5000	131.171	643.278	528.439	574.197	-834.665	820.440	-8.571
5100	131.240	645.876	530.716	587.317	-837.267	853.568	-8.742
5200	131.305	648.425	532.955	600.445	-840.080	886.748	-8.907
5300	131.367	650.927	535.157	613.578	-843.106	919.985	-9.067
5400	131.426	653.383	537.324	626.718	-846.343	953.281	-9.221
5500	131.482	655.795	539.456	639.864	-849.793	986.638	-9.370
5600	131.535	658.165	541.555	653.014	-853.452	1020.061	-9.515
5700	131.585	660.493	543.621	666.170	-856.629	1053.455	-9.654
5800	131.632	662.782	545.656	679.331	-860.614	1087.001	-9.789
5900	131.678	665.033	547.660	692.497	-864.792	1120.614	-9.921
6000	131.721	667.246	549.635	705.667	-869.158	1154.302	-10.049

PREVIOUS: December 1970 (1 atm) CURRENT: December 1970 (1 bar)

Potassium Hydroxide ((KOH)$_2$) $H_2K_2O_2(g)$

LITHIUM HYDROXIDE $((LiOH)_2)$ IDEAL GAS M_r = 47.89668

$\Delta_f H°(0\ K)$ = -699.39 ± 33.5 kJ mol^{-1}

S°(298.15 K) = [269.794 ± 8.4] J K^{-1} mol^{-1} $\Delta_f H°(298.15\ K)$ = -711.28 ± 33.5 kJ mol^{-1}

Vibrational Frequencies and Degeneracies (ν, cm^{-1}, all singly degenerate)

[400] [400] [500] [500] [450] [350] [3700] [3700] [1250] [1250] [1250] [1250]

Ground State Quantum Weight: [1] σ =2

Point Group: [C_{2h}]

Bond Distances: Li-O = [1.74] Å O-H = [0.97] Å

Bond Angles: Li-O-Li = [90]° Li-O-H = [110]°

Product of the Moments of Inertia: $I_A I_B I_C$ = [4.293246 x 10^{-115}] g^3 cm^6

Enthalpy of Formation

The reaction between water vapor and condensed lithium oxide has been investigated as a function of temperature and water pressure by means of such techniques as mass spectrometry (1, 2) and transpiration (3). At low water pressures (<0.01 Torr) in the temperature range 780-900 K, Schoonmaker and Porter (1) observed only Li^+ and $LiOH^+$ ion species in the mass spectrum of the reaction products. The precursor of these two ions was assumed to be monomeric LiOH. Berkowitz et al. (2) in a more extensive mass spectrometric study of the $Li_2O(cr)$-$H_2O(g)$ equilibrium detected not only monomeric and dimeric LiOH in the vapor phase, but also trace amounts of trimer ($Li_3(OH)_3$). Their work covered the temperature range 1100-1400 K at water pressures of 0.04 and 0.16 Torr. Absolute equilibrium constants were determined for the dimerization reaction, 2 LiOH(g) = $Li_2(OH)_2$(g), by calibrating the mass spectrometer in units of pressure. This was accomplished by measuring the water pressure externally and simultaneously monitoring the intensity of the H_2O^+ peak. In an attempt to investigate the same equilibrium at water pressures inaccessible to mass spectrometry, Berkowitz-Mattuck and Buchler (3) used the transpiration method. Their work covered water pressures of 4.58 and 19.5 Torr at 1095 and 1145 K. An analysis of these equilibrium data are summarized below.

Source	Reaction[a]	Method	T/K	Data Points	$\Delta_r H°$(298.15 K), kcal mol^{-1} 2nd law	3rd law	Drift cal K^{-1} mol^{-1}	$\Delta_f H°$(298.15 K) kcal mol^{-1}
Berkowitz, et al (2)	A	Mass Spec.	1119-1352	9[b]	-57.2	-63.8±1.4	-5.4±2.8	-175.8±4.0
Berkowitz-Mattuck et al (3)	B	Transp.	1095, 1145	2		12.7		-188.2 ±4.0

[a]Reactions: A) 2 LiOH(g) = $Li_2(OH)_2$(g) B) H_2O(g) + Li_2O(cr) = $Li_2(OH)_2$(g)

[b]One point rejected due to failure of a statistical test.

It seems most likely that the equilibrium data of Berkowitz-Mattuck and Buchler (3) are in error, since no satisfactory compromise in the flow rates of the inert carrier gas could be found such that saturation was achieved while eliminating thermal diffusion effects. Under these conditions the simple equations relating partial pressures to the masses transported during the transpiration experiment are now longer valid. Furthermore, the authors assumed that under the conditions of their experiment only dimer LiOH would be formed. The work of Berkowitz et al. (2) clearly establishes the existence of a trimer at water pressures some 100 times lower than those employed by Berkowitz-Mattuck and Buchler (3).

Trends in the dimerization energies for the higher alkali metal hydroxides, as well as for the dimeric alkali fluorides and chlorides, suggest that the enthalpy of dimerization for LiOH from the work of Berkowitz et al. (2) may be slightly high. Such a comparison results in $\Delta_f H°$(dimerization, 298.15 K) values in the range -(52-61) kcal mol^{-1}. Further support for a lower value comes from the mass spectral work of Porter and Schoonmaker (1). They investigated the reaction of H_2O(g) with a mixture of condensed Li_2O-Na_2O. From the reported ion current intensities for the various ions found in the reaction products we calculate, $\Delta_{dim}H°$(NaOH, g, 298.15 K) $-\Delta_{dim}H°$(LiOH, g, 298.15 K) = 5.3 kcal mol^{-1}. This value yields $\Delta_{dim}H°$(298.15 K) = -56.0 ± 5 kcal mol^{-1} for LiOH(g) when used in combination with the adopted value for NaOH(g) (4). A high enthalpy of dimerization from the mass spectral work of Berkowitz et al. (2) is easily reconciled in terms of nonequilibrium measurements. Based upon these findings, we adopt $\Delta_{dim}H°$(298.15 K) = -58.0 ± 5.0 kcal mol for LiOH(g). Combining this result with the enthalpy of formation for LiOH(g) (4), that for the dimer is $\Delta_f H°(Li_2(OH)_2$, g, 298.15 K) = -170.0 ± 8 kcal mol^{-1}

Heat Capacity and Entropy

The molecular structure of $Li_2(OH)_2$(g) is assumed identical to that adopted for the higher alkali metal hydroxide dimers. It consists of a planar configuration for the lithium and oxygen atoms with the O-Li-O bond angle equal to 90°. The hydrogen atoms are placed in a trans-configuration above and below the plane of the rhombus formed by the lithium and oxygen atoms. The Li-O-H bond angle is assumed equal to 110°. The Li-O bond distance is estimated as being 10% longer than in LiOH(g). The O-H bond distance is estimated to be the same as in H_2O(g). The principal moments of inertia are: I_A = [3.7044 x 10^{-39}], I_B = [9.2519 x 10^{-39}], and I_C = [12.5268 x 10^{-39}] g cm^2.

The first six vibrational frequencies are estimated by comparison with calculated (5) and experimental (4, 6) frequencies for Li_2F_2(g). In the estimation it is assumed that the bulkier OH group would tend to reduce the frequencies of $Li_2(OH)_2$(g) relative to those for Li_2F_2(g). The remaining OH stretching and bending frequencies are estimates from the work of Berkowitz et al. (2).

References

1. R. C. Schoonmaker and R. F. Porter, J. Phys. Chem. 64, 457 (1960).
2. J. Berkowitz, D. J. Meschi and W. A. Chupka, J. Chem. Phys. 33, 533 (1960).
3. J. B. Berkowitz-Mattuck and A. Buchler, J. Phys. Chem. 67, 1386 (1963).
4. JANAF Thermochemical Tables: LiOH(g), 6-31-71; Li_2O(cr), 3-31-64; $Na_2(OH)_2$(g), 12-31-70; Li_2F_2(g), 12-31-68.
5. J. Berkowitz, J. Chem. Phys. 32, 1519 (1960).
6. a) A. Snelson, J. Chem. Phys. 46, 3652 (1967); b) A. Snelson, J. Phys. Chem. 73, 1919 (1969).

Lithium Hydroxide $((LiOH)_2)$ $H_2Li_2O_2$(g)

Enthalpy Reference Temperature = T_r = 298.15 K Standard State Pressure = p° = 0.1 MPa

T/K	$C_p°$ (J K^{-1}mol^{-1})	S°	$-[G°-H°(T_r)]/T$	$H°-H°(T_r)$ (kJ mol^{-1})	$\Delta_f H°$	$\Delta_f G°$	Log K_f
0	0.	0.	INFINITE	-14.502	-699.387	-699.387	INFINITE
100	37.588	212.012	323.024	-11.101	-702.880	-695.209	363.140
200	57.261	244.155	275.895	-6.348	-707.415	-685.736	179.096
250	65.023	257.797	270.933	-3.284	-709.461	-680.076	142.094
298.15	71.223	269.794	269.794	0.	-711.280	-674.249	118.126
300	71.443	270.235	269.795	0.132	-711.347	-674.019	117.357
350	76.958	281.672	270.686	3.845	-713.085	-667.659	99.643
400	81.765	292.270	272.730	7.816	-714.741	-661.056	86.325
450	85.956	302.148	275.456	12.011	-716.370	-654.247	75.943
500	89.602	311.397	278.592	16.402	-724.002	-646.643	67.554
600	95.542	328.283	285.495	25.673	-726.787	-630.900	54.925
700	100.130	343.370	292.704	35.466	-729.036	-614.734	45.872
800	103.799	356.988	299.902	45.669	-730.916	-598.274	39.063
900	106.840	369.395	306.945	56.205	-732.542	-581.594	33.755
1000	109.432	380.789	313.767	67.022	-733.963	-564.744	29.499
1100	111.684	391.327	320.345	78.080	-735.215	-547.760	26.011
1200	113.662	401.131	326.674	89.349	-736.324	-530.668	23 099
1300	115.412	410.300	332.758	100.805	-737.317	-513.489	20 632
1400	116.966	418.911	338.607	112.425	-738.207	-496.238	18.515
1500	118.349	427.029	344.234	124.192	-739.004	-478.926	16.678
1600	119.583	434.707	349.650	136.090	-739.723	-461.564	15.069
1700	120.685	441.990	354.870	148.105	-1030.849	-429.806	13.206
1800	121.671	448.917	359 904	160.223	-1029.941	-394.477	11.447
1900	122.554	455.519	364.764	172.435	-1029.005	-359.199	9.875
2000	123.348	461.826	369.460	184.731	-1028.049	-323.971	8.461
2100	124.062	467.862	374.004	197.102	-1027.081	-288.791	7.183
2200	124.705	473.648	378.402	209.541	-1026.110	-253.656	6.023
2300	125.287	479.205	382.665	222.041	-1025.142	-218.567	4.964
2400	125.813	484.548	386.799	234.597	-1024.183	-183.519	3.994
2500	126.291	489.694	390.813	247.202	-1023.240	-148.511	3.103
2600	126.725	494.656	394.712	259.854	-1022.319	-113.540	2.281
2700	127.121	499.446	398.503	272.546	-1021.425	-78.604	1.521
2800	127.482	504.076	402.191	285.277	-1020.564	-43.700	0.815
2900	127.813	508.555	405.782	298.042	-1019.741	-8.827	0.159
3000	128.116	512.893	409.281	310.838	-1018.961	26.019	-0.453
3100	128.395	517.099	412.691	323.664	-1018.227	60.840	-1.025
3200	128.651	521.179	416.018	336.517	-1017.546	95.636	-1.561
3300	128.888	525.142	419.265	349.394	-1016.919	130.413	-2.064
3400	129.106	528.993	422.436	362.293	-1016.352	165.173	-2.538
3500	129.308	532.738	425.534	375.214	-1015.849	199.916	-2.984
3600	129.495	536.384	428.563	388.155	-1015.412	234.646	-3.405
3700	129.669	539.934	431.525	401.113	-1015.046	269.364	-3.803
3800	129.831	543.394	434.424	414.088	-1014.753	304.074	-4.180
3900	129.981	546.769	437.261	427.079	-1014.538	338.778	-4.537
4000	130.122	550.061	440.040	440.084	-1014.403	373.476	-4.877
4100	130.253	553.276	442.763	453.103	-1014.351	408.172	-5.200
4200	130.376	556.416	445.432	466.134	-1014.387	442.868	-5.508
4300	130.491	559.485	448.049	479.178	-1014.513	477.566	-5.801
4400	130.598	562.487	450.616	492.232	-1014.690	512.264	-6.081
4500	130.700	565.423	453.134	505.297	-1014.986	546.970	-6.349
4600	130.795	568.296	455.607	518.372	-1015.385	581.684	-6.605
4700	130.884	571.110	458.034	531.456	-1015.884	616.408	-6.851
4800	130.968	573.867	460.419	544.549	-1016.487	651.144	-7.086
4900	131.048	576.568	462.762	557.649	-1017.197	685.894	-7.312
5000	131.123	579.216	465.065	570.758	-1018.017	720.660	-7.529
5100	131.194	581.813	467.328	583.874	-1018.940	755.441	-7.737
5200	131.261	584.362	469.555	596.997	-1019.988	790.242	-7.938
5300	131.325	586.863	471.744	610.126	-1021.157	825.066	-8.132
5400	131.385	589.318	473.899	623.261	-1022.450	859.912	-8.318
5500	131.442	591.729	476.020	636.403	-1023.870	894.783	-8.498
5600	131.496	594.098	478.107	649.550	-1025.421	929.682	-8.672
5700	131.548	596.426	480.162	662.702	-1027.107	964.609	-8.840
5800	131.597	598.714	482.187	675.859	-1028.930	999.568	-9.002
5900	131.643	600 964	484.181	689.021	-1030.895	1034.557	-9.159
6000	131.688	603.177	486.146	702.188	-1033.005	1069.583	-9.312

PREVIOUS: June 1971 (1 atm) CURRENT: June 1971 (1 bar)

Lithium Hydroxide $((LiOH)_2)$ $H_2Li_2O_2$(g)

MAGNESIUM HYDRIDE (MgH_2) CRYSTAL M_r = 26.32088

$S°(298.15\ K) = 31.091 \pm 0.84\ J\ K^{-1}\ mol^{-1}$

$\Delta_f H°(0\ K) = -68.00 \pm 9.2\ kJ\ mol^{-1}$
$\Delta_f H°(298.15\ K) = -76.15 \pm 9.2\ kJ\ mol^{-1}$

Enthalpy of Formation

The enthalpy of hydrolysis of MgH_2(cr) was measured by Sinke and Stull (1). The value of $\Delta_f H°$(298.15 K) was calculated to be -19.1±2 kcal mol^{-1}. From the decomposition pressure measurements for MgH_2(cr) reported by Ellinger (2) and Stampfer et al. (3), the values of $\Delta_f H°$(298.15 K) for MgH_2(cr) were found to be -17.9±1.0 and -17.7 kcal mol^{-1}, respectively. The value used was the average of these three.

Heat Capacity and Entropy

The low temperature heat capacities (25-300 K) were measured by Sinke and Hildenbrand (4). Above 300 K the $C_p°$ values were estimated by comparison with those for MgF_2(cr). S°(298.15 K) is obtained from Sinke and Hildenbrand (4) using S°(25 K) = 0.037 cal K^{-1} mol^{-1} (extrap.).

Decomposition Data

T_{dcm} = 560 K is estimated from the value of $\Delta_f G°$ in the table.

References

1. G. C. Sinke and D. R. Stull, The Dow Chemical Company, personal communication, July, 1956.
2. F. H. Ellinger, et al., J. Amer. Chem. Soc. 77, 2647 (1955).
3. J. F. Stampfer, Jr., C. E. Holley, Jr. and J. F. Suttle, J. Amer. Chem. Soc. 82, 3504 (1960).
4. G. C. Sinke and D. L. Hildenbrand, The Dow Chemical Company, personal communication, February, 1958.

Magnesium Hydride (MgH_2) H_2Mg_1(cr)

Enthalpy Reference Temperature = T_r = 298.15 K Standard State Pressure = p° = 0.1 MPa

T/K	$C_p°$ (J K⁻¹mol⁻¹)	S° (J K⁻¹mol⁻¹)	$-[G°-H°(T_r)]/T$ (J K⁻¹mol⁻¹)	$H°-H°(T_r)$ (kJ mol⁻¹)	$\Delta_f H°$ (kJ mol⁻¹)	$\Delta_f G°$ (kJ mol⁻¹)	Log K_f
0	0.	0.	INFINITE	-5.314	-67.997	-67.997	INFINITE
100	12.774	6.389	55.049	-4.866	-71.191	-60.806	31.762
200	25.284	19.372	33.974	-2.920	-73.946	-49.309	12.878
298.15	35.334	31.091	31.091	0.	-76.149	-36.716	6.432
300	35.510	31.310	31.092	0.066	-76.183	-36.471	6.350
400	43.304	42.620	32.581	4.015	-77.694	-22.989	3.002
500	49.999	53.029	35.644	8.693	-78.606	-9.195	0.961
600	55.229	62.621	39.352	13.962	-79.033	4.734	-0.412
700	59.538	71.468	43.316	19.706	-79.099	18.705	-1.396
800	63.095	79.659	47.353	25.844	-78.901	32.665	-2.133
900	65.898	87.259	51.370	32.300	-78.539	46.591	-2.704
1000	67.990	94.315	55.316	38.998	-86.701	61.184	-3.196
1100	69.486	100.867	59.163	45.874	-86.295	75.953	-3.607
1200	70.751	106.969	62.896	52.888	-85.790	90.682	-3.947
1300	71.761	112.673	66.508	60.015	-85.214	105.365	-4.234
1400	72.592	118.023	69.998	67.234	-212.000	123.169	-4.595
1500	73.204	123.052	73.369	74.525	-209.996	147.039	-5.120
1600	73.722	127.794	76.624	81.872	-207.978	170.776	-5.575
1700	74.088	132.274	79.767	89.262	-205.960	194.386	-5.973
1800	74.475	136.520	82.803	96.690	-203.945	217.878	-6.323
1900	74.883	140.557	85.737	104.158	-201.929	241.257	-6.633
2000	75.312	144.409	88.575	111.667	-199.908	264.531	-6.909

PREVIOUS: June 1963 CURRENT: September 1963

Magnesium Hydride (MgH_2) H_2Mg_1(cr)

MAGNESIUM HYDROXIDE ($Mg(OH)_2$) CRYSTAL M_r = 58.31968

$\Delta_f H°(0\ K) = -915.64 \pm 2.1\ kJ\ mol^{-1}$

S°(298.15 K) = 63.243 J K^{-1} mol^{-1} $\Delta_f H°(298.15\ K) = -924.66 \pm 2.1\ kJ\ mol^{-1}$

Enthalpy of Formation

The adopted value is an average of -221.10 and -220.86 kcal mol^{-1} obtained from $\Delta_r H°$(298.15 K) = 9.08 and 8.84 kcal mol^{-1} for the dehydration reaction $Mg(OH)_2$(cr) = MgO(cr) + H_2O(ℓ), using auxiliary data (13, 14). The two values correspond to natually occurring brucite and to synthetic $Mg(OH)_2$ prepared from MgO by hydration at 177°C under a steam pressure of about 9 atm. The enthalpies of reaction are derived from studies of Taylor and Wells (1) on the enthalpys of solution of various samples of $Mg(OH)_2$ and MgO in aqueous HCl. These studies show that the enthalpy of dehydration is quite dependent on the nature of the MgO sample; e.g., values for synthetic $Mg(OH)_2$ vary monotonically from 9.79 to 8.84 kcal mol^{-1} for MgO ignited at temperatures from 450 to 1425°C, respectively. The authors ascribe this difference primarily to increase in particle size at the higher ignition temperatures. More recent work (15) suggests that the difference is caused by crystalline disorder rather than particle size alone. We have adopted the results at 1425°C since they are reasonably consistent with the high temperature samples on which the JANAF enthalpy of formation for MgO is based. The uncertainty is estimated as ±0.5 kcal mol^{-1}.

Shomate and Huffman (2) have confirmed the enthalpy of solution of MgO (ignited at 1000°C) within about 0.3 kcal mol^{-1}. Their value may be combined with the enthalpy of solution for synthetic $Mg(OH)_2$, determined under similar conditions by Torgeson and Sahama (3), to obtain $\Delta_r H°$ = 8.85 kcal mol^{-1} for the dehydration reaction. At the other extreme, the enthalpy of solution data of Giauque and Archibald (4) lead to $\Delta_r H°$ = 9.74 kcal mol^{-1}; however, this value corresponds to MgO ignited at 350°C. The MgO sample was microcrystalline and had significant excess $C_p°$, as suggested by Giauque and Archibald (4) and later confirmed by data of Barron et al. (5). Giauque and Archibald (4) intentionally used the microcrystalline form which would be similar to the product in their decomposition pressure measurements (190 and 212°C). Thus, it is not surprising that 3rd law analysis of their decomposition pressures yields essentially the same enthalpy of formation as their solution data (see summary below). Decomposition data at higher pressures have been reported by Fyfe (6), Roy and Roy (7), Kennedy (8), Zhabrova and Kadenatsi (9), and Berg and Rassonskaya (10). These data do not significantly affect the enthalpy of formation, however, because of uncertainties in the attainment of equilibrium and in reduction to standard state values.

Source	$Mg(OH)_2$ Sample	Temp. of MgO Formation, °C	Enthalpy of Dehydration $\Delta_r H°$(298.15 K), kcal mol^{-1}	$\Delta_f H°$(298.15 K)
Taylor & Wells (1)	Brucite	1425	9.08	-221.10
	Synthetic	1425	8.84	-220.86
	Synthetic	1000	9.18	-221.20
	Synthetic	450	9.79	>-221.81
Torgeson (3) - Shomate (2)	Synthetic	1000	8.85	-220.87
Giauque & Archibald (4)	Synthetic[a]	350	9.74	>-221.74
K_p data	Synthetic[a]	190, 212	20.10[b]	>-221.59

[a]Prepared by reaction of KOH with $MgCl_2$(aq) at 210°C, rather than by hydration of MgO.

[b]For $Mg(OH)_2$(cr) = MgO(cr) + H_2O(g) rather than $Mg(OH)_2$(cr) = MgO(cr) + H_2O(ℓ).

Heat Capacity and Entropy

The low temperature heat capacities are from the measurements (22-321 K) of Giauque and Archibald (4) who used a synthetic, macrocrystalline sample of magnesium dihydroxide prepared by reacting KOH with aqueous $MgCl_2$ at 210°C. The entropy is derived from the heat capacities starting with S°(20 K) = 0.027 cal K^{-1} mol^{-1} from a T^3 extrapolation. S°(298.15 K) = 15.11 cal K^{-1} mol^{-1} compares favorably with 15.09 cal K^{-1} mol^{-1} given by Giauque and Archibald (4). The drop calorimeter enthalpy measurements (350-699 K) of King et al. (11) are used to derive heat capacities above 298.15 K. They used a ground sample of natural brucite, which analyzed 98.77% $Mg(OH)_2$. Their published data are corrected for impurities and to one atmosphere total pressure. The low and high temperature heat capacities were joined smoothly by mathematical curve fitting techniques. Heat capacities above 700 K are obtained by extrapolation.

The reference temperature for the enthalpy measurements (393-667 K) of Lashchenko and Kompanskii (12) is uncertain. Their values appear to be near those of King et al. (11) up to around 550 K; above 550 K their values average about 5% lower than those of King et al. (11).

Decomposition Data

T_{dcm} = 542.2 K is calculated as the temperature at which the fugacity of H_2O(g) is 1 atm for the reaction $Mg(OH)_2$(cr) = MgO(cr) + H_2O(g).

References

1. K. Taylor and L. S. Wells, J. Res. Nat. Bur. Stand. 21, 133 (1938).
2. C. H. Shomate and E. H. Huffman, J. Amer. Chem. Soc. 65, 1625 (1943).
3. D. R. Torgeson and Th. G. Sahama, J. Amer. Chem. Soc. 70, 2156 (1948).
4. W. F. Giauque and R. C. Archibald, J. Amer. Chem. Soc. 59, 561 (1937).
5. T. H. K. Barron, W. T. Berg, and J. A. Morrison, Proc. Roy. Soc. (London) A250, 70 (1959).
6. W. S. Fyfe, Amer. J. Sci. 256, 729 (1958).
7. D. M. Roy and R. Roy, Amer. J. Sci. 255, 573 (1957).
8. G. C. Kennedy, Amer. J. Sci. 254, 567 (1956).
9. G. M. Zhabrova and B. M. Kadenatsi, J. Gen. Chem. U.S.S.R. 24, 1129 (1954).
10. L. G. Berg and I. S. Rassonskaya, Dokl. Akad. Nauk S.S.S.R. 81, 855 (1951).
11. E. G. King, M. J. Ferrante, and L. B. Pankratz, U. S. Bur. Mines RI 8041, 1975.
12. P. N. Lashchenko and D. I. Kompanskii, Zh. Prikl. Khim. 8, 628 (1935).
13. JANAF Thermochemical Tables: MgO(cr), 12-31-74; H_2O(g), 3-31-61.
14. U. S. Nat. Bur. Stand. Tech. Note 270-3, 1968: H_2O(ℓ).
15. M. Natarajan, T. S. Sarma, J. C. Ahluwalia, and C. N. R. Rao, Trans. Faraday Soc. 65, 3088 (1969).

Magnesium Hydroxide ($Mg(OH)_2$) $H_2Mg_1O_2$(cr)

Enthalpy Reference Temperature = T_r = 298.15 K   Standard State Pressure = p° = 0.1 MPa

T/K	$C_p°$ (J $K^{-1}mol^{-1}$)	S° (J $K^{-1}mol^{-1}$)	$-[G°-H°(T_r)]/T$ (J $K^{-1}mol^{-1}$)	$H°-H°(T_r)$ (kJ mol^{-1})	$\Delta_f H°$ (kJ mol^{-1})	$\Delta_f G°$ (kJ mol^{-1})	Log K_f
0	0.	0.	INFINITE	-13.127	-915.642	-915.642	INFINITE
100	21.618	9.770	116.934	-10.716	-919.778	-892.401	466.142
200	56.868	36.341	69.736	-6.679	-923.352	-863.412	225.500
298.15	77.249	63.242	63.242	0.	-924.664	-833.652	146.052
300	77.555	63.721	63.244	0.143	-924.675	-833.087	145.053
400	91.705	88.120	66.474	8.659	-924.591	-802.538	104.801
500	99.617	109.525	72.994	18.266	-923.632	-772.123	80.663
600	104.441	128.143	80.669	28.484	-922.270	-741.945	64.592
700	107.880	144.513	88.644	39.108	-920.710	-712.012	53.131
800	110.579	159.101	96.556	50.036	-919.060	-682.311	44.550
900	112.847	172.259	104.248	61.210	-917.385	-652.817	37.889
1000	114.851	184.254	111.658	72.597	-924.320	-622.798	32.532

PREVIOUS: March 1967   CURRENT: December 1975

Magnesium Hydroxide ($Mg(OH)_2$) $H_2Mg_1O_2$(cr)

MAGNESIUM HYDROXIDE ($Mg(OH)_2$) IDEAL GAS M_r = 58.31968

$\Delta_f H°(0\ K) = -564.70 \pm 33.5\ kJ\ mol^{-1}$

$S°(298.15\ K) = [267.270 \pm 12.6]\ J\ K^{-1}\ mol^{-1}$ $\Delta_f H°(298.15\ K) = -572.37 \pm 33.5\ kJ\ mol^{-1}$

Vibrational Frequencies and Degeneracies

ν, cm^{-1}
[508](1)
[215](1)
[875](1)
[3650](2)
[492](4)

Ground State Quantum Weight: [1] $\sigma = 2$

Point Group: [C_{2v}]

Bond Distances: Mg–O = [1.79] Å O–H = [0.96] Å

Bond Angles: O–Mg–O = [158]° Mg–O–H = [180]°

Product of the Moments of Inertia: $I_A I_B I_C = [1.092990 \times 10^{-115}]\ g^3\ cm^6$

Enthalpy of Formation

Alexander, et al. (1) determined the temperature dependence of the equilibrium constant for the reaction $MgO(cr) + H_2O(g) = Mg(OH)_2(g)$ in the range of 1650 to 2020 K by measuring vapor densities using a transpiration technique. The data are presented graphically and are represented by a linear equation. With auxiliary data (2), analysis of the equation yields a 2nd law $\Delta_r H°(298.15\ K) = 67.1\ kcal\ mol^{-1}$ and a 3rd law $\Delta_r H°(298.15\ K) = 64.70\ kcal\ mol^{-1}$ with a drift of $-1.3\ cal\ K^{-1}\ mol^{-1}$. From the 3rd law heat of reaction, the adopted $\Delta_f H°(298.15\ K) = -136.80\ kcal\ mol^{-1}$ is calculated. An uncertainty of $\pm 8.0\ kcal\ mol^{-1}$ is assigned to allow for the uncertainty shown in the vapor density measurements and for the uncertainty in the entropy.

The enthalpy of dissociation listed by Jackson (6) leads to $\Delta_f H°(Mg(OH)_2,\ g,\ 298.15\ K) = -140.74\ kcal\ mol^{-1}$. Another recent compilation (7) lists $\Delta_f H°(298.15\ K) = -134\ kcal\ mol^{-1}$.

Based on the adopted $\Delta_f H°(298.15\ K) = -136.80 \pm 8\ kcal\ mol^{-1}$, $D_0^0 = 188.6\ kcal\ mol^{-1}$ is calculated for the reaction $Mg(OH)_2(g) = Mg(g) + 2\ OH(g)$.

Heat Capacity and Entropy

The analogy between gaseous mono- and dihydroxides and gaseous mono- and dihalides, particularly the mono- and difluorides, has been recognized (3-6). The O-Mg-O bond angle is assumed to be the same as the F-Mg-F bond angle (2); the Mg-O-H bond is considered to be linear as in MgOH (2). The Mg-O bond distance is estimated to be slightly larger, 0.02 Å, than the Mg-F distance in MgF_2 (2) after noting the close similarity in the bond distance of the alkali metal fluorides and hydroxides. The O-H bond distance is taken to be the same as in water (2).

The vibrational frequencies are estimated to be the same as in MgF_2 (2) (O-Mg-O symmetrical and asymmetrical stretch, and bend) and as in MgOH (2) (O-H stretch and Mg-O-H bend). The principal moments of inertia are: $I_A = 0.3029 \times 10^{-39}$, $I_B = 18.8449 \times 10^{-39}$, and $I_C = 19.1478 \times 10^{-39}\ g\ cm^2$.

Jackson (6) has used a different molecular configuration and different vibrational frequencies to estimate $S°(298.15\ K) = 63.829\ cal\ K^{-1}\ mol^{-1}$. We assign an uncertainty of $\pm 3.0\ cal\ K^{-1}\ mol^{-1}$ to the adopted entropy.

References

1. C. A. Alexander, J. S. Ogden, and A. Levy, J. Chem. Phys. 39, 3057 (1963).
2. JANAF Thermochemical Tables: MgO(cr), 12-31-74; $H_2O(g)$, 3-31-61; $Mg(OH)_2(g)$, MgOH(g), 12-31-75, $MgF_2(g)$, 6-30-75.
3. R. C. Schoonmaker and R. F. Porter, J. Chem. Phys. 31, 830 (1959).
4. J. Berkowitz, D. J. Meschi, and W. A. Chupka, J. Chem. Phys. 33, 533 (1960).
5. F. E. Stafford and J. Berkowitz, J. Chem. Phys. 40, 2963 (1964).
6. D. D. Jackson, Lawrence Livermore Laboratory, Univ. of Calif., Report UCRL-51137, Dec. 8, 1971.
7. U. S. Nat. Bur. Stand. Tech. Note 270-6, 1971.

Magnesium Hydroxide ($Mg(OH)_2$) $H_2Mg_1O_2(g)$

Enthalpy Reference Temperature = T_r = 298.15 K Standard State Pressure = p° = 0.1 MPa

T/K	$C_p°$ (J K⁻¹mol⁻¹)	S° (J K⁻¹mol⁻¹)	$-[G°-H°(T_r)]/T$ (J K⁻¹mol⁻¹)	$H°-H°(T_r)$ (kJ mol⁻¹)	$\Delta_f H°$ (kJ mol⁻¹)	$\Delta_f G°$ (kJ mol⁻¹)	Log K_f
0	0.	0.	INFINITE	-14.472	-564.695	-564.695	INFINITE
100	38.926	209.778	319.782	-11.000	-567.769	-560.393	292.719
200	56.563	242.056	273.257	-6.240	-570.621	-551.824	144.122
250	63.982	255.509	268.385	-3.219	-571.624	-547.003	114.290
298.15	69.476	267.270	267.270	0.	-572 371	-542.190	94.989
300	69.659	267.700	267.271	0.129	-572.396	-542.003	94.371
350	73.933	278.775	268.136	3.724	-572.989	-536.888	80.126
400	77.165	288.868	270.106	7.505	-573.453	-531.698	69.433
450	79.646	298.106	272.711	11.428	-573.828	-526.455	61.109
500	81.588	306.602	275.681	15.460	-574.145	-521.174	54.447
600	84.438	321.746	282.128	23.771	-574.691	-510.528	44.445
700	86.510	334.924	288.749	32.322	-575.204	-499.793	37.295
800	88.203	346.589	295.264	41.060	-575.743	-488.984	31.927
900	89.704	357.066	301.559	49.957	-576.345	-478.104	27.748
1000	91.089	366.590	307.593	58.997	-585.627	-466.440	24.364
1100	92.385	375.333	313.359	68.172	-586.431	-454.482	21.582
1200	93.596	383.424	318.865	77.471	-587.190	-442.452	19.259
1300	94.723	390.961	324.124	86.888	-587.908	-430.362	17.292
1400	95.765	398.020	329.153	96.413	-716.000	-415.050	15.486
1500	96.722	404.660	333.968	106.038	-715.303	-393 578	13.706
1600	97.597	410.930	338.584	115.755	-714.584	-372.153	12.150
1700	98.395	416.871	343.016	125.555	-713.847	-350.773	10.778
1800	99.121	422.516	347.277	135.432	-713.099	-329.438	9.560
1900	99.781	427.894	351.379	145.377	-712.344	-308.144	8.471
2000	100.380	433.027	355.334	155.386	-711.587	-286 889	7.493
2100	100.925	437.938	359.152	165.451	-710.831	-265.673	6.608
2200	101.419	442.645	362.841	175.569	-710.080	-244.492	5.805
2300	101.869	447.163	366.409	185.734	-709.337	-223.346	5.072
2400	102.279	451.507	369.865	195.942	-708.604	-202.232	4.401
2500	102.653	455.690	373.215	206.188	-707.884	-181.148	3.785
2600	102.994	459.723	376.465	216.471	-707.179	-160.093	3.216
2700	103.307	463.616	379.621	226.786	-706.491	-139.064	2.690
2800	103.593	467.379	382.689	237.131	-705.822	-118.060	2.202
2900	103.856	471.018	385.672	247.504	-705 174	-97.081	1.749
3000	104.097	474.543	388.576	257.902	-704.549	-76.122	1.325
3100	104.319	477.960	391.405	268.323	-703.948	-55.184	0.930
3200	104.524	481.276	394.162	278.765	-703.374	-34.266	0.559
3300	104.714	484.495	396.850	289.227	-702.829	-13.365	0.212
3400	104.889	487.624	399.474	299.708	-702.314	7.520	-0.116
3500	105.052	490.667	402.037	310.205	-701.832	28.390	-0.424
3600	105.203	493.628	404.540	320.718	-701.385	49.247	-0.715
3700	105.343	496.512	406.987	331.245	-700.975	70.092	-0.990
3800	105.474	499.324	409.380	341.786	-700.605	90.928	-1.250
3900	105.595	502.065	411.721	352.339	-700.276	111.753	-1.497
4000	105.709	504.740	414.014	362.905	-699 992	132.570	-1.731
4100	105.815	507.351	416.258	373.481	-699.754	153.381	-1.954
4200	105.915	509.902	418.458	384.067	-699.566	174.187	-2.166
4300	106.008	512.396	420.613	394.664	-699.429	194.989	-2.369
4400	106.096	514.834	422.727	405.269	-699.347	215.788	-2.562
4500	106.178	517.219	424.801	415.883	-699.321	236.586	-2.746
4600	106.255	519.554	426.835	426.504	-699.354	257.384	-2.923
4700	106.328	521.839	428.832	437.134	-699.448	278.184	-3.092
4800	106.397	524.079	430.793	447.770	-699.606	298.986	-3.254
4900	106.461	526.273	432.720	458.413	-699.830	319.793	-3.409
5000	106.523	528.425	434.612	469.062	-700.122	340.604	-3.558
5100	106.581	530.535	436.473	479.717	-700.485	361.422	-3.702
5200	106.635	532.605	438.301	490.378	-700 919	382.247	-3.840
5300	106.687	534.637	440.100	501.044	-701 427	403.083	-3.973
5400	106.736	536.631	441.869	511.715	-702.012	423.928	-4.101
5500	106.783	538.590	443.610	522.391	-702.674	444.784	-4.224
5600	106.827	540.515	445.323	533.072	-703.416	465 654	-4.343
5700	106.869	542.406	447.010	543.757	-704.239	486.537	-4.459
5800	106.910	544.265	448.671	554.446	-705.145	507.437	-4.570
5900	106.948	546.093	450.306	565.139	-706.136	528.350	-4.678
6000	106.984	547.890	451.918	575.835	-707.212	549.283	-4.782

PREVIOUS: December 1975 (1 atm) CURRENT: December 1975 (1 bar)

Magnesium Hydroxide ($Mg(OH)_2$) $H_2Mg_1O_2(g)$

MOLYBDIC ACID (H_2MoO_4) IDEAL GAS M_r = 161.95348

$S°(298.15 K) = [355.620 \pm 8.4] J K^{-1} mol^{-1}$

$\Delta_f H°(0 K) = -842.89 \pm 4.2 kJ mol^{-1}$

$\Delta_f H°(298.15 K) = -851.03 \pm 4.2 kJ mol^{-1}$

Vibrational Frequencies and Degeneracies

ν, cm^{-1}	ν, cm^{-1}	ν, cm^{-1}
[2000](1)	[100](1)	[830](1)
[1000](1)	[150](1)	[200](1)
[700](1)	[100](1)	[950](1)
[800](1)	[3600](1)	[200](1)
[250](1)	[600](1)	[250](1)

Ground State Quantum Weight: [1] σ = 2

Point Group: [C_{2v}]

Bond Distances: Mo-O = [1.83] Å O-H = [0.958] Å

Bond Angles: O-Mo-O = [109.5]° H-O-Mo = [104.5]°

HO-Mo-OH plane perpendicular to O-M-O plane

Product of the Moments of Inertia: $I_A I_B I_C = [1.536748 \times 10^{-113}] g^3 cm^6$

Enthalpy of Formation

The enhanced volatility of metallic molybdenum and molybdenum trioxide in the presence of water vapor at high temperatures has been investigated by several workers (1, 2, 3) using transpiration techniques. Glemser and Haeseler (1) and Bulten (2) have shown that the major volatilization reaction of the oxide is:

A. $MoO_3(cr) + H_2O(g) = MoO_2(OH_2)(g)$.

For the reaction of metallic molybdenum with water vapor, Belton and Jordan (3) reported the existence of $MoO_2(OH)_2(g)$ according to the equilibrium:

B. $Mo(cr) + 4H_2O(g) = MoO_2(OH)_2(g) + 3H_2(g)$.

2nd and 3rd law analyses of these equilibrium data are summarized below.

Source	Reaction	T/K	Data Points	$\Delta_r H°$(298.15 K), kcal mol^{-1} 2nd law	3rd law	Drift cal K^{-1} mol^{-1}	$\Delta_f H°$(298.15 K) kcal mol^{-1}
Glemser and Haeseler (1)	A	873-963	Equation	36.2	31.9±0.3	-4.2	-204.0±0.5
Buiten (2)	A	673-773	Equation	30.0	32.2±0.3	3.6	-203.7±0.5
Belton and Jordan (3)	B	1473-1773	Equation	30.1	28.6±0.1	-0.1	-202.6±0.3

The adopted value, $\Delta_f H°(MoO_2(OH)_2, g) = -203.4 \pm 1.0$ kcal mol^{-1}, is an average of these three results.

Heat Capacity and Entropy

The assumed tetrahedral structure for $MoO_2(OH)_2$ is from Jordan (4). The O-H bond length and H-O-Mo bond angle are assumed to be the same as for $H_2O(g)$. The Mo-O bond length is from the compilation of Sutton (5) and assumed to be the same as for Ag_2MoO_4. Principal moments of inertia are: $I_A = 23.7897 \times 10^{-39}$, $I_B = 25.3843 \times 10^{-39}$, and $I_C = 25.4477 \times 10^{-39}$ g cm^2.

The vibrational frequencies are estimated by comparison with those for $H_2SO_4(g)$ (6, 7) and are adjusted to give $S°(MoO_2(OH)_2$, g) = 137.4 cal K^{-1} mol^{-1} at 1600 K. This entropy value is derived from a 2nd law analysis of the equilibrium data for reaction B reported by Belton and Jordan (3).

References

1. V. O. Glemser and R. V. Haeseler, Z. Anorg. Allg. Chem. 316, 168 (1962).
2. J. Buiten, J. Catalysis 10, 188 (1968).
3. G. R. Belton and A. S. Jordan, J. Phys. Chem. 69, 2065 (1965).
4. A. S. Jordan, Ph.D. Thesis, Univ. Penn., 1965.
5. L. E. Sutton, "Tables of Interatomic Distances and Configuration in Molecules and Ions," The Chemical Society, London (1958).
6. P. A. Giguere and R. Savoie, J. Amer. Chem. Soc. 85, 287 (1963).
7. S. M. Chackalachal and F. E. Stafford, J. Amer. Chem. Soc. 88, 723 (1966).

Molybdic Acid ($O_2Mo(OH)_2$) $H_2Mo_1O_4(g)$

Enthalpy Reference Temperature = T_r = 298.15 K Standard State Pressure = p° = 0.1 MPa

T/K	$C_p°$ (J $K^{-1}mol^{-1}$)	S°	$-[G°-H°(T_r)]/T$	$H°-H°(T_r)$ (kJ mol^{-1})	$\Delta_f H°$	$\Delta_f G°$	Log K_f
0	0.	0.	INFINITE	-22.282	-842.889	-842.889	INFINITE
100	68.112	262.488	437.713	-17.523	-847.429	-828.241	432.629
200	89.457	317.041	364.698	-9.531	-849.796	-808.032	211.036
250	97.403	337.873	357.298	-4.856	-850.551	-797.499	166.628
298.15	104.154	355.620	355.620	0.	-851.026	-787.234	137.920
300	104.395	356.265	355.622	0.193	-851.039	-786.838	137.001
350	110.374	372.819	356.916	5.566	-851.282	-776.115	115.829
400	115.405	387.896	359.860	11.214	-851.321	-765.372	99.947
450	119.623	401.740	363.755	17.093	-851.202	-754.635	87.596
500	123.177	414.532	368.201	23.166	-850.959	-743.917	77.716
600	128.789	437.512	377.882	35.778	-850.212	-722.574	62.906
700	133.009	457.696	387.871	48.877	-849.232	-701.376	52.337
800	136.310	475.681	397.744	62.350	-848.107	-680.329	44.421
900	138.977	491.895	407.319	76.118	-846.891	-659.429	38.272
1000	141.185	506.656	416.526	90.130	-845.620	-638.668	33.361
1100	143.043	520.202	425.344	104.344	-844.325	-618.035	29.348
1200	144.627	532.718	433.777	118.729	-843.034	-597.520	26.009
1300	145.988	544.349	441.840	133.262	-841.770	-577.112	23.189
1400	147.167	555.212	449.554	147.921	-840.552	-556.800	20.774
1500	148.192	565.401	456.941	162.690	-839.398	-536.573	18.685
1600	149.089	574.995	464.023	177.555	-838.326	-516.420	16.859
1700	149.876	584.057	470.820	192.504	-837.349	-496.331	15.250
1800	150.570	592.644	477.351	207.527	-836.483	-476.297	13.822
1900	151 184	600.802	483.636	222.615	-835.741	-456.308	12.545
2000	151.729	608 571	489.690	237.762	-835.136	-436.353	11.396
2100	152.214	615.985	495.529	252.959	-834.682	-416.426	10.358
2200	152.648	623.077	501.166	268.203	-834.389	-396.516	9.414
2300	153.037	629.871	506.615	283.487	-834.271	-376.617	8.553
2400	153.386	636.391	511.888	298.809	-834.340	-356.718	7.764
2500	153.701	642.659	516.994	314.163	-834.609	-336.812	7.037
2600	153.987	648.693	521.944	329.548	-835.100	-316.892	6.366
2700	154.245	654.510	526.747	344.960	-835.854	-296.947	5.745
2800	154.480	660.124	531.411	360.396	-836.934	-276.968	5.167
2900	154.694	665.548	535.943	375.855	-874.304	-256.896	4.627
3000	154.890	670.796	540.351	391.334	-874.242	-235.607	4.102
3100	155.069	675.878	544.642	406.833	-874.221	-214.319	3.611
3200	155.233	680.804	548.820	422.348	-874.241	-193.033	3.151
3300	155.384	685.583	552.892	437.879	-874.301	-171.745	2.718
3400	155.524	690.224	556.864	453.424	-874.401	-150.453	2.311
3500	155.652	694.734	560 739	468.983	-874.540	-129.159	1.928
3600	155.771	699.120	564.522	484.554	-874.718	-107.860	1.565
3700	155.882	703.390	568.218	500.137	-874.933	-86.556	1.222
3800	155.984	707 548	571.830	515.730	-875.187	-65.243	0.897
3900	156.079	711.601	575.362	531.334	-875.477	-43.925	0.588
4000	156.168	715.554	578.818	546.946	-875.803	-22.600	0.295
4100	156.250	719.411	582.200	562.567	-876.166	-1.265	0.016
4200	156.327	723.177	585.512	578.196	-876.564	20.078	-0.250
4300	156.399	726.857	588.756	593.832	-876.998	41.433	-0.503
4400	156.467	730.453	591.936	609.476	-877.468	62.796	-0.745
4500	156.530	733.970	595.053	625.125	-877.972	84.170	-0.977
4600	156.590	737.411	598.111	640.782	-878.513	105.557	-1.199
4700	156.646	740.779	601.111	656.443	-879.089	126.956	-1.411
4800	156.699	744.078	604.055	672 111	-879.701	148.368	-1.615
4900	156.748	747.309	606.945	687.783	-880.350	169.794	-1.810
5000	156.795	750.477	609.785	703.460	-1472.420	196.964	-2.058
5100	156.839	753.582	612.574	719.142	-1474.822	230.376	-2.360
5200	156.881	756.628	615.315	734.828	-1477.409	263.834	-2.650
5300	156.921	759.617	618.009	750.518	-1480.178	297.347	-2.931
5400	156.958	762.550	620.659	766.212	-1483.131	330.912	-3.201
5500	156.994	765.430	623.265	781.910	-1486.263	364.532	-3.462
5600	157.027	768.260	625.829	797.611	-1489.576	398.215	-3.714
5700	157.059	771.039	628.352	813.315	-1493.065	431.955	-3.958
5800	157.090	773.771	630.836	829.023	-1496.731	465.761	-4.195
5900	157.119	776.457	633.281	844.733	-1500.570	499.626	-4.423
6000	157.146	779.098	635.690	860.446	-1504.583	533.563	-4.645

PREVIOUS: June 1970 (1 atm) CURRENT: June 1970 (1 bar)

Molybdic Acid ($O_2Mo(OH)_2$) $H_2Mo_1O_4(g)$

AMIDOGEN (NH_2) IDEAL GAS M_r = 16.02258

$S°(298.15\ K) = 194.707\ J\ K^{-1}\ mol^{-1}$

$\Delta_f H°(0\ K) = 193.246 \pm 6.3\ kJ\ mol^{-1}$

$\Delta_f H°(298.15\ K) = 190.372 \pm 6.3\ kJ\ mol^{-1}$

Electronic Levels (Quantum Weights) and Vibrational Frequencies (Degeneracies)

State	ε, cm^{-1}	ν_1, cm^{-1}	ν_2, cm^{-1}	ν_3, cm^{-1}
$\tilde{X}^2B_1$	0(2)	[3173](1)	1497.2(1)	3220 (1)
$\tilde{A}^2A_1(\pi)$	10249(2)	3325 (1)	633 (1)	[3220](1)

Point Group: C_{2v} σ = 2

Bond Distance: N-H = 1.024 Å

Bond Angle: H-N-H = 103°

Product of the Moments of Inertia: $I_AI_BI_C = 0.0085 \times 10^{-117}\ g^3\ cm^6$

Enthalpy of Formation

Direct kinetic studies (1 - 5) onthe decomposition of hydrazine, $N_2H_4(g) = 2\ NH_2(g)$, give activation energies which suggest $\Delta_f H°(NH_2,\ 298.15\ K) \leq 40\ kcal\ mol^{-1}$. [The kinetic rates are discribed by k_d for the the forward reaction and k_r for the reverse reaction]. The pre-exponential factor suggests $k_r \sim 10^7 - 10^8$ 1/(mol s). This is exceptionally slow for a radical combination reaction (6) and is at variance with the most recent direct measurements of $k_r = 10^{10.8}$ 1/(mol s) (300 K, 1500 mmHg) (7) and $k_r = 10^{9.4}$ 1/(mol s) (300 K, 10 mmHg) (8). A possible explanation for this discrepancy is that the hydrazine decomposition studies have been carried out in the energy dependent region (9, 10). This has been confirmed on the basis of RRKM calculations. With $\Delta_f H°(298.15\ K) = 40\ kcal\ mol^{-1}$, it is not possible to simultaneously reproduce the measured forward (k_d) and backward (k_r) rates by orders of magnitude. With a value of $\Delta_f H°(298.15\ K) = 45.5\ kcal\ mol^{-1}$, all of the rate data can be fitted to a factor of three. For higher values of $\Delta_f H°(298.15\ K)$, the discrepancy increases.

Data on the thermal decomposition of benzylamine from toluene-carrier studies (2, 11) yield $\Delta_f H°(298.15\ K) = 36\ kcal\ mol^{-1}$. This technique, however, has consistently yielded erroneous rate paramters (12). Using their very low pressure pyrolysis technique, Golden et al. (13) find $\Delta_f H°(NH_2,\ 298.15\ K) = 47.2\ kcal\ mol^{-1}$. This is in reasonable agreement with the recommended value.

Heat Capacity and Entropy

The bond distance and angle are from the electronic absorption spectrum as summarized by Herzberg (14). The vibrational frequencies are obtained from Herzberg (14) and from matrix-isolation studies of Milligan and Jacox (15).

The National Bureau of Standards prepared this table (16) by critical analysis of data existing in 1972. Using S°, $C_p°$ and $\Delta_f H°$ selected by NBS (16), we recalculate the table in terms of current JANAF reference states for the elements.

References

1. M. Szwarc, Proc. Roy. Soc. A198, 267 (1949).
2. J. A. Kerr, R. C. Sekhar, and A. F. Trotman-Dickenson, J. Chem. Soc. 3217 (1963).
3. E. T. McHale, B. E. Knox, and H. B. Palmer, "Tenth Symposium on Combustion (International)," p. 341, The Combustion Institute, Pittsburgh, Pa., 1965.
4. K. W. Michel and H. Gg. Wagner, "Tenth Symposium on Combustion (International)," p. 353, The Combustion Institute, Pittsburgh, Pa., 1965.
5. E. Meyer, H. A. Olscheusek, T. Troe, and H. Gg. Wagner, "Eleventh Symposium on Combustion (International)," p. 345, The Combustion Institute, Pittsburgh, Pa., 1967.
6. S. W. Benson and H. E. O'Neal, NSRDS-NBS 21, p. 34, 1970
7. S. Gordon, W. Mulac, and P. Nangia, J. Phys. Chem. 75, 2087 (1971).
8. J. D. Saltzman and E. J. Bair, J. Chem. Phys. 41, 3564 (1964).
9. M. Gilbert, Combustion and Flame 2, 137, 149 (1958).
10. D. W. Setzer and W. C. Richardson, Can. J. Chem. 47, 2593 (1969).
11. M. Szwarc, Proc. Roy. Soc. A198, 285 (1949).
12. S. W. Benson, J. Chem. Education 42, 505 (1965).
13. D. M. Golden, R. K. Solly, N. A. Gac, and S. W. Benson, J. Amer. Chem. Soc. 94, 363 (1972).
14. G. Herzberg, "Electronic Spectra of Polyatomic Molecules," D. Van Nostrand Co., Inc., New York, 1966.
15. D. E. Milligan and M. E. Jacox, J. Chem. Phys. 43, 4487 (1965).
16. S. Abramowitz et al., U. S. Nat. Bur. Stand. Report 10904, 239, July, 1972.

Amidogen (NH_2) $H_2N_1(g)$

Enthalpy Reference Temperature = T_r = 298.15 K Standard State Pressure = p° = 0.1 MPa

T/K	$C_p°$ (J K⁻¹mol⁻¹)	S°	$-[G°-H°(T_r)]/T$	$H°-H°(T_r)$ (kJ mol⁻¹)	$\Delta_f H°$	$\Delta_f G°$	Log K_f
0	0.	0.	INFINITE	-9.929	193.246	193.246	INFINITE
100	33.259	158.327	224.351	-6.602	192.122	194.352	-101.519
200	33.280	181.381	197.762	-3.276	191.299	196.904	-51.426
298.15	33.572	194.707	194.707	0.	190.372	199.846	-35.012
300	33.585	194.915	194.708	0.062	190.354	199.905	-34.807
400	34.395	204.680	196.034	3.458	189.386	203.236	-26.540
500	35.533	212.474	198.567	6.953	188.488	206.804	-21.605
600	36.838	219.066	201.448	10.571	187.685	210.545	-18.330
700	38.254	224.851	204.386	14.325	186.980	214.412	-16.000
800	39.715	230.054	207.274	18.224	186.371	218.373	-14.258
900	41.179	234.817	210.074	22.269	185.853	222.405	-12.908
1000	42.597	239.229	212.771	26.458	185.419	226.491	-11.831
1100	43.945	243.353	215.366	30.786	185.059	230.616	-10.951
1200	45.196	247.231	217.861	35.244	184.764	234.771	-10.219
1300	46.355	250.895	220.263	39.822	184.525	238.948	-9.601
1400	47.417	254.370	222.576	44.511	184.333	243.142	-9.072
1500	48.392	257.675	224.807	49.303	184.182	247.348	-8.613
1600	49.292	260.827	226.960	54.187	184.066	251.563	-8.213
1700	50.128	263.841	229.042	59.159	183.982	255.784	-7.859
1800	50.907	266.728	231.056	64.211	183.925	260.010	-7.545
1900	51.647	269.501	233.007	69.339	183.895	264.238	-7.264
2000	52.346	272.168	234.899	74.539	183.891	268.467	-7.012
2100	53.020	274.738	236.735	79.807	183.912	272.695	-6.783
2200	53.664	277.220	238.519	85.142	183.957	276.922	-6.575
2300	54.287	279.619	240.254	90.539	184.027	281.146	-6.385
2400	54.894	281.942	241.943	95.999	184.123	285.366	-6.211
2500	55.484	284.195	243.588	101.518	184.243	289.583	-6.050
2600	56.053	286.383	245.192	107.095	184.389	293.793	-5.902
2700	56.605	288.508	246.757	112.728	184.560	297.998	-5.765
2800	57.145	290.577	248.286	118.415	184.757	302.196	-5.638
2900	57.664	292.591	249.779	124.156	184.977	306.386	-5.519
3000	58.162	294.555	251.239	129.947	185.222	310.569	-5.407
3100	58.643	296.470	252.667	135.788	185.489	314.742	-5.303
3200	59.103	298.339	254.065	141.675	185.778	318.907	-5.206
3300	59.542	300.164	255.435	147.608	186.088	323.063	-5.114
3400	59.961	301.948	256.776	153.583	186.417	327.209	-5.027
3500	60.359	303.692	258.092	159.599	186.763	331.345	-4.945
3600	60.735	305.398	259.383	165.654	187.126	335.471	-4.868
3700	61.087	307.067	260.649	171.746	187.503	339.586	-4.794
3800	61.417	308.700	261.892	177.871	187.892	343.691	-4.724
3900	61.725	310.299	263.113	184.028	188.291	347.786	-4.658
4000	62.011	311.866	264.312	190.215	188.700	351.870	-4.595
4100	62.274	313.400	265.491	196.430	189.115	355.945	-4.535
4200	62.517	314.904	266.649	202.669	189.535	360.009	-4.477
4300	62.740	316.378	267.789	208.932	189.958	364.063	-4.422
4400	62.944	317.822	268.909	215.217	190.383	368.107	-4.370
4500	63.128	319.239	270.012	221.521	190.807	372.141	-4.320
4600	63.291	320.628	271.098	227.842	191.230	376.166	-4.271
4700	63.438	321.991	272.166	234.178	191.650	380.182	-4.225
4800	63.568	323.328	273.218	240.529	192.066	384.189	-4.181
4900	63.684	324.640	274.254	246.891	192.476	388.187	-4.138
5000	63.785	325.928	275.275	253.265	192.879	392.177	-4.097
5100	63.871	327.192	276.280	259.648	193.275	396.159	-4.057
5200	63.944	328.433	277.271	266.039	193.663	400.133	-4.019
5300	64.003	329.651	278.248	272.436	194.042	404.100	-3.983
5400	64.052	330.848	279.211	278.839	194.412	408.060	-3.947
5500	64.091	332.024	280.161	285.246	194.772	412.013	-3.913
5600	64.116	333.179	281.097	291.657	195.122	415.960	-3.880
5700	64.134	334.314	282.021	298.069	195.462	419.901	-3.848
5800	64.143	335.429	282.932	304.483	195.791	423.835	-3.817
5900	64.144	336.526	283.831	310.898	196.110	427.764	-3.787
6000	64.137	337.604	284.718	317.312	196.419	431.688	-3.758

PREVIOUS: June 1977 (1 atm) CURRENT: June 1977 (1 bar)

Amidogen (NH_2) $H_2N_1(g)$

DIAZENE, CIS-(N_2H_2) IDEAL GAS M_r = 30.02928

$S°(298.15\ K) = [218.603]\ J\ K^{-1}\ mol^{-1}$

$\Delta_fH°(0\ K) = 220.00 \pm 20.9\ kJ\ mol^{-1}$
$\Delta_fH°(298.15\ K) = 212.97 \pm 10.9\ kJ\ mol^{-1}$

Vibrational Frequencies and Degeneracies

ν, cm^{-1}	ν, cm^{-1}
[3095](1)	[3205](1)
[1406](1)	[1495](1)
[1360](1)	[910](1)

Ground State Quantum Weight: 1 $\sigma = 2$
Point Group: C_{2v}
Bond Distances: N-H = [1.014] Å N-N = [1.23] Å
Bond Angle: H-N-N = [100]°
Product of the Moments of Inertia: $I_AI_BI_C = [1.3981 \times 10^{-117}]\ g^3\ cm^6$

Enthalpy of Formation

The standard enthalpy of formation, $\Delta_fH°(N_2H_2, g, 298.15\ K) = 50.9\ kcal\ mol^{-1}$, was calculated from the appearance and ionization potentials of di-imide reported by Foner and Hudson (1) using the JANAF $\Delta_fH°(N_2H_4, g, 0\ K) = 26.1\ kcal\ mol^{-1}$ in calculation. For example:

	$\Delta_rH°(0\ K)$
$N_2H_4(g) \rightarrow N_2H_2^+ + H_2(g) + e$	$11.0 \pm 0.2\ cal\ K^{-1}\ mol^{-1}$
$N_2H_2(g) \rightarrow N_2H_2^+ + e$	$9.85 \pm 0.1\ cal\ K^{-1}\ mol^{-1}$
$N_2H_4(g) \rightarrow N_2H_2(g) + H_2$	$1.15 \pm 0.2\ cal\ K^{-1}\ mol^{-1}$
	$(26.5 \pm 5\ kcal\ mol^{-1}$

From the above enthalpy of formation and that for NH(g) in the JANAF Tables 12-31-65, one can calculate a N-N bond energy of $109.4 \pm 5\ kcal\ mol^{-1}$ at 0 K. This is in agreement with the N-N bond in molecules such as N_2F_2.

Heat Capacity and Entropy

The bond distances and angle were obtained from a quantum-mechanical calculation by Wheland and Chen (2). The principal moments of inertia are: $I_A = [0.3114] \times 10^{-39}$, $I_B = [1.9689] \times 10^{-39}$ and $I_C = [2.2803] \times 10^{-39}\ g\ cm^2$.

Since there are 12 valency electrons in N_2H_2, Walsh (3) has predicted a bent ground state, existing in either a cis or a trans configuration for diimide. Blau and Hochheimer (4) have measured the infrared spectrum of diimide which was frozen in a cold trap as one of the decomposition products of N_2H_4. The spectrum of the solid N_2H_2 at liquid nitrogen temperature agrees with the predicted non-linear planar structure, and the molecule N_2H_2 appears mainly in cis form. They have also assigned six fundamental vibrational frequencies for cis H_2H_4(s) which have been adopted here for gaseous cis N_2H_2. Blau and Hochheimer also observed two gas phase bands (1277 - 1314 and 3050 - 3190 cm^{-1}). Rosengren and Pimentel (5) have observed these bands in matrix isolation studies. They assigned ν_1 or $\nu_4 = 3074\ cm^{-1}$ and $\nu_3 = 1279\ cm^{-1}$ which are in satisfactory agreement with the selected frequencies.

References

1. S. N. Foner and R. L. Hudson, J. Chem. Phys. 29, 442 (1958).
2. W. Wheland and P. S. K. Chen, J. Chem. Phys. 24, 67 (1956).
3. A. D. Walsh, J. Chem. Soc. 2288 (1953).
4. E. J. Blau and B. F. Hochheimer, J. Chem. Phys. 41, 1174 (1964).
5. K. Rosengren and G. C. Pimentel, J. Chem. Phys. 43, 507 (1965).

Diazene, Cis (HNNH) $H_2N_2(g)$

Enthalpy Reference Temperature = T_r = 298.15 K Standard State Pressure = p° = 0.1 MPa

T/K	$C_p°$ (J $K^{-1}mol^{-1}$)	S° (J $K^{-1}mol^{-1}$)	$-[G°-H°(T_r)]/T$ (J $K^{-1}mol^{-1}$)	$H°-H°(T_r)$ (kJ mol^{-1})	$\Delta_fH°$ (kJ mol^{-1})	$\Delta_fG°$ (kJ mol^{-1})	Log K_f
0	0.	0.	INFINITE	-10.107	219.996	219.996	INFINITE
100	33.261	181.498	249.315	-6.782	217.420	225.324	-117.697
200	33.870	204.651	221.843	-3.438	215.159	234.108	-61.143
250	34.966	212.315	219.195	-1.720	214.026	238.977	-49.932
298.15	36.554	218.603	218.603	0.	212.966	243.880	-42.727
300	36.624	218.829	218.604	0.068	212.926	244.071	-42.497
350	38.678	224.625	219.057	1.949	211.901	249.345	-37.213
400	40.937	229.936	220.089	3.939	210.974	254.758	-33.268
450	43.247	234.891	221.461	6.044	210.153	260.282	-30.213
500	45.508	239.565	223.040	8.263	209.436	265.891	-27.777
600	49.713	248.242	226.529	13.028	208.288	277.296	-24.141
700	53.426	256.190	230.206	18.189	207.469	288.866	-21.555
800	56.683	263.541	233.919	23.698	206.916	300.535	-19.623
900	59.540	270.386	237.595	29.512	206.579	312.259	-18.123
1000	62.049	276.792	241.198	35.594	206.417	324.012	-16.925
1100	64.250	282.812	244.710	41.912	206.398	335.773	-15.945
1200	66.179	288.487	248.124	48.435	206.495	347.531	-15.128
1300	67.871	293.852	251.437	55.140	206.685	359.277	-14.436
1400	69.353	298.937	254.650	62.002	206.950	371.005	-13.842
1500	70.655	303.767	257.765	69.004	207.275	382.712	-13.327
1600	71.799	308.365	260.785	76.128	207.649	394.396	-12.876
1700	72.807	312.748	263.713	83.359	208.061	406.055	-12.477
1800	73.697	316.936	266.555	90.686	208.504	417.689	-12.121
1900	74.486	320.942	269.313	98.095	208.972	429.298	-11.802
2000	75.186	324.781	271.991	105.580	209.457	440.882	-11.515
2100	75.809	328.464	274.593	113.130	209.957	452.441	-11.254
2200	76.366	332.004	277.122	120.739	210.468	463.976	-11.016
2300	76.865	335.410	279.583	128.401	210.986	475.487	-10.799
2400	77.313	338.691	281.978	136.111	211.508	486.976	-10.599
2500	77.718	341.855	284.310	143.863	212.034	498.443	-10.414
2600	78.083	344.911	286.583	151.653	212.560	509.889	-10.244
2700	78.413	347.864	288.798	159.478	213.085	521.315	-10.085
2800	78.714	350.721	290.959	167.335	213.608	532.721	-9.938
2900	78.987	353.488	293.067	175.220	214.127	544.109	-9.800
3000	79.237	356.170	295.126	183.131	214.642	555.479	-9.672
3100	79.466	358.772	297.138	191.067	215.151	566.832	-9.551
3200	79.675	361.298	299.103	199.024	215.653	578.168	-9.438
3300	79.868	363.753	301.025	207.001	216.149	589.489	-9.331
3400	80.045	366.140	302.906	214.997	216.636	600.795	-9.230
3500	80.209	368.463	304.746	223.010	217.114	612.087	-9.135
3600	80.360	370.724	306.547	231.038	217.583	623.365	-9.045
3700	80.500	372.928	308.312	239.081	218.041	634.631	-8.959
3800	80.630	375.077	310.040	247.138	218.490	645.884	-8.878
3900	80.751	377.173	311.735	255.207	218.927	657.125	-8.801
4000	80.863	379.219	313.397	263.288	219.352	668.356	-8.728
4100	80.968	381.217	315.026	271.379	219.766	679.576	-8.658
4200	81.066	383.169	316.626	279.481	220.168	690.786	-8.591
4300	81.157	385.077	318.196	287.592	220.557	701.986	-8.527
4400	81.243	386.944	319.737	295.712	220.933	713.178	-8.466
4500	81.323	388.771	321.251	303.841	221.296	724.361	-8.408
4600	81.399	390.559	322.738	311.977	221.646	735.536	-8.352
4700	81.470	392.310	324.200	320.120	221.984	746.704	-8.299
4800	81.536	394.026	325.637	328.271	222.308	757.865	-8.247
4900	81.599	395.708	327.050	336.427	222.619	769.019	-8.198
5000	81.658	397.357	328.439	344.590	222.917	780.168	-8.150
5100	81.714	398.975	329.807	352.759	223.202	791.309	-8.105
5200	81.767	400.562	331.152	360.933	223.475	802.446	-8.061
5300	81.817	402.120	332.476	369.112	223.736	813.578	-8.018
5400	81.865	403.650	333.780	377.296	223.986	824.705	-7.977
5500	81.909	405.153	335.064	385.485	224.224	835.826	-7.938
5600	81.952	406.629	336.329	393.678	224.452	846.945	-7.900
5700	81.992	408.080	337.575	401.875	224.670	858.059	-7.863
5800	82.031	409.506	338.803	410.077	224.878	869.169	-7.828
5900	82.067	410.909	340.013	418.282	225.076	880.275	-7.793
6000	82.102	412.288	341.207	426.490	225.267	891.379	-7.760

PREVIOUS: December 1965 (1 atm) CURRENT: December 1965 (1 bar)

Diazene, Cis (HNNH) $H_2N_2(g)$

SODIUM HYDROXIDE ($(NaOH)_2$) IDEAL GAS M_r = 79.99422

$S°(298.15\ K) = [307.396 \pm 8.4]\ J\ K^{-1}\ mol^{-1}$

$\Delta_f H°(0\ K) = -594.77 \pm 25.1\ kJ\ mol^{-1}$
$\Delta_f H°(298.15\ K) = -607.52 \pm 25.1\ kJ\ mol^{-1}$

Vibrational Frequencies and Degeneracies

ν, cm^{-1}	ν, cm^{-1}
[350](1)	[3700](1)
[209](1)	[3700](1)
[270](1)	[1250](1)
[201](1)	[1250](1)
[294](1)	[1250](1)
[313](1)	[1250](1)

Ground State Quantum Weight: 1 σ = 2
Point Group: C_{2h}
Bond Distances: Na-O = [2.17] Å O-H = [0.96] Å
Bond Angles: Na-O-Na = [90]° Na-O-H = [110]°
Product of the Moments of Inertia: $I_A I_B I_C = [8.159128 \times 10^{-114}]\ g^3\ cm^6$

Enthalpy of Formation

Mass spectrometric studies of the vapors in equilibrium with NaOH(ℓ) (1, 2) and mixed NaOH-KOH condensed phases (3, 4), along with a study of the vapor composition by a molecular beam velocity analysis technique (5), have established the importance of monomeric and dimeric species in the temperature range 573-1050 K. Absolute partial pressures for the monomer and dimer calculated from peak intensity data by Porter and Schoonmaker (3) are analyzed by the 3rd law method with JANAF Gibbs energy functions (6) to give a enthalpy of dimerization of -50.0 cal mol^{-1} at 298.15 K. Schoonmaker and Porter (4) also calculated differences in the Gibbs energies of dimerization for several alkali metal hydroxide pairs from peak intensity data on mixed systems. A 3rd law analysis of the Gibbs energy data for the NaOH-KOH system leads to a difference in their enthalpies of dimerization of -6.0 kcal mol^{-1} at 298.15 K. Based upon the adopted value for KOH(g), $\Delta_{dim}H°$(= -45.3 kcal mol^{-1} (6), we derive $\Delta_{dim}H°$(= -51.3 kcal mol^{-1} for NaOH(g). For the reaction 2 NaOH(g) = $Na_2(OH)_2$(g), we adopt a mean of these two results, $\Delta_{dim}H°$(298.15 K) = -50.7±3.0 kcal mol^{-1}. Combining this result with the adopted enthalpy of formation for NaOH(g) (6), that for the dimer is $\Delta_f H°(Na_2(OH)_2$, g, 298.15 K) = -145.2±6.0 kcal mol^{-1}.

Heat Capacity and Entropy

A key to the molecular structure of $Na_2(OH)_2$(g) is provided by the work of Buchler et al. (7) which involves the electric deflection of molecular beams of $Cs_2(OH)_2$(g) and $Na_2(OH)_2$(g). Their results indicate that these dimers are nonpolar which implies a planar structure for at least the M_2O_2 part of the molecule. The three models proposed thus far for the dimeric alkali metal hydroxides each incorporate this planar configuration for the alkali and oxygen atoms. Bauer et al. (8) proposed a model consisting of a square planar configuration for the alkali and oxygen atoms with two hydrogen bonded bridges between the oxygens and hydrogens. Schoonmaker and Porter (4) adopted the square planar model for the M_2O_2 configuration but did not allow for hydrogen bonding. Berkowitz et al. (9) proposed a model for $Li_2(OH)_2$(g) which consisted of a trans-configuration with hydrogens above and below the plane of a rhombus formed by the lithium and oxygen atoms, the O-Li-O bond angle being 100°. The adopted molecular structure for $Na_2(OH)_2$(g) is similar to the model of Berkowitz et al. (9) but with the O-Na-O bond angle equal to 90°. The Na-O-H bond angle is assumed equal to 110°. The Na-O bond distance is estimated as being 12% longer than that in NaOH(g). The O-H bond distance is estimated to be the same as in H_2O(g). The principal moments of inertia are: $I_A = [14.0843] \times 10^{-39}$, $I_B = [18.1898] \times 10^{-39}$, and $I_C = [31.8478] \times 10^{-39}\ g\ cm^2$.

The first six vibrational frequencies are estimated by comparison with those for other alkali metal fluorides and hydroxides. The remaining O-H stretching and bending modes are estimates taken from the work of Berkowitz et al. (9).

References

1. R. F. Porter and R. C. Schoonmaker, J. Chem. Phys. 28, 168 (1958).
2. R. C. Schoonmaker and R. F. Porter, J. Chem. Phys. 28, 454 (1958).
3. R. F. Porter and R. C. Schoonmaker, J. Phys. Chem. 62, 486 (1958).
4. R. C. Schoonmaker and R. F. Porter, J. Chem. Phys. 31, 830 (1959).
5. V. S. Rao and R. C. Schoonmaker, J. Chem. Phys. 33, 1718 (1960).
6. JANAF Thermochemical Tables: NaOH(g) and $Na_2(OH)_2$(g), 12-31-70; KOH(g), 12-31-70.
7. A. Buchler, J. L. Stauffer, and W. Klemperer, J. Chem. Phys. 46, 605 (1967).
8. S. H. Bauer, R. H. Diner, and R. F. Porter, J. Chem. Phys. 29, 991 (1958).
9. J. Berkowitz, D. J. Meschi, and W. A. Chupka, J. Chem. Phys. 33, 533 (1960).

Sodium Hydroxide ($(NaOH)_2$) $H_2Na_2O_2$(g)

Enthalpy Reference Temperature = T_r = 298.15 K Standard State Pressure = p° = 0.1 MPa

T/K	$C_p°$ (J K⁻¹mol⁻¹)	S° (J K⁻¹mol⁻¹)	-[G°-H°(T_r)]/T (J K⁻¹mol⁻¹)	H°-H°(T_r) (kJ mol⁻¹)	$\Delta_f H°$ (kJ mol⁻¹)	$\Delta_f G°$ (kJ mol⁻¹)	Log K_f
0	0.	0.	INFINITE	-17.301	-594.774	-594.774	INFINITE
100	49.910	235.707	370.856	-13.515	-599.571	-590.975	308.694
200	70.007	277.594	314.373	-7.356	-603.928	-580.589	151.634
250	75.172	293.797	308.679	-3.720	-605.796	-574.535	120.043
298.15	79.284	307.396	307.396	0.	-607.517	-568.358	99.574
300	79.433	307.887	307.398	0.147	-607.582	-568.115	98.918
350	83.306	320.426	308.379	4.217	-609.343	-561.398	83.784
400	86.884	331.787	310.606	8.473	-616.377	-554.016	72.347
450	90.149	342.212	313.546	12.900	-618.052	-546.119	63.392
500	93.089	351.866	316.901	17.482	-619.551	-538.044	56.209
600	98.048	369.293	324.213	27.048	-622.106	-521.494	45.400
700	102.010	384.716	331.775	37.058	-624.193	-504.554	37.650
800	105.258	398.556	339.273	47.427	-625.932	-487.341	31.820
900	108.004	411.117	346.568	58.093	-627.420	-469.926	27.274
1000	110.381	422.622	353.606	69.015	-628.738	-452.355	23.629
1100	112.472	433.242	360.369	80.160	-629.958	-434.656	20.640
1200	114.327	443.110	366.858	91.502	-824.665	-411.970	17.933
1300	115.980	452.327	373.082	103.019	-824.009	-377.604	15.172
1400	117.457	460.977	379.055	114.692	-823.270	-343.293	12.808
1500	118.778	469.127	384.790	126.505	-822.464	-309.037	10.762
1600	119.960	476.831	390.304	138.443	-821.602	-274.836	8.972
1700	121.020	484.136	395.611	150.493	-820.694	-240.691	7.396
1800	121.970	491.081	400.723	162.643	-819.752	-206.600	5.995
1900	122.823	497.699	405.655	174.884	-818.783	-172.562	4.744
2000	123.590	504.019	410.416	187.205	-817.794	-138.576	3.619
2100	124.282	510.066	415.018	199.599	-816.793	-104.640	2.603
2200	124.906	515.862	419.471	212.059	-815.785	-70.751	1.680
2300	125.471	521.427	423.784	224.578	-814.776	-36.909	0.838
2400	125.982	526.778	427.965	237.151	-813.771	-3.110	0.068
2500	126.447	531.930	432.021	249.773	-812.775	30.647	-0.640
2600	126.869	536.898	435.960	262.439	-811.792	64.364	-1.293
2700	127.255	541.693	439.787	275.146	-810.828	98.044	-1.897
2800	127.607	546.328	443.510	287.889	-809.885	131.689	-2.457
2900	127.929	550.811	447.133	300.666	-808.968	165.300	-2.977
3000	128.225	555.153	450.662	313.474	-808.082	198.881	-3.463
3100	128.496	559.362	454.101	326.310	-807.230	232.432	-3.916
3200	128.746	563.446	457.454	339.173	-806.417	265.956	-4.341
3300	128.977	567.411	460.727	352.059	-805.646	299.455	-4.740
3400	129.190	571.265	463.921	364.967	-804.921	332.933	-5.115
3500	129.388	575.012	467.042	377.896	-804.248	366.389	-5.468
3600	129.571	578.660	470.092	390.845	-803.630	399.827	-5.801
3700	129.740	582.212	473.075	403.810	-803.073	433.248	-6.116
3800	129.898	585.675	475.992	416.792	-802.580	466.656	-6.415
3900	130.046	589.051	478.848	429.789	-802.156	500.052	-6.697
4000	130.183	592.345	481.645	442.801	-801.808	533.436	-6.966
4100	130.311	595.561	484.384	455.826	-801.540	566.814	-7.221
4200	130.431	598.703	487.069	468.863	-801.354	600.186	-7.464
4300	130.543	601.773	489.701	481.912	-801.262	633.555	-7.696
4400	130.649	604.775	492.282	494.971	-801.184	666.915	-7.917
4500	130.748	607.713	494.815	508.041	-801.246	700.280	-8.129
4600	130.841	610.587	497.300	521.121	-801.428	733.648	-8.331
4700	130.928	613.402	499.741	534.209	-801.719	767.023	-8.525
4800	131.011	616.160	502.137	547.306	-802.123	800.405	-8.710
4900	131.089	618.862	504.492	560.411	-802.648	833.796	-8.888
5000	131.162	621.511	506.806	573.524	-803.283	867.199	-9.060
5100	131.231	624.109	509.081	586.643	-804.061	900.616	-9.224
5200	131.297	626.658	511.317	599.770	-804.976	934.048	-9.383
5300	131.359	629.159	513.517	612.903	-806.033	967.502	-9.535
5400	131.418	631.615	515.682	626.042	-807.238	1000.975	-9.683
5500	131.474	634.027	517.811	639.186	-808.595	1034.472	-9.825
5600	131.527	636.397	519.908	652.336	-810.111	1067.997	-9.962
5700	131.578	638.725	521.972	665.492	-811.790	1101.549	-10.095
5800	131.626	641.014	524.005	678.652	-813.636	1135.133	-10.223
5900	131.671	643.264	526.007	691.817	-815.655	1168.748	-10.347
6000	131.715	645.478	527.980	704.986	-817.534	1202.364	-10.468

PREVIOUS: December 1970 (1 atm) CURRENT December 1970 (1 bar)

Sodium Hydroxide ($(NaOH)_2$) $H_2Na_2O_2$(g)

WATER (H_2O) LIQUID, p = 1 BAR M_r = 18.01528

Water (H_2O) $H_2O_1(l)$

$S°(298.15\ K) = 69.950 \pm 0.079\ J\ K^{-1}\ mol^{-1}$ $\Delta_f H°(298.15\ K) = -285.830 \pm 0.042\ kJ\ mol^{-1}$

Enthalpy of Formation

The adopted enthalpy of formation and uncertainty are taken from the CODATA recommended values (1) which are, in turn, based on calorimetric measurements by Rossini (2) and King and Armstrong (3).

Heat Capacity and Entropy

The adopted $C_p°$ data from 273.15 to 373.15 K are taken from the very accurate calorimetric measurements of Osborne et al. (4). Heat capacity data taken from the recent equation of state formulation of Haar et al. (5) agree with the adopted data to within 0.06% above 320 K with deviations up to 0.12% being noted near 305 K. In the latter region the experimental $C_p°$ data go through a single smooth minimum while data derived from the equation of state exhibit an incipient double minimum behavior. These deviations are very small and would lead to nearly negligible differences in the thermochemical functions.

We have made an approximate extrapolation of $C_p°$ (ℓ, p=1 bar) to temperatures where water is unstable with respect to vaporization by using data derived from Haar's equation of state formulation (5). From 373.15 to 390 K the adopted $C_p°$(1 bar) data were taken directly from Haar et al. (5), while from 400 to 450 K the adopted $C_p°$ data were derived by taking advantage of the constant difference of 0.0090 cal K^{-1} mol^{-1} between C_p(10 bar) and $C_p°$(1 bar) observed between 320 to 390 K using the data of Haar et al. (5). Above 450 K the C_p(100 bar) data (5) were adjusted by a similar procedure although the difference is no longer constant. The $C_p°$ adjustment from 10 bar to 1 bar is less than 0.05% while that from 100 bar to 1 bar ranges from 0.9 to 1.2%.

The adopted value of S°(298.15 K) = 16.718±0.019 cal K^{-1} mol^{-1} is taken from the CODATA recommended value (1). This was calculated by CODATA from the entropy of the ideal gas with appropriate corrections for real gas behavior and vaporization.

Vaporization Data

The boiling point and enthalpy of vaporization to the real gas at p=1 bar are taken from the work of Haar et al. (5); see the H_2O (liquid-real gas at p=1 bar) JANAF Table (6) for further details. The enthalpy of vaporization, $\Delta_{vap}H°$, to the ideal gas is calculated as $\Delta_{vap}H° = \Delta_f H°(g) - \Delta_f H°(\ell, p=1\ bar)$ using JANAF data (6). The temperature and enthalpy of vaporization at f=1 bar represent the calculated values at the point where $\Delta_r G° = 0$ for the process $H_2O(\ell) \rightarrow H_2O$(ideal gas). Thus, these values represent boiling to the (hypothetical) ideal vapor at a fugacity of 1 bar.

References

1. ICSU-CODATA Task Group, J. Chem. Thermodyn. 4, 331 (1972); see also, CODATA Bulletin #17, "CODATA Recommended Key Values for Thermodynamics - 1975", 1976.
2. F. D. Rossini, J. Res. Nat. Bur. Stand. 6, 1 (1931); ibid 6, 36 (1931); ibid 22, 407 (1939).
3. R. King and G. T. Armstrong, J. Res. Nat. Bur. Stand. 72A, 113 (1968).
4. N. S. Osborne, H. F. Stimson and D. C. Ginnings, J. Res. Nat. Bur. Stand. 23, 197 (1939).
5. L. Haar, J. Gallagher and G. S. Kell, Ninth International Conference on Properties of Steam, Sept. 1979, Munich; and personal communication from L. Haar and J. Gallagher, U. S. Nat. Bur. Stand., May, 1980.
6. JANAF Thermochemical Tables: H_2O(liquid-real gas at p=1 bar), H_2O(ideal gas) 3-31-79.

Enthalpy Reference Temperature = T_r = 298.15 K Standard State Pressure = p° = 0.1 MPa

T/K	$C_p°$ (J K⁻¹mol⁻¹)	S° (J K⁻¹mol⁻¹)	$-[G°-H°(T_r)]/T$ (J K⁻¹mol⁻¹)	$H°-H°(T_r)$ (kJ mol⁻¹)	$\Delta_f H°$ (kJ mol⁻¹)	$\Delta_f G°$ (kJ mol⁻¹)	Log K_f
0							
100							
200							
280	75.563	65.215	70.102	-1.368	-286.410	-240.123	44.796
298.15	75.351	69.950	69.950	0.	-285.830	-237.141	41.546
300	75.349	70.416	69.952	0.139	-285.771	-236.839	41.237
320	75.344	75.279	70.134	1.646	-285.137	-233.598	38.131
340	75.388	79.847	70.573	3.153	-284.506	-230.396	35.396
360	75.679	84.164	71.209	4.664	-283.874	-227.231	32.970
372.780	75.962	86.808	71.699	5.633	----- LIQUID <--> REAL GAS -----		
380	76.154	88.267	72.000	6.182	-283.237	-224.102	30.805
400	76.770	92.189	72.912	7.711	-282.591	-221.006	28.860
420	77.547	95.952	73.920	9.254	-281.934	-217.943	27.105
440	78.543	99.582	75.004	10.814	-281.262	-214.912	25.513
460	79.793	103.100	76.150	12.397	-280.569	-211.911	24.063
480	81.463	106.530	77.344	14.009	-279.850	-208.941	22.737
500	83.694	109.898	78.579	15.659	-279.095	-206.002	21.521

PREVIOUS: March 1979 (1 atm) CURRENT: March 1979 (1 bar)

Water (H_2O) $H_2O_1(l)$

WATER (H_2O) IDEAL GAS M_r = 18.01528

Water (H_2O) $H_2O_1(g)$

$\Delta_{at}H°(0\ K)$ = 917.769 ± 0.109 kJ mol^{-1}
$S°(298.15\ K)$ = 188.834 ± 0.042 J K^{-1} mol^{-1}

$\Delta_fH°(0\ K)$ = -238.921 ± 0.042 kJ mol^{-1}
$\Delta_fH°(298.15\ K)$ = -241.826 ± 0.042 kJ mol^{-1}

Vibrational Frequencies and Degeneracies[a]

ν, cm^{-1}
3651.1(1)
1594.7(1)
3755.9(1)

Point Group: C_{2v} σ = 2
Rotational Constants[a]: A_0 = 27.8847 cm^{-1} B_0 = 14.5118 cm^{-1} C_0 = 9.2806 cm^{-1}

[a]See Wooley (4) for detailed molecular constants on which this table is based.

Enthalpy of Formation

The adopted value of $\Delta_fH°(298.15\ K)$ is that recommended by CODATA (1) and was calculated by CODATA from the heat of formation of liquid water at 298.15 K using vaporization data given by Keenan et al. (2). The enthalpy of atomization, $\Delta_{at}H°(0\ K)$ = 219.352±0.026 kcal mol^{-1}, is calculated using auxiliary JANAF data (3).

Heat Capacity and Entropy

The adopted heat capacity and entropy data below 2000 K, which are for water with the natural isotopic abundance, are taken from recent calculations by Woolley (4). Wooley has done detailed calculations on the thermodynamic functions of light isotopic water ($^1H_2{}^{16}O$) and then corrected those to functions appropriate for water of natural isotopic abundance. Below 240 K, his calculation was a direct summation over rotational levels of the group vibrational state, with small additions above 130 K for the v_2=1 vibrational level. Above 240 K, individual vibrational partition functions were summed to obtain the total partition function. During this summation the molecule was treated as a rigid rotator with modified corrections of Stripp-Kirkwood and centrifugal effect types, guided by the behavior found for the vibrational ground state. Details are given by Woolley (4).

Between 2000 and 5000 K we adopt $C_p°$ data calculated by Friedman and Haar (5) for light isotopic water, after a small correction. Our correction consists of adding 0.021 cal K^{-1} mol^{-1} to Friedman and Haar's $C_p°$ data. This correction arises from the constant difference of 0.021±0.001 cal K^{-1} mol^{-1} between Woolley's $C_p°$ data (natural abundance) and Friedman and Haar's $C_p°$ data (light isotopic water) in the temperature range 1300-2000 K. Above 5000 K, we have linearly extrapolated Friedman and Haar's corrected $C_p°$ data.

Below 2000 K there is good agreement among thermodynamic functions given by Woolley (4), Friedman and Haar (5) and Glushko et al. (6). Above 2000 K the $C_p°$ data from references (5) and (6) begin to deviate significantly; at 4000 and 6000 K the differences [$C_p°$ (6) - $C_p°$ (5)] are 0.57 and 1.07 cal K^{-1} mol^{-1}, respectively. Woolley is working on extending his calculations to higher temperature (7) and his initial results (4) yield $C_p°$ values slightly higher than (5) but not nearly so high as (6).

Values of $S°(298.15\ K)$ and $H°(298.15\ K)-H°(0\ K)$ in the present table are in excellent agreement with CODATA values (1) of 45.106±0.01 cal K^{-1} mol^{-1} K) and 2.368±0.002 kcal mol^{-1}, respectively.

References

1. ICSU-CODATA Task Group, J. Chem. Thermodyn. 4, 331 (1972); see also, CODATA Bulletin #17, "CODATA Recommended Key Values for Thermodynamics - 1975," 1976.
2. J. H. Keenan, F. G. Keyes, P. G. Hill and J. G. Moore, "Steam Tables," Wiley, New York, 1969.
3. JANAF Thermochemical Tables: H(g), O(g), 3-31-77.
4. H. W. Woolley, U. S. Nat. Bur. Stand., paper presented at the 9th International Conference on the Properties of Steam, Munich, Sept. 10-14, 1979.
5. A. S. Friedman and L. Haar, J. Chem. Phys. 22, 2051 (1954).
6. V. P. Glushko, L. V. Gurvich et al., "Thermodynamic Properties of Individual Substances," Third edition, Volume I, Book 1, Nauka, Moscow, 1978.
7. H. W. Woolley, U. S. Nat. Bur. Stand., personal communication, August, 1980.

Enthalpy Reference Temperature = T_r = 298.15 K Standard State Pressure = p° = 0.1 MPa

T/K	$C_p°$ (J $K^{-1}mol^{-1}$)	S° (J $K^{-1}mol^{-1}$)	$-[G°-H°(T_r)]/T$ (J $K^{-1}mol^{-1}$)	$H°-H°(T_r)$ (kJ mol^{-1})	$\Delta_fH°$ (kJ mol^{-1})	$\Delta_fG°$ (kJ mol^{-1})	Log K_f
0	0.	0.	INFINITE	-9.904	-238.921	-238.921	INFINITE
100	33.299	152.388	218.534	-6.615	-240.083	-236.584	123.579
200	33.349	175.485	191.896	-3.282	-240.900	-232.766	60.792
298.15	33.590	188.834	188.834	0.	-241.826	-228.582	40.047
300	33.596	189.042	188.835	0.062	-241.844	-228.500	39.785
400	34.262	198.788	190.159	3.452	-242.846	-223.901	29.238
500	35.226	206.534	192.685	6.925	-243.826	-219.051	22.884
600	36.325	213.052	195.550	10.501	-244.758	-214.007	18.631
700	37.495	218.739	198.465	14.192	-245.632	-208.812	15.582
800	38.721	223.825	201.322	18.002	-246.443	-203.496	13.287
900	39.987	228.459	204.084	21.938	-247.185	-198.083	11.496
1000	41.268	232.738	206.738	26.000	-247.857	-192.590	10.060
1100	42.536	236.731	209.285	30.191	-248.460	-187.033	8.881
1200	43.768	240.485	211.730	34.506	-248.997	-181.425	7.897
1300	44.945	244.035	214.080	38.942	-249.473	-175.774	7.063
1400	46.054	247.407	216.341	43.493	-249.894	-170.089	6.346
1500	47.090	250.620	218.520	48.151	-250.265	-164.376	5.724
1600	48.050	253.690	220.623	52.908	-250.592	-158.639	5.179
1700	48.935	256.630	222.655	57.758	-250.881	-152.883	4.698
1800	49.749	259.451	224.621	62.693	-251.138	-147.111	4.269
1900	50.496	262.161	226.526	67.706	-251.368	-141.325	3.885
2000	51.180	264.769	228.374	72.790	-251.575	-135.528	3.540
2100	51.823	267.282	230.167	77.941	-251.762	-129.721	3.227
2200	52.408	269.706	231.909	83.153	-251.934	-123.905	2.942
2300	52.947	272.048	233.604	88.421	-252.092	-118.082	2.682
2400	53.444	274.312	235.253	93.741	-252.239	-112.252	2.443
2500	53.904	276.503	236.860	99.108	-252.379	-106.416	2.223
2600	54.329	278.625	238.425	104.520	-252.513	-100.575	2.021
2700	54.723	280.683	239.952	109.973	-252.643	-94.729	1.833
2800	55.089	282.680	241.443	115.464	-252.771	-88.878	1.658
2900	55.430	284.619	242.899	120.990	-252.897	-83.023	1.495
3000	55.748	286.504	244.321	126.549	-253.024	-77.163	1.344
3100	56.044	288.337	245.711	132.139	-253.152	-71.298	1.201
3200	56.323	290.120	247.071	137.757	-253.282	-65.430	1.068
3300	56.583	291.858	248.402	143.403	-253.416	-59.558	0.943
3400	56.828	293.550	249.705	149.073	-253.553	-53.681	0.825
3500	57.058	295.201	250.982	154.768	-253.696	-47.801	0.713
3600	57.276	296.812	252.233	160.485	-253.844	-41.916	0.608
3700	57.480	298.384	253.459	166.222	-253.997	-36.027	0.509
3800	57.675	299.919	254.661	171.980	-254.158	-30.133	0.414
3900	57.859	301.420	255.841	177.757	-254.326	-24.236	0.325
4000	58.033	302.887	256.999	183.552	-254.501	-18.334	0.239
4100	58.199	304.322	258.136	189.363	-254.684	-12.427	0.158
4200	58.357	305.726	259.252	195.191	-254.876	-6.516	0.081
4300	58.507	307.101	260.349	201.034	-255.078	-0.600	0.007
4400	58.650	308.448	261.427	206.892	-255.288	5.320	-0.063
4500	58.787	309.767	262.486	212.764	-255.508	11.245	-0.131
4600	58.918	311.061	263.528	218.650	-255.738	17.175	-0.195
4700	59.044	312.329	264.553	224.548	-255.978	23.111	-0.257
4800	59.164	313.574	265.562	230.458	-256.229	29.052	-0.316
4900	59.275	314.795	266.554	236.380	-256.491	34.998	-0.373
5000	59.390	315.993	267.531	242.313	-256.763	40.949	-0.428
5100	59.509	317.171	268.493	248.258	-257.046	46.906	-0.480
5200	59.628	318.327	269.440	254.215	-257.338	52.869	-0.531
5300	59.746	319.464	270.373	260.184	-257.639	58.838	-0.580
5400	59.864	320.582	271.293	266.164	-257.950	64.811	-0.627
5500	59.982	321.682	272.199	272.157	-258.268	70.791	-0.672
5600	60.100	322.764	273.092	278.161	-258.595	76.777	-0.716
5700	60.218	323.828	273.973	284.177	-258.930	82.769	-0.758
5800	60.335	324.877	274.841	290.204	-259.272	88.767	-0.799
5900	60.453	325.909	275.698	296.244	-259.621	94.770	-0.839
6000	60.571	326.926	276.544	302.295	-259.977	100.780	-0.877

PREVIOUS: March 1961 (1 atm) CURRENT: March 1979 (1 bar)

Water (H_2O) $H_2O_1(g)$

WATER (H_2O) LIQUID-REAL GAS, p=1 BAR M_r = 18.01528

$S°(298.15\ K) = 69.950 \pm 0.079\ J\ K^{-1}\ mol^{-1}$ $\Delta_f H°(298.15\ K) = -285.830 \pm 0.042\ kJ\ mol^{-1}$

Enthalpy of Formation

See the $H_2O(\ell$, p=1 bar) table for details (1).

Heat Capacity and Entropy

The adopted heat capacity data in the liquid region are taken from the very accurate calorimetric measurements of Osborne et al. (2). The adopted heat capacity data for the real gas at one bar pressure are taken from the recent equation of state formulation of Haar et al. (3). See the JANAF Table for $H_2O(\ell$, p=1 bar) (1) for details concerning the entropy.

Phase Data

Our table refers to the real fluid existing at the standard-state pressure of 1 bar. Since this pressure is below the critical pressure of water [p_c = 217.6 bar at T_c = 647.14 K (4)], there is a first order transition at the normal boiling point. Therefore, this is a two-phase (liquid-real gas) table for the real fluid at p=1 bar.

Fugacity and Fugacity Coefficient Data

At temperature T the fugacity f for the real gas at p=1 bar is given by $f = \exp[\Delta_f G(\text{real gas}) - \Delta_f G°(\text{ideal gas})/(RT)]$ using real gas data from this table and ideal gas data from (1). Since p=1 bar, the fugacity coefficient (f/P) is equal to the fugacity. The fugacity coefficient is 0.986 at 400 K and approaches 1 as temperature increases, indicating nearly ideal behavior at all temperatures.

References

1. JANAF Thermochemical Tables: $H_2O(\ell$, p=1 bar), H_2O(ideal gas), 3-31-79.
2. N. S. Osborne, H. F. Stimson and D. C. Ginnings, J. Res. Nat. Bur. Stand. 23, 197 (1939).
3. L. Haar, J. Gallagher and G. S. Kell, Ninth International Conference on Properties of Steam, Sept. 1979, Munich; and personal communication from L. Haar and J. Gallagher, U. S. Nat. Bur. Stand., May, 1980.
4. J. F. Matthews, Chem. Rev. 72, 71 (1972).

Water, 1 Bar (H_2O) $H_2O_1(l,g)$

Enthalpy Reference Temperature = T_r = 298.15 K Standard State Pressure = p° = 0.1 MPa

	J K⁻¹mol⁻¹			kJ mol⁻¹			
T/K	$C_p°$	S°	$-[G°-H°(T_r)]/T$	$H°-H°(T_r)$	$\Delta_f H°$	$\Delta_f G°$	Log K_f
0							
100							
200							
280	75.563	65.215	70.102	-1.368	-286.410	-240.123	44.796
298.15	75.351	69.950	69.950	0.	-285.830	-237.141	41.546
300	75.349	70.416	69.952	0.139	-285.771	-236.839	41.237
320	75.344	75.279	70.134	1.646	-285.137	-233.598	38.131
340	75.388	79.847	70.573	3.153	-284.506	-230.396	35.396
360	75.679	84.164	71.209	4.664	-283.874	-227.231	32.970
372.780	75.962	86.808	71.699	5.633	LIQUID <--> REAL GAS		
372.780	36.800	195.911	71.699	46.304	PRESSURE = 1 bar		
400	35.982	198.473	80.240	47.293	-243.009	-223.937	29.243
500	35.699	206.428	104.712	50.858	-243.896	-219.069	22.886
600	36.521	213.003	122.227	54.466	-244.797	-214.018	18.632
700	37.596	218.712	135.612	58.170	-245.658	-208.819	15.582
800	38.780	223.809	146.323	61.988	-246.461	-203.501	13.287
900	40.023	228.448	155.194	65.928	-247.198	-198.086	11.497
1000	41.292	232.730	162.736	69.994	-247.868	-192.593	10.060
1100	42.554	236.725	169.283	74.186	-248.468	-187.035	8.882
1200	43.781	240.481	175.061	78.504	-249.004	-181.426	7.897
1300	44.954	244.032	180.231	82.941	-249.479	-175.775	7.063
1400	46.062	247.404	184.910	87.492	-249.899	-170.090	6.346
1500	47.097	250.618	189.184	92.151	-250.269	-164.376	5.724
1600	48.057	253.689	193.120	96.909	-250.595	-158.639	5.179
1700	48.941	256.629	196.770	101.760	-250.884	-152.883	4.698
1800	49.753	259.450	200.175	106.695	-251.141	-147.111	4.269

SINCE THE STANDARD STATE FOR A GAS IS DEFINED AS THE IDEAL GAS AT ONE BAR PRESSURE, THE GAS PHASE PROPERTIES IN THIS TABLE ARE NOT STANDARD STATE PROPERTIES. THE SUPERSCRIPT ZERO, REPRESENTING STANDARD STATE, REFERS ONLY TO LIQUID-PHASE PROPERTIES AND NOT TO GAS-PHASE PROPERTIES WHICH ARE FOR THE REAL GAS AT P = 1 BAR.

PREVIOUS: March 1979 (1 atm) CURRENT: March 1979 (1 bar)

Water, 1 Bar (H_2O) $H_2O_1(l,g)$

WATER, p=10 BAR (H_2O) LIQUID-REAL GAS, p=10 BAR M_r = 18.01528

S(298.15 K, 10 bar) = 69.946 ± 0.084 J K^{-1} mol^{-1} Δ_fH(298.15 K, 10 bar) = -285.815 ± 0.042 kJ mol^{-1}

Enthalpy of Formation

The enthalpy of formation of H_2O at 298.15 K and p=10 bar [Δ_fH(298.15 K, 10 bar)] is derived from the CODATA (1) enthalpy of formation of the liquid at 298.15 K and 1 bar [Δ_fH°(ℓ, 298.15 K) = -68.315±0.010 kcal mol^{-1}] by addition of the enthalpy increment [H(298.15 K, 10 bar)-H°(298.15 K)]. The enthalpy increment is taken from the recent equation of state formulation of Haar et al. (2) which yields enthalpy data at any temperature and pressure referenced to the triple point of water. The values of Δ_fH*, Δ_fG* and log K* in this table refer to formation of H_2O at p=10 bar from H_2 and O_2 in their standard reference states (ideal gas at a pressure of 1 bar).

Heat Capacity and Entropy

The adopted heat capacity data are taken from the recent equation of state formulation of Haar et al. (2). The entropy and enthalpy are obtained by appropriate integration of the C_p data and differ from entropy and enthalpy data calculated directly from the equation of state formulation by less than 0.0001 cal K^{-1} mol^{-1} and 0.0001 kcal mol^{-1}, respectively. The starting point for the entropy integration [S(298.15 K, 10 bar)] is obtained from the CODATA (1) entropy at standard conditions [S°(ℓ, 298.15 K) = 16.7184±0.019 cal K^{-1} mol^{-1})] by addition of the entropy increment [S(298.15 K, 10 bar)-S°(298.15 K)]. The entropy increment is also taken from the equation of state formulation (2).

Phase Data

Our table refers to the real fluid phase existing at the nonstandard-state pressure of 10 bar. Since this pressure is below the critical pressure of water [p_c = 217.6 bar at T_c = 647.14 K (3)] there is a first order transition at the boiling temperature. Therefore this is a two-phase (liquid-real gas) table for the real fluid at p=10 bar.

Vaporization Data

The adopted boiling temperature (at 10 bar) and enthalpy of vaporization are taken from the equation of state formulation of Haar et al. (2).

Fugacity and Fugacity Coefficients Data

At temperature T the fugacity f of water at p=10 bar is given by f=K/K* or f=exp[(Δ_fG*-Δ_fG°)/(RT)], where K and Δ_fG° are values for H_2O(ideal gas, T) (4). The fugacity coefficient (f/P) at p=10 bar is given by f/10. At p=10 bar the fugacity coefficient is 0.99 at 800 K and the gas behaves in a nearly ideal manner above that temperature.

References

1. ICSU-CODATA Task Group, J. Chem. Thermodyn. 4, 331 (1972); see also, CODATA Bulletin #17, "CODATA Recommended Key Values for Thermodynamics - 1975," 1976.
2. L. Haar, J. Gallagher and G. S. Kell, Ninth International Conference on Properties of Steam, Sept. 1979, Munich; and personal communication from L. Haar and J. Gallagher, U. S. Nat. Bur. Stand., May, 1980.
3. J. F. Matthews, Chem. Rev. 72, 71 (1972).
4. JANAF Thermochemical Tables: H_2O(ideal gas), 3-31-79.

Water, 10 Bar (H_2O) H_2O_1(l,g)

Enthalpy Reference Temperature = T_r = 298.15 K Standard State Pressure = p° = 0.1 MPa

T/K	C_p	S	-[G - H (T_r)]/T	H - H (T_r)	Δ_fH*	Δ_fG*	Log K_f^*
	J $K^{-1}mol^{-1}$ (at P = 10 bar)			kJ mol^{-1}			
0							
100							
200							
298.15	75.313	69.946	69.946	0.	-285.815	-237.125	41.543
300	75.314	70.412	69.948	0.139	-285.756	-236.823	41.235
400	76.732	92.173	72.906	7.707	-282.580	-220.989	28.858
453.070	79.322	101.872	75.740	11.840	LIQUID <--> REAL GAS		
453.070	46.063	181.988	75.740	48.138	PRESSURE = 10 bar		
500	41.002	186.240	85.919	50.160	-244.579	-209.657	21.903
600	38.430	193.409	103.261	54.089	-245.159	-202.623	17.640
700	38.527	199.326	116.572	57.927	-245.886	-195.477	14.587
800	39.310	204.517	127.247	61.816	-246.618	-188.225	12.290
900	40.357	209.206	136.097	65.798	-247.314	-180.884	10.498
1000	41.517	213.517	143.626	69.891	-247.955	-173.468	9.061
1100	42.714	217.531	150.164	74.103	-248.537	-165.990	7.882
1200	43.900	221.298	155.937	78.434	-249.059	-158.462	6.898
1300	45.045	224.858	161.103	82.881	-249.524	-150.893	6.063
1400	46.134	228.236	165.778	87.441	-249.935	-143.291	5.346
1500	47.155	231.454	170.050	92.106	-250.299	-135.660	4.724
1600	48.104	234.528	173.985	96.869	-250.620	-128.007	4.179
1700	48.981	237.471	177.633	101.724	-250.904	-120.335	3.697
1800	49.786	240.294	181.037	106.663	-251.157	-112.647	3.269

VALUES OF $\Delta_F H^*$, $\Delta_F G^*$ AND LOG K_F^* IN THIS TABLE REFER TO FORMATION OF H_2O AT P = 10 BAR FROM H_2 AND O_2 IN THEIR STANDARD REFERENCE STATES (IDEAL GAS AT A PRESSURE OF 1 BAR).

PREVIOUS: March 1979 (1 atm) CURRENT: March 1979 (1 bar)

Water, 10 Bar (H_2O) H_2O_1(l,g)

WATER, p=100 BAR (H_2O) LIQUID-REAL GAS, p=100 BAR M_r = 18.01528

S(298.15 K, 100 bar) = 69.903 ± 0.105 J K^{-1} mol^{-1} Δ_fH(298.15 K, 100 bar) = -285.666 ± 0.084 kJ mol^{-1}

Enthalpy of Formation

The enthalpy of formation of H_2O at 298.15 K and p=100 bar [Δ_fH(298.15 K, 100 bar)] is derived from the CODATA (1) enthalpy of formation of the liquid at 298.15 K and 1 bar [Δ_fH°(ℓ, 298.15 K) = -68.315±0.10 kcal mol^{-1}] by addition of the enthalpy increment [H(298.15 K, 100 bar) - H°(298.15 K)]. The enthalpy increment is taken from the recent equation of state formulation of Haar et al. (2) which yields enthalpy data at any temperature and pressure referenced to the triple point of water. The values of Δ_fH*, Δ_fG* and log K* in this table refer to formation of H_2O at p=100 bar from H_2 and O_2 in their standard reference states (ideal gas at a pressure of 1 bar).

Heat Capacity and Entropy

The adopted heat capacity data are taken from the recent equation of state formulation of Haar et al. (2). The entropy and enthalpy are obtained by appropriate integration of the C_p data and differ from entropy and enthalpy data calculated directly from the equation of state formulation by less than 0.0012 cal K^{-1} mol^{-1} and 0.0008 kcal mol^{-1}, respectively. The starting point for the entropy integration [S(298.15 K, 100 bar)] is obtained from the CODATA (1) entropy at standard conditions [S°(ℓ, 298.15 K) = 16.7184±0.019 cal K^{-1} mol^{-1}] by addition of the entropy increment [S(298.15 K, 100 bar) - S°(298.15 K)] The entropy increment is also taken from the equation of state formulation (2).

Phase Data

Our table refers to the real fluid phase existing at the nonstandard-state pressure of 100 bar. Since this pressure is below the critical pressure of water [p_c = 217.6 bar at T_c = 647.14 K (3)], there is a first order transition at the boiling temperature. Therefore this is a two-phase (liquid-real gas) table for the real fluid at p=100 bar.

Vaporization Data

The adopted boiling temperature (at 100 bar) and enthalpy of vaporization are taken from the equation of state formulation of Haar et al. (2).

Fugacity and Fugacity Coefficient Data

At temperature T the fugacity f of water at p=100 bar is given by f=K/K* or f=exp[Δ_fG*-Δ_fG°)/(RT)], where K and Δ_fG° are values for H_2O(ideal gas) (4). The fugacity coefficient (f/p) at p=100 bar is given by f/100. At p=100 bar the fugacity coefficient is 0.99 at 1200 K and the gas behaves in a nearly ideal manner above that temperature.

References

1. ICSU-CODATA Task Group, J. Chem. Thermodyn. 4, 331 (1972); see also, CODATA Bulletin #17, "CODATA Recommended Key Values for Thermodynamics - 1975," 1976.
2. L. Haar, J. Gallagher and G. S. Kell, Ninth International Conference on Properties of Steam, Sept. 1979, Munich; and personal communication from L. Haar and J. Gallagher, U. S. Nat. Bur. Stand., May, 1980.
3. J. F. Matthews, Chem. Rev. 72, 71 (1972).
4. JANAF Thermochemical Tables: H_2O(ideal gas), 3-31-79.

Water, 100 Bar (H_2O) H_2O_1(l,g)

Enthalpy Reference Temperature = T_r = 298.15 K Standard State Pressure = p° = 0.1 MPa

T/K	C_p (J $K^{-1}mol^{-1}$)	S	-[G - H (T_r)]/T	H - H (T_r) (kJ mol^{-1})	$\Delta_f H^*$	$\Delta_f G^*$	Log K_f^*
0	at P = 100 bar						
100							
200							
298.15	74.864	69.903	69.903	0.	-285.666	-236.963	41.515
300	74.878	70.366	69.905	0.139	-285.608	-236.661	41.206
400	76.313	92.021	72.848	7.669	-282.468	-220.816	28.836
500	82.696	109.578	78.481	15.548	-279.041	-205.789	21.499
584.180	110.330	123.855	83.972	23.299	LIQUID <--> REAL GAS		
584.180	124.260	164.476	83.972	47.029	PRESSURE = 100 bar		
600	92.684	167.317	86.134	48.710	-250.389	-192.197	16.732
700	51.828	177.340	98.532	55.165	-248.498	-182.699	13.633
800	45.622	183.761	108.802	59.968	-248.317	-173.320	11.317
900	44.071	189.023	117.430	64.434	-248.528	-163.934	9.514
1000	43.949	193.653	124.825	68.828	-248.869	-154.517	8.071
1100	44.414	197.861	131.276	73.244	-249.247	-145.064	6.889
1200	45.145	201.756	136.989	77.721	-249.623	-135.576	5.901
1300	45.991	205.403	142.113	82.277	-249.978	-126.057	5.065
1400	46.872	208.844	146.758	86.920	-250.306	-116.513	4.347
1500	47.745	212.108	151.007	91.651	-250.603	-106.945	3.724
1600	48.586	215.216	154.923	96.468	-250.871	-97.359	3.178
1700	49.380	218.186	158.558	101.367	-251.112	-87.757	2.696
1800	50.123	221.029	161.950	106.343	-251.328	-78.142	2.268

VALUES OF $\Delta_F H^*$, $\Delta_F G^*$ AND LOG K_F^* IN THIS TABLE REFER TO FORMATION OF H_2O AT P = 100 BAR FROM H_2 AND O_2 IN THEIR STANDARD REFERENCE STATES (IDEAL GAS AT A PRESSURE OF 1 BAR).

PREVIOUS: March 1979 (1 atm) CURRENT: March 1979 (1 bar)

Water, 100 Bar (H_2O) H_2O_1(l,g)

WATER, p=500 BAR (H_2O) REAL FLUID, p=500 BAR M_r = 18.01528

S(298.15 K, 500 bar) = 69.696 ± 0.167 J K^{-1} mol^{-1} Δ_fH(298.15 K, 500 bar) = -285.014 ± 0.167 kJ mol^{-1}

Enthalpy of Formation

The enthalpy of formation of H_2O at 298.15 K and p=500 bar [$\Delta_fH°$(298.15 K, 500 bar)] is derived from the CODATA (1) enthalpy of formation of the liquid at 298.15 K and 1 bar [$\Delta_fH°$(ℓ, 298.15 K) = -68.315±0.010 kcal mol^{-1}] by addition of the enthalpy increment [H(298.15 K, 500 bar) - H°(298.15 K)]. The enthalpy increment is taken from the recent equation of state formulation of Haar et al. (2) which yields enthalpy data at any temperature and pressure referenced to the triple point of water. The values of Δ_fH*, Δ_fG* and log K* in this table refer to formation of H_2O (real fluid at p=500 bar) from H_2 and O_2 in their standard reference states (ideal gas at a pressure of 1 bar).

Heat Capacity and Entropy

The adopted heat capacity data are taken from the recent equation of state formulation of Haar et al. (2). The entropy and enthalpy are obtained by appropriate integration of the C_p data and differ from entropy and enthalpy data calculated directly from the equation of state formulation by less than 0.0004 cal K^{-1} mol^{-1} and 0.0002 kcal mol^{-1}, respectively. The starting point for the entropy integration [S(298.15 K, 500 bar)] is obtained from the CODATA (1) entropy at standard condtions [S°(ℓ, 298.15 K) = 16.7184±0.019 cal K^{-1} mol^{-1}] by addition of the entropy increment [S(298.15 K, 500 bar) - S°(298.15 K)]. The entropy increment is also taken from the equation of state formulation (2).

There is a significant maximum in the C_p(500 bar) data which rises to 40.2 cal K^{-1} mol^{-1} near 730 K. In this temperature region the density is varying rapidly but continuously with temperature; the maximum in the C_p data is due to a change in sign of the derivative of the density-temperature curve. These same density changes are also responsible for large effects on intermolecular interaction as measured by the fugacity coefficient which changes from 0.22 at 600 K to 0.69 at 800 K (see below).

Phase Data

Our table refers to the fluid phase existing at the nonstandard-state pressure of 500 bar. Since 500 bar exceeds the critical pressure of water (p_c = 217.6 bar at T_c = 647.14 K (3)) the properties of the fluid vary continuously with temperature. Thus our table is a single-phase table for the compressed fluid. No first-order phase transition occurs, even though labels such as compressed liquid and dense gas are sometimes used for the regions below and above T_c.

Fugacity and Fugacity Coefficient Data

At temperature T the fugacity f of water at p=500 bar is given by f=K/K* or f=exp[(Δ_fG*-$\Delta_fG°$)/(RT)], where K and $\Delta_fG°$ are values for H_2O(ideal gas) (4). The fugacity coefficient (f/p) at p=500 bar is given by f/500.

References

1. ICSU-CODATA Task Group, J. Chem. Thermodyn. 4, 331 (1972); see also, CODATA Bulletin #17, "CODATA Recommended Key Values for Thermodynamics - 1975," 1976.
2. L. Haar, J. Gallagher and G. S. Kell, Ninth International Conference on Properties of Steam, Sept. 1979, Munich; and prersonal communication from L. Haar and J. Gallagher, U. S. Nat. Bur. Stand., May, 1980.
3. J. F. Matthews, Chem. Rev. 72, 71 (1972).
4. JANAF Thermochemical Tables: H_2O(ideal gas), 3-31-79.

Water, 500 Bar (H_2O) H_2O_1(fl)

Enthalpy Reference Temperature = T_r = 298.15 K Standard State Pressure = p° = 0.1 MPa

T/K	C_p (J $K^{-1}mol^{-1}$)	S (J $K^{-1}mol^{-1}$)	-[G - H (T_r)]/T (J $K^{-1}mol^{-1}$)	H - H (T_r) (kJ mol^{-1})	Δ_fH* (kJ mol^{-1})	Δ_fG* (kJ mol^{-1})	Log K_f*
0	at P = 500 bar						
100							
200							
298.15	73.198	69.696	69.696	0.	-285.014	-236.249	41.390
300	73.249	70.149	69.698	0.135	-284.959	-235.947	41.082
400	74.726	91.395	72.585	7.524	-281.962	-220.060	28.737
500	78.518	108.401	78.098	15.152	-278.786	-204.945	21.410
600	90.964	123.595	84.422	23.504	-274.943	-190.519	16.586
700	149.910	140.724	91.172	34.686	-268.326	-176.895	13.200
800	104.930	160.611	98.728	49.507	-258.126	-164.609	10.748
900	67.740	170.291	106.191	57.690	-254.620	-153.167	8.890
1000	57.261	176.794	112.941	63.853	-253.192	-141.981	7.416
1100	53.101	182.029	118.990	69.342	-252.497	-130.897	6.216
1200	51.274	186.560	124.436	74.549	-252.143	-119.860	5.217
1300	50.530	190.631	129.374	79.634	-251.970	-108.845	4.373
1400	50.349	194.367	133.885	84.675	-251.899	-97.838	3.650
1500	50.479	197.844	138.034	89.715	-251.888	-86.835	3.024
1600	50.781	201.111	141.875	94.778	-251.910	-75.831	2.476
1700	51.174	204.201	145.451	99.875	-251.952	-64.824	1.992
1800	51.611	207.138	148.797	105.014	-252.005	-53.816	1.562

VALUES OF Δ_FH^*, Δ_FG^* AND LOG K_F^* IN THIS TABLE REFER TO FORMATION OF H_2O AT P = 500 BAR FROM H_2 AND O_2 IN THEIR STANDARD REFERENCE STATES (IDEAL GAS AT A PRESSURE OF 1 BAR).

PREVIOUS: March 1979 (1 atm) CURRENT: March 1979 (1 bar)

Water, 500 Bar (H_2O) H_2O_1(fl)

WATER, p=5000 BAR (H_2O) REAL FLUID, p=5000 BAR M_r = 18.01528

S(298.15 K, 5000 bar) = 66.927 ± 0.209 $J\ K^{-1}\ mol^{-1}$ Δ_fH(298.15 K, 5000 bar) = -278.394 ± 0.418 $kJ\ mol^{-1}$

Enthalpy of Formation

The enthalpy of formation of H_2O at 298.15 K and p=5000 bar [Δ_fH(298.15 K, 5000 bar)] is derived from the CODATA (1) heat of formation of the liquid at 298.15 K and 1 bar [$\Delta_fH°$(ℓ, 298.15 K) = -68.315±0.010 $kcal\ mol^{-1}$] by addition of the enthalpy increment [H(298.15, 5000 bar)-H°(298.15 K)]. The enthalpy increment is taken from the recent equation of state formulation of Haar et al. (2) which yields enthalpy data at any temperature and pressure referenced to the triple point of water. The values of $\Delta_fH°$, $\Delta_fG°$ and log K in this table refer to formation of H_2O (real fluid at p=5000 bar) from H_2 and O_2 in their standard reference states (ideal gas at a pressure of 1 bar).

Heat Capacity and Entropy

The adopted heat capacity data are taken from the recent equation of state formulation of Haar et al. (2). The entropy and enthalpy are obtained by appropriate integration of the C_p data and differ from entropy and enthalpy data calculated directly from the equation of state formulation by less than 0.0002 $cal\ K^{-1}\ mol^{-1}$ and 0.0001 $kcal\ mol^{-1}$, respectively. The starting point for the entropy integration [S(298.15 K, 5000 bar)] is obtained from the CODATA (1) entropy at standard condtions [S°(ℓ, 298.15 K) = 16.7184±0.019 $cal\ K^{-1}\ mol^{-1}$] by addition of the entropy increment [S(298.15 K, 5000 bar)-S°(298.15 K)]. The entropy increment is also taken from the equation of state formulation (2).

There are two rather small maxima in the C_p(5000 bar) data near 420 and 780 K.

Phase Data

Our table refers to the fluid phase existing at the nonstandard-state pressure of 5000 bar. Since 5000 bar exceeds the critical pressure of water (P_c = 217.6 bar at T_c = 647.14 K (3)) the properties of the fluid vary continuously with temperature. Thus our table is a single-phase table for the compressed fluid. No first-order phase transition occurs, even though labels such as compressed liquid and dense gas are sometimes used for the regions below and above T_c.

Fugacity and Fugacity Coefficient Data

At temperature T the fugacity f of water at p=5000 bar is given by f=K/K or f=exp[(Δ_fG-$\Delta_fG°$)/(RT)], where K and $\Delta_fG°$ are values for H_2O(ideal gas) (4). The fugacity coefficient (f/p) at p=5000 bar is given by f/5000.

References

1. ICSU-CODATA Task Group: J. Chem. Thermodyn. 4, 331 (1972); see also, CODATA Bulletin #17, "CODATA Recommended Key Values for Thermodynamics - 1975," 1976.
2. L. Haar, J. Gallagher and G. S. Kell, Ninth International Conference on Properties of Steam, Sept. 1979, Munich; and personal communication from L. Haar and J. Gallagher, U. S. Nat. Bur. Stand., May, 1980.
3. J. F. Matthews, Chem. Rev. 72, 71 (1972).
4. JANAF Thermochemical Tables: H_2O(ideal gas), 3-31-79.

Water, 5000 Bar (H_2O) H_2O_1(fl)

Enthalpy Reference Temperature = T_r = 298.15 K Standard State Pressure = p° = 0.1 MPa

T/K	C_p (J K⁻¹mol⁻¹)	S	-[G - H (T_r)]/T	H - H (T_r) (kJ mol⁻¹)	Δ_fH^*	Δ_fG^*	Log K_f^*
0	at P = 5000 bar						
100							
200							
298.15	62.350	66.927	66.927	0.	-278.394	-228.804	40.086
300	63.153	67.316	66.929	0.116	-278.358	-228.496	39.785
400	67.912	86.697	69.552	6.858	-276.008	-212.226	27.714
500	67.526	101.838	74.553	13.642	-273.676	-196.553	20.534
600	67.186	114.105	80.154	20.370	-271.457	-181.338	15.787
700	67.489	124.482	85.764	27.102	-269.290	-166.490	12.424
800	67.633	133.510	91.180	33.863	-267.150	-151.951	9.921
900	67.178	141.455	96.334	40.609	-265.082	-137.676	7.991
1000	66.153	148.483	101.205	47.278	-263.148	-123.625	6.458
1100	64.807	154.725	105.792	53.826	-261.393	-109.759	5.212
1200	63.396	160.303	110.106	60.236	-259.836	-96.045	4.181
1300	62.089	165.324	114.164	66.508	-258.475	-82.452	3.313
1400	60.968	169.883	117.984	72.659	-257.295	-68.957	2.573
1500	60.056	174.057	121.585	78.709	-256.274	-55.541	1.934
1600	59.344	177.910	124.986	84.678	-255.391	-42.188	1.377
1700	58.804	181.490	128.206	90.584	-254.624	-28.887	0.888
1800	58.408	184.840	131.260	96.443	-253.956	-15.628	0.454

VALUES OF Δ_FH^*, Δ_FG^* AND LOG K_F^* IN THIS TABLE REFER TO FORMATION OF H_2O AT P = 5000 BAR FROM H_2 AND O_2 IN THEIR STANDARD REFERENCE STATES (IDEAL GAS AT A PRESSURE OF 1 BAR).

PREVIOUS: March 1979 (1 atm) CURRENT: March 1979 (1 bar)

Water, 5000 Bar (H_2O) H_2O_1(fl)

HYDROGEN PEROXIDE (H_2O_2) IDEAL GAS M_r = 34.01468

Hydrogen Peroxide (HOOH) $H_2O_2(g)$

S°(298.15 K) = 232.991 J K^{-1} mol^{-1}

$\Delta_fH°$(0 K) = -129.808 kJ mol^{-1}
$\Delta_fH°$(298.15 K) = -136.106 kJ mol^{-1}

Point Group: C_2

Heat Capacity and Entropy

$C_p^°$ from Giguere and Lin (1). $C_p^°$ below 298.15 K estimated.

Reference

1. P. A. Giguere and I. D. Lin, J. Amer. Chem. Soc. 77, 6477 (1955).

Enthalpy Reference Temperature = T_r = 298.15 K; Standard State Pressure = p° = 0.1 MPa

	J $K^{-1}mol^{-1}$			kJ mol^{-1}			
T/K	$C_p^°$	S°	$-[G°-H°(T_r)]/T$	$H°-H°(T_r)$	$\Delta_fH°$	$\Delta_fG°$	Log K_f
0	0.	0.	INFINITE	-10.853	-129.808	-129.808	INFINITE
100	33.526	186.540	261.768	-7.523	-132.382	-123.632	64.579
200	36.874	210.632	228.978	-3.669	-134.133	-113.680	29.690
298.15	43.116	232.991	232.991	0.	-136.106	-105.445	18.474
300	43.221	233.258	232.992	0.080	-136.133	-105.255	18.326
400	48.451	246.435	234.753	4.673	-137.417	-94.757	12.374
500	52.551	257.705	238.243	9.731	-138.341	-83.978	8.773
600	55.689	267.576	242.326	15.150	-139.011	-73.039	6.359
700	57.990	276.339	246.571	20.838	-139.515	-62.002	4.627
800	59.831	284.205	250.792	26.731	-139.912	-50.901	3.324
900	61.463	291.349	254.908	32.797	-140.225	-39.755	2.307
1000	62.844	297.897	258.884	39.013	-140.475	-28.578	1.493
1100	64.141	303.949	262.709	45.364	-140.673	-17.378	0.825
1200	65.312	309.580	266.383	51.837	-140.827	-6.162	0.268
1300	66.400	314.852	269.911	58.423	-140.944	5.065	-0.203
1400	67.404	319.810	273.300	65.114	-141.031	16.299	-0.608
1500	68.325	324.492	276.558	71.901	-141.093	27.539	-0.959

PREVIOUS: December 1960 (1 atm) CURRENT: December 1960 (1 bar)

Hydrogen Peroxide (HOOH) $H_2O_2(g)$

STRONTIUM HYDROXIDE ($Sr(OH)_2$) CRYSTAL M_r = 121.63468

$\Delta_f H°$(0 K) = Unknown
$S°$(298.15 K) = [97.069 ± 8.4] J K^{-1} mol^{-1} $\Delta_f H°$(298.15 K) = -968.89 ± 9.2 kJ mol^{-1}
T_{fus} = 783.15 ± 15 K $\Delta_{fus}H°$ = 21.020 ± 2.1 kJ mol^{-1}

Enthalpy of Formation

The adopted $\Delta_f H°$(298.15 K) = -231.57±2.2 kcal mol^{-1} is obtained from the enthalpy of formation adopted for the liquid (1) minus enthalpy of melting and the enthalpy difference of the liquid and the crystal between the melting point and 298.15 K. The dissociation pressure measurements by Johnston (2) and by Tamaru and Shiomi (3) span the melting point, but there is not a definite break in the log P vs 1/T curve at the melting point. The temperatures of the dissociation pressures determined by Sano (4) are all below the melting point. For 2nd and 3rd law analysis, using auxiliary data from the JANAF Thermochemical Tables (1), the dissociation pressures below the melting point are considered to pertain to the equilibrium $Sr(OH)_2$(cr) = SrO(cr) + H_2O(g).

Source	Data Points	T/K	$\Delta_r H°$(298.15 K), kcal mol^{-1} 2nd law	3rd law	Drift cal K^{-1} mol^{-1}	$\Delta_f H°$(298.15 K) kcal mol^{-1}
Johnston (2)	4	703-776	18.7	31.65±1.06	17.5±1.4	-230.9±1.9
Tamaru and Shiomi (3)	10	667-782	28.0	31.50±0.42	4.9±0.3	-230.8±1.2
Sano (4)	5	613-743	26.0	31.27±0.75	7.6±0.2	-230.6±1.6

These three 3rd law values for $\Delta_f H°$(cr, 298.15 K) are in excellent agreement, are in reasonable agreement with the adopted value, and are closer to another recent evaluation of -229.2 kcal mol^{-1} (5) than is the adopted value. The route taken to derive the adopted value is presently preferred because dissociation pressure measurements of the liquid are judged to be better than those of the solid.

Heat Capacity and Entropy

The heat capacities are derived from a curve fit of the drop calorimeter enthalpy measurements of Powers and Blalock (6, 545-783 K), extrapolated and forced through zero relative enthalpy at 273.15 K. The change in heat capacity with temperature is larger than it is for the other alkaline earth dihydroxides (1). There appears to be no definite explanation for this. The purity of the $Sr(OH)_2$ sample, as indicated by total alkalinity, did change more during the enthalpy measurement than other samples in the same report (6). The total alkalinity of the $Sr(OH)_2$ sample changed from 99.80 to 94.1% while for $Ba(OH)_2$ the change was 100.4 to 99.81%, for NaOH the change was 99.97 to 99.46%, and for KOH the change was 100.00 to 98.68% (6). The known existence of crystalline allotropy in some of the alkaline earth halides and hydroxides suggests this possibility for $Sr(OH)_2$ crystal also.

The adopted entropy, $S°$(298.15 K) = [23.2±2] kcal mol^{-1}, is calculated from Kelley's additive entropy constants for cations and anions (7).

Fusion Data

From their drop calorimetry, Powers and Blalock (6) selected a melting point of 783.15 K where they derived $\Delta_{fus}H°$ = 5.23 kcal mol^{-1}. Our smoothing of their data leads to the adopted $\Delta_{fus}H°$ = 5.024±0.50 kcal mol^{-1}. Brcic and Jernejcic (8) determined a melting point of 771 K and Berggren and Brown (9), 723 K. The latter dihydroxide sample contained 5 mole percent $SrCO_3$. T_{fus} = 783.15±15 K is adopted.

References

1. JANAF Thermochemical Tables: SrO(cr), 12-31-72; H_2O(g), 3-31-61; $Be(OH)_2$ (β-c), $Mg(OH)_2$(cr), $Ca(OH)_2$(cr), $Ba(OH)_2$(cr), $Sr(OH)_2$(ℓ), 12-31-75.
2. J. Johnston, Z. Physik. Chem. 62, 330 (1908).
3. S. Tamaru and K. Shiomi, Z. Physik. Chem. A171, 221 (1935).
4. K. Sano, J. Chem. Soc. Japan 60, 758 (1939).
5. U. S. Nat. Bur. Stand. Tech. Note 270-6, 1971.
6. W. D. Powers an G. C. Blalock, U. S. Atomic Energy Comm., ORNL-1653 (1954).
7. K. K. Kelley in Chap. 13, D. R. Stull and H. Prophet, "The Characterization of High Temperature Vapors," J. L. Margrave, ed., John Wiley and Sons, New York, 1967.
8. B. S. Brcic and J. Jernejcic, Vestn. Sloven. Kem. Drustva 9, 65 (1962).
9. G. Berggren and A. Brown, Acta Chem. Scand. 25, 1377 (1971).

Strontium Hydroxide ($Sr(OH)_2$) $H_2O_2Sr_1$(cr)

Enthalpy Reference Temperature = T_r = 298.15 K; Standard State Pressure = p° = 0.1 MPa

T/K	$C_p°$ (J K^{-1}mol^{-1})	S°	-[G°-H°(T_r)]/T	H°-H°(T_r) (kJ mol^{-1})	$\Delta_f H°$	$\Delta_f G°$	Log K_f
0							
100							
200							
298.15	74.894	97.069	97.069	0.	-968.889	-881.098	154.365
300	75.145	97.533	97.070	0.139	-968.907	-880.553	153.318
400	88.450	120.970	100.176	8.318	-969.335	-851.006	111.130
500	101.755	142.137	106.481	17.828	-968.634	-821.482	85.820
600	115.018	161.859	114.083	28.666	-966.815	-792.204	68.967
700	128.323	180.589	122.256	40.833	-963.874	-763.319	56.960
783.150	139.370	195.601	129.251	51.963	------ CRYSTAL <--> LIQUID -----		
800	141.628	198.592	130.680	54.330	-959.806	-734.933	47.986
900	154.934	216.041	139.199	69.158	-955.232	-707.052	41.036
1000	168.239	233.054	147.737	85.316	-948.573	-679.822	35.510

PREVIOUS: CURRENT: December 1975

Strontium Hydroxide ($Sr(OH)_2$) $H_2O_2Sr_1$(cr)

STRONTIUM HYDROXIDE ($Sr(OH)_2$) LIQUID M_r = 121.63468

$S°$(298.15 K) = [108.739] J K^{-1} mol^{-1} $\Delta_f H°$(298.15 K) = [-957.304] kJ mol^{-1}

T_{fus} = 783.15 ± 15 K $\Delta_{fus}H°$ = 21.020 ± 2.1 kJ mol^{-1}

Enthalpy of Formation

The adopted $\Delta_f H°$(298.15 K) = -228.8±1.7 kcal mol^{-1} is the average of values derived from the 3rd law analysis of the dissociation pressure measurements of Johnston (1) and of Tamaru and Shiomi (2). The data from both sources (1, 2) span the melting point; only those data above the melting point are used for the 2nd and 3rd law analysis of the reaction $Sr(OH)_2(\ell)$ = SrO(cr) + H_2O(g).

Source	Data Points	T/K	$\Delta_r H°$(298.15 K), kcal mol^{-1} 2nd law	3rd law	Drift cal K^{-1} mol^{-1}	$\Delta_f H°$(298.15 K) kcal mol^{-1}
Johnston (1)	11[a]	806-1038	28.5	30.06±0.28[b]	1.6±0.4	-229.36±1.1
Tamaru and Shioni (2)	6	880-942	30.9	29.93±0.26[b]	-2.3±0.2	-228.23±1.1

[a]One point rejected due to failure of a statistical test.
[b]Considering all of the data as liquid data (733-1038 K, 1; 667-942 K, 2) yields 3rd law enthalpies of reaction of 29.97±0.48 kcal mol^{-1} for (1) and 29.08±0.16 kcal mol^{-1} for (2) leading to standard state enthalpies of formation at 298.15 K of -229.27±1.3 and -228.38±1.0 kcal mol^{-1}, respectively.

Heat Capacity and Entropy

$C°_p$ is derived from the drop ice calorimeter measurements of Powers and Blalock (4, 789-1187 K) and is assumed constant at 27.7 cal K^{-1} mol^{-1} over the range of 520 to 1600 K. A glass transition is assumed at 520 K below which $C°_p$ is that of the crystal. The entropy of 298.15 K is calculated from the value adopted for the crystal (3).

Decomposition Data

T_{dcm} = 1017 K is calculated as the temperature at which the fugacity of H_2O(g) is 1 atm for the reaction $Sr(OH)_2(\ell)$ = SrO(cr) + H_2O(g). Auxiliary data used in the calculations are from the JANAF Thermochemical Tables (3).

References

1. J. Johnston, Z. physik. Chem. 62, 330 (1908).
2. S. Tamaru and K. Shiomi, Z. physik. Chem. A171, 221 (1935).
3. JANAF Thermochemical Tables: SrO(cr), 12-31-72; H_2O(g), 3-31-61; $Sr(OH)_2$(cr), 12-31-75.
4. W. D. Powers and G. C. Blalock, U. S. Atomic Energy Comm. ORNL-1653 (1954).

Strontium Hydroxide ($Sr(OH)_2$) $H_2O_2Sr_1(l)$

Enthalpy Reference Temperature = T_r = 298.15 K Standard State Pressure = p° = 0.1 MPa

T/K	$C°_p$ (J $K^{-1}mol^{-1}$)	S°	$-[G°-H°(T_r)]/T$	$H°-H°(T_r)$ (kJ mol^{-1})	$\Delta_f H°$	$\Delta_f G°$	Log K_f
0							
100							
200							
298.15	74.894	108.739	108.739	0.	-957.304	-872.993	152.945
300	75.145	109.203	108.740	0.139	-957.323	-872.470	151.910
400	88.450	132.640	111.846	8.318	-957.750	-844.090	110.227
500	101.755	153.807	118.151	17.828	-957.049	-815.732	85.219
520.000	104.392	157.850	119.600	19.890	GLASS <--> LIQUID TRANSITION		
520.000	157.737	157.850	119.600	19.890			
600	157.737	180.422	126.241	32.509	-951.387	-787.914	68.594
700	157.737	204.737	135.762	48.282	-944.839	-761.189	56.801
783.150	157.737	222.442	144.043	61.398	------ CRYSTAL <--> LIQUID -----		
800	157.737	225.800	145.730	64.056	-938.495	-735.388	48.016
900	157.737	244.379	155.679	79.830	-932.976	-710.299	41.225
1000	157.737	260.998	165.395	95.603	-926.701	-685.895	35.827
1100	157.737	276.032	174.780	111.377	-928.412	-661.749	31.424
1200	157.737	289.757	183.798	127.151	-923.211	-637.737	27.760
1300	157.737	302.382	192.441	142.924	-918.088	-614.156	24.677
1400	157.737	314.072	200.716	158.698	-913.038	-590.966	22.049
1500	157.737	324.955	208.640	174.472	-908.060	-568.135	19.784
1600	157.737	335.135	216.231	190.245	-903.151	-545.634	17.813
1700	157.737	344.697	223.510	206.019	-1035.223	-522.258	16.047
1800	157.737	353.713	230.495	221.793	-1028.585	-492.276	14.285
1900	157.737	362.242	237.207	237.566	-1022.015	-462.660	12.719
2000	157.737	370.333	243.663	253.340	-1015.513	-433.389	11.319

PREVIOUS: CURRENT: December 1975

Strontium Hydroxide ($Sr(OH)_2$) $H_2O_2Sr_1(l)$

STRONTIUM HYDROXIDE ($Sr(OH)_2$)   CRYSTAL–LIQUID M_r = 121.63468

0 to 783.15 K crystal
above 783.15 K liquid

Refer to the individual tables for details.

Strontium Hydroxide ($Sr(OH)_2$) $H_2O_2Sr_1$(cr,l)

Enthalpy Reference Temperature = T_r = 298.15 K   Standard State Pressure = p° = 0.1 MPa

T/K	C_p° (J K⁻¹mol⁻¹)	S° (J K⁻¹mol⁻¹)	$-[G^\circ-H^\circ(T_r)]/T$ (J K⁻¹mol⁻¹)	$H^\circ-H^\circ(T_r)$ (kJ mol⁻¹)	Δ_fH° (kJ mol⁻¹)	Δ_fG° (kJ mol⁻¹)	Log K_f
0							
100							
200							
298.15	74.894	97.069	97.069	0.	-968.889	-881.098	154.365
300	75.145	97.533	97.070	0.139	-968.907	-880.553	153.318
400	88.450	120.970	100.176	8.318	-969.335	-851.006	111.130
500	101.755	142.137	106.481	17.828	-968.634	-821.482	85.820
600	115.018	161.859	114.083	28.666	-966.815	-792.204	68.967
700	128.323	180.589	122.256	40.833	-963.874	-763.319	56.960
783.150	139.370	195.601	129.251	51.963	CRYSTAL <--> LIQUID		
783.150	157.737	222.442	129.251	72.983	TRANSITION		
800	157.737	225.800	131.249	75.641	-938.495	-735.388	48.016
900	157.737	244.379	142.807	91.415	-932.976	-710.299	41.225
1000	157.737	260.998	153.810	107.188	-926.701	-685.895	35.827
1100	157.737	276.032	164.248	122.962	-928.412	-661.749	31.424
1200	157.737	289.757	174.144	138.736	-923.211	-637.737	27.760
1300	157.737	302.382	183.529	154.509	-918.088	-614.156	24.677
1400	157.737	314.072	192.441	170.283	-913.038	-590.966	22.049
1500	157.737	324.955	200.917	186.057	-908.060	-568.135	19.784
1600	157.737	335.135	208.991	201.830	-903.151	-545.634	17.813
1700	157.737	344.697	216.695	217.604	-1035.223	-522.258	16.047
1800	157.737	353.713	224.059	233.378	-1028.585	-492.276	14.285
1900	157.737	362.242	231.110	249.151	-1022.015	-462.660	12.719
2000	157.737	370.333	237.870	264.925	-1015.513	-433.389	11.319

PREVIOUS: CURRENT: December 1975

Strontium Hydroxide ($Sr(OH)_2$) $H_2O_2Sr_1$(cr,l)

STRONTIUM HYDROXIDE ($Sr(OH)_2$) IDEAL GAS M_r = 121.63468

$S°(298.15 K) = [305.042 \pm 8.4]\ J\ K^{-1}\ mol^{-1}$

$\Delta_f H°(0 K) = -588.31 \pm 41.8\ kJ\ mol^{-1}$

$\Delta_f H°(298.15 K) = -595.80 \pm 41.8\ kJ\ mol^{-1}$

Vibrational Frequencies and Degeneracies

ν, cm^{-1}
[442](1)
[82](1)
[443](1)
[3650](2)
[438](4)

Ground State Quantum Weight: [1] [σ = 2]

Point Group: [C_{2v}]

Bond Distances: Sr-O = [2.22] Å O-H = [0.96] Å

Bond Angles: O-Sr-O = [108]° Sr-O-H = [180]°

Product of the Moments of Inertia: $I_A I_B I_C = [3.824302] \times 10^{-114}\ g^3\ cm^6$

Enthalpy of Formation

Dissociation energies for the reaction $Sr(OH)_2(g) = Sr(g) + 2\ OH(g)$ have been derived from flame spectral measurements (1-3). Ryabova and Gurvich (1) believed the dominant reaction to be $Sr(g) + H_2O(g) = SrOH(g) + H(g)$ but they also considered the possibility that the reaction $Sr(g) + 2\ H_2O(g) = Sr(OH)_2 + 2\ H(g)$ was dominant and derived D_0^0 = 180±20 kcal mol^{-1}. Sugden and Schofield (2) considered the dihydroxide to be the dominant product and derived D_0^0 = 215±12 kcal mol^{-1}. Cotton and Jenkins (3) found both SrOH and $Sr(OH)_2$ to be present in significant amounts in fuel-rich hydrogen-oxygen-nitrogen flames and derived D_0^0 = 202.2±5 kcal mol^{-1}. Cotton and Jenkins (3) recalculated the work of Ryabova and Gurvich (1) and of Sugden and Schofield (2) considering both SrOH and $Sr(OH)_2$ to be present and obtained the recalculated D_0^0 values of 196 and 200 kcal mol^{-1}, respectively.

A 3rd law analysis of the experimental equilibrium constants tabulated by Cotton and Jenkins (3) using current JANAF auxiliary data (4) leads to D_0^0 = 204.8 kcal mol^{-1} which is 2.6 kcal mol^{-1} higher than the 202.2 kcal mol^{-1} derived by Cotton and Jenkins (3). Applying this difference to the data of Ryabova and Gurvich (1) and of Sugden and Schofield (2) as recalculated by Cotton and Jenkins (3) gives D_0^0 = 198.6 and 202 kcal mol^{-1}, respectively.

For $Ba(OH)_2(g)$ (4) the corrected dissociation energy of Ryabova and Gurvich (4), D_0^0 = 208.8 kcal mol^{-1}, is in better agreement with the "adopted" value of 209.6 kcal mol^{-1}, based on good Knudsen cell mass spectrometric measurements, than are the corrected dissociation energies of Sugden and Schofield or Cotton and Jenkins (4). We adopt D_0^0 = 198.6 kcal mol^{-1} for the dissociation of $Sr(OH)_2$ from which $\Delta_f H°(Sr(OH)_2$ = -140.66±6.0 kcal mol^{-1} is calculated.

The enthalpy of dissociation listed by Jackson (8) leads to $\Delta_f H°(Sr(OH)_2$, g, 298.15 K) = -143.82 kcal mol^{-1}. Another recent compilation (9) lists $\Delta_f H°$(298.15 K) = -135 kcal mol^{-1}.

Heat Capacity and Entropy

The analogy between gaseous mono- and dihydroxides and gaseous mono- and dihalides, particularly the mono- and difluorides, has been recognized (5-8). The O-Sr-O bond angle is assumed to be the same as the F-Sr-F bond angle (4); the Sr-O-H bond angle is considered to be linear as in SrOH (4). The Sr-O bond distance is estimated to be slightly larger, 0.02 Å, than the Sr-F distance in SrF_2 (4) after noting the close similarity in the bond distance of the alkali metal fluorides and hydroxides. The O-H bond distance is taken to be the same as in water (4).

The vibrational frequencies are assumed to be the same as in SrF_2(4) (O-Sr-O symmetrical and asymmetrical stretch, and bend) and as in SrOH(4) (O-H stretch and Sr-O-H bend). The principal moments of inertia are: $I_A = 7.3880 \times 10^{-39}$, $I_B = 19.3556 \times 10^{-39}$, and $I_C = 26.7436 \times 10^{-39}\ g\ cm^2$.

Jackson (8) has used a different molecular configuration and different vibrational frequencies to estimate S°(298.15 K) = 74.057 cal K^{-1} mol^{-1}. We assign an uncertainty of ±2.0 cal K^{-1} mol^{-1} to the adopted entropy.

References

1. V. G. Ryabova and L. V. Gurvich, Teplofiz. Vysokikh Temperatur, Akad. Nauk SSSR 3, 318 (1965).
2. T. M. Sugden and K. Schofield, Trans. Faraday Soc. 62, 566 (1966).
3. D. H. Cotton and D. R. Jenkins, Trans. Faraday Soc. 64, 2988 (1968).
4. JANAF Thermochemical Tables: Sr(g), HO(g), 12-31-70; $H_2O(g)$, 31-31-61; H(g), 6-30-74; SrOH(g), $Sr(OH)_2(g)$, $Ba(OH)_2(g)$, 12-31-75; $SrF_2(g)$, 31-31-72.
5. R. C. Schoonmaker and R. F. Porter, J. Chem. Phys. 31, 830 (1959).
6. J. Berkowitz, D. J. Meschi and W. A. Chupka, J. Chem. Phys. 33, 533 (1960).
7. F. E. Stafford and J. Berkowitz, J. Chem. Phys. 33, 2963 (1964).
8. D. D. Jackson, Lawrence Livermore Laboratory, Univ. of Calif., UCRL-51137, Dec. 8, 1971.
9. U. S. Nat. Bur. Stand. Tech. Note 270-6, 1971.

Strontium Hydroxide ($Sr(OH)_2$) $H_2O_2Sr_1(g)$

Enthalpy Reference Temperature = T_r = 298.15 K Standard State Pressure = p° = 0.1 MPa

	J $K^{-1}mol^{-1}$			kJ mol^{-1}			
T/K	$C_p°$	S°	$-[G°-H°(T_r)]/T$	$H°-H°(T_r)$	$\Delta_f H°$	$\Delta_f G°$	Log K_f
0	0.	0.	INFINITE	-16.235	-588.313	-588.313	INFINITE
100	44.263	239.906	364.080	-12.417	-591.898	-585.681	305.928
200	64.366	276.853	311.694	-6.968	-594.542	-578.292	151.034
250	71.457	292.025	306.273	-3.562	-595.293	-574.142	119.960
298.15	76.209	305.042	305.042	0.	-595.797	-570.013	99.864
300	76.360	305.514	305.044	0.141	-595.813	-569.853	99.220
350	79.769	317.557	305.987	4.049	-596.183	-565.496	84.396
400	82.195	328.375	308.121	8.102	-596.458	-561.092	73.271
450	83.972	338.164	310.924	12.258	-596.681	-556.658	64.615
500	85.321	347.085	314.101	16.492	-596.878	-552.200	57.688
600	87.261	362.823	320.945	25.127	-597.261	-543.229	47.292
700	88.697	376.386	327.918	33.928	-597.687	-534.190	39.862
800	89.938	388.313	334.737	42.860	-598.183	-525.087	34.285
900	91.109	398.974	341.293	51.913	-599.385	-515.844	29.939
1000	92.247	408.632	347.551	61.081	-599.716	-506.544	26.459
1100	93.355	417.477	353.512	70.361	-607.920	-496.846	23.593
1200	94.420	425.645	359.187	79.751	-609.104	-486.696	21.185
1300	95.430	433.243	364.594	89.244	-610.261	-476.448	19.144
1400	96.378	440.351	369.754	98.835	-611.394	-466.112	17.391
1500	97.259	447.030	374.686	108.517	-612.507	-455.696	15.869
1600	98.071	453.334	379.406	118.284	-613.605	-445.206	14.534
1700	98.817	459.302	383.932	128.129	-751.606	-433.468	13.319
1800	99.498	464.970	388.278	138.045	-750.825	-414.777	12.037
1900	100.120	470.366	392.457	148.027	-750.047	-396.129	10.890
2000	100.687	475.516	396.483	158.068	-749.279	-377.521	9.860
2100	101.203	480.442	400.364	168.162	-748.526	-358.952	8.928
2200	101.674	485.161	404.112	178.307	-747.797	-340.418	8.083
2300	102.102	489.690	407.735	188.496	-747.100	-321.917	7.311
2400	102.493	494.044	411.241	198.726	-746.443	-303.445	6.604
2500	102.851	498.235	414.638	208.993	-745.836	-284.999	5.955
2600	103.177	502.275	417.931	219.295	-745.288	-266.577	5.356
2700	103.476	506.175	421.128	229.628	-744.810	-248.174	4.801
2800	103.751	509.943	424.233	239.989	-744.412	-229.787	4.287
2900	104.003	513.588	427.251	250.377	-744.104	-211.414	3.808
3000	104.235	517.118	430.188	260.789	-743.899	-193.049	3.361
3100	104.449	520.540	433.048	271.224	-743.805	-174.689	2.943
3200	104.646	523.859	435.834	281.679	-743.835	-156.331	2.552
3300	104.828	527.082	438.551	292.152	-743.996	-137.969	2.184
3400	104.997	530.214	441.201	302.644	-744.300	-119.600	1.837
3500	105.153	533.260	443.788	313.151	-744.755	-101.221	1.511
3600	105.299	536.224	446.315	323.674	-745.367	-82.826	1.202
3700	105.434	539.111	448.784	334.211	-746.145	-64.412	0.909
3800	105.560	541.924	451.198	344.760	-747.092	-45.974	0.632
3900	105.677	544.668	453.560	355.322	-748.214	-27.509	0.368
4000	105.787	547.345	455.871	365.896	-749.514	-9.014	0.118
4100	105.890	549.958	458.134	376.480	-750.993	9.517	-0.121
4200	105.986	552.511	460.351	387.073	-752.648	28.086	-0.349
4300	106.076	555.006	462.523	397.676	-754.486	46.696	-0.567
4400	106.160	557.446	464.653	408.288	-756.500	65.351	-0.776
4500	106.239	559.832	466.741	418.908	-758.688	84.053	-0.976
4600	106.314	562.168	468.791	429.536	-760.982	102.801	-1.167
4700	106.385	564.455	470.802	440.171	-763.487	121.606	-1.351
4800	106.451	566.696	472.776	450.813	-766.147	140.466	-1.529
4900	106.513	568.891	474.716	461.461	-768.957	159.383	-1.699
5000	106.573	571.044	476.621	472.115	-771.903	178.358	-1.863
5100	106.629	573.155	478.493	482.775	-774.963	197.391	-2.022
5200	106.681	575.226	480.333	493.441	-778.168	216.487	-2.175
5300	106.732	577.258	482.143	504.112	-781.489	235.648	-2.322
5400	106.779	579.254	483.923	514.787	-784.918	254.870	-2.465
5500	106.824	581.214	485.674	525.467	-788.447	274.157	-2.604
5600	106.867	583.139	487.397	536.152	-792.068	293.511	-2.738
5700	106.908	585.031	489.094	546.841	-795.772	312.929	-2.868
5800	106.947	586.890	490.764	557.534	-799.552	332.414	-2.994
5900	106.984	588.719	492.409	568.230	-803.189	351.937	-3.116
6000	107.019	590.517	494.029	578.930	-807.067	371.549	-3.235

PREVIOUS: December 1975 (1 atm) CURRENT: December 1975 (1 bar)

Strontium Hydroxide ($Sr(OH)_2$) $H_2O_2Sr_1(g)$

SULFURIC ACID (H_2SO_4) CRYSTAL–LIQUID M_r = 98.07348

$S°(298.15\ K) = 156.895 \pm 0.08\ J\ K^{-1}\ mol^{-1}$

$T_{fus} = 283.46 \pm 0.05\ K$

$\Delta_fH°(0\ K) = -811.969 \pm 0.67\ kJ\ mol^{-1}$

$\Delta_fH°(298.15\ K) = -813.989 \pm 0.67\ kJ\ mol^{-1}$

$\Delta_{fus}H° = 10.711 \pm 0.13\ kJ\ mol^{-1}$

Enthalpy of Formation

The enthalpy of combustion, $\Delta_cH°(298.15\ K)$, of rhombic sulfur to aqueous sulfuric acid, S(cr, rhombic) + 3/2 O_2(g) + H_2O(ℓ) = H_2SO_4 (n H_2O), has been determined by many investigators. Based on the reported combustion data, values of $\Delta_fH°(298.15\ K)$ for liquid H_2SO_4 are derived using tabulated enthalpy of solution and enthalpy of dilution data (1, 2). The value adopted, $\Delta_fH°(H_2SO_4$, ℓ, 298.15 K) = -194.548±0.16 kcal mol^{-1}, is that recommended by NBS (2). The uncertainty of ±0.16 kcal mol^{-1} not only accounts for the scatter in the combustion data but also the uncertainty arising from conversion of these data to be consistent with the current atomic weights and fundamental constants (~tens of calories/mol). An uncertainty of -0.08 kcal mol^{-1} may exist due to the difference in the heat of formation of $SO_4^{=}$(aq) as reported by CODATA (11) and NBS (2).

Source	Product	$\Delta_cH°$(298.15 K), kcal mol^{-1}	$\Delta_fH°$(298.15 K), kcal mol^{-1}
McCullough et al. (3)	H_2SO_4 (70 H_2O)	-143.58±0.09	-194.400
Scott et al. (4)	H_2SO_4 (75 H_2O)	-143.98±0.22	-194.775
Good et al. (5)	H_2SO_4 (115 H_2O)	-143.85±0.06	-194.521
Mansson and Sunner (6)	H_2SO_4 (115 H_2O)	-143.93±0.07	-194.601

Heat Capacity and Entropy

The heat capacity has been measured experimentally (7) in the range 15 - 300 K and has been summarized by Giauque et al. (1). Above 300 K, the heat capacity values are derived from a nomograph constructed by Kharbanda (8) based on experimental data (0 - 300°C). These temperature-dependent values are corrected to current atomic weights, converted from IPTS-48 to IPTS-68, and adjusted graphically in the region 290-350 K so as to provide smoothly varying heat capacity values. In the region above 350 K, a linear dependence is assumed (as suggested from a least squares fit of the nomograph data); this linear dependence is used for extrapolation of the heat capacity values to 1000 K.

Other heat capacity studies on sulfuric acid and its aqueous solutions are too numerous to mention. However, a recent compilation (9) summarizes much of the work and shows that our adopted heat capacity values for sulfuric acid and its hydrates from room temperature to ~80°C are consistent with the available literature.

The entropy at 298.15 K is calculated by the appropriate integration of the crystal and liquid heat capacity values, S°(15 K) = 0.269 cal K^{-1} mol^{-1} and addition of the entropy of melting.

Fusion Data

The melting point and enthalpy of melting have been measured calorimetrically by Rubin and Giauque (7). The value for T_{fus} used by Giauque et al. (1) in their re-evaluation and correlation of aqueous sulfuric acid properties was slightly different; the change was 0.07 K. Many other studies have also reported melting data but they are too numerous to mention.

Vaporization Data

T_{vap} = 610 K is the temperature at which the fugacity of H_2SO_4(g) is one atm for the reaction H_2SO_4(ℓ) = H_2SO_4(g). $\Delta_{vap}H°$ = 13.9 kcal mol^{-1} is calculated as the difference in the respective $\Delta_fH°$ values at T_{vap}.

According to Luchinskii (10), the total pressure over H_2SO_4(ℓ) reaches one atmosphere at 553 K with a vapor composition of 76.6 mole% SO_3, 22.6 mole% H_2SO_4, and 0.8 mole% H_2O.

References

1. W. F. Giauque, E. W. Horning, J. E. Kunzler, and T. R. Rubin, J. Amer. Chem. Soc. 82, 62 (1960).
2. U. S. Nat. Bur. Stand. Tech. Note 270-3, 1968.
3. J. P. McCullough, S. Sunner, H. L. Finke, W. N. Hubbard, J. E. Gross, R. E. Pennington, J. F. Messerly, W. D. Good, and G. Waddington, J. Amer. Chem. Soc. 75, 5075 (1953).
4. D. W. Scott, J. P. McCullough, W. N. Hubbard, J. F. Messerly, I. A. Hossenlopp, F. R. Frow, and G. Waddington, J. Amer. Chem. Soc. 78, 5463 (1956).
5. W. D. Good, J. L. Lacina, and J. P. McCullough, J. Amer. Chem. Soc. 82, 5589 (1960).
6. M. Mansson and S. Sunner, Acta Chem. Scand. 17, 723 (1963).
7. T. R. Rubin and W. F. Giauque, J. Amer. Chem. Soc. 74, 800 (1952).
8. Om. P. Kharbanda, Ind. Chemist 31, 538 (1955).
9. Landolt-Bornstein, "Numerical Data and Functional Relationships in Science and Technology," New Series, Group IV, Vol. 1, Part 6, Springer-Verlag, New York, 1977.
10. G. P. Lucinskii, Zh. Foz. Khim. 30, 1207 (1956).
11. ICSU-CODATA Task Group, J. Chem. Thermodyn. 4, 331 (1972).

Sulfuric Acid ($O_2S(OH)_2$) $H_2O_4S_1$(cr,l)

Enthalpy Reference Temperature = T_r = 298.15 K | Standard State Pressure = p° = 0.1 MPa

T/K	$C_p°$ (J $K^{-1}mol^{-1}$)	S° (J $K^{-1}mol^{-1}$)	-[G°-H°(T_r)]/T (J $K^{-1}mol^{-1}$)	H°-H°(T_r) (kJ mol^{-1})	$\Delta_fH°$ (kJ mol^{-1})	$\Delta_fG°$ (kJ mol^{-1})	Log K_f
0	0.	0.	INFINITE	-28.226	-811.969	-811.969	INFINITE
100	45.142	42.586	300.474	-25.789	-819.030	-777.302	406.021
200	73.587	82.684	181.887	-19.841	-823.240	-733.773	191.642
283.456	96.902	112.175	157.073	-12.727	CRYSTAL <--> LIQUID TRANSITION		
283.456	135.770	149.962	157.073	-2.016			
298.15	138.584	156.895	156.895	0.	-813.989	-689.918	120.871
300	138.940	157.753	156.897	0.257	-813.936	-689.149	119.991
400	158.205	200.356	162.576	15.112	-812.526	-647.968	84.616
500	177.619	237.744	173.938	31.903	-808.704	-607.248	63.439
600	197.033	271.844	187.451	50.636	-802.804	-567.476	49.403
700	216.447	303.674	201.804	71.310	-794.924	-528.851	39.463
800	235.860	333.844	216.438	93.925	-785.146	-491.486	32.091
900	255.274	362.745	231.099	118.482	-826.632	-454.377	26.371
1000	274.688	390.646	245.666	144.980	-811.918	-413.788	21.614

PREVIOUS: December 1966 CURRENT: September 1977

Sulfuric Acid ($O_2S(OH)_2$) $H_2O_4S_1$(cr,l)

SULFURIC ACID (H_2SO_4) IDEAL GAS M_r = 98.07348

Sulfuric Acid ($O_2S(OH)_2$) $H_2O_4S_1(g)$

$S°(298.15\ K) = [298.796 \pm 2.1]\ J\ K^{-1}\ mol^{-1}$

$\Delta_fH°(0\ K) = -721.22 \pm 8.4\ kJ\ mol^{-1}$

$\Delta_fH°(298.15\ K) = -735.13 \pm 8.4\ kJ\ mol^{-1}$

Vibrational Frequencies and Degeneracies

ν, cm^{-1}	ν, cm^{-1}	ν, cm^{-1}
[3500](1)	[380](1)	883 (1)
1223 (1)	[265](1)	[400](1)
1138 (1)	[390](1)	1450 (1)
834 (1)	3610 (1)	[265](1)
550 (1)	1159 (1)	568 (1)

Ground State Quantum Weight: [1] σ = 2

Point Group: C_{2v}

Bond Distances: S-O = [1.42] Å S-OH = [1.55] Å O-H = [0.97] Å

Bond Angles: O-S-O = [109.47122]° S-O-H = [105]°

Product of the Moments of Inertia: $I_AI_BI_C = [4.510247 \times 10^{-114}]\ g^3\ cm^6$

Enthalpy of Formation

The vapor phase equilibria of the H_2SO_4, H_2O and SO_3 system have been determined by PVT studies by Bodenstein and Katayama (5) and Suvorov et al. (4) and by vapor pressure measurements by Luchinskii (3). Using the reported equilibrium values (after correcting a few typographical errors) for the reaction $H_2SO_4(g) = H_2O(g) + SO_3(g)$, we calculate $\Delta_fH°(298.15\ K)$ using a 2nd and 3rd law analysis and auxiliary data (6). The results are tabulated below. We adopt $\Delta_fH°(H_2SO_4, g, 298.15\ K) = -175.7\pm2$ kcal mol^{-1} which is the median of the three values derived from the 3rd law $\Delta_rH°(298.15\ K)$. The analysis of vaporization data of Thomas and Ramsey (1) and Abel (2) leads to $\Delta_fH°(298.15\ K)$ values within 2 kcal mol^{-1} of our adopted value.

Source	T/K	Data Points	Reaction*	$\Delta_rH°(298.15\ K)$, kcal mol^{-1} 2nd law	3rd law	Drift cal K^{-1} mol^{-1}	$\Delta_fH°(298.15\ K)$ kcal^{-1} mol^{-1}
(1)	463-493	4	A	18.1±0.1	20.5±0.1	4.9±0.2	-174.0
(2)	523-673	7	A	17.7±0.2	19.7±0.4	3.3±0.3	-174.8
(2)	298-603	9	B	24.2±0.3	23.4±0.5	-1.6±0.1	-175.8
(3)	333-473	8	B	23.7±0.1	23.5±0.1	-0.6±0.3	-175.9
(4)	583-785	40	B	25.1±0.4	23.3±0.3	-2.7±0.7	-175.7
(5)	596-756	53	B	23.3±0.6	23.0±0.5	-0.4±0.9	-175.4

*Reactions: A) $H_2SO_4(\ell) = H_2SO_4(g)$ B) $H_2SO_4(g) = H_2O(g) + SO_3(g)$

Heat Capacity and Entropy

The molecular structure of $H_2SO_4(g)$ is assumed to be tetrahedral with differing bond lengths for S-O and S-OH. This adopted structure is consistent with the X-ray and electron diffraction studies on $H_2SO_4(cr)$ by Pascard (7) and Pascard-Billy (8), respectively, and the X-ray scattering study on $H_2SO_4(\ell)$ by Weidner et al. (9). The latter study gave results which could be interpreted by models having point symmetry C_2 and C_{2v}. In both models, the sulfur-oxygen bond distance are equivalent within the experimental uncertainty. For the C_{2v} model, the O-S-O bond angles coincide with the tetrahedral angle. In the C_2 model these angles are split, but not as extremely as in $H_2SO_4(cr)$ (7, 8). The OH distance and the S-O-H angle are taken from those for $H_2O(g)$. The principal moments of inertia are: $I_A = 14.9802 \times 10^{-39}$, $I_B = 17.0080 \times 10^{-39}$ and $I_C = 17.7023 \times 10^{-39}$ g cm^2.

The vibrational frequencies are from the infrared studies of Chackalackal and Stafford (10) and the comparisons of Giguere and Savoie (11). The infrared studies led to the assignement of nine vibrational frequencies while the remaining six were estimated (10) by comparison with the infrared frequencies for the crystal and Raman frequencies for the liquid.

References

1. J. S. Thomas and A. G. Ramsey, J. Chem. Soc. 123, 3256 (1923).
2. E. ABel, J. Phys. Chem. 50, 260 (1946); J. Phys. Chem. 52, 908 (1948).
3. G. P. Luchinskii, Zh. Fiz. Khim. 30, 1207 (1956).
4. A. V. Suvorov, R. B. Dobrotin, and S. M. Gadzhiev, Russ. J. Inorg. Chem. 10, 710 (1965).
5. B. Bodenstein and M. Katayama, Z. Elektrochem. 15, 244 (1909); Z. Physik. Chem. 69, 26 (1909).
6. JANAF Thermochemical Tables: $H_2SO_4(\ell)$, 9-30-77; $SO_3(g)$, 9-30-65; $H_2O(g)$, 3-31-61.
7. R. Pascard, Compt. Rend. 240, 2162 (1955).
8. C. Pascard-Billy, Acta Cryst. 18, 829 (1965).
9. J. U. Weidner, H. Geisenfelder, and H. Zimmerman, Ber. Bunsenges; Phys. Chem. 75, 800 (1971).
10. S. M. Shackalackal and F. E. Stafford, J. Amer. Chem. Soc. 88, 723 (1966).
11. P. A. Giguere and R. Savoie, J. Amer. Chem. Soc. 85, 287 (1963).

Enthalpy Reference Temperature = T_r = 298.15 K Standard State Pressure = p° = 0.1 MPa

T/K	$C_p°$ (J K^{-1}mol^{-1})	S°	$-[G°-H°(T_r)]/T$	$H°-H°(T_r)$ (kJ mol^{-1})	$\Delta_fH°$	$\Delta_fG°$	Log K_f
0	0.	0.	INFINITE	-16.340	-721.223	-721.223	INFINITE
100	42.100	232.305	360.452	-12.815	-727.196	-704.440	367.962
200	65.850	269.002	305.914	-7.382	-731.922	-679.718	177.524
250	75.588	284.768	300.131	-3.841	-733.711	-666.453	139.248
298.15	83.761	298.796	298.796	0.	-735.129	-653.366	114.467
300	84.054	299.315	298.798	0.155	-735.178	-652.859	113.673
350	91.435	312.840	299.849	4.547	-736.354	-639.042	95.372
400	97.833	325.477	302.271	9.282	-739.496	-624.986	81.615
450	103.344	337.327	305.515	14.315	-740.883	-610.596	70.876
500	108.076	348.466	309.259	19.604	-742.143	-596.048	62.269
600	115.661	368.875	317.527	30.809	-743.771	-566.661	49.332
700	121.411	387.155	326.191	42.675	-744.699	-537.062	40.076
800	125.932	403.674	334.860	55.051	-745.161	-507.364	33.127
900	129.611	418.725	343.355	67.833	-798.420	-476.548	27.658
1000	132.690	432.545	351.592	80.953	-797.084	-440.854	23.028
1100	135.316	445.318	359.540	94.356	-795.609	-405.301	19.246
1200	137.583	457.192	367.188	108.004	-794.025	-369.888	16.101
1300	139.557	468.284	374.543	121.863	-792.355	-334.610	13.445
1400	141.285	478.691	381.614	135.907	-790.620	-299.464	11.173
1500	142.803	488.492	388.416	150.113	-788.838	-264.444	9.209
1600	144.143	497.751	394.963	164.462	-787.023	-229.543	7.494
1700	145.328	506.526	401.269	178.937	-785.188	-194.756	5.984
1800	146.380	514.863	407.351	193.523	-783.344	-160.079	4.645
1900	147.315	522.803	413.220	208.209	-781.499	-125.503	3.450
2000	148.150	530.381	418.890	222.983	-779.661	-91.024	2.377
2100	148.897	537.628	424.373	237.836	-777.837	-56.638	1.409
2200	149.567	544.570	429.680	252.760	-776.031	-22.336	0.530
2300	150.169	551.232	434.821	267.747	-774.248	11.882	-0.270
2400	150.713	557.635	439.805	282.792	-772.491	46.023	-1.002
2500	151.204	563.798	444.643	297.888	-770.764	80.092	-1.673
2600	151.649	569.737	449.340	313.031	-769.067	114.093	-2.292
2700	152.054	575.468	453.906	328.216	-767.402	148.030	-2.864
2800	152.422	581.005	458.347	343.441	-765.771	181.905	-3.393
2900	152.759	586.359	462.670	358.700	-764.174	215.721	-3.886
3000	153.067	591.543	466.879	373.991	-762.612	249.484	-4.344
3100	153.349	596.567	470.982	389.312	-761.086	283.196	-4.772
3200	153.608	601.440	474.983	404.660	-759.595	316.858	-5.172
3300	153.847	606.170	478.887	420.033	-758.140	350.474	-5.548
3400	154.067	610.766	482.699	435.429	-756.720	384.048	-5.900
3500	154.270	615.235	486.422	450.846	-755.336	417.579	-6.232
3600	154.459	619.584	490.061	466.283	-753.987	451.072	-6.545
3700	154.633	623.818	493.619	481.737	-752.674	484.527	-6.840
3800	154.796	627.944	497.100	497.209	-751.396	517.949	-7.120
3900	154.947	631.967	500.507	512.696	-750.154	551.337	-7.384
4000	155.087	635.892	503.842	528.198	-748.947	584.692	-7.635
4100	155.219	639.723	507.110	543.713	-747.776	618.019	-7.874
4200	155.341	643.465	510.312	559.241	-746.641	651.317	-8.100
4300	155.456	647.121	513.451	574.781	-745.542	684.590	-8.316
4400	155.564	650.697	516.530	590.332	-744.480	717.835	-8.522
4500	155.665	654.194	519.551	605.894	-743.454	751.057	-8.718
4600	155.759	657.616	522.515	621.465	-742.467	784.256	-8.906
4700	155.849	660.967	525.425	637.045	-741.517	817.437	-9.085
4800	155.933	664.249	528.283	652.634	-740.607	850.596	-9.256
4900	156.012	667.465	531.091	668.232	-739.736	883.738	-9.421
5000	156.086	670.618	533.850	683.837	-738.906	916.863	-9.578
5100	156.157	673.709	536.562	699.449	-738.117	949.970	-9.730
5200	156.224	676.742	539.229	715.068	-737.371	983.061	-9.875
5300	156.287	679.718	541.852	730.693	-736.668	1016.141	-10.015
5400	156.347	682.640	544.432	746.325	-736.010	1049.206	-10.149
5500	156.403	685.510	546.971	761.963	-735.398	1082.258	-10.278
5600	156.457	688.328	549.470	777.606	-734.833	1115.303	-10.403
5700	156.508	691.098	551.931	793.254	-734.316	1148.336	-10.523
5800	156.557	693.820	554.354	808.907	-733.848	1181.363	-10.639
5900	156.603	696.497	556.740	824.565	-733.432	1214.376	-10.751
6000	156.647	699.129	559.091	840.228	-733.068	1247.389	-10.859

PREVIOUS: September 1977 (1 atm) CURRENT: September 1977 (1 bar)

Sulfuric Acid ($O_2S(OH)_2$) $H_2O_4S_1(g)$

TUNGSTIC ACID ($WO_3 \cdot H_2O$) CRYSTAL M_r = 249.86348 Tungstic Acid ($O_2W(OH)_2$) $H_2O_4W_1(cr)$

$\Delta_f H°(0\ K)$ = Unknown

$S°(298.15\ K) = [144.766 \pm 25.1]\ J\ K^{-1}\ mol^{-1}$ $\Delta_f H°(298.15\ K) = -1131.77 \pm 1.7\ kJ\ mol^{-1}$

Enthalpy of Formation

Spitsyn and Patsukova (1) have measured calorimetrically the enthalpy of reaction (1) and (2) at 298.15 K as -13.80±0.01 and -13.05±0.02 kcal mol^{-1}, respectively. The reactions are given as follows:

(1) $WO_3(cr) + 2\ NaOH(66H_2O) \rightarrow Na_2WO_4(4200\ H_2O) + H_2O(\ell)$

(2) $H_2WO_4(cr) + 2\ NaOH(66H_2O) \rightarrow Na_2WO_4(4200\ H_2O) + 2\ H_2O(\ell)$

Combination of reactions (1) and (2) gives $\Delta_r H°(298.15\ K) = -0.75\pm0.04$ kcal mol^{-1} for $WO_3(cr) + H_2O(\ell) \rightarrow H_2WO_4(cr)$, which leads to $\Delta_f H°(H_2WO_4$, cr, 298.15 K) = -270.5±0.4 kcal mol^{-1}.

To obtain $WO_3(cr)$ in a form soluble in the calorimetric solution, Spitsyn and Patsukova dehydrated $H_2WO_4(cr)$ at the relatively low temperature of 250°C. The resulting $WO_3(cr)$ was a very fine lemon-yellow powder which may have excess surface free energy compared with the JANAF standard state. The latter is WO_3 formed at high temperature in a calorimetric bomb. The actual enthalpy of formation of tungstic acid should be less negative if this effect is significant.

The previously accepted value (2, 3, 4) -280.2±0.4 kcal mol^{-1}, was based on the pressure-temperature-composition studies of Huttig and Kurre (5) for the WO_3-H_2O system. We suspect that the data do not correspond to equilibrium (see entropy section) for the reaction $H_2WO_4(cr) = WO_3(cr) + H_2O(g)$.

Heat Capacity and Entropy

$C_p°(300\ K) = 27.0$ cal K^{-1} mol^{-1} is estimated using Kopp's rule. Since the decomposition temperature of $H_2WO_4(cr)$ is approximately 400 K, which is relatively low for hydrogen and oxygen to reach their maximum $C_p°$ contribution, we estimate $C_p°(400\ K)$ = 31.0 cal K^{-1} mol^{-1} for $H_2WO_4(cr)$.

The entropy, S°(298.15 K) = 34.6 cal K^{-1} mol^{-1}, is estimated so that the decomposition pressure of water is one atmosphere at 120°C. This value is quite uncertain as suggested by comparison with the following estimates. A lower limit may be obtained from S°(283.5 K) = 26.8 cal K^{-1} mol^{-1} for $H_2SO_4(cr)$. Summation of the entropies for $WO_3(cr)$ and $H_2O(cr)$ [10.7 cal K^{-1} mol^{-1}] yields 28.8 cal K^{-1} mol^{-1}, and estimate which is somewhat larger than the lower limit. An upper limit may be obtained by assuming that the pressure of about 14 mm at 190°C corresonds to equilibrium for the reaction $H_2WO_4(cr) = WO_3(cr) + H_2O(g)$. This pressure, based on the plateau in the temperature-composition data of Huttig and Kurre (5), yields S°(298.15 K) = 46.5 cal K^{-1} mol^{-1} when combined with the adopted $\Delta_f H°$(298.15 K) and JANAF auxiliary data. Since the estimate appears to be unreasonably large, we suggest that this region of the curve may correspond to the reaction $WO_3 \cdot 0.5\ H_2O(cr) = WO_3(cr) + 0.5\ H_2O(g)$. In contrast, the observed pressures (5) for samples having mole ratios (H_2O/WO_3) of 1.2 - 1.35 leads to S°(298.15 K) = 34 cal K^{-1} mol^{-1}, if the reaction is $WO_3 \cdot H_2O(cr) = WO_3(cr) + H_2O(g)$.

Decomposition Data

T_{dcm} = 393 K is calculated as the temperature at which the fugucity of water is 1 atmosphere for the reaction $H_2WO_4(cr) \rightarrow WO_3(cr) + H_2O(g)$ approaches zero.

Spitsyn and Patsukova (1) observed that some dehydration of stoichiometric $H_2WO_4(cr)$ occurs at temperatures as low as 90°C. Thermal studies (6, 7) on tungstic acid containing some exces water indicated the start of dehydration somewhere in the region of 60 - 120°C. These temperatures yield a lower limit for T_{dec} on the presumption that the experiments correspond to a partial pressure of water of less than one atmosphere.

References

1. V. I. Spitsyn and N. N. Patsukova, Russ. J. Inorg. Chem. 10, 1304 (1965).
2. D. D. Wagman, U. S. Nat. Bur. Stand. Report 8919, 1965.
3. M. F. Kohler, L. B. Parkratz and R. Barany, U. S. Bur. Mines RI 5973, 1962.
4. U. S. Nat. Bur. Stand. Circ. 500, 1952.
5. G. F. Huttig and B. Kurre, Z. Anorg. Chem. 122, 44 (1922); 126, 167 (1923).
6. W. Wanek, Silikaty 6, 70 (1962).
7. P. F. Kerr and F. Young, Amer. Mineral. 29, 192 (1944).

Enthalpy Reference Temperature = T_r = 298.15 K — Standard State Pressure = p° = 0.1 MPa

T/K	$C_p°$ (J K⁻¹mol⁻¹)	S°	$-[G°-H°(T_r)]/T$	$H°-H°(T_r)$ (kJ mol⁻¹)	$\Delta_f H°$	$\Delta_f G°$	Log K_f
0							
100							
200							
298.15	112.633	144.766	144.766	0.	-1131.772	-1003.905	175.880
300	112.968	145.464	144.769	0.209	-1131.770	-1003.112	174.657
400	129.704	180.255	149.399	12.342	-1130.949	-960.310	125.404
500	146.440	211.003	158.697	26.153	-1128.693	-917.882	95.890
600	159.829	238.972	169.783	41.513	-1125.138	-876.036	76.266
700	170.707	264.509	181.518	58.094	-1120.606	-834.865	62.298
800	176.565	287.724	193.367	75.485	-1115.486	-794.391	51.868
900	182.422	308.889	205.044	93.461	-1109.985	-754.580	43.795
1000	186.606	328.326	216.414	111.912	-1104.195	-715.399	37.369

PREVIOUS: June 1963 CURRENT: March 1967

Tungstic Acid ($O_2W(OH)_2$) $H_2O_4W_1(cr)$

TUNGSTIC ACID ($O_2W(OH)_2$) IDEAL GAS M_r = 249.86348

$S°(298.15 K) = [351.999] J K^{-1} mol^{-1}$

$\Delta_fH°(0 K) = -896.09 \pm 4.2 kJ mol^{-1}$
$\Delta_fH°(298.15 K) = -905.84 \pm 4.2 kJ mol^{-1}$

Vibrational Frequencies and Degeneracies

ν, cm^{-1}	ν, cm^{-1}	ν, cm^{-1}
[2200]	[100]	[570]
[1000]	[225]	[120]
[620]	[100]	[1100]
[560]	[2500]	[300]
[500]	[630]	[540]

Ground State Quantum Weight: 1 σ = [2]
Point Group: [C_{2v}]
Bond Distances: W-O = [1.81] Å O-H = [0.97] Å
Bond Angles: O-W-O = [109.47]° W-O-H = [105]°
Product of the Moments of Inertia: $I_AI_BI_C = [1.442326 \times 10^{-113}] g^3 cm^6$

Enthalpy of Formation

The adopted enthalpy of formation, $\Delta_fH°(298.15 K) = -216.5\pm1.0 kcal mol^{-1}$, is derived from equilibrium data (1, 2, 3, 4, 5) obtained by the transpiration method. The data are in reasonable agreement and they are summarized below. Belton and McCarron (1) have concluded that $WO_3 \cdot H_2O$ is the species responsible for the increased volatilization of W(cr) and WO_3(cr) in the presence of water vapor. Based on their data for W(cr) and the earlier data for WO_3(cr), Belton and McCarron (1) ruled out the possibility of significant amounts of $(WO_3)_2 \cdot H_2O$ and $WO_3 \cdot 2 H_2O$.

Source	Reaction*	T/K	Data Points	$\Delta_rH°$(298.15 K), kcal mol^{-1} 2nd law	3rd law	Drift cal K^{-1} mol^{-1}	$\Delta_fH°$(298.15 K) kcal mol^{-1}
Belton & McCarron (1)	A	1473-1773	Equation	14.70	14.63	-0.1±1.1	-216.56
Glemser & Ackermann(2)	B	673-1173	6	38.56	41.67	3.2±0.9	-217.59
Glemser & Haeseler (3)	B	1173-1373	Equation	38.27	42.27	1.6	-216.99
Glemser & Volz (4)	B	1373	1	-	43.44	-	-215.82
Meyer et al. (5)	B	1393-1518	Equation	26.14	43.51	11.9	-215.75

*Reactions: A) $W(cr) + 4H_2O(g) \rightarrow H_2WO_4(g) + 3H_2(g)$ B) $WO_3(cr) + H_2O(g) \rightarrow H_2WO_4(g)$

Heat Capacity and Entropy

The vibrational frequencies are estimated by comparison with those for H_2SO_4(g) reported by Giguere and Savoie (6), and Chackalackal and Stafford (7). These frequencies are also adjusted so that $S°(H_2WO_4$, g, 1600 K) is 136.7 cal K^{-1} mol^{-1} as reported by Belton and McCarron (1). This entropy was derived a 2nd law analysis of the equilibrium constants for reaction A.

The configuration of WO_4^- group is assumed to be tetrahedral. The bond angle W-O-H is estimated to be the same as the angle of H_2O(g). The bond distances are assumed to be the same as those in WO_3(g) and H_2O(g). The principal moments of inertia are: $I_A = 23.2705 \times 10^{-39}$, $I_B = 24.8660 \times 10^{-39}$ and $I_C = 24.9259 \times 10^{-39}$ g cm^2.

References

1. G. R. Belton and R. L. McCarron, J. Phys. Chem. 68, 1852 (1964).
2. O. Glemser and H. Ackermann, Z. Anorg. Allg. Chem. 325, 281 (1963).
3. O. Glemser and R. V. Haeseler, Z. Anorg. Allg. Chem. 316, 168 (1962).
4. O. Glemser and H. G. Volz, Naturwissenschaften 43, 33 (1956).
5. G. Meyer, J. F. Oosterom and W. J. V. Oeveren, Rev. Trav. Chim. 78, 417 (1959).

Tungstic Acid ($O_2W(OH)_2$) $H_2O_4W_1$(g)

Enthalpy Reference Temperature = T_r = 298.15 K; Standard State Pressure = p° = 0.1 MPa

T/K	$C_p°$ (J $K^{-1}mol^{-1}$)	S°	$-[G°-H°(T_r)]/T$	$H°-H°(T_r)$ (kJ mol^{-1})	$\Delta_fH°$	$\Delta_fG°$	Log K_f
0	0.	0.	INFINITE	-21.062	-896.091	-896.091	INFINITE
100	60.677	265.197	430.296	-16.510	-901.000	-881.824	460.618
200	84.476	314.672	360.846	-9.235	-904.248	-861.251	224.936
250	94.555	334.639	353.644	-4.751	-905.233	-850.380	177.677
298.15	102.527	351.999	351.999	0.	-905.836	-839.756	147.122
300	102.802	352.634	352.001	0.190	-905.853	-839.346	146.143
350	109.449	368.999	353.278	5.502	-906.167	-828.233	123.607
400	114.836	383.977	356.193	11.114	-906.241	-817.091	106.701
450	119.269	397.767	360.056	16.970	-906.131	-805.953	93.553
500	122.986	410.531	364.473	23.029	-905.881	-794.834	83.036
600	128.916	433.504	374.107	35.638	-905.077	-772.694	67.269
700	133.483	453.733	384.066	48.767	-903.996	-750.713	56.019
800	137.120	471.804	393.924	62.304	-902.731	-728.900	47.592
900	140.070	488.130	403.499	76.168	-901.341	-707.253	41.048
1000	142.491	503.017	412.717	90.300	-899.871	-685.766	35.821
1100	144.492	516.695	421.557	104.652	-898.355	-664.429	31.551
1200	146.159	529.341	430.018	119.187	-896.820	-643.230	27.999
1300	147.555	541.097	438.116	133.875	-895.289	-622.160	24.999
1400	148.733	552.076	445.868	148.691	-893.780	-601.207	22.431
1500	149.733	562.373	453.296	163.616	-892.309	-580.361	20.210
1600	150.587	572.064	460.419	178.633	-890.887	-559.611	18.269
1700	151.320	581.216	467.258	193.729	-889.525	-538.948	16.560
1800	151.954	589.884	473.832	208.894	-888.231	-518.364	15.043
1900	152.504	598.114	480.158	224.117	-887.011	-497.850	13.687
2000	152.985	605.949	486.253	239.392	-885.872	-477.397	12.468
2100	153.407	613.424	492.132	254.712	-884.818	-457.000	11.367
2200	153.779	620.569	497.809	270.072	-883.853	-436.649	10.367
2300	154.109	627.412	503.297	285.467	-882.979	-416.342	9.455
2400	154.402	633.978	508.606	300.892	-882.199	-396.072	8.620
2500	154.664	640.286	513.748	316.346	-881.513	-375.831	7.853
2600	154.898	646.357	518.732	331.824	-880.921	-355.615	7.144
2700	155.109	652.207	523.568	347.325	-880.448	-335.420	6.489
2800	155.300	657.851	528.263	362.846	-880.131	-315.240	5.881
2900	155.472	663.304	532.826	378.384	-880.008	-295.069	5.315
3000	155.628	668.577	537.264	393.939	-880.106	-274.897	4.786
3100	155.771	673.683	541.583	409.509	-880.467	-254.718	4.292
3200	155.901	678.630	545.789	425.093	-881.123	-234.524	3.828
3300	156.020	683.429	549.887	440.689	-882.098	-214.304	3.392
3400	156.129	688.089	553.884	456.297	-883.426	-194.048	2.981
3500	156.229	692.616	557.783	471.915	-885.171	-173.748	2.593
3600	156.321	697.018	561.590	487.542	-887.428	-153.391	2.226
3700	156.407	701.303	565.308	503.179	-925.062	-132.773	1.874
3800	156.486	705.475	568.942	518.823	-925.055	-111.359	1.531
3900	156.559	709.540	572.495	534.476	-925.087	-89.945	1.205
4000	156.627	713.505	575.971	550.135	-925.157	-68.531	0.895
4100	156.690	717.373	579.373	565.801	-925.266	-47.114	0.600
4200	156.749	721.150	582.704	581.473	-925.412	-25.694	0.320
4300	156.804	724.839	585.967	597.151	-925.595	-4.269	0.052
4400	156.856	728.444	589.164	612.834	-925.816	17.158	-0.204
4500	156.904	731.970	592.299	628.522	-926.073	38.591	-0.448
4600	156.949	735.419	595.372	644.214	-926.368	60.031	-0.682
4700	156.991	738.795	598.388	659.911	-926.699	81.479	-0.906
4800	157.031	742.100	601.348	675.612	-927.068	102.933	-1.120
4900	157.068	745.339	604.254	691.317	-927.475	124.397	-1.326
5000	157.103	748.512	607.107	707.026	-927.920	145.869	-1.524
5100	157.137	751.624	609.910	722.738	-928.403	167.349	-1.714
5200	157.168	754.675	612.665	738.453	-928.926	188.838	-1.897
5300	157.198	757.669	615.373	754.172	-929.488	210.340	-2.073
5400	157.226	760.608	618.035	769.893	-930.092	231.851	-2.243
5500	157.252	763.493	620.654	785.617	-930.737	253.373	-2.406
5600	157.278	766.327	623.230	801.343	-931.425	274.911	-2.564
5700	157.301	769.111	625.765	817.072	-932.156	296.458	-2.717
5800	157.324	771.847	628.260	832.803	-932.933	318.021	-2.864
5900	157.346	774.536	630.716	848.537	-933.755	339.592	-3.007
6000	157.366	777.181	633.136	864.272	-934.625	361.184	-3.144

PREVIOUS: March 1967 (1 atm) CURRENT March 1967 (1 bar)

Tungstic Acid ($O_2W(OH)_2$) $H_2O_4W_1$(g)

PHOSPHINO (PH_2) IDEAL GAS M_r = 32.98964

$S°(298.15\ K)$ = [212.652] $J\ K^{-1}\ mol^{-1}$

$\Delta_f H°(0\ K)$ = [128.63 ± 96.2] $kJ\ mol^{-1}$
$\Delta_f H°(298.15\ K)$ = [125.94 ± 96.2] $kJ\ mol^{-1}$

Vibrational Frequencies and Degeneracies

ν, cm^{-1}
[2327](1)
[1020](1)
[2322](1)

Ground State Quantum Weight: [2] σ = [2]
Point Group = [C_{2v}]
Bond Distance: P-H = [1.419] Å
Bond Angle: H-P-H = [93.7]°
Product of the Moments of Inertia: $I_A I_B I_C$ = [0.0695×10^{-117}] $g^3\ cm^6$

Enthalpy of Formation

$\Delta_f H°(298.15\ K)$ was estimated by DiStefano et al. (1). The $\Delta_f H°(298.15\ K)$ was estimated by taking 63% of the dissociation energy of PH_3 and using it as the dissociation energy of PH_2. Willes and Winkler (2) have estimated $\Delta_f H°(PH_2$, g, 298.15 K) to be 30 kcal mol^{-1}.

Heat Capacity and Entropy

The vibratonal spectra of PH_2 from the flash photolysis of PH_3 has been reported by Ramsey (3) and in (4). Similarities between the spectra of NH_2 and PH_2 were found. For NH_2 and NH_3 bond angles and bond distances were reported to be about the same. By analogy then the bond angles and bond distances of PH_2 and PH_3 should be nearly the same. The moments of inertia for PH_2 were therefore calculated from the molecular constants for PH_3. The principal moments of inertia are: $I_A = 0.2960 \times 10^{-39}$, $I_B = 0.3588 \times 10^{-39}$ and $I_C = 0.6548\ g\ cm^2 \times 10^{-39}$.

DiStefano et al. (1) have assumed the ground state degeneracy to be 2 because of the unpaired electron. The vibrational frequencies reported by DiStefano et al. have been used. They used a simple force field and the force constants of PH_3 to obtain the vibrational frequencies.

References

1. V. N. DiStefano, R. L. Potter and S. N. Fox, "The Thermodynamic Functions of Some Combustion Products Containing Phosphorus-I.", American Cyanamid Company.
2. D. Willes and C. Winkler, J. Phys. Chem. 61, 902 (1957).
3. D. A. Ramsey, Nature 178, 375 (1956).
4. N. Y. Acad. Sci. 67, 485 (1957).

Phosphino (PH_2) $H_2P_1(g)$

Enthalpy Reference Temperature = T_r = 298.15 K Standard State Pressure = p° = 0.1 MPa

	J $K^{-1}mol^{-1}$			kJ mol^{-1}			
T/K	$C_p°$	S°	$-[G°-H°(T_r)]/T$	$H°-H°(T_r)$	$\Delta_f H°$	$\Delta_f G°$	Log K_f
0	0.	0.	INFINITE	-10.006	128.633	128.633	INFINITE
100	33.258	175.957	242.760	-6.680	127.460	124.974	-65.280
200	33.549	199.055	215.787	-3.347	126.753	122.766	-32.063
250	34.075	206.593	213.221	-1.657	126.348	121.815	-25.452
298.15	34.774	212.652	212.652	0.	125.938	120.979	-21.195
300	34.804	212.868	212.653	0.064	125.923	120.949	-21.059
350	35.641	218.294	213.080	1.825	125.510	120.152	-17.932
400	36.535	223.112	214.038	3.629	125.129	119.414	-15.594
450	37.468	227.468	215.292	5.479	124.788	118.720	-13.781
500	38.430	231.466	216.712	7.377	124.492	118.062	-12.334
600	40.400	238.646	219.783	11.318	124.040	116.822	-10.170
700	42.343	245.020	222.941	15.456	123.771	115.642	-8.629
800	44.169	250.796	226.067	19.783	123.669	114.490	-7.475
900	45.820	256.095	229.113	24.284	123.708	113.342	-6.578
1000	47.278	261.000	232.060	28.940	123.859	112.183	-5.860
1100	48.548	265.567	234.901	33.733	124.093	111.004	-5.271
1200	49.645	269.839	237.636	38.644	124.387	109.802	-4.780
1300	50.590	273.851	240.269	43.657	124.719	108.573	-4.363
1400	51.403	277.631	242.804	48.758	125.073	107.318	-4.004
1500	52.104	281.202	245.246	53.934	125.437	106.037	-3.693
1600	52.711	284.584	247.600	59.175	125.802	104.732	-3.419
1700	53.238	287.796	249.871	64.473	126.160	103.404	-3.177
1800	53.696	290.852	252.063	69.821	126.506	102.056	-2.962
1900	54.097	293.767	254.182	75.211	126.837	100.688	-2.768
2000	54.450	296.551	256.231	80.638	127.150	99.304	-2.594
2100	54.760	299.215	258.215	86.099	127.442	97.904	-2.435
2200	55.035	301.769	260.137	91.589	127.714	96.491	-2.291
2300	55.280	304.221	262.001	97.105	127.964	95.066	-2.159
2400	55.498	306.578	263.810	102.644	128.191	93.631	-2.038
2500	55.693	308.848	265.566	108.204	128.396	92.187	-1.926
2600	55.869	311.036	267.273	113.782	128.577	90.735	-1.823
2700	56.027	313.147	268.933	119.377	128.736	89.276	-1.727
2800	56.169	315.187	270.549	124.987	128.872	87.812	-1.638
2900	56.299	317.161	272.122	130.611	128.984	86.344	-1.555
3000	56.417	319.071	273.656	136.247	129.074	84.872	-1.478
3100	56.524	320.923	275.151	141.894	129.142	83.397	-1.405
3200	56.622	322.719	276.609	147.551	129.186	81.920	-1.337
3300	56.712	324.463	278.033	153.218	129.208	80.443	-1.273
3400	56.795	326.157	279.424	158.893	129.207	78.965	-1.213
3500	56.871	327.804	280.783	164.577	129.183	77.488	-1.156
3600	56.941	329.408	282.111	170.267	129.136	76.011	-1.103
3700	57.005	330.969	283.411	175.965	129.066	74.537	-1.052
3800	57.065	332.490	284.682	181.668	128.974	73.064	-1.004
3900	57.121	333.973	285.927	187.377	128.858	71.594	-0.959
4000	57.172	335.419	287.146	193.092	128.719	70.128	-0.916
4100	57.220	336.832	288.341	198.812	128.558	68.665	-0.875
4200	57.265	338.211	289.512	204.536	128.373	67.207	-0.836
4300	57.307	339.559	290.660	210.265	128.165	65.753	-0.799
4400	57.346	340.877	291.787	215.997	127.935	64.304	-0.763
4500	57.383	342.166	292.892	221.734	127.682	62.860	-0.730
4600	57.417	343.428	293.977	227.474	127.407	61.422	-0.697
4700	57.449	344.663	295.042	233.217	127.109	59.991	-0.667
4800	57.480	345.873	296.089	238.964	126.790	58.567	-0.637
4900	57.508	347.058	297.117	244.713	126.450	57.149	-0.609
5000	57.535	348.220	298.127	250.465	126.089	55.739	-0.582
5100	57.560	349.360	299.121	256.220	125.708	54.335	-0.557
5200	57.584	350.478	300.098	261.977	125.309	52.939	-0.532
5300	57.607	351.575	301.059	267.737	124.890	51.552	-0.508
5400	57.628	352.652	302.004	273.499	124.455	50.172	-0.485
5500	57.649	353.710	302.935	279.263	124.003	48.800	-0.463
5600	57.668	354.749	303.851	285.028	123.535	47.437	-0.442
5700	57.686	355.769	304.753	290.796	123.054	46.083	-0.422
5800	57.703	356.773	305.641	296.566	122.559	44.737	-0.403
5900	57.720	357.759	306.516	302.337	122.053	43.399	-0.384
6000	57.736	358.730	307.378	308.110	121.537	42.070	-0.366

PREVIOUS: March 1963 (1 atm) CURRENT: March 1963 (1 bar)

Phosphino (PH_2) $H_2P_1(g)$

HYDROGEN SULFIDE (H_2S) IDEAL GAS M_r = 34.07588

$S°(298.15\ K) = 205.757\ J\ K^{-1}\ mol^{-1}$

$\Delta_fH°(0\ K) = -17.584 \pm 0.8\ kJ\ mol^{-1}$
$\Delta_fH°(298.15\ K) = -20.502 \pm 0.8\ kJ\ mol^{-1}$

Vibrational Frequencies and Degeneracies

ν, cm^{-1}
2614.6(1)
1182.7(1)
2626.5(1)

Ground State Quantum Weight: 1 $\sigma = 2$

Point Group: C_{2v}

Bond Distance: S-H = 1.328 Å

Bond Angle: H-S-H = 92.2°

Product of the Moments of Inertia: $I_AI_BI_C = 0.0469 \times 10^{-117}\ g^3\ cm^6$

Enthalpy of Formation

The enthalpy of formation of $H_2S(g)$ determined by various workers including those reviewed in the previous JANAF analysis (1) were recalculated with the results shown below. No weight was given to the work performed prior to 1934 and less weight was given to the work of Kapustinskii and Kankovskii (7) because of the difficulty of establishing the state of the products.

Source	Reaction*	[from (1)]	$\Delta_fH°(298.15\ K)$, kcal mol^{-1} Recalculated
Thomsen (2)	$H_2S(g) + HI_3(aq) = 3\ HI(aq) + S(amorph)$	----	-4.4, -5.1 if S(rhombic)
Thomsen (3)	$^*H_2S(g) + O_2(g) = SO_2(g) + H_2SO_4 \cdot H_2O$	----	-2.0
Pollitzer (4)	$H_2S(g) + I_2(cr) = 2\ HI(g) + S(rh)$	----	-4.7, -4.0 if S(amorph)
Lewis and Randall (5)	$3\ S(\ell) + 2\ H_2O(g) = SO_2(g) + 2\ H_2S(g)$	----	-3.6±0.2
Zeumer and Roth (6)	$^*H_2S(g) + O_2(g) = SO_2(g) + H_2SO_4 \cdot H_2O$	-4.80±0.15	-4.89±0.24
Kapustinskii and Kankovskii (7)	$^*H_2S(g) + 3/2\ O_2(g) = SO_2(g) + H_2O(\ell) + S(rh)$	-4.9 ±0.8	-4.92±0.08

*Equations which are not balanced involve two reactions occurring in various proportions.

Heat Capacity and Entropy

The thermodynamic functions were taken from the JANAF table for $H_2S(g)$ dated Dec. 31, 1965 (1). These in turn were taken from Gordon (8) except below 298 K were they were calculated by the rigid-rotor, harmonic-oscillator approximation. Gordon had calculated $C_p°$ from 298 K to 6000 K by a method which takes into account second-order corrections for vibrational anharmonicity, vibration-rotation interaction, and centrifugal stretching. The spectroscopic constants used were taken from Allen and Plyler (9).

The National Bureau of Standards prepared this table (10) by critical analysis of data existing in 1972. Using $\Delta_fH°$ selected by NBS (10), we recalculate the table in terms of current JANAF reference states for the elements.

References

1. JANAF Thermochemical Tables, 2nd ed., NSRDS-NBS 37, 1971.
2. J. Thomsen, "Thermochemical Investigations," Vol. II, p. 60, J. A. Barth, Leipzig, 1882.
3. J. Thomsen, "Thermochemical Investigations," Vol. IV, p. 188, J. A. Barth, Leipzig, 1882.
4. F. Pollitzer, Z. Anorg. Chem. 64, 121 (1909).
5. G. N. Lewis and M. Randall, J. Amer. Chem. Soc. 40, 362 (1918).
6. H. Zeumer and W. A. Roth, Z. Elektrochem. 40, [11], 777 (1934).
7. A. F. Kapustinskii and R. T. Kankovskii, Zhur. Fiz. Khim. 32, 2810 (1958).
8. J. S. Gordon, Thiokol Chem. Corp., personal communication to D. R. Stull, Feb. 7, 1961.
9. H. C. Allen and E. K. Plyler, J. Chem. Phys. 25, 1132 (1956); 24, 35 (1956).
10. S. Abramowitz et al., U. S. Nat. Bur. Stand. Rept. 10904, 239, July, 1972.

Hydrogen Sulfide (H_2S) $H_2S_1(g)$

Enthalpy Reference Temperature = T_r = 298.15 K Standard State Pressure = p° = 0.1 MPa

T/K	$C_p°$ (J $K^{-1}mol^{-1}$)	S° (J $K^{-1}mol^{-1}$)	$-[G°-H°(T_r)]/T$ (J $K^{-1}mol^{-1}$)	$H°-H°(T_r)$ (kJ mol^{-1})	$\Delta_fH°$ (kJ mol^{-1})	$\Delta_fG°$ (kJ mol^{-1})	Log K_f
0	0.	0.	INFINITE	-9.962	-17.584	-17.584	INFINITE
100	33.259	168.972	235.330	-6.636	-17.947	-23.519	12.285
200	33.380	192.038	208.586	-3.310	-18.957	-28.754	7.510
298.15	34.192	205.757	205.757	0.	-20.502	-33.329	5.839
300	34.208	205.969	205.758	0.063	-20.534	-33.408	5.817
400	35.581	215.992	207.115	3.551	-24.549	-37.343	4.876
500	37.192	224.102	209.726	7.188	-27.762	-40.179	4.197
600	38.936	231.037	212.713	10.994	-30.470	-42.399	3.691
700	40.740	237.174	215.777	14.978	-32.771	-44.201	3.298
800	42.518	242.731	218.804	19.142	-34.771	-45.694	2.984
900	44.212	247.838	221.750	23.479	-89.665	-45.870	2.662
1000	45.786	252.579	224.599	27.980	-90.024	-40.984	2.141
1100	47.200	257.011	227.346	32.631	-90.284	-36.066	1.713
1200	48.467	261.173	229.993	37.415	-90.465	-31.129	1.355
1300	49.593	265.098	232.544	42.320	-90.584	-26.179	1.052
1400	50.593	268.810	235.003	47.330	-90.656	-21.222	0.792
1500	51.476	272.331	237.375	52.434	-90.693	-16.261	0.566
1600	52.262	275.679	239.666	57.622	-90.705	-11.298	0.369
1700	52.961	278.869	241.879	62.883	-90.699	-6.336	0.195
1800	53.589	281.914	244.019	68.212	-90.681	-1.374	0.040
1900	54.145	284.827	246.091	73.599	-90.656	3.587	-0.099
2000	54.656	287.617	248.098	79.039	-90.627	8.547	-0.223
2100	55.107	290.295	250.044	84.528	-90.597	13.505	-0.336
2200	55.522	292.868	251.932	90.059	-90.566	18.461	-0.438
2300	55.898	295.345	253.766	95.631	-90.538	23.416	-0.532
2400	56.250	297.731	255.549	101.238	-90.512	28.370	-0.617
2500	56.568	300.034	257.282	106.879	-90.490	33.323	-0.696
2600	56.861	302.258	258.970	112.551	-90.472	38.275	-0.769
2700	57.133	304.409	260.613	118.251	-90.458	43.227	-0.836
2800	57.384	306.492	262.214	123.977	-90.450	48.178	-0.899
2900	57.618	308.510	263.776	129.727	-90.447	53.129	-0.957
3000	57.840	310.467	265.300	135.500	-90.450	58.080	-1.011
3100	58.049	312.367	266.788	141.294	-90.459	63.031	-1.062
3200	58.245	314.213	268.241	147.109	-90.473	67.982	-1.110
3300	58.430	316.008	269.662	152.943	-90.495	72.935	-1.154
3400	58.605	317.755	271.051	158.795	-90.524	77.887	-1.197
3500	58.773	319.456	272.409	164.664	-90.560	82.841	-1.236
3600	58.932	321.114	273.739	170.549	-90.604	87.796	-1.274
3700	59.086	322.731	275.042	176.450	-90.656	92.752	-1.309
3800	59.229	324.308	276.317	182.366	-90.717	97.710	-1.343
3900	59.371	325.849	277.568	188.296	-90.788	102.670	-1.375
4000	59.509	327.354	278.794	194.240	-90.868	107.631	-1.406
4100	59.639	328.825	279.996	200.197	-90.958	112.595	-1.434
4200	59.768	330.263	281.176	206.168	-91.058	117.561	-1.462
4300	59.890	331.671	282.334	212.151	-91.169	122.529	-1.488
4400	60.011	333.050	283.471	218.146	-91.290	127.500	-1.514
4500	60.128	334.399	284.588	224.153	-91.423	132.474	-1.538
4600	60.237	335.722	285.685	230.171	-91.568	137.451	-1.561
4700	60.346	337.019	286.763	236.200	-91.725	142.431	-1.583
4800	60.459	338.291	287.824	242.240	-91.893	147.416	-1.604
4900	60.563	339.538	288.866	248.292	-92.073	152.403	-1.625
5000	60.668	340.763	289.892	254.353	-92.265	157.394	-1.644
5100	60.764	341.965	290.901	260.425	-92.468	162.389	-1.663
5200	60.869	343.146	291.895	266.506	-92.682	167.388	-1.681
5300	60.965	344.306	292.873	272.598	-92.907	172.392	-1.699
5400	61.061	345.447	293.836	278.699	-93.143	177.400	-1.716
5500	61.153	346.568	294.785	284.810	-93.389	182.412	-1.732
5600	61.250	347.671	295.719	290.930	-93.644	187.429	-1.748
5700	61.342	348.756	296.640	297.060	-93.908	192.451	-1.764
5800	61.429	349.823	297.548	303.198	-94.180	197.477	-1.778
5900	61.526	350.874	298.443	309.346	-94.459	202.507	-1.793
6000	61.609	351.909	299.325	315.503	-94.745	207.543	-1.807

PREVIOUS: June 1977 (1 atm) CURRENT: June 1977 (1 bar)

TITANIUM HYDRIDE (TiH_2) CRYSTAL M_r = 49.89588

$S°(298.15\ K) = 29.711\ J\ K^{-1}\ mol^{-1}$

$\Delta_fH°(0\ K) = [-136.00 \pm 8.4]\ kJ\ mol^{-1}$
$\Delta_fH°(298.15\ K) = [-144.35 \pm 8.4]\ kJ\ mol^{-1}$

Enthalpy of Formation

The enthalpy of combustion of $TiH_{1.73}$(cr), the stable species of titanium hydride, was reported to be -247.8 kcal mol^{-1} by Sieverts and Gotta (1). From this data the value of $\Delta_fH°$(298.15 K) for $TiH_{1.73}$(cr) was calculated to be -29.40 kcal mol^{-1}. The enthalpy of dissociation of titanium hydride having compositions between the rough limits $TiH_{1.75}$-$TiH_{1.93}$ was given as 19.15 kcal mol^{-1} of H_2 dissociated by Gibb (2). Assuming this value is applicable to compositions between the limits $TiH_{1.73}$-TiH_2, the difference in the enthalpy of formation of titanium hydride between $TiH_{1.73}$ and TiH_2 was estimated to be -5.17 kcal mol^{-1}. Thus the value of $\Delta_fH°$(298.15 K) for TiH_2(cr) was calculated.

Heat Capacity and Entropy

The low temperature $C_p°$ for $TiH_{1.97}$(cr) were measured by Stalinski and Bieganski (3, 24-363 K). This data were adopted for TiH_2(cr) due to lack of the measured $C_p°$ values. A small but well marked λ-type anomaly with a maximal value of $C_p°$ at 288.5 K in the $C_p°$ vs T curve was observed. S°(298.15 K) was reported by Stalinski and Bieganski (3), based on S°(extrap 25 K) = 0.65 cal K^{-1} mol^{-1}.

References

1. A. Sieverts and A. Gotta, Z. Anorg. Allg. Chem. 199, 384 (1931).
2. T. R. P. Gibb, Jr., J. J. McSharry, and R. W. Bragdon, J. Amer. Chem. Soc. 73, 1751 (1951).
3. B. Stalinski and Z. Bieganski, Bull. Acad. Polon. Sci., Ser. Sci. Chim. 8, No. 5, 243 (1960).

Titanium Hydride (TiH_2) H_2Ti_1(cr)

Enthalpy Reference Temperature = T_r = 298.15 K Standard State Pressure = p° = 0.1 MPa

	J K⁻¹mol⁻¹			kJ mol⁻¹			
T/K	$C_p°$	S°	$-[G°-H°(T_r)]/T$	$H°-H°(T_r)$	$\Delta_fH°$	$\Delta_fG°$	Log K_f
0	0.	0.	INFINITE	-4.945	-135.996	-135.996	INFINITE
100	12.678	6.745	51.513	-4.477	-139.087	-128.863	67.311
200	22.719	19.041	32.284	-2.648	-141.870	-117.550	30.701
298.15	30.087	29.711	29.711	0.	-144.348	-105.073	18.408
300	30.259	29.897	29.711	0.056	-144.392	-104.829	18.252
400	39.330	39.847	31.006	3.536	-146.431	-91.315	11.924
500	47.279	49.504	33.747	7.879	-147.752	-77.368	8.083
600	53.806	58.721	37.148	12.944	-148.441	-63.217	5.504
700	58.994	67.418	40.857	18.593	-148.618	-48.993	3.656
800	63.137	75.577	44.693	24.707	-148.381	-34.773	2.270
900	66.159	83.194	48.553	31.178	-147.877	-20.600	1.196
1000	68.534	90.294	52.376	37.918	-147.261	-6.491	0.339
1100	70.057	96.899	56.127	44.849	-146.684	7.557	-0.359
1200	71.337	103.051	59.784	51.921	-150.185	21.677	-0.944
1300	72.251	108.798	63.336	59.101	-149.106	35.955	-1.445
1400	73.046	114.182	66.777	66.367	-148.063	50.151	-1.871
1500	73.722	119.246	70.108	73.707	-147.082	64.274	-2.238
1600	74.279	124.022	73.330	81.108	-146.187	78.335	-2.557
1700	74.716	128.539	76.446	88.558	-145.402	92.343	-2.837
1800	75.034	132.819	79.460	96.047	-144.751	106.308	-3.085
1900	75.232	136.882	82.376	103.561	-144.257	120.242	-3.306
2000	75.312	140.743	85.199	111.089	-158.631	134.607	-3.516

PREVIOUS: CURRENT: June 1963

Titanium Hydride (TiH_2) H_2Ti_1(cr)

IODOSILANE (SiH_3I) IDEAL GAS M_r = 158.01382

$\Delta_f H°(0\ K) = [8.35 \pm 16.7]\ kJ\ mol^{-1}$

$S°(298.15\ K) = 271.022 \pm 0.25\ J\ K^{-1}\ mol^{-1}$ $\Delta_f H°(298.15\ K) = [-2.09 \pm 16.7]\ kJ\ mol^{-1}$

Vibrational Frequencies and Degeneracies

ν, cm^{-1}	ν, cm^{-1}
2192(1)	2206(2)
903(1)	941(2)
362(1)	592(2)

Ground State Quantum Weight: [1] σ = 3

Point Group: C_{3v}

Bond Distances: Si-H = 1.486 ± 0.008 Å Si-H = 2.437 Å

Bond Angles: H-Si-H = 110.42° ± 0.5° H-Si-I = 108.5° ± 0.5°

Product of the Moments of Inertia: $I_A I_B I_C = 6.831704 \times 10^{-115}\ g^3\ cm^6$

Enthalpy of Formation

$\Delta_f H$ (298.15 K) is estimated by linear interpolation between the values (1) of SiI_4(g) and SiH_4(g). There are no experimental $\Delta_f H°$ data for SiH_3I, SiH_2I_2 and $SiHI_3$. It is difficult to justify a nonlinear interpolation scheme such as that adopted for the chlorosilanes (1). A cubic interpolation fits $\Delta_f H°$ data for the chloromethanes and has been proposed for bromomethanes and iodomethanes (2). Experimental $\Delta_f H°$ values do not seem to eixst for CI_4, CBr_4 and $CHBr_3$, while published values for CBr_4 and $CHBr_3$ are the subject of controversy (1, SiH_3Br, g). Moreover, the cubic scheme proposed (2) for CH_nI_{4-n} is surprisingly different from those of the other halomethanes (2); i.e., the first differences in $\Delta_f H°$ exhibit a minimum instead of a maximum. These uncertainties preclude use of the iodomethanes for predicting $\Delta_f H°$ of the iodosilanes. Hunt and Sirtl (3) also concluded that linear interpolation of $\Delta_f H°$ is all that is justified by available data.

Heat Capacity and Entropy

The molecular structure is based on microwave data of Kewley et al. (4) for isotopic forms of the four monohalosilanes. The authors concluded that the silyl group is effectively the same in all four halides. We adopt effective (r_o) structural parameters rather than substitutional (r_s) parameters. These are in reasonable agreement with previous structural determinations (5). The principal moments of inertia are $I_A = 0.9972 \times 10^{-39}$ and $I_B = I_C = 26.1748 \times 10^{-39}\ g\ cm^2$.

Vibrational frequencies are from gas-phase infrared spectra (6, 7). Assignments are consistent with liquid-phase Raman spectra (8, 9) which yield slightly lower values for most of the frequencies.

We neglect excited states and assume the electronic ground state to be $^1A_1$ by analogy with CH_3I (10). Comparison of the halosilanes (11) and halomethanes (10, 11) suggests that the lowest excited state of SiH_3I is at ≥30000 cm^{-1}. Contributions from this state should be unimportant.

References

1. JANAF Thermochemical Tables: ClH_3Si(g), BrH_3Si(g) 12-31-76; H_4Si(g), I_4Si(g) 6-30-76.
2. S. A. Kudchadker and A. P. Kudchadker, J. Phys. Chem. Ref. Data 4, 457 (1975).
3. L. P. Hunt and E. Sirtl, J. Electrochem. Soc. 120, 806 (1973).
4. R. Kewley, P. M. McKinney and A. G. Robiette, J. Mol. Spectrosc. 34, 390 (1970).
5. L. E. Sutton, "Interatomic Distances," Spec. Publ. No. 11, 1958; "Interatomic Distances Supplement," Spec. Publ. No. 18, The Chemical Society, London, 1965.
6. R. N. Dixon and N. Sheppard, Trans. Faraday Soc. 53, 282 (1957); J. Chem. Phys. 23, 215 (1955).
7. H. R. Linton and E. R. Nixon, Spectrochim. Acta 12, 41 (1958).
8. D. F. Ball, M. J. Buttler and D. C. McKean, Spectrochim. Acta 21, 451 (1965); unpublished data from R. Taylor, Dissert., Oxford Univ., 1961.
9. F. Feher, A. G. Wronka and B. Mostert, Monatsh. Chem. 104, 360 (1973).
10. G. Herzberg, "Electronic Spectra of Polyatomic Molecules," D. Van Nostrand Co., Inc. New York, 1966.
11. G. C. Causley, J. B. Clark and B. R. Russell, Chem. Phys. Lett. 38, 602, (1976).

Iodosilane (SiH_3I) $H_3I_1Si_1$(g)

Enthalpy Reference Temperature = T_r = 298.15 K Standard State Pressure = p° = 0.1 MPa

T/K	$C_p°$ (J K⁻¹mol⁻¹)	S° (J K⁻¹mol⁻¹)	−[G°−H°(T_r)]/T (J K⁻¹mol⁻¹)	H°−H°(T_r) (kJ mol⁻¹)	$\Delta_f H°$ (kJ mol⁻¹)	$\Delta_f G°$ (kJ mol⁻¹)	Log K_f
0	0.	0.	INFINITE	-12.078	8.348	8.348	INFINITE
100	34.754	224.891	312.147	-8.726	5.410	1.463	-0.764
200	43.918	251.491	275.668	-4.835	1.615	-1.031	0.269
250	49.420	261.884	271.890	-2.501	-0.309	-1.469	0.307
298.15	54.401	271.022	271.022	0.	-2.092	-1.529	0.268
300	54.582	271.359	271.023	0.101	-2.159	-1.525	0.266
350	59.178	280.125	271.704	2.947	-3.927	-1.279	0.191
400	63.215	288.296	273.273	6.009	-13.550	-0.514	0.067
450	66.784	295.952	275.372	9.261	-15.629	1.242	-0.144
500	69.979	303.157	277.794	12.681	-37.638	5.122	-0.535
600	75.514	316.420	283.146	19.964	-39.001	13.809	-1.202
700	80.152	328.419	288.770	27.754	-39.950	22.691	-1.693
800	84.055	339.385	294.421	35.970	-40.561	31.685	-2.069
900	87.331	349.480	299.986	44.545	-40.906	40.739	-2.364
1000	90.075	358.827	305.408	53.419	-41.049	49.820	-2.602
1100	92.372	367.523	310.664	62.545	-41.046	58.908	-2.797
1200	94.299	375.646	315.745	71.881	-40.942	67.991	-2.960
1300	95.922	383.260	320.648	81.394	-40.778	77.062	-3.096
1400	97.295	390.420	325.379	91.057	-40.587	86.120	-3.213
1500	98.462	397.173	329.942	100.847	-40.398	95.163	-3.314
1600	99.460	403.561	334.346	110.744	-40.232	104.195	-3.402
1700	100 318	409.617	338.597	120.734	-90.285	113.665	-3.492
1800	101.059	415.372	342.704	130.804	-89.986	125.653	-3.646
1900	101.703	420.854	346.674	140.942	-89.711	137.625	-3.784
2000	102.265	426 085	350.514	151.141	-89.467	149.584	-3.907
2100	102.759	431.087	354.233	161.393	-89.257	161.531	-4.018
2200	103.194	435.877	357.836	171.691	-89.082	173.469	-4.119
2300	103.579	440 473	361.330	182.030	-88.941	185.400	-4.211
2400	103.922	444.889	364.720	192.406	-88.833	197.325	-4.295
2500	104.228	449.138	368.012	202.814	-88.752	209.247	-4.372
2600	104.502	453.231	371.212	213.250	-88.694	221.166	-4.443
2700	104.748	457.179	374.323	223.713	-88.656	233.082	-4.509
2800	104.970	460.993	377.351	234.199	-88.632	244.998	-4.570
2900	105.171	464.680	380.299	244.706	-88.617	256.913	-4.628
3000	105.354	468.249	383.171	255.233	-88.607	268.828	-4.681
3100	105.520	471.706	385.972	265.776	-88.600	280.742	-4.730
3200	105.672	475.059	388.704	276.336	-88.591	292.656	-4.777
3300	105.811	478.312	391.370	286.910	-88.578	304.570	-4.821
3400	105.938	481.473	393.974	297.498	-88.559	316.483	-4.862
3500	106.055	484.546	396.518	308.098	-88.533	328.396	-4.901
3600	106.163	487.535	399.005	318.709	-472.649	350.769	-5.090
3700	106.262	490.445	401.437	329.330	-472.190	373.635	-5.275
3800	106.354	493.280	403.817	339.961	-471.724	396.489	-5.450
3900	106.440	496.044	406.146	350.601	-471.249	419.331	-5.616
4000	106.519	498.740	408.427	361.248	-470.765	442.160	-5.774
4100	106.593	501.371	410.662	371.904	-470.273	464.977	-5.924
4200	106.661	503.940	412.853	382.567	-469.773	487.782	-6.066
4300	106.725	506.451	415.000	393.236	-469.265	510.575	-6.202
4400	106.785	508.905	417.107	403.912	-468.750	533.356	-6.332
4500	106.841	511.305	419.174	414.593	-468.228	556.125	-6.455
4600	106.894	513.654	421.202	425.280	-467.699	578.882	-6.573
4700	106.943	515.954	423.194	435.972	-467.164	601.628	-6.686
4800	106.990	518.206	425.150	446.669	-466.624	624.364	-6.794
4900	107.033	520.412	427.071	457.370	-466.079	647.086	-6.898
5000	107.074	522.575	428.960	468.075	-465.530	669.799	-6.997
5100	107.113	524.696	430.816	478.784	-464.976	692.500	-7.093
5200	107.149	526.776	432.642	489.498	-464.419	715.190	-7.184
5300	107.184	528.817	434.437	500.214	-463.857	737.869	-7.272
5400	107.217	530.821	436.204	510.934	-463.292	760.538	-7.357
5500	107.248	532.789	437.942	521.658	-462.722	783.196	-7.438
5600	107.277	534.721	439.653	532.384	-462.150	805.844	-7.517
5700	107.305	536.620	441.338	543.113	-461.573	828.482	-7.592
5800	107.331	538.487	442.996	553.845	-460.992	851.110	-7.665
5900	107.356	540.322	444.631	564.579	-460.407	873.727	-7.735
6000	107.380	542.126	446.240	575.316	-459.817	896.334	-7.803

PREVIOUS: December 1976 (1 atm) CURRENT: December 1976 (1 bar)

Iodosilane (SiH_3I) $H_3I_1Si_1$(g)

AMMONIA (NH_3) IDEAL GAS M_r = 17.03052

$\Delta_f H°(0\ K) = -38.907 \pm 0.4\ kJ\ mol^{-1}$

$S°(298.15\ K) = 192.774 \pm 0.025\ J\ K^{-1}\ mol^{-1}$ $\Delta_f H°(298.15\ K) = -45.898 \pm 0.4\ kJ\ mol^{-1}$

Vibrational Frequencies and Degeneracies

ν, cm^{-1}	ν, cm^{-1}	ν, cm^{-1}	ν, cm^{-1}
3506(1)	1022(1)	3577(2)	1691(2)

Ground State Quantum Weight: 1 σ(internal) = 2

Point Group: C_{3v} σ(external) = 3

Bond Length: N-H = 1.0124 Å

Bond Angle: 106.67°

Product of the Moments of Inertia: $I_A I_B I_C = 0.0348 \times 10^{-117}\ g^3\ cm^6$

Enthalpy of Formation

2nd and 3rd law analyses of equilibrium data for the reaction $1/2N_2(g)+3/2H_2(g) = NH_3(g)$ cited in the previous JANAF evaluation (1) plus more recent work of Schulz and Schaefer (6) were made using the revised thermal functions for $NH_3(g)$. All of the previously cited work in reaction calorimetry plus the early work of Berthelot (7, 8) and Thomsen (9) were reevaluated. No significant differences in the 3rd law calculations of the equilibrium data or in the corrections to the flow calorimetry data of Haber and Tamaru (12) and Wittig and Schmatz (13) were found. Thus, the 0.1 kcal discrepancy between the results of the equilibrium and reaction calorimetry measurements remains unresolved. The previous JANAF selection (1) for $\Delta_f H°(298.15\ K)$ of $NH_3(g)$ was adopted. A recent evaluation (14) which includes new indirect calorimetry (unpublished) further confirms this selection.

Source	Method	$\Delta_f H°(298.15\ K)$ kcal mol^{-1}	$\Delta_f H°(298.15\ K)$ kcal mol^{-1}*	$\Delta S°$(obs.-calc. 298.15 K)* cal K^{-1} mol^{-1}
Larson, Dodge (2)	$K_p°(4)$ from $K_p°$(10-1,000 atm, 600-800 K)	-10.88	-10.70±0.11	+0.24±0.15[a]
Haber et al. (3)	$K_p°(4)$ from $K_p°$(30 atm, 800-1200 K)	-10.86	-10.88±0.15	-0.02±0.15[b]
Haber, Maschke (5)	$K_p°$(1 atm, 900-1400 K)	-10.85	-10.62±0.22	0.20±0.19[c]
Schulz, Schaefer (6)	$K_p°$(1 atm, 567-673 K)	-10.87	-10.78±0.20	0.14±0.3[d]
Berthelot (7)	Indirect; Reaction of $Br_2(aq)$ and $NH_3(aq)$	-11.4		
Berthelot (8)	Indirect; Reaction of $O_2(g)$ with $NH_3(g)$	-12.1		
Thomsen (9)	Indirect; Reaction of $O_2(g)$ with $NH_3(g)$	-11.9		
Becker, Roth (10)	Indirect; Heat of combustion oxalates	-11.00±0.15		
Haber et al. (11)	Flow calorimetry at 298 K	-11.10±0.05		
Haber, Tamaru (12)	Flow calorimetry (739-932 K)	-10.97±0.008		
Wittig, Schmatz (13)	Flow calorimetry at 832 K	-10.99±0.05		

*2nd law analysis assuming $\Delta C_p°(cal\ K^{-1}\ mol^{-1})$ equals (a) -2.672+0.00591(T-700), (b) -1.236+0.00404(T-1000), (c) -0.855+0.00305(T-1100), (d) 3.287+0.00651(T-600).

Heat Capacity and Entropy

The thermodynamic functions differ from those of the 1965 JANAF table (1) in being taken directly form the later and more complete work of Haar (15). Haar treated in detail the contribution of the highly anharmonic out-of-plane vibrational mode, including its large coupling with rotation and its coupling with the other vibrational modes. Haar's values of $C_p°$ pass through a shallow maximum between 4000 and 5000 K; they were extrapolated from 5000 to 6000 K by assuming a constant value (19.300 cal K^{-1} mol^{-1}). A summary of Haar's estimated uncertainties and of the differences of the 1965 table from the present table (in cal K^{-1} mol^{-1}) is as follows:

	Uncertainties (Haar, 15)		1965 Table minus This Table	
T, K	$C_p°$	S°	$C_p°$	S°
1000	0.006	0.006	-0.034	-0.033
3000	0.10	0.06	+0.142	-0.122
5000	0.6	0.04	+1.775	+0.265

The National Bureau of Standards prepared this table (16) by critical analysis of data existing in 1972. Using the results of Haar (15) and $\Delta_f H°$ selected by NBS (16), we recalculate the table in terms of R=1.987192 cal K^{-1} mol^{-1} (17) and current JANAF reference states for the elements.

References

1. JANAF Thermochemical Tables, 2nd ed., NSRDS-NBS 37, 1971.
2. A. T. Larson and R. L. Dodge, J. Amer. Chem. Soc. 45, 2918 (1923).
3. F. Haber, S. Tamaru, and Ch. Ponnaz, Z. Elektrochem. 21, 89 (1914).
4. C. C. Stephenson and H. O. McMahan, J. Amer. Chem. Soc. 61, 437 (1939).
5. F. Haber and A. Maschke, Z. Electrochem. 21, 128 (1915).
6. G. Schulz and H. Schaefer, Ber. Bunsenges. Physik. Chem. 70, 21 (1966).
7. M. Berthelot, Compt. Rend. 89, 877 (1879); Ann. Chim. Phys. [5] 20, 247 (1880).
8. M. Berthelot, Ann. Chim. Phys. [5] 20, 244 (1880).
9. J. Thomsen, "Thermochemical Investigations," Vol. II, p. 68, Johann A. Barth, Leipzig, 1882.
10. G. Becker and W. A. Rogh, Z. Elektrochem. 40, 836 (1934).
11. F. Haber, S. Tamaru, and L. W. Oeholm. Z. Elektrochem. 21, 206 (1915).
12. F. Haber and S. Tamaru, Z. Elektrochem. 21, 191 (1915).
13. E. Wittig and W. Schmatz, Z. Elektrochem. 63, 475 (1959).
14. W. H. Evans, Personal Communication, U. S. Nat. Bur. Stand, July 1972; evaluation of ICSU-CODATA Task Group on Key Values for Thermodynamics, June (1972).
15. L. Haar, J. Res. Nat. Bur. Stand. 72A, 207 (1968).
16. S. Abramowitz et al., U. S. Nat. Bur. Stand., Rept. 10904, 239, July, 1972.
17. CODATA Task Group on Fundamental Constants, CODATA Bulletin 11, December, 1973.

Ammonia (NH_3) $H_3N_1(g)$

Enthalpy Reference Temperature = T_r = 298.15 K Standard State Pressure = p° = 0.1 MPa

T/K	$C_p°$ (J $K^{-1}mol^{-1}$)	S° (J $K^{-1}mol^{-1}$)	$-[G°-H°(T_r)]/T$ (J $K^{-1}mol^{-1}$)	$H°-H°(T_r)$ (kJ mol^{-1})	$\Delta_f H°$ (kJ mol^{-1})	$\Delta_f G°$ (kJ mol^{-1})	Log K_f
0	0.	0.	INFINITE	-10.045	-38.907	-38.907	INFINITE
100	33.284	155.840	223.211	-6.737	-41.550	-34.034	17.777
200	33.757	178.990	195.962	-3.394	-43.703	-25.679	6.707
298.15	35.652	192.774	192.774	0.	-45.898	-16.367	2.867
300	35.701	192.995	192.775	0.066	-45.939	-16.183	2.818
400	38.716	203.663	194.209	3.781	-48.041	-5.941	0.776
500	42.048	212.659	197.021	7.819	-49.857	4.800	-0.501
600	45.293	220.615	200.302	12.188	-51.374	15.879	-1.382
700	48.354	227.829	203.727	16.872	-52.618	27.190	-2.029
800	51.235	234.476	207.160	21.853	-53.621	38.662	-2.524
900	53.948	240.669	210.543	27.113	-54.411	50.247	-2.916
1000	56.491	246.486	213.849	32.637	-55.013	61.910	-3.234
1100	58.859	251.983	217.069	38.406	-55.451	73.625	-3.496
1200	61.048	257.199	220.197	44.402	-55.746	85.373	-3.716
1300	63.057	262.166	223.236	50.609	-55.917	97.141	-3.903
1400	64.893	266.907	226.187	57.008	-55.982	108.918	-4.064
1500	66.564	271.442	229.054	63.582	-55.954	120.696	-4.203
1600	68.079	275.788	231.840	70.315	-55.847	132.469	-4.325
1700	69.452	279.957	234.549	77.193	-55.672	144.234	-4.432
1800	70.695	283.962	237.184	84.201	-55.439	155.986	-4.527
1900	71.818	287.815	239.748	91.328	-55.157	167.725	-4.611
2000	72.833	291.525	242.244	98.561	-54.833	179.447	-4.687
2100	73.751	295.101	244.677	105.891	-54.473	191.152	-4.755
2200	74.581	298.552	247.048	113.309	-54.084	202.840	-4.816
2300	75.330	301.884	249.360	120.805	-53.671	214.509	-4.872
2400	76.009	305.104	251.616	128.372	-53.238	226.160	-4.922
2500	76.626	308.220	253.818	136.005	-52.789	237.792	-4.968
2600	77.174	311.236	255.969	143.695	-52.329	249.406	-5.011
2700	77.672	314.158	258.070	151.438	-51.860	261.003	-5.049
2800	78.132	316.991	260.124	159.228	-51.386	272.581	-5.085
2900	78.529	319.740	262.132	167.062	-50.909	284.143	-5.118
3000	78.902	322.409	264.097	174.933	-50.433	295.689	-5.148
3100	79.228	325.001	266.020	182.840	-49.959	307.218	-5.177
3200	79.521	327.521	267.903	190.778	-49.491	318.733	-5.203
3300	79.785	329.972	269.747	198.744	-49.030	330.233	-5.227
3400	80.011	332.358	271.554	206.734	-48.578	341.719	-5.250
3500	80.216	334.680	273.324	214.745	-48.139	353.191	-5.271
3600	80.400	336.942	275.060	222.776	-47.713	364.652	-5.291
3700	80.550	339.147	276.763	230.824	-47.302	376.101	-5.310
3800	80.684	341.297	278.433	238.886	-46.908	387.539	-5.327
3900	80.793	343.395	280.072	246.960	-46.534	398.967	-5.344
4000	80.881	345.441	281.680	255.043	-46.180	410.385	-5.359
4100	80.956	347.439	283.260	263.136	-45.847	421.795	-5.374
4200	81.006	349.391	284.811	271.234	-45.539	433.198	-5.388
4300	81.048	351.297	286.335	279.337	-45.254	444.593	-5.401
4400	81.065	353.161	287.833	287.442	-44.996	455.981	-5.413
4500	81.073	354.983	289.305	295.550	-44.764	467.364	-5.425
4600	81.057	356.765	290.752	303.656	-44.561	478.743	-5.436
4700	81.032	358.508	292.175	311.761	-44.387	490.117	-5.447
4800	80.990	360.213	293.575	319.862	-44.242	501.488	-5.457
4900	80.931	361.882	294.952	327.958	-44.129	512.856	-5.467
5000	80.856	363.517	296.307	336.048	-44.047	524.223	-5.477
5100	80.751	365.117	297.641	344.127	-43.999	535.587	-5.486
5200	80.751	366.685	298.954	352.202	-43.979	546.951	-5.494
5300	80.751	368.223	300.246	360.277	-43.982	558.315	-5.503
5400	80.751	369.732	301.519	368.352	-44.006	569.680	-5.511
5500	80.751	371.214	302.773	376.428	-44.049	581.044	-5.518
5600	80.751	372.669	304.008	384.503	-44.112	592.410	-5.526
5700	80.751	374.098	305.225	392.578	-44.193	603.778	-5.533
5800	80.751	375.503	306.425	400.653	-44.291	615.147	-5.540
5900	80.751	376.883	307.607	408.728	-44.404	626.516	-5.547
6000	80.751	378.240	308.773	416.803	-44.531	637.889	-5.553

PREVIOUS: June 1977 (1 atm) CURRENT: June 1977 (1 bar)

Ammonia (NH_3) $H_3N_1(g)$

HYDRONIUM, ION (H_3O^+) IDEAL GAS M_r = 19.02267

$S°(298.15\ K) = [192.257]\ J\ K^{-1}\ mol^{-1}$

$\Delta_f H°(0\ K) = 581.99 \pm 12.6\ kJ\ mol^{-1}$
$\Delta_f H°(298.15\ K) = 581.16 \pm 12.6\ kJ\ mol^{-1}$

Vibrational Frequencies and Degeneracies

ν, cm^{-1}	ν, cm^{-1}
[3760](1)	[3870](2)
[1050](1)	[1550](2)

Ground State Quantum Weight: 1 σ = [3]
Point Group: [C_{3v}]
Bond Distance: H–O = [0.95] Å
Bond Angle: H–O–H = [109]°
Product of the Moments of Inertia: $I_A I_B I_C = [0.0240 \times 10^{-117}]\ g^3\ cm^6$

Enthalpy of Formation

The proton affinity of water has been measured by the ion impact method as −169 kcal mol^{-1} by Tal'rose and Frankevitch (1). The selected enthalpy of formation of hydronium ion, $\Delta_f H°(H_3O^+, g, 0\ K) = 139.1\ kcal\ mol^{-1}$, was derived from the $\Delta_r H°(0\ K) = -169\ kcal\ mol^{-1}$ for $H_2O(g) + H^+(g) + H_3O^+(g)$, using the JANAF $\Delta_f H°(H_2O, g, 0\ K) = -57.1$ and $\Delta_f H°(H^+, g, 0\ K) = 365.2\ kcal\ mol^{-1}$.

The values of 195 and 207 kcal mol^{-1} for the enthalpy of formation of $H_3O^+(g)$ have also been reported by Field and Franklin (2) and Bernecker and Long (3), respectively. These two values were based on the electron impact studies which may involve unaccounted excitations and excess energy in the fragments. Other literature values have been reviewed by Bishop (4).

Green and Sugden (5), based on the studies of the positive ions in flames by mass spectrometry, prefer the value derived from the proton affinity of water obtained by Tal'rose and Frankevitch, rather than the electron impact data. From the approximate equilibrium constant calculations for the reaction $HCO^+(g) + H_2O(g) = H_3O^+(g) + CO(g)$, Clifton (6) has also found that the heat of formation of $H_3O^+(g)$ derived from the proton affinity of water, not from the electron impact data, leads to agreement with the experimental data given by Green and Sugden.

Heat Capacity and Entropy

Since the molecule $H_3O^+(g)$ has the same number of valence electrons as $NH_3(g)$, a pyramidal molecular configuration (C_{3v}) is assumed for $H_3O^+(g)$ in the tabulation. Clifton (7) has obtained a bond distance O-H of 0.95 Å from an application of covalent bond radii and the effect of electric charge, and a bond angle H-O-H of 109° by extrapolating a plot of bond length versus bond angle for $NH_3(g)$ and $PH_3(g)$ to 0.95 Å. Both values are adopted in the tabulation. The principal moments of inertia are: $I_A = I_B = 0.2446 \times 10^{-39}$, $I_C = 0.4005 \times 10^{-39}\ g\ cm^2$. Bishop (4) has calculated theoretically the bond distance OH of 0.95 Å and the bond angle H-O-H of 114° 26' with a planar D_{3h} molecular configuration. The vibrational frequencies were calculated by the valence force method from the estimated stretching and bending force constants of Clifton.

References

1. V. L. Tal'rose and E. L. Frankevitch, Doklady Akad. Nauk SSSR 111, 376 (1956); J. Amer. Chem. Soc. 80, 2344 (1958).
2. F. M. Field and J. L. Franklin, "Electron Impact Phenomena," Academic Press, New York, 1957.
3. R. R. Bernecker and F. A. Long, J. Phys. Chem. 65, 1565 (1961).
4. D. M. Bishop, J. Chem. Phys. 43, 4453 (1965).
5. J. A. Green and T. M. Sugden, Ninth Symposium (International) on Combustion, Academic Press, New York, 1963.
6. D. G. Clifton, AD444078, G. M. Defense Research Laboratory, General Motors Corp., Santa Barbara, Calif., July 1964.
7. D. G. Clifton, J. Chem. Phys. 41, 3656 (1964).

Hydronium, Ion (H_3O^+) $H_3O_1{}^+(g)$

Enthalpy Reference Temperature = T_r = 298.15 K Standard State Pressure = p° = 0.1 MPa

T/K	$C_p°$ (J $K^{-1}mol^{-1}$)	S° (J $K^{-1}mol^{-1}$)	−[G°−H°(T_r)]/T (J $K^{-1}mol^{-1}$)	H°−H°(T_r) (kJ mol^{-1})	$\Delta_f H°$ (kJ mol^{-1})	$\Delta_f G°$ (kJ mol^{-1})	Log K_f
0	0.	0.	INFINITE	-10.016	581.986		
100	33.258	155.526	222.432	-6.691			
200	33.536	178.619	195.407	-3.358			
250	34.159	186.161	192.829	-1.667			
298.15	35.146	192.257	192.257	0.	581.158	606.607	-106.275
300	35.190	192.474	192.257	0.065	581.154	606.765	-105.647
350	36.540	197.997	192.691	1.857	581.074	611.040	-91.193
400	38.081	202.975	193.670	3.722	581.045	615.324	-80.353
450	39.697	207.554	194.961	5.667	581.080	619.607	-71.922
500	41.309	211.820	196.436	7.692	581.180	623.883	-65.177
600	44.368	219.626	199.663	11.978	581.571	632.391	-55.055
700	47.158	226.678	203.026	16.556	582.194	640.815	-47.818
800	49.727	233.145	206.392	21.402	583.021	649.135	-42.384
900	52.131	239.142	209.702	26.496	584.029	657.341	-38.151
1000	54.401	244.753	212.929	31.824	585.199	665.426	-34.758
1100	56.542	250.040	216.065	37.372	586.513	673.386	-31.976
1200	58.553	255.047	219.107	43.128	587.955	681.221	-29.653
1300	60.428	259.808	222.056	49.078	589.512	688.930	-27.682
1400	62.164	264.351	224.916	55.209	591.168	696.517	-25.987
1500	63.763	268.695	227.691	61.506	592.911	703.981	-24.515
1600	65.229	272.858	230.385	67.957	594.730	711.327	-23.222
1700	66.569	276.853	233.001	74.548	596.614	718.557	-22.079
1800	67.790	280.693	235.545	81.267	598.552	725.674	-21.059
1900	68.903	284.389	238.019	88.102	600.537	732.682	-20.143
2000	69.915	287.949	240.427	95.044	602.561	739.585	-19.316
2100	70.837	291.383	242.772	102.082	604.618	746.386	-18.565
2200	71.676	294.698	245.058	109.209	606.700	753.088	-17.881
2300	72.440	297.901	247.286	116.415	608.803	759.695	-17.253
2400	73.138	300.999	249.460	123.694	610.923	766.210	-16.676
2500	73.775	303.998	251.582	131.041	613.055	772.637	-16.143
2600	74.358	306.903	253.654	138.448	615.196	778.977	-15.650
2700	74.891	309.719	255.678	145.911	617.343	785.236	-15.191
2800	75.381	312.452	257.657	153.424	619.492	791.415	-14.764
2900	75.831	315.105	259.593	160.985	621.642	797.517	-14.365
3000	76.245	317.683	261.486	168.589	623.790	803.545	-13.991
3100	76.626	320.189	263.340	176.233	625.935	809.502	-13.640
3200	76.979	322.628	265.155	183.914	628.074	815.389	-13.310
3300	77.304	325.001	266.932	191.628	630.207	821.210	-12.999
3400	77.606	327.314	268.674	199.374	632.331	826.966	-12.705
3500	77.886	329.567	270.382	207.149	634.445	832.659	-12.427
3600	78.146	331.765	272.057	214.950	636.548	838.293	-12.163
3700	78.387	333.910	273.700	222.777	638.639	843.868	-11.913
3800	78.612	336.003	275.312	230.627	640.718	849.387	-11.676
3900	78.823	338.048	276.894	238.499	642.782	854.852	-11.449
4000	79.019	340.046	278.448	246.391	644.832	860.263	-11.234
4100	79.202	342.000	279.975	254.302	646.866	865.624	-11.028
4200	79.374	343.910	281.474	262.231	648.884	870.935	-10.832
4300	79.535	345.780	282.948	270.177	650.885	876.198	-10.644
4400	79.687	347.610	284.397	278.138	652.869	881.415	-10.464
4500	79.829	349.402	285.822	286.114	654.835	886.586	-10.291
4600	79.962	351.158	287.223	294.103	656.782	891.715	-10.126
4700	80.088	352.880	288.602	302.106	658.712	896.801	-9.967
4800	80.207	354.567	289.958	310.121	660.622	901.847	-9.814
4900	80.319	356.222	291.294	318.147	662.513	906.853	-9.667
5000	80.425	357.846	292.609	326.185	664.386	911.821	-9.526
5100	80.526	359.439	293.904	334.232	666.239	916.751	-9.389
5200	80.621	361.004	295.179	342.290	668.074	921.645	-9.258
5300	80.710	362.540	296.435	350.356	669.890	926.504	-9.131
5400	80.796	364.050	297.674	358.431	671.687	931.329	-9.009
5500	80.877	365.533	298.894	366.515	673.467	936.120	-8.891
5600	80.953	366.991	300.097	374.607	675.228	940.881	-8.776
5700	81.027	368.425	301.283	382.706	676.972	945.609	-8.666
5800	81.096	369.834	302.453	390.812	678.699	950.307	-8.558
5900	81.162	371.221	303.607	398.925	680.410	954.974	-8.455
6000	81.225	372.586	304.745	407.044	682.104	959.614	-8.354

PREVIOUS. June 1966 (1 atm) CURRENT: June 1966 (1 bar)

Hydronium, Ion (H_3O^+) $H_3O_1{}^+(g)$

$S°(298.15\ K) = 110.544 \pm 0.4\ J\ K^{-1}\ mol^{-1}$

$T_{fus} = 315.5 \pm 5\ K$

$\Delta_fH°(0\ K) = -1265.9 \pm 2.1\ kJ\ mol^{-1}$

$\Delta_fH°(298.15\ K) = -1284.4 \pm 2.1\ kJ\ mol^{-1}$

$\Delta_{fus}H° = 13.389 \pm 0.8\ kJ\ mol^{-1}$

Enthalpy of Formation

Recent calorimetric measurements (1, 2, 3) indicate the National Bureau of Standards' selected values quoted in reference (4), $\Delta_fH°(H_3PO_4$, cr, 298.15 K) = -305.7 ± 0.3 kcal mol^{-1}, is probably in error. The value was taken from Holmes' work (5). The data are summarized below, with heat values based on white phosphorus reference state.

Source	Method	n Moles of H_2O	$\Delta_fH°(H_3PO_4$, aq, 298.15 K) kcal mol^{-1}	$\Delta_fH°(298.15\ K)$ kcal mol^{-1}
Head (1)	a & b	40	-309.34 ± 0.38	-307.0 ± 0.4
Birley (2)	a	100	-309.8 ± 0.35	-307.18 ± 0.5
Irving (3)	a	0.6416	-305.9 ± 0.5	-307.36 ± 0.5
Holmes (5)	a & b	---	---	-307.4 ± 0.5*

a = solution calorimetry; b = combustion calorimetry

* see text

Head and Lewis (1) determined the enthalpy of combustion of phosphorus and enthalpy of hydrolysis of combustion products which led to $\Delta_fH°(H_3PO_4 \cdot 40\ H_2O$, aq, 298.15 K) = -309.34 ± 0.38 kcal mol^{-1}. Birley and Skinner (2) measured the enthalpy of hydrolysis of PCl_5 and derived $\Delta_fH°(H_3PO_4 \cdot 100\ H_2O$, aq, 298.15 K) = -309.8 ± 0.35 kcal mol^{-1} based on $\Delta_fH°(PCl_5$, cr, 298.15 K) = -106.0 kcal mol^{-1} obtained by direct chlorination of phosphorus (4). Irving and McKerrell (3) measured the enthalpy of solution of P_4O_{10} from which they derived $\Delta_fH°(H_3PO_4 \cdot 0.6416\ H_2O$, aq, 298.15 K) = -305.9 ± 0.5 kcal mol^{-1}. Using enthalpy of solution data for $H_3PO_4(cr) + H_3PO_4(363\ H_2O$, aq, (5) and enthalpy of dilution data for $H_3PO_4(363\ H_2O, aq) + H_3PO_4(n\ H_2O, aq)$ (6) we obtain enthalpy of formation of $H_3PO_4(cr)$ listed in the last column of the table. (These values are based on white phosphorus as the reference state.) Holmes (5) determined $\Delta_fH°(H_3PO_4$, cr, 298.15 K) = -305.7 ± 0.3 kcal mol^{-1} based on his enthalpy of combustion of white phosphorus and related enthalpy of solution data which were combined to form a thermal cycle. Using the white phosphorus based $\Delta_fH°(P_4O_{10}$, cr, 298.15 K) = -719.4 kcal mol^{-1} (7) (instead of his value -713.2 kcal mol^{-1}, we have recalculated his thermal cycle and find $\Delta_fH°(H_3PO_4$, cr, 298.15 K) = -307.4 ± 0.5 kcal mol^{-1} which is in good agreement with the recent calorimetric measurements (1, 2, 3).

The value, -307.0 ± 0.5 kcal mol^{-1}, is tentatively selected for the enthalpy of formation of $H_3PO_4(cr)$, based on white phosphorus as reference state.

Heat Capacity and Entropy

Low temperature heat capacity from 10 to 300 K were measured by Egan and Wakefield (8). Their smooth $C_p°$ are adopted to the tabulation. Heat capacities above 300 K are estimated graphically based on the measured $C_p°$ curve.

The entropy S°(298.15 K) is derived from heat capacity data, using S°(10 K) = 0.04 cal k^{-1} mol^{-1}.

Fusion Data

Egan and Wakefield (8) measured calorimetrically the heat of fusion $\Delta_{fus}H°$ = 3.1 ± 0.2 kcal mol^{-1}, but failed to determine the exact melting point due to continuous fusion between 301 to 319 K in their measurements. They suggested the melting point 315.5 K which was found by Ross and Jones (9).

Holmes (5) derived the enthalpy of fusion $\Delta_{fus}H°(298.15\ K)$ = 3.2 ± 0.5 kcal mol^{-1} from his enthalpy of solution data for crystalline and liquid H_3PO_4 in water and in aq. $HClO_4$ solution, which corresponds to $\Delta_{fus}H°$ = 3.36 kcal mol^{-1}. We adopt an intermediate value of $\Delta_{fus}H°$ = 3.2 ± 0.2 kcal mol^{-1}.

References

1. A. J. Head and G. B. Lewis, J. Chem. Thermodyn. 2, 701 (1970).
2. G. I. Birley and H. A. Skinner, Trans. Faraday Soc. 64, 3232 (1968).
3. R. J. Irving and H. McKerrell, Trans. Faraday Soc. 63, 2582 (1967).
4. U. S. Nat. Bur. Stand. Tech. Note 270-3, 1968; personal communication D. D. Wagman, Oct., 1971.
5. W. S. Holmes, Trans. Faraday Soc. 58, 1916 (1962).
6. E. P. Egan and B. B. Luff, J. Phys. Chem. 65, 523 (1961).
7. JANAF Thermochemical Table: P_4O_{10}(cr), 12-31-65.
8. E. P. Egan, Jr., and Z. T. Wakefield, J. Phys. Chem. 61, 1500 (1957).
9. W. H. Ross and R. M. Jones, J. Amer. Chem. Soc. 47, 2165 (1925).

Enthalpy Reference Temperature = T_r = 298.15 K; Standard State Pressure = p° = 0.1 MPa

	J $K^{-1}mol^{-1}$			kJ mol^{-1}			
T/K	$C_p°$	S°	$-[G°-H°(T_r)]/T$	$H°-H°(T_r)$	$\Delta_fH°$	$\Delta_fG°$	Log K_f
0	0.	0.	INFINITE	-16.980	-1265.927	-1265.927	INFINITE
100	42.844	34.528	183.578	-14.905	-1275.048	-1226.961	640.899
200	75.856	74.598	119.270	-8.935	-1281.203	-1176.480	307.265
298.15	106.064	110.544	110.544	0.	-1284.375	-1124.314	196.975
300	106.608	111.202	110.546	0.197	-1284.411	-1123.321	195.588
315.500	111.303	116.688	110.713	1.885	------ CRYSTAL <--> LIQUID -----		
400	136.900	146.019	115.089	12.372	-1285.789	-1069.271	139.633
500	167.193	179.822	124.668	27.577	-1283.719	-1015.323	106.070
600	197.485	212.982	136.631	45.811	-1278.830	-962.056	83.754
700	227.777	245.700	149.880	67.074	-1271.116	-909.831	67.892
800	258.069	278.092	163.885	91.366	-1260.559	-858.906	56.081
900	288.361	310.238	178.363	118.688	-1247.143	-809.475	46.981
1000	318.653	342.187	193.149	149.038	-1230.854	-761.692	39.787

PREVIOUS: June 1962 CURRENT: December 1971

PHOSPHORIC ACID (H_3PO_4) LIQUID M_r = 97.99518

S°(298.15 K) = [150.775] J K^{-1} mol^{-1}

T_{fus} = 315.5 K

$\Delta_f H°$(298.15 K) = [-1271.663] kJ mol^{-1}

$\Delta_{fus} H°$ = 13.389 ± 0.8 kJ mol^{-1}

Enthalpy of Formation

$\Delta_f H°$ (H_3PO_4, ℓ, 298.15 K) is calculated from that of the crystal by adding $\Delta_{fus}H°$ and the difference in enthalpy, H°(315.5 K) - H°(298.15 K), between the crystal and liquid.

Heat Capacity and Entropy

Heat capacities of the liquid are estimated by comparison with those of H_2SO_4(cr, ℓ) (1) and H_3PO_4(cr). S°(298.15 K) is calculated in a manner similar to that used for the enthalpy of formation.

Fusion Data

See the crystal table for details (2).

Decomposition Data

The phosphoric acid tends to be dehydrated to pyro- and meta-phosphoric acids when it is heated.

References

1. W. F. Giauque, E. W. Hornung, J. E. Kunzler and T. R. Rubin, J. Amer. Chem. Soc. 82, 62 (1960).
2. JANAF Thermochemical Tables: H_3PO_4(cr), 12-30-71.

Phosphoric Acid (H_3PO_4) $H_3O_4P_1$(l)

Enthalpy Reference Temperature = T_r = 298.15 K Standard State Pressure = p° = 0.1 MPa

T/K	$C_p°$ (J K⁻¹mol⁻¹)	S° (J K⁻¹mol⁻¹)	-[G°-H°(T_r)]/T (J K⁻¹mol⁻¹)	H°-H°(T_r) (kJ mol⁻¹)	$\Delta_f H°$ (kJ mol⁻¹)	$\Delta_f G°$ (kJ mol⁻¹)	Log K_f
0							
100							
200							
298.15	145.047	150.775	150.775	0.	-1271.663	-1123.597	196.849
300	145.603	151.673	150.777	0.269	-1271.627	-1122.678	195.476
315.500	150.273	159.125	151.005	2.562	------ CRYSTAL <--> LIQUID -----		
400	175.728	197.687	156.848	16.335	-1269.114	-1073.263	140.154
500	205.853	240.135	169.306	35.414	-1263.169	-1024.930	107.074
600	235.978	280.330	184.486	57.506	-1254.423	-978.057	85.147
700	266.102	318.968	200.954	82.610	-1242.867	-932.871	69.612
800	296.227	356.468	218.059	110.726	-1228.486	-889.534	58.081
900	326.352	393.097	235.480	141.855	-1211.263	-848.169	49.226
1000	356.477	429.041	253.044	175.997	-1191.183	-808.875	42.251

PREVIOUS: June 1962 CURRENT: December 1971

Phosphoric Acid (H_3PO_4) $H_3O_4P_1$(l)

PHOSPHORIC ACID (H_3PO_4) CRYSTAL-LIQUID M_r = 97.99518 Phosphoric Acid (H_3PO_4) $H_3O_4P_1$(cr,l)

0 to 315.5 K crystal
above 315.5 K liquid

Refer to the individual tables for details.

Enthalpy Reference Temperature = T_r = 298.15 K Standard State Pressure = $p°$ = 0.1 MPa

T/K	$C_p°$	$S°$	$-[G°-H°(T_r)]/T$	$H°-H°(T_r)$	$\Delta_f H°$	$\Delta_f G°$	Log K_f
	J K⁻¹mol⁻¹			kJ mol⁻¹			
0	0.	0.	INFINITE	-16.980	-1265.927	-1265.927	INFINITE
100	42.844	34.528	183.578	-14.905	-1275.048	-1226.961	640.899
200	75.856	74.598	119.270	-8.935	-1281.203	-1176.480	307.265
298.15	106.064	110.544	110.544	0.	-1284.375	-1124.314	196.975
300	106.608	111.202	110.546	0.197	-1284.411	-1123.321	195.588
315.500	111.303	116.688	110.713	1.885	CRYSTAL <--> LIQUID		
315.500	150.273	159.125	110.713	15.274	TRANSITION		
400	175.728	197.687	125.068	29.048	-1269.114	-1073.263	140.154
500	205.853	240.135	143.882	48.127	-1263.169	-1024.930	107.074
600	235.978	280.330	163.299	70.218	-1254.423	-978.057	85.147
700	266.102	318.968	182.794	95.322	-1242.867	-932.871	69.612
800	296.227	356.468	202.169	123.439	-1228.486	-889.534	58.081
900	326.352	393.097	221.356	154.568	-1211.263	-848.169	49.226
1000	356.477	429.041	240.332	188.709	-1191.183	-808.875	42.251

PREVIOUS: CURRENT: December 1971

Phosphoric Acid (H_3PO_4) $H_3O_4P_1$(cr,l)

PHOSPHINE (PH_3) IDEAL GAS M_r = 33.99758

$S°(298.15\ K) = 210.243\ J\ K^{-1}\ mol^{-1}$

$\Delta_f H°(0\ K) = 30.81 \pm 1.7\ kJ\ mol^{-1}$
$\Delta_f H°(298.15\ K) = 22.89 \pm 1.7\ kJ\ mol^{-1}$

Vibrational Frequencies and Degeneracies

ν, cm^{-1}	ν, cm^{-1}
2322.9(1)	2327.7(2)
992 (1)	1122.4(2)

Point Group: C_{3v} $\sigma = 3$
Bond Distance: P-H = 1.42 Å
Bond Angle: H-P-H = 93° 50'
Product of the Moments of Inertia: $I_A I_B I_C = 0.2826 \times 10^{-117}\ g^3\ cm^6$

Enthalpy of Formation

The enthalpy of formation was derived by Gunn and Green (1). In order to obtain the $\Delta_f H°(298.15\ K)$ they measured the enthalpy of explosive decomposition of phosphine gas. There is a good deal of uncertainty about the enthalpy of formation of $PH_3(g)$, for instance:

$\Delta_f H°(298.15\ K)$	Source
+3.6 kcal mol^{-1}	The enthalpy of reaction between gaseous PH_3 and saturated bromine water (2)
+6.3 kcal mol^{-1}	The enthalpy of combustion of $PH_3(g)$ in $O_2(g)$ to form H_3PO_4 (3).
-2.34 kcal mol^{-1}	Equilibrium data of (4). Their data showed a wide variation in $\Delta_f H°$, as much as 11 kcal mol^{-1}.
+1.6 kcal mol^{-1}	U.S. Nat. Bur. Stand. (5).

The enthalpy of formation have been reviewed by Cottrell (6) and by Bichowsky and Rossini (7).
The enthalpy of formation selected was, +1.3±0.4 kcal mol^{-1}, as reported by Gunn and Green (1).

Heat Capacity and Entropy

The fundamental wave number and degeneracy assignments are from the infrared spectral data by McConaghie and Nielsen (8). The molecular constants are those obtained from infrared data by Nielsen (9). Nielsen (9) has reported the molecular constants after calculating the Coriolis coupling effect between the wave numbers 2 and 4, or (992 cm^{-1}) and (1122.4 cm^{-1}).

The principal moments of inertia are: $I_A = I_B = 0.6264 \times 10^{-39}\ g\ cm^2$ and $I_C = 0.7201 \times 10^{-39}\ g\ cm^2$.

References

1. S. R. Gunn and L. G. Green, J. Phys. Chem. 65, 779 (1961).
2. M. J. Ogier, Ann. Chim. Phys. 20, 5 (1880).
3. M. P. Lemoult, Compt. Rend. 145, 374 (1907).
4. W. N. Ipatiew, and A. W. Frost, Ber. dtsch. Chem. Ges. 63B, 1104 (1930).
5. U.S. Nat. Bur. Stand. Report No. 7437, pp. 80-91, 1962.
6. T. L. Cottrell, "The Strength of Chemical Bonds," Butterworth's Scientific Publications, London, (1958).
7. F. R. Bichowsky and F. D. Rossini, "The Thermochemistry of the Chemical Substances," Reinhold Publishing Corp., New York (1936).
8. V. N. McConaghie and H. H. Nielsen, J. Chem. Phys. 21, 1836, (1953).
9. H. H. Nielsen, J. Chem. Phys. 20, 759 (1952).

Phosphine (PH_3) $H_3P_1(g)$

Enthalpy Reference Temperature = T_r = 298.15 K Standard State Pressure = p° = 0.1 MPa

T/K	$C_p°$	S°	$-[G°-H°(T_r)]/T$	$H°-H°(T_r)$	$\Delta_f H°$	$\Delta_f G°$	Log K_f
	J $K^{-1}mol^{-1}$			kJ mol^{-1}			
0	0.	0.	INFINITE	-10.136	30.811	30.811	INFINITE
100	33.259	173.027	241.134	-6.811	28.751	28.326	-14.796
200	33.933	196.180	213.516	-3.467	25.790	28.803	-7.523
250	35.259	203.881	210.843	-1.740	24.331	29.723	-6.210
298.15	37.102	210.243	210.243	0.	22.886	30.893	-5.412
300	37.180	210.473	210.244	0.069	22.831	30.943	-5.388
350	39.422	216.370	210.704	1.983	20.636	32.483	-4.848
400	41.782	221.787	211.755	4.013	19.164	34.277	-4.476
450	44.151	226.845	213.154	6.161	17.806	36.249	-4.208
500	46.478	231.618	214.764	8.427	16.562	38.366	-4.008
600	50.905	240.489	218.323	13.299	14.408	42.937	-3.738
700	54.935	248.644	222.080	18.595	12.664	47.835	-3.569
800	58.511	256.218	225.879	24.271	11.279	52.958	-3.458
900	61.623	263.294	229.648	30.282	10.195	58.235	-3.380
1000	64.300	269.929	233.348	36.581	9.356	63.620	-3.323
1100	66.585	276.167	236.960	43.128	8.712	69.079	-3.280
1200	68.532	282.047	240.474	49.887	-55.351	75.668	-3.294
1300	70.191	287.599	243.888	56.825	-54.952	86.571	-3.478
1400	71.608	292.854	247.199	63.917	-54.469	97.440	-3.636
1500	72.822	297.837	250.411	71.140	-53.925	108.272	-3.770
1600	73.867	302.571	253.524	78.476	-53.336	119.066	-3.887
1700	74.770	307.077	256.543	85.909	-52.717	129.822	-3.989
1800	75.554	311.374	259.470	93.426	-52.077	140.541	-4.078
1900	76.238	315.477	262.311	101.016	-51.424	151.224	-4.157
2000	76.837	319.404	265.068	108.671	-50.766	161.873	-4.228
2100	77.365	323.165	267.746	116.381	-50.107	172.489	-4.290
2200	77.831	326.775	270.347	124.142	-49.451	183.073	-4.347
2300	78.244	330.244	272.877	131.946	-48.800	193.628	-4.397
2400	78.612	333.582	275.337	139.789	-48.158	204.155	-4.443
2500	78.942	336.798	277.732	147.667	-47.526	214.655	-4.485
2600	79.237	339.900	280.063	155.576	-46.906	225.130	-4.523
2700	79.504	342.896	282.335	163.514	-46.299	235.581	-4.558
2800	79.744	345.792	284.550	171.476	-45.707	246.010	-4.589
2900	79.962	348.594	286.711	179.462	-45.129	256.418	-4.619
3000	80.160	351.308	288.819	187.468	-44.567	266.807	-4.646
3100	80.340	353.939	290.877	195.493	-44.022	277.177	-4.670
3200	80.505	356.493	292.888	203.535	-43.495	287.530	-4.693
3300	80.655	358.972	294.853	211.593	-42.984	297.866	-4.715
3400	80.794	361.382	296.775	219.666	-42.493	308.188	-4.735
3500	80.921	363.726	298.654	227.752	-42.020	318.495	-4.753
3600	81.039	366.007	300.494	235.850	-41.566	328.789	-4.771
3700	81.147	368.229	302.294	243.959	-41.132	339.070	-4.787
3800	81.247	370.395	304.058	252.079	-40.718	349.341	-4.802
3900	81.340	372.506	305.786	260.208	-40.324	359.600	-4.816
4000	81.427	374.567	307.480	268.347	-39.951	369.850	-4.830
4100	81.507	376.578	309.141	276.494	-39.599	380.090	-4.842
4200	81.582	378.544	310.770	284.648	-39.268	390.323	-4.854
4300	81.652	380.464	312.369	292.810	-38.958	400.548	-4.866
4400	81.718	382.342	313.938	300.978	-38.670	410.765	-4.876
4500	81.779	384.179	315.478	309.153	-38.403	420.976	-4.887
4600	81.836	385.977	316.991	317.334	-38.157	431.182	-4.896
4700	81.890	387.738	318.478	325.520	-37.932	441.383	-4.905
4800	81.941	389.462	319.939	333.712	-37.728	451.579	-4.914
4900	81.988	391.152	321.375	341.908	-37.544	461.770	-4.923
5000	82.033	392.809	322.787	350.110	-37.380	471.959	-4.931
5100	82.076	394.434	324.176	358.315	-37.236	482.144	-4.938
5200	82.116	396.028	325.543	366.525	-37.110	492.327	-4.945
5300	82.153	397.593	326.887	374.738	-37.002	502.508	-4.953
5400	82.189	399.129	328.211	382.955	-36.911	512.686	-4.959
5500	82.223	400.637	329.514	391.176	-36.837	522.863	-4.966
5600	82.255	402.119	330.798	399.400	-36.777	533.039	-4.972
5700	82.286	403.575	332.062	407.627	-36.732	543.214	-4.978
5800	82.315	405.006	333.307	415.857	-36.699	553.388	-4.984
5900	82.342	406.414	334.534	424.090	-36.679	563.561	-4.989
6000	82.368	407.798	335.744	432.325	-36.668	573.735	-4.995

PREVIOUS: June 1962 (1 atm) CURRENT: June 1962 (1 bar)

Phosphine (PH_3) $H_3P_1(g)$

AMMONIUM IODIDE (NH_4I) CRYSTAL M_r = 144.94296

$\Delta_fH°$(0 K) = Unknown
S°(298.15 K) = 112.968 ± 6.3 kJ mol^{-1} $\Delta_fH°$(298.15 K) = -202.087 kJ mol^{-1}
T_{trs} = 260 K $\Delta_{trs}H°$ = 2.929 kJ mol^{-1}
T_{fus} = 824 K $\Delta_{fus}H°$ = [20.920] kJ mol^{-1}

Enthalpy of Formation

The Δ_fH, T_{trs} and T_{fus} was taken from U. S. Nat. Bur. Stand. (1). Sublimation data from Luft (3); $\Delta_{sub}H°$(798 K) = 40.28 kcal mol^{-1}.

Heat Capacity and Entropy

Data extrapolated from Simon et al. (4). S°(298.15 K) was taken from Kelley (2).

References

1. U. S. Nat. Bur. Stand. Circ. 500, 1952.
2. K. K. Kelley, U. S. Bur. Mines Bull. 584, 1960.
3. N. W. Luft, Ind. Chemist, 31, 502 (1955).
4. Simon, Simson, and Ruhemann, Z. Phys. Chemie, 129, 339 (1927).

Ammonium Iodide (NH_4I) $H_4I_1N_1$(cr)

Enthalpy Reference Temperature = T_r = 298.15 K Standard State Pressure = p° = 0.1 MPa

T/K	$C_p°$ (J $K^{-1}mol^{-1}$)	S° (J $K^{-1}mol^{-1}$)	$-[G°-H°(T_r)]/T$ (J $K^{-1}mol^{-1}$)	$H°-H°(T_r)$ (kJ mol^{-1})	$\Delta_fH°$ (kJ mol^{-1})	$\Delta_fG°$ (kJ mol^{-1})	Log K_f
0							
100							
200							
298.15	81.755	112.968	112.968	0.	-202.087	-111.966	19.616
300	81.881	113.474	112.970	0.151	-202.120	-111.407	19.398
400	88.998	138.004	116.266	8.695	-211.666	-80.646	10.531
500	96.148	158.632	122.727	17.953	-233.822	-45.736	4.778
600	103.261	176.789	130.252	27.922	-233.079	-8.177	0.712
700	110.458	193.247	138.093	38.608	-231.674	29.204	-2.179
800	117.696	208.468	145.949	50.015	-229.616	66.340	-4.332
900	124.976	222.753	153.697	62.150	-226.914	103.179	-5.988
1000	132.089	236.288	161.285	75.003	-223.590	139.683	-7.296
1100	139.327	249.217	168.695	88.574	-219.652	175.825	-8.349
1200	146.440	261.656	175.927	102.875	-215.099	211.581	-9.210
1300	152.030	273.609	182.984	117.812	-210.029	246.935	-9.922
1400	156.098	285.034	189.869	133.231	-204.598	281.884	-10.517
1500	158.643	295.899	196.579	148.980	-198.962	316.437	-11.019
1600	159.666	306.178	203.110	164.909	-193.275	350.611	-11.446
1700	159.166	315.851	209.461	180.863	-187.688	384.432	-11.812
1800	157.144	324.898	215.626	196.691	-182.352	417.931	-12.128
1900	153.599	333.307	221.601	212.241	-177.418	451.143	-12.403
2000	148.532	341.063	227.383	227.360	-173.034	484.109	-12.644

PREVIOUS: CURRENT: March 1962

Ammonium Iodide (NH_4I) $H_4I_1N_1$(cr)

HYDRAZINE (N_2H_4) LIQUID M_r = 32.04516

$S°(298.15\ K) = 121.544 \pm 0.4\ J\ K^{-1}\ mol^{-1}$ $\Delta_fH°(298.15\ K) = 50.626 \pm 0.4\ kJ\ mol^{-1}$

$T_{fus} = 274.69\ K$ $\Delta_{fus}H° = 12.657\ kJ\ mol^{-1}$

Enthalpy of Formation

Hughes et al. (1), measured the enthalpies of combustion of hydrazine and hydrazine monohydrate. The data, adjusted as suggested by Cole and Gilbert (2), are summarized below. Enthalpies of solution determined by Bushnell et al. (3) interrelate $N_2H_4(\ell)$. An intermediate value was selected from the resulting enthalpies of formation.

Source	Reaction	$\Delta_rH°(298.15\ K)$ kcal mol^{-1}	$\Delta_fH°(298.15\ K)$ kcal mol^{-1}
Hughes (1)	$N_2H_4(\ell) + O_2(g) \rightarrow N_2(g) + 2H_2O(\ell)$	-148.68±0.06	12.05
Hughes (1)	$N_2H_4 \cdot H_2O(\ell) + O_2(g) \rightarrow N_2(g) + 3H_2O(\ell)$	-147.00±0.06	-
Bushnell (3)	$N_2H_4(\ell) + H_2O(\ell) \rightarrow N_2H_4 \cdot H_2O(\ell)$	-1.80	12.17*

*From combination of second and third reactions.

Heat Capacity and Entropy

Heat capacities (12-340 K) of hydrazine crystal and liquid were determined by Scott et al. (4). Values beyond the experimental range for the liquid were extrapolated linearly. The entropy was obtained from the enthalpy of melting and the smoothed heat capacities of the crystal using $S°(12) = 0.023\ cal\ K^{-1}\ mol^{-1}$.

Fusion Data

Taken from Scott et al. (4).

Vaporization Data

The normal boiling point and enthalpy of vaporization are those calculated for the ideal gas from these tables. The boiling point found for the real gas (387.3 K) by Hieber and Woerner (5) is in good agreement. The selected enthalpy of vaporization, $\Delta H°(298.15\ K) = 10.69\ kcal\ mol^{-1}$, is based on the vapor pressure data of Scott et al., (4), series III. Vapor pressure measurements of Hieber and Woerner (5), are in satisfactory agreement as shown below.

Source	T/K	$\Delta_{vap}H°(298.15\ K)$, kcal mol^{-1} 2nd law	3rd law	Drift cal K^{-1} mol^{-1}
Scott, et al. (4)	273-343	10.683±0.003	10.696	0.05±0.01
Chang, Gockcen (6)	276-325	10.40±0.03	10.68	0.91±0.09
Hieber, Woerner (5)	293-387	10.82±0.03	10.71	-0.33±0.07

The 2nd law-3rd law consistency of Scott et al. (4) reflects the use of these data in selection of the gas phase functions (see H_4N_2 gas).

References

1. A. M. Hughes, R. J. Corruccini and E. C. Gilbert, J. Amer. Chem. Soc. 61, 2639 (1939).
2. L. G. Cole and E. C. Gilbert, J. Amer. Chem. Soc. 73, 5423 (1951).
3. V. C. Bushnell, A. M. Hughes and E. C. Gilbert, J. Amer. Chem. Soc. 59, 2142 (1937).
4. D. W. Scott, G. D. Oliver, M. E. Gross, W. N. Hubbard and H. M. Huffman, J. Amer. Chem. Soc. 71, 2293 (1949).
5. W. Hieber and A. Woerner, Z. Elektrochem. 40, 252 (1934).
6. E. T. Chang and N. A. Gokcen, AD-458288, Defense Documentation Center, January, 1964.

Hydrazine (N_2H_4) $H_4N_2(l)$

Enthalpy Reference Temperature = T_r = 298.15 K — Standard State Pressure = p° = 0.1 MPa

T/K	$C_p°$ (J $K^{-1}mol^{-1}$)	S° (J $K^{-1}mol^{-1}$)	$-[G°-H°(T_r)]/T$ (J $K^{-1}mol^{-1}$)	$H°-H°(T_r)$ (kJ mol^{-1})	$\Delta_fH°$ (kJ mol^{-1})	$\Delta_fG°$ (kJ mol^{-1})	Log K_f
0							
100	87.236	21.037	205.395	-18.436	48.895	82.918	-43.312
200	93.094	83.302	130.399	-9.419	49.613	116.715	-30.483
298.15	98.840	121.544	121.544	0.	50.626	149.440	-26.181
300	98.952	122.156	121.546	0.183	50.649	150.054	-26.127
400	107.361	151.693	125.516	10.471	52.208	182.975	-23.894
500	116.566	176.639	133.305	21.667	54.619	215.406	-22.503
600	125.771	198.705	142.398	33.784	57.895	247.270	-21.527
700	134.976	218.784	151.896	46.822	62.013	278.518	-20.783
800	144.181	237.409	161.435	60.779	66.957	309.120	-20.183

PREVIOUS: CURRENT: December 1965

Hydrazine (N_2H_4) $H_4N_2(l)$

$\Delta_fH°$(0 K) = 109.43 kJ mol^{-1}

S°(298.15 K) = 238.719 ± 1.7 J K^{-1} mol^{-1} $\Delta_fH°$(298.15 K) = 95.35 ± 0.8 kJ mol^{-1}

Vibrational Frequencies and Degeneracies

ν, cm^{-1}	ν, cm^{-1}	ν, cm^{-1}
3350	875	3280
3261	780	1275
1312	377	[933]
1098	3330	966

Ground State Quantum Weight: 1 σ = 2

Point Group: C_2

Bond Distances: N-H = 1.022 Å N-N = 1.449 Å

Bond Angles: H-N-H = 109.47° H-N-N = 112.0°

Torsional angle between two NH_2 groups ≥ 90° from eclipsed position.

Product of the Moments of Inertia: $I_AI_BI_C$ = 7.3095 x 10^{-117} g^3 cm^6

Enthalpy of Formation

This was obtained from the enthalpy of formation of the liquid and the selected enthalpy of vaporization, $\Delta_vH°$(298.15 K) = 10.69 kcal mol^{-1}, derived from the vapor pressure data (Series III) of Scott et al. (1). Other vapor pressure data from Hieber and Woerner (2) and Chang and Gokcen (3) are in satisfactory agreement (see N_2H_4 liquid table).

Heat Capacity and Entropy

Functions were calculated from the constants given above using the rigid rotator harmonic oscillator method. The entropy was increased by R ℓn 2 because two rotameric forms are implied by use of the torsional frequency. Small but arbitrary adjustments were made in the assignment of the bending mode frequencies in order to reproduce the vapor pressure data of Scott et al. (1) as closely as possible. Calculated values of S°(298.15 K) = 57.03 and S°(340 K) = 58.70 cal K^{-1} mol^{-1} may be compared with 56.99 and 58.69, respectively, derived from the data of Scott et al. Internal rotation calculations would require a complex potential function in order to fit the data. The barrier to inversion (990 cm^{-1} ~ 2.8 kcal mol^{-1}) is slightly less than the barrier to internal rotation at the trans position (1110 cm^{-1} ~ 3.2 kcal mol^{-1}) according to analyses of the microwave data by Kasuya (4) and Kasuya and Kojima (5).

From matrix isolation and gas phase infrared studies, Catalano et al. (6) have suggested possible uncertainties in the assignment of the bending mode frequencies in the region from 1630 to 740 cm^{-1}. The question hinges on the uncertain existence of a gas phase fundamental at about 1600 cm^{-1} as observed by Giguere and Liu (7). No corresponding matrix band was found and the gas phase band appeared to be due to absorption on the cell windows. A vapor phase Raman frequency at 1564 cm^{-1} was observed, however, by Kotov and Tatevskii (8). Catalano et al. suggest one assignment based on the uncertain fundamental and a tentative alternative assignment excluding it. The two assignments give entropies that are 0.26 cal K^{-1} mol^{-1} lower and 0.12 higher than the 3rd law value. The lower calculated entropy is probably more reasonable, but the higher value could be more easily modified to agree with the 3rd law value. Thus the alternative assignment was adopted with ν_{11} arbitrarily increased from ~810 to 933 cm^{-1}.

The structural constants are based primarily on the electron diffraction study of Morino et al. (9). Related analyses of the infrared fine structure of the torsional frequency (377 cm^{-1}), Yamaguchi et al. (10) and of the microwave spectra, Kasuya and Kojima (5) are in close agreement. The microwave data give an H-N-H angle of 106° and a torsional angle of 90°2'. Principal moments of inertia of the adopted structure are: I_A = 0.5741 x 10^{-39}, I_B = 3.5633 x 10^{-39} and I_C = 3.5734 x 10^{-39} g cm^2.

References

1. D. W. Scott, G. D. Oliver, M. E. Gross, W. N. Hubbard and H. M. Huffman, J. Amer. Chem. Soc. 71, 2293 (1949).
2. W. Hieber and A. Woerner, Z. Elektrochem. 40, 252 (1934).
3. E. T. Chang and N. A. Gokcen, AD-458288, Defense Documentation Center, January, 1964.
4. T. Kasuya, Sci. Papers Inst. Phys. Chem. Res. (Tokyo) 56, No. 1, 1-39 (1962).
5. T. Kasuya and T. Kojima, J. Phys. Soc. Japan 18, 364-8 (1963).
6. E. Catalano, R. H. Sanborn and J. W. Frazer, J. Chem. Phys. 38, 2265 (1963).
7. P. A. Giguere and I. D. Liu, J. Chem. Phys. 20, 136 (1952).
8. Yu. I. Kotov and V. M. Tatevskii, Opt. i Spektroskopiya 15, 65 (1963).
9. Y. Morino, T. Iijima and Y. Murata, Bull. Chem. Soc. Japan 33, 46 (1960).
10. A. Yamaguchi, I. Ichishima, T. Shimanouchi and S. Mizushima, Spectrochim. Acta 16, 1471 (1960).

Enthalpy Reference Temperature = T_r = 298.15 K Standard State Pressure = p° = 0.1 MPa

T/K	$C_p°$	S°	-[G°-H°(T_r)]/T	H°-H°(T_r)	$\Delta_fH°$	$\Delta_fG°$	Log K_f
	J K^{-1}mol^{-1}			kJ mol^{-1}			
0	0.	0.	INFINITE	-11.528	109.430	109.430	INFINITE
100	34.368	195.185	277.002	-8.182	103.876	120.484	-62.934
200	40.658	220.673	243.020	-4.469	99.290	138.917	-36.281
250	45.534	230.252	239.526	-2.318	97.193	149.070	-31.146
298.15	50.813	238.719	238.719	0.	95.353	159.232	-27.897
300	51.019	239.034	238.720	0.094	95.287	159.628	-27.794
350	56.531	247.315	239.362	2.784	93.621	170.487	-25.444
400	61.702	255.207	240.853	5.742	92.205	181.567	-23.710
450	66.379	262.749	242.870	8.946	91.023	192.811	-22.381
500	70.544	269.963	245.221	12.371	90.050	204.175	-21.330
600	77.562	283.469	250.487	19.789	88.626	227.144	-19.775
700	83.302	295.869	256.096	27.841	87.760	250.305	-18.678
800	88.202	307.320	261.793	36.422	87.326	273.560	-17.862
900	92.511	317.962	267.449	45.462	87.240	296.848	-17.229
1000	96.356	327.912	273.004	54.908	87.439	320.130	-16.722
1100	99.803	337.260	278.424	64.720	87.875	343.379	-16.306
1200	102.891	346.079	283.698	74.857	88.507	366.581	-15.957
1300	105.653	354.426	288.821	85.287	89.302	389.722	-15.659
1400	108.119	362.348	293.792	95.978	90.231	412.797	-15.402
1500	110.318	369.884	298.616	106.902	91.270	435.802	-15.176
1600	112.279	377.067	303.296	118.034	92.401	458.734	-14.976
1700	114.026	383.928	307.839	129.350	93.606	481.594	-14.798
1800	115.586	390.490	312.250	140.833	94.871	504.380	-14.637
1900	116.980	396.778	316.534	152.462	96.185	527.095	-14.491
2000	118.227	402.810	320.698	164.224	97.538	549.740	-14.358
2100	119.346	408.606	324.747	176.103	98.922	572.316	-14.236
2200	120.352	414.182	328.687	188.089	100.330	594.826	-14.123
2300	121.257	419.552	332.521	200.170	101.756	617.271	-14.019
2400	122.075	424.730	336.256	212.338	103.195	639.654	-13.922
2500	122.815	429.729	339.896	224.583	104.644	661.977	-13.831
2600	123.486	434.559	343.444	236.898	106.097	684.241	-13.747
2700	124.096	439.231	346.906	249.278	107.553	706.450	-13.667
2800	124.652	443.754	350.284	261.716	109.007	728.604	-13.592
2900	125.159	448.137	353.583	274.207	110.459	750.707	-13.522
3000	125.624	452.388	356.806	286.746	111.904	772.760	-13.455
3100	126.050	456.515	359.956	299.330	113.342	794.765	-13.392
3200	126.441	460.523	363.037	311.955	114.770	816.723	-13.332
3300	126.802	464.419	366.050	324.617	116.187	838.638	-13.275
3400	127.134	468.210	368.999	337.314	117.591	860.509	-13.220
3500	127.441	471.899	371.887	350.043	118.980	882.339	-13.168
3600	127.726	475.493	374.715	362.802	120.354	904.130	-13.119
3700	127.989	478.997	377.487	375.588	121.711	925.883	-13.071
3800	128.234	482.413	380.203	388.399	123.050	947.599	-13.026
3900	128.462	485.747	382.867	401.234	124.370	969.281	-12.982
4000	128.675	489.002	385.480	414.091	125.669	990.928	-12.940
4100	128.873	492.182	388.043	426.968	126.948	1012.544	-12.900
4200	129.058	495.290	390.560	439.865	128.205	1034.129	-12.861
4300	129.232	498.329	393.031	452.780	129.440	1055.684	-12.824
4400	129.394	501.302	395.458	465.711	130.652	1077.210	-12.788
4500	129.546	504.211	397.843	478.658	131.841	1098.709	-12.753
4600	129.690	507.060	400.186	491.620	133.007	1120.182	-12.720
4700	129.824	509.851	402.490	504.596	134.149	1141.630	-12.688
4800	129.951	512.585	404.755	517.585	135.268	1163.055	-12.657
4900	130.071	515.266	406.983	530.586	136.364	1184.455	-12.626
5000	130.183	517.895	409.175	543.599	137.436	1205.835	-12.597
5100	130.290	520.474	411.332	556.622	138.487	1227.191	-12.569
5200	130.391	523.005	413.456	569.656	139.515	1248.529	-12.542
5300	130.486	525.489	415.546	582.700	140.522	1269.846	-12.515
5400	130.576	527.929	417.605	595.753	141.509	1291.145	-12.489
5500	130.662	530.326	419.632	608.815	142.477	1312.425	-12.464
5600	130.743	532.681	421.630	621.886	143.426	1333.689	-12.440
5700	130.820	534.996	423.599	634.964	144.359	1354.935	-12.417
5800	130.894	537.272	425.539	648.050	145.275	1376.166	-12.394
5900	130.964	539.510	427.452	661.142	146.177	1397.379	-12.371
6000	131.030	541.712	429.338	674.242	147.065	1418.579	-12.350

PREVIOUS: December 1965 (1 atm) CURRENT: December 1965 (1 bar)

SULFURIC ACID, MONOHYDRATE ($H_2SO_4 \cdot H_2O$) CRYSTAL-LIQUID M_r = 116.08876

$\Delta_f H°(0\ K) = -1126.433 \pm 0.8\ kJ\ mol^{-1}$

$S°(298.15\ K) = 211.510 \pm 0.21\ J\ K^{-1}\ mol^{-1}$ $\Delta_f H°(298.15\ K) = -1127.621 \pm 0.8\ kJ\ mol^{-1}$

$T_{fus} = 281.63 \pm 0.05\ K$ $\Delta_{fus}H° = 19.464 \pm 0.042\ kJ\ mol^{-1}$

Enthalpy of Formation

The enthalpy of formation of the monohydrate of sulfuric acid is derived from the enthalpy of solution data of Giauque et al. (1). We calculate $\Delta_f H°(298.15\ K)$ using $\Delta_r H°(298.15\ K) = -6.645\ kcal\ mol^{-1}$ for the solution process $H_2SO_4(\ell) + H_2O(\ell) = H_2SO_4 \cdot H_2O(\ell)$ and auxiliary data for H_2SO_4 (2) and H_2O (3).

Heat Capacity and Entropy

The heat capacity has been measured experimentally (4) in the range 15 - 300 K and has been re-analyzed and summarized by Giauque et al. (1). Above 300 K, the heat capacity values are derived from a nomograph constructed by Kharbanda (5) based on experimental data (273 - 573 K). These temperature-dependent values are corrected to current atomic weights, converted from IPTS-48 to IPTS-68, and adjusted graphically in the region 290 - 390 K so as to provide smoothly varying heat capacity values. In the region above 390 K, a linear dependence is assumed (as suggested from a least squares fit of the nomograph data); this linear dependence is used to obtain heat capacity data by extrapolation to 1000 K.

The entropy of 298.15 K is calculated by the appropriate integration of the crystal and liquid heat capacity values, addition of the entropy of melting, and $S°(15\ K) = 0.190\ cal\ K^{-1}\ mol^{-1}$.

Fusion Data

The melting point and enthalpy of melting have been measured calorimetrically by Rubin and Giauque (4). The values used by Giauque et al. (1) in their re-evaluation and correlation of aqueous sulfuric acid properties were slightly different from those reported by Rubin and Giauque (4); the changes were 0.01 K in the melting point and 6 cal $K^{-1}\ mol^{-1}$ in the enthalpy of melting. In addition, the value for T_{fus} is corrected from IPTS-48 to IPTS-68.

References

1. W. F. Giauque, E. W. Hornung, J. E. Kinzler, and T. R. Rubin, J. Amer. Chem. Soc. 82, 62 (1960).
2. JANAF Thermochemical Tables: $H_2SO_4(\ell)$, 9-30-77.
3. ICSU-CODATA Task Group, J. Chem. Thermodyn. 4, 331 (1972).
4. T. R. Rubin and W. F. Giauque, J. Amer. Chem. Soc. 74, 800 (1952).
5. Om. P. Kharbanda, Ind. Chemist 31, 538 (1955).

Sulfuric Acid, Monohydrate ($H_2SO_4 \cdot H_2O$) $H_4O_5S_1$(cr,l)

Enthalpy Reference Temperature = T_r = 298.15 K Standard State Pressure = p° = 0.1 MPa

T/K	$C_p°$ (J K⁻¹mol⁻¹)	S° (J K⁻¹mol⁻¹)	$-[G°-H°(T_r)]/T$ (J K⁻¹mol⁻¹)	$H°-H°(T_r)$ (kJ mol⁻¹)	$\Delta_f H°$ (kJ mol⁻¹)	$\Delta_f G°$ (kJ mol⁻¹)	Log K_f
0	0.	0.	INFINITE	-41.866	-1126.433	-1126.433	INFINITE
100	54.220	43.699	436.019	-39.232	-1137.748	-1077.394	562.773
200	92.914	93.433	252.619	-31.837	-1144.661	-1014.113	264.859
281.626	123.973	130.247	211.864	-22.985	CRYSTAL <--> LIQUID		
281.626	211.961	199.360	211.864	-3.521	TRANSITION		
298.15	214.264	211.510	211.510	0.	-1127.621	-950.290	166.487
300	214.523	212.836	211.514	0.397	-1127.510	-949.190	165.269
400	228.538	276.470	220.096	22.550	-1123.193	-890.620	116.303
500	243.601	329.085	236.772	46.157	-1117.007	-833.180	87.042
600	258.663	374.830	256.047	71.270	-1109.236	-777.118	67.654
700	273.726	415.834	275.992	97.889	-1099.975	-722.477	53.912
800	288.788	453.369	295.850	126.015	-1089.308	-669.261	43.698
900	303.850	488.253	315.312	155.647	-1130.396	-616.382	35.774
1000	318.913	521.046	334.261	186.785	-1115.776	-560.042	29.254

PREVIOUS: CURRENT: September 1977

Sulfuric Acid, Monohydrate ($H_2SO_4 \cdot H_2O$) $H_4O_5S_1$(cr,l)

SILANE (SiH_4) IDEAL GAS M_r = 32.11726

$S°(298.15 K) = 204.653 J K^{-1} mol^{-1}$

$\Delta_f H°(0 K) = 43.92 \pm 2.1 kJ mol^{-1}$
$\Delta_f H°(298.15 K) = 34.31 \pm 2.1 kJ mol^{-1}$

Vibrational Frequencies and Degeneracies

ν, cm^{-1}
2185.7 (1)
972.1 (2)
2189.08(3)
913.28(3)

Ground State Quantum Weight: [1] $\sigma = 12$
Point Group: T_d
Bond Distance: Si-H = 1.4806 ± 0.001 Å
Bond Angle: H-Si-H = 109.47122
Product of the Moments of Inertia: $I_A I_B I_C = 0.9367 \times 10^{-117} g^3 cm^6$

Enthalpy of Formation

The adopted $\Delta_f H°(SiH_4, g, 298.15 K) = 8.2\pm0.5 kcal mol^{-1}$ is derived from the enthalpy of decomposition measurements by Gunn and Green (1). We have taken the average of the three measurements, $\Delta E = -7.83 kcal mol^{-1}$, to obtain $\Delta H = -7.24 kcal mol^{-1}$ for the decomposition. Some question exists about the final state of the silicon; we choose to consider it as amorphous and apply a correction of 1 kcal mol^{-1} to convert to the crystalline state (2) (the authors chose to neglect this correction). The results of Gunn and Green (1) are in good agreement with the value of $\Delta_f H°(SiH_4, g, 298.15 K) = 7.8\pm3.5 kcal mol^{-1}$ determined by Brimm and Humphreys (3) from high temperature (680°C) enthalpy of decomposition measurements. We question the earlier negative values for the enthalpy of formation of $SiH_4(g)$ (4-6) which range from -8.7 to -14.8 kcal mol^{-1} and the later result of Feher, et al. (7), -11.3 kcal mol^{-1}, because of suspected uncertainty in the final states and/or corrections for the final states.

Heat Capacity and Entropy

The four fundamental vibrational frequencies and the Si-H bond distance are the result of gas-phase infrared and Raman spectral studies by Kattenberg and Oskam (8). They are in excellent agreement with values selected from earlier work as given in the compilations of Janz and Mikawa (9) and Shimanouchi (10). The heat capacity and entropy calculations are based on a tetrahedral structure. The $S°(298.15 K) = 48.89\pm0.01 cal K^{-1} mol^{-1}$ is in good agreement with another evaluation (2).

The principal moments of inertia are $I_A = I_B = I_C = 0.9784 \times 10^{-39} g cm^2$.

References
1. S. R. Gunn and L. H. Green, J. Phys. Chem. 65, 779 (1961).
2. U. S. Nat. Bur. Stand. Tech. Note 270-3, 1968.
3. E. O. Brimm and H. M. Humphreys, J. Phys. Chem. 61, 829 (1957).
4. H. Von Wartenberg, Z. Anorg. Allgem. Chem. 79, 71 (1913).
5. F. R. Bichowsky and F. D. Rossini, "The Thermochemistry of Chemical Substances," Reinhold, New York, 1936.
6. U. S. Nat. Bur. Stand. Circ. 500, 1952.
7. F. Feher, G. Jansen, and H. Rohmer, Z. Anorg. Allg. Chem. 329, 31 (1964).
8. H. W. Kattenberg and A. Oskam, J. Mol. Spectrosc. 49, 52 (1974).
9. G. J. Janz and Y. Mikawa, Bull. Chem. Soc. Japan 34, 1495 (1961).
10. T. Shimanouchi, NSRDS-NBS 39, 1972.

Silane (SiH_4) $H_4Si_1(g)$

Enthalpy Reference Temperature = T_r = 298.15 K Standard State Pressure = p° = 0.1 MPa

	J $K^{-1}mol^{-1}$			kJ mol^{-1}			
T/K	$C_p°$	S°	$-[G°-H°(T_r)]/T$	$H°-H°(T_r)$	$\Delta_f H°$	$\Delta_f G°$	Log K_f
0	0.	0.	INFINITE	-10.538	43.923	43.923	INFINITE
100	33.269	165.774	237.899	-7.212	40.984	44.936	-23.472
200	35.520	189.213	208.310	-3.819	37.811	50.067	-13.076
250	38.857	197.475	205.335	-1.965	36.023	53.336	-11.144
298.15	42.827	204.653	204.653	0.	34.309	56.827	-9.956
300	42.985	204.918	204.654	0.079	34.244	56.967	-9.919
350	47.291	211.868	205.193	2.336	32.568	60.888	-9.087
400	51.473	218.458	206.442	4.806	31.038	65.040	-8.493
450	55.431	224.752	208.130	7.480	29.667	69.374	-8.053
500	59.148	230.786	210.096	10.345	28.454	73.853	-7.715
600	65.877	242.179	214.505	16.604	26.478	83.128	-7.237
700	71.715	252.784	219.225	23.491	25.042	92.690	-6.917
800	76.713	262.695	224.046	30.919	24.056	102.426	-6.688
900	80.948	271.982	228.862	38.808	23.433	112.263	-6.516
1000	84.513	280.701	233.614	47.086	23.093	122.153	-6.381
1100	87.507	288.900	238.271	55.692	22.965	132.067	-6.271
1200	90.024	296.625	242.815	64.572	22.989	141.986	-6.180
1300	92.146	303.917	247.238	73.683	23.119	151.897	-6.103
1400	93.942	310.814	251.535	82.990	23.315	161.796	-6.037
1500	95.470	317.348	255.706	92.463	23.548	171.680	-5.978
1600	96.777	323.553	259.755	102.077	23.793	181.547	-5.927
1700	97.901	329.454	263.682	111.812	-26.143	191.846	-5.895
1800	98.872	335.078	267.494	121.652	-25.691	204.656	-5.939
1900	99.716	340.447	271.193	131.582	-25.226	217.440	-5.978
2000	100.454	345.581	274.785	141.592	-24.756	230.200	-6.012
2100	101.101	350.498	278.274	151.670	-24.288	242.937	-6.043
2200	101.671	355.215	281.665	161.809	-23.826	255.651	-6.070
2300	102.176	359.746	284.962	172.002	-23.375	268.344	-6.094
2400	102.625	364.104	288.170	182.243	-22.936	281.018	-6.116
2500	103.026	368.302	291.291	192.526	-22.513	293.674	-6.136
2600	103.385	372.350	294.332	202.847	-22.108	306.313	-6.154
2700	103.708	376.258	297.294	213.202	-21.720	318.937	-6.170
2800	104.000	380.035	300.182	223.587	-21.353	331.548	-6.185
2900	104.264	383.689	302.999	234.001	-21.007	344.145	-6.199
3000	104.503	387.228	305.748	244.439	-20.683	356.731	-6.211
3100	104.721	390.658	308.432	254.900	-20.381	369.306	-6.223
3200	104.920	393.986	311.054	265.383	-20.102	381.872	-6.233
3300	105.102	397.217	313.616	275.884	-19.847	394.430	-6.243
3400	105.269	400.357	316.121	286.402	-19.617	406.980	-6.252
3500	105.422	403.411	318.572	296.937	-19.412	419.524	-6.261
3600	105.564	406.383	320.970	307.486	-403.383	442.523	-6.421
3700	105.694	409.277	323.318	318.049	-402.815	466.013	-6.579
3800	105.815	412.097	325.617	328.625	-402.275	489.488	-6.728
3900	105.927	414.847	327.870	339.212	-401.764	512.949	-6.870
4000	106.031	417.530	330.078	349.810	-401.280	536.396	-7.005
4100	106.127	420.150	332.243	360.418	-400.824	559.833	-7.132
4200	106.217	422.708	334.367	371.035	-400.397	583.258	-7.254
4300	106.302	425.209	336.450	381.661	-399.997	606.674	-7.370
4400	106.380	427.653	338.495	392.295	-399.625	630.080	-7.480
4500	106.454	430.045	340.503	402.937	-399.280	653.478	-7.585
4600	106.522	432.385	342.475	413.586	-398.963	676.869	-7.686
4700	106.587	434.677	344.413	424.242	-398.672	700.254	-7.782
4800	106.648	436.922	346.317	434.903	-398.408	723.633	-7.875
4900	106.705	439.121	348.188	445.571	-398.170	747.006	-7.963
5000	106.759	441.277	350.029	456.244	-397.957	770.375	-8.048
5100	106.810	443.392	351.839	466.923	-397.768	793.739	-8.130
5200	106.858	445.467	353.619	477.606	-397.603	817.101	-8.208
5300	106.903	447.502	355.372	488.294	-397.460	840.459	-8.283
5400	106.946	449.501	357.096	498.986	-397.339	863.815	-8.356
5500	106.986	451.464	358.794	509.683	-397.239	887.168	-8.426
5600	107.025	453.392	360.466	520.384	-397.158	910.521	-8.493
5700	107.061	455.287	362.113	531.088	-397.095	933.872	-8.558
5800	107.096	457.149	363.736	541.796	-397.050	957.222	-8.621
5900	107.129	458.980	365.335	552.507	-397.019	980.570	-8.681
6000	107.160	460.781	366.910	563.222	-397.003	1003.919	-8.740

PREVIOUS: June 1976 (1 atm) CURRENT: June 1976 (1 bar)

Silane (SiH_4) $H_4Si_1(g)$

SULFURIC ACID, DIHYDRATE ($H_2SO_4 \cdot 2\ H_2O$) CRYSTAL–LIQUID M_r = 134.10404

$\Delta_f H°(0\ K) = -1422.274 \pm 0.8\ kJ\ mol^{-1}$
$S°(298.15\ K) = 276.363 \pm 0.42\ J\ K^{-1}\ mol^{-1}$ $\Delta_f H°(298.15\ K) = -1427.100 \pm 0.8\ kJ\ mol^{-1}$
$T_{fus} = 233.70 \pm 0.05\ K$ $\Delta_{fus} H° = 18.242 \pm 0.17\ kJ\ mol^{-1}$

Enthalpy of Formation

The enthalpy of formation of the dihydrate of sulfuric acid is derived from the enthalpy of solution data of Giauque et al. (1). We calculate $\Delta_f H°(298.15\ K)$ for the dihydrate using $\Delta_r H°(298.15\ K) = -9.907\ kcal\ mol^{-1}$ for the solution process $H_2SO_4(\ell) + 2\ H_2O(\ell) = H_2SO_4 \cdot 2\ H_2O(\ell)$ and auxiliary data for H_2SO_4 (2) and H_2O (3).

Heat Capacity and Entropy

The heat capacity has been measured experimentally (4) in the range 15 - 300 K and has been summarized by Giauque et al. (1). Above 300 K, the heat capacity values are derived from a nomograph constructed by Kharbanda (5) based on experimental data (273 - 573 K). These temperature-dependent values are corrected to current atomic weights, converted from IPTS-48 to IPTS-68, and adjusted graphically in the region 290 - 310 K so as to provide smoothly varying heat capacity values. In the region above 310 K, a linear dependence is assumed (as suggested from a least squares fit of the nomograph data); this linear dependence is used to obtain heat capacity data by extrapolation to 1000 K.

The entropy at 298.15 K is calculated by the appropriate integration of the crystal and liquid heat capacity values, addition of the entropy of melting, and $S°(15\ K) = 0.363\ cal\ K^{-1}\ mol^{-1}$.

Fusion Data

The melting point and enthalpy of melting have been measured calorimetrically by Rubin and Giauque (4). The melting point adopted in this table differs from that reported by Rubin and Giauque (4) by 0.02 K. This is due to the re-analysis of the data by Giauque et al. (1) and a conversion of the resulting T_{fus} value from IPTS-48 to IPTS-68.

References

1. W. F. Giauque, E. W. Horning, J. E. Kunzler, and T. R. Rubin, J. Amer. Chem. Soc. 82, 62 (1960).
2. JANAF Thermochemical Tables: $H_2SO_4(\ell)$, 9-30-77.
3. ICSU-CODATA Task Group, J. Chem. Thermodyn. 4, 331 (1972).
4. T. R. Rubin and W. F. Giauque, J. Amer. Chem. Soc. 74, 800 (1952).
5. Om. P. Kharbanda, Ind. Chemist 31, 538 (1955).

Sulfuric Acid, Dihydrate ($H_2SO_4 \cdot 2H_2O$) $H_6O_6S_1$(cr,l)

Enthalpy Reference Temperature = T_r = 298.15 K Standard State Pressure = p° = 0.1 MPa

T/K	$C_p°$ (J K⁻¹mol⁻¹)	S° (J K⁻¹mol⁻¹)	$-[G°-H°(T_r)]/T$ (J K⁻¹mol⁻¹)	$H°-H°(T_r)$ (kJ mol⁻¹)	$\Delta_f H°$ (kJ mol⁻¹)	$\Delta_f G°$ (kJ mol⁻¹)	Log K_f
0	0.	0.	INFINITE	-51.038	-1422.274	-1422.274	INFINITE
100	66.143	56.938	534.191	-47.725	-1437.362	-1359.594	710.179
200	113.413	117.668	311.144	-38.695	-1446.789	-1277.857	333.742
233.700	128.910	136.505	284.609	-34.612	CRYSTAL <--> LIQUID TRANSITION		
233.700	248.446	214.564	284.609	-16.370			
298.15	261.453	276.363	276.363	0.	-1427.100	-1199.560	210.158
300	262.055	277.982	276.368	0.484	-1426.981	-1198.148	208.616
400	294.604	357.829	287.036	28.317	-1421.376	-1122.886	146.634
500	327.155	427.065	308.254	59.405	-1412.161	-1049.281	109.618
600	359.707	489.589	333.342	93.748	-1399.668	-977.824	85.127
700	392.258	547.483	359.845	131.347	-1383.994	-908.713	67.809
800	424.810	601.987	386.737	172.200	-1365.221	-842.061	54.981
900	457.361	653.901	413.559	216.309	-1396.509	-776.864	45.088
1000	489.913	703.774	440.102	263.672	-1370.399	-709.387	37.055

PREVIOUS: CURRENT: September 1977

Sulfuric Acid, Dihydrate ($H_2SO_4 \cdot 2H_2O$) $H_6O_6S_1$(cr,l)

SULFURIC ACID, TRIHYDRATE ($H_2SO_4 \cdot 3\ H_2O$) CRYSTAL–LIQUID M_r = 152.11932

$S°(298.15\ K) = 345.373 \pm 0.42\ J\ K^{-1}\ mol^{-1}$

$T_{fus} = 236.78 \pm 0.05\ K$

$\Delta_f H°(0\ K) = -1716.249 \pm 0.8\ kJ\ mol^{-1}$

$\Delta_f H°(298.15\ K) = -1720.402 \pm 0.8\ kJ\ mol^{-1}$

$\Delta_{fus} H° = 23.999 \pm 0.21\ kJ\ mol^{-1}$

Enthalpy of Formation

The enthalpy of formation of the trihydrate of sulfuric acid is derived from the enthalpy of solution data of Giauque et al. (1). We calculate $\Delta_f H°(298.15\ K)$ for the trihydrate using $\Delta_r H°(298.15\ K) = -11.693\ kcal\ mol^{-1}$ for the solution process $H_2SO_4(\ell) + 3\ H_2O(\ell) = H_2SO_4 \cdot 3\ H_2O(\ell)$ and auxiliary data for H_2SO_4 (2) and H_2O (3).

Heat Capacity and Entropy

The heat capacity has been measured experimentally (4, 5) in the range 15-300 K and has been summarized by Giauque et al. (1). Above 300 K, the heat capacity values are derived from a nomograph constructed by Kharbanda (6) based on experimental data (273-573 K). These temperature-dependent values are corrected to current atomic weights and converted from IPTS-48 to IPTS-68. Of the five hydrates of sulfuric acid (mono-, di-, tri-, tetra-, and hemihexa-) only in the case of the trihydrate is there a serious discrepancy in the heat capacity values (at ambient temperature) between Giauque et al. (6) and Kharbanda (7). Assuming that the data of Giauque et al. (1) are correct and that the linear dependence suggested by the nomograph of Kharbanda (6) is correct, we have shifted the data derived from the nomograph by 0.78 $cal\ K^{-1}\ mol^{-1}$ to obtain a smooth junction of the low temperature and high temperature heat capacity data. The data are extrapolated linearly to 1000 K.

The entropy at 298.15 K is calculated by the appropriate integration of the crystal and liquid heat capacity values, addition of the entropy of melting, and $S°(15\ K) = 0.398\ cal\ K^{-1}\ mol^{-1}$.

Fusion Data

The melting point and enthalpy of melting have been measured calorimetrically by Hornung and Giauque (5). The melting point adopted in this table differs from that reported by Hornung and Giauque (5) by 0.02 K. This is due to the re-analysis of the data by Giauque et al. (1) and a conversion of the resulting T_{fus} value from IPTS-48 to IPTS-68.

References

1. W. F. Giauque, E. W. Hornung, J. E. Kunzler, and T. R. Rubin, J. Amer. Chem. Soc. 82, 62 (1960).
2. JANAF Thermochemical Tables: $H_2SO_4(\ell)$, 9-30-77.
3. ICSU-CODATA Task Group, J. Chem. Thermodyn. 4, 331 (1972).
4. J. E. Kunzler and W. F. Giauque, J. Amer. Chem. Soc. 74, 797 (1952).
5. E. W. Hornung and W. F. Giauque, J. Amer. Chem. Soc. 77, 2983 (1955).
6. Om. P. Kharbanda, Ind. Chemist 31, 538 (1955).

Sulfuric Acid, Trihydrate ($H_2SO_4 \cdot 3H_2O$) $H_8O_7S_1(cr,l)$

Enthalpy Reference Temperature = T_r = 298.15 K Standard State Pressure = $p°$ = 0.1 MPa

T/K	$C_p°$ (J K⁻¹mol⁻¹)	$S°$	$-[G°-H°(T_r)]/T$	$H°-H°(T_r)$ (kJ mol⁻¹)	$\Delta_f H°$	$\Delta_f G°$	Log K_f
0	0.	0.	INFINITE	-64.520	-1716.249	-1716.249	INFINITE
100	82.711	68.614	673.238	-60.462	-1735.045	-1639.705	856.494
200	145.372	145.125	390.475	-49.070	-1746.258	-1539.587	402.099
236.780	169.746	171.661	354.430	-43.276	CRYSTAL <--> LIQUID TRANSITION		
236.780	310.829	273.019	354.430	-19.276			
298.15	319.090	345.373	345.373	0.	-1720.402	-1443.893	252.964
300	319.320	347.348	345.379	0.591	-1720.258	-1442.178	251.106
400	347.779	442.970	358.216	33.902	-1713.566	-1350.672	176.380
500	376.606	523.671	383.430	70.121	-1703.672	-1261.054	131.741
600	405.434	594.883	412.845	109.223	-1690.930	-1173.680	102.178
700	434.262	659.546	443.535	151.208	-1675.434	-1088.659	81.237
800	463.089	719.415	474.321	196.075	-1657.267	-1006.044	65.688
900	491.917	775.624	504.706	243.825	-1689.591	-924.781	53.673
1000	520.745	828.944	534.486	294.459	-1664.947	-841.100	43.935

PREVIOUS: CURRENT: September 1977

SULFURIC ACID, TETRAHYDRATE ($H_2SO_4 \cdot 4H_2O$) CRYSTAL–LIQUID M_r = 170.13460

$\Delta_fH°(0\ K)$ = -2007.121 ± 0.8 kJ mol^{-1}

$S°(298.15\ K)$ = 414.529 ± 0.42 J K^{-1} mol^{-1} $\Delta_fH°(298.15\ K)$ = -2011.199 ± 0.8 kJ mol^{-1}

T_{fus} = 244.90 ± 0.05 K $\Delta_{fus}H°$ = 30.644 ± 0.25 kJ mol^{-1}

Enthalpy of Formation

The enthalpy of formation of the tetrahydrate of sulfuric acid is derived from the enthalpy of solution data of Giauque et al. (1). We calculate $\Delta_fH°(298.15\ K)$ for the tetrahydrate using $\Delta_rH°(298.15\ K)$ = -12.880 kcal mol^{-1} for the solution process $H_2SO_4(\ell) + 4H_2O(\ell) = H_2SO_4 \cdot 4\ H_2O(\ell)$ and auxiliary data for H_2SO_4 (2) and H_2O (3).

Heat Capacity and Entropy

The heat capacity has been measured experimentally (4) in the range 15-300 K and has been summarized by Giauque et al. (1). Above 300 K, the heat capacity values are derived from a nomograph constructed by Kharbanda (5) based on experimental data (273-573 K). These temperature-dependent values are corrected to current atomic weights, converted from IPTS-48 to IPTS-68, and adjusted graphically in the region 300-330 K so as to provide smoothly varying heat capacity values. In the region above 350 K, a linear dependence is assumed (as suggested from a least squares fit of the nomograph data); this linear dependence is used to obtain heat capacity data by extrapolation to 1000 K.

The entropy at 298.15 K is calculated by the appropriate integration of the crystal and liquid heat capacity values, addition of the entropy of melting, and $S°(15\ K)$ = 0.687 cal K^{-1} mol^{-1}.

Fusion Data

The melting point and enthalpy of melting have been measured calorimetrically by Hornung and Giauque (4). The values adopted in this table for T_{fus} and $\Delta_{fus}H°$ differ from those reported by Hornung and Giauque (4) by 0.02 K and 2 cal mol^{-1}, respectively. This is due to the re-analysis of this data by Giauque et al. (1) and a conversion of the resulting T_{fus} value from IPTS-48 to IPTS-68.

References

1. W. F. Giauque, E. W. Hornung, J. E. Kunzler, and T. R. Rubin, J. Amer. Chem. Soc. 82, 62 (1960).
2. JANAF Thermochemical Tables: $H_2SO_4(\ell)$, 9-30-77.
3. ICSU-CODATA Task Group, J. Chem. Thermodyn. 4, 331 (1972).
4. E. W. Hornung and W. F. Giauque, J. Amer. Chem. Soc. 77, 2983 (1955).
5. Om. P. Kharbanda, Ind. Chemist 31, 538 (1955).

Sulfuric Acid, Tetrahydrate ($H_2SO_4 \cdot 4H_2O$) $H_{10}O_8S_1(cr,l)$

Enthalpy Reference Temperature = T_r = 298.15 K Standard State Pressure = p° = 0.1 MPa

T/K	$C_p°$ (J $K^{-1}mol^{-1}$)	S° (J $K^{-1}mol^{-1}$)	$-[G°-H°(T_r)]/T$ (J $K^{-1}mol^{-1}$)	$H°-H°(T_r)$ (kJ mol^{-1})	$\Delta_fH°$ (kJ mol^{-1})	$\Delta_fG°$ (kJ mol^{-1})	Log K_f
0	0.	0.	INFINITE	-77.404	-2007.121	-2007.121	INFINITE
100	98.358	88.906	811.941	-72.303	-2029.325	-1917.277	1001.483
200	166.040	177.734	473.245	-59.102	-2042.879	-1799.498	469.981
244.900	197.695	214.443	422.447	-50.940	CRYSTAL <--> LIQUID	TRANSITION	
244.900	376.392	339.570	422.447	-20.297			
298.15	386.382	414.529	414.529	0.	-2011.199	-1685.764	295.339
300	386.771	416.920	414.536	0.715	-2011.010	-1683.745	293.166
400	410.342	531.343	429.977	40.546	-2002.190	-1576.184	205.829
500	434.525	625.506	459.927	82.789	-1990.724	-1470.981	153.672
600	458.709	706.867	494.448	127.451	-1976.931	-1368.290	119.120
700	482.892	779.393	530.063	174.531	-1960.905	-1268.086	94.626
800	507.076	845.453	565.416	224.030	-1942.729	-1170.328	76.415
900	531.259	906.574	599.967	275.946	-1975.563	-1073.893	62.327
1000	555.443	963.799	633.518	330.281	-1951.951	-974.955	50.926

PREVIOUS: CURRENT: September 1977

Sulfuric Acid, Tetrahydrate ($H_2SO_4 \cdot 4H_2O$) $H_{10}O_8S_1(cr,l)$

SULFURIC ACID, HEMIHEXAHYDRATE ($H_2SO_4 \cdot 6.5H_2O$) CRYSTAL–LIQUID $M_r = 215.1780$

$\Delta_f H°(0\ K) = -2727.58 \pm 0.8\ kJ\ mol^{-1}$

$S°(298.15\ K) = 587.819 \pm 0.63\ J\ K^{-1}\ mol^{-1}$ $\Delta_f H°(298.15\ K) = -2733.26 \pm 0.8\ kJ\ mol^{-1}$

$T_{fus} = 220.30 \pm 0.05\ K$ $\Delta_{fus}H° = 34.116 \pm 0.25\ kJ\ mol^{-1}$

Enthalpy of Formation

The enthalpy of formation of the hemihexahydrate of sulfuric acid is derived from the enthalpy of solution data of Giauque et al. (1). We calculate $\Delta_f H°(298.15\ K)$ for the hemihexahydrate using $\Delta_r H°(298.15\ K) = -14.669\ kcal\ mol^{-1}$ for the solution process $H_2SO_4(l) + 6.5H_2O(l) = H_2SO_4 \cdot 6.5H_2O(l)$ and auxiliary data for H_2SO_4 (2) and H_2O (3).

Heat Capacity and Entropy

The heat capacity has been measured experimentally (4) in the range 15–300 K and has been summarized by Giauque et al. (1). Above 300 K, the heat capacity values are derived from a nomograph constructed by Kharbanda (5) based on experimental data (273–573 K). These temperature-dependent values are corrected to current atomic weights, converted from IPTS-48 to IPTS-68, and adjusted graphically in the region 300–320 K so as to provide smoothly varying heat capacity values. In the region above 350 K, a linear dependence is assumed (as suggested from a least squares fit of the nomograph data); this linear dependence is used to obtain heat capacity data by extrapolation to 1000 K.

The entropy at 298.15 K is calculated by the appropriate integration of the crystal and liquid heat capacity values, addition of the entropy of melting, and $S°(15\ K) = 0.721\ cal\ K^{-1}\ mol^{-1}$.

Fusion Data

The melting point and enthalpy of melting have been measured calorimetrically by Hornung et al. These data were reevaluated by Giauque et al. (1). We have adopted the revised values and have also corrected T_{fus} to the IPTS-68.

References

1. W. F. Giauque, E. W. Hornung, J. E. Kunzler, and T. R. Rubin, J. Amer. Chem. Soc. 82, 62 (1960).
2. JANAF Thermochemical Tables: $H_2SO_4(l)$, 9-30-77.
3. ICSU-CODATA Task Group, J. Chem. Thermodyn. 4, 331 (1972).
4. E. W. Hornung, T. E. Brackett, and W. F. Giauque, J. Amer. Chem. Soc. 78, 5747 (1956).
5. Om. P. Kharbanda, Ind. Chemist 31, 538 (1955).

Sulfuric Acid, Hemihexahydrate ($H_2SO_4 \cdot 6.5H_2O$) $H_{15}O_{10.5}S_1(cr,l)$

Enthalpy Reference Temperature = T_r = 298.15 K Standard State Pressure = $p°$ = 0.1 MPa

T/K	$C_p°$ (J K⁻¹mol⁻¹)	$S°$ (J K⁻¹mol⁻¹)	$-[G°-H°(T_r)]/T$ (J K⁻¹mol⁻¹)	$H°-H°(T_r)$ (kJ mol⁻¹)	$\Delta_f H°$ (kJ mol⁻¹)	$\Delta_f G°$ (kJ mol⁻¹)	Log K_f
0	0.	0.	INFINITE	-107.821	-2727.578	-2727.578	INFINITE
100	136.554	115.659	1125.761	-101.010	-2759.200	-2602.982	1359.658
200	239.876	242.053	652.984	-82.186	-2777.504	-2438.910	636.978
220.300	261.042	266.230	616.234	-77.106	CRYSTAL <--> LIQUID	TRANSITION	
220.300	521.282	421.093	616.234	-42.990			
298.15	570.279	587.819	587.819	0.	-2733.261	-2285.631	400.433
300	570.615	591.348	587.830	1.055	-2732.933	-2282.855	397.480
400	583.250	757.240	610.364	58.750	-2717.228	-2135.430	278.859
500	595.843	888.741	653.331	117.705	-2700.180	-1991.951	208.098
600	608.437	998.489	701.957	177.919	-2682.107	-1851.986	161.230
700	621.031	1093.226	751.237	239.393	-2663.102	-1715.123	127.984
800	633.625	1176.975	799.319	302.125	-2643.243	-1581.044	103.232
900	646.219	1252.333	845.536	366.118	-2675.696	-1448.421	84.064
1000	658.813	1321.071	889.701	431.369	-2653.004	-1313.267	68.598

PREVIOUS: CURRENT: September 1977

Sulfuric Acid, Hemihexahydrate ($H_2SO_4 \cdot 6.5H_2O$) $H_{15}O_{10.5}S_1(cr,l)$

0 to 6000 K Ideal Monatomic Gas

IP(He, g) = 198310.76 ± 0.02 cm^{-1}
S°(298.15 K) = 126.152 ± 0.002 $J\ K^{-1}\ mol^{-1}$

$\Delta_f H°$(0 K) = 0 kJ mol^{-1}
$\Delta_f H°$(298.15 K) = 0 kJ mol^{-1}

Electronic Levels and Quantum Weights

State	ϵ_i, cm^{-1}	g_i
$^1S_0$	0	1

Enthalpy of Formation

Zero by definition.

Heat Capacity and Entropy

Information on the electronic energy levels and quantum weights is taken from Moore (1) and Martin (9). All predicted levels have been observed through n=4 but above that many predicted levels are missing. Our calculations indicate that any reasonable method of filling in these missing levels and cutting off the summation in the partition function (2) has no effect on the thermodynamic properties to 6000 K. This is undoubtedly a result of the high energy of these levels; the first excited level is over 159000 cm^{-1} above the ground state. Therefore, we list the ground state only. Extension to higher temperatures may require consideration of excited states and utilization of different fill and cutoff procedures (2).

The thermodynamic functions at 298.15 K agree exactly with recent CODATA recommendations (3) except for two minor differences. First, the entropy differs by 0.1094 $J\ K^{-1}\ mol^{-1}$ because this table uses a standard-state pressure of 1 bar, whereas the CODATA recommendations are based on 1 atm. Second, entropy differences of the order of 0.001-0.004 $J\ K^{-1}\ mol^{-1}$ for the rare gases arise due to the use of slightly different values for R; this table uses R = 8.31441 $J\ K^{-1}\ mol^{-1}$. Considering these minor changes, this table agrees within the estimated uncertainty with those by Hultgren et al. (4), McBride et al. (5), Gurvich et al. (8) and Wagman et al. (12). The estimated uncertainty is due to uncertainties in the relative atomic mass and fundamental constants, which in this tabulation are based on the 1981 scale (6) and the 1973 values (7), respectively.

The thermodynamic functions reported here are calculated assuming that helium gas obeys Boltzmann statistics. In fact helium is a Bose-Einstein gas (4). However, the differences in thermal properties between the two are very small except at extremely low temperature (10, 11). At 298.15 K the difference in $C_p°$ amounts to only 0.004 $J\ K^{-1}\ mol^{-1}$ and will be less at higher temperatures. Since $C_p°$ is the most sensitive of the thermodynamic properties, the differences are well within the stated uncertainty. At temperatures below approximately 50 K the differences may become significant and Bose-Einstein statistics should be used.

Phase Data

The boiling point, 4.215 K (1 atm), is taken from Hultgren et al. (4). This value should prove reasonably accurate, although it has not been evaluated by the present authors, and is furnished for the convenience of the reader. As a result of this low value, the reference state for helium is chosen to be the ideal gas at all temperatures. This may differ from the choice of other authors.

References

1. C. E. Moore, U. S. Nat. Bur. Stand., NSRDS-NBS-35, Volume I, 1970 [Reprint of NBS Circular 467, Volume I, 1949].
2. J. R. Downey, Jr., The Dow Chemical Co., AFOSR-TR-78-0960, Contract No. F44620-75-1-0048, 1978.
3. J. D. Cox, chairman, ICSU-CODATA Task Group on Key Values for Thermodynamics, J. Chem. Thermodynamics 10, 903 (1978).
4. R. Hultgren, P. D. Desai et al., "Selected Values of the Thermodynamic Properties of the Elements", American Society for Metals, Metals Park, Ohio, 1973.
5. B. J. McBride, S. Heimel, J. G. Ehlers and S. Gordon, NASA SP-3001, 1973.
6. N. E. Holden and R. L. Martin, Pure Appl. Chem. 55, 1101 (1983).
7. E. R. Cohen and B. N. Taylor, J. Phys. Chem. Ref. Data 2, 663 (1973).
8. L. V. Gurvich, I. V. Veits et al., "Thermodynamic Properties of Individual Substances," 3rd ed., Vol. I, Nauka, Moscow, 1978.
9. W. C. Martin, J. Phys. Chem. Ref. Data 2, 257 (1973).
10. N. Davidson, "Statistical Mechanics," McGraw-Hill, New York, 1962, p. 95.
11. N. V. Tsederberg, V. N. Popov and N. A. Morozova, "Thermodynamic and Thermophysical Properties of Helium," Israel Program for Scientific Translations, Jerusalem, 1971, avail. NTIS.
12. D. D. Wagman, W. H. Evans et al., J. Phys. Chem. Ref. Data 11, Supp. 2 (1982).

Enthalpy Reference Temperature = T_r = 298.15 K — Standard State Pressure = p° = 0.1 MPa

T/K	$C_p°$ (J $K^{-1}mol^{-1}$)	S° (J $K^{-1}mol^{-1}$)	$-[G°-H°(T_r)]/T$ (J $K^{-1}mol^{-1}$)	$H°-H°(T_r)$ (kJ mol^{-1})	$\Delta_f H°$ (kJ mol^{-1})	$\Delta_f G°$ (kJ mol^{-1})	Log K_f
0	0.	0.	INFINITE	-6.197	0.	0.	0.
100	20.786	103.445	144.632	-4.119	0.	0.	0.
200	20.786	117.853	128.053	-2.040	0.	0.	0.
250	20.786	122.491	126.494	-1.001	0.	0.	0.
298.15	20.786	126.152	126.152	0.	0.	0.	0.
300	20.786	126.281	126.152	0.038	0.	0.	0.
350	20.786	129.485	126.405	1.078	0.	0.	0.
400	20.786	132.260	126.968	2.117	0.	0.	0.
450	20.786	134.709	127.694	3.156	0.	0.	0.
500	20.786	136.899	128.507	4.196	0.	0.	0.
600	20.786	140.688	130.231	6.274	0.	0.	0.
700	20.786	143.893	131.960	8.353	0.	0.	0.
800	20.786	146.668	133.629	10.431	0.	0.	0.
900	20.786	149.116	135.216	12.510	0.	0.	0.
1000	20.786	151.306	136.718	14.589	0.	0.	0.
1100	20.786	153.288	138.135	16.667	0.	0.	0.
1200	20.786	155.096	139.475	18.746	0.	0.	0.
1300	20.786	156.760	140.741	20.824	0.	0.	0.
1400	20.786	158.300	141.941	22.903	0.	0.	0.
1500	20.786	159.734	143.080	24.982	0.	0.	0.
1600	20.786	161.076	144.163	27.060	0.	0.	0.
1700	20.786	162.336	145.196	29.139	0.	0.	0.
1800	20.786	163.524	146.181	31.217	0.	0.	0.
1900	20.786	164.648	147.124	33.296	0.	0.	0.
2000	20.786	165.714	148.027	35.375	0.	0.	0.
2100	20.786	166.728	148.893	37.453	0.	0.	0.
2200	20.786	167.695	149.726	39.532	0.	0.	0.
2300	20.786	168.619	150.528	41.610	0.	0.	0.
2400	20.786	169.504	151.300	43.689	0.	0.	0.
2500	20.786	170.352	152.045	45.768	0.	0.	0.
2600	20.786	171.168	152.765	47.846	0.	0.	0.
2700	20.786	171.952	153.461	49.925	0.	0.	0.
2800	20.786	172.708	154.135	52.004	0.	0.	0.
2900	20.786	173.438	154.789	54.082	0.	0.	0.
3000	20.786	174.142	155.422	56.161	0.	0.	0.
3100	20.786	174.824	156.037	58.239	0.	0.	0.
3200	20.786	175.484	156.634	60.318	0.	0.	0.
3300	20.786	176.123	157.215	62.397	0.	0.	0.
3400	20.786	176.744	157.781	64.475	0.	0.	0.
3500	20.786	177.346	158.331	66.554	0.	0.	0.
3600	20.786	177.932	158.867	68.632	0.	0.	0.
3700	20.786	178.501	159.390	70.711	0.	0.	0.
3800	20.786	179.056	159.901	72.790	0.	0.	0.
3900	20.786	179.596	160.399	74.868	0.	0.	0.
4000	20.786	180.122	160.885	76.947	0.	0.	0.
4100	20.786	180.635	161.361	79.025	0.	0.	0.
4200	20.786	181.136	161.826	81.104	0.	0.	0.
4300	20.786	181.625	162.280	83.183	0.	0.	0.
4400	20.786	182.103	162.726	85.261	0.	0.	0.
4500	20.786	182.570	163.161	87.340	0.	0.	0.
4600	20.786	183.027	163.588	89.418	0.	0.	0.
4700	20.786	183.474	164.007	91.497	0.	0.	0.
4800	20.786	183.912	164.417	93.576	0.	0.	0.
4900	20.786	184.340	164.819	95.654	0.	0.	0.
5000	20.786	184.760	165.214	97.733	0.	0.	0.
5100	20.786	185.172	165.601	99.811	0.	0.	0.
5200	20.786	185.575	165.981	101.890	0.	0.	0.
5300	20.786	185.971	166.355	103.969	0.	0.	0.
5400	20.786	186.360	166.722	106.047	0.	0.	0.
5500	20.786	186.741	167.082	108.126	0.	0.	0.
5600	20.786	187.116	167.437	110.204	0.	0.	0.
5700	20.786	187.484	167.785	112.283	0.	0.	0.
5800	20.786	187.845	168.128	114.362	0.	0.	0.
5900	20.786	188.201	168.465	116.440	0.	0.	0.
6000	20.786	188.550	168.797	118.519	0.	0.	0.

PREVIOUS: March 1977 (1 atm) CURRENT: March 1982 (1 bar)

HELIUM, ION (He^+) IDEAL GAS M_r = 4.00205

$IP(He^+, g) = 438908.85 \pm 0.02\ cm^{-1}$

$S°(298.15\ K) = 131.913 \pm 0.002\ J\ K^{-1}\ mol^{-1}$

$\Delta_f H°(0\ K) = 2372.324 \pm 0.001\ kJ\ mol^{-1}$

$\Delta_f H°(298.15\ K) = [2378.522]\ kJ\ mol^{-1}$

Electronic Levels and Quantum Weights

State	ε_i, cm^{-1}	g_i
$^2S_{1/2}$	0	4

Enthalpy of Formation

The ionization limit of neutral helium (198310.76±0.02 cm^{-1}) reported by Moore (1) and Martin (10), is adopted as $\Delta_f H°(0\ K)$ for $He^+(g)$. The ionization limit is converted from cm^{-1} to kJ mol^{-1} using the factor, 1 cm^{-1} = 0.01196266 kJ mol^{-1}, which is derived from the 1973 CODATA fundamental constants (2). The uncertainty in the ionization limit corresponds to an uncertainty of less than 0.001 kJ mol^{-1} in the enthalpy of formation. Rosenstock et al. (9) have summarized additional ionization potential and appearance potential data. Gurvich et al. (8) and Wagman et al. (11) adopted the same ionization potential, but the use of slightly different fundamental constants by Wagman et al. (11) resulted in an enthalpy of formation difference of 0.021 kJ mol^{-1}.

$\Delta_f H°(He^+, g, 298.15\ K)$ is obtained from $\Delta_f H°(He, g, 0\ K)$ by using IP(He) with JANAF (3) enthalpies H°(0 K)-H°(298.15 K) for $He^+(g)$, He(r), and $e^-(r)$. $\Delta_r H°(He \rightarrow He^+ + e^-, 298.15\ K)$ differs from a room temperature threshold energy due to inclusion of these enthalpies and to threshold effects discussed by Rosenstock et al. (9). $\Delta_f H°(298.15\ K)$ should be changed by -6.197 kJ mol^{-1} if it is to be used in the ion convention that excludes the enthalpy of the electron.

Heat Capacity and Entropy

The information on electronic energy levels and quantum weights given by Moore (4) is complete to n=14. Our calculations indicate that any reasonable method of including these levels and cutting off the summation in the partition function (5) has no effect on the thermodynamic functions to 6000 K. This is a result of the high energy of these levels; the first excited level is over 329000 cm^{-1} above the ground state. Since inclusion of these upper levels has no effect on the thermodynamic functions (to 6000 K) we list only the ground state. The reported uncertainty in S°(298.15 K) is due to uncertainties in the relative ionic mass and fundamental constants. Extension of these calculations above 6000 K may require consideration of the higher excited states and use of different fill and cutoff procedures (5).

The thermodynamic functions reported here agree with those of Green et al. (6), Hilsenrath et al. (7), and Gurvich et al. (8) except for one or two minor differences. First, the entropy differs by 0.1094 J K^{-1} mol^{-1} because this table uses a standard-state pressure of 1 bar, whereas the cited references used a pressure of 1 atm. Second, smaller differences arise from the use of slightly different values for the fundamental constants and the relative ionic mass.

References

1. C. E. Moore, U. S. Nat. Bur. Stand., NSRDS-NBS-34, 1970.
2. E. R. Cohen and B. N. Taylor, J. Phys. Chem. Ref. Data 2, 663 (1973).
3. JANAF Thermochemical Tables: He(r), 3-31-82; e^-(r), 3-31-82.
4. C. E. Moore, U. S. Nat. Bur. Stand., NSRDS-NBS-35, Volume I, 1970 [Reprint of NBS Circular 467, Volume I, 1949].
5. J. R. Downey, Jr., The Dow Chemical Co., AFOSR-TR-78-0960, Contract No. F44620-75-1-0048, 1978.
6. J. W. Green, D. E. Poland, and J. L. Margrave, University of Wisconsin, Report AD-275542, 1961.
7. J. Hilsenrath, C. G. Messina, and W. H. Evans, U.S. Nat. Bur. Stand., Report AD-606163, 1964.
8. L. V. Gurvich, I. V. Veits et al., "Thermodynamic Properties of Individual Substances", 3rd ed., Vol. 1, Nauka, Moscow, 1978.
9. H. M. Rosenstock, K. Draxl et al., J. Phys. Chem. Ref. Data 6, Supp. 1 (1977).
10. W. C. Martin, J. Phys. Chem. Ref. Data 2, 257 (1973).
11. D. D. Wagman, W. H. Evans, et al., J. Phys. Chem. Ref. Data 11, Supp. 2 (1982).

Helium, Ion (He^+) $He_1^+(g)$

Enthalpy Reference Temperature = T_r = 298.15 K Standard State Pressure = p° = 0.1 MPa

T/K	$C_p°$ (J $K^{-1}mol^{-1}$)	S° (J $K^{-1}mol^{-1}$)	$-[G°-H°(T_r)]/T$ (J $K^{-1}mol^{-1}$)	$H°-H°(T_r)$ (kJ mol^{-1})	$\Delta_f H°$ (kJ mol^{-1})	$\Delta_f G°$ (kJ mol^{-1})	Log K_f
0	0.	0.	INFINITE	-6.197	2372.324		
100	20.786	109.206	150.394	-4.119			
200	20.786	123.614	133.815	-2.040			
250	20.786	128.252	132.256	-1.001			
298.15	20.786	131.913	131.913	0.	2378.522	2370.549	-415.310
300	20.786	132.042	131.914	0.038	2378.560	2370.499	-412.741
350	20.786	135.246	132.167	1.078	2379.599	2369.074	-353.565
400	20.786	138.022	132.729	2.117	2380.639	2367.499	-309.164
450	20.786	140.470	133.456	3.156	2381.678	2365.794	-274.614
500	20.786	142.660	134.269	4.196	2382.717	2363.974	-246.963
600	20.786	146.450	135.993	6.274	2384.796	2360.030	-205.459
700	20.786	149.654	137.721	8.353	2386.874	2355.738	-175.787
800	20.786	152.430	139.390	10.431	2388.953	2351.148	-153.514
900	20.786	154.878	140.978	12.510	2391.032	2346.298	-136.176
1000	20.786	157.068	142.479	14.589	2393.110	2341.216	-122.293
1100	20.786	159.049	143.897	16.667	2395.189	2335.926	-110.924
1200	20.786	160.858	145.236	18.746	2397.267	2330.446	-101.442
1300	20.786	162.521	146.502	20.824	2399.346	2324.794	-93.411
1400	20.786	164.062	147.702	22.903	2401.425	2318.981	-86.522
1500	20.786	165.496	148.841	24.982	2403.503	2313.020	-80.547
1600	20.786	166.837	149.925	27.060	2405.582	2306.920	-75.313
1700	20.786	168.097	150.957	29.139	2407.660	2300.690	-70.692
1800	20.786	169.286	151.943	31.217	2409.739	2294.337	-66.580
1900	20.786	170.409	152.885	33.296	2411.818	2287.869	-62.898
2000	20.786	171.476	153.788	35.375	2413.896	2281.292	-59.581
2100	20.786	172.490	154.655	37.453	2415.975	2274.611	-56.578
2200	20.786	173.457	155.488	39.532	2418.053	2267.830	-53.845
2300	20.786	174.381	156.289	41.610	2420.132	2260.955	-51.348
2400	20.786	175.265	157.062	43.689	2422.211	2253.990	-49.057
2500	20.786	176.114	157.807	45.768	2424.289	2246.938	-46.947
2600	20.786	176.929	158.527	47.846	2426.368	2239.803	-44.998
2700	20.786	177.714	159.223	49.925	2428.446	2232.588	-43.192
2800	20.786	178.469	159.897	52.004	2430.525	2225.296	-41.513
2900	20.786	179.199	160.550	54.082	2432.604	2217.930	-39.949
3000	20.786	179.904	161.183	56.161	2434.682	2210.492	-38.488
3100	20.786	180.585	161.798	58.239	2436.761	2202.985	-37.120
3200	20.786	181.245	162.396	60.318	2438.839	2195.410	-35.836
3300	20.786	181.885	162.977	62.397	2440.918	2187.771	-34.630
3400	20.786	182.505	163.542	64.475	2442.997	2180.069	-33.493
3500	20.786	183.108	164.092	66.554	2445.075	2172.305	-32.420
3600	20.786	183.693	164.629	68.632	2447.154	2164.482	-31.406
3700	20.786	184.263	165.152	70.711	2449.232	2156.602	-30.446
3800	20.786	184.817	165.662	72.790	2451.311	2148.665	-29.535
3900	20.786	185.357	166.160	74.868	2453.390	2140.673	-28.671
4000	20.786	185.883	166.647	76.947	2455.468	2132.629	-27.849
4100	20.786	186.397	167.122	79.025	2457.547	2124.532	-27.067
4200	20.786	186.898	167.587	81.104	2459.625	2116.384	-26.321
4300	20.786	187.387	168.042	83.183	2461.704	2108.187	-25.609
4400	20.786	187.864	168.487	85.261	2463.783	2099.942	-24.929
4500	20.786	188.332	168.923	87.340	2465.861	2091.650	-24.279
4600	20.786	188.788	169.350	89.418	2467.940	2083.311	-23.657
4700	20.786	189.235	169.768	91.497	2470.018	2074.927	-23.060
4800	20.786	189.673	170.178	93.576	2472.097	2066.499	-22.488
4900	20.786	190.102	170.580	95.654	2474.176	2058.027	-21.939
5000	20.786	190.522	170.975	97.733	2476.254	2049.513	-21.411
5100	20.786	190.933	171.362	99.811	2478.333	2040.958	-20.904
5200	20.786	191.337	171.743	101.890	2480.411	2032.362	-20.415
5300	20.786	191.733	172.116	103.969	2482.490	2023.725	-19.945
5400	20.786	192.121	172.483	106.047	2484.569	2015.050	-19.492
5500	20.786	192.503	172.844	108.126	2486.647	2006.336	-19.055
5600	20.786	192.877	173.198	110.204	2488.726	1997.584	-18.633
5700	20.786	193.245	173.546	112.283	2490.804	1988.796	-18.225
5800	20.786	193.607	173.889	114.362	2492.883	1979.970	-17.832
5900	20.786	193.962	174.226	116.440	2494.962	1971.109	-17.451
6000	20.786	194.311	174.558	118.519	2497.040	1962.213	-17.083

PREVIOUS: March 1977 (1 atm) CURRENT: March 1982 (1 bar)

Helium, Ion (He^+) $He_1^+(g)$

HAFNIUM (Hf)

REFERENCE STATE

0 to 2054 K	Crystal, Alpha
2054 to 2500 K	Crystal, Beta
2500 to 4963.793 K	Liquid
4963.793 to 6000 K	Ideal Monatomic Gas

Refer to the individual tables for details.

A_r = 178.49

Hafnium (Hf) Hf_1(ref)

Enthalpy Reference Temperature = T_r = 298.15 K Standard State Pressure = p° = 0.1 MPa

T/K	C_p° (J K⁻¹mol⁻¹)	S° (J K⁻¹mol⁻¹)	$-[G^\circ-H^\circ(T_r)]/T$ (J K⁻¹mol⁻¹)	$H^\circ-H^\circ(T_r)$ (kJ mol⁻¹)	$\Delta_f H^\circ$ (kJ mol⁻¹)	$\Delta_f G^\circ$ (kJ mol⁻¹)	Log K_f
0	0.	0.	INFINITE	-5.842	0.	0.	0.
100	20.580	17.753	-65.394	-4.764	0.	0.	0.
200	24.418	33.561	-45.871	-2.462	0.	0.	0.
298.15	25.687	43.560	43.560	0.	0.	0.	0.
300	25.708	43.719	43.560	0.048	0.	0.	0.
400	26.751	51.261	44.581	2.672	0.	0.	0.
500	27.692	57.332	46.543	5.394	0.	0.	0.
600	28.583	62.460	48.780	8.208	0.	0.	0.
700	29.445	66.932	51.060	11.110	0.	0.	0.
800	30.288	70.919	53.298	14.097	0.	0.	0.
900	31.120	74.534	55.459	17.167	0.	0.	0.
1000	31.943	77.856	57.535	20.320	0.	0.	0.
1100	32.759	80.938	59.524	23.556	0.	0.	0.
1200	33.572	83.824	61.430	26.872	0.	0.	0.
1300	34.372	86.543	63.258	30.270	0.	0.	0.
1400	35.064	89.115	65.014	33.741	0.	0.	0.
1500	35.765	91.558	66.703	37.283	0.	0.	0.
1600	36.466	93.889	68.330	40.894	0.	0.	0.
1700	37.167	96.121	69.899	44.576	0.	0.	0.
1800	37.868	98.265	71.416	48.328	0.	0.	0.
1900	38.569	100.331	72.884	52.149	0.	0.	0.
2000	39.269	102.327	74.306	56.041	0.	0.	0.
2054.000	39.648	103.378	75.057	58.172	ALPHA <--> BETA		
2054.000	33.959	106.255	75.057	64.080	TRANSITION		
2100	34.384	107.011	75.749	65.652	0.	0.	0.
2200	35.307	108.632	77.207	69.136	0.	0.	0.
2300	36.231	110.222	78.608	72.713	0.	0.	0.
2400	37.154	111.784	79.957	76.383	0.	0.	0.
2500	38.077	113.353	81.262	80.228	0.	0.	0.
2500.000	38.077	113.353	81.262	80.228	BETA <--> LIQUID		
2500.000	37.656	125.068	81.262	109.516	TRANSITION		
2600	37.656	126.545	82.976	113.281	0.	0.	0.
2700	37.656	127.966	84.616	117.047	0.	0.	0.
2800	37.656	129.336	86.189	120.813	0.	0.	0.
2900	37.656	130.657	87.699	124.578	0.	0.	0.
3000	37.656	131.934	89.153	128.344	0.	0.	0.
3100	37.656	133.169	90.553	132.109	0.	0.	0.
3200	37.656	134.364	91.903	135.875	0.	0.	0.
3300	37.656	135.523	93.208	139.641	0.	0.	0.
3400	37.656	136.647	94.469	143.406	0.	0.	0.
3500	37.656	137.739	95.690	147.172	0.	0.	0.
3600	37.656	138.799	96.872	150.937	0.	0.	0.
3700	37.656	139.831	98.020	154.703	0.	0.	0.
3800	37.656	140.835	99.133	158.469	0.	0.	0.
3900	37.656	141.814	100.215	162.234	0.	0.	0.
4000	37.656	142.767	101.267	166.000	0.	0.	0.
4100	37.656	143.697	102.291	169.765	0.	0.	0.
4200	37.656	144.604	103.287	173.531	0.	0.	0.
4300	37.656	145.490	104.258	177.297	0.	0.	0.
4400	37.656	146.356	105.205	181.062	0.	0.	0.
4500	37.656	147.202	106.129	184.828	0.	0.	0.
4600	37.656	148.030	107.031	188.593	0.	0.	0.
4700	37.656	148.840	107.912	192.359	0.	0.	0.
4800	37.656	149.632	108.773	196.125	0.	0.	0.
4900	37.656	150.409	109.615	199.890	0.	0.	0.
4963.793	37.656	150.896	110.142	202.292	LIQUID <--> IDEAL GAS		
4963.793	37.996	263.754	110.142	762.496	FUGACITY = 1 bar		
5000	38.154	264.031	111.256	763.875	0.	0.	0.
5200	39.044	265.544	117.161	771.594	0.	0.	0.
5400	39.964	267.035	122.684	779.494	0.	0.	0.
5600	40.910	268.505	127.866	787.581	0.	0.	0.
5800	41.879	269.958	132.741	795.859	0.	0.	0.
6000	42.870	271.394	137.339	804.334	0.	0.	0.

PREVIOUS: March 1979 (1 atm) CURRENT: March 1979 (1 bar)

Hafnium (Hf) Hf_1(ref)

HAFNIUM, ALPHA (α-Hf) CRYSTAL A_r = 178.49

$S°(298.15\ K) = 43.560\ J\ K^{-1}\ mol^{-1}$

$T_{trs} = 2054 \pm 50\ K\ (\alpha-\beta)$

$\Delta_f H°(0\ K) = 0\ kJ\ mol^{-1}$

$\Delta_f H°(298.15\ K) = 0\ kJ\ mol^{-1}$

$\Delta_{trs} H° = 5.908 \pm 0.2\ kJ\ mol^{-1}$

Enthalpy of Formation

Zero by definition.

Heat Capacity and Entropy

The adopted thermal functions for α-Hf(cr) are derived from the studies of Collings and Ho (1), Westrum (2), Hawkins et al. (3) and Cezairliyan and McClure (4). The mathematical and graphical treatment of these four studies yields a continuous and smooth heat capacity curve.

Roberts (5) surveyed the superconductive properties of the elements and reported a critical temperature of 0.128 K for α-Hf(cr). Since this temperature is so low, the effects of superconductivity on the thermal functions are not considered.

There are four studies (1, 6, 7, 8) on the heat capaicty of hafnium in the region 1-5 K. Only Wolcott (6) reported the experimental data (56 data points). In the other three states, an equation (with two constants) was given to describe the entire set of experimental data. The second set of constants given by Betterton and Scarbrough (8) are stated to be an average of the two studies of Kneip et al. (7) and the more recent study by Betterton and Scarbrough (8). However, the values appear suspect and may contain a typographical error. Nevertheless the maximum difference in the calculated values for S°(5 K) is 0.00024 cal K^{-1} mol^{-1} with the adopted value being S°(5 K) = 0.0038 cal K^{-1} mol^{-1}. Similarly for Cp°(5 K) the maximum difference is 0.00024 cal K^{-1} mol^{-1}, with the adopted value being Cp°(5 K) = 0.00638 cal K^{-1} mol^{-1}

In the region above 5 K, and below 350 K, there are four heat capacity studied (2, 9-11). Cristescu and Simon (9) studies Hf from 13 to 210 K and reported 15 heat capacity values with a peak near 75 K. Burk and Darnell (10) made heat capacity measurements (40-190 K) and reported that the peak is completely absent. The heat capacity measurements of Burk et al. (10-200 K, 11) and Westrum (5.82-348.55 K, 1, 12) support this latter conclusion, i.e., no peak. The values of Burk et al. (11), after correction for the ~2% Zr in this sample, agree with the data of Westrum (2, 12) between 10 and 40 K but deviate above 40 K, becoming ~5% higher at 100 K and decreasing to ~1% higher at 200 K.

Above 350 K, there are four heat capacity studies and four enthalpy studies:

		Source (Year)	Data Points	T/K	Wt % Zr in Hf Sample
Enthalpy	(13)	Adenstedt (1952)	1	373	0.7
	(14)	Fieldhouse and Lang (1961)	14	534-1884	1.0
	(3)	Hawkins et al. (1963)	50	338-1346	2.8
	(15)	Golutvin and Maslennikova (1970)	12	551-1309	0.79
	(15)	Golutvin and Maslennikova (1970)	9	695-137	0.79
Heat Capacity	(16)	Peletskii et al. (1971)	13 (smoothed)	1200-2200	0.66
	(17)	Arutyunov et al. (1972)	16 (smoothed)	1100-2050	0.65
	(4)	Cezairliyand and McClure (1975)	26	1500-2400	3.12
	(18)	Rumyantsev et al. (1979)	graph	1000-1900	-

The enthalpy studies show an unusually large scatter within each study as well as a lack of agreement between the studies. The heat capacity studies all have a similar temperature dependence in comparison with the adopted values. However, the studies of Arutyunov et al. (17, 19) and Rumyantsev et al. (18) lie roughly 10% higher than the adopted values, whereas the study of Peletskii et al. (16) lies 5-24% below. The studies by Arutyunov et al. (17, 19) and Rumyantsev et al. (18) suggest a sharp decrease in the heat capacity values within 50 K of the α-β transition temperature. The experimental results were corrected for this zirconium content.

Transition Data

Refer to the β-crystal table for details.

Sublimation Data

There are no sublimation studies involving α-Hf(cr). The enthalpy of sublimation is calculated as the difference between the enthalpy of formation of the ideal gas and the enthalpy of formation of α-Hf(cr).

References

1. E. W. Collins and J. C. Ho, Phys. Rev. 4B, 349 (1971).
2. E. F. Westrum, University of Michigan, personal communication, 1979.
3. D. T. Hawkins, M. Onillon and R. L. Orr, J. Chem. Eng. Data 8, 628 (1963).
4. A. Cezairliyan and J. L. McClure, J. Res. Nat. Bur. Stand. 79A, 431 (1975).
5. B. W. Roberts, J. Phys. Chem. Ref. Data 5, 581 (1976).
6. N. M. Wolcott, Phil. Mag. 2, 1246 (1957).
7. G. D. Kneip, Jr., J. O. Betterton, Jr. and J. O. Scarbrough, Phys. Rev. 130, 1687 (1963).
8. J. O. Betterton, Jr. and J. O. Scarbrough, Phys. Rev. 168, 715 (1968).
9. S. Cristescu and F. Simon, Z. physik. chem. B25, 273 (1934).
10. D. L. Burk and F. J. Darnell, Phys. Rev. 86, 628 (1952).
11. D. L. Burk, I. Estermann and S. A. Friedberg, Z. Physik. Chem. Neue Folge, Bd. 16, 183 (1958).
12. L. A. McClaine and P. E. Blackburn, Rept. ASD-TRD-62-204, Part III, 1964.
13. H. K. Adenstedt, Trans. A. S. M. 44, 949 (1952).
14. I. B. Fieldhouse and J. I. Lang, WADD-TR-60-904, 1961.
15. Yu. M. Golutvin and E. G. Maslennikova, Izv. Akad. Nauk SSSR, Metally (5), 174 (1970).
16. V. E. Peletskii and V. P. Druzhinin, Teplofiz. Vys. Temp. 9, 539 (1971).
17. A. V. Arutyunov, S. N. Banchila and L. P. Filippov, Teplofiz. Vys. Temp. 10, 425 (1972).
18. A. V. Rumyantsev, I. N. Makarenko and S. N. Banchila, Inzh. -Fiz. Zh. 36, 581 (1979).
19. L. P. Filippov, Int. J. Heat Mass Transfer 16, 865 (1973).

Hafnium, Alpha (Hf) $Hf_1(cr)$

Enthalpy Reference Temperature = T_r = 298.15 K — Standard State Pressure = p° = 0.1 MPa

T/K	$C_p°$ (J K⁻¹mol⁻¹)	S° (J K⁻¹mol⁻¹)	$-[G°-H°(T_r)]/T$ (J K⁻¹mol⁻¹)	$H°-H°(T_r)$ (kJ mol⁻¹)	$\Delta_f H°$ (kJ mol⁻¹)	$\Delta_f G°$ (kJ mol⁻¹)	Log K_f
0	0.	0.	INFINITE	-5.842	0.	0.	0.
100	20.580	17.753	-65.394	-4.764	0.	0.	0.
200	24.418	33.561	-45.871	-2.462	0.	0.	0.
250	25.110	39.086	-43.979	-1.223	0.	0.	0.
298.15	25.687	43.560	43.560	0.	0.	0.	0.
300	25.708	43.719	43.560	0.048	0.	0.	0.
350	26.247	47.723	43.875	1.347	0.	0.	0.
400	26.751	51.261	44.581	2.672	0.	0.	0.
450	27.230	54.439	45.503	4.021	0.	0.	0.
500	27.692	57.332	46.543	5.394	0.	0.	0.
600	28.583	62.460	48.780	8.208	0.	0.	0.
700	29.445	66.932	51.060	11.110	0.	0.	0.
800	30.288	70.919	53.298	14.097	0.	0.	0.
900	31.120	74.534	55.459	17.167	0.	0.	0.
1000	31.943	77.856	57.535	20.320	0.	0.	0.
1100	32.759	80.938	59.524	23.556	0.	0.	0.
1200	33.572	83.824	61.430	26.872	0.	0.	0.
1300	34.372	86.543	63.258	30.270	0.	0.	0.
1400	35.064	89.115	65.014	33.741	0.	0.	0.
1500	35.765	91.558	66.703	37.283	0.	0.	0.
1600	36.466	93.889	68.330	40.894	0.	0.	0.
1700	37.167	96.121	69.899	44.576	0.	0.	0.
1800	37.868	98.265	71.416	48.328	0.	0.	0.
1900	38.569	100.331	72.884	52.149	0.	0.	0.
2000	39.269	102.327	74.306	56.041	0.	0.	0.
2054.000	39.648	103.378	75.057	58.172	-------- ALPHA <--> BETA	-------	
2100	39.970	104.260	75.687	60.003	-5.648	0.129	-0.003
2200	40.671	106.136	77.029	64.035	-5.101	0.392	-0.009
2300	41.372	107.959	78.334	68.138	-4.576	0.629	-0.014
2400	42.073	109.734	79.605	72.310	-4.073	0.845	-0.018
2500	42.774	111.466	80.845	76.552	-3.676	1.042	-0.022

PREVIOUS: CURRENT: March 1979

Hafnium, Alpha (Hf) $Hf_1(cr)$

HAFNIUM, BETA (β-Hf) CRYSTAL A_r = 178.49

$S°(298.15\ K) = 61.104\ J\ K^{-1}\ mol^{-1}$ $\Delta_f H°(298.15\ K) = [18.686]\ kJ\ mol^{-1}$

$T_{trs} = 2054 \pm 50\ K\ (\alpha-\beta) J\ K^{-1}$ $\Delta_{trs}H° = 5.908 \pm 0.2\ kJ\ mol^{-1}$

$T_{fus} = 2500 \pm 20\ K\ (\beta-\ell) J\ K^{-1}$ $\Delta_{fus}H° = [29.288 \pm 4.2]\ kJ\ mol^{-1}$

Enthalpy of Formation

The enthalpy of formation is calculated from that of the α-crystal by addition of $\Delta_{trs}H°$ and the difference in enthalpy, H°(2054 K)-H°(298.15 K), between the α-crsytal and β-crystal.

Heat Capacity and Entropy

The only heat capacity study for the β-crystal is that of Cezairliyand McClure (1). Using a pulse heating technique, they studied the heat capacity of a sample containing 3.12% (wt) Zr in the range 2054 to 2400 K. These results were corrected by Cezairliyan and McClure (1) to represent pure β-Hf and are extrapolated linearly to the melting point, 2500 K, and to 298.15 K. The entropy at 298.15 is calculated in a manner similar to that for the enthalpy of formation.

Phase Data

Hafnium, at ambient pressures, exists in two crystal modifications. The low temperature form, α-Hf, is hexagonal close packed, an hcp(A3) structure isotypic with Mg. The high temperature form, β-Hf, is body centered cubic, a bcc(A2) structure isotypic with W. The wide variation in the observed transition temperature is undoubtedly due to the effects of various impurities, in particular, zirconium.

Transition Data

Cezairliyan and McClure (2) have summarized the reported transition temperatures for the α to β transition of hafnium. Their summary is given below. All temperatures were corrected to IPTS-68. The pure hafnium value reported by the cited author was corrected for the zirconium content of the alloy. Hultgren et al. (19), based on data available through 1965, recommended T_{trs} = 2013±20 K. We adopt T_{trs} = 2054±50 K based on the study of Cezairliyan and McClure (2) since we have also adopted their measured heat capacity values. In addition there are many studies of binary phase diagrams which lend support to this adopted value.

	Source	Zirconium Content (W/%)	Transformation Temperature (K) Hafnium Alloy	Pure Hafnium
(3)	Duwez	unknown	1586±10	
(4)	Fast (1952)	3	1960-2075	
(5)	Gibson et al. (1958)	0.01	2011	
(6)	Deardorff and Kato (1959)	1.2-8.8		2026±20
(7)	Peterson and Beernsten (1960)	0.03	1966-2096	
(8)	Giessen et al. (1963)	2.3	1991±15	2031±20
(9)	Ross and Hume-Rothery (1963)	1.6	2236-2266 50	2271±70
(10)	Krikorian and Wallace (1964)	0.02-3.45	2007-2100	2016±14
(11)	Taylor and Doyle (1964)	2.3	2226±50	2270±70
(12)	Romans et al. (1965)	1.5	2028	2053
(13)	Bates and Barnes (1967)	3.6	2033	
(14)	Peletskii and Druzhinin (1971)	0.66	1970±5	
(15)	Carlson et al. (1973)	3.3	1990-2003	
(2)	Cezairliyan and McClure (1976)	3.12	2012±10	2054

Cezairliyand and McClure (2), using a millisecond-resolution pulse-heating technique, measured the transformation enthalpy to be 33.1 $J\ g^{-1}$ (1.412±0.05 kcal mol^{-1}). Peletskii and Druzhinin (14) measured 29 $J\ g^{-1}$ (1.237 kcal mol^{-1}), a value which is roughly 12% lower than that of Cezairliyan and McClure (2). Martynyuk and Tsapkov (80), using a continuous-pulse heating technique, measured a transformation enthalpy of 1.24 kcal mol^{-1}. The samples used in all three studies contained significant amounts of zirconium: 3.12 wt % (2), 0.66 wt % (14), and 0.7 wt % (80), respectively. Hoerster et al. (70) measured an enthalpy of transition of 0.52 kcal mol^{-1} at 2090 K. We adopted $\Delta_{trs}H°$ = 1.412±0.05 kcal mol^{-1} as derived from the study of Cezairliyan and McClure (2).

Fusion Data

Refer to the liquid table.

Sublimation Data

The sublimation studies for β-Hf(cr) are summarized on the Hf(g) table. The enthalpy of sublimation for Hf(β,cr) is calculated as the difference in the enthalpies of formation at 298.15 K of Hf(g) and Hf(β,cr).

References for β-Hf(cr) are listed on the Hf(ℓ) table.

Hafnium, Beta (Hf) $Hf_1(cr)$

Enthalpy Reference Temperature = T_r = 298.15 K Standard State Pressure = p° = 0.1 MPa

T/K	$C_p°$ (J K⁻¹mol⁻¹)	S°	$-[G°-H°(T_r)]/T$	$H°-H°(T_r)$ (kJ mol⁻¹)	$\Delta_f H°$	$\Delta_f G°$	Log K_f
0							
100							
200							
250							
298.15	17.747	61.104	61.104	0.	18.686	13.455	-2.357
300	17.764	61.214	61.105	0.033	18.671	13.422	-2.337
350	18.226	63.987	61.323	0.933	18.272	12.579	-1.877
400	18.687	66.451	61.812	1.855	17.869	11.793	-1.540
450	19.149	68.679	62.454	2.801	17.466	11.058	-1.284
500	19.611	70.720	63.180	3.770	17.062	10.368	-1.083
600	20.534	74.377	64.748	5.778	16.255	9.105	-0.793
700	21.457	77.612	66.359	7.877	15.453	7.976	-0.595
800	22.381	80.538	67.951	10.069	14.658	6.963	-0.455
900	23.304	83.227	69.501	12.353	13.872	6.048	-0.351
1000	24.227	85.730	71.000	14.730	13.095	5.221	-0.273
1100	25.151	88.082	72.447	17.199	12.329	4.470	-0.212
1200	26.074	90.310	73.844	19.760	11.573	3.789	-0.165
1300	26.997	92.434	75.193	22.414	10.829	3.171	-0.127
1400	27.921	94.468	76.497	25.160	10.104	2.609	-0.097
1500	28.844	96.426	77.761	27.998	9.401	2.098	-0.073
1600	29.767	98.317	78.987	30.928	8.720	1.634	-0.053
1700	30.691	100.150	80.178	33.951	8.061	1.211	-0.037
1800	31.614	101.930	81.338	37.067	7.424	0.827	-0.024
1900	32.537	103.664	82.467	40.274	6.810	0.477	-0.013
2000	33.461	105.356	83.569	43.574	6.218	0.159	-0.004
2054.000	33.959	106.255	84.154	45.394	-------- ALPHA <--> BETA -------		
2100	34.384	107.011	84.646	46.966	0.	0.	0.
2200	35.307	108.632	85.700	50.451	0.	0.	0.
2300	36.231	110.222	86.732	54.028	0.	0.	0.
2400	37.154	111.784	87.743	57.697	0.	0.	0.
2500	38.077	113.353	88.736	61.542	0.	0.	0.
2500.000	38.077	113.353	88.736	61.542	------- BETA <--> LIQUID -------		
2600	39.001	114.865	89.712	65.396	-29.200	1.170	-0.024
2700	39.924	116.354	90.672	69.342	-29.019	2.335	-0.045
2800	40.847	117.822	91.615	73.381	-28.746	3.492	-0.065
2900	41.771	119.272	92.544	77.512	-28.381	4.637	-0.084
3000	42.694	120.704	93.459	81.735	-27.923	5.768	-0.100

PREVIOUS: CURRENT: March 1979

Hafnium, Beta (Hf) $Hf_1(cr)$

HAFNIUM (Hf) LIQUID A_r = 178.49

$S°(298.15\ K) = [71.938]\ J\ K^{-1}\ mol^{-1}$

$T_{fus} = 2500 \pm 20\ K$

$\Delta_f H°(298.15\ K) = [46.209]\ kJ\ mol^{-1}$

$\Delta_{fus} H° = [29.288 \pm 4.2]\ kJ\ mol^{-1}$

Enthalpy of Formation

The enthalpy of formation is calculated from that of the β-crystal by addition of $\Delta_{fus}H°$ and the difference in enthalpy, H°(2500 K)-H°(298.15 K), between the β-crystal and liquid.

Heat Capacity and Entropy

There are no heat capacity or enthalpy studies covering the liquid region of hafnium. A constant value of Cp° = 9.0 cal K^{-1} mol^{-1} is assumed for the region 1820-5500 K. A glass transition is assumed at 1820 K below which extrapolated β-Hf heat capacity values are used. The entropy is calculated in a manner similar to that used for the enthalpy of formation.

Fusion Data

Garg and Ackermann (1), using pyrometric techniques, measured the melting point of hafnium as 2501±3 K. Hultgren (2) reviewed nine studies available through 1955 and recommended T_{fus} = 2500±20 K. We adopted T_{fus} = 2500±20 K.

The studies (and the reported melting temperatures in K) examined by Hultgren (2) are: DeBoer and Fast (3) (1930), 2503±50; Skinner et al. (4) (1950), 2498; Litton (5) (1971), 2403; Adenstedt (6) (1952), 2248; Spedding (7) (1955), 2508±5; Deardoff and Hayes (8) (1956, 2495±20; Gregorovitch (9) (1960), 2116; Taylor et al. (10) (1964), 2503±20.

Ackermann and Rauh (11) measured a melting point of 2467±4 K in a HfO_2 container. However, they state that the melting point may be as high as 2495 K since their sample contained 0.7 wt % Zr.

The enthalpy melting for Hf has not been measured. Various estimates and calculations have been made. Martynyuk et al. (12) correlated graphically $\Delta_{fus}S°$ versus T_{fus} for many metals and used the resulting linear relationship to estimate $\Delta_{fus}H°$ = 7.0 kcal mol^{-1}. We adopt this value. Ackermann and Rauh (13) measured vapor pressures over Hf(β,cr) and Hf(ℓ), adjusted the vapor pressures for Hf(β,cr) to match those of Hf(ℓ) at T_{fus}, and derived $\Delta_{fus}H°$ = 6.2 kcal mol^{-1}.

Vaporization Data

The vaporization studies are summarized on the ideal gas table.

The boiling point is calculated as that temperature for which ΔG° = 0 for Hf(ℓ)=Hf(g). $\Delta_{vap}H°$ is the corresponding enthalpy change. T_{vap} is the temperature at which the fugacity is one bar. The normal boiling point (p=1 bar) would be a slightly lower temperature.

References - Hf(ℓ)

1. S. P. Garg and R. J. Ackermann, Met. Trans. 8A, 239 (1977).
2. R. Hultgren, P. D. Desai, et al., "Selected Values of the Thermodynamic Properties of the Elements", American Society for Metals, Metals Park, Ohio, 1973.
3. J. H. deBoer and J. D. Fast, Z. Anorg. Allgem. Chem. 187, 193 (1930).
4. G. B. Skinner, C. W. Beckett and H. L. Johnston, Ohio State University, Tech. Rept. No. 102-AC49/1-100-3, 1950.
5. F. B. Litton, J. Electrohem. Soc. 98, 488 (1951).
6. H. K. Adenstedit, Trans. A. S. M. 44, 949 (1952).
7. F. H. Spedding, U. S. AEC Rept. ISC-531, 1955.
8. D. K. Deardorff and E. T. Hayes, Trans. AIME 206, 509 (1956).
9. V. K. Gregorovitch, Izvest. Akad. Nauk SSSR, Otdel. Tekhn. Nauk Met. i Toplivo (6), 93 (1960).
10. A. Taylor, B. J. Kagle and N. J. Doyle, J. Less-Common Metals 5, 26 (1963).
11. R. J. Ackermann and E. G. Rauh, High Temp. Sci. 4, 272 (1972).
12. M. M. Martynyuk, Zh. Fiz. Khim. 51, 1199 (1977).
13. R. J. Ackermann and E. G. Rauh, J. Chem. Thermodynamics 4, 521 (1972).

References - β-Hf(cr)

1. A. Cezairliyan and J. L. McClure, J. Res. Nat. Bur. Stand. 79A, 431 (1975).
2. A. Cezairliyan and J. L. McClure, High Temp.-High Pressures 8, 467 (1976).
3. P. Duwez, J. Appl. Phys. 22, 1174 (1951).
4. J. D. Fast, J. Appl. Phys. 23, 350 (1952).
5. E. D. Gibson, B. A. Loomis and O. N. Carlson, Trans. A. S. M. 50, 348 (1958).
6. D. K. Deardorff and H. Kato, AIME 215, 876 (1959); Trans. AIME 227, 264 (1963).
7. D. T. Peterson and D. J. Beernsten, Trans. A. S. M. 52, 763 (1960).
8. B. C. Geissen, I. Rump and N. J. Grant, Trans. AIME 224, 60 (1962); Trans. SIME 227, 264 (1963).
9. R. G. Ross and W. Hume-Rothery, J. Less-Common Metals 5, 258 (1963).
10. N. N. Krikorian and T. C. Wallace, J. Electrochem. Soc. 111, 1431 (1964).
11. A. Taylor and N. J. Doyle, J. Less-Common Metals 7, 37 (1964).
12. P. A. Romans, O. G. Paasche and H. Kato, J. Less-Common Metals 8, 213 (1965).
13. G. N. Bates and G. Barnes, Appl. Phys. Lett. 11, 75 (1967).
14. V. E. Peletskii and V. P. Druzhinin, Teplofiz. Vys. Temp. 9, 539 (1971).
15. O. N. Carlson, F. A. Schmidt, and J. C. Sever, Met. Trans. 4, 2407 (1973).
16. M. M. Martynyuk and V. I. Tsapkov, Izv. Akad. Nauk SSSR, Metal. (2), 181 (1974).
17. H. Hoerster, E. Kauer, F. Kettel and A. Rabemai, Colloq. Int. Cent. Nat. Rech. Sci. 205, 39 (1972).
18. R. Hultgren, P. D. Desai, et al., "Selected Values of the Thermodynamic Properties of the Elements", American Society for Metals, Metals Park, Ohio, 1973.

Hafnium (Hf) $Hf_1(l)$

Enthalpy Reference Temperature = T_r = 298.15 K Standard State Pressure = p° = 0.1 MPa

T/K	$C_p°$ (J K⁻¹mol⁻¹)	S° (J K⁻¹mol⁻¹)	$-[G°-H°(T_r)]/T$ (J K⁻¹mol⁻¹)	$H°-H°(T_r)$ (kJ mol⁻¹)	$\Delta_f H°$ (kJ mol⁻¹)	$\Delta_f G°$ (kJ mol⁻¹)	Log K_f
0							
100							
200							
250							
298.15	17.747	71.938	71.938	0.	46.209	37.748	-6.613
300	17.764	72.048	71.939	0.033	46.194	37.695	-6.563
350	18.226	74.821	72.157	0.933	45.795	36.311	-5.419
400	18.687	77.285	72.646	1.855	45.393	34.983	-4.568
450	19.149	79.513	73.288	2.801	44.989	33.706	-3.912
500	19.611	81.554	74.014	3.770	44.585	32.474	-3.393
600	20.534	85.211	75.582	5.778	43.778	30.128	-2.623
700	21.457	88.446	77.193	7.877	42.976	27.916	-2.083
800	22.381	91.372	78.785	10.069	42.181	25.819	-1.686
900	23.304	94.061	80.335	12.353	41.395	23.821	-1.383
1000	24.227	96.564	81.834	14.730	40.619	21.910	-1.144
1100	25.151	98.916	83.281	17.199	39.852	20.076	-0.953
1200	26.074	101.144	84.678	19.760	39.097	18.312	-0.797
1300	26.997	103.268	86.027	22.414	38.353	16.610	-0.667
1400	27.921	105.302	87.331	25.160	37.627	14.965	-0.558
1500	28.844	107.260	88.595	27.998	36.924	13.371	-0.466
1600	29.767	109.151	89.821	30.928	36.243	11.823	-0.386
1700	30.691	110.984	91.012	33.951	35.584	10.317	-0.317
1800	31.614	112.764	92.172	37.067	34.948	8.849	-0.257
1820.000	31.799	113.114	92.400	37.701	GLASS <--> LIQUID		
1820.000	37.656	113.114	92.400	37.701	TRANSITION		
1900	37.656	114.734	93.306	40.713	34.773	7.406	-0.204
2000	37.656	116.666	94.426	44.479	34.646	5.969	-0.156
2100	37.656	118.503	95.529	48.244	28.802	4.669	-0.116
2200	37.656	120.255	96.614	52.010	29.083	3.513	-0.083
2300	37.656	121.929	97.678	55.776	29.271	2.346	-0.053
2400	37.656	123.531	98.722	59.541	29.368	1.173	-0.026
2500	37.656	125.068	99.746	63.307	29.288	0.	0.
2500.000	37.656	125.068	99.746	66.307	------- BETA <--> LIQUID -------		
2600	37.656	126.545	100.748	67.072	0.	0.	0.
2700	37.656	127.966	101.730	70.838	0.	0.	0.
2800	37.656	129.336	102.692	74.604	0.	0.	0.
2900	37.656	130.657	103.633	78.369	0.	0.	0.
3000	37.656	131.934	104.556	82.135	0.	0.	0.
3100	37.656	133.169	105.459	85.900	0.	0.	0.
3200	37.656	134.364	106.344	89.666	0.	0.	0.
3300	37.656	135.523	107.210	93.432	0.	0.	0.
3400	37.656	136.647	108.060	97.197	0.	0.	0.
3500	37.656	137.739	108.892	100.963	0.	0.	0.
3600	37.656	138.799	109.708	104.728	0.	0.	0.
3700	37.656	139.831	110.508	108.494	0.	0.	0.
3800	37.656	140.835	111.293	112.260	0.	0.	0.
3900	37.656	141.814	112.063	116.025	0.	0.	0.
4000	37.656	142.767	112.819	119.791	0.	0.	0.
4100	37.656	143.697	113.561	123.556	0.	0.	0.
4200	37.656	144.604	114.289	127.322	0.	0.	0.
4300	37.656	145.490	115.005	131.088	0.	0.	0.
4400	37.656	146.356	115.707	134.853	0.	0.	0.
4500	37.656	147.202	116.398	138.619	0.	0.	0.
4600	37.656	148.030	117.077	142.384	0.	0.	0.
4700	37.656	148.840	117.744	146.150	0.	0.	0.
4800	37.656	149.632	118.400	149.916	0.	0.	0.
4900	37.656	150.409	119.045	153.681	0.	0.	0.
4963.793	37.656	150.896	119.452	156.083	------- FUGACITY = 1 bar -------		
5000	37.656	151.170	119.680	157.447	-560.219	4.086	-0.043
5100	37.656	151.915	120.305	161.212	-560.290	15.373	-0.157
5200	37.656	152.646	120.920	164.978	-560.407	26.662	-0.268
5300	37.656	153.364	121.525	168.744	-560.568	37.953	-0.374
5400	37.656	154.068	122.121	172.509	-560.776	49.248	-0.476
5500	37.656	154.759	122.709	176.275	-561.030	60.547	-0.575

PREVIOUS: CURRENT: March 1979

Hafnium (Hf) $Hf_1(l)$

HAFNIUM (Hf) ALPLHA-BETA-LIQUID A_r = 178.49

0 to 2054 K crystal, alpha
2054 to 2500 K crystal, beta
above 2500 K liquid

Refer to the individual tables for details.

Hafnium (Hf) $Hf_1(cr,l)$

Enthalpy Reference Temperature = T_r = 298.15 K					Standard State Pressure = $p°$ = 0.1 MPa		
	J K⁻¹mol⁻¹			kJ mol⁻¹			
T/K	$C_p°$	S°	$-[G°-H°(T_r)]/T$	$H°-H°(T_r)$	$\Delta_f H°$	$\Delta_f G°$	Log K_f
0	0.	0.	INFINITE	-5.842	0.	0.	0.
100	20.580	17.753	-65.394	-4.764	0.	0.	0.
200	24.418	33.561	-45.871	-2.462	0.	0.	0.
250	25.110	39.086	-43.979	-1.223	0.	0.	0.
298.15	25.687	43.560	43.560	0.	0.	0.	0.
300	25.708	43.719	43.560	0.048	0.	0.	0.
350	26.247	47.723	43.875	1.347	0.	0.	0.
400	26.751	51.261	44.581	2.672	0.	0.	0.
450	27.230	54.439	45.503	4.021	0.	0.	0.
500	27.692	57.332	46.543	5.394	0.	0.	0.
600	28.583	62.460	48.780	8.208	0.	0.	0.
700	29.445	66.932	51.060	11.110	0.	0.	0.
800	30.288	70.919	53.298	14.097	0.	0.	0.
900	31.120	74.534	55.459	17.167	0.	0.	0.
1000	31.943	77.856	57.535	20.320	0.	0.	0.
1100	32.759	80.938	59.524	23.556	0.	0.	0.
1200	33.572	83.824	61.430	26.872	0.	0.	0.
1300	34.372	86.543	63.258	30.270	0.	0.	0.
1400	35.064	89.115	65.014	33.741	0.	0.	0.
1500	35.765	91.558	66.703	37.283	0.	0.	0.
1600	36.466	93.889	68.330	40.894	0.	0.	0.
1700	37.167	96.121	69.899	44.576	0.	0.	0.
1800	37.868	98.265	71.416	48.328	0.	0.	0.
1900	38.569	100.331	72.884	52.149	0.	0.	0.
2000	39.269	102.327	74.306	56.041	0.	0.	0.
2054.000	39.648	103.378	75.057	58.172	ALPHA <--> BETA		
2054.000	33.959	106.255	75.057	64.080	TRANSITION		
2100	34.384	107.011	75.749	65.652	0.	0.	0.
2200	35.307	108.632	77.207	69.136	0.	0.	0.
2300	36.231	110.222	78.608	72.713	0.	0.	0.
2400	37.154	111.784	79.957	76.383	0.	0.	0.
2500	38.077	113.353	81.262	80.228	0.	0.	0.
2500.000	38.077	113.353	81.262	80.228	BETA <--> LIQUID		
2500.000	37.656	125.068	81.262	109.516	TRANSITION		
2600	37.656	126.545	82.976	113.281	0.	0.	0.
2700	37.656	127.966	84.616	117.047	0.	0.	0.
2800	37.656	129.336	86.189	120.813	0.	0.	0.
2900	37.656	130.657	87.699	124.578	0.	0.	0.
3000	37.656	131.934	89.153	128.344	0.	0.	0.
3100	37.656	133.169	90.553	132.109	0.	0.	0.
3200	37.656	134.364	91.903	135.875	0.	0.	0.
3300	37.656	135.523	93.208	139.641	0.	0.	0.
3400	37.656	136.647	94.469	143.406	0.	0.	0.
3500	37.656	137.739	95.690	147.172	0.	0.	0.
3600	37.656	138.799	96.872	150.937	0.	0.	0.
3700	37.656	139.831	98.020	154.703	0.	0.	0.
3800	37.656	140.835	99.133	158.469	0.	0.	0.
3900	37.656	141.814	100.215	162.234	0.	0.	0.
4000	37.656	142.767	101.267	166.000	0.	0.	0.
4100	37.656	143.697	102.291	169.765	0.	0.	0.
4200	37.656	144.604	103.287	173.531	0.	0.	0.
4300	37.656	145.490	104.258	177.297	0.	0.	0.
4400	37.656	146.356	105.205	181.062	0.	0.	0.
4500	37.656	147.202	106.129	184.828	0.	0.	0.
4600	37.656	148.030	107.031	188.593	0.	0.	0.
4700	37.656	148.840	107.912	192.359	0.	0.	0.
4800	37.656	149.632	108.773	196.125	0.	0.	0.
4900	37.656	150.409	109.615	199.890	0.	0.	0.
4963.793	37.656	150.896	110.142	202.292	------- FUGACITY = 1 bar -------		
5000	37.656	151.170	110.438	203.656	-560.219	4.086	-0.043
5100	37.656	151.915	111.244	207.421	-560.290	15.373	-0.157
5200	37.656	152.646	112.034	211.187	-560.407	26.662	-0.268
5300	37.656	153.364	112.807	214.953	-560.568	37.953	-0.374
5400	37.656	154.068	113.564	218.718	-560.776	49.248	-0.476
5500	37.656	154.759	114.307	222.484	-561.030	60.547	-0.575

PREVIOUS: CURRENT: March 1979

Hafnium (Hf) $Hf_1(cr,l)$

HAFNIUM (Hf) IDEAL GAS A_r = 178.49

IP(Hf, g) = 54700 ± 600 cm^{-1}
S°(298.15 K) = 186.897 ± 0.2 J K^{-1} mol^{-1}

$\Delta_f H°(0\ K)$ = 618.0± 6.3 kJ mol^{-1}
$\Delta_f H°$ = (298.15 K) = 618.4 ± 6.3 kJ mol^{-1}

Electronic Levels and Quantum Weights

State	ϵ_i, cm^{-1}	g_i
$^3F_2$	0.00	5
$^3F_3$	2356.68	7
$^3F_4$	4567.64	9
.	.	.
	53853.61	
IP	54700	

Enthalpy of Formation

There are six sublimation studies and two vaporization studies from which an enthalpy of formation for Hf(g) could be derived. A plot of these vapor pressure data, log p vs 1/T, reveals considerable discrepancy between the various studies. In the liquid region, the vapor pressure values of Koch et al. (1) and Ackermann and Rauh (2) are in close agreement. The pressures of the former study are 5% higher at 2500 K but become 10% lower at 2800 K. The measured temperatures in the former study (1) were adjusted (i.e., the pyrometer reading) by Koch et al. (1) so that at the observed melting point the temperature would be 2500 K. The sample purity was not stated. In the latter study (2) the reported vapor pressures were corrected pressures (assuming ideal solution) from a eutectic of Hf and W. The Hf sample had a measured melting point of 2464 K; the purity was not reported.

The sublimation studies of Golubtsov and Mikul'skaya (3) and Ackermann and Rauh (2) spanned a temperature range which involved the α and β phases of hafnium, yet neither study mentioned the characterization of the condensed phase. The study of Golubtsov and Mikul'skaya (3) is not considered further since their reported vapor pressure equation is ambiguous and cannot be easily modified to yield reasonable vapor pressures. In general terms, the sublimation pressures of Ackermann and Rauh (2), Kibler et al. (4) and McClaine (5) are in reasonable agreement. The pressures of Panish and Reif (6) are 60% above these three studies (2, 4, 5); there is considerable scatter with one distinctly low pressure point at 2113 K. On the other hand, the results of Kebler and Aspinal (7) are substantially lower by 50% than the three studies (2, 4, 5) with two distinctly lower points.

A 2nd and 3rd law analysis is given in the following table. We adopt $\Delta_f H°$ (Hf, g, 298.15 K) = 147.8±1.5 kcal mol^{-1}. This result is based on the reasonable consistency of the two vapor pressure studies (2, 4) with each other and with our adopted thermodynamic functions as evidenced by the small values of the drift. Hultgren (8), using different thermodynamic functions and data available through 1966, recommended an enthalpy of formation of 148.0±1.0 kcal mol^{-1} based heavily on the study by Kibler et al. (4). Other values for $\Delta_f H°$(298.15 K), also derived from vapor pressure studies, have been reported by Krupka (9), Krikorian (10) and Lowrie (11).

Source (Year)	Reaction[a]	Pts	T/K	Method	$\Delta_r H°$(298.15 K), kcal mol^{-1} 2nd law	3rd law	Drift cal K^{-1} mol^{-1}	$\Delta_f H°$(298.15 K) kcal mol^{-1}
(11) Kibler et al. (1963)	A	8	2035-2325	Langmuir	141.5± 2.4	143.4±0.5	0.9±1.1	147.8
(6) Panish and Reif (1963)	A	6	2066-2274	Langmuir	160.3±15.5	140.9±2.7	-9.0±7.2	145.4
(5) McClaine (1964)	A	5	2200-2363	Langmuir	124.3± 5.7	144.5±1.3	8.8±2.5	149.0
(7) Kebler and Aspinall (1965)	A	15[b]	2075-2273	Knudsen	145.1± 4.2	145.9±0.7	0.4±1.9	150.4
(2) Ackermann and Rauh (1972)	A	eqn	1950-2464	mass eff.	145.4	143.6	-0.8	148.0
(1) Koch et al. (1968)	B	17	2500-2810	Langmuir	131.6± 3.4	136.9±1.2	2.0±1.3	147.9
(2) Ackermann and Rauh (1972)	B	eqn	1464-2800	mass eff.	140.1	136.7	-1.3	147.8

[a]Reaction: A) Hf(β,cr) = Hf(g) B) Hf(ℓ) = Hf(g)

[b]Two points discarded due to a statistical test.

Heat Capacity and Entropy

The electronic energy levels are given in the compilation by Meggers and Moore (12). Although we have listed only the ground state, the lowest two excited states, the highest observed excited state (below the ionization potential), and the ionization potential for Hf(g), all levels listed by Meggers and Moore (12), as well as estimated levels (for n = 5,6,7), are used in the calculation. The 233 observed levels, two of which lie above the ionization potential, are too numerous to list completely. In our calculations, the missing levels for n = 6 and 7 have been arbitrarily added near the ionization limit. The calculations indicate that for Hf(g) the thermodynamic functions up to at least 3000 K are independent of the estimated missing levels (for n = 6 and 7), the cut-off procedure, and the inclusion of n = 7 levels. At this temperature there are no experimental heat capacity enthalpy, or vapor pressure data. The Gibbs energy function is essentially unaffected up to 5500 K.

References

1. R. K. Koch, W. E. Anable and R. A. Beall, U. S. Bur. Mines RI-7125, 1968.
2. R. J. Ackerman and E. G. Rauh, J., Chem. Thermodynamics 4, 521 (1972).
3. I. V. Golubtsov and G. Yu. Mikul'skaya, Svoistva Metal Splavov 75, 62 (1975).
4. G. M. Kibler, T. F. Lyon, M. J. Linevsky and V. J. DeSantis, AEC Accesion No. 4661, Rept. No. WADD-TR-60-646, Part IV, 1963.
5. L. A. McClaine and P. E. Blackburn, ASD-TRD-62-204, Part III, 1964.
6. M. B. Panish and L. Reif, J. Chem. Phys. 38, 253 (1963).
7. O. C. Trulson, P. O. Schissel and H. W. Goldstein, AEC Accession No. 46956, Rept. No. NP-15514, Final Rept., Volume I, Appendix XII, 1965.
8. R. Hultgren, P. D. Desai, et al., "Selected Values of the Thermodynamic Properties of the Elements", American Society for Metals, Metals Park, Ohio, 1973.
9. M. C. Krupka, U. S. AEC LS-2611, 1962.
10. O. H. Krikorian, J. Phys. Chem. 67, 1586 (1963).
11. G. M. Kibler, T. F. Lyon, M. J. Linevsky and V. J. DeSantis, U. S. AEC WADD-TR-60-646, Part 3, Volume II, 1962.
12. W. F. Meggers and C. E. Moore, U. S. Nat. Bur. Stand. Monograph 153, 1976.

Hafnium (Hf) $Hf_1(g)$

Enthalpy Reference Temperature = T_r = 298.15 K Standard State Pressure = p° = 0.1 MPa

T/K	$C_p°$ (J $K^{-1}mol^{-1}$)	S°	-[G°-H°(T_r)]/T	H°-H°(T_r) (kJ mol^{-1})	$\Delta_f H°$	$\Delta_f G°$	Log K_f
0	0.	0.	INFINITE	-6.198	618.040	618.040	INFINITE
100	20.786	164.188	205.380	-4.119	619.040	604.397	-315.704
200	20.786	178.596	188.799	-2.041	618.816	589.809	-154.042
250	20.789	183.235	187.240	-1.001	618.617	582.580	-121.723
298.15	20.803	186.897	186.897	0.	618.395	575.659	-100.853
300	20.804	187.026	186.898	0.038	618.386	575.394	-100.185
350	20.854	190.236	187.151	1.080	618.128	568.249	-84.806
400	20.960	193.027	187.715	2.125	617.848	561.142	-73.278
450	21.140	195.505	188.445	3.177	617.551	554.071	-64.315
500	21.397	197.745	189.265	4.240	617.241	547.034	-57.148
600	22.113	201.706	191.017	6.414	616.600	533.053	-46.406
700	23.012	205.181	192.797	8.669	615.954	519.180	-38.742
800	23.991	208.317	194.544	11.019	615.317	505.398	-32.999
900	24.978	211.200	196.236	13.467	614.695	491.696	-28.537
1000	25.925	213.881	197.868	16.013	614.088	478.062	-24.971
1100	26.806	216.394	199.439	18.650	613.490	464.489	-22.057
1200	27.603	218.761	200.952	21.371	612.894	450.969	-19.630
1300	28.305	220.999	202.409	24.167	612.293	437.500	-17.579
1400	28.912	223.119	203.813	27.029	611.683	424.077	-15.823
1500	29.424	225.132	205.168	29.947	611.059	410.698	-14.302
1600	29.847	227.045	206.476	32.911	610.412	397.362	-12.973
1700	30.190	228.865	207.740	35.913	609.733	384.067	-11.801
1800	30.465	230.599	208.962	38.947	609.014	370.813	-10.761
1900	30.681	232.252	210.144	42.004	608.250	357.600	-9.831
2000	30.852	233.830	211.290	45.081	607.435	344.429	-8.996
2100	30.986	235.339	212.399	48.174	600.917	331.429	-8.244
2200	31.095	236.783	213.475	51.278	600.537	318.605	-7.565
2300	31.187	238.167	214.519	54.392	600.074	305.800	-6.945
2400	31.270	239.496	215.532	57.515	599.528	293.017	-6.377
2500	31.351	240.775	216.516	60.646	598.813	280.260	-5.856
2600	31.436	242.006	217.473	63.785	568.899	268.702	-5.398
2700	31.528	243.194	218.404	66.933	568.282	257.168	-4.975
2800	31.632	244.342	219.310	70.091	567.674	245.656	-4.583
2900	31.750	245.454	220.192	73.260	567.077	234.166	-4.218
3000	31.884	246.533	221.052	76.442	566.493	222.696	-3.877
3100	32.037	247.581	221.891	79.638	565.924	211.246	-3.559
3200	32.208	248.601	222.710	82.850	565.370	199.814	-3.262
3300	32.399	249.595	223.510	86.080	564.835	188.398	-2.982
3400	32.610	250.565	224.291	89.330	564.319	176.999	-2.719
3500	32.841	251.513	225.056	92.603	563.826	165.614	-2.472
3600	33.090	252.442	225.803	95.899	563.357	154.243	-2.238
3700	33.359	253.352	226.536	99.221	562.914	142.885	-2.017
3800	33.645	254.246	227.253	102.571	562.498	131.539	-1.808
3900	33.948	255.124	227.957	105.951	562.112	120.203	-1.610
4000	34.267	255.987	228.647	109.362	561.757	108.877	-1.422
4100	34.602	256.837	229.324	112.805	561.435	97.559	-1.243
4200	34.951	257.675	229.989	116.282	561.147	86.248	-1.073
4300	35.313	258.502	230.642	119.796	560.894	74.944	-0.910
4400	35.688	259.318	231.285	123.345	560.678	63.645	-0.756
4500	36.074	260.124	231.917	126.933	560.501	52.351	-0.608
4600	36.471	260.921	232.539	130.561	560.362	41.061	-0.466
4700	36.878	261.710	233.151	134.228	560.264	29.773	-0.331
4800	37.294	262.491	233.754	137.936	560.207	18.486	-0.201
4900	37.720	263.264	234.348	141.687	560.192	7.201	-0.077
4963.793	37.996	263.754	234.723	144.102	------- FUGACITY = 1 bar -------		
5000	38.154	264.031	234.934	145.481	0.	0.	0.
5100	38.595	264.790	235.512	149.318	0.	0.	0.
5200	39.044	265.544	236.083	153.200	0.	0.	0.
5300	39.501	266.292	236.646	157.127	0.	0.	0.
5400	39.964	267.035	237.202	161.100	0.	0.	0.
5500	40.434	267.773	237.751	165.120	0.	0.	0.
5600	40.910	268.505	238.293	169.187	0.	0.	0.
5700	41.392	269.234	238.830	173.302	0.	0.	0.
5800	41.879	269.958	239.360	177.465	0.	0.	0.
5900	42.372	270.678	239.885	181.678	0.	0.	0.
6000	42.870	271.394	240.404	185.940	0.	0.	0.

PREVIOUS: March 1979 (1 atm) CURRENT: March 1979 (1 bar)

Hafnium (Hf) $Hf_1(g)$

HAFNIUM, ION (Hf^+) IDEAL GAS M_r = 178.48945

IP(Hf^+, g) = 120000 ± 1000 cm^{-1}
S°(298.15 K) = 185.040 ± 0.01 J K mol^{-1}

Δ_fH°(0 K) = 1272.4 ± 4.0 kJ mol^{-1}
Δ_fH°(298.15 K) = [1278.950] kJ mol^{-1}

Electronic Levels and Quantum Weights

State	ε_i, cm^{-1}	g_i
$^2D_{3/2}$	0.00	4
$^2D_{5/2}$	3050.88	6
.	.	.
.	.	.
.	.	.
	83129.8	10
IP	120000	

Enthalpy of Formation

Δ_fH°(Hf^+, g 0 K) is calculated from Δ_fH°(Hf, g, 0 K) (1) using the spectroscopic value of IP(Hf) = 54700 ± 600 cm^{-1} (654.358 ± 1.18 kJ mol^{-1}) from Meggers and Moore (2). The ionization limit is converted from cm^{-1} to kJ mol^{-1} using the factor, 1 cm^{-1} = 0.01196266 kJ cm^{-1}, which is derived from the 1973 CODATA fundamental constants (3). Rosenstock et al. (4) and Levin and Lias (5) have summarized additional ionization and appearance potential data.

Δ_fH°(Hf^+, g, 298.15 K) is calculated from Δ_fH°(Hf, g, 0 K) by using IP (Hf) with JANAF (1) enthalpies, H°(0 K) - H°(298.15 K), for Hf(g), Hf^+(g), and e^-(g). Δ_fH°(Hf + Hf^+ + e^-, 298.15 K) differs from a room temperature threshold energy due to inclusion of these enthalpies and to threshold effects discussed by Rosenstock et al. (4). Δ_fH°(298.15 K) should be changed by -6.197 kJ mol^{-1} if it is to be used in the ion convention that excludes the enthalpy of the electron.

Heat Capacity and Entropy

The electronic energy levels are given in the compilation by Moore (6). Although we have listed only the ground state, the first excited state, the highest observed excited state, and the ionization potential for Hf^+(g), all levels listed by Moore (6), as well as estimated levels (for n=6 and 7), are used in the calculation. The 121 observed levels are too numerous to list completely. In our calculations the missing levels for n=6 and 7 have been arbitrarily added near the ionization limit. The calculations indicate that for Hf^+(g) the thermodyanmic functions up to 5000 K are independent of the estimated missing levels (for n=6 and 7), the cut-off procedure, and the inclusion of n=7 levels. The free energy function is unaffected up to 10000 K. The reported uncertainty in S°_{298} is due to uncertainties in the gram formula weight and the fundamental constants. Extension of these calculations above 6000 K may require consideration of the higher excited states (n>7), more precise estimation of the missing levels (for all n), and utilization of proper fill and cut-off procedures (7).

References

1. JANAF Thermochemical Tables: Hf(g), 3-31-84, e^-(g), 3-31-82.
2. W. F. Meggers and C. E. Moore, NBS Monograph 153 (1976).
3. E. R. Cohen and B. N. Taylor, J. Phys. Chem. Ref. Data 2, 663 (1973).
4. H. M. Rosenstock, K. Draxl, B. W. Steiner, and J. T. Herron, J. Phys. Chem. Ref. Data 6, Supp. 1 (1977).
5. R. D. Levin and S. G. Lias, U.S. Nat. Bur. Stand., NSRDS-NBS-71, 1982.
6. C. E. Moore, U.S. Nat. Bur. Stand., NSRDS-NBS 35, Volume III, 1971 [Reprint of NBS Circualr 476, Volume III, 1958].
7. J. R. Downey, Jr., The Dow Chemical Company, Thermal Research, AFOSR-TR-78-0960, March, 1978.

Hafnium, Ion (Hf^+) Hf_1^+(g)

Enthalpy Reference Temperature = T_r = 298.15 K   Standard State Pressure = p° = 0.1 MPa

	J $K^{-1}mol^{-1}$			kJ mol^{-1}			
T/K	C_p°	S°	-[G°-H°(T_r)]/T	H°-H°(T_r)	Δ_fH°	Δ_fG°	Log K_f
0	0.	0.	INFINITE	-6.197	1272.397		
100	20.786	162.333	203.521	-4.119			
200	20.786	176.741	186.942	-2.040			
250	20.786	181.379	185.382	-1.001			
298.15	20.786	185.040	185.040	0.	1278.950	1230.512	-215.581
300	20.786	185.169	185.041	0.038	1278.979	1230.212	-214.199
350	20.786	188.373	185.294	1.078	1279.758	1222.022	-182.377
400	20.786	191.149	185.856	2.117	1280.512	1213.722	-158.496
450	20.786	193.597	186.583	3.156	1281.241	1205.329	-139.911
500	20.786	195.787	187.396	4.196	1281.946	1196.857	-125.035
600	20.786	199.577	189.120	6.274	1283.290	1179.711	-102.703
700	20.786	202.781	190.848	8.353	1284.545	1162.347	-86.735
800	20.786	205.556	192.517	10.431	1285.716	1144.810	-74.748
900	20.786	208.005	194.105	12.510	1286.802	1127.130	-65.417
1000	20.786	210.195	195.606	14.589	1287.806	1109.334	-57.946
1100	20.787	212.176	197.024	16.667	1288.729	1091.442	-51.828
1200	20.789	213.985	198.363	18.746	1289.569	1073.469	-46.727
1300	20.794	215.649	199.629	20.825	1290.329	1055.429	-42.408
1400	20.804	217.190	200.829	22.905	1291.016	1037.334	-38.703
1500	20.823	218.626	201.968	24.986	1291.635	1019.192	-35.491
1600	20.857	219.971	203.052	27.070	1292.186	1001.010	-32.680
1700	20.910	221.237	204.085	29.158	1292.671	982.797	-30.198
1800	20.991	222.434	205.071	31.253	1293.092	964.556	-27.991
1900	21.107	223.572	206.015	33.358	1293.454	946.294	-26.015
2000	21.264	224.658	206.920	35.476	1293.759	928.015	-24.237
2100	21.472	225.701	207.790	37.612	1288.363	909.850	-22.631
2200	21.737	226.705	208.627	39.772	1289.117	891.808	-21.174
2300	22.065	227.679	209.434	41.962	1289.808	873.733	-19.843
2400	22.462	228.626	210.214	44.187	1290.444	855.629	-18.622
2500	22.931	229.552	210.970	46.456	1290.946	837.501	-17.499
2600	23.477	230.462	211.702	48.776	1262.291	820.523	-16.485
2700	24.100	231.359	212.413	51.154	1262.982	803.519	-15.545
2800	24.800	232.248	213.106	53.599	1263.739	786.488	-14.672
2900	25.576	233.132	213.781	56.117	1264.571	769.429	-13.859
3000	26.424	234.013	214.441	58.716	1265.483	752.340	-13.099
3100	27.341	234.894	215.086	61.404	1266.484	735.219	-12.388
3200	28.320	235.778	215.719	64.187	1267.579	718.064	-11.721
3300	29.355	236.665	216.340	67.070	1268.776	700.873	-11.094
3400	30.437	237.557	216.951	70.059	1270.078	683.645	-10.503
3500	31.559	238.455	217.553	73.159	1271.490	666.376	-9.945
3600	32.711	239.360	218.146	76.372	1273.017	649.066	-9.418
3700	33.883	240.273	218.732	79.702	1274.659	631.712	-8.918
3800	35.066	241.192	219.311	83.149	1276.420	614.312	-8.444
3900	36.249	242.118	219.884	86.715	1278.298	596.863	-7.994
4000	37.424	243.051	220.451	90.399	1280.295	579.366	-7.566
4100	38.580	243.989	221.014	94.199	1282.409	561.816	-7.158
4200	39.710	244.932	221.572	98.114	1284.636	544.214	-6.768
4300	40.805	245.880	222.126	102.140	1286.975	526.558	-6.396
4400	41.857	246.830	222.677	106.273	1289.422	508.846	-6.041
4500	42.862	247.782	223.224	110.510	1291.971	491.077	-5.700
4600	43.812	248.734	223.768	114.844	1294.618	473.250	-5.374
4700	44.704	249.686	224.310	119.270	1297.358	455.365	-5.061
4800	45.534	250.636	224.848	123.783	1300.183	437.420	-4.760
4900	46.299	251.583	225.384	128.375	1303.088	419.416	-4.471
5000	46.999	252.526	225.918	133.040	745.848	405.438	-4.236
5100	47.631	253.463	226.449	137.772	748.821	398.601	-4.083
5200	48.197	254.393	226.977	142.564	751.810	391.704	-3.935
5300	48.696	255.316	227.503	147.409	754.807	384.751	-3.792
5400	49.130	256.231	228.027	152.301	757.804	377.740	-3.654
5500	49.501	257.136	228.548	157.233	760.795	370.675	-3.520
5600	49.811	258.030	229.066	162.199	763.772	363.555	-3.391
5700	50.063	258.914	229.582	167.194	766.730	356.382	-3.266
5800	50.260	259.787	230.095	172.210	769.662	349.157	-3.144
5900	50.405	260.647	230.606	177.244	772.562	341.882	-3.027
6000	50.502	261.495	231.114	182.290	775.424	334.558	-2.913

PREVIOUS: March 1979 (1 atm)   CURRENT: March 1984 (1 bar)

Hafnium, Ion (Hf^+) Hf_1^+(g)

HAFNIUM, ION (Hf^-) IDEAL GAS M_r = 178.49055

Hafnium, Ion (Hf^-) $Hf_1^-(g)$

EA(Hf, g) = 0.0 eV

S°(298.15 K) = 185.040 ± 0.002 J K^{-1} mol^{-1}

$\Delta_f H°(0\ K)$ = 618.040 kJ mol^{-1}

$\Delta_f H°(298.15\ K)$ = [612.197] kJ mol^{-1}

Electonic Level and Quantum Weight

State	ε_i, cm^{-1}	g_i
$^4F_{3/2}$	0.0	4

Enthalpy of Formation

$\Delta_f H°(Hf^-, g, 0\ K)$ is calculated from $\Delta_f H°(Hf, g, 0\ K)$ (1) using the adopted electron affinity of EA(Hf) = ~ 0.0 eV. This value, recommended by Hotop and Lineberger (2), is based on a semiempirical extrapolation (3, 4). Additional information on $Hf^-(g)$ and may be obtained in the critical discussions of Hotop and Lineberger (2, 4), Rosenstock et al. (5) and Massey (6).

$\Delta_f H°(Hf^-, g, 298.15\ K)$ is obtained from $\Delta_f H°(Hf, g, 0\ K)$ by using EA(Hf) with JANAF (1) enthalpies, H°(0 K) - H°(298.15 K), for $Hf^-(g)$, Hf(g), and $e^-(g)$. $\Delta_r H°(Hf^- \rightarrow Hf + e^-, 298.15\ K)$ differs from a room-temperature threshold energy due to inclusion of these enthalpies and to threshold effects discussed by Rosenstock et al. (5). $\Delta_f H°(298.15\ K)$ should be changed by +6.197 kJ mol^{-1} if it is to be used in the ion convention that excludes the enthalpy of the electron.

Heat Capacity and Entropy

The ground state electronic configuration for $Hf^-(g)$ is given by Hotop and Lineberger (2, 4) and Rosenstock et al. (5). Lacking any experimental evidence as to the stability of any excited states, we assume that no stable excited states exist.

References

1. JANAF Thermochemical Tables: Hf(g), 3-31-79; $e^-(g)$, 3-31-82.
2. H. Hotop and W. C. Lineberger, J. Phys. Chem. Ref. Data, to be published, 1985.
3. R. J. Zollweg, J. Chem. Phys. 50, 4251 (1969).
4. H. Hotop and W. C. Lineberger, J. Phys. Chem. Ref. Data 4, 539 (1975).
5. H. M. Rosenstock, K. Draxl et al., J. Phys. Chem. Ref. Data 6, Supp. 1 (1977).
6. H. S. W. Massey, "Negative Ions", 3rd ed., Cambridge University Press, Cambridge, 1976.

Enthalpy Reference Temperature = T_r = 298.15 K; Standard State Pressure = p° = 0.1 MPa

T/K	$C_p°$ (J K⁻¹mol⁻¹)	S° (J K⁻¹mol⁻¹)	$-[G°-H°(T_r)]/T$ (J K⁻¹mol⁻¹)	$H°-H°(T_r)$ (kJ mol⁻¹)	$\Delta_f H°$ (kJ mol⁻¹)	$\Delta_f G°$ (kJ mol⁻¹)	Log K_f
0	0.	0.	INFINITE	-6.197	618.040		
100	20.786	162.333	203.521	-4.119			
200	20.786	176.741	186.942	-2.040			
250	20.786	181.379	185.383	-1.001			
298.15	20.786	185.040	185.040	0.	612.197	576.270	-100.960
300	20.786	185.169	185.041	0.038	612.150	576.047	-100.299
350	20.786	188.373	185.294	1.078	610.851	570.132	-85.088
400	20.786	191.149	185.856	2.117	609.526	564.405	-73.704
450	20.786	193.597	186.583	3.156	608.176	558.846	-64.869
500	20.786	195.787	187.396	4.196	606.803	553.438	-57.817
600	20.786	199.577	189.120	6.274	603.989	543.028	-47.275
700	20.786	202.781	190.848	8.353	601.087	533.096	-39.780
800	20.786	205.556	192.517	10.431	598.101	523.586	-34.187
900	20.786	208.005	194.105	12.510	595.030	514.455	-29.858
1000	20.786	210.195	195.606	14.589	591.877	505.671	-26.414
1100	20.786	212.176	197.024	16.667	588.642	497.206	-23.610
1200	20.786	213.984	198.363	18.746	585.325	489.040	-21.287
1300	20.786	215.648	199.629	20.824	581.928	481.153	-19.333
1400	20.786	217.189	200.829	22.903	578.456	473.531	-17.668
1500	20.786	218.623	201.968	24.982	574.915	466.159	-16.233
1600	20.786	219.964	203.052	27.060	571.303	459.026	-14.986
1700	20.786	221.224	204.084	29.139	567.621	452.121	-13.892
1800	20.786	222.412	205.069	31.217	563.870	445.435	-12.926
1900	20.786	223.536	206.012	33.296	560.048	438.959	-12.068
2000	20.786	224.602	206.915	35.375	556.156	432.687	-11.301
2100	20.786	225.617	207.782	37.453	546.546	426.740	-10.615
2200	20.786	226.584	208.615	39.532	543.061	421.116	-9.999
2300	20.786	227.508	209.416	41.610	539.484	415.653	-9.440
2400	20.786	228.392	210.188	43.689	535.815	410.347	-8.931
2500	20.786	229.241	210.934	45.768	531.970	405.198	-8.466
2600	20.786	230.056	211.654	47.846	498.916	401.373	-8.064
2700	20.786	230.840	212.350	49.925	495.150	397.693	-7.694
2800	20.786	231.596	213.024	52.004	491.385	394.153	-7.353
2900	20.786	232.326	213.677	54.082	487.619	390.747	-7.038
3000	20.786	233.030	214.310	56.161	483.854	387.470	-6.746
3100	20.786	233.712	214.925	58.239	480.088	384.320	-6.476
3200	20.786	234.372	215.523	60.318	476.322	381.290	-6.224
3300	20.786	235.012	216.104	62.397	472.557	378.379	-5.989
3400	20.786	235.632	216.669	64.475	468.791	375.581	-5.770
3500	20.786	236.235	217.219	66.554	465.026	372.895	-5.565
3600	20.786	236.820	217.756	68.632	461.260	370.316	-5.373
3700	20.786	237.390	218.279	70.711	457.494	367.841	-5.193
3800	20.786	237.944	218.789	72.790	453.729	365.469	-5.024
3900	20.786	238.484	219.287	74.868	449.963	363.195	-4.864
4000	20.786	239.010	219.774	76.947	446.198	361.018	-4.714
4100	20.786	239.524	220.249	79.025	442.432	358.936	-4.573
4200	20.786	240.024	220.714	81.104	438.666	356.945	-4.439
4300	20.786	240.514	221.169	83.183	434.901	355.043	-4.313
4400	20.786	240.991	221.614	85.261	431.135	353.230	-4.193
4500	20.786	241.459	222.050	87.340	427.370	351.501	-4.080
4600	20.786	241.915	222.477	89.418	423.604	349.857	-3.973
4700	20.786	242.362	222.895	91.497	419.838	348.294	-3.871
4800	20.786	242.800	223.305	93.576	416.073	346.812	-3.774
4900	20.786	243.229	223.707	95.654	412.307	345.408	-3.682
5000	20.786	243.649	224.102	97.733	-151.677	348.167	-3.637
5100	20.786	244.060	224.489	99.811	-155.514	358.202	-3.669
5200	20.786	244.464	224.870	101.890	-159.396	368.313	-3.700
5300	20.786	244.860	225.243	103.969	-163.324	378.499	-3.730
5400	20.786	245.248	225.610	106.047	-167.297	388.759	-3.760
5500	20.786	245.630	225.970	108.126	-171.317	399.093	-3.790
5600	20.786	246.004	226.325	110.204	-175.384	409.501	-3.820
5700	20.786	246.372	226.673	112.283	-179.499	419.982	-3.849
5800	20.786	246.734	227.016	114.362	-183.662	430.535	-3.877
5900	20.786	247.089	227.353	116.440	-187.874	441.161	-3.906
6000	20.786	247.438	227.685	118.519	-192.136	451.858	-3.934

PREVIOUS: March 1979 (1 atm) CURRENT: March 1984 (1 bar)

Hafnium, Ion (Hf^-) $Hf_1^-(g)$

MERCURY (Hg) REFERENCE STATE A_r = 200.59

0 to 234.29	K	crystal
234.29 to 629.839	K	liquid
above 629.839	K	ideal monatomic gas

Refer to the individual tables for details.

Mercury (Hg) Hg_1(ref)

Enthalpy Reference Temperature = T_r = 298.15 K — Standard State Pressure = p° = 0.1 MPa

T/K	C_p° (J K⁻¹mol⁻¹)	S° (J K⁻¹mol⁻¹)	$-[G^\circ-H^\circ(T_r)]/T$ (J K⁻¹mol⁻¹)	$H^\circ-H^\circ(T_r)$ (kJ mol⁻¹)	Δ_fH° (kJ mol⁻¹)	Δ_fG° (kJ mol⁻¹)	Log K_f
0	0.	0.	INFINITE	-9.343	0.	0.	0.
100	24.255	37.219	113.592	-7.637	0.	0.	0.
200	27.275	55.022	80.284	-5.052	0.	0.	0.
234.290	28.485	59.428	76.915	-4.097	CRYSTAL <--> LIQUID TRANSITION		
234.290	28.476	69.225	76.915	-1.802			
298.15	27.978	76.028	76.028	0.	0.	0.	0.
300	27.963	76.201	76.028	0.052	0.	0.	0.
400	27.414	84.161	77.118	2.817	0.	0.	0.
500	27.175	90.248	79.158	5.545	0.	0.	0.
600	27.139	95.197	81.432	8.259	0.	0.	0.
629.839	27.164	96.514	82.115	9.069	LIQUID <--> IDEAL GAS FUGACITY = 1 bar		
629.839	20.786	190.515	82.115	68.274			
700	20.786	192.710	93.092	69.733	0.	0.	0.
800	20.786	195.486	105.722	71.811	0.	0.	0.
900	20.786	197.934	115.834	73.890	0.	0.	0.
1000	20.786	200.124	124.156	75.969	0.	0.	0.
1100	20.786	202.105	131.153	78.047	0.	0.	0.
1200	20.786	203.914	137.143	80.126	0.	0.	0.
1300	20.786	205.578	142.344	82.204	0.	0.	0.
1400	20.786	207.118	146.916	84.283	0.	0.	0.
1500	20.786	208.552	150.978	86.362	0.	0.	0.
1600	20.786	209.894	154.619	88.440	0.	0.	0.
1700	20.786	211.154	157.908	90.519	0.	0.	0.
1800	20.786	212.342	160.899	92.597	0.	0.	0.
1900	20.786	213.466	163.636	94.676	0.	0.	0.
2000	20.786	214.532	166.155	96.755	0.	0.	0.
2100	20.786	215.546	168.483	98.833	0.	0.	0.
2200	20.786	216.513	170.644	100.912	0.	0.	0.
2300	20.786	217.437	172.659	102.990	0.	0.	0.
2400	20.786	218.322	174.543	105.069	0.	0.	0.
2500	20.786	219.170	176.311	107.148	0.	0.	0.
2600	20.786	219.986	177.975	109.226	0.	0.	0.
2700	20.786	220.770	179.546	111.305	0.	0.	0.
2800	20.786	221.526	181.032	113.383	0.	0.	0.
2900	20.786	222.255	182.441	115.462	0.	0.	0.
3000	20.786	222.960	183.780	117.541	0.	0.	0.
3100	20.786	223.642	185.055	119.619	0.	0.	0.
3200	20.786	224.302	186.271	121.698	0.	0.	0.
3300	20.787	224.941	187.433	123.776	0.	0.	0.
3400	20.787	225.562	188.546	125.855	0.	0.	0.
3500	20.787	226.164	189.612	127.934	0.	0.	0.
3600	20.788	226.750	190.635	130.013	0.	0.	0.
3700	20.789	227.319	191.619	132.091	0.	0.	0.
3800	20.790	227.874	192.566	134.170	0.	0.	0.
3900	20.791	228.414	193.478	136.249	0.	0.	0.
4000	20.793	228.940	194.358	138.329	0.	0.	0.
4100	20.795	229.454	195.208	140.408	0.	0.	0.
4200	20.798	229.955	196.029	142.488	0.	0.	0.
4300	20.802	230.444	196.824	144.568	0.	0.	0.
4400	20.807	230.923	197.594	146.648	0.	0.	0.
4500	20.813	231.390	198.339	148.729	0.	0.	0.
4600	20.820	231.848	199.063	150.811	0.	0.	0.
4700	20.829	232.296	199.765	152.893	0.	0.	0.
4800	20.839	232.734	200.448	154.977	0.	0.	0.
4900	20.851	233.164	201.111	157.061	0.	0.	0.
5000	20.866	233.586	201.756	159.147	0.	0.	0.
5100	20.882	233.999	202.384	161.234	0.	0.	0.
5200	20.902	234.405	202.996	163.324	0.	0.	0.
5300	20.924	234.803	203.593	165.415	0.	0.	0.
5400	20.949	235.194	204.174	167.508	0.	0.	0.
5500	20.978	235.579	204.742	169.605	0.	0.	0.
5600	21.010	235.957	205.296	171.704	0.	0.	0.
5700	21.046	236.329	205.837	173.807	0.	0.	0.
5800	21.087	236.696	206.366	175.913	0.	0.	0.
5900	21.132	237.057	206.883	178.024	0.	0.	0.
6000	21.182	237.412	207.389	180.140	0.	0.	0.

PREVIOUS: December 1961 (1 atm) CURRENT: December 1961 (1 bar)

Mercury (Hg) Hg_1(ref)

MERCURY (Hg) CRYSTAL–LIQUID A_r = 200.59

$S°(298.15\ K) = 76.03 \pm 0.2\ J\ K^{-1}\ mol^{-1}$

$T_{fus} = 234.29\ K$

$\Delta_f H°(0\ K) = \ kJ\ mol^{-1}$

$\Delta_f H°(298.15\ K) = 0\ kJ\ mol^{-1}$

$\Delta_{fus} H° = 2.2953 \pm 0.0004\ kJ\ mol^{-1}$

Enthalpy of Formation

Zero by definition.

Heat Capacity and Entropy

The heat capacities of Smith and Wolcott, (1.2-20.2 K, 1), Busey and Giaugue (15-330 K, 2), and Douglas et al. (300-770 K, 3) have been smoothly joined. There is fair agreement between the selected valaues and those of Pickard and Simon (4), Carpenter and Oakley (5), Carpenter and Stoodley (6), Cohen et al. (7), Dixon and Rodebush (8), Simon (9), Bronsted (10), and Pollitzer (11).

Fusion Data

The results of Busey and Giauque (2) are adopted.

Vaporization Data

See the gas phase tables for details.

References

1. P. L. Smith and N. M. Wolcott, Phil. Mag. 1[8], 854(1956).
2. R. H. Busey and W. F. Giauque, J. Am. Chem. Soc. 75, 806(1953).
3. T. B. Douglas, A. T. Ball, and d. C. Ginnings, J. Res. Nat. Bur. Stands. 46, 334(1951).
4. G. L. Pickard and T. E. Simon, Proc, Phys. Soc. A61, 1(1948).
5. L. G. Carpenter and F. H. Oakley, Phil. Mag. 12, 511(1931).
6. L. G. Carpenter and L. G. Stoodley, Phil. Mag. 10, 249 (1930).
7. E. Cohen, C. L. Kruisheer, and A. L. Morsvold, Z. Phys. Chem. A96, 437(1920).
8. A. L. Dixon and W. H. Rodebush, J. Am. Chem. Soc. 49, 1162 (1927).
9. F. Simon, Z. Phys. Chem. 107, 282(1923); Ann. Physik 68, 241(1922).
10. J. N. Bronsted, Z. Elektrochem. 18, 714(1912).
11. F. Pollitzer, Z. Elektrochem. 17, 5(1911).

Mercury (Hg) $Hg_1(cr,l)$

Enthalpy Reference Temperature = T_r = 298.15 K Standard State Pressure = p° = 0.1 MPa

T/K	$C_p°$ (J K⁻¹mol⁻¹)	S° (J K⁻¹mol⁻¹)	$-[G°-H°(T_r)]/T$ (J K⁻¹mol⁻¹)	$H°-H°(T_r)$ (kJ mol⁻¹)	$\Delta_f H°$ (kJ mol⁻¹)	$\Delta_f G°$ (kJ mol⁻¹)	Log K_f
0	0.	0.	INFINITE	-9.343	0.	0.	0.
100	24.255	37.219	113.592	-7.637	0.	0.	0.
200	27.275	55.022	80.284	-5.052	0.	0.	0.
234.290	28.485	59.428	76.915	-4.097	CRYSTAL <--> LIQUID	TRANSITION	
234.290	28.476	69.225	76.915	-1.802			
250	28.351	71.069	76.490	-1.355	0.	0.	0.
298.15	27.978	76.028	76.028	0.	0.	0.	0.
300	27.963	76.201	76.028	0.052	0.	0.	0.
350	27.639	80.486	76.368	1.441	0.	0.	0.
400	27.414	84.161	77.118	2.817	0.	0.	0.
450	27.267	87.381	78.083	4.184	0.	0.	0.
500	27.175	90.248	79.158	5.545	0.	0.	0.
600	27.139	95.197	81.432	8.259	0.	0.	0.
629.839	27.164	96.514	82.115	9.069	------- FUGACITY	= 1 bar	-------
700	27.291	99.389	83.705	10.979	-58.754	6.571	-0.490
800	27.582	103.052	85.899	13.722	-58.089	15.858	-1.035
900	27.896	106.318	87.990	16.496	-57.394	25.060	-1.454
1000	28.210	109.274	89.973	19.301	-56.667	34.183	-1.786
1100	28.523	111.977	91.852	22.138	-55.909	43.232	-2.053
1200	28.837	114.472	93.634	25.006	-55.120	52.210	-2.273
1300	29.151	116.793	95.328	27.905	-54.299	61.121	-2.456
1400	29.465	118.965	96.939	30.836	-53.447	69.968	-2.611
1500	29.778	121.008	98.476	33.798	-52.563	78.752	-2.742
1600	30.092	122.940	99.945	36.792	-51.649	87.477	-2.856
1700	30.406	124.774	101.352	39.816	-50.702	96.144	-2.954
1800	30.719	126.521	102.703	42.873	-49.725	104.754	-3.040
1900	31.033	128.190	104.000	45.960	-48.716	113.308	-3.115
2000	31.347	129.790	105.250	49.079	-47.675	121.809	-3.181

PREVIOUS: CURRENT: December 1961

Mercury (Hg) $Hg_1(cr,l)$

MERCURY (Hg) IDEAL GAS A_r = 200.59

IP(Hg, g) = 84184.1 ± 0.5 cm^{-1}
S°(298.15 K) = 174.970 ± 0.020 J K^{-1} mol^{-1}

$\Delta_f H°(0\ K)$ = 64.53 ± 0.04 kJ mol^{-1}
$\Delta_f H°(298.15\ K)$ = 61.38 ± 0.04 kJ mol^{-1}

Electronic Levels and Quantum Weights

State	ε_i, cm^{-1}	g_i
$^1S_0$	0.000	1
$^3P_0$	37645.080	1
$^3P_1$	39412.300	3
$^3P_2$	44042.977	5

Enthalpy of Formation

The value adopted for the enthalpy of formation of Hg(g) is that recommended by CODATA (1). This value was derived from the vapor pressure studies summarized by Hultren et al. (2). Of the 24 studies examined, most weight was given to the following studies: spedding and Dye (533 - 630 K, 3), Menzies (395 625 K, 4), Beattie, Blaisdell, and Kalinsky (623 - 635 K, 5), Burlingame (343 - 474) K, 6), Rodebush and Dixon (444 - 476 K, 7) and Mayer (262 - 298 K, 8).

Heat Capacity and Entropy

The information on electronic energy levels and quantum weights, given by Moore (9, 10), is incomplete because many theoretically predicted levels have not been observed. Our calculations indicate that any reasonable method of filling in these missing levels and cutting off the summation in the partition function (11) has no effect on the thermodynamic functions to 6000 K. This is a result of the high energy of all levels other than the ground state and the three levels below 45000 cm^{-1}; the next excited state is approximately 54000 cm^{-1} above the ground state. Since inclusion of these excited states has no effect on the thermodynamic functions (to 6000 K), we list only the ground state and the three P levels. The reported uncertainty in S° (298.15 K) is due to uncertainties in the relative ionic mass and the fundamental constants. Extension of these calculations above 6000 K may require consideration of the excited states and use of different fill and cutoff procedures (11).

The thermal functions at 298.15 K agree with the CODATA recommendations (1) except that the entropy differs by 0.1094 J K^{-1} mol^{-1}, since this table uses a standard state pressure of 1 bar (CODATA recommendations are based on 1 atm).

References

1. J. D. Cox, chairman, CODATA Task Group on Key Values for Thermodynamics, J. Chem. Thermodynamics 10, 903 (1978).
2. R. Hultgren, P. D. Desai, et al., "Selected Values of the Thermodynamic Properties of the Elements," American Society for Metals, Metals Park, Ohio, 1973.
3. F. H. Spedding and J. L. Dye, J. Phys. Chem. 59, 581 (1955).
4. A. W. C. Menzies, Z. Physik. Chem. 130, 90 (1927).
5. J. A. Beattie, B. E. Blaisdell, and J. Kaminsky, Proc. Amer. Acad. Arts and Sci. 71, 375 (1937).
6. J. W. Burlingame, University of Pennsylvania, Ph. D. Thesis, 1968.
7. W. H. Rodebush and A. L. Dixon, Phys. Rev. 26, 851 (1925).
8. H. Mayer, Z. Physik. 67, 240 (1931).
9. C. E. Moore, U. S. Nat. Bur. Stand., NSRDS-NBS-34, 1970.
10. C. E. Moore, U. S. Nat. Bur. Stand., NSRDS-NBS-35, Volume III, 1971 [Reprint of NBS Circular 467, Volume III, 1958].
11. J. R. Downey, Jr., The Dow Chemical Company, AFOSR-TR-78-0960, Contract No. F44670-75-1-0048, 1978.

Mercury (Hg) $Hg_1(g)$

Enthalpy Reference Temperature = T_r = 298.15 K — Standard State Pressure = p° = 0.1 MPa

T/K	$C_p°$ (J $K^{-1}mol^{-1}$)	S°	$-[G°-H°(T_r)]/T$	$H°-H°(T_r)$ (kJ mol^{-1})	$\Delta_f H°$	$\Delta_f G°$	Log K_f
0	0.	0.	INFINITE	-6.197	64.526	64.526	INFINITE
100	20.786	152.263	193.450	-4.119	64.899	53.394	-27.890
200	20.786	166.670	176.871	-2.040	64.392	42.062	-10.986
250	20.786	171.309	175.312	-1.001	61.735	36.675	-7.663
298.15	20.786	174.970	174.970	0.	61.380	31.880	-5.585
300	20.786	175.098	174.970	0.038	61.367	31.697	-5.519
350	20.786	178.303	175.223	1.078	61.016	26.781	-3.997
400	20.786	181.078	175.786	2.117	60.680	21.913	-2.862
450	20.786	183.526	176.512	3.156	60.352	17.087	-1.983
500	20.786	185.717	177.325	4.196	60.031	12.297	-1.285
600	20.786	189.506	179.049	6.274	59.395	2.810	-0.245
629.839	20.786	190.515	179.569	6.894	------- FUGACITY = 1 bar -------		
700	20.786	192.710	180.778	8.353	0.	0.	0.
800	20.786	195.486	182.447	10.431	0.	0.	0.
900	20.786	197.934	184.034	12.510	0.	0.	0.
1000	20.786	200.124	185.536	14.589	0.	0.	0.
1100	20.786	202.105	186.953	16.667	0.	0.	0.
1200	20.786	203.914	188.292	18.746	0.	0.	0.
1300	20.786	205.578	189.559	20.824	0.	0.	0.
1400	20.786	207.118	190.759	22.903	0.	0.	0.
1500	20.786	208.552	191.898	24.982	0.	0.	0.
1600	20.786	209.894	192.981	27.060	0.	0.	0.
1700	20.786	211.154	194.013	29.139	0.	0.	0.
1800	20.786	212.342	194.999	31.217	0.	0.	0.
1900	20.786	213.466	195.942	33.296	0.	0.	0.
2000	20.786	214.532	196.845	35.375	0.	0.	0.
2100	20.786	215.546	197.711	37.453	0.	0.	0.
2200	20.786	216.513	198.544	39.532	0.	0.	0.
2300	20.786	217.437	199.346	41.610	0.	0.	0.
2400	20.786	218.322	200.118	43.689	0.	0.	0.
2500	20.786	219.170	200.863	45.768	0.	0.	0.
2600	20.786	219.986	201.583	47.846	0.	0.	0.
2700	20.786	220.770	202.279	49.925	0.	0.	0.
2800	20.786	221.526	202.953	52.004	0.	0.	0.
2900	20.786	222.255	203.606	54.082	0.	0.	0.
3000	20.786	222.960	204.240	56.161	0.	0.	0.
3100	20.786	223.642	204.855	58.239	0.	0.	0.
3200	20.786	224.302	205.452	60.318	0.	0.	0.
3300	20.787	224.941	206.033	62.397	0.	0.	0.
3400	20.787	225.562	206.598	64.475	0.	0.	0.
3500	20.787	226.164	207.149	66.554	0.	0.	0.
3600	20.788	226.750	207.685	68.633	0.	0.	0.
3700	20.789	227.319	208.208	70.712	0.	0.	0.
3800	20.790	227.874	208.719	72.790	0.	0.	0.
3900	20.791	228.414	209.217	74.869	0.	0.	0.
4000	20.793	228.940	209.703	76.949	0.	0.	0.
4100	20.795	229.454	210.179	79.028	0.	0.	0.
4200	20.798	229.955	210.644	81.108	0.	0.	0.
4300	20.802	230.444	211.098	83.188	0.	0.	0.
4400	20.807	230.923	211.544	85.268	0.	0.	0.
4500	20.813	231.390	211.979	87.349	0.	0.	0.
4600	20.820	231.848	212.406	89.431	0.	0.	0.
4700	20.829	232.296	212.825	91.513	0.	0.	0.
4800	20.839	232.734	213.235	93.597	0.	0.	0.
4900	20.851	233.164	213.637	95.681	0.	0.	0.
5000	20.866	233.586	214.032	97.767	0.	0.	0.
5100	20.882	233.999	214.420	99.855	0.	0.	0.
5200	20.902	234.405	214.800	101.944	0.	0.	0.
5300	20.924	234.803	215.174	104.035	0.	0.	0.
5400	20.949	235.194	215.541	106.129	0.	0.	0.
5500	20.978	235.579	215.902	108.225	0.	0.	0.
5600	21.010	235.957	216.256	110.324	0.	0.	0.
5700	21.046	236.329	216.605	112.427	0.	0.	0.
5800	21.087	236.696	216.949	114.534	0.	0.	0.
5900	21.132	237.057	217.286	116.644	0.	0.	0.
6000	21.182	237.412	217.619	118.760	0.	0.	0.

PREVIOUS: December 1961 (1 atm) CURRENT: September 1984 (1 bar)

Mercury (Hg) $Hg_1(g)$

MERCURY, ION (Hg^+) IDEAL GAS M_r = 200.58945

IP(Hg^+, g) = 151280 ± 10 cm^{-1}

S°(298.15 K) = 224.617 ± 0.025 J K^{-1} mol^{-1}

$\Delta_f H°$(0 K) = 1071.57 ± 4.0 kJ mol^{-1}

$\Delta_f H°$(298.15 K) = [1074.623] kJ mol^{-1}

Electronic Levels and Quantum Weights

State	ϵ_i, cm^{-1}	g_i
$^2S_{1/2}$	0.0	2
$^2D_{5/2}$	35514	6
$^2D_{3/2}$	50552	4
$^2P_{1/2}$	51485	2
$^2P_{3/2}$	60608	4

Enthalpy of Formation

$\Delta_f H°$(Hg^+, g, 0 K) is calculated from $\Delta_f H°$(Hg, g, 0 K) (1) using the spectroscopic value of IP(Hg) = 84184.1 ± 0.5 cm^{-1} (1007.066 ± 0.006 kJ mol^{-1}) from Moore (2). The ionization limit is converted from cm^{-1} to kJ mol^{-1} using the factor, 1 cm^{-1} = 0.01196266 kJ cm^{-1}, which is derived from the 1973 CODATA fundamental constants (3). Rosenstock et al. (4) and Levin and Lias (5) have summarized additional ionization and appearance potential data.

$\Delta_f H°$(Hg^+, g, 298.15 K) is calculated from $\Delta_f H°$(Hg, g, 0 K) by using IP(Hg) with JANAF (1) enthalpies, H°(0 K) - H°(298.15 K), for Hg(g), Hg^+(g), and e^-(g). $\Delta_f H°$(Hg + Hg^+ + e^-, 298.15 K) differs from a room temperature threshold energy due to inclusion of these enthalpies and to threshold effects discussed by Rosenstock et al. (4). $\Delta_f H°$(298.15 K) should be changed by -6.197 kJ mol^{-1} if it is to be used in the ion convention that excludes the enthalpy of the electron.

Heat Capacity and Entropy

The information on electronic energy levels and quantum weights, given by Moore (2, 6), is incomplete because many theoretically predicted levels have not been observed. Our calculations indicae that any reasonable method of filling in these missing levels and cutting of the summation in th partition functions (7) has no effect on the thermodynamic functions to 6000 K. This is a result of the high energy of all levels other than the ground state and the first four excited states; the next excied state is approximately 79704 cm-1 above the ground state. Since inclusion of these excited states has no effect on the thermodynamic functions (to 6000), we list only the ground state and the first four excited states. The reported uncertainty in S°(298.15 K) is due to uncertain ties in the relative ionic mass and the fundamental constants. Extension of these calculations above 6000 K may require consideration of the excited states and use of different fill and cutoff procedures (7).

References

1. JANAF Thermochemical Tables = Hg(g), 9-30-84; e-(g), 3-31-82.
2. C. E. Moore, U.S. Nat. Bur. Stand., NSRDS-NBS-34, 1970.
3. E. R. Cohen and B. N. Taylor, J. Phys. Chem. Ref. Data 2, 663(1973).
4. H. M. Rosenstock, K. Draxel et al., J. Phys. Chem. Ref. Data 6, Supp. 1 (1977).
5. R. D. Levin and S. G. Lias, U.S. Nat. Bur. Stand., NSRDS-NBS-71, 1982.
6. C. E. Moore, U.S. Nat. Bur. Stand., NSRDS-NBS-35, Volume III, 1971 [Reprinted from NBS Circular 467, Volume III, 1958]. 7.
7. J. R. Downey, Jr., The Dow Chemcial Company, AFOSR-TR-78-0960, Contract No. F44620-75-1-0048, 1978.

Mercury, Ion (Hg^+) Hg_1^+(g)

Enthalpy Reference Temperature = T_r = 298.15 K Standard State Pressure = p° = 0.1 MPa

T/K	$C_p°$ (J $K^{-1}mol^{-1}$)	S°	-[G°-H°(T_r)]/T	H°-H°(T_r) (kJ mol^{-1})	$\Delta_f H°$	$\Delta_f G°$	Log K_f
0	0.	0.	INFINITE	-6.197	1071.571		
100	20.786	201.910	243.098	-4.119			
200	20.786	216.318	226.519	-2.040			
250	20.786	220.956	224.960	-1.001			
298.15	20.786	224.617	224.617	0.	1074.623	1024.066	-179.412
300	20.786	224.746	224.618	0.038	1074.648	1023.753	-178.251
350	20.786	227.950	224.871	1.078	1075.337	1015.216	-151.513
400	20.786	230.726	225.433	2.117	1076.040	1006.579	-131.446
450	20.786	233.174	226.160	3.156	1076.752	997.854	-115.828
500	20.786	235.364	226.973	4.196	1077.470	989.049	-103.325
600	20.786	239.154	228.697	6.274	1078.913	971.229	-84.553
700	20.786	242.358	230.425	8.353	1021.596	959.739	-71.617
800	20.786	245.133	232.094	10.431	1023.675	950.761	-62.078
900	20.786	247.582	233.682	12.510	1025.753	941.522	-54.645
1000	20.786	249.772	235.183	14.589	1027.832	932.051	-48.685
1100	20.786	251.753	236.601	16.667	1029.911	922.373	-43.800
1200	20.786	253.561	237.940	18.746	1031.989	912.505	-39.720
1300	20.786	255.225	239.206	20.824	1034.068	902.464	-36.261
1400	20.786	256.766	240.406	22.903	1036.146	892.262	-33.291
1500	20.786	258.200	241.545	24.982	1038.225	881.912	-30.711
1600	20.786	259.541	242.629	27.060	1040.304	871.424	-28.449
1700	20.786	260.801	243.661	29.139	1042.382	860.805	-26.449
1800	20.786	261.989	244.646	31.217	1044.461	850.064	-24.668
1900	20.786	263.113	245.589	33.296	1046.539	839.207	-23.071
2000	20.786	264.180	246.492	35.375	1048.618	828.241	-21.631
2100	20.786	265.194	247.359	37.453	1050.697	817.172	-20.326
2200	20.786	266.161	248.192	39.532	1052.775	806.003	-19.137
2300	20.786	267.085	248.993	41.610	1054.854	794.739	-18.049
2400	20.786	267.969	249.765	43.689	1056.932	783.385	-17.050
2500	20.786	268.818	250.511	45.768	1059.011	771.945	-16.129
2600	20.786	269.633	251.231	47.846	1061.090	760.421	-15.277
2700	20.786	270.418	251.927	49.925	1063.168	748.817	-14.487
2800	20.786	271.173	252.601	52.004	1065.247	737.137	-13.751
2900	20.786	271.903	253.254	54.082	1067.325	725.382	-13.066
3000	20.786	272.608	253.887	56.161	1069.404	713.555	-12.424
3100	20.786	273.289	254.502	58.239	1071.483	701.659	-11.823
3200	20.786	273.949	255.100	60.318	1073.561	689.696	-11.258
3300	20.786	274.589	255.681	62.397	1075.640	677.668	-10.727
3400	20.786	275.209	256.246	64.475	1077.718	665.578	-10.225
3500	20.786	275.812	256.796	66.554	1079.797	653.425	-9.752
3600	20.786	276.397	257.333	68.632	1081.875	641.214	-9.304
3700	20.786	276.967	257.856	70.711	1083.954	628.945	-8.879
3800	20.786	277.521	258.366	72.790	1086.032	616.619	-8.476
3900	20.786	278.061	258.864	74.868	1088.110	604.239	-8.093
4000	20.786	278.587	259.351	76.947	1090.188	591.806	-7.728
4100	20.786	279.101	259.826	79.025	1092.266	579.321	-7.381
4200	20.786	279.601	260.291	81.104	1094.343	566.785	-7.049
4300	20.786	280.091	260.746	83.183	1096.421	554.199	-6.732
4400	20.786	280.568	261.191	85.261	1098.497	541.566	-6.429
4500	20.786	281.036	261.627	87.340	1100.574	528.885	-6.139
4600	20.786	281.492	262.054	89.418	1102.649	516.157	-5.861
4700	20.786	281.939	262.472	91.497	1104.724	503.385	-5.595
4800	20.786	282.377	262.882	93.576	1106.798	490.569	-5.338
4900	20.786	282.806	263.284	95.654	1108.870	477.709	-5.092
5000	20.787	283.226	263.679	97.733	1110.942	464.808	-4.856
5100	20.787	283.637	264.066	99.812	1113.012	451.864	-4.628
5200	20.787	284.041	264.447	101.890	1115.080	438.880	-4.409
5300	20.787	284.437	264.820	103.969	1117.146	425.857	-4.197
5400	20.787	284.825	265.187	106.048	1119.210	412.794	-3.993
5500	20.787	285.207	265.547	108.126	1121.271	399.694	-3.796
5600	20.787	285.581	265.902	110.205	1123.329	386.555	-3.606
5700	20.787	285.949	266.250	112.284	1125.383	373.381	-3.422
5800	20.788	286.311	266.593	114.363	1127.434	360.170	-3.244
5900	20.788	286.666	266.930	116.441	1129.480	346.923	-3.071
6000	20.788	287.016	267.262	118.520	1131.522	333.642	-2.905

PREVIOUS: CURRENT: September 1984 (1 bar)

Mercury, Ion (Hg^+) Hg_1^+(g)

MERCURY IODIDE (HgI) IDEAL GAS M_r = 327.4945

S°(298.15 K) = 280.752 J K^{-1} mol^{-1}

$\Delta_f H°$(0 K) = 138.82 kJ mol^{-1}
$\Delta_f H°$(298.15 K) = 133.47 ± 4.6 kJ mol^{-1}

Electronic Level and Quantum weight

State	ε, cm^{-1}	g_i
$^2\Sigma$	0	2

ω_e = 125.0 cm^{-1} $\omega_e x_e$ = 1.00 cm^{-1} σ = 1
B_e = [0.0351] cm^{-1} α_e = [0.0003] cm^{-1} r_e = [2.49] Å

Enthalpy of Formation

Wieland and Herczog (1) gave 8.3±1.1 kcal mol^{-1} for the enthalpy of dissociation into gaseous atoms.

Heat Capacity and Entropy

Vibrational constants were taken from Wieland (2). Rotational constants are estimated using relationship in Herzberg (3). The bond length was estimated to be 0.965 of the mercuric bond length by analogy with the mercury chlorides.

References

1. K. Wieland and A Herczog, Helv. Chim. Acta, 32, 889, (1949).
2. K. Wieland, Zeits, F. Electrochemie 64, 761 (1960).
3. G. Herzberg, "Spectra of Diatomic Molecules," D. van Nostrand, New York, 123, p. 108, 1950.

Mercury Iodide (HgI) Hg_1I_1(g)

Enthalpy Reference Temperature = T_r = 298.15 K Standard State Pressure = p° = 0.1 MPa

T/K	$C_p°$ (J $K^{-1}mol^{-1}$)	S° (J $K^{-1}mol^{-1}$)	$-[G°-H°(T_r)]/T$ (J $K^{-1}mol^{-1}$)	$H°-H°(T_r)$ (kJ mol^{-1})	$\Delta_f H°$ (kJ mol^{-1})	$\Delta_f G°$ (kJ mol^{-1})	Log K_f
0	0.	0.	INFINITE	-10.595	138.817	138.817	INFINITE
100	35.757	240.288	314.043	-7.375	138.805	121.548	-63.490
200	37.382	265.715	284.205	-3.698	137.432	104.789	-27.368
250	37.696	274.092	281.375	-1.821	134.298	96.877	-20.241
298.15	37.919	280.752	280.752	0.	133.470	89.745	-15.723
300	37.926	280.986	280.752	0.070	133.438	89.474	-15.579
350	38.116	286.847	281.215	1.971	132.543	82.216	-12.270
400	38.284	291.948	282.245	3.881	123.664	75.362	-9.841
450	38.438	296.466	283.578	5.799	122.199	69.412	-8.057
500	38.585	300.524	285.074	7.725	100.682	65.524	-6.845
600	38.864	307.583	288.254	11.598	99.964	58.561	-5.098
700	39.132	313.594	291.455	15.497	40.506	58.287	-4.349
800	39.394	318.837	294.557	19.424	40.464	60.830	-3.972
900	39.653	323.492	297.518	23.376	40.443	63.377	-3.678
1000	39.910	327.683	300.329	27.354	40.441	65.926	-3.444
1100	40.166	331.499	302.992	31.358	40.459	68.474	-3.252
1200	40.420	335.005	305.515	35.387	40.493	71.019	-3.091
1300	40.674	338.250	307.910	39.442	40.541	73.561	-2.956
1400	40.928	341.274	310.186	43.522	40.599	76.099	-2.839
1500	41.181	344.106	312.354	47.628	40.663	78.633	-2.738
1600	41.434	346.772	314.423	51.759	40.726	81.162	-2.650
1700	41.687	349.291	316.400	55.915	40.786	83.687	-2.571
1800	41.940	351.681	318.295	60.096	40.838	86.209	-2.502
1900	42.192	353.956	320.112	64.303	40.881	88.729	-2.439
2000	42.444	356.126	321.859	68.534	40.914	91.246	-2.383
2100	42.697	358.203	323.541	72.791	40.938	93.762	-2.332
2200	42.949	360.195	325.162	77.074	40.957	96.277	-2.286
2300	43.201	362.110	326.727	81.381	40.973	98.791	-2.244
2400	43.453	363.954	328.240	85.714	40.992	101.305	-2.205
2500	43.705	365.733	329.704	90.072	41.020	103.817	-2.169
2600	43.957	367.452	331.123	94.455	41.061	106.328	-2.136
2700	44.209	369.116	332.500	98.863	41.122	108.838	-2.106
2800	44.461	370.728	333.836	103.297	41.209	111.344	-2.077
2900	44.713	372.293	335.136	107 756	41.327	113.847	-2.051
3000	44.965	373.813	336.400	112.239	41.481	116.345	-2.026
3100	45.217	375.291	337.630	116.749	41.674	118.837	-2.002
3200	45.469	376.731	338.830	121.283	41.912	121.323	-1.980
3300	45.721	378.134	340.000	125.842	42.196	123.800	-1.960
3400	45.973	379.502	341.142	130.427	42.530	126.268	-1.940
3500	46.225	380.839	342.257	135.037	42.915	128.726	-1.921
3600	46.476	382.144	343.347	139.672	43.352	131.171	-1.903
3700	46.728	383.421	344.413	144.332	43.844	133.604	-1.886
3800	46.980	384.671	345.456	149.018	44.389	136.023	-1.870
3900	47.232	385.894	346.477	153.728	44.990	138.427	-1.854
4000	47.484	387.093	347.477	158.464	45.645	140.814	-1.839
4100	47.736	388.269	348.458	163.225	46.355	143.185	-1.824
4200	47.988	389.422	349.420	168.011	47.119	145.537	-1.810
4300	48.239	390.554	350.363	172.823	47.936	147.871	-1 796
4400	48.491	391.666	351.289	177.659	48.806	150.185	-1.783
4500	48.743	392.759	352.199	182.521	49.729	152.479	-1.770
4600	48.995	393.833	353.092	187.408	50.702	154.751	-1.757
4700	49.247	394.889	353.970	192.320	51.726	157.003	-1.745
4800	49.498	395.929	354.834	197.257	52.799	159.231	-1.733
4900	49.750	396.952	355.683	202.219	53.920	161.437	-1.721
5000	50.002	397.960	356.518	207.207	55.088	163.619	-1.709
5100	50.254	398.952	357.341	212.220	56.302	165.778	-1.698
5200	50.506	399.931	358.150	217.258	57.561	167.913	-1.687
5300	50.757	400.895	358.948	222.321	58.864	170.022	-1.676
5400	51.009	401.846	359.733	227.409	60.210	172.107	-1.665
5500	51.261	402.784	360.507	232.523	61.598	174.166	-1.654
5600	51.513	403.710	361.271	237.661	63.025	176.201	-1.644
5700	51.765	404.624	362.023	242.825	64.493	178.208	-1.633
5800	52.016	405.527	362.766	248.014	65.998	180.190	-1.623
5900	52.268	406.418	363.498	253.229	67.541	182.146	-1.613
6000	52.520	407.299	364.221	258.468	69.120	184.075	-1.603

PREVIOUS: December 1961 (1 atm) CURRENT: December 1961 (1 bar)

Mercury Iodide (HgI) Hg_1I_1(g)

MERCURY IODIDE (HgI_2) CRYSTAL M_r = 454.3990

$S°(298.15\ K)$ = [181.326 ± 6.3] $J\ K^{-1}\ mol^{-1}$
T_{trs} = 402 K
T_{fus} = 530 K

$\Delta_f H°(298.15\ K)$ = -105.437 ± 1.7 $kJ\ mol^{-1}$
$\Delta_{trs} H°$ = 2.515 ± 0.17 $kJ\ mol^{-1}$
$\Delta_{fus} H°$ = 18.966 ± 0.21 $kJ\ mol^{-1}$

Enthalpy of Formation

The value recommended by NBS (1) is adopted.

Heat Capacity and Entropy

The heat capacity is calculated from the enthalpy data of Guinchant (2) and Ewald (3). The entropy is estimated by adjusting its value to provide the best fit of the melting, sublimation and vaporization phenomena.

Transition and Fusion Data

The enthalpy of transition, $\Delta_{trs}H°$ = 0.601±0.04 $kcal^{-1}$, and the temperature of transition are taken from NBS (1). The enthalpy of melting, $\Delta_{fus}H°$ = 4.533±0.05 $kcal\ mol^{-1}$ was taken from Guinchant (2), the temperature of melting from NBS (1).

References

1. U. S. Nat. Bur. Stand. Circ. 500, 1952.
2. M. Guinchant, Comp. Rend. 68, 145 (1907).
3. R. Ewald, Ann. d. Physik (4) 44, 1213 (1914).

Mercury Iodide (HgI_2) $Hg_1I_2(cr)$

Enthalpy Reference Temperature = T_r = 298.15 K Standard State Pressure = p° = 0.1 MPa

T/K	$C_p°$ (J K⁻¹mol⁻¹)	S°	$-[G°-H°(T_r)]/T$	$H°-H°(T_r)$ (kJ mol⁻¹)	$\Delta_f H°$	$\Delta_f G°$	Log K_f
0							
100							
200							
298.15	77.747	181.326	181.326	0.	-105.437	-102.204	17.906
300	77.822	181.807	181.328	0.144	-105.445	-102.184	17.792
400	82.006	204.768	184.430	8.135	-121.858	-100.476	13.121
402.000	82.090	205.178	184.532	8.299	I <--> II	TRANSITION	
402.000	84.119	211.433	184.532	10.814			
500	84.119	229.784	191.669	19.058	-161.859	-91.667	9.576
530.000	84.119	234.686	193.966	21.581	-------- II <-->	LIQUID --------	
600	84.119	245.121	199.338	27.470	-159.916	-77.812	6.774
700	84.119	258.088	206.828	35.882	-216.745	-57.709	4.306
800	84.119	269.320	213.954	44.294	-214.191	-35.164	2.296
900	84.119	279.228	220.667	52.705	-211.648	-12.939	0.751
1000	84.119	288.091	226.974	61.117	-209.116	9.004	-0.470
1100	84.119	296.109	232.900	69.529	-206.598	30.694	-1.458
1200	84.119	303.428	238.477	77.941	-204.098	52.156	-2.270
1300	84.119	310.161	243.735	86.353	-201.620	73.410	-2.950
1400	84.119	316.395	248.706	94.765	-199.174	94.474	-3.525
1500	84.119	322.199	253.414	103.177	-196.768	115.364	-4.017

PREVIOUS: CURRENT: March 1962

Mercury Iodide (HgI_2) $Hg_1I_2(cr)$

MERCURY IODIDE (HgI_2) LIQUID M_r = 454.3990

$S°(298.15\ K) = [215.663]\ J\ K^{-1}mol^{-1}$ $\Delta_f H°(298.15\ K) = [-87.287]\ kJ\ mol^{-1}$

$T_{fus} = 530\ K$ $\Delta_{fus}H° = 18.966 \pm 0.21\ kJ\ mol^{-1}$

Enthalpy of Formation

$\Delta_f H°(HgI_2, \ell, 298.15\ K)$ is calculated from $\Delta_f H°(HgI_2, cr, 298.15\ K)$ by adding the enthalpy of fusion, $\Delta_{fus}H°$, and the difference in enthalpy, H°(530 K) - H°(298.15 K), between the crystal and liquid.

Heat Capacity and Entropy

Heat capacity measurements were obtained from the data of Guinchant (1) in the range 540-600 K and were assumed constant above and below this range. A glass type transition is assumed at 353 K below which the heat capacity is assumed to be that of $HgI_2(cr)$. $S°(\ell, 298.15\ K)$ is calculated in a manner analogous to that used for the enthalpy of formation.

Fusion and Vaporization Data

T_{vap} = 627 K (1 atm) is taken from NBS (2), $\Delta_{vap}H° = 14.141 \pm 0.25\ kcal\ mol^{-1}$ (for T_{vap} = 627 K) was obtained from the data of Prideaux (3) and Johnson (4).

References

1. M. Guinchant, Compt. Rend. 68, 145 (1907).
2. U. S. Nat. Bur. Stand. Circ. 500, 1952.
3. E. B. R. Prideaux, J. Chem. Soc. (London) 97, 2032 (1910).
4. F. M. G. Johnson, J. Amer. Chem. Soc. 33, 777 (1911).

Mercury Iodide (HgI_2) $Hg_1I_2(l)$

Enthalpy Reference Temperature = T_r = 298.15 K Standard State Pressure = p° = 0.1 MPa

T/K	$C_p°$ (J K⁻¹mol⁻¹)	S° (J K⁻¹mol⁻¹)	$-[G°-H°(T_r)]/T$ (J K⁻¹mol⁻¹)	$H°-H°(T_r)$ (kJ mol⁻¹)	$\Delta_f H°$ (kJ mol⁻¹)	$\Delta_f G°$ (kJ mol⁻¹)	Log K_f
0							
100							
200							
298.15	77.747	215.663	215.663	0.	-87.287	-94.291	16.519
300	77.822	216.144	215.665	0.144	-87.295	-94.335	16.425
353.000	80.040	228.980	216.722	4.327	GLASS <--> LIQUID TRANSITION		
353.000	102.090	228.980	216.722	4.327			
400	102.090	241.741	218.928	9.125	-102.717	-96.125	12.553
500	102.090	264.522	225.853	19.334	-143.432	-90.609	9.466
530.000	102.090	270.471	228.212	22.397	-------- II <--> LIQUID --------		
600	102.090	283.135	233.896	29.543	-139.692	-80.396	6.999
700	102.090	298.872	242.083	39.752	-194.724	-64.237	4.793
800	102.090	312.504	250.053	49.961	-190.373	-45.893	2.997
900	102.090	324.529	257.673	60.170	-186.033	-28.094	1.631
1000	102.090	335.285	264.906	70.379	-181.704	-10.778	0.563
1100	102.090	345.015	271.753	80.588	-177.389	6.106	-0.290
1200	102.090	353.898	278.234	90.797	-173.091	22.597	-0.984
1300	102.090	362.070	284.373	101.006	-168.817	38.731	-1.556
1400	102.090	369.635	290.196	111.215	-164.574	54.537	-2.035
1500	102.090	376.679	295.729	121.424	-160.371	70.041	-2.439

PREVIOUS: CURRENT: March 1962

Mercury Iodide (HgI_2) $Hg_1I_2(l)$

MERCURY IODIDE (HgI_2) CRYSTAL–CRYSTAL–LIQUID M_r = 454.3990

0 to 402 K crystal, I
402 to 530 K crystal, II
above 530 K liquid

Refer to the individual tables for details.

Mercury Iodide (HgI_2) Hg_1I_2(cr,l)

Enthalpy Reference Temperature = T_r = 298.15 K Standard State Pressure = $p°$ = 0.1 MPa

T/K	$C_p°$ (J K⁻¹mol⁻¹)	$S°$ (J K⁻¹mol⁻¹)	$-[G°-H°(T_r)]/T$ (J K⁻¹mol⁻¹)	$H°-H°(T_r)$ (kJ mol⁻¹)	$\Delta_fH°$ (kJ mol⁻¹)	$\Delta_fG°$ (kJ mol⁻¹)	Log K_f
0							
100							
200							
298.15	77.747	181.326	181.326	0.	-105.437	-102.204	17.906
300	77.822	181.807	181.328	0.144	-105.445	-102.184	17.792
400	82.006	204.768	184.430	8.135	-121.858	-100.476	13.121
402.000	82.090	205.178	184.532	8.299	I <--> II	TRANSITION	
402.000	84.119	211.433	184.532	10.814			
500	84.119	229.784	191.669	19.058	-161.859	-91.667	9.576
530.000	84.119	234.686	193.966	21.581	II <--> LIQUID	TRANSITION	
530.000	102.090	270.471	193.966	40.547			
600	102.090	283.135	203.646	47.694	-139.692	-80.396	6.999
700	102.090	298.872	216.154	57.903	-194.724	-64.237	4.793
800	102.090	312.504	227.365	68.112	-190.373	-45.893	2.997
900	102.090	324.529	237.506	78.321	-186.033	-28.094	1.631
1000	102.090	335.285	246.756	88.529	-181.704	-10.778	0.563
1100	102.090	345.015	255.253	98.738	-177.389	6.106	-0.290
1200	102.090	353.898	263.109	108.947	-173.091	22.597	-0.984
1300	102.090	362.070	270.411	119.156	-168.817	38.731	-1.556
1400	102.090	369.635	277.232	129.365	-164.574	54.537	-2.035
1500	102.090	376.679	283.629	139.574	-160.371	70.041	-2.439

PREVIOUS: CURRENT: March 1962

Mercury Iodide (HgI_2) Hg_1I_2(cr,l)

MERCURY IODIDE (HgI_2) IDEAL GAS M_r = 454.3990

$S°(298.15 K) = 336.206 J K^{-1} mol^{-1}$

$\Delta_f H°(0 K) = -9.84 kJ mol^{-1}$
$\Delta_f H°(298.15 K) = -16.13 \pm 2.1 kJ mol^{-1}$

Vibrational Frequencies and Degeneracies

ν, cm^{-1}
156 (1)
33 (2)
[233](1)

Point Group: $D_{\infty h}$ $\sigma = 2$
Bond Distance: Hg-I = 2.58 Å
Bond Angle: I-Hg-I = 180°
Rotational Constant: $B_0 = 0.009978 cm^{-1}$

Enthalpy of Formation

The enthalpy of formation of the crystal as given in National Bureau of Standards (1) was combined with the 3rd law heat of sublimation at 298.15 K derived from the data of Niwa and Shibata (2), Magee (3) and Johnson (4).

Heat Capacity and Entropy

The vibrational constants were given by Klemperer and Lindeman (5). The anti-symmetric stretching frequency is estimated by analogy with the bromide and chloride. The bond length is an average of the values given by Braune and Knoke (6), Gregg et al. (7) and Akishin et al. (8).

References
1. U. S. Nat. Bur. Stand. Circ. 500, 1952.
2. K. Niwa and Z. Shibata, J. Fac. Sci. Hokkaido Imp. Univ. Ser. III 2, 183 (1938).
3. D. W. Magee, Univ. Microfilms No. 14474 Ann Arbor, Michigan.
4. F. M. G. Johnson, J. Amer. Chem. Soc. 33, 777 (1911).
5. W. Klemperer and L. Lindeman, J. Chem. Phys. 25, 397 (1956).
6. H. Braune and S. Knoke, Naturwiss. 21, 349 (1933).
7. A. H. Gregg, G. C. Hampson, G. I. Jenkins, P. L. F. Jones and L. E. Sutton, Trans. Faraday Soc. 33, 852 (1937).
8. P. A. Akishin, V. P. Spiridinov and A. N. Khodchenkov, Zhur. Fiz. Khim. 33, 20 (1959).

Mercury Iodide (HgI_2) $Hg_1I_2(g)$

Enthalpy Reference Temperature = T_r = 298.15 K Standard State Pressure = p° = 0.1 MPa

T/K	$C_p°$ (J K⁻¹mol⁻¹)	S° (J K⁻¹mol⁻¹)	$-[G°-H°(T_r)]/T$ (J K⁻¹mol⁻¹)	$H°-H°(T_r)$ (kJ mol⁻¹)	$\Delta_f H°$ (kJ mol⁻¹)	$\Delta_f G°$ (kJ mol⁻¹)	Log K_f
0	0.	0.	INFINITE	-16.253	-9.841	-9.841	INFINITE
100	54.486	272.205	389.418	-11.721	-10.065	-27.465	14.346
200	59.759	312.043	341.763	-5.944	-11.805	-44.218	11.549
250	60.630	325.481	337.210	-2.932	-15.118	-52.052	10.876
298.15	61.118	336.206	336.206	0.	-16.129	-59.074	10.349
300	61.132	336.584	336.207	0.113	-16.169	-59.340	10.332
350	61.446	346.033	336.952	3.178	-17.305	-66.448	9.917
400	61.654	354.252	338.612	6.256	-34.430	-72.842	9.512
450	61.798	361.523	340.762	9.342	-36.743	-77.504	8.996
500	61.903	368.040	343.169	12.435	-79.175	-78.110	8.160
600	62.040	379.339	348.284	18.633	-79.445	-77.872	6.779
700	62.124	388.909	353.422	24.841	-138.478	-71.017	5.299
800	62.178	397.209	358.388	31.057	-138.120	-61.404	4.009
900	62.216	404.534	363.116	37.276	-137.769	-51.836	3.008
1000	62.243	411.091	367.592	43.499	-137.427	-42.306	2.210
1100	62.263	417.024	371.820	49.725	-137.095	-32.810	1.558
1200	62.278	422.443	375.816	55.952	-136.780	-23.344	1.016
1300	62.290	427.428	379.597	62.180	-136.486	-13.903	0.559
1400	62.299	432.044	383.180	68.410	-136.222	-4.484	0.167
1500	62.307	436.343	386.583	74.640	-135.997	4.918	-0.171
1600	62.313	440.364	389.820	80.871	-135.822	14.307	-0.467
1700	62.318	444.142	392.905	87.103	-135.705	23.686	-0.728
1800	62.322	447.704	395.852	93.335	-135.652	33.060	-0.959
1900	62.326	451.074	398.670	99.567	-135.669	42.433	-1.167
2000	62.329	454.271	401.371	105.800	-135.755	51.808	-1.353
2100	62.332	457.312	403.963	112.033	-135.909	61.190	-1.522
2200	62.334	460.212	406.454	118.266	-136.124	70.580	-1.676
2300	62.336	462.983	408.852	124.500	-136.395	79.982	-1.816
2400	62.338	465.636	411.164	130.733	-136.710	89.396	-1.946
2500	62.340	468.181	413.394	136.967	-137.058	98.824	-2.065
2600	62.341	470.626	415.548	143.201	-137.429	108.267	-2.175
2700	62.342	472.978	417.632	149.435	-137.811	117.724	-2.278
2800	62.343	475.246	419.649	155.670	-138.191	127.195	-2.373
2900	62.344	477.433	421.604	161.904	-138.560	136.680	-2.462
3000	62.345	479.547	423.501	168.139	-138.907	146.177	-2.545
3100	62.346	481.591	425.342	174.373	-139.225	155.685	-2.623
3200	62.347	483.571	427.131	180.608	-139.505	165.203	-2.697
3300	62.347	485.489	428.870	186.842	-139.742	174.729	-2.766
3400	62.348	487.350	430.563	193.077	-139.931	184.261	-2.831
3500	62.349	489.158	432.211	199.312	-140.067	193.799	-2.892
3600	62.349	490.914	433.818	205.547	-140.149	203.339	-2.950
3700	62.350	492.622	435.384	211.782	-140.173	212.881	-3.005
3800	62.350	494.285	436.912	218.017	-140.138	222.422	-3.057
3900	62.350	495.905	438.404	224.252	-140.044	231.962	-3.107
4000	62.351	497.483	439.862	230.487	-139.891	241.499	-3.154
4100	62.351	499.023	441.286	236.722	-139.679	251.031	-3.198
4200	62.352	500.526	442.679	242.957	-139.409	260.558	-3.241
4300	62.352	501.993	444.041	249.192	-139.082	270.077	-3.281
4400	62.352	503.426	445.374	255.428	-138.698	279.588	-3.319
4500	62.352	504.827	446.680	261.663	-138.261	289.090	-3.356
4600	62.353	506.198	447.959	267.898	-137.771	298.581	-3.390
4700	62.353	507.539	449.213	274.133	-137.230	308.061	-3.424
4800	62.353	508.852	450.441	280.369	-136.640	317.529	-3.455
4900	62.353	510.137	451.647	286.604	-136.003	326.985	-3.486
5000	62.353	511.397	452.829	292.839	-135.320	336.426	-3.515
5100	62.354	512.632	453.990	299.075	-134.595	345.854	-3.542
5200	62.354	513.842	455.129	305.310	-133.828	355.268	-3.569
5300	62.354	515.030	456.248	311.545	-133.021	364.665	-3.594
5400	62.354	516.196	457.347	317.781	-132.178	374.048	-3.618
5500	62.354	517.340	458.428	324.016	-131.298	383.414	-3.641
5600	62.354	518.463	459.490	330.252	-130.385	392.765	-3.664
5700	62.355	519.567	460.534	336.487	-129.440	402.098	-3.685
5800	62.355	520.651	461.561	342.722	-128.465	411.415	-3.705
5900	62.355	521.717	462.572	348.958	-127.462	420.715	-3.725
6000	62.355	522.765	463.567	355.193	-126.432	429.997	-3.743

PREVIOUS: March 1962 (1 atm) CURRENT: March 1962 (1 bar)

Mercury Iodide (HgI_2) $Hg_1I_2(g)$

MERCURY OXIDE (HgO) CRYSTAL M_r = 216.5894

$S°(298.15\ K) = 70.270 \pm 0.13\ J\ K^{-1}\ mol^{-1}$

$\Delta_f H°(0\ K) = -86.208 \pm 0.100\ kJ\ mol^{-1}$
$\Delta_f H°(298.15\ K) = -90.789 \pm 0.100\ kJ\ mol^{-1}$

Enthalpy of Formation

Calculated from the entropy obtained by Bauer and Johnston (1) and the Gibbs energy of formation as determined from the numerous cell measurements described by Bauer and Johnston. The value adopted for $\Delta_f H°(298.15\ K)$ is -42707±14 cal mol^{-1}.

Heat Capacity and Entropy

The heat capacity has been measured from 15 to 298.15 K by Bauer and Johnston (1). The extrapolation below 15 K was done assuming a T^3 law and gave $S°(15\ K) = 0.264\ cal\ K^{-1}\ mol^{-1}$. Above 298.15 the data was extrapolated smoothly and adjusted to give the observed decomposition vapor pressures of Taylor and Hulett (2).

Decomposition Data

According to Taylor and Hulett (2), the vapor pressure reaches 1.89 atmospheres at 749 K. This corresponds to the decomposition to 2 Hg(g) + O_2(g), the equilibrium constant being unity at 1.89 atm.

References

1. T. W. Bauer and H. L. Johnston, J. Amer. Chem. Soc. 75, 2217 (1953).
2. G. B. Taylor and G. A. Hulett, J. Phys. Chem. 17, 565 (1913).

Mercury Oxide (HgO) $Hg_1O_1(cr)$

Enthalpy Reference Temperature = T_r = 298.15 K Standard State Pressure = p° = 0.1 MPa

	J K⁻¹mol⁻¹			kJ mol⁻¹			
T/K	$C_p°$	S°	$-[G°-H°(T_r)]/T$	$H°-H°(T_r)$	$\Delta_f H°$	$\Delta_f G°$	Log K_f
0	0.	0.	INFINITE	-9.104	-86.208	-86.208	INFINITE
100	28.434	30.840	105.148	-7.431	-87.693	-78.390	40.947
200	38.171	53.856	74.128	-4.054	-88.357	-68.775	17.962
298.15	44.062	70.270	70.270	0.	-90.789	-58.490	10.247
300	44.145	70.543	70.271	0.082	-90.786	-58.289	10.149
400	48.325	83.836	72.055	4.712	-90.406	-47.502	6.203
500	51.631	94.990	75.556	9.717	-89.659	-36.856	3.850
600	54.120	104.632	79.617	15.009	-88.660	-26.386	2.297
700	56.066	113.127	83.809	20.523	-146.248	-9.526	0.711
800	57.509	120.711	87.956	26.204	-144.314	9.874	-0.645
900	58.626	127.553	91.982	32.013	-142.285	29.027	-1.685
1000	59.413	133.773	95.855	37.918	-140.190	47.950	-2.505

PREVIOUS: CURRENT: June 1962

Mercury Oxide (HgO) $Hg_1O_1(cr)$

MERCURY OXIDE (HgO) IDEAL GAS M_r = 216.5894

$\Delta_f H°(0\ K)$ = [46.514] kJ mol^{-1}
$\Delta_f H°(298.15\ K)$ = [41.840] kJ mol^{-1}

S°(298.15 K) = [239.161] J K^{-1} mol^{-1}

Electronic Level and Quantum Weight

State	ε_i, cm^{-1}	g_i
$^1\Sigma$	0	1

ω_e = [680] cm^{-1} $\omega_e x_e$ = [4.7] cm^{-1} σ = 1
B_e = [0.336] cm^{-1} α_e = [0.0027] cm^{-1} r_e = [1.84] Å

Enthalpy of Formation

The enthalpy of formation, $\Delta_f H°(H_gO, g, 298.15\ K)$, is estimated to be 10 ± 15 kcal mol^{-1}. This value is obtained from the enthalpy of formation of PbO(g) and comparison of the enthalpy of formation of analogous lead and mercury compounds using the method of Karapet'yants (1). The corresponding dissociation energy, D_0^0, of HgO(g) is 64 ± 15 kcal mol^{-1}.

Heat Capacity and Entropy

The equilibrium internuclear separation (r_e) of HgO(g) is estimated from the corresponding quantity for PbO(g) and comparison of Hg-X and Pb-X bond distances for cases in which values of both distances are known. The rotational constant B_e is calculated from r_e. The fundamental vibrational frequency ω_e is estimated from Guggenheimer's relation for multiple bonded molecules (2). The anharmonic vibrational term $\omega_e x_e$ is calculated from $\omega_e x_e = \omega_e^2/(4D_0 + 2\ \omega_e)$. The value of α_e is calculated from the Morse potential function. The ground state configuration is estimated from the building-up principle (3), the united atom theory (3), and from analogy with BeO.

References

1. M. Kh. Karapet'yants, Russ. J. Inorg. Chem. 10, 837 (1965).
2. K. M. Guggenheimer, Proc. Phys. Soc. (London) 58, 456 (1946).
3. G. Herzberg, "Spectra of Diatomic Molecules," D. Van Nostrand Co., Inc., New York, 1950.

Mercury Oxide (HgO) $Hg_1O_1(g)$

Enthalpy Reference Temperature = T_r = 298.15 K Standard State Pressure = p° = 0.1 MPa

T/K	$C_p°$ (J K^{-1}mol^{-1})	S°	$-[G°-H°(T_r)]/T$	$H°-H°(T_r)$ (kJ mol^{-1})	$\Delta_f H°$	$\Delta_f G°$	Log K_f
0	0.	0.	INFINITE	-9.010	46.514	46.514	INFINITE
100	29.158	205.913	266.918	-6.101	46.266	38.062	-19.882
200	30.709	226.480	242.102	-3.124	45.202	30.259	-7.903
250	31.874	233.459	239.697	-1.560	42.341	26.742	-5.587
298.15	32.883	239.161	239.161	0.	41.840	23.784	-4.167
300	32.918	239.365	239.162	0.061	41.822	23.672	-4.122
350	33.776	244.506	239.566	1.729	41.362	20.684	-3.087
400	34.460	249.063	240.474	3.436	40.945	17.759	-2.319
450	35.002	253.154	241.659	5 173	40.557	14.884	-1.728
500	35.435	256.865	242.997	6.934	40.187	12.052	-1.259
600	36.068	263.385	245.867	10.511	39.470	6.492	-0.565
700	36.501	268.980	248.778	14.141	-20.001	7.624	-0.569
800	36.811	273.875	251.616	17.807	-20.082	11.575	-0.756
900	37.045	278.225	254.335	21.501	-20.170	15.538	-0.902
1000	37.229	282.138	256.923	25.215	-20.265	19.510	-1.019
1100	37.379	285.693	259.380	28.945	-20.368	23.493	-1.116
1200	37.504	288.951	261.710	32.690	-20.476	27.485	-1.196
1300	37.613	291.958	263.923	36.446	-20.591	31.486	-1.265
1400	37.708	294.749	266.026	40.212	-20.710	35.496	-1.324
1500	37.794	297.353	268.029	43.987	-20.834	39.515	-1.376
1600	37.873	299.795	269.938	47.770	-20.963	43.543	-1.422
1700	37.946	302.093	271.763	51.561	-21.096	47.579	-1.462
1800	38.014	304.264	273.509	55.359	-21.235	51.622	-1.498
1900	38.079	306.321	275.182	59.164	-21.378	55.674	-1.531
2000	38.141	308.276	276.788	62.975	-21.527	59.733	-1.560
2100	38.201	310.138	278.333	66.792	-21.681	63.800	-1.587
2200	38.258	311.917	279.819	70.615	-21.841	67.874	-1.612
2300	38.314	313.619	281.252	74.444	-22.006	71.956	-1.634
2400	38.369	315.250	282.635	78.278	-22.177	76.044	-1.655
2500	38.422	316.818	283.971	82.117	-22.354	80.141	-1.674
2600	38.474	318.326	285.263	85.962	-22.536	84.244	-1.692
2700	38.526	319.779	286.515	89.812	-22.723	88.355	-1.709
2800	38.577	321.181	287.728	93.667	-22.915	92.472	-1.725
2900	38.627	322.535	288.905	97.528	-23.113	96.596	-1.740
3000	38.676	323.846	290.048	101.393	-23.315	100.728	-1.754
3100	38.725	325.115	291.159	105.263	-23.521	104.866	-1.767
3200	38.774	326.345	292.239	109.138	-23.731	109.011	-1.779
3300	38.822	327.539	293.291	113.018	-23.946	113.162	-1.791
3400	38.870	328.698	294.315	116.902	-24.164	117.320	-1.802
3500	38.918	329.826	295.314	120.792	-24.385	121.485	-1.813
3600	38.965	330.923	296.288	124.686	-24.609	125.656	-1.823
3700	39.013	331.991	297.239	128.585	-24.836	129.832	-1.833
3800	39.060	333.032	298.167	132.488	-25.066	134.016	-1.842
3900	39.106	334.047	299.074	136.397	-25.297	138.206	-1.851
4000	39.153	335.038	299.961	140.310	-25.531	142.401	-1.860
4100	39.199	336.005	300.828	144.227	-25.767	146.602	-1.868
4200	39.246	336.951	301.677	148.150	-26.005	150.809	-1.876
4300	39.292	337.875	302.508	152.076	-26.245	155.022	-1.883
4400	39.338	338.778	303.322	156.008	-26.487	159.240	-1.890
4500	39.384	339.663	304.120	159.944	-26.731	163.464	-1.897
4600	39.430	340.529	304.902	163.885	-26.977	167.693	-1.904
4700	39.476	341.378	305.669	167.830	-27.226	171.927	-1.911
4800	39.521	342.209	306.422	171.780	-27.477	176 167	-1.917
4900	39.567	343.025	307.160	175.734	-27.731	180.413	-1.923
5000	39.613	343.824	307.886	179.693	-27.988	184.663	-1.929
5100	39.658	344.609	308.598	183.657	-28.249	188.919	-1.935
5200	39.704	345.380	309.298	187.625	-28.514	193.179	-1.941
5300	39.749	346.137	309.986	191.598	-28.784	197.446	-1.946
5400	39.794	346.880	310.662	195.575	-29.060	201.717	-1.951
5500	39.840	347.611	311.328	199.556	-29.342	205.993	-1.956
5600	39.885	348.329	311.982	203.543	-29.631	210.275	-1.961
5700	39.930	349.035	312.626	207.533	-29.927	214.561	-1.966
5800	39.975	349.730	313.260	211.529	-30.232	218.853	-1.971
5900	40.021	350.414	313.883	215.529	-30.547	223.150	-1.976
6000	40.066	351.087	314.498	219.533	-30.872	227.453	-1.980

PREVIOUS: June 1967 (1 atm) CURRENT: June 1967 (1 bar)

Mercury Oxide (HgO) $Hg_1O_1(g)$

MERCURY IODIDE (Hg_2I_2) CRYSTAL M_r = 654.9890

Mercury Iodide (Hg_2I_2) Hg_2I_2(cr)

S°(298.15 K) = 241.29 ± 8.4 J K^{-1} mol^{-1}

T_{fus} = 563 K

$\Delta_f H°$(298.15 K) = -119.085 ± 2.1 kJ mol^{-1}

$\Delta_{fus}H°$ = [27.196 ± 8.4] kJ mol^{-1}

Enthalpy of Formation

The average value of the cell measurements of Vosburgh (1), Cohen (2), Yoshida (3), Oholm (4) and the calorimetric values of Varet (5) and Nernst (6) is adopted.

Heat Capacity and Entropy

The heat capacity at 298.15 K is taken from NBS (7) and is estimated above this by analogy with mercurous chloride. The entropy is obtained from the enthalpy and Gibbs energy of formation obtained from the cell measurements listed under enthalpy of formation.

Fusion Data

Yvon (8) gives 563 K for the melting point. The enthalpy of melting is estimated by assuming the entropy of melting per atom to be the same as that for mercuric iodide.

References

1. W. C. Vosburgh, J. Amer. Chem. Soc. 50, 2386 (1928).
2. E. Cohen, Zeits. physik. Chemie 94, 210 (1920).
3. T. Yoshida, J. Chem. Soc. Japan 48, 435 (1927).
4. L. W. Oholm, Acta. Soc. Scientarum Fennicae 41, 1 (1913).
5. R. Varet, Ann. Chim. Phys. (7) 8, 91 (1896).
6. W. Nernst, Zeits. physik. Chemie 2, 23 (1888).
7. U. S. Nat. Bur. Stand. Circ. 500, 1952.
8. P. Yvon, Comp. Rend. 76, 1607 (1873).

Enthalpy Reference Temperature = T_r = 298.15 K; Standard State Pressure = p° = 0.1 MPa

T/K	$C_p°$ (J $K^{-1}mol^{-1}$)	S° (J $K^{-1}mol^{-1}$)	$-[G°-H°(T_r)]/T$ (J $K^{-1}mol^{-1}$)	$H°-H°(T_r)$ (kJ mol^{-1})	$\Delta_f H°$ (kJ mol^{-1})	$\Delta_f G°$ (kJ mol^{-1})	Log K_f
0							
100							
200							
298.15	105.855	241.291	241.291	0.	-119.085	-111.063	19.458
300	105.868	241.946	241.293	0.196	-119.093	-111.013	19.329
400	110.416	273.066	245.506	11.024	-135.435	-107.707	14.065
500	113.679	298.073	253.599	22.237	-177.873	-96.702	10.102
563.000	115.244	311.658	259.349	29.450	------ CRYSTAL <--> LIQUID -----		
600	116.064	319.018	262.804	33.729	-175.564	-80.680	7.024
700	117.905	337.052	272.152	45.430	-290.578	-51.920	3.874
800	119.411	352.898	281.275	57.298	-286.646	-18.092	1.181
900	120.625	367.034	290.033	69.301	-282.590	15.235	-0.884
1000	121.713	379.800	298.381	81.419	-278.432	48.104	-2.513
1100	122.687	391.447	306.320	93.640	-274.183	80.553	-3.825
1200	123.554	402.161	313.866	105.953	-269.860	112.611	-4.902
1300	124.236	412.076	321.044	118.341	-265.485	144.307	-5.798
1400	125.143	421.314	327.880	130.808	-261.062	175.664	-6.554
1500	126.273	429.986	334.401	143.377	-256.578	206.702	-7.198

PREVIOUS: CURRENT: March 1962

Mercury Iodide (Hg_2I_2) Hg_2I_2(cr)

MERCURY IODIDE (Hg_2I_2) LIQUID M_r = 654.9890

$S°(298.15\ K)$ = [273.258] J K^{-1} mol^{-1} $\Delta_fH°(298.15\ K)$ = [-98.564] kJ mol^{-1}

T_{fus} = 563 K $\Delta_{fus}H°$ = [27.196 ± 8.4] kJ mol^{-1}

Enthalpy of Formation

$\Delta_fH°(Hg_2I_2$, ℓ, 298.15 K) is calculated from $\Delta_fH°(Hg_2I_2$, cr, 298.15 K) by adding the enthalpy of fusion, $\Delta_{fus}H°$, and the difference in enthalpy, H°(563 K) - H°(298.15 K), between the crystal and liquid.

Heat Capacity and Entropy

$C_p°$ is estimated as 1.33 times the heat capacity of mercuric iodide. S°(ℓ, 298.15 K) is calculated in a manner analogous to that used for the enthalpy of formation.

Vaporization Data

Mercurous iodide decomposes to Hg(g) and HgI_2(g) at the boiling point which was estimated from the Gibbs energy change of the reaction. For 1 atm, T_{vap} = [630] K (decomposition).

Reference

1. P. Yvon, Comp. Rend. 76, 1607 (1873).

Mercury Iodide (Hg_2I_2) Hg_2I_2(l)

Enthalpy Reference Temperature = T_r = 298.15 K Standard State Pressure = p° = 0.1 MPa

T/K	$C_p°$	S°	$-[G°-H°(T_r)]/T$	$H°-H°(T_r)$	$\Delta_fH°$	$\Delta_fG°$	Log K_f
	J $K^{-1}mol^{-1}$			kJ mol^{-1}			
0							
100							
200							
298.15	136.398	273.258	273.258	0.	-98.564	-100.073	17.532
300	136.398	274.101	273.260	0.252	-98.516	-100.082	17.426
400	136.398	313.341	278.610	13.892	-112.046	-100.428	13.115
500	136.398	343.777	288.713	27.532	-152.057	-93.738	9.793
563.000	136.398	359.964	295.798	36.125	------ CRYSTAL <--> LIQUID -----		
600	136.398	368.646	300.026	41.172	-147.600	-82.493	7.182
700	136.398	389.672	311.369	54.812	-260.675	-58.850	4.391
800	136.398	407.885	322.321	68.452	-254.971	-30.408	1.985
900	136.398	423.950	332.738	82.091	-249.279	-2.679	0.155
1000	136.398	438.321	342.590	95.731	-243.598	24.416	-1.275
1100	136.398	451.322	351.893	109.371	-237.931	50.943	-2.419
1200	136.398	463.190	360.681	123.011	-232.281	76.955	-3.350
1300	136.398	474.107	368.992	136.651	-226.654	102.496	-4.118
1400	136.398	484.216	376.865	150.291	-221.059	127.605	-4.761
1500	136.398	493.626	384.339	163.930	-215.503	152.316	-5.304

PREVIOUS. CURRENT: March 1962

Mercury Iodide (Hg_2I_2) Hg_2I_2(l)

MERCURY IODIDE (Hg_2I_2) CRYSTAL-LIQUID M_r = 654.9890 Mercury Iodide (Hg_2I_2) Hg_2I_2(cr,l)

0 to 563 K crystal
above 563 K liquid

Refer to the individual tables for details.

Enthalpy Reference Temperature = T_r = 298.15 K Standard State Pressure = p° = 0.1 MPa

T/K	C_p° (J K⁻¹mol⁻¹)	S° (J K⁻¹mol⁻¹)	−[G°−H°(T_r)]/T (J K⁻¹mol⁻¹)	H°−H°(T_r) (kJ mol⁻¹)	Δ_fH° (kJ mol⁻¹)	Δ_fG° (kJ mol⁻¹)	Log K_f
0							
100							
200							
298.15	105.855	241.291	241.291	0.	-119.085	-111.063	19.458
300	105.868	241.946	241.293	0.196	-119.093	-111.013	19.329
400	110.416	273.066	245.506	11.024	-135.435	-107.707	14.065
500	113.679	298.073	253.599	22.237	-177.873	-96.702	10.102
563.000	115.244	311.658	259.349	29.450	CRYSTAL <--> LIQUID		
563.000	136.398	359.964	259.349	56.646	TRANSITION		
600	136.398	368.646	265.824	61.693	-147.600	-82.493	7.182
700	136.398	389.672	282.053	75.333	-260.675	-58.850	4.391
800	136.398	407.885	296.669	88.972	-254.971	-30.408	1.985
900	136.398	423.950	309.937	102.612	-249.279	-2.679	0.155
1000	136.398	438.321	322.069	116.252	-243.598	24.416	-1.275
1100	136.398	451.322	333.238	129.892	-237.931	50.943	-2.419
1200	136.398	463.190	343.580	143.532	-232.281	76.955	-3.350
1300	136.398	474.107	353.206	157.172	-226.654	102.496	-4.118
1400	136.398	484.216	362.207	170.811	-221.059	127.605	-4.761
1500	136.398	493.626	370.659	184.451	-215.503	152.316	-5.304

PREVIOUS: CURRENT: March 1962

Mercury Iodide (Hg_2I_2) Hg_2I_2(cr,l)

IODINE (I) IDEAL GAS A_r = 126.9045

IP(I, g) = 84295.1 ± 0.2 cm^{-1}

S°(298.15 K) = 180.786 ± 0.002 J K^{-1} mol^{-1}

$\Delta_fH°(0\ K)$ = 107.16 ± 0.04 kJ mol^{-1}

$\Delta_fH°(298.15\ K)$ = 106.76 ± 0.04 kJ mol^{-1}

Electronic Levels and Quantum Weights

State	ϵ_i, cm^{-1}	g_i
$^2P°_{3/2}$	0.0	4
$^2P°_{1/2}$	7603.15	2

Enthalpy of Formation

The adopted value for the enthalpy of formation of I(g) is derived from the spectroscopically determined dissociation energy of $I_2(g)$ as reported by Barrow et al. (1, 2). This value is also recommended by Brewer and Winn (3) and Huber and Herzberg (4). The adopted value is $D_0^0(I_2, g)$ = 12440.06±0.03 cm^{-1} (148.816±0.001 kJ mol^{-1}). The convergence limit in the absorption spectrum of the $^3\Pi_{0^+u}$ - $^1\Sigma^+_g$ system corresponds to dissociation to a ground state iodine atom ($^3P_{3/2}$) and an excited state iodine atom ($^3P_{1/2}$).

Using earlier data, the CODATA recommended value (5) was calculated using $D_0^0(I_2, g)$ = 12440.9±1.1 cm^{-1} from LeRoy (6), Brown (7), LeRoy and Bernstein (8), and Sitterley (9). LeRoy (6), LeRoy and Bernstein (8), and a recent study by LeRoy and Bernstein (10) obtained the dissociation energy from the analysis of the distribution of the uppermost vibrational levels. Their calculations were based on the data of Brown (7), with reference to the study by Verma (11).

Heat Capacity and Entropy

Information on the electronic energy levels and quantum weights is taken from Moore (12, 13). Only two levels are included in the calculation. All other levels (observed and predicted) lie above 54633 cm^{-1}. Our calculations indicate that any reasonable method of filling in the missing levels and cutting off the summation in the partition function (14) has no effect on the thermodynamic properties to 6000 K. This is undoubtedly a result of the high energy of these levels. Therefore, we list only the ground state and the first excited state. Extension to higher temperatures may require consideration of the higher excited states and utilization of different fill and cutoff procedures.

The thermodynamic functions at 298.15 K are in agreement with recent CODATA recommendations (15) except for two minor differences. First, the entropy differs by 0.1094 J K^{-1} mol^{-1} because this table uses a standard-state pressure of 1 bar, whereas the CODATA recommendations are based on 1 atm. Second, entropy differences of the order of 0.001-0.004 J K^{-1} mol^{-1} for the monatomic halogens arise due to the use of slightly different values for the relative atomic mass and for R; this table uses R 8.31441 J K^{-1} mol^{-1}. Considering these minor changes, this table agrees within the estimated uncertainty with those by Hultgren et al. (16), Gurvich et al. (17) and Wagman et al. (20). The estimated uncertainty is due to uncertainties in the relative aton mass and fundamental constants which are based on the 1981 scale (18) and the 1973 values (19), respectively.

References

1. R. F. Barrow, D. F. Broyd, L. B. Pederson and K. K. Yee, Chem. Phys. Letters 18, 357 (1973).
2. R. F. Barrow, Clark, Coxon and K. K. Yee, J. Mol. Spectrosc. 51, 428 (1974).
3. L. Brewer and J. S. Winn, Faraday Symp. Chem. Soc. (GB) 14, 126 (1980).
4. K. P. Huber and G. Herzberg "Constants of Diatomic Molecules", Van Nostrand Reinhold Company, New York, N.Y. (1979).
5. J. D. Cox, J. Chem. Thermodynamics 4, 331 (1972).
6. R. J. LeRoy, J. Chem. Phys. 52, 2678 (1970).
7. W. G. Brown, Phys. Rev. 38, 709 (1931).
8. R. J. LeRoy and R. B. Bernstein, J. Chem. Phys. 52, 3869 (1970).
9. C. M. Sitterly, U. S. Nat. Bur. Stand., personal communication to ISCU-CODATA Task Group on Key Values for Thermodynamics.
10. R. J. LeRoy and R. B. Bernstein, Chem. Phys. Lett. 5, 42 (1970); J. Mol. Spectrosc. 37 109 (1971).
11. R. D. Varma, J. Chem. Phys. 32 738 (1960).
12. C. E. Moore, U. S. Nat. Bur. Stand., NSRDS-NBS-34, 1970.
13. C. E. Moore, U. S. Nat. Bur. Stand. NSRDS-NBS-35, Vol. II, 1970 [Reprint of NBS Circular 467, Volume II, 1952].
14. J. R. Downey, Jr., The Dow Chemical Co., AFOSR-TR-78-0960, Contract No. F44620-75-1-0048, 1978.
15. J. D. Cox, chairman, ICSU-CODATA Task Group on Key Values for Thermodynamics, J. Chem. Thermodynamics, 10, 903 (1978).
16. R. Hultgren, P. D. Desai et al., "Selected Values of the Thermodynamic Properties of the Elements", American Society for Metals, Metals Park, Ohio, 1973.
17. L. V. Gurvich, I. V. Veits et al., "Thermodynamic Properties of Individual Substances," 3rd ed., Vol. I, Nauka, Moscow, 1978.
18. N. E. Holden and R. L. Martin, Pure App. Chem. 55, 1101 (1983).
19. E. R. Cohen and B. N. Taylor, J. Phys. Chem. Ref. Data 2, 663 (1973).
20. D. D. Wayman, W. H. Evans et al., J. Phys. Chem. Ref. Data 11, Supp. 2 (1982).

Iodine (I) $I_1(g)$

Enthalpy Reference Temperature = T_r = 298.15 K   Standard State Pressure = p° = 0.1 MPa

T/K	$C_p°$ (J $K^{-1}mol^{-1}$)	S° (J $K^{-1}mol^{-1}$)	-[G°-H°(T_r)]/T (J $K^{-1}mol^{-1}$)	H°-H°(T_r) (kJ mol^{-1})	$\Delta_fH°$ (kJ mol^{-1})	$\Delta_fG°$ (kJ mol^{-1})	Log K_f
0	0.	0.	INFINITE	-6.197	107.164	107.164	INFINITE
100	20.786	158.079	199.267	-4.119	107.717	94.959	-49.601
200	20.786	172.487	182.688	-2.040	107.330	82.328	-21.502
250	20.786	177.125	181.128	-1.001	107.055	76.109	-15.902
298.15	20.786	180.786	180.786	0.	106.762	70.174	-12.294
300	20.786	180.915	180.787	0.038	106.750	69.947	-12.179
350	20.786	184.119	181.040	1.078	106.383	63.841	-9.528
400	20.786	186.895	181.602	2.117	98.010	58.064	-7.582
450	20.786	189.343	182.329	3.156	97.032	53.130	-6.167
500	20.786	191.533	183.142	4.196	75.990	50.203	-5.245
600	20.786	195.323	184.866	6.274	76.191	45.027	-3.920
700	20.786	198.527	186.594	8.353	76.386	39.818	-2.971
800	20.787	201.302	188.263	10.432	76.575	34.580	-2.258
900	20.789	203.751	189.851	12.510	76.759	29.320	-1.702
1000	20.795	205.942	191.352	14.589	76.937	24.039	-1.256
1100	20.806	207.924	192.770	16.669	77.110	18.741	-0.890
1200	20.824	209.735	194.109	18.751	77.275	13.428	-0.584
1300	20.851	211.403	195.376	20.835	77.430	8.101	-0.325
1400	20.889	212.949	196.577	22.921	77.574	2.762	-0.103
1500	20.936	214.392	197.717	25.013	77.701	-2.586	0.090
1600	20.994	215.745	198.802	27.109	77.809	-7.942	0.259
1700	21.062	217.020	199.836	29.212	77.894	-13.304	0.409
1800	21.137	218.226	200.825	31.322	77.954	-18.671	0.542
1900	21.220	219.371	201.771	33.440	77.986	-24.040	0.661
2000	21.308	220.461	202.679	35.566	77.992	-29.410	0.768
2100	21.399	221.503	203.550	37.701	77.974	-34.780	0.865
2200	21.493	222.501	204.389	39.846	77.933	-40.148	0.953
2300	21.588	223.458	205.198	42.000	77.875	-45.514	1.034
2400	21.682	224.379	205.978	44.163	77.803	-50.877	1.107
2500	21.775	225.266	206.732	46.336	77.724	-56.237	1.175
2600	21.865	226.122	207.461	48.518	77.643	-61.594	1.237
2700	21.953	226.949	208.168	50.709	77.565	-66.948	1.295
2800	22.036	227.749	208.853	52.909	77.497	-72.299	1.349
2900	22.116	228.523	209.518	55.116	77.442	-77.648	1.399
3000	22.191	229.274	210.164	57.332	77.406	-82.995	1.445
3100	22.261	230.003	210.792	59.554	77.392	-88.342	1.489
3200	22.327	230.711	211.404	61.784	77.403	-93.688	1.529
3300	22.387	231.399	211.999	64.019	77.442	-99.035	1.568
3400	22.443	232.068	212.580	66.261	77.511	-104.384	1.604
3500	22.494	232.720	213.146	68.508	77.612	-109.735	1.638
3600	22.540	233.354	213.698	70.760	77.745	-115.090	1.670
3700	22.581	233.972	214.238	73.016	77.911	-120.448	1.700
3800	22.618	234.575	214.765	75.276	78.110	-125.812	1.729
3900	22.651	235.163	215.281	77.539	78.342	-131.181	1.757
4000	22.679	235.736	215.785	79.806	78.608	-136.557	1.783
4100	22.704	236.297	216.279	82.075	78.905	-141.940	1.808
4200	22.725	236.844	216.762	84.346	79.234	-147.330	1.832
4300	22.743	237.379	217.235	86.620	79.593	-152.729	1.855
4400	22.757	237.902	217.699	88.895	79.983	-158.136	1.877
4500	22.769	238.414	218.153	91.171	80.400	-163.552	1.898
4600	22.777	238.914	218.599	93.448	80.846	-168.979	1.919
4700	22.783	239.404	219.037	95.726	81.318	-174.415	1.938
4800	22.787	239.884	219.466	98.005	81.816	-179.861	1.957
4900	22.788	240.354	219.888	100.284	82.338	-185.318	1.976
5000	22.787	240.814	220.302	102.562	82.883	-190.786	1.993
5100	22.784	241.265	220.708	104.841	83.450	-196.265	2.010
5200	22.780	241.708	221.108	107.119	84.039	-201.755	2.027
5300	22.774	242.142	221.501	109.397	84.648	-207.257	2.043
5400	22.766	242.567	221.887	111.674	85.276	-212.770	2.058
5500	22.757	242.985	222.267	113.950	85.922	-218.296	2.073
5600	22.747	243.395	222.640	116.225	86.586	-223.833	2.088
5700	22.736	243.797	223.008	118.500	87.266	-229.382	2.102
5800	22.724	244.193	223.370	120.773	87.962	-234.943	2.116
5900	22.711	244.581	223.726	123.044	88.673	-240.517	2.129
6000	22.697	244.963	224.077	125.315	89.399	-246.102	2.143

PREVIOUS: June 1974 (1 atm)   CURRENT: June 1982 (1 bar)

Iodine (I) $I_1(g)$

IODINE, ION (I^+) IDEAL GAS M_r = 126.90395

$IP(I^+, g) = 154304 \pm 1\ cm^{-1}$
$S°(298.15\ K) = 182.642 \pm 0.005\ J\ K^{-1}\ mol^{-1}$

$\Delta_f H°(0\ K) = 1115.557 \pm 0.05\ kJ\ mol^{-1}$
$\Delta_f H°(298.15\ K) = [1121.353]\ kJ\ mol^{-1}$

Electronic Levels and Quantum Weights

State	ε_i, cm^{-1}	g_i
$^3P_2$	0.0	5
$^3P_1$	7087.0	3
$^3P_0$	6447.9	1
$^1D_2$	13731	5
$^1S_0$	32629	1

Enthalpy of Formation

$\Delta_f H°(I^+, g, 0\ K)$ is calculated from $\Delta_f H°(I, g, 0\ K)$ (1) using the spectroscopic value of IP(I) = 84295.1±0.2 cm^{-1} (1008.39±0.01 kJ mol^{-1}) from Moore (2). The ionization limit is converted from cm^{-1} to kJ mol^{-1} using the factor, 1 cm^{-1} = 0.01196266 kJ mol^{-1}, which is derived from the 1973 CODATA fundamental constants (3). Rosenstock et al. (4) and Levin and Lias (5) have summarized additional ionization potential and appearance potential data.

$\Delta_f H°(I^+, g, 298.15\ K)$ is calculated from $\Delta_f H°(I, g, 0\ K)$ by using IP(I) with JANAF (1) enthalpies, H°(0 K)-H°(298.15 K), for I(g), $I^+(g)$, and $e^-(g)$. $\Delta_f H°(I \rightarrow I^+ + e^-, 298.15\ K)$ differs from a room temperature threshold energy due to inclusion of these enthalpies and to threshold effects discussed by Rosenstock et al. (4). $\Delta_f H°(298.15\ K)$ should be changed by -6.197 kJ mol^{-1} if it is to be used in the ion convention that excludes the enthalpy of the electron.

Heat Capacity and Entropy

The information on electronic energy levels and quantum weights, given by Moore (2, 6), is incomplete because many theoretically predicted levels have not been observed. Our calculations indicate that any reasonable method of filling in these missing levels and cutting off the summation in the partition function (7) has no effect on the thermodynamic functions to 6000 K. This is a result of the high energy of all levels other than the ground state and the lowest four excited states; the next lowest level is approximately 81040 cm^{-1} above the ground state. Since inclusion of these upper levels has no effect on the thermodynamic functions (to 6000 K), we list only the ground state and four excited states, with the energy of these states taken from recent study by Moore (2, 6). The reported uncertainty in S°(298.15 K) is due to uncertainties in the relative ionic mass, fundamental constants, and the position of the four lowest excited states. Extension of these calculations above 6000 K may require consideration of the higher excited states and use of different fill and cutoff procedures (7).

References

1. JANAF Thermochemical Tables: I(g), 3-31-82; e^-(r), 3-31-82.
2. C. E. Moore, U. S. Nat. Bur. Stand., NSRDS-NBS-34, 1970.
3. E. R. Cohen and B. N. Taylor, J. Phys. Chem. Ref. Data 2, 663 (1973).
4. H. M. Rosenstock, K. Draxl et al., J. Phys. Chem. Ref. Data 6, Supp. 1 (1977).
5. R. D. Levin and S. G. Lias, U. S. Nat. Bur. Stand., NSRDS-NBS-71, 1982.
6. C. E. Moore, U. S. Nat. Bur. Stand., NSRDS-NBS-35, Volume III, 1970 [reprint of NBS Circular 467, Volume III, 1958].
7. J. R. Downey, Jr., The Dow Chemical Company, AFOSR-TR-78-0960, Contract No. F44620-75-1-0048, 1978.

Iodine, Ion (I^+) $I_1^+(g)$

Enthalpy Reference Temperature = T_r = 298.15 K Standard State Pressure = p° = 0.1 MPa

T/K	$C_p°$ (J $K^{-1}mol^{-1}$)	S°	-[G°-H°(T_r)]/T	H°-H°(T_r) (kJ mol^{-1})	$\Delta_f H°$	$\Delta_f G°$	Log K_f
0	0.	0.	INFINITE	-6.197	1115.557		
100	20.786	159.934	201.122	-4.119			
200	20.786	174.342	184.543	-2.040			
250	20.786	178.980	182.984	-1.001			
298.15	20.786	182.642	182.642	0.	1121.353	1077.957	-188.854
300	20.786	182.770	182.642	0.038	1121.379	1077.688	-187.642
350	20.786	185.974	182.895	1.078	1122.052	1070.352	-159.741
400	20.786	188.750	183.457	2.117	1114.718	1063.195	-138.839
450	20.786	191.198	184.184	3.156	1114.780	1056.751	-122.664
500	20.786	193.388	184.997	4.196	1094.777	1052.200	-109.923
600	20.786	197.178	186.721	6.274	1097.057	1043.470	-90.842
700	20.787	200.382	188.449	8.353	1099.330	1034.359	-77.185
800	20.790	203.158	190.118	10.432	1101.598	1024.923	-66.921
900	20.800	205.607	191.706	12.511	1103.861	1015.203	-58.921
1000	20.819	207.800	193.208	14.592	1106.119	1005.231	-52.508
1100	20.852	209.785	194.626	16.675	1108.374	995.032	-47.250
1200	20.903	211.602	195.966	18.763	1110.624	984.629	-42.860
1300	20.974	213.277	197.234	20.857	1112.868	974.039	-39.137
1400	21.065	214.835	198.436	22.959	1115.105	963.276	-35.940
1500	21.176	216.292	199.578	25.070	1117.332	952.353	-33.164
1600	21.305	217.663	200.666	27.194	1119.546	941.282	-30.730
1700	21.450	218.959	201.705	29.332	1121.744	930.073	-28.578
1800	21.608	220.189	202.698	31.485	1123.925	918.735	-26.661
1900	21.776	221.362	203.649	33.654	1126.088	907.277	-24.943
2000	21.952	222.483	204.563	35.840	1128.233	895.705	-23.393
2100	22.134	223.559	205.442	38.045	1130.361	884.026	-21.989
2200	22.320	224.593	206.289	40.267	1132.477	872.246	-20.710
2300	22.506	225.589	207.107	42.509	1134.585	860.370	-19.540
2400	22.693	226.551	207.897	44.769	1136.688	848.403	-18.465
2500	22.879	227.481	208.662	47.047	1138.793	836.348	-17.475
2600	23.062	228.382	209.403	49.344	1140.906	824.208	-16.559
2700	23.241	229.256	210.122	51.659	1143.031	811.987	-15.709
2800	23.416	230.104	210.821	53.992	1145.175	799.687	-14.918
2900	23.587	230.929	211.500	56.343	1147.341	787.310	-14.181
3000	23.751	231.731	212.161	58.710	1149.535	774.858	-13.491
3100	23.910	232.513	212.805	61.093	1151.760	762.332	-12.845
3200	24.062	233.274	213.433	63.491	1154.019	749.734	-12.238
3300	24.208	234.017	214.046	65.905	1156.315	737.065	-11.667
3400	24.346	234.741	214.644	68.333	1158.649	724.325	-11.128
3500	24.478	235.449	215.228	70.774	1161.022	711.517	-10.619
3600	24.603	236.140	215.799	73.228	1163.436	698.639	-10.137
3700	24.720	236.816	216.358	75.694	1165.891	685.695	-9.680
3800	24.830	237.477	216.905	78.172	1168.387	672.683	-9.247
3900	24.933	238.123	217.441	80.660	1170.922	659.605	-8.834
4000	25.029	238.756	217.966	83.158	1173.498	646.462	-8.442
4100	25.118	239.375	218.481	85.666	1176.112	633.254	-8.068
4200	25.200	239.981	218.986	88.182	1178.764	619.981	-7.711
4300	25.276	240.575	219.481	90.705	1181.452	606.645	-7.369
4400	25.345	241.157	219.967	93.236	1184.176	593.246	-7.043
4500	25.407	241.727	220.444	95.774	1186.934	579.785	-6.730
4600	25.464	242.286	220.913	98.318	1189.725	566.262	-6.430
4700	25.514	242.834	221.373	100.867	1192.546	552.678	-6.142
4800	25.559	243.372	221.826	103.420	1195.398	539.034	-5.866
4900	25.598	243.899	222.271	105.978	1198.277	525.330	-5.600
5000	25.632	244.417	222.709	108.540	1201.184	511.567	-5.344
5100	25.661	244.925	223.140	111.104	1204.116	497.746	-5.098
5200	25.685	245.423	223.563	113.672	1207.072	483.866	-4.860
5300	25.705	245.913	223.980	116.241	1210.052	469.930	-4.631
5400	25.720	246.393	224.391	118.813	1213.052	455.937	-4.410
5500	25.732	246.865	224.795	121.385	1216.074	441.889	-4.197
5600	25.739	247.329	225.194	123.959	1219.115	427.786	-3.990
5700	25.743	247.785	225.586	126.533	1222.174	413.628	-3.790
5800	25.743	248.233	225.973	129.107	1225.249	399.416	-3.597
5900	25.740	248.673	226.354	131.682	1228.342	385.151	-3.410
6000	25.734	249.105	226.729	134.255	1231.449	370.833	-3.228

PREVIOUS: CURRENT: June 1982 (1 bar)

Iodine, Ion (I^+) $I_1^+(g)$

IODINE, ION (I^-) IDEAL GAS M_r = 126.90505

EA(I, g) = 3.0591 ± 0.0004 eV
S°(298.15 K) = 169.260 ± 0.005 J K^{-1} mol^{-1}

$\Delta_f H°$(0 K) = -187.992 ± 0.4 kJ mol^{-1}
$\Delta_f H°$(298.15 K) = [-194.591] kJ mol^{-1}

Electronic Level and Quantum Weight

State	ε_i, cm^{-1}	g_i
$^1S_0$	0.0	1

Enthalpy of Formation

$\Delta_f H°(I^-$, g, 0 K) is calculated from $\Delta_f H°$(I, g, 0 K) (1) using the adopted electron affinity of EA(I) = 3.0591 ± 0.0004 eV (295.156±0.039 kJ mol^{-1}). This value, recommended by Hotop and Lineberger (6), is based on laser optogalvanic spectroscopy (5). Additional information on I^-(g) may be obtained in the critical discussions of Hotop and Lineberger (2, 6), Rosenstock et al. (3) and Massey (4).

$\Delta_f H°(I^-$, g, 298.15 K) is obtained from $\Delta_f H°$(I, g, 0 K) by using EA(I) with JANAF (1) enthalpies, H°(0 K) - H°(298.15 K), for I^-(g), I(g), and e^-(g). $\Delta_r H°(I^- \rightarrow I + e^-$, 298.15 K) differs from a room-temperature threshold energy due to inclusion of these enthalpies and to threshold effects discussed by Rosenstock et al. (3). $\Delta_f H°$(298.15 K) should be changed by +6.197 kJ mol^{-1} if it is to be used in the ion convention that excludes the enthalpy of the electron.

Heat Capacity and Entropy

The ground state configuration for I^-(g) is given by Hotop and Lineberger (2, 6), Rosenstock et al. (3) and Massey (4). Lacking any experimental evidence as to the stability of any excited states, we assume that no stable excited states exist.

The entropy at all temperatures (0-6000 K) agrees within 0.001 J K^{-1} mol^{-1} with the values of Gurvich et al. (8) except for one minor difference. The entropy differs by 0.1094 J K^{-1} mol^{-1} because this table uses a standard-state pressure of 1 bar, whereas the tabulation of Gurvich, et al (8) is based on 1 atm. The estimated uncertainty is due to uncertainties in the relative ionic mass and the fundamental constants which are based on the 1981 scale (9) and the 1973 values (7), respectively.

References

1. JANAF Thermochemical Tables: I(g), 6-30-82; e^-(g), 3-31-82.
2. H. Hotop and W. C. Lineberger, J. Phys. Chem. Ref. Data 4, 539 (1975).
3. H. M. Rosenstock, K. Draxl et al. J. Phys. Chem. Ref. Data 6, Supp. 1 (1977).
4. H. S. W. Massey, "Negative Ions," 3rd ed., Cambridge University Press, Cambridge, 1976.
5. C. R. Webster, I. S. McDermid, and C. T. Rettner, J. Chem. Phys. 78, 646 (1983).
6. H. Hotop and W. C. Lineberger, J. Phys. Chem. Ref. Data, to be published, 1985.
7. E. R. Cohen and B. N. Taylor, J. Phys. Chem. Ref. Data 2, 663 (1973).
8. L. V. Gurvich, I. V. Veits et al., "Thermodynamic Properties of Individual Substances", 3rd ed., Vol. 1, Nauka, Moscow, 1978.
9. N. E. Holden and R. L. Martin, Pure Appl. Chem. 55, 1101 (1983).

Iodine, Ion (I^-) I_1^-(g)

Enthalpy Reference Temperature = T_r = 298.15 K Standard State Pressure = p° = 0.1 MPa

T/K	$C_p°$ (J $K^{-1}mol^{-1}$)	S° (J $K^{-1}mol^{-1}$)	-[G°-H°(T_r)]/T (J $K^{-1}mol^{-1}$)	H°-H°(T_r) (kJ mol^{-1})	$\Delta_f H°$ (kJ mol^{-1})	$\Delta_f G°$ (kJ mol^{-1})	Log K_f
0	0.	0.	INFINITE	-6.197	-187.992		
100	20.786	146.553	187.740	-4.119			
200	20.786	160.961	171.161	-2.040			
250	20.786	165.599	169.602	-1.001			
298.15	20.786	169.260	169.260	0.	-194.591	-221.488	38.804
300	20.786	169.389	169.260	0.038	-194.642	-221.654	38.593
350	20.786	172.593	169.514	1.078	-196.048	-226.047	33.736
400	20.786	175.368	170.076	2.117	-205.461	-229.961	30.030
450	20.786	177.817	170.803	3.156	-207.477	-232.902	27.035
500	20.786	180.007	171.615	4.196	-229.559	-233.720	24.417
600	20.786	183.796	173.339	6.274	-231.436	-234.376	20.404
700	20.786	187.001	175.068	8.353	-233.320	-234.717	17.515
800	20.786	189.776	176.737	10.431	-235.209	-234.788	15.330
900	20.786	192.224	178.324	12.510	-237.104	-234.621	13.617
1000	20.786	194.414	179.826	14.589	-239.005	-234.243	12.236
1100	20.786	196.396	181.244	16.667	-240.913	-233.675	11.096
1200	20.786	198.204	182.583	18.746	-242.829	-232.932	10.139
1300	20.786	199.868	183.849	20.824	-244.757	-232.029	9.323
1400	20.786	201.408	185.049	22.903	-246.701	-230.977	8.618
1500	20.786	202.842	186.188	24.982	-248.664	-229.786	8.002
1600	20.786	204.184	187.271	27.060	-250.653	-228.463	7.459
1700	20.786	205.444	188.304	29.139	-252.671	-227.014	6.975
1800	20.786	206.632	189.289	31.217	-254.721	-225.446	6.542
1900	20.786	207.756	190.232	33.296	-256.807	-223.763	6.152
2000	20.786	208.822	191.135	35.375	-258.927	-221.969	5.797
2100	20.786	209.836	192.001	37.453	-261.081	-220.068	5.474
2200	20.786	210.803	192.834	39.532	-263.266	-218.064	5.178
2300	20.786	211.727	193.636	41.610	-265.479	-215.960	4.905
2400	20.786	212.612	194.408	43.689	-267.714	-213.759	4.652
2500	20.786	213.461	195.153	45.768	-269.966	-211.465	4.418
2600	20.786	214.276	195.873	47.846	-272.229	-209.080	4.200
2700	20.786	215.060	196.570	49.925	-274.497	-206.608	3.997
2800	20.786	215.816	197.243	52.004	-276.765	-204.053	3.807
2900	20.786	216.546	197.897	54.082	-279.027	-201.416	3.628
3000	20.786	217.250	198.530	56.161	-281.279	-198.701	3.460
3100	20.786	217.932	199.145	58.239	-283.516	-195.912	3.301
3200	20.786	218.592	199.742	60.318	-285.734	-193.050	3.151
3300	20.786	219.231	200.323	62.397	-287.931	-190.120	3.009
3400	20.786	219.852	200.889	64.475	-290.103	-187.123	2.875
3500	20.786	220.454	201.439	66.554	-292.249	-184.063	2.747
3600	20.786	221.040	201.975	68.632	-294.368	-180.942	2.625
3700	20.786	221.610	202.498	70.711	-296.458	-177.762	2.510
3800	20.786	222.164	203.009	72.790	-298.519	-174.527	2.399
3900	20.786	222.704	203.507	74.868	-300.550	-171.237	2.293
4000	20.786	223.230	203.993	76.947	-302.551	-167.896	2.192
4100	20.786	223.743	204.469	79.025	-304.523	-164.505	2.096
4200	20.786	224.244	204.934	81.104	-306.466	-161.067	2.003
4300	20.786	224.733	205.389	83.183	-308.380	-157.582	1.914
4400	20.786	225.211	205.834	85.261	-310.265	-154.053	1.829
4500	20.786	225.678	206.269	87.340	-312.124	-150.482	1.747
4600	20.786	226.135	206.696	89.418	-313.956	-146.870	1.668
4700	20.786	226.582	207.115	91.497	-315.762	-143.218	1.592
4800	20.786	227.020	207.525	93.576	-317.543	-139.528	1.518
4900	20.786	227.448	207.927	95.654	-319.299	-135.801	1.448
5000	20.786	227.868	208.322	97.733	-321.033	-132.039	1.379
5100	20.786	228.280	208.709	99.811	-322.744	-128.241	1.313
5200	20.786	228.684	209.089	101.890	-324.434	-124.411	1.250
5300	20.786	229.079	209.463	103.969	-326.103	-120.549	1.188
5400	20.786	229.468	209.830	106.047	-327.752	-116.655	1.128
5500	20.786	229.849	210.190	108.126	-329.381	-112.731	1.071
5600	20.786	230.224	210.545	110.204	-330.993	-108.777	1.015
5700	20.786	230.592	210.893	112.283	-332.587	-104.795	0.960
5800	20.786	230.953	211.236	114.362	-334.164	-100.784	0.908
5900	20.786	231.309	211.573	116.440	-335.724	-96.747	0.857
6000	20.786	231.658	211.905	118.519	-337.269	-92.684	0.807

PREVIOUS: CURRENT: June 1982 (1 bar)

Iodine, Ion (I^-) I_1^-(g)

POTASSIUM IODIDE (KI) CRYSTAL M_r = 166.0028

$S°(298.15\ K) = 106.387\ J\ K^{-1}\ mol^{-1}$
$T_{fus} = 954\ K$

$\Delta_f H°(0\ K) = -326.93 \pm 0.4\ kJ\ mol^{-1}$
$\Delta_f H°(298.15\ K) = -327.90 \pm 0.4\ kJ\ mol^{-1}$
$\Delta_{fus} H° = 24.016\ kJ\ mol^{-1}$

Enthalpy of Formation

The enthalpies of solution of KI(cr) have been critically reviewed by Parker (1). Six pertinent results are quoted in the table below. Adopting the best value, $\Delta_\infty H°(298.15\ K) = 4.86 \pm 0.03\ kcal\ mol^{-1}$, reported by Parker, and $\Delta_f H°(298.15\ K) = -60.32$ and $-13.19\ kcal\ mol^{-1}$ for $K^+(aq, \infty)$ and $I^-(aq, \infty)$, respectively, obtained from (2), we obtain $\Delta_f H°(KI, cr, 298.15\ K) = -78.37 \pm 0.1\ kcal\ mol^{-1}$, which is adopted.

$\Delta_f H°$, kcal mol^{-1}(a)	T/°C	n(b)	Source
4.830±0.05	19	200	Thomsen (3)
4.897±0.04	25	18	Wust and Lange (4)
4.873	25	500	Lange and Martin (5)
4.790±0.10	20.5	600	Popov et al. (6)
5.020	24.5	18.5	Bobtelsky and Lairsch (7)
4.830±0.07	25	75000	Kapustinskii and Drakin (8)

a Values are adjusted to 298.15 K and to infinite dilution.
b n is the number of moles of H_2O per mole of KI(cr) in solution.

Heat Capacity and Entropy

The low temperature heat capacities are based on those measured by Berg and Morrison (9), 2.86-268.0 K. The adopted values are in good agreement with the $C°_p$ values reported by Scales (10), 2-7 K. The data of Clusius et al. (11), 10.27-269.9 K, appear to be too low in the temperature range 70-270 K and are not used. Low temperature enthalpies, 83-273 K, have been measured by Koref (12). The value of S°(298.15 K) is derived based on the adopted low temperature heat capacities, using $S°(2.86\ K) = 0.0033\ cal\ K^{-1}\ mol^{-1}$.

The high temperature enthalpies were determined by Skuratov and Lapushkin (13), 623-923 K; Dworkin (14), 854-954 K; and Cooper (15), 297-973 K. Heat capacities are derived from the enthalpies by a curve fitting method which requires a smooth joint with the low temperature $C°_p$. Enthalpy values used in the curve fit include smooth values calculated from the equation of Skuratov and the value $H°(954)-H°(298.15) = 9.2\ kcal\ mol^{-1}$ from Dworkin. Deviation of the input values from the selected enthalpies are all less than 0.3 percent, while the deviations of the data of Cooper range from +1.0 at 570 K to -1.6 percent at 974 K. Since the latter point is 20 above the melting point, there appears to be an error in Cooper's measurement of temperature. $C°_p$ above the melting point is obtained by graphical extrapolation.

Fusion Data

T_{fus} has been reported to be 953, 952, and 954 K by Ray and Dayal (16), Phipps and Partridge (17), and Johnson and Bredig (18, respectively. The latter value is adopted. The enthalpy of fusion was determined by Dworkin and Bredig (19), using drop calorimetry

Sublimation Data

The difference between $\Delta_f H°(298.15\ K)$ for KI(g) and KI(cr) is $\Delta_{sub}H°$(to monomer, 298.15 K). The $\Delta_{sub}H°$(to dimer, 298.15 K) is calculated as the enthalpy change for the following reaction: $2KI(cr) = K_2I_2(g)$.

References

1. V. B. Parker, NSRDS-NBS 2, 1965.
2. U. S. Nat. Bur. Stand. Tech. Note 270-1, 1965.
3. J. Thomsen, Thermochemische Untersuchungen, J. Barth Verlag, Leipzig, 1882-86.
4. J. Wust and E. Lange, Z. Physik. Chem. 116, 161 (1925).
5. E. Lange and W. Martin, Z. Physik. Chem. A180, 233 (1937).
6. M. M. Popov, S. M. Skuratov, and M. M. Strel'tsova, Zhur. Obshch. Khim. 10, 2023 (1940).
7. M. Bobtelsky and R. D. Lairsch, J. Chem. Soc. 1950, 3612 (1950).
8. A. F. Kapustinskii and S. I. Drakin, Zhur. Fiz. Khim. 26, 581 (1952).
9. W. T. Berg and J. A. Morrison, Proc. Roy. Soc. (London) A242, 467 (1957).
10. W. S. Scales, Phys. Rev. 112, 49 (1958).
11. K. Clusius, J. Goldmann, and A. Perlick, Z. Naturforsch. 4A, 424 (1949).
12. F. Koref, Ann. Physik 4, 49 (1911).
13. S. M. Skuratov and S. A. Lapushkin, Russ. J. Gen. Chem. 21, 2485 (1951).
14. A. S. Dworkin, Oak Ridge National Laboratory, personal communication, Dec. 1, 1964.
15. C. B. Cooper, J. Chem. Phys. 21, 777 (1953).
16. R. C. Ray and V. Dayal, Trans. Faraday Soc. 32, 741 (1936).
17. T. E. Phipps and E. G. Partridge, J. Amer. Chem. Soc. 51, 1331 (1929).
18. J. W. Johnson and M. A. Bredig, J. Phys. Chem. 62, 604 (1958).
19. A. S. Dworkin and M. A. Bredig, J. Phys. Chem. 64, 269 (1960).

Potassium Iodide (KI) $I_1K_1(cr)$

Enthalpy Reference Temperature = T_r = 298.15 K — Standard State Pressure = p° = 0.1 MPa

T/K	$C°_p$ (J K⁻¹mol⁻¹)	S° (J K⁻¹mol⁻¹)	-[G°-H°(T_r)]/T (J K⁻¹mol⁻¹)	H°-H°(T_r) (kJ mol⁻¹)	$\Delta_f H°$ (kJ mol⁻¹)	$\Delta_f G°$ (kJ mol⁻¹)	Log K_f
0	0.	0.	INFINITE	-12.711	-326.930	-326.930	INFINITE
100	45.003	52.413	151.616	-9.920	-327.397	-326.028	170.300
200	50.518	85.772	111.148	-5.075	-327.612	-324.572	84.770
298.15	52.777	106.387	106.387	0.	-327.900	-323.024	56.592
300	52.806	106.713	106.388	0.098	-327.907	-322.994	56.238
400	53.932	122.054	108.472	5.433	-338.870	-320.549	41.859
500	55.363	134.235	112.446	10.894	-360.614	-313.870	32.790
600	57.300	144.490	116.953	16.522	-359.903	-304.584	26.516
700	59.777	153.503	121.543	22.372	-358.933	-295.437	22.046
800	62.639	161.668	126.055	28.490	-357.681	-286.448	18.703
900	65.793	169.225	130.437	34.909	-356.140	-277.633	16.113
954.000	67.594	173.111	132.743	38.510	------ CRYSTAL <--> LIQUID -----		
1000	69.036	176.329	134.674	41.654	-354.308	-269.006	14.051
1100	71.546	183.033	138.769	48.691	-431.198	-255.964	12.155
1200	73.136	189.332	142.723	55.930	-427.954	-240.176	10.455
1300	74.090	195.226	146.537	63.295	-424.595	-224.663	9.027
1400	74.601	200.737	150.214	70.732	-421.182	-209.412	7.813
1500	74.927	205.895	153.756	78.209	-417.748	-194.405	6.770
1600	75.145	210.739	157.168	85.713	-414.313	-179.628	5.864
1700	75.234	215.297	160.454	93.232	-410.894	-165.065	5.072
1800	75.291	219.599	163.622	100.759	-407.503	-150.702	4.373
1900	75.317	223.671	166.676	108.290	-404.146	-136.527	3.753
2000	75.312	227.534	169.623	115.821	-400.829	-122.528	3.200

PREVIOUS: December 1961 CURRENT: June 1967

Potassium Iodide (KI) $I_1K_1(cr)$

POTASSIUM IODIDE (KI) LIQUID M_r = 166.0028

$S°(298.15\ K) = 114.098\ J\ K^{-1}\ mol^{-1}$ $\Delta_f H°(298.15\ K) = -312.846\ kJ\ mol^{-1}$

$T_{fus} = 954\ K$ $\Delta_{fus}H° = 24.016\ kJ\ mol^{-1}$

Enthalpy of Formation

$\Delta_f H°$(KI, ℓ, 298.15 K), is calculated from $\Delta_f H°$(KI, cr, 298.15 K) by adding the enthalpy of fusion, $\Delta_{fus}H°$, and the difference in enthalpy, H°(954 K)−H°(298.15 K), between the crystal and liquid.

Heat Capacity and Entropy

The heat capacity is derived as 17.3 cal K^{-1} mol^{-1} from the enthalpy measurements, 954-1014 K, by Dworkin (1). This $C_p°$ value is adopted in the temperature range 298.15 - 2500 K. S°(298.15 K) is calculated in a manner analogous to that used for the enthalpy of formation.

Fusion Data

See the crystal table for details.

Vaporization Data

The boiling point, T_{vap} = 1618 K (1 atm), is calculated as the temperature at which the sum of the partial pressures of KI(g) and K_2I_2(g) over KI(ℓ) equals one atmosphere. The value of $\Delta_{vap}H°$ is calculated as the heat required to produce one mole of vapor mixture, which contains 78 percent monomer (KI) and 22 percent dimer (K_2I_2) at T_{vap}.

T_{vap} was reported as 1603 and 1590 K by Wartenberg and Albrecht (2), and Ruff and Mugdan (3), respectively.

References

1. A. S. Dworkin, Oak Ridge National Laboratory, personal communication, Dec. 1, 1964.
2. H. von Wartenberg and P. Albrecht, Z. Elektrochem. 27, 162 (1921).
3. O. Ruff and S. Mugdan, Z. Anorg. Chem. 117, 147 (1921).

Potassium Iodide (KI) $I_1K_1(l)$

Enthalpy Reference Temperature = T_r = 298.15 K Standard State Pressure = p° = 0.1 MPa

	J K⁻¹mol⁻¹			kJ mol⁻¹			
T/K	$C_p°$	S°	−[G°−H°(T_r)]/T	H°−H°(T_r)	$\Delta_f H°$	$\Delta_f G°$	Log K_f
0							
100							
200							
298.15	72.383	114.098	114.098	0.	-312.846	-310.269	54.358
300	72.383	114.546	114.100	0.134	-312.817	-310.253	54.020
400	72.383	135.369	116.939	7.372	-321.876	-308.881	40.336
500	72.383	151.521	122.300	14.611	-341.844	-303.743	31.732
600	72.383	164.718	128.303	21.849	-339.522	-296.340	25.799
700	72.383	175.876	134.323	29.087	-337.163	-289.329	21.590
800	72.383	185.541	140.135	36.326	-334.792	-282.657	18.456
900	72.383	194.067	145.663	43.564	-332.431	-276.282	16.035
954.000	72.383	198.285	148.523	47.473	------ CRYSTAL <--> LIQUID -----		
1000	72.383	201.693	150.891	50.802	-330.106	-270.169	14.112
1100	72.383	208.592	155.828	58.040	-406.794	-259.675	12.331
1200	72.383	214.890	160.491	65.279	-403.551	-246.444	10.727
1300	72.383	220.684	164.902	72.517	-400.320	-233.483	9.381
1400	72.383	226.048	169.080	79.755	-397.104	-220.770	8.237
1500	72.383	231.042	173.046	86.994	-393.909	-208.286	7.253
1600	72.383	235.714	176.819	94.232	-390.740	-196.015	6.399
1700	72.383	240.102	180.413	101.470	-387.602	-183.941	5.652
1800	72.383	244.239	183.845	108.709	-384.499	-172.051	4.993
1900	72.383	248.153	187.128	115.947	-381.435	-160.332	4.408
2000	72.383	251.866	190.273	123.185	-378.411	-148.773	3.886
2100	72.383	255.397	193.291	130.424	-375.428	-137.365	3.417
2200	72.383	258.764	196.191	137.662	-372.485	-126.098	2.994
2300	72.383	261.982	198.982	144.900	-369.581	-114.963	2.611
2400	72.383	265.063	201.671	152.139	-366.713	-103.955	2.263
2500	72.383	268.017	204.267	159.377	-363.879	-93.065	1.944

PREVIOUS: December 1961 CURRENT: June 1967

Potassium Iodide (KI) $I_1K_1(l)$

POTASSIUM IODIDE (KI) CRYSTAL-LIQUID M_r = 166.0028

0 to 954 K crystal
above 954 K liquid

Refer to the individual tables for details.

Potassium Iodide (KI) $I_1K_1(cr,l)$

Enthalpy Reference Temperature = T_r = 298.15 K Standard State Pressure = $p°$ = 0.1 MPa

T/K	$C_p°$ (J K⁻¹mol⁻¹)	S° (J K⁻¹mol⁻¹)	$-[G°-H°(T_r)]/T$ (J K⁻¹mol⁻¹)	$H°-H°(T_r)$ (kJ mol⁻¹)	$\Delta_fH°$ (kJ mol⁻¹)	$\Delta_fG°$ (kJ mol⁻¹)	Log K_f
0	0.	0.	INFINITE	-12.711	-326.930	-326.930	INFINITE
100	45.003	52.413	151.616	-9.920	-327.397	-326.028	170.300
200	50.518	85.772	111.148	-5.075	-327.612	-324.572	84.770
298.15	52.777	106.387	106.387	0.	-327.900	-323.024	56.592
300	52.806	106.713	106.388	0.098	-327.907	-322.994	56.238
400	53.932	122.054	108.472	5.433	-338.870	-320.549	41.859
500	55.363	134.235	112.446	10.894	-360.614	-313.870	32.790
600	57.300	144.490	116.953	16.522	-359.903	-304.584	26.516
700	59.777	153.503	121.543	22.372	-358.933	-295.437	22.046
800	62.639	161.668	126.055	28.490	-357.681	-286.448	18.703
900	65.793	169.225	130.437	34.909	-356.140	-277.633	16.113
954.000	67.594	173.111	132.743	38.510	CRYSTAL <--> LIQUID		
954.000	72.383	198.285	132.743	62.527	TRANSITION		
1000	72.383	201.693	135.837	65.856	-330.106	-270.169	14.112
1100	72.383	208.592	142.143	73.095	-406.794	-259.675	12.331
1200	72.383	214.890	147.946	80.333	-403.551	-246.444	10.727
1300	72.383	220.684	153.322	87.571	-400.320	-233.483	9.381
1400	72.383	226.048	158.327	94.810	-397.104	-220.770	8.237
1500	72.383	231.042	163.010	102.048	-393.909	-208.286	7.253
1600	72.383	235.714	167.410	109.286	-390.740	-196.015	6.399
1700	72.383	240.102	171.558	116.524	-387.602	-183.941	5.652
1800	72.383	244.239	175.482	123.763	-384.499	-172.051	4.993
1900	72.383	248.153	179.205	131.001	-381.435	-160.332	4.408
2000	72.383	251.866	182.746	138.239	-378.411	-148.773	3.886
2100	72.383	255.397	186.122	145.478	-375.428	-137.365	3.417
2200	72.383	258.764	189.348	152.716	-372.485	-126.098	2.994
2300	72.383	261.982	192.437	159.954	-369.581	-114.963	2.611
2400	72.383	265.063	195.399	167.193	-366.713	-103.955	2.263
2500	72.383	268.017	198.245	174.431	-363.879	-93.065	1.944

PREVIOUS: CURRENT: June 1967

Potassium Iodide (KI) $I_1K_1(cr,l)$

POTASSIUM IODIDE (KI) IDEAL GAS M_r = 166.0028

$S°(298.15\ K) = 258.283\ J\ K^{-1}\ mol^{-1}$

$\Delta_f H°(0\ K) = -122.08 \pm 2.1\ kJ\ mol^{-1}$
$\Delta_f H°(298.15\ K) = -125.52 \pm 2.1\ kJ\ mol^{-1}$

Electronic Level and Quantum Weight

State	ε_i, cm^{-1}	g_i
$^1\Sigma^+$	0	1

$\omega_e = 186.287 \pm 0.04\ cm^{-1}$ $\omega_e x_e = 0.5725 \pm 0.006\ cm^{-1}$ $\sigma = 1$
$B_e = 0.06072\ cm^{-1}$ $\alpha_e = 0.000267\ cm^{-1}$ $r_e = 3.0478$ Å

Enthalpy of Formation

The enthalpy of formation is derived from the sublimation and vaporization data analyzed below. Many investigators have determined the total pressures of KI(g) and K_2I_2(g) over KI(cr) or KI(ℓ) by static methods or "apparent" pressures by Knudsen effusion or transpiration. These pressures are converted to monomer pressures by use of functions [see K_2I_2(g) table] which are consistent with the dimer-monomer equilibrium data of Datz (9). 2nd and 3rd law analyses of the monomer pressures show reasonable agreement. The selected value of $\Delta_f H°(298.15\ K) = -30.0 \pm 0.5$ kcal mol is also in good agreement with -30.23 ± 1.15 and -30.46 kcal mol^{-1} derived from $D_0^0 = 3.32 \pm 0.05$ and 3.33 eV reported by Gaydon (10) and Herzberg (11), respectively.

Source	Reaction*	T/K	Data Points	$\Delta_r H°(298.15\ K)$, kcal mol^{-1} 2nd law	3rd law	Drift cal K^{-1} mol^{-1}	$\Delta_f H°(298.15\ K)$ kcal mol^{-1}
Wartenberg-Albrecht (1)	(B)	1336.2-1606.2	9	47.79±0.38	44.76	-2.1±0.2	-30.01
Ruff-Mugdan (2)	(B)	1319.2-1590.2	12	49.38±1.03	44.59	-3.3±0.7	-30.18
Flock-Rodebush (3)	(B)	1116.1-1275.7	8	45.43±0.03	44.75	-0.6±0.1	-30.02
Greiner-Jellinek (4)	(B)	1453.2	1	-	45.24	-	-29.53
Niwa (5)	(A)	803.2-873.2	8	46.11±0.35	48.30	2.6±0.4	-30.07
Zimm-Mayer (6)	(A)	618.4-892.9	10	48.66±0.25	48.47	-0.4±0.3	-29.90
Cogin-Kimball (7)	(A)	721.5-897.7	29	48.52±0.40	47.73	-1.1±0.5	-30.64
Bridgers (8)	(A)	697.5-891.7	25	48.48±0.22	47.72	-1.1±0.3	-30.65

*Reactions: (A) KI(cr) = KI(g); (B) KI(ℓ) = KI(g).

Heat Capacity and Entropy

The ground state configuration is obtained from Herzberg (11). The adopted values of ω_e, $\omega_e x_e$, B_e, α_e, and r_e were determined using millimeter wave molecular beam spectroscopy by Rusk and Gordy (12), and corrected to the average isotopic species. Similar values were reported by Honig et al. (13) from microwave spectra, except ω_e which was given as 199.74 cm^{-1}. The value of ω_e was derived to be 200 and 173 cm^{-1} by Barrow and Caunt (14) and Rice and Klemperer (15). The moment of inertia is $4.5979 \times 10^{-38}\ g\ cm^2$.

References

1. H. von Wartenberg and P. Albrecht, Z. Elektrochem. 27, 162 (1921).
2. O. Ruff and S. Mugdan, Z. Anorg. Chem. 117, 147 (1921).
3. E. F. Flock and W. H. Rodebush, J. Amer. Chem. Soc. 48, 2522 (1926).
4. B. Greiner and K. Jellinek, Z. Physik. Chem. (Leipzig) A165, 97 (1933).
5. K. Niwa, J. Fac. Sci. Hokkaido Imperial Univ., Ser. III, 2, 201 (1938).
6. B. H. Zimm and J. E. Mayer, J. Chem. Phys. 12, 362 (1944).
7. G. E. Cogin and G. E. Kimball, J. Chem. Phys. 16, 1035 (1948).
8. H. E. Bridgers, Ph.D. Thesis, The Ohio State University, 1953.
9. S. Datz, Oak Ridge National Laboratory, ORNL-2933, 1960.
10. A. G. Gaydon, "Dissociation Energies," Chapman and Hall Ltd., London, 1953.
11. G. Herzberg, "Spectra of Diatomic Molecules," D. Van Nostrand Co., Inc., New York, 1950.
12. J. R. Rusk and W. Gordy, Phys. Rev. 127, 817 (1962).
13. A. Honig, M. Mandel, M. L. Stitch and C. H. Townes, Phys. Rev. 96, 629 (1954).
14. R. F. Barrow and A. D. Caunt, Proc. Roy. Soc. (London) A219, 120 (1953).
15. S. A. Rice and W. Klemperer, J. Chem. Phys. 27, 573 (1957).

Potassium Iodide (KI) I_1K_1(g)

Enthalpy Reference Temperature = T_r = 298.15 K Standard State Pressure = p° = 0.1 MPa

T/K	$C_p°$ (J $K^{-1}mol^{-1}$)	S°	$-[G°-H°(T_r)]/T$	$H°-H°(T_r)$ (kJ mol^{-1})	$\Delta_f H°$	$\Delta_f G°$	Log K_f
0	0.	0.	INFINITE	-10.242	-122.081	-122.081	INFINITE
100	33.908	219.083	290.729	-7.165	-122.261	-137.559	71.854
200	36.441	243.584	261.662	-3.615	-123.772	-152.295	39.775
250	36.869	251.766	258.893	-1.782	-124.627	-159.328	33.290
298.15	37.122	258.283	258.283	0.	-125.520	-165.932	29.071
300	37.130	258.513	258.284	0.069	-125.556	-166.182	28.935
350	37.308	264.251	258.737	1.930	-128.992	-172.775	25.785
400	37.439	269.241	259.745	3.799	-138.124	-178.678	23.333
450	37.543	273.657	261.050	5.673	-139.830	-183.644	21.317
500	37.629	277.617	262.512	7.553	-161.576	-186.523	19.486
600	37.771	284.491	265.619	11.323	-162.722	-191.403	16.663
700	37.889	290.322	268.742	15.106	-163.818	-196.097	14.633
800	37.994	295.389	271.763	18.900	-164.891	-200.634	13.100
900	38.093	299.870	274.642	22.705	-165.964	-205.038	11.900
1000	38.186	303.888	277.369	26.519	-167.064	-209.321	10.934
1100	38.277	307.532	279.948	30.342	-247.167	-208.881	9.919
1200	38.366	310.866	282.388	34.174	-247.330	-205.394	8.941
1300	38.453	313.940	284.698	38.015	-247.496	-201.892	8.112
1400	38.539	316.793	286.890	41.864	-247.669	-198.378	7.402
1500	38.624	319.455	288.973	45.723	-247.854	-194.851	6.785
1600	38.709	321.951	290.957	49.589	-248.056	-191.311	6.246
1700	38.793	324.300	292.850	53.464	-248.282	-187.757	5.769
1800	38.877	326.520	294.660	57.348	-248.534	-184.190	5.345
1900	38.961	328.624	296.392	61.240	-248.816	-180.608	4.965
2000	39.044	330.624	298.054	65.140	-249.130	-177.010	4.623
2100	39.128	332.531	299.651	69.049	-249.477	-173.396	4.313
2200	39.211	334.353	301.187	72.966	-249.856	-169.764	4.031
2300	39.294	336.098	302.667	76.891	-250.265	-166.114	3.773
2400	39.377	337.772	304.095	80.824	-250.702	-162.446	3.536
2500	39.459	339.381	305.475	84.766	-251.163	-158.760	3.317
2600	39.542	340.931	306.809	88.716	-251.647	-155.054	3.115
2700	39.625	342.425	308.101	92.675	-252.149	-151.329	2.928
2800	39.707	343.867	309.352	96.641	-252.668	-147.586	2.753
2900	39.790	345.262	310.567	100.616	-253.201	-143.824	2.591
3000	39.872	346.612	311.746	104.599	-253.746	-140.043	2.438
3100	39.955	347.921	312.892	108.590	-254.303	-136.243	2.296
3200	40.037	349.191	314.006	112.590	-254.872	-132.426	2.162
3300	40.119	350.424	315.091	116.598	-255.445	-128.591	2.035
3400	40.202	351.623	316.148	120.614	-256.037	-124.738	1.916
3500	40.284	352.790	317.179	124.638	-256.642	-120.868	1.804
3600	40.366	353.925	318.184	128.671	-257.266	-116.980	1.697
3700	40.449	355.033	319.165	132.712	-257.912	-113.074	1.596
3800	40.531	356.112	320.123	136.761	-258.583	-109.151	1.500
3900	40.613	357.166	321.059	140.818	-259.283	-105.209	1.409
4000	40.696	358.196	321.975	144.883	-260.011	-101.250	1.322
4100	40.778	359.201	322.871	148.957	-260.778	-97.272	1.239
4200	40.860	360.185	323.747	153.039	-261.593	-93.274	1.160
4300	40.942	361.148	324.606	157.129	-262.457	-89.257	1.084
4400	41.025	362.090	325.447	161.227	-263.098	-85.246	1.012
4500	41.107	363.013	326.272	165.334	-264.006	-81.193	0.942
4600	41.189	363.917	327.080	169.449	-264.968	-77.120	0.876
4700	41.271	364.804	327.874	173.572	-265.965	-73.028	0.812
4800	41.353	365.673	328.652	177.703	-266.911	-68.925	0.750
4900	41.436	366.527	329.416	181.842	-268.021	-64.789	0.691
5000	41.518	367.365	330.167	185.990	-269.192	-60.630	0.633
5100	41.600	368.188	330.904	190.146	-270.426	-56.447	0.578
5200	41.682	368.996	331.629	194.310	-271.726	-52.238	0.525
5300	41.764	369.791	332.342	198.482	-273.091	-48.005	0.473
5400	41.846	370.573	333.042	202.663	-274.523	-43.744	0.423
5500	41.929	371.341	333.732	206.852	-276.022	-39.457	0.375
5600	42.011	372.097	334.410	211.048	-277.589	-35.142	0.328
5700	42.093	372.842	335.078	215.254	-278.876	-30.846	0.283
5800	42.175	373.574	335.735	219.467	-280.531	-26.480	0.238
5900	42.257	374.296	336.383	223.689	-282.246	-22.085	0.196
6000	42.339	375.007	337.021	227.918	-284.019	-17.661	0.154

PREVIOUS: June 1967 (1 atm) CURRENT: June 1967 (1 bar)

Potassium Iodide (KI) I_1K_1(g)

LITHIUM IODIDE (LiI) CRYSTAL M_r = 133.8455

$S°(298.15\ K) = [85.772]\ J\ K^{-1}\ mol^{-1}$

$T_{fus} = 742\ K$

$\Delta_f H°(0\ K)$ = Unknown

$\Delta_f H°(298.15\ K) = -270.08 \pm 0.4\ kJ\ mol^{-1}$

$\Delta_{fus} H° = 14.644\ kJ\ mol^{-1}$

Enthalpy of Formation

The enthalpy of formation. $\Delta_f H°$(LiI, cr, 298.15 K) = -64.55 kcal mol^{-1}, was calculated from the heat of solution of lithium iodide (cr) at infinite dilution and the ionic heats of formation of Li^+(aq, ∞) and I^+(aq, ∞).

Parker (1) reviewed the literature heat of solution data and gave the "best" value, $\Delta_{sol} H°$(aq, ∞) = -15.130 ± 75 cal K^{-1} mol^{-1} for LiI(cr) + LiI(aq, ∞) at 298.15 K. The ionic heat of formation, $\Delta_f H°(Li^+$, aq, ∞, 298.15 K) = -66.49 kcal mol^{-1} was calculated from ΔH°(LiOH·aq, ∞, 298.15 K) = -121.46 kcal mol^{-1} (see JANAF LiOH(cr) table, Mar. 31, 1966) with $\Delta_f H°(OH^-$·aq, ∞, 298.15 K) = -54.97 kcal mol^{-1} from (2). The value $\Delta_f H°(I^-$·aq ∞, 298.15 K) = -13.19 kcal mol^{-1} was also obtained from (2). Combination of the ionic heats of formation of Li^+(aq, ∞) and I^-(aq, ∞) gives the heat of formation of LiI(aq, ∞), $\Delta_f H°$(298.15 K) = -79.68 kcal mol^{-1}.

Heat Capacity and Entropy

Dworkin (3) has measured the enthalpy changes by the drop method (642-802 K), yielding H°(742 K)-H°(298.15 K) = 6.0 kcal mol^{-1} for the crystal at the melting point. Heat capacities derived from his data were $C_p°$(692 K) = 14.8 and $C_p°$ (772 K) = 15.1 cal $K^{-1} mol^{-1}$ for the crystal and liquid, respectively. The tabulated heat capacities were estimated based on these values and on the heat capacities of LiCl(cr), NaCl(cr) and NaI(cr).

The entropy, S°(298.15 K) = 20.5 cal K^{-1} mol^{-1}, was estimated by adding the entropy difference at 298.15 K between NaI(cr) and NaCl(cr) to the entropy of LiCl(cr). Comparisons with other alkali halides give results within ±0.5 cal K^{-1} mol^{-1} of this value.

Kelley (4) has estimated the heat capacities ($C_p° = 12.30 + 2.44 \times 10^{-3}T$ cal K^{-1} mol^{-1}) and the entropy, S°(298.15 K) = 17.50 cal K^{-1} mol^{-1}.

Fusion Data

The selected enthalpy of fusion, ΔH°(823 K) = 350 kcal mol^{-1}) was obtained from enthalpy measurements in a drop calorimeter by Dworkin and Bredig (5).

References

1. V. B. Parker, NSRDS-NBS 2, 1965.
2. U. S. Nat. Bur. Stand. Tech. Note 270-1, 1965.
3. A. S. Dworkin, Oak Ridge National Laboratory, personal communication, Dec. 1964.
4. K. K. Kelley, U. S. Bur. Mines Bull. 584, 1960; U.S. Bur. Mines Bull. 592, 1961.
5. A. S. Dworkin and M. A. Bredig, J. Phys. Chem. 64, 269 (1960).

Lithium Iodide (LiI) $I_1Li_1(cr)$

Enthalpy Reference Temperature = T_r = 298.15 K; Standard State Pressure = p° = 0.1 MPa

T/K	$C_p°$ (J K⁻¹mol⁻¹)	S° (J K⁻¹mol⁻¹)	-[G°-H°(T_r)]/T (J K⁻¹mol⁻¹)	H°-H°(T_r) (kJ mol⁻¹)	$\Delta_f H°$ (kJ mol⁻¹)	$\Delta_f G°$ (kJ mol⁻¹)	Log K_f
0							
100							
200							
298.15	50.082	85.772	85.772	0.	-270.077	-269.665	47.244
300	50.187	86.082	85.773	0.093	-270.080	-269.662	46.952
400	53.430	100.875	87.767	5.243	-278.350	-269.208	35.155
500	55.948	113.073	91.645	10.714	-302.910	-264.545	27.637
600	58.660	123.499	96.104	16.437	-302.047	-256.948	22.369
700	62.258	132.807	100.693	22.480	-300.814	-249.523	18.620
800	66.107	141.363	105.248	28.891	-299.188	-242.302	15.821
900	69.517	149.355	109.710	35.680	-297.185	-235.308	13.657
1000	71.965	156.815	114.052	42.763	-294.889	-228.553	11.938
1100	73.534	163.751	118.259	50.042	-292.399	-222.039	10.544
1200	74.559	170.195	122.321	57.448	-289.785	-215.757	9.392
1300	75.375	176.196	126.237	64.946	-287.087	-209.697	8.426
1400	75.877	181.801	130.008	72.510	-284.333	-203.847	7.606
1500	76.191	187.049	133.638	80.116	-281.548	-198.195	6.902
1600	76.191	191.966	137.132	87.735	-278.767	-192.729	6.292
1700	76.191	196.585	140.494	95.354	-421.243	-180.261	5.539
1800	76.191	200.940	143.733	102.974	-417.755	-166.186	4.823
1900	76.191	205.060	146.853	110.593	-414.303	-152.304	4.187
2000	76.191	208.968	149.862	118.212	-410.888	-138.603	3.620
2100	76.191	212.685	152.766	125.831	-407.509	-125.072	3.111
2200	76.191	216.229	155.570	133.450	-404.165	-111.702	2.652
2300	76.191	219.616	158.282	141.069	-400.854	-98.482	2.237
2400	76.191	222.859	160.906	148.688	-397.571	-85.406	1.859
2500	76.191	225.969	163.446	156.307	-394.313	-72.466	1.514

PREVIOUS: September 1961 CURRENT: June 1966

Lithium Iodide (LiI) $I_1Li_1(cr)$

LITHIUM IODIDE (LiI) LIQUID M_r = 133.8455

$S°(298.15\ K)$ = [98.611] J K^{-1} mol^{-1}

T_{fus} = 742 K

$\Delta_f H°(298.15\ K)$ = -258.349 kJ mol^{-1}

$\Delta_{fus} H°$ = 14.644 kJ mol^{-1}

Enthalpy of Formation

$\Delta_f H°$(LiI, ℓ, 298.15 K) is calculated from $\Delta_f H°$(LiI, cr, 298.15 K) by adding the enthalpy of fusion, $\Delta_{fus} H°$, and the difference in enthalpy, H°(742 K)-H°(298.15 K), between the crystal and liquid.

Heat Capacity and Entropy

Dworkin (1), has derived the heat capacity ($C_p°$ = 15.1 cal K^{-1} mol^{-1}) from enthalpy measurements (742-802 K) by the drop method. The liquid heat capacity was assumed to be constant as 15.1 cal K^{-1} mol^{-1}. The entropy at 298.15 K is calculated in a manner analogous to that used for the enthalpy of formation.

Fusion Data

See the crystal table for details.

Vaporization Data

T_{vap}(to monomer only) = 1497 K is determined as the temperature at which the fugacity is 1 atm for the reaction LiI(ℓ) = LiI(g). The corresponding enthalpy change ($\Delta_r H°$) at boiling point is the enthalpy of vaporization $\Delta_{vap} H°$(to monomer only).

T_{vap}(to equilibrium mixture) = 1447 K is the temperature at which the sum of the calculated partial vapor pressures of LiI(g) and Li_2I_2(g) reaches one atmosphere (trimer and higher polymer have been neglected in calculation). This value is in good agreement with the boiling point of 1444 K from the least squares fit of the total vapor pressure data measured by Ruff and Mugdan (2). $\Delta_{vap} H°$(to equilibrium mixture) at the boiling point is calculated as the heat of vaporization of one mole of liquid to vapor containing 30.44 mole percent of dimer. For detailed information see LiI(g) and Li_2I_2(g) tables.

References

1. A. S. Dworkin, Oak Ridge National Labortory, personal communication, Dec. 1964.
2. O. Ruff and S. Mugdan, Z. Anorg. Chem. 117, 147 (1921).

Lithium Iodide (LiI) $I_1Li_1(l)$

Enthalpy Reference Temperature = T_r = 298.15 K Standard State Pressure = p° = 0.1 MPa

	J $K^{-1}mol^{-1}$			kJ mol^{-1}			
T/K	$C_p°$	S°	$-[G°-H°(T_r)]/T$	$H°-H°(T_r)$	$\Delta_f H°$	$\Delta_f G°$	Log K_f
0							
100							
200							
298.15	63.178	98.611	98.611	0.	-258.349	-261.764	45.860
300	63.178	99.002	98.612	0.117	-258.328	-261.785	45.581
400	63.178	117.177	101.090	6.435	-265.430	-262.809	34.319
500	63.178	131.275	105.770	12.753	-289.143	-259.879	27.149
600	63.178	142.794	111.010	19.070	-287.685	-254.163	22.127
700	63.178	152.533	116.264	25.388	-286.177	-248.694	18.558
742.000	63.178	156.214	118.422	28.042	------ CRYSTAL <--> LIQUID -----		
800	63.178	160.969	121.336	31.706	-284.644	-243.444	15.895
900	63.178	168.410	126.162	38.024	-283.113	-238.386	13.836
1000	63.178	175.067	130.725	44.342	-281.582	-233.498	12.197
1100	63.178	181.088	135.034	50.660	-280.053	-228.764	10.863
1200	63.178	186.586	139.104	56.977	-278.527	-224.169	9.758
1300	63.178	191.643	142.954	63.295	-277.010	-219.700	8.828
1400	63.178	196.325	146.601	69.613	-275.502	-215.348	8.035
1500	63.178	200.683	150.063	75.931	-274.005	-211.104	7.351
1600	63.178	204.761	153.355	82.249	-272.525	-206.959	6.757
1700	63.178	208.591	156.493	88.567	-416.302	-195.730	6.014
1800	63.178	212.202	159.489	94.884	-414.116	-182.818	5.305
1900	63.178	215.618	162.354	101.202	-411.965	-170.027	4.674
2000	63.178	218.859	165.099	107.520	-409.851	-157.349	4.110
2100	63.178	221.941	167.733	113.838	-407.773	-144.775	3.601
2200	63.178	224.880	170.264	120.156	-405.731	-132.299	3.141
2300	63.178	227.689	172.700	126.474	-403.721	-119.916	2.723
2400	63.178	230.378	175.048	132.792	-401.739	-107.619	2.342
2500	63.178	232.957	177.313	139.109	-399.782	-95.404	1.993

PREVIOUS: September 1961 CURRENT: June 1966

Lithium Iodide (LiI) $I_1Li_1(l)$

LITHIUM IODIDE (LiI) CRYSTAL-LIQUID M_r = 133.8455

0 to 742 K crystal
above 742 K liquid

Refer to the individual tables for details.

Lithium Iodide (LiI) $I_1Li_1(cr,l)$

Enthalpy Reference Temperature = T_r = 298.15 K; Standard State Pressure = $p°$ = 0.1 MPa

T/K	$C_p°$ (J K⁻¹mol⁻¹)	$S°$ (J K⁻¹mol⁻¹)	$-[G°-H°(T_r)]/T$ (J K⁻¹mol⁻¹)	$H°-H°(T_r)$ (kJ mol⁻¹)	$\Delta_fH°$ (kJ mol⁻¹)	$\Delta_fG°$ (kJ mol⁻¹)	Log K_f
0							
100							
200							
298.15	50.082	85.772	85.772	0.	-270.077	-269.665	47.244
300	50.187	86.082	85.773	0.093	-270.080	-269.662	46.952
400	53.430	100.875	87.767	5.243	-278.350	-269.208	35.155
500	55.948	113.073	91.645	10.714	-302.910	-264.545	27.637
600	58.660	123.499	96.104	16.437	-302.047	-256.948	22.369
700	62.258	132.807	100.693	22.480	-300.814	-249.523	18.620
742.000	63.764	136.478	102.615	25.126	CRYSTAL <--> LIQUID TRANSITION		
742.000	63.178	156.214	102.615	39.770			
800	63.178	160.969	106.676	43.435	-284.644	-243.444	15.895
900	63.178	168.410	113.130	49.752	-283.113	-238.386	13.836
1000	63.178	175.067	118.997	56.070	-281.582	-233.498	12.197
1100	63.178	181.088	124.372	62.388	-280.053	-228.764	10.863
1200	63.178	186.586	129.331	68.706	-278.527	-224.169	9.758
1300	63.178	191.643	133.932	75.024	-277.010	-219.700	8.828
1400	63.178	196.325	138.224	81.342	-275.502	-215.348	8.035
1500	63.178	200.684	142.244	87.659	-274.005	-211.104	7.351
1600	63.178	204.761	146.025	93.977	-272.525	-206.959	6.757
1700	63.178	208.591	149.594	100.295	-416.302	-195.730	6.014
1800	63.178	212.202	152.973	106.613	-414.116	-182.818	5.305
1900	63.178	215.618	156.181	112.931	-411.965	-170.027	4.674
2000	63.178	218.859	159.234	119.249	-409.851	-157.349	4.110
2100	63.178	221.941	162.148	125.566	-407.773	-144.775	3.601
2200	63.178	224.880	164.933	131.884	-405.731	-132.299	3.141
2300	63.178	227.689	167.601	138.202	-403.721	-119.916	2.723
2400	63.178	230.378	170.161	144.520	-401.739	-107.619	2.342
2500	63.178	232.957	172.622	150.838	-399.782	-95.404	1.993

PREVIOUS: CURRENT: June 1966

Lithium Iodide (LiI) $I_1Li_1(cr,l)$

LITHIUM IODIDE (LiI) IDEAL GAS M_r = 133.8455

$S°(298.15\ K) = 232.262\ J\ K^{-1}\ mol^{-1}$

$\Delta_f H°(0\ K) = -89.08 \pm 8.4\ kJ\ mol^{-1}$
$\Delta_f H°(298.15\ K) = -91.00 \pm 8.4\ kJ\ mol^{-1}$

Electronic Level and Quantum Weight

State	ε_i, cm^{-1}	g_i
$^1\Sigma^+$	0	1

$\omega_e = 498.16\ cm^{-1}$ $\omega_e x_e = 3.39\ cm^{-1}$ $\sigma = 1$
$B_e = 0.44837\ cm^{-1}$ $\alpha_e = 0.00416\ cm^{-1}$ $r_e = 2.3919$ Å

Enthalpy of Formation

The enthalpy of formation, $\Delta_f H°(LiI, g, 298.15\ K) = -21.75\pm2\ kcal\ mol^{-1}$, was calculated from the selected heat of sublimation and the enthalpy of formation for lithium iodide (cr). The heat of sublimation has been determined mass-spectrometrically from the ion intensity measurement (2nd law method) by Gorokhov (1) as $\Delta_{sub}H°(cr \rightarrow monomer, 298.15\ K) = 42.9\pm0.6$ and $\Delta_{sub}H°(cr \rightarrow dimer, 298.15\ K) = 43.2\pm0.4\ kcal\ mol^{-1}$ in the temperature range from 626-722 K. Gorokhov also reported the partial vapor pressure of monomeric lithium iodide ($P_{monomer} = 3 \times 10^{-3}$ mm Hg) and the ratio of dimer to monomer ($P_{dimer}/P_{monomer} = 6$) which have been used for 3rd law calculation. Using the same technique, Friedman (2) reported the heats of sublimation as $\Delta_{sub}H°(cr \rightarrow monomer, 727\ K) = 41.9\pm0.5$ and $\Delta_{sub}H°(cr \rightarrow dimer, 727\ K) = 40.8\pm0.5\ kcal\ mol^{-1}$ in the temperature range from 630 to 730 K. Berkowitz et al. (3) have also studied the lithium iodide vapor in the mass spectrometer and reported the 2nd law and 3rd law values of the heat of dimerization at 800 K as 37.9 and 40.4 $kcal\ mol^{-1}$, respectively. Using the JANAF thermodynamic functions, all the heats of sublimation and dimerization at different temperatures have been reduced to 298.15 K and summarized in the following table. The total vapor pressure over the liquid has been measured by a boiling point method in the temperature range 1223 to 1413 K by Ruff and Mugdan (4). (Their data have been quoted by Wartenberg and Schulz (5). In order to have good agreement between the calculated vapor pressure and the observed total vapor pressure, the heats of sublimation have been so chosen as $\Delta_{sub}H°(298.15\ K) = 42.8\ kcal\ mol^{-1}$ and $\Delta_{sub}H°(cr \rightarrow dimer, 298.15\ K) = 42.6\ kcal\ mol^{-1}$ and the heat of dissociation of dimer, $\Delta_d H°(cr \rightarrow monomer, 298.15\ K) = 43.0\ kcal\ mol^{-1}$. The calculated boiling point (to equilibrium mixture) is 1449 K which agrees with 1443 K reported by Wartenberg and Schulz, and also 1462 K by Ruff and Mugdan.

Reaction	Source	$\Delta_r H°(298.15\ K)$, $kcal\ mol^{-1}$ 2nd Law	3rd Law
$LiI(cr) \rightarrow LiI(g)$	Gorokhov	44.7	42.2
	Friedman	43.9	--
$2\ LiI(cr) \rightarrow Li_2I_2(g)$	Gorokhov	46.0	41.3
	Friedman	44.1	--
$Li_2I_2(g) \rightarrow 2\ LiI(g)$	Berkowitz et al.	38.8	41.3
	Friedman	43.7	--
	Gorokov	43.2	43.1

Heat Capacity and Entropy

The bond distance and the rotational constants (B_e and ν_e), corrected to the isotopic abundance of 7.4% Li^6 and 92.6% Li^7, were obtained from the microwave studies by Honig et al. (6). The vibrational constants (ω_e and $\omega_e x_e$) were determined from the infrared spectrum by Klemperer et al. (7).

The tabulated thermodynamic functions are in reasonable agreement with those calculated by Wilkins (8) who used slightly different molecular constants.

References

1. L. N. Gorokhov, Dokl. Akad. Nauk SSSR 142, 113 (1962).
2. L. Friedman, J. Chem. Phys. 23, 477 (1955).
3. J. Berkowitz, H. A. Tasman and W. A. Chupka, J. Chem. Phys. 36, 2170 (1962).
4. O. Ruff and S. Mugdan, Z. Anorg. Chem. 117, 147 (1921).
5. H. von Wartenberg and H. Schulz, Z. Elektrochem. 27, 568 (1921).
6. A. Honig, M. Mandel, M. L. Stitch and C. H. Townes, Phys. Rev. 96, 629 (1954).
7. W. Klemperer, W. G. Norris, A. Buchler and A. G. Emslie, J. Chem. Phys. 33, 1534 (1960).
8. R. L. Wilkins, J. Chem. Eng. Data 5, 337 (1960).

Lithium Iodide (LiI) $I_1Li_1(g)$

Enthalpy Reference Temperature = T_r = 298.15 K Standard State Pressure = p° = 0.1 MPa

T/K	$C_p°$ (J K⁻¹mol⁻¹)	S°	$-[G°-H°(T_r)]/T$	$H°-H°(T_r)$ (kJ mol⁻¹)	$\Delta_f H°$	$\Delta_f G°$	Log K_f
0	0.	0.	INFINITE	-9.297	-89.078	-89.078	INFINITE
100	29.475	197.607	261.437	-6.383	-88.184	-104.161	54.408
200	32.394	218.879	235.363	-3.297	-89.405	-119.720	31.268
250	33.663	226.251	232.826	-1.644	-90.194	-127.209	26.579
298.15	34.572	232.262	232.262	0.	-91.002	-134.265	23.523
300	34.602	232.476	232.262	0.064	-91.034	-134.534	23.424
350	35.289	237.864	232.686	1.812	-91.955	-141.713	21.149
400	35.798	242.611	233.636	3.590	-100.928	-148.480	19.390
450	36.185	246.851	234.873	5.390	-102.568	-154.327	17.914
500	36.485	250.679	236.265	7.207	-127.342	-157.780	16.483
600	36.921	257.372	239.241	10.879	-128.530	-163.755	14.256
700	37.223	263.088	242.250	14.587	-129.631	-169.537	12.651
800	37.448	268.074	245.173	18.321	-130.683	-175.166	11.437
900	37.627	272.495	247.967	22.075	-131.715	-180.664	10.485
1000	37.777	276.467	250.622	25.845	-132.732	-186.048	9.718
1100	37.906	280.074	253.138	29.629	-133.736	-191.331	9.086
1200	38.022	283.377	255.522	33.426	-134.732	-196.523	8.554
1300	38.128	286.425	257.784	37.234	-135.725	-201.632	8.102
1400	38.227	289.254	259.932	41.051	-136.717	-206.665	7.711
1500	38.321	291.895	261.976	44.879	-137.710	-211.627	7.369
1600	38.410	294.371	263.924	48.715	-138.712	-216.522	7.069
1700	38.496	296.702	265.784	52.561	-284.962	-214.178	6.581
1800	38.580	298.905	267.563	56.415	-285.239	-210.006	6.094
1900	38.662	300.993	269.268	60.277	-285.543	-205.819	5.658
2000	38.742	302.978	270.905	64.147	-285.877	-201.614	5.266
2100	38.821	304.870	272.477	68.025	-286.239	-197.392	4.910
2200	38.899	306.678	273.991	71.911	-286.629	-193.152	4.586
2300	38.976	308.409	275.450	75.805	-287.043	-188.894	4.290
2400	39.052	310.069	276.858	79.706	-287.478	-184.617	4.018
2500	39.128	311.665	278.219	83.615	-287.930	-180.322	3.768
2600	39.203	313.201	279.535	87.532	-288.394	-176.009	3.536
2700	39.277	314.682	280.810	91.456	-288.867	-171.677	3.321
2800	39.352	316.112	282.045	95.387	-289.345	-167 328	3.122
2900	39.426	317.494	283.244	99.326	-289.824	-162.962	2.935
3000	39.499	318.832	284.408	103.272	-290.300	-158.580	2.761
3100	39.572	320.128	285.539	107.226	-290.771	-154.181	2.598
3200	39.645	321.386	286.640	111.187	-291.236	-149.768	2.445
3300	39.718	322.607	287.711	115.155	-291.693	-145.340	2.301
3400	39.791	323.794	288.755	119.131	-292.140	-140.898	2.165
3500	39.863	324.948	289.773	123.113	-292.578	-136.443	2.036
3600	39.936	326.072	290.766	127.103	-293.007	-131.976	1.915
3700	40.008	327.167	291.735	131.100	-293.426	-127.497	1.800
3800	40.080	328.235	292.681	135.105	-293.838	-123.007	1.691
3900	40.152	329.277	293.606	139.116	-294.242	-118.506	1.587
4000	40.224	330.295	294.511	143.135	-294.641	-113.995	1.489
4100	40.296	331.289	295.396	147.161	-295.035	-109.474	1.395
4200	40.368	332.261	296.262	151.194	-295.428	-104.943	1.305
4300	40.439	333.211	297.110	155.235	-295.821	-100.404	1.220
4400	40.511	334.142	297.941	159.282	-296.194	-95.856	1.138
4500	40.583	335.053	298.756	163.337	-296.583	-91.299	1.060
4600	40.654	335.946	299.555	167.399	-296.980	-86.732	0.985
4700	40.726	336.821	300.338	171.468	-297.383	-82.158	0.913
4800	40.797	337.679	301.107	175.544	-297.796	-77.574	0.844
4900	40.869	338.521	301.862	179.627	-298.222	-72.981	0.778
5000	40.940	339.347	302.604	183.718	-298.661	-68.380	0.714
5100	41.011	340.159	303.332	187.815	-299.111	-63.771	0.653
5200	41.083	340.956	304.048	191.920	-299.585	-59.152	0.594
5300	41.154	341.739	304.752	196.032	-300.081	-54.523	0.537
5400	41.225	342.509	305.444	200.151	-300.600	-49.885	0.483
5500	41.296	343.266	306.125	204.277	-301.145	-45.238	0.430
5600	41.368	344.011	306.795	208.410	-301.719	-40.579	0.379
5700	41.439	344.744	307.454	212.550	-302.324	-35.911	0.329
5800	41.510	345.465	308.103	216.698	-302.962	-31.231	0.281
5900	41.581	346.175	308.743	220.852	-303.636	-26.541	0.235
6000	41.652	346.875	309.372	225.014	-304.347	-21.838	0.190

PREVIOUS: June 1966 (1 atm) CURRENT: June 1966 (1 bar)

Lithium Iodide (LiI) $I_1Li_1(g)$

MAGNESIUM IODIDE (MgI) IDEAL GAS M_r = 151.2095

$\Delta_f H°(0\ K) = [26.45 \pm 41.8]\ kJ\ mol^{-1}$
$\Delta_f H°(298.15\ K) = [24.61 \pm 41.8]\ kJ\ mol^{-1}$

$S°(298.15\ K) = 252.647 \pm 0.21\ J\ K^{-1}\ mol^{-1}$

Electronic Levels and Quantum Weights

State	ε_i, cm^{-1}	g_i
$X^2\Sigma^+$	0	2
$A[^2\Pi]$	24319	2
	25294	2

$\omega_e = 312\ cm^{-1}$ $\omega_e x_e = [1.09]\ cm^{-1}$ $\sigma = 1$
$B_e = [0.13014]\ cm^{-1}$ $\alpha_e = [0.00062]\ cm^{-1}$ $r_e = [2.52]\ Å$

Enthalpy of Formation

The adopted $\Delta_f H°(0\ K) = 6.309 \pm 10\ kcal\ mol^{-1}$ is calculated from a D_0^0 value of 54.3 kcal mol^{-1} (2.35 cal $K^{-1}\ mol^{-1}$). An approximate value of 312 cm^{-1} has been given for ω_e (1). A value of $\omega_e x_e$, 1.09 cm^{-1}, is calculated from an average value of $x_e \mu^{1/2}$ calculated from data for CaI, SrI, BaI, MgBr, and MgCl (1, 2). A linear Birge-Sponer extrapolation of these ω_e and $\omega_e x_e$ data is corrected for the ionic character of the molecule as described by Hildenbrand (3) to give the adopted value of D_0^0 = 54.3 kcal mol^{-1}.

From a consideration of ionic bonding forces, Krasnov and Karaseva (4) calculated $D_0^0 \geq 55 \pm 10$ kcal mol^{-1}.

Using JANAF data, $D_0^0(MgI)\ D_0^0(MgI_2) = 0.44$, in reasonable agreement with the ratios found by Blue et al. (5) for the alkaline earth fluorides and by Hildenbrand (6) for the alkaline with fluorides and chlorides.

$\Delta_f H°(298.15\ K) = 5.883 \pm 10\ kcal\ mol^{-1}$ is calculated from the adopted $\Delta_f H°(0\ K) = 6.309\ kcal\ mol^{-1}$.

Heat Capacity and Entropy

The ground state vibrational constants are as given in the Enthalpy of Formation section above. The adopted value of r_e = 2.52 Å is the bond distance in MgI_2 (7); Margrave (8) has estimated 2.72 Å from additive ionic radii. B_e is calculated from the adopted r_e, and α_e is calculated assuming a Morse potential function.

The electronic levels and their designation are those given by Rosen (1). The possibility of additional electronic levels near 15000 cm^{-1} is suggested by a comparison with CaI, SrI, and BaI (2). Including these levels does not change the entropy at 298.15 K and increases the entropy at 6000 K by only 0.69 cal $K^{-1}\ mol^{-1}$.

References

1. B. Rosen, Ed. "Spectroscopic Data Relative to Diatomic Molecules," Pergamon Press, New York, 1970.
2. JANAF Thermochemical Tables: CaI(g), SrI(g), BaI(g), 6-30-74; MgBr(g), 6-30-66; MgCl(g) 3-31-66.
3. D. L. Hildenbrand in "Advances in High Temperature Chemistry," Vol. 1, L. Eyring, Ed., Academic Press, New York, pp. 193-215, 1967.
4. K. S. Krasnov and N. V. Karaseva, Optics and Spectrosc. 19, 14 (1965).
5. G. D. Blue, J. W. Green, T. C. Ehlert, and J. L. Margrave, Nature 199, 804 (1963).
6. D. L. Hildenbrand, Stanford Research Institute, personal communication, July 2, 1974.
7. P. A. Akishin and V. P. Spiridinov, Zhur. Fiz. Khim. 32, 1682 (1958).
8. J. L. Margrave, J. Phys. Chem. 58, 258 (1954).

Magnesium Iodide (MgI) $I_1Mg_1(g)$

Enthalpy Reference Temperature = T_r = 298.15 K Standard State Pressure = p° = 0.1 MPa

T/K	$C_p°$ (J K⁻¹mol⁻¹)	S° (J K⁻¹mol⁻¹)	-[G°-H°(T_r)]/T (J K⁻¹mol⁻¹)	H°-H°(T_r) (kJ mol⁻¹)	$\Delta_f H°$ (kJ mol⁻¹)	$\Delta_f G°$ (kJ mol⁻¹)	Log K_f
0	0.	0.	INFINITE	-9.765	26.446	26.446	INFINITE
100	31.085	215.601	283.716	-6.811	27.233	9.673	-5.053
200	34.773	238.459	255.920	-3.492	26.079	-7.488	1.956
250	35.645	246.320	253.240	-1.730	25.356	-15.798	3.301
298.15	36.178	252.647	252.647	0.	24.614	-23.658	4.145
300	36.195	252.871	252.648	0.067	24.585	-23.957	4.171
350	36.562	258.480	253.090	1.886	23.736	-31.982	4.773
400	36.821	263.380	254.076	3.721	14.865	-39.608	5.172
450	37.013	267.728	255.356	5.567	13.374	-46.327	5.378
500	37.160	271.636	256.792	7.422	-8.199	-50.980	5.326
600	37.376	278.431	259.849	11.149	-9.116	-59.452	5.176
700	37.530	284.205	262.926	14.895	-10.126	-67.763	5.057
800	37.651	289.224	265.907	18.654	-11.245	-75.922	4.957
900	37.751	293.665	268.749	22.425	-12.488	-83.933	4.871
1000	37.840	297.647	271.443	26.204	-22.466	-91.086	4.758
1100	37.920	301.257	273.992	29.992	-24.016	-97.873	4.648
1200	37.994	304.560	276.404	33.788	-25.568	-104.519	4.550
1300	38.064	307.604	278.688	37.591	-27.123	-111.035	4.461
1400	38.132	310.427	280.855	41.401	-156.097	-114.264	4.263
1500	38.197	313.060	282.916	45.217	-156.323	-111.268	3.875
1600	38.261	315.528	284.878	49.040	-156.567	-108.257	3.534
1700	38.324	317.849	286.750	52.869	-156.835	-105.229	3.233
1800	38.385	320.041	288.539	56.705	-157.128	-102.185	2.965
1900	38.446	322.118	290.252	60.546	-157.451	-99.124	2.725
2000	38.506	324.092	291.895	64.394	-157.802	-96.045	2.508
2100	38.565	325.972	293.473	68.247	-158.181	-92.948	2.312
2200	38.625	327.768	294.992	72.107	-158.586	-89.833	2.133
2300	38.684	329.486	296.454	75.972	-159.013	-86.698	1.969
2400	38.743	331.133	297.865	79.844	-159.458	-83.544	1.818
2500	38.802	332.716	299.228	83.721	-159.914	-80.372	1.679
2600	38.861	334.239	300.545	87.604	-160.378	-77.181	1.551
2700	38.921	335.707	301.821	91.493	-160.843	-73.972	1.431
2800	38.981	337.123	303.056	95.388	-161.305	-70.747	1.320
2900	39.042	338.492	304.255	99.289	-161.759	-67.504	1.216
3000	39.104	339.817	305.418	103.197	-162.202	-64.247	1.119
3100	39.168	341.100	306.549	107.110	-162.630	-60.974	1.027
3200	39.232	342.345	307.648	111.030	-163.041	-57.688	0.942
3300	39.299	343.553	308.718	114.957	-163.433	-54.390	0.861
3400	39.368	344.727	309.760	118.890	-163.806	-51.080	0.785
3500	39.439	345.869	310.775	122.831	-164.158	-47.759	0.713
3600	39.512	346.982	311.765	126.778	-164.491	-44.429	0.645
3700	39.588	348.065	312.732	130.733	-164.805	-41.090	0.580
3800	39.667	349.122	313.676	134.696	-165.101	-37.742	0.519
3900	39.750	350.153	314.598	138.667	-165.380	-34.387	0.461
4000	39.835	351.161	315.499	142.646	-165.646	-31.024	0.405
4100	39.925	352.146	316.381	146.634	-165.899	-27.656	0.352
4200	40.017	353.109	317.244	150.631	-166.142	-24.281	0.302
4300	40.114	354.052	318.089	154.637	-166.378	-20.901	0.254
4400	40.214	354.975	318.917	158.654	-166.609	-17.515	0.208
4500	40.318	355.880	319.729	162.680	-166.837	-14.124	0.164
4600	40.427	356.767	320.524	166.718	-167.066	-10.728	0.122
4700	40.539	357.638	321.305	170.766	-167.298	-7.326	0.081
4800	40.654	358.492	322.070	174.826	-167.535	-3.920	0 043
4900	40.774	359.332	322.822	178.897	-167.780	-0.509	0.005
5000	40.897	360.157	323.561	182.980	-168.035	2.907	-0.030
5100	41.024	360.968	324.286	187.076	-168.303	6.329	-0.065
5200	41.155	361.766	324.999	191.185	-168.586	9.756	-0.098
5300	41.289	362.551	325.701	195.308	-168.885	13.188	-0.130
5400	41.427	363.324	326.390	199.443	-169.204	16.626	-0.161
5500	41.567	364.086	327.069	203.593	-169.544	20.071	-0.191
5600	41.711	364.836	327.736	207.757	-169.908	23.521	-0.219
5700	41.857	365.575	328.394	211.935	-170.296	26.979	-0.247
5800	42.006	366.305	329.041	216.128	-170.711	30.443	-0.274
5900	42.158	367.024	329.679	220.337	-171.155	33.915	-0.300
6000	42.312	367.734	330.307	224.560	-171.628	37.395	-0.326

PREVIOUS: December 1974 (1 atm) CURRENT: December 1974 (1 bar)

Magnesium Iodide (MgI) $I_1Mg_1(g)$

MOLYBDENUM IODIDE (MoI) IDEAL GAS $M_r = 222.8445$

$S°(298.15\ K) = [278.974 \pm 8.4]\ J\ K^{-1}\ mol^{-1}$

$\Delta_f H°(0\ K) = [499.74 \pm 41.8]\ kJ\ mol^{-1}$
$\Delta_f H°(298.15\ K) = [498.73 \pm 41.8]\ kJ\ mol^{-1}$

Electronic Levels and Quantum Weights

ε_i, cm^{-1}	g_i	ε_i, cm^{-1}	g_i	ε_i, cm^{-1}	g_i
0	[6]	[15199]	[6]	[15890]	[2]
[11783]	[2]	[15331]	[8]	[16796]	[2]
[12034]	[4]	[15428]	[10]	[16947]	[8]
[12417]	[6]	[15447]	[12]	[17174]	[4]
[12900]	[8]	[15691]	[6]	[17344]	[6]
[13461]	[10]	[15699]	[4]		

$\omega_e = [200]\ cm^{-1}$ $\omega_e x_e = [0.41]\ cm^{-1}$ $\sigma = 1$
$B_e = [0.04361]\ cm^{-1}$ $\alpha_e = [0.00013]\ cm^{-1}$ $r_e = [2.66]$ Å

Enthalpy of Formation

The adopted value of $\Delta_f H°$ at 298.15 K is that estimated by Brewer (1). The estimated value is consistent with the low volatility of solid molybdenum in iodine vapor reported by Schafer et al. (2). Brewer's estimated value is given as $\Delta_f H°(298.15\ K)/R = 60000 \pm 5000$. The assigned uncertainty (± 10 kcal mol^{-1}) is determined from the product 500xR which has been rounded to the nearest kcal mol^{-1}. The value of $\Delta_f H°(0\ K)$ combined with JANAF data (3) for Mo(g) and I(g) gives $D_0^0(Mo-I) = 63 \pm 11$ kcal mol^{-1}.

Heat Capacity and Entropy

We use values of r_e and ω_e estimated by Brewer (1). The electronic contribution is taken to be the same as that for the free gaseous Mo^+ ion, as suggested by Brewer (1). The levels listed by Moore (4) below 17334 cm^{-1} are used in the calculations. Our thermodynamic functions differ slightly from those reported by Brewer (1) due to the use of non-zero values for $\omega_e x_e$ and α_e. Estimates for α_e and ω_e are based on the empirical correlations of Calder and Ruedenberg (5).

References

1. L. Brewer, Materials and Molecular Research Division, Lawrence Berkeley Laboratory, University of California, Berkeley; personal communication; Sept. 28, 1978, preliminary draft of review to be submitted for publication in Atomic Energy Review, International Atomic Energy Agency, Vienna, Austria.
2. H. Schafer, T. Grofe, and M. Trenkel, J. Solid State Chem. 8, 14 (1973).
3. JANAF Thermochemical Tables: Mo(g), 3-31-78; I(g) and SrI(g), 6-30-74.
4. C. E. Moore, NSRDS-NBS 34, 1970; NSRDS-NBS 35, Vol. III, Reissued 1971.
5. G. V. Calder and K. Ruedenberg, J. Chem. Phys. 49, 5399 (1968).

Molybdenum Iodide (MoI) $I_1Mo_1(g)$

Enthalpy Reference Temperature = T_r = 298.15 K Standard State Pressure = $p°$ = 0.1 MPa

T/K	$C_p°$ (J K^{-1}mol^{-1})	$S°$ (J K^{-1}mol^{-1})	$-[G°-H°(T_r)]/T$ (J K^{-1}mol^{-1})	$H°-H°(T_r)$ (kJ mol^{-1})	$\Delta_f H°$ (kJ mol^{-1})	$\Delta_f G°$ (kJ mol^{-1})	Log K_f
0	0.	0.	INFINITE	-10.173	499.744	499.744	INFINITE
100	33.511	240.056	311.234	-7.118	500.783	480.529	-251.003
200	36.229	264.348	282.337	-3.598	499.994	460.516	-120.274
250	36.692	272.487	279.581	-1.774	499.386	450.715	-94.172
298.15	36.962	278.974	278.974	0.	498.733	441.399	-77.331
300	36.970	279.203	278.975	0.068	498.707	441.043	-76.792
350	37.154	284.917	279.426	1.922	497.940	431.492	-64.397
400	37.285	289.887	280.429	3.783	489.146	422.328	-55.150
450	37.384	294.284	281.729	5.650	487.732	414.060	-48.063
500	37.463	298.227	283.185	7.521	466.237	407.851	-42.608
600	37.586	305.069	286.279	11.274	465.497	396.242	-34.496
700	37.682	310.870	289.389	15.037	464.704	384.762	-28.711
800	37.764	315.907	292.395	18.810	463.865	373.398	-24.380
900	37.837	320.360	295.260	22.590	462.984	362.142	-21.018
1000	37.905	324.350	297.973	26.377	462.057	350.987	-18.334
1100	37.970	327.966	300.538	30.171	461 081	339.926	-16.142
1200	38.035	331.272	302.963	33.971	460.046	328.958	-14.319
1300	38.100	334.319	305.259	37.778	458.944	318.078	-12.781
1400	38.170	337.145	307.437	41.591	457.764	307.286	-11.465
1500	38.249	339.781	309.507	45.412	456.496	296.581	-10.328
1600	38.342	342.253	311 477	49.241	455.130	285.964	-9.336
1700	38.456	344.580	313.356	53.081	453.656	275.436	-8.463
1800	38.599	346.782	315.153	56.933	452.067	264.997	-7.690
1900	38.780	348.874	316.873	60.802	450.356	254.651	-7.001
2000	39.007	350.869	318.523	64.691	448.518	244.397	-6.383
2100	39.288	352.778	320.109	68.605	446.552	234.239	-5.826
2200	39.630	354.614	321.636	72.550	444.456	224.177	-5.323
2300	40.040	356.384	323.109	76.533	442.232	214.214	-4.865
2400	40.523	358.098	324.531	80.561	439.881	204.350	-4.448
2500	41.082	359.763	325.907	84.640	437.405	194.587	-4.066
2600	41.718	361.387	327.241	88.780	434.796	184.925	-3.715
2700	42.430	362.974	328.535	92.987	432.027	175.367	-3.393
2800	43.218	364.531	329.793	97.268	429.050	165 915	-3.095
2900	44.075	366.063	331.017	101.632	389.911	156.623	-2.821
3000	44.997	367.572	332.210	106.086	388.347	148.605	-2.587
3100	45.975	369.064	333.375	110.634	386.893	140.638	-2.370
3200	47.001	370.539	334.514	115.282	385.557	132.716	-2.166
3300	48.065	372.002	335.628	120.035	384.348	124.834	-1.976
3400	49.156	373.453	336.719	124.896	383.271	116.986	-1.797
3500	50.263	374.894	337.789	129.867	382.330	109.168	-1.629
3600	51.375	376.325	338.839	134.949	381.528	101.375	-1.471
3700	52.481	377.748	339.872	140.142	380.865	93.603	-1.321
3800	53.568	379.162	340.887	145.444	380.341	85.846	-1.180
3900	54.628	380.567	341.887	150.854	379.955	78.102	-1.046
4000	55.649	381.963	342.871	156.369	379.702	70.366	-0.919
4100	56.625	383.350	343.842	161.983	379.579	62.634	-0.798
4200	57.546	384.725	344.799	167.692	379.579	54.904	-0.683
4300	58.407	386.090	345.743	173.490	379.698	47.172	-0.573
4400	59.202	387.441	346.675	179.371	379.928	39.437	-0.468
4500	59.928	388.780	347.596	185.328	380.261	31.695	-0.368
4600	60.581	390.105	348.506	191.354	380.689	23.944	-0.272
4700	61.161	391.414	349.405	197.442	381.206	16.183	-0.180
4800	61.665	392.707	350.294	203.584	381.801	8.411	-0.092
4900	62.096	393.983	351.172	209.772	382.467	0.625	-0.007
5000	62.454	395.241	352.041	216.001	-208.189	-1.444	0.015
5100	62.742	396.481	352.900	222.261	-209.085	2.700	-0.028
5200	62.961	397.701	353.750	228.547	-210.080	6.862	-0.069
5300	63.116	398.902	354.591	234.851	-211.179	11.044	-0.109
5400	63.210	400.083	355.422	241.168	-212.390	15.249	-0.148
5500	63.247	401.243	356.245	247.491	-213.714	19.476	-0.185
5600	63.231	402.383	357.059	253.815	-215.157	23.729	-0.221
5700	63.166	403.502	357.864	260.136	-216.720	28.008	-0.257
5800	63.057	404.599	358.660	266.447	-218.408	32.316	-0.291
5900	62.908	405.676	359.448	272.746	-220.219	36.654	-0.325
6000	62.723	406.732	360.227	279.028	-222.160	41.024	-0.357

PREVIOUS: September 1978 (1 atm) CURRENT: September 1978 (1 bar)

Molybdenum Iodide (MoI) $I_1Mo_1(g)$

NITROSYL IODIDE (NOI) IDEAL GAS $M_r = 156.9106$

$\Delta_f H°(0\ K) = [115.30 \pm 20.9]\ kJ\ mol^{-1}$

$S°(298.15\ K) = [282.960]\ J\ K^{-1}\ mol^{-1}$ $\Delta_f H°(298.15\ K) = [112.13 \pm 20.9]\ kJ\ mol^{-1}$

Vibrational Frequencies and Degeneracies

$\nu,\ cm^{-1}$

[1780](1)
[440](1)
[200](1)

Ground State Quantum Weight: 1 $\sigma = 1$

Point Group: C_s

Bond Distances: O-N = [1.15] Å N-I = [2.3] Å

Bond Angle: O-N-I = [120]°

Product of the Moments of Inertia: $I_A I_B I_C = [7.908120 \times 10^{-115}]\ g^3\ cm^6$

Enthalpy of Formation

The enthalpy of formation $\Delta_f H°(298.15\ K)$ for INO(g) is estimated by comparison with those for FNO(g), ClNO(g), BrNO(g), and the related cyanogen halides.

Heat Capacity and Entropy

The molecular structure and bond angle are estimated by comparison with those for ClNO(g) and BrNO(g). The N-O bond distance is assumed to be the same as that in NO(g). The N-I bond distance, r(N-I) is calculated based on an assumption that r(ON-Cl)-r(ON-I) = r(I-Cl-r(I-I). The vibrational frequencies are estimated by comparison with those for FNO(g), ClNO(g) and BrNO(g). The three principal moments of inertia are: $I_A = 0.8962 \times 10^{-39}$, $I_B = 29.2605 \times 10^{-39}$ and $I_C = 30.1568 \times 10^{-39}\ g\ cm^2$.

Nitrosyl Iodide (ONI) $I_1N_1O_1(g)$

Enthalpy Reference Temperature = T_r = 298.15 K; Standard State Pressure = p° = 0.1 MPa

T/K	$C_p°$ (J K⁻¹mol⁻¹)	S° (J K⁻¹mol⁻¹)	$-[G°-H°(T_r)]/T$ (J K⁻¹mol⁻¹)	$H°-H°(T_r)$ (kJ mol⁻¹)	$\Delta_f H°$ (kJ mol⁻¹)	$\Delta_f G°$ (kJ mol⁻¹)	Log K_f
0	0.	0.	INFINITE	-12.107	115.300	115.300	INFINITE
100	38.203	236.247	322.541	-8.629	114.349	110.430	-57.683
200	44.111	264.773	287.171	-4.480	113.122	107.010	-27.948
250	45.744	274.803	283.726	-2.231	112.600	105.543	-22.052
298.15	46.856	282.960	282.960	0.	112.131	104.227	-18.260
300	46.893	283.250	282.961	0.087	112.113	104.178	-18.139
350	47.790	290.548	283.535	2.455	111.608	102.894	-15.356
400	48.566	296.981	284.822	4.864	103.127	101.957	-13.314
450	49.279	302.743	286.498	7.310	102.065	101.875	-11.825
500	49.950	307.970	288.388	9.791	80.957	103.810	-10.845
600	51.181	317.188	292.440	14.849	81.066	108.371	-9.435
700	52.248	325.160	296.558	20.022	81.207	112.911	-8.425
800	53.146	332.197	300.581	25.293	81.365	117.429	-7.667
900	53.890	338.501	304.451	30.646	81.532	121.927	-7.076
1000	54.503	344.212	308.146	36.066	81.701	126.407	-6.603
1100	55.006	349.431	311.665	41.543	81.866	130.869	-6.214
1200	55.422	354.236	315.015	47.065	82.023	135.317	-5.890
1300	55.767	358.686	318.205	52.625	82.166	139.752	-5.615
1400	56.055	362.829	321.246	58.216	82.291	144.177	-5.379
1500	56.297	366.705	324.149	63.834	82.390	148.594	-5.174
1600	56.502	370.345	326.924	69.474	82.459	153.005	-4.995
1700	56.677	373.776	329.580	75.133	82.492	157.413	-4.837
1800	56.827	377.020	332.126	80.809	82.484	161.820	-4.696
1900	56.956	380.096	334.571	86.498	82.434	166.229	-4.570
2000	57.069	383.020	336.921	92.199	82.339	170.641	-4.457
2100	57.167	385.807	339.183	97.911	82.202	175.059	-4.354
2200	57.253	388.469	341.363	103.632	82.024	179.486	-4.262
2300	57.329	391.015	343.467	109.362	81.808	183.920	-4.177
2400	57.397	393.457	345.499	115.098	81.561	188.365	-4.100
2500	57.457	395.801	347.465	120.841	81.286	192.820	-4.029
2600	57.511	398.056	349.367	126.589	80.990	197.288	-3.964
2700	57.559	400.227	351.211	132.343	80.678	201.767	-3.903
2800	57.602	402.321	352.999	138.101	80.357	206.258	-3.848
2900	57.641	404.343	354.735	143.863	80.032	210.759	-3.796
3000	57.677	406.298	356.422	149.629	79.708	215.273	-3.748
3100	57.709	408.190	358.061	155.398	79.390	219.797	-3.704
3200	57.738	410.022	359.656	161.171	79.081	224.331	-3.662
3300	57.765	411.799	361.210	166.946	78.785	228.875	-3.623
3400	57.790	413.524	362.723	172.724	78.504	233.427	-3.586
3500	57.812	415.200	364.199	178.504	78.241	237.988	-3.552
3600	57.833	416.829	365.638	184.286	77.997	242.555	-3.519
3700	57.852	418.414	367.043	190.070	77.775	247.129	-3.489
3800	57.870	419.957	368.415	195.856	77.574	251.710	-3.460
3900	57.886	421.460	369.756	201.644	77.395	256.295	-3.433
4000	57.902	422.926	371.067	207.434	77.239	260.884	-3.407
4100	57.916	424.356	372.350	213.225	77.105	265.476	-3.382
4200	57.929	425.751	373.605	219.017	76.994	270.072	-3.359
4300	57.941	427.115	374.833	224.810	76.904	274.671	-3.337
4400	57.953	428.447	376.037	230.605	76.836	279.270	-3.315
4500	57.964	429.749	377.216	236.401	76.788	283.872	-3.295
4600	57.974	431.023	378.372	242.198	76.760	288.474	-3.276
4700	57.983	432.270	379.505	247.996	76.752	293.076	-3.257
4800	57.992	433.491	380.617	253.794	76.761	297.679	-3.239
4900	58.000	434.687	381.709	259.594	76.786	302.282	-3.222
5000	58.008	435.859	382.780	265.394	76.828	306.883	-3.206
5100	58.015	437.008	383.832	271.195	76.884	311.484	-3.190
5200	58.022	438.134	384.866	276.997	76.954	316.083	-3.175
5300	58.029	439.240	385.881	282.800	77.037	320.681	-3.160
5400	58.035	440.324	386.879	288.603	77.130	325.277	-3.146
5500	58.041	441.389	387.861	294.407	77.234	329.871	-3.133
5600	58.047	442.435	388.826	300.211	77.346	334.464	-3.120
5700	58.052	443.463	389.775	306.016	77.466	339.054	-3.107
5800	58.057	444.472	390.710	311.822	77.592	343.642	-3.095
5900	58.062	445.465	391.629	317.628	77.723	348.228	-3.083
6000	58.066	446.441	392.535	323.434	77.858	352.812	-3.072

PREVIOUS: December 1966 (1 atm) CURRENT: December 1966 (1 bar)

Nitrosyl Iodide (ONI) $I_1N_1O_1(g)$

SODIUM IODIDE (NaI) CRYSTAL M_r = 149.89427

$S°(298.15\ K) = 98.500\ J\ K^{-1}\ mol^{-1}$

$T_{fus} = 933\ K$

$\Delta_f H°(0\ K) = -287.34 \pm 0.8\ kJ\ mol^{-1}$

$\Delta_f H°(298.15\ K) = -287.86 \pm 0.8\ kJ\ mol^{-1}$

$\Delta_{fus} H° = 23.598\ kJ\ mol^{-1}$

Enthalpy of Formation

The enthalpy of formation was obtained as follows:

	Reaction	$\Delta_f H°$(298.15 K) kcal mol^{-1}	Ref.
(1)	$HI\cdot100H_2O + NaOH\cdot100H_2O \rightarrow NaI\cdot201H_2O$	-13.51±0.02	(1)
(2)	$Na(cr) + 1001H_2O(\ell) \rightarrow NaOH\cdot1000H_2O + 1/2H_2(g)$	-44.069±0.006	(2)
(3)	$NaOH\cdot1000H_2O \rightarrow NaOH\cdot100H_2O + 900H_2O(\ell)$	+0.031±0.004	(3)
(4)	$1/2H_2(g) + 1/2I_2(s) \rightarrow HI(g)$	+6.30±0.05	(4)
(5)	$HI(g) + 100H_2O(\ell) \rightarrow HI\cdot100H_2O$	-19.30±0.05	(5)
(6)	$NaI\cdot201H_2O(\ell) \rightarrow NaI(cr) + 201\ H_2O(\ell)$	+1.75±0.06	(6, 7, 8)
	(1) + (2) + (3) + (4) + (5) + (6) gives $Na(cr) + 1/2I_2(s) \rightarrow NaI(cr)$	-68.80±0.2	

Heat Capacity and Entropy

Smoothed heat capacities between 3 and 270 K have been presented by Berg and Morrison (9). Above 270 K the heat capacity was obtained by comparing with KI(cr) on a reduced temperature plot of C_p/T versus T. The values so obtained were increased by 1.4% to bring them into line with the low temperature data.

Fusion Data

Dworkin and Bredig (10) have measured both T_{fus} and $\Delta_{fus}H$.

References

1. T. W. Richards and A. W. Rowe, J. Amer. Chem. Soc. 44, 684 (1922).
2. S. R. Gunn and L. G. Green, J. Amer. Chem. Soc. 80, 4782 (1958).
3. U. S. Nat. Bur. Stand. Circ. 500, 1952.
4. JANAF Thermochemical Tables.
5. W. A. Roth, Z. Elektrochem. 50, 107 (1944).
6. J. Wust and E. Lange, Z. Physik. Chem. 116, 161 (1925).
7. E. Lange and A. L. Robinson, Chem. Revs. 9, 89 (1931).
8. Askew, Bullock, Smith, Tinkler, Gatty and Wolfenden, J. Chem. Soc., 1368 (1934).
9. W. T. Berg and J. A. Morrison, Proc. Roy. Soc. (London) 242A, 467 (1957).
10. A. Dworkin and M. Bredig, J. Phys. Chem. 64, 269 (1960).

Sodium Iodide (NaI) $I_1Na_1(cr)$

Enthalpy Reference Temperature = T_r = 298.15 K Standard State Pressure = p° = 0.1 MPa

T/K	$C_p°$ (J K⁻¹mol⁻¹)	S° (J K⁻¹mol⁻¹)	$-[G°-H°(T_r)]/T$ (J K⁻¹mol⁻¹)	$H°-H°(T_r)$ (kJ mol⁻¹)	$\Delta_f H°$ (kJ mol⁻¹)	$\Delta_f G°$ (kJ mol⁻¹)	Log K_f
0	0.	0.	INFINITE	-12.523	-287.336	-287.336	INFINITE
100	43.032	45.601	143.089	-9.749	-287.427	-286.556	149.681
200	49.852	78.103	103.207	-5.021	-287.621	-285.606	74.593
298.15	52.225	98.500	98.500	0.	-287.859	-284.572	49.856
300	52.258	98.823	98.501	0.097	-287.865	-284.551	49.545
400	53.806	114.077	100.570	5.403	-299.000	-282.898	36.943
500	55.020	126.213	104.526	10.844	-320.758	-276.772	28.914
600	56.233	136.353	109.008	16.407	-320.088	-268.035	23.335
700	57.363	145.106	113.554	22.087	-319.244	-259.424	19.358
800	58.492	152.840	117.990	27.880	-318.250	-250.945	16.385
900	59.580	159.793	122.255	33.784	-317.128	-242.598	14.080
933.000	59.939	161.945	123.621	35.756	------ CRYSTAL <--> LIQUID -----		
1000	60.675	166.127	126.330	39.796	-315.904	-234.381	12.243
1100	61.764	171.961	130.217	45.918	-314.598	-226.292	10.746
1200	62.844	177.381	133.924	52.149	-409.994	-215.884	9.397
1300	63.914	182.454	137.464	58.487	-407.663	-199.802	8.028
1400	64.974	187.229	140.850	64.931	-405.241	-183.903	6.861
1500	66.024	191.748	144.093	71.481	-402.733	-168.180	5.857
1600	67.064	196.042	147.207	78.136	-400.146	-152.627	4.983
1700	68.095	200.139	150.201	84.894	-397.484	-137.238	4.217
1800	69.116	204.060	153.085	91.754	-394.753	-122.008	3.541
1900	70.127	207.824	155.868	98.717	-391.956	-106.931	2.940
2000	71.128	211.446	158.557	105.779	-389.093	-92.004	2.403

PREVIOUS: CURRENT: September 1963

Sodium Iodide (NaI) $I_1Na_1(cr)$

SODIUM IODIDE (NaI) LIQUID M_r = 149.89427

$S°(298.15\ K) = 120.754\ J\ K^{-1}\ mol^{-1}$ $\Delta_f H°(298.15\ K) = -266.508\ kJ\ mol^{-1}$

$T_{fus} = 933\ K$ $\Delta_{fus}H° = 23.598\ kJ\ mol^{-1}$

Enthalpy of Formation

$\Delta_f H°$(NaI,ℓ, 298.15 K) is calculated from that of the crystal by adding $\Delta_{fus}H°$ and the difference in enthalpy, H°(933 K)-H°(298.15 K), between the crystal and liquid.

Heat Capacity and Entropy

$C°_p$ is estimated assuming each atom contributed 7.75 cal K^{-1} mol^{-1}. S°(NaI, ℓ, 298.15 K) is calculated in a manner similar to that used for the enthalpy of formation.

Fusion Data

See the crystal table for details.

Vaporization Data

The boiling point, T_{vap} = 1577 K (ℓ + equilibrium mixture, 1 atm), is that given by Kelley (1) from examination of the high temperature vapor pressure.

Reference

1. K. K. Kelley, U. S. Bur. of Mines Bull. 383, 1935.

Sodium Iodide (NaI) $I_1Na_1(l)$

Enthalpy Reference Temperature = T_r = 298.15 K Standard State Pressure = p° = 0.1 MPa

T/K	$C°_p$ (J K⁻¹mol⁻¹)	S° (J K⁻¹mol⁻¹)	−[G°−H°(T_r)]/T (J K⁻¹mol⁻¹)	H°−H°(T_r) (kJ mol⁻¹)	$\Delta_f H°$ (kJ mol⁻¹)	$\Delta_f G°$ (kJ mol⁻¹)	Log K_f
0							
100							
200							
298.15	52.225	120.754	120.754	0.	-266.508	-269.856	47.278
300	52.258	121.077	120.755	0.097	-266.514	-269.876	46.990
400	53.806	136.331	122.824	5.403	-277.649	-270.448	35.317
500	55.020	148.468	126.780	10.844	-299.407	-266.548	27.846
600	56.233	158.606	131.262	16.407	-298.737	-260.036	22.638
600.000	56.233	158.606	131.262	16.407	GLASS <--> LIQUID TRANSITION		
600.000	64.852	158.606	131.262	16.407			
700	64.852	168.603	135.901	22.892	-297.088	-253.716	18.933
800	64.852	177.263	140.542	29.377	-295.401	-247.635	16.169
900	64.852	184.902	145.055	35.862	-293.699	-241.766	14.032
933.000	64.852	187.237	146.506	38.002	CRYSTAL <--> LIQUID		
1000	64.852	191.734	149.387	42.347	-292.002	-236.087	12.332
1100	64.852	197.915	153.522	48.833	-290.333	-230.576	10.949
1200	64.852	203.558	157.460	55.318	-385.474	-222.776	9.697
1300	64.852	208.749	161.208	61.803	-382.995	-209.318	8.411
1400	64.852	213.555	164.778	68.288	-380.532	-196.051	7.315
1500	64.852	218.030	168.181	74.773	-378.089	-182.959	6.371
1600	64.852	222.215	171.428	81.259	-375.672	-170.030	5.551
1700	64.852	226.147	174.533	87.744	-373.283	-157.250	4.832
1800	64.852	229.854	177.504	94.229	-370.927	-144.611	4.196
1900	64.852	233.360	180.352	100.714	-368.607	-132.101	3.632
2000	64.852	236.686	183.087	107.199	-366.322	-119.713	3.127
2100	64.852	239.851	185.715	113.685	-364.071	-107.438	2.672
2200	64.852	242.867	188.245	120.170	-361.854	-95.269	2.262
2300	64.852	245.750	190.683	126.655	-359.666	-83.201	1.890
2400	64.852	248.510	193.035	133.140	-357.503	-71.227	1.550
2500	64.852	251.158	195.308	139.625	-355.360	-59.343	1.240
2600	64.852	253.701	197.505	146.111	-353.233	-47.544	0.955
2700	64.852	256.149	199.632	152.596	-351.116	-35.827	0.693
2800	64.852	258.507	201.693	159.081	-349.005	-24.188	0.451
2900	64.852	260.783	203.691	165.566	-346.897	-12.625	0.227
3000	64.852	262.982	205.631	172.051	-344.787	-1.134	0.020

PREVIOUS: CURRENT: September 1963

Sodium Iodide (NaI) $I_1Na_1(l)$

SODIUM IODIDE (NaI) CRYSTAL-LIQUID M_r = 149.89427 Sodium Iodide (NaI) I_1Na_1(cr,l)

0 to 933 K crystal
above 933 K liquid

Refer to the individual tables for details.

Enthalpy Reference Temperature = T_r = 298.15 K   Standard State Pressure = p° = 0.1 MPa

T/K	C_p° (J K⁻¹mol⁻¹)	S° (J K⁻¹mol⁻¹)	−[G°−H°(T_r)]/T (J K⁻¹mol⁻¹)	H°−H°(T_r) (kJ mol⁻¹)	Δ_fH° (kJ mol⁻¹)	Δ_fG° (kJ mol⁻¹)	Log K_f
0	0.	0.	INFINITE	-12.523	-287.336	-287.336	INFINITE
100	43.032	45.601	143.089	-9.749	-287.427	-286.556	149.681
200	49.852	78.103	103.207	-5.021	-287.621	-285.606	74.593
298.15	52.225	98.500	98.500	0.	-287.859	-284.572	49.856
300	52.258	98.823	98.501	0.097	-287.865	-284.551	49.545
400	53.806	114.077	100.570	5.403	-299.000	-282.898	36.943
500	55.020	126.213	104.526	10.844	-320.758	-276.772	28.914
600	56.233	136.353	109.008	16.407	-320.088	-268.035	23.335
700	57.363	145.106	113.554	22.087	-319.244	-259.424	19.358
800	58.492	152.840	117.990	27.880	-318.250	-250.945	16.385
900	59.580	159.793	122.255	33.784	-317.128	-242.598	14.080
933.000	59.939	161.945	123.621	35.756	CRYSTAL <--> LIQUID		
933.000	64.852	187.237	123.621	59.354	TRANSITION		
1000	64.852	191.734	128.036	63.699	-292.002	-236.087	12.332
1100	64.852	197.915	134.112	70.184	-290.333	-230.576	10.949
1200	64.852	203.558	139.667	76.669	-385.474	-222.776	9.697
1300	64.852	208.749	144.784	83.154	-382.995	-209.318	8.411
1400	64.852	213.555	149.527	89.639	-380.532	-196.051	7.315
1500	64.852	218.030	153.946	96.125	-378.089	-182.959	6.371
1600	64.852	222.215	158.084	102.610	-375.672	-170.030	5.551
1700	64.852	226.147	161.973	109.095	-373.283	-157.250	4.832
1800	64.852	229.854	165.642	115.580	-370.927	-144.611	4.196
1900	64.852	233.360	169.115	122.065	-368.607	-132.101	3.632
2000	64.852	236.686	172.411	128.551	-366.322	-119.713	3.127
2100	64.852	239.850	175.548	135.036	-364.071	-107.438	2.672
2200	64.852	242.867	178.540	141.521	-361.854	-95.269	2.262
2300	64.852	245.750	181.400	148.006	-359.666	-83.201	1.890
2400	64.852	248.510	184.139	154.491	-357.503	-71.227	1.550
2500	64.852	251.158	186.767	160.977	-355.360	-59.343	1.240
2600	64.852	253.701	189.293	167.462	-353.233	-47.544	0.955
2700	64.852	256.149	191.724	173.947	-351.116	-35.827	0.693
2800	64.852	258.507	194.067	180.432	-349.005	-24.188	0.451
2900	64.852	260.783	196.329	186.917	-346.897	-12.625	0.227
3000	64.852	262.982	198.514	193.403	-344.787	-1.134	0.020

PREVIOUS: CURRENT: September 1963

Sodium Iodide (NaI) I_1Na_1(cr,l)

LEAD IODIDE (PbI) IDEAL GAS M_r = 334.1045

$S°(298.15\ K) = 280.094 \pm 0.21\ J\ K^{-1}\ mol^{-1}$

$\Delta_f H°(0\ K) = 110.88 \pm 37.7\ kJ\ mol^{-1}$
$\Delta_f H°(298.15\ K) = 107.74 \pm 37.7\ kJ\ mol^{-1}$

Electronic Levels and Quantum Weights

State	ϵ_i, cm^{-1}	g_i
X $^2\Pi_{1/2}$	0	2
$^2\Pi_{3/2}$	[8300]	2
A $[^2\Delta]$	20528.5	[4]
B $[^2\Sigma]$	33488	[2]

$\omega_e = 160.5\ cm^{-1}$ $\omega_e x_e = 0.25\ cm^{-1}$ $\sigma = 1$
$B_e = [0.02861]\ cm^{-1}$ $\alpha_e = [0.00006]\ cm^{-1}$ $r_e = [2.736]$ Å

Enthalpy of Formation

$\Delta_f H°$(PbI, g, 0 K) = 26.50±9 kcal mol^{-1} is calculated from D_0^0 = 46±9 kcal mol^{-1} with $\Delta_f H°$(Pb, g) = 46.91±0.13 kcal mol^{-1} (1) and $\Delta_f H°$(I, g, 0 K) = 25.633 kcal mol^{-1} (1). The dissociation energy is uncertain. Both Rosen (2) and Gaydon (3) referred to the most probable value of D_0^0 = 1.98 eV determined by Wieland and Newburgh (4) from emission and absorption spectra. Interpretation is difficult and Wieland and Newburgh (4, 5) have given possible values of 2.84, 1.98, and 1.36 eV with 1.98 eV given as the most probable value. The linear Birge-Sponer extrapolation of the adopted ground state vibrational constants, derived from emission and absorption spectra (v, 0-36) (5), gives a dissociation energy of 3.18 eV. Correction for the ionic character of PbI as suggested by Hildenbrand (6) gives D_0^0 = 2.7 eV (62.3 kcal mol^{-1}). This is in better agreement with the original D_0^0 = 2.84 eV given by Wieland and Newburgh (5) than it is with D_0^0 = 1.98 eV recommended later (4). An intercomparison of the dissociaiton energies of lead (1), mercury (2), and potassium (1) halides suggests a D_0^0 value near 45 kcal mol . The rounded value D_0^0 = 2.0±0.4 eV (46±9 kcal mol^{-1}) listed by Gaydon (3) is adopted. $\Delta_f H°$(PbI, g, 298.15 K) = 25.75±9 kcal mol^{-1} is calculated from the selected $\Delta_f H°$(PbI, g, 0 K) = 26.5±9 kcal mol^{-1}.

Heat Capacity and Entropy

The vibrational constants are taken from the compilation of Rosen (2). The electronic states and levels are from Rosen (2) except that the $^2\Pi_{3/2}$ state at 8300 cm^{-1} is added in analogy to PbF(1) and PbCl (1). The value of r_e = 2.736 Å is estimated from a comparison of the sums of the ionic radii of the lead halides with the adopted r_e values for PbF, PbCl, and PbBr (1). B_e is calculated from the adopted r_e. The value for α_e is estimated assuming a Morse potential function.

References

1. JANAF Thermochemical Tables: Pb(g), 3-31-62; I(g), 12-31-61; PbF(g), 12-31-73; PbCl(g), 6-30-73; PbBr(g), 12-31-73; KF(g), 6-30-69; KCl(g), 3-31-66; KBr(g), 3-31-67; KI(g), 6-30-67.
2. B. Rosen, Ed., "Spectroscopic Data Relative to Diatomic Molecules," Pergamon Press, New York, 1970.
3. A. G. Gaydon, "Dissociation Energies and Spectra of Diatomic Molecules," 3rd ed., Chapman and Hall, London, 1968.
4. K. Wieland and R. Newburgh, Helv. Phys. Acta 25, 87 (1952).
5. K. Wieland and R. Newburgh, Helv. Phys. Acta 22, 590 (1949).
6. D. L. Hildenbrand in "Advances in High Temperature Chemistry," Vol. 1, L. Eyring, Ed., Academic Press, New York, pp. 193-215, 1967.

Lead Iodide (PbI) $I_1Pb_1(g)$

Enthalpy Reference Temperature = T_r = 298.15 K Standard State Pressure = p° = 0.1 MPa

T/K	$C_p°$ (J K^{-1}mol^{-1})	S° (J K^{-1}mol^{-1})	$-[G°-H°(T_r)]/T$ (J K^{-1}mol^{-1})	$H°-H°(T_r)$ (kJ mol^{-1})	$\Delta_f H°$ (kJ mol^{-1})	$\Delta_f G°$ (kJ mol^{-1})	Log K_f
0	0.	0.	INFINITE	-10.339	110.877	110.877	INFINITE
100	34.581	240.551	312.703	-7.215	110.709	93.387	-48.780
200	36.646	265.350	283.479	-3.626	109.308	76.588	-20.003
250	36.968	273.565	280.704	-1.785	108.530	68.497	-14.312
298.15	37.155	280.094	280.094	0.	107.738	60.858	-10.662
300	37.160	280.323	280.094	0.069	107.707	60.567	-10.546
350	37.288	286.062	280.547	1.930	106.809	52.780	-7.877
400	37.380	291.047	281.555	3.797	97.887	45.399	-5.929
450	37.451	295.454	282.859	5.668	96.345	38.930	-4.519
500	37.508	299.403	284.320	7.542	74.720	34.534	-3.608
600	37.601	306.250	287.421	11.297	73.710	26.591	-2.315
700	37.676	312.052	290.536	15.061	67.768	19.620	-1.464
800	37.742	317.087	293.547	18.832	66.632	12.820	-0.837
900	37.803	321.536	296.415	22.610	65.529	6.160	-0.358
1000	37.865	325.522	299.130	26.393	64.457	-0.379	0.020
1100	37.931	329.134	301.696	30.183	63.414	-6.812	0.323
1200	38.004	332.438	304.122	33.979	62.401	-13.151	0.572
1300	38.088	335.483	306.419	37.784	61.407	-19.407	0.780
1400	38.187	338.309	308.597	41.597	60.414	-25.586	0.955
1500	38.301	340.948	310.666	45.422	59.410	-31.694	1.104
1600	38.433	343.424	312.637	49.258	58.386	-37.734	1.232
1700	38.582	345.758	314.518	53.109	57.333	-43.709	1.343
1800	38.747	347.968	316.315	56.975	56.247	-49.622	1.440
1900	38.926	350.068	318.037	60.859	55.126	-55.473	1.525
2000	39.119	352.069	319.689	64.761	53.968	-61.264	1.600
2100	39.322	353.983	321.276	68.683	-124.446	-59.882	1.489
2200	39.533	355.817	322.805	72.625	-125.282	-56.788	1.348
2300	39.750	357.579	324.279	76.590	-126.211	-53.654	1.219
2400	39.972	359.275	325.702	80.576	-127.232	-50.478	1.099
2500	40.196	360.911	327.078	84.584	-128.338	-47.258	0.987
2600	40.421	362.492	328.410	88.615	-129.524	-43.991	0.884
2700	40.645	364.022	329.700	92.668	-130.780	-40.678	0.787
2800	40.867	365.504	330.953	96.744	-132.099	-37.317	0.696
2900	41.086	366.942	332.169	100.842	-133.470	-33.908	0.611
3000	41.303	368.339	333.352	104.961	-134.882	-30.450	0.530
3100	41.515	369.696	334.502	109.102	-136.327	-26.946	0.454
3200	41.724	371.018	335.623	113.264	-137.795	-23.394	0.382
3300	41.928	372.305	336.715	117.447	-139.275	-19.796	0.313
3400	42.128	373.560	337.780	121.649	-140.759	-16.153	0.248
3500	42.323	374.784	338.820	125.872	-142.238	-12.467	0.186
3600	42.514	375.978	339.836	130.114	-143.704	-8.738	0.127
3700	42.702	377.146	340.828	134.375	-145.151	-4.969	0.070
3800	42.885	378.287	341.799	138.654	-146.572	-1.161	0.016
3900	43.064	379.403	342.749	142.951	-147.962	2.684	-0.036
4000	43.240	380.496	343.679	147.267	-149.314	6.564	-0.086
4100	43.412	381.566	344.590	151.599	-150.626	10.477	-0.133
4200	43.580	382.614	345.483	155.949	-151.892	14.422	-0.179
4300	43.746	383.641	346.359	160.315	-153.111	18.396	-0.223
4400	43.908	384.649	347.217	164.698	-154.280	22.398	-0.266
4500	44.068	385.637	348.060	169.097	-155.397	26.426	-0.307
4600	44.224	386.608	348.888	173.511	-156.460	30.478	-0.346
4700	44.378	387.560	349.701	177.942	-157.468	34.553	-0.384
4800	44.529	388.496	350.499	182.387	-158.421	38.649	-0.421
4900	44.678	389.416	351.284	186.847	-159.318	42.764	-0.456
5000	44.824	390.320	352.056	191.322	-160.160	46.897	-0.490
5100	44.967	391.209	352.815	195.812	-160.947	51.046	-0.523
5200	45.108	392.084	353.561	200.316	-161.679	55.210	-0.555
5300	45.247	392.944	354.296	204.834	-162.357	59.387	-0.585
5400	45.383	393.791	355.020	209.365	-162.982	63.577	-0.615
5500	45.516	394.625	355.732	213.910	-163.556	67.778	-0.644
5600	45.647	395.447	356.434	218.468	-164.080	71.989	-0.671
5700	45.776	396.256	357.126	223.039	-164.555	76.209	-0.698
5800	45.902	397.053	357.807	227.623	-164.983	80.436	-0.724
5900	46 025	397.839	358.479	232.220	-165.364	84.671	-0.750
6000	46.147	398.613	359.142	236.828	-165.702	88.912	-0.774

PREVIOUS: December 1973 (1 atm) CURRENT: December 1973 (1 bar)

Lead Iodide (PbI) $I_1Pb_1(g)$

IODOSILYLIDYNE (SiI) IDEAL GAS M_r = 154.9900

$\Delta_f H°(0\ K) = 313.51 \pm 41.8\ kJ\ mol^{-1}$

$S°(298.15\ K) = 253.854\ J\ K^{-1}\ mol^{-1}$ $\Delta_f H°(298.15\ K) = 313.63 \pm 41.8\ kJ\ mol^{-1}$

Electronic Levels and Molecular Constants (σ = 1)

State	ε_i, cm^{-1}	g_i	ω_e, cm^{-1}	$\omega_e x_e$, cm^{-1}	B_e, cm^{-1}	α_e, cm^{-1}	r_e, Å	Source
$X\ ^2\pi_{1/2}$	0	2	363.2	1.25	0.123	[0.001]	2.45	(1)
$^2\pi_{3/2}$	650	2	358.5	1.1	0.123	[0.001]	2.45	(1, 6)
$A'\ ^4\Sigma_{1/2}$	20939.1	4	275.3	5.6	0.118	[0.001]	2.50	(1)
$A\ ^2\Sigma^+$	21204.9	2	208.3	1.66	0.085	[0.001]	2.95	(1)
$B\ ^2\Sigma^+$	32380.3	2	471.0	0.9	[0.123]	[0.001]	[2.45]	(1)
$C\ ^2\pi$	42710	4	485	3.5	[0.123]	[0.001]	[2.45]	(6)

Enthalpy of Formation

The adopted value is $\Delta_f H°(298.15\ K) = 75.0 \pm 10\ kcal\ mol^{-1}$. The value is based on a Birge-Sponer extrapolation of the $A\ ^2\Sigma$ state by Billingsley (1) who obtains a dissociation limit of 69.76 $kcal\ mol^{-1}$. Correction for the ionic character of the bond according to Hildenbrand (2) yields $D_0^0 = 57.3\ kcal\ mol^{-1}$ which, with auxiliary JANAF data (3), gives the adopted value. An uncertainty of ±10 $kcal\ mol^{-1}$ is assigned to $\Delta_f H°(298.15\ K)$ due to the approximate nature of Birge-Sponer extrapolation based on data from the first few vibrational levels only. The adopted value of $D_0^0 = 57.3 \pm 8\ kcal\ mol^{-1}$ is in good agreement with D_0^0 = 56.9 and 61.8 $kcal\ mol^{-1}$ obtained in a similar manner from the $A'\ ^4\Sigma$ state and ground state data, respectively. The value is also in reasonable agreement with D_0^0 = 72.4 and 65±10 $kcal\ mol^{-1}$ obtained by theoretical and empirical procedures, respectively (4, 5). The upper limit for D_0^0 = 4.02 ev (92.7 $kcal\ mol^{-1}$) is based on the observed predissociation of the $B\ ^2\Sigma$ state (1). The adopted value of D_0^0 is also in good agreement with the average (per bond) heat of atomization of $SiI_4(g)$ of 58.6 $kcal\ mol^{-1}$ (3).

Heat Capacity and Entropy

Electronic levels and molecular constants are taken from the studies by Oldershaw and Robinson (6) and Billingsley (1). Several systems above 40,000 cm^{-1} have been observed by Oldershaw and Robinson (7) but are not included since their states and degeneracies are not known. They would make only a minor contribution to the thermodynamic properties at 6000 K and a negligible contribution at lower temperatures. Of more importance is the uncertainty in the value of the ground state spin splitting constant and the energy of the $A'\ ^4\Sigma^-$ state. Billingsley (1) interprets the observed A'-X emission spectrum as terminating at the upper level of the split ground state and calculates the ground state splitting of 646.4 cm^{-1} from the observation of one absorption line. Lakshminarayana and Haranath (8) observe a second emission series 650.3 cm^{-1} to the red of that observed by Billingsley, and although the ground state splittings are in agreement, this would place the A' state approximately 650 cm^{-1} lower than reported. We favor Billingsley's interpretation while recognizing the need for additional work to resolve this paradox. In addition the $B'\ ^2\Delta$ state observed for the other silicon monohalides (3) is not observed for SiI. Its energy is difficult to predict due to the rapidly dropping energy of this state across the series and the possible misidentification of this state for SiBr. Accordingly, we have assigned an uncertainty of ±0.1 $cal\ K^{-1}\ mol^{-1}$ to $S°(298.15\ K)$ to account for this uncertainty. Since no information is available on the rotational constants for the B and C states, the ground state B_e = .123 cm^{-1} was used in our calculations. Likewise, values of α_e = .001 cm^{-1} are estimated for all states. The ground state is treated as two distinct levels due to the splitting of this state as expressed by the spin coupling constant (A = 650 cm^{-1}). This leads to slightly biased results at low temperature; the stated uncertainty in $S°(298.15\ K)$ should account for this also. All molecular constants have been corrected to reflect the natural abundance of Si.

References

1. J. Billingsley, J. Mol. Spectrosc. 43, 128 (1972).
2. D. L. Hildenbrand, "Advances in High Temperature Chemistry," Vol. I, L. Eyring, ed., Academic Press, NY, 1967.
3. JANAF Thermochemical Tables: Si(g), 3-31-67; I(g), 6-30-74; $SiI_4(g)$, 6-30-76; SiF(g), SiCl(g), SiBr(g), 12-31-76.
4. S. P. Tandon and K. Tandon, J. Phys. 38, 460 (1964).
5. O. G. Polyachenok, Russ. J. Phys. Chem. 40, 1420 (1966).
6. G. A. Oldershaw and K. Robinson, Trans. Faraday Soc. 64, 2256 (1968).
7. G. A. Oldershaw and K. Robinson, J. Mol. Spectrosc. 44, 602 (1972).
8. A. Lakshminarayana and P. B. V. Haranath, Proc. Phys. Soc., London (At. Mol. Phys.) 3, 576 (1970).

Iodosilylidyne (SiI) $I_1Si_1(g)$

Enthalpy Reference Temperature = T_r = 298.15 K Standard State Pressure = p° = 0.1 MPa

T/K	$C_p°$ (J K⁻¹mol⁻¹)	S° (J K⁻¹mol⁻¹)	−[G°−H°(T_r)]/T (J K⁻¹mol⁻¹)	H°−H°(T_r) (kJ mol⁻¹)	$\Delta_f H°$ (kJ mol⁻¹)	$\Delta_f G°$ (kJ mol⁻¹)	Log K_f
0	0.	0.	INFINITE	-9.936	313.510	313.510	INFINITE
100	30.446	216.152	286.167	-7.001	314.654	296.471	-154.861
200	35.736	238.883	257.338	-3.691	314.319	278.371	-72.703
250	37.748	247.087	254.491	-1.851	313.995	269.420	-56.292
298.15	39.038	253.854	253.854	0.	313.629	260.868	-45.703
300	39.076	254.096	253.855	0.072	313.614	260.540	-45.364
350	39.850	260.184	254.334	2.047	313.148	251.730	-37.569
400	40.238	265.534	255.407	4.051	304.652	243.264	-31.767
450	40.381	270.283	256.801	6.067	303.527	235.658	-27.354
500	40.380	274.539	258.366	8.087	282.312	230.077	-24.036
600	40.179	281.886	261.692	12.116	282.088	219.650	-19.122
700	39.905	288.059	265.030	16.120	281.761	209.268	-15.616
800	39.652	293.370	268.248	20.098	281.339	198.940	-12.989
900	39.442	298.028	271.304	24.052	280.837	188.669	-10.950
1000	39.276	302.175	274.187	27.987	280.260	178.458	-9.322
1100	39.151	305.912	276.904	31.909	279.618	168.309	-7.992
1200	39.057	309.314	279.465	35.819	278.913	158.220	-6.887
1300	38.989	312.438	281.883	39.721	278.147	148.193	-5.954
1400	38.943	315.325	284.170	43.617	277.317	138.228	-5.157
1500	38.913	318.011	286.338	47.510	276.422	128.324	-4.469
1600	38.897	320.522	288.397	51.400	275.458	118.481	-3.868
1700	38.891	322.880	290.356	55.290	224.244	109.149	-3.354
1800	38.895	325.103	292.226	59.179	223.363	102.403	-2.972
1900	38.906	327.206	294.012	63.069	222.449	95.708	-2.631
2000	38.924	329.202	295.722	66.960	221.500	89.062	-2.326
2100	38.948	331.102	297.362	70.854	220.520	82.464	-2.051
2200	38.976	332.914	298.937	74.750	219.512	75.914	-1.802
2300	39.010	334.647	300.452	78.649	218.479	69.409	-1.576
2400	39.049	336.308	301.912	82.552	217.427	62.951	-1.370
2500	39.093	337.903	303.320	86.459	216.362	56.536	-1.181
2600	39.143	339.438	304.679	90.371	215.292	50.164	-1.008
2700	39.199	340.916	305.994	94.288	214.221	43.833	-0.848
2800	39.262	342.343	307.267	98.211	213.156	37.542	-0.700
2900	39.333	343.722	308.501	102.141	212.104	31.289	-0.564
3000	39.412	345.056	309.697	106.078	211.070	25.072	-0.437
3100	39.500	346.350	310.859	110.024	210.059	18.888	-0.318
3200	39.598	347.606	311.987	113.978	209.076	12.737	-0.208
3300	39.708	348.826	313.085	117.944	208.125	6.617	-0.105
3400	39.829	350.013	314.154	121.920	207.210	0.524	-0.008
3500	39.963	351.169	315.195	125.910	206.333	-5.542	0.083
3600	40.110	352.297	316.210	129.913	-178.653	-1.123	0.016
3700	40.271	353.398	317.200	133.932	-179.030	3.814	-0.054
3800	40.448	354.475	318.167	137.968	-179.362	8.760	-0.120
3900	40.640	355.528	319.112	142.022	-179.648	13.714	-0.184
4000	40.848	356.559	320.035	146.097	-179.885	18.675	-0.244
4100	41.072	357.571	320.938	150.193	-180.071	23.642	-0.301
4200	41.314	358.563	321.822	154.312	-180.206	28.612	-0.356
4300	41.572	359.538	322.688	158.456	-180.287	33.585	-0.408
4400	41.848	360.497	323.537	162.627	-180.313	38.559	-0.458
4500	42.141	361.441	324.368	166.826	-180.283	43.533	-0.505
4600	42.452	362.370	325.185	171.056	-180.196	48.506	-0.551
4700	42.781	363.287	325.985	175.317	-180.051	53.477	-0.594
4800	43.127	364.191	326.772	179.612	-179.847	58.443	-0.636
4900	43.491	365.084	327.545	183.943	-179.583	63.405	-0.676
5000	43.872	365.967	328.304	188.311	-179.258	68.361	-0.714
5100	44.269	366.839	329.051	192.718	-178.871	73.310	-0.751
5200	44.683	367.703	329.786	197.166	-178.422	78.250	-0.786
5300	45.113	368.558	330.510	201.655	-177.910	83.181	-0.820
5400	45.559	369.406	331.222	206.189	-177.334	88.102	-0 852
5500	46.020	370.246	331.924	210.768	-176.693	93.012	-0.883
5600	46.496	371.079	332.616	215.393	-175.988	97.909	-0.913
5700	46.985	371.906	333.298	220.067	-175.217	102.793	-0.942
5800	47.488	372.728	333.971	224.791	-174.380	107.664	-0.970
5900	48.003	373.544	334.635	229.565	-173.478	112.519	-0.996
6000	48.531	374.355	335.290	234.392	-172.508	117.358	-1.022

PREVIOUS: December 1976 (1 atm) CURRENT: December 1976 (1 bar)

Iodosilylidyne (SiI) $I_1Si_1(g)$

STRONTIUM IODIDE (SrI) IDEAL GAS M_r = 214.5245

$\Delta_f H°(0\ K) = -27.61 \pm 83.7\ kJ\ mol^{-1}$

$S°(298.15\ K) = 272.213 \pm 0.21\ J\ K^{-1}\ mol^{-1}$ $\Delta_f H°(298.15\ K) = -30.49 \pm 83.7\ kJ\ mol^{-1}$

Electronic Levels and Quantum Weights

State	ε_i, cm^{-1}	g_i
$X\ ^2\Sigma^+$	0	2
$A\ ^2\pi$	14422.7	2
	14815.9	
$B\ ^2\Sigma$	14748.8	2
$C\ ^2\pi$	22666.1	2
	23223.4	2
$D\ ^2\Sigma$	28928.3	2

$\omega_e = 173.9\ cm^{-1}$ $\omega_e x_e = 0.42\ cm^{-1}$ $\sigma = 1$

$B_e = [0.0354]\ cm^{-1}$ $\alpha_e = [0.00011]\ cm^{-1}$ $r_e = [3.03]$ Å

Enthalpy of Formation

The adopted $\Delta_f H°(0\ K) = -6.65 \pm 20\ kcal\ mol^{-1}$ is obtained from $D_0^0 = 71.5\ kcal\ mol^{-1}$ calculated from the ratio $D_0^0(SrI)/D_0^0(SrI_2) = 0.46$. This value for the ratio was found for a series of mono and difluorides (1) and for other alkaline earth halide systems (2). Other values for D_0^0 in $kcal\ mol^{-1}$ are 54.6 obtained from a linear Birge-Sponer extrapolation of ω_e and $\omega_e x_e$ data (3) with a correction for the ionic character of the molecule as described by Hildenbrand (4), 67.3 obtained by Krasnov and Karaseva (5) as a lower bound from a consideration of ionic bonding forces, and 65 obtained by Mims et al. (6) as a lower bound from a consideration of energy conservation and measured reaction threshold relative kinetic energy from crossed molecular beam experiments. $\Delta_f H°(298.15\ K) = -7.29 \pm 20\ kcal\ mol^{-1}$ from the adopted $\Delta_f H°(0\ K) = -6.65\ kcal\ mol^{-1}$.

Heat Capacity and Entropy

The ground state vibrational constants are from the compilation of Rosen (3). The value of r_e = 3.03 Å is assumed the same as the bond distance in SrI_2 (7). B_e is calculated from the adopted r_e. α_e is calculated assuming a Morse potential function.

The electronic levels and their designations are from Ashrafunnisa et al. (8, 9) and Reddy et al. (10). The data are similar to the values in Rosen (2) but are more complete. The $D\ ^2\Sigma$ level is confirmed by Shah et al. (11).

References

1. G. D. Blue, J. W. Green, T. C. Ehlert and J. L. Margrave, Nature 199, 804 (1963).
2. JANAF Thermochemical Tables: all alkaline earth mono- and di-halides.
3. B. Rosen, Ed., "Spectroscopic Data Relative to Diatomic Molecules," Pergamon Press, New York, 1970.
4. D. L. Hildenbrand in "Advances in High Temperature Chemistry," Vol. 1, L. Eyring, Ed., Academic Press, New York, pp. 193-215, 1967.
5. K. S. Krasnov and N. V. Karaseva, Optics and Spectrosc. 19, 14 (1965).
6. C. A. Mims, S. Lin, and R. R. Herm, J. Chem. Phys. 57, 3099 (1972).
7. P. A. Akishin, V. P. Spiridonov, G. A. Sobolev, and V. A. Naumov, Zhur. Fiz. Khim. 31, 1871 (1957).
8. Ashrafunnisa, D. V. K. Rao and P. T. Rao, J. Phys. B 6, 503 (1973).
9. Ashrafunnisa, D. V. K. Rao and P. T. Rao, J. Phys. B 6, 2653 (1973).
10. B. R. K. Reddy, Y. P. Reddy, and P. T. Rao, J. Phys. B 4 574 (1971).
11. S. G. Shah, M. M. Patel and A. B. Darji, J. Phys. B 6, 1344 (1973).

Strontium Iodide (SrI) $I_1Sr_1(g)$

Enthalpy Reference Temperature = T_r = 298.15 K Standard State Pressure = p° = 0.1 MPa

	J K⁻¹mol⁻¹			kJ mol⁻¹			
T/K	$C_p°$	S°	$-[G°-H°(T_r)]/T$	$H°-H°(T_r)$	$\Delta_f H°$	$\Delta_f G°$	Log K_f
0	0.	0.	INFINITE	-10.287	-27.614	-27.614	INFINITE
100	34.228	232.849	304.737	-7.189	-27.540	-44.971	23.490
200	36.539	257.493	275.595	-3.620	-28.926	-61.887	16.163
250	36.915	265.691	272.824	-1.783	-29.705	-70.044	14.635
298.15	37.135	272.213	272.213	0.	-30.494	-77.736	13.619
300	37.141	272.443	272.214	0.069	-30.526	-78.029	13.586
350	37.294	278.181	272.667	1.930	-31 423	-85.877	12.816
400	37.405	283.168	273.675	3.797	-40.345	-93.318	12.186
450	37.492	287.579	274.979	5.670	-41.891	-99.847	11.590
500	37.563	291.533	276.440	7.546	-63.522	-104.304	10.897
600	37.679	298.392	279.544	11.309	-64.567	-112.364	9.782
700	37.774	304.208	282.663	15.081	-65.712	-120.241	8.972
800	37.859	309.257	285.678	18.863	-66.959	-127.946	8.354
900	37.936	313.721	288.551	22.653	-68.939	-135.418	7.859
1000	38.010	317.722	291.272	26.450	-70.076	-142.744	7.456
1100	38.082	321.348	293.843	30.255	-79.115	-149.590	7.103
1200	38.152	324.664	296.276	34.067	-81.166	-155.906	6.786
1300	38.220	327.721	298.579	37.885	-83.221	-162.051	6.511
1400	38.289	330.556	300.763	41.711	-85.286	-168.038	6.270
1500	38.358	333.200	302.838	45.543	-87.363	-173.876	6.055
1600	38.428	335.678	304.814	49.382	-89.459	-179.575	5.863
1700	38.502	338.010	306.699	53.229	-228.491	-183.963	5.653
1800	38.580	340.212	308.500	57.083	-228.773	-181.336	5.262
1900	38.663	342.301	310.224	60.945	-229.088	-178.692	4.913
2000	38.755	344.286	311.878	64.816	-229.437	-176.031	4.597
2100	38.857	346.179	313.467	68.696	-229.822	-173.351	4.312
2200	38.970	347.989	314.995	72.587	-230.244	-170.652	4.052
2300	39.097	349.725	316.468	76.491	-230.704	-167.934	3.814
2400	39.238	351.391	317.888	80.407	-231.201	-165.194	3.595
2500	39.395	352.996	319.261	84.339	-231.737	-162.433	3.394
2600	39.570	354.545	320.588	88.287	-232.312	-159.649	3.207
2700	39.762	356.042	321.874	92.253	-232.927	-156.843	3.034
2800	39.972	357.491	323.120	96.240	-233.585	-154.013	2.873
2900	40.200	358.898	324.330	100.248	-234.288	-151.159	2.723
3000	40.447	360.265	325.505	104.281	-235.039	-148.280	2.582
3100	40.711	361.596	326.648	108.338	-235.843	-145.375	2.450
3200	40.992	362.892	327.760	112.423	-236.705	-142.443	2.325
3300	41.289	364.158	328.844	116.537	-237.629	-139.483	2.208
3400	41.601	365.396	329.901	120.682	-238.620	-136.494	2.097
3500	41.928	366.606	330.932	124.858	-239.683	-133.475	1.992
3600	42.267	367.792	331.940	129.068	-240.824	-130.425	1.892
3700	42.618	368.955	332.925	133.312	-242.045	-127.342	1.798
3800	42.980	370.096	333.888	137.592	-243.350	-124.224	1.708
3900	43.350	371.217	334.831	141.908	-244.744	-121.071	1.622
4000	43.727	372.320	335.754	146.262	-246.226	-117.882	1.539
4100	44.110	373.404	336.659	150.654	-247.800	-114.654	1.461
4200	44.498	374.472	337.547	155.084	-249.461	-111.386	1.385
4300	44.888	375.523	338.418	159.553	-251.215	-108.079	1.313
4400	45.280	376.560	339.273	164.062	-253.057	-104.729	1.243
4500	45.673	377.582	340.113	168.609	-254.985	-101.336	1.176
4600	46.064	378.590	340.938	173.196	-256.931	-97.905	1.112
4700	46.454	379.585	341.750	177.822	-259.001	-94.425	1.049
4800	46.840	380.567	342.549	182.487	-261.141	-90.901	0.989
4900	47.221	381.536	343.334	187.190	-263.345	-87.332	0.931
5000	47.598	382.494	344.108	191.931	-265.601	-83.718	0.875
5100	47.967	383.440	344.870	196.709	-267.890	-80.059	0.820
5200	48.330	384.375	345.621	201.524	-270.242	-76.353	0.767
5300	48.684	385.299	346.361	206.375	-272.630	-72.602	0.716
5400	49.030	386.213	347.090	211.261	-275.048	-68.805	0.666
5500	49.367	387.115	347.810	216.181	-277.489	-64.964	0.617
5600	49.693	388.008	348.520	221.134	-279.946	-61.077	0.570
5700	50.010	388.890	349.220	226.119	-282.413	-57.147	0.524
5800	50.316	389.763	349.912	231.135	-284.882	-53.173	0.479
5900	50.611	390.625	350.595	236.182	-287.138	-49.181	0.435
6000	50.894	391.478	351.269	241.257	-289.565	-45.128	0.393

PREVIOUS: June 1974 (1 atm) CURRENT: June 1974 (1 bar)

Strontium Iodide (SrI) $I_1Sr_1(g)$

TITANIUM IODIDE (TiI) IDEAL GAS M_r = 174.7845

$S°(298.15\ K) = 268.827 \pm 8.4\ J\ K^{-1}\ mol^{-1}$

$\Delta_f H°(0\ K) = [275.34 \pm 41.8]\ kJ\ mol^{-1}$
$\Delta_f H°(298.15\ K) = [274.05 \pm 41.8]\ kJ\ mol^{-1}$

Electronic Levels and Quantum Weights

State	ε_i, cm^{-1}	g_i
[$^4\Sigma$]	0	[4]
	[1000]	[4]
	[2000]	[4]
	[4000]	[4]
	[6000]	[4]
	[8000]	[4]
	[10000]	[4]

$\omega_e = [240]\ cm^{-1}$ $\omega_e x_e = [1.67]\ cm^{-1}$ $\sigma = 1$
$B_e = [0.062]\ cm^{-1}$ $\alpha_e = [0.00043]\ cm^{-1}$ $r_e = [2.8]\ Å$

Enthalpy of Formation

The enthalpy of formation, $\Delta_f H°(298.15\ K)$, of TiI(g) is calculated from the dissocation energy, D_0^0, which is estimated as 73 kcal mol^{-1}. This estimate is obtained from the relation $\Delta_{at}H°(TiI_4,\ 0\ K)/4 < D_0^0(TiI) < \Delta_{at}H°(TiI_3,\ 0\ K)/3)$, which is valid for the titanium flourides. The dissociation energy of TiF(g) from which the relation is derived was estimated relative to that of TiF_2(g) by Zmbov and Margrave (1).

Heat Capacity and Entropy

The vibrational frequency, ω_e, and the anharmonic vibrational term, $\omega_e x_e$, are estimated from those of TiCl(g) and comparisons of the mercury and alkali monohalides. The interatomic distance is estimated from Guggenheimer's relation (2). B_e is calculated from r_e. The ground state term and electronic levels are estimated from the ground state multiplet of Ti^+ (3). α_e is estimated from the Morse potential function.

References

1. K. F. Zmbov and J. L. Margrave, J. Phys. Chem. 71, 2893 (1967).
2. K. M. Guggenheimer, Proc. Phys. Soc. (London) 58, 456 (1946).
3. C. E. Moore, U. S. Nat. Bur. Stand. Circ. 467, 1949.

Enthalpy Reference Temperature = T_r = 298.15 K   Standard State Pressure = p° = 0.1 MPa

	J K^{-1}mol^{-1}			kJ mol^{-1}			
T/K	$C_p°$	S°	$-[G°-H°(T_r)]/T$	$H°-H°(T_r)$	$\Delta_f H°$	$\Delta_f G°$	Log K_f
0	0.	0.	INFINITE	-10.138	275.343	275.343	INFINITE
100	32.578	230.074	301.372	-7.130	276.266	257.134	-134.313
200	36.180	253.948	272.273	-3.665	275.347	238.299	-62.237
250	37.373	262.151	269.455	-1.826	274.709	229.109	-47.870
298.15	38.459	268.827	268.827	0.	274.052	220.386	-38.611
300	38.500	269.065	268.828	0.071	274.026	220.053	-38.315
350	39.542	275.080	269.301	2.023	273.284	211.115	-31.507
400	40.464	280.421	270.363	4.023	264.546	202.559	-26.451
450	41.251	285.234	271.752	6.067	263.215	194.890	-22.622
500	41.902	289.615	273.323	8.146	241.830	189.269	-19.773
600	42.842	297.345	276.700	12.387	241.368	178.800	-15.566
700	43.407	303.995	280.135	16.702	240.911	168.409	-12.567
800	43.728	309.815	283.489	21.060	240.455	158.082	-10.322
900	43.903	314.976	286.706	25.443	239.952	147.815	-8.579
1000	44.002	319.607	289.769	29.838	239.325	137.609	-7.188
1100	44.066	323.804	292.675	34.242	238.506	127.476	-6.053
1200	44.119	327.641	295.431	38.651	233.504	117.545	-5.117
1300	44.175	331.174	298.047	43.066	233 011	107.902	-4.336
1400	44.238	334.450	300.531	47.487	232.429	98.299	-3.668
1500	44.312	337.505	302.896	51.914	231.743	88.742	-3.090
1600	44.395	340.367	305.149	56.349	230.934	79.234	-2.587
1700	44.487	343.062	307.301	60.793	229.988	69.781	-2.144
1800	44.585	345.607	309.359	65.247	228.887	60.388	-1.752
1900	44.687	348.020	311.331	69.710	227.618	51.061	-1.404
2000	44.791	350.315	313.223	74.184	211.480	42.258	-1.104
2100	44.896	352.503	315.042	78.669	209.086	33.856	-0.842
2200	45.001	354.594	316.792	83.163	206.672	25.568	-0.607
2300	45.103	356.597	318.480	87.669	204.241	17.391	-0.395
2400	45.203	358.518	320.108	92.184	201.797	9.319	-0.203
2500	45.299	360.366	321.682	96.709	199.347	1.349	-0.028
2600	45.391	362.144	323.204	101.244	196.895	-6.522	0.131
2700	45.479	363.859	324.678	105.787	194.446	-14.299	0.277
2800	45.562	365.514	326.107	110.339	192.006	-21.986	0.410
2900	45.641	367.115	327.494	114.900	189.581	-29.585	0.533
3000	45.716	368.663	328.841	119.468	187.173	-37.102	0.646
3100	45.787	370.163	330.150	124.043	184.788	-44.538	0.750
3200	45.854	371.618	331.423	128.625	182.428	-51.898	0.847
3300	45.917	373.030	332.662	133.213	180.096	-59.185	0.937
3400	45.977	374.402	333.870	137.808	177.795	-66.401	1.020
3500	46.034	375.735	335.047	142.409	175.525	-73.550	1.098
3600	46.089	377.033	336.195	147.015	173.289	-80.635	1.170
3700	46.141	378.296	337.316	151.626	-238.012	-79.870	1.128
3800	46.191	379.528	338.411	156.243	-238.945	-75.583	1.039
3900	46.240	380.728	339.481	160.864	-239.900	-71.271	0.955
4000	46.287	381.899	340.527	165.491	-240.874	-66.935	0.874
4100	46.333	383.043	341.550	170.122	-241.866	-62.575	0.797
4200	46.378	384.160	342.551	174.757	-242.875	-58.190	0.724
4300	46.422	385.252	343.531	179.397	-243.901	-53.780	0.653
4400	46.466	386.319	344.492	184.042	-244.941	-49.347	0.586
4500	46.510	387.364	345.433	188.691	-245.996	-44.890	0.521
4600	46.554	388.387	346.356	193.344	-247.063	-40.409	0.459
4700	46.597	389.388	347.261	198.001	-248.142	-35.905	0.399
4800	46.641	390.370	348.148	202.663	-249.231	-31.378	0.341
4900	46.686	391.332	349.020	207.330	-250.329	-26.828	0.286
5000	46.731	392.276	349.876	212.000	-251.435	-22.255	0.232
5100	46.776	393.202	350.716	216.676	-252.548	-17.660	0.181
5200	46.823	394.110	351.542	221.356	-253.667	-13.044	0.131
5300	46.869	395.003	352.354	226.040	-254.789	-8.406	0.083
5400	46.917	395.879	353.152	230.730	-255.915	-3.747	0.036
5500	46.966	396.741	353.936	235.424	-257.043	0.933	-0.009
5600	47.015	397.587	354.708	240.123	-258.172	5.634	-0.053
5700	47.066	398.420	355.468	244.827	-259.300	10.355	-0.095
5800	47.117	399.239	356.215	249.536	-260.427	15.096	-0.136
5900	47.170	400.045	356.951	254.250	-261.551	19.856	-0.176
6000	47.223	400.838	357.676	258.970	-262.672	24.635	-0.214

PREVIOUS: December 1968 (1 atm)   CURRENT: December 1968 (1 bar)

ZIRCONIUM IODIDE (ZrI) IDEAL GAS M_r = 218.1245

$S°(298.15\ K) = [275.805]\ J\ K^{-1}\ mol^{-1}$

$\Delta_f H°(0\ K) = [593.20 \pm 41.8]\ kJ\ mol^{-1}$
$\Delta_f H°(298.15\ K) = [591.20 \pm 41.8]\ kJ\ mol^{-1}$

Electronic Level and Quantum Weight

State	ε_i, cm^{-1}	g_i
$^4\Sigma$	0	4

$\omega_e = [223]\ cm^{-1}$ $\omega_e x_e = [1.2]\ cm^{-1}$ $\sigma = 1$
$B_e = [0.03789]\ cm^{-1}$ $\alpha_e = [0.00020]\ cm^{-1}$ $r_e = [2.9]$ Å

Enthalpy of Formation

The dissociation energy was estimated from a correlation of the JANAF values for TiCl(g), TiI(g), and ZrCl(g). The resulting value, 29.7 kcal mol^{-1}, was employed with auxiliary JANAF enthalpies of formation for Zr(g) and I(g) in calculating $\Delta_f H°$(298.15 K).

Heat Capacity and Entropy

The vibrational frequency was estimated from those of TiI(g), TiCl(g), ZrCl(g), and the alkali halides. The ground state configuration was assumed to be analogous to the ground term of TiCl(g) as given by Shenyavskaya et al. (1). The interatomic distance was estiamted from those of TiBr(g), TiI_4(g) and ZrI_4(g).

Reference

1. E. A. Shenyavskaya et al., Optics and Spectroscopy 12, 359 (1962).

Zirconium Iodide (ZrI) $I_1Zr_1(g)$

Enthalpy Reference Temperature = T_r = 298.15 K Standard State Pressure = p° = 0.1 MPa

T/K	$C°_p$ (J K⁻¹mol⁻¹)	S° (J K⁻¹mol⁻¹)	$-[G°-H°(T_r)]/T$ (J K⁻¹mol⁻¹)	$H°-H°(T_r)$ (kJ mol⁻¹)	$\Delta_f H°$ (kJ mol⁻¹)	$\Delta_f G°$ (kJ mol⁻¹)	Log K_f
0	0.	0.	INFINITE	-10.097	593.198	593.198	INFINITE
100	32.964	237.198	307.908	-7.071	593.813	574.549	-300.113
200	36.028	261.223	279.160	-3.587	592.637	555.699	-145.134
250	36.593	269.329	276.411	-1.771	591.925	546.545	-114.194
298.15	36.931	275.805	275.805	0.	591.199	537.871	-94.233
300	36.941	276.033	275.806	0.068	591.171	537.540	-93.594
350	37.177	281.747	276.256	1.922	590.347	528.665	-78.899
400	37.349	286.723	277.260	3.785	581.508	520.185	-67.929
450	37.483	291.130	278.561	5.656	580.058	512.607	-59.502
500	37.592	295.085	280.019	7.533	558.533	507.090	-52.975
600	37.767	301.955	283.120	11.301	557.733	496.875	-43.257
700	37.910	307.788	286.237	15.085	556.867	486.800	-36.325
800	38.034	312.858	289.255	18.882	555.925	476.853	-31.135
900	38.148	317.345	292.132	22.692	554.892	467.031	-27.106
1000	38.256	321.370	294.858	26.512	553.755	457.328	-23.888
1100	38.360	325.021	297.436	30.343	552.507	447.745	-21.262
1200	38.460	328.363	299.876	34.184	547.430	438.502	-19.088
1300	38.559	331.445	302.188	38.035	546.484	429.463	-17.256
1400	38.656	334.306	304.381	41.895	545.491	420.498	-15.689
1500	38.751	336.976	306.466	45.766	544.434	411.607	-14.333
1600	38.847	339.480	308.452	49.646	543.298	402.788	-13.150
1700	38.941	341.838	310.347	53.535	542.069	394.043	-12.107
1800	39.035	344.067	312.159	57.434	540.732	385.374	-11.183
1900	39.128	346.180	313.894	61.342	539.274	376.782	-10.358
2000	39.221	348.189	315.559	65.260	537.685	368.270	-9.618
2100	39.314	350.105	317.159	69.186	535.954	359.842	-8.951
2200	39.407	351.936	318.699	73.122	512.752	352.244	-8.363
2300	39.499	353.690	320.182	77.068	510.300	345.003	-7.835
2400	39.592	355.373	321.613	81.022	507.836	337.869	-7.354
2500	39.684	356.991	322.996	84.986	505.364	330.838	-6.912
2600	39.776	358.549	324.334	88.959	502.889	323.906	-6.507
2700	39.868	360.052	325.629	92.941	500.419	317.069	-6.134
2800	39.960	361.503	326.885	96.933	497.959	310.323	-5.789
2900	40.052	362.907	328.103	100.933	495.513	303.665	-5.470
3000	40.143	364.267	329.286	104.943	493.087	297.091	-5.173
3100	40.235	365.584	330.435	108.962	490.685	290.598	-4.897
3200	40.327	366.863	331.554	112.990	488.311	284.181	-4.639
3300	40.418	368.106	332.643	117.027	485.968	277.838	-4.398
3400	40.510	369.314	333.704	121.074	483.658	271.566	-4.172
3500	40.602	370.489	334.738	125.129	481.383	265.361	-3.960
3600	40.693	371.634	335.747	129.194	479.145	259.221	-3.761
3700	40.785	372.751	336.732	133.268	476.945	253.142	-3.574
3800	40.876	373.839	337.694	137.351	474.783	247.123	-3.397
3900	40.968	374.902	338.635	141.443	472.660	241.159	-3.230
4000	41.059	375.941	339.555	145.544	470.576	235.250	-3.072
4100	41.150	376.956	340.454	149.655	468.531	229.392	-2.922
4200	41.242	377.948	341.335	153.774	466.524	223.584	-2.781
4300	41.333	378.920	342.198	157.903	464.555	217.823	-2.646
4400	41.425	379.871	343.044	162.041	462.623	212.107	-2.518
4500	41.516	380.803	343.872	166.188	460.728	206.435	-2.396
4600	41.607	381.717	344.685	170.344	458.868	200.805	-2.280
4700	41.699	382.612	345.483	174.510	457.043	195.215	-2.170
4800	41.790	383.491	346.265	178.684	-105.684	201.280	-2.190
4900	41.881	384.354	347.034	182.868	-107.093	207.690	-2.214
5000	41.973	385.201	347.789	187.060	-108.486	214.128	-2.237
5100	42.064	386.033	348.531	191.262	-109.864	220.594	-2.259
5200	42.155	386.851	349.260	195.473	-111.225	227.087	-2.281
5300	42.247	387.655	349.977	199.693	-112.570	233.606	-2.302
5400	42.338	388.445	350.682	203.923	-113.897	240.150	-2.323
5500	42.429	389.223	351.375	208.161	-115.206	246.718	-2.343
5600	42.521	389.988	352.058	212.408	-116.496	253.311	-2.363
5700	42.612	390.742	352.730	216.665	-117.768	259.926	-2.382
5800	42.703	391.483	353.392	220.931	-119.020	266.563	-2.401
5900	42.795	392.214	354.044	225.206	-120.252	273.221	-2.419
6000	42.886	392.934	354.686	229.490	-121.463	279.901	-2.437

PREVIOUS: September 1964 (1 atm) CURRENT: September 1964 (1 bar)

Zirconium Iodide (ZrI) $I_1Zr_1(g)$

IODINE (I_2)

REFERENCE STATE

0 to 386.75 K		crystal
386.75 to 457.666 K		liquid
above 457.666 K		diatomic ideal gas

Refer to the individual tables for details.

M_r = 253.8090

Iodine (I_2) I_2(ref)

Enthalpy Reference Temperature = T_r = 298.15 K   Standard State Pressure = $p°$ = 0.1 MPa

T/K	$C_p°$ (J K⁻¹mol⁻¹)	$S°$ (J K⁻¹mol⁻¹)	$-[G°-H°(T_r)]/T$ (J K⁻¹mol⁻¹)	$H°-H°(T_r)$ (kJ mol⁻¹)	$\Delta_f H°$ (kJ mol⁻¹)	$\Delta_f G°$ (kJ mol⁻¹)	Log K_f
0	0.	0.	INFINITE	-13.198	0.	0.	0.
100	45.650	60.992	162.472	-10.148	0.	0.	0.
200	51.566	94.958	121.036	-5.216	0.	0.	0.
298.15	54.436	116.142	116.142	0.	0.	0.	0.
300	54.506	116.479	116.143	0.101	0.	0.	0.
386.750	63.545	131.224	117.899	5.153	CRYSTAL <--> LIQUID		
386.750	80.669	171.344	117.899	20.670	TRANSITION		
400	80.669	174.062	119.714	21.739	0.	0.	0.
457.666	80.669	184.926	127.262	26.391	LIQUID <--> IDEAL GAS		
457.666	37.386	276.609	127.262	68.351	FUGACITY = 1 bar		
500	37.464	279.920	140.049	69.935	0.	0.	0.
600	37.612	286.764	163.948	73.690	0.	0.	0.
700	37.735	292.571	181.918	77.457	0.	0.	0.
800	37.847	297.618	196.072	81.236	0.	0.	0.
900	37.959	302.082	207.608	85.027	0.	0.	0.
1000	38.081	306.087	217.259	88.828	0.	0.	0.
1100	38.232	309.724	225.502	92.644	0.	0.	0.
1200	38.431	313.058	232.661	96.476	0.	0.	0.
1300	38.699	316.144	238.966	100.332	0.	0.	0.
1400	39.055	319.025	244.583	104.219	0.	0.	0.
1500	39.507	321.734	249.637	108.146	0.	0.	0.
1600	40.051	324.301	254.224	112.124	0.	0.	0.
1700	40.674	326.747	258.418	116.159	0.	0.	0.
1800	41.352	329.091	262.280	120.260	0.	0.	0.
1900	42.054	331.346	265.856	124.431	0.	0.	0.
2000	42.748	333.521	269.185	128.671	0.	0.	0.
2100	43.404	335.622	272.299	132.979	0.	0.	0.
2200	43.994	337.655	275.224	137.349	0.	0.	0.
2300	44.494	339.622	277.981	141.775	0.	0.	0.
2400	44.890	341.525	280.589	146.245	0.	0.	0.
2500	45.171	343.363	283.064	150.749	0.	0.	0.
2600	45.335	345.139	285.417	155.275	0.	0.	0.
2700	45.382	346.851	287.661	159.812	0.	0.	0.
2800	45.319	348.500	289.805	164.348	0.	0.	0.
2900	45.154	350.088	291.856	168.872	0.	0.	0.
3000	44.897	351.615	293.823	173.376	0.	0.	0.
3100	44.561	353.082	295.711	177.849	0.	0.	0.
3200	44.157	354.490	297.526	182.286	0.	0.	0.
3300	43.697	355.842	299.273	186.679	0.	0.	0.
3400	43.193	357.139	300.956	191.024	0.	0.	0.
3500	42.654	358.384	302.579	195.316	0.	0.	0.
3600	42.091	359.577	304.146	199.554	0.	0.	0.
3700	41.510	360.723	305.660	203.734	0.	0.	0.
3800	40.919	361.822	307.123	207.855	0.	0.	0.
3900	40.325	362.877	308.539	211.917	0.	0.	0.
4000	39.731	363.891	309.911	215.920	0.	0.	0.
4100	39.143	364.864	311.239	219.864	0.	0.	0.
4200	38.563	365.801	312.527	223.749	0.	0.	0.
4300	37.995	366.701	313.777	227.577	0.	0.	0.
4400	37.439	367.568	314.989	231.348	0.	0.	0.
4500	36.899	368.404	316.167	235.065	0.	0.	0.
4600	36.374	369.209	317.311	238.729	0.	0.	0.
4700	35.866	369.986	318.424	242.341	0.	0.	0.
4800	35.375	370.736	319.506	245.903	0.	0.	0.
4900	34.901	371.460	320.559	249.416	0.	0.	0.
5000	34.444	372.161	321.584	252.883	0.	0.	0.
5100	34.005	372.838	322.582	256.306	0.	0.	0.
5200	33.582	373.495	323.555	259.685	0.	0.	0.
5300	33.176	374.130	324.503	263.023	0.	0.	0.
5400	32.785	374.747	325.428	266.321	0.	0.	0.
5500	32.410	375.345	326.330	269.580	0.	0.	0.
5600	32.051	375.926	327.211	272.803	0.	0.	0.
5700	31.705	376.490	328.071	275.991	0.	0.	0.
5800	31.374	377.038	328.910	279.145	0.	0.	0.
5900	31.056	377.572	329.730	282.266	0.	0.	0.
6000	30.751	378.091	330.532	285.356	0.	0.	0.

PREVIOUS: September 1961 (1 atm)   CURRENT: June 1982 (1 bar)

Iodine (I_2) I_2(ref)

IODINE (I_2) CRYSTAL M_r = 253.8090

$S°(298.15\ K) = 116.142 \pm 0.080\ J\ K^{-1}\ mol^{-1}$

$T_{fus} = 386.75 \pm 0.3\ K$

$\Delta_f H°(0\ K) = 0\ kJ\ mol^{-1}$

$\Delta_f H°(298.15\ K) = 0\ kJ\ mol^{-1}$

$\Delta_{fus} H° = 15.517 \pm 0.16\ kJ\ mol^{-1}$

Enthalpy of Formation

Zero by definition.

Heat Capacity and Entropy

The heat capacity measurements of Shirley and Giauque (1) over the temperature range 13 to 327 K are adopted. No weight is given to the earlier work of Lange (2) which deviate by 1% or more. Carpenter and Harle (3) measured the heat capacity of crystalline and liquid iodine by an adiabatic method. Their data are in poor agreement with Shirley and Giauque (1) in the range 275–330 K and are given no weight. Frederick and Hildebrand (4) measured the enthalpy of crystalline and liquid iodine relative to 298.15 K by the method of mixtures. For calibration they used a slug of copper. The enthalpy of copper which they used appears to be too high (5). The results are corrected by a factor which varies from 0 at 300 K to 0.995 at 400 K, and 0.990 at 500 K. Using these corrected enthalpies and the heat capacities of Shirley and Giaugue (1) avove 298.15 K, the following equation is derived for I_2(cr). $H°(T) - H°(298.15\ K) = -\ 50.646T + 0.12345T^2 - 27.97 \times 10^5\ T^{-1} + 13508.7$.

The CODATA recommendation (6) for S°(298.15 K) is within 0.003 $J\ K^{-1}\ mol^{-1}$ of our calculated value and is based on the same heat capacity study (1).

Fusion Data

T_{fus} = 113.6 °C is given by Frederick and Hildebrand (4). The difference in the enthalpy equations at T_{fus} is the enthalpy of fusion.

Sublimation Data

See the I_2(g) table for details.

References

1. D. A. Shirley and W. F. Giauque, J. Am. Chem. Soc. 81, 4778 (1959).
2. E. Lange, Z. physik. Chem. 110, 343 (1924).
3. L. G. Carpenter and T. F. Harle, Phil. Mag. 23, 193 (1937).
4. K. J. Frederick and J. H. Hildebrand, J. Am. Chem. Soc. 60, 1436 (1939).
5. JANAF Thermochemical Tables: Cu(cr), 6-30-77.
6. J. D. Cox, Chairman, CODATA Task Group on Key Values for Chemical Thermodynamics, J. Chem. Thermodynamics 10, 903 (1978).

Iodine (I_2) I_2(cr)

Enthalpy Reference Temperature = T_r = 298.15 K Standard State Pressure = p° = 0.1 MPa

T/K	$C_p°$	S°	$-[G°-H°(T_r)]/T$	$H°-H°(T_r)$	$\Delta_f H°$	$\Delta_f G°$	Log K_f
	J K⁻¹mol⁻¹			kJ mol⁻¹			
0	0.	0.	INFINITE	-13.198	0.	0.	0.
100	45.650	60.992	162.472	-10.148	0.	0.	0.
200	51.566	94.958	121.036	-5.216	0.	0.	0.
250	53.244	106.678	117.030	-2.588	0.	0.	0.
298.15	54.436	116.142	116.142	0.	0.	0.	0.
300	54.506	116.479	116.143	0.101	0.	0.	0.
350	58.605	125.140	116.818	2.913	0.	0.	0.
386.750	63.545	131.224	117.899	5.153	------ CRYSTAL <--> LIQUID -----		
400	65.598	133.398	118.376	6.009	-15.730	0.535	-0.070
450	73.402	141.575	120.500	9.484	-16.288	2.606	-0.303
500	81.207	149.713	123.015	13.349	-56.586	8.517	-0.890
600	96.816	165.898	128.815	22.250	-51.439	21.080	-1.835
700	112.425	181.995	135.263	32.712	-44.745	32.659	-2.437

PREVIOUS: CURRENT: June 1982

Iodine (I_2) I_2(cr)

IODINE (I_2) LIQUID M_r = 253.8090

S°(298.15 K) = [150.356) J K^{-1} mol^{-1}

T_{fus} = 386.75 ± 0.03 K

$\Delta_f H°$(298.15 K) = [13.523] kJ mol^{-1}

$\Delta_{fus}H°$ = 15.517 ± 0.16 kJ mol^{-1}

Enthalpy of Formation

$\Delta_f H°(I_2$, ℓ, 298.15 K), is calculated from that of the crystal by adding $\Delta_{fus}H°$ and the difference in enthalpy, H°(386.75 K) - H°(298.15 K), between the crystal and liquid.

Heat Capacity and Entropy

The heat capacity of the liquid is adopted to be a constant value, 80.672 J K^{-1} mol^{-1}, based on the data of Frederick and Hildebrand (1). Their enthalpy data was corrected as discussed in the I_2(cr) table. The enthalpy equation is H°(T) - H°(298.15 K) = 80.672 T - 10529.

The entropy is calcualted in a manner analogous to that used for the enthalpy of formation.

Vaporization Data

See the I_2(g) table for details.

References

1. K. J. Frederick and J. H. Hidebrand, J. Am., Chem. Soc. 60, 1436 (1938).

Iodine (I_2) I_2(l)

Enthalpy Reference Temperature = T_r = 298.15 K Standard State Pressure = p° = 0.1 MPa

	J K⁻¹mol⁻¹			kJ mol⁻¹			
T/K	$C_p°$	S°	$-[G°-H°(T_r)]/T$	$H°-H°(T_r)$	$\Delta_f H°$	$\Delta_f G°$	Log K_f
0							
100							
200							
250							
298.15	80.669	150.356	150.356	0.	13.523	3.322	-0.582
300	80.669	150.855	150.357	0.149	13.571	3.259	-0.567
350	80.669	163.290	151.339	4.183	14.793	1.440	-0.215
386.750	80.669	171.344	152.864	7.147	------ CRYSTAL <--> LIQUID -----		
400	80.669	174.062	153.521	8.216	0.	0.	0.
450	80.669	183.563	156.342	12.250	0.	0.	0.
457.666	80.669	184.926	156.809	12.868	------- FUGACITY = 1 bar -------		
500	80.669	192.062	159.496	16.283	-40.130	3.799	-0.397
600	80.669	206.770	166.187	24.350	-35.817	12.179	-1.060
700	80.669	219.205	172.895	32.417	-31.518	19.839	-1.480
800	80.669	229.977	179.372	40.484	-27.230	26.883	-1.755
900	80.669	239.478	185.533	48.550	-22.953	33.390	-1.938
1000	80.669	247.978	191.360	56.617	-18.688	39.421	-2.059

PREVIOUS: September 1961 CURRENT: June 1982

Iodine (I_2) I_2(l)

IODINE (I_2) CRYSTAL–LIQUID M_r = 253.8090

0 to 386.75 K crystal
above 386.75 K liquid

Refer to the individual tables for details.

Iodine (I_2) I_2(cr,l)

Enthalpy Reference Temperature = T_r = 298.15 K Standard State Pressure = p° = 0.1 MPa

T/K	C_p° (J K⁻¹mol⁻¹)	S° (J K⁻¹mol⁻¹)	$-[G^\circ-H^\circ(T_r)]/T$ (J K⁻¹mol⁻¹)	$H^\circ-H^\circ(T_r)$ (kJ mol⁻¹)	Δ_fH° (kJ mol⁻¹)	Δ_fG° (kJ mol⁻¹)	Log K_f
0	0.	0.	INFINITE	-13.198	0.	0.	0.
100	45.650	60.992	162.472	-10.148	0.	0.	0.
200	51.566	94.958	121.036	-5.216	0.	0.	0.
250	53.244	106.678	117.030	-2.588	0.	0.	0.
298.15	54.436	116.142	116.142	0.	0.	0.	0.
300	54.506	116.479	116.143	0.101	0.	0.	0.
350	58.605	125.140	116.818	2.913	0.	0.	0.
386.750	63.545	131.224	117.899	5.153	CRYSTAL <--> LIQUID TRANSITION		
386.750	80.669	171.344	117.899	20.670			
400	80.669	174.062	119.714	21.739	0.	0.	0.
450	80.669	183.563	126.291	25.772	0.	0.	0.
457.666	80.669	184.926	127.262	26.391	------- FUGACITY = 1 bar -------		
500	80.669	192.062	132.451	29.806	-40.130	3.799	-0.397
600	80.669	206.770	143.649	37.873	-35.817	12.179	-1.060
700	80.669	219.205	153.577	45.940	-31.518	19.839	-1.480
800	80.669	229.977	162.469	54.006	-27.230	26.883	-1.755
900	80.669	239.478	170.508	62.073	-22.953	33.390	-1.938
1000	80.669	247.978	177.837	70.140	-18.688	39.421	-2.059

PREVIOUS: CURRENT: June 1982

Iodine (I_2) I_2(cr,l)

IODINE (I_2) IDEAL GAS M_r = 253.8090

D_0^0 = 148.816 ± 0.001 kJ mol^{-1} $\Delta_fH°(0\ K)$ = 65.504 ± 0.092 kJ mol^{-1}

S°(298.15 K) = 260.685 ± 0.062 J K^{-1} mol^{-1} $\Delta_fH°(298.15\ K)$ = 62.421 ± 0.079 kJ mol^{-1}

Spectroscopic Data for I_2 in cm^{-1}

State	T_e	D_0^0(j)	ω_e	ω_ex_e	B_e	$10^3\ \alpha_e$	$10^6\ D_e$	r_e, Å	References
$^1\Sigma_g^+$	0	12440	214.5481	0.616259(a)	0.037395	0.12435(b)	0.00454(c)	2.6657	(1) (4, D_0^0)
$^3\Pi_{2u}$	10200	12440	110.366	1.3927(d)	0.028	0.1	0.04	3.104	(2)
$^3\Pi_{1u}$	11059	12440	90.6	2.57(e)	0.02763	0.542(f)	0.03	3.101	(3)
$^3\Pi_{0^+u}$	15769.01	20043.15	125.697	0.7642(g)	0.029039	0.1582(h)	0.00543(i)	3.0247	(4)

(a) $+7.507\times10^{-5}(v+1/2)^3 - 1.263643\times10^{-4}(v+1/2)^4 + 6.198129\times10^{-6}(v+1/2)^5 - 2.0255975\times10^{-7}(v+1/2)^6 + 3.9662824\times10^{-9}(v+1/2)^7$

(b) $+4.4498\times10^{-7}(v+1/2)^2 - 1.482\times10^{-8}(v+1/2)^3$

(c) $+1.7\times10^{-11}(v+1/2) + 7\times10^{-12}(v+1/2)^2$

(d) $-3.45\times10^{-3}(v+1/2)^3$

(e) $+3.36\times10^{-2}(v+1/2)^3$

(f) $+2.46\times10^{-6}(v+1/2)^2$

(g) $-1.78\times10^{-3}(v+1/2)^3 + 1.03\times10^{-6}(v+1/2)^4$

(h) $-3.36\times10^{-7}(v+1/2)^2 - 4.78\times10^{-8}(v+1/2)^3$

(i) $+3.0\times10^{-10}(v+1/2)$

(j) The dissociation energy of each state is relative to the lowest level of the ground state; the $^3\Pi_{0^+u}$ state dissociates to $I(^2P_{3/2}) + I(^2P_{1/2})$.

Enthalpy of Formation

We adopt the enthalpy of formation values of CODATA (5) as follows: $\Delta_fH°(298.15\ K)$ = 62.421 ± 0.079 kJ mol^{-1} and $\Delta_fH°$ (0 K) = 65.500 ± 0.092 kJ mol^{-1}. These values were based on the heat capacity study of Shirley and Giauque (8) and the vapor pressure data of Baxter et al (9), Baxter and Grose (10), Gillespie and Fraser (11), and Stern (12). Note that the latter value of $\Delta_fH°$ is deduced from the former using the CODATA values of H°(298.15 K) - H°(0 K) for I_2(g) and I_2(cr). Our calculated value of H°(298.15) - H°(0 K) for I_2(g) is only 1.5 J mol^{-1} smaller than that recommended by CODATA. We adopt the dissociation energy for the $^1\Sigma_g^+$, $^3\Pi_{2u}$, and $^3\Pi_{1u}$ states as given by Huber et al. (4). The D_0^0 value for the $^3\Pi_0+$ state was deduced from the dissociative state energies given by Gurvich et al. (2, table 10.1). Refer to the monatomic iodine gas table for additional discussion of the dissociation energy.

Heat Capacity and Entropy

The thermal functions are calculated using a direct summation technique. Included in the calculation are the ground state and the three lowest lying excited electronic states. Spectroscopic constants for the ground state are from Leroy (1); however, we truncate the B_v expression at the $(v+1/2)^3$ term. The $^3\Pi_{2u}$ constants are taken from Gurvich et. al. (2). We adopt the spectroscopic constants of Ashby (3) for the $^3\Pi_{1u}$ state and of Huber and Herzberg (1) for the $^3\Pi_{0^+u}$ state. For the ground state and the $^3\Pi_{0^+u}$ state, the calculation is performed including the quasi-bound rotational levels above the dissociation limit (see Cl_2(g) table). The values of J_{lim} and v_{max} for the ground state are those given by Gurvich et al. (2). The corresponding values for the $^3\Pi_{0^+u}$ state are estimated as J_{lim} = 500 and v_{max} = 69 based on the spectroscopic constants given by Gurvich et al. (2) and Ashby (3).

Splitting of the rotational levels due to the rotational-electronic interaction is taken into account only insofar as the statistical weight of the rotational levels is increased appropriately. The rotational levels are extrapolated to high J values according to the method of Khachkurozov (6) who proposed a simpler form of Woolley's method (7). Our calculated value of S°(298.15 K) is only 0.112 J K^{-1} mol^{-1} larger than that given by Gurvich (2) and CODATA (5).

References

1. R. J. Leroy, J. Chem. Phys. 52, 2683 (1970).
2. L. V. Gurvich, I. V. Veits et al., "Thermodynamic Properties of Individual Substances," 3rd ed., Vol. I, Table 10.3, Nauka, Moscow, 1978.
3. R. A. Ashby, Can. J. Phys. 57, 698 (1979).
4. K. P. Huber and G. Herzberg, "Constants of Diatomic Molecules," Van Nostrand-Reinhold Co., New York, 1979.
5. J. D. Cox, chairman, ICSU-CODATA Task Group on Key Values for Thermodynamics, J. Chem. Thermodynamics 10, 903 (1978).
6. G. A. Khachkurozov, Opt. Spectrosc. 30, 455 (1971).
7. H. W. Woolley, R. B. Scott and F. G. Brickwedde, J. Res. Nat. Bur. Stand. 41, 379 (1948).
8. D. A. Shirley and W. F. Giauque, J. Am. Chem. Soc. 81, 4778 (1959).
9. G. P. Baxter, C. H. Hickley, and W. C. Holmes, J. Am. Chem. Soc. 29, 127 (1907).
10. G. P. Baxter and M. R. Grose, J. Am. Chem. Soc. 37, 1061 (1915).
11. L. J. Gillespie and L. H. D. Fraser, J. Am. Chem. Soc. 58, 2260 (1936).
12. T. H. Stern, Ph. D. Thesis, Univ. of Washington (1958).

Iodine (I_2) I_2(g)

Enthalpy Reference Temperature = T_r = 298.15 K Standard State Pressure = p° = 0.1 MPa

T/K	$C_p°$	S°	$-[G°-H°(T_r)]/T$	$H°-H°(T_r)$	$\Delta_fH°$	$\Delta_fG°$	Log K_f
	J K^{-1}mol^{-1}			kJ mol^{-1}			
0	0.	0.	INFINITE	-10.116	65.504	65.504	INFINITE
100	33.135	222.004	292.808	-7.080	65.489	49.387	-25.797
200	36.062	246.105	264.038	-3.587	64.050	33.821	-8.833
250	36.582	254.213	261.291	-1.769	63.240	26.356	-5.507
298.15	36.887	260.685	260.685	0.	62.421	19.325	-3.386
300	36.897	260.913	260.686	0.068	62.388	19.058	-3.318
350	37.105	266.617	261.136	1.919	61.427	11.910	-1.777
400	37.256	271.582	262.138	3.778	44.460	5.452	-0.712
450	37.371	275.977	263.436	5.644	42.292	0.706	-0.082
457.666	37.386	276.609	263.651	5.930	------- FUGACITY = 1 bar -------		
500	37.464	279.920	264.891	7.515	0.	0.	0.
600	37.612	286.764	267.983	11.269	0.	0.	0.
700	37.735	292.571	271.091	15.036	0.	0.	0.
800	37.847	297.618	274.098	18.815	0.	0.	0.
900	37.959	302.082	276.964	22.606	0.	0.	0.
1000	38.081	306.087	279.680	26.407	0.	0.	0.
1100	38.232	309.724	282.248	30.223	0.	0.	0.
1200	38.431	313.058	284.679	34.055	0.	0.	0.
1300	38.699	316.144	286.982	37.911	0.	0.	0.
1400	39.055	319.025	289.169	41.798	0.	0.	0.
1500	39.507	321.734	291.251	45.726	0.	0.	0.
1600	40.051	324.301	293.237	49.703	0.	0.	0.
1700	40.674	326.747	295.136	53.738	0.	0.	0.
1800	41.352	329.091	296.958	57.839	0.	0.	0.
1900	42.054	331.346	298.709	62.010	0.	0.	0.
2000	42.748	333.521	300.396	66.250	0.	0.	0.
2100	43.404	335.622	302.023	70.558	0.	0.	0.
2200	43.994	337.655	303.597	74.929	0.	0.	0.
2300	44.494	339.622	305.121	79.354	0.	0.	0.
2400	44.890	341.525	306.598	83.824	0.	0.	0.
2500	45.171	343.363	308.032	88.328	0.	0.	0.
2600	45.335	345.139	309.425	92.854	0.	0.	0.
2700	45.382	346.851	310.780	97.391	0.	0.	0.
2800	45.319	348.500	312.098	101.927	0.	0.	0.
2900	45.154	350.088	313.381	106.451	0.	0.	0.
3000	44.897	351.615	314.630	110.955	0.	0.	0.
3100	44.561	353.082	315.847	115.428	0.	0.	0.
3200	44.157	354.490	317.033	119.865	0.	0.	0.
3300	43.697	355.842	318.188	124.258	0.	0.	0.
3400	43.193	357.139	319.315	128.603	0.	0.	0.
3500	42.654	358.384	320.414	132.895	0.	0.	0.
3600	42.091	359.577	321.485	137.133	0.	0.	0.
3700	41.510	360.723	322.530	141.313	0.	0.	0.
3800	40.919	361.822	323.550	145.434	0.	0.	0.
3900	40.325	362.877	324.545	149.496	0.	0.	0.
4000	39.731	363.891	325.516	153.499	0.	0.	0.
4100	39.143	364.864	326.464	157.443	0.	0.	0.
4200	38.563	365.801	327.389	161.328	0.	0.	0.
4300	37.995	366.701	328.293	165.156	0.	0.	0.
4400	37.439	367.568	329.176	168.928	0.	0.	0.
4500	36.899	368.404	330.038	172.644	0.	0.	0.
4600	36.374	369.209	330.881	176.308	0.	0.	0.
4700	35.866	369.986	331.705	179.920	0.	0.	0.
4800	35.375	370.736	332.510	183.482	0.	0.	0.
4900	34.901	371.460	333.298	186.995	0.	0.	0.
5000	34.444	372.161	334.068	190.462	0.	0.	0.
5100	34.005	372.838	334.822	193.885	0.	0.	0.
5200	33.582	373.495	335.559	197.264	0.	0.	0.
5300	33.176	374.130	336.281	200.602	0.	0.	0.
5400	32.785	374.747	336.988	203.900	0.	0.	0.
5500	32.410	375.345	337.680	207.159	0.	0.	0.
5600	32.051	375.926	338.357	210.382	0.	0.	0.
5700	31.705	376.490	339.022	213.570	0.	0.	0.
5800	31.374	377.038	339.672	216.724	0.	0.	0.
5900	31.056	377.572	340.310	219.845	0.	0.	0.
6000	30.751	378.091	340.936	222.935	0.	0.	0.

PREVIOUS: September 1961 (1 atm) CURRENT: June 1982 (1 bar)

Iodine (I_2) I_2(g)

POTASSIUM IODIDE ((KI)$_2$) IDEAL GAS M_r = 332.0056

$\Delta_f H°(0\ K)$ = -416.48 ± 4.2 kJ mol^{-1}

S°(298.15 K) = [395.773] J K^{-1} mol^{-1} $\Delta_f H°(298.15\ K)$ = -422.17 ± 4.2 kJ mol^{-1}

Vibrational Frequencies and Degeneracies

ν, cm^{-1}	ν, cm^{-1}
[90](1)	[80](1)
[60](1)	[105](1)
[110](1)	[120](1)

Ground State Quantum Weight: [1] σ = [4]

Point Group: [D_{2h}]

Bond Distance: K-I = [3.29] Å K-K = [4.00] Å

Bond Angles: K-I-K = [75°] I-K-I = [105°]

Product of the Moments of Inertia: $I_A I_B I_C$ = [5.073451 x 10^{-111}] g^3 cm^6

Enthalpy of Formation

The chemical equilibrium for the reaction $K_2I_2(g) = 2\ KI(g)$ has been studied by Datz (1) and Datz et al. (2). They measured the temperature dependence of the molecular weight of KI(g) which was determined by measurement of the absolute pressure exerted by a known weight of completely vaporized salt contained in an isothermal fused silica bulb of known volume. Based on the reported P-V-T measurements, the enthalpy change of the reaction at 298.15 K is evaluated by the 2nd and 3rd law methods. The results are given in the table below.

Miller and Kusch (3) determined the molecular composition of KI vapor by measurement of the velocity distribution of the molecules in the beam produced as the vapor effused through a small slit in a source. The analysis was based on an assumption that the velocity distribution within the oven is Maxwellian and that the vapor effuses through the ideal slit of kinetic theory. The velocity distributions of potassium and thallium atomic beams were found to be in excellent agreement with the theoretical distributions so the determination of the molecular composition of KI beams was tried. Using the derived equilibrium constants, we calculate the enthalpy change of the dissociation reaction by the 2nd and 3rd law methods. The results are presented in the following table.

The value of $\Delta_f H°(K_2I_2$, g, 298.15 K) is adopted as -100.9±1 kcal mol^{-1}.

Source	T/K	Data Points	$\Delta_r H°$(298.15 K), kcal mol^{-1} 2nd law	3rd law	Drift cal K^{-1} mol^{-1}	$\Delta_f H°$(298.15 K) kcal mol^{-1}
Datz (1)	1226-1385	10	40.09±0.82	40.66	0.4±0.6	-100.66
	1292-1408	6	41.37±1.91	41.01	-0.2±1.4	-101.01
Miller-Kusch (2)	816-921	10	54.62±2.46	36.28	-21.4±2.8	-96.28

Heat Capacity

The molecular structure and bond distance and angles were estimated by Berkowitz (4). The vibrational frequences are estimated by comparison with those calculated by Berkowitz (5) and adjusted to give good agreement between 2nd and 3rd law evaluations of the equilibrium data. The three principal moments of inertia are: I_A = 52.0871 x 10^{-39}, I_B = 287.1362 x 10^{-39} and I_C = 339.2233 x 10^{-39} g cm^2.

References

1. S. Datz, Oak Ridge National Laboratory, ORNL-2933, 1960.
2. S. Datz, W. T. Smith, Jr., and E. H. Taylor, J. Chem. Phys. 34, 558 (1961).
3. R. C. Miller and P. Kusch, J. Chem. Phys. 25, 860 (1956).
4. J. Berkowitz, J. Chem. Phys. 29, 1386 (1958).
5. J. Berkowitz, J. Chem. Phys. 32, 1519 (1960).

Potassium Iodide ((KI)$_2$) $I_2K_2(g)$

Enthalpy Reference Temperature = T_r = 298.15 K Standard State Pressure = p° = 0.1 MPa

T/K	$C_p°$ (J K^{-1}mol^{-1})	S°	-[G°-H°(T_r)]/T	H°-H°(T_r) (kJ mol^{-1})	$\Delta_f H°$	$\Delta_f G°$	Log K_f
0	0.	0.	INFINITE	-21.676	-416.480	-416.480	INFINITE
100	75.971	308.279	467.994	-15.972	-417.290	-434.898	227.167
200	81.203	363.113	403.271	-8.032	-419.471	-451.705	117.973
250	81.890	381.315	397.125	-3.953	-420.768	-459.616	96.032
298.15	82.258	395.772	395.772	0.	-422.166	-466.974	81.812
300	82.268	396.281	395.774	0.152	-422.223	-467.252	81.356
350	82.499	408.982	396.777	4.272	-428.698	-474.433	70.805
400	82.649	420.008	399.007	8.401	-446.570	-480.288	62.719
450	82.752	429.749	401.892	12.536	-449.596	-484.321	56.218
500	82.826	438.472	405.121	16.675	-492.707	-484.220	50.586
600	82.923	453.583	411.977	24.963	-494.252	-482.375	41.995
700	82.982	466.370	418.857	33.259	-495.716	-480.279	35.839
800	83.020	477.453	425.504	41.559	-497.149	-477.976	31.209
900	83.046	487.233	431.831	49.862	-498.601	-475.493	27.597
1000	83.064	495.984	437.816	58.168	-500.123	-472.845	24.699
1100	83.078	503.901	443.470	66.475	-659.668	-460.819	21.882
1200	83.089	511.131	448.811	74.783	-659.350	-442.756	19.273
1300	83.097	517.782	453.864	83.093	-659.054	-424.718	17.065
1400	83.103	523.940	458.652	91.403	-658.789	-406.703	15.174
1500	83.109	529.674	463.198	99.713	-658.565	-388.705	13.536
1600	83.113	535.038	467.522	108.025	-658.393	-370.720	12.103
1700	83.116	540.076	471.644	116.336	-658.282	-352.744	10.839
1800	83.119	544.827	475.579	124.648	-658.241	-334.773	9.715
1900	83.122	549.321	479.343	132.960	-658.277	-316.802	8.709
2000	83.124	553.585	482.949	141.272	-658.394	-298.827	7.805
2100	83.126	557.641	486.410	149.585	-658.592	-280.844	6.986
2200	83.128	561.508	489.736	157.897	-658.871	-262.850	6.241
2300	83.129	565.203	492.938	166.210	-659.227	-244.841	5.561
2400	83.130	568.741	496.023	174.523	-659.655	-226.815	4.937
2500	83.131	572.135	499.000	182.836	-660.149	-208.771	4.362
2600	83.132	575.395	501.876	191.149	-660.703	-190.704	3.831
2700	83.133	578.533	504.657	199.463	-661.311	-172.616	3.339
2800	83.134	581.556	507.350	207.776	-661.968	-154.505	2.882
2900	83.135	584.473	509.960	216.089	-662.670	-136.369	2.456
3000	83.135	587.292	512.491	224.403	-663.413	-118.208	2.058
3100	83.136	590.018	514.948	232.717	-664.196	-100.022	1.685
3200	83.136	592.657	517.335	241.030	-665.020	-81.809	1.335
3300	83.137	595.215	519.657	249.344	-665.868	-63.572	1.006
3400	83.137	597.697	521.916	257.657	-666.770	-45.307	0.696
3500	83.138	600.107	524.115	265.971	-667.714	-27.016	0.403
3600	83.138	602.449	526.259	274.285	-668.714	-8.696	0.126
3700	83.138	604.727	528.349	282.599	-669.773	9.652	-0.136
3800	83.139	606.944	530.388	290.913	-670.899	28.030	-0.385
3900	83.139	609.104	532.379	299.227	-672.101	46.438	-0.622
4000	83.139	611.209	534.324	307.540	-673.373	64.877	-0.847
4100	83.139	613.262	536.224	315.854	-674.741	83.350	-1.062
4200	83.140	615.265	538.082	324.168	-676.221	101.857	-1.267
4300	83.140	617.221	539.900	332.482	-677.815	120.402	-1.463
4400	83.140	619.133	541.679	340.796	-678.979	138.930	-1.649
4500	83.140	621.001	543.421	349.110	-680.696	157.538	-1.829
4600	83.140	622.829	545.128	357.424	-682.534	176.185	-2.001
4700	83.140	624.617	546.800	365.738	-684.461	194.870	-2.166
4800	83.141	626.367	548.439	374.052	-686.302	213.574	-2.324
4900	83.141	628.081	550.047	382.366	-688.486	232.343	-2.477
5000	83.141	629.761	551.625	390.681	-690.809	251.159	-2.624
5100	83.141	631.407	553.173	398.995	-693.275	270.023	-2.766
5200	83.141	633.022	554.693	407.309	-695.888	288.936	-2.902
5300	83.141	634.605	556.186	415.623	-698.649	307.901	-3.035
5400	83.141	636.160	557.653	423.937	-701.560	326.919	-3.162
5500	83.141	637.685	559.094	432.251	-704.622	345.993	-3.286
5600	83.142	639.183	560.511	440.565	-707.835	365.124	-3.406
5700	83.142	640.655	561.904	448.879	-710.506	384.218	-3.521
5800	83.142	642.101	563.274	457.194	-713.928	403.454	-3.633
5900	83.142	643.522	564.622	465.508	-717.486	422.749	-3.743
6000	83.142	644.919	565.949	473.822	-721.178	442.106	-3.849

PREVIOUS: June 1967 (1 atm) CURRENT: June 1967 (1 bar)

Potassium Iodide ((KI)$_2$) $I_2K_2(g)$

LITHIUM IODIDE $((LiI)_2)$ IDEAL GAS M_r = 267.6910

$S°(298.15\ K) = [330.628]\ J\ K^{-1}\ mol^{-1}$

$\Delta_f H°(0\ K) = -356.74 \pm 16.7\ kJ\ mol^{-1}$

$\Delta_f H°(298.15\ K) = -361.92 \pm 16.7\ kJ\ mol^{-1}$

Vibrational Frequencies and Degeneracies

ν, cm^{-1}	ν, cm^{-1}
[166](1)	[310](1)
[160](1)	248 (1)
[383](1)	375 (1)

Ground State Quantum Weight: 1 $\sigma = 4$

Point Group: D_{2h}

Bond Distances: Li-I = 2.54 Å Li-Li = 2.71 Å

Bond Angles: I-Li-I = 116° ± 4° Li-I-Li = 64°

Product of the Moments of Inertia: $I_A I_B I_C = 1.631229 \times 10^{-112}\ g^3\ cm^6$

Enthalpy of Formation

This was obtained from the enthalpy of formation of the crystal and the enthalpy of sublimation to the dimer, which derivation has been given in the LiI(g) table.

Heat Capacity and Entropy

Berkowitz (1) has calculated the molecular structure and vibrational frequencies based on the ionic model. The planar rhombic structure proposed by Berkowitz has been confirmed by the lack of polarity in electric deflection by Buchler et al. (2). The selected bond distances and angle were obtained from the electron diffraction studies of monomer-dimer vapor by Akischin and Rambidi (3). The bond distances, r(Li-Li) = 2.712 Å and r(I-I) = 4.520 Å, calculated by Berkowitz are in good agreement with those from Akischin and Rambidi (3). The principal moments of inertia are: $I_A = 4.1763 \times 10^{-39}$, $I_B = 195.5561 \times 10^{-39}$ and $I_C = 199.7324 \times 10^{-39}\ g\ cm^2$.

Bauer et al. (4) have estimated six vibrational frequencies (501, 501, 148, 200, 294, 360 cm^{-1}) for $Li_2I_2(g)$ in the electron diffraction studies of the $Li_2Cl_2(g)$. Klemperer and Norris (5) have observed two fundamental vibrational frequencies (248 and 375 cm^{-1}) in the infrared spectrum and tentatively assigned them as B_{2u} and B_{3u} modes; these have been adopted in the tabulation. The remaining four vibrational frequencies were obtained from Berkowitz (1) because his model and derivation are self-consistent.

References

1. J. Berkowitz, J. Chem. Phys. 29, 1386 (1958), 32, 1519 (1960).
2. A. Buchler, J. L. Stauffer and W. Klemperer, J. Amer. Chem. Soc. 86, 4544 (1964).
3. P. A. Akischin and N. G. Rambidi, Z. Physik Chem. 213, 111 (1960).
4. S. H. Bauer, T. Ino and R. F. Porter, J. Chem. Phys. 33, 685 (1960).
5. W. Klemperer and W. G. Norris, J. Chem. Phys. 34, 1071 (1961).

Lithium Iodide $((LiI)_2)$ $I_2Li_2(g)$

Enthalpy Reference Temperature = T_r = 298.15 K Standard State Pressure = p° = 0.1 MPa

T/K	$C_p°$ (J K⁻¹mol⁻¹)	S° (J K⁻¹mol⁻¹)	$-[G°-H°(T_r)]/T$ (J K⁻¹mol⁻¹)	$H°-H°(T_r)$ (kJ mol⁻¹)	$\Delta_f H°$ (kJ mol⁻¹)	$\Delta_f G°$ (kJ mol⁻¹)	Log K_f
0	0.	0.	INFINITE	-17.267	-356.740	-356.740	INFINITE
100	51.234	259.423	393.015	-13.359	-356.873	-375.247	196.009
200	69.370	301.482	337.413	-7.186	-359.315	-392.689	102.560
250	73.525	317.446	331.867	-3.605	-360.618	-400.883	83.760
298.15	76.044	330.628	330.628	0.	-361.916	-408.522	71.571
300	76.121	331.098	330.629	0.141	-361.967	-408.811	71.180
350	77.820	342.969	331.563	3.992	-363.454	-416.504	62.160
400	78.982	353.441	333.656	7.914	-381.033	-423.426	55.294
450	79.808	362.794	336.383	11.885	-383.942	-428.552	49.745
500	80.414	371.235	339.453	15.891	-433.119	-428.934	44.810
600	81.222	385.974	346.014	23.976	-434.754	-427 940	37.256
700	81.720	398.535	352.642	32.125	-436.223	-426.686	31.840
800	82.048	409.470	359.076	40.315	-437.605	-425.229	27.765
900	82.275	419.147	365.224	48.531	-438.961	-423.601	24.585
1000	82.438	427.825	371.057	56.767	-440.299	-421.822	22.034
1100	82.559	435.688	376.581	65.018	-441.626	-419.910	19.940
1200	82.652	442.876	381.810	73.278	-442.950	-417.877	18.190
1300	82.724	449.494	386.766	81.547	-444.281	-415.734	16.704
1400	82.782	455.627	391.468	89.823	-445.625	-413.488	15.427
1500	82.828	461.340	395.938	98.103	-446.987	-411.145	14.317
1600	82.866	466.687	400.194	106.388	-448.378	-408.710	13.343
1700	82.898	471.712	404.255	114.676	-740.280	-391.835	12.040
1800	82.924	476.451	408.135	122.968	-740.251	-371.340	10.776
1900	82.947	480.935	411.850	131.261	-740.291	-350.844	9.645
2000	82.966	485.190	415.412	139.557	-740.403	-330.344	8.628
2100	82.983	489.238	418.832	147.854	-740.587	-309.837	7.707
2200	82.997	493.099	422.120	156.153	-740.839	-289.320	6.869
2300	83.009	496.789	425.287	164.454	-741.154	-268.789	6.104
2400	83.020	500.322	428.340	172.755	-741.525	-248.244	5.403
2500	83.030	503.711	431.288	181.058	-741.944	-227.682	4.757
2600	83.039	506.968	434.136	189.361	-742.403	-207.103	4.161
2700	83.046	510.102	436.892	197.665	-742.893	-186.505	3.608
2800	83.053	513.122	439.561	205.970	-743.406	-165.889	3.095
2900	83.059	516.037	442.148	214.276	-743.936	-145.254	2.616
3000	83.065	518.852	444.658	222.582	-744.475	-124.600	2.169
3100	83.070	521.576	447.096	230.889	-745.018	-103.929	1.751
3200	83.074	524.214	449.465	239.196	-745.562	-83.240	1.359
3300	83.079	526.770	451.769	247.504	-746.104	-62.534	0.990
3400	83.082	529.250	454.012	255.812	-746.642	-41.812	0.642
3500	83.086	531.659	456.196	264.120	-747.175	-21.074	0.315
3600	83.089	533.999	458.325	272.429	-747.703	-0.320	0.005
3700	83.092	536.276	460.401	280.738	-748.227	20.448	-0.289
3800	83.095	538.492	462.427	289.047	-748.750	41.230	-0.567
3900	83.097	540.650	464.405	297.357	-749.272	62.025	-0.831
4000	83.099	542.754	466.338	305.667	-749.797	82.834	-1.082
4100	83.102	544.806	468.226	313.977	-750.329	103.657	-1.321
4200	83.104	546.809	470.074	322.287	-750.870	124.492	-1.548
4300	83.105	548.764	471.881	330.597	-751.425	145.341	-1.766
4400	83.107	550.675	473.650	338.908	-751.957	166.199	-1.973
4500	83.109	552.542	475.383	347.219	-752.533	187.072	-2.171
4600	83.110	554.369	477.080	355.530	-753.139	207.959	-2.361
4700	83.112	556.157	478.744	363.841	-753.773	228.859	-2.543
4800	83.113	557.906	480.375	372.152	-754.441	249.773	-2.718
4900	83.114	559.620	481.974	380.464	-755.146	270.702	-2.886
5000	83.116	561.299	483.544	388.775	-755.894	291.645	-3.047
5100	83.117	562.945	485.085	397.087	-756.679	312.603	-3.202
5200	83.118	564.559	486.598	405.398	-757.524	333.577	-3.351
5300	83.119	566.142	488.084	413.710	-758.427	354.568	-3.494
5400	83.120	567.696	489.544	422.022	-759.391	375.577	-3.633
5500	83.120	569.221	490.979	430.334	-760.422	396.604	-3.767
5600	83.121	570.719	492.389	438.646	-761.525	417.651	-3.896
5700	83.122	572.190	493.776	446.958	-762.703	438.718	-4.020
5800	83.123	573.636	495.141	455.271	-763.962	459.806	-4.141
5900	83.124	575.057	496.483	463.583	-765.306	480.917	-4.258
6000	83.124	576.454	497.805	471.895	-766.738	502.052	-4.371

PREVIOUS: June 1966 (1 atm) CURRENT: June 1966 (1 bar)

Lithium Iodide $((LiI)_2)$ $I_2Li_2(g)$

MAGNESIUM IODIDE (MgI_2) CRYSTAL M_r = 278.1140

$S°(298.15\ K) = [129.704 \pm 4.2]\ J\ K^{-1}\ mol^{-1}$

$T_{fus} = 907 \pm 15\ K$

$\Delta_f H°(0\ K)$ = Unknown

$\Delta_f H°(298.15\ K) = -366.94 \pm 6.3\ kJ\ mol^{-1}$

$\Delta_{fus} H° = 29.288 \pm 8.4\ kJ\ mol^{-1}$

Enthalpy of Formation

Finch et al. (1) measured the enthalpy of solution of MgO(cr) in aqueous HI and of MgI_2(cr) in the same solvent. Combining and correcting for a concentration change gives $\Delta_r H°(298.15\ K) = 14.53\ kcal\ mol^{-1}$ for the process MgO(cr) + 2 HI (0.5m) = MgI_2(cr) + H_2O(ℓ) from which $\Delta_f H°(MgI_2$, cr, 298.15 K) = -87.70 ± 1.5 kcal mol^{-1} is calculated. The auxiliary data used are $\Delta_f H°$(MgO, cr, 298.15 K) = -143.7 kcal mol^{-1} (2), $\Delta_f H°(H_2O$, ℓ, 298.15 K) = -68.315 kcal mol^{-1} (3), and $\Delta_f H°$(HI, 0.5m, 298.15 K) = -13.42 kcal mol^{-1}. This last value was calculated from the CODATA value of -13.60 for $\Delta_f H°(I^-$, aq, std. state, 298.15 K) (3) = $\Delta_f H°$(HI, aq, std. state, 298.15 K) and the enthalpy of dilution of HI (4).

Combination of $\Delta_f H°(Mg^{+2}$, aq, std. state, 298.15 K) = -111.58 kcal mol^{-1} selected by Parker (4, 5) with the CODATA value of $\Delta_f H°(I^-$, aq, std. state, 298.15 K) = -13.60 kcal mol^{-1} gives $\Delta_f H°(MgI_2$, aq, std. state, 298.15 K) = -138.78 kcal mol^{-1}. Further combination with the enthalpy of solution, -50.96 kcal mol^{-1} (4), gives $\Delta_f H°(MgI_2$, cr, 298.15 K) = -87.82 kcal mol^{-1}. We adopt $\Delta_f H°(MgI_2$, cr, 298.15 K) = -87.7 ± 1.5 kcal mol^{-1} to retain internal consistency.

Heat Capacity and Entropy

$C_p°$(298.15 K) = 17.88 cal K^{-1} mol^{-1} is calculated from the reaction $MgCl_2$(cr) + 2 NaI(cr) = MgI_2(cr) + 2 NaCl(cr) assuming no net change in heat capacity. $C_p°$ = 21.0 cal k^{-1} mol^{-1} at T_{fus} = 907 K estimated from a Kopp's rule value of 7.0 cal k^{-1} g-atom appears reasonable in comparison with a $C_p°$ vs. T plot for $MgBr_2$(cr), $MgCl_2$(cr) and CaI_2(cr) (2). Values of $C_p°$ between 298.15 and 907 K are read from the graphical comparison curve. The adopted values lead to H°(500 K)-H°(298.15 K) = 3.77 kcal mol^{-1}, in reasonable agreement with the estimate of 4.0 by Brewer et al. (3).

S°(298.15 K) = 31.0±1.0 cal K^{-1} mol^{-1} is adopted from Kelley and King (7).

Fusion Data

The melting point of 650°C (923 K) quoted in the literature appears to be based on the assumed value of Devoto and Jeny (8). In a study of molten electrolytes, Bockris et al. (9), found MgI_2 to be molten at 910 K; a graph in the same paper infers that a T_{fus} = 907 K was chosen. We adopt T_{fus} = 907±15 K.

From the KI/MgI_2 phase diagram by Klemm et al. (10), values of $\Delta_{fus}H°(MgI_2)$ ranging from 7 to 9 kcal mol^{-1} are calculated. Brewer (11) earlier estimated 5.3 kcal mol^{-1}. We adopt 7.0±2.0 kcal mol^{-1}.

Sublimation Data

By a mass spectrometric - Knudsen cell technique, Berkowitz and Marguart (12) found the sublimation pressure of MgI_2(cr) to be 5.18×10^{-6} atm at 734 K and 2.02×10^{-5} atm at 757 K. A 3rd law analysis of these two points gives $\Delta_{sub}H°$(298.15 K) = 49.4±0.8 kcal mol^{-1}. The 2nd law analysis of these two points gives 67.5 kcal mol^{-1} but other data covering an approximate range of 550 to 750 K are shown graphically by Berkowitz and Marguardt (12) from which they calculate $\Delta_{sub}H°$(550-750 K) = 45 kcal mol^{-1}. We adopt $\Delta_{sub}H°$(298.15 K) = 49.4±0.8 kcal mol^{-1}.

References

1. A. Finch, P. J. Gardner, and C. J. Steadman, Can. J. Chem. 46, 3447 (1968).
2. JANAF Thermochemical Tables: MgO(cr), 12-31-74; $MgBr_2$(cr), 6-30-74; $MgCl_2$(cr), 12-31-65; $CaCl_2$(cr), 6-30-70.
3. J. D. Cox, ICSU-CODATA Task Group on Key Values for Thermodynamics, J. Chem. Thermody. 4, 331 (1972).
4. U. S. Nat. Bur. Stand. Tech. Note 270-3, 1968.
5. V. B. Parker, U. S. Nat. Bur. Stand. Report 9803, 1968.
6. L. Brewer, L. A. Bromley, P. W. Gilles, and N. L. Lofgren in "The Chemistry and Metallurgy of Miscellaneous Materials, Thermodynamics," L. L. Quill, Ed., McGraw-Hill, New York, 1950.
7. K. K. Kelley and E. G. King, U. S. Bur. Mines Bull. 592, 1961.
8. G. Devoto and G. Jeny, Gazz. Chim., Ital. 61, 305 (1931).
9. J. O'M. Bockris, E. H. Crook, H. Bloom and N. E. Richards, Proc. Roy. Soc. (London) A255, 558 (1960).
10. W. Klemm, K. Beyersdorfer, and J. Oryschkewitsch, Z. Anorg. Allg. Chem. 256, 34 (1948).
11. L. Brewer in "The Chemistry and Metallurgy of Miscellaneous Materials, Thermodynamics," L. L. Quill, Ed., McGraw-Hill, New York, 1950.
12. J. Berkowitz and J. R. Marquart, J. Chem. Phys. 37, 1853 (1962).

Magnesium Iodide (MgI_2) I_2Mg_1(cr)

Enthalpy Reference Temperature = T_r = 298.15 K Standard State Pressure = p° = 0.1 MPa

	J $K^{-1}mol^{-1}$			kJ mol^{-1}			
T/K	$C_p°$	S°	$-[G°-H°(T_r)]/T$	$H°-H°(T_r)$	$\Delta_f H°$	$\Delta_f G°$	Log K_f
0							
100							
200							
298.15	74.810	129.704	129.704	0.	-366.937	-361.240	63.288
300	74.977	130.167	129.705	0.139	-366.945	-361.204	62.891
400	78.408	152.229	132.690	7.815	-383.462	-358.661	46.836
500	80.919	170.002	138.432	15.785	-426.355	-348.339	36.391
600	83.011	184.946	144.971	23.985	-424.677	-332.893	28.981
700	84.726	197.873	151.626	32.373	-422.928	-317.733	23.710
800	86.274	209.290	158.134	40.925	-421.144	-302.826	19.773
900	87.655	219.502	164.395	49.596	-419.381	-288.143	16.723
907.000	87.864	220.182	164.823	50.210	------ CRYSTAL <--> LIQUID -----		
1000	88.994	228.812	170.378	58.434	-426.201	-272.943	14.257
1100	90.249	237.354	176.084	67.396	-424.485	-257.700	12.237
1200	91.462	245.259	181.523	76.482	-422.663	-242.617	10.561
1300	92.634	252.626	186.713	85.687	-420.744	-227.690	9.149
1400	93.763	259.532	191.670	95.008	-546.151	-209.748	7.826
1500	94.851	266.039	196.413	104.439	-542.726	-185.838	6.471

PREVIOUS: CURRENT: December 1974

Magnesium Iodide (MgI_2) I_2Mg_1(cr)

MAGNESIUM IODIDE (MgI_2) LIQUID M_r = 278.1140

$S°(298.15\ K)$ = [155.737] J K^{-1} mol^{-1}

T_{fus} = 907 ± 15 K

$\Delta_f H°(298.15\ K)$ = [-342.252] kJ mol^{-1}

$\Delta_{fus} H°$ = 29.288 ± 8.4 kJ mol^{-1}

Enthalpy of Formation

$\Delta_f H°(MgI_2$, ℓ, 298.15 K) is calculated from $\Delta_f H°(MgI_2$, cr, 298.15 K) by adding the enthalpy of fusion, $\Delta_{fus} H°$, and the difference in enthalpy, H°(907 K)-H°(298.15 K), between the crystal and the liquid.

Heat Capacity and Entropy

The liquid heat capacity of 24.0 cal K^{-1} mol^{-1} (8 cal K^{-1} g-atom) follows the trend of the liquid heat capacities of the alkaline earth chlorides and bromides and the other alkaline earth iodides. This value is adopted and assumed constant over a 600-2000 K range. A glass transition is assumed at 600 K below which the heat capacity is that of the crystal.

$S°(MgI_2$, ℓ, 298.15 K) is calculated in a manner similar to that used for the enthalpy of formation.

Fusion Data

See the crystal table for details.

Vaporization Data

T_{vap} = 1255 K is calculated as the temperature at which the fugacity is 1 atm for the reaction MgI_2(ℓ) = MgI_2(g). $\Delta_{vap} H°$ = 36.126 kcal mol^{-1} is the calculated difference between the enthalpies of formation of the ideal gas and the liquid at T_{vap}.

Magnesium Iodide (MgI_2) I_2Mg_1(l)

Enthalpy Reference Temperature = T_r = 298.15 K Standard State Pressure = p° = 0.1 MPa

T/K	$C_p°$	S°	$-[G°-H°(T_r)]/T$	$H°-H°(T_r)$	$\Delta_f H°$	$\Delta_f G°$	Log K_f
	J $K^{-1}mol^{-1}$			kJ mol^{-1}			
0							
100							
200							
298.15	74.810	155.737	155.737	0.	-342.252	-344.316	60.323
300	74.977	156.200	155.738	0.139	-342.260	-344.329	59.953
400	78.408	178.261	158.723	7.815	-358.776	-344.389	44.973
500	80.919	196.035	164.464	15.785	-401.669	-336.671	35.172
600	83.011	210.980	171.004	23.985	-399.991	-323.827	28.192
600.000	83.011	210.980	171.004	23.985	GLASS <--> LIQUID TRANSITION		
600.000	100.416	210.980	171.004	23.985			
700	100.416	226.459	177.849	34.027	-396.589	-311.403	23.237
800	100.416	239.867	184.782	44.068	-393.315	-299.459	19.553
900	100.416	251.695	191.572	54.110	-390.182	-287.917	16.710
907.000	100.416	252.473	192.040	54.813	------ CRYSTAL <--> LIQUID -----		
1000	100.416	262.275	198.123	64.152	-395.798	-276.003	14.417
1100	100.416	271.845	204.397	74.193	-393.003	-264.159	12.544
1200	100.416	280.583	210.387	84.235	-390.225	-252.568	10.994
1300	100.416	288.620	216.100	94.276	-387.470	-241.209	9.692
1400	100.416	296.062	221.549	104.318	-512.156	-226.893	8.465
1500	100.416	302.990	226.750	114.360	-508.120	-206.658	7.196
1600	100.416	309.471	231.720	124.401	-504.134	-186.691	6.095
1700	100.416	315.558	236.474	134.443	-500.207	-166.972	5.130
1800	100.416	321.298	241.029	144.484	-496.345	-147.481	4.280
1900	100.416	326.727	245.398	154.526	-492.552	-128.204	3.525
2000	100.416	331.878	249.594	164.568	-488.830	-109.124	2.850

PREVIOUS: CURRENT: December 1974

Magnesium Iodide (MgI_2) I_2Mg_1(l)

MAGNESIUM IODIDE (MgI_2) CRYSTAL-LIQUID M_r = 278.1140

0 to 907 K crystal
above 907 K liquid

Refer to the individual tables for details.

Magnesium Iodide (MgI_2) I_2Mg_1(cr,l)

Enthalpy Reference Temperature = T_r = 298.15 K Standard State Pressure = $p°$ = 0.1 MPa

T/K	$C_p°$ (J K⁻¹mol⁻¹)	$S°$ (J K⁻¹mol⁻¹)	$-[G°-H°(T_r)]/T$ (J K⁻¹mol⁻¹)	$H°-H°(T_r)$ (kJ mol⁻¹)	$\Delta_fH°$ (kJ mol⁻¹)	$\Delta_fG°$ (kJ mol⁻¹)	Log K_f
0							
100							
200							
298.15	74.810	129.704	129.704	0.	-366.937	-361.240	63.288
300	74.977	130.167	129.705	0.139	-366.945	-361.204	62.891
400	78.408	152.229	132.690	7.815	-383.462	-358.661	46.836
500	80.919	170.002	138.432	15.785	-426.355	-348.339	36.391
600	83.011	184.946	144.971	23.985	-424.677	-332.893	28.981
700	84.726	197.873	151.626	32.373	-422.928	-317.733	23.710
800	86.274	209.290	158.134	40.925	-421.144	-302.826	19.773
900	87.655	219.502	164.395	49.596	-419.381	-288.143	16.723
907.000	87.864	220.182	164.823	50.210	CRYSTAL <--> LIQUID		
907.000	100.416	252.473	164.823	79.498	TRANSITION		
1000	100.416	262.275	173.438	88.837	-395.798	-276.003	14.417
1100	100.416	271.845	181.956	98.879	-393.003	-264.159	12.544
1200	100.416	280.583	189.816	108.920	-390.225	-252.568	10.994
1300	100.416	288.620	197.111	118.962	-387.470	-241.209	9.692
1400	100.416	296.062	203.917	129.003	-512.156	-226.893	8.465
1500	100.416	302.990	210.293	139.045	-508.120	-206.658	7.196
1600	100.416	309.471	216.291	149.087	-504.134	-186.691	6.095
1700	100.416	315.558	221.953	159.128	-500.207	-166.972	5.130
1800	100.416	321.298	227.315	169.170	-496.345	-147.481	4.280
1900	100.416	326.727	232.405	179.211	-492.552	-128.204	3.525
2000	100.416	331.878	237.251	189.253	-488.830	-109.124	2.850

PREVIOUS: CURRENT: December 1974

Magnesium Iodide (MgI_2) I_2Mg_1(cr,l)

MAGNESIUM IODIDE (MgI_2) IDEAL GAS M_r = 278.1140

$S°(298.15\ K) = 317.489\ J\ K^{-1}\ mol^{-1}$

$\Delta_f H°(0\ K) = -157.56 \pm 10.5\ kJ\ mol^{-1}$
$\Delta_f H°(298.15\ K) = -160.25 \pm 10.5\ kJ\ mol^{-1}$

Vibrational Frequencies and Degeneracies

ν, cm^{-1}
[125](1)
[60](2)
[423](1)

Ground State Quantum Weight: [1] σ = 2
Point Group: $D_{\infty h}$
Bond Distance: Mg-I = 2.52 ± 0.03 Å
Bond Angle: I-Mg-I = 180°
Rotational Constant: B_o = [0.010459] cm^{-1}

Enthalpy of Formation

$\Delta_f H°(MgI_2, g, 298.15\ K)$ is obtained from that of the crystal by addition of $\Delta_{sub}H°(298.15\ K) = 49.4 \pm 0.8\ kcal\ mol^{-1}$. The enthalpy of sublimation is derived from the sublimation pressure measurements of Berkowitz and Marquart (1). See MgI_2(cr) table.

Heat Capacity and Entropy

The bond distance is taken from the electron diffraction study of Akishin and Spiridonov (2). From an electric deflection study of mass spectrometrically detected molecular beams, Buechler et al. (3) concluded that MgI_2 is linear.

The equality of the stretching force constant of the alkaline earth monohalide and the stretching force constant of the alkaline earth dihalide has been indicated as a reasonable assumption (4, 5). The stretching force constant for MgI(g) is calculated, using the valence force model, from the ground state vibrational frequency given by Rosen (6). The bending force constant for MgI_2(g) is taken as 0.01 of the stretching force constant. Brewer et al. (4) used a factor of 0.1 for MgI_2; however, a factor of 0.01 appears more reasonable (7). The adopted vibrational frequencies are calculated from the estimated force constants. Other estimates for ν_1, ν_2, and ν_3 are 132, 199, 446 (4) and 137, 140, 464 (8).

We assign an uncertainty of ±2 cal K^{-1} mol^{-1} to the entropy to allow for error in estimating the vibrational frequencies.

References

1. J. Berkowitz and J. R. Marquart, J. Chem. Phys. 37, 1853 (1962).
2. P. A. Akishin and V. P. Spiridonov, Zhur. Fiz. Khim. 32, 1682 (1958).
3. A. Buechler, J. L. Stauffer and W. Klemperer, J. Amer. Chem. Soc. 86, 4544 (1964).
4. L. Brewer, G. R. Somyajulu, and E. Brackett, Chem. Rev. 63, 111 (1963).
5. JANAF Thermochemcial Tables: BeF_2(g), 6-30-70; MgF_2(g), 3-31-66; CaF_2(g), 12-31-68; SrF_2(g), 12-31-72; BaF_2(g), 12-31-72; CaI_2(g), 6-30-74; SrI_2(g), 6-30-74; BaI_2(g), 6-30-74.
6. B. Rosen, Ed., "Spectroscopic Data Relative to Diatomic Molecules," Pergamon Press, New York, 1970.
7. JANAF Thermochemical Table: $MgBr_2$(g), 6-30-74.
8. K. S. Krasnov and V. I. Svettsov, Izv. Vysshikh Uchebn. Zavedenii, Khim. i Khim. Tekhnol. 6, 167 (1963).

Magnesium Iodide (MgI_2) I_2Mg_1(g)

Enthalpy Reference Temperature = T_r = 298.15 K — Standard State Pressure = p° = 0.1 MPa

	J K^{-1}mol^{-1}			kJ mol^{-1}			
T/K	$C_p°$	S°	$-[G°-H°(T_r)]/T$	$H°-H°(T_r)$	$\Delta_f H°$	$\Delta_f G°$	Log K_f
0	0.	0.	INFINITE	-15.512	-157.563	-157.563	INFINITE
100	51.832	256.256	368.725	-11.247	-156.990	-175.566	91.706
200	57.298	294.115	322.882	-5.753	-158.436	-193.639	50.573
250	58.718	307.064	318.466	-2.850	-159.332	-202.337	42.276
298.15	59.623	317.489	317.489	0.	-160.247	-210.538	36.885
300	59.651	317.858	317.490	0.110	-160.284	-210.850	36.712
350	60.281	327.104	318.219	3.110	-161.359	-219.196	32.713
400	60.721	335.183	319.845	6.135	-178.452	-226.834	29.621
450	61.038	342.355	321.955	9.180	-180.762	-232.743	27.016
500	61.273	348.798	324.323	12.238	-223.213	-234.595	24.508
600	61.590	360.000	329.363	18.382	-223.590	-236.838	20.619
700	61.787	369.510	334.436	24.552	-224.060	-239.010	17.835
800	61.917	377.770	339.348	30.737	-224.641	-241.108	15.743
900	62.008	385.068	344.030	36.934	-225.354	-243.125	14.111
1000	62.073	391.605	348.467	43.138	-234.808	-244.342	12.763
1100	62.122	397.523	352.662	49.348	-235.844	-245.245	11.646
1200	62.159	402.930	356.629	55.562	-236.893	-246.054	10.710
1300	62.189	407.907	360.384	61.780	-237.962	-246.774	9.916
1400	62.212	412.517	363.945	68.000	-366.470	-244.244	9.113
1500	62.230	416.809	367.328	74.222	-366.253	-235.521	8.202
1600	62.246	420.826	370.548	80.446	-366.085	-226.812	7.405
1700	62.259	424.600	373.617	86.671	-365.974	-218.111	6.702
1800	62.269	428.159	376.550	92.897	-365.928	-209.414	6.077
1900	62.278	431.526	379.355	99.125	-365.949	-200.719	5.518
2000	62.286	434.721	382.044	105.353	-366.040	-192.021	5.015
2100	62.293	437.760	384.626	111.582	-366.198	-183.316	4.560
2200	62.298	440.658	387.107	117.811	-366.418	-174.602	4.146
2300	62.304	443.427	389.496	124.042	-366.693	-165.878	3.767
2400	62.308	446.079	391.799	130.272	-367.014	-157.140	3.420
2500	62.312	448.623	394.021	136.503	-367.368	-148.388	3.100
2600	62.315	451.067	396.169	142.734	-367.747	-139.621	2.805
2700	62.318	453.418	398.246	148.966	-368.138	-130.840	2.531
2800	62.321	455.685	400.257	155.198	-368.531	-122.044	2.277
2900	62.324	457.872	402.206	161.430	-368.916	-113.234	2.040
3000	62.326	459.985	404.097	167.663	-369.285	-104.411	1.818
3100	62.328	462.028	405.933	173.896	-369.631	-95.576	1.610
3200	62.330	464.007	407.717	180.128	-369.947	-86.730	1.416
3300	62.332	465.925	409.452	186.362	-370.230	-77.875	1.233
3400	62.333	467.786	411.141	192.595	-370.475	-69.012	1.060
3500	62.334	469.593	412.785	198.828	-370.680	-60.142	0.898
3600	62.336	471.349	414.387	205.062	-370.846	-51.268	0.744
3700	62.337	473.057	415.950	211.295	-370.971	-42.388	0.598
3800	62.338	474.719	417.475	217.529	-371.057	-33.507	0.461
3900	62.339	476.339	418.964	223.763	-371.104	-24.623	0.330
4000	62.340	477.917	420.418	229.997	-371.117	-15.739	0.206
4100	62.341	479.456	421.839	236.231	-371.095	-6 855	0.087
4200	62.342	480.959	423.229	242.465	-371.044	2.029	-0.025
4300	62.342	482.426	424.589	248.699	-370.966	10.911	-0.133
4400	62.343	483.859	425.919	254.934	-370.865	19.790	-0.235
4500	62.344	485.260	427.223	261.168	-370.744	28.668	-0.333
4600	62.344	486.630	428.499	267.402	-370.608	37.542	-0.426
4700	62.345	487.971	429.750	273.637	-370.459	46.413	-0.516
4800	62.346	489.283	430.977	279.871	-370.302	55.281	-0.602
4900	62.346	490.569	432.180	286.106	-370.140	64.146	-0.684
5000	62.347	491.829	433.360	292.341	-369.978	73.007	-0.763
5100	62.347	493.063	434.519	298.575	-369.819	81.865	-0.838
5200	62.347	494.274	435.657	304.810	-369.665	90.721	-0.911
5300	62.348	495.461	436.774	311.045	-369.521	99.572	-0.981
5400	62.348	496.627	437.871	317.279	-369.390	108.422	-1.049
5500	62.349	497.771	438.950	323.514	-369.275	117.269	-1.114
5600	62.349	498.894	440.011	329.749	-369.179	126.115	-1.176
5700	62.349	499.998	441.053	335.984	-369.105	134.959	-1.237
5800	62.349	501.082	442.079	342.219	-369.055	143.801	-1.295
5900	62.350	502.148	443.088	348.454	-369.032	152.643	-1.351
6000	62.350	503.196	444.081	354.689	-369.039	161.485	-1.406

PREVIOUS: December 1974 (1 atm) CURRENT: December 1974 (1 bar)

Magnesium Iodide (MgI_2) I_2Mg_1(g)

MOLYBDENUM IODIDE (α-MoI_2) CRYSTAL M_r = 349.7490

$\Delta_fH°$(0 K) = Unknown

S°(298.15 K) = [149.662 ± 16.7] J K^{-1} mol^{-1} $\Delta_fH°$(298.15 K) = -103.93 ± 12.6 kJ mol^{-1}

T_{dcm} = 1305 K (total p = 1 atm)

Enthalpy of Formation

Barnes et al. (1) have measured calorimetrically the enthalpy of formation of $MoI_{2.76}$ by reacting iodine vapor with $Mo(CO)_6$ at 539 K. The products were shown to consist of MoI_3, MoI_2, and Mo by chemical analysis. Assuming a difference in the enthalpies of formation of MoI_3 and MoI_2 of -1.5 kcal mol^{-1}, Barnes et al. (1) determined $\Delta_fH°$(MoI_2, cr, 298.15 K) = -24.7±2 kcal mol^{-1}. Brewer (2) has estimated that the difference $\Delta_fH°(MoI_3) - \Delta_fH°(MoI_2)$ may be as large as -5 kcal mol^{-1}, and he reanalyzed the enthalpy of iodination data with the result $\Delta_fH°$(MoI_2, cr, 298.15 K)/R = -12500±1500 K which is adopted. The assigned uncertainty (±3 kcal mol^{-1}) is determined from the product 1500xR which has been rounded to the nearest kcal mol^{-1}.

Heat Capacity and Entropy

Values of S°(298.15 K) and $C_p°$ over the temperature range 298-1500 K are estimates reported by Brewer (2). We treat his reported value of S°(298.15 K)/R in a manner similar to that for the heat of formation.

Phase Data

The crystal structure (2) of α-MoI_2 in the Pearson classification system of oC72. Glicksman and Walton (3) have reported preparing a new molybdenum diiodide phase (β-MoI_2) by reacting $Mo_2(O_2CCH_3)_4$ with HI gas at 573 K. We assume that this new phase is unstable with respect to α-MoI_2 at all temperatures.

Decomposition Data

T_{dcm} is calculated as the temperature at which the partial pressures of I_2(g) and I(g) reach one atmosphere over crystalline MoI_2 and Mo. At T_{dcm} = 1305 K, we calculate the vapor composition to consist of 0.63 atm of I_2 and 0.37 atm of I. Saturated vapor pressure measurements by Drobot et al. (4) on MoI_2(cr) prepared by thermal decomposition of MoI_3(cr) show that initial decomposition to the metal and iodine occurs at 1063 K and 0.33 atm.

References

1. D. S. Barnes, G. Pilcher, D. A. Pittam, H. A. Skinner, D. Todd, and Y. Virmani, J. Less-Common Met. 36, 177 (1974).
2. L. Brewer, Materials and Molecular Research Division, Lawrence Berkeley Laboratory, University of California, Berkeley; personal communication, Sept. 28, 1978, preliminary draft of review to be submitted for publication in Atomic Energy Review, International Atomic Energy Agency, Vienna, Austria.
3. H. D. Glicksman and R. A. Walton, Inorg. Chem. 17, 200 (1978).
4. D. V. Drobot, L. G. Mikhailova, K. A. Bol'shakov, and V. L. Sbitnev, Russ. J. Inorg. Chem. 23, 643 (1978).

Molybdenum Iodide, Alpha (MoI_2) I_2Mo_1(cr)

Enthalpy Reference Temperature = T_r = 298.15 K Standard State Pressure = p° = 0.1 MPa

T/K	$C_p°$ (J $K^{-1}mol^{-1}$)	S° (J $K^{-1}mol^{-1}$)	$-[G°-H°(T_r)]/T$ (J $K^{-1}mol^{-1}$)	$H°-H°(T_r)$ (kJ mol^{-1})	$\Delta_fH°$ (kJ mol^{-1})	$\Delta_fG°$ (kJ mol^{-1})	Log K_f
0							
100							
200							
298.15	82.705	149.662	149.662	0.	-103.931	-105.396	18.465
300	82.730	150.173	149.663	0.153	-103.923	-105.405	18.353
400	83.977	174.144	152.923	8.488	-119.681	-105.390	13.763
500	85.224	193.017	159.120	16.948	-161.966	-97.768	10.214
600	86.471	208.665	166.110	25.533	-159.752	-85.135	7.412
700	87.718	222.089	173.171	34.243	-157.483	-72.878	5.438
800	88.964	233.883	180.037	43.077	-155.149	-60.950	3.980
900	90.211	244.433	186.616	52.035	-152.747	-49.318	2.862
1000	91.458	254.003	192.884	61.119	-150.278	-37.958	1.983
1100	92.705	262.778	198.844	70.327	-147.748	-26.848	1.275
1200	93.952	270.898	204.515	79.660	-145.166	-15.971	0.695
1300	95.199	278.467	209.915	89.117	-142.546	-5.310	0.213
1400	96.445	285.568	215.068	98.700	-139.900	5.147	-0.192
1500	97.696	292.264	219.993	108.407	-137.246	15.415	-0.537

PREVIOUS: CURRENT: September 1978

Molybdenum Iodide, Alpha (MoI_2) I_2Mo_1(cr)

MOLYBDENUM IODIDE (MoI_2) IDEAL GAS M_r = 349.7490

$S°(298.15\ K) = [339.290 \pm 8.4]\ J\ K^{-1}\ mol^{-1}$

$\Delta_f H°(0\ K) = [257.38 \pm 41.8]\ kJ\ mol^{-1}$

$\Delta_f H°(298.15\ K) = [257.73 \pm 41.8]\ kJ\ mol^{-1}$

Electronic Levels and Quantum Weights

ϵ_i, cm^{-1}	g_i
0	[1]
[243.1]	[3]
[669.6]	[5]
[1225.2]	[7]
[1873.0]	[9]
[11271.0]	[1]
[12510.0]	[3]
[12630.0]	[9]
[13201.0]	[11]
[13742.0]	[13]

Vibrational Frequencies and Degeneracies

ν, cm^{-1}
[200](1)
[50](2)
[200](1)

Point Group: $[D_{\infty h}]$ σ = [2]

Bond Distance: Mo-I = [2.66] Å

Bond Angle: I-Mo-I = [180]°

Rotational Constant: $B_0 = [0.009387]\ cm^{-1}$

Enthalpy of Formation

The adopted value of $\Delta_f H°$ at 298.15 K is that estimated by Brewer (1). The estimated value is consistent with the low volatility of solid molybdenum in iodine vapor reported by Schafer et al. (2). Brewer (1) has given no weight to the equilibrium data reported by Allen and Yost (3) for the formation of $MoI_2(g)$ from Mo(cr) and I(g) since the product formed may have been MoO_2I_2 rather than the diiodide (2).

Brewer's estimated value is given as $\Delta_f H°(298.15\ K/F)$ = 31000±5000 K. The assigned uncertainty (±10 kcal mol^{-1}) is determined from the product 5000xR which has been rounded to the nearest kcal mol^{-1}. The value of $\Delta_f H°(0\ K)$ combined with JANAF data (4) for Mo(g) and I(g) gives $\Delta_{at}H°(0\ K)$ = 147±11 kcal mol^{-1} and an average Mo-I bond energy of $\Delta_{at}H°(0\ K/2)$ = 73.4±5.5 kcal mol^{-1}.

Heat Capacity and Entropy

The bond length and vibrational frequencies are values estimated by Brewer (1). A linear structure has been predicted for MoI_2 by Drake and Rosenblatt (5). The electronic contribution is taken to be the same as that for the free gaseous Mo^{+2} ion, as estimated by Brewer (1). The levels listed by Rico (6) below 1374 cm^{-1} are used in the calculations.

Our heat capacity and enthalpy values up to 3000 K are in agreement with those determined by Brewer (1). However, his entropies and free energy functions differ from our values by about 1 cal K^{-1} mol^{-1} at all temperatures over the range 298-3000 K. We believe that his results are erroneous due to the use of an incorrect value for the molecular weight of MoI_2. We reproduce his entropies and free energy functions if we use a molecular weight of 476.65 which corresponds to MoI_3.

References

1. L. Brewer, Materials and Molecular Research Division, Lawrence Berkeley Laboratory, University of California, Berkeley: personal communication, Sept. 20, 1978, preliminary draft of review to be submitted for publication in Atomic Energy Review, International Atomic Energy Agency, Vienna, Austria.
2. H. Schafer, T. Grofe, and M. Trenkel, J. Solid State Chem. 8, 14 (1973).
3. T. L. Allen and D. M. Yost, J. Chem. Phys. 22, 855 (1954).
4. JANAF Thermochemical Tables: Mo(g), 3-31-78; I(g), 6-30-74.
5. M. C. Drake and G. M. Rosenblatt, Paper 388, The Electrochemical Society, Atlanta, Georgia, October, 1977.
6. F. R. Rico, An. Real Soc. Espan. Fis. Quim., Ser. A 61, 103 (1965).

Molybdenum Iodide (MoI_2) $I_2Mo_1(g)$

Enthalpy Reference Temperature = T_r = 298.15 K Standard State Pressure = p° = 0.1 MPa

	J $K^{-1}mol^{-1}$			kJ mol^{-1}			
T/K	$C_p°$	S°	$-[G°-H°(T_r)]/T$	$H°-H°(T_r)$	$\Delta_f H°$	$\Delta_f G°$	Log K_f
0	0.	0.	INFINITE	-18.140	257.377	257.377	INFINITE
100	61.708	265.272	400.919	-13.565	258.411	238.686	-124.677
200	69.265	311.318	345.722	-6.881	258.321	218.944	-57.182
250	70.171	326.880	340.452	-3.393	258.063	209.128	-43.695
298.15	70.717	339.290	339.290	0.	257.734	199.731	-34.992
300	70.733	339.728	339.291	0.131	257.720	199.372	-34.714
350	71.082	350.660	340.154	3.677	257.241	189.682	-28.308
400	71.254	360.164	342.074	7.236	240.732	180.614	-23.586
450	71.268	368.559	344.560	10.800	238.997	173.204	-20.105
500	71.146	376.063	347.342	14.361	197.111	169.786	-17.737
600	70.599	388.991	353.240	21.451	197.830	164.252	-14.299
700	69.827	399.818	359.142	28.473	198.412	158.607	-11.835
800	68.995	409.087	364.820	35.414	198.853	152.889	-9.983
900	68.198	417.167	370.197	42.273	199.156	147.124	-8.539
1000	67.478	424.314	375.258	49.056	199.324	141.332	-7.382
1100	66.847	430.715	380.014	55.772	199.362	135.530	-6.436
1200	66.305	436.508	384.484	62.429	199.267	129.731	-5.647
1300	65.842	441.796	388.692	69.035	199.037	123.944	-4.980
1400	65.449	446.661	392.662	75.599	198.664	118.181	-4.409
1500	65.118	451.165	396.414	82.127	198.140	112.450	-3.916
1600	64.840	455.358	399.968	88.625	197.453	106.758	-3.485
1700	64.609	459.282	403.343	95.097	196.594	101.115	-3.107
1800	64.422	462.970	406.554	101.548	195.553	95.528	-2.772
1900	64.272	466.449	409.616	107.982	194.322	90.004	-2.474
2000	64.157	469.742	412.540	114.403	192.897	84.549	-2.208
2100	64.075	472.870	415.339	120.815	191.274	79.171	-1.969
2200	64.022	475.850	418.023	127.219	189.452	73.875	-1.754
2300	63.996	478.695	420.599	133.620	187.433	68.666	-1.559
2400	63.995	481.419	423.077	140.019	185.219	63.549	-1.383
2500	64.017	484.031	425.463	146.420	182.812	58.528	-1.223
2600	64.059	486.543	427.765	152.823	180.204	53.607	-1.077
2700	64.120	488.962	429.987	159.232	177.369	48.791	-0.944
2800	64.198	491.295	432.135	165.648	174.257	44.085	-0.822
2900	64.289	493.549	434.214	172.072	134.916	39.545	-0.712
3000	64.393	495.730	436.228	178.506	133.082	36.288	-0.632
3100	64.507	497.844	438.182	184.951	131.287	33.091	-0.558
3200	64.630	499.894	440.079	191.408	129.542	29.952	-0.489
3300	64.760	501.884	441.921	197.878	127.853	26.866	-0.425
3400	64.895	503.820	443.714	204.360	126.225	23.830	-0.366
3500	65.034	505.703	445.458	210.857	124.663	20.841	-0.311
3600	65.175	507.537	447.157	217.367	123.171	17.896	-0.260
3700	65.317	509.324	448.813	223.892	121.750	14.992	-0.212
3800	65.458	511.068	450.429	230.430	120.401	12.125	-0.167
3900	65.599	512.770	452.005	236.983	119.126	9.292	-0.124
4000	65.738	514.433	453.545	243.550	117.925	6.491	-0.085
4100	65.873	516.058	455.050	250.131	116.796	3.719	-0.047
4200	66.005	517.647	456.522	256.725	115.739	0.974	-0.012
4300	66.133	519.201	457.962	263.332	114.753	-1.746	0.021
4400	66.257	520.723	459.371	269.951	113.835	-4.445	0.053
4500	66.375	522.214	460.751	276.583	112.984	-7.123	0.083
4600	66.487	523.674	462.103	283.226	112.198	-9.784	0.111
4700	66.594	525.105	463.428	289.880	111.475	-12.428	0.138
4800	66.695	526.508	464.728	296.545	110.812	-15.056	0.164
4900	66.790	527.884	466.003	303.219	110.207	-17.672	0.188
5000	66.879	529.234	467.254	309.902	-481.727	-14.545	0.152
5100	66.962	530.559	468.482	316.595	-483.902	-5.180	0.053
5200	67.038	531.860	469.688	323.295	-486.173	4.229	-0.042
5300	67.109	533.138	470.874	330.002	-488.538	13.681	-0.135
5400	67.173	534.393	472.038	336.716	-491.000	23.180	-0.224
5500	67.232	535.626	473.183	343.436	-493.557	32.725	-0.311
5600	67.284	536.838	474.309	350.162	-496.210	42.318	-0.395
5700	67.331	538.029	475.417	356.893	-498.956	51.959	-0.476
5800	67.373	539.201	476.506	363.628	-501.797	61.649	-0.555
5900	67.409	540.353	477.579	370.368	-504.729	71.388	-0.632
6000	67.440	541.486	478.634	377.110	-507.755	81.178	-0.707

PREVIOUS: September 1978 (1 atm) CURRENT: September 1978 (1 bar)

Molybdenum Iodide (MoI_2) $I_2Mo_1(g)$

LEAD IODIDE (PbI_2) CRYSTAL M_r = 461.0090

$S°(298.15 K) = 174.836 \pm 0.21\ J\ K^{-1}\ mol^{-1}$
$T_{fus} = 683 \pm 2\ K$

$\Delta_fH°(0\ K) = -174.817 \pm 0.42\ kJ\ mol^{-1}$
$\Delta_fH°(298.15\ K) = -175.393 \pm 0.42\ kJ\ mol^{-1}$
$\Delta_{fus}H° = 23.430 \pm 0.8\ kJ\ mol^{-1}$

Enthalpy of Formation

Koref and Braune (1) determined $\Delta_fH°(PbI_2$, cr, 298.15 K) = -41.850 kcal mol^{-1} by enthalpy of solution calorimetry. 3rd law analyses of emf data at 298.15 K reported by Gerke (2) and Cann and Taylor (3) for the reaction Pb(cr) + I_2(cr) = PbI_2(cr) lead to $\Delta_fH°(PbI_2$, cr, 298.15 K) = -41.923 and -41.925 kcal mol^{-1}, respectively. $\Delta_fH°(PbI_2$, cr, 298.15 K) = -41.92 ± 0.10 kcal mol^{-1} is adopted.

Heat Capacity and Entropy

McBride (4) measured the low temperature heat capacity (4.52-291.00 K) of PbI_2. Enthalpies have been measured by Ehrhardt (5) (448-776 K), Magnus (6) (373-523 K), and Linsey (7) (323-377 K). The enthalpy data of Linsey (7) is subjected to a curve fitting procedure subject to the constraint that it join smoothly in the region of 298.15 K with the enthalpy derived from the heat capacity data of McBride (4). S°(10 K) = 0.662 cal K^{-1} mol^{-1} is adopted from McBride (4). The data of Linsey (7) did not indicate any transitions other than the solid-liquid transition at 683 K.

Fusion Data

Germann and Metz (8), van Klooster and Stearns (9) and Popovkin, Odin, and Novoselova (10) determined T_{fus} = 685±1 K. Blanc and Petit (11) found T_{fus} = 680 K. Sumarokova and Modestova (12) found PbI_2 to exist in two crystalline forms with the α-β transition at 645 K. On a DTA cooling curve, Popovkin et al. (10) observed a peak at 656 K which they attribute to the β-α transition. Modestova and Sumarokova (13) observed the β-form to melt at 669 K. Ehrhardt's (5) drop calorimetry data and his interpretation of the enthalpy increase at 648 K as indicating the melting points are consistent with the existence of a β-phase. The observation by Novoselova et al. (14) that only the α-form crystallizes on cooling melts of the PbI_2 rich region of the PbI_2-PbSe system is consistent with the interpretation that the reference state for Ehrhardt's work is α-PbI_2. The enthalpy measurements by Linsey (7) led to a reported value of T_{fus} = 683 K. This value is adopted.

From a plot of Ehrhardt's enthalpy data, $\Delta_{fus}H°$ = 5.43 kcal mol^{-1} at 685 K is obtained. Kelley (15) derived 5.57, 5.67, and 5.80 kcal mol^{-1} from phase diagram information and also gives 6.01 kcal mol^{-1} (16). Another analysis (17) has resulted in $\Delta_{fus}H°$ = 5.20 kcal mol^{-1}. Blanc and Petit's (11) value of 3.87 kcal mol^{-1} is too low. The enthalpy data of Linsey (7) was analyzed to yield $\Delta_{fus}H°$ = 5.6 kcal mol^{-1}. $\Delta_{fus}H°$ = 5.6±0.2 kcal mol^{-1} is adopted.

Sublimation Data

$\Delta_{sub}H°$(298.15 K) is calculated from the difference between the adopted values for $\Delta_fH°$(298.15 K) of the gas and crystal. Refer to PbI_2(g) table.

References

1. F. Koref and H. Braune, Z. Elektrochem. 18, 818 (1912).
2. R. H. Gerke, J. Amer. Chem. Soc. 44, 1684 (1922).
3. J. Y. Cann and A. C. Taylor, J. Amer. Chem. Soc. 59, 1987 (1937).
4. J. J. McBride, Ph.D. Dissertation, University of Michigan, 1955.
5. O. Ehrhardt, Wied. Ann. 24, 215 (1885).
6. A. Magnus, Ann. Physik 31, 597 (1910).
7. C. W. Linsey, Ph.D. Dissertation, North Texas State University, 1970.
8. F. E. E. Germann and C. F. Metz, J. Phys. Chem. 35, 1944 (1931).
9. H. S. van Klooster and E. E. Stearns, J. Amer. Chem. Soc. 55, 4121 (1933).
10. B. A. Popovkin, I. N. Odin, A. V. Novoselova, Vestn. Mosk. Univ., Khim. 12, 316 (1971).
11. M. Blanc and G. Petit, Compt. rend. 248, 1305 (1959).
12. T. N. Sumarokova, and T. Modestova, Zhur. Neorg. Khim. 1, 2027 (1956).
13. T. Modestova and T. N. Sumarokova, Zhur. Neorg. Khim. 3, 1355 (1958).
14. A. V. Novoselova, I. N. Odin, and B. A. Popovkin, Isv. Akad. Nauk SSSR, Neorg. Materialy 2, 1397 (1966).
15. K. K. Kelley, U. S. Bur. Mines Bull. 393, 1936.
16. K. K. Kelley, U. S. Bur. Mines Bull. 584, 1960.
17. U. S. Nat. Bur. Stand. Circ. 500, 1952.

Lead Iodide (PbI_2) I_2Pb_1(cr)

Enthalpy Reference Temperature = T_r = 298.15 K — Standard State Pressure = p° = 0.1 MPa

T/K	$C_p°$ (J $K^{-1}mol^{-1}$)	S° (J $K^{-1}mol^{-1}$)	$-[G°-H°(T_r)]/T$ (J $K^{-1}mol^{-1}$)	$H°-H°(T_r)$ (kJ mol^{-1})	$\Delta_fH°$ (kJ mol^{-1})	$\Delta_fG°$ (kJ mol^{-1})	Log K_f
0	0.	0.	INFINITE	-19.501	-174.817	-174.817	INFINITE
100	69.789	93.717	242.109	-14.839	-174.972	-174.562	91.182
200	75.396	144.332	181.860	-7.506	-175.096	-174.116	45.474
298.15	77.573	174.836	174.836	0.	-175.393	-173.577	30.410
300	77.613	175.316	174.838	0.144	-175.400	-173.566	30.221
400	78.856	197.846	177.901	7.978	-191.933	-172.328	22.504
500	80.391	215.574	183.723	15.925	-234.996	-163.287	17.059
600	83.713	230.498	190.305	24.116	-233.448	-149.087	12.979
683.000	87.780	241.591	195.872	31.226	------ CRYSTAL <--> LIQUID -----		
700	88.743	243.761	197.009	32.726	-236.428	-134.371	10.027
800	95.102	256.015	203.627	41.910	-234.040	-119.947	7.832
900	102.428	267.637	210.098	51.784	-230.941	-105.864	6.144
1000	110.064	278.821	216.415	62.406	-227.076	-92.166	4.814

PREVIOUS: March 1962 CURRENT: December 1973

Lead Iodide (PbI_2) I_2Pb_1(cr)

LEAD IODIDE (PbI_2) LIQUID M_r = 461.0090

$S°(298.15\ K) = [198.902]\ J\ K^{-1}\ mol^{-1}$

$T_{fus} = 683 \pm 2\ K$

$\Delta_f H°(298.15\ K) = [-157.679]\ kJ\ mol^{-1}$

$\Delta_{fus}H° = 23.430 \pm 0.8\ kJ\ mol^{-1}$

Enthalpy of Formation

$\Delta_f H°(PbI_2$, ℓ, 298.15 K) is calculated from $\Delta_f H°(PbI_2$, cr, 298.15 K) by adding $\Delta_{fus}H°$ and the difference in the enthalpy, H°(683 K)-H°(298.15 K), between the crystal and the liquid.

$\Delta_f H°$(ℓ) can also be obtained from a 2nd and 3rd law analysis of emf data for the cell reaction Pb(ℓ) + I_2(g) = PbI_2(ℓ) as reported by Murgulescu et al. (1) in the form of an equation covering a range of 723 to 923 K. The enthalpy changes for the reaction are $\Delta_r H°$(298.15 K) = -57.65±0.18 kcal mol^{-1} (2nd law) and -52.74±1.08 kcal mol^{-1} (3rd law) with a drift of 5.8±0.2 cal K^{-1} mol^{-1}. With auxiliary data $\Delta_f H°$(Pb, ℓ, 298.15 K) = 1.025 kcal mol^{-1} (2) and $\Delta_f H°(I_2$, g, 298.15 K) = 14.924 kcal mol^{-1} (2), $\Delta_f H°(PbI_2$, ℓ, 298.15 K) = -41.70 kcal mol^{-1} and -36.79 kcal mol^{-1} from the 2nd and 3rd law calculations, respectively.

Heat Capacity and Entropy

The heat capacity is assumed to be constant at 25.95 cal K^{-1} mol^{-1} from 460 to 2000 K and is based on a linear least squares fit of the liquid range enthalpy measurements (683-877 K) of Linsey (4). The enthalpy measurements of Ehrhardt (3) for the liquid range covered a smaller temperature region (687-776 K) and are considered less reliable. A least squares fit of the Ehrhardt data (3) leads to a heat capacity value of 32.5 cal K^{-1} mol^{-1}. A glass transition is assumed at 460 K below which the heat capacity is that of the crystal.

$S°(PbI_2$, ℓ, 298.15 K) is calculated in a manner analogous to that used for the enthalpy of formation.

Vaporization Data

T_{vap} = 1105 K is calculated as the temperature at which the fugacity is 1 atm for the reaction PbI_2(cr) = PbI_2(g). $\Delta_{vap}H°$ = 28.34 kcal mol^{-1} is calculated as the difference between $\Delta_f H°$ at T_{vap} for the gas and the liquid.

References

1. I. G. Murgulescu, S. Sternberg, and M. Terzi, Electrochim. Acta 12, 1121 (1967).
2. JANAF Thermochemical Tables: Pb(ℓ), 3-31-62; I_2(g), 9-30-61.
3. O. Ehrhardt, Wied. Ann. 24, 215 (1885).
4. C. W. Linsey, Ph.D. Dissertation, North Texas State University, 1970.

Lead Iodide (PbI_2) I_2Pb_1(l)

Enthalpy Reference Temperature = T_r = 298.15 K Standard State Pressure = p° = 0.1 MPa

T/K	$C_p°$ (J K⁻¹mol⁻¹)	S°	$-[G°-H°(T_r)]/T$	$H°-H°(T_r)$ (kJ mol⁻¹)	$\Delta_f H°$	$\Delta_f G°$	Log K_f
0							
100							
200							
298.15	77.571	198.902	198.902	0.	-157.679	-163.038	28.564
300	77.613	199.382	198.904	0.144	-157.686	-163.072	28.393
400	78.856	221.912	201.967	7.978	-174.219	-164.241	21.448
460.000	79.592	232.980	205.306	12.730	GLASS <--> LIQUID TRANSITION		
460.000	108.575	232.980	205.306	12.730			
500	108.575	242.033	207.887	17.073	-216.134	-157.655	16.470
600	108.575	261.829	215.278	27.930	-211.920	-146.357	12.741
683.000	108.575	275.896	221.808	36.942	------ CRYSTAL <--> LIQUID -----		
700	108.575	278.565	223.154	38.788	-212.652	-134.959	10.071
800	108.575	293.064	231.007	49.645	-208.591	-124.137	8.105
900	108.575	305.852	238.627	60.503	-204.509	-113.825	6.606
1000	108.575	317.291	245.931	71.360	-200.408	-103.969	5.431
1100	108.575	327.640	252.896	82.218	-196.290	-94.524	4.489
1200	108.575	337.087	259.524	93.075	-192.159	-85.455	3.720
1300	108.575	345.778	265.829	103.933	-188.028	-76.730	3.083
1400	108.575	353.824	271.831	114.790	-183.920	-68.323	2.549
1500	108.575	361.315	277.550	125.648	-179.855	-60.208	2.097

PREVIOUS: March 1962 CURRENT: December 1973

Lead Iodide (PbI_2) I_2Pb_1(l)

LEAD IODIDE (PbI_2) CRYSTAL–LIQUID M_r = 461.0090

0 to 683 K crystal
above 683 K liquid

Refer to the individual tables for details.

Lead Iodide (PbI_2) $I_2Pb_1(cr,l)$

Enthalpy Reference Temperature = T_r = 298.15 K Standard State Pressure = $p°$ = 0.1 MPa

T/K	$C_p°$ (J K⁻¹mol⁻¹)	$S°$ (J K⁻¹mol⁻¹)	$-[G°-H°(T_r)]/T$ (J K⁻¹mol⁻¹)	$H°-H°(T_r)$ (kJ mol⁻¹)	$\Delta_f H°$ (kJ mol⁻¹)	$\Delta_f G°$ (kJ mol⁻¹)	Log K_f
0	0.	0.	INFINITE	-19.501	-174.817	-174.817	INFINITE
100	69.789	93.717	242.109	-14.839	-174.972	-174.562	91.182
200	75.396	144.332	181.860	-7.506	-175.096	-174.116	45.474
298.15	77.573	174.836	174.836	0.	-175.393	-173.577	30.410
300	77.613	175.316	174.838	0.144	-175.400	-173.566	30.221
400	78.856	197.846	177.901	7.978	-191.933	-172.328	22.504
500	80.391	215.574	183.723	15.925	-234.996	-163.287	17.059
600	83.713	230.498	190.305	24.116	-233.448	-149.087	12.979
683.000	87.780	241.591	195.872	31.226	CRYSTAL <--> LIQUID TRANSITION		
683.000	108.575	275.896	195.872	54.656			
700	108.575	278.565	197.848	56.502	-212.652	-134.959	10.071
800	108.575	293.064	208.864	67.359	-208.591	-124.137	8.105
900	108.575	305.852	218.944	78.217	-204.509	-113.825	6.606
1000	108.575	317.291	228.217	89.074	-200.408	-103.969	5.431
1100	108.575	327.640	236.792	99.932	-196.290	-94.524	4.489
1200	108.575	337.087	244.762	110.789	-192.159	-85.455	3.720
1300	108.575	345.778	252.203	121.647	-188.028	-76.730	3.083
1400	108.575	353.824	259.178	132.504	-183.920	-68.323	2.549
1500	108.575	361.315	265.740	143.362	-179.855	-60.208	2.097

PREVIOUS: CURRENT: December 1973

Lead Iodide (PbI_2) $I_2Pb_1(cr,l)$

LEAD IODIDE (PbI_2) IDEAL GAS M_r = 461.0090

$S°(298.15\ K) = [359.552 \pm 8.4]\ J\ K^{-1}\ mol^{-1}$

$\Delta_f H°(0\ K) = 1.16 \pm 4.2\ kJ\ mol^{-1}$
$\Delta_f H°(298.15\ K) = -3.18 \pm 4.2\ kJ\ mol^{-1}$

Vibrational Frequencies and Degeneracies

ν, cm^{-1}
[100](1)
[40](1)
[160](1)

Ground State Quantum Weight: [1] $\sigma = 2$
Point Group: C_{2v}
Bond Distance: Pb-I = 2.79 ± 0.02 Å
Bond Angle: Br-Pb-Br = [95]°
Product of the Moments of Inertia: $I_A I_B I_C = [2.948087 \times 10^{-111}]\ g^3\ cm^6$

Enthalpy of Formation

The adopted value of $\Delta_f H°(298.15\ K) = -0.76 \pm 1.0\ kcal\ mol^{-1}$ is the average of results from 3rd law analyses of two sets (1, 2) of sublimation data and one set (3) of vaporization data. Auxiliary data $\Delta_f H°(PbI_2, cr, 298.15\ K) = -41.92\ kcal\ mol^{-1}$ (4) and $\Delta_f H°(PbI_2, \ell, 298.15\ K) = -37.686\ kcal\ mol^{-1}$ (4) are used in the calculations. The adopted enthalpy of formation for $PbI_2(g)$ yields an atomization energy of $\Delta_{at} H°(0\ K)$= 4.29 eV for the process $PbI_2(g) = Pb(g) + 2\ I(g)$.

Source	Reaction	Data Points	T/K	$\Delta_r H°$, kcal mol^{-1} 2nd law	3rd law	Drift cal K^{-1} mol^{-1}	$-\Delta_f H°(PbI_2, g, 298.15\ K)$ kcal mol^{-1}
Niwa et al. (1)	A	7	579-650	39.77±0.37	41.54±0.16	2.9±0.60	0.38
Duncan and Thomas (2)	A	13	563-613	39.44±0.40	40.24±0.09	1.4±0.7	1.69
Jellinek and Rudat (3)	B	4	923-1073	35.19±0.52	37.49±0.32	2.3±0.6	0.20

*Reactions: (A) $PbI_2(cr) = PbI_2(g)$ (B) $PbI_2(\ell) = PbI_2(g)$

Heat Capacity and Entropy

Molecular dimensions are those given by Sutton (5). Vibrational frequencies are estimated by comparison with the corresponding values for $HgI_2(g)$ (4) and with the lead dihalide series (4). The electronic ground state is assumed to be A_1 based on analogy with $PbCl_2(g)$ (4). The principal moments of inertia are: $I_A = 67.3005 \times 10^{-39}$, $I_B = 178.3336 \times 10^{-39}$ and $I_C = 245.6341 \times 10^{-39}\ g\ cm^2$.

References

1. K. Niva, M. Sato, and M. Yosiyama, J. Chem. Soc. Japan 60, 918 (1939).
2. J. F. Duncan and F. G. Thomas, J. Chem. Soc. 1964, 360 (1964).
3. K. Jellinek and A. Rudat, Z. physik. Chem. 143, 55 (1929).
4. JANAF Thermochemical Tables: $PbI_2(cr)$, 12-31-73; $PbI_2(\ell)$, 12-31-73; $HgI_2(g)$, 3-31-62; $PbF_2(g)$, 12-31-73; $PbCl_2(g)$, 6-30-73; $PbBr_2(g)$, 12-31-73.
5. L. E. Sutton, Ed., "Tables of Interatomic Distances and Configuration in Molecules and Ions," The Chemical Society, London, 1958.

Lead Iodide (PbI_2) $I_2Pb_1(g)$

Enthalpy Reference Temperature = T_r = 298.15 K Standard State Pressure = p° = 0.1 MPa

T/K	$C_p°$ (J $K^{-1}mol^{-1}$)	S°	$-[G°-H°(T_r)]/T$	$H°-H°(T_r)$ (kJ mol^{-1})	$\Delta_f H°$	$\Delta_f G°$	Log K_f
0	0.	0.	INFINITE	-15.736	1.161	1.161	INFINITE
100	53.807	298.094	410.195	-11.210	0.870	-19.157	10.007
200	56.934	336.667	364.806	-5.628	-1.005	-38.492	10.053
250	57.375	349.424	360.499	-2.769	-2.078	-47.741	9.975
298.15	57.615	359.552	359.552	0.	-3.180	-56.437	9.888
300	57.622	359.908	359.553	0.107	-3.224	-56.767	9.884
350	57.773	368 803	360.255	2.992	-4.504	-65.593	9.789
400	57.872	376.524	361.817	5.883	-21.814	-73.681	9.622
450	57.940	383.345	363.837	8.778	-24.349	-80.012	9.288
500	57.989	389.452	366.099	11.677	-67.031	-82.262	8.594
600	58.054	400.031	370.899	17.479	-67.872	-85.230	7.420
700	58.092	408.983	375.717	23.287	-73.654	-87.253	6.511
800	58.118	416.742	380.371	29.097	-74.639	-89.128	5.819
900	58.135	423.588	384.800	34.910	-75.602	-90 882	5.275
1000	58.148	429.714	388.990	40.724	-76.545	-92.529	4.833
1100	58.157	435.257	392.948	46.539	-77.469	-94.082	4.468
1200	58.164	440.317	396.688	52.355	-78.380	-95.552	4.159
1300	58.169	444.973	400.226	58.172	-79.290	-96 946	3.895
1400	58.174	449.284	403.578	63.989	-80.222	-98.270	3.666
1500	58.177	453.298	406.760	69.807	-81.196	-99.525	3.466
1600	58.180	457.053	409.787	75.625	-82.228	-100.714	3.288
1700	58.182	460.580	412.672	81.443	-83.331	-101.836	3.129
1800	58.184	463.906	415.427	87.261	-84.515	-102.890	2.986
1900	58.186	467.052	418.062	93.080	-85.787	-103.877	2.856
2000	58.188	470.036	420.587	98.898	-87.149	-104.794	2.737
2100	58.189	472.875	423.010	104.717	-265.820	-98.526	2.451
2200	58.190	475.582	425.338	110.536	-266.964	-90.534	2.150
2300	58.191	478.169	427.580	116.355	-268.251	-82.486	1.873
2400	58.192	480.645	429.739	122.174	-269.674	-74.379	1.619
2500	58.192	483.021	431.823	127.993	-271.222	-66.210	1.383
2600	58.193	485.303	433.837	133.813	-272.882	-57.977	1.165
2700	58.194	487.499	435.784	139.632	-274.641	-49.678	0.961
2800	58.194	489.616	437.669	145.451	-276.484	-41.313	0.771
2900	58.195	491.658	439.496	151.271	-278.395	-32.881	0.592
3000	58.195	493.631	441.267	157.090	-280.359	-24.381	0.425
3100	58.195	495.539	442.987	162.910	-282.362	-15.816	0.266
3200	58.196	497.387	444.659	168.729	-284.390	-7.185	0.117
3300	58.196	499.177	446.284	174.549	-286.430	1.509	-0.024
3400	58.196	500.915	447.865	180.368	-288.470	10.265	-0.158
3500	58.197	502.602	449.405	186.188	-290.498	19.081	-0.285
3600	58.197	504.241	450.906	192.008	-292.506	27.954	-0.406
3700	58.197	505.836	452.369	197.827	-294.484	36.884	-0.521
3800	58.197	507.388	453.796	203.647	-296.425	45.866	-0.630
3900	58.197	508.899	455.190	209.467	-298.323	54.898	-0.735
4000	58.198	510.373	456.551	215.287	-300.172	63.979	-0.835
4100	58.198	511.810	457.882	221.106	-301.969	73.105	-0.931
4200	58.198	513.212	459.182	226.926	-303.708	82.274	-1.023
4300	58.198	514.582	460.455	232.746	-305.387	91.484	-1.111
4400	58.198	515.920	461.700	238.566	-307.005	100.732	-1.196
4500	58.198	517.228	462.920	244.386	-308.559	110.017	-1.277
4600	58.198	518.507	464.114	250.205	-310.048	119.335	-1.355
4700	58.198	519 758	465.285	256.025	-311.473	128.685	-1.430
4800	58.199	520.984	466.433	261.845	-312.832	138.065	-1.502
4900	58.199	522.184	467.558	267.665	-314.127	147.472	-1.572
5000	58.199	523.359	468.662	273.485	-315.357	156.904	-1.639
5100	58.199	524.512	469.746	279.305	-316.525	166.362	-1.704
5200	58.199	525.642	470.810	285.125	-317.630	175.841	-1.766
5300	58.199	526.751	471.855	290.945	-318.675	185.340	-1.827
5400	58.199	527.838	472.882	296.764	-319.661	194.860	-1.885
5500	58.199	528.906	473.891	302.584	-320.590	204.396	-1.941
5600	58.199	529.955	474.883	308.404	-321.463	213.950	-1.996
5700	58.199	530.985	475.858	314.224	-322.284	223.518	-2.048
5800	58.199	531.997	476.817	320.044	-323.052	233.100	-2.099
5900	58.199	532.992	477.761	325.864	-323.771	242.695	-2.149
6000	58.199	533.970	478.690	331.684	-324.442	252.302	-2.196

PREVIOUS: December 1973 (1 atm) CURRENT: December 1973 (1 bar)

Lead Iodide (PbI_2) $I_2Pb_1(g)$

DIIODOSILYLENE (SiI_2) IDEAL GAS M_r = 281.8945

$\Delta_f H°(0\ K)$ = 95.09 ± 8.4 kJ mol^{-1}

$S°(298.15\ K)$ = [320.972 ± 4.2] J K^{-1} mol^{-1} $\Delta_f H(298.15\ K)$ = 92.47 ± 8.4 kJ mol^{-1}

Electronic Levels and Quantum Weights

State	ε_i, cm^{-1}	g_i
$[^1A_1]$	0	[1]
$[^3B_1]$	[20000]	[3]

Vibrational Frequencies and Degeneracies

ν, cm^{-1}
[350](1)
[90](1)
[350](1)

Point Group: $[C_{2v}]$ σ = [2]

Bond Distance: Si-I = [2.44] Å

Bond Angle: I-Si-I = [113]°

Product of the Moments of Inertia: $I_A I_B I_C$ = [2.419819 x 10^{-112}] g^3 cm^6

Enthalpy of Formation

An equilibrium system involving $SiI_2(g)$ has been studied by Schafer et al. (1) and Wolf and Herbst (2). They both used a flow technique in an argon atmosphere. A 2nd and 3rd law analysis of their results (reported only in equation form) is summarized below. As in the case of $SiCl_2(g)$ and $SiBr_2(g)$ (3), there is acceptable agreement between the 3rd law results of these two studies. Using the mean $\Delta_f H°(298.15\ K)$, we calculate and adopt $\Delta_f H°(SiI_2, g, 298.15\ K)$ = 22.1 kcal mol^{-1}. This corresponds to $\Delta_{at} H°$ = 135.2 kcal mol^{-1} and an average bond energy of 67.6 kcal mol^{-1}. In comparison, the average bond energy of $SiI_4(g)$ is 58.6 kcal mol^{-1}. This implies that the average bond energy is 15% stronger in SiI_2 than in SiI_4.

Source	Reaction	T/K	$\Delta_r H°(298.15\ K)$, kcal mol^{-1} 2nd law	3rd law	Drift cal K^{-1} mol^{-1}	$\Delta_f H°(g, 298.15\ K)$* kcal mol^{-1}
(1)	$SiI_4(g) = SiI_2(g) + 2I(g)$	1361-1526	98.92	101.04	1.46	23.60
(2)	$Si(cr) + I_2(g) = SiI_2(g)$	1300-1470	3.21	5.14	1.38	20.60
(4)	$SiI_4(g) + Si(cr) = 2SiI_2(g)$	1173-1373	83.90	68.09	-12.42	20.84

*Based on 3rd law $\Delta_r H°(298.15\ K)$ value and auxiliary data (3).

Uchimura et al. (4) also studied $SiI_2(g)$. An analysis of their equilibrium data yields $\Delta_f H°(SiI_2, g, 298.15\ K)$ = 20.84 kcal mol^{-1} which is in good agreement with our adopted value. However, because of the large drift (-12.42 cal K^{-1} mol^{-1}), no weight was given to this study.

Heat Capacity and Entropy

The Si-I bond distance is assumed to be the same as in SiH_2I_2 (3). The bond angle is estimated to be 113° based on the trend observed in SiF_2, $SiCl_2$, and $SiBr_2$ (3). C_{2v} symmetry is also assumed to be consistent with the other three silicon dihalides. The principal moments of inertia are: I_A = 7.6158 x 10^{-39}, I_B = 174.4841 x 10^{-39}, and I_C = 182.0999 x 10^{-39} g cm^2.

The vibrational frequencies are estimated based on trends observed in the SiX_2 and SiH_2X_2 species, where X = F, Cl, Br, and I (3). As suggested by SiF_2 and $SiCl_2$, an electronic excited state is included at 20000 cm^{-1}.

References

1. H. Schafer, H. Bruderrick, and B. Morcher, Z. Anorg. Allg. Chem. 352, 122 (1967).
2. E. Wolf and C. Herbst, Z. Chem. 7, 34 (1967).
3. JANAF Thermochemical Tables: Si(g), 3-31-67; I(g), 6-30-74; $SiH_2F_2(g)$ and $SiI_4(g)$, 6-30-76, $SiBr_4(g)$, $SiH_2X_2(g)$, X = Cl, Br, I, 12-31-76; $SiF_2(g)$, $SiCl_2(g)$, and $SiBr_2(g)$, 12-31-77.
4. K. Uchimura, T. Takuma, M. Yuizumi, and T. Kunugi, Denki Kagaku 35, 876 (1967).

Diiodosilylene (SiI_2) $I_2Si_1(g)$

Enthalpy Reference Temperature = T_r = 298.15 K Standard State Pressure = p° = 0.1 MPa

T/K	$C_p°$ (J $K^{-1}mol^{-1}$)	S°	$-[G°-H°(T_r)]/T$	$H°-H°(T_r)$ (kJ mol^{-1})	$\Delta_f H°$	$\Delta_f G°$	Log K_f
0	0.	0.	INFINITE	-13.792	95.090	95.090	INFINITE
100	43.279	266.948	367.003	-10.006	95.561	75.348	-39.358
200	51.344	299.763	325.883	-5.224	94.231	55.603	-14.522
250	53.375	311.456	321.865	-2.602	93.375	46.042	-9.620
298.15	54.624	320.972	320.972	0.	92.466	37.008	-6.484
300	54.663	321.310	320.973	0.101	92.430	36.664	-6.384
350	55.512	329.804	321.641	2.857	91.339	27.452	-4.097
400	56.096	337.257	323.137	5.648	74.217	18.952	-2.475
450	56.512	343.889	325.081	8.464	71.875	12.184	-1.414
500	56.817	349.860	327.266	11.297	29.392	9.477	-0.990
600	57.226	360.259	331.923	17.001	28.966	5.533	-0.482
700	57.478	369.100	336.618	22.737	28.486	1.665	-0.124
800	57.644	376.787	341.170	28.494	27.955	-2.131	0.139
900	57.759	383.584	345.512	34.265	27.373	-5.858	0.340
1000	57.842	389.674	349.629	40.045	26.741	-9.517	0.497
1100	57.904	395.190	353.524	45.832	26.057	-13.110	0.623
1200	57.951	400.230	357.209	51.625	25.318	-16.638	0.724
1300	57.988	404.870	360.699	57.422	24.519	-20.103	0.808
1400	58.017	409.169	364.010	63.222	23.650	-23.503	0.877
1500	58.040	413.172	367.155	69.025	22.701	-26.838	0.935
1600	58.060	416.919	370.150	74.830	21.663	-30.108	0.983
1700	58.076	420.439	373.005	80.637	-29.651	-32.862	1.010
1800	58.090	423.759	375.734	86.445	-30.663	-33.022	0.958
1900	58.102	426.900	378.345	92.255	-31.743	-33.124	0.911
2000	58.113	429.881	380.848	98.066	-32.892	-33.167	0.866
2100	58.124	432.716	383.251	103.878	-34.108	-33.151	0.825
2200	58.135	435.420	385.561	109.691	-35.385	-33.076	0.785
2300	58.147	438.005	387.785	115.505	-36.716	-32.942	0.748
2400	58.160	440.480	389.930	121.320	-38.090	-32.748	0.713
2500	58.176	442.854	392.000	127.137	-39.497	-32.497	0.679
2600	58.195	445.136	394.000	132.955	-40.925	-32.189	0.647
2700	58.218	447.333	395.935	138.776	-42.360	-31.825	0.616
2800	58.245	449.451	397.808	144.599	-43.793	-31.409	0.586
2900	58.278	451.495	399.625	150.425	-45.211	-30.942	0.557
3000	58.317	453.472	401.387	156.255	-46.604	-30.426	0.530
3100	58.363	455.385	403.098	162.089	-47.963	-29.864	0.503
3200	58.416	457.238	404.761	167.928	-49.280	-29.259	0.478
3300	58.477	459.037	406.379	173.772	-50.548	-28.614	0.453
3400	58.546	460.784	407.953	179.623	-51.762	-27.931	0.429
3500	58.624	462.482	409.487	185.482	-52.916	-27.213	0.406
3600	58.710	464.134	410.982	191.349	-438.158	-16.002	0.232
3700	58.805	465.744	412.440	197.224	-438.768	-4.267	0.060
3800	58.909	467.314	413.864	203.110	-439.311	7.484	-0.103
3900	59.022	468.846	415.254	209.006	-439.786	19.248	-0.258
4000	59.143	470.341	416.613	214.915	-440.190	31.023	-0.405
4100	59.273	471.803	417.941	220.835	-440.523	42.808	-0.545
4200	59.411	473.233	419.241	226.769	-440.785	54.600	-0.679
4300	59.557	474.633	420.513	232.718	-440.976	66.397	-0.807
4400	59.710	476.004	421.758	238.681	-441.096	78.198	-0.928
4500	59.870	477.348	422.979	244.660	-441.145	90.001	-1.045
4600	60.036	478.665	424.175	250.655	-441.124	101.804	-1.156
4700	60.208	479.958	425.348	256.667	-441.034	113.606	-1.263
4800	60.385	481.228	426.499	262.697	-440.877	125.405	-1.365
4900	60.567	482.475	427.629	268.745	-440.652	137.201	-1.463
5000	60.753	483.700	428.738	274.811	-440.363	148.990	-1.556
5100	60.943	484.905	429.827	280.896	-440.010	160.774	-1.647
5200	61.135	486.090	430.898	286.999	-439.594	172.551	-1.733
5300	61.330	487.257	431.950	293.123	-439.117	184.318	-1.817
5400	61.526	488.405	432.985	299.265	-438.581	196.076	-1.897
5500	61.723	489.536	434.003	305.428	-437.986	207.823	-1.974
5600	61.920	490.650	435.005	311.610	-437.335	219.560	-2.048
5700	62.118	491.747	435.991	317.812	-436.630	231.284	-2.119
5800	62.315	492.829	436.961	324.034	-435.873	242.995	-2.188
5900	62.511	493.896	437.917	330.275	-435.064	254.693	-2.255
6000	62.705	494.948	438.859	336.536	-434.206	266.376	-2.319

PREVIOUS: December 1977 (1 atm) CURRENT: December 1977 (1 bar)

Diiodosilylene (SiI_2) $I_2Si_1(g)$

STRONTIUM IODIDE (SrI_2) CRYSTAL M_r = 341.4290

$S°(298.15 K) = 159.120 \pm 0.8 J K^{-1} mol^{-1}$

$T_{fus} = 811 \pm 10 K$

$\Delta_f H°(0 K) = -560.65 \pm 2.1 kJ mol^{-1}$

$\Delta_f H°(298.15 K) = -561.49 \pm 2.1 kJ mol^{-1}$

$\Delta_{fus} H° = 19.665 \pm 0.38 kJ mol^{-1}$

$\Delta_{sub} H°(298.15 K) = 286.604 kJ mol^{-1}$

Enthalpy of Formation

Ehrlich et al. (1) derived $\Delta_f H°(SrI_2$, cr, 298.15 K) = -133.8 ± 0.4 kcal mol^{-1} from enthalpy of solution measurements of Sr(cr) and SrI_2(cr) in 0.1 N HI. An auxiliary HI enthalpy of solution value, -13.22 kcal mol^{-1}, calculated from data in reference (2) was used in the derivation. This auxiliary value will be changed by incorporation of the accepted CODATA key value of $\Delta_f H°(I^-$, aq. std. state) = -13.60 kcal mol^{-1} (3) in the $\Delta_f H°$(HI, aq, 298.15 K) table so that the above derived enthalpy of formation of SrI_2 (cr) will be approximately -134.3 kcal mol^{-1}.

Combining $\Delta_f H°(Sr^{+2}$, aq, std. state, 298.15 K) = -130.45 kcal mol^{-1} selected by Parker (4, 5) with the CODATA value for I^-(aq, std. state) (3) gives $\Delta_f H°(SrI_2$, aq, std. state, 298.15 K) = -157.65 kcal mol^{-1}. Further combination with the enthalpy of solution, -23.43 kcal mol^{-1} (5) gives $\Delta_f H°(SrI_2$, cr, 298.15 K) = -134.22 kcal mol^{-1}. This enthalpy of solution value is based solely on the work of Ehrlich et al. (6).

To be consistent with the selection of the enthalpies of formation of BaI_2(cr) and CaI_2(cr), we adopt $\Delta_f H°(SrI_2$, cr, 298.15 K) = -134.2 ± 0.5 kcal mol^{-1} calculated via the combination of aqueous ions.

Heat Capacity and Entropy

$C_p°$(13.9 - 300.9 K) has been measured by Paukov et al. (7) and smoothed values have been published. Starting values of entropy and enthalpy, S°(14 K) and H°(14 K)-H°(0 K), were not given. Our T^3 extrapolation leads to S°(8 K) = 0.148 cal K^{-1} mol^{-1} and H°(8 K)-H(0 K) = 0.891 cal K^{-1} mol^{-1}, which, when combined with our $C_p°$ extrapolation from 14 K to 8 K and with the Paukov et al. (7) data, gives S°(298.15 K) = 38.03 cal K^{-1} mol^{-1} and H°(298.15 K)-H°(0 K) = 4522 cal K^{-1} mol^{-1}. Paukov et al. (7) gave 37.42 ± 0.06 cal K^{-1} mol^{-1} and 4156 ± 7 kcal mol^{-1} for the entropy and enthalpy at 298.15 K. Inasmuch as our integration of their data gives S°(298.15 K)-S°(14 K) = 37.26 cal K^{-1} mol^{-1} it appears unlikely that their published value for S°(298.15 K) is correct.

The low temperature heat capacity joins smoothly with a linear extrapolation from $C_p°$ = 18.45 at 270 K to $C_p°$ = 22.0 at the adopted T_{fus} of 811 K. The linearly extrapolated heat capacity at 298 K is 0.6% higher than the published smoothed value (7). Dworkin and Bredig (8) determined the heat capacity of the crystal near the melting point to be 22.9 ± 5%. The graphically selected $C_p°$(811 K) = 22.0 is within the experimental error and gives H°(811 K)-H°(298.15 K) = 10.42 kcal mol^{-1} in agreement with Dworkin and Bredig's measured value of 10.4 kcal mol^{-1} (8).

Fusion Data

Emons and Loeffelholz (9) determined $\Delta_{fus}H°$ = 4.580 kcal mol^{-1} (±5%) and T_{fus} = 802 K by high temperature cryoscopy, in good agreement with the drop calorimetry $\Delta_{fus}H°$ = 4.70 kcal mol^{-1} (±2%) and T_{fus} = 811 K by Dworkin and Bredig (8). Hutchison (10) found T_{fus} = 793 K, noting that his melting point data were only approximate.

Because drop calorimetry is a more direct measure of the enthalpy of melting, we adopt $\Delta_{fus}H° = 4.70\pm0.09$ kcal mol^{-1} at T_{fus} = 811 ± 10 K.

Sublimation Data

$\Delta_{sub}H°$(298.15 K) is derived as the difference between the enthalpies of formation of the ideal gas and the crystal state at 298.15 K.

References

1. P. Ehrlich, K. Peik, and E. Koch, Z. Anorg. Allg. Chem. 324, 113 (1963).
2. U. S. Nat. Bur. Stand. Circ. 500, 1952.
3. J. D. Cox, ICSU-CODATA Task Group on Key Values for Thermodynamics, J. Chem. Thermodyn. 4, 331 (1972).
4. V. B. Parker, U. S. Nat. Bur. Stand. Report 10074, 164 (1969).
5. U. S. Nat. Bur. Stand. Tech. Note 270-6, 1971.
6. V. B. Parker, U. S. Nat. Bur. Stand., personal communication, May 1974.
7. I. E. Paukov, F. S. Rakmenkulov, and S. S. Vrublevskaya, Russ. J. Phys. Chem. 47, 1226 (1973).
8. A. S. Dworkin and M. A. Bredig, J. Phys. Chem. 67, 697 (1963).
9. H. H. Emons and B. Loeffelholz, Wiss. Z. Hochsch. Chem. Leuna-Merseburg 6, 261 (1964).
10. J. F. Hutchison, U. S. At. Energy Comm. IS-T-50, 1965.

Strontium Iodide (SrI_2) I_2Sr_1(cr)

Enthalpy Reference Temperature = T_r = 298.15 K — Standard State Pressure = p° = 0.1 MPa

	J $K^{-1}mol^{-1}$			kJ mol^{-1}			
T/K	$C_p°$	S°	$-[G°-H°(T_r)]/T$	$H°-H°(T_r)$	$\Delta_f H°$	$\Delta_f G°$	Log K_f
0	0.	0.	INFINITE	-18.919	-560.645	-560.645	INFINITE
100	67.404	78.914	226.165	-14.725	-561.001	-559.988	292.508
200	75.019	128.568	166.164	-7.519	-561.216	-558.896	145.969
298.15	77.948	159.120	159.120	0.	-561.493	-557.702	97.707
300	78.032	159.603	159.122	0.144	-561.499	-557.678	97.100
400	80.751	182.421	162.214	8.083	-577.928	-555.790	72.579
500	83.513	200.737	168.145	16.296	-620.739	-546.142	57.055
600	86.274	216.207	174.898	24.785	-618.933	-531.390	46.262
700	88.994	229.710	181.784	33.548	-616.972	-516.953	38.575
800	91.755	241.776	188.542	42.587	-614.852	-502.807	32.830
811.000	92.048	243.031	189.272	43.598	------ CRYSTAL <--> LIQUID -----		
900	94.492	252.741	195.075	51.899	-613.204	-488.864	28.373
1000	97.236	262.839	201.353	61.486	-610.453	-475.194	24.822
1100	99.971	272.234	207.374	71.346	-615.344	-461.446	21.912
1200	102.717	281.051	213.150	81.480	-612.989	-447.557	19.482
1300	105.493	289.382	218.697	91.891	-610.380	-433.876	17.433
1400	108.240	297.301	224.031	102.578	-607.527	-420.404	15.685
1500	110.960	304.861	229.169	113.538	-604.440	-407.144	14.178

PREVIOUS: CURRENT: June 1974

Strontium Iodide (SrI_2) I_2Sr_1(cr)

STRONTIUM IODIDE (SrI_2) LIQUID M_r = 341.4290

$S°$(298.15 K) = [172.440] J K^{-1} mol^{-1} $\Delta_f H°$(298.15 K) = [-548.748] kJ mol^{-1}

T_{fus} = 811 ± 10 K $\Delta_{fus}H°$ = 19.665 ± 0.38 kJ mol^{-1}

Enthalpy of Formation

$\Delta_f H°$(SrI_2, ℓ, 298.15 K) is calculated from $\Delta_f H°$(SrI_2, cr, 298.15 K) by adding the enthalpy of fusion, $\Delta_{fus}H°$, and the difference in enthalpy, H°(811 K)-H°(298.15 K), between the crystal and the liquid.

Heat Capacity and Entropy

The liquid heat capacity near the melting point, 26.3 cal K^{-1} mol^{-1}, was determined by Dworkin and Bredig (1) from drop calorimeter measurements. This value is adopted and assumed constant over a 500-2600 K range. A glass transition is assumed at 500 K below which the heat capacity is that of the crystal.

$S°$(SrI_2, ℓ, 298.15 K) is calculated in a manner similar to that used for the enthalpy of formation.

Vaporization Data

The temperature at which the fugacity is 1 atm for the reaction SrI_2(ℓ) = SrI_2(g) is 2181 K, the adopted T_{vap}. Peterson and Hutchison (2) have extrapolated Knudsen effusion cell measurements in the 1061-1308 K range to obtain a normal boiling point of 2046 K.

$\Delta_{vap}H°$ = 45.35 kcal mol^{-1} is the calculated difference between the enthalpies of formation of the ideal gas and the liquid at T_{vap}. See the ideal gas table for details.

References

1. A. S. Dworkin and M. A. Bredig, J. Phys. Chem. 67, 697 (1963).
2. D. T. Peterson and J. F. Hutchison, J. Chem. Eng. Data 15, 320 (1970).

Strontium Iodide (SrI_2) I_2Sr_1(l)

Enthalpy Reference Temperature = T_r = 298.15 K Standard State Pressure = p° = 0.1 MPa

T/K	$C_p°$ (J K⁻¹mol⁻¹)	S° (J K⁻¹mol⁻¹)	-[G°-H°(T_r)]/T (J K⁻¹mol⁻¹)	H°-H°(T_r) (kJ mol⁻¹)	$\Delta_f H°$ (kJ mol⁻¹)	$\Delta_f G°$ (kJ mol⁻¹)	Log K_f
0							
100							
200							
298.15	77.948	172.440	172.440	0.	-548.748	-548.928	96.170
300	78.032	172.923	172.442	0.144	-548.754	-548.929	95.577
400	80.751	195.741	175.534	8.083	-565.183	-548.373	71.610
499.990	83.513	214.055	181.464	16.295	GLASS <--> LIQUID		
499.990	110.039	214.055	181.464	16.295	TRANSITION		
500	110.039	214.057	181.465	16.296	-607.994	-540.057	56.419
600	110.039	234.120	188.620	27.300	-603.674	-526.878	45.869
700	110.039	251.082	196.362	38.304	-599.471	-514.413	38.386
800	110.039	265.776	204.141	49.308	-595.386	-502.542	32.813
811.000	110.039	267.279	204.987	50.518	------ CRYSTAL <--> LIQUID -----		
900	110.039	278.737	211.724	60.312	-592.047	-491.103	28.503
1000	110.039	290.330	219.015	71.316	-587.878	-480.111	25.078
1100	110.039	300.818	225.982	82.320	-591.626	-469.170	22.279
1200	110.039	310.393	232.623	93.323	-588.401	-458.180	19.944
1300	110.039	319.201	238.949	104.327	-585.199	-447.459	17.979
1400	110.039	327.356	244.976	115.331	-582.028	-436.982	16.304
1500	110.039	334.948	250.724	126.335	-578.898	-426.731	14.860
1600	110.039	342.049	256.212	137.339	-575.818	-416.688	13.603
1700	110.039	348.720	261.460	148.343	-709.710	-405.655	12.464
1800	110.039	355.010	266.484	159.347	-704.893	-387.909	11.257
1900	110.039	360.960	271.301	170.351	-700.151	-370.429	10.184
2000	110.039	366.604	275.926	181.355	-695.487	-353.196	9.225
2100	110.039	371.973	280.373	192.359	-690.902	-336.194	8.362
2200	110.039	377.092	284.654	203.363	-686.397	-319.410	7.584
2300	110.039	381.983	288.780	214.367	-681.969	-302.828	6.877
2400	110.039	386.666	292.762	225.370	-677.614	-286.437	6.234
2500	110.039	391.158	296.609	236.374	-673.330	-270.226	5.646
2600	110.039	395.474	300.329	247.378	-669.112	-254.185	5.107

PREVIOUS: CURRENT: June 1974

Strontium Iodide (SrI_2) I_2Sr_1(l)

STRONTIUM IODIDE (SrI_2) CRYSTAL–LIQUID M_r = 341.4290

0 to 811 K crystal
above 811 K liquid

Refer to the individual tables for details.

Strontium Iodide (SrI_2) I_2Sr_1(cr,l)

Enthalpy Reference Temperature = T_r = 298.15 K Standard State Pressure = p° = 0.1 MPa

T/K	C_p° (J K⁻¹mol⁻¹)	S° (J K⁻¹mol⁻¹)	$-[G^\circ-H^\circ(T_r)]/T$ (J K⁻¹mol⁻¹)	$H^\circ-H^\circ(T_r)$ (kJ mol⁻¹)	Δ_fH° (kJ mol⁻¹)	Δ_fG° (kJ mol⁻¹)	Log K_f
0	0.	0.	INFINITE	-18.919	-560.645	-560.645	INFINITE
100	67.404	78.914	226.165	-14.725	-561.001	-559.988	292.508
200	75.019	128.568	166.164	-7.519	-561.216	-558.896	145.969
298.15	77.948	159.120	159.120	0.	-561.493	-557.702	97.707
300	78.032	159.603	159.122	0.144	-561.499	-557.678	97.100
400	80.751	182.421	162.214	8.083	-577.928	-555.790	72.579
500	83.513	200.737	168.145	16.296	-620.739	-546.142	57.055
600	86.274	216.207	174.898	24.785	-618.933	-531.390	46.262
700	88.994	229.710	181.784	33.548	-616.972	-516.953	38.575
800	91.755	241.776	188.542	42.587	-614.852	-502.807	32.830
811.000	92.048	243.031	189.272	43.598	CRYSTAL <--> LIQUID		
811.000	110.039	267.279	189.272	63.263	TRANSITION		
900	110.039	278.737	197.563	73.057	-592.047	-491.103	28.503
1000	110.039	290.330	206.270	84.061	-587.878	-480.111	25.078
1100	110.039	300.818	214.396	95.064	-591.626	-469.170	22.279
1200	110.039	310.393	222.003	106.068	-588.401	-458.180	19.944
1300	110.039	319.201	229.145	117.072	-585.199	-447.459	17.979
1400	110.039	327.356	235.873	128.076	-582.028	-436.982	16.304
1500	110.039	334.948	242.227	139.080	-578.898	-426.731	14.860
1600	110.039	342.049	248.247	150.084	-575.818	-416.688	13.603
1700	110.039	348.720	253.963	161.088	-709.710	-405.655	12.464
1800	110.039	355.010	259.403	172.092	-704.893	-387.909	11.257
1900	110.039	360.960	264.593	183.096	-700.151	-370.429	10.184
2000	110.039	366.604	269.554	194.100	-695.487	-353.196	9.225
2100	110.039	371.973	274.304	205.104	-690.902	-336.194	8.362
2200	110.039	377.092	278.861	216.108	-686.397	-319.410	7.584
2300	110.039	381.983	283.239	227.111	-681.969	-302.828	6.877
2400	110.039	386.666	287.452	238.115	-677.614	-286.437	6.234
2500	110.039	391.158	291.511	249.119	-673.330	-270.226	5.646
2600	110.039	395.474	295.427	260.123	-669.112	-254.185	5.107

PREVIOUS: CURRENT: June 1974

Strontium Iodide (SrI_2) I_2Sr_1(cr,l)

STRONTIUM IODIDE (SrI_2) IDEAL GAS M_r = 341.4290

$S°(298.15\ K)$ = [339.554 ± 8.4] J K^{-1} mol^{-1}

$\Delta_f H°(0\ K)$ = -271.66 ± 6.3 kJ mol^{-1}
$\Delta_f H°(298.15\ K)$ = -274.89 ± 6.3 kJ mol^{-1}

Vibrational Frequencies and Degeneracies

ν, cm^{-1}
[111](1)
[31](2)
[219](1)

Ground State Quantum Weight: [1] σ = 2
Point Group: $D_{\infty h}$
Bond Distance: Sr-I = 3.03 ± 0.03 Å
Bond Angle: I-Sr-I = [180]°
Rotational Constant: B_0 = 0.007234 cm^{-1}

Enthalpy of Formation

Peterson and Hutchison (1, 2) used a weight loss Knudsen effusion technique to observe vapor pressures of the liquid at 20 temperatures ranging from 1061 to 1308 K. Our 2nd and 3rd law analyses of these data give $\Delta_{vap}H°(298.15\ K)$ (2nd law) = 65.72 ± 1.48 kcal mol^{-1} and $\Delta_{vap}H°(298.15\ K)$ (3rd law) = 65.48 ± 0.68 kcal mol^{-1} with a drift of -0.2 ± 1.2 cal K^{-1} mol^{-1}. We adopt $\Delta_{vap}H°(298.15\ K)$ = 65.5 ± 0.7 kcal mol^{-1} and combine this with the adopted enthalpy of formation of the liquid to obtain $\Delta_f H°(SrI_2,\ g,\ 298.15\ K)$ = -65.7 ± 1.5 kcal mol^{-1}.

Heat Capacity and Entropy

The bond distance, as determined from an electron diffraction study, is taken from Akishin et al. (3) who also judged that the bond angle was 180° ± 10°. We have assumed a linear structure.

An analysis of the trend of the ratio of the stretching force constants, k(monohalide)/k_1(dihalide), has indicated that k(SrI)/$k_1(SrI_2)$ = 1 is a reasonable approximation (4). The bending force constant is taken as 0.01 times the stretching force constant. These approximations have been used by Brewer et al. (5). Our vibrational frequencies calculated from the force constants, using k for SrI calculated from the ground state vibrational frequency given by Rosen (6), are in exact agreement with those calculated by Brewer et al. (5), which are adopted. Other estimates of ν_1, ν_2, and ν_3 are 111, 49, and 220 (7).

We assign an uncertainty of ±2 cal K^{-1} mol^{-1} to the entropy to allow for error in bond angle and vibrational frequency estimates.

References

1. J. F. Hutchison, U. S. At. Energy Comm. IS-T-50, 1965.
2. D. T. Peterson and J. F. Hutchison, J. Chem. Eng. Data 15, 320 (1970).
3. P. A. Akishin, V. P. Spiridonov, G. A. Sobolev, and V. A. Naumov, Zhur. Fiz. Khim. 31, 1871 (1957).
4. JANAF Thermochemical Table: $SrBr_2$(g), 6-30-74.
5. L. Brewer, G. R. Somayajulu, and E. Brackett, Chem. Rev. 63, 111 (1963).
6. B. Rosen, Ed., "Spectroscopic Data Relative to Diatomic Molecules," Pergamon Press, New York, 1970.
7. K. S. Krasnov and V. I. Svettsov, Ziv. Vysshikh Uchebn. Zavedenii, Khim. i Khim. Tekhnol. 6, 167 (1963).

Strontium Iodide (SrI_2) I_2Sr_1(g)

Enthalpy Reference Temperature = T_r = 298.15 K   Standard State Pressure = p° = 0.1 MPa

T/K	$C_p°$ (J $K^{-1}mol^{-1}$)	S°	$-[G°-H°(T_r)]/T$	$H°-H°(T_r)$ (kJ mol^{-1})	$\Delta_f H°$	$\Delta_f G°$	Log K_f
0	0.	0.	INFINITE	-16.542	-271.664	-271.664	INFINITE
100	56.065	274.673	393.256	-11.858	-271.530	-290.093	151.529
200	60.336	315.227	345.144	-5.983	-273.076	-308.088	80.464
250	61.019	328.771	340.563	-2.948	-273.971	-316.744	66.180
298.15	61.399	339.554	339.554	0.	-274.889	-324.894	56.920
300	61.410	339.934	339.555	0.114	-274.926	-325.204	56.623
350	61.653	349.420	340.304	3.191	-276.013	-333.501	49.772
400	61.814	357.664	341.970	6.278	-293.129	-341.088	44.542
450	61.926	364.951	344.126	9.371	-295.470	-346.943	40.272
500	62.007	371.480	346.541	12.470	-337.961	-348.736	36.432
600	62.113	382.796	351.669	18.676	-338.439	-350.848	30.544
700	62.177	392.376	356.817	24.891	-339.025	-352.872	26.332
800	62.220	400.681	361.793	31.111	-339.724	-354.804	23.166
900	62.248	408.011	366.529	37.334	-341.165	-356.569	20.695
1000	62.269	414.571	371.011	43.560	-341.774	-358.248	18.713
1100	62.285	420.507	375.245	49.788	-350.298	-359.500	17.071
1200	62.296	425.927	379.246	56.017	-351.848	-360.268	15.682
1300	62.305	430.914	383.031	62.247	-353.420	-360.906	14.501
1400	62.313	435.531	386.618	68.478	-355.022	-361.422	13.485
1500	62.318	439.830	390.024	74.710	-356.664	-361.822	12.600
1600	62.323	443.853	393.264	80.942	-358.356	-362.111	11.822
1700	62.327	447.631	396.352	87.174	-497.020	-361.113	11.096
1800	62.331	451.194	399.301	93.407	-496.974	-353.120	10.247
1900	62.333	454.564	402.121	99.640	-497.002	-345.128	9.488
2000	62.336	457.761	404.824	105.874	-497.108	-337.132	8.805
2100	62.338	460.803	407.418	112.108	-497.294	-329.129	8.187
2200	62.340	463.703	409.911	118.341	-497.559	-321.115	7.624
2300	62.341	466.474	412.310	124.576	-497.901	-313.088	7.110
2400	62.343	469.127	414.623	130.810	-498.316	-305.044	6.639
2500	62.344	471.672	416.854	137.044	-498.801	-296.981	6.205
2600	62.345	474.117	419.010	143.278	-499.353	-288.897	5.804
2700	62.346	476.470	421.095	149.513	-499.968	-280.792	5.432
2800	62.347	478.737	423.113	155.748	-500.646	-272.662	5.087
2900	62.347	480.925	425.069	161.982	-501.384	-264.507	4.764
3000	62.348	483.039	426.967	168.217	-502.185	-256.325	4.463
3100	62.349	485.083	428.808	174.452	-503.049	-248.116	4.181
3200	62.349	487.063	430.598	180.687	-503.979	-239.877	3.916
3300	62.350	488.981	432.338	186.922	-504.978	-231.609	3.666
3400	62.350	490.843	434.032	193.157	-506.051	-223.309	3.431
3500	62.351	492.650	435.681	199.392	-507.202	-214.976	3.208
3600	62.351	494.407	437.288	205.627	-508.436	-206.610	2.998
3700	62.352	496.115	438.855	211.862	-509.756	-198.208	2.798
3800	62.352	497.778	440.384	218.097	-511.167	-189.769	2.609
3900	62.352	499.397	441.876	224.333	-512.672	-181.292	2.428
4000	62.352	500.976	443.334	230.568	-514.275	-172.775	2.256
4100	62.353	502.516	444.759	236.803	-515.977	-164.216	2.092
4200	62.353	504.018	446.152	243.038	-517.776	-155.615	1.935
4300	62.353	505.485	447.515	249.274	-519.677	-146.970	1.785
4400	62.353	506.919	448.849	255.509	-521.678	-138.280	1.642
4500	62.354	508.320	450.155	261.744	-523.777	-129.543	1.504
4600	62.354	509.691	451.434	267.980	-525.906	-120.763	1.371
4700	62.354	511.032	452.688	274.215	-528.173	-111.931	1.244
4800	62.354	512.344	453.917	280.451	-530.523	-103.050	1.121
4900	62.354	513.630	455.123	286.686	-532.952	-94.120	1.003
5000	62.355	514.890	456.306	292.921	-535.447	-85.140	0.889
5100	62.355	516.125	457.466	299.157	-537.990	-76.110	0.780
5200	62.355	517.335	458.606	305.392	-540.611	-67.028	0.673
5300	62.355	518.523	459.725	311.628	-543.283	-57.896	0.571
5400	62.355	519.689	460.825	317.863	-546.001	-48.712	0.471
5500	62.355	520.833	461.906	324.099	-548.756	-39.477	0.375
5600	62.355	521.956	462.968	330.334	-551.542	-30.192	0.282
5700	62.355	523.060	464.013	336.570	-554.352	-20.858	0.191
5800	62.355	524.145	465.040	342.805	-557.179	-11.474	0.103
5900	62.356	525.210	466.051	349.041	-559.807	-2.064	0.018
6000	62.356	526.259	467.046	355.276	-562.618	7.413	-0.065

PREVIOUS: June 1974 (1 atm)   CURRENT: June 1974 (1 bar)

Strontium Iodide (SrI_2) I_2Sr_1(g)

TITANIUM IODIDE (TiI_2) CRYSTAL M_r = 301.6890

$S°(298.15\ K) = [122.591 \pm 12.6]\ J\ K^{-1}\ mol^{-1}$

$\Delta_f H°(0\ K)$ = Unknown
$\Delta_f H°(298.15\ K) = -266.10 \pm 12.6\ kJ\ mol^{-1}$

Enthalpy of Formation

The enthalpy of formation of TiI_2(cr) is calculated from the enthalpy of reaction for the process $2TiI_2(cr) = Ti(cr) + TiI_4(g)$, and the enthalpy of formation of TiI_4(cr). Vapor pressure data for this reaction were reported by Herczog and Pidgeon (1) over the temperature range 750 to 902 K. 2nd and 3rd law analyses of these data give values for $\Delta_r H°(298.15\ K)$ of 60.5±0.8 and 60.85 kcal mol^{-1}, respectively, the 3rd law drift being 0.4±1.0 cal $K^{-1}\ mol^{-1}$. The 3rd law $\Delta\ H°(298.15\ K)$ is used to obtain the adopted value of $\Delta_f H°(298.15\ K)$.

Heat Capacity and Entropy

The estimated heat capacity is that reported by Kelley (2). The value of S°(298.15 K) is estimated from that of $TiCl_2$(cr) and the difference between ionic entropy contributions of Cl^- and I^-.

Sublimation Data

The enthalpy of sublimation of TiI_2(cr) is taken as the difference in the enthalpies of formation of TiI_2(cr) and TiI_2(g) at the sublimation temperature. The sublimation temperature is taken as the point at which $\Delta_r G° = 0$ for the reaction $TiI_2(cr) = TiI_2(g)$.

References

1. A. Herczog and L. M. Pidgeon, Can. J. Chem. 34, 1687 (1956).
2. K. K. Kelley, U. S. Bur. Mines Bull. 584, 1961.

Titanium Iodide (TiI_2) I_2Ti_1(cr)

Enthalpy Reference Temperature = T_r = 298.15 K Standard State Pressure = p° = 0.1 MPa

T/K	$C_p°$ (J K⁻¹mol⁻¹)	S° (J K⁻¹mol⁻¹)	$-[G°-H°(T_r)]/T$ (J K⁻¹mol⁻¹)	$H°-H°(T_r)$ (kJ mol⁻¹)	$\Delta_f H°$ (kJ mol⁻¹)	$\Delta_f G°$ (kJ mol⁻¹)	Log K_f
0							
100							
200							
298.15	86.228	122.591	122.591	0.	-266.102	-258.854	45.350
300	86.241	123.125	122.593	0.160	-266.090	-258.810	45.063
400	86.969	148.034	125.984	8.820	-281.682	-255.901	33.417
500	87.697	167.519	132.412	17.553	-323.885	-245.418	25.639
600	88.425	183.572	139.640	26.359	-321.659	-229.934	20.018
700	89.153	197.258	146.917	35.238	-319.435	-214.822	16.030
800	89.881	209.210	153.973	44.190	-317.188	-200.030	13.061
900	90.609	219.838	160.711	53.214	-314.944	-185.521	10.767
1000	91.337	229.423	167.111	62.312	-312.770	-171.258	8.946
1100	92.065	238.162	173.179	71.482	-310.730	-157.207	7.465
1200	92.793	246.204	178.933	80.725	-312.815	-143.215	6.234
1300	93.521	253.660	184.398	90.040	-310.336	-129.182	5.191
1400	94.249	260.617	189.597	99.429	-307.893	-115.339	4.303
1500	94.977	267.145	194.551	108.890	-305.509	-101.669	3.540
1600	95.705	273.298	199.282	118.424	-303.207	-88.155	2.878
1700	96.433	279.121	203.809	128.031	-301.008	-74.782	2.298
1800	97.161	284.654	208.148	137.711	-298.933	-61.535	1.786
1900	97.889	289.927	212.315	147.463	-296.999	-48.400	1.331
2000	98.617	294.966	216.322	157.288	-309.906	-34.909	0.912

PREVIOUS: June 1964 CURRENT: December 1968

TITANIUM IODIDE (TiI_2) IDEAL GAS M_r = 301.6890

$S°(298.15\ K) = [323.673 \pm 16.7]\ J\ K^{-1}\ mol^{-1}$

$\Delta_f H°(0\ K) = [-17.15 \pm 33.5]\ kJ\ mol^{-1}$
$\Delta_f H°(298.15\ K) = [-19.66 \pm 33.5]\ kJ\ mol^{-1}$

Electronic Levels and Quantum Weights

ε_i, cm^{-1}	g_i
0	[3]
[7000]	[6]
[17000]	[6]
[22000]	[15]

Vibrational Frequencies and Degeneracies

ν, cm^{-1}
[134](1)
[92](2)
[290](1)

Ground State Quantum Weight: [3] $\sigma = 2$
Point Group: [$D_{\infty h}$]
Bond Distance: Ti-I = [2.7] Å
Bond Angle: I-Ti-I = [180]°
Rotational Constant: B_0 = [0.009111] cm^{-1}

Enthalpy of Formation

The enthalpy of formation of $TiI_2(g)$ is calculated from that of $TiI_2(cr)$ and the enthalpy of vaporization. Vapor pressure data over the temperature range 750 K to 902 K have been reported by Herczog and Pidgeon (1). 2nd and 3rd law analyses of these data give values for $\Delta_{vap}H°(298.15\ K)$ of 51.9 and 58.9 kcal mol^{-1}, respectively. The 3rd law drift for these data is 8.3±1.1 cal $K^{-1}\ mol^{-1}$, with one point rejected due to failure of a statistical test. The adopted value of $\Delta_f H°(298.15\ K)$ is calculated from the 3rd law enthalpy of vaporization.

Heat Capacity and Entropy

The interatomic distances are estimated from those of $TiCl_2(g)$, $TiCl_4(g)$, $TiBr_2(g)$, $TiBr_4(g)$ and $TiI_4(g)$. The vibrational frequencies are estimated from a valence force field model. The force constant K is estimated as 1.0 millidynes/Å and the constant K_δ/ℓ^2 is assumed to be 0.05 millidynes/Å.

The electronic levels are assumed to be the same as $TiCl_2(g)$. The levels of $TiCl_2$ are estimated by assuming they correspond to the inverted states of $NiCl_2(g)$ (2). The linear configuration is adopted, since experimental evidence indicates that other transition metal dihalides are linear (3, 4, 5, 6).

References

1. A. Herczog and L. M. Pidgeon, Can. J. Chem. 34, 1687 (1956).
2. C. W. DeKock and D. M. Gruen, J. Chem. Phys. 44, 4387 (1966).
3. R. A. Berg and O. Sinanoghi, J. Chem. Phys. 32, 1082 (1960).
4. J. T. Houghen, G. E. Leroi and T. C. James, J. Chem. Phys. 34, 1670 (1961).
5. G. E. Leroi, T. C. James, J. T. Hougen and W. Klemperer, J. Chem. Phys. 36, 2879 (1962).
6. A. Buchler, J. L. Stauffer, and W. Klemperer, J. Chem. Phys. 40, 3471 (1964).

Titanium Iodide (TiI_2) $I_2Ti_1(g)$

Enthalpy Reference Temperature = T_r = 298.15 K; Standard State Pressure = p° = 0.1 MPa

T/K	$C_p°$ (J $K^{-1}mol^{-1}$)	S° (J $K^{-1}mol^{-1}$)	$-[G°-H°(T_r)]/T$ (J $K^{-1}mol^{-1}$)	$H°-H°(T_r)$ (kJ mol^{-1})	$\Delta_f H°$ (kJ mol^{-1})	$\Delta_f G°$ (kJ mol^{-1})	Log K_f
0	0.	0.	INFINITE	-15.514	-17.151	-17.151	INFINITE
100	51.957	261.203	375.991	-11.479	-16.726	-35.921	18.763
200	58.692	299.824	329.167	-5.869	-17.965	-54.693	14.284
250	59.887	313.062	324.667	-2.901	-18.789	-63.782	13.327
298.15	60.570	323.673	323.673	0.	-19.665	-72.369	12.679
300	60.591	324.048	323.674	0.112	-19.700	-72.696	12.658
350	61.036	333.424	324.414	3.154	-20.758	-81.449	12.156
400	61.333	341.595	326.062	6.213	-37.851	-89.495	11.687
450	61.542	348.832	328.197	9.285	-40.170	-95.811	11.121
500	61.693	355.324	330.591	12.366	-82.634	-98.070	10.245
600	61.893	366.591	335.680	18.547	-83.034	-101.120	8.803
700	62.016	376.142	340.795	24.742	-83.493	-104.099	7.768
800	62.103	384.429	345.743	30.949	-83.991	-107.009	6.987
900	62.178	391.748	350.456	37.163	-84.558	-109.853	6.376
1000	62.260	398.303	354.919	43.384	-85.260	-112.628	5.883
1100	62.365	404.242	359.137	49.615	-86.159	-115.323	5.476
1200	62.505	409.674	363.125	55.859	-91.243	-117.807	5.128
1300	62.688	414.684	366.901	62.118	-91.820	-119.998	4.822
1400	62.916	419.338	370.482	68.398	-92.486	-122.141	4.557
1500	63.188	423.688	373.886	74.703	-93.259	-124.233	4.326
1600	63.500	427.775	377.127	81.037	-94.157	-126.269	4.122
1700	63.845	431.635	380.221	87.404	-95.198	-128.245	3.940
1800	64.215	435.295	383.180	93.806	-96.400	-130.155	3.777
1900	64.603	438.777	386.015	100.247	-97.777	-131.994	3.629
2000	65.000	442.101	388.737	106.727	-114.029	-133.302	3.481
2100	65.400	445.282	391.355	113.247	-116.541	-134.204	3.338
2200	65.795	448.334	393.876	119.807	-119.076	-134.985	3.205
2300	66.182	451.267	396.308	126.406	-121.626	-135.652	3.081
2400	66.555	454.091	398.657	133.043	-124.183	-136.207	2.964
2500	66.913	456.816	400.929	139.717	-126.737	-136.655	2.855
2600	67.254	459.447	403.129	146.425	-129.278	-137.001	2.752
2700	67.576	461.991	405.263	153.167	-131.797	-137.251	2.655
2800	67.879	464.454	407.333	159.940	-134.284	-137.407	2.563
2900	68.164	466.841	409.344	166.742	-136.730	-137.475	2.476
3000	68.431	469.157	411.299	173.572	-139.127	-137.460	2.393
3100	68.681	471.405	413.202	180.428	-141.469	-137.366	2.315
3200	68.916	473.589	415.055	187.308	-143.749	-137.197	2.240
3300	69.137	475.713	416.861	194.211	-145.963	-136.958	2.168
3400	69.346	477.780	418.623	201.135	-148.107	-136.652	2.099
3500	69.543	479.793	420.342	208.079	-150.179	-136.285	2.034
3600	69.730	481.755	422.021	215.043	-152.176	-135.860	1.971
3700	69.908	483.668	423.661	222.025	-563.197	-127.591	1.801
3800	70.078	485.534	425.265	229.024	-563.808	-115.810	1.592
3900	70.242	487.357	426.834	236.040	-564.399	-104.013	1.393
4000	70.399	489.137	428.369	243.072	-564.969	-92.201	1.204
4100	70.551	490.877	429.872	250.120	-565.516	-80.375	1.024
4200	70.697	492.579	431.345	257.182	-566.042	-68.536	0.852
4300	70.839	494.244	432.789	264.259	-566.544	-56.684	0.689
4400	70.977	495.875	434.204	271.350	-567.024	-44.822	0.532
4500	71.110	497.471	435.592	278.455	-567.481	-32.948	0.382
4600	71.240	499.035	436.955	285.572	-567.916	-21.065	0.239
4700	71.365	500.569	438.292	292.702	-568.328	-9.172	0.102
4800	71.486	502.073	439.605	299.845	-568.717	2.729	-0.030
4900	71.603	503.548	440.895	306.999	-569.084	14.638	-0.156
5000	71.716	504.996	442.162	314.165	-569.428	26.554	-0.277
5100	71.825	506.417	443.408	321.343	-569.751	38.477	-0.394
5200	71.928	507.813	444.634	328.530	-570.051	50.406	-0.506
5300	72.028	509.184	445.839	335.728	-570.330	62.340	-0.614
5400	72.122	510.531	447.024	342.936	-570.586	74.280	-0.719
5500	72.212	511.855	448.191	350.152	-570.822	86.224	-0.819
5600	72.296	513.157	449.339	357.378	-571.035	98.173	-0.916
5700	72.375	514.437	450.470	364.611	-571.228	110.125	-1.009
5800	72.449	515.697	451.584	371.853	-571.400	122.080	-1.099
5900	72.517	516.936	452.681	379.101	-571.550	134.037	-1.187
6000	72.580	518.155	453.762	386.356	-571.681	145.998	-1.271

PREVIOUS: December 1968 (1 atm) CURRENT: December 1968 (1 bar)

Titanium Iodide (TiI_2) $I_2Ti_1(g)$

ZIRCONIUM IODIDE (ZrI_2) CRYSTAL M_r = 345.0290

$S°(298.15\ K)$ = [150.206] J K^{-1} mol^{-1} $\Delta_f H°(298.15\ K)$ = [-259.408] kJ mol^{-1}

T_{fus} = [700] K $\Delta_{fus}H°$ = [25.104] kJ mol^{-1}

Enthalpy of Formation

Estimated from a consideration of $\Delta_f H°(ZrI_4, cr)$ and of the disproportionation data given by Rolsten (1).

Heat Capacity and Entropy

Heat capacity estimated by comparison to the other zirconium halides and titanium halides. Entropy estimated from additive constants.

Fusion and Vaporization Data

$\Delta_{fus}H°$ was estimated. T_{fus}, T_{vap} and $\Delta_{vap}H°$ from Brewer (2).

References

1. R. F. Rolsten, "Iodide Metals and Metal Iodides", John Wiley and Sons, Inc., New York, (1961).
2. L. Brewer, National Nuclear Energy Series, Div. IV, 19B, paper 7, McGraw-Hill Book Co., Inc., New York, 1950.

Zirconium Iodide (ZrI_2) $I_2Zr_1(cr)$

Enthalpy Reference Temperature = T_r = 298.15 K; Standard State Pressure = $p°$ = 0.1 MPa

T/K	$C_p°$ (J K⁻¹mol⁻¹)	$S°$ (J K⁻¹mol⁻¹)	$-[G°-H°(T_r)]/T$ (J K⁻¹mol⁻¹)	$H°-H°(T_r)$ (kJ mol⁻¹)	$\Delta_f H°$ (kJ mol⁻¹)	$\Delta_f G°$ (kJ mol⁻¹)	Log K_f
0							
100							
200							
298.15	94.140	150.206	150.206	0.	-259.408	-257.975	45.196
300	94.349	150.789	150.207	0.174	-259.381	-257.966	44.916
400	95.019	178.019	153.915	9.641	-274.112	-257.141	33.579
500	95.855	199.313	160.942	19.186	-315.389	-248.966	26.009
600	96.609	216.855	168.841	28.808	-312.211	-235.980	20.544
700	97.445	231.809	176.794	38.510	-309.043	-223.526	16.680
700.000	97.445	231.809	176.794	38.510	------ CRYSTAL <--> LIQUID -----		
800	98.157	244.862	184.504	48.286	-305.897	-211.524	13.811
900	99.621	256.520	191.870	58.185	-302.735	-199.917	11.603
1000	99.830	267.022	198.869	68.153	-299.625	-188.661	9.855
1100	100.583	276.571	205.505	78.173	-296.592	-177.712	8.439
1200	101.420	285.359	211.798	88.274	-297.326	-166.815	7.261
1300	102.215	293.509	217.774	98.455	-293.868	-156.079	6.271
1400	103.010	301.113	223.458	108.717	-290.405	-145.609	5.433
1500	103.805	308.247	228.875	119.057	-286.955	-135.387	4.715

PREVIOUS: CURRENT: June 1962

Zirconium Iodide (ZrI_2) $I_2Zr_1(cr)$

ZIRCONIUM IODIDE (ZrI_2)　　LIQUID　　M_r = 345.0290

$S°(298.15\ K)$ = [186.168] J K^{-1} mol^{-1}　　$\Delta_fH°(298.15\ K)$ = [-234.239] kJ mol^{-1}

T_{fus} = [700] K　　$\Delta_{fus}H°$ = [25.104] kJ mol^{-1}

Enthalpy of Formation

$\Delta_fH°(ZrI_2, \ell, 298.15\ K)$ is calculated from that of the crystal by adding the enthalpy of fusion, $\Delta_{fus}H°$, and the difference in enthalpy, H°(700 K)-H°(298.15 K), between the crystal and liquid.

Heat Capacity and Entropy

The liquid phase heat capacity is estimated. $S°(ZrI_2, \ell, 298.15\ K)$ is calculated in a manner analogous to that used for the enthalpy of formation.

Fusion and Vaporization Data

$\Delta_{fus}H°$ was estimated. T_{fus}, T_{vap} and $\Delta_{vap}H°$ from Brewer (1).

Reference

1. L. Brewer, National Nuclear Energy Series, Div. IV, 19B, paper 7, McGraw-Hill Book Co., Inc., 1950.

Zirconium Iodide (ZrI_2)　　$I_2Zr_1(l)$

Enthalpy Reference Temperature = T_r = 298.15 K　　Standard State Pressure = p° = 0.1 MPa

T/K	$C_p°$ (J K⁻¹mol⁻¹)	S° (J K⁻¹mol⁻¹)	$-[G°-H°(T_r)]/T$ (J K⁻¹mol⁻¹)	$H°-H°(T_r)$ (kJ mol⁻¹)	$\Delta_fH°$ (kJ mol⁻¹)	$\Delta_fG°$ (kJ mol⁻¹)	Log K_f
0							
100							
200							
298.15	94.140	186.168	186.168	0.	-234.239	-243.529	42.665
300	94.349	186.751	186.170	0.174	-234.212	-243.586	42.412
400	95.019	213.982	189.878	9.641	-248.943	-246.357	32.171
500	95.855	235.276	196.904	19.186	-290.220	-241.779	25.258
600	96.609	252.818	204.804	28.808	-287.043	-232.389	20.231
700	97.445	267.672	212.750	38.445	-283.939	-223.526	16.680
700.000	97.445	267.672	212.750	38.445	------ CRYSTAL <--> LIQUID -----		
800	106.148	281.251	220.471	48.625	-280.389	-215.128	14.046
900	114.893	294.259	227.951	59.677	-276.075	-207.222	12.027
1000	123.595	306.815	235.213	71.602	-271.008	-199.836	10.438
1100	132.298	319.003	242.279	84.396	-265.200	-192.994	9.165
1200	141.001	330.887	249.170	98.061	-262.369	-186.492	8.118
1300	149.704	342.517	255.905	112.596	-254.559	-180.480	7.252
1400	158.448	353.931	262.500	128.004	-245.949	-175.099	6.533
1500	167.151	365.160	268.971	144.284	-236.560	-170.362	5.933
1600	175.854	376.225	275.329	161.434	-226.413	-166.275	5.428
1700	184.556	387.148	281.586	179.455	-215.529	-162.846	5.004
1800	193.259	397.943	287.751	198.345	-203.925	-160.078	4.645
1900	202.004	408.626	293.832	218.108	-191.613	-157.975	4.343
2000	210.706	419.209	299.837	238.744	-178.604	-156.538	4.088

PREVIOUS:　　CURRENT: June 1962

Zirconium Iodide (ZrI_2)　　$I_2Zr_1(l)$

ZIRCONIUM IODIDE (ZrI_2) CRYSTAL-LIQUID M_r = 345.0290

0 to 700 K crystal
above 700 K liquid

Refer to the individual tables for details.

Zirconium Iodide (ZrI_2) I_2Zr_1(cr,l)

Enthalpy Reference Temperature = T_r = 298.15 K Standard State Pressure = p° = 0.1 MPa

T/K	C_p° (J K⁻¹mol⁻¹)	S° (J K⁻¹mol⁻¹)	$-[G°-H°(T_r)]/T$ (J K⁻¹mol⁻¹)	$H°-H°(T_r)$ (kJ mol⁻¹)	$\Delta_fH°$ (kJ mol⁻¹)	$\Delta_fG°$ (kJ mol⁻¹)	Log K_f
0							
100							
200							
298.15	94.140	150.206	150.206	0.	-259.408	-257.975	45.196
300	94.349	150.789	150.207	0.174	-259.381	-257.966	44.916
400	95.019	178.019	153.915	9.641	-274.112	-257.141	33.579
500	95.855	199.313	160.942	19.186	-315.389	-248.966	26.009
600	96.609	216.855	168.841	28.808	-312.211	-235.980	20.544
700	97.445	231.809	176.794	38.510	-309.043	-223.526	16.680
700.000	97.445	267.672	176.794	63.614	CRYSTAL <--> LIQUID		
700.000	97.445	267.672	176.794	63.614	TRANSITION		
800	106.148	281.251	189.009	73.794	-280.389	-215.128	14.046
900	114.893	294.259	199.986	84.846	-276.075	-207.222	12.027
1000	123.595	306.815	210.044	96.770	-271.008	-199.836	10.438
1100	132.298	319.003	219.398	109.565	-265.200	-192.994	9.165
1200	141.001	330.887	228.196	123.230	-262.369	-186.492	8.118
1300	149.704	342.517	236.544	137.765	-254.559	-180.480	7.252
1400	158.448	353.931	244.522	153.173	-245.949	-175.099	6.533
1500	167.151	365.160	252.191	169.453	-236.560	-170.362	5.933
1600	175.854	376.225	259.598	186.603	-226.413	-166.275	5.428
1700	184.556	387.148	266.781	204.623	-215.529	-162.846	5.004
1800	193.259	397.943	273.768	223.514	-203.925	-160.078	4.645
1900	202.004	408.626	280.585	243.277	-191.613	-157.975	4.343
2000	210.706	419.209	287.252	263.913	-178.604	-156.538	4.088

PREVIOUS: CURRENT: June 1962

Zirconium Iodide (ZrI_2) I_2Zr_1(cr,l)

ZIRCONIUM IODIDE (ZrI_2) IDEAL GAS M_r = 345.0290

$\Delta_fH°$(0 K) = [-63.262] kJ mol^{-1}

S°(298.15 K) = [344.782] J K^{-1} mol^{-1} $\Delta_fH°$(298.15 K) = [-66.643] kJ mol^{-1}

Vibrational Frequencies and Degeneracies

ν, cm^{-1}
[60](1)
[120](1)
[210](1)

Ground State Quantum Weight: [1] σ = 2

Point Group: C_{2v}

Bond Distance: Zr-I = [2.66] Å

Bond Angle: I-Zr-I = 120°

Product of the Moments of Inertia: $I_AI_BI_C = 1.072892 \times 10^{-111}\ g^3\ cm^6$

Enthalpy of Formation

Based upon the estimated value of the $\Delta_fH°$ for the crystal and the appropriate crystal and gas functions.

Heat Capacity and Entropy

Molecular constants were all estimated by comparison to similar molecules of zirconium and titanium. The principal moments of inertia are: $I_A = 19.7106 \times 10^{-39}$, $I_B = 223.6596 \times 10^{-39}$ and $I_C = 243.3703 \times 10^{-39}\ g\ cm^2$.

Zirconium Iodide (ZrI_2) $I_2Zr_1(g)$

Enthalpy Reference Temperature = T_r = 298.15 K; Standard State Pressure = p° = 0.1 MPa

T/K	$C_p°$ (J $K^{-1}mol^{-1}$)	S° (J $K^{-1}mol^{-1}$)	-[G°-H°(T_r)]/T (J $K^{-1}mol^{-1}$)	H°-H°(T_r) (kJ mol^{-1})	$\Delta_fH°$ (kJ mol^{-1})	$\Delta_fG°$ (kJ mol^{-1})	Log K_f
0	0.	0.	INFINITE	-15.315	-63.262	-63.262	INFINITE
100	51.686	284.507	394.767	-11.026	-62.910	-83.856	43.802
200	56.159	322.118	349.992	-5.575	-64.584	-104.206	27.216
250	56.854	334.733	345.723	-2.748	-65.600	-113.996	23.818
298.15	57.238	344.782	344.782	0.	-66.643	-123.223	21.588
300	57.249	345.136	344.783	0.106	-66.684	-123.574	21.516
350	57.495	353.981	345.481	2.975	-67.898	-132.963	19.844
400	57.656	361.670	347.035	5.854	-85.134	-141.624	18.494
450	57.769	368.468	349.046	8.740	-87.586	-148.538	17.242
500	57.850	374.558	351.298	11.630	-130.179	-151.379	15.814
600	57.956	385.116	356.081	17.421	-130.833	-155.559	13.543
700	58.020	394.055	360.883	23.220	-131.568	-159.623	11.911
800	58.062	401.806	365.525	29.024	-132.393	-163.576	10.680
900	58.091	408.646	369.944	34.832	-133.323	-167.419	9.717
1000	58.112	414.768	374.125	40.642	-134.371	-171.152	8.940
1100	58.127	420.307	378.076	46.454	-135.545	-174.774	8.299
1200	58.139	425.366	381.809	52.268	-140.566	-178.062	7.751
1300	58.148	430.020	385.341	58.082	-141.476	-181.151	7.279
1400	58.156	434.329	388.688	63.897	-142.459	-184.166	6.871
1500	58.161	438.342	391.866	69.713	-143.534	-187.108	6.516
1600	58.166	442.095	394.889	75.530	-144.721	-189.975	6.202
1700	58.170	445.622	397.771	81.346	-146.041	-192.764	5.923
1800	58.173	448.947	400.523	87.164	-147.510	-195.471	5.672
1900	58.176	452.092	403.155	92.981	-149.144	-198.092	5.446
2000	58.179	455.076	405.677	98.799	-150.953	-200 622	5.240
2100	58.181	457.915	408.097	104.617	-152.947	-203.057	5.051
2200	58.182	460.621	410.424	110.435	-176.452	-204.647	4.859
2300	58.184	463.208	412.663	116.253	-179.243	-205.866	4.675
2400	58.185	465.684	414.821	122.072	-182.079	-206.963	4.504
2500	58.187	468.059	416.903	127.890	-184.948	-207.941	4.345
2600	58.188	470.342	418.915	133.709	-187.840	-208.804	4.195
2700	58.189	472.538	420.861	139.528	-190.742	-209.555	4.054
2800	58.190	474.654	422.744	145.347	-193.643	-210.199	3.921
2900	58.190	476.696	424.570	151.166	-196.532	-210.739	3.796
3000	58.191	478.668	426.340	156.985	-199.401	-211.180	3.677
3100	58.192	480.577	428.059	162.804	-202.239	-211.526	3.564
3200	58.192	482.424	429.729	168.623	-205.040	-211.780	3.457
3300	58.193	484.215	431.353	174.442	-207.798	-211.948	3.355
3400	58.193	485.952	432.934	180.262	-210.508	-212.033	3.257
3500	58.194	487.639	434.473	186.081	-213.165	-212.039	3.165
3600	58.194	489.278	435.973	191.900	-215.767	-211.970	3.076
3700	58.194	490.873	437.435	197.720	-218.312	-211.829	2.990
3800	58.195	492.425	438.862	203.539	-220.798	-211.621	2.909
3900	58.195	493.936	440.255	209.359	-223.225	-211.347	2.831
4000	58.195	495.410	441.615	215.178	-225.592	-211.013	2.756
4100	58.196	496.847	442.945	220.998	-227.900	-210.620	2.683
4200	58.196	498.249	444.245	226.817	-230.150	-210.171	2.614
4300	58.196	499.618	445.517	232.637	-232.342	-209.669	2.547
4400	58.196	500.956	446.762	238.457	-234.478	-209.117	2.483
4500	58.196	502.264	447.981	244.276	-236.559	-208.517	2.420
4600	58.197	503.543	449.175	250.096	-238.587	-207.872	2.360
4700	58.197	504.795	450.345	255.916	-240.563	-207.182	2.303
4800	58.197	506.020	451.492	261.735	-803.426	-194.834	2.120
4900	58.197	507.220	452.617	267.555	-804.955	-182.139	1.942
5000	58.197	508.396	453.721	273.375	-806.456	-169.414	1.770
5100	58.197	509.548	454.804	279.195	-807.927	-156.658	1.605
5200	58.198	510.678	455.868	285.014	-809.369	-143.874	1.445
5300	58.198	511.787	456.913	290.834	-810.782	-131.063	1.292
5400	58.198	512.875	457.939	296.654	-812.168	-118.225	1.144
5500	58.198	513.943	458.947	302.474	-813.525	-105.362	1.001
5600	58.198	514.991	459.939	308.293	-814.855	-92.474	0.863
5700	58.198	516.021	460.914	314.113	-816.157	-79.562	0.729
5800	58.198	517.034	461.873	319.933	-817.432	-66.629	0.600
5900	58.198	518.028	462.816	325.753	-818.680	-53.673	0.475
6000	58.198	519.007	463.744	331.573	-819.901	-40.696	0.354

PREVIOUS: June 1962 (1 atm) CURRENT June 1962 (1 bar)

Zirconium Iodide (ZrI_2) $I_2Zr_1(g)$

MOLYBDENUM IODIDE (MoI_3) CRYSTAL M_r = 476.6535

$S°(298.15\ K)$ = [195.393 ± 16.7] J K^{-1} mol^{-1}

$\Delta_f H°(0\ K)$ = Unknown

$\Delta_f H°(298.15\ K)$ = -124.68 ± 8.4 kJ mol^{-1}

Enthalpy of Formation

Barnes et al. (1) have measured calorimetrically the enthalpy of formation of $MoI_{2.76}$ by reacting iodine vapor with $Mo(CO)_6$ at 539 K. The products were found to consist of MoI_3, MoI_2, and Mo by chemical analysis. Assuming a difference in the enthalpies of formation of MoI_3 and MoI_2 of 1.5 kcal mol^{-1}, Barnes et al. (1) determined that the $\Delta_f H°(MoI_3$, cr, 298.15 K) = -26.2±2.0 kcal mol^{-1}. Brewer (2) has estimated that the difference $\Delta_f H°(MoI_3)-\Delta_f H°(MoI_2)$ may be as large as -5 kcal mol^{-1} and reanalyzed the heat of iodination data with the result $\Delta_f H°(MoI_3$, cr, 298.15 K)/R = -15000±1000 K which is adopted. We adopt the estimate of Brewer(2). The assigned uncertainty (±2.0 kcal mol^{-1}) is determined from the product 1000xR which has been rounded to the nearest kcal mol^{-1}.

Heat Capacity and Entropy

The values of S°(298.15 K) and $C_p°$ over the temperature range 298-1000 K are estimates reported by Brewer (2). We treat the estimated entropy value given as S°(298.15 K)/R in a manner similar to that for the enthalpy of formation.

Phase Data

Drobot et al. (3) have studied the thermal decomposition of MoI_3 in evacuated sealed ampoules at various temperatures. X-ray diffraction and chemical analyses of the condensed decomposition products showed the existence of a homogeneity range in which the atomic ratio of I/Mo varied from 3.00 to 2.5. The crystal structure of MoI_3 is oP16 (2) in the Pearson classification system.

Decomposition Data

T_{dcm} of 629.5 K is calculated as the decomposition temperature at which the fugacity of iodine is 1 atm for the dissociation reaction $MoI_3(cr) = MoI_2(cr) + 0.5\ I_2(g)$. Saturated vapor pressure measurements by Drobot et al. (3) indicate that MoI_3(cr) decomposes to a non-stoichiometric triiodide phase beginning at a temperature near 363.2 K.

References

1. D. S. Barnes, G. Pilcher, D. A. Pittam, H. A. Skinner, D. Todd, and Y. Virmani, J. Less-Common Met. 36, 177 (1974).
2. L. Brewer, Materials and Molecular Research Division, Lawrence Berkeley Laboratory, University of California, Berkeley; personal communication, Sept. 29, 1978, preliminary draft of review to be submitted for publication in Atomic Energy Review, International Atomic Energy Agency, Vienna, Austria.
3. D. V. Drobot, L. G. Mikhailova, K. A. Bol'shakov, and V. L. Sbitnev, Russ. J. Inorg. Chem. 23, (5), 643 (1978).

Molybdenum Iodide (MoI_3) $I_3Mo_1(cr)$

Enthalpy Reference Temperature = T_r = 298.15 K — Standard State Pressure = p° = 0.1 MPa

T/K	$C_p°$	S°	$-[G°-H°(T_r)]/T$	$H°-H°(T_r)$	$\Delta_f H°$	$\Delta_f G°$	Log K_f
	J $K^{-1}mol^{-1}$			kJ mol^{-1}			
0							
100							
200							
298.15	108.889	195.393	195.393	0.	-124.683	-122.469	21.456
300	108.918	196.066	195.395	0.201	-124.677	-122.456	21.321
400	110.583	227.628	199.687	11.177	-148.615	-120.905	15.789
500	112.244	252.483	207.847	22.318	-212.317	-107.872	11.269
600	113.909	273.094	217.052	33.626	-209.257	-87.269	7.597
700	115.570	290.778	226.350	45.100	-206.107	-67.185	5.013
800	117.231	306.319	235.394	56.740	-202.857	-47.559	3.105
900	118.897	320.223	244.061	68.546	-199.502	-28.347	1.645
1000	120.558	332.836	252.317	80.519	-196.045	-9.515	0.497

PREVIOUS: CURRENT: September 1978

Molybdenum Iodide (MoI_3) $I_3Mo_1(cr)$

MOLYBDENUM IODIDE (MoI_3)   IDEAL GAS M_r = 476.6535

$S°(298.15 K) = [403.532 \pm 8.4] J K^{-1} mol^{-1}$

$\Delta_f H°(0 K) = [186.15 \pm 41.8] kJ mol^{-1}$
$\Delta_f H°(298.15 K) = [182.84 \pm 41.8] kJ mol^{-1}$

Electronic Levels and Quantum Weights

ϵ_i, cm^{-1}	g_i
[0]	2
[3505.4]	2
[10861.0]	4
[11062.0]	4

Vibrational Frequencies and Degeneracies

ν, cm^{-1}
[110](1)
[75](1)
[200](2)
[60](2)

Point Group: $[D_{3h}]$ $\sigma = [6]$
Bond Distance: Mo–I = [2.66] Å
Bond Angle: I–Mo–I = [120]°
Product of the Moments of Inertia: $I_A I_B I_C = [2.237654 \times 10^{-110}] g^3 cm^6$

Enthalpy of Formation

The adopted value of $\Delta_f H°$ at 298.15 K is that estimated by Brewer (1). The estimated value is consistent with the low volatility of solid molybdenum in iodine vapor reported by Schafer et al. (2). The estimated value given by Brewer (1) is $\Delta_f H°(298.15 K)/R = 22000 \pm 5000$ K. The assigned uncertainty (± 10 kcal mol^{-1}) is determined from the product 5000xR which has been rounded to the nearest kcal mol^{-1}. The value of $\Delta_f H°(0 K)$ combined with JANAF data (3) for Mo(g) and I(g) gives $\Delta_{at} H°(0 K)$ = 189 ± 11 kcal mol^{-1} and an average Mo-I bond energy of $\Delta_{at} H°(0 K)/3 = 63 \pm 4$ kcal mol^{-1}.

Heat Capacity and Entropy

The bond length and vibrational frequencies are values estimated by Brewer (1). A planar structure has been predicted for MoI_3 by Drak and Rosenblatt (4). The electronic contribution is taken to be the same as that for TaO, as estimated by Brewer (1). The levels are taken from the latest JANAF table (3), however, we only include four levels. The principal moments of inertia are: $I_A = I_B = 223.6596 \times 10^{-39}$, and $I_C = 447.3193 \times 10^{-39}$ g cm^2.

References

1. L. Brewer, Materials and Molecular Research Division, Lawrence Berkeley Laboratory, University of California, Berkeley; personal communication, Sept. 29, 1978, preliminary draft of review to be submitted for publication in Atomic Energy Review, International Atomic Energy Agency, Vienna, Austria.
2. H. Schafer, T. Grofe, and M. Trenkel, J. Solid State Chem. 8, 14 (1973).
3. JANAF Thermochemical Tables: Mo(g), 3-31-78; I(g), 3-30-74; TaO(g), 12-31-73.
4. M. C. Drake and G. M. Rosenblatt, Paper 388, The Electrochemical Society, Atlanta, Georgia, October, 1977.

Molybdenum Iodide (MoI_3) $I_3Mo_1(g)$

Enthalpy Reference Temperature = T_r = 298.15 K   Standard State Pressure = $p°$ = 0.1 MPa

T/K	$C_p°$ (J $K^{-1}mol^{-1}$)	$S°$	$-[G°-H°(T_r)]/T$	$H°-H°(T_r)$ (kJ mol^{-1})	$\Delta_f H°$	$\Delta_f G°$	Log K_f
0	0.	0.	INFINITE	-21.071	186.152	186.152	INFINITE
100	71.920	318.351	474.458	-15.611	186.546	164.562	-85.958
200	79.674	371.309	410.943	-7.927	184.989	143.110	-37.377
250	80.859	389.229	404.871	-3.910	183.946	132.758	-27.738
298.15	81.513	403.532	403.532	0.	182.841	122.998	-21.549
300	81.532	404.036	403.533	0.151	182.796	122.627	-21.351
350	81.949	416.638	404.528	4.239	181.452	112.700	-16.820
400	82.227	427.600	406.742	8.343	156.076	103.797	-13.554
450	82.427	437.297	409.608	12.460	152.878	97.454	-11.312
500	82.585	445.990	412.819	16.585	89.474	97 166	-10.151
600	82.861	461.072	419.642	24.858	89.499	98.701	-8.593
700	83.159	473.867	426.497	33.159	89.476	100.236	-7.480
800	83.512	484.994	433.129	41.492	89.419	101.777	-6.645
900	83.914	494.853	439.450	49.863	89.338	103.326	-5.997
1000	84.342	503.716	445.441	58.275	89.236	104.886	-5.479
1100	84.772	511.775	451.110	66.731	89.106	106.457	-5.055
1200	85.183	519.169	456.478	75.229	88.936	108.041	-4.703
1300	85.562	526.002	461.566	83.767	88.709	109.642	-4.405
1400	85.906	532.356	466.399	92.340	88.402	111.264	-4.151
1500	86.214	538.294	470.996	100.947	87.992	112.910	-3.932
1600	86.492	543.867	475.378	109.582	87.455	114.588	-3.741
1700	86.745	549.118	479.563	118.244	86.768	116.304	-3.574
1800	86.980	554.083	483.566	126.931	85.912	118.065	-3.426
1900	87.205	558.792	487.402	135.640	84.871	119.879	-3.296
2000	87.425	563.270	491.085	144.371	83.636	121.753	-3.180
2100	87.644	567.541	494.625	153.125	82.201	123.693	-3.077
2200	87.865	571.624	498.032	161.900	80.565	125.706	-2.985
2300	88.090	575.534	501.318	170.698	78.730	127.798	-2.902
2400	88.318	579.288	504.489	179.519	76.702	129.974	-2.829
2500	88.550	582.898	507.553	188.362	74.486	132.239	-2.763
2600	88.784	586.376	510.519	197.229	72.078	134.596	-2.704
2700	89.018	589.731	513.391	206.119	69.455	137.050	-2.651
2800	89.250	592.972	516.175	215.032	66.574	139.605	-2.604
2900	89.479	596.108	518.878	223.969	27.483	142.318	-2.563
3000	89.702	599.146	521.503	232.928	25.922	146.305	-2.547
3100	89.917	602.090	524.055	241.909	24.427	150.343	-2.533
3200	90.122	604.948	526.539	250.911	23.008	154.427	-2.521
3300	90.316	607.725	528.957	259.933	21.675	158.555	-2.510
3400	90.498	610.424	531.314	268.973	20.433	162.721	-2.500
3500	90.666	613.049	533.612	278.032	19.287	166.924	-2.491
3600	90.819	615.606	535.854	287.106	18.239	171.157	-2.483
3700	90.957	618.096	538.043	296.195	17.293	175.418	-2.476
3800	91.081	620.523	540.182	305.297	16.447	179.703	-2.470
3900	91.188	622.890	542.272	314.411	15.701	184.009	-2.465
4000	91.280	625.200	544.317	323.534	15.055	188.333	-2.459
4100	91.357	627.455	546.317	332.666	14 506	192.672	-2.455
4200	91.418	629.658	548.275	341.805	14.052	197.023	-2.450
4300	91.466	631.809	550.193	350.949	13.689	201.384	-2.446
4400	91.499	633.912	552.072	360.098	13.414	205.752	-2.443
4500	91.519	635.969	553.914	369.249	13.224	210.126	-2.439
4600	91.527	637.980	555.719	378.401	13.116	214.503	-2.436
4700	91.522	639.949	557.491	387.554	13.085	218.881	-2.433
4800	91.507	641.876	559.229	396.705	13.128	223.260	-2.430
4900	91.481	643.762	560.935	405.855	13.241	227.637	-2.427
5000	91.446	645.610	562.610	415.001	-577.964	237.741	-2.484
5100	91.402	647.420	564.255	424.144	-579.400	254.070	-2.602
5200	91.349	649.195	565.871	433.281	-580.922	270.427	-2.716
5300	91.290	650.934	567.460	442.413	-582.532	286.814	-2.827
5400	91.223	652.640	569.022	451.539	-584.231	303.233	-2.933
5500	91.151	654.313	570.557	460.658	-586.019	319.683	-3.036
5600	91.073	655.955	572.068	469.769	-587.898	336.168	-3.136
5700	90.991	657.566	573.554	478.872	-589.866	352.686	-3.232
5800	90.904	659.148	575.016	487.967	-591.925	369.240	-3.325
5900	90.813	660.701	576.455	497.053	-594.070	385.830	-3.416
6000	90.720	662.227	577.872	506.129	-596.307	402.457	-3.504

PREVIOUS: September 1978 (1 atm)   CURRENT: September 1978 (1 bar)

Molybdenum Iodide (MoI_3) $I_3Mo_1(g)$

TRIIODO SILYL (SiI_3) IDEAL GAS M_r = 408.7990

$\Delta_f H°(0\ K)$ = [39.75 ± 62.8] kJ mol^{-1}

S°(298.15 K) = [378.305 ± 8.4] J K^{-1} mol^{-1} $\Delta_f H°(298.15\ K)$ = [35.31 ± 62.8] kJ mol^{-1}

Vibrational Frequencies and Degeneracies

ν, cm^{-1}
[286](1)
[122](1)
[399](2)
[80](2)

Ground State Quantum Weight: [2] σ = [3]

Point Group: [C_{3v}]

Bond Distance: Si-I = [2.435] Å

Bond Angle: I-Si-I = [112]°

Product of the Moments of Inertia: $I_A I_B I_C$ = [1.038908 x 10^{-110}] $g^3\ cm^6$

Enthalpy of Formation

The enthalpy of formation of SiI_3(g) is based on an assumed average bond energy of 58±5 kcal mol^{-1} (1). This average bond energy is that of SiI (g), i.e. $\Delta_{at}H°(SiI_4$, g, 0 K)/4. The rationale for this assumption is based on the same relationship existing for the silicon chloride and fluoride species (1).

Heat Capacity and Entropy

The molecular structure is assumed to be identical to the SiI_3 group in $SiHI_3$ (1). From this structure we estimate the following principal moments of inertia: $I_A = I_B$ = 173.9077 x 10^{-39}, and I_C = 343.5099 x 10^{-39} g cm^2. The vibrational frequencies are assumed to be those of the SiI_3 group in $SiHI_3$(g).

Reference

1. JANAF Thermochemical Tables: SiI_4(g), $SiCl_3$(g) and SiF_3(g), 12-31-77; $SiHI_3$(g), 12-31-76.

Triiodosilyl (SiI_3) I_3Si_1(g)

Enthalpy Reference Temperature = T_r = 298.15 K Standard State Pressure = p° = 0.1 MPa

T/K	$C_p°$ (J K^{-1}mol^{-1})	S° (J K^{-1}mol^{-1})	$-[G°-H°(T_r)]/T$ (J K^{-1}mol^{-1})	$H°-H°(T_r)$ (kJ mol^{-1})	$\Delta_f H°$ (kJ mol^{-1})	$\Delta_f G°$ (kJ mol^{-1})	Log K_f
0	0.	0.	INFINITE	-18.578	39.751	39.751	INFINITE
100	58.786	303.097	442.837	-13.974	39.514	18.736	-9.787
200	71.873	348.463	385.228	-7.353	37.557	-1.315	0.343
250	75.152	364.881	379.565	-3.671	36.448	-10.907	2.279
298.15	77.191	378.305	378.305	0.	35.314	-19.925	3.491
300	77.254	378.783	378.306	0.143	35.268	-20.268	3.529
350	78.652	390.803	379.253	4.043	33.916	-29.421	4.391
400	79.619	401.373	381.370	8.001	8.548	-37.551	4.904
450	80.310	410.793	384.126	12.000	5.373	-43.123	5.006
500	80.820	419.282	387.224	16.029	-57.997	-42.643	4.455
600	81.503	434.083	393.836	24.148	-57.885	-39.583	3.446
700	81.926	446.681	400.508	32.321	-57.811	-36.539	2.727
800	82.206	457.640	406.979	40.529	-57.781	-33.503	2.188
900	82.399	467.334	413.157	48.759	-57.798	-30.467	1.768
1000	82.539	476.023	419.017	57.007	-57.864	-27.428	1.433
1100	82.643	483.895	424.563	65.266	-57.984	-24.379	1.158
1200	82.722	491.090	429.811	73.534	-58.163	-21.316	0.928
1300	82.784	497.714	434.783	81.810	-58.412	-18.236	0.733
1400	82.833	503.850	439.500	90.091	-58.744	-15.134	0.565
1500	82.873	509.567	443.983	98.376	-59.174	-12.004	0.418
1600	82.906	514.916	448.251	106.665	-59.717	-8.843	0.289
1700	82.933	519.943	452.321	114.957	-110.563	-5.196	0.160
1800	82.955	524.684	456.211	123.252	-111.140	1.018	-0.030
1900	82.975	529.170	459.934	131.548	-111.818	7.267	-0.200
2000	82.991	533.426	463.503	139.846	-112.600	13.554	-0.354
2100	83.005	537.476	466.930	148.146	-113.482	19.883	-0.495
2200	83.018	541.338	470.225	156.447	-114.456	26.256	-0.623
2300	83.028	545.028	473.398	164.750	-115.511	32.676	-0.742
2400	83.038	548.562	476.456	173.053	-116.632	39.142	-0.852
2500	83.046	551.952	479.409	181.357	-117.804	45.657	-0.954
2600	83.053	555.209	482.262	189.662	-119.008	52.219	-1.049
2700	83.060	558.344	485.022	197.968	-120.227	58.828	-1.138
2800	83.066	561.365	487.695	206.274	-121.444	65.482	-1.222
2900	83.071	564.280	490.286	214.581	-122.644	72.179	-1.300
3000	83.076	567.096	492.800	222.888	-123.811	78.917	-1.374
3100	83.080	569.820	495.241	231.196	-124.933	85.693	-1.444
3200	83.084	572.458	497.613	239.505	-125.999	92.505	-1.510
3300	83.088	575.014	499.920	247.813	-127.000	99.349	-1.573
3400	83.091	577.495	502.165	256.122	-127.928	106.222	-1.632
3500	83.094	579.904	504.352	264.431	-128.777	113.121	-1.688
3600	83.097	582.244	506.483	272.741	-513.695	130.504	-1.894
3700	83.099	584.521	508.562	281.051	-513.961	148.403	-2.095
3800	83.102	586.737	510.590	289.361	-514 140	166.307	-2.286
3900	83.104	588.896	512.570	297.671	-514.232	184.215	-2.467
4000	83.106	591.000	514.505	305.981	-514.236	202.124	-2.639
4100	83.108	593.052	516.396	314 292	-514.151	220.032	-2.803
4200	83.109	595.055	518.245	322.603	-513.979	237.937	-2.959
4300	83.111	597.010	520.054	330.914	-513.721	255.837	-3.108
4400	83.112	598.921	521.825	339.225	-513.378	273.730	-3.250
4500	83.114	600.789	523.559	347.536	-512.953	291.614	-3.385
4600	83.115	602.616	525.257	355.848	-512.448	309.488	-3.514
4700	83.116	604.403	526.922	364.160	-511.865	327.350	-3.638
4800	83.117	606.153	528.555	372.471	-511.206	345.199	-3.757
4900	83.119	607.867	530.156	380.783	-510.475	363.033	-3.870
5000	83.120	609.546	531.727	389.095	-509.673	380.851	-3.979
5100	83.121	611.192	533.269	397.407	-508.804	398.654	-4.083
5200	83.121	612.806	534.783	405.719	-507.869	416.438	-4.183
5300	83.122	614.390	536.270	414.031	-506.872	434.203	-4.279
5400	83.123	615.943	537.732	422.343	-505.815	451.950	-4.372
5500	83.124	617.469	539.167	430.656	-504.700	469.676	-4.461
5600	83.125	618.966	540.579	438.968	-503.531	487.382	-4.546
5700	83.125	620.438	541.967	447.281	-502.310	505.066	-4.628
5800	83.126	621.883	543.333	455.593	-501.038	522.728	-4.708
5900	83.126	623.304	544.676	463.906	-499.719	540.368	-4.784
6000	83.127	624.701	545.998	472.219	-498.353	557.985	-4.858

PREVIOUS: December 1977 (1 atm) CURRENT: December 1977 (1 bar)

Triiodosilyl (SiI_3) I_3Si_1(g)

TITANIUM IODIDE (TiI_3) CRYSTAL M_r = 428.5935

$S°(298.15\ K)$ = [192.464 ± 8.4] J K^{-1} mol^{-1}

T_{sub} = [1000] K

$\Delta_f H°(0\ K)$ = Unknown

$\Delta_f H°(298.15\ K)$ = [-322.17 ± 20.9] kJ mol^{-1}

Enthalpy of Formation

The enthalpy of formation of $TiI_3(cr)$ is calculated from the free energy of formation at 623 K combined with the increment $\Delta_f G°(623\ K)-\Delta_f G°(298.15\ K)$ and the entropies, $S°(298.15\ K)$, of $TiI_2(cr)$, $TiI_3(cr)$ and $TiI_4(g)$. The free energy of formation, $\Delta_f G°(623\ K)$, is estimated such that $\Delta_r G°(623\ K) = 0$ for the process $2TiI_3(cr) = TiI_2(cr) + TiI_4(g)$. This assumption is based on the observations of Hercog and Pidgeon (1) and Fast (2).

Heat Capacity and Entropy

The heat capacities of $TiI_3(cr)$ and the value of $S°(298.15\ K)$ estimated by Kelley (3) have been adopted.

Sublimation Data

The enthalpy of sublimation of $TiI_3(cr)$ is taken as the difference in the enthalpies of formation of $TiI_3(cr)$ and $TiI_3(g)$ at the sublimation temperature. The sublimation temperature is estimated as the point at which $\Delta_r G° = 0$ for the process $TiI_3(cr) = TiI_3(g)$.

References

1. A. Herczog and L. M. Pidgeon, Can. J. Chem. 34, 1687 (1956).
2. J. P. Fast, Rec. Trav. Chim. 58, 174 (1939).
3. K. K. Kelley, U. S. Bur. Mines Bull. 584, 1960; U. S. Bur. Mines Bull. 592, 1961.

Titanium Iodide (TiI_3) $I_3Ti_1(cr)$

Enthalpy Reference Temperature = T_r = 298.15 K   Standard State Pressure = p° = 0.1 MPa

T/K	$C_p°$ (J $K^{-1}mol^{-1}$)	S° (J $K^{-1}mol^{-1}$)	$-[G°-H°(T_r)]/T$ (J $K^{-1}mol^{-1}$)	$H°-H°(T_r)$ (kJ mol^{-1})	$\Delta_f H°$ (kJ mol^{-1})	$\Delta_f G°$ (kJ mol^{-1})	Log K_f
0							
100							
200							
298.15	116.771	192.464	192.464	0.	-322.168	-318.439	55.789
300	116.784	193.186	192.466	0.216	-322.150	-318.416	55.441
400	117.512	226.883	197.056	11.931	-345.506	-316.453	41.324
500	118.240	253.183	205.746	23.718	-408.753	-303.138	31.669
600	118.968	274.805	215.507	35.579	-405.350	-282.335	24.579
700	119.696	293.199	225.324	47.512	-401.956	-262.101	19.558
800	120.424	309.229	234.832	59.518	-398.543	-242.354	15.824
900	121.152	323.455	243.903	71.597	-395.141	-223.035	12.945
1000	121.880	336.257	252.509	83.748	-391.813	-204.092	10.661
1100	122.608	347.908	260.660	95.973	-388.627	-185.476	8.808
1200	123.336	358.607	268.383	108.270	-389.573	-167.022	7.270
1300	124.064	368.508	275.708	120.640	-385.968	-148.623	5.972
1400	124.792	377.729	282.670	133.083	-382.414	-130.499	4.869
1500	125.520	386.364	289.298	145.598	-378.940	-112.627	3.922

PREVIOUS: June 1964   CURRENT: December 1968

Titanium Iodide (TiI_3) $I_3Ti_1(cr)$

TITANIUM IODIDE (TiI_3) IDEAL GAS M_r = 428.5935

$\Delta_f H°(0\ K)$ = [-144.51 ± 33.5] kJ mol^{-1}

S°(298.15 K) = [382.164 ± 12.6] J K^{-1} mol^{-1} $\Delta_f H°(298.15\ K)$ = [-150.21 ± 33.5] kJ mol^{-1}

Electronic Levels and Quantum Weights

ϵ_i, cm^{-1}	g_i	ϵ_i, cm^{-1}	g_i
0	[2]	4000	[2]
600	[2]	10000	[2]
1500	[2]		

Vibrational Frequencies and Degeneracies

ν, cm^{-1}	ν, cm^{-1}
[341](1)	[383](2)
[123](1)	[74](2)

Ground State Quantum Weight: [2] σ = 3

Point Group: [C_{3v}]

Bond Distance: Ti-I = [2.7] Å

Bond Angle: I-Ti-I = [100]°

Product of the Moments of Inertia: $I_A I_B I_C$ = 1.322449 x 10^{-110}] g^3 cm^6

Enthalpy of Formation

The enthalpy of formation, $\Delta_f H°(TiI_3$, g, 298.15 K), is estimated from the average Ti-I bond energy. The bond energy is estimated from the corresponding quantities for $TiBr_4$(g), $TiBr_3$(g) and TiI_4(g).

Heat Capacity and Entropy

The interatomic distance is estimated from those of TiI_4, $TiCl_4$ and $TiCl_3$. The pyramidal bond angle is estimated by assuming that TiI_3(g) is similar to the group V trihalides. The principal moments of inertia are: $I_A = I_B$ = 191.5028 x 10^{-39} and I_C = 360.6023 x 10^{-39} g cm^2. The vibrational frequencies are estimated from valence force field predictions and comparisons with group V trihalides. The electronic levels are estimated from the levels of Ti^{+3} reported by Moore (1).

Reference

1. C. E. Moore, U. S. Nat. Bur. Stand. Circ. 467, 1949.

Titanium Iodide (TiI_3) I_3Ti_1(g)

Enthalpy Reference Temperature = T_r = 298.15 K Standard State Pressure = p° = 0.1 MPa

T/K	$C_p°$ (J $K^{-1}mol^{-1}$)	S° (J $K^{-1}mol^{-1}$)	$-[G°-H°(T_r)]/T$ (J $K^{-1}mol^{-1}$)	$H°-H°(T_r)$ (kJ mol^{-1})	$\Delta_f H°$ (kJ mol^{-1})	$\Delta_f G°$ (kJ mol^{-1})	Log K_f
0	0.	0.	INFINITE	-18.933	-144.511	-144.511	INFINITE
100	58.519	305.563	448.602	-14.304	-145.018	-165.600	86.500
200	73.632	351.232	389.367	-7.627	-147.657	-185.170	48.362
250	78.018	368.168	383.480	-3.828	-148.963	-194.397	40.617
298.15	80.800	382.164	382.164	0.	-150.206	-203.035	35.571
300	80.886	382.664	382.166	0.150	-150.254	-203.363	35.409
350	82.769	395.284	383.158	4.244	-151.664	-212.107	31.655
400	84.024	406.424	385.384	8.416	-177.058	-219.822	28.706
450	84.870	416.373	388.284	12.640	-180.242	-224.976	26.114
500	85.440	425.346	391.549	16.899	-243.611	-224.077	23.409
600	86.063	440.987	398.523	25.478	-243.488	-220.183	19.169
700	86.303	454.275	405.563	34.099	-243.407	-216.305	16.141
800	86.363	465.804	412.388	42.733	-243.366	-212.437	13.871
900	86.344	475.976	418.900	51.369	-243.407	-208.570	12.105
1000	86.297	485.071	425.070	60.001	-243.598	-204.690	10.692
1100	86.242	493.293	430.904	68.628	-244.010	-200.782	9.534
1200	86.187	500.795	436.420	77.249	-248.632	-196.705	8.562
1300	86.133	507.691	441.641	85.865	-248.780	-192.373	7.730
1400	86.083	514.073	446.590	94.476	-249.058	-188.024	7.015
1500	86.034	520.010	451.289	103.082	-249.494	-183.651	6.395
1600	85.988	525.561	455.759	111.683	-250.113	-179.242	5.852
1700	85.943	530.773	460.020	120.279	-250.943	-174.788	5.371
1800	85.900	535.684	464.088	128.871	-252.006	-170.279	4.941
1900	85.860	540.327	467.980	137.459	-253.321	-165.704	4.556
2000	85.822	544.730	471.708	146.044	-269.590	-160.599	4.194
2100	85.787	548.917	475.286	154.624	-272.195	-155.086	3.858
2200	85.754	552.907	478.724	163.201	-274.898	-149.447	3.548
2300	85.723	556.718	482.033	171.775	-277.685	-143.683	3.263
2400	85.695	560.366	485.221	180.346	-280.543	-137.795	2.999
2500	85.669	563.863	488.298	188.914	-283.455	-131.788	2.754
2600	85.644	567.223	491.269	197 479	-286.402	-125.663	2.525
2700	85.622	570.455	494.142	206.043	-289.368	-119.424	2.310
2800	85.600	573.568	496.924	214.604	-292.335	-113.076	2.109
2900	85.580	576.571	499.619	223.163	-295.286	-106.621	1.920
3000	85.561	579.472	502.232	231.720	-298.208	-100.066	1.742
3100	85.543	582.278	504.770	240.275	-301.086	-93.414	1.574
3200	85.525	584.993	507.234	248.829	-303.911	-86.669	1.415
3300	85.507	587.625	509.631	257.380	-306.673	-79.838	1.264
3400	85.490	590.177	511.962	265.930	-309.364	-72.923	1.120
3500	85.473	592.655	514.233	274.478	-311.979	-65.931	0.984
3600	85.456	595.063	516.445	283.025	-314.512	-58.865	0.854
3700	85.439	597.404	518.601	291.569	-726.060	-43.941	0.620
3800	85.422	599.682	520.705	300.112	-727.189	-25.490	0.350
3900	85.404	601.901	522.759	308.654	-728.285	-7.010	0.094
4000	85.386	604.063	524.764	317.193	-729.349	11.498	-0.150
4100	85.369	606.171	526.724	325.731	-730.378	30.032	-0.383
4200	85.350	608.228	528.641	334.267	-731.372	48.590	-0.604
4300	85.332	610.236	530.515	342.801	-732.332	67.172	-0.816
4400	85.313	612.198	532.349	351.333	-733.256	85.776	-1.018
4500	85.294	614.115	534.145	359.864	-734.146	104.400	-1.212
4600	85.275	615.989	535.904	368.392	-735.001	123.044	-1.397
4700	85.255	617.823	537.627	376.919	-735.823	141.706	-1.575
4800	85.235	619.617	539.317	385.443	-736.611	160.386	-1.745
4900	85.215	621.375	540.974	393.966	-737.367	179.081	-1.909
5000	85.195	623.096	542.599	402.486	-738.090	197.791	-2.066
5100	85.174	624.783	544.194	411.005	-738.783	216.516	-2.218
5200	85.154	626.437	545.760	419.521	-739.444	235.254	-2.363
5300	85.133	628.059	547.297	428.035	-740.075	254.004	-2.503
5400	85.112	629.650	548.808	436.548	-740.676	272.766	-2.638
5500	85.091	631.211	550.292	445.058	-741.247	291.538	-2.769
5600	85.070	632.744	551.750	453.566	-741.790	310.322	-2.895
5700	85.049	634.250	553.185	462.072	-742.304	329.114	-3.016
5800	85.028	635.729	554.595	470.576	-742.790	347.915	-3.133
5900	85.007	637.182	555.983	479.077	-743.248	366.724	-3.247
6000	84.986	638.611	557.348	487.577	-743.678	385.541	-3.356

PREVIOUS: December 1968 (1 atm) CURRENT: December 1968 (1 bar)

Titanium Iodide (TiI_3) I_3Ti_1(g)

ZIRCONIUM IODIDE (ZrI_3) CRYSTAL M_r = 471.9335

S°(298.15 K) = [204.585] J K^{-1} mol^{-1} $\Delta_f H°$(0 K) = [-396.545] kJ mol^{-1}

T_{sub} = [970] K $\Delta_f H°$(298.15 K) = [-397.480] kJ mol^{-1}

Enthalpy of Formation

Larsen and Leddy (1) studied the reaction Zr(cr) + 3 ZrI_4(g) = 4 ZrI_3(cr) in the temperature range 473 to 973 K and pressure range 5 to 15 atm. They present a plot of the fractional reaction, at a number of temperatures, as a function of reaction time; however, they express doubt concerning the attainment of thermodynamic equilibrium.

The Gibbs energy function change for this reaction was calculated at 100 K from 500 to 1100 K. At each temperature the equilibrium constant was assumed to be 1, and $\Delta_r H°$(298.15 K) accordingly calculated. These values and the enthalpy of formation of ZrI_4(g) [see ZrI_4 table] were used to compute a series of values for the enthalpy of formation of ZrI_3(cr) at 298.15 K. The following table gives representative values at four temperatures.

T/K	K_p	$\Delta_r H°$(298.15 K), kcal mol^{-1}	$\Delta_f H°$(298.15 K), kcal mol^{-1}
500	1	-66.0	-80.2
700	1	-90.3	-86.2
900	1	-113.5	-92.1
1100	1	-135.9	-97.7

If one assumes an initial pressure of from 5 to 10 atm. for ZrI_4 and stoichiometric amounts of reactants, then, on the basis of zirconium, the reaction must proceed to the extent of 80 to 90% for the pressure of ZrI_4 to attain a value of 1 atm. and hence an equilibrium constant of 1. The data of Larsen and Leddy indicate that the reaction proceeds to the extent of 88% at 973 K. From this it was assumed that at around 1000 K the equilibrium constant attains a value of 1 giving -95 kcal mol^{-1} for the enthalpy of formation of ZrI_3 at 298.15 K.

Their data were subjected to a 2nd law calculation but the results are of doubtful value in view of the uncertainty in the attainment of thermodynamic equilibrium at the lower temperature. The limits of error assigned to the enthalpy of formation are ±15 kcal mol^{-1} corresponding to a temperature spread of ±500 in the above table.

Heat Capacity and Entropy

The heat capacity was estimated in the same manner as for $ZrBr_4$(cr) [see $ZrBr_4$(cr) table]. The value θ_D = 60 K and θ_E = 115 K were taken to be the same as those estimated for ZrI_4(cr). The internal contribution was obtained from the estimated ZrI_3 vibrational frequencies and the anharmonicity factor "a" was taken to be 2.5 x 10^{-3}. The specific heat above 300 K was obtained by graphical extrapolation.

For the above estimation, it was assumed that the crystalline lattice is made up of ZrI_3 molecules. However, Holze (2) came to the conclusion that crystalline ZrI_3 is composed of a chain lattice of $(Zr_2I_6)_n$ units. The results of an analysis for a crystalline lattice composed of Zr_2I_6 units would probably not differ significantly from that for a crystalline lattice composed of ZrI_3 units since both are approaching the classical harmonic heat capacity of 12R calories per formula weight of ZrI_3 at relatively low temperatures.

Until more quantitative information becomes available, it is felt that the above analysis gives a fair approximation to the heat capacity of ZrI_3.

Decomposition Data

As described in the "Enthalpy of Formation" section, the temperature of decomposition was assumed to be 1000 ± 500 K.

Liquid Data

It is assumed that the liquid phase is thermodynamically unstable under ordinary conditions.

Sublimation Data

The enthalpy of sublimation at 298.15 K was obtained from the difference in the enthalpies of formation of the gas and solid at 298.15 K. The sublimation point was obtained from the free energy crossover between gas and solid.

References

1. E. M. Larsen and J. J. Leddy, J. Amer. Chem. Soc. 78, 5983 (1956).
2. E. Holze: see R. F. Rolsten, "Iodide Metals and Metal Iodides," John Wiley & Sons, Inc., New York, pp. 46, 1961.

Zirconium Iodide (ZrI_3) I_3Zr_1(cr)

Enthalpy Reference Temperature = T_r = 298.15 K Standard State Pressure = p° = 0.1 MPa

T/K	$C_p°$ (J $K^{-1}mol^{-1}$)	S° (J $K^{-1}mol^{-1}$)	-[G°-H°(T_r)]/T (J $K^{-1}mol^{-1}$)	H°-H°(T_r) (kJ mol^{-1})	$\Delta_f H°$ (kJ mol^{-1})	$\Delta_f G°$ (kJ mol^{-1})	Log K_f
0	0.	0.	INFINITE	-24.359	-396.545	-396.545	INFINITE
100	83.366	100.663	292.625	-19.196	-396.843	-396.355	207.035
200	98.328	164.189	213.936	-9.950	-397.188	-395.728	103.354
298.15	103.818	204.585	204.585	0.	-397.480	-394.947	69.193
300	103.897	205.227	204.587	0.192	-397.486	-394.931	68.764
400	105.855	235.422	208.690	10.693	-422.002	-393.180	51.344
500	106.274	259.088	216.490	21.299	-486.315	-379.800	39.677
600	106.692	278.504	225.255	31.949	-483.987	-358.717	31.229
700	106.901	294.966	234.068	42.629	-481.726	-338.018	25.223
800	107.110	309.254	242.593	53.329	-479.544	-317.638	20.740
900	107.320	321.883	250.715	64.051	-477.454	-297.526	17.268
1000	107.445	333.197	258.407	74.790	-475.474	-277.642	14.503
1100	107.529	343.442	265.679	85.539	-473.620	-257.949	12.249
1200	107.612	352.802	272.555	96.296	-475.614	-238.198	10.368
1300	107.696	361.419	279.064	107.062	-473.500	-218.500	8.779
1400	107.738	369.401	285.235	117.833	-471.470	-198.961	7.423
1500	107.780	376.836	291.097	128.609	-469.548	-179.564	6.253
1600	107.822	383.793	296.675	139.389	-467.761	-160.291	5.233
1700	107.864	390.331	301.994	150.174	-466.131	-141.124	4.336
1800	107.905	396.498	307.074	160.962	-464.679	-122.049	3.542
1900	107.947	402.333	311.936	171.755	-463.423	-103.050	2.833
2000	107.989	407.871	316.595	182.551	-462.373	-84.111	2.197

PREVIOUS: June 1962 CURRENT: June 1964

Zirconium Iodide (ZrI_3) I_3Zr_1(cr)

ZIRCONIUM IODIDE (ZrI_3) IDEAL GAS M_r = 471.9335

$\Delta_f H°(0\ K)$ = [-216.860] kJ mol^{-1}

$S°(298.15\ K)$ = [397.776] J K^{-1} mol^{-1} $\Delta_f H°(298.15\ K)$ = [-221.752] kJ mol^{-1}

Vibrational Frequencies and Degeneracies

ν, cm^{-1}
[195](1)
[88](1)
[220](2)
[65](2)

Ground State Quantum Weight: [2] σ = [3]

Point Group: [C_{3v}]

Bond Distance: Zr-I = [2.59] Å

Bond Angle: I-Zr-I = [101]°

Product of the Moments of Inertia: $I_A I_B I_C$ = [1.155142 x 10^{-110}] g^3 cm^6

Enthalpy of Formation

The enthalpies of formation, from the gaseous atoms, of the gaseous zirconium tetrahalides were computed from data issued in these tables. The zirconium-halide bond energy, taken as 1/4 of this enthalpy of formation, was found to be linear with internuclear separation. From gaseous titanium tri- and tetrachloride, it was found that the bond energy of the trichloride was around 7 kcal mol^{-1} greater than that for titanium tetrachloride. This amount was added to the bond energy of zirconium tetrachloride to get that of zirconium trichloride which, when used with its estimated internuclear distance, was found to lie almost exactly on the bond energy versus internuclear distance curve for the tetrahalides. The bond energy for each of the gaseous trihalides of zirconium was determined from this curve and their estimated bond distances.

For zirconium triiodide, the enthalpy of formation from the gaseous atoms was -275 kcal mol^{-1} and from the elements in their standard state, -53 kcal mol^{-1}.

Heat Capacity and Entropy

The measured internuclear distances for the triiodides of phosphorus, arsenic, and antimony were plotted as a function of the atomic weight of these elements and a smooth curve was drawn through the points. The bond distance in ZrI_3 was taken from this plot. The bond angle was assumed to be 101°. The principal moments of inertia are: $I_A = I_B$ = 185.2317 x 10^{-39} and I_C = 336.6703 x 10^{-39} g cm^2.

Force constants for ZrI_3 were estimated from those of the triiodides of phosphorus and arsenic. These were used in a valence force field calculation giving ν_1 = 195, ν_2 = 88, ν_3 = 181, and ν_4 = 65 cm^{-1}. For the fluorides and chlorides of P, As, and Sb, ν_1 is greater than ν_3 but there appears to be a reversal as one goes to the bromides and iodides with ν_3 becoming greater than ν_1. An empirical correlation procedure for ZrI_3 gave ν_1 = 195, ν_2 = 90, ν_3 = 220, and ν_4 = 65 cm^{-1} in which $\nu_3 > \nu_1$. Since $\nu_3 > \nu_1$ for PI_3 and AsI_3, the correlation value for ν_3 was used to represent this frequency in ZrI_3.

Zirconium Iodide (ZrI_3) $I_3Zr_1(g)$

Enthalpy Reference Temperature = T_r = 298.15 K Standard State Pressure = p° = 0.1 MPa

T/K	$C_p°$ (J $K^{-1}mol^{-1}$)	S° (J $K^{-1}mol^{-1}$)	$-[G°-H°(T_r)]/T$ (J $K^{-1}mol^{-1}$)	$H°-H°(T_r)$ (kJ mol^{-1})	$\Delta_f H°$ (kJ mol^{-1})	$\Delta_f G°$ (kJ mol^{-1})	Log K_f
0	0.	0.	INFINITE	-20.403	-216.860	-216.860	INFINITE
100	68.161	314.735	467.501	-15.277	-217.195	-238.115	124.378
200	78.257	365.960	405.106	-7.829	-219.340	-258.234	67.444
250	79.902	383.617	399.102	-3.871	-220.539	-267.821	55.958
298.15	80.820	397.776	397.776	0.	-221.752	-276.819	48.497
300	80.847	398.276	397.778	0.150	-221.800	-277.160	48.258
350	81.436	410.786	398.764	4.208	-223.231	-286.278	42.725
400	81.826	421.688	400.963	8.290	-248.677	-294.361	38.440
450	82.097	431.342	403.812	12.388	-251.933	-299.876	34.809
500	82.292	440.002	407.006	16.498	-315.388	-299.330	31.271
600	82.549	455.031	413.795	24.741	-315.467	-296.112	25.779
700	82.706	467.768	420.619	33.005	-315.621	-292.875	21.855
800	82.808	478.819	427.218	41.281	-315.864	-289.610	18.910
900	82.878	488.577	433.504	49.565	-316.212	-286.309	16.617
1000	82.928	497.312	439.456	57.856	-316.681	-282.962	14.780
1100	82.966	505.218	445.081	66.151	-317.280	-279.563	13.275
1200	82.994	512.438	450.397	74.449	-321.733	-275.881	12.009
1300	83.016	519.082	455.428	82.749	-322.085	-272.046	10.931
1400	83.034	525.235	460.198	91.052	-322.523	-268.181	10.006
1500	83.048	530.964	464.727	99.356	-323.074	-264.281	9.203
1600	83.060	536.324	469.036	107.661	-323.761	-260.340	8.499
1700	83.069	541.360	473.143	115.968	-324.609	-256.351	7.877
1800	83.077	546.108	477.066	124.275	-325.638	-252.307	7.322
1900	83.084	550.600	480.819	132.583	-326.866	-248.201	6.824
2000	83.090	554.862	484.416	140.892	-328.305	-244.024	6.373
2100	83.095	558.916	487.868	149.201	-329.961	-239.770	5.964
2200	83.099	562.782	491.186	157.511	-353.161	-234.687	5.572
2300	83.103	566.476	494.380	165.821	-355.672	-229.245	5.206
2400	83.107	570.013	497.458	174.131	-358.251	-223.694	4.869
2500	83.109	573.405	500.428	182.442	-360.880	-218.033	4.556
2600	83.112	576.665	503.298	190.753	-363.542	-212.267	4.264
2700	83.114	579.802	506.074	199.065	-366.220	-206.397	3.993
2800	83.116	582.824	508.761	207.376	-368.897	-200.429	3.739
2900	83.118	585.741	511.366	215.688	-371.556	-194.365	3.501
3000	83.120	588.559	513.892	224.000	-374.183	-188.211	3.277
3100	83.122	591.284	516.345	232.312	-376.765	-181.969	3.066
3200	83.123	593.923	518.728	240.624	-379.291	-175.645	2.867
3300	83.124	596.481	521.046	248.937	-381.753	-169 242	2.679
3400	83.125	598.963	523.301	257.249	-384.141	-162.767	2.501
3500	83.126	601.372	525.498	265.562	-386.452	-156.222	2.331
3600	83.127	603.714	527.638	273.874	-388.679	-149.612	2.171
3700	83.128	605.992	529.725	282.187	-390.821	-142.942	2.018
3800	83.129	608.209	531.761	290.500	-392.874	-136.214	1.872
3900	83.130	610.368	533.749	298.813	-394.838	-129.434	1.734
4000	83.131	612.473	535.691	307.126	-396.714	-122.605	1.601
4100	83.131	614.525	537.589	315.439	-398.500	-115.730	1.474
4200	83.132	616.529	539.445	323.752	-400.199	-108.813	1.353
4300	83.132	618.485	541.260	332.065	-401.811	-101.856	1.237
4400	83.133	620.396	543.037	340.379	-403.339	-94.862	1.126
4500	83.133	622.264	544.777	348.692	-404.785	-87.835	1.020
4600	83.134	624.091	546.482	357.005	-406.151	-80.777	0.917
4700	83.134	625.879	548.152	365.319	-407.439	-73.689	0.819
4800	83.135	627.630	549.789	373.632	-969.589	-54.957	0.598
4900	83.135	629.344	551.396	381.946	-970.382	-35.894	0.383
5000	83.135	631.023	552.971	390.259	-971.122	-16.816	0.176
5100	83.136	632.670	554.518	398.573	-971.810	2.278	-0.023
5200	83.136	634.284	556.037	406.886	-972.448	21.384	-0.215
5300	83.136	635.868	557.528	415.200	-973.037	40.501	-0.399
5400	83.137	637.422	558.993	423.514	-973.577	59.630	-0.577
5500	83.137	638.947	560.433	431.827	-974.071	78.767	-0.748
5600	83.137	640.445	561.848	440.141	-974.518	97.915	-0.913
5700	83.137	641.917	563.240	448.455	-974.920	117.069	-1.073
5800	83.138	643.362	564.609	456.769	-975.278	136.229	-1.227
5900	83.138	644.784	565.956	465.082	-975.593	155.396	-1.376
6000	83.138	646.181	567.282	473.396	-975.865	174.568	-1.520

PREVIOUS: June 1964 (1 atm) CURRENT: June 1964 (1 bar)

Zirconium Iodide (ZrI_3) $I_3Zr_1(g)$

MOLYBDENUM IODIDE (MoI_4) CRYSTAL M_r = 603.5580

$\Delta_f H°(0\ K)$ = Unknown

$S°(298.15\ K) = [266.061 \pm 16.7]\ J\ K^{-1}\ mol^{-1}$ $\Delta_f H°(298.15\ K) = [-123.010 \pm 8.4]\ kJ\ mol^{-1}$

T_{dcm} = [293] K

Enthalpy of Formation

We adopt the $\Delta_f H°(298.15\ K)$ value estimated by Brewer (1). This value predicts that MoI_4(cr) is unstable with respect to MoI_3(cr) at temperatures above 293 K (1). This is consistent with the observations of Lewis et al. (2) who found that only the crystalline triiodide is formed by heating pure Mo metal in excess iodine in a sealed tube to 573 K. Brewer's estimated value is given as $\Delta_f H°(298.15\ K)/R = -14800 \pm 1000$ K. The assigned uncertainty ($\pm 2\ kcal\ mol^{-1}$) is determined from the product 1000xR which has been rounded to the nearest $kcal\ mol^{-1}$.

Heat Capacity and Entropy

The values of $C_p°$ and S° at 298.15 K are estimates reported by Brewer (1). We assume a constant $C_p°$ of 32.79 cal $K^{-1}\ mol^{-1}$ over the temperature range 298-600 K. The value of S°(298.15 K)/R reported by Brewer (1) is treated in a manner similar to that for the enthalpy of formation.

Phase Data

The preparation of MoI_4(cr) has been reported by Guichard (3) but the sample was impure. The stability range of the tetraiodide appears to be small based on the observations of Lewis et al. (2) who found that MoI_4 could not be formed by heating Mo in excess iodine in a sealed tube to 573 K.

Decomposition Data

Brewer (1) estimated a T_{dcm} of 293 K from free energy data for the dissociation reaction $MoI_4(cr) = MoI_3(cr) + 0.5\ I_2(cr)$. We adopt this value since T_{dcm} appears to lie outside the temperature range selected for our MoI_4(cr) phase table.

References

1. L. Brewer, Materials and Molecular Research Division, Lawrence Berkeley Laboratory, University of California, Berkeley; personal communication, Sept. 29, 1978, preliminary draft of review to be submitted for publication in Atomic Energy Review, International Atomic Energy Agency, Vienna, Austria.
2. J. Lewis, D. J. Machin, R. S. Nyholm, P. Pauling, and P. W. Smith, Chemistry and Industry, 1960, 259 (1960).
3. M. Guichard, Annales de Chimie et de Physique, 23, 498 (1901).

Molybdenum Iodide (MoI_4) I_4Mo_1(cr)

Enthalpy Reference Temperature = T_r = 298.15 K Standard State Pressure = p° = 0.1 MPa

T/K	$C_p°$ (J K⁻¹mol⁻¹)	S° (J K⁻¹mol⁻¹)	$-[G°-H°(T_r)]/T$ (J K⁻¹mol⁻¹)	$H°-H°(T_r)$ (kJ mol⁻¹)	$\Delta_f H°$ (kJ mol⁻¹)	$\Delta_f G°$ (kJ mol⁻¹)	Log K_f
0							
100							
200							
298.15	137.193	266.061	266.061	0.	-123.010	-124.552	21.821
300	137.193	266.909	266.063	0.254	-123.002	-124.561	21.688
400	137.193	306.377	271.444	13.973	-155.014	-123.992	16.192
500	137.193	336.991	281.606	27.692	-240.237	-108.065	11.290
600	137.193	362.004	292.985	41.412	-236.642	-81.970	7.136

PREVIOUS: CURRENT: September 1978

Molybdenum Iodide (MoI_4) I_4Mo_1(cr)

MOLYBDENUM IODIDE (MoI_4) IDEAL GAS M_r = 603.5580

S°(298.15 K) = [452.812 ± 12.6] J K^{-1} mol^{-1}

$\Delta_f H°$(0 K) = [129.32 ± 41.8] kJ mol^{-1}
$\Delta_f H°$(298.15 K) = [124.68 ± 41.8] kJ mol^{-1}

Electronic Levels and Quantum Weights

ϵ_i, cm^{-1}	g_i
0	[3]
[15000]	[11]
[20000]	[9]

Vibrational Frequencies and Degeneracies

ν, cm^{-1}	ν, cm^{-1}
[150](1)	[240](3)
[50](2)	[60](3)

Point Group: [T_d] σ = [12]
Bond Distance: Mo-I = [2.66] Å
Bond Angle: I-Mo-I = [109°28']
Product of the Moments of Inertia: $I_A I_B I_C$ = [6.286307 x 10^{-110}] g^3 cm^6

Enthalpy of Formation

We adopt the $\Delta_f H°$(298.15 K) value estimated by Brewer (1). This estimated value is consistent with the low volatility of metallic molybdenum in iodine vapor reported by Schafer et al. (2). Brewer's estimated value is given as $\Delta_f H°$(298.15 K)/R = 15000±5000 K. The assigned uncertainty (10 kcal mol^{-1}) is determined from the product 5000xR which has been rounded to the nearest kcal mol^{-1}. Combining the $\Delta_f H°$(0 K) value with JANAF (3) data on Mo(g) and I(g), we calculate a heat of atomization $\Delta_{at} H°$(0 K) of 228.7±11 kcal mol^{-1} and a mean bond dissociation energy $\Delta_{at} H°$(0 K)/4=57.2±2.5 kcal mol^{-1}.

Heat Capacity and Entropy

The vibrational frequencies and molecular structural data are estimates taken from Brewer (1). The principal moments of inertia are: $I_A = I_B = I_C$ = 397.6172 x 10^{-39} g cm^2. The electronic levels and degeneracies are those predicted by Brewer (1) for MoF_4(g) from crystal field theory. The upper levels contribute 0.23 cal K^{-1} mol^{-1} to the Gibbs energy function at 6000 K.

References

1. L. Brewer, Materials and Molecular Research Division, Lawrence Berkeley Laboratory, University of California, Berkeley; personal communication, Sept. 29, 1978, preliminary draft of review to be submitted for publication in Atomic Energy Review, International Atomic Energy Agency, Vienna, Austria.
2. H. Schafer, T. Grofe, and M. Trenkel, J. Solid State Chem. 8, 14 (1973).
3. JANAF Thermochemical Tables: Mo(g), 3-31-78, I(g), 6-30-74.

Molybdenum Iodide (MoI_4) I_4Mo_1(g)

Enthalpy Reference Temperature = T_r = 298.15 K Standard State Pressure = p° = 0.1 MPa

T/K	$C_p°$ (J $K^{-1}mol^{-1}$)	S° (J $K^{-1}mol^{-1}$)	-[G°-H°(T_r)]/T (J $K^{-1}mol^{-1}$)	H°-H°(T_r) (kJ mol^{-1})	$\Delta_f H°$ (kJ mol^{-1})	$\Delta_f G°$ (kJ mol^{-1})	Log K_f
0	0.	0.	INFINITE	-26.344	129.320	129.320	INFINITE
100	88.395	345.225	543.185	-19.796	129.277	107.655	-56.233
200	101.392	411.558	462.319	-10.152	127.214	86.781	-22.665
250	103.615	434.446	454.532	-5.022	125.971	76.814	-16.049
298.15	104.869	452.812	452.812	0.	124.683	67.461	-11.819
300	104.906	453.461	452.814	0.194	124.631	67.106	-11.684
350	105.716	469.698	454.095	5.461	123.061	57.638	-8.602
400	106.254	483.851	456.948	10.761	89.467	49.499	-6.464
450	106.629	496.389	460.647	16.084	85.458	44.744	-5.194
500	106.901	507.639	464.794	21.422	1.186	48.034	-5.018
600	107.258	527.163	473.610	32.132	1.771	57.347	-4.993
700	107.476	543.715	482.473	42.869	2.301	66.567	-4.967
800	107.618	558.076	491.045	53.625	2.776	75.715	-4.944
900	107.716	570.758	499.212	64.392	3.196	84.807	-4.922
1000	107.786	582.111	506.944	75.167	3.555	93.855	-4.902
1100	107.838	592.386	514.252	85.948	3.843	102.870	-4.885
1200	107.878	601.771	521.159	96.734	4.045	111.863	-4.869
1300	107.909	610.407	527.697	107.524	4.142	120.843	-4.856
1400	107.935	618.405	533.894	118.316	4.110	129.820	-4.844
1500	107.957	625.853	539.779	129.110	3.925	138.805	-4.834
1600	107.977	632.821	545.379	139.907	3.560	147.808	-4.825
1700	107.998	639.367	550.717	150.706	2.992	156.840	-4.819
1800	108.022	645.541	555.815	161.507	2.200	165.911	-4.815
1900	108.051	651.382	560.693	172.310	1.169	175.033	-4.812
2000	108.088	656.925	565.367	183.117	-0.111	184.216	-4.811
2100	108.135	662.200	569.853	193.928	-1.643	193.469	-4.812
2200	108.195	667.232	574.166	204.745	-3.424	202.800	-4.815
2300	108.270	672.043	578.318	215.568	-5.445	212.219	-4.820
2400	108.363	676.653	582.320	226.399	-7.697	221.730	-4.826
2500	108.476	681.079	586.182	237.241	-10.167	231.339	-4.834
2600	108.611	685.336	589.914	248.095	-12.851	241.052	-4.843
2700	108.768	689.438	593.525	258.964	-15.763	250.872	-4.853
2800	108.950	693.396	597.022	269.850	-18.940	260.805	-4.865
2900	109.156	697.223	600.411	280.755	-58.325	270.905	-4.880
3000	109.388	700.928	603.700	291.682	-60.170	282.290	-4.915
3100	109.644	704.518	606.895	302.633	-61.931	293.735	-4.949
3200	109.924	708.004	610.000	313.611	-63.591	305.234	-4.982
3300	110.228	711.391	613.022	324.619	-65.136	316.784	-5.014
3400	110.554	714.686	615.964	335.658	-66.552	328.379	-5.045
3500	110.901	717.896	618.830	346.730	-67.830	340.014	-5.074
3600	111.267	721.025	621.626	357.838	-68.963	351.683	-5.103
3700	111.650	724.079	624.354	368.984	-69.943	363.382	-5.130
3800	112.049	727.062	627.017	380.169	-70.767	375.105	-5.156
3900	112.460	729.978	629.620	391.394	-71.431	386.847	-5.181
4000	112.882	732.830	632.165	402.661	-71.935	398.604	-5.205
4100	113.313	735.623	634.654	413.971	-72.279	410.372	-5.228
4200	113.749	738.359	637.091	425.324	-72.462	422.147	-5.250
4300	114.189	741.040	639.477	436.721	-72.486	433.924	-5.271
4400	114.630	743.671	641.816	448.162	-72.354	445.699	-5.291
4500	115.071	746.252	644.108	459.647	-72.068	457.470	-5.310
4600	115.508	748.786	646.356	471.176	-71.632	469 233	-5.328
4700	115.940	751.274	648.562	482.748	-71.048	480.985	-5.346
4800	116.365	753.720	650.727	494.364	-70.323	492.724	-5.362
4900	116.780	756.123	652.854	506.021	-69.458	504.445	-5.377
5000	117.185	758.487	654.943	517.719	-659.845	521.878	-5.452
5100	117.578	760.811	656.996	529.458	-660.396	545.519	-5.587
5200	117.957	763.098	659.014	541.234	-660.969	569.170	-5.717
5300	118.321	765.348	661.000	553.048	-661.566	592.831	-5.843
5400	118.669	767.563	662.953	564.898	-662.190	616.505	-5.964
5500	119.000	769.744	664.874	576.782	-662.843	640.190	-6.080
5600	119.314	771.891	666.766	588.697	-663.529	663.889	-6.192
5700	119.609	774.005	668.629	600.644	-664.248	687.599	-6.301
5800	119.886	776.088	670.464	612.619	-665.003	711.321	-6.406
5900	120.144	778.140	672.272	624.620	-665.794	735.058	-6.508
6000	120.383	780.161	674.053	636 647	-666.625	758.808	-6.606

PREVIOUS: September 1978 (1 atm) CURRENT: September 1978 (1 bar)

Molybdenum Iodide (MoI_4) I_4Mo_1(g)

LEAD IODIDE (PbI_4) IDEAL GAS M_r = 714.8180

$S°(298.15\ K) = [466.266 \pm 12.6]\ J\ K^{-1}\ mol^{-1}$

$\Delta_f H°(0\ K) = [-218.82 \pm 83.7]\ kJ\ mol^{-1}$

$\Delta_f H°(298.15\ K) = [-224.47 \pm 83.7]\ kJ\ mol^{-1}$

Vibrational Frequencies and Degeneracies

ν_i, cm^{-1}
[137](1)
[37](2)
[168](3)
[48](3)

Ground State Quantum Weight: [1]
Point Group: $[T_d]$
Bond Distance: Pb-I = [2.77] Å
Bond Angle: I-Pb-I = [109°28']
Product of the Moments of Inertia: $I_A I_B I_C = [8.016506 \times 10^{-110}]\ g^3\ cm^6$

σ = [12]

Enthalpy of Formation

The adopted $\Delta_f H°(0\ K) = -52.3\pm20\ kcal\ mol^{-1}$ is derived from the atomization energy for the process $PbI_4(g) = Pb(g) + 4\ I(g)$. $\Delta_{at}H°(0\ K) = 8.75\ eV$ is estimated from an intercomparison of the lead mono-, di-, and tetra- halides (1). This adopted value for $PbI_4(g)$ is greater than the $\Delta_{at}H°(0\ K)$ value for PbI (g) by a factor of 2.04. The auxiliary values $\Delta_f H°(Pb, g, 0\ K) = 46.91\ kcal\ mol^{-1}$ (1) and $\Delta_f H°(I, g, 0\ K) = 25.633\ kcal\ mol^{-1}$ (1) are used. $\Delta_f H°(PbI_4, g, 298.15\ K) = -53.6\pm20\ kcal\ mol^{-1}$ is calculated from $\Delta_f H°(0\ K)$.

Heat Capacity and Entropy

The molecular constants and vibrational frequencies are those calculated by Aleksandrovskaya et al. (2). Later works (3, 4, 5) mistakenly cite this work as observed data.

The principal moments of inertia are: $I_A = I_B = I_C = 431.1831 \times 10^{-39}\ g\ cm^2$.

References

1. JANAF Thermochemical Tables: PbF(g), PbF_2(g), PbF_4(g), PbBr(g), $PbBr_2$(g), PbI(g) and PbI_2(g), 12-31-73; PbCl(g) and $PbCl_2$(g), 6-30-73; Pb(g), 3-31-62; I(g), 12-31-61.
2. A. M. Aleksandrovskaya, I. V. Rigina, and I. N. Godnev, Optics and Spectrosc. 7, 495 (1959).
3. G. Nagarajan, Bull. Soc. Chim. Belges, 71, 119 (1962).
4. M. Radhakrishnan, Indian J. Pure Appl. Phys. 1, 402 (1963).
5. S. P. Singh, Labdev, Part A 7, 185 (1969).

Lead Iodide (PbI_4) $I_4Pb_1(g)$

Enthalpy Reference Temperature = T_r = 298.15 K; Standard State Pressure = p° = 0.1 MPa

T/K	$C_p°$ (J K⁻¹mol⁻¹)	S° (J K⁻¹mol⁻¹)	$-[G°-H°(T_r)]/T$ (J K⁻¹mol⁻¹)	$H°-H°(T_r)$ (kJ mol⁻¹)	$\Delta_f H°$ (kJ mol⁻¹)	$\Delta_f G°$ (kJ mol⁻¹)	Log K_f
0	0.	0.	INFINITE	-27.627	-218.820	-218.820	INFINITE
100	95.214	354.506	559.058	-20.455	-219.515	-239.084	124.885
200	104.275	424.175	475.941	-10.353	-221.802	-257.799	67.330
250	105.593	447.600	468.013	-5.103	-223.112	-266.650	55.713
298.15	106.313	466.266	466.266	0.	-224.467	-274.914	48.164
300	106.334	466.924	466.268	0.197	-224.522	-275.226	47.921
350	106.790	483 352	467.565	5.526	-226.170	-283.553	42.318
400	107.089	497.633	470.450	10.873	-259.851	-290.536	37.940
450	107.296	510.259	474.185	16.233	-263.954	-294.126	34.141
500	107.445	521.572	478.368	21.602	-348.329	-289.659	30.261
600	107.640	541.180	487.252	32.357	-347.971	-277.961	24.199
700	107.758	557.782	496.172	43.127	-352.558	-265.517	19.813
800	107.835	572.177	504.793	53.907	-352.353	-253.096	16.525
900	107.888	584.881	513.000	64.693	-352.133	-240.702	13.970
1000	107.926	596.250	520.766	75.484	-351.901	-228.333	11.927
1100	107.954	606.538	528.103	86.278	-351.662	-215.988	10.256
1200	107.975	615.932	535.037	97.075	-351.424	-203.664	8.865
1300	107.991	624.576	541.596	107.873	-351.208	-191.360	7.689
1400	108.005	632.579	547.813	118.673	-351.045	-179.070	6.681
1500	108.015	640.031	553.715	129.474	-350.963	-166.790	5.808
1600	108.024	647.002	559.330	140.276	-350.988	-154.512	5.044
1700	108.031	653.552	564.682	151.079	-351.142	-142.228	4.370
1800	108.037	659.727	569.792	161.882	-351.442	-129.931	3.770
1900	108.042	665.568	574.681	172.686	-351.899	-117.613	3.233
2000	108.047	671.110	579.365	183.491	-352.515	-105.267	2.749
2100	108.051	676.382	583.860	194.295	-530.508	-85.771	2.133
2200	108.054	681.408	588.181	205.101	-531.036	-64.582	1.533
2300	108.057	686.212	592.339	215.906	-531.762	-43.364	0.985
2400	108.059	690.810	596.347	226.712	-532.669	-22.110	0.481
2500	108.061	695.222	600.215	237.518	-533.733	-0.816	0.017
2600	108.063	699.460	603.951	248.324	-534.933	20.525	-0.412
2700	108.065	703.538	607.564	259.131	-536.242	41.913	-0.811
2800	108.067	707.468	611.062	269.937	-537.633	63.351	-1.182
2900	108.068	711.261	614.452	280.744	-539.081	84.840	-1.528
3000	108.069	714.924	617.741	291.551	-540.562	106.380	-1.852
3100	108.070	718.468	620.933	302.358	-542.051	127.969	-2.156
3200	108.071	721.899	624.035	313.165	-543.528	149.606	-2.442
3300	108.072	725.225	627.051	323.972	-544.973	171.289	-2.711
3400	108.073	728.451	629.986	334.779	-546.370	193.015	-2.965
3500	108.074	731.584	632.845	345.587	-547.703	214.782	-3.205
3600	108.075	734.628	635.630	356.394	-548.960	236.585	-3.433
3700	108.075	737.589	638.346	367.202	-550.131	258.422	-3.648
3800	108.076	740.472	640.995	378.009	-551.206	280.290	-3.853
3900	108.077	743.279	643.582	388.817	-552.178	302.184	-4.047
4000	108.077	746.015	646.109	399.625	-553.042	324.102	-4.232
4100	108.078	748.684	648.578	410.432	-553.794	346.040	-4.409
4200	108.078	751.288	650.993	421.240	-554.430	367.995	-4.577
4300	108.079	753.831	653.355	432.048	-554.950	389.964	-4.737
4400	108.079	756.316	655.667	442.856	-555.351	411.943	-4.890
4500	108.079	758.745	657.931	453.664	-555.633	433.931	-5.037
4600	108.080	761.120	660.148	464.472	-555.798	455.923	-5.177
4700	108.080	763.445	662.322	475.280	-555.847	477.917	-5.311
4800	108.080	765.720	664.452	486.088	-555.780	499.913	-5.440
4900	108.081	767.949	666.542	496.896	-555.600	521.905	-5.564
5000	108.081	770.132	668.592	507.704	-555.309	543.891	-5.682
5100	108.081	772.273	670.604	518.512	-554.911	565.872	-5.796
5200	108.081	774.371	672.579	529.320	-554.407	587.843	-5.905
5300	108.082	776.430	674.519	540.128	-553.802	609.803	-6.010
5400	108.082	778.450	676.425	550.936	-553.097	631.752	-6.111
5500	108.082	780.434	678.298	561.745	-552.298	653.686	-6.208
5600	108.082	782.381	680.140	572.553	-551.406	675.606	-6.302
5700	108.082	784.294	681.950	583.361	-550.425	697.508	-6.392
5800	108.083	786.174	683.731	594.169	-549.359	719.392	-6.479
5900	108.083	788.022	685.483	604.978	-548.211	741.257	-6.563
6000	108.083	789.838	687.207	615.786	-546.984	763.103	-6.643

PREVIOUS: December 1973 (1 atm) CURRENT: December 1973 (1 bar)

Lead Iodide (PbI_4) $I_4Pb_1(g)$

TETRAIODOSILANE (SiI_4) CRYSTAL M_r = 535.7035

$\Delta_f H°$(0 K) = Unknown

$S°$(298.15 K) = [258.153 ± 3.3] J K^{-1} mol^{-1} $\Delta_f H°$(298.15 K) = -189.54 ± 16.7 kJ mol^{-1}

T_{fus} = 393.65 ± 3.0 K $\Delta_{fus} H°$ = 19.665 ± 2.1 kJ mol^{-1}

Enthalpy of Formation

Schafer and Heine (1) measured the enthalpies of solution of Si(cr) and SiI_4(cr) in an HF solution containing AgF. For the overall reaction SiI_4(cr) + 4Ag(cr) = Si(cr) + 4 AgI(cr), they reported $\Delta_r H°$(298.15 K) = -13.80±0.44 kcal mol^{-1}. Using auxiliary data for AgI (2), we calculate $\Delta_f H°$(298.15 K) = -45.3 kcal mol^{-1} for SiI_4(cr). We adopt this value and assign an uncertainty of ±4.0 kcal mol^{-1} due to the overall inconsistency in the SiI_4 thermochemical data. Refer to the SiI_4(g) table for more details.

Wolf (3) studied the enthalpies of solution of SiI_4(cr) and Na_2SiO_3(cr) in caustic solution. The net reaction of interest SiI_4(cr) + 6NaOH(cr) = Na_2SiO_3(cr) + 4NaI(cr) + $3H_2O$(ℓ) yielded a enthalpy of reaction of $\Delta_r H°$(298.15 K) = -191.5±1.5 kcal mol^{-1} based on the appropriate combination of results from five solution studies. Using current auxiliary data (4, 5), we calculate $\Delta_f H°$(298.15 K) = -51.0 kcal mol^{-1} for SiI_4(cr). This data is suspect due to uncertainties in how Wolf converted his data from Na_2SiO_3(calorimetric solution) to Na_2SiO_3(cr).

Heat Capacity and Entropy

The adopted heat capacity values are from the study of Kurosawa, et al. (6). The data was obtained from an adiabatic calorimeter of a type used for specific heat measurements and differential thermal analysis. Similar studies on Cr(cr) and Si(cr) would suggest uncertainties of the order of 3% near room temperature and 0-10% at 600-70 K.

There is no low temperature heat capacity data for T<298.15 K. The entropy at 298.15 K is chosen from a consideration of the sublimation and vaporization data. Refer to the SiI_4(g) table for additional information. However, the discrepancies in the various vapor pressure studies are such that a precise estimate of the entropy is not possible.

Fusion Data

Kurosawa et al. (6), using differential thermal analysis (DTA) techniques, determined $\Delta_{fus} H°$ = 4.7 kcal mol^{-1}, an average of eight measurements with a standard deviation of 0.08 kcal mol^{-1}. The same study reported T_{fus} = 120.5°C as determined from the cooling curve. McCarty et al. (7) reported T_{fus} = 123.68±0.06°C for 99.998% (mole) pure SiI_4. Other studies have listed melting points in the range 120-127 C as reported by Kurosawa et al. (6). The sublimation and vaporization studies of Bartsch and Wolf (8) suggest a melting point of 123.5 C and a heat of melting of 4.1±0.6 kcal mol^{-1}. Due to the short temperature range of measurement in each condensed phase, precise melting data is not expected. In general, the vapor pressure studies, as discussed in the SiI_4(g) table (5), show poor agreement in the vicinity of T_{fus}.

We adopt T_{fus} = 120.5°C and $\Delta_{fus} H°$ = 4.7 kcal mol^{-1} as derived from Kurosawa et al. (6). We assign respective uncertainties of ±3.0 K and 0.5 kcal mol^{-1} due to the variances among the various studies.

Sublimation Data

The enthalpy of sublimation, $\Delta_{sub} H°$(298.15 K), is the difference between the $\Delta_f H°$(298.15 K) values for SiI_4(g) and SiI_4(cr). Three sublimation studies are summarized in the SiI_4(g) table.

References

1. H. Schafer and H. Heine, Z. Anorg. Allg. Chem. 332, 25 (1964).
2. U. S. Nat. Bur. Stand. Tech. Note 270-4, 1969.
3. E. Wolf, Z. Chem. 7, 283 (1967).
4. J. D. Cox, ICSU-CODATA Task Group on Key Values for Thermodynamics, J. Chem. Thermodynamics 4, 331 (1972).
5. JANAF Thermochemical Tables: NaI(cr), 9-30-63; NaOH(cr), 12-31-70; Na_2SiO_3(cr), 9-30-67; $SiCl_4$(g), 12-31-70; SiI_4(cr), 6-30-76.
6. T. Kurosawa, R. Hasegawa, and T. Yagihashi, Trans. Japan Inst. Metals. 6, 229 (1965); Trans. Natl. Res. Inst. Metals. (Tokyo) 7, 222 (1965).
7. L. V. McCarty, L. C. Landauer, and J. M. Binkowski, J. Chem. Eng. Data 5, 365-6 (1960).
8. K. Bartsch and E. Wolf, Z. Anorg. Allg. Chem. 383, 66 (1971).

Tetraiodosilane (SiI_4) I_4Si_1(cr)

Enthalpy Reference Temperature = T_r = 298.15 K Standard State Pressure = p° = 0.1 MPa

T/K	$C_p°$ (J $K^{-1}mol^{-1}$)	$S°$ (J $K^{-1}mol^{-1}$)	$-[G°-H°(T_r)]/T$ (J $K^{-1}mol^{-1}$)	$H°-H°(T_r)$ (kJ mol^{-1})	$\Delta_f H°$ (kJ mol^{-1})	$\Delta_f G°$ (kJ mol^{-1})	Log K_f
0							
100							
200							
298.15	108.035	258.153	258.153	0.	-189.535	-191.637	33.574
300	108.198	258.822	258.155	0.200	-189.574	-191.650	33.369
393.650	116.388	289.279	262.056	10.716	------ CRYSTAL <--> LIQUID -----		
400	116.943	291.146	262.503	11.457	-223.715	-190.911	24.930
500	125.687	318.180	271.003	23.589	-310.254	-174.369	18.216

PREVIOUS: CURRENT: June 1976

Tetraiodosilane (SiI_4) I_4Si_1(cr)

TETRAIODOSILANE (SiI_4) LIQUID M_r = 535.7035

S°(298.15 K) = [294.309] J K^{-1} mol^{-1} $\Delta_f H°$(298.15 K) = [-174.603] kJ mol^{-1}

T_{fus} = 393.65 ± 3.0 K $\Delta_{fus}H°$ = 19.665 ± 2.1 kJ mol^{-1}

Enthalpy of Formation

$\Delta_f H°$(SiI_4, ℓ, 298.15 K) is calculated from that of the crystal by adding the enthalpy of melting, $\Delta_{fus}H°$, and the difference in enthalpy, H°(393.65 K)-H°(298.15 K), between the crystal and liquid.

Heat Capacity and Entropy

Kurosawa et al. (1) measured the heat capacity of SiI_4 from the melting point (unspecified). No data was presented but rather smoothed results were represented by an equation, linear in T. This equation is extrapolated to 700 K and to 298.15 K. S°(298.15 K) is calculated in a manner analogous to that used for $\Delta_f H°$(298.15 K).

Vaporization Data

T_{vap} = ~576 K is calculated as that temperature at which the fugacity is one atmosphere for the process SiI_4(ℓ) = SiI_4(g). $\Delta_{vap}H°$ is calculated as the difference between the $\Delta_f H°$ values for the ideal gas and the liquid at T_{vap}. The vaporization studies near T_{vap} are discussed in the SiI_4(g) table; they yielded normal boiling points (p = 760 mm Hg) of 574.7 K (2) and 573.7 K (3). The normal boiling point should be slightly lower than our calculated T_{vap} which corresponds to f = 760 mm Hg.

References

1. T. Kurosawa, R. Hasegawa, and T. Yagihanshi, Trans. Japan Inst. Metals. 6, 229 (1965); Trans. Natl. Res. Metals (Toyko) 7, 222 (1965).
2. H. C. Andersen and L. H. Belz, J. Amer. Chem. Soc. 75, 4828 (1953).
3. K. Uchimura, T. Takuma, M. Yuizumi and T. Kunugi, Denki Kagaku 35, 876 (1967).

Tetraiodosilane (SiI_4) I_4Si_1(l)

Enthalpy Reference Temperature = T_r = 298.15 K Standard State Pressure = p° = 0.1 MPa

	J $K^{-1}mol^{-1}$			kJ mol^{-1}			
T/K	$C_p°$	S°	$-[G°-H°(T_r)]/T$	$H°-H°(T_r)$	$\Delta_f H°$	$\Delta_f G°$	Log K_f
0							
100							
200							
298.15	159.800	294.309	294.309	0.	-174.603	-187.485	32.847
300	159.875	295.298	294.312	0.296	-174.546	-187.565	32.658
393.650	163.744	339.234	299.988	15.449	------ CRYSTAL <--> LIQUID -----		
400	164.004	341.856	300.632	16.490	-203.750	-191.230	24.972
500	168.134	378.897	312.703	33.097	-285.814	-180.287	18.834
600	172.264	409.916	326.389	50.117	-278.678	-159.847	13.916
700	176.393	436.781	340.282	67.549	-271.228	-140.627	10.494

PREVIOUS: June 1976 CURRENT: December 1977

Tetraiodosilane (SiI_4) I_4Si_1(l)

TETRAIODOSILANE (SiI_4) CRYSTAL–LIQUID M_r = 535.7035

0 to 393.65 K crystal
above 393.65 K liquid

Refer to the individual tables for details.

Tetraiodosilane (SiI_4) I_4Si_1(cr,l)

Enthalpy Reference Temperature = T_r = 298.15 K Standard State Pressure = p° = 0.1 MPa

T/K	C_p° (J K⁻¹mol⁻¹)	S° (J K⁻¹mol⁻¹)	$-[G°-H°(T_r)]/T$ (J K⁻¹mol⁻¹)	$H°-H°(T_r)$ (kJ mol⁻¹)	$\Delta_f H°$ (kJ mol⁻¹)	$\Delta_f G°$ (kJ mol⁻¹)	Log K_f
0							
100							
200							
298.15	108.035	258.153	258.153	0.	-189.535	-191.637	33.574
300	108.198	258.822	258.155	0.200	-189.574	-191.650	33.369
393.650	116.388	289.279	262.056	10.716	CRYSTAL <--> LIQUID TRANSITION		
393.650	163.744	339.234	262.056	30.381			
400	164.004	341.856	263.302	31.422	-203.750	-191.230	24.972
500	168.134	378.897	282.839	48.029	-285.814	-180.287	18.834
600	172.264	409.916	301.502	65.049	-278.678	-159.847	13.916
700	176.393	436.781	318.950	82.481	-271.228	-140.627	10.494

PREVIOUS: CURRENT: December 1977

Tetraiodosilane (SiI_4) I_4Si_1(cr,l)

$S°(298.15\ K) = [416.449 \pm 0.21]\ J\ K^{-1}\ mol^{-1}$

$\Delta_f H°(0\ K) = -105.04 \pm 16.7\ kJ\ mol^{-1}$
$\Delta_f H°(298.15\ K) = -110.46 \pm 16.7\ kJ\ mol^{-1}$

Vibrational Frequencies and Degeneracies

ν, cm^{-1}	ν, cm^{-1}
165.9(1)	404 (3)
57.4(2)	88.8(3)

Ground State Quantum Weight: [1] σ = 12
Point Group: T_d
Bond Distance: Si-I = 2.43 ± 0.02 Å
Bond Angle: I-Si-I = 109° 28'
Product of the Moments of Intertia: $I_A I_B I_C = [3.653782 \times 10^{-110}]\ g^3\ cm^6$

Enthalpy of Formation

The sublimation, vaporization, and decomposition studies on SiI_4 are summarized in the following table. The vapor pressure data from the sublimation and vaporization studies are corrected for vapor non-ideality by means of the equation $\Delta G°/T = -R\ln p - Bp/T$. The Berthelot equation of state and critical constants Tc = 850 K and Pc = 42.1 atm as suggested by Lapidus et al. (1) are used to calculate B.

Source	Reaction*	Data Points	T/K	$\Delta_r H°(298.15\ K)$, kcal mol^{-1} 2nd law	3rd law	Drift cal K^{-1} mol^{-1}	$\Delta_f H°(g, 298.15\ K)$ kcal mol^{-1}
(2)	A	13	343-396	16.23±0.32	18.82±0.28	6.9±0.9	-26.48
(3)	A	eqn (flow)	373-392	18.75	19.08	0.9	-26.22
(3)	A	eqn (static)	370-395	18.33	19.08	2.0	-26.22
(3)	B	eqn (static)	398-425	15.58	15.49	-0.5	-26.24
(4)	B	10	399-572	15.50±0.06	15.38±0.05	-0.2±0.1	-26.35
(5)	B	eqn	393-573	14.82	15.27	0.9	-26.46
(6)	C	6	1313-1513	59.09±1.14	59.91±0.26	0.6±0.8	-30.06
(7)	C	eqn	1300-1470	59.44	59.84	0.3	-29.99

Reactions: A) $SiI_4(cr) = SiI_4(g)$ B) $SiI_4(\ell) = SiI_4(g)$ C) $SiI_4(g) = Si(cr) + 2I_2(g)$

The 3rd law results of the sublimation and vaporization studies (reactions A and B) perhaps imply a better agreement among the studies than actually exists. For example, the vaporization studies of Andersen and Belz (4) and Uchimura et al. (5) agree very well near T_{vap} but diverge by up to 33% as T decreases to T_{fus}. The vapor pressure results of Bartsch and Wolf (3) cover a very small temperature range and lie 10-15% lower than those of Andersen and Belz (4) and Krichevskii et al. (2). In addition, the decomposition studies (reaction C) suggest a $\Delta_f H°(g, 298.15\ K)$ value 3.6 kcal mol^{-1} more negative than the vapor pressure studies.

We adopt $\Delta_f H°(g, 298.15\ K) = -26.4$ kcal mol^{-1} based primarily on Andersen and Belz (4) and Krichevskii et al. (2). Anderson and Belz (4) are also given primary weight in our selection of S°(cr, 298.15 K). We assign an uncertainty of ±4 kcal mol^{-1} to include the possibility that the decomposition studies (6, 7) may be correct.

Heat Capacity and Entropy

The adopted vibrational frequencies are from the work of Clark and Rippon (8), who recorded the Raman spectra in the vapor phase (210°C). The spectral data were interpreted in terms of a tetrahedral structure. This structure is consistent with the electron diffraction data of Liston and Sutton (9) which suggested a tetrahedral structure with a Si-I bond distance of 2.43±0.02 Å. We adopt this bond distance. The principal moments of inertia are: $I_A = I_B = I_C = 331.8289 \times 10^{-39}\ g\ cm^2$. Shimanouchi, in a recent compilation of molecular vibrational frequencies (10), suggested somewhat different values (168, 63, 405, 94 cm^{-1}), based on earlier infrared and Raman studies by Delwaulle (11) and Delwaulle and Francois (12). Much literature has been published on the inter-relationships between force constants and vibrational frequencies. Since the majority of this work was published prior to the Clark and Rippon study (8), it will not be further discussed or referenced. The same situation exists for temperature dependent thermochemical tabulations for $SiI_4(g)$.

References

1. I. I. Lapidus, L. A. Nisel'son and A. L. Seifer in "Thermophysical Properties of Gases and Liquids," Vol. 1, V. A. Rabinovich, Ed., Isreal Program for Scientific Translations, Jerusalem, pp. 102-135, 1970.
2. I. R. Krichevskii, G. F. Ivanovskii, and E. K. Safronov, Zh. Fiz. Khim 39, 2594 (1965).
3. K. Bartsch and E. Wolf, Z. Anorg. Allg. Chem. 383, 66 (1971).
4. H. C. Andersen and L. H. Belz, J. Amer. Chem. Soc. 75, 4828 (1953).
5. K. Uchimura, T. Takuma, M. Yuizumi, and T. Kunugi, Kenki Kagaku 35, 876 (1967).
6. H. Schafer and B. Morcher, A. Anorg. Allg. Chem. 352, 122 (1967).
7. E. Wolf and C. Herbst, Z. Chem. 7, 34 (1967).
8. R. J. H. Clark and D. M. Rippon, J. Mol. Spectrosc. 44, 479 (1972).
9. M. W. Liston and L. E. Sutton, Trans. Faraday Soc. 37, 393 (1941).
10. T. Shimanouchi, NSRDS-NBS 39, 1972.
11. M. L. Delwaulle, J. Phys. Chem. 56, 355 (1952).
12. M. L. Delwaulle and F. Francois, J. Phys. Radium 15, 206 (1954).

Enthalpy Reference Temperature = T_r = 298.15 K Standard State Pressure = p° = 0.1 MPa

T/K	$C_p°$ (J $K^{-1}mol^{-1}$)	S°	$-[G°-H°(T_r)]/T$	$H°-H°(T_r)$ (kJ mol^{-1})	$\Delta_f H°$	$\Delta_f G°$	Log K_f
0	0.	0.	INFINITE	-24.199	-105.043	-105.043	INFINITE
100	78.597	317.592	500.874	-18.328	-105.538	-124.716	65.145
200	94.049	377.499	425.478	-9.596	-107.849	-143.032	37.356
250	98.051	398.948	418.092	-4.786	-109.144	-151.681	31.692
298.15	100.574	416.449	416.449	0.	-110.458	-159.755	27.988
300	100.653	417.071	416.451	0.186	-110.510	-160.061	27.869
350	102.399	432.726	417.683	5.265	-112.090	-168.200	25.102
400	103.613	446.484	420.441	10.417	-145.677	-175.008	22.854
450	104.485	458.741	424.028	15.621	-149.664	-178.435	20.712
500	105.129	469.784	428.060	20.862	-233.903	-173.820	18.159
600	105.996	489.035	436.665	31.422	-233.227	-161.867	14.092
700	106.534	505.418	445.346	42.050	-232.582	-150.026	11.195
800	106.889	519.668	453.765	52.723	-231.977	-138.274	9.028
900	107.136	532.273	461.801	63.425	-231.417	-126.595	7.347
1000	107.314	543.570	469.423	74.147	-230.909	-114.976	6.006
1100	107.447	553.805	476 636	84.886	-230.457	-103.405	4.910
1200	107.548	563.159	483.462	95.636	-230.071	-91.872	3.999
1300	107.627	571.770	489.928	106.395	-229.765	-80.368	3.229
1400	107.690	579.749	496.062	117.161	-229.555	-68.885	2.570
1500	107.741	587.180	501.892	127.932	-229.462	-57.412	1.999
1600	107.782	594.135	507.442	138.708	-229.506	-45.942	1.500
1700	107.817	600.670	512.736	149.488	-279.883	-34.017	1.045
1800	107.846	606.834	517.794	160.272	-280.021	-19.551	0.567
1900	107.871	612.665	522.635	171.058	-280.295	-5.074	0.139
2000	107.892	618.199	527.276	181.846	-280.707	9.422	-0.246
2100	107.910	623.464	531.732	192.636	-281.253	23.941	-0.596
2200	107.926	628.484	536.017	203.428	-281.922	38.489	-0.914
2300	107.939	633.282	540.142	214.221	-282.699	53.070	-1.205
2400	107.951	637.876	544.119	225.015	-283 564	67.687	-1.473
2500	107.962	642.283	547.958	235.811	-284.496	82.342	-1 720
2600	107.971	646.517	551.668	246.608	-285.471	97.035	-1.949
2700	107.980	650.592	555.257	257.405	-286.467	111.765	-2.162
2800	107.987	654.520	558.732	268.204	-287.460	126.533	-2.360
2900	107.994	658.309	562.101	279.003	-288.430	141.335	-2.546
3000	108.000	661.970	565.369	289.803	-289.356	156.171	-2.719
3100	108.006	665.512	568.543	300.603	-290.222	171.036	-2.882
3200	108.011	668.941	571.627	311.404	-291.014	185.928	-3.035
3300	108.015	672.265	574.627	322.205	-291.719	200.844	-3.179
3400	108.019	675.489	577.546	333.007	-292.326	215.779	-3.315
3500	108.023	678.621	580.389	343.809	-292.829	230.731	-3.443
3600	108.027	681.664	583.161	354.611	-677.373	256.156	-3.717
3700	108.030	684.624	585.863	365.414	-677.236	282.087	-3.982
3800	108.033	687.505	588.500	376.217	-676.983	308.011	-4.234
3900	108.036	690.311	591.075	387.021	-676.612	333.928	-4.472
4000	108.038	693.046	593.590	397.825	-676.124	359.833	-4.699
4100	108.041	695.714	596.048	408.628	-675.518	385.724	-4.914
4200	108.043	698.317	598.452	419.433	-674.795	411.600	-5.119
4300	108.045	700.860	600.805	430.237	-673.958	437.456	-5.314
4400	108.047	703.344	603.107	441.042	-673.008	463.292	-5.500
4500	108.049	705.772	605.361	451.846	-671.948	489.106	-5.677
4600	108.050	708.147	607.570	462.651	-670.781	514.894	-5.847
4700	108.052	710.470	609.735	473.456	-669.510	540.656	-6.009
4800	108.053	712.745	611.857	484.262	-668.139	566.391	-6.164
4900	108.055	714.973	613.939	495.067	-666.670	592.095	-6.312
5000	108.056	717.156	615.982	505.873	-665.108	617.768	-6.454
5100	108.057	719.296	617.987	516.678	-663.456	643.410	-6.590
5200	108.058	721.394	619.955	527.484	-661.718	669.018	-6.720
5300	108.059	723.453	621.888	538.290	-659.896	694.591	-6.846
5400	108.060	725.473	623.788	549.096	-657.994	720.130	-6.966
5500	108.061	727.455	625.655	559.902	-656.016	745.632	-7.081
5600	108.062	729.402	627.490	570.708	-653.964	771.099	-7.193
5700	108.063	731.315	629.295	581.514	-651.843	796.528	-7.299
5800	108.064	733.195	631.070	592.321	-649.654	821.918	-7.402
5900	108.065	735.042	632.817	603.127	-647.402	847.271	-7.501
6000	108.066	736.858	634.536	613.934	-645.088	872.585	-7.597

PREVIOUS: December 1977 (1 atm) CURRENT: December 1977 (1 bar)

TITANIUM IODIDE (TiI_4) CRYSTAL M_r = 555.4980

$S°(298.15\ K) = 246.150 \pm 6.7\ J\ K^{-1}\ mol^{-1}$
$T_{trs} = 379\ K$
$T_{fus} = 428\ K$

$\Delta_f H°(0\ K) = -373.64 \pm 8.4\ kJ\ mol^{-1}$
$\Delta_f H°(298.15\ K) = -375.72 \pm 8.4\ kJ\ mol^{-1}$
$\Delta_{trs} H° = 9.916 \pm 0.63\ kJ\ mol^{-1}$
$\Delta_{fus} H° = 19.832 \pm 0.63\ kJ\ mol^{-1}$

Enthalpy of Formation

The enthalpy of formation of TiI_4(cr) was measured by Johnson et al. (1), based on the enthalpies of hydrolysis of $TiBr_4$(cr) and TiI_4(cr) in sulfuric acid. These data result in a enthalpy of reaction of -57.88 kcal mol^{-1} for the process $TiI_4(cr) + 2Br_2(\ell) = TiBr_4(cr) + 2I_2(cr)$, when recalculated with revised values of $\Delta_f H°(298.15\ K)$ for HBr(aq) and HI(aq) (2). Combination of $\Delta_r H°(298.15\ K)$ with auxiliary JANAF data gives the adopted value of $\Delta_f H°(298.15\ K)$.

Heat Capacity and Entropy

King et al. (3) reported the heat capacity and enthalpy of TiI_4(cr) over the temperature range 51 to 428 K. The value of S°(298.15 K) is calculated from these data based on S°(51 K) = 14.62 cal $K^{-1}\ mol^{-1}$. The value of H°(51 K)-H°(0 K) is calculated to be 0.356 kcal mol^{-1} from a Debye-Einstein extrapolation of the measured heat capacities.

Transition Data

A first order transition was observed by King et al. (2) at 379 K. They reported the measured value of 2.37 kcal mol^{-1}, noting that this value had an uncertainty of 0.15 kcal mol^{-1} due to hysteresis in the transition.

Fusion Data

The melting temperature and enthalpy of fusion were reported by King et al. (2). The large uncertainty of $\Delta_{fus}H°$ is due to the uncertainty and close proximity of the transition at 379 K.

References

1. W. H. Johnson, A. A. Gilliland, and E. J. Prosen, J. Res. Nat. Bur. Stand. 63A, 161 (1959).
2. U. S. Nat. Bur. Stand. Tech. Note 270-3, 1968.
3. E. G. King, W. W. Weller, A. U. Christensen and K. K. Kelley, U. S. Bur. Mines RI 5799, 1961.

Titanium Iodide (TiI_4) I_4Ti_1(cr)

Enthalpy Reference Temperature = T_r = 298.15 K   Standard State Pressure = p° = 0.1 MPa

T/K	$C_p°$ (J K⁻¹mol⁻¹)	S° (J K⁻¹mol⁻¹)	$-[G°-H°(T_r)]/T$ (J K⁻¹mol⁻¹)	$H°-H°(T_r)$ (kJ mol⁻¹)	$\Delta_f H°$ (kJ mol⁻¹)	$\Delta_f G°$ (kJ mol⁻¹)	Log K_f
0	0.	0.	INFINITE	-29.141	-373.638	-373.638	INFINITE
100	100.671	120.059	352.958	-23.290	-374.448	-373.429	195.059
200	119.349	197.192	257.466	-12.055	-374.995	-372.204	97.210
298.15	125.629	246.150	246.150	0.	-375.723	-370.687	64.943
300	125.947	246.928	246.153	0.233	-375.739	-370.655	64.537
379.000	139.608	277.904	249.613	10.722		I <--> II	
379.000	148.114	304.068	249.613	20.639		TRANSITION	
400	148.114	312.056	252.683	23.749	-398.112	-368.316	48.097
428.000	148.114	322.077	256.899	27.896	-------- II <--> LIQUID --------		
500	148.114	345.106	267.986	38.560	-482.434	-352.801	36.857
600	148.114	372.111	283.158	53.372	-477.957	-327.296	28.494
700	148.114	394.942	297.538	68.183	-473.568	-302.534	22.575
800	148.114	414.720	310.977	82.994	-469.240	-278.397	18.177
900	148.114	432.166	323.492	97.806	-465.000	-254.797	14.788
1000	148.114	447.771	335.154	112.617	-460.914	-231.662	12.101
1100	148.114	461.888	346.044	127.428	-457.048	-208.926	9.921
1200	148.114	474.775	356.242	142.240	-457.397	-186.412	8.114
1300	148.114	486.631	365.822	157.051	-453.278	-163.998	6.590
1400	148.114	497.607	374.848	171.863	-449.299	-141.896	5.294
1500	148.114	507.826	383.376	186.674	-445.492	-120.072	4.181

PREVIOUS: June 1964   CURRENT: December 1968

Titanium Iodide (TiI_4) I_4Ti_1(cr)

TITANIUM IODIDE (TiI_4) LIQUID M_r = 555.4980

$\Delta_f H°$(0 K) = Unknown

S°(298.15 K) = [311.841] J K^{-1} mol^{-1} $\Delta_f H°$(298.15 K) = [-348.314] kJ mol^{-1}

T_{fus} = 428 K $\Delta_{fus} H°$ = 19.832 ± 0.63 kJ mol^{-1}

Enthalpy of Formation

$\Delta_f H°(TiI_4$, ℓ, 298.15 K) is calculated from $\Delta_f H°(TiI_4$, cr, 298.15 K) by adding the enthalpy of fusion, $\Delta_{fus} H°$, and the difference in enthalpy, H°(428 K)-H°(298.15 K), between the crystal and liquid.

Heat Capacity and Entropy

King et al. (1) reported the enthalpy of TiI_4(ℓ) above the melting temperature. The heat capacity derived from these data is constant. S°(TiI_4, ℓ, 298.15 K) is calculated in a manner analogous to that used for the enthalpy of formation.

Fusion Data

The melting temperature and enthalpy of fusion were reported by King et al. (1).

Vaporization Data

The boiling temperature, T_{vap}, is taken as the temperature at which the fugacity is 1 bar for the reaction TiI_4(ℓ) = TiI_4(g). The enthalpy of vaporization is calculated as the difference between $\Delta_f H°$ of the liquid and gas at the boiling temperature. The vapor pressure data are discussed in the table for TiI_4(g).

Reference

1. E. G. King, W. W. Weller, A. U. Christensen, and K. K. Kelley, U. S. Bur. Mines RI 5799, 1961.

Titanium Iodide (TiI_4) I_4Ti_1(l)

Enthalpy Reference Temperature = T_r = 298.15 K Standard State Pressure = p° = 0.1 MPa

	J $K^{-1}mol^{-1}$			kJ mol^{-1}			
T/K	$C_p°$	S°	-[G°-H°(T_r)]/T	H°-H°(T_r)	$\Delta_f H°$	$\Delta_f G°$	Log K_f
0							
100							
200							
298.15	156.482	311.841	311.841	0.	-348.314	-362.863	63.572
300	156.482	312.809	311.844	0.289	-348.273	-362.954	63.196
400	156.482	357.826	317.982	15.938	-378.515	-367.027	47.929
428.000	156.482	368.414	320.939	20.319	-------- II <--> LIQUID --------		
500	156.482	392.744	329.573	31.586	-462.000	-356.185	37.210
600	156.482	421.274	342.551	47.234	-456.685	-335.523	29.210
700	156.482	445.396	355.564	62.882	-451.460	-315.744	23.561
800	156.482	466.291	368.128	78.530	-446.295	-296.709	19.373
900	156.482	484.722	380.079	94.178	-441.219	-278.316	16.153
1000	156.482	501.209	391.382	109.827	-436.295	-260.482	13.606
1100	156.482	516.123	402.055	125.475	-431.593	-243.130	11.545
1200	156.482	529.739	412.137	141.123	-431.105	-226.076	9.841
1300	156.482	542.264	421.671	156.771	-426.149	-209.193	8.405
1400	156.482	553.861	430.704	172.419	-421.333	-192.685	7.189
1500	156.482	564.657	439.279	188.067	-416.690	-176.516	6.147

PREVIOUS: June 1964 CURRENT: December 1968

Titanium Iodide (TiI_4) I_4Ti_1(l)

TITANIUM IODIDE (TiI_4) CRYSTAL–CRYSTAL–LIQUID M_r = 555.4980

0 to 379 K crystal, I
379 to 428 K crystal, II
above 428 K liquid

Refer to the individual tables for details.

Titanium Iodide (TiI_4) I_4Ti_1(cr,l)

Enthalpy Reference Temperature = T_r = 298.15 K Standard State Pressure = $p°$ = 0.1 MPa

T/K	$C_p°$ (J K⁻¹mol⁻¹)	$S°$ (J K⁻¹mol⁻¹)	$-[G°-H°(T_r)]/T$ (J K⁻¹mol⁻¹)	$H°-H°(T_r)$ (kJ mol⁻¹)	$\Delta_f H°$ (kJ mol⁻¹)	$\Delta_f G°$ (kJ mol⁻¹)	Log K_f
0	0.	0.	INFINITE	-29.141	-373.638	-373.638	INFINITE
100	100.671	120.059	352.958	-23.290	-374.448	-373.429	195.059
200	119.349	197.192	257.466	-12.055	-374.995	-372.204	97.210
298.15	125.629	246.150	246.150	0.	-375.723	-370.687	64.943
300	125.947	246.928	246.153	0.233	-375.739	-370.655	64.537
379.000	139.608	277.904	249.613	10.722		I <--> II	
379.000	148.114	304.068	249.613	20.639		TRANSITION	
400	148.114	312.056	252.683	23.749	-398.112	-368.316	48.097
428.000	148.114	322.077	256.899	27.896		II <--> LIQUID	
428.000	156.482	368.414	256.899	47.728		TRANSITION	
500	156.482	392.744	274.754	58.995	-462.000	-356.185	37.210
600	156.482	421.274	296.869	74.643	-456.685	-335.523	29.210
700	156.482	445.396	316.408	90.291	-451.460	-315.744	23.561
800	156.482	466.291	333.867	105.939	-446.295	-296.709	19.373
900	156.482	484.722	349.625	121.588	-441.219	-278.316	16.153
1000	156.482	501.209	363.973	137.236	-436.295	-260.482	13.606
1100	156.482	516.123	377.138	152.884	-431.593	-243.130	11.545
1200	156.482	529.739	389.296	168.532	-431.105	-226.076	9.841
1300	156.482	542.264	400.587	184.180	-426.149	-209.193	8.405
1400	156.482	553.861	411.126	199.828	-421.333	-192.685	7.189
1500	156.482	564.657	421.006	215.477	-416.690	-176.516	6.147

PREVIOUS: CURRENT: December 1968

Titanium Iodide (TiI_4) I_4Ti_1(cr,l)

TITANIUM IODIDE (TiI_4) IDEAL GAS M_r = 555.4980

$S°(298.15\ K) = 433.072 \pm 8.4\ J\ K^{-1}\ mol^{-1}$

$\Delta_fH°(0\ K) = -271.86 \pm 8.4\ kJ\ mol^{-1}$

$\Delta_fH°(298.15\ K) = -277.27 \pm 8.4\ kJ\ mol^{-1}$

Vibrational Frequencies and Degeneracies

ν, cm^{-1}	ν, cm^{-1}
[175](1)	[260](3)
[55](2)	[69](3)

Ground State Quantum Weight: 1 σ = 12

Point Group: T_d

Bond Distance: Ti-I = [2.5] Å

Bond Angle: I-Ti-I = [109°28']

Product of the Moments of Inertia: $I_AI_BI_C = [4.332573 \times 10^{-110}]\ g^3\ cm^6$

Enthalpy of Formation

The enthalpy of formation of $TiI_4(g)$ is calculated from that of $TiI_4(\ell)$ and the enthalpy of vaporization of the liquid. Blocher and Campbell (1) reported vapor pressure data for $TiI_4(\ell)$ over the temperature range 473 to 655 K. 2nd and 3rd law analyses of these data give values for $\Delta_{vap}H°(298.15\ K)$ of 17.2±0.1 and 16.98 kcal mol^{-1}, respectively, the 3rd law drift being -0.4±0.2 cal K^{-1} mol^{-1}. The adopted value of $\Delta_fH°(298.15\ K)$ is based on the 3rd law enthalpy of vaporization.

Heat Capacity and Entropy

The interatomic distances are estimated from those of $TiBr_4(cr)$, $TiBr_4(g)$ and $TiI_4(cr)$ (2). The tetrahedral model is assumed by analogy with $TiCl_4(g)$ and $TiBr_4(g)$. The principal moments of inertia are: $I_A = I_B = I_C = 351.2222 \times 10^{-39}\ g\ cm^2$.

The vibrational frequencies of $TiI_4(g)$ have been estimated by Ushanova et al. (3). The given estimates are based on their predictions and valence force field calculations.

References

1. J. M. Blocher, Jr. and I. E. Campbell, J. Amer. Chem. Soc. 69, 2100 (1947).
2. O. Hasoel and H. Kringstand, Z. Phys. Chem. 1315, 274 (1932).
3. N. I. Ushanova, I. N. Godnev and I. V. Orlova, Opt. Spektrosk. 5, 567 (1958).

Titanium Iodide (TiI_4) $I_4Ti_1(g)$

Enthalpy Reference Temperature = T_r = 298.15 K Standard State Pressure = p° = 0.1 MPa

T/K	$C_p°$ (J K⁻¹mol⁻¹)	S° (J K⁻¹mol⁻¹)	-[G°-H°(T_r)]/T (J K⁻¹mol⁻¹)	H°-H°(T_r) (kJ mol⁻¹)	$\Delta_fH°$ (kJ mol⁻¹)	$\Delta_fG°$ (kJ mol⁻¹)	Log K_f
0	0.	0.	INFINITE	-25.812	-271.860	-271.860	INFINITE
100	85.757	327.171	522.445	-19.527	-272.236	-291.928	152.488
200	100.193	392.170	442.507	-10.068	-274.558	-310.763	81.163
250	102.779	414.832	434.781	-4.987	-275.896	-319.662	66.790
298.15	104.254	433.072	433.072	0.	-277.274	-327.968	57.459
300	104.298	433.717	433.074	0.193	-277.329	-328.282	57.159
350	105.256	449.872	434.347	5.433	-278.999	-336.648	50.242
400	105.895	463.971	437.188	10.713	-312.699	-343.668	44.878
450	106.342	476.471	440.871	16.020	-316.816	-347.292	40.313
500	106.666	487.693	445.002	21.345	-401.200	-342.859	35.818
600	107.093	507.181	453.789	32.035	-400.844	-331.226	28.836
700	107.354	523.711	462.627	42.759	-400.543	-319.647	23.852
800	107.524	538.058	471.179	53.503	-400.282	-308.109	20.117
900	107.641	550.730	479.328	64.262	-400.095	-296.600	17.214
1000	107.726	562.075	487.045	75.030	-400.051	-285.104	14.892
1100	107.788	572.346	494.340	85.806	-400.221	-273.603	12.992
1200	107.836	581.727	501.237	96.587	-404.600	-261.957	11.403
1300	107.873	590.360	507.765	107.373	-404.507	-250.074	10.048
1400	107.902	598.355	513.954	118.162	-404.550	-238.194	8.887
1500	107.926	605.800	519.832	128.953	-404.764	-226.305	7.881
1600	107.945	612.766	525.425	139.747	-405.179	-214.396	6.999
1700	107.962	619.311	530.757	150.542	-405.828	-202.453	6.221
1800	107.975	625.482	535.850	161.339	-406.737	-190.465	5.527
1900	107.987	631.320	540.722	172.137	-407.927	-178.419	4.905
2000	107.996	636.860	545.392	182.936	-424.101	-165.849	4.332
2100	108.005	642.129	549.874	193.736	-426.640	-152.874	3.803
2200	108.012	647.154	554.182	204.537	-429.304	-139.776	3.319
2300	108.019	651.955	558.330	215.339	-432.077	-126.554	2.874
2400	108.024	656.552	562.327	226.141	-434.939	-113.210	2.464
2500	108.029	660.962	566.185	236.943	-437.868	-99.744	2.084
2600	108.034	665.199	569.912	247.747	-440.841	-86.160	1.731
2700	108.037	669.277	573.517	258.550	-443.835	-72.462	1.402
2800	108.041	673.206	577.008	269.354	-446.827	-58.653	1.094
2900	108.044	676.997	580.391	280.158	-449.795	-44.737	0.806
3000	108.047	680.660	583.672	290.963	-452.721	-30.719	0.535
3100	108.049	684.203	586.859	301.768	-455.587	-16.605	0.280
3200	108.052	687.633	589.954	312.573	-458.378	-2.400	0.039
3300	108.054	690.958	592.965	323.378	-461.083	11.891	-0.188
3400	108.056	694.184	595.895	334.184	-463.691	26.263	-0.403
3500	108.058	697.316	598.748	344.989	-466.194	40.711	-0.608
3600	108.059	700.361	601.529	355.795	-468.587	55.227	-0.801
3700	108.061	703.321	604.240	366.601	-879.963	77.598	-1.095
3800	108.062	706.203	606.885	377.407	-880.890	103.491	-1.423
3900	108.063	709.010	609.468	388.213	-881.752	129.407	-1.733
4000	108.065	711.746	611.991	399.020	-882.550	155.344	-2.029
4100	108.066	714.414	614.457	409.826	-883.283	181.301	-2.310
4200	108.067	717.019	616.868	420.633	-883.949	207.274	-2.578
4300	108.068	719.561	619.227	431.440	-884.549	233.263	-2.834
4400	108.069	722.046	621.535	442.247	-885.085	259.265	-3.078
4500	108.069	724.475	623.796	453.053	-885.557	285.278	-3.311
4600	108.070	726.850	626.011	463.860	-885.965	311.301	-3.535
4700	108.071	729.174	628.181	474.667	-886.312	337.333	-3.749
4800	108.072	731.449	630.309	485.475	-886.599	363.372	-3.954
4900	108.072	733.678	632.396	496.282	-886.827	389.415	-4.151
5000	108.073	735.861	634.443	507.089	-886.997	415.462	-4.340
5100	108.073	738.001	636.453	517.896	-887.112	441.513	-4.522
5200	108.074	740.100	638.426	528.704	-887.172	467.565	-4.697
5300	108.074	742.158	640.364	539.511	-887.178	493.617	-4.865
5400	108.075	744.178	642.268	550.319	-887.133	519.670	-5.027
5500	108.075	746.162	644.139	561.126	-887.037	545.721	-5.183
5600	108.076	748.109	645.978	571.934	-886.892	571.770	-5.333
5700	108.076	750.022	647.786	582.741	-886.698	597.816	-5.478
5800	108.077	751.901	649.565	593.549	-886.457	623.858	-5.618
5900	108.077	753.749	651.316	604.356	-886.170	649.896	-5.754
6000	108.077	755.565	653.038	615.164	-885.838	675.928	-5.884

PREVIOUS: December 1968 (1 atm) CURRENT: December 1968 (1 bar)

Titanium Iodide (TiI_4) $I_4Ti_1(g)$

ZIRCONIUM IODIDE (ZrI_4) CRYSTAL M_r = 598.8380

$S°(298.15\ K) = [260.270 \pm 4.2]\ J\ K^{-1}\ mol^{-1}$

$T_{fus} = 772 \pm 2\ K$

$T_{sub} = 705.6\ K$

$\Delta_f H°(0\ K) = [-487.24 \pm 8.4]\ kJ\ mol^{-1}$

$\Delta_f H°(298.15\ K) = -488.69 \pm 6.3\ kJ\ mol^{-1}$

$\Delta_{fus} H°$ = Unknown

$\Delta_{sub} H°(298.15\ K) = 126.357 \pm 2.1\ kJ\ mol^{-1}$

Enthalpy of Formation

Turnbull (1) measured the enthalpies of reaction for the dissolution of the zirconium tetrahalides in caustic and in water. The reported enthalpies of reaction and the corresponding reactions may be combined to yield the following:

$ZrI_4(cr) + 4\ NaCl(aq) = ZrCl_4(cr) + 4\ NaI(aq),\ \Delta_r H°(298.15\ K) = -12.50\ kcal\ mol^{-1}$

$ZrI_4(cr) + 4\ Cl^-(aq) = ZrCl_4(cr) + 4\ I^-(aq),\ \Delta_r H°(298.15\ K) = -11.97\ kcal\ mol^{-1}$

Using auxiliary data (2, 4), we calculate $\Delta_f H°(298.15\ K) = -116.52$ and $-117.05\ kcal\ mol^{-1}$ for $ZrI_4(cr)$ from these two reactions. We adopt a mean of these two values, $\Delta_f H°(298.15\ K) = -116.8\ kcal\ mol^{-1}$, and assign an uncertainty of $\pm$ 1.5 kcal mol^{-1}. The value suggested by NBS (3) is 1.7 kcal mol^{-1} more positive.

Baev and Shelton (5) reference an enthalpy of formation value for $ZrI_4(cr)$. The reported $\Delta_f H°(298.15\ K) = -105.9\ kcal\ mol^{-1}$ as supposedly extracted from Karapet'yants and Karapet'yants (6). We tentatively discard this value due to the large number of inconsistencies in the reported tabular results of Baev and Shelton (5).

Heat Capacity and Entropy

There is no heat capacity and enthalpy data reported in the literature for $ZrI_4(cr)$. The adopted heat capacity values are estimated such they they parallel those adopted for $ZrCl_4(cr)$ and $ZrBr_4(cr)$ (2) and are consistent with the sublimation data.

The crystal data compilation of Donnay and Ondik (8) does not list the crystal structure of $ZrI_4(cr)$ but does tabulate both $ZrCl_4$ and $ZrBr_4$ as cubic structures. The literature data related to the crystal structure of ZrI_4 is not definitive. Assuming $ZrI_4(cr)$ also has a cubic structure, the adopted heat capacity values are estimated so as to parallel those for $ZrCl_4$ and $ZrBr_4$. The values below 300 K are calculated in the same manner as for $ZrBr_4(cr)$ (2). The high temperature heat capacities are obtained graphically.

Fusion Data

The melting point was observed by Rahlfs and Fischer (9) to be 772±2 K and by Nisel'son (10) to be 773 K.

Sublimation Data

The sublimation data are treated in the $ZrI_4(g)$ table (2). The sublimation temperature, T_{sub}, is calculated from the Gibbs energy crossover between the crystal and gas. Since T_{sub} is less than T_{fus}, the liquid phase is thermodynamically unstable at one atmosphere pressure.

References

1. A. G. Turnbull, J. Phys. Chem. 65, 1652 (1961).
2. JANAF Thermochemical Tables: $ZrCl_4(cr)$, $ZrBr_4(cr)$ and $ZrI_4(g)$, 6-30-75.
3. U. S. Nat. Bur. Stand. Tech. Note 270-5, 1971.
4. ICSU-CODATA Task Group, J. Chem. Thermodyn. 4, 331 (1972).
5. A. K. Baev and R. A. J. Shelton, Russ. J. Phys. Chem. 47, 1340 (1973).
6. M. K. Karapet'yants and M. L. Karapet'yants, "Osnovnye Termokhimicheskie Konstanty Neorganisheskekh i Organicheskikh Veshchestv," Izd. Khimiya, Moscow, 1968 (as referenced by Baev and Shelton (13)).
7. F. R. Sale and R. A. J. Shelton, J. Less-Common Metals 9, 54 (1965).
8. J. D. H. Donnay and H. M. Ondik, "Crystal Data, Determinative Tables," 3rd edition, Vol. 2, U. S. Nat. Bur. Stand. and Joint Committee on Powder Diffraction Standards, 1973.
9. O. Rahlfs and W. Fischer, Z. Anorg. Allg. Chem. 211, 349 (1933).
10. L. A. Nisel'son, Russ. J. Inorg. Chem. 7, 354 (1962).

Zirconium Iodide (ZrI_4) $I_4Zr_1(cr)$

Enthalpy Reference Temperature = T_r = 298.15 K; Standard State Pressure = p° = 0.1 MPa

T/K	$C_p°$ (J K⁻¹mol⁻¹)	S° (J K⁻¹mol⁻¹)	$-[G°-H°(T_r)]/T$ (J K⁻¹mol⁻¹)	$H°-H°(T_r)$ (kJ mol⁻¹)	$\Delta_f H°$ (kJ mol⁻¹)	$\Delta_f G°$ (kJ mol⁻¹)	Log K_f
0	0.	0.	INFINITE	-30.439	-487.236	-487.236	INFINITE
100	104.207	131.545	369.071	-23.753	-487.537	-487.087	254.428
200	121.579	210.426	271.784	-12.272	-488.114	-486.406	127.036
298.15	127.800	260.270	260.270	0.	-488.691	-485.446	85.048
300	127.880	261.061	260.272	0.236	-488.703	-485.426	84.520
400	131.001	298.303	265.329	13.190	-521.586	-483.104	63.087
500	133.009	327.759	274.971	26.394	-607.399	-465.240	48.603
600	134.599	352.154	285.860	39.776	-604.216	-437.106	38.053
700	135.938	373.005	296.856	53.304	-600.990	-409.509	30.558
800	137.152	391.236	307.537	66.959	-597.743	-382.376	24.967
900	138.365	407.460	317.755	80.735	-594.496	-355.650	20.641
1000	139.578	422.101	327.469	94.632	-591.258	-329.285	17.200

PREVIOUS: June 1964 CURRENT: June 1975

ZIRCONIUM IODIDE (ZrI_4) IDEAL GAS M_r = 598.8380

$S°(298.15\ K) = 446.563 \pm 0.13\ J\ K^{-1}\ mol^{-1}$

$\Delta_f H°(0\ K) = -356.75 \pm 8.4\ kJ\ mol^{-1}$

$\Delta_f H°(298.15\ K) = -362.33 \pm 8.4\ kJ\ mol^{-1}$

Vibrational Frequencies and Degeneracies

ν, cm^{-1}
158 (1)
43 (2)
254 (3)
55 (3)

Ground State Quantum Weight: [1] $\sigma = 12$

Point Group: T_d

Bond Distance: Zr-I = 2.660 ± 0.005 Å

Bond Angle: I-Zr-I = 109° 28'

Product of the Moments of Inertia: $I_A I_B I_C = 6.286307 \times 10^{-110}\ g^3\ cm^6$

Enthalpy of Formation

The enthalpy of formation for ZrI_4(g) is calculated from the enthalpies of formation and sublimation at 298.15 K. The adopted value for the heat of sublimation, $\Delta_{sub}H°(298.15\ K) = 30.2 \pm 0.5\ kcal\ mol^{-1}$, is based on the 2nd and 3rd law analyses of the sublimation data of Sale and Shelton (1) and Rahlfs and Fischer (2). Using Knudsen effusion techniques, Sale and Shelton (1) studied effusion in a nickel effusion cell for three different orifice areas. The results were then extrapolated to zero orifice area (refer to last line in the following table). Other sublimation studies have been conducted by Rahlfs and Fischer (2), using a liquid tin tensimeter, and by Gerlach et al. (3), using the oscillating quartz fiber method. It should be noted that the work of Rahlfs and Fischer (2) covered an observed pressure range of 4.0 - 272 mm Hg, which is much wider than range covered by Sale and Shelton (1). The adopted $\Delta_{sub}H°(298.15\ K)$ is based on the mean of the 3rd law results of Sale and Shelton (30.01 kcal mol^{-1}, 1) and Rahlfs and Fischer (30.32 kcal mol^{-1}, 2).

The effusion studies (membrane method) by Baev and Shelton (4) are discarded due to the large number of discrepancies in their tabulated results.

In analyzing the vapor pressures for the three sublimation studies (1, 2, 3), corrections were made for non-ideality by means of the equation $\Delta G°/T = -R\ln p - Bp/T$. The Bertholet equation of state and the critical constants $T_c = 959$ K and $p_c = 40.7$ atm as determined by Nisel'son and Sokolova (5) are used to calculate B.

Source	T/K	Data Points	$\Delta_{sub}H°(298.15\ K)$, kcal mol^{-1} 2nd law	3rd law	Drift cal K^{-1} mol^{-1}
(2)	558-671	15*	31.33±0.24	30.32±0.12	-1.62±0.38
(3)	393-453	eqn	26.91	29.98	7.25
(1)	426-474A**	12*	29.84±1.28	31.22±0.30	3.09±2.84
	432-500B**	14*	30.81±0.57	30.86±0.15	0.10±1.24
	436-480C**	12	30.24±0.76	30.51±0.17	0.61±1.66
	423-503	eqn	29.97	30.01	0.08

*One point rejected due to a statistical test

**Orifice area ($cm^2\ 10^2$) A = 0.5364 B = 0.0935

Heat Capacity and Entropy

The adopted vibrational frequencies are from the work of Clark et al. (6, 7) who recorded the Raman spectra of ZrI_4 in the vapor phase (380-420°C). These studies indicated that ZrI_4 is a tetrahedral monomer in the vapor phase. Shimanouchi, in his compilation of molecular vibrational frequencies (8), also adopted the values of Clark et al. (6, 7) for ZrI_4(g). Rahlfs and Fischer (2), though vapor density measurements had earlier concluded that ZrI_4 was monomeric in the vapor phase.

An electron diffraction study of ZrI_4(g) at 250-270°C was conducted by Zasorin et al. (9). This study confirmed the tetrahedral structure and determined the Zr-I bond length to be 2.660±0.005 Å. We adopt this value. Other interatomic distances reported in the literature are estimates. The principal moments of inertia are: $I_A = I_B = I_C = 397.6172 \times 10^{-39}\ g\ cm^2$.

Much literature has been published on the inter-relationships between force constants and vibrational frequencies. Since the majority of these are based on estimated frequencies, they will not be further discussed or referenced. The same situation is true for thermodynamic tabulations of ZrI_4(g). One exception is that Clark et al. (6) calculated thermodynamic properties based on their experimental vibrational frequencies. Their tabulation is very similar to ours: the differences in the entropy being 0.04 cal K^{-1} mol^{-1} in the range 100-1000 K.

References

1. F. R. Sale and R. A. J. Shelton, J. Less Common Metals 3, 54 (1965).
2. O. Rahlfs and W. Fischer, Z. Anorg. Allg. Chem. 211, 349 (1933).
3. J. Gerlach, J. P. Krumme, F. Pawlek, and H. Probst, Z. Phys. Chem. (Frankfurt am Main) 53, 135 (1967).
4. A. K. Baev and R. A. J. Shelton, Russ. J. Phys. Chem. 47, 1340 (1973).
5. L. A. Nisel'son and T. D. Sokolova, Russ. J. Inorg. Chem. 7, 1382 (1962).
6. R. J. H. Clark, B. K. Hunter and D. M. Rippon, Chem. Ind. (London) 1971, 787 (1971); Inorg. Chem. 11, 56 (1972).
7. R. J. H. Clark and D. M. Rippon, J. Mol. Spectrosc. 44, 479 (1972); Advan. Raman Spectrosc. 1, 443 (1972).
8. T. Shimanouchi, J. Phys. Chem. Ref. Data 3, 269 (1971).
9. E. Z. Zasoren, G. V. Girichev, V. P. Spiridonov, K. S. Karasnov, and V. I. Tsirel'nikov, Izv. Vyssh. Ucheb. Zaved., Khim. Khim. Teknol. 16, 802 (1973).

Zirconium Iodide (ZrI_4) I_4Zr_1(g)

Enthalpy Reference Temperature = T_r = 298.15 K Standard State Pressure = p° = 0.1 MPa

T/K	$C_p°$ (J $K^{-1}mol^{-1}$)	S°	$-[G°-H°(T_r)]/T$	$H°-H°(T_r)$ (kJ mol^{-1})	$\Delta_f H°$	$\Delta_f G°$	Log K_f
0	0.	0.	INFINITE	-26.311	-356.752	-356.752	INFINITE
100	87.629	339.627	536.514	-19.689	-357.116	-377.475	197.173
200	100.873	405.465	456.038	-10.115	-359.600	-396.900	103.660
250	103.241	428.254	448.278	-5.006	-360.962	-406.069	84.843
298.15	104.589	446.563	446.563	0.	-362.334	-414.633	72.642
300	104.629	447.210	446.565	0.194	-362.389	-414.957	72.250
350	105.504	463.409	447.843	5.448	-364.029	-423.594	63.218
400	106.088	477.538	450.690	10.739	-397.679	-430.892	56.269
450	106.496	490.058	454.382	16.054	-401.736	-434.800	50.470
500	106.791	501.295	458.521	21.387	-486.049	-430.658	44.991
600	107.181	520.803	467.324	32.087	-485.549	-419.628	36.532
700	107.418	537.344	476.175	42.818	-485.119	-408.676	30.496
800	107.574	551.698	484.738	53.568	-484.777	-397.780	25.972
900	107.681	564.375	492.896	64.331	-484.542	-386.921	22.456
1000	107.758	575.725	500.622	75.103	-484.430	-376.081	19.644
1100	107.814	585.998	507.924	85.882	-484.453	-365.246	17.344
1200	107.858	595.381	514.826	96.666	-488.336	-354.181	15.417
1300	107.892	604.016	521.359	107.453	-488.129	-343.011	13.782
1400	107.919	612.012	527.553	118.244	-488.023	-331.852	12.382
1500	107.940	619.459	533.434	129.037	-488.048	-320.697	11.168
1600	107.958	626.426	539.031	139.832	-488.235	-309.536	10.105
1700	107.973	632.971	544.366	150.628	-488.610	-298.356	9.167
1800	107.985	639.143	549.462	161.426	-489.200	-287.149	8.333
1900	107.996	644.982	554.337	172.225	-490.022	-275.903	7.585
2000	108.005	650.521	559.009	183.025	-491.089	-264.607	6.911
2100	108.012	655.791	563.493	193.826	-492.408	-253.251	6.299
2200	108.019	660.816	567.803	204.628	-515.301	-241.082	5.724
2300	108.025	665.618	571.953	215.430	-517.533	-228.567	5.191
2400	108.030	670.215	575.952	226.233	-519.855	-215.954	4.700
2500	108.034	674.625	579.811	237.036	-522.243	-203.243	4.247
2600	108.038	678.863	583.540	247.839	-524.676	-190.435	3.826
2700	108.042	682.940	587.146	258.643	-527.130	-177.532	3.435
2800	108.045	686.869	590.638	269.448	-529.582	-164.540	3.070
2900	108.048	690.661	594.022	280.252	-532.010	-151.460	2.728
3000	108.050	694.324	597.305	291.057	-534.395	-138.296	2.408
3100	108.053	697.867	600.492	301.863	-536.721	-125.054	2.107
3200	108.055	701.298	603.589	312.668	-538.973	-111.739	1.824
3300	108.057	704.623	606.600	323.474	-541.137	-98.354	1.557
3400	108.059	707.848	609.531	334.279	-543.205	-84.905	1.304
3500	108.060	710.981	612.385	345.085	-545.169	-71.397	1.066
3600	108.062	714.025	615.166	355.891	-547.021	-57.834	0.839
3700	108.063	716.986	617.878	366.698	-548.759	-44.221	0.624
3800	108.064	719.868	620.525	377.504	-550.380	-30.563	0.420
3900	108.066	722.675	623.108	388.310	-551.882	-16.864	0.226
4000	108.067	725.411	625.631	399.117	-553.265	-3.128	0.041
4100	108.068	728.079	628.098	409.924	-554.529	10.641	-0.136
4200	108.069	730.683	630.509	420.731	-555.677	24.441	-0.304
4300	108.069	733.226	632.869	431.537	-556.710	38.265	-0.465
4400	108.070	735.711	635.178	442.344	-557.630	52.113	-0.619
4500	108.071	738.139	637.439	453.152	-558.441	65.980	-0.766
4600	108.072	740.515	639.654	463.959	-559.144	79.863	-0.907
4700	108.072	742.839	641.825	474.766	-559.745	93.762	-1.042
4800	108.073	745.114	643.953	485.573	-1121.182	119.289	-1.298
4900	108.074	747.343	646.040	496.380	-1121.238	145.133	-1.547
5000	108.074	749.526	648.088	507.188	-1121.218	170.977	-1.786
5100	108.075	751.666	650.098	517.995	-1121.123	196.821	-2.016
5200	108.075	753.765	652.072	528.803	-1120.957	222.661	-2.237
5300	108.076	755.823	654.010	539.610	-1120.720	248.497	-2.449
5400	108.076	757.844	655.914	550.418	-1120.416	274.329	-2.654
5500	108.076	759.827	657.786	561.225	-1120.045	300.154	-2.851
5600	108.077	761.774	659.625	572.033	-1119.610	325.973	-3.041
5700	108.077	763.687	661.434	582.841	-1119.112	351.782	-3.224
5800	108.077	765.567	663.213	593.649	-1118.553	377.582	-3.400
5900	108.078	767.414	664.964	604.456	-1117.934	403.372	-3.571
6000	108.078	769.231	666.687	615.264	-1117.257	429.152	-3.736

PREVIOUS: June 1975 (1 atm) CURRENT: June 1975 (1 bar)

Zirconium Iodide (ZrI_4) I_4Zr_1(g)

POTASSIUM (K) REFERENCE STATE A_r = 39.0983

0 to 336.35	K	crystal
336.35 to 1039.540	K	liquid
above 1039.540	K	ideal monatomic gas

Refer to the individual tables for details.

Enthalpy Reference Temperature = T_r = 298.15 K — Standard State Pressure = p° = 0.1 MPa

T/K	C_p° (J K⁻¹mol⁻¹)	S° (J K⁻¹mol⁻¹)	$-[G^\circ - H^\circ(T_r)]/T$ (J K⁻¹mol⁻¹)	$H^\circ - H^\circ(T_r)$ (kJ mol⁻¹)	$\Delta_f H^\circ$ (kJ mol⁻¹)	$\Delta_f G^\circ$ (kJ mol⁻¹)	Log K_f
0	0.	0.	INFINITE	-7.082	0.	0.	0.
100	24.631	35.603	89.101	-5.350	0.	0.	0.
200	27.004	53.493	67.270	-2.755	0.	0.	0.
298.15	29.497	64.670	64.670	0.	0.	0.	0.
300	29.573	64.853	64.671	0.055	0.	0.	0.
336.350	34.567	68.370	64.880	1.174	CRYSTAL <--> LIQUID		
336.350	32.139	75.310	64.880	3.508	TRANSITION		
400	31.499	80.825	66.993	5.533	0.	0.	0.
500	30.695	87.763	70.482	8.641	0.	0.	0.
600	30.136	93.306	73.840	11.680	0.	0.	0.
700	29.824	97.925	76.960	14.676	0.	0.	0.
800	29.757	101.901	79.834	17.653	0.	0.	0.
900	29.936	105.413	82.485	20.636	0.	0.	0.
1000	30.361	108.587	84.939	23.648	0.	0.	0.
1039.540	30.604	109.769	85.861	24.854	LIQUID <--> IDEAL GAS		
1039.540	20.786	186.300	85.861	104.410	FUGACITY = 1 bar		
1100	20.786	187.475	91.414	105.667	0.	0.	0.
1200	20.787	189.284	99.496	107.746	0.	0.	0.
1300	20.789	190.948	106.467	109.825	0.	0.	0.
1400	20.793	192.488	112.557	111.904	0.	0.	0.
1500	20.801	193.923	117.934	113.983	0.	0.	0.
1600	20.814	195.266	122.726	116.064	0.	0.	0.
1700	20.836	196.529	127.031	118.146	0.	0.	0.
1800	20.868	197.720	130.925	120.231	0.	0.	0.
1900	20.914	198.850	134.471	122.320	0.	0.	0.
2000	20.975	199.924	137.717	124.415	0.	0.	0.
2100	21.056	200.949	140.703	126.516	0.	0.	0.
2200	21.157	201.931	143.464	128.627	0.	0.	0.
2300	21.281	202.874	146.027	130.748	0.	0.	0.
2400	21.431	203.783	148.415	132.884	0.	0.	0.
2500	21.606	204.661	150.647	135.035	0.	0.	0.
2600	21.809	205.512	152.741	137.206	0.	0.	0.
2700	22.043	206.340	154.711	139.398	0.	0.	0.
2800	22.307	207.146	156.569	141.615	0.	0.	0.
2900	22.605	207.934	158.327	143.861	0.	0.	0.
3000	22.936	208.706	159.993	146.137	0.	0.	0.
3100	23.306	209.464	161.577	148.449	0.	0.	0.
3200	23.714	210.210	163.085	150.800	0.	0.	0.
3300	24.123	210.943	164.524	153.184	0.	0.	0.
3400	24.599	211.670	165.900	155.619	0.	0.	0.
3500	25.107	212.390	167.218	158.102	0.	0.	0.
3600	25.666	213.105	168.483	160.640	0.	0.	0.
3700	26.270	213.816	169.698	163.236	0.	0.	0.
3800	26.920	214.526	170.869	165.896	0.	0.	0.
3900	27.618	215.234	171.997	168.622	0.	0.	0.
4000	28.345	215.940	173.087	171.414	0.	0.	0.
4100	29.124	216.649	174.141	174.283	0.	0.	0.
4200	29.962	217.361	175.161	177.237	0.	0.	0.
4300	30.848	218.076	176.151	180.277	0.	0.	0.
4400	31.145	218.726	177.106	183.131	0.	0.	0.
4500	32.003	219.436	178.039	186.288	0.	0.	0.
4600	32.892	220.149	178.946	189.532	0.	0.	0.
4700	33.767	220.861	179.830	192.846	0.	0.	0.
4800	34.464	221.553	180.689	196.143	0.	0.	0.
4900	35.385	222.273	181.531	199.635	0.	0.	0.
5000	36.319	222.997	182.353	203.220	0.	0.	0.
5100	37.263	223.725	183.157	206.899	0.	0.	0.
5200	38.212	224.458	183.944	210.673	0.	0.	0.
5300	39.163	225.195	184.715	214.542	0.	0.	0.
5400	40.111	225.936	185.472	218.506	0.	0.	0.
5500	41.052	226.681	186.214	222.564	0.	0.	0.
5600	41.980	227.429	186.944	226.716	0.	0.	0.
5700	42.477	228.111	187.652	230.615	0.	0.	0.
5800	43.337	228.857	188.356	234.906	0.	0.	0.
5900	44.172	229.605	189.049	239.281	0.	0.	0.
6000	44.978	230.354	189.731	243.739	0.	0.	0.

PREVIOUS: December 1961 (1 atm) CURRENT: December 1961 (1 bar)

POTASSIUM (K) CRYSTAL $A_r = 39.0983$

$S°(298.15\ K) = 64.67\ J\ K^{-1}\ mol^{-1}$

$T_{fus} = 336.35\ K$

$\Delta_f H°(0\ K) = 0\ kJ\ mol^{-1}$

$\Delta_f H°(298.15\ K) = 0\ kJ\ mol^{-1}$

$\Delta_{fus} H° = 2.334\ kJ\ mol^{-1}$

Enthalpy of Formation

Zero by definition.

Heat Capacity and Entropy

Low temperature heat capacity measurements have been reported by Roberts (1.5-20 K, 1), Krier, Craig, and Wallace (12-320 K, 2), Dauphinee, Martin, and Preston-Thomas (20-330 K, 3), Simon and Zeidler (15-300 K, 4), and Eastman and Rodebush, (70-290 K, 5). A smooth curve was drawn through a large scale plot of the data, giving the most weight to the first three references. Above 300 K, the heat capacity data were adjusted to join smoothly with the enthalpy measurements of Douglas et al. (6).

Fusion Data

The enthalpy of fusion is from the enthalpy measurements of Douglas et al. (6).

Sublimation Data

See the K(g) and $K_2(g)$ tables for details.

References

1. L. M. Roberts, Proc. Phys. Soc. (London) 70B, 744 (1957).
2. C. A. Krier, R. S. Craig, and W. E. Wallace, J. Phys. Chem. 61, 522 (1957).
3. T. M. Dauphinee, D. L. Martin, and H. Preston-Thomas, Proc. Roy. Soc. (London) A233, 214 (1955).
4. F. Simon and W. Zeidler, Z. Physik. Chem. 123, 383 (1926).
5. E. D. Eastman and W. H. Rodebush, J. Am. Chem. Soc. 40, 489 (1918).
6. T. B. Douglas, A. F. Ball, D. C. Ginnings, and W. D. Davis, J. Am. Chem. Soc. 74, 2472 (1952).

Potassium (K) $K_1(cr)$

Enthalpy Reference Temperature = T_r = 298.15 K Standard State Pressure = $p°$ = 0.1 MPa

T/K	$C_p°$	$S°$	$-[G°-H°(T_r)]/T$	$H°-H°(T_r)$	$\Delta_f H°$	$\Delta_f G°$	Log K_f
	J K⁻¹mol⁻¹			kJ mol⁻¹			
0	0.	0.	INFINITE	-7.082	0.	0.	0.
100	24.631	35.603	89.101	-5.350	0.	0.	0.
200	27.004	53.493	67.270	-2.755	0.	0.	0.
250	28.008	59.623	65.146	-1.381	0.	0.	0.
298.15	29.497	64.670	64.670	0.	0.	0.	0.
300	29.573	64.853	64.671	0.055	0.	0.	0.
336.350	34.567	68.370	64.880	1.174	------ CRYSTAL <--> LIQUID -----		
350	38.484	69.819	65.044	1.671	-2.275	0.094	-0.014
400	52.287	75.858	66.008	3.940	-1.593	0.394	-0.051

PREVIOUS: CURRENT: December 1961

Potassium (K) $K_1(cr)$

POTASSIUM (K) LIQUID A_r = 39.0983

$S°(298.15\ K) = [71.403]\ J\ K^{-1}\ mol^{-1}$

$T_{fus} = 336.35\ K$

$\Delta_f H°(298.15\ K) = [2.270]\ kJ\ mol^{-1}$

$\Delta_{fus} H° = 2.334\ kJ\ mol^{-1}$

Enthalpy of Formation

$\Delta_f H°$(K, ℓ, 298.15 K) is calculated from that of the crystal by adding the enthalpy of fusion and the difference in enthalpy, H°(336.35 K) - H°(298.15 K), between the crystal and liquid.

Heat Capacity and Entropy

The liquid heat capacity values are derived from the enthalpy measurements of Douglas et al. (1). The heat capactiy curve was extropolated to 298.15 K and 200 K in a reasonable manner.

The entropy is calculated in a manner analogous to that used for the enthalpy of formation.

Vaporization Data

Potassium vaporizes to a mixture of monatomic and diatomic gas. The total vapor pressure reaches 1 atm. at 1037 K. The vapor composition at this temperature is 94.5 mole % monatomic and 5.5 mole % diatomic gas. The enthalpy of vaporization of 39.1 grams of liquid potassium to this equilibrium mixture is 18.38 kcal.

Reference

1. T. B. Douglas, A. F. Ball, D. C. Ginnings, and W. D. Davis, J. Am. Chem. Soc. 74, 2472 (1952).

Potassium (K) $K_1(l)$

Enthalpy Reference Temperature = T_r = 298.15 K — Standard State Pressure = p° = 0.1 MPa

T/K	$C_p°$ (J K⁻¹mol⁻¹)	S° (J K⁻¹mol⁻¹)	$-[G°-H°(T_r)]/T$ (J K⁻¹mol⁻¹)	$H°-H°(T_r)$ (kJ mol⁻¹)	$\Delta_f H°$ (kJ mol⁻¹)	$\Delta_f G°$ (kJ mol⁻¹)	Log K_f
0							
100							
200							
250							
298.15	32.702	71.403	71.403	0.	2.270	0.263	-0.046
300	32.677	71.605	71.404	0.060	2.276	0.250	-0.044
336.350	32.139	75.310	71.630	1.238	------ CRYSTAL <--> LIQUID -----		
350	31.993	76.586	71.799	1.676	0.	0.	0.
400	31.499	80.825	72.669	3.263	0.	0.	0.
450	31.067	84.510	73.784	4.827	0.	0.	0.
500	30.695	87.763	75.023	6.370	0.	0.	0.
600	30.136	93.306	77.623	9.410	0.	0.	0.
700	29.824	97.925	80.203	12.406	0.	0.	0.
800	29.757	101.901	82.672	15.383	0.	0.	0.
900	29.936	105.413	85.007	18.365	0.	0.	0.
1000	30.361	108.587	87.209	21.378	0.	0.	0.
1039.540	30.604	109.769	88.045	22.583	------- FUGACITY = 1 bar -------		
1100	31.031	111.511	89.287	24.446	-78.951	4.610	-0.219
1200	31.882	114.247	91.254	27.591	-77.884	12.160	-0.529
1300	32.803	116.835	93.123	30.825	-76.729	19.617	-0.788
1400	33.723	119.300	94.906	34.152	-75.482	26.982	-1.007
1500	34.644	121.658	96.611	37.570	-74.143	34.255	-1.193

PREVIOUS: CURRENT: December 1961

Potassium (K) $K_1(l)$

POTASSIUM (K) CRYSTAL-LIQUID A_r = 39.0983

0 to 336.35 K crystal
above 336.35 K liquid

Refer to the individual tables for details.

Potassium (K) K_1(cr,l)

Enthalpy Reference Temperature = T_r = 298.15 K Standard State Pressure = p° = 0.1 MPa

T/K	C_p° (J K⁻¹mol⁻¹)	S° (J K⁻¹mol⁻¹)	$-[G°-H°(T_r)]/T$ (J K⁻¹mol⁻¹)	$H°-H°(T_r)$ (kJ mol⁻¹)	$\Delta_f H°$ (kJ mol⁻¹)	$\Delta_f G°$ (kJ mol⁻¹)	Log K_f
0	0.	0.	INFINITE	-7.082	0.	0.	0.
100	24.631	35.603	89.101	-5.350	0.	0.	0.
200	27.004	53.493	67.270	-2.755	0.	0.	0.
250	28.008	59.623	65.146	-1.381	0.	0.	0.
298.15	29.497	64.670	64.670	0.	0.	0.	0.
300	29.573	64.853	64.671	0.055	0.	0.	0.
336.350	34.567	68.370	64.880	1.174	CRYSTAL <--> LIQUID	TRANSITION	
336.350	32.139	75.310	64.880	3.508			
350	31.993	76.586	65.312	3.946	0.	0.	0.
400	31.499	80.825	66.993	5.533	0.	0.	0.
450	31.067	84.510	68.739	7.097	0.	0.	0.
500	30.695	87.763	70.482	8.641	0.	0.	0.
600	30.136	93.306	73.840	11.680	0.	0.	0.
700	29.824	97.925	76.960	14.676	0.	0.	0.
800	29.757	101.901	79.834	17.653	0.	0.	0.
900	29.936	105.413	82.485	20.636	0.	0.	0.
1000	30.361	108.587	84.939	23.648	0.	0.	0.
1039.540	30.604	109.769	85.861	24.854	------- FUGACITY = 1 bar -------		
1100	31.031	111.511	87.223	26.716	-78.951	4.610	-0.219
1200	31.882	114.247	89.362	29.862	-77.884	12.160	-0.529
1300	32.803	116.835	91.377	33.096	-76.729	19.617	-0.788
1400	33.723	119.300	93.284	36.422	-75.482	26.982	-1.007
1500	34.644	121.658	95.098	39.840	-74.143	34.255	-1.193

PREVIOUS: CURRENT: December 1961

Potassium (K) K_1(cr,l)

POTASSIUM (K) IDEAL GAS A_r = 39.0983

IP(K, g) = 35009.78 ± 0.02 cm^{-1}
S°(298.15 K) = 160.340 ± 0.025 J K^{-1} mol^{-1}

$\Delta_f H°$(0 K) = 89.9 ± 0.4 kJ mol^{-1}
$\Delta_f H°$(298.15 K) = 89.0 ± 0.4 kJ mol^{-1}

Electronic Levels and Quantum Weights

State	ε_i, cm^{-1}	g_i
$^2S_{1/2}$	0	2
$^2P_{1/2}$	12985.170	2
$^2P_{3/2}$	13042.876	4
$^2S_{1/2}$	21026.551	2
.	.	.
$^2P_{3/2}$	34991.2	4
IP	35009.78	

Enthalpy of Formation

The enthalpy of formation of potassium gas, $\Delta_f H°$(298.15 K) = 89.0 ± 0.4 kJ mol^{-1}, is chosen to be the value recommended by CODATA (1). This value was obtained from 2nd and 3rd law treatments of vapor pressure data. The reported vapor pressures were corrected for the effects of K_2(g), using D_0^0= 49.01 kJ mol^{-1}. At higher pressures an alternate virial treatment was employed.

As stated by CODATA (1), the selected value is the weighted average of the 3rd law values of $\Delta_{sub}H°$(298.15 K) calculated from the following studies:

Source	$\Delta_{sub}H°$(298.15 K), kJ mol^{-1}	Source	$\Delta_{sub}H°$(298.15 K), kJ mol^{-1}
Haycock and Lamplough (2)	89.31 ± 0.65	Venogradov and Volyak (13)	89.27 ± 0.65
Fiock and Rodenbush (3)	89.41 ± 0.65	Achener (14)	88.97 ± 0.69
Edmondson and Egerton (4)	89.71 ± 0.25	Ahpil'rain et al. (15)	88.94 ± 0.47
Weiler (5)	89.71 ± 0.65	Bowles (16)	88.50 ± 1.35
Mayer (6)	88.58 ± 1.00	Gerasimov et al. (17)	89.43 ± 0.50
Neumann and volker (7)	89.56 ± 0.25	Sphil'rain and Nikanorov (18)	88.70 ± 0.85
Walling (8)	88.59 ± 0.75	Schins et al. (19)	88.94 ± 0.50
Makansi et al. (9)	88.73 ± 0.72	Bonilla (20)	88.75 ± 0.65
Buck and Pauly (10)	89.72 ± 0.35	Roeder and Morawitz (21)	89.59 ± 0.38
Stone et al. (11)	88.81 ± 0.90	Kondrat'ev (22)	88.27 ± 0.55
Rigney et al. (12)	88.90 ± 0.93		

The 2nd law calculations from the above sources led to somewhat more positive values.

Heat Capacity and Entropy

The thermal functions for the five alkali metal monatomic gases are calculated by the same procedure. Observed and estimated atomic energy levels are included in the partition function calculation, using an ionization potential lowering (IP-kT) technique as the cutoff procedure in the energy level summation (24).

The lowest lying levels for these metals [14904 cm^{-1} (Li), 16956 cm^{-1} (Na), 12985 cm^{-1} (K), 12578 cm^{-1} (Cs)] do not contribute to the thermal functions below ~1000 K; there is only a translational contribution below this temperature. Above this approximate temperature, the thermal functions become increasingly sensitive to the partition function cutoff procedure used, due to the combined effect of the observation of atomic energy levels of high principal quantum number and a low ionization potential. Energy levels have been observed up to n = 42 (Li), 59(Na), 79(K), 77(Rb), and 73(Cs). However, not all predicted levels have been observed for each of these principal quantum numbers. The ionization potentials vary from 43487.29 cm^{-1} for Li to 31406.1 cm^{-1} for Cs. In calculating the thermal functions with the inclusion of missing levels up to the high principal quantum numbers just mentioned, the Gibbs energy functions show significant differences (depending on the cutoff procedure) above 3000 K (24).

The atomic energy levels have been taken from the compilation of Corliss and Sugar (14). Our calculated values for the thermal functions are similar to those recommended by CODATA (1). They do differ for two reasons, however. First, the entropy differs by 0.1094 J K^{-1} mol^{-1} because this table uses a reference pressure of 1 bar, whereas the CODATA recommendations are based on 1 atm. Second, the entropies at 298.15 K for all alkali metal gases differs by ~0.004 J K^{-1} mol^{-1}, presumably due to the use of slightly different values for auxiliary data.

References

1. J. D. Cox, chairman, CODATA Task Group on Key Values for Thermodynamics, J. Chem. Thermodynamics 10, 903 (1978).
2. C. T. Heycock and F. E. E. Lamplough, proc. Chem. Soc. 28, 3 (1912).
3. E. F. Fiock and W. H. Rodebush, J. Am. Chem. Soc. 48, 2522 (1926).
4. W. Edmondson and C. A. Egerton, Proc. Roy. Soc. (London) A113, 520 (1927).
5. J. Weiler, Ann. Physik [5] 1, 361, 33 (1932).
6. H. Mayer, Z. Physik 69, 786 (1931).
7. K. Neumann and E. Volker, Z. physik. Chem. 161, 33 (1932).
8. J. F. Walling, J. Phys. Chem. 67, 1380 (1963).
9. M. M. Makansi, M. Madsen, et al., J. Phys. Chem. 60, 128 (1956).
10. U. Buck and H. Pauly, Z. physik. Chem. (Frankfurt) 44, 345 (1965).
11. J. P. Stone, C. P. Ewing, et al., J. Chem. Eng. Data 11, 315 (1966).
12. D. V. Rigney, S. M. Kapelner, and r. E. Cleary, U. S. AEC Rept. TIM-810 (1965).
13. Yu. K. Vinogradov and L. D. volyak, Teplofiz. vys. Temp. 4, 50 (1966).
14. P. Y. Achneer, U. S. AEC Rept. AGN-8141, 1965.
15. E. E.Shpil'rain, E. E. Totsku, and V. A. Sherishevsku, Teplofiz. Vep. Temp. 6, 924 (1968).
16. K. J. Bowles, NASA Tech. Note D-4535, 1968.
17. Ya. I Gerasimov, G. F. Voronin and N. T. Shiu, J. Chem. Thermodynamics 1, 425 (1969,; Zhur. Fiz. Khim. 41, 1468 (1968).
18. E. E. Shpil'rain and E. V. Nikanorov, Teplofiz. Svoistva Gasov Mater. Vsis. Teplofiz. Konf. Svoistvam. Vishchister Vyz. Temp. (1968), Mosk. "Nauka", 141 (1970).
19. H. E. J. Schins, R. W. M. van Wijk and b. Dorpema, Z. Metallkunde 62, 330 (1971).
20. C. F. Bonilla, U. S. AEC Rept. ORNL - 3605, 1, 286 (1964).
21. A. Roeder and W. Morawitz, Z. Elektrochem. 60, 431 (1956).
22. N. S. Kondrat'ev and I. F. Parfent'eva, Teplo-massoperenmos v odno-z-dvukhfaznykh sudakh, Moskov, "Nauke", 19 (1971).
23. C. Corliss and J. Sugar, J. Phys. Chem. Ref. Data 8, 1109 (1979).
24. J. R. Downey, Jr., The Dow Chemical Company, Rept. AFOSR-TR-78-0960, Contract No. F44620-75-0048, 1978.

Potassium (K) K_1(g)

Enthalpy Reference Temperature = T_r = 298.15 K — Standard State Pressure = p° = 0.1 MPa

T/K	$C_p°$ (J $K^{-1}mol^{-1}$)	S° (J $K^{-1}mol^{-1}$)	$-[G°-H°(T_r)]/T$ (J $K^{-1}mol^{-1}$)	$H°-H°(T_r)$ (kJ mol^{-1})	$\Delta_f H°$ (kJ mol^{-1})	$\Delta_f G°$ (kJ mol^{-1})	Log K_f
0	0.	0.	INFINITE	-6.197	89.885	89.885	INFINITE
100	20.786	137.632	178.820	-4.119	90.231	80.028	-41.802
200	20.786	152.040	162.241	-2.040	89.715	70.006	-18.284
250	20.786	156.678	160.682	-1.001	89.380	65.116	-13.605
298.15	20.786	160.340	160.340	0.	89.000	60.476	-10.595
300	20.786	160.468	160.340	0.038	88.984	60.299	-10.499
350	20.786	163.672	160.593	1.078	86.132	55.652	-8.306
400	20.786	166.448	161.155	2.117	85.584	51.335	-6.704
450	20.786	168.896	161.882	3.156	85.060	47.086	-5.466
500	20.786	171.086	162.695	4.196	84.555	42.894	-4.481
600	20.786	174.876	164.419	6.274	83.594	34.652	-3.017
700	20.786	178.080	166.147	8.353	82.677	26.568	-1.983
800	20.786	180.856	167.816	10.431	81.778	18.614	-1.215
900	20.786	183.304	169.404	12.510	80.875	10.773	-0.625
1000	20.786	185.494	170.905	14.589	79.940	3.034	-0.158
1039.540	20.786	186.300	171.476	15.411	------- FUGACITY = 1 bar -------		
1100	20.786	187.475	172.323	16.667	0.	0.	0.
1200	20.787	189.284	173.662	18.746	0.	0.	0.
1300	20.789	190.948	174.929	20.825	0.	0.	0.
1400	20.793	192.488	176.129	22.904	0.	0.	0.
1500	20.801	193.923	177.268	24.983	0.	0.	0.
1600	20.814	195.266	178.351	27.064	0.	0.	0.
1700	20.836	196.529	179.384	29.147	0.	0.	0.
1800	20.868	197.720	180.369	31.232	0.	0.	0.
1900	20.914	198.850	181.313	33.321	0.	0.	0.
2000	20.975	199.924	182.217	35.415	0.	0.	0.
2100	21.056	200.949	183.084	37.516	0.	0.	0.
2200	21.157	201.931	183.919	39.627	0.	0.	0.
2300	21.281	202.874	184.723	41.748	0.	0.	0.
2400	21.431	203.783	185.498	43.884	0.	0.	0.
2500	21.606	204.661	186.247	46.035	0.	0.	0.
2600	21.809	205.512	186.972	48.206	0.	0.	0.
2700	22.043	206.340	187.674	50.398	0.	0.	0.
2800	22.307	207.146	188.355	52.615	0.	0.	0.
2900	22.605	207.934	189.017	54.861	0.	0.	0.
3000	22.936	208.706	189.660	57.137	0.	0.	0.
3100	23.306	209.464	190.287	59.449	0.	0.	0.
3200	23.714	210.210	190.898	61.800	0.	0.	0.
3300	24.123	210.943	191.494	64.184	0.	0.	0.
3400	24.599	211.670	192.077	66.619	0.	0.	0.
3500	25.107	212.390	192.647	69.102	0.	0.	0.
3600	25.666	213.105	193.205	71.640	0.	0.	0.
3700	26.270	213.816	193.752	74.237	0.	0.	0.
3800	26.920	214.526	194.290	76.896	0.	0.	0.
3900	27.618	215.234	194.818	79.622	0.	0.	0.
4000	28.345	215.940	195.337	82.414	0.	0.	0.
4100	29.124	216.649	195.848	85.283	0.	0.	0.
4200	29.962	217.361	196.352	88.237	0.	0.	0.
4300	30.848	218.076	196.849	91.277	0.	0.	0.
4400	31.145	218.726	197.333	94.131	0.	0.	0.
4500	32.003	219.436	197.816	97.288	0.	0.	0.
4600	32.892	220.149	198.294	100.532	0.	0.	0.
4700	33.767	220.861	198.766	103.847	0.	0.	0.
4800	34.464	221.553	199.231	107.143	0.	0.	0.
4900	35.385	222.273	199.694	110.635	0.	0.	0.
5000	36.319	222.997	200.153	114.220	0.	0.	0.
5100	37.263	223.725	200.608	117.899	0.	0.	0.
5200	38.212	224.458	201.059	121.673	0.	0.	0.
5300	39.163	225.195	201.508	125.542	0.	0.	0.
5400	40.111	225.936	201.953	129.506	0.	0.	0.
5500	41.052	226.681	202.396	133.564	0.	0.	0.
5600	41.980	227.429	202.837	137.716	0.	0.	0.
5700	42.477	228.111	203.266	141.615	0.	0.	0.
5800	43.337	228.857	203.701	145.906	0.	0.	0.
5900	44.172	229.605	204.134	150.281	0.	0.	0.
6000	44.978	230.354	204.564	154.739	0.	0.	0.

PREVIOUS: June 1962 (1 atm) CURRENT: December 1983 (1 bar)

Potassium (K) K_1(g)

POTASSIUM, ION (K^+) IDEAL GAS M_r = 39.09775

$IP(K^+, g) = 255100 \pm 100\ cm^{-1}$
$S°(298.15\ K) = 154.576 \pm 0.025\ J\ K^{-1}\ mol^{-1}$

$\Delta_fH°(0\ K) = 508.7 \pm 0.4\ kJ\ mol^{-1}$
$\Delta_fH°(298.15\ K) = [514.007]\ kJ\ mol^{-1}$

Electronic Level and Quantum Weight

State	ε_i, cm^{-1}	g_i
$^1S_0$	0	1

Enthalpy of Formation

$\Delta_fH°(K^+, g, 0\ K)$ is calculated from $\Delta_fH°(K, g, 0\ K)$ (1) using the spectroscopic value of $IP(K) = 35009.78 \pm 0.02\ cm^{-1}$ ($418.810 \pm 0.000\ kJ\ mol^{-1}$) from Corliss and Sugar. The ionization limit is converted from cm^{-1} to $kJ\ mol^{-1}$ using the factor, $1\ cm^{-1} = 0.01196266\ kJ\ cm^{-1}$, which is derived from the 1973 CODATA fundamental constants (3). Rosenstock et al. (4) and Levin and Lias (5) have summarized additional ionization and appearance potential data.

$\Delta_fH°(K^+, g, 298.15\ K)$ is calculated from $\Delta_fH°(K, g, 0\ K)$ by using IP(K) with JANAF (1) enthalpies, H°(0 K) - H°(298.15 K), for K(g), $K^+(g)$, and $e^-(g)$. $\Delta_fH°(K + K^+ + e^-, 298.15\ K)$ differs from a room temperature threshold energy due to inclusion of these enthalpies and to threshold effects discussed by Rosenstock et al. (4). $\Delta_fH°(298.15\ K)$ should be changed by $-6.197\ kJ\ mol^{-1}$ if it is to be used in the ion convention that excludes the enthalpy of the electron.

Heat Capacity and Entropy

The information on electronic energy levels and quantum weights, given by Corliss and Sugar (2), is incomplete because many theoretically predicted levels have not been oberved. Our calculations indicate that any reasonable method of filling in these missing levels and cutting off the summation in the partition function (6) has no effect on the thermodynamic functions to 6000 K. This is a result of the high energy of all levels other than the ground state; the first excited state is approximately 162507 cm^{-1} above the ground state. Since inclusion of these excited states has no effect on the thermodynamic functions (to 6000 K), we list only the ground state. The reported uncertainty in S°(298.15 K) is due to uncertainties in the relative ionic mass and the fundamental constants. Extension of these calculations above 6000 K may require consideration of the excited states and use of different fill and cutoff procedures (6).

References

1. JANAF Thermochemical Tables: K(g), 12-31-83; $e^-(g)$, 3-31-82.
2. C. Corliss and J. Sugar, J. Phys. Chem. Ref. Data 8(4), 1109-1145 (1979).
3. E. R. Cohen and B. N. Taylor, J. Phys. Chem. Ref. Data 2, 663 (1973).
4. H. M. Rosenstock, K. Draxl et al., J. Phys. Chem. Ref. Data 6, Supp. 1 (1977).
5. R. D. Levin and S. G. Lias, U. S. Nat. Bur. Stand., NSRDS-NBS-71, 1982.
6. J. R. Downey, Jr., The Dow Chemical Company, AFOSR-TR-78-0960, Contract No. F4460-75-1-0048, 1978.

Potassium, Ion (K^+) $K_1^+(g)$

Enthalpy Reference Temperature = T_r = 298.15 K   Standard State Pressure = p° = 0.1 MPa

T/K	$C_p°$ (J K⁻¹mol⁻¹)	S° (J K⁻¹mol⁻¹)	-[G°-H°(T_r)]/T (J K⁻¹mol⁻¹)	H°-H°(T_r) (kJ mol⁻¹)	$\Delta_fH°$ (kJ mol⁻¹)	$\Delta_fG°$ (kJ mol⁻¹)	Log K_f
0	0.	0.	INFINITE	-6.197	508.695		
100	20.786	131.869	173.057	-4.119			
200	20.786	146.277	156.478	-2.040			
250	20.786	150.915	154.919	-1.001			
298.15	20.786	154.576	154.576	0.	514.007	480.947	-84.260
300	20.786	154.705	154.577	0.038	514.030	480.742	-83.705
350	20.786	157.909	154.830	1.078	512.217	475.245	-70.926
400	20.786	160.685	155.392	2.117	512.709	469.930	-61.367
450	20.786	163.133	156.119	3.156	513.223	464.552	-53.924
500	20.786	165.323	156.932	4.196	513.758	459.116	-47.963
600	20.786	169.113	158.656	6.274	514.876	448.083	-39.009
700	20.786	172.317	160.384	8.353	516.037	436.860	-32.599
800	20.786	175.092	162.053	10.431	517.217	425.468	-27.780
900	20.786	177.541	163.641	12.510	518.392	413.929	-24.024
1000	20.786	179.731	165.142	14.589	519.536	402.260	-21.012
1100	20.786	181.712	166.560	16.667	441.675	395.089	-18.761
1200	20.786	183.520	167.899	18.746	443.753	390.762	-17.009
1300	20.786	185.184	169.165	20.824	445.832	386.262	-15.520
1400	20.786	186.725	170.365	22.903	447.910	381.602	-14.238
1500	20.786	188.159	171.504	24.982	449.988	376.793	-13.121
1600	20.786	189.500	172.588	27.060	452.064	371.845	-12.140
1700	20.786	190.760	173.620	29.139	454.139	366.768	-11.269
1800	20.786	191.948	174.605	31.217	456.211	361.569	-10.492
1900	20.786	193.072	175.548	33.296	458.279	356.255	-9.794
2000	20.786	194.138	176.451	35.375	460.342	350.832	-9.163
2100	20.786	195.153	177.318	37.453	462.398	345.305	-8.589
2200	20.786	196.120	178.151	39.532	464.445	339.682	-8.065
2300	20.786	197.044	178.952	41.610	466.480	333.965	-7.585
2400	20.786	197.928	179.724	43.689	468.502	328.160	-7.142
2500	20.786	198.777	180.470	45.768	470.508	322.271	-6.733
2600	20.786	199.592	181.190	47.846	472.494	316.302	-6.355
2700	20.786	200.376	181.886	49.925	474.459	310.258	-6.002
2800	20.786	201.132	182.560	52.004	476.399	304.141	-5.674
2900	20.786	201.862	183.213	54.082	478.311	297.955	-5.367
3000	20.786	202.566	183.846	56.161	480.192	291.703	-5.079
3100	20.786	203.248	184.461	58.239	482.037	285.390	-4.809
3200	20.786	203.908	185.059	60.318	483.844	279.017	-4.554
3300	20.786	204.548	185.640	62.397	485.617	272.589	-4.315
3400	20.786	205.168	186.205	64.475	487.339	266.107	-4.088
3500	20.786	205.771	186.755	66.554	489.013	259.576	-3.874
3600	20.786	206.356	187.292	68.632	490.632	252.997	-3.671
3700	20.786	206.926	187.815	70.711	492.193	246.375	-3.478
3800	20.786	207.480	188.325	72.790	493.691	239.711	-3.295
3900	20.786	208.020	188.823	74.868	495.122	233.008	-3.121
4000	20.786	208.546	189.310	76.947	496.487	226.269	-2.955
4100	20.786	209.060	189.785	79.025	497.775	219.497	-2.796
4200	20.786	209.560	190.250	81.104	498.978	212.695	-2.645
4300	20.786	210.050	190.705	83.183	500.095	205.866	-2.501
4400	20.786	210.527	191.150	85.261	501.399	198.984	-2.362
4500	20.786	210.995	191.586	87.340	502.399	192.099	-2.230
4600	20.786	211.451	192.013	89.418	503.312	185.194	-2.103
4700	20.786	211.898	192.431	91.497	504.155	178.267	-1.981
4800	20.786	212.336	192.841	93.576	505.016	171.311	-1.864
4900	20.786	212.765	193.243	95.654	505.680	164.352	-1.752
5000	20.786	213.185	193.638	97.733	506.253	157.380	-1.644
5100	20.786	213.596	194.025	99.811	506.731	150.398	-1.540
5200	20.786	214.000	194.406	101.890	507.114	143.407	-1.441
5300	20.786	214.396	194.779	103.969	507.403	136.410	-1.344
5400	20.786	214.784	195.146	106.047	507.596	129.408	-1.252
5500	20.786	215.166	195.506	108.126	507.695	122.403	-1.162
5600	20.786	215.540	195.861	110.204	507.701	115.398	-1.076
5700	20.786	215.908	196.209	112.283	507.959	108.345	-0.993
5800	20.786	216.270	196.552	114.362	507.825	101.336	-0.913
5900	20.786	216.625	196.889	116.440	507.607	94.329	-0.835
6000	20.786	216.974	197.221	118.519	507.306	87.327	-0.760

PREVIOUS: March 1965 (1 atm)   CURRENT: December 1983 (1 bar)

Potassium, Ion (K^+) $K_1^+(g)$

POTASSIUM, ION (K^-) IDEAL GAS M_r = 39.09885

EA(K, g) = 0.50147 ± 0.00010 eV
S°(298.15 K) = 154.577 ± 0.005 J K^{-1} mol^{-1}

$\Delta_f H°$(0 K) = 41.5 ± 1 kJ mol^{-1}
$\Delta_f H°$(298.15 K) = [34.419] kJ mol^{-1}

Electonic Level and Quantum Weight

State	ε_i, cm^{-1}	g_i
$^1S_0$	0.0	1

Enthalpy of Formation

The enthalpy of formation, at 0 K, for K^-(g) is calculated from the adopted electron affinity, EA(K, g) = 0.50147 ± 0.00010 eV (48.384±0.010 kJ mol^{-1}). This value, recommended by Hotop and Lineberger (6), was measured by a tunable laser photodetachment threshold technique (2). Additional discussion on K^-(g) may be obtained in the critical discussions of Hotop and Lineberger (1, 6), Rosenstock et al. (3), and Massey (4).

$\Delta_f H°(K^-$, g, 298.15 K) is obtained from $\Delta_f H°$(K, g, 0 K) by using EA(K, g) with JANAF (5) enthalpies, H°(0 K) - H°(298.15 K), for K^-(g), K(g), and e^-(g). $\Delta_r H°(K^- \rightarrow K + e^-$, 298.15 K) differs from a room-temperature threshold energy due to inclusion of these enthalpies and to threshold effects discussed by Rosenstock et al. (3). $\Delta_f H°$(298.15 K) should be changed by +6.197 kJ mol^{-1} if it is to be used in the ion convention which excludes the enthalpy of the electron.

Capacity and Entropy

The ground state configuration for K^-(g) is given by Hotop and Lineberger (1, 6) and Rosenstock et al. (3). Lacking any experimental evidence as to the stability of any excited states, we assume that no stable excited states exist.

References

1. H. Hotop and W. C. Lineberger, J. Phys. Chem. Ref. Data 4, 539 (1975).
2. J. Slater, F. H. Read, S. E. Novick, and W. C. Lineberger, Phys. Rev. A 17, 201 (1978).
3. H. M. Rosenstock, K. Draxl et al., J. Phys. Chem. Ref. Data 6, Supp. 1 (1977).
4. H. S. W. Massey, "Negative Ions," 3rd ed., Cambridge University Press, Cambridge, 1976.
5. JANAF Thermochemical Tables: K(g), 12-31-83; e^-(g), 3-31-82.
6. H. Hotop and W. C. Lineberger, J. Phys. Chem. Ref. Data, to be published, 1985.

Potassium, Ion (K^-) K_1^-(g)

Enthalpy Reference Temperature = T_r = 298.15 K Standard State Pressure = p° = 0.1 MPa

T/K	$C_p°$ (J $K^{-1}mol^{-1}$)	S° (J $K^{-1}mol^{-1}$)	-[G°-H°(T_r)]/T (J $K^{-1}mol^{-1}$)	H°-H°(T_r) (kJ mol^{-1})	$\Delta_f H°$ (kJ mol^{-1})	$\Delta_f G°$ (kJ mol^{-1})	Log K_f
0	0.	0.	INFINITE	-6.197	41.501		
100	20.786	131.869	173.057	-4.119			
200	20.786	146.277	156.478	-2.040			
250	20.786	150.916	154.919	-1.001			
298.15	20.786	154.577	154.577	0.	34.419	13.868	-2.430
300	20.786	154.705	154.577	0.038	34.364	13.740	-2.392
350	20.786	157.909	154.830	1.078	30.473	10.519	-1.570
400	20.786	160.685	155.392	2.117	28.886	7.777	-1.016
450	20.786	163.133	156.119	3.156	27.322	5.232	-0.607
500	20.786	165.323	156.932	4.196	25.778	2.861	-0.299
600	20.786	169.113	158.656	6.274	22.738	-1.437	0.125
700	20.786	172.317	160.385	8.353	19.743	-5.228	0.390
800	20.786	175.093	162.053	10.431	16.766	-8.592	0.561
900	20.786	177.541	163.641	12.510	13.783	-11.583	0.672
1000	20.786	179.731	165.142	14.589	10.770	-14.241	0.744
1100	20.786	181.712	166.560	16.667	-71.249	-11.984	0.569
1200	20.786	183.521	167.899	18.746	-73.327	-6.504	0.283
1300	20.786	185.185	169.166	20.824	-75.406	-0.851	0.034
1400	20.786	186.725	170.366	22.903	-77.485	4.961	-0.185
1500	20.786	188.159	171.505	24.982	-79.565	10.923	-0.380
1600	20.786	189.501	172.588	27.060	-81.645	17.023	-0.556
1700	20.786	190.761	173.620	29.139	-83.728	23.254	-0.715
1800	20.786	191.949	174.606	31.217	-85.813	29.607	-0.859
1900	20.786	193.073	175.548	33.296	-87.902	36.076	-0.992
2000	20.786	194.139	176.451	35.375	-89.996	42.656	-1.114
2100	20.786	195.153	177.318	37.453	-92.098	49.340	-1.227
2200	20.786	196.120	178.151	39.532	-94.208	56.125	-1.333
2300	20.786	197.044	178.952	41.610	-96.330	63.005	-1.431
2400	20.786	197.929	179.725	43.689	-98.465	69.979	-1.523
2500	20.786	198.777	180.470	45.768	-100.617	77.041	-1.610
2600	20.786	199.592	181.190	47.846	-102.787	84.190	-1.691
2700	20.786	200.377	181.886	49.925	-104.980	91.423	-1.769
2800	20.786	201.133	182.560	52.004	-107.197	98.738	-1.842
2900	20.786	201.862	183.213	54.082	-109.442	106.132	-1.912
3000	20.786	202.567	183.847	56.161	-111.719	113.605	-1.978
3100	20.786	203.248	184.462	58.239	-114.030	121.153	-2.041
3200	20.786	203.908	185.059	60.318	-116.381	128.777	-2.102
3300	20.786	204.548	185.640	62.397	-118.765	136.475	-2.160
3400	20.786	205.168	186.205	64.475	-121.201	144.246	-2.216
3500	20.786	205.771	186.756	66.554	-123.683	152.089	-2.270
3600	20.786	206.357	187.292	68.632	-126.221	160.004	-2.322
3700	20.786	206.926	187.815	70.711	-128.818	167.990	-2.372
3800	20.786	207.480	188.325	72.790	-131.477	176.047	-2.420
3900	20.786	208.020	188.823	74.868	-134.204	184.176	-2.467
4000	20.786	208.547	189.310	76.947	-136.995	192.374	-2.512
4100	20.786	209.060	189.785	79.025	-139.865	200.643	-2.556
4200	20.786	209.561	190.250	81.104	-142.819	208.984	-2.599
4300	20.786	210.050	190.705	83.183	-145.859	217.396	-2.641
4400	20.786	210.528	191.150	85.261	-148.712	225.852	-2.681
4500	20.786	210.995	191.586	87.340	-151.869	234.401	-2.721
4600	20.786	211.452	192.013	89.418	-155.114	243.020	-2.760
4700	20.786	211.899	192.431	91.497	-158.428	251.709	-2.797
4800	20.786	212.336	192.841	93.576	-161.724	260.457	-2.834
4900	20.786	212.765	193.244	95.654	-165.217	269.289	-2.871
5000	20.786	213.185	193.638	97.733	-168.802	278.192	-2.906
5100	20.786	213.596	194.026	99.811	-172.481	287.169	-2.941
5200	20.786	214.000	194.406	101.890	-176.254	296.218	-2.976
5300	20.786	214.396	194.779	103.969	-180.123	305.341	-3.009
5400	20.786	214.785	195.146	106.047	-184.087	314.537	-3.043
5500	20.786	215.166	195.507	108.126	-188.145	323.808	-3.075
5600	20.786	215.541	195.861	110.204	-192.297	333.154	-3.108
5700	20.786	215.908	196.210	112.283	-196.196	342.527	-3.139
5800	20.786	216.270	196.552	114.362	-200.487	352.015	-3.170
5900	20.786	216.625	196.890	116.440	-204.863	361.579	-3.201
6000	20.786	216.975	197.221	118.519	-209.320	371.217	-3.232

PREVIOUS: CURRENT: December 1983 (1 bar)

Potassium, Ion (K^-) K_1^-(g)

POTASSIUM OXIDE (KO) IDEAL GAS M_r = 55.0977

$S°(298.15\ K) = [238.027 \pm 2.1]\ J\ K^{-1}\ mol^{-1}$

$\Delta_f H°(0\ K) = [72.89 \pm 41.8]\ kJ\ mol^{-1}$
$\Delta_f H°(298.15\ K) = [71.13 \pm 41.8]\ kJ\ mol^{-1}$

Electronic Level and Quantum Weight

State	ε_i, cm^{-1}	g_i
$[^2\Pi]$	0	[4]

$\omega_e = [350]\ cm^{-1}$ $\omega_e x_e = [2.2]\ cm^{-1}$ $\sigma = 1$
$B_e = [0.3068]\ cm^{-1}$ $\alpha_e = [0.0027]\ cm^{-1}$ $r_e = [2.2]$ Å

Enthalpy of Formation

Brewer and Margrave (1) suggested that the $\Delta_f H°$ value for MO(g) from the gaseous ions can be taken as the mean of the corresponding $\Delta_f H°$ values for the alkali fluoride and chloride with a maximum uncertainty of 10 kcal mol^{-1}. We obtain the values $\Delta_r H°(0\ K) = -137.94$ and -117.7 kcal mol^{-1} for the reactions $K^+(g) + F^-(g) = KF(g)$ and $K^+(g) + Cl^-(g) = KCl(g)$, respectively. Assuming $\Delta_r H°(0\ K) = -127.83$ kcal mol^{-1} for the reaction $K^+(g) + O^-(g) = KO(g)$, we calculate $\Delta_f H°$(KO, g, 298.15 K) = 19.0 kcal mol^{-1} with the JANAF auxiliary values for the gaseous ions.

Based on the assumption that $D_0^0(KO) = 1/2[D_0^0(K_2) + D_0^0(O_2)]$, or $\Delta_r H°(0\ K) = 64.9$ kcal mol^{-1} for the reaciton KO(g) = K(g) + O(g), we evaluate $\Delta_f H°$(KO, g, 0 K) = 15.6 kcal mol^{-1} or $\Delta_f H°$(KO, g, 298.15 K) = 15.2 kcal mol^{-1}.

According to Somayajulu (2), in a sequence of similar diatomic molecules, $k_e r_e / D_0^0$ = constant, where k_e = force constant and r_e = equilibrium bond distance. Using r_e = 1.62 Å, D_0^0 = 77.9 kcal mol^{-1} for LiO(g), r_e = 2.2 Å and $k_e(KO)/k_e(LiO) = 0.3015$, we calculate D_0^0 = 55 kcal mol^{-1} for KO(g), yielding $\Delta_f H°$(KO, g, 298.15 K) = 26.1 kcal mol^{-1}. The value k_e is derived from the relation $k_e = 4\pi^2 \nu^2 \mu$ where ν is the vibrational frequency for the diatomic molecule and μ is the reduced mass.

The value of $\Delta_f H°$(KO, g, 298.15 K) is tentatively adopted as 17±10 kcal mol^{-1}.

Heat Capacity and Entropy

The ground state configuration is assumed to be the same as that for the OH(g) molecule, which has the same number of valence electrons. The values of ω_e and $\omega_e x_e$ are estimated by comparison with those for LiF(g) and KF(g). The bond distance is estimated from those for HO(g), HF(g) and KF(g). B_e and α_e are derived from r_e, ω_e and $\omega_e x_e$ by the method suggested by Herzberg (3).

References

1. L. Brewer and J. Margrave, J. Phys. Chem. 59, 421 (1955).
2. G. R. Somayajulu, J. Chem. Phys. 33, 1541 (1960).
3. G. Herzberg, "Spectra of Diatomic Molecules," D. Van Nostrand Co., Inc., New York, 1950.

Potassium Oxide (KO) $K_1O_1(g)$

Enthalpy Reference Temperature = T_r = 298.15 K   Standard State Pressure = p° = 0.1 MPa

	J K^{-1}mol^{-1}			kJ mol^{-1}			
T/K	$C_p°$	S°	$-[G°-H°(T_r)]/T$	$H°-H°(T_r)$	$\Delta_f H°$	$\Delta_f G°$	Log K_f
0	0.	0.	INFINITE	-9.666	72.886	72.886	INFINITE
100	30.583	201.486	268.756	-6.727	72.640	64.717	-33.805
200	34.332	223.966	241.275	-3.462	71.855	57.109	-14.915
250	35.340	231.744	238.616	-1.718	71.496	53.464	-11.171
298.15	35.981	238.027	238.027	0.	71.128	50.024	-8.764
300	36.001	238.249	238.028	0.067	71.113	49.893	-8.687
350	36.455	243.835	238.468	1.879	68.295	46.487	-6.938
400	36.782	248.725	239.451	3.710	67.792	43.407	-5.668
450	37.029	253.073	240.727	5.555	67.315	40.387	-4.688
500	37.223	256.984	242.160	7.412	66.857	37.420	-3.909
600	37.514	263.798	245.215	11.150	65.976	31.616	-2.752
700	37.731	269.598	248.295	14.912	65.115	25.957	-1.937
800	37.906	274.648	251.280	18.694	64.252	20.422	-1.333
900	38.056	279.121	254.130	22.493	63.365	14.996	-0.870
1000	38.191	283.138	256.833	26.305	62.433	9.671	-0.505
1100	38.316	286.784	259.393	30.130	-17.515	9.053	-0.430
1200	38.433	290.123	261.816	33.968	-17.530	11.469	-0.499
1300	38.546	293.204	264.114	37.817	-17.552	13.886	-0.558
1400	38.655	296.064	266.295	41.677	-17.577	16.305	-0.608
1500	38.761	298.735	268.370	45.548	-17.607	18.727	-0.652
1600	38.865	301.240	270.347	49.429	-17.640	21.150	-0.690
1700	38.968	303.599	272.234	53.321	-17.676	23.575	-0.724
1800	39.070	305.829	274.039	57.223	-17.717	26.003	-0.755
1900	39.170	307.944	275.768	61.135	-17.764	28.433	-0.782
2000	39.270	309.956	277.428	65.057	-17.818	30.866	-0.806
2100	39.369	311.874	279.023	68.989	-17.880	33.302	-0.828
2200	39.467	313.708	280.558	72.930	-17.953	35.741	-0.849
2300	39.565	315.465	282.038	76.882	-18.038	38.183	-0.867
2400	39.663	317.151	283.466	80.843	-18.139	40.629	-0.884
2500	39.760	318.772	284.846	84.815	-18.256	43.080	-0.900
2600	39.857	320.333	286.181	88.795	-18.394	45.536	-0.915
2700	39.954	321.839	287.474	92.786	-18.555	47.998	-0.929
2800	40.050	323.294	288.727	96.786	-18.741	50.467	-0.941
2900	40.147	324.701	289.944	100.796	-18.955	52.942	-0.954
3000	40.243	326.064	291.125	104.815	-19.201	55.425	-0.965
3100	40.339	327.385	292.274	108.845	-19.481	57.918	-0.976
3200	40.435	328.667	293.391	112.883	-19.800	60.419	-0.986
3300	40.531	329.913	294.479	116.932	-20.151	62.930	-0.996
3400	40.627	331.124	295.539	120.990	-20.552	65.454	-1.006
3500	40.723	332.303	296.573	125.057	-20.999	67.990	-1.015
3600	40.819	333.452	297.581	129.134	-21.500	70.539	-1.024
3700	40.914	334.571	298.566	133.221	-22.057	73.103	-1.032
3800	41.010	335.664	299.528	137.317	-22.674	75.684	-1.040
3900	41.105	336.730	300.468	141.423	-23.356	78.281	-1.048
4000	41.201	337.772	301.388	145.538	-24.101	80.896	-1.056
4100	41.296	338.791	302.288	149.663	-24.919	83.530	-1.064
4200	41.392	339.787	303.169	153.797	-25.819	86.186	-1.072
4300	41.487	340.762	304.032	157.941	-26.802	88.865	-1.079
4400	41.582	341.717	304.877	162.095	-27.595	91.539	-1.087
4500	41.678	342.652	305.706	166.258	-28.688	94.259	-1.094
4600	41.773	343.570	306.519	170.430	-29.865	97.004	-1.102
4700	41.868	344.469	307.317	174.612	-31.109	99.773	-1.109
4800	41.964	345.351	308.101	178.804	-32.331	102.558	-1.116
4900	42.059	346.218	308.870	183.005	-33.746	105.383	-1.123
5000	42.154	347.068	309.625	187.216	-35.251	108.238	-1.131
5100	42.249	347.904	310.368	191.436	-36.847	111.123	-1.138
5200	42.344	348.725	311.097	195.665	-38.535	114.040	-1.146
5300	42.440	349.533	311.815	199.905	-40.316	116.992	-1.153
5400	42.535	350.327	312.521	204.153	-42.191	119.977	-1.161
5500	42.630	351.108	313.215	208.412	-44.158	122.998	-1.168
5600	42.725	351.877	313.899	212.679	-46.217	126.056	-1.176
5700	42.820	352.634	314.572	216.957	-48.024	129.103	-1.183
5800	42.915	353.380	315.234	221.243	-50.222	132.230	-1.191
5900	43.010	354.114	315.887	225.540	-52.505	135.395	-1.199
6000	43.105	354.838	316.530	229.845	-54.871	138.600	-1.207

PREVIOUS: December 1967 (1 atm)   CURRENT: December 1967 (1 bar)

Potassium Oxide (KO) $K_1O_1(g)$

$\Delta_f H°(0\ K)$ = [-130.09 ± 83.7] kJ mol^{-1}

$S°(298.15\ K)$ = [226.308 ± 1.3] J K^{-1} mol^{-1} $\Delta_f H°(298.15\ K)$ = [-138.07 ± 83.7] kJ mol^{-1}

Electronic Level and Quantum Weight

State	ε_i, cm^{-1}	g_i
$[^1\Sigma]$	0	[1]

ω_e = [360] cm^{-1} $\omega_e x_e$ = [2.3] cm^{-1} σ = 1

B_e = [0.3068] cm^{-1} α_e = [0.0027] cm^{-1} r_e = [2.2] Å

Enthalpy of Formation

Since the electron affinity of KO(g) is unavailable, the value of $\Delta_f H°(KO^-, g, 298.15\ K)$ is estimated. Assuming that the enthalpy change $\Delta_r H°(0\ K)$ of the reaction (1) $KO^-(g) = K(g) + O^-(g)$ is approximately the average of the $\Delta_r H°(0\ K)$ value for the reaction (2) KO(g) = K(g) + O(g) and (3) KF(g) = K(g) + F(g), we obtain $\Delta_r H°(0\ K)$ = 1/2(63.5 + 117.3) = 90.4 kcal mol^{-1} for reaction (1). However, we compare the $\Delta_r H°(0\ K)$ values for the reactions (4) HO(g) = H(g) + O(g), (5) $HO^-(g) = H(g) + O^-(g)$, and (6) HF(g) = H(g) + F(g), and find that $\Delta_r H°(0\ K)$ for reaction (5) is experimentally determined as 109.7 kcal mol^{-1} which is about 8 kcal mol^{-1} more negative than the average of $\Delta_r H°(0\ K)$ values for reactions (4) and (6). Based on this fact we estimated $\Delta_r H°(0\ K)$ = 80 kcal mol^{-1} for reaction (1), yielding $\Delta_f H°(KO^-, g, 298.15\ K)$ = -33 kcal mol^{-1}. The value of $\Delta_f H°(KO^-, g, 298.15\ K)$ is tentatively adopted as -33±20 kcal mol^{-1}. The electronic affinity of KO(g) is derived to be 2.2±0.9 eV. The value of EA for HO(g) is determined as 1.8±0.1 eV.

Heat Capacity and Entropy

The ground state configuration is assumed to be the same as that of KF(g) which is isoelectronic with $KO^-(g)$. The values of ω_e, $\omega_e x_e$ and r_e are estimated by comparison with those for KF(g). The values of B_e and α_e are calculated from r_e, ω_e and $\omega_e x_e$ using the method recommended by Herzberg (1).

Reference

1. G. Herzberg, "Spectra of Diatomic Molecules," D. Van Nostrand Co., Inc. New York, 1950.

Enthalpy Reference Temperature = T_r = 298.15 K; Standard State Pressure = p° = 0.1 MPa

T/K	$C_p°$ (J $K^{-1}mol^{-1}$)	S° (J $K^{-1}mol^{-1}$)	$-[G°-H°(T_r)]/T$ (J $K^{-1}mol^{-1}$)	$H°-H°(T_r)$ (kJ mol^{-1})	$\Delta_f H°$ (kJ mol^{-1})	$\Delta_f G°$ (kJ mol^{-1})	Log K_f
0	0.	0.	INFINITE	-9.636	-130.087		
100	30.460	189.920	256.923	-6.700			
200	34.190	212.295	229.546	-3.450			
250	35.226	220.044	226.896	-1.713			
298.15	35.888	226.308	226.308	0.	-138.072	-149.427	26.179
300	35.909	226.530	226.309	0.066	-138.126	-149.498	26.030
350	36.379	232.103	226.748	1.874	-141.987	-151.180	22.562
400	36.717	236.984	227.729	3.702	-143.533	-152.387	19.900
450	36.973	241.324	229.003	5.545	-145.052	-153.402	17.806
500	37.173	245.230	230.433	7.398	-146.552	-154.250	16.114
600	37.473	252.036	233.483	11.132	-149.517	-155.510	13.538
700	37.693	257.830	236.558	14.890	-152.460	-156.276	11.661
800	37.870	262.875	239.539	18.669	-155.405	-156.621	10.226
900	38.021	267.344	242.385	22.464	-158.375	-156.595	9.089
1000	38.155	271.357	245.085	26.272	-161.388	-156.236	8.161
1100	38.279	275.000	247.641	30.094	-243.418	-150.962	7.169
1200	38.395	278.335	250.062	33.928	-245.516	-142.465	6.201
1300	38.506	281.413	252.357	37.773	-247.620	-133.792	5.376
1400	38.614	284.271	254.535	41.629	-249.728	-124.957	4.662
1500	38.718	286.938	256.608	45.496	-251.841	-115.971	4.038
1600	38.820	289.440	258.582	49.373	-253.957	-106.844	3.488
1700	38.921	291.797	260.467	53.260	-256.076	-97.584	2.998
1800	39.020	294.024	262.270	57.157	-258.201	-88.200	2.559
1900	39.118	296.137	263.998	61.064	-260.331	-78.697	2.164
2000	39.215	298.145	265.655	64.980	-262.469	-69.082	1.804
2100	39.312	300.061	267.248	68.907	-264.615	-59.360	1.477
2200	39.408	301.892	268.782	72.843	-266.772	-49.535	1.176
2300	39.504	303.646	270.260	76.788	-268.942	-39.613	0.900
2400	39.599	305.329	271.686	80.743	-271.128	-29.595	0.644
2500	39.694	306.948	273.065	84.708	-273.331	-19.486	0.407
2600	39.788	308.506	274.398	88.682	-275.554	-9.289	0.187
2700	39.883	310.010	275.689	92.666	-277.800	0.995	-0.019
2800	39.977	311.462	276.941	96.659	-280.071	11.362	-0.212
2900	40.071	312.866	278.156	100.661	-282.372	21.811	-0.393
3000	40.165	314.226	279.336	104.673	-284.704	32.340	-0.563
3100	40.258	315.545	280.482	108.694	-287.071	42.947	-0.724
3200	40.352	316.825	281.598	112.725	-289.477	53.631	-0.875
3300	40.445	318.068	282.685	116.764	-291.915	64.390	-1.019
3400	40.538	319.277	283.743	120.814	-294.403	75.224	-1.156
3500	40.632	320.453	284.775	124.872	-296.938	86.132	-1.285
3600	40.725	321.599	285.782	128.940	-299.527	97.114	-1.409
3700	40.818	322.716	286.765	133.017	-302.172	108.167	-1.527
3800	40.911	323.806	287.726	137.103	-304.877	119.294	-1.640
3900	41.004	324.870	288.665	141.199	-307.648	130.493	-1.748
4000	41.097	325.909	289.583	145.304	-310.481	141.762	-1.851
4100	41.190	326.925	290.481	149.419	-313.389	153.104	-1.951
4200	41.283	327.919	291.361	153.542	-316.378	164.518	-2.046
4300	41.376	328.891	292.222	157.675	-319.451	176.004	-2.138
4400	41.468	329.843	293.067	161.817	-322.334	187.535	-2.226
4500	41.561	330.776	293.894	165.969	-325.517	199.158	-2.312
4600	41.654	331.691	294.706	170.130	-328.784	210.854	-2.394
4700	41.747	332.588	295.503	174.300	-332.118	222.619	-2.474
4800	41.839	333.467	296.284	178.479	-335.432	234.444	-2.551
4900	41.932	334.331	297.052	182.667	-338.938	246.353	-2.626
5000	42.025	335.179	297.806	186.865	-342.534	258.334	-2.699
5100	42.117	336.012	298.547	191.072	-346.222	270.388	-2.769
5200	42.210	336.831	299.276	195.289	-350.002	282.515	-2.838
5300	42.302	337.636	299.992	199.514	-353.875	294.716	-2.905
5400	42.395	338.428	300.696	203.749	-357.842	306.990	-2.970
5500	42.488	339.206	301.389	207.993	-361.902	319.339	-3.033
5600	42.580	339.973	302.071	212.247	-366.054	331.763	-3.095
5700	42.673	340.727	302.743	216.509	-369.954	344.213	-3.154
5800	42.765	341.470	303.404	220.781	-374.245	356.780	-3.213
5900	42.858	342.202	304.056	225.062	-378.622	369.421	-3.271
6000	42.950	342.923	304.698	229.353	-383.082	382.137	-3.327

PREVIOUS: December 1967 (1 atm) CURRENT: December 1967 (1 bar)

POTASSIUM SUPEROXIDE (KO_2) CRYSTAL M_r = 71.0971

$S°(298.15\ K) = 122.499 \pm 4.2\ J\ K^{-1}\ mol^{-1}$
$T_{trs1} = 193.5\ K$
$T_{trs2} = 230.9\ K$
$T_{fus} = 782 \pm 20\ K$

$\Delta_f H°(0\ K) = -285.48 \pm 21\ kJ\ mol^{-1}$
$\Delta_f H°(298.15\ K) = -284.51 \pm 2.1\ kJ\ mol^{-1}$
$\Delta_{trs1}H° = 0.302\ kJ\ mol^{-1}$
$\Delta_{trs2}H° = 0.157\ kJ\ mol^{-1}$
$\Delta_{fus}H°$ = Unknown

Enthalpy of Formation

D'Orazio and Wood (1) measured the enthalpy of solution, $\Delta_{sln}H°(25°C) = -13.0 \pm 0.4\ kcal\ mol^{-1}$ for $KO_2(cr) + 1/2\ H_2O(\ell)$. $KOH(aq,\ 12{,}000,\ H_2O) + 3/4\ O_2(g)$ which leads to $\Delta_f H°(KO_2,\ cr,\ 298.15\ K) = -68.09 \pm 0.4\ kcal\ mol^{-1}$, using the following auxiliary data. $\Delta_f H°(KOH{\cdot}aq,\ 12{,}000\ H_2O,\ 298.15\ K) = -115.243\ kcal\ mol^{-1}$ (2) and $\Delta_f H°(H_2O,\ \ell,\ 298.15\ K) = -68.315\ kcal\ mol^{-1}$ (3). Giles and Margrave (4) also measured the enthalpy of solution as $-13.2 \pm 0.8\ kcal\ mol^{-1}$ from which we obtain $\Delta_f H°(KO_2,\ cr,\ 298.15\ K) = -67.88 \pm 0.8\ kcal\ mol^{-1}$. Both $\Delta_{sol}H°$ values have been corrected for impurites present in their samples. The value $\Delta_f H°(KO_2,\ cr,\ 298.15\ K) = -68.0 \pm 0.5\ kcal\ mol^{-1}$ is adopted in the tabulation.

Kazarnovskaya and Kazarnovskii (5) and deForcrand (6) determined the enthalpy of solution in $H_2SO_4(aq)$ and derived the heat of formation of $KO_2(cr)$ as $-67.9\ kcal\ mol^{-1}$ and $-66.8\ kcal\ mol^{-1}$, respectively. The former is reliable since their sample was well characterized and corrections for the impurities were included in the $\Delta_{sln}H°$ data (5). The latter did not give enough experimental details to check their derivation.

Heat Capacity and Entropy

The low temperature heat capacities in the temperature range from 52-298 K are obtained from Todd's measurements (7). Two peaks at 193.5 K and 230.9 K were found in his heat capacity data. Below 50 K, the heat capacities were extrapolated using a combination of 1 Debye (θ = 139) and 2 Einstein (θ = 260) functions as suggested by Todd. This extrapolation yields the entropy from lattice contribution as $3.278\ cal\ K^{-1}\ mol^{-1}$ at 50 K. By neutron diffraction, Smith et al. (8) found an antiferromagnetic transition at 7 K which indicates the existence of an unpaired electron in $KO_2(cr)$. We tentatively adopt $S°(50\ K) = 4.656 \pm 1\ cal\ K^{-1}\ mol^{-1}$ which includes both lattice ($3.278\ cal\ K^{-1}\ mol^{-1}$) and unpaired electron (Rℓn2) contributions. Heat capacities above 298 K are estimated graphically.

The entropy at 298 K is calculated from the $C_p°$ data and is based on an extrapolation of $S°(50\ K) = 4.656\ cal\ K^{-1}\ mol^{-1}$.

Transition Data

Todd (7) observed two anomalies in the heat capacity at 193.5 and 230.9 K which are adopted as two phase transitions in the tabulation. The heats of transition are calculated from the total heats of absorption given by Todd between 180 and 208 K for transition 1 and 217-240 K for transition 2 with extrapolated $C_p°$ below 176 K and above 250 K.

Smith et al. (8) confirmed these two transitions in the neutron diffraction measurements. In addition, they found an antiferromagnetic transition at 7.1 ± 0.5 K and an unusually low temperature phase transition at 12 K. Neuman (9) observed a color change at -75.7 ± 15°C in the warming and cooling curves for $KO_2(cr)$. This color change represents a phase transition which is in good agreement with Todd's observation.

Fusion Data

Firsova et al. (10) studied the thermograms of $KO_2(cr)$ and found the melting point 509 ± 20°C under atmospheric pressure which is adopted in the tabulation. Earlier melting point data were reported as 380°C by Bluemental (11) and 440°C by deForcrand (6).

References

1. L. A. D'Orazio and R. H. Wood, J. Phys. Chem. 69, 2550 (1965).
2. This value is derived from JANAF $\Delta_f H°(KOH{\cdot}aq,\ \infty,\ 298.15\ K) = -115.273\ kcal\ mol^{-1}$ with the heat of dilution data from V. B. Parker, NSRDS-NBS 2, 1965.
3. U. S. Nat. Bur. Stand. Tech. Note 270-3, 1968.
4. P. W. Gilles and J. L. Margrave, J. Phys. Chem. 60, 1333 (1956).
5. L. I. Kazarnovskaya and I. A. Kazarnovskii, Zh. Fiz. Khim. 25, 293 (1951).
6. M. R. deForcrand, Comp. Rend. 158, 843 (1914).
7. S. S. Todd, J. Amer. Chem. Soc. 75, 1229 (1953).
8. H. G. Smith, R. M. Nicklow, L. J. Raubenheimer and M. K. Wilkinson, J. Appl. Phys. 37, 1047 (1966).
9. E. W. Neuman, J. Chem. Phys. 3, 243 (1935).
10. T. P. Firsova, A. N. Molodkina, T. G. Morozova and I. V. Aksenova, Izv. Akad. Nauk SSSR, Ser. Khim. No. 9 (English transl.) 1639 (1965).
11. M. Blumental, Roczniki Chem. 12, 126 (1932); M. Centnerszwer and M. Blumental, Bull. Intern. Acad. Polonaise, Classe Sci. Math. Nat. 1933A, 499.

Potassium Superoxide (KO_2) $K_1O_2(cr)$

Enthalpy Reference Temperature = T_r = 298.15 K; Standard State Pressure = p° = 0.1 MPa

T/K	$C_p°$ (J K⁻¹mol⁻¹)	S° (J K⁻¹mol⁻¹)	$-[G°-H°(T_r)]/T$ (J K⁻¹mol⁻¹)	$H°-H°(T_r)$ (kJ mol⁻¹)	$\Delta_f H°$ (kJ mol⁻¹)	$\Delta_f G°$ (kJ mol⁻¹)	Log K_f
0	0.	0.	INFINITE	-16.734	-285.480	-285.480	INFINITE
100	52.400	46.225	188.849	-14.262	-287.646	-271.377	141.753
200	69.454	92.567	129.436	-7.374	-286.263	-255.381	66.699
298.15	77.530	122.499	122.499	0.	-284.512	-240.589	42.150
300	77.613	122.979	122.500	0.144	-284.478	-240.316	41.843
400	83.889	146.226	125.627	8.240	-284.831	-225.443	29.440
500	87.655	165.373	131.719	16.827	-282.410	-210.868	22.029
600	90.165	181.588	138.714	25.724	-279.712	-196.810	17.134
700	91.922	195.625	145.864	34.833	-276.854	-183.217	13.672
800	93.220	207.987	152.872	44.092	-273.908	-170.041	11.103
900	94.224	219.028	159.620	53.467	-270.921	-157.237	9.126
1000	94.893	228.992	166.067	62.924	-267.939	-144.765	7.562
1100	95.395	238.061	172.206	72.440	-343.951	-127.980	6.077
1200	95.730	246.376	178.045	81.997	-340.021	-108.520	4.724
1300	95.981	254.049	183.600	91.583	-336.097	-89.388	3.592
1400	96.148	261.169	188.890	101.191	-332.182	-70.557	2.633
1500	96.232	267.806	193.932	110.810	-328.284	-52.006	1.811

PREVIOUS: June 1962 CURRENT: June 1971

Potassium Superoxide (KO_2) $K_1O_2(cr)$

POTASSIUM (K_2) IDEAL GAS M_r = 78.1966

D_0^0 = 53.48 ± 1.20 kJ mol^{-1} (natural abundance)
S°(298.15 K) = 249.690 ± 0.042 J K^{-1} mol^{-1}

$\Delta_fH°$(0 K) = 127.11 ± 2.0 kJ mol^{-1}
$\Delta_fH°$(298.15 K) = 123.68 ± 2.0 kJ mol^{-1}

Spectroscopic Data for $^{39}K_2$ (Ground State: $^1\Sigma_g^+$) in cm^{-1}

ω_e = 92.021 ω_ex_e = 0.2829 ω_ey_e = -2.055 x 10^{-3}
B_e = 0.056743 α_e = 1.65 x 10^{-4} (a)
D_e = 8.63 x 10^{-8} r_e = 3.9051 Å

(a) $-7.2 \times 10^{-6} (v + 1/2)^2 + 1.5 \times 10^{-7} (v + 1/2)^3$.

Enthalpy of Formation

Using the adopted value of the dissociation energy for natural abundance K_2(g), D_0^0 = 53.477 kJ mol^{-1} [D_0^0 for $^{39}K_2$ = 53.474 kJ mol^{-1} as determined by (2)], and the recommended $\Delta_fH°$(K, g, 0 K) = 90.14 kJ mol^{-1} from NBS (4), we calculate $\Delta_fH°$(K_2, g, 0 K) = 126.80 kJ mol^{-1}. Since this value is only 0.31 kJ mol^{-1} smaller than the NBS recommended value of 127.1 kJ mol^{-1}, we adopt the NBS result. Correcting this value to 298.15 K using the current H°(298.15 K)-H°(0 K) = 1.0738 kJ mol^{-1} for K_2(g) and the NBS value of H°(298.15 K)-H°(0 K) = 7.092 kJ mol^{-1} for K(cr) (4), we obtain $\Delta_fH°$(K_2, g, 298.15 K) = 123.68 kJ mol^{-1}.

Heat Capacity and Entropy

The heat capacity and entropy were calculated using a direct summation technique similar to the Li_2(g) species (see Li_2(g) table) using data given above (1,2). Since insufficient spectroscopic data exist for the lower lying excited electronic states, we performed the calculation in the ground state only. Based on the results of Li_2 and Na_2, we expect that the neglect of these excited states results in only ca. 1-2% lower free energy functions near 6000 K. The heat capacity, however, above 2000 K will be seriously deficient. The actual Cp will probably exhibit temperature dependence similar to that calculated for Li_2 and Na_2. Likewise, the actual entropy will be somewhat larger than given here but these effects will not manifest themselves at temperatures less than ca. 3000 K (for Na_2(g) the entropy of the ground state only at 3000 K is only 1.3 J K^{-1} less than that calculated including the excited states). Separate calculations were performed on the isotopic species $^{39}K_2$, $^{41}K_2$, and $^{39}K^{41}K$ and the results combined according to the natural atomic isotopic abundances (3): ^{39}K, 93.258%; ^{41}K, 6.730% (the diatomic species $^{40}K_2$ was neglected due to the small isotopic abundance of .011%). The spectroscopic input data for the calculation of the isotopic species were calculated from the $^{39}K_2$(g) data using mass reduced scaling. The rotational levels were weighted according to the nuclear spins of the individual atoms (^{39}K, I=3/2; ^{41}K, I=4) as follows: $^{39}K_2$, even J weight = 0.625, odd J weight = 0.375; $^{41}K_2$, even J weight = 0.556, odd J weight = 0.444. The adopted value of S°(298.15 K) is only 0.15 J K^{-1} mol^{-1} lower than that adopted by NBS (4).

References

1. W. C. Stwalley, Director: Iowa Laser Facility and Professor of Chemistry and Physics, University of Iowa, personal communication, 1982.
2. W. J. Tango, J. K. Link and R. N. Zare, J. Chem. Phys. 49, 4264 (1968).
3. Pure Appl. Chem. 47, 75 (1976).
4. D. D. Wagman, et al., J. Phys. Chem. Ref. Data 11, Supp. 2 (1982).

Potassium (K_2) K_2(g)

Enthalpy Reference Temperature = T_r = 298.15 K   Standard State Pressure = p° = 0.1 MPa

T/K	$C_p°$ (J K^{-1}mol^{-1})	S°	-[G°-H°(T_r)]/T	H°-H°(T_r) (kJ mol^{-1})	$\Delta_fH°$	$\Delta_fG°$	Log K_f
0	0.	0.	INFINITE	-10.738	127.110	127.110	INFINITE
100	36.488	208.920	283.115	-7.420	126.963	113.192	-59.125
200	37.533	234.619	253.150	-3.706	125.488	99.961	-26.107
298.15	37.962	249.690	249.690	0.	123.683	87.801	-15.382
300	37.969	249.925	249.691	0.070	123.644	87.579	-15.249
400	38.347	260.900	251.185	3.886	116.504	76.804	-10.030
500	38.698	269.496	254.018	7.739	114.141	67.157	-7.016
600	38.895	276.572	257.204	11.621	111.944	57.968	-5.047
700	38.769	282.563	260.411	15.507	109.838	49.139	-3.667
800	38.242	287.710	263.509	19.361	107.738	40.611	-2.652
900	37.358	292.167	266.452	23.143	105.556	32.350	-1.878
1000	36.231	296.046	269.222	26.824	103.211	24.340	-1.271
1100	34.982	299.441	271.817	30.386	-57.265	25.795	-1.225
1200	33.710	302.430	274.247	33.820	-57.988	33.377	-1.453
1300	32.482	305.079	276.519	37.129	-58.837	41.024	-1.648
1400	31.337	307.444	278.644	40.319	-59.805	48.742	-1.819
1500	30.292	309.570	280.637	43.400	-60.884	56.532	-1.969
1600	29.352	311.494	282.506	46.381	-62.064	64.397	-2.102
1700	28.514	313.248	284.264	49.274	-63.336	72.339	-2.223
1800	27.771	314.856	285.919	52.087	-64.693	80.359	-2.332
1900	27.114	316.340	287.482	54.831	-66.127	88.456	-2.432
2000	26.533	317.716	288.960	57.512	-67.634	96.631	-2.524
2100	26.020	318.998	290.360	60.140	-69.209	104.883	-2.609
2200	25.566	320.197	291.689	62.718	-70.851	113.211	-2.688
2300	25.162	321.325	292.953	65.254	-72.559	121.616	-2.762
2400	24.804	322.388	294.158	67.752	-74.332	130.096	-2.831
2500	24.485	323.394	295.307	70.217	-76.171	138.651	-2.897
2600	24.199	324.348	296.406	72.651	-78.078	147.282	-2.959
2700	23.944	325.257	297.458	75.057	-80.055	155.986	-3.018
2800	23.715	326.124	298.466	77.440	-82.107	164.766	-3.074
2900	23.508	326.952	299.434	79.801	-84.237	173.621	-3.127
3000	23.321	327.746	300.365	82.143	-86.449	182.549	-3.178
3100	23.153	328.508	301.261	84.466	-88.749	191.553	-3.228
3200	22.999	329.240	302.124	86.774	-91.142	200.633	-3.275
3300	22.860	329.946	302.956	89.066	-93.617	209.788	-3.321
3400	22.733	330.626	303.760	91.346	-96.209	219.021	-3.365
3500	22.617	331.284	304.537	93.613	-98.907	228.331	-3.408
3600	22.510	331.919	305.289	95.870	-101.727	237.720	-3.449
3700	22.412	332.535	306.017	98.116	-104.674	247.189	-3.490
3800	22.322	333.131	306.723	100.352	-107.755	256.740	-3.529
3900	22.240	333.710	307.407	102.580	-110.980	266.375	-3.568
4000	22.163	334.272	308.072	104.801	-114.344	276.091	-3.605
4100	22.092	334.818	308.718	107.013	-117.870	285.896	-3.642
4200	22.026	335.350	309.345	109.219	-121.572	295.788	-3.679
4300	21.966	335.868	309.956	111.419	-125.452	305.769	-3.714
4400	21.909	336.372	310.551	113.612	-128.965	315.791	-3.749
4500	21.856	336.864	311.130	115.801	-133.091	325.944	-3.783
4600	21.807	337.343	311.695	117.984	-137.397	336.192	-3.818
4700	21.761	337.812	312.245	120.162	-141.847	346.532	-3.851
4800	21.718	338.270	312.783	122.336	-146.267	356.944	-3.884
4900	21.677	338.717	313.308	124.506	-151.081	367.477	-3.917
5000	21.640	339.155	313.820	126.672	-156.086	378.110	-3.950
5100	21.604	339.583	314.321	128.834	-161.281	388.847	-3.983
5200	21.571	340.002	314.811	130.993	-166.670	399.684	-4.015
5300	21.539	340.412	315.290	133.148	-172.252	410.629	-4.047
5400	21.510	340.815	315.759	135.301	-178.027	421.681	-4.079
5500	21.482	341.209	316.218	137.450	-183.994	432.841	-4.111
5600	21.456	341.596	316.668	139.597	-190.151	444.111	-4.142
5700	21.431	341.976	317.109	141.741	-195.805	455.399	-4.173
5800	21.407	342.348	317.541	143.883	-202.245	466.878	-4.205
5900	21.385	342.714	317.964	146.023	-208.856	478.472	-4.236
6000	21.364	343.073	318.380	148.160	-215.634	490.178	-4.267

PREVIOUS: June 1962 (1 atm)   CURRENT: December 1983 (1 bar)

Potassium (K_2) K_2(g)

POTASSIUM OXIDE (K_2O) CRYSTAL M_r = 94.1960

Potassium Oxide (K_2O) $K_2O_1(cr)$

$\Delta_fH°(0\ K)$ = Unknown

$S°(298.15\ K) = [94.14 \pm 6.3]\ J\ K^{-1}\ mol^{-1}$ $\Delta_fH°(298.15\ K) = -363.17 \pm 2.1\ kJ\ mol^{-1}$

Enthalpy of Formation

$\Delta_fH°(298.15\ K)$ was reported by Rengade (1) based on the measurements of the enthalpy of solution of K(cr) and $K_2O(cr)$ in H_2O. The same value of $\Delta_fH°(298.15\ K)$ was also given by de Forcrand (2).

Heat Capacity and Entropy

Both $C_p°$ and S°(298.15 K) were estimated by comparison with those of the $Na_2O(cr)$.

Decomposition Data

T_{dcm} = 1154 K(600mmHg) was taken from (3).

References

1. M. Rengade, Ann. Chim. et phys. [8] 14, 540 (1908).
2. R. de Forcrand, Compt. rend. 158, 991 (1914).
3. "Data on Chemicals for Ceramic Use," National Research Council Bulleton 118 (1949).

Enthalpy Reference Temperature = T_r = 298.15 K							Standard State Pressure = p° = 0.1 MPa
	J K⁻¹mol⁻¹			kJ mol⁻¹			
T/K	$C_p°$	S°	$-[G°-H°(T_r)]/T$	$H°-H°(T_r)$	$\Delta_fH°$	$\Delta_fG°$	Log K_f
0							
100							
200							
298.15	83.680	94.140	94.140	0.	-363.171	-322.094	56.429
300	83.889	94.658	94.142	0.155	-363.153	-321.839	56.037
400	91.546	119.958	97.539	8.968	-366.782	-307.331	40.133
500	94.266	140.671	104.161	18.255	-365.240	-292.638	30.572
600	97.404	158.126	111.736	27.834	-363.320	-278.292	24.227
700	101.384	173.439	119.478	37.773	-361.000	-264.299	19.722
800	105.395	187.239	127.099	48.111	-358.283	-250.665	16.367
900	109.455	199.887	134.494	58.854	-355.209	-237.394	13.778
1000	113.512	211.629	141.627	70.002	-351.817	-224.482	11.726
1100	117.549	222.637	148.496	81.555	-506.056	-202.704	9.626
1200	121.592	233.038	155.112	93.512	-500.031	-175.390	7.635
1300	125.642	242.931	161.489	105.874	-493.618	-148.594	5.971
1400	129.699	252.390	167.647	118.641	-486.816	-122.306	4.563
1500	133.762	261.477	173.601	131.814	-479.623	-96.519	3.361
1600	137.833	270.240	179.369	145.394	-472.038	-71.223	2.325
1700	141.910	278.718	184.965	159.381	-464.062	-46.415	1.426
1800	145.993	286.945	190.403	173.776	-455.695	-22.087	0.641
1900	150.083	294.948	195.696	188.580	-446.939	1.766	-0.049
2000	154.180	302.751	200.854	203.793	-437.796	25.147	-0.657

PREVIOUS: CURRENT: June 1963

Potassium Oxide (K_2O) $K_2O_1(cr)$

POTASSIUM PEROXIDE (K_2O_2) CRYSTAL M_r = 110.1954

$S°(298.15\ K)$ = [112.968 ± 6.3] $J\ K^{-1}\ mol^{-1}$

T_{fus} = 763 K

$\Delta_f H°(0\ K)$ = Unknown

$\Delta_f H°(298.15\ K)$ = [495.80 ± 4.2] $kJ\ mol^{-1}$

$\Delta_{fus} H°$ = Unknown

Enthalpy of Formation

$\Delta_f H°(298.15\ K)$ was reported by NBS (1). The value was based on the measured decomposition pressure of $K_2O_2(cr)$ by Centnerzwer and Blumenthal (2).

Heat Capacity and Entropy

The heat capacities and $S°(298.15\ K)$ were estimated by comparison with those for $Na_2O_2(cr)$, $Na_2O(cr)$, and $K_2O(cr)$.

Fusion Data

T_{fus} was reported by Centnerzwer and Blumenthal, (2).

References

1. D. D. Wagman, U. S. Nat. Bur. Stand., Report 7437, Table 4, Chap. B4, p 95, January 1, 1962.
2. M. Centnerzwer and M. Blumenthal, IX Congr. IUPAC, 3A, 201 (1935).

Potassium Peroxide (K_2O_2) $K_2O_2(cr)$

Enthalpy Reference Temperature = T_r = 298.15 K Standard State Pressure = p° = 0.1 MPa

T/K	$C_p°$ (J K⁻¹mol⁻¹)	S° (J K⁻¹mol⁻¹)	$-[G°-H°(T_r)]/T$ (J K⁻¹mol⁻¹)	$H°-H°(T_r)$ (kJ mol⁻¹)	$\Delta_f H°$ (kJ mol⁻¹)	$\Delta_f G°$ (kJ mol⁻¹)	Log K_f
0							
100							
200							
298.15	100.165	112.968	112.968	0.	-495.804	-429.758	75.292
300	100.374	113.588	112.970	0.185	-495.782	-429.348	74.756
400	107.110	143.348	116.982	10.546	-499.349	-406.479	53.081
500	115.478	168.206	124.802	21.702	-497.468	-383.461	40.060
600	120.708	189.704	133.869	33.501	-494.907	-360.891	31.418
700	127.952	208.862	143.236	45.938	-491.716	-338.798	25.281
800	134.725	226.395	152.550	59.076	-487.870	-317.208	20.712
900	140.798	242.614	161.667	72.852	-483.464	-296.134	17.187
1000	146.858	257.762	170.527	87.235	-478.569	-275.578	14.395
1100	152.917	272.043	179.112	102.224	-631.126	-246.314	11.696
1200	158.959	285.608	187.426	117.818	-623.239	-211.675	9.214
1300	164.983	298.569	195.481	134.015	-614.782	-177.717	7.141
1400	170.992	311.016	203.292	150.814	-605.755	-144.431	5.389
1500	176.983	323.018	210.876	168.213	-596.156	-111.812	3.894
1600	182.958	334.631	218.249	186.210	-585.988	-79.850	2.607
1700	188.916	345.901	225.428	204.804	-575.251	-48.542	1.492
1800	194.857	356.868	232.427	223.993	-563.948	-17.883	0.519
1900	200.782	367.562	239.259	243.775	-552.083	12.133	-0.334
2000	206.690	378.011	245.937	264.148	-539.660	41.509	-1.084

PREVIOUS: June 1963 CURRENT: September 1963

Potassium Peroxide (K_2O_2) $K_2O_2(cr)$

POTASSIUM SILICATE (K_2SiO_3) CRYSTAL M_r = 154.2803

$S°(298.15\ K) = 146.147 \pm 0.8\ J\ K^{-1}\ mol^{-1}$

$T_{fus} = 1249\ K$

$\Delta_f H°(0\ K) = -1539.54 \pm 8.4\ kJ\ mol^{-1}$

$\Delta_f H°(298.15\ K) = -1548.08 \pm 8.4\ kJ\ mol^{-1}$

$\Delta_{fus} H° = [50.208 \pm 12.6]\ kJ\ mol^{-1}$

Enthalpy of Formation

Hatton et al. (1) conducted enthalpy of solution measurements on K_2SiO_3(cr) and KOH(cr) in HF(aq) at 50°C to determine the enthalpy of formation of the silicate. The results of this work gave a $\Delta_r H°(298.15\ K) = -17.5 \pm 0.5$ kcal mol^{-1} for the process 2 KOH(cr) + SiO_2(cr, quartz) → K_2SiO_3(cr) + H_2O(ℓ). When combined with the auxiliary data for H_2O(ℓ) KOH(cr) and SiO_2(cr) (2, 3), we calculate $\Delta_f H°(K_2SiO_3$, cr, 298.15 K) = -369.9 ± 1 kcal mol^{-1}.

Urusov (4) summarized the various methods of estimating the enthalpies of formation of silicates using the Born-Haber cycle and energies of atomization. His estimate for $\Delta_f H°(K_2SiO_3$, cr, 298.15 K) is -369 kcal mol^{-1}.

An earlier estimate by Wilcox and Bromley (5) based upon ionic contributions to crystal lattice energies gave $\Delta_f H°(K_2SiO_3$, cr, 298.15 K) = -367 ± 15 kcal mol^{-1}.

An attempt to calculate the $\Delta_f H°(K_2SiO_3$, cr, 298.15 K) from HF solution experiments on higher potassium silicates ($K_2Si_2O_5$, $K_2Si_3O_7$) and sodium silicates (6) yielded a value of -383.8 kcal mol^{-1} which is discrepant with the former values.

A value for $\Delta_f H°(K_2SiO_3$, cr, 298.15 K) = -370 ± 2 kcal mol^{-1} has been selected.

Heat Capacity and Entropy

The low temperature heat capacity of K_2SiO_3(cr) has been measured and reported by Stull et al. (7) in the range 15-298 K. They report S°(298.15 K) = 34.93 cal K^{-1} mol^{-1} which is derived from $C_p°$ and an extrapolation of S°(15 K) = 0.18 cal k^{-1} mol^{-1}.

Heat capacities above 298.15 K have been estimated by analogy with other alkali metal silicates and borates.

Fusion Data

A melting temperature of 1249 K for K_2SiO_3 was measured by Morey and Bowen (8). The enthalpy of fusion, $\Delta_{fus}H°$ = 12 ± 3 kcal mol^{-1}, was estimated by comparison with other silicates.

References

1. W. E. Hatton, D. L. Hildenbrand, G. C. Sinke and D. R. Stull, unpublished data, The Dow Chemical Company, Midland, MI (1959).
2. U. S. Nat. Bur. Stand. Tech. Note 270-3, 1968.
3. JANAF Thermochemical Tables: KOH(cr), 12-31-70; SiO_2(cr), 6-30-67.
4. V. S. Urusov, Geokhimiya, No. 5, 551 (1965).
5. D. E. Wilcox and L. A. Bromley, Ind. Eng. Chem. 55, No. 7, 32 (1963).
6. G. W. Morey, Geophysical Laboratory, Carnegie Institute of Washington, Paper No. 1215 (1952-1953).
7. D. R. Stull, D. L. Hildenbrand, F. L. Oetting and G. C. Sinke, J. Chem. Eng. Data 15, 15 (1970).
8. G. W. Morey and H. L. Bowen, J. Phys. Chem. 28, 1167 (1924).

Potassium Silicate (K_2SiO_3) $K_2O_3Si_1$(cr)

Enthalpy Reference Temperature = T_r = 298.15 K Standard State Pressure = p° = 0.1 MPa

T/K	$C_p°$ (J K⁻¹mol⁻¹)	S°	-[G°-H°(T_r)]/T	H°-H°(T_r) (kJ mol⁻¹)	$\Delta_f H°$	$\Delta_f G°$	Log K_f
0	0.	0.	INFINITE	-21.870	-1539.542	-1539.542	INFINITE
100	63.262	46.944	236.438	-18.949	-1544.710	-1515.904	791.827
200	98.198	103.052	156.335	-10.657	-1547.151	-1485.986	388.100
298.15	118.407	146.147	146.147	0.	-1548.080	-1455.733	255.038
300	118.758	146.881	146.149	0.219	-1548.089	-1455.160	253.366
400	135.562	183.448	151.021	12.971	-1552.872	-1423.255	185.858
500	148.114	215.101	160.743	27.179	-1551.745	-1390.957	145.312
600	157.737	242.994	172.173	42.492	-1549.626	-1358.985	118.310
700	164.850	267.859	184.099	58.632	-1546.808	-1327.426	99.054
800	170.707	290.268	195.992	75.420	-1543.488	-1296.307	84.640
900	175.310	310.648	207.617	92.728	-1539.816	-1265.627	73.455
1000	179.075	329.321	218.867	110.455	-1535.919	-1235.369	64.529
1100	182.004	346.523	229.700	128.505	-1689.825	-1196.287	56.807
1200	185.351	362.515	240.110	146.887	-1683.623	-1151.692	50.132
1249.000	186.376	369.955	245.059	155.996	------ CRYSTAL <--> LIQUID -----		
1300	187.443	377.434	250.106	165.526	-1677.256	-1107.622	44.505
1400	189.535	391.401	259.705	184.375	-1670.768	-1064.047	39.700
1500	191.627	404.549	268.927	203.434	-1664.155	-1020.940	35.552
1600	193.719	416.983	277.795	222.701	-1657.416	-978.278	31.937
1700	195.811	428.790	286.333	242.177	-1700.727	-935.594	28.747
1800	197.903	440.042	294.562	261.863	-1693.505	-890.795	25.850
1900	199.995	450.798	302.504	281.758	-1686.116	-846.400	23.269
2000	202.087	461.110	310.179	301.862	-1678.564	-802.399	20.957
2100	204.179	471.020	317.603	322.175	-1670.852	-758.780	18.874
2200	206.271	480.567	324.795	342.698	-1662.982	-715.531	16.989

PREVIOUS: CURRENT: June 1972

Potassium Silicate (K_2SiO_3) $K_2O_3Si_1$(cr)

POTASSIUM SILICATE (K_2SiO_3) LIQUID M_r = 154.2803

S°(298.15 K) = [184.256] J K^{-1} mol^{-1}

T_{fus} = 1249 K

$\Delta_f H°$(298.15 K) = [-1499.290] kJ mol^{-1}

$\Delta_{fus}H°$ = [50.208 ± 12.6] kJ mol^{-1}

Enthalpy of Formation

$\Delta_f H°(K_2SiO_3$, ℓ, 298.15 K) is calculated from that of the crystal by adding the enthalpy of fusion, $\Delta_{fus}H°$, and the difference in enthalpy, H°(1249 K) - H°(298.15 K), between the crystal and liquid.

Heat Capacity and Entropy

The heat capacity of K_2SiO_3(ℓ) has been estimated. A constant $C_p°$ of 43 cal K^{-1} mol^{-1} for K_2SiO_3(ℓ) was selected above a hypothetical glass transition temperature of 700 K. Below this temperature the heat capacity was taken as that of the crystal.

S°(K_2SiO_3, ℓ, 298.15 K) is calculated in a manner analogous to that used for the enthalpy of formation.

Fusion Data

See the crystal table for details.

Potassium Silicate (K_2SiO_3) $K_2O_3Si_1$(l)

Enthalpy Reference Temperature = T_r = 298.15 K Standard State Pressure = p° = 0.1 MPa

T/K	$C_p°$	S°	-[G°-H°(T_r)]/T	H°-H°(T_r)	$\Delta_f H°$	$\Delta_f G°$	Log K_f
	J $K^{-1}mol^{-1}$			kJ mol^{-1}			
0							
100							
200							
298.15	118.407	184.256	184.256	0.	-1499.291	-1418.306	248.481
300	118.758	184.990	184.259	0.219	-1499.299	-1417.803	246.861
400	135.562	221.557	189.130	12.971	-1504.082	-1389.709	181.477
500	148.114	253.210	198.852	27.179	-1502.956	-1361.222	142.206
600	157.737	281.103	210.282	42.492	-1500.837	-1333.061	116.053
700	164.850	305.985	222.209	58.643	-1498.008	-1305.314	97.404
700.000	164.850	305.985	222.209	58.643	GLASS <--> LIQUID	TRANSITION	
700.000	179.912	305.985	222.209	58.643			
800	179.912	330.008	234.216	76.634	-1493.485	-1278.097	83.451
900	179.912	351.199	246.060	94.625	-1489.129	-1251.437	72.632
1000	179.912	370.155	257.538	112.616	-1484.968	-1225.251	64.001
1100	179.912	387.302	268.568	130.607	-1638.934	-1190.252	56.520
1200	179.912	402.956	279.124	148.599	-1633.122	-1149.720	50.046
1249.000	179.912	410.154	284.121	157.414	------ CRYSTAL <--> LIQUID -----		
1300	179.912	417.357	289.211	166.590	-1627.403	-1109.670	44.587
1400	179.912	430.690	298.846	184.581	-1621.773	-1070.056	39.924
1500	179.912	443.103	308.055	202.572	-1616.227	-1030.842	35.897
1600	179.912	454.714	316.862	220.563	-1610.764	-991.994	32.385
1700	179.912	465.621	325.295	238.555	-1655.560	-953.040	29.283
1800	179.912	475.905	333.379	256.546	-1650.032	-911.875	26.462
1900	179.912	485.632	341.139	274.537	-1644.548	-871.016	23.946
2000	179.912	494.860	348.596	292.528	-1639.109	-830.445	21.689
2100	179.912	503.638	355.772	310.519	-1633.718	-790.144	19.654
2200	179.912	512.008	362.685	328.511	-1628.380	-750.098	17.810
2300	179.912	520.005	369.352	346.502	-1623.098	-710.296	16.131
2400	179.912	527.662	375.790	364.493	-1617.876	-670.721	14.598
2500	179.912	535.006	382.013	382.484	-1612.720	-631.363	13.192
2600	179.912	542.063	388.034	400.475	-1607.634	-592.209	11.898
2700	179.912	548.853	393.865	418.467	-1602.623	-553.249	10.703
2800	179.912	555.396	399.518	436.458	-1597.692	-514.474	9.598
2900	179.912	561.709	405.002	454.449	-1592.847	-475.874	8.571
3000	179.912	567.808	410.328	472.440	-1588.095	-437.438	7.616

PREVIOUS: CURRENT: June 1972

Potassium Silicate (K_2SiO_3) $K_2O_3Si_1$(l)

POTASSIUM SILICATE (K_2SiO_3) CRYSTAL-LIQUID M_r = 154.2803

0 to 1249 K crystal
above 1249 K liquid

Refer to the individual tables for details.

Potassium Silicate (K_2SiO_3) $K_2O_3Si_1$(cr,l)

Enthalpy Reference Temperature = T_r = 298.15 K Standard State Pressure = p° = 0.1 MPa

T/K	C_p° (J K⁻¹mol⁻¹)	S° (J K⁻¹mol⁻¹)	$-[G^\circ - H^\circ(T_r)]/T$ (J K⁻¹mol⁻¹)	$H^\circ - H^\circ(T_r)$ (kJ mol⁻¹)	$\Delta_f H^\circ$ (kJ mol⁻¹)	$\Delta_f G^\circ$ (kJ mol⁻¹)	Log K_f
0	0.	0.	INFINITE	-21.870	-1539.542	-1539.542	INFINITE
100	63.262	46.944	236.438	-18.949	-1544.710	-1515.904	791.827
200	98.198	103.052	156.335	-10.657	-1547.151	-1485.986	388.100
298.15	118.407	146.147	146.147	0.	-1548.080	-1455.733	255.038
300	118.758	146.881	146.149	0.219	-1548.089	-1455.160	253.366
400	135.562	183.448	151.021	12.971	-1552.872	-1423.255	185.858
500	148.114	215.101	160.743	27.179	-1551.745	-1390.957	145.312
600	157.737	242.994	172.173	42.492	-1549.626	-1358.985	118.310
700	164.850	267.859	184.099	58.632	-1546.808	-1327.426	99.054
800	170.707	290.268	195.992	75.420	-1543.488	-1296.307	84.640
900	175.310	310.648	207.617	92.728	-1539.816	-1265.627	73.455
1000	179.075	329.321	218.867	110.455	-1535.919	-1235.369	64.529
1100	182.004	346.523	229.700	128.505	-1689.825	-1196.287	56.807
1200	185.351	362.515	240.110	146.887	-1683.623	-1151.692	50.132
1249.000	186.376	369.955	245.059	155.996	CRYSTAL <--> LIQUID		
1249.000	179.912	410.154	245.059	206.204	TRANSITION		
1300	179.912	417.354	251.677	215.379	-1627.403	-1109.665	44.587
1400	179.912	430.687	263.994	233.370	-1621.773	-1070.051	39.924
1500	179.912	443.099	275.525	251.362	-1616.227	-1030.837	35.897
1600	179.912	454.711	286.365	269.353	-1610.764	-991.989	32.385
1700	179.912	465.618	296.592	287.344	-1655.560	-953.034	29.283
1800	179.912	475.901	306.271	305.335	-1650.032	-911.869	26.462
1900	179.912	485.629	315.457	323.326	-1644.548	-871.010	23.946
2000	179.912	494.857	324.198	341.318	-1639.109	-830.438	21.689
2100	179.912	503.635	332.535	359.309	-1633.718	-790.137	19.654
2200	179.912	512.004	340.504	377.300	-1628.380	-750.091	17.809
2300	179.912	520.002	348.136	395.291	-1623.098	-710.288	16.131
2400	179.912	527.659	355.458	413.282	-1617.876	-670.713	14.598
2500	179.912	535.003	362.494	431.274	-1612.720	-631.354	13.191
2600	179.912	542.059	369.265	449.265	-1607.634	-592.200	11.897
2700	179.912	548.849	375.791	467.256	-1602.623	-553.240	10.703
2800	179.912	555.392	382.090	485.247	-1597.692	-514.464	9.597
2900	179.912	561.706	388.175	503.238	-1592.847	-475.865	8.571
3000	179.912	567.805	394.062	521.230	-1588.095	-437.429	7.616

PREVIOUS: CURRENT: June 1972

Potassium Silicate (K_2SiO_3) $K_2O_3Si_1$(cr,l)

$\Delta_fH°(0\ K) = -1427.20 \pm 0.4\ kJ\ mol^{-1}$

$S°(298.15\ K) = 175.544 \pm 0.8\ J\ K^{-1}\ mol^{-1}$ $\Delta_fH°(298.15\ K) = -1437.71 \pm 0.4\ kJ\ mol^{-1}$

$T_{trs}(\alpha \rightarrow \beta) = 857 \pm 1\ K$ $\Delta_{trs}H°(\alpha \rightarrow \beta) = 8.452 \pm 0.418\ kJ\ mol^{-1}$

Enthalpy of Formation

Shibata et al. (1) measured the emf of the cell: K (amalgam, 0.2201%)|K_2SO_4(sat. soln.), Hg_2SO_4|Hg at 298-303 K. From their results at 298 K, correcting the observed voltage for the formation of the K-amalgam, and converting to absolute volts, we have E_{cell} = 3.59172 volts for the reaction at 298 K, 2 K(cr) + Hg_2SO_4(cr) = 2 Hg(ℓ) + K_2SO_4(cr). From this voltage, we calculate $\Delta_rG°$(298.15 K) = -165.656 kcal mol^{-1}. Taking $\Delta_fG°$(Hg_2SO_4, cr, 298.15 K) = -149.658 kcal mol^{-1} (2) and reference entropies from (3), we calculate $\Delta_fG°$(K_2SO_4, cr, 298.15 K) = -315.314 kcal mol^{-1} and $\Delta_fH°$(K_2SO_4, cr, 298.15 K) = -343.525±0.2 kcal mol^{-1}.

The enthalpy of solution at infinite dilution, $\Delta_{sol}H°$(K_2SO_4, cr, 298.15 K) = 5.677±0.020 kcal mol^{-1}, adopted in this tabulation is based on our reanalysis of the enthalpy of solution measurements of Mishchenko and Pronina (4), Tsvetkov and Rabinovich (5), Romanova and Samoilov (6), Cohen and Kooy (7), Bronsted (8), Samilov et al. (9), and Gritsus et al. (10). Enthalpies of dilution are taken from Lange and Streeck (11). When this result is combined with enthalpy of formation of the infinitely dilute ions from CODATA (2), we obtain $\Delta_fH°$(K_2SO_4, aq, ∞, 298.15 K) = -337.940±0.10 kcal mol^{-1} and $\Delta_fH°$(K_2SO_4, cr, 298.15 K) = -343.617±0.10 kcal mol^{-1}.

The solubility of K_2SO_4(cr) in water at 298 K has been measured by many workers (4, 12, 14, 15, 16). When the solubility of 0.6919 molal is combined with the activity coefficient data of Stokes (13), we obtain $\Delta_rG°$(298.15 K) = 2423±20 cal mol^{-1} for the reaction K_2SO_4(cr) = 2 K^+(aq, ∞) + SO_4^{-2}(aq, ∞). When this result is combined with enthalpies of formation and entropies from (3), we obtain $\Delta_fH°$(K_2SO_4, cr, 298.15 K) = -343.595±0.2 kcal mol^{-1}.

We adopt the value of $\Delta_fH°$(298.15 K) obtained from the enthalpy of solution measurements.

Heat Capacity and Entropy

Low temperature heat capacities of K_2SO_4(cr, α) have been measured by Moore and Kelley (17) from 52.7-295.4 K and by Paukov (18) from 12.5-299.5 K. The heat capacities reported by Moore and Kelley are systematically lower by about 0.5% than those of Paukov. Our adopted value of S°(298.15 K) = 41.956±0.2 cal K^{-1} mol^{-1} is based on S°(12.5 K) = 0.068 cal K^{-1} mol^{-1} (18) and H°(12.5 K) - H°(0 K) = 0.640 cal K^{-1} mol^{-1} derived from a Debye T^3-law extrapolation.

Heat capacity data measured by adiabatic calorimetry in the range 298.5-773.7 K have been reportd by Shmidt (19). In addition, drop calorimetric measurements have been reported by the following workers; Shomate and Naylor (20), Dworkin and Bredig (21), Denielou et al. (22), and Rubinchik et al. (23). The heat capacity and enthalpy curves between 298 K and 857 K are obtained by orthogonal polynomial curve-fitting of all the data listed above. These data are in good agreement.

Transition Data

K_2SO_4(cr, α) is the low-temperature form (sometimes denoted as K_2SO_4, II). The mineral is known as arcanite and exists in the form of orthorhombic crystals (space group D_{2h}^{16} -Pnma, 24). The orthorhombic-to-hexagonal transition temperature for K_2SO_4 is selected as 857±1 K from the calorimetric work of Dworkin and Bredig (21) and Rubinchik et al. (23) and the polymorphism study of Pistorius and Rapoport (25). The enthalpy of transition is determined by taking the difference between the adopted enthalpy curves of α- and β-K_2SO_4 extrapolated to 857 K.

References

1. F. L. E. Shibata, S. Oda, and S. Furukawa, J. Sci. Hiroshima Univ. (Japan) 3A, 227 (1933).
2. ICSU-CODATA Task Group on Key Values for Thermodynamics, J. Chem. Thermodyn. 10, 903 (1978).
3. JANAF Thermochemical Tables: K(ref. st.), 12-31-61; O_2(ref. st.), 3-31-77; S(ref. st.), 9-30-77; K_2SO_4(cr, β), 9-30-78.
4. K. P. Mishchenko and M. S. Pronina, J. Gen. Chem. 6, 85 (1936).
5. V. G. Tsvetkov and I. B. Rabinovich, Russ. J. Phys. Chem. 43, 675 (1969).
6. I. I. Romanova and O. Ya Samoilov, Russ. J. Inorg. Chem. 14, 131 (1969).
7. E. Cohen and J. Kooy, Z. Phys. Chem. 139, 273 (1928).
8. J. N. Bronsted, Z. Phys. Chem. 77, 315 (1911).
9. O. Ya Samoilov, H. K'O-yuan, and T. A. Nosova, J. Struct. Chem. 1, 379 (1960).
10. B. V. Gritsus, E. I. Akhumov, and L. P. Zhilina, J. Appl. Chem. USSR 45, 1969 (1972).
11. E. Lange and H. Streeck, Z. Phys. Chem. 152, 1 (1931).
12. T. Ishii and S. Fujita, J. Chem. Eng. Data 23, 19 (1978).
13. R. H. Stokes, Trans. Faraday Soc. 44, 295 (1948).
14. S. Z. Makarov and I. G. Druzhinin, Izv. Akad. Nauk SSSR, Ser. Khim. 1291 (1937).
15. A. P. Perova, Zh. Prikl. Khim. 11, 1045 (1930).
16. J. N Pearce and H. C. Eckstrom, J. Amer. Chem. Soc. 59, 2689 (1937).
17. G. E. Moore and K. K. Kelley, J. Amer. Chem. Soc. 64, 2949 (1942).
18. I. E. Paukov, Russ. J. Phys. Chem. 43, 1134 (1969).
19. N. E. Shmidt, Russ. J. Inorg. Chem. 12, 929 (1967).
20. C. H. Shomate and B. F. Naylor, J. Amer. Chem. Soc. 67, 72 (1945).
21. A. S. Dworkin and M. A. Bredig, J. Phys. Chem. 74, 3403 (1970).
22. L. Denielou, Y. Fournier, J. P. Petitet, and C. Tequi, C. R. Acad. Sci., Paris 270, 1854 (1970).
23. S. M. Rubinchik, E. I. Banashek, V. A. Sokolov, and A. I. Fomin, Russ. J. Phys. Chem. 45, 599 (1971).
24. M. T. Robinson, J. Phys. Chem. 62, 925 (1958).
25. C. W. F. T. Pistorius and E. Rapoport, J. Phys. Chem. Solids 30, 195.

Enthalpy Reference Temperature = T_r = 298.15 K Standard State Pressure = p° = 0.1 MPa

T/K	$C_p°$ (J K⁻¹mol⁻¹)	S° (J K⁻¹mol⁻¹)	-[G°-H°(T_r)]/T (J K⁻¹mol⁻¹)	H°-H°(T_r) (kJ mol⁻¹)	$\Delta_fH°$ (kJ mol⁻¹)	$\Delta_fG°$ (kJ mol⁻¹)	Log K_f
0	0.	0.	INFINITE	-25.436	-1427.199	-1427.199	INFINITE
100	79.195	61.609	277.387	-21.578	-1433.305	-1396.431	729.421
200	110.667	127.344	186.920	-11.915	-1436.295	-1358.245	354.737
298.15	131.319	175.544	175.544	0.	-1437.706	-1319.595	231.188
300	131.666	176.357	175.546	0.243	-1437.723	-1318.862	229.634
400	147.649	216.494	180.911	14.233	-1445.229	-1278.152	166.909
500	160.733	250.870	191.544	29.663	-1446.060	-1236.272	129.153
600	172.527	281.186	204.005	46.309	-1445.397	-1194.353	103.978
700	186.368	308.829	217.031	64.259	-1443.295	-1152.659	86.012
800	199.619	334.580	230.132	83.558	-1439.835	-1111.361	72.564
857.000	207.169	348.575	237.547	95.151	--------	ALPHA <--> BETA	-------
900	212.865	358.856	243.098	104.182	-1488.244	-1069.491	62.072
1000	226.116	381.969	255.838	126.131	-1481.100	-1023.335	53.454
1100	239.365	404.142	268.319	149.405	-1630.753	-968.729	46.001
1200	252.613	425.538	280.534	174.004	-1619.296	-909.041	39.570

PREVIOUS: December 1971 CURRENT: June 1978

POTASSIUM SULFATE (β-K_2SO_4) CRYSTAL M_r = 174.2542

$S°(298.15\ K) = [192.2\&(]\ J\ K^{-1}\ mol^{-1}$

$T_{trs}(\alpha \rightarrow \beta) = 857 \pm 1\ K$

$T_{fus}H°(\beta \rightarrow \ell) = 1342 \pm 1\ K$

$\Delta_fH°(298.15\ K) = [-1424.243]\ kJ\ mol^{-1}$

$\Delta_{trs}H°(\alpha \rightarrow \beta) = 8.452 \pm 0.418\ kJ\ mol^{-1}$

$\Delta_{fus}H°(\beta \rightarrow \ell) = 34.392\ kJ\ mol^{-1}$

Enthalpy of Formation

$\Delta_fH°(K_2SO_4$, cr, β, 298.15 K) is calculated from $\Delta_fH°(K_2SO_4$, cr, α, 298.15 K) by adding the enthalpy of transition ($\alpha \rightarrow \beta$), $\Delta_{trs}H°$, and the difference in enthalpy, H°(857 K)-H°(298.15 K), between the alpha and beta crystalline forms (1).

Heat Capacity and Entropy

High temperature enthalpy measurements by drop calorimetry were carried out by Shomate and Naylor (2, 872.4-1258.8 K), Rubinchik et al. (3, 859.0-1340.7 K) Dworkin and Bredig (4, 868.5-979.9 K), and Denielou et al. (5, 864-1300 K). The data of (2) are systematically higher by about 0.5% than the data from (3, 4, 5). $C_p°$ values are derived from the smoothed curves and extrapolated to 298.15 K and 1500 K. S°(K_2SO_4, β, 298.15 K) is calculated in a manner analogous to that used for the enthalpy of formation.

Transition Data

The high temperature β-modification of K_2SO_4 is of hexagonal structure, space group D^3_{3d}-P3ml (6). The transition from the low temperature orthorhombic pseudohexagonal modification occurs at 857±1 K (1).

Fusion Data

The adopted melting point of 1342 K was determined by Shomate and Naylor (2) and Rubinchik et al. (3). The heat of fusion, $\Delta_{fus}H° = 8.220 \pm 0.400\ kcal\ mol^{-1}$, is calculated from the difference between the smoothed relative enthalpy of the liquid (see $K_2SO_4(\ell)$ table) and the adopted enthalpy value for the β-crystalline form at T_{fus}.

References

1. JANAF Thermochemical Tables: K_2SO_4(cr, α), $K_2SO_4(\ell)$ 6-30-78.
2. C. H. Shomate and B. F. Naylor, J. Amer. Chem. Soc. 67, 72 (1945).
3. S. M. Rubinchik, E. I. Banashek, V. A. Sokolov, and A. I. Fomin, Russ. J. Phys. Chem. 45, 599 (1971).
4. A. S. Dworkin and M. A. Bredig, J. Phys. Chem. 74, 3403 (1970).
5. L. Denielou, Y. Fournier, J. P. Petitet, and C. Tequi, C. R. Acad. Sci., Paris 270, 1854 (1970).
6. M. Bernard and R. Hocart, Bull. Soc. Franc. Miner. Crist. 84, 396 (1961); G. Pannetier and M. Gaultier, Bull. Chim. Fr. 188 (1966).
7. H. S. Roberts, J. Amer. Chem. Sci. 35A, 273 (1938).

Potassium Sulfate, Beta (K_2SO_4) $K_2O_4S_1(cr)$

Enthalpy Reference Temperature = T_r = 298.15 K Standard State Pressure = p° = 0.1 MPa

T/K	$C_p°$	S°	-[G°-H°(T_r)]/T	H°-H°(T_r)	$\Delta_fH°$	$\Delta_fG°$	Log K_f
	J K⁻¹mol⁻¹			kJ mol⁻¹			
298.15	138.574	192.279	192.279	0.	-1424.243	-1311.121	229.703
300	138.741	193.137	192.282	0.257	-1424.246	-1310.419	228.164
400	146.858	234.162	197.820	14.537	-1431.462	-1271.452	166.035
500	154.975	267.803	208.548	29.628	-1432.632	-1231.311	128.634
600	163.134	296.781	220.891	45.534	-1432.709	-1191.022	103.688
700	171.251	322.537	233.605	62.253	-1431.838	-1150.797	85.874
800	179.368	345.935	246.205	79.784	-1430.146	-1110.755	72.525
857.000	184.000	358.437	253.257	90.140	-------- ALPHA <--> BETA -------		
900	187.485	367.530	258.500	98.127	-1480.836	-1069.890	62.095
1000	195.602	387.703	270.423	117.281	-1476.487	-1024.456	53.512
1100	203.761	406.729	281.957	137.249	-1629.446	-970.268	46.074
1200	211.878	424.807	293.114	158.031	-1621.806	-910.674	39.641
1300	219.995	442.087	303.914	179.625	-1613.438	-851.747	34.224
1342.000	223.384	449.136	308.349	188.936	------- BETA <--> LIQUID -------		
1400	228.112	458.687	314.380	202.030	-1604.337	-793.492	29.606
1500	236.229	474.702	324.537	225.247	-1594.495	-735.914	25.627

PREVIOUS: CURRENT: June 1978

Potassium Sulfate, Beta (K_2SO_4) $K_2O_4S_1(cr)$

POTASSIUM SULFATE (K_2SO_4) LIQUID M_r = 174.2542

Potassium Sulfate (K_2SO_4) $K_2O_4S_1(l)$

$S°(298.15 K) = [211.505] J K^{-1} mol^{-1}$ $\Delta_fH°(298.15 K) = [-1393.668] kJ mol^{-1}$

$T_{fus}(\beta \rightarrow \ell) = 1342 \pm 1 K$ $\Delta_{fus}H°(\beta \rightarrow \ell) = 34.392 kJ mol^{-1}$

$T_{dcm} = 2567 K$

Enthalpy of Formation

$\Delta_fH°(K_2SO_4, \ell, 298.15 K)$ is calculated from $\Delta_fH°(K_2SO_4, \beta, 298.15 K)$ by adding the enthalpy of fusion, $\Delta_{fus}H°$, and the difference in enthalpy, H°(1342 K)-H°(298.15 K), between the β-crystal and liquid (1).

Heat Capacity and Entropy

Enthalpies for $K_2SO_4(\ell)$ have been measured by drop calorimetry by Shomate and Naylor (1358-1698 K, 2), Denielou et al. (1350-1550 K, 3), and Rubinchik et al. (single point at 1344.84 K, 4). A constant heat capacity of 46.15 cal $K^{-1} mol^{-1}$ is selected above a hypothetical glass transition at 800 K. Below the glass transition, the heat capacity is taken as that of the orthorhombic crystal.

$S°(K_2SO_4, \ell, 298.15 K)$ is calculated in a manner analogous to that used for the enthalpy of formation.

Fusion Data

See K_2SO_4(cr, β) table for details.

Vaporization and Decomposition Data

$K_2SO_4(\ell)$ can vaporize to the gaseous monomer $K_2SO_4(g)$ or decompose according to $K_2SO_4(\ell) = 2 K(g) + SO_2(g) + O_2(g)$. Vaporization pressures from the two processes are of comparable magnitude in the vicinity of the melting point (6). Recent experimental data for vaporization of K_2SO_4(cr) (5, 6) are in good agreement with dissociation pressures and the heat of reaction calculated from the JANAF Thermochemical Tables (1). Molecular vaporization data are discussed on the $K_2SO_4(g)$ table (1). T_{vap} = 2975 K is the calculated temperature at which the Gibbs energy difference is zero for the reaction $K_2SO_4(\ell) = 2 K(g) + SO_2(g) + O_2(g)$. Decomposition to the oxides of potassium produces significantly lower decomposition pressures than decomposition to the metal. Significant decomposition can occur in the presence of water vapor according to $K_2SO_4(\ell) + H_2O(g) = 2 KOH(g) + SO_2(g) + 1/2 O_2(g)$.

References - $K_2SO_4(\ell)$

1. JANAF Thermochemical Tables: K_2SO_4(cr, α), K_2SO_4(cr, β), K_2SO_4(g), 6-30-78.
2. C. H. Shomate and B. F. Naylor, J. Amer. Chem. Soc. 67, 72 (1945).
3. L. Denielou, Y. Fournier, J. P. Petitet, and C. Tequi, C. R. Acad. Sci., Paris 270, 1854 (1975).
4. S. M. Rubinchik, E. I. Banashek, V. A. Sokolov, and A. I. Fomin, Russ. J. Phys. Chem. 45, 599 (1971).
5. P. J. Ficalora, D. M. Uy, D. W. Muenow, and J. L. Margrave, J. Amer. Ceram. Soc. 51, 574 (1968).
6. K. H. Lau, D. Cubicciotti, and D. L. Hildenbrand, J. Electrochem. Soc. 126, 490 (1979).

References - $K_2SO_4(g)$

1. A. G. Effimova and L. N. Gorokhov, Teplofiz. Vys. Temp. 16, 1195 (1978); L. N. Gorokhov, High Temperature Technology, Proc. Third International Symposium on High Temperature Technology (1967) Butterworths, London; L. N. Gorokhov and A. G. Efimova Annual Report, Institute for High Temperatures, Academy of Sciences, USSR, 1975, "Nauka," 1976.
2. E. R. Plante, C. D. Olson, and T. Negas, U. S. Nat. Bur. Stand., personal communication, 1978.
3. K. H. Lau, D. Cubicciotti, and D. L. Hildenbrand, J. Electrochem. Soc. 126, 490 (1979).
4. V. V. Ugarov, Yu. S. Ezhov, and N. G. Rambidi, J. Mol. Struct. 25, 357 (1975).
5. V. P. Spiridonov and B. I. Lutoshkin, Vestn. Mosk. Univ. Khim. 11, 509 (1970)(English trans. NASA TT F 16199).
6. L. V. Gurvich, O. V. Dorofeeva, and V. S. Yungman, Proceedings of the Conference on High Temperature Science related to Open Cycle Coal-Fuel MHD Systems, Argonne National Lab. Report ANL-77-21 (1977).
7. R. M. Atkins and K. A. Ginerich, Chem. Phys. Lett. 53, 347 (1978).
8. A. A. Belyaeva, M. I. Dworkin, and L. D. Shcherba, Opt. Spectroscopy 38, 170 (1975); 38, 291 (1975); 43, 114 (1977).
9. J. E. D. Davies and W. F. Sandford, J. Chem. Soc. 1912 (1975).
10. I. Eliezer and R. A. Howald, J. Chem. Phys. 65, 3053 (1976).
11. T. Kosugi, Bull. Chem. Soc. Japan 45, 15 (1972); Kogyo Kagaku Zasshi 70, 2089 (1967).
12. P. J. Ficalora, D. M. Uy, D. W. Muenow, and J. L. Margrave, J. Amer. Ceram. Soc. 51, 574 (1968).
13. W. D. Halstead, Trans. Faraday Soc. 66, 1966 (1970).
14. J. Dubois and J. Millet, C. R. Acad. Sci. 266, 852 (1968).

Enthalpy Reference Temperature = T_r = 298.15 K; Standard State Pressure = p° = 0.1 MPa

	J K⁻¹mol⁻¹			kJ mol⁻¹			
T/K	$C_p°$	S°	$-[G°-H°(T_r)]/T$	$H°-H°(T_r)$	$\Delta_fH°$	$\Delta_fG°$	Log K_f
298.15	131.319	211.505	211.505	0.	-1393.668	-1286.278	225.351
300	131.666	212.319	211.508	0.243	-1393.685	-1285.612	223.845
400	147.649	252.455	216.872	14.233	-1401.191	-1248.498	163.037
500	160.733	286.831	227.505	29.663	-1402.022	-1210.215	126.430
600	172.527	317.148	239.966	46.309	-1401.359	-1171.892	102.022
700	186.368	344.790	252.992	64.259	-1399.257	-1133.794	84.605
800	199.619	370.547	266.094	83.563	-1395.792	-1096.092	71.567
800.000	199.619	370.547	266.094	83.563	GLASS <--> LIQUID		
800.000	201.460	370.547	266.094	83.563	TRANSITION		
900	201.460	394.276	279.044	103.709	-1444.679	-1057.804	61.393
1000	201.460	415.502	291.647	123.855	-1439.338	-1015.106	53.024
1100	201.460	434.703	303.793	144.001	-1592.120	-963.712	45.763
1200	201.460	452.232	315.443	164.147	-1585.116	-906.894	39.476
1300	201.460	468.358	326.594	184.293	-1578.196	-850.657	34.180
1342.000	201.460	474.763	331.132	192.754	------- BETA <--> LIQUID -------		
1400	201.460	483.287	337.260	204.439	-1571.353	-794.950	29.660
1500	201.460	497.187	347.464	224.585	-1564.582	-739.730	25.760
1600	201.460	510.189	357.232	244.731	-1557.880	-684.958	22.362
1700	201.460	522.402	366.592	264.877	-1551.245	-630.604	19.376
1800	201.460	533.917	375.571	285.022	-1544.678	-576.639	16.734
1900	201.460	544.809	384.195	305.168	-1538.178	-523.036	14.379
2000	201.460	555.143	392.486	325.314	-1531.747	-469.774	12.269
2100	201.460	564.972	400.467	345.460	-1525.387	-416.833	10.368
2200	201.460	574.344	408.159	365.606	-1519.101	-364.191	8.647
2300	201.460	583.299	415.581	385.752	-1512.892	-311.835	7.082
2400	201.460	591.873	422.749	405.898	-1506.763	-259.747	5.653
2500	201.460	600.097	429.680	426.044	-1500.719	-207.913	4.344
2600	201.460	607.999	436.387	446.190	-1494.762	-156.318	3.140
2700	201.460	615.602	442.885	466.336	-1488.898	-104.951	2.030
2800	201.460	622.929	449.185	486.482	-1483.129	-53.799	1.004
2900	201.460	629.998	455.299	506.628	-1477.463	-2.853	0.051
3000	201.460	636.828	461.237	526.774	-1471.903	47.901	-0.834

PREVIOUS: December 1971 CURRENT: June 1978

POTASSIUM SULFATE (K_2SO_4) ALPHA-BETA-LIQUID M_r = 174.2542

0 to 857 K		crystal, alpha
857 to 1342 K		crystal, beta
above 1342 K		liquid

Refer to the individual tables for details.

Potassium Sulfate (K_2SO_4) $K_2O_4S_1$(cr,l)

Enthalpy Reference Temperature = T_r = 298.15 K   Standard State Pressure = p° = 0.1 MPa

T/K	C_p° (J K⁻¹mol⁻¹)	S°	-[G°-H°(T_r)]/T	H°-H°(T_r) (kJ mol⁻¹)	Δ_fH°	Δ_fG°	Log K_f
0	0.	0.	INFINITE	-25.436	-1427.199	-1427.199	INFINITE
100	79.195	61.609	277.387	-21.578	-1433.305	-1396.431	729.421
200	110.667	127.344	186.920	-11.915	-1436.295	-1358.245	354.737
298.15	131.319	175.544	175.544	0.	-1437.706	-1319.595	231.188
300	131.666	176.357	175.546	0.243	-1437.723	-1318.862	229.634
400	147.649	216.494	180.911	14.233	-1445.229	-1278.152	166.909
500	160.733	250.870	191.544	29.663	-1446.060	-1236.272	129.153
600	172.527	281.186	204.005	46.309	-1445.397	-1194.353	103.978
700	186.368	308.829	217.031	64.259	-1443.295	-1152.659	86.012
800	199.619	334.580	230.132	83.558	-1439.835	-1111.361	72.564
857.000	207.169	348.575	237.547	95.151	ALPHA <--> BETA		
857.000	184.000	358.437	237.547	103.603	TRANSITION		
900	187.485	367.530	243.541	111.590	-1480.836	-1069.890	62.095
1000	195.602	387.703	256.959	130.744	-1476.487	-1024.456	53.512
1100	203.761	406.729	269.718	150.713	-1629.446	-970.268	46.074
1200	211.878	424.807	281.895	171.494	-1621.806	-910.674	39.641
1300	219.995	442.087	293.557	193.089	-1613.438	-851.747	34.224
1342.000	223.384	449.136	298.316	202.400	BETA <--> LIQUID		
1342.000	201.460	474.763	298.316	236.792	TRANSITION		
1400	201.460	483.287	305.804	248.477	-1571.353	-794.950	29.660
1500	201.460	497.187	318.105	268.623	-1564.582	-739.730	25.760
1600	201.460	510.189	329.708	288.769	-1557.880	-684.958	22.362
1700	201.460	522.402	340.688	308.915	-1551.245	-630.604	19.376
1800	201.460	533.917	351.106	329.061	-1544.678	-576.639	16.734
1900	201.460	544.809	361.017	349.207	-1538.178	-523.037	14.379
2000	201.460	555.143	370.467	369.353	-1531.747	-469.774	12.269
2100	201.460	564.972	379.497	389.499	-1525.387	-416.833	10.368
2200	201.460	574.344	388.142	409.644	-1519.101	-364.191	8.647
2300	201.460	583.299	396.434	429.790	-1512.892	-311.835	7.082
2400	201.460	591.873	404.400	449.936	-1506.763	-259.747	5.653
2500	201.460	600.097	412.064	470.082	-1500.719	-207.913	4.344
2600	201.460	607.999	419.449	490.228	-1494.762	-156.318	3.140
2700	201.460	615.602	426.574	510.374	-1488.898	-104.951	2.030
2800	201.460	622.929	433.457	530.520	-1483.129	-53.799	1.004
2900	201.460	629.998	440.113	550.666	-1477.463	-2.853	0.051
3000	201.460	636.828	446.557	570.812	-1471.903	47.901	-0.834

PREVIOUS: CURRENT: June 1978

Potassium Sulfate (K_2SO_4) $K_2O_4S_1$(cr,l)

POTASSIUM SULFATE (K_2SO_4) IDEAL GAS M_r = 174.2542

$\Delta_fH°(0\ K) = -1080.64 \pm 16.7\ kJ\ mol^{-1}$

$S°(298.15\ K) = 366.152 \pm 16.7\ J\ K^{-1}\ mol^{-1}$ $\Delta_fH°(298.15\ K) = -1094.12 \pm 16.7\ kJ\ mol^{-1}$

Vibrational Frequencies, Symmetries, and Degeneracies

	ν, cm^{-1}		ν, cm^{-1}		ν, cm^{-1}
A_1	962 (1)	B_1	[450](1)	E	1098 (2)
	[458](1)	B_2	1124 (1)		613 (2)
	[190](1)		639 (1)		[230](2)
			[262](1)		[62](2)

Ground State Quantum Weight: [1] σ = 4

Point Group: D_{2h}

Bond Distances: S-O = 1.47 Å K-O = 1.45 Å

Bond Angles: O-S-O = 109.47° O-K-O = 58.67°

Product of the Moments of Inertia: $I_AI_BI_C = 2.626307 \times 10^{-112}\ g^3\ cm^6$

Enthalpy of Formation

The adopted value of the enthalpy of formation is based on JANAF analyses of the vaporization data given below. Earlier data on the vaporizaiton of K_2SO_4 is extremely discordant (for a review, see 10). Dissociation and other reactions play an important role in K_2SO_4 vaporization (see $K_2SO_4(\ell)$ table).

Torsion-effusion studies on the vaporization of $K_2SO_4(g)$ by Lau et al. (3) showed an orifice dependence which was reproduced in four separate runs. Plante et al. (2) did not observe this orifice size dependence in their effusion study, obtaning similar total pressures in two separate experiments. Calculated dissociation pressures were subtracted in both studies to obtain $K_2SO_4(g)$ pressures. Similar results were obtained by Efimova and Gorokhov (1). We have corrected the effusion data of Kosugi (11) for dissociation. The 3rd law enthalpy of sublimation from the three studies (1, 2, 3) are in very good agreement indicating similar $K_2SO_4(g)$ pressures. The studies of Lau et al. (3) and Efimova and Gorokhov (1) yield 2nd law enthalply of sublimation (from $K_2SO_4(cr, \infty)$) some 6 kcal mol^{-1} greater than our calculated 3rd law enthalpy.

Lau et al. (3) and Efimova and Gorokhov (1) also carried out mass spectrometric measurements of the $K_2SO_4^+$ ion intensity as a function of temperature. The 2nd law enthalpy of sublimation of Lau et al. (3) obtained in two separate mass spectrometric experiments tend to corroborate their 2nd law enthalpy from the torsion-effusion measurements. The 2nd law enthalpy of Efimova and Gorokhov (1) as well as that of Ficalora et al. (12) are more in line with our 3rd law effusion values. Kosugi (11) did not observe the $K_2SO_4^+$ ion in his mass spectrometric experiments. The appearance potential of $K_2SO_4^+$ has been measured as 7.40ϕ0.5 eV (12), 8.40±0.3 eV (3) and 8.0 eV (1).

We have discounted the transpiration experiments of Halstead (13) and Dubois and Millet (14). The calculated 3rd law $\Delta_rH°(298.15\ K)$ may have an uncertainty of 3 kcal mol^{-1} since the JANAF Gibbs energy functions are partially based on the estimated molecular constants of $K_2SO_4(g)$. Our adopted $\Delta_fH°$ is based on a weighted average of the 3rd law results from (1, 2, 3).

Source	Method	Reaction*	T/K	Data Points	δS cal K^{-1} mol^{-1}	$\Delta_rH°$(298.15 K), kcal mol^{-1} 2nd law	3rd law	$\Delta_fH°(K_2SO_r$, g, 298.15 K) kcal mol^{-1}
Efimova (1)	effusion	A	1171-1331	equation	4.38	84.77	79.02±0.11	-261.38±1.5
Lau et al. (3)	torsion-effusion	A	1180-1274	37	5.18	85.36	78.93±0.45	-261.47±1.5
Plante (2)	effusion	A	1166-1322	26	-1.42±0.8	77.03±0.9	78.78±0.42	-261.62±1.5
Plante (2)	effusion	A	1196-1321	15	-0.76±2.0	77.87±2.0	78.83±0.51	-261.57±1.0
Halstead	effusion	A	1180-1340	equation	1.07	75.75	74.39±0.15	-266.02±1.0
Kosugi (11)	effusion	B	1409-1530	12	2.70±2.8	73.67±4.2	69.66±0.76	-263.43±2.0
Efimova (1)	mass spec	A	1171-1331	?	--	78.0 ±1.0	--	-262.4 ±2
Lau et al. (3)	mass spec	A	1183-1248	9	--	84.12±2.02	--	-256.28±2.0
Lau et al. (3)	mass spec	A	1180-1314	10	--	81.96±0.40	--	-258.44±1.5
Ficalora (12)	mass spec	A	1080-1230	?	--	79.8 ±2.0	--	-260.6 ±4

*Reactions: A) $K_2SO_r(cr, \beta) = K_2SO_r(g)$ B) $K_2SO_4(\ell) = K_2SO_4(g)$

Heat Capacity and Entropy

The adopted structure (D_{2h} symmetry) is based on the high temperature electron diffraction study of Spiridonov and Lutoshkin (5). This symmetry is supported by the similar work of Ugarov et al. (4) on Cs_2SO_4 and the IR and Raman study of Atkins and Gingerich (7) on K_2SO_4 and Na_2SO_4. The molecule can be pictured as having a central sulfur atom surrounded at the corners of a tetrahedron by four oxygen atoms with bridging potassium atoms between two pairs of oxygens forming two planar mutually perpendicular KO_2S rings. The sulphate core is fairly rigid while the metal atoms undergo large amplitude motions. The internuclear distances of 1.47 Å for S-O and 2.45 Å for K-O are from (5) as are the O-S-O and O-K-O angles. The principal moments of inertia are $I_A = 15.3096 \times 10^{-39}$ and $I_B = I_C = 130.9758 \times 10^{-39}\ g\ cm^2$.

Atkins and Gingerich (7) have observed the infrared and Raman spectra of K_2SO_4 and Na_2SO_4 isolated in oxygen and nitrogen matrices at 12 K. Five of the eleven fundamental frequencies were observed in this work. Good agreement exists between these observations and the frequencies adopted by Gurvich et al. (6). The two other sulphate-ion group fundamentals are taken from the Raman study of Davies and Sandford (9). The low-frequency fundamentals involving the metal atoms are taken from the infrared study of Belyaeva et al. (8). These authors observed the spectra of K_2SO_4, Rb_2SO_4 and Cs_2SO_4 isolated in Ar, Kr, and Xe matrices at 4.2 K. Uncertainties in the location of the gas phase frequencies may contribute 4 cal K^{-1} mol^{-1} to S°(298.15 K).

References

References are listed on the liquid table.

Potassium Sulfate (K_2SO_4) $K_2O_4S_1(g)$

Enthalpy Reference Temperature = T_r = 298.15 K Standard State Pressure = p° = 0.1 MPa

T/K	$C_p°$ (J $K^{-1}mol^{-1}$)	S°	$-[G°-H°(T_r)]/T$	$H°-H°(T_r)$ (kJ mol^{-1})	$\Delta_fH°$	$\Delta_fG°$	Log K_f
0	0.	0.	INFINITE	-22.463	-1080.636	-1080.636	INFINITE
100	64.770	273.209	449.682	-17.647	-1085.784	-1070.071	558.948
200	90.384	326.456	375.546	-9.818	-1090.608	-1052.380	274.854
250	100.439	347.734	367.897	-5.041	-1092.515	-1042.597	217.839
298.15	108.730	366.152	366.152	0.	-1094.116	-1032.835	180.948
300	109.023	366.826	366.154	0.201	-1094.175	-1032.454	179.766
350	116.277	384.193	367.509	5.839	-1100.439	-1021.853	152.503
400	122.342	400.128	370.604	11.809	-1104.063	-1010.439	131.950
450	127.380	414.838	374.713	18.056	-1105.903	-998.632	115.918
500	131.555	428.481	379.416	24.533	-1107.600	-986.618	103.071
600	137.903	453.063	389.688	38.025	-1110.091	-962.173	83.765
700	142.345	474.674	400.317	52.050	-1111.914	-937.369	69.947
800	145.525	493.901	410.835	66.452	-1113.351	-912.333	59.569
900	147.859	511.183	421.041	81.127	-1167.708	-886.050	51.425
1000	149.612	526.856	430.851	96.005	-1167.636	-854.758	44.648
1100	150.956	541.181	440.239	111.036	-1325.533	-814.251	38.666
1200	152.006	554.363	449.208	126.186	-1323.524	-767.860	33.424
1300	152.842	566.564	457.772	141.430	-1321.507	-721.636	28.996
1400	153.515	577.917	465.953	156.749	-1319.491	-675.568	25.206
1500	154.067	588.527	473.775	172.129	-1317.486	-629.644	21.926
1600	154.523	598.486	481.261	187.559	-1315.500	-583.853	19.061
1700	154.904	607.865	488.435	203.031	-1313.539	-538.185	16.536
1800	155.226	616.729	495.319	218.538	-1311.610	-492.632	14.296
1900	155.500	625.129	501.932	234.075	-1309.720	-447.185	12.294
2000	155.736	633.111	508.293	249.637	-1307.873	-401.836	10.495
2100	155.940	640.715	514.419	265.221	-1306.075	-356.579	8.869
2200	156.117	647.973	520.326	280.824	-1304.332	-311.405	7.394
2300	156.272	654.916	526.028	296.444	-1302.649	-266.310	6.048
2400	156.409	661.570	531.538	312.078	-1301.032	-221.288	4.816
2500	156.529	667.957	536.868	327.725	-1299.487	-176.330	3.684
2600	156.637	674.099	542.028	343.383	-1298.017	-131.433	2.641
2700	156.733	680.012	547.030	359.052	-1296.630	-86.591	1.675
2800	156.819	685.714	551.882	374.729	-1295.330	-41.798	0.780
2900	156.896	691.218	556.592	390.415	-1294.124	2.948	-0.053
3000	156.966	696.538	561.169	406.108	-1293.017	47.656	-0.830
3100	157.029	701.686	565.619	421.808	-1292.015	92.329	-1.556
3200	157.087	706.673	569.950	437.514	-1291.126	136.970	-2.236
3300	157.139	711.507	574.166	453.225	-1290.336	181.583	-2.874
3400	157.187	716.199	578.275	468.942	-1289.682	226.178	-3.475
3500	157.231	720.756	582.281	484.663	-1289.155	270.753	-4.041
3600	157.272	725.186	586.189	500.388	-1288.769	315.316	-4.575
3700	157.309	729.496	590.005	516.117	-1288.530	359.870	-5.080
3800	157.343	733.691	593.731	531.850	-1288.445	404.421	-5.559
3900	157.375	737.779	597.372	547.586	-1288.523	448.972	-6.013
4000	157.404	741.764	600.932	563.325	-1288.761	493.523	-6.445
4100	157.432	745.651	604.415	579.066	-1289.181	538.085	-6.855
4200	157.457	749.445	607.823	594.811	-1289.798	582.659	-7.246
4300	157.481	753.150	611.160	610.558	-1290.615	627.252	-7.620
4400	157.503	756.771	614.428	626.307	-1291.086	671.809	-7.975
4500	157.523	760.310	617.631	642.058	-1292.192	716.432	-8.316
4600	157.543	763.773	620.770	657.812	-1293.501	761.081	-8.642
4700	157.561	767.161	623.849	673.567	-1294.978	805.758	-8.955
4800	157.578	770.479	626.869	689.324	-1296.449	850.445	-9.255
4900	157.594	773.728	629.834	705.082	-1298.342	895.192	-9.543
5000	157.609	776.912	632.743	720.842	-1300.452	939.979	-9.820
5100	157.623	780.033	635.601	736.604	-1302.781	984.812	-10.087
5200	157.636	783.094	638.408	752.367	-1305.334	1029.689	-10.343
5300	157.649	786.097	641.166	768.131	-1308.111	1074.621	-10.591
5400	157.660	789.044	643.878	783.897	-1311.116	1119.605	-10.830
5500	157.672	791.937	646.543	799.663	-1314.347	1164.647	-11.061
5600	157.682	794.778	649.165	815.431	-1317.805	1209.752	-11.284
5700	157.693	797.569	651.744	831.200	-1320.800	1254.823	-11.499
5800	157.702	800.311	654.282	846.969	-1324.621	1300.044	-11.708
5900	157.711	803.007	656.780	862.740	-1328.659	1345.328	-11.911
6000	157.720	805.658	659.239	878.512	-1332.909	1390.687	-12.107

PREVIOUS: June 1978 (1 atm) CURRENT: June 1978 (1 bar)

Potassium Sulfate (K_2SO_4) $K_2O_4S_1(g)$

POTASSIUM SULFIDE (K_2S) CRYSTAL M_r = 110.2566

$S°(298.15\ K) = 115.06 \pm 16.7\ J\ K^{-1}\ mol^{-1}$
$T_{trs} = 1050 \pm 10\ K$ (diffuse lambda transition)
$T_{fus} = 1221 \pm 10\ K$

$\Delta_fH°(0\ K)$ = Unknown
$\Delta_fH°(298.15\ K) = -376.56 \pm 12.6\ kJ\ mol^{-1}$
$\Delta_{trs}H° = 0\ kJ\ mol^{-1}$
$\Delta_{fus}H° = 16.150 \pm 0.17\ kJ\ mol^{-1}$

Enthalpy of Formation

Experimental measurements leading to the enthalpy of formation of $K_2S(cr)$ are summarized below.

Source	Reaction	$\Delta_fH°(298.15\ K)$, kcal mol^{-1}
Sabatier (1)	$\Delta_{sln}H°$ $K_2S(cr)$ and K(cr); $\Delta_3H°$ KOH(aq) + $H_2S(aq)$	-103.4
Rengade and Costeanu (2)		-87.3
Letoffe et al. (3)	H_2SO_4(aq, 1100 H_2O) + K_2S(cr) = K SO (soln.) + H S(g)	-92.1[a]

a. The authors' value of -91.1 ± 0.9 kcal mol^{-1} is changed by new $\Delta_fH°(298.15\ K)$ values for H_2SO_4(aq, 1100 H_2O), $K_2SO_4(cr)$ and $H_2S(g)$ (4, 5).

Since Sabatier himself has expressed reservations because of impure material, we eliminate his value and adopt the average of the other two values, $\Delta_fH°(K_2S, cr, 298.15\ K) = -90\pm3$ kcal mol^{-1}.

Heat Capacity and Entropy

The heat capacity is taken from the drop calorimetry of Dworkin and Bredig (384 to 1260 K) (6). Between 298 and 820 K, the observed enthalpy differences and the constraint of passing through zero at 298.15 K are fit by a linear least squares technique; $C_p° = 15.99 + 6.22 \times 10^{-3}$ T cal K^{-1} mol^{-1} (298–820 K). The heat capacity in the observed diffuse lambda transition region, 820–1100 K, was adjusted to properly reproduce the observed enthalpies. The sharp heat capacity maximum occurs at 1050 K (6) and 46.60 cal K^{-1} mol^{-1}. From 1100 K to the melting point, 1221 K (7), the heat capacity was found to be constant at 34.02 cal K^{-1} mol^{-1}.

Freeman (8) has estimated $S°(K_2S, cr, 298.15\ K)$ = 27 cal K^{-1} mol^{-1}; Voronin (9) also has estimated 27 cal K^{-1} mol^{-1}. Kelley's additive entropy constants give 28.0 We adopt $S°(K_2S, cr, 298.15\ K) = 27.5\pm4$ cal K^{-1} mol^{-1}.

Transition and Fusion Data

A crystal II to crystal I transition at 419.6 K with a enthalpy change of 0.085 kcal mol^{-1} has been listed (10) from Bridgman (11). However, West (12) found no change in the x-ray powder diffraction pattern over the 298–473 K range. Although not definitive because of the small heat change listed, the drop calorimetric enthalpy measurements of Dworkin and Bredig (6) do not show this transition. Therefore, we do not indicate any solid state transition at 419.6 K.

The diffuse lambda transition has been discussed above.

A melting point of 1221 K was determined by thermal analysis (cooling curves) of the $K\text{-}K_2S$ system by Dworkin and Bredig (7). This is considerably higher than earlier values of 1108 K (13) and 1185 K (14), but it is confirmed by their drop calometric measurements (6). A melting point of 1221 ± 10 K is adopted with $\Delta_{fus}H° = 3.86\pm0.04$ kcal mol^{-1} (6).

References

1. P. Sabatier, Annls. Chim. Phys. Ser. 5, 22, 5 (1881).
2. E. Rengade and N. Costeanu, Compt. rend. 158, 946 (1914), and Bull. Soc. Chim. Fr. 15, 717 (1914).
3. J. M. Letoffe, R. D. Joly, J. Thourey, G. Perachon, and J. Bousquet, J. Chem. Phys. Physicochim. Biol. 71, 427 (1974), and J. M. Letoffe, Doctorate Thesis, Universite Claude Bernard-Lyon I, 1975.
4. U. S. Nat. Bur. Stand. Tech. Note 270-3, 1968.
5. JANAF Thermochemical Tables: $K_2SO_4(cr)$, 12-31-71, $H_2S(g)$, 6-30-77.
6. A. S. Dworkin and M. A. Bredig, J. Phys. Chem. 72, 1277 (1968).
7. A. S. Dworkin and M. A. Bredig, J. Phys. Chem. 71, 764 (1967).
8. R. D. Freeman, Oklahoma State University, Department of Chemistry, Research Foundation Report No. 60, 1962.
9. G. F. Voronin, Russ. J. Phys. Chem. 44, 1717 (1970).
10. U. S. Nat. Bur. Stand. Circ. 500, 1952.
11. P. W. Bridgman, Proc. Am. Acad. Arts Sci. 51, 53 (1915).
12. C. D. West, Z. Krist. 88, 97 (1934).
13. O. Kubaschewski, E. LL. Evans and C. B. Alcock, "Metallurgical Thermochemistry," 4th ed., Pergamon Press, Oxford, 1967.
14. J. Goubeau, H. Kolb, and H. G. Krall, Z. Anorg. Allg. Chem. 236, 45 (1938).

Potassium Sulfide (K_2S) $K_2S_1(cr)$

Enthalpy Reference Temperature = T_r = 298.15 K; Standard State Pressure = p° = 0.1 MPa

T/K	$C_p°$ (J $K^{-1}mol^{-1}$)	S°	$-[G°-H°(T_r)]/T$	$H°-H°(T_r)$ (kJ mol^{-1})	$\Delta_fH°$	$\Delta_fG°$	Log K_f
0							
100							
200							
298.15	74.684	115.060	115.060	0.	-376.560	-362.745	63.551
300	74.726	115.522	115.061	0.138	-376.573	-362.659	63.145
400	77.320	137.375	118.023	7.741	-384.524	-356.897	46.606
500	79.914	154.907	123.702	15.602	-386.806	-349.730	36.536
600	82.508	169.706	130.167	23.723	-388.349	-342.158	29.787
700	85.144	182.621	136.757	32.105	-389.306	-334.377	24.952
800	87.738	194.161	143.224	40.749	-389.826	-326.490	21.318
900	103.554	205.100	149.486	50.053	-442.745	-317.488	18.427
1000	156.063	218.274	155.672	62.602	-438.078	-303.773	15.867
1050.000	194.982	226.816	158.850	71.364	Cp HUMP MAXIMUM TRANSITION		
1050.000	194.985	226.816	158.850	71.364			
1100	142.340	234.558	162.127	79.675	-586.914	-281.577	13.371
1200	142.340	246.943	168.686	93.908	-578.724	-254.180	11.064
1221.000	142.340	249.413	170.053	96.898	CRYSTAL <--> LIQUID		
1300	142.340	258.336	175.150	108.142	-570.550	-227.467	9.140
1400	142.340	268.885	181.473	122.376	-562.393	-201.382	7.514
1500	142.340	278.705	187.632	136.610	-554.251	-175.879	6.125
1600	142.340	287.892	193.614	150.844	-546.127	-150.920	4.927
1700	142.340	296.521	199.416	165.078	-538.021	-126.467	3.886
1800	142.340	304.657	205.039	179.312	-529.933	-102.492	2.974
1900	142.340	312.353	210.486	193.546	-521.866	-78.965	2.171
2000	142.340	319.654	215.764	207.780	-513.823	-55.863	1.459

PREVIOUS: CURRENT: March 1978

Potassium Sulfide (K_2S) $K_2S_1(cr)$

POTASSIUM SULFIDE (K_2S) LIQUID M_r = 110.2566

$S°(298.15\ K) = [141.172]\ J\ K^{-1}\ mol^{-1}$

$T_{fus} = 1221 \pm 10\ K$

$\Delta_f H°(298.15\ K) = [-346.506]\ kJ\ mol^{-1}$

$\Delta_{fus}H° = 16.150 \pm 0.17\ kJ\ mol^{-1}$

Enthalpy of Formation

$\Delta_f H°(K_2S,\ \ell,\ 298.15\ K)$ is calculated from that of the crystal by adding the enthalpy of fusion, $\Delta_{fus}H°$, and the difference in enthalpy, H°(1221 K) - H°(298.15 K), between the crystal and liquid.

Heat Capacity and Entropy

The heat capacity was derived from drop calorimetric enthalpy measurements by Dworkin and Bredig (2, 1221-1260 K) and is constant at 24.13 cal $K^{-1}\ mol^{-1}$ over the range measured. A glass transition temperature is assumed at 820 K with $C_p°$ = 24.13 above this temperature and $C_p°$ equal to that of the crystal below this temperature.

$S°(K_2S,\ \ell,\ 298.15\ K)$ is calculated in a manner analogous to that used for the enthalpy of formation.

References

1. JANAF Thermochemical Tables: $K_2S(cr)$, 3-31-78.
2. A. S. Dworkin and M. A. Bredig, J. Phys. Chem. 72, 1277 (1968).

Potassium Sulfide (K_2S) $K_2S_1(l)$

Enthalpy Reference Temperature = T_r = 298.15 K Standard State Pressure = p° = 0.1 MPa

T/K	$C_p°$ (J K⁻¹mol⁻¹)	S° (J K⁻¹mol⁻¹)	$-[G°-H°(T_r)]/T$ (J K⁻¹mol⁻¹)	$H°-H°(T_r)$ (kJ mol⁻¹)	$\Delta_f H°$ (kJ mol⁻¹)	$\Delta_f G°$ (kJ mol⁻¹)	Log K_f
0							
100							
200							
298.15	74.684	141.172	141.172	0.	-346.506	-340.476	59.650
300	74.726	141.634	141.174	0.138	-346.519	-340.439	59.276
400	77.320	163.487	144.136	7.741	-354.470	-337.288	44.045
500	79.914	181.019	149.815	15.602	-356.752	-332.732	34.760
600	82.508	195.819	156.280	23.723	-358.295	-327.771	28.535
700	85.144	208.734	162.869	32.105	-359.252	-322.602	24.073
800	87.738	220.273	169.336	40.749	-359.773	-317.326	20.719
820.010	88.241	222.447	170.606	42.510	GLASS <--> LIQUID		
820.010	100.960	222.447	170.606	42.510	TRANSITION		
900	100.960	231.844	175.637	50.586	-412.158	-310.970	18.048
1000	100.960	242.481	181.799	60.682	-409.944	-299.847	15.662
1100	100.960	252.103	187.760	70.778	-565.757	-279.720	13.283
1200	100.960	260.888	193.493	80.874	-561.705	-253.895	11.052
1221.000	100.960	262.640	194.667	82.994	------ CRYSTAL <--> LIQUID -----		
1300	100.960	268.969	198.992	90.970	-557.669	-228.408	9.178
1400	100.960	276.451	204.261	101.066	-553.649	-203.231	7.583
1500	100.960	283.417	209.309	111.162	-549.646	-178.341	6.210
1600	100.960	289.932	214.146	121.258	-545.660	-153.717	5.018
1700	100.960	296.053	218.786	131.354	-541.691	-129.343	3.974
1800	100.960	301.824	223.241	141.450	-537.742	-105.201	3.053
1900	100.960	307.282	227.521	151.546	-533.813	-81.278	2.234
2000	100.960	312.461	231.640	161.642	-529.907	-57.562	1.503
2100	100.960	317.387	235.607	171.738	-526.026	-34.040	0.847
2200	100.960	322.083	239.432	181.834	-522.174	-10.702	0.254
2300	100.960	326.571	243.124	191.930	-518.353	12.461	-0.283
2400	100.960	330.868	246.691	202.026	-514.569	35.458	-0.772
2500	100.960	334.990	250.141	212.122	-510.825	58.299	-1.218
2600	100.960	338.949	253.481	222.218	-507.125	80.991	-1.627
2700	100.960	342.760	256.717	232.314	-503.476	103.541	-2.003
2800	100.960	346.431	259.856	242.410	-499.882	125.957	-2.350
2900	100.960	349.974	262.903	252.506	-496.350	148.246	-2.670
3000	100.960	353.397	265.863	262.602	-492.887	170.414	-2.967

PREVIOUS: CURRENT: March 1978

Potassium Sulfide (K_2S) $K_2S_1(l)$

POTASSIUM SULFIDE (K_2S) CRYSTAL–CRYSTAL–LIQUID M_r = 110.2566

0 to 1050 K* crystal
1050 to 1221 K crystal
above 1221 K liquid

Refer to the individual tables for details.

*There is a diffuse transition at 1030 K.

Potassium Sulfide (K_2S) K_2S_1(cr,l)

Enthalpy Reference Temperature = T_r = 298.15 K; Standard State Pressure = $p°$ = 0.1 MPa

T/K	$C_p°$ (J K⁻¹mol⁻¹)	S° (J K⁻¹mol⁻¹)	−[G°−H°(T_r)]/T (J K⁻¹mol⁻¹)	H°−H°(T_r) (kJ mol⁻¹)	$\Delta_fH°$ (kJ mol⁻¹)	$\Delta_fG°$ (kJ mol⁻¹)	Log K_f
0							
100							
200							
298.15	74.684	115.060	115.060	0.	-376.560	-362.745	63.551
300	74.726	115.522	115.061	0.138	-376.573	-362.659	63.145
400	77.320	137.375	118.023	7.741	-384.524	-356.897	46.606
500	79.914	154.907	123.702	15.602	-386.806	-349.730	36.536
600	82.508	169.706	130.167	23.723	-388.349	-342.158	29.787
700	85.144	182.621	136.757	32.105	-389.306	-334.377	24.952
800	87.738	194.161	143.224	40.749	-389.826	-326.490	21.318
900	103.554	205.100	149.486	50.053	-442.745	-317.488	18.427
1000	156.063	218.274	155.672	62.602	-438.078	-303.773	15.867
1050.000	194.982	226.816	158.850	71.364	Cp HUMP MAXIMUM		
1050.000	194.985	226.816	158.850	71.364	TRANSITION		
1100	142.340	234.558	162.127	79.675	-586.914	-281.577	13.371
1200	142.340	246.943	168.686	93.908	-578.724	-254.180	11.064
1221.000	142.340	249.413	170.053	96.898	CRYSTAL <--> LIQUID		
1221.000	100.960	262.640	170.053	113.048	TRANSITION		
1300	100.960	268.969	175.874	121.024	-557.669	-228.408	9.178
1400	100.960	276.451	182.794	131.120	-553.649	-203.231	7.583
1500	100.960	283.417	189.273	141.216	-549.646	-178.341	6.210
1600	100.960	289.932	195.363	151.312	-545.660	-153.717	5.018
1700	100.960	296.053	201.107	161.408	-541.691	-129.343	3.974
1800	100.960	301.824	206.544	171.504	-537.742	-105.201	3.053
1900	100.960	307.282	211.704	181.600	-533.813	-81.278	2.234
2000	100.960	312.461	216.613	191.696	-529.907	-57.562	1.503
2100	100.960	317.387	221.296	201.792	-526.026	-34.040	0.847
2200	100.960	322.084	225.771	211.888	-522.174	-10.702	0.254
2300	100.960	326.571	230.057	221.984	-518.353	12.461	-0.283
2400	100.960	330.868	234.168	232.080	-514.569	35.458	-0.772
2500	100.960	334.990	238.119	242.176	-510.825	58.299	-1.218
2600	100.960	338.949	241.922	252.272	-507.125	80.991	-1.627
2700	100.960	342.760	245.586	262.368	-503.476	103.541	-2.003
2800	100.960	346.431	249.123	272.464	-499.882	125.957	-2.350
2900	100.960	349.974	252.540	282.560	-496.350	148.246	-2.670
3000	100.960	353.397	255.845	292.656	-492.887	170.414	-2.967

PREVIOUS: CURRENT: March 1978

Potassium Sulfide (K_2S) K_2S_1(cr,l)

KRYPTON (Kr) REFERENCE STATE A_r = 83.80

0 to 6000 K Ideal Monatomic Gas

IP(Kr, g) = 112914.5 ± 0.1 cm^{-1}
S°(298.15 K) = 164.084 ± 0.003 J K^{-1} mol^{-1}

$\Delta_f H°$(0 K) = 0 kJ mol^{-1}
$\Delta_f H°$(298.15 K) = 0 kJ mol^{-1}

Electronic Levels and Quantum Weights

State	ϵ_i, cm^{-1}	g_i
$^1S_0$	0	1

Ethalpy of Formation

Zero by definition.

Heat Capacity and Entropy

Information on the electronic energy levels and quantum weights is taken from Moore (1, 9). All predicted levels have been observed for n=4 but above that many predicted levels are missing. Our calculations indicate that any reasonable method of filling in these missing levels and cutting off the summation in the partition function (2) has no effect on the thermodynamic properties to 6000 K. This is undoubtedly a result of the high energy of these levels; the first excited level is nearly 80000 cm^{-1} above the ground state. Therefore, we list the ground state only. Extension to higher temperatures may require consideration of excited states and utilization of different fill and cutoff procedures (2).

The thermodynamic functions at 298.15 K agree exactly with recent CODATA recommendations (3) except for two minor differences First, the entropy differs by 0.1094 J K^{-1} mol^{-1} because this table uses a standard-state pressure of 1 bar, whereas the CODATA recommendations are based on 1 atm. Second, entropy differences of the order of 0.001-0.004 J K^{-1} mol^{-1} for the rare gases arise due to the use of slightly different values for R; this table uses R = 8.31441 J K^{-1} mol^{-1}. Considering these minor changes, this table agrees within the estimated uncertainty with those by Hultgren et al. (4), Hilsenrath et al. (5), Gurvich et al. (8), and Wagman et al. (12). The estimated uncertainty is due to uncertainties in the relative atomic mass and fundamental constants which in this tabulation are based on the 1981 scale (6) and the 1973 values (7), respectively.

Phase Data

The triple point, 115.770 K, and boiling point, 119.800 K, are secondary fixed points of IPTS-68 (10,11). Hultgren et al. (4) had recommended a triple point of 115.78 K (0.7220 atm) and a boiling point of 119.86 K (1 atm). These values are provided for the convenience of the reader and have not been evaluated by the present authors. As a result of these low values, the reference state for krypton is chosen to be the ideal gas at all temperatures. This may differ from the choice of other authors.

References

1. C. E. Moore, U. S. Nat. Bur. Stand., NSRDS-NBS-35, Volume II, 1970 [Reprint of NBS Circular 467, Volume II, 1952].
2. J. R. Downey, Jr., The Dow Chemical Co., AFOSR-TR-78-0960, Contract No. F44620-75-1-0048, 1978.
3. J. D. Cox, chairman, ICSU CODATA Task Group on Key Values for Thermodynamics, J. Chem. Thermodynamics 10, 903 (1978).
4. R. Hultgren, P. D. Desai et al., "Selected Values of the Thermodynamic Properties of the Elements", American Society for Metals, Metals Park, Ohio, 1973.
5. J. Hilsenrath, C. G. Messina and W. H. Evans, U.S. Nat. Bur. Stand., Report AD-606163 (avail. NTIS), 1964.
6. N. E. Holden and R. L. Martin, Pure Appl. Chem. 55, 1101 (1983).
7. E. R. Cohen and B. N. Taylor, J. Phys. Chem. Ref. Data 2, 663 (1973).
8. L. V. Gurvich, I. V. Veits et al., "Thermodynamic Properties of Individual Substances," 3rd ed., Vol. I, Nauka, Moscow, 1978.
9. C. E. Moore, U. S. Nat. Bur. Stand., NSRDS-NBS-34, 1970.
10. H. Preston-Thomas, Metrologia 12, 7 (1976).
11. L. Crovini, R. E. Bedford, and A. Moser, Metrologia 13, 197 (1977).
12. D. D. Wagman, W. H. Evans et al., J. Phys. Chem. Ref. Data 11, Supp. 2 (1982).

Krypton (Kr) Kr_1(ref)

Enthalpy Reference Temperature = T_r = 298.15 K Standard State Pressure = p° = 0.1 MPa

	J $K^{-1}mol^{-1}$			kJ mol^{-1}			
T/K	$C_p°$	S°	$-[G°-H°(T_r)]/T$	$H°-H°(T_r)$	$\Delta_f H°$	$\Delta_f G°$	Log K_f
0	0.	0.	INFINITE	-6.197	0.	0.	0.
100	20.786	141.377	182.565	-4.119	0.	0.	0.
200	20.786	155.785	165.986	-2.040	0.	0.	0.
250	20.786	160.423	164.427	-1.001	0.	0.	0.
298.15	20.786	164.084	164.084	0.	0.	0.	0.
300	20.786	164.213	164.085	0.038	0.	0.	0.
350	20.786	167.417	164.338	1.078	0.	0.	0.
400	20.786	170.193	164.900	2.117	0.	0.	0.
450	20.786	172.641	165.627	3.156	0.	0.	0.
500	20.786	174.831	166.440	4.196	0.	0.	0.
600	20.786	178.621	168.164	6.274	0.	0.	0.
700	20.786	181.825	169.892	8.353	0.	0.	0.
800	20.786	184.600	171.561	10.431	0.	0.	0.
900	20.786	187.049	173.149	12.510	0.	0.	0.
1000	20.786	189.239	174.650	14.589	0.	0.	0.
1100	20.786	191.220	176.068	16.667	0.	0.	0.
1200	20.786	193.028	177.407	18.746	0.	0.	0.
1300	20.786	194.692	178.673	20.824	0.	0.	0.
1400	20.786	196.233	179.873	22.903	0.	0.	0.
1500	20.786	197.667	181.012	24.982	0.	0.	0.
1600	20.786	199.008	182.096	27.060	0.	0.	0.
1700	20.786	200.268	183.128	29.139	0.	0.	0.
1800	20.786	201.456	184.113	31.217	0.	0.	0.
1900	20.786	202.580	185.056	33.296	0.	0.	0.
2000	20.786	203.646	185.959	35.375	0.	0.	0.
2100	20.786	204.661	186.826	37.453	0.	0.	0.
2200	20.786	205.628	187.659	39.532	0.	0.	0.
2300	20.786	206.552	188.460	41.610	0.	0.	0.
2400	20.786	207.436	189.232	43.689	0.	0.	0.
2500	20.786	208.285	189.978	45.768	0.	0.	0.
2600	20.786	209.100	190.698	47.846	0.	0.	0.
2700	20.786	209.884	191.394	49.925	0.	0.	0.
2800	20.786	210.640	192.068	52.004	0.	0.	0.
2900	20.786	211.370	192.721	54.082	0.	0.	0.
3000	20.786	212.074	193.354	56.161	0.	0.	0.
3100	20.786	212.756	193.969	58.239	0.	0.	0.
3200	20.786	213.416	194.567	60.318	0.	0.	0.
3300	20.786	214.056	195.148	62.397	0.	0.	0.
3400	20.786	214.676	195.713	64.475	0.	0.	0.
3500	20.786	215.279	196.263	66.554	0.	0.	0.
3600	20.786	215.864	196.800	68.632	0.	0.	0.
3700	20.786	216.434	197.323	70.711	0.	0.	0.
3800	20.786	216.988	197.833	72.790	0.	0.	0.
3900	20.786	217.528	198.331	74.868	0.	0.	0.
4000	20.786	218.054	198.818	76.947	0.	0.	0.
4100	20.786	218.567	199.293	79.025	0.	0.	0.
4200	20.786	219.068	199.758	81.104	0.	0.	0.
4300	20.786	219.557	200.213	83.183	0.	0.	0.
4400	20.786	220.035	200.658	85.261	0.	0.	0.
4500	20.786	220.502	201.094	87.340	0.	0.	0.
4600	20.786	220.959	201.521	89.418	0.	0.	0.
4700	20.786	221.406	201.939	91.497	0.	0.	0.
4800	20.786	221.844	202.349	93.576	0.	0.	0.
4900	20.786	222.273	202.751	95.654	0.	0.	0.
5000	20.786	222.692	203.146	97.733	0.	0.	0.
5100	20.786	223.104	203.533	99.811	0.	0.	0.
5200	20.786	223.508	203.914	101.890	0.	0.	0.
5300	20.786	223.904	204.287	103.969	0.	0.	0.
5400	20.786	224.292	204.654	106.047	0.	0.	0.
5500	20.786	224.674	205.014	108.126	0.	0.	0.
5600	20.786	225.048	205.369	110.204	0.	0.	0.
5700	20.786	225.416	205.717	112.283	0.	0.	0.
5800	20.786	225.778	206.060	114.362	0.	0.	0.
5900	20.786	226.133	206.397	116.440	0.	0.	0.
6000	20.786	226.482	206.729	118.519	0.	0.	0.

PREVIOUS: March 1977 (1 atm) CURRENT: March 1982 (1 bar)

Krypton (Kr) Kr_1(ref)

KRYPTON, ION (Kr^+) IDEAL GAS M_r = 83.79945

$IP(Kr^+, g) = 196474.8 \pm 0.1\ cm^{-1}$
$S°(298.15\ K) = 175.610 \pm 0.003\ J\ K^{-1}\ mol^{-1}$

$\Delta_f H°(0\ K) = 1350.758 \pm 0.012\ kJ\ mol^{-1}$
$\Delta_f H°(298.15\ K) = [1356.955]\ kJ\ mol^{-1}$

Electronic Levels and Quantum Weights

State	ε_i, cm^{-1}	g_i
$^2P_{3/2}$	0	4
$^2P_{1/2}$	5371.00	2

Enthalpy of Formation

The ionization limit of neutral krypton (112914.5±0.1 cm^{-1}) reported by Moore (1) is adopted as $\Delta_f H°(0\ K)$ for $Kr^+(g)$. The ionization limit is converted from cm^{-1} to kJ mol^{-1} using the factor, 1 cm^{-1} = 0.01196266 kJ mol^{-1}, which is derived from the 1973 CODATA fundamental constants (2). The uncertainty in the ionization limit is estimated to be ± 0.1 cm^{-1} which corresponds to an uncertainty of ±0.012 kJ mol^{-1} in the enthalpy of formation. Rosenstock et al. (9) and Levin and Lias (11) have summarized additional ionization potential and appearance potential data. The recent spectroscopic study by Yoshino et al. (12) is in agreement with our adopted value. Gurvich et al. (8) and Wagman et al. (10) adopted the same ionization potential, but the use of slightly different fundamental constants by Wagman et al. (10) resulted in an enthalpy of formation difference of 0.012 kJ mol^{-1}. However, the study by Chaghtai et al. (13) suggests an ionization potential which is larger by 10 cm^{-1}.

$\Delta_f H°(Kr^+, g, 298.15\ K)$ is obtained from $\Delta_f H°(Kr, g, 0\ K)$ by using IP(Kr) with JANAF (3) enthalpies $H°(0\ K)-H°(298.15\ K)$ for $Kr^+(g)$, Kr(r), and $e^-(r)$. $\Delta_r H°(Kr \rightarrow Kr^+ + e^-, 298.15\ K)$ differs from a room temperature threshold energy due to inclusion of these enthalpies and to threshold effects discussed by Rosenstock et al. (9). $\Delta_f H°(298.15\ K)$ should be changed by -6.197 kJ mol^{-1} if it is to be used in the ion convention that excludes the enthalpy of the electron.

Heat Capacity and Entropy

The information on electronic energy levels and quantum weights given by Moore (4) is incomplete because many theoretically predicted levels have not been observed. Our calculations indicate that any reasonable method of filling in these missing levels and cutting off the summation in the partition function (5) has no effect on the thermodynamic functions to 6000 K. This is a result of the high energy of all levels other than the ground state and the $^2P_{1/2}$ level; the next lowest level is over 109000 cm^{-1} above the ground state. Since inclusion of these upper levels has no effect on the thermodynamic functions (to 6000 K) we list only the ground state and the $^2P_{1/2}$ state, with the energy of the latter state taken from a more recent study by Moore (1). The reported uncertainty in S°(298.15 K) is due to uncertainties in the relative ionic mass and fundamental constants. Extension of these calculations above 6000 K may require consideration of the higher excited states and use of different fill and cutoff procedures (5).

The thermodynamic functions reported here agree with those of Green et al. (6), Hilsenrath et al. (7), and Gurvich et al. (8) except for one or two minor changes. First, the entropy differs by 0.1094 J K^{-1} mol^{-1} because this table uses a standard-state pressure of 1 bar, whereas the cited references used a pressure of 1 atm. Second, smaller differences arise from the use of slightly different values for the fundamental constants, the relative ionic mass, and the position of the $^2P_{1/2}$ electronic level.

References

1. C. E. Moore, U. S. Nat. Bur. Stand., NSRDS-NBS-34, 1970.
2. E. R. Cohen and B. N. Taylor, J. Phys. Chem. Ref. Data 2, 663 (1973).
3. JANAF Thermochemical Tables: Kr(r), 3-31-82; e⁻(r), 3-31-82.
4. C. E. Moore, U. S. Nat. Bur. Stand., NSRDS-NBS-35, Volume II, 1970 [Reprint of NBS Circular 467, Volume II, 1952].
5. J. R. Downey, Jr., The Dow Chemical Co., AFOSR-TR-78-0960, Contract No. F44620-75-1-0048, 1978.
6. J. W. Green, D. E. Poland and J. L. Margrave, University of Wisconsin, Report AD-275542, 1961.
7. J. Hilsenrath, C. G. Messina and W. H. Evans, U.S. Nat. Bur. Stand., Report AD-606163, 1964.
8. L. V. Gurvich, I. V. Veits et al., "Thermodynamic Properties of Individual Substances", 3rd ed., Vol. 1, Nauka, Moscow, 1978.
9. H. M. Rosenstock, K. Draxl et al., J. Phys. Chem. Ref. Data 6, Supp. 1 (1977).
10. D. D. Wagman, W. H. Evans et al., J. Phys. Chem. Ref. Data 11, Supp. 2 (1982).
11. R. D. Levine and S. G. Lias, U. S. Nat. Bur. Stand., NSRDS-NBS-71, 1982.
12. K Yoshino and Y. Tanaka, J. Opt. Soc. Am. 69, 159 (1974).
13. M. S. Z. Chaghtai and V. Hassan, J. Phys. B 6, 433 (1973).

Krypton, Ion (Kr^+) $Kr_1^+(g)$

Enthalpy Reference Temperature = T_r = 298.15 K; Standard State Pressure = p° = 0.1 MPa

T/K	$C_p°$ (J K⁻¹mol⁻¹)	S° (J K⁻¹mol⁻¹)	-[G°-H°(T_r)]/T (J K⁻¹mol⁻¹)	H°-H°(T_r) (kJ mol⁻¹)	$\Delta_f H°$ (kJ mol⁻¹)	$\Delta_f G°$ (kJ mol⁻¹)	Log K_f
0	0.	0.	INFINITE	-6.197	1350.758		
100	20.786	152.903	194.091	-4.119			
200	20.786	167.311	177.512	-2.040			
250	20.786	171.949	175.953	-1.001			
298.15	20.786	175.610	175.610	0.	1356.955	1347.264	-236.035
300	20.786	175.739	175.611	0.038	1356.994	1347.204	-234.569
350	20.786	178.943	175.864	1.078	1358.033	1345.490	-200.803
400	20.786	181.719	176.426	2.117	1359.072	1343.627	-175.460
450	20.786	184.167	177.153	3.156	1360.111	1341.634	-155.733
500	20.786	186.357	177.966	4.196	1361.151	1339.525	-139.939
600	20.788	190.147	179.690	6.274	1363.229	1335.005	-116.222
700	20.794	193.352	181.418	8.353	1365.308	1330.136	-99.256
800	20.811	196.129	183.087	10.434	1367.389	1324.970	-86.512
900	20.843	198.582	184.675	12.516	1369.471	1319.542	-76.584
1000	20.895	200.781	186.178	14.603	1371.558	1313.883	-68.630
1100	20.968	202.776	187.598	16.696	1373.651	1308.014	-62.112
1200	21.061	204.604	188.940	18.797	1375.752	1301.954	-56.673
1300	21.170	206.294	190.210	20.909	1377.864	1295.719	-52.063
1400	21.291	207.867	191.416	23.032	1379.987	1289.321	-48.105
1500	21.421	209.340	192.562	25.167	1382.122	1282.770	-44.670
1600	21.555	210.727	193.655	27.316	1384.271	1276.077	-41.660
1700	21.688	212.038	194.698	29.478	1386.433	1269.248	-38.999
1800	21.819	213.281	195.696	31.653	1388.609	1262.292	-36.631
1900	21.944	214.464	196.653	33.842	1390.797	1255.215	-34.508
2000	22.062	215.593	197.572	36.042	1392.997	1248.022	-32.595
2100	22.171	216.672	198.456	38.254	1395.209	1240.719	-30.861
2200	22.271	217.706	199.308	40.476	1397.431	1233.311	-29.283
2300	22.361	218.698	200.129	42.707	1399.663	1225.801	-27.839
2400	22.441	219.651	200.923	44.948	1401.903	1218.194	-26.513
2500	22.512	220.569	201.691	47.195	1404.151	1210.493	-25.292
2600	22.573	221.453	202.434	49.450	1406.405	1202.702	-24.163
2700	22.626	222.306	203.154	51.710	1408.665	1194.824	-23.115
2800	22.669	223.130	203.853	53.975	1410.930	1186.863	-22.141
2900	22.705	223.926	204.531	56.243	1413.199	1178.821	-21.233
3000	22.734	224.696	205.191	58.515	1415.471	1170.700	-20.384
3100	22.756	225.442	205.832	60.790	1417.745	1162.503	-19.588
3200	22.771	226.164	206.456	63.066	1420.021	1154.233	-18.841
3300	22.782	226.865	207.064	65.344	1422.299	1145.892	-18.138
3400	22.787	227.546	207.657	67.623	1424.578	1137.482	-17.475
3500	22.788	228.206	208.234	69.901	1426.856	1129.005	-16.849
3600	22.785	228.848	208.798	72.180	1429.135	1120.463	-16.257
3700	22.778	229.472	209.348	74.458	1431.413	1111.857	-15.697
3800	22.768	230.080	209.886	76.735	1433.690	1103.190	-15.164
3900	22.755	230.671	210.411	79.012	1435.967	1094.463	-14.659
4000	22.740	231.247	210.925	81.286	1438.241	1085.677	-14.177
4100	22.723	231.808	211.428	83.560	1440.515	1076.835	-13.719
4200	22.704	232.355	211.919	85.831	1442.786	1067.938	-13.282
4300	22.684	232.889	212.401	88.100	1445.056	1058.986	-12.864
4400	22.662	233.411	212.873	90.368	1447.323	1049.981	-12.465
4500	22.640	233.920	213.335	92.633	1449.588	1040.925	-12.083
4600	22.616	234.417	213.788	94.896	1451.851	1031.819	-11.717
4700	22.591	234.903	214.232	97.156	1454.111	1022.663	-11.366
4800	22.566	235.378	214.667	99.414	1456.369	1013.460	-11.029
4900	22.541	235.844	215.095	101.669	1458.624	1004.209	-10.705
5000	22.515	236.299	215.514	103.922	1460.877	994.912	-10.394
5100	22.489	236.744	215.926	106.172	1463.127	985.571	-10.094
5200	22.463	237.181	216.331	108.420	1465.375	976.185	-9.806
5300	22.436	237.608	216.728	110.665	1467.620	966.756	-9.528
5400	22.410	238.027	217.119	112.907	1469.862	957.285	-9.260
5500	22.384	238.438	217.503	115.147	1472.102	947.772	-9.001
5600	22.357	238.841	217.880	117.384	1474.339	938.219	-8.751
5700	22.332	239.237	218.251	119.618	1476.573	928.625	-8.510
5800	22.306	239.625	218.617	121.850	1478.805	918.992	-8.276
5900	22.280	240.006	218.976	124.079	1481.034	909.321	-8.051
6000	22.255	240.380	219.329	126.306	1483.261	899.613	-7.832

PREVIOUS: March 1977 (1 atm) CURRENT: March 1982 (1 bar)

Krypton, Ion (Kr^+) $Kr_1^+(g)$

LITHIUM (Li) REFERENCE STATE A_r = 6.941

0 to 453.69 K		crystal
453.69 to 1620.120 K		liquid
above 1620.120 K		ideal monatomic gas

Refer to the individual tables for details.

Lithium (Li) Li_1(ref)

Enthalpy Reference Temperature = T_r = 298.15 K   Standard State Pressure = p° = 0.1 MPa

T/K	C_p° (J K⁻¹mol⁻¹)	S° (J K⁻¹mol⁻¹)	$-[G^\circ-H^\circ(T_r)]/T$ (J K⁻¹mol⁻¹)	$H^\circ-H^\circ(T_r)$ (kJ mol⁻¹)	Δ_fH° (kJ mol⁻¹)	Δ_fG° (kJ mol⁻¹)	Log K_f
0	0.	0.	INFINITE	-4.622	0.	0.	0.
100	13.360	7.342	48.615	-4.127	0.	0.	0.
107.250	16.491	8.370	45.859	-4.021	Cp HUMP MAXIMUM		
107.250	16.494	8.370	45.859	-4.021	TRANSITION		
200	21.573	19.826	31.255	-2.286	0.	0.	0.
298.15	24.623	29.085	29.085	0.	0.	0.	0.
300	24.669	29.237	29.085	0.046	0.	0.	0.
400	27.610	36.698	30.083	2.646	0.	0.	0.
453.690	29.387	40.288	31.080	4.178	CRYSTAL <--> LIQUID		
453.690	30.392	46.902	31.080	7.178	TRANSITION		
500	30.125	49.843	32.684	8.579	0.	0.	0.
600	29.539	55.283	36.013	11.562	0.	0.	0.
700	28.987	59.794	39.097	14.487	0.	0.	0.
800	28.937	63.661	41.931	17.384	0.	0.	0.
900	28.886	67.066	44.539	20.275	0.	0.	0.
1000	28.836	70.107	46.946	23.161	0.	0.	0.
1100	28.786	72.853	49.179	26.042	0.	0.	0.
1200	28.736	75.356	51.257	28.918	0.	0.	0.
1300	28.702	77.655	53.201	31.790	0.	0.	0.
1400	28.619	79.779	55.024	34.656	0.	0.	0.
1500	28.535	81.750	56.741	37.514	0.	0.	0.
1600	28.451	83.589	58.362	40.363	0.	0.	0.
1620.120	28.431	83.945	58.678	40.936	LIQUID <--> IDEAL GAS		
1620.120	20.794	173.965	58.678	186.779	FUGACITY = 1 bar		
1700	20.799	174.966	64.118	188.441	0.	0.	0.
1800	20.810	176.155	70.310	190.521	0.	0.	0.
1900	20.826	177.281	75.911	192.603	0.	0.	0.
2000	20.849	178.349	81.006	194.687	0.	0.	0.
2100	20.882	179.367	85.666	196.773	0.	0.	0.
2200	20.925	180.340	89.947	198.863	0.	0.	0.
2300	20.980	181.271	93.898	200.958	0.	0.	0.
2400	21.049	182.165	97.557	203.060	0.	0.	0.
2500	21.132	183.026	100.959	205.169	0.	0.	0.
2600	21.231	183.857	104.131	207.287	0.	0.	0.
2700	21.347	184.660	107.099	209.415	0.	0.	0.
2800	21.479	185.439	109.883	211.556	0.	0.	0.
2900	21.628	186.195	112.502	213.712	0.	0.	0.
3000	21.794	186.931	114.970	215.883	0.	0.	0.
3100	21.977	187.649	117.303	218.071	0.	0.	0.
3200	22.177	188.350	119.513	220.278	0.	0.	0.
3300	22.391	189.035	121.609	222.506	0.	0.	0.
3400	22.622	189.707	123.602	224.757	0.	0.	0.
3500	22.868	190.366	125.500	227.031	0.	0.	0.
3600	23.130	191.014	127.311	229.331	0.	0.	0.
3700	23.406	191.652	129.041	231.658	0.	0.	0.
3800	23.697	192.280	130.697	234.013	0.	0.	0.
3900	24.003	192.899	132.284	236.398	0.	0.	0.
4000	24.323	193.511	133.807	238.814	0.	0.	0.
4100	24.658	194.115	135.271	241.263	0.	0.	0.
4200	25.009	194.714	136.679	243.746	0.	0.	0.
4300	25.375	195.307	138.036	246.265	0.	0.	0.
4400	25.696	195.889	139.344	248.800	0.	0.	0.
4500	26.069	196.470	140.607	251.386	0.	0.	0.
4600	26.464	197.047	141.827	254.012	0.	0.	0.
4700	26.873	197.621	143.008	256.679	0.	0.	0.
4800	27.297	198.191	144.152	259.387	0.	0.	0.
4900	27.736	198.758	145.261	262.139	0.	0.	0.
5000	28.191	199.323	146.336	264.935	0.	0.	0.
5200	29.134	200.446	148.396	270.661	0.	0.	0.
5400	30.152	201.564	150.344	276.588	0.	0.	0.
5600	31.233	202.680	152.193	282.726	0.	0.	0.
5800	32.377	203.796	153.954	289.086	0.	0.	0.
6000	33.581	204.914	155.634	295.681	0.	0.	0.

PREVIOUS: June 1962 (1 atm)   CURRENT: June 1962 (1 bar)

Lithium (Li) Li_1(ref)

LITHIUM (Li) CRYSTAL $A_r = 6.941$

$S°(298.15\ K) = 29.085\ J\ K^{-1}\ mol^{-1}$

$T_{fus} = 453.69\ K$

$\Delta_f H°(0\ K) = 0\ kJ\ mol^{-1}$

$\Delta_f H°(298.15\ K) = 0\ kJ\ mol^{-1}$

$\Delta_{fus} H° = 3.000 \pm 0.015\ kJ\ mol^{-1}$

Enthalpy of Formation

Zero by definition.

Heat Capacity and Entropy

Heat capacities have been reported by the following authors for the temperature ranges indicated: Roberts (1), 1.5 to 20 K (Li 99.5% pure); Simon and Swain (2) 15 to 300 K; Martin (3) 22 to 300 K(Li 99.95% pure); Douglas et al. (4) 289.15 K to 1200 K (Li 99.98% pure). Martin (3) in a series of determinations, found that C_p depended on the thermal history of the sample; a peak at roughly 107 K was ascribed to a martensitic transformation. There is excellent continuty between the results of Roberts (1) and of Martin (3), while those of Martin (3) and Douglas et al. (4) differ by 1% at 300 K and have been smoothly joined. The values of Simon and Swain (2) were not used as they were 5% lower than Martin's (3) and did not show the peak at 107 K; also their values for Al_2O_3, reported at the same time, are lower than recent measurements. Heat capacities between 1200 K and the normal boiling point were obtained from the enthalpy equation given by Douglas et al. (4) between 420° and 900°C.

Fusion Data

Douglas et al. (4) report the triple point to be 180.54°C, which is here adopted as the melting point since the densities of solid and liquid lithium at 180°C are not available for calculating dT_{fus}/dP. The same authors (4) give the enthalpy of melting as $432.3 \pm 2.2\ J\ g^{-1}$.

Sublimation Data

Lithium vapor contains an appreciable amount of dimer, whose enthalpy of dissociation has been selected by Evans (5), from spectroscopic and molecular beam measuresments to be $25.76 \pm 0.10\ kcal\ mol^{-1}$ at 0 K. This enthalpy of dissociation, together with the thermodynamic functions calculated in this work, has been used to find the partial pressures of Li(g) and Li_2(g) from the measured total vapor pressures. Hartmann and Schneider (6), report values from 1204 to 1353 K; while Mancherat (7) reports effusion measurements from 735 to 915 K. Mancherat's (7) pressures are calculated on the assumption of monatomic vapor and have been recalculated to fine the true total pressure. Effusion measurements by Lewis (8) and Bogros (9) have been disregarded. Mancherat (7) considers them to be inaccurate because of impurities in the lithium used, and Lewis (8) used a doubtful calibration method. Enthalpy of sublimation to monatomic vapor calculated from the vapor pressures of Hartmann and Schneider (6) and of Mancherat (7) agree to within 2% and the average value has been adopted. The enthalpy of sublimation of the dimer was then calculated using this value.

References

1. L. M. Roberts, Proc. Phys. Soc. 70, 744 (1957).
2. F. Simon and R. C. Swain, Z. phys. Chem. B28, 189 (1935).
3. D. L. Martin, Proc. Roy. Soc. A254, 444 (1960).
4. T. B. Douglas, L. F. Epstein, J. L. Dever, and W. H. Howland, J. Am. Chem. Soc. 77, 2144 (1955).
5. W. H. Evans, R. Jacobson, T. R. Munson, and D. D. Wagman, J. Res. Natl. Bur. of Stand. 55, 83 (1955).
6. H. Hartmann and R. Schneider, Z. anorg. allgem. Chem. 180, 275 (1929).
7. M. Mancherat, C. Rendus 208, 499 (1939).
8. L. C. Lewis, Z. Physik 69, 786 (1931).
9. A. Bogros, C. Rendus 191, 560 (1930); Ann. Phys. 17, 199 (1932).

Lithium (Li) Li_1(cr)

Enthalpy Reference Temperature = T_r = 298.15 K; Standard State Pressure = p° = 0.1 MPa

T/K	$C_p°$	S°	$-[G°-H°(T_r)]/T$	$H°-H°(T_r)$	$\Delta_f H°$	$\Delta_f G°$	Log K_f
	J K⁻¹mol⁻¹			kJ mol⁻¹			
0	0.	0.	INFINITE	-4.622	0.	0.	0.
100	13.360	7.342	48.615	-4.127	0.	0.	0.
107.250	16.491	8.370	45.859	-4.021	Cp HUMP MAXIMUM TRANSITION		
107.250	16.494	8.370	45.859	-4.021			
200	21.573	19.826	31.255	-2.286	0.	0.	0.
250	23.406	24.854	29.484	-1.157	0.	0.	0.
298.15	24.623	29.085	29.085	0.	0.	0.	0.
300	24.669	29.237	29.085	0.046	0.	0.	0.
350	25.930	33.128	29.390	1.308	0.	0.	0.
400	27.610	36.698	30.083	2.646	0.	0.	0.
450	29.280	40.049	31.006	4.069	0.	0.	0.
453.690	29.387	40.288	31.080	4.178	CRYSTAL <--> LIQUID		
500	30.217	43.188	32.069	5.559	-3.020	0.308	-0.032
600	30.380	48.714	34.396	8.591	-2.972	0.970	-0.084
700	30.384	53.397	36.785	11.629	-2.859	1.619	-0.121
800	30.384	57.455	39.121	14.667	-2.716	2.249	-0.147
900	30.384	61.033	41.360	17.706	-2.569	2.860	-0.166
1000	30.384	64.235	43.491	20.744	-2.417	3.456	-0.180

PREVIOUS: CURRENT: June 1962

Lithium (Li) Li_1(cr)

LITHIUM (Li) LIQUID A_r = 6.941

S°(298.15 K) = [33.938] J K^{-1} mol^{-1}

T_{fus} = 1843 K

$\Delta_f H°$(298.15 K) = [2.380] kJ mol^{-1}

$\Delta_{fus} H°$ = 3.000 ± 0.015 kJ mol^{-1}

Enthalpy of Formation

$\Delta_f H°$(Li, ℓ, 298.15 K) is calculated from that of the crystal by adding the enthalpy of fusion and the difference in enthalpy, H°(453.69 K) - H°(298.15 K), between the crystal and liquid.

Heat Capacity and Entropy

The liquid heat capacity values are derived from the enthalpy measurements of Douglas et al. (1). The heat capacity curve was extrapolated to 298.15 K and 2000 K in a reasonable manner.

The entropy is calculated in a manner analogous to that used for the enthalpy of formation.

Fusion Data

The enthalpy of fusion and the melting point were measured by Douglas et al. (1).

Vaporization Data

Lithium vaporizes to a mixture of monatomic and diatomic gas. The total vapor pressure reaches 1 atm at 1620 K; the vapor pressure of the monatomic gas reaches 1 atm at 1638 K and the enthalpy of vaporization to monatomic gas is 35.16 kcal mol^{-1}.

Reference

1. T. B. Douglas, L. F. Epstein, J. L. Dever, and W. H. Howland, J. Am. Chem. Soc. 77, 2144 (1955).

Lithium (Li) Li_1(l)

Enthalpy Reference Temperature = T_r = 298.15 K Standard State Pressure = p° = 0.1 MPa

T/K	$C_p°$	S°	$-[G°-H°(T_r)]/T$	$H°-H°(T_r)$	$\Delta_f H°$	$\Delta_f G°$	Log K_f
	J K^{-1} mol^{-1}			kJ mol^{-1}			
0							
100							
200							
250							
298.15	31.300	33.938	33.938	0.	2.380	0.933	-0.163
300	31.284	34.131	33.938	0.058	2.392	0.924	-0.161
350	31.000	38.933	34.318	1.615	2.686	0.655	-0.098
400	30.711	43.053	35.159	3.158	2.891	0.349	-0.046
450	30.414	46.653	36.240	4.686	2.996	0.024	-0.003
453.690	30.392	46.902	36.335	4.794	------ CRYSTAL <--> LIQUID -----		
500	30.125	49.843	37.444	6.199	0.	0.	0.
600	29.539	55.283	39.979	9.182	0.	0.	0.
700	28.987	59.793	42.497	12.108	0.	0.	0.
800	28.937	63.661	44.906	15.004	0.	0.	0.
900	28.886	67.066	47.183	17.895	0.	0.	0.
1000	28.836	70.107	49.326	20.781	0.	0.	0.
1100	28.786	72.853	51.342	23.662	0.	0.	0.
1200	28.736	75.356	53.240	26.538	0.	0.	0.
1300	28.702	77.655	55.031	29.410	0.	0.	0.
1400	28.619	79.779	56.724	32.276	0.	0.	0.
1500	28.535	81.750	58.327	35.134	0.	0.	0.
1600	28.451	83.589	59.850	37.983	0.	0.	0.
1620.120	28.431	83.945	60.147	38.556	------- FUGACITY = 1 bar -------		
1700	28.368	85.311	61.297	40.824	-145.237	7.176	-0.220
1800	28.326	86.931	62.677	43.658	-144.483	16.120	-0.468
1900	28.326	88.463	63.994	46.491	-143.732	25.022	-0.688
2000	28.326	89.916	65.254	49.323	-142.983	33.884	-0.885
2100	28.326	91.298	66.462	52.156	-142.237	42.709	-1.062
2200	28.326	92.615	67.621	54.989	-141.495	51.499	-1.223

PREVIOUS: CURRENT: June 1962

Lithium (Li) Li_1(l)

LITHIUM (Li) CRYSTAL-LIQUID A_r = 6.941 Lithium (Li) Li_1(cr,l)

0 to 453.69 K crystal
above 453.69 K liquid

Refer to the individual tables for details.

Enthalpy Reference Temperature = T_r = 298.15 K Standard State Pressure = p° = 0.1 MPa

T/K	C_p° (J K⁻¹mol⁻¹)	S° (J K⁻¹mol⁻¹)	$-[G^\circ-H^\circ(T_r)]/T$ (J K⁻¹mol⁻¹)	$H^\circ-H^\circ(T_r)$ (kJ mol⁻¹)	$\Delta_f H^\circ$ (kJ mol⁻¹)	$\Delta_f G^\circ$ (kJ mol⁻¹)	Log K_f
0	0.	0.	INFINITE	-4.622	0.	0.	0.
100	13.360	7.342	48.615	-4.127	0.	0.	0.
107.250	16.491	8.370	45.859	-4.021	Cp HUMP MAXIMUM TRANSITION		
107.250	16.494	8.370	45.859	-4.021			
200	21.573	19.826	31.255	-2.286	0.	0.	0.
250	23.406	24.854	29.484	-1.157	0.	0.	0.
298.15	24.623	29.085	29.085	0.	0.	0.	0.
300	24.669	29.237	29.085	0.046	0.	0.	0.
350	25.930	33.128	29.390	1.308	0.	0.	0.
400	27.610	36.698	30.083	2.646	0.	0.	0.
450	29.280	40.049	31.006	4.069	0.	0.	0.
453.690	29.387	40.288	31.080	4.178	CRYSTAL <--> LIQUID TRANSITION		
453.690	30.392	46.902	31.080	7.178			
500	30.125	49.843	32.684	8.579	0.	0.	0.
600	29.539	55.283	36.013	11.562	0.	0.	0.
700	28.987	59.794	39.097	14.487	0.	0.	0.
800	28.937	63.661	41.931	17.384	0.	0.	0.
900	28.886	67.066	44.539	20.275	0.	0.	0.
1000	28.836	70.107	46.946	23.161	0.	0.	0.
1100	28.786	72.853	49.179	26.042	0.	0.	0.
1200	28.736	75.356	51.257	28.918	0.	0.	0.
1300	28.702	77.655	53.201	31.790	0.	0.	0.
1400	28.619	79.779	55.024	34.656	0.	0.	0.
1500	28.535	81.750	56.741	37.514	0.	0.	0.
1600	28.451	83.589	58.362	40.363	0.	0.	0.
1620.120	28.431	83.945	58.678	40.936	------- FUGACITY = 1 bar -------		
1700	28.368	85.311	59.897	43.204	-145.237	7.176	-0.220
1800	28.326	86.931	61.355	46.038	-144.483	16.120	-0.468
1900	28.326	88.463	62.741	48.871	-143.732	25.022	-0.688
2000	28.326	89.916	64.064	51.703	-142.983	33.884	-0.885
2100	28.326	91.298	65.328	54.536	-142.237	42.709	-1.062
2200	28.326	92.615	66.539	57.368	-141.495	51.499	-1.223

PREVIOUS: CURRENT: June 1962

Lithium (Li) Li_1(cr,l)

LITHIUM (Li) IDEAL GAS A_r = 6.941

IP(Li, g) = 43487.150 ± 0.005 cm^{-1} $\Delta_f H°(298.15\ K)$ = 157.7 ± 10 kJ mol^{-1}

S°(298.15 K) = 138.781 ± 0.025 J K^{-1} mol^{-1} $\Delta_f H°(298.15\ K)$ = 159.3 ± 1.0 kJ mol^{-1}

Electronic Levels and Quantum Weights

State	ε_i, cm^{-1}	g_i
$^2S_{1/2}$	0	2
$^2P_{1/2}$	14903.66	2
$^2P_{3/2}$	14904.00	4
$^2S_{1/2}$	27206.12	2
.	.	.
.	.	.
.	.	.
$^2P_{3/2}$	43424.3	4
IP	43487.150	

Enthalpy of Formation

The enthalpy of formation of lithium gas is chosen to be the value recommended by CODATA (1). This value was obtained from 2nd and 3rd law treatments of vapor pressure data. The reported vapor pressures were corrected for the effects of Li_2(g), using D_0^0= 101.02 kJ mol^{-1}.

As stated by CODATA (2), the selected value is the weighted average of the 3rd law values of $\Delta_{sub}H°(298.15\ K)$ calculated from the following data:

Source	$\Delta_{sub}H°(298.15\ K)$, kJ mol^{-1}
Hartmann and Sckneider (3)	159.04 ± 1.56
Bohdansky and Schins (4)	159.54 ± 1.85
Schins et al. (5)	159.22 ± 1.81
Rigney et al. (6)	159.34 ± 1.85
Anisimov and Volyak (7)	159.60 ± 1.58
Shpilrain and Belova (8)	159.06 ± 1.99

The weighted average is $\Delta_{sub}H°(298.15\ K)$ = 159.3 ± 1.0 kJ mol^{-1}.

The 2nd law values calculated from the above sources lead to more positive values. 3rd law values in reasonable agreement with the selected value were calculated from the data of Bogros (9), Maucherat (10), Achener (11), Wu(12), and Bonilla (13).

Heat Capacity and Entropy

The thermal functions for the five alkali metal monatomic gases are calculated by the same procedure. Observed and estimated atomic energy levels are included in the partition function calculation, using an ionization potential lowering (IP-kT) technique as the cutoff procedure in the energy level summation (16).

The lowest lying levels for these metals [14904 cm^{-1} (Li), 16956 cm^{-1} (Na), 12985 cm^{-1} (K), 12578 cm^{-1} (Cs)] do not contribute to the thermal functions below ~1000 K; there is only a translational contribution below this temperature. Above this approximate temperature, the thermal functions become increasingly sensitive to the partition function cutoff procedure used, due to the combined effect of the observation of atomic energy levels of high principal quantum number and a low ionization potential. Energy levels have been observed up to n = 42 (Li), 59(Na), 79(K), 77(Rb), and 73(Cs). However, not all predicted levels have been observed for each of these principal quantum numbers. The ionization potentials vary from 43487.29 cm^{-1} for Li to 31406.1 cm^{-1} for Cs. In calculating the thermal functions with the inclusion of missing levels up to the high principal quantum numbers just mentioned, the Gibbs energy functions show significant differences (depending on the cutoff procedure) above 3000 K (16).

The atomic energy levels have been taken from the compilation of Moore (14, 15). Our calculated values for the thermal functions are similar to those recommended by CODATA (1). They do differ for two reasons, however. First, the entropy differs by 0.1094 J K^{-1} mol^{-1} because this table uses a reference pressure of 1 bar, whereas the CODATA recommendations are based on 1 atm. Second, the entropies at 298.15 K for all alkali metal gases differs by ~0.004 J K^{-1} mol^{-1}, presumably due to the use of slightly different values for auxiliary data.

References

1. J. D. Cox, chairman, CODATA Task Group on Key Values for Thermodynamics, J. Chem. Thermodynamics 10, 903 (1978).
2. J. D. Cox, chairman, CODATA Task Group on Key Values for Thermodynamics, CODATA Special Report No. 8, 1980.
3. H. Hartmann and R. Schneider, Z. anorg allgem. Chem. 180, 275 (1929).
4. J. Bohdansky and H. E. J. Schins, J. Appl. Phys. 36, 3683 (1965); J. Phys. Chem. 71, 215 (1967).
5. H. E. J. Schins, R. W. M. van Wijik and B. Dorpima, Z. Metallkundi 62, 330 (1971).
6. D. V. Rigney, S. M. Kapelner, and R. E. Cleary, U. S. AEC Report TIM-844, 1965.
7. V. M. Anisimov and L. D. Volyak, Teplofiz. Vussh. Temp. 7, 317 (1969).
8. E. E. Shpil'rain and A. M. Belova, Teplofiz. Vyssh. Temp. 6. 342 (1968).
9. A. Bogros, Ann. Phys. [10] 17, 199 (1929); Compt. rend. 191, 322 (1930).
10. M. Maucherat, Compt. rend 208, 499 (1939); J. Phys. Radium 10, 441 (1939).
11. P. Y. Achener and D. L. Fisher, U. S. AEC Report AGN-8191, Volume 2, 1967.
12. C. H. J. Wu, J. Chem. Phys. 65, 3181 (1976).
13. C. F. Bonilla, U. S. DOE Report COO-3027-30, 1977.
14. C. E. Moore, U. S. Nat. Bur. Stand., NSRDS-NBS 34, 1970.
15. C. E. Moore, U. S. Nat. Bur. Stand., NSRDS-NBS 35, Volume I, 1970 [Reprint of NBS Circular 467, Volume I, 1949].
16. J. R. Downey, Jr., The Dow Chemical company, Rept. AFOSR-TR-78-0960, Contract No. F44620-75-1-0048, 1978.

Lithium (Li) Li_1(g)

Enthalpy Reference Temperature = T_r = 298.15 K Standard State Pressure = p° = 0.1 MPa

T/K	$C_p°$	S°	$-[G°-H°(T_r)]/T$	$H°-H°(T_r)$	$\Delta_f H°$	$\Delta_f G°$	Log K_f
	J $K^{-1}mol^{-1}$			kJ mol^{-1}			
0	0.	0.	INFINITE	-6.197	157.725	157.725	INFINITE
100	20.786	116.074	157.261	-4.119	159.308	148.435	-77.535
200	20.786	130.481	140.682	-2.040	159.545	137.415	-35.889
250	20.786	135.120	139.123	-1.001	159.457	131.890	-27.557
298.15	20.786	138.781	138.781	0.	159.300	126.594	-22.179
300	20.786	138.909	138.781	0.038	159.293	126.391	-22.007
350	20.786	142.114	139.034	1.078	159.069	120.924	-18.047
400	20.786	144.889	139.596	2.117	158.771	115.495	-15.082
450	20.786	147.337	140.323	3.156	158.387	110.107	-12.781
500	20.786	149.527	141.136	4.196	154.917	105.074	-10.977
600	20.786	153.317	142.860	6.274	154.012	95.192	-8.287
700	20.786	156.521	144.589	8.353	153.165	85.456	-6.377
800	20.786	159.297	146.258	10.431	152.348	75.839	-4.952
900	20.786	161.745	147.845	12.510	151.535	66.324	-3.849
1000	20.786	163.935	149.346	14.589	150.728	56.900	-2.972
1100	20.786	165.916	150.764	16.667	149.925	47.556	-2.258
1200	20.786	167.725	152.103	18.746	149.128	38.285	-1.666
1300	20.786	169.389	153.370	20.825	148.334	29.080	-1.168
1400	20.787	170.929	154.570	22.903	147.547	19.936	-0.744
1500	20.789	172.363	155.709	24.982	146.768	10.848	-0.378
1600	20.793	173.705	156.792	27.061	145.998	1.812	-0.059
1620.120	20.794	173.965	157.004	27.479	------- FUGACITY = 1 bar -------		
1700	20.799	174.966	157.824	29.141	0.	0.	0.
1800	20.810	176.155	158.810	31.221	0.	0.	0.
1900	20.826	177.281	159.753	33.303	0.	0.	0.
2000	20.849	178.349	160.656	35.386	0.	0.	0.
2100	20.882	179.367	161.523	37.473	0.	0.	0.
2200	20.925	180.340	162.356	39.563	0.	0.	0.
2300	20.980	181.271	163.159	41.658	0.	0.	0.
2400	21.049	182.165	163.932	43.760	0.	0.	0.
2500	21.132	183.026	164.679	45.868	0.	0.	0.
2600	21.231	183.857	165.401	47.986	0.	0.	0.
2700	21.347	184.660	166.099	50.115	0.	0.	0.
2800	21.479	185.439	166.776	52.256	0.	0.	0.
2900	21.628	186.195	167.433	54.412	0.	0.	0.
3000	21.794	186.931	168.070	56.582	0.	0.	0.
3100	21.977	187.649	168.690	58.771	0.	0.	0.
3200	22.177	188.350	169.294	60.978	0.	0.	0.
3300	22.391	189.035	169.882	63.206	0.	0.	0.
3400	22.622	189.707	170.455	65.457	0.	0.	0.
3500	22.868	190.366	171.014	67.731	0.	0.	0.
3600	23.130	191.014	171.561	70.031	0.	0.	0.
3700	23.406	191.652	172.095	72.358	0.	0.	0.
3800	23.697	192.280	172.618	74.713	0.	0.	0.
3900	24.003	192.899	173.130	77.098	0.	0.	0.
4000	24.323	193.511	173.632	79.514	0.	0.	0.
4100	24.658	194.115	174.125	81.963	0.	0.	0.
4200	25.009	194.714	174.608	84.446	0.	0.	0.
4300	25.375	195.307	175.082	86.965	0.	0.	0.
4400	25.696	195.889	175.548	89.500	0.	0.	0.
4500	26.069	196.470	176.007	92.086	0.	0.	0.
4600	26.464	197.047	176.458	94.712	0.	0.	0.
4700	26.873	197.621	176.902	97.379	0.	0.	0.
4800	27.297	198.191	177.339	100.087	0.	0.	0.
4900	27.736	198.758	177.771	102.839	0.	0.	0.
5000	28.191	199.323	178.196	105.635	0.	0.	0.
5100	28.650	199.885	178.616	108.472	0.	0.	0.
5200	29.134	200.446	179.030	111.361	0.	0.	0.
5300	29.635	201.006	179.440	114.299	0.	0.	0.
5400	30.152	201.564	179.844	117.288	0.	0.	0.
5500	30.684	202.122	180.244	120.330	0.	0.	0.
5600	31.233	202.680	180.640	123.426	0.	0.	0.
5700	31.798	203.238	181.031	126.577	0.	0.	0.
5800	32.377	203.796	181.419	129.786	0.	0.	0.
5900	32.972	204.354	181.803	133.053	0.	0.	0.
6000	33.581	204.914	182.184	136.381	0.	0.	0.

PREVIOUS: June 1962 (1 atm) CURRENT: December 1983 (1 bar)

Lithium (Li) Li_1(g)

LITHIUM, ION (Li^+) IDEAL GAS M_r = 6.94045

$IP(Li^+, g) = 610079.0 \pm 0.1\ cm^{-1}$

$S°(298.15\ K) = 133.017 \pm 0.02\ J\ K^{-1}\ mol^{-1}$

$\Delta_f H°(0\ K) = 677.947 \pm 0.002\ kJ\ mol^{-1}$

$\Delta_f H°(298.15\ K) = [685.719]\ kJ\ mol^{-1}$ u

Electronic Level and Quantum Weight

State	ε_i, cm^{-1}	g_i
$^1S_0$	0	1

Enthalpy of Formation

$\Delta_f H°(Li^+, g, 0\ K)$ is calculated from $\Delta_f H°(Li, g, 0\ K)$ (1) using the spectroscopic value of $IP(Li) = 43487.150 \pm 0.005\ cm^{-1}$ ($520.222 \pm .000\ kJ\ mol^{-1}$) from Moore (2). The ionization limit is converted from cm^{-1} to $kJ\ mol^{-1}$ using the factor, $1\ cm^{-1} = 0.01196266\ kJ\ cm^{-1}$, which is derived from the 1973 CODATA fundamental constants (3). Rosenstock et al. (4) and Levin and Lias (5) have summarized additional ionization and appearance potential data.

$\Delta_f H°(Li^+, g, 298.15\ K)$ is calculated from $\Delta_f H°(Li, g, 0\ K)$ by using IP(Li) with JANAF (1) enthalpies, $H°(0\ K) - H°(298.15\ K)$, for Li(g), $Li^+(g)$, and $e^-(g)$. $\Delta_f H°(Li \rightarrow Li^+ + e^-, 298.15\ K)$ differs from a room temperature threshold energy due to inclusion of these enthalpies and to threshold effects discussed by Rosenstock et al. (4). $\Delta_f H°(298.15\ K)$ should be changed by $-6.197\ kJ\ mol^{-1}$ if it is to be used in the ion convention that excludes the enthalpy of the electron.

Heat Capacity and Entropy

The information on electronic energy levels and quantum weights, given by Moore (6), is incomplete because many theoretically predicted levels have not been observed. Our calculations indicate that any reasonable method of filling in these missing levels and cutting off the summation in the partition function (7) has no effect on the thermodynamic functions to 6000 K. This is a result of the high energy of all levels other than the ground state; the first excited state is approximately 476046 cm^{-1} above the ground state. Since inclusion of these excited states has no effect on the thermodynamic functions (to 6000 K), we list only the ground state. The reported uncertainty in S°(298.15 K) is due to uncertainties in the relative ionic mass and the fundamental constants. Extension of these calculations above 6000 K may require consideration of the excited states and use of different fill and cutoff procedures (7).

References

1. JANAF Thermochemical Tables: Li(g), 6-30-62; e^-(g), 3-31-82.
2. C. E. Moore, U. S. Nat. Bur. Stand., NSRDS-NBS-34, 1970.
3. E. R. Cohen and B. N. Taylor, J. Phys. Chem. Ref. Data 2, 663 (1973).
4. H. M. Rosenstock, K. Draxl et al., J. Phys. Chem. Ref. Data 6, Supp. 1 (1977).
5. R. D. Levin and S. G. Lias, U. S. Nat. Bur. Stand., NSRDS-NBS-71, 1982.
6. C. E. Moore, U. S. Nat. Bur. Stand., NSRDS-NBS-35, Volume 1, 1970 (Reprint of NBS Circular 467, Volume 1, 1949).
7. J. R. Downey, Jr., The Dow Chemical Company, AFOSR-TR-78-0960, Contract No. F44620-75-1-0048, 1978.

Lithium, Ion (Li^+) $Li_1^+(g)$

Enthalpy Reference Temperature = T_r = 298.15 K — Standard State Pressure = p° = 0.1 MPa

	J K⁻¹mol⁻¹			kJ mol⁻¹			
T/K	$C_p°$	S°	$-[G°-H°(T_r)]/T$	$H°-H°(T_r)$	$\Delta_f H°$	$\Delta_f G°$	Log K_f
0	0.	0.	INFINITE	-6.197	677.947		
100	20.786	110.309	151.497	-4.119			
200	20.786	124.717	134.918	-2.040			
250	20.786	129.356	133.359	-1.001			
298.15	20.786	133.017	133.017	0.	685.719	648.477	-113.611
300	20.786	133.145	133.017	0.038	685.751	648.246	-112.870
350	20.786	136.349	133.270	1.078	686.566	641.930	-95.803
400	20.786	139.125	133.832	2.117	687.307	635.502	-82.988
450	20.786	141.573	134.559	3.156	687.963	628.986	-73.011
500	20.786	143.763	135.372	4.196	685.532	622.709	-65.054
600	20.786	147.553	137.096	6.274	686.706	610.035	-53.108
700	20.786	150.757	138.825	8.353	687.938	597.160	-44.561
800	20.786	153.533	140.493	10.431	689.199	584.105	-38.138
900	20.786	155.981	142.081	12.510	690.465	570.893	-33.134
1000	20.786	158.171	143.582	14.589	691.736	557.539	-29.123
1100	20.786	160.152	145.000	16.667	693.012	544.057	-25.835
1200	20.786	161.961	146.339	18.746	694.293	530.460	-23.090
1300	20.786	163.625	147.606	20.824	695.578	516.755	-20.763
1400	20.786	165.165	148.806	22.903	696.869	502.951	-18.765
1500	20.786	166.599	149.945	24.982	698.169	489.054	-17.030
1600	20.786	167.941	151.028	27.060	699.477	475.070	-15.509
1700	20.786	169.201	152.060	29.139	555.556	468.181	-14.385
1800	20.786	170.389	153.046	31.217	557.633	462.981	-13.435
1900	20.786	171.513	153.988	33.296	559.709	457.666	-12.582
2000	20.786	172.579	154.891	35.375	561.782	452.242	-11.811
2100	20.786	173.593	155.758	37.453	563.853	446.714	-11.111
2200	20.786	174.560	156.591	39.532	565.920	441.087	-10.473
2300	20.786	175.484	157.392	41.610	567.982	435.367	-9.887
2400	20.786	176.369	158.165	43.689	570.038	429.557	-9.349
2500	20.786	177.217	158.910	45.768	572.086	423.661	-8.852
2600	20.786	178.032	159.630	47.846	574.125	417.684	-8.391
2700	20.786	178.817	160.326	49.925	576.154	411.629	-7.963
2800	20.786	179.573	161.000	52.004	578.170	405.498	-7.565
2900	20.786	180.302	161.653	54.082	580.172	399.296	-7.192
3000	20.786	181.007	162.287	56.161	582.158	393.025	-6.843
3100	20.786	181.688	162.902	58.239	584.127	386.688	-6.516
3200	20.786	182.348	163.499	60.318	586.077	380.288	-6.208
3300	20.786	182.988	164.080	62.397	588.006	373.827	-5.917
3400	20.786	183.608	164.645	64.475	589.913	367.308	-5.643
3500	20.786	184.211	165.196	66.554	591.795	360.733	-5.384
3600	20.786	184.797	165.732	68.632	593.653	354.105	-5.138
3700	20.786	185.366	166.255	70.711	595.483	347.426	-4.905
3800	20.786	185.920	166.765	72.790	597.286	340.697	-4.683
3900	20.786	186.460	167.263	74.868	599.058	333.922	-4.472
4000	20.786	186.987	167.750	76.947	600.799	327.101	-4.272
4100	20.786	187.500	168.225	79.025	602.507	320.238	-4.080
4200	20.786	188.001	168.690	81.104	604.181	313.333	-3.897
4300	20.786	188.490	169.145	83.183	605.819	306.388	-3.722
4400	20.786	188.968	169.590	85.261	607.441	299.405	-3.554
4500	20.786	189.435	170.026	87.340	609.013	292.386	-3.394
4600	20.786	189.892	170.453	89.418	610.544	285.333	-3.240
4700	20.786	190.339	170.871	91.497	612.034	278.247	-3.092
4800	20.786	190.776	171.281	93.576	613.483	271.129	-2.950
4900	20.786	191.205	171.684	95.654	614.889	263.982	-2.814
5000	20.786	191.625	172.078	97.733	616.250	256.807	-2.683
5100	20.786	192.036	172.466	99.811	617.570	249.605	-2.556
5200	20.786	192.440	172.846	101.890	618.838	242.377	-2.435
5300	20.786	192.836	173.219	103.969	620.057	235.126	-2.317
5400	20.786	193.225	173.586	106.047	621.225	227.852	-2.204
5500	20.786	193.606	173.947	108.126	622.341	220.557	-2.095
5600	20.786	193.981	174.301	110.204	623.402	213.242	-1.989
5700	20.786	194.348	174.650	112.283	624.408	205.909	-1.887
5800	20.786	194.710	174.992	114.362	625.357	198.558	-1.788
5900	20.786	195.065	175.330	116.440	626.246	191.192	-1.693
6000	20.786	195.415	175.661	118.519	627.076	183.811	-1.600

PREVIOUS: March 1965 (1 atm) CURRENT: December 1983 (1 bar)

Lithium, Ion (Li^+) $Li_1^+(g)$

LITHIUM, ION (Li^-) IDEAL GAS M_r = 6.94155

EA(Li, g) = 0.6180 ± 0.0005 eV
S°(298.15 K) = 133.019 ± 0.005 J K^{-1} mol^{-1}

$\Delta_f H°$(0 K) = 98.10 ± 0.1 kJ mol^{-1}
$\Delta_f H°$(298.15 K) = [93.475] kJ mol^{-1}

Electonic Level and Quantum Weight

State	ε_i, cm^{-1}	g_i
$^1S_0$	0.0	1

Enthalpy of Formation

The enthalpy of formation, at 0 K, for Li^-(g) is calculated from the adopted electron affinity, EA(Li, g) = 0.6180 ± 0.0005 eV (59.627 ± 0.048 kJ mol^{-1}). This value, recommended by Hotop and Lineberger (1), was measured by a tunable laser photodetachment threshold technique (2, 3). Additional discussion on Li^-(g) may be obtained in the critical discussions of Hotop and Lineberger (1, 4), Rosenstock et al. (5), and Massey (6).

$\Delta_f H°$(Li^-, g, 298.15 K) is obtained from $\Delta_f H°$(Li, g, 0 K) by using EA(Li, g) with JANAF (7) enthalpies, H°(0 K) - H°(298.15 K), for Li^-(g), Li(g), and e^-(g). $\Delta_r H°$(Li^- + Li + e^-, 298.15 K) differs from a room-temperature threshold energy due to inclusion of these enthalpies and to threshold effects discussed by Rosenstock et al. (5). $\Delta_f H°$(298.15 K) should be changed by +6.197 kJ mol^{-1} if it is to be used in the ion convention which excludes the enthalpy of the electron.

Heat Capacity and Entropy

The ground state configuration for Li^-(g) is given by Hotop and Lineberger (1, 4) and Rosenstock et al. (5). Lacking any experimental evidence as to the stability of any excited states, we assume that no stable excited states exist.

References

1. H. Hotop and W. C. Lineberger, J. Phys. Chem. Ref. Data, to be published, 1985.
2. D. Feldman, Z. Physik A 277, 19 (1976).
3. Y. K. Bae and J. R. Peterson, Phys. Rev. A, to be published.
4. H. Hotop and W. C. Lineberger, J. Phys. Chem. Ref. Data 4, 539 (1975).
5. H. M. Rosenstock, K. Draxl et al., J. Phys. Chem. Ref. Data 6, Supp. 1 (1977).
6. H. S. W. Massey, "Negative Ions," 3rd ed., Cambridge University Press, Cambridge, 1976.
7. JANAF Thermochemical Tables: Li(g), 3-31-82; e^-(g), 3-31-82.

Lithium, Ion (Li^-) Li_1^-(g)

Enthalpy Reference Temperature = T_r = 298.15 K   Standard State Pressure = p° = 0.1 MPa

T/K	$C_p°$ (J $K^{-1}mol^{-1}$)	S° (J $K^{-1}mol^{-1}$)	-[G°-H°(T_r)]/T (J $K^{-1}mol^{-1}$)	H°-H°(T_r) (kJ mol^{-1})	$\Delta_f H°$ (kJ mol^{-1})	$\Delta_f G°$ (kJ mol^{-1})	Log K_f
0	0.	0.	INFINITE	-6.197	98.097		
100	20.786	110.311	151.499	-4.119			
200	20.786	124.719	134.920	-2.040			
250	20.786	129.357	133.361	-1.001			
298.15	20.786	133.019	133.019	0.	93.475	68.742	-12.043
300	20.786	133.147	133.019	0.038	93.430	68.589	-11.942
350	20.786	136.351	133.272	1.078	92.167	64.548	-9.633
400	20.786	139.127	133.834	2.117	90.829	60.692	-7.926
450	20.786	141.575	134.561	3.156	89.406	57.010	-6.618
500	20.786	143.765	135.374	4.196	84.896	53.797	-5.620
600	20.786	147.555	137.098	6.274	81.913	47.859	-4.166
700	20.786	150.759	138.827	8.353	78.988	42.415	-3.165
800	20.786	153.535	140.495	10.431	76.092	37.388	-2.441
900	20.786	155.983	142.083	12.510	73.200	32.724	-1.899
1000	20.786	158.173	143.584	14.589	70.314	28.381	-1.482
1100	20.786	160.154	145.002	16.667	67.433	24.328	-1.155
1200	20.786	161.963	146.341	18.746	64.557	20.536	-0.894
1300	20.786	163.627	147.608	20.824	61.685	16.984	-0.682
1400	20.786	165.167	148.808	22.903	58.819	13.653	-0.509
1500	20.786	166.601	149.947	24.982	55.961	10.527	-0.367
1600	20.786	167.943	151.030	27.060	53.112	7.591	-0.248
1700	20.786	169.203	152.062	29.139	-94.965	12.009	-0.369
1800	20.786	170.391	153.048	31.217	-97.046	18.361	-0.533
1900	20.786	171.515	153.990	33.296	-99.128	24.829	-0.683
2000	20.786	172.581	154.893	35.375	-101.211	31.407	-0.820
2100	20.786	173.595	155.760	37.453	-103.298	38.090	-0.947
2200	20.786	174.562	156.593	39.532	-105.388	44.871	-1.065
2300	20.786	175.486	157.394	41.610	-107.483	51.748	-1.175
2400	20.786	176.371	158.167	43.689	-109.584	58.716	-1.278
2500	20.786	177.219	158.912	45.768	-111.693	65.772	-1.374
2600	20.786	178.034	159.632	47.846	-113.811	72.913	-1.465
2700	20.786	178.819	160.328	49.925	-115.940	80.135	-1.550
2800	20.786	179.575	161.002	52.004	-118.081	87.436	-1.631
2900	20.786	180.304	161.655	54.082	-120.236	94.814	-1.708
3000	20.786	181.009	162.289	56.161	-122.407	102.266	-1.781
3100	20.786	181.690	162.904	58.239	-124.596	109.791	-1.850
3200	20.786	182.350	163.501	60.318	-126.803	117.387	-1.916
3300	20.786	182.990	164.082	62.397	-129.031	125.053	-1.979
3400	20.786	183.610	164.647	64.475	-131.282	132.786	-2.040
3500	20.786	184.213	165.198	66.554	-133.556	140.586	-2.098
3600	20.786	184.799	165.734	68.632	-135.856	148.451	-2.154
3700	20.786	185.368	166.257	70.711	-138.183	156.380	-2.208
3800	20.786	185.922	166.767	72.790	-140.538	164.373	-2.259
3900	20.786	186.462	167.265	74.868	-142.922	172.428	-2.309
4000	20.786	186.989	167.752	76.947	-145.339	180.544	-2.358
4100	20.786	187.502	168.227	79.025	-147.788	188.722	-2.404
4200	20.786	188.003	168.692	81.104	-150.271	196.959	-2.450
4300	20.786	188.492	169.147	83.183	-152.790	205.256	-2.493
4400	20.786	188.970	169.592	85.261	-155.325	213.611	-2.536
4500	20.786	189.437	170.028	87.340	-157.910	222.024	-2.577
4600	20.786	189.894	170.455	89.418	-160.537	230.496	-2.617
4700	20.786	190.341	170.873	91.497	-163.204	239.026	-2.656
4800	20.786	190.778	171.283	93.576	-165.912	247.612	-2.695
4900	20.786	191.207	171.686	95.654	-168.664	256.256	-2.732
5000	20.786	191.627	172.080	97.733	-171.460	264.956	-2.768
5100	20.786	192.038	172.468	99.811	-174.297	273.712	-2.803
5200	20.786	192.442	172.848	101.890	-177.186	282.524	-2.838
5300	20.786	192.838	173.221	103.969	-180.124	291.393	-2.872
5400	20.786	193.227	173.588	106.047	-183.113	300.317	-2.905
5500	20.786	193.608	173.949	108.126	-186.155	309.298	-2.937
5600	20.786	193.982	174.303	110.204	-189.251	318.334	-2.969
5700	20.786	194.350	174.652	112.283	-192.402	327.426	-3.001
5800	20.786	194.712	174.994	114.362	-195.611	336.573	-3.031
5900	20.786	195.067	175.332	116.440	-198.878	345.777	-3.061
6000	20.786	195.417	175.663	118.519	-202.206	355.036	-3.091

PREVIOUS: CURRENT: December 1983 (1 bar)

Lithium, Ion (Li^-) Li_1^-(g)

LITHIUM NITRIDE (LiN) IDEAL GAS $M_r = 20.9477$

$\Delta_f H°(0\ K) = 334.7 \pm 167\ kJ\ mol^{-1}$

$S°(298.15\ K) = [208.245]\ J\ K^{-1}\ mol^{-1}$ $\Delta_f H°(298.15\ K) = 334.7 \pm 167\ kJ\ mol^{-1}$

Electronic Level and Quantum Weight

State	ε_i, cm^{-1}	g_i
$[^3\Sigma]$	0	[3]

$\omega_e = [700]\ cm^{-1}$ $\omega_e x_e = [7]\ cm^{-1}$ $\sigma = 1$

$B_e = [1.25]\ cm^{-1}$ $\alpha_e = [0.0183]\ cm^{-1}$ $r_e = [1.71]$ Å

Enthalpy of Formation

The adopted $\Delta_f H°(298.15\ K) = 80 \pm 40\ kcal\ mol^{-1}$ is calculated from the following estimates:

Method	D_0^0, $kcal\ mol^{-1}$	$\Delta_f H°(298.15\ K)$ $kcal\ mol^{-1}$
Estimate D_0^0 from geometric means for Li_2 and N_2	76	75
Estimate from D_0^0 for series $N_2(g)$, CN(g), BN(g), and BeN(g)	40 - 100	110 - 50
Estimate from D_0^0 for series HN(g), HO(g), HF(g), LiF(g) and LiO(g)	20 - 60	130 - 90

Margrave and Sthapitanonda (1) report estimated dissociation energies in the range 111-178 $kcal\ mol^{-1}$ based on approximate ionic binding energies calculated from ionic radii for N^- in the range 0.70 to 1.08 Å.

Heat Capacity and Entropy

The molecular constants, ω_e, r_e and $\omega_e x_e$ are estimated by comparison with those of CN, BN, CO, BO and LiO. The values of α_e and B_e are calculated from the above constants. The ground state configuration is assumed to be $^3\Sigma$ by analogy with NH and OH^+.

Reference

1. J. L. Margrave and P. Sthapitanonda, J. Phys. Chem. 59, 1231 (1955).

Lithium Nitride (LiN) $Li_1N_1(g)$

Enthalpy Reference Temperature = T_r = 298.15 K Standard State Pressure = p° = 0.1 MPa

T/K	$C_p°$ (J $K^{-1}mol^{-1}$)	S° (J $K^{-1}mol^{-1}$)	$-[G°-H°(T_r)]/T$ (J $K^{-1}mol^{-1}$)	$H°-H°(T_r)$ (kJ mol^{-1})	$\Delta_f H°$ (kJ mol^{-1})	$\Delta_f G°$ (kJ mol^{-1})	Log K_f
0	0.	0.	INFINITE	-8.999	334.678	334.678	INFINITE
100	29.167	175.034	235.957	-6.092	335.639	326.860	-170.734
200	30.651	195.587	211.181	-3.119	335.316	318.162	-83.095
250	31.812	202.552	208.780	-1.557	335.021	313.907	-65.587
298.15	32.838	208.245	208.245	0.	334.720	309.868	-54.288
300	32.875	208.448	208.246	0.061	334.708	309.713	-53.926
350	33.762	213.585	208.649	1.727	334.383	305.573	-45.604
400	34.479	218.142	209.556	3.434	334.022	301.481	-39.369
450	35.056	222.237	210.742	5.173	333.605	297.438	-34.526
500	35.521	225.956	212.080	6.938	330.123	293.752	-30.688
600	36.215	232.497	214.952	10.527	329.238	286.562	-24.947
700	36.702	238.118	217.870	14.174	328.438	279.514	-20.858
800	37.064	243.044	220.715	17.863	327.676	272.577	-17.797
900	37.345	247.426	223.444	21.584	326.918	265.735	-15.423
1000	37.574	251.373	226.043	25.330	326.158	258.977	-13.528
1100	37.767	254.964	228.511	29.098	325.396	252.296	-11.981
1200	37.935	258.257	230.855	32.883	324.631	245.684	-10.694
1300	38.084	261.300	233.081	36.684	323.862	239.137	-9.609
1400	38.221	264.127	235.199	40.499	323.095	232.648	-8.680
1500	38.347	266.768	237.216	44.328	322.332	226.214	-7.877
1600	38.466	269.247	239.142	48.169	321.573	219.831	-7.177
1700	38.578	271.582	240.982	52.021	175.586	220.671	-6.780
1800	38.686	273.791	242.744	55.884	175.594	223.323	-6.481
1900	38.790	275.885	244.434	59.758	175.601	225.975	-6.212
2000	38.890	277.877	246.056	63.642	175.607	228.626	-5.971
2100	38.989	279.777	247.617	67.536	175.612	231.277	-5.753
2200	39.085	281.593	249.121	71.440	175.616	233.927	-5.554
2300	39.179	283.333	250.570	75.353	175.617	236.578	-5.373
2400	39.273	285.002	251.971	79.275	175.616	239.228	-5.207
2500	39.364	286.607	253.324	83.207	175.611	241.879	-5.054
2600	39.455	288.153	254.634	87.148	175.600	244.530	-4.913
2700	39.545	289.643	255.903	91.098	175.584	247.181	-4.782
2800	39.634	291.083	257.134	95.057	175.559	249.833	-4.661
2900	39.722	292.476	258.329	99.025	175.526	252.486	-4.548
3000	39.810	293.824	259.490	103.002	175.482	255.141	-4.442
3100	39.898	295.131	260.619	106.987	175.426	257.797	-4.344
3200	39.984	296.399	261.717	110.981	175.356	260.455	-4.251
3300	40.071	297.630	262.787	114.984	175.271	263.116	-4.165
3400	40.157	298.828	263.829	118.995	175.170	265.779	-4.083
3500	40.243	299.993	264.846	123.015	175.051	268.446	-4.006
3600	40.328	301.128	265.838	127.044	174.912	271.116	-3.934
3700	40.414	302.234	266.807	131.081	174.753	273.791	-3.865
3800	40.499	303.313	267.753	135.127	174.571	276.470	-3.800
3900	40.584	304.366	268.679	139.181	174.366	279.154	-3.739
4000	40.668	305.395	269.584	143.243	174.136	281.844	-3.681
4100	40.753	306.400	270.470	147.314	173.880	284.540	-3.625
4200	40.837	307.383	271.337	151.394	173.596	287.242	-3.572
4300	40.921	308.345	272.186	155.482	173.282	289.952	-3.522
4400	41.006	309.287	273.019	159.578	172.959	292.667	-3.474
4500	41.090	310.209	273.835	163.683	172.592	295.392	-3.429
4600	41.173	311.113	274.636	167.796	172.191	298.125	-3.385
4700	41.257	311.999	275.421	171.918	171.757	300.867	-3.344
4800	41.341	312.869	276.192	176.048	171.287	303.619	-3.304
4900	41.425	313.722	276.950	180.186	170.781	306.381	-3.266
5000	41.508	314.560	277.693	184.332	170.236	309.154	-3.230
5100	41.592	315.383	278.424	188.487	169.658	311.938	-3.195
5200	41.675	316.191	279.143	192.651	169.034	314.733	-3.162
5300	41.759	316.986	279.849	196.822	168.367	317.542	-3.130
5400	41.842	317.767	280.544	201.003	167.657	320.363	-3.099
5500	41.925	318.536	281.228	205.191	166.900	323.198	-3.069
5600	42.009	319.292	281.901	209.388	166.096	326.047	-3.041
5700	42.092	320.036	282.564	213.593	165.243	328.910	-3.014
5800	42.175	320.769	283.216	217.806	164.339	331.789	-2.988
5900	42.258	321.491	283.859	222.028	163.383	334.685	-2.963
6000	42.341	322.201	284.492	226.258	162.373	337.597	-2.939

PREVIOUS: December 1966 (1 atm) CURRENT: December 1966 (1 bar)

Lithium Nitride (LiN) $Li_1N_1(g)$

LITHIUM OXYNITRIDE (LiON) IDEAL GAS M_r = 36.9471

$S°(298.15\ K) = 245.332\ J\ K^{-1}\ mol^{-1}$

$\Delta_fH°(0\ K) = [181.91 \pm 41.8]\ kJ\ mol^{-1}$
$\Delta_fH°(298.15\ K) = [179.91 \pm 41.8]\ kJ\ mol^{-1}$

Vibrational Frequencies and Degeneracies

ν, cm^{-1}
1350(1)
330(1)
650(1)

Ground State Quantum Weight: [1] $\sigma = 1$
Point Group: C_s
Bond Distances: O-Li = 1.63 ± 0.07 Å O-N = 1.30 ± 0.05 Å
Bond Angle: Li-O-N = 100 ± 10°
Product of the Moments of Inertia: $I_AI_BI_C = 2.65370 \times 10^{-116}\ g^3\ cm^6$

Enthalpy of Formation

In order to estimate the $\Delta_fH°(298.15\ K)$ of LiON(g) a comparison with the known $\Delta_fH°(LiOH, g, 298.15\ K) = -59\ kcal\ mol^{-1}$ was made. This was estimated from bond energies using D(Li-O) = 88 kcal mol^{-1}, the average value in $Li_2O(g)$, and D(O-H) = 111 kcal mol^{-1}, the average value in $H_2O(g)$, which yields an estimated $\Delta_fH°(LiOH, g, 298.15\ K) = -49\ kcal\ mol^{-1}$ or a difference of -10 kcal mol^{-1}. Using D(N=O) = 69 kcal mol^{-1} from $NO_2Cl(g)$(equation 1) and D(Li-O) = 88 kcal mol^{-1} from $Li_2O(g)$, the estimated $\Delta_fH°(298.15\ K) = 53\ kcal\ mol^{-1}$ of LiON(g) was obtained (equation 2). Thus adding the -10 kcal mol^{-1} correction from the above comparison a $\Delta_fH°(298.15\ K) = 43\ kcal\ mol^{-1}$ for LiON(g) was obtained.

The justification for using a N=O bond energy is from the 1.7 bond order reported by Andrews and Pimentel (1) and a comparison of bond lengths.

Equations: 1. $NO_2Cl(g) \rightarrow NOCl(g) + O(g)$
2. $LiON(g) \rightarrow Li(g) + O(g) + N(g)$

Heat Capacity and Entropy

The vibration frequencies, bond distances and bond angle were obtained from Andrews and Pimentel (1). The principal moments of inertia are: $I_A = 1.2707 \times 10^{-39}$, $I_B = 3.9785 \times 10^{-39}$ and $I_C = 5.2492 \times 10^{-39}\ g\ cm^2$.

Reference

1. W. L. S. Andrews and G. C. Pimentel, J. Chem. Phys. 44, 2361 (1966).

Lithium Oxynitride (LiON) $Li_1N_1O_1(g)$

Enthalpy Reference Temperature = T_r = 298.15 K Standard State Pressure = p° = 0.1 MPa

T/K	$C_p°$ (J K⁻¹mol⁻¹)	S° (J K⁻¹mol⁻¹)	$-[G°-H°(T_r)]/T$ (J K⁻¹mol⁻¹)	$H°-H°(T_r)$ (kJ mol⁻¹)	$\Delta_fH°$ (kJ mol⁻¹)	$\Delta_fG°$ (kJ mol⁻¹)	Log K_f
0	0.	0.	INFINITE	-11.301	181.910	181.910	INFINITE
100	34.974	202.507	281.906	-7.940	181.873	179.012	-93.506
200	40.335	228.423	249.272	-4.170	180.890	176.518	-46.102
250	42.576	237.671	246.053	-2.096	180.380	175.484	-36.665
298.15	44.424	245.332	245.332	0.	179.912	174.584	-30.586
300	44.490	245.607	245.333	0.082	179.895	174.551	-30.392
350	46.144	252.592	245.881	2.349	179.432	173.698	-25.923
400	47.585	258.850	247.117	4.693	178.961	172.910	-22.580
450	48.838	264.529	248.741	7.104	178.457	172.184	-19.987
500	49.925	269.732	250.584	9.574	174.909	171.823	-17.950
600	51.676	278.997	254.566	14.658	173.939	171.299	-14.913
700	52.982	287.066	258.645	19.895	173.101	170.927	-12.755
800	53.962	294.208	262.653	25.244	172.332	170.669	-11.144
900	54.706	300.608	266.521	30.679	171.585	170.506	-9.896
1000	55.280	306.403	270.224	36.180	170.848	170.426	-8.902
1100	55.729	311.694	273.757	41.731	170.115	170.419	-8.093
1200	56.085	316.559	277.124	47.322	169.382	170.479	-7.421
1300	56.371	321.060	280.332	52.946	168.644	170.601	-6.855
1400	56.605	325.246	283.393	58.595	167.904	170.779	-6.372
1500	56.797	329.158	286.315	64.265	167.162	171.010	-5.955
1600	56.957	332.829	289.108	69.953	166.417	171.291	-5.592
1700	57.092	336.286	291.783	75.656	20.434	178.795	-5.494
1800	57.206	339.553	294.347	81.371	20.436	188.110	-5.459
1900	57.304	342.649	296.808	87.097	20.425	197.425	-5.428
2000	57.388	345.590	299.174	92.831	20.401	206.742	-5.400
2100	57.461	348.392	301.452	98.574	20.362	216.060	-5.374
2200	57.525	351.066	303.647	104.323	20.307	225.380	-5.351
2300	57.581	353.625	305.765	110.079	20.235	234.703	-5.330
2400	57.630	356.076	307.810	115.839	20.145	244.029	-5.311
2500	57.674	358.430	309.788	121.604	20.036	253.360	-5.294
2600	57.713	360.693	311.703	127.374	19.906	262.695	-5.278
2700	57.747	362.871	313.558	133.147	19.754	272.036	-5.263
2800	57.779	364.972	315.357	138.923	19.578	281.383	-5.249
2900	57.807	367.000	317.103	144.702	19.377	290.737	-5.237
3000	57.832	368.960	318.799	150.484	19.150	300.098	-5.225
3100	57.855	370.857	320.448	156.269	18.895	309.467	-5.214
3200	57.876	372.694	322.052	162.055	18.611	318.845	-5.205
3300	57.895	374.475	323.614	167.844	18.296	328.232	-5.195
3400	57.913	376.204	325.135	173.634	17.950	337.629	-5.187
3500	57.929	377.883	326.618	179.426	17.571	347.037	-5.179
3600	57.943	379.515	328.065	185.220	17.158	356.456	-5.172
3700	57.957	381.103	329.477	191.015	16.709	365.887	-5.165
3800	57.970	382.649	330.856	196.811	16.224	375.331	-5.159
3900	57.981	384.155	332.204	202.609	15.702	384.788	-5.154
4000	57.992	385.623	333.521	208.408	15.140	394.258	-5.148
4100	58.002	387.055	334.809	214.207	14.538	403.744	-5.144
4200	58.011	388.453	336.070	220.008	13.894	413.244	-5.139
4300	58.020	389.818	337.304	225.809	13.207	422.761	-5.136
4400	58.028	391.152	338.513	231.612	12.497	432.292	-5.132
4500	58.035	392.456	339.697	237.415	11.730	441.841	-5.129
4600	58.042	393.732	340.858	243.219	10.915	451.407	-5.126
4700	58.049	394.980	341.996	249.023	10.052	460.993	-5.123
4800	58.055	396.202	343.113	254.829	9.140	470.597	-5.121
4900	58.061	397.399	344.208	260.634	8.177	480.220	-5.119
5000	58.067	398.572	345.284	266.441	7.162	489.864	-5.118
5100	58.072	399.722	346.340	272.248	6.099	499.529	-5.116
5200	58.077	400.850	347.378	278.055	4.975	509.214	-5.115
5300	58.081	401.956	348.397	283.863	3.793	518.923	-5.114
5400	58.086	403.042	349.399	289.672	2.551	528.653	-5.114
5500	58.090	404.108	350.384	295.480	1.248	538.407	-5.113
5600	58.094	405.154	351.353	301.289	-0.119	548.187	-5.113
5700	58.097	406.183	352.306	307.099	-1.552	557.990	-5.113
5800	58.101	407.193	353.243	312.909	-3.053	567.820	-5.114
5900	58.104	408.186	354.166	318.719	-4.624	577.675	-5.114
6000	58.107	409.163	355.075	324.530	-6.268	587.559	-5.115

PREVIOUS: September 1966 (1 atm) CURRENT: September 1966 (1 bar)

Lithium Oxynitride (LiON) $Li_1N_1O_1(g)$

LITHIUM SODIUM OXIDE (LiONa) IDEAL GAS $M_r = 45.93017$

Lithium Sodium Oxide (LiONa) $Li_1Na_1O_1(g)$

$S°(298.15\ K) = [256.459]\ J\ K^{-1}\ mol^{-1}$

$\Delta_f H°(0\ K) = [-100.16 \pm 125.5]\ kJ\ mol^{-1}$
$\Delta_f H°(298.15\ K) = [-104.60 \pm 125.5]\ kJ\ mol^{-1}$

Vibrational Frequencies and Degeneracies

ν, cm^{-1}
[400](1)
[1000](1)
[900](1)

Ground State Quantum Weight: [1] $\sigma = 1$
Point Group: C_s
Bond Distances: O-Li = [1.82] Å O-Na = [2.21] Å
Bond Angle: Li-O-Na = [105]°
Product of the Moments of Inertia: $I_A I_B I_C = [3.182324 \times 10^{-115}]\ g^3\ cm^6$

Enthalpy of Formation

Estimated from bond energies derived from enthalpies of formation used in the JANAF Thermochemical Tables.

Heat Capacity and Entropy

The molecular constants have been estimated by comparison of similar molecules. The principal moments of inertia are: $I_A = 1.9618 \times 10^{-39}$, $I_B = 11.7930 \times 10^{-39}$, and $I_C = 13.7548 \times 10^{-39}\ g\ cm^2$.

Enthalpy Reference Temperature = T_r = 298.15 K Standard State Pressure = $p°$ = 0.1 MPa

T/K	$C_p°$ (J K⁻¹mol⁻¹)	S° (J K⁻¹mol⁻¹)	$-[G°-H°(T_r)]/T$ (J K⁻¹mol⁻¹)	$H°-H°(T_r)$ (kJ mol⁻¹)	$\Delta_f H°$ (kJ mol⁻¹)	$\Delta_f G°$ (kJ mol⁻¹)	Log K_f
0	0.	0.	INFINITE	-10.966	-100.156	-100.156	INFINITE
100	34.139	215.303	291.557	-7.625	-100.102	-109.851	57.380
200	38.470	240.232	260.252	-4.004	-102.233	-118.826	31.034
250	40.849	249.071	257.156	-2.021	-103.432	-122.836	25.665
298.15	43.079	256.459	256.459	0.	-104.600	-126.468	22.157
300	43.161	256.726	256.460	0.080	-104.645	-126.604	22.044
350	45.267	263.540	256.993	2.292	-105.887	-130.166	19.426
400	47.099	269.708	258.203	4.602	-109.831	-133.340	17.412
450	48.652	275.348	259.799	6.997	-111.180	-136.198	15.809
500	49.952	280.543	261.617	9.463	-115.534	-138.598	14.479
600	51.941	289.837	265.564	14.564	-118.012	-142.973	12.447
700	53.334	297.955	269.624	19.831	-120.249	-146.954	10.966
800	54.331	305.145	273.623	25.217	-122.337	-150.625	9.835
900	55.060	311.589	277.490	30.689	-124.346	-154.040	8.940
1000	55.607	317.419	281.196	36.223	-126.316	-157.234	8.213
1100	56.026	322.740	284.734	41.806	-128.278	-160.230	7.609
1200	56.353	327.629	288.108	47.425	-227.019	-160.608	6.991
1300	56.613	332.151	291.324	53.074	-228.112	-155.029	6.229
1400	56.822	336.354	294.392	58.746	-229.192	-149.367	5.573
1500	56.993	340.280	297.322	64.437	-230.258	-143.628	5.002
1600	57.134	343.963	300.123	70.144	-231.313	-137.818	4.499
1700	57.252	347.431	302.805	75.863	-377.595	-124.767	3.834
1800	57.352	350.706	305.376	81.594	-377.882	-109.886	3.189
1900	57.437	353.809	307.844	87.333	-378.174	-94.990	2.611
2000	57.510	356.757	310.217	93.081	-378.471	-80.078	2.091
2100	57.573	359.565	312.500	98.835	-378.777	-65.151	1.621
2200	57.628	362.244	314.701	104.595	-379.094	-50.208	1.192
2300	57.676	364.807	316.824	110.360	-379.423	-35.252	0.801
2400	57.718	367.262	318.875	116.130	-379.768	-20.281	0.441
2500	57.755	369.619	320.858	121.904	-380.132	-5.295	0.111
2600	57.788	371.885	322.777	127.681	-380.515	9.706	-0.195
2700	57.818	374.067	324.637	133.461	-380.922	24.723	-0.478
2800	57.844	376.170	326.440	139.244	-381.356	39.755	-0.742
2900	57.868	378.200	328.190	145.030	-381.819	54.802	-0.987
3000	57.890	380.162	329.890	150.818	-382.314	69.866	-1.216
3100	57.910	382.061	331.542	156.608	-382.845	84.948	-1.431
3200	57.927	383.900	333.150	162.400	-383.414	100.047	-1.633
3300	57.943	385.683	334.715	168.193	-384.024	115.164	-1.823
3400	57.958	387.413	336.240	173.988	-384.680	130.301	-2.002
3500	57.972	389.093	337.726	179.785	-385.383	145.457	-2.171
3600	57.984	390.726	339.175	185.583	-386.138	160.635	-2.331
3700	57.996	392.315	340.590	191.382	-386.947	175.834	-2.482
3800	58.006	393.862	341.972	197.182	-387.814	191.056	-2.626
3900	58.016	395.369	343.322	202.983	-388.743	206.301	-2.763
4000	58.025	396.838	344.641	208.785	-389.738	221.571	-2.893
4100	58.034	398.271	345.932	214.588	-390.802	236.867	-3.018
4200	58.041	399.669	347.195	220.392	-391.937	252.189	-3.136
4300	58.049	401.035	348.431	226.196	-393.151	267.540	-3.250
4400	58.056	402.369	349.642	232.002	-394.384	282.914	-3.359
4500	58.062	403.674	350.828	237.807	-395.733	298.322	-3.463
4600	58.068	404.950	351.991	243.614	-397.179	313.761	-3.563
4700	58.073	406.199	353.131	249.421	-398.715	329.232	-3.659
4800	58.079	407.422	354.249	255.229	-400.344	344.738	-3.752
4900	58.084	408.620	355.347	261.037	-402.073	360.278	-3.841
5000	58.088	409.793	356.424	266.845	-403.897	375.854	-3.927
5100	58.093	410.943	357.482	272.654	-405.828	391.468	-4.009
5200	58.097	412.072	358.521	278.464	-407.876	407.120	-4.090
5300	58.101	413.178	359.542	284.274	-410.040	422.814	-4.167
5400	58.104	414.264	360.545	290.084	-412.325	438.549	-4.242
5500	58.108	415.330	361.531	295.895	-414.735	454.328	-4.315
5600	58.111	416.378	362.502	301.705	-417.275	470.153	-4.385
5700	58.114	417.406	363.456	307.517	-419.949	486.023	-4.454
5800	58.117	418.417	364.395	313.328	-422.762	501.942	-4.520
5900	58.120	419.410	365.319	319.140	-425.716	517.909	-4.585
6000	58.123	420.387	366.228	324.952	-428.659	533.911	-4.648

PREVIOUS: June 1961 (1 atm) CURRENT: June 1961 (1 bar)

Lithium Sodium Oxide (LiONa) $Li_1Na_1O_1(g)$

LITHIUM OXIDE (LiO) IDEAL GAS M_r = 22.9404

$S°(298.15\ K) = [210.963]\ J\ K^{-1}\ mol^{-1}$

$\Delta_f H°(0\ K) = 84.12 \pm 20.9\ kJ\ mol^{-1}$
$\Delta_f H°(298.15\ K) = 84.10 \pm 20.9\ kJ\ mol^{-1}$

Electronic Level and Quantum Weight

State	ε_i, cm^{-1}	g_i
$[^2\Pi]$	0	4

$\omega_e = 745\ cm^{-1}$ $\omega_e x_e = [4.835]\ cm^{-1}$ $\sigma = 1$
$B_e = [1.333]\ cm^{-1}$ $\alpha_e = [0.01288]\ cm^{-1}$ $r_e = [1.62]$ Å

Enthalpy of Formation

White et al. (1) calculated the enthalpy of formation from their mass spectral data by two different methods.

For the first calculation (a) they combined the slope of the log IT vs 1/T curve for LiO^+ with the slopes for Li^+ and Li_2O^+ for the reaction $Li_2O(g) \rightarrow Li(g) + LiO(g)$. For the second calculation (b) they assumed a cross section σ LiO/σ Li_2O = 1 and from this they calculated a partial pressure for LiO(g) at 1500 K and an equilibrium constant for the reaction $Li_2O(g) \rightarrow Li(g) + LiO(g)$. They obtained (a) $\Delta_f H°(298.15\ k) = 18.3\ kcal\ mol^{-1}$ and (b) $\Delta_f H°(298.15\ K) = 16.3\ kcal\ mol^{-1}$, and used the average value from the two methods for the enthalpy of formation of LiO(g). The JANAF enthalpy of formation, 20.1±5 $kcal\ mol^{-1}$, was obtained in the same way using their data and the present tables i.e. (a) $\Delta_f H°(298.15\ K) = 22.0\ kcal\ mol^{-1}$ and (b) $\Delta_f H°(298.15\ K) = 18.3\ kcal\ mol^{-1}$.

Heat Capacity and Entropy

The molecular constants, $\omega_e x_e$, B_e and α_e, were calculated from ω_e and r_e, reported by White et al. (1). The ground state configuration, $^2\Pi$, was estimated by Gordon (2), Buchler and Stauffer (3) and White et al. (1).

References

1. D. White, K. S. Seshadri, D. F. Dever, and D. E. Mann, and M. J. Linevsky, J. Chem. Phys. 39, 2463 (1963).
2. J. S. Gordon, Thiokol Chemical Corp., Reaction Motor Division, Denville, N.J., "Thermodynamic Data for Combustion Products", Janaury, 1960.
3. A. Buchler and J. L. Stauffer, J. Chem. Phys. 39, 2299 (1963).

Lithium Oxide (LiO) $Li_1O_1(g)$

Enthalpy Reference Temperature = T_r = 298.15 K Standard State Pressure = p° = 0.1 MPa

T/K	$C_p°$ ($J\ K^{-1}mol^{-1}$)	S°	$-[G°-H°(T_r)]/T$	$H°-H°(T_r)$ ($kJ\ mol^{-1}$)	$\Delta_f H°$	$\Delta_f G°$	Log K_f
0	0.	0.	INFINITE	-8.945	84.118	84.118	INFINITE
100	29.147	178.000	238.387	-6.039	85.076	76.676	-40.051
200	30.343	198.459	213.861	-3.080	84.737	68.359	-17.854
250	31.411	205.343	211.491	-1.537	84.424	64.300	-13.435
298.15	32.406	210.963	210.963	0.	84.098	60.454	-10.591
300	32.442	211.163	210.963	0.060	84.086	60.307	-10.500
350	33.332	216.233	211.362	1.705	83.729	56.372	-8.413
400	34.067	220.733	212.257	3.391	83.330	52.490	-6.855
450	34.664	224.782	213.427	5.109	82.867	48.662	-5.649
500	35.148	228.460	214.750	6.855	79.332	45.197	-4.722
600	35.872	234.936	217.589	10.408	78.323	38.466	-3.349
700	36.376	240.506	220.474	14.022	77.384	31.898	-2.380
800	36.744	245.389	223.290	17.679	76.476	25.462	-1.663
900	37.024	249.733	225.991	21.368	75.571	19.140	-1.111
1000	37.246	253.646	228.564	25.082	74.668	12.918	-0.675
1100	37.428	257.205	231.008	28.816	73.766	6.787	-0.322
1200	37.582	260.468	233.329	32.567	72.867	0.738	-0.032
1300	37.716	263.482	235.534	36.332	71.968	-5.237	0.210
1400	37.834	266.281	237.632	40.109	71.073	-11.142	0.416
1500	37.942	268.895	239.630	43.898	70.183	-16.983	0.591
1600	38.041	271.347	241.536	47.697	69.300	-22.766	0.743
1700	38.133	273.656	243.358	51.506	-76.815	-21.317	0.655
1800	38.219	275.838	245.103	55.324	-76.936	-18.049	0.524
1900	38.302	277.907	246.775	59.150	-77.061	-14.774	0.406
2000	38.381	279.874	248.381	62.984	-77.192	-11.492	0.300
2100	38.457	281.748	249.926	66.826	-77.329	-8.204	0.204
2200	38.530	283.539	251.414	70.675	-77.474	-4.909	0.117
2300	38.602	285.253	252.848	74.532	-77.628	-1.607	0.036
2400	38.672	286.897	254.233	78.396	-77.792	1.702	-0.037
2500	38.741	288.477	255.571	82.266	-77.967	5.017	-0.105
2600	38.808	289.998	256.866	86.144	-78.156	8.340	-0.168
2700	38.875	291.464	258.120	90.028	-78.359	11.671	-0.226
2800	38.941	292.879	259.337	93.919	-78.579	15.010	-0.280
2900	39.005	294.247	260.517	97.816	-78.815	18.356	-0.331
3000	39.070	295.570	261.664	101.720	-79.071	21.711	-0.378
3100	39.133	296.852	262.778	105.630	-79.347	25.075	-0.423
3200	39.196	298.096	263.862	109.547	-79.645	28.448	-0.464
3300	39.259	299.303	264.918	113.469	-79.966	31.831	-0.504
3400	39.321	300.476	265.947	117.398	-80.311	35.224	-0.541
3500	39.383	301.616	266.950	121.334	-80.682	38.627	-0.576
3600	39.445	302.727	267.928	125.275	-81.080	42.042	-0.610
3700	39.506	303.808	268.883	129.223	-81.506	45.467	-0.642
3800	39.567	304.863	269.816	133.176	-81.962	48.906	-0.672
3900	39.628	305.891	270.728	137.136	-82.448	52.356	-0.701
4000	39.689	306.895	271.620	141.102	-82.966	55.819	-0.729
4100	39.749	307.876	272.492	145.074	-83.517	59.295	-0.755
4200	39.810	308.835	273.346	149.052	-84.103	62.785	-0.781
4300	39.870	309.772	274.182	153.036	-84.725	66.290	-0.805
4400	39.930	310.689	275.002	157.026	-85.364	69.808	-0.829
4500	39.990	311.587	275.805	161.022	-86.052	73.342	-0.851
4600	40.049	312.467	276.592	165.024	-86.781	76.891	-0.873
4700	40.109	313.329	277.365	169.031	-87.551	80.458	-0.894
4800	40.169	314.174	278.123	173.045	-88.364	84.041	-0.915
4900	40.228	315.003	278.867	177.065	-89.219	87.642	-0.934
5000	40.287	315.816	279.598	181.091	-90.120	91.260	-0.953
5100	40.347	316.615	280.316	185.123	-91.062	94.897	-0.972
5200	40.406	317.399	281.022	189.160	-92.058	98.552	-0.990
5300	40.465	318.169	281.715	193.204	-93.104	102.228	-1.008
5400	40.524	318.926	282.397	197.253	-94.203	105.924	-1.025
5500	40.583	319.670	283.068	201.309	-95.357	109.640	-1.041
5600	40.642	320.402	283.728	205.370	-96.567	113.379	-1.058
5700	40.701	321.122	284.378	209.437	-97.835	117.139	-1.073
5800	40.760	321.830	285.018	213.510	-99.165	120.922	-1.089
5900	40.819	322.527	285.648	217.589	-100.557	124.728	-1.104
6000	40.878	323.214	286.268	221.674	-102.013	128.559	-1.119

PREVIOUS: March 1964 (1 atm) CURRENT: March 1964 (1 bar)

Lithium Oxide (LiO) $Li_1O_1(g)$

$\Delta_fH°(0\ K)$ = [-60.5 ± 62.8] kJ mol^{-1}

S°(298.15 K) = [199.241 ± 1.3] J K^{-1} mol^{-1} $\Delta_fH°(298.15\ K)$ = [-66.944] kJ mol^{-1}

Electronic Level and Quantum Weight

State	ε_i, cm^{-1}	g_i
[$^1\Sigma$]	0	[1]

ω_e = [750] cm^{-1} ω_ex_e = [4.5] cm^{-1} σ = 1
B_e = [1.360] cm^{-1} α_e = [0.0121] cm^{-1} r_e = [1.6] Å

Enthalpy of Formation

Data on the electron affinity of LiO(g) are unavailable, therefore the value of $\Delta_fH°(LiO^-$, g, 298.15 K) is estimated. We first assume that the enthalpy change $\Delta_rH°(0\ K)$ of the reaction (1) $LiO^-(g) = Li(g) + O^-(g)$ is close to the average of the $\Delta_rH°(0\ K)$ values for the reactions (2) LiO(g) = Li(g) + O(g) and (3) LiF(g) = Li(g) + F(g). The atom F(g) is isoelectronic with $O^-(g)$. In other words, the $\Delta_rH°(0\ K)$ for reaction (1) is approximately 1/2(76.91 + 135.84) = 106.4 kcal mol^{-1}. Then we compare the $\Delta_rH°(0\ K)$ values for the reactions (4) HO(g) = H(g) + O(g), (5) $HO^-(g) = H(g) + O^-(g)$, and (6) HF(g) + H(g) + F(g), and find that $\Delta_rH°(0\ K)$ for reaction (5) is experimentally determined as 109.7 kcal mol^{-1} which is about 8 kcal mol^{-1} more negative than the average of $\Delta_rH°(0\ K)$ values for reactions (4) and (6). Based on this fact we estimate $\Delta_rH°(0\ K)$ = 96.4 kcal mol^{-1} for reaction (1), yielding $\Delta_fH°(LiO^-$, g, 298.15 K) = -15.5 kcal mol^{-1}.

Applying the difference in electron affinity between H(g) and Li(g) to OH(g) gives an approximate value, EA = 1.7 eV (39.2 kcal mol^{-1}) for LiO(g). The values of EA for H(g), Li(g) and OH(g) are taken from (1). Based on $\Delta_rH°(0\ K)$ = 39.2 kcal mol^{-1} for the reaction $LiO^-(g) = LiO(g) + e^-(g)$, we derive $\Delta_fH°(298.15\ K)$ = -19.1 kcal mol^{-1} for $LiO^-(g)$, which is in reasonable agreement with the above estimated value.

The value of $\Delta_fH°(298.15\ K)$ for $LiO^-(g)$ is tentatively adopted as -16±15 kcal mol^{-1}. Using this $\Delta_fH°(298.15\ K)$ value, we derive EA = 1.6±0.7 eV for LiO(g). The electron affinity of OH(g) is 1.8±0.1 eV.

Heat Capacity and Entropy

The ground state configuration is assumed to be the same as that of LiF(g) which is isoelectronic with $LiO^-(g)$. The values of ω_e, ω_ex_e and r_e are estimated by comparison with the observed data for LiF(g). The values of B_e and α_e are calculated from r_e, ω_e and ω_ex_e by use of the method suggested by Herzberg (2).

References

1. U. S. Nat. Bur. Stand. Report 8628, 1965.
2. G. Herzberg, "Spectra of Diatomic Molecules," D. Van Nostrand Co., Inc., New York, 1950.

Enthalpy Reference Temperature = T_r = 298.15 K   Standard State Pressure = p° = 0.1 MPa

T/K	$C_p°$ (J K^{-1}mol^{-1})	S° (J K^{-1}mol^{-1})	$-[G°-H°(T_r)]/T$ (J K^{-1}mol^{-1})	$H°-H°(T_r)$ (kJ mol^{-1})	$\Delta_fH°$ (kJ mol^{-1})	$\Delta_fG°$ (kJ mol^{-1})	Log K_f
0	0.	0.	INFINITE	-8.939	-60.473		
100	29.145	166.304	226.636	-6.033			
200	30.312	186.754	202.136	-3.076			
250	31.369	193.630	199.769	-1.535			
298.15	32.359	199.241	199.241	0.	-66.944	-71.753	12.571
300	32.395	199.442	199.242	0.060	-67.009	-71.782	12.498
350	33.284	204.504	199.640	1.703	-68.820	-72.436	10.811
400	34.020	208.998	200.534	3.386	-73.391	-72.616	9.483
450	34.619	213.041	201.703	5.102	-75.419	-72.398	8.404
500	35.106	216.715	203.023	6.846	-80.495	-71.645	7.485
600	35.833	223.184	205.859	10.395	-84.525	-69.494	6.050
700	36.339	228.748	208.740	14.005	-88.419	-66.679	4.976
800	36.707	233.625	211.552	17.658	-92.240	-63.312	4.134
900	36.986	237.965	214.250	21.343	-96.035	-59.468	3.451
1000	37.207	241.874	216.821	25.054	-99.830	-55.202	2.883
1100	37.388	245.429	219.262	28.784	-103.644	-50.555	2.401
1200	37.541	248.689	221.580	32.530	-204.258	-43.119	1.877
1300	37.672	251.699	223.783	36.291	-207.240	-29.570	1.188
1400	37.789	254.495	225.878	40.064	-210.218	-15.791	0.589
1500	37.894	257.106	227.874	43.848	-213.191	-1.800	0.063
1600	37.991	259.555	229.778	47.643	-216.158	12.390	-0.404
1700	38.080	261.861	231.598	51.446	-364.356	33.941	-1.043
1800	38.165	264.040	233.341	55.259	-366.561	57.433	-1.667
1900	38.244	266.106	235.011	59.079	-368.772	81.049	-2.228
2000	38.321	268.069	236.615	62.908	-370.988	104.781	-2.737
2100	38.394	269.941	238.158	66.743	-373.213	128.624	-3.199
2200	38.465	271.728	239.644	70.586	-375.447	152.574	-3.623
2300	38.534	273.440	241.076	74.436	-377.691	176.625	-4.011
2400	38.602	275.081	242.459	78.293	-379.949	200.774	-4.370
2500	38.668	276.658	243.796	82.157	-382.223	225.017	-4.701
2600	38.732	278.176	245.089	86.027	-384.514	249.352	-5.010
2700	38.796	279.639	246.342	89.903	-386.825	273.775	-5.296
2800	38.859	281.051	247.556	93.786	-389.159	298.285	-5.565
2900	38.921	282.416	248.735	97.675	-391.518	322.877	-5.816
3000	38.982	283.736	249.880	101.570	-393.906	347.552	-6.051
3100	39.043	285.016	250.993	105.471	-396.326	372.308	-6.273
3200	39.104	286.256	252.075	109.379	-398.779	397.141	-6.483
3300	39.163	287.460	253.129	113.292	-401.270	422.052	-6.681
3400	39.223	288.630	254.156	117.211	-403.801	447.039	-6.868
3500	39.282	289.768	255.158	121.137	-406.375	472.101	-7.046
3600	39.341	290.876	256.135	125.068	-408.997	497.237	-7.215
3700	39.399	291.954	257.088	129.005	-411.668	522.447	-7.376
3800	39.458	293.006	258.020	132.948	-414.392	547.730	-7.529
3900	39.516	294.032	258.930	136.896	-417.174	573.085	-7.676
4000	39.573	295.033	259.820	140.851	-420.016	598.512	-7.816
4100	39.631	296.011	260.691	144.811	-422.923	624.012	-7.950
4200	39.689	296.966	261.543	148.777	-425.896	649.582	-8.079
4300	39.746	297.901	262.378	152.749	-428.943	675.225	-8.202
4400	39.803	298.815	263.196	156.726	-432.004	700.934	-8.321
4500	39.860	299.710	263.997	160.709	-435.176	726.717	-8.436
4600	39.917	300.587	264.783	164.698	-438.439	752.572	-8.546
4700	39.974	301.446	265.554	168.693	-441.787	778.500	-8.652
4800	40.030	302.288	266.311	172.693	-445.224	804.500	-8.755
4900	40.087	303.114	267.053	176.699	-448.755	830.572	-8.854
5000	40.144	303.925	267.783	180.710	-452.376	856.717	-8.950
5100	40.200	304.720	268.499	184.727	-456.099	882.935	-9.043
5200	40.257	305.501	269.203	188.750	-459.934	909.228	-9.133
5300	40.313	306.269	269.895	192.779	-463.879	935.596	-9.221
5400	40.369	307.023	270.576	196.813	-467.940	962.038	-9.306
5500	40.425	307.764	271.245	200.853	-472.121	988.557	-9.389
5600	40.481	308.493	271.904	204.898	-476.427	1015.154	-9.469
5700	40.538	309.210	272.552	208.949	-480.861	1041.829	-9.547
5800	40.594	309.915	273.190	213.005	-485.428	1068.583	-9.624
5900	40.650	310.610	273.819	217.068	-490.133	1095.415	-9.698
6000	40.706	311.294	274.438	221.135	-494.820	1122.312	-9.771

PREVIOUS: June 1961 (1 atm)   CURRENT: June 1961 (1 bar)

LITHIUM (Li_2) IDEAL GAS M_r = 13.882

D_0^0 = 99.833 ± 0.096 kJ mol^{-1} (natural abundance)
S°(298.15 K) = 196.998 ± 0.030 J K^{-1} mol^{-1}

$\Delta_f H°$(0 K) = 215.5 ± 3.0 kJ mol^{-1}
$\Delta_f H°$(298.15 K) = 215.9 ± 3.0 kJ mol^{-1}

Spectroscopic Data for $^7Li_2$ in cm^{-1}

State	T_e	D_0^0	ω_e	$\omega_e x_e$	B_e	$10^3 \alpha_e$	$10^6 D_e$	r_e, Å	References
$^1\Sigma_g^+$	0	8346.45	351.39	2.5779[a]	0.67257	7.0173[b]	9.8478[c]	2.6730	(9(D°), 5)
$^3\Sigma_u^+$	8229.5	8346.45	61.0	3.2	0.26806	[17.35][d]	[20.706][d]	4.234	(10)
$^3\Pi_u^+$	11942.5	23250.4	339.0	2.5	0.69368	[7.65][d]	[11.618][d]	2.632	(10)
$^1\Sigma_u^+$	14068.3	23250.4	255.47	1.58	0.4975	5.4	7.54	3.107	(9)
$^3\Sigma_g^+$	16614.5	23250.4	245.0	2.2	0.50134	[6.74][d]	[8.385][d]	3.096	(10)

(a) $-6.4724 \times 10^{-3} (v + 1/2)^3 - 9.712 \times 10^{-5} (v + 1/2)^4$.
(b) $-3.6071 \times 10^{-5} (v + 1/2)^2 - 4.2570 \times 10^{-7} (v + 1/2)^3$.
(c) $+5.672 \times 10^{-8} (v + 1/2) - 1.9002 \times 10^{-9} (v + 1/2)^2$.
(d) These values of α_e and D_e were approximated: equations III,118 and III,123, (4).

Enthalpy of Formation

We adopt the $\Delta_f H°(Li_2$, g, 298.15 K) of 215.9 kJ mol^{-1} as recommended by NBS (12). The value adopted for $\Delta_f H°(Li_2$, g, 0 K) was obtained using the values of H°(298.15 K)-H°(0 K) for Li_2(g) and Li(cr), from JANAF and NBS (12), respectively. This results in $\Delta_f H°(Li_2$, g, 0 K) = 215.474 kJ mol^{-1}. Note that this adoption of the NBS value results in a minor inconsistency in the value of D_0^0 for the ground state. In the calculation of $C_p°$ and S° described below, we used D_0^0 = 99.846 kJ mol^{-1} (for $^7Li_2$) whereas the above procedure results in D_0^0 = 100.125 kJ mol^{-1} based on $\Delta_f H°$(Li, g, 0 K) from NBS (12). This results in uncertainties in the calculated $C_p°$ and S° of less than 0.002 J K^{-1} mol^{-1} at 6000 K.

Heat Capacity and Entropy

The thermal functions were calculated using a direct summation technique. Included in the calculation were the ground state plus the four lowest lying excited electronic states. Of these excited states, only the $^1\Sigma_u^+$ has been observed spectroscopically. All spectroscopic data for $^7Li_2$ used in this calculation are those recommended by Stwalley (1,2). Separate calculations were performed for each of the three isotopic species $^6Li_2$, $^7Li_2$, and $^6Li^7Li$ and the results were combined according to the natural abundances of the atoms, ^{7}Li: 92.3191%, ^{6}Li: 7.6809% (3). Calculation of the $^6Li_2$ and $^6Li^7Li$ species were carried out by adjusting the spectroscopic constants for $^7Li_2$ using a standard reduced mass scaling routine. The summation was extended to the dissociation limit of each state, and thus did not include any of the so-called quasi-bound levels above D° that owe their existence to rotational barriers. A detailed study of the effects of the quasi-bound levels on the thermodynamic properties of diatomic molecules has been recently performed by Mies and Julienne (11). According to their approach, the metastable and "free" contribution to the Gibbs energy function for Li_2(g) at 6000 K is only ca. 4 J K^{-1} mol^{-1} (13).

The excited states contribute significantly to the heat capacity above 2000 K but the affect on the Gibbs energy function is considerably less, only 0.5 J K^{-1} mol^{-1} larger than the ground state contribution at 6000 K. The ground state and $^1\Sigma_u^+$ excited state rotational levels were weighted in accordance with the nuclear spin-rotation interaction (4) as follows: even J weight = 0.625, odd J weight = 0.375 for the $^7Li_2$ species; even J weight = 0.667, odd J weight = 0.333 for the $^6Li_2$ species. Note also that although the Dunham coefficients for the ground state are known to higher order than used here (5), the neglect of these data leads to small errors in the computed functions as follows: S°(6000 K) is 0.4 J K^{-1} mol^{-1} larger than the nontruncated rovibrational constant calculation (based on a ground state calculation). The rotational levels were extrapolated to high J values according to the method of Khachkuruzov (6) who proposed a simpler form of Wooley's method (7). Splitting of the rotational levels in the $^3\Sigma$ and $^3\Pi$ states due to rotational-electronic coupling was taken into account only insofar as the electronic degeneracy of these states was increased accordingly. The errors invoked by this approximation are much less than the other uncertainties. The adopted value of S°(298.15 K) is only 0.013 J K^{-1} mol^{-1} lower than that adopted by NBS (12).

References

1. W. C. Stwalley, Director, Iowa Laser Facility and Professor of Chemistry and Physics, University of Iowa, personal communication, 1982.
2. Y. C. Chan and W. C. Stwalley, to be published in J. Chem. Thermodynamics, (1983).
3. Pure Appl. Chem. 47, 75 (1976).
4. G. Herzberg, "Spectra of Diatomic Molecules," 2nd Ed., D. Van Nostrand Co., Inc., New York, Chapter V, 1950.
5. N. M. Hessel and C. R. Vidal, J. Chem. Phys. 70, 4439 (1979).
6. G. A. Khachkuruzov, Opt. Spectrosk. 30, 455 (1971).
7. H. W. Wooley, R. B. Scott and F. G. Brickwedde, J. Res. Nat. Bur. Stand. 41, 379 (1948).
8. K. P. Huber and G. Herzberg, "Constants of Diatomic Molecules," Van Nostrand Reinhold Co., New York, 1979.
9. K. K. Verma, M. E. Koch, and W. C. Stwalley, J. Chem. Phys., 78, 3614 (1983).
10. M. L. Olson and D. D. Konowalow, Chem. Phys. 21, 393 (1977); 22, 29 (1977).
11. F. H. Mies and P. S. Julienne, J. Chem. Phys. 77, 6162 (1982).
12. D. D. Wagman, et al., J. Phys. Chem. Ref. Data 11, Supp. 2 (1982).
13. F. M. Mies and P. S. Julienne, U.S. Nat. Bur. Stand., personal communication, 1982.

Lithium (Li_2) Li_2(g)

Enthalpy Reference Temperature = T_r = 298.15 K Standard State Pressure = p° = 0.1 MPa

T/K	$C_p°$ (J K^{-1}mol^{-1})	S° (J K^{-1}mol^{-1})	-[G°-H°(T_r)]/T (J K^{-1}mol^{-1})	H°-H°(T_r) (kJ mol^{-1})	$\Delta_f H°$ (kJ mol^{-1})	$\Delta_f G°$ (kJ mol^{-1})	Log K_f
0	0.	0.	INFINITE	-9.675	215.469	215.469	INFINITE
100	30.583	160.414	227.785	-6.737	217.417	202.844	-105.955
200	34.380	182.905	200.255	-3.470	217.001	188.351	-49.192
250	35.426	190.697	197.589	-1.723	216.492	181.244	-37.869
298.15	36.103	196.998	196.998	0.	215.900	174.508	-30.573
300	36.125	197.222	196.999	0.067	215.876	174.251	-30.340
350	36.617	202.830	197.441	1.886	215.169	167.368	-24.978
400	36.985	207.744	198.428	3.726	214.334	160.595	-20.972
450	37.274	212.118	199.711	5.583	213.344	153.935	-17.868
500	37.513	216.058	201.151	7.453	206.195	148.009	-15.462
600	37.902	222.933	204.225	11.225	204.000	136.580	-11.890
700	38.225	228.800	207.327	15.031	201.956	125.507	-9.365
800	38.499	233.923	210.338	18.868	200.001	114.720	-7.490
900	38.714	238.471	213.216	22.729	198.080	104.175	-6.046
1000	38.843	242.557	215.949	26.608	196.186	93.843	-4.902
1100	38.859	246.261	218.539	30.494	194.310	83.700	-3.975
1200	38.744	249.638	220.992	34.375	192.439	73.727	-3.209
1300	38.498	252.731	223.316	38.238	190.558	63.910	-2.568
1400	38.133	255.571	225.520	42.071	188.658	54.239	-2.024
1500	37.675	258.187	227.612	45.862	186.734	44.705	-1.557
1600	37.155	260.602	229.599	49.604	184.777	35.300	-1.152
1700	36.607	262.838	231.490	53.292	-107.689	40.370	-1.240
1800	36.067	264.915	233.290	56.926	-108.217	49.095	-1.425
1900	35.569	266.851	235.006	60.507	-108.799	57.850	-1.590
2000	35.142	268.665	236.644	64.042	-109.431	66.637	-1.740
2100	34.814	270.371	238.209	67.539	-110.107	75.457	-1.877
2200	34.610	271.985	239.708	71.009	-110.818	84.310	-2.002
2300	34.547	273.522	241.145	74.465	-111.551	93.196	-2.117
2400	34.643	274.993	242.525	77.924	-112.296	102.114	-2.222
2500	34.909	276.412	243.853	81.400	-113.037	111.063	-2.321
2600	35.350	277.790	245.131	84.911	-113.762	120.041	-2.412
2700	35.970	279.135	246.366	88.476	-114.455	129.047	-2.497
2800	36.765	280.457	247.560	92.111	-115.102	138.077	-2.576
2900	37.728	281.763	248.717	95.834	-115.689	147.130	-2.650
3000	38.846	283.061	249.840	99.662	-116.203	156.202	-2.720
3100	40.103	284.355	250.933	103.608	-116.634	165.289	-2.785
3200	41.479	285.649	251.997	107.686	-116.971	174.389	-2.847
3300	42.948	286.948	253.037	111.907	-117.206	183.498	-2.905
3400	44.483	288.253	254.053	116.278	-117.336	192.612	-2.959
3500	46.057	289.565	255.049	120.805	-117.358	201.729	-3.011
3600	47.640	290.884	256.026	125.490	-117.272	210.845	-3.059
3700	49.201	292.211	256.986	130.332	-117.083	219.957	-3.105
3800	50.713	293.543	257.931	135.328	-116.797	229.062	-3.149
3900	52.149	294.879	258.861	140.472	-116.423	238.159	-3.190
4000	53.485	296.217	259.778	145.755	-115.973	247.246	-3.229
4100	54.702	297.553	260.683	151.165	-115.460	256.320	-3.266
4200	55.781	298.884	261.577	156.691	-114.901	265.381	-3.300
4300	56.711	300.208	262.460	162.317	-114.314	274.428	-3.334
4400	57.483	301.521	263.333	168.028	-113.673	283.459	-3.365
4500	58.092	302.820	264.196	173.808	-113.063	292.477	-3.395
4600	58.537	304.102	265.049	179.640	-112.484	301.483	-3.423
4700	58.822	305.364	265.894	185.510	-111.948	310.476	-3.451
4800	58.950	306.604	266.729	191.400	-111.475	319.459	-3.476
4900	58.930	307.820	267.555	197.295	-111.083	328.432	-3.501
5000	58.772	309.009	268.373	203.181	-110.789	337.399	-3.525
5100	58.487	310.170	269.181	209.045	-110.599	346.360	-3.547
5200	58.086	311.302	269.980	214.875	-110.547	355.319	-3.569
5300	57.582	312.404	270.770	220.659	-110.640	364.279	-3.590
5400	56.988	313.475	271.551	226.388	-110.889	373.242	-3.610
5500	56.317	314.514	272.323	232.054	-111.306	382.210	-3.630
5600	55.580	315.523	273.085	237.649	-111.903	391.188	-3.649
5700	54.789	316.499	273.838	243.168	-112.686	400.179	-3.667
5800	53.955	317.445	274.582	248.606	-113.666	409.185	-3.685
5900	53.087	318.360	275.316	253.958	-114.849	418.209	-3.703
6000	52.196	319.245	276.041	259.222	-116.239	427.256	-3.720

PREVIOUS: June 1962 (1 atm) CURRENT: December 1983 (1 bar)

Lithium (Li_2) Li_2(g)

LITHIUM OXIDE (Li_2O) CRYSTAL M_r = 29.8814

$S°(298.15\ K) = 37.890\ J\ K^{-1}\ mol^{-1}$

T_{fus} = 1843 K

$\Delta_f H°(0\ K) = -592.39 \pm 2.1\ kJ\ mol^{-1}$

$\Delta_f H°(298.15\ K) = -598.73 \pm 2.1\ kJ\ mol^{-1}$

$\Delta_{fus} H° = [58.576]\ kJ\ mol^{-1}$

Enthalpy of Formation

The selected enthalpy of formation was calculated as follows:

	Reaction	$\Delta_r H°(298.15\ K)$ kcal mol^{-1}	Source
(A)	$Li_2O(cr) + H_2O(\ell) = 2\ LiOH(aq, \infty)$	-31.74 ± 0.08	(1)
(B)	$Li(cr) + 1/2\ O_2(g) = LiOH(aq, \infty)$	-121.572 ± 0.019	(2)
(C)	$H_2(g) + 1/2\ O_2(g) = H_2O(\ell)$	-68.3174	(3)
(D)	$2\ Li(cr) + 1/2\ O_2(g) = Li_2O(cr)$*	-143.1 ± 0.5	

* Reaction (D) is the sum of -(A) + 2(B) - (C).

Kolesov et al. (1) measured the enthalpy of the reaction of Li_2O with water at 20° in 5 experiments. The mean $\Delta_r H°$ was -31.41 ± 0.08 kcal mol^{-1}. This was brought to 298.15 K by Kolesov using enthalpy data of Johnston and Bauer (4) for Li_2O, Osborne et al. (5) for H_2O, Gucker and Schminke (6) for $C°_p$ of LiOH(aq) and Rossini et al. (3) for aqueous LiOH. In addition to the above, Beketov (7) and Forcrand (8) have reported experimental enthalpies for Li_2O. Rossini et al. and Johnston and Bauer have reported enthalpies of formation for Li_2O recalculated from older data. Fasolino (9) reported a $\Delta_f H°(298.15\ K)$ of -141.1 ± 4.2 kcal mol^{-1}. He measured the enthalpy of combustion of Li(cr) in O (g) in a Parr bomb.

Heat Capacity and Entropy

The low temperature $C°_p$ data of Johnston and Bauer (4) from [179 - 298 K] was joined graphically to the high temperature $C°_p$ data of Shomate and Cohen (10, 425 - 1045 K) and Rodigina and Gomelskii (11, 370 - 1125 K). S°(298.15 K) was reported by Johnston and Bauer using S°(16 K) = 0.003 cal K^{-1} mol^{-1} (extrap.).

Fusion Data

The melting point of Li_2O is uncertain. The selected melting point, 1843 K, was measured by van Arkel et al. (12). Brewer and Margrave (13) reported the melting point 1700 K. They also stated that van Kooster and Jaeger (14) had reported 1700 (1973 K) as the melting point. The heat of melting, [14] kcal mol^{-1}, was estimated by Glassner (15).

Sublimation Data

$Li_2O(cr)$ vaporizes to $Li_2O(g)$, LiO(g), $Li_2O_2(g)$ and $O_2(g)$ (16 - 20). The major species (0 - 1700 K) are reported to be $Li_2O(g)$, Li(g), and $O_2(g)$ with lesser amounts of LiO(g) and $Li_2O_2(g)$. Hildenbrand et al. (18), U-2298, have reported about equal amounts of Li_2O^+ and Li^+ (1300 to 1550 K) and about 1/30th of this amount of LiO^+. White et al. (17) calculated $\Delta_{sub}H°$ from their mass spectra data. They made two different calculations: (I) for $Li_2O(cr) \rightarrow Li_2O(g)$, and (II) for $Li_2O(g) \rightarrow 2\ LI(g) + 1/2\ O_2(g)$, and found: (a) 102.5 kcal mol^{-1} and (b) 99.0 kcal mol^{-1}. They used a weighted average of (a) and (b) and obtained 101.6 kcal mol^{-1} at 0 K. The selected $\Delta_{sub}H°(298.15\ K)$ of 102.3 ± 3 kcal mol^{-1} was obtained in the same manner using their data and the present JANAF functions, i.e., (a) 104.2 kcal mol^{-1} and (b) 100.3 kcal mol^{-1}. Knudson-effusion data of White et al., corrected for dissociation gave 3rd law heats at 298 K of 101.9 kcal mol^{-1} and a 2nd law heat of 94 kcal mol^{-1}. Recalculation of Hildenbrand's dissociation pressure with JANAF values give Li_2O pressures slightly different than those reported. The recalculated pressures give a 3rd law $\Delta_{sub}H°(298.15\ K)$ of 101.7 kcal mol^{-1} for both cells. The corresponding 2nd law values are 85 and 95.5 kcal mol^{-1} for cell 7 and 8. The 3rd law treatment of both sets of data show relatively large trends of $\Delta_{sub}H°$ with T which are probably too large to be only due to errors in the function.

A sample of $LiAlO_2$ was melted 7 times in argon in an arc-image furnace, Prophet (21). The sample decomposed on fusion and successive melting points approached that of Al_2O_3. The residue was found to be 95% Al_2O_3 and 5% $LiAl_5O_8$. The sample and residue were characterized by X-ray diffraction. A reasonable assumption from this is that $\Delta_r F°(2000\ K) = 0$ for $2LiAlO_2(cr) = Al_2O_3(cr) + Li_2O(g)$. From this ΔF, a $\Delta_{sub}H°(298.15\ K)$ for Li_2O of 50 kcal mol^{-1} was calculated. A vapor pressure study of the molecular species over the decomposing $LiAlO_2$ might help resolve the discrepancies in lithium oxide $\Delta_{sub}H°$ values.

References

1. V. P. Kolesov, S. M. Skuratov, and I. D. Zaikin, Russian J. Inorganic Chemistry 4, 558 (1959).
2. S. R. Gunn and L. G. Green, J. Amer. Chem. Soc. 80, 4782 (1958).
3. U. S. Nat. Bur. Stand. Circ. 500, 1952.
4. H. L. Johnston and T. W. Bauer, J. Amer. Chem. Soc. 73, 1119 (1951).
5. N. S. Osborne, J. F. Stimson, and D. C. Ginnings, J. Res. Nat. Bur. Stand. A 23, 197 (1939).
6. F. T. Gucker and K. N. Schminke, J. Amer. Chem. Soc. 55, 1013 (1933).
7. N. N. Beketov, Bull. Acad. Sci. Russ. 32, 186 (1888).
8. H. R. Forcrand, Compt. Rend. 144, 1403 (1907).
9. L. G. Fasolino, in a Nat. Research Corp. Quarterly Report, July 1 to Sept. 30, 1962.
10. C. H. Shomate and A. J. Cohen, J. Amer. Chem. Soc. 77, 285 (1955).
11. E. N. Rodigina and K. A. Gomelskii, Russian J. Phys. Chem. 35, 898 (1961).
12. A. E. van Arkel, E. A. Flood and N. F. H. Bright, Can. J. Chem. 31, 1009 (1953).
13. L. Brewer and J. L. Margrave, J. Phys. Chem. 59, 421 (1955).
14. H. van Kooster and F. Jaeger, Proc. Roy. Acad. Sco., Amsterdam 16, 857 (1914).
15. A. Glassner, Argonne National Laboratory, ANL-5750, 1957.
16. J. Berkowitz, W. A. Chupka, G. D. Blue and J. L. Margrave, J. Phys. Chem. 63, 644 (1959).
17. D. White, K. S. Seshadri, D. F. Dever, D. E. Mann and M. J. Linevsky, J. Chem. Phys. 39, 2463 (1963).
18. D. L. Hildenbrand, W. F. Hall and N. D. Potter, J. Chem. Phys. 39, 296 (1963).
19. D. L. Hildenbrand, L. P. Theard, W. F. Hall and N. D. Potter, Fifth Quarterly Technical Report, U-2289, Aeronutronic Division of Philco Corporation, September 15, 1963.
20. L. P. Firsova and A. N. Nesmeyanov, Russian J. Phys. Chem. 34, 1232 (1960).
21. H. Prophet, Dow Chemical Co., ARPA-1-150, March 9, 1961.

Lithium Oxide (Li_2O) $Li_2O_1(cr)$

Enthalpy Reference Temperature = T_r = 298.15 K; Standard State Pressure = p° = 0.1 MPa

T/K	$C°_p$ (J K^{-1}mol^{-1})	S° (J K^{-1}mol^{-1})	$-[G°-H°(T_r)]/T$ (J K^{-1}mol^{-1})	$H°-H°(T_r)$ (kJ mol^{-1})	$\Delta_f H°$ (kJ mol^{-1})	$\Delta_f G°$ (kJ mol^{-1})	Log K_f
0	0.	0.	INFINITE	-7.248	-592.392	-592.392	INFINITE
100	10.447	3.632	73.417	-6.979	-594.565	-584.794	305.465
200	37.204	19.581	42.362	-4.556	-597.282	-573.919	149.892
298.15	54.087	37.890	37.890	0.	-598.730	-562.102	98.478
300	54.379	38.226	37.891	0.100	-598.748	-561.875	97.831
400	64.003	55.315	40.156	6.064	-599.472	-549.465	71.753
500	69.580	70.232	44.714	12.759	-606.172	-536.272	56.024
600	73.822	83 284	50.077	19.924	-606.553	-522.248	45.466
700	77.404	94.940	55.668	27.490	-606.464	-508.198	37.922
800	80.563	105.484	61.246	35.390	-606.026	-494.187	32.267
900	83.429	115.137	66.705	43.589	-605.311	-480.247	27.873
1000	86.211	124.073	72.001	52.073	-604.331	-466.401	24.362
1100	88.805	132.415	77.118	60.827	-603.094	-452.666	21.495
1200	91.211	140.244	82.055	69.826	-601.621	-439.054	19.112
1300	93.199	147.625	86.818	79.049	-599.934	-425.574	17.100
1400	94.872	154.595	91.413	88.455	-598.066	-412.230	15.381
1500	96.232	161.189	95.847	98.013	-596.045	-399.026	13.895
1600	97.320	167.434	100.127	107.691	-593.899	-385.961	12.600
1700	98.332	173.364	104.262	117.473	-882.118	-358.681	11.021
1800	99.454	179.017	108.260	127.363	-878.247	-328.002	9.518
1843.000	99.914	181.370	109.938	131.649	------ CRYSTAL <--> LIQUID -----		
1900	100.542	184.423	112.127	137.363	-874.280	-297.541	8.180
2000	101.420	189.603	115.872	147.461	-870.230	-267.290	6.981
2100	102.287	194.572	119.502	157.646	-866.111	-237.244	5.901
2200	103.143	199.350	123.024	167.918	-861.924	-207.396	4.924
2300	103.988	203.954	126.443	178.274	-857.673	-177.740	4.037
2400	104.821	208.397	129.766	188.715	-853.361	-148.270	3.227
2500	105.642	212.693	132.997	199.238	-848.993	-118.981	2.486
2600	106.452	216.852	136.143	209.843	-844.572	-89.867	1.805
2700	107.251	220.884	139.207	220.528	-840.103	-60.925	1.179
2800	108.038	224.799	142.195	231.293	-835.590	-32.149	0.600
2900	108.814	228.604	145.109	242.136	-831.036	-3.535	0.064
3000	109.579	232.306	147.954	253.055	-826.447	24.922	-0.434

PREVIOUS: December 1960 CURRENT: March 1964

Lithium Oxide (Li_2O) $Li_2O_1(cr)$

LITHIUM OXIDE (Li_2O) LIQUID M_r = 29.8814

$S°(298.15\ K) = [55.083]\ J\ K^{-1}\ mol^{-1}$

T_{fus} = 1843 K

$\Delta_f H°(298.15\ K) = [-553.247]\ kJ\ mol^{-1}$

$\Delta_{fus}H° = [58.576]\ kJ\ mol^{-1}$

Enthalpy of Formation

$\Delta_f H°(Li_2O$, ℓ, 298.15 K) is calculated from $\Delta_f H°(Li_2O$, cr, 298.15 K) by adding the enthalpy of fusion, $\Delta_{fus}H°$, and the difference in enthalpy, H°(1843 K) - H°(298.15 K), between the crystal and liquid.

Heat Capacity and Entropy

The heat capacity was estimated at 8 cal K^{-1} mol^{-1} from comparisons with similar compounds and from Kopp's rule. $S°(Li_2O$, ℓ, 298.15 K) is calculated in a manner analogous to that used for the enthalpy of formation.

Fusion Data

See the crystal table for details.

Vaporization Data

The vapor over Li_2O(cr) at 1700 K is composed of Li_2O(g), Li(g), O_2(g), LiO(g), and Li_2O_2(g). Thus a boiling point becomes almost meaningless. The value, T_{vap} = 2836 K, listed above is the temperature at which the fugacity of Li_2O(g) reaches one atm. Brewer (1) estimated a boiling point of 2600 K.

Reference

1. L. Brewer, Chem. Rev. 52, (1953).

Lithium Oxide (Li_2O) Li_2O_1(l)

Enthalpy Reference Temperature = T_r = 298.15 K Standard State Pressure = p° = 0.1 MPa

T/K	$C_p°$ (J K⁻¹mol⁻¹)	S° (J K⁻¹mol⁻¹)	$-[G°-H°(T_r)]/T$ (J K⁻¹mol⁻¹)	$H°-H°(T_r)$ (kJ mol⁻¹)	$\Delta_f H°$ (kJ mol⁻¹)	$\Delta_f G°$ (kJ mol⁻¹)	Log K_f
0							
100							
200							
298.15	54.087	55.083	55.083	0.	-553.247	-521.744	91.407
300	54.379	55.419	55.084	0.100	-553.265	-521.549	90.810
400	64.003	72.477	57.344	6.053	-553.999	-510.857	66.711
500	69.580	87.383	61.896	12.743	-560.704	-499.380	52.170
600	73.822	100.464	67.256	19.925	-561.069	-487.072	42.403
600.000	73.822	100.464	67.256	19.925	GLASS <--> LIQUID	TRANSITION	
600.000	100.416	100.464	67.256	19.925			
700	100.416	115.943	73.134	29.966	-558.505	-474.941	35.441
800	100.416	129.352	79.342	40.008	-555.924	-463.180	30.243
900	100.416	141.179	85.569	50.049	-553.368	-451.741	26.218
1000	100.416	151.759	91.668	60.091	-550.829	-440.585	23.014
1100	100.416	161.330	97.573	70.133	-548.304	-429.683	20.404
1200	100.416	170.067	103.255	80.174	-545.789	-419.010	18.239
1300	100.416	178.105	108.708	90.216	-543.284	-408.547	16.416
1400	100.416	185.546	113.934	100.257	-540.781	-398.277	14.860
1500	100.416	192.474	118.942	110.299	-538.275	-388.185	13.518
1600	100.416	198.955	123.742	120.341	-535.766	-378.261	12.349
1700	100.416	205.043	128.347	130.382	-823.725	-354.142	10.881
1800	100.416	210.782	132.769	140.424	-819.702	-326.636	9.479
1843.000	100.416	213.153	134.617	144.742	------ CRYSTAL <--> LIQUID -----		
1900	100.416	216.212	137.019	150.465	-815.694	-299.352	8.230
2000	100.416	221.362	141.109	160.507	-811.701	-272.280	7.111
2100	100.416	226.262	145.048	170.549	-807.725	-245.407	6.104
2200	100.416	230.933	148.847	180.590	-803.768	-218.722	5.193
2300	100.416	235.397	152.513	190.632	-799.832	-192.218	4.365
2400	100.416	239.670	156.056	200.673	-795.919	-165.884	3.610
2500	100.416	243.770	159.483	210.715	-792.033	-139.712	2.919
2600	100.416	247.708	162.801	220.757	-788.175	-113.696	2.284
2700	100.416	251.498	166.017	230.798	-784.350	-87.827	1.699
2800	100.416	255.150	169.135	240.840	-780.559	-62.099	1.158
2900	100.416	258.673	172.162	250.881	-776.807	-36.506	0.658
3000	100.416	262.078	175.103	260.923	-773.096	-11.042	0.192
3100	100.416	265.370	177.962	270.965	-769.429	14.299	-0.241
3200	100.416	268.558	180.744	281.006	-765.809	39.523	-0.645
3300	100.416	271.648	183.452	291.048	-762.239	64.634	-1.023
3400	100.416	274.646	186.090	301.089	-758.722	89.638	-1.377
3500	100.416	277.557	188.662	311.131	-755.261	114.539	-1.709
3600	100.416	280.386	191.171	321.173	-751.859	139.342	-2.022
3700	100.416	283.137	193.619	331.214	-748.518	164.051	-2.316
3800	100.416	285.815	196.011	341.256	-745.240	188.672	-2.593
3900	100.416	288.423	198.347	351.297	-742.029	213.207	-2.856
4000	100.416	290.965	200.631	361.339	-738.888	237.660	-3.104

PREVIOUS: December 1960 CURRENT: March 1964

Lithium Oxide (Li_2O) Li_2O_1(l)

LITHIUM OXIDE (Li_2O) CRYSTAL-LIQUID M_r = 29.8814

0 to 1843 K crystal
above 1843 K liquid

Refer to the individual tables for details.

Lithium Oxide (Li_2O) Li_2O_1(cr,l)

Enthalpy Reference Temperature = T_r = 298.15 K — Standard State Pressure = p° = 0.1 MPa

T/K	C_p° (J K⁻¹mol⁻¹)	S° (J K⁻¹mol⁻¹)	−[G°−H°(T_r)]/T (J K⁻¹mol⁻¹)	H°−H°(T_r) (kJ mol⁻¹)	$\Delta_f H^\circ$ (kJ mol⁻¹)	$\Delta_f G^\circ$ (kJ mol⁻¹)	Log K_f
0	0.	0.	INFINITE	-7.248	-592.392	-592.392	INFINITE
100	10.447	3.632	73.417	-6.979	-594.565	-584.794	305.465
200	37.204	19.581	42.362	-4.556	-597.282	-573.919	149.892
298.15	54.087	37.890	37.890	0.	-598.730	-562.102	98.478
300	54.379	38.226	37.891	0.100	-598.748	-561.875	97.831
400	64.003	55.315	40.156	6.064	-599.472	-549.465	71.753
500	69.580	70.232	44.714	12.759	-606.172	-536.272	56.024
600	73.822	83.284	50.077	19.924	-606.553	-522.248	45.466
700	77.404	94.940	55.668	27.490	-606.464	-508.198	37.922
800	80.563	105.484	61.246	35.390	-606.026	-494.187	32.267
900	83.429	115.137	66.705	43.589	-605.311	-480.247	27.873
1000	86.211	124.073	72.001	52.073	-604.331	-466.401	24.362
1100	88.805	132.415	77.118	60.827	-603.094	-452.666	21.495
1200	91.211	140.244	82.055	69.826	-601.621	-439.054	19.112
1300	93.199	147.625	86.818	79.049	-599.934	-425.574	17.100
1400	94.872	154.595	91.413	88.455	-598.066	-412.230	15.381
1500	96.232	161.189	95.847	98.013	-596.045	-399.026	13.895
1600	97.320	167.434	100.127	107.691	-593.899	-385.961	12.600
1700	98.332	173.364	104.262	117.473	-882.118	-358.681	11.021
1800	99.454	179.017	108.260	127.363	-878.247	-328.002	9.518
1843.000	99.914	181.370	109.938	131.649	CRYSTAL <--> LIQUID		
1843.000	100.416	213.153	109.938	190.225	TRANSITION		
1900	100.416	216.212	113.081	195.949	-815.694	-299.352	8.230
2000	100.416	221.362	118.367	205.990	-811.701	-272.280	7.111
2100	100.416	226.262	123.389	216.032	-807.725	-245.407	6.104
2200	100.416	230.933	128.172	226.074	-803.768	-218.722	5.193
2300	100.416	235.397	132.738	236.115	-799.832	-192.218	4.365
2400	100.416	239.670	137.105	246.157	-795.919	-165.884	3.610
2500	100.416	243.769	141.290	256.198	-792.033	-139.712	2.919
2600	100.416	247.708	145.308	266.240	-788.175	-113.696	2.284
2700	100.416	251.498	149.171	276.282	-784.350	-87.827	1.699
2800	100.416	255.150	152.891	286.323	-780.559	-62.099	1.158
2900	100.416	258.673	156.478	296.365	-776.807	-36.506	0.658
3000	100.416	262.077	159.942	306.406	-773.096	-11.042	0.192
3100	100.416	265.370	163.290	316.448	-769.429	14.299	-0.241
3200	100.416	268.558	166.530	326.490	-765.809	39.523	-0.645
3300	100.416	271.648	169.669	336.531	-762.239	64.634	-1.023
3400	100.416	274.646	172.713	346.573	-758.722	89.638	-1.377
3500	100.416	277.557	175.667	356.614	-755.261	114.539	-1.709
3600	100.416	280.386	178.537	366.656	-751.859	139.342	-2.022
3700	100.416	283.137	181.327	376.698	-748.518	164.052	-2.316
3800	100.416	285.815	184.041	386.739	-745.240	188.672	-2.593
3900	100.416	288.423	186.684	396.781	-742.029	213.207	-2.856
4000	100.416	290.965	189.260	406.822	-738.888	237.660	-3.104

PREVIOUS: CURRENT: March 1964

Lithium Oxide (Li_2O) Li_2O_1(cr,l)

LITHIUM OXIDE (Li_2O) IDEAL GAS M_r = 29.8814

Lithium Oxide (Li_2O) $Li_2O_1(g)$

$S°(298.15\ K) = 229.108\ J\ K^{-1}\ mol^{-1}$

$\Delta_f H°(0\ K) = -165.84 \pm 10.5\ kJ\ mol^{-1}$
$\Delta_f H°(298.15\ K) = -166.94 \pm 1.5\ kJ\ mol^{-1}$

Vibrational Frequencies and Degeneracies

ν, cm^{-1}
[760](1)
[140](2)
987 (1)

Ground State Quantum Weight: [1] σ = 2
Point Group: $D_{\infty h}$
Bond Distance: Li-O = [1.59] Å
Bond Angle: Li-O-Li = 180°
Rotational Constant: B_0 = 0.480342 cm^{-1}

Enthalpy of Formation

The enthalpy of formation was calculated from the enthalpy of formation of the crystal and the enthalpy of sublimation.

Heat Capacity and Entropy

White et al. (1) made an estimate of the bond distance Li-O = 1.59 Å based on an analysis of their vibrational spectra. They found that Li_2O was a linear molecule. Buchler and Stauffer (2) have shown from electric deflection experiments that Li_2O is very probably linear. They also report from the results of an effusion experiment evidence that the electron diffraction data and Li-O bond distance, 1.82 Å reported by Akishin and Rambidi (3) are for Mo-O in Li_2MoO_4 rather than for Li-O in Li_2O. Buchler and Stauffer (2) estimate an Li-O distance of 1.55 Å. White, et al., (1) have estimated the vibrational frequencies ν_1 = [760] cm^{-1} and ν_2 = [140] cm^{-1} and assigned ν_3 = 987 cm^{-1} from their infrared matrix spectra.

References

1. D. White, K. S. Seshadri, D. F. Dever, D. E. Mann, and M. J. Linevsky, J. Chem. Phys. 39, 2463 (1963).
2. A. Buchler and J. L. Stauffer, J. Chem. Phys. 39, 2299 (1963).
3. P. A. Akishin and N. G. Rambidi, Dokl. Akad. Nauk. SSSR 118, 973 (1958) [English Translation, Proc. Acad. Sci. USSR, Phys. Chem. Sect. 118, 83 (1958)].

Enthalpy Reference Temperature = T_r = 298.15 K Standard State Pressure = p° = 0.1 MPa

T/K	$C_p°$ (J K⁻¹mol⁻¹)	S°	-[G°-H°(T_r)]/T	H°-H°(T_r) (kJ mol⁻¹)	$\Delta_f H°$	$\Delta_f G°$	Log K_f
0	0.	0.	INFINITE	-12.485	-165.840	-165.840	INFINITE
100	41.106	179.969	270.575	-9.061	-164.858	-172.721	90.220
200	45.797	210.079	233.532	-4.691	-165.627	-180.364	47.106
250	47.836	220.517	229.917	-2.350	-166.272	-183.975	38.440
298.15	49.758	229.108	229.108	0.	-166.942	-187.325	32.818
300	49.829	229.416	229.109	0.092	-166.968	-187.451	32.638
350	51.633	237.236	229.723	2.630	-167.694	-190.808	28.477
400	53.190	244.235	231.107	5.251	-168.495	-194.056	25.341
450	54.500	250.578	232.923	7.945	-169.407	-197.198	22.890
500	55.590	256.378	234.983	10.698	-176.444	-199.618	20.854
600	57.243	266.669	239.428	16.344	-178.344	-204.070	17.766
700	58.393	275.584	243.971	22.130	-180.036	-208.221	15.538
800	59.211	283.438	248.423	28.012	-181.615	-212.139	13.851
900	59.808	290.448	252.710	33.964	-183.147	-215.863	12.528
1000	60.254	296.774	256.805	39.968	-184.646	-219.417	11.461
1100	60.595	302.533	260.705	46.012	-186.120	-222.822	10.581
1200	60.861	307.818	264.414	52.085	-187.573	-226.095	9.842
1300	61.072	312.698	267.942	58.182	-189.012	-229.247	9.211
1400	61.242	317.230	271.303	64.298	-190.435	-232.288	8.667
1500	61.380	321.460	274.507	70.429	-191.840	-235.228	8.191
1600	61.495	325.425	277.567	76.573	-193.228	-238.076	7.772
1700	61.591	329.156	280.493	82.728	-485.074	-226.484	6.959
1800	61.672	332.679	283.295	88.891	-484.930	-211.278	6.131
1900	61.740	336.016	285.983	95.062	-484.792	-196.078	5.391
2000	61.799	339.184	288.565	101.239	-484.664	-180.886	4.724
2100	61.850	342.200	291.047	107.421	-484.547	-165.700	4.122
2200	61.895	345.079	293.439	113.608	-484.444	-150.519	3.574
2300	61.933	347.831	295.744	119.800	-484.358	-135.343	3.074
2400	61.968	350.467	297.970	125.995	-484.292	-120.170	2.615
2500	61.998	352.998	300.120	132.193	-484.249	-104.999	2.194
2600	62.025	355.430	302.201	138.394	-484.232	-89.830	1.805
2700	62.049	357.771	304.216	144.598	-484.245	-74.660	1.444
2800	62.070	360.028	306.170	150.804	-484.290	-59.490	1.110
2900	62.089	362.207	308.065	157.012	-484.371	-44.317	0.798
3000	62.107	364.312	309.905	163.222	-484.492	-29.141	0.507
3100	62.123	366.349	311.693	169.433	-484.655	-13.960	0.235
3200	62.137	368.321	313.432	175.646	-484.864	1.227	-0.020
3300	62.150	370.233	315.124	181.861	-485.121	16.421	-0.260
3400	62.162	372.089	316.772	188.076	-485.430	31.624	-0.486
3500	62.173	373.891	318.379	194.293	-485.794	46.836	-0.699
3600	62.183	375.643	319.945	200.511	-486.215	62.060	-0.900
3700	62.192	377.346	321.474	206.730	-486.697	77.296	-1.091
3800	62.201	379.005	322.966	212.949	-487.241	92.547	-1.272
3900	62.209	380.621	324.424	219.170	-487.852	107.813	-1.444
4000	62.216	382.196	325.848	225.391	-488.531	123.094	-1.607
4100	62.223	383.732	327.241	231.613	-489.281	138.394	-1.763
4200	62.229	385.232	328.604	237.836	-490.105	153.713	-1.912
4300	62.235	386.696	329.938	244.059	-491.007	169.053	-2.054
4400	62.241	388.127	331.245	250.283	-491.947	184.411	-2.189
4500	62.246	389.526	332.524	256.507	-492.992	199.793	-2.319
4600	62.251	390.894	333.778	262.732	-494.125	215.201	-2.444
4700	62.255	392.233	335.008	268.957	-495.345	230.634	-2.563
4800	62.259	393.544	336.214	275.183	-496.653	246.094	-2.678
4900	62.263	394.827	337.397	281.409	-498.054	261.583	-2.789
5000	62.267	396.085	338.558	287.636	-499.550	277.100	-2.895
5100	62.271	397.318	339.698	293.862	-501.134	292.648	-2.997
5200	62.274	398.528	340.818	300.090	-502.830	308.228	-3.096
5300	62.277	399.714	341.918	306.317	-504.630	323.843	-3.192
5400	62.280	400.878	342.999	312.545	-506.540	339.492	-3.284
5500	62.283	402.021	344.062	318.773	-508.562	355.178	-3.373
5600	62.285	403.143	345.107	325.002	-510.701	370.902	-3.460
5700	62.288	404.245	346.135	331.230	-512.959	386.664	-3.543
5800	62.290	405.329	347.146	337.459	-515.342	402.468	-3.625
5900	62.293	406.394	348.141	343.688	-517.851	418.313	-3.703
6000	62.295	407.441	349.121	349.918	-520.490	434.203	-3.780

PREVIOUS: March 1964 (1 atm) CURRENT: March 1964 (1 bar)

Lithium Oxide (Li_2O) $Li_2O_1(g)$

LITHIUM PEROXIDE (Li_2O_2) CRYSTAL M_r = 45.8808

$\Delta_f H°(0\ K)$ = Unknown

S°(298.15 K) = [56.484 ± 4.2] J K^{-1} mol^{-1} $\Delta_f H°(298.15\ K)$ = [632.62 ± 8.4] kJ mol^{-1}

Enthalpy of Formation

$\Delta_f H°(298.15\ K)$ was reported by Forcrand (1), based on the measured enthalpies of the reactions of Li_2O_2(cr) with HCl(aq) and LiOH(aq) with H_2O_2(aq).

Heat Capacity and Entropy

Both $C_p°$ and S°(298.15 K) were estimated by comparison with those of the Na_2O_2(cr), respectively.

Decomposition Data

T_{dcm} was reported to be about 300°C by Wiederhorn et al. (2). The value adopted, T_{dcm} = 468 K, was reported by (3) obtained from Centnerzwer and Blumenthal (4).

References

1. R. de Forcrand, Compt. Rend. 130, 1465 (1900).
2. N. M. Wiederhorn, N. F. Surprenant, and A. J. Leffler, "Research on New Methods of Separation of Air into O_2 and N_2 Components," C-63825, First-Quarterly Progress Report, September 1961, Arthur D. Little, Inc.
3. Bulletin of the National Research Council, No. 118, "Data on Chemicals for Ceramic Use", June, 1949.
4. M. Centnerzwer and M. Blementhal, Bull. Intern. Acad. Polonaise, Classe. Sci. Math. Nat., A 499 (1933).

Lithium Peroxide (Li_2O_2) Li_2O_2(cr)

Enthalpy Reference Temperature = T_r = 298.15 K Standard State Pressure = p° = 0.1 MPa

T/K	J K⁻¹mol⁻¹ $C_p°$	S°	$-[G°-H°(T_r)]/T$	kJ mol⁻¹ $H°-H°(T_r)$	$\Delta_f H°$	$\Delta_f G°$	Log K_f
0							
100							
200							
298.15	70.626	56.484	56.484	0.	-632.621	-570.954	100.029
300	70.919	56.922	56.485	0.131	-632.635	-570.571	99.345
400	82.676	79.021	59.419	7.841	-633.098	-549.799	71.796
500	90.793	98.369	65.316	16.527	-639.337	-528.332	55.194
600	97.445	115.520	72.280	25.944	-639.045	-506.147	44.064
700	103.428	130.995	79.580	35.991	-638.104	-484.063	36.121
800	109.077	145.178	86.905	46.618	-636.605	-462.154	30.176
900	114.432	158.336	94.120	57.794	-634.617	-440.462	25.564
1000	119.704	170.666	101.164	69.502	-632.144	-419.017	21.887
1100	124.892	182.319	108.017	81.733	-629.184	-397.844	18.892
1200	129.997	193.406	114.674	94.478	-625.740	-376.961	16.409
1300	134.980	204.008	121.141	107.727	-621.818	-356.385	14.320
1400	139.905	214.192	127.426	121.472	-617.419	-336.129	12 541
1500	144.766	224.010	133.540	135.706	-612.541	-316.205	11.011
1600	149.570	233.507	139.492	150.423	-607.190	-296.621	9.684
1700	154.318	242.717	145.295	165.618	-891.841	-263.034	8.082
1800	159.005	251.671	150.957	181.285	-884.052	-226.268	6.566
1900	163.628	260.392	156.488	197.417	-875.822	-189.947	5.222
2000	168.197	268.902	161.897	214.009	-867.160	-154.070	4.024

PREVIOUS: June 1963 CURRENT: September 1963

Lithium Peroxide (Li_2O_2) Li_2O_2(cr)

LITHIUM OXIDE $((LiO)_2)$ IDEAL GAS M_r = 45.8808

$S°(298.15\ K)$ = [273.562] J K^{-1} mol^{-1}

$\Delta_f H°(0\ K)$ = [-240.60 ± 25.1] kJ mol^{-1}
$\Delta_f H°(298.15\ K)$ = [-242.67 ± 25.1] kJ mol^{-1}

Vibrational Frequencies and Degeneracies

ν, cm^{-1}	ν, cm^{-1}
[400](1)	[270](1)
[250](1)	330 (1)
[300](1)	530 (1)

Ground State Quantum Weight: [1] σ = 4
Point Group: $[D_{2h}]$
Bond Distance: Li-O = [1.90] Å
Bond Angle: O-Li-O = [116]°
Product of the Moments of Inertia: $I_A I_B I_C$ = [5.200811 x 10^{-115}] g^3 cm^6

Enthalpy of Formation

The enthalpy of formation, -58 kcal mol^{-1}, was calculated from an equilibrium constant for the reaction 1/2 $O_2(g)$ + $Li_2O(g)$ → $Li_2O_2(g)$ and the current JANAF functions. The equilibrium constant, 0.39, was estimated by White et al. (1), from their mass spectrometric data.

Heat Capacity and Entropy

White, et al., (1) from their spectral data have assigned two observed frequencies and estimated the remaining frequencies using the ionic model calculations of Berkowitz (2) as a guide. They proposed a planar rhombic structure of D_{2h} symmetry, similar to the alkali halide dimers, for the previously undetected Li_2O_2 molecule, and estimated the angle, 116°, and bond distance, 1.90 Å, from their matrix spectra. The principal moments calculated from these molecular constants are: I_A = 2.3369 x 10^{-39}, I_B = 13.7955 x 10^{-39}, and I_C = 16.1324 x 10^{-39} g cm^2.

References

1. D. White, K. S. Seshadri, D. F. Dever, and D. E. Mann, and M. J. Linevsky, J. Chem. Phys. 39, 2463 (1963).
2. J. Berkowitz, J. Chem. Phys. 29, 1386 (1958); 32, 1519 (1960).

Lithium Oxide $((LiO)_2)$ $Li_2O_2(g)$

Enthalpy Reference Temperature = T_r = 298.15 K Standard State Pressure = p° = 0.1 MPa

T/K	$C_p°$ (J $K^{-1}mol^{-1}$)	S° (J $K^{-1}mol^{-1}$)	$-[G°-H°(T_r)]/T$ (J $K^{-1}mol^{-1}$)	$H°-H°(T_r)$ (kJ mol^{-1})	$\Delta_f H°$ (kJ mol^{-1})	$\Delta_f G°$ (kJ mol^{-1})	Log K_f
0	0.	0.	INFINITE	-15.854	-240.598	-240.598	INFINITE
100	43.954	208.779	331.528	-12.275	-240.914	-242.992	126.926
200	63.946	246.153	279.984	-6.766	-241.999	-244.602	63.884
250	69.312	261.042	274.742	-3.425	-242.372	-245.208	51.233
298.15	72.730	273.562	273.562	0.	-242.672	-245.727	43.050
300	72.837	274.012	273.563	0.135	-242.683	-245.746	42.788
350	75.223	285.431	274.460	3.840	-242.980	-246.233	36.748
400	76.893	295.591	276.478	7.645	-243.345	-246.674	32.212
450	78.098	304.720	279.117	11.521	-243.832	-247.062	28.678
500	78.993	312.998	282.098	15.450	-250.465	-246.774	25.780
600	80.201	327.516	288.491	23.415	-251.626	-245.925	21.410
700	80.955	339.940	294.975	31.475	-252.670	-244.891	18.274
800	81.454	350.784	301.288	39.597	-253.678	-243.711	15.913
900	81.801	360.399	307.332	47.761	-254.702	-242.404	14.069
1000	82.052	369.032	313.078	55.954	-255.743	-240.982	12.588
1100	82.239	376.861	318.526	64.169	-256.799	-239.455	11.371
1200	82.381	384.023	323.690	72.400	-257.868	-237.831	10.353
1300	82.493	390.622	328.588	80.644	-258.952	-236.117	9.487
1400	82.582	396.739	333.240	88.898	-260.044	-234.320	8.743
1500	82.654	402.439	337.665	97.160	-261.138	-232.444	8.094
1600	82.713	407.775	341.882	105.429	-262.236	-230.496	7.525
1700	82.762	412.791	345.907	113.702	-553.809	-214.126	6.579
1800	82.803	417.523	349.756	121.981	-553.407	-194.157	5.634
1900	82.838	422.001	353.441	130.263	-553.028	-174.209	4.789
2000	82.867	426.250	356.976	138.548	-552.672	-154.280	4.029
2100	82.893	430.294	360.372	146.836	-552.343	-134.369	3.342
2200	82.915	434.151	363.639	155.127	-552.041	-114.472	2.718
2300	82.935	437.837	366.785	163.419	-551.769	-94.589	2.148
2400	82.952	441.367	369.820	171.713	-551.531	-74.717	1.626
2500	82.967	444.754	372.750	180.009	-551.327	-54.854	1.146
2600	82.980	448.008	375.582	188.307	-551.162	-34.999	0.703
2700	82.992	451.140	378.323	196.605	-551.038	-15.148	0.293
2800	83.003	454.158	380.978	204.905	-550.959	4.698	-0.088
2900	83.012	457.071	383.552	213.206	-550.926	24.541	-0.442
3000	83.021	459.886	386.050	221.507	-550.943	44.386	-0.773
3100	83.029	462.608	388.476	229.810	-551.013	64.231	-1.082
3200	83.036	465.244	390.834	238.113	-551.139	84.079	-1.372
3300	83.042	467.799	393.127	246.417	-551.322	103.932	-1.645
3400	83.048	470.278	395.360	254.722	-551.566	123.792	-1.902
3500	83.053	472.686	397.535	263.027	-551.874	143.660	-2.144
3600	83.058	475.026	399.656	271.332	-552.247	163.538	-2.373
3700	83.063	477.301	401.724	279.638	-552.688	183.426	-2.590
3800	83.067	479.517	403.742	287.945	-553.200	203.328	-2.795
3900	83.071	481.674	405.712	296.252	-553.785	223.245	-2.990
4000	83.075	483.778	407.638	304.559	-554.446	243.176	-3.176
4100	83.078	485.829	409.520	312.867	-555.184	263.126	-3.352
4200	83.081	487.831	411.361	321.175	-556.004	283.094	-3.521
4300	83.084	489.786	413.162	329.483	-556.908	303.084	-3.682
4400	83.087	491.696	414.925	337.791	-557.856	323.091	-3.836
4500	83.089	493.563	416.652	346.100	-558.915	343.124	-3.983
4600	83.092	495.390	418.344	354.409	-560.069	363.181	-4.124
4700	83.094	497.177	420.002	362.719	-561.316	383.265	-4.260
4800	83.096	498.926	421.628	371.028	-562.659	403.377	-4.390
4900	83.098	500.639	423.224	379.338	-564.100	423.518	-4.515
5000	83.100	502.318	424.789	387.648	-565.643	443.689	-4.635
5100	83.101	503.964	426.325	395.958	-567.281	463.891	-4.751
5200	83.103	505.577	427.834	404.268	-569.038	484.126	-4.863
5300	83.105	507.160	429.316	412.578	-570.907	504.398	-4.971
5400	83.106	508.714	430.772	420.889	-572.893	524.705	-5.076
5500	83.107	510.239	432.203	429.199	-575.000	545.049	-5.176
5600	83.109	511.736	433.609	437.510	-577.232	565.435	-5.274
5700	83.110	513.207	434.993	445.821	-579.593	585.859	-5.369
5800	83.111	514.653	436.354	454.132	-582.086	606.328	-5.461
5900	83.112	516.073	437.693	462.443	-584.717	626.839	-5.550
6000	83.113	517.470	439.011	470.755	-587.489	647.398	-5.636

PREVIOUS: March 1964 (1 atm) CURRENT: March 1964 (1 bar)

Lithium Oxide $((LiO)_2)$ $Li_2O_2(g)$

LITHIUM SILICATE (Li_2SiO_3) CRYSTAL M_r = 89.9657

$\Delta_fH°(0\ K) = -1638.57 \pm 4.2\ kJ\ mol^{-1}$
$S°(298.15\ K) = 80.295 \pm 1.3\ J\ K^{-1}\ mol^{-1}$ $\Delta_fH°(298.15\ K) = -1649.50 \pm 4.2\ kJ\ mol^{-1}$
$T_{fus} = 1474 \pm 1\ K$ $\Delta_{fus}H° = 28.033 \pm 2.1\ kJ\ mol^{-1}$

Enthalpy of Formation

The enthalpies of solution of LiCl(cr), Li_2SO_4(cr), SiO_2(quartz) and Li_2SO_3(cr) in 20 percent hydrofluoric acid at 74.7°C were measured by Kracek (1). The authors derived two values of $\Delta_{sln}H°$ of Li_2O(cr) from those of LiCl(cr) and Li_2SO_4(cr), presumably employing auxiliary data from (2). We recalculate $\Delta_{sln}H°$ of Li_2O(cr) as -58.74 (chloride scheme, originally -59.36) and -59.83 kcal mol^{-1} (sulfate scheme, originally -60.86), using recent $\Delta_fH°$(298.15 K) values for LiCl(cr), Li_2SO_4(cr), HCl(aq), H_2SO_4(aq) and Li_2O(cr). The corresponding enthalpy changes, $\Delta_rH°$(298.15 K), for the reaction Li_2O(cr) + SiO_2(quartz) = Li_2SiO_3(cr) are derived as -33.08 and -34.17 kcal mol^{-1}. Adopting the weighted average $\Delta_rH°$(298.15 K) = -33.44 kcal mol^{-1} with $\Delta_fH°$(Li_2O, cr, 298.15 K) = -143.1 kcal mol^{-1} and $\Delta_fH°$(SiO_2, quartz, 298.15 K) = -217.7 kcal mol^{-1}, we obtain $\Delta_fH°$(298.15 K) = -394.24 kcal mol^{-1} for Li_2SiO_3(cr). This $\Delta_fH°$(Li_2SiO_3, cr, 298.15 K) value is independent of future changes in $\Delta_fH°$(298.15 K) of Li_2O(cr) but the values of $\Delta_{sln}H°$ and $\Delta_rH°$ are not independent.

The enthalpies of solution of LiOH(cr) and Li_2SiO_3(cr) in 20 percent HF(aq) at 50°C were measured by Hatton et al. (2). The Li_2SiO_3 sample was prepared from lithium carbonate and silica by fusion under vacuum at 1500 K. Analysis gave 66.74 percent SiO_2 and 15.12 percent Li. The corresponding calculated values are 66.79 and 15.43. Also present were 0.5 percent K and 0.06 percent Na. Corrections were made for impurities in the enthalpy of formation measurements. The results are given as follows:

	$\Delta_rH°$(298.15 K) kcal mol^{-1}
2 LiOH(cr, 25°C) + 2 HF(sol, 50°C) → 2 LiF(sol, 50°C) + 2 H_2O(sol, 50°C)	-38.728
SiO_2(cr, 25°C) + 6 HF(sol, 50°C) → H_2SiF_6(sol, 50°C) + 2 H_2O(sol, 50°C)	-32.810
3 H_2O(sol, 50°C) + 2 LiF(sol, 50°C) + H_2SiF_6(sol, 50°C) → Li_2SiO_3(cr, 25°C) + 8 HF(sol, 50°C)	+58.850
H_2O(sol, 50°C) → H_2O(ℓ, 25°C)	-0.450
2 LiOH(cr, 25°C) + SiO_2(cr, 25°C) → Li_2SO_3(cr, 25°C) + H_2O(ℓ, 25°C)	-13.14±0.70

The $\Delta_rH°$(298.15 K) for the second reaction was determined by Torgeon and Sahama (4). Based on the calculated enthalpy change for the overall reaction, $\Delta_rH°$(298.15 K) = -13.14 kcal mol^{-1} and $\Delta_fH°$(298.15 K) = -115.84, -217.7 and -68.33 kcal mol^{-1} for LiOH(cr), SiO_2(low quartz), and H_2O(ℓ), respectively, we derived $\Delta_fH°$(Li_2SO_3, cr, 298.15 K) = -394.21±1.0 kcal mol^{-1}, which is in agreement with the adopted value.

Kroger and Fingas (5) measured the equilibrium pressure of CO_2 over a mixture of Li_2CO_3(cr), SiO_2(quartz) and Li_2SiO_3(cr) from 585 to 781 K. The enthalpy change $\Delta_rH°$(298.15 K) of the reaction Li_2CO_3(cr) + SiO_2(cr) → Li_2SiO_3(cr) + CO_2(g) is evaluated by the second and third law methods to be 30.94 and 29.47 kcal mol^{-1}, respectively. Using $\Delta_fH°$(298.15 K) = -290.64, -217.7 and -94.05 kcal mol^{-1} for Li_2CO_3(cr), SiO_2(cr), and CO_2(g), respectively, and 3rd law $\Delta_rH°$(298.15 K), we obtain $\Delta_fH°$(Li_2SiO_3, cr, 298.15 K) = -384.82±5 kcal mol^{-1}, which is not used.

Heat Capacity and Entropy

The low temperature heat capacities, 25 - 340 K, were measured by Hatton et al. (3) with an adiabatic vacuum calorimeter, using a nickel-plated copper sample container. During the filling and sealing of the calorimeter, the compound was handled in an anhydrous, CO_2-free atmosphere so as to avoid contamination. The purity of the sample is not well established. The high temperature heat capacities are estimated by summation of the $C_p°$'s of the constituent oxides, Li_2O and SiO_2, and a subsequent correction based upon a comparison of the heat capacities of Na_2SiO_3(cr) and its constituent oxides. These estimated $C_p°$ values are plotted and the resulting curve is joined smoothly with the low temperature $C_p°$ curve at 298 K. The entropy, S°(298.15 K), is derived using the measured low temperature heat capacities, based on S°(25 K) = 0.023 cal K^{-1} mol^{-1}.

Fusion Data

See the liquid table for details.

References

1. F. C. Kracek, Ann. Rept. Director of the Geophysical Laboratory, No. 1215, 69 (1953).
2. U. S. Nat. Bur. Stand. Circ. 500, 1952.
3. W. E. Hatton, D. L. Hildenbrand, G. C. Sinke and D. R. Stull, unpublished work, The Dow Chemical Co., Midland, Michigan, 1959.
4. D. R. Torgeson and T. G. Sahama, J. Amer. Chem. Soc. 70, 2156 (1948).
5. C. Kroger and E. Fingas, Z. Anorg. Allg. Chem. 213, 12 (1933).

Lithium Silicate (Li_2SiO_3) $Li_2O_3Si_1$(cr)

Enthalpy Reference Temperature = T_r = 298.15 K; Standard State Pressure = p° = 0.1 MPa

T/K	$C_p°$	S°	-[G°-H°(T_r)]/T	H°-H°(T_r)	$\Delta_fH°$	$\Delta_fG°$	Log K_f
	J $K^{-1}mol^{-1}$			kJ mol^{-1}			
0	0.	0.	INFINITE	-14.554	-1638.567	-1638.567	INFINITE
100	27.669	12.276	148.950	-13.667	-1643.293	-1616.673	844.463
200	72.036	45.798	88.665	-8.573	-1647.428	-1588.278	414.816
298.15	100.567	80.295	80.295	0.	-1649.500	-1558.739	273.085
300	101.010	80.919	80.297	0.186	-1649.524	-1558.176	271.303
400	118.826	112.628	84.499	11.252	-1650.237	-1527.594	199.483
500	127.809	140.155	92.946	23.604	-1656.617	-1496.277	156.315
600	134.306	164.049	102.850	36.720	-1656.584	-1464.203	127.470
700	139.649	185.161	113.129	50.422	-1656.061	-1432.175	106.870
800	144.432	204.128	123.338	64.632	-1655.158	-1400.249	91.427
900	148.565	221.382	133.288	79.285	-1653.958	-1368.455	79.423
1000	152.302	237.232	142.900	94.331	-1652.487	-1336.808	69.828
1100	155.576	251.903	152.151	109.728	-1650.773	-1305.321	61.985
1200	158.574	265.571	161.040	125.437	-1648.837	-1274.001	55.456
1300	161.293	278.371	169.578	141.431	-1646.703	-1242 850	49.938
1400	163.942	290.422	177.783	157.694	-1644.375	-1211 870	45.215
1474.000	165.856	298.915	183.653	169.896	------ CRYSTAL <--> LIQUID -----		
1500	166.523	301.821	185.676	174.217	-1641.853	-1181.064	41.128
1600	169.034	312.649	193.276	190.996	-1639.140	-1150.432	37.558
1700	171.502	322.971	200.604	208.023	-1976.890	-1105.177	33.958
1800	173.887	332.841	207.678	225.293	-1972.074	-1054.038	30.587
1900	176.188	342.305	214.517	242.798	-1967.061	-1003.172	27.579
2000	178.406	351.399	221.135	260.528	-1961.862	-952.574	24.879
2100	180.537	360.155	227.548	278.476	-1956.485	-902.241	22.442
2200	182.590	368.602	233.768	296.633	-1950.940	-852.168	20.233
2300	184.565	376.762	239.809	314.991	-1945.238	-802.351	18.222
2400	186.461	384.657	245.681	333.543	-1939.387	-752.786	16.384
2500	188.280	392.306	251.394	352.281	-1933.399	-703.467	14.698

PREVIOUS: March 1964 CURRENT: June 1967

Lithium Silicate (Li_2SiO_3) $Li_2O_3Si_1$(cr)

LITHIUM SILICATE (Li_2SiO_3) LIQUID M_r = 89.9657

S°(298.15 K) = [96.064] J K^{-1} mol^{-1} $\Delta_f H°$(298.15 K) = [-1625.231] kJ mol^{-1}

T_{fus} = 1474 ± 1 K $\Delta_{fus}H°$ = 28.033 ± 2.1 kJ mol^{-1}

Enthalpy of Formation

$\Delta_f H°$(Li_2SiO_3, ℓ, 298.15 K) is calculated from $\Delta_f H°$(Li_2SiO_3, cr, 298.15 K) by adding the enthalpy of fusion, $\Delta_{fus}H°$, and the difference in enthalpy, H°(1474 K)-H°(298.15 K), between the crystal and liquid.

Tschernobaeff (1) measured the enthalpy of reaction between SiO_2 and Li_2CO_3 in a bomb calorimeter, using carbon as the auxiliary combustible material. Based on the derived $\Delta_r H°$(298.15 K) = 24.99 kcal mol^{-1} for the reaction Li_2CO_3(cr) + SiO_2(cr) → Li_2SiO_3(ℓ) + CO_2(g), the enthalpy of formation for Li_2SiO_3(ℓ) is evaluated as -389.30 kcal mol^{-1}, which is in fair agreement with the adopted value.

Heat Capacity and Entropy

The heat capacity is estimated by comparison with those for Na_2SiO_3(ℓ), Na_2O(ℓ) and Li_2O(ℓ). A glass transition is assumed at 1000 K; i.e., the heat capacities below 1000 K are taken to be the same as those for Li_2SiO_3(cr).

S°(Li_2SiO_3, ℓ, 298.15 K) is calculated in a manner analogous to that used for the enthalpy of formation.

Fusion Data

The melting point has been reported as 1474±1 and 1461 K by Fracek (2) and Van Klooster (3), respectively. The value reported by Kracek is adopted.

Schwartz and Sturm (4) used a cooling-curve method to compare the heats of fusion of Li_2SiO_3(cr) and NaCl(cr). From the reported ratio, ε(Li_2SiO_3)/ε(NaCl) = 0.261/0.402, where ε = heat of fusion in kcal g^{-1} and heat of fusion of NaCl(cr), the value $\Delta_{fus}H°$ = 6.7 kcal mol^{-1} is derived.

References

1. Tschernobaeff, Rev. Met. 21, 729 (1905).
2. F. C. Kracek, J. Phys. Chem. 34, 2641 (1930).
3. H. S. Van Klooster, Z. Anorg. Chem. 69, 135 (1911).
4. R. Schwarz and H. Sturm, Ber. 47, 1730 (1914).

Lithium Silicate (Li_2SiO_3) $Li_2O_3Si_1$(l)

Enthalpy Reference Temperature = T_r = 298.15 K Standard State Pressure = p° = 0.1 MPa

T/K	$C_p°$ (J $K^{-1}mol^{-1}$)	S° (J $K^{-1}mol^{-1}$)	$-[G°-H°(T_r)]/T$ (J $K^{-1}mol^{-1}$)	$H°-H°(T_r)$ (kJ mol^{-1})	$\Delta_f H°$ (kJ mol^{-1})	$\Delta_f G°$ (kJ mol^{-1})	Log K_f
0							
100							
200							
298.15	100.567	96.064	96.064	0.	-1625.231	-1539.171	269.656
300	101.010	96.688	96.066	0.186	-1625.254	-1538.637	267.901
400	118.826	128.398	100.268	11.252	-1625.968	-1509.633	197.138
500	127.809	155.924	108.716	23.604	-1632.348	-1479.892	154.603
600	134.306	179.818	118.619	36.720	-1632.314	-1449.396	126.181
700	139.649	200.930	128.898	50.422	-1631.792	-1418.944	105.883
800	144.432	219.897	139.108	64.632	-1630.889	-1388.595	90.666
900	148.565	237.151	149.057	79.285	-1629.689	-1358.378	78.838
1000	152.302	253.001	158.670	94.331	-1628.218	-1328.309	69.384
1000.000	152.302	253.001	158.670	94.331	GLASS <--> LIQUID TRANSITION		
1000.000	167.360	253.001	158.670	94.331			
1100	167.360	268.952	167.982	111.067	-1625.164	-1298.466	61.659
1200	167.360	283.514	177.012	127.803	-1622.201	-1268.898	55.234
1300	167.360	296.910	185.726	144.539	-1619.325	-1239.573	49.807
1400	167.360	309.313	194.116	161.275	-1616.524	-1210.467	45.163
1474.000	167.360	317.933	200.118	173.660	------ CRYSTAL <--> LIQUID -----		
1500	167.360	320.860	202.185	178.011	-1613.789	-1181.559	41.146
1600	167.360	331.661	209.944	194.747	-1611.119	-1152.831	37.636
1700	167.360	341.807	217.405	211.483	-1949.160	-1109.469	34.090
1800	167.360	351.373	224.584	228.219	-1944.879	-1060.199	30.766
1900	167.360	360.422	231.498	244.955	-1940.635	-1011.166	27.799
2000	167.360	369.006	238.160	261.691	-1936.429	-962.356	25.134
2100	167.360	377.172	244.587	278.427	-1932.264	-913.755	22.728
2200	167.360	384.957	250.792	295.163	-1928.141	-865.350	20.546
2300	167.360	392.397	256.788	311.899	-1924.060	-817.134	18.558
2400	167.360	399.519	262.588	328.635	-1920.026	-769.094	16.739
2500	167.360	406.351	268.203	345.371	-1916.040	-721.220	15.069
2600	167.360	412.915	273.643	362.107	-1912.104	-673.505	13.531
2700	167.360	419.232	278.919	378.843	-1908.221	-625.940	12.110
2800	167.360	425.318	284.040	395.579	-1904.393	-578.518	10.792
2900	167.360	431.191	289.013	412.315	-1900.623	-531.233	9.569
3000	167.360	436.865	293.848	429.051	-1896.915	-484.076	8.429
3100	167.360	442.352	298.550	445.787	-1893.268	-437.041	7.364
3200	167.360	447.666	303.127	462.523	-1889.688	-390.124	6.368
3300	167.360	452.816	307.586	479.259	-1886.174	-343.318	5.434
3400	167.360	457.812	311.931	495.995	-1882.730	-296.616	4.557
3500	167.360	462.663	316.169	512.731	-1879.358	-250.015	3.731
3600	167.360	467.378	320.304	529.467	-2260.212	-193.048	2.801
3700	167.360	471.964	324.341	546.203	-2256.576	-135.678	1.915
3800	167.360	476.427	328.285	562.939	-2253.020	-78.403	1.078
3900	167.360	480.774	332.139	579.675	-2249.546	-21.221	0.284
4000	167.360	485.011	335.908	596.411	-2246.155	35.871	-0.468

PREVIOUS: March 1964 CURRENT: June 1967

Lithium Silicate (Li_2SiO_3) $Li_2O_3Si_1$(l)

LITHIUM SILICATE (Li_2SiO_3) CRYSTAL-LIQUID M_r = 89.9657

0 to 1474 K crystal
above 1474 K liquid

Refer to the individual tables for details.

Lithium Silicate (Li_2SiO_3) $Li_2O_3Si_1$(cr,l)

Enthalpy Reference Temperature = T_r = 298.15 K   Standard State Pressure = p° = 0.1 MPa

T/K	C_p° (J K⁻¹mol⁻¹)	S° (J K⁻¹mol⁻¹)	-[G°-H°(T_r)]/T (J K⁻¹mol⁻¹)	H°-H°(T_r) (kJ mol⁻¹)	Δ_fH° (kJ mol⁻¹)	Δ_fG° (kJ mol⁻¹)	Log K_f
0	0.	0.	INFINITE	-14.554	-1638.567	-1638.567	INFINITE
100	27.669	12.276	148.950	-13.667	-1643.293	-1616.673	844.463
200	72.036	45.798	88.665	-8.573	-1647.428	-1588.278	414.816
298.15	100.567	80.295	80.295	0.	-1649.500	-1558.739	273.085
300	101.010	80.919	80.297	0.186	-1649.524	-1558.176	271.303
400	118.826	112.628	84.499	11.252	-1650.237	-1527.594	199.483
500	127.809	140.155	92.946	23.604	-1656.617	-1496.277	156.315
600	134.306	164.049	102.850	36.720	-1656.584	-1464.203	127.470
700	139.649	185.161	113.129	50.422	-1656.061	-1432.175	106.870
800	144.432	204.128	123.338	64.632	-1655.158	-1400.249	91.427
900	148.565	221.382	133.288	79.285	-1653.958	-1368.455	79.423
1000	152.302	237.232	142.900	94.331	-1652.487	-1336.808	69.828
1100	155.576	251.903	152.151	109.728	-1650.773	-1305.321	61.985
1200	158.574	265.571	161.040	125.437	-1648.837	-1274.001	55.456
1300	161.293	278.371	169.578	141.431	-1646.703	-1242.850	49.938
1400	163.942	290.422	177.783	157.694	-1644.375	-1211.870	45.215
1474.000	165.856	298.915	183.653	169.896	CRYSTAL <--> LIQUID		
1474.000	167.360	317.933	183.653	197.929	TRANSITION		
1500	167.360	320.860	186.006	202.281	-1613.789	-1181.559	41.146
1600	167.360	331.661	194.775	219.017	-1611.119	-1152.831	37.636
1700	167.360	341.807	203.129	235.753	-1949.160	-1109.469	34.090
1800	167.360	351.373	211.102	252.489	-1944.879	-1060.199	30.766
1900	167.360	360.422	218.725	269.225	-1940.635	-1011.166	27.799
2000	167.360	369.006	226.026	285.961	-1936.429	-962.356	25.134
2100	167.360	377.172	233.030	302.697	-1932.264	-913.755	22.728
2200	167.360	384.957	239.761	319.433	-1928.141	-865.350	20.546
2300	167.360	392.397	246.236	336.169	-1924.060	-817.134	18.558
2400	167.360	399.519	252.476	352.905	-1920.026	-769.094	16.739
2500	167.360	406.351	258.495	369.641	-1916.040	-721.220	15.069
2600	167.360	412.915	264.309	386.377	-1912.104	-673.505	13.531
2700	167.360	419.232	269.931	403.113	-1908.221	-625.940	12.110
2800	167.360	425.318	275.372	419.849	-1904.393	-578.518	10.792
2900	167.360	431.191	280.645	436.585	-1900.623	-531.233	9.569
3000	167.360	436.865	285.758	453.321	-1896.915	-484.076	8.429
3100	167.360	442.352	290.721	470.057	-1893.268	-437.041	7.364
3200	167.360	447.666	295.543	486.793	-1889.688	-390.124	6.368
3300	167.360	452.816	300.231	503.529	-1886.174	-343.318	5.434
3400	167.360	457.812	304.793	520.265	-1882.730	-296.616	4.557
3500	167.360	462.663	309.235	537.001	-1879.358	-250.015	3.731
3600	167.360	467.378	313.562	553.737	-2260.212	-193.048	2.801
3700	167.360	471.964	317.782	570.473	-2256.576	-135.678	1.915
3800	167.360	476.427	321.898	587.209	-2253.020	-78.403	1.078
3900	167.360	480.774	325.916	603.945	-2249.546	-21.221	0.284
4000	167.360	485.011	329.841	620.681	-2246.155	35.871	-0.468

PREVIOUS: CURRENT: June 1967

Lithium Silicate (Li_2SiO_3) $Li_2O_3Si_1$(cr,l)

LITHIUM TITANIUM OXIDE (Li_2TiO_3) CRYSTAL (ALPHA-BETA) M_r = 109.7602

$S°(298.15\ K) = 91.755 \pm 0.4\ J\ K^{-1}\ mol^{-1}$
T_{trs} = 1485 K
T_{fus} = 1820 K

$\Delta_f H°(0\ K) = -1660.065\ kJ\ mol^{-1}$
$\Delta_f H°(298.15\ K) = -1670.671\ kJ\ mol^{-1}$
$\Delta_{trs} H° = 11.506\ kJ\ mol^{-1}$
$\Delta_{fus} H° = 110.165\ kJ\ mol^{-1}$

Enthalpy of Formation

$\Delta_f H°(298.15\ K)$ was calculated from measurements of solution calorimetry by Todd and Kelley (1) according to the following equation:

$TiO_2(cr) + Li_2SO_4(cr) + 13.312\ H_2O(\ell) = Li_2TiO_3(cr) + H_2SO_4(aq,\ 12.312\ H_2O)$ $\Delta_r H°(298.15\ K) = 27.04\ kcal\ mol^{-1}$

$\Delta_f H°(TiO_2,\ cr,\ 298.15\ K)$ value for rutile was obtained from the JANAF tables and the other $\Delta_f H(298.15\ K)$ values were obtained from (2).

Heat Capacity and Entropy

The low temperature heat capacities were taken from King (3). These low temperature heat capacities were fit to the high temperature enthalpies from Christensen et al. (4).

$S°(298.15\ K) = 21.9\ cal\ K^{-1}\ mol^{-1}$ was obtained from King (3) based on the measured heat capacities and the extrapolated S°(51 K) value.

Transition Data

Christensen et al. (4) reported that lithium metatitanate undergoes a transition from the α to the β form at 1,485 K with an absorption of 2.750 kcal mol^{-1}.

Fusion Data

The T_{fus} and $\Delta_{fus}H°$ were obtained from Christensen et al. (4).

References

1. S. S. Todd and K. K. Kelley, U. S. Bur. RI 5193, 1956.
2. U. S. Nat. Bur. Stand. Circ. 500, 1952.
3. E. G. King, J. Amer. Chem. Soc. 77, 2150 (1955).
4. A. U. Christensen, K. C. Conway and K. K. Kelley, U. S. Bur. Mines RI 5565, 1960.

Lithium Titanium Oxide (Li_2TiO_3) $Li_2O_3Ti_1(cr)$

Enthalpy Reference Temperature = T_r = 298.15 K Standard State Pressure = p° = 0.1 MPa

T/K	$C_p°$	S°	$-[G°-H°(T_r)]/T$	$H°-H°(T_r)$	$\Delta_f H°$	$\Delta_f G°$	Log K_f
	J K⁻¹mol⁻¹			kJ mol⁻¹			
0	0.	0.	INFINITE	-16.493	-1660.065	-1660.065	INFINITE
100	31.142	14.129	169.105	-15.498	-1664.977	-1638.099	855.655
200	82.663	52.886	101.316	-9.686	-1669.132	-1609.488	420.355
298.15	109.855	91.755	91.755	0.	-1670.671	-1579.767	276.769
300	110.750	92.437	91.757	0.204	-1670.687	-1579.203	274.964
400	127.361	126.619	96.309	12.124	-1671.038	-1548.635	202.231
500	135.645	156.002	105.388	25.307	-1677.049	-1517.421	158.524
600	141.503	181.283	115.979	39.182	-1676.706	-1485.520	129.326
700	145.854	203.436	126.924	53 559	-1675.949	-1453.710	108.477
800	149.034	223.130	137.741	68.311	-1674.920	-1422.030	92.849
900	151.670	240.839	148.229	83.349	-1673.763	-1390.488	80.702
1000	153.888	256.936	158.307	98.628	-1672.570	-1359.077	70.991
1100	155.812	271.696	167.954	114.117	-1671.422	-1327.783	63.051
1200	157.318	285.322	177.174	129.778	-1674.331	-1296.471	56.434
1300	158.490	297.965	185.985	145.574	-1672.635	-1265 052	50.830
1400	159.494	309.751	194.410	161.478	-1670.942	-1233.763	46.032
1485.000	160.344	319.174	201.284	175.067	ALPHA <--> BETA	TRANSITION	
1485.000	176.063	326.922	201.284	186.573			
1500	176.565	328.694	202.549	189.218	-1657.529	-1202.712	41.882
1600	179.912	340.196	210.795	207.042	-1654.160	-1172.500	38.278
1700	183.259	351.204	218.733	225.200	-1941.066	-1128.152	34.664
1800	186.606	361.773	226.388	243.693	-1935.811	-1080.485	31.355
1820.000	187.276	363.839	227.887	247.432	BETA <--> LIQUID		
1900	189.954	371.952	233.783	262.521	-1930.404	-1033.113	28.402
2000	193.301	381.781	240.939	281.684	-1939.544	-985.576	25.741
2100	196.648	391.293	247.873	301.182	-1934.622	-937.998	23.331
2200	199.995	400.518	254.603	321.014	-1929.406	-890.660	21.147
2300	203.342	409.482	261.143	341.181	-1923.899	-843.567	19.158
2400	206.690	418.207	267.506	361.682	-1918.103	-796.720	17.340
2500	210.037	426.712	273.705	382.519	-1912.021	-750.119	15.673
2600	213.384	435.015	279.750	403.690	-1905.654	-703.768	14.139
2700	216.731	443.131	285.652	425.195	-1899.005	-657.666	12.723
2800	220.078	451.074	291.418	447.036	-1892.077	-611.816	11.414
2900	223.426	458.855	297.058	469.211	-1884.873	-566.220	10.199
3000	226.773	466.486	302.579	491.721	-1877.394	-520.876	9.069

PREVIOUS: June 1961 CURRENT: March 1964

LITHIUM TITANIUM OXIDE (Li_2TiO_3) LIQUID M_r = 109.7602

S°(298.15 K) = [147.656] J K^{-1} mol^{-1} $\Delta_f H°$(298.15 K) = [-1566.760] kJ mol^{-1}

T_{fus} = 1820 K $\Delta_{fus}H°$ = 110.165 kJ mol^{-1}

Enthalpy of Formation

$\Delta_f H°$(Li_2TiO_3, ℓ, 298.15 K) is calculated from that of the crystal by adding the enthalpy of fusion, $\Delta_{fus}H°$, and the difference in enthalpy, H°(1820 K) - H°(298.15 K), between the crystal and liquid.

Heat Capacity and Entropy

The heat capacities were reported by Christensen et al. (1). The measured heat capacity is extrapolated to an assumed glass transition temperature of 1214 K, below which the heat capacity is taken to be equal to that of the crystal.

S°(Li_2TiO_3, ℓ, 298.15 K) is calculated in a manner analogous to that used for the enthalpy of formation.

Fusion Data

The T_{fus} and $\Delta_{fus}H°$ were obtained from Christensen et al. (1).

Reference

1. A. U. Christensen, K. C. Conway and K. K. Kelley, U. S. Bur. Mines RI 5565, 1960.

Lithium Titanium Oxide (Li_2TiO_3) $Li_2O_3Ti_1$(l)

Enthalpy Reference Temperature = T_r = 298.15 K Standard State Pressure = p° = 0.1 MPa

T/K	$C_p°$ (J $K^{-1}mol^{-1}$)	S° (J $K^{-1}mol^{-1}$)	-[G°-H°(T_r)]/T (J $K^{-1}mol^{-1}$)	H°-H°(T_r) (kJ mol^{-1})	$\Delta_f H°$ (kJ mol^{-1})	$\Delta_f G°$ (kJ mol^{-1})	Log K_f
0							
100							
200							
298.15	109.855	147.656	147.656	0.	-1566.760	-1492.523	261.484
300	110.750	148.338	147.658	0.204	-1566.775	-1492.062	259.791
400	127.361	182.520	152.209	12.124	-1567.127	-1467.084	191.582
500	135.645	211.903	161.289	25.307	-1573.138	-1441.460	150.588
600	141.503	237.183	171.880	39.182	-1572.795	-1415.149	123.200
700	145.854	259.337	182.824	53.559	-1572.038	-1388.930	103.643
800	149.034	279.031	193.642	68.311	-1571.009	-1362.840	88.984
900	151.670	296.740	204.130	83.349	-1569.852	-1336.887	77.591
1000	153.888	312.837	214.208	98.628	-1568.659	-1311.066	68.483
1100	155.812	327.597	223.855	114.117	-1567.511	-1285.363	61.037
1200	157.318	341.223	233.075	129.778	-1570.420	-1259.641	54.831
1214.000	157.569	343.049	234.333	131.982	------- GLASS <--> LIQUID ----- TRANSITION		
1214.000	200.832	343.049	234.333	131.982			
1300	200.832	356.795	241.984	149.253	-1565.044	-1233.940	49.580
1400	200.832	371.678	250.723	169.336	-1559.172	-1208.691	45.097
1500	200.832	385.534	259.254	189.420	-1553.416	-1183.859	41.226
1600	200.832	398.495	267.556	209.503	-1547.787	-1159.406	37.851
1700	200.832	410.671	275.620	229.586	-1832.769	-1120.948	34.443
1800	200.832	422.150	283.445	249.669	-1825.924	-1079.275	31.320
1820.000	200.832	424.369	284.965	253.716	------- BETA <--> LIQUID -------		
1900	200.832	433.008	291.033	269.752	-1819.262	-1037.977	28.536
2000	200.832	443.310	298.392	289.836	-1827.482	-996.572	26.028
2100	200.832	453.108	305.528	309.919	-1821.974	-955.162	23.758
2200	200.832	462.451	312.450	330.002	-1816.507	-914.012	21.701
2300	200.832	471.378	319.167	350.085	-1811.084	-873.112	19.829
2400	200.832	479.926	325.689	370.168	-1805.706	-832.447	18.118
2500	200.832	488.124	332.023	390.252	-1800.377	-792.004	16.548
2600	200.832	496.001	338.180	410.335	-1795.098	-751.773	15.103
2700	200.832	503.580	344.166	430.418	-1789.872	-711.744	13.770
2800	200.832	510.884	349.991	450.501	-1784.701	-671.908	12.535
2900	200.832	517.931	355.661	470.584	-1779.588	-632.257	11.388
3000	200.832	524.740	361.184	490.668	-1774.536	-592.780	10.321
3100	200.832	531.325	366.567	510.751	-1769.547	-553.470	9.326
3200	200.832	537.701	371.816	530.834	-1764.623	-514.321	8.395
3300	200.832	543.881	376.937	550.917	-1759.766	-475.325	7.524
3400	200.832	549.877	381.935	571.000	-1754.979	-436.474	6.706
3500	200.832	555.698	386.817	591.084	-1750.265	-397.763	5.936
3600	200.832	561.356	391.587	611.167	-1745.624	-359.186	5.212
3700	200.832	566.859	396.250	631.250	-2150.157	-312.949	4.418
3800	200.832	572.214	400.811	651.333	-2144.436	-263.371	3.620
3900	200.832	577.431	405.273	671.416	-2138.850	-213.942	2.865
4000	200.832	582.516	409.641	691.500	-2133.402	-164.655	2.150

PREVIOUS: June 1961 CURRENT: March 1964

Lithium Titanium Oxide (Li_2TiO_3) $Li_2O_3Ti_1$(l)

LITHIUM TITANIUM OXIDE (Li_2TiO_3) ALPHA-BETA-LIQUID M_r = 109.7602

0 to 1485 K crystal, alpha
1485 to 1820 K crystal, beta
above 1820 K liquid

Refer to the individual tables for details.

Lithium Titanium Oxide (Li_2TiO_3) $Li_2O_3Ti_1$(cr,l)

Enthalpy Reference Temperature = T_r = 298.15 K Standard State Pressure = $p°$ = 0.1 MPa

T/K	$C_p°$	$S°$	$-[G°-H°(T_r)]/T$	$H°-H°(T_r)$	$\Delta_f H°$	$\Delta_f G°$	Log K_f
	J K⁻¹mol⁻¹			kJ mol⁻¹			
0	0.	0.	INFINITE	-16.493	-1660.065	-1660.065	INFINITE
100	31.142	14.129	169.105	-15.498	-1664.977	-1638.099	855.655
200	82.663	52.886	101.316	-9.686	-1669.132	-1609.488	420.355
298.15	109.855	91.755	91.755	0.	-1670.671	-1579.767	276.769
300	110.750	92.437	91.757	0.204	-1670.687	-1579.203	274.964
400	127.361	126.619	96.309	12.124	-1671.038	-1548.635	202.231
500	135.645	156.002	105.388	25.307	-1677.049	-1517.421	158.524
600	141.503	181.283	115.979	39.182	-1676.706	-1485.520	129.326
700	145.854	203.436	126.924	53.559	-1675.949	-1453.710	108.477
800	149.034	223.130	137.741	68.311	-1674.920	-1422.030	92.849
900	151.670	240.839	148.229	83.349	-1673.763	-1390.488	80.702
1000	153.888	256.936	158.307	98.628	-1672.570	-1359.077	70.991
1100	155.812	271.696	167.954	114.117	-1671.422	-1327.783	63.051
1200	157.318	285.322	177.174	129.778	-1674.331	-1296.471	56.434
1300	158.490	297.965	185.985	145.574	-1672.635	-1265.052	50.830
1400	159.494	309.751	194.410	161.478	-1670.942	-1233.763	46.032
1485.000	160.344	319.174	201.284	175.067		ALPHA <--> BETA	
1485.000	176.063	326.922	201.284	186.573		TRANSITION	
1500	176.565	328.694	202.549	189.218	-1657.529	-1202.712	41.882
1600	179.912	340.196	210.795	207.042	-1654.160	-1172.500	38.278
1700	183.259	351.204	218.733	225.200	-1941.066	-1128.152	34.664
1800	186.606	361.773	226.388	243.693	-1935.811	-1080.485	31.355
1820.000	187.276	363.839	227.887	247.432		BETA <--> LIQUID	
1820.000	200.832	424.369	227.887	357.597		TRANSITION	
1900	200.832	433.008	236.343	373.664	-1819,262	-1037.977	28.536
2000	200.832	443.310	246.436	393.747	-1827.482	-996.572	26.028
2100	200.832	453.108	256.046	413.830	-1821.974	-955.162	23.758
2200	200.832	462.451	265.218	433.913	-1816.507	-914.012	21.701
2300	200.832	471.378	273.989	453.996	-1811.084	-873.112	19.829
2400	200.832	479.926	282.392	474.080	-1805.706	-832.447	18.118
2500	200.832	488.124	290.459	494.163	-1800.377	-792.004	16.548
2600	200.832	496.001	298.214	514.246	-1795.098	-751.773	15.103
2700	200.832	503.580	305.681	534.329	-1789.872	-711.744	13.770
2800	200.832	510.884	312.880	554.412	-1784.701	-671.908	12.535
2900	200.832	517.931	319.830	574.496	-1779.588	-632.257	11.388
3000	200.832	524.740	326.547	594.579	-1774.536	-592.780	10.321
3100	200.832	531.325	333.047	614.662	-1769.547	-553.470	9.326
3200	200.832	537.701	339.344	634.745	-1764.623	-514.321	8.395
3300	200.832	543.881	345.449	654.828	-1759.766	-475.325	7.524
3400	200.832	549.877	351.373	674.912	-1754.979	-436.474	6.706
3500	200.832	555.698	357.128	694.995	-1750.265	-397.763	5.936
3600	200.832	561.356	362.723	715.078	-1745.624	-359.186	5.212
3700	200.832	566.859	368.166	735.161	-2150.157	-312.949	4.418
3800	200.832	572.214	373.466	755.244	-2144.436	-263.371	3.620
3900	200.832	577.431	378.629	775.328	-2138.850	-213.942	2.865
4000	200.832	582.516	383.663	795.411	-2133.402	-164.655	2.150

PREVIOUS: CURRENT: March 1964

Lithium Titanium Oxide (Li_2TiO_3) $Li_2O_3Ti_1$(cr,l)

LITHIUM SULFATE (α-Li_2SO_4) CRYSTAL M_r = 109.9396

$S°(298.15\ K) = 113.947 \pm 0.4\ J\ K^{-1}\ mol^{-1}$

$T_{trs}(\alpha \to \beta) = 848 \pm 1\ K$

$\Delta_f H°(0\ K) = -1423.980 \pm 1.7\ kJ\ mol^{-1}$

$\Delta_f H°(298.15\ K) = -1436.367 \pm 1.7\ kJ\ mol^{-1}$

$\Delta_{trs}H°(\alpha \to \beta) = 28.451 \pm 1.7\ kJ\ mol^{-1}$

Enthalpy of Formation

The enthalpy of solution of Li_2SO_4(cr, α) has been measured by Ueda (1), Thomsen (2), Pickering (3), Romanova and Samoilov (4) and Tsvetkov and Rabinovich (5). Adopting the results from (1, 4, 5) we have $\Delta_{sln}H°(298.15\ K) = -7.300\pm0.1\ kcal\ mol^{-1}$ for the enthalpy of solution at infinite dilution. We have used data from Thompson et al. (6) to correct the enthalpy of solution data to infinite dilution. When this result is combined with enthalpies of formation of the infinitely dilute ions from CODATA (7) we obtain $\Delta_f H°(Li_2SO_4$, aq, ∞, 298.15 K) = -350.504±0.2 kcal mol^{-1} and $\Delta_f H°(Li_2SO_4$, cr, 298.15 K) = -343.20±0.3 kcal mol^{-1}.

Lithium sulfate was studied by hydrochloric acid solution calorimetry by Barany and Adami (8) who obtained $\Delta_r H°(298.15\ K)$ = -16.020±0.120 kcal mol^{-1} for the reaction 2 LiCl(cr) + $H_2SO_4 \cdot 7.068\ H_2O$(ℓ) + 18.394 H_2O(ℓ) = Li_2SO_4(cr) + 2 HCL·12.731 H_2O(ℓ). When this result is combined with auxiliary data from (9, 10), we obtain $\Delta_f H°(Li_2SO_4$, cr, 298.15 K) = -343.30±0 kcal mol^{-1}. Auxiliary data used in the analysis are (in kcal mol^{-1}): $\Delta_f H°(H_2SO_4 \cdot 7.068\ H_2O$, 298.15 K) = -209.566±0.10 and $\Delta_f H°$(HCl·12.731 H_2O, 298.15 K) = -38.792±0.010.

Ueda (1) measured the emf of the cell: Li (amalgam, 0.035%)|$Li_2SO_4 \cdot H_2O$ (sat. soln.), Hg_2SO_4|Hg at 298 K as E_{cell} = 2.67489 volts. Kelley (17) used this value along with other data to determine $\Delta_f H°(298.15\ K)$ for Li_2SO_4(cr). Adopting results given in Kelley (17), we calculate $\Delta_r G°(298.15\ K) = -165.780\ kcal\ mol^{-1}$ for the reaction 2 Li(cr) + Hg_2SO_4(cr) = 2 Hg(ℓ) + Li_2SO_4(cr). Taking $\Delta_f G°(Hg_2SO_4$, cr, 298.15 K) = -149.658 kcal mol^{-1} (7) and reference entropies from (7), we calculate $\Delta_f G°(Li_2SO_4$, cr, 298.15 K) = -315.438 and $\Delta_f H°(Li_2SO_4$, cr, 298.15 K) = -343.0±0.5 kcal mol^{-1}.

We adopt a value for the enthalpy of reaction based on the solution work of Barany and Adami (8).

Heat Capacity and Entropy

Low temperature heat capacities of Li_2SO_4(cr, α) have been measured by Paukov (11) in an adiabatic calorimeter from 12.00 - 311.11 K. We have smoothed the experimental $C_p°$ data by fitting to orthogonal polynomials over selected temperature intervals. Our adopted value of $S°(298.15\ K) = 27.234\pm0.10\ kcal\ mol^{-1}$ is based on $S°(14\ K) = 0.051\ kcal\ mol^{-1}$ and $H°(14;0) = 0.540\ cal\ K^{-1}\ mol^{-1}$ derived from a Debye T^3-law extrapolation.

Heat capacity data measured by adiabatic calorimetry have been reported by Shmidt (12, 298.5 - 774.3) and by Sjoblom (13, 430 - 845 K). In addition, drop calorimetric enthalpy measurements have been performed by Denielou et al. (14, 400 - 848 K), by Clark (15, 488.0 - 848.7 K) and by Voskresenskaya and Banashek (16, 360 - 859 K). The data of Clark is systematically low by 15% while the data from Sjoblom is too high by 10%. There is good agreement among the other three studies (12, 14, 16). We adopt heat capacities above 300 K based on a smooth fit to the combined data from (11, 12, 14, and 16).

Transition Data

Li_2SO_4(cr, α) is the low-temperature form of lithium sulfate (sometimes denoted Li_2SO_4 II). The crystals are monoclinic (space group C_{2h}, $P2_1/c$). The transformation to the high-temperature β-phase (cubic form) occurs at 848±1 K (14, 15). The enthalpy of transition is determined by taking the difference between the adopted enthalpy curves of α- and β-Li_2SO_4 extrapolated to 848 K (see the discussion on the Li_2SO_4(cr, β) table, 18).

References

1. Y. Ueda, Science Repts. Tohuku Imp. Univ., First Ser. 22, 879 (1933), see Chem. Abstr. 28, 954 (1934) and also J. Chem. Soc. Japan 53, 559 (1932) and Chem. Abstr. 26, 5249 (1932).
2. J. Thomsen, "Thermochemische Untersuchungen," Vol. III, p. 189, J. Barth (Leipzig) 1883.
3. S. U. Pickering, J. Chem. Soc. 47, 98 (1885); J. Chem. Soc. 49, 260 (1886).
4. I. I. Romanova and O. Ya. Samoilov, Russ. J. Inorg. Chem. 14, 131 (1969).
5. V. G. Tsvetkov and I. B. Rabinovich, Russ. J. Phys. Chem. 43, 675 (1969).
6. P. T. Thompson, D. E. Smith and R. H. Wood, J. Chem. Eng. Data 19, 386 (1974).
7. ICSU-CODATA Task Group on Key Values for Thermodynamics, J. Chem. Thermodyn. 10, 903 (1978).
8. R. Barany and L. H. Adami, U. S. Bur. Mines RI 6873 (1966).
9. U. S. Nat. Bur. Stand. Tech. Note 270-3, 1968.
10. V. B. Parker, D. D. Wagman and D. Garvin, U. S. Nat. Bur. Stand. Interim Report 75-968, 1976.
11. I. E. Paukov, Russ. J. Phys. Chem. 43, 1134 (1969).
12. N. E. Shmidt, Russ. J. Inorg. Chem. 12, 929 (1967).
13. C. A. Sjoblom, High Temp. - High Pressures 8, 499 (1976).
14. L. Denielou, Y. Fournier, J. P. Petitet, and C. Tequi, Rev. Int. Hautes Temp. 8, 119 (1971).
15. R. P. Clark, J. Chem. Eng. Data 20, 17 (1975).
16. N. K. Voskresenskaya and E. I. Banashek, Izv. Sekt. Fiz. Khim. Anal., Akad. Nauk SSSR 25, 150 (1954).
17. K. K. Kelley, U. S. Bur. Mines Bull. 406, 1937.
18. JANAF Thermochemical Tables: Li_2SO_4(cr, β), 12-31-78.

Lithium Sulfate, Alpha (Li_2SO_4) $Li_2O_4S_1$(cr)

Enthalpy Reference Temperature = T_r = 298.15 K   Standard State Pressure = p° = 0.1 MPa

T/K	$C_p°$ (J $K^{-1}mol^{-1}$)	S° (J $K^{-1}mol^{-1}$)	$-[G°-H°(T_r)]/T$ (J $K^{-1}mol^{-1}$)	$H°-H°(T_r)$ (kJ mol^{-1})	$\Delta_f H°$ (kJ mol^{-1})	$\Delta_f G°$ (kJ mol^{-1})	Log K_f
0	0.	0.	INFINITE	-18.636	-1423.980	-1423.980	INFINITE
100	43.869	28.932	196.627	-16.769	-1429.603	-1395.114	728.733
200	86.872	73.028	123.821	-10.159	-1434.139	-1358.693	354.854
298.15	117.608	113.947	113.947	0.	-1436.367	-1321.111	231.453
300	118.026	114.676	113.949	0.218	-1436.391	-1320.395	229.901
400	139.244	151.679	118.850	13.131	-1439.218	-1281.517	167.349
500	154.473	184.409	128.754	27.827	-1446.434	-1241.336	129.682
600	168.469	213.812	140.519	43.975	-1446.156	-1200.315	104.497
700	182.360	240.827	152.943	61.518	-1444.320	-1159.466	86.520
800	196.062	266.069	165.522	80.437	-1441.078	-1118.979	73.062
848.000	202.732	277.686	171.544	90.008	--------	ALPHA <--> BETA	-------
900	209.702	289.957	178.033	100.732	-1489.633	-1077.896	62.559
1000	223.384	312.760	190.373	122.387	-1482.531	-1032.517	53.933
1100	237.051	334.692	202.502	145.408	-1474.161	-987.910	46.912
1200	250.705	355.903	214.406	169.796	-1464.510	-944.121	41.097

PREVIOUS: CURRENT: December 1978

Lithium Sulfate, Alpha (Li_2SO_4) $Li_2O_4S_1$(cr)

LITHIUM SULFATE (β-Li_2SO_4) CRYSTAL M_r = 109.9396

$S°(298.15\ K) = [113.802]\ J\ K^{-1}\ mol^{-1}$

$T_{trs}(\alpha \to \beta) = 848 \pm 1\ K$

$T_{fus}(\beta \to \ell) = 1132 \pm 1\ K$

$\Delta_f H°(298.15\ K) = [-1425.373]\ kJ\ mol^{-1}$

$\Delta_{trs}H°(\alpha \to \beta) = 28.451 \pm 1.7\ kJ\ mol^{-1}$

$\Delta_{fus}H°(\beta \to \ell) = 8.577 \pm 0.8\ kJ\ mol^{-1}$

Enthalpy of Formation

$\Delta_f H°(Li_2SO_4$, cr, β, 298.15 K) is calculated from $\Delta_f H°(Li_2SO_4$, cr, α, 298.15 K) by adding the enthalpy of transition ($\alpha \to \beta$), $\Delta_{trs}H°$, and the difference in enthalpy, H°(848 K)-H°(298.15 K), between the alpha and beta crystalline forms (1).

Heat Capacity and Entropy

High temperature enthalpy measurements by drop calorimetry were carried out by Denielou et al. (2, 860-1110 K), by Clark (3, 874-1103 K), and by Voskresenskaya and Banashek (4, 859-1133 K). In addition, heat capacities have been measured in an adiabatic calorimeter by Sjoblom (5, 873-1073 K). All of the data indicate that $C_p°$ is nearly constant over the temperature interval 848-1132 K. The data of Clark (3) is systematically low by 15% while the other studies (2, 4, 5) are in substantial agreement. We have adopted enthalpies from Denielou et al. (2) for our curve fits, but have neglected points near T_{trs} and T_{fus} where the data shows pronounced curvature. This leads to a constant heat capacity of 51.0 cal K^{-1} mol^{-1} over the range 909-1119 K. Our adopted enthalpies are 1.5% higher than those from Denielou et al. (2) at 848 K and 0.5% higher at 1132 K. We have extrapolated a constant heat capacity below T_{trs} to 700 K at which point we slowly bend $C_p°$ until it is only 5% higher than $C_p°$ for the α-form at 298.15 K. We also extrapolate a constant heat capacity above T_{fus}.

Transition Data

The high temperature β-modification of lithium sulfate (sometimes denoted Li_2SO_4 I) is of cubic structure, space group T_d^2 (6). The α-β transition at 848 K is somewhat gradual and shows a second order rather than first order behavior. We have treated the transition as a first order transition, taking some of the curvature in the enthalpy data (see above) and lumping it into the heat of transition. Reasonable $C_p°$ curves are obtained by omitting from the curve fits those enthalpy data in the vicinity of T_{trs}. T_{trs} = 848 K is selected as the peak of the transition. $\Delta_{trs}H°$ is calculated as the difference in the enthalpy curve fits at T_{trs}.

Fusion Data

The adopted melting point, T_{fus} = 1132±1 K, was determined from the work of Denielou et al. (2, 1130±3 K), Voskresenskaya and Banashek (4), 1133±1 K), and Rea (7, 1132 K). The heat of fusion, $\Delta_{fus}H°$ = 2.050±0.2 kcal mol^{-1}, is calculated from the difference between the smoothed relative enthalpy of the liquid and the adopted enthalpy value for the β-crystalline form at T_{fus}.

References

1. JANAF Thermochemical Tables: Li_2SO_4(cr, α), 12-31-78.
2. L. Denielou, Y. Fournier, J. P. Petitet, and C. Tequi, Rev. Int. Hautes Temp. 8, 119 (1971).
3. R. P. Clark, J. Chem. Eng. Data 20, 17 (1975).
4. N. K. Voskresenskaya and E. I. Banashek, Izv. Sekt. Fiz. Khim. Anal., Akad. Nauk SSSR 25, 150 (1954).
5. C. A. Sjoblom, High Temp. - High Pressures 8, 499 (1976).
6. T. Forland and J. Krogh-Moe, Acta. Chem. Scand. 11, 565 (1957).
7. R. F. Rea, J. Amer. Ceram. Soc. 21, 98 (1938).

Lithium Sulfate, Beta (Li_2SO_4) $Li_2O_4S_1$(cr)

Enthalpy Reference Temperature = T_r = 298.15 K Standard State Pressure = p° = 0.1 MPa

T/K	$C_p°$ (J K⁻¹mol⁻¹)	S° (J K⁻¹mol⁻¹)	-[G°-H°(T_r)]/T (J K⁻¹mol⁻¹)	H°-H°(T_r) (kJ mol⁻¹)	$\Delta_f H°$ (kJ mol⁻¹)	$\Delta_f G°$ (kJ mol⁻¹)	Log K_f
0							
100							
200							
298.15	129.704	113.802	113.802	0.	-1425.373	-1310.073	229.519
300	132.214	114.612	113.804	0.242	-1425.373	-1309.358	227.979
400	176.565	159.302	119.593	15.883	-1425.472	-1270.820	165.952
500	196.648	201.029	131.788	34.621	-1428.646	-1231.859	128.691
600	207.526	237.933	146.469	54.878	-1424.259	-1192.891	103.850
700	212.547	270.354	161.901	75.917	-1418.927	-1154.743	86.168
800	213.384	298.803	177.274	97.223	-1413.298	-1117.386	72.958
848.000	213.384	311.237	184.509	107.465	-------- ALPHA <--> BETA -------		
900	213.384	323.936	192.201	118.561	-1460.810	-1079.654	62.661
1000	213.384	346.418	206.518	139.900	-1454.024	-1037.668	54.202
1100	213.384	366.756	220.176	161.238	-1447.337	-996.357	47.313
1132.000	213.384	372.875	224.406	168.066	------- BETA <--> LIQUID -------		
1200	213.384	385.323	233.176	182.576	-1440.735	-955.650	41.598
1300	213.384	402.402	245.545	203.915	-1434.210	-915.492	36.785
1400	213.384	418.216	257.321	225.253	-1427.749	-875.833	32.678
1500	213.384	432.938	268.543	246.592	-1421.342	-836.635	29.134
1600	213.384	446.709	279.253	267.930	-1414.985	-797.862	26.048
1700	213.384	459.646	289.488	289.268	-1699.147	-745.133	22.895
1800	213.384	471.842	299.283	310.607	-1691.378	-689.240	20.001
1900	213.384	483.379	308.671	331.945	-1683.671	-633.776	17.424
2000	213.384	494.325	317.683	353.284	-1676.027	-578.716	15.115

PREVIOUS: CURRENT: December 1978

Lithium Sulfate, Beta (Li_2SO_4) $Li_2O_4S_1$(cr)

LITHIUM SULFATE (Li_2SO_4) LIQUID M_r = 109.9396

$S°(298.15 K) = [154.332] J K^{-1} mol^{-1}$

$T_{fus}(\beta \rightarrow \ell) = 1132 \pm 1 K$

$\Delta_f H°(298.15 K) = [-1399.296] kJ mol^{-1}$

$\Delta_f H°(\beta \rightarrow \ell) = 8.577 \pm 0.8 kJ mol^{-1}$

Enthalpy of Formation

$\Delta_f H°(Li_2SO_4, \ell, 298.15 K)$ is calculated from $\Delta_f H°(Li_2SO_4, \beta, 298.15 K)$ by adding the enthalpy of fusion, $\Delta_{fus}H°$, and the difference in enthalpy, H°(1132 K) - H°(298.15 K), between the β crystal and liquid.

Heat Capacity and Entropy

Enthalpies for $Li_2SO_4(\ell)$ have been measured by Clark (2, 1145-1179 K), by Voskresenskaya and Banashek (3, 1133-1270 K), and by Denielou et al. (4, 1138-1530 K). The data of Clark is systematically low by 15% while the other two studies (3, 4) are in substantial agreement. We adopt enthalpies based on the work of Denielou et al. (4). We adopt a constant heat capacity of 49.00 cal K^{-1} mol^{-1} at temperatures above a hypothetical glass transition at 680 K. Below the glass transition, the heat capacity is taken as that of the alpha phase. S°(298.15 K) is calculated in a manner analogous to that used for the enthalpy of formation.

Fusion Data

See the β-crystal table for details.

Vaporization and Decomposition Data

$Li_2SO_4(\ell)$ can vaporize to the gaseous monomer, $Li_2SO_4(g)$, or decompose according to $Li_2SO_4(\ell) = Li_2O(cr) + SO_2(g) + 1/2\ O_2(g)$ or according to $Li_2SO_4(\ell) = 2\ Li(g) + SO_2(g) + O_2(g)$. Ficalora et al. (5) did not observe ions from Li_2O or Li_2SO_4 in their effusion mass-spectrometric experiments, indicating decomposition occurs to the metal. They determined $\Delta_{dcm}H°(1080 K) = 324\pm12$ kcal mol^{-1}. This is in substantial agreement with JANAF thermodynamic data (1) which indicates $\Delta_{dcm}H°(1080 K) = 332.3$ kcal mol^{-1}. Effusion and transpiration results of Jagannathan and Wyatt (6) also indicate that decomposition to the oxide is less important then decomposition to the metal. T_{dcm} = ~2466 K is the temperature at which the Gibbs energy change is zero for the reaction $Li_2SO_4(\ell) = Li_2O(\ell) + SO_2(g) + 1/2\ O_2(g)$ and T_{dcm} = 2800 K for the reaction $Li_2SO_4(\ell) = 2\ Li(g) + SO_2(g) + O_2(g)$.

References

1. JANAF Thermochemical Tables: Li_2SO_4(cr, β), 12-31-78.
2. R. P. Clark, J. Chem. Eng. Data 20, 17 (1975).
3. N. K. Voskresenskaya and E. I. Banashek, Izv. Sekt. Fiz. Khim. Anal., Akad. Nauk SSSR. 25, 150 (1974).
4. L. Denielou, Y. Fournier, J. P. Petitet, and C. Tequi, Rev. Int. Hautes Temp. 8, 119 (1971).
5. P. J. Ficalora, O. M. Uy, D. W. Muenow, and J. L. Margrave, J. Amer. Ceram. Soc. 51, 574 (1968).
6. G. V. Jagannathan and P. A. H. Wyatt, J. Chem. Res. (M) 3744 (1978).

Lithium Sulfate (Li_2SO_4) $Li_2O_4S_1(l)$

Enthalpy Reference Temperature = T_r = 298.15 K; Standard State Pressure = p° = 0.1 MPa

T/K	$C_p°$	S°	$-[G°-H°(T_r)]/T$	$H°-H°(T_r)$	$\Delta_f H°$	$\Delta_f G°$	Log K_f
	J K⁻¹mol⁻¹			kJ mol⁻¹			
0							
100							
200							
298.15	117.608	154.332	154.332	0.	-1399.296	-1296.080	227.068
300	118.026	155.061	154.334	0.218	-1399.320	-1295.439	225.556
400	139.244	192.064	159.235	13.131	-1402.147	-1260.599	164.617
500	154.473	224.794	169.139	27.827	-1409.362	-1224.457	127.918
600	168.469	254.196	180.904	43.975	-1409.085	-1187.475	103.379
680.000	179.619	275.966	190.820	57.899	GLASS <--> LIQUID TRANSITION		
680.000	205.016	275.966	190.820	57.899			
700	205.016	281.909	193.338	61.999	-1406.767	-1150.671	85.864
800	205.016	309.285	206.159	82.501	-1401.943	-1114.416	72.764
900	205.016	333.432	218.985	103.002	-1450.291	-1077.682	62.547
1000	205.016	355.033	231.529	123.504	-1444.342	-1036.601	54.146
1100	205.016	374.573	243.659	144.006	-1438.492	-996.110	47.301
1132.000	205.016	380.452	247.443	150.566	------- BETA <--> LIQUID -------		
1200	205.016	392.412	255.322	164.507	-1432.727	-956.149	41.620
1300	205.016	408.822	266.507	185.009	-1427.039	-916.665	36.832
1400	205.016	424.015	277.222	205.510	-1421.414	-877.617	32.744
1500	205.016	438.160	287.485	226.012	-1415.844	-838.970	29.216
1600	205.016	451.391	297.320	246.514	-1410.324	-800.692	26.140
1700	205.016	463.820	306.752	267.015	-1695.323	-748.405	22.996
1800	205.016	475.538	315.807	287.517	-1688.391	-692.905	20.108
1900	205.016	486.623	324.508	308.018	-1681.521	-637.788	17.534
2000	205.016	497.139	332.879	328.520	-1674.713	-583.031	15.227
2100	205.016	507.142	340.941	349.022	-1667.968	-528.613	13.149
2200	205.016	516.679	348.714	369.523	-1661.285	-474.513	11.266
2300	205.016	525.792	356.216	390.025	-1654.667	-420.718	9.555
2400	205.016	534.518	363.465	410.526	-1648.115	-367.210	7.992
2500	205.016	542.887	370.476	431.028	-1641.630	-313.972	6.560
2600	205.016	550.928	377.263	451.530	-1635.212	-260.993	5.243
2700	205.016	558.665	383.839	472.031	-1628.865	-208.258	4.029
2800	205.016	566.121	390.217	492.533	-1622.589	-155.758	2.906
2900	205.016	573.315	396.407	513.034	-1616.386	-103.482	1.864
3000	205.016	580.266	402.420	533.536	-1610.259	-51.417	0.895

PREVIOUS: CURRENT: December 1978

Lithium Sulfate (Li_2SO_4) $Li_2O_4S_1(l)$

LITHIUM SULFATE (Li_2SO_4) ALPHA-BETA-LIQUID M_r = 109.9396

0 to 848 K	crystal, alpha	
848 to 1132 K	crystal, beta	
above 1132 K	liquid	

Refer to the individual tables for details.

Lithium Sulfate (Li_2SO_4) $Li_2O_4S_1$(cr,l)

Enthalpy Reference Temperature = T_r = 298.15 K Standard State Pressure = p° = 0.1 MPa

	J K⁻¹mol⁻¹			kJ mol⁻¹			
T/K	C_p°	S°	−[G°−H°(T_r)]/T	H°−H°(T_r)	$\Delta_f H^\circ$	$\Delta_f G^\circ$	Log K_f
0	0.	0.	INFINITE	-18.636	-1423.980	-1423.980	INFINITE
100	43.869	28.932	196.627	-16.769	-1429.603	-1395.114	728.733
200	86.872	73.028	123.821	-10.159	-1434.139	-1358.693	354.854
298.15	117.608	113.947	113.947	0.	-1436.367	-1321.111	231.453
300	118.026	114.676	113.949	0.218	-1436.391	-1320.395	229.901
400	139.244	151.679	118.850	13.131	-1439.218	-1281.517	167.349
500	154.473	184.409	128.754	27.827	-1446.434	-1241.336	129.682
600	168.469	213.812	140.519	43.975	-1446.156	-1200.315	104.497
700	182.360	240.827	152.943	61.518	-1444.320	-1159.466	86.520
800	196.062	266.069	165.522	80.437	-1441.078	-1118.979	73.062
848.000	202.732	277.686	171.544	90.008	ALPHA <--> BETA		
848.000	213.384	311.237	171.544	118.459	TRANSITION		
900	213.384	323.936	179.985	129.555	-1460.810	-1079.654	62.661
1000	213.384	346.418	195.524	150.894	-1454.024	-1037.668	54.202
1100	213.384	366.756	210.181	172.232	-1447.337	-996.357	47.313
1132.000	213.384	372.875	214.694	179.060	BETA <--> LIQUID		
1132.000	205.016	380.452	214.694	187.638	TRANSITION		
1200	205.016	392.412	224.429	201.579	-1432.727	-956.149	41.620
1300	205.016	408.822	237.991	222.080	-1427.039	-916.665	36.832
1400	205.016	424.015	250.742	242.582	-1421.414	-877.617	32.744
1500	205.016	438.159	262.770	263.084	-1415.844	-838.970	29.216
1600	205.016	451.391	274.150	283.585	-1410.324	-800.691	26.140
1700	205.016	463.820	284.945	304.087	-1695.323	-748.405	22.996
1800	205.016	475.538	295.211	324.588	-1688.391	-692.905	20.108
1900	205.016	486.623	304.997	345.090	-1681.521	-637.788	17.534
2000	205.016	497.139	314.343	365.592	-1674.713	-583.031	15.227
2100	205.016	507.142	323.288	386.093	-1667.968	-528.613	13.149
2200	205.016	516.679	331.863	406.595	-1661.285	-474.513	11.266
2300	205.016	525.792	340.098	427.096	-1654.667	-420.718	9.555
2400	205.016	534.518	348.019	447.598	-1648.115	-367.210	7.992
2500	205.016	542.887	355.647	468.100	-1641.630	-313.972	6.560
2600	205.016	550.928	363.004	488.601	-1635.212	-260.993	5.243
2700	205.016	558.665	370.109	509.103	-1628.865	-208.258	4.029
2800	205.016	566.121	376.977	529.604	-1622.589	-155.758	2.906
2900	205.016	573.315	383.624	550.106	-1616.386	-103.482	1.864
3000	205.016	580.266	390.063	570.608	-1610.259	-51.417	0.895

PREVIOUS: CURRENT: December 1978

Lithium Sulfate (Li_2SO_4) $Li_2O_4S_1$(cr,l)

LITHIUM SULFATE (Li_2SO_4) IDEAL GAS M_r = 109.9396

$\Delta_fH°(0\ K)$ = [-1030.445 ± 83.7] kJ mol^{-1}

S°(298.15 K) = [322.820 ± 41.8] J K^{-1} mol^{-1} $\Delta_fH°(298.15\ K)$ = [-1041.816 ± 83.7] kJ mol^{-1}

Vibrational Frequencies, Symmetries, and Degeneracies

	ν, cm^{-1}		ν, cm^{-1}		ν, cm^{-1}
A_1	[1020](1)	B_1	[460](1)	E	[1115](2)
	[500](1)	B_2	[1150](1)		[61[(2)
	[380](1)		[660](1)		[410](2)
			[420](1)		[100](2)

Ground State Quantum Weight: [1] σ = [4]

Point Group: [D_{2d}]

Bond Distances: S-O = [1.48] Å Li-O = [1.90] Å

Bond Angles: O-S-O = [109.47°] O-Li-O = [79.0°]

Product of the Moments of Inertia: $I_AI_BI_C$ = [1.210866 x 10^{-113}] g^3 cm^6

Enthalpy of Formation

Vaporization experiments on Li_2SO_4 have been carried out by Ficalora et al. (1), by Vasilev and Dmitrievskaya (2, 973-1223 K), and by Jagannathan and Wyatt (3, 1227-1371 K). Ficalora et al. (1) did not observe the $Li_2SO_4^+$ ion nor ions from Li_2O in their Knudsen effusion mass-spectrometric experiments. Decomposition to Li(g) was the dominant process observed. Vasilev and Dmitrievskaya (2) carried out Langmuir method experiments on solid and liquid lithium sulfate. Their experimental pressures are less than JANAF calculated dissociation pressures from the two reactions Li_2SO_4(ℓ) = Li_2O(cr) + SO_2(g) + 1/2 O_2(g) and Li_2SO_4(ℓ) = 2 Li(g) + SO_2(g) + O_2(g). We have discounted their results.

Jagannathan and Wyatt (3) carried out Knudsen effusion weight-loss experiments and transpiration experiments (in N_2) on molten lithium sulfate. Their transpiration experiments identified high lithium concentrations in the vapor, so that Li_2O(cr) formation cannot be the major reaction. The effusion experiments yield apparent Li_2SO_4(g) pressures of the same order of magnitude as our calculated decomposition pressures (to Li_2O(cr). The data from (3) leads to $\Delta_rH°(1300\ K)$ = 58±20 kcal mol^{-1} for molecular vaporization. When this result is combined with auxiliary JANAF data (7), we obtain $\Delta_fH°(Li_2SO_4$, g, 298.15 K) = -266±20 kcal mol^{-1}.

From trends in the enthalpies and entropies of sublimation for the other alkali sulfates (4), we would estimate $\Delta_rH°(1200\ K)$ = 75±15 kcal mol^{-1} and $\Delta_rS°(1200\ K)$ = 35±5 cal K^{-1} mol^{-1} for Li_2SO_4(ℓ) = Li_2SO_4(g). This estimate leads to $\Delta_fH°(Li_2SO_4$, g, 298.15 K) = -249±28 kcal mol^{-1}. We adopt this estimate for this tabulation.

Heat Capacity and Entropy

The adopted structure (D_{2d} symmetry) is based on the similar symmetry assigned to Cs_2SO_4(g) and K_2SO_4(g) in the high-temperature electron diffraction studies of Ugarov et al. (5) and Spiridonov and Lutoshkin (6). The internuclear distance of 1.48 Å for S-O is taken from data on other sulfates (7) while Li-O = 1.90 Å is an estimate based on trends among lithium and sodium compounds. The Li-O distance of 1.96 Å found by Alcock et al. (8) seems a bit long. The principal moments of inertia are: I_A = 15.5185 x 10^{-39} g cm^2 and $I_B = I_C$ = 27.9333 x 10^{-39} g cm^2.

Estimates of the eleven fundamental frequencies of Li_2SO_4 are based on comparisons with Na_2SO_4(g) and K_2SO_4(g) (7), data on crystalline Li_2SO_4 (9), and on a normal coordinate analysis of Devlin (10). Uncertainties in the estimates of the molecular constants and gas-phase frequencies may contribute 10 cal K^{-1} mol^{-1} to S°(298.15 K).

References

1. P. J. Ficalora, O. M. Uy, D. W. Muenow, and J. L. Margrave, J. Amer. Ceram. Soc. 51, 574 (1968).
2. V. G. Vasilev and E. V. Dmitrievskaya, Russ. J. Phys. Chem. 48, 1696 (1974).
3. G. V. Jagannathan and P. A. H. Wyatt, J. Chem. Res. (M) 2744 (1978).
4. K. H. Lau, D. Cubicciotti, and D. L. Hildenbrand, J. Electrochem. Soc. 126, 490 (1979).
5. V. V. Ugarov, Yu. S. Ezhov, and N. G. Rambidi, J. Molec. Struct. 25, 257 (1975).
6. V. P. Spiridonov and B. I. Lutoshkin, Vestn. Mosk. Univ. Khim. 11, 509 (1970).
7. JANAF Thermochemical Tables: K_2SO_4(g), Na_2SO_4(g), 6-30-78; Li_2SO_4(ℓ), 12-31-78.
8. N. W. Alcock, D. A. Evans, and H. D. B. Jenkins, Acta. Cryst. B29, 360 (1973).
9. H. Takahashi, S. Meshitsuka, and K. Higasi, Spectrochim. Acta. 31A, 1617 (1975).
10. J. P. Devlin, Oklahoma State University, personal communication, July, 1979.

Lithium Sulfate (Li_2SO_4) $Li_2O_4S_1$(g)

Enthalpy Reference Temperature = T_r = 298.15 K — Standard State Pressure = p° = 0.1 MPa

T/K	$C_p°$	S°	-[G°-H°(T_r)]/T	H°-H°(T_r)	$\Delta_fH°$	$\Delta_fG°$	Log K_f
	J K^{-1}mol^{-1}			kJ mol^{-1}			
0	0.	0.	INFINITE	-19.652	-1030.445	-1030.445	INFINITE
100	51.737	242.404	397.499	-15.509	-1033.791	-1020.650	533.133
200	79.551	286.652	331.470	-8.963	-1038.393	-1005.671	262.654
250	91.859	305.764	324.443	-4.670	-1040.290	-997.264	208.367
298.15	101.804	322.820	322.820	0.	-1041.816	-988.835	173.240
300	102.151	323.451	322.822	0.189	-1041.869	-988.506	172.114
350	110.669	339.858	324.098	5.516	-1043.187	-979.505	146.183
400	117.688	355.110	327.033	11.231	-1046.568	-970.239	126.700
450	123.463	369.316	330.951	17.264	-1048.340	-960.600	111.504
500	128.218	382.578	335.458	23.560	-1056.150	-950.137	99.260
600	135.407	406.631	345.359	36.763	-1058.817	-928.668	80.848
700	140.417	427.903	355.661	50.569	-1060.718	-906.818	67.668
800	143.997	446.899	365.900	64.799	-1062.165	-884.730	57.767
900	146.621	464.019	375.867	79.337	-1116.477	-861.396	49.994
1000	148.589	479.574	385.472	94.102	-1116.264	-833.064	43.515
1100	150.099	493.810	394.683	109.040	-1115.979	-804.757	38.215
1200	151.278	506.922	403.497	124.111	-1115.644	-776.479	33.799
1300	152.216	519.070	411.926	139.287	-1115.281	-748.230	30.064
1400	152.972	530.379	419.987	154.548	-1114.897	-720.009	26.864
1500	153.591	540.954	427.703	169.877	-1114.499	-691.817	24.091
1600	154.102	550.884	435.095	185.262	-1114.095	-663.651	21.666
1700	154.530	560.239	442.184	200.695	-1404.164	-621.159	19.086
1800	154.891	569.083	448.990	216.166	-1402.261	-575.156	16.691
1900	155.199	577.466	455.533	231.671	-1400.388	-529.256	14.550
2000	155.463	585.433	461.831	247 205	-1398.549	-483.455	12.627
2100	155.692	593.024	467.899	262.763	-1396.747	-437.744	10.888
2200	155.891	600.271	473.752	278.342	-1394.987	-392.118	9.310
2300	156.065	607.205	479.405	293.940	-1393.273	-346.572	7.871
2400	156.218	613.850	484.869	309.554	-1391.607	-301.100	6.553
2500	156.354	620.230	490.157	325.183	-1389.995	-255.696	5.342
2600	156.474	626.365	495.279	340.825	-1388.438	-210.355	4.226
2700	156.582	632.272	500.244	356.477	-1386.939	-165.072	3.194
2800	156.678	637.969	505.061	372.141	-1385.501	-119.843	2.236
2900	156.765	643.468	509.740	387.813	-1384.128	-74.667	1.345
3000	156.843	648.784	514.286	403.493	-1382.822	-29.535	0.514
3100	156.915	653.928	518.708	419.181	-1381.586	15.555	-0.262
3200	156.979	658.911	523.012	434.876	-1380.422	60.604	-0.989
3300	157.038	663.742	527.204	450.577	-1379.331	105.618	-1.672
3400	157.092	668.431	531.289	466.283	-1378.317	150.602	-2.314
3500	157.141	672.986	535.273	481.995	-1377.382	195.556	-2.919
3600	157.186	677.413	539.160	497.711	-1376.528	240.485	-3.489
3700	157.228	681.721	542.955	513.432	-1375.757	285.390	-4.029
3800	157.266	685.914	546.662	529.157	-1375.072	330.279	-4.540
3900	157.302	690.000	550.285	544.885	-1374.475	375.149	-5.025
4000	157.335	693.983	553.828	560.617	-1373.968	420.003	-5.485
4100	157.366	697.868	557.294	576.352	-1373.554	464.847	-5.922
4200	157.394	701.660	560.687	592.090	-1373.236	509.682	-6.339
4300	157.421	705.364	564.008	607.831	-1373.017	554.512	-6.736
4400	157.445	708.984	567.262	623.574	-1372.858	599.332	-7.115
4500	157.468	712.522	570.451	639.320	-1372.826	644.153	-7.477
4600	157.490	715.983	573.577	655.068	-1372.904	688.974	-7.824
4700	157.510	719.371	576.643	670.818	-1373.092	733.801	-8.155
4800	157.529	722.687	579.651	686.570	-1373.391	778.631	-8.473
4900	157.547	725.935	582.604	702.324	-1373.808	823.470	-8.778
5000	157.564	729.118	585.502	718.079	-1374.344	868.318	-9.071
5100	157.580	732.239	588.349	733.837	-1374.993	913.177	-9.353
5200	157.595	735.299	591.146	749.595	-1375.781	958.048	-9.624
5300	157.609	738.301	593.894	765.356	-1376.701	1002.940	-9.885
5400	157.622	741.247	596.596	781.117	-1377.761	1047.848	-10.136
5500	157.635	744.139	599.252	796.880	-1378.962	1092.776	-10.378
5600	157.647	746.980	601.865	812.644	-1380.312	1137.731	-10.612
5700	157.658	749.770	604.435	828.409	-1381.815	1182.708	-10.838
5800	157.669	752.512	606.965	844.176	-1383.476	1227.715	-11.057
5900	157.679	755.207	609.454	859.943	-1385.299	1272.748	-11.268
6000	157.689	757.858	611.906	875.712	-1387.292	1317.818	-11.473

PREVIOUS: December 1978 (1 atm) CURRENT: December 1978 (1 bar)

Lithium Sulfate (Li_2SO_4) $Li_2O_4S_1$(g)

LITHIUM SILICATE ($Li_2Si_2O_5$) CRYSTAL(I-II) M_r = 150.0500

$S°$(298.15 K) = [125.520] J K^{-1} mol^{-1}
T_{trs} = 1209 K
T_{fus} = 1307 ± 1 K

$\Delta_f H°$(0 K) = Unknown
$\Delta_f H°$(298.15 K) = -2560.90 ± 4.2 kJ mol^{-1}
$\Delta_{trs} H°$ = 0.941 kJ mol^{-1}
$\Delta_{fus} H°$ = [53.806] kJ mol^{-1}

Enthalpy of Formation

The enthalpies of solution of LiCl(cr), Li_2SO_4(cr), SiO_2(quartz) and $Li_2Si_2O_5$(cr) in 20 per cent hydrofluoric acid at 74.7°C were measured by Kracek (1). The authors derived two values of $\Delta_{soln}H$ of Li_2O(cr) from those of LiCl(cr) and Li_2SO_4(cr), presumably employing auxiliary data from (2). We recalculate $\Delta_{soln}H$ of Li_2O(cr) as -58.74 (chloride scheme, originally -59.36) and -59.83 kcal mol^{-1} (sulfate scheme, originally -60.86), using recent $\Delta_f H°$(298.15 K) values for LiCl(cr), Li_2SO_4(cr), HCl(aq), H_2SO_4(aq), and Li_2O(cr). The corresponding enthalpy changes, $\Delta_r H°$(298.15 K), for the reaction Li_2O(cr) + $2SiO_2$(quartz) = $Li_2Si_2O_5$(cr) are derived as -33.21 and -34.30 kcal mol^{-1}. Adopting the weighted average $\Delta_r H°$(298.15 K) = -33.57 kcal mol^{-1} with $\Delta_f H°$(Li_2O, cr, 298.15 K) = -143.1 kcal mol^{-1} and $\Delta_f H°$(SiO_2, quartz, 298.15 K) = -217.7 kcal mol^{-1}, we obtain $\Delta_f H°$(298.15 K) = -612.07 kcal mol^{-1} for $Li_2Si_2O_5$(cr). This value is independent of future changes in $\Delta_f H°$(298.15 K) of Li_2O(cr) but the values of $\Delta_{soln}H$ and $\Delta_r H°$ are not independent.

The enthalpies of solution of LiOH(cr) and $Li_2Si_2O_6$(cr) in 20 per cent HF(aq) at 50°C were measured by Hatton et al. (3). By use of a reaction scheme similar to that described on the Li_2SiO_3(cr) table, the enthalpy change at 298.15 K for the reaction 2LiOH(cr) + $2SiO_2$(cr) + $Li_2Si_2O_5$(cr) + H_2O(ℓ) was calculated to be -13.44 kcal mol^{-1}. Incorporating this value with $\Delta_f H°$ (298.15 K) = -115.84, -217.7 and -68.315 kcal mol^{-1} for LiOH(cr), SiO_2(cr), and H_2O(ℓ), respectively, we obtain $\Delta_f H°$($Li_2Si_2O_5$, cr, 298.15 K) = -612.20±2.0 kcal mol^{-1}, which is in excellent agreement with the adopted value.

Heat Capacity and Entropy

Both heat capacities and $S°$(298.15 K) are estimated by comparison with those for $Na_2Si_2O_5$(cr), Na_2O(cr) and Li_2O(cr).

Transition Data

The transition temperature, T_{trs} = 1209 K, is taken from Kracek (4). The author estimated a enthalpy of transition of 1 to 2 cal K^{-1} mol^{-1} based on the arrest in the cooling and heating curves. The value of $\Delta_{trs}H°$ is calculated assuming the enthalpy of transition at 1209 K is 1.5 cal K^{-1} mol^{-1}.

Fusion Data

Incongruent melting of $Li_2Si_2O_5$(cr) at 1033° has been reported by Kracek (5) and Kracek (4). The liquidus temperature of 1034 at the composition 19.92 per cent Li_2O and 80.08 per cent SiO_2, reported by Kracek (4), is adopted as T_{fus}. The enthalpy of melting is evaluated such that $\Delta_{fus}H°$(298.15 K) + $\Delta_{trs}H°$(298.15 K) = 11.43 kcal mol^{-1}, where the value 11.43 kcal mol^{-1} is the difference between $\Delta_f H°$(298.15 K) for $Li_2Si_2O_5$(cr) and $Li_2Si_2O_5$(ℓ). The latter were determined by solution calorimetry by Kracek (see Enthalpy of Formation sections in crystal and liquid tables).

References

1. R. C. Kracek, Ann. Rept. Director of the Geophysical Laboratory, No. 1215, 69 (1953).
2. U. S. Nat. Bur. Stand. Circ. 500, 1952.
3. W. E. Hatton, G. C. Sinke, D. L. Hildenbrand, and D. R. Stull, The Dow Chemical Co., Midland, Michigan, 1959.
4. F. C. Kracek, J. Amer. Chem. Soc. 61, 2863 (1939).
5. F. C. Kracek, J. Phys. Chem. 34, 2641 (1930).

Lithium Silicate ($Li_2Si_2O_5$) $Li_2O_5Si_2$(cr)

Enthalpy Reference Temperature = T_r = 298.15 K Standard State Pressure = p° = 0.1 MPa

T/K	$C_p°$	S°	-[G°-H°(T_r)]/T	H°-H°(T_r)	$\Delta_f H°$	$\Delta_f G°$	Log K_f
	J $K^{-1}mol^{-1}$			kJ mol^{-1}			
0							
100							
200							
298.15	138.072	125.520	125.520	0.	-2560.901	-2416.848	423.422
300	138.909	126.377	125.523	0.256	-2560.946	-2415.954	420.655
400	174.891	171.701	131.457	16.098	-2561.976	-2367.401	309.151
500	192.673	212.726	143.694	34.516	-2567.627	-2318.171	242.178
600	205.685	249.062	158.287	54.466	-2566 295	-2268.388	197 481
700	214.890	281.479	173.613	75.507	-2564.136	-2218.898	165.576
800	222.589	310.685	188.951	97.387	-2561.408	-2169.759	141.671
900	229.492	337.312	203.978	120.000	-2558.215	-2120 990	123.099
1000	235.350	361.800	218.552	143.248	-2554.617	-2072.599	108.261
1100	240.617	384.483	232.618	167.052	-2550.660	-2024 585	96.140
1200	245.182	405.621	246.164	191.347	-2546.385	-1976.947	86.054
1209.000	245.559	407.454	247.358	193.556		I <--> II	
1209.000	245.559	408.233	247.358	194.497		TRANSITION	
1300	248.195	426.145	259.253	216.960	-2540.956	-1929.746	77.538
1307.000	248.417	427.478	260.150	218.698	--------	II <--> LIQUID	--------
1400	251.207	444.648	271.841	241.929	-2536.318	-1882.902	70.252
1500	254.868	462.109	283.949	267.239	-2531.474	-1836.398	63.949
1600	257.734	478.653	295.606	292.876	-2526.436	-1790.223	58.445
1700	259.617	494.337	306.839	318.747	-2912.119	-1729.121	53.129
1800	261.082	509.220	317.672	344.786	-2904.970	-1659.740	48.164
1900	262.075	523.364	328.129	370.946	-2897.761	-1590.756	43.733
2000	262.755	536.825	338.230	397.190	-2890.530	-1522.153	39.755
2100	262.978	549.650	347.995	423.477	-2883.319	-1453.912	36.164
2200	263.174	561.889	357.441	449.785	-2876.151	-1386.012	32.908
2300	263.341	573.591	366.586	476.111	-2869.031	-1318.439	29.943
2400	263.480	584.802	375.447	502.452	-2861.964	-1251.174	27.231
2500	263.592	595.560	384.038	528.806	-2854.954	-1184.202	24.743

PREVIOUS: CURRENT: June 1967

Lithium Silicate ($Li_2Si_2O_5$) $Li_2O_5Si_2$(cr)

LITHIUM SILICATE ($Li_2Si_2O_5$) LIQUID M_r = 150.0500

S°(298.15 K) = [160.243] J K^{-1} mol^{-1} $\Delta_fH°$(298.15 K) = [-2513.068] kJ mol^{-1}

T_{fus} = 1307 K $\Delta_{fus}H°$ = [53.806] kJ mol^{-1}

Enthalpy of Formation

The enthalpies of solution of LiCl(cr), Li_2SO_4(cr), SiO_2(quartz) and $Li_2Si_2O_5$(gl) in 20 percent hydrofluoric acid at 74.7°C were measured by Kracek (1). Following the same procedure as described in $Li_2Si_2O_5$(cr) table, we derived the enthalpy changes, $\Delta_rH°$(298.15 K), for the reaction Li_2O(cr) + $2SiO_2$(quartz) = $Li_2Si_2O_5$(gl) as -21.78 and -22.87 kcal mol^{-1}. Adopting the weighted average $\Delta_rH°$(298.15 K) = -22.14 kcal mol^{-1} with $\Delta_fH°$(Li_2O, cr, 298.15 K) = -143.1 kcal mol^{-1} and $\Delta_fH°$(SiO_2, quartz, 298.15 K) = -217.7 kcal mol^{-1}, we obtain $\Delta_fH°$(298.15 K) = -600.640 kcal mol^{-1} for $Li_2Si_2O_5$(gl). This $\Delta_fH°$($Li_2Si_2O_5$, gl, 298.15 K) value is independent of future changes in $\Delta_fH°$(298.15 K) of Li_2O(cr) but the values of $\Delta_{sol}H°$ and $\Delta_rH°$ are not independent.

Heat Capacity and Entropy

The heat capacity is estimated by comparison with those for $Na_2Si_2O_5$(ℓ), Na_2SiO_3(ℓ) and Li_2SiO_3(ℓ). A glass transition temperature at 800 K is assumed, i.e. the heat capacities below 800 K are taken from those for $Li_2Si_2O_5$(cr) and above 800 K are assumed to be constant.

S°($Li_2Si_2O_5$, ℓ, 298.15 K) is calculated from S°($Li_2Si_2O_5$, cr, 298.15 K) by adding the entropy of fusion, $\Delta_{fus}S°$(1307 K), and the difference in entropy, S°(1307 K)-S°(298.15 K), between the crystal and liquid.

Fusion Data

See the crystal table for details.

Reference

1. F. C. Kracek, Ann. Rept. Director of the Geophysical Laboratory, No. 1215, 69 (1953).

Lithium Silicate ($Li_2Si_2O_5$) $Li_2O_5Si_2$(l)

Enthalpy Reference Temperature = T_r = 298.15 K Standard State Pressure = p° = 0.1 MPa

T/K	$C_p°$	S°	$-[G°-H°(T_r)]/T$	$H°-H°(T_r)$	$\Delta_fH°$	$\Delta_fG°$	Log K_f
	J $K^{-1}mol^{-1}$			kJ mol^{-1}			
0							
100							
200							
298.15	138.072	160.243	160.243	0.	-2513.068	-2379.367	416.855
300	138.909	161.099	160.245	0.256	-2513.113	-2378.537	414.140
400	174.891	206.424	166.179	16.098	-2514.143	-2333.457	304.718
500	192.673	247.448	178.417	34.516	-2519.794	-2287.699	238.994
600	205.685	283.785	193.009	54.466	-2518.462	-2241.388	195.130
700	214.890	316.202	208.335	75.507	-2516.303	-2195.371	163.821
800	222.589	345.416	223.674	97.393	-2513.569	-2149.704	140.361
800.000	222.589	345.416	223.674	97.393	GLASS <--> LIQUID TRANSITION		
800.000	251.040	345.416	223.674	97.393			
900	251.040	374.984	238.876	122.497	-2507.885	-2104.565	122.146
1000	251.040	401.434	253.833	147.601	-2502.430	-2060.046	107.606
1100	251.040	425.361	268.356	172.705	-2497.173	-2016.063	95.735
1200	251.040	447.204	282.363	197.809	-2492.090	-1972.552	85.863
1300	251.040	467.298	295.826	222.913	-2487.169	-1929.458	77.527
1307.000	251.040	468.646	296.748	224.671	-------- II <--> LIQUID --------		
1400	251.040	485.902	308.747	248.017	-2482.396	-1886.737	70.395
1500	251.040	503.222	321.141	273.121	-2477.759	-1844.353	64.226
1600	251.040	519.424	333.033	298.225	-2473.253	-1802.273	58.838
1700	251.040	534.643	344.449	323.329	-2859.704	-1745.226	53.624
1800	251.040	548.992	355.418	348.433	-2853.489	-1679.849	48.748
1900	251.040	562.565	365.966	373.537	-2847.336	-1614.814	44.394
2000	251.040	575.442	376.121	398.641	-2841.245	-1550.102	40.485
2100	251.040	587.690	385.906	423.745	-2835.217	-1485.694	36.955
2200	251.040	599.368	395.346	448.849	-2829.253	-1421.569	33.752
2300	251.040	610.527	404.461	473.953	-2823.356	-1357.717	30.835
2400	251.040	621.212	413.271	499.057	-2817.526	-1294.119	28.166
2500	251.040	631.460	421.795	524.161	-2811.766	-1230.763	25.715
2600	251.040	641.306	430.050	549.265	-2806.078	-1167.635	23.458
2700	251.040	650.780	438.050	574.369	-2800.463	-1104.724	21.372
2800	251.040	659.910	445.812	599.473	-2794.925	-1042.020	19.439
2900	251.040	668.719	453.347	624.577	-2789.465	-979.515	17.643
3000	251.040	677.230	460.669	649.681	-2784.085	-917.194	15.970

PREVIOUS: CURRENT: June 1967

Lithium Silicate ($Li_2Si_2O_5$) $Li_2O_5Si_2$(l)

LITHIUM SILICATE ($Li_2Si_2O_5$)

CRYSTAL–CRYSTAL–LIQUID

0 to 1209 K crystal, I
1209 to 1307 K crystal, II
above 1307 K liquid

Refer to the individual tables for details.

M_r = 150.0500

Lithium Silicate ($Li_2Si_2O_5$) $Li_2O_5Si_2$(cr,l)

Enthalpy Reference Temperature = T_r = 298.15 K   Standard State Pressure = p° = 0.1 MPa

	J K⁻¹mol⁻¹			kJ mol⁻¹			
T/K	C_p°	S°	$-[G^\circ - H^\circ(T_r)]/T$	$H^\circ - H^\circ(T_r)$	$\Delta_f H^\circ$	$\Delta_f G^\circ$	Log K_f
0							
100							
200							
298.15	138.072	125.520	125.520	0.	-2560.901	-2416.848	423.422
300	138.909	126.377	125.523	0.256	-2560.946	-2415.954	420.655
400	174.891	171.701	131.457	16.098	-2561.976	-2367.401	309.151
500	192.673	212.726	143.694	34.516	-2567.627	-2318.171	242.178
600	205.685	249.062	158.287	54.466	-2566.295	-2268.388	197.481
700	214.890	281.479	173.613	75.507	-2564.136	-2218.898	165.576
800	222.589	310.685	188.951	97.387	-2561.408	-2169.759	141.671
900	229.492	337.312	203.978	120.000	-2558.215	-2120.990	123.099
1000	235.350	361.800	218.552	143.248	-2554.617	-2072.599	108.261
1100	240.617	384.483	232.618	167.052	-2550.660	-2024.585	96.140
1200	245.182	405.621	246.164	191.347	-2546.385	-1976.947	86.054
1209.000	245.559	407.454	247.358	193.556		I <--> II	
1209.000	245.559	408.233	247.358	194.497		TRANSITION	
1300	248.195	426.145	259.253	216.960	-2540.956	-1929.746	77.538
1307.000	248.417	427.478	260.150	218.698		II <--> LIQUID	
1307.000	251.040	468.646	260.150	272.504		TRANSITION	
1400	251.040	485.902	274.580	295.851	-2482.396	-1886.737	70.395
1500	251.040	503.222	289.252	320.955	-2477.759	-1844.353	64.226
1600	251.040	519.424	303.137	346.059	-2473.253	-1802.273	58.838
1700	251.040	534.643	316.312	371.163	-2859.704	-1745.226	53.624
1800	251.040	548.992	328.844	396.267	-2853.489	-1679.849	48.748
1900	251.040	562.565	340.791	421.371	-2847.336	-1614.814	44.394
2000	251.040	575.442	352.204	446.475	-2841.245	-1550.102	40.485
2100	251.040	587.690	363.129	471.579	-2835 217	-1485.694	36.955
2200	251.040	599.368	373.603	496.683	-2829.253	-1421.569	33.752
2300	251.040	610.527	383.664	521.787	-2823.356	-1357.717	30.835
2400	251.040	621.212	393.341	546.891	-2817.526	-1294.119	28.166
2500	251.040	631.460	402.662	571.995	-2811.766	-1230.763	25.715
2600	251.040	641.306	411.652	597.099	-2806.078	-1167.635	23.458
2700	251.040	650.780	420.334	622.203	-2800.463	-1104.724	21.372
2800	251.040	659.910	428.729	647.307	-2794.925	-1042.020	19.439
2900	251.040	668.719	436.853	672.411	-2789.465	-979.515	17.643
3000	251.040	677.230	444.725	697.515	-2784.085	-917.194	15.970

PREVIOUS: CURRENT. June 1967

Lithium Silicate ($Li_2Si_2O_5$) $Li_2O_5Si_2$(cr,l)

LITHIUM NITRIDE (Li_3N) CRYSTAL M_r = 34.8297

$\Delta_fH°(0\ K) = -157.57 \pm 1.3\ kJ\ mol^{-1}$

$S°(298.15\ K) = 62.593 \pm 0.13\ J\ K^{-1}\ mol^{-1}$ $\Delta_fH°(298.15\ K) = -164.56 \pm 1.1\ kJ\ mol^{-1}$

T_{fus} = 1086 K

Enthalpy of Formation

The adopted enthalpy of formation for Li_3N(cr) is derived from the solution calorimetric study of O'Hare and Johnson (1). The reported values have been revised by Osborne and Flotow (2) to correct for the current atomic weight of lithium to give $\Delta_fH°(298.15\ K) = -39.33\pm0.26\ kcal\ mol^{-1}$. This uncertainty is derived from the experimental studies.

Previous studies on the measurement of the enthalpy of reaction of lithium with nitrogen (4) and the enthalpy of solution of Li_3N in water (3) are currently thought to be less reliable (1). Both these studies (3, 4) yielded $\Delta_fH°(298.15\ K)$ values of approximately -47 kcal mol^{-1}. Yonco et al. (5) studied the solubility of nitrogen in liquid lithium (468-714 K) and the equilibrium pressure of nitrogen over solid Li_3N (933-1051 K). A 2nd and 3rd law analysis of their reported $\Delta_fG°(T)$ values supports our adopted S°(298.15 K) value, to within 0.1 cal K^{-1} mol^{-1}, but suggests a 3rd law $\Delta_fH°(298.15\ k)$ value which is more positive by roughly 1.5 kcal mol^{-1} than our adopted value. Emf measurements (823-973 K) of Bonomi et al. (6) gave $\Delta_fG°(T)$ values which are also more positive than our adopted values by roughly 2 kcal mol^{-1} at 900 K. These are suspect since $Li_9N_2Cl_3$ may be involved in the reaction (6). A 2nd and 3rd law analysis of this latter data suggests a S°(298.15 K) value which is 1.8 cal K^{-1} mol^{-1} smaller than our adopted value and a 3rd law $\Delta_fH°(298.15\ K)$ value which is more positive by 2 kcal mol^{-1} than our adopted value. The S°(298.15 K) discrepancy is very difficult to reconcile without a unreasonable change in the heat capacity values.

Heat Capacity and Entropy

The thermal functions adopted are those tabulated by Osborne and Flotow (2). These authors determined the heat capacity of a well characterized sample of Li_3N between 5 and 350 K. They used the results of Satoh (7), ice calorimetric data from 273 to 773 K, to extend the heat capacity results to 773 K and then extrapolate to 1086 K, the melting point of Li_3N (1, 5).

As a check, the low temperature experimental heat capacity values are smoothed by fitting the data with orthogonal polynomials over selected overlapping temperature intervals. This fitting procedure gave an S°(298.15 K) identical to that reported by Osborne and Flotow (2). The ice calorimetric data of Satoh (7) are not extensive enough to reliably define the entire heat capacity dependence from 350 to 1086 K, however the three data prints at 373 K, 578 K, and 773 K are sufficient to define a reasonable temperature dependence.

References

1. P. A. G. O'Hare and G. K. Johnson, J. Chem. Thermodyn. 7, 13 (1975).
2. D. W. Osborne and H. E. Flotow, J. Chem. Thermodyn. 10, 675 (1978).
3. A. Guntz, Compt. Rend. 123, 995 (1896).
4. B. Neumann, C. Kroger, and H. Z. Haebler, Anorg. Allg. Chem. 204, 81 (1932).
5. R. M. Yonco, E. Veleckis, and V. A. Maroni, Nucl. Mat. 57, 317 (1975).
6. A. Bonomi, M. Hadate, and C. Gentaz, "Proceedings of the International Symposium on Molten Salts," J. P. Pemsler, ed., Electrochemical Society, Princeton, N.J., May, pp. 78-92, 1961.
7. S. Satoh, Inst. Phys. Chem. Res. Sci. Papers (Toyko) 35, 182 (1939).

Lithium Nitride (Li_3N) Li_3N_1(cr)

Enthalpy Reference Temperature = T_r = 298.15 K Standard State Pressure = p° = 0.1 MPa

	J K⁻¹mol⁻¹			kJ mol⁻¹			
T/K	$C_p°$	S°	$-[G°-H°(T_r)]/T$	$H°-H°(T_r)$	$\Delta_fH°$	$\Delta_fG°$	Log K_f
0	0.	0.	INFINITE	-11.212	-157.567	-157.567	INFINITE
100	22.204	10.114	114.995	-10.488	-159.779	-150.597	78.664
200	55.417	36.403	68.917	-6.503	-162.774	-140.160	36.606
298.15	75.266	62 593	62.593	0.	-164.557	-128.640	22.537
300	75.544	63.059	62.594	0.139	-164.581	-128.417	22.359
400	87.100	86.504	65.713	8.316	-165.664	-116.192	15.173
500	96.940	107.004	71.961	17.521	-175.728	-102.781	10.737
600	106.430	125.522	79.367	27.693	-175.998	-88.148	7.674
700	115.580	142.620	87.197	38.796	-175.191	-73.556	5.489
800	124.380	158.632	95.135	50.797	-173.434	-59.146	3.862
900	132.840	173.773	103.039	63.661	-170.832	-45.008	2.612
1000	140.950	188.192	110.839	77.353	-167.418	-31.203	1.630
1100	148.693	201.995	118.503	91.841	-163.222	-17.779	0.844
1200	156.108	215.253	126.016	107.084	-158.281	-4.769	0.208
1300	163.177	228.030	133.375	123.051	-152.628	7.799	-0.313

PREVIOUS: September 1966 CURRENT: March 1978

Lithium Nitride (Li_3N) Li_3N_1(cr)

MAGNESIUM (Mg)

REFERENCE STATE

A_r = 24.305

0 to 923	K	crystal
923 to 1366.104	K	liquid
above 1366.104	K	ideal monatomic gas

Refer to the individual tables for details.

Magnesium (Mg) Mg_1(ref)

Enthalpy Reference Temperature = T_r = 298.15 K   Standard State Pressure = p° = 0.1 MPa

T/K	C_p° (J K⁻¹mol⁻¹)	S° (J K⁻¹mol⁻¹)	$-[G°-H°(T_r)]/T$ (J K⁻¹mol⁻¹)	$H°-H°(T_r)$ (kJ mol⁻¹)	$\Delta_f H°$ (kJ mol⁻¹)	$\Delta_f G°$ (kJ mol⁻¹)	Log K_f
0	0.	0.	INFINITE	-4.998	0.	0.	0.
100	15.762	9.505	53.066	-4.356	0.	0.	0.
200	22.724	23.143	34.888	-2.349	0.	0.	0.
298.15	24.869	32.671	32.671	0.	0.	0.	0.
300	24.897	32.825	32.671	0.046	0.	0.	0.
400	26.144	40.167	33.664	2.601	0.	0.	0.
500	27.171	46.113	35.578	5.268	0.	0.	0.
600	28.184	51.156	37.764	8.035	0.	0.	0.
700	29.279	55.581	39.999	10.907	0.	0.	0.
800	30.507	59.569	42.200	13.895	0.	0.	0.
900	31.895	63.241	44.336	17.014	0.	0.	0.
923.000	32.238	64.050	44.818	17.751	CRYSTAL <--> LIQUID		
923.000	34.309	73.234	44.818	26.228	TRANSITION		
1000	34.309	75.983	47.113	28.870	0.	0.	0.
1100	34.309	79.253	49.888	32.301	0.	0.	0.
1200	34.309	82.238	52.462	35.732	0.	0.	0.
1300	34.309	84.984	54.859	39.163	0.	0.	0.
1366.104	34.309	86.686	56.358	41.431	LIQUID <--> IDEAL GAS		
1366.104	20.786	180.286	56.359	169.298	FUGACITY = 1 bar		
1400	20.786	180.796	59.365	170.003	0.	0.	0.
1500	20.786	182.230	67.509	172.082	0.	0.	0.
1600	20.786	183.571	74.721	174.160	0.	0.	0.
1700	20.786	184.832	81.162	176.239	0.	0.	0.
1800	20.787	186.020	86.954	178.317	0.	0.	0.
1900	20.787	187.144	92.198	180.396	0.	0.	0.
2000	20.789	188.210	96.972	182.475	0.	0.	0.
2100	20.791	189.224	101.341	184.554	0.	0.	0.
2200	20.795	190.192	105.358	186.633	0.	0.	0.
2300	20.802	191.116	109.067	188.713	0.	0.	0.
2400	20.812	192.002	112.504	190.794	0.	0.	0.
2500	20.826	192.851	115.701	192.875	0.	0.	0.
2600	20.846	193.669	118.684	194.959	0.	0.	0.
2700	20.874	194.456	121.476	197.045	0.	0.	0.
2800	20.909	195.216	124.096	199.134	0.	0.	0.
2900	20.956	195.950	126.561	201.227	0.	0.	0.
3000	21.014	196.661	128.886	203.326	0.	0.	0.
3100	21.085	197.352	131.084	205.430	0.	0.	0.
3200	21.172	198.022	133.165	207.543	0.	0.	0.
3300	21.275	198.675	135.140	209.665	0.	0.	0.
3400	21.396	199.312	137.019	211.799	0.	0.	0.
3500	21.537	199.934	138.807	213.945	0.	0.	0.
3600	21.697	200.543	140.514	216.107	0.	0.	0.
3700	21.879	201.140	142.144	218.285	0.	0.	0.
3800	22.083	201.726	143.705	220.483	0.	0.	0.
3900	22.310	202.303	145.200	222.703	0.	0.	0.
4000	22.559	202.871	146.634	224.946	0.	0.	0.
4100	22.832	203.431	148.013	227.215	0.	0.	0.
4200	23.128	203.985	149.339	229.513	0.	0.	0.
4300	23.447	204.533	150.616	231.842	0.	0.	0.
4400	23.789	205.076	151.848	234.203	0.	0.	0.
4500	24.152	205.614	153.037	236.600	0.	0.	0.
4600	24.537	206.149	154.185	239.034	0.	0.	0.
4700	24.944	206.681	155.297	241.508	0.	0.	0.
4800	25.372	207.211	156.373	244.023	0.	0.	0.
4900	25.820	207.739	157.416	246.583	0.	0.	0.
5000	26.287	208.265	158.427	249.188	0.	0.	0.
5100	26.773	208.790	159.410	251.841	0.	0.	0.
5200	27.276	209.315	160.364	254.543	0.	0.	0.
5300	27.794	209.839	161.293	257.296	0.	0.	0.
5400	28.329	210.364	162.197	260.102	0.	0.	0.
5500	28.878	210.888	163.077	262.962	0.	0.	0.
5600	29.442	211.414	163.936	265.878	0.	0.	0.
5700	30.020	211.940	164.773	268.851	0.	0.	0.
5800	30.610	212.467	165.591	271.882	0.	0.	0.
5900	31.213	212.996	166.390	274.973	0.	0.	0.
6000	31.827	213.525	167.171	278.125	0.	0.	0.

PREVIOUS: September 1962 (1 atm)   CURRENT: September 1983 (1 bar)

Magnesium (Mg) Mg_1(ref)

MAGNESIUM (Mg) CRYSTAL A_r = 24.305

Pearson Notation: hP2

$S°(298.15\ K) = 32.671 \pm 0.10\ J\ K^{-1}\ mol^{-1}$

$T_{fus} = 923 \pm 1\ K$

$\Delta_f H°(0\ K) = 0\ kJ\ mol^{-1}$

$\Delta_f H°(298.15\ K) = 0\ kJ\ mol^{-1}$

$\Delta_{fus} H°(923\ K) = 8.477 \pm 0.418\ kJ\ mol^{-1}$

Enthalpy of Formation

Zero by definition.

Heat Capacity and Entropy

The adopted heat capacities are derived from heat capacity (1, 2) and enthalpy data (3) by fitting the experimental data with orthogonal polynomials over selected overlapping temperature intervals. The temperature region of 1-14 K is described by the equation: $C_p° = 0.3x10^{-4}\ T + 0.8x10^{-6}\ T^3\ cal\ K^{-1}\ mol^{-1}$. The S°(298.15 K) value calculated from the polynomials agrees with those of Hultgren et al. (4), Gurvich et al. (5), and CODATA (6) to within 0.01 $J\ K^{-1}\ mol^{-1}$.

Craig et al. (1) measured the heat capacity in the region 12-320 K. Only smoothed values at 10 K intervals were reported and these were based on an incorrect relative atomic mass. Furukawa et al. (4) stated that the heat capacity values should be multiplied by 24.32/24.22 = 1.0041 to be consistent with the atomic mass of 1954. The current value for the relative atomic mass of magnesium requires a multiplier of 24.305/24.22 = 1.0035. Our analysis of this smoothed data (with orthogonal polynomials) suggests that the reported values at 18 K and 200 K may be in error.

The heat capacity study of Saba et al. (2) consisted of 37 determinations covering the range 293-548 K. Again only smoothed values at 10 K intervals were given. Using the current relative atomic mass for magnesium, this data should be multiplied by 24.305/24.32 = 0.99938. The enthalpy data of McDonald (3), 6 values in the range 404-914 K, also need to be corrected by a factor of 24.305/24.312 = 0.99971. In the region 300-320 K, the corrected results of Craig et al. (1) lie 0.1% higher and Saba et al. (2] 0.2% lower than the adopted values. The maximum deviation in the enthalpy data of McDonald (3) in comparison with our derived enthalpy values is 0.1%, except for the point at 403 K which is 0.7%. At the melting point, the entropies of JANAF, Hultgren et al. (4), and Gurvich et al. (5) are 15.308, 15.34, and 15.313 $cal\ K^{-1}\ mol^{-1}$, respectively.

Other studies (8-13) are considered for comparison with and confirmation of the adopted values, but they are not included in the polynomial fitting procedure. The heat capacity results of Mannchen and Bornkessel (8), 12-300 K, were presented only graphically and support the results of Craig et al. (1). The quantitative agreement is uncertain since it is not clear if the authors (8) were aware of the correction required for the data of Craig et al. (1). Clusius and Vaughn (9) made experimental heat capacity determinations in the range 11-228 K. These data scatter about the selected values within ±2-4%.

Phase Data

According to Pearson (15), magnesium has a hexagonal-close-packed structure, designated as hP2. In the alkaline earth metal series, only α-Be has a similar structure.

Melting Data

Refer to the Mg(ℓ) single phase table for details.

Sublimation Data

The enthalpy of sublimation of magnesium is the enthalpy of formation of the gas. It is chosen to be that adopted by CODATA (6) based on numerous experimental studies which are discussed in the Mg(g) single phase table (14).

References

1. R. S. Craig, C. A. Krier, L. W. Coffee, E. A. Bates, and W. E. Wallace, J. Amer. Chem. Soc. 76, 238 (1958).
2. W. G. Saba, K. F. Sterrett, R. S. Craig and W. E. Wallace, J. Amer. Chem. Soc., 79, 3637 (1957).
3. R. A. McDonald, J. Chem. Eng. Data 12, 131 (1967).
4. R. Hultgren et al., "Selected Values of the Thermodynamic Properties of the Elements," American Society for Metals, Metal Park, Ohio, 1973.
5. L. V. Gurvich et al., "Thermodynamic Properties of Individual Substances," Volume III, Nauka, Moscow, 1981.
6. ICSU-CODATA Task Group on Key Values for Thermodynamics, J. Chem. Thermodynamics 10, 903 (1978).
7. U. S. Nat. Bur. Stand. Rept. No. 6297, 1959.
8. W. Mannchen and K. Bornkessel, Z. Naturforsch. 14a, 925 (1959).
9. K. Clusius and J. V. Vaughn, J. Amer. Chem. Soc. 52, 4684 (1930).
10. J. K. Logan, J. R. Clement, and H. R. Jeffers, Phys. Rev. 105, 1435 (1957).
11. P. L. Smith, Phil. Mag. 46, 744 (1955).
12. J. A. Rayne, J. Phys. Chem. Solids 7, 268 (1958).
13. D. L. Martin, Proc. Phys. Soc. 78, 1482 (1961).
14. JANAF Thermochemical Tables: Mg(g), 9-30-83.
15. W. B. Pearson, "A Handbook of Lattice Spacings and Structures of Metals and Alloys," Pergamon Press, London, 1958, 1967.

Magnesium (Mg) $Mg_1(cr)$

Enthalpy Reference Temperature = T_r = 298.15 K; Standard State Pressure = p° = 0.1 MPa

T/K	$C_p°$ (J K⁻¹mol⁻¹)	S° (J K⁻¹mol⁻¹)	$-[G°-H°(T_r)]/T$ (J K⁻¹mol⁻¹)	$H°-H°(T_r)$ (kJ mol⁻¹)	$\Delta_f H°$ (kJ mol⁻¹)	$\Delta_f G°$ (kJ mol⁻¹)	Log K_f
0	0.	0.	INFINITE	-4.998	0.	0.	0.
100	15.762	9.505	53.066	-4.356	0.	0.	0.
200	22.724	23.143	34.888	-2.349	0.	0.	0.
250	24.018	28.364	33.076	-1.178	0.	0.	0.
298.15	24.869	32.671	32.671	0.	0.	0.	0.
300	24.897	32.825	32.671	0.046	0.	0.	0.
350	25.568	36.715	32.977	1.308	0.	0.	0.
400	26.144	40.167	33.664	2.601	0.	0.	0.
450	26.668	43.277	34.562	3.922	0.	0.	0.
500	27.171	46.113	35.578	5.268	0.	0.	0.
600	28.184	51.156	37.764	8.035	0.	0.	0.
700	29.279	55.581	39.999	10.907	0.	0.	0.
800	30.507	59.569	42.200	13.895	0.	0.	0.
900	31.895	63.241	44.336	17.014	0.	0.	0.
923.000	32.238	64.050	44.818	17.751	------ CRYSTAL	<--> LIQUID	-----
1000	33.460	66.681	46.401	20.280	-8.590	0.712	-0.037
1100	35.214	69.951	48.394	23.712	-8.589	1.644	-0.078
1200	37.164	73.097	50.322	27.330	-8.402	2.567	-0.112
1300	39.316	76.155	52.192	31.152	-8.011	3.467	-0.139
1400	41.673	79.154	54.011	35.200	-134.803	7.496	-0.280
1500	44.238	82.115	55.786	39.493	-132.588	17.584	-0.612
1600	47.013	85.058	57.524	44.054	-130.106	27.516	-0.898
1700	50.000	87.996	59.230	48.903	-127.336	37.284	-1.146
1800	53.199	90.944	60.910	54.061	-124.256	46.881	-1.360
1900	56.613	93.911	62.568	59.550	-120.846	56.297	-1.548
2000	60.241	96.906	64.210	65.391	-117.084	65.524	-1.711

PREVIOUS: September 1962 CURRENT: September 1983

Magnesium (Mg) $Mg_1(cr)$

MAGNESIUM (Mg) LIQUID A_r = 24.305

$S°(298.15\ K) = [34.464]\ J\ K^{-1}\ mol^{-1}$

$T_{fus} = 923 \pm 1\ K$

$\Delta_f H°(298.15\ K) = [4.790]\ kJ\ mol^{-1}$

$\Delta_{fus} H°(923\ K) = 8.477 \pm 0.418\ kJ\ mol^{-1}$

Enthalpy of Formation

$\Delta_f H°$(Mg, ℓ, 298.15 K) is calculated from $\Delta_f H°$(Mg, cr, 298.15 K) by adding the enthalpy of fusion, $\Delta_{fus} H°$(923 K), and the difference in enthalpy, H°(923 K) - H°(298.15 K), between the crystal and the liquid. This value is dependent on the arbitrary choice of the liquid phase heat capacity values for temperatures less than 923 K.

Heat Capacity and Entropy

The enthalpy measurements of McDonald (1), 5 observations in the temperature range 954-1263 K, are used to derive a constant liquid phase heat capacity of 8.2 cal K^{-1} mol^{-1}. This constant value is adopted for the entire liquid region and the region between 298.15 and 923 K. Gurvich et al. (13) have adopted the same value. Kubaschewski (2) quoted an unpublished study by Reinartz which gave $C_p°(\ell) = 8.1 \pm 0.3$ cal K^{-1} mol^{-1}. The enthalpies measured by McDonald deviate by -0.2 to +0.3% from the adopted values. The enthalpy study of Awbery and Griffiths (3) extended only 100 K into the liquid region (3 data points) and cannot be used to reliably predict the heat capacity; in addition, their reported enthalpies are 15% lower than the adopted values.

The entropy is calculated in a manner analogous to that used for the enthalpy of formation.

Fusion Data

The melting point is adopted as 923±1 K. Charlesworth [4] and Geschneider [5] both adopted this value after critically reviewing the available data in 1964 and 1970, respectively. The primary studies are by Chadwick [6 (923 K)], Jones [7 (920.5 K)], and Haughton and Payne [8 (922±0.5 K)].

The adopted enthalpy of melting, $\Delta_{fus}H° = 2.026$ kcal mol^{-1}, is derived from the enthalpy study of McDonald (1). Although McDonald (1) reports an uncertainty of ±50 cal mol^{-1}, we feel a value of ±100 cal mol^{-1} reflects better the scatter of the various studies. Kubaschewski (2) quoted values for $\Delta_{fus}H°$ of 2.030±0.080 and 1.830±0.090 kcal mol^{-1} from the unpublished work of Reinartz and a thesis of Wittig, respectively. The calorimetric studies of Ginsberg and Wrigge (9) and the DTA study of Chiotti et al. (10) led to values of 90.6±2 cal/g (2.202±0.049 kcal mol^{-1}) and 1.970±0.016 kcal mol^{-1}, respectively. An enthalpy study of Stull and McDonald (11) led to a heat of melting of 2.14 kcal mol^{-1}; however, this value was thought to be in error due to a container reaction. These reported values all support the adopted value. The enthalpy study of Awbery and Griffiths (3) led to a much lower value of 1.130 kcal mol^{-1}.

Vaporization Data

T_{vap} is calculated as the temperature for which $\Delta_{vap}G° = 0$ for the liquid vaporizing to the ideal monatomic gas. T_{vap} corresponds to a fugacity of one bar. The enthalpy of vaporization is calculated from the enthalpy of formation of the liquid and the gas at T_{vap}. Vaporization studies are discussed in the ideal gas table (12).

References

1. R. A. McDonald, J. Chem. Eng. Data 12, 131 (1967).
2. O. Kubaschewski, Z. Metallkunde 41, 445 (1950).
3. J. H. Awbery and E. Griffiths, Proc. Phys. Soc. (London) 38, 378 (1926).
4. J. H. Charlesworth, Wright-Patterson Air Force Base, AFML-TR-70-137, 1970.
5. K. A. Gschneider, Jr., Solid State Phys. 16, 275 (1964).
6. R. Chadwick, J. Inst. Metals 39, 285 (1928).
7. W. R. D. Jones, J. Inst. Metals 46, 395 (1931).
8. J. L. Haughton and R. J. M. Payne, J. Inst. Metals 54, 278 (1934).
9. H. Ginsberg and H. Ch. Wrigge, Metall. 24, 40 (1970).
0. P. Chiotti, G. J. Gartner, E. R. Stevens and Y. Sato, J. Chem. Eng. Data 11, 571 (1966).
11. D. R. Stull and R. A. McDonald, J. Amer. Chem. Soc. 77, 5293 (1955).
12. JANAF Thermochemical Tables: Mg(g), 9-30-83.
13. L. V. Gurvich, I. V. Veits et al., "Thermodynamic Properties of Individual Substances," 3rd ed., Vol. III, Nauka, Moscow, 1981.

Magnesium (Mg) $Mg_1(l)$

Enthalpy Reference Temperature = T_r = 298.15 K — Standard State Pressure = p° = 0.1 MPa

T/K	$C_p°$ (J K⁻¹mol⁻¹)	S° (J K⁻¹mol⁻¹)	-[G°-H°(T_r)]/T (J K⁻¹mol⁻¹)	H°-H°(T_r) (kJ mol⁻¹)	$\Delta_f H°$ (kJ mol⁻¹)	$\Delta_f G°$ (kJ mol⁻¹)	Log K_f
0							
100							
200							
250							
298.15	34.309	34.464	34.464	0.	4.790	4.256	-0.746
300	34.309	34.676	34.465	0.063	4.808	4.252	-0.740
350	34.309	39.965	34.882	1.779	5.261	4.124	-0.615
400	34.309	44.546	35.810	3.494	5.684	3.932	-0.513
450	34.309	48.587	37.010	5.210	6.078	3.689	-0.428
500	34.309	52.202	38.352	6.925	6.448	3.403	-0.356
600	34.309	58.457	41.197	10.356	7.111	2.731	-0.238
700	34.309	63.746	44.050	13.787	7.670	1.955	-0.146
800	34.309	68.327	46.805	17.218	8.113	1.107	-0.072
900	34.309	72.368	49.425	20.649	8.425	0.211	-0.012
923.000	34.309	73.234	50.008	21.438	------ CRYSTAL <--> LIQUID -----		
1000	34.309	75.983	51.903	24.080	0.	0.	0.
1100	34.309	79.253	54.243	27.511	0.	0.	0.
1200	34.309	82.238	56.454	30.941	0.	0.	0.
1300	34.309	84.984	58.544	34.372	0.	0.	0.
1366.104	34.309	86.686	59.865	36.640	------- FUGACITY = 1 bar -------		
1400	34.309	87.527	60.525	37.803	-127.409	3.167	-0.118
1500	34.309	89.894	62.405	41.234	-126.057	12.447	-0.433
1600	34.309	92.108	64.193	44.665	-124.705	21.636	-0.706
1700	34.309	94.188	65.897	48.096	-123.352	30.741	-0.945
1800	34.309	96.149	67.523	51.527	-122.000	39.766	-1.154
1900	34.309	98.004	69.079	54.958	-120.648	48.717	-1.339
2000	34.309	99.764	70.570	58.388	-119.296	57.596	-1.504

PREVIOUS: September 1962 CURRENT: September 1983

Magnesium (Mg) $Mg_1(l)$

MAGNESIUM (Mg) CFYSTAL–LIQUID A_r = 24.305

0 tc 923 K crystal
above 923 K liquid

Refer to the individual tables for details.

Magnesium (Mg) $Mg_1(cr,l)$

Enthalpy Reference Temperature = T_r = 298.15 K   Standard State Pressure = p° = 0.1 MPa

T/K	C_p° (J K⁻¹mol⁻¹)	S° (J K⁻¹mol⁻¹)	$-[G°-H°(T_r)]/T$ (J K⁻¹mol⁻¹)	$H°-H°(T_r)$ (kJ mol⁻¹)	$\Delta_fH°$ (kJ mol⁻¹)	$\Delta_fG°$ (kJ mol⁻¹)	Log K_f
0	0.	0.	INFINITE	-4.998	0.	0.	0.
100	15.762	9.505	53.066	-4.356	0.	0.	0.
200	22.724	23.143	34.888	-2.349	0.	0.	0.
250	24.018	28.364	33.076	-1.178	0.	0.	0.
298.15	24.869	32.671	32.671	0.	0.	0.	0.
300	24.897	32.825	32.671	0.046	0.	0.	0.
350	25.568	36.715	32.977	1.308	0.	0.	0.
400	26.144	40.167	33.664	2.601	0.	0.	0.
450	26.668	43.277	34.562	3.922	0.	0.	0.
500	27.171	46.113	35.578	5.268	0.	0.	0.
600	28.184	51.156	37.764	8.035	0.	0.	0.
700	29.279	55.581	39.999	10.907	0.	0.	0.
800	30.507	59.569	42.200	13.895	0.	0.	0.
900	31.895	63.241	44.336	17.014	0.	0.	0.
923.000	32.238	64.050	44.818	17.751	CRYSTAL <--> LIQUID		
923.000	34.309	73.234	44.818	26.228	TRANSITION		
1000	34.309	75.983	47.113	28.870	0.	0.	0.
1100	34.309	79.253	49.888	32.301	0.	0.	0.
1200	34.309	82.238	52.462	35.732	0.	0.	0.
1300	34.309	84.984	54.859	39.163	0.	0.	0.
1366.104	34.309	86.686	56.358	41.431	------- FUGACITY = 1 bar -------		
1400	34.309	87.527	57.103	42.594	-127.409	3.167	-0.118
1500	34.309	89.894	59.211	46.024	-126.057	12.447	-0.433
1600	34.309	92.108	61.199	49.455	-124.705	21.636	-0.706
1700	34.309	94.188	63.079	52.886	-123.352	30.741	-0.945
1800	34.309	96.149	64.862	56.317	-122.000	39.766	-1.154
1900	34.309	98.004	66.558	59.748	-120.648	48.717	-1.339
2000	34.309	99.764	68.175	63.179	-119.296	57.596	-1.504

PREVIOUS: CURRENT: September 1983

Magnesium (Mg) $Mg_1(cr,l)$

MAGNESIUM (Mg) IDEAL GAS $A_r = 24.305$

$IP(Mg, g) = 61671.02 \pm 0.10\ cm^{-1}$ $\Delta_f H°(0\ K) = 145.90 \pm 0.80\ kJ\ mol^{-1}$

$S°(298.15\ K) = 148.648 \pm 0.020\ J\ K^{-1}\ mol^{-1}$ $\Delta_f H°(298.15\ K) = 147.10 \pm 0.80\ kJ\ mol^{-1}$

Electronic Levels and Quantum Weights

State	ε_i, cm^{-1}	g_i
$^1S_0$	0.00	1
$^3P_0$	21850.405	1
$^3P_1$	21870.464	3
$^3P_2$	21911.178	5
$^1P_1$	35051.264	3

Enthalpy of Formation

The adopted enthalpy of formation values for Mg(g), $\Delta_f H°(298.15\ K) = 147.10 \pm 0.80\ kJ\ mol^{-1}$, is that recommended by CODATA (1). This value was calculated, according to CODATA, from the vapor pressure measurements by Priselkov (673-773 K, 2), Gilbreath (495-658 K, 3), McCreary and Thorn (716-800 K, 4), Pahlman and Smith (650-930 K, 5), and others cited in Hultgren et al. (6).

Heat Capacity and Entropy

The information on electronic energy levels and quantum weights, given by Martin and Zulubas (7), is incomplete because many theoretically predicted levels have not been observed. Our calculations indicate that any reasonable method of fitting in these missing levels (for n < 60) and cutting off the summation in the partition function (8) has no effect on the thermodynamic functions to 4000 K. This is a result of the high energy of all levels other than the ground state and the levels listed above; the next excited state is approximately 41197 cm^{-1} above the ground state. Although we list only a few levels, all levels listed by Martin and Zulubas (7) and estimated levels (for n < 60) are considered in the calculation. The reported uncertainty in S°(298.15 K) is due to uncertainties in the relative atomic mass and the fundamental constants. Extension of these calculations above 6000 K may require consideration of the additional states and use of different fill and cutoff procedures (8).

The thermal functions at 298.15 K differ from the CODATA recommendations (1) for two reasons. First, the entropy differs by 0.1094 $J\ K^{-1}\ mol^{-1}$ because this table uses a reference pressure of 1 bar, whereas CODATA recommendations are based on 1 atm. Second, small changes, ~0.002 $J\ K^{-1}\ mol^{-1}$, arise due to the use of slightly different values for the fundamental constants.

References

1. J. D. Cox, Chairman, CODATA Task Group on Key Values for Thermodynamics, J. Chem. Thermodynamics 10, 903 (1978); CODATA Bulletin No. 22, 1977.
2. Yu. A. Priselkov, Moscow State University, Dissertation, 1954; cited by A. N. Nesmeyanov, "Vapor Pressures of the Chemical Elements", Moscow, 1961.
3. W. P. Gilbreath, NASA TN-D-2723, 1965.
4. J. R. McCreary and R. J. Thorn, High Temp. Sci. 3, 300 (1971).
5. J. E. Pahlman and J. F. Smith, Met. Trans. 3, 2423 (1972).
6. R. Hultgren, P. D. Desai, et al., "Selected Values of the Thermodynamic Properties of the Elements", American Society of Metals, Metals Park, Ohio, 1973.
7. W. C. Martin and R. Zalubas, J. Phys. Chem. Ref. Data 9, 1 (1980).
8. J. R. Downey, Jr., The Dow Chemical Company, Report AFOSR-TR-78-0960, Contract No. F44670-75-1-0048, 1978

Magnesium (Mg) $Mg_1(g)$

Enthalpy Reference Temperature = T_r = 298.15 K — Standard State Pressure = p° = 0.1 MPa

T/K	$C_p°$ (J K⁻¹mol⁻¹)	S° (J K⁻¹mol⁻¹)	−[G°−H°(T_r)]/T (J K⁻¹mol⁻¹)	H°−H°(T_r) (kJ mol⁻¹)	$\Delta_f H°$ (kJ mol⁻¹)	$\Delta_f G°$ (kJ mol⁻¹)	Log K_f
0	0.	0.	INFINITE	-6.197	145.901	145.901	INFINITE
100	20.786	125.940	167.128	-4.119	147.337	135.694	-70.879
200	20.786	140.348	150.549	-2.040	147.409	123.968	-32.377
250	20.786	144.986	148.990	-1.001	147.277	118.122	-24.680
298.15	20.786	148.648	148.648	0.	147.100	112.522	-19.713
300	20.786	148.776	148.648	0.038	147.092	112.307	-19.554
350	20.786	151.980	148.901	1.078	146.870	106.527	-15.898
400	20.786	154.756	149.463	2.117	146.616	100.780	-13.161
450	20.786	157.204	150.190	3.156	146.335	95.067	-11.035
500	20.786	159.394	151.003	4.196	146.028	89.387	-9.338
600	20.786	163.184	152.727	6.274	145.339	78.122	-6.801
700	20.786	166.388	154.455	8.353	144.546	66.981	-4.998
800	20.786	169.164	156.124	10.431	143.636	55.961	-3.654
900	20.786	171.612	157.712	12.510	142.596	45.062	-2.615
1000	20.786	173.802	159.213	14.589	132.819	35.000	-1.828
1100	20.786	175.783	160.631	16.667	131.466	25.283	-1.201
1200	20.786	177.592	161.970	18.746	130.114	15.690	-0.683
1300	20.786	179.255	163.237	20.824	128.762	6.209	-0.249
1366.104	20.786	180.286	164.037	22.199	------- FUGACITY = 1 bar -------		
1400	20.786	180.796	164.437	22.903	0.	0.	0.
1500	20.786	182.230	165.575	24.982	0.	0.	0.
1600	20.786	183.571	166.659	27.060	0.	0.	0.
1700	20.786	184.832	167.691	29.139	0.	0.	0.
1800	20.787	186.020	168.677	31.218	0.	0.	0.
1900	20.787	187.144	169.619	33.296	0.	0.	0.
2000	20.789	188.210	170.522	35.375	0.	0.	0.
2100	20.791	189.224	171.389	37.454	0.	0.	0.
2200	20.795	190.192	172.222	39.533	0.	0.	0.
2300	20.802	191.116	173.023	41.613	0.	0.	0.
2400	20.812	192.002	173.796	43.694	0.	0.	0.
2500	20.826	192.851	174.541	45.776	0.	0.	0.
2600	20.846	193.669	175.261	47.859	0.	0.	0.
2700	20.874	194.456	175.958	49.945	0.	0.	0.
2800	20.909	195.216	176.632	52.034	0.	0.	0.
2900	20.956	195.950	177.285	54.127	0.	0.	0.
3000	21.014	196.661	177.920	56.226	0.	0.	0.
3100	21.085	197.352	178.535	58.331	0.	0.	0.
3200	21.172	198.022	179.134	60.443	0.	0.	0.
3300	21.275	198.675	179.716	62.566	0.	0.	0.
3400	21.396	199.312	180.283	64.699	0.	0.	0.
3500	21.537	199.934	180.836	66.845	0.	0.	0.
3600	21.697	200.543	181.375	69.007	0.	0.	0.
3700	21.879	201.140	181.901	71.186	0.	0.	0.
3800	22.083	201.726	182.415	73.384	0.	0.	0.
3900	22.310	202.303	182.918	75.603	0.	0.	0.
4000	22.559	202.871	183.409	77.846	0.	0.	0.
4100	22.832	203.431	183.891	80.116	0.	0.	0.
4200	23.128	203.985	184.363	82.413	0.	0.	0.
4300	23.447	204.533	184.825	84.742	0.	0.	0.
4400	23.789	205.076	185.279	87.103	0.	0.	0.
4500	24.152	205.614	185.725	89.500	0.	0.	0.
4600	24.537	206.149	186.164	91.934	0.	0.	0.
4700	24.944	206.681	186.594	94.408	0.	0.	0.
4800	25.372	207.211	187.018	96.923	0.	0.	0.
4900	25.820	207.739	187.436	99.483	0.	0.	0.
5000	26.287	208.265	187.847	102.088	0.	0.	0.
5100	26.773	208.790	188.253	104.741	0.	0.	0.
5200	27.276	209.315	188.653	107.443	0.	0.	0.
5300	27.794	209.839	189.048	110.196	0.	0.	0.
5400	28.329	210.364	189.437	113.002	0.	0.	0.
5500	28.878	210.888	189.823	115.862	0.	0.	0.
5600	29.442	211.414	190.204	118.778	0.	0.	0.
5700	30.020	211.940	190.580	121.751	0.	0.	0.
5800	30.610	212.467	190.953	124.782	0.	0.	0.
5900	31.213	212.996	191.322	127.873	0.	0.	0.
6000	31.827	213.525	191.688	131.025	0.	0.	0.

PREVIOUS: September 1962 (1 atm) CURRENT: September 1983 (1 bar)

Magnesium (Mg) $Mg_1(g)$

MAGNESIUM, ION (Mg^+) IDEAL GAS M_r = 24.30445

IP(Mg^+, g) = 121267.61 ± 0.05 cm^{-1}

S°(298.15 K) = 154.410 ± 0.02 J K^{-1} mol^{-1}

$\Delta_f H°$(0 K) = 883.65 ± 1.3 kJ mol^{-1}

$\Delta_f H°$(298.15 K) = [891.047] kJ mol^{-1}

Electronic Levels and Quantum Weights

State	ϵ_i cm^{-1}	g_i
$^2S_{1/2}$	0	2
$^2P_{1/2}$	35669.31	2
$^2P_{3/2}$	35760.88	4

Enthalpy of Formation

$\Delta_f H°$(Mg^+, g, 0 K) is calculated from $\Delta_f H°$(Mg, g, 0 K) (1) using the spectroscopic value of IP(Mg) = 61671.02 ± 0.10 cm^{-1} (737.749 ± 0.001 kJ mol^{-1}) from Martin and Zolubas (2). The ionization limit is converted from cm^{-1} to kJ mol^{-1} using the factor, 1 cm^{-1} = 0.01196266 kJ cm^{-1}, which is derived from the 1973 CODATA fundamental constants (3). Rosenstock et al. (4) and Levin and Lias (5) have summarized additional ionization and appearance potential data.

$\Delta_f H°$(Mg^+, g, 298.15 K) is calculated from $\Delta_f H°$(Mg, g, 0 K) by using IP(Mg) with JANAF (1) enthalpies, H°(0 K) - H°(298.15 K), for Mg(g), Mg^+(g), and e^-(g). $\Delta_r H°$(Mg + Mg^+ + e^-, 298.15 K) differs from a room temperature threshold energy due to inclusion of these enthalpies and to threshold effects discussed by Rosenstock et al. (4). $\Delta_f H°$(298.15 K) should be changed by -6.197 kJ mol^{-1} if it is to be used in the ion convention that excludes the enthalpy of the electron.

Heat Capacity and Entropy

The information on electronic energy levels and quantum weights, given by Martin and Zolubas (2), is incomplete because many theoretically predicted levels have not been observed. Our calculations indicate that any reasonable method of filling in these missing levels and cutting off the summation in the partition function (6) has no effect on the thermodynamic functions to 6000 K. This is a result of the high energy of all levels other than the ground state and the first two excited states; the next excited state is 69804.95 cm^{-1} above the ground state. Since inclusion of these higher excited states has no effect on the thermodynamic functions (to 6000 K), we list only the ground state and the first two excited states. The reported uncertainty in S°(298.15 K) is due to uncertainties in the relative ionic mass and the fundamental constants. Extension of these calculations above 6000 K may require consideration of the excited states and use of different fill and cutoff procedures (6).

References

1. JANAF Thermochemical Tables: Mg(g), 9-30-83; e^-(g), 3-31-82.
2. W. C. Martin and R. Zolubas, J. Phys. Chem. Ref. Data 9, 1 (1980).
3. E. R. Cohen and B. N. Taylor, J. Phys. Chem. Ref. Data 2, 663 (1973).
4. H. M. Rosenstock, K. Draxl, et al., J. Phys. Chem. Ref. Data 6, Supp. 1 (1977).
5. R. D. Levin and S. G. Lias, U. S. Nat. Bur. Stand., NSRDS-NBS-71, 1982.
6. J. R. Downey, Jr., The Dow Chemical Company, AFOSRS-TR-78-0960, Contract No. F44620-75-1-0048, 1978.

Magnesium, Ion (Mg^+) Mg_1^+(g)

Enthalpy Reference Temperature = T_r = 298.15 K Standard State Pressure = p° = 0.1 MPa

	J $K^{-1}$$mol^{-1}$			kJ mol^{-1}			
T/K	$C_p°$	S°	$-[G°-H°(T_r)]/T$	$H°-H°(T_r)$	$\Delta_f H°$	$\Delta_f G°$	Log K_f
0	0.	0.	INFINITE	-6.197	883.650		
100	20.786	131.703	172.891	-4.119			
200	20.786	146.111	156.312	-2.040			
250	20.786	150.749	154.753	-1.001			
298.15	20.786	154.410	154.410	0.	891.047	848.495	-148.653
300	20.786	154.539	154.411	0.038	891.078	848.231	-147.690
350	20.786	157.743	154.664	1.078	891.894	841.025	-125.516
400	20.786	160.519	155.226	2.117	892.680	833.704	-108.871
450	20.786	162.967	155.953	3.156	893.438	826.287	-95.913
500	20.786	165.157	156.766	4.196	894.170	818.786	-85.538
600	20.786	168.947	158.490	6.274	895.560	803.577	-69.958
700	20.786	172.151	160.218	8.353	896.845	788.143	-58.812
800	20.786	174.926	161.887	10.431	898.014	772.533	-50.441
900	20.786	177.375	163.475	12.510	899.053	756.784	-43.923
1000	20.786	179.565	164.976	14.589	891.354	741.639	-38.739
1100	20.786	181.546	166.394	16.667	892.080	726.633	-34.505
1200	20.786	183.354	167.733	18.746	892.807	711.560	-30.973
1300	20.786	185.018	168.999	20.824	893.533	696.427	-27.983
1400	20.786	186.559	170.199	22.903	766.850	684.404	-25.535
1500	20.786	187.993	171.338	24.982	768.929	678.443	-23.625
1600	20.786	189.334	172.422	27.060	771.007	672.343	-21.950
1700	20.786	190.594	173.454	29.139	773.086	666.113	-20.467
1800	20.786	191.783	174.439	31.217	775.164	659.760	-19.146
1900	20.786	192.906	175.382	33.296	777.243	653.292	-17.960
2000	20.786	193.973	176.285	35.375	779.321	646.714	-16.890
2100	20.786	194.987	177.152	37.453	781.400	640.033	-15.920
2200	20.786	195.954	177.985	39.532	783.477	633.253	-15.035
2300	20.786	196.878	178.786	41.610	785.555	626.378	-14.225
2400	20.786	197.762	179.558	43.689	787.631	619.412	-13.481
2500	20.786	198.611	180.304	45.768	789.707	612.361	-12.795
2600	20.786	199.426	181.024	47.846	791.780	605.226	-12.159
2700	20.786	200.211	181.720	49.925	793.852	598.011	-11.569
2800	20.786	200.966	182.394	52.004	795.920	590.720	-11.020
2900	20.786	201.696	183.047	54.082	797.984	583.355	-10.507
3000	20.786	202.401	183.680	56.161	800.043	575.919	-10.028
3100	20.786	203.082	184.295	58.239	802.095	568.414	-9.578
3200	20.787	203.742	184.893	60.318	804.140	560.844	-9.155
3300	20.787	204.382	185.474	62.397	806.175	553.209	-8.757
3400	20.788	205.002	186.039	64.475	808.199	545.513	-8.381
3500	20.788	205.605	186.589	66.554	810.210	537.758	-8.026
3600	20.789	206.191	187.126	68.633	812.205	529.945	-7.689
3700	20.790	206.760	187.649	70.712	814.184	522.077	-7.370
3800	20.792	207.315	188.159	72.791	816.144	514.156	-7.068
3900	20.794	207.855	188.657	74.871	818.083	506.184	-6.780
4000	20.797	208.381	189.144	76.950	819.998	498.162	-6.505
4100	20.800	208.895	189.619	79.030	821.887	490.093	-6.244
4200	20.804	209.396	190.084	81.110	823.748	481.978	-5.994
4300	20.809	209.886	190.539	83.191	825.578	473.818	-5.756
4400	20.815	210.364	190.984	85.272	827.377	465.617	-5.528
4500	20.822	210.832	191.420	87.354	829.141	457.375	-5.309
4600	20.830	211.290	191.847	89.436	830.868	449.095	-5.100
4700	20.839	211.738	192.265	91.520	832.556	440.777	-4.899
4800	20.850	212.177	192.676	93.604	834.203	432.424	-4.706
4900	20.862	212.607	193.078	95.690	835.808	424.037	-4.520
5000	20.876	213.028	193.473	97.777	837.368	415.617	-4.342
5100	20.892	213.442	193.860	99.865	838.882	407.167	-4.170
5200	20.910	213.848	194.241	101.955	840.349	398.688	-4.005
5300	20.930	214.246	194.615	104.047	841.767	390.181	-3.845
5400	20.952	214.638	194.982	106.141	843.133	381.647	-3.692
5500	20.976	215.022	195.343	108.238	844.448	373.089	-3.543
5600	21.003	215.400	195.697	110.337	845.710	364.508	-3.400
5700	21.032	215.772	196.046	112.438	846.918	355.904	-3.261
5800	21.063	216.138	196.390	114.543	848.069	347.279	-3.128
5900	21.098	216.499	196.727	116.651	849.165	338.636	-2.998
6000	21.134	216.854	197.060	118.763	850.203	329.974	-2.873

PREVIOUS: December 1970 (1 atm) CURRENT: September 1983 (1 bar)

Magnesium, Ion (Mg^+) Mg_1^+(g)

MAGNESIUM NITRIDE (MgN) IDEAL GAS M_r = 38.3117

$\Delta_f H°(0\ K)$ = [289.04 ± 25.1] kJ mol^{-1}

S°(298.15 K) = [224.835] J K^{-1} mol^{-1} $\Delta_f H°(298.15\ K)$ = [288.70 ± 25.1] kJ mol^{-1}

Electronic Level and Quantum Weight

State	ε_i, cm^{-1}	g_i
[$^2\Pi$]	0	[4]

ω_e = [700] cm^{-1} $\omega_e x_e$ = [4.8] cm^{-1} σ = 1

B_e = [0.5542] cm^{-1} α_e = [0.0052] cm^{-1} r_e = [1.85] Å

Enthalpy of Formation

$\Delta_f H°$(MgN, g, 298.15 K) was estimated from bond strengths using MgO and comparison with analogous pairs of compounds such as, PN and PO, and also the bond strengths of C-N and C-O in organic compounds.

Heat Capacity and Entropy

The r_e value for MgN was estimated from MgO by comparison with similar pairs PO and PN, SiO and SiN, and AlO and AlN. Using Guggenheimer (1). ω for MgO was found to lie between the single and multiple bonding cases. Magnesium nitride was assumed to be similar to MgO and ω_e 700 cm^{-1} was obtained. It was assumed that the $x_e\mu^{1/2}$ value for MgN was equal to that for MgO thus giving $\omega_e x_e$ = 4.8 cm^{-1}. B_e was calculated from r_e. The α_e value was calculated from B_e, ω_e and $\omega_e x_e$.

Reference

1. Guggenheimer's Relation, Proc. Phys. Soc. (London) 58, 456 (1946).

Magnesium Nitride (MgN) $Mg_1N_1(g)$

Enthalpy Reference Temperature = T_r = 298.15 K Standard State Pressure = p° = 0.1 MPa

T/K	$C_p°$ (J K^{-1}mol^{-1})	S° (J K^{-1}mol^{-1})	$-[G°-H°(T_r)]/T$ (J K^{-1}mol^{-1})	$H°-H°(T_r)$ (kJ mol^{-1})	$\Delta_f H°$ (kJ mol^{-1})	$\Delta_f G°$ (kJ mol^{-1})	Log K_f
0	0.	0.	INFINITE	-8.989	289.040	289.040	INFINITE
100	29.151	191.683	252.482	-6.080	289.856	279.629	-146.063
200	30.587	212.211	227.762	-3.110	289.363	269.548	-70.399
250	31.725	219.159	225.368	-1.552	289.023	264.634	-55.292
298.15	32.733	224.835	224.835	0.	288.696	259.966	-45.545
300	32.768	225.037	224.835	0.061	288.684	259.788	-45.233
350	33.639	230.156	225.238	1.722	288.354	254.999	-38.056
400	34.340	234.696	226.141	3.422	288.031	250.256	-32.680
450	34.901	238.774	227.322	5.153	287.709	245.553	-28.503
500	35.350	242.475	228.655	6.910	287.383	240.886	-25.165
600	36.013	248.983	231.516	10.480	286.694	231.651	-20.167
700	36.468	254.570	234.420	14.106	285.926	222.536	-16.606
800	36.797	259.463	237.251	17.770	285.047	213.539	-13.943
900	37.045	263.812	239.965	21.462	284.033	204.660	-11.878
1000	37.241	267.725	242.548	25.177	274.271	196.614	-10.270
1100	37.401	271.282	245.001	28.909	272.924	188.914	-8.971
1200	37.536	274.543	247.329	32.656	271.566	181.337	-7.893
1300	37.653	277.552	249.540	36.416	270.198	173.873	-6.986
1400	37.756	280.346	251.642	40.186	141.411	169.682	-6.331
1500	37.849	282.954	253.643	43.967	141.379	171.702	-5.979
1600	37.935	285.400	255.552	47.756	141.340	173.725	-5.672
1700	38.014	287.702	257.376	51.553	141.296	175.751	-5.400
1800	38.089	289.877	259.122	55.359	141.248	177.779	-5.159
1900	38.159	291.938	260.795	59.171	141.197	179.809	-4.943
2000	38.227	293.897	262.402	62.990	141.143	181.843	-4.749
2100	38.292	295.764	263.947	66.816	141.088	183.879	-4.574
2200	38.355	297.547	265.434	70.649	141.031	185.918	-4.414
2300	38.417	299.253	266.867	74.487	140.973	187.960	-4.269
2400	38.477	300.889	268.251	78.332	140.914	190.004	-4.135
2500	38.535	302.461	269.588	82.183	140.855	192.051	-4.013
2600	38.593	303.974	270.882	86.039	140.795	194.100	-3.900
2700	38.649	305.431	272.134	89.901	140.733	196.151	-3.795
2800	38.705	306.838	273.349	93.769	140.669	198.205	-3.698
2900	38.761	308.197	274.527	97.642	140.603	200.261	-3.607
3000	38.815	309.512	275.672	101.521	140.534	202.319	-3.523
3100	38.870	310.786	276.784	105.405	140.460	204.380	-3.444
3200	38.923	312.021	277.866	109.295	140.381	206.443	-3.370
3300	38.977	313.219	278.919	113.190	140.294	208.509	-3.300
3400	39.030	314.383	279.945	117.090	140.199	210.577	-3.235
3500	39.082	315.516	280.945	120.996	140.093	212.649	-3.174
3600	39.135	316.617	281.921	124.907	139.975	214.724	-3.116
3700	39.187	317.690	282.873	128.823	139.843	216.802	-3.061
3800	39.239	318.736	283.803	132.744	139.694	218.884	-3.009
3900	39.290	319.756	284.712	136.670	139.527	220.970	-2.960
4000	39.342	320.751	285.601	140.602	139.339	223.060	-2.913
4100	39.393	321.723	286.470	144.539	139.128	225.156	-2.869
4200	39.444	322.673	287.321	148.481	138.891	227.257	-2.826
4300	39.496	323.602	288.154	152.428	138.627	229.364	-2.786
4400	39.546	324.511	288.970	156.380	138.334	231.477	-2.748
4500	39.597	325.400	289.769	160.337	138.008	233.598	-2.712
4600	39.648	326.271	290.553	164.299	137.649	235.726	-2.677
4700	39.699	327.124	291.323	168.267	137.253	237.862	-2.644
4800	39.749	327.960	292.077	172.239	136.818	240.008	-2.612
4900	39.800	328.780	292.818	176.216	136.343	242.162	-2.581
5000	39.850	329.585	293.545	180.199	135.826	244.327	-2.552
5100	39.901	330.375	294.260	184.186	135.264	246.502	-2.525
5200	39.951	331.150	294.962	188.179	134.656	248.689	-2.498
5300	40.001	331.911	295.652	192.177	134.001	250.889	-2.473
5400	40.051	332.659	296.330	196.179	133.296	253.101	-2.448
5500	40.101	333.395	296.997	200.187	132.541	255.326	-2.425
5600	40.152	334.118	297.654	204.200	131.733	257.566	-2.402
5700	40.202	334.829	298.300	208.217	130.871	259.820	-2.381
5800	40.252	335.529	298.936	212.240	129.953	262.090	-2.360
5900	40.302	336.217	299.562	216.268	128.979	264.377	-2.341
6000	40.351	336.895	300.178	220.300	127.947	266.680	-2.322

PREVIOUS: March 1964 (1 atm) CURRENT: March 1964 (1 bar)

Magnesium Nitride (MgN) $Mg_1N_1(g)$

MAGNESIUM OXIDE (MgO) CRYSTAL M_r = 40.3044

$S°(298.15\ K) = 26.924 \pm 0.08\ J\ K^{-1}\ mol^{-1}$

$T_{fus} = 3105 \pm 30\ K$

$\Delta_f H°(0\ K) = -597.060 \pm 0.63\ kJ\ mol^{-1}$

$\Delta_f H°(298.15\ K) = -601.241 \pm 0.63\ kJ\ mol^{-1}$

$\Delta_{fus} H° = [77.822 \pm 16.7]\ kJ\ mol^{-1}$

Enthalpy of Formation

$\Delta_f H°$ is based on oxygen-bomb calorimetry of Holley and Huber (1) and HCl-solution calorimetry of Shomate and Huffman (2). The reported values of -143.70 ± 0.12 (1) and -143.84 ± 0.05 (2) kcal mol^{-1} become $\Delta_f H°(298.15\ K)$ = -143.61 ± 0.12 and -143.78 ± 0.08 kcal mol^{-1} when adjusted to 1969 atomic weights. We adopt the median value of -143.70 ± 0.15 kcal mol^{-1}.

NBS (3) adopted -143.81 kcal mol^{-1} for macrocrystalline periclase, based on the data of Shomate and Huffman (2) converted to 1961 atomic weights. Parker (3) preferred the data of Shomate and Huffman because of the inference that their sample corresponded more closely to the macrocrystalline standard state. New $\Delta_{sln}H°$ data (4) suggest that excess energy of up to ~1.1 kcal mol^{-1} arises mainly from disorder (defects) in MgO rather than from high surface area and small particle size. Since the excess energy is eliminated at T>~1000°C (4), it is unlikely that this effect would bias the combustion data of Holley and Huber (1).

Other data for $\Delta_f H°$ were reviewed by Parker (3). New measurements exist for $\Delta_{sln}H°$ of MgO (4) and Mg (5), but we have not attempted to derive $\Delta_f H°$ because of difference in the final states of the solutions.

Heat Capacity and Entropy

$C_p°$ below 270 K is based on data (3-270 K) for single crystals measured by Barron et al. (6). Values above 270 K are from a constrained fit of $C_p°$ (6) and enthalpies (373-1173 K) of fused MgO measured in a Bunsen ice calorimeter by Victor & Douglas (7). S°(298.15 K) is obtained from $C_p°$ based on a negligible extrapolation to absolute zero.

Maximum deviatons of the $C_p°$ data (6) from the adopted curve are ±1.7% near 15 K, ±1.0% near 20 K, ±0.4% near 40 K, and ~0.2% at higher temperatures. Gmelin (8) later reported smoothed $C_p°$ values (2-320 K) derived from data for MgO sintered at high temperature. These values are in good agreement with the adopted curve except from 70 to 140 K, where deviations are all positive and reach a maximum of almost 7% at 90 K. Gmelin's values gave S°(298.15 K) = 6.50, i.e., higher by ~0.06 cal $K^{-1}\ mol^{-1}$. Larger positive deviations in $C_p°$ exist at all temperatures (20-301 K) for the non-standard state MgO used by Giauque and Archibald (9). Their sample was prepared from hydroxide in vacuo at ~350°C and was not annealed at high temperature. The microcrystalline product presumably was disordered, leading to excess $C_p°$ and S°(298.15 K) = 6.66 cal $K^{-1}\ mol^{-1}$. Limited data of Parks and Kelley (10) for fused MgO tend to confirm the adopted curve.

Deviations of the enthalpy data from the adopted curve are ±0.15% (7, 373-1173 K), -0.3 ± 0.3% (11, 402-1799 K), +0.8 ± 0.9% (12, 370-1591 K) and -2.8 to +1.0% (13, equation for 1200-2500 K). Microcalorimetric $C_p°$ data (14, 305-455 K) agree with the adopted curve at the mid-temperature but have a temperature derivative which is too small.

Fusion Data

See the liquid table for details.

References

1. C. E. Holley and E. J. Huber, J. Amer. Chem. Soc. 73, 5577 (1951).
2. C. H. Shomate and E. H. Huffman, J. Amer. Chem. Soc. 65, 1625 (1943).
3. U. S. Nat. Bur. Stand. Tech. Note 270-6, 1971; V. B. Parker, U. S. Nat. Bur. Stand. Report 9803, pp. 91-2 1968.
4. M. Natarajan, T. S. Sarma, J. C. Ahluwalia and C. N. R. Rao, Trans. Faraday Soc. 65, 3088 (1969), and references cited therein.
5. A. S. Monaenkova, E. B. Pashlova and A. F. Vorob'ev, Dokl. Akad. Nauk SSSR 199, 1332 (1971).
6. T. H. K. Barron, W. T. Berg and J. A. Morrison, Proc. Roy. Soc. (London) 250A, 70 (1959).
7. A. C. Victor and T. B. Douglas, J. Res. Nat. Bur. Stand. 67A, 325 (1963).
8. E. Gmelin, Z. Naturforsch. 24a, 1794 (1969); Cryogenics 7, 225 (1967).
9. W. F. Giauque and R. C. Archibald, J. Amer. Chem. Soc. 59, 561 (1937).
10. G. S. Parks and K. K. Kelley, J. Phys. Chem. 30, 47 (1926).
11. L. B. Pankratz and K. K. Kelley, U. S. Bur. Mines RI 6295, 1963.
12. D. S. Tsagareishvili and G. G. Gvelesiani, Teplofiz. Vys. Temp. 12, 208 (1974).
13. V. P. Bondarenko, E. N. Fomichev and V. V. Kandyba, High Temp. High Press. 5, 5 (1973).
14. L. Malaspino, R. Gigli, and G. Bardi, Calore 42, 317 (1971).

Magnesium Oxide (MgO) $Mg_1O_1(cr)$

Enthalpy Reference Temperature = T_r = 298.15 K Standard State Pressure = p° = 0.1 MPa

T/K	$C_p°$	S°	$-[G°-H°(T_r)]/T$	$H°-H°(T_r)$	$\Delta_f H°$	$\Delta_f G°$	Log K_f
	J $K^{-1}mol^{-1}$			kJ mol^{-1}			
0	0.	0.	INFINITE	-5.159	-597.060	-597.060	INFINITE
100	7.802	2.548	52.212	-4.966	-598.962	-589.601	307.976
200	26.681	14.096	30.037	-3.188	-600.646	-579.488	151.347
298.15	37.106	26.924	26.924	0.	-601.241	-568.945	99.677
300	37.244	27.154	26.925	0.069	-601.245	-568.745	99.027
400	42.561	38.678	28.460	4.087	-601.268	-557.898	72.854
500	45.544	48.523	31.513	8.505	-601.046	-547.078	57.153
600	47.430	57.006	35.072	13.160	-600.738	-536.312	46.690
700	48.748	64.420	38.746	17.972	-600.426	-525.600	39.221
800	49.740	70.996	42.374	22.898	-600.156	-514.930	33.621
900	50.539	76.902	45.888	27.913	-599.962	-504.289	29.268
1000	51.208	82.262	49.262	33.001	-608.462	-492.952	25.749
1100	51.794	87.171	52.488	38.151	-608.496	-481.399	22.860
1200	52.325	91.701	55.569	43.358	-608.495	-469.844	20.452
1300	52.810	95.908	58.512	48.615	-608.461	-458.291	18.414
1400	53.262	99.839	61.325	53.918	-735.804	-443.575	16.550
1500	53.693	103.528	64.017	59.266	-734.355	-422.752	14.722
1600	54.107	107.007	66.596	64.656	-732.877	-402.026	13.125
1700	54.509	110.299	69.071	70.087	-731.371	-381.394	11.719
1800	54.898	113.426	71.449	75.558	-729.837	-360.851	10.472
1900	55.278	116.404	73.737	81.067	-728.277	-340.395	9.358
2000	55.651	119.249	75.942	86.613	-726.690	-320.021	8.358
2100	56.019	121.973	78.070	92.197	-725.079	-299.727	7.455
2200	56.379	124.587	80.125	97.816	-723.442	-279.510	6.636
2300	56.738	127.101	82.114	103.472	-721.781	-259.368	5.890
2400	57.094	129.524	84.039	109.164	-720.097	-239.300	5.208
2500	57.445	131.862	85.905	114.891	-718.389	-219.301	4.582
2600	57.797	134.121	87.716	120.653	-716.659	-199.372	4.005
2700	58.145	136.309	89.476	126.450	-714.906	-179.509	3.473
2800	58.491	138.430	91.187	132.282	-713.132	-159.712	2.979
2900	58.836	140.489	92.851	138.148	-711.338	-139.979	2.521
3000	59.177	142.489	94.473	144.049	-709.524	-120.308	2.095
3100	59.516	144.436	96.053	149.987	-707.689	-100.697	1.697
3105.000	59.523	144.532	96.131	150.284	------ CRYSTAL <--> LIQUID -----		
3200	59.859	146.331	97.595	155.955	-705.840	-81.147	1.325
3300	60.201	148.178	99.100	161.958	-703.975	-61.654	0.976
3400	60.541	149.980	100.570	167.995	-702.095	-42.218	0.649
3500	60.878	151.740	102.007	174.066	-700.203	-22.837	0.341
3600	61.214	153.460	103.412	180.171	-698.299	-3.511	0.051
3700	61.551	155.142	104.788	186.309	-696.387	15.763	-0.223
3800	61.888	156.787	106.135	192.481	-694.467	34.985	-0.481
3900	62.225	158.399	107.454	198.687	-692.542	54.156	-0.725
4000	62.561	159.979	108.748	204.926	-690.613	73.277	-0.957

PREVIOUS: December 1965 CURRENT: December 1974

Magnesium Oxide (MgO) $Mg_1O_1(cr)$

MAGNESIUM OXIDE (MgO) LIQUID M_r = 40.3044

$S°(298.15\ K) = [48.366]\ J\ K^{-1}\ mol^{-1}$

$T_{fus} = 3105 \pm 30\ K$

$\Delta_f H°(298.15\ K) = [-532.609]\ kJ\ mol^{-1}$

$\Delta_{fus}H° = [77.822 \pm 16.7]\ kJ\ mol^{-1}$

Enthalpy of Formation

$\Delta_f H°$(MgO, ℓ, 298.15 K) is calculated from that of the crystal by adding $\Delta_{fus}H°$ and the difference in enthalpy, H°(3105 K)-H°(298.15 K), between the crystal and liquid.

Heat Capacity and Entropy

$C_p°$ is assumed to be 16 cal K^{-1} mol^{-1} except below the glass transition (2100 K) where $C_p°$(ℓ) is taken equal to $C_p°$(cr). Enthalpy data yield $C_p° \geq 19$ cal K^{-1} mol^{-1} for liquid BeO (1) but there may be a positive bias due to volatilization. S°(298.15 K) is calculated in a manner analogous to that used for the enthalpy of formation.

Fusion Data

Documented values for T_{fus} (in °C, IPTS-48) include 2800±20 (2), 2827±20 (3) and 2852 (4, 5). We adopt the median value (3) which becomes 2832°C = 3105 K on IPTS-68.

Calorimetric data for $\Delta_{fus}H°$ are not available. Kelley (6) derived an apparent $\Delta_{fus}H°$ = 18.5 kcal mol^{-1} at T_{fus} = 2915 K from liquidus data for the MgO-ZrO_2 binary (7). The old (7) and new (8, 9) data for MgO-ZrO_2 imply gross non-ideality which makes the system unsuitable for obtaining the enthalpy of fusion. Apparent $\Delta_{fus}H°$ values ranging from 8 to 30 kcal mol^{-1} are obtained from more suitable binary systems (9, 10) involving CaO, Al_2O_3, Cr_2O_3, PuO_2 and UO_2. These values are so uncertain that we assume $\Delta_{fus}H°$ = 18.6 kcal mol^{-1} based on Δ_{fus} = 6 cal k^{-1} mol^{-1}. $\Delta_{fus}H°$ = 20 kcal mol^{-1} was measured calorimetrically for BeO (1), but the crystal is hexagonal or tetragonal rather than cubic as in the case of MgO.

Vaporization Data

Vaporization rates based on weight loss at T_{fus} were repoted by Noguchi (8) for MgO and six other refractory oxides. Vaporization of MgO near 2000 K has been studied by Langmuir-torsion (11), Langmuir-weight-loss (12, 13), Knudsen-mass-spectrometric (14), Knudsen-effusion (15) and transpiration (15, 16) methods. Vaporization near 2000 K is primarily to the elements (14), Mg(g) and presumably O(g) and O_2(g), rather than to MgO(g). Quantitative interpretation of the vaporization species is uncertain. The adopted tables (1) predict that MgO(g) is insignificant near 2000 K but should become one of the significant components of the vapor near T_{fus}.

References

1. JANAF Thermochemical Tables: BeO(cr, ℓ) and MgO(cr, g), 12-31-74.
2. B. Riley, Rev. Int. Hautes Temp. Refract. 3, 327 (1966).
3. R. N. McNally, F. I. Peters and P. H. Ribbe, J. Amer. Ceram. Soc. 44, 491 (1961).
4. C. W. Kanolt, Bull. Bur. Stand. 10, 295 (1914); J. Wash. Acad. Sci. 3, 315 (1913).
5. S. J. Schneider, U. S. Nat. Bur. Stand. Monograph 68, 1963.
6. K. K. Kelley, U. S. Bur. Mines Bull. 393, 1936.
7. F. Ebert and E. Cohn, Z. Anorg. Allg. Chem. 213, 321 (1933).
8. T. Noguchi, Advan. High Temp. Chem. 2, 235 (1969).
9. N. A. Toropov, V. P. Barzakovskii, V. A. Lapin and N. N. Kurtseva, "Handbook of Phase Diagrams of Silicate Systems, Vol. I, Binary Systems," Second revised edition (1969), English transl. (avail. NTIS), Israel Program for Scientific Translations, Jerusalem, 1972.
10. E. M. Levin, C. R. Robbins and H. F. McMurdie, "Phase Diagrams for Ceramists," Third edition and 1969 Supplement. The American Ceramic Society, Columbus, Ohio, 1964 and 1969.
11. C. B. Alcock and M. Peleg, Trans. Brit. Ceram. Soc. 66, 217 (1967); High Temp. Sci. 6, 52 (1974); Bull. Soc. Br. Ceram. 72, 25 (1966).
12. W. Kroenert, Glas-Email-Keramo-Tech. 23, 319 (1972).
13. A. G. Karaulov et al., Refractories (USSR), No. 2, 121, 1971.
14. R. F. Porter, W. A. Chupka, and M. G. Inghram, J. Chem. Phys. 23, 1347 (1955).
15. R. L. Altman, J. Phys. Chem. 67, 366 (1963).
16. C. A. Alexander, J. S. Ogden, and A. Levy, J. Chem. Phys. 39, 3057 (1963).

Magnesium Oxide (MgO) Mg_1O_1(l)

Enthalpy Reference Temperature = T_r = 298.15 K   Standard State Pressure = p° = 0.1 MPa

T/K	$C_p°$ (J K⁻¹mol⁻¹)	S°	$-[G°-H°(T_r)]/T$	$H°-H°(T_r)$ (kJ mol⁻¹)	$\Delta_f H°$	$\Delta_f G°$	Log K_f
0							
100							
200							
298.15	37.106	48.366	48.366	0.	-532.609	-506.707	88.773
300	37.244	48.596	48.367	0.069	-532.614	-506.546	88.197
400	42.561	60.120	49.902	4.087	-532.636	-497.843	65.012
500	45.544	69.966	52.956	8.505	-532.414	-489.167	51.103
600	47.430	78.448	56.514	13.160	-532.106	-480.546	41.835
700	48.748	85.862	60.188	17.972	-531.794	-471.978	35.219
800	49.740	92.439	63.816	22.898	-531.525	-463.452	30.260
900	50.539	98.344	67.330	27.913	-531.331	-454.955	26.405
1000	51.208	103.705	70.704	33.001	-539.830	-445.763	23.284
1100	51.794	108.613	73.930	38.151	-539.865	-436.354	20.721
1200	52.325	113.143	77.012	43.358	-539.864	-426.943	18.584
1300	52.810	117.351	79.955	48.615	-539.829	-417.535	16.777
1400	53.262	121.281	82.768	53.918	-667.172	-404.962	15.109
1500	53.693	124.970	85.459	59.266	-665.724	-386.284	13.452
1600	54.107	128.449	88.039	64.656	-664.246	-367.702	12.004
1700	54.509	131.741	90.513	70.087	-662.740	-349.214	10.730
1800	54.898	134.868	92.891	75.558	-661.206	-330.816	9.600
1900	55.278	137.846	95.180	81.067	-659.645	-312.503	8.591
2000	55.651	140.691	97.385	86.613	-658.059	-294.274	7.686
2100	56.019	143.415	99.512	92.197	-656.447	-276.124	6.868
2100.001	56.019	143.415	99.512	92.197	GLASS <--> LIQUID TRANSITION		
2100.001	66.944	143.415	99.512	92.197			
2200	66.944	146.530	101.579	98.891	-653.736	-258.077	6.128
2300	66.944	149.505	103.599	105.585	-651.037	-240.153	5.454
2400	66.944	152.354	105.571	112.280	-648.350	-222.346	4.839
2500	66.944	155.087	107.498	118.974	-645.674	-204.651	4.276
2600	66.944	157.713	109.379	125.669	-643.012	-187.062	3.758
2700	66.944	160.239	111.216	132.363	-640.362	-169.576	3.281
2800	66.944	162.674	113.011	139.057	-637.725	-152.188	2.839
2900	66.944	165.023	114.764	145.752	-635.103	-134.894	2.430
3000	66.944	167.293	116.477	152.446	-632.496	-117.690	2.049
3100	66.944	169.488	118.152	159.140	-629.904	-100.572	1.695
3105.000	66.944	169.595	118.235	159.475	------ CRYSTAL <--> LIQUID -----		
3200	66.944	171.613	119.790	165.835	-627.329	-83.538	1.364
3300	66.944	173.673	121.391	172.529	-624.772	-66.584	1.054
3400	66.944	175.671	122.959	179.224	-622.235	-49.708	0.764
3500	66.944	177.612	124.493	185.918	-619.719	-32.905	0.491
3600	66.944	179.498	125.994	192.612	-617.226	-16.174	0.235
3700	66.944	181.332	127.465	199.307	-614.757	0.487	-0.007
3800	66.944	183.117	128.906	206.001	-612.315	17.083	-0.235
3900	66.944	184.856	130.319	212.696	-609.901	33.615	-0.450
4000	66.944	186.551	131.704	219.390	-607.518	50.084	-0.654
4100	66.944	188.204	133.062	226.084	-605.167	66.495	-0.847
4200	66.944	189.817	134.394	232.779	-602.851	82.849	-1.030
4300	66.944	191.393	135.701	239.473	-600.572	99.149	-1.204
4400	66.944	192.932	136.984	246.168	-598.332	115.395	-1.370
4500	66.944	194.436	138.244	252.862	-596.133	131.591	-1.527
4600	66.944	195.907	139.482	259.556	-593.978	147.739	-1.678
4700	66.944	197.347	140.698	266.251	-591.869	163.840	-1.821
4800	66.944	198.756	141.893	272.945	-589.807	179.897	-1.958
4900	66.944	200.137	143.067	279.640	-587.796	195.912	-2.088
5000	66.944	201.489	144.222	286.334	-585.838	211.886	-2.214

PREVIOUS: December 1965   CURRENT: December 1974

Magnesium Oxide (MgO) Mg_1O_1(l)

MAGNESIUM OXIDE (MgO) CRYSTAL-LIQUID M_r = 40.3044 Magnesium Oxide (MgO) $Mg_1O_1(cr,l)$

0 to 3105 K crystal
above 3105 K liquid

Refer to the individual tables for details.

Enthalpy Reference Temperature = T_r = 298.15 K Standard State Pressure = $p°$ = 0.1 MPa

T/K	$C_p°$ (J K⁻¹mol⁻¹)	S° (J K⁻¹mol⁻¹)	$-[G°-H°(T_r)]/T$ (J K⁻¹mol⁻¹)	$H°-H°(T_r)$ (kJ mol⁻¹)	$\Delta_f H°$ (kJ mol⁻¹)	$\Delta_f G°$ (kJ mol⁻¹)	Log K_f
0	0.	0.	INFINITE	-5.159	-597.060	-597.060	INFINITE
100	7.802	2.548	52.212	-4.966	-598.962	-589.601	307.976
200	26.681	14.096	30.037	-3.188	-600.646	-579.488	151.347
298.15	37.106	26.924	26.924	0.	-601.241	-568.945	99.677
300	37.244	27.154	26.925	0.069	-601.245	-568.745	99.027
400	42.561	38.678	28.460	4.087	-601.268	-557.898	72.854
500	45.544	48.523	31.513	8.505	-601.046	-547.078	57.153
600	47.430	57.006	35.072	13.160	-600.738	-536.312	46.690
700	48.748	64.420	38.746	17.972	-600.426	-525.600	39.221
800	49.740	70.996	42.374	22.898	-600.156	-514.930	33.621
900	50.539	76.902	45.888	27.913	-599.962	-504.289	29.268
1000	51.208	82.262	49.262	33.001	-608.462	-492.952	25.749
1100	51.794	87.171	52.488	38.151	-608.496	-481.399	22.860
1200	52.325	91.701	55.569	43.358	-608.495	-469.844	20.452
1300	52.810	95.908	58.512	48.615	-608.461	-458.291	18.414
1400	53.262	99.839	61.325	53.918	-735.804	-443.575	16.550
1500	53.693	103.528	64.017	59.266	-734.355	-422.752	14.722
1600	54.107	107.007	66.596	64.656	-732.877	-402.026	13.125
1700	54.509	110.299	69.071	70.087	-731.371	-381.394	11.719
1800	54.898	113.426	71.449	75.558	-729.837	-360.851	10.472
1900	55.278	116.404	73.737	81.067	-728.277	-340.395	9.358
2000	55.651	119.249	75.942	86.613	-726.690	-320.021	8.358
2100	56.019	121.973	78.070	92.197	-725.079	-299.727	7.455
2200	56.379	124.587	80.125	97.816	-723.442	-279.510	6.636
2300	56.738	127.101	82.114	103.472	-721.781	-259.368	5.890
2400	57.094	129.524	84.039	109.164	-720.097	-239.300	5.208
2500	57.445	131.862	85.905	114.891	-718.389	-219.301	4.582
2600	57.797	134.121	87.716	120.653	-716.659	-199.372	4.005
2700	58.145	136.309	89.476	126.450	-714.906	-179.509	3.473
2800	58.491	138.430	91.187	132.282	-713.132	-159.712	2.979
2900	58.836	140.489	92.851	138.148	-711.338	-139.979	2.521
3000	59.177	142.489	94.473	144.049	-709.524	-120.308	2.095
3100	59.516	144.436	96.053	149.987	-707.689	-100.697	1.697
3105.000	59.523	144.532	96.131	150.284	CRYSTAL <--> LIQUID		
3105.000	66.944	169.595	96.131	228.107	TRANSITION		
3200	66.944	171.613	98.342	234.466	-627.329	-83.538	1.364
3300	66.944	173.673	100.594	241.161	-624.772	-66.584	1.054
3400	66.944	175.671	102.773	247.855	-622.235	-49.708	0.764
3500	66.944	177.612	104.884	254.550	-619.719	-32.905	0.491
3600	66.944	179.498	106.930	261.244	-617.226	-16.174	0.235
3700	66.944	181.332	108.916	267.938	-614.757	0.487	-0.007
3800	66.944	183.117	110.846	274.633	-612.315	17.083	-0.235
3900	66.944	184.856	112.721	281.327	-609.901	33.615	-0.450
4000	66.944	186.551	114.546	288.022	-607.518	50.084	-0.654
4100	66.944	188.204	116.322	294.716	-605.167	66.495	-0.847
4200	66.944	189.817	118.053	301.410	-602.851	82.849	-1.030
4300	66.944	191.393	119.740	308.105	-600.572	99.149	-1.204
4400	66.944	192.932	121.386	314.799	-598.332	115.395	-1.370
4500	66.944	194.436	122.993	321.494	-596.133	131.591	-1.527
4600	66.944	195.907	124.562	328.188	-593.978	147.739	-1.678
4700	66.944	197.347	126.095	334.882	-591.869	163.840	-1.821
4800	66.944	198.756	127.595	341.577	-589.807	179.897	-1.958
4900	66.944	200.137	129.061	348.271	-587.796	195.912	-2.088
5000	66.944	201.489	130.496	354.966	-585.838	211.886	-2.214

PREVIOUS: CURRENT: December 1974

Magnesium Oxide (MgO) $Mg_1O_1(cr,l)$

MAGNESIUM OXIDE (MgO) IDEAL GAS M_r = 40.3044

$\Delta_f H°(0\ K)$ = 58.59 ± 25.1 kJ mol^{-1}

S°(298.15 K) = 213.269 ± 3.3 J K^{-1} mol^{-1} $\Delta_f H°(298.15\ K)$ = 58.16 ± 25.1 kJ mol^{-1}

Electronic States and Molecular Constants (σ=1)

State	ε_i, cm^{-1}	g_i	ω_e, cm^{-1}	$\omega_e x_e$, cm^{-1}	B_e, cm^{-1}	α_e, cm^{-1}	r_e, Å	Source
$X^1\Sigma^+$	0.0	1	785.1	5.18	0.5743	0.0050	1.749	(1)
$^3\Pi$	[2300]	6	[664.4]	[3.9]	[0.5050]	[0.0040]	[1.864]	(2-7)
$A^1\Pi$	3503.3	2	664.4	3.9	0.5050	0.0040	1.864	(1)
$^3\Sigma^+$	[14000]	3	[824.1]	[4.8]	[0.5822]	[0.0045]	[1.737]	(2-5)
$B^1\Sigma^+$	20004.	1	824.1	4.76	0.5822	0.0045	1.737	(1)
$^3\Sigma^+$	[28000]	3	[632.5]	[5.3]	[0.501]	[0.0048]	[1.872]	(4, 3)
$^3\Sigma$	[29000]	6	[632.5]	[5.3]	[0.501]	[0.0048]	[1.872]	(4, 3)
$D^1\Delta$	29775.	2	632.5	5.3	0.5014	0.0048	1.872	(1, 8, 9)
$^3\Sigma^-$	[30000]	3	[632.5]	[5.3]	[0.501]	[0.0048]	[1.872]	(4, 3)
$C^1\Sigma^-$	30004.	1	632.4	5.2	0.5008	0.0048	1.873	(1)
$E^1\Sigma^+$	37684.	1	[632.4]	[5.2]	0.5273	[0.0048]	1.825	(10, 3, 4)
$^3\Pi$	[37000]	6	[710]	[5.]	[0.5615]	[0.005]	[1.768]	(4, 3)
$F^1\Pi$	37879.	2	710	[5.]	0.5615	[0.005]	1.768	(10, 3, 4)
$^3\Pi$	[39000]	6	[790]	[5.]	[0.5249]	[0.005]	[1.83]	(4, 3)
$G^1\Pi$	39868.	2	790	[5.]	0.5249	[0.005]	1.83	(10, 3, 4)

Enthalpy of Formation

We adopt D_0^0 = 80±6 kcal mol^{-1} and $\Delta_f H°(298.15\ K)$ = 13.9±6 kcal mol^{-1} based on equilibria analyzed below. Greatest weight is given to mass-spectrometric data (11) for reactions A and B. We give "<" or ">" for $\Delta_f H°$ and D_0^0 values which may have a systematic bias. The JANAF differences in $\Delta_f H°(298.15\ K)$ for WO_3-WO_2 (3) may be biased by as much as -3.5 kcal mol^{-1}, leading to a positive bias in D_0^0 from reaction B. Mass spectra (12) and transpiration data (13) yield an upper limit to D_0^0; we neglect $Mg(OH)_2$ and MgOH which probably are important in (13). Cotton and Jenkins (15) concluded that hydroxides were negligible in their H_2-O_2-N_2 flames; we doubt this conclusion and presume that reaction D yields an upper limit to D_0^0. Our adopted value is comparable with those of Richards et al. (7, 1) and Brewer & Rosenblatt (16), considering the difference in functions for MgO.

Other published values of D_0^0 are dismissed for reasons cited in critical reviews (17, 18, 11). Schofield's criticism (17) of flame studies is supported by new evidence for importance of hydroxides (3, CaO, g). Inadequate allowance for hydroxides may bias a recent flame result of D_0^0 = 88±2.3 kcal mol^{-1} (19); furthermore, we cannot firm this D_0^0 from the published information. D_0^0 = 83 kcal mol^{-1} was derived (20) by fitting an electronegativity potential function to X'Σ. This state is predicted (18, 4, 11) to dissociate to excited state $O(^1D)$ at 45.4 kcal mol^{-1}. Theoretical calculations (21) support this prediction. Linear Birge-Sponer extrapolation yields a very low estimate for D_0^0, just as it does for the alkali halides (18), unless $X^1\Sigma$ fails to dissociate to $O^1(D)$ as predicted.

Source	Method	Reaction[a]	T/K	Data Points	δS cal K^{-1} mol^{-1}	$\Delta_r H°(298.15\ K)$ kcal mol^{-1} 2nd law	3rd law	$\Delta_f H°(298.15\ K)$ kcal mol^{-1}	D_0^0
(11)	Knudsen mass spec.	A	2026-2274	8	-13±5	12±11	39.5±3	15.3	78.6±3
	Knudsen mass spec.	B	2106-2225	6	-18±8	23±16	63.1±3	>10.1	<83.8±5
(12)	Knudsen mass spec.	C	1950	1	-	-	>148.7	>5.0	<86.9±5
(13)	Transpiration	C	1780-2010	7	-8±3	132±5	>148.3±2	>4.7	<89.2±2
(14)	Transpiration	C	2033-2175	8	-52±32	53±68	162.8±8	19.1	74.7±8
(15)	Flame spec.	D	1570-2370	4	-9±1	-11±2	>5.7±6	>-1.6	<95.4±6

[a]Reactions: A) Mg(g) + O_2(g) = MgO(g) + O(g) C) MgO(cr) = MgO(g)
B) Mg(g) + WO_3(g) = MgO(g) + WO_2(g) D) Mg(g) + OH(g) = MgO(g) + H(g)

Heat Capacity and Entropy

Electronic levels (T_0) and vibrational-rotational constants of observed states are from Rosen (1) and other recent sources (3-10). States are listed in the isoconfigurational order given by Schamps and Lefebvre-Brion (4). Designations (3) of the new E, F and G states are based on theoretical calculations (4). Field (2) concluded that calculations (4-6) of the isoconfigurational $A^1\Pi$-$^3\Pi$ separation should be adequate for estimating the low-lying $^3\Pi$ state. The other potentially low-lying state ($^3\Sigma^+$) is estimated (4) similarly from isoconfigurational $B^1\Sigma^+$. Higher estimated levels are also based on calculations (4). Vibrational-rotational constants are estimated in isoconfigurational groups by comparison with BeO, CaO, SrO and BaO (3). Low-lying levels and their vibrational-rotational constants cause marked changes in the thermodynamic functions, which are calculated using first-order anharmonic corrections to Q_r^i and Q_v^i in the partition function $Q = Q_t \Sigma Q_r^i Q_v^i g_i \exp(-c_2\varepsilon_i/T)$.

References

1. B. Rosen, "Spectroscopic Data Relative to Diatomic Molecules," Pergamon Press, New York, p. 255. 1970.
2. R. W. Field, J. Chem. Phys. 60, 2400 (1974).
3. JANAF Thermochemical Tables: MgO(g), BeO(g), CaO(g), 12-31-74; SrO(g), BaO(g), 6-30-74; WO_2(g), WO_3(g), 9-30-66.
4. J. Schamps and H. Lefebvre-Brion, J. Chem. Phys. 56, 573 (1972).
5. B. Huron and P. Rancurel, Chem. Phys. Lett. 13, 515 (1972).
6. W. M. Huo, K. F. Freed and W. Klemperer, J. Chem. Phys. 46, 3556 (1967).
7. W. G. Richards, G. Verhaegen and C. M. Moser, J. Chem. Phys. 45, 3226 (1966).
8. D. W. Green, Ph.D. Thesis, Univ. California, Berkeley, UCRL-17878-Rev., 1968.
9. J. Schamps, G. Gandara and M. Becart, Can. Spectrosc. 14, 13 (1969).
10. M. Singh, J. Phys. B4, 565 (1971); B6, 1339, 1917 (1973).
11. J. Drowart, G. Exsteen and G. Verhaegen, Trans. Faraday Soc. 60, 1920 (1964).
12. R. F. Porter, W. A. Chupka and M. G. Inghram, J. Chem. Phys. 23, 1347 (1955).
13. C. A. Alexander, J. S. Ogden and A. Levy, J. Chem. Phys. 39, 3057 (1963).
14. R. L. Altman, J. Phys. Chem. 67, 366 (1963).
15. D. H. Cotton and D. R. Jenkins, Trans. Faraday Soc. 65, 376 (1969).
16. L. Brewer and G. Rosenblatt, Advan. High Temp. Chem. 2, 1 (1969).
17. K. Schofield, Chem. Rev. 67, 707 (1967).
18. A. G. Gaydon, "Dissociation Energies and Spectra of Diatomic Molecules," 3rd ed., Chapman and Hall Ltd., London, pp. 236-244, 109, 1968.
19. P. J. T. Zeegers, W. P. Townsend and J. D. Winefordner, Spectrochim. Acta 24B, 243 (1969).
20. B. Rai and S. N. Rai, Indian J. Pure Appl. Phys. 10, 401 (1972).
21. N. J. Stag and W. G. Richards, Mol. Phys. 27, 787 (1974).

Magnesium Oxide (MgO) $Mg_1O_1(g)$

Enthalpy Reference Temperature = T_r = 298.15 K Standard State Pressure = p° = 0.1 MPa

T/K	$C_p°$ (J $K^{-1}mol^{-1}$)	S°	-[G°-H°(T_r)]/T	H°-H°(T_r) (kJ mol^{-1})	$\Delta_f H°$	$\Delta_f G°$	Log K_f
0	0.	0.	INFINITE	-8.909	58.588	58.588	INFINITE
100	29.124	180.485	240.493	-6.001	59.402	50.970	-26.624
200	30.117	200.873	216.143	-3.054	58.886	42.689	-11.149
250	31.121	207.699	213.793	-1.524	58.517	38.682	-8.082
298.15	32.173	213.269	213.269	0.	58.158	34.895	-6.113
300	32.215	213.468	213.270	0.060	58.144	34.750	-6.051
350	33.401	218.522	213.666	1.699	57.783	30.880	-4.609
400	34.783	223.069	214.562	3.403	57.446	27.060	-3.534
450	36.452	227.259	215.743	5.182	57.147	23.280	-2.702
500	38.434	231.200	217.093	7.053	56.901	19.531	-2.040
600	43.078	238.608	220.068	11.124	56.624	12.088	-1.052
700	47.785	245.608	223.221	15.671	56.672	4.666	-0.348
800	51.560	252.250	226.439	20.649	56.994	-2.783	0.182
900	53.890	258.472	229.657	25.934	57.457	-10.282	0.597
1000	54.795	264.208	232.829	31.379	49.315	-17.121	0.894
1100	54.608	269.429	235.923	36.857	49.607	-23.779	1.129
1200	53.729	274.147	238.915	42.278	49.823	-30.461	1.326
1300	52.496	278.400	241.792	47.591	49.914	-37.156	1.493
1400	51.139	282.241	244.546	52.773	-77.551	-40.685	1.518
1500	49.802	285.723	247.177	57.820	-76.404	-38.093	1.327
1600	48.558	288.897	249.687	62.737	-75.399	-35.573	1.161
1700	47.439	291.807	252.080	67.535	-74.525	-33.111	1.017
1800	46.454	294.490	254.363	72.229	-73.768	-30.697	0.891
1900	45.599	296.978	256.541	76.831	-73.114	-28.323	0.779
2000	44.862	299.298	258.621	81.353	-72.552	-25.980	0.679
2100	44.230	301.471	260.611	85.806	-72.070	-23.664	0.589
2200	43.691	303.516	262.515	90.202	-71.658	-21.368	0.507
2300	43.231	305.447	264.340	94.547	-71.308	-19.091	0.434
2400	42.841	307.279	266.091	98.850	-71.012	-16.827	0.366
2500	42.509	309.021	267.774	103.117	-70.764	-14.574	0.305
2600	42.229	310.682	269.392	107.354	-70.559	-12.331	0.248
2700	41.993	312.272	270.951	111.565	-70.393	-10.095	0.195
2800	41.795	313.795	272.454	115.754	-70.262	-7.864	0.147
2900	41.629	315.259	273.905	119.925	-70.163	-5.637	0.102
3000	41.493	316.668	275.307	124.081	-70.094	-3.414	0.059
3100	41.381	318.026	276.664	128.224	-70.053	-1.191	0.020
3200	41.291	319.339	277.977	132.357	-70.040	1.029	-0.017
3300	41.220	320.608	279.250	136.483	-70.052	3.250	-0.051
3400	41.167	321.838	280.484	140.602	-70.090	5.472	-0.084
3500	41.128	323.030	281.683	144.717	-70.154	7.695	-0.115
3600	41.103	324.189	282.848	148.828	-70.243	9.921	-0.144
3700	41.091	325.315	283.980	152.938	-70.359	12.149	-0.172
3800	41.090	326.411	285.082	157.047	-70.503	14.381	-0.198
3900	41.099	327.478	286.156	161.156	-70.674	16.617	-0.223
4000	41.117	328.519	287.202	165.267	-70.874	18.858	-0.246
4100	41.144	329.534	288.222	169.380	-71.105	21.104	-0.269
4200	41.179	330.526	289.218	173.496	-71.367	23.356	-0.290
4300	41.222	331.496	290.190	177.616	-71.662	25.615	-0.311
4400	41.272	332.444	291.139	181.741	-71.992	27.881	-0.331
4500	41.329	333.372	292.067	185.871	-72.358	30.154	-0.350
4600	41.393	334.281	292.975	190.007	-72.761	32.437	-0.368
4700	41.463	335.172	293.864	194.149	-73.203	34.729	-0.386
4800	41.539	336.046	294.733	198.300	-73.686	37.030	-0.403
4900	41.621	336.903	295.585	202.458	-74.212	39.342	-0.419
5000	41.709	337.745	296.420	206.624	-74.781	41.665	-0.435
5100	41.802	338.572	297.238	210.800	-75.395	44.000	-0.451
5200	41.901	339.384	298.041	214.985	-76.057	46.347	-0.466
5300	42.004	340.183	298.829	219.180	-76.766	48.709	-0.480
5400	42.113	340.970	299.602	223.386	-77.525	51.083	-0.494
5500	42.227	341.743	300.361	227.603	-78.335	53.471	-0.508
5600	42.345	342.505	301.107	231.831	-79.198	55.876	-0.521
5700	42.467	343.256	301.840	236.072	-80.115	58.296	-0.534
5800	42.594	343.995	302.560	240.325	-81.087	60.733	-0.547
5900	42.725	344.725	303.269	244.591	-82.116	63.186	-0.559
6000	42.860	345.444	303.966	248.870	-83.202	65.658	-0.572

PREVIOUS: December 1974 (1 atm) CURRENT: December 1974 (1 bar)

Magnesium Oxide (MgO) $Mg_1O_1(g)$

MAGNESIUM SILICATE ($MgSiO_3$) CRYSTAL (I-II-III) M_r = 100.3887

$S°(298.15\ K) = 67.77 \pm 0.8\ J\ K^{-1}\ mol^{-1}$
$T_{trs1} = 903\ K$
$T_{trs2} = 1258\ K$
$T_{fus} = 1850 \pm 2\ K$

$\Delta_f H°(0\ K) = -1539.79 \pm 4.2\ kJ\ mol^{-1}$
$\Delta_f H°(298.15\ K) = -1548.92 \pm 4.2\ kJ\ mol^{-1}$
$\Delta_{trs1} H° = 0.67\ kJ\ mol^{-1}$
$\Delta_{trs2} H° = 1.632\ kJ\ mol^{-1}$
$\Delta_{fus} H° = [75.312 \pm 20.9]\ kJ\ mol^{-1}$

Enthalpy of Formation

Torgeson et al. (1) report a $\Delta_r H°(298.15\ K) = -8.69 \pm 0.15\ kcal\ mol^{-1}$ for reaction (a). They used a hydrofluoric acid solution calorimeter and measured the enthalpies of solution of each component. They appear to have made the water-absorption correction to the enthalpy of solution of SiO_2(cr) in the wrong direction. We reverse the correction so that the new $\Delta_{sol}H$ almost overlaps the range observed by King (11). We derive $\Delta_f H°(MgSiO_3$, cr, 298.15 K) = $-370.20 \pm 1.0\ kcal\ mol^{-1}$ using the $\Delta_f H°$(298.15 K) of MgO(cr) and SiO_2(cr, low quartz) from the JANAF tables.

Reesman et al. (2) calculated a $\Delta_r G°(298.15\ K) = 23.72\ kcal\ mol^{-1}$ for reaction (b) using aqueous solubility data. We use the $\Delta_f G°$(298.15 K) of $H_2O(\ell)$ and OH^-(aq) from source 3, $\Delta_f G°$(298.15 K) of H_4SiO_4(aq) from source 4 and $\Delta_f G°$(298.15 K) of Mg^{++}(aq) from source 5 and we obtain a $\Delta_f G°(MgSiO_3$, cr, 298.15 K) = $-352.534\ kcal\ mol^{-1}$. We derive $\Delta_f H°(MgSiO_3$, cr, 298.15 K) = $-373.293\ kcal\ mol^{-1}$ using the above $\Delta_f G°$(298.15 K) of $MgSiO_3$(cr) and the entropies of Mg(cr), Si(cr) and O_2(g) from the JANAF tables.

We choose the enthalpy of formation derived from the result of Torgeson et al. (1) which is considered to be better since it relates directly to the oxides. Also the $\Delta_f G°(MgSiO_3$, ℓ, 1873 K) thus calculated is in very good agreement with that derived from the activities of SiO_2 and MgO (in the SiO_2 - MgO system) measured by Rein et al. (6). The $\Delta_f H°(MgSiO_3$, cr, 298.15 K) derived from the work of Reesman et al. is less accurate and involves uncertainties in the auxiliary data used to calculate the value.

Source	Method	Reaction	$\Delta_r H°$(298.15 K) kcal mol^{-1}	$\Delta_r G°$(298.15 K) kcal mol^{-1}	$\Delta_f H°$(298.15 K) kcal mol^{-1}
(1)	HF solution calorimeter	(a) $MgO(cr) + SiO_2(cr) = MgSiO_3(cr)$	-8.7±0.15		-370.20
(2)	Aqueous solubility	(b) $MgSiO_3(cr) + 3H_2O(\ell) = Mg^{++}(aq) + H_4SiO_4(aq) + 2OH^-(aq)$		23.72	-373.3

Heat Capacity and Entropy

Low temperature heat capacities are from the data (53 - 295 K) of Kelley (7). Wagner (8) measured the average heat capacities in the temperature range 580 - 1570 K. Using his data we derive the high temperature heat capacities which are joined smoothly with the low temperature values by a constrained fitting technique. The $C_p°$ values at 903 = 1258 K and above 1258 K are derived as 28.762 and 29.262 cal $K^{-1}\ mol^{-1}$, respectively, by assuming that the data of Wagner (8) pertain to the stable phase in these regions. The entropy is based on S°(50 K) = 0.49 cal $K^{-1}\ mol^{-1}$.

Transition Data

T_{trs1} = 903 K is obtained from Boyd et al. (9). The authors determined the phase diagram boundary at high pressures and extrapolated to one atmosphere. T_{trs2} = 1258 K is taken from Atlas (10). Below 903 K the phase is known as clinoenstatite. Between 903 and 1258 K it is rhombic enstatite and above 1258 K it is protoenstatite. We calculate $\Delta_{trs1}H°$ and $\Delta_{trs2}H°$ (from dT/dP (slopes of the inversion) = 2.6/kbar and 84/kbar reported by Boyd et al. (9, 13) and from the densities of clinoenstatite (3.210 gm/cc) and rhombic enstatite (3.208 gm/cc) reported by Stephenson et al. (11) and the density of protoenstatite (3.10 gm/cc) reported by Smith (12).

Fusion Data

See the liquid table for details.

References

1. D. R. Torgeson and Th. G. Sahama, J. Amer. Chem. Soc. 70, 2156 (1948).
2. A. L. Reesman and W. D. Keller, Am. Mineralogist 50, 1729-39 (1965).
3. U. S. Nat. Bur. Stand. Tech. Note 270-1, 1965.
4. U. S. Nat. Bur. Stand. Tech. Note 270-2, 1966.
5. U. S. Nat. Bur. Stand. Circ. 500, 1952.
6. R. H. Rein and J. Chipman, Trans. AIME 233, 415-425 (1965).
7. K. K. Kelley, J. Amer. Chem. Soc. 65, 339-41 (1943).
8. H. Wagner, Z. Anorg. Allg. Chem. 208, 1-22 (1932).
9. F. R. Boyd and J. L. England, Ann. Rept. Director of the Geophysical Laboratory, No. 1455, 117-120 (1965).
10. L. Atlas, J. Geol. 60, 125-47 (1952).
11. D. A. Stephenson, C. B. Sclar and J. V. Smith, Mineral. Mag. 35, 839-46 (1966).
12. J. V. Smith, Acta Cryst. 12, 515 (1959).
13. F. R. Boyd, J. L. England and B. T. C. Davis, J. Geophys. Res. 69, (10), 2101 (1964).

Magnesium Silicate ($MgSiO_3$) $Mg_1O_3Si_1$(cr)

Enthalpy Reference Temperature = T_r = 298.15 K — Standard State Pressure = p° = 0.1 MPa

T/K	$C_p°$ (J $K^{-1}mol^{-1}$)	S° (J $K^{-1}mol^{-1}$)	$-[G°-H°(T_r)]/T$ (J $K^{-1}mol^{-1}$)	$H°-H°(T_r)$ (kJ mol^{-1})	$\Delta_f H°$ (kJ mol^{-1})	$\Delta_f G°$ (kJ mol^{-1})	Log K_f
0	0.	0.	INFINITE	-12.113	-1539.788	-1539.788	INFINITE
100	23.912	11.117	124.420	-11.330	-1544.271	-1518.053	792.950
200	59.643	39.430	74.617	-7.037	-1547.531	-1490.410	389.255
298.15	81.927	67.768	67.768	0.	-1548.917	-1462.023	256.140
300	82.228	68.276	67.770	0.152	-1548.930	-1461.484	254.467
400	94.194	93.738	71.160	9.031	-1549.184	-1432.276	187.036
500	101.512	115.590	77.915	18.838	-1548.910	-1403.073	146.578
600	106.997	134.600	85.812	29.273	-1548.358	-1373.955	119.613
700	111.621	151.449	94.008	40.208	-1547.624	-1344.943	100.361
800	115.780	166.629	102.152	51.581	-1546.754	-1316.047	85.929
900	119.675	180.492	110.097	63.356	-1545.768	-1287.267	74.711
903.000	119.788	180.891	110.332	63.715		I <--> II	
903.000	120.340	181.632	110.332	64.384		TRANSITION	
1000	120.340	193.911	117.854	76.057	-1552.726	-1257.958	65.709
1100	120.340	205.381	125.298	88.091	-1552.043	-1228.515	58.337
1200	120.340	215.852	132.414	100.125	-1551.461	-1199.129	52.197
1258.000	120.340	221.532	136.393	107.105		II <--> III	
1258.000	122.432	222.829	136.393	108.737		TRANSITION	
1300	122.432	226.850	139.250	113.879	-1549.254	-1169.845	47.005
1400	122.432	235.923	145.836	126.122	-1676.053	-1137.516	42.441
1500	122.432	244.370	152.126	138.365	-1674.175	-1099.115	38.275
1600	122.432	252.271	158.141	150.609	-1672.377	-1060.837	34.633
1700	122.432	259.694	163.899	162.852	-1720.835	-1022.222	31.409
1800	122.432	266.692	169.417	175.095	-1718.964	-981.182	28.473
1850.000	122.432	270.046	172.091	181.217	--------	III <--> LIQUID	-------
1900	122.432	273.311	174.712	187.338	-1717.128	-940.244	25.849
2000	122.432	279.591	179.801	199.581	-1715.327	-899.402	23.490
2100	122.432	285.565	184.696	211.825	-1713.561	-858.649	21.358
2200	122.432	291.260	189.411	224.068	-1711.829	-817.979	19.421
2300	122.432	296.703	193.959	236.311	-1710.131	-777.388	17.655
2400	122.432	301.913	198.349	248.554	-1708.467	-736.870	16.038
2500	122.432	306.911	202.592	260.798	-1706.838	-696.421	14.551

PREVIOUS: September 1964 CURRENT: December 1967

Magnesium Silicate ($MgSiO_3$) $Mg_1O_3Si_1$(cr)

MAGNESIUM SILICATE ($MgSiO_3$) LIQUID M_r = 100.3887

$S°(298.15\ K) = [92.515]\ J\ K^{-1}\ mol^{-1}$

$T_{trs1} = 903\ K$

$T_{trs2} = 1258\ K$

$T_{fus} = 1850 \pm 2\ K$

$\Delta_f H°(298.15\ K) = [-1494.862 \pm 20.9]\ kJ\ mol^{-1}$

$\Delta_{trs1}H° = 0.67\ kJ\ mol^{-1}$

$\Delta_{trs2}H° = 1.632\ kJ\ mol^{-1}$

$\Delta_{fus}H° = [75.312 \pm 20.9]\ kJ\ mol^{-1}$

Enthalpy of Formation

$\Delta_f H°(MgSiO_3, \ell, 298.15\ K)$ is calculated from $\Delta_f H°(MgSiO_3, cr, 298.15\ K)$ by adding $\Delta_{fus}H°$ and the difference in enthalpy, $H°(1850\ K)-H°(298.15\ K)$, between the crystal and liquid.

Heat Capacity and Entropy

A glass transition is assumed at 900 K. Below 900 K the heat capacity is obtained from the heat capacity of the crystal. Above 900 K the heat capacity is assumed constant and estimated as 35 cal K^{-1} mol^{-1} or 7 cal K^{-1} g-atom^{-1}. $S°(\ell, 298.15\ K)$ is calculated in a manner analogous to that used for the enthalpy of formation.

Transition Data

See the crystal table for details.

Fusion Data

T_{fus} is taken from Bowen and Andersen (1). $MgSiO_3$(cr) melts incongruently in the temperature range 1830 - 1850 K at atmospheric pressure. $\Delta_{fus}H°$ is estimated from the phase diagrams of the $MgSiO_3$ - TiO_2 system reported by MacGregor (2). The phase diagrams were determined at 10 and 20 kbars pressure where $MgSiO_3$(cr) melts congruently.

References

1. N. L. Bowen and O. Andersen, J. Amer. Chem. Sci. 37, 487 (1914).
2. I. D. MacGregor, Ann. Rept. Director of the Geophysical Laboratory, No. 1455, 135-9, 1965.

Magnesium Silicate ($MgSiO_3$) $Mg_1O_3Si_1(l)$

Enthalpy Reference Temperature = T_r = 298.15 K Standard State Pressure = p° = 0.1 MPa

T/K	$C_p°$	S°	$-[G°-H°(T_r)]/T$	$H°-H°(T_r)$	$\Delta_f H°$	$\Delta_f G°$	Log K_f
	J K⁻¹mol⁻¹			kJ mol⁻¹			
0							
100							
200							
298.15	81.927	92.515	92.515	0.	-1494.862	-1415.346	247.963
300	82.228	93.022	92.516	0.152	-1494.875	-1414.853	246.348
400	94.194	118.484	95.906	9.031	-1495.129	-1388.120	181.270
500	101.512	140.337	102.661	18.838	-1494.855	-1361.392	142.224
600	106.997	159.347	110.559	29.273	-1494.303	-1334.747	116.200
700	111.621	176.195	118.754	40.208	-1493.569	-1308.211	97.620
800	115.780	191.375	126.899	51.581	-1492.699	-1281.789	83.692
900	119.675	205.239	134.844	63.356	-1491.713	-1255.483	72.866
900.000	119.675	205.239	134.844	63.356	GLASS <--> LIQUID TRANSITION		
900.000	146.440	205.239	134.844	63.356			
1000	146.440	220.668	142.668	78.000	-1496.729	-1228.717	64.182
1100	146.440	234.625	150.404	92.644	-1493.435	-1202.076	57.082
1200	146.440	247.367	157.961	107.288	-1490.244	-1175.731	51.178
1300	146.440	259.088	165.295	121.932	-1487.146	-1149.648	46.193
1400	146.440	269.941	172.387	136.576	-1611.545	-1120.632	41.811
1500	146.440	280.044	179.231	151.220	-1607.265	-1085.717	37.808
1600	146.440	289.495	185.830	165.864	-1603.067	-1051.085	34.314
1700	146.440	298.373	192.192	180.508	-1649.124	-1016.266	31.226
1800	146.440	306.743	198.326	195.152	-1644.852	-979.163	28.415
1850.000	146.440	310.756	201.310	202.474	-------- III <--> LIQUID -------		
1900	146.440	314.661	204.242	209.796	-1640.616	-942.295	25.906
2000	146.440	322.172	209.952	224.440	-1636.414	-905.650	23.653
2100	146.440	329.317	215.468	239.084	-1632.247	-869.215	21.621
2200	146.440	336.129	220.799	253.728	-1628.114	-832.976	19.777
2300	146.440	342.639	225.956	268.372	-1624.015	-796.926	18.099
2400	146.440	348.871	230.948	283.016	-1619.951	-761.053	16.564
2500	146.440	354.849	235.786	297.660	-1615.921	-725.349	15.155
2600	146.440	360.593	240.476	312.304	-1611.924	-689.805	13.858
2700	146.440	366.120	245.028	326.948	-1607.962	-654.414	12.660
2800	146.440	371.445	249.448	341.592	-1604.033	-619.169	11.551
2900	146.440	376.584	253.744	356.236	-1600.138	-584.065	10.520
3000	146.440	381.549	257.922	370.880	-1596.278	-549.093	9.561

PREVIOUS: September 1964 CURRENT: December 1967

Magnesium Silicate ($MgSiO_3$) $Mg_1O_3Si_1(l)$

MAGNESIUM SILICATE ($MgSiO_3$) CRYSTAL-CRYSTAL-CRYSTAL-LIQUID M_r = 100.3887

0 to 903 K	crystal, I
903 to 1258 K	crystal, II
1258 to 1850 K	crystal, III
above 1850 K	liquid

Refer to the individual tables for details.

Magnesium Silicate ($MgSiO_3$) $Mg_1O_3Si_1$(cr,l)

Enthalpy Reference Temperature = T_r = 298.15 K Standard State Pressure = $p°$ = 0.1 MPa

T/K	$C_p°$ (J K⁻¹mol⁻¹)	S°	$-[G°-H°(T_r)]/T$	$H°-H°(T_r)$ (kJ mol⁻¹)	$\Delta_fH°$	$\Delta_fG°$	Log K_f
0	0.	0.	INFINITE	-12.113	-1539.788	-1539.788	INFINITE
100	23.912	11.117	124.420	-11.330	-1544.271	-1518.053	792.950
200	59.643	39.430	74.617	-7.037	-1547.531	-1490.410	389.255
298.15	81.927	67.768	67.768	0.	-1548.917	-1462.023	256.140
300	82.228	68.276	67.770	0.152	-1548.930	-1461.484	254.467
400	94.194	93.738	71.160	9.031	-1549.184	-1432.276	187.036
500	101.512	115.590	77.915	18.838	-1548.910	-1403.073	146.578
600	106.997	134.600	85.812	29.273	-1548.358	-1373.955	119.613
700	111.621	151.449	94.008	40.208	-1547.624	-1344.943	100.361
800	115.780	166.629	102.152	51.581	-1546.754	-1316.047	85.929
900	119.675	180.492	110.097	63.356	-1545.768	-1287.267	74.711
903.000	119.788	180.891	110.332	63.715		I <--> II	
903.000	120.340	181.632	110.332	64.384		TRANSITION	
1000	120.340	193.911	117.854	76.057	-1552.726	-1257.958	65.709
1100	120.340	205.381	125.298	88.091	-1552.043	-1228.515	58.337
1200	120.340	215.852	132.414	100.125	-1551.461	-1199.129	52.197
1258.000	120.340	221.532	136.393	107.105		II <--> III	
1258.000	122.432	222.829	136.393	108.737		TRANSITION	
1300	122.432	226.850	139.250	113.879	-1549.254	-1169.845	47.005
1400	122.432	235.923	145.836	126.122	-1676.053	-1137.516	42.441
1500	122.432	244.370	152.126	138.365	-1674.175	-1099.115	38.275
1600	122.432	252.271	158.141	150.609	-1672.377	-1060.837	34.633
1700	122.432	259.694	163.899	162.852	-1720.835	-1022.222	31.409
1800	122.432	266.692	169.417	175.095	-1718.964	-981.182	28.473
1850.000	122.432	270.046	172.091	181.217		III <--> LIQUID	
1850.000	146.440	310.756	172.091	256.529		TRANSITION	
1900	146.440	314.661	175.792	263.851	-1640.616	-942.295	25.906
2000	146.440	322.172	182.925	278.495	-1636.414	-905.650	23.653
2100	146.440	329.317	189.727	293.139	-1632.247	-869.215	21.621
2200	146.440	336.129	196.228	307.783	-1628.114	-832.976	19.777
2300	146.440	342.639	202.453	322.427	-1624.015	-796.926	18.099
2400	146.440	348.871	208.425	337.071	-1619.951	-761.053	16.564
2500	146.440	354.849	214.164	351.715	-1615.921	-725.349	15.155
2600	146.440	360.593	219.686	366.359	-1611.924	-689.805	13.858
2700	146.440	366.120	225.007	381.003	-1607.962	-654.414	12.660
2800	146.440	371.445	230.143	395.647	-1604.033	-619.169	11.551
2900	146.440	376.584	235.104	410.291	-1600.138	-584.065	10.520
3000	146.440	381.549	239.904	424.935	-1596.278	-549.093	9.561

PREVIOUS: CURRENT: December 1967

Magnesium Silicate ($MgSiO_3$) $Mg_1O_3Si_1$(cr,l)

MAGNESIUM TITANIUM OXIDE ($MgTiO_3$) CRYSTAL M_r = 120.1832

$S°(298.15\ K) = 74.56 \pm 0.4\ J\ K^{-1}\ mol^{-1}$

$T_{fus} = 1953 \pm 20\ K$

$\Delta_f H°(0\ K) = -1563.26 \pm 6.3\ kJ\ mol^{-1}$

$\Delta_f H°(298.15\ K) = -1572.56 \pm 6.3\ kJ\ mol^{-1}$

$\Delta_{fus} H° = [90.374]\ kJ\ mol^{-1}$

Enthalpy of Formation

The adopted enthalpy of formation is calculated from $\Delta_r H°(298.15\ K) = -6.35 \pm 0.25\ kcal\ mol^{-1}$ for the reaction MgO(cr) + TiO_2(rutile) → $MgTiO_3$(cr) measured by Kelley et al. (1), using a solution calorimetric method. For the same reaction a value of $\Delta_r H°(298.15\ K) = -5.5 \pm 0.1\ kcal\ mol^{-1}$ was obtained with a calorimetric bomb method by Ronfilov and Feodos'ev (2). These authors used the oxidation of lampblack in order to initiate the formation of the titanate from the oxides. Unfortunately the desired enthalpy of reaction was only 1.2 to 1.5 percent of the total measured heat, and consequently this result appears to be more uncertain than that measured by the solution calorimetric method.

Heat Capacity and Entropy

Low temperature heat capacities are from the data (52 - 296 K) of Shomate (3). High temperature enthalpies have been measured (402 - 1720 K) by Naylor and Cook (4). High temperature heat capacities are derived from the enthalpies by a fitting technique which constrains the curve to join smoothly with the low temperature values. The entropy is based on $S°(52\ K) = 0.45\ cal\ K^{-1}\ mol^{-1}$.

Fusion Data

See the liquid table for details.

References

1. K. K. Kelley, S. S. Todd and E. G. King, U. S. Bur. Mines RI 5059, 1954.
2. B. I. Ronfilov and N. N. Freodos'ev, Russ. J. Inorg. Chem. 9, 1452 (1964).
3. C. H. Shomate, J. Amer. Chem. Soc. 67, 964 (1946).
4. B. F. Naylor and O. A. Cook, J. Amer. Chem. Soc. 68, 1003 (1946).

Magnesium Titanium Oxide ($MgTiO_3$) $Mg_1O_3Ti_1$(cr)

Enthalpy Reference Temperature = T_r = 298.15 K; Standard State Pressure = p° = 0.1 MPa

T/K	$C_p°$ (J K⁻¹mol⁻¹)	S° (J K⁻¹mol⁻¹)	$-[G°-H°(T_r)]/T$ (J K⁻¹mol⁻¹)	$H°-H°(T_r)$ (kJ mol⁻¹)	$\Delta_f H°$ (kJ mol⁻¹)	$\Delta_f G°$ (kJ mol⁻¹)	Log K_f
0	0.	0.	INFINITE	-13.556	-1563.260	-1563.260	INFINITE
100	25.510	10.857	138.595	-12.774	-1568.037	-1541.350	805.119
200	67.873	42.539	82.308	-7.954	-1571.507	-1513.096	395.180
298.15	91.881	74.559	74.559	0.	-1572.556	-1484.128	260.013
300	92.203	75.128	74.561	0.170	-1572.560	-1483.579	258.314
400	105.244	103.609	78.355	10.102	-1572.255	-1453.939	189.865
500	113.123	128.001	85.906	21.047	-1571.304	-1424.461	148.812
600	118.453	149.124	94.722	32.641	-1570.043	-1395.208	121.464
700	122.353	167.690	103.847	44.690	-1568.636	-1366.179	101.946
800	125.382	184.233	112.880	57.082	-1567.162	-1337.357	87.320
900	127.846	199.147	121.650	69.747	-1565.714	-1308.719	75.956
1000	129.938	212.728	130.089	82.639	-1572.993	-1279.523	66.835
1100	131.775	225.200	138.177	95.726	-1571.915	-1250.229	59.368
1200	133.432	236.738	145.915	108.987	-1574.903	-1220.909	53.145
1300	134.972	247.480	153.319	122.408	-1573.268	-1191.476	47.874
1400	136.423	257.536	160.408	135.979	-1699.017	-1159.002	43.243
1500	137.821	266.996	167.202	149.691	-1695.995	-1120.536	39.021
1600	139.185	275.934	173.721	163.542	-1692.979	-1082.270	35.333
1700	140.532	284.413	179.985	177.527	-1689.982	-1044.193	32.084
1800	141.879	292.483	186.012	191.648	-1687.017	-1006.291	29.202
1900	143.231	300.191	191.820	205.903	-1684.097	-968.553	26.627
1953.000	143.949	304.141	194.815	213.514	------ CRYSTAL <--> LIQUID -----		
2000	144.595	307.572	197.425	220.294	-1695.921	-930.513	24.303
2100	145.980	314.660	202.840	234.823	-1693.874	-892.293	22.195
2200	147.394	321.484	208.078	249.491	-1691.721	-854.172	20.281
2300	148.833	328.067	213.153	264.302	-1689.459	-816.152	18.535
2400	150.310	334.433	218.074	279.259	-1687.086	-778.233	16.938
2500	151.825	340.599	222.853	294.366	-1684.598	-740.415	15.470
2600	153.381	346.584	227.497	309.626	-1681.989	-702.699	14.117
2700	154.975	352.402	232.016	325.043	-1679.257	-665.085	12.867
2800	156.615	358.068	236.417	340.622	-1676.398	-627.574	11.708
2900	158.302	363.593	240.708	356.368	-1673.405	-590.170	10.630
3000	160.038	368.989	244.894	372.284	-1670.277	-552.869	9.626

PREVIOUS: June 1961 CURRENT: June 1967

Magnesium Titanium Oxide ($MgTiO_3$) $Mg_1O_3Ti_1$(cr)

MAGNESIUM TITANIUM OXIDE ($MgTiO_3$) LIQUID M_r = 120.1832

$S°(298.15\ K) = [111.082]\ J\ K^{-1}\ mol^{-1}$

$T_{fus} = 1953 \pm 20\ K$

$\Delta_f H°(298.15\ K) = [-1497.631 \pm 6.3]\ kJ\ mol^{-1}$

$\Delta_{fus} H° = [90.374]\ kJ\ mol^{-1}$

Enthalpy of Formation

$\Delta_f H°(MgTiO_3$, ℓ, 298.15 K) is calculated from $\Delta_f H°(MgTiO_3$, cr, 298.15 K) by adding $\Delta_{fus}H°$ and the difference in enthalpy, H°(1953 K)-H°(298.15 K), between the crystal and liquid.

Heat Capacity and Entropy

A glass transition is assumed at 1300 K. The heat capacity below 1300 K is obtained from the heat capacity of the crystal. Above 1300 K the heat capacity is assumed constant and estimated as 39 cal K^{-1} mol^{-1} or 7.8 cal K^{-1} g-atom^{-1}. S°(ℓ, 298.15 K) is calculated in a manner analogous to that used for the enthalpy of formation.

Fusion Data

The incongruent melting point is obtained from Massazza (1). Previous work by Coughanour (2) indicates congruent melting point at 1903 K. The $\Delta_{fus}H°$ is estimated from those of $Na_2O \cdot 2TiO_2$ and $Na_2O \cdot TiO_2$ observed by Naylor (3) and $CaO \cdot TiO_2 \cdot SiO_2$ observed by King (4).

References

1. F. Massazza and E. Sirchia, 16th Int. Congress of Pure and Appl. Chem., Inorg. Sect., Paris (1957), pp. 161-8, Butterworths Scientific Publications, London, 1958.
2. L. W. Coughanour and V. A. DeProsse, J. Res. Nat. Bur. Stand. 51, 87 (1953).
3. B. F. Naylor, J. Amer. Chem. Soc. 67, 2120 (1945).
4. E. G. King, R. L. Orr and K. R. Bonnickson, J. Amer. Chem. Soc. 76, 4320 (1954).

Magnesium Titanium Oxide ($MgTiO_3$) $Mg_1O_3Ti_1(l)$

Enthalpy Reference Temperature = T_r = 298.15 K Standard State Pressure = p° = 0.1 MPa

T/K	$C_p°$ (J K⁻¹mol⁻¹)	S° (J K⁻¹mol⁻¹)	$-[G°-H°(T_r)]/T$ (J K⁻¹mol⁻¹)	$H°-H°(T_r)$ (kJ mol⁻¹)	$\Delta_f H°$ (kJ mol⁻¹)	$\Delta_f G°$ (kJ mol⁻¹)	Log K_f
0							
100							
200							
298.15	91.881	111.082	111.082	0.	-1497.631	-1420.091	248.794
300	92.203	111.651	111.083	0.170	-1497.635	-1419.610	247.176
400	105.244	140.132	114.878	10.102	-1497.329	-1393.623	181.988
500	113.123	164.524	122.429	21.047	-1496.378	-1367.797	142.893
600	118.453	185.647	131.245	32.641	-1495.117	-1342.196	116.849
700	122.353	204.213	140.370	44.690	-1493.710	-1316.819	98.262
800	125.382	220.756	149.403	57.082	-1492.236	-1291.649	84.336
900	127.846	235.670	158.173	69.747	-1490.789	-1266.664	73.515
1000	129.938	249.251	166.612	82.639	-1498.067	-1241.121	64.829
1100	131.775	261.723	174.699	95.726	-1496.990	-1215.479	57.718
1200	133.432	273.261	182.438	108.987	-1499.977	-1189.811	51.791
1300	134.972	284.003	189.842	122.409	-1498.342	-1164.030	46.771
1300.000	134.972	284.003	189.842	122.409	GLASS <--> LIQUID		
1300.000	163.176	284.003	189.842	122.409	TRANSITION		
1400	163.176	296.095	197.005	138.726	-1621.344	-1135.313	42.359
1500	163.176	307.353	203.991	155.044	-1615.717	-1100.794	38.333
1600	163.176	317.884	210.784	171.361	-1610.233	-1066.645	34.822
1700	163.176	327.777	217.378	187.679	-1604.905	-1032.835	31.735
1800	163.176	337.104	223.772	203.997	-1599.743	-999.334	29.000
1900	163.176	345.926	229.971	220.314	-1594.761	-966.115	26.560
1953.000	163.176	350.416	233.179	228.963	------ CRYSTAL <--> LIQUID -----		
2000	163.176	354.296	235.980	236.632	-1604.658	-932.699	24.360
2100	163.176	362.258	241.805	252.949	-1600.822	-899.195	22.366
2200	163.176	369.848	247.454	269.267	-1597.019	-865.873	20.558
2300	163.176	377.102	252.935	285.585	-1593.251	-832.724	18.912
2400	163.176	384.047	258.254	301.902	-1589.518	-799.738	17.406
2500	163.176	390.708	263.420	318.220	-1585.818	-766.907	16.024
2600	163.176	397.108	268.439	334.537	-1582.152	-734.223	14.751
2700	163.176	403.266	273.320	350.855	-1578.520	-701.679	13.575
2800	163.176	409.200	278.067	367.173	-1574.922	-669.269	12.485
2900	163.176	414.926	282.688	383.490	-1571.357	-636.988	11.473
3000	163.176	420.458	287.189	399.808	-1567.828	-604.829	10.531
3100	163.176	425.809	291.575	416.125	-1564.332	-572.786	9.651
3200	163.176	430.989	295.851	432.443	-1560.871	-540.857	8.829
3300	163.176	436.011	300.023	448.761	-1557.446	-509.035	8.057
3400	163.176	440.882	304.094	465.078	-1554.057	-477.315	7.333
3500	163.176	445.612	308.070	481.396	-1550.706	-445.695	6.652
3600	163.176	450.209	311.955	497.713	-1547.392	-414.170	6.009
3700	163.176	454.680	315.752	514.031	-1953.217	-374.949	5.293
3800	163.176	459.031	319.466	530.349	-1948.749	-332.352	4.568
3900	163.176	463.270	323.099	546.666	-1944.379	-289.872	3.882
4000	163.176	467.401	326.655	562.984	-1940.107	-247.504	3.232

PREVIOUS: CURRENT: June 1967

Magnesium Titanium Oxide ($MgTiO_3$) $Mg_1O_3Ti_1(l)$

MAGNESIUM TITANIUM OXIDE ($MgTiO_3$) CRYSTAL–LIQUID M_r = 120.1832

0 to 1953 K crystal
above 1953 K liquid

Refer to the individual tables for details.

Magnesium Titanium Oxide ($MgTiO_3$) $Mg_1O_3Ti_1$(cr,l)

Enthalpy Reference Temperature = T_r = 298.15 K Standard State Pressure = p° = 0.1 MPa

T/K	C_p°	S°	$-[G^\circ-H^\circ(T_r)]/T$	$H^\circ-H^\circ(T_r)$	Δ_fH°	Δ_fG°	Log K_f
	J K⁻¹mol⁻¹			kJ mol⁻¹			
0	0.	0.	INFINITE	-13.556	-1563.260	-1563.260	INFINITE
100	25.510	10.857	138.595	-12.774	-1568.037	-1541.350	805.119
200	67.873	42.539	82.308	-7.954	-1571.507	-1513.096	395.180
298.15	91.881	74.559	74.559	0.	-1572.556	-1484.128	260.013
300	92.203	75.128	74.561	0.170	-1572.560	-1483.579	258.314
400	105.244	103.609	78.355	10.102	-1572.255	-1453.939	189.865
500	113.123	128.001	85.906	21.047	-1571.304	-1424.461	148.812
600	118.453	149.124	94.722	32.641	-1570.043	-1395.208	121.464
700	122.353	167.690	103.847	44.690	-1568.636	-1366.179	101.946
800	125.382	184.233	112.880	57.082	-1567.162	-1337.357	87.320
900	127.846	199.147	121.650	69.747	-1565.714	-1308.719	75.956
1000	129.938	212.728	130.089	82.639	-1572.993	-1279.523	66.835
1100	131.775	225.200	138.177	95.726	-1571.915	-1250.229	59.368
1200	133.432	236.738	145.915	108.987	-1574.903	-1220.909	53.145
1300	134.972	247.480	153.319	122.408	-1573.268	-1191.476	47.874
1400	136.423	257.536	160.408	135.979	-1699.017	-1159.002	43.243
1500	137.821	266.996	167.202	149.691	-1695.995	-1120.536	39.021
1600	139.185	275.934	173.721	163.542	-1692.979	-1082.270	35.333
1700	140.532	284.413	179.985	177.527	-1689.982	-1044.193	32.084
1800	141.879	292.483	186.012	191.648	-1687.017	-1006.291	29.202
1900	143.231	300.191	191.820	205.903	-1684.097	-968.553	26.627
1953.000	143.949	304.141	194.815	213.514	CRYSTAL <--> LIQUID		
1953.000	163.176	350.416	194.815	303.888	TRANSITION		
2000	163.176	354.296	198.517	311.557	-1604.658	-932.699	24.360
2100	163.176	362.258	206.127	327.875	-1600.822	-899.195	22.366
2200	163.176	369.848	213.397	344.193	-1597.019	-865.873	20.558
2300	163.176	377.102	220.358	360.510	-1593.251	-832.724	18.912
2400	163.176	384.047	227.035	376.828	-1589.518	-799.739	17.406
2500	163.176	390.708	233.450	393.145	-1585.818	-766.907	16.024
2600	163.176	397.108	239.622	409.463	-1582.152	-734.223	14.751
2700	163.176	403.266	245.570	425.781	-1578.520	-701.679	13.575
2800	163.176	409.200	251.308	442.098	-1574.922	-669.269	12.485
2900	163.176	414.926	256.852	458.416	-1571.357	-636.988	11.473
3000	163.176	420.458	262.214	474.733	-1567.828	-604.829	10.531
3100	163.176	425.809	267.405	491.051	-1564.332	-572.786	9.651
3200	163.176	430.989	272.437	507.369	-1560.871	-540.857	8.829
3300	163.176	436.011	277.318	523.686	-1557.446	-509.035	8.057
3400	163.176	440.882	282.057	540.004	-1554.057	-477.315	7.333
3500	163.176	445.612	286.663	556.321	-1550.706	-445.695	6.652
3600	163.176	450.209	291.142	572.639	-1547.392	-414.170	6.009
3700	163.176	454.680	295.502	588.957	-1953.217	-374.949	5.293
3800	163.176	459.031	299.749	605.274	-1948.749	-332.352	4.568
3900	163.176	463.270	303.887	621.592	-1944.379	-289.872	3.882
4000	163.176	467.401	307.924	637.909	-1940.107	-247.504	3.232

PREVIOUS: CURRENT: June 1967

Magnesium Titanium Oxide ($MgTiO_3$) $Mg_1O_3Ti_1$(cr,l)

MAGNESIUM SULFATE ($MgSO_4$) CRYSTAL M_r = 120.3626

$S°(298.15\ K) = 91.40 \pm 0.8\ J\ K^{-1}\ mol^{-1}$

$T_{fus} = 1400\ K$

$\Delta_f H°(0\ K) = -1250.40 \pm 20.9\ kJ\ mol^{-1}$

$\Delta_f H°(298.15\ K) = -1261.79 \pm 20.9\ kJ\ mol^{-1}$

$\Delta_{fus} H° = 14.644\ kJ\ mol^{-1}$

Enthalpy of Formation

The adopted $\Delta_f H°(298.15\ K) = -301.6 \pm 5\ kcal\ mol^{-1}$ was calculated from the $\Delta_r H°(298.15\ K) = 63.3 \pm 5\ kcal\ mol^{-1}$ for the reaction $MgO(cr) + SO_3(g) = MgSO_4(cr)$. The value of $\Delta_r H°(298.15\ K)$ was calculated by the 3rd law method using the equilibrium data obtained from a transpiration study by Knopf and Staude (1). In addition Marchal (2) measured the total pressure of $SO_3(g)$, $SO_2(g)$ and $O_2(g)$ above $MgSO_4(cr)$ and partial pressures were obtained by application of the SO_3-SO_2 equilibrium data. Another value of $\Delta_f H°(MgSO_4$, cr, 298.15 K) may be derived from the measured enthalpy of solution of $MgSO_4(cr)$ by Thomsen (3) and the measured enthalpies of mixing of $MgCl_2(aq, 200\ H_2O)$ with $H_2SO_4(aq, 200\ H_2O)$ and $MgSO_4(aq, 200\ H_2O)$ with $2HCl(100\ H_2O)$ by Thomsen (4).

Source	Reaction	T/K	Drift cal K^{-1} mol^{-1}	$\Delta_r H°(298.15\ K)$ kcal mol^{-1}	$\Delta_f H°(298.15\ K)$ kcal mol^{-1}
(1)	$MgO(cr) + SO_3(g) = MgSO_4(cr)$	1144-1313	0.514	63.284	-301.574
(2)	$MgO(cr) + SO_3(g) = MgSO_4(cr)$	1223-1428	0.525	72.394	-310.867
(3)	$MgSO_4(cr) = MgSO_4(aq, 200\ H_2O)$	298.15		-20.84	
(4)	$MgCl_2(aq, 200\ H_2O) + H_2SO_4(aq, 200\ H_2O)$ $= MgSO_4(aq, 200\ H_2O) + 2HCl(aq, 100\ H_2O)$	298.15		- 3.52	-305.1*

*Combination of the reactions 3 and 4 with the following $\Delta_f H°(298.15\ K)$:

$H_2SO_4(aq, 200\ H_2O) = -212.387$, $HCl(aq, 100\ H_2O) = -39.74$ and $MgCl_2(aq, 200\ H_2O) = -189.5\ kcal\ mol^{-1}$.

Heat Capacity and Entropy

The low temperature heat capacities 53.3-295.4 K were measured by Moore and Kelley (4). The heat capacities in the temperature range 298-1400 K were estimated from the equation $MgSO_4(cr) = CaSO_4(cr) + MgO(cr) - CaO(cr)$. $CaSO_4(cr)$ and $CaO(cr)$ $C_p°$ were calculated using the equation $C_p° = 16.78 + 23.60 \times 10^{-3}T$ and $C_p° = 11.67 + 1.08 \times 10^{-3}T - 1.56 \times 10^5 T^{-2}$ respectively, Kelley (5). The heat capacity of $MgO(cr)$ was calculated from the correlation of the low temperature heat capacities. The values from the two sources join smoothly at 298.15 K. Above the melting point (1400 K) the heat capacity was graphically extrapolated. The entropy was calculated at 50.12 K using the Debye and Einstein function D(242/T)+2E(378/T)+2E(866/T)+E(1683/T) given by Moore and Kelley (4). The value of $S°(50.12\ K) = 1.08\ cal\ K^{-1}\ mol^{-1}$.

Fusion Data

T_{fus} and $\Delta_{fus}H°$ from Kelley (6).

References

1. H. J. Knopf and H. Staude, Z. Physik. Chem. (Leipzig) 204, 265-275 (1955).
2. G. Marchal, J. Chim. Phys. 22, 493 (1925).
3. J. Thomsen, "Thermochemische Untersuchungen," Vols. I-IV, Barth, Leipzig (1882-1886).
4. G. E. Moore and K. K. Kelley, J. Amer. Chem. Soc. 64, 2949 (1942).
5. K. K. Kelley, U. S. Bur. Mines Bull. 584, 1960.
6. K. K. Kelley, U. S. Bur. Mines Bull. 393, 1936.

Magnesium Sulfate ($MgSO_4$) $Mg_1O_4S_1(cr)$

Enthalpy Reference Temperature = T_r = 298.15 K   Standard State Pressure = p° = 0.1 MPa

T/K	$C_p°$ (J $K^{-1}mol^{-1}$)	S° (J $K^{-1}mol^{-1}$)	$-[G°-H°(T_r)]/T$ (J $K^{-1}mol^{-1}$)	$H°-H°(T_r)$ (kJ mol^{-1})	$\Delta_f H°$ (kJ mol^{-1})	$\Delta_f G°$ (kJ mol^{-1})	Log K_f
0	0.	0.	INFINITE	-15.390	-1250.399	-1250.399	INFINITE
100	36.313	19.979	160.297	-14.032	-1256.182	-1221.316	637.950
200	73.309	57.532	99.534	-8.400	-1260.022	-1184.778	309.432
298.15	96.232	91.396	91.396	0.	-1261.786	-1147.408	201.021
300	96.441	91.992	91.398	0.178	-1261.804	-1146.698	199.658
400	109.997	121.692	95.355	10.535	-1264.542	-1108.138	144.708
500	119.411	147.271	103.240	22.015	-1265.774	-1068.894	111.667
600	127.612	169.790	112.492	34.379	-1266.082	-1029.474	89.624
700	134.306	189.970	122.143	47.479	-1265.710	-990.061	73.879
800	140.499	208.314	131.785	61.223	-1264.839	-950.737	62.077
900	146.231	225.196	141.238	75.562	-1316.686	-910.467	52.842
1000	151.670	240.888	150.428	90.461	-1322.424	-864.769	45.171
1100	156.691	255.582	159.326	105.881	-1319.324	-819.150	38.898
1200	161.419	269.424	167.930	121.793	-1315.827	-773.831	33.684
1300	165.352	282.504	176.245	138.137	-1311.983	-728.819	29.284
1400	168.619	294.881	184.281	154.840	-1435.265	-680.946	25.406
1400.000	168.619	294.882	184.281	154.841	------ CRYSTAL <--> LIQUID -----		
1500	171.389	306.611	192.049	171.843	-1429.556	-627.265	21.843
1600	173.787	317.750	199.560	189.105	-1423.656	-573.970	18.738
1700	175.895	328.350	206.826	206.591	-1417.595	-521.049	16.010
1800	177.778	338.458	213.861	224.276	-1411.397	-468.490	13.595
1900	179.477	348.117	220.675	242.140	-1405.080	-416.278	11.444
2000	181.033	357.363	227.279	260.166	-1398.658	-364.401	9.517
2100	182.468	366.230	233.686	278.342	-1392.145	-312.848	7.782
2200	183.807	374.750	239.906	296.657	-1385.549	-261.606	6.211
2300	185.067	382.949	245.948	315.101	-1378.878	-210.668	4.784
2400	186.259	390.850	251.822	333.668	-1372.138	-160.021	3.483
2500	187.393	398.477	257.537	352.351	-1365.335	-109.656	2.291

PREVIOUS: December 1960   CURRENT: March 1966

Magnesium Sulfate ($MgSO_4$) $Mg_1O_4S_1(cr)$

MAGNESIUM SULFATE ($MgSO_4$) LIQUID M_r = 120.3626

$S°$(298.15 K) = [102.125] J K^{-1} mol^{-1}

T_{fus} = 1400 K

$\Delta_f H°$(298.15 K) = [-1246.587] kJ mol^{-1}

$\Delta_{fus} H°$ = 14.644 kJ mol^{-1}

Enthalpy of Formation

$\Delta_f H°$($MgSO_4$, ℓ, 298.15 K) is calculated from $\Delta_f H°$($MgSO_4$, cr, 298.15 K) by adding $\Delta_{fus} H°$ and the difference in enthalpy, H°(1400 K)-H°(298.15 K), between the crystal and liquid.

Heat Capacity and Entropy

A glass transition was assumed at 1000 K. The heat capacity below 1000 K was obtained from the heat capacity of the crystal. Above 100 K the heat capacity was assumed constant and estimated as 38.0 cal K^{-1} mol^{-1} by comparison with those for $MgCl_2$(ℓ), KCl(ℓ) and K_2SO_4(ℓ). S°(ℓ, 298.15 K) is calculated in a manner analogous to that used for the enthalpy of formation.

Fusion Data

See the crystal table for details.

Magnesium Sulfate ($MgSO_4$) $Mg_1O_4S_1$(l)

Enthalpy Reference Temperature = T_r = 298.15 K Standard State Pressure = p° = 0.1 MPa

T/K	$C_p°$	S°	$-[G°-H°(T_r)]/T$	$H°-H°(T_r)$	$\Delta_f H°$	$\Delta_f G°$	Log K_f
	J $K^{-1}mol^{-1}$			kJ mol^{-1}			
0							
100							
200							
298.15	96.232	102.125	102.125	0.	-1246.587	-1135.408	198.919
300	96.441	102.721	102.127	0.178	-1246.605	-1134.718	197.572
400	109.997	132.421	106.084	10.535	-1249.343	-1097.231	143.284
500	119.411	158.000	113.970	22.015	-1250.575	-1059.059	110.639
600	127.612	180.519	123.221	34.379	-1250.883	-1020.713	88.861
700	134.306	200.700	132.873	47.479	-1250.511	-982.373	73.306
800	140.499	219.043	142.514	61.223	-1249.640	-944.122	61.645
900	146.231	235.858	151.964	75.505	-1301.545	-904.921	52.520
1000	158.574	251.845	161.156	90.690	-1306.996	-860.299	44.937
1000.000	158.574	251.845	161.156	90.690	GLASS <--> LIQUID		
1000.000	158.992	251.845	161.156	90.690	TRANSITION		
1100	158.992	266.999	170.100	106.589	-1303.417	-815.802	38.739
1200	158.992	280.833	178.760	122.488	-1299.933	-771.628	33.588
1300	158.992	293.559	187.108	138.387	-1296.534	-727.741	29.241
1400	158.992	305.342	195.137	154.286	-1420.620	-680.946	25.406
1400.000	158.992	305.342	195.137	154.286	------ CRYSTAL <--> LIQUID -----		
1500	158.992	316.311	202.854	170.186	-1416.015	-628.274	21.878
1600	158.992	326.572	210.269	186.085	-1411.477	-575.906	18.801
1700	158.992	336.211	217.397	201.984	-1407.003	-523.820	16.095
1800	158.992	345.299	224.253	217.883	-1402.590	-471.996	13.697
1900	158.992	353.895	230.852	233.782	-1398.238	-420.415	11.558
2000	158.992	362.050	237.209	249.682	-1393.944	-369.062	9.639
2100	158.992	369.808	243.340	265.581	-1389.707	-317.923	7.908
2200	158.992	377.204	249.258	281.480	-1385.526	-266.982	6.339
2300	158.992	384.271	254.976	297.379	-1381.400	-216.233	4.911
2400	158.992	391.038	260.505	313.278	-1377.328	-165.662	3.606
2500	158.992	397.528	265.857	329.178	-1373.310	-115.258	2.408

PREVIOUS: December 1960 CURRENT: March 1966

Magnesium Sulfate ($MgSO_4$) $Mg_1O_4S_1$(l)

MAGNESIUM SULFATE ($MgSO_4$) CRYSTAL-LIQUID M_r = 120.3626

0 to 1400 K crystal
above 1400 K liquid

Refer to the individual tables for details.

Magnesium Sulfate ($MgSO_4$) $Mg_1O_4S_1$(cr,l)

Enthalpy Reference Temperature = T_r = 298.15 K Standard State Pressure = p° = 0.1 MPa

	J K⁻¹mol⁻¹			kJ mol⁻¹			
T/K	C_p°	S°	$-[G°-H°(T_r)]/T$	$H°-H°(T_r)$	$\Delta_fH°$	$\Delta_fG°$	Log K_f
0	0.	0.	INFINITE	-15.390	-1250.399	-1250.399	INFINITE
100	36.313	19.979	160.297	-14.032	-1256.182	-1221.316	637.950
200	73.309	57.532	99.534	-8.400	-1260.022	-1184.778	309.432
298.15	96.232	91.396	91.396	0.	-1261.786	-1147.408	201.021
300	96.441	91.992	91.398	0.178	-1261.804	-1146.698	199.658
400	109.997	121.692	95.355	10.535	-1264.542	-1108.138	144.708
500	119.411	147.271	103.240	22.015	-1265.774	-1068.894	111.667
600	127.612	169.790	112.492	34.379	-1266.082	-1029.474	89.624
700	134.306	189.970	122.143	47.479	-1265.710	-990.061	73.879
800	140.499	208.314	131.785	61.223	-1264.839	-950.737	62.077
900	146.231	225.196	141.238	75.562	-1316.686	-910.467	52.842
1000	151.670	240.888	150.428	90.461	-1322.424	-864.769	45.171
1100	156.691	255.582	159.326	105.881	-1319.324	-819.150	38.898
1200	161.419	269.424	167.930	121.793	-1315.827	-773.831	33.684
1300	165.352	282.504	176.245	138.137	-1311.983	-728.819	29.284
1400	168.619	294.882	184.281	154.841	-1435.264	-680.946	25.406
1400.000	168.619	305.342	184.281	169.485	CRYSTAL <--> LIQUID		
1400.000	158.992	305.342	184.281	169.485	TRANSITION		
1500	158.992	316.311	192.721	185.385	-1416.015	-628.274	21.878
1600	158.992	326.572	200.770	201.284	-1411.477	-575.906	18.801
1700	158.992	336.211	208.456	217.183	-1407.003	-523.820	16.095
1800	158.992	345.299	215.809	233.082	-1402.590	-471.996	13.697
1900	158.992	353.895	222.852	248.981	-1398.238	-420.415	11.558
2000	158.992	362.050	229.610	264.881	-1393.944	-369.062	9.639
2100	158.992	369.808	236.103	280.780	-1389.707	-317.923	7.908
2200	158.992	377.204	242.350	296.679	-1385.526	-266.982	6.339
2300	158.992	384.271	248.368	312.578	-1381.400	-216.233	4.911
2400	158.992	391.038	254.172	328.477	-1377.328	-165.662	3.606
2500	158.992	397.528	259.778	344.377	-1373.310	-115.258	2.408

PREVIOUS: CURRENT: March 1966

Magnesium Sulfate ($MgSO_4$) $Mg_1O_4S_1$(cr,l)

MAGNESIUM TUNGSTEN OXIDE ($MgWO_4$) CRYSTAL M_r = 272.1526

$S°(298.15\ K) = 101.19 \pm 0.8\ J\ K^{-1}\ mol^{-1}$

T_{fus} = Unknown

$\Delta_f H°(0\ K) = -1505.74 \pm 33.5\ kJ\ mol^{-1}$

$\Delta_f H°(298.15\ k) = -1515.86 \pm 33.5\ kJ\ mol^{-1}$

$\Delta_{fus}H°$ = Unknown

Enthalpy of Formation

Rezukhina and Levitskii (1) measured the emf, E = 0.2433 - 0.0001264T, for the cell reaction MgO(cr) + W(cr) + $3Fe_{0.95}O$(cr) = $MgWO_4$(cr) + 2.85 Fe(cr), in the temperature range 1220 - 1370 K. Incorporating the results with auxiliary thermal data, the authors obtained $\Delta_r G° = -224{,}370 + 65.67T$ (cal K^{-1} mol^{-1}) for MgO(cr) + W(cr) + $3/2O_2$(g) → $MgWO_4$(cr). 3rd law analysis of the $\Delta_r G°$ equation gives $\Delta_r H°(298.15\ K) = -218.6$ kcal mol^{-1}, from which the adopted $\Delta_f H°(MgWO_4$, cr, 298.15 K), -362.3 kcal mol^{-1}, is calculated. The 2nd law analysis gives $\Delta_f H°(MgWO_4$, cr, 298.15 K) = -372.7 kcal mol^{-1}. The drift is 8.0 cal K^{-1} mol^{-1}.

Rezukhina, et. al. (2) determined the Gibbs energy changes $\Delta_r G° = 51{,}300 - 27.67T$ (cal K^{-1} mol^{-1}) for the reaction $MgWO_4$(cr) + $3H_2$(g) → MgO(cr) + W(cr) + $3H_2O$(g) in the temperature range 1073 K - 1273 K by a circulation method. This leads to heats of formation of -366.1 kcal mol^{-1} by the 3rd law method and -377.2 kcal mol^{-1} by the 2nd law method. Both determinations were performed by Rezukhina and his co-workers, but they considered the data obtained by the emf method to be more reliable than that derived from heterogeneous equilibria.

Heat Capacity and Entropy

Low temperature heat capacities, 52.92 - 296.09 K, were measured calorimetrically by King and Weller (3). The heat capacities above 300 K are estimated by comparison with those of $CaWO_4$(cr) (4), CaO(cr) (5) and MgO(cr) (6). These two sets of data are joined smoothly at 298.15 K by a graphical method.

The entropy, S°(298.15 K) = 21.184 cal K^{-1} mol^{-1}, is taken from King and Weller (3), based on an extrapolation of S°(51 K) = 1.27 cal K^{-1} mol^{-1}. No spin-magnetic entropy is predicted below 51 K based on the diamagnetism observed by Gerasimov (7).

References

1. T. N. Rezukhina and V. A. Levitskii, Russ. J. Phys. Chem. 37, 1277 (1963).
2. T. N. Rezukhina, Ya. I. Gerasimov and Yu. P. Simanov, Vestnik Moskov. Univ. 4, No. 6, 103 (1949).
3. E. G. King and W. W. Weller, U. S. Bur. Mines RI 5791, 1961.
4. R. A. Yakovleva and T. N. Rezukhina, Russ. J. Phys. Chem. 34, 390 (1960).
5. K. K. Kelley, U.S. Bur. Mines Bull. 584, 1960.
6. JANAF Thermochemical Tables: MgO, 12-31-65.
7. Ya. I. Gerasimov, Papers Presented to Section on Inorganic Chemistry, 16th Intern. Congress Pure and Applied Chem. Paris (1957), pp. 227-35, Butterworths Scientific Publications, London, 1958.

Magnesium Tungsten Oxide ($MgWO_4$) $Mg_1O_4W_1$(cr)

Enthalpy Reference Temperature = T_r = 298.15 K Standard State Pressure = p° = 0.1 MPa

T/K	$C_p°$ (J K⁻¹mol⁻¹)	S°	-[G°-H°(T_r)]/T	H°-H°(T_r) (kJ mol⁻¹)	$\Delta_f H°$	$\Delta_f G°$	Log K_f
0	0.	0.	INFINITE	-17.218	-1505.744	-1505.744	INFINITE
100	38.221	21.150	179.146	-15.800	-1511.429	-1476.970	771.490
200	83.220	62.706	110.418	-9.542	-1515.008	-1440.872	376.317
298.15	109.077	101.186	101.186	0.	-1515.863	-1404.224	246.014
300	109.621	101.862	101.188	0.202	-1515.861	-1403.532	244.377
400	123.428	135.424	105.670	11.902	-1515.122	-1366.171	178.404
500	131.587	163.893	114.543	24.675	-1513.648	-1329.093	138.849
600	137.026	188.377	124.856	38.112	-1511.854	-1292.348	112.509
700	141.670	209.851	135.496	52.049	-1509.900	-1255.916	93.718
800	146.147	229.063	146.011	66.441	-1507.815	-1219.774	79.643
900	150.415	246.522	156.223	81.268	-1505.606	-1183.900	68.712
1000	154.808	262.595	166.067	96.528	-1511.860	-1147.567	59.943
1100	159.410	277.566	175.530	112.239	-1509.378	-1111.254	52.769
1200	164.013	291.633	184.625	128.410	-1506.559	-1075.183	46.801
1300	168.615	304.945	193.373	145.043	-1503.393	-1039.361	41.762
1400	173.008	317.600	201.798	162.122	-1627.297	-1000.628	37.334
1500	177.611	329.694	209.924	179.655	-1622.088	-956.046	33.292
1600	182.004	341.296	217.775	197.634	-1616.532	-911.823	29.768
1700	186.606	352.469	225.371	216.066	-1610.619	-867.958	26.669
1800	191.000	363.259	232.734	234.945	-1604.355	-824.451	23.925
1900	195.602	373.710	239.880	254.277	-1597.733	-781.302	21.479
2000	199.995	383.855	246.826	274.058	-1590.756	-738.511	19.288

PREVIOUS: June 1963 CURRENT: March 1967

Magnesium Tungsten Oxide ($MgWO_4$) $Mg_1O_4W_1$(cr)

MAGNESIUM TITANIUM OXIDE ($MgTi_2O_5$) CRYSTAL M_r = 200.0620

$S°(298.15\ K) = 135.60 \pm 6.3\ J\ K^{-1}\ mol^{-1}$

$T_{fus} = 1963 \pm 20\ K$

$\Delta_f H°(0\ K) = -2495.43 \pm 10.5\ kJ\ mol^{-1}$

$\Delta_f H°(298.15\ K) = -2509.35 \pm 10.5\ kJ\ mol^{-1}$

$\Delta_{fus} H° = [146.440]\ kJ\ mol^{-1}$

Enthalpy of Formation

$\Delta_f H°(298.15\ K)$ is calculated from $\Delta_r H°(298.15\ K) = -4.45 \pm 0.45\ kcal\ mol^{-1}$ for the reaction $MgO(cr) + 2TiO_2(rutile) = MgTi_2O_5(cr)$ measured by Kelley et al. (1), using a solution calorimetric method. The uncertainty of ±2.5 kcal mol^{-1} assigned to the enthalpy of formation is due primarily to TiO_2(rutile).

Heat Capacity and Entropy

Low temperature heat capacities are from the data (52-296 K) of Todd (2). High temperature enthalpies have been measured (397-1812 K) by Orr and Coughlin (3). High temperature heat capacities are derived from the enthalpies by a fitting technique which constrains the curve to join smoothly with the low temperature values. The entropy is based on $S°(51\ K) = 1.11\ cal\ K^{-1}\ mol^{-1}$.

Kelley et al. (1), suggested that there may be a zero-point entropy due to randomness of the cations. The authors estimated S°(0 K) from the following alternative assumptions: (a) all the Mg and Ti ions are randomly distributed among the latice sites giving 3.79 cal $K^{-1}\ mol^{-1}$ and (b) all the Mg and one-half of the Ti ions are randomly distributed giving 2.75 cal $K^{-1}\ mol^{-1}$. These assumptions were proposed in order to make $MgTi_2O_5$ stable relative to $MgTiO_3$ at high temperatures. Phase data of Massazza and Sirchia (4), indicate that $MgTi_2O_5$ melts congruently at 1963±20 K and that there is a temperature range. To make $\Delta_r G°(1800\ K) < 0$ would require that $S°(0\ K) \geq 0.35\ cal\ K^{-1}\ mol^{-1}$. We adopt a compromise zero-point entropy of 2 cal $K^{-1}\ mol^{-1}$ between the upper (3.79 cal $K^{-1}\ mol^{-1}$) and the lower (0.35 cal $K^{-1}\ mol^{-1}$) limits

Fusion Data

See the liquid table for details.

References

1. K. K. Kelley, S. S. Todd and E. G. King, U. S. Bur. Mines RI 5059, 1954.
2. S. S. Todd, J. Amer. Chem. Soc. 74, 4669 (1952).
3. R. L. Orr and J. P. Coughlin, J. Amer. Chem. Soc. 74, 3186 (1952).
4. F. Massazza and E. Sirchia, 16th Int. Congress of Pure and Appl. Chem., Inorg. Sect., Paris, 1975, pp. 161-168, Butterworths Scientific publications, London, 1958.

Magnesium Titanium Oxide ($MgTi_2O_5$) $Mg_1O_5Ti_2$(cr)

Enthalpy Reference Temperature = T_r = 298.15 K   Standard State Pressure = p° = 0.1 MPa

T/K	$C_p°$	S°	$-[G°-H°(T_r)]/T$	$H°-H°(T_r)$	$\Delta_f H°$	$\Delta_f G°$	Log K_f
	J $K^{-1}mol^{-1}$			kJ mol^{-1}			
0	0.	0.	INFINITE	-22.439	-2495.427	-2495.427	INFINITE
100	46.129	22.217	231.124	-20.891	-2502.903	-2459.195	1284.552
200	110.533	75.638	139.695	-12.811	-2507.943	-2413.208	630.265
298.15	146.858	135.603	135.603	0.	-2509.354	-2368.790	415.002
300	147.335	136.513	135.606	0.272	-2509.357	-2367.918	412.291
400	165.753	181.688	141.640	16.019	-2508.821	-2320.819	303.068
500	176.665	219.925	153.575	33.175	-2507.459	-2273.965	237.559
600	184.498	252.858	167.441	51.250	-2505.701	-2227.426	193.915
700	190.807	281.785	181.751	70.024	-2503.712	-2181.202	162.763
800	196.276	307.628	195.899	89.383	-2501.533	-2135.276	139.419
900	201.238	331.036	209.634	109.262	-2499.268	-2089.630	121.279
1000	205.870	352.481	222.861	129.620	-2505.664	-2043.521	106.743
1100	210.284	372.311	235.557	150.429	-2503.688	-1997.403	94.849
1200	214.539	390.791	247.732	171.671	-2509.739	-1951.222	84.934
1300	218.681	408.127	259.410	193.332	-2506.427	-1904.812	76.536
1400	222.739	424.482	270.622	215.404	-2630.346	-1855.496	69.229
1500	226.731	439.986	281.401	237.878	-2625.354	-1800.324	62.693
1600	230.672	454.745	291.777	260.748	-2620.240	-1745.488	56.984
1700	234.572	468.847	301.781	284.011	-2615.031	-1690.975	51.957
1800	238.438	482.364	311.441	307.662	-2609.756	-1636.771	47.498
1900	242.279	495.359	320.781	331.698	-2604.443	-1582.861	43.516
1963.000	244.685	503.301	326.512	347.037	------ CRYSTAL <--> LIQUID -----		
2000	246.099	507.883	329.825	356.116	-2628.493	-1528.324	39.916
2100	249.902	519.982	338.593	380.917	-2624.683	-1473.408	36.649
2200	253.688	531.695	347.106	406.096	-2620.551	-1418.679	33.684
2300	257.458	543.055	355.380	431.654	-2616.098	-1364.148	30.981
2400	261.220	554.092	363.431	457.588	-2611.324	-1309.817	28.507
2500	264.973	564.832	371.273	483.897	-2606.231	-1255.691	26.236
2600	268.713	575.297	378.919	510.582	-2600.818	-1201.776	24.144
2700	272.450	585.508	386.383	537.640	-2595.087	-1148.073	22.211
2800	276.177	595.484	393.673	565.071	-2589.036	-1094.589	20.420
2900	279.901	605.241	400.801	592.875	-2582.667	-1041.330	18.756
3000	283.621	614.792	407.775	621.051	-2575.979	-988.292	17.208

PREVIOUS: June 1961   CURRENT: June 1967

Magnesium Titanium Oxide ($MgTi_2O_5$) $Mg_1O_5Ti_2$(cr)

MAGNESIUM TITANIUM OXIDE ($MgTi_2O_5$) LIQUID M_r = 200.0620

$S°(298.15\ K) = [197.783]\ J\ K^{-1}\ mol^{-1}$ $\Delta_fH°(298.15\ K) = [-2382.31 \pm 8.4]\ kJ\ mol^{-1}$

$T_{fus} = 1963 \pm 20\ K$ $\Delta_{fus}H° = [146.440]\ kJ\ mol^{-1}$

Enthalpy of Formation

$\Delta_fH°(MgTi_2O_5, \ell, 298.15\ K)$ is calculated from $\Delta_fH°(MgTi_2O_5, cr, 298.15\ K)$ by adding $\Delta_{fus}H°$ and the difference in enthalpy, H°(1963 K)-H°(298.15 K), between the crystal and liquid.

Heat Capacity and Entropy

A glass transition is assumed at 1300 K. Below 1300 K the heat capacity is obtained from the heat capacity of the crystal. Above 1300 K the heat capacity is assumed constant and estimated as 62.4 cal $K^{-1}\ mol^{-1}$ or 7.8 cal K^{-1} g-atom^{-1}. $S°(\ell, 298.15\ K)$ is calculated in a manner analogous to that used for the enthalpy of formation.

Fusion Data

The congruent melting point is obtained from Massazza (1). Previous work by Coughanour (2) indicates congruent melting at 1925 K. The $\Delta_{fus}H°$ is estimated from those of $Na_2O\cdot 2TiO_2$ and $Na_2O\cdot TiO_2$ observed by Naylor (3) and $CaO\cdot TiO_2\cdot SiO_2$ observed by King (4).

References

1. F. Massazza and E. Sirchia, 16th Int. Congress of Pure and Appl. Chem., Inorg. Sect., Paris (1957), pp. 161-8, Butterworths Scientific Publications, London, 1958.
2. L. W. Coughanour and V. A. DeProsse, J. Res. Nat. Bur. Stand. 51, 87 (1953).
3. B. V. Naylor, J. Amer. Chem. Soc. 67, 2120 (1945).
4. E. G. King, R. L. Orr and K. R. Bonnickson, J. Amer. Chem. Soc. 76, 4320 (1954).

Magnesium Titanium Oxide ($MgTi_2O_5$) $Mg_1O_5Ti_2(l)$

Enthalpy Reference Temperature = T_r = 298.15 K Standard State Pressure = p° = 0.1 MPa

T/K	$C_p°$ (J K⁻¹mol⁻¹)	S° (J K⁻¹mol⁻¹)	$-[G°-H°(T_r)]/T$ (J K⁻¹mol⁻¹)	$H°-H°(T_r)$ (kJ mol⁻¹)	$\Delta_fH°$ (kJ mol⁻¹)	$\Delta_fG°$ (kJ mol⁻¹)	Log K_f
0							
100							
200							
298.15	146.858	197.783	197.783	0.	-2382.307	-2260.282	395.992
300	147.335	198.692	197.785	0.272	-2382.310	-2259.525	393.418
400	165.753	243.867	203.819	16.019	-2381.773	-2218.643	289.725
500	176.665	282.105	215.754	33.175	-2380.411	-2178.007	227.535
600	184.498	315.037	229.620	51.250	-2378.654	-2137.686	186.102
700	190.807	343.964	243.931	70.024	-2376.665	-2097.680	156.531
800	196.276	369.807	258.078	89.383	-2374.486	-2057.972	134.372
900	201.238	393.215	271.813	109.262	-2372.221	-2018.544	117.153
1000	205.870	414.660	285.040	129.620	-2378.616	-1978.653	103.354
1100	210.284	434.490	297.736	150.429	-2376.641	-1938.752	92.064
1200	214.539	452.970	309.911	171.671	-2382.691	-1898.789	82.652
1300	218.681	470.307	321.589	193.333	-2379.379	-1858.597	74.679
1300.000	218.681	470.307	321.589	193.333	GLASS <--> LIQUID TRANSITION		
1300.000	261.082	470.307	321.589	193.333			
1400	261.082	489.655	332.911	219.441	-2499.262	-1815.654	67.743
1500	261.082	507.668	343.968	245.549	-2490.636	-1767.128	61.537
1600	261.082	524.518	354.732	271.657	-2482.284	-1719.168	56.125
1700	261.082	540.345	365.189	297.765	-2474.229	-1671.721	51.366
1800	261.082	555.268	375.339	323.873	-2466.497	-1624.740	47.149
1900	261.082	569.384	385.184	349.982	-2459.112	-1578.179	43.387
1963.000	261.082	577.901	391.233	366.430	------ CRYSTAL <--> LIQUID -----		
2000	261.082	582.776	394.731	376.090	-2481.473	-1531.089	39.988
2100	261.082	595.514	403.992	402.198	-2476.355	-1483.696	36.905
2200	261.082	607.660	412.975	428.306	-2471.294	-1436.544	34.108
2300	261.082	619.265	421.694	454.414	-2466.290	-1389.624	31.559
2400	261.082	630.377	430.159	480.522	-2461.342	-1342.919	29.228
2500	261.082	641.035	438.383	506.631	-2456.451	-1296.419	27.087
2600	261.082	651.275	446.375	532.739	-2451.614	-1250.113	25.115
2700	261.082	661.128	454.148	558.847	-2446.832	-1203.992	23.293
2800	261.082	670.623	461.710	584.955	-2442.105	-1158.047	21.604
2900	261.082	679.785	469.073	611.063	-2437.431	-1112.272	20.034
3000	261.082	688.636	476.245	637.171	-2432.812	-1066.655	18.572
3100	261.082	697.196	483.235	663.280	-2428.245	-1021.190	17.207
3200	261.082	705.485	490.052	689.388	-2423.732	-975.875	15.930
3300	261.082	713.519	496.702	715.496	-2419.271	-930.700	14.732
3400	261.082	721.313	503.195	741.604	-2414.863	-885.656	13.606
3500	261.082	728.882	509.535	767.712	-2410.508	-840.744	12.547
3600	261.082	736.236	515.731	793.820	-2406.207	-795.954	11.549
3700	261.082	743.390	521.788	819.928	-3220.157	-735.708	10.386
3800	261.082	750.352	527.711	846.037	-3213.497	-668.648	9.191
3900	261.082	757.134	533.507	872.145	-3207.003	-601.763	8.060
4000	261.082	763.744	539.181	898.253	-3200.675	-535.044	6.987

PREVIOUS: June 1961 CURRENT: June 1967

Magnesium Titanium Oxide ($MgTi_2O_5$) $Mg_1O_5Ti_2(l)$

MAGNESIUM TITANIUM OXIDE ($MgTi_2O_5$) CRYSTAL-LIQUID M_r = 200.0620

0 to 1963 K crystal
above 1963 K liquid

Refer to the individual tables for details.

Magnesium Titanium Oxide ($MgTi_2O_5$) $Mg_1O_5Ti_2$(cr,l)

Enthalpy Reference Temperature = T_r = 298.15 K Standard State Pressure = p° = 0.1 MPa

T/K	C_p° (J K⁻¹mol⁻¹)	S° (J K⁻¹mol⁻¹)	−[G°−H°(T_r)]/T (J K⁻¹mol⁻¹)	H°−H°(T_r) (kJ mol⁻¹)	Δ_fH° (kJ mol⁻¹)	Δ_fG° (kJ mol⁻¹)	Log K_f
0	0.	0.	INFINITE	-22.439	-2495.427	-2495.427	INFINITE
100	46.129	22.217	231.124	-20.891	-2502.903	-2459.195	1284.552
200	110.533	75.638	139.695	-12.811	-2507.943	-2413.208	630.265
298.15	146.858	135.603	135.603	0.	-2509.354	-2368.790	415.002
300	147.335	136.513	135.606	0.272	-2509.357	-2367.918	412.291
400	165.753	181.688	141.640	16.019	-2508.821	-2320.819	303.068
500	176.665	219.925	153.575	33.175	-2507.459	-2273.965	237.559
600	184.498	252.858	167.441	51.250	-2505.701	-2227.426	193.915
700	190.807	281.785	181.751	70.024	-2503.712	-2181.202	162.763
800	196.276	307.628	195.899	89.383	-2501.533	-2135.276	139.419
900	201.238	331.036	209.634	109.262	-2499.268	-2089.630	121.279
1000	205.870	352.481	222.861	129.620	-2505.664	-2043.521	106.743
1100	210.284	372.311	235.557	150.429	-2503.688	-1997.403	94.849
1200	214.539	390.791	247.732	171.671	-2509.739	-1951.222	84.934
1300	218.681	408.127	259.410	193.332	-2506.427	-1904.812	76.536
1400	222.739	424.482	270.622	215.404	-2630.346	-1855.496	69.229
1500	226.731	439.986	281.401	237.878	-2625.354	-1800.324	62.693
1600	230.672	454.745	291.777	260.748	-2620.240	-1745.488	56.984
1700	234.572	468.847	301.781	284.011	-2615.031	-1690.975	51.957
1800	238.438	482.364	311.441	307.662	-2609.756	-1636.771	47.498
1900	242.279	495.359	320.781	331.698	-2604.443	-1582.861	43.516
1963.000	244.685	503.301	326.512	347.037	CRYSTAL <--> LIQUID		
1963.000	261.082	577.901	326.512	493.477	TRANSITION		
2000	261.082	582.776	331.208	503.137	-2481.473	-1531.089	39.988
2100	261.082	595.514	343.493	529.245	-2476.355	-1483.696	36.905
2200	261.082	607.660	355.227	555.353	-2471.294	-1436.544	34.108
2300	261.082	619.265	366.456	581.462	-2466.290	-1389.624	31.559
2400	261.082	630.377	377.223	607.570	-2461.342	-1342.919	29.228
2500	261.082	641.035	387.564	633.678	-2456.451	-1296.419	27.087
2600	261.082	651.275	397.511	659.786	-2451.614	-1250.113	25.115
2700	261.082	661.128	407.093	685.894	-2446.832	-1203.992	23.293
2800	261.082	670.623	416.336	712.002	-2442.105	-1158.047	21.604
2900	261.082	679.785	425.264	738.110	-2437.431	-1112.272	20.034
3000	261.082	688.636	433.896	764.219	-2432.812	-1066.655	18.572
3100	261.082	697.196	442.252	790.327	-2428.245	-1021.190	17.207
3200	261.082	705.485	450.350	816.435	-2423.732	-975.875	15.930
3300	261.082	713.519	458.203	842.543	-2419.271	-930.700	14.732
3400	261.082	721.313	465.828	868.651	-2414.863	-885.656	13.606
3500	261.082	728.882	473.236	894.759	-2410.508	-840.744	12.547
3600	261.082	736.236	480.440	920.868	-2406.207	-795.954	11.549
3700	261.082	743.390	487.450	946.976	-3220.157	-735.708	10.386
3800	261.082	750.352	494.278	973.084	-3213.497	-668.648	9.191
3900	261.082	757.134	500.931	999.192	-3207.003	-601.763	8.060
4000	261.082	763.744	507.419	1025.300	-3200.675	-535.044	6.987

PREVIOUS: CURRENT: June 1967

Magnesium Titanium Oxide ($MgTi_2O_5$) $Mg_1O_5Ti_2$(cr,l)

MAGNESIUM SULFIDE (MgS) CRYSTAL M_r = 56.365

$\Delta_f H°(0\ K) = -344.65 \pm 4.2\ kJ\ mol^{-1}$

$S°(298.15\ K) = 50.33 \pm 0.4\ J\ K^{-1}\ mol^{-1}$ $\Delta_f H°(298.15\ K) = -345.72 \pm 4.2\ kJ\ mol^{-1}$

Enthalpy of Formation

Sabatier (1) determined $\Delta_r H°(286\ K) = -37.8\pm1.0\ kcal\ mol^{-1}$ while Mourlot (2) determined $\Delta_r H°(291\ K) = -38.0\pm1.0\ kcal\ mol^{-1}$ for $MgS(cr) + 2\ HCl(aq,\ 100\ H_2O) + MgCl_2(aq,\ 100\ H_2O) + H_2S(aq)$. We derive $\Delta_f H°(MgS,\ cr,\ 298.15\ K) = -83.0\pm2.0\ kcal\ mol^{-1}$ and $\Delta_f H°(MgS,\ cr,\ 298.15\ K) = -82.8\pm2.0\ kcal\ mol^{-1}$ respectively based on the following auxiliary data: $\Delta_f H°(HCl\cdot100\ H_2O,\ aq,\ 298.15\ K) = -39.657\ kcal\ mol^{-1}$ (3), $\Delta_f H°(H_2S,\ aq,\ 298.15\ K) = -9.5\ kcal\ mol^{-1}$ (3), and $\Delta_f H°(MgCl_2\cdot100\ H_2O,\ aq,\ 298.15\ K) = -190.59\ kcal\ mol^{-1}$(4). The uncertainty reflects the correction to 298.15 K and the ambiguity of the exact solution composition. Von Wartenberg (5) measured $\Delta_r H°(293\ K) = -33.95\pm0.79\ kcal\ mol^{-1}$ for $MgS(cr) + 2\ HCl(30\ H_2O,\ aq) + MgCl_2(30\ H_2O,\ aq) + H_2S(g)$. We derive $\Delta_f H°(MgS,\ cr,\ 298.15\ K) = -82.0\pm1.0\ kcal\ mol^{-1}$ based on the following auxiliary data: $\Delta_f H°(HCl\cdot30\ H_2O,\ aq,\ 298.15\ K) = -39.357\ kcal\ mol^{-1}$ (3), $\Delta_f H°(H_2S,\ g,\ 298.15\ K) = -4.90\ kcal\ mol^{-1}$ (3), and $\Delta_f H°(MgCl_2\cdot30\ H_2O,\ aq,\ 298.15\ K) = -189.78\ kcal\ mol^{-1}$ (4).

Kapustinskii and Korshunov (6) performed calorimetric measurements of the enthalpy of formation of MgS(cr) from the pure elements. They determined $\Delta_f H°(298.15\ K) = -84.33\pm0.27\ kcal\ mol^{-1}$ for $Mg(cr) + S(cr,\ rhombic) = MgS(cr)$.

Our analyses of equilibrium studies of Curlock and Pidgeon (7) and of Dewing and Richardson (8) are given below. The two studies are in reasonable agreement. The 3rd law $\Delta_r H°(298.15\ K)$ may be in error by as much as 0.5 kcal mol^{-1} since the Gibbs energy functions employed in the calculations are partially based on estimated high temperature $C_p°$ values.

Source	Method	Data Points	Reaction[a]	T/K	δS cal K^{-1} mol^{-1}	$\Delta_r H°(298.15\ K)$, kcal mol^{-1} 2nd law	3rd law	$\Delta_f H°(298.15\ K)$ kcal mol^{-1}
Curlock (7)	Equilibrium	6	A	1180-1483	-0.60±0.45	-8.49±0.59	-9.19±0.24	-81.59±0.4
Dewing (8)	Equilibrium	6	A	1267-1768	-2.23±0.22	-5.53±0.38	-8.73±1.01	-82.05±1.2

[a]Reaction: A) $MgS(cr) + H_2O(g) = MgO(cr) + H_2S(g)$

Our adopted value is based on a weighted average of all the data.

Heat Capacity and Entropy

Stull et al. (9) measured the low temperature heat capacities in an adiabatic calorimeter from 15 - 320 K. Their smooth values are adopted in the tabulation. The entropy $S°(298.15\ K) = 12.03\pm0.1\ cal\ K^{-1}\ mol^{-1}$ is based on $S° = 0.010\ cal\ K^{-1}\ mol^{-1}$ at 15 K (9). The $C_p°$ values above 320 K are estimated by graphical extrapolation combined with method B of Kubaschewski et al. (10).

Fusion Data

MgS melts above 2300 K (11) and no other literature melting data are available.

References

1. M. Sabatier, Ann. Chim. Phys. 22, 5 (1881).
2. M. A. Mourlot, Ann. Chim. Phys. 17, 510 (1899).
3. U. S. Nat. Bur. Stand. Tech. Note 270-3, 1968.
4. U. S. Nat. Bur. Stand. Tech. Note 270-6, 1971.
5. H. Von Wartenberg, Z. Anorg. Chem. 252, 136 (1943).
6. A. F. Kapustinskii and I. A. Korshunov, Acta. Physicochim. URSS. 10, 259 (1939); data taken from Chem. Abstr. 33, 61325.
7. W. Curlock and L. M. Pidgeon, Trans. AIME 212, 671 (1958).
8. E. W. Dewing and F. D. Richardson, J. Iron Steel Inst. 195, 56 (1964).
9. D. R. Stull, D. L. Hildenbrand, F. L. Oetting, and G. C. Sinke, J. Chem. Eng. Data 15, 52 (1970).
10. O. Kubaschewski, E. L. Evans, and C. B. Alcock, "Metallurgical Thermochemistry," 4th ed., Pergamon Press, Oxford, 1967.
11. E. Tiede and A. Schleede, Berichte 53, 1721 (1920).
12. JANAF Thermochemical Tables: MgO(cr) 12-13-74; $H_2O(g)$ 3-31-61; $H_2S(g)$ 6-30-77.

Magnesium Sulfide (MgS) $Mg_1S_1(cr)$

Enthalpy Reference Temperature = T_r = 298.15 K Standard State Pressure = p° = 0.1 MPa

T/K	$C_p°$ (J $K^{-1}mol^{-1}$)	S° (J $K^{-1}mol^{-1}$)	$-[G°-H°(T_r)]/T$ (J $K^{-1}mol^{-1}$)	$H°-H°(T_r)$ (kJ mol^{-1})	$\Delta_f H°$ (kJ mol^{-1})	$\Delta_f G°$ (kJ mol^{-1})	Log K_f
0	0.	0.	INFINITE	-8.333	-344.647	-344.647	INFINITE
100	22.677	10.920	86.467	-7.555	-345.200	-344.090	179.734
200	40.125	33.095	54.342	-4.249	-345.544	-342.807	89.532
298.15	45.564	50.334	50.334	0.	-345.724	-341.433	59.818
300	45.647	50.616	50.334	0.084	-345.728	-341.406	59.444
400	47.614	64.032	52.150	4.753	-348.212	-339.840	44.379
500	48.953	74.807	55.639	9.584	-349.975	-337.556	35.264
600	49.915	83.823	59.606	14.530	-351.381	-334.934	29.159
700	50.543	91.566	63.631	19.555	-352.575	-332.097	24.781
800	51.003	98.346	67.556	24.632	-353.697	-329.094	21.488
900	51.463	104.380	71.318	29.756	-407.949	-324.871	18.855
1000	51.882	109.823	74.901	34.923	-416.494	-314.931	16.450
1100	52.347	114.790	78.305	40.134	-416.585	-304.770	14.472
1200	52.802	119.365	81.538	45.392	-416.645	-294.602	12.824
1300	53.252	123.609	84.613	50.694	-416.676	-284.431	11.429
1400	53.681	127.571	87.542	56.041	-544.087	-271.091	10.115
1500	54.073	131.288	90.335	61.429	-542.712	-251.639	8.763
1600	54.476	134.791	93.005	66.856	-541.311	-232.279	7.583
1700	54.910	138.106	95.562	72.326	-539.883	-213.008	6.545
1800	55.312	141.256	98.014	77.837	-538.427	-193.822	5.625
1900	55.647	144.256	100.369	83.385	-536.947	-174.718	4.803
2000	56.024	147.120	102.636	88.968	-535.444	-155.692	4.066
2100	56.468	149.864	104.820	94.593	-533.911	-136.742	3.401
2200	56.902	152.501	106.927	100.261	-532.344	-117.866	2.798
2300	57.326	155.039	108.964	105.973	-530.744	-99.062	2.250
2400	57.739	157.488	110.935	111.726	-529.112	-80.328	1.748
2500	58.131	159.853	112.845	117.520	-527.450	-61.662	1.288
2600	58.534	162.141	114.697	123.353	-525.756	-43.064	0.865
2700	58.953	164.358	116.496	129.227	-524.029	-24.532	0.475
2800	59.371	166.509	118.244	135.143	-522.270	-6.064	0.113
2900	59.789	168.600	119.944	141.101	-520.479	12.341	-0.222
3000	60.208	170.634	121.600	147.101	-518.656	30.683	-0.534

PREVIOUS: December 1971 CURRENT: September 1977

Magnesium Sulfide (MgS) $Mg_1S_1(cr)$

MAGNESIUM SULFIDE (MgS) IDEAL GAS $M_r = 56.365$

$S°(298.15\ K) = 225.483 \pm 4.2\ J\ K^{-1}\ mol^{-1}$

$\Delta_f H°(0\ K) = [145.39 \pm 66.9]\ kJ\ mol^{-1}$
$\Delta_f H°(298.15\ K) = [145.23 \pm 66.9]\ kJ\ mol^{-1}$

Electronic States and Molecular Constants

State	ε_i, cm^{-1}	g_i	ω_e, cm^{-1}	$\omega_e x_e$, cm^{-1}	B_e, cm^{-1}	α_e, cm^{-1}	r_e, Å	Source
$X^1\Sigma^+$	0.0	1	526.47	2.681	0.26570	0.00174	2.1425	(1)
$^3\Pi$	[2000]	6	[430]	[2.2]	[0.237]	[0.0015]	[2.27]	(2)
$^1\Pi$	[3000]	2	[430]	[2.2]	[0.237]	[0.0015]	[2.27]	(3)
$^3\Sigma^+$	[17000]	3	[493.20]	[2.313]	[0.2530]	[0.00153]	[2.1956]	(1, 2)
$A^1\Sigma^+$	23036.98	1	495.20	2.313	0.2530	0.00153	2.1956	(1)

Enthalpy of Formation

Colin et al. (1) searched unsuccessfully for gaseous MgS in a Knudsen effusion mass-spectrometric experiment. They set the upper limit $D_0^0(MgS, g) = 55\ kcal\ mol^{-1}$. There are no other equilibrium or positive mass spectrometric data involving MgS(g); consequently $D_0^0(MgS, g)$ must be estimated. The Birge-Sponer extrapolation is of little utility in this case. Gaydon (2) has gone into great detail concerning the problems inherent in linear-Birge-Sponer extrapolations (LBX) on the oxides and sulfides of the alkaline earths. The normal LBX yields $D_0^0(MgS, g) = 73\ kcal\ mol^{-1}$. When adjusted for ionic character with the Hildenbrand correction (3), $D_0^0(MgS, g) = 65\ kcal\ mol^{-1}$. The ground state of MgS cannot dissociate into ground state atoms. After correcting for the excited atomic products, $D_0^0(MgS, g) = 42\ kcal\ mol^{-1}$. A more reliable method of estimation is that of Hauge and Margrave (4) based on the observation that the ratio $D_{MS}K_{MO}/D_{MO}K_{MS}$ is very similar for elements within a given group. Based on data involving known bond energies and force constants of the gaseous diatomic oxides and sulfides, we calculate $D_0^0(MgS, g) = 78 \pm 10\ kcal\ mol^{-1}$. Dissociation energies of the alkaline earth monoxides were taken from Srivastava (5), dissociation energies for the alkaline earth sulfides were taken the JANAF Thermochemical Tables (6) and force constants were calculated from spectroscopic data of Barrow and Cousins (7). Further comparisons of vapor pressures and enthalpies of sublimation for the other alkaline earth oxides and sulfides (6) suggest lower values of $D_0^0(MgS, g)$ than the Hauge and Margrave method leads to. We adopt $D_0^0(MgS, g) \leq 66 \pm 16\ kcal\ mol^{-1}$. Using auxiliary JANAF data (6), we calculate $\Delta_f H°(MgS, g, 298.15\ K) = 34.71 \pm 16\ kcal\ mol^{-1}$.

Heat Capacity and Entropy

The spectroscopic constants of the observed states, taken from Marcano and Barrow (8) for the isotopic species $Mg^{24}S^{32}$, are corrected to account for the natural abundances of the elements. Other low-lying electronic states and their vibrational-rotational constants are estimated in isoconfigurational groups by analogy with MgO (6) and from trends observed in the known states of the other alkaline-earth oxides and sulfides (6). Uncertainty in the energy and molecular constants for the estimated states may contribute significantly, perhaps as much as $1.0\ cal\ K^{-1}\ mol^{-1}$ to the entropy at 298.15 K.

The ground state of gaseous MgS is taken as $^1\Sigma^+$. There should be low-lying triplet states. A $^1\Pi$ state is predicted for MgS at 3000 cm^{-1} based on experimentally observed $^1\Pi$ states for MgO (9), BeO (6), and BeS (10). The corresponding triplet, $^3\Pi$, is assumed to lie 1000 cm^{-1} lower in energy based on trends in the triplet-singlet splitting observed by Field (11) for the alkaline-earth oxides. Similar trends in the triplet-singlet splitting of the isoconfigurational Σ states are used to place the $^3\Sigma^+$ state 6000 cm^{-1} below the experimentally observed (8) $A^1\Sigma^+$ state. By analogy with MgO (6) we do not feel that a $^1\Delta$ state exists below 25000 cm^{-1}. The thermodynamic functions are calculated using first-order anharmonic corrections to Q_r^i and Q_v^i in the partition function $Q = Q_t \sum_i Q_r^i Q_v^i g_i \exp(-c_2\varepsilon_i/T)$.

References

1. R. Colin, P. Goldfinger, and M. Jeunehomme, Trans. Faraday Soc. 60, 314 (1964).
2. A. G. Gaydon, "Dissociation Energies and Spectra of Diatomic Molecules," 3rd ed., Chapman and Hall, London, 1968.
3. D. L. Hildenbrand "Advances in High Temperature Chemistry," Vol. 1, L. Eyring (ed.) pp. 198-206, Academic Press, New York, 1967.
4. R. H. Hauge and J. L. Margrave, High Temp. Sci. 4, 170 (1972).
5. R. D. Srivastava, High Temp. Sci. 8, 225 (1976).
6. JANAF Thermochemical Tables: Mg(g) 9-30-62; S(g) 6-30-71; S_2(g) 12-31-65; BaO(g), SrO(g) 6-30-74; MgO(g), BeO(g) CaO(g), 12-31-74; MgS(g) 6-30-71; BaS(g), SrS(g), CaS(g) 9-30-77.
7. R. F. Barrow and C. Cousins, Adv. High Temp. Chem. 4, 161 (1971).
8. M. Marcano and R. F. Barrow, Trans. Faraday Soc. 66, 2936 (1970).
9. B. Rosen, "Spectroscopic Data Relative to Diatomic Molecules," Pergamon Press, New York, 1970.
10. C. J. Cheetham, W. J. M. Gissane, and R. F. Barrow, Trans. Faraday Soc. 61, 1308 (1965).
11. R. W. Field, J. Chem. Phys. 60, 2400 (1974).

Magnesium Sulfide (MgS) $Mg_1S_1(g)$

Enthalpy Reference Temperature = T_r = 298.15 K; Standard State Pressure = p° = 0.1 MPa

T/K	$C_p°$ (J K⁻¹mol⁻¹)	S° (J K⁻¹mol⁻¹)	$-[G°-H°(T_r)]/T$ (J K⁻¹mol⁻¹)	$H°-H°(T_r)$ (kJ mol⁻¹)	$\Delta_f H°$ (kJ mol⁻¹)	$\Delta_f G°$ (kJ mol⁻¹)	Log K_f
0	0.	0.	INFINITE	-9.245	145.392	145.392	INFINITE
100	29.371	191.105	254.427	-6.332	146.973	130.065	-67.939
200	32.039	212.204	228.567	-3.273	146.383	113.299	-29.591
250	33.380	219.500	226.046	-1.636	145.829	105.090	-21.957
298.15	34.599	225.483	225.483	0.	145.227	97.297	-17.046
300	34.648	225.697	225.484	0.064	145.203	97.000	-16.889
350	36.111	231.144	225.911	1.832	144.542	89.018	-13.285
400	37.972	236.082	226.878	3.682	141.668	81.219	-10.606
450	40.277	240.684	228.159	5.636	140.377	73.732	-8.559
500	42.922	245.062	229.632	7.715	139.107	66.399	-6.937
600	48.397	253.377	232.903	12.284	137.324	52.038	-4.530
700	52.741	261.186	236.392	17.356	136.177	37.921	-2.830
800	55.100	268.406	239.950	22.765	135.386	23.941	-1.563
900	55.594	274.940	243.481	28.313	81.559	11.133	-0.646
1000	54.817	280.766	246.924	33.842	73.375	3.996	-0.209
1100	53.380	285.926	250.240	39.255	73.486	-2.948	0.140
1200	51.717	290.500	253.408	44.510	73.424	-9.896	0.431
1300	50.079	294.574	256.421	49.599	73.179	-16.830	0.676
1400	48.585	298.230	259.279	54.531	-54.647	-20.572	0.768
1500	47.278	301.536	261.988	59.323	-53.867	-18.166	0.633
1600	46.161	304.551	264.555	63.993	-53.224	-15.808	0.516
1700	45.217	307.320	266.990	68.560	-52.698	-13.486	0.414
1800	44.426	309.881	269.303	73.042	-52.272	-11.192	0.325
1900	43.765	312.265	271.502	77.450	-51.931	-8.920	0.245
2000	43.213	314.496	273.596	81.798	-51.663	-6.663	0.174
2100	42.752	316.592	275.594	86.096	-51.457	-4.419	0.110
2200	42.368	318.572	277.503	90.351	-51.304	-2.183	0.052
2300	42.048	320.448	279.330	94.572	-51.195	0.048	-0.001
2400	41.781	322.232	281.081	98.763	-51.126	2.274	-0.049
2500	41.559	323.933	282.761	102.929	-51.090	4.498	-0.094
2600	41.375	325.559	284.376	107.076	-51.082	6.721	-0.135
2700	41.225	327.118	285.931	111.205	-51.101	8.945	-0.173
2800	41.102	328.615	287.428	115.321	-51.141	11.169	-0.208
2900	41.003	330.055	288.874	119.426	-51.203	13.396	-0.241
3000	40.925	331.444	290.270	123.523	-51.284	15.625	-0.272
3100	40.865	332.785	291.620	127.612	-51.383	17.857	-0.301
3200	40.820	334.081	292.926	131.696	-51.499	20.092	-0.328
3300	40.790	335.337	294.193	135.776	-51.633	22.331	-0.353
3400	40.772	336.554	295.421	139.854	-51.784	24.575	-0.378
3500	40.764	337.736	296.613	143.931	-51.955	26.823	-0.400
3600	40.766	338.885	297.771	148.008	-52.144	29.077	-0.422
3700	40.776	340.002	298.898	152.085	-52.354	31.335	-0.442
3800	40.793	341.089	299.994	156.163	-52.587	33.600	-0.462
3900	40.817	342.149	301.061	160.243	-52.843	35.872	-0.480
4000	40.847	343.183	302.101	164.327	-53.125	38.150	-0.498
4100	40.882	344.192	303.116	168.413	-53.434	40.436	-0.515
4200	40.921	345.178	304.105	172.503	-53.773	42.729	-0.531
4300	40.965	346.141	305.072	176.597	-54.143	45.031	-0.547
4400	41.012	347.083	306.016	180.696	-54.548	47.342	-0.562
4500	41.062	348.005	306.939	184.800	-54.987	49.663	-0.576
4600	41.115	348.909	307.841	188.909	-55.466	51.993	-0.590
4700	41.170	349.793	308.725	193.023	-55.984	54.335	-0.604
4800	41.228	350.661	309.589	197.143	-56.545	56.688	-0.617
4900	41.287	351.511	310.436	201.269	-57.150	59.053	-0.630
5000	41.348	352.346	311.266	205.400	-57.801	61.431	-0.642
5100	41.410	353.166	312.080	209.538	-58.500	63.823	-0.654
5200	41.473	353.970	312.878	213.682	-59.250	66.229	-0.665
5300	41.537	354.761	313.660	217.833	-60.050	68.649	-0.677
5400	41.602	355.538	314.429	221.990	-60.905	71.085	-0.688
5500	41.668	356.302	315.183	226.153	-61.814	73.538	-0.698
5600	41.734	357.053	315.924	230.323	-62.779	76.007	-0.709
5700	41.801	357.792	316.652	234.500	-63.803	78.495	-0.719
5800	41.868	358.520	317.368	238.683	-64.886	81.001	-0.729
5900	41.935	359.236	318.071	242.874	-66.029	83.526	-0.739
6000	42.002	359.942	318.763	247.070	-67.234	86.070	-0.749

PREVIOUS: September 1977 (1 atm) CURRENT: September 1977 (1 bar)

Magnesium Sulfide (MgS) $Mg_1S_1(g)$

MAGNESIUM (Mg_2) IDEAL GAS M_r = 48.610

D_0^0 = 4.84 ± 0.01 kJ mol^{-1} (natural abundance)
$S°$(298.15 K) = 240.19 ± 0.50 J K^{-1} mol^{-1}

$\Delta_fH°$(0 K) = 288.16 ± 0.80 kJ mol^{-1}
$\Delta_fH°$(298.15 K) = 287.63 ± 0.80 kJ mol^{-1}

Spectroscopic Constants for $^{24}Mg_2$ in cm^{-1}

State	T_e	D°	ω_e	ω_ex_e	B_e	$10^3\alpha_e$	10^6D_e	r_e, Å	References
$^1\Sigma_g^+$	0	404	51.121	1.6448[a]	0.092866	3.7758[b]	1.2166[c]	3.8905	1
$^3\Pi_g$	[15645]	[21740]	[220.1]	[1.95]	[0.171]	[1.89]	[0.413]	[2.91]	4[d]
$^3\Sigma_u^+$	[16773]	[21740]	[184.2]	[1.67]	[0.145]	[1.64]	[0.359]	[3.18]	4[d]
$^1\Pi_g$	[20267]	[34941]	[263.1]	[1.17]	[0.181]	[1.15]	[0.343]	[2.91]	4[d]

(a) $+1.6244 \times 10^{-2} (v + 1/2)^3 - 6.106 \times 10^{-4} (v + 1/2)^4$.
(b) $+1.0679 \times 10^{-4} (v + 1/2)^2 - 9.6831 \times 10^{-6} (v + 1/2)^3$.
(c) $-3.0768 \times 10^{-8} (v + 1/2) + 2.1344 \times 10^{-8} (v + 1/2)^2$.
(d) These values are estimated based on the theoretical potential curves of Stevens and Krauss (4).

Enthalpy of Formation

The ground state dissociation energy for $^{24}Mg_2$ was determined by Li and Stwalley (2) to be 404.1±0.5 cm^{-1}. This value was corrected to natural abundance Mg_2 with the result D_0^0= 404.3 cm^{-1}. Using this value and the recommended value for $\Delta_fH°$(Mg, g, 0 K) = 146.499 kJ mol^{-1} from CODATA (13), we calculate the adopted value of $\Delta_fH°(Mg_2$, g, 0) = 288.16 kJ mol^{-1}. Our adopted D° is in essential agreement with the value adopted by Brewer (12) and by Wagman et al. (3). It should be noted that a value of D_0^0 = 11.3 ± 4.1 kJ mol^{-1} was determined by a mass spectrometric study of Wu, Ihle and Gingerich (7). In light of the spectroscopic evidence, this value appears to be too large.

Heat Capacity and Entropy

The thermal functions are calculated using a direct summation technique analogous to the alkali dimers and diatomic halogens. Included in the calculation are the ground state and the three lowest excited states. Spectrocopic constants for the $^1\Sigma_g^+$ ground state are taken from an experimental study from Balfour and Douglas (1). No known spectroscopic data exists for the three lowest excited electronic states of Mg_2(g). In order to generate the spectroscopic constants of these states, a Morse potential was least squares fit to the calculated potential curves of Stevens and Krause (4), an ab initio investigation which included configuration interaction. Although this procedure for estimating the spectroscopic data results in substantial uncertainty in the generated constants, the inclusion of the excited states results in thermal functions which are more realistic at temperatures greater than 1000 K. The large maximum in the heat capacity at 5000 K, for example, would not be present if the excited states were not included. Also, it should be noted that Mg_2(g) is an example of a molecule where the traditional integration methods of calculating the thermal functions fail due to the shallow ground state.

The rotational levels are extrapolated to high J values according to the method of Khachkuruzov (5) who proposed a simpler form of Wooley's method (6). For the ground state only, we included the so-called quasi-bound rotational levels above the dissociation limit (9). The values used for J_{lim} and v_{max} are 93 and 17, respectively, as determined by (8). Separate calculations were performed for each of the six isotopic dimeric species (made up from the three atomic isotopes ^{24}Mg, ^{25}Mg, ^{26}Mg) by adjusting the spectroscopic constants of ^{24}Mg using a standard reduced mass scaling routine (10). The results were combined according to the natural abundance of the atoms, ^{24}Mg: 78.99%, ^{25}Mg: 10.00%, ^{26}Mg: 11.01% (11). Splitting of the rotational levels in the $^3\Pi_g$ and $^3\Sigma_u^+$ states was taken into account only insofar as the degeneracy of these states were increased accordingly. The adopted value of S°(298.15 K) is 0.64 J K^{-1} mol^{-1} smaller than that adopted by (8).

References

1. W. J. Balfour and A. E. Douglas, Can. J. Phys. 48, 901 (1970).
2. K. C. Li and W. C. Stwalley, J. Chem. Phys. 59, 4423 (1973).
3. D. D. Wagman, W. H. Evans, et al., J. Phys. Chem. Ref. Data 11, Supplement 2 (1982).
4. W. J. Stevens and M. Krauss, J. Chem. Phys. 67, 1977 (1977).
5. G. A. Khachkuruzov, Opt. Spectrosk. 30, 455 (1971).
6. H. W. Wooley, R. B. Scott and F. G. Brickwedde, J. Res. Nat. Bur. Stand. 41, 379 (1948).
7. C. H. Wu, H. R. Ihle, and K. A. Gingerich, Int. J. Mass. Spec. Ion Phys. 47, 235 (1983).
8. L. V. Gurvich, I. V. Veits, et al., "Thermodynamic Properties of Individual Substances", Vol. 3, Nauka, Moscow, 1981.
9. JANAF Thermochemical Tables: F_2(r) (1983).
10. G. Herzberg, "Spectra of Diatomic Molecules", 2nd Ed., D. Van Nostrand Co., Inc., New York, 1950.
11. N. E. Holden, R. L. Martin, and I. L. Barnes, Pure Appl. Chem. 55, 1119 (1983).
12. L. Brewer and J. S. Winn, Faraday Symp. Chem. Soc., (14), 126 (1980).
13. J. D. Cox, J. Chem. Thermodynamics 10, 903 (1977).

Magnesium (Mg_2) Mg_2(g)

Enthalpy Reference Temperature = T_r = 298.15 K   Standard State Pressure = p° = 0.1 MPa

T/K	$C_p°$ (J K^{-1}mol^{-1})	S°	$-[G°-H°(T_r)]/T$	$H°-H°(T_r)$ (kJ mol^{-1})	$\Delta_fH°$	$\Delta_fG°$	Log K_f
0	0.	0.	INFINITE	-9.463	288.158	288.158	INFINITE
100	38.343	206.636	264.861	-5.822	290.515	271.752	-141.949
200	28.193	229.790	242.505	-2.543	289.780	253.079	-66.098
250	25.711	235.789	240.594	-1.201	288.780	244.015	-50.984
298.15	24.293	240.186	240.186	0.	287.625	235.495	-41.258
300	24.250	240.337	240.187	0.045	287.578	235.172	-40.947
350	23.340	244.001	240.478	1.233	286.242	226.541	-33.809
400	22.742	247.076	241.116	2.384	284.806	218.110	-28.482
450	22.329	249.730	241.929	3.510	283.292	209.863	-24.360
500	22.034	252.066	242.828	4.619	281.708	201.788	-21.081
600	21.649	256.046	244.711	6.801	278.356	186.115	-16.203
700	21.418	259.365	246.574	8.954	274.764	171.022	-12.762
800	21.268	262.214	248.355	11.088	270.922	156.461	-10.216
900	21.166	264.713	250.036	13.209	266.806	142.398	-8.265
1000	21.093	266.939	251.617	15.322	245.207	130.233	-6.803
1100	21.039	268.947	253.103	17.428	240.451	118.966	-5.649
1200	20.999	270.776	254.501	19.530	235.691	108.132	-4.707
1300	20.967	272.455	255.818	21.628	230.928	97.696	-3.925
1400	20.944	274.008	257.063	23.724	-28.657	93.960	-3.506
1500	20.927	275.452	258.241	25.817	-30.721	102.790	-3.579
1600	20.919	276.803	259.359	27.909	-32.786	111.758	-3.649
1700	20.923	278.071	260.423	30.001	-34.851	120.856	-3.713
1800	20.946	279.267	261.437	32.095	-36.915	130.074	-3.775
1900	20.998	280.401	262.406	34.192	-38.976	139.408	-3.833
2000	21.094	281.480	263.333	36.296	-41.029	148.850	-3.888
2100	21.253	282.513	264.221	38.413	-43.070	158.394	-3.940
2200	21.497	283.507	265.076	40.549	-45.092	168.035	-3.990
2300	21.853	284.470	265.898	42.716	-47.085	177.767	-4.037
2400	22.351	285.410	266.692	44.925	-49.038	187.585	-4.083
2500	23.024	286.336	267.459	47.192	-50.934	197.484	-4.126
2600	23.904	287.255	268.203	49.536	-52.757	207.457	-4.168
2700	25.023	288.177	268.925	51.980	-54.485	217.498	-4.208
2800	26.412	289.112	269.629	54.550	-56.093	227.601	-4.246
2900	28.096	290.067	270.318	57.273	-57.557	237.760	-4.283
3000	30.094	291.052	270.992	60.179	-58.847	247.966	-4.317
3100	32.417	292.076	271.656	63.302	-59.934	258.211	-4.351
3200	35.067	293.146	272.310	66.674	-60.788	268.488	-4.383
3300	38.032	294.270	272.959	70.326	-61.380	278.788	-4.413
3400	41.290	295.453	273.603	74.290	-61.683	289.101	-4.442
3500	44.805	296.700	274.245	78.593	-61.673	299.419	-4.469
3600	48.526	298.014	274.887	83.258	-61.331	309.732	-4.494
3700	52.393	299.396	275.530	88.303	-60.643	320.031	-4.518
3800	56.336	300.845	276.177	93.739	-59.603	330.306	-4.540
3900	60.277	302.360	276.829	99.570	-58.211	340.549	-4.561
4000	64.134	303.935	277.487	105.792	-56.475	350.753	-4.580
4100	67.825	305.564	278.152	112.391	-54.414	360.908	-4.598
4200	71.274	307.240	278.824	119.349	-52.053	371.010	-4.614
4300	74.410	308.955	279.505	126.636	-49.423	381.053	-4.629
4400	77.176	310.698	280.194	134.218	-46.563	391.031	-4.642
4500	79.525	312.460	280.891	142.057	-43.517	400.942	-4.654
4600	81.426	314.229	281.597	150.108	-40.335	410.785	-4.665
4700	82.864	315.996	282.310	158.327	-37.064	420.556	-4.674
4800	83.838	317.752	283.030	166.666	-33.756	430.258	-4.682
4900	84.358	319.487	283.756	175.079	-30.462	439.891	-4.689
5000	84.450	321.193	284.488	183.523	-27.228	449.457	-4.695
5100	84.145	322.863	285.224	191.956	-24.101	458.959	-4.701
5200	83.481	324.491	285.964	200.340	-21.121	468.402	-4.705
5300	82.502	326.072	286.706	208.641	-18.326	477.789	-4.709
5400	81.254	327.603	287.449	216.831	-15.748	487.125	-4.712
5500	79.780	329.081	288.193	224.885	-13.414	496.415	-4.715
5600	78.126	330.504	288.936	232.781	-11.349	505.665	-4.717
5700	76.331	331.871	289.677	240.505	-9.571	514.882	-4.718
5800	74.435	333.182	290.416	248.044	-8.095	524.070	-4.720
5900	72.470	334.438	291.151	255.390	-6.931	533.234	-4.721
6000	70.466	335.639	291.883	262.537	-6.088	542.382	-4.722

PREVIOUS: December 1974 (1 atm)   CURRENT: September 1983 (1 bar)

Magnesium (Mg_2) Mg_2(g)

$S°(298.15 K) = 95.14 \pm 0.8\ J\ K^{-1}\ mol^{-1}$

$T_{fus} = 2171 \pm 11\ K$

$\Delta_f H°(0\ K) = -2163.63 \pm 4.2\ kJ\ mol^{-1}$

$\Delta_f H°(298.15\ K) = -2176.94 \pm 4.2\ kJ\ mol^{-1}$

$\Delta_{fus} H° = [71.128 \pm 20.9]\ kJ\ mol^{-1}$

Enthalpy of Formation

Torgeson et al. (1) have reported a $\Delta_r H°(298.15\ K) = -15.1 \pm 0.2\ kcal\ mol^{-1}$ for reaction (a). They used an HF solution calorimeter and measured the enthalpies of solution of each component. We derive $\Delta_f H°(Mg_2SiO_4$, cr, 298.15 K) = -520.30 ± 1.0 kcal mol^{-1} using the $\Delta_f H°$(298.15 K) of MgO(cr) and SiO_2(cr, low quartz) from JANAF tables.

Kapustinskii et al. (2) used the same method but measured only $\Delta_r H°(298.15\ K) = 95.00 \pm 0.47\ kcal\ mol^{-1}$ for reaction (b). We calculate $\Delta_f H°(Mg_2SiO_4$, cr, 298.15 K) = −522.27 kcal mol^{-1} using the $\Delta_f H°$(298.15 K) of H_2SiF_6(aq) from source 3 and the $\Delta_f H°$(298.15 K) of $H_2O(\ell)$ and HF(aq) from source 4.

Reesman et al. (5) calculated a $\Delta_r G°(298.15\ K) = 42.46\ kcal\ mol^{-1}$ for reaction (c) using aqueous solubility data. We use the $\Delta_f G°$(298.15 K) of $H_2O(\ell)$ and OH^-(aq) from source 4, $\Delta_f G°$(298.15 K) of H_4SiO_4(aq) from source 3 and $\Delta_f G°$(298.15 K) of Mg^{++}(aq) from source 6 and we obtain a $\Delta_f G°(Mg_2SiO_4$, cr, 298.15 K) = −498.764 kcal mol^{-1}. We derive a $\Delta_f H°(Mg_2SiO_4$, cr, 298.15 K) = −524.848 kcal mol^{-1} using the above $\Delta_f G°$(298.15 K) of Mg_2SiO_4(cr) and the entropies of Mg(cr), Si(cr) and O_2(g) from the JANAF tables.

The adopted value is from the work of Torgeson et al. (1) which is considered to be the most reliable since it relates directly to the oxides. The other reported values are less accurate and involve uncertainties in the auxiliary data used to derive the heat of formation.

Source	Method	Reaction	$\Delta_r H°$(298.15 K) kcal mol^{-1}	$\Delta_f G°$(298.15 K) kcal mol^{-1}	$\Delta_f H°$(298.15 K) kcal mol^{-1}
(1)	hydrofluoric acid solution calorimeter	(a) 2 MgO(cr) + SiO_2(cr) = Mg_2SiO_4(cr)	−15.10 ±0.2		−520.3
(2)	hydrofluoric acid solution calorimeter	(b) Mg_2SiO_4(cr) + 10 HF(aq, 4.44 H_2O) = 2 MgF_2(cr) + H_2SiF_6(714 HF(aq) + 322 OH_2O) + 4 $H_2O(\ell)$	95.0 ±0.47	−522.27	
(5)	aqueous solubility	(c) Mg_2SiO_4(cr) + 4 $H_2O(\ell)$ = 2 Mg^{++}(aq) + H_4SiO_4(aq) + 4 OH^-(aq)		42.46	−524.848

Heat Capacity and Entropy

Low temperature heat capacities are from the data (53-295 K) of Kelley (7). High temperature enthalpies have been measured (398-1808 K) by Orr (8). High temperature heat capacities are derived from the enthalpies by a fitting technique which constrains the curve to join smoothly with the low temperature values. The entropy is based on S°(50 K) = 0.60 cal K^{-1} mol^{-1}.

Fusion Data

T_{fus} is obtained from Davis et al. (9). $\Delta_{fus}H°$ is estimated from MacGregor's (10) phase diagrams of the Mg_2SiO_4 − TiO_2 system. The phase diagrams were determined at 10 and 20 kbars pressure.

References

1. D. R. Torgeson and Th. G. Sahama, J. Amer. Chem. Soc. 70, 2156 (1948).
2. A. F. Kapustinskii and K. K. Samplavskaya, Russ. J. Inorg. Chem. 6, 1141 (1961).
3. U. S. Nat. Bur. Stand. Tech. Note 270-2, 1966.
4. U. S. Nat. Bur. Stand. Tech. Note 270-1, 1965.
5. A. L. Reesman and W. D. Keller, Am. Mineralogist 50, 1729-39 (1965).
6. U. S. Nat. Bur. Stand. Circ. 500, 1952.
7. K. K. Kelley, J. Amer. Chem. Soc. 65, 339-41 (1943).
8. R. L. Orr, J. Amer. Chem. Soc. 75, 528-9 (1953).
9. B. T. C. Davis and J. L. England, J. Geophys. Res. 69, 1113-1116 (1965).
10. I. D. MacGregor, Ann. Rept. Director of the Geophysical Laboratory, No. 1455, 135-9 (1965).

Enthalpy Reference Temperature = T_r = 298.15 K; Standard State Pressure = p° = 0.1 MPa

T/K	$C_p°$ (J $K^{-1}mol^{-1}$)	S° (J $K^{-1}mol^{-1}$)	−[G°−H°(T_r)]/T (J $K^{-1}mol^{-1}$)	H°−H°(T_r) (kJ mol^{-1})	$\Delta_f H°$ (kJ mol^{-1})	$\Delta_f G°$ (kJ mol^{-1})	Log K_f
0	0.	0.	INFINITE	-17.280	-2163.634	-2163.634	INFINITE
100	32.317	14.050	176.849	-16.280	-2169.994	-2134.453	1114.924
200	86.144	54.162	105.060	-10.180	-2174.908	-2096.757	547.617
298.15	118.688	95.140	95.140	0.	-2176.935	-2057.879	360.532
300	119.148	95.876	95.142	0.220	-2176.953	-2057.141	358.180
400	137.557	132.920	100.068	13.141	-2177.207	-2017.131	263.410
500	148.695	164.894	109.913	27.491	-2176.585	-1977.171	206.554
600	156.431	192.724	121.447	42.767	-2175.539	-1937.382	168.664
700	162.310	217.297	133.419	58.715	-2174.292	-1897.786	141.615
800	167.063	239.291	145.302	75.191	-2172.975	-1858.375	121.339
900	171.088	259.206	156.869	92.104	-2171.673	-1819.128	105.579
1000	174.607	277.418	168.026	109.392	-2187.632	-1778.598	92.904
1100	177.770	294.210	178.744	127.013	-2186.547	-1737.746	82.519
1200	180.665	309.804	189.024	144.936	-2185.281	-1697.000	73.868
1300	183.360	324.372	198.881	163.139	-2183.847	-1656.367	66.554
1400	185.895	338.054	208.338	181.602	-2437.073	-1609.516	60.052
1500	188.309	350.962	217.420	200.314	-2432.626	-1550.560	53.995
1600	190.619	363.190	226.152	219.261	-2428.036	-1491.904	48.706
1700	192.845	374.813	234.558	238.434	-2473.488	-1433.093	44.034
1800	195.008	385.897	242.660	257.828	-2468.404	-1372.040	39.816
1900	197.108	396.497	250.480	277.434	-2463.154	-1311.273	36.049
2000	199.163	406.660	258.036	297.248	-2457.742	-1250.787	32.667
2100	201.175	416.426	265.348	317.264	-2452.175	-1190.575	29.614
2171.000	202.673	423.139	270.399	331.600	------ CRYSTAL <--> LIQUID -----		
2200	203.154	425.833	272.430	337.487	-2446.445	-1130.631	26.845
2300	205.100	434.907	279.298	357.900	-2440.573	-1070.953	24.322
2400	207.020	443.676	285.965	378.506	-2434.554	-1011.534	22.015
2500	208.915	452.166	292.445	399.303	-2428.390	-952.367	19.899
2600	210.790	460.396	298.747	420.289	-2422.084	-893.451	17.950
2700	212.648	468.386	304.883	441.461	-2415.638	-834.778	16.150
2800	214.490	476.153	310.861	462.818	-2409.054	-776.348	14.483
2900	216.315	483.712	316.692	484.358	-2402.334	-718.156	12.935
3000	218.124	491.076	322.382	506.080	-2395.483	-660.197	11.495

PREVIOUS: September 1964 CURRENT: December 1967

MAGNESIUM SILICATE (Mg_2SiO_4) LIQUID M_r = 140.6931

$S°(298.15\ K) = [123.044]\ J\ K^{-1}\ mol^{-1}$

$T_{fus} = 2171 \pm 11\ K$

$\Delta_fH°(298.15\ K) = [-2113.88 \pm 20.9]\ kJ\ mol^{-1}$

$\Delta_{fus}H° = [71.128 \pm 20.9]\ kJ\ mol^{-1}$

Enthalpy of Formation

$\Delta_fH°(Mg_2SiO_4$, ℓ, 298.15 K) is calculated from $\Delta_fH°(Mg_2SiO_4$, cr, 298.15 K) by adding $\Delta_{fus}H°$ and the difference in enthalpy, H°(2171 K)-H°(298.15 K), between the crystal and liquid.

Heat Capacity and Entropy

A glass transition is assumed at 1400 K. Below 1400 K the heat capacity is obtained from the heat capacity of the crystal. Above 1400 K the heat capacity is assumed constant and estimated as 49 cal $K^{-1}\ mol^{-1}$ or 7 cal K^{-1} g-atom^{-1}. S°(ℓ, 298.15 K) is calculated in a manner analogous to that used for the enthalpy of formation.

Fusion Data

See the crystal table for details.

Magnesium Silicate (Mg_2SiO_4) $Mg_2O_4Si_1(l)$

Enthalpy Reference Temperature = T_r = 298.15 K Standard State Pressure = p° = 0.1 MPa

T/K	$C_p°$ (J K⁻¹mol⁻¹)	S° (J K⁻¹mol⁻¹)	$-[G°-H°(T_r)]/T$ (J K⁻¹mol⁻¹)	$H°-H°(T_r)$ (kJ mol⁻¹)	$\Delta_fH°$ (kJ mol⁻¹)	$\Delta_fG°$ (kJ mol⁻¹)	Log K_f
0							
100							
200							
298.15	118.688	123.044	123.044	0.	-2113.878	-2003.141	350.942
300	119.148	123.780	123.046	0.220	-2113.895	-2002.454	348.658
400	137.557	160.824	127.972	13.141	-2114.149	-1965.234	256.633
500	148.695	192.798	137.817	27.491	-2113.528	-1928.065	201.424
600	156.431	220.628	149.351	42.767	-2112.482	-1891.067	164.632
700	162.310	245.201	161.323	58.715	-2111.235	-1854.261	138.367
800	167.063	267.195	173.206	75.191	-2109.918	-1817.640	118.680
900	171.088	287.110	184.773	92.104	-2108.615	-1781.184	103.377
1000	174.607	305.322	195.930	109.392	-2124.574	-1743.445	91.068
1100	177.770	322.114	206.648	127.013	-2123.489	-1705.382	80.982
1200	180.665	337.708	216.928	144.936	-2122.224	-1667.427	72.581
1300	183.360	352.276	226.785	163.139	-2120.789	-1629.585	65.478
1400	185.895	365.958	236.242	181.603	-2374.015	-1585.524	59.157
1400.000	185.895	365.958	236.242	181.603	GLASS <--> LIQUID TRANSITION		
1400.000	205.016	365.958	236.242	181.603			
1500	205.016	380.103	245.367	202.104	-2367.777	-1529.422	53.259
1600	205.016	393.334	254.206	222.606	-2361.633	-1473.732	48.112
1700	205.016	405.763	262.759	243.108	-2405.758	-1417.978	43.569
1800	205.016	417.482	271.032	263.609	-2399.565	-1360.052	39.468
1900	205.016	428.567	279.035	284.111	-2393.419	-1302.469	35.807
2000	205.016	439.082	286.776	304.612	-2387.319	-1245.209	32.522
2100	205.016	449.085	294.269	325.114	-2381.266	-1188.252	29.556
2171.000	205.016	455.902	299.444	339.670	CRYSTAL <--> LIQUID		
2200	205.016	458.623	301.525	345.616	-2375.259	-1131.582	26.867
2300	205.016	467.736	308.555	366.117	-2369.298	-1075.186	24.418
2400	205.016	476.461	315.370	386.619	-2363.384	-1019.047	22.179
2500	205.016	484.830	321.982	407.120	-2357.515	-963.154	20.124
2600	205.016	492.871	328.401	427.622	-2351.693	-907.495	18.232
2700	205.016	500.609	334.637	448.124	-2345.917	-852.058	16.484
2800	205.016	508.065	340.699	468.625	-2340.189	-796.834	14.865
2900	205.016	515.259	346.595	489.127	-2334.508	-741.816	13.362
3000	205.016	522.209	352.333	509.628	-2328.877	-686.991	11.962
3100	205.016	528.932	357.922	530.130	-2323.296	-632.353	10.655
3200	205.016	535.441	363.368	550.632	-2317.767	-577.896	9.433
3300	205.016	541.749	368.679	571.133	-2312.291	-523.610	8.288
3400	205.016	547.870	373.860	591.635	-2306.872	-469.488	7.213
3500	205.016	553.813	378.917	612.136	-2301.510	-415.527	6.201
3600	205.016	559.588	383.855	632.638	-2680.362	-351.255	5.097
3700	205.016	565.205	388.681	653.140	-2674.711	-286.637	4.047
3800	205.016	570.673	393.399	673.641	-2669.130	-222.167	3.054
3900	205.016	575.998	398.013	694.143	-2663.620	-157.845	2.114
4000	205.016	581.189	402.528	714.644	-2658.186	-93.665	1.223

PREVIOUS: September 1964 CURRENT: December 1967

Magnesium Silicate (Mg_2SiO_4) $Mg_2O_4Si_1(l)$

MAGNESIUM ORTHOSILICATE (Mg_2SiO_4) CRYSTAL-LIQUID M_r = 140.6931

0 to 2171 K crystal
above 2171 K liquid

Refer to the individual tables for details.

Magnesium Silicate (Mg_2SiO_4) $Mg_2O_4Si_1$(cr,l)

Enthalpy Reference Temperature = T_r = 298.15 K Standard State Pressure = p° = 0.1 MPa

T/K	C_p° (J K⁻¹mol⁻¹)	S° (J K⁻¹mol⁻¹)	$-[G^\circ-H^\circ(T_r)]/T$ (J K⁻¹mol⁻¹)	$H^\circ-H^\circ(T_r)$ (kJ mol⁻¹)	Δ_fH° (kJ mol⁻¹)	Δ_fG° (kJ mol⁻¹)	Log K_f
0	0.	0.	INFINITE	-17.280	-2163.634	-2163.634	INFINITE
100	32.317	14.050	176.849	-16.280	-2169.994	-2134.453	1114.924
200	86.144	54.162	105.060	-10.180	-2174.908	-2096.757	547.617
298.15	118.688	95.140	95.140	0.	-2176.935	-2057.879	360.532
300	119.148	95.876	95.142	0.220	-2176.953	-2057.141	358.180
400	137.557	132.920	100.068	13.141	-2177.207	-2017.131	263.410
500	148.695	164.894	109.913	27.491	-2176.585	-1977.171	206.554
600	156.431	192.724	121.447	42.767	-2175.539	-1937.382	168.664
700	162.310	217.297	133.419	58.715	-2174.292	-1897.786	141.615
800	167.063	239.291	145.302	75.191	-2172.975	-1858.375	121.339
900	171.088	259.206	156.869	92.104	-2171.673	-1819.128	105.579
1000	174.607	277.418	168.026	109.392	-2187.632	-1778.598	92.904
1100	177.770	294.210	178.744	127.013	-2186.547	-1737.746	82.519
1200	180.665	309.804	189.024	144.936	-2185.281	-1697.000	73.868
1300	183.360	324.372	198.881	163.139	-2183.847	-1656.367	66.554
1400	185.895	338.054	208.338	181.602	-2437.073	-1609.516	60.052
1500	188.309	350.962	217.420	200.314	-2432.626	-1550.560	53.995
1600	190.619	363.190	226.152	219.261	-2428.036	-1491.904	48.706
1700	192.845	374.813	234.558	238.434	-2473.488	-1433.093	44.034
1800	195.008	385.897	242.660	257.828	-2468.404	-1372.040	39.816
1900	197.108	396.497	250.480	277.434	-2463.154	-1311.273	36.049
2000	199.163	406.660	258.036	297.248	-2457.742	-1250.787	32.667
2100	201.175	416.426	265.348	317.264	-2452.175	-1190.575	29.614
2171.000	202.673	423.139	270.399	331.600	CRYSTAL <--> LIQUID		
2171.000	205.016	455.902	270.399	402.728	TRANSITION		
2200	205.016	458.623	272.862	408.673	-2375.259	-1131.582	26.867
2300	205.016	467.736	281.138	429.175	-2369.298	-1075.186	24.418
2400	205.016	476.461	289.096	449.676	-2363.384	-1019.047	22.179
2500	205.016	484.830	296.759	470.178	-2357.515	-963.154	20.124
2600	205.016	492.871	304.148	490.680	-2351.693	-907.495	18.232
2700	205.016	500.609	311.282	511.181	-2345.917	-852.058	16.484
2800	205.016	508.065	318.178	531.683	-2340.189	-796.834	14.865
2900	205.016	515.259	324.850	552.184	-2334.508	-741.816	13.362
3000	205.016	522.209	331.314	572.686	-2328.877	-686.991	11.962
3100	205.016	528.932	337.581	593.188	-2323.296	-632.353	10.655
3200	205.016	535.441	343.663	613.689	-2317.767	-577.896	9.433
3300	205.016	541.749	349.570	634.191	-2312.291	-523.610	8.288
3400	205.016	547.870	355.313	654.692	-2306.872	-469.488	7.213
3500	205.016	553.813	360.900	675.194	-2301.510	-415.527	6.201
3600	205.016	559.588	366.339	695.696	-2680.362	-351.255	5.097
3700	205.016	565.205	371.639	716.197	-2674.711	-286.637	4.047
3800	205.016	570.673	376.805	736.699	-2669.130	-222.167	3.054
3900	205.016	575.998	381.844	757.200	-2663.620	-157.845	2.114
4000	205.016	581.189	386.763	777.702	-2658.186	-93.665	1.223

PREVIOUS: CURRENT: December 1967

Magnesium Silicate (Mg_2SiO_4) $Mg_2O_4Si_1$(cr,l)

MAGNESIUM TITANIUM OXIDE (Mg_2TiO_4) CRYSTAL M_r = 160.4876

$S°(298.15\ K) = 115.10 \pm 6.3\ J\ K^{-1}\ mol^{-1}$

$T_{fus} = 2013 \pm 20\ K$

$\Delta_f H°(0\ K) = -2151.03 \pm 6.3\ kJ\ mol^{-1}$

$\Delta_f H°(298.15\ K) = -2164.38 \pm 6.3\ kJ\ mol^{-1}$

$\Delta_{fus} H° = [129.704]\ kJ\ mol^{-1}$

Enthalpy of Formation

The enthalpy of formation is calculated from $\Delta_r H°(298.15\ K) = -4.1 \pm 0.25\ kcal\ mol^{-1}$ for the reaction 2 MgO(cr) + TiO_2(rutile) → Mg_2TiO_4(cr) measured by Kelley (1), using a solution calorimetric method. The uncertainty of ±1.5 kcal mol^{-1} assigned to the enthalpy of formation is due primarily to TiO_2(rutile).

Heat Capacity and Entropy

Low temperature heat capacities are from the data (52 - 296 K) of Todd (2).

High temperature enthalpies have been measured (392 - 1817 K) by Orr (3). High temperature heat capacities are derived from the enthalpies by a fitting technique which constrains the curve to join smoothly with the low temperature values.

The entropy is based on $S°(51\ K) = 0.66 + 2.75 = 3.41\ cal\ K^{-1}\ mol^{-1}$, where the first term is a lattice contribution and the second term is a zero-point entropy of 2Rℓn2 corresponding to random distribution of Ti and Mg on the octahedral sites of the spinel lattice. The adopted value is $S°(298.15\ K) = 27.51 \pm 1.5\ cal\ K^{-1}\ mol^{-1}$ instead of $24.76 \pm 0.15\ cal\ K^{-1}\ mol^{-1}$ as given by Todd.

Barth (4) showed from intensities of the X-ray diffraction lines that Mg_2TiO_4 is a spinel of the inverse (or variate) class. Thus, half of the Mg ions occupy the tetrahedral interstices in the oxygen lattice, while the remaining Mg ions and the Ti ions are (presumably) randomly located in the octahedral interstices. This conclusion was confirmed by Romeijn (5).

King and Kelley (6) suggested that the degree of structural randomness is not established and that it may vary with the sample of Mg_2TiO_4; thus, they recommended an intermediate value of Rℓn2 for the zero-point entropy. In a previous review, however, Kelley (1) used the theoretical value of 2Rℓn2 in order to make Mg_2TiO_4 stable relative to $MgTiO_3$ at high temperatures. We choose to adopt 2Rℓn2, although the evidence is not conclusive. Recent studies of solid state reactions by Bautsch and Werner (7) indicate that $\Delta_r G°$ is negative for $MgTiO_3$(cr) + MgO(cr) → Mg_2TiO_4(cr) at temperatures of 1673 K and above. Both Rℓn2 and 2Rℓn2 are consistent with this evidence.

Fusion Data

See the liquid table for details.

References

1. K. K. Kelley, S. S. Todd and E. G. King, U. S. Bur. Mines RI 5059, 1954.
2. S. S. Todd, J. Amer. Chem. Soc. 74, 4669 (1952).
3. R. L. Orr and J. P. Coughlin, J. Amer. Chem. Soc. 74, 3186 (1952).
4. T. F. W. Barth and E. Posnjak, Z. Krist. 82, 325 (1932).
5. F. C. Romeijn, Philips Res. Rep. 8, 304 (1953).
6. E. G. King and K. K. Kelley, U. S. Bur. Mines RI 5502, 1959.
7. H. J. Bautsch and I. Werner, Silikat Tech. 14, 203 (1963).

Magnesium Titanium Oxide (Mg_2TiO_4) $Mg_2O_4Ti_1$(cr)

Enthalpy Reference Temperature = T_r = 298.15 K; Standard State Pressure = p° = 0.1 MPa

T/K	$C_p°$	S°	$-[G°-H°(T_r)]/T$	$H°-H°(T_r)$	$\Delta_f H°$	$\Delta_f G°$	Log K_f
	J K⁻¹mol⁻¹			kJ mol⁻¹			
0	0.	0.	INFINITE	-18.836	-2151.027	-2151.027	INFINITE
100	34.639	26.740	204.267	-17.753	-2157.597	-2122.882	1108.880
200	94.320	70.337	125.901	-11.113	-2162.710	-2085.881	544.776
298.15	128.574	115.102	115.102	0.	-2164.383	-2047.719	358.752
300	129.026	115.899	115.104	0.238	-2164.393	-2046.995	356.414
400	146.285	155.637	120.405	14.093	-2164.204	-2007.859	262.199
500	156.410	189.441	130.920	29.260	-2163.228	-1968.875	205.687
600	163.686	218.628	143.161	45.280	-2161.887	-1930.126	168.032
700	169.582	244.316	155.813	61.952	-2160.357	-1891.618	141.154
800	174.707	267.301	168.337	79.171	-2158.713	-1853.339	121.011
900	179.360	288.151	180.509	96.877	-2157.045	-1815.267	105.355
1000	183.686	307.275	192.243	115.032	-2172.648	-1775.953	92.766
1100	187.778	324.975	203.515	133.607	-2171.269	-1736.349	82.452
1200	191.690	341.483	214.332	152.581	-2173.748	-1696.756	73.858
1300	195.456	356.976	224.714	171.940	-2171.398	-1657.101	66.583
1400	199.091	371.594	234.689	191.668	-2423.636	-1611.300	60.118
1500	202.614	385.451	244.282	211.754	-2418.140	-1553.467	54.097
1600	206.041	398.637	253.520	232.188	-2412.452	-1496.007	48.840
1700	209.367	411.229	262.429	252.959	-2406.595	-1438.908	44.212
1800	212.610	423.288	271.033	274.058	-2400.588	-1382.158	40.109
1900	215.765	434.868	279.353	295.478	-2394.453	-1325.745	36.447
2000	218.840	446.014	287.410	317.209	-2402.896	-1269.205	33.148
2013.000	219.233	447.433	288.438	320.056	------ CRYSTAL <--> LIQUID -----		
2100	221.831	456.764	295.220	339.243	-2397.315	-1212.657	30.163
2200	224.752	467.151	302.800	361.573	-2391.484	-1156.379	27.456
2300	227.593	477.205	310.165	384.191	-2385.411	-1100.373	24.990
2400	230.358	486.950	317.329	407.089	-2379.103	-1044.637	22.736
2500	233.049	496.408	324.304	430.260	-2372.570	-989.168	20.668
2600	235.668	505.600	331.101	453.696	-2365.816	-933.964	18.764
2700	238.212	514.542	337.730	477.391	-2358.852	-879.024	17.006
2800	240.685	523.250	344.201	501.336	-2351.684	-824.346	15.378
2900	243.086	531.738	350.522	525.525	-2344.320	-769.929	13.868
3000	245.417	540.018	356.701	549.951	-2336.769	-715.767	12.463

PREVIOUS: June 1961 CURRENT: June 1967

Magnesium Titanium Oxide (Mg_2TiO_4) $Mg_2O_4Ti_1$(cr)

MAGNESIUM TITANIUM OXIDE (Mg_2TiO_4) LIQUID M_r = 160.4876

$S°(298.15\ K) = [172.412]\ J\ K^{-1}\ mol^{-1}$

$T_{fus} = 2013 \pm 20\ K$

$\Delta_f H°(298.15\ K) = [-2046.33 \pm 6.3]\ kJ\ mol^{-1}$

$\Delta_{fus}H° = [129.704]\ kJ\ mol^{-1}$

Enthalpy of Formation

$\Delta_f H°(Mg_2TiO_4$, ℓ, 298.15 K) is calculated from $\Delta_f H°(Mg_2TiO_4$, cr, 298.15 K) by adding $\Delta_{fus}H°$ and the difference in enthalpy, H°(2013 K)-H°(298.15 K), between the crystal and liquid.

Heat Capacity and Entropy

A glass transition is assumed at 1400 K. Below 1400 K the heat capacity is obtained from the heat capacity of the crystal. Above 1400 K the heat capacity is assumed constant and estimated as 54.6 cal K^{-1} mol^{-1} or 7.8 cal K^{-1} g-atom^{-1}. S°(ℓ, 298.15 K) is calculated in a manner analogous to that used for the enthalpy of formation.

Fusion Data

The incongruent melting point is obtained from Massazza (1). Previous work by Coughanour (2) suggested congruent melting at 2005 K. $\Delta_{fus}H°$ is estimated from those of $Na_2O \cdot 2TiO_2$ and $Na_2O \cdot TiO_2$ observed by Naylor (3) and $CaO \cdot TiO_2 \cdot SiO_2$ observed by King (4).

References

1. F. Massazza and E. Sirchia, 16th Int. Congress of Pure and Appl. Chem., Inorg. Sect., Paris (1957), pp. 161-8, Butterworths Scientific Publications, London, 1958.
2. L. W. Coughanour and V. A. DeProsse, J. Res. Nat. Bur. Stand. 51, 87 (1953).
3. B. F. Naylor, J. Amer. Chem. Soc. 67, 2120 (1945).
4. E. G. King, R. L. Orr and K. R. Bonnickson, J. Amer. Chem. Soc. 76, 4320 (1954).

Magnesium Titanium Oxide (Mg_2TiO_4) $Mg_2O_4Ti_1$(l)

Enthalpy Reference Temperature = T_r = 298.15 K Standard State Pressure = p° = 0.1 MPa

T/K	$C_p°$	S°	$-[G°-H°(T_r)]/T$	$H°-H°(T_r)$	$\Delta_f H°$	$\Delta_f G°$	Log K_f
	J K⁻¹mol⁻¹			kJ mol⁻¹			
0							
100							
200							
298.15	128.574	172.412	172.412	0.	-2046.329	-1946.752	341.063
300	129.026	173.209	172.415	0.238	-2046.338	-1946.134	338.852
400	146.285	212.948	177.716	14.093	-2046.150	-1912.729	249.777
500	156.410	246.751	188.230	29.260	-2045.173	-1879.476	196.347
600	163.686	275.939	200.472	45.280	-2043.833	-1846.458	160.748
700	169.582	301.626	213.124	61.952	-2042.303	-1813.681	135.339
800	174.707	324.612	225.648	79.171	-2040.658	-1781.133	116.296
900	179.360	345.462	237.820	96.877	-2038.991	-1748.792	101.497
1000	183.686	364.585	249.553	115.032	-2054.594	-1715.209	89.593
1100	187.778	382.286	260.825	133.607	-2053.214	-1681.336	79.840
1200	191.690	398.793	271.642	152.581	-2055.694	-1647.475	71.713
1300	195.456	414.286	282.025	171.940	-2053.344	-1613.550	64.833
1400	199.091	428.905	291.999	191.668	-2305.581	-1573.481	58.707
1400.000	199.091	428.905	291.999	191.668	GLASS <--> LIQUID		
1400.000	228.446	428.905	291.999	191.668	TRANSITION		
1500	228.446	444.666	301.658	214.513	-2297.326	-1521.477	52.983
1600	228.446	459.410	311.061	237.357	-2289.228	-1470.019	47.991
1700	228.446	473.259	320.199	260.202	-2281.297	-1419.062	43.602
1800	228.446	486.317	329.069	283.047	-2273.545	-1368.568	39.715
1900	228.446	498.668	337.673	305.891	-2265.985	-1318.498	36.248
2000	228.446	510.386	346.018	328.736	-2273.315	-1268.368	33.126
2013.000	228.446	511.866	347.084	331.706	------ CRYSTAL <--> LIQUID -----		
2100	228.446	521.532	354.113	351.580	-2266.923	-1218.278	30.303
2200	228.446	532.159	361.966	374.425	-2260.577	-1168.490	27.743
2300	228.446	542.314	369.588	397.270	-2254.277	-1118.992	25.413
2400	228.446	552.037	376.989	420.114	-2248.023	-1069.766	23.283
2500	228.446	561.362	384.179	442.959	-2241.816	-1020.800	21.328
2600	228.446	570.322	391.167	465.804	-2235.654	-972.081	19.529
2700	228.446	578.944	397.963	488.648	-2229.540	-923.598	17.868
2800	228.446	587.252	404.576	511.493	-2223.473	-875.340	16.330
2900	228.446	595.268	411.014	534.338	-2217.453	-827.300	14.901
3000	228.446	603.013	417.286	557.182	-2211.483	-779.465	13.572
3100	228.446	610.504	423.398	580.027	-2205.563	-731.828	12.331
3200	228.446	617.757	429.359	602.872	-2199.695	-684.383	11.171
3300	228.446	624.786	435.175	625.716	-2193.881	-637.120	10.085
3400	228.446	631.606	440.853	648.561	-2188.122	-590.031	9.065
3500	228.446	638.228	446.398	671.405	-2182.422	-543.113	8.106
3600	228.446	644.664	451.817	694.250	-2176.783	-496.355	7.202
3700	228.446	650.923	457.114	717.095	-2580.306	-441.966	6.239
3800	228.446	657.015	462.294	739.939	-2573.563	-384.261	5.282
3900	228.446	662.949	467.364	762.784	-2566.947	-326.734	4.376
4000	228.446	668.733	472.326	785.629	-2560.458	-269.377	3.518

PREVIOUS: June 1961 CURRENT: June 1967

Magnesium Titanium Oxide (Mg_2TiO_4) $Mg_2O_4Ti_1$(l)

MAGNESIUM TITANIUM OXIDE (Mg_2TiO_4) CRYSTAL–LIQUID M_r = 160.4876

0 to 2013 K crystal
above 2013 K liquid

Refer to the individual tables for details.

Magnesium Titanium Oxide (Mg_2TiO_4) $Mg_2O_4Ti_1$(cr,l)

Enthalpy Reference Temperature = T_r = 298.15 K Standard State Pressure = p° = 0.1 MPa

T/K	C_p° (J K⁻¹mol⁻¹)	S° (J K⁻¹mol⁻¹)	$-[G^\circ-H^\circ(T_r)]/T$ (J K⁻¹mol⁻¹)	$H^\circ-H^\circ(T_r)$ (kJ mol⁻¹)	Δ_fH° (kJ mol⁻¹)	Δ_fG° (kJ mol⁻¹)	Log K_f
0	0.	0.	INFINITE	-18.836	-2151.027	-2151.027	INFINITE
100	34.639	26.740	204.267	-17.753	-2157.597	-2122.882	1108.880
200	94.320	70.337	125.901	-11.113	-2162.710	-2085.881	544.776
298.15	128.574	115.102	115.102	0.	-2164.383	-2047.719	358.752
300	129.026	115.899	115.104	0.238	-2164.393	-2046.995	356.414
400	146.285	155.637	120.405	14.093	-2164.204	-2007.859	262.199
500	156.410	189.441	130.920	29.260	-2163.228	-1968.875	205.687
600	163.686	218.628	143.161	45.280	-2161.887	-1930.126	168.032
700	169.582	244.316	155.813	61.952	-2160.357	-1891.618	141.154
800	174.707	267.301	168.337	79.171	-2158.713	-1853.339	121.011
900	179.360	288.151	180.509	96.877	-2157.045	-1815.267	105.355
1000	183.686	307.275	192.243	115.032	-2172.648	-1775.953	92.766
1100	187.778	324.975	203.515	133.607	-2171.269	-1736.349	82.452
1200	191.690	341.483	214.332	152.581	-2173.748	-1696.756	73.858
1300	195.456	356.976	224.714	171.940	-2171.398	-1657.101	66.583
1400	199.091	371.594	234.689	191.668	-2423.636	-1611.300	60.118
1500	202.614	385.451	244.282	211.754	-2418.140	-1553.467	54.097
1600	206.041	398.637	253.520	232.188	-2412.452	-1496.007	48.840
1700	209.367	411.229	262.429	252.959	-2406.595	-1438.908	44.212
1800	212.610	423.288	271.033	274.058	-2400.588	-1382.158	40.109
1900	215.765	434.868	279.353	295.478	-2394.453	-1325.745	36.447
2000	218.840	446.014	287.410	317.209	-2402.896	-1269.205	33.148
2013.000	219.233	447.433	288.438	320.056	CRYSTAL <--> LIQUID		
2013.000	228.446	511.866	288.438	449.760	TRANSITION		
2100	228.446	521.532	297.896	469.635	-2266.923	-1218.278	30.303
2200	228.446	532.159	308.305	492.480	-2260.577	-1168.490	27.743
2300	228.446	542.314	318.260	515.324	-2254.277	-1118.992	25.413
2400	228.446	552.037	327.800	538.169	-2248.023	-1069.766	23.283
2500	228.446	561.362	336.957	561.013	-2241.816	-1020.800	21.328
2600	228.446	570.322	345.761	583.858	-2235.654	-972.081	19.529
2700	228.446	578.944	354.239	606.703	-2229.540	-923.598	17.868
2800	228.446	587.252	362.414	629.547	-2223.473	-875.340	16.330
2900	228.446	595.268	370.306	652.392	-2217.453	-827.300	14.901
3000	228.446	603.013	377.934	675.237	-2211.483	-779.465	13.572
3100	228.446	610.504	385.316	698.081	-2205.563	-731.828	12.331
3200	228.446	617.757	392.467	720.926	-2199.695	-684.383	11.171
3300	228.446	624.786	399.401	743.771	-2193.881	-637.120	10.085
3400	228.446	631.606	406.131	766.615	-2188.122	-590.031	9.065
3500	228.446	638.228	412.668	789.460	-2182.422	-543.113	8.106
3600	228.446	644.664	419.024	812.305	-2176.783	-496.355	7.202
3700	228.446	650.923	425.207	835.149	-2580.306	-441.966	6.239
3800	228.446	657.015	431.227	857.994	-2573.563	-384.261	5.282
3900	228.446	662.949	437.093	880.838	-2566.947	-326.734	4.376
4000	228.446	668.733	442.812	903.683	-2560.458	-269.377	3.518

PREVIOUS: CURRENT: June 1967

Magnesium Titanium Oxide (Mg_2TiO_4) $Mg_2O_4Ti_1$(cr,l)

MAGNESIUM SILICIDE (Mg_2Si) CRYSTAL M_r = 76.6955

$S°(298.15\ K) = [81.588]\ J\ K^{-1}\ mol^{-1}$ $\Delta_f H°(298.15\ K) = [-77.822]\ kJ\ mol^{-1}$

$T_{fus} = 1375\ K$ $\Delta_{fus} H° = [85.772]\ kJ\ mol^{-1}$

Enthalpy of Formation

$\Delta_f H°(298.15\ K)$ and T_{fus} from (1).

Heat Capacity and Entropy

$C_p°$ and $\Delta_{fus} H°$ from Kubaschewski and Evans (2). S°(298.15 K) estimated.

References

1. U. S. Nat. Bur. Stand., Circ. 500, 1952.
2. O. Kubaschewski and E. L. Evans, "Metallurgical Thermochemistry", 3rd Edn., Pergamon Press Ltd., New York, (1958).

Magnesium Silicide (Mg_2Si) $Mg_2Si_1(cr)$

Enthalpy Reference Temperature = T_r = 298.15 K Standard State Pressure = p° = 0.1 MPa

T/K	$C_p°$ (J K⁻¹mol⁻¹)	S° (J K⁻¹mol⁻¹)	$-[G°-H°(T_r)]/T$ (J K⁻¹mol⁻¹)	$H°-H°(T_r)$ (kJ mol⁻¹)	$\Delta_f H°$ (kJ mol⁻¹)	$\Delta_f G°$ (kJ mol⁻¹)	Log K_f
0							
100							
200							
298.15	67.864	81.588	81.588	0.	-77.822	-77.055	13.500
300	67.994	82.008	81.589	0.126	-77.826	-77.050	13.416
400	73.764	102.413	84.332	7.232	-77.951	-76.770	10.025
500	77.278	119.268	89.683	14.793	-78.002	-76.468	7.989
600	79.831	133.589	95.837	22.651	-78.054	-76.157	6.630
700	82.006	146.062	102.140	30.745	-78.152	-75.833	5.659
800	83.889	157.137	108.335	39.041	-78.342	-75.491	4.929
900	85.688	167.122	114.321	47.520	-78.661	-75.117	4.360
1000	87.404	176.239	120.064	56.175	-96.329	-73.274	3.827
1100	89.077	184.649	125.558	65.000	-97.022	-70.933	3.368
1200	90.667	192.468	130.812	73.987	-97.596	-68.536	2.983
1300	92.257	199.788	135.839	83.134	-98.052	-66.095	2.656
1375.000	93.450	204.996	139.470	90.098	------ CRYSTAL <--> LIQUID -----		
1400	93.805	206.684	140.655	92.439	-353.209	-57.289	2.137
1500	95.312	213.207	145.277	101.896	-350.734	-36.237	1.262
1600	96.776	219.405	149.718	111.500	-348.152	-15.354	0.501
1700	98.198	225.315	153.992	121.249	-395.645	5.811	-0.179
1800	99.579	230.967	158.113	131.139	-392.633	29.339	-0.851
1900	100.918	236.387	162.091	141.164	-389.485	52.697	-1.449
2000	102.215	241.597	165.937	151.321	-386.205	75.886	-1.982
2100	103.470	246.615	169.660	161.606	-382.798	98.907	-2.460
2200	104.684	251.456	173.268	172.014	-379.268	121.763	-2.891
2300	105.855	256.136	176.770	182.541	-375.620	144.456	-3.281
2400	106.985	260.665	180.172	193.183	-371.859	166.988	-3.634
2500	108.073	265.054	183.480	203.936	-367.989	189.360	-3.956

PREVIOUS: CURRENT: December 1960

Magnesium Silicide (Mg_2Si) $Mg_2Si_1(cr)$

MAGNESIUM SILICIDE (Mg_2Si) LIQUID M_r = 76.6955

$S°(298.15\ K) = [123.472]\ J\ K^{-1}\ mol^{-1}$

$T_{fus} = 1375\ K$

$\Delta_f H°(298.15\ k) = [-3.327]\ kJ\ mol^{-1}$

$\Delta_{fus} H° = [85.772]\ kJ\ mol^{-1}$

Enthalpy of Formation

$\Delta_f H°(Mg_2Si, \ell, 298.15\ K)$ is calculated from $\Delta_f H°(Mg_2Si, cr, 298.15\ K)$ by adding $\Delta_{fus}H°$ and the difference in enthalpy, H°(1375 K)-H°(298.15 K), between the crystal and liquid.

Fusion Data

T_{fus} from (1). $\Delta_{fus}H°$ from Kubaschewski and Evans (2). Other data estimated.

References

1. U. S. Nat. Bur. Stand., Circ. 500, 1952.
2. O. Kubaschewski and E. L. Evans, "Metallurgical Thermochemistry", 3rd Edn., Pergamon Press. Ltd., New York, (1958).

Magnesium Silicide (Mg_2Si) $Mg_2Si_1(l)$

Enthalpy Reference Temperature = T_r = 298.15 K Standard State Pressure = p° = 0.1 MPa

T/K	$C_p°$	S°	$-[G°-H°(T_r)]/T$	$H°-H°(T_r)$	$\Delta_f H°$	$\Delta_f G°$	Log K_f
	J K⁻¹mol⁻¹			kJ mol⁻¹			
0							
100							
200							
298.15	94.140	123.472	123.472	0.	-3.327	-15.048	2.636
300	94.140	124.054	123.474	0.174	-3.282	-15.121	2.633
400	94.140	151.136	127.166	9.588	-1.100	-19.408	2.534
500	94.140	172.143	134.139	19.002	0.703	-24.201	2.528
600	94.140	189.307	141.947	28.416	2.206	-29.327	2.553
700	94.140	203.819	149.776	37.830	3.428	-34.683	2.588
800	94.140	216.389	157.334	47.244	4.357	-40.195	2.624
900	94.140	227.477	164.524	56.658	4.971	-45.804	2.658
1000	94.140	237.396	171.324	66.072	-11.938	-50.039	2.614
1100	94.140	246.369	177.745	75.486	-12.041	-53.844	2.557
1200	94.140	254.560	183.810	84.900	-12.188	-57.638	2.509
1300	94.140	262.095	189.546	94.314	-12.376	-61.418	2.468
1375.000	94.140	267.375	193.648	101.375	------ CRYSTAL <--> LIQUID -----		
1400	94.140	269.072	194.980	103.728	-267.425	-58.848	2.196
1500	94.140	275.567	200.138	113.142	-264.992	-44.035	1.533
1600	94.140	281.642	205.045	122.556	-262.602	-29.383	0.959
1700	94.140	287.349	209.720	131.970	-310.430	-14.432	0.443
1800	94.140	292.730	214.184	141.384	-307.893	2.906	-0.084
1900	94.140	297.820	218.453	150.798	-305.356	20.104	-0.553
2000	94.140	302.649	222.543	160.212	-302.819	37.168	-0.971
2100	94.140	307.242	226.468	169.626	-300.282	54.105	-1.346
2200	94.140	311.621	230.240	179.040	-297.747	70.921	-1.684
2300	94.140	315.806	233.870	188.454	-295.212	87.622	-1.990
2400	94.140	319.813	237.368	197.868	-292.679	104.212	-2.268
2500	94.140	323.656	240.743	207.282	-290.148	120.697	-2.522
2600	94.140	327.348	244.003	216.696	-287.621	137.081	-2.754
2700	94.140	330.901	247.156	226.110	-285.098	153.368	-2.967
2800	94.140	334.324	250.209	235.524	-282.582	169.562	-3.163
2900	94.140	337.628	253.167	244.938	-280.074	185.665	-3.344
3000	94.140	340.819	256.035	254.352	-277.576	201.683	-3.512
3100	94.140	343.906	258.820	263.766	-275.091	217.617	-3.667
3200	94.140	346.895	261.526	273.180	-272.623	233.471	-3.811
3300	94.140	349.792	264.157	282.594	-270.173	249.249	-3.945
3400	94.140	352.602	266.718	292.008	-267.745	264.952	-4.070
3500	94.140	355.331	269.211	301.422	-265.344	280.585	-4.188
3600	94.140	357.983	271.640	310.836	-647.124	306.610	-4.449
3700	94.140	360.563	274.008	320.250	-644.373	333.065	-4.702
3800	94.140	363.073	276.319	329.664	-641.662	359.445	-4.941
3900	94.140	365.518	278.575	339.078	-638.996	385.756	-5.167
4000	94.140	367.902	280.779	348.492	-636.378	411.998	-5.380

PREVIOUS: CURRENT: December 1960

MAGNESIUM SILICIDE (Mg_2Si) CRYSTAL-LIQUID M_r = 76.6955

0 to 1375 K crystal
above 1375 K liquid

Refer to the individual tables for details.

Magnesium Silicide (Mg_2Si) Mg_2Si_1(cr,l)

Enthalpy Reference Temperature = T_r = 298.15 K Standard State Pressure = p° = 0.1 MPa

T/K	C_p° (J K⁻¹mol⁻¹)	S° (J K⁻¹mol⁻¹)	$-[G^\circ-H^\circ(T_r)]/T$ (J K⁻¹mol⁻¹)	$H^\circ-H^\circ(T_r)$ (kJ mol⁻¹)	Δ_fH° (kJ mol⁻¹)	Δ_fG° (kJ mol⁻¹)	Log K_f
0							
100							
200							
298.15	67.864	81.588	81.588	0.	-77.822	-77.055	13.500
300	67.994	82.008	81.589	0.126	-77.826	-77.050	13.416
400	73.764	102.413	84.332	7.232	-77.951	-76.770	10.025
500	77.278	119.268	89.683	14.793	-78.002	-76.468	7.989
600	79.831	133.589	95.837	22.651	-78.054	-76.157	6.630
700	82.006	146.062	102.140	30.745	-78.152	-75.833	5.659
800	83.889	157.137	108.335	39.041	-78.342	-75.491	4.929
900	85.688	167.122	114.321	47.520	-78.661	-75.117	4.360
1000	87.404	176.239	120.064	56.175	-96.329	-73.274	3.827
1100	89.077	184.649	125.558	65.000	-97.022	-70.933	3.368
1200	90.667	192.468	130.812	73.987	-97.596	-68.536	2.983
1300	92.257	199.788	135.839	83.134	-98.052	-66.095	2.656
1375.000	93.450	204.996	139.470	90.098	CRYSTAL <--> LIQUID		
1375.000	94.140	267.375	139.470	175.870	TRANSITION		
1400	94.140	269.072	141.769	178.223	-267.425	-58.848	2.196
1500	94.140	275.567	150.475	187.637	-264.992	-44.035	1.533
1600	94.140	281.642	158.485	197.051	-262.602	-29.383	0.959
1700	94.140	287.349	165.899	206.465	-310.430	-14.432	0.443
1800	94.140	292.730	172.797	215.879	-307.893	2.906	-0.084
1900	94.140	297.820	179.245	225.293	-305.356	20.104	-0.553
2000	94.140	302.649	185.295	234.707	-302.819	37.168	-0.971
2100	94.140	307.242	190.994	244.121	-300.282	54.105	-1.346
2200	94.140	311.621	196.378	253.535	-297.747	70.921	-1.684
2300	94.140	315.806	201.480	262.949	-295.212	87.622	-1.990
2400	94.140	319.813	206.328	272.363	-292.679	104.212	-2.268
2500	94.140	323.656	210.945	281.777	-290.148	120.697	-2.522
2600	94.140	327.348	215.351	291.191	-287.621	137.081	-2.754
2700	94.140	330.901	219.566	300.605	-285.098	153.368	-2.967
2800	94.140	334.324	223.603	310.019	-282.582	169.562	-3.163
2900	94.140	337.628	227.479	319.433	-280.074	185.665	-3.344
3000	94.140	340.819	231.204	328.847	-277.576	201.683	-3.512
3100	94.140	343.906	234.790	338.261	-275.091	217.617	-3.667
3200	94.140	346.895	238.247	347.675	-272.623	233.471	-3.811
3300	94.140	349.792	241.583	357.089	-270.173	249.249	-3.945
3400	94.140	352.602	244.807	366.503	-267.745	264.952	-4.070
3500	94.140	355.331	247.926	375.917	-265.344	280.585	-4.188
3600	94.140	357.983	250.947	385.331	-647.124	306.610	-4.449
3700	94.140	360.563	253.875	394.745	-644.373	333.065	-4.702
3800	94.140	363.073	256.715	404.159	-641.662	359.445	-4.941
3900	94.140	365.518	259.474	413.573	-638.996	385.756	-5.167
4000	94.140	367.902	262.155	422.987	-636.378	411.998	-5.380

PREVIOUS: CURRENT: December 1960

Magnesium Silicide (Mg_2Si) Mg_2Si_1(cr,l)

MAGNESIUM NITRIDE (Mg_3N_2) CRYSTAL M_r = 100.9284

$S°(298.15\ K) = [87.864 \pm 8.4]\ J\ K^{-1}\ mol^{-1}$ $\Delta_f H°(298.15\ K) = -461.08 \pm 4.2\ kJ\ mol^{-1}$

Enthalpy of Formation

Data from (1).

Heat Capacity and Entropy

Data from (1).

Reference

1. F. B. Douglas, A. C. Victor, and A. R. Beaudoin, U. S. Nat. Bur. Stand., Report 6928, Appendix B, 1960.

Magnesium Nitride (Mg_3N_2) Mg_3N_2(cr)

Enthalpy Reference Temperature = T_r = 298.15 K Standard State Pressure = p° = 0.1 MPa

	J K⁻¹mol⁻¹			kJ mol⁻¹			
T/K	$C_p°$	S°	$-[G°-H°(T_r)]/T$	$H°-H°(T_r)$	$\Delta_f H°$	$\Delta_f G°$	Log K_f
0							
100							
200							
298.15	104.516	87.864	87.864	0.	-461.077	-400.923	70.240
300	104.600	88.511	87.866	0.193	-461.075	-400.549	69.742
400	107.654	119.021	92.005	10.806	-461.046	-380.381	49.673
500	110.709	143.371	99.922	21.724	-461.067	-360.213	37.631
600	113.763	163.826	108.913	32.948	-461.128	-340.037	29.603
700	116.817	181.592	118.053	44.477	-461.258	-319.846	23.867
800	119.872	197.390	127.001	56.311	-461.498	-299.630	19.564
823.000	120.574	200.797	129.016	59.076		I <--> II	
823.000	123.846	201.916	129.016	59.997		TRANSITION	
900	123.846	212.993	135.733	69.533	-460.808	-279.470	16.220
1000	123.846	226.041	144.123	81.918	-487.232	-257.154	13.432
1061.000	123.846	233.374	149.046	89.472		II <--> III	
1061.000	123.595	234.399	149.046	90.560		TRANSITION	
1100	123.595	238.861	152.152	95.380	-487.359	-234.128	11.118
1200	123.595	249.615	159.832	107.740	-488.641	-211.050	9.187
1300	123.595	259.508	167.124	120.100	-489.968	-187.864	7.548
1400	123.595	268.668	174.054	132.459	-873.563	-155.073	5.786
1500	123.595	277.195	180.649	144.819	-870.908	-103.845	3.616
1600	123.595	285.171	186.935	157.178	-868.283	-52.793	1.724
1700	123.595	292.664	192.936	169.538	-865.684	-1.905	0.059
1800	123.595	299.729	198.675	181.897	-863.110	48.831	-1.417
1900	123.595	306.411	204.171	194.257	-860.556	99.425	-2.733
2000	123.595	312.751	209.443	206.616	-858.022	149.885	-3.915
2100	123.595	318.781	214.507	218.976	-855.504	200.218	-4.980
2200	123.595	324.531	219.378	231.335	-853.002	250.432	-5.946
2300	123.595	330.025	224.071	243.695	-850.516	300.532	-6.825
2400	123.595	335.285	228.596	256.054	-848.043	350.524	-7.629
2500	123.595	340.330	232.965	268.414	-845.586	400.414	-8.366

PREVIOUS: CURRENT: December 1960

Magnesium Nitride (Mg_3N_2) Mg_3N_2(cr)

MAGNESIUM PHOSPHATE ($Mg_3P_2O_8$) CRYSTAL M_r = 262.85772

$S°(298.15\ K) = [188.280]\ J\ K^{-1}\ mol^{-1}$

T_{fus} = 1621 K

$\Delta_fH°(0\ K)$ = Unknown

$\Delta_fH°(298.15\ K) = -3745.10 \pm 10.5\ kJ\ mol^{-1}$

$\Delta_{fus}H° = [121.336 \pm 41.8]\ kJ\ mol^{-1}$

Enthalpy of Formation

The selected $\Delta_fH°$(298.15 K) is a weighted average of a revised calorimetric determination and a 3rd law calculation using equilibrium data.

A calorimetric determination of the enthalpy of formation reported by Stevens and Turkdogan (1) was revised three ways:

1. The enthalpy of reaction for the reaction 3 MgO(cr) + P_2O_5(orthorohombic) + $Mg_3P_2O_8$(cr) had been evaluated at 348 K. A small temperature correction to 298.15 K was made with thermal data not available to Stevens and Turkdogan (1) on MgO (Giauque (2)), P_2O_5 JANAF Table, Dec. 31, 1962) and $Mg_3P_2O_8$ (this table).
2. A new $\Delta_fH°$(298.15 K) for P_2O_5(hexagonal) as reported by Holmes (3) was combined with the enthalpy of reaction, -6.85 ± 1.0 kcal mol^{-1}, for the reaction: P_2O_5(hexagonal) + P_2O_5(orthorhombic) as reported by Hill et al. (4).
3. A correction of +8.35 kcal mol^{-1} was made for the change in phosphorus reference state from white (α) + red (V). (The correction is +8.35 kcal mol^{-1} of $Mg_3P_2O_8$ or +4.173 kcal mol^{-1} of P.)

This revised $\Delta_fH°$(298.15 K), -897.60±2.0 kcal mol^{-1}, would appear to be more reliable than the original value reported by Stevens and Turkdogan (1).

For the reaction $Mg_3P_2O_8$(cr) + 5 H_2(g) + 3 MgO(cr) + F_2(g) + 5 H_2O(g), the 3rd law treatment of the equilibrium data of Bookey (5) and the JANAF table free energy functions gave a $\Delta_fH°$(298.15 K) = -893.11 kcal mol^{-1}.

The quantity, -895.10±2.50 kcal mol^{-1}, is taken as the most probable value for the enthalpy of formation of magnesium orthophosphate.

Heat Capacity and Entropy

The entropy at 298.15 was estimated by considering the additive entropy constants of Kelley (6) and Evans and Kubaschewski (7). The heat capacity was estimated in a similar manner by considering the heat capacities of the constituent elements and related compounds such as Mg, O_2, MgO, P, P_2O_5, $Ca_2P_2O_7$, and $Ca_3P_2O_8$.

Fusion Data

The melting point, 1621 K, as reported by Stevens and Turkdogan (1) was selected. There is disagreement on the melting point in the literature. From the work of Winter (8), Kelley (9) has quoted a melting point of 1457 K and has calculated a "rather uncertain value" of 11.300 kcal mol^{-1} for the heat of fusion. Bookey (5) has found a melting point of 1703 K. Berak (10) gives 1630 K as the melting point.

References

1. C. G. Stevens and E. T. Turkdogan, Trans. Faraday Soc. 50, 370 (1954).
2. W. F. Giauque, J. Amer. Chem. Soc. 71, 3192 (1949).
3. W. S. Holmes, Trans. Faraday Soc. 58, 1916 (1962).
4. W. L. Hill, G. T. Faust, and S. B. Hendricks, J. Amer. Chem. Soc. 65, 794 (1943).
5. J. B. Bookey, J. Iron Steel Institute (London) 172, 66 (1952).
6. K. K. Kelley, U. S. Bur. Mines, personal communication, June 1960.
7. E. L. Evans and O. Kubaschewski, "Metallurgical Thermochemistry, 1958.
8. H. Winter, Diss. Universitats zu Leipzig, 1913.
9. K. K. Kelley, U. S. Bur. Mines Bull. 393, 1936.
10. J. Berak, Roczniki Chim. 32, 17 (1958).

Magnesium Phosphate ($Mg_3P_2O_8$) $Mg_3O_8P_2$(cr)

Enthalpy Reference Temperature = T_r = 298.15 K; Standard State Pressure = p° = 0.1 MPa

T/K	$C_p°$ (J K⁻¹mol⁻¹)	S° (J K⁻¹mol⁻¹)	$-[G°-H°(T_r)]/T$ (J K⁻¹mol⁻¹)	$H°-H°(T_r)$ (kJ mol⁻¹)	$\Delta_fH°$ (kJ mol⁻¹)	$\Delta_fG°$ (kJ mol⁻¹)	Log K_f
0							
100							
200							
298.15	212.756	188.280	188.280	0.	-3745.098	-3502.859	613.686
300	213.384	189.598	188.284	0.394	-3745.148	-3501.356	609.640
400	240.162	254.762	196.988	23.110	-3748.487	-3419.399	446.528
500	262.337	310.777	214.270	48.254	-3748.843	-3337.054	348.620
600	282.211	360.377	234.563	75.488	-3747.814	-3254.772	283.353
700	301.248	405.305	255.793	104.658	-3745.544	-3172.761	236.754
800	320.620	446.840	277.109	135.785	-3741.994	-3091.163	201.832
900	335.975	485.488	298.140	168.613	-3737.409	-3010.075	174.700
1000	351.540	521.754	318.706	203.048	-3757.656	-2927.411	152.912
1100	360.075	555.683	338.726	238.653	-3751.645	-2844.671	135.082
1200	365.682	587.236	358.136	274.920	-3872.275	-2760.346	120.155
1300	373.757	616.851	376.909	311.925	-3863.611	-2668.032	107.203
1400	377.899	644.714	395.053	349.525	-4236.710	-2566.904	95.772
1500	379.907	670.862	412.578	387.425	-4223.344	-2448.098	85.250
1600	380.744	695.407	429.496	425.458	-4209.957	-2330.185	76.073
1621.000	380.920	700.373	432.973	433.455	------ CRYSTAL <--> LIQUID -----		
1700	381.581	718.517	445.824	463.578	-4196.587	-2213.109	68.001
1800	381.581	740.328	461.585	501.736	-4183.281	-2096.819	60.848
1900	381.581	760.959	476.804	539.894	-4170.074	-1981.265	54.469
2000	381.581	780.531	491.505	578.052	-4156.965	-1866.404	48.745
2100	381.581	799.149	505.715	616.210	-4143.953	-1752.196	43.583
2200	381.581	816.900	519.460	654.368	-4131.036	-1638.604	38.905
2300	381.581	833.862	532.763	692.526	-4118.215	-1525.600	34.647
2400	381.581	850.102	545.650	730.685	-4105.489	-1413.152	30.756
2500	381.581	865.679	558.142	768.843	-4092.858	-1301.231	27.188
2600	381.581	880.644	570.260	807.001	-4080.321	-1189.814	23.904
2700	381.581	895.045	582.024	845.159	-4067.878	-1078.876	20.872
2800	381.581	908.923	593.452	883.317	-4055.529	-968.398	18.066
2900	381.581	922.313	604.563	921.475	-4043.276	-858.362	15.461
3000	381.581	935.249	615.371	959.633	-4031.118	-748.744	13.037

PREVIOUS: December 1960 CURRENT: March 1963

Magnesium Phosphate ($Mg_3P_2O_8$) $Mg_3O_8P_2$(cr)

MAGNESIUM PHOSPHATE ($Mg_3P_2O_8$) LIQUID M_r = 262.85772

$S°(298.15\ K) = [235.102]\ J\ K^{-1}\ mol^{-1}$ $\Delta_f H°(298.15\ K) = [-3663.190]\ kJ\ mol^{-1}$

T_{fus} = 1621 K $\Delta_{fus}H° = [121.336 \pm 41.8]\ kJ\ mol^{-1}$

Enthalpy of Formation

$\Delta_f H°(Mg_3P_2O_8$, ℓ, 298.15 K) is calculated from $\Delta_f H°(Mg_3P_2O_8$, cr, 298.15 K) by adding $\Delta_{fus}H°$ and the difference in enthalpy, H°(1621 K)−H°(298.15 K), between the crystal and liquid.

Heat Capacity and Entropy

The heat capacity was estimated as 13/11 of the corresponding heat capacity of $Ca_2P_2O_7$ as given in Kelley (1). S°(ℓ, 298.15 K) is calculated in a manner similar to that used for the enthalpy of formation.

Fusion Data

The enthalpy of fusion was estimated by taking 13/11 of the $\Delta_{fus}H°$ of $Ca_2P_2O_7$ as given in Kelley (1).

Reference

1. K. K. Kelley, U. S. Bur. Mines Bull. 5841, 1960.

Magnesium Phosphate ($Mg_3P_2O_8$) $Mg_3O_8P_2(l)$

Enthalpy Reference Temperature = T_r = 298.15 K Standard State Pressure = p° = 0.1 MPa

T/K	$C_p°$ (J K⁻¹mol⁻¹)	S° (J K⁻¹mol⁻¹)	$-[G°-H°(T_r)]/T$ (J K⁻¹mol⁻¹)	$H°-H°(T_r)$ (kJ mol⁻¹)	$\Delta_f H°$ (kJ mol⁻¹)	$\Delta_f G°$ (kJ mol⁻¹)	Log K_f
0							
100							
200							
298.15	212.756	235.102	235.102	0.	-3663.190	-3434.910	601.782
300	213.384	236.420	235.106	0.394	-3663.240	-3433.494	597.824
400	240.162	301.584	243.810	23.110	-3666.578	-3356.219	438.277
500	262.337	357.600	261.092	48.254	-3666.934	-3278.557	342.508
600	282.211	407.199	281.385	75.488	-3665.905	-3200.957	278.668
700	301.248	452.127	302.615	104.658	-3663.635	-3123.628	233.088
800	320.620	493.663	323.931	135.785	-3660.086	-3046.712	198.930
900	335.975	532.311	344.962	168.613	-3655.500	-2970.306	172.392
1000	351.540	568.576	365.529	203.048	-3675.747	-2892.324	151.080
1100	360.075	602.505	385.548	238.653	-3669.736	-2814.266	133.638
1200	365.682	634.097	404.960	274.965	-3790.321	-2734.626	119.035
1220.000	366.451	640.148	408.766	282.287	GLASS <--> LIQUID		
1220.000	475.302	640.148	408.766	282.287	TRANSITION		
1300	475.302	670.336	423.944	320.311	-3773.316	-2647.268	106.369
1400	475.302	705.560	442.817	367.841	-4136.486	-2551.863	95.211
1500	475.302	738.353	461.439	415.371	-4113.490	-2439.480	84.950
1600	475.302	769.028	479.715	462.901	-4090.605	-2328.626	76.022
1621.000	475.302	775.226	483.503	472.883	------ CRYSTAL <--> LIQUID -----		
1700	475.302	797.843	497.589	510.432	-4067.825	-2219.200	68.188
1800	475.302	825.011	515.032	557.962	-4045.147	-2111.114	61.263
1900	475.302	850.709	532.029	605.492	-4022.568	-2004.284	55.102
2000	475.302	875.089	548.577	653.022	-4000.086	-1898.640	49.587
2100	475.302	898.279	564.682	700.553	-3977.702	-1794.118	44.626
2200	475.302	920.390	580.352	748.083	-3955.413	-1690.659	40.141
2300	475.302	941.518	595.599	795.613	-3933.220	-1588.214	36.069
2400	475.302	961.747	610.437	843.143	-3911.122	-1486.733	32.358
2500	475.302	981.149	624.880	890.674	-3889.119	-1386.169	28.962
2600	475.302	999.791	638.943	938.204	-3867.209	-1286.483	25.846
2700	475.302	1017.729	652.642	985.734	-3845.394	-1187.638	22.976
2800	475.302	1035.015	665.992	1033.264	-3823.673	-1089.599	20.327
2900	475.302	1051.694	679.006	1080.795	-3802.047	-992.339	17.874
3000	475.302	1067.807	691.699	1128.325	-3780.518	-895.818	15.598
3100	475.302	1083.392	704.084	1175.855	-3759.085	-800.013	13.480
3200	475.302	1098.482	716.175	1223.385	-3737.751	-704.902	11.506
3300	475.302	1113.108	727.982	1270.916	-3716.518	-610.456	9.663
3400	475.302	1127.298	739.519	1318.446	-3695.387	-516.648	7.937
3500	475.302	1141.075	750.797	1365.976	-3674.363	-423.464	6.320
3600	475.302	1154.465	761.824	1413.506	-3653.448	-330.876	4.801
3700	475.302	1167.488	772.613	1461.036	-3632.647	-238.871	3.372
3800	475.302	1180.163	783.172	1508.567	-3611.963	-147.421	2.026
3900	475.302	1192.510	793.510	1556.097	-3591.401	-56.516	0.757
4000	475.302	1204.543	803.636	1603.627	-3570.967	33.860	-0.442
4100	475.302	1216.280	813.558	1651.157	-3550.666	123.729	-1.576
4200	475.302	1227.733	823.284	1698.688	-3530.503	213.104	-2.650
4300	475.302	1238.917	832.820	1746.218	-3510.485	302.002	-3.669
4400	475.302	1249.844	842.174	1793.748	-3490.617	390.433	-4.635
4500	475.302	1260.526	851.353	1841.278	-3470.904	478.414	-5.553

PREVIOUS: December 1960 CURRENT: March 1963

Magnesium Phosphate ($Mg_3P_2O_8$) $Mg_3O_8P_2(l)$

MAGNESIUM PHOSPHATE ($Mg_3P_2O_8$) CRYSTAL-LIQUID M_r = 262.85772

0 to	1621 K	crystal
above	1621 K	liquid

Refer to the individual tables for details.

Magnesium Phosphate ($Mg_3P_2O_8$) $Mg_3O_8P_2$(cr,l)

Enthalpy Reference Temperature = T_r = 298.15 K Standard State Pressure = p° = 0.1 MPa

T/K	C_p° (J K⁻¹mol⁻¹)	S° (J K⁻¹mol⁻¹)	$-[G^\circ-H^\circ(T_r)]/T$ (J K⁻¹mol⁻¹)	$H^\circ-H^\circ(T_r)$ (kJ mol⁻¹)	Δ_fH° (kJ mol⁻¹)	Δ_fG° (kJ mol⁻¹)	Log K_f
0							
100							
200							
298.15	212.756	188.280	188.280	0.	-3745.098	-3502.859	613.686
300	213.384	189.598	188.284	0.394	-3745.148	-3501.356	609.640
400	240.162	254.762	196.988	23.110	-3748.487	-3419.399	446.528
500	262.337	310.777	214.270	48.254	-3748.843	-3337.054	348.620
600	282.211	360.377	234.563	75.488	-3747.814	-3254.772	283.353
700	301.248	405.305	255.793	104.658	-3745.544	-3172.761	236.754
800	320.620	446.840	277.109	135.785	-3741.994	-3091.163	201.832
900	335.975	485.488	298.140	168.613	-3737.409	-3010.075	174.700
1000	351.540	521.754	318.706	203.048	-3757.656	-2927.411	152.912
1100	360.075	555.683	338.726	238.653	-3751.645	-2844.671	135.082
1200	365.682	587.236	358.136	274.920	-3872.275	-2760.346	120.155
1300	373.757	616.851	376.909	311.925	-3863.611	-2668.032	107.203
1400	377.899	644.714	395.053	349.525	-4236.710	-2566.904	95.772
1500	379.907	670.862	412.578	387.425	-4223.344	-2448.098	85.250
1600	380.744	695.407	429.496	425.458	-4209.957	-2330.185	76.073
1621.000	380.920	700.373	432.973	433.455	CRYSTAL <--> LIQUID		
1621.000	475.302	775.226	432.973	554.791	TRANSITION		
1700	475.302	797.843	449.408	592.340	-4067.825	-2219.200	68.188
1800	475.302	825.011	469.527	639.870	-4045.147	-2111.114	61.263
1900	475.302	850.709	488.919	687.401	-4022.568	-2004.284	55.102
2000	475.302	875.089	507.623	734.931	-4000.086	-1898.640	49.587
2100	475.302	898.279	525.678	782.461	-3977.702	-1794.118	44.626
2200	475.302	920.390	543.121	829.991	-3955.413	-1690.659	40.141
2300	475.302	941.518	559.987	877.522	-3933.220	-1588.214	36.069
2400	475.302	961.747	576.308	925.052	-3911.122	-1486.733	32.358
2500	475.302	981.149	592.116	972.582	-3889.119	-1386.169	28.962
2600	475.302	999.791	607.440	1020.112	-3867.209	-1286.483	25.846
2700	475.302	1017.729	622.306	1067.643	-3845.394	-1187.638	22.976
2800	475.302	1035.015	636.739	1115.173	-3823.673	-1089.599	20.327
2900	475.302	1051.694	650.762	1162.703	-3802.047	-992.339	17.874
3000	475.302	1067.807	664.396	1210.233	-3780.518	-895.818	15.598
3100	475.302	1083.392	677.662	1257.764	-3759.085	-800.013	13.480
3200	475.302	1098.482	690.578	1305.294	-3737.751	-704.902	11.506
3300	475.302	1113.108	703.162	1352.824	-3716.518	-610.456	9.663
3400	475.302	1127.298	715.429	1400.354	-3695.387	-516.648	7.937
3500	475.302	1141.075	727.394	1447.885	-3674.363	-423.464	6.320
3600	475.302	1154.465	739.072	1495.415	-3653.448	-330.876	4.801
3700	475.302	1167.488	750.476	1542.945	-3632.647	-238.871	3.372
3800	475.302	1180.163	761.617	1590.475	-3611.963	-147.421	2.026
3900	475.302	1192.510	772.508	1638.006	-3591.401	-56.516	0.757
4000	475.302	1204.543	783.159	1685.536	-3570.967	33.860	-0.442
4100	475.302	1216.280	793.581	1733.066	-3550.666	123.729	-1.576
4200	475.302	1227.733	803.782	1780.596	-3530.503	213.104	-2.650
4300	475.302	1238.917	813.772	1828.126	-3510.485	302.002	-3.669
4400	475.302	1249.844	823.559	1875.657	-3490.617	390.433	-4.635
4500	475.302	1260.526	833.151	1923.187	-3470.904	478.414	-5.553

PREVIOUS: CURRENT: March 1963

Magnesium Phosphate ($Mg_3P_2O_8$) $Mg_3O_8P_2$(cr,l)

MANGANESE (Mn)

REFERENCE STATE

0 to 980	K	crystal, alpha
980 to 1361	K	crystal, beta
1361 to 1412	K	crystal, gamma
1412 to 1519	K	crystal, delta
1519 to 2334.526	K	liquid
above 2334.526		ideal monatomic gas

Refer to the individual tables for details.

A_r = 54.9380 Manganese (Mn) Mn_1(ref)

Enthalpy Reference Temperature = T_r = 298.15 K Standard State Pressure = p° = 0.1 MPa

T/K	C_p° (J K⁻¹mol⁻¹)	S° (J K⁻¹mol⁻¹)	$-[G^\circ-H^\circ(T_r)]/T$ (J K⁻¹mol⁻¹)	$H^\circ-H^\circ(T_r)$ (kJ mol⁻¹)	Δ_fH° (kJ mol⁻¹)	Δ_fG° (kJ mol⁻¹)	Log K_f
0	0.	0.	INFINITE	-4.994	0.	0.	0.
100	14.723	8.875	52.876	-4.400	0.	0.	0.
200	23.054	22.133	34.325	-2.438	0.	0.	0.
298.15	26.299	32.010	32.010	0.	0.	0.	0.
300	26.347	32.173	32.010	0.049	0.	0.	0.
400	28.527	40.066	33.072	2.798	0.	0.	0.
500	30.292	46.696	35.152	5.772	0.	0.	0.
600	31.899	52.362	37.559	8.882	0.	0.	0.
700	33.426	57.395	40.039	12.149	0.	0.	0.
800	34.915	61.956	42.498	15.566	0.	0.	0.
900	36.384	66.153	44.896	19.131	0.	0.	0.
980.000	37.545	69.300	46.761	22.088	ALPHA <--> BETA TRANSITION		
980.000	37.572	71.572	46.761	24.314			
1000	37.698	72.332	47.265	25.067	0.	0.	0.
1100	38.116	75.945	49.710	28.858	0.	0.	0.
1200	38.535	79.279	52.037	32.690	0.	0.	0.
1300	38.953	82.380	54.253	36.565	0.	0.	0.
1361.000	39.204	84.172	55.554	38.948	BETA <--> GAMMA TRANSITION		
1361.000	43.095	85.731	55.554	41.070			
1400	43.430	86.953	56.412	42.757	0.	0.	0.
1412.000	43.514	87.324	56.673	43.279	GAMMA <--> DELTA TRANSITION		
1412.000	45.229	88.655	56.673	45.158			
1500	45.982	91.413	58.631	49.173	0.	0.	0.
1519.000	46.108	91.993	59.045	50.048	DELTA <--> LIQUID TRANSITION		
1519.000	46.024	99.931	59.045	62.106			
1600	46.024	102.322	61.176	65.834	0.	0.	0.
1700	46.024	105.112	63.679	70.436	0.	0.	0.
1800	46.024	107.743	66.055	75.039	0.	0.	0.
1900	46.024	110.231	68.315	79.641	0.	0.	0.
2000	46.024	112.592	70.470	84.244	0.	0.	0.
2100	46.024	114.837	72.530	88.846	0.	0.	0.
2200	46.024	116.978	74.502	93.448	0.	0.	0.
2300	46.024	119.024	76.393	98.051	0.	0.	0.
2334.526	46.024	119.710	77.029	99.640	LIQUID <--> IDEAL GAS FUGACITY = 1 bar		
2334.526	20.951	216.509	77.029	325.620			
2400	20.997	217.089	80.842	326.993	0.	0.	0.
2500	21.085	217.948	86.309	329.097	0.	0.	0.
2600	21.198	218.777	91.388	331.211	0.	0.	0.
2700	21.341	219.580	96.121	333.338	0.	0.	0.
2800	21.515	220.359	100.545	335.480	0.	0.	0.
2900	21.726	221.118	104.689	337.642	0.	0.	0.
3000	21.976	221.858	108.583	339.827	0.	0.	0.
3100	22.269	222.583	112.248	342.039	0.	0.	0.
3200	22.606	223.296	115.708	344.282	0.	0.	0.
3300	22.990	223.997	118.978	346.561	0.	0.	0.
3400	23.424	224.690	122.077	348.882	0.	0.	0.
3500	23.909	225.376	125.019	351.248	0.	0.	0.
3600	24.445	226.056	127.816	353.665	0.	0.	0.
3700	25.034	226.734	130.480	356.139	0.	0.	0.
3800	25.675	227.410	133.022	358.674	0.	0.	0.
3900	26.370	228.086	135.451	361.275	0.	0.	0.
4000	27.118	228.763	137.776	363.949	0.	0.	0.
4200	28.772	230.125	142.141	369.535	0.	0.	0.
4400	30.635	231.506	146.171	375.472	0.	0.	0.
4600	32.606	232.907	149.912	381.777	0.	0.	0.
4800	34.798	234.340	153.399	388.514	0.	0.	0.
5000	37.153	235.808	156.666	395.707	0.	0.	0.
5200	39.664	237.313	159.739	403.386	0.	0.	0.
5400	42.327	238.860	162.641	411.582	0.	0.	0.
5600	43.514	240.321	165.382	419.658	0.	0.	0.
5800	45.714	241.887	167.993	428.580	0.	0.	0.
6000	47.832	243.472	170.483	437.936	0.	0.	0.

PREVIOUS: CURRENT: September 1967 (1 bar)

Manganese (Mn) Mn_1(ref)

MANGANESE (Mn) ALPHA-BETA-GAMMA-DELTA A_r = 54.9380

$S°(298.15\ K) = 32.01 \pm 0.08\ J\ K^{-1}\ mol^{-1}$
$T_{trs}(\alpha \rightarrow \beta) = 980 \pm 20\ K$
$T_{trs}(\beta \rightarrow \gamma) = 1361 \pm 10\ K$
$T_{trs}(\gamma - \delta) = 1412 \pm 5\ K$
$T_{fus}(\delta - \ell) = 1519 \pm 5\ K$

$\Delta_f H°(0\ K) = 0\ kJ\ mol^{-1}$
$\Delta_f H°(298.15\ K) = 0\ kJ\ mol^{-1}$
$\Delta_{trs} H°(\alpha \rightarrow \beta) = 2.226 \pm 0.21\ kJ\ mol^{-1}$
$\Delta_{trs} H°(\beta \rightarrow \gamma) = 2.122 \pm 0.33\ kJ\ mol^{-1}$
$\Delta_{trs} H°(\gamma \rightarrow \delta) = 1.879 \pm 0.33\ kJ\ mol^{-1}$
$\Delta_{fus} H°(\delta \rightarrow \ell) = 12.058 \pm 1.0\ kJ\ mol^{-1}$

Enthalpy of Formation

Zero by definition.

Heat Capacity and Entropy

A thorough re-examination of the manganese thermal functions is in process. Temporarily, the recommendations of Hultgren et al. (1) are adotped. These results are based on the heat capacity studies of Scurlock and Stevens (0.3-1.0 K, 2), Gaumer (0.64-3 K, 3) , Guthrie et al. (1.7-4.2 K, 4), Booth et al. (12-20 K, 5), Kelley (52-298 K, 6), and Shomate (52-298 K, 7) and the enthalpy study of Naylor (373-1436 K, 8). As Hultgren et al. (1) states, the study of Naylor was chosen because it covers all forms of Mn in a self consistant manner. It is not stated why Hultgren, et al. (1) altered Naylor's equations (8), which appear more appropriate than the current values.

Phase and Fusion Data

As we are temporarily adopting the analysis of Hultgren, et al (1), we rely on his analysis of the transition and fusion temperatures. Ten studies are analyzed by Hultgren et al. (1). The temperatures were adjusted to IPTS-68.

Sublimation Data

See the ideal gas table for details.

References

1. R. Hultgren, P. D. Desai, et al., "Selected Values of the Thermodynamic Properties of the Elements," American Society for Metals," Metals Park, Ohio, 1973.
2. R. G. Scurlock and W. N. R. Stevens, Proc. Phys. Soc. 86, 331 (1965).
3. R. E. Gaumer, Ohio State University, Ph. D. Thesis, 1959.
4. G. L. Guthrie, S. A. Friedburg, and J. E. Goldman, Phys. Rev. 139, 1200 (1965).
5. G. L. Booth, F. E. Hoare and B. T. Murphy, Proc. Phys. Soc. 68, 830 (1955).
6. K. K. Kelley, J. Am. Chem. Soc. 61, 203 (1939).
7. C. H. Shomate, U. S. Bur. Mines Tech. Paper 686, 1946.
8. B. F. Naylor, J. Chem. Phys. 13, 329 (1945).

Manganese (Mn) $Mn_1(cr)$

Enthalpy Reference Temperature = T_r = 298.15 K; Standard State Pressure = p° = 0.1 MPa

T/K	$C_p°$ (J K⁻¹mol⁻¹)	S° (J K⁻¹mol⁻¹)	$-[G°-H°(T_r)]/T$ (J K⁻¹mol⁻¹)	$H°-H°(T_r)$ (kJ mol⁻¹)	$\Delta_f H°$ (kJ mol⁻¹)	$\Delta_f G°$ (kJ mol⁻¹)	Log K_f
0	0.	0.	INFINITE	-4.994	0.	0.	0.
100	14.723	8.875	52.876	-4.400	0.	0.	0.
200	23.054	22.133	34.325	-2.438	0.	0.	0.
250	24.948	27.493	32.436	-1.236	0.	0.	0.
298.15	26.299	32.010	32.010	0.	0.	0.	0.
300	26.347	32.173	32.010	0.049	0.	0.	0.
350	27.516	36.325	32.336	1.396	0.	0.	0.
400	28.527	40.066	33.072	2.798	0.	0.	0.
450	30.020	43.524	34.043	4.266	0.	0.	0.
500	30.292	46.696	35.152	5.772	0.	0.	0.
600	31.899	52.362	37.559	8.882	0.	0.	0.
700	33.426	57.395	40.039	12.149	0.	0.	0.
800	34.915	61.956	42.498	15.566	0.	0.	0.
900	36.384	66.153	44.896	19.131	0.	0.	0.
980.000	37.545	69.300	46.761	22.088	ALPHA <--> BETA TRANSITION		
980.000	37.572	71.572	46.761	24.314			
1000	37.698	72.332	47.265	25.067	0.	0.	0.
1100	38.116	75.945	49.710	28.858	0.	0.	0.
1200	38.535	79.279	52.037	32.690	0.	0.	0.
1300	38.953	82.380	54.253	36.565	0.	0.	0.
1361.000	39.204	84.172	55.554	38.948	BETA <--> GAMMA TRANSITION		
1361.000	43.095	85.731	55.554	41.070			
1400	43.430	86.953	56.412	42.757	0.	0.	0.
1412.000	43.514	87.324	56.673	43.279	GAMMA <--> DELTA TRANSITION		
1412.000	45.229	88.655	56.673	45.158			
1500	45.982	91.413	58.631	49.173	0.	0.	0.
1519.000	46.108	91.993	59.045	50.048	DELTA <--> LIQUID		
1600	46.803	94.406	60.774	53.811	-12.023	0.642	-0.021
1700	47.624	97.268	62.837	58.532	-11.904	1.431	-0.044
1800	48.446	100.013	64.827	63.336	-11.703	2.210	-0.064
1900	49.267	102.655	66.749	68.221	-11.420	2.975	-0.082

PREVIOUS: CURRENT: September 1967

Manganese (Mn) $Mn_1(cr)$

MANGANESE (Mn) LIQUID A_r = 54.9380

$S°(298.15\ K) = [43.499]\ J\ K^{-1}\ mol^{-1}$ $\Delta_f H°(298.15\ K) = [16.289]\ kJ\ mol^{-1}$

$T_{fus} = 1519 \pm 5\ K$ $\Delta_{fus}H° = 12.058 \pm 1.0\ kJ\ mol^{-1}$

Enthalpy of Formation

The enthalpy of formation of Mn(ℓ), $\Delta_f H°(298.15\ K)$ is calculated from that of the crystal by adding $\Delta_{fus}H°$ and the difference in enthalpy, H°(1519 K) - H°(298.15 K), between the crystal and liquid.

Heat Capacity and Entropy

Hultgren (1) adopted $Cp°(ℓ) = 11.0\ cal\ K^{-1}\ mol^{-1}$ based on the recommendation of Kelly (2). A glass transition temperature is assumed at 1125 K, below which the heat capacity values of α-Mn(cr) are adopted. a re-examination of Mn(ℓ) is in progress. The entropy at 298.15 K is calculated in a manner analogous to that used for the enthalpy of formation.

Vaporization Data

T_{vap}(1 bar) is calculated as the temperature at which $\Delta_r G° = 0$ for the process Mn(ℓ) = Mn(g). $\Delta_{vap}H°$ is calculated as the difference in the enthalpies of formation of Mn(ℓ) and Mn(g) at T_{vap}.

References

1. R. Hultgren, P. D. Desai, et al., "Selected Values of the Thermodynamic Properties of the Elements," American Society for Metals, Metals Park, Ohio, 1973.
2. K. K. Kelley, U. S. Bur. Mines Bull. 584, 1960.

Manganese (Mn) $Mn_1(l)$

Enthalpy Reference Temperature = T_r = 298.15 K Standard State Pressure = p° = 0.1 MPa

	J K⁻¹mol⁻¹			kJ mol⁻¹			
T/K	$C_p°$	S°	$-[G°-H°(T_r)]/T$	$H°-H°(T_r)$	$\Delta_f H°$	$\Delta_f G°$	Log K_f
0							
100							
200							
250							
298.15	26.299	43.499	43.499	0.	16.289	12.863	-2.254
300	26.347	43.662	43.500	0.049	16.289	12.842	-2.236
350	27.516	47.814	43.825	1.396	16.289	12.267	-1.831
400	28.527	51.555	44.561	2.798	16.289	11.693	-1.527
450	30.020	55.013	45.533	4.266	16.289	11.118	-1.291
500	30.292	58.185	46.642	5.772	16.289	10.544	-1.102
600	31.899	63.852	49.048	8.882	16.289	9.395	-0.818
700	33.426	68.884	51.529	12.149	16.289	8.246	-0.615
800	34.915	73.445	53.988	15.566	16.289	7.097	-0.463
900	36.384	77.643	56.386	19.131	16.289	5.948	-0.345
1000	37.834	81.551	58.709	22.842	14.064	4.845	-0.253
1100	39.275	85.225	60.954	26.698	14.129	3.921	-0.186
1125.000	39.634	86.111	61.503	27.684	GLASS <--> LIQUID TRANSITION		
1125.000	46.024	86.111	61.503	27.684			
1200	46.024	89.082	63.135	31.136	14.734	2.971	-0.129
1300	46.024	92.766	65.275	35.738	15.462	1.961	-0.079
1400	46.024	96.176	67.362	40.341	13.872	0.960	-0.036
1500	46.024	99.352	69.390	44.943	12.059	0.151	-0.005
1519.000	46.024	99.931	69.768	45.817	------- DELTA <--> LIQUID ------		
1600	46.024	102.322	71.356	49.545	0.	0.	0.
1700	46.024	105.112	73.260	54.148	0.	0.	0.
1800	46.024	107.743	75.104	58.750	0.	0.	0.
1900	46.024	110.231	76.888	63.353	0.	0.	0.
2000	46.024	112.592	78.614	67.955	0.	0.	0.
2100	46.024	114.837	80.286	72.557	0.	0.	0.
2200	46.024	116.978	81.906	77.160	0.	0.	0.
2300	46.024	119.024	83.476	81.762	0.	0.	0.
2334.526	46.024	119.710	84.006	83.351	------- FUGACITY = 1 bar -------		
2400	46.024	120.983	84.998	86.365	-224.340	6.315	-0.137
2500	46.024	122.862	86.475	90.967	-221.841	15.874	-0.332
2600	46.024	124.667	87.909	95.569	-219.353	25.334	-0.509
2700	46.024	126.404	89.303	100.172	-216.877	34.698	-0.671
2800	46.024	128.078	90.658	104.774	-214.417	43.970	-0.820
2900	46.024	129.693	91.977	109.377	-211.977	53.155	-0.957
3000	46.024	131.253	93.260	113.979	-209.559	62.256	-1.084

PREVIOUS: CURRENT: September 1967

Manganese (Mn) $Mn_1(l)$

MANGANESE (Mn)

ALPHA-BETA-GAMMA-DELTA-LIQUID

0 to 980 K	crystal, alpha
980 to 1361 K	crystal, beta
1361 to 1412 K	crystal, gamma
1412 to 1519 K	crystal, delta
above 1519 K	liquid

Refer to the individual tables for details.

A_r = 54.9380 Manganese (Mn) Mn_1(cr,l)

Enthalpy Reference Temperature = T_r = 298.15 K Standard State Pressure = p° = 0.1 MPa

T/K	C_p° (J K⁻¹mol⁻¹)	S° (J K⁻¹mol⁻¹)	$-[G^\circ-H^\circ(T_r)]/T$ (J K⁻¹mol⁻¹)	$H^\circ-H^\circ(T_r)$ (kJ mol⁻¹)	$\Delta_f H^\circ$ (kJ mol⁻¹)	$\Delta_f G^\circ$ (kJ mol⁻¹)	Log K_f
0	0.	0.	INFINITE	-4.994	0.	0.	0.
100	14.723	8.875	52.876	-4.400	0.	0.	0.
200	23.054	22.133	34.325	-2.438	0.	0.	0.
250	24.948	27.493	32.436	-1.236	0.	0.	0.
298.15	26.299	32.010	32.010	0.	0.	0.	0.
300	26.347	32.173	32.010	0.049	0.	0.	0.
350	27.516	36.325	32.336	1.396	0.	0.	0.
400	28.527	40.066	33.072	2.798	0.	0.	0.
450	30.020	43.524	34.043	4.266	0.	0.	0.
500	30.292	46.696	35.152	5.772	0.	0.	0.
600	31.899	52.362	37.559	8.882	0.	0.	0.
700	33.426	57.395	40.039	12.149	0.	0.	0.
800	34.915	61.956	42.498	15.566	0.	0.	0.
900	36.384	66.153	44.896	19.131	0.	0.	0.
980.000	37.545	69.300	46.761	22.088		ALPHA <--> BETA	
980.000	37.572	71.572	46.761	24.314		TRANSITION	
1000	37.698	72.332	47.265	25.067	0.	0.	0.
1100	38.116	75.945	49.710	28.858	0.	0.	0.
1200	38.535	79.279	52.037	32.690	0.	0.	0.
1300	38.953	82.380	54.253	36.565	0.	0.	0.
1361.000	39.204	84.172	55.554	38.948		BETA <--> GAMMA	
1361.000	43.095	85.731	55.554	41.070		TRANSITION	
1400	43.430	86.953	56.412	42.757	0.	0.	0.
1412.000	43.514	87.324	56.673	43.279		GAMMA <--> DELTA	
1412.000	45.229	88.655	56.673	45.158		TRANSITION	
1500	45.982	91.413	58.631	49.173	0.	0.	0.
1519.000	46.108	91.993	59.045	50.048		DELTA <--> LIQUID	
1519.000	46.024	99.931	59.045	62.106		TRANSITION	
1600	46.024	102.322	61.176	65.834	0.	0.	0.
1700	46.024	105.112	63.679	70.436	0.	0.	0.
1800	46.024	107.743	66.055	75.039	0.	0.	0.
1900	46.024	110.231	68.315	79.641	0.	0.	0.
2000	46.024	112.592	70.470	84.244	0.	0.	0.
2100	46.024	114.837	72.530	88.846	0.	0.	0.
2200	46.024	116.978	74.502	93.448	0.	0.	0.
2300	46.024	119.024	76.393	98.051	0.	0.	0.
2334.526	46.024	119.710	77.029	99.640	-------	FUGACITY = 1 bar	-------
2400	46.024	120.983	78.211	102.653	-224.340	6.315	-0.137
2500	46.024	122.862	79.960	107.256	-221.841	15.874	-0.332
2600	46.024	124.667	81.645	111.858	-219.353	25.334	-0.509
2700	46.024	126.404	83.270	116.460	-216.877	34.698	-0.671
2800	46.024	128.078	84.841	121.063	-214.417	43.970	-0.820
2900	46.024	129.693	86.360	125.665	-211.977	53.155	-0.957
3000	46.024	131.253	87.830	130.268	-209.559	62.256	-1.084

PREVIOUS: CURRENT: September 1967

Manganese (Mn) Mn_1(cr,l)

MANGANESE (Mn) IDEAL GAS A_r = 54.9380

IP (Mn, g) = 59981 ± 1 cm^{-1} $\Delta_f H°$(0 K) = 282.1 ± 4.2 kJ mol^{-1}

S°(298.15 K) = 173.716 J K^{-1} mol^{-1} $\Delta_f H°$ = (298.15 K) = 283.3 ± 4.2 kJ mol^{-1}

Electronic Levels and Quantum Weights

State	ε_i, cm^{-1}	g_i
$^6S_{5/2}$	0.00	6
$^6D_{9/2}$	17052.29	10
$^6D_{7/2}$	17282.00	8
$^6D_{5/2}$	17451.52	6
$^6D_{3/2}$	17568.48	4
$^6D_{1/2}$	17637.15	2

Enthalpy of Formation

The value adopted for the enthalpy of formation of Mn(g), $\Delta_f H°$(298.15 K) = 283.3 ± 4.2 kJ mol^{-1}, is that recommended by Hultgren et al (1). This value is a weighted mean of sublimation and vaporization studies. The studies are: Bogatyrev and Golubtsov (1043-1168 K), 2), Pratt and Spencer (1248-1620 K, 3), Wiedemeier (1269-1448 K, 4), Butler et al. (1075-1335 K, 5), Woolf, et al. (1523-1823 K, 6), McCabe and Hudson (1075-1235 K, 7) and Baur and Brunner (1587-1987 K, 8).

A reanalysis of the vapor pressure data is in progress.

Heat Capacity and Entropy

The information on electronic energy levels and quantum weights, given by Corliss and Sugar (9), is incomplete because many theoretically predicted levels have not been observed. Although we have listed only a few of the lowest lying levels for Mn(g), all levels listed by Corliss and Sugar (9), as well as estimated levels, are used in the calculation. The observed levels are too numerous to list completely. The calculations indicate that for Mn(g), the thermodynamic functions are independent of the estimated missing levels (for n = 4-8), the cut-off procedure, and the inclusion of n<9 levels up to 4000 K; the Gibbs energy function showing no significant variations at this temperature. The reported uncertainty in S°(298.15 K) is due to uncertainties in the relative ionic mass, and the fundamental constants. Extension of these calculations above 6000 K may require consideration of the higher excited states (n>8), and use of different fill and cut-off procedures (10).

References

1. R. Hultgren, P. D. Desai, et al., "Selected Values for Thermodynamic Properties of Metals", American Society of Metals, Metals Park, Ohio, 1973.
2. V. L. Bogatyrev and I. V. Golubtsev, Zh. Fiz. Khim. 40, 938 (1966).
3. J. N. Pratt and P. J. Spencer, Q.T.S.R. No. 1, U.S. Dept. of the Army, Contract No. DA-91-591-EVC-3762, 1966.
4. H. Z. Wiedemeier, Z. Anorg. Allgem. Chem. 320, 225 (1964).
5. J. F. Butler, C. L. McCabe, and H. W. Paxton, Trans. AIME 221, 479 (1961).
6. P. L. Woolf, G. R. Zellars, et al., U. S. Bur. Mines RI-5634, 1959.
7. C. L. McCabe and R. G. Hudson, Trans. AIME, 209, 17 (1957).
8. E. Baur and R. Brunner, Helv. Chim. Acta 17, 958 (1934).
9. C. Corliss and J. Sugar, J. Phys. Chem. Ref. Data 6, 1253 (1977).
10. J. R. Downey, Jr., The Dow Chemical Company, AFOSR-TR-78-0960, Contract No. F44620-75-1-0048, 1978.

Manganese (Mn) Mn_1(g)

Enthalpy Reference Temperature = T_r = 298.15 K   Standard State Pressure = p° = 0.1 MPa

T/K	$C_p°$ (J $K^{-1}mol^{-1}$)	S° (J $K^{-1}mol^{-1}$)	$-[G°-H°(T_r)]/T$ (J $K^{-1}mol^{-1}$)	$H°-H°(T_r)$ (kJ mol^{-1})	$\Delta_f H°$ (kJ mol^{-1})	$\Delta_f G°$ (kJ mol^{-1})	Log K_f
0	0.	0.	INFINITE	-6.197	282.053	282.053	INFINITE
100	20.786	151.009	192.196	-4.119	283.538	269.325	-140.681
200	20.786	165.416	175.617	-2.040	283.655	254.998	-66.599
250	20.786	170.055	174.058	-1.001	283.492	247.851	-51.786
298.15	20.786	173.716	173.716	0.	283.257	241.007	-42.223
300	20.786	173.844	173.716	0.038	283.247	240.745	-41.917
350	20.786	177.049	173.969	1.078	282.938	233.685	-34.876
400	20.786	179.824	174.532	2.117	282.576	226.673	-29.600
450	20.786	182.272	175.258	3.156	282.147	219.710	-25.503
500	20.786	184.462	176.071	4.196	281.681	212.797	-22.231
600	20.786	188.252	177.795	6.274	280.649	199.115	-17.335
700	20.786	191.456	179.524	8.353	279.461	185.618	-13.851
800	20.786	194.232	181.193	10.431	278.122	172.301	-11.250
900	20.786	196.680	182.780	12.510	276.636	159.161	-9.237
1000	20.786	198.870	184.282	14.589	272.778	146.240	-7.639
1100	20.786	200.851	185.699	16.667	271.066	133.669	-6.347
1200	20.786	202.660	187.038	18.746	269.312	121.255	-5.278
1300	20.786	204.324	188.305	20.824	267.517	108.990	-4.379
1400	20.786	205.864	189.505	22.903	263.403	96.927	-3.616
1500	20.787	207.298	190.644	24.982	259.066	85.238	-2.968
1600	20.788	208.640	191.727	27.061	244.483	74.375	-2.428
1700	20.791	209.900	192.759	29.139	241.960	63.820	-1.961
1800	20.797	211.089	193.745	31.219	239.437	53.414	-1.550
1900	20.806	212.213	194.688	33.299	236.914	43.148	-1.186
2000	20.822	213.281	195.591	35.380	234.393	33.015	-0.862
2100	20.845	214.297	196.458	37.464	231.874	23.008	-0.572
2200	20.880	215.268	197.291	39.550	229.358	13.121	-0.312
2300	20.930	216.197	198.093	41.640	226.846	3.348	-0.076
2334.526	20.951	216.509	198.363	42.363	------- FUGACITY = 1 bar -------		
2400	20.997	217.089	198.866	43.736	0.	0.	0.
2500	21.085	217.948	199.612	45.840	0.	0.	0.
2600	21.198	218.777	200.333	47.954	0.	0.	0.
2700	21.341	219.580	201.031	50.081	0.	0.	0.
2800	21.515	220.359	201.708	52.223	0.	0.	0.
2900	21.726	221.118	202.364	54.385	0.	0.	0.
3000	21.976	221.858	203.002	56.570	0.	0.	0.
3100	22.269	222.583	203.622	58.782	0.	0.	0.
3200	22.606	223.296	204.225	61.025	0.	0.	0.
3300	22.990	223.997	204.814	63.305	0.	0.	0.
3400	23.424	224.690	205.388	65.625	0.	0.	0.
3500	23.909	225.376	205.950	67.991	0.	0.	0.
3600	24.445	226.056	206.499	70.408	0.	0.	0.
3700	25.034	226.734	207.036	72.882	0.	0.	0.
3800	25.675	227.410	207.564	75.417	0.	0.	0.
3900	26.370	228.086	208.081	78.019	0.	0.	0.
4000	27.118	228.763	208.590	80.693	0.	0.	0.
4100	27.919	229.442	209.090	83.444	0.	0.	0.
4200	28.772	230.125	209.583	86.278	0.	0.	0.
4300	29.678	230.813	210.068	89.200	0.	0.	0.
4400	30.635	231.506	210.548	92.216	0.	0.	0.
4500	31.574	232.201	211.021	95.311	0.	0.	0.
4600	32.606	232.907	211.489	98.520	0.	0.	0.
4700	33.681	233.619	211.952	101.834	0.	0.	0.
4800	34.798	234.340	212.411	105.258	0.	0.	0.
4900	35.956	235.069	212.866	108.795	0.	0.	0.
5000	37.153	235.808	213.318	112.450	0.	0.	0.
5100	38.389	236.556	213.766	116.227	0.	0.	0.
5200	39.664	237.313	214.211	120.129	0.	0.	0.
5300	40.976	238.081	214.655	124.161	0.	0.	0.
5400	42.327	238.860	215.096	128.325	0.	0.	0.
5500	43.716	239.649	215.535	132.627	0.	0.	0.
5600	43.514	240.321	215.964	136.401	0.	0.	0.
5700	44.624	241.101	216.398	140.808	0.	0.	0.
5800	45.714	241.887	216.831	145.324	0.	0.	0.
5900	46.784	242.677	217.262	149.948	0.	0.	0.
6000	47.832	243.472	217.692	154.679	0.	0.	0.

PREVIOUS: CURRENT: June 1984 (1 bar)

Manganese (Mn) Mn_1(g)

MANGANESE, ION (MN^+) IDEAL GAS M_r = 54.93745

IP(Mn^+, g) = 126145.0 ± 0.6 cm^{-1}
S°(298.15 K) = 174.997 ± 0.05 J K^{-1} mol^{-1}

$\Delta_f H°$(0 K) = 999.6 ± 0.5 kJ mol^{-1}
$\Delta_f H°$(298.15 K) = [1006.986] kJ mol^{-1}

Electronic Levels and Quantum Weights

State	ϵ_i cm^{-1}	g_i
$^7S_3$	0	7
$^5S_2$	9472.97	5
$^5D_4$	14325.86	9
.	.	.
.	.	.
.	.	.
$^7H_8$	120721.9	17
IP	126145.0	

Enthalpy of Formation

$\Delta_f H°$(Mn^+, g, 0 K) is calculated from $\Delta_f H°$(Mn, g, 0 K) (1) using the spectroscopic value of IP(Mn) = 59981 ± 1 cm^{-1} (717.532 ± 0.012 mol^{-1}) from Corliss and Sugar (2). The ionization limit is converted from cm^{-1} to kJ mol^{-1} using the factor, 1 cm^{-1} = 0.01196266 kJ cm^{-1}, which is derived from the 1973 CODATA fundamental constants (3). Rosenstock et al. (4) and Levin and Lias (5) have summarized additional ionization and appearance potential data.

$\Delta_f H°$(Mn^+, g, 298.15 K) is calculated from $\Delta_f H°$(Mn, g, 0 K) by using IP(Mn) with JANAF (1) enthalpies, H°(0 K) - H°(298.15 K), for Mn(g), Mn^+(g), and e^-(g). $\Delta_f H°$(Mn → Mn^+ + e^-, 298.15 K) differs from a room temperature threshold energy due to inclusion of these enthalpies and to threshold effects discussed by Rosenstock et al. (4). $\Delta_f H°$(298.15 K) should be changed by -6.197 kJ mol^{-1} if it is to be used in the ion convention that excludes the enthalpy of the electron.

Heat Capacity and Entropy

The information on electronic energy levels and quantum weights, given by Corliss and Sugar (2), is incomplete because many theoretically predicted levels have not been observed. Although we have listed only the ground, the first two excited state, the highest observed excited state, and the ionization potential for Mn^+(g), all levels listed by Corliss and Sugar (2), as well as estimated levels, are used in the calculation. The observed levels are too numerous to list completely. The calculations indicate that for Mn^+(g), the thermodynamic functions are independent of the estimated missing levels (for n=4-7), the cut-off procedure, and the inclusion of n<8 levels up to 6000 K; the Gibbs energy function showing, no significant at this temperature. The reported uncertainty in S°(298.15 K) is due to uncertainties in the relative ionic mass, and the fundamental constants. Extension of these calculations above 6000 K may require consideration of the higher excited states (n>7), and use of different fill and cut-off procedures (6).

References

1. JANAF Thermochemical Tables: Mn(g), 6-30-84; e^-(g), 3-31-82.
2. C. Corliss and J. Sugar, J. Phys. Chem. Ref. Data 6, 1253 (1977).
3. E. R. Cohen and B. N. Taylor, J. Phys. Chem. Ref. Data 2, 663 (1973).
4. H. M. Rosenstock, K. Draxl et al., J. Phys. Chem. Ref. Data 6, Supp. 1 (1977).
5. R. D. Levin and S. G. Lias, U. S. Natl. Bur. Stand., NSRDS-NBS-71, 1982.
6. J. R. Downey, Jr., The Dow Chemical Company, AFOSR-TR-78-0960, Contract No. F44620-75-1-0048, 1978.

Manganese, Ion (Mn^+) Mn_1^+(g)

Enthalpy Reference Temperature = T_r = 298.15 K Standard State Pressure = p° = 0.1 MPa

T/K	$C_p°$ (J $K^{-1}mol^{-1}$)	S° (J $K^{-1}mol^{-1}$)	-[G°-H°(T_r)]/T (J $K^{-1}mol^{-1}$)	H°-H°(T_r) (kJ mol^{-1})	$\Delta_f H°$ (kJ mol^{-1})	$\Delta_f G°$ (kJ mol^{-1})	Log K_f
0	0.	0.	INFINITE	-6.197	999.586		
100	20.786	152.290	193.478	-4.119			
200	20.786	166.698	176.899	-2.040			
250	20.786	171.336	175.340	-1.001			
298.15	20.786	174.997	174.997	0.	1006.986	958.100	-167.855
300	20.786	175.126	174.998	0.038	1007.015	957.797	-166.767
350	20.786	178.330	175.251	1.078	1007.746	949.535	-141.710
400	20.786	181.106	175.813	2.117	1008.423	941.172	-122.904
450	20.786	183.554	176.540	3.156	1009.033	932.728	-108.268
500	20.786	185.744	177.353	4.196	1009.606	924.219	-96.553
600	20.786	189.534	179.077	6.274	1010.653	907.041	-78.965
700	20.786	192.738	180.805	8.353	1011.543	889.700	-66.390
800	20.786	195.514	182.474	10.431	1012.283	872.241	-56.952
900	20.786	197.962	184.062	12.510	1012.875	854.699	-49.605
1000	20.787	200.152	185.563	14.589	1011.097	837.144	-43.728
1100	20.790	202.133	186.981	16.668	1011.464	819.731	-38.926
1200	20.795	203.942	188.320	18.747	1011.789	802.286	-34.923
1300	20.805	205.607	189.587	20.827	1012.073	784.815	-31.534
1400	20.821	207.150	190.787	22.908	1010.040	767.387	-28.632
1500	20.846	208.587	191.926	24.991	1007.787	750.184	-26.124
1600	20.882	209.933	193.010	27.078	995.290	733.668	-23.952
1700	20.932	211.201	194.043	29.168	994.857	717.330	-22.041
1800	20.996	212.399	195.030	31.265	994.430	701.017	-20.343
1900	21.078	213.536	195.974	33.368	994.009	684.728	-18.824
2000	21.179	214.620	196.880	35.481	993.598	668.461	-17.458
2100	21.301	215.656	197.749	37.605	993.198	652.214	-16.223
2200	21.445	216.650	198.586	39.742	992.812	635.985	-15.100
2300	21.611	217.607	199.392	41.894	992.440	619.774	-14.076
2400	21.800	218.531	200.171	44.065	767.747	609.894	-13.274
2500	22.011	219.425	200.923	46.255	769.912	603.272	-12.605
2600	22.246	220.293	201.651	48.468	772.090	596.564	-11.985
2700	22.503	221.137	202.357	50.705	774.279	589.771	-11.410
2800	22.782	221.960	203.043	52.969	776.479	582.898	-10.874
2900	23.080	222.765	203.709	55.262	778.689	575.945	-10.374
3000	23.398	223.553	204.358	57.586	780.906	568.916	-9.906
3100	23.735	224.325	204.989	59.942	783.130	561.813	-9.466
3200	24.088	225.085	205.605	62.333	785.356	554.638	-9.054
3300	24.457	225.831	206.207	64.761	787.582	547.394	-8.665
3400	24.841	226.567	206.795	67.225	789.805	540.082	-8.297
3500	25.238	227.293	207.370	69.729	792.021	532.705	-7.950
3600	25.648	228.010	207.934	72.273	794.227	525.265	-7.621
3700	26.070	228.718	208.486	74.859	796.418	517.763	-7.310
3800	26.504	229.419	209.028	77.488	798.590	510.203	-7.013
3900	26.948	230.113	209.559	80.160	800.739	502.586	-6.731
4000	27.402	230.801	210.082	82.878	802.861	494.914	-6.463
4100	27.866	231.484	210.595	85.641	804.952	487.189	-6.207
4200	28.341	232.161	211.101	88.451	807.007	479.414	-5.962
4300	28.825	232.833	211.598	91.310	809.021	471.590	-5.729
4400	29.319	233.502	212.089	94.217	810.992	463.720	-5.505
4500	29.822	234.166	212.572	97.174	812.932	455.805	-5.291
4600	30.336	234.827	213.049	100.181	814.809	447.848	-5.085
4700	30.859	235.485	213.519	103.241	816.634	439.851	-4.888
4800	31.391	236.140	213.983	106.354	818.401	431.815	-4.699
4900	31.934	236.793	214.442	109.520	820.108	423.744	-4.517
5000	32.485	237.444	214.896	112.741	821.753	415.638	-4.342
5100	33.046	238.093	215.344	116.017	823.331	407.500	-4.174
5200	33.616	238.740	215.788	119.350	824.841	399.331	-4.011
5300	34.195	239.386	216.227	122.741	826.278	391.135	-3.855
5400	34.781	240.030	216.662	126.189	827.641	382.911	-3.704
5500	35.375	240.674	217.093	129.697	828.926	374.664	-3.558
5600	35.976	241.317	217.520	133.265	830.798	366.345	-3.417
5700	36.583	241.959	217.943	136.893	832.097	358.040	-3.281
5800	37.195	242.600	218.362	140.581	833.349	349.712	-3.149
5900	37.810	243.241	218.778	144.332	834.553	341.363	-3.022
6000	38.429	243.882	219.192	148.144	835.713	332.994	-2.899

PREVIOUS: CURRENT: June 1984 (1 bar)

Manganese, Ion (Mn^+) Mn_1^+(g)

MOLYBDENUM (Mo)

REFERENCE STATE

0 to 2896 K crystal
2896 to 4951.969 K liquid
above 4951.969 K ideal monatomic gas

Refer to the individual tables for details.

A_r = 95.94

Molybdenum (Mo) Mo_1(ref)

Enthalpy Reference Temperature = T_r = 298.15 K — Standard State Pressure = $p°$ = 0.1 MPa

T/K	$C_p°$ (J K⁻¹mol⁻¹)	$S°$ (J K⁻¹mol⁻¹)	$-[G°-H°(T_r)]/T$ (J K⁻¹mol⁻¹)	$H°-H°(T_r)$ (kJ mol⁻¹)	$\Delta_fH°$ (kJ mol⁻¹)	$\Delta_fG°$ (kJ mol⁻¹)	Log K_f
0	0.	0.	INFINITE	-4.585	0.	0.	0.
100	13.495	7.026	47.961	-4.094	0.	0.	0.
200	21.507	19.478	30.735	-2.251	0.	0.	0.
298.15	23.933	28.605	28.605	0.	0.	0.	0.
300	23.958	28.753	28.606	0.044	0.	0.	0.
400	25.080	35.810	29.560	2.500	0.	0.	0.
500	25.850	41.494	31.397	5.049	0.	0.	0.
600	26.460	46.263	33.487	7.665	0.	0.	0.
700	26.980	50.382	35.613	10.338	0.	0.	0.
800	27.440	54.015	37.691	13.059	0.	0.	0.
900	27.890	57.272	39.689	15.825	0.	0.	0.
1000	28.370	60.235	41.597	18.638	0.	0.	0.
1100	28.900	62.963	43.417	21.501	0.	0.	0.
1200	29.490	65.502	45.153	24.419	0.	0.	0.
1300	30.140	67.888	46.811	27.400	0.	0.	0.
1400	30.860	70.148	48.398	30.450	0.	0.	0.
1500	31.650	72.304	49.920	33.575	0.	0.	0.
1600	32.500	74.373	51.384	36.782	0.	0.	0.
1700	33.420	76.371	52.796	40.078	0.	0.	0.
1800	34.420	78.309	54.160	43.469	0.	0.	0.
1900	35.490	80.198	55.481	46.964	0.	0.	0.
2000	36.650	82.048	56.763	50.570	0.	0.	0.
2100	37.900	83.866	58.010	54.297	0.	0.	0.
2200	39.240	85.659	59.226	58.152	0.	0.	0.
2300	40.670	87.435	60.414	62.147	0.	0.	0.
2400	42.210	89.198	61.577	66.290	0.	0.	0.
2500	43.890	90.954	62.717	70.594	0.	0.	0.
2600	45.880	92.713	63.837	75.079	0.	0.	0.
2700	48.370	94.490	64.939	79.786	0.	0.	0.
2800	51.570	96.304	66.027	84.777	0.	0.	0.
2896.000	54.835	98.098	67.060	89.885	CRYSTAL <--> LIQUID TRANSITION		
2896.000	37.656	110.522	67.060	125.867			
2900	37.656	110.574	67.120	126.018	0.	0.	0.
3000	37.656	111.851	68.590	129.784	0.	0.	0.
3100	37.656	113.086	70.005	133.549	0.	0.	0.
3200	37.656	114.281	71.370	137.315	0.	0.	0.
3300	37.656	115.440	72.688	141.080	0.	0.	0.
3400	37.656	116.564	73.962	144.846	0.	0.	0.
3500	37.656	117.656	75.195	148.612	0.	0.	0.
3600	37.656	118.716	76.389	152.377	0.	0.	0.
3700	37.656	119.748	77.547	156.143	0.	0.	0.
3800	37.656	120.752	78.671	159.908	0.	0.	0.
3900	37.656	121.731	79.763	163.674	0.	0.	0.
4000	37.656	122.684	80.824	167.440	0.	0.	0.
4100	37.656	123.614	81.856	171.205	0.	0.	0.
4200	37.656	124.521	82.861	174.971	0.	0.	0.
4300	37.656	125.407	83.841	178.736	0.	0.	0.
4400	37.656	126.273	84.795	182.502	0.	0.	0.
4500	37.656	127.119	85.726	186.268	0.	0.	0.
4600	37.656	127.947	86.635	190.033	0.	0.	0.
4700	37.656	128.757	87.523	193.799	0.	0.	0.
4800	37.656	129.549	88.390	197.564	0.	0.	0.
4900	37.656	130.326	89.238	201.330	0.	0.	0.
4951.969	37.656	130.723	89.671	203.287	LIQUID <--> IDEAL GAS FUGACITY = 1 bar		
4951.969	53.001	249.995	89.671	793.918			
5000	53.713	250.510	91.214	796.481	0.	0.	0.
5100	55.185	251.588	94.348	801.926	0.	0.	0.
5200	56.640	252.674	97.382	807.517	0.	0.	0.
5300	58.072	253.766	100.322	813.252	0.	0.	0.
5400	59.485	254.865	103.174	819.130	0.	0.	0.
5500	60.874	255.969	105.942	825.148	0.	0.	0.
5600	62.239	257.078	108.631	831.303	0.	0.	0.
5700	63.577	258.191	111.245	837.593	0.	0.	0.
5800	64.887	259.308	113.788	844.015	0.	0.	0.
5900	66.164	260.428	116.264	850.565	0.	0.	0.
6000	67.413	261.550	118.676	857.243	0.	0.	0.

PREVIOUS: March 1978 (1 atm) CURRENT: March 1978 (1 bar)

Molybdenum (Mo) Mo_1(ref)

MOLYBDENUM (Mo) CRYSTAL A_r = 95.94

$\Delta_f H°(0\ K) = 0\ kJ\ mol^{-1}$
$\Delta_f H°(298.15\ K) = 0\ kJ\ mol^{-1}$
$\Delta_{fus} H° = 36.0 \pm 5.4\ kJ\ mol^{-1}$

$S°(298.15\ K) = 28.605 \pm 0.050\ J\ K^{-1}\ mol^{-1}$
$T_{fus} = 2896 \pm 8\ K$

Enthalpy of Formation

Zero by definition.

Heat Capacity and Entropy

The low temperature heat capacities of Mo(cr) have been measured by Clusius and Franzosini (1) between 16 and 256 K (63 measured Cp points). The reported Cp values are smoothed by fitting the data with orthogonal polynomials over selected overlapping temperature intervals. This fitting procedure also includes the smoothed Cp° values (275-335 K) reported in the critical evaluation by Ditmars et al. (2) so as to provide smoothly varying heat capacity values in the range 256-275 K. The data of Simon and Zeidler (3), 15-238 K, are as much as 2.5% high below 78 K and as much as 1% low above 78 K.

Below 20 K several investigators (4, 5, 6, 7) have determined the heat capacity. Heiniger et al. (8) have summarized and critically reviewed this work as well as other studies. These sources give S°(20 K) values in the range 0.021-0.025 cal K^{-1} mol^{-1}. Adopting S°(20 K) = 0.024 cal K^{-1} mol^{-1} and combining this with our analysis of the Clusius and Franzosoni work (1), we calculate S°(298.15 K) = 6.836 ± 0.012 cal K^{-1} mol^{-1}. This result agrees with the review of the heat capacity data (T <298.15 K) by Reilly and Furukawa (9) for Brewer (10).

The high temperature heat capacity values, 273-2800 K, are those selected by Ditmars et al. (2). This study reports three recent NBS heat capacity/enthalpy studies covering the 270-2800 K on NBS Standard Reference Material No. 781. In addition, the report contains a bibliography of publications on the high temperature enthalpy and heat capacity of crystalline molybdenum. A graphical comparison of the NBS results with those of all the principal investigators and critical evaluation is also given in the publication. These results are linearly extrapolated from 2800 K to 3300 K. The high temperature heat capacity values adopted by Brewer (10), based on IPTS-48, are within 0.1% of the NBS results between 298 and 900 K and are 0.2 to 0.3% higher between 1000 and 1400 K. Between 1500 and 1900 K, the difference is less than 0.1%. From 2000 to 2300 K, the values are 0.2 to 0.3% lower and are 0.1% low to 0.2% high between 2400 to 2700 K. The value at 2800 K is 0.5% lower than our adopted value. At 2800 K, the entropy difference is 0.004 cal K^{-1} mol^{-1} with the JANAF value being smaller.

Fusion Data

Refer to the liquid table for details.

Sublimation Data

Eight sublimation studies are discussed on the ideal gas table.

References

1. K. Clusius and P. Franzosini, Z. Naturforsch 14a, 99 (1959).
2. D. A. Ditmars, A. Cezairliyan, S. Ishihara, and T. B. Douglas, NBS special Publication 260-55 (1977).
3. F. Simon and W. Zeidler, Z. Phys. Chem. 123, 383 (1926).
4. C. A. Bryant and P. H. Keesom, J. Chem. Phys. 35, 1149 (1961).
5. M. Horowitz and J. G. Daunt, Phys. Rev. 91, 1099 (1953).
6. F. J. Morin and J. P. Maita, Phys. Rev. 129, 1115 (1963).
7. D. C. Rohrer, D. H. Onn, and H. Meyer, Phys. Rev. 138, A1661 (1965).
8. F. Heiniger, E. Bucher, and J. Muller, Phys. kondens. Materie 5, 243 (1966).
9. M. L. Reilly and G. T. Furukawa, U.S. Natl. Bur. Stand. personal communication to L. Brewer, February, 1976.
10. L. Brewer, Materials and Molecular Research Division, Lawrence Berkeley Laboratory, University of California, Berkeley; personal communication of priliminary draft of review to be submitted for publication in Atomic Energy Review, International Atomic Energy Agency, Vienna, Austria.

Molybdenum (Mo) $Mo_1(cr)$

Enthalpy Reference Temperature = T_r = 298.15 K — Standard State Pressure = p° = 0.1 MPa

T/K	$C_p°$ (J K⁻¹mol⁻¹)	S° (J K⁻¹mol⁻¹)	$-[G°-H°(T_r)]/T$ (J K⁻¹mol⁻¹)	$H°-H°(T_r)$ (kJ mol⁻¹)	$\Delta_f H°$ (kJ mol⁻¹)	$\Delta_f G°$ (kJ mol⁻¹)	Log K_f
0	0.	0.	INFINITE	-4.585	0.	0.	0.
100	13.495	7.026	47.961	-4.094	0.	0.	0.
200	21.507	19.478	30.735	-2.251	0.	0.	0.
250	23.074	24.462	28.995	-1.133	0.	0.	0.
298.15	23.933	28.605	28.605	0.	0.	0.	0.
300	23.958	28.753	28.606	0.044	0.	0.	0.
350	24.580	32.494	28.900	1.258	0.	0.	0.
400	25.080	35.810	29.560	2.500	0.	0.	0.
450	25.500	38.789	30.423	3.765	0.	0.	0.
500	25.850	41.494	31.397	5.049	0.	0.	0.
600	26.460	46.263	33.487	7.665	0.	0.	0.
700	26.980	50.382	35.613	10.338	0.	0.	0.
800	27.440	54.015	37.691	13.059	0.	0.	0.
900	27.890	57.272	39.689	15.825	0.	0.	0.
1000	28.370	60.235	41.597	18.638	0.	0.	0.
1100	28.900	62.963	43.417	21.501	0.	0.	0.
1200	29.490	65.502	45.153	24.419	0.	0.	0.
1300	30.140	67.888	46.811	27.400	0.	0.	0.
1400	30.860	70.148	48.398	30.450	0.	0.	0.
1500	31.650	72.304	49.920	33.575	0.	0.	0.
1600	32.500	74.373	51.384	36.782	0.	0.	0.
1700	33.420	76.371	52.796	40.078	0.	0.	0.
1800	34.420	78.309	54.160	43.469	0.	0.	0.
1900	35.490	80.198	55.481	46.964	0.	0.	0.
2000	36.650	82.048	56.763	50.570	0.	0.	0.
2100	37.900	83.866	58.010	54.297	0.	0.	0.
2200	39.240	85.659	59.226	58.152	0.	0.	0.
2300	40.670	87.435	60.414	62.147	0.	0.	0.
2400	42.210	89.198	61.577	66.290	0.	0.	0.
2500	43.890	90.954	62.717	70.594	0.	0.	0.
2600	45.880	92.713	63.837	75.079	0.	0.	0.
2700	48.370	94.490	64.939	79.786	0.	0.	0.
2800	51.570	96.304	66.027	84.777	0.	0.	0.
2896.000	54.835	98.098	67.060	89.885	------ CRYSTAL <--> LIQUID -----		
2900	54.970	98.173	67.103	90.104	-35.913	0.050	-0.001
3000	58.370	100.094	68.170	95.771	-34.012	1.258	-0.022
3100	61.770	102.064	69.232	101.778	-31.771	2.398	-0.040
3200	65.170	104.078	70.289	108.125	-29.189	3.460	-0.056
3300	68.570	106.136	71.344	114.812	-26.268	4.436	-0.070

PREVIOUS: December 1966 CURRENT: March 1978

Molybdenum (Mo) $Mo_1(cr)$

MOLYBDENUM (Mo) LIQUID A_r = 95.94

$S°(298.15\ K)$ = [43.146] J K^{-1} mol^{-1}

T_{fus} = 2896 ± 8 K

$\Delta_f H°(298.15\ K)$ = [41.568] kJ mol^{-1}

$\Delta_{fus}H°$ = 36.0 ± 5.4 kJ mol^{-1}

Enthalpy of Formation

The enthalpy of formation is calculated from that of the crystal by addition of $\Delta_{fus}H°$ and the difference in enthalpy, H°(2896 K) - H°(298.15 K), between the crystal and liquid.

Heat Capacity and Entropy

Enthalpy measurements via levitation calorimetry by Treverton and Margrave (3) were over a sufficient range of temperature, 2692-3112 K, to establish Cp°(ℓ) = 8.19 cal K^{-1} mol^{-1}. The enthalpy measurements of Berezin et al. (2) in the liquid region (14 pts., 2890-2925 K) are insufficient to define a liquid phase heat capacity value. Shaner et al. (5) and Dikhter and Lebedev (6) have reported Cp°(ℓ) values of 16.7 cal K^{-1} mol^{-1} and 14.0 cal K^{-1} mol^{-1}, respectively, using pulse heating techniques.

We adopt Cp°(ℓ) = 9.0 cal K^{-1} mol^{-1} as did Brewer (9). This value implies a heavier weighting of the drop calorimetric studies rather than the pulse heating techniques. The magnitude of the difference of the Cp°(ℓ) values derived from the various studies (3, 5, 6) suggests further study in this area. A glass transition is assumed at 2150 K below which the heat capacity is assumed to be that of the crystal. The entropy is calculated in a manner analogous to the enthalpy of formation.

Fusion Data

The adopted melting point, T_{fus} = 2896 ± 8 K, is derived from the recent experimental study and critical evaluation of earlier data by Kenisarin et al. (1).

Bererzin et al (2) have measured the enthalpy of crystal and liquid molybdenum by levitation calorimetry in the range 1962-2869 K and 2890-2925 K, respectively. They reported $\Delta_{fus}H°$ = 8.741 ± 0.314 kcal mol^{-1} at an assumed melting temperature of 2890 K. Correcting these results to T_{fus} = 2896 K yields 8.693 ± 0.314 kcal mol^{-1}. Shaner et al. (5), using a pulse-heating technique, reported enthalpy measurements (1978-4450 K) which gave $\Delta_{fus}H°$ = 8.553 kcal mol^{-1} at a melting point of 2883 K.

Treverton and Margrave (3) also used levitation calorimetry to measure the liquid phase enthalpy of molybdenum in the range 2693-3115 K. Using the enthalpy value for the liquid at the melting point as determined by Treverton and Margrave (3) and the enthalpy value of the crystal at the melting point as determined by our adopted functions, we calculate $\Delta_{fus}H°$ = 7.264 kcal mol^{-1}. Lebedev et al. (7) reported $\Delta_{fus}H°$ = 9.864 kcal mol^{-1} (exploding wire technique) while Kostekov et al. (8) found $\Delta_{fus}H°$ = 6.095 kcal mol^{-1} (drop technique).

We adopt $\Delta_{fus}H°$ = 8.6 ± 1.3 kcal mol^{-1}, which corresponds to $\Delta_{fus}S°$ = 2.97 cal K^{-1} mol^{-1}. This $\Delta_{fus}S°$ appears high for a metal with a bcc (A2) structure; however, this value is supported by the studies of Berezin et al. (2) and Shaner et al. (5). Brewer (9) selected a value of $\Delta_{fus}H°$ = 8.54 ± 0.3 kcal mol^{-1} (IPTS-48).

Vaporization Data

The boiling point and enthalpy of vaporization are calculated from the adopted functions and enthalpy of sublimation in order to maintain proper themodynamic consistency.

Koch and Anable (4) have measured the vapor pressure of liquid molybdenum by a Langmuir free-evaporation method. They measured 16 data points in the range 2895-2995 K. A 2nd and 3th law analysis gives $\Delta_{vap}H°(298.15\ K)$ values of 136.8 ± 18.4 and 147.7 ± 1.7 kcal mol^{-1}, respectively. This corresponds to our adopted value of $\Delta_{vap}H°(298.15\ K)$ = 147.565 kcal mol^{-1}. Note, however, that there is a significant difference between the 2nd and 3rd law results.

References

1. M. M. Kenisarin, B. Ya. Berezin, and V. Ya. Chekhovskoi, High Temperatures - High Pressures 4, 707 (1972).
2. B. Ya Berezin, V. Ya. Chekhovskoi, and A. E. Sheindlin, High Temperatures - High Pressures 3, 287 (1971).
3. J. A. Treverton and J. L. Margrave, Proc. 5th Symp. Thermophysical Properties, p. 489 (1970).
4. R. K. Koch and W. E. Anable, U.S. Bur. Mines RI 7063, 1968.
5. J. W. Shaner, G. R. Gathers, and C. Minichino, High Temperatures - High Pressures 9, 331 (1977).
6. I. Ya Dikhter and S. V. Lebedev, High Temp. 9, 845 (1971).
7. S. V. Lebedev, A. I. Savvatimskii, and Yu. B. Smirnov, High Temp. (USSR) 9, 578 (1971).
8. V. I. Kostikov, M. A. Maurakh, B. S. Mitin, I. A. Pen'kov, and G. M. Sverdlov, in the collection of papers "Nekotorye Fizicheskie Svoistva Zhidkikh Tugoplavkikh Metallov i Okislov," ed. V. P. Elyutin (Metallorgiya, Moscow), pp 106-113 (1968). [ref 2].
9. L. Brewer, Materials and Molecular Research Division, Lawrence Berkeley Laboratory, University of California, Berkeley; personal communication of preliminary draft of review to be sumbitted for publication in Atomic Energy Review, International Atomic Energy Agency, Vienna, Austria.

Molybdenum (Mo) $Mo_1(l)$

Enthalpy Reference Temperature = T_r = 298.15 K — Standard State Pressure = p° = 0.1 MPa

T/K	$C_p°$ (J $K^{-1}mol^{-1}$)	S° (J $K^{-1}mol^{-1}$)	$-[G°-H°(T_r)]/T$ (J $K^{-1}mol^{-1}$)	$H°-H°(T_r)$ (kJ mol^{-1})	$\Delta_f H°$ (kJ mol^{-1})	$\Delta_f G°$ (kJ mol^{-1})	Log K_f
0							
100							
200							
298.15	23.932	43.146	43.146	0.	41.568	37.233	-6.523
300	23.958	43.294	43.146	0.044	41.568	37.206	-6.478
350	24.580	47.035	43.440	1.258	41.568	36.479	-5.444
400	25.080	50.351	44.101	2.500	41.568	35.752	-4.669
450	25.500	53.330	44.964	3.765	41.568	35.025	-4.066
500	25.850	56.035	45.938	5.049	41.568	34.298	-3.583
600	26.460	60.804	48.028	7.665	41.568	32.844	-2.859
700	26.980	64.922	50.154	10.338	41.568	31.390	-2.342
800	27.440	68.555	52.232	13.059	41.568	29.935	-1.955
900	27.890	71.813	54.230	15.825	41.568	28.481	-1.653
1000	28.370	74.776	56.138	18.638	41.568	27.027	-1.412
1100	28.900	77.504	57.958	21.501	41.568	25.573	-1.214
1200	29.490	80.043	59.694	24.419	41.568	24.119	-1.050
1300	30.140	82.429	61.352	27.400	41.568	22.665	-0.911
1400	30.860	84.689	62.939	30.450	41.568	21.211	-0.791
1500	31.650	86.845	64.461	33.575	41.568	19.757	-0.688
1600	32.500	88.914	65.925	36.782	41.568	18.303	-0.598
1700	33.420	90.912	67.337	40.078	41.568	16.849	-0.518
1800	34.420	92.850	68.701	43.469	41.568	15.395	-0.447
1900	35.490	94.739	70.021	46.964	41.568	13.940	-0.383
2000	36.650	96.589	71.304	50.570	41.568	12.486	-0.326
2100	37.900	98.407	72.551	54.297	41.568	11.032	-0.274
2150.000	38.550	99.306	73.163	56.208	GLASS <--> LIQUID TRANSITION		
2150.000	37.656	99.306	73.163	56.208			
2200	37.656	100.172	73.767	58.091	41.506	9.579	-0.227
2300	37.656	101.846	74.952	61.856	41.277	8.132	-0.185
2400	37.656	103.448	76.106	65.622	40.900	6.699	-0.146
2500	37.656	104.985	77.231	69.387	40.362	5.284	-0.110
2600	37.656	106.462	78.327	73.153	39.642	3.894	-0.078
2700	37.656	107.884	79.395	76.919	38.700	2.537	-0.049
2800	37.656	109.253	80.437	80.684	37.475	1.219	-0.023
2896.000	37.656	110.522	81.414	84.299	CRYSTAL <--> LIQUID		
2900	37.656	110.574	81.454	84.450	0.	0.	0.
3000	37.656	111.851	82.446	88.215	0.	0.	0.
3100	37.656	113.086	83.414	91.981	0.	0.	0.
3200	37.656	114.281	84.360	95.747	0.	0.	0.
3300	37.656	115.440	85.285	99.512	0.	0.	0.
3400	37.656	116.564	86.188	103.278	0.	0.	0.
3500	37.656	117.656	87.072	107.043	0.	0.	0.
3600	37.656	118.716	87.936	110.809	0.	0.	0.
3700	37.656	119.748	88.782	114.575	0.	0.	0.
3800	37.656	120.752	89.610	118.340	0.	0.	0.
3900	37.656	121.731	90.421	122.106	0.	0.	0.
4000	37.656	122.684	91.216	125.871	0.	0.	0.
4100	37.656	123.614	91.995	129.637	0.	0.	0.
4200	37.656	124.521	92.759	133.403	0.	0.	0.
4300	37.656	125.407	93.508	137.168	0.	0.	0.
4400	37.656	126.273	94.243	140.934	0.	0.	0.
4500	37.656	127.119	94.964	144.699	0.	0.	0.
4600	37.656	127.947	95.672	148.465	0.	0.	0.
4700	37.656	128.757	96.367	152.231	0.	0.	0.
4800	37.656	129.549	97.050	155.996	0.	0.	0.
4900	37.656	130.326	97.721	159.762	0.	0.	0.
4951.969	37.656	130.723	98.066	161.719	FUGACITY = 1 bar		
5000	37.656	131.087	98.381	163.527	-591.385	5.731	-0.060
5100	37.656	131.832	99.030	167.293	-593.064	17.690	-0.181
5200	37.656	132.564	99.668	171.059	-594.890	29.683	-0.298
5300	37.656	133.281	100.295	174.824	-596.860	41.713	-0.411
5400	37.656	133.985	100.912	178.590	-598.972	53.781	-0.520
5600	37.656	135.354	102.118	186.121	-603.614	78.040	-0.728
5800	37.656	136.676	103.287	193.652	-608.795	102.474	-0.923
6000	37.656	137.952	104.422	201.183	-614.491	127.096	-1.106

PREVIOUS: December 1966 CURRENT: March 1978

Molybdenum (Mo) $Mo_1(l)$

MOLYBDENUM (Mo) CRYSTAL–LIQUID A_r = 95.94

0 to 2896 K crystal
above 2896 K liquid

Refer to the individual tables for details.

Molybdenum (Mo) $Mo_1(cr,l)$

Enthalpy Reference Temperature = T_r = 298.15 K Standard State Pressure = p° = 0.1 MPa

T/K	C_p° (J K⁻¹mol⁻¹)	S° (J K⁻¹mol⁻¹)	$-[G^\circ-H^\circ(T_r)]/T$ (J K⁻¹mol⁻¹)	$H^\circ-H^\circ(T_r)$ (kJ mol⁻¹)	Δ_fH° (kJ mol⁻¹)	Δ_fG° (kJ mol⁻¹)	Log K_f
0	0.	0.	INFINITE	-4.585	0.	0.	0.
100	13.495	7.026	47.961	-4.094	0.	0.	0.
200	21.507	19.478	30.735	-2.251	0.	0.	0.
250	23.074	24.462	28.995	-1.133	0.	0.	0.
298.15	23.933	28.605	28.605	0.	0.	0.	0.
300	23.958	28.753	28.606	0.044	0.	0.	0.
350	24.580	32.494	28.900	1.258	0.	0.	0.
400	25.080	35.810	29.560	2.500	0.	0.	0.
450	25.500	38.789	30.423	3.765	0.	0.	0.
500	25.850	41.494	31.397	5.049	0.	0.	0.
600	26.460	46.263	33.487	7.665	0.	0.	0.
700	26.980	50.382	35.613	10.338	0.	0.	0.
800	27.440	54.015	37.691	13.059	0.	0.	0.
900	27.890	57.272	39.689	15.825	0.	0.	0.
1000	28.370	60.235	41.597	18.638	0.	0.	0.
1100	28.900	62.963	43.417	21.501	0.	0.	0.
1200	29.490	65.502	45.153	24.419	0.	0.	0.
1300	30.140	67.888	46.811	27.400	0.	0.	0.
1400	30.860	70.148	48.398	30.450	0.	0.	0.
1500	31.650	72.304	49.920	33.575	0.	0.	0.
1600	32.500	74.373	51.384	36.782	0.	0.	0.
1700	33.420	76.371	52.796	40.078	0.	0.	0.
1800	34.420	78.309	54.160	43.469	0.	0.	0.
1900	35.490	80.198	55.481	46.964	0.	0.	0.
2000	36.650	82.048	56.763	50.570	0.	0.	0.
2100	37.900	83.866	58.010	54.297	0.	0.	0.
2200	39.240	85.659	59.226	58.152	0.	0.	0.
2300	40.670	87.435	60.414	62.147	0.	0.	0.
2400	42.210	89.198	61.577	66.290	0.	0.	0.
2500	43.890	90.954	62.717	70.594	0.	0.	0.
2600	45.880	92.713	63.837	75.079	0.	0.	0.
2700	48.370	94.490	64.939	79.786	0.	0.	0.
2800	51.570	96.304	66.027	84.777	0.	0.	0.
2896.000	54.835	98.098	67.060	89.885	CRYSTAL <--> LIQUID	TRANSITION	
2896.000	37.656	110.522	67.060	125.867			
2900	37.656	110.574	67.120	126.018	0.	0.	0.
3000	37.656	111.851	68.590	129.784	0.	0.	0.
3100	37.656	113.086	70.005	133.549	0.	0.	0.
3200	37.656	114.281	71.370	137.315	0.	0.	0.
3300	37.656	115.440	72.688	141.080	0.	0.	0.
3400	37.656	116.564	73.962	144.846	0.	0.	0.
3500	37.656	117.656	75.195	148.612	0.	0.	0.
3600	37.656	118.716	76.389	152.377	0.	0.	0.
3700	37.656	119.748	77.547	156.143	0.	0.	0.
3800	37.656	120.752	78.671	159.908	0.	0.	0.
3900	37.656	121.731	79.763	163.674	0.	0.	0.
4000	37.656	122.684	80.824	167.440	0.	0.	0.
4100	37.656	123.614	81.856	171.205	0.	0.	0.
4200	37.656	124.521	82.861	174.971	0.	0.	0.
4300	37.656	125.407	83.841	178.736	0.	0.	0.
4400	37.656	126.273	84.795	182.502	0.	0.	0.
4500	37.656	127.119	85.726	186.268	0.	0.	0.
4600	37.656	127.947	86.635	190.033	0.	0.	0.
4700	37.656	128.757	87.523	193.799	0.	0.	0.
4800	37.656	129.549	88.390	197.564	0.	0.	0.
4900	37.656	130.326	89.238	201.330	0.	0.	0.
4951.969	37.656	130.723	89.671	203.287	------- FUGACITY = 1 bar -------		
5000	37.656	131.087	90.068	205.096	-591.385	5.731	-0.060
5100	37.656	131.832	90.879	208.861	-593.064	17.690	-0.181
5200	37.656	132.564	91.674	212.627	-594.890	29.683	-0.298
5300	37.656	133.281	92.452	216.392	-596.860	41.713	-0.411
5400	37.656	133.985	93.215	220.158	-598.972	53.781	-0.520
5500	37.656	134.676	93.962	223.924	-601.224	65.889	-0.626
5600	37.656	135.354	94.695	227.689	-603.614	78.040	-0.728
5800	37.656	136.676	96.120	235.220	-608.795	102.475	-0.923
6000	37.656	137.952	97.494	242.752	-614.491	127.096	-1.106

PREVIOUS: CURRENT: March 1978

Molybdenum (Mo) $Mo_1(cr,l)$

MOLYBDENUM (Mo) IDEAL GAS A_r = 95.94

IP(Mo, g) = 57260 ± 50 cm^{-1}
S°(298.15 K) = 181.951 ± 0.04 J K^{-1} mol^{-1}

$\Delta_fH°$(0 K) = 657.37 ± 3.8 kJ mol^{-1}
$\Delta_fH°$(298.15 K) = 658.98 ± 3.8 kJ mol^{-1}

Electronic Levels and Quantum Weights

State	ϵ_i, cm^{-1}	g_i
$^7S_3$	0	7
$^5S_2$	10768.33	5
$^5D_0$	10965.97	1
.	.	.
.	.	.
.	.	.
.	.	.
IP	57260	

Enthalpy of Formation

The enthalpy of formation at 298.15 K for Mo(g) is simply the enthalpy of sublimation at 298.15 K; i.e. Mo(cr) → Mo(g). The following table summarizes the analysis of the numerous sublimation studies by a 2nd and 3rd law calculation. All temperatures have been corrected to IPTS-68. All studies are based on the Langmuir effusion method.

Source	Data Points	T/K	$\Delta_{sub}H°$(298.15 K), kcal mol^{-1} 2nd Law	3rd Law	Drift cal K^{-1} mol^{-1}
1	eqn	1201-2396	159.8	158.5	-0.7
2	9	2068-2500	162.2±3.8	156.2±1.4	-2.6±1.6
3	12	2144-2537	158.4±2.6	157.8±0.8	-0.3±1.1
4	9	2154-2466	146.8±1.4	158.4±1.2	5.1±0.6
5	eqn	1601-2297	131.7	148.8	8.8
6	12	2089-2493	156.8±1.6	156.9±0.6	0.1±0.7
7	10	2293-2673	152.4±4.1	158.2±1.4	2.3±1.6

A mass spectrometric study of the enthalpy of sublimation of molybdenum (8) yielded , $\Delta_{sub}H°$(2500 K) = 144.1±2.2 kcal mol^{-1}; correcting to 298.15 K, we calculate $\Delta_{sub}H°$(298.15 K) = 149.8 kcal mol^{-1}.

We adopt $\Delta_{sub}H°$(298.15 K) = 157.5±0.9 kcal mol^{-1} from a consideration of five studies (1, 2, 3, 6, 7). Brewer (12) has adopted a value 0.2 kcal mol^{-1} less positive; however his vapor pressure analysis is based on the 1948 temperature scale.

Heat Capacity and Entropy

The electronic energy levels are given in the compilations by Moore (9, 10). Although we have only listed the ground state, first excited state, highest observed excited state, and the ionization limit for Mo(g), all levels listed by Moore (9, 10), as well as estimated levels, are used in our calculations. The observed levels are too numerous to list completely. An examination of the estimates for missing levels suggests that they all lie above 20000 cm^{-1}. Our calculations indicate that for Mo(g) the thermochemical functions are independent of the estimated missing levels (for n = 5, 6, 7), the cut-off procedure, and the inclusion of n = 6, 7 levels up to 3000 K. The Gibbs energy function is essentially unaffected up to 4000 K. The reported uncertainty in S°(298.15 K) is due to uncertainties in the relative atomic gas and the fundamental constants. Extension of these calculations above 6000 K may require consideration of the higher excited states and utilization of proper fill and cut-off procedures (11). In fact, the inclusion of some higher states (n = 5, 6, 7) and consideration of various cut-off procedures leads to calculational differences in the Gibbs energy function of roughly 0.1 cal K^{-1} mol^{-1} or greater at temperatures in excess of 7600 K. The thermal functions adopted here are the same as those of Brewer (12).

References

1. C. Zwikker, Physica 7, 71 (1927).
2. H. Z. Jones, I. Langmuir, and G. M. J. Mackay, Phys. Rev. 30, 201 (1927).
3. P. A. Vozella, A. D. Miller, and M. A. DeCrescente, J. Chem. Phys. 41, 589 (1964).
4. J. W. Edwards, H. L. Johnston, and P. E. Blackburn, J. Am. Chem. Soc. 74, 1539 (1952).
5. L. Norris and A. G. Worthing, Phys. Rev. 44, 323 (1933).
6. R. J. Fries, AEC Accession No. 13029, Rept. No. LA-3423, January, 1966.
7. I. Ya. Dekhtyar, V. N. Kolesnik, V. I. Patoka, and J. I. Silant'ev, High Temperatures 12, 870 (1974).
8. T. P. J. H. Babeliowsky, Physica 28, 1160 (1962).
9. C. E. Moore, U. S. Nat. Bur. Stand., NSRDS-NBS 35, Volume III, 1970 [Reprint of NBS Circular 497, Volume III, 1958].
10. C. E. Moore, U. S. Nat. Bur. Stand,, NSRDS-NBS 34, 1970.
11. J. R. Downey, The Dow Chemical Company, Report AFOSR-TR-78-0960, Contract No. F44620-75-1-0048, 1978.
12. L. Brewer, Materials and Molecular Research Division. Lawrence Berkeley Laboratory, University of California, Berkeley; personal communication of preliminary draft of review to be submitted for publication in Atomic Energy Review, International Atomic Energy Agency, Vienna, Austria.

Molybdenum (Mo) $Mo_1(g)$

Enthalpy Reference Temperature = T_r = 298.15 K Standard State Pressure = p° = 0.1 MPa

T/K	$C_p°$ (J $K^{-1}mol^{-1}$)	S°	$-[G°-H°(T_r)]/T$	$H°-H°(T_r)$ (kJ mol^{-1})	$\Delta_fH°$	$\Delta_fG°$	Log K_f
0	0.	0.	INFINITE	-6.197	657.367	657.367	INFINITE
100	20.786	159.243	200.431	-4.119	658.955	643.733	-336.252
200	20.786	173.651	183.852	-2.040	659.191	628.357	-164.110
250	20.786	178.290	182.293	-1.001	659.112	620.655	-129.679
298.15	20.786	181.951	181.951	0.	658.980	613.260	-107.441
300	20.786	182.079	181.951	0.038	658.974	612.976	-106.729
350	20.786	185.283	182.204	1.078	658.800	605.323	-90.340
400	20.786	188.059	182.766	2.117	658.597	597.697	-78.051
450	20.786	190.507	183.493	3.156	658.372	590.098	-68.497
500	20.786	192.697	184.306	4.196	658.127	582.525	-60.856
600	20.786	196.487	186.030	6.274	657.589	567.455	-49.401
700	20.786	199.691	187.759	8.353	656.995	552.478	-41.226
800	20.786	202.467	189.427	10.431	656.352	537.591	-35.101
900	20.786	204.915	191.015	12.510	655.665	522.786	-30.342
1000	20.787	207.105	192.516	14.589	654.931	508.061	-26.538
1100	20.788	209.086	193.934	16.667	654.147	493.411	-23.430
1200	20.793	210.895	195.273	18.746	653.307	478.836	-20.843
1300	20.803	212.560	196.540	20.826	652.406	464.333	-18.657
1400	20.823	214.102	197.740	22.907	651.437	449.901	-16.786
1500	20.856	215.540	198.879	24.991	650.396	435.542	-15.167
1600	20.910	216.888	199.963	27.079	649.277	421.254	-13.753
1700	20.989	218.158	200.996	29.174	648.076	407.039	-12.507
1800	21.100	219.360	201.983	31.278	646.789	392.897	-11.402
1900	21.250	220.505	202.928	33.395	645.412	378.829	-10.415
2000	21.443	221.600	203.835	35.530	643.940	364.836	-9.529
2100	21.687	222.652	204.706	37.686	642.369	350.919	-8.729
2200	21.985	223.667	205.545	39.869	640.697	337.079	-8.003
2300	22.342	224.652	206.354	42.085	638.918	323.318	-7.343
2400	22.761	225.611	207.137	44.339	637.029	309.636	-6.739
2500	23.248	226.550	207.895	46.639	635.026	296.036	-6.185
2600	23.802	227.473	208.630	48.991	632.892	282.518	-5.676
2700	24.428	228.382	209.345	51.402	630.596	269.085	-5.206
2800	25.127	229.283	210.041	53.879	628.082	255.741	-4.771
2900	25.898	230.178	210.720	56.430	589.392	242.541	-4.369
3000	26.742	231.070	211.383	59.061	588.258	230.600	-4.015
3100	27.659	231.962	212.033	61.781	587.212	218.696	-3.685
3200	28.647	232.855	212.669	64.595	586.261	206.823	-3.376
3300	29.704	233.753	213.295	67.512	585.412	194.979	-3.086
3400	30.827	234.656	213.910	70.538	584.672	183.160	-2.814
3500	32.012	235.567	214.515	73.680	584.048	171.360	-2.557
3600	33.256	236.486	215.113	76.943	583.546	159.576	-2.315
3700	34.554	237.415	215.703	80.333	583.170	147.805	-2.087
3800	35.900	238.354	216.287	83.855	582.927	136.042	-1.870
3900	37.288	239.304	216.865	87.514	582.820	124.283	-1.665
4000	38.713	240.266	217.438	91.314	582.855	112.526	-1.469
4100	40.169	241.240	218.006	95.258	583.033	100.765	-1.284
4200	41.649	242.226	218.571	99.349	583.358	88.999	-1.107
4300	43.147	243.223	219.133	103.588	583.832	77.223	-0.938
4400	44.658	244.232	219.692	107.978	584.456	65.435	-0.777
4500	46.176	245.253	220.248	112.520	585.233	53.630	-0.623
4600	47.696	246.284	220.803	117.214	586.161	41.807	-0.475
4700	49.214	247.327	221.356	122.059	587.240	29.962	-0.333
4800	50.725	248.378	221.908	127.056	588.472	18.092	-0.197
4900	52.226	249.440	222.459	132.204	589.854	6.195	-0.066
4951.969	53.001	249.995	222.746	134.938	------- FUGACITY = 1 bar -------		
5000	53.713	250.510	223.010	137.501	0.	0.	0.
5100	55.185	251.588	223.560	142.946	0.	0.	0.
5200	56.640	252.674	224.109	148.537	0.	0.	0.
5300	58.072	253.766	224.658	154.272	0.	0.	0.
5400	59.485	254.865	225.207	160.150	0.	0.	0.
5500	60.874	255.969	225.757	166.168	0.	0.	0.
5600	62.239	257.078	226.306	172.323	0.	0.	0.
5700	63.577	258.191	226.856	178.613	0.	0.	0.
5800	64.887	259.308	227.406	185.035	0.	0.	0.
5900	66.164	260.428	227.956	191.585	0.	0.	0.
6000	67.413	261.550	228.506	198.263	0.	0.	0.

PREVIOUS: March 1978 (1 atm) CURRENT: March 1978 (1 bar)

Molybdenum (Mo) $Mo_1(g)$

MOLYBDENUM, ION (Mo^+) IDEAL GAS M_r = 95.93945

$IP(Mo^+, g) = 130300 \pm 500\ cm^{-1}$

$S°(298.15\ K) = 180.669 \pm 0.05\ J\ K^{-1}\ mol^{-1}$

$\Delta_f H°(0\ K) = 1342.35 \pm 4.0\ kJ\ mol^{-1}$

$\Delta_f H°(298.15\ K) = [1350.159]\ kJ\ mol^{-1}$

Electronic Levels and Quantum Weights

State	ϵ_i, cm^{-1}	g_i
$^6S_{5/2}$	0.00	6
$^6D_{1/2}$	11783.36	2
$^6D_{3/2}$	12034.06	4
$^6D_{5/2}$	12417.28	6
$^6D_{7/2}$	12900.33	8
$^6D_{9/2}$	13460.70	10

Enthalpy of Formation

$\Delta_f H°(Mo^+, g, 0\ K)$ is calculated from $\Delta_f H°(Mo, g, 0\ K)$ (1) using the spectroscopic value of IP(Mo) = 57260 ± 50 cm^{-1} (684.982 ± 0.60 kJ mol^{-1}) from Moore (2). The ionization limit is converted from cm^{-1} to kJ mol^{-1} using the factor, 1 cm^{-1} = 0.01196266 kJ cm^{-1}, which is derived from the 1973 CODATA fundamental constants (3). Rosenstock et al. (4) and Levin and Lias (5) have summarized additional ionization and appearance potential data.

$\Delta_f H°(Mo^+, g, 298.15\ K)$ is calculated from $\Delta_f H°(Mo, g, 0\ K)$ by using IP(Mo) with JANAF (1) enthalpies, H°(0 K) − H°(298.15 K), for Mo(g), $Mo^+(g)$, and $e^-(g)$. $\Delta_f H°(Mo \rightarrow Mo^+ + e^-, 298.15\ K)$ differs from a room temperature threshold energy due to inclusion of these enthalpies and to threshold effects discussed by Rosenstock et al. (4). $\Delta_f H°(298.15\ K)$ should be changed by −6.197 kJ mol^{-1} if it is to be used in the ion convention that excludes the enthalpy of the electron.

Heat Capacity and Entropy

The information on electronic energy levels and quantum weights, given by Moore (6), is incomplete because many theoretically predicted levels have not been observed. Although we have listed only the ground, the first excited state, the highest observed excited state, and the ionization potential for $Mo^+(g)$, all levels listed by Moore, as well as estimated levels, are used in the calculation. The observed levels are too numerous to list completely. The calculations indicate that for $Mo^+(g)$, the thermodynamic functions are independent of the estimated missing levels (for n=4, 5), the cut-off procedure, and the inclusion of n=5 levels up to 4000 K; the Gibbs energy function showing variations of 0.4 J K^{-1} mol^{-1} at this temperature. The reported uncertainty in S°(298.15 K) is due to uncertainties in the relative ionic mass, and the fundamental constants. Extension of these calculations above 6000 K may require consideration of the higher excited states (n>5), and use of different fill and cut-off procedures (7).

References

1. JANAF Thermochemical Tables: Mo(g), 3-31-78; e^-(g), 3-31-82.
2. C. E. Moore, U. S. Nat. Bur. Stand., NSRDS-NBS-34, 1970.
3. E. R. Cohen and B. N. Taylor, J. Phys. Chem. Ref. Data 2, 663 (1973).
4. H. M. Rosenstock, K. Draxl et al., J. Phys. Chem. Ref. Data 6, Supp. 1 (1977).
5. R. D. Levin and S. G. Lias, U. S. Natl. Bur. Stand. NSRDS-NBS-71, 1982.
6. C. E. Moore, U.S. Nat. Bur. Stand., NSRDS-NBS-35, Volume III, 1970 [Reprint of NBS Circular 467, Volume III, 1958].
7. J. R. Downey, Jr., The Dow Chemical Company, AFOSR-TR-78-0960, Contract No. F44620-75-1-0048, 1978.

Molybdenum, Ion (Mo^+) $Mo_1^+(g)$

Enthalpy Reference Temperature = T_r = 298.15 K Standard State Pressure = p° = 0.1 MPa

T/K	$C_p°$ (J K^{-1}mol^{-1})	S° (J K^{-1}mol^{-1})	−[G°−H°(T_r)]/T (J K^{-1}mol^{-1})	H°−H°(T_r) (kJ mol^{-1})	$\Delta_f H°$ (kJ mol^{-1})	$\Delta_f G°$ (kJ mol^{-1})	Log K_f
0	0.	0.	INFINITE	-6.197	1342.349		
100	20.786	157.962	199.149	-4.119			
200	20.786	172.369	182.570	-2.040			
250	20.786	177.008	181.011	-1.001			
298.15	20.786	180.669	180.669	0.	1350.159	1298.567	-227.503
300	20.786	180.798	180.669	0.038	1350.192	1298.246	-226.045
350	20.786	184.002	180.922	1.078	1351.057	1289.520	-192.450
400	20.786	186.777	181.485	2.117	1351.893	1280.672	-167.239
450	20.786	189.226	182.211	3.156	1352.707	1271.720	-147.617
500	20.786	191.416	183.024	4.196	1353.502	1262.679	-131.911
600	20.786	195.205	184.748	6.274	1355.043	1244.368	-108.332
700	20.786	198.409	186.477	8.353	1356.527	1225.804	-91.471
800	20.786	201.185	188.146	10.431	1357.963	1207.031	-78.811
900	20.786	203.633	189.733	12.510	1359.354	1188.081	-68.954
1000	20.786	205.823	191.235	14.589	1360.699	1168.978	-61.061
1100	20.787	207.804	192.652	16.667	1361.993	1149.742	-54.597
1200	20.789	209.613	193.992	18.746	1363.232	1130.392	-49.205
1300	20.794	211.277	195.258	20.825	1364.409	1110.940	-44.638
1400	20.804	212.819	196.458	22.905	1365.517	1091.401	-40.721
1500	20.824	214.255	197.597	24.986	1366.552	1071.784	-37.323
1600	20.860	215.600	198.681	27.070	1367.507	1052.102	-34.348
1700	20.917	216.866	199.714	29.159	1368.379	1032.362	-31.721
1800	21.004	218.064	200.700	31.255	1369.162	1012.573	-29.384
1900	21.128	219.202	201.644	33.361	1369.853	992.743	-27.292
2000	21.300	220.290	202.549	35.482	1370.446	972.879	-25.409
2100	21.526	221.335	203.419	37.623	1370.939	952.988	-23.704
2200	21.816	222.343	204.257	39.789	1371.328	933.076	-22.154
2300	22.176	223.320	205.064	41.988	1371.611	913.149	-20.738
2400	22.613	224.273	205.845	44.227	1371.785	893.212	-19.440
2500	23.131	225.206	206.601	46.514	1371.847	873.270	-18.246
2600	23.734	226.125	207.334	48.856	1371.783	853.327	-17.144
2700	24.423	227.033	208.047	51.263	1371.561	833.390	-16.123
2800	25.198	227.935	208.741	53.744	1371.129	813.465	-15.175
2900	26.060	228.834	209.418	56.306	1334.529	793.610	-14.294
3000	27.004	229.733	210.081	58.958	1335.495	774.941	-13.493
3100	28.028	230.635	210.729	61.709	1336.559	756.239	-12.743
3200	29.127	231.542	211.365	64.566	1337.729	737.500	-12.038
3300	30.296	232.456	211.991	67.537	1339.013	718.724	-11.376
3400	31.528	233.379	212.606	70.628	1340.416	699.906	-10.753
3500	32.817	234.311	213.213	73.844	1341.946	681.046	-10.164
3600	34.156	235.254	213.812	77.193	1343.607	662.140	-9.607
3700	35.541	236.209	214.404	80.677	1345.405	643.186	-9.080
3800	36.963	237.176	214.991	84.302	1347.343	624.181	-8.580
3900	38.419	238.155	215.572	88.071	1349.425	605.124	-8.105
4000	39.904	239.146	216.149	91.987	1351.653	586.011	-7.653
4100	41.412	240.150	216.722	96.053	1354.032	566.841	-7.222
4200	42.942	241.166	217.292	100.270	1356.563	547.611	-6.811
4300	44.491	242.195	217.859	104.642	1359.247	528.319	-6.418
4400	46.057	243.235	218.424	109.169	1362.087	508.962	-6.042
4500	47.638	244.288	218.987	113.853	1365.085	489.540	-5.682
4600	49.235	245.352	219.549	118.697	1368.242	470.049	-5.338
4700	50.847	246.429	220.109	123.701	1371.559	450.487	-5.007
4800	52.473	247.516	220.669	128.867	1375.037	430.853	-4.689
4900	54.112	248.615	221.228	134.196	1378.680	411.146	-4.383
5000	55.765	249.725	221.787	139.690	791.101	397.093	-4.148
5100	57.429	250.845	222.346	145.349	793.394	389.190	-3.986
5200	59.102	251.977	222.904	151.176	795.708	381.242	-3.830
5300	60.781	253.118	223.464	157.170	798.046	373.249	-3.679
5400	62.463	254.270	224.024	163.332	800.409	365.212	-3.533
5500	64.142	255.432	224.584	169.662	802.800	357.131	-3.392
5600	65.814	256.603	225.145	176.160	805.221	349.006	-3.255
5700	67.470	257.782	225.708	182.825	807.674	340.837	-3.123
5800	69.103	258.970	226.271	189.654	810.159	332.625	-2.996
5900	70.706	260.165	226.835	196.644	812.679	324.370	-2.872
6000	72.268	261.366	227.401	203.793	815.229	316.072	-2.752

PREVIOUS: March 1978 (1 atm) CURRENT: March 1984 (1 bar)

Molybdenum, Ion (Mo^+) $Mo_1^+(g)$

MOLYBDENUM, ION (Mo^-) IDEAL GAS M_r = 95.94055

EA(Mo, g) = 0.746 ± 0.010 eV

S°(298.15 K) = 180.669 ± 0.003 J K^{-1} mol^{-1}

$\Delta_f H°$(0 K) = 585.390 ± 1.5 kJ mol^{-1}

$\Delta_f H°$(298.15 K) = [580.805] kJ mol^{-1}

Electonic Level and Quantum Weight

State	ε_i, cm^{-1}	g_i
$^6S_{5/2}$	0.0	6

Enthalpy of Formation

$\Delta_f H°(Mo^-$, g, 0 K) is calculated from $\Delta_f H°$(Mo, g 0 K) (1) using the adopted electron affinity of EA(Mo) = 0.746 ± 0.010 eV (71.978 ± 0.965 kJ mol^{-1}). This value, recommended by Hotop and Lineberger (2), is based on a laser photodetachment electron spectrometry study (3). Additional information on Mo^-(g) may be obtained in the critical discussions of Hotop and Lineberger (2, 4), Rosenstock et al. (5) and Massey (6).

$\Delta_f H°(Mo^-$, g, 298.15 K) is obtained from $\Delta_f H°$(Mo, g, 0 K) by using EA(Mo) with JANAF (1) enthalpies, H°(0 K) - H°(298.15 K), for Mo^-(g), Mo(g), and e^-(g). $\Delta_r H°(Mo^- \rightarrow Mo + e^-$, 298.15 K) differs from a room-temperature threshold energy due to inclusion of these enthalpies and to threshold effects discussed by Rosenstock et al. (5). $\Delta_f H°$(298.15 K) should be changed by +6.197 kJ mol^{-1} if it is to be used in the ion convention that excludes the enthalpy of the electron.

Heat Capacity and Entropy

The ground state electronic configuration for Mo^-(g) is given by Hotop and Lineberger (2, 4) and Rosenstock et al. (5). Lacking any experimental evidence as to the stability of any excited states, we assume that no stable excited states exist.

References

1. JANAF Thermochemical Tables: Mo(g), 3-31-78; e^-(g), 3-31-82.
2. H. Hotop and W. kC. Lineberger, J. Phys. Chem. Ref. kData, to be published, 1985.
3. C. S. Feigerle, R. R. Cordermann, S. V. Bobashev and W. C. Lineberger, J. Chem. Phys. 74, 1580 (1981).
4. H. Hotop and W. C. Lineberger, J. Phys. Chem. Ref. Data 4, 539 (1975).
5. H. M. Rosenstock, K. Draxl et al., J. Phys. Chem. Ref. Data 6, Supp.1 (1977).
6. H. S. W. Massey, "Negative Ions", 3rd ed., Cambridge University Press, Cambridge, 1976.

Molybdenum, Ion (Mo^-) Mo_1^-(g)

Enthalpy Reference Temperature = T_r = 298.15 K Standard State Pressure = p° = 0.1 MPa

T/K	$C_p°$ (J $K^{-1}mol^{-1}$)	S° (J $K^{-1}mol^{-1}$)	$-[G°-H°(T_r)]/T$ (J $K^{-1}mol^{-1}$)	$H°-H°(T_r)$ (kJ mol^{-1})	$\Delta_f H°$ (kJ mol^{-1})	$\Delta_f G°$ (kJ mol^{-1})	Log K_f
0	0.	0.	INFINITE	-6.197	585.390		
100	20.786	157.962	199.149	-4.119			
200	20.786	172.370	182.570	-2.040			
250	20.786	177.008	181.011	-1.001			
298.15	20.786	180.669	180.669	0.	580.805	541.722	-94.907
300	20.786	180.798	180.669	0.038	580.761	541.480	-94.280
350	20.786	184.002	180.923	1.078	579.547	535.028	-79.849
400	20.786	186.777	181.485	2.117	578.305	528.753	-69.048
450	20.786	189.226	182.212	3.156	577.040	522.635	-60.666
500	20.786	191.416	183.024	4.196	575.757	516.658	-53.975
600	20.786	195.205	184.748	6.274	573.140	505.083	-43.971
700	20.786	198.410	186.477	8.353	570.467	493.951	-36.859
800	20.786	201.185	188.146	10.431	567.746	483.205	-31.550
900	20.786	203.633	189.733	12.510	564.980	472.804	-27.441
1000	20.786	205.823	191.235	14.589	562.167	462.712	-24.170
1100	20.786	207.805	192.653	16.667	559.304	452.905	-21.507
1200	20.786	209.613	193.992	18.746	556.386	443.360	-19.299
1300	20.786	211.277	195.258	20.824	553.405	434.062	-17.441
1400	20.786	212.817	196.458	22.903	550.355	424.996	-15.857
1500	20.786	214.251	197.597	24.982	547.230	416.150	-14.492
1600	20.786	215.593	198.680	27.060	544.023	407.516	-13.304
1700	20.786	216.853	199.713	29.139	540.727	399.084	-12.262
1800	20.786	218.041	200.698	31.217	537.336	390.850	-11.342
1900	20.786	219.165	201.641	33.296	533.842	382.806	-10.524
2000	20.786	220.231	202.544	35.375	530.235	374.950	-9.793
2100	20.786	221.245	203.410	37.453	526.509	367.277	-9.136
2200	20.786	222.212	204.243	39.532	522.653	359.784	-8.542
2300	20.786	223.136	205.045	41.610	518.658	352.470	-8.005
2400	20.786	224.021	205.817	43.689	514.515	345.333	-7.516
2500	20.786	224.869	206.562	45.768	510.212	338.371	-7.070
2600	20.786	225.685	207.282	47.846	505.726	331.585	-6.662
2700	20.786	226.469	207.979	49.925	501.019	324.976	-6.287
2800	20.786	227.225	208.652	52.004	496.028	318.547	-5.943
2900	20.786	227.955	209.306	54.082	454.787	312.351	-5.626
3000	20.786	228.659	209.939	56.161	451.022	307.503	-5.354
3100	20.786	229.341	210.554	58.239	447.256	302.781	-5.102
3200	20.786	230.001	211.151	60.318	443.490	298.181	-4.867
3300	20.786	230.640	211.732	62.397	439.725	293.698	-4.649
3400	20.786	231.261	212.298	64.475	435.959	289.330	-4.445
3500	20.786	231.863	212.848	66.554	432.194	285.072	-4.254
3600	20.786	232.449	213.384	68.632	428.428	280.922	-4.076
3700	20.786	233.018	213.907	70.711	424.662	276.876	-3.909
3800	20.786	233.573	214.418	72.790	420.897	272.932	-3.752
3900	20.786	234.113	214.916	74.868	417.131	269.088	-3.604
4000	20.786	234.639	215.402	76.947	413.366	265.340	-3.465
4100	20.786	235.152	215.878	79.025	409.600	261.686	-3.334
4200	20.786	235.653	216.343	81.104	405.834	258.123	-3.210
4300	20.786	236.142	216.797	83.183	402.069	254.651	-3.093
4400	20.786	236.620	217.243	85.261	398.303	251.266	-2.983
4500	20.786	237.087	217.678	87.340	394.538	247.967	-2.878
4600	20.786	237.544	218.105	89.418	390.772	244.751	-2.779
4700	20.786	237.991	218.524	91.497	387.006	241.617	-2.685
4800	20.786	238.429	218.934	93.576	383.241	238.564	-2.596
4900	20.786	238.857	219.336	95.654	379.475	235.589	-2.511
5000	20.786	239.277	219.731	97.733	-215.675	238.422	-2.491
5100	20.786	239.689	220.118	99.811	-221.120	247.557	-2.536
5200	20.786	240.093	220.498	101.890	-226.712	256.802	-2.580
5300	20.786	240.488	220.872	103.969	-232.447	266.154	-2.623
5400	20.786	240.877	221.239	106.047	-238.325	275.617	-2.666
5500	20.786	241.258	221.599	108.126	-244.342	285.189	-2.709
5600	20.786	241.633	221.954	110.204	-250.498	294.873	-2.750
5700	20.786	242.001	222.302	112.283	-256.788	304.667	-2.792
5800	20.786	242.362	222.645	114.362	-263.210	314.573	-2.833
5900	20.786	242.718	222.982	116.440	-269.760	324.590	-2.874
6000	20.786	243.067	223.314	118.519	-276.437	334.720	-2.914

PREVIOUS: March 1978 (1 atm) CURRENT: March 1984 (1 bar)

Molybdenum, Ion (Mo^-) Mo_1^-(g)

MOLYBDENUM OXIDE (MoO) IDEAL GAS M_r = 111.9394

$S°(298.15\ K) = 241.764 \pm 8.4\ J\ K^{-1}\ mol^{-1}$

$\Delta_f H°(0\ K) = 310.25 \pm 33.5\ kJ\ mol^{-1}$
$\Delta_f H°(298.15\ K) = 310.96 \pm 33.5\ kJ\ mol^{-1}$

Electronic Levels and Quantum Weights

ϵ_i, cm^{-1}	g_i
[0]	[3]
[450]	[5]

$\omega_e = [1000 \pm 20]\ cm^{-1}$ $\omega_e x_e = [3.99]\ cm^{-1}$ $\sigma = 1$
$B_e = [0.40604]\ cm^{-1}$ $\alpha_e = [0.00309]\ cm^{-1}$ $r_e = [1.74]$ Å

Enthalpy of Formation

The enthalpy of formation of MoO(g) is taken from the critical analysis of the Mo-O system by Brewer (1). The adopted value is $\Delta_f H°(298.15\ K) = 74.32 \pm 8\ kcal\ mol^{-1}$. All the current JANAF Thermochemical Tables for molybdenum-oxygen species are based on the internally consistent values of Brewer (1).

Choudary et al. (2) have reviewed earlier data and evaluations of $\Delta_f H°$ for MoO(g). They have also presented new data between 1462 and 1908 K. For the reaction 2 $MoO_2(g)$ = MoO(g) + $MoO_3(g)$, the 3rd law values are mostly in the range -2.4 to -4.6 kcal mol^{-1} compared to 2nd law values of -7.8 kcal mol^{-1} or more negative. $\Delta_r H°/R = -2200 \pm 2000$ K was selected by Brewer (1) for the reaction as written.

Heat Capacity and Entropy

The ground state configuration is assumed to be similar to that for gaseous WO_3 (3). Swaminathan and Krisnamurthy (4) observed ten band systems which they attributed to molybdenum oxide with lower state values of $\omega_e = 950 \pm 10\ cm^{-1}$. Iorns and Stafford (5) also observed a band with a peak at 969 cm^{-1} in the vapor above MoO_3(cr). It is surprising that the observed vibrational frequency of MoO(g) is lower than 1055 cm^{-1} observed for WO(g). The latter comes not only from electronic spectra of the gas but also from infrared observation in matrix isolation; the former comes only from tentative assignments of electronic spectra presumably arising from MoO(g). It is possible that the observed states could be low lying excited states. We tentatively assume the fundamental vibrational frequency to be 1000±20 cm^{-1}. The anharmonicity correction x_e is estimated from the relation $x_e \mu^{1/2} = 0.014$ given by Barrow and Caunt (6). The constant (0.014) is assumed to be the same as that for WO(g). The value of α_e is calculated from the Morse potential function. Schroder (7) gave a relationship between bond order and Mo-O internuclear distance which yields values of 1.74 Å for a bond order of two and 1.62 Å for a bond order of three. A value 1.74±0.05 Å was used.

The suggestion of Hewett et al., (8) that the electronic structure of MoO should be closer to that of CrO than WO, was followed in using the same treatment as for MoF_2 (1) where g=3 was taken for the ground state and g=5 for a low-lying excited state at 450 cm^{-1}.

References

1. L. Brewer, Materials and Research Division, Lawrence Berkeley Laboratory, University of California, Berkeley, personal communication, Sept. 29, 1978, preliminary draft of review to be submitted for publication in Atomic Energy Review, International Atomic Energy Agency, Vienna, Austria.
2. U. V. Choudary, K. A. Gingerich and J. E. Kingcade, J. Less-Common Metals 42, 111 (1975).
3. JANAF Thermochemical Tables: WO(g) and WO_3(g), 9-30-66.
4. T. M. Swaminathan and A. G. Kirshnamurthy, Current Sci. 23, 258 (1954).
5. T. V. Iorns and F. E. Stafford, J. Amer. Chem. Soc. 88, 4819 (1966).
6. R. F. Barrow and A. D. Caunt, Proc. Roy. Soc. (London) 219A, 120 (1953).
7. F. A. Schroder, Acta Cryst. B31, 2294 (1975).
8. W. D. Hewett, Jr., J. H. Newton, and W. Weltner, Jr., J. Phys. Chem. 79, 2640 (1975).

Molybdenum Oxide (MoO) $Mo_1O_1(g)$

Enthalpy Reference Temperature = T_r = 298.15 K   Standard State Pressure = p° = 0.1 MPa

T/K	$C_p°$ (J K⁻¹mol⁻¹)	S° (J K⁻¹mol⁻¹)	$-[G°-H°(T_r)]/T$ (J K⁻¹mol⁻¹)	$H°-H°(T_r)$ (kJ mol⁻¹)	$\Delta_f H°$ (kJ mol⁻¹)	$\Delta_f G°$ (kJ mol⁻¹)	Log K_f
0	0.	0.	INFINITE	-9.636	310.249	310.249	INFINITE
100	29.997	205.390	272.530	-6.714	311.228	300.056	-156.733
200	34.472	227.640	245.025	-3.477	311.167	288.883	-75.449
250	35.531	235.461	242.355	-1.724	311.073	283.322	-59.197
298.15	36.002	241.764	241.764	0.	310.959	277.988	-48.702
300	36.014	241.987	241.765	0.067	310.954	277.783	-48.366
350	36.242	247.558	242.205	1.874	310.809	272.266	-40.633
400	36.376	252.407	243.183	3.689	310.635	266.771	-34.837
450	36.478	256.697	244.451	5.511	310.433	261.300	-30.331
500	36.567	260.545	245.871	7.337	310.205	255.852	-26.729
600	36.732	267.227	248.890	11.002	309.674	245.030	-21.332
700	36.879	272.901	251.925	14.683	309.054	234.304	-17.484
800	37.006	277.834	254.862	18.377	308.359	223.672	-14.604
900	37.114	282.199	257.662	22.083	307.596	213.131	-12.370
1000	37.207	286.114	260.314	25.799	306.769	202.679	-10.587
1100	37.287	289.664	262.824	29.524	305.876	192.313	-9.132
1200	37.357	292.911	265.198	33.256	304.916	182.031	-7.924
1300	37.418	295.904	267.446	36.995	303.882	171.832	-6.904
1400	37.473	298.679	269.579	40.740	302.770	161.715	-6.034
1500	37.522	301.266	271.606	44.490	301.574	151.681	-5.282
1600	37.568	303.689	273.537	48.244	300.288	141.730	-4.627
1700	37.610	305.968	275.378	52.003	298.905	131.862	-4.052
1800	37.649	308.119	277.138	55.766	297.419	122.078	-3.543
1900	37.686	310.155	278.822	59.533	295.821	112.380	-3.090
2000	37.721	312.089	280.438	63.303	294.104	102.769	-2.684
2100	37.754	313.931	281.989	67.077	292.259	93.247	-2.319
2200	37.786	315.688	283.481	70.854	290.276	83.816	-1.990
2300	37.817	317.368	284.918	74.634	288.146	74.479	-1.691
2400	37.846	318.978	286.304	78.417	285.859	65.237	-1.420
2500	37.875	320.524	287.642	82.203	283.405	56.094	-1.172
2600	37.903	322.010	288.936	85.992	280.760	47.054	-0.945
2700	37.931	323.441	290.187	89.784	277.886	38.119	-0.737
2800	37.958	324.821	291.400	93.578	274.720	29.295	-0.547
2900	37.985	326.153	292.575	97.376	235.298	20.640	-0.372
3000	38.011	327.441	293.716	101.175	233.344	13.272	-0.231
3100	38.036	328.688	294.824	104.978	231.383	5.969	-0.101
3200	38.062	329.896	295.901	108.783	229.415	-1.272	0.021
3300	38.087	331.068	296.949	112.590	227.441	-8.450	0.134
3400	38.112	332.205	297.970	116.400	225.462	-15.568	0.239
3500	38.136	333.310	298.964	120.212	223.477	-22.629	0.338
3600	38.161	334.385	299.933	124.027	221.486	-29.632	0.430
3700	38.185	335.431	300.878	127.844	219.491	-36.581	0.516
3800	38.209	336.449	301.801	131.664	217 491	-43.474	0.598
3900	38.233	337.442	302.702	135.486	215.486	-50.315	0.674
4000	38.257	338.410	303.583	139.311	213.477	-57.106	0.746
4100	38.280	339.355	304.444	143.138	211.464	-63.845	0.813
4200	38.304	340.278	305.286	146.967	209.447	-70.536	0.877
4300	38.327	341.180	306.110	150.798	207.426	-77.178	0.938
4400	38.350	342.061	306.917	154.632	205 402	-83.773	0.995
4500	38.374	342.923	307.708	158.468	203.373	-90.323	1.048
4600	38.397	343.767	308.483	162.307	201.341	-96.827	1.100
4700	38.420	344.593	309.242	166.148	199.305	-103.287	1.148
4800	38.443	345.402	309.987	169.991	197.265	-109.704	1.194
4900	38.466	346.195	310.718	173.836	195.221	-116.077	1.237
5000	38.489	346.972	311.435	177.684	-398.212	-116.678	1.219
5100	38.512	347.735	312.140	181.534	-401.945	-111.011	1.137
5200	38.534	348.483	312.831	185.386	-405.828	-105.269	1.057
5300	38.557	349.217	313.511	189.241	-409.860	-99.450	0.980
5400	38.580	349.938	314.179	193.098	-414.040	-93.555	0.905
5500	38.602	350.646	314.836	196.957	-418.366	-87.581	0.832
5600	38.625	351.342	315.481	200.818	-422.836	-81.526	0.760
5700	38.648	352.026	316.116	204.682	-427.446	-75.391	0.691
5800	38.670	352.698	316.741	208.548	-432.196	-69.173	0.623
5900	38.693	353.359	317.356	212.416	-437.081	-62.873	0.557
6000	38.715	354.010	317.962	216.286	-442.103	-56.488	0.492

PREVIOUS: September 1978 (1 atm)   CURRENT: September 1978 (1 bar)

Molybdenum Oxide (MoO) $Mo_1O_1(g)$

MOLYBDENUM OXIDE (MoO_2) CRYSTAL M_r = 127.9388

Pearson Notation: mP12

$S°(298.15\ K) = 46.459 \pm 0.42\ J\ K^{-1}\ mol^{-1}$

$\Delta_f H°(0\ K) = -582.90 \pm 2.1\ kJ\ mol^{-1}$

$\Delta_f H°(298.15\ K) = -587.85 \pm 2.9\ kJ\ mol^{-1}$

Enthalpy of Formation

The adopted enthalpy of formation of MoO_2(cr) at 298.15 K is -140.5±0.2 kcal mol^{-1}. This value is chosen so as to be in agreement with Brewer (1) who has performed a critical analysis of all data pertaining to the molybdenum-oxygen system. However, Brewer (1) chose to adopt the value previously used by JANAF (14); a value which was derived as a weighted average of two calorimetric determinations and seven equilibrium studies, the latter results based on S°(298.15 K) = 11.954 cal K^{-1} mol^{-1}. In revising the S°(298.15 K) value to 11.04 cal K^{-1} mol^{-1} in this tabulation, the enthalpy of formation should also have been revised. With the exception of the data from Barbi (5) the following equilibrium studies suggest $\Delta_f H°$(298.15 K) = -141.20 kcal mol^{-1} in comparison with the calorimetric mean of $\Delta_f H°$(298.15 K) = -140.8 kcal mol^{-1} for MoO_2(cr). In addition, eight recent emf studies (published since the analysis by Brewer) also yield a value of $\Delta_f H°$(298.15 K) = -141.2 kcal mol^{-1}. Although there is some overlap in the values for $\Delta_f H°$ derived from equilibrium and calorimetric studies, the equilibrium studies appear to be more negative by 0.4 kcal mol^{-1}. Brewer (1) noted (in reference to $\Delta_f H°$(298.15 K) = -140.5 kcal mol^{-1}, however, that a better overall fit of all Mo-O data would be obtained if the enthalpy of formation of MoO_2(cr) were made more negative by 0.5 kcal mol^{-1} with a corresponding decrease of ~0.5 kcal mol^{-1} for MoO_2(g) and MoO(g) and ~0.8 kcal mol^{-1} for MoO_3 (g). Such a change could be accommodated in the MoO_2 tables.

Ignition of the bomb products from the calorimetric studies of Staskiewicz et al. (2) indicated that oxidation was only about 70 percent complete. The authors also digested the products with NH_4OH to remove the trioxide. The residue was the dioxide, as indicated by both oxidation and reduction. This evidence does not necessarily eliminate the possibility of known intermediate oxides in the combustion products. Mah (3) found 72-75 percent conversion to trioxide by weighing the total combustion products; no additional analysis was performed.

Source	Method	Reaction	T/K	Data Points	$\Delta_r H°$(298.15 K), kcal mol^{-1} 2nd Law	3rd Law	Drift cal K^{-1} mol^{-1}	$\Delta_f H°$(298.15 K) kcal mol^{-1}
Staskiewicz et al. (2)	Calorimetric	A	-	-	-37.12		-	-140.86±0.13**
Mah (3)	Calorimetric	A	-	-	-37.52		-	-140.64±0.13**
Rapp (4)	emf	B	1023-1323	35	-13.26	-14.18	-0.8±0.1	-141.46
		C	1023-1273	22	-25.64	-26.65	-0.9±0.1	-141.25
Barbi (5)	emf	F	873-1073	Eqn.	-140.4	-139.45	0.9	-139.45
Gleiser and Chipman (6)	Equilibrium	D	1296-1341	9	+3.02	-6.57	-3.6±1.5	-141.84
Gokcen (7)	Equilibrium	E	949-1344	Eqn.	-19.88	-25.50	-3.4	-141.11
Collins (8)	Equilibrium	E	973-1173	Eqn.	-22.18	-25.46	-1.5	-141.07
Tonosaki (9)	Equilibrium	E	981-1096	Eqn.	-18.86	-25.86	-3.0	-140.47

**Auxiliary data for MoO_3(cr) are taken from corresponding measurements of the same authors. See also (13).

Reaction:

A. $MoO_2(cr) + 1/2\ O_2(g) = MoO_3(cr)$

B. $2\ Fe_xO(cr) + Mo(cr) = MoO_2(cr) + 2\ xFe(cr)$

C. $2\ NiO(cr) + Mo(cr) = MoO_2(cr) + 2\ Ni(cr)$

D. $Mo(cr) + 2\ CO_2(g) = MoO_2(cr) + 2\ CO(g)$

E. $Mo(cr) + 2\ H_2O(g) = MoO_2(cr) + 2\ H_2(g)$

F. $Mo(cr) + O_2(g) = MoO_2(cr)$ derived by the author from the cell reaction $2\ Fe_xO(cr) + Mo(cr) = MoO_2(cr) + 2\ xFe(cr)$ using $\Delta_f G°$ data for $Fe_xO(cr)$

Heat Capacity and Entropy

King (10) measured low temperature heat capacities from 53.3 to 296 K. King et al. (11) measured the high temperature enthalpy changes from 399 to 1801 K by drop calorimetry. The adopted heat capacities are derived from these two sets of data, subject to the constraint that they join smoothly at 298.15 K. Deviations of the enthalpy data from the adopted values are -1.3 to +0.7%, excluding the point at 1801 K. This point at the highest temperature deviates by +2.3%, which suggests that the apparent heat capacity is rising very rapidly. No explanation for this rapid rise is available.

The entropy S°(298.15 K) = 11.104 cal K^{-1} mol^{-1} is obtained from the low temperature heat capacity data, based on S°(51 K) = 0.36 cal K^{-1} mol^{-1}. As discussed by Brewer (1), Ghose et al. (12) used neutron diffraction measurements to demonstrate the absence of a permanent magnetic moment. Thus no additional heat capacity or entropy contribution below 50 K due to a possible antiferromagnetic effect is necessary.

References

1. L. Brewer, Materials and Research Division, Lawrence Berkeley Laboratory, University of California, Berkeley, personal communication, Sept. 29, 1978, preliminary draft of review to be submitted for publication in Atomic Energy Review, International Atomic Energy Agency, Vienna, Austria.
2. H. A. Staskiewicz, J. R. Tucker and P. E. Snyder, J. Amer. Chem. Soc. 77, 2987 (1955).
3. A. D. Mah, J. Phys. Chem. 61, 1572 (1957).
4. R. A. Rapp, Trans. AIME 227, 371 (1963).
5. G. B. Barbi, J. Phys. Chem. 68, 1025 (1964).
6. M. Gleiser and J. Chipman, J. Phys. Chem. 66, 1539 (1962).
7. N. A. Gokcen, Trans. AIME, 197, 1019 (1953).
8. B. T. Collins, Sc. D. Thesis, Massachusetts Institute of Technology, 1949.
9. K. Tonosaki, Bull. Inst. Phys. Chem. Res. (Tokyo) 19, 126 (1940).
10. E. G. King, J. Amer. Chem. Soc. 80, 1799 (1958).
11. E. G. King, W. W. Weller and A. U. Christensen, U. S. Bur. Mines, RI 5664, 1960.
12. J. Ghose, N. N. Greenwood, G. C. Hallam, and D. A. Read, J. Solid State Chem. 19, 365 (1976).
13. JANAF Thermochemical Tables: MoO_3(cr), 9-30-78.
14. JANAF Thermochemical Tables, 2nd ed., NSRDS-NBS 37, 1971; MoO_2(cr), 6-30-67.

Molybdenum Oxide (MoO_2) Mo_1O_2(cr)

Enthalpy Reference Temperature = T_r = 298.15 K   Standard State Pressure = p° = 0.1 MPa

T/K	$C_p°$ (J $K^{-1}mol^{-1}$)	S°	$-[G°-H°(T_r)]/T$	$H°-H°(T_r)$ (kJ mol^{-1})	$\Delta_f H°$	$\Delta_f G°$	Log K_f
0	0.	0.	INFINITE	-8.319	-582.903	-582.903	INFINITE
100	16.175	7.463	85.470	-7.801	-585.780	-568.493	296.951
200	41.363	26.960	51.165	-4.841	-587.574	-550.373	143.743
298.15	55.898	46.459	46.459	0.	-587.852	-532.011	93.206
300	56.095	46.806	46.460	0.104	-587.847	-531.664	92.571
400	63.551	64.076	48.764	6.125	-587.253	-513.011	66.992
500	67.936	78.758	53.333	12.713	-586.272	-494.558	51.666
600	71.174	91.441	58.651	19.674	-585.087	-476.324	41.468
700	73.935	102.623	64.150	26.932	-583.757	-458.300	34.199
800	76.488	112.664	69.597	34.454	-582.293	-440.476	28.760
900	78.965	121.816	74.898	42.227	-580.692	-422.843	24.541
1000	81.437	130.264	80.017	50.247	-578.946	-405.397	21.176
1100	83.948	138.143	84.947	58.515	-577.049	-388.132	18.431
1200	86.513	145.557	89.692	67.038	-574.994	-371.048	16.151
1300	89.148	152.585	94.262	75.820	-572.776	-354.141	14.230
1400	91.872	159.291	98.669	84.870	-570.389	-337.411	12.589
1500	94.680	165.725	102.926	94.197	-567.829	-320.858	11.173
1600	97.588	171.927	107.046	103.810	-565.090	-304.482	9.940
1700	100.588	177.933	111.040	113.718	-562.170	-288.282	8.858
1800	103.688	183.770	114.919	123.931	-559.064	-272.260	7.901
1900	106.893	189.461	118.694	134.459	-555.770	-256.416	7.049
2000	110.198	195.028	122.372	145.313	-552.285	-240.749	6.288
2100	113.608	200.487	125.962	156.502	-548.608	-225.262	5.603
2200	117.123	205.852	129.472	168.038	-544.736	-209.954	4.985
2300	120.742	211.138	132.908	179.930	-540.669	-194.828	4.425
2400	124.470	216.355	136.276	192.190	-536.405	-179.882	3.915
2500	128.302	221.514	139.583	204.827	-531.946	-165.118	3.450
2600	132.244	226.622	142.832	217.854	-527.301	-150.536	3.024
2700	136.290	231.689	146.030	231.280	-522.500	-136.135	2.634
2800	140.440	236.720	149.179	245.115	-517.593	-121.915	2.274
2900	144.704	241.722	152.284	259.371	-548.535	-107.822	1.942
3000	149.072	246.701	155.348	274.059	-541.590	-92.742	1.615

PREVIOUS: June 1967   CURRENT: September 1978

Molybdenum Oxide (MoO_2) Mo_1O_2(cr)

MOLYBDENUM OXIDE (MoO_2) IDEAL GAS M_r = 127.9388

$\Delta_{at}H°(0\ K) = 1157.294 \pm 20.9\ kJ\ mol^{-1}$

$S°(298.15\ K) = [276.990 \pm 8.4]\ J\ K^{-1}\ mol^{-1}$

$\Delta_f H°(0\ K) = -6.32 \pm 12.6\ kJ\ mol^{-1}$

$\Delta_f H°(298.15\ K) = -8.31 \pm 12.6\ kJ\ mol^{-1}$

Vibrational Frequencies and Degeneracies

ν, cm^{-1}
900 (1)
[300](1)
950 (1)

Electronic Levels and Quantum Weights

ε_i, cm^{-1}	g_i
0	[3]
[13000]	[6]
[25000]	[6]

Point Group: C_{2v} σ = 2

Bond Distance: Mo–O = 1.73 Å

Bond Angle: O–Mo–O = 116°

Product of the Moments of Inertia: $I_A I_B I_C = 5.663388 \times 10^{-115}\ g^3\ cm^6$

Enthalpy of Formation

The adopted enthalpy of formation for MoO_2(g) at 298.15 K is −1.99±3 kcal mol^{-1}. This value was adopted by Brewer (1) who has performed a critical analysis of all data pertaining to the molybdenum-oxygen system. Brewer (1) adopted $\Delta_{sub}H°(298.15\ K)/R$ = 69700±1500 K for the sublimation process MoO_2(cr) = MoO_2(g). The vapor pressures of MoO_2 calculated from the adopted tabulation slightly exceed those of Burns et al. (2) and are 50% lower than those of Chizhikov et al. (6). The adopted ΔG values for the formation of MoO_2(g) from Mo(cr) and O_2(g) average 1 kcal mol^{-1} more negative than those determined by Franklin and Stickney (7). The only recent determination that deviates seriously is that of Belton and Jordan (8) who determined the volatility of molybdenum in steam-hydrogen mixtures at 1200 and 1500°C. The variation of their results with water partial pressure at 1383 to 1500°C indicate an intercept corresponding to non-hydrated species which they attribute to MoO_3 and MoO_2. They assumed the ratio of MoO_2 to MoO_3 to be that observed by Burns et al. (2) and obtained partial pressures of MoO_2 about six times higher than calculated from the adopted values. Their values would require that $\Delta_f H°(298.15\ K)/R$ of Mo(cr) + O_2(g) = MoO_2(g) be made more negative by 3200 K. If both $\Delta_{sub}H°$ and $\Delta_f H°$(cr) were shifted by their assigned uncertainties indicated, a change of 2000 K could be accomodated.

Burns et al. (2) have investigated mass-spectrometrically the vapor in equilibrium with powdered MoO_2(cr) and also reported the partial pressures of MoO_2(g), MoO_3(g), $(MoO_3)_2$(g) and $(MoO_3)_3$(g). DeMaria et al. (3) have measured the partial pressures of gaseous oxides MoO, MoO_2, and O(g) in the Mo-Al_2O_3 system by mass spectrometry. Blackburn et al. (4) and Plante (5) have studied the vapor pressures above MoO_2 (cr) by the Knudsen effusion method. More recently there have been mass spectroscopic Knudsen cell studies of Chizhikov et al. (6) of the vapor over MoO_2(cr) between 1500 and 1730 K and the Langmuir type studies of Franklin and Stickney (7) on the vapor resulting from oxygen on a molybdenum surface.

Source	Reaction	T/K	Data Points	$\Delta_r H°(298.15\ K)$, kcal mol^{-1} 2nd law	3rd law	Drift cal K^{-1} mol^{-1}	$\Delta_f H°(298.15\ K)$ kcal mol^{-1}
Burns et al. (2)	MoO_2(cr) → MoO_2(g)	1598-1777	11*	137.6±3.0	137.38	-0.7±1.8	-1.67
DeMaria et al. (3)	Mo(cr)+2O(g) → MoO_2(g)	2262-2466	6	-61.3±32.0	138.83	-20.0±13.6	+10.74
Blackburn et al. (4)	MoO_2(cr) → MoO_2(g)	1818-2028	7	107.6±6.8	126.55	-9.9±3.5	-13.95
Plante (5)	MoO_2(cr) → MoO_2(g)	1611-1802	6	151.1±13.4	142.89	-5.0±7.7	+1.89
Chizhikov et al. (6)	MoO_2(cr) = MoO_2(g)	1500-1750	Eqn	131.3±0.2	136.88	3.4±0.1	-3.6
Franklin and Stickney (7)	Mo(cr)+O_2(g) = MoO_2(g)	1900-2500	Eqn	-0.2±1.3	-0.80	-0.3±0.6	-0.80
Belton and Jordan (8)	Mo(cr)+2H_2O(g) = MoO_2(g)+2H_2(g)	1473-1773	Eqn	-118.1±0.1	109.91	-6.5±0.0	-5.7

*2 points rejected due to failure of statistical test

Heat Capacity and Entropy

Hewett et al. (9) have reviewed the spectroscopic studies on MoO_2(g). In addition they have studied spectroscopically MoO_2 in neon and argon matrices at 4 K. The adopted bond distance and angle and two stretching frequencies are derived from this study. Brewer (1) adopted the same values. The bending frequency (300 cm^{-1}) is estimated by comparison with other transition metal dioxides.

The electronic levels and quantum weights are estimated from the data reported by Weltner and McLeod (10) on WO_2. The ground state is assumed to be $^3A_1$ and the first excited level is assumed to lie close to the observed transitions at 7890 Å and 7806 Å in WO_2. Since both these states are presumed to be triplet, we have assigned a total degeneracy of 6. In addition, we expect from the electronic levels of Mo^{+4}(g) (11) that other electronic states will also be found and these are arbitrarily estimated in the level at 25000 cm^{-1}. The principal moments of inertia are: $I_A = 3.3489 \times 10^{-39}$, $I_B = 11.4373 \times 10^{-39}$, and $I_C = 14.7862 \times 10^{-39}$ g cm^2.

References

1. L. Brewer, Materials and Research Division, Lawrence Berkeley Laboratory, University of California, Berkeley, personal communication, Sept. 29, 1978, preliminary draft of review to be submitted for publication in Atomic Energy Review, International Atomic Energy Agency, Vienna, Austria.
2. R. P. Burns, G. DeMaria, J. Drowart and R. T. Grimley, J. Chem. Phys. 32, 1363 (1960).
3. G. DeMaria, R. P. Burns, J. Drowart and M. G. Inghram, J. Chem. Phys. 32, 1373 (1960).
4. P. E. Blackburn, M. Hoch and H. L. Johnston, J. Phys. Chem. 62, 769 (1958).
5. E. R. Plante, Ph. D. Thesis, Univ. of Kansas, 1960.
6. D. M. Chizhikov, E. Kazenas, and I. O. Ermilova, Russ. J. Phys. Chem. 50, 878 (1976).
7. J. E. Franklin and R. E. Stickney, High Temp. Sci. 3, 401 (1971).
8. G. R. Belton and A. S. Jordan, J. Phys. Chem. 69, 2065 (1965).
9. W. D. Hewett, Jr., J. H. Newton, and W. Weltner Jr., J. Phys. Chem. 79, 2640 (1975).
10. W. Weltner, Jr., and D. McLeod, Jr., J. Mol. Spectrosc. 17, 276 (1965).
11. C. E. Moore, NSRDS-NBS 35, Vol. III, reissued 1971.

Molybdenum Oxide (MoO_2) Mo_1O_2(g)

Enthalpy Reference Temperature = T_r = 298.15 K; Standard State Pressure = p° = 0.1 MPa

T/K	$C_p°$ (J $K^{-1}mol^{-1}$)	S°	−[G°−H°(T_r)]/T	H°−H°(T_r) (kJ mol^{-1})	$\Delta_f H°$	$\Delta_f G°$	Log K_f
0	0.	0.	INFINITE	-11.278	-6.324	-6.324	INFINITE
100	35.387	234.266	313.300	-7.903	-6.346	-11.739	6.132
200	39.935	260.251	280.895	-4.129	-7.324	-16.782	4.383
250	42.110	269.395	277.706	-2.078	-7.849	-19.085	3.988
298.15	44.174	276.990	276.990	0.	-8.314	-21.206	3.715
300	44.250	277.264	276.991	0.082	-8.331	-21.286	3.706
350	46.212	284.235	277.537	2.344	-8.759	-23.410	3.494
400	47.922	290.521	278.774	4.699	-9.141	-25.477	3.327
450	49.371	296.251	280.402	7.132	-9.490	-27.498	3.192
500	50.583	301.518	282.254	9.632	-9.815	-29.481	3.080
600	52.431	310.914	286.267	14.788	-10.435	-33.355	2.904
700	53.723	319.099	290.385	20.100	-11.051	-37.127	2.770
800	54.644	326.336	294.435	25.521	-11.688	-40.809	2.665
900	55.317	332.813	298.346	31.020	-12.360	-44.409	2.577
1000	55.821	338.668	302.090	36.578	-13.077	-47.932	2.504
1100	56.206	344.008	305.662	42.180	-13.847	-51.380	2.440
1200	56.507	348.912	309.064	47.817	-14.678	-54.756	2 383
1300	56.747	353.444	312.306	53.480	-15.579	-58.060	2.333
1400	56.942	357.657	315.397	59.165	-16.557	-61.292	2.287
1500	57.104	361.591	318.347	64.867	-17.621	-64.451	2.244
1600	57.243	365.281	321.166	70.585	-18.778	-67.536	2.205
1700	57.365	368.755	323.864	76.315	-20.035	-70.545	2.168
1800	57.479	372.037	326.450	82.057	-21.400	-73.478	2.132
1900	57.587	375.148	328.932	87.811	-22.880	-76.331	2.098
2000	57.694	378.105	331.317	93.575	-24.485	-79.103	2.066
2100	57.805	380.922	333.613	99.350	-26.222	-81.792	2.034
2200	57.920	383.614	335.825	105.136	-28.100	-84.394	2.004
2300	58.043	386.191	337.959	110.934	-30.127	-86.908	1.974
2400	58.176	388.664	340.021	116.745	-32.313	-89.331	1.944
2500	58.318	391.042	342.014	122.569	-34.666	-91.659	1.915
2600	58.471	393.332	343.944	128.409	-37.208	-93.890	1.886
2700	58.634	395.542	345.815	134.264	-39.978	-96.017	1.858
2800	58.809	397.677	347.629	140.136	-43.035	-98.038	1.829
2900	58.994	399.744	349.391	146.026	-82.343	-99.894	1.799
3000	59.188	401.748	351.103	151.935	-84.176	-100.468	1.749
3100	59.392	403.692	352.768	157.864	-86.009	-100.980	1.702
3200	59.603	405.581	354.389	163.814	-87.838	-101.435	1.656
3300	59.822	407.418	355.968	169.785	-89.664	-101.832	1.612
3400	60.046	409.207	357.508	175.778	-91.484	-102.173	1.570
3500	60.276	410.951	359.010	181.794	-93.297	-102.461	1.529
3600	60.510	412.652	360.476	187.833	-95.103	-102.697	1.490
3700	60.746	414.313	361.909	193.896	-96.900	-102.884	1.452
3800	60.985	415.937	363.310	199.983	-98.687	-103.021	1.416
3900	61.225	417.524	364.679	206.093	-100.464	-103.111	1.381
4000	61.465	419.077	366.020	212.228	-102.231	-103.157	1.347
4100	61.704	420.598	367.333	218.386	-103.987	-103.158	1.314
4200	61.941	422.087	368.619	224.568	-105.731	-103.117	1.282
4300	62.177	423.548	369.879	230.774	-107.465	-103.034	1.252
4400	62.409	424.980	371.115	237.004	-109.187	-102.911	1.222
4500	62.638	426.385	372.328	243.256	-110.898	-102.750	1.193
4600	62.863	427.764	373.518	249.531	-112.599	-102.550	1.164
4700	63.083	429.118	374.687	255.828	-114.290	-102.313	1.137
4800	63.298	430.449	375.835	262.148	-115.971	-102.041	1.110
4900	63.507	431.756	376.963	268.488	-117.644	-101.733	1.084
5000	63.711	433.041	378.071	274.849	-710.695	-95.660	0.999
5100	63.909	434.305	379.162	281.230	-714.033	-83.326	0.853
5200	64.100	435.548	380.234	287.630	-717.513	-70.926	0.712
5300	64.285	436.770	381.289	294.050	-721.131	-58.457	0.576
5400	64.462	437.974	382.328	300.487	-724.890	-45.919	0.444
5500	64.633	439.158	383.350	306.942	-728.787	-33.310	0.316
5600	64.798	440.324	384.357	313.413	-732.823	-20.628	0.192
5700	64.954	441.472	385.349	319.901	-736.994	-7.874	0.072
5800	65.104	442.603	386.327	326.404	-741.301	4.956	-0.045
5900	65.247	443.717	387.290	332.922	-745.740	17.858	-0.158
6000	65.383	444.815	388.240	339.453	-750.314	30.840	-0.268

PREVIOUS: September 1978 (1 atm) CURRENT: September 1978 (1 bar)

Molybdenum Oxide (MoO_2) Mo_1O_2(g)

MOLYBDENUM OXIDE ($MoO_{2.750}$) CRYSTAL M_r = 139.9384

Pearson Notation: mP60

$S°(298.15\ K) = [69.936 \pm 1.7]\ J\ K^{-1}\ mol^{-1}$

$T_{trs} = 875 \pm 50\ K$

$\Delta_f H°(0\ K)$ = Unknown

$\Delta_f H°(298.15\ K) = -708.41 \pm 3.3\ kJ\ mol^{-1}$

$\Delta_{trs} H° = [0]\ kJ\ mol^{-1}$

Enthalpy of Formation

For this oxide, the atomic ratio of oxygen to molybdenum is 2.750 or 11/4. The formula $MoO_{2.750}$ (upon which this tabulation is based) is more indicative of the oxygen content relative to MoO_2 and MoO_3, as opposed to Mo_4O_{11}. The thermochemical values describing $MoO_{2.750}$ are numerically 1/4 those of Mo_4O_{11}.

The high-temperature emf measurements (1, 2, 3) on compositions between MoO_2 and MoO_3 gave widely divergent results, some of which are contradicted by the Mo-O phase diagram recommended by Brewer (4). Brewer (4) found that it was more practical to determine a $\Delta_f H°(298.15\ K)$ value for Mo_4O_{11}(cr) by fixing the thermodynamic properties from the peritectic equilibrium together with the thermodynamic data for the liquid. From the phase diagram (4), MoO_2 and Mo_4O_{11} solids are in equilibrium with the peritectic liquid of 74.2±0.1 atomic % oxygen at 1091±7 K. For Mo_4O_{11}(cr) = MoO_2(cr) + 3 MoO_3(ℓ), Brewer (4) obtained $\Delta_r G°/RT$ = 0.863±0.05 at 1091 K. This yielded, using auxiliary data (5), $\Delta Hr°/R$ = 1280±400 K for the disproportionation of Mo_4O_{11} to crystalline MoO_2 and MoO_3 at 298.15 K. This heat of reaction yields $\Delta_r H°(298.15\ K)/R$ = -85203±200 K for Mo_4O_{11}, from which we calculate $\Delta_f H°(MoO_{2.750}$, cr, 298.15 K) = -169.315±0.4 kcal mol^{-1}. We increase the uncertainty in $\Delta_f H°$ to ±0.8 kcal mol^{-1} which better represents the effect of errors in reducing a $\Delta_r G°$ value at 1091 K to a $\Delta_f H°$ value at 298.15 K.

Heat Capacity and Entropy

Bousquet et al. (6) used differential enthalpy analysis to obtain enthalpies between 379 and 971 K for Mo_4O_{11}(cr), Mo_8O_{23}(cr), and Mo_9O_{26}(cr). Brewer (1) compared their values of MoO_2(cr) and MoO_3(cr) mixtures with the same composition. For Mo_4O_{11}(cr), the enthalpies were higher by 3% at 400 K, decreasing to 1.5% at 700 K. As these differences were within the scatter of the experimental measurements, Brewer (1) assumed $\Delta C_p°$ = 0 and $\Delta S°$ = 0 for the formation of Mo_4O_{11} from 3 MoO_3(cr) and MoO_2(cr).

Bousquet et al. (6) also measured the high temperature enthalpy data of MoO_3 fom 379 to 971 K. Their values agree with the adopted JANAF values for MoO_3 (5) within 2% below 500 K, but the deviations increase rapidly at higher temperatures becoming 10% high at 950 K. Thus, the adopted heat capacity values for Mo_4O_{11}(cr), based on the assumption $\Delta C_p°$ = 0 as discussed in the preceding paragraph, may be less reliable at higher temperatures.

Phase Data

Mo_4O_{11}(oP60) disproportionates to MoO_2 and liquid at 1091±7 K, according to the phase diagram of Phillips and Chang (7, 8). Mo_4O_{11}(oP60) transforms on cooling below 875±50 K to Mo_4O_{11}(mP60). Bousquet et al. (6) did not interpret their enthalpy data to include a transition at 875 K. However, a graphical examination of their data suggests two comments. First, the data point at 670 K may be a typographical error; perhaps 62.58 cal g^{-1} should be 52.58 cal g^{-1}. Second, a transition at approximately 850 K with a corresponding heat of ~0.7 kcal mol^{-1} could be easily derived. To be consistent with Brewer (4), we assume a transition at 875±50 K with a corresponding zero heat of transition.

In addition to the discussion of the molybdenum-oxygen system by Brewer (4), Gmelin Handbuch der Anorganischen Chemie (9) has recently presented a thorough summary of the literature dealing with the phase diagram and properties of the oxide phases. Ekstrom and Tilley (10) have reviewed the work of Ekstrom as well as the earlier work of Kihlborg (11) on the phases in the Mo-O system.

References

1. H. LeBresq, J. J. Oehlig, and F. Marion, C. R. Acad. Sci. (Paris), Ser. C. 268, 1047 (1969).
2. S. Zador and C. B. Alcock, J. Chem. Thermodyn. 2, 9 (1970).
3. V. M. Zhukovskii et al., Russ. J. Phys. Chem. 46, 1542 (1972).
4. L. Brewer, Materials and Molecular Research Division, Lawrence Berkeley Laboratory, University of California, Berkeley; personal communication, Sept. 29, 1978, preliminary draft of review to be submitted for publication in Atomic Energy Review, International Atomic Energy Agency, Vienna, Austria.
5. JANAF Thermochemical Tables: MoO_2(cr) and MoO_3(ℓ), 9-30-78.
6. J. Bousquet, J. C. David, M. Diot. and A. Guillon, Bull. Soc. Chim. Fr. (6), 2390 (1968).
7. L. L. Y. Chang and B. Phillips, J. Amer. Ceram. Soc. 52, 527 (1969).
8. B. Phillips and L. L. Y. Chang, Trans. AIME 233, 1433 (1965).
9. Gmelin Handbuch der Anorganischen Chemie, Main Series, 8th Ed., Molybdan, System No. 53, Eganzungsband Teil B1, Springer-Verlag, Berlin, 1975.
10. T. Ekstrom and R. J. D. Tilley, J. Solid State Chem. 19, 125 (1976).
11. L. Kihlborg, Acta Chem. Scand. 13, 954 (1959).

Molybdenum Oxide ($MoO_{2.750}$) $Mo_1O_{2.750}$(cr)

Enthalpy Reference Temperature = T_r = 298.15 K Standard State Pressure = $p°$ = 0.1 MPa

T/K	$C_p°$ (J K⁻¹mol⁻¹)	$S°$ (J K⁻¹mol⁻¹)	$-[G°-H°(T_r)]/T$ (J K⁻¹mol⁻¹)	$H°-H°(T_r)$ (kJ mol⁻¹)	$\Delta_f H°$ (kJ mol⁻¹)	$\Delta_f G°$ (kJ mol⁻¹)	Log K_f
0							
100							
200							
298.15	70.135	69.936	69.936	0.	-708.412	-636.633	111.535
300	70.354	70.370	69.937	0.130	-708.401	-636.188	110.770
400	78.208	91.763	72.800	7.585	-707.487	-612.239	79.950
500	82.797	109.722	78.438	15.642	-706.185	-588.572	61.488
600	86.626	125.158	84.968	24.114	-704.674	-565.189	49.204
700	90.344	138.790	91.701	32.962	-702.973	-542.073	40.450
800	94.145	151.100	98.369	42.185	-701.059	-519.215	33.901
900	98.072	162.414	104.864	51.795	-698.899	-496.612	28.823
1000	102.139	172.956	111.152	61.804	-696.462	-474.263	24.773
1100	106.331	182.887	117.226	72.227	-693.727	-452.173	21.472
1200	110.638	192.323	123.094	83.074	-690.678	-430.346	18.732
1300	115.050	201.352	128.769	94.358	-687.302	-408.786	16.425
1400	119.553	210.043	134.266	106.087	-683.591	-387.498	14.458
1500	124.139	218.447	139.599	118.271	-679.539	-366.489	12.762

PREVIOUS: CURRENT: September 1978

Molybdenum Oxide ($MoO_{2.750}$) $Mo_1O_{2.750}$(cr)

MOLYBDENUM OXIDE ($MoO_{2.875}$) CRYSTAL M_r = 141.9383

Pearson Notation: mP62

$S°(298.15\ K)$ = [73.849 ± 1.7] J K^{-1} mol^{-1}

$\Delta_f H°(0\ K)$ = Unknown

$\Delta_f H°(298.15\ K)$ = -727.13 ± 2.5 kJ mol^{-1}

Enthalpy of Formation

For this oxide, the atomic ratio of oxygen to molybdenum is 2.875 or 23/8. The formula $MoO_{2.875}$ (upon which this tabulation is based) is more indicative of the oxygen content relative to MoO_2 and MoO_3, as opposed to Mo_8O_{23}. The thermochemical values describing $MoO_{2.875}$ are numerically 1/8 those of Mo_8O_{23}.

The high temperature emf measurements (1, 2, 3) on compositions between MoO_2 and MoO_3 gave widely divergent results, some of which are contradicted by the Mo-O phase diagram recommended by Brewer (4). Brewer (4) found that it was more practical to determine a $\Delta_f H°(298.15\ K)$ value for $MoO_{2.875}$ by fixing the thermodynamic properties from the peritectic equilibria together with the thermodynamic data for the liquid. From the Mo-O phase diagram (4), Mo_8O_{23} and Mo_4O_{11} solids are in equilibrium with the peritectic liquid of 74.45±0.1 atomic % oxygen at 1058±5 K. For the equilibrium $Mo_8O_{23}(cr) = Mo_4O_{11}(cr) + 4\ MoO_3(\ell)$, Brewer (4) calculated a $\Delta_r G°/RT$ value at 1058 K. This yielded, using auxiliary data (5), $\Delta_r H°/R$ = 380±200 K for the disproportionation of Mo_8O_{23} to crystalline Mo_4O_{11} and MoO_3 at 298.15 K. This enthalpy of reaction yields $\Delta_f H°(298.15\ K)/R$ = -87454±150 K from which we calculate $\Delta_f H°(MoO_{2.875}$, cr, 298.15 K) = -173.788±0.3 kcal mol^{-1}. We increase the uncertainty in $\Delta_f H°$ to ±0.6 kcal mol^{-1}; a value which better represents the effect of uncertainties in reducing a $\Delta_r G°$ value at 1058 K to a $\Delta_f H°$ value at 298.15 K.

Heat Capacity and Entropy

Bousquet et al. (6) used differential enthalpy analysis to obtain enthalpies between 379 and 971 K for $Mo_4O_{11}(cr)$, $Mo_8O_{23}(cr)$, and $Mo_9O_{26}(cr)$. Brewer (4) compared their values with values of $MoO_2(cr)$ and $MoO_3(cr)$ mixtures with the same composition. For $Mo_8O_{23}(cr)$, the enthalpies were lower by 15% at 400 K, lower by 3% at 500 K, and higher at higher temperatures, becoming 6% high at 700 K. As these differences were within the scalter of the expeirmental measurements, Brewer (4) assumed $\Delta C_p°$ = 0 and $\Delta S°$ = 0 for the formation of $Mo_4O(cr)$ from $3MoO_2(cr)$ and $MoO_2(cr)$.

Bousquet et al. (6) also measured high temperature enthalpy data of MoO_3 from 379 to 971 K. Their values agree with the adopted JANAF values (5) for $MoO_3(cr)$ within 2% below 500 K, but the deviations increase rapidly at higher temperatures becoming 10% high at 950 K. Thus, the adopted heat capacity values for $Mo_8O_{23}(cr)$, based on the assumption $\Delta C_p°$ = 0 as discussed in the preceding paragraph, may be less reliable at higher temperatures.

Phase Data

The phase Mo_8O_{23}(mP62) was found by Kihlborg (7) and Bousquet and Guillon (8) at temperatures up to 1058-1063 K. Bousquet and Guillon (8) observed Mo_8O_{23} to disproportionate to $Mo_4O_{11}(cr)$ and $MoO_3(g)$ at 1053 K by differential thermal analysis. Brewer (4) adopted Mo_8O_{23} as a stable phase which disproportionates to Mo_4O_{11} and liquid at a temperature, T_{dec} = 1058±5 K, which is slightly above the disproportionation temperature of Mo_9O_{26}.

In addition to the discussion of the molybdenum-oxygen system by Brewer (4), Gmeli Handbuch der Anorganischen Chemie (9) has recently presented a thorough summary of the literature dealing with the phase diagram and properties of the oxide phases. Ekstrom and Tilley (10) have reviewed the work of Ekstrom as well as the earlier work of Kihlborg (11) on the phases in the Mo-O system.

References

1. H. Le Bresq, J. J. Oehlig, and F. Marion, C. R. Acad. Sci. (Paris), Ser. C, 268, 1047 (1969).
2. S. Zador and C. B. Alcock, J. Chem. Thermodyn. 2, 9 (1970).
3. V. M. Zhukovskii, et al., Russ. J. Phys. Chem. 46, 1542 (1972).
4. L. Brewer, Materials and Molecular Research Division, Lawrence Berkeley Laboratory, University of California, Berkeley; personal communication, Sept. 29, 1978, preliminary draft of review to be submitted for publication to Atomic Energy Review, International Atomic Energy Agency, Vienna, Austria.
5. JANAF Thermochemical Tables: $MoO_2(cr)$, $MoO_3(cr)$, and $MoO_{2.750}(cr)$, 9-30-78.
6. J. Bousquet, J. C. David, M. Diot, and A. Guillon, Bull. Soc. Chim. Fr. (6), 2390 (1968).
7. L. Kihlborg, Acta Chem. Scand. 13, 954 (1959).
8. J. Bousquet and A. Guillon, Bull. Soc. Chim. Fr. (7), 2250 (1966).
9. Gmelin Handbuch der Anorganische Chemie, Main Series, 8th Ed., Molybdan, System No. 53, Eganzungsband Teil B1, Springer-Verlag, Berlin, 1975.
10. T. Ekstrom and R. J. D. Tilley, J. Solid State Chem. 19, 125 (1976).

Molybdenum Oxide ($MoO_{2.875}$) $Mo_1O_{2.875}(cr)$

Enthalpy Reference Temperature = T_r = 298.15 K — Standard State Pressure = p° = 0.1 MPa

T/K	$C_p°$ (J $K^{-1}mol^{-1}$)	S° (J $K^{-1}mol^{-1}$)	$-[G°-H°(T_r)]/T$ (J $K^{-1}mol^{-1}$)	$H°-H°(T_r)$ (kJ mol^{-1})	$\Delta_f H°$ (kJ mol^{-1})	$\Delta_f G°$ (kJ mol^{-1})	Log K_f
0							
100							
200							
298.15	72.508	73.849	73.849	0.	-727.128	-652.662	114.344
300	72.730	74.298	73.850	0.134	-727.116	-652.200	113.558
400	80.651	96.385	76.807	7.831	-726.147	-627.358	81.925
500	85.274	114.891	82.625	16.133	-724.793	-602.813	62.975
600	89.201	130.786	89.358	24.857	-723.229	-578.561	50.368
700	93.079	144.826	96.297	33.970	-721.468	-554.586	41.384
800	97.088	157.515	103.169	43.477	-719.481	-530.878	34.663
900	101.256	169.189	109.864	53.393	-717.229	-507.435	29.451
1000	105.589	180.081	116.347	63.734	-714.679	-484.259	25.295
1100	110.062	190.353	122.612	74.516	-711.806	-461.353	21.908
1200	114.658	200.126	128.667	85.751	-708.592	-438.723	19.097
1300	119.366	209.489	134.527	97.451	-705.025	-416.376	16.730
1400	124.166	218.510	140.205	109.627	-701.096	-394.318	14.712
1500	129.049	227.243	145.718	122.287	-696.797	-372.553	12.973

PREVIOUS: CURRENT: September 1978

Molybdenum Oxide ($MoO_{2.875}$) $Mo_1O_{2.875}(cr)$

MOLYBDENUM OXIDE ($MoO_{2.889}$) CRYSTAL M_r = 142.1623

Pearson Notation: mP70

$S°(298.15\ K) = [74.281 \pm 1.7]\ J\ K^{-1}\ mol^{-1}$

$\Delta_f H°(0\ K)$ = Unknown

$\Delta_f H°(298.15\ K) = -729.16 \pm 2.5\ kJ\ mol^{-1}$

Enthalpy of Formation

For this oxide, the atomic ratio of oxygen to molybdenum is 2.889 or, more exactly, 26/9. The formula $MoO_{2.889}$ (upon which this tabulation is based) is more indicative of the oxygen content relative to MoO_2 and MoO_3, as opposed to Mo_9O_{26}. The thermochemical values describing $MoO_{2.889}$ are numerically 1/9 those of Mo_9O_{26}.

The high temperature emf measurements (1, 2, 3) on compositions between MoO_2 and MoO_3 gave widely divergent results, some of which are contradicted by the Mo-O phase diagram recommended by Brewer (4). Brewer (4) found that it was more practical to determine a $\Delta_f H°(298.15\ K)$ value for $MoO_{2.889}$ by fixing the thermodynamic properties from the peritectic equilibria together with the thermodynamic data for the liquid. From the Mo-O phase diagram (4), Mo_9O_{26} and Mo_8O_{23} solids are in equilibrium with the peritectic liquid of 74.6±0.1 atomic % oxygen at 1053±5 K. For the equilibrium $Mo_9O_{26}(cr) = Mo_8O_{23}(cr) + MoO_3(\ell)$, Brewer (4) calculated a $\Delta_r G°/RT$ value at 1053 K. This yielded, using auxiliary data (5), $\Delta_r H°/R = 50 \pm 50$ K for the disproportionation of Mo_9O_{26} to crystalline Mo_8O_{23} and MoO_3 at 298.15 K. This enthalpy of reaction yields $\Delta_f H°(298.15\ K)/R = -87699 \pm 150$ K from which we calculate $\Delta_f H°(MoO_{2.889}, cr, 298.15\ K) = -174.275 \pm 0.3\ kcal\ mol^{-1}$. We increase the uncertainty in $\Delta_f H°$ to $\pm 0.6\ kcal\ mol^{-1}$; a value which better represents the effects of uncertainties in reducing a $\Delta_r G°$ value of 1053 K to a $\Delta_f H°$ value at 298.15 K.

Heat Capacity and Entropy

Bousquet et al. (6) used differential enthalpy analysis to obtain enthalpies between 379 and 971 K for $Mo_4O_{11}(cr)$, $Mo_8O_{23}(cr)$, and $Mo_9O_{26}(cr)$. Brewer (4) compared their values with values of $MoO_2(cr)$ and $MoO_3(cr)$ mixtures with the same composition. For $Mo_9O_{26}(cr)$, the enthalpies were lower by 10% at 400 K but crossed over between 500 and 600 K and were 3% higher at 700 K. As these differences were within the scatter of the experimental measurements, Brewer (4) assumed $\Delta C_p° = 0$ and $\Delta S° = 0$ for the formation of $Mo_9O_{25}(cr)$ from $MoO_2(cr)$ and 8 $MoO_3(cr)$.

Bousquet et al. (6) also measured high temperature enthalpy data of MoO_3 from 379 to 971 K. Their values agree with the adopted JANAF values (5) for $MoO_3(cr)$ within 2% below 500 K, but the deviations increase rapidly at higher temperatures becoming 10% high at 950 K. Thus, the adopted heat capacity values for $Mo_9O_{26}(cr)$, based on the assumption $\Delta C_p° = 0$ as discussed in the preceding paragraph, may be less reliable at hgher temperatures.

Phase Data

Mo_9O_{26} (mP70) disproportionates at 1053±5 K to Mo_4O_{11} and liquid according to the phase diagram of Phillips and Chang (7, 8). Brewer (4) adopted a phase diagram which included Mo_8O_{23} which Phillips and Chang (7, 8) did not have.

In addition to the discussion of the molybdenum-oxygen system by Brewer (4), Gmelin Handbuch der Anorganischen Chemie (9) has recently presented a thorough summary of the literature dealing with the phase diagram and properties of the oxide phases. Ekstrom and Tilley (10) have reviewed the work of Ekstrom as well as the earlier work of Kihlborg (11) on the phases in the Mo-O system.

References

1. H. LeBresq, J. J. Oehlig, and F. Marion, C. R. Acad. Sci. (Paris), Ser. C 268, 1047 (1969).
2. S. Zador and C. B. Alcock, J. Chem. Thermodyn. 2, 9 (1970).
3. V. M. Zhukovskii et al., Russ. J. Phys. Chem. 46, 1542 (1972).
4. L. Brewer, Materials and Molecular Research Division, Lawrence Berkeley Laboratory, University of California, Berkeley: personal communication, Sept. 29, 1978, preliminary draft of review to be submitted for publication in Atomic Energy Review, International Atomic Energy Agency, Vienna, Austria.
5. JANAF Thermochemical Tables: $MoO_2(cr)$, $MoO_3(cr)$, and $Mo_8O_{23}(cr)$, 9-30-78.
6. J. Bousquet, J. C. David, M. Diot, and A. Guillon, Bull. Soc. Chim. Fr. (6), 2390 (1968).
7. L. L. Y. Chang and B. Phillips, J. Amer. Chem. Soc. 52, 527 (1969).
8. B. Phillips and L. L. Y. Chang, AIME 233, 1433 (1965).
9. Gmelin Handbuch der Anorganischen Chemie, Main Series, 8th Ed., Molybdan, System No. 53, Eganzungsband Teil Bl, Springer-Verlag, Berlin, 1975.
10. T. Ekstrom and R. J. D. Tilley, J. Solid State Chem. 19, 125 (1976).
11. L. Kihlborg, Acta Chem. Scand. 13, 954 (1959).

Molybdenum Oxide ($MoO_{2.889}$) $Mo_1O_{2.889}(cr)$

Enthalpy Reference Temperature = T_r = 298.15 K; Standard State Pressure = p° = 0.1 MPa

T/K	$C_p°$ (J K⁻¹mol⁻¹)	S° (J K⁻¹mol⁻¹)	$-[G°-H°(T_r)]/T$ (J K⁻¹mol⁻¹)	$H°-H°(T_r)$ (kJ mol⁻¹)	$\Delta_f H°$ (kJ mol⁻¹)	$\Delta_f G°$ (kJ mol⁻¹)	Log K_f
0							
100							
200							
298.15	72.772	74.281	74.281	0.	-729.165	-654.461	114.659
300	72.995	74.732	74.282	0.135	-729.153	-653.998	113.871
400	80.923	96.896	77.250	7.858	-728.175	-629.078	82.149
500	85.550	115.463	83.088	16.188	-726.812	-604.456	63.147
600	89.487	131.409	89.843	24.940	-725.239	-580.129	50.505
700	93.383	145.494	96.806	34.082	-723.469	-556.082	41.495
800	97.415	158.225	103.700	43.621	-721.470	-532.303	34.756
900	101.611	169.940	110.417	53.571	-719.204	-508.790	29.529
1000	105.973	180.870	116.922	63.949	-716.637	-485.545	25.362
1100	110.476	191.181	123.208	74.770	-713.746	-462.573	21.966
1200	115.105	200.991	129.284	86.048	-710.511	-439.879	19.147
1300	119.846	210.391	135.164	97.795	-706.919	-417.470	16.774
1400	124.679	219.449	140.863	110.021	-702.961	-395.351	14.751
1500	129.595	228.218	146.396	122.734	-698.631	-373.528	13.007

PREVIOUS: CURRENT: September 1978

Pearson Notation: oP16

$S°(298.15\ K) = 77.760 \pm 1.3\ J\ K^{-1}\ mol^{-1}$

$T_{fus} = 1074 \pm 5\ K$

$\Delta_f H°(0\ K) = -740.15 \pm 0.4\ kJ\ mol^{-1}$

$\Delta_f H°(298.15\ K) = -745.17 \pm 0.4\ kJ\ mol^{-1}$

$\Delta_{fus} H° = 48.534 \pm 4.2\ kJ\ mol^{-1}$

Enthalpy of Formation

The adopted enthalpy of formation, $-178.1\ kcal\ mol^{-1}$, is the weighted average of two calorimetric determinations (1, 2). The result of these and other calorimetric determinations, all for the reaction $Mo(cr) + 3/2\ O_2(g) \rightarrow MoO_3(cr)$, are given below. Staskiewicz et al. (1) found that oxidation was only about 80% complete and tentatively attributed this to formation of $MoO_2(cr)$. Using their analytical results and their value of enthalpy of combustion for the dioxide, we obtain $\Delta_f H°(MoO_3,\ cr,\ 298.15\ K) = -177.98\ kcal\ mol^{-1}$. Subsequent discovery of oxides intermediate between MoO_2 and MoO_3 raises a possible uncertainty concerning the incomplete oxidation. Mah (2) also reported that combustion of the Mo(cr) sample ranged from 84 to 93 percent complete, as determined only by weight gain of the total combustion products upon prolonged low temperature ignition. In contrast, the earlier workers reported almost complete combustion. Neumann et al. (3) found about 97% completion using a wet analysis involving aqua regia. Moose and Parr (4) found that in several combustions, they obtained over 99% completion, based on the weight of the oxide formed.

Source	Year	$\Delta_f H°(298.15\ K)$, kcal mol^{-1}
(1)	1955	-177.98
(2)	1957	-178.16
(3)	1934	-180.4
(4)	1924	-175.6

Heat Capacity and Entropy

Smith et al. (5) measured low temperature heat capacities from 18.3 to 299.8 K. King et al. (6) measured high temperature enthalpy data from 399.3 to 1063.9 K by drop calorimetry. The adopted heat capacities are derived from these two sets of data, subject to the constraint that they join smoothly near 300 K. Deviations of the enthalpy data from the adopted values are -0.65% to +0.17%. Seltz et al. (7) have also measured low temperature heat capacities from 70 to 298.7 K which are not in agreement with the values adopted; the discrepancy has been discussed by Smith. Cosgrove and Snyder (8) measured high temperture enthalpy data from 273 to 1068 K by drop calorimetry. Their smooth values are about 2% higher than the values adopted in the tabulation. Bousquet et al. (9) also measured high temperature enthalpy data from 379 to 971 K. Their values agree with the adopted values within 2% below 500 K, but the deviations increase rapidly at higher temperature becoming +10% at 950 K.

The entropy, $S°(298.15\ K) = 18.585\ cal\ K^{-1}\ mol^{-1}$, is obtained from the low temperature heat capacity data of Smith et al. (5), based on $S°(20\ K) = 0.197\ cal\ K^{-1}\ mol^{-1}$. This starting entropy was obtained by the authors from a T^2 extrapolation of the data. Recently, McBride and Westrum (10) have shown that the T^2 limiting law is inappropriate for MoS_2 and a T^3 law is suggested for this and other compounds with lamellar lattices. They have also shown that the heat capacity data of Smith et al. for MoS_2 (5) was positively biased by 0.07 cal $K^{-1}\ mol^{-1}$ below 100 K. If the same bias is present in the $C°_p$ data of Smith et al. for MoO_3, and a T^3 extrapolation is used, our analysis suggests a reduction of ~0.12 cal $K^{-1}\ mol^{-1}$ in S°(298.15 K). Other treatments result in estimated decreases in S°(298.15 K) ranging from 0.08 to 0.17 cal $K^{-1}\ mol^{-1}$. The increasing scatter in the $C°_p$ data below 40 K (indicative of a loss of calorimeter sensitivity) ccupled with uncertainty due to bias suggests an uncertainty of ±0.3 cal $K^{-1}\ mol^{-1}$ for S°(298.15 K). Therefore, we adopt the author's original interpretation while recognizing the need for new low temperature $C°_p$ measurements. It is surprising that the entropy of $MoO_3(cr)$ is larger than that of $WO_3(cr)$ by about 0.4 cal $K^{-1}\ mol^{-1}$. This may be due to the difference in the crystal structures, MoO_3 being orthorhomibc (oP16) and WO_3 monoclinic, and also to uncertainty in the entropy extrapolation below 50 K for $WO_3(cr)$.

Fusion Data

The adopted melting point, 1074 Δ 5 K, is obtained from King et al. (6). The enthalpy of fusion is calculated from the adopted enthalpies of the crystal and liquid based on the enthalpy data of King. Cosgrove and Snyder (8) found the melting point 1068 K by the cooling curve method and derived $\Delta_{fus} H° = 12.54\ kcal\ mol^{-1}$ from their smooth enthalpy values at the melting point.

References

1. B. A. Staskiewicz, J. R. Tucker and P. E. Snyder, J. Amer. Chem. Soc. 77, 2987 (1955).
2. A. D. Mah, J. Phys. Chem. 61, 1572 (1957).
3. B. Neumann, C. Kroger and H. Kunz, Z. Anorg. Allg. Chem. 218, 387 (1934).
4. J. E. Moose and S. W. Parr, J. Amer. Chem. Soc. 46, 2856 (1924).
5. D. F. Smith, D. B. Brown, A. S. Dworkin, D. J. Sasmore and E. R. Van Artsdalen, J. Amer. Chem. Soc. 78, 1533 (1956).
6. E. G. King, W. W. Weller and A. U. Christensen, U. S. Bur. Mines RI 5664, 1960.
7. H. Seltz, F. J. Dunkerley and B. J. DeWitt, J. Amer. Chem. Soc. 65, 600 (1943).
8. L. A. Cosgrove and P. E. Snyder, J. Amer. Chem. Soc. 75, 1227 (1953).
9. J. Bousquet, J. C. David, M. Diot and A. Guillon, Bull. Soc. Chim. France, 2390 (1968).
10. J. J. McBride and E. F. Westrum, Jr., J. Chem. Thermodyn. 8, 37 (1976).

Enthalpy Reference Temperature = T_r = 298.15 K; Standard State Pressure = p° = 0.1 MPa

T/K	$C°_p$ (J K⁻¹mol⁻¹)	S° (J K⁻¹mol⁻¹)	$-[G°-H°(T_r)]/T$ (J K⁻¹mol⁻¹)	$H°-H°(T_r)$ (kJ mol⁻¹)	$\Delta_f H°$ (kJ mol⁻¹)	$\Delta_f G°$ (kJ mol⁻¹)	Log K_f
0	0.	0.	INFINITE	-12.590	-740.150	-740.150	INFINITE
100	31.648	19.553	132.814	-11.326	-743.735	-718.992	375.563
200	59.279	50.977	84.136	-6.632	-745.249	-693.504	181.125
298.15	74.881	77.760	77.760	0.	-745.170	-668.079	117.045
300	75.107	78.224	77.761	0.139	-745.158	-667.601	116.240
400	83.094	101.065	80.821	8.098	-744.111	-641.890	83.822
500	87.751	120.130	86.830	16.650	-742.696	-616.494	64.405
600	91.776	136.487	93.775	25.627	-741.074	-591.403	51.486
700	95.814	150.936	100.927	35.006	-739.250	-566.599	42.280
800	100.031	164.003	108.007	44.796	-737.186	-542.072	35.394
900	104.441	176.037	114.906	55.018	-734.839	-517.821	30.054
1000	109.039	187.278	121.587	65.691	-732.172	-493.847	25.796
1074.000	112.562	195.186	126.387	73.889	------ CRYSTAL <--> LIQUID -----		
1100	113.792	197.894	128.046	76.833	-729.156	-470.158	22.326
1200	118.679	208.004	134.291	88.455	-725.775	-446.759	19.447
1300	123.683	217.700	140.336	100.573	-722.014	-423.657	17.023
1400	128.779	227.052	146.198	113.195	-717.862	-400.860	14.956
1500	133.959	236.113	151.892	126.331	-713.312	-378.374	13.176

PREVIOUS: June 1967 CURRENT: September 1978

MOLYBDENUM OXIDE (MoO_3) LIQUID M_r = 143.9382

$S°(298.15\ K)$ = [114.935] $J\ K^{-1}\ mol^{-1}$ $\Delta_f H°(298.15\ K)$ = [-703.727] $kJ\ mol^{-1}$

T_{fus} = 1074 ± 5 K $\Delta_{fus}H°$ = 48.534 ± 4.2 $kJ\ mol^{-1}$

Enthalpy of Formation

$\Delta_f H°(MoO_3, \ell, 298.15\ K)$ is calculated from that of the crystal by adding the enthalpy of fusion, $\Delta_{fus}H°$, and the difference in enthalpy, H°(1074 K)-H°(298.15 K), between the crystal and liquid.

Heat Capacity and Entropy

The heat capacity is derived from the enthalpy data (1074 - 1392 K) measured by King et al. (1). The heat capacity is assumed constant above and below the measured range. At 750 K, a glass transition is assumed, below which the heat capacity is taken to be the same as that of the crystal. $S°(MoO_3, \ell, 298.15\ K)$ is calculated in a manner analogous to that used for the enthalpy of formation.

Cosgrove and Snyder (2) also measured enthalpy data for $MoO_3(\ell)$ in the temperature range from 1068 to 1300 K. Their smooth values are 2.5 to 4.0 percent higher than our tabulated values.

Fusion Data

See the crystal table for details.

Vaporization Data

Effusion (3) and transpiration (4) measurements indicate that $MoO_3(g)$ is not the major species found in the vapor above MoO_3 condensed phases. Both sets of measurements indicate the presence of polymeric species $(MoO_3)_x$ where the predominant species near the normal boiling point are those for x=3 and 4. At higher temperatures the equilibrium shifts toward lower x but $MoO_3(\ell)$ will reach 1 atm at 1380±30 K based primarily on the transpiration measurements of Ackermann et al. (4). Feiser (6), using indirect methods, reported a normal boiling point of 1438 K. We adopt Brewer's estimate (5) and emphasize that this represents vaporization to a mixture of gases, $(MoO_3)_x$.

References

1. E. G. King, W. W. Weller and A. U. Christensen, U. S. Bur. Mines RI 5664, 1960.
2. L. A. Cosgrove and P. E. Snyder, J. Amer. Chem. Soc. 75, 1227 (1953).
3. J. Berkowitz, M. G. Inghram and W. A. Chupka, J. Chem. Phys. 26, 842 (1957).
4. R. J. Ackermann, R. J. Thorn, C. Alexander and M. Tetenbaum, J. Phys. Chem. 64, 350 (1960).
5. L. Brewer, Materials and Research Division, Lawrence Berkeley Laboratory, University of California, Berkeley, personal communication, Sept. 29, 1978, of preliminary draft of review to be submitted for publication in Atomic Energy Review, International Atomic Energy Agency, Vienna, Austria.
6. J. Feiser, Metall Erz. 28, 297 (1931).

Molybdenum Oxide (MoO_3) $Mo_1O_3(l)$

Enthalpy Reference Temperature = T_r = 298.15 K — Standard State Pressure = p° = 0.1 MPa

T/K	$C_p°$ (J K⁻¹mol⁻¹)	S° (J K⁻¹mol⁻¹)	-[G°-H°(T_r)]/T (J K⁻¹mol⁻¹)	H°-H°(T_r) (kJ mol⁻¹)	$\Delta_f H°$ (kJ mol⁻¹)	$\Delta_f G°$ (kJ mol⁻¹)	Log K_f
0							
100							
200							
298.15	74.881	114.935	114.935	0.	-703.727	-637.719	111.726
300	75.107	115.399	114.936	0.139	-703.714	-637.310	110.966
400	83.094	138.240	117.996	8.098	-702.668	-615.317	80.352
500	87.751	157.305	124.005	16.650	-701.252	-593.638	62.017
600	91.776	173.661	130.949	25.627	-699.631	-572.264	49.820
700	95.814	188.110	138.102	35.006	-697.807	-551.178	41.129
750.000	97.898	194.791	141.660	39.848	GLASS <--> LIQUID		
750.000	126.951	194.791	141.660	39.848	TRANSITION		
800	126.951	202.985	145.240	46.196	-694.343	-530.415	34.633
900	126.951	217.937	152.503	58.891	-689.523	-510.214	29.612
1000	126.951	231.313	159.727	71.586	-684.833	-490.544	25.623
1074.000	126.951	240.376	164.975	80.980	CRYSTAL <--> LIQUID		
1100	126.951	243.413	166.794	84.281	-680.265	-471.337	22.382
1200	126.951	254.459	173.645	96.976	-675.811	-452.541	19.699
1300	126.951	264.620	180.258	109.671	-671.472	-434.112	17.443
1400	126.951	274.028	186.624	122.366	-667.247	-416.012	15.522
1500	126.951	282.787	192.746	135.061	-663.139	-398.212	13.867
1600	126.951	290.980	198.633	147.757	-659.152	-380.680	12.428
1700	126.951	298.677	204.293	160.452	-655.290	-363.395	11.166
1800	126.951	305.933	209.740	173.147	-651.560	-346.333	10.050
1900	126.951	312.797	214.985	185.842	-647.968	-329.475	9.058
2000	126.951	319.309	220.040	198.537	-644.523	-312.801	8.170

PREVIOUS: June 1967 CURRENT: September 1978

Molybdenum Oxide (MoO_3) $Mo_1O_3(l)$

MOLYBDENUM OXIDE (MoO_3) CRYSTAL-LIQUID M_r = 143.9382

0 to 1074 K	crystal (oP16)
above 1074 K	liquid

Refer to the individual tables for details.

Molybdenum Oxide (MoO_3) $Mo_1O_3(cr,l)$

Enthalpy Reference Temperature = T_r = 298.15 K Standard State Pressure = $p°$ = 0.1 MPa

	J K⁻¹mol⁻¹			kJ mol⁻¹			
T/K	$C_p°$	$S°$	$-[G°-H°(T_r)]/T$	$H°-H°(T_r)$	$\Delta_fH°$	$\Delta_fG°$	Log K_f
0	0.	0.	INFINITE	-12.590	-740.150	-740.150	INFINITE
100	31.648	19.553	132.814	-11.326	-743.735	-718.992	375.563
200	59.279	50.977	84.136	-6.632	-745.249	-693.504	181.125
298.15	74.881	77.760	77.760	0.	-745.170	-668.079	117.045
300	75.107	78.224	77.761	0.139	-745.158	-667.601	116.240
400	83.094	101.065	80.821	8.098	-744.111	-641.890	83.822
500	87.751	120.130	86.830	16.650	-742.696	-616.494	64.405
600	91.776	136.487	93.775	25.627	-741.074	-591.403	51.486
700	95.814	150.936	100.927	35.006	-739.250	-566.599	42.280
800	100.031	164.003	108.007	44.796	-737.186	-542.072	35.394
900	104.441	176.037	114.906	55.018	-734.839	-517.821	30.054
1000	109.039	187.278	121.587	65.691	-732.172	-493.847	25.796
1074.000	112.562	195.186	126.387	73.889	CRYSTAL <--> LIQUID		
1074.000	126.951	240.376	126.387	122.424	TRANSITION		
1100	126.951	243.413	129.118	125.724	-680.265	-471.337	22.382
1200	126.951	254.459	139.109	138.420	-675.811	-452.541	19.699
1300	126.951	264.620	148.378	151.115	-671.472	-434.112	17.443
1400	126.951	274.028	157.022	163.810	-667.247	-416.012	15.522
1500	126.951	282.787	165.117	176.505	-663.139	-398.212	13.867
1600	126.951	290.980	172.730	189.200	-659.152	-380.680	12.428
1700	126.951	298.677	179.915	201.895	-655.290	-363.395	11.166
1800	126.951	305.933	186.716	214.590	-651.560	-346.333	10.050
1900	126.951	312.797	193.173	227.285	-647.968	-329.475	9.058
2000	126.951	319.309	199.319	239.980	-644.523	-312.801	8.170

PREVIOUS: CURRENT: September 1978

Molybdenum Oxide (MoO_3) $Mo_1O_3(cr,l)$

MOLYBDENUM OXIDE (MoO_3) IDEAL GAS M_r = 143.9382

$\Delta_{at}H°(0\ K)$ = 1740.126 ± 21.3 kJ mol^{-1}
S°(298.15 K) = [283.898 ± 12.6] J K^{-1} mol^{-1}

$\Delta_f H°(0\ K)$ = -342.22 ± 20.9 kJ mol^{-1}
$\Delta_f H°(298.15\ K)$ = -346.44 ± 20.9 kJ mol^{-1}

Vibrational Frequencies and Degeneracies

ν, cm^{-1}	ν, cm^{-1}
978 (1)	924 (2)
[340](1)	[310](2)

Ground State Quantum Weight: [1] σ = 3
Point Group: C_{3v}
Bond Distance: Mo-O = [1.73] Å
Bond Angle: O-Mo-O = 99°
Product of the Moments of Inertia: $I_A I_B I_C$ = [3.0307222 x 10^{-114}] g^3 cm^6

Enthalpy of Formation

Burns et al. (1) determined the partial pressures of gaseous oxides MoO_2, MoO_3, $(MoO_3)_2$ and $(MoO_3)_3$ in equilibrium with powdered MoO_2 by mass spectrometry. Plante (2), Blackburn et al. (3) and Chizhikov et al. (4) also investigated the vapor pressures above molybdenum dioxide by the Knudsen effusion method. These studies (1-4) involve the disproportionation of MoO_2(cr) to Mo(cr) and MoO_3(g). DeMaria et al. (5) reported the partial pressures of MoO(g), MoO_2(g), MoO_3(g) and O(g) in the Mo-Al_2O_2 system from mass spectrometry.

Source	Reaction	T/K	Data Points	$\Delta_r H°$(298.15 K), kcal mol^{-1} 2nd law	3rd law	Drift cal K^{-1} mol^{-1}	$\Delta_f H°$(298.15 K) kcal mol^{-1}
(1)	1.5 MoO_2(cr) → MoO_3(g) + 0.5 Mo(cr)	1481-1777	34	131.7±1.5	128.0±0.8	-2.2±0.9	-82.8
(2)	1.5 MoO_2(cr) → MoO_3(g) + 0.5 Mo(cr)	1592-1918	119	133.5±1.4	124.0±1.3	-6.3±0.8	-86.8
(3)	1.5 MoO_2(cr) → MoO_3(g) + 0.5 Mo(cr)	1818-2028	7	108.8±6.4	121.2±1.6	6.3±3.4	-89.5
(4)	1.5 MoO_2(cr) → MoO_3(g) + 0.5 Mo(cr)	1500-1730	Eqn	137.9	133.1±0.5	-3.0	-77.6
(5)	Mo(cr) + 3 O(g) → MoO_3(g)	2281-2416	6	-210.5±28.0	-253.7±4.0	-18.6±11.8	-75.0

The selected enthalpy of formation is -82.8 kcal mol^{-1}, based on the 3rd law value of $\Delta_f H°$(298.15 K) of Burns et al. (1). The choice is made because there is serious doubt about the data of Blackburn (3) as discussed by Burns et al. (1). In addition, the assumption made by Plante (2) that they only vapor species of any importance is MoO_3 may be erroneous. The mass spectrometric work of Burns et al. (1) is thus preferred. Our adopted value differs slightly from the value of -82.57 kcal mol^{-1} adopted by Brewer (7), which is also based on Burns et al. (1). The difference is a result of slight differences in Gibbs energy functions in the respective auxiliary data for Mo(cr) (6, 7).

The heat of atomization calculated from the JANAF tables (6) is $\Delta_{at}H°(0\ K)$ = 415.9±5.1 kcal mol^{-1}.

Heat Capacity and Entropy

In a recent IR matrix isolation study, Hewett et al. (8) observed peaks at 976 and 922 cm^{-1} which they assigned to stretching modes of MoO_3(g). We add 2 cm^{-1} to these frequencies to correct for the matrix effect. The observation of two stretching modes in the infrared spectrum suggests a pyramidal C_{3v} structure for MoO_3(g). Hewett et al. performed a normal coordinate analysis using a bond distance of 1.8 Å and an estimated bending force constant (f_α) of 0.296 mdyn/Å. Variations of 20% in f_α changed the calculated stretching frequencies by only a few cm^{-1}. The calculated stretching frequencies were more sensitive to the estimated pyramidal angle (β) which was adjusted to obtain the best fit to the observed stretching frequencies of oxide molecules containing ^{16}O, ^{18}O, and mixed ^{16}O-^{18}O. The final results yielded β = 61.5±2° and values of 261, 275 and 275 cm^{-1} for the bending frequencies of $Mo^{16}O_3$. Due to the insensitivity of the normal coordinate analysis to f_α these bending frequencies are useful only for comparing changes due to isotopic substitution. In contrast, we adopt bending frequencies of 310, 310 and 340 cm^{-1} and a bond distance of 1.73 Å as selected by Brewer (7) for consistency with other metal oxides. The pyramidal angle of 61.5° is adopted; the possible effect of using a shorter bond distance during the normal coordinate analysis is unknown. The pyramidal angle of 61.5° corresponds to an O-Mo-O bond angle of 99°. Based on the adopted structure, the principal moments of inertia are $I_A=I_B$=[12.8372 x 10^{-39}] and I_C=[18.3909 x 10^{-39}] g cm^2. The thermochemical functions are consistent with those of Brewer (7).

References

1. R. P. Burns, G. DeMaria, J. Drowart and R. T. Grimley, J. Chem. Phys. 32, 1363 (1960).
2. E. R. Plante, Ph.D. Thesis, Univ. of Kansas, 1960.
3. P. E. Blackburn, M. Hoch and H. L. Johnston, J. Phys. Chem. 62, 769 (1958).
4. D. M. Chizhikov, E. K. Kazenas and I. D. Ermilova, Russ. J. Phys. Chem. 50, 878 (1976).
5. G. DeMaria, R. P. Burns, J. Drowart and M. G. Inghram, J. Chem. Phys. 32, 1373 (1960).
6. JANAF Thermochemical Tables: Mo(cr) and Mo(g), 3-31-78; MoO_2(cr), 9-30-78; O(g), 3-31-77; WO_3(g), 9-30-66.
7. L. Brewer, Materials and Molecular Research Division, Lawrence Berkeley Laboratory, University of California, Berkeley, personal communication of preliminary draft of review to be submitted for publication in Atomic Energy Review, International Atomic Energy Agency, Vienna, Austria; Sept. 29, 1978.
8. W. D. Hewett, J. H. Newton and W. Weltner, Jr., J. Phys. Chem. 79, 2640 (1975).

Molybdenum Oxide (MoO_3) Mo_1O_3(g)

Enthalpy Reference Temperature = T_r = 298.15 K Standard State Pressure = p° = 0.1 MPa

T/K	$C_p°$ (J K^{-1}mol^{-1})	S°	-[G°-H°(T_r)]/T	H°-H°(T_r) (kJ mol^{-1})	$\Delta_f H°$	$\Delta_f G°$	Log K_f
0	0.	0.	INFINITE	-13.399	-342.225	-342.225	INFINITE
100	38.694	231.037	330.596	-9.956	-343.630	-340.035	177.616
200	50.906	261.923	289.070	-5.429	-345.312	-335.755	87.690
250	55.482	273.785	284.854	-2.767	-345.954	-333.289	69.637
298.15	59.378	283.897	283.897	0.	-346.435	-330.804	57.955
300	59.518	284.265	283.899	0.110	-346.451	-330.707	57.581
350	63.017	293.710	284.637	3.176	-346.814	-328.052	48.959
400	65.979	302.324	286.317	6.403	-347.071	-325.353	42.487
450	68.449	310.242	288.541	9.765	-347.249	-322.627	37.450
500	70.494	317.563	291.082	13.241	-347.370	-319.885	33.418
600	73.589	330.707	296.616	20.454	-347.512	-314.373	27.369
700	75.738	342.222	302.326	27.927	-347.594	-308.842	23.046
800	77.266	352.440	307.964	35.581	-347.666	-303.302	19.804
900	78.381	361.609	313.424	43.366	-347.756	-297.751	17.281
1000	79.214	369.912	318.664	51.248	-347.880	-292.189	15.262
1100	79.851	377.493	323.672	59.203	-348.051	-286.612	13.610
1200	80.348	384.463	328.452	67.214	-348.282	-281.017	12.232
1300	80.742	390.911	333.011	75.269	-348.582	-275.400	11.066
1400	81.059	396.906	337.364	83.359	-348.962	-269.756	10.065
1500	81.318	402.508	341.522	91.479	-349.430	-264.083	9.196
1600	81.532	407.763	345.500	99.622	-349.995	-258.376	8.435
1700	81.711	412.711	349.309	107.784	-350.665	-252.629	7.762
1800	81.862	417.386	352.962	115.963	-351.452	-246.841	7.163
1900	81.990	421.816	356.471	124.156	-352.363	-241.005	6.626
2000	82.100	426.024	359.844	132.360	-353.408	-235.117	6.141
2100	82.196	430.032	363.092	140.575	-354.598	-229.174	5.700
2200	82.278	433.858	366.222	148.799	-355.942	-223.170	5.299
2300	82.351	437.517	369.243	157.030	-357.451	-217.102	4.931
2400	82.415	441.023	372.161	165.269	-359.136	-210.965	4.592
2500	82.471	444.389	374.983	173.513	-361.007	-204.754	4.278
2600	82.521	447.624	377.716	181.763	-363.087	-198.463	3.987
2700	82.566	450.740	380.363	190.017	-365.416	-192.088	3.716
2800	82.606	453.743	382.930	198.276	-368.055	-185.621	3.463
2900	82.642	456.642	385.422	206.538	-406.969	-179.005	3.224
3000	82.675	459.445	387.843	214.804	-408.435	-171.118	2.979
3100	82.704	462.156	390.197	223.073	-409.925	-163.183	2.750
3200	82.731	464.782	392.487	231.345	-411.439	-155.200	2.533
3300	82.756	467.328	394.716	239.619	-412.977	-147.169	2.329
3400	82.778	469.799	396.889	247.896	-414.537	-139.090	2.137
3500	82.798	472.199	399.006	256.175	-416.120	-130.967	1.955
3600	82.817	474.532	401.072	264.456	-417.724	-122.796	1.782
3700	82.835	476.801	403.088	272.738	-419.348	-114.582	1.618
3800	82.851	479.010	405.057	281.023	-420.992	-106.322	1.462
3900	82.865	481.163	406.981	289.308	-422.655	-98.019	1.313
4000	82.879	483.261	408.862	297.596	-424.337	-89.675	1.171
4100	82.892	485.308	410.702	305.884	-426.037	-81.287	1.036
4200	82.904	487.305	412.502	314.174	-427.754	-72.858	0.906
4300	82.915	489.256	414.264	322.465	-429.489	-64.387	0.782
4400	82.925	491.162	415.990	330.757	-431.242	-55.877	0.663
4500	82.934	493.026	417.682	339.050	-433.011	-47.327	0.549
4600	82.943	494.849	419.339	347.344	-434.798	-38.737	0.440
4700	82.952	496.633	420.965	355.639	-436.603	-30.106	0.335
4800	82.960	498.379	422.560	363.934	-438.426	-21.438	0.233
4900	82.967	500.090	424.125	372.230	-440.267	-12.731	0.136
5000	82.974	501.766	425.661	380.528	-1033.511	1.744	-0.018
5100	82.981	503.409	427.169	388.825	-1037.070	22.485	-0.230
5200	82.987	505.021	428.651	397.124	-1040.796	43.295	-0.435
5300	82.993	506.602	430.107	405.423	-1044.686	64.181	-0.633
5400	82.998	508.153	431.538	413.722	-1048.742	85.140	-0.824
5500	83.004	509.676	432.945	422.022	-1052.961	106.176	-1.008
5600	83.009	511.172	434.328	430.323	-1057.343	127.292	-1.187
5700	83.013	512.641	435.689	438.624	-1061.885	148.485	-1.361
5800	83.018	514.085	437.028	446.926	-1066.587	169.761	-1.529
5900	83.022	515.504	438.347	455.228	-1071.446	191.116	-1.692
6000	83.026	516.899	439.644	463.530	-1076.463	212.559	-1.850

PREVIOUS: September 1978 (1 atm) CURRENT: September 1978 (1 bar)

Molybdenum Oxide (MoO_3) Mo_1O_3(g)

MOLYBDENUM SULFIDE (MoS_2) CRYSTAL M_r = 160.06

Pearson Notation: hP6

$S°(298.15\ K) = 62.572 \pm 0.08\ J\ K^{-1}\ mol^{-1}$

$T_{fus} = 2023 \pm 50\ K\ [p = 1\ atm\ S_2(g)]$

$\Delta_f H°(0\ K) = -273.31 \pm 2.5\ kJ\ mol^{-1}$

$\Delta_f H°(298.15\ K) = -276.14 \pm 2.5\ kJ\ mol^{-1}$

$\Delta_{fus} H°$ = Unknown

Enthalpy of Formation

The adopted enthalpy of formation is from Brewer (1) and is a weighted average of calorimetric (2) and equilibrium (3) values. O'Hare et al. (2) obtained $\Delta_f H°(298.15\ K) = -65.8 \pm 1.2\ kcal\ mol^{-1}$ using a fluorine bomb calorimeter. Larson and Elliott (3) studied the equilibrium $MoO_2(cr) + 2\ SO_2(g) = MoS_2(cr) + 3\ O_2(g)$ via emf technique (1051-1372 K); using auxiliary data (1, 4) Brewer calculated $\Delta_f H°(MoS_2, 298.15\ K) = -66.64 \pm 0.6\ kcal\ mol^{-1}$ from their results. Thermodynamic analysis of other calorimetric and equilibrium measurements (5-7) are summarized by O'Hare et al. (2) and Mills (8).

Heat Capacity and Entropy

The adopted values of $C_p°$ are from Brewer (1) and are based on experimental $C_p°$ data (6-346 K) of McBride and Westrum (9) and high temperature enthalpy data (523-1204 K) of Fredrickson and Chasanov (10). $C_p°$ values above 1200 K are extrapolated. The recent enthalpy data (500-1700 K) of Volovik et al. (11) agrees with enthalpies in this table to within ±2% to 2000 K. Older $C_p°$ data of Smith et al. (12) and Anderson (13) contain considerable scatter and are given no weight.

S°(298.15 K) is obtained from the appropriate integration of the adopted $C_p°$ data based on $S°(10\ K) = 0.012\ cal\ K^{-1}\ mol^{-1}$.

Fusion Data

The crystal structure of MoS_2 is (hP6). Mering and Levialdi (14) have shown that nonstoichiometric sulfur-rich modifications (which have essentially the hP6 structure with expansion and distortion of the lattice) can be prepared as metastable compositions up to $MoS_{2.59}$; amorphous metastable phases are obtained with higher sulfur contents. For information on other metastable phases, see (1) and (15). According to Brewer's (1) phase diagram the compositional range of stability decreases as the temperature increases.

Based on Cannon's evidence (16) of a melting point in excess of 2073 K, Brewer (1) adopted a congruent melting point of 2173 K. However, congruent melting is observed only at high sulfur pressures; the melting point of 2173 K corresponds to a $S_2(g)$ pressure of ~4 atm. Incongruent melting points up to 350 K lower may be obtained depending upon the pressure of $S_2(g)$. The adopted melting point is that selected by Brewer (1) for an $S_2(g)$ pressure of 1 atm. Mills (8) adopted a 1 atm value of 1458 K which appears much too low.

References

1. L. Brewer, Materials and Molecular Research Division, Lawrence Berkeley Laboratory, University of California, Berkeley; personal communication of preliminary draft of review to be submitted for publication in Atomic Energy Review, International Atomic Energy Agency, Vienna, Austria; September 29, 1978.
2. P. A. G. O'Hare, E. Benn, F. Y. Cheng and G. Kuzmycz, J. Chem. Thermodyn. 2, 797 (1970).
3. H. R. Larson and J. F. Elliott, Trans. AIME 239, 1713 (1967).
4. JANAF Thermochemical Tables: $O_2(g)$, 3-31-77; $SO_2(g)$, 6-30-61.
5. R. L. Nuttall and D. R. Fredrickson, U. S. Atomic Energy Report, ANL-6029 (1959).
6. I. A. Makolkin, Acta Physicochim. USSR 13, 361 (1940).
7. J. R. Stubbles and F. D. Richardson, Trans. Faraday Soc. 56, 1460 (1960).
8. K. C. Mills, "Thermodynamic Data for Inorganic Sulphides, Selenides and Tellurides," Butterworths, London, 1974.
9. J. J. McBride and E. F. Westrum, J. Chem. Thermodyn. 8, 37 (1976).
10. D. R. Fredrickson and M. G. Chasanov, J. Chem. Thermodyn. 3, 693 (1971).
11. L. S. Volovik, V. V. Fesenko, A. S. Bolgar, S. V. Drozdova, L. A. Klochkov and V. F. Primachenko, Chem. Abstr. 90, 13088t (1979).
12. D. F. Smith, D. Brown, A. S. Dworkin, D. J. Sasmar and E. R. VanArtsdalen, J. Amer. Chem. Soc. 78, 1533 (1956).
13. T. Anderson, J. Amer. Chem. Soc. 59, 486 (1937).
14. J. Mering and A. Levialdi, CR Acad. Sci. Paris 213, 798 (1941).
15. M. Hansen, "Constitution of Binary Alloys," McGraw-Hill, NY, 1958: R. P. Elliott, ibid., first supplement, 1965: F. A. Shunk, ibid., second supplement, 1969.
16. P. Cannon, Nature 183, 1612 (1959).

Molybdenum Sulfide (MoS_2) $Mo_1S_2(cr)$

Enthalpy Reference Temperature = T_r = 298.15 K — Standard State Pressure = p° = 0.1 MPa

T/K	$C_p°$ (J K⁻¹mol⁻¹)	S° (J K⁻¹mol⁻¹)	$-[G°-H°(T_r)]/T$ (J K⁻¹mol⁻¹)	$H°-H°(T_r)$ (kJ mol⁻¹)	$\Delta_f H°$ (kJ mol⁻¹)	$\Delta_f G°$ (kJ mol⁻¹)	Log K_f
0	0.	0.	INFINITE	-10.579	-273.314	-273.314	INFINITE
100	24.050	12.991	110.098	-9.711	-274.317	-272.409	142.292
200	51.840	39.359	68.080	-5.744	-275.476	-269.998	70.516
298.15	63.555	62.572	62.572	0.	-276.144	-267.156	46.805
300	63.680	62.965	62.573	0.118	-276.155	-267.100	46.506
400	68.910	82.099	65.146	6.781	-281.141	-263.822	34.452
500	71.756	97.805	70.155	13.825	-284.502	-259.125	27.071
600	73.597	111.057	75.897	21.096	-287.017	-253.800	22.095
700	75.019	122.511	81.756	28.528	-288.951	-248.105	18.514
800	76.232	132.610	87.494	36.092	-290.530	-242.159	15.811
900	77.278	141.650	93.018	43.769	-398.134	-233.867	13.573
1000	78.241	149.842	98.297	51.545	-396.882	-215.682	11.266
1100	79.119	157.341	103.329	59.413	-395.620	-197.623	9.384
1200	79.998	164.263	108.122	67.370	-394.356	-179.680	7.821
1300	80.835	170.699	112.691	75.411	-393.101	-161.841	6.503
1400	81.672	176.721	117.052	83.537	-391.861	-144.098	5.376
1500	82.467	182.383	121.220	91.744	-390.646	-126.444	4.403
1600	83.262	187.730	125.211	100.030	-389.463	-108.869	3.554
1700	84.057	192.802	129.039	108.396	-388.317	-91.367	2.807
1800	84.810	197.628	132.717	116.839	-387.219	-73.931	2.145
1900	85.605	202.235	136.255	125.361	-386.171	-56.555	1.555
2000	86.358	206.645	139.665	133.959	-385.182	-39.233	1.025
2100	87.111	210.876	142.956	142.632	-384.261	-21.958	0.546
2200	87.906	214.947	146.137	151.383	-383.410	-4.726	0.112
2300	88.659	218.871	149.214	160.211	-382.640	12.469	-0.283
2400	89.412	222.660	152.196	169.115	-381.961	29.633	-0.645

PREVIOUS: CURRENT: June 1978

Molybdenum Sulfide (MoS_2) $Mo_1S_2(cr)$

MOLYBDENUM SULFIDE (Mo_2S_3) CRYSTAL M_r = 288.06

Pearson notation: mP10

S°(298.15 K) = 114.976 ± 8.4 J K^{-1} mol^{-1}

T_{fus} = 2080 ± 20 K

$\Delta_f H°$(0 K) = Unknown

$\Delta_f H°$(298.15 K) = -407.103 ± 8.4 kJ mol^{-1}

$\Delta_{fus} H°$ = -129.704 ± 41.8 kJ mol^{-1}

Enthalpy of Formation

The adopted enthalpy of formation is taken from the critical review by Brewer (1). His linear regression analysis of equilibrium data, which is based on Cubicciotti's sigma-plot method (2), yields both $\Delta_f H°$(298.15 K) and S°(298.15 K). The data used in Brewer's analysis are outlined below. Auxiliary data are taken from the JANAF tables (3).

Source	Reaction	T/K
McCabe (4)	A	1303 - 1425
Isakova (5)	B	1523 - 1738
Hager and Elliott (6)	C	1366 - 1612
Stubbles and Richardson (7)	C	1119 - 1467
Pouillard and Perrot (8)	C	1030 - 1280
Stubbles and Richardson (7)	D	1270 - 1470
Pouillard and Perrot (8)	D	1030 - 1280

Reactions:
(A) 2 Mo(cr) + 3/2 S_2(g) = Mo_2S_3(cr)
(B) 2 MoS_2(cr) = Mo_2S_3(cr) + 1/2S_2(g)
(C) 2 Mo(cr) + 3 H_2S(g) = Mo_2S_3(cr) + 3 H_2(g)
(D) 2 MoS_2(cr) + H_2(g) = Mo_2S_3(cr) + H_2S(g)

Heat Capacity and Entropy

There are no experimental high temperature heat capacity or enthalpy studies. The adopted heat capacity, which is assumed to represent the high temperature crystal form, is based on the estimate of Gronvold and Westrum (9), $C_p°$ = 26.36 + 7.88 x 10^{-3}T = 2.30 x $10^5/T^2$. The value of S°(298.15 K) is taken from Brewer (1), see above.

Rastogi and Ray (10) measured the low temperature $C_p°$ (80-350 K) but report only $C_p°$(298.15 K) = 33.2 cal K^{-1} mol^{-1}. This value apparently refers to the low temperature superstructure crystal form but is not expected to differ significantly from the high temperature form. The reason for the 7 cal K^{-1} mol^{-1} discrepancy between the Rastogi-Ray value of $C_p°$(298.15 K) and that adopted here is not known but it should be mentioned that the Rastogi-Ray value is well above the classical limit of 15R. Insufficient information on sample preparation and calorimetric technique is available to assess the validity of the Rastogi-Ray data.

Phase Data

The crystal structure of Mo_2S_3 is monoclinic and belongs to space group $P2_1/m$ above 310 K (11). Below 310 K all axes are doubled and a superstructure is formed. The heat of transition is reported to be 5 cal K^{-1} mol^{-1} (10).

The stability region for Mo_2S_3 is from 900±40 K (1) to the melting point. Below 900 K the material is metastable with respect to disproportionation to Mo(cr) and MoS_2(cr). However equilibrium is established very slowly below 1100 K (11, 12) and, once prepared, Mo_2S_3 may be quenched and stored indefinitely at room temperature. The enthalpy of decomposition at 900 K ($\Delta_{dcm}H°$) for the reaction Mo_2S_3(cr) + 1/2 Mo(cr) + 3/2 MoS_2(cr) is calculated from the JANAF tables (3).

Fusion Data

We adopt the congruent melting point from Brewer (1). The enthalpy of fusion is estimated assuming an entropy of fusion of ~3 cal K^{-1} g-atom^{-1} as suggested by Kubaschewski et al. (13). This gives a enthalpy of fusion of 31 kcal mol^{-1} and the uncertainty is estimated to be 10 kcal mol^{-1}.

References

1. L. Brewer, Materials and Research Division, Lawrence Berkeley Laboratory, University of California, Berkeley; personal communication of preliminary draft of review to be submitted for publication in Atomic Energy Review, International Atomic Energy Agency, Vienna, Austria; Sept. 29, 1978.
2. D. Cubicciotti, J. Phys. Chem. 70, 2410 (1966).
3. JANAF Thermochemical Tables: Mo(cr) 3-31-78; MoS_2(cr) 6-30-78; S_2(g) 9-30-77; H_2S(g) 6-30-77.
4. C. L. McCabe, Trans. AIME, 203, 61 (1955).
5. R. A. Isakova, Izv. Akad. Nauk Kazakh, SSR, Ser. Metallurgi Obogasch i. Ogneuporov, #3, 3 (1961).
6. J. P. Hager and J. R. Elliott, Trans. AIME 239, 513 (1967).
7. J. R. Stubbles and F. D. Richardson, Trans. Faraday Soc. 56, 1460 (1960).
8. G. Pouillard and P. Perrot, C. R. Acad. Sci. Paris 281C, 143 (1975).
9. F. Gronvold and E. F. Westrum, Jr., Inorg. Chem. 1, 36 (1962).
10. A. K. Rastogi and R. K. Ray, Indian J. Cryogenics 1, 145 (1978).
11. R. DeJonge, T. J. A. Popma, G. A. Wiegers and F. Jellinek, J. Sol. State Chem. 2, 188 (1970).
12. N. Morimoto and G. Kullerud, Carnegie Inst. Washington Yearbook 61, 143 (1962).
13. O. Kubaschewski, E. L. Evans and C. B. Alcock, "Metallurgical Thermochemistry," 4th ed., Pergamon Press, Oxford, pp. 210-215, 1967.

Molybdenum Sulfide (Mo_2S_3) Mo_2S_3(cr)

Enthalpy Reference Temperature = T_r = 298.15 K; Standard State Pressure = p° = 0.1 MPa

T/K	$C_p°$ (J K⁻¹mol⁻¹)	S°	-[G°-H°(T_r)]/T	H°-H°(T_r) (kJ mol⁻¹)	$\Delta_f H°$	$\Delta_f G°$	Log K_f
0							
100							
200							
298.15	109.294	114.976	114.976	0.	-407.103	-395.654	69.317
300	109.489	115.653	114.978	0.202	-407.116	-395.583	68.877
400	117.464	148.339	119.381	11.583	-414.437	-391.374	51.108
500	122.926	175.164	127.934	23.615	-419.287	-385.077	40.229
600	127.399	197.981	137.754	36.136	-422.753	-377.885	32.898
700	131.405	217.925	147.812	49.079	-425.196	-370.204	27.625
800	135.162	235.719	157.707	62.409	-426.942	-362.222	23.651
900	138.775	251.848	167.285	76.107	-587.548	-350.811	20.361
1000	142.297	266.652	176.491	90.161	-584.687	-324.658	16.958
1100	145.762	280.378	185.319	104.564	-581.623	-298.802	14.189
1200	149.186	293.208	193.781	119.312	-578.373	-273.233	11.894
1300	152.582	305.283	201.898	134.401	-574.954	-247.942	9.962
1400	155.957	316.714	209.695	149.828	-571.381	-222.920	8.317
1500	159.317	327.589	217.195	165.591	-567.668	-198.159	6.901
1600	162.666	337.978	224.421	181.691	-563.828	-173.650	5.669
1700	166.006	347.940	231.396	198.124	-559.873	-149.384	4.590
1800	169.339	357.523	238.138	214.892	-555.817	-125.354	3.638
1900	172.666	366.768	244.667	231.992	-551.674	-101.552	2.792
2000	175.989	375.709	250.996	249.425	-547.458	-77.970	2.036
2080.000	178.645	382.663	255.927	263.610	------ CRYSTAL <--> LIQUID -----		
2100	179.309	384.376	257.142	267.190	-543.184	-54.601	1.358
2200	182.625	392.794	263.118	285.286	-538.867	-31.436	0.746
2300	185.939	400.985	268.935	303.715	-534.523	-8.468	0.192
2400	189.251	408.968	274.604	322.474	-530.172	14.311	-0.311

PREVIOUS: CURRENT: June 1978

Molybdenum Sulfide (Mo_2S_3) Mo_2S_3(cr)

MOLYBDENUM SULFIDE (Mo_2S_3) LIQUID M_r = 288.06

S°(298.15 K) = [181.170] J K^{-1} mol^{-1} $\Delta_fH°$(298.15 K) = [-270.304] kJ mol^{-1}

T_{fus} = 2080 ± 20 K $\Delta_{fus}H°$ = 129.704 ± 41.8 kJ mol^{-1}

Enthalpy of Formation

$\Delta_fH°$(Mo_2S_3, ℓ, 298.15 K) is calculated from that of the crystal by adding the enthalpy fusion, of $\Delta_{fus}H°$, and the difference in enthalpy, H°(2080 K)-H°(298.15 K), between the crystal and liquid.

Heat Capacity and Entropy

A constant liquid $C_p°$ of 7.5 cal K^{-1} g-$atom^{-1}$ is assumed which yields a $C_p°$ value of 37.5 cal K^{-1} mol^{-1}. This value is adopted above an assumed glass transition at 1450 K. The heat capacity of the crystal (1) is assumed below the glass transition.

S°(Mo_2S_3, ℓ, 298.15 K) in calculated by a manner analogous to that used for the enthalpy of formation.

Fusion Data

See the crystal table for details (1).

Decomposition Data

According to Brewer's phase diagram (2), Mo_2S_3 has a very short liquid range at 1 atm S_2(g). We adopt a decomposition temperature of 2140±30 K based on the phase diagram (2). Above this temperature Mo_2S_3(ℓ) decomposes to S_2(g) plus a liquid poorer in sulfur than Mo_2S_3. At higher sulfur pressures, the liquid range is much more extensive.

References

1. JANAF Thermochemical Tables: Mo_2S_3(cr) 6-30-78.
2. L. Brewer, Materials Research Division, Lawrence Berkeley Laboratory, University of California, Berkeley; personal communication of preliminary draft of review to be submitted for publication in Atomic Energy Review, International Atomic Energy Agency, Vienna, Austria; Sept. 29, 1978.

Molybdenum Sulfide (Mo_2S_3) Mo_2S_3(l)

Enthalpy Reference Temperature = T_r = 298.15 K Standard State Pressure = p° = 0.1 MPa

T/K	$C_p°$ (J $K^{-1}mol^{-1}$)	S°	-[G°-H°(T_r)]/T	H°-H°(T_r) (kJ mol^{-1})	$\Delta_fH°$	$\Delta_fG°$	Log K_f
0							
100							
200							
298.15	109.294	181.170	181.170	0.	-270.304	-278.590	48.808
300	109.489	181.847	181.172	0.202	-270.316	-278.641	48.516
400	117.464	214.533	185.575	11.583	-277.637	-281.052	36.702
500	122.926	241.358	194.129	23.615	-282.487	-281.374	29.395
600	127.399	264.175	203.948	36.136	-285.953	-280.802	24.446
700	131.405	284.119	214.006	49.079	-288.396	-279.740	20.874
800	135.162	301.913	223.901	62.409	-290.142	-278.377	18.176
900	138.775	318.042	233.479	76.107	-450.748	-273.586	15.878
1000	142.297	332.846	242.685	90.161	-447.887	-254.052	13.270
1100	145.762	346.572	251.513	104.564	-444.824	-234.816	11.150
1200	149.186	359.402	259.975	119.312	-441.574	-215.866	9.396
1300	152.582	371.477	268.092	134.401	-438.155	-197.195	7.923
1400	155.957	382.908	275.889	149.828	-434.582	-178.792	6.671
1450.000	157.639	388.410	279.674	157.668	GLASS <--> LIQUID		
1450.000	156.900	388.410	279.674	157.668	TRANSITION		
1500	156.900	393.730	283.388	165.513	-430.947	-160.649	5.594
1600	156.900	403.856	290.604	181.203	-427.516	-142.743	4.660
1700	156.900	413.368	297.548	196.893	-424.305	-125.044	3.842
1800	156.900	422.336	304.234	212.583	-421.327	-107.527	3.120
1900	156.900	430.819	310.676	228.273	-418.594	-90.170	2.479
2000	156.900	438.867	316.886	243.963	-416.121	-72.949	1.905
2080.000	156.900	445.021	321.696	256.515	------ CRYSTAL <--> LIQUID -----		
2100	156.900	446.522	322.878	259.653	-413.922	-55.846	1.389
2200	156.900	453.821	328.665	275.343	-412.011	-38.841	0.922
2300	156.900	460.796	334.260	291.033	-410.406	-21.915	0.498
2400	156.900	467.473	339.672	306.723	-409.124	-5.053	0.110
2500	156.900	473.878	344.913	322.413	-408.188	11.763	-0.246
2600	156.900	480.032	349.992	338.103	-407.636	28.548	-0.574
2700	156.900	485.953	354.919	353.793	-407.546	45.321	-0.877
2800	156.900	491.659	359.701	369.483	-408.042	62.101	-1.159
2900	156.900	497.165	364.347	385.173	-481.053	79.010	-1.423
3000	156.900	502.484	368.863	400.863	-479.131	98.290	-1.711
3100	156.900	507.629	373.257	416.553	-477.222	117.506	-1.980
3200	156.900	512.610	377.535	432.243	-475.327	136.660	-2.231
3300	156.900	517.439	381.701	447.933	-473.444	155.756	-2.465
3400	156.900	522.122	385.763	463.623	-471.573	174.794	-2.685
3500	156.900	526.671	389.724	479.313	-469.716	193.778	-2.892

PREVIOUS: CURRENT: June 1978

Molybdenum Sulfide (Mo_2S_3) Mo_2S_3(l)

MOLYBDENUM SULFIDE (Mo_2S_3) CRYSTAL-LIQUID M_r = 288.06

298 to 2080 K* crystal
above 2080 K liquid

Refer to the individual tables for details.

*This material is reported to be metastable with respect to disproportionation to Mo(cr) and MoS_2(cr) below ~900 K. However, disproportionation is very slow and the material may be quenched and stored at room temperature indefinitely.

Molybdenum Sulfide (Mo_2S_3) Mo_2S_3(cr,l)

Enthalpy Reference Temperature = T_r = 298.15 K   Standard State Pressure = p° = 0.1 MPa

T/K	C_p°	S°	$-[G^\circ-H^\circ(T_r)]/T$	$H^\circ-H^\circ(T_r)$	Δ_fH°	Δ_fG°	Log K_f
	J K⁻¹mol⁻¹			kJ mol⁻¹			
0							
100							
200							
298.15	109.294	114.976	114.976	0.	-407.103	-395.654	69.317
300	109.489	115.653	114.978	0.202	-407.116	-395.583	68.877
400	117.464	148.339	119.381	11.583	-414.437	-391.374	51.108
500	122.926	175.164	127.934	23.615	-419.287	-385.077	40.229
600	127.399	197.981	137.754	36.136	-422.753	-377 885	32.898
700	131.405	217.925	147.812	49.079	-425.196	-370.204	27.625
800	135.162	235.719	157.707	62.409	-426.942	-362.222	23.651
900	138.775	251.848	167.285	76.107	-587.548	-350.811	20.361
1000	142.297	266.652	176.491	90.161	-584.687	-324.658	16.958
1100	145.762	280.378	185.319	104.564	-581.623	-298.802	14.189
1200	149.186	293.208	193.781	119.312	-578.373	-273.233	11.894
1300	152.582	305.283	201.898	134.401	-574.954	-247.942	9.962
1400	155.957	316.714	209.695	149.828	-571.381	-222.920	8.317
1500	159.317	327.589	217.195	165.591	-567.668	-198.159	6.901
1600	162.666	337.978	224.421	181.691	-563.828	-173.650	5.669
1700	166.006	347.940	231.396	198.124	-559.873	-149.384	4.590
1800	169.339	357.523	238.138	214.892	-555.817	-125.354	3.638
1900	172.666	366.768	244.667	231.992	-551.674	-101.552	2.792
2000	175.989	375.709	250.996	249.425	-547.458	-77.970	2.036
2080.000	178.645	382.663	255.927	263.610	CRYSTAL <--> LIQUID		
2080.000	156.900	445.021	255.927	393.314	TRANSITION		
2100	156.900	446.522	257.735	396.452	-413.922	-55.846	1.389
2200	156.900	453.821	266.484	412.142	-412.011	-38.841	0.922
2300	156.900	460.796	274.782	427.832	-410.406	-21.915	0.498
2400	156.900	467.473	282.672	443.522	-409.124	-5.053	0.110
2500	156.900	473.878	290.193	459.212	-408.187	11.763	-0.246
2600	156.900	480.032	297.377	474.902	-407.636	28.548	-0.574
2700	156.900	485.953	304.252	490.592	-407.546	45.321	-0.877
2800	156.900	491.659	310.844	506.282	-408.042	62.101	-1.159
2900	156.900	497.165	317.175	521.972	-481.053	79.010	-1.423
3000	156.900	502.484	323.264	537.662	-479.131	98.290	-1.711
3100	156.900	507.629	329.128	553.352	-477.222	117.506	-1.980
3200	156.900	512.610	334.785	569.042	-475.327	136.660	-2.231
3300	156.900	517.438	340.247	584.732	-473.443	155.756	-2.465
3400	156.900	522.122	345.528	600.422	-471.573	174.794	-2.685
3500	156.900	526.671	350.639	616.112	-469.716	193.778	-2.892

PREVIOUS: CURRENT: June 1978

Molybdenum Sulfide (Mo_2S_3) Mo_2S_3(cr,l)

VANADIUM SUBNITRIDE ($VN_{0.465}$) CRYSTAL M_r = 57.4546

$S°(298.15\ K) = 26.712 \pm 0.042\ J\ K^{-1}\ mol^{-1}$

$\Delta_f H°(0\ K) = -130.187 \pm 6.3\ kJ\ mol^{-1}$
$\Delta_f H°(298.15\ K) = -132.214 \pm 6.3\ kJ\ mol^{-1}$

Enthalpy of Formation

Pankratz et al. (1) measured the enthalpy of combustion of vanadium subnitride at 303.15 K. The composition of their sample was $VN_{0.465}$ which corresponds to the nitrogen-rich boundary of the subnitride phase (see Phase Data section). Chemical analysis also showed that the sample contained 0.74 weight percent VO and trace amounts of CaO and Mo as impurities. The composition of the combustion products fell between V_2O_4 and V_2O_5. JANAF data (2) for these two oxides are used to correct their combustion values to correspond to the formation of the pentoxide as the final state. Our values for this correction are roughly 3 cal K^{-1} mol^{-1} subnitride greater than those used by Pankratz et al. (1). Impurity corrections for VO and Mo are also based on JANAF (2) data and amount to 9.03 and 0.37 cal g^{-1} subnitride, respectively. For the process $VN_{0.465}(cr) + 1.25\ O_2(g) = 0.5\ V_2O_5(cr) + 0.2325\ N_2(g)$, we obtain $\Delta_c H° = -153.69 \pm 0.75$ kcal mol^{-1} at unit fugacities of oxygen and nitrogen. When this value is combined with $\Delta_f H°(V_2O_5$, cr, 298.15 K) (2), that for vanadium subnitride is $\Delta_f H°(VN_{0.465}$, cr, 298.15 K) = -31.6 ± 1.5 kcal mol^{-1}. We adopt this value and include in the overall uncertainty (+1.5 kcal mol^{-1}) the error associated with correcting the combustion values to correspond to the pentoxide as the final state. The only other value for $\Delta_f H°$ is derived from dissociation pressure data for $VN_{0.43}$ which were reported by Brauer and Schnell (3) for the temperature range 1655-1879 K (corrected to IPTS-68). JANAF functions for $VN_{0.465}$ are used to analyze these pressures by the 3rd law method. These calculations lead to $\Delta_f H°$ for the subnitride phase of -25.2 ± 1.7 kcal mol^{-1}. The 2nd law value is $\Delta_f H°(298.15\ K) = -10.7$ kcal mol^{-1}; the 3rd law drift is 8.3 ± 1.1 cal K^{-1} mol^{-1}. These results show considerable scatter and are believed to be less reliable than those obtained by combustion calorimetry.

Heat Capacity and Entropy

Pankratz et al. (1) have measured low temperature heat capacities (9 to 307 K) for $VN_{0.465}$ in an adiabatic calorimeter. The composition of the subnitride sample was similar to that employed in their combustion work. Their $C_p°$ data are adopted and used to obtain a value (6.384 cal K^{-1} mol^{-1}) for S°(298.15 K). This latter value is based on S°(9 K) = 0.004 cal K^{-1} mol^{-1}.

In the same paper, Pankratz et al. (1) reported measurements of the high temperature enthalpies (401-1794 K) for $VN_{0.465}$ which were obtained in a copper-block drop calorimeter. The subnitride sample was the same as that used in their combustion work and was contained in Pt-Rh alloy capsules during the "drop" experiments. Temperature measurements were based on the IPTS-68 scale. A technique employing orthogonal polynomials is used to fit their experimental enthalpies by computer. The curve is constrained to join smoothly with the low temperature $C_p°$ data near 298.15 K. The average deviation of the smoothed enthalpies from the experimental values is +0.46%; the maximum deviation is +0.73% at 702 K. $C_p°$ data above 1800 K are obtained by graphical extrapolation. No anomalies are observed in either the low temperature $C_p°$ data or the high temperature enthalpies.

Phase Data

In a review of vanadium binary systems, Rostoker and Yamamoto (4) have questioned the existence of a subnitride phase, since Muthmann et al. (5) were unable to recognize its presence. However, several recent investigations (1, 3, 6-8) have reported the preparation of the subnitride at temperatures in excess of 1300 K. Using x-ray techniques, Brauer and Schnell (3) and Hahn (6) found that the phase had an extensive homogeneity range which extends from about $VN_{0.37}$ to $VN_{0.49}$ (27 to 33 atomic percent nitrogen). Pankratz et al. (1) reported the composition of the nitrogen-rich phase boundary as $VN_{0.465}$. These results are in agreement with the earlier x-ray work (3, 6). The structure of vanadium subnitride has been determined to be hexagonal (6, 7). Further information on the vanadium-vanadium nitride system has been reviewed by Storms (9).

Decomposition Data

No information is available on the melting behavior of $VN_{0.465}$ at elevated temperatures. Mass spectrometric evidence (10) has been presented which indicates that vanadium mononitride decomposes to its elements rather than melting. By analogy with VN, we assume the process $VN_{0.465}(cr) = V(\ell) + 0.2325\ N_2(g)$ to be the mode of decomposition upon heating the solid. T_{dcm} = 2875 K is the temperature at which the fugacity for the above process approaches 1 atm. $\Delta_{dcm}H°$ = 34.3 kcal mol^{-1} is the negative of the enthalpy of formation of $VN_{0.465}$ at T_{dcm}.

References

1. L. B. Pankratz, J. M. Stuve, H. O. Poppleton, L. L. Oder, and A. D. Mah, U. S. Bur. Mines RI 7585, 1971.
2. JANAF Thermochemical Tables: V_2O_4(cr) and VO(cr), 12-31-73; V_2O_5(cr), 6-30-73; MoO_3(cr), 6-30-67.
3. G. Brauer and W. D. Schnell, J. Less-Common Metals 6, 326 (1964).
4. W. Rostoker and A. Yamamoto, Trans. ASM 46, 1136 (1954).
5. W. Muthmann, L. Weiss, and R. Riedelbauch, Liebigs Ann. Chem. 355, 58 (1907).
6. H. Hahn, Z. Anorg. Chem. 58, 258 (1949).
7. D. Potter and R. Geils, Sci. Met. 6, 395 (1972).
8. E. T. Turkdogan and E. M. Penn, J. Iron Steel Inst. 181, 343 (1955).
9. E. K. Storms, Los Alamos Scientific Laboratory, LA-2942, August 1964.
10. M. Farber and R. D. Srivastava, J. Chem. Soc. Faraday Trans. 1, 69, 390 (1973).

Vanadium Nitride ($VN_{0.465}$) $N_{0.465}V_1$(cr)

Enthalpy Reference Temperature = T_r = 298.15 K; Standard State Pressure = p° = 0.1 MPa

T/K	$C_p°$ (J $K^{-1}mol^{-1}$)	S° (J $K^{-1}mol^{-1}$)	$-[G°-H°(T_r)]/T$ (J $K^{-1}mol^{-1}$)	$H°-H°(T_r)$ (kJ mol^{-1})	$\Delta_f H°$ (kJ mol^{-1})	$\Delta_f G°$ (kJ mol^{-1})	Log K_f
0	0.	0.	INFINITE	-4.623	-130.187	-130.187	INFINITE
100	10.647	4.977	47.673	-4.270	-130.995	-127.066	66.373
200	22.457	16.425	29.173	-2.550	-131.786	-122.808	32.074
298.15	28.877	26.712	26.712	0.	-132.214	-118.297	20.725
300	28.971	26.891	26.712	0.054	-132.219	-118.211	20.582
400	32.677	35.771	27.898	3.149	-132.365	-113.513	14.823
500	35.016	43.324	30.247	6.538	-132.318	-108.802	11.367
600	36.815	49.871	32.984	10.132	-132.138	-104.114	9.064
700	38.359	55.664	35.818	13.892	-131.860	-99.464	7.422
800	39.765	60.878	38.630	17.799	-131.508	-94.860	6.194
900	41.095	65.639	41.370	21.842	-131.104	-90.303	5.241
1000	42.376	70.036	44.020	26.016	-130.653	-85.793	4.481
1100	43.627	74.133	46.573	30.316	-130.165	-81.330	3.862
1200	44.852	77.982	49.032	34.740	-129.652	-76.913	3.348
1300	46.066	81.620	51.400	39.286	-129.119	-72.540	2.915
1400	47.262	85.078	53.683	43.953	-128.577	-68.208	2.545
1500	48.455	88.379	55.887	48.739	-128.026	-63.915	2.226
1600	49.643	91.544	58.017	53.644	-127.465	-59.659	1.948
1700	50.823	94.589	60.080	58.667	-126.904	-55.438	1.703
1800	51.999	97.529	62.079	63.810	-126.344	-51.251	1.487
1900	52.928	100.365	64.020	69.055	-125.814	-47.093	1.295
2000	53.974	103.106	65.906	74.401	-125.322	-42.963	1.122
2100	54.936	105.763	67.741	79.846	-124.880	-38.856	0.966
2200	55.940	108.342	69.528	85.390	-147.378	-34.665	0.823
2300	56.902	110.850	71.270	91.032	-147.199	-29.545	0.671
2400	57.907	113.293	72.971	96.774	-146.923	-24.435	0.532
2500	58.743	115.674	74.631	102.606	-146.560	-19.339	0.404
2600	59.622	117.995	76.255	108.524	-146.113	-14.258	0.286
2700	60.584	120.263	77.843	114.534	-145.576	-9.197	0.178
2800	61.505	122.483	79.398	120.639	-144.947	-4.157	0.078
2900	62.425	124.657	80.921	126.836	-144.227	0.859	-0.015
3000	63.304	126.789	82.414	133.123	-143.419	5.848	-0.102
3100	64.141	128.878	83.880	139.494	-142.527	10.809	-0.182
3200	65.019	130.928	85.318	145.952	-141.551	15.740	-0.257
3300	65.898	132.942	86.731	152.498	-140.488	20.639	-0.327
3400	66.777	134.923	88.119	159.132	-139.339	25.505	-0.392
3500	67.655	136.871	89.484	165.854	-138.103	30.336	-0.453

PREVIOUS: CURRENT: December 1973

Vanadium Nitride ($VN_{0.465}$) $N_{0.465}V_1$(cr)

NITROGEN (N) IDEAL GAS A_r = 14.0067

$S°(298.15\ K) = 153.300 \pm 0.02\ J\ K^{-1}\ mol^{-1}$

$\Delta_f H°(0\ K) = 470.82 \pm 0.10\ kJ\ mol^{-1}$
$\Delta_f H°(298.15\ K) = 472.68 \pm 0.10\ kJ\ mol^{-1}$

Electronic Levels and Quantum Weights

State	ϵ_i, cm^{-1}	g_i
$^4S°_{3/2}$	0.000	4
$^2D°_{5/2}$	19224.464	6
$^2D°_{3/2}$	19233.177	4
$^2P°_{1/2}$	28838.920	2
$^2P°_{3/2}$	28839.306	4

Enthalpy of Formation

$\Delta_f H°$ is based on the D_0^0 value for $N_2(g)$ adopted by the Committee on Data for Science and Technology (CODATA) of the International Council of Scientific Unions (1). The adopted value for $\Delta_f H°$(N, 298.15, 0 K) is derived using $D_0^0(N_2, g) = 78715 \pm 50\ cm^{-1}$ (225.057 ± 0.143 kcal mol^{-1}) from Buttenbender and Herzberg (2), Tanaka, Ogawa, and Jursa (3), and Telford, Vanderslice, and Wilkinson (4). This is discussed further by Gaydon(5) and Lofthus and Krupenie (6).

Heat Capacity and Entropy

The electronic levels for N(g) are given in the compilation by Moore (7). Our calculations indicate that, except for the five states lying below 29000 cm^{-1}, the inclusion of levels up to n = 12 has no effect on the thermodynamic functions to 6000 K. This is a result of the high energy of these levels; the fifth excited state lies at 83284.070 cm^{-1} above the ground state. Since the inclusion of these upper levels has no effect on the thermodynamic function (to 6000 K) we list only the lowest lying states (ϵ_i < 29000 cm^{-1}). The reported uncertainty in S°(298.15 K) is due to uncertainties in the relative atomic mass and the fundamental constants. Extension of these calculations above 6000 K may require consideration of the higher excited states and utilization of proper fill and cut off procedures (8).

References

1. ICSU-CODATA Task Group, J. Chem. Thermodynamics 4, 331 (1972).
2. G. Buttenbender, and G. Herzberg, Ann. Phys. Leipzig [5] 21, 577 (1935).
3. Y. Tanaka, M. Ogawa, and A. S. Jursa, J. Chem. Phys. 40. 3690 (1964).
4. S. G. Telford, J. T. Vanderslice, and P. G. Wilkinson, Astrophys. J. 142, 1203 91965).
5. A. G. Gaydon, "Dissociation Energies and Spectra of Diatomic Molecules", 3rd ed., Chapman and Hall Ltd., London, 1968.
6. A. Lofthus and P. H. Krupenie, J. Phys. Chem. Ref. Data 6, 113 (1977).
7. C. E. Moore, NSRDS-NBS3, Section 5 (1975).
8. J. R. Downey, AFOSR-TR 78-0960, March 9, 1978.

Nitrogen (N) $N_1(g)$

Enthalpy Reference Temperature = T_r = 298.15 K   Standard State Pressure = p° = 0.1 MPa

T/K	$C_p°$ (J K⁻¹mol⁻¹)	S° (J K⁻¹mol⁻¹)	$-[G°-H°(T_r)]/T$ (J K⁻¹mol⁻¹)	$H°-H°(T_r)$ (kJ mol⁻¹)	$\Delta_f H°$ (kJ mol⁻¹)	$\Delta_f G°$ (kJ mol⁻¹)	Log K_f
0	0.	0.	INFINITE	-6.197	470.820	470.820	INFINITE
100	20.786	130.593	171.780	-4.119	471.448	466.379	-243.611
200	20.786	145.001	155.201	-2.040	472.071	461.070	-120.419
250	20.786	149.639	153.642	-1.001	472.383	458.283	-95.753
298.15	20.786	153.300	153.300	0.	472.683	455.540	-79.809
300	20.786	153.429	153.300	0.038	472.694	455.434	-79.298
350	20.786	156.633	153.554	1.078	473.005	452.533	-67.537
400	20.786	159.408	154.116	2.117	473.314	449.587	-58.710
450	20.786	161.857	154.843	3.156	473.621	446.603	-51.840
500	20.786	164.047	155.655	4.196	473.923	443.584	-46.341
600	20.786	167.836	157.379	6.274	474.510	437.461	-38.084
700	20.786	171.041	159.108	8.353	475.067	431.242	-32.180
800	20.786	173.816	160.777	10.431	475.591	424.945	-27.746
900	20.786	176.264	162.364	12.510	476.081	418.584	-24.294
1000	20.786	178.454	163.866	14.589	476.540	412.171	-21.530
1100	20.786	180.436	165.284	16.667	476.970	405.713	-19.266
1200	20.786	182.244	166.623	18.746	477.374	399.217	-17.377
1300	20.786	183.908	167.889	20.824	477.756	392.688	-15.778
1400	20.786	185.448	169.089	22.903	478.118	386.131	-14.407
1500	20.786	186.882	170.228	24.982	478.462	379.548	-13.217
1600	20.786	188.224	171.311	27.060	478.791	372.943	-12.175
1700	20.786	189.484	172.344	29.139	479.107	366.318	-11.256
1800	20.787	190.672	173.329	31.218	479.411	359.674	-10.437
1900	20.788	191.796	174.272	33.296	479.705	353.014	-9.705
2000	20.790	192.863	175.175	35.375	479.990	346.339	-9.045
2100	20.793	193.877	176.042	37.454	480.266	339.650	-8.448
2200	20.797	194.844	176.874	39.534	480.536	332.947	-7.905
2300	20.804	195.769	177.676	41.614	480.799	326.233	-7.409
2400	20.813	196.655	178.448	43.695	481.057	319.507	-6.954
2500	20.826	197.504	179.194	45.777	481.311	312.770	-6.535
2600	20.843	198.322	179.914	47.860	481.561	306.024	-6.148
2700	20.864	199.109	180.610	49.945	481.809	299.268	-5.790
2800	20.891	199.868	181.285	52.033	482.054	292.502	-5.457
2900	20.924	200.601	181.938	54.124	482.299	285.728	-5.147
3000	20.963	201.311	182.572	56.218	482.543	278.946	-4.857
3100	21.010	202.000	183.188	58.317	482.789	272.155	-4.586
3200	21.064	202.667	183.786	60.420	483.036	265.357	-4.332
3300	21.126	203.317	184.368	62.530	483.286	258.550	-4.093
3400	21.197	203.948	184.935	64.646	483.540	251.736	-3.867
3500	21.277	204.564	185.487	66.769	483.799	244.915	-3.655
3600	21.365	205.164	186.025	68.902	484.064	238.086	-3.455
3700	21.463	205.751	186.550	71.043	484.335	231.249	-3.265
3800	21.569	206.325	187.063	73.194	484.614	224.405	-3.085
3900	21.685	206.887	187.564	75.357	484.903	217.554	-2.914
4000	21.809	207.437	188.054	77.532	485.201	210.695	-2.751
4100	21.941	207.977	188.534	79.719	485.510	203.829	-2.597
4200	22.082	208.508	189.003	81.920	485.830	196.955	-2.449
4300	22.231	209.029	189.463	84.136	486.164	190.073	-2.309
4400	22.388	209.542	189.913	86.367	486.510	183.183	-2.175
4500	22.551	210.047	190.355	88.614	486.871	176.285	-2.046
4600	22.722	210.544	190.788	90.877	487.247	169.379	-1.923
4700	22.899	211.035	191.214	93.158	487.638	162.465	-1.806
4800	23.081	211.519	191.632	95.457	488.046	155.542	-1.693
4900	23.269	211.997	192.043	97.775	488.471	148.610	-1.584
5000	23.461	212.469	192.447	100.111	488.912	141.670	-1.480
5100	23.658	212.935	192.844	102.467	489.372	134.721	-1.380
5200	23.858	213.397	193.235	104.843	489.849	127.762	-1.283
5300	24.061	213.853	193.619	107.238	490.345	120.794	-1.190
5400	24.266	214.305	193.998	109.655	490.860	113.817	-1.101
5500	24.474	214.752	194.371	112.092	491.394	106.829	-1.015
5600	24.682	215.195	194.739	114.550	491.947	99.832	-0.931
5700	24.892	215.633	195.102	117.028	492.519	92.825	-0.851
5800	25.102	216.068	195.460	119.528	493.110	85.808	-0.773
5900	25.312	216.499	195.813	122.049	493.720	78.780	-0.697
6000	25.521	216.926	196.161	124.590	494.349	71.742	-0.625

PREVIOUS: March 1977 (1 atm)   CURRENT: December 1982 (1 bar)

Nitrogen (N) $N_1(g)$

NITROGEN, ION (N^+) IDEAL GAS M_r = 14.00615 Nitrogen, Ion (N^+) $N_1^+(g)$

$IP(N^+, g)$ = 238750.50 ± 1.3 cm^{-1}

$S°(298.15\ K)$ = 159.797 ± 0.02 $JK^{-1}\ mol^{-1}$

$\Delta_f H°(0\ K)$ = 1873.152 ± 0.42 kJ mol^{-1}

$\Delta_f H°(298.15\ K)$ = [1882.130] kJ mol^{-1}

Electronic Levels and Quantum Weights

State	ε_i cm^{-1}	g_i
$^3P_0$	0	1
$^3P_1$	48.7	3
$^3P_2$	130.8	5
$^1D_2$	15316.2	5
$^1S_0$	32688.8	1
$^5S_2$	46784.6	5

Enthalpy of Formation

$\Delta_f H°(N^+, g, 0\ K)$ is calculated from $\Delta_f H°(N, g, 0\ K)$ (1) using the spectroscopic value of IP(N) = 117225.7 ± 0.3 cm^{-1} (1402.331 ± 0.004 kJ mol^{-1}) from Moore (2). The ionization limit is converted from cm^{-1} to kJ mol^{-1} using the factor, 1 cm^{-1} = 0.01196266 kJ mol^{-1}, which is derived from the 1973 CODATA fundamental constants (3). Rosenstock et al. (4) and Levin and Lias (5) have summarized additional ionization and appearance potential data.

$\Delta_f H°(N^+, g, 298.15\ K)$ is calculated from $\Delta_f H°(N, g, 0\ K)$ by using IP(N) with JANAF (1) enthalpies, H°(0 K) - H°(298.15 K), for N(g), $N^+(g)$, and $e^-(g)$. $\Delta_f H°(N \rightarrow N^+ + e^-$, 298.15 K) differs from a room temperature threshold energy due to inclusion of these enthalpies and to threshold effects discussed by Rosenstock et al. (4). $\Delta_f H°(298.15\ K)$ should be changed by -6.197 kJ mol^{-1} if it is to be used in the ion convention that excludes the enthalpy of the electron.

Heat Capacity and Entropy

The information on electronic energy levels and quantum weights, given by Moore (2) is incomplete because many theoretically predicted levels have not been observed. Our calculations indicate that any reasonable method of filling in these missing levels and cutting off the summation in the partition function (6) has no effect on the thermodynamic functions to 6000 K. This is a result of the high energy of all levels other than the ground state and the first five excited states; the next excited state is 92237.2 cm^{-1} above the ground state. Since inclusion of these excited states has no effect on the thermodynamic functions (to 6000 K), we list only the ground state and the first five excited states. The reported uncertainty in S°(298.15 K) is due to uncertainties in the relative ionic mass and the fundamental constants. Extension of these calculations above 6000 K may require consideration of the excited states and use of different fill and cutoff procedures (6).

References

1. JANAF Thermochemical Tables: N(g), 12-31-82 ; $e^-(g)$, 3-31-82.
2. C. E. Moore, U. S. Nat. Bur. Stand., NSRDS-NBS-3, Section 5, 1975.
3. E. R. Cohen and B. N. Taylor, J. Phys. Chem. Ref. Data 2, 663 (1973).
4. H. M. Rosenstock, K. Draxl et al., J. Phys. Chem. Ref. Data 6, Supp. 1 (1977).
5. R. D. Levin and S. G. Lias, U. S. Nat. Bur. Stand., NSRDS-NBS-71, 1982.
6. J. R. Downey, Jr., The Dow Chemical Company, AFOSR-TR-0960, Contract No. F44620-75-1-0048, 1978.

Enthalpy Reference Temperature = T_r = 298.15 K Standard State Pressure = p° = 0.1 MPa

T/K	$C_p°$ (J $K^{-1}mol^{-1}$)	S° (J $K^{-1}mol^{-1}$)	-[G°-H°(T_r)]/T (J $K^{-1}mol^{-1}$)	H°-H°(T_r) (kJ mol^{-1})	$\Delta_f H°$ (kJ mol^{-1})	$\Delta_f G°$ (kJ mol^{-1})	Log K_f
0	0.	0.	INFINITE	-7.116	1873.152		
100	24.729	135.201	179.245	-4.404			
200	21.896	151.192	161.758	-2.113			
250	21.497	156.030	160.148	-1.030			
298.15	21.284	159.797	159.797	0.	1882.130	1856.796	-325.303
300	21.278	159.929	159.797	0.039	1882.181	1856.639	-323.270
350	21.146	163.198	160.056	1.100	1883.552	1852.273	-276.437
400	21.060	166.016	160.629	2.155	1884.917	1847.712	-241.286
450	21.002	168.492	161.368	3.206	1886.274	1842.979	-213.927
500	20.960	170.703	162.193	4.255	1887.626	1838.096	-192.025
600	20.906	174.519	163.939	6.348	1890.306	1827.938	-159.136
700	20.874	177.739	165.686	8.437	1892.952	1817.334	-135.611
800	20.853	180.525	167.371	10.523	1895.562	1806.353	-117.943
900	20.839	182.980	168.972	12.608	1898.137	1795.047	-104.182
1000	20.829	185.175	170.484	14.691	1900.679	1783.456	-93.158
1100	20.821	187.160	171.911	16.774	1903.191	1771.612	-84.127
1200	20.815	188.972	173.259	18.855	1905.677	1759.540	-76.591
1300	20.811	190.638	174.532	20.937	1908.140	1747.262	-70.206
1400	20.808	192.180	175.738	23.018	1910.583	1734.795	-64.726
1500	20.805	193.615	176.883	25.098	1913.008	1722.154	-59.971
1600	20.803	194.958	177.971	27.179	1915.418	1709.351	-55.805
1700	20.802	196.219	179.008	29.259	1917.814	1696.399	-52.124
1800	20.803	197.408	179.997	31.339	1920.198	1683.306	-48.848
1900	20.804	198.533	180.943	33.420	1922.572	1670.080	-45.914
2000	20.806	199.600	181.850	35.500	1924.937	1656.730	-43.269
2100	20.810	200.615	182.719	37.581	1927.294	1643.261	-40.874
2200	20.816	201.583	183.555	39.662	1929.644	1629.681	-38.694
2300	20.824	202.509	184.359	41.744	1931.988	1615.994	-36.700
2400	20.835	203.395	185.134	43.827	1934.327	1602.205	-34.871
2500	20.848	204.246	185.882	45.911	1936.661	1588.318	-33.186
2600	20.864	205.064	186.604	47.997	1938.992	1574.339	-31.629
2700	20.883	205.852	187.302	50.084	1941.320	1560.269	-30.185
2800	20.904	206.612	187.978	52.174	1943.646	1546.114	-28.843
2900	20.929	207.346	188.634	54.265	1945.970	1531.875	-27.592
3000	20.956	208.056	189.269	56.359	1948.293	1517.556	-26.423
3100	20.987	208.743	189.886	58.457	1950.616	1503.160	-25.328
3200	21.020	209.410	190.486	60.557	1952.938	1488.689	-24.300
3300	21.056	210.058	191.069	62.661	1955.261	1474.145	-23.334
3400	21.094	210.687	191.637	64.768	1957.585	1459.531	-22.423
3500	21.135	211.299	192.190	66.880	1959.910	1444.849	-21.563
3600	21.178	211.895	192.729	68.995	1962.237	1430.100	-20.750
3700	21.223	212.476	193.255	71.115	1964.566	1415.286	-19.980
3800	21.269	213.042	193.768	73.240	1966.897	1400.410	-19.250
3900	21.317	213.595	194.270	75.369	1969.231	1385.472	-18.556
4000	21.366	214.136	194.760	77.503	1971.567	1370.474	-17.897
4100	21.417	214.664	195.239	79.642	1973.906	1355.417	-17.268
4200	21.468	215.180	195.707	81.787	1976.248	1340.304	-16.669
4300	21.520	215.686	196.166	83.936	1978.594	1325.135	-16.097
4400	21.573	216.182	196.615	86.091	1980.943	1309.911	-15.551
4500	21.625	216.667	197.056	88.251	1983.296	1294.634	-15.028
4600	21.679	217.143	197.487	90.416	1985.652	1279.304	-14.527
4700	21.732	217.610	197.910	92.586	1988.011	1263.923	-14.047
4800	21.784	218.068	198.326	94.762	1990.374	1248.492	-13.586
4900	21.837	218.517	198.733	96.943	1992.741	1233.012	-13.144
5000	21.889	218.959	199.133	99.129	1995.111	1217.483	-12.719
5100	21.941	219.393	199.526	101.321	1997.485	1201.907	-12.310
5200	21.992	219.820	199.912	103.518	1999.862	1186.284	-11.916
5300	22.042	220.239	200.292	105.719	2002.242	1170.616	-11.537
5400	22.092	220.652	200.665	107.926	2004.626	1154.902	-11.171
5500	22.141	221.057	201.032	110.138	2007.013	1139.145	-10.819
5600	22.189	221.457	201.393	112.354	2009.404	1123.344	-10.478
5700	22.236	221.850	201.749	114.575	2011.797	1107.500	-10.149
5800	22.281	222.237	202.099	116.801	2014.192	1091.614	-9.831
5900	22.326	222.618	202.443	119.032	2016.591	1075.687	-9.523
6000	22.370	222.994	202.783	121.267	2018.991	1059.719	-9.226

PREVIOUS: March 1977 (1 atm) CURRENT: December 1982 (1 bar)

NITROGEN, ION (N^-) IDEAL GAS M_r = 14.00725

EA(N, g) = -0.07 ± 0.02 eV
S°(298.15 K) = 159.927 ± 0.001 J K^{-1} mol^{-1}

$\Delta_f H°$(0 K) = 477.574 kJ mol^{-1}
$\Delta_f H°$(298.15 K) = [473.540] kJ mol^{-1}

Electonic Levels and Quantum Weights

State	ε_i, cm^{-1}	g_i
$^3P_2$	0.0	5
$^3P_1$	[63]	3
$^3P_0$	[90]	1
$^1D_2$	[10800]	5
$^1S_0$	[22390]	1

Enthalpy of Formation

$\Delta_f H°(N^-$, g, 0 K) is calculated from $\Delta_f H°$(N, g, 0 K) (1) using the adopted electron affinity of EA(N) = -0.07 ± 0.02 eV (-6.754 ± 1.930 kJ mol^{-1}). This value, recommended by Hotop and Lineberger (2), is based on a dissociative attachment of electrons study (3). Additional information on $N^-(g)$ may be obtained in the critical discussions of Hotop and Lineberger (2, 4), Rosenstock et al. (5) and Massey (6).

$\Delta_f H°(N^-$, g, 298.15 K) is obtained from $\Delta_f H°$(N, g, 0 K) by using EA(N) with JANAF (1) enthalpies, H°(0 K) - H°(298.15 K), for $N^-(g)$, N(g), and $e^-(g)$. $\Delta_r H°(N^- \rightarrow N + e^-$, 298.15 K) differs from a room-temperature threshold energy due to inclusion of these enthalpies and to threshold effects discussed by Rosenstock et al. (5). $\Delta_f H°$(298.15 K) should be changed by +6.197 kJ mol^{-1} if it is to be used in the ion convention that excludes the enthalpy of the electron.

Heat Capacity and Entropy

The ground state for $N^-(g)$ is reported to be $2p^4$ ^3P. We estimate the fine structure separations in the ground state by comparison of isoelectronic series based on $N^-(g)$ [N^-, O, F^+, Ne^{++}, Na^{+++}] and C(g) [C, N^+, O^{++}, F^{+++}]. The uncerataintv in S°(298.15 K) in part reflects this estimation. Massey (6) listed the hyperfine splitting of the ground state as 97 and 137 cm^{-1} based on a quadratic isoelectronic extrapolation. This technique led to high values in the case of $O^-(g)$ (5) and is assumed to give high results for $N^-(g)$.

A comparison of the isoelectronic sequence - $N^-(g)$, O(g), $F^+(g)$ - would suggest that stable electronic states may exist at low wave numbers. These states would greatly affect the entropy. We estimate two of these levels.

References

1. JANAF Thermochemical Tables: N(g), 12-31-82; $e^-(g)$, 3-31-82; $O^-(g)$, 9-30-82.
2. H. Hotop and W. C. Lineberger, J. Phys. Chem. Ref. Data, to be published, 1985.
3. J. Mazeau, F. Gresteau, R. I. Hall and A. Huetz, J. Phys. B 11, L557 (1978).
4. H. Hotop and W. C. Lineberger, J. Phys. Chem. Ref. Data 4, 539 (1975).
5. H. M. Rosenstock, K. Draxl et al., J. Phys. Chem. Ref. Data 6, Supp. 1 (1977).
6. H. S. W. Massey, "Negative Ions", 3rd ed., Cambridge University Press, Cambridge, 1976.

Nitrogen, Ion (N^-) $N_1^-(g)$

Enthalpy Reference Temperature = T_r = 298.15 K   Standard State Pressure = p° = 0.1 MPa

T/K	$C_p°$ (J $K^{-1}mol^{-1}$)	S° (J $K^{-1}mol^{-1}$)	-[G°-H°(T_r)]/T (J $K^{-1}mol^{-1}$)	H°-H°(T_r) (kJ mol^{-1})	$\Delta_f H°$ (kJ mol^{-1})	$\Delta_f G°$ (kJ mol^{-1})	Log K_f
0	0.	0.	INFINITE	-6.498	477.574		
100	22.234	136.475	178.809	-4.233			
200	21.252	151.496	161.855	-2.072			
250	21.096	156.220	160.274	-1.014			
298.15	21.009	159.927	159.927	0.	473.540	460.677	-80.709
300	21.007	160.057	159.928	0.039	473.514	460.597	-80.197
350	20.951	163.291	160.183	1.088	472.795	458.501	-68.428
400	20.913	166.086	160.750	2.134	472.072	456.508	-59.614
450	20.888	168.548	161.483	3.179	471.345	454.606	-52.769
500	20.869	170.747	162.301	4.223	470.612	452.786	-47.302
600	20.844	174.550	164.035	6.309	469.128	449.360	-39.120
700	20.829	177.762	165.773	8.392	467.611	446.184	-33.295
800	20.819	180.542	167.449	10.475	466.060	443.229	-28.940
900	20.812	182.994	169.043	12.556	464.475	440.470	-25.564
1000	20.807	185.187	170.550	14.637	462.857	437.889	-22.873
1100	20.804	187.170	171.972	16.718	461.210	435.471	-20.679
1200	20.801	188.980	173.315	18.798	459.538	433.205	-18.857
1300	20.799	190.645	174.585	20.878	457.842	431.079	-17.321
1400	20.797	192.186	175.787	22.958	456.126	429.085	-16.009
1500	20.796	193.621	176.929	25.037	454.393	427.214	-14.877
1600	20.794	194.963	178.015	27.117	452.645	425.459	-13.890
1700	20.793	196.223	179.049	29.196	450.883	423.814	-13.022
1800	20.793	197.412	180.037	31.275	449.109	422.272	-12.254
1900	20.792	198.536	180.981	33.355	447.325	420.830	-11.569
2000	20.791	199.602	181.886	35.434	445.531	419.482	-10.956
2100	20.791	200.617	182.754	37.513	443.729	418.224	-10.403
2200	20.790	201.584	183.588	39.592	441.920	417.052	-9.902
2300	20.790	202.508	184.390	41.671	440.103	415.962	-9.447
2400	20.790	203.393	185.164	43.750	438.281	414.951	-9.031
2500	20.789	204.242	185.910	45.829	436.453	414.017	-8.650
2600	20.789	205.057	186.631	47.908	434.620	413.156	-8.300
2700	20.789	205.842	187.328	49.987	432.783	412.365	-7.978
2800	20.789	206.598	188.003	52.066	430.941	411.642	-7.679
2900	20.789	207.327	188.657	54.145	429.095	410.986	-7.403
3000	20.788	208.032	189.291	56.223	427.246	410.393	-7.146
3100	20.788	208.714	189.906	58.302	425.393	409.862	-6.906
3200	20.788	209.374	190.504	60.381	423.536	409.390	-6.683
3300	20.788	210.013	191.086	62.460	421.677	408.977	-6.474
3400	20.788	210.634	191.652	64.539	419.815	408.620	-6.278
3500	20.788	211.236	192.203	66.617	417.951	408.318	-6.094
3600	20.788	211.822	192.740	68.696	416.083	408.069	-5.921
3700	20.788	212.392	193.263	70.775	414.214	407.872	-5.758
3800	20.788	212.946	193.774	72.854	412.342	407.726	-5.605
3900	20.787	213.486	194.272	74.932	410.468	407.629	-5.460
4000	20.787	214.012	194.759	77.011	408.591	407.580	-5.322
4100	20.787	214.526	195.235	79.090	406.713	407.578	-5.193
4200	20.787	215.026	195.701	81.169	404.833	407.622	-5.070
4300	20.787	215.516	196.156	83.247	402.950	407.710	-4.953
4400	20.787	215.993	196.601	85.326	401.066	407.843	-4.842
4500	20.787	216.461	197.037	87.405	399.180	408.018	-4.736
4600	20.787	216.917	197.465	89.484	397.293	408.235	-4.636
4700	20.787	217.365	197.883	91.562	395.403	408.493	-4.540
4800	20.787	217.802	198.294	93.641	393.512	408.792	-4.449
4900	20.787	218.231	198.696	95.720	391.619	409.130	-4.361
5000	20.787	218.651	199.091	97.798	389.724	409.506	-4.278
5100	20.787	219.062	199.479	99.877	387.828	409.921	-4.198
5200	20.787	219.466	199.859	101.956	385.930	410.373	-4.122
5300	20.787	219.862	200.233	104.034	384.030	410.861	-4.049
5400	20.787	220.251	200.600	106.113	382.129	411.385	-3.979
5500	20.787	220.632	200.961	108.192	380.226	411.944	-3.912
5600	20.787	221.006	201.315	110.270	378.321	412.538	-3.848
5700	20.787	221.374	201.664	112.349	376.414	413.166	-3.786
5800	20.787	221.736	202.007	114.428	374.506	413.827	-3.727
5900	20.787	222.091	202.344	116.506	372.595	414.522	-3.670
6000	20.787	222.441	202.676	118.585	370.682	415.248	-3.615

PREVIOUS: March 1977 (1 atm)   CURRENT: December 1982 (1 bar)

Nitrogen, Ion (N^-) $N_1^-(g)$

$S°(298.15\ K) = 210.758\ J\ K^{-1}\ mol^{-1}$

$\Delta_f H°(0\ K) = 89.775 \pm 0.17\ kJ\ mol^{-1}$
$\Delta_f H°(298.15\ K) = 90.291 \pm 0.17\ kJ\ mol^{-1}$

Electronic Level and Quantum Weight

State	ε_i, cm^{-1}	g_i
$^2\Pi$	0	2
	121.1	2

$\omega_e = 1903.60\ cm^{-1}$ $\quad \omega_e x_e = 13.97\ cm^{-1}$ $\quad \sigma = 1$
$B_e = 1.7042\ cm^{-1}$ $\quad \alpha_e = 0.0178\ cm^{-1}$ $\quad r_e = 1.1508$ Å

Enthalpy of Formation

Herzberg (1), lists two D_0^0(NO) values, 5.296 and 6.487 eV. Gaydon (2) lists D_0^0(NO) = 6.49±0.05 eV. A Birge-Sponer extrapolation of the vibrational levels, up to v = 18, of the ground state leads to a D_0^0 of about 6.5 eV. Gaydon's (2) analysis on the predissociation of NO also leads to the higher value for D°(NO).

Tanaka (3), has observed and analyzed the emmission bands of NO in the vacuum ultraviolet region. Tanaka's (3), observation and interpretation of the abrupt cutoff (predissociation) of certain v' values in the β, γ, and δ bands and of the complete non appearance of ε and β' bands strongly favors a D°(NO) = 6.48 eV. $\Delta_f H°$(NO, g, 298.15 K) was calculated on this basis to be 22.033±1 kcal mol^{-1}.

More recently Frisch (4), has calorimetrically determined the enthalpy of reaction for:

$$NO(g) + CO(g) \rightarrow CO_2(g) + 1/2N_2(g)$$

at 298.15 K, from which the author derives a $\Delta_f H°$(NO, g, 298.15 K) = 21.58±0.04 kcal mol^{-1}. The auxiliary values used by Frisch to calculate the enthalpy of formation of nitric oxide were based upon the 1961 atomic weight scale. Recalculating $\Delta_f H°$(NO, g, 298.15 K) using auxiliary functions based on the 1956 atomic weight scale had a negligible effect on the final result.

The calorimetrically determined heat of formation for nitric oxide, 21.58±0.04 kcal mol^{-1}, was selected as the best value.

Heat Capacity and Entropy

The spectroscopic constants were taken from Gillette and Eyster (5), who measured and analyzed the fundamental rotation-vibration band of nitric oxide using a grating spectrometer. The constants used in this calculation were adjusted to correspond to the naturally occurring isotopic mixture.

References

1. G. Herzberg, "Molecular Spectra and Molecular Structure, I. Spectra of Diatomic Molecules", D. Van Nostrand Co., Inc., New York, 1950.
2. A. G. Gaydon, "Dissociation Energies and Spectra of Diatomic Molecules", Chapman and Hall, Ltd., London, 1953.
3. Y. Tanaka, J. Chem. Phys. 22, 2045 (1954).
4. M. A. Frisch, Thesis, University of Wisconsin, Madison, Wisconsin, 1962; Diss. Absts. 23, 1940 (1961).
5. R. H. Gillette and E. H. Eyster, Phys. Revs. 56, 1113 (1939).

Enthalpy Reference Temperature = T_r = 298.15 K — Standard State Pressure = p° = 0.1 MPa

T/K	$C_p°$ (J K⁻¹mol⁻¹)	S° (J K⁻¹mol⁻¹)	$-[G°-H°(T_r)]/T$ (J K⁻¹mol⁻¹)	$H°-H°(T_r)$ (kJ mol⁻¹)	$\Delta_f H°$ (kJ mol⁻¹)	$\Delta_f G°$ (kJ mol⁻¹)	Log K_f
0	0.	0.	INFINITE	-9.192	89.775	89.775	INFINITE
100	32.302	177.031	237.757	-6.073	89.991	88.944	-46.460
200	30.420	198.747	213.501	-2.951	90.202	87.800	-22.931
250	30.025	205.488	211.251	-1.441	90.256	87.193	-18.218
298.15	29.845	210.758	210.758	0.	90.291	86.600	-15.172
300	29.841	210.943	210.759	0.055	90.292	86.577	-15.074
350	29.823	215.540	211.122	1.546	90.316	85.955	-12.828
400	29.944	219.529	211.929	3.040	90.332	85.331	-11.143
450	30.175	223.068	212.974	4.542	90.343	84.705	-9.832
500	30.486	226.263	214.145	6.059	90.352	84.079	-8.784
600	31.238	231.886	216.646	9.144	90.366	82.822	-7.210
700	32.028	236.761	219.179	12.307	90.381	81.564	-6.086
800	32.767	241.087	221.652	15.548	90.398	80.303	-5.243
900	33.422	244.985	224.031	18.858	90.417	79.041	-4.587
1000	33.987	248.536	226.307	22.229	90.437	77.775	-4.063
1100	34.468	251.799	228.478	25.653	90.457	76.508	-3.633
1200	34.877	254.816	230.549	29.120	90.476	75.239	-3.275
1300	35.226	257.621	232.525	32.626	90.493	73.969	-2.972
1400	35.524	260.243	234.412	36.164	90.508	72.697	-2.712
1500	35.780	262.703	236.217	39.729	90.518	71.425	-2.487
1600	36.002	265.019	237.945	43.319	90.525	70.151	-2.290
1700	36.195	267.208	239.603	46.929	90.526	68.878	-2.116
1800	36.364	269.282	241.195	50.557	90.522	67.605	-1.962
1900	36.514	271.252	242.725	54.201	90.511	66.332	-1.824
2000	36.647	273.128	244.199	57.859	90.494	65.060	-1.699
2100	36.767	274.919	245.619	61.530	90.469	63.788	-1.587
2200	36.874	276.632	246.990	65.212	90.438	62.519	-1.484
2300	36.971	278.273	248.315	68.904	90.398	61.251	-1.391
2400	37.060	279.849	249.596	72.606	90.350	59.984	-1.306
2500	37.141	281.363	250.837	76.316	90.295	58.720	-1.227
2600	37.216	282.822	252.039	80.034	90.231	57.458	-1.154
2700	37.285	284.227	253.205	83.759	90.160	56.199	-1.087
2800	37.350	285.585	254.338	87.491	90.081	54.943	-1.025
2900	37.410	286.896	255.438	91.229	89.994	53.689	-0.967
3000	37.466	288.165	256.508	94.973	89.899	52.439	-0.913
3100	37.519	289.395	257.549	98.722	89.798	51.192	-0.863
3200	37.570	290.587	258.563	102.477	89.689	49.948	-0.815
3300	37.617	291.744	259.551	106.236	89.574	48.708	-0.771
3400	37.663	292.867	260.514	110.000	89.451	47.472	-0.729
3500	37.706	293.960	261.454	113.768	89.323	46.239	-0.690
3600	37.747	295.022	262.372	117.541	89.189	45.010	-0.653
3700	37.787	296.057	263.269	121.318	89.049	43.784	-0.618
3800	37.825	297.065	264.145	125.098	88.903	42 563	-0.585
3900	37.862	298.048	265.002	128.883	88.752	41.346	-0.554
4000	37.898	299.008	265.840	132.671	88.596	40.132	-0.524
4100	37.933	299.944	266.660	136.462	88.434	38.922	-0.496
4200	37.966	300.858	267.464	140.257	88.268	37.717	-0.469
4300	37.999	301.752	268.251	144.056	88.097	36.515	-0.444
4400	38.031	302.626	269.022	147.857	87.922	35.318	-0.419
4500	38.062	303.481	269.778	151.662	87.741	34.124	-0.396
4600	38.092	304.318	270.520	155.469	87.556	32.934	-0.374
4700	38.122	305.137	271.248	159.280	87.366	31.749	-0.353
4800	38.151	305.940	271.962	163.094	87.171	30.568	-0.333
4900	38.180	306.727	272.664	166.910	86.970	29.391	-0.313
5000	38.208	307.499	273.353	170.730	86.765	28.218	-0.295
5100	38.235	308.256	274.030	174.552	86.553	27.049	-0.277
5200	38.262	308.998	274.695	178.377	86.336	25.884	-0.260
5300	38.289	309.728	275.349	182.204	86.112	24.724	-0.244
5400	38.316	310.443	275.993	186.034	85.881	23.568	-0.228
5500	38.342	311.147	276.625	189.867	85.644	22.416	-0.213
5600	38.367	311.838	277.248	193.703	85.399	21.269	-0.198
5700	38.393	312.517	277.861	197.541	85.146	20.125	-0.184
5800	38.418	313.185	278.464	201.381	84.884	18.987	-0.171
5900	38.443	313.842	279.058	205.224	84.613	17.853	-0.158
6000	38.468	314.488	279.643	209.070	84 331	16.724	-0.146

PREVIOUS: June 1963 (1 atm) CURRENT: June 1963 (1 bar)

NITROGEN OXIDE, ION (NO^+) IDEAL GAS M_r = 30.00555

$S°(298.15\ K) = 198.219\ J\ K^{-1}\ mol^{-1}$

$\Delta_fH°(0\ K) = 983.995 \pm 0.8\ kJ\ mol^{-1}$
$\Delta_fH°(298.15\ K) = 990.185 \pm 0.8\ kJ\ mol^{-1}$

Electronic Level and Quantum Weight

State	ε_i, cm^{-1}	g_i
$^1\Sigma^+$	0	1

$\omega_e = 2377.1\ cm^{-1}$ $\omega_ex_e = 16.35\ cm^{-1}$ $\sigma = 1$
$B_e = 2.002\ cm^{-1}$ $\alpha_e = 0.0202\ cm^{-1}$ $r_e = 1.0619$ Å

Enthalpy of Formation

Watanabe (1) and Walker and Weissler (2) have measured the ionization potential of NO(g) by photoionization, yielding IP = 9.25±0.02 and 9.20±0.03 eV, respectively. Hagstrum (3), reported IP = 9.4±0.2 eV by electron impact. Huber (4) determined spectroscopically the ionization potential for NO(g), IP = 9.267±0.005 eV (213.72±0.1 kcal mol^{-1}). The last value has been selected, and yields the enthalpy of formation of NO^+(g), $\Delta_fH°(NO^+, g, 0\ K) = 235.18\ kcal\ mol^{-1}$, using $\Delta_fH°(NO, g, 0\ K) = 21.46\pm0.04\ kcal\ mol^{-1}$.

Heat Capacity and Entropy

The selected molecular constants were obtained from the ultraviolet spectra by Miescher (5).

The tabulated thermodynamic functions are in reasonable agreement with those calculated by Hilsenrath and Messina (6), who have used slightly different molecular constants.

References

1. K. Watanabe, J. Chem. Phys. 22, 1564 (1954).
2. W. C. Walker and G. L. Weissler, J. Chem. Phys. 23, 1962 (1955).
3. H. D. Hagstrum, J. Chem. Phys. 23, 1178 (1955).
4. K. P. Huber, Helv. Phys. Acta 34, 929 (1961).
5. E. Miescher, Can. J. Phys. 33, 355 (1955), Helv. Phys. Acta 29, 135 (1956).
6. J. Hilsenrath and C. G. Messina, U. S. Nat. Bur. Stand. Report 8504, 1964.

Nitrogen Oxide, Ion (NO^+) $N_1O_1^+(g)$

Enthalpy Reference Temperature = T_r = 298.15 K Standard State Pressure = p° = 0.1 MPa

T/K	$C_p°$ (J K⁻¹mol⁻¹)	S° (J K⁻¹mol⁻¹)	$-[G°-H°(T_r)]/T$ (J K⁻¹mol⁻¹)	$H°-H°(T_r)$ (kJ mol⁻¹)	$\Delta_fH°$ (kJ mol⁻¹)	$\Delta_fG°$ (kJ mol⁻¹)	Log K_f
0	0.	0.	INFINITE	-8.670	983.995		
100	29.104	166.421	224.100	-5.768			
200	29.107	186.595	200.882	-2.857			
250	29.111	193.091	198.698	-1.402			
298.15	29.123	198.219	198.219	0.	990.185	983.978	-172.389
300	29.124	198.399	198.219	0.054	990.224	983.939	-171.319
350	29.163	202.891	198.574	1.511	991.253	982.811	-146.676
400	29.244	206.790	199.363	2.971	992.275	981.535	-128.175
450	29.378	210.242	200.383	4.436	993.288	980.131	-113.771
500	29.568	213.346	201.527	5.910	994.293	978.615	-102.235
600	30.089	218.781	203.962	8.891	996.282	975.293	-84.907
700	30.728	223.466	206.421	11.931	998.252	971.639	-72.505
800	31.403	227.613	208.816	15.038	1000.214	967.703	-63.185
900	32.059	231.350	211.115	18.211	1002.175	963.521	-55.921
1000	32.666	234.760	213.312	21.448	1004.139	959.121	-50.099
1100	33.213	237.899	215.406	24.742	1006.109	954.524	-45.327
1200	33.697	240.810	217.403	28.088	1008.085	949.747	-41.341
1300	34.124	243.525	219.309	31.480	1010.066	944.805	-37.963
1400	34.497	246.067	221.131	34.911	1012.053	939.711	-35.061
1500	34.825	248.459	222.874	38.378	1014.043	934.474	-32.541
1600	35.112	250.716	224.544	41.875	1016.036	929.104	-30.332
1700	35.365	252.852	226.147	45.399	1018.030	923.610	-28.379
1800	35.588	254.880	227.687	48.947	1020.024	917.998	-26.640
1900	35.786	256.810	229.170	52.516	1022.017	912.276	-25.080
2000	35.963	258.650	230.598	56.103	1024.008	906.449	-23.674
2100	36.121	260.408	231.976	59.708	1025.995	900.522	-22.399
2200	36.263	262.092	233.307	63.327	1027.979	894.501	-21.238
2300	36.391	263.707	234.594	66.960	1029.959	888.389	-20.176
2400	36.507	265.258	235.839	70.605	1031.933	882.191	-19.200
2500	36.612	266.750	237.046	74.261	1033.902	875.911	-18.301
2600	36.709	268.188	238.216	77.927	1035.865	869.553	-17.470
2700	36.798	269.575	239.352	81.602	1037.823	863.119	-16 698
2800	36.880	270.915	240.456	85.286	1039.774	856.613	-15.980
2900	36.955	272.211	241.529	88.978	1041.720	850.037	-15.311
3000	37.025	273.465	242.572	92.677	1043.659	843.395	-14.685
3100	37.091	274.680	243.589	96.383	1045.593	836.687	-14.098
3200	37.152	275.858	244.579	100.095	1047.520	829.917	-13.547
3300	37.209	277.003	245.544	103.813	1049.442	823.087	-13.028
3400	37.263	278.114	246.486	107.537	1051.358	816.200	-12.539
3500	37.313	279.195	247.405	111.266	1053.269	809.255	-12.077
3600	37.361	280.247	248.303	114.999	1055.174	802.257	-11.640
3700	37.407	281.271	249.180	118.738	1057.074	795.205	-11.226
3800	37.450	282.269	250.038	122.481	1058.969	788.102	-10.833
3900	37.491	283.243	250.877	126.228	1060.860	780.950	-10.460
4000	37.531	284.192	251.698	129.979	1062.745	773.748	-10.104
4100	37.569	285.120	252.502	133.734	1064.626	766.500	-9.765
4200	37.605	286.025	253.289	137.492	1066.502	759.206	-9.442
4300	37.640	286.911	254.061	141.255	1068.374	751.867	-9.133
4400	37.673	287.776	254.817	145.020	1070.241	744.485	-8.838
4500	37.705	288.623	255.559	148.789	1072.103	737.060	-8.556
4600	37.737	289.452	256.287	152.561	1073.961	729.594	-8.285
4700	37.767	290.264	257.001	156.337	1075.814	722.088	-8.025
4800	37.796	291.060	257.702	160.115	1077.662	714.542	-7.776
4900	37.825	291.839	258.391	163.896	1079.505	706.959	-7.536
5000	37.853	292.604	259.068	167.680	1081.342	699.337	-7.306
5100	37.880	293.354	259.733	171.466	1083.174	691.679	-7.084
5200	37.906	294.089	260.386	175.256	1084.999	683.984	-6.871
5300	37.932	294.812	261.029	179.048	1086.819	676.255	-6.665
5400	37.957	295.521	261.661	182.842	1088.631	668.492	-6.466
5500	37.982	296.218	262.283	186.639	1090.436	660.694	-6 275
5600	38.006	296.902	262.895	190.438	1092.234	652.865	-6.090
5700	38 030	297.575	263.498	194.240	1094.023	645.003	-5.911
5800	38.053	298.237	264.091	198.044	1095.803	637.110	-5.738
5900	38.076	298.887	264.675	201.851	1097.574	629.186	-5.570
6000	38.098	299.528	265.251	205.660	1099.334	621.233	-5.408

PREVIOUS: June 1966 (1 atm) CURRENT: June 1966 (1 bar)

Nitrogen Oxide, Ion (NO^+) $N_1O_1^+(g)$

NITROGEN OXIDE (NO_2) IDEAL GAS M_r = 46.0055

$S°(298.15\ K) = 240.034 \pm 0.13\ J\ K^{-1}\ mol^{-1}$

$\Delta_fH°(0\ K) = 35.93 \pm 0.8\ kJ\ mol^{-1}$
$\Delta_fH°(298.15\ K) = 33.10 \pm 0.8\ kJ\ mol^{-1}$

Vibrational Frequencies and Degeneracies

ν, cm^{-1}
1357.8(1)
756.8(1)
1665.5(1)

Ground State Quantum Weight: 2 $\sigma = 2$
Point Group: C_{2v}
Bond Distance: N–O = 1.197 Å
Bond Angle: O–N–O = 134° 15'
Product of the Moments of Inertia: $I_AI_BI_C = 1.54229 \times 10^{-116}\ g^3\ cm^6$

Enthalpy of Formation

The equilibrium constants for NO(g) + 1/2 O_2 → NO_2(g) of Bodenstein and Lindner (1), as given by Giauque and Kemp (2) were subjected to 2nd and 3rd law analysis. 3rd law analysis gave $\Delta_fH°(298.15\ K) = -13.67\ kcal\ mol^{-1}$ and 2nd law $\Delta_fH°(298.15\ K) = -13.52\pm0.01\ kcal\ mol^{-1}$. The 3rd law value had a very slight trend with temperature which was consistent with an entropy error of 0.25 cal K^{-1} mol^{-1}, the 2nd law entropy was also consistent with an entropy error of 0.25 cal K^{-1} mol^{-1}. However the functions for all three gases are well established to about 0.03 cal K^{-1} mol^{-1} or better and so this must be interpreted as due to errors in the data. On this basis the 3rd law value was chosen and combined with JANAF auxiliary data to give $\Delta_fH°(298.15\ K) = 7.91\pm0.2$ kcal mol^{-1}, where the uncertainty includes that due to the possible error in the equilibrium constants.

Heat Capacity and Entropy

Bird et al. have given a complete review of the microwave spectrum of NO_2. They also give a corrected infra-red vibrational assignment, adopted here, which is in excellent agreement with calculations from the microwave data. They report $I_A = 0.34981 \times 10^{-39}$, g cm^2 and $I_B = 6.45446 \times 10^{-39}$ g cm^2 and $I_C = 6.81994 \times 10^{-39}$ g cm^2. The bond length and angle were chosen to give the best fit of the moments of inertia.

References

1. M. Bodenstein and Lindner, Z. physik. Chem. 100, 82 (1922).
2. W. F. Giauque and J. D. Kemp, J. Chem. Phys. 6, 40 (1938).
3. G. R. Bird, J. C. Baird, A. W. Jache, J. A. Hodgeson, R. F. Curl Jr., A. C. Kunkle, J. W. Bransford, J. Rastrup-Andersen and J. Rosenthal, J. Chem. Phys. 40, 3378 (1964).

Nitrogen Oxide (NO_2) $N_1O_2(g)$

Enthalpy Reference Temperature = T_r = 298.15 K Standard State Pressure = p° = 0.1 MPa

T/K	$C_p°$ (J $K^{-1}mol^{-1}$)	S° (J $K^{-1}mol^{-1}$)	$-[G°-H°(T_r)]/T$ (J $K^{-1}mol^{-1}$)	$H°-H°(T_r)$ (kJ mol^{-1})	$\Delta_fH°$ (kJ mol^{-1})	$\Delta_fG°$ (kJ mol^{-1})	Log K_f
0	0.	0.	INFINITE	-10.186	35.927	35.927	INFINITE
100	33.276	202.563	271.168	-6.861	34.898	39.963	-20.874
200	34.385	225.852	243.325	-3.495	33.897	45.422	-11.863
250	35.593	233.649	240.634	-1.746	33.460	48.355	-10.103
298.15	36.974	240.034	240.034	0.	33.095	51.258	-8.980
300	37.029	240.262	240.034	0.068	33.083	51.371	-8.944
350	38.583	246.086	240.491	1.958	32.768	54.445	-8.125
400	40.171	251.342	241.524	3.927	32.512	57.560	-7.517
450	41.728	256.164	242.886	5.975	32.310	60.703	-7.046
500	43.206	260.638	244.440	8.099	32.154	63.867	-6.672
600	45.834	268.755	247.830	12.555	31.959	70.230	-6.114
700	47.986	275.988	251.345	17.250	31.878	76.616	-5.717
800	49.708	282.512	254.840	22.138	31.874	83.008	-5.420
900	51.076	288.449	258.250	27.179	31.923	89.397	-5.188
1000	52.166	293.889	261.545	32.344	32.005	95.779	-5.003
1100	53.041	298.903	264.717	37.605	32.109	102.152	-4.851
1200	53.748	303.550	267.761	42.946	32.226	108.514	-4.724
1300	54.326	307.876	270.683	48.351	32.351	114.867	-4.615
1400	54.803	311.920	273.485	53.808	32.478	121.209	-4.522
1500	55.200	315.715	276.175	59.309	32.603	127.543	-4.441
1600	55.533	319.288	278.759	64.846	32.724	133.868	-4.370
1700	55.815	322.663	281.244	70.414	32.837	140.186	-4.307
1800	56.055	325.861	283.634	76.007	32.940	146.497	-4.251
1900	56.262	328.897	285.937	81.624	33.032	152.804	-4.201
2000	56.441	331.788	288.158	87.259	33.111	159.106	-4.155
2100	56.596	334.545	290.302	92.911	33.175	165.404	-4.114
2200	56.732	337.181	292.373	98.577	33.223	171.700	-4.077
2300	56.852	339.706	294.377	104.257	33.255	177.993	-4.042
2400	56.958	342.128	296.316	109.947	33.270	184.285	-4.011
2500	57.052	344.455	298.196	115.648	33.268	190.577	-3.982
2600	57.136	346.694	300.018	121.357	33.248	196.870	-3.955
2700	57.211	348.852	301.787	127.075	33.210	203.164	-3.930
2800	57.278	350.934	303.505	132.799	33.155	209.460	-3.908
2900	57.339	352.945	305.176	138.530	33.082	215.757	-3.886
3000	57.394	354.889	306.800	144.267	32.992	222.058	-3.866
3100	57.444	356.772	308.382	150.009	32.885	228.363	-3.848
3200	57.490	358.597	309.923	155.756	32.761	234.670	-3.831
3300	57.531	360.366	311.425	161.507	32.622	240.981	-3.814
3400	57.569	362.084	312.890	167.262	32.467	247.298	-3.799
3500	57.604	363.754	314.319	173.020	32.297	253.618	-3.785
3600	57.636	365.377	315.715	178.783	32.113	259.945	-3.772
3700	57.666	366.957	317.079	184.548	31.914	266.276	-3.759
3800	57.693	368.495	318.412	190.316	31.701	272.613	-3.747
3900	57.719	369.994	319.715	196.086	31.475	278.956	-3.736
4000	57.742	371.455	320.991	201.859	31.236	285.305	-3.726
4100	57.764	372.881	322.239	207.635	30.985	291.659	-3.716
4200	57.784	374.274	323.461	213.412	30.720	298.020	-3.706
4300	57.803	375.634	324.659	219.191	30.444	304.388	-3.698
4400	57.821	376.963	325.833	224.973	30.155	310.762	-3.689
4500	57.837	378.262	326.983	230.756	29.854	317.142	-3.681
4600	57.853	379.534	328.112	236.540	29.540	323.530	-3.674
4700	57.867	380.778	329.219	242.326	29.214	329.925	-3.667
4800	57.881	381.996	330.306	248.114	28.875	336.326	-3.660
4900	57.894	383.190	331.373	253.902	28.523	342.736	-3.654
5000	57.906	384.360	332.421	259.692	28.158	349.152	-3.648
5100	57.917	385.507	333.451	265.483	27.778	355.576	-3.642
5200	57.928	386.631	334.463	271.276	27.384	362.006	-3.636
5300	57.938	387.735	335.458	277.069	26.974	368.446	-3.631
5400	57.948	388.818	336.436	282.863	26.548	374.892	-3.626
5500	57.957	389.881	337.398	288.658	26.106	381.347	-3.622
5600	57.965	390.926	338.344	294.455	25.646	387.811	-3.617
5700	57.973	391.952	339.276	300.251	25.167	394.281	-3.613
5800	57.981	392.960	340.193	306.049	24.669	400.762	-3.609
5900	57.988	393.951	341.096	311.848	24.150	407.249	-3.606
6000	57.995	394.926	341.985	317.647	23.608	413.748	-3.602

PREVIOUS: September 1964 (1 atm) CURRENT: September 1964 (1 bar)

Nitrogen Oxide (NO_2) $N_1O_2(g)$

NITROGEN OXIDE, ION (NO_2^-) IDEAL GAS M_r = 46.006048

$S°(298.15\ K) = 236.598 \pm 2.1\ J\ K^{-1}\ mol^{-1}$

$\Delta_fH°(0\ K) = -193.73 \pm 5.8\ kJ\ mol^{-1}$

$\Delta_fH°(298.15\ K) = -202.72 \pm 5.8\ kJ\ mol^{-1}$

Electronic Levels and Quantum Weights

State	ε_i, cm^{-1}	g_i
[$^1A_1$]	[0]	[1]
[$^3B_1$]	[22000]	[3]
[$^1B_1$]	[27000]	[1]

Vibrational Frequencies and Degeneracies

ν, cm^{-1}
[1320](1)
[750](1)
1244 (1)

Point Group: C_{2v} σ = 2

Bond Distance: N-O = [1.23 ± 0.04] Å

Bond Angle: O-N-O = 115° ± 3°

Product of the Moments of Inertia: $I_AI_BI_C = [2.59592 \times 10^{-116}]\ g^3\ cm^6$

Enthalpy of Formation

The recent literature concerning the electron affinity of NO_2(g) is summarized in the following table.

Source	Technique	electron affinity NO_2(g), eV
(1)	magnetron technique, direct electron-capture process	3.99 ± 0.16
(2)	endothermic charge transfer reaction	>2.04, >3.063
(3)	endothermic charge transfer reaction	2.30 ± 0.15
(4)	photodetachment of e^- from NO_2^- in argon matrix	<3.9
(5)	reaction of NO_2^- with HCl and HBr in flowing after glow system	<2.6
(6)	correlation between heats of hydration and heterolytic bond dissociation energies	2.5
(7)	charge transfer reactions, flowing after glow system	2.38 ± 0.06

The early literature on the electron affinity of NO_2(g) is best summarized and referenced by Berkowitz et al. (2), Ferguson, et al. (5) and Dunkin, et al. (7). Much of these early data are not only widely scattered but also conflicting.

The value chosen for the electron affinity of NO_2(g) is 2.38 ± 0.06 eV, based on the work of Dunkin et al. (7). This leads to $\Delta_fH°(NO_2^-$, g, 298.15 K) = -48.45 ± 1.38 kcal mol^{-1}, using JANAF auxiliary data.

Heat Capacity and Entropy

The infrared spectrum of the NO_2^- ion isolated in an argon matrix was studied by Milligan et al. (4) and Milligan and Jacox (8). They reported that the antisymmetric stretching fundamental was ν_3 = 1244 cm^{-1}.

Hochstrasser and Marchetti (9), in combination with data from Sidman (10), reported ν_1 = 1325 cm^{-1}, ν_2 = 831 cm^{-1}, and ν_3 = 1243 cm^{-1} as the vibrational frequencies of the nitrite ion in solution. These frequencies are also very similar to the values reported for NO_2(g) by Herzberg (12). The ν_1 and ν_2 vibrational frequencies of NO_2^-(g) were assumed to be the same as those for nitrite ion in solution (11). The value adopted for ν_3 was that of Milligan, et al. (4, 8).

Milligan, et al (4, 8), via studies of the infrared spectrum of NO_2^- ion isolated in an argon matrix, concluded that the isotopic data were consistent with a O-N-O angle of 115°. Hochstrasser and Marchetti (9), relying on crystallographic data for $NaNO_2$(cr) collected by other investigators, reported the N-O bond distance as 1.23 ± 0.04 Å and the O-N-O bond angle as 115° ± 3°. The N-O bond distance in NO_2^-(g) is assumed to be the same as that reported for $NaNO_2$(cr).

The electronic states are estimated by analogy with the electronic levels of the isoelectronic molecule SO_2 (12).

References

1. A. L. Farragher, F. M. Page and R. C. Wheeler, Disc. Faraday Soc. 37, 203 (1964).
2. J. Berkowitz, W. A. Chupka and D. Gutman, J. Chem. Phys. 55, 2733 (1971).
3. C. Lifshitz, B. M. Hughes, and T. O. Tiernan, Chem. Phys. Letters 7, 469 (1970).
4. D. E. Milligan, H. E. Jacox and W. A. Guillory, J. Chem. Phys. 52, 3864 (1970).
5. E. E. Ferguson, D. B. Dunkin and F. C. Fehsenfeld, J. Chem. Phys. 57, Aug. 15, 1972.
6. J. D. Payzant, R. Yamdagni and P. Kebarle, Can. J. Chem. 49, 3308 (1971).
7. D. B. Dunkin, F. C. Fehsenfeld, and E. E. Ferguson, Chem. Phys. Letters, in press.
8. D. E. Milligan and H. E. Jacox, J. Chem. Phys. 55, 3404 (1971).
9. R. M. Hochstrasser and A. P. Marchetti, J. Chem. Phys. 50, 1727 (1969).
10. J. W. Sidman, J. Amer. Chem. Soc. 79, 2669 (1957).
11. D. Williams, J. Amer. Chem. Soc. 61, 2987 (1939).
12. G. Herzberg, "Electronic Spectra of Polyatomic Molecules," D. Van Nostrand, Company, Inc., Princeton, New Jersey (1966).

Nitrogen Oxide, Ion (ONO^-) $N_1O_2^-$(g)

Enthalpy Reference Temperature = T_r = 298.15 K Standard State Pressure = p° = 0.1 MPa

T/K	$C_p°$ (J $K^{-1}mol^{-1}$)	S° (J $K^{-1}mol^{-1}$)	$-[G°-H°(T_r)]/T$ (J $K^{-1}mol^{-1}$)	$H°-H°(T_r)$ (kJ mol^{-1})	$\Delta_fH°$ (kJ mol^{-1})	$\Delta_fG°$ (kJ mol^{-1})	Log K_f
0	0.	0.	INFINITE	-10.227	-193.726		
100	33.278	198.965	267.973	-6.901			
200	34.509	222.275	239.931	-3.531			
250	35.955	230.122	237.208	-1.771			
298.15	37.663	236.598	236.598	0.	-202.715	-177.273	31.057
300	37.733	236.831	236.599	0.070	-202.765	-177.115	30.838
350	39.645	242.791	237.065	2.004	-204.075	-172.735	25.779
400	41.538	248.209	238.124	4.034	-205.309	-168.173	21.961
450	43.317	253.206	239.526	6.156	-206.476	-163.460	18.974
500	44.933	257.855	241.129	8.363	-207.587	-158.621	16.571
600	47.641	266.297	244.635	12.997	-209.683	-148.628	12.939
700	49.714	273.804	248.275	17.870	-211.665	-138.295	10.320
800	51.291	280.550	251.895	22.924	-213.581	-127.682	8.337
900	52.496	286.664	255.424	28.116	-215.462	-116.831	6.781
1000	53.429	292.245	258.831	33.414	-217.324	-105.773	5.525
1100	54.160	297.373	262.105	38.795	-219.179	-94.527	4.489
1200	54.741	302.111	265.244	44.241	-221.035	-83.113	3.618
1300	55.208	306.512	268.251	49.739	-222.896	-71.545	2.875
1400	55.590	310.618	271.132	55.279	-224.764	-59.832	2.232
1500	55.904	314.464	273.894	60.855	-226.643	-47.986	1.671
1600	56.166	318.081	276.544	66.459	-228.534	-36.014	1.176
1700	56.387	321.493	279.089	72.087	-230.439	-23.923	0.735
1800	56.574	324.721	281.535	77.735	-232.360	-11.720	0.340
1900	56.734	327.784	283.889	83.400	-234.297	0.591	-0.016
2000	56.872	330.698	286.157	89.081	-236.252	13.004	-0.340
2100	56.992	333.476	288.345	94.774	-238.226	25.515	-0.635
2200	57.098	336.129	290.457	100.479	-240.218	38.121	-0.905
2300	57.191	338.670	292.498	106.193	-242.229	50.817	-1.154
2400	57.275	341.105	294.473	111.917	-244.260	63.602	-1.384
2500	57.351	343.445	296.386	117.648	-246.310	76.471	-1.598
2600	57.421	345.696	298.239	123.387	-248.380	89.424	-1.797
2700	57.487	347.864	300.037	129.132	-250.468	102.456	-1.982
2800	57.550	349.956	301.783	134.884	-252.574	115.566	-2.156
2900	57.612	351.977	303.479	140.642	-254.699	128.750	-2.319
3000	57.673	353.931	305.129	146.407	-256.840	142.010	-2.473
3100	57.735	355.823	306.733	152.177	-258.997	155.341	-2.617
3200	57.799	357.657	308.296	157.954	-261.169	168.740	-2.754
3300	57.866	359.436	309.819	163.737	-263.354	182.208	-2.884
3400	57.937	361.165	311.304	169.527	-265.553	195.744	-3.007
3500	58.012	362.845	312.753	175.324	-267.763	209.343	-3.124
3600	58.092	364.481	314.167	181.130	-269.983	223.006	-3.236
3700	58.178	366.074	315.548	186.943	-272.212	236.731	-3.342
3800	58.271	367.626	316.899	192.765	-274.449	250.517	-3.444
3900	58.370	369.141	318.219	198.597	-276.692	264.361	-3.541
4000	58.476	370.620	319.510	204.440	-278.940	278.262	-3.634
4100	58.590	372.066	320.775	210.293	-281.193	292.220	-3.723
4200	58.711	373.479	322.013	216.158	-283.448	306.233	-3.809
4300	58.840	374.862	323.226	222.035	-285.705	320.300	-3.891
4400	58.977	376.216	324.415	227.926	-287.963	334.419	-3.970
4500	59.121	377.543	325.581	233.831	-290.221	348.589	-4.046
4600	59.273	378.844	326.725	239.751	-292.478	362.809	-4.120
4700	59.432	380.121	327.847	245.686	-294.733	377.079	-4.191
4800	59.598	381.374	328.949	251.637	-296.987	391.397	-4.259
4900	59.771	382.604	330.032	257.606	-299.238	405.762	-4.325
5000	59.951	383.814	331.095	263.592	-301.486	420.173	-4.390
5100	60.136	385.003	332.141	269.596	-303.731	434.629	-4.452
5200	60.327	386.172	333.169	275.619	-305.973	449.127	-4.512
5300	60.523	387.323	334.179	281.662	-308.212	463.671	-4.570
5400	60.724	388.456	335.174	287.724	-310.448	478.255	-4.626
5500	60.930	389.573	336.153	293.807	-312.681	492.881	-4.681
5600	61.139	390.672	337.117	299.910	-314.913	507.548	-4.734
5700	61.351	391.756	338.066	306.035	-317.142	522.254	-4.786
5800	61.566	392.825	339.001	312.181	-319.371	537.001	-4.836
5900	61.783	393.879	339.922	318.348	-321.600	551.783	-4.885
6000	62.002	394.920	340.830	324.537	-323.831	566.606	-4.933

PREVIOUS: June 1972 (1 atm) CURRENT: June 1972 (1 bar)

Nitrogen Oxide, Ion (ONO^-) $N_1O_2^-$(g)

NITROGEN OXIDE (NO_3) IDEAL GAS $M_r = 62.0049$

$S°(298.15\ K) = [252.619]\ J\ K^{-1}\ mol^{-1}$

$\Delta_fH°(0\ K) = 77.53 \pm 20.9\ kJ\ mol^{-1}$

$\Delta_fH°(298.15\ K) = 71.13 \pm 20.9\ kJ\ mol^{-1}$

Vibrational Frequencies and Degeneracies

$\nu,\ cm^{-1}$

[1158](2)

[940](1)

[704](2)

[765](1)

Ground State Quantum Weight: 2 $\sigma = [6]$

Point Group: $[D_{3h}]$

Bond Distance: O-N = [1.2] Å

Bond Angle: O-N-O = [120]°

Product of the Moments of Inertia: $I_AI_BI_C = [5.311350 \times 10^{-115}]\ g^3\ cm^6$

Enthalpy of Formation

The enthalpy of formation is based on equilibrium constants for the reaction $N_2O_5(g) \rightarrow NO_2(g) + NO_3(g)$ obtained from shock wave studies by Schott and Davidson (1). These equilibria give a 2nd law value of $\Delta_rH°(495\ K) = 21.5 \pm 1.4\ kcal\ mol^{-1}$, corresponding to $\Delta_rH°(298.15\ K) = 22.2\ kcal\ mol^{-1}$, and a 3rd law value of $\Delta_rH°(298.15\ K) = 22.2\ kcal\ mol^{-1}$. The resulting enthalpy of formation is 17.0 $kcal\ mol^{-1}$.

Guillory and Johnston (2) report an equilibrium constant based on estimated Beer's law constants for NO_3 and attributed to the reaction $O_2 + NO \rightarrow NO_3$. This gives a 3rd law heat of formation of 12 ± 2 $kcal\ mol^{-1}$ which was not considered. A tentative value of 23 $kcal\ mol^{-1}$ (or greater) may be derived from the D° given by Ramsay (3).

Heat Capacity and Entropy

The structure and frequencies are those estimated by Hisatsune (4). A symmetry of D_{3h} was chosen on the basis of Walsh's (5) correlation, and the bond distance was taken as 2% longer than in the nitrate ion. The planar mode frequencies were calculated from estimated Urey-Bradley force constants, while the out-of-plane wagging frequency (765 cm^{-1}) was assumed the same as in $HNO_3(g)$.

There are several conflicting bits of information concerning NO_3 radicals. Chantry et al. (6) studied the electron resonance and optical absorption spectra of what appears to be NO_3 radicals trapped in crystals of urea nitrate. The authors suggest a planar structure without a threefold axis (thus C_s or C_{2v} symmetry), at least for the trapped radicals. Guillory and Johnston (2) report for gaseous NO_3 radicals an infrared frequency at 1840 cm^{-1} which they assign as the N-O stretch. The similarity of this frequency to those in N-O and X-N-O molecules led the authors to suggest an OONO structure (thus C_s symmetry) rather than the nitrate structure. The remaining frequences for such a structure may be crudely estimated by analogy with those measured for $(NO)_2$ by Smith et al. (7). Based on such frequencies and reasonable bond distances, the room temperature entropy for the OONO structure would be 68-72 $cal\ K^{-1}\ mol^{-1}$. Such an entropy is quite inconsistent with the 2nd law $\Delta_rS°(495\ K) = 33.2 \pm 3\ cal\ K^{-1}\ mol^{-1}$ for $N_2O_5(g) \rightarrow NO_2(g) + NO_3(g)$ obtained from the shock wave studies of Schott and Davidson. The Hisatsune functions based on the nitrate structure were selected because they are consistent with the experimental entropy. The principal moments of inertia for this structure are $I_A = I_B = 6.4277 \times 10^{-39}$ and $I_C = 12.8555 \times 10^{-39}\ g\ cm^2$. A structure of C_{2v} symmetry would result if the nitrate structure were distorted to give one non-equivalent oxygen. Ramsay (3) suggests that such distortion can only be small, based on the high dispersion optical spectra of the gas phase.

References

1. G. Schott and N. Davidson, J. Amer. Chem. Soc. 80, 1941 (1958).
2. W. Guillory and H. S. Johnston, J. Amer. Chem. Soc. 85, 1695 (1963).
3. D. A. Ramsay, Proc. 10th Colloq. Spect. Intnl. (1962), Univ. Maryland, pp. 583-596, 1963.
4. I. C. Hisatsune, J. Phys. Chem. 65, 2249 (1961).
5. A. D. Walsh, J. Chem. Soc. 1953, 2301 (1953).
6. G. W. Chantry, A. Horsfield, J. R. Morton and D. H. Whiffen, Mol. Phys. 5, 589 (1962).
7. A. L. Smith, W. E. Keller and H. L. Johnston, J. Chem. Phys. 19, 189 (1951).

Nitrogen Oxide (NO_3) $N_1O_3(g)$

Enthalpy Reference Temperature = T_r = 298.15 K Standard State Pressure = p° = 0.1 MPa

	$J\ K^{-1}mol^{-1}$			$kJ\ mol^{-1}$			
T/K	$C_p°$	S°	$-[G°-H°(T_r)]/T$	$H°-H°(T_r)$	$\Delta_fH°$	$\Delta_fG°$	Log K_f
0	0.	0.	INFINITE	-10.959	77.529	77.529	INFINITE
100	33.345	211.871	288.193	-7.632	75.048	87.847	-45.887
200	37.739	235.888	256.603	-4.143	72.715	101.581	-26.530
250	42.240	244.778	253.366	-2.147	71.797	108.909	-22.755
298.15	46.934	252.619	252.619	0.	71.128	116.121	-20.344
300	47.113	252.910	252.620	0.087	71.106	116.400	-20.267
350	51.765	260.527	253.211	2.561	70.637	123.989	-18.504
400	55.929	267.717	254.579	5.255	70.359	131.631	-17.189
450	59.533	274.519	256.420	8.144	70.240	139.299	-16.169
500	62.601	280.954	258.555	11.200	70.246	146.973	-15.354
600	67.380	292.815	263.295	17.712	70.527	162.297	-14.129
700	70.793	303.472	268.286	24.630	71.042	177.554	-13.249
800	73.266	313.095	273.296	31.840	71.691	192.726	-12.584
900	75.095	321.836	278.211	39.262	72.417	207.813	-12.061
1000	76.476	329.822	282.979	46.844	73.186	222.816	-11.639
1100	77.539	337.163	287.576	54.547	73.977	237.741	-11.289
1200	78.373	343.947	291.994	62.344	74.776	252.593	-10.995
1300	79.037	350.248	296.236	70.216	75.576	267.379	-10.743
1400	79.573	356.125	300.306	78.147	76.371	282.103	-10.525
1500	80.013	361.631	304.213	86.127	77.155	296.771	-10.334
1600	80.377	366.807	307.965	94.147	77.925	311.387	-10.166
1700	80.682	371.689	311.571	102.200	78.678	325.955	-10.015
1800	80.939	376.308	315.040	110.282	79.411	340.479	-9.880
1900	81.159	380.690	318.381	118.387	80.122	354.963	-9.759
2000	81.348	384.858	321.602	126.513	80.809	369.411	-9.648
2100	81.511	388.831	324.709	134.656	81.472	383.825	-9.547
2200	81.653	392.626	327.711	142.814	82.108	398.208	-9.455
2300	81.777	396.259	330.613	150.986	82.717	412.562	-9.370
2400	81.887	399.741	333.421	159.169	83.298	426.890	-9.291
2500	81.984	403.086	336.141	167.363	83.851	441.195	-9.218
2600	82.070	406.303	338.778	175.565	84.376	455.478	-9.151
2700	82.147	409.402	341.337	183.776	84.873	469.743	-9.088
2800	82.216	412.391	343.822	191.994	85.343	483.989	-9.029
2900	82.278	415.277	346.236	200.219	85.785	498.217	-8.974
3000	82.334	418.067	348.584	208.450	86.200	512.432	-8.922
3100	82.385	420.768	350.869	216.686	86.590	526.634	-8.874
3200	82.431	423.384	353.095	224.927	86.954	540.822	-8.828
3300	82.473	425.922	355.263	233.172	87.293	555.000	-8.785
3400	82.512	428.384	357.378	241.421	87.608	569.169	-8.744
3500	82.547	430.777	359.441	249.674	87.901	583.327	-8.706
3600	82.580	433.102	361.455	257.931	88.171	597.479	-8.669
3700	82.609	435.365	363.422	266.190	88.419	611.622	-8.635
3800	82.637	437.569	365.345	274.452	88.647	625.761	-8.602
3900	82.663	439.716	367.224	282.717	88.854	639.893	-8.570
4000	82.686	441.809	369.063	290.985	89.042	654.019	-8.541
4100	82.708	443.851	370.862	299.255	89.210	668.142	-8.512
4200	82.728	445.844	372.624	307.526	89.360	682.260	-8.485
4300	82.747	447.791	374.349	315.800	89.491	696.376	-8.459
4400	82.765	449.694	376.040	324.076	89.603	710.487	-8.435
4500	82.782	451.554	377.697	332.353	89.698	724.597	-8.411
4600	82.797	453.373	379.323	340.632	89.774	738.704	-8.388
4700	82.812	455.154	380.917	348.913	89.831	752.811	-8.367
4800	82.825	456.898	382.482	357.194	89.869	766.917	-8.346
4900	82.838	458.606	384.018	365.478	89.887	781.023	-8.326
5000	82.850	460.279	385.527	373.762	89.886	795.127	-8.307
5100	82.862	461.920	387.009	382.048	89.863	809.233	-8.288
5200	82.872	463.529	388.465	390.334	89.819	823.337	-8.271
5300	82.883	465.108	389.896	398.622	89.753	837.445	-8.254
5400	82.892	466.657	391.303	406.911	89.662	851.552	-8.237
5500	82.901	468.178	392.687	415.201	89.547	865.662	-8.221
5600	82.910	469.672	394.049	423.491	89.406	879.776	-8.206
5700	82.918	471.140	395.388	431.782	89.237	893.890	-8.192
5800	82.926	472.582	396.707	440.075	89.040	908.010	-8.178
5900	82.933	474.000	398.005	448.368	88.811	922.129	-8.164
6000	82.940	475.393	399.283	456.661	88.550	936.257	-8.151

PREVIOUS: December 1964 (1 atm) CURRENT: December 1964 (1 bar)

Nitrogen Oxide (NO_3) $N_1O_3(g)$

PHOSPHORUS NITRIDE (PN) IDEAL GAS M_r = 44.98046

S°(298.15 K) = 211.137 J K^{-1} mol^{-1}

$\Delta_f H°$(0 K) = 105.77 ± 5.0 kJ mol^{-1}
$\Delta_f H°$(298.15 K) = 104.78 ± 5.0 kJ mol^{-1}

Electronic Level and Multiplicity

State	ε_i, cm^{-1}	g_i
$^1\Sigma^+$	0	1

ω_e = 1337.24 cm^{-1} $\omega_e x_e$ = 6.983 cm^{-1} σ = 1
B_e = 0.7862 cm^{-1} α_e = 0.00557 cm^{-1} r_e = 1.4910 Å

Enthalpy of Formation

The selected $\Delta_f H°$(298.15 K) was calculated from the D_0^0 = 165.8±1.2 kcal mol^{-1} reported by Potter and DiStefano (1). D_0^0 = 163.7±1.2 kcal mol^{-1} was experimentally determined by Huffman et al. (2). The D_0^0 reported by Huffman et al. (2) was derived using a pressure-composition method, i.e., measurement of equilibrium vapor pressure, at 1173 K for the reaction [PN(g) → P_2(g) + N_2(g)]. Potter and DiStefano (1) recalculated Huffman's results to obtain their D_0^0 value. Gaydon (3) lists a D_0^0 = 138.4±19 kcal mol^{-1}. The corresponding enthalpy of formation are:

Source	$\Delta_f H°$(298.15 K), kcal mol^{-1}
Potter and DiStefano (1)	25.0 ± 1.2
Huffman et al. (2)	27.1 ± 1.2
Gaydon (3)	53 ± 19

Heat Capacity and Entropy

The molecular constants are taken from Herzburg (4).

References

1. R. L. Potter and V. N. DiStefano, J. Phys. Chem. 65, 849 (1961).
2. E. O. Huffman, G. Tarbutton, K. L. Elmore, W. E. Cate, H. K. Walters, Jr., and G. V. Elmore, J. Amer. Chem. Soc. 76, 6329 (1954).
3. A. G. Gaydon, "Dissociation Energies and Spectra of Diatomic Molecules," Chapman and Hall (1953).
4. G. Herzburg, "Diatomic Molecules," D. Van Nostrand Co., New York, N. Y., 2nd Ed. (1960).

Phosphorus Nitride (PN) N_1P_1(g)

Enthalpy Reference Temperature = T_r = 298.15 K Standard State Pressure = p° = 0.1 MPa

T/K	$C_p°$ (J $K^{-1}mol^{-1}$)	S° (J $K^{-1}mol^{-1}$)	$-[G°-H°(T_r)]/T$ (J $K^{-1}mol^{-1}$)	$H°-H°(T_r)$ (kJ mol^{-1})	$\Delta_f H°$ (kJ mol^{-1})	$\Delta_f G°$ (kJ mol^{-1})	Log K_f
0	0.	0.	INFINITE	-8.703	105.772	105.772	INFINITE
100	29.105	179.230	237.189	-5.796	106.341	98.177	-51.282
200	29.168	199.412	213.832	-2.884	105.534	90.075	-23.525
250	29.361	205.938	211.624	-1.422	105.177	86.251	-18.021
298.15	29.710	211.137	211.137	0.	104.780	82.641	-14.478
300	29.726	211.321	211.138	0.055	104.764	82.503	-14.365
350	30.230	215.940	211.501	1.553	103.598	78.899	-11.775
400	30.809	220.013	212.315	3.079	103.077	75.407	-9.847
450	31.409	223.677	213.378	4.635	102.584	71.978	-8.355
500	31.991	227.017	214.577	6.220	102.116	68.603	-7.167
600	33.033	232.944	217.157	9.473	101.244	61.983	-5.396
700	33.879	238.102	219.788	12.820	100.437	55.504	-4.142
800	34.548	242.672	222.369	16.242	99.673	49.138	-3.208
900	35.075	246.772	224.856	19.725	98.934	42.866	-2.488
1000	35.494	250.490	227.237	23.254	98.211	36.675	-1.916
1100	35.830	253.890	229.507	26.821	97.496	30.556	-1.451
1200	36.103	257.019	231.671	30.418	33.214	25.580	-1.113
1300	36.328	259.918	233.734	34.040	33.281	24.941	-1.002
1400	36.516	262.618	235.702	37.682	33.344	24.297	-0.907
1500	36.675	265.143	237.581	41.342	33.403	23.649	-0.824
1600	36.812	267.514	239.379	45.016	33.458	22.996	-0.751
1700	36.930	269.749	241.100	48.704	33.509	22.341	-0.686
1800	37.033	271.863	242.751	52.402	33.556	21.683	-0.629
1900	37.124	273.868	244.336	56.110	33.600	21.022	-0.578
2000	37.206	275.774	245.861	59.826	33.642	20.359	-0.532
2100	37.279	277.591	247.329	63.551	33.680	19.694	-0.490
2200	37.346	279.327	248.744	67.282	33.716	19.027	-0.452
2300	37.407	280.988	250.110	71.020	33.750	18.358	-0.417
2400	37.463	282.582	251.430	74.763	33.781	17.688	-0.385
2500	37.515	284.112	252.707	78.512	33.811	17.017	-0.356
2600	37.563	285.584	253.944	82.266	33.839	16.345	-0.328
2700	37.608	287.003	255.142	86.024	33.866	15.672	-0.303
2800	37.651	288.371	256.304	89.787	33.891	14.997	-0.280
2900	37.692	289.693	257.433	93.555	33.915	14.322	-0.258
3000	37.730	290.972	258.530	97.326	33.937	13.646	-0.238
3100	37.767	292.210	259.596	101.101	33.959	12.969	-0.219
3200	37.802	293.409	260.634	104.879	33.979	12.292	-0.201
3300	37.836	294.573	261.645	108.661	33.999	11.614	-0.184
3400	37.869	295.703	262.630	112.446	34.018	10.935	-0.168
3500	37.900	296.801	263.591	116.235	34.036	10.256	-0.153
3600	37.931	297.869	264.529	120.026	34.053	9.577	-0.139
3700	37.961	298.909	265.444	123.821	34.069	8.896	-0.126
3800	37.990	299.922	266.338	127.618	34.085	8.216	-0.113
3900	38.018	300.909	267.212	131.419	34.100	7.535	-0.101
4000	38.046	301.872	268.066	135.222	34.115	6.854	-0.089
4100	38.073	302.811	268.902	139.028	34.129	6.172	-0.079
4200	38.099	303.729	269.721	142.836	34.143	5.490	-0.068
4300	38.125	304.626	270.522	146.648	34.156	4.807	-0.058
4400	38.151	305.503	271.307	150.461	34.168	4.125	-0.049
4500	38.176	306.360	272.076	154.278	34.180	3.442	-0.040
4600	38.201	307.200	272.831	158.097	34.192	2.759	-0.031
4700	38.226	308.022	273.571	161.918	34.203	2.075	-0.023
4800	38.250	308.827	274.297	165.742	34.214	1.391	-0.015
4900	38.274	309.616	275.010	169.568	34.224	0.708	-0.008
5000	38.298	310.389	275.710	173.397	34.234	0.023	-0.000
5100	38.321	311.148	276.397	177.228	34.243	-0.661	0.007
5200	38.345	311.892	277.073	181.061	34.251	-1.345	0.014
5300	38.368	312.623	277.737	184.897	34.259	-2.030	0.020
5400	38.391	313.340	278.389	188.735	34.267	-2.715	0.026
5500	38.414	314.045	279.031	192.575	34.274	-3.400	0.032
5600	38.436	314.737	279.663	196.417	34.280	-4.085	0.038
5700	38.459	315.418	280.284	200.262	34.285	-4.770	0.044
5800	38.481	316.087	280.895	204.109	34.290	-5.455	0.049
5900	38.503	316.745	281.497	207.958	34.294	-6.140	0.054
6000	38.525	317.392	282.090	211.810	34.296	-6.826	0.059

PREVIOUS: September 1962 (1 atm) CURRENT: September 1962 (1 bar)

Phosphorus Nitride (PN) N_1P_1(g)

SULFUR NITRIDE (SN) IDEAL GAS M_r = 46.0667

$\Delta_f H°(0\ K) = 262.9 \pm 105\ kJ\ mol^{-1}$

$S°(298.15\ K) = 222.087\ J\ K^{-1}\ mol^{-1}$ $\Delta_f H°(298.15\ K) = 263.6 \pm 105\ kJ\ mol^{-1}$

Electronic Levels and Multiplicities

State	ε_i, cm^{-1}	g_i
$^2\Pi_r$	0	2
	223	2

$\omega_e = 1220.0\ cm^{-1}$ $\omega_e x_e = 7.75\ cm^{-1}$ $\sigma = 1$

$B_e = [0.7762]\ cm^{-1}$ $\alpha_e = [0.0064]\ cm^{-1}$ $r_e = 1.495(7)$ Å

Enthalpy of Formation

The dissociation energy is listed by Gaydon (1).

Heat Capacity and Entropy

The spectroscopic constants other than B_e and α_e are from Herzberg (2). r_e is from (3) and has been used, together with the equation $\equiv_e = 6\ (f_e x_e B_e^3 - 6B_e^2)/f_e$ to calculate B_e and $\equiv_e$ (2).

References

1. A. G. Gaydon, "Dissociation Energies and Spectra of Diatomic Molecules," 2nd Ed., Chapman and Hall, London (1953).
2. G. Herzberg, Spectra of Diatomic Molecules," 2nd Ed., D. Van Nostrand Company, New York (1950).
3. "Tables of Interatomic Distances and Configuration in Mclecules and Ions, The Chemical Society, London (1958).

Nitrogen Sulfide (NS) $N_1S_1(g)$

Enthalpy Reference Temperature = T_r = 298.15 K Standard State Pressure = p° = 0.1 MPa

	J $K^{-1}mol^{-1}$			kJ mol^{-1}			
T/K	$C_p°$	S°	$-[G°-H°(T_r)]/T$	$H°-H°(T_r)$	$\Delta_f H°$	$\Delta_f G°$	Log K_f
0	0.	0.	INFINITE	-9.397	262.942	262.942	INFINITE
100	32.301	186.768	250.632	-6.386	263.812	254.378	-132.873
200	32.201	209.348	224.999	-3.130	263.971	244.827	-63.942
250	31.833	216.488	222.611	-1.531	263.823	240.056	-50.157
298.15	31.793	222.087	222.087	0.	263.592	235.498	-41.258
300	31.797	222.284	222.088	0.059	263.582	235.324	-40.974
350	32.020	227.200	222.476	1.653	263.282	230.637	-34.421
400	32.391	231.498	223.340	3.263	260.731	226.084	-29.524
450	32.824	235.338	224.464	4.894	259.703	221.806	-25.747
500	33.267	238.820	225.728	6.546	258.615	217.656	-22.738
600	34.081	244.959	228.435	9.914	256.907	209.632	-18.250
700	34.751	250.265	231.183	13.357	255.482	201.868	-15.064
800	35.284	254.941	233.866	16.860	254.219	194.297	-12.686
900	35.704	259.122	236.444	20.410	199.924	187.958	-10.909
1000	36.039	262.902	238.904	23.998	200.036	186.622	-9.748
1100	36.308	266.350	241.245	27.616	200.133	185.276	-8.798
1200	36.529	269.519	243.471	31.258	200.214	183.922	-8.006
1300	36.713	272.450	245.589	34.920	200.277	182.562	-7.335
1400	36.867	275.177	247.606	38.600	200.322	181.197	-6.761
1500	36.999	277.725	249.530	42.293	200.348	179.830	-6.262
1600	37.114	280.117	251.367	45 999	200.356	178.462	-5.826
1700	37.214	282.370	253.125	49.715	200.347	177.094	-5.441
1800	37.303	284.499	254.810	53.441	200.322	175.726	-5.099
1900	37.383	286.519	256.426	57.176	200.282	174.361	-4.794
2000	37.455	288.438	257.979	60.918	200.228	172.998	-4.518
2100	37.521	290.267	259.473	64.667	200.162	171.638	-4.269
2200	37.582	292.014	260.913	68.422	200.085	170.282	-4.043
2300	37.639	293.686	262.302	72.183	199.997	168.929	-3.836
2400	37.691	295.289	263.643	75.949	199.900	167.580	-3.647
2500	37.741	296.828	264.940	79.721	199.795	166.236	-3.473
2600	37.788	298.310	266.195	83.498	199.683	164.895	-3.313
2700	37.833	299.737	267.411	87.279	199.564	163.560	-3.164
2800	37.875	301.113	268.590	91.064	199.439	162.228	-3.026
2900	37.916	302.443	269.735	94.854	199.309	160.902	-2.898
3000	37.955	303.729	270.847	98.647	199.174	159.580	-2.779
3100	37.993	304.974	271.928	102.445	199.035	158.262	-2.667
3200	38.030	306.181	272.979	106.246	198.892	156.949	-2.562
3300	38.066	307.352	274.003	110.051	198.746	155.641	-2 464
3400	38.101	308.489	275.001	113.859	198.596	154.337	-2.371
3500	38.135	309.594	275.974	117.671	198.442	153.037	-2.284
3600	38.168	310.669	276.922	121.486	198.286	151.742	-2.202
3700	38.201	311.715	277.849	125.304	198.126	150.451	-2.124
3800	38.233	312.734	278.753	129.126	197.963	149.165	-2.050
3900	38.265	313.727	279.637	132.951	197.796	147.883	-1.981
4000	38.296	314.697	280.502	136.779	197.626	146.605	-1.914
4100	38.326	315.643	281.347	140.610	197.452	145.332	-1.852
4200	38.357	316.567	282.175	144.444	197.275	144.063	-1.792
4300	38.386	317.469	282.985	148.281	197.093	142.798	-1.735
4400	38.416	318.352	283.779	152.122	196.907	141.538	-1.680
4500	38.445	319.216	284.557	155.965	196.717	140.281	-1.628
4600	38.474	320.061	285.320	159.811	196.523	139.029	-1.579
4700	38.503	320.889	286.068	163.660	196.323	137.782	-1.531
4800	38.532	321.700	286.802	167.511	196.119	136.538	-1.486
4900	38.560	322.495	287.522	171.366	195.909	135.299	-1.442
5000	38.588	323 274	288.229	175.223	195.694	134.064	-1.401
5100	38.616	324.038	288.924	179.083	195.474	132.834	-1.360
5200	38.644	324.789	289.607	182.946	195.247	131.608	-1.322
5300	38.671	325.525	290.277	186.812	195.015	130.386	-1.285
5400	38.699	326.248	290.937	190.681	194.776	129.169	-1.249
5500	38.726	326.958	291.585	194.552	194.532	127.956	-1.215
5600	38.753	327.656	292.223	198.426	194.281	126.748	-1.182
5700	38.780	328.343	292.851	202.303	194.023	125.544	-1.150
5800	38.807	329.017	293.469	206.182	193.759	124.345	-1.120
5900	38.834	329.681	294.077	210.064	193.488	123.151	-1.090
6000	38.861	330.334	294.676	213.949	193.211	121.961	-1.062

PREVIOUS: June 1961 (1 atm) CURRENT June 1961 (1 bar)

Nitrogen Sulfide (NS) $N_1S_1(g)$

SILICON NITRIDE (SiN) IDEAL GAS M_r = 42.0922

$S°(298.15\ K) = 216.812\ J\ K^{-1}\ mol^{-1}$

$\Delta_f H°(0\ K) = 371.20 \pm 62.8\ kJ\ mol^{-1}$
$\Delta_f H°(298.15\ K) = 372.38 \pm 62.8\ kJ\ mol^{-1}$

Electronic Levels and Quantum Weights

State	ε_i, cm^{-1}	g_i
$^2\Sigma^+$	0	2
	[6000]	4
	24299.4	2
	[31000]	4

$\omega_e = 1150.8\ cm^{-1}$ $\omega_e x_e = 6.55\ cm^{-1}$ $\sigma = 1$
$B_e = 0.7299\ cm^{-1}$ $\alpha_e = 0.00567\ cm^{-1}$ $r_e = 1.5718$ Å

Enthalpy of Formation

A linear Birge-Sponer extrapolation of the ground state gives a dissociation energy of 142.9 kcal mol^{-1}. The ground state can correlate with normal atoms $Si(^3P) + N(^4S)$ or with $Si(^3P) + N(^2D)$; in this latter case D_0^0 would be 88 kcal mol^{-1}. The uncertainty in the extrapolation is quite large, ±30 kcal mol^{-1}, since only 6 levels, of the estimated total of 87 levels, are observed.

A dissociation energy can also be calculated from the $B^2\Sigma^+$ state by a much shorter extrapolation. This extrapolation is done graphically since some curvature is evident in the first six levels. A value of $D_0^0(B^2\Sigma)$ = 50.4 kcal mol^{-1} is obtained. This corresponds to $D_0^0(X^2\Sigma)$ = 119.9 kcal mol^{-1} for normal products or $D_0^0(X^2\Sigma)$ = 65 kcal mol^{-1} for excited nitrogen as a product. The uncertainty in this extrapolation is less than the ground state and is probably ±20 kcal mol^{-1}.

An estimate of $D_0^0(SiN)$ may be obtained by equating $D_0^0(SiN)/D_0^0(SiO) = D_0^0(CN)/D_0^0(CO)$; this yields $D_0^0(SiN)$ = 135 kcal mol^{-1}.

A comparison of the isoelectronic molecules CP, BS and BeCl and the related molecules CN, BO, BeF indicates $D_0^0(SiN)$ > 115 kcal mol^{-1}.

A value of $D_0^0(SiN)$ = 130 ± 15 kcal mol^{-1} is in agreement within the uncertainties of both ground state and B state extrapolations assuming normal atoms as products, and is in agreement with the other estimates. This value is adopted and yields $\Delta_f H°$(SiN, g, 298.15 K) = 89 kcal mol^{-1}.

Heat Capacity and Entropy

The B-X transition has been reported by Jenkins and de Laszlo (1). The rotational and vibrational constants are derived from their work and adjusted to normal isotopic abundance. The $A^2\Pi - C^2\Pi$ transition has been observed by Shetlar (2). The height of the A state is estimated by correlation with CP(g). The level of the B state is taken from Herzberg (3).

References

1. F. A. Jenkins and H. de Laszlo, Proc. Roy. Soc. (London) 122, 103 (1929).
2. M. D. Shetlar, Ph.D. Thesis, Univ. of California, Berkeley (1965).
3. G. Herzberg, "Spectra of Diatomic Molecules," D. Van Nostrand Co., Inc., 2nd Edition, New York, 1950.

Silicon Nitride (SiN) $N_1Si_1(g)$

Enthalpy Reference Temperature = T_r = 298.15 K Standard State Pressure = p° = 0.1 MPa

T/K	$C_p°$ (J K⁻¹mol⁻¹)	S° (J K⁻¹mol⁻¹)	$-[G°-H°(T_r)]/T$ (J K⁻¹mol⁻¹)	$H°-H°(T_r)$ (kJ mol⁻¹)	$\Delta_f H°$ (kJ mol⁻¹)	$\Delta_f G°$ (kJ mol⁻¹)	Log K_f
0	0.	0.	INFINITE	-8.733	371.197	371.197	INFINITE
100	29.106	184.787	243.038	-5.825	372.387	362.282	-189.237
200	29.267	204.984	219.537	-2.910	372.667	352.002	-91.934
250	29.630	211.550	217.306	-1.439	372.561	346.845	-72.469
298.15	30.168	216.812	216.812	0.	372.376	341.909	-59.901
300	30.191	216.999	216.813	0.056	372.368	341.720	-59.499
350	30.859	221.703	217.183	1.582	372.130	336.630	-50.239
400	31.551	225.868	218.013	3.142	371.874	331.576	-43.299
450	32.213	229.623	219.098	4.736	371.611	326.554	-37.905
500	32.818	233.049	220.324	6.362	371.347	321.562	-33.593
600	33.832	239.126	222.964	9.697	370.814	311.655	-27.132
700	34.616	244.403	225.658	13.121	370.269	301.838	-22.523
800	35.234	249.067	228.298	16.615	369.698	292.101	-19.072
900	35.752	253.248	230.842	20.165	369.098	282.437	-16.392
1000	36.222	257.039	233.275	23.764	368.466	272.841	-14.252
1100	36.683	260.513	235.596	27.409	367.807	263.311	-12.504
1200	37.156	263.725	237.808	31.101	367.125	253.841	-11.049
1300	37.652	266.718	239.918	34.841	366.428	244.429	-9.821
1400	38.173	269.528	241.933	38.632	365.720	235.071	-8.771
1500	38.713	272.180	243.862	42.476	365.006	225.764	-7.862
1600	39.264	274.696	245.711	46.375	364.289	216.504	-7.068
1700	39.814	277.093	247.487	50.329	313.395	207.736	-6.383
1800	40.355	279.384	249.196	54.338	312.910	201.535	-5.848
1900	40.877	281.580	250.843	58.399	312.467	195.360	-5.371
2000	41.372	283.689	252.433	62.512	312.066	189.207	-4.942
2100	41.834	285.719	253.970	66.673	311.705	183.073	-4.554
2200	42.258	287.675	255.458	70.878	311.380	176.956	-4.201
2300	42.642	289.562	256.900	75.123	311.089	170.852	-3.880
2400	42.984	291.384	258.299	79.405	310.829	164.760	-3.586
2500	43.285	293.145	259.658	83.719	310.595	158.679	-3.315
2600	43.544	294.848	260.979	88.060	310.384	152.607	-3.066
2700	43.765	296.496	262.264	92.426	310.192	146.542	-2.835
2800	43.948	298.091	263.515	96.812	310.016	140.484	-2.621
2900	44.097	299.636	264.734	101.215	309.853	134.433	-2.421
3000	44.215	301.133	265.923	105.630	309.699	128.386	-2.235
3100	44.304	302.584	267.082	110.057	309.553	122.345	-2.061
3200	44.368	303 992	268.214	114.490	309.411	116.308	-1.899
3300	44.409	305.358	269.319	118.929	309.271	110.276	-1.746
3400	44.430	306.684	270.398	123.372	309.131	104.248	-1.602
3500	44.435	307.972	271.453	127.815	308.990	98.224	-1.466
3600	44.425	309.224	272.485	132.258	-75.306	102.665	-1.490
3700	44.402	310.440	273.495	136.699	-75.040	107.605	-1.519
3800	44.370	311.624	274.483	141.138	-74.781	112.538	-1.547
3900	44.330	312.776	275.450	145.573	-74.529	117.464	-1.573
4000	44.283	313 898	276.397	150.004	-74.284	122.383	-1.598
4100	44.231	314.991	277.325	154.430	-74.047	127.297	-1.622
4200	44.176	316.056	278.235	158.850	-73.819	132.205	-1 644
4300	44.118	317.095	279.126	163.265	-73.598	137.108	-1.666
4400	44.059	318.108	280.001	167.673	-73.384	142.006	-1.686
4500	44.000	319.098	280.859	172.076	-73.178	146.898	-1.705
4600	43.941	320.064	281.700	176.473	-72.980	151.787	-1.724
4700	43.883	321.009	282.527	180.865	-72.789	156.671	-1.741
4800	43.827	321.932	283.338	185.250	-72.605	161.551	-1.758
4900	43.772	322.835	284.135	189.630	-72.428	166.428	-1.774
5000	43.721	323.719	284.918	194.005	-72.257	171.300	-1.790
5100	43.672	324.584	285.687	198.374	-72.093	176.170	-1.804
5200	43.626	325.432	286.443	202.739	-71.935	181.036	-1.819
5300	43.584	326.262	287.187	207.100	-71.783	185.900	-1.832
5400	43.544	327.077	287.918	211.456	-71.637	190.760	-1.845
5500	43.508	327.875	288.637	215.808	-71.496	195.618	-1.858
5600	43.476	328.659	289.345	220.158	-71.360	200.474	-1.870
5700	43.447	329.428	290.042	224.504	-71.230	205.327	-1.882
5800	43.422	330.184	290.727	228.847	-71.106	210.177	-1.893
5900	43.400	330.926	291.402	233.188	-70.986	215.026	-1.904
6000	43.382	331.655	292.067	237.528	-70.872	219.873	-1.914

PREVIOUS: March 1967 (1 atm) CURRENT: March 1967 (1 bar)

Silicon Nitride (SiN) $N_1Si_1(g)$

SILICON NITRIDE (Si_2N) IDEAL GAS M_r = 70.1777 Silicon Nitride (Si_2N) $N_1Si_2(g)$

$\Delta_fH°(0\ K) = 396.43 \pm 20.9\ kJ\ mol^{-1}$

$S°(298.15\ K) = [256.487 \pm 8.4]\ J\ K^{-1}\ mol^{-1}$ $\Delta_fH°(298.15\ K) = 397.48 \pm 20.9\ kJ\ mol^{-1}$

Electronic Levels and Quantum Weights

ε_i, cm^{-1}	g_i
0	[4]
[24000]	[4]
[28000]	[2]

Vibrational Frequencies and Degeneracies

ν, cm^{-1}
[600](1)
[240](2)
[1000](1)

Point Group: $[D_{\infty h}]$ σ = 2

Bond Distance: Si-N = [170] Å

Bond Angle: Si-N-Si = [180]°

Rotational Constant: $B_0 = [0.103845]\ cm^{-1}$

Enthalpy of Formation

Zmbov and Margrave (1) have identified $Si_2N(g)$ in a mass spectrometer. They also determined partial pressures of $Si_2N(g)$ in equilibrium with Si(g) and $N_2(g)$. From four of these measurements, in the temperature range 1742 - 1846 K, we obtain a 2nd law $\Delta_rH°(298.15\ K) = 99.6\pm6\ kcal\ mol^{-1}$ for the reaction $Si_2N(g) = 2\ Si(g) + 0.5\ N_2(g)$; and a 3rd law $\Delta_rH°(298.15\ K) = 120.2\ kcal\ mol^{-1}$ with a drift of $-11.4\pm3.4\ cal\ K^{-1}\ mol^{-1}$. Considering the small temperature range and the uncertainties in the free energy functions, the data appear consistent. We adopt the 3rd law enthalpy of reaction which yields $\Delta_fH°(Si_2N, g, 298.15\ K) = 95\ kcal\ mol^{-1}$.

Heat Capacity and Entropy

$Si_2N(g)$ is considered to be closely related to $C_2N(g)$, and its structure and electronic configuration are adopted by analogy to this species. The bond length was obtained by adding the difference between the bond distances of SiN(g) and CN(g) to the bond length in CNC(g). The vibrational frequencies were estimated by comparison with several molecules, CNC(g); $^1\pi$ state of $C_3(g)$; NCN(g); and $Si_2C(g)$. It should be noted that there is a wide variation in observed frequencies for these molecules and ν_1 and ν_3 may be significantly different from the chosen values.

The electronic ground state is considered to be $^2\pi g$ with excited states of $^2\Delta_u$ and $^2\Sigma_u^-$ as in the case of CNC(g). The levels are obtained from those in CNC(g) by use of the shift between $C_3(g)$ and SiCSi(g).

Reference

1. K. F. Zmbov and J. L. Margrave, Rice University, personal communication, March 1967.

Enthalpy Reference Temperature = T_r = 298.15 K Standard State Pressure = p° = 0.1 MPa

	J K⁻¹mol⁻¹			kJ mol⁻¹			
T/K	$C_p°$	S°	$-[G°-H°(T_r)]/T$	$H°-H°(T_r)$	$\Delta_fH°$	$\Delta_fG°$	Log K_f
0	0.	0.	INFINITE	-11.822	396.430	396.430	INFINITE
100	35.904	209.763	296.993	-8.723	397.545	385.325	-201.273
200	44.596	237.627	260.893	-4.653	397.802	372.941	-97.402
250	47.543	247.906	257.295	-2.347	397.680	366.736	-76.625
298.15	49.884	256.487	256.487	0.	397.480	360.795	-63.210
300	49.965	256.795	256.488	0.092	397.471	360.567	-62.780
350	51.951	264.652	257.104	2.642	397.223	354.436	-52.897
400	53.570	271.699	258.495	5.281	396.959	348.341	-45.489
450	54.886	278.087	260.323	7.994	396.690	342.280	-39.731
500	55.958	283.927	262.395	10.766	396.418	336.249	-35.128
600	57.554	294.280	266.869	16.447	395.855	324.268	-28.230
700	58.648	303.239	271.439	22.260	395.251	312.384	-23.310
800	59.420	311.124	275.917	28.166	394.584	300.590	-19.627
900	59.981	318.156	280.226	34.137	393.843	288.885	-16.766
1000	60.399	324.499	284.342	40.157	393.021	277.266	-14.483
1100	60.718	330.271	288.259	46.214	392.117	265.734	-12.619
1200	60.966	335.565	291.983	52.298	391.130	254.287	-11.069
1300	61.163	340.453	295.526	58.405	390.059	242.926	-9.761
1400	61.321	344.992	298.899	64.529	388.902	231.651	-8.643
1500	61.450	349.227	302.115	70.668	387.658	220.462	-7.677
1600	61.557	353.196	305.185	76.819	386.327	209.359	-6.835
1700	61.646	356.931	308.120	82.979	284.554	199.235	-6.122
1800	61.721	360.457	310.930	89.147	283.509	194.246	-5.637
1900	61.785	363.795	313.626	95.323	282.460	189.316	-5.205
2000	61.840	366.966	316.214	101.504	281.408	184.440	-4.817
2100	61.887	369.984	318.703	107.690	280.353	179.618	-4.468
2200	61.928	372.864	321.100	113.881	279.295	174.846	-4.151
2300	61.965	375.618	323.411	120.076	278.233	170.122	-3.864
2400	61.997	378.256	325.642	126.274	277.170	165.444	-3.601
2500	62.025	380.787	327.797	132.475	276.103	160.811	-3.360
2600	62.051	383.220	329.882	138.679	275.035	156.220	-3.139
2700	62.075	385.563	331.902	144.885	273.964	151.671	-2.934
2800	62.097	387.821	333.859	151.094	272.891	147.161	-2.745
2900	62.118	390.000	335.757	157.305	271.817	142.690	-2.570
3000	62.137	392.106	337.601	163.517	270.741	138.255	-2.407
3100	62.156	394.144	339.392	169.732	269.663	133.857	-2.255
3200	62.175	396.118	341.134	175.949	268.584	129.493	-2.114
3300	62.194	398.031	342.829	182.167	267.504	125.163	-1.981
3400	62.214	399.888	344.480	188.388	266.423	120.866	-1.857
3500	62.234	401.692	346.089	194.610	265.341	116.601	-1.740
3600	62.255	403.445	347.658	200.834	-504.045	133.288	-1.934
3700	62.278	405.151	349.189	207.061	-504.299	150.995	-2.132
3800	62.302	406.813	350.684	213.290	-504.557	168.709	-2.319
3900	62.327	408.431	352.144	219.521	-504.817	186.430	-2.497
4000	62.355	410.010	353.571	225.755	-505.079	204.158	-2.666
4100	62.384	411.550	354.966	231.992	-505.342	221.892	-2.827
4200	62.416	413.053	356.331	238.232	-505.604	239.633	-2.980
4300	62.450	414.522	357.668	244.476	-505.866	257.379	-3.127
4400	62.486	415.959	358.976	250.722	-506.126	275.132	-3.266
4500	62.524	417.363	360.258	256.973	-506.384	292.891	-3.400
4600	62.566	418.738	361.515	263.227	-506.639	310.656	-3.528
4700	62.609	420.084	362.746	269.486	-506.891	328.426	-3.650
4800	62.655	421.402	363.955	275.749	-507.139	346.201	-3.767
4900	62.704	422.695	365.140	282.017	-507.384	363.981	-3.880
5000	62.755	423.962	366.304	288.290	-507.625	381.767	-3.988
5100	62.809	425.205	367.447	294.568	-507.861	399.557	-4.092
5200	62.865	426.426	368.569	300.852	-508.093	417.352	-4.192
5300	62.923	427.624	369.672	307.141	-508.321	435.151	-4.289
5400	62.984	428.800	370.756	313.437	-508.544	452.954	-4.381
5500	63.047	429.957	371.822	319.738	-508.761	470.762	-4.471
5600	63.112	431.093	372.871	326.046	-508.976	488.573	-4.557
5700	63.179	432.211	373.902	332.361	-509.187	506.388	-4.641
5800	63.248	433.310	374.917	338.682	-509.395	524.207	-4.721
5900	63.318	434.392	375.916	345.010	-509.600	542.030	-4.799
6000	63.391	435.457	376.899	351.346	-509.801	559.856	-4.874

PREVIOUS: March 1967 (1 atm) CURRENT: March 1967 (1 bar)

Silicon Nitride (Si_2N) $N_1Si_2(g)$

TITANIUM NITRIDE (TiN) CRYSTAL M_r = 61.8867

$S°(298.15\ K) = 30.235 \pm 0.21\ J\ K^{-1}\ mol^{-1}$

$T_{fus} = 3220 \pm 50\ K$

$\Delta_f H°(0\ K) = -333.971 \pm 4.2\ kJ\ mol^{-1}$

$\Delta_f H°(298.15\ K) = -337.649 \pm 4.2\ kJ\ mol^{-1}$

$\Delta_{fus} H° = [66.944]\ kJ\ mol^{-1}$

Enthalpy of Formation

Humphrey (1) and Neumann et al. (2) determined the enthalpy of combustion of titanium nitride by burning the material in an oxygen bomb calorimeter. For reaction (a) Humphrey obtained $\Delta_r H°(298.15\ K) = -145.05 \pm 0.14\ kcal\ mol^{-1}$ and Neumann et al., $\Delta_r H°(298.15\ K) = -145.0\ kcal\ mol^{-1}$. From an examination of the combustion products Humphrey reported that the TiO_2 was >95 percent rutile and <5 percent anatase, but made no correction for the anatase. The adopted $\Delta_f H°(298.15\ K) = -80.7\ kcal\ mol^{-1}$ is the average of three values, two from Humphrey (-80.75, -80.61) one as reported and one corrected for anatase and one from Neumann et al. (-80.8). The uncertainty in the adopted enthalpy is due mainly to the uncertainty in the enthalpy of formation of rutile. Hoch et al. (3) determined the vapor pressure of reaction (b), in the temperature range 1987-2241 K, using a Knudsen effusion method. 3rd law analysis of the data yields $\Delta_f H°(298.15\ K) = -79.5\ kcal\ mol^{-1}$. Linevsky (4) obtained the titanium partial pressures of reaction (C) at fixed nitrogen pressures using atomic absorption spectroscopy. 3rd law analysis of the data gives a $\Delta_f H°(298.15\ K) = -81 \pm 18\ kcal\ mol^{-1}$. These equilibrium measurements were not considered in deriving the heat of formation due to their large uncertainties in comparison to the calorimetric data.

Source	Method	Reaction	T/K	$\Delta_r H°(298.15\ K)$ kcal mol^{-1} 2nd Law	3rd Law	Drift cal K^{-1} mol^{-1}	$\Delta_f H°(298.15\ K)$ kcal mol^{-1}
(1)	Calorimeter	(a) $TiN(cr) + O_2(g) \rightarrow TiO_2(rutile) + 1/2\ N_2(g)$	298	-145.05±.14			-80.75
(1)	Calorimeter	(a) $TiN(cr) + O_2(g) \rightarrow TiO_2(rutile) + 1/2\ N_2(g)$	298	-145.05±.14			-80.61*
(2)	Calorimeter	(a) $TiN(cr) + O_2(g) \rightarrow TiO_2(rutile) + 1/2\ N_2(g)$	298	-145.0			-80.80
(3)	Equilibrium	(b) $TiN(cr) \rightarrow Ti(g) + 1/2\ N_2(g)$	1987-2241	212±9	192.5	-9.3±4.3	-79.5
(4)	Equilibrium	(c) $TiN(cr) \rightarrow Ti(g) + 1/2\ N_2(g)$	2088-2141	230±17	194.2	-17±8	-81.2

*Assumes that the product TiO_2 is 95 percent rutile and 5 percent antase.

Heat Capacity and Entropy

Low temperature heat capacities are from the data (52-296 K) of Shomate (5). Naylor (6) measured high temperature heat contents (388-1738 K). High temperature heat capacities are derived from the enthalpies by a fitting technique which constrains the curve to join smoothly with the low temperature values. Above 1738 K the heat capacity is graphically extrapolated. The entropy is based on $S°(50;0) = 0.138\ cal\ K^{-1}\ mol^{-1}$.

Fusion Data

See the liquid table for details.

References

1. G. L. Humphrey, J. Amer. Chem. Soc. 73, 2261 (1951).
2. B. Neumann, C. Kroger and H. Kunz, Z. Anorg. Allg. Chem. 218, 379-401 (1934).
3. M. Hoch, D. P. Dingledy and H. L. Johnston, J. Amer. Chem. Soc. 77, 304-6 (1955).
4. M. J. Linevsky, General Electric Co., AFML-TR-64-420, February 1965.
5. C. H. Shomate, J. Amer. Chem. Soc. 68, 310 (1946).
6. B. F. Naylor, J. Amer. Chem. Soc. 68, 370 (1946).

Titanium Nitride (TiN) $N_1Ti_1(cr)$

Enthalpy Reference Temperature = T_r = 298.15 K Standard State Pressure = p° = 0.1 MPa

T/K	$C_p°$ (J $K^{-1}mol^{-1}$)	S°	$-[G°-H°(T_r)]/T$	$H°-H°(T_r)$ (kJ mol^{-1})	$\Delta_f H°$	$\Delta_f G°$	Log K_f
0	0.	0.	INFINITE	-5.487	-333.971	-333.971	INFINITE
100	10.761	4.422	56.044	-5.162	-335.658	-327.283	170.955
200	27.330	17.338	33.350	-3.202	-337.070	-318.294	83.130
298.15	37.079	30.235	30.235	0.	-337.649	-308.929	54.123
300	37.238	30.465	30.236	0.069	-337.654	-308.750	53.758
400	43.811	42.171	31.787	4.154	-337.641	-299.104	39.059
500	47.057	52.338	34.906	8.716	-337.289	-289.506	30.245
600	48.723	61.074	38.557	13.510	-336.812	-279.993	24.376
700	49.781	68.667	42.328	18.437	-336.294	-270.564	20.190
800	50.618	75.370	46.048	23.458	-335.753	-261.211	17.055
900	51.384	81.376	49.645	28.558	-335.232	-251.925	14.621
1000	52.145	86.829	53.095	33.734	-334.797	-242.693	12.677
1100	52.919	91.835	56.392	38.987	-334.508	-233.498	11.088
1200	53.718	96.474	59.542	44.319	-338.345	-224.203	9.759
1300	54.538	100.806	62.551	49.731	-337.610	-214.721	8.628
1400	55.384	104.878	65.430	55.227	-336.890	-205.295	7.660
1500	56.246	108.729	68.190	60.809	-336.193	-195.920	6.823
1600	57.128	112.387	70.839	66.477	-335.528	-186.591	6.092
1700	58.020	115.877	73.386	72.235	-334.906	-177.301	5.448
1800	58.927	119.219	75.840	78.082	-334.337	-168.047	4.877
1900	59.848	122.430	78.208	84.020	-333.831	-158.823	4.366
2000	60.777	125.523	80.497	90.052	-348.087	-149.170	3.896
2100	61.710	128.511	82.713	96.176	-348.489	-139.214	3.463
2200	62.655	131.403	84.861	102.394	-348.804	-129.240	3.069
2300	63.605	134.209	86.946	108.707	-349.031	-119.255	2.708
2400	64.559	136.937	88.972	115.115	-349.170	-109.261	2.378
2500	65.517	139.591	90.944	121.619	-349.218	-99.264	2.074
2600	66.480	142.180	92.865	128.219	-349.175	-89.266	1.793
2700	67.446	144.707	94.738	134.915	-349.040	-79.272	1.534
2800	68.413	147.177	96.567	141.708	-348.813	-69.284	1.293
2900	69.387	149.595	98.354	148.598	-348.493	-59.307	1.068
3000	70.362	151.964	100.102	155.586	-348.079	-49.341	0.859
3100	71.337	154.287	101.812	162.670	-347.571	-39.392	0.664
3200	72.316	156.567	103.488	169.853	-346.969	-29.460	0.481
3220.000	72.513	157.018	103.819	171.301	------ CRYSTAL <--> LIQUID -----		
3300	73.295	158.807	105.130	177.134	-346.271	-19.548	0.309
3400	74.279	161.010	106.742	184.512	-345.478	-9.659	0.148
3500	75.262	163.177	108.323	191.989	-344.590	0.205	-0.003
3600	76.244	165.311	109.877	199.565	-343.606	10.043	-0.146
3700	77.228	167.414	111.403	207.238	-751.624	27.640	-0.390
3800	78.215	169.486	112.905	215.010	-749.214	48.669	-0.669
3900	79.203	171.531	114.382	222.881	-746.762	69.634	-0.933
4000	80.193	173.549	115.836	230.851	-744.267	90.536	-1.182
4100	81.182	175.541	117.268	238.920	-741.729	111.375	-1.419
4200	82.171	177.509	118.679	247.087	-739.144	132.151	-1.644
4300	83.161	179.454	120.069	255.354	-736.511	152.865	-1.857
4400	84.152	181.377	121.441	263.720	-733.829	173.517	-2.060
4500	85.144	183.280	122.794	272.185	-731.095	194.108	-2.253

PREVIOUS: December 1960 CURRENT: June 1968

Titanium Nitride (TiN) $N_1Ti_1(cr)$

TITANIUM NITRIDE (TiN) LIQUID M_r = 61.8867

$S°(298.15\ K) = [52.733]\ J\ K^{-1}\ mol^{-1}$ $\Delta_f H°(298.15\ K) = [-265.813]\ kJ\ mol^{-1}$

$T_{fus} = 3220 \pm 50\ K$ $\Delta_{fus}H° = [66.944]\ kJ\ mol^{-1}$

Enthalpy of Formation

$\Delta_f H°$(TiN, ℓ, 298.15 K) is calculated from $\Delta_f H°$(TiN, cr, 298.15 K) by adding $\Delta_{fus}H°$ and the difference in enthalpy, H°(3220 K)−H°(298.15 K), between the crystal and liquid.

Heat Capacity and Entropy

A glass transition is assumed at 2200 K. Below 2200 K the heat capacity is obtained from the heat capacity of the crystal. Above 2200 K it is assumed constant and estimated as 15 cal $K^{-1}\ mol^{-1}$ or 7.5 cal K^{-1} g-atom. S°(ℓ, 298.15 K) is calculated in a manner analogous to that used for the enthalpy of formation.

Fusion Data

T_{fus} is taken from Agte and Moers (1). $\Delta_{fus}H°$ is derived from an estimated $\Delta_m S° = 2.5$ cal K^{-1} g-atom as suggested by Kubaschewski et al. (2).

References

1. C. Agte and K. Moers, Z. Anorg. Allg. Chem. 198, 233-43 (1931).
2. O. Kubaschewski, E. Ll. Evans and C. B. Alcock, "Metallurgical Thermochemistry," Pergamon Press, New York, 1967.

Titanium Nitride (TiN) $N_1Ti_1(l)$

Enthalpy Reference Temperature = T_r = 298.15 K; Standard State Pressure = p° = 0.1 MPa

T/K	$C_p°$ (J K⁻¹mol⁻¹)	S° (J K⁻¹mol⁻¹)	−[G°−H°(T_r)]/T (J K⁻¹mol⁻¹)	H°−H°(T_r) (kJ mol⁻¹)	$\Delta_f H°$ (kJ mol⁻¹)	$\Delta_f G°$ (kJ mol⁻¹)	Log K_f
0							
100							
200							
298.15	37.079	52.733	52.733	0.	-265.813	-243.800	42.713
300	37.238	52.963	52.734	0.069	-265.818	-243.664	42.426
400	43.811	64.670	54.285	4.154	-265.805	-236.267	30.853
500	47.057	74.836	57.404	8.716	-265.453	-228.919	23.915
600	48.723	83.572	61.056	13.510	-264.976	-221.656	19.297
700	49.781	91.165	64.827	18.437	-264.458	-214.477	16.004
800	50.618	97.868	68.546	23.458	-263.917	-207.374	13.540
900	51.384	103.874	72.143	28.558	-263.396	-200.337	11.627
1000	52.145	109.327	75.593	33.734	-262.961	-193.355	10.100
1100	52.919	114.333	78.891	38.987	-262.672	-186.410	8.852
1200	53.718	118.972	82.040	44.319	-266.509	-179.365	7.808
1300	54.538	123.304	85.049	49.731	-265.774	-172.133	6.916
1400	55.384	127.377	87.929	55.227	-265.054	-164.957	6.155
1500	56.246	131.227	90.688	60.809	-264.357	-157.831	5.496
1600	57.128	134.885	93.337	66.477	-263.692	-150.751	4.922
1700	58.020	138.375	95.884	72.235	-263.070	-143.712	4.416
1800	58.927	141.717	98.338	78.082	-262.501	-136.708	3.967
1900	59.848	144.928	100.706	84.020	-261.995	-129.733	3.567
2000	60.777	148.021	102.995	90.052	-276.251	-122.330	3.195
2100	61.710	151.009	105.211	96.176	-276.653	-114.624	2.851
2200	62.655	153.901	107.359	102.394	-276.968	-106.900	2.538
2200.000	62.655	153.901	107.359	102.394	GLASS <--> LIQUID TRANSITION		
2200.000	62.760	153.901	107.359	102.394			
2300	62.760	156.691	109.443	108.670	-277.232	-99.164	2.252
2400	62.760	159.362	111.468	114.946	-277.503	-91.416	1.990
2500	62.760	161.924	113.436	121.222	-277.779	-83.657	1.748
2600	62.760	164.386	115.348	127.498	-278.060	-75.886	1.525
2700	62.760	166.754	117.208	133.774	-278.345	-68.105	1.318
2800	62.760	169.037	119.019	140.050	-278.635	-60.313	1.125
2900	62.760	171.239	120.782	146.326	-278.929	-52.511	0.946
3000	62.760	173.367	122.499	152.602	-279.227	-44.698	0.778
3100	62.760	175.425	124.174	158.878	-279.527	-36.876	0.621
3200	62.760	177.417	125.807	165.154	-279.832	-29.043	0.474
3220.000	62.760	177.808	126.128	166.409	------ CRYSTAL <--> LIQUID -----		
3300	62.760	179.348	127.400	171.430	-280.139	-21.201	0.336
3400	62.760	181.222	128.956	177.706	-280.449	-13.350	0.205
3500	62.760	183.041	130.475	183.982	-280.761	-5.490	0.082
3600	62.760	184.809	131.960	190.258	-281.076	2.379	-0.035
3700	62.760	186.529	133.412	196.534	-690.492	18.046	-0.255
3800	62.760	188.203	134.832	202.810	-689.578	37.183	-0.511
3900	62.760	189.833	136.221	209.086	-688.721	56.297	-0.754
4000	62.760	191.422	137.581	215.362	-687.920	75.390	-0.984
4100	62.760	192.971	138.913	221.638	-687.174	94.463	-1.203
4200	62.760	194.484	140.219	227.914	-686.481	113.519	-1.412
4300	62.760	195.961	141.498	234.190	-685.839	132.559	-1.610
4400	62.760	197.403	142.752	240.466	-685.247	151.585	-1.800
4500	62.760	198.814	143.982	246.742	-684.701	170.597	-1.980

PREVIOUS: December 1960 CURRENT: June 1968

Titanium Nitride (TiN) $N_1Ti_1(l)$

TITANIUM NITRIDE (TiN) CRYSTAL-LIQUID M_r = 61.8867

0 to 3220 K crystal
above 3220 K liquid

Refer to the individual tables for details.

Titanium Nitride (TiN) $N_1Ti_1(cr,l)$

Enthalpy Reference Temperature = T_r = 298.15 K Standard State Pressure = $p°$ = 0.1 MPa

T/K	$C_p°$ (J K⁻¹mol⁻¹)	$S°$ (J K⁻¹mol⁻¹)	$-[G°-H°(T_r)]/T$ (J K⁻¹mol⁻¹)	$H°-H°(T_r)$ (kJ mol⁻¹)	$\Delta_f H°$ (kJ mol⁻¹)	$\Delta_f G°$ (kJ mol⁻¹)	Log K_f
0	0.	0.	INFINITE	-5.487	-333.971	-333.971	INFINITE
100	10.761	4.422	56.044	-5.162	-335.658	-327.283	170.955
200	27.330	17.338	33.350	-3.202	-337.070	-318.294	83.130
298.15	37.079	30.235	30.235	0.	-337.649	-308.929	54.123
300	37.238	30.465	30.236	0.069	-337.654	-308.750	53.758
400	43.811	42.171	31.787	4.154	-337.641	-299.104	39.059
500	47.057	52.338	34.906	8.716	-337.289	-289.506	30.245
600	48.723	61.074	38.557	13.510	-336.812	-279.993	24.376
700	49.781	68.667	42.328	18.437	-336.294	-270.564	20.190
800	50.618	75.370	46.048	23.458	-335.753	-261.211	17.055
900	51.384	81.376	49.645	28.558	-335.232	-251.925	14.621
1000	52.145	86.829	53.095	33.734	-334.797	-242.693	12.677
1100	52.919	91.835	56.392	38.987	-334.508	-233.498	11.088
1200	53.718	96.474	59.542	44.319	-338.345	-224.203	9.759
1300	54.538	100.806	62.551	49.731	-337.610	-214.721	8.628
1400	55.384	104.878	65.430	55.227	-336.890	-205.295	7.660
1500	56.246	108.729	68.190	60.809	-336.193	-195.920	6.823
1600	57.128	112.387	70.839	66.477	-335.528	-186.591	6.092
1700	58.020	115.877	73.386	72.235	-334.906	-177.301	5.448
1800	58.927	119.219	75.840	78.082	-334.337	-168.047	4.877
1900	59.848	122.430	78.208	84.020	-333.831	-158.823	4.366
2000	60.777	125.523	80.497	90.052	-348.087	-149.170	3.896
2100	61.710	128.511	82.713	96.176	-348.489	-139.214	3.463
2200	62.655	131.403	84.861	102.394	-348.804	-129.240	3.069
2300	63.605	134.209	86.946	108.707	-349.031	-119.255	2.708
2400	64.559	136.937	88.972	115.115	-349.170	-109.261	2.378
2500	65.517	139.591	90.944	121.619	-349.218	-99.264	2.074
2600	66.480	142.180	92.865	128.219	-349.175	-89.266	1.793
2700	67.446	144.707	94.738	134.915	-349.040	-79.272	1.534
2800	68.413	147.177	96.567	141.708	-348.813	-69.284	1.293
2900	69.387	149.595	98.354	148.598	-348.493	-59.307	1.068
3000	70.362	151.964	100.102	155.586	-348.079	-49.341	0.859
3100	71.337	154.287	101.812	162.670	-347.571	-39.392	0.664
3200	72.316	156.567	103.488	169.853	-346.969	-29.460	0.481
3220.000	72.513	157.018	103.819	171.301	CRYSTAL <--> LIQUID		
3220.000	62.760	177.808	103.819	238.245	TRANSITION		
3300	62.760	179.348	105.631	243.266	-280.139	-21.201	0.336
3400	62.760	181.222	107.827	249.542	-280.449	-13.350	0.205
3500	62.760	183.041	109.950	255.818	-280.761	-5.490	0.082
3600	62.760	184.809	112.005	262.094	-281.076	2.379	-0.035
3700	62.760	186.529	113.996	268.370	-690.492	18.046	-0.255
3800	62.760	188.203	115.927	274.646	-689.578	37.183	-0.511
3900	62.760	189.833	117.801	280.922	-688.721	56.297	-0.754
4000	62.760	191.422	119.622	287.198	-687.920	75.390	-0.984
4100	62.760	192.971	121.392	293.474	-687.174	94.463	-1.203
4200	62.760	194.484	123.115	299.750	-686.481	113.519	-1.412
4300	62.760	195.961	124.792	306.026	-685.839	132.559	-1.610
4400	62.760	197.403	126.426	312.302	-685.247	151.585	-1.800
4500	62.760	198.814	128.019	318.578	-684.701	170.597	-1.980

PREVIOUS: CURRENT: June 1968

Titanium Nitride (TiN) $N_1Ti_1(cr,l)$

VANADIUM NITRIDE (VN) CRYSTAL M_r = 64.9482

$\Delta_f H°(0\ K) = -214.403 \pm 5.0\ kJ\ mol^{-1}$

$S°(298.15\ K) = 37.264 \pm 0.42\ J\ K^{-1}\ mol^{-1}$ $\Delta_f H°(298.15\ K) = -217.150 \pm 5.0\ kJ\ mol^{-1}$

Enthalpy of Formation

Mah (1) measured the enthalpy of combustion of vanadium mononitride at 303.15 K. The composition of the sample was $VN_{1.00}$ which corresponds to the nitrogen-rich boundary of the mononitride phase (see Phase Data section). Chemical analysis also showed that the sample contained 0.07 weight percent of unreacted vanadium. The composition of the combustion products ranged from $VO_{2.231}$ to $VO_{2.296}$. X-ray diffraction patterns showed the solid combustion products to be a mixture of V_2O_4 and V_2O_5. JANAF enthalpy of formation data (2) for these two oxides are used to correct Mah's combustion values to correspond to the formation of the pentoxide as the final state. Our values for this correction agree with those of Mah (1) to within 0.1 cal $K^{-1}\ mol^{-1}$ mononitride. For the process VN(cr) + 1.25 O_2(g) = 0.5 V_2O_5(cr) + 0.5 N_2(g), we obtain $\Delta_c H° = -133.41 \pm 0.43$ kcal mol^{-1} at unit fugacities of oxygen and nitrogen. When this value is combined with the enthalpy of formation of V_2O_5 (2), that for vanadium mononitride is $\Delta_f H°$(VN, cr, 298.15 K) = -51.9 ± 1.2 kcal mol^{-1}. We adopt this value and include in the overall uncertainty (±1.2 kcal mol^{-1}) the error associated with correcting the combustion values for the V_2O_4(cr) formed to V_2O_5(cr) as the final state.

Other values for $\Delta_f H°$ can be derived from the results of dissociation pressure measurements (3), emf data (4), and equilibrium data (5) obtained by an effusion-mass spectrometric technique. Analyses of these data are summarized below.

Source	Method	Reaction	Data Points	T/K	$\Delta_r H°$(298.15 K), kcal mol^{-1} 2nd law	3rd law	Drift cal $K^{-1}\ mol^{-1}$	$-\Delta_f H°$(298.15 K) kcal mol^{-1}
Slade and Higson (3)	Static	A	2	1476-1544	-68.8	-41.1±1.8	-18.3	41.1
Voleinik and Shabdenov (4)	emf	A	Equation	363-403	-54.4	-55.5±0.1	-2.7	55.5±0.1
Farber and Srivastava (5)	Mass Spec.	B	13	1900-2412	173.5	181.7±1.6	3.8±0.8	58.5±3.6

Reactions: A) V(cr) + 0.5 N_2(g) = VN(cr) B) VN(cr) = V(g) + 0.5 N_2(g)

Results obtained from the dissociation pressures of Slade and Higson (3) are obviously not reliable. Two other more extensive sets of dissociation pressure data (6, 7) have been reported for the mononitride phase. Brauer and Schnell (6) measured the nitrogen dissociation pressures for $VN_{0.72}$ between 1573 and 1873 K, while Kozheurov et al. (7) determined equilibrium pressures for VN_x (x = 0.55-0.9) at temperatures in the range 1573-1923 K. Unfortunately, these compositions are significantly different from $VN_{1.00}$, and thus the data are probably of no use in defining the enthalpy of formation of the mononitride at its nitrogen-rich phase boundary. The 3rd law values of $\Delta_f H°$ obtained from the emf (4) and mass spectral (5) data are 3.6 and 6.6 kcal mol^{-1}, respectively, more negative than our adopted value. However, since the thermal functions used in the analysis of the mass spectral data (5) are based on extrapolated data, the 2nd law value, $\Delta_f H°$(298.15 K) = -50.3 kcal mol^{-1}, is probably more reliable. This value is in much better agreement with that obtained by combustion calorimetry.

Heat Capacity and Entropy

Shomate and Kelley (8, 52.6-296.3 K) measured the low temperature heat capacities for VN. The purity of the mononitride sample based on an analysis for vanadium by permanganate titration was found to be 99.76 weight percent. Their $C_p°$ data corrected to 1969 Atomic Weight Scale are adopted. S°(298.15 K) is obtained from the adopted $C_p°$ data and is based on S°(52 K) = 0.47 cal $K^{-1}\ mol^{-1}$. This latter value is calculated from the Debye function D(351/T) which was suggested by Shomate and Kelley (8).

High temperature enthalpies for VN have been reported by King (9) for the temperature range 408-1611 K. The sample was identical to that employed by Shomate and Kelley (8) and was contained in platinum-rhodium alloy capsules for the "drop" experiments. His experimental enthalpies are corrected to refer to the 1969 Atomic Weight Scale. Also, his reported temperatures are corrected to the IPTS-68 scale. A technique based on orthogonal polynomials is used to fit the corrected enthalpies by computer. The curve is constrained to join smoothly with the low temperature $C_p°$ data (8) near 298 K. Our smoothed enthalpies show an average deviation of ±0.4% from the experimental values (9). The maximum deviation is +2.1% at 408 K. No anomalies are observed in either the low temperature $C_p°$ data (8) or the high temperature enthalpies (9). $C_p°$ data above 1613 K are obtained by graphical extrapolation. Satoh (10) has measured the mean specific heats of VN over three temperature intervals by means of an ice calorimeter. However, the sample used in his study was impure (97.93% VN), and we believe the results are probably not reliable. They do show considerable scatter from our adopted results; deviations are in the range of ±1-5%.

Phase Data

The mononitride phase has been prepared by several workers (1, 5-8, and 11-13) at temperatures as high as 1600 K. X-ray data (6, 11) and dissociation pressure measurements (7) have shown that the phase has an extensive homogeneity range. Although Kozeurov et al. (7) were unable to obtain $VN_{1.00}$ at 1273 K, Brauer and Schnell (6) and Hahn (11) reported that the composition of the mononitride phase extends from $VN_{0.72}$ to $VN_{1.00}$ (42 to 50 atomic percent nitrogen) at 1273-1873 K. The dissociation pressure data of Kozheurov et al. (7) indicate that the vanadium-rich phase boundary decreases from $VN_{0.7}$ at 1573 K to $VN_{0.56}$ at 1923 K. The structure of the mononitride is face-centered cubic (6, 11). Further information on the vanadium-vanadium nitride system has been reviewed by Storms (14).

Decomposition Data

The melting point of VN has been reported to be approximately 2323 K by Rostoker and Yamamoto (15). However, Farber and Srivastava (5) in their mass spectrometric studies observed solid nitride at temperatures as high as 2412 K. Liquid vanadium metal, presumably formed from decomposition of the nitride, was deposited in the sample. We assume that VN decomposes to its elements prior to melting. T_{dcm} = 2619 K is the temperature at which the fugacity is 1 atm for the decomposition reaction VN(cr) + V(ℓ) + 0.5 N_2(g) approaches zero. $\Delta_{dcm}H°$ is the negative of the heat of formation of VN(cr) at T_{dcm}.

References

References are listed on the ideal gas table.

Vanadium Nitride (VN) N_1V_1(cr)

Enthalpy Reference Temperature = T_r = 298.15 K Standard State Pressure = p° = 0.1 MPa

T/K	$C_p°$ (J $K^{-1}mol^{-1}$)	S°	$-[G°-H°(T_r)]/T$	$H°-H°(T_r)$ (kJ mol^{-1})	$\Delta_f H°$	$\Delta_f G°$	Log K_f
0	0.	0.	INFINITE	-6.228	-214.403	-214.403	INFINITE
100	15.111	8.354	64.935	-5.658	-215.773	-207.899	108.595
200	29.675	23.719	40.504	-3.357	-216.763	-199.597	52.129
298.15	37.981	37.264	37.264	0.	-217.150	-191.068	33.474
300	38.200	37.499	37.264	0.070	-217.152	-190.907	33.240
400	43.296	49.239	38.830	4.164	-217.082	-182.158	23.787
500	46.183	59.227	41.937	8.645	-216.731	-173.464	18.122
600	48.208	67.832	45.553	13.368	-216.221	-164.856	14.352
700	49.819	75.387	49.286	18.271	-215.615	-156.342	11.666
800	51.212	82.132	52.978	23.324	-214.951	-147.919	9.658
900	52.480	88.238	56.561	28.509	-214.256	-139.582	8.101
1000	53.668	93.829	60.013	33.817	-213.540	-131.323	6.860
1100	54.806	98.998	63.325	39.241	-212.812	-123.137	5.847
1200	55.907	103.814	66.500	44.777	-212.084	-115.017	5.007
1300	56.986	108.332	69.546	50.421	-211.362	-106.957	4.298
1400	58.045	112.594	72.470	56.173	-210.655	-98.952	3.692
1500	59.091	116.634	75.281	62.030	-209.963	-90.998	3.169
1600	60.124	120.481	77.987	67.990	-209.284	-83.089	2.713
1700	61.170	124.158	80.595	74.057	-208.625	-75.222	2.311
1800	62.049	127.679	83.114	80.217	-207.999	-67.393	1.956
1900	62.969	131.058	85.549	86.468	-207.420	-59.598	1.638
2000	63.932	134.313	87.906	92.813	-206.890	-51.832	1.354
2100	64.852	137.454	90.191	99.252	-206.420	-44.091	1.097
2200	65.772	140.493	92.409	105.784	-228.900	-36.266	0.861
2300	66.651	143.436	94.564	112.405	-228.716	-27.514	0.625
2400	67.530	146.291	96.660	119.114	-228.450	-18.771	0.409
2500	68.408	149.065	98.701	125.911	-228.102	-10.041	0.210
2600	69.287	151.766	100.690	132.796	-227.671	-1.327	0.027
2700	70.166	154.397	102.631	139.768	-227.156	7.369	-0.143
2800	71.044	156.965	104.526	146.829	-226.558	16.044	-0.299
2900	71.923	159.473	106.377	153.977	-225.877	24.697	-0.445
3000	72.802	161.926	108.188	161.214	-225.110	33.325	-0.580
3100	73.638	164.327	109.961	168.536	-224.262	41.925	-0.706
3200	74.475	166.678	111.696	175.941	-223.333	50.497	-0.824
3300	75.312	168.983	113.398	183.430	-222.323	59.039	-0.935
3400	76.191	171.244	115.066	191.006	-221.231	67.549	-1.038
3500	77.027	173.465	116.703	198.667	-220.055	76.026	-1.135

PREVIOUS: CURRENT: December 1973

Vanadium Nitride (VN) N_1V_1(cr)

VANADIUM NITRIDE (VN) IDEAL GAS M_r = 64.9482

$S°(298.15\ K)$ = [233.384 ± 8.4] J K^{-1} mol^{-1}

$\Delta_f H°(0\ K)$ = 523.18 ± 20.9 kJ mol^{-1}
$\Delta_f H°(298.15\ K)$ = 523.00 ± 20.9 kJ mol^{-1}

Electronic Levels and Quantum Weights

ε_i, cm^{-1}	g_i	ε_i, cm^{-1}	g_i	ε_i, cm^{-1}	g_i	ε_i, cm^{-1}	g_i
0	[6]	[12000]	[2]	[16000]	[6]	[21000]	[2]
[1000]	[2]	[13000]	[3]	[18000]	[2]	[22000]	[6]
[3000]	[1]	[14000]	[6]	[19000]	[2]	[24000]	[2]
[10000]	[6]	[15000]	[1]	[20000]	[6]	[25000]	[6]
						[28000]	[6]

Ground State Configuration: $[^3\Delta]$

ω_e = [1025] cm^{-1} $\omega_e x_e$ = [4.41] cm^{-1} σ = 1

B_e = [0.5905] cm^{-1} α_e = [0.00354] cm^{-1} r_e = [1.612] Å

Enthalpy of Formation

Dissociation energies for the diatomic group III-VI transition-metal nitrides have been estimated by Gingerich (1) and provide support for the existence of these gaseous molecules. Very recently, mass-spectrometric evidence (2) has been presented which verifies the existence of VN(g). Using an effusion-mass spectrometric technique, Farber and Srivastava (2) observed the VN^+ ion as one of the vapor species in equilibrium with solid VN in the temperature range 1900-2412 K. The molecular precursor of this ion is assumed to be VN(g). Results of effusion weight loss and mass spectrometric experiments (2) were combined to yield partial pressures for the species involved in the following two reactions: (A) VN(cr) = VN(g) and (B) VN(g) = V(g) + 0.5 N_2(g). During the course of these experiments, the ion intensities for each species did not show any variation with time. This implies that the mononitride phase remained at near unit activity during their measurements.

Source	Reaction	T/K	Data Points	$\Delta_r H°(298.15\ K)$, kcal mol^{-1} 2nd law	3rd law	Drift cal K^{-1} mol^{-1}	$\Delta_f H°(298.15\ K)$ kcal mol^{-1}
(2)	A	1900-2412	13	174.3	185.4±2.4	5.2±1.4	133.5±3.6
(2)	B	1900-2412	12*	-1.9	-3.6±1.6	-0.8±1.4	126.8±3.6

*One point rejected due to failure of a statistical test.

The 3rd law value of 126.8±3.6 kcal mol^{-1} for $\Delta_f H°$ is considered to be more reliable than that (133.5 kcal mol^{-1}) derived from the sublimation data, since the results for the gas phase reaction (B) show the least drift. Also, Farber and Srivastava (2) reported a 2nd law value determined from ion intensity plots of $\Delta_r H°$ = 3±5 kcal mol^{-1} for reaction (B) at 2162 K. Using auxiliary data (3), this value leads to $\Delta_f H°$(VN, g, 298.15 K) = 122.0±7.0 kcal mol^{-1}. We adopt $\Delta_f H°$(VN, g, 298.15 K) = 125.0±5.0 kcal mol^{-1} which is a weighted average of the 3rd law value derived from reaction (B) and the 2nd law value based on ion intensity plots. The adopted $\Delta_f H°$ value corresponds to a bond dissociation energy, D°(298.5 K), of 111.2 kcal mol^{-1}. This compares favorably with the value (116 kcal mol^{-1}) estimated by Gingerich (1).

Heat Capacity and Entropy

The value of r_e(VN) is obtained from the expression r_e(MO)-r_e(MN) = -0.023 A with r_e(VO) = 1.589 A (3). Differences in the bond lengths (4) for the oxides and nitrides of As, B, P, Si, and Ti are used to establish the value -0.023. B_e is calculated from the estimated value for r_e. The value of α_e is obtained from the other constants assuming a Morse potential function. The values of ω_e and $\omega_e x_e$ are assumed to lie between those for TiO(g) and AsN(g) (4). VN is isoelectronic with TiO. Farber and Srivastava (2) have estimated a somewhat higher value (1065 cm^{-1}) for ω_e. However, their estimate is based partially on estimated JANAF data (3) for ZrN(g).

The ground state electronic configuration ($^3\Delta$), upper electronic levels and quantum weights are estimated by analogy with those for TiO (3, 5). It is possible, however, that the ground state for VN is $^1\Delta$. VN is also isoelectronic with ScF and ZrO which both have singlet ground states (5). If the adopted ground state and first excited state are interchanged, then the value of S°(298.15 K) is reduced by 1.94 cal K^{-1} mol^{-1}. Our adopted upper levels and quantum weights contribute 0.61 and 1.11 cal K^{-1} mol^{-1} to the entropy of VN(g) at 2000 and 4000 K, respectively.

References - VN(g)

1. K. A. Gingerich, J. Chem. Phys. 49, 19 (1968).
2. M. Farber and R. D. Srivastava, J. Chem. Soc. Faraday Trans. 1, 69, 390 (1973).
3. JANAF Thermochemical Tables: VN(cr), VO(g) and TiO(g), 12-31-73; V(g), 6-30-73; N_2(g), 9-30-65; ZrN(g), 6-30-63.
4. B. Rosen, Ed., "Spectroscopic Data Relative to Diatomic Molecules," Pergamon Press, New York, 1970.
5. L. Brewer and D. W. Green, High Temp. Sci. 1, 26 (1969).

References - VN(cr)

1. A. D. Mah, U. S. Bur. Mines RI 6177, 1963.
2. JANAF Thermochemical Tables: V_2O_5(cr) and V(g), 6-30-73; V_2O_4(cr), 12-31-73.
3. R. E. Slade and G. I. Higson, J. Chem. Soc. 115, 215 (1919).
4. V. V. Voleinik and B. A. Shaddenov, Fiz. Khim. Elektrokhim. Rasplov. Salei Shlakov, No. 1, 157 (1969).
5. M. Farber and R. D. Srivastava, J. Chem. Soc. Faraday Trans. 1, 69, 390 (1973).
6. G. Brauer and W. D. Schnell, J. Less-Common Metals 6, 326 (1964).
7. V. A. Kozheurov, V. M. Zhikharev, V. I. Shishkov, and G. V. Gritsishina, Izv. Vyssh. Ucheb. Zaved., Chem. Met. No. 8, 10 (1972).
8. C. H. Shomate and K. K. Kelley, J. Amer. Chem. Soc. 71, 314 (1949).
9. E. G. King, J. Amer. Chem. Soc. 71, 316 (1949).
10. S. Satoh, Sci. Papers Inst. Phys. Chem. Research (Tokoyo) 34, 241 (1938).
11. H. Hahn, Z. Anorg. Chem. 258, 58 (1949).
12. E. T. Turkdogan and E. M. Fenn, J. Iron Steel Inst. 181, 343 (1955).
13. D. Potter and R. Geils, Scripta Met. 6, 395 (1972).
14. E. K. Storms, Los Alamos Scientific Laboratory, LA-2942, August, 1964.
15. W. Rostoker and A. Yamamoto, Trans. ASM 46, 1136 (1954).

Vanadium Nitride (VN) N_1V_1(g)

Enthalpy Reference Temperature = T_r = 298.15 K Standard State Pressure = p° = 0.1 MPa

T/K	$C_p°$ (J $K^{-1}mol^{-1}$)	S° (J $K^{-1}mol^{-1}$)	$-[G°-H°(T_r)]/T$ (J $K^{-1}mol^{-1}$)	$H°-H°(T_r)$ (kJ mol^{-1})	$\Delta_f H°$ (kJ mol^{-1})	$\Delta_f G°$ (kJ mol^{-1})	Log K_f
0	0.	0.	INFINITE	-8.800	523.175	523.175	INFINITE
100	29.107	201.085	260.002	-5.892	524.143	512.744	-267.830
200	29.518	221.325	236.173	-2.970	523.774	501.419	-130.957
250	30.234	227.982	233.891	-1.477	523.395	495.873	-103.607
298.15	31.147	233.384	233.384	0.	523.000	490.608	-85.952
300	31.184	233.577	233.384	0.058	522.985	490.407	-85.387
350	32.193	238.460	233.768	1.642	522.574	485.010	-72.384
400	33.143	242.821	234.632	3.276	522.180	479.671	-62.639
450	33.983	246.775	235.765	4.955	521.804	474.380	-55.065
500	34.699	250.393	237.049	6.672	521.446	469.130	-49.010
600	35.799	256.823	239.823	10.200	520.761	458.732	-39.936
700	36.560	262.402	242.659	13.821	520.084	448.447	-33.464
800	37.092	267.321	245.440	17.505	519.380	438.260	-28.615
900	37.472	271.713	248.120	21.234	518.619	428.165	-24.850
1000	37.750	275.676	250.680	24.996	517.789	418.159	-21.842
1100	37.959	279.284	253.119	28.782	516.879	408.239	-19.386
1200	38.118	282.594	255.439	32.586	515.875	398.406	-17.342
1300	38.245	285.651	257.647	36.404	514.771	388.661	-15.617
1400	38.349	288.489	259.750	40.234	513.556	379.005	-14.141
1500	38.441	291.138	261.755	44.074	512.231	369.440	-12.865
1600	38.527	293.622	263.670	47.922	510.798	359.967	-11.752
1700	38.614	295.960	265.501	51.779	509.248	350.587	-10.772
1800	38.706	298.169	267.256	55.645	507.579	341.301	-9.904
1900	38.807	300.265	268.938	59.521	505.783	332.112	-9.130
2000	38.921	302.258	270.555	63.407	503.853	323.021	-8.436
2100	39.048	304.160	272.110	67.305	501.783	314.029	-7.811
2200	39.192	305.980	273.609	71.217	476.683	305.245	-7.247
2300	39.353	307.726	275.054	75.144	474.173	297.508	-6.757
2400	39.531	309.404	276.451	79.088	471.674	289.881	-6.309
2500	39.727	311.022	277.801	83.051	469.188	282.357	-5.900
2600	39.940	312.584	279.109	87.034	466.717	274.933	-5.523
2700	40.171	314.096	280.377	91.040	464.265	267.603	-5.177
2800	40.417	315.561	281.608	95.069	461.831	260.364	-4.857
2900	40.678	316.984	282.803	99.124	459.419	253.211	-4.561
3000	40.954	318.367	283.966	103.205	457.031	246.142	-4.286
3100	41.242	319.715	285.097	107.315	454.667	239.151	-4.030
3200	41.541	321.029	286.200	111.454	452.329	232.236	-3.791
3300	41.850	322.312	287.275	115.623	450.019	225.394	-3.568
3400	42.167	323.566	288.324	119.824	447.737	218.622	-3.359
3500	42.492	324.793	289.348	124.057	445.485	211.916	-3.163
3600	42.821	325.995	290.350	128.323	443.263	205.274	-2.978
3700	43.154	327.173	291.329	132.621	-5.736	199.894	-2.822
3800	43.490	328.328	292.287	136.954	-6.210	205.458	-2.824
3900	43.827	329.462	293.226	141.320	-6.703	211.035	-2.826
4000	44.163	330.576	294.146	145.719	-7.216	216.624	-2.829
4100	44.498	331.670	295.048	150.152	-7.752	222.227	-2.831
4200	44.829	332.747	295.933	154.618	-8.309	227.843	-2.834
4300	45.157	333.805	296.801	159.118	-8.891	233.472	-2.836
4400	45.479	334.847	297.654	163.650	-9.497	239.115	-2.839
4500	45.795	335.873	298.492	168.213	-10.129	244.773	-2.841
4600	46.104	336.883	299.316	172.808	-10.788	250.445	-2.844
4700	46.405	337.877	300.126	177.434	-11.475	256.131	-2.847
4800	46.698	338.858	300.922	182.089	-12.192	261.832	-2.849
4900	46.981	339.823	301.706	186.773	-12.939	267.549	-2.852
5000	47.254	340.775	302.478	191.485	-13.718	273.281	-2.855
5100	47.517	341.714	303.238	196.224	-14.531	279.029	-2.858
5200	47.770	342.639	303.987	200.988	-15.378	284.793	-2.861
5300	48.011	343.551	304.725	205.777	-16.262	290.574	-2.864
5400	48.241	344.451	305.452	210.590	-17.184	296.372	-2.867
5500	48.459	345.338	306.170	215.425	-18.145	302.187	-2.870
5600	48.665	346.213	306.877	220.281	-19.148	308.020	-2.873
5700	48.860	347.076	307.575	225.158	-20.194	313.872	-2.876
5800	49.043	347.927	308.263	230.053	-21.285	319.742	-2.880
5900	49.214	348.767	308.942	234.966	-22.422	325.631	-2.883
6000	49.374	349.596	309.613	239.895	-23.608	331.541	-2.886

PREVIOUS: December 1973 (1 atm) CURRENT: December 1973 (1 bar)

ZIRCONIUM NITRIDE (ZrN) CRYSTAL M_r = 105.2267

$S°(298.15 K) = 38.861 \pm 0.21 J K^{-1} mol^{-1}$

T_{fus} = [3225 K]

$\Delta_f H°(0 K) = -362.020 \pm 8.4 kJ mol^{-1}$

$\Delta_f H°(298.15 K) = -365.263 \pm 8.4 kJ mol^{-1}$

$\Delta_{fus} H°$ = [67.362] $kJ mol^{-1}$

Enthalpy of Formation

The enthalpy of reaction I, -174.25 ± 0.30 $kcal mol^{-1}$, was measured by Mah and Gelbert (7). The $\Delta_f H°$(298.15 K) of ZrN was obtained using -261.5 ± 0.2 $kcal mol^{-1}$ for $\Delta_f H°$(298.15 K) of ZrO_2. For the same reaction, Neumann et al. (1) obtain -176.0 kcal mol^{-1} on a sample of unspecified purity. Vapor pressure measurements of Hoch et al. (2) lead to -81 $kcal mol^{-1}$ for the $\Delta_f H°$(298.15 K) of ZrN. However, this value is uncertain by 10% as evidenced by the work on ZrO_2 by Chupka et al. (3).

$$\text{(I)} \quad ZrN(cr) + O_2(g) = ZrO_2(cr) + 1/2\, N_2(g)$$

Heat Capacity and Entropy

Low temperature measurements, 53-297 K, of Todd (4) were smoothly joined to those of Coughlin and King (5), 298-1073 K.

Fusion Data

Estimated by Henderson and Scheffee (6).

References

1. B. Neumann, C. Kroger and H. Kunz, Z. Anorg. Chem. 218, 379 (1934).
2. M. Hoch, D. P. Dingledy and H. L. Johnston, J. Amer. Chem. Soc. 77, 304 (1955).
3. W. A. Chupka, J. Berkowitz and M. G. Inghram, J. Chem. Phys. 26, 1207 (1957).
4. S. S. Todd, J. Amer. Chem. Soc. 72, 2914 (1950).
5. J. P. Coughlin and E. G. King, J. Amer. Chem. Soc. 72, 2262 (1950).
6. C. B. Henderson and R. S. Scheffee, Alexandria, Va., "Survey of Thermochemical Data," Atlantic Research Corp., January, 1960.
7. A. D. Mah and N. L. Gelbert, J. Am. Chem. Soc. 78, 3261 (1956).

Zirconium Nitride (ZrN) $N_1Zr_1(cr)$

Enthalpy Reference Temperature = T_r = 298.15 K Standard State Pressure = p° = 0.1 MPa

T/K	$C_p°$	S°	$-[G°-H°(T_r)]/T$	$H°-H°(T_r)$	$\Delta_f H°$	$\Delta_f G°$	Log K_f
	J K⁻¹mol⁻¹			kJ mol⁻¹			
0	0.	0.	INFINITE	-6.590	-362.020	-362.020	INFINITE
100	15.548	8.134	68.383	-6.025	-363.793	-355.210	185.543
200	31.765	24.376	42.325	-3.590	-365.006	-346.073	90.385
298.15	40.443	38.861	38.861	0.	-365.263	-336.697	58.988
300	40.556	39.112	38.862	0.075	-365.262	-336.519	58.593
400	44.756	51.399	40.508	4.356	-364.999	-326.969	42.698
500	47.078	61.649	43.741	8.954	-364.495	-317.516	33.171
600	48.660	70.378	47.471	13.744	-363.888	-308.176	26.829
700	49.894	77.974	51.298	18.673	-363.247	-298.942	22.307
800	50.940	84.706	55.061	23.716	-362.609	-289.799	18.922
900	51.882	90.760	58.697	28.857	-362.003	-280.734	16.293
1000	52.752	96.272	62.183	34.089	-361.447	-271.735	14.194
1100	53.580	101.339	65.515	39.406	-360.950	-262.788	12.479
1200	54.379	106.036	68.698	44.804	-364.228	-253.662	11.042
1300	55.153	110.419	71.741	50.281	-363.317	-244.485	9.824
1400	55.915	114.534	74.652	55.835	-362.391	-235.379	8.782
1500	56.668	118.418	77.442	61.464	-361.459	-226.339	7.882
1600	57.409	122.099	80.119	67.168	-360.532	-217.362	7.096
1700	58.141	125.601	82.692	72.945	-359.618	-208.442	6.405
1800	58.873	128.945	85.169	78.796	-358.727	-199.575	5.792
1900	59.597	132.147	87.558	84.720	-357.869	-190.757	5.244
2000	60.321	135.223	89.865	90.716	-357.054	-181.983	4.753
2100	60.973	138.181	92.096	96.780	-356.296	-173.248	4.309
2200	61.756	141.036	94.256	102.917	-376.922	-163.802	3.889
2300	62.471	143.797	96.350	109.128	-376.712	-154.119	3.500
2400	63.187	146.471	98.383	115.411	-376.435	-144.447	3.144
2500	63.898	149.065	100.359	121.765	-376.093	-134.788	2.816
2600	64.614	151.585	102.281	128.191	-375.685	-125.143	2.514
2700	65.321	154.037	104.152	134.688	-375.210	-115.516	2.235
2800	66.032	156.425	105.977	141.255	-374.669	-105.907	1.976
2900	66.743	158.755	107.757	147.894	-374.060	-96.319	1.735
3000	67.450	161.029	109.495	154.604	-373.384	-86.754	1.511
3100	68.157	163.252	111.193	161.384	-372.641	-77.211	1.301
3200	68.864	165.428	112.854	168.235	-371.830	-67.694	1.105
3225.000	69.041	165.964	113.264	169.959	------ CRYSTAL <--> LIQUID -----		
3300	69.572	167.557	114.480	175.157	-370.952	-58.203	0.921
3400	70.279	169.645	116.072	182.149	-370.005	-48.740	0.749
3500	70.986	171.692	117.632	189.213	-368.991	-39.306	0.587
3600	71.693	173.702	119.161	196.347	-367.908	-29.901	0.434
3700	72.396	175.676	120.662	203.551	-366.758	-20.528	0.290
3800	73.103	177.616	122.135	210.826	-365.539	-11.186	0.154
3900	73.810	179.524	123.582	218.172	-364.252	-1.878	0.025
4000	74.517	181.402	125.005	225.588	-362.896	7.397	-0.097

PREVIOUS: CURRENT: June 1961

Zirconium Nitride (ZrN) $N_1Zr_1(cr)$

ZIRCONIUM NITRIDE (ZrN) LIQUID M_r = 105.2267

$S°(298.15\ K)$ = [62.342] $J\ K^{-1}\ mol^{-1}$

$\Delta_f H°(298.15\ K)$ = [-290.749] $kJ\ mol^{-1}$

T_{fus} = [3225] K

$\Delta_{fus}H°$ = [67.362] $kJ\ mol^{-1}$

Enthalpy of Formation

$\Delta_f H°$(ZrN, ℓ, 298.15 K) is calculated from that of the crystal by adding the enthalpy of fusion, $\Delta_{fus}H°$, and the difference in enthalpy, H°(3225 K)-H°(298.15 K), between the crystal and liquid.

Heat Capacity and Entropy

Calculated from the solid. The heat capacity function of the solid was assumed to approximate those of the liquid up to the estimated glass point of 2150 K. Above the glass point the heat capacity was estimated.

S°(ZrN, ℓ, 298.15 K) is calculated in a manner analogous to that used for the enthalpy of formation.

Zirconium Nitride (ZrN) $N_1Zr_1(l)$

Enthalpy Reference Temperature = T_r = 298.15 K; Standard State Pressure = p° = 0.1 MPa

T/K	$C_p°$ (J K⁻¹mol⁻¹)	S° (J K⁻¹mol⁻¹)	$-[G°-H°(T_r)]/T$ (J K⁻¹mol⁻¹)	$H°-H°(T_r)$ (kJ mol⁻¹)	$\Delta_f H°$ (kJ mol⁻¹)	$\Delta_f G°$ (kJ mol⁻¹)	Log K_f
0							
100							
200							
298.15	40.443	62.342	62.342	0.	-290.749	-269.183	47.160
300	40.556	62.592	62.342	0.075	-290.747	-269.049	46.846
400	44.756	74.880	63.989	4.356	-290.484	-261.846	34.194
500	47.078	85.130	67.222	8.954	-289.981	-254.742	26.613
600	48.660	93.858	70.952	13.744	-289.374	-247.750	21.569
700	49.894	101.454	74.778	18.673	-288.732	-240.864	17.973
800	50.940	108.186	78.542	23.716	-288.094	-234.069	15.283
900	51.882	114.241	82.177	28.857	-287.489	-227.353	13.195
1000	52.752	119.753	85.663	34.089	-286.932	-220.701	11.528
1100	53.580	124.820	88.996	39.406	-286.436	-214.103	10.167
1200	54.379	129.516	92.179	44.804	-289.714	-207.325	9.025
1300	55.153	133.900	95.222	50.281	-288.802	-200.496	8.056
1400	55.915	138.015	98.133	55.835	-287.876	-193.738	7.228
1500	56.668	141.898	100.922	61.464	-286.945	-187.046	6.514
1600	57.409	145.579	103.599	67.168	-286.017	-180.416	5.890
1700	58.141	149.082	106.173	72.945	-285.103	-173.844	5.342
1800	58.873	152.426	108.650	78.796	-284.213	-167.326	4.856
1900	59.597	155.628	111.039	84.720	-283.355	-160.856	4.422
2000	60.321	158.704	113.346	90.716	-282.539	-154.430	4.033
2100	60.973	161.663	115.577	96.781	-281.780	-148.043	3.682
2150.000	61.273	163.101	116.665	99.838	GLASS <--> LIQUID		
2150.000	58.576	163.101	116.665	99.838	TRANSITION		
2200	58.576	164.448	117.736	102.766	-302.558	-140.944	3.346
2300	58.576	167.052	119.824	108.624	-302.701	-133.594	3.034
2400	58.576	169.545	121.844	114.482	-302.850	-126.239	2.748
2500	58.576	171.936	123.800	120.339	-303.005	-118.877	2.484
2600	58.576	174.233	125.696	126.197	-303.164	-111.509	2.240
2700	58.576	176.444	127.535	132.054	-303.329	-104.134	2.015
2800	58.576	178.574	129.320	137.912	-303.497	-96.753	1.805
2900	58.576	180.630	131.054	143.770	-303.670	-89.367	1.610
3000	58.576	182.615	132.740	149.627	-303.846	-81.974	1.427
3100	58.576	184.536	134.380	155.485	-304.025	-74.575	1.257
3200	58.576	186.396	135.976	161.342	-304.208	-67.171	1.096
3225.000	58.576	186.852	136.369	162.807	------ CRYSTAL <--> LIQUID -----		
3300	58.576	188.198	137.532	167.200	-304.394	-59.760	0.946
3400	58.576	189.947	139.048	173.058	-304.583	-52.345	0.804
3500	58.576	191.645	140.526	178.915	-304.774	-44.923	0.670
3600	58.576	193.295	141.969	184.773	-304.968	-37.496	0.544
3700	58.576	194.900	143.378	190.630	-305.164	-30.063	0.424
3800	58.576	196.462	144.755	196.488	-305.362	-22.626	0.311
3900	58.576	197.984	146.100	202.346	-305.563	-15.183	0.203
4000	58.576	199.467	147.416	208.203	-305.766	-7.734	0.101
4100	58.576	200.913	148.703	214.061	-305.971	-0.281	0.004
4200	58.576	202.325	149.963	219.918	-306.178	7.177	-0.089
4300	58.576	203.703	151.197	225.776	-306.387	14.641	-0.178
4400	58.576	205.050	152.406	231.634	-306.597	22.109	-0.262
4500	58.576	206.366	153.590	237.491	-306.810	29.582	-0.343
4600	58.576	207.653	154.752	243.349	-307.024	37.059	-0.421
4700	58.576	208.913	155.891	249.206	-307.240	44.542	-0.495
4800	58.576	210.146	157.008	255.064	-868.394	63.646	-0.693
4900	58.576	211.354	158.105	260.922	-868.265	83.062	-0.885
5000	58.576	212.538	159.182	266.779	-868.155	102.476	-1.071
5100	58.576	213.698	160.239	272.637	-868.062	121.888	-1.248
5200	58.576	214.835	161.278	278.494	-867.986	141.298	-1.419
5300	58.576	215.951	162.299	284.352	-867.923	160.706	-1.584
5400	58.576	217.046	163.303	290.210	-867.875	180.114	-1.742
5500	58.576	218.120	164.290	296.067	-867.838	199.521	-1.895
5600	58.576	219.176	165.261	301.925	-867.811	218.927	-2.042
5700	58.576	220.213	166.216	307.782	-867.795	238.333	-2.184
5800	58.576	221.231	167.156	313.640	-867.787	257.739	-2.321
5900	58.576	222.233	168.081	319.498	-867.786	277.144	-2.454
6000	58.576	223.217	168.991	325.355	-867.792	296.550	-2.582

PREVIOUS: CURRENT: June 1961

Zirconium Nitride (ZrN) $N_1Zr_1(l)$

ZIRCONIUM NITRIDE (ZrN) CRYSTAL-LIQUID M_r = 105.2267 Zirconium Nitride (ZrN) $N_1Zr_1(cr,l)$

0 to 3225 K crystal
above 3225 K liquid

Refer to the individual tables for details.

Enthalpy Reference Temperature = T_r = 298.15 K Standard State Pressure = p° = 0.1 MPa

T/K	C_p° (J K⁻¹mol⁻¹)	S° (J K⁻¹mol⁻¹)	−[G°−H°(T_r)]/T (J K⁻¹mol⁻¹)	H°−H°(T_r) (kJ mol⁻¹)	Δ_fH° (kJ mol⁻¹)	Δ_fG° (kJ mol⁻¹)	Log K_f
0	0.	0.	INFINITE	-6.590	-362.020	-362.020	INFINITE
100	15.548	8.134	68.383	-6.025	-363.793	-355.210	185.543
200	31.765	24.376	42.325	-3.590	-365.006	-346.073	90.385
298.15	40.443	38.861	38.861	0.	-365.263	-336.697	58.988
300	40.556	39.112	38.862	0.075	-365.262	-336.519	58.593
400	44.756	51.399	40.508	4.356	-364.999	-326.969	42.698
500	47.078	61.649	43.741	8.954	-364.495	-317.516	33.171
600	48.660	70.378	47.471	13.744	-363.888	-308.176	26.829
700	49.894	77.974	51.298	18.673	-363.247	-298.942	22.307
800	50.940	84.706	55.061	23.716	-362.609	-289.799	18.922
900	51.882	90.760	58.697	28.857	-362.003	-280.734	16.293
1000	52.752	96.272	62.183	34.089	-361.447	-271.735	14.194
1100	53.580	101.339	65.515	39.406	-360.950	-262.788	12.479
1200	54.379	106.036	68.698	44.804	-364.228	-253.662	11.042
1300	55.153	110.419	71.741	50.281	-363.317	-244.485	9.824
1400	55.915	114.534	74.652	55.835	-362.391	-235.379	8.782
1500	56.668	118.418	77.442	61.464	-361.459	-226.339	7.882
1600	57.409	122.099	80.119	67.168	-360.532	-217.362	7.096
1700	58.141	125.601	82.692	72.945	-359.618	-208.442	6.405
1800	58.873	128.945	85.169	78.796	-358.727	-199.575	5.792
1900	59.597	132.147	87.558	84.720	-357.869	-190.757	5.244
2000	60.321	135.223	89.865	90.716	-357.054	-181.983	4.753
2100	60.973	138.181	92.096	96.780	-356.296	-173.248	4.309
2200	61.756	141.036	94.256	102.917	-376.922	-163.802	3.889
2300	62.471	143.797	96.350	109.128	-376.712	-154.119	3.500
2400	63.187	146.471	98.383	115.411	-376.435	-144.447	3.144
2500	63.898	149.065	100.359	121.765	-376.093	-134.788	2.816
2600	64.614	151.585	102.281	128.191	-375.685	-125.143	2.514
2700	65.321	154.037	104.152	134.688	-375.210	-115.516	2.235
2800	66.032	156.425	105.977	141.255	-374.669	-105.907	1.976
2900	66.743	158.755	107.757	147.894	-374.060	-96.319	1.735
3000	67.450	161.029	109.495	154.604	-373.384	-86.754	1.511
3100	68.157	163.252	111.193	161.384	-372.641	-77.211	1.301
3200	68.864	165.428	112.854	168.235	-371.830	-67.694	1.105
3225.000	69.041	165.964	113.264	169.959	CRYSTAL <--> LIQUID		
3225.000	58.576	186.852	113.264	237.321	TRANSITION		
3300	58.576	188.198	114.952	241.715	-304.394	-59.760	0.946
3400	58.576	189.947	117.132	247.572	-304.583	-52.345	0.804
3500	58.576	191.645	119.236	253.430	-304.774	-44.923	0.670
3600	58.576	193.295	121.271	259.287	-304.968	-37.496	0.544
3700	58.576	194.900	123.239	265.145	-305.164	-30.063	0.424
3800	58.576	196.462	125.146	271.003	-305.362	-22.626	0.311
3900	58.576	197.984	126.994	276.860	-305.563	-15.183	0.203
4000	58.576	199.467	128.787	282.718	-305.766	-7.734	0.101
4100	58.576	200.913	130.529	288.575	-305.971	-0.281	0.004
4200	58.576	202.325	132.222	294.433	-306.178	7.177	-0.089
4300	58.576	203.703	133.868	300.291	-306.387	14.641	-0.178
4400	58.576	205.050	135.470	306.148	-306.597	22.109	-0.262
4500	58.576	206.366	137.031	312.006	-306.810	29.582	-0.343
4600	58.576	207.653	138.553	317.863	-307.024	37.059	-0.421
4700	58.576	208.913	140.036	323.721	-307.240	44.542	-0.495
4800	58.576	210.146	141.484	329.579	-868.394	63.646	-0.693
4900	58.576	211.354	142.898	335.436	-868.265	83.062	-0.885
5000	58.576	212.538	144.279	341.294	-868.155	102.476	-1.071
5100	58.576	213.698	145.629	347.151	-868.062	121.888	-1.248
5200	58.576	214.835	146.949	353.009	-867.986	141.298	-1.419
5300	58.576	215.951	148.240	358.867	-867.923	160.706	-1.584
5400	58.576	217.046	149.504	364.724	-867.875	180.114	-1.742
5500	58.576	218.120	150.742	370.582	-867.838	199.521	-1.895
5600	58.576	219.176	151.955	376.439	-867.811	218.927	-2.042
5700	58.576	220.213	153.143	382.297	-867.795	238.333	-2.184
5800	58.576	221.231	154.308	388.155	-867.787	257.739	-2.321
5900	58.576	222.233	155.451	394.012	-867.786	277.144	-2.454
6000	58.576	223.217	156.572	399.870	-867.792	296.550	-2.582

PREVIOUS: CURRENT: June 1961

Zirconium Nitride (ZrN) $N_1Zr_1(cr,l)$

ZIRCONIUM NITRIDE (ZrN) IDEAL GAS M_r = 105.2267

$S°(298.15\ K) = [233.487]\ J\ K^{-1}\ mol^{-1}$

$\Delta_f H°(0\ K) = [714.342]\ kJ\ mol^{-1}$
$\Delta_f H°(298.15\ K) = [713.372]\ kJ\ mol^{-1}$

Electronic Level and Quantum Weight

State	ε_i, cm^{-1}	g_i
$[^2\Sigma]$	0	[2]

$\omega_e = [840]\ cm^{-1}$ $\omega_e x_e = [5]\ cm^{-1}$ $\sigma = 1$
$B_e = [0.4146]\ cm^{-1}$ $\alpha_e = [0.0025]\ cm^{-1}$ $r_e = [1.83]$ Å

Enthalpy of Formation

$\Delta_f H°(0\ K)$ was estimated as 170.7 kcal mol^{-1} by Gordon (1). $\Delta_f H°(298.15\ K)$ was then calculated.

Heat Capacity and Entropy

Molecular constants were estimated by Gordon (1). r_e was calculated from the relationship $r_e = (16.858778/\mu B_e)^{1/2}$ Å.

Reference

1. J. S. Gordon, Astrosystems International, Livingston, New Jersey, personal communication, Jan. 10, 1963.

Zirconium Nitride (ZrN) $N_1Zr_1(g)$

Enthalpy Reference Temperature = T_r = 298.15 K — Standard State Pressure = p° = 0.1 MPa

T/K	$C_p°$ (J K⁻¹mol⁻¹)	S° (J K⁻¹mol⁻¹)	$-[G°-H°(T_r)]/T$ (J K⁻¹mol⁻¹)	$H°-H°(T_r)$ (kJ mol⁻¹)	$\Delta_f H°$ (kJ mol⁻¹)	$\Delta_f G°$ (kJ mol⁻¹)	Log K_f
0	0.	0.	INFINITE	-8.863	714.342	714.342	INFINITE
100	29.114	200.910	260.451	-5.954	714.913	704.218	-367.846
200	29.877	221.235	236.323	-3.018	714.201	693.763	-181.192
250	30.752	227.993	234.003	-1.503	713.772	688.703	-143.897
298.15	31.661	233.487	233.487	0.	713.372	683.911	-119.818
300	31.695	233.683	233.488	0.059	713.357	683.728	-119.048
350	32.569	238.636	233.877	1.666	712.965	678.821	-101.309
400	33.326	243.036	234.752	3.314	712.594	673.969	-88.011
450	33.961	246.999	235.896	4.996	712.239	669.162	-77.674
500	34.488	250.605	237.189	6.708	711.893	664.395	-69.409
600	35.286	256.968	239.970	10.199	711.202	654.960	-57.019
700	35.845	262.452	242.799	13.757	710.472	645.643	-48.178
800	36.248	267.266	245.562	17.363	709.673	636.435	-41.555
900	36.549	271.554	248.217	21.003	708.778	627.333	-36.409
1000	36.781	275.417	250.747	24.670	707.769	618.336	-32.299
1100	36.964	278.931	253.151	28.358	706.637	609.447	-28.940
1200	37.114	282.154	255.436	32.062	701.665	600.888	-26.156
1300	37.239	285.130	257.607	35.780	700.817	592.524	-23.808
1400	37.345	287.894	259.673	39.509	699.919	584.227	-21.798
1500	37.437	290.474	261.641	43.248	698.960	575.996	-20.058
1600	37.519	292.892	263.520	46.996	697.932	567.832	-18.538
1700	37.592	295.169	265.315	50.752	696.824	559.734	-17.199
1800	37.659	297.320	267.034	54 515	695.626	551.704	-16.010
1900	37.720	299.358	268.682	58.283	694.330	543.743	-14.949
2000	37.776	301.294	270.265	62.058	692.923	535.853	-13.995
2100	37.829	303.138	271.786	65.839	691.398	528.037	-13.134
2200	37.880	304.899	273.252	69.624	668.420	521.042	-12.371
2300	37.927	306.584	274.665	73.414	666.210	514.392	-11.682
2400	37.973	308.199	276.029	77.209	663.998	507.839	-11.053
2500	38.017	309.750	277.347	81.009	661.786	501.378	-10.476
2600	38.059	311.242	278.622	84.813	659.572	495.005	-9.945
2700	38.100	312.679	279.857	88.621	657.358	488.717	-9.455
2800	38.140	314.066	281.054	92.433	655.144	482.512	-9.001
2900	38.179	315.405	282.215	96.249	652.930	476.386	-8.581
3000	38.218	316.700	283.343	100.069	650.716	470.336	-8.189
3100	38.255	317.953	284.440	103.892	648.503	464.359	-7.824
3200	38.292	319.169	285.506	107.720	646.290	458.455	-7.484
3300	38.328	320.347	286.544	111.551	644.077	452.619	-7.164
3400	38.364	321.492	287.555	115.385	641.866	446.850	-6.865
3500	38.399	322.605	288.541	119.224	639.655	441.147	-6.584
3600	38.434	323.687	289.502	123.065	637.445	435.507	-6.319
3700	38.469	324.740	290.440	126.910	635.237	429.927	-6.069
3800	38.503	325.767	291.357	130.759	633.029	424.408	-5.834
3900	38.537	326.767	292.252	134.611	630.823	418.947	-5.611
4000	38.571	327.744	293.127	138.466	628.618	413.542	-5.400
4100	38.604	328.696	293.983	142.325	626.414	408.193	-5.200
4200	38.638	329.627	294.821	146.187	624.212	402.897	-5.011
4300	38.671	330.537	295.641	150.053	622.011	397.653	-4.831
4400	38.704	331.426	296.444	153.921	619.811	392.461	-4.659
4500	38.736	332.296	297.231	157.793	617.613	387.319	-4.496
4600	38.769	333.148	298.003	161.669	615.417	382.226	-4.340
4700	38.801	333.982	298.759	165.547	613.222	377.180	-4.192
4800	38.834	334.799	299.502	169.429	50.092	383.798	-4.177
4900	38.866	335.600	300.230	173.314	48.248	390.769	-4.166
5000	38.898	336.386	300.945	177.202	46.389	397.778	-4.156
5100	38.930	337.156	301.648	181.093	44.515	404.825	-4.146
5200	38.962	337.913	302.338	184.988	42.629	411.908	-4.138
5300	38.994	338.655	303.016	188.886	40.731	419.028	-4.130
5400	39.026	339.384	303.683	192.787	38.823	426.183	-4.123
5500	39.057	340.101	304.339	196.691	36.907	433.374	-4.116
5600	39.089	340.805	304.984	200.598	34.983	440.600	-4.110
5700	39.121	341.497	305.618	204.509	33.052	447.860	-4.104
5800	39.152	342.178	306.243	208.423	31.116	455.155	-4.099
5900	39.184	342.847	306.857	212.339	29.176	462.482	-4.095
6000	39.215	343.506	307.463	216.259	27.233	469.843	-4.090

PREVIOUS: June 1963 (1 atm) CURRENT: June 1963 (1 bar)

Zirconium Nitride (ZrN) $N_1Zr_1(g)$

NITROGEN (N_2) REFERENCE STATE - IDEAL GAS M_r = 28.0134

$\Delta_f H°(0\ K) = 0\ kJ\ mol^{-1}$

$S°(298.15\ K) = 191.609 \pm 0.020\ J\ K^{-1}\ mol^{-1}$ $\Delta_f H°(298.15\ K) = 0\ kJ\ mol^{-1}$

Vibrational and Rotational Levels (cm^{-1})

Direct Summation of Electronic Ground State:

$$E = G - G_0 + F = G - G_0 + BZ - DZ^2 + \ldots \approx G - G_0 + B^2Z/(B + DZ),$$

where $Z = J(J+1)$, $Y = v + 1/2$, and we omit subscript v on G, F, B, and D

$$G = 2358.583Y - 14.33594Y^2 - 3.767755 \times 10^{-3}Y^3 + 2.753652 \times 10^{-4}Y^4 - 2.701198 \times 10^{-5}Y^5 + 1.749062 \times 10^{-7}Y^6$$

$$B = 1.998197 - 0.017279Y - 3.283 \times 10^{-5}Y^2,\ D = 5.74 \times 10^{-6}$$

$$v_{max} = 58,\ J_{max} = 260 - 250v/v_{max}$$

Ground State Configuration, $^1\Sigma_g^+$ $\sigma = 2$

Excited Electronic States - Contributions at 4800 - 6000 K Calculated Using First-Order Corrections:

State	ϵ_i/cm^{-1}	g_i	B_e/cm^{-1}	α_e/cm^{-1}	ω_e/cm^{-1}	$\omega_e x_e/cm^{-1}$
$A^3\Sigma_u^+$	49754.78	3	1.4546	0.01799	1460.638	13.8723
$B^3\Pi_g$	59306.81	6	1.63745	0.017906	1733.391	14.1221
$W^3\Delta_g$	59380.	6	[1.47327]	[0.016656]	1501.4	11.6

Enthalpy of Formation

Zero by definition.

Heat Capacity and Entropy

These are calculated by combining direct summation of vibrational-rotational energy levels of the electronic ground state with an approximate correction for the contribution of three excited states. All spectroscopic constants are from Lofthus and Krupenie (1) except the vibrational polynomial G used in direct summation of the ground state. We revise the G equation (1) so that is converges near $D_0^0 = 78715\ cm^{-1}$ (2, 1). Experimental data are limited to $v \leq 27$ and there is uncertainty in the extrapolation to high v values. Methods proposed by Khachkuruzov (3) are used to approximate the infinite-series (F) equation for rotational levels and the limiting values (J_{max}) of rotational quantum number. Values in the J_{max} equation are estimated by comparison with Khachkuruzov (3) and Gurvich et al. (4).

Contributions of excited states begin to appear in Cp° at ~4800 K. We obtain these contributions using the difference between approximate calculations for four states (X, A, B, and W) and one state (ground state). These approximate calculations used first-order corrections for anharmonicity and vibrational-rotational interaction. The excited state contribution to Cp° is $0.019\ cal^{-1}\ K^{-1}\ mol^{-1}$ at 5000 K.

Values and uncertainties at 298.15 K are the same as those selected by CODATA (2). Uncertainties at 6000 K are considerably larger due to uncertainity in the extrapolation of G and F. Comparing our approximate calculation (a) with our adopted values (b) and those (c) of Gurvich et al. (4), we have S°(600 K) = 69.981, 69.999 and 70.005 $cal^{-1}\ mol^{-1}$ and H(T) - H(0 K) = 51.196, 51.271, and 51.289 $J\ K\ mol^{-1}$.

References

1. A. Lofthus and P. H. Krupenie, J. Phys. Chem. Ref. Data 6, 113 (1977).
2. ICSU-CODATA Task Group, J. Chem. Thermodynamics 4, 331 (1972).
3. G. A. Khachkuruzov, Opt. Spectrosc. 30, 455 (1971).
4. L. V. Gurvich, G. A. Khachkuruzov et al., "Thermodynamic Properties of Individual Substances. Volume I. Calculation of the Thermodynamic Properties," English Translation, Reports AD-659660, AD-659659, AD-659679 (avail. NTIS), 1967; Volume II, Russian Edition, Moscow, 1962.

Nitrogen (N_2) N_2(ref)

Enthalpy Reference Temperature = T_r = 298.15 K Standard State Pressure = p° = 0.1 MPa

T/K	$C_p°$ (J K⁻¹mol⁻¹)	S° (J K⁻¹mol⁻¹)	-[G°-H°(T_r)]/T (J K⁻¹mol⁻¹)	H°-H°(T_r) (kJ mol⁻¹)	$\Delta_f H°$ (kJ mol⁻¹)	$\Delta_f G°$ (kJ mol⁻¹)	Log K_f
0	0.	0.	INFINITE	-8.670	0.	0.	0.
100	29.104	159.811	217.490	-5.768	0.	0.	0.
200	29.107	179.985	194.272	-2.857	0.	0.	0.
250	29.111	186.481	192.088	-1.402	0.	0.	0.
298.15	29.124	191.609	191.609	0.	0.	0.	0.
300	29.125	191.789	191.610	0.054	0.	0.	0.
350	29.165	196.281	191.964	1.511	0.	0.	0.
400	29.249	200.181	192.753	2.971	0.	0.	0.
450	29.387	203.633	193.774	4.437	0.	0.	0.
500	29.580	206.739	194.917	5.911	0.	0.	0.
600	30.110	212.176	197.353	8.894	0.	0.	0.
700	30.754	216.866	199.813	11.937	0.	0.	0.
800	31.433	221.017	202.209	15.046	0.	0.	0.
900	32.090	224.757	204.510	18.223	0.	0.	0.
1000	32.697	228.170	206.708	21.463	0.	0.	0.
1100	33.241	231.313	208.804	24.760	0.	0.	0.
1200	33.723	234.226	210.802	28.109	0.	0.	0.
1300	34.147	236.943	212.710	31.503	0.	0.	0.
1400	34.518	239.487	214.533	34.936	0.	0.	0.
1500	34.843	241.880	216.277	38.405	0.	0.	0.
1600	35.128	244.138	217.948	41.904	0.	0.	0.
1700	35.378	246.275	219.552	45.429	0.	0.	0.
1800	35.600	248.304	221.094	48.978	0.	0.	0.
1900	35.796	250.234	222.577	52.548	0.	0.	0.
2000	35.971	252.074	224.006	56.137	0.	0.	0.
2100	36.126	253.833	225.385	59.742	0.	0.	0.
2200	36.268	255.517	226.717	63.361	0.	0.	0.
2300	36.395	257.132	228.004	66.995	0.	0.	0.
2400	36.511	258.684	229.250	70.640	0.	0.	0.
2500	36.616	260.176	230.458	74.296	0.	0.	0.
2600	36.713	261.614	231.629	77.963	0.	0.	0.
2700	36.801	263.001	232.765	81.639	0.	0.	0.
2800	36.883	264.341	233.869	85.323	0.	0.	0.
2900	36.959	265.637	234.942	89.015	0.	0.	0.
3000	37.030	266.891	235.986	92.715	0.	0.	0.
3100	37.096	268.106	237.003	96.421	0.	0.	0.
3200	37.158	269.285	237.993	100.134	0.	0.	0.
3300	37.216	270.429	238.959	103.852	0.	0.	0.
3400	37.271	271.541	239.901	107.577	0.	0.	0.
3500	37.323	272.622	240.821	111.306	0.	0.	0.
3600	37.373	273.675	241.719	115.041	0.	0.	0.
3700	37.420	274.699	242.596	118.781	0.	0.	0.
3800	37.465	275.698	243.454	122.525	0.	0.	0.
3900	37.508	276.671	244.294	126.274	0.	0.	0.
4000	37.550	277.622	245.115	130.027	0.	0.	0.
4100	37.590	278.549	245.919	133.784	0.	0.	0.
4200	37.629	279.456	246.707	137.545	0.	0.	0.
4300	37.666	280.341	247.479	141.309	0.	0.	0.
4400	37.702	281.208	248.236	145.078	0.	0.	0.
4500	37.738	282.056	248.978	148.850	0.	0.	0.
4600	37.773	282.885	249.706	152.625	0.	0.	0.
4700	37.808	283.698	250.420	156.405	0.	0.	0.
4800	37.843	284.494	251.122	160.187	0.	0.	0.
4900	37.878	285.275	251.811	163.973	0.	0.	0.
5000	37.912	286.041	252.488	167.763	0.	0.	0.
5100	37.947	286.792	253.153	171.556	0.	0.	0.
5200	37.981	287.529	253.807	175.352	0.	0.	0.
5300	38.013	288.253	254.451	179.152	0.	0.	0.
5400	38.046	288.964	255.083	182.955	0.	0.	0.
5500	38.080	289.662	255.705	186.761	0.	0.	0.
5600	38.116	290.348	256.318	190.571	0.	0.	0.
5700	38.154	291.023	256.921	194.384	0.	0.	0.
5800	38.193	291.687	257.515	198.201	0.	0.	0.
5900	38.234	292.341	258.099	202.023	0.	0.	0.
6000	38.276	292.984	258.675	205.848	0.	0.	0.

PREVIOUS: March 1977 (1 atm) CURRENT: March 1977 (1 bar)

Nitrogen (N_2) N_2(ref)

NITROGEN, ION (N_2^+) IDEAL GAS M_r = 28.01285 Nitrogen, Ion (N_2^+) $N_2^+(g)$

$S°(298.15\ K) = 197.660 \pm 0.04\ kJ\ mol^{-1}$

$\Delta_f H°(0\ K) = 1503.303 \pm 0.04\ kJ\ mol^{-1}$
$\Delta_f H°(298.15\ K) = [1509.499]\ kJ\ mol^{-1}$

Electronic and Molecular Constants (σ = 2)

State	g_i	T_o, cm^{-1}	ω_e, cm^{-1}	$\omega_e x_e$, cm^{-1}	$\omega_e y_e$, cm^{-1}	B_e, cm^{-1}	α_e, cm^{-1}	r_e, Å
$X^2\Sigma_g^+$	2	0	2207.00	16.10	-0.040	1.9319	0.0190	1.116384
$A^2\Pi_{ui}$	4	9016.4	1903.53	15.011		1.748	0.020	1.17364
$B^2\Sigma_u^+$	2	25566.0	2419.84	23.19		2.073	0.020	1.07772

Enthalpy of Formation

The adopted value for the enthalpy of formation of $N_2^+(g)$, $\Delta_f H°(0\ K) = 359.298 \pm 0.01$ kcal mol^{-1} is recommended in a critical review on the energetics of gaseous ions by Rosenstock et al. (1). This value is derived from ionization potentials determined in the spectroscopic studies of Worley (2) and Ogawa and Tanaka (3); the IP values are 125665.8 and 125666.8 cm^{-1} respectively.

$\Delta_f H°(N_2^+, 298.15\ K)$ is obtained from $\Delta_f H°(N_2, g, 0\ K)$ by using IP (N_2, g, 0 K) with JANAF (7) enthalpies H°(0 K) - H°(298.15) for $N_2(g)$, $N_2(g)$, and $e^-(g)$. $\Delta_r H°(298.15)$ for the reaction $N_2(g) = N_2^+(g) + e^-(g)$ differs from a room temperature threshold energy due to the inclusion of these enthalpies and to threshold effects discussed by Rosenstock et al. (1). $\Delta_f H°(298.15)$ should be changed by -6.197 kJ mol^{-1} if it is to be used in the ion convention which excludes the enthalpy of the electron.

Heat Capacity and Entropy

The thermal functions are calculated using the program of McBride and Gordon (4). The contributions of vibrational anharmonicity and rotation-vibration interaction are calculated via the procedures given by Pennington and Kobe (5). In this procedure the energy levels are calculated as follows: $\epsilon_i = T_o + G(v) - G(o) + B_v J(J+1) - D_e J^2(J+1)^2$ where $G(v) = \omega_e(v + 1/2) - \omega_e x_e(v + 1/2)^2 + \omega_e y_e(v + 1/2)^3$ and $B_v = B_e - \alpha_e(v + 1/2)$. When a value of D_e is not available, it is calculated to be $D_e = 4B_e^3/\omega_e^2$. Molecular constant data in the table are from the critical review of Lofthus and Krupenie (6). Electronic levels above 5000 cm^{-1} given by Lofthus and Krupenie are not included since they make negligible contributions to the thermodynamic properties at 6000 K.

References

1. H. M. Rosenstock, H. Draxl, B. W. Steiner, and J. T. Herron, J. Phys. Chem. Ref. Data 6, Supp. 1 (1977).
2. R. E. Worley, Phys. Rev. 64, 207 (1943).
3. M. Ogawa and Y. Tanaka, Can. J. Phys. 40, 1593 (1962).
4. B. J. McBride and S. Gordon, NASA TN D-4097, 1967.
5. R. E. Pennington and K. A. Kobe, J. Chem. Phys. 22, 1442 (1954).
6. A. Lofthus and P. H. Krupenie, J. Phys. Chem. Ref. Data 6, 113 (1977).
7. JANAF Thermochemical Tables: $N_2(g)$, 3-31-77; $e^-(g)$, 3-31-77.

Enthalpy Reference Temperature = T_r = 298.15 K Standard State Pressure = p° = 0.1 MPa

T/K	$C_p°$ (J K^{-1}mol^{-1})	S° (J K^{-1}mol^{-1})	-[G°-H°(T_r)]/T (J K^{-1}mol^{-1})	H°-H°(T_r) (kJ mol^{-1})	$\Delta_f H°$ (kJ mol^{-1})	$\Delta_f G°$ (kJ mol^{-1})	Log K_f
0	0.	0.	INFINITE	-8.669	1503.303		
100	29.104	165.859	223.544	-5.768			
200	29.137	197.660	211.948	-2.858			
298.15	29.137	197.660	197.660	0.	1509.499	1501.440	-263.046
300	29.139	197.840	197.660	0.054	1509.538	1501.390	-261.415
400	29.321	206.242	198.804	2.975	1511.620	1498.361	-195.666
500	29.752	212.827	200.973	5.927	1513.711	1494.804	-156.161
600	30.379	218.305	203.418	8.933	1515.812	1490.826	-129.788
700	31.097	223.042	205.890	12.006	1517.922	1486.495	-110.924
800	31.819	227.242	208.302	15.153	1520.037	1481.861	-96.756
900	32.495	231.030	210.620	18.369	1522.155	1476.962	-85.721
1000	33.108	234.486	212.836	21.650	1524.275	1471.826	-76.880
1100	33.657	237.668	214.951	24.988	1526.395	1466.479	-69.637
1200	34.148	240.617	216.968	28.379	1528.515	1460.939	-63.593
1300	34.597	243.369	218.894	31.816	1530.637	1455.221	-58.472
1400	35.018	245.948	220.736	35.297	1532.763	1449.340	-54.076
1500	35.422	248.378	222.498	38.819	1534.895	1443.307	-50.260
1600	35.822	250.677	224.188	42.381	1537.037	1437.131	-46.918
1700	36.225	252.860	225.811	45.984	1539.193	1430.821	-43.964
1800	36.636	254.943	227.372	49.627	1541.365	1424.384	-41.335
1900	37.059	256.935	228.876	53.311	1543.559	1417.825	-38.979
2000	37.496	258.847	230.327	57.039	1545.776	1411.150	-36.855
2100	37.944	260.687	231.729	60.811	1548.022	1404.364	-34.932
2200	38.403	262.463	233.086	64.628	1550.298	1397.470	-33.180
2300	38.869	264.180	234.401	68.492	1552.607	1390.472	-31.579
2400	39.339	265.844	235.676	72.402	1554.951	1383.372	-30.108
2500	39.809	267.459	236.916	76.360	1557.330	1376.174	-28.754
2600	40.275	269.030	238.121	80.364	1559.746	1368.880	-27.501
2700	40.733	270.559	239.294	84.414	1562.200	1361.493	-26.340
2800	41.181	272.048	240.437	88.510	1564.690	1354.014	-25.259
2900	41.615	273.501	241.552	92.650	1567.216	1346.445	-24.252
3000	42.032	274.919	242.641	96.833	1569.778	1338.789	-23.310
3100	42.433	276.304	243.705	101.056	1572.374	1331.047	-22.428
3200	42.811	277.657	244.745	105.318	1575.002	1323.220	-21.599
3300	43.165	278.980	245.762	109.617	1577.661	1315.311	-20.820
3400	43.496	280.273	246.758	113.951	1580.348	1307.321	-20.085
3500	43.804	281.538	247.734	118.316	1583.062	1299.251	-19.390
3600	44.084	282.776	248.690	122.710	1585.801	1291.103	-18.733
3700	44.343	283.988	249.628	127.132	1588.561	1282.879	-18.111
3800	44.581	285.174	250.548	131.578	1591.342	1274.580	-17.520
3900	44.798	286.335	251.451	136.048	1594.141	1266.208	-16.959
4000	44.994	287.471	252.337	140.537	1596.957	1257.764	-16.425
4100	45.158	288.584	253.207	145.045	1599.786	1249.249	-15.916
4200	45.307	289.674	254.063	149.568	1602.627	1240.665	-15.430
4300	45.439	290.742	254.903	154.106	1605.478	1232.013	-14.966
4400	45.556	291.788	255.730	158.656	1608.338	1223.295	-14.522
4500	45.656	292.813	256.543	163.216	1611.205	1214.512	-14.098
4600	45.730	293.817	257.342	167.786	1614.078	1205.665	-13.691
4700	45.794	294.801	258.129	172.362	1616.954	1196.755	-13.300
4800	45.846	295.766	258.903	176.944	1619.832	1187.784	-12.926
4900	45.889	296.712	259.665	181.531	1622.711	1178.753	-12.566
5000	45.920	297.639	260.415	186.122	1625.591	1169.664	-12.219
5100	45.933	298.549	261.154	190.714	1628.469	1160.517	-11.886
5200	45.939	299.441	261.881	195.308	1631.345	1151.313	-11.565
5300	45.939	300.316	262.598	199.902	1634.218	1142.054	-11.256
5400	45.932	301.174	263.305	204.495	1637.087	1132.741	-10.957
5500	45.919	302.017	264.001	209.088	1639.952	1123.375	-10.669
5600	45.900	302.844	264.687	213.679	1642.812	1113.957	-10.391
5700	45.875	303.657	265.364	218.268	1645.666	1104.488	-10.121
5800	45.843	304.454	266.031	222.854	1648.513	1094.969	-9.861
5900	45.805	305.237	266.689	227.436	1651.353	1085.400	-9.609
6000	45.761	306.007	267.338	232.015	1654.184	1075.784	-9.366

PREVIOUS: September 1977 (1 atm) CURRENT: September 1977 (1 bar)

Nitrogen, Ion (N_2^+) $N_2^+(g)$

NITROGEN, ION (N_2^-) IDEAL GAS M_r = 28.01395

Nitrogen, Ion (N_2^-) $N_2^-(g)$

$S°(298.15\ K) = 204.54 \pm 0.4\ J\ K^{-1}\ mol^{-1}$

$\Delta_f H°(0\ K) = 155 \pm 96\ kJ\ mol^{-1}$
$\Delta_f H°(298.15\ K) = [148.615]\ kJ\ mol^{-1}$

Electronic Level and Quantum Weight

State	ε_i, cm^{-1}	g_i
$^2\Pi_g$	0	4

$\omega_e = [1900]\ cm^{-1}$ $\omega_e x_e = [12.0]\ cm^{-1}$ $\sigma = 2$
$B_e = [1.691]\ cm^{-1}$ $\alpha_e = [0.015]\ cm^{-1}$ $r_e = [1.193 \pm 0.003]$ Å

Enthalpy of Formation

The enthalpy of formation at 0 K for $N_2^-(g)$ is based on the electron affinity value derived from the Rydberg-Klein-type calculations of Gilmore (1). Use of this adopted value, $EA(N_2) = -1.6 \pm 1.0$ eV and auxiliary data (2) leads to $\Delta_f H° = 37 \pm 23$ kJ mol^{-1} and $D_0^0(N_2^-) = 190 \pm 30\ kJ\ mol^{-1}$. Additional discussion on $N_2^-(g)$ may be found in the reviews by Lofthus and Krupenie (3) and Massey (4).

The adopted EA is the only tabulated value for the ground state given in a recent compilation by Rosenstock et al. (5). This compilation contains results through the end of 1973. In support of the adopted value there are two additional studies to consider. First, Massey (4) discusses qualitatively the electronic and molecular structure as follows. The ground state of N_2 has a closed shell configuration. The corresponding configuration for N_2^- is obtained by adding an additional electron in an antibonding Πg 2p orbital. According to Mulliken's rule this reduces the binding energy by about 1.25 eV (~29 kJ mol^{-1}). In contrast to $O_2^-(g)$, in the separated atom limit the ion N_2^- is not stable so that we expect the lowest level of N_2^- will be at least 1.25 eV above that for N_2 as well as occurring at a larger nuclear separation. Second, Birtwistle and Herzenberg (6) developed a theory of vibrational excitation through an intermediate state which predicts that the lowest energies of N_2^- and N_2 are separated by 1.925 eV (~44 kJ mol^{-1}). Comparable results were obtained by Krauss and Mies (7) using a variational method.

$\Delta_f H°(0\ K)$ is converted to $\Delta_f H°(298.15\ 0\ K)$ by use of JANAF (2) enthalpies $H°(0\ K) - H°(298.15\ K)$ for $N_2(g)$, $N_2^-(g)$, and $e^-(g)$. $\Delta_f H°(298.15\ K)$ should be changed by +1.481 kJ mol^{-1} if it is to be used in the ion convention which excludes the enthalpy of the electron.

Heat Capacity and Entropy

The vibrational constants are estimated based on trends implied in isoelectronic sequences involving (N_2^-, NO, and O_2^+) and (N_2 and NO^+). The value adopted for ω_e is ~35 cm^{-1} and ~65 cm^{-1} lower than those calculated by Krauss and Mies (7) and Birtwistle and Herzenberg (6), respectively.

We adopt the $r_e = 1.193 \pm 0.003$ Å value calculated by Birtwistle and Herzenberg (6) as it agrees with the value suggested from the above isoelectronic series. The calculations of Krauss and Mies (7) suggest r_e = 1.22 Å which, in our estimation, is too high. B_e is calculated from r_e while α_e is estimated from the aforementioned isoelectronic series. This α_e value is the same as one would calculate assuming a Morse potential for the electronic ground state.

No low lying excited states are assumed to exist (3, 4).

References

1. F. R. Gilmore, J. Quant. Spectry. Radiative Transfer 5, 369 (1965).
2. JANAF Thermochemcial Tables: $N_2(g)$, $e^-(g)$, N(g), and $N^-(g)$, 3-31-77.
3. A. Lofthus and P. H. Krupenie, J. Phys. Chem. Ref. Data 6, 113 (1977).
4. H. S. W. Massey, "Negative Ions," 3rd ed., Canbridge University Press, Cambridge, 1976.
5. H. M. Rosenstock, K. Draxl, B. W. Steiner, and J. T. Herron, J. Phys. Chem. Ref. Data 6, Supp. 1 (1977).
6. D. F. Birtwistle and A. Herzenberg, J. Phys. B 4, 53 (1971).
7. M. Krauss and F. H. Mies, Phyus. Rev. A 1, 1592 (1970).

Enthalpy Reference Temperature = T_r = 298.15 K — Standard State Pressure = p° = 0.1 MPa

T/K	$C_p°$ (J K⁻¹mol⁻¹)	S° (J K⁻¹mol⁻¹)	-[G°-H°(T_r)]/T (J K⁻¹mol⁻¹)	H°-H°(T_r) (kJ mol⁻¹)	$\Delta_f H°$ (kJ mol⁻¹)	$\Delta_f G°$ (kJ mol⁻¹)	Log K_f
0	0.	0.	INFINITE	-8.674	154.808		
100	29.105	172.727	230.434	-5.771			
200	29.111	192.903	207.202	-2.860			
250	29.131	199.400	205.016	-1.404			
298.15	29.194	204.536	204.536	0.	148.615	151.015	-26.457
300	29.198	204.716	204.536	0.054	148.576	151.030	-26.297
350	29.339	209.227	204.892	1.517	147.543	151.521	-22.613
400	29.567	213.158	205.685	2.989	146.516	152.160	-19.870
450	29.874	216.658	206.713	4.475	145.497	152.926	-17.751
500	30.240	219.824	207.869	5.978	144.486	153.806	-16.068
600	31.062	225.409	210.339	9.042	142.489	155.858	-13.569
700	31.895	230.260	212.845	12.190	140.516	158.243	-11.808
800	32.661	234.570	215.297	15.419	138.556	160.909	-10.506
900	33.333	238.457	217.657	18.719	136.601	163.820	-9.508
1000	33.909	241.999	219.917	22.082	134.646	166.950	-8.721
1100	34.398	245.255	222.075	25.498	132.686	170.275	-8.086
1200	34.813	248.266	224.133	28.959	130.719	173.779	-7.564
1300	35.165	251.067	226.099	32.459	128.746	177.447	-7.130
1400	35.465	253.684	227.977	35.991	126.766	181.268	-6.763
1500	35.723	256.140	229.773	39.550	124.779	185.230	-6.450
1600	35.946	258.453	231.494	43.134	122.785	189.325	-6.181
1700	36.139	260.638	233.145	46.739	120.785	193.545	-5.947
1800	36.308	262.709	234.730	50.361	118.780	197.883	-5.742
1900	36.457	264.676	236.255	54.000	116.770	202.332	-5.562
2000	36.590	266.549	237.723	57.652	114.755	206.887	-5.403
2100	36.708	268.337	239.139	61.317	112.737	211.544	-5.262
2200	36.814	270.047	240.505	64.993	110.715	216.296	-5.136
2300	36.910	271.686	241.825	68.680	108.689	221.141	-5.022
2400	36.997	273.259	243.103	72.375	106.661	226.073	-4.920
2500	37.077	274.771	244.339	76.079	104.629	231.090	-4.828
2600	37.150	276.226	245.538	79.790	102.596	236.189	-4.745
2700	37.218	277.630	246.701	83.509	100.560	241.366	-4.670
2800	37.280	278.984	247.830	87.234	98.522	246.618	-4.601
2900	37.339	280.294	248.927	90.964	96.482	251.943	-4.538
3000	37.393	281.560	249.993	94.701	94.441	257.339	-4.481
3100	37.444	282.787	251.032	98.443	92.397	262.803	-4.428
3200	37.493	283.977	252.043	102.190	90.353	268.332	-4.380
3300	37.538	285.131	253.028	105.941	88.307	273.926	-4.336
3400	37.582	286.253	253.989	109.697	86.260	279.581	-4.295
3500	37.623	287.343	254.926	113.458	84.212	285.297	-4.258
3600	37.662	288.403	255.841	117.222	82.163	291.071	-4.223
3700	37.700	289.436	256.735	120.990	80.113	296.902	-4.192
3800	37.736	290.441	257.609	124.762	78.062	302.789	-4.162
3900	37.771	291.422	258.464	128.537	76.010	308.730	-4.135
4000	37.805	292.379	259.300	132.316	73.957	314.723	-4.110
4100	37.837	293.313	260.118	136.098	71.904	320.767	-4.087
4200	37.869	294.225	260.919	139.884	69.849	326.862	-4.065
4300	37.899	295.116	261.704	143.672	67.795	333.006	-4.045
4400	37.929	295.988	262.474	147.463	65.739	339.197	-4.027
4500	37.958	296.841	263.228	151.258	63.683	345.435	-4.010
4600	37.986	297.675	263.968	155.055	61.626	351.719	-3.994
4700	38.014	298.492	264.693	158.855	59.568	358.048	-3.979
4800	38.041	299.293	265.406	162.658	57.510	364.420	-3.966
4900	38.067	300.078	266.106	166.463	55.450	370.835	-3.953
5000	38.093	300.847	266.793	170.271	53.390	377.293	-3.942
5100	38.118	301.602	267.468	174.082	51.329	383.791	-3.931
5200	38.143	302.342	268.131	177.895	49.267	390.330	-3.921
5300	38.168	303.069	268.784	181.710	47.205	396.909	-3.912
5400	38.192	303.782	269.425	185.528	45.141	403.526	-3.903
5500	38.216	304.483	270.056	189.349	43.077	410.182	-3.896
5600	38.239	305.172	270.677	193.171	41.011	416.875	-3.888
5700	38.263	305.849	271.289	196.996	38.944	423.606	-3.882
5800	38.286	306.515	271.890	200.824	36.875	430.372	-3.876
5900	38.308	307.170	272.483	204.654	34.805	437.174	-3.870
6000	38.331	307.814	273.066	208.486	32.733	444.012	-3.865

PREVIOUS: September 1977 (1 atm) CURRENT: September 1977 (1 bar)

Nitrogen, Ion (N_2^-) $N_2^-(g)$

NITROGEN OXIDE (N_2O) IDEAL GAS M_r = 44.0128

$\Delta_f H°$(0 K) = 85.48 ± 0.4 kJ mol^{-1}

S°(298.15 K) = 219.957 J K^{-1} mol^{-1} $\Delta_f H°$(298.15 K) = 82.05 ± 0.4 kJ mol^{-1}

Vibrational Frequencies and Degeneracies

ν, cm^{-1}
1276.5(1)
589.2(2)
2223.7(1)

Ground State Quantum Weight: 1 σ = 1

Point Group: $C_{\infty v}$

Bond Distances: N-N = 1.1282 Å N–O = 1.1842 Å

Bond Angle: N-N-O = 180°

Rotational Constant: B_o = 0.421019 cm^{-1}

Enthalpy of Formation

An analysis of the enthalpy of reaction data is summarized below. Carlton-Sutton et al. (1) studied the direct decomposition at pressures of 40 to 50 atmospheres and made P-V-T measurements in order to reduce the data. They obtained a similar but less certain enthalpy of formation by combustion of N_2O in CO. Flame calorimetric measurements of the combustion by Awberry and Griffiths (2) are in good agreement, but the bomb calorimetry of Fenning and Cotton (3) gives a enthalpy of formation about 0.2 kcal mol^{-1} larger. The selected enthalpy of formation is a weighted average of these values.

Source	Method	Reaction	$\Delta_r H°$(298.15 K), kcal mol^{-1}	$\Delta_f H°$(298.15 K), kcal mol^{-1}
(1)	Bomb calorimetry	$N_2O(g) \rightarrow N_2(g) + 0.5\ O_2(g)$	-19.52 ± 0.10	19.52
(1)	Bomb calorimetry	$N_2O(g) + CO(g) \rightarrow CO_2(g) + N_2(g)$	-87.10 ± 0.30	19.46*
(2)	Flame calorimetry	$N_2O(g) + CO(g) \rightarrow CO_2(g) + N_2(g)$	-87.09 ± 0.22	19.45*
(3)	Bomb calorimetry	$N_2O(g) + CO(g) \rightarrow CO_2(g) + N_2(g)$	-87.37 ± 0.12	19.73*
(3)	Bomb calorimetry	$N_2O(g) + H_2(g) \rightarrow H_2O(\ell) + N_2(g)$	-88.03 ± 0.12	19.71*

*Based on $\Delta_f H°$(298.15 K) = -68.32 (H_2O, ℓ), -94.054 (CO_2, g), -26.417 (CO, g)

Heat Capacity and Entropy

The functions are based on the rotational constants of Pliva (4) and the frequencies determined by Tidwell et al. (5). The bond distances reported by Pliva were obtained from measurements of isotopic shifts in the infrared. They include higher-order corrections for vibration-rotation interaction and are in satisfactory agreement with the infrared and microwave studies of Douglas and Moller (6) and the electron diffraction data of Schomaker and Spurr (7).

References

1. T. Carlton-Sutton, H. R. Ambler, G. W. Williams, Proc. Phys. Soc. (London) 48, 189 (1936).
2. J. H. Awberry, E. Giffiths, Proc. Roy. Soc. (London) A141, 17 (1933).
3. R. W. Fenning, F. T. Cotton, Proc. Roy. Soc. (London) A141, 1 (1933).
4. Pliva, J. Mol. Spectrosc. 12, 360 (1964).
5. Tidwell, Plyler and Benedict, J. Opt. Soc. Am. 50, 1243 (1960).
6. Douglas and Moller, J. Chem. Phys. 22, 275 (1954).
7. Schomaker and Spurr, J. Amer. Chem. Soc. 64, 1184 (1942).

Nitrogen Oxide (N_2O) $N_2O_1(g)$

Enthalpy Reference Temperature = T_r = 298.15 K Standard State Pressure = p° = 0.1 MPa

T/K	$C_p°$ (J $K^{-1}mol^{-1}$)	S° (J $K^{-1}mol^{-1}$)	$-[G°-H°(T_r)]/T$ (J $K^{-1}mol^{-1}$)	$H°-H°(T_r)$ (kJ mol^{-1})	$\Delta_f H°$ (kJ mol^{-1})	$\Delta_f G°$ (kJ mol^{-1})	Log K_f
0	0.	0.	INFINITE	-9.579	85.481	85.481	INFINITE
100	29.349	184.150	250.829	-6.668	84.038	90.269	-47.152
200	33.610	205.571	223.335	-3.553	82.787	97.018	-25.338
250	36.290	213.362	220.580	-1.805	82.351	100.629	-21.025
298.15	38.617	219.957	219.957	0.	82.048	104.179	-18.252
300	38.701	220.196	219.958	0.072	82.039	104.316	-18.163
350	40.817	226.324	220.437	2.061	81.832	108.046	-16.125
400	42.680	231.899	221.526	4.149	81.713	111.800	-14.600
450	44.338	237.023	222.967	6.325	81.665	115.565	-13.414
500	45.826	241.773	224.613	8.580	81.675	119.331	-12.466
600	48.389	250.363	228.204	13.295	81.827	126.851	-11.043
700	50.500	257.986	231.924	18.243	82.105	134.334	-10.024
800	52.241	264.846	235.618	23.383	82.467	141.772	-9.257
900	53.678	271.085	239.217	28.681	82.886	149.160	-8.657
1000	54.865	276.804	242.694	34.110	83.344	156.500	-8.175
1100	55.848	282.081	246.038	39.647	83.829	163.792	-7.778
1200	56.665	286.976	249.248	45.274	84.333	171.039	-7.445
1300	57.349	291.540	252.328	50.976	84.849	178.244	-7.162
1400	57.924	295.811	255.283	56.740	85.373	185.408	-6.918
1500	58.410	299.825	258.120	62.557	85.902	192.535	-6.705
1600	58.825	303.608	260.846	68.420	86.432	199.627	-6.517
1700	59.180	307.185	263.467	74.320	86.961	206.685	-6.351
1800	59.486	310.577	265.991	80.254	87.488	213.713	-6.202
1900	59.752	313.800	268.423	86.216	88.010	220.711	-6.068
2000	59.983	316.871	270.769	92.203	88.527	227.682	-5.946
2100	60.186	319.803	273.035	98.212	89.038	234.627	-5.836
2200	60.365	322.607	275.225	104.240	89.542	241.548	-5.735
2300	60.523	325.294	277.344	110.284	90.038	248.446	-5.642
2400	60.663	327.873	279.396	116.344	90.526	255.323	-5.557
2500	60.788	330.352	281.385	122.417	91.005	262.179	-5.478
2600	60.900	332.738	283.314	128.501	91.474	269.017	-5.405
2700	61.001	335.038	285.188	134.596	91.935	275.837	-5.336
2800	61.091	337.258	287.008	140.701	92.387	282.640	-5.273
2900	61.173	339.404	288.778	146.814	92.829	289.427	-5.213
3000	61.248	341.479	290.500	152.935	93.262	296.199	-5.157
3100	61.315	343.488	292.177	159.063	93.686	302.956	-5.105
3200	61.377	345.436	293.811	165.198	94.101	309.700	-5.055
3300	61.434	347.325	295.405	171.339	94.508	316.431	-5.009
3400	61.486	349.160	296.959	177.485	94.905	323.150	-4.965
3500	61.533	350.943	298.476	183.636	95.295	329.857	-4.923
3600	61.577	352.677	299.957	189.791	95.676	336.554	-4.883
3700	61.618	354.365	301.405	195.951	96.049	343.240	-4.846
3800	61.655	356.009	302.821	202.115	96.414	349.916	-4.810
3900	61.690	357.611	304.205	208.282	96.772	356.583	-4.776
4000	61.722	359.173	305.560	214.453	97.121	363.240	-4.743
4100	61.752	360.697	306.886	220.626	97.464	369.888	-4.712
4200	61.780	362.186	308.185	226.803	97.799	376.529	-4.683
4300	61.806	363.640	309.458	232.982	98.127	383.162	-4.654
4400	61.830	365.061	310.705	239.164	98.447	389.786	-4.627
4500	61.853	366.451	311.929	245.348	98.760	396.404	-4.601
4600	61.874	367.810	313.129	251.534	99.066	403.015	-4.576
4700	61.894	369.141	314.307	257.723	99.364	409.619	-4.552
4800	61.913	370.445	315.463	263.913	99.654	416.217	-4.529
4900	61.931	371.721	316.598	270.106	99.937	422.810	-4.507
5000	61.947	372.973	317.713	276.299	100.211	429.396	-4.486
5100	61.963	374.200	318.808	282.495	100.476	435.977	-4.465
5200	61.978	375.403	319.885	288.692	100.732	442.553	-4.445
5300	61.992	376.584	320.944	294.891	100.980	449.124	-4.426
5400	62.005	377.743	321.985	301.090	101.218	455.690	-4.408
5500	62.018	378.880	323.009	307.292	101.445	462.252	-4.390
5600	62.030	379.998	324.017	313.494	101.662	468.811	-4.373
5700	62.041	381.096	325.009	319.697	101.868	475.365	-4.356
5800	62.052	382.175	325.985	325.902	102.061	481.916	-4.340
5900	62.062	383.236	326.946	332.108	102.242	488.463	-4.325
6000	62.072	384.279	327.893	338.314	102.409	495.009	-4.309

PREVIOUS: December 1964 (1 atm) CURRENT: December 1964 (1 bar)

Nitrogen Oxide (N_2O) $N_2O_1(g)$

NITROGEN OXIDE, ION (N_2O^+) IDEAL GAS M_r = 44.012251

S°(298.15 K) = 233.856 ± 0.04 J K^{-1} mol^{-1}

$\Delta_f H°$(0 K) = 1329.592 ± 0.63 kJ mol^{-1}
$\Delta_f H°$(298.15 K) = 1333.399 ± 0.63 kJ mol^{-1}

Electronic Levels and Quantum Weights

State	ε_i, cm^{-1}	g_i
$X^2\pi_{3/2}$	0	2
$X^2\pi_{1/2}$	132.4	2
$A^2\Sigma^+$	28229	2

Vibrational Frequencies and Degeneracies

ν, cm^{-1}
1736.6(1)
461.2(2)
1126.4(1)

Point Group: $C_{\infty v}$ σ = 1
Bond Distances: N-N = 1.155 Å N-O = 1.185 Å
Bond Angle: N-N-O = 180°
Rotational Constant: B_0 = 0.411407 cm^{-1}

Enthalpy of Formation

The ionization potential of 104000 cm^{-1} (297.35 kcal mol^{-1}) has been obtained by Tanaka et al. (1) from a Rydberg series terminating on $X^2\pi_{3/2}$. Other Rydberg series terminating on $X^2\pi_{1/2}$ and $A^2\Sigma^+$ were also obtained and confirm the adopted value. This value in conjunction with the enthalpy of formation of N_2O(g) (2) was used to calculate the adopted enthalpy of formation.

Heat Capacity and Entropy

The electronic, vibrational and rotational constants are those reported by Herzberg (3) from the work of Callomon and Creutzberg (4).

References

1. Y. Tanaka, A. S. Jura and F. L. LeBlanc, J. Chem. Phys. 28, 350 (1958).
2. JANAF Thermochemical Tables: N_2O(g), 12-31-64.
3. G. Herzberg, "Electronic Spectra of Polyatomic Molecules," D. Von Nostrand Co., Inc., New York, 1966.
4. J. H. Callomon and F. C. Creutzberg, Proc. Int. Conf. Spectrosc., 1st Bombay 1, 171 (1967).

Nitrogen Oxide, Ion (N_2O^+) $N_2O_1^+$(g)

Enthalpy Reference Temperature = T_r = 298.15 K Standard State Pressure = p° = 0.1 MPa

	J $K^{-1}mol^{-1}$			kJ mol^{-1}			
T/K	$C_p°$	S°	$-[G°-H°(T_r)]/T$	$H°-H°(T_r)$	$\Delta_f H°$	$\Delta_f G°$	Log K_f
0	0.	0.	INFINITE	-10.622	1329.592		
100	33.467	193.445	268.380	-7.494			
200	37.926	217.872	237.588	-3.943			
250	40.258	226.590	234.540	-1.988			
298.15	42.262	233.856	233.856	0.	1333.399	1345.131	-235.661
300	42.335	234.117	233.856	0.078	1333.435	1345.203	-234.221
350	44.212	240.787	234.379	2.243	1334.443	1347.086	-201.042
400	45.924	246.804	235.562	4.497	1335.529	1348.819	-176.138
450	47.485	252.304	237.121	6.833	1336.680	1350.412	-156.752
500	48.903	257.382	238.896	9.243	1337.885	1351.874	-141.229
600	51.329	266.520	242.756	14.259	1340.416	1354.436	-117.914
700	53.267	274.584	246.738	19.492	1343.058	1356.565	-101.228
800	54.801	281.801	250.678	24.899	1345.765	1358.310	-88.689
900	56.016	288.329	254.504	30.442	1348.508	1359.714	-78.916
1000	56.983	294.283	258.189	36.094	1351.267	1360.811	-71.081
1100	57.759	299.751	261.722	41.832	1354.032	1361.632	-64.659
1200	58.387	304.805	265.104	47.641	1356.796	1362.201	-59.295
1300	58.900	309.499	268.341	53.506	1359.554	1362.539	-54.747
1400	59.324	313.880	271.439	59.418	1362.305	1362.666	-50.842
1500	59.677	317.985	274.407	65.368	1365.045	1362.596	-47.450
1600	59.973	321.847	277.252	71.351	1367.774	1362.343	-44.476
1700	60.225	325.490	279.984	77.361	1370.491	1361.921	-41.847
1800	60.439	328.939	282.608	83.395	1373.196	1361.338	-39.505
1900	60.623	332.212	285.134	89.448	1375.889	1360.606	-37.406
2000	60.783	335.325	287.566	95.519	1378.568	1359.732	-35.513
2100	60.922	338.294	289.912	101.604	1381.234	1358.725	-33.796
2200	61.043	341.131	292.176	107.702	1383.887	1357.591	-32.233
2300	61.150	343.847	294.364	113.812	1386.527	1356.336	-30.803
2400	61.245	346.452	296.480	119.932	1389.154	1354.967	-29.490
2500	61.329	348.954	298.529	126.061	1391.767	1353.489	-28.280
2600	61.404	351.361	300.515	132.197	1394.368	1351.907	-27.160
2700	61.471	353.679	302.442	138.341	1396.956	1350.224	-26.122
2800	61.532	355.916	304.312	144.491	1399.532	1348.447	-25.156
2900	61.587	358.076	306.129	150.647	1402.095	1346.576	-24.254
3000	61.636	360.165	307.895	156.809	1404.647	1344.619	-23.412
3100	61.681	362.187	309.614	162.975	1407.187	1342.576	-22.622
3200	61.723	364.146	311.288	169.145	1409.717	1340.451	-21.881
3300	61.761	366.046	312.919	175.319	1412.235	1338.247	-21.183
3400	61.796	367.890	314.508	181.497	1414.743	1335.968	-20.525
3500	61.828	369.682	316.059	187.678	1417.242	1333.614	-19.903
3600	61.859	371.424	317.573	193.862	1419.730	1331.190	-19.315
3700	61.887	373.119	319.052	200.050	1422.209	1328.696	-18.758
3800	61.914	374.770	320.496	206.240	1424.680	1326.136	-18.229
3900	61.939	376.378	321.908	212.433	1427.141	1323.511	-17.726
4000	61.964	377.947	323.290	218.628	1429.594	1320.822	-17.248
4100	61.987	379.477	324.642	224.825	1432.039	1318.073	-16.792
4200	62.009	380.971	325.965	231.025	1434.476	1315.263	-16.358
4300	62.031	382.431	327.261	237.227	1436.905	1312.397	-15.942
4400	62.053	383.857	328.532	243.431	1439.326	1309.473	-15.545
4500	62.073	385.252	329.777	249.638	1441.740	1306.494	-15.165
4600	62.094	386.616	330.997	255.846	1444.147	1303.462	-14.801
4700	62.115	387.952	332.195	262.056	1446.545	1300.378	-14.452
4800	62.135	389.260	333.370	268.269	1448.936	1297.243	-14.117
4900	62.155	390.541	334.524	274.483	1451.319	1294.058	-13.795
5000	62.176	391.797	335.657	280.700	1453.695	1290.824	-13.485
5100	62.197	393.028	336.770	286.919	1456.062	1287.544	-13.187
5200	62.217	394.236	337.863	293.139	1458.420	1284.216	-12.900
5300	62.238	395.422	338.938	299.362	1460.771	1280.844	-12.623
5400	62.260	396.585	339.995	305.587	1463.112	1277.427	-12.357
5500	62.281	397.728	341.034	311.814	1465.444	1273.966	-12.099
5600	62.303	398.850	342.057	318.043	1467.767	1270.464	-11.850
5700	62.326	399.953	343.063	324.275	1470.079	1266.920	-11.610
5800	62.348	401.037	344.053	330.508	1472.380	1263.336	-11.378
5900	62.371	402.103	345.028	336.744	1474.670	1259.712	-11.153
6000	62.395	403.152	345.988	342.983	1476.947	1256.050	-10.935

PREVIOUS: December 1970 (1 atm) CURRENT: December 1970 (1 bar)

Nitrogen Oxide, Ion (N_2O^+) $N_2O_1^+$(g)

NITROGEN OXIDE (N_2O_3) IDEAL GAS M_r = 76.0116

$S°(298.15\ K) = 308.54 \pm 2.1\ J\ K^{-1}\ mol^{-1}$

$\Delta_f H°(0\ K)$ = Unknown
$\Delta_f H°(298.15\ K) = 82.843 \pm 0.8\ kJ\ mol^{-1}$

Vibrational Frequencies and Degeneracies

ν, cm^{-1}	ν, cm^{-1}	ν, cm^{-1}
1863(1)	783(1)	313(1)
1589(1)	627(1)	253(1)
1297(1)	407(1)	[Rotation]

Ground State Quantum Weight: 1 $\sigma = 1$
Point Group: C_s
Bond Distances: N-N = 2.08 Å N-O (nitroso) = 1.12 Å
N-O(nitro) = 1.18 Å
Bond Angles: O-N-N (nitroso) = 110° O-N-O = 134°
Product of Moments of Inertia: $I_AI_BI_C = 4.6201 \times 10^{-114}\ g^3\ cm^6$

Enthalpy of Formation

Equilibria for the reaction $N_2O_3(g) \rightarrow NO(g) + NO_2(g)$ have been studied by Beattie and Bell (1), Verhoek and Daniels (2), and Abel and Proisl (3). The results of Beattie and Bell are the most extensive, but they depend on the early analysis of Giauque and Kemp (4) for the simultaneous equilibrium $N_2O_4(g) \rightarrow 2\ NO_2(g)$. The JANAF analysis of the tetroxide dissociation has been used to recalculate the data of Beattie and Bell. Non-ideality of N_2O_4 and NO_2 was allowed for by use of the equations of state of Giauque and Kemp, while non-ideality of NO and N_2O_3 was removed by extrapolation of the equilibrium constants to zero pressure in a manner similar to that of Beattie and Bell. The data of Verhoek and Daniels and of Abel and Proisl have not been recalculated. 2nd and 3rd law analyses of the results are summarized below. The enthalpy of formation was obtained from $\Delta_r H°(298.15\ K) = 9.70\ kcal\ mol^{-1}$.

Source	Temp. Range	T/K	$\Delta_r S°$, cal K^{-1} mol^{-1} 2nd law	3rd law	$\Delta_r H°(298.15\ K)$, kcal mol^{-1} 2nd law	3rd law
(1)[a]	278-318	298	33.25 ± 0.35	--	9.527 ± 0.096	--
(1)[b]	278-318	298	33.04 ± 0.31	33.77	9.476 ± 0.093	9.695
(1)[c]	278-318	298	33.56 ± 0.52	33.77	9.642 ± 0.155	9.706
(2)	298-318	308	38.8 ± 1.9	33.78	11.1 ± 0.6	9.59
(3)	291-308	294	32.9	33.77	9.60	9.86

a) given by authors b) from author's K_p values c) from recalculated K_p values

Heat Capacity and Entropy

The functions are similar to those calculated by Hisatsune (5) based on the analysis of Devlin and Hisatsune (6) of the Raman and infrared spectra in terms of Urey - Bradley force constants. Devlin and Hisatsune used observed isotopic shifts in the spectra along with force constants transferred from nitrogen oxides and oxyhalides in order to select the N-N bond distance and the O-N-N bond angle.

The unobserved mode was assumed to be free internal rotation around the N-N bond, leading to $S°(298.15\ K) = 73.91\ cal\ K^{-1}\ mol^{-1}$ which is in good agreement with $74.13 \pm 0.52\ cal\ K^{-1}\ mol^{-1}$ obtained from the recalculated 2nd law of Beattie and Bell. The principal moments of inertia are $I_A = 22.694 \times 10^{-39}$, $I_B = 29.577 \times 10^{-39}$, and $I_C = 6.883 \times 10^{-39}\ g\ cm^2$, while the reduced moment for internal rotation was calculated as $I_r = 0.990 \times 10^{-39}\ g\ cm^2$.

References

1. Beattie and Bell, J. Chem. Soc. 1957, 1681 (1957).
2. Verhoek and Daniels, J. Amer. Chem. Soc. 53, 1250 (1931).
3. Abel and Proisl, Z. Elektrochem. 35, 712 (1929).
4. Giauque and Kemp, J. Chem. Phys. 6, 40 (1938).
5. I. C. Hisatsune, J. Phys. Chem. 65, 2249 (1961).
6. Devlin and Hisatsune, Spectrochim. Acta 17, 218 (1961).

Nitrogen Oxide (N_2O_3) $N_2O_3(g)$

Enthalpy Reference Temperature = T_r = 298.15 K Standard State Pressure = p° = 0.1 MPa

T/K	$C_p°$ (J $K^{-1}mol^{-1}$)	S°	$-[G°-H°(T_r)]/T$	$H°-H°(T_r)$ (kJ mol^{-1})	$\Delta_f H°$	$\Delta_f G°$	Log K_f
0	4.157	0.	INFINITE	-14.920	89.618	89.618	INFINITE
100	43.290	249.899	360.299	-11.040	86.239	103.226	-53.920
200	56.497	284.180	314.268	-6.018	83.984	121.191	-31.652
250	61.524	297.344	309.597	-3.063	83.297	130.577	-27.283
298.15	65.615	308.539	308.539	0.	82.843	139.727	-24.480
300	65.761	308.946	308.541	0.122	82.829	140.080	-24.390
350	69.450	319.366	309.355	3.504	82.540	149.647	-22.334
400	72.738	328.858	311.208	7.060	82.394	159.246	-20.795
450	75.698	337.599	313.661	10.772	82.365	168.855	-19.600
500	78.370	345.715	316.466	14.625	82.431	178.462	-18.644
600	82.931	360.422	322.592	22.698	82.781	197.640	-17.206
700	86.579	373.491	328.947	31.181	83.339	216.741	-16.173
800	89.478	385.248	335.262	39.989	84.033	235.752	-15.393
900	91.780	395.925	341.418	49.056	84.815	254.671	-14.781
1000	93.617	405.694	347.364	58.330	85.656	273.499	-14.286
1100	95.095	414.688	353.081	67.768	86.533	292.242	-13.877
1200	96.294	423.016	358.567	77.339	87.432	310.903	-13.533
1300	97.276	430.764	363.826	87.019	88.344	329.489	-13.239
1400	98.087	438.003	368.868	96.789	89.259	348.004	-12.984
1500	98.764	444.794	373.706	106.632	90.173	366.453	-12.761
1600	99.333	451.187	378.351	116.538	91.079	384.843	-12.564
1700	99.816	457.224	382.815	126.496	91.974	403.175	-12.388
1800	100.228	462.941	387.109	136.499	92.854	421.455	-12.230
1900	100.583	468.370	391.244	146.540	93.716	439.688	-12.088
2000	100.890	473.537	395.230	156.614	94.557	457.875	-11.958
2100	101.157	478.466	399.078	166.716	95.377	476.021	-11.840
2200	101.391	483.178	402.794	176.844	96.172	494.129	-11.732
2300	101.598	487.689	406.388	186.994	96.943	512.200	-11.632
2400	101.780	492.017	409.866	197.163	97.687	530.238	-11.540
2500	101.942	496.175	413.236	207.349	98.405	548.246	-11.455
2600	102.087	500.177	416.503	217.551	99.095	566.226	-11.376
2700	102.216	504.032	419.674	227.766	99.759	584.180	-11.302
2800	102.333	507.751	422.754	237.993	100.396	602.110	-11.233
2900	102.438	511.344	425.747	248.232	101.006	620.017	-11.168
3000	102.533	514.819	428.658	258.481	101.589	637.904	-11.107
3100	102.619	518.182	431.492	268.738	102.147	655.773	-11.050
3200	102.698	521.441	434.253	279.004	102.680	673.622	-10.996
3300	102.770	524.603	436.943	289.278	103.188	691.456	-10.945
3400	102.836	527.672	439.566	299.558	103.672	709.275	-10.897
3500	102.896	530.653	442.126	309.845	104.133	727.080	-10.851
3600	102.952	533.553	444.626	320.137	104.572	744.872	-10.808
3700	103.003	536.374	447.068	330.435	104.989	762.652	-10.767
3800	103.050	539.122	449.454	340.738	105.385	780.423	-10.728
3900	103.094	541.799	451.788	351.045	105.760	798.182	-10.690
4000	103.135	544.410	454.071	361.356	106.115	815.931	-10.655
4100	103.173	546.957	456.305	371.672	106.451	833.673	-10.621
4200	103.208	549.444	458.494	381.991	106.767	851.406	-10.589
4300	103.240	551.873	460.637	392.313	107.064	869.133	-10.558
4400	103.271	554.246	462.738	402.639	107.342	886.851	-10.528
4500	103.300	556.568	464.797	412.967	107.602	904.564	-10.500
4600	103.326	558.838	466.817	423.299	107.842	922.271	-10.473
4700	103.352	561.061	468.798	433.633	108.063	939.974	-10.447
4800	103.375	563.237	470.743	443.969	108.265	957.672	-10.422
4900	103.397	565.369	472.653	454.308	108.446	975.367	-10.398
5000	103.418	567.458	474.528	464.648	108.606	993.057	-10.374
5100	103.438	569.506	476.370	474.991	108.744	1010.746	-10.352
5200	103.456	571.515	478.181	485.336	108.860	1028.429	-10.331
5300	103.474	573.485	479.960	495.682	108.952	1046.114	-10.310
5400	103.491	575.420	481.710	506.031	109.020	1063.794	-10.290
5500	103.506	577.319	483.432	516.380	109.062	1081.474	-10.271
5600	103.521	579.184	485.125	526.732	109.077	1099.156	-10.253
5700	103.535	581.016	486.791	537.085	109.063	1116.835	-10.235
5800	103.549	582.817	488.431	547.439	109.018	1134.517	-10.217
5900	103.561	584.587	490.046	557.794	108.942	1152.196	-10.201
6000	103.574	586.328	491.636	568.151	108.831	1169.881	-10.185

PREVIOUS: December 1964 (1 atm) CURRENT: December 1964 (1 bar)

Nitrogen Oxide (N_2O_3) $N_2O_3(g)$

NITROGEN OXIDE (N_2O_4) CRYSTAL M_r = 92.0110

$S°(298.15\ K) = 150.290\ J\ K^{-1}\ mol^{-1}$

$T_{fus} = 261.95 \pm 0.05\ K$

$\Delta_f H°(0\ K) = -30.399 \pm 1.7\ kJ\ mol^{-1}$

$\Delta_f H°(298.15\ K) = -35.046 \pm 1.7\ kJ\ mol^{-1}$

$\Delta_{fus} H° = 14.652 \pm 0.013\ kJ\ mol^{-1}$

Enthalpy of Formation

The enthalpy of formation was calculated from that of the liquid using the measured enthalpy of melting and the enthalpy of the crystal and liquid.

Heat Capacity and Entropy

Giauque and Kemp (1) measured the low temperature heat capacity from 15-295 K. The entropy was obtained by integration of these data using $S°(15\ K) = 0.34\ cal\ K^{-1}\ mol^{-1}$.

Fusion Data

The temperature and enthalpy of fusion were also measured by Giauque and Kemp (1).

Reference

1. W. F. Giauque and J. D. Kemp, J. Chem. Phys. 6, 40 (1938).

Nitrogen Oxide (N_2O_4) $N_2O_4(cr)$

Enthalpy Reference Temperature = T_r = 298.15 K Standard State Pressure = p° = 0.1 MPa

T/K	$C_p°$ (J K⁻¹mol⁻¹)	S° (J K⁻¹mol⁻¹)	$-[G°-H°(T_r)]/T$ (J K⁻¹mol⁻¹)	$H°-H°(T_r)$ (kJ mol⁻¹)	$\Delta_f H°$ (kJ mol⁻¹)	$\Delta_f G°$ (kJ mol⁻¹)	Log K_f
0	0.	0.	INFINITE	-21.390	-30.399	-30.399	INFINITE
100	60.702	56.130	237.659	-18.153	-35.874	9.156	-4.783
200	91.701	108.002	160.472	-10.494	-36.947	54.843	-14.324
261.950	110.918	135.224	151.325	-4.218	------ CRYSTAL <--> LIQUID -----		
298.15	122.103	150.290	150.290	0.	-35.046	99.603	-17.450
300	122.675	151.047	150.292	0.226	-34.982	100.438	-17.488
400	153.636	190.991	155.526	14.186	-29.882	144.891	-18.921
500	167.360	227.068	166.310	30.379	-22.746	187.782	-19.617

PREVIOUS: June 1961 CURRENT: September 1964

Nitrogen Oxide (N_2O_4) $N_2O_4(cr)$

NITROGEN OXIDE (N_2O_4) LIQUID M_r = 92.0110

$S°(298.15\ K) = 209.198\ J\ K^{-1}\ mol^{-1}$

$T_{fus} = 261.95 \pm 0.05\ K$

$\Delta_f H°(298.15\ K) = -19.564 \pm 1.7\ kJ\ mol^{-1}$

$\Delta_{fus} H° = 14.652 \pm 0.013\ kJ\ mol^{-1}$

Enthalpy of Formation

The enthalpy of formation was calculated from that of the gas using the calorimetrically measured enthalpy of vaporization to the equilibrium mixture, in conjunction with the fraction of NO_2 dimerized α = 0.161.

Heat Capacity and Entropy

Giauque and Kemp (1) measured the low temperature heat capacity from 15-295 K. Above 295 K the heat capacity was smoothly extrapolated. S°(ℓ, 298.15 K) is calculated from $C_p°$.

Fusion and Vaporization Data

Giauque and Kemp (1) measured the temperature and enthalpy of fusion and also the boiling point and the calorimetric value of the enthalpy of vaporization to the equilibrium mixture. The enthalpy of vaporization (f = 1 atm) to a hypothetical pure N_2O_4 gas at 294.25 K was calculated to be 6.91 kcal mol^{-1}.

References

1. W. F. Giauque and J. D. Kemp, J. Chem. Phys. 6, 40 (1938).

Nitrogen Oxide (N_2O_4) N_2O_4(l)

Enthalpy Reference Temperature = T_r = 298.15 K Standard State Pressure = p° = 0.1 MPa

T/K	$C_p°$ (J K⁻¹mol⁻¹)	S° (J K⁻¹mol⁻¹)	$-[G°-H°(T_r)]/T$ (J K⁻¹mol⁻¹)	$H°-H°(T_r)$ (kJ mol⁻¹)	$\Delta_f H°$ (kJ mol⁻¹)	$\Delta_f G°$ (kJ mol⁻¹)	Log K_f
0							
100	112.884	72.843	325.324	-25.248	-27.487	15.871	-8.290
200	127.528	155.582	221.720	-13.227	-24.199	58.075	-15.168
261.950	136.843	191.160	210.426	-5.047	------ CRYSTAL <--> LIQUID -----		
298.15	142.509	209.198	209.198	0.	-19.564	97.521	-17.085
300	142.842	210.081	209.201	0.264	-19.463	98.247	-17.106
400	160.833	253.638	215.019	15.448	-13.139	136.576	-17.835
500	178.824	291.459	226.598	32.431	-5.213	173.120	-18.086

PREVIOUS: September 1964 CURRENT: December 1964

Nitrogen Oxide (N_2O_4) N_2O_4(l)

NITROGEN OXIDE (N_2O_4) CRYSTAL–LIQUID M_r = 92.0110 Nitrogen Oxide (N_2O_4) N_2O_4(cr,l)

0 to 261.95 K crystal
above 261.95 K liquid

Refer to the individual tables for details.

Enthalpy Reference Temperature = T_r = 298.15 K   Standard State Pressure = p° = 0.1 MPa

T/K	C_p° (J K⁻¹mol⁻¹)	S° (J K⁻¹mol⁻¹)	$-[G°-H°(T_r)]/T$ (J K⁻¹mol⁻¹)	$H°-H°(T_r)$ (kJ mol⁻¹)	$\Delta_f H°$ (kJ mol⁻¹)	$\Delta_f G°$ (kJ mol⁻¹)	Log K_f
0	0.	0.	INFINITE	-36.872	-30.399	-30.399	INFINITE
100	60.702	56.130	392.476	-33.635	-35.874	9.156	-4.783
200	91.701	108.002	237.881	-25.976	-36.947	54.843	-14.324
261.950	110.918	135.224	210.426	-19.699	CRYSTAL <--> LIQUID TRANSITION		
261.950	136.843	191.160	210.426	-5.047			
298.15	142.509	209.198	209.198	0.	-19.564	97.521	-17.085
300	142.842	210.081	209.201	0.264	-19.463	98.247	-17.106
400	160.833	253.638	215.019	15.448	-13.139	136.576	-17.835
500	178.824	291.459	226.598	32.431	-5.213	173.120	-18.086

PREVIOUS: CURRENT: December 1964

Nitrogen Oxide (N_2O_4) N_2O_4(cr,l)

NITROGEN OXIDE (N_2O_4) IDEAL GAS M_r = 92.0110

S°(298.15 K) = 304.376 J K^{-1} mol^{-1}

$\Delta_f H°$(0 K) = 18.72 ± 1.7 kJ mol^{-1}
$\Delta_f H°$(298.15 K) = 9.08 ± 1.7 kJ mol^{-1}

Vibrational Frequencies and Degeneracies

ν, cm^{-1}	ν, cm^{-1}	ν, cm^{-1}
1373 (1)	1710(1)	1748(1)
823 (1)	480(1)	385(1)
260 (1)	430(1)	1261(1)
[50](1)	675(1)	750(1)

Ground State Quantum Weight: 1 σ = 4
Point Group: D_{2h}
Bond Distances: N–N = 1.750 Å N–O = 1.180 Å
Bond Angle: O–N–O = 133.7°
Product of the Moments of Inertia: $I_A I_B I_C$ = 9.935917 x 10^{-114} g^3 cm^6

Enthalpy of Formation

Giauque and Kemp (1) calculated idealized equilibrium constants for the reaction N_2O_4(g) = 2 NO_2(g) from the work of Bodenstein and Boes (2), Verhoek and Daniels (3) and Wourtzel (4). The 2nd and 3rd law analysis of these equilibrium constants has been repeated using more recent functions. The results are shown below.

Source	$\Delta_f H°$(298.15 K), kcal mol^{-1} 2nd law	3rd law
(2)	13.634 ± 0.015*	13.644
(3)	14.429 ± 0.126	13.672
(4)	13.599 ± 0.019**	13.657

*Ten points rejected because of statistically significant scatter.
**One point rejected.

The 2nd law value of Verhoek and Daniels (3) was not considered significant due to the extremely small temperature range and the use of only three temperatures. The adopted value was 13.65 ± 0.03 kcal mol^{-1} which leads to 2.17 kcal mol^{-1} for the enthalpy of formation of N_2O_4(g).

Heat Capacity and Entropy

The structural parameters were taken from the electron diffraction measurements of Smith and Hedberg (5). The vibrational frequencies are those chosen by Hitsatsune et al. (6). The torsional frequency of 50 cm^{-1} was estimated in order to bring the entropy of the gas into agreement with that determined by Giauque and Kemp (1) from low temperature heat capacities and heats of fusion and vaporization. The principal moments of inertia are: I_A = 12.5103 x 10^{-39}, I_B = 22.6126 x 10^{-39} and I_C = 35.1229 x 10^{-39} g cm^2.

References

1. W. F. Guauque and J. D. Kemp, J. Chem. Phys. 6, 40 (1938).
2. Bodenstein and Boes, Zeits. fur physik. Chem. 100, 75 (1922).
3. Verhoek and Daniels, J. Amer. Chem. Soc. 53, 1250 (1931).
4. Wourtzel, Compt. Rend. 169, 1397 (1919).
5. D. W. Smith and K. Hedberg, J. Chem. Phys. 25, 1282 (1956).
6. I. C. Hitsatsune, J. P. Devlin and V. Wada, J. Chem. Phys. 33, 714 (1960).

Nitrogen Oxide (N_2O_4) N_2O_4(g)

Enthalpy Reference Temperature = T_r = 298.15 K Standard State Pressure = p° = 0.1 MPa

T/K	$C_p°$ (J $K^{-1}mol^{-1}$)	S° (J $K^{-1}mol^{-1}$)	–[G°–H°(T_r)]/T (J $K^{-1}mol^{-1}$)	H°–H°(T_r) (kJ mol^{-1})	$\Delta_f H°$ (kJ mol^{-1})	$\Delta_f G°$ (kJ mol^{-1})	Log K_f
0	0.	0.	INFINITE	-16.398	18.718	18.718	INFINITE
100	46.262	239.166	363.098	-12.393	14.011	40.737	-21.279
200	63.206	276.398	311.015	-6.923	10.749	68.860	-17.984
250	70.825	291.339	305.613	-3.568	9.732	83.513	-17.449
298.15	77.256	304.376	304.376	0.	9.079	97.787	-17.132
300	77.487	304.854	304.377	0.143	9.060	98.338	-17.122
350	83.340	317.248	305.343	4.167	8.673	113.251	-16.902
400	88.521	328.722	307.557	8.466	8.523	128.204	-16.742
450	93.121	339.419	310.509	13.009	8.567	143.163	-16.618
500	97.204	349.446	313.907	17.769	8.769	158.109	-16.518
600	104.012	367.795	321.388	27.844	9.542	187.911	-16.359
700	109.314	384.244	329.213	38.522	10.667	217.555	-16.234
800	113.439	399.122	337.037	49.668	12.030	247.019	-16.129
900	116.665	412.677	344.700	61.180	13.554	276.303	-16.036
1000	119.208	425.106	352.127	72.978	15.189	305.410	-15.953
1100	121.235	436.566	359.290	85.004	16.899	334.350	-15.877
1200	122.867	447.187	366.177	97.212	18.661	363.132	-15.807
1300	124.196	457.076	372.794	109.567	20.456	391.765	-15.741
1400	125.290	466.321	379.148	122.044	22.272	420.259	-15.680
1500	126.198	474.997	385.251	134.619	24.097	448.623	-15.622
1600	126.959	483.167	391.118	147.278	25.922	476.865	-15.568
1700	127.603	490.884	396.762	160.007	27.742	504.994	-15.517
1800	128.152	498.193	402.196	172.796	29.550	533.014	-15.468
1900	128.623	505.135	407.432	185.635	31.340	560.936	-15.421
2000	129.030	511.743	412.484	198.518	33.110	588.764	-15.377
2100	129.385	518.047	417.362	211.439	34.855	616.503	-15.335
2200	129.694	524.074	422.077	224.394	36.573	644.162	-15.294
2300	129.967	529.845	426.638	237.377	38.262	671.741	-15.256
2400	130.208	535.381	431.054	250.386	39.920	699.247	-15.219
2500	130.421	540.701	435.334	263.418	41.545	726.685	-15.183
2600	130.612	545.820	439.486	276.469	43.138	754.059	-15.149
2700	130.783	550.753	443.516	289.539	44.698	781.373	-15.117
2800	130.936	555.512	447.431	302.625	46.224	808.630	-15.085
2900	131.074	560.109	451.238	315.726	47.718	835.831	-15.055
3000	131.200	564.555	454.942	328.840	49.178	862.983	-15.026
3100	131.313	568.859	458.547	341.966	50.606	890.087	-14.998
3200	131.417	573.029	462.060	355.102	52.002	917.144	-14.971
3300	131.511	577.075	465.484	368.249	53.368	944.157	-14.945
3400	131.597	581.002	468.824	381.404	54.703	971.132	-14.920
3500	131.677	584.818	472.084	394.568	56.010	998.066	-14.895
3600	131.750	588.528	475.268	407.739	57.288	1024.964	-14.872
3700	131.817	592.139	478.378	420.918	58.538	1051.826	-14.849
3800	131.879	595.655	481.418	434.102	59.762	1078.657	-14.827
3900	131.936	599.082	484.391	447.293	60.960	1105.455	-14.806
4000	131.990	602.423	487.300	460.490	62.132	1132.220	-14.785
4100	132.039	605.683	490.148	473.691	63.279	1158.958	-14.765
4200	132.085	608.865	492.937	486.897	64.402	1185.669	-14.746
4300	132.128	611.973	495.669	500.108	65.501	1212.353	-14.727
4400	132.168	615.011	498.347	513.323	66.575	1239.011	-14.709
4500	132.206	617.982	500.973	526.542	67.626	1265.644	-14.691
4600	132.241	620.888	503.548	539.764	68.653	1292.255	-14.674
4700	132.274	623.733	506.075	552.990	69.654	1318.845	-14.657
4800	132.305	626.518	508.555	566.219	70.630	1345.413	-14.641
4900	132.334	629.246	510.991	579.451	71.581	1371.962	-14.625
5000	132.361	631.920	513.383	592.685	72.504	1398.491	-14.610
5100	132.387	634.541	515.733	605.923	73.400	1425.003	-14.595
5200	132.411	637.112	518.042	619.163	74.267	1451.494	-14.580
5300	132.434	639.635	520.313	632.405	75.104	1477.973	-14.566
5400	132.456	642.110	522.545	645.649	75.908	1504.434	-14.553
5500	132.476	644.541	524.742	658.896	76.680	1530.880	-14.539
5600	132.496	646.928	526.902	672.145	77.417	1557.315	-14.526
5700	132.514	649.273	529.029	685.395	78.116	1583.734	-14.513
5800	132.532	651.578	531.122	698.648	78.776	1610.145	-14.501
5900	132.549	653.844	533.183	711.902	79.394	1636.539	-14.489
6000	132.564	656.072	535.212	725.157	79.968	1662.928	-14.477

PREVIOUS: September 1964 (1 atm) CURRENT: September 1964 (1 bar)

Nitrogen Oxide (N_2O_4) N_2O_4(g)

NITROGEN OXIDE (N_2O_5) IDEAL GAS M_r = 108.0104

$S°(298.15\ K) = 346.548 \pm 4.2\ J\ K^{-1}\ mol^{-1}$

$\Delta_f H°(0\ K)$ = Unknown

$\Delta_f H°(298.15\ K) = 11.297 \pm 1.3\ kJ\ mol^{-1}$

Vibrational Frequencies and Degeneracies

ν, cm^{-1}	ν, cm^{-1}	ν, cm^{-1}	ν, cm^{-1}	ν, cm^{-1}
1728(1)	614 (1)	[614](1)	[Rotation)	860(1)
1338(1)	353 (1)	[Rotation]	1728(1)	743(1)
743(1)	[85](1)	577 (1)	1247(1)	353(1)

Ground State Quantum Weight: 1 σ = 2

Point Group: C_{2v}

Bond Distances: O-N = 1.21 Å N-O' = 1.46 Å

Bond Angles: O-N-O = 134° N-O'-N = 95°

O-N-O planes perpendicular to N-O'-N plane

Product of the Moments of Inertia: $I_A I_B I_C = 2.066 \times 10^{-113}\ g^3\ cm^6$

Enthalpy of Formation

The enthalpy of formation was obtained by recalculation of the calorimetric data of Ray and Ogg for the reaction of N_2O_5(g) with NO. The stoichiometry of the reaction products, originally based on the results for N_2O_4 + 2 NO_2 from Giauque and Kemp (1) was based on the JANAF analysis of this simultaneous equilibrium. A correction of 0.29 kcal mol^{-1} was made in the enthalpy of reaction because of the additional formation of N_2O_3 from NO and NO_2.

Confirmation of the enthalpy of formation was obtained by combination of the enthalpies of hydrolysis and sublimation of N_2O_5(cr). The results are summarized below. 2nd law values are given for the enthalpies of sublimation, but 3rd law analyses suggest that the value from Russ and Pokorny (4) is more reliable. Most of the uncertainty in this path, however, arises from the enthalpy of hydrolysis.

Source	Method	Reaction	$\Delta_r H°$(298.15 K) kcal mol^{-1}	$\Delta_f H°$(298.15 K) kcal mol^{-1}
(2)	Calorimetric	$N_2O_5(g) + NO(g) \rightarrow 0.724\ NO_2(g) + 1.138\ N_2O_4(g)$	-16.05 ± 0.20	2.67
(3)	Calorimetric	$N_2O_5(cr) + H_2O(\ell) \rightarrow 2\ HNO_3(aq)$	-20.2 ± 0.5	--
(4)	Vapor Pressure	$N_2O_5(cr) \rightarrow N_2O_5(g)$	13.37 ± 0.06	2.79*
(5)	Vapor Pressure	$N_2O_5(cr) \rightarrow N_2O_5(g)$	13.25 ± 0.12	2.67*

*Using reaction 2 and $\Delta_f H°(298.15\ K) = -68.32(H_2O,\ \ell),\ -49.55(HNO_3,\ aq)$.

Heat Capacity and Entropy

Ray and Ogg (6) reported S°(298.15 K) = 85 ± 0.5 cal K^{-1} mol^{-1} from combination of kinetic and thermochemical values. This entropy was recalculated as follows. The rate data of Ray and Ogg (6) and of Daniels and Johnston (7) give for the reaction $N_2O_5(g) + NO(g) + NO_2(g) + O_2(g)$ at 25°C the equilibrium constant $K = 5.15 \times 10^{-7}\ moles^2\ liter^{-2} = 3.08 \times 10^{-4}\ atm^2$. The resulting value of $\Delta_r H°$(298.15 K) = 4.79 kcal mol^{-1} leads to S°(298.15 K) = 82.8 cal K^{-1} mol^{-1} for N_2O_5(g) when combined with the JANAF entropies and heats of formation of the components of the reaction.

The molecular structure is the nonplanar configuration from the electron diffraction study of Akishin et al. (8). A planar model (also point group C_{2v}) was assumed by Hisatsune et al. (9) in their approximate normal coordinate analysis of the infrared and Raman spectra. The frequency assignments of these authors are listed above in the order for the planar model, although the vibrations for the nonplanar form will separate differently into the species $5A_1$, $3A_2$, $3B_1$ and $4B_2$. Hisatsune et al. (9) estimated the N-O'-N deformation frequency (8 cm^{-1}) from combination bands in the solid and gas phase spectra. The JANAF thermodynamic functions were obtained using these frequencies and assuming the two NO_2 groups to be hindered internal rotators. Barriers to internal rotation of 4.4 kcal mol^{-1} (corresponding to torsional vibratons of about 55 cm^{-1}) were selected in order to give the observed entropy at room temperature. The principal moments of inertia for the nonplanar model were taken as $I_A = 30.60 \times 10^{-39}$, $I_B = 40.15 \times 10^{-39}$ and $I_C = 16.82 \times 10^{-39}\ g\ cm^2$, while $I_r = 4.693 \times 10^{-39}\ g\ cm^2$ was used for the reduced moment of inertia of the O_2 tops.

References

1. Giauque and Kemp, J. Chem. Phys. 6, 40 (1938).
2. J. D. Ray and R. A. Ogg, J. Phys. Chem. 61, 1087 (1957).
3. R. A. Ogg, J. Chem. Phys. 15, 337 (1947).
4. Russ and Pokorny, Monatsh. 34, 1027 (1913).
5. F. Daniels and A. C. Bright, J. Amer. Chem. Soc. 42, 1131 (1920).
6. J. D. Ray and R. A. Ogg, J. Chem. Phys. 26, 984 (1957).
7. F. Daniels and Johnston, J. Amer. Chem. Soc. 43, 53 (1921).
8. Akishin, Vilkov and Rosolovskii, Zhur. Strukt. Khim. 1, 1 (1960).
9. Hisatsune, Devlin and Wada, Spectrochim. Acta 18, 1641 (1962).

Nitrogen Oxide (N_2O_5) N_2O_5(g)

Enthalpy Reference Temperature = T_r = 298.15 K   Standard State Pressure = p° = 0.1 MPa

T/K	$C_p°$ (J $K^{-1}mol^{-1}$)	S° (J $K^{-1}mol^{-1}$)	$-[G°-H°(T_r)]/T$ (J $K^{-1}mol^{-1}$)	$H°-H°(T_r)$ (kJ mol^{-1})	$\Delta_f H°$ (kJ mol^{-1})	$\Delta_f G°$ (kJ mol^{-1})	Log K_f
0							
100							
200	78.203	311.842	354.795	-8.591	12.733	83.104	-21.704
298.15	96.303	346.548	346.548	0.	11.297	118.013	-20.675
300	96.604	347.145	346.550	0.178	11.285	118.675	-20.663
400	110.851	376.991	350.523	10.587	11.349	154.497	-20.175
500	121.190	402.904	358.466	22.219	12.395	190.178	-19.868
600	128.382	425.677	367.809	34.720	14.013	225.590	-19.639
700	133.432	445.869	377.546	47.826	15.940	260.703	-19.454
800	137.001	463.930	387.235	61.356	18.018	295.528	-19.296
900	139.599	480.224	396.677	75.192	20.164	330.089	-19.158
1000	141.444	495.031	405.784	89.248	22.324	364.409	-19.035
1100	142.829	508.579	414.522	103.463	24.470	398.514	-18.924
1200	143.971	521.058	422.886	117.805	26.591	432.424	-18.823
1300	144.816	532.616	430.888	132.246	28.680	466.158	-18.730
1400	145.528	543.375	438.543	146.764	30.731	499.734	-18.645
1500	146.078	553.434	445.871	161.345	32.741	533.163	-18.566
1600	146.553	562.877	452.892	175.977	34.706	566.460	-18.493
1700	146.879	571.772	459.626	190.649	36.623	599.636	-18.425
1800	147.179	580.176	466.091	205.352	38.487	632.699	-18.360
1900	147.454	588.141	472.307	220.084	40.300	665.662	-18.300
2000	147.704	595.711	478.290	234.842	42.064	698.530	-18.244
2100	147.846	602.921	484.054	249.620	43.773	731.311	-18.190
2200	147.981	609.802	489.615	264.411	45.424	764.013	-18.140
2300	148.109	616.383	494.985	279.216	47.018	796.640	-18.092
2400	148.230	622.689	500.175	294.033	48.558	829.198	-18.047
2500	148.344	628.742	505.198	308.861	50.043	861.694	-18.004
2600	148.450	634.562	510.062	323.701	51.475	894.131	-17.963
2700	148.549	640.167	514.777	338.551	52.857	926.516	-17.925
2800	148.641	645.571	519.353	353.411	54.188	958.850	-17.888
2900	148.726	650.788	523.796	368.279	55.470	991.134	-17.852
3000	148.804	655.832	528.113	383.156	56.704	1023.378	-17.819
3100	148.851	660.712	532.312	398.038	57.892	1055.582	-17.786
3200	148.896	665.438	536.399	412.926	59.032	1087.746	-17.756
3300	148.939	670.021	540.379	427.818	60.127	1119.875	-17.726
3400	148.980	674.468	544.258	442.714	61.180	1151.974	-17.698
3500	149.019	678.787	548.040	457.614	62.190	1184.041	-17.671
3600	149.057	682.985	551.730	472.517	63.161	1216.080	-17.645
3700	149.092	687.070	555.333	487.425	64.094	1248.092	-17.620
3800	149.125	691.046	558.853	502.336	64.989	1280.082	-17.596
3900	149.156	694.920	562.292	517.250	65.849	1312.047	-17.573
4000	149.185	698.697	565.655	532.167	66.674	1343.989	-17.551
4100	149.206	702.381	568.945	547.086	67.465	1375.912	-17.529
4200	149.227	705.977	572.165	562.008	68.223	1407.816	-17.509
4300	149.247	709.488	575.318	576.932	68.948	1439.704	-17.489
4400	149.267	712.920	578.407	591.858	69.640	1471.572	-17.470
4500	149.285	716.274	581.433	606.785	70.301	1503.426	-17.451
4600	149.303	719.556	584.400	621.715	70.929	1535.266	-17.433
4700	149.319	722.767	587.310	636.646	71.525	1567.093	-17.416
4800	149.335	725.911	590.165	651.578	72.088	1598.908	-17.400
4900	149.350	728.990	592.967	666.513	72.616	1630.713	-17.384
5000	149.365	732.007	595.718	681.448	73.111	1662.506	-17.368
5100	149.378	734.965	598.419	696.386	73.569	1694.290	-17.353
5200	149.391	737.866	601.073	711.324	73.990	1726.062	-17.339
5300	149.402	740.712	603.681	726.264	74.373	1757.831	-17.324
5400	149.413	743.505	606.244	741.204	74.714	1789.590	-17.311
5500	149.423	746.246	608.765	756.146	75.014	1821.343	-17.298
5600	149.432	748.939	611.244	771.089	75.269	1853.094	-17.285
5700	149.441	751.584	613.683	786.033	75.477	1884.838	-17.273
5800	149.448	754.183	616.083	800.977	75.635	1916.581	-17.261
5900	149.455	756.738	618.446	815.922	75.741	1948.317	-17.249
6000	149.461	759.250	620.772	830.868	75.791	1980.057	-17.238

PREVIOUS: December 1964 (1 atm)   CURRENT: December 1964 (1 bar)

Nitrogen Oxide (N_2O_5) N_2O_5(g)

AZIDE (N_3) IDEAL GAS M_r = 42.0201

$S°(298.15\ K) = [226.469 \pm 2.1]\ J\ K^{-1}\ mol^{-1}$

$\Delta_f H°(0\ K) = 416.97 \pm 20.9\ kJ\ mol^{-1}$
$\Delta_f H°(298.15\ K) = 414.22 \pm 20.9\ kJ\ mol^{-1}$

Electronic Levels and Quantum Weights

State	ϵ_i, cm^{-1}	g_i
$X^2\Pi_g$ 3/2	0	2
$X^2\Pi_g$ 1/2	71.3	2
$A^2\Sigma_u^+$	36811	2

Vibrational Frequencies and Degeneracies

ν, cm^{-1}
[1800](1)
[500](2)
[1200](1)

Point Group: $D_{\infty h}$ $\sigma = 2$
Bond Distance: N-N = 1.1815 Å
Bond Angle: N-N-N = 180°
Rotational Constant: $B_0 = 0.431085\ cm^{-1}$

Enthalpy of Formation

Okabe and Mele (1) have measured the photodissociation onset of the process

$CN{\cdot}N_3(g) \rightarrow CN(g) + N_3(g)$ $\Delta H°(0\ K) = 96 \pm 2.5\ kcal\ mol^{-1}$

Using a value of $\Delta_f H°(CN{\cdot}N_3,\ g,\ 298.15\ K) = 107 \pm 3\ kcal\ mol^{-1}$ and $\Delta_f H°(CN,\ g,\ 298.15\ K) = 104 \pm 2.5\ kcal\ mol^{-1}$ (2), we derive $\Delta_f H°(N_3,\ g,\ 298.15\ K) = 99 \pm 5\ kcal\ mol^{-1}$. The enthalpy of formation of $CN{\cdot}N_3$ is obtained from $\Delta_f H°(CN{\cdot}N_3,\ cr,\ 298.15\ K) = 92.6 \pm 3\ kcal\ mol^{-1}$ (3) and an estimated $\Delta_{sub}H°(298.15\ K) = 14.4 \pm 3\ kcal\ mol^{-1}$. The enthalpy of sublimation is similar to that for ICN(cr) (2). A rough confirmation of $\Delta_{sub}H°$ can be made by assuming a boiling point of 227°C (500 K), from the facts that $CN{\cdot}N_3$ sublimes in vacuo at 40°C and can be heated as a liquid to 170°C. Thus, using the Trouton rule, $\Delta_{vap}H° = 11\ kcal\ mol^{-1}$, in addition the enthalpy of melting would be expected to be about 3.6 $kcal\ mol^{-1}$ for this 5 atom molecule, thus $\Delta_{sub}H° = 14.6\ kcal\ mol^{-1}$.

Heat Capacity and Entropy

The electronic levels and rotational constants have been reported by Douglas and Jones (4) and Thrush (5). The vibrational frequencies were estimated from the isoelectronic species NCO, NNO^+, and the related molecules NCN and NNO (2).

References

1. H. Okabe and A. Mele, J. Chem. Phys. 51, 2100 (1959).
2. JANAF Thermochemical Tables: CN(g), 6-30-69; ICN(g), 6-30-69; NCO(g), 12-31-70; NNO^+(g), 12-31-70; NCN(g), 12-31-70; N_2O(g), 12-31-64.
3. U. S. Nat. Bur. Stand. Tech. Note 270-3, 1968.
4. A. E. Douglas and W. J. Jones, Can. J. Phys. 43, 2216 (1965).
5. B. A. Thrush, Proc. Roy. Soc. (London) 235A, 143 (1956).

Azide (N_3) N_3(g)

Enthalpy Reference Temperature = T_r = 298.15 K Standard State Pressure = p° = 0.1 MPa

T/K	$C_p°$ (J K⁻¹mol⁻¹)	S° (J K⁻¹mol⁻¹)	$-[G°-H°(T_r)]/T$ (J K⁻¹mol⁻¹)	$H°-H°(T_r)$ (kJ mol⁻¹)	$\Delta_f H°$ (kJ mol⁻¹)	$\Delta_f G°$ (kJ mol⁻¹)	Log K_f
0	0.	0.	INFINITE	-10.253	416.969	416.969	INFINITE
100	31.448	188.244	259.348	-7.110	415.757	420.905	-219.858
200	35.960	211.175	230.051	-3.775	414.727	426.487	-111.387
250	38.562	219.485	227.128	-1.911	414.408	429.467	-89.732
298.15	40.762	226.469	226.469	0.	414.216	432.387	-75.752
300	40.842	226.721	226.470	0.075	414.211	432.499	-75.305
350	42.864	233.172	226.974	2.169	414.119	435.556	-65.003
400	44.686	239.016	228.120	4.359	414.118	438.620	-57.278
450	46.339	244.377	229.633	6.635	414.196	441.679	-51.269
500	47.838	249.338	231.358	8.990	414.340	444.725	-46.460
600	50.407	258.295	235.117	13.907	414.782	450.764	-39.242
700	52.470	266.226	239.006	19.054	415.365	456.716	-34.081
800	54.113	273.344	242.861	24.387	416.033	462.578	-30.203
900	55.421	279.796	246.612	29.866	416.748	468.354	-27.183
1000	56.467	285.691	250.229	35.462	417.484	474.048	-24.762
1100	57.308	291.114	253.703	41.152	418.228	479.669	-22.778
1200	57.992	296.131	257.032	46.919	418.971	485.222	-21.121
1300	58.552	300.796	260.221	52.747	419.709	490.712	-19.717
1400	59.015	305.152	263.277	58.626	420.437	496.147	-18.511
1500	59.402	309.238	266.206	64.547	421.156	501.530	-17.465
1600	59.727	313.082	269.017	70.504	421.865	506.865	-16.547
1700	60.003	316.711	271.717	76.491	422.563	512.156	-15.737
1800	60.239	320.148	274.313	82.503	423.252	517.406	-15.015
1900	60.442	323.410	276.812	88.538	423.932	522.618	-14.368
2000	60.617	326.515	279.220	94.591	424.602	527.795	-13.785
2100	60.770	329.477	281.543	100.660	425.264	532.938	-13.256
2200	60.904	332.307	283.787	106.744	425.918	538.050	-12.775
2300	61.022	335.017	285.956	112.841	426.565	543.132	-12.335
2400	61.127	337.616	288.054	118.948	427.204	548.187	-11.931
2500	61.219	340.113	290.087	125.066	427.837	553.215	-11.559
2600	61.302	342.516	292.058	131.192	428.464	558.217	-11.215
2700	61.376	344.831	293.970	137.326	429.084	563.196	-10.896
2800	61.443	347.064	295.826	143.467	429.699	568.152	-10.599
2900	61.503	349.221	297.630	149.614	430.308	573.086	-10.322
3000	61.558	351.307	299.385	155.767	430.912	577.999	-10.064
3100	61.607	353.327	301.093	161.926	431.510	582.892	-9.822
3200	61.652	355.283	302.756	168.089	432.104	587.766	-9.594
3300	61.693	357.181	304.376	174.256	432.693	592.621	-9.380
3400	61.731	359.024	305.957	180.427	433.278	597.459	-9.179
3500	61.766	360.813	307.499	186.602	433.858	602.279	-8.989
3600	61.798	362.554	309.004	192.780	434.434	607.083	-8.809
3700	61.827	364.248	310.474	198.962	435.006	611.871	-8.638
3800	61.855	365.897	311.911	205.146	435.574	616.643	-8.476
3900	61.880	367.504	313.316	211.332	436.138	621.401	-8.323
4000	61.904	369.071	314.690	217.522	436.697	626.144	-8.177
4100	61.926	370.600	316.035	223.713	437.253	630.874	-8.037
4200	61.946	372.092	317.352	229.907	437.806	635.589	-7.905
4300	61.966	373.550	318.642	236.102	438.354	640.292	-7.778
4400	61.984	374.975	319.907	242.300	438.899	644.982	-7.657
4500	62.001	376.368	321.146	248.499	439.440	649.660	-7.541
4600	62.018	377.731	322.361	254.700	439.978	654.325	-7.430
4700	62.033	379.065	323.553	260.903	440.512	658.979	-7.324
4800	62.048	380.371	324.724	267.107	441.042	663.622	-7.222
4900	62.063	381.650	325.872	273.312	441.568	668.253	-7.124
5000	62.077	382.904	327.001	279.519	442.091	672.874	-7.029
5100	62.090	384.134	328.109	285.728	442.610	677.485	-6.939
5200	62.103	385.340	329.198	291.937	443.125	682.085	-6.852
5300	62.116	386.523	330.268	298.148	443.637	686.676	-6.768
5400	62.128	387.684	331.321	304.360	444.144	691.256	-6.687
5500	62.141	388.824	332.356	310.574	444.649	695.828	-6.608
5600	62.153	389.944	333.374	316.788	445.149	700.390	-6.533
5700	62.165	391.044	334.377	323.004	445.644	704.944	-6.460
5800	62.177	392.125	335.363	329.222	446.135	709.489	-6.390
5900	62.189	393.188	336.334	335.440	446.622	714.025	-6.321
6000	62.202	394.234	337.290	341.659	447.103	718.553	-6.256

PREVIOUS: December 1970 (1 atm) CURRENT: December 1970 (1 bar)

Azide (N_3) N_3(g)

SILICON NITRIDE, ALPHA (α-Si_3N_4) CRYSTAL M_r = 140.2833

S°(298.15 K) = [112.967 ± 16.7] J K^{-1} mol^{-1}

$\Delta_f H^\circ$(0 K) = Unknown
$\Delta_f H^\circ$(298.15 K) = -744.752 ± 29.3 kJ mol^{-1}

Enthalpy of Formation

The equilibria (A) Si_3N_4(cr) = 3 Si(ℓ) + 2 N_2(g) and (B) Si_3N_4(cr) = 3 Si(cr) + 2 N_2(g) have been investigated by three groups. Pehlke and Elliott (1) made four series of measurements which mainly concerned reaction (A). Hincke and Brantley (2) made a single series of measurements which covered reactions (A) and (B). Matignon (3) also gave a single equilibrium point for reaction (C) 3 SiO_2(cr) + 6 C + 2 N_2 = Si_3N_4(cr) + 6 CO(g). Below are shown 2nd and 3rd law analyses of the data. It should be noted that these analyses are based on the estimated enthalpy capacities above 900 K and on an estimated entropy. The uncertainty due to these estimates is estimated as ± 7 kcal mol^{-1} in $\Delta_f E^\circ$(298.15 K).

Source	Data Points	T/K	Reaction	$\Delta_r H^\circ$(298.15 K), kcal mol^{-1} 2nd law	3rd law	Drift cal K^{-1} mol^{-1}	$\Delta_f H^\circ$(298.15 K) kcal mol^{-1}
Pehlke - 1	9	1699-1934	A	190.1 ± 7.6	211.79	11.8 ± 4.3	-177.04
Pehlke - 2	9	1691-1939	A	210.7 ± 8.6	213.29	1.8 ± 4.9	-178.53
Pehlke - 3	10	1692-1966	A	212.3 ± 3.3	212.61	0.1 ± 1.9	-177.85
Pehlke - 4	6	1690-1872	A	207.0 ± 7.5	212.47	2.6 ± 4.3	-177.71
Pehlke	30**	1690-1966	A	212.8 ± 3.2	212.67	-0.2 ± 1.8	-177.91
Hinke	4	1709-1802	A	196.0 ±18.5	213.43	10.3 ±10.4	-178.67
Pehlke	4	1659-1666	B		177.41		-177.41
Hinke	3	1606-1675	B		177.82	22.8 ±11.8	-177.82
Matignon	1	1700	C		318.9		-175.1

**Four points rejected due to failure of a statistical test; this set is a combination of the first four sets.

The adopted $\Delta_f H^\circ$(α-Si_3N_4, 298.15 K) = -178 ± 7 kcal mol^{-1}.

McClaine and Coppel (4) have investigated the decomposition reaction (B). The conclude that at 1618 K the equilibrium pressure is roughly 0.15 mm. This corresponds to $\Delta_f H^\circ$(298.15 K) = 182.3 kcal mol^{-1}, and was not given any weight in the adopted value.

Heat Capacity and Entropy

The enthalpy of α-Si_3N_4 has been measured at three temperatures, in the range 373 - 858 K, by Sato (5). The heat capacity is Calculated from the data assuming that a limit of 45 cal K^{-1} mol^{-1} is attained by 3000 K. The entropy is obtained from the equilibrium data of Pehlke and Elliott (1).

Decomposition Data

The temperature of decomposition, T_{dcm} = 2151 K, is calculated as the point at which the decomposition pressure of nitrogen reaches a fugacity of 1 atm.

References

1. R. D. Pehlke and J. F. Elliott, Trans. Met. Soc. AIME 215, 781 (1959).
2. W. B. Hincke and L. R. Brantley, J. Amer. Chem. Soc. 52, 48 (1930).
3. C. Matignon, Bull. Soc. Chim. France 13, 791 (1913).
4. L. A. McClaine and C. P. Coppel, A. D. Little, Inc., "Equilibrium Studies of Refractory Nitrides," Tech. Rept. AFML-TR-65-99, Part 11, Dec. 1966.
5. S. Sato, Bull. Chem. Soc. Japan 13, 41 (1938).

Silicon Nitride, Alpha (Si_3N_4) N_4Si_3(cr)

Enthalpy Reference Temperature = T_r = 298.15 K; Standard State Pressure = p° = 0.1 MPa

T/K	C_p° (J $K^{-1}mol^{-1}$)	S°	$-[G^\circ - H^\circ(T_r)]/T$	$H^\circ - H^\circ(T_r)$ (kJ mol^{-1})	$\Delta_f H^\circ$	$\Delta_f G^\circ$	Log K_f
0							
100							
200							
298.15	99.533	112.967	112.967	0.	-744.752	-647.343	113.412
300	99.788	113.583	112.969	0.184	-744.787	-646.739	112.607
400	110.667	143.790	117.014	10.711	-746.459	-613.792	80.153
500	120.708	169.582	125.007	22.288	-747.595	-580.482	60.643
600	129.704	192.417	134.374	34.826	-748.151	-546.998	47.620
700	138.072	213.048	144.162	48.220	-748.186	-513.462	38.315
800	145.812	232.003	153.973	62.424	-747.729	-479.956	31.338
900	152.298	249.566	163.631	77.342	-746.850	-446.533	25.916
1000	158.155	265.920	173.051	92.869	-745.635	-413.227	21.585
1100	163.678	281.258	182.198	108.966	-744.100	-380.058	18.047
1200	168.490	295.712	191.061	125.581	-742.280	-347.041	15.106
1300	172.531	309.362	199.641	142.637	-740.233	-314.187	12.624
1400	175.920	322.276	207.944	160.065	-738.019	-281.496	10.503
1500	178.657	334.510	215.977	177.799	-735.694	-248.968	8.670
1600	180.548	346.102	223.751	195.762	-733.327	-216.597	7.071
1700	182.171	357.098	231.274	213.900	-881.495	-183.033	5.624
1800	183.452	367.548	238.557	232.183	-878.469	-142.034	4.122
1900	184.539	377.496	245.610	250.584	-875.367	-101.205	2.782
2000	185.435	386.986	252.444	269.084	-872.202	-60.541	1.581
2100	185.780	396.041	259.068	287.645	-869.010	-20.037	0.498
2200	186.111	404.692	265.492	306.240	-865.814	20.316	-0.482
2300	186.429	412.972	271.725	324.867	-862.612	60.523	-1.375
2400	186.734	420.913	277.777	343.525	-859.403	100.590	-2.189
2500	187.025	428.541	283.656	362.213	-856.187	140.524	-2.936
2600	187.303	435.882	289.371	380.930	-852.962	180.329	-3.623
2700	187.567	442.956	294.929	399.673	-849.729	220.009	-4.256
2800	187.818	449.782	300.338	418.443	-846.487	259.570	-4.842
2900	188.056	456.377	305.606	437.236	-843.236	299.015	-5.386
3000	188.280	462.756	310.738	456.053	-839.977	338.347	-5.891

PREVIOUS: December 1960 CURRENT: March 1967

Silicon Nitride, Alpha (Si_3N_4) N_4Si_3(cr)

PHOSPHORUS NITRIDE (P_3N_5) CRYSTAL M_r = 162.95478

$\Delta_fH°$(0 K) = Unknown

S°(298.15 K) = [184.096 ± 25.1] J K^{-1} mol^{-1}

$\Delta_fH°$(298.15 K) = -320.155 ± 20.9 kJ mol^{-1}

Enthalpy of Formation

The enthalpy of formation is based on the work of Stock and Wrede (1) who burned P_3N_5 in oxygen. They report a $\Delta_rH°$(298.15 K) = -949.4 kcal mol^{-1} for the reaction 2 P_3N_5(cr) + 7.5 O_2(g) = 1.5 P_4O_{10}(amorph) + 5 N_2(g). In (2) a value of -947.7 kcal mol^{-1} is used for the above reaction and we adopt this value. Using the JANAF value of P_4O_{10}(cr) and correcting this to P_4O_{10}(amorph) by adding -13.8 kcal mol^{-1}, from (2) we obtain $\Delta_fH°$(298.15 K) P_4O_{10}(amorph) = -716.5 kcal mol^{-1}. Thus, we derive $\Delta_fH°$(P_3N_5, cr, 298.15 K) = -63.525 kcal mol^{-1}; however, because of the uncertainties in the enthalpy of reaction and correction for amorphous P_4O_{10}, we round the value to -64 ± 5 kcal mol^{-1}.

Heat Capacity and Entropy

Sato (3) gave an equation for the heat capacity in the range 273 - 578 K. Judging from his results for Be_3N_2(cr) in the same paper, we believe his values will be slightly high at the upper end of the range. Thus, taking 8.2 cal K^{-1} g-atom as the limiting value, a curve was drawn through Sato's values at the low end and approaching the limit asympototically. The entropy was estimated from that of Si_3N_4(cr) by assuming that the ratio $C_p°$(298.15 K)/S°(298.15 K) on a gram atom basis would be the same for P_3N_5(cr).

Decomposition Data

T_{dcm} = 730 K is calculated as the point at which the total pressure reaches a fugacity of 1 atm. This corresponds to log $K_p°$ = -0.76 for the reaction P_3N_5(cr) = 0.75 P_4(g) + 2.5 N_2(g). The temperature of decomposition has been reported by Postnikov and Kuzmin (4) as 1073 K. This has been confirmed by Moreau and Rocquet (5) and Huffman et al. (6). However, it was also noted that P_3N_5(cr) is not reformed on cooling the decomposition products; thus the decomposition is non-equilibrium. It is known that the decomposition of AlN, BN and Be_3N_2 is kinetically limited and the equilibrium pressure is not attained until the average energy is greater than the activation energy. It appears that this is also the case for P_3N_5 and a high activation energy is expected.

References

1. A. Stock and F. Wrede, Chem. Ber. 40, 2923 (1907).
2. U. S. Nat. Bur. Stand. Tech. Note 270-1, 1965.
3. S. Sato, Sci. Papers Inst. Phys. Chem. Res. (Tokyo) 34, 584 (1938).
4. V. F. Postnikov and L. L. Kuzmin, J. Appl. Chem. (USSR) 8, 429 (1935).
5. H. Moreau and P. Rocquet, Compt. Rend. 198, 1691 (1934); Bull. Soc. Chim. France 3, 1801 (1936).
6. E. O. Huffman, G. Tarbutton, G. V. Elmore, A. J. Smith and M. G. Rountree, J. Amer. Chem. Soc. 79, 1765 (1957).

Phosphorus Nitride (P_3N_5) N_5P_3(cr)

Enthalpy Reference Temperature = T_r = 298.15 K Standard State Pressure = p° = 0.1 MPa

T/K	$C_p°$ (J K⁻¹mol⁻¹)	S° (J K⁻¹mol⁻¹)	-[G°-H°(T_r)]/T (J K⁻¹mol⁻¹)	H°-H°(T_r) (kJ mol⁻¹)	$\Delta_fH°$ (kJ mol⁻¹)	$\Delta_fG°$ (kJ mol⁻¹)	Log K_f
0							
100							
200							
298.15	148.950	184.096	184.096	0.	-320.155	-195.482	34.248
300	149.787	185.020	184.099	0.276	-320.146	-194.708	33.902
400	192.464	234.376	190.538	17.535	-319.936	-152.610	19.929
500	217.568	280.175	203.962	38.107	-314.612	-111.344	11.632
600	234.304	321.385	220.160	60.735	-307.339	-71.348	6.211
700	246.856	358.500	237.317	84.828	-298.751	-32.680	2.439
800	255.224	392.032	254.596	109.949	-289.301	4.694	-0.306
900	261.500	422.480	271.584	135.807	-279.283	40.846	-2.371
1000	265.266	450.236	288.083	162.153	-268.934	75.864	-3.963
1100	267.888	475.646	303.995	188.816	-258.412	109.835	-5.216
1200	269.868	499.045	319.287	215.709	-438.505	146.079	-6.359
1300	271.207	520.703	333.958	242.768	-425.504	194.267	-7.806
1400	271.904	540.831	348.025	269.929	-412.515	241.454	-9.009
1500	271.960	559.596	361.511	297.128	-399.588	287.715	-10.019

PREVIOUS: September 1962 CURRENT: March 1967

Phosphorus Nitride (P_3N_5) N_5P_3(cr)

SODIUM (Na) REFERENCE STATE A_r = 22.98977

0 to 370.98 K		crystal
370.98 to 1170.525 K		liquid
above 1170.525 K		ideal monatomic gas

Refer to the individual tables for details.

Sodium (Na) Na_1(ref)

Enthalpy Reference Temperature = T_r = 298.15 K Standard State Pressure = p° = 0.1 MPa

T/K	C_p° (J K⁻¹mol⁻¹)	S° (J K⁻¹mol⁻¹)	$-[G°-H°(T_r)]/T$ (J K⁻¹mol⁻¹)	$H°-H°(T_r)$ (kJ mol⁻¹)	$\Delta_f H°$ (kJ mol⁻¹)	$\Delta_f G°$ (kJ mol⁻¹)	Log K_f
0	0.	0.	INFINITE	-6.447	0.	0.	0.
55.120	18.800	11.738	121.072	-6.027	Cp HUMP MAXIMUM		
55.120	18.791	11.738	121.072	-6.027	TRANSITION		
100	22.451	23.814	74.885	-5.107	0.	0.	0.
200	25.987	40.697	53.954	-2.651	0.	0.	0.
298.15	28.154	51.455	51.455	0.	0.	0.	0.
300	28.204	51.629	51.455	0.052	0.	0.	0.
370.980	31.583	57.898	52.098	2.152	CRYSTAL <--> LIQUID		
370.980	31.827	64.915	52.098	4.755	TRANSITION		
400	31.510	67.301	53.116	5.674	0.	0.	0.
500	30.552	74.225	56.675	8.775	0.	0.	0.
600	29.807	79.726	60.074	11.791	0.	0.	0.
700	29.271	84.278	63.216	14.743	0.	0.	0.
800	28.945	88.163	66.097	17.652	0.	0.	0.
900	28.836	91.564	68.742	20.540	0.	0.	0.
1000	28.945	94.606	71.179	23.427	0.	0.	0.
1100	29.259	97.377	73.436	26.335	0.	0.	0.
1170.525	29.613	99.206	74.934	28.411	LIQUID <--> IDEAL GAS		
1170.525	20.786	182.094	74.934	125.433	FUGACITY = 1 bar		
1200	20.786	182.611	77.573	126.046	0.	0.	0.
1300	20.786	184.275	85.717	128.124	0.	0.	0.
1400	20.786	185.815	92.813	130.203	0.	0.	0.
1500	20.787	187.249	99.061	132.282	0.	0.	0.
1600	20.787	188.591	104.616	134.360	0.	0.	0.
1700	20.789	189.851	109.593	136.439	0.	0.	0.
1800	20.792	191.039	114.085	138.518	0.	0.	0.
1900	20.797	192.164	118.165	140.598	0.	0.	0.
2000	20.805	193.231	121.892	142.678	0.	0.	0.
2100	20.816	194.246	125.313	144.759	0.	0.	0.
2200	20.833	195.215	128.469	146.841	0.	0.	0.
2300	20.856	196.141	131.391	148.925	0.	0.	0.
2400	20.886	197.029	134.108	151.012	0.	0.	0.
2500	20.924	197.883	136.642	153.103	0.	0.	0.
2600	20.973	198.704	139.013	155.198	0.	0.	0.
2700	21.032	199.497	141.238	157.298	0.	0.	0.
2800	21.104	200.263	143.333	159.405	0.	0.	0.
2900	21.189	201.005	145.309	161.519	0.	0.	0.
3000	21.289	201.725	147.177	163.643	0.	0.	0.
3100	21.403	202.425	148.948	165.777	0.	0.	0.
3200	21.535	203.106	150.630	167.924	0.	0.	0.
3300	21.680	203.771	152.230	170.084	0.	0.	0.
3400	21.845	204.421	153.756	172.260	0.	0.	0.
3500	22.030	205.057	155.213	174.454	0.	0.	0.
3600	22.234	205.680	156.606	176.667	0.	0.	0.
3700	22.460	206.292	157.941	178.901	0.	0.	0.
3800	22.709	206.894	159.221	181.160	0.	0.	0.
3900	22.981	207.488	160.451	183.444	0.	0.	0.
4000	23.278	208.073	161.634	185.757	0.	0.	0.
4100	23.602	208.652	162.774	188.100	0.	0.	0.
4200	23.945	209.224	163.873	190.475	0.	0.	0.
4300	24.324	209.792	164.934	192.889	0.	0.	0.
4400	24.611	210.345	165.959	195.298	0.	0.	0.
4500	24.992	210.901	166.952	197.769	0.	0.	0.
4600	25.421	211.454	167.913	200.289	0.	0.	0.
4700	25.876	212.006	168.846	202.854	0.	0.	0.
4800	26.359	212.556	169.750	205.466	0.	0.	0.
4900	26.868	213.104	170.630	208.127	0.	0.	0.
5000	27.388	213.651	171.484	210.832	0.	0.	0.
5200	28.537	214.747	173.127	216.423	0.	0.	0.
5400	29.789	215.847	174.689	222.254	0.	0.	0.
5600	31.141	216.955	176.179	228.346	0.	0.	0.
5800	32.583	218.073	177.604	234.717	0.	0.	0.
6000	33.880	219.173	178.969	241.226	0.	0.	0.

PREVIOUS: June 1962 (1 atm) CURRENT: June 1962 (1 bar)

Sodium (Na) Na_1(ref)

$S°(298.15\ K) = 51.455\ J\ K^{-1}\ mol^{-1}$

$T_{fus} = 370.98 \pm 0.02\ K$

$\Delta_f H°(0\ K) = 0\ kJ\ mol^{-1}$

$\Delta_f H°(298.15\ K) = 0\ kJ\ mol^{-1}$

$\Delta_{fus} H° = 2.602\ kJ\ mol^{-1}$

Enthalpy of Formation

Zero by definition.

Heat Capacity and Entropy

Martin (1) reports that the low temperature heat capacity of sodium depends on the thermal history of the sample and ascribes this to a martensitic transformation. The effect is most pronounced in the range 40-80 K, where the graph of C_p against T increases continuously for some samples and rises to a peak at 55 K for others. Martin (1) lists heat capacities obtained in several runs on "high purity" sodium at 21 to 300 K and the higher values, those defining the peak at 55 K, have been selected here. The heat capacities reported by Dauphinee et al. (2), have not been used, since one of the authors in a later publication, Martin (1) questions the accuracy of the measurements above 100 K, and at lower temperatures there is no rise to a peak at 55 K. In the range 1.5 to 20 K, Roberts (3) and Parkinson and Quarrington (4) have measured the heat capacity of very high purity sodium. The values of Parkinson and Quarrington (4) are somewhat higher than those of Roberts (3) and have been given greater weight, since they show better continuity with Martin's (1) results. The results of Pickard and Simon (5) have been rejected below 10 K, since they show a peak between 5 and 8 K, a feature not verified by other workers; between 10 and 25 K they agree exactly with the results of Parkinson and Quarrington (4). Rayne (6) states that the graph of C_p against T has a small peak at about 0.9 K. This peak has not been observed in the work of Lien and Phillips (7) from 0.15 to 1 K; Gaumer and Heer from 0.4 to 2 K and Martin (9) from 0.4 to 1.5 K.

Douglas et al. (10) report heat capacities in the range 273 to 1173 K on a sample of not less than 99.9% purity. These values agree to within 1% with those of Martin (1) in the range 273 to 300 K, and the two sets of dates have been smoothly joined.

Fusion Data

Douglas et al. (10) have determined the triple point to be 97.82 ± 0.02°C, from which the melting point is calculated to be 97.83 ± 0.02°C, and have measured the enthalpy of fusion.

Sublimation Data

At the boiling point and higher temperatures sodium vapor contains an appreciable proportion of diatomic moleculses. Thompson and Garelis (11) have made a careful analysis of the available vapor pressure data. Their results are consistent with the Gibbs energy functions calculated in the present work and have been adopted.

References

1. D. L. Martin, Proc. Roy. Soc. A254, 433 (1960).
2. T. M. Dauphinee, D. L. Martin, and H. Preston-Thomas, Proc. Roy. Soc. A233, 214 (1955).
3. L. M. Roberts, Proc. Phys. Soc. B70, 744 (1957).
4. D. H. Parkinson and J. E. Quarrington Proc. Phys. Soc. A68, 762 (1955).
5. G. L. Pickard and F. E. Simon, Proc. Phys. Soc. 61, 1 (1948).
6. J. Rayne, Phys. Rev. 95, 1428 (1954).
7. W. H. Lien and N. E. Phillips, Phys. Rev. 118, 958 (1960).
8. R. E. Gaumer and C. V. Heer, Phys. Rev. 118, 955 (1960).
9. D. L. Martin, Phys. Rev. 124, 438 (1961).
10. T. B. Douglas, A. F. Ball, D. C. Ginnings, and W. D. Davis, J. Am. Chem. Soc. 74, 2472 (1952).
11. G. W. Thompson and E. Garelis "Sodium - its Manufacture, Properties and Uses," Ed. M. Sittig, Chapter 3, Reinhold Publishing Corporation, Inc., New York (1956).

Enthalpy Reference Temperature = T_r = 298.15 K   Standard State Pressure = p° = 0.1 MPa

T/K	$C_p°$ (J K⁻¹mol⁻¹)	S° (J K⁻¹mol⁻¹)	$-[G°-H°(T_r)]/T$ (J K⁻¹mol⁻¹)	$H°-H°(T_r)$ (kJ mol⁻¹)	$\Delta_f H°$ (kJ mol⁻¹)	$\Delta_f G°$ (kJ mol⁻¹)	Log K_f
0	0.	0.	INFINITE	-6.447	0.	0.	0.
55.120	18.800	11.738	121.072	-6.027	Cp HUMP MAXIMUM TRANSITION		
55.120	18.791	11.738	121.072	-6.027			
100	22.451	23.814	74.885	-5.107	0.	0.	0.
200	25.987	40.697	53.954	-2.651	0.	0.	0.
250	27.008	46.604	51.911	-1.327	0.	0.	0.
298.15	28.154	51.455	51.455	0.	0.	0.	0.
300	28.204	51.629	51.455	0.052	0.	0.	0.
350	30.142	56.104	51.805	1.505	0.	0.	0.
370.980	31.583	57.898	52.098	2.152	------ CRYSTAL <--> LIQUID -----		
400	32.489	60.315	52.608	3.083	-2.591	0.203	-0.027
450	33.103	64.180	53.683	4.724	-2.513	0.548	-0.064
500	33.472	67.689	54.910	6.389	-2.386	0.882	-0.092
600	33.472	73.792	57.564	9.736	-2.055	1.506	-0.131
700	33.472	78.951	60.260	13.084	-1.660	2.069	-0.154
800	33.472	83.421	62.882	16.431	-1.222	2.572	-0.168
900	33.472	87.363	65.388	19.778	-0.762	3.019	-0.175
1000	33.472	90.890	67.765	23.125	-0.302	3.414	-0.178

PREVIOUS: CURRENT: June 1962

SODIUM (Na) LIQUID A_r = 22.98977

$S°(298.15\ K) = [57.862]\ J\ K^{-1}\ mol^{-1}$

$T_{fus} = 370.98 \pm 0.02\ K$

$\Delta_f H°(298.15\ K) = [2.406]\ kJ\ mol^{-1}$

$\Delta_{fus} H° = 2.602\ kJ\ mol^{-1}$

Enthalpy of Formation

$\Delta_f H°$(Na, ℓ, 298.15 K) is calculated from that of the crystal by adding the enthalpy of fusion and the difference in enthalpy, H°(370.98 K) - H°(298.15 K), between the crystal and liquid.

Heat Capacity and Entropy

The liquid heat capacity values are derived from the enthalpy measurements of Douglas et al. (1). The heat capacity curve was extrapolated to 298.15 K and 2000 K in a reasonable manner.

The entropy is calculated in a manner anologous to that used for the enthalpy of formation.

Fusion Data

The enthalpy of fusion and the melting point were measured by Douglas et al. (1).

Vaporization Data

Sodium vaporizes to a mixture of monatomic and diatomic gas. The total vapor pressure reaches 1 atm at 1156 K. The vapor pressure of the monatomic gas reaches 1 atm at 1176.9 K and the enthalpy of vaporization to monatomic gas is 23.285 kcal mol^{-1}.

Reference

1. T. B. Douglas, A. F. Ball, D. C. Ginnings and W. D. Davis, J. Am. Chem. Soc. 74, 2472 (1952).

Sodium (Na) $Na_1(l)$

Enthalpy Reference Temperature = T_r = 298.15 K   Standard State Pressure = p° = 0.1 MPa

T/K	$C_p°$	S°	$-[G°-H°(T_r)]/T$	$H°-H°(T_r)$	$\Delta_f H°$	$\Delta_f G°$	Log K_f
	J K⁻¹mol⁻¹			kJ mol⁻¹			
0						00	INFINITE
100							
200							
250							
298.15	32.706	57.862	57.862	0.	2.406	0.495	-0.087
300	32.685	58.064	57.863	0.060	2.414	0.483	-0.084
350	32.068	63.056	58.258	1.679	2.580	0.147	-0.022
370.980	31.827	64.915	58.583	2.349	------ CRYSTAL <--> LIQUID -----		
400	31.510	67.301	59.130	3.268	0.	0.	0.
450	31.004	70.982	60.247	4.831	0.	0.	0.
500	30.552	74.225	61.486	6.370	0.	0.	0.
600	29.807	79.726	64.083	9.386	0.	0.	0.
700	29.271	84.278	66.652	12.338	0.	0.	0.
800	28.945	88.163	69.104	15.247	0.	0.	0.
900	28.836	91.564	71.415	18.134	0.	0.	0.
1000	28.945	94.606	73.584	21.021	0.	0.	0.
1100	29.259	97.377	75.623	23.930	0.	0.	0.
1170.525	29.613	99.206	76.989	26.005	------- FUGACITY = 1 bar -------		
1200	29.790	99.945	77.544	26.881	-96.760	2.440	-0.106
1300	30.514	102.357	79.361	29.895	-95.824	10.669	-0.429
1400	31.351	104.649	81.086	32.988	-94.809	18.823	-0.702
1500	32.188	106.840	82.730	36.165	-93.711	26.902	-0.937
1600	33.024	108.944	84.303	39.426	-92.529	34.905	-1.140

PREVIOUS: CURRENT: June 1962

Sodium (Na) $Na_1(l)$

SODIUM (Na) CRYSTAL–LIQUID A_r = 22.98977

0 to 370.98 K crystal
above 370.98 K liquid

Refer to the individual tables for details.

Sodium (Na) $Na_1(cr,l)$

Enthalpy Reference Temperature = T_r = 298.15 K — Standard State Pressure = $p°$ = 0.1 MPa

T/K	$C_p°$ (J K⁻¹mol⁻¹)	$S°$ (J K⁻¹mol⁻¹)	$-[G°-H°(T_r)]/T$ (J K⁻¹mol⁻¹)	$H°-H°(T_r)$ (kJ mol⁻¹)	$\Delta_f H°$ (kJ mol⁻¹)	$\Delta_f G°$ (kJ mol⁻¹)	Log K_f
0	0.	0.	INFINITE	-6.447	0.	0.	0.
55.120	18.800	11.738	121.072	-6.027	Cp HUMP MAXIMUM TRANSITION		
55.120	18.791	11.738	121.072	-6.027			
100	22.451	23.814	74.885	-5.107	0.	0.	0.
200	25.987	40.697	53.954	-2.651	0.	0.	0.
250	27.008	46.604	51.911	-1.327	0.	0.	0.
298.15	28.154	51.455	51.455	0.	0.	0.	0.
300	28.204	51.629	51.455	0.052	0.	0.	0.
350	30.142	56.104	51.805	1.505	0.	0.	0.
370.980	31.583	57.898	52.098	2.152	CRYSTAL <--> LIQUID TRANSITION		
370.980	31.827	64.915	52.098	4.755			
400	31.510	67.301	53.116	5.674	0.	0.	0.
450	31.004	70.982	54.901	7.236	0.	0.	0.
500	30.552	74.225	56.675	8.775	0.	0.	0.
600	29.807	79.726	60.074	11.791	0.	0.	0.
700	29.271	84.278	63.216	14.743	0.	0.	0.
800	28.945	88.163	66.097	17.652	0.	0.	0.
900	28.836	91.564	68.742	20.540	0.	0.	0.
1000	28.945	94.606	71.179	23.427	0.	0.	0.
1100	29.259	97.377	73.436	26.335	0.	0.	0.
1170.525	29.613	99.206	74.934	28.411	------- FUGACITY = 1 bar -------		
1200	29.790	99.945	75.539	29.286	-96.760	2.440	-0.106
1300	30.514	102.357	77.510	32.300	-95.824	10.669	-0.429
1400	31.351	104.649	79.368	35.394	-94.809	18.823	-0.702
1500	32.188	106.840	81.127	38.571	-93.711	26.902	-0.937
1600	33.024	108.944	82.800	41.831	-92.529	34.905	-1.140

PREVIOUS: CURRENT: June 1962

Sodium (Na) $Na_1(cr,l)$

SODIUM (Na) IDEAL GAS A_r = 22.98977

IP(Na, g) = 41449.44 ± 0.03 cm^{-1}
S°(298.15 K) = 153.667 ± 0.025 J K^{-1} mol^{-1}

$\Delta_fH°$(0 K) = 107.5 ± 0.7 kJ mol^{-1}
$\Delta_fH°$(298.15 K) = 107.3 ± 0.7 kJ mol^{-1}

Electronic Levels and Quantum Weights

State	ϵ_i, cm^{-1}	g_i
$^2S_{1/2}$	0	2
$^2P_{1/2}$	16956.172	2
$^2P_{3/2}$	16973.368	4
$^2S_{1/2}$	25739.991	2
.	.	.
.	.	.
.	.	.
$^2P_{3/2}$	41404.006	4
IP	41449.44	

Enthalpy of Formation

The enthalpy of formation of sodium gas, $\Delta_fH°$(298.15 K) = 107.5 ± 0.7 kJ mol^{-1}, is chosen to be the value recommended by CODATA (1). This value was obtained from 2nd and 3rd law treatment of vapor pressure data. The reported vapor pressures were corrected for the effects of Na_2(g), using D_0^0= 70.63 kJ mol^{-1}. At higher pressures, an alternate virial treatment was employed.

As stated by CODATA (1), the selected value is the weighted average of the 3rd law values of $\Delta_{sub}H°$(298.15 K) calculated from the following:

Source	$\Delta_{sub}H°$(298.15 K), kJ mol^{-1}
Haycock and Lamplough (2)	107.52 ± 0.68
Rodebush and Devries (3)	107.37 ± 0.60
Thiele (4)	107.99 ± 0.60
Kis Tiakowsky (5)	107.38 ± 0.70
Makansi et al. (6)	107.18 ± 0.95
Shpil'rain and Belova (7)	107.26 ± 0.78
Sowa (8)	107.60 ± 1.35
Stone et al. (9)	107.68 ± 1.33
Achener and Jouthas (10)	107.85 ± 0.80
Vinogradov and Volyak (11)	107.44 ± 0.60
Bohdansky and Schins (12)	107.47 ± 1.00
Schins et al. (13)	107.36 ± 1.22

The 2nd law calculations from the above sources led to somewhat more positive values(1).

Heat Capacity and Entropy

The thermal functions for the five alkali metal monatomic gases are calculated by the same procedure. Oberved and estimated atomic energy levels are included in the partition function calculation, using an ionization potential lowering (IP-kT) technique as the cutoff procedure in the energy level summation (15).

The lowest lying levels for these metals [14904 cm^{-1} (Li), 16956 cm^{-1} (Na), 12985 cm^{-1} (K), 12578 cm^{-1} (Cs)] do not contribute to the thermal functions below ~1000 K; there is only a translational contribution below this temperature. Above this approximate temperature, the thermal functions become increasingly sensitive to the partition function cutoff procedure used, due to the combined effect of the observation of atomic energy levels of high principal quantum number and a low ionization potential. Energy levels have been observed up to n = 42 (Li), 59(Na), 79(K), 77(Rb), and 73(Cs). However, not all predicted levels have been observed for each of these principal quantum numbers. The ionization potentials vary from 43487.29 cm^{-1} for Li to 31406.1 cm^{-1} for Cs. In calculating the thermal functions with the inclusion of missing levels up to the high principal quantum numbers just mentioned, the Gibbs energy functions show significant differences (depending on the cutoff procedure) above 3000 K (15).

The atomic energy levels have been taken from the compilation of Martin and Zalubas (14). Our calculated values for the thermal functions are similar to those recommended by CODATA (1). They do differ for two reasons, however. First, the entropy differs by 0.1094 J K^{-1} mol^{-1} because this table uses a reference pressure of 1 bar, whereas the CODATA recommendations are based on 1 atm. Second, the entropies at 298.15 K for all alkali metal gases differs by ~0.004 J K^{-1} mol^{-1}, presumably due to the use of slightly different values for auxiliary data.

References
1. J. D. Cox, chairman, CODATA Task Group on Key Values for Thermodynamics, J. Chem. Thermodynamics 10, 908 (1978); CODATA Special Report No. 8, 1980.
2. C. T. Heycock and F. E. E. Lamplough, Proc. Chem. Soc. 28, 3, (1912).
3. W. H. Rodebush and T. DeVries, J. Am,. Chem. Soc. 47, 2488 (1925).
4. E. Thiele, Ann. Physik [5] 14, 937 (1932).
5. G. B. Kistiakowshy, Naval Res. Lab. Rept. NRL-2958, 1941.
6. M. M. Makansi, C. H. Nuendel and W. A. Selke, J. Phys. Chem. 59, 40 (1955).
7. E. E. Shpil'rain and A. M. Belova, Nauch. Issled. Inst. Vys. Temp. Rept. No. 264, 1862.
8. E. S. Sowa, Nucleonics 21, 76 (1963).
9. J. P. Stone, C. P. Ewing, et al., J. Chem. Eng. Data 11, 315 (1965).
10. P. Y. Achener and J. T. Jouthas, U. S. AEC rept. AGN-1891, 1966.
11. Yu. K. Vonogradov and L. D. Volyak, Teplofiz. Vys. Temp., 4, 50 (1966).
12. J. Bohdansky and H. E. J. Schins, J. Apl. Phys. 36, 3683 (1965).
13. H. E. J. Schins, R. W. M. van Wijk and B. Dorpema, Z. Meallkunde 62, 330 (1971).
14. W. C. Martin and R. Zalubas, J. Phys. Chem. Ref. Data 10, 153 (1981).
15. J. R. Downey, Jr., The Dow Chemical Company, Rept. AFOSR-TR-78-0960, Contract No. F44620-75-1-0048, 1978.

Sodium (Na) Na_1(g)

Enthalpy Reference Temperature = T_r = 298.15 K Standard State Pressure = p° = 0.1 MPa

T/K	$C_p°$ (J $K^{-1}mol^{-1}$)	S° (J $K^{-1}mol^{-1}$)	-[G°-H°(T_r)]/T (J $K^{-1}mol^{-1}$)	H°-H°(T_r) (kJ mol^{-1})	$\Delta_fH°$ (kJ mol^{-1})	$\Delta_fG°$ (kJ mol^{-1})	Log K_f
0	0.	0.	INFINITE	-6.197	107.549	107.549	INFINITE
100	20.786	130.960	172.147	-4.119	108.288	97.574	-50.967
200	20.786	145.367	155.568	-2.040	107.911	86.977	-22.716
250	20.786	150.006	154.009	-1.001	107.626	81.776	-17.086
298.15	20.786	153.667	153.667	0.	107.300	76.825	-13.460
300	20.786	153.795	153.667	0.038	107.286	76.636	-13.344
350	20.786	157.000	153.920	1.078	106.873	71.560	-10.680
400	20.786	159.775	154.482	2.117	103.743	66.753	-8.717
450	20.786	162.223	155.209	3.156	103.220	62.161	-7.216
500	20.786	164.413	156.022	4.196	102.721	57.626	-6.020
600	20.786	168.203	157.746	6.274	101.783	48.697	-4.239
700	20.786	171.407	159.475	8.353	100.909	39.919	-2.979
800	20.786	174.183	161.144	10.431	100.079	31.263	-2.041
900	20.786	176.631	162.731	12.510	99.270	22.710	-1.318
1000	20.786	178.821	164.232	14.589	98.462	14.246	-0.744
1100	20.786	180.802	165.650	16.667	97.632	5.865	-0.278
1170.525	20.786	182.094	166.602	18.133	------- FUGACITY = 1 bar -------		
1200	20.786	182.611	166.989	18.746	0.	0.	0.
1300	20.786	184.275	168.256	20.824	0.	0.	0.
1400	20.786	185.815	169.456	22.903	0.	0.	0.
1500	20.787	187.249	170.595	24.982	0.	0.	0.
1600	20.787	188.591	171.678	27.060	0.	0.	0.
1700	20.789	189.851	172.710	29.139	0.	0.	0.
1800	20.792	191.039	173.696	31.218	0.	0.	0.
1900	20.797	192.164	174.639	33.298	0.	0.	0.
2000	20.805	193.231	175.542	35.378	0.	0.	0.
2100	20.816	194.246	176.408	37.459	0.	0.	0.
2200	20.833	195.215	177.241	39.541	0.	0.	0.
2300	20.856	196.141	178.043	41.626	0.	0.	0.
2400	20.886	197.029	178.816	43.713	0.	0.	0.
2500	20.924	197.883	179.562	45.803	0.	0.	0.
2600	20.973	198.704	180.282	47.898	0.	0.	0.
2700	21.032	199.497	180.979	49.998	0.	0.	0.
2800	21.104	200.263	181.654	52.105	0.	0.	0.
2900	21.189	201.005	182.309	54.219	0.	0.	0.
3000	21.289	201.725	182.944	56.343	0.	0.	0.
3100	21.403	202.425	183.561	58.477	0.	0.	0.
3200	21.535	203.106	184.161	60.624	0.	0.	0.
3300	21.680	203.771	184.746	62.784	0.	0.	0.
3400	21.845	204.421	185.315	64.960	0.	0.	0.
3500	22.030	205.057	185.870	67.154	0.	0.	0.
3600	22.234	205.680	186.411	69.367	0.	0.	0.
3700	22.460	206.292	186.941	71.601	0.	0.	0.
3800	22.709	206.894	187.458	73.860	0.	0.	0.
3900	22.981	207.488	187.964	76.144	0.	0.	0.
4000	23.278	208.073	188.459	78.457	0.	0.	0.
4100	23.602	208.652	188.945	80.800	0.	0.	0.
4200	23.945	209.224	189.421	83.175	0.	0.	0.
4300	24.324	209.792	189.888	85.589	0.	0.	0.
4400	24.611	210.345	190.346	87.998	0.	0.	0.
4500	24.992	210.901	190.796	90.469	0.	0.	0.
4600	25.421	211.454	191.239	92.990	0.	0.	0.
4700	25.876	212.006	191.675	95.554	0.	0.	0.
4800	26.359	212.556	192.105	98.166	0.	0.	0.
4900	26.868	213.104	192.528	100.827	0.	0.	0.
5000	27.388	213.651	192.944	103.533	0.	0.	0.
5100	27.949	214.199	193.356	106.299	0.	0.	0.
5200	28.537	214.747	193.762	109.123	0.	0.	0.
5300	29.150	215.297	194.163	112.007	0.	0.	0.
5400	29.789	215.847	194.560	114.954	0.	0.	0.
5500	30.453	216.400	194.952	117.966	0.	0.	0.
5600	31.141	216.955	195.340	121.046	0.	0.	0.
5700	31.851	217.512	195.724	124.195	0.	0.	0.
5800	32.583	218.073	196.104	127.417	0.	0.	0.
5900	33.334	218.636	196.481	130.712	0.	0.	0.
6000	33.880	219.173	196.852	133.926	0.	0.	0.

PREVIOUS: June 1962 (1 atm) CURRENT: December 1983 (1 bar)

Sodium (Na) Na_1(g)

SODIUM, ION (Na^+) IDEAL GAS M_r = 22.98922

IP(Na^+, g) = 381390 ± 50 eV
S°(298.15 K) = 147.953 ± 0.025 J K^{-1} mol^{-1}

Δ_fH°(O K) = 603.4 ± 0.7 kJ mol^{-1}
Δ_fH°(298.15 K) = [609.343] kJ mol^{-1}

Electronic Level and Quantum Weight

State	ε_i, cm^{-1}	g_i
$^1S_0$	0	1

Enthalpy of Formation

Δ_fH°(Na^+, g, O K) is calculated from Δ_fH°(Na, g, O K) (1) using the spectroscopic value of IP(Na) = 41449.44 ± 0.03 cm^{-1} (495.846 ± 0.000 kJ mol^{-1}) from Martin and Zolubas (2). The ionization limit is converted from cm^{-1} to kJ mol^{-1} using the factor, 1 cm^{-1} = 0.01196266 kJ cm^{-1}, which is derived from the 1973 CODATA fundamental constants (3). Rosenstock et al. (4) and Levin and Lias (5) have summarized additional ionization and appearance potential data.

Δ_fH°(Na^+, g, 298.15 K) is calculated from Δ_fH°(Na, g, O K) by using IP(Na) with JANAF (1) enthalpies, H°(O K) - H°(298.15 K), for Na(g), Na^+(g), and e^-(g). Δ_rH°(Na → Na^+ + e^-, 298.15 K) differs from a room temperature threshold energy due to inclusion of these enthalpies and to threshold effects discussed by Rosenstock et al. (4). Δ_fH°(298.15 K) should be changed by -6.197 kJ mol^{-1} if it is to be used in the ion convention that excludes the enthalpy of the electron.

Heat Capacity and Entropy

The information on electronic energy levels and quantum weights, given by Martin and Zolubas (2), is incomplete because many theoretically predicted levels have not been observed. Our calculations indicate that any reasonable method of filling in these missing levels and cutting off the summation in the partition function (6) has no effect on the thermodynamic functions to 6000 K. This is a result of the high energy of all levels other than the ground state; the first excited state is approximately 264924 cm^{-1} above the ground state. Since inclusion of these excited states has no effect on the thermodynamic functions (to 6000 K), we list only the ground state. The reported uncertainty in S°(298.15 K) is due to uncertainties in the relative ionic mass and the fundamental constants. Extension of these calculations above 6000 K may require consideration of the excited states and use of different fill and cutoff procedures (6).

References

1. JANAF Thermochemical Tables: Na(g), 12-31-83; e^-(g), 3-31-82.
2. W. C. Martin and R. Zolubas, J. Phys. Chem. Ref. Data 10, 153 (1981).
3. E. R. Cohen and B. N. Taylor, J. Phys. Chem. Ref. Data 2, 663 (1973).
4. H. M. Rosenstock, K. Draxl et al., J. Phys. Chem. Ref. Data 6, Supp. 1 (1977).
5. R. D. Levin and S. G. Lias, U. S. Nat. Bur. Stand., NSRDS-NBS-71, 1982.
6. J. R. Downey, The Dow Chemical Company, AFOSR-TR-78-960, Contract No. F44620-75-1-0048, 1978.

Sodium, Ion (Na^+) Na_1^+(g)

Enthalpy Reference Temperature = T_r = 298.15 K Standard State Pressure = p° = 0.1 MPa

T/K	$C_p°$ (J $K^{-1}mol^{-1}$)	S° (J $K^{-1}mol^{-1}$)	-[G°-H°(T_r)]/T (J $K^{-1}mol^{-1}$)	H°-H°(T_r) (kJ mol^{-1})	Δ_fH° (kJ mol^{-1})	Δ_fG° (kJ mol^{-1})	Log K_f
0	0.	0.	INFINITE	-6.197	603.395		
100	20.786	125.246	166.434	-4.119			
200	20.786	139.654	149.855	-2.040			
250	20.786	144.292	148.296	-1.001			
298.15	20.786	147.953	147.953	0.	609.343	574.317	-100.618
300	20.786	148.082	147.954	0.038	609.368	574.100	-99.960
350	20.786	151.286	148.207	1.078	609.994	568.171	-84.795
400	20.786	154.062	148.769	2.117	607.903	562.364	-73.437
450	20.786	156.510	149.496	3.156	608.419	556.641	-64.613
500	20.786	158.700	150.309	4.196	608.959	550.859	-57.548
600	20.786	162.490	152.033	6.274	610.100	539.133	-46.936
700	20.786	165.694	153.761	8.353	611.305	527.211	-39.341
800	20.786	168.470	155.430	10.431	612.553	515.112	-33.633
900	20.786	170.918	157.018	12.510	613.823	502.856	-29.185
1000	20.786	173.108	158.519	14.589	615.093	490.458	-25.619
1100	20.786	175.089	159.937	16.667	616.342	477.934	-22.695
1200	20.786	176.898	161.276	18.746	520.789	467.738	-20.360
1300	20.786	178.561	162.542	20.824	522.867	463.232	-18.613
1400	20.786	180.102	163.742	22.903	524.946	458.567	-17.109
1500	20.786	181.536	164.881	24.982	527.025	453.753	-15.801
1600	20.786	182.877	165.965	27.060	529.103	448.801	-14.652
1700	20.786	184.137	166.997	29.139	531.181	443.718	-13.634
1800	20.786	185.326	167.982	31.217	533.260	438.513	-12.725
1900	20.786	186.449	168.925	33.296	535.337	433.193	-11.909
2000	20.786	187.516	169.828	35.375	537.415	427.763	-11.172
2100	20.786	188.530	170.695	37.453	539.491	422.229	-10.502
2200	20.786	189.497	171.528	39.532	541.566	416.597	-9.891
2300	20.786	190.421	172.329	41.610	543.638	410.870	-9.331
2400	20.786	191.305	173.102	43.689	545.709	405.053	-8.816
2500	20.786	192.154	173.847	45.768	547.775	399.150	-8.340
2600	20.786	192.969	174.567	47.846	549.838	393.164	-7.899
2700	20.786	193.754	175.263	49.925	551.895	387.099	-7.489
2800	20.786	194.509	175.937	52.004	553.945	380.958	-7.107
2900	20.786	195.239	176.590	54.082	555.988	374.744	-6.750
3000	20.786	195.944	177.223	56.161	558.021	368.460	-6.415
3100	20.786	196.625	177.838	58.239	560.044	362.107	-6.101
3200	20.786	197.285	178.436	60.318	562.055	355.690	-5.806
3300	20.786	197.925	179.017	62.397	564.052	349.210	-5.528
3400	20.786	198.545	179.582	64.475	566.033	342.670	-5.264
3500	20.786	199.148	180.132	66.554	567.997	336.072	-5.016
3600	20.786	199.733	180.669	68.632	569.941	329.418	-4.780
3700	20.786	200.303	181.192	70.711	571.863	322.710	-4.556
3800	20.786	200.857	181.702	72.790	573.762	315.951	-4.343
3900	20.786	201.397	182.200	74.868	575.635	309.142	-4.140
4000	20.786	201.923	182.687	76.947	577.480	302.285	-3.947
4100	20.786	202.437	183.162	79.025	579.293	295.383	-3.763
4200	20.786	202.937	183.627	81.104	581.075	288.437	-3.587
4300	20.786	203.427	184.082	83.183	582.819	281.448	-3.419
4400	20.786	203.904	184.527	85.261	584.567	274.417	-3.258
4500	20.786	204.372	184.963	87.340	586.253	267.348	-3.103
4600	20.786	204.828	185.390	89.418	587.890	260.243	-2.955
4700	20.786	205.275	185.808	91.497	589.483	253.103	-2.813
4800	20.786	205.713	186.218	93.576	591.028	245.929	-2.676
4900	20.786	206.142	186.620	95.654	592.524	238.724	-2.545
5000	20.786	206.562	187.015	97.733	593.976	231.488	-2.418
5100	20.786	206.973	187.402	99.811	595.366	224.225	-2.297
5200	20.786	207.377	187.783	101.890	596.700	216.934	-2.179
5300	20.786	207.773	188.156	103.969	597.973	209.619	-2.066
5400	20.786	208.161	188.523	106.047	599.183	202.280	-1.957
5500	20.786	208.543	188.883	108.126	600.328	194.919	-1.851
5600	20.786	208.917	189.238	110.204	601.406	187.539	-1.749
5700	20.786	209.285	189.586	112.283	602.414	180.139	-1.651
5800	20.786	209.647	189.929	114.362	603.349	172.722	-1.556
5900	20.786	210.002	190.266	116.440	604.211	165.290	-1.463
6000	20.786	210.351	190.598	118.519	605.155	157.826	-1.374

PREVIOUS: March 1965 (1 atm) CURRENT: December 1983 (1 bar)

Sodium, Ion (Na^+) Na_1^+(g)

SODIUM, ION (Na^-) IDEAL GAS M_r = 22.99032

EA(Na, g) = 0.547930 ± 0.000025 eV

S°(298.15 K) = 147.954 ± 0.005 J K^{-1} mol^{-1}

$\Delta_fH°$(0 K) = 54.68 ± 0.1 kJ mol^{-1}

$\Delta_fH°$(298.15 K) = [48.236] kJ mol^{-1}

Electonic Level and Quantum Weight

State	ε_i, cm^{-1}	g_i
$^1S_0$	0.0	1

Enthalpy of Formation

The enthalpy of formation, at 0 K, for Na^-(g) is calculated from the adopted electron affinity, EA(Na, g) = 0.547930 ± 0.000025 eV (52.867 ± 0.002 kJ mol^{-1}). This value, recommended by Hotop and Lineberger (6), was measured by a tunable laser detachment threshold technique (2). Additional discussion on Na^-(g) may be obtained in the critical discussions of Hotop and Lineberger (1, 6), Rosenstock et al. (3), and Massey (4).

$\Delta_fH°$(Na^-, g, 298.15 K) is obtained from $\Delta_fH°$(Na, g, 0 K) by using EA(Na, g) with JANAF (5) enthalpies, H°(0 K)−H°(298.15 K), for Na^-(g), Na(g), and e^-(g). $\Delta_rH°$(Na^- → Na + e^-, 298.15 K) differs from a room-temperature threshold energy due to inclusion of these enthalpies and to threshold effects discussed by Rosenstock et al. (3). $\Delta_fH°$(298.15 K) should be changed by +6.197 kJ mol^{-1} if it is to be used in the ion convention which excludes the enthalpy of the electron.

Heat Capacity and Entropy

The ground state configuration for Na^-(g) is given by Hotop and Lineberger (1, 6) and Rosenstock et al. (3). Lacking any experimental evidence as to the stability of any excited states, we assume that no stable excited states exist.

References

1. H. Hotop and W. C. Lineberger, J. Phys. Chem. Ref. Data 4, 539 (1975).
2. P. A. Schulz, R. D. Mead, and W. C. Lineberger, Phys. Rev. A, in preparation.
3. H. M. Rosenstock, K. Draxl et al., J. Phys. Chem. Ref. Data 6, Supp. 1 (1977).
4. H. S. W. Massey, "Negative Ions," 3rd ed., Cambridge University Press, Cambridge, 1976.
5. JANAF Thermochemical Tables: Na(g), 3-31-82; e^-(g), 3-31-82.
6. H. Hotop and W. C. Lineberger, J. Phys. Chem. Ref. Data, to be published, 1985.

Sodium, Ion (Na^-) Na_1^-(g)

Enthalpy Reference Temperature = T_r = 298.15 K   Standard State Pressure = p° = 0.1 MPa

T/K	$C_p°$ (J $K^{-1}mol^{-1}$)	S° (J $K^{-1}mol^{-1}$)	−[G°−H°(T_r)]/T (J $K^{-1}mol^{-1}$)	H°−H°(T_r) (kJ mol^{-1})	$\Delta_fH°$ (kJ mol^{-1})	$\Delta_fG°$ (kJ mol^{-1})	Log K_f
0	0.	0.	INFINITE	-6.197	54.683		
100	20.786	125.247	166.434	-4.119			
200	20.786	139.655	149.855	-2.040			
250	20.786	144.293	148.296	-1.001			
298.15	20.786	147.954	147.954	0.	48.236	25.719	-4.506
300	20.786	148.083	147.954	0.038	48.184	25.580	-4.454
350	20.786	151.287	148.207	1.078	46.731	21.926	-3.272
400	20.786	154.062	148.770	2.117	42.562	18.692	-2.441
450	20.786	156.511	149.496	3.156	40.999	15.803	-1.834
500	20.786	158.701	150.309	4.196	39.461	13.086	-1.367
600	20.786	162.490	152.033	6.274	36.445	8.095	-0.705
700	20.786	165.695	153.762	8.353	33.492	3.604	-0.269
800	20.786	168.470	155.431	10.431	30.583	-0.467	0.030
900	20.786	170.918	157.018	12.510	27.696	-4.174	0.242
1000	20.786	173.108	158.520	14.589	24.809	-7.561	0.395
1100	20.786	175.090	159.937	16.667	21.900	-10.657	0.506
1200	20.786	176.898	161.277	18.746	-77.810	-11.047	0.481
1300	20.786	178.562	162.543	20.824	-79.889	-5.400	0.217
1400	20.786	180.102	163.743	22.903	-81.967	0.408	-0.015
1500	20.786	181.536	164.882	24.982	-84.046	6.365	-0.222
1600	20.786	182.878	165.965	27.060	-86.125	12.460	-0.407
1700	20.786	184.138	166.998	29.139	-88.203	18.685	-0.574
1800	20.786	185.326	167.983	31.217	-90.282	25.033	-0.726
1900	20.786	186.450	168.926	33.296	-92.362	31.496	-0.866
2000	20.786	187.516	169.829	35.375	-94.442	38.069	-0.994
2100	20.786	188.530	170.695	37.453	-96.523	44.745	-1.113
2200	20.786	189.497	171.528	39.532	-98.605	51.521	-1.223
2300	20.786	190.421	172.330	41.610	-100.690	58.392	-1.326
2400	20.786	191.306	173.102	43.689	-102.777	65.353	-1.422
2500	20.786	192.154	173.847	45.768	-104.867	72.401	-1.513
2600	20.786	192.970	174.567	47.846	-106.962	79.533	-1.598
2700	20.786	193.754	175.263	49.925	-109.062	86.746	-1.678
2800	20.786	194.510	175.937	52.004	-111.169	94.037	-1.754
2900	20.786	195.239	176.590	54.082	-113.283	101.403	-1.826
3000	20.786	195.944	177.224	56.161	-115.407	108.842	-1.895
3100	20.786	196.626	177.839	58.239	-117.541	116.352	-1.961
3200	20.786	197.286	178.436	60.318	-119.688	123.931	-2.023
3300	20.786	197.925	179.017	62.397	-121.848	131.578	-2.083
3400	20.786	198.546	179.583	64.475	-124.024	139.290	-2.140
3500	20.786	199.148	180.133	66.554	-126.218	147.066	-2.195
3600	20.786	199.734	180.669	68.632	-128.431	154.906	-2.248
3700	20.786	200.303	181.192	70.711	-130.665	162.807	-2.298
3800	20.786	200.858	181.703	72.790	-132.924	170.769	-2.347
3900	20.786	201.398	182.201	74.868	-135.208	178.790	-2.395
4000	20.786	201.924	182.687	76.947	-137.521	186.871	-2.440
4100	20.786	202.437	183.163	79.025	-139.865	195.010	-2.484
4200	20.786	202.938	183.628	81.104	-142.239	203.206	-2.527
4300	20.786	203.427	184.082	83.183	-144.653	211.459	-2.569
4400	20.786	203.905	184.528	85.261	-147.062	219.766	-2.609
4500	20.786	204.372	184.963	87.340	-149.533	228.130	-2.648
4600	20.786	204.829	185.390	89.418	-152.054	236.550	-2.686
4700	20.786	205.276	185.809	91.497	-154.618	245.026	-2.723
4800	20.786	205.714	186.219	93.576	-157.230	253.556	-2.759
4900	20.786	206.142	186.621	95.654	-159.891	262.142	-2.794
5000	20.786	206.562	187.016	97.733	-162.597	270.781	-2.829
5100	20.786	206.974	187.403	99.811	-165.363	279.476	-2.862
5200	20.786	207.377	187.783	101.890	-168.187	288.226	-2.895
5300	20.786	207.773	188.157	103.969	-171.071	297.031	-2.927
5400	20.786	208.162	188.524	106.047	-174.018	305.890	-2.959
5500	20.786	208.543	188.884	108.126	-177.030	314.805	-2.990
5600	20.786	208.918	189.239	110.204	-180.110	323.775	-3.020
5700	20.786	209.286	189.587	112.283	-183.259	332.801	-3.050
5800	20.786	209.647	189.930	114.362	-186.481	341.883	-3.079
5900	20.786	210.003	190.267	116.440	-189.776	351.020	-3.108
6000	20.786	210.352	190.599	118.519	-192.990	360.197	-3.136

PREVIOUS: CURRENT: December 1983 (1 bar)

Sodium, Ion (Na^-) Na_1^-(g)

SODIUM OXIDE (NaO) IDEAL GAS M_r = 38.98917

Sodium Oxide (NaO) $Na_1O_1(g)$

$S°(298.15\ K) = [229.13 \pm 2.1]\ J\ K^{-1}\ mol^{-1}$

$\Delta_f H°(0\ K) = [85.04 \pm 41.8]\ kJ\ mol^{-1}$
$\Delta_f H°(298.15\ K) = [83.68 \pm 41.8]\ kJ\ mol^{-1}$

Electronic Level and Quantum Weight

State	ε_i, cm^{-1}	g_i
$[^2\Pi]$	0	[4]

$\omega_e = [440]\ cm^{-1}$ $\omega_e x_e = [3.0]\ cm^{-1}$ $\sigma = 1$
$B_e = [0.4467]\ cm^{-1}$ $\alpha_e = [0.0043]\ cm^{-1}$ $r_e = [2.0]$ Å

Enthalpy of Formation

Bawn and Evans (1) studied the reaction rate of sodium vapor with nitrous oxide, nitrogen peroxide, nitromethane, ethyl nitrate and amyl nitrite, using the diffusion flame method. On the evidence of reaction rate measurements they estimated the value, $\Delta_r H°(298.15\ K) < 0$, for the reaction $Na(g) + NO_2(g) = NaO(g) + NO(g)$. Based on this result, we calculate $\Delta_f H°(NaO, g, 298.15\ K) > 12\ kcal\ mol^{-1}$, using $\Delta_f H°(298.15\ K)$ = 25.8, 7.9 and 21.6 kcal mol^{-1} for Na(g), $NO_2(g)$ and NO(g), respectively.

Assuming that $D_0^0(NaO) = 1/2\ [D_0^0(Na_2) + D_0^0(O_2)]$, or $\Delta_r H°(0\ K) = 68.1\ kcal\ mol^{-1}$ for the reaction $NaO(g) = Na(g) + O(g)$, we derive $\Delta_f H°$ = 16.7 and $\Delta_f H°(298.15\ K) = 16.4\ kcal\ mol^{-1}$, using $\Delta_r H°(0\ K)$ = 25.8 and 59.0 kcal mol^{-1} for Na(g) and O(g), respectively.

Somayajulu (2) found that in a sequence of similar diatomic molecules the force constant (k_e), dissociation energy (D_0^0) and equilibrium bond distance (r_e) are related by the expression $k_e r_e (D_0^0)$ = constant. Using r_e = 1.62 Å, $D_0^0 = 77.9\ kcal\ mol^{-1}$ for LiO(g), r_e = 2.0 Å for NaO(g) and $k_e(NaO)/k_e(LiO) = 0.4827$, we evaluate $D_0^0 = 65.4\ kcal\ mol^{-1}$ for NaO(g), yielding $\Delta_f H°(NaO, g, 298.15\ K) = 19.4\ kcal\ mol^{-1}$. The value k_e is calculated from the relation $k_e = 4\pi^2\nu^2\mu$ where ν is the vibrational frequency of the diatomic molecule and μ is the reduced mass.

The value of $\Delta_f H°(NaO, g, 298.15\ K)$ is tentatively adopted as 20 ± 10 kcal mol^{-1}. The corresponding $D_0^0(NaO)$ is 65 ± 10 kcal mol^{-1}.

Heat Capacity and Entropy

The ground state configuration is assumed to be the same as that for OH(g) which has the same number of valence electrons as NaO(g). The values of ω_e and $\omega_e x_e$ are estimated by comparison with those for LiO(g), LiF(g) and NaF(g). The value of r_e is estimated from those for OH(g), HF(g) and NaF(g). B_e and α_e are derived from r_e, ω_e and $\omega_e x_e$ by the method suggested by Herzberg (3).

References

1. C. E. H. Bawn and A. G. Evans, Trans. Faraday Soc. 33, 1571 (1937).
2. G. R. Somayajulu, J. Chem. Phys. 33, 1541 (1960).
3. G. Herzberg, "Spectra of Diatomic Molecules," D. Van Nostrand Co., Inc., New York, 1950.

Enthalpy Reference Temperature = T_r = 298.15 K Standard State Pressure = p° = 0.1 MPa

T/K	$C_p°$ (J K⁻¹mol⁻¹)	S° (J K⁻¹mol⁻¹)	$-[G°-H°(T_r)]/T$ (J K⁻¹mol⁻¹)	$H°-H°(T_r)$ (kJ mol⁻¹)	$\Delta_f H°$ (kJ mol⁻¹)	$\Delta_f G°$ (kJ mol⁻¹)	Log K_f
0	0.	0.	INFINITE	-9.427	85.042	85.042	INFINITE
100	29.760	193.820	258.890	-6.507	85.169	76.834	-40.134
200	33.116	215.486	232.293	-3.361	84.404	68.794	-17.967
250	34.324	223.013	229.708	-1.674	84.038	64.935	-13.567
298.15	35.147	229.133	229.133	0.	83.680	61.287	-10.737
300	35.173	229.351	229.134	0.065	83.666	61.149	-10.647
350	35.777	234.821	229.565	1.840	83.249	57.428	-8.571
400	36.218	239.628	230.528	3.640	80.133	53.977	-7.049
450	36.550	243.914	231.782	5.460	79.632	50.738	-5.889
500	36.809	247.779	233.191	7.294	79.157	47.553	-4.968
600	37.189	254.526	236.201	10.995	78.262	41.317	-3.597
700	37.457	260.280	239.240	14.728	77.415	35.227	-2.629
800	37.664	265.295	242.190	18.484	76.594	29.256	-1.910
900	37.833	269.742	245.009	22.259	75.779	23.388	-1.357
1000	37.979	273.736	247.685	26.050	74.952	17.611	-0.920
1100	38.109	277.362	250.221	29.855	74.093	11.918	-0.566
1200	38.228	280.683	252.623	33.672	-23.575	8.745	-0.381
1300	38.339	283.747	254.901	37.500	-23.616	11.440	-0.460
1400	38.445	286.592	257.064	41.339	-23.662	14.139	-0.528
1500	38.546	289.248	259.122	45.189	-23.712	16.840	-0.586
1600	38.645	291.739	261.084	49.049	-23.765	19.546	-0.638
1700	38.740	294.085	262.956	52.918	-23.820	22.254	-0.684
1800	38.834	296.302	264.748	56.797	-23.878	24.966	-0.724
1900	38.926	298.404	266.464	60.685	-23.940	27.681	-0.761
2000	39.017	300.403	268.112	64.582	-24.004	30.400	-0.794
2100	39.107	302.309	269.695	68.488	-24.071	33.122	-0.824
2200	39.196	304.130	271.219	72.403	-24.142	35.847	-0.851
2300	39.285	305.874	272.688	76.327	-24.218	38.575	-0.876
2400	39.373	307.548	274.106	80.260	-24.299	41.307	-0.899
2500	39.460	309.157	275.476	84.202	-24.385	44.043	-0.920
2600	39.547	310.706	276.802	88.152	-24.477	46.781	-0.940
2700	39.633	312.201	278.085	92.111	-24.577	49.524	-0.958
2800	39.720	313.643	279.330	96.079	-24.685	52.271	-0.975
2900	39.806	315.039	280.537	100.055	-24.802	55.021	-0.991
3000	39.891	316.390	281.710	104.040	-24.930	57.776	-1.006
3100	39.977	317.699	282.850	108.033	-25.068	60.535	-1.020
3200	40.062	318.970	283.959	112.035	-25.220	63.298	-1.033
3300	40.147	320.204	285.038	116.046	-25.385	66.067	-1.046
3400	40.233	321.404	286.090	120.065	-25.566	68.841	-1.058
3500	40.317	322.571	287.116	124.092	-25.764	71.621	-1.069
3600	40.402	323.708	288.117	128.128	-25.981	74.406	-1.080
3700	40.487	324.816	289.094	132.173	-26.218	77.198	-1.090
3800	40.572	325.897	290.048	136.226	-26.478	79.996	-1.100
3900	40.656	326.952	290.981	140.287	-26.761	82.802	-1.109
4000	40.741	327.982	291.893	144.357	-27.072	85.615	-1.118
4100	40.825	328.989	292.786	148.435	-27.412	88.436	-1.127
4200	40.909	329.974	293.659	152.522	-27.781	91.266	-1.135
4300	40.994	330.938	294.515	156.617	-28.186	94.106	-1.143
4400	41.078	331.881	295.354	160.721	-28.585	96.951	-1.151
4500	41.162	332.805	296.176	164.833	-29.043	99.808	-1.159
4600	41.246	333.711	296.982	168.953	-29.548	102.677	-1.166
4700	41.330	334.599	297.773	173.082	-30.095	105.557	-1.173
4800	41.415	335.470	298.549	177.219	-30.687	108.450	-1.180
4900	41.499	336.325	299.311	181.365	-31.326	111.355	-1.187
5000	41.583	337.164	300.060	185.519	-32.008	114.273	-1.194
5100	41.667	337.988	300.796	189.681	-32.749	117.206	-1.200
5200	41.751	338.798	301.519	193.852	-33.547	120.154	-1.207
5300	41.835	339.594	302.230	198.032	-34.403	123.118	-1.213
5400	41.918	340.377	302.929	202.219	-35.321	126.098	-1.220
5500	42.002	341.147	303.617	206.415	-36.304	129.096	-1.226
5600	42.086	341.904	304.294	210.620	-37.355	132.114	-1.232
5700	42.170	342.650	304.960	214.832	-38.476	135.149	-1.239
5800	42.254	343.384	305.616	219.054	-39.670	138.206	-1.245
5900	42.338	344.107	306.263	223.283	-40.940	141.283	-1.251
6000	42.422	344.819	306.899	227.521	-42.130	144.365	-1.257

PREVIOUS: December 1967 (1 atm) CURRENT: December 1967 (1 bar)

Sodium Oxide (NaO) $Na_1O_1(g)$

SODIUM OXIDE, ION (NaO^-) IDEAL GAS M_r = 38.98972

$\Delta_f H°(0\ K)$ = [-113.76 ± 83.7] kJ mol^{-1}

S°(298.15 K) = [217.49 ± 2.1] J K^{-1} mol^{-1} $\Delta_f H°(298.15\ K)$ = [-121.34 ± 83.7] kJ mol^{-1}

Electronic Level and Quantum Weight

State	ε_i, cm^{-1}	g_i
[$^1\Sigma$]	0	[1]

ω_e = [450] cm^{-1} $\omega_e x_e$ = [3.2] cm^{-1} σ = 1

B_e = [0.4467] cm^{-1} α_e = [0.00495] cm^{-1} r_e = [2.0] Å

Enthalpy of Formation

The electron affinity of $NaO^-(g)$ is unavailable, therefore the value of $\Delta_f H°(NaO^-$, g, 298.15 K) is estimated. We first assume that the enthalpy change $\Delta_r H°(0\ K)$ of the reaction (1) $NaO^{-1}(g) = Na(g) + O^-(g)$ is close to the average of the $\Delta_r H°(0\ K)$ values for the reactions (2) $NaO(g) = Na(g) + O(g)$ and (3) $NaF(g) = Na(g) + F(g)$. The atom F(g) is isoelectronic with $O^-(g)$. In other words, the $\Delta_r H°(0\ K)$ for reaction (1) is approximately 1/2 (64.8 + 113.9) = 89.3 kcal mol^{-1}. Then we compare the $\Delta_r H°(0\ K)$ values for the reactions (4) $HO(g) = H(g) + O(g)$, (5) $HO^-(g) = H(g) + O^-(g)$, and (6) $HF(g) = H(g) + F(g)$, and find that $\Delta_r H°(0\ K)$ for reaction (5) is experimentally determined as 109.7 kcal mol^{-1} which is about 8 kcal mol^{-1} more negative than the average of $\Delta_r H°(0\ K)$ values for reactions (4) and (6). Based on this fact we estimate $\Delta_r H°(0\ K)$ = 80 kcal mol^{-1} for reaction (1), yielding $\Delta_f H°(NaO^-$, g, 298.15 K) = -29 kcal mol^{-1} which is tenatively adopted. The uncertainty of this value may be ±20 kcal mol^{-1}. Based on this $\Delta_f H°(298.15\ K)$ value we calculate E. A. = 2.1 ± 0.9 eV for NaO(g). The corresponding value for HO(g) is 1.8 ± 0.1 eV.

Heat Capacity and Entropy

The ground state configuration is assumed to be the same as that of NaF(g) which is isoelectronic with $NaO^-(g)$. The values of ω_e, $\omega_e x_e$ and r_e are estimated by comparison with the observed data for NaF(g). The values of B_e and α_e are calculated from r_e, ω_e and $\omega_e x_e$ using the method recommended by Herzberg (1).

Reference

1. G. Herzberg, "Spectra of Diatomic Molecules," D. Van Nostrand Co., Inc., New York, 1950.

Sodium Oxide, Ion (NaO^-) $Na_1O_1^-(g)$

Enthalpy Reference Temperature = T_r = 298.15 K Standard State Pressure = p° = 0.1 MPa

T/K	$C_p°$ (J K⁻¹mol⁻¹)	S° (J K⁻¹mol⁻¹)	-[G°-H°(T_r)]/T (J K⁻¹mol⁻¹)	H°-H°(T_r) (kJ mol⁻¹)	$\Delta_f H°$ (kJ mol⁻¹)	$\Delta_f G°$ (kJ mol⁻¹)	Log K_f
0	0.	0.	INFINITE	-9.405	-113.755		
100	29.702	182.286	247.150	-6.486			
200	32.996	203.881	220.638	-3.351			
250	34.222	211.383	218.060	-1.669			
298.15	35.064	217.487	217.487	0.	-121.336	-134.001	23.476
300	35.091	217.704	217.488	0.065	-121.389	-134.080	23.345
350	35.712	223.163	217.917	1.836	-122.848	-136.081	20.309
400	36.168	227.963	218.879	3.633	-127.006	-137.662	17.977
450	36.513	232.243	220.131	5.451	-128.549	-138.901	16.123
500	36.781	236.105	221.538	7.283	-130.066	-139.970	14.623
600	37.175	242.848	224.544	10.982	-133.041	-141.670	12.333
700	37.455	248.601	227.580	14.715	-135.967	-142.876	10.662
800	37.670	253.617	230.527	18.471	-138.866	-143.665	9.380
900	37.846	258.064	233.345	22.247	-141.759	-144.091	8.363
1000	37.997	262.059	236.020	26.040	-144.663	-144.195	7.532
1100	38.132	265.687	238.554	29.846	-147.598	-144.006	6.838
1200	38.256	269.011	240.956	33.666	-247.342	-141.109	6.142
1300	38.371	272.077	243.233	37.497	-249.460	-132.170	5.311
1400	38.481	274.925	245.397	41.340	-251.581	-123.068	4.592
1500	38.586	277.584	247.455	45.193	-253.705	-113.815	3.963
1600	38.688	280.077	249.417	49.057	-255.833	-104.419	3.409
1700	38.788	282.426	251.290	52.931	-257.962	-94.891	2.916
1800	38.885	284.645	253.082	56.814	-260.094	-85.237	2.474
1900	38.981	286.750	254.799	60.708	-262.228	-75.465	2.075
2000	39.075	288.752	256.447	64.611	-264.366	-65.580	1.713
2100	39.168	290.661	258.031	68.523	-266.506	-55.588	1.383
2200	39.261	292.485	259.556	72.444	-268.649	-45.493	1.080
2300	39.352	294.232	261.026	76.375	-270.797	-35.302	0.802
2400	39.443	295.909	262.445	80.315	-272.949	-25.017	0.544
2500	39.534	297.521	263.816	84.264	-275.107	-14.642	0.306
2600	39.624	299.074	265.142	88.221	-277.270	-4.181	0.084
2700	39.714	300.571	266.427	92.188	-279.441	6.364	-0.123
2800	39.803	302.017	267.672	96.164	-281.619	16.989	-0.317
2900	39.892	303.415	268.881	100.149	-283.806	27.692	-0.499
3000	39.981	304.769	270.055	104.143	-286.004	38.471	-0.670
3100	40.070	306.081	271.196	108.145	-288.212	49.323	-0.831
3200	40.158	307.355	272.306	112.157	-290.433	60.246	-0.983
3300	40.247	308.592	273.387	116.177	-292.667	71.239	-1.128
3400	40.335	309.795	274.440	120.206	-294.916	82.301	-1.264
3500	40.423	310.965	275.467	124.244	-297.182	93.428	-1.394
3600	40.511	312.105	276.469	128.290	-299.467	104.621	-1.518
3700	40.598	313.216	277.447	132.346	-301.772	115.877	-1.636
3800	40.686	314.300	278.403	136.410	-304.099	127.196	-1.748
3900	40.774	315.358	279.337	140.483	-306.450	138.577	-1.856
4000	40.861	316.391	280.250	144.565	-308.827	150.018	-1.959
4100	40.949	317.402	281.144	148.655	-311.233	161.519	-2.058
4200	41.036	318.389	282.019	152.755	-313.668	173.079	-2.153
4300	41.124	319.356	282.876	156.863	-316.139	184.697	-2.244
4400	41.211	320.302	283.716	160.979	-318.603	196.370	-2.331
4500	41.298	321.229	284.539	165.105	-321.126	208.101	-2.416
4600	41.385	322.138	285.347	169.239	-323.696	219.890	-2.497
4700	41.473	323.029	286.139	173.382	-326.308	231.735	-2.575
4800	41.560	323.903	286.917	177.534	-328.964	243.636	-2.651
4900	41.647	324.761	287.681	181.694	-331.667	255.594	-2.725
5000	41.734	325.603	288.431	185.863	-334.413	267.606	-2.796
5100	41.821	326.431	289.168	190.041	-337.217	279.674	-2.864
5200	41.908	327.243	289.892	194.227	-340.078	291.797	-2.931
5300	41.995	328.043	290.604	198.422	-342.997	303.977	-2.996
5400	42.082	328.828	291.305	202.626	-345.978	316.211	-3.059
5500	42.169	329.601	291.994	206.839	-349.023	328.502	-3.120
5600	42.256	330.362	292.673	211.060	-352.135	340.849	-3.179
5700	42.343	331.111	293.340	215.290	-355.318	353.252	-3.237
5800	42.430	331.848	293.998	219.529	-358.573	365.711	-3.294
5900	42.517	332.574	294.646	223.776	-361.903	378.227	-3.349
6000	42.604	333.289	295.284	228.032	-365.154	390.782	-3.402

PREVIOUS: December 1967 (1 atm) CURRENT: December 1967 (1 bar)

Sodium Oxide, Ion (NaO^-) $Na_1O_1^-(g)$

SODIUM SUPEROXIDE (NaO_2) CRYSTAL M_r = 54.98857

$S°(298.15\ K) = 115.90 \pm 1.3\ J\ K^{-1}\ mol^{-1}$

$T_{trs1} = 196.5\ K$

$T_{trs2} = 223.3\ K$

$T_{fus} = 825 \pm 10\ K$

$\Delta_f H°(0\ K) = -263.88 \pm 2.9\ kJ\ mol^{-1}$

$\Delta_f H°(298.15\ K) = -260.66\ kJ\ mol^{-1}$

$\Delta_{trs1} H° = 1464.400\ kJ\ mol^{-1}$

$\Delta_{trs2} H° = 1548.080\ kJ\ mol^{-1}$

$\Delta_{fus} H°$ = Unknown

Enthalpy of Formation

$\Delta_f H°(298.15\ K)$ was obtained from (1). Giles and Margrave (2) measured the enthalpy of decomposition of NaO_2(cr) in H_2O, yielding $\Delta_f H°(298.15\ K) = -62.1 \pm 0.7\ kcal\ mol^{-1}$. From the original experimental data by de Forcrand (3), the value of $\Delta_f H°(298.15\ K)$ was recalculated to be $-65.0\ kcal\ mol^{-1}$ by Gilles and Margrave (2).

Heat Capacity and Entropy

The low temperature heat capacities, 52.13 ~ 296.46 K, were measured by Todd (4). The heat capacities above 298.15 K were estimated by comparison with those for Na_2O(cr). S°(298.15 K) was given by Todd (4) using $S°(51\ K) = 2.23\ cal\ K^{-1}\ mol^{-1}$ (extrap.).

Transition Data

T_{trs1}, T_{trs2}, $\Delta_{trs1}H°$ and $\Delta_{trs2}H°$ were calculated from the data reported by Todd (4).

Fusion Data

The melting temperature was obtained from Margrave (5).

References

1. U. S. Nat. Bur. Stand. Report 7437, 1962.
2. P. W. Giles and J. L. Margrave, J. Phys. Chem. 60, 1333 (1956).
3. R. de Forcrand, Compt. Rend. 127, 514 (1898); 158, 843, 991 (1914).
4. S. S. Todd, J. Amer. Chem. Soc. 75, 1229 (1953).
5. J. L. Margrave, University of Wisconsin, personal communication, May 21, 1963.

Sodium Superoxide (NaO_2) Na_1O_2(cr)

Enthalpy Reference Temperature = T_r = 298.15 K Standard State Pressure = p° = 0.1 MPa

T/K	$C_p°$ (J K⁻¹mol⁻¹)	S° (J K⁻¹mol⁻¹)	$-[G°-H°(T_r)]/T$ (J K⁻¹mol⁻¹)	$H°-H°(T_r)$ (kJ mol⁻¹)	$\Delta_f H°$ (kJ mol⁻¹)	$\Delta_f G°$ (kJ mol⁻¹)	Log K_f
0	0.	0.	INFINITE	-18.351	-263.884	-263.884	INFINITE
100	45.564	30.585	194.891	-16.431	-266.208	-249.554	130.354
200	78.450	78.241	118.679	-8.088	-263.232	-232.044	60.604
298.15	72.141	115.897	115.897	0.	-260.663	-218.712	38.317
300	72.216	116.343	115.898	0.134	-260.636	-218.452	38.036
400	76.274	137.673	118.779	7.558	-261.805	-204.405	26.693
500	80.375	155.134	124.354	15.390	-260.133	-190.240	19.874
600	84.475	170.151	130.763	23.633	-258.066	-176.450	15.361
700	88.502	183.474	137.358	32.281	-255.625	-163.035	12.166
800	92.634	195.560	143.889	41.337	-252.814	-149.996	9.794
900	96.962	206.721	150.258	50.817	-249.627	-137.331	7.970
1000	101.211	217.158	156.431	60.726	-246.067	-125.041	6.531
1100	105.186	226.991	162.403	71.046	-242.164	-113.124	5.372
1200	109.177	236.314	168.177	81.764	-334.706	-99.137	4.315
1300	113.186	245.211	173.763	92.882	-329.249	-79.725	3.203
1400	117.211	253.746	179.173	104.402	-323.422	-60.747	2.267
1500	121.252	261.971	184.421	116.325	-317.219	-42.199	1.470
1600	125.311	269.925	189.517	128.653	-310.637	-24.077	0.786
1700	129.386	277.645	194.475	141.388	-303.672	-6.378	0.196
1800	133.478	285.156	199.305	154.531	-296.324	10.899	-0.316
1900	137.587	292.483	204.017	168.084	-288.590	27.758	-0.763
2000	141.712	299.645	208.620	182.048	-280.468	44.200	-1.154

PREVIOUS: CURRENT: June 1963

Sodium Superoxide (NaO_2) Na_1O_2(cr)

SODIUM (Na_2) IDEAL GAS M_r = 45.975

D_0^0 = 71.113 ± 0.072 kJ mol^{-1}
S°(298.15 K) = 230.243 ± 0.030 J K^{-1} mol^{-1}

$\Delta_fH°$(0 K) = 144.56 ± 1.20 kJ mol^{-1}
$\Delta_fH°$(298.15 K) = 142.07 ± 1.20 kJ mol^{-1}

Spectroscopic Data for Na_2 in cm^{-1}

State	T_e	D_0^0	ω_e	ω_ex_e	B_e	$10^3\ \alpha_e$	$10^6\ D_e$	r_e, Å	Reference
$^1\Sigma_g^+$	0	6024	159.11	0.72142[a]	0.15474	0.86827[b]	0.58972[c]	3.07857	(2)[d], (6)[e]
$^1\Sigma_u^+$	14681	22900[f]	117.32	0.35758[g]	0.11078	0.54884[h]	0.38816	3.6384	(5)

(a) $-1.5354 \times 10^{-3}(v + 1/2)^3 - 2.3576 \times 10^{-5}\ (v + 1/2)^4$
(b) $-4.2855 \times 10^{-6}(v + 1/2)^2 - 1.4784 \times 10^{-7}(v + 1/2)^3$
(c) $-3.9068 \times 10^{-9}(v + 1/2) + 9.1188 \times 10^{-11}(v + 1/2)^2$
(d) Reference for D_0^0
(e) Reference for constants
(f) This value was estimated based on the estimate of Konowalow, et al (4), D_e = 8275 cm^{-1}
(g) $+5.1671 \times 10^{-5}(v + 1/2)^3 + 9.2770 \times 10^{-6}(v + 1/2)^4$
(h) $+1.6252 \times 10^{-8}(v + 1/2)^2 + 3.1645 \times 10^{-8}(v + 1/2)^3$

Enthalpy of Formation

Using the adopted value of the dissociation energy for Na_2(g), D_0^0 = 71.113 kJ mol^{-1} (2), and the recommended $\Delta_fH°$(Na, g, 0 K) = 107.32 kJ mol^{-1} from NBS (7), we calculate $\Delta_fH°$(Na_2, g, 0 K) = 143.51 kJ mol^{-1}. Since this value is only 1.05 kJ mol^{-1} smaller than the NBS recommended value of 144.56 kJ mol^{-1} (7) we adopt the NBS result. Correcting this value to 298.15 K using the current value of H°(298.15 K)−H°(0 K) = 10403.27 J K^{-1} mol^{-1} for Na_2 and the NBS value of H°(298.15 K)−H°(0 K) for Na(cr) (1) we obtain $\Delta_fH°$(Na_2, g, 298.15 K) = 142.05 kJ mol^{-1}.

Heat Capacity and Entropy

The heat capacity and entropy were calculated using a direct summation technique similar to the Li_2(g) calculation (see Li_2(g) table) using data recommended by Stwalley (1, 2) for the ground state. Reliable spectroscopic data for the lower lying excited states of Na_2 is sparse. Only the $^1\Sigma_u^+$ state at T_e = 14680.6 cm^{-1} has been experimentally characterized with any degree of confidence (3, 6). Theoretical calculations by Konowalow, Rosenkrantz and Olson (4) show that the relative ordering of states is equivalent to that of Li_2(g) (see Li_2(g) table). Since the excited state contribution to all thermodynamic functions (except for the heat capacity) for Li_2(g) was small (~0.2% at 6000 K) we decided to treat the excited states for Na_2(g) in an approximate way as follows. First, the $^3\Sigma_u^+$ state was neglected entirely due to the extremely shallow potential well (The $^3\Sigma_u^+$ state of Li_2(g) contributes insignificantly to the thermodynamic functions.) Second, the $^3\Pi_u$, $^1\Sigma_u^+$, and $^3\Sigma_g^+$ states were combined as a single $^1\Sigma_u^+$ state using the spectroscopic constants of Kamimsky (5) and an artifical degeneracy of 10 (6 for the $^3\Pi_u$, 1 for the $^1\Sigma_u^+$, and 3 for the $^3\Sigma_g^+$ state). This procedure certainly overestimates the $^3\Sigma_g^+$ contribution (which lies above the $^1\Sigma_u^+$ state by ca. 3300 cm^{-1}) but at the same time underestimates the $^3\Pi_u$ contribution (which lies below the $^1\Sigma_u^+$ state by ca 1100 cm^{-1}) (4). In any case the contribution to the Gibbs energy function is very small (< 0.01 J K^{-1} mol^{-1}) at temperatures below 2500 K and only contributes ca. 3 J mol^{-1} K^{-1} at 6000 K. Note also that the Dunham coefficients for the ground state are reported to higher order in (6) than used in the calculation. The neglect of these higher order terms results in uncertainties similar to the Li_2(g) calculation (the Gibbs energy function at 6000 K is probably only 0.3 J K^{-1} mol^{-1} smaller than the truncated rovib expression method). The rotational levels of the ground state were weighted in accordance with the nuclear spin of 3/2 (even J weight = 0.625, odd J weight = 0.375). The adopted value of S°(298.15 K) is only 0.1 J K^{-1} mol^{-1} lower than that adopted by NBS (7).

Reference

1. W. C. Stwalley, Director, Iowa Laser Facility and Professor of Chemistry and Physics, University of Iowa, personal communication, 1982.
2. K. K. Verma, J. T. Bahns, A. R. Rizi, W. C. Stwalley, and W. T. Zemke, J. Chem. Phys. 78, 3599 (1983).
3. K. P. Huber and G. Herzberg, "Constants of Diatomic Molecules," Van Nostrand Reinhold Co., New York, 1979.
4. D. D. Konowalow, M. E. Rosenkrantz, and M. L. Olson, J. Chem. Phys. 72, 2612 (1980).
5. M. E. Kaminsky, J. Chem. Phys, 66, 4951 (1977).
6. P. Kusch and M. M. Hessel, J. Chem. Phys. 68, 2591 (1978).
7. D. D. Wagman, et al., J. Phys. Chem. Ref. Data 11, Supp. 2 (1982).

Sodium (Na_2) Na_2(g)

Enthalpy Reference Temperature = T_r = 298.15 K Standard State Pressure = p° = 0.1 MPa

T/K	$C_p°$ (J K^{-1}mol^{-1})	S° (J K^{-1}mol^{-1})	−[G°−H°(T_r)]/T (J K^{-1}mol^{-1})	H°−H°(T_r) (kJ mol^{-1})	$\Delta_fH°$ (kJ mol^{-1})	$\Delta_fG°$ (kJ mol^{-1})	Log K_f
0	0.	0.	INFINITE	-10.403	144.560	144.560	INFINITE
100	34.747	190.419	263.127	-7.271	145.013	130.734	-68.288
200	36.922	215.363	233.663	-3.660	143.713	116.919	-30.536
250	37.315	223.647	230.860	-1.803	142.920	110.310	-23.048
298.15	37.577	230.243	230.243	0.	142.070	104.105	-18.239
300	37.585	230.476	230.244	0.070	142.035	103.870	-18.085
350	37.799	236.286	230.702	1.954	141.014	97.587	-14.564
400	37.984	241.346	231.723	3.849	134.571	91.873	-11.997
450	38.157	245.830	233.046	5.752	133.349	86.610	-10.053
500	38.327	249.859	234.529	7.665	132.184	81.480	-8.512
600	38.667	256.876	237.686	11.514	130.002	71.547	-6.229
700	38.998	262.862	240.865	15.398	127.981	61.966	-4.624
800	39.269	268.088	243.949	19.312	126.077	52.667	-3.439
900	39.411	272.723	246.893	23.247	124.238	43.601	-2.531
1000	39.369	276.875	249.687	27.188	122.404	34.740	-1.815
1100	39.123	280.618	252.332	31.114	120.513	26.064	-1.238
1200	38.680	284.004	254.833	35.006	-75.016	22.445	-0.977
1300	38.074	287.077	257.197	38.845	-75.334	30.579	-1.229
1400	37.347	289.873	259.432	42.617	-75.720	38.741	-1.445
1500	36.546	292.423	261.548	46.312	-76.182	46.932	-1.634
1600	35.714	294.755	263.552	49.925	-76.726	55.157	-1.801
1700	34.891	296.895	265.451	53.455	-77.354	63.418	-1.949
1800	34.110	298.867	267.253	56.904	-78.062	71.719	-2.081
1900	33.399	300.692	268.966	60.279	-78.846	80.061	-2.201
2000	32.782	302.389	270.595	63.587	-79.698	88.447	-2.310
2100	32.280	303.976	272.147	66.840	-80.608	96.876	-2.410
2200	31.908	305.468	273.628	70.048	-81.565	105.350	-2.501
2300	31.681	306.881	275.044	73.226	-82.555	113.868	-2.586
2400	31.610	308.227	276.398	76.389	-83.566	122.430	-2.665
2500	31.700	309.519	277.698	79.553	-84.583	131.034	-2.738
2600	31.957	310.767	278.946	82.735	-85.591	139.678	-2.806
2700	32.381	311.980	280.147	85.950	-86.575	148.361	-2.870
2800	32.968	313.168	281.305	89.217	-87.523	157.080	-2.930
2900	33.712	314.337	282.424	92.549	-88.419	165.832	-2.987
3000	34.601	315.495	283.507	95.964	-89.252	174.614	-3.040
3100	35.624	316.646	284.557	99.474	-90.011	183.422	-3.091
3200	36.761	317.794	285.578	103.092	-90.686	192.253	-3.138
3300	37.996	318.944	286.572	106.830	-91.269	201.105	-3.183
3400	39.305	320.098	287.541	110.694	-91.756	209.972	-3.226
3500	40.666	321.257	288.488	114.692	-92.145	218.852	-3.266
3600	42.055	322.422	289.414	118.828	-92.435	227.742	-3.304
3700	43.449	323.593	290.322	123.103	-92.629	236.639	-3.341
3800	44.823	324.770	291.213	127.517	-92.732	245.540	-3.375
3900	46.156	325.952	292.088	132.067	-92.751	254.442	-3.408
4000	47.426	327.136	292.950	136.746	-92.697	263.344	-3.439
4100	48.616	328.322	293.798	141.549	-92.582	272.243	-3.468
4200	49.710	329.507	294.634	146.466	-92.414	281.139	-3.496
4300	50.695	330.689	295.459	151.488	-92.220	290.031	-3.523
4400	51.560	331.864	296.273	156.601	-91.925	298.911	-3.549
4500	52.299	333.032	297.077	161.795	-91.673	307.789	-3.573
4600	52.908	334.188	297.871	167.057	-91.452	316.664	-3.596
4700	53.386	335.331	298.656	172.373	-91.266	325.534	-3.618
4800	53.733	336.459	299.432	177.730	-91.132	334.401	-3.639
4900	53.952	337.569	300.199	183.115	-91.069	343.265	-3.659
5000	54.050	338.660	300.957	188.516	-91.079	352.128	-3.679
5100	54.032	339.731	301.707	193.921	-91.207	360.993	-3.697
5200	53.906	340.779	302.448	199.319	-91.458	369.862	-3.715
5300	53.681	341.804	303.181	204.699	-91.846	378.737	-3.733
5400	53.366	342.804	303.906	210.052	-92.386	387.620	-3.749
5500	52.971	343.780	304.622	215.370	-93.093	396.516	-3.766
5600	52.504	344.731	305.330	220.644	-93.977	405.425	-3.782
5700	51.976	345.655	306.029	225.868	-95.052	414.353	-3.797
5800	51.395	346.554	306.720	231.037	-96.326	423.300	-3.812
5900	50.769	347.428	307.403	236.146	-97.809	432.272	-3.827
6000	50.107	348.275	308.077	241.190	-99.192	441.234	-3.841

PREVIOUS: June 1962 (1 atm) CURRENT: December 1983 (1 bar)

Sodium (Na_2) Na_2(g)

$S°(298.15\ K) = 75.042\ J\ K^{-1}\ mol^{-1}$
$T_{trs}(\gamma \rightarrow \beta) = 1023.2\ K$
$T_{trs}(\beta \rightarrow \alpha) = 1243.2\ K$
$T_{fus} = 1405.2\ K$

$\Delta_f H°(0\ K) = -413.15 \pm 4.2\ kJ\ mol^{-1}$
$\Delta_f H°(298.15\ K) = -417.98 \pm 4.2\ kJ\ mol^{-1}$
$\Delta_{trs} H°(\gamma \rightarrow \beta) = 1.757\ kJ\ mol^{-1}$
$\Delta_{trs} H°(\beta \rightarrow \alpha) = 11.924\ kJ\ mol^{-1}$
$\Delta_{fus} H° = 47.698\ kJ\ mol^{-1}$

Enthalpy of Formation

The enthalpy change of the reaction $Na_2O(cr) + H_2O(\ell) = 2\ NaOH(aq)$ has been determined by Matsui (1, 2), Roth (3, 4) and Rengade (5). Based on their reported $\Delta_r H°$ values, the corresponding $\Delta_f H°(298.15\ K)$ for $Na_2O(cr)$ are calculated, using $\Delta_f H°(H_2O, \ell, 298.15\ K) = -68.315\ kcal\ mol^{-1}$ obtained from (6) and $\Delta_f H°(NaOH, nH_2O, 298.15\ K)$ derived from JANAF $\Delta_f H°(NaOH, cr, 298.15\ K)$ and $\Delta_{sln} H°$ and $\Delta_{dil} H°$ reported by Parker (7). The sample employed by Matsui was prepared by the method of Rengade (9) and contains $Na_2O(cr)$ and $Na(cr)$ as impurities. The sample used by Roth also contains 3.96% (by weight) $Na_2O_2(cr)$. Corrections in the reported results for such impurities have been made by the authors. However, in order to obtain better results, more measurements using purer samples are necessary. The enthalpy change of the reaciton $2\ Na(cr) + H_2O(\ell) = Na_2O(cr) + H_2(g)$ was calculated by Rengade (5) from his measurements (5) incorporating with $\Delta_r H° = -85.2\ kcal\ mol^{-1}$ for the reaction $2\ Na(cr) + 2\ H_2O(\ell) = 2\ NaOH(aq) + H_2(g)$ determined by Joannis (10). The value $\Delta_f H°(298.15\ K) = -94.9\ kcal\ mol^{-1}$ is derived from the data of Forcrand (8) using $\Delta_f H°(Na_2O_2, cr, 298.15\ K) = -122.66\ kcal\ mol^{-1}$. The results are presented in the table below.

Brewer (13) determined the equilibrium constants, 918-1467 K, for the vaporization of $Na_2O(cr)$, using an effusion method. The Na_2O vapor consists predominately of Na(g) and $O_2(g)$ molecules with any oxide molecules being not important. The Na/O ratio was not measured and assumed to be 2/1 in all calculations on Na_2O. Based on the reported data, we derive the equilibrium constants for the reaction $Na_2O(cr) = 2\ Na(g) + 1/2\ O_2(g)$ and evaluate the 3rd law $\Delta_r H°(298.15\ K)$ as $117.1 \pm 10\ kcal\ mol^{-1}$. From this $\Delta_r H°(298.15\ K)$ value, we obtain $\Delta_f H°(Na_2O, cr, 298.15\ K) = -65.6 \pm 10\ kcal\ mol^{-1}$, employing $\Delta_f H°(298.15\ K)$ for Na(g). The value $-65.6\ kcal\ mol^{-1}$ is not consistent with the other values described above. This may be caused by an invalid assumption used in the calculation.

The enthalpy of formation at 298.15 K for $Na_2O(cr)$ is tentatively selected as $-99.90 \pm 1.0\ kcal\ mol^{-1}$.

Source	Reaction	$\Delta_r H°(298.15\ K)$ kcal mol^{-1}	$\Delta_f H°(298.15\ K)$ kcal mol^{-1}
(1)	$Na_2O(cr) + H_2O(\ell) = 2\ NaOH(7890\ H_2O)$	-56.61 ± 0.16	-100.08
(2)	$Na_2O(cr) + H_2O(\ell) = 2\ NaOH(4600\ H_2O)$	-56.70 ± 0.07	-99.89
(3)	$Na_2O(cr) + H_2O(\ell) = 2\ NaOH(825\ H_2O)$	-56.03 ± 0.13	-100.55
(4)	$Na_2O(cr) + H_2O(\ell) = 2\ NaOH(825\ H_2O)$	-55.98 ± 0.13	-100.60
(5)	$Na_2O(cr) + H_2O(\ell) = 2\ NaOH(aq)$	-56.50	-100.09
	$2\ Na(cr) + H_2O(\ell) = Na_2O(cr) + H_2(g)$	-28.70	-97.02
(8)	$Na_2O(cr) + 1/2\ O_2(g) = Na_2O_2(cr)$	-27.76	-94.90

Heat Capacity and Entropy

The low temperature heat capacities, 15-380 K, were obtained from Furukawa (11). We make the corrections in the reported $C_p°$ values for impurity $Na_2CO_3(cr)$ 2.33% (by weight), analyzed by Grimley (12). The effect on heat capacity due to the presence of Na_2O_2(0.91%) impurity is insignificant. Using the smoothed low temperature $C_p°$ data, the S°(298.15 K) is calculated, based on $S°(15\ K) = 0.0122\ cal\ K^{-1}\ mol^{-1}$.

Grimley (12) measured the high temperature enthalpies, 380.1-1174.6 K, with a copper block drop-type calorimeter. The enthalpies for the γ phase, 380.1-980.4 K, are joined smoothly with the low temperature data and extrapolated to 1023.2 K (the $\gamma \rightarrow \beta$ transition temperature). The enthalpies above 1023.2 K are obtained by extrapolation. The enthalpies measured at 1078.3 and 1174.6 K by Grimley are unreasonably large, and are not adopted.

Transition Data

The transition temperature and heats of transition were determined calorimetrically by Bouaziz (14) who used a very pure sample. The adopted $\Delta_{trs} H° = 0.42$ and $2.85\ kcal\ mol^{-1}$, are averages of the derived values, 0.36 - 0.48 and 2.7 - 3.0 kcal mol^{-1}, respectively.

Fusion Data

T_{fus} and $\Delta_{fus} H°$ were obtained from Bouaziz (14). The adopted $\Delta_{fus} H°$ is an average of the reported 10.8-12.0 kcal mol^{-1}. T_{fus} has been reported as 1193 and 1190 ± 10 K by Bunzel (15) and Brewer (13), respectively, which are not adopted.

References

1. M. Matsui and S. Oka, J. Soc. Chem. Ind. (Japan) 32, 79 (1929).
2. M. Matsui and S. Oka, J. Soc. Chem. Ind. (Japan) 32, 83 (1929).
3. W. A. Roth and H. L. Kaule, Z. Anorg. Chem. 253, 352 (1947).
4. W. A. Roth, Z. Anorg. Chem. 255, 324 (1948).
5. E. Rengade, Compt. Rend. 145, 236 (1907).
6. U. S. Nat. Bur. Stand. Tech. Note 270-1, 1965.
7. V. B. Parker, NSRDS-NBS 2, 1965.
8. de Forcrand, Compt. Rend. 130, 1465 (1900).
9. E. Rengade, Compt. Rend. 143, 1152 (1906).
10. Joannis, Ann. Chim. Phys. 12, 376 (1887).
11. G. T. Furukawa, U. S. Nat. Bur. Stand., personal communication, Nov. 13, 1967.
12. R. T. Grimley and J. L. Margrave, J. Phys. Chem. 64, 1763 (1960).
13. L. Brewer and J. Margrave, J. Phys. Chem. 59, 421 (1955).
14. R. Bouaziz, G. Papin and A. P. Rollet, Compt. Rend. Ser. C 262, 1051 (1966).
15. E. G. Bunzel and E. J. Kohlmeyer, Z. Anorg. Chem. 254, 1 (1947).

Enthalpy Reference Temperature = T_r = 298.15 K; Standard State Pressure = p° = 0.1 MPa

T/K	$C_p°$ (J $K^{-1}mol^{-1}$)	S° (J $K^{-1}mol^{-1}$)	$-[G°-H°(T_r)]/T$ (J $K^{-1}mol^{-1}$)	$H°-H°(T_r)$ (kJ mol^{-1})	$\Delta_f H°$ (kJ mol^{-1})	$\Delta_f G°$ (kJ mol^{-1})	Log K_f
0	0.	0.	INFINITE	-12.399	-413.146	-413.146	INFINITE
100	32.752	16.649	128.981	-11.233	-416.111	-404.348	211.210
200	59.551	49.209	81.114	-6.381	-417.626	-391.841	102.338
298.15	69.103	75.042	75.042	0.	-417.982	-379.090	66.415
300	69.245	75.470	75.043	0.128	-417.985	-378.849	65.964
400	75.776	96.304	77.840	7.386	-423.456	-365.363	47.712
500	81.496	113.855	83.332	15.262	-423.312	-350.842	36.652
600	85.709	129.103	89.717	23.631	-422.555	-336.410	29.287
700	88.851	142.560	96.324	32.365	-421.353	-322.142	24.039
800	91.291	154.589	102.869	41.376	-419.828	-308.070	20.115
900	93.266	165.458	109.229	50.607	-418.075	-294.204	17.075
1000	94.918	175.373	115.355	60.018	-416.169	-280.541	14.654
1023.150	95.264	177.549	116.737	62.219		GAMMA <--> BETA	
1023.150	95.265	179.266	116.737	63.977		TRANSITION	
1100	96.353	186.205	121.351	71.340	-412.418	-267.206	12.689
1200	97.629	194.645	127.111	81.040	-603.913	-249.215	10.848
1243.150	98.140	198.103	129.516	85.264		BETA <--> ALPHA	
1243.150	98.142	207.695	129.516	97.188		TRANSITION	
1300	98.788	212.098	133.031	102.786	-588.116	-220.358	8.854
1400	99.860	219.459	138.945	112.719	-584.147	-192.218	7.172
1405.200	99.913	219.829	139.243	113.239	-------	ALPHA <--> LIQUID	------
1500	100.868	226.383	144.545	122.756	-580.088	-164.364	5.724
1600	101.822	232.923	149.866	132.891	-575.944	-136.784	4.466
1700	102.734	239.124	154.936	143.119	-571.720	-109.465	3.363
1800	103.617	245.021	159.778	153.437	-567.418	-82.398	2.391
1900	104.470	250.646	164.414	163.841	-563.042	-55.571	1.528
2000	105.307	256.026	168.861	174.330	-558.594	-28.977	0.757

PREVIOUS: June 1962 CURRENT: June 1968

SODIUM OXIDE (Na_2O) LIQUID M_r = 61.97894

S°(298.15 K) = [91.607] J K^{-1} mol^{-1} $\Delta_f H°$(298.15 K) = [-372.843] kJ mol^{-1}

T_{fus} = 1405.2 K $\Delta_{fus}H°$ = 47.698 kJ mol^{-1}

Enthalpy of Formation

$\Delta_f H°$(Na_2O, ℓ, 298.15 K) is calculated from $\Delta_f H°$(Na_2O, cr, 298.15 K) by adding $\Delta_{fus}H°$ and the difference in enthalpy, H°(1405.2 K)-H°(298.15 K), between Na_2O(cr) and Na_2O(ℓ).

Heat Capacity and Entropy

The heat capacity is estimated by comparison with those for Na_2SiO_3(ℓ) and SiO_2(ℓ). S°(ℓ, 298.15 K) is calculated in a manner analogous to that used for the enthalpy of formation.

Fusion Data

See the crystal table for details.

Decomposition Data

T_{dcm} = 2223 K is the temperature at which the fugacity of oxygen is one atmosphere for the decomposition reaction Na_2O(ℓ) = 2 Na(g) + 1/2 O_2(g).

Sodium Oxide (Na_2O) Na_2O_1(l)

Enthalpy Reference Temperature = T_r = 298.15 K Standard State Pressure = p° = 0.1 MPa

T/K	$C_p°$ (J $K^{-1}mol^{-1}$)	S° (J $K^{-1}mol^{-1}$)	$-[G°-H°(T_r)]/T$ (J $K^{-1}mol^{-1}$)	$H°-H°(T_r)$ (kJ mol^{-1})	$\Delta_f H°$ (kJ mol^{-1})	$\Delta_f G°$ (kJ mol^{-1})	Log K_f
0							
100							
200							
298.15	104.600	91.607	91.607	0.	-372.843	-338.890	59.372
300	104.600	92.254	91.609	0.194	-372.780	-338.680	58.969
400	104.600	122.346	95.712	10.654	-375.049	-327.373	42.751
500	104.600	145.687	103.460	21.114	-372.321	-315.766	32.988
600	104.600	164.757	112.135	31.574	-369.474	-304.721	26.528
700	104.600	180.882	120.834	42.034	-366.545	-294.160	21.950
800	104.600	194.849	129.232	52.494	-363.572	-284.022	18.545
900	104.600	207.169	137.221	62.954	-360.589	-274.257	15.917
1000	104.600	218.190	144.776	73.414	-357.635	-264.824	13.833
1100	104.600	228.159	151.911	83.874	-354.746	-255.683	12.141
1200	104.600	237.261	158.649	94.334	-545.481	-241.922	10.531
1300	104.600	245.633	165.023	104.794	-540.970	-216.808	8.711
1400	104.600	253.385	171.061	115.254	-536.474	-192.041	7.165
1405.200	104.600	253.773	171.366	115.797	------- ALPHA <--> LIQUID ------		
1500	104.600	260.601	176.792	125.714	-531.992	-167.596	5.836
1600	104.600	267.352	182.244	136.174	-527.523	-143.448	4.683
1700	104.600	273.693	187.438	146.634	-523.066	-119.580	3.674
1800	104.600	279.672	192.398	157.094	-518.622	-95.974	2.785
1900	104.600	285.328	197.142	167.554	-514.191	-72.614	1.996
2000	104.600	290.693	201.686	178.014	-509.772	-49.488	1.292
2100	104.600	295.796	206.047	188.474	-505.367	-26.582	0.661
2200	104.600	300.662	210.238	198.934	-500.976	-3.886	0.092
2300	104.600	305.312	214.271	209.394	-496.600	18.611	-0.423
2400	104.600	309.764	218.158	219.854	-492.240	40.918	-0.891
2500	104.600	314.034	221.908	230.314	-487.899	63.044	-1.317
2600	104.600	318.136	225.531	240.774	-483.576	84.996	-1.708
2700	104.600	322.084	229.034	251.234	-479.275	106.783	-2.066
2800	104.600	325.888	232.426	261.694	-474.998	128.410	-2.396
2900	104.600	329.559	235.712	272.154	-470.745	149.885	-2.700
3000	104.600	333.105	238.900	282.614	-466.522	171.214	-2.981

PREVIOUS: June 1962 CURRENT: June 1968

Sodium Oxide (Na_2O) Na_2O_1(l)

SODIUM OXIDE (Na_2O)

GAMMA-BETA-ALPHA-LIQUID

M_r = 61.67894

0 to 1023.2 K		crystal, gamma
1023.2 to 1243.2 K		crystal, beta
1243.2 to 1405.2 K		crystal, alpha
above 1405.2 K		liquid

Refer to the individual tables for details.

Sodium Oxide (Na_2O) Na_2O_1(cr,l)

Enthalpy Reference Temperature = T_r = 298.15 K   Standard State Pressure = p° = 0.1 MPa

T/K	C_p° (J K⁻¹mol⁻¹)	S° (J K⁻¹mol⁻¹)	$-[G°-H°(T_r)]/T$ (J K⁻¹mol⁻¹)	$H°-H°(T_r)$ (kJ mol⁻¹)	$\Delta_f H°$ (kJ mol⁻¹)	$\Delta_f G°$ (kJ mol⁻¹)	Log K_f
0	0.	0.	INFINITE	-12.399	-413.146	-413.146	INFINITE
100	32.752	16.649	128.981	-11.233	-416.111	-404.348	211.210
200	59.551	49.209	81.114	-6.381	-417.626	-391.841	102.338
298.15	69.103	75.042	75.042	0.	-417.982	-379.090	66.415
300	69.245	75.470	75.043	0.128	-417.985	-378.849	65.964
400	75.776	96.304	77.840	7.386	-423.456	-365.363	47.712
500	81.496	113.855	83.332	15.262	-423.312	-350.842	36.652
600	85.709	129.103	89.717	23.631	-422.555	-336.410	29.287
700	88.851	142.560	96.324	32.365	-421.353	-322.142	24.039
800	91.291	154.589	102.869	41.376	-419.828	-308.070	20.115
900	93.266	165.458	109.229	50.607	-418.075	-294.204	17.075
1000	94.918	175.373	115.355	60.018	-416.169	-280.541	14.654
1023.150	95.264	177.549	116.737	62.219	GAMMA <--> BETA	TRANSITION	
1023.150	95.265	179.266	116.737	63.977			
1100	96.353	186.205	121.351	71.340	-412.418	-267.206	12.689
1200	97.629	194.645	127.111	81.040	-603.913	-249.215	10.848
1243.150	98.140	198.103	129.516	85.264	BETA <--> ALPHA	TRANSITION	
1243.150	98.142	207.695	129.516	97.188			
1300	98.788	212.098	133.031	102.786	-588.116	-220.358	8.854
1400	99.860	219.459	138.945	112.719	-584.147	-192.218	7.172
1405.200	99.913	219.829	139.243	113.239	ALPHA <--> LIQUID	TRANSITION	
1405.200	104.600	253.773	139.243	160.936			
1500	104.600	260.601	146.700	170.853	-531.992	-167.596	5.836
1600	104.600	267.352	154.032	181.313	-527.523	-143.448	4.683
1700	104.600	273.693	160.886	191.773	-523.066	-119.580	3.674
1800	104.600	279.672	167.321	202.233	-518.622	-95.974	2.785
1900	104.600	285.328	173.384	212.693	-514.191	-72.614	1.996
2000	104.600	290.693	179.117	223.153	-509.772	-49.488	1.292
2100	104.600	295.796	184.552	233.613	-505.367	-26.582	0.661
2200	104.600	300.662	189.720	244.073	-500.976	-3.886	0.092
2300	104.600	305.312	194.646	254.533	-496.600	18.611	-0.423
2400	104.600	309.764	199.350	264.993	-492.240	40.918	-0.891
2500	104.600	314.034	203.853	275.453	-487.899	63.044	-1.317
2600	104.600	318.136	208.170	285.913	-483.576	84.996	-1.708
2700	104.600	322.084	212.316	296.373	-479.275	106.783	-2.066
2800	104.600	325.888	216.305	306.833	-474.998	128.410	-2.396
2900	104.600	329.559	220.147	317.293	-470.745	149.885	-2.700
3000	104.600	333.105	223.854	327.753	-466.522	171.214	-2.981

PREVIOUS: CURRENT: June 1968

Sodium Oxide (Na_2O) Na_2O_1(cr,l)

SODIUM OXIDE (Na_2O_2) CRYSTAL (ALPHA-BETA) M_r = 77.97834

Sodium Peroxide (Na_2O_2) Na_2O_2(cr)

$S°(298.15 K) = 94.80 \pm 1.3 J K^{-1} mol^{-1}$
T_{trs} = 785 ± 1 K
T_{fus} = 948 K

$\Delta_f H°(0 K) = -507.34 \pm 5.0 kJ mol^{-1}$
$\Delta_f H°(298.15 K) = -513.21 \pm 5.0 kJ mol^{-1}$
$\Delta_{trs}H° = 5.732 kJ mol^{-1}$
$\Delta_{fus}H°$ = Unknown

Enthalpy of Formation

The enthalpy change of the reaction $Na_2O_2(cr) + H_2O(\ell) = 2 NaOH(aq, 1200 H_2O) + 1/2 O_2(g)$ was measured at 18.5°C by Roth (1). He employed a sample with composition 96.81% Na_2O_2 and 3.4% Na_2O by weight. Using $\Delta_{sln}H°(Na_2O_2, cr) = -56.02 kcal mol^{-1}$, Roth derived the value $\Delta_{sln}H°(Na_2O_2, cr) = -34.85 kcal mol^{-1}$. This value was later corrected to be $-35.75 kcal mol^{-1}$ (2). From these data we calculate the enthalpy of formation (298.15 K) of Na_2O_2(cr) as $-121.0 kcal mol^{-1}$. Forcrand obtained $\Delta_{sln}H° = -36.65 kcal mol^{-1}$ for the same reaction, quoted by Roth (1), yielding $\Delta_f H°(Na_2O_2, cr, 298.15 K) = -120.0 kcal mol^{-1}$. The auxiliary data $\Delta_f H°(H_2O, \ell, 298.15 K) = -68.315 kcal mol^{-1}$ is taken from (3); and $\Delta_f H°(NaOH\cdot aq, 1200 H_2O, 298.15 K) = -112.458 kcal mol^{-1}$ is derived based on $\Delta_f H°(NaOH, cr, 298.15 K) = -101.90 kcal mol^{-1}$ from JANAF Table and $\Delta_{sol}H°(aq, \infty) = -10.637 kcal mol^{-1}$ and $\Delta_{dil}H°(aq, 1200 H_2O \to aq, \infty) = -0.079 kcal mol^{-1}$ from Parker (4).

Gilles (5) determined the enthalpy change of the reaction $Na_2O_2(cr) + H_2O(\ell) = 2 NaOH(aq, 3800 H_2O) + 1/2 O_2(g)$, using a purer sample ($Na_2O_2$, 99.2%; Na_2O, 0.8%). Based on the reported value $\Delta_r H°(298.15 K) = -34.0 \pm 0.3 kcal mol^{-1}$ and derived $\Delta_f H°(NaOH\cdot aq, 3800 H_2O, 298.15 K) = -112.487 kcal mol^{-1}$, we calculate $\Delta_f H°(298.15 K) = -122.66 \pm 1.2 kcal mol^{-1}$ for Na_2O_2(cr).

Leffler (6) studied the oxygen pressure-melt composition for the liquid sodium-oxygen system between 780 and 980°C. These results, together with the thermodynamic data for Na_2O_2, were used to calculate the enthalpies and Gibbs energies of formation of melt compositions between $NaO_{1.0}$ and $NaO_{0.5}$. However, there are no numerical results reported. Therefore these data are not used for evaluation.

The enthalpy of formation at 298.15 K is adopted as $-122.66 \pm 1.2 kcal mol^{-1}$.

Heat Capacity and Entropy

The low temperature heat capacities, 52.31-296.22 K, were measured by Todd (7). His sample contains (percent) 94 Na_2O_2(cr), 3.6 Na_2O(cr) and 2.4 Na_2CO_3. Corrections in the reported results have been made for Na_2CO_3 content, but not for Na_2O content which was estimated to be small. The $C_p°$ values below 52.31 K are calculated using the equation $C_p° = D(252/T)+3E(399/T)$ reported by Todd. S°(298.15 K) is derived from the adopted $C_p°$, based on $S°(50 K) = 1.012 cal K^{-1} mol^{-1}$.

The high temperature enthalpies, 375.4-869.2 K, were determined by Chandrasekharaiah (8), using drop calorimetry. Corrections were made for the contributions from impurities (Na_2CO_3, 1.5 and Na_2O, 0.2%). A discontinuity was observed in the enthalpies between 773 and 793 K, indicating some kind of transition (see Transition Data). The $C_p°$ for $Na_2O_2(\alpha)$ at temperatures 298-769.5 K are derived from the enthalpy measurements, 375.4-769.5 K, and joined smoothly with the low temperature data at 298.15 K, and extrapolated to 785 K(T_{trs}). The $C_p°$ for $Na_2O_2(\beta)$ is derived from the enthalpy data, 794-869.2 K, as constant, 27.15 cal $K^{-1} mol^{-1}$, and extrapolated to 2500 K.

Vedeneev (9) measured the high temperature specific heat of Na_2O_2(cr) in the temperature interval 292-373 K using adiabatic calorimetry. The reported average value, $21.46 cal K^{-1} mol^{-1}$, is in fair agreement with the adopted ones.

Transition Data

T_{trs} was reported by Tallman (10), who measured the X-ray power diffraction pattern by a Geiger-counter diffractometer and found that at about 510°C the pattern changes entirely and a new set of peaks indicates a new phase. Separate thermal analysis measurements (10) confirm the transition temperature ($\alpha \to \beta$) is 512 ± 1°C or 785 ± 1 K, which is adopted here. Foppl (11) found a thermal anomaly at 485°C in both heating and cooling curves. $\Delta_{trs}H°$ is derived from the adopted enthalpy and $C_p°$ data on $Na_2O_2(\alpha)$ and $Na_2O_2(\beta)$.

Fusion Data

T_{fus} is obtained from Bunzel (12). T_{fus} = 733 and 783 K were reported by Blumenthal (13) and Rode (14), respectively, which are not adopted.

References

1. W. A. Roth and H. L. Kaule, Z. Anorg. Chem. 253, 352 (1947).
2. W. A. Roth, Z. Anorg. Chem. 255, 324 (1948).
3. U. S. Nat. Bur. Stand. Tech. Note 270-1, 1965.
4. V. B. Parker, NSRDS-NBS 2, 1965.
5. P. W. Gilles and J. L. Margrave, J. Phys. Chem. 68, 1333 (1956).
6. A. J. Leffler and N. M. Wiederhorn, J. Phys. Chem. 68, 2882 (1964).
7. S. S. Todd, J. Amer. Chem. Soc. 75, 1229 (1953).
8. M. C. Chandrasekharaiah, R. T. Grimley and J. L. Margrave, J. Phys. Chem. 63, 1505 (1959).
9. A. V. Vedeneev and S. M. Skuratov, Zh. Fiz. Khim. 25, 837 (1951).
10. R. L. Tallman and J. L. Margrave, J. Inorg. Nucl. Chem. 21, 40 (1961).
11. H. Foppl, Z. Anorg. Chem. 291, 12 (1957).
12. E. G. Bunzel and E. J. Kohlmeyer, Z. Anorg. Chem. 254, 1 (1947).
13. M. Blumenthal, Roczniki Chem. 11, 865 (1931).
14. T. V. Rode and G. A. Gol'der, Izv. Akad. Nauk SSSR, Otd. Khim. Nauk 3, 299 (1956).

Enthalpy Reference Temperature = T_r = 298.15 K; Standard State Pressure = p° = 0.1 MPa

T/K	$C_p°$ (J K⁻¹mol⁻¹)	S° (J K⁻¹mol⁻¹)	$-[G°-H°(T_r)]/T$ (J K⁻¹mol⁻¹)	$H°-H°(T_r)$ (kJ mol⁻¹)	$\Delta_f H°$ (kJ mol⁻¹)	$\Delta_f G°$ (kJ mol⁻¹)	Log K_f
0	0.	0.	INFINITE	-15.707	-507.339	-507.339	INFINITE
100	40.689	20.882	163.515	-14.263	-511.480	-491.475	256.720
200	75.479	61.785	102.600	-8.163	-513.202	-470.583	122.904
298.15	89.266	94.801	94.801	0.	-513.209	-449.627	78.773
300	89.408	95.354	94.803	0.165	-513.203	-449.233	78.218
400	97.721	122.268	98.416	9.541	-518.042	-427.560	55.834
500	103.767	144.748	105.495	19.626	-517.218	-405.020	42.312
600	108.445	164.093	113.687	30.244	-515.792	-382.706	33.318
700	112.332	181.108	122.127	41.287	-513.908	-360.668	26.913
785.000	115.253	194.149	129.230	50.961	ALPHA <--> BETA	TRANSITION	
785.000	113.596	201.451	129.230	56.694			
800	113.596	203.601	130.604	58.397	-505.952	-339.036	22.137
900	113.596	216.981	139.473	69.757	-503.773	-318.303	18.474
1000	113.596	228.949	147.833	81.117	-501.650	-297.810	15.556
1100	113.596	239.776	155.707	92.476	-499.616	-277.525	13.179
1200	113.596	249.660	163.130	103.836	-691.226	-252.540	10.993
1300	113.596	258.753	170.141	115.195	-687.607	-216.130	8.684
1400	113.596	267.171	176.775	126.555	-684.018	-179.997	6.716
1500	113.596	275.008	183.065	137.914	-680.457	-144.121	5.019
1600	113.596	282.340	189.043	149.274	-676.922	-108.480	3.542
1700	113.596	289.226	194.736	160.634	-673.412	-73.060	2.245
1800	113.596	295.719	200.168	171.993	-669.926	-37.846	1.098
1900	113.596	301.861	205.360	183.353	-666.465	-2.825	0.078
2000	113.596	307.688	210.332	194.712	-663.028	32.014	-0.836
2100	113.596	313.230	215.101	206.072	-659.616	66.682	-1.659
2200	113.596	318.515	219.682	217.431	-656.229	101.189	-2.403
2300	113.596	323.564	224.090	228.791	-652.869	135.542	-3.078
2400	113.596	328.399	228.336	240.150	-649.537	169.748	-3.694
2500	113.596	333.036	232.432	251.510	-646.233	203.817	-4.259

PREVIOUS: June 1963 CURRENT: June 1968

SODIUM SILICATE (Na_2SiO_3) CRYSTAL M_r = 122.06324

Sodium Silicate (Na_2SiO_3) $Na_2O_3Si_1$(cr)

$S°(298.15\ K) = 113.847 \pm 1.3\ J\ K^{-1}\ mol^{-1}$

$T_{fus} = 1362 \pm 0.5\ K$

$\Delta_f H°(0\ K) = -1551.20 \pm 4.2\ kJ\ mol^{-1}$

$\Delta_f H°(298.15\ K) = -1561.43 \pm 4.2\ kJ\ mol^{-1}$

$\Delta_{fus} H° = 51.798\ kJ\ mol^{-1}$

Enthalpy of Formation

The enthalpies of solution of NaCl(cr), Na_2SO_4(cr), SiO_2(quartz), Na_2SiO_3(cr), HCl(aq) and H_2SO_4(aq) in 20% hydrofluoric acid at 74.7°C were measured by Kracek (1). The author derived two values of $\Delta_{sln}H°$ of Na_2O(cr) from those of NaCl(cr), Na_2SO_4(cr), HCl(aq) and H_2SO_4(aq) presumably employing auxiliary data from (2). We recalculate $\Delta_{sln}H°$ of Na_2O(cr) in the calorimetric solution as -87.50 (chloride scheme, originally -87.66) and -88.74 kcal mol^{-1} (sulfate scheme, originally -89.19), using recent $\Delta_f H°$(298.15 K) values for NaCl(cr), Na_2SO_4(cr), HCl(aq), H_2SO_4(aq), and Na_2O(cr). The corresponding enthalpy changes, $\Delta_r H°$(298.15 K), for the reaction $Na_2O(cr) + SiO_2(quartz) = Na_2SiO_3(cr)$ are derived as -55.20 and -56.38 kcal mol^{-1}. Adopting the weighted average $\Delta_r H°$(298.15 K) = -55.59 kcal mol^{-1} with $\Delta_f H°$(Na_2O, cr, 298.15 K) = -99.9 kcal mol^{-1} and $\Delta_f H°$(SiO_2, quartz, 298.15 K) = -217.7 kcal mol^{-1}, we obtain $\Delta_f H°$(298.15 K) = -373.19 kcal mol^{-1} for Na_2SiO_3(cr). This $\Delta_f H°$(Na_2SiO_3, cr, 298.15 K) value is independent of future changes in $\Delta_f H°$(298.15 K) of Na_2O(cr) but the values of $\Delta_{sln}H°$ and $\Delta_r H°$ are not independent. The adopted value of $\Delta_f H°$(Na_2, cr, 298.15 K) is the average of -100.1 and -99.7 kcal mol^{-1} reported by Roth and Kaule (3) and Matsui and Oka (4), respectively.

Roth and Troitzsch (5) have measured the enthalpies of solution of SiO_2(quartz), Na_2SiO_3(cr) and Na_2O(cr) in 10.3% HF at 22°C to be -32.85, -67.96 and -89.9 kcal mol^{-1}, respectively. Using these results, we calculate the enthalpy change of the reaction $Na_2O(cr) + SiO_2(quartz) = Na_2SiO_3(cr)$ as -53.79 kcal mol^{-1} and $\Delta_f H°$(Na_2SiO_3, cr, 298.15 K) = -371.39 kcal mol^{-1}.

By use of high temperature reaction calorimetry, Kroger and Janetzko (6) have determined the enthalpy change of the reaction $Na_2C_3(cr) + Na_2Si_2O_5(g, \ell) = 2\ Na_2SiO_3(cr) + CO_2(g)$ as 17.99 kcal mol^{-1}. Based on $\Delta_f H°$(298.15 K) = -270.26, -582.92 and -94.05 kcal mol^{-1} for Na_2SiO_3(cr).

The enthalpies of solution of Na_2CO_3(cr), SiO_2(quartz) and Na_2SiO_3(cr) in 39% HF at 26.5°C were measured to be -244.1, -564.5 and -673.6 cal g^{-1} by Hummel and Schwiete (7). From these results we calculate the enthalpy change to be 22.43 kcal mol^{-1} for the reaction $Na_2CO_3(cr) + SiO_2(quartz) = Na_2SiO_3(cr) + CO_2(g)$, yielding $\Delta_f H°$(Na_2SiO_3, cr, 298.15 K) = -371.48 kcal mol^{-1}.

Many other older calorimetric measurements have been made in order to derive the enthalpies of formation for Na_2SiO_2(cr), but most of these are of questionable accuracy and in some cases the products of the reaction are uncertain. For further details, see the review by McCready (8).

Heat Capacity and Entropy

The low temperature heat capacities, 53.6 - 294.5 K, were measured by Kelley (9), and high temperature enthalpies, 360.5 - 1747 K were determined by Naylor (10). The high temperature heat capacities are derived from the measured enthalpy data and joined smoothly with the low temperature $C_p°$ values. The entropy at 298.15 K is calculated using the adopted low temperature heat capacities, based on S°(50 K) = 1.195 cal K^{-1} mol^{-1}.

Fusion Data

See the liquid table for details.

References

1. F. C. Kracek, Ann. Rept. Director of the Geophysical Laboratory, No. 1215, 69 (1953).
2. U. S. Nat. Bur. Stand. Circ. 500, 1952.
3. W. A. Roth and H. L. Kaule, Z. Anorg. Chem. 253, 352 (1947); ibid. 255, 324 (1948).
4. M. Matsui and S. Oka, J. Soc. Chem. Ind. (Japan) 32, 79 (1929).
5. W. A. Roth and H. Troitzsch, Z. Anorg. Chem. 260, 337 (1949).
6. D. C. Kroger and W. Janetzko, Z. Anorg. Allg. Chem. 284, 83 (1956).
7. C. Hummel and H. E. Schwiete, Glastech. Ber. 32, 327 (1959).
8. N. W. McCready, J. Phys. Colloid. Chem. 52, 1277 (1948).
9. K. K. Kelley, J. Amer. Chem. Soc. 61, 471 (1939).
10. B. F. Naylor, J. Amer. Chem. Soc. 67, 466 (1945).

Enthalpy Reference Temperature = T_r = 298.15 K Standard State Pressure = p° = 0.1 MPa

	J $K^{-1}mol^{-1}$			kJ mol^{-1}			
T/K	$C_p°$	S°	$-[G°-H°(T_r)]/T$	$H°-H°(T_r)$	$\Delta_f H°$	$\Delta_f G°$	Log K_f
0	0.	0.	INFINITE	-18.913	-1551.203	-1551.203	INFINITE
100	48.308	25.568	196.995	-17.143	-1556.736	-1528.150	798.224
200	89.743	73.542	123.451	-9.982	-1560.031	-1498.083	391.259
298.15	111.880	113.847	113.847	0.	-1561.427	-1467.330	257.070
300	112.395	114.540	113.849	0.207	-1561.442	-1466.746	255.383
400	127.779	149.118	118.459	12.264	-1567.208	-1434.679	187.350
500	138.716	178.865	127.635	25.615	-1566.925	-1401.558	146.420
600	147.114	204.926	138.390	39.922	-1565.766	-1368.581	119.146
700	153.946	228.133	149.583	54.985	-1563.937	-1335.853	99.683
800	159.749	249.078	160.732	70.677	-1561.578	-1303.425	85.105
900	164.841	268.193	171.625	86.911	-1558.788	-1271.319	73.785
1000	169.423	285.801	182.174	103.628	-1555.650	-1239.544	64.747
1100	173.628	302.149	192.346	120.783	-1552.231	-1208.096	57.368
1200	177.540	317.425	202.140	138.343	-1742.114	-1172.091	51.020
1300	181.230	331.783	211.566	156.283	-1736.446	-1124.817	45.196
1362.000	183.407	340.278	217.233	167.587	------ CRYSTAL <--> LIQUID -----		
1400	184.736	345.343	220.641	174.582	-1730.507	-1077.989	40.220
1500	188.100	358.203	229.387	193.224	-1724.308	-1031.596	35.923
1600	191.347	370.441	237.824	212.187	-1717.869	-985.624	32.177
1700	195.748	382.186	245.972	231.564	-1761.273	-939.617	28.871
1800	197.560	393.420	253.854	251.219	-1754.069	-891.492	25.870
1900	200.556	404.182	261.485	271.125	-1746.650	-843.772	23.197
2000	203.489	414.544	268.880	291.328	-1738.971	-796.449	20.801
2100	206.376	424.543	276.056	311.822	-1731.037	-749.517	18.643
2200	209.213	434.209	283.026	332.601	-1722.854	-702.968	16.691
2300	212.016	443.571	289.804	353.663	-1714.427	-656.798	14.916
2400	214.781	452.653	296.401	375.003	-1705.760	-611.000	13.298
2500	217.518	461.476	302.829	396.618	-1696.857	-565.567	11.817

PREVIOUS: June 1965 CURRENT: September 1967

Sodium Silicate (Na_2SiO_3) $Na_2O_3Si_1$(cr)

SODIUM SILICATE (Na_2SiO_3) LIQUID M_r = 122.06324

S°(298.15 K) = [150.496] J K^{-1} mol^{-1} $\Delta_fH°$(298.15 K) = [-1510.875] kJ mol^{-1}

T_{fus} = 1362 ± 0.5 K $\Delta_{fus}H°$ = 51.798 kJ mol^{-1}

Enthalpy of Formation

$\Delta_fH°$(Na_2SiO_3, ℓ, 298.15 K) is calculated from $\Delta_fH°$(Na_2SiO_3, cr, 298.15 K) by adding $\Delta_{fus}H°$ and the difference in enthalpy, H°(1362 K)-H°(298.15 K), between the crystal and liquid.

Heat Capacity and Entropy

The heat capacity above 900 K is derived as constant from the high temperature enthalpy data, 1414 - 1747 K, measured by Naylor (1). A glass transition temperature is assumed at 900 K, i.e., the heat capacities below 900 K are taken to be those for Na_2SiO_3(cr).

S°(ℓ, 298.15 K) is calculated in a manner analogous to that used for the enthalpy of formation. The low temperature heat capacities, 63.74 -162.48 K, have been measured by Tarasov and Savitskaya (2); however, these are not adequate to be used to evaluate S°(298.15 K).

Fusion Data

The adopted T_{fus} = 1362 K, determined by Kracek (3), is in good agreement with the value 1261 K reported by Jaeger (4) and later confirmed by Morey and Bowen (5).

Other reported T_{fus} values are 1280, 1291, and 1329 K by Kultascheff (6), Wallace (7) and van Klooster (8), respectively. These values are all lower than the adopted T_{fus} value, probably because the sample compositions were not the exact composition of metasilicate.

The value of $\Delta_{fus}H°$ is evaluated from the adopted heat capacities for Na_2SiO_3(cr) and Na_2SiO_3(ℓ), and the enthalpy data reported by Naylor (1), assuming the final state of the calorimetric sample to be Na_2SiO_3(cr). From the phase diagrams of Na_2SiO_3 with other compounds, Kelley (9) derived the values of $\Delta_{fus}H°$ of 9.8 - 10.5 kcal mol^{-1}.

References

1. B. F. Naylor, J. Amer. Chem. Soc. 67, 466 (1945).
2. V. V. Tarasov and Y. S. Savitskaya, Dokl. Akad. Nauk SSSR 88, 1019 (1953).
3. F. C. Kracek, J. Phys. Chem. 34, 1583 (1930).
4. F. M. Jaeger, J. Wash. Acad. Sci. 1, 49 (1911).
5. G. W. Morey and N. L. Bowen, J. Phys. Chem. 28, 1167 (1924).
6. N. V. Kultascheff, Z. Anorg. Chem. 35, 186 (1903).
7. R. C. Wallace, Z. Anorg. Chem. 63, 1 (1909).
8. H. S. van Klooster, Z. Anorg. Chem. 69, 135 (1910).
9. K. K. Kelley, U. S. Bur. Mines Bull. 393, 1936.

Sodium Silicate (Na_2SiO_3) $Na_2O_3Si_1$(l)

Enthalpy Reference Temperature = T_r = 298.15 K Standard State Pressure = p° = 0.1 MPa

	J $K^{-1}mol^{-1}$			kJ mol^{-1}			
T/K	$C_p°$	S°	$-[G°-H°(T_r)]/T$	$H°-H°(T_r)$	$\Delta_fH°$	$\Delta_fG°$	Log K_f
0							
100							
200							
298.15	111.880	150.496	150.496	0.	-1510.875	-1427.704	250.128
300	112.395	151.189	150.498	0.207	-1510.890	-1427.188	248.496
400	127.779	185.767	155.108	12.264	-1516.655	-1398.786	182.663
500	138.716	215.514	164.284	25.615	-1516.373	-1369.330	143.053
600	147.114	241.575	175.039	39.922	-1515.214	-1340.018	116.659
700	153.946	264.782	186.232	54.985	-1513.385	-1310.955	97.825
800	159.749	285.727	197.381	70.677	-1511.025	-1282.192	83.719
900	164.842	304.843	208.274	86.912	-1508.235	-1253.751	72.766
900.000	164.842	304.843	208.274	86.912	GLASS <--> LIQUID TRANSITION		
900.000	177.318	304.843	208.274	86.912			
1000	177.318	323.525	218.882	104.644	-1504.081	-1225.699	64.024
1100	177.318	340.425	229.175	122.375	-1500.086	-1198.055	56.891
1200	177.318	355.854	239.098	140.107	-1689.797	-1165.888	50.750
1300	177.318	370.047	248.632	157.839	-1684.338	-1122.451	45.101
1362.000	177.318	378.308	254.349	168.833	------ CRYSTAL <--> LIQUID -----		
1400	177.318	383.188	257.780	175.571	-1678.966	-1079.431	40.274
1500	177.318	395.421	266.553	193.303	-1673.677	-1036.792	36.104
1600	177.318	406.865	274.969	211.034	-1668.470	-994.504	32.467
1700	177.318	417.615	283.047	228.766	-1713.518	-952.091	29.254
1800	177.318	427.750	290.807	246.498	-1708.238	-907.454	26.334
1900	177.318	437.337	298.269	264.230	-1702.994	-863.110	23.729
2000	177.318	446.433	305.452	281.962	-1697.785	-819.040	21.391
2100	177.318	455.084	312.373	299.693	-1692.613	-775.230	19.283
2200	177.318	463.333	319.049	317.425	-1687.478	-731.665	17.372
2300	177.318	471.215	325.495	335.157	-1682.381	-688.333	15.633
2400	177.318	478.762	331.725	352.889	-1677.322	-645.223	14.043
2500	177.318	486.000	337.752	370.621	-1672.303	-602.322	12.585

PREVIOUS: June 1965 CURRENT: September 1967

Sodium Silicate (Na_2SiO_3) $Na_2O_3Si_1$(l)

SODIUM SILICATE (Na_2SiO_3) CRYSTAL–LIQUID M_r = 122.06324 Sodium Silicate (Na_2SiO_3) $Na_2O_3Si_1$(cr,l)

0 to 1362 K crystal
above 1362 K liquid

Refer to the individual tables for details.

Enthalpy Reference Temperature = T_r = 298.15 K Standard State Pressure = p° = 0.1 MPa

T/K	C_p° (J K⁻¹mol⁻¹)	S° (J K⁻¹mol⁻¹)	$-[G^\circ-H^\circ(T_r)]/T$ (J K⁻¹mol⁻¹)	$H^\circ-H^\circ(T_r)$ (kJ mol⁻¹)	Δ_fH° (kJ mol⁻¹)	Δ_fG° (kJ mol⁻¹)	Log K_f
0	0.	0.	INFINITE	-18.913	-1551.203	-1551.203	INFINITE
100	48.308	25.568	196.995	-17.143	-1556.736	-1528.150	798.224
200	89.743	73.542	123.451	-9.982	-1560.031	-1498.083	391.259
298.15	111.880	113.847	113.847	0.	-1561.427	-1467.330	257.070
300	112.395	114.540	113.849	0.207	-1561.442	-1466.746	255.383
400	127.779	149.118	118.459	12.264	-1567.208	-1434.679	187.350
500	138.716	178.865	127.635	25.615	-1566.925	-1401.558	146.420
600	147.114	204.926	138.390	39.922	-1565.766	-1368.581	119.146
700	153.946	228.133	149.583	54.985	-1563.937	-1335.853	99.683
800	159.749	249.078	160.732	70.677	-1561.578	-1303.425	85.105
900	164.841	268.193	171.625	86.911	-1558.788	-1271.319	73.785
1000	169.423	285.801	182.174	103.628	-1555.650	-1239.544	64.747
1100	173.628	302.149	192.346	120.783	-1552.231	-1208.096	57.368
1200	177.540	317.425	202.140	138.343	-1742.114	-1172.091	51.020
1300	181.230	331.783	211.566	156.283	-1736.446	-1124.817	45.196
1362.000	183.407	340.278	217.233	167.587	CRYSTAL <--> LIQUID		
1362.000	177.318	378.308	217.233	219.385	TRANSITION		
1400	177.318	383.188	221.671	226.123	-1678.966	-1079.431	40.274
1500	177.318	395.421	232.851	243.855	-1673.677	-1036.792	36.104
1600	177.318	406.865	243.374	261.587	-1668.470	-994.504	32.467
1700	177.318	417.615	253.310	279.319	-1713.518	-952.091	29.254
1800	177.318	427.750	262.722	297.050	-1708.238	-907.454	26.334
1900	177.318	437.337	271.663	314.782	-1702.994	-863.110	23.729
2000	177.318	446.433	280.176	332.514	-1697.785	-819.040	21.391
2100	177.318	455.084	288.300	350.246	-1692.613	-775.230	19.283
2200	177.318	463.333	296.070	367.978	-1687.478	-731.665	17.372
2300	177.318	471.215	303.515	385.709	-1682.381	-688.334	15.633
2400	177.318	478.762	310.661	403.441	-1677.322	-645.223	14.043
2500	177.318	486.000	317.531	421.173	-1672.303	-602.322	12.585

PREVIOUS: CURRENT: September 1967

Sodium Silicate (Na_2SiO_3) $Na_2O_3Si_1$(cr,l)

SODIUM SULFATE, δ (Na_2SO_4) CRYSTAL M_r = 142.03714

$S°(298.15\ K) = [163.212]\ J\ K^{-1}\ mol^{-1}$

$T_{trs}(I \rightarrow \delta) = [986]\ K$

$T_{fus}(\delta \rightarrow \ell) = [1170]\ K$

$\Delta_f H°(298.15\ K) = [-1379.294]\ kJ\ mol^{-1}$

$\Delta_{trs} H°(I \rightarrow \delta) = [0.502]\ kJ\ mol^{-1}$

$\Delta_{fus} H°(\delta \rightarrow \ell) = [21.338]\ kJ\ mol^{-1}$

Enthalpy of Formation

The enthalpy of formation adopted in this tabulation is obtained from $\Delta_f H°(298.15\ K)$ for Na_2SO_4(cr, I) by adding $\Delta_{trs}H°(I \rightarrow \delta)$ and the difference between H°(986 K)-H°(298.15 K) for Na_2SO_4(cr, I) and Na_2SO_4(cr, δ) (2).

Heat Capacity and Entropy

Heat capacities in the range 900 - 1010 K were measured by Shmidt and Sokolov (1) in an adiabatic calorimeter. We have graphically extrapolated these data down to 298.15 K, paralleling the $C_p°$ data of Na_2SO_4(cr, I), and extended the data smoothly to 1500 K. Our calculated enthalpies are biased by some 300 cal K^{-1} mol^{-1} above the experimental results (for the I phase) of Denielou et al. (3) and Coughlin (4). See the discussion in the Na_2SO_4(cr, I) table (2).

Transition Data

The existence of Na_2SO_4(cr, δ) is based on the work of Shmidt and Sokolov (1) who found an anomolous region in the heat capacity curve of Na_2SO_4(cr, I) between 963 - 993 K. Popov and Ginzburg (5) also observed a sharp change in the heat capacity above 837.6 K. Simanov and Kirkina (6) have observed an orthorhombic modification of Na_2SO_4 at 993 K by X-ray diffraction. As noted above, our calculated enthalpies do not agree with measured enthalpies above 980 K. The evidence for the existence of the δ-phase is not conclusive. We have deleted the δ-phase from our combined phase table for Na_2SO_4. The heat of transition $\Delta_{trs}H°(I \rightarrow \delta$, 986 K) = 120 ± 10 cal mol^{-1} is obtained by graphical integration of the area between the measured $C_p°$ curve from (1) and the adopted $C_p°$ curve for Na_2SO_4(cr, I) (2). The temperature of the transition is assumed to be 986 K.

Fusion Data

Based on our selected $C_p°$ values for Na_2SO_4(cr, δ) and Na_2SO_4(ℓ), we derive $\Delta_{fus}H°(\delta \rightarrow \ell)$ = 5.10 ± 0.10 kcal mol^{-1} using the experimental liquid enthalpies from Denielou et al. (3) and Coughlin (4). $T_{fus}(\delta \rightarrow \ell)$ = 1170 K is the calculated temperature at which $\Delta_r G° = 0$ for Na_2SO_4(cr, δ) → Na_2SO_4(ℓ).

References

1. N. E. Shmidt and V. A. Sokolov, Russ. J. Inorg. Chem. 6, 1321 (1961).
2. JANAF Thermochemical Tables: Na_2SO_4(cr, I), Na_2SO_4(ℓ), 6-30-78.
3. L. Denielou, V. Fournier, J. P. Petitet, and C. Tequi, C. R. Acad. Sci., Paris, Ser. C269, 1577 (1969).
4. J. P. Coughlin, J. Amer. Chem. Soc. 77, 868 (1955).
5. M. M. Popov and D. M. Ginzburg, J. Gen. Chem. USSR 26, 1107 (1956).
6. Y. P. Simanov and D. R. Kirkina, Russ. J. Inorg. Chem. 2, 364 (1957).

Sodium Sulfate, Delta (Na_2SO_4) $Na_2O_4S_1$(cr)

Enthalpy Reference Temperature = T_r = 298.15 K Standard State Pressure = p° = 0.1 MPa

T/K	$C_p°$ (J K⁻¹mol⁻¹)	S° (J K⁻¹mol⁻¹)	-[G°-H°(T_r)]/T (J K⁻¹mol⁻¹)	H°-H°(T_r) (kJ mol⁻¹)	$\Delta_f H°$ (kJ mol⁻¹)	$\Delta_f G°$ (kJ mol⁻¹)	Log K_f
0							
100							
200							
298.15	155.812	163.212	163.212	0.	-1379.294	-1265.387	221.690
300	155.896	164.176	163.215	0.288	-1379.261	-1264.680	220.200
400	163.050	210.099	169.427	16.269	-1385.063	-1226.248	160.132
500	166.251	246.782	181.355	32.714	-1384.866	-1186.573	123.960
600	171.816	277.574	194.889	49.611	-1383.905	-1146.990	99.854
700	178.113	304.516	208.664	67.097	-1382.180	-1107.631	82.652
800	185.707	328.766	222.184	85.266	-1379.714	-1068.569	69.770
900	195.979	351.212	235.287	104.333	-1429.489	-1028.786	59.709
1000	208.112	372.490	247.951	124.539	-1423.838	-984.557	51.428
1100	220.413	392.907	260.207	145.970	-1417.113	-940.945	44.682
1200	232.086	412.592	272.090	168.602	-1602.887	-893.112	38.876
1300	242.881	431.600	283.633	192.357	-1592.358	-834.384	33.526
1400	252.923	449.977	294.862	217.160	-1580.856	-776.504	28.972
1500	261.333	467.723	305.798	242.887	-1568.503	-719.477	25.054

PREVIOUS: December 1966 CURRENT: June 1978

Sodium Sulfate, Delta (Na_2SO_4) $Na_2O_4S_1$(cr)

SODIUM SULFATE, I (Na_2SO_4) CRYSTAL M_r = 142.03714

$S°(298.15\ K) = [160.132 \pm 4.2]\ J\ K^{-1}\ mol^{-1}$
$T_{trs}(III \to I) = 509 \pm 1\ K$
$T_{trs}(IV \to I) = 514 \pm 1\ K$
$T_{fus}(I \to \ell) = 1157 \pm 1\ K$

$\Delta_f H°(298.15\ K) = [-1380.899 \pm 0.8]\ kJ\ mol^{-1}$
$\Delta_{trs}H°(III \to I) = 6.887 \pm 0.084\ kJ\ mol^{-1}$
$\Delta_{trs}H°(IV \to I) = 10.908 \pm 0.084\ kJ\ mol^{-1}$
$\Delta_{fus}H°(I \to \ell) = 23.849 \pm 0.42\ kJ\ mol^{-1}$

Enthalpy of Formation

The enthalpy of formation adopted in this tabulation is obtained from $\Delta_f H°(Na_2SO_4$, cr, III, 298.15 K) by adding $\Delta_{trs}H°$(III → I, 509 K) = 1646 cal $K^{-1}\ mol^{-1}$ (1, 2) and the difference between H°(509 K)-H°(298.15 K) for Na_2SO_4(cr, III) and Na_2SO_4(cr, I) (2).

Heat Capacity and Entropy

Heat capacities of Na_2SO_4(cr, I) have been measured by Shmidt and Sokolov (3) in an adiabatic calorimeter over the temperature range 538.65 - 1021.75 K. The direct heating measurements of Popov and Galchenko (4) using a heat-flow calorimeter are in substantial agreement. Enthalpy measurements by drop calorimetry have been carried out by Coughlin (5, 515 - 1143 K), Popov and Ginzburg (6, 1173 - 1290 K), May (7, 520 - 1157 K), and by Denielou et al. (8, 516 - 1154 K).

The data of May (7) are not in agreement with the other enthalpy results. The enthalpy data of references (5, 6, 8) are in general agreement but our derived $C_p°$ values are lower than the calorimetric $C_p°$ (3) results below 800 K but show the proper slope. Derived $C_p°$ values from Denielou et al. (8) are in agreement with the calorimetric results above 930 K. All of the experiments seem to show a change in the slope of the $C_p°$ curve near 900 K. We adopt $C_p°$ from 530 - 880 K from (3) and extrapolate smoothly to 1500 K. $C_p°$ values below 530 K are obtained by graphical extrapolation. The entropy, S°(298.15 K), is calculated in a manner analogous to that of the heat of formation.

Transition Data

Na_2SO_4(cr, I) is the hexagonal modification (space group D^3_{3d} or C3ml (9, 10)) of anhydrous sodium sulfate stable from 514 - 1157 K. All the low-temperature modifications (III, IV, V) are converted to the high temperature form between 500 - 530 K (2); $\Delta_{trs}H°$(IV → I, 514 K) = 2607 cal mol^{-1} (1, 2), $\Delta_{trs}H°$(V → I, 512 K) = 2663 cal mol^{-1} (1, 2), and $\Delta_{trs}H°$(III → I, 509 K) = 1646 cal mol^{-1} (1, 2).

A small anomalous region between 963.15 - 993.15 K on the $C_p°$ curve for Na_2SO_4(cr, I) was reported by Shmidt and Sokolov (3), who attributed the anomaly to a transition to another phase of Na_2SO_4 which they referred to as the δ-phase. There is not sufficient evidence to warrant our inclusion of the δ-phase.

Fusion Data

The enthalpy of melting of Na_2SO_4(cr, I) → Na_2SO_4(ℓ) has been derived from drop calorimetry data by Denielou et al. (8, 5.58 ± 0.13 kcal mol^{-1}), Coughlin (5, 5.67 ± 0.1 kcal mol^{-1}), Popov and Ginzburg (6, 5.77 ± 0.2 kcal mol^{-1}), and May (7, 5.8 ± 0.2 kcal mol^{-1}). No mention was made of the δ phase. Based on our selected $C_p°$ values for Na_2SO_4(cr, ℓ) and Na_2SO_4(ℓ), we derive $\Delta_{fus}H°$(I → ℓ) = 5.70 ± 0.10 kcal mol^{-1}, using the enthalpy data of Coughlin (5). There is good agreement on the melting temperature which we take as T_{fus}(I → ℓ) = 1157 ± 1 K.

References

1. G. E. Brodale and W. F. Giauque, J. Phys. Chem. 76, 737 (1972).
2. JANAF Thermochemical Tables: Na_2SO_4(cr, V), Na_2SO_4(cr, IV), Na_2SO_4(cr, III), 6-30-78.
3. N. E. Shmidt and V. A. Sokolov, Russ. J. Inorg. Chem. 6, 1321 (1961).
4. M. M. Popov and G. L. Galchenko, J. Gen. Chem. USSR 21, 2489 (1951).
5. J. P. Coughlin, J. Amer. Chem. Soc. 77, 868 (1955).
6. M. M. Popov and D. M. Ginzberg, J. Gen. Chem. USSR 26, 1107 (1956).
7. M. N. May, TAPPI 35, 511 (1952).
8. L. Denielou, Y. Fournier, J. P. Petitet, and C. Tequi, C. R. Acad. Soc. Paris 269, 1577 (1969).
9. Y. P. Simanov and D. F. Kirkina, Russ. J. Inorg. Chem. 2, 364 (1957).
10. H. F. Fischmesiter, Monatsh. Chem. 93, 420 (1962).

Sodium Sulfate, I (Na_2SO_4) $Na_2O_4S_1$(cr)

Enthalpy Reference Temperature = T_r = 298.15 K Standard State Pressure = p° = 0.1 MPa

T/K	$C_p°$ (J K⁻¹mol⁻¹)	S° (J K⁻¹mol⁻¹)	$-[G°-H°(T_r)]/T$ (J K⁻¹mol⁻¹)	$H°-H°(T_r)$ (kJ mol⁻¹)	$\Delta_f H°$ (kJ mol⁻¹)	$\Delta_f G°$ (kJ mol⁻¹)	Log K_f
0							
100							
200							
298.15	159.620	160.132	160.132	0.	-1380.899	-1266.074	221.811
300	159.724	161.120	160.135	0.295	-1380.859	-1265.361	220.319
400	164.758	207.754	166.460	16.518	-1386.419	-1226.666	160.186
500	170.017	245.088	178.571	33.258	-1385.927	-1186.787	123.983
514.000	170.732	249.793	180.447	35.644	-----------	IV <--> I	----------
600	175.331	276.549	192.345	50.522	-1384.599	-1147.069	99.861
700	181.042	303.999	206.375	68.337	-1382.545	-1107.634	82.653
800	187.276	328.577	220.140	86.750	-1379.835	-1068.539	69.768
900	193.845	351.015	233.452	105.807	-1429.621	-1028.740	59.707
1000	200.309	371.773	246.259	125.515	-1424.467	-984.470	51.423
1100	206.731	391.167	258.560	145.867	-1418.821	-940.739	44.672
1157.000	210.367	401.702	265.354	157.755	---------	I <--> LIQUID	--------
1200	213.133	409.428	270.379	166.860	-1606.234	-892.663	38.857
1300	219.447	426.739	281.746	188.490	-1597.830	-833.536	33.492
1400	225.598	443.227	292.696	210.744	-1588.878	-775.076	28.918
1500	231.584	458.997	303.261	233.604	-1579.391	-717.276	24.978

PREVIOUS: December 1966 CURRENT: June 1978

Sodium Sulfate, I (Na_2SO_4) $Na_2O_4S_1$(cr)

SODIUM SULFATE, III (Na_2SO_4) CRYSTAL M_r = 142.03714

$S°(298.15\ K) = 154.908 \pm 0.4\ J\ K^{-1}\ mol^{-1}$
$T_{trs}(III \rightarrow I) = 509 \pm 1\ K$

$\Delta_f H°(0\ K) = -1373.71 \pm 0.63\ kJ\ mol^{-1}$
$\Delta_f H°(298.15\ K) = -1384.85 \pm 0.63\ kJ\ mol^{-1}$
$\Delta_{trs} H°(III \rightarrow I) = 6.887 \pm 0.084\ kJ\ mol^{-1}$

Enthalpy of Formation

Brodale and Giauque (1) determined the enthalpy of solution at infinite dilution $\Delta_{sln}H°(Na_2SO_4$, cr, III, 298.15 K) = -1273 ± 4 cal K^{-1} mol^{-1} for the process Na_2SO_4(cr, III) → Na_2SO_4(aq, ∞). When this result is combined with the enthalpies of formation of the infinitely dilute ions from CODATA (2), we obtain $\Delta_f H°(Na_2SO_4$, aq, ∞, 298.15 K) = -332.266 ± 0.13 kcal mol^{-1} and $\Delta_f H°(Na_2SO_4$, cr, III, 298.15 K) = -330.993 ± 0.15 kcal mol^{-1}. Coughlin (3) also measured the enthalpy of solution, $\Delta_{sln}H°(Na_2SO_4$, cr, III, 303 K) = -1363 ± 4 cal K^{-1} mol^{-1} for the process Na_2SO_4(cr, III) → Na_2SO_4(aq, 1068 H_2O). Using $\Delta C_p°$ as determined by Coughlin to extrapolate the heat of solution to 298 K gives $\Delta_{sln}H°(Na_2SO_4$, cr, III, 298.15 K) = -1037 ± 20 cal K^{-1} mol^{-1}. Extrapolating to infinite dilution using the data of Lange and Streeck (4) and Wallace and Robinson (5) gives $\Delta_{sln}H°(Na_2SO_4$, cr, III, 298.15 K) = -1282 ± 20 cal K^{-1} mol^{-1}. Using the same auxiliary data as above (2) $\Delta_f H°(Na_2SO_4$, cr, III, 298.15 K) = -330.984 ± 0.15 kcal mol^{-1}. The enthalpy of solution data of Coughlin (3) as corrected by Brodale and Giauque (1) leads to $\Delta_r H°$(V → III, 298.15 K) = 716 ± 8 cal K^{-1} mol^{-1}. When this is combined with auxiliary data for Na_2SO_4(cr, V) (6), we obtain $\Delta_f H°(Na_2SO_4$, cr, III, 298.15 K) = -330.980±0.2 kcal mol^{-1}. We adopt an intermediate value $\Delta_f H°(Na_2SO_4$, cr, III, 298.15 K) = -330.988 ± 0.15 kcal mol^{-1} for this tabulation.

Heat Capacity and Entropy

Heat capacities of Na_2SO_4(cr, III) have been measured by Brodale and Giauque (1) over the ragne 13.16 - 297.71 K and by Shmidt and Sokolov (7) in an adiabatic calorimeter over the temperature range 319.03 - 520.56 K. Paukov and Lavrent'eva (8) also measured the low temperature heat capacity of Na_2SO_4(cr, III) but their results are intermediate between Na_2SO_4(cr, V) and Na_2SO_4(cr, III) indicating that their sample may have been converted to as much as two-thirds of the V modification. Coughlin (3) measured enthalpies of Na_2SO_4(cr, III) from 370.2 - 514.9 K by drop calorimetry, but his data appear high by 2% at 450 K. Enthalpy measurements of Denielou et al. (9) are in very good agreement with the $C_p°$ measurements of Shmidt and Sokolov (7). We combine the three sets of experimental data (1, 7, 9) by graphical smoothing and by fitting the data with orthogonal polynomials over selected overlapping temperature intervals. $C_p°$ is smoothly extrapolated to 1000 K. S°(298.15 K) is derived from our combined smooth fit based on S°(14 K) = 0.100 cal K^{-1} mol^{-1} and H°(14 K)-H°(0 K) = 1.051 cal K^{-1} mol^{-1} derived from a Debye T^3-law extrapolation.

Transition Data

Na_2SO_4(cr, III) is the meltable low-temperature orthorhombic modification of anhydrous sodium sulfate which is obtained on cooling, in the absence of moisture, the high-temperature hexagonal form Na_2SO_4(cr, I). Na_2SO_4(cr, III) is metastable with respect to the two other low-temperature orthorhombic forms Na_2SO_4(cr, V) and Na_2SO_4(cr, IV). The space group of Na_2SO_4(cr, III) has been given as Cmcm (10).

The second-order transformation from Na_2SO_4(cr, III) to Na_2SO_4(cr, I) in which both phases are metastable, has been observed by Shmidt and Sokolov (7) at 521.5 K, $\Delta_{trs}H°$(III → I) = 1642 ± 20 cal mol^{-1}; by Coughlin (3) at 514 K, $\Delta_{trs}H°$(III → I) = 1680 cal mol^{-1}, by Denielou et al. (9) at 515 ± 1 K, $\Delta_{trs}H°$(III → I) = 1650 ± 90 cal mol^{-1}; and by May (11) at 520 ± 5 K, $\Delta_{trs}H°$(III → I) = 1150 cal mol^{-1}. We adopt the transition temperature from the analysis of Brodale and Giauque (1). This is consistent with our analysis which yields $\Delta_{trs}H°$(III → I, 298.15 K) = 1646 ± 20 cal mol^{-1}. The heat of transition from metastable Na_2SO_4(cr, V) to metastable Na_2SO_4(cr, III) has been measured by heat of solution measurements on the separate crystalline forms by Coughlin (3), Brodale and Giauque (1) and Pickering (12). $\Delta_r H°$(V → III, 298.15 K) = 716 ± 8 cal mol^{-1} (1, 3), 699 ± 10 cal mol^{-1} (1), 709 ± 10 cal mol^{-1} (1, 13) and $\Delta_r H°$(V → III, 294 K) = 708 ± 20 cal mol^{-1} (12). Using the heat of transition value from (1) and the difference between H°(517 K)-H°(298.15 K) for Na_2SO_4(cr, III) and Na_2SO_4(cr, V) (6), we obtain $\Delta_{trs}H°$(V → III, 517 K) = 1041 ± 20 cal mol^{-1}. The heat of transition from metastable Na_2SO_4(cr, IV) to metastable Na_2SO_4(cr, III) is discussed on the table for Na_2SO_4(cr, IV). Our analysis yields $\Delta_{trs}H°$(IV → III, 519 K) = 988 ± 20 cal mol^{-1}.

References

1. G. E. Brodale and W. F. Giauque, J. Phys. Chem. 76, 737 (1972).
2. ICSU-CODATA Task Group on Key Values for Thermodynamics, CODATA Bulletin No. 28, April 1978.
3. J. P. Coughlin, J. Amer. Chem. Soc. 77, 868 (1955).
4. E. Lange and H. Streeck, Z. Phys. Chem. A157, 1 (1931).
5. W. E. Wallace and A. L. Robinson, J. Amer. Chem. Soc. 63, 958 (1941).
6. JANAF Thermochemical Tables: Na_2SO_4(cr, V), Na_2SO_4(cr, IV), Na_2SO_4(cr, I), 6-30-78.
7. N. E. Shmidt and V. A. Sokolov, Russ. J. Inorg. Chem. 6, 1321 (1961).
8. I. E. Paukov and M. N. Lavrent'eva, Russ. J. Phys. Chem. 43, 1185 (1969).
9. L. Denielou, Y. Fournier, J. P. Petitet, and C. Tequi, C. R. Acad. Sci., Paris, Ser. C 269, 1577 (1969).
10. H. Fischmeister, Acta Crystallogr. 7, 776 (1954).
11. M. N. May, TAPPI 35, 511 (1952).
12. S. U. Pickering, J. Chem. Soc. 65, 686 (1884).
13. K. S. Pitzer and L. V. Coulter, J. Amer. Chem. Soc. 60, 1310 (1938).

Sodium Sulfate, III (Na_2SO_4) $Na_2O_4S_1$(cr)

Enthalpy Reference Temperature = T_r = 298.15 K Standard State Pressure = p° = 0.1 MPa

	J K⁻¹mol⁻¹			kJ mol⁻¹			
T/K	$C_p°$	S°	$-[G°-H°(T_r)]/T$	$H°-H°(T_r)$	$\Delta_f H°$	$\Delta_f G°$	Log K_f
0	0.	0.	INFINITE	-23.527	-1373.708	-1373.708	INFINITE
100	68.216	47.861	252.668	-20.481	-1379.841	-1343.950	702.008
200	105.621	108.169	165.992	-11.565	-1383.300	-1306.534	341.232
298.15	129.168	154.908	154.908	0.	-1384.854	-1268.471	222.231
300	129.570	155.709	154.911	0.239	-1384.870	-1267.748	220.735
400	150.143	195.743	160.231	14.205	-1392.687	-1228.129	160.377
500	173.167	231.587	170.966	30.311	-1392.829	-1186.938	123.999
600	200.091	265.539	183.916	48.973	-1390.103	-1145.967	99.765
700	227.011	298.405	197.935	70.329	-1384.508	-1105.681	82.507
800	253.935	330.475	212.505	94.376	-1376.164	-1066.386	69.628
900	280.859	361.939	227.366	121.116	-1418.266	-1027.217	59.618
1000	307.779	392.924	242.376	150.548	-1403.389	-984.542	51.427

PREVIOUS: December 1966 CURRENT: June 1978

Sodium Sulfate, III (Na_2SO_4) $Na_2O_4S_1$(cr)

SODIUM SULFATE, IV (Na_2SO_4) CRYSTAL M_r = 142.03714

$S°(298.15 K) = [150.157 \pm 0.4]\ J\ K^{-1}\ mol^{-1}$
$T_{trs}(V \to IV) = 458 \pm 1\ K$
$T_{trs}(IV \to I) = 514 \pm 1\ K$

$\Delta_f H°(0\ K) = [-1380.09]\ kJ\ mol^{-1}$
$\Delta_f H°(298.15\ K) = -1387.559 \pm 0.63\ kJ\ mol^{-1}$
$\Delta_{trs} H°(V \to IV) = 0.255 \pm 0.063\ kJ\ mol^{-1}$
$\Delta_{trs} H°(IV \to I) = 10.908 \pm 0.084\ kJ\ mol^{-1}$

Enthalpy of Formation

Brodale and Giauque (1) determined the enthalpy of solution at infinite dilution $\Delta_{sln}H°(Na_2SO_4$, cr, IV, 298.15 K) = -635 ± 1 kcal mol^{-1} for the process Na_2SO_4(cr, IV) → Na_2SO_4(aq, ∞). When this result is combined with the enthalpies of formation of the infinitely dilute ions from CODATA (2), we obtain $\Delta_f H°(Na_2SO_4$, aq, ∞, 298.15 K) = -332.266 ± 0.13 kcal mol^{-1} and $\Delta_f H°(Na_2SO_4$, cr, IV, 298.15 K) = -331.631 ± 0.15 kcal mol^{-1}. Brodale and Giauque (1) have also determined the difference in $\Delta_f H°$(298.15 K) between Na_2SO_4(cr, IV) and Na_2SO_4(cr, V) by direct heat of solution experiments. We prefer to give complete weight to these results using the same calorimeter. $\Delta_{sln}H°(Na_2SO_4$, cr, V, 298.15 K) = -574 cal K^{-1} mol^{-1} (1) which leads to $\Delta_r H°$(V → IV, 298.15 K) = 61 ± 15 cal K^{-1} mol^{-1}. When this is combined with auxiliary data for Na_2SO_4(cr, V) (3), we obtain $\Delta_f H°(Na_2SO_4$, cr, IV, 298.15 K) = -331.635 ± 0.15 kcal mol^{-1}. This value of $\Delta_f H°$(298.15 K) is adopted in this tabulation.

Heat Capacity and Entropy

The heat capacity of Na_2SO_4(cr, IV) has been taken equal to the heat capacity of the V form from 298-514 K (1, 3). The heat of transition (V → IV) adopted here leads to a calculated entropy of transition of 0.133 cal K^{-1} mol^{-1} and $S°(Na_2SO_4$, cr, IV, 458 K) = 50.304 cal K^{-1} mol^{-1}. Brodale and Giauque (1) state that the small difference in entropy (between IV and V) means that on the average the heat capacity of Na_2SO_4(cr, IV) should exceed the heat capacity of Na_2SO_4(cr, V) by only about 0.3% over the range 298-514 K. We calculate $S°(Na_2SO_4$, cr, IV, 298.15 K) = 35.887 cal K^{-1} mol^{-1}.

Transition Data

Na_2SO_4(cr, IV) is the orthorhombic modification of anhydrous sodium sulfate stable from 458-514 K (1, 4). This form can persist down to low temperatures. The presence of moisture enhances the achievement of equilibrium. A recent Raman and DSC study has failed to observe the IV modification (8). The (V → IV) transition at 458 K has been discussed previously (see (3) and above), $\Delta_r H°$(V → IV, 298.15 K) = 61 ± 15 cal mol^{-1}. Since $\Delta C°_p$ = 0 (1, 3), $\Delta_{trs}H°$(V → IV, 458 K) = 61 ± 15 cal mol^{-1}. The temperature of 458 K for the transition is taken from Kracek (4) and Brodale and Giauque (1).

The transition from Na_2SO_4(cr, IV) to Na_2SO_4(cr, I) has been found to occur at 514 K (1, 4). The heat of the transition has not been directly measured although Shmidt and Sokolov (5) obtained $\Delta_{trs}H°$(516 K) = 2584 ± 15 cal mol^{-1} for a sample of ground Na_2SO_4 which could have been Na_2SO_4(cr, IV) (see 1). We adopt the value of the heat of transition $\Delta_{trs}H°$(IV → I, 514 K) = 2607 ± 20 cal mol^{-1}. This is obtained by summation of $\Delta_{trs}H°$(IV → III) (1, 3) and $\Delta_{trs}H°$(III → I) (1, 3), all reduced to 514 K. Other measurements include that of Kreidl and Simon (6) who found $\Delta_{trs}H°$(IV → I, 511 K) = 2300 cal mol^{-1} and Popov and Galchenko (7) who found $\Delta_{trs}H°$(518 K) = 2586 ± 64 cal mol^{-1} on first heating. Later heating apparently converted some sample over to the III modification. Brodale and Giauque (1) give $\Delta_{trs}H°$(IV → I, 514 K) = 2611 cal mol^{-1}.

The heat of transition from metastable Na_2SO_4(cr, IV) to metastable Na_2SO_4(cr, III) is obtained from heat of solution measurements on the separate crystalline forms. Brodale and Giauque (1) obtained $\Delta_{sln}H°(Na_2SO_4$, cr, IV, 298.15 K) = -635 ± 1 cal mol^{-1} and $\Delta_{sln}H°(Na_2SO_4$, cr, III, 298.15 K) = -1273 ± 4 cal mol^{-1} from which $\Delta_r H°$(IV → III, 298.15 K) = 638 ± 10 cal mol^{-1}. Combining this result with the difference between H°(519 K)-H°(298.15 K) for Na_2SO_4(cr, IV) and Na_2SO_4(cr, III) we obtain $\Delta_{trs}H°$(IV → III, 519 K) = 988 ± 20 cal mol^{-1}. Shmidt and Sokolov (5) have measured the heats of transformation of Na_2SO_4(thenardite) → Na_2SO_4,cr, I) and Na_2SO_4(cr, III) → Na_2SO_4(cr, I) calorimetrically from which we obtain $\Delta_{trs}H°$(thenardite → III, 516 K) = 954 ± 20 cal mol^{-1}. Brodale and Giauque (1) have speculated that the "ground thenardite" of (5) was actually the IV modification of Na_2SO_4.

References

1. G. E. Bordale and W. F. Giauque, J. Phys. Chem. 76, 737 (1972).
2. ICSU-CODATA Task Group on Key Values for Thermodynamics, CODATA Bulletin No. 28, April, 1978.
3. JANAF Thermochemical Tables: Na_2SO_4(cr, V), Na_2SO_4(cr, III), Na_2SO_4(cr, I), 6-30-78.
4. F. C. Kracek, J. Phys. Chem. 33, 1281 (1929); F. C. Kracek and R. E. Gibson, J. Phys. Chem. 33, 1304 (1929); 34, 188 (1930).
5. N. E. Shmidt and V. A. Sokolov, Russ. J. Inorg. Chem. 6, 1321 (1961).
6. E. L. Kreidl and I. Simon, Nature 181, 1529 (1958).
7. M. M. Popov and G. L. Galchenko, J. Gen. Chem. USSR 21, 2489 (1951).
8. J. E. D. Davies and W. F. Sandford, J. Chem. Soc., Dalton Trans. 1912 (1975).

Sodium Sulfate, IV (Na_2SO_4) $Na_2O_4S_1$(cr)

Enthalpy Reference Temperature = T_r = 298.15 K   Standard State Pressure = p° = 0.1 MPa

T/K	$C°_p$ (J K⁻¹mol⁻¹)	S°	$-[G°-H°(T_r)]/T$	$H°-H°(T_r)$ (kJ mol⁻¹)	$\Delta_f H°$	$\Delta_f G°$	Log K_f
0							
100							
200							
298.15	128.143	150.157	150.157	0.	-1387.559	-1269.759	222.456
300	128.495	150.950	150.159	0.237	-1387.577	-1269.028	220.957
400	145.101	190.263	155.407	13.943	-1395.654	-1228.904	160.478
458.000	153.331	210.462	161.117	22.600	-----------	V <--> IV	----------
500	158.908	224.159	165.842	29.158	-1396.687	-1187.082	124.014
514.000	160.712	228.572	167.491	31.396	-----------	IV <--> I	----------
600	171.322	254.242	178.113	45.677	-1396.104	-1145.190	99.698
700	182.933	281.530	190.967	63.394	-1394.148	-1103.508	82.345
800	194.046	306.686	203.879	82.246	-1390.999	-1062.190	69.354
900	204.828	330.166	216.620	102.191	-1439.895	-1020.251	59.214
1000	215.371	352.295	229.091	123.203	-1433.438	-973.962	50.875

PREVIOUS: CURRENT: June 1978

Sodium Sulfate, IV (Na_2SO_4) $Na_2O_4S_1$(cr)

SODIUM SULFATE, V (Na_2SO_4) CRYSTAL M_r = 142.03714

$S°(298.15 K) = 149.595 \pm 0.4 J K^{-1} mol^{-1}$

$T_{trs}(V \rightarrow IV) = 458 \pm 1 K$

$\Delta_f H°(0 K) = -1376.361 \pm 0.63 kJ mol^{-1}$

$\Delta_f H°(298.15 K) = -1387.816 \pm 0.63 kJ mol^{-1}$

$\Delta_{trs} H°(V \rightarrow IV) = 0.255 \pm 0.063 kJ mol^{-1}$

Enthalpy of Formation

Shibata et al. (1) measured the emf of the cell: Na(amalgam, 0.2077%)|Na_2SO_4(sat. soln), Hg_2SO_4|Hg at 306-310 K. Extrapolating their results to 298 K, correcting the observed voltage for the formation of the Na-amalgam, and converting to absolute volts, we have E_{cell} = 3.33404 volts for the reaction at 298 K, 2 Na(cr) + Hg_2SO_4(cr) = 2 Hg(ℓ) + Na_2SO_4(cr, V). From this voltage we calculate $\Delta_r G°$(298.15 K) = -153.77 kcal mol^{-1}. Taking $\Delta_f G°$(Hg_2SO_4, cr, 298.15 K) = -149.66 kcal mol^{-1} (2) and reference entropies from (3), we calculate $\Delta_f G°$(Na_2SO_4, cr, V, 298.15 K) = -303.43 kcal mol^{-1} and $\Delta_f H°$(Na_2SO_4, cr, V, 298.15 K) = -331.60 ± 0.2 kcal mol^{-1}.

The enthalpy of solution at infinite dilution, $\Delta_{sln}H°$(Na_2SO_4, cr, V, 298.15 K) = -570 ± 10 cal K^{-1} mol^{-1} adopted in this tabulation is based on our reanalysis of the enthalpy of solution measurements of Gardner et al. (4), Brodale and Giauque (5), Pitzer and Coulter (6), Coughlin (7) as corrected in (5), and Readnour and Cobble (8). When this result is combined with heats of formation of the infinitely dilute ions from CODATA (2), we obtain $\Delta_f H°$(Na_2SO_4, aq, ∞, 298.15 K) = -332.266 ± 0.13 kcal mol^{-1} and $\Delta_f H°$(Na_2SO_4, cr, V, 298.15 K) = -331.696 ± 0.15 kcal mol^{-1}. The experimental $\Delta_{sln}H°$ values and the number of determinations are -582 ± 5 cal K^{-1} mol^{-1}, 6 (4), -574 cal K^{-1} mol^{-1}, 1 (5), -564 ± 10 cal K^{-1} mol^{-1}, 2 (6), -565 ± 30 cal K^{-1} mol^{-1}, 5 (7, 5), and -567 ± 12 cal K^{-1} mol^{-1}, 9 (8).

We adopt $\Delta_f H°$(Na_2SO_4, cr, V, 298.15 K) = -331.696 ± 0.5 kcal mol^{-1} from the heat of solution results for consistency with the other modifications of Na_2SO_4(cr) (3).

Heat Capacity and Entropy

Heat capacities of Na_2SO_4(cr, V) have been measured by Pitzer and Coulter (6) over the range 13.74-313.44 K and by Shmidt and Sokolov (9) in an adiabatic calorimeter over the temperature range 324.62-503.06 K. Shmidt (10) later carried out additional measurements which he combined with the earlier measurements of Shmidt and Sokolov (9) and reported $C_p°$ values extending from 298.5-503.1 K. We have smoothed the experimental data (6, 9, 10) by fitting the data with orthogonal polynomials over selected overlapping temperature intervals. The data have been smoothly extrapolated to 1000 K. We have given no weight to the calorimetry enthalpy differences measured by Coughlin (7), May (11), and Denielou et al. (12). These measured enthalpies are higher than our calculated enthalpies by 6%, 9% and 2%, respectively at 400 K. The $C_p°$ measurements of Popov and Galchenko (13) from 373-468 K appear to be biased, being too low at 373 K and too high at 468 K. S°(298.15 K) is derived from our combined smooth fit based on S°(14 K) = 0.059 cal K^{-1} mol^{-1} and H°(14 K)-H°(0 K) = 0.620 cal mol^{-1} derived from a T^3-law extrapolation.

Transition Data

Na_2SO_4(cr, V) is the stable low-temperature modification of anhydrous sodium sulfate which is obtained by crystallization from aqueous solution above 305.65 K under ordinary pressure. The mineral is known as thenardite and exists in the form of orthorhombic crystals (space group Fddd) (15).

Polymorphism studies by Kracek (14) indicated five distinct modifications of anhydrous sodium sulfate, i.e., V, IV, III, II, I. Calorimetric studies by Shmidt and Sokolov (9) as well as the X-ray work of Simanov and Kirkina (15) suggested another form, stable at high temperatures which is labeled as δ. Brodale and Giauque (5) have recently reviewed the relationships among the various crystalline forms of Na_2SO_4(cr). Their analysis shows that Na_2SO_4(cr, V) is the stable form from 0 to 458 K; Na_2SO_4(cr, IV) rather than Na_2SO_4 (cr, III) is the stable form from 458 to 514 K; Na_2SO_4(cr, I) is the form stable from 514 K to the melting point. Na_2SO_4(cr, III) is metastable at all temperatures from 0 to 517 K. Na_2SO_4(cr, II) is a meta-stable phase intermediate between forms III and I. The phase transitions are sluggish and subject to hysteresis. Forms V, IV and III can all persist far beyond their stable regions.

The heat of transition $\Delta_{trs}H°$(V → IV, 458 K) = 61 ± 15 cal mol^{-1} is discussed on the table for Na_2SO_4(cr, IV) (3). The temperature of the transition is taken from the work of Kracek (14) and Brodale and Giauque (5). The heat of transition from metastable Na_2SO_4(cr, V) to metastable Na_2SO_4(cr, III) $\Delta_{trs}H°$(V → III, 517 K) = 1041 ± 20 cal mol^{-1} has been discussed on the table for Na_2SO_4(cr, III) (3). This compares to $\Delta_{trs}H°$(V → III, 521 K) = 1030 ± 25 cal mol^{-1} from Shmidt and Sokolov (9) and $\Delta_{trs}H°$(V → III, 514 K) = 1040 ± 20 cal mol^{-1} from (5). The heat of transition from metastable Na_2SO_4(cr, V) to metastable Na_2SO_4(cr, I) at 512 ± 1 K is obtained by summation of $\Delta_{trs}H°$(V → III) = 1021 cal mol^{-1} and $\Delta_{trs}H°$(III → I) = 1642 cal mol^{-1} (5) to yield $\Delta_{trs}H°$(V → I, 512 K) = 2663 ± 25 cal mol^{-1}, in good agreement with the value measured by Shmidt and Sokolov (9), $\Delta_{trs}H°$(V → I, 521 K) = 2673 ± 5 cal mol^{-1}. Brodale and Giauque (5) give $\Delta_{trs}H°$(V → I, 512 K) = 2681 ± 20 cal mol^{-1}.

References

1. F. L. E. Shibata, S. Oda, and Furukawa, J. Sci. Hiroshima Univ. (Japan) 3A, 227 (1933).
2. ICSU-CODATA Task Group on Key Values for Thermodynamics, CODATA Bulletin No. 28, April, 1978.
3. JANAF Thermochemical Tables: Na(ref st), 6-30-62; O_2(ref st), 3-31-77; S(ref st), 9-30-77; Na_2SO_4(cr, IV), Na_2SO_4(cr, III), Na_2SO_4(cr, I), 6-30-78.
4. W. L. Gardner, E. C. Jekel, and J. W. Cobble, J. Phys. Chem. 73, 2017 (1969).
5. G. E. Brodale and W. F. Giauque, J. Phys. Chem. 76, 737 (1972).
6. K. S. Pitzer and L. V. Coulter, J. Amer. Chem. Soc. 60, 1310 (1938).
7. J. P. Coughlin, J. Amer. Chem. Soc. 77, 868 (1955).
8. J. M. Readnour and J. W. Cobble, Inorg. Chem. 8, 2174 (1969).
9. N. E. Shmidt and V. A. Sokolov, Russ. J. Inorg. Chem. 6, 1321 (1961).
10. N. E. Shmidt, Russ. J. Inorg. Chem. 12, 929 (1967).
11. M. N. May, TAPPI 35, 511 (1952).
12. L. Denielous, Y. Fournier, J. P. Petitet, and C. Tequi, C. R. Acad. Sci., Paris, Ser. C 269, 1577 (1969).
13. M. M. Popov and G. L. Galchenko, J. Gen. Chem. USSR 21, 2489 (1951).
14. F. C. Kracek, J. Phys. Chem. 33, 1281 (1929); F. C. Kracek and R. E. Gibson, J. Phys. Chem. 33, 1304 (1929); 34, 188 (1930).
15. Y. P. Simanov and D. F. Kirkina, Russ. J. Inorg. Chem. 2, 334 (1957).

Sodium Sulfate, V (Na_2SO_4) $Na_2O_4S_1$(cr)

Enthalpy Reference Temperature = T_r = 298.15 K — Standard State Pressure = p° = 0.1 MPa

T/K	$C_p°$ (J K⁻¹mol⁻¹)	S° (J K⁻¹mol⁻¹)	-[G°-H°(T_r)]/T (J K⁻¹mol⁻¹)	H°-H°(T_r) (kJ mol⁻¹)	$\Delta_f H°$ (kJ mol⁻¹)	$\Delta_f G°$ (kJ mol⁻¹)	Log K_f
0	0.	0.	INFINITE	-23.217	-1376.361	-1376.361	INFINITE
100	66.542	43.120	247.007	-20.389	-1382.711	-1346.347	703.259
200	105.667	103.010	160.624	-11.523	-1386.221	-1308.422	341.725
298.15	128.143	149.595	149.595	0.	-1387.816	-1269.849	222.472
300	128.495	150.388	149.597	0.237	-1387.834	-1269.117	220.973
400	145.101	189.702	154.845	13.943	-1395.911	-1228.937	160.483
458.000	153.331	209.905	160.555	22.602	-----------	V <--> IV	----------
500	158.908	223.596	165.280	29.158	-1396.944	-1187.058	124.011
600	171.322	253.678	177.551	45.676	-1396.362	-1145.110	99.691
700	182.933	280.966	190.404	63.393	-1394.406	-1103.372	82.335
800	194.046	306.122	203.316	82.245	-1391.257	-1061.997	69.341
900	204.828	329.602	216.057	102.190	-1440.154	-1020.002	59.199
1000	215.371	351.731	228.529	123.202	-1433.696	-973.656	50.859

PREVIOUS: December 1966 CURRENT: June 1978

Sodium Sulfate, V (Na_2SO_4) $Na_2O_4S_1$(cr)

SODIUM SULFATE (Na_2SO_4) LIQUID M_r = 142.03714

S°(298.15 K) = [181.172] J K^{-1} mol^{-1}

T_{fus}(I → ℓ) = 1157 ± 1 K

T_{dcm} = 2500 K

$\Delta_f H°$(298.15 K) = [-1356.385] kJ mol^{-1}

$\Delta_{fus}H°$(I → ℓ) = 238.49 ± 0.42 kJ mol^{-1}

Enthalpy of Formation

$\Delta_f H°$(ℓ, 298.15 K) is calculated from $\Delta_f H°$(Na_2SO_4, cr, I, 298.15 K) by adding $\Delta_{fus}H°$ and the difference in enthalpy, H°(1157 K)-H°(298.15 K), between Na_2SO_4(cr, I) and Na_2SO_4(ℓ) (5).

Heat Capacity and Entropy

The enthalpies for Na_2SO_4(ℓ) have been measured by drop calorimetry in the temperature range 1164.7-1826.2 K by Coughlin (1), 1173.3-1502.6 K by Denielou (2), 1157-1500 K by May (3) and 1173.4-1290.2 K by Popov and Ginzburg (4). The more recent studies (1, 2, 4) are in very good agreement. The study of May (4) appears biased, being 1% low at 1200 K and 2% high at 1500 K. We have adopted heat capacities derived from the enthalpy data of Coughlin (1) over the range 1157-1800 K. The data have been extended to 800 K and 3000 K. Below an assumed glass transition temperature of 800 K, the heat capacities are taken to be identical to those of the I phase. The entropy at 298.15 K is calculated in a manner analogous to that used for the enthalpy of formation.

Fusion Data

The enthalpy of melting of Na_2SO_4(cr, I) → Na_2SO_4(ℓ) has been discussed in the Na_2SO_4(cr, I) table (5). Kelley (6) had previously reported $\Delta_{fus}H°$(1157 K) = 5.83 kcal mol^{-1} from data on binary systems. We prefer the results from direct calorimetric measurements.

Vaporization and Decomposition Data

Vaporization data on Na_2SO_4(ℓ) is discordant due primarily to the extent to which the vapor is dissociated. Na_2SO_4(ℓ) can decompose according to various mechanisms, the most significant of which is the reaction Na_2SO_4(ℓ) = 2 Na(g) + SO_2(g) + O_2(g). Recent experimental data (7, 8, 9) are in good agreement with dissociation pressures and the heat of reaction calculated from the JANAF Thermochemical Tables (5). Decomposition to the oxides of sodium produces lower decomposition pressures than decomposition to the metal. In the presence of water vapor, significant decomposition can occur according to Na_2SO_4(ℓ) + H_2O(g) = 2 NaOH(g) + SO_2(g) + 1/2 O_2(g) (10). Molecular vaporization data are discussed in the Na_2SO_4(g) table (5). T_{dec} = 2500 K is the calculated temperature at which the Gibbs energy difference is zero for the reaction Na_2SO_4(ℓ) = 2 Na(g) + SO_2(g) + O_2(g).

References - Na_2SO_4(ℓ)

1. J. P. Coughlin, J. Amer. Chem. Soc. 77, 868 (1955).
2. L. Denielou, V. Fournier, J. P. Petitet, and C. Tequi, C. R. Acad. Sci., Paris, Ser. C 269, 1577 (1969).
3. M. N. May, TAPPI 35, 511 (1952).
4. M. M. Popov and D. M. Ginzburg, J. Gen. Chem. USSR 26, 1107 (1956).
5. JANAF Thermochemical Tables: Na_2SO_4(cr, I), Na_2SO_4(cr, δ), Na_2SO_4(g), 6-30-78.
6. K. K. Kelley, U. S. Bur. Mines Bull. 393, 1936.
7. D. Cubicciotti and F. J. Kenesha, High Temp. Sci. 4, 32 (1972).
8. P. J. Ficalora, O. M. Uy, D. W. Muenow, and J. J. Margrave, J. Amer. Ceram. Soc. 51, 574 (1968).
9. K. H. Lau, D. Cubicciotti, and D. L. Hildenbrand, J. Electrochem. Soc. 126, 490 (1979).
10. B. J. Kirkbride, Chem. Soc. Special Publ. No. 30, 65 (1977).

References - Na_2SO_4(g)

1. D. Cubicciotti and F. J. Kenesha, High Temp. Sci. 4, 32 (1972).
2. P. J. Ficalora, O. M. Uy, D. W. Muenow and J. L. Margrave, J. Amer. Ceram. Soc. 51, 574 (1968).
3. T. Kosugi, Kogyo Kogaku Zasshi 73, 1087 (1970).
4. R. E. Fryxell, C. A. Trythall, and R. J. Perkins, Corrosion 29, 423 (1973).
5. F. J. Kohl, C. A. Stearns, and G. C. Fryburg, NASA TM X-71641, May 1975. Article also appeared in "Metal-Slag-Gas Reactions and Processes," ed. by Z. A. Foroulis and W. W. Smeltzer, Electrochemical Society, p. 649, 1975.
6. V. V. Ugarov, Yu S. Ezhov, and N. G. Rambidi, J. Molec. Struct. 25, 357 (1975).
7. V. P. Spiridonov and B. I. Lutoshkin, Vestn. Mosk. Univ. Khim. 11, 509 (1970); English trans. NASA TT-F-16199, March 1975.
8. A. A. Belyawva, M. I. Dvorkin, and L. D. Shcherba, Opt. Spektrosk. 38, 170 (1975); 38, 291 (1975).
9. R. M. Atkins and K. A. Gingerich, Chem. Phys. Lett. 53, 347 (1978).
10. L. V. Gurvich, O. V. Dorofeeva, and V. S. Yungman, ANL-77-21, "Conference on High Temperature Sciences Related to Open-Cycle, Coal-Fired MHD Systems," Argonne Natl. Lab., 1977.
11. S. Montero, Spectrochim. Acta 32A, 843 (1976).
12. L. Andrews, J. Phys. Chem. 73, 3922 (1969).
13. G. V. Jagannathan and P. A. W. Wyatt, J. Chem. Res. (s) 208, (1978).
14. D. Bonnell and J. Hastie, U. S. Nat. Bur. Stand., personal communication.
15. JANAF Thermochemical Tables: Na_2SO_4(ℓ), 12-31-66; Na(g), 6-30-62; SO_2(g), 6-30-61; O_2(g), 3-31-77; Na_2SO_4(cr, I), 6-30-78.

Sodium Sulfate (Na_2SO_4) $Na_2O_4S_1$(l)

Enthalpy Reference Temperature = T_r = 298.15 K Standard State Pressure = p° = 0.1 MPa

T/K	$C_p°$ (J $K^{-1}mol^{-1}$)	S° (J $K^{-1}mol^{-1}$)	$-[G°-H°(T_r)]/T$ (J $K^{-1}mol^{-1}$)	$H°-H°(T_r)$ (kJ mol^{-1})	$\Delta_f H°$ (kJ mol^{-1})	$\Delta_f G°$ (kJ mol^{-1})	Log K_f
0							
100							
200							
298.15	159.620	181.172	181.172	0.	-1356.385	-1247.832	218.615
300	159.724	182.160	181.175	0.295	-1356.345	-1247.159	217.150
400	164.758	228.794	187.500	16.518	-1361.905	-1210.568	158.084
500	170.017	266.128	199.611	33.258	-1361.412	-1172.792	122.521
600	175.331	297.589	213.385	50.522	-1360.085	-1135.179	98.826
700	181.042	325.038	227.415	68.337	-1358.031	-1097.848	81.922
800	187.276	349.614	241.179	86.748	-1355.323	-1060.856	69.267
800.000	187.276	349.614	241.179	86.748	GLASS <--> LIQUID TRANSITION		
800.000	197.033	349.614	241.179	86.748			
900	197.033	372.822	254.542	106.451	-1404.461	-1023.207	59.385
1000	197.033	393.581	267.426	126.155	-1399.313	-981.123	51.249
1100	197.033	412.360	279.762	145.858	-1394.316	-939.547	44.615
1157.000	197.033	422.314	286.541	157.089	--------- I <--> LIQUID --------		
1200	197.033	429.504	291.537	165.561	-1583.018	-893.538	38.895
1300	197.033	445.275	302.764	185.265	-1576.541	-836.345	33.605
1400	197.033	459.877	313.471	204.968	-1570.140	-779.647	29.089
1500	197.033	473.471	323.690	224.671	-1563.809	-723.405	25.191
1600	197.033	486.187	333.453	244.374	-1557.546	-667.583	21.794
1700	197.033	498.132	342.792	264.078	-1551.347	-612.150	18.809
1800	197.033	509.394	351.738	283.781	-1545.210	-557.082	16.166
1900	197.033	520.047	360.319	303.484	-1539.134	-502.351	13.811
2000	197.033	530.154	368.560	323.188	-1533.117	-447.939	11.699
2100	197.033	539.767	376.486	342.891	-1527.159	-393.827	9.796
2200	197.033	548.933	384.118	362.594	-1521.259	-339.997	8.073
2300	197.033	557.692	391.475	382.298	-1515.418	-286.435	6.505
2400	197.033	566.077	398.577	402.001	-1509.635	-233.125	5.074
2500	197.033	574.120	405.439	421.704	-1503.912	-180.055	3.762
2600	197.033	581.848	412.076	441.407	-1498.246	-127.213	2.556
2700	197.033	589.284	418.503	461.111	-1492.640	-74.587	1.443
2800	197.033	596.450	424.731	480.814	-1487.093	-22.168	0.414
2900	197.033	603.364	430.772	500.517	-1481.607	30.053	-0.541
3000	197.033	610.044	436.637	520.221	-1476.185	82.087	-1.429

PREVIOUS: December 1966 CURRENT: June 1978

Sodium Sulfate (Na_2SO_4) $Na_2O_4S_1$(l)

SODIUM SULFATE (Na_2SO_4) V-IV-I-LIQUID M_r = 142.03714

0 to 458 K	crystal, V
458 to 514 K	crystal, IV
514 to 1157 K	crystal, I
above 1157 K	liquid

Refer to the individual tables for details.

Sodium Sulfate (Na_2SO_4) $Na_2O_4S_1$(cr,l)

Enthalpy Reference Temperature = T_r = 298.15 K Standard State Pressure = $p°$ = 0.1 MPa

T/K	$C_p°$ (J K⁻¹mol⁻¹)	$S°$ (J K⁻¹mol⁻¹)	$-[G°-H°(T_r)]/T$ (J K⁻¹mol⁻¹)	$H°-H°(T_r)$ (kJ mol⁻¹)	$\Delta_f H°$ (kJ mol⁻¹)	$\Delta_f G°$ (kJ mol⁻¹)	Log K_f
0	0.	0.	INFINITE	-23.217	-1376.361	-1376.361	INFINITE
100	66.542	43.120	247.007	-20.389	-1382.711	-1346.347	703.259
200	105.667	103.010	160.624	-11.523	-1386.221	-1308.422	341.725
298.15	128.143	149.595	149.595	0.	-1387.816	-1269.849	222.472
300	128.495	150.388	149.597	0.237	-1387.834	-1269.117	220.973
400	145.101	189.702	154.845	13.943	-1395.911	-1228.937	160.483
458.000	153.331	209.905	160.555	22.602		V <--> IV	
458.000	153.331	210.462	160.555	22.857		TRANSITION	
500	158.908	224.159	165.328	29.415	-1396.687	-1187.082	124.014
514.000	160.712	228.572	166.990	31.653		IV <--> I	
514.000	170.732	249.793	166.990	42.560		TRANSITION	
600	175.331	276.549	180.817	57.439	-1384.599	-1147.069	99.861
700	181.042	303.999	196.494	75.253	-1382.545	-1107.634	82.653
800	187.276	328.577	211.494	93.666	-1379.835	-1068.539	69.768
900	193.845	351.015	225.767	112.723	-1429.621	-1028.740	59.707
1000	200.309	371.773	239.342	132.431	-1424.467	-984.470	51.423
1100	206.731	391.166	252.272	152.784	-1418.821	-940.739	44.672
1157.000	210.367	401.702	259.376	164.671		I <--> LIQUID	
1157.000	197.033	422.314	259.376	188.520		TRANSITION	
1200	197.033	429.504	265.344	196.992	-1583.018	-893.538	38.895
1300	197.033	445.275	278.586	216.696	-1576.541	-836.345	33.605
1400	197.033	459.877	291.021	236.399	-1570.140	-779.647	29.089
1500	197.033	473.471	302.736	256.102	-1563.809	-723.405	25.191
1600	197.033	486.187	313.809	275.806	-1557.546	-667.583	21.794
1700	197.033	498.132	324.303	295.509	-1551.347	-612.150	18.809
1800	197.033	509.394	334.276	315.212	-1545.210	-557.081	16.166
1900	197.033	520.047	343.776	334.916	-1539.134	-502.351	13.811
2000	197.033	530.154	352.844	354.619	-1533.117	-447.939	11.699
2100	197.033	539.767	361.518	374.322	-1527.159	-393.827	9.796
2200	197.033	548.933	369.831	394.025	-1521.259	-339.997	8.073
2300	197.033	557.692	377.809	413.729	-1515.418	-286.435	6.505
2400	197.033	566.077	385.480	433.432	-1509.635	-233.125	5.074
2500	197.033	574.120	392.866	453.135	-1503.912	-180.055	3.762
2600	197.033	581.848	399.987	472.839	-1498.246	-127.213	2.556
2700	197.033	589.284	406.861	492.542	-1492.640	-74.587	1.443
2800	197.033	596.450	413.505	512.245	-1487.093	-22.168	0.414
2900	197.033	603.364	419.934	531.948	-1481.607	30.053	-0.541
3000	197.033	610.044	426.160	551.652	-1476.185	82.087	-1.429

PREVIOUS: CURRENT: June 1978

Sodium Sulfate (Na_2SO_4) $Na_2O_4S_1$(cr,l)

SODIUM SULFATE (Na_2SO_4) IDEAL GAS M_r = 142.03714

$\Delta_fH°$(0 K) = -1020.106 ± 25.1 kJ mol^{-1}

S°(298.15 K) = [346.854 ± 16.7] J K^{-1} mol^{-1} $\Delta_fH°$(298.15 K) = -1033.615 ± 25.1 kJ mol^{-1}

Vibrational Frequencies, Symmetries and Degeneracies

	ν, cm^{-1}	ν, cm^{-1}	ν, cm^{-1}
A_1	961 (1)	B_1[460](1)	E 1101 (2)
	[510](1)	$B_2$1131 (1)	610 (2)
	[160](1)	640 (1)	[333](2)
		[391](1)	[81](2)

Ground State Quantum Weight: [1] σ = [4]

Point Group: [D_{2d}]

Bond Distances: S–O = [1.48] Å Na–O = [2.21] Å

Bond Angles: O–S–O = [109.47]° O–Na–O = [66.205]°

Product of the Moments of Inertia: $I_AI_BI_C$ = [7.9153 x 10^{-113}] g^3 cm^6

Enthalpy of Formation

The adopted value of the enthalpy of formation, $\Delta_fH°(Na_2SO_4$, g, 298.15 K) = -247.04 ± 6.0 kcal mol^{-1}, is based on JANAF analyses of the vaporization data given below. Early studies of the vaporization of sodium sulfate failed to take dissociation into account. The many of the effusion and transpiration results in the literature are in error. The dissociation is retarded in the presence of SO_2, O_2 mixtures as in the study of Cubicciotti and Kenesha (1) or in the presence of air (O_2) as in the study of Fryxell et al. (4). We have "corrected" some of the original data by subtracting the calculated dissociation pressures (15) from the total pressure to give molecular vaporization pressures. These "corrected data sets" are explicitly noted. Mass spectrometric experiments (2, 3, 5, and 14) can lead to direct molecular vaporization data provided the $Na_2SO_4^+$ ion is directly measured. Results from the Knudsen effusion experiments (5, 14) are in reasonable agreement with the most reliable transpiration experiments (1, 4). The calculated 3rd law $\Delta_rH°$(298.15 K) may have an uncertainty of 4 kcal mol^{-1} since the JANAF free energy functions are partially based on the estimated molecular constants of Na_2SO_4(g). Vapor pressures from (1, 4 and 5) are in general agreement. Our adopted heat of formation is based on a weighted average of the 2nd or 3rd law results from (1, 4, 5, 14).

Source	Method	Reaction[a]	T/K	Data Points	δS[b] cal K^{-1} mol^{-1}	$\Delta_rH°$(298.15 K), kcal mol^{-1} 2nd law	3rd law	$\Delta_fH°$(298.15 K) kcal mol^{-1}
(1)[b]	Transpiration (N_2)	A	1400-1624	8	9.24 ± 2.8	90.36 ± 4.2	76.31 ± 2.0	-247.86 ± 6
(1)	Transpiration (SO_2 + O_2)	A	1305-1663	12	4.69 ± 1.2	84.26 ± 1.8	77.06 ± 1.1	-247.11 ± 4
(4)[b]	Transpiration (air)	A	1283-1477	20	0.23 ± 1.1	78.51 ± 1.5	78.20 ± 0.6	-245.97 ± 4
(4)	Transpiration (air)	A	1339-1477	16	3.70 ± 1.1	83.18 ± 1.5	78.05 ± 0.6	-246.12 ± 3
(13)	Transpiration (SO_2 + O_2)	A	1289-1412	6	5.60 ± 1.4	81.39 ± 1.8	73.81 ± 0.6	-250.36 ± 6
(13)[b]	Transpiration (N_2)	A	1301-1426	4	17.3 ± 2.5	98.78 ± 3.5	75.05 ± 2.0	-249.12 ± 6
(5)	Knudsen mass spec	A	1180-1370	2	----	75.76 ± 4.0	----	-248.41 ± 4
(5)	Knudsen mass spec	A	1196-1349	19	-1.89± 0.9	78.36 ± 0.9	78.76 ± 0.4	-245.41 ± 2
(14)	Knudsen mass spec	A,B	1069-1281	?	----	74.93 ± 2.5	----	-249.24 ± 4
(14)	Knudsen mass spec	A	1440-1540	?	----	77.41 ± 4.0	----	-246.76 ± 4

[a] A) $Na_2SO_4(\ell) = Na_2SO_4(g)$ B) $Na_2SO_4(cr) = Na_2SO_4(g)$

[b]data corrected for dissociation.

Heat Capacity and Entropy

The adopted structure (D_{2d} symmetry) is based on the similar symmetry assigned to Cs_2SO_4(g) and to K_2SO_4(g) in the high temperature electron diffraction studies of Ugarov et al. (6) and Spiridonov and Lutoshkin (7). This symmetry is supported by the IR and Raman study of Atkins and Gingerich (9) on Na_2SO_4 and K_2SO_4. The molecule can be pictured as having a central sulfur atom surrounded at the corners of a tetrahedron by four oxygen atoms with bridging sodium atoms between two pairs of oxygens. The internuclear distance of 1.48 Å for S–O is taken from data on other sulphates (6, 7, 8) while Na–O = 2.21 Å is an estimate from (7). The principal moments of inertia are I_A = 15.518 x 10^{-39} and $I_B = I_C$ = 71.418 x 10^{-39} g cm^2.

Atkins and Gingerich (9) have observed the infrared and Raman spectra of Na_2SO_4 and K_2SO_4 isolated in oxygen and nitrogen matrices at 12 K. Five of the eleven fundamental frequencies were observed in this work. The two other sulphate-ion group fundamentals are estimates taken from Gurvich et al. (10). The four fundamentals involving the metal atoms are based on the Raman single-crystal study of Montero (11) and comparisons involving K_2SO_4, Cs_2SO_4 (10), LiO_2 and NaO_2 (12). Uncertainties in the estimates of the gas phase frequencies may contribute an uncertainty of 4 cal K^{-1} mol^{-1} to S°(298.15 K).

References

References are listed on the liquid table.

Sodium Sulfate (Na_2SO_4) $Na_2O_4S_1$(g)

Enthalpy Reference Temperature = T_r = 298.15 K Standard State Pressure = p° = 0.1 MPa

T/K	$C_p°$ (J K^{-1}mol^{-1})	S°	-[G°-H°(T_r)]/T	H°-H°(T_r) (kJ mol^{-1})	$\Delta_fH°$	$\Delta_fG°$	Log K_f
0	0.	0.	INFINITE	-21.163	-1020.106	-1020.106	INFINITE
100	58.797	259.777	426.257	-16.648	-1024.770	-1010.071	527.607
200	85.280	308.790	355.907	-9.423	-1029.920	-993.278	259.418
250	96.480	329.057	348.544	-4.872	-1031.953	-983.875	205.569
298.15	105.619	346.854	346.854	0.	-1033.615	-974.461	170.721
300	105.940	347.509	346.856	0.196	-1033.675	-974.094	169.605
350	113.834	364.451	348.176	5.696	-1035.199	-964.041	143.875
400	120.370	380.092	351.201	11.556	-1044.097	-953.279	124.485
450	125.760	394.591	355.226	17.714	-1046.024	-941.817	109.323
500	130.203	408.078	359.845	24.117	-1047.785	-930.140	97.171
600	136.924	432.449	369.960	37.494	-1050.344	-906.354	78.905
700	141.606	453.929	380.452	51.434	-1052.164	-882.204	65.831
800	144.949	473.068	390.854	65.771	-1053.531	-857.827	56.010
900	147.397	490.289	400.962	80.394	-1107.749	-832.215	48.301
1000	149.234	505.918	410.688	95.230	-1107.468	-801.615	41.872
1100	150.641	520.211	420.005	110.227	-1107.178	-771.044	36.614
1200	151.741	533.367	428.911	125.348	-1300.462	-735.618	32.021
1300	152.614	545.549	437.420	140.567	-1298.469	-688.628	27.669
1400	153.319	556.885	445.553	155.865	-1296.473	-641.792	23.946
1500	153.895	567.483	453.332	171.227	-1294.484	-595.099	20.723
1600	154.371	577.431	460.781	186.641	-1292.511	-548.538	17.908
1700	154.769	586.802	467.921	202.098	-1290.557	-502.099	15.428
1800	155.106	595.658	474.774	217.592	-1288.629	-455.776	13.226
1900	155.392	604.052	481.359	233.118	-1286.731	-409.558	11.260
2000	155.638	612.029	487.695	248.670	-1284.865	-363.439	9.492
2100	155.851	619.628	493.798	264.244	-1283.036	-317.413	7.895
2200	156.036	626.883	499.683	279.839	-1281.245	-271.472	6.446
2300	156.198	633.822	505.366	295.451	-1279.496	-225.613	5.124
2400	156.340	640.473	510.858	311.078	-1277.789	-179.830	3.914
2500	156.467	646.858	516.171	326.718	-1276.128	-134.116	2.802
2600	156.579	652.997	521.316	342.371	-1274.513	-88.467	1.777
2700	156.679	658.908	526.303	358.034	-1272.947	-42.879	0.830
2800	156.769	664.608	531.141	373.706	-1271.431	2.652	-0.049
2900	156.849	670.111	535.839	389.387	-1269.968	48.127	-0.867
3000	156.922	675.429	540.404	405.076	-1268.560	93.555	-1.629
3100	156.988	680.576	544.843	420.771	-1267.208	138.938	-2.341
3200	157.049	685.561	549.163	436.473	-1265.915	184.275	-3.008
3300	157.103	690.394	553.370	452.181	-1264.681	229.573	-3.634
3400	157.153	695.085	557.469	467.894	-1263.512	274.837	-4.222
3500	157.199	699.641	561.467	483.611	-1262.410	320.066	-4.777
3600	157.241	704.070	565.367	499.333	-1261.377	365.265	-5.300
3700	157.280	708.379	569.174	515.059	-1260.416	410.435	-5.794
3800	157.316	712.574	572.893	530.789	-1259.533	455.583	-6.262
3900	157.349	716.661	576.527	546.523	-1258.729	500.708	-6.706
4000	157.380	720.645	580.080	562.259	-1258.011	545.811	-7.128
4100	157.408	724.531	583.556	577.998	-1257.383	590.899	-7.528
4200	157.435	728.325	586.958	593.741	-1256.844	635.972	-7.909
4300	157.459	732.030	590.289	609.485	-1256.410	681.034	-8.273
4400	157.482	735.650	593.552	625.232	-1255.996	726.079	-8.620
4500	157.504	739.189	596.749	640.982	-1255.731	771.121	-8.951
4600	157.524	742.651	599.883	656.733	-1255.594	816.159	-9.268
4700	157.543	746.039	602.957	672.486	-1255.573	861.198	-9.571
4800	157.560	749.356	605.972	688.242	-1255.676	906.236	-9.862
4900	157.577	752.605	608.932	703.999	-1255.908	951.280	-10.141
5000	157.593	755.789	611.837	719.757	-1256.261	996.327	-10.409
5100	157.608	758.910	614.691	735.517	-1256.767	1041.384	-10.666
5200	157.621	761.970	617.494	751.278	-1257.422	1086.450	-10.914
5300	157.635	764.973	620.248	767.041	-1258.231	1131.533	-11.152
5400	157.647	767.919	622.955	782.805	-1259.203	1176.631	-11.382
5500	157.659	770.812	625.618	798.571	-1260.343	1221.748	-11.603
5600	157.670	773.653	628.236	814.337	-1261.658	1266.891	-11.817
5700	157.680	776.444	630.811	830.105	-1263.155	1312.055	-12.024
5800	157.690	779.186	633.346	845.873	-1264.839	1357.250	-12.223
5900	157.700	781.882	635.841	861.643	-1266.717	1402.471	-12.417
6000	157.709	784.533	638.297	877.413	-1268.480	1447.695	-12.603

PREVIOUS: June 1978 (1 atm) CURRENT: June 1978 (1 bar)

Sodium Sulfate (Na_2SO_4) $Na_2O_4S_1$(g)

SODIUM TUNGTEN OXIDE (Na_2WO_4) CRYSTAL M_r = 293.82714

$S°(298.15 K) = 160.331 \pm 2.1 J K^{-1} mol^{-1}$
T_{trs1} = 862.0 K
T_{trs2} = 860.8 K
T_{fus} = 968.7 K

$\Delta_f H°(0 K) = -1534.79 \pm 8.4 kJ mol^{-1}$
$\Delta_f H°(298.15 K) = -1544.73 \pm 8.4 kJ mol^{-1}$
$\Delta_{trs1} H° = 4.113 kJ mol^{-1}$
$\Delta_{trs2} H° = 30.849 kJ mol^{-1}$
$\Delta_{fus} H° = 23.799 kJ mol^{-1}$

Enthalpy of Formation

The adopted enthalpy of formation, $\Delta_f H°(Na_2WO_4$, cr, 298.15 K) = $-369.2 kcal mol^{-1}$, is calculated from $\Delta_r H°(303.15 K) = 19.44 \pm 0.08 kcal mol^{-1}$ for the reaction of $H_2WO_4(cr) + 2 NaCl(cr) = Na_2WO_4(cr) + 2 HCl(aq, 12.731 H_2O)$, using the JANAF value of $\Delta_f H°(H_2WO_4$, cr, 298.15 K) = $-270.5 kcal mol^{-1}$ and $\Delta_f H°$(NaCl, cr, 298.15 K) = $-98.26 kcal mol^{-1}$, and auxiliary data for HCl(aq) (11). The value of $\Delta_r H°(303.15 K)$ was determined by solution calorimetry by Koehler et al. (1).

Sherfey and Brenner (2) measured by electrochemical calorimetry the enthalpy of reaction $\Delta_r H°(298.15 K) = -7 \pm 1.5 kcal mol^{-1}$ for $W(cr) + 2 NaOH(aq, 55.55 H_2O) + 2 H_2O(\ell) = Na_2WO_4(aq, \infty) + 3 H_2(g)$. This value, combined with the heat of solution (1, 3, 9), $\Delta_{sln} H° = 1.7 \pm 0.1 kcal mol^{-1}$ for $Na_2WO_4(cr) = Na_2WO_4(aq, \infty)$, gives $\Delta_f H°(298.15 K)$ (Na_2WO_4, cr) = $-368.4 kcal mol^{-1}$, which is in good agreement with the value adopted.

The previously accepted value (1, 3), $\Delta_f H°(Na_2WO_4$, cr, 298.15 K) = $-379 \pm 0.5 kcal mol^{-1}$, was based on the unreliable $\Delta_f H°(H_2WO_4$, cr, 298.15 K) = $-280.2 \pm 0.4 kcal mol^{-1}$. (See JANAF H_2WO_4, cr, table of March 31, 1967.) The other quoted value (4), $\Delta_f H°(Na_2WO_4$, cr, 298.15 K) = $-395 kcal mol^{-1}$, was obtained from Mixter's (5) measurement of the heat of reaction of tungsten powder with excess Na_2O_2. This value was probably in error because the complex tungstate and peroxytungstates were formed in the reaction.

Heat Capacity and Entropy

Low temperature heat capacities (52-299.87 K) were measured calorimetrically by King and Weller (6). The heat capacities above 300 K are estimated by extrapolation of the low temperature heat capacity curve to $C°_p(860 K) = 49.0 cal K^{-1} mol^{-1}$. The latter is obtained from an estimate of $C°_p$ equal to 7 cal K^{-1} g-atom at the first transition temperature.

The entropy, $S°(298.15 K) = 38.32 \pm 0.5 cal K^{-1} mol^{-1}$, is calculated from low temperature heat capacity data of King and Weller (6), based on an extrapolation of $S°(51 K) = 2.77 \pm 0.5 cal K^{-1} mol^{-1}$.

Transition Data

Transition temperatures and heats were obtained from differential heating and cooling curves both at atmospheric and higher pressures by Goranson and Kracek (7). Existence of three phases is confirmed by the earlier birefringence studies of Boeke (8).

Riccardi and Sinistri (10) found only one transition at 864 K with $\Delta_{trs} H° = 8.23 kcal mol^{-1}$ by differential thermal analysis. This heat is apparently the sum of the two adopted heats of transition.

Fusion Data

The adopted melting data were measured by the differential heating and cooling curve method by Goranson and Kracek (7). Riccardi and Sinistri (10) found the melting point at 971 K with $\Delta_{fus} H° = 7.52 kcal mol^{-1}$ by differential thermal analysis.

References

1. M. F. Koehler, L. B. Pankratz and R. Barany, U. S. Bur. Mines RI 5973, 1962.
2. J. M. Sherfey and A. Brenner, J. Electrochem. Soc. 105, 665 (1958).
3. R. L. Graham and L. G. Hepler, J. Amer. Chem. Soc. 80, 3538 (1958).
4. U. S. Nat. Bur. Stand. Circ. 500, 1952.
5. W. G. Mixter, Amer. J. Sci. 26, 125 (1908).
6. E. G. King and W. W. Weller, U. S. Bur. Mines RI 5791, 1961.
7. R. W. Goranson and F. C. Kracek, J. Chem. Phys. 3, 87 (1935).
8. H. E. Boeke, Z. Anorg. Allg. Chem. 50, 359 (1906).
9. S. A. Shchukarev and G. A. Kokovin, Zh. Neorg. Khim. 9, 1309 (1964).
10. R. Riccardi and C. Sinistri, Ric. Sci. Rend. Sez. A 8, 1026 (1965).
11. V. B. Parker, NSRDS-NBS 2, 1965; U. S. Nat. Bur. Stand. Tech. Note 270-1, 1965.

Sodium Tungsten Oxide (Na_2WO_4) $Na_2O_4W_1$(cr)

Enthalpy Reference Temperature = T_r = 298.15 K; Standard State Pressure = p° = 0.1 MPa

T/K	$C°_p$ (J K⁻¹mol⁻¹)	S° (J K⁻¹mol⁻¹)	$-[G°-H°(T_r)]/T$ (J K⁻¹mol⁻¹)	$H°-H°(T_r)$ (kJ mol⁻¹)	$\Delta_f H°$ (kJ mol⁻¹)	$\Delta_f G°$ (kJ mol⁻¹)	Log K_f
0	0.	0.	INFINITE	-25.286	-1534.786	-1534.786	INFINITE
100	69.860	43.482	267.673	-22.419	-1541.060	-1505.022	786.143
200	117.244	108.962	172.462	-12.700	-1544.082	-1467.547	383.284
298.15	139.775	160.331	160.331	0.	-1544.733	-1429.786	250.493
300	140.122	161.197	160.334	0.259	-1544.732	-1429.073	248.824
400	155.335	203.750	166.032	15.087	-1549.553	-1390.158	181.536
500	167.360	239.748	177.264	31.242	-1548.233	-1350.438	141.079
600	178.238	271.237	190.355	48.529	-1545.854	-1311.086	114.140
700	188.698	299.500	203.961	66.877	-1542.521	-1272.209	94.933
800	198.740	325.351	217.541	86.248	-1538.287	-1233.871	80.564
860.800	205.101	340.137	225.682	98.524		I <--> II	
860.800	205.099	375.975	225.682	129.372		TRANSITION	
862.000	205.226	376.260	225.891	129.618		II <--> III	
862.000	205.224	381.032	225.891	133.731		TRANSITION	
900	209.200	389.971	232.631	141.606	-1498.203	-1197.704	69.513
1000	219.242	412.530	249.504	163.026	-1492.216	-1164.628	60.834
1100	229.492	433.907	265.304	185.463	-1485.393	-1132.192	53.763
1200	239.743	454.314	280.211	208.923	-1671.275	-1095.535	47.687
1300	250.120	473.913	294.362	233.417	-1660.975	-1047.968	42.108
1400	260.454	492.828	307.866	258.946	-1649.746	-1001.228	37.356
1500	270.705	511.147	320.811	285.505	-1637.590	-955.325	33.267

PREVIOUS: CURRENT: June 1967

Sodium Tungsten Oxide (Na_2WO_4) $Na_2O_4W_1$(cr)

SODIUM SILICATE ($Na_2Si_2O_5$) CRYSTAL M_r = 182.14754

$S°(298.15\ K) = 164.055 \pm 4.2\ J\ K^{-1}\ mol^{-1}$
$\Delta_fH°(0\ K) = -2455.46 \pm 4.2\ kJ\ mol^{-1}$
$\Delta_fH°(298.15\ K) = -2470.07 \pm 4.2\ kJ\ mol^{-1}$
$T_{trs}(B \rightarrow A\beta) = 951\ K$ $\Delta_{trs}H° = 0.418\ kJ\ mol^{-1}$
$T_{trs}(A\beta \rightarrow A\alpha) = 980\ K$ $\Delta_{trs}H° = 0.628\ kJ\ mol^{-1}$
$T_{fus} = 1147\ K$ $\Delta_{fus}H° = 35.564\ kJ\ mol^{-1}$

Enthalpy of Formation

The enthalpies of solution of NaCl(cr), Na_2SO_4(cr) SiO_2(quartz), $Na_2Si_2O_5$(cr), HCl(aq) and H_2SO_4(aq), in 20% hydrofluoric acid at 74.7°C were measured by Kracek (1). The author derived two values of $\Delta_{sln}H°$ of Na_2O(cr) from those of NaCl(cr), Na_2SO_4(cr), HCl(aq) and H_2SO_4(aq), presumably employing auxiliary data from (2). We recalculate $\Delta_{sln}H°$ of Na_2O(cr) in the calorimetric solution as -87.50 (chlorine scheme, originally -87.66) and -88.74 kcal mol^{-1} (sulfate scheme, originally -89.19), using recent $\Delta_fH°$(298.15 K) values (16, 17, 18) for NaCl(cr), HCl(aq), H_2SO_4(aq), Na_2SO_4(cr), and Na_2O(cr). The corresponding enthalpy changes, $\Delta_rH°$(298.15 K), for the reaction $Na_2O(cr) + 2\ SiO_2(quartz) = Na_2Si_2O_5(cr)$ are derived as -54.65 and -55.89 kcal mol^{-1}. Adopting the weighted average $\Delta_rH°$(298.15 K) = -55.06 kcal mol^{-1} with $\Delta_fH°(Na_2O$, cr, 298.15 K) = -99.9 kcal mol^{-1} and $\Delta_fH°(SiO_2$, quartz, 298.15 K) = -217.7 kcal mol^{-1}, we obtain $\Delta_fH°$(298.15 K) = -590.36 kcal mol^{-1} for $Na_2Si_2O_5$(cr). This $\Delta_fH°(Na_2Si_2O_5$, cr, 298.15 K) value is independent of future changes in $\Delta_fH°$(298.15 K) of Na_2O(cr), but the values of $\Delta_{sln}H°$ and $\Delta_rH°$ are not independent.

Stevens et al. (3) measured the enthalpies of solution of $Na_2Si_2O_5$(cr), Na_2O(cr) and SiO_2(quartz) in HF(aq) at 25°C. Based on the reported results, the enthalpy change for the reaction $Na_2O(cr) + 2\ SiO_2(quartz) = Na_2Si_2O_5(cr)$ is calculated as -53.25 ± 0.32 kcal mol^{-1}, yielding $\Delta_fH°(Na_2Si_2O_5$, cr, 298.15 K) = -588.55 kcal mol^{-1}.

Hummel (4) determined the enthalpies of solution of $Na_2Si_2O_5$(cr), SiO_2(quartz) and Na_2CO_3(cr) in 39% HF at 26.5°C. From the data reported we derive the enthalpy change to be 20.88 kcal mol^{-1} for the reaction $Na_2CO_3(cr) + 2\ SiO_2(quartz) = Na_2Si_2O_5(cr) + CO_2(g)$. Employing $\Delta_fH°$(298.15 K) = -270.26, -217.7 and -94.05 kcal mol^{-1} for Na_2CO_3(cr), SiO_2(quartz) and CO_2(g), respectively, we obtain $\Delta_fH°$(298.15 K) = -590.78 kcal mol^{-1} for $Na_2Si_2O_5$(cr) which is in good agreement with the adopted one.

Heat Capacity and Entropy

The low temperature heat capacities, 54.3 - 294.7 K, have been measured by Kelley (5) and the high temperature enthalpies, 376.8 - 1100.2 K, by Naylor (6). The derived high temperature heat capacities are joined smoothly with the low temperature ones at 298.15 K. An approximate $C_p°$ of 70 cal $K^{-1}\ mol^{-1}$ is derived from the enthalpies above 980 K; this value is extrapolated to 2000 K. The value of S°(298.15 K) is calculated from the adopted low temperature heat capacities based on S°(50 K) = 2.945 cal $K^{-1}\ mol^{-1}$. The entropy extrapolation derives from the Debye-Einstein extrapolation of Kelley (5) which yields a $C_p°$ curve with an extra inflection in the region from 15 to 50 K. A more normal extrapolation would lead to a value of S°(50 K) which is lower by 0.5 cal $K^{-1}\ mol^{-1}$.

Transition Data

There are six slowly inverting polymorphous phases (1, 7-9), namely A, B, C, D, E and F, of which only the first two phases are stable. Two rapid transitions have been reported at 951 and 980 K (1, 10). Phase A, which has two subphases, Aα and Aβ, is stable above 951 K. The stable phase below 951 K is known as phase B, previously called β-$Na_2Si_2O_5$. The $\Delta_{trs}H°$ values are derived from the enthalpy data of Naylor (6) using the adopted $C_p°$ values. Three more rapid transitions have been reported at 822, 846 and 866 K (1, 10) in E, previously called γ-$Na_2Si_2O_5$. Willgallis and Range (10) have examined these transitions.

Fusion Data

T_{fus} is obtained from Morey (11). The selected value of $\Delta_{fus}H°$ is based on the following $\Delta_{fus}H°$ values derived from four different kinds of measurements.

Source	$\Delta_{fus}H°$, kcal mol^{-1} 298.15 K	1147 K	Method**
(12)	(7.51)	8.46	I
(1)	5.25	(6.20)	II
(13)	8.81	(9.76)	III
(14)	8.83	(9.78)	II
(15)	(7.50)	8.45	IV
(4)	8.09	(9.04)	II

* The numbers in parentheses are derived from the measured ones at the other listed temperatures.
**I = phase diagram, II = solution calorimetry; III = high temperature reaction calorimetry; and IV = dynamic-differential calorimetric method.

References

1. F. C. Kracek, Ann. Rept. Director of the Geophysical Laboratory, No. 1215, 58, 69 (1953).
2. U. S. Nat. Bur. Stand. Circ. 500, 1952.
3. C. G. Stevens and E. T. Turkdogan, Trans. Faraday Soc. 51, 356 (1955).
4. C. Hummel and H. E. Schwiete, Glastech. Ber. 32, 327 (1959).
5. K. K. Kelley, J. Amer. Chem. Soc. 61, 471 (1939).
6. B. F. Naylor, J. Amer. Chem. Soc. 67, 466 (1945).
7. G. W. Morey and N. L. Bowen, J. Soc. Glass Technol. 9, 226 (1925).
8. F. C. Kracek, J. Phys. Chem. 34, 1583 (1930).
9. J. d'Ans and J. Loffler, Z. Anorg. Allg. Chem. 191, 1 (1930).
10. A. Willgallis and K. J. Range, Glastech. Ber. 37, 194 (1964).
11. G. W. Morey and N. L. Bowen, J. Phys. Chem. 28, 1167 (1924).
12. K. K. Kelley, U. S. Bur. Mines Bull. 393, 1936.
13. C. Kroger and W. Janetzko, Z. Anorg. Allg. Chem. 284, 84 (1956).
14. C. Kroger and G. Kreitlow, Glastech. Ber. 29, 393 (1956).
15. H. E. Schwiete and G. Ziegler, Ber. Deut. Keram. Ges. 35, 193 (1958).
16. JANAF Thermochemical Tables: NaCl(cr), 9-30-64; Na_2SO_4(cr, V), 12-31-66.
17. U. S. Nat. Bur. Stand. Tech. Note 270-1, 1965.
18. The value of $\Delta_fH°(Na_2O$, cr, 298.15 K) is selected based on the heat of solution data reported by W. A. Roth and H. L. Kaule, Z. Anorg. Chem. 253, 352 (1947), and M. Matsui and S. Oka, J. Soc. Chem. Ind. (Japan) 32, 79 (1929).

Sodium Silicate ($Na_2Si_2O_5$) $Na_2O_5Si_2$(cr)

Enthalpy Reference Temperature = T_r = 298.15 K Standard State Pressure = p° = 0.1 MPa

T/K	$C_p°$ (J K⁻¹mol⁻¹)	S° (J K⁻¹mol⁻¹)	$-[G°-H°(T_r)]/T$ (J K⁻¹mol⁻¹)	$H°-H°(T_r)$ (kJ mol⁻¹)	$\Delta_fH°$ (kJ mol⁻¹)	$\Delta_fG°$ (kJ mol⁻¹)	Log K_f
0	0.	0.	INFINITE	-26.430	-2455.459	-2455.459	INFINITE
100	66.906	42.058	279.416	-23.736	-2463.238	-2418.587	1263.340
200	124.081	107.976	177.452	-13.895	-2467.943	-2371.851	619.464
298.15	156.984	164.055	164.055	0.	-2470.066	-2324.163	407.184
300	157.494	165.027	164.058	0.291	-2470.090	-2323.258	404.515
400	183.401	214.007	170.560	17.379	-2475.916	-2273.781	296.925
500	203.246	257.178	183.653	36.762	-2474.938	-2223.325	232.269
600	217.576	295.572	199.168	57.843	-2472.541	-2173.208	189.195
700	227.840	329.925	215.438	80.141	-2469.179	-2123.575	158.463
800	235.162	360.855	231.715	103.312	-2465.186	-2074.470	135.449
900	240.325	388.869	247.645	127.102	-2460.809	-2025.890	117.580
951.000	242.313	402.170	255.578	139.409		I <--> II	
951.000	242.316	402.610	255.578	139.828		TRANSITION	
980.000	243.275	409.903	260.037	146.869		II <--> III	
980.000	292.880	410.544	260.037	147.497		TRANSITION	
1000	292.880	416.461	263.107	153.354	-2454.208	-1977.854	103.312
1100	292.880	444.375	278.337	182.642	-2444.821	-1930.674	91.680
1147.000	292.880	456.629	285.393	196.408	--------	III <--> LIQUID	-------
1200	292.880	469.859	293.251	211.930	-2629.223	-1879.459	81.811
1300	292.880	493.302	307.750	241.218	-2618.531	-1817.413	73.025
1400	292.880	515.007	321.788	270.506	-2607.999	-1756.184	65.524
1500	292.880	535.213	335.351	299.794	-2597.620	-1695.704	59.050
1600	292.880	554.115	348.439	329.082	-2587.389	-1635.911	53.407
1700	292.880	571.871	361.065	358.370	-2677.658	-1575.859	48.420
1800	292.880	588.612	373.246	387.658	-2667.257	-1511.349	43.858
1900	292.880	604.447	385.002	416.946	-2656.916	-1447.414	39.792
2000	292.880	619.470	396.353	446.234	-2646.633	-1384.022	36.147

PREVIOUS: June 1965 CURRENT: September 1967

Sodium Silicate ($Na_2Si_2O_5$) $Na_2O_5Si_2$(cr)

SODIUM SILICATE ($Na_2Si_2O_5$) LIQUID M_r = 182.14754

$S°(298.15\ K) = [188.705]\ J\ K^{-1}\ mol^{-1}$

$T_{fus} = 1147\ K$

$\Delta_f H°(298.15\ K) = [-2438.485]\ kJ\ mol^{-1}$

$\Delta_{fus} H° = 35.564\ kJ\ mol^{-1}$

Enthalpy of Formation

$\Delta_f H°$(ℓ, 298.15 K) is calculated from $\Delta_f H°$(cr, 298.15 K) by adding $\Delta_{fus}H°$ and the difference in enthalpy, H°(1147 K)-H°(298.15 K), between the crystal and liquid.

The enthalpies of solution of NaCl(cr), Na_2SO_4(cr), HCl(aq), H_2SO_4(aq), SiO_2(quartz) and $Na_2Si_2O_5$(g, ℓ) in 20% hydrofluoric acid at 74.7°C were measured by Kracek (1). Following the same procedure as described in the $Na_2Si_2O_5$(cr) table, we derive the enthalpy changes, $\Delta_r H°$(298.15 K), for the reaction Na_2O(cr) + 2 SiO_2(quartz) = $Na_2Si_2O_5$(g, ℓ) as -49.40 and -50.64 kcal mol^{-1} using the heats of solution of Na_2O(cr) as -87.50 (chloride scheme) and -88.74 kcal mol^{-1} (sulfate scheme), respectively. Adopting the weighted average $\Delta_r H°$(298.15 K) = -49.81 kcal mol^{-1} with $\Delta_f H°$(Na_2O, cr, 298.15 K) = -99.9 kcal mol^{-1} and $\Delta_f H°$(SiO_2, quartz, 298.15 K) = -217.7 kcal mol^{-1}, we obtain $\Delta_f H°$(298.15 K) = -585.113 kcal mol^{-1} for $Na_2Si_2O_5$(g, ℓ).

Using high temperature reaction calorimetry, Kroger (2) determined the enthalpy changes as 25.28 ± 0.22 and -18.58 ± 0.36 kcal mol^{-1} for the reactions Na_2CO_3(cr) + 2 SiO_2(quartz) = $Na_2Si_2O_5$(g, ℓ) + CO_2(g) and 2 NaOH(cr) + 2 SiO_2(quartz) = $Na_2Si_2O_5$(g, ℓ) + H_2O(ℓ), respectively. From the given results, we calculate the corresponding values of $\Delta_f H°$($Na_2Si_2O_5$, g, ℓ, 298.15 K) to be -586.32 and -589.46 kcal mol^{-1}, using $\Delta_f H°$(298.15 K) = -270.26, -217.70, -94.05, -101.90 and -68.32 kcal mol^{-1} for Na_2CO_3(cr), SiO_2(quartz, CO_2(g), NaOH(cr) and H_2O(ℓ), respectively. These two $\Delta_f H°$($Na_2Si_2O_5$, ℓ, 298.15 K) values, as well as the value -585.113 kcal mol^{-1} based on Kracek's data, are not adopted because incorporating these $\Delta_f H°$(g, ℓ, 298.15 K) values with the value of $\Delta_f H°$(cr, 298.15 K), gives values of the enthalpy of melting at 298.15 K which are too low in comparison with the other values (see the $Na_2Si_2O_5$, cr, table).

Hummel (3) measured the enthalpies of solution of Na_2CO_3(cr), SiO_2(quartz) and $Na_2Si_2O_5$(g, ℓ) in 39% hydrofluoric acid at 26.5°C, as -244.1, -564.5 and -673.7 cal g^{-1}, respectively. Based on these results, we derive the enthalpy change for the reaction Na_2CO_3(cr) + 2 SiO_2(quartz) = $Na_2Si_2O_5$(g, ℓ) + CO_2(g) to be 29.01 kcal mol^{-1}, yielding $\Delta_f H°$($Na_2Si_2O_5$, g, ℓ, 298.15 K) = -582.6 kcal mol^{-1} which is in good agreement with the adopted value.

Heat Capacity and Entropy

Using drop calorimetry, Naylor (4) has measured the enthalpies of $Na_2Si_2O_5$(g, ℓ), 445.5 - 1120.7 K, and $Na_2Si_2O_5$(ℓ), 1173.2 - 1744 K. Based on the liquid enthalpy data, we evaluate the heat capacity of $Na_2Si_2O_5$(ℓ) to be 62.43 cal K^{-1} mol^{-1}. The $C_p°$ values below 1147 K are derived from the enthalpies measured at 445.5 - 1120.7 K by Naylor (4). Below 600 K, $C_p°$ is identical with that of the crystal. The adopted curve rises rapidly to a maximum in the glass transition region, 650-950 K, then smoothly approaches the value of the real liquid near T_{fus}. The heat capacity of $Na_2Si_2O_5$(ℓ) is assumed to be constant in the temperature range 1147 to 2500 K.

S°(ℓ, 298.15 K) is calculated in a manner similar to that used for the enthalpy of formation.

Fusion Data

See the crystal table for details.

References

1. F. C. Kracek, Ann. Rept. Director of the Geophysical Laboratory, No. 1215, 69 (1953).
2. C. Kroger and W. Janetzko, Z. Anorg. Allg. Chem. 284, 84 (1956).
3. C. Hummel and H. E. Schwiete, Glastech. Ber. 32, 327 (1959).
4. B. F. Naylor, J. Amer. Chem. Soc. 67, 466 (1945).

Sodium Silicate ($Na_2Si_2O_5$) $Na_2O_5Si_2$(l)

Enthalpy Reference Temperature = T_r = 298.15 K Standard State Pressure = p° = 0.1 MPa

T/K	$C_p°$ (J K⁻¹mol⁻¹)	S° (J K⁻¹mol⁻¹)	-[G°-H°(T_r)]/T (J K⁻¹mol⁻¹)	H°-H°(T_r) (kJ mol⁻¹)	$\Delta_f H°$ (kJ mol⁻¹)	$\Delta_f G°$ (kJ mol⁻¹)	Log K_f
0							
100							
200							
298.15	156.984	188.705	188.705	0.	-2438.485	-2299.931	402.938
300	157.494	189.678	188.708	0.291	-2438.508	-2299.071	400.304
400	183.401	238.657	195.210	17.379	-2444.334	-2252.060	294.089
500	203.246	281.828	208.304	36.762	-2443.356	-2204.069	230.258
600	217.576	320.191	223.817	57.825	-2440.978	-2156.416	187.733
700	245.810	355.525	240.121	80.783	-2436.956	-2109.271	157.396
800	272.169	390.353	256.734	106.895	-2430.022	-2062.904	134.694
900	276.228	422.809	273.415	134.455	-2421.874	-2017.501	117.093
1000	267.776	451.569	289.825	161.744	-2414.237	-1972.990	103.058
1100	261.207	476.706	305.694	188.113	-2407.769	-1929.186	91.609
1147.000	261.207	487.635	312.927	200.390	--------	III <--> LIQUID	-------
1200	261.207	499.434	320.906	214.234	-2595.338	-1881.064	81.881
1300	261.207	520.342	335.454	240.355	-2587.813	-1821.847	73.203
1400	261.207	539.700	349.360	266.475	-2580.448	-1763.204	65.786
1500	261.207	557.721	362.657	292.596	-2573.236	-1705.082	59.376
1600	261.207	574.579	375.381	318.717	-2566.173	-1647.437	53.783
1700	261.207	590.415	387.569	344.838	-2659.609	-1589.334	48.834
1800	261.207	605.345	399.257	370.958	-2652.376	-1526.587	44.300
1900	261.207	619.468	410.479	397.079	-2645.202	-1464.239	40.255
2000	261.207	632.866	421.266	423.200	-2638.086	-1402.267	36.623
2100	261.207	645.610	431.648	449.320	-2631.031	-1340.650	33.347
2200	261.207	657.761	441.652	475.441	-2624.034	-1279.366	30.376
2300	261.207	669.373	451.302	501.562	-2617.098	-1218.401	27.671
2400	261.207	680.489	460.622	527.683	-2610.223	-1157.736	25.197
2500	261.207	691.152	469.631	553.803	-2603.410	-1097.356	22.928

PREVIOUS: June 1965 CURRENT: September 1967

Sodium Silicate ($Na_2Si_2O_5$) $Na_2O_5Si_2$(l)

SODIUM SILICATE ($Na_2Si_2O_5$) CRYSTAL–CRYSTAL–CRYSTAL–LIQUID M_r = 182.14754

0 to 951 K	crystal, B
951 to 980 K	crystal, Aβ
980 to 1147 K	crystal, Aα
above 1147 K	liquid

Refer to the individual tables for details.

Sodium Silicate ($Na_2Si_2O_5$) $Na_2O_5Si_2$(cr,l)

Enthalpy Reference Temperature = T_r = 298.15 K Standard State Pressure = $p°$ = 0.1 MPa

T/K	$C_p°$ (J K⁻¹mol⁻¹)	S° (J K⁻¹mol⁻¹)	$-[G°-H°(T_r)]/T$ (J K⁻¹mol⁻¹)	$H°-H°(T_r)$ (kJ mol⁻¹)	$\Delta_f H°$ (kJ mol⁻¹)	$\Delta_f G°$ (kJ mol⁻¹)	Log K_f
0	0.	0.	INFINITE	-26.430	-2455.459	-2455.459	INFINITE
100	66.906	42.058	279.416	-23.736	-2463.238	-2418.587	1263.340
200	124.081	107.976	177.452	-13.895	-2467.943	-2371.851	619.464
298.15	156.984	164.055	164.055	0.	-2470.066	-2324.163	407.184
300	157.494	165.027	164.058	0.291	-2470.090	-2323.258	404.515
400	183.401	214.007	170.560	17.379	-2475.916	-2273.781	296.925
500	203.246	257.178	183.653	36.762	-2474.938	-2223.325	232.269
600	217.576	295.572	199.168	57.843	-2472.541	-2173.208	189.195
700	227.840	329.925	215.438	80.141	-2469.179	-2123.575	158.463
800	235.162	360.855	231.715	103.312	-2465.186	-2074.470	135.449
900	240.325	388.869	247.645	127.102	-2460.809	-2025.890	117.580
951.000	242.313	402.170	255.578	139.409		I <--> II	
951.000	242.316	402.610	255.578	139.828		TRANSITION	
980.000	243.275	409.903	260.037	146.869		II <--> III	
980.000	292.880	410.544	260.037	147.497		TRANSITION	
1000	292.880	416.461	263.107	153.354	-2454.208	-1977.854	103.312
1100	292.880	444.375	278.337	182.642	-2444.821	-1930.674	91.680
1147.000	292.880	456.629	285.393	196.408		III <--> LIQUID	
1147.000	261.207	487.635	285.393	231.972		TRANSITION	
1200	261.207	499.434	294.588	245.816	-2595.338	-1881.064	81.881
1300	261.207	520.342	311.160	271.936	-2587.813	-1821.847	73.203
1400	261.207	539.700	326.802	298.057	-2580.448	-1763.204	65.786
1500	261.207	557.721	341.603	324.178	-2573.236	-1705.082	59.376
1600	261.207	574.579	355.643	350.298	-2566.173	-1647.437	53.783
1700	261.207	590.415	368.992	376.419	-2659.609	-1589.334	48.834
1800	261.207	605.345	381.712	402.540	-2652.376	-1526.587	44.300
1900	261.207	619.468	393.857	428.661	-2645.202	-1464.239	40.255
2000	261.207	632.866	405.475	454.781	-2638.086	-1402.267	36.623
2100	261.207	645.610	416.609	480.902	-2631.031	-1340.650	33.347
2200	261.207	657.761	427.297	507.023	-2624.035	-1279.366	30.376
2300	261.207	669.373	437.571	533.143	-2617.098	-1218.401	27.671
2400	261.207	680.489	447.463	559.264	-2610.223	-1157.736	25.197
2500	261.207	691.152	456.999	585.385	-2603.410	-1097.356	22.928

PREVIOUS: CURRENT: September 1967

Sodium Silicate ($Na_2Si_2O_5$) $Na_2O_5Si_2$(cr,l)

SODIUM SULFIDE (Na_2S) CRYSTAL M_r = 78.03954

$S°(298.15 K) = 96.232 \pm 16.7 J K^{-1} mol^{-1}$

T_{trs} = [1276 ± 20 K] (lambda transition)

T_{fus} = 1445 ± 10 K

$\Delta_f H°(0 K)$ = Unknown

$\Delta_f H°(298.15 K) = -366.10 \pm 12.6 kJ mol^{-1}$

$\Delta_{trs} H° = 0$

$\Delta_{fus} H° = [19.246 \pm 4.2] kJ mol^{-1}$

Enthalpy of Formation

We adopt the average value, $\Delta_f H°(298.15 K) = -87.5 kcal mol^{-1}$, derived from the experimental results summarized below.

Source	Reaction	$\Delta_f H°(298.15 K)$, kcal mol^{-1}
(1)	$\Delta_{sln}H°$ Na_2S(cr) and Na(cr), $\Delta_n H°$ NaOH(aq) + H_2S(aq)	-88.2
(2)		-89.7
(3)	Na (in NH_3) + S(cr) = Na_2S(cr)	-87.0
(4)	H_2SO_4(aq, 1100 H_2O) + Na_2S(cr) = Na_2SO_4(soln.) + H_2S(g)	-85.1*

* The author's value of -84.35 ± 0.9 kcal mol^{-1} is corrected for newer values for $\Delta_f H°(298.15 K)$ of Na_2SO_4(c, V) and H_2S(g) (5, 11).

The adopted value is reasonably consistent with -86.6 ± 1 kcal mol^{-1} calculated from the 2nd law analysis of Uusitalo's equilibrium data (10) described in the following section.

Heat Capacity and Entropy

The heat capacity is based on the drop calorimetry of May (6) (400-1500 K). The pre-melt S-shaped enthalpy curve is reinterpreted as incorporating a lambda transition in view of the enthalpy measurements on K_2S by Dworkin and Bredig (7) and the occurrence of lambda transitions in other materials having the fluorite or anti-flourite type of structure (8). The adopted heat capacity shows the maximum of the lambda transition at 50.65 cal K^{-1} mol^{-1} and 1276 K. The heat capacity then falls to a constant value of 32 cal K^{-1} mol^{-1} at 1420 K. This interpretation along with the adopted melting point, heat of melting, and liquid heat capacity (22 cal K^{-1} mol^{-1}, 9) leads to an enthalpy of the liquid consistent with May's measurement.

May (6, 648-656 K) and Uusitalo (10, 462-763 K) have measured equilibrium pressures for the reaction Na_2SO_4(cr) + 2 C(solid) = Na_2S(cr) + 2 CO_2(g) and have derived S°(Na_2S, cr, 298.15 K) = 18.6 and 18.5 cal K^{-1} mol^{-1}. These derivations are open to question because of the uncertainty of the crystalline form of Na_2SO_4 used in the reaction (forward direction) or resulting from the reaction (reverse direction). Five crystalline forms of Na_2SO_4 have been identified (11, 12). S°(298.15 K) for form V, stable at room temperature, has erroneously been used to derive S°(Na_2S, cr, 298.15 K) rather than S°(298.15 K) for form I which is the form stable at the equilibrium measurement temperature. Furthermore, Uusitalo (10) used charcoal which probably deviates from graphite, the standard reference state. Our 2nd law treatment of Uusitalo's data (10) with auxiliary data for Na_2SO_4(cr, I) and C(graphite) (11), which is still questionable, leads to S°(Na_2S, cr, 298.15 K) = 23.2 cal K^{-1} mol^{-1}. The agreement of this derivation with the estimate below is almost surely fortuitous.

Kubaschewski et al. (13) have estimated the entropy as 23.5 ± 2.5 cal K^{-1} mol^{-1}. Voronin's (14) estimate is 21.6 cal K^{-1} mol^{-1}. The summation of Kelley's additive entropy constants (15) gives 23.0 cal K^{-1} mol^{-1}. We adopt S°(Na_2S, cr, 298.15 K) = 23.0 ± 4 cal K^{-1} mol^{-1}.

Fusion Data

Tegman (16) indicated that many of the early measurements of the melting point of Na_2S were made on impure brown material. Several measurements on white or light pink material are in reasonable agreement: in °C, ~1200 (17), 1180 ± 10 (18), 1180 (19), 1160 ± 5 (20, 21), 1175 ± 10 (22), and 1168 ± 10 (23). We adopt a melting point of 1445 ± 10 K (1172 ± 10°C). This is consistent with May's (6) enthalpy measurements.

In an examination of May's enthalpy data, Wagman (24) has used an entropy of fusion of 3.21 cal K^{-1} mol^{-1} --the entropy of fusion of K_2S is 3.16 cal K^{-1} mol^{-1} --which leads to an entahlpy of fusion of 4.8 kcal mol^{-1} at the adopted melting point of 1445 K. From the Na_2S-S phase diagram work of Rosen and Tegman (23), Wagman (24) calculates an average enthalpy of fusion of 5.7 kcal mol^{-1} and recommends $\Delta_{fus}H°$ = 5±1 kcal mol^{-1}. We adopt $\Delta_{fus}E°$ = 4.8 kcal mol^{-1} to remain consistent with our incorporation of a lambda transition in the evaluation of May's (6) enthalpy measurements.

References

1. P. Sabatier, Annls. Chim. Phys. Ser. 5, 22, 5 (1881).
2. E. Rengade and N. Costeanu, Compt. rend. 158, 946 (1914).
3. C. A. Kraus and J. A. Ridderhof, J. Amer. Chem. Soc. 56, 79 (1934).
4. J. M. Letoffe, J. Thourey, G. Perachon, and J. Bousquet, Bull. Soc. Chim. Fr. 1976, 424 (1976).
5. U. S. Nat. Bur. Stand. Tech. Note 270-3, 1968; U. S. Nat. Bur. Stand. Report PB-254460, 73 pp., 1976.
6. M. N. May, Tapp 35, 511 (1952).
7. A. S. Dworkin and M. A. Bredig, J. Phys. Chem. 72, 1277 (1968).
8. M. A. Bredig, Colloq. Int. Centre Nat. Rech. Sci. No. 205, 183 (1972).
9. I. Barin, O. Knacke, and O. Kubaschewski, "Thermochemical Properties of Inorganic Substances," Supplement, Springer-Verlag, Berlin, 1977.
10. E. Uusitalo, Soumen Kemstilehti 31B, 228 (1958).
11. JANAF Thermochemical Tables: Na_2SO_4(cr, V), Na_2SO_4(cr, I), 12-31-66; C(ref st.), 3-31-78; H_2S(g), 6-30-77.
12. G. E. Brodale and W. F. Giauque, J. Phys. Chem. 76, 737 (1972).
13. O. Kubaschewski, E. L. Evans, C. B. Alcock, "Metallurgical Thermochemistry," 4th ed., Pergamon Press, Oxford, 1967.
14. G. F. Voronin, Russ. J. Phys. Chem. 44, 1717 (1970).
15. Quoted by D. R. Stull and H. Prophet, Chap. 13, J. L. Margrave, Ed., "The Characterization of High Temperature Vapors," Wiley, New York 1967.
16. R. Tegman, Univ. Umea, Umea, Sweden, personal communication, March 7, 1978.
17. P. Ley, Chem.-Ztg. 58, 859 (1934).
18. G. Courtois, Compt. rend. 207, 1220 (1938).
19. E. J. Kohlmeyer and G. Lohrke, Z. Anorg. Allg. Chem. 281, 54 (1955).
20. E. K. Ovechkin, L. N. Shevtsova, A. E. Voitsekhovskii, and L. V. Kuznetsova, Russ. J. Inorg. Chem. 16, 1672 (1971).
21. E. K. Ovechkin, L. N. Shevtsova, A. E. Voitsekhovskii, L. I. Oboznaya, and L. V. Kuznetsova, Russ. J. Inorg. Chem. 18, 571 (1973).
22. R. Tegman and B. Warngvist, Acta Chem. Scand. 26, 413 (1972).
23. E. Rosen and R. Tegman, Chem. Scr. 2, 221 (1972).
24. D. D. Wagman, U. S. Nat. Bur. Stand., personal communication, March 9, 1978.

Sodium Sulfide (Na_2S) Na_2S_1(cr)

Enthalpy Reference Temperature = T_r = 298.15 K; Standard State Pressure = p° = 0.1 MPa

T/K	$C_p°$ (J K⁻¹mol⁻¹)	S°	-[G°-H°(T_r)]/T	H°-H°(T_r) (kJ mol⁻¹)	$\Delta_f H°$	$\Delta_f G°$	Log K_f
0							
100							
200							
298.15	82.801	96.232	96.232	0.	-366.100	-354.552	62.116
300	82.843	96.744	96.234	0.153	-366.093	-354.480	61.720
400	84.308	120.780	99.501	8.512	-373.575	-350.129	45.722
500	85.772	139.753	105.719	17.017	-375.200	-344.086	35.946
600	87.236	155.522	112.742	25.668	-376.166	-337.761	29.405
700	88.701	169.081	119.844	34.466	-376.619	-331.319	24.723
800	90.165	181.023	126.760	43.411	-376.704	-324.838	21.210
900	91.337	191.717	133.394	52.491	-429.655	-317.282	18.415
1000	92.843	201.418	139.718	61.700	-428.077	-304.880	15.925
1100	110.458	210.823	145.749	71.581	-425.884	-292.653	13.897
1200	159.494	222.269	151.627	84.771	-614.002	-275.864	12.008
1276.000	211.921	233.718	156.162	98.962	Cp LAMBDA MAXIMUM	TRANSITION	
1276.000	211.939	233.718	156.162	98.962			
1300	187.652	237.411	157.629	103.718	-601.115	-248.179	9.972
1400	137.152	249.134	163.773	119.505	-591.403	-221.426	8.262
1445.000	133.888	253.390	166.498	125.559	------ CRYSTAL <--> LIQUID -----		
1500	133.888	258.392	169.777	132.923	-584.076	-195.256	6.799
1600	133.888	267.033	175.588	146.312	-576.793	-169.572	5.536
1700	133.888	275.150	181.208	159.700	-569.524	-144.343	4.435
1800	133.888	282.802	186.642	173.089	-562.270	-119.542	3.469
1900	133.888	290.041	191.895	186.478	-555.029	-95.144	2.616
2000	133.888	296.909	196.976	199.867	-547.802	-71.126	1.858

PREVIOUS: December 1960 CURRENT: March 1978

Sodium Sulfide (Na_2S) Na_2S_1(cr)

SODIUM SULFIDE (Na_2S) LIQUID M_r = 78.03954

$S°(298.15\ K) = [127.656]\ J\ K^{-1}\ mol^{-1}$ $\Delta_f H°(298.15\ K) = [-323.940]\ kJ\ mol^{-1}$

$T_{fus} = 1445 \pm 10\ K$ $\Delta_{fus}H° = [19.246 \pm 4.2]\ kJ\ mol^{-1}$

Enthalpy of Formation

$\Delta_f H°(Na_2S,\ \ell,\ 298.15\ K)$ is calculated from $\Delta_f H°(Na_2S,\ cr,\ 298.15\ K)$ by adding $\Delta_{fus}H°$ and the difference in enthalpy, H°(1445 K)-H°(298.15 K), between the crystal and liquid.

Heat Capacity and Entropy

The heat capacity of 22 cal $K^{-1}\ mol^{-1}$ is from the estimated of Barin et al. (1) and is assumed to be constant above the assumed glass transition at 970 K. Below the glass transition, the heat capacity is that of the crystal.

The entropy at 298.15 K is calculated in a manner similar to that used for the enthalpy of formation.

Reference

1. I. Barin, D. Knacke, and O. Kubaschewski, "Thermochemical Properties of Inorganic Substances," Supplement, Springer-Verlag, Berlin, 1977.

Enthalpy Reference Temperature = T_r = 298.15 K Standard State Pressure = p° = 0.1 MPa

T/K	$C_p°$ (J K⁻¹mol⁻¹)	S° (J K⁻¹mol⁻¹)	$-[G°-H°(T_r)]/T$ (J K⁻¹mol⁻¹)	$H°-H°(T_r)$ (kJ mol⁻¹)	$\Delta_f H°$ (kJ mol⁻¹)	$\Delta_f G°$ (kJ mol⁻¹)	Log K_f
0							
100							
200							
298.15	82.801	127.656	127.656	0.	-323.940	-321.761	56.371
300	82.843	128.169	127.658	0.153	-323.933	-321.747	56.021
400	84.308	152.205	130.925	8.512	-331.415	-320.539	41.858
500	85.772	171.177	137.143	17.017	-333.040	-317.638	33.183
600	87.236	186.947	144.166	25.668	-334.006	-314.455	27.376
700	88.701	200.506	151.268	34.466	-334.459	-311.156	23.219
800	90.165	212.447	158.184	43.411	-334.544	-307.817	20.098
900	91.337	223.141	164.818	52.491	-387.495	-303.404	17.609
970.010	92.425	230.023	169.278	58.923	GLASS <--> LIQUID	TRANSITION	
970.010	92.048	230.023	169.278	58.923			
1000	92.048	232.826	171.142	61.684	-385.933	-294.144	15.365
1100	92.048	241.599	177.155	70.888	-384.416	-285.039	13.535
1200	92.048	249.608	182.864	80.093	-576.519	-271.189	11.805
1300	92.048	256.976	188.285	89.298	-573.374	-245.872	9.879
1400	92.048	263.798	193.438	98.503	-570.245	-220.797	8.238
1445.000	92.048	266.710	195.675	102.645	------ CRYSTAL <--> LIQUID -----		
1500	92.048	270.148	198.343	107.708	-567.131	-195.945	6.823
1600	92.048	276.089	203.019	116.912	-564.031	-171.301	5.592
1700	92.048	281.669	207.483	126.117	-560.947	-146.849	4.512
1800	92.048	286.930	211.752	135.322	-557.877	-122.579	3.557
1900	92.048	291.907	215.841	144.527	-554.820	-98.480	2.707
2000	92.048	296.629	219.763	153.732	-551.777	-74.541	1.947
2100	92.048	301.120	223.531	162.936	-548.747	-50.753	1.262
2200	92.048	305.402	227.156	172.141	-545.729	-27.110	0.644
2300	92.048	309.494	230.647	181.346	-542.725	-3.604	0.082
2400	92.048	313.411	234.015	190.551	-539.735	19.772	-0.430
2500	92.048	317.169	237.266	199.756	-536.760	43.024	-0.899
2600	92.048	320.779	240.409	208.960	-533.800	66.157	-1.329
2700	92.048	324.253	243.451	218.165	-530.858	89.176	-1.725
2800	92.048	327.600	246.397	227.370	-527.934	112.087	-2.091
2900	92.048	330.830	249.253	236.575	-525.032	134.893	-2.430
3000	92.048	333.951	252.024	245.780	-522.153	157.600	-2.744

PREVIOUS: December 1960 CURRENT: March 1978

SODIUM SULFIDE (Na_2S) CRYSTAL-CRYSTAL - LIQUID M_r = 78.03954 Sodium Sulfide (Na_2S) Na_2S_1(cr,l)

0 to 1276 K* crystal
1276 to 1445 K crystal
above 1445 K liquid

*There is a lambda transition at 1276 K.

Refer to the individual tables for details.

Enthalpy Reference Temperature = T_r = 298.15 K; Standard State Pressure = p° = 0.1 MPa

	J K⁻¹mol⁻¹			kJ mol⁻¹			
T/K	C_p°	S°	$-[G°-H°(T_r)]/T$	$H°-H°(T_r)$	$\Delta_f H°$	$\Delta_f G°$	Log K_f
0							
100							
200							
298.15	82.801	96.232	96.232	0.	-366.100	-354.552	62.116
300	82.843	96.744	96.234	0.153	-366.093	-354.480	61.720
400	84.308	120.780	99.501	8.512	-373.575	-350.129	45.722
500	85.772	139.753	105.719	17.017	-375.200	-344.086	35.946
600	87.236	155.522	112.742	25.668	-376.166	-337.761	29.405
700	88.701	169.081	119.844	34.466	-376.619	-331.319	24.723
800	90.165	181.023	126.760	43.411	-376.704	-324.838	21.210
900	91.337	191.717	133.394	52.491	-429.655	-317.282	18.415
1000	92.843	201.418	139.718	61.700	-428.077	-304.880	15.925
1100	110.458	210.823	145.749	71.581	-425.884	-292.653	13.897
1200	159.494	222.269	151.627	84.771	-614.002	-275.864	12.008
1276.000	211.921	233.718	156.162	98.962	Cp LAMBDA MAXIMUM		
1276.000	211.939	233.718	156.162	98.962	TRANSITION		
1300	187.652	237.411	157.629	103.718	-601.115	-248.179	9.972
1400	137.152	249.134	163.773	119.505	-591.403	-221.426	8.262
1445.000	133.888	253.390	166.498	125.559	CRYSTAL <--> LIQUID		
1445.000	92.048	266.710	166.498	144.805	TRANSITION		
1500	92.048	270.148	170.236	149.868	-567.131	-195.945	6.823
1600	92.048	276.089	176.668	159.073	-564.031	-171.301	5.592
1700	92.048	281.669	182.682	168.278	-560.947	-146.849	4.512
1800	92.048	286.931	188.329	177.482	-557.877	-122.579	3.557
1900	92.048	291.907	193.651	186.687	-554.820	-98.480	2.707
2000	92.048	296.629	198.683	195.892	-551.777	-74.541	1.947
2100	92.048	301.120	203.455	205.097	-548.747	-50.753	1.262
2200	92.048	305.402	207.992	214.302	-545.729	-27.110	0.644
2300	92.048	309.494	212.317	223.506	-542.725	-3.604	0.082
2400	92.048	313.411	216.448	232.711	-539.735	19.772	-0.430
2500	92.048	317.169	220.402	241.916	-536.760	43.024	-0.899
2600	92.048	320.779	224.194	251.121	-533.800	66.157	-1.329
2700	92.048	324.253	227.836	260.326	-530.858	89.176	-1.725
2800	92.048	327.600	231.339	269.530	-527.934	112.087	-2.091
2900	92.048	330.830	234.715	278.735	-525.032	134.893	-2.430
3000	92.048	333.951	237.971	287.940	-522.153	157.600	-2.744

PREVIOUS: CURRENT: March 1978

Sodium Sulfide (Na_2S) Na_2S_1(cr,l)

SODIUM SULFIDE, BETA (β-Na_2S_2) CRYSTAL M_r = 110.09954

$S°(298.15\ K)$ = 150.6 ± 24 J K^{-1} mol^{-1}
T_{trs} = 423 - 523 K
T_{fus} = 753 ± 5 K

$\Delta_f H°(0\ K)$ = Unknown
$\Delta_f H°(298.15\ K)$ = -397.1 ± 8 kJ mol^{-1}
$\Delta_{trs} H°$ = Unknown
$\Delta_{fus} H°$ = [18.91 ± 1.7] kJ mol^{-1}

Enthalpy of Formation

Kraus and Ridderhof (1) determined $\Delta_f H°(Na_2S_2$, cr, 298.15 K) = -96.2 kcal mol^{-1} from the enthalpy of reaction of Na(in NH_3) and S(cr) to form Na_2S_2(cr) in liquid ammonia. Letoffe et al. (2) derived $\Delta_f H°(Na_2S_2$, cr, 298.15 K) = -92.7 kcal mol^{-1} from the enthalpy of reaction of Na_2S_2(cr) and H_2SO_4(1100 H_2O) to form Na_2SO_4(acid solution), H_2S(g), and S(cr). With newer values for the enthalpies of formation of H_2SO_4(1100 H_2O) (3), Na_2SO_4(cr) (4), and H_2S(g) (4), the value of Letoffe et al. corrects to -93.6 kcal mol^{-1}. We adopt the average of $\Delta_f H°(Na_2S_2$, cr, 298.15 K) = -94.9±2 kcal mol^{-1}.

Heat Capacity and Entropy

The heat capacities are those estimated by Barin et al. (5).

Mills (6) gives S°(298.15 K) = 21.4±5 cal K^{-1} mol^{-1} derived from the dissociation pressure measurements of Teder and Tiberg (7); the path of the derivation is not clear. Tegman (8) has measured the vapor pressure of sulfur over sodium polysulfide melts which he has supplemented with other measurements from the literature to form an extensive data set covering a composition range of $Na_2S_{1.3}$ - $Na_2S_{6.0}$ and a temperature range of 523 - 1273 K. His analysis of this data set via computer modeling gave, among other things, the entropy changes for the reactions: $Na_2S(\ell)$ + 1/2 S_2(g) = $Na_2S_2(\ell)$; ΔS° = -9.00 cal K^{-1} mol^{-1} and Na_2S_2(cr) = $Na_2S_2(\ell)$; ΔS° = 4.43 cal K^{-1} mol^{-1}. It had been assumed these values are independent of temperature, but for our calculation of S°(Na_2S_2, cr, 298.15 K) we have used a temperature of 900 K, near the mid-point of the experimental range. With auxiliary data for $Na_2S(\ell)$ and S_2(g) (4) and S°(900 K)-S°(298.15 K) for Na_2S_2(cr) from a preliminary table, we derive S°(Na_2S_2, cr, 298.15 K) = 42.1 cal K^{-1} mol^{-1}.

A graphical comparison of the standard entropies of K_2S(cr) (5), K_2S_3(cr) (19), K_2S_5(cr) (19), and Na_2S(cr) (5) indicates that 21.4 cal K^{-1} mol^{-1} is probably too low and 42.1 cal K^{-1} mol^{-1} is probably too high for S°(Na_2S_2, cr, 298.15 K). We adopt S°(Na_2S_2, cr, 298.15 K) = 36±6 cal K^{-1} mol^{-1}.

Transition and Fusion Data

The existence of two forms of Na_2S_2 is well established (9-13), but the α-β transition is sluggish. Some indicate the transition is irreversible (9-12) while others find it reversible (13). On heating the α-form, it changes to the β-form in the 150-250° range (9, 12, 13) (423-523 K).

The melting point of the β-form has been determined as 473.9 (14), ~490 (9), 482±2 (15), 478±5 (16), 475 (17), 475 (12), 455 (13), and 484°C (18). We adopt 480±5°C (753±5 K).

The adopted $\Delta_{fus}H°$ = 4.52 kcal mol^{-1} is taken from the tabulation of Barin et al. (5), referenced to private communications from Mills (6). It is not known if this is a measured or an estimated value. Tegman (8) derived a $\Delta_{fus}H°$ = 4.62 kcal mol^{-1} from liquid-vapor and solid-liquid equilibria data via complex computer modeling. This indirectly calculated figure apparently corresponds to an average temperature of approximately 1000 K so that $\Delta_{fus}H°(753\ K)$ ~ 4.8 kcal mol^{-1}; this value tends to confirm the adopted value and is used in estimating an uncertainty of ±0.4 kcal mol^{-1}.

References

1. C. A. Kraus and J. A. Ridderhof, J. Amer. Chem. Soc. 56, 79 (1934).
2. J. M. Letoffe, J. Thourey, G. Perachon, and J. Bousquet, Bull. Soc. Chim. Fr. 1976, 424 (1976).
3. U. S. Nat. Bur. Stand. Tech. Note 270-3, 1968.
4. JANAF Thermochemical Tables: Na_2SO_4, V(cr), 6-30-78; H_2S(g), 6-30-77; Na_2S(cr), $Na_2S(\ell)$, K_2S(cr), 3-31-78; S_2(g), 9-30-77.
5. I. Barin, O. Knacke, and O. Kubaschewski, "Thermochemical Properties of Inorganic Substances", Supplement, Springer-Verlag, Berlin, 1977.
6. K. C. Mills, "Thermodynamic Data for Inorganic Sulphides, Selenides and Tellurides", Butterworths, London, 1974.
7. A. Teder and J. Tiberg, Acta Chem. Scand. 24, 991 (1970).
8. R. Tegman, Chem. Scr. 9, 158 (1976).
9. F. Feher and H. J. Berthold, Z. Anorg. Allgem. Chem. 273, 144 (1953).
10. O. Erametsa and K. Karlsson, Acta Polytech. Scand. 1961, No. 15 (1961).
11. H. Foeppl, E. Busmann, and F. K. Frorath, Z. Anorg. Allgem. Chem. 314, 12 (1962).
12. D. G. Oei, Inorg. Chem. 12, 435 (1973).
13. J. M. Letoffe, J. M. Blanchard, and J. Bousquet, Bull. Soc. Chim. Fr. 1976, 395 (1976).
14. T. G. Pearson and P. L. Robinson, J. Chem. Soc. 1473 (1930).
15. E. Rosen and R. Tegman, Acta Chem. Scand. 25, 3329 (1971).
16. E. Rosen and R. Tegman, Chem. Scr. 2, 221 (1972).
17. N. K. Gupta and P. P. Tischer, J. Electrochem. Soc. 119, 1033 (1972).
18. G. J. Janz, J. R. Downey, E. Roduner, C. J. Wasilczyk, J. M. Coutts, and A. Eluard, Inorg. Chem. 15, 1757 (1976).
19. J. M. Letoffe, Doctorate Thesis, Universite Claude Bernard-Lyon I, 1975.

Sodium Sulfide, Beta (Na_2S_2) Na_2S_2(cr)

Enthalpy Reference Temperature = T_r = 298.15 K   Standard State Pressure = p° = 0.1 MPa

	J K⁻¹mol⁻¹			kJ mol⁻¹			
T/K	$C_p°$	S°	$-[G°-H°(T_r)]/T$	$H°-H°(T_r)$	$\Delta_f H°$	$\Delta_f G°$	Log K_f
0							
100							
200							
298.15	98.625	150.624	150.624	0.	-397.062	-392.173	68.707
300	98.726	151.234	150.626	0.183	-397.067	-392.142	68.278
400	104.299	180.399	154.565	10.334	-407.354	-389.839	50.908
500	109.872	204.272	162.187	21.042	-410.704	-385.082	40.229
600	115.445	224.796	170.949	32.308	-412.640	-379.752	33.060
700	121.018	243.011	179.966	44.131	-413.415	-374.196	27.923
753.000	123.972	251.950	184.720	50.624	------- BETA <--> LIQUID -------		
800	126.591	259.534	188.894	56.512	-413.275	-368.593	24.067
900	132.164	274.766	197.600	69.450	-518.625	-360.893	20.946
1000	137.737	288.979	206.035	82.945	-514.617	-343.578	17.947

PREVIOUS: CURRENT: December 1983

Sodium Sulfide, Beta (Na_2S_2) Na_2S_2(cr)

SODIUM SULFIDE (Na_2S_2) LIQUID M_r = 110.09954

$S°(298.15\ K)$ = [172.365] $J\ K^{-1}\ mol^{-1}$

T_{fus} = 753 ± 5 K

$\Delta_f H°(298.15\ K)$ = [-380.114] $kJ\ mol^{-1}$

$\Delta_{fus}H°$ = [18.91 ± 1.7] $kJ\ mol^{-1}$

Enthalpy of Formation

$\Delta_f H°(Na_2S_2, \ell, 298.15\ K)$ is calculated from that of the crystal by adding $\Delta_{fus}H°$ and the difference in enthalpy, H°(753 K)-H°(298.15 K), between the crystal and liquid.

Heat Capacity and Entropy

The heat capacity of 29.8 cal $K^{-1}\ mol^{-1}$ is from the estimate of Barin et al. (1) and is assumed to be constant above an assumed glass transition at 500 K. Below the glass transition, the heat capacity is that of the crystal.

The entropy is calculated in a manner analogous to that used for the enthalpy of formation.

Fusion Data

See the β-crystal table for details.

References

1. I. Barin, O. Knacke, and O. Kubaschewski, "Thermochemical Properties of Inorganic Substances", Supplement, Springer-Verlag, Berlin, 1977.

Sodium Sulfide (Na_2S_2) $Na_2S_2(l)$

Enthalpy Reference Temperature = T_r = 298.15 K; Standard State Pressure = p° = 0.1 MPa

T/K	$C_p°$ (J K⁻¹mol⁻¹)	S° (J K⁻¹mol⁻¹)	−[G°−H°(T_r)]/T (J K⁻¹mol⁻¹)	H°−H°(T_r) (kJ mol⁻¹)	$\Delta_f H°$ (kJ mol⁻¹)	$\Delta_f G°$ (kJ mol⁻¹)	Log K_f
0							
100							
200							
298.15	98.625	172.365	172.365	0.	-380.114	-381.707	66.873
300	98.726	172.975	172.367	0.183	-380.119	-381.717	66.463
400	104.299	202.140	176.305	10.334	-390.405	-381.587	49.830
500	109.872	226.012	183.928	21.042	-393.756	-379.005	39.594
500.000	109.872	226.012	183.928	21.042	GLASS <--> LIQUID TRANSITION		
500.000	124.683	226.012	183.928	21.042			
600	124.683	248.745	192.894	33.511	-394.489	-375.971	32.731
700	124.683	267.965	202.280	45.979	-394.619	-372.868	27.824
753.000	124.683	277.065	215.768	46.157	BETA <--> LIQUID		
800	124.683	284.614	211.555	58.447	-394.391	-369.773	24.144
900	124.683	299.299	220.504	70.916	-500.211	-364.559	21.158
1000	124.683	312.436	229.052	83.384	-497.229	-349.647	18.264
1100	124.683	324.320	237.181	95.852	-494.321	-335.030	15.909
1200	124.683	335.168	244.901	108.321	-685.047	-315.794	13.746
1300	124.683	345.148	252.234	120.789	-680.541	-285.206	11.460
1400	124.683	354.388	259.205	133.257	-676.066	-254.963	9.513
1500	124.683	362.991	265.840	145.726	-671.622	-225.040	7.837

PREVIOUS: CURRENT: December 1983

Sodium Sulfide (Na_2S_2) $Na_2S_2(l)$

SODIUM SULFIDE (Na_2S_2)   CRYSTAL-LIQUID M_r = 110.09954

298 to 753 K   crystal, beta
above 753 K   liquid

Refer to the individual tables for details.

Sodium Sulfide (Na_2S_2) Na_2S_2(cr,l)

Enthalpy Reference Temperature = T_r = 298.15 K   Standard State Pressure = p° = 0.1 MPa

T/K	C_p° (J K⁻¹mol⁻¹)	S° (J K⁻¹mol⁻¹)	−[G°−H°(T_r)]/T (J K⁻¹mol⁻¹)	H°−H°(T_r) (kJ mol⁻¹)	Δ_fH° (kJ mol⁻¹)	Δ_fG° (kJ mol⁻¹)	Log K_f
0							
100							
200							
298.15	98.625	150.624	150.624	0.	-397.062	-392.173	68.707
300	98.726	151.234	150.626	0.183	-397.067	-392.142	68.278
400	104.299	180.399	154.565	10.334	-407.354	-389.839	50.908
500	109.872	204.272	162.187	21.042	-410.704	-385.082	40.229
600	115.445	224.796	170.949	32.308	-412.640	-379.752	33.060
700	121.018	243.011	179.966	44.131	-413.415	-374.196	27.923
753.000	123.972	251.950	184.720	50.624	BETA <--> LIQUID		
753.000	124.683	277.065	184.720	69.535	TRANSITION		
800	124.683	284.614	190.370	75.395	-394.391	-369.773	24.144
900	124.683	299.299	201.673	87.864	-500.211	-364.559	21.158
1000	124.683	312.436	212.104	100.332	-497.229	-349.647	18.264
1100	124.683	324.320	221.774	112.800	-494.321	-335.030	15.909
1200	124.683	335.168	230.778	125.269	-685.047	-315.794	13.746
1300	124.683	345.148	239.197	137.737	-680.541	-285.206	11.460
1400	124.683	354.388	247.099	150.205	-676.066	-254.963	9.513
1500	124.683	362.991	254.542	162.674	-671.622	-225.040	7.837

PREVIOUS: CURRENT: December 1983

Sodium Sulfide (Na_2S_2) Na_2S_2(cr,l)

NIOBIUM (Nb) REFERENCE STATE A_r = 92.9064 Niobium (Nb) Nb_1(ref)

0 to 2750 K		crystal
2750 to 5130.824 K		liquid
above 5130.824 K		ideal monatomic gas

Refer to the individual tables for details.

Enthalpy Reference Temperature = T_r = 298.15 K — Standard State Pressure = p° = 0.1 MPa

	J K⁻¹mol⁻¹			kJ mol⁻¹			
T/K	$C_p°$	S°	$-[G°-H°(T_r)]/T$	$H°-H°(T_r)$	$\Delta_fH°$	$\Delta_fG°$	Log K_f
0	0.	0.	INFINITE	-5.241	0.	0.	0.
100	17.439	12.558	57.153	-4.460	0.	0.	0.
200	23.091	26.878	38.685	-2.361	0.	0.	0.
298.15	24.694	36.464	36.464	0.	0.	0.	0.
300	24.711	36.616	36.464	0.046	0.	0.	0.
400	25.393	43.826	37.442	2.553	0.	0.	0.
500	25.899	49.548	39.311	5.118	0.	0.	0.
600	26.347	54.309	41.424	7.731	0.	0.	0.
700	26.769	58.403	43.564	10.387	0.	0.	0.
800	27.183	62.004	45.649	13.084	0.	0.	0.
900	27.593	65.230	47.648	15.823	0.	0.	0.
1000	27.999	68.158	49.555	18.603	0.	0.	0.
1100	28.405	70.846	51.370	21.423	0.	0.	0.
1200	28.798	73.334	53.098	24.284	0.	0.	0.
1300	29.179	75.654	54.745	27.182	0.	0.	0.
1400	29.589	77.831	56.317	30.120	0.	0.	0.
1500	30.062	79.888	57.820	33.102	0.	0.	0.
1600	30.606	81.846	59.261	36.135	0.	0.	0.
1700	31.221	83.719	60.645	39.226	0.	0.	0.
1800	31.903	85.523	61.977	42.381	0.	0.	0.
1900	32.639	87.267	63.263	45.608	0.	0.	0.
2000	33.430	88.961	64.506	48.911	0.	0.	0.
2100	34.275	90.612	65.710	52.296	0.	0.	0.
2200	35.187	92.228	66.878	55.768	0.	0.	0.
2300	36.192	93.813	68.015	59.336	0.	0.	0.
2400	37.317	95.377	69.123	63.010	0.	0.	0.
2500	38.635	96.926	70.204	66.805	0.	0.	0.
2600	40.233	98.471	71.261	70.745	0.	0.	0.
2700	42.283	100.027	72.298	74.868	0.	0.	0.
2750.000	43.423	100.813	72.809	77.011	CRYSTAL <--> LIQUID TRANSITION		
2750.000	33.472	110.596	72.809	103.914			
2800	33.472	111.199	73.489	105.587	0.	0.	0.
2900	33.472	112.374	74.810	108.935	0.	0.	0.
3000	33.472	113.508	76.081	112.282	0.	0.	0.
3100	33.472	114.606	77.306	115.629	0.	0.	0.
3200	33.472	115.669	78.489	118.976	0.	0.	0.
3300	33.472	116.699	79.631	122.323	0.	0.	0.
3400	33.472	117.698	80.736	125.671	0.	0.	0.
3500	33.472	118.668	81.806	129.018	0.	0.	0.
3600	33.472	119.611	82.843	132.365	0.	0.	0.
3700	33.472	120.528	83.849	135.712	0.	0.	0.
3800	33.472	121.421	84.826	139.059	0.	0.	0.
3900	33.472	122.290	85.776	142.407	0.	0.	0.
4000	33.472	123.138	86.699	145.754	0.	0.	0.
4100	33.472	123.964	87.598	149.101	0.	0.	0.
4200	33.472	124.771	88.474	152.448	0.	0.	0.
4300	33.472	125.558	89.327	155.795	0.	0.	0.
4400	33.472	126.328	90.159	159.143	0.	0.	0.
4500	33.472	127.080	90.971	162.490	0.	0.	0.
4600	33.472	127.816	91.764	165.837	0.	0.	0.
4700	33.472	128.536	92.539	169.184	0.	0.	0.
4800	33.472	129.240	93.296	172.531	0.	0.	0.
4900	33.472	129.930	94.037	175.879	0.	0.	0.
5000	33.472	130.607	94.762	179.226	0.	0.	0.
5100	33.472	131.270	95.471	182.573	0.	0.	0.
5130.824	33.472	131.471	95.687	183.605	LIQUID <--> IDEAL GAS FUGACITY = 1 bar		
5130.824	36.434	266.042	95.686	874.064			
5200	36.642	266.531	97.956	876.592	0.	0.	0.
5300	36.943	267.232	101.143	880.271	0.	0.	0.
5400	37.115	267.914	104.225	883.923	0.	0.	0.
5600	37.678	269.274	110.095	891.403	0.	0.	0.
5800	38.251	270.606	115.607	898.995	0.	0.	0.
6000	38.839	271.913	120.795	906.704	0.	0.	0.

PREVIOUS: December 1973 (1 atm) CURRENT: December 1973 (1 bar)

Niobium (Nb) Nb_1(ref)

NIOBIUM (Nb) CRYSTAL A_r = 92.9064

$S°(298.15\ K) = 36.46 \pm 0.4\ J\ K^{-1}\ mol^{-1}$

$T_{fus} = 2750 \pm 10\ K$

$\Delta_f H°(0\ K) = 0\ kJ\ mol^{-1}$

$\Delta_f H°(298.15\ K) = 0\ kJ\ mol^{-1}$

$\Delta_{fus} H° = 26.90 \pm 0.8\ kJ\ mol^{-1}$

Enthalpy of Formation

Zero by definition.

Heat Capacity and Entropy

The heat capacity values for T < 10 are chosen to be the same as those adopted by Hultgren et al. (1). A graphical integration of these Cp° data yields $S°(10\ K) = 0.029\ J\ K^{-1}\ mol^{-1}$ and $H°(10\ K) = 0.19\ cal\ mol^{-1}$. The work by DaSilva et al. (2, 3) on the low temperature (1-10 K) heat capacity of annealed and unannealed niobium wires in magnetic fields supports this choice. Clusius et al. (4) measured the heat capacity (61 pts.) in the region 11-274 K. This data is smoothed graphically and is constrained to join smoothly with the Cp° values for T ≤ 10 K. Deviations up to 4% occur between the observed and adopted Cp° values for T < 80 K. The deviations here parallel the deviations obtained in our analysis of the heat capacity for Ta (5). The deviations decrease steadily with increasing temperature up to 274 K.

There are numerous high temperature heat capacity and enthalpy measurements for Nb(cr). The various studies are listed below.

Source	T/K	Method
(6)	273-1873	drop
(7)	454-1882	drop
(8)	433-1840	drop
(9)	1400-2350	modul
(10)	1300-2700	modul
(11)	358-1415	drop
(12)	600-2600	drop
(13)	1273-2593	drop
(14)	1100-2400	pulse
(15)	1500-2700	pulse
(16)	1650-2707	drop

The adopted Cp° values for T > 298.15 K are obtained by a combination of graphical and polynomial curve fitting techniques. A polynomial curve fit procedure is used on the enthalpy data of Hawkins and Orr (11) with the constraint that it join smoothly in the 298.15 K region with the enthalpy derived from the Clusius et al. (4) heat capacity data. Above 1300 K, a heat capacity curve is adopted through graphical procedures. The adopted curve is chosen so as to be intermediate between the Cp° values of Cezairliyan (15) and the higher Cp° values implied from the enthalpy data of Kirillin et al. (12) and Sheindlin et al. (16). In general there is good agreement between the various sets of data.

The heat capacity values near T_{fus} are the smoothed values suggested by Cezairliyan (15). In comparison with the adopted Cp° values, Cezairliyan's smoothed data lies low by 1.9% at 1500 K with the difference steadily decreasing to zero at 2600 K. The smoothed results of Cezairliyan (15) are used for extrapolation to 3200 K. The Cp° values reported by Makarenko and Trukhanov (14) are linear in the region 1100-2400 K whereas the adopted Cp° values are nonlinear; the Cp° - T curve being concave upward in this region. The Cp° values of Makarenko and Trukhanov (14) are within -0.4% to +0.8% of the adopted values in the region 1400-2100 K but drift to -3% at 1100 K and 2400 K. In comparing enthalpies, Kirillin et al. (12) reported smoothed values (12) which are 15 $cal\ mol^{-1}$ or 0.8% lower than our tabulated values at 600 K and drift to 849 $cal\ mol^{-1}$ or 4.5% higher at 2700 K. The enthalpy results of Sheindlin et al. (16) lie above the JANAF values by 0.9% in the region 1700-2700 K (82 $cal\ mol^{-1}$ at 1700 K and 181 $cal\ mol^{-1}$ at 2700 K).

Fusion Data

See the liquid table for details.

Sublimation Data

$\Delta_{sub}H°(298.15\ K)$ for Nb(cr) = Nb(g) is simply $\Delta_f H°(298.15\ K)$ for Nb(g). See the ideal gas table for details.

References

1. R. Hultgren, R. L. Orr, and K. K. Kelley, Supplement to Selected Values of Thermodynamic Properties of Metals and Alloys, Nb table, Jan., 1966.
2. J. F. DaSilva, N. W. J. van Duykeren, and Z. Dokoupil. Physica 32, 1253 (1966).
3. J. F. DaSilva, E. A. Burgemeister, and Z. Dokoupil. Physica 41, 409 (1969).
4. K. Clusius, P. Franzosini, and U. Piesbergen, Z. Naturforsch. 15, 728 (1960).
5. JANAF Thermochemical Tables: Ta(cr), 12-31-72.
6. F. M. Jaeger and W. A. Veenstra, Rec. Trav. Chem. 53, 677 (1934); Proc. Roy. Acad. Amsterdam 37, 61 (1934).
7. I. B. Fieldhouse, J. C. Hedge, and J. I. Lang, WADC TR 58-274, 1958 [AD 206 892].
8. P. V. Gel'd and F. G. Kusenko, Izv. Akad. Nauk SSSR, Otdel. Tekh. Nauk, Met. i Toplivv, No. 2, 79 (1960).
9. G. C. Lowenthal, Aust. J. Phys. 16, 47 (1963).
10. Ya. A. Kraftmakher, Solid State (USSR) 5, 696 (1963).
11. D. T. Hawkins and R. L. Orr, J. Chem. Eng. Data 9, 505 (1964).
12. V. A. Kirillin, A. E. Sheindlin, V. Ya. Chekhovskoi, and I. A. Zhukova, High Temp. 3, 357 (1965); High Temp. 3, 801 (1965).
13. J. B. Conway and R. A. Hein, in "Avances in Thermophysical Properties at Extreme Temperatures and Pressures," S. Gratch, Ed., ASME, New York, 1965, p. 131.
14. I. N. Makarenko, L. N. Trukhanov, and L. P. Filippov, High Temp. 8, 667 (1970).
15. A. Cezairliyan, J. Res. Nat. Bur. Stand. 75A, 565 (1971).
16. A. E. Sheindlin, B. Ya. Berezin, and V. Ya. Chekhovskoi, High Temp.-High Press. 4, 611 (1972).

Niobium (Nb) $Nb_1(cr)$

Enthalpy Reference Temperature = T_r = 298.15 K   Standard State Pressure = p° = 0.1 MPa

T/K	$C_p°$ (J K⁻¹mol⁻¹)	S° (J K⁻¹mol⁻¹)	$-[G°-H°(T_r)]/T$ (J K⁻¹mol⁻¹)	$H°-H°(T_r)$ (kJ mol⁻¹)	$\Delta_f H°$ (kJ mol⁻¹)	$\Delta_f G°$ (kJ mol⁻¹)	Log K_f
0	0.	0.	INFINITE	-5.241	0.	0.	0.
100	17.439	12.558	57.153	-4.460	0.	0.	0.
200	23.091	26.878	38.685	-2.361	0.	0.	0.
250	24.154	32.159	36.867	-1.177	0.	0.	0.
298.15	24.694	36.464	36.464	0.	0.	0.	0.
300	24.711	36.616	36.464	0.046	0.	0.	0.
350	25.088	40.455	36.766	1.291	0.	0.	0.
400	25.393	43.826	37.442	2.553	0.	0.	0.
450	25.653	46.832	38.322	3.830	0.	0.	0.
500	25.899	49.548	39.311	5.118	0.	0.	0.
600	26.347	54.309	41.424	7.731	0.	0.	0.
700	26.769	58.403	43.564	10.387	0.	0.	0.
800	27.183	62.004	45.649	13.084	0.	0.	0.
900	27.593	65.230	47.648	15.823	0.	0.	0.
1000	27.999	68.158	49.555	18.603	0.	0.	0.
1100	28.405	70.846	51.370	21.423	0.	0.	0.
1200	28.798	73.334	53.098	24.284	0.	0.	0.
1300	29.179	75.654	54.745	27.182	0.	0.	0.
1400	29.589	77.831	56.317	30.120	0.	0.	0.
1500	30.062	79.888	57.820	33.102	0.	0.	0.
1600	30.606	81.846	59.261	36.135	0.	0.	0.
1700	31.221	83.719	60.645	39.226	0.	0.	0.
1800	31.903	85.523	61.977	42.381	0.	0.	0.
1900	32.639	87.267	63.263	45.608	0.	0.	0.
2000	33.430	88.961	64.506	48.911	0.	0.	0.
2100	34.275	90.612	65.710	52.296	0.	0.	0.
2200	35.187	92.228	66.878	55.768	0.	0.	0.
2300	36.192	93.813	68.015	59.336	0.	0.	0.
2400	37.317	95.377	69.123	63.010	0.	0.	0.
2500	38.635	96.926	70.204	66.805	0.	0.	0.
2600	40.233	98.471	71.261	70.745	0.	0.	0.
2700	42.283	100.027	72.298	74.868	0.	0.	0.
2750.000	43.423	100.813	72.809	77.011	------ CRYSTAL <--> LIQUID -----		
2800	44.643	101.606	73.316	79.212	-26.375	0.484	-0.009
2900	47.342	103.219	74.319	83.808	-25.126	1.423	-0.026
3000	50.410	104.874	75.310	88.692	-23.589	2.313	-0.040
3100	53.879	106.583	76.291	93.903	-21.725	3.146	-0.053
3200	57.777	108.354	77.265	99.483	-19.494	3.914	-0.064

PREVIOUS: June 1972   CURRENT: December 1973

Niobium (Nb) $Nb_1(cr)$

NIOBIUM (Nb) LIQUID A_r = 92.9064

$S°(298.15\ K)$ = [47.302] J K^{-1} mol^{-1} $\Delta_f H°(298.15\ K)$ = [29.647] kJ mol^{-1}

T_{fus} = 2750 ± 10 K $\Delta_{fus}H°$ = 26.90 ± 0.8 kJ mol^{-1}

Enthalpy of Formation

The enthalpy of formation of Nb(ℓ) at 298.15 K is calculated from that of the crystal by adding $\Delta_{fus}H°$ and the difference in enthalpy, H°(2750 K) - H°(298.15 K), between the crystal and liquid.

Heat Capacity and Entropy

The enthalpy measurements on Nb(ℓ) by Sheindlin et al. (7) and Margrave (8) do not extend over a sufficiently wide temperature range such that a heat capacity value could be derived. Therefore, the heat capacity for Nb(ℓ) is estimated as 8.0 cal K^{-1} mol^{-1} by analogy with other monatomic metals. The entropy at 298.15 K is calculated in a manner analogous to that used in calculating the enthalpy of formation. A glass transition is assumed at 1750 K. The heat capacity values used below 1750 K are those of Nb(cr).

Fusion Data

Melting point determinations for Nb have been numerous, with most reported values in the range 2680-2795 K (IPTS-68 scale). A prime cause for diversity in the results is that the melting point is quite sensitive to impurities, especially gases such as oxygen and nitrogen. These effects have been documented by Schofield (1) and Pemsler (2). The value chosen is 2750 ± 10 K as reported by Cezairliyan (3) using a subsecond pulse heating method. This value is consistent with the heat capacity values near T_{fus} which are also based on related work by Cezairliyan (4). Other values in this range (corrected to IPTS-68 scale) are 2746 K by Schofield (1), 2745 K by Pemsler (2), and 2738 K by Buehler (5). Kocherzhinskii and Kobzenko (6) have reported a T_{fus} value of 2793 K, using a sample purified by electron zone fusion.

The value chosen for $\Delta_{fus}H°$ is 6.43 ± 0.20 kcal mol^{-1} as determined by Cezairliyan (3) using a pulse heating technique. Adopting this value gives consistency between the Cp° values near T_{fus}, $\Delta_{fus}H°$, and T_{fus} as all these values are dependent upon the same heating technique. Sheindlin et al. (7) reported $\Delta_{fus}H°$ = 6.597 ± 0.170 kcal mol^{-1} based on enthalpy measurements on Nb(cr) and Nb(ℓ) obtained from levitation calorimetry techniques. In the vicinity of T_{fus}, the adopted enthalpy vlaues for Nb(cr) and Nb(ℓ) lie roughly 200 cal mol^{-1} lower than those reported by Sheindlin et al. (7). Margrave (8), also using levitation techniques, reported a value $\Delta_{fus}H°$ = 7.9 kcal mol^{-1}. It is felt that Margrave's value was calculated from a measured enthalpy of Nb(ℓ) at T_{fus} and the enthalpy of Nb(cr) at T_{fus} as suggested by Hultgren et al. (9). Our enthalpy vlaues at T_{fus} for Nb(cr) would reduce Margrave's value to roughly 7.4 kcal mol^{-1} for $\Delta_{fus}H°$.

Martynyuk and Tsapkov (10) measured the enthalpies of Nb(cr) and Nb(ℓ) at the melting point using the method of pulse heating. They reported enthalpy of melting value of 7.98 kcal mol^{-1} for an Nb sample which was reported to be 99.94% pure. No T_{fus} value was reported.

Vaporization Data

T_{vap} is the temperature at which the Gibbs energy change for the reaction Nb(ℓ) = Nb(g) is zero. The difference between $\Delta_f H°$(Nb, g) and $\Delta_f H°$(Nb, ℓ) at T_{vap} is $\Delta_{vap}H°$. The uncertainty in the boiling point is probably of the order of ±100 K. Also refer to Nb(g) table.

References

1. T. H. Schofield, J. Inst. Metals 85, 372 (1956).
2. J. P. Pemsler, J. Electrochem. Soc. 108, 744 (1961).
3. A. Cezairliyan, U. S. Nat. Bur. Stand. Report. 10904, July 1972, Ch. 8 and 9.
4. A. Cezairliyan, U. S. Nat. Bur. Stand. 75A, 565 (1971).
5. E. Buehler, Trans. AIME 212, 694 (1958).
6. Y. A. Kocherzhinskii and G. F. Kobzenko, FTD-TT-65-1883, 1 (1966); as reported by J. H. Charlesworth, Melting Points of Metallic Elements and Selected Compounds, AFML-TR-70-137, Oct., 1970.
7. A. E. Sheindlin, B. Ya. Berezin, and V. Ya. Chekhovskoi, High Temperatures-High Pressures 4, 611 (1972).
8. J. L. Margrave, High Temperatures-High Pressures 2, 583 (1970).
9. R. Hultgren, R. L. Orr, and K. K. Kelley, Supplement to Selected Values of Thermodynamic Properties of Metals and Alloys, Nb table, Jan., 1966.
10. M. M. Martynyuk and V. I. Tsapkov, Russ. J. Phys. Chem. 47, 741 (1973).

Niobium (Nb) $Nb_1(l)$

Enthalpy Reference Temperature = T_r = 298.15 K   Standard State Pressure = p° = 0.1 MPa

T/K	$C_p°$ (J $K^{-1}mol^{-1}$)	S° (J $K^{-1}mol^{-1}$)	$-[G°-H°(T_r)]/T$ (J $K^{-1}mol^{-1}$)	$H°-H°(T_r)$ (kJ mol^{-1})	$\Delta_f H°$ (kJ mol^{-1})	$\Delta_f G°$ (kJ mol^{-1})	Log K_f
0							
100							
200							
298.15	24.694	47.302	47.302	0.	29.647	26.415	-4.628
300	24.711	47.455	47.302	0.046	29.647	26.395	-4.596
350	25.088	51.293	47.605	1.291	29.647	25.853	-3.858
400	25.393	54.664	48.281	2.553	29.647	25.311	-3.305
450	25.653	57.670	49.160	3.830	29.647	24.770	-2.875
500	25.899	60.386	50.149	5.118	29.647	24.228	-2.531
600	26.347	65.148	52.263	7.731	29.647	23.144	-2.015
700	26.769	69.241	54.403	10.387	29.647	22.060	-1.646
800	27.183	72.842	56.487	13.084	29.647	20.976	-1.370
900	27.593	76.068	58.486	15.823	29.647	19.892	-1.155
1000	27.999	78.996	60.393	18.603	29.647	18.808	-0.982
1100	28.405	81.684	62.208	21.423	29.647	17.725	-0.842
1200	28.798	84.172	63.936	24.284	29.647	16.641	-0.724
1300	29.179	86.492	65.583	27.182	29.647	15.557	-0.625
1400	29.589	88.669	67.155	30.120	29.647	14.473	-0.540
1500	30.062	90.727	68.658	33.102	29.647	13.389	-0.466
1600	30.606	92.684	70.099	36.135	29.647	12.306	-0.402
1700	31.221	94.557	71.483	39.226	29.647	11.222	-0.345
1750.000	31.559	95.467	72.156	40.795	GLASS <--> LIQUID	TRANSITION	
1750.000	33.472	95.467	72.156	40.795			
1800	33.472	96.410	72.816	42.469	29.734	10.137	-0.294
1900	33.472	98.220	74.106	45.816	29.854	9.044	-0.249
2000	33.472	99.937	75.355	49.163	29.899	7.948	-0.208
2100	33.472	101.570	76.565	52.510	29.861	6.851	-0.170
2200	33.472	103.127	77.737	55.857	29.736	5.757	-0.137
2300	33.472	104.615	78.874	59.205	29.515	4.672	-0.106
2400	33.472	106.039	79.976	62.552	29.188	3.599	-0.078
2500	33.472	107.406	81.046	65.899	28.740	2.541	-0.053
2600	33.472	108.718	82.085	69.246	28.148	1.504	-0.030
2700	33.472	109.982	83.095	72.593	27.372	0.494	-0.010
2750.000	33.472	110.596	83.590	74.267	CRYSTAL <--> LIQUID		
2800	33.472	111.199	84.077	75.941	0.	0.	0.
2900	33.472	112.374	85.033	79.288	0.	0.	0.
3000	33.472	113.508	85.963	82.635	0.	0.	0.
3100	33.472	114.606	86.870	85.982	0.	0.	0.
3200	33.472	115.669	87.753	89.329	0.	0.	0.
3300	33.472	116.699	88.615	92.677	0.	0.	0.
3400	33.472	117.698	89.456	96.024	0.	0.	0.
3500	33.472	118.668	90.276	99.371	0.	0.	0.
3600	33.472	119.611	91.078	102.718	0.	0.	0.
3700	33.472	120.528	91.862	106.065	0.	0.	0.
3800	33.472	121.421	92.628	109.413	0.	0.	0.
3900	33.472	122.290	93.377	112.760	0.	0.	0.
4000	33.472	123.138	94.111	116.107	0.	0.	0.
4100	33.472	123.964	94.829	119.454	0.	0.	0.
4200	33.472	124.771	95.532	122.801	0.	0.	0.
4300	33.472	125.558	96.221	126.149	0.	0.	0.
4400	33.472	126.328	96.897	129.496	0.	0.	0.
4500	33.472	127.080	97.559	132.843	0.	0.	0.
4600	33.472	127.816	98.209	136.190	0.	0.	0.
4700	33.472	128.536	98.847	139.537	0.	0.	0.
4800	33.472	129.240	99.473	142.885	0.	0.	0.
4900	33.472	129.931	100.087	146.232	0.	0.	0.
5000	33.472	130.607	100.691	149.579	0.	0.	0.
5100	33.472	131.270	101.284	152.926	0.	0.	0.
5130.824	33.472	131.471	101.465	153.958	FUGACITY = 1 bar		
5200	33.472	131.920	101.867	156.273	-690.672	9.309	-0.094
5300	33.472	132.557	102.440	159.621	-691.004	22.774	-0.224
5400	33.472	133.183	103.004	162.968	-691.309	36.240	-0.351
5600	33.472	134.400	104.103	169.662	-692.094	63.200	-0.590
5800	33.472	135.575	105.168	176.357	-692.992	90.191	-0.812
6000	33.472	136.709	106.201	183.051	-694.007	117.214	-1.020

PREVIOUS: June 1972   CURRENT: December 1973

Niobium (Nb) $Nb_1(l)$

NIOBIUM (Nb) CRYSTAL-LIQUID A_r = 92.9064 Niobium (Nb) $Nb_1(cr,l)$

0 to 2750 K crystal
above 2750 K liquid

Refer to the individual tables for details.

Enthalpy Reference Temperature = T_r = 298.15 K Standard State Pressure = p° = 0.1 MPa

T/K	C_p° (J K⁻¹mol⁻¹)	S° (J K⁻¹mol⁻¹)	$-[G^\circ-H^\circ(T_r)]/T$ (J K⁻¹mol⁻¹)	$H^\circ-H^\circ(T_r)$ (kJ mol⁻¹)	Δ_fH° (kJ mol⁻¹)	Δ_fG° (kJ mol⁻¹)	Log K_f
0	0.	0.	INFINITE	-5.241	0	0.	0.
100	17.439	12.558	57.153	-4.460	0.	0.	0.
200	23.091	26.878	38.685	-2.361	0.	0.	0.
250	24.154	32.159	36.867	-1.177	0.	0.	0.
298.15	24.694	36.464	36.464	0.	0.	0.	0.
300	24.711	36.616	36.464	0.046	0.	0.	0.
350	25.088	40.455	36.766	1.291	0.	0.	0.
400	25.393	43.826	37.442	2.553	0.	0.	0.
450	25.653	46.832	38.322	3.830	0.	0.	0.
500	25.899	49.548	39.311	5.118	0.	0.	0.
600	26.347	54.309	41.424	7.731	0.	0.	0.
700	26.769	58.403	43.564	10.387	0.	0.	0.
800	27.183	62.004	45.649	13.084	0.	0.	0.
900	27.593	65.230	47.648	15.823	0.	0.	0.
1000	27.999	68.158	49.555	18.603	0.	0.	0.
1100	28.405	70.846	51.370	21.423	0.	0.	0.
1200	28.798	73.334	53.098	24.284	0.	0.	0.
1300	29.179	75.654	54.745	27.182	0.	0.	0.
1400	29.589	77.831	56.317	30.120	0.	0.	0.
1500	30.062	79.888	57.820	33.102	0.	0.	0.
1600	30.606	81.846	59.261	36.135	0.	0.	0.
1700	31.221	83.719	60.645	39.226	0.	0.	0.
1800	31.903	85.523	61.977	42.381	0.	0.	0.
1900	32.639	87.267	63.263	45.608	0.	0.	0.
2000	33.430	88.961	64.506	48.911	0.	0.	0.
2100	34.275	90.612	65.710	52.296	0.	0.	0.
2200	35.187	92.228	66.878	55.768	0.	0.	0.
2300	36.192	93.813	68.015	59.336	0.	0.	0.
2400	37.317	95.377	69.123	63.010	0.	0.	0.
2500	38.635	96.926	70.204	66.805	0.	0.	0.
2600	40.233	98.471	71.261	70.745	0.	0.	0.
2700	42.283	100.027	72.298	74.868	0.	0.	0.
2750.000	43.423	100.813	72.809	77.011	CRYSTAL <--> LIQUID		
2750.000	33.472	110.596	72.809	103.914	TRANSITION		
2800	33.472	111.199	73.489	105.587	0.	0.	0.
2900	33.472	112.374	74.810	108.935	0.	0.	0.
3000	33.472	113.508	76.081	112.282	0.	0.	0.
3100	33.472	114.606	77.306	115.629	0.	0.	0.
3200	33.472	115.669	78.489	118.976	0.	0.	0.
3300	33.472	116.699	79.631	122.323	0.	0.	0.
3400	33.472	117.698	80.736	125.671	0.	0.	0.
3500	33.472	118.668	81.806	129.018	0.	0.	0.
3600	33.472	119.611	82.843	132.365	0.	0.	0.
3700	33.472	120.528	83.849	135.712	0.	0.	0.
3800	33.472	121.421	84.826	139.059	0.	0.	0.
3900	33.472	122.290	85.776	142.407	0.	0.	0.
4000	33.472	123.138	86.699	145.754	0.	0.	0.
4100	33.472	123.964	87.598	149.101	0.	0.	0.
4200	33.472	124.771	88.474	152.448	0.	0.	0.
4300	33.472	125.558	89.327	155.795	0.	0.	0.
4400	33.472	126.328	90.159	159.143	0.	0.	0.
4500	33.472	127.080	90.971	162.490	0.	0.	0.
4600	33.472	127.816	91.764	165.837	0.	0.	0.
4700	33.472	128.536	92.539	169.184	0.	0.	0.
4800	33.472	129.240	93.296	172.531	0.	0.	0.
4900	33.472	129.930	94.037	175.879	0.	0.	0.
5000	33.472	130.607	94.762	179.226	0.	0.	0.
5100	33.472	131.270	95.471	182.573	0.	0.	0.
5130.824	33.472	131.471	95.687	183.605	------- FUGACITY = 1 bar -------		
5200	33.472	131.920	96.166	185.920	-690.672	9.310	-0.094
5300	33.472	132.557	96.846	189.267	-691.004	22.774	-0.224
5400	33.472	133.183	97.513	192.615	-691.309	36.240	-0.351
5600	33.472	134.400	98.809	199.309	-692.094	63.200	-0.590
5800	33.472	135.575	100.057	206.003	-692.992	90.191	-0.812
6000	33.472	136.709	101.260	212.698	-694.007	117.214	-1.020

PREVIOUS: CURRENT: December 1973

Niobium (Nb) $Nb_1(cr,l)$

NIOBIUM (Nb) IDEAL GAS A_r = 92.9064

IP(Nb, g) = 55511 ± 1 cm^{-1}

S°(298.15 K) = 186.26 ± 0.4 J K^{-1} mol^{-1}

$\Delta_f H°$(0 K) = 729.9 ± 8 kJ mol^{-1}

$\Delta_f H°$(298.15 K) = 733.0 ± 8 kJ mol^{-1}

Electronic Levels and Quantum Weights

State	ε_i, cm^{-1}	g_i
$^6D_{1/2}$	0.00	2
$^6D_{3/2}$	154.19	4
$^6D_{5/2}$	391.99	6
$^6D_{7/2}$	695.25	8
$^6D_{9/2}$	1050.26	10

Enthalpy of Formation

A 2nd and 3rd law analysis of six reported vapor pressure studies is tabulated below. Five studies (1-5) used the Langmuir free evaporation method while one study (6) used the integral variation of the Knudsen method.

Source	Reaction	T/K (IPTS-68)	Data Points	$\Delta_r H°$(298.15 K), kcal mol^{-1} 2nd law	3rd law	Drift cal K^{-1} mol^{-1}	$\Delta_f H°$(298.15 K) kcal mol^{-1}
(1)	A	2744-3146	15*	146.2	168.4	7.5±1.5	175.5
(2)	B	2307-2600	14**	175.8	172.5	-1.3±1.6	172.5
(3)	B	2456-2691	8	182.8	175.7	-2.8±5.3	175.7
(4)	B	2253-2604	eqn	183.4	174.8	-3.6	174.8
(5)	B	2370-2685	eqn	175.0	177.7	1.1	177.7
(6)	B	2553-2684	4	143.1	191.4	18.4±4.8	191.4

* 2939.9 K point deleted due to a statistical test. A = vaporization

** 2307, 2328, 2584 K points deleted due to a statistical test. B = sublimation

We adopt $\Delta_f H°$(298.15 K) = 175.2 kcal mol^{-1} for Nb(g) which is a rounded average of four results (2, 3, 4, 5). With a drift of 7.5 cal K^{-1} mol^{-1} in that data, Shchukarev et al. (7) reported an enthalpy of sublimation of Nb(cr), determined at 2243-2393 K, of 172 ± 5 kcal mol^{-1}. Our tabulations give $\Delta_{sub}H°$(2300 K) = 173.5 kcal mol^{-1}. The same workers (7) later reported $\Delta_{sub}H°$(298.15 K) = 173 kcal mol^{-1} based on a 2nd law analysis of Nb^+ ion currents at temperatures above 2573 K.

Heat Capacity and Entropy

The electronic levels and quantum weights are obtained from Moore (8). Only a few of the levels are listed above. All levels reported by Moore (4) are considered in the calculation. There are predicted electronic levels which have not been observed (8). These levels are assumed to lie above 20,000 cm^{-1} and thus will not significantly affect the entropy at temperatures below 3000 K. The heat capacity and entropy values are very similar to those adopted by Hultgren (9), being the same at 298.15 K and differing by 0.39 cal K^{-1} mol^{-1} in Cp at 5500 K.

References

1. R. K. Koch, W. E. Anable, and R. A. Beall, U.S. Bur. Mines RI-7125, 1968.
2. R. Speiser, P. Blackburn, and H. L. Johnston, J. Electrochem. Soc. 106, 52 (1959).
3. L. Kaufman, Semi-Annual Report No. 2, Manlabs, Inc., Cambridge, Mass., April, 1963.
4. P. F. Woerner and G. F. Wakefield, Rev. Sci. Instr. 33, 1456 (1982).
5. A. L. Reimann and C. K. Grant, Phil. Mag. 22, 34 (1936).
6. V. N. Vlasov, Dissertation, Dept. of Chemistry, Moscow University, as quoted by An. N. Nesmeyanov, "Vapor Pressure of the Chemical Elements", R. Gary (ed.), Elsevier Publishing Co., New York (1963).
7. S. A. Shchukarev, G. A. Semenov, and K. E. Frantseva, Izv. Vysshikh Uchebn. Zavendenii, Khim. i Khim. Tekhnol. 5, 691 (1962); Russ. J. Inorg. Chem. 11, 129 (1966).
8. C. E. Moore, U.S. Nat. Bur. Stand., NSRDS-NBS 35, Volume II, 1970 [Reprint of NBS Circular 467, Volume II, 1952].
9. R. Hultgren, R. L. Orr, and K. K. Kelley, Supplement to Selected Values of Thermodynamic Properties of Metals and Alloys, Nb table, Jan., 1966.

Niobium (Nb) $Nb_1(g)$

Enthalpy Reference Temperature = T_r = 298.15 K — Standard State Pressure = p° = 0.1 MPa

T/K	$C_p°$ (J $K^{-1}mol^{-1}$)	S°	$-[G°-H°(T_r)]/T$	$H°-H°(T_r)$ (kJ mol^{-1})	$\Delta_f H°$	$\Delta_f G°$	Log K_f
0	0.	0.	INFINITE	-8.354	729.923	729.923	INFINITE
100	28.819	153.823	212.891	-5.907	731.590	717.463	-374.764
200	29.932	174.240	189.016	-2.955	732.443	702.971	-183.597
250	30.144	180.945	186.756	-1.453	732.761	695.565	-145.330
298.15	30.159	186.259	186.259	0.	733.037	688.375	-120.600
300	30.155	186.446	186.260	0.056	733.047	688.098	-119.809
350	29.969	191.082	186.626	1.560	733.305	680.586	-101.572
400	29.646	195.064	187.438	3.050	733.534	673.039	-87.890
450	29.253	198.534	188.482	4.523	733.730	665.464	-77.245
500	28.840	201.594	189.644	5.975	733.894	657.870	-68.727
600	28.050	206.781	192.083	8.819	734.125	642.642	-55.947
700	27.365	211.052	194.497	11.589	734.239	627.384	-46.816
800	26.785	214.668	196.798	14.296	734.248	612.117	-39.967
900	26.294	217.793	198.961	16.949	734.162	596.855	-34.641
1000	25.880	220.542	200.984	19.557	733.991	581.607	-30.380
1100	25.533	222.991	202.876	22.127	733.741	566.380	-26.895
1200	25.248	225.200	204.646	24.666	733.419	551.179	-23.992
1300	25.023	227.212	206.305	27.179	733.033	536.008	-21.537
1400	24.857	229.060	207.866	29.672	732.589	520.869	-19.434
1500	24.748	230.771	209.336	32.152	732.087	505.763	-17.612
1600	24.696	232.366	210.726	34.624	731.526	490.693	-16.019
1700	24.700	233.863	212.044	37.093	730.904	475.659	-14.615
1800	24.759	235.276	213.296	39.566	730.221	460.664	-13.368
1900	24.872	236.618	214.488	42.047	729.475	445.709	-12.253
2000	25.034	237.898	215.627	44.542	728.667	430.794	-11.251
2100	25.244	239.124	216.717	47.055	727.796	415.922	-10.345
2200	25.496	240.304	217.762	49.592	726.860	401.092	-9.523
2300	25.788	241.444	218.767	52.156	725.856	386.307	-8.773
2400	26.115	242.548	219.735	54.751	724.777	371.567	-8.087
2500	26.472	243.621	220.669	57.380	723.611	356.873	-7.456
2600	26.854	244.667	221.572	60.046	722.338	342.229	-6.875
2700	27.256	245.688	222.446	62.751	720.920	327.635	-6.338
2800	27.675	246.686	223.294	65.498	692.947	313.582	-5.850
2900	28.105	247.665	224.118	68.287	692.389	300.044	-5.404
3000	28.544	248.625	224.919	71.119	691.874	286.523	-4.989
3100	28.986	249.568	225.699	73.995	691.403	273.020	-4.600
3200	29.430	250.496	226.459	76.916	690.977	259.530	-4.236
3300	29.871	251.408	227.202	79.881	690.595	246.054	-3.895
3400	30.309	252.306	227.927	82.890	690.257	232.588	-3.573
3500	30.740	253.191	228.636	85.943	689.962	219.131	-3.270
3600	31.162	254.063	229.330	89.038	689.710	205.683	-2.984
3700	31.576	254.922	230.010	92.175	689.500	192.241	-2.714
3800	31.979	255.770	230.677	95.353	689.330	178.804	-2.458
3900	32.371	256.606	231.331	98.570	689.201	165.370	-2.215
4000	32.751	257.430	231.973	101.827	689.110	151.940	-1.984
4100	33.120	258.243	232.604	105.120	689.056	138.512	-1.765
4200	33.479	259.046	233.224	108.450	689.039	125.084	-1.556
4300	33.826	259.838	233.834	111.816	689.057	111.656	-1.356
4400	34.164	260.619	234.434	115.215	689.110	98.228	-1.166
4500	34.493	261.391	235.024	118.648	689.195	84.798	-0.984
4600	34.814	262.152	235.606	122.114	689.313	71.365	-0.810
4700	35.127	262.904	236.179	125.611	689.463	57.930	-0.644
4800	35.436	263.647	236.743	129.139	689.644	44.492	-0.484
4900	35.740	264.381	237.300	132.698	689.856	31.049	-0.331
5000	36.041	265.106	237.849	136.287	690.098	17.601	-0.184
5100	36.342	265.823	238.390	139.906	690.370	4.149	-0.042
5130.824	36.434	266.042	238.556	141.028	------- FUGACITY = 1 bar -------		
5200	36.642	266.531	238.925	143.555	0.	0.	0.
5300	36.943	267.232	239.452	147.234	0.	0.	0.
5400	37.115	267.914	239.972	150.887	0.	0.	0.
5500	37.396	268.598	240.486	154.612	0.	0.	0.
5600	37.678	269.274	240.994	158.366	0.	0.	0.
5700	37.963	269.943	241.496	162.148	0.	0.	0.
5800	38.251	270.606	241.993	165.959	0.	0.	0.
5900	38.543	271.263	242.483	169.798	0.	0.	0.
6000	38.839	271.913	242.968	173.667	0.	0.	0.

PREVIOUS: December 1973 (1 atm) CURRENT: December 1973 (1 bar)

Niobium (Nb) $Nb_1(g)$

IP(Nb^+, g) = 115500 ± 100 cm^{-1}

S°(298.15 K) = 183.333 ± 0.4 J K^{-1} mol^{-1}

$\Delta_fH°$(0 K) = 1394.0 ± 8 kJ mol^{-1}

$\Delta_fH°$(298.15 K) = [1402.954] kJ mol^{-1}

Electronic Levels and Quantum Weights

State	ε_i, cm^{-1}	g_i
$^5D_0$	0.00	1
$^5D_1$	158.38	3
$^5D_2$	438.99	5
$^5D_3$	801.38	7
$^5D_4$	1224.87	9

Enthalpy of Formation

$\Delta_fH°$(Nb^+, g, 0 K) is calculated from $\Delta_fH°$(Nb, g, 0 K) (1) using the spectroscopic value of IP(Nb) = 55511 ± 1 cm^{-1} (55511 ± 1 cm^{-1} (664.059 ± 0.01 kJ mol^{-1}) from Moore (2). The ionization limit is converted from cm^{-1} to kJ mol^{-1} using the factor, 1 cm^{-1} = 0.01196266 kJ cm^{-1}, which is derived from the 1973 CODATA fundamental constants (3). Rosenstock et al. (4) and Levin and Lias (5) have summarized additional ionization and appearance potential data.

$\Delta_fH°$(Nb^+, g, 298.15 K) is calculated from $\Delta_fH°$(Nb, g, 0 K) by using IP(Nb) with JANAF (1) enthalpies, H°(0 K) - H°(298.15 K), for Nb(g), Nb^+(g), and e^-(g). $\Delta_fH°$(Nb → Nb^+ + e^-, 298.15 K) differs from a room temperature threshold energy due to inclusion of these enthalpies and to threshold effects discussed by Rosenstock et al. (4). $\Delta_fH°$(298.15 K) should be changed by -6.197 kJ mol^{-1} if it is to be used in the ion convention that excludes the enthalpy of the electron.

Heat Capacity and Entropy

The information on electronic energy levels and quantum weights, given by Moore (6), is incomplete because many theoretically predicted levels have not been observed. Although we have listed only the ground, the first excited state, the highest observed excited state, and the ionization potential for Nb^+(g), all levels listed by Moore (6), as well as estimated levels, are used in the calculation. The observed levels are too numerous to list completely. The calculations indicate that for Nb^+(g), the thermodynamic functions are independent of the estimated missing levels (for n=4, 5), the cut-off procedure, and the inclusion of n=5 levels up to 6000 K; the Gibbs energy function showing variations of 0.2 % at this temperature. The reported uncertainty in S°(298.15 K) is due to uncertainties in the relative ionic mass, and the fundamental constants. Extension of these calculations above 6000 K may require consideration of the higher excited states (n>5), and use of different fill and cut-off procedures (7).

References

1. JANAF Thermochemical Tables: Nb(g), 3-31-84; e^-(g), 3-31-82.
2. C. E. Moore, U. S. Nat. Bur. Stand., NSRDS-NBS-34, 1970.
3. E. R. Cohen and B. N. Taylor, J. Phys. Chem. Ref. Data 2, 663 (1973).
4. H. M. Rosenstock, K. Draxl et al., J. Phys. Chem. Ref. Data 6, Supp. 1 (1977).
5. R. D. Levin and S. G. Lias, U. S. Natl. Bur. Stand. NSRDS-NBS-71, 1982.
6. C. E. Moore, U. S. Nat. Bur. Stand. 35, Volume II, 1970 [Reprint of NBS Circular 497, Volume II, 1952].
7. J. R. Downey, Jr., The Dow Chemical Company, AFOSR-TR-78-0960, Contract No. F44620-75-1-0048, 1978.

Enthalpy Reference Temperature = T_r = 298.15 K Standard State Pressure = p° = 0.1 MPa

	J $K^{-1}mol^{-1}$			kJ mol^{-1}			
T/K	$C_p°$	S°	-[G°-H°(T_r)]/T	H°-H°(T_r)	$\Delta_fH°$	$\Delta_fG°$	Log K_f
0	0.	0.	INFINITE	-8.015	1393.983		
100	26.964	152.351	209.002	-5.665			
200	28.783	171.684	186.012	-2.866			
250	29.250	178.162	183.816	-1.414			
298.15	29.421	183.333	183.333	0.	1402.954	1352.911	-237.024
300	29.423	183.515	183.333	0.054	1403.002	1352.600	-235.509
350	29.331	188.046	183.691	1.524	1404.265	1344.099	-200.596
400	29.059	191.947	184.485	2.985	1405.503	1335.419	-174.388
450	28.698	195.349	185.507	4.429	1406.710	1326.586	-153.986
500	28.323	198.353	186.645	5.854	1407.886	1317.620	-137.651
600	27.697	203.458	189.036	8.653	1410.151	1299.353	-113.119
700	27.316	207.696	191.407	11.402	1412.322	1280.714	-95.568
800	27.143	211.330	193.676	14.123	1414.425	1261.768	-82.385
900	27.099	214.524	195.819	16.835	1416.476	1242.562	-72.116
1000	27.118	217.380	197.834	19.545	1418.485	1223.131	-63.890
1100	27.159	219.966	199.731	22.259	1420.457	1203.499	-57.149
1200	27.203	222.331	201.517	24.977	1422.394	1183.690	-51.525
1300	27.242	224.510	203.203	27.700	1424.296	1163.721	-46.759
1400	27.281	226.530	204.798	30.426	1426.163	1143.606	-42.668
1500	27.324	228.414	206.310	33.156	1427.990	1123.360	-39.119
1600	27.378	230.179	207.747	35.891	1429.771	1102.993	-36.009
1700	27.448	231.841	209.116	38.632	1431.500	1082.516	-33.262
1800	27.537	233.412	210.423	41.381	1433.171	1061.939	-30.817
1900	27.647	234.904	211.672	44.140	1434.782	1041.271	-28.627
2000	27.780	236.325	212.870	46.911	1436.329	1020.519	-26.653
2100	27.934	237.684	214.019	49.697	1437.809	999.692	-24.866
2200	28.108	238.988	215.125	52.499	1439.217	978.796	-23.240
2300	28.300	240.242	216.190	55.319	1440.548	957.838	-21.753
2400	28.507	241.450	217.217	58.159	1441.793	936.823	-20.389
2500	28.726	242.618	218.210	61.021	1442.938	915.759	-19.134
2600	28.956	243.750	219.171	63.905	1443.961	894.651	-17.974
2700	29.193	244.847	220.101	66.812	1444.823	873.507	-16.899
2800	29.437	245.913	221.004	69.744	1419.114	852.819	-15.910
2900	29.686	246.950	221.881	72.700	1420.802	832.564	-14.996
3000	29.939	247.961	222.734	75.681	1422.514	812.251	-14.143
3100	30.196	248.947	223.563	78.688	1424.252	791.880	-13.343
3200	30.457	249.909	224.372	81.720	1426.016	771.453	-12.593
3300	30.723	250.851	225.160	84.779	1427.807	750.970	-11.887
3400	30.996	251.772	225.929	87.865	1429.624	730.433	-11.222
3500	31.276	252.674	226.680	90.979	1431.469	709.842	-10.594
3600	31.566	253.560	227.415	94.121	1433.343	689.197	-10.000
3700	31.868	254.429	228.133	97.292	1435.246	668.500	-9.438
3800	32.183	255.283	228.837	100.495	1437.179	647.752	-8.904
3900	32.513	256.123	229.525	103.729	1439.145	626.952	-8.397
4000	32.861	256.950	230.201	106.998	1441.145	606.101	-7.915
4100	33.229	257.766	230.863	110.302	1443.181	585.200	-7.456
4200	33.617	258.571	231.513	113.644	1445.255	564.248	-7.017
4300	34.027	259.367	232.152	117.027	1447.368	543.247	-6.599
4400	34.461	260.154	232.779	120.451	1449.524	522.196	-6.199
4500	34.917	260.934	233.396	123.919	1451.724	501.096	-5.817
4600	35.398	261.707	234.003	127.435	1453.971	479.946	-5.450
4700	35.902	262.473	234.601	131.000	1456.267	458.747	-5.098
4800	36.428	263.235	235.190	134.616	1458.615	437.498	-4.761
4900	36.977	263.991	235.770	138.286	1461.016	416.200	-4.437
5000	37.546	264.744	236.342	142.012	1463.473	394.853	-4.125
5100	38.133	265.493	236.906	145.796	1465.989	373.456	-3.825
5200	38.737	266.240	237.463	149.639	777.892	361.318	-3.629
5300	39.356	266.983	238.013	153.544	780.196	353.285	-3.482
5400	39.985	267.725	238.556	157.511	782.589	345.204	-3.339
5500	40.624	268.464	239.093	161.541	784.972	337.082	-3.201
5600	41.268	269.202	239.624	165.636	787.392	328.917	-3.068
5700	41.915	269.938	240.150	169.795	789.847	320.708	-2.939
5800	42.561	270.673	240.670	174.019	792.339	312.456	-2.814
5900	43.203	271.406	241.184	178.307	794.866	304.160	-2.693
6000	43.839	272.137	241.694	182.659	797.428	295.822	-2.575

PREVIOUS: CURRENT: March 1984 (1 bar)

NIOBIUM, ION (Nb^-) IDEAL GAS M_r = 92.90695

EA(Nb, g) = 0.893 ± 0.025 eV

S°(298.15 K) = 186.095 ± 0.002 J K^{-1} mol^{-1}

$\Delta_f H°$(0 K) = 643.763 ± 4.5 kJ mol^{-1}

$\Delta_f H°$(298.15 K) = [640.978] kJ mol^{-1}

Electonic Levels and Quantum Weights

State	ϵ_i, cm^{-1}	g_i
$^5D_0$	0.0	1
$^5D_1$	110	3
$^5D_2$	310	5
$^5D_3$	560	7
$^5D_4$	870	9

Enthalpy of Formation

$\Delta_f H°$(Nb^-, g, 0 K) is calculated from $\Delta_f H°$(Nb, g, 0 K) (1) using the adopted electron affinity of EA(Nb) = 0.893 ± 0.025 eV (86.161 ± 2.412 kJ mol^{-1}). This value, recommended by Hotop and Lineberger (2), is based on a laser photodetachment electron spectrometry study (3). Additional information on Nb^-(g) may be obtained in the critical discussions of Hotop and Lineberger (2, 4), Rosenstock et al. (5) and Massey (6).

$\Delta_f H°$(Nb^-, g, 298.15 K) is obtained from $\Delta_f H°$(Nb, g, 0 K) by using EA(Nb) with JANAF (1) enthalpies, H°(0 K) - H°(298.15 K), for Nb^-(g), Nb(g), and e^-(g). $\Delta_r H°$(Nb^- → Nb + e^-, 298.15 K) differs from a room-temperature threshold energy due to inclusion of these enthalpies and to threshold effects discussed by Rosenstock et al. (5). $\Delta_f H°$(298.15 K) should be changed by +6.197 kJ mol^{-1} if it is to be used in the ion convention that excludes the enthalpy of the electron.

Heat Capacity and Entropy

The ground state electronic configuration for Nb^-(g) is given by Hotop and Lineberger (2, 4), Rosenstock et al (5), and Massey(6). The fine structure separation has been calculated by an isoelectronic extrapolation of ratios of fine structure separations (3) and is that recommended by Hotop and Lineberger (2).

References

1. JANAF Thermochemical Tables: Nb(g), 12-31-73; e^-(g), 3-31-82.
2. H. Hotop and W. C. Lineberger, J. Phys. Chem. Ref. Data, to be published, 1985.
3. C. S. Feigerle, R. R. Cordermann, S. V. Bobashev and W. C. Lineberger, J. Chem. Phys. 74, 1580 (1981).
4. H. Hotop and W. C. Lineberger, J. Phys. Chem. Ref. Data 4, 539 (1975).
5. H. M. Rosenstock, K. Draxl et al. J. Phys. Chem. Ref. Data 6, Supp. 1 (1977).
6. H. S. W. Massey, "Negative Ions", 3rd ed., Cambridge University Press, Cambrige, 1976.

Niobium, Ion (Nb^-) Nb_1^-(g)

Enthalpy Reference Temperature = T_r = 298.15 K Standard State Pressure = p° = 0.1 MPa

	J $K^{-1}mol^{-1}$			kJ mol^{-1}			
T/K	$C_p°$	S°	-[G°-H°(T_r)]/T	H°-H°(T_r)	$\Delta_f H°$	$\Delta_f G°$	Log K_f
0	0.	0.	INFINITE	-8.654	643.763		
100	30.251	153.166	212.710	-5.954			
200	30.317	174.205	188.796	-2.918			
250	29.768	180.919	186.576	-1.414			
298.15	28.948	186.095	186.095	0.	640.978	602.621	-105.577
300	28.913	186.274	186.095	0.054	640.948	602.383	-104.884
350	27.956	190.659	186.444	1.475	640.085	596.023	-88.952
400	27.029	194.331	187.207	2.850	639.158	589.790	-77.019
450	26.196	197.465	188.177	4.180	638.172	583.678	-67.752
500	25.474	200.187	189.245	5.471	637.135	577.678	-60.350
600	24.342	204.725	191.461	7.959	634.932	565.991	-49.274
700	23.538	208.414	193.627	10.350	632.589	554.685	-41.391
800	22.962	211.517	195.675	12.674	630.136	543.722	-35.501
900	22.541	214.196	197.587	14.948	627.593	533.072	-30.939
1000	22.227	216.554	199.368	17.186	624.972	522.709	-27.304
1100	21.988	218.661	201.028	19.396	622.284	512.613	-24.342
1200	21.803	220.566	202.578	21.585	619.534	502.764	-21.885
1300	21.656	222.305	204.030	23.758	616.729	493.146	-19.815
1400	21.539	223.905	205.393	25.917	613.872	483.746	-18.049
1500	21.443	225.388	206.677	28.066	610.961	474.553	-16.525
1600	21.364	226.769	207.890	30.206	607.990	465.556	-15.199
1700	21.299	228.062	209.039	32.339	604.953	456.746	-14.034
1800	21.243	229.278	210.130	34.466	601.846	448.117	-13.004
1900	21.197	230.425	211.168	36.588	598.663	439.663	-12.087
2000	21.157	231.512	212.159	38.706	595.399	431.379	-11.266
2100	21.122	232.543	213.105	40.820	592.049	423.260	-10.528
2200	21.092	233.525	214.011	42.931	588.609	415.303	-9.861
2300	21.066	234.462	214.880	45.038	585.070	407.504	-9.255
2400	21.043	235.358	215.715	47.144	581.423	399.862	-8.703
2500	21.023	236.217	216.518	49.247	577.653	392.374	-8.198
2600	21.005	237.041	217.291	51.349	573.736	385.039	-7.736
2700	20.989	237.833	218.038	53.448	569.634	377.859	-7.310
2800	20.975	238.596	218.758	55.546	538.934	371.319	-6.927
2900	20.962	239.332	219.455	57.643	535.605	365.391	-6.581
3000	20.950	240.043	220.130	59.739	532.275	359.579	-6.261
3100	20.940	240.729	220.783	61.833	528.944	353.877	-5.963
3200	20.930	241.394	221.417	63.927	525.611	348.283	-5.685
3300	20.922	242.038	222.032	66.019	522.278	342.793	-5.426
3400	20.914	242.662	222.630	68.111	518.944	337.404	-5.184
3500	20.906	243.268	223.211	70.202	515.609	332.113	-4.957
3600	20.900	243.857	223.776	72.293	512.274	326.918	-4.743
3700	20.894	244.430	224.327	74.382	508.938	321.815	-4.543
3800	20.888	244.987	224.863	76.471	505.601	316.802	-4.355
3900	20.883	245.530	225.386	78.560	502.264	311.877	-4.177
4000	20.878	246.058	225.896	80.648	498.926	307.038	-4.010
4100	20.874	246.574	226.394	82.735	495.588	302.282	-3.851
4200	20.869	247.077	226.881	84.823	492.249	297.608	-3.701
4300	20.866	247.568	227.356	86.909	488.910	293.013	-3.559
4400	20.862	248.047	227.821	88.996	485.570	288.496	-3.425
4500	20.859	248.516	228.276	91.082	482.231	284.054	-3.297
4600	20.855	248.974	228.721	93.167	478.891	279.687	-3.176
4700	20.853	249.423	229.156	95.253	475.550	275.393	-3.061
4800	20.850	249.862	229.583	97.338	472.210	271.169	-2.951
4900	20.847	250.292	230.001	99.423	468.869	267.016	-2.846
5000	20.845	250.713	230.412	101.507	465.527	262.930	-2.747
5100	20.842	251.126	230.814	103.592	462.186	258.911	-2.652
5200	20.840	251.530	231.208	105.676	-231.828	264.268	-2.655
5300	20.838	251.927	231.595	107.760	-235.501	273.843	-2.699
5400	20.836	252.317	231.975	109.844	-239.149	283.483	-2.742
5500	20.834	252.699	232.349	111.927	-242.869	293.196	-2.785
5600	20.833	253.075	232.716	114.010	-246.618	302.977	-2.826
5700	20.831	253.443	233.076	116.094	-250.396	312.824	-2.867
5800	20.830	253.806	233.430	118.177	-254.202	322.738	-2.907
5900	20.828	254.162	233.779	120.260	-258.037	332.718	-2.946
6000	20.827	254.512	234.121	122.342	-261.902	342.764	-2.984

PREVIOUS: CURRENT: March 1984 (1 bar)

Niobium, Ion (Nb^-) Nb_1^-(g)

NIOBIUM OXIDE (NbO) CRYSTAL M_r = 108.9058

$S°(298.15\ K) = [46.024 \pm 8.4]\ J\ K^{-1}\ mol^{-1}$

$T_{fus} = 2210 \pm 15\ K$

$\Delta_f H°(0\ K)$ = Unknown

$\Delta_f H°(298.15\ K) = -419.66 \pm 12.6\ kJ\ mol^{-1}$

$\Delta_{fus} H° = 85.354 \pm 20.9\ kJ\ mol^{-1}$

$\Delta_{sub} H°(298.15\ K) = 618.395 \pm 20.9\ kJ\ mol^{-1}$

Enthalpy of Formation

There are two reported oxygen bomb calorimetric studies for NbO(cr). The results must be interpreted in light of impure samples, incomplete combustion to Nb_2O_5(cr), and the possible nonstoichiometry of the reactants and products. Morozova and Getskina (1) reported $\Delta_c H°(291\ K) = 255.0\ kcal\ mol^{-1}$ while Kusenko and Gel'd (2) reported $\Delta_c H°(298.15\ K) = 263.2\ kcal\ mol^{-1}$ for the combustion of NbO(cr). Using auxiliary data (3), we calculate $\Delta_f H°(298.15\ K) = -99.5$ and $-95.4\ kcal\ mol^{-1}$ for NbO(cr). Schaefer and Liedmeier (4), using Cl_2 bomb calorimetry, reported a value of $-97 \pm 1\ kcal\ mol^{-1}$, based on $\Delta_f H°(NbOCl_3$, cr, 298.15 K) = $-210.2\ kcal\ mol^{-1}$.

A 2nd and 3rd law analysis of the majority of the reported smoothed emf data is tabulated below. Note that the enthalpy of formation for NbO(cr) calculated from the 3rd law $\Delta_r H°$ values are roughly 3 kcal mol^{-1} more negative than the adopted value. There is, however, excellent agreement within the equilibrium data; the 3rd law $\Delta_f H°(298.15\ K)$ values average $-100.3\ kcal\ mol^{-1}$ with a spread of only 1.8 kcal mol^{-1}. We adopt $\Delta_f H°(298.15\ K) = 100.3 \pm 3.0\ kcal\ mol^{-1}$ based on an average of the equilibrium studies (5-11). The reason for the discrepancy between the equilibrium and combustion results is not known at this time (8).

Source	T/K	Reaction	$\Delta_r H°(298.15\ K)$, kcal mol^{-1} 2nd law	3rd law	Drift cal $K^{-1}\ mol^{-1}$	$\Delta_f H°(298.15\ K)$ kcal mol^{-1}
(5)	1245-1379	A	-35.56	-36.38	-0.63	-100.02
(6)	1115-1347	A	-35.34	-37.60	-1.80	-101.24
(7)	1123-1323	A	-38.62	-36.20	1.98	-99.84
(8)	1030-1300	B	-11.17	3.19	6.62	-100.99
(9)	1000-1400	A	-37.82	-35.91	1.60	-99.55
(10)	1073-1373	A	-36.95	-36.46	0.40	-100.10
(11)	1177-1388	C	- 9.69	-9.72	-0.03	-100.12

A) $Nb(cr) + Fe_{0.95}O = NbO(cr) + 0.95\ Fe(cr)$

B) $Nb(cr) + 0.2\ Ta_2O_5(cr) = 0.4\ Ta(cr) + NbO(cr)$

C) $Nb(cr) + 1/3\ Cr_2O_3(cr) = NbO(cr) + 2/3\ Cr(cr)$

Heat Capacity and Entropy

Gel'd and Kusenko (12) measured the enthalpy of NbO(cr) and reported fourteen data points in the range 420-1702 K (IPTS-68). A Shomate-type equation is used to represent the data. The average deviation between the data and calculated values is 0.31%; the maximum deviation of 0.62% occurs at 1518 K. This equation is used to obtain extrapolated values of $C°_p$ up to T_{fus}.

There is no low temperature heat capacity data reported in the literature for T < 298 K. In order to have the 3rd law results of the equilibrium data agree with the combustion data, an entropy value of the order of 13-14 cal $K^{-1}\ mol^{-1}$ would be necessary. At this point, however, the 3rd law drifts would be all positive. $S°(298.15\ K) = 10.2\ cal\ K^{-1}\ mol^{-1}$ would lead to a more satisfying variation in the 3rd law drifts for the condensed phase equilibrium data but an intermediate value S°(298.15 K) = 11.0 cal $K^{-1}\ mol^{-1}$ gives better consistency with the vapor pressure data. For more details refer to the NbO(g) and NbO_2(g) tables (3).

Phase Data

Brauer (13) reported a homogeneity range of x = 0.89-1.04 for NbO_x. NbO(cr) has a cubic structure, a NaCl type with ordered vacancies (13, 14). Further information may be found in the review of the Nb-O system by Elliott (15).

Fusion Data

See the liquid table for details.

Sublimation Data

The mass spectrometric study by Shchukarev et al. (16) indicated that NbO(cr) at elevated temperatures yields NbO(g) and NbO_2(g). In the range 1873-2473 K, the concentration of NbO in the vapor phase over NbO(cr, ℓ) varied in the range 5-30% (16). $\Delta_{sub} H°(298.15\ K) = 147.8 \pm 5.0\ kcal\ mol^{-1}$ is calculated as the difference between the $\Delta_f H°(298.15\ K)$ values for NbO(g) and NbO(cr).

References

1. M. P. Morozova and L. L. Getskina, J. Gen. Chem. USSR 29, 1019 (1959).
2. F. G. Kusenko and P. V. Gel'd, Izv. Sibir. Otdel. Akad. Nauk SSSR, No. 2, 46 (1960); Chem. Abstr. 54, 16160 (1960).
3. JANAF Thermochemical Tables: Nb_2O_5(cr), 6-30-72; O_2(g), 9-30-65, NbO(g) and NbO_2(g), 12-31-73.
4. H. Schaefer and F. Liedmeier, Z. Anorg. Chem. 329, 225 (1964).
5. V. N. Drobyshev and T. N. Rezukhina, Russ. J. Phys. Chem. 39, 75 (1965).
6. V. I. Lavrent'ev, Ya. I. Genasimov, and T. N. Rezukhina, Dokl. Akad. Nauk SSSR 136, 1372 (1960).
7. B. C. H. Steel and C. B. Alcock, Trans. AIME 233, 800 (1968).
8. W. L. Worrell, Thermodynamics, IAEA, Vienna, pp. 131, 1966.
9. G. B. Barbi, Z. Naturforsch. A23, 800 (1968).
10. S. Ignatowicz and M. W. Davies, J. Less-Common Metals 15, 100 (1968).
11. T. Hiraoka, N. Sano, and Y. Matsushita, Trans. Iron Steel Inst. Jap. 11, 102 (1971).
12. P. V. Gel'd and F. G. Kusenko, Izv. Akad. Nauk SSSR, Otdel Tekh. Nauk, Met. i Toplivo, No. 2, 79 (1960).
13. G. Brauer, Z. Anorg. Chem. 248, 1 (1941).
14. D. Adler, Solid State Phys. 21, 1 (1968).
15. R. P. Elliott, Trans. Amer. Soc. Metals 52, 990 (1960).
16. S. A. Shchukarev, G. A. Semenov, and K. E. Frantseva, Russ. J. Inorg. Chem. 11, 129 (1966).

Niobium Oxide (NbO) Nb_1O_1(cr)

T/K	$C°_p$ (J $K^{-1}mol^{-1}$)	S° (J $K^{-1}mol^{-1}$)	$-[G°-H°(T_r)]/T$ (J $K^{-1}mol^{-1}$)	$H°-H°(T_r)$ (kJ mol^{-1})	$\Delta_f H°$ (kJ mol^{-1})	$\Delta_f G°$ (kJ mol^{-1})	Log K_f
0							
100							
200							
298.15	41.112	46.024	46.024	0.	-419.655	-391.923	68.663
300	41.187	46.279	46.025	0.076	-419.652	-391.751	68.210
400	44.032	58.555	47.679	4.350	-419.371	-382.488	49.948
500	45.827	68.584	50.888	8.848	-418.968	-373.313	39.000
600	47.204	77.064	54.562	13.501	-418.507	-364.224	31.709
700	48.369	84.429	58.314	18.280	-418.011	-355.216	26.507
800	49.467	90.960	61.994	23.173	-417.485	-346.281	22.610
900	50.481	96.845	65.545	28.170	-416.929	-337.414	19.583
1000	51.467	102.215	68.947	33.268	-416.342	-328.610	17.165
1100	52.418	107.165	72.199	38.462	-415.722	-319.866	15.189
1200	53.359	111.766	75.307	43.751	-415.068	-311.181	13.545
1300	54.287	116.074	78.279	49.134	-414.376	-302.551	12.157
1400	55.208	120.131	81.125	54.608	-413.646	-293.976	10.968
1500	56.117	123.971	83.855	60.175	-412.882	-285.455	9.940
1600	57.024	127.622	86.477	65.832	-412.091	-276.986	9.043
1700	57.929	131.106	89.000	71.579	-411.280	-268.566	8.252
1800	58.831	134.443	91.433	77.417	-410.456	-260.195	7.551
1900	59.730	137.647	93.781	83.345	-409.624	-251.870	6.924
2000	60.626	140.734	96.052	89.363	-408.791	-243.589	6.362
2100	61.526	143.714	98.251	95.471	-407.960	-235.349	5.854
2200	62.413	146.596	100.384	101.668	-407.140	-227.148	5.393
2210.000	62.501	146.880	100.593	102.293	------ CRYSTAL <--> LIQUID -----		
2300	63.306	149.391	102.454	107.954	-406.337	-218.985	4.973
2400	64.199	152.104	104.466	114.329	-405.562	-210.856	4.589
2500	65.091	154.743	106.425	120.794	-404.831	-202.759	4.236
2600	65.982	157.313	108.333	127.347	-404.165	-194.689	3.911
2700	66.871	159.820	110.194	133.990	-403.604	-186.644	3.611
2800	67.760	162.268	112.010	140.722	-429.560	-178.131	3.323

Enthalpy Reference Temperature = T_r = 298.15 K; Standard State Pressure = p° = 0.1 MPa

PREVIOUS: CURRENT: December 1973

Niobium Oxide (NbO) Nb_1O_1(cr)

NIOBIUM OXIDE (NbO) LIQUID M_r = 108.9058

$S°(298.15\ K) = 83.233\ J\ K^{-1}\ mol^{-1}$

$T_{fus} = 2210 \pm 15\ K$

$\Delta_f H°(298.15\ K) = -336.743\ kJ\ mol^{-1}$

$\Delta_{fus} H° = 85.354 \pm 20.9\ kJ\ mol^{-1}$

Enthalpy of Formation

$\Delta_f H°$(NbO, ℓ, 298.15 K) is calculated from $\Delta_f H°$(NbO, cr, 298.15K) by adding $\Delta_{fus} H°$ and the difference in enthalpy, H°(2210 K)-H°(298.15 K), between the crystal and liquid.

Heat Capacity and Entropy

There is no data reported in the literature concerning the heat capacity or enthalpy of NbO(ℓ). We estimate a constant value of $C_p° = 15.0\ cal\ K^{-1}\ mol^{-1}$ for the liquid phase. A glass transition is also assumed at 1500 K, so as to insure the proper thermodynamic relationship between the crystal values and extrapolated liquid values. At temperatures below 1500 K, $C_p°$ values of the crystal are used. The entropy at 298.15 K is calculated in a manner analogous to that used for the enthalpy of formation.

Fusion Data

Elliott (1) reported the melting point of NbO(cr) as T_{fus} = 2222 K (IPTS-68). Measurements by Kolchin and Sumarokova (2) gave T_{fus} = 2212 K (IPTS-68). The vapor pressure study of NbO(cr, ℓ) by Shchukarev et al. (3) gave T_{fus} = 2207 K. We adopt T_{fus} = 2210 ± 15 K in order to maintain good consistency with the vapor pressure data.

Since the crystal structure of NbO(cr) is a defect NaCl structure, we can estimate $\Delta_{fus} H° = 13.9\ kcal\ mol^{-1}$ based on $\Delta_{fus} S° = 6.27\ cal\ K^{-1} mol^{-1}$ as observed for NaCl (4). The data of Shchukarev et al. (3) are more consistent with $\Delta_{fus} H° = 20.4\ kcal\ mol^{-1}$, as determined from our analysis of the vapor pressures over NbO(cr, ℓ) (4). We adopt $\Delta_{fus} H° = 20.4 \pm 5.0\ kcal\ mol^{-1}$. This leads to $\Delta_{fus} S° = 9.23\ cal\ K^{-1}\ mol^{-1}$ which is roughly 3 cal $K^{-1}\ mol^{-1}$ greater than the NaCl value (4).

Vaporization Data

The vapors over NbO(ℓ) have been shown by Shchukarev et al. (3) to be NbO(g) and NbO_2(g). At 2300 K, the rates of the vapor pressures NbO_2(g):NbO(g) is 1.00:0.23 (3).

References

1. R. P. Elliott, Trans. Amer. Soc. Metals 52, 999 (1960).
2. O. Kolchin and N. Sumarokova, Atomnaya Energ. 10, 168 (1961).
3. S. A. Shchukarev, G. A. Semenoo, and K. E. Frantseva, Russ. J. Inorg. Chem. 11, 129 (1966).
4. JANAF Thermochemical Tables: NaCl(cr), 9-30-64; NbO(g), 12-31-73.

Niobium Oxide (NbO) Nb_1O_1(l)

Enthalpy Reference Temperature = T_r = 298.15 K; Standard State Pressure = p° = 0.1 MPa

T/K	$C_p°$ (J K⁻¹mol⁻¹)	S° (J K⁻¹mol⁻¹)	-[G°-H°(T_r)]/T (J K⁻¹mol⁻¹)	H°-H°(T_r) (kJ mol⁻¹)	$\Delta_f H°$ (kJ mol⁻¹)	$\Delta_f G°$ (kJ mol⁻¹)	Log K_f
0							
100							
200							
298.15	41.112	83.233	83.233	0.	-336.743	-320.105	56.081
300	41.187	83.488	83.234	0.076	-336.740	-320.002	55.717
400	44.032	95.764	84.888	4.350	-336.459	-314.460	41.064
500	45.827	105.793	88.097	8.848	-336.056	-309.005	32.282
600	47.204	114.273	91.771	13.501	-335.595	-303.638	26.434
700	48.369	121.638	95.523	18.280	-335.099	-298.351	22.263
800	49.467	128.169	99.203	23.173	-334.573	-293.136	19.140
900	50.481	134.054	102.754	28.170	-334.017	-287.990	16.714
1000	51.467	139.424	106.156	33.268	-333.430	-282.907	14.778
1100	52.418	144.374	109.408	38.462	-332.810	-277.884	13.196
1200	53.359	148.975	112.516	43.751	-332.156	-272.920	11.880
1300	54.287	153.283	115.488	49.134	-331.464	-268.011	10.769
1400	55.208	157.340	118.334	54.608	-330.734	-263.157	9.819
1500	56.120	161.180	121.063	60.175	-329.970	-258.357	8.997
1500.000	56.120	161.180	121.063	60.175	GLASS <--> LIQUID	TRANSITION	
1500.000	62.760	161.180	121.063	60.175			
1600	62.760	165.231	123.699	66.451	-328.560	-253.629	8.280
1700	62.760	169.035	126.255	72.727	-327.221	-248.987	7.650
1800	62.760	172.623	128.732	79.003	-325.959	-244.422	7.093
1900	62.760	176.016	131.132	85.279	-324.779	-239.925	6.596
2000	62.760	179.235	133.458	91.555	-323.687	-235.487	6.150
2100	62.760	182.297	135.711	97.831	-322.689	-231.102	5.748
2200	62.760	185.217	137.895	104.107	-321.789	-226.762	5.384
2210.000	62.760	185.501	138.110	104.734	------ CRYSTAL	<--> LIQUID	-----
2300	62.760	188.006	140.014	110.383	-320.997	-222.461	5.052
2400	62.760	190.677	142.070	116.659	-320.321	-218.192	4.749
2500	62.760	193.239	144.066	122.935	-319.778	-213.948	4.470
2600	62.760	195.701	146.004	129.211	-319.389	-209.723	4.213
2700	62.760	198.070	147.889	135.487	-319.195	-205.510	3.976
2800	62.760	200.352	149.722	141.763	-345.607	-200.815	3.746
2900	62.760	202.554	151.506	148.039	-344.657	-195.661	3.524
3000	62.760	204.682	153.244	154.315	-343.717	-190.539	3.318

PREVIOUS: CURRENT: December 1973

Niobium Oxide (NbO) Nb_1O_1(l)

NIOBIUM OXIDE (NbO) CRYSTAL–LIQUID M_r = 108.9058

0 to 2210 K crystal
above 2210 K liquid

Refer to the individual tables for details.

Niobium Oxide (NbO) $Nb_1O_1(cr,l)$

Enthalpy Reference Temperature = T_r = 298.15 K Standard State Pressure = $p°$ = 0.1 MPa

T/K	$C_p°$ (J K⁻¹mol⁻¹)	S° (J K⁻¹mol⁻¹)	−[G°−H°(T_r)]/T (J K⁻¹mol⁻¹)	H°−H°(T_r) (kJ mol⁻¹)	$\Delta_fH°$ (kJ mol⁻¹)	$\Delta_fG°$ (kJ mol⁻¹)	Log K_f
0							
100							
200							
298.15	41.112	46.024	46.024	0.	-419.655	-391.923	68.663
300	41.187	46.279	46.025	0.076	-419.652	-391.751	68.210
400	44.032	58.555	47.679	4.350	-419.371	-382.488	49.948
500	45.827	68.584	50.888	8.848	-418.968	-373.313	39.000
600	47.204	77.064	54.562	13.501	-418.507	-364.224	31.709
700	48.369	84.429	58.314	18.280	-418.011	-355.216	26.507
800	49.467	90.960	61.994	23.173	-417.485	-346.281	22.610
900	50.481	96.845	65.545	28.170	-416.929	-337.414	19.583
1000	51.467	102.215	68.947	33.268	-416.342	-328.610	17.165
1100	52.418	107.165	72.199	38.462	-415.722	-319.866	15.189
1200	53.359	111.766	75.307	43.751	-415.068	-311.181	13.545
1300	54.287	116.074	78.279	49.134	-414.376	-302.551	12.157
1400	55.208	120.131	81.125	54.608	-413.646	-293.976	10.968
1500	56.117	123.971	83.855	60.175	-412.882	-285.455	9.940
1600	57.024	127.622	86.477	65.832	-412.091	-276.986	9.043
1700	57.929	131.106	89.000	71.579	-411.280	-268.566	8.252
1800	58.831	134.443	91.433	77.417	-410.456	-260.195	7.551
1900	59.730	137.647	93.781	83.345	-409.624	-251.870	6.924
2000	60.626	140.734	96.052	89.363	-408.791	-243.589	6.362
2100	61.526	143.714	98.251	95.471	-407.960	-235.349	5.854
2200	62.413	146.596	100.384	101.668	-407.140	-227.148	5.393
2210.000	62.501	146.880	100.593	102.293	CRYSTAL <--> LIQUID		
2210.000	62.760	185.501	100.593	187.646	TRANSITION		
2300	62.760	188.006	103.965	193.295	-320.997	-222.461	5.052
2400	62.760	190.677	107.523	199.571	-320.321	-218.192	4.749
2500	62.760	193.239	110.901	205.847	-319.778	-213.948	4.470
2600	62.760	195.701	114.115	212.123	-319.389	-209.723	4.213
2700	62.760	198.070	117.181	218.399	-319.195	-205.510	3.976
2800	62.760	200.352	120.111	224.675	-345.607	-200.815	3.746
2900	62.760	202.554	122.916	230.951	-344.657	-195.661	3.524
3000	62.760	204.682	125.606	237.227	-343.717	-190.539	3.318

PREVIOUS: CURRENT: December 1973

Niobium Oxide (NbO) $Nb_1O_1(cr,l)$

NIOBIUM OXIDE (NbO) IDEAL GAS M_r = 108.9058

$S°(298.15 K) = 238.976 \pm 3.56 J K^{-1} mol^{-1}$

$\Delta_f H°(0 K) = 199.54 \pm 20.9 kJ mol^{-1}$
$\Delta_f H°(298.15 K) = 198.74 \pm 20.9 kJ mol^{-1}$

Electronic Levels and Quantum Weights

State	ε_i, cm^{-1}	g_i
$X[^4\Sigma^-]$	0	4
$[^4\Pi]$	15000	8
$[^4\Sigma^-]$	21316.2	4

$\omega_e = 989.03 cm^{-1}$ $\omega_e x_e = 3.83 cm^{-1}$ $\sigma = 1$
$B_e = 0.432 cm^{-1}$ $\alpha_e = 2.1 \times 10^{-3} cm^{-1}$ $r_e = 1.691$ Å

Enthalpy of Formation

Using the adopted vibrational constants [ω_e and $\omega_e x_e$], we calculate D° = 7.855 eV as the dissociation energy of NbO(g) based on the linear Birge-Sponer extrapolation of the ground state. Applying a correction for the ionicity in the Nb-O bond as suggested by Hildenbrand and Murad (1), we recalculate D° = 8.032 eV. This latter value corresponds to $\Delta_f H°$(NbO, g, 298.15 K) = 48.03 kcal mol^{-1}, assuming dissociation to normal atoms, while the former yields 52.11 kcal mol^{-1}.

Shchukarev et al. (2, 3) have investigated mass spectrometrically the composition of the vapors over NbO(cr, ℓ) and NbO_2(cr, ℓ). In their first work (2), the authors reported that, besides NbO sublimation at 1773-1923 K, the reaction 2 NbO(cr) = Nb(cr) + NbO_2(g) occurs. A 2nd law $\Delta_{sub}H°$ = 138 ± 3 kcal mol^{-1} was reported for the range 1773-1923 K. Using auxiliary data (5), we calculate $\Delta_f H°$(NbO, g, 298.15 K) = 47.2 ± 3.0 kcal mol^{-1}, assuming the $\Delta_{sub}H°$ reported value refers to 1900 K.

Shchukarev et al. (3) measured the vapor pressures of NbO and NbO_2(cr, ℓ) and NbO_2(cr, ℓ) by the effusion method coupled with a mass spectrometer. The results are summarized in the following table.

Reaction	T/K	$\Delta_r H°$(298.15 K), kcal mol^{-1} 2nd law	3rd law	Drift cal K^{-1} mol^{-1}	$\Delta_f H°$(298.15 K), kcal mol^{-1} 2nd law	3rd law
NbO(cr) = NbO(g)	1773-2210	148.0	147.5	-0.24	47.7	47.2
NbO(ℓ) = NbO(g)	2210-2473	128.1	128.0	-0.04	47.6	47.5

We adopt $\Delta_f H°$(298.15 K) = 47.5 ± 5.0 kcal mol^{-1} for NbO(g). The adopted value is an average value based on the 2nd and 3rd law results tabulated above. It is interesting to note that the adjusted Birge-Sponer result shows excellent agreement with the adopted $\Delta_f H°$(298.15 K). The adopted $\Delta_f H°$(298.15 K) leads to a dissociation energy D° = 8.05 eV.

Heat Capacity and Entropy

The spectroscopic work involving NbO(g) has been reviewed and referenced by Rosen (4). The adopted vibrational and rotational constants as well as the electronic levels are those tabulated by Rosen (4). As indicated in Rosen (4), the data suggest a $^4\Sigma^-$ ground state as in VO(g) (5). However, a possible ground state $^2\Delta$ as in TaO(g) (5) is not definitely excluded. The recent reviews by Cheetham and Barrow (6) and Weltner (7) favor a $^2\Delta$ ground state. Using a $^2\Delta$ ground state and $^2\Delta$-$^2\Delta$ transition at 21316.2 cm^{-1} produces the same thermodynamic functions as does a $^4\Sigma^-$ ground state and a $^4\Sigma$-$^4\Sigma$ transition at 21316.2 cm^{-1}. Green et al. (9) studied the infrared absorption spectra of isotopic NbO isolated in an Ar matrix. Green et al. (9) reported that a partial analysis of the observed bands supported the assignment of a $^4\Sigma^-$ electronic ground state for NbO.

The free energy functions used here are approximately 0.85 cal K^{-1} mol^{-1} at 298 K and 2.90 cal K^{-1} mol^{-1} at 3000 K lower than those proposed by Brewer and Rosenblatt (8). The difference arises from the fact that Brewer and Rosenblatt (8) approximated the NbO electronic levels by using Nb^{++} electronic levels (which included low lying levels) whereas we have used the observed NbO electronic levels.

References

1. D. L. Hildenbrand and E. Murad, J. Chem. Phys. 51, 807 (1969).
2. S. A. Shchukarev, G. A. Semenov and K. E. Frantseva, Izv. Vyssh Ucheb. Zavend, Khim. i Khim. Tekhnol. 5, 691 (1962).
3. S. A. Shchukarev, G. A. Semenov and K. E. Frantseva, Russ. J. Inorg. Chem. 11, 129 (1966).
4. B. Rosen, Ed., "Spectroscopic Data Relative to Diatomic Molecules," Pergamon Press, New York, 1970.
5. JANAF Thermochemical Tables: VO(g), TaO(g), and NbO(cr), 12-31-73.
6. C. J. Cheetham and R. F. Barrow in "Advances in High Temperature Chemistry," Vol. 1, L. Eyring, Ed., Academic Press, New York, pp. 7-42, 1967.
7. W. Weltner, Jr., in "Advances in High Temperature Chemistry," Vol. 2, L. Eyring, Ed., Academic Press, New York, pp. 85-105, 1967.
8. L. Brewer and G. M. Rosenblatt, in "Advances in High Temperature Chemistry," Vol. 2, L. Eyring, Ed., Academic Press, New York, pp. 1-83, 1967.
9. D. W. Green, W. Korfmacher and D. M. Gruen, J. Chem. Phys. 58, 404 (1973).

Niobium Oxide (NbO) $Nb_1O_1(g)$

Enthalpy Reference Temperature = T_r = 298.15 K Standard State Pressure = p° = 0.1 MPa

T/K	$C_p°$ (J $K^{-1}mol^{-1}$)	S°	-[G°-H°(T_r)]/T	H°-H°(T_r) (kJ mol^{-1})	$\Delta_f H°$	$\Delta_f G°$	Log K_f
0	0.	0.	INFINITE	-8.781	199.541	199.541	INFINITE
100	29.106	206.753	265.477	-5.872	200.216	189.462	-98.965
200	29.467	226.986	241.744	-2.952	199.584	178.910	-46.727
250	30.057	233.620	239.478	-1.464	199.157	173.791	-36.312
298.15	30.788	238.976	238.976	0.	198.740	168.943	-29.598
300	30.818	239.166	238.976	0.057	198.724	168.758	-29.383
350	31.614	243.977	239.354	1.618	198.301	163.798	-24.445
400	32.364	248.248	240.204	3.218	197.891	158.897	-20.750
450	33.033	252.099	241.315	4.853	197.492	154.047	-17.881
500	33.613	255.610	242.572	6.519	197.099	149.241	-15.591
600	34.531	261.824	245.276	9.929	196.316	139.742	-12.166
700	35.197	267.200	248.033	13.417	195.521	130.376	-9.729
800	35.687	271.934	250.730	16.963	194.700	121.125	-7.909
900	36.054	276.159	253.325	20.550	193.847	111.979	-6.499
1000	36.336	279.973	255.803	24.171	192.956	102.930	-5.377
1100	36.557	283.447	258.160	27.816	192.026	93.972	-4.462
1200	36.735	286.636	260.402	31.481	191.056	85.101	-3.704
1300	36.880	289.582	262.535	35.161	190.047	76.312	-3.066
1400	37.002	292.320	264.566	38.856	188.997	67.602	-2.522
1500	37.106	294.876	266.502	42.561	187.900	58.969	-2.053
1600	37.196	297.274	268.351	46.277	186.749	50.411	-1.646
1700	37.277	299.531	270.120	50.000	185.536	41.926	-1.288
1800	37.352	301.664	271.813	53.732	184.254	33.515	-0.973
1900	37.423	303.686	273.438	57.471	182.896	25.177	-0.692
2000	37.493	305.607	274.999	61.216	181.457	16.913	-0.442
2100	37.564	307.438	276.500	64.969	179.933	8.723	-0.217
2200	37.638	309.187	277.947	68.729	178.316	0.608	-0.014
2300	37.717	310.862	279.342	72.497	176.601	-7.432	0.169
2400	37.802	312.469	280.689	76.273	174.776	-15.394	0.335
2500	37.895	314.014	281.991	80.057	172.828	-23.279	0.486
2600	37.997	315.502	283.251	83.852	170.735	-31.082	0.624
2700	38.108	316.938	284.473	87.657	168.458	-38.802	0.751
2800	38.230	318.326	285.657	91.474	139.587	-45.948	0.857
2900	38.362	319.670	286.807	95.303	138.091	-52.549	0.947
3000	38.506	320.973	287.924	99.147	136.598	-59.097	1.029
3100	38.660	322.238	289.011	103.005	135.111	-65.595	1.105
3200	38.825	323.468	290.068	106.879	133.631	-72.046	1.176
3300	39.000	324.665	291.099	110.770	132.160	-78.451	1.242
3400	39.185	325.832	292.103	114.679	130.698	-84.811	1.303
3500	39.380	326.971	293.083	118.607	129.247	-91.128	1.360
3600	39.583	328.083	294.040	122.556	127.808	-97.404	1.413
3700	39.794	329.171	294.975	126.524	126.383	-103.641	1.463
3800	40.012	330.235	295.889	130.515	124.972	-109.838	1.510
3900	40.236	331.277	296.783	134.527	123.576	-115.999	1.554
4000	40.466	332.299	297.658	138.562	122.196	-122.125	1.595
4100	40.700	333.301	298.515	142.620	120.833	-128.216	1.633
4200	40.937	334.284	299.355	146.702	119.487	-134.274	1.670
4300	41.177	335.250	300.179	150.808	118.158	-140.300	1.704
4400	41.418	336.200	300.987	154.938	116.848	-146.296	1.737
4500	41.661	337.133	301.780	159.091	115.556	-152.262	1.767
4600	41.903	338.052	302.558	163.270	114.281	-158.199	1.796
4700	42.144	338.955	303.323	167.472	113.025	-164.109	1.824
4800	42.384	339.845	304.075	171.699	111.787	-169.992	1.850
4900	42.622	340.722	304.814	175.949	110.566	-175.850	1.875
5000	42.856	341.585	305.540	180.223	109.363	-181.683	1.898
5100	43.087	342.436	306.256	184.520	108.175	-187.492	1.920
5200	43.313	343.275	306.959	188.840	-583.668	-183.969	1.848
5300	43.535	344.102	307.652	193.182	-585.156	-176.268	1.737
5400	43.752	344.918	308.335	197.547	-586.603	-168.543	1.630
5500	43.963	345.722	309.007	201.933	-588.110	-160.788	1.527
5600	44.168	346.516	309.670	206.339	-589.633	-153.004	1.427
5700	44.366	347.300	310.323	210.766	-591.173	-145.194	1.331
5800	44.559	348.073	310.968	215.212	-592.731	-137.356	1.237
5900	44.744	348.837	311.603	219.677	-594.309	-129.492	1.146
6000	44.922	349.590	312.230	224.161	-595.909	-121.600	1.059

PREVIOUS: December 1973 (1 atm) CURRENT: December 1973 (1 bar)

Niobium Oxide (NbO) $Nb_1O_1(g)$

NIOBIUM OXIDE (NbO_2) CRYSTAL (I-II-III) M_r = 124.9052

$S°(298.15 K) = 54.506 \pm 0.29\ J\ K^{-1}\ mol^{-1}$

$T_{trs1} = 1090 \pm 20$ K

$T_{trs2} = 1200$ K

$T_{fus} = 2175 \pm 15$ K

$\Delta_f H°(0 K) = -790.31 \pm 8.4\ kJ\ mol^{-1}$

$\Delta_f H°(298.15 K) = -794.96 \pm 8.4\ kJ\ mol^{-1}$

$\Delta_{trs1} H° = 3.423\ kJ\ mol^{-1}$

$\Delta_{trs2} H° = 0.000\ kJ\ mol^{-1}$

$\Delta_{fus} H° = 92.048 \pm 20.9\ kJ\ mol^{-1}$

Enthalpy of Formation

There are three oxygen bomb calorimetric studies for NbO_2(cr). As with NbO(cr), the combustion results must be interpreted in light of impure samples, possible incomplete combustion, and possible nonstoichiometry. The following $\Delta_c H°$ results have been reported for the oxygen combustion of one mole of NbO_2(cr) to yield Nb_2O_5(cr): 37.0 ± 0.4 kcal mol^{-1} (291 K) by Morozova and Getskina (1), 37.6 ± 2.6 kcal mol^{-1} (298 K) by Kusenko and Gel'd (2), and 36.67 ± 0.10 kcal mol^{-1} (298 K) by Mah (3). Using the value $\Delta_c H°(298.15 K) = 37.0 \pm 2.0\ kcal\ mol^{-1}$ and JANAF auxiliary data (4), we calculate and adopt $\Delta_f H°(NbO_2, cr, 298.15 K) = -190.0 \pm 2.0\ kcal\ mol^{-1}$.

There are numerous equilibrium studies involving NbO_2. Early work involved the reduction of niobium oxides with hydrogen by Grube et al. (5), Sue (6), and Schaefer and Breil (7), and the carbon reduction of Nb_2O_5 by Sue (8). The more recent emf equilibrium studies are analyzed and tabulated in the following table.

Source	T/K	Reaction	$\Delta_r H°(298.15 K)$, kcal mol^{-1} 2nd law	3rd law	Drift cal K^{-1} mol^{-1}	$\Delta_f H°(298.15 K)$ kcal mol^{-1}
(9)	1673-1823	$NbO_2(cr)+H_2(g)=NbO(cr)+H_2O(g)$	36.35	32.30	-2.3	-190.4
(10)	1100-1300	$NbO_2(cr)+Nb(cr)=2\ NbO(cr)$	-11.96	-9.59	2.0	-187.8
(11)	1074-1175	$Nb_2O_5(cr)+C(cr)=2\ NbO_2(cr) + CO(g)$	38.43	45.87	6.6	-190.9
(12)	1177-1361	$NbO(cr)+1/3\ Cr_2O_3(cr)=NbO_2(cr)+2/3\ Cr(cr)$	2.32	1.78	-0.4	-188.9

Heat Capacity and Entropy

King (13, 53-296 K) measured the heat capacity of NbO_2(cr). Using the combination of Debye and Einstein functions suggested by King (13), we calculate $S°(51 K) = 0.616\ cal\ K^{-1}mol^{-1}$ and $H°(51 K)-H°(0 K) = 23\ cal\ K^{-1}\ mol^{-1}$. The enthalpy data of Gel'd and Kusenko (14), and King and Christensen (15), indicated a transition in the region 1000-1200 K. We adopt the reported enthalpy equations of King and Christensen (15) for the ranges 1090-1200 K and 1200-1800 K. The latter equation is extrapolated to T_{fus}. For the region 298-1090 K, we use a five term polynomial to fit the enthalpy data of King and Christensen (15) with constraints to join smoothly with the enthalpy derived from the King data (13): the deviations are +0.34 to -0.67% in this range 298-1090 K. In the same range, the Gel'd and Kusenko data (14) lies approximately 2-3% higher; for T > 1200 K, it is 0.6-2.5% high.

Phase Data

Brauer (16) reported a homogeneity range of x = 1.95 to 2.05 for NbO_2. A recent study by Janninck and Whitmore (17), using isopiestic reduction techniques at 1373 K, reported x = 1.9975 to 2.003.

Brauer (16), Magneli et al. (18), Terao (19), and Marinder (20) have shown that at room temperature NbO_2 has a deformed rutile structure. NbO_2 is similar to V_2O_4 in terms of crystal structure (21).

Transition Data

Electric conductivity and thermoelectric power measurements and a DTA and x-ray diffraction study indicated a phase transition in the region 1068-1173 K (17, 21, 22). In this region, the structure of NbO_2 transforms from a deformed rutile to a rutile structure. The evidence suggested that the transition is not sharp but is actually spread over at least 50° (21).

The enthalpy data of King and Christensen (15) and Gel'd and Kusenko (14) are consistent with the above interpretation. We treat this transition as first order with $T_{trs1} = 1090 \pm 20$ K and add an artificial transition T_{trs2} = 1200 K as in King and Christensen (15). This combination reduces the reported enthalpy data (15). The value $\Delta_{trs1} H° = 0.818\ kcal\ mol^{-1}$ is calculated as the difference between the enthalpies of the two "phases" at 1090 K associated with the transition at T_{trs1}.

Fusion Data

See liquid table for details.

Sublimation Data

$\Delta_{sub} H°(298.15 K) = 142.1 \pm 5.0$ kcal/mol is calculated as the $\Delta_f H°(298.15 K)$ differences for NbO_2(g) and NbO_2(cr).

References

1. M. P. Morozova and L. Getskina, Russ. J. Gen. Chem. 29, 1019 (1959).
2. F. G. Kusenko and P. V. Gel'd, Izvest. Sibir. Otdel. Akad. Nauk SSSR, No. 2, 46 (1960).
3. A. D. Mah, J. Amer. Chem. Soc. 80, 3872 (1958).
4. JANAF Thermochemical Tables: Nb_2O_5(cr), 12-31-72; NbO(cr), 12-31-73.
5. G. Grube, O. Kubaschewski, and K. Zwiauer, Z. Elektrochem. 45, 885 (1939).
6. P. Sue, Comp. rend. 208, 1088 (1939).
7. H. Schaefer and G. Breil, Z. Anorg. Chem. 267, 265 (1952).
8. P. Sue, J. Chim. Phys. 36, 280 (1939).
9. V. I. Lavrent'ev, Ya. I. Gerasimov, and T. N. Rezukhina, Dokl. Akad. Nauk SSSR 136, 1372 (1961).
10. W. L. Worrell, Thermodynamics, IAEA, Vienna, p. 131, 1966.
11. W. L. Worrell, J. Phys. Chem. 68, 952 (1964).
12. T. Hiraoka, N. Sano, and Y. Matsushita, Trans. Iron Steel Inst. Jap. 11, 102 (1971).
13. E. G. King, J. Amer. Chem. Soc. 80, 1799 (1958).
14. P. V. Gel'd and F. G. Kusenko, Izvest. Akad. Nauk SSSR, Otdel. Tekh. Nauk, Met. i Toplivo, No. 2, 79 (1960).
15. E. G. King and A. U. Christensen, U. S. Bur. Mines RI 5789, 1961.
16. G. Brauer, Z. Anorg. Chem. 248, 1 (1941).
17. R. F. Janninck and D. H. Whitmore, J. Phys. Chem. Solids 27, 1183 (1966).
18. A. Magneli, G. Anderson and G. Sundkvist, Acta Chem. Scand. 9, 1402 (1955).
19. N. Terao, Japan J. Appl. Phys. 2, 565 (1963).
20. B. A. Marinder, Acta Chem. Scand. 15, 707 (1961).
21. K. Sakata, J. Phys. Soc. Japan 26, 582 (1969).
22. T. Sakata, K. Sakata, and I. Nishida, Phys. Status Solidi 20, K155 (1967).

Niobium Oxide (NbO_2) Nb_1O_2(cr)

Enthalpy Reference Temperature = T_r = 298.15 K Standard State Pressure = p° = 0.1 MPa

T/K	$C_p°$ (J $K^{-1}mol^{-1}$)	S°	$-[G°-H°(T_r)]/T$	$H°-H°(T_r)$ (kJ mol^{-1})	$\Delta_f H°$	$\Delta_f G°$	Log K_f
0	0.	0.	INFINITE	-9.271	-790.308	-790.308	INFINITE
100	21.143	11.275	96.297	-8.502	-793.224	-775.765	405.218
200	44.918	33.967	59.417	-5.090	-794.821	-757.542	197.850
298.15	57.452	54.506	54.506	0.	-794.960	-739.175	129.500
300	57.605	54.862	54.507	0.106	-794.954	-738.829	128.642
400	63.547	72.327	56.848	6.191	-794.348	-720.199	94.048
500	67.739	86.967	61.447	12.760	-793.403	-701.766	73.313
600	71.651	99.665	66.781	19.730	-792.205	-683.547	59.508
700	75.555	111.003	72.302	27.090	-790.755	-665.549	49.664
800	79.496	121.348	77.795	34.843	-789.037	-647.776	42.295
900	83.479	130.941	83.173	42.991	-787.033	-630.236	36.578
1000	87.500	139.945	88.405	51.540	-784.726	-612.934	32.016
1090.000	91.144	147.640	92.980	59.579		I <--> II	
1090.000	92.885	150.780	92.980	63.001		TRANSITION	
1100	92.885	151.628	93.510	63.930	-778.665	-595.911	28.297
1200	92.885	159.710	98.694	73.219	-775.786	-579.425	25.222
1200.000	92.885	159.710	98.694	73.219		II <--> III	
1200.000	83.052	159.710	98.694	73.219		TRANSITION	
1300	83.052	166.358	103.647	81.524	-773.962	-563.136	22.627
1400	83.052	172.513	108.349	89.829	-772.208	-546.984	20.408
1500	83.052	178.243	112.820	98.135	-770.526	-530.956	18.490
1600	83.052	183.603	117.078	106.440	-768.921	-515.038	16.814
1700	83.052	188.638	121.141	114.745	-767.398	-499.217	15.339
1800	83.052	193.385	125.024	123.050	-765.965	-483.484	14.030
1900	83.052	197.875	128.741	131.355	-764.626	-467.827	12.861
2000	83.052	202.135	132.305	139.661	-763.386	-452.239	11.811
2100	83.052	206.188	135.728	147.966	-762.251	-436.710	10.863
2175.000	83.052	209.102	138.208	154.195	--------	III <--> LIQUID	-------
2200	83.052	210.051	139.019	156.271	-761.226	-421.232	10.001
2300	83.052	213.743	142.188	164.576	-760.320	-405.799	9.216
2400	83.052	217.278	145.244	172.882	-759.541	-390.402	8.497
2500	83.052	220.668	148.193	181.187	-758.906	-375.035	7.836
2600	83.052	223.925	151.044	189.492	-758.437	-359.690	7.226
2700	83.052	227.060	153.802	197.797	-758.172	-344.359	6.662
2800	83.052	230.080	156.472	206.103	-784.523	-328.549	6.129

PREVIOUS: CURRENT: December 1973

Niobium Oxide (NbO_2) Nb_1O_2(cr)

NIOBIUM OXIDE (NbO_2) LIQUID M_r = 124.9052

S°(298.15 K) = [92.835] J K^{-1} mol^{-1} $\Delta_fH°$(298.15 K) = [-710.872] kJ mol^{-1}

T_{fus} = 2175 ± 15 K $\Delta_{fus}H°$ = 92.048 ± 20.9 kJ mol^{-1}

Enthalpy of Formation

$\Delta_fH°(NbO_2$, ℓ, 298.15 K) is calculated from $\Delta_fH°(NbO_2$, cr, 298.15 K) by adding $\Delta_{fus}H°$ and the difference in enthalpy, H°(2175 K)-H(298.15 K), between the crystal and the liquid.

Heat Capacity and Entropy

There is no data reported in the literature concerning the heat capacity or enthalpy of NbO_2(ℓ). We estimate a constant value of $C_p°$ = 7.5 cal K^{-1}g-$atom^{-1}$ or 22.5 cal $K^{-1}mol^{-1}$ for the liquid phase. A glass transition is assumed at 1000 K below which the $C_p°$ values of the crystal are used. The entropy at 298.15 K is calculated in a manner analogous to that used in calculating the enthalpy of formation.

Fusion Data

Elliott (1) reported the melting point of NbO_2(cr) as T_{fus} = 2191 K (IPTS-68) and stated that NbO_2(cr) melts without decomposition. The vapor pressure study on NbO_2(cr, ℓ) by Shchukarev et al. (2) gave T_{fus} = 2173 K (IPTS-68) with congruent melting. We adopt T_{fus} = 2175 ± 15 K in order to maintain good consistency with the vapor pressure data. A value of T_{fus} = 2357 K reported by Kolchin and Sumarokova (3) appears too high.

The vapor pressure data of Shchukarev et al. (2) suggests $\Delta_{fus}H°$ = 22.0 kcal mol^{-1} which we adopt. This leads to $\Delta_{fus}S°$ = 10.11 cal K^{-1} mol^{-1}.

Vaporization Data

The vapors over NbO_2(ℓ) have been shown by Shchukarev et al. (2) to be NbO(g), NbO_2(g), and O(g). At 2300 K, the ratio of the vapor pressures NbO_2(g):NbO(g) is 1.00:0.07.

References

1. R. P. Elliott, Trans. Amer. Soc. Metals 52, 990 (1960).
2. S. A. Shchukarev, G. A. Semenov, and K. E. Frantseva, Russ. J. Inorg. Chem. 11, 129 (1966).
3 O. Kolchin and N. Sumarokova, Atomnaya Energ. 10, 168 (1961).

Niobium Oxide (NbO_2) Nb_1O_2(l)

Enthalpy Reference Temperature = T_r = 298.15 K Standard State Pressure = p° = 0.1 MPa

T/K	$C_p°$ (J $K^{-1}mol^{-1}$)	S° (J $K^{-1}mol^{-1}$)	$-[G°-H°(T_r)]/T$ (J $K^{-1}mol^{-1}$)	$H°-H°(T_r)$ (kJ mol^{-1})	$\Delta_fH°$ (kJ mol^{-1})	$\Delta_fG°$ (kJ mol^{-1})	Log K_f
0							
100							
200							
298.15	57.450	92.835	92.835	0.	-710.872	-666.514	116.771
300	57.605	93.191	92.836	0.106	-710.865	-666.239	116.003
400	63.547	110.656	95.177	6.191	-710.259	-651.442	85.070
500	67.739	125.296	99.776	12.760	-709.314	-636.842	66.530
600	71.651	137.993	105.109	19.730	-708.116	-622.456	54.190
700	75.555	149.331	110.631	27.090	-706.667	-608.290	45.391
800	79.496	159.677	116.124	34.843	-704.949	-594.351	38.807
900	83.479	169.270	121.502	42.991	-702.945	-580.643	33.700
1000	87.500	178.273	126.733	51.540	-700.637	-567.175	29.626
1000.000	87.500	178.273	126.733	51.540	GLASS <--> LIQUID TRANSITION		
1000.000	94.140	178.273	126.733	51.540			
1100	94.140	187.246	131.833	60.954	-697.553	-553.978	26.306
1200	94.140	195.437	136.797	70.368	-694.548	-541.059	23.552
1300	94.140	202.972	141.602	79.782	-691.616	-528.388	21.231
1400	94.140	209.949	146.237	89.196	-688.753	-515.940	19.250
1500	94.140	216.444	150.704	98.610	-685.962	-503.694	17.540
1600	94.140	222.520	155.005	108.024	-683.248	-491.632	16.050
1700	94.140	228.227	159.146	117.438	-680.617	-479.737	14.741
1800	94.140	233.608	163.134	126.852	-678.075	-467.994	13.581
1900	94.140	238.698	166.979	136.266	-675.627	-456.390	12.547
2000	94.140	243.526	170.686	145.680	-673.278	-444.913	11.620
2100	94.140	248.119	174.265	155.094	-671.034	-433.550	10.784
2175.000	94.140	251.423	176.869	162.154	-------- III <--> LIQUID -------		
2200	94.140	252.499	177.722	164.508	-668.901	-422.291	10.026
2300	94.140	256.683	181.065	173.922	-666.886	-411.128	9.337
2400	94.140	260.690	184.300	183.336	-664.998	-400.049	8.707
2500	94.140	264.533	187.433	192.750	-663.254	-389.046	8.129
2600	94.140	268.225	190.470	202.164	-661.676	-378.109	7.596
2700	94.140	271.778	193.416	211.578	-660.303	-367.229	7.104
2800	94.140	275.202	196.276	220.992	-685.546	-355.912	6.640
2900	94.140	278.505	199.055	230.406	-683.437	-344.177	6.199
3000	94.140	281.697	201.757	239.820	-681.347	-332.514	5.790

PREVIOUS: CURRENT: December 1973

NIOBIUM OXIDE (NbO_2) CRYSTAL–CRYSTAL–CRYSTAL–LIQUID M_r = 124.9052

0 to 1090 K	crystal, I
1090 to 1200 K	crystal, II
1200 to 2175 K	crystal, III
above 2175 K	liquid

Refer to the individual tables for details.

Enthalpy Reference Temperature = T_r = 298.15 K Standard State Pressure = p° = 0.1 MPa

T/K	C_p° (J K⁻¹mol⁻¹)	S° (J K⁻¹mol⁻¹)	−[G°−H°(T_r)]/T (J K⁻¹mol⁻¹)	H°−H°(T_r) (kJ mol⁻¹)	Δ_fH° (kJ mol⁻¹)	Δ_fG° (kJ mol⁻¹)	Log K_f
0	0.	0.	INFINITE	-9.271	-790.308	-790.308	INFINITE
100	21.143	11.275	96.297	-8.502	-793.224	-775.765	405.218
200	44.918	33.967	59.417	-5.090	-794.821	-757.542	197.850
298.15	57.452	54.506	54.506	0.	-794.960	-739.175	129.500
300	57.605	54.862	54.507	0.106	-794.954	-738.829	128.642
400	63.547	72.327	56.848	6.191	-794.348	-720.199	94.048
500	67.739	86.967	61.447	12.760	-793.403	-701.766	73.313
600	71.651	99.665	66.781	19.730	-792.205	-683.547	59.508
700	75.555	111.003	72.302	27.090	-790.755	-665.549	49.664
800	79.496	121.348	77.795	34.843	-789.037	-647.776	42.295
900	83.479	130.941	83.173	42.991	-787.033	-630.236	36.578
1000	87.500	139.945	88.405	51.540	-784.726	-612.934	32.016
1090.000	91.144	147.640	92.980	59.579		I <--> II	
1090.000	92.885	150.780	92.980	63.001		TRANSITION	
1100	92.885	151.628	93.510	63.930	-778.665	-595.911	28.297
1200	92.885	159.710	98.694	73.219	-775.786	-579.425	25.222
1200.000	92.885	159.710	98.694	73.219		II <--> III	
1200.000	83.052	159.710	98.694	73.219		TRANSITION	
1300	83.052	166.358	103.647	81.524	-773.962	-563.136	22.627
1400	83.052	172.513	108.349	89.829	-772.208	-546.984	20.408
1500	83.052	178.243	112.820	98.135	-770.526	-530.956	18.490
1600	83.052	183.603	117.078	106.440	-768.921	-515.038	16.814
1700	83.052	188.638	121.141	114.745	-767.398	-499.217	15.339
1800	83.052	193.385	125.024	123.050	-765.965	-483.484	14.030
1900	83.052	197.875	128.741	131.355	-764.626	-467.827	12.861
2000	83.052	202.135	132.305	139.661	-763.386	-452.239	11.811
2100	83.052	206.188	135.728	147.966	-762.251	-436.710	10.863
2175.000	83.052	209.102	138.208	154.195		III <--> LIQUID	
2175.000	94.140	251.423	138.208	246.243		TRANSITION	
2200	94.140	252.499	139.500	248.596	-668.901	-422.291	10.026
2300	94.140	256.683	144.505	258.010	-666.886	-411.128	9.337
2400	94.140	260.690	149.263	267.424	-664.998	-400.049	8.707
2500	94.140	264.533	153.798	276.838	-663.254	-389.046	8.129
2600	94.140	268.225	158.128	286.252	-661.676	-378.109	7.596
2700	94.140	271.778	162.272	295.666	-660.303	-367.229	7.104
2800	94.140	275.202	166.245	305.080	-685.546	-355.912	6.640
2900	94.140	278.505	170.059	314.494	-683.437	-344.177	6.199
3000	94.140	281.697	173.727	323.908	-681.347	-332.514	5.790

PREVIOUS: CURRENT: December 1973

NIOBIUM OXIDE (NbO_2) IDEAL GAS M_r = 124.9052

$S°(298.15\ K)$ = [272.921 ± 8.4] $J\ K^{-1}\ mol^{-1}$

$\Delta_fH°(0\ K)$ = -197.33 ± 20.9 $kJ\ mol^{-1}$

$\Delta_fH°(298.15\ K)$ = -200.00 ± 20.9 $kJ\ mol^{-1}$

Vibrational Frequencies and Degeneracies

ν, cm^{-1}
[970](1)
[300](1)
[930](1)

Ground State Quantum Weight: [2] σ = [2]

Point Group: [C_{2v}]

Bond Distance: Nb-O = [1.691] Å

Bond Angle: O-Nb-O = [110]°

Product of the Moments of Inertia: $I_AI_BI_C$ = [5.274350 x 10^{-115}] $g^3\ cm^6$

Enthalpy of Formation

Shchukarev et al. (1-4) measured the vapor pressures of NbO and NbO_2 over NbO(cr, ℓ) and NbO_2(cr, ℓ) by the effusion method coupled with a mass spectrometer. An analysis of the reported smoothed results is presented in the following table (4).

Reaction	T/K	$\Delta_rH°$(298.15 K), kcal mol^{-1} 2nd law	$\Delta_rH°$(298.15 K), kcal mol^{-1} 3rd law	Drift cal K^{-1} mol^{-1}	$\Delta_fH°$(298.15 K), kcal mol^{-1} 2nd law	$\Delta_fH°$(298.15 K), kcal mol^{-1} 3rd law
NbO_2(cr) = NbO_2(g)	1773-2175	142.0	143.4	0.69	-48.0	-46.6
NbO_2(ℓ) = NbO_2(g)	2175-2473	124.2	123.7	-0.22	-46.8	-46.3
2 NbO(cr) = NbO_2(g) + Nb(cr)	1773-2210	147.8	149.9	1.03	-52.8	-50.7
2 NbO(ℓ) = NbO_2(g) + Nb(cr)	2210-2473	119.7	111.1	-3.73	-41.3	-49.9

We adopt $\Delta_fH°(NbO_2$, g, 298.15 K) = -47.8 ± 5.0 kcal mol^{-1}. This value is an average value of the 2nd and 3rd law results for the sublimation and vaporization of NbO_2. This adopted $\Delta_fH°$(298.15 K) value leads to D_0^0 = 14.73 eV for the process NbO_2(g) = Nb(g) + 2 O(g). The D_0^0 value is 1.83 times as great as the dissociation energy of NbO(g) (8.05 eV, 2).

The earlier less complete studies by Shchukarev et al. (1-3) are in fair agreement with the above conclusions. Using a platinum strip to vaporize NbO_2(cr) in a mass spectrometer at temperatures of 1500 to 1880 K, Shchukarev et al. (1) reported a 2nd law sublimation heat, $\Delta_{sub}H°$(1700 K) = 142 ± 3 kcal mol^{-1}. Using auxiliary data (6), we calculate $\Delta_fH°(NbO_2$, g, 298.15 K) = -38.8 kcal mol^{-1}. The mass spectrometric investigation of the NbO(cr) evaporation by Shchukarev et al. (2) led to a reported value $\Delta_rH°$(1800 K) = 140 ± 3 kcal mol^{-1} for the process 2 NbO(cr) = Nb(cr) + NbO_2(g). Using auxiliary data (6), we calculate $\Delta_fH°(NbO_2$, g, 298.15 K) = -53.3 kcal mol^{-1}. NbO_2 pressures were obtained (1938-2122 K) by the differential effusion method using the radioactive tracer ^{95}Nb (3). The pressures agree within 5% of those obtained in the more complete study (4). Golubtsov et al. (5) also measured the vapor pressure of NbO_2(g) over NbO_2(cr) in the range 1489-1905 K by using labelled atoms in a Knudsen cell. The same authors also studied the dissociation of Nb_2O_5 in vacuo in the range 1432-1756 K, the process being Nb_2O_5(cr) → 2 NbO_2(g) + 1/2 O_2(g). These latter data (5) are not considered.

Heat Capacity and Entropy

The vibrational frequencies and geometry of NbO_2(g) are estimated based on existing and estimated data for related metal dioxides such as TiO_2(g), TaO_2(g), and WO_2(g) (6). The Nb-O bond distance is assumed to be the same as in NbO(g) (6). The ground state quantum weight of 2 is adopted so as to be consistent with TaO_2(g) (6). The principal moments of inertia are calculated as I_A = 3.7181 x 10^{-39}, I_B = 10.1954 x 10^{-39}, and I_C = 13.9136 x 10^{-39} g cm^2.

References

1. S. A. Shchukarev, G. A. Semenov, and K. E. Frantseva, Russ. J. Inorg. Chem. 4, 1217 (1959).
2. S. A. Shchukarev, G. A. Semenov, and K. E. Frantseva, Izv. Vyssh. Uchebn. Zaved. Khim. i Khim. Tekhnol. 5, 691 (1962).
3. S. A. Shchukarev, G. A. Semenov, and K. E. Frantseva, Dokl. Akad. Nauk SSSR 145, 119 (1962).
4. S. A. Shchukarev, G. A. Semenov, and K. E. Frantseva, Russ. J. Inorg. Chem. 11, 129 (1966).
5. I. V. Golubtsov, A. V. Lapitskii, and V. K. Shiryaev, Izv. Vyssh. Uchebn. Zavend. Khim. i Khim. Tekhnol. 3, 571 (1960); Chem. Abstr. 55, 1116b (1961).
6. JANAF Thermochemical Tables: NbO_2(cr), NbO(cr, g), Nb(cr), TaO_2(g) and TiO_2(g), 12-31-73; WO_2(g), 9-30-66.

Niobium Oxide (NbO_2) Nb_1O_2(g)

Enthalpy Reference Temperature = T_r = 298.15 K; Standard State Pressure = p° = 0.1 MPa

T/K	$C_p°$ (J $K^{-1}mol^{-1}$)	S° (J $K^{-1}mol^{-1}$)	-[G°-H°(T_r)]/T (J $K^{-1}mol^{-1}$)	H°-H°(T_r) (kJ mol^{-1})	$\Delta_fH°$ (kJ mol^{-1})	$\Delta_fG°$ (kJ mol^{-1})	Log K_f
0	0.	0.	INFINITE	-11.254	-197.325	-197.325	INFINITE
100	35.385	230.299	309.095	-7.880	-197.637	-202.080	105.556
200	39.819	256.260	276.805	-4.109	-198.875	-206.055	53.816
250	41.903	265.368	273.633	-2.066	-199.475	-207.779	43.413
298.15	43.907	272.921	272.921	0.	-199.995	-209.330	36.674
300	43.982	273.193	272.922	0.081	-200.014	-209.388	36.458
350	45.916	280.121	273.465	2.330	-200.488	-210.912	31.477
400	47.622	286.366	274.693	4.669	-200.905	-212.373	27.733
450	49.082	292.062	276.311	7.088	-201.280	-213.783	24.815
500	50.311	297.299	278.152	9.573	-201.625	-215.154	22.477
600	52.201	306.649	282.141	14.705	-202.266	-217.799	18.961
700	53.531	314.801	286.237	19.995	-202.886	-220.338	16.442
800	54.485	322.015	290.267	25.398	-203.517	-222.789	14.547
900	55.184	328.475	294.160	30.884	-204.176	-225.159	13.068
1000	55.708	334.317	297.888	36.429	-204.872	-227.453	11.881
1100	56.110	339.647	301.446	42.021	-205.609	-229.676	10.906
1200	56.424	344.543	304.836	47.649	-206.391	-231.829	10.091
1300	56.674	349.070	308.067	53.304	-207.217	-233.916	9.399
1400	56.875	353.277	311.147	58.982	-208.091	-235.937	8.803
1500	57.039	357.207	314.088	64.678	-209.018	-237.895	8.284
1600	57.175	360.893	316.900	70.388	-210.007	-239.788	7.828
1700	57.288	364.362	319.591	76.112	-211.067	-241.617	7.424
1800	57.384	367.640	322.170	81.846	-212.205	-243.382	7.063
1900	57.466	370.744	324.645	87.588	-213.428	-245.081	6.738
2000	57.536	373.694	327.025	93.338	-214.743	-246.713	6.443
2100	57.597	376.503	329.314	99.095	-216.157	-248.277	6.176
2200	57.649	379.183	331.521	104.857	-217.675	-249.771	5.930
2300	57.695	381.747	333.649	110.625	-219.306	-251.194	5.705
2400	57.736	384.203	335.705	116.396	-221.062	-252.544	5.496
2500	57.772	386.561	337.692	122.172	-222.956	-253.817	5.303
2600	57.804	388.827	339.616	127.951	-225.013	-255.012	5.123
2700	57.832	391.009	341.479	133.732	-227.272	-256.123	4.955
2800	57.858	393.113	343.286	139.517	-256.144	-256.662	4.788
2900	57.881	395.144	345.039	145.304	-257.662	-256.655	4.623
3000	57.902	397.106	346.742	151.093	-259.197	-256.593	4.468
3100	57.921	399.005	348.398	156.884	-260.749	-256.480	4.322
3200	57.938	400.845	350.008	162.677	-262.317	-256.318	4.184
3300	57.953	402.628	351.576	168.472	-263.901	-256.107	4.054
3400	57.967	404.358	353.103	174.268	-265.500	-255.846	3.931
3500	57.981	406.038	354.591	180.065	-267.113	-255.539	3.814
3600	57.992	407.672	356.043	185.864	-268.741	-255.185	3.703
3700	58.004	409.261	357.460	191.664	-270.382	-254.787	3.597
3800	58.014	410.808	358.844	197.465	-272.037	-254.342	3.496
3900	58.023	412.315	360.196	203.266	-273.705	-253.854	3.400
4000	58.032	413.784	361.517	209.069	-275.385	-253.325	3.308
4100	58.040	415.217	362.809	214.873	-277.077	-252.752	3.220
4200	58.047	416.616	364.074	220.677	-278.781	-252.138	3.136
4300	58.055	417.982	365.312	226.482	-280.497	-251.483	3.055
4400	58.061	419.317	366.524	232.288	-282.224	-250.789	2.977
4500	58.067	420.622	367.712	238.094	-283.963	-250.056	2.903
4600	58.073	421.898	368.876	243.901	-285.713	-249.283	2.831
4700	58.078	423.147	370.017	249.709	-287.476	-248.472	2.761
4800	58.083	424.370	371.137	255.517	-289.250	-247.623	2.695
4900	58.088	425.567	372.236	261.326	-291.036	-246.737	2.630
5000	58.092	426.741	373.314	267.135	-292.835	-245.815	2.568
5100	58.097	427.891	374.373	272.944	-294.647	-244.856	2.508
5200	58.101	429.020	375.413	278.754	-987.145	-234.553	2.356
5300	58.104	430.126	376.435	284.564	-989.316	-220.059	2.169
5400	58.108	431.212	377.439	290.375	-991.477	-205.529	1.988
5500	58.111	432.279	378.427	296.186	-993.725	-190.954	1.814
5600	58.114	433.326	379.398	301.997	-996.019	-176.337	1.645
5700	58.117	434.354	380.353	307.809	-998.359	-161.679	1.482
5800	58.120	435.365	381.293	313.621	-1000.745	-146.979	1.324
5900	58.123	436.359	382.218	319.433	-1003.180	-132.240	1.171
6000	58.126	437.336	383.128	325.245	-1005.664	-117.456	1.023

PREVIOUS: December 1973 (1 atm) CURRENT: December 1973 (1 bar)

Niobium Oxide (NbO_2) Nb_1O_2(g)

NIOBIUM OXIDE (Nb_2O_5) CRYSTAL M_r = 265.8098

$\Delta_f H°(0\ K) = -1889.616 \pm 4.2$ kJ mol^{-1}

$S°(298.15\ K) = 137.30 \pm 1.3$ J K^{-1} mol^{-1} $\Delta_f H°(298.15\ K) = -1899.536 \pm 4.2$ kJ mol^{-1}

$T_{fus} = 1785 \pm 30$ K $\Delta_{fus} H° = 104.265 \pm 2.1$ kJ mol^{-1}

Enthalpy of Formation

The $\Delta_f H°(Nb_2O_5$, cr, 298.15 K) investigations suffer from uncertainties concerning the polymorphic state of the samples employed and the oftentimes incomplete impurity analysis of the samples (1). There is considerable scatter in the following tabulation for the enthalpy of formation values, all of which are based on enthalpy of combustion studies.

Source	Nb(cr) purity, %	% completion*	$\Delta_f H°$(298.15 K), kcal mol^{-1}
(2)			-442.8
(3)	98.8	96.34 - 96.94	-459.9 ± 0.7**
(4)	99.69	97.63 - 99.38	-454.0 ± 0.6**
(5)			-472.6 ± 1.0
(6)	99.49		-454.8 ± 0.8
(7)			-458.6 ± 5.0
(8)	99.01	99.1 - 99.33	-455.1 ± 0.5
(9)	99.45	98.78 - 100	-454.4 ± 1.6
(10)	98.47 - 99.35	99.48 - 99.96	-453.5 ± 0.4
(11)	98.47 - 99.35	99.48 - 99.96	-453.5 ± 0.4
(12)			-456.9

* Percent completion refers to extent of reaction 2 Nb(cr) + 2.5 O_2(g) = Nb_2O_5(cr).

** Values adjusted by Kubaschewski and Catterall (13) to more satisfactorily correct for incomplete reaction.

The value chosen for $\Delta_f H°(Nb_2O_5$, cr, 298.15 K) is -454.0 ± 1 kcal mol^{-1}. This value is representative of the work by Humphrey (4), Kornilov et al. (10, 11), Huber et al. (9), Kusenko and Gel'd (8) and Morozova and Stolyarova (6). These works have discussed sample purity and have percentage conversions of Nb(cr) to Nb_2O_5(cr) of the order of 97% or better. Based on the discussion by Reisman and Holtzberg (1), this $\Delta_f H°$(298.15 K) value is for the high temperature α-phase of Nb_2O_5(cr). The heat capacity and enthalpy work, also dealing with Nb_2O_5(cr, α), indicated no phase changes for 53-297 K and 381-1785 K (14-16).

$\Delta_f G°$ values for Nb_2O_5(cr) in the range 1050-1300 K were calculated from measured emf values in an oxygen concentration cell and the known $\Delta_f G°$ values of Cr_2O_3 (17). The reported smoothed $\Delta_f G°$ values were treated by a 3rd law analysis, yielding $\Delta_f H°(Nb_2O_5$, cr, 298.15 K) = -451.6 kcal mol^{-1} with a drift equivalent to an entropy error of -4.8 cal K^{-1} mol^{-1}.

Heat Capacity and Entropy

King (14) measured the heat capacity of the high temperature α-phase Nb_2O_5 from 53.24 to 296.64 K and fitted the data (29 data points) with a combination of Debye and Einstein functions. These functions fit the data over the entire measured temperature range with a maximum deviation of 0.6% (14) and are used to calculate S°(50 K) = 2.42 cal K^{-1} mol^{-1}.

Orr (15), using the same material as King (14), measured the enthalpy from 381.4 to 1809.2 K. Gel'd and Kusenko (16) also measured the enthalpy from 461 to 1828 K. Combining these data (15, 16) with a portion of the King data (a $C_p°$ fit, then integrated), a six term polynomial fit is used to represent the enthalpy data, yielding a deviation of -0.99% to +0.81% for the Orr data and -0.64% to +1.10% for the Gel'd and Kusenko data. Two points (644 K, 734 K) of the Gel'd and Kusenko data are not used as they deviate from the polynomial-fitted values by 2-3%. Entahlpy values corresponding to temperatures greater than 1750 K but less than T_{fus} are considered to be involved in premelting (15). Enthalpy values for 1785 $\leq$ T $\leq$ 2300 K are obtained by extrapolating the values obtained from the polynomial fit. The chosen heat capacity values are obtained by differentiating the enthalpy polynomial fit for the range 298-2300 K. For temperatures 50-298 K, the King data were fit to a six term polynomial.

The existing experimental data (14, 15, 16) did not indicate any transitions other than a crystal-liquid transition. Lacking thermodynamic information for the other crystal phases of Nb_2O_5 (1), we assume that there is little difference in the heat capacity and enthalpy values for the various crystal phases. Thus the values tabulated here apply to all crystal Nb_2O_5 phases. The data for Ta_2O_5(cr) presented in these tables are treated in the same manner as discussed here so as to maintain consistency between Nb_2O_5(cr, ℓ) and Ta_2O_5(cr, ℓ).

Fusion Data

See the liquid table for details.

References

1. A. Reisman and F. Holtzberg, High Temp. Oxides 2, 217 (1970).
2. W. Muthmann, L. Weiss, and R. Reidelbauch, Ann. Chem. 355, 67, 82 (1907).
3. G. Becker and W. A. Roth, Z. Physik. Chem. 167A, 16 (1933).
4. G. L. Humphrey, J. Amer. Chem. Soc. 76, 978 (1954).
5. M. P. Morozova and L. L. Getskina, Zhur. Obshchei Khim. 29, 1049 (1959); J. Gen. Chem. USSR 29, 1019 (1959).
6. M. P. Morozova and T. A. Stolyarova, Zhur. Obshchei Khim. 30, 3848 (1960); J. Gen. Chem. USSR 30, 3808 (1960).
7. F. G. Kusenko and P. V. Gel'd, Izvest. Sib. Otdel. Akad. Nauk SSSR, No. 2, 46 (1960); cited by H. L. Shick, "Thermodynamics of Certain Refractory Compounds," Vol. I, Academic Press, New York, 1966.
8. F. G. Kusenko and P. V. Gel'd, Zhur. Obshchei Khim. 30, 3847 (1960); J. Gen. Chem. USSR 30, 3806 (1960).
9. E. J. Huber, Jr., E. L. Head, C. E. Holley, Jr., E. K. Storms, and N. H. Krikorian, J. Phys. Chem. 65, 1846 (1961).
10. A. N. Kornilov, V. Ya. Leonidov, and S. M. Skuratov, Vest. Moskov Univ. Khim., No. 6, 48 (1962); cited by H. L. Shick (7).
11. A. N. Kornilov, U. Ya. Leonidov, and S. M. Skuratov, Dokl. Akad. Nauk SSSR 144, 355 (1962); Proc. Acad. Sci. USSR 144, 427 (1962).
12. V. I. Lavrent'ev, Y. I. Gerasimov, and J. N. Rezukhina, Dokl. Akad. Nauk SSSR 136, 1372 (1961).
13. O. Kubaschewski and J. A. Catterall, "Thermochemical Data of Alloys," Pergamon Press, New York (1956).
14. E. G. King, J. Amer. Chem. Soc. 76, 3289 (1954).
15. R. L. Orr, J. Amer. Chem. Soc. 75, 2808 (1953).
16. P. V. Gel'd and F. G. Kusenko, Izvest. Akad. Nauk SSSR O. T. N. Met. i Top. 1960 (2), 79 (1960); C.A. 55, 16114 (1961), data extracted from the following: Ya. I. Gerasimov, A. N. Krestovnikov and A. S. Shakhov, "Chemical Thermodynamics in Nonferrous Metallurgy," Vol. III, Israel Program for Scientific Translations, Jerusalem, 1965.
17. T. Hiraoka, N. Sano, and Y. Matsushita, Trans. Iron Steel Inst. Jap. 11, 102 (1971); C.A. 75, 81056w (1971).

Niobium Oxide (Nb_2O_5) Nb_2O_5(cr)

Enthalpy Reference Temperature = T_r = 298.15 K Standard State Pressure = p° = 0.1 MPa

T/K	$C_p°$ (J $K^{-1}mol^{-1}$)	S° (J $K^{-1}mol^{-1}$)	$-[G°-H°(T_r)]/T$ (J $K^{-1}mol^{-1}$)	$H°-H°(T_r)$ (kJ mol^{-1})	$\Delta_f H°$ (kJ mol^{-1})	$\Delta_f G°$ (kJ mol^{-1})	Log K_f
0	0.	0.	INFINITE	-22.287	-1889.634	-1889.634	INFINITE
100	54.455	34.317	235.233	-20.092	-1896.262	-1853.855	968.355
200	105.922	89.529	148.666	-11.827	-1899.472	-1809.884	472.693
298.15	131.988	137.298	137.298	0.	-1899.536	-1765.817	309.364
300	132.320	138.115	137.300	0.244	-1899.519	-1764.987	307.312
400	144.992	178.096	142.666	14.172	-1898.035	-1720.341	224.653
500	153.841	211.441	153.178	29.131	-1895.852	-1676.159	175.107
600	160.737	240.126	165.334	44.875	-1893.233	-1632.460	142.118
700	166.013	265.318	177.855	61.225	-1890.331	-1589.225	118.589
800	170.038	287.760	190.216	78.035	-1887.259	-1546.418	100.971
900	173.117	307.972	202.195	95.199	-1884.086	-1504.003	87.290
1000	175.502	326.340	213.706	112.634	-1880.865	-1461.944	76.364
1100	177.360	343.157	224.720	130.280	-1877.632	-1420.207	67.440
1200	178.828	358.654	235.244	148.092	-1874.413	-1378.766	60.016
1300	179.991	373.016	245.296	166.035	-1871.225	-1337.591	53.745
1400	180.920	386.390	254.902	184.082	-1868.087	-1296.661	48.379
1500	181.665	398.898	264.090	202.213	-1865.024	-1255.953	43.736
1600	182.263	410.642	272.886	220.410	-1862.060	-1215.445	39.680
1700	182.745	421.707	281.318	238.661	-1859.220	-1175.119	36.107
1785.000	183.075	430.631	288.217	254.209	------ CRYSTAL <--> LIQUID -----		
1800	183.129	432.163	289.410	256.956	-1856.527	-1134.957	32.936
1900	183.435	442.073	297.187	275.284	-1854.000	-1094.939	30.102
2000	183.678	451.489	304.668	293.640	-1851.656	-1055.050	27.555
2100	183.862	460.455	311.875	312.018	-1849.512	-1015.273	25.254
2200	184.004	469.012	318.825	330.411	-1847.584	-975.592	23.164
2300	184.109	477.193	325.534	348.817	-1845.891	-935.996	21.257

PREVIOUS: CURRENT: December 1972

Niobium Oxide (Nb_2O_5) Nb_2O_5(cr)

NIOBIUM OXIDE (Nb_2O_5) LIQUID M_r = 265.8098

$S°(298.15\ K) = [171.488]\ J\ K^{-1}\ mol^{-1}$ $\Delta_f H°(298.15\ K) = [-1830.873]\ kJ\ mol^{-1}$

$T_{fus} = 1785 \pm 30\ K$ $\Delta_{fus}H° = 104.265 \pm 2.1\ kJ\ mol^{-1}$

Enthalpy of Formation

$\Delta_f H°(Nb_2O_5, \ell, 298.15\ K)$ is calculated from that of the crystal by adding $\Delta_{fus}H°$ and the difference in enthalpy, H°(1785 K)-H°(298.15 K), between the crystal and liquid.

Heat Capacity and Entropy

Orr (1) and Gel'd and Kusenko (2) reported enthalpy measurements at temperatures greater than T_{fus}. The work of Orr (1) indicates a $C_p°$ value of 57.90 cal K^{-1} mol^{-1} based on two enthalpy points in the liquid region. The three enthalpy points of Gel'd and Kusenko (2) deviate by +0.06%, -0.86%, and -0.37% from the linear representation of the Orr data. Thus, the constant $C_p°$ value of 57.90 cal K^{-1} mol^{-1} is chosen for the liquid phase. A glass transition is also chosen at 1200 K so as to ensure the proper thermodynamic relationship between crystal values and extrapolated liquid value. At temperatures below 1200 K, the heat capacity values of the crystal are used. The entropy at 298.15 K is calculated in a manner analogous to that used in calculating the enthalpy of formation.

Fusion Data

The enthalpy work of Orr (1) and Gel'd and Kusenko (2) provide information as to the heat of fusion. The heat capacity discussion for Nb_2O_5(cr) and $Nb_2O_5(\ell)$ provide details as to the fitting of the data in the crystal and liquid regions. The heat of fusion is then calculated to be $\Delta_{fus}H°$ = 24.31 ± 0.5 kcal mol^{-1} at a melting point of 1785 K [H°(1785 K) -H°(298.15 K) = -60.759 kcal mol^{-1} for the crystal]. This melting point, although it is consistent with the enthalpy work (1, 2), is 21 K greater than the value reported by Holtzberg et al. (3) and is greater than most T_{fus} values cited by Charlesworth (4) and Schneider (5). Reisman and Holtzberg (6) attributed this high value of 1785 K to an impure sample. The melting point, T_{fus}, is chosen to be 1785 K strictly on the basis of being consistent with the two sets of enthalpy data. More data are necessary to accurately define consistent values for T_{fus}, $\Delta_{fus}H°$, and enthalpy values in the liquid and crystal regions.

References

1. R. L. Orr, J. Amer. Chem. Soc. 75, 2808 (1953).
2. P. V. Gel'd and F. G. Kusenko, Izvest. Akad. Nauk SSSR O.T.N. Met. i Top. 1960 (2), 79 (1960); Chem. Abstr. 55, 16117 (1961), data extracted from the following: Ya. I. Gerasimov, A. N. Kustovnikov, and A. S. Shakhov, "Chemical Thermodynamics in Nonferrous Metallurgy," Vol. III, Israel Program for Scientific Translations, Jerusalem, 1965.
3. F. Holtzberg, A. Reisman, M. Berry, and M. Berkenblit, J. Amer. Chem. Soc. 79, 2039 (1957).
4. J. H. Charlesworth, AFML-TR-70-137; October, 1970
5. S. J. Schneider, U.S. Nat. Bur. Stand. Monograph 68, 1963.
6. A. Reisman and F. Holtzberg, High Temp. Oxides 2, 217 (1970).

Niobium Oxide (Nb_2O_5) $Nb_2O_5(l)$

Enthalpy Reference Temperature = T_r = 298.15 K Standard State Pressure = p° = 0.1 MPa

T/K	$C_p°$ (J K⁻¹mol⁻¹)	S° (J K⁻¹mol⁻¹)	-[G°-H°(T_r)]/T (J K⁻¹mol⁻¹)	H°-H°(T_r) (kJ mol⁻¹)	$\Delta_f H°$ (kJ mol⁻¹)	$\Delta_f G°$ (kJ mol⁻¹)	Log K_f
0							
100							
200							
298.15	131.988	171.488	171.488	0.	-1830.873	-1707.347	299.120
300	132.320	172.306	171.491	0.244	-1830.856	-1706.581	297.142
400	144.992	212.286	176.857	14.172	-1829.371	-1665.354	217.473
500	153.841	245.631	187.368	29.131	-1827.189	-1624.591	169.720
600	160.737	274.316	199.524	44.875	-1824.569	-1584.311	137.927
700	166.013	299.509	212.045	61.225	-1821.668	-1544.495	115.252
800	170.038	321.950	224.406	78.035	-1818.595	-1505.107	98.273
900	173.117	342.162	236.386	95.199	-1815.423	-1466.111	85.091
1000	175.502	360.530	247.896	112.634	-1812.202	-1427.470	74.563
1100	177.360	377.347	258.910	130.280	-1808.969	-1389.153	65.965
1200	178.828	392.845	269.434	148.093	-1805.749	-1351.131	58.813
1200.000	178.828	392.845	269.434	148.093	GLASS <--> LIQUID		
1200.000	242.254	392.845	269.434	148.093	TRANSITION		
1300	242.254	412.236	279.683	172.318	-1796.279	-1313.631	52.782
1400	242.254	430.189	289.800	196.544	-1786.963	-1276.855	47.640
1500	242.254	446.902	299.723	220.769	-1777.804	-1240.739	43.206
1600	242.254	462.537	309.416	244.994	-1768.813	-1205.229	39.347
1700	242.254	477.224	318.859	269.220	-1759.998	-1170.276	35.958
1785.000	242.254	489.043	326.684	289.811	CRYSTAL <--> LIQUID		
1800	242.254	491.070	328.045	293.445	-1751.374	-1135.837	32.961
1900	242.254	504.168	336.973	317.670	-1742.951	-1101.871	30.293
2000	242.254	516.594	345.646	341.896	-1734.738	-1068.343	27.902
2100	242.254	528.414	354.071	366.121	-1726.746	-1035.221	25.750
2200	242.254	539.684	362.253	390.347	-1718.985	-1002.472	23.802
2300	242.254	550.452	370.204	414.572	-1711.473	-970.074	22.031
2400	242.254	560.762	377.930	438.797	-1704.227	-937.996	20.415
2500	242.254	570.652	385.443	463.023	-1697.280	-906.214	18.934
2600	242.254	580.153	392.750	487.248	-1690.674	-874.703	17.573
2700	242.254	589.296	399.861	511.473	-1684.489	-843.437	16.317
2800	242.254	598.106	406.785	535.699	-1731.546	-811.424	15.137
2900	242.254	606.607	413.530	559.924	-1723.909	-778.699	14.026
3000	242.254	614.820	420.103	584.149	-1716.320	-746.234	12.993
3100	242.254	622.763	426.513	608.375	-1708.778	-714.021	12.031
3200	242.254	630.455	432.767	632.600	-1701.282	-682.053	11.133
3300	242.254	637.909	438.871	656.826	-1693.829	-650.319	10.294
3400	242.254	645.141	444.832	681.051	-1686.417	-618.807	9.507
3500	242.254	652.163	450.656	705.276	-1679.046	-587.515	8.768
3600	242.254	658.988	456.349	729.502	-1671.713	-556.432	8.074
3700	242.254	665.625	461.915	753.727	-1664.417	-525.554	7.419
3800	242.254	672.086	467.362	777.952	-1657.157	-494.869	6.802
3900	242.254	678.379	472.692	802.178	-1649.932	-464.376	6.220
4000	242.254	684.512	477.911	826.403	-1642.740	-434.070	5.668
4100	242.254	690.494	483.023	850.628	-1635.580	-403.942	5.146
4200	242.254	696.331	488.033	874.854	-1628.452	-373.988	4.651
4300	242.254	702.032	492.944	899.079	-1621.355	-344.203	4.181
4400	242.254	707.601	497.759	923.304	-1614.289	-314.585	3.735
4500	242.254	713.045	502.483	947.530	-1607.253	-285.126	3.310
4600	242.254	718.370	507.119	971.755	-1600.248	-255.824	2.905
4700	242.254	723.580	511.669	995.981	-1593.273	-226.671	2.519
4800	242.254	728.680	516.137	1020.206	-1586.330	-197.668	2.151
4900	242.254	733.675	520.526	1044.431	-1579.418	-168.808	1.800
5000	242.254	738.569	524.838	1068.657	-1572.540	-140.089	1.464

PREVIOUS: CURRENT: December 1972

Niobium Oxide (Nb_2O_5) $Nb_2O_5(l)$

NIOBIUM OXIDE (Nb_2O_5) CRYSTAL-LIQUID M_r = 265.8098

0 to 1785 K crystal
above 1785 K liquid

Refer to the individual tables for details.

Niobium Oxide (Nb_2O_5) Nb_2O_5(cr,l)

Enthalpy Reference Temperature = T_r = 298.15 K Standard State Pressure = p° = 0.1 MPa

T/K	C_p° (J K⁻¹mol⁻¹)	S° (J K⁻¹mol⁻¹)	$-[G^\circ - H^\circ(T_r)]/T$ (J K⁻¹mol⁻¹)	$H^\circ - H^\circ(T_r)$ (kJ mol⁻¹)	$\Delta_f H^\circ$ (kJ mol⁻¹)	$\Delta_f G^\circ$ (kJ mol⁻¹)	Log K_f
0	0.	0.	INFINITE	-22.287	-1889.634	-1889.634	INFINITE
100	54.455	34.317	235.233	-20.092	-1896.262	-1853.855	968.355
200	105.922	89.529	148.666	-11.827	-1899.472	-1809.884	472.693
298.15	131.988	137.298	137.298	0.	-1899.536	-1765.817	309.364
300	132.320	138.115	137.300	0.244	-1899.519	-1764.987	307.312
400	144.992	178.096	142.666	14.172	-1898.035	-1720.341	224.653
500	153.841	211.441	153.178	29.131	-1895.852	-1676.159	175.107
600	160.737	240.126	165.334	44.875	-1893.233	-1632.460	142.118
700	166.013	265.318	177.855	61.225	-1890.331	-1589.225	118.589
800	170.038	287.760	190.216	78.035	-1887.259	-1546.418	100.971
900	173.117	307.972	202.195	95.199	-1884.086	-1504.003	87.290
1000	175.502	326.340	213.706	112.634	-1880.865	-1461.944	76.364
1100	177.360	343.157	224.720	130.280	-1877.632	-1420.207	67.440
1200	178.828	358.654	235.244	148.092	-1874.413	-1378.766	60.016
1300	179.991	373.016	245.296	166.035	-1871.225	-1337.591	53.745
1400	180.920	386.390	254.902	184.082	-1868.087	-1296.661	48.379
1500	181.665	398.898	264.090	202.213	-1865.024	-1255.953	43.736
1600	182.263	410.642	272.886	220.410	-1862.060	-1215.445	39.680
1700	182.745	421.707	281.318	238.661	-1859.220	-1175.119	36.107
1785.000	183.075	430.631	288.217	254.209	CRYSTAL <--> LIQUID	TRANSITION	
1785.000	242.254	489.043	288.217	358.475			
1800	242.254	491.070	289.899	362.108	-1751.374	-1135.837	32.961
1900	242.254	504.168	300.835	386.334	-1742.951	-1101.871	30.293
2000	242.254	516.594	311.315	410.559	-1734.738	-1068.343	27.902
2100	242.254	528.414	321.374	434.784	-1726.746	-1035.221	25.750
2200	242.254	539.684	331.043	459.010	-1718.985	-1002.472	23.802
2300	242.254	550.452	340.350	483.235	-1711.473	-970.074	22.031
2400	242.254	560.762	349.321	507.461	-1704.227	-937.996	20.415
2500	242.254	570.652	357.977	531.686	-1697.280	-906.214	18.934
2600	242.254	580.153	366.341	555.911	-1690.674	-874.703	17.573
2700	242.254	589.296	374.430	580.137	-1684.489	-843.437	16.317
2800	242.254	598.106	382.262	604.362	-1731.546	-811.424	15.137
2900	242.254	606.607	389.853	628.587	-1723.909	-778.699	14.026
3000	242.254	614.820	397.216	652.813	-1716.320	-746.234	12.993
3100	242.254	622.763	404.364	677.038	-1708.778	-714.021	12.031
3200	242.254	630.455	411.310	701.263	-1701.282	-682.053	11.133
3300	242.254	637.909	418.064	725.489	-1693.829	-650.319	10.294
3400	242.254	645.141	424.637	749.714	-1686.417	-618.807	9.507
3500	242.254	652.163	431.038	773.940	-1679.046	-587.515	8.768
3600	242.254	658.988	437.275	798.165	-1671.713	-556.432	8.074
3700	242.254	665.625	443.358	822.390	-1664.417	-525.554	7.419
3800	242.254	672.086	449.292	846.616	-1657.157	-494.869	6.802
3900	242.254	678.379	455.086	870.841	-1649.932	-464.376	6.220
4000	242.254	684.512	460.745	895.066	-1642.740	-434.070	5.668
4100	242.254	690.494	466.276	919.292	-1635.580	-403.942	5.146
4200	242.254	696.331	471.685	943.517	-1628.452	-373.988	4.651
4300	242.254	702.032	476.975	967.742	-1621.355	-344.203	4.181
4400	242.254	707.601	482.154	991.968	-1614.289	-314.585	3.735
4500	242.254	713.045	487.225	1016.193	-1607.253	-285.126	3.310
4600	242.254	718.370	492.192	1040.418	-1600.248	-255.824	2.905
4700	242.254	723.580	497.060	1064.644	-1593.273	-226.671	2.519
4800	242.254	728.680	501.832	1088.869	-1586.330	-197.668	2.151
4900	242.254	733.675	506.513	1113.095	-1579.418	-168.808	1.800
5000	242.254	738.569	511.105	1137.320	-1572.540	-140.089	1.464

PREVIOUS: CURRENT: December 1972

Niobium Oxide (Nb_2O_5) Nb_2O_5(cr,l)

NEON (Ne) REFERENCE STATE $A_r = 20.179$

0 to 6000 K Ideal Monatomic Gas

IP(Ne, g) = 173929.70 ± 0.1 cm^{-1}
S°(298.15 K) = 146.327 ± 0.003 $J\ K^{-1}\ mol^{-1}$

$\Delta_f H°(0\ K) = 0\ kJ\ mol^{-1}$
$\Delta_f H°(298.15\ K) = 0\ kJ\ mol^{-1}$

Electronic Levels and Quantum Weights

State	ε_i, cm^{-1}	g_i
$^1S_0$	0	1

Enthalpy of Formation

Zero by definition.

Heat Capacity and Entropy

Information on the electronic energy levels and quantum weights is taken from Moore (1, 9). All predicted levels have been observed for n=2 and 3 but above that many predicted levels are missing. Our calculations indicate that any reasonable method of filling in these missing levels and cutting off the summation in the partition function (2) has no effect on the thermodynamic properties to 6000 K. This is undoubtedly a result of the high energy of these levels; the first excited level is over 134000 cm^{-1} above the ground state. Therefore, we list the ground state only. Extension to higher temperatures may require consideration of excited states and utilization of different fill and cutoff procedures (2).

The thermodynamic functions at 298.15 K agree exactly with recent CODATA recommendations (3) except for two minor differences. First, the entropy differs by 0.1094 $J\ K^{-1}\ mol^{-1}$ because this table uses a standard-state pressure of 1 bar, whereas the CODATA recommendations are based on 1 atm. Second, entropy differences of the order of 0.001-0.004 $J\ K^{-1}\ mol^{-1}$ for the rare gases arise due to the use of slightly different values for R; this table uses R = 8.31441 $J\ K^{-1}\ mol^{-1}$. Considering these minor changes, this table agrees within the estimated uncertainty with those by Hultgren et al. (4), McBride et al. (5), Gurvich et al. (8) and Wagman et al. (12). The estimated uncertainty is due to uncertainties in the relative atomic mass and fundamental constants which in this tabulation are based on the 1981 scale (6) and the 1973 values (7), respectively.

Phase Data

The triple point, 24.561 K, and boiling point, 27.102 K, are a secondary reference point and a defining fixed-point, respectively, of IPTS-68 (10, 11). Hultgren et al. (4) had recommended a triple point of 24.533 K (0.4277 atm) and a boiling point of 27.096 K (1 atm). These values are provided for the convenience of the reader and have not been evaluated by the present authors. As a result of these low values, the reference state for neon is chosen to be the ideal gas at all temperatures. This may differ from the choice of other authors.

References

1. C. E. Moore, U. S. Nat. Bur. Stand., NSRDS-NBS-35, Volume I, 1970 [Reprint of NBS Circular 467, Volume I, 1949].
2. J. R. Downey, Jr., The Dow Chemical Co., AFOSR-TR-78-0960, Contract No. F44620-75-1-0048, 1978.
3. J. D. Cox, chairman, ICSU-CODATA Task Group on Key Values for Thermodynamics, J. Chem. Thermodynamics 10, 903 (1978).
4. R. Hultgren, P. D. Desai et al., "Selected Values of the Thermodynamic Properties of the Elements", American Society for Metals, Metals Park, Ohio, 1973.
5. B. J. McBride, S. Heimel, J. G. Ehlers and S. Gordon, NASA SP-3001, 1973.
6. N. E. Holden and R. L. Martin, Pure Appl. Chem. 51, 405 (1979).
7. E. R. Cohen and B. N. Taylor, J. Phys. Chem. Ref. Data 2, 663 (1973).
8. L. V. Gurvich, I. V. Veits et al., "Thermodynamic Properties of Individual Substances," 3rd ed., Vol. I, Nauka, Moscow, 1978.
9. C. E. Moore, U. S. Nat. Bur. Stand., NSRDS-NBS-34, 1970.
10. H. Preston-Thomas, Metrologia 12, 7 (1976).
11. L. Crovini, R. E. Bedford and A. Moser, Metrologia 13, 197 (1977).
12. D. D. Wagman, W. H. Evans et al., J. Phys. Chem. Ref. Data 11, Supp. 2 (1982).

Neon (Ne) Ne_1(ref)

Enthalpy Reference Temperature = T_r = 298.15 K Standard State Pressure = p° = 0.1 MPa

	J $K^{-1}mol^{-1}$			kJ mol^{-1}			
T/K	$C_p°$	S°	$-[G°-H°(T_r)]/T$	$H°-H°(T_r)$	$\Delta_f H°$	$\Delta_f G°$	Log K_f
0	0.	0.	INFINITE	-6.197	0.	0.	0.
100	20.786	123.620	164.808	-4.119	0.	0.	0.
200	20.786	138.028	148.229	-2.040	0.	0.	0.
250	20.786	142.666	146.670	-1.001	0.	0.	0.
298.15	20.786	146.327	146.327	0.	0.	0.	0.
300	20.786	146.456	146.328	0.038	0.	0.	0.
350	20.786	149.660	146.581	1.078	0.	0.	0.
400	20.786	152.436	147.143	2.117	0.	0.	0.
450	20.786	154.884	147.870	3.156	0.	0.	0.
500	20.786	157.074	148.683	4.196	0.	0.	0.
600	20.786	160.864	150.407	6.274	0.	0.	0.
700	20.786	164.068	152.135	8.353	0.	0.	0.
800	20.786	166.843	153.804	10.431	0.	0.	0.
900	20.786	169.292	155.392	12.510	0.	0.	0.
1000	20.786	171.482	156.893	14.589	0.	0.	0.
1100	20.786	173.463	158.311	16.667	0.	0.	0.
1200	20.786	175.271	159.650	18.746	0.	0.	0.
1300	20.786	176.935	160.916	20.824	0.	0.	0.
1400	20.786	178.476	162.116	22.903	0.	0.	0.
1500	20.786	179.910	163.255	24.982	0.	0.	0.
1600	20.786	181.251	164.339	27.060	0.	0.	0.
1700	20.786	182.511	165.371	29.139	0.	0.	0.
1800	20.786	183.699	166.356	31.217	0.	0.	0.
1900	20.786	184.823	167.299	33.296	0.	0.	0.
2000	20.786	185.889	168.202	35.375	0.	0.	0.
2100	20.786	186.904	169.069	37.453	0.	0.	0.
2200	20.786	187.871	169.902	39.532	0.	0.	0.
2300	20.786	188.795	170.703	41.610	0.	0.	0.
2400	20.786	189.679	171.475	43.689	0.	0.	0.
2500	20.786	190.528	172.221	45.768	0.	0.	0.
2600	20.786	191.343	172.941	47.846	0.	0.	0.
2700	20.786	192.127	173.637	49.925	0.	0.	0.
2800	20.786	192.883	174.311	52.004	0.	0.	0.
2900	20.786	193.613	174.964	54.082	0.	0.	0.
3000	20.786	194.317	175.597	56.161	0.	0.	0.
3100	20.786	194.999	176.212	58.239	0.	0.	0.
3200	20.786	195.659	176.810	60.318	0.	0.	0.
3300	20.786	196.299	177.391	62.397	0.	0.	0.
3400	20.786	196.919	177.956	64.475	0.	0.	0.
3500	20.786	197.522	178.506	66.554	0.	0.	0.
3600	20.786	198.107	179.043	68.632	0.	0.	0.
3700	20.786	198.677	179.566	70.711	0.	0.	0.
3800	20.786	199.231	180.076	72.790	0.	0.	0.
3900	20.786	199.771	180.574	74.868	0.	0.	0.
4000	20.786	200.297	181.061	76.947	0.	0.	0.
4100	20.786	200.811	181.536	79.025	0.	0.	0.
4200	20.786	201.311	182.001	81.104	0.	0.	0.
4300	20.786	201.801	182.456	83.183	0.	0.	0.
4400	20.786	202.278	182.901	85.261	0.	0.	0.
4500	20.786	202.745	183.337	87.340	0.	0.	0.
4600	20.786	203.202	183.764	89.418	0.	0.	0.
4700	20.786	203.649	184.182	91.497	0.	0.	0.
4800	20.786	204.087	184.592	93.576	0.	0.	0.
4900	20.786	204.516	184.994	95.654	0.	0.	0.
5000	20.786	204.936	185.389	97.733	0.	0.	0.
5100	20.786	205.347	185.776	99.811	0.	0.	0.
5200	20.786	205.751	186.157	101.890	0.	0.	0.
5300	20.786	206.147	186.530	103.969	0.	0.	0.
5400	20.786	206.535	186.897	106.047	0.	0.	0.
5500	20.786	206.917	187.257	108.126	0.	0.	0.
5600	20.786	207.291	187.612	110.204	0.	0.	0.
5700	20.786	207.659	187.960	112.283	0.	0.	0.
5800	20.786	208.021	188.303	114.362	0.	0.	0.
5900	20.786	208.376	188.640	116.440	0.	0.	0.
6000	20.786	208.725	188.972	118.519	0.	0.	0.

PREVIOUS: March 1977 (1 atm) CURRENT: March 1982 (1 bar)

Neon (Ne) Ne_1(ref)

NEON, ION (Ne^+) IDEAL GAS $M_r = 20.17845$

$IP(Ne^+, g) = 330391.0 \pm 0.1\ cm^{-1}$
$S°(298.15\ K) = 158.307 \pm 0.003\ J\ K^{-1}\ mol^{-1}$

$\Delta_f H°(0\ K) = 2080.662 \pm 0.001\ kJ\ mol^{-1}$
$\Delta_f H°(298.15\ K) = [2086.966]\ kJ\ mol^{-1}$

Electronic Levels and Quantum Weights

State	ε_i, cm^{-1}	g_i
$^2P_{3/2}$	0	4
$^2P_{1/2}$	780.45	2

Enthalpy of Formation

The ionization limit of neutral neon (173929.70±0.1 cm^{-1}) reported by Moore (1] is adopted as $\Delta_f H°(0\ K)$ for $Ne^+(g)$. The ionization limit is converted from cm^{-1} to kJ mol^{-1} using the factor, 1 cm^{-1} = 0.01196266 kJ mol^{-1}, which is derived from the 1973 CODATA fundamental constants (2). The uncertainty in the ionization limit is estimated to be ±0.1 cm^{-1} which corresponds to an uncertainty of ±0.001 kJ mol^{-1} in the enthalpy of formation. Rosenstock et al. (9) and Levin and Lias (11) have summarized additional ionization potential and appearance potential data. Gurvich et al. (8) and Wagman et al. (10) adopted the same ionization potential, but the use of slightly different fundamental constants by Wagman et al. (10) resulted in a enthalpy of formation difference of 0.017 kJ mol^{-1}.

$\Delta_f H°(Ne^+, g, 298.15\ K)$ is obtained from $\Delta_f H°(Ne, g, 0\ K)$ by using IP(Ne) with JANAF (3) enthalpies $H°(0\ K)-H°(298.15\ K)$ for $Ne^+(g)$, Ne(r), and $e^-(r)$. $\Delta_r H°(Ne \rightarrow Ne^+ + e^-, 298.15\ K)$ differs from a room temperature threshold energy due to inclusion of these enthalpies and to threshold effects discussed by Rosenstock et al. (9). $\Delta_f H°(298.15\ K)$ should be changed by -6.197 kJ mol^{-1} if it is to be used in the ion convention that excludes the enthalpy of the electron.

Heat Capacity and Entropy

The information on electronic energy levels and quantum weights given by Moore (4) is incomplete because many theoretically predicted levels have not been observed. Our calculations indicate that any reasonable method of filling in these missing levels and cutting off the summation in the partition function (5) has no effect on the thermodynamic functions to 6000 K. This is a result of the high energy of all levels other than the ground state and the $^2P_{1/2}$ level; the next lowest level is over 217000 cm^{-1} above the ground state. Since inclusion of these upper levels has no effect on the thermodynamic functions (to 6000 K), we list only the ground state and the $^2P_{1/2}$ state, with the energy of the latter state taken from a more recent study by Moore (1). The reported uncertainty in S°(298.15 K) is due to uncertainties in the relative ionic mass and fundamental constants. Extension of these calculations above 6000 K may require consideration of the higher excited states and use of different fill and cutoff procedures (5).

The thermodynamic functions reported here agree with those of Green et al. (6), Hilsenrath et al. (7), and Gurvich et al. (8) except for one or two minor differences. First, the entropy differs by 0.1094 J K^{-1} mol^{-1} because this table uses a standard-state pressure of 1 bar, whereas the cited references used a pressure of 1 atm. Second, smaller differences arise from the use of slightly different values for the fundamental constants, the relative ionic mass, and the position of the $^2P_{1/2}$ electronic level.

References

1. C. E. Moore, U. S. Nat. Bur. Stand., NSRDS-NBS-34, 1970.
2. E. R. Cohen and B. N. Taylor, J. Phys. Chem. Ref. Data 2, 663 (1973).
3. JANAF Thermochemical Tables: Ne(r), 3-31-82; e⁻(r), 3-31-82.
4. C. E. Moore, U. S. Nat. Bur. Stand., NSRDS-NBS-35, Volume I, 1970 [Reprint of NBS Circular 467, Volume I, 1949].
5. J. R. Downey, Jr., The Dow Chemical Co., AFOSR-TR-78-0960, Contract No. F44620-75-1-0048, 1978.
6. J. W. Green, D. E. Poland, and J. L. Margrave, University of Wisconsin, Report AD-275542, 1961.
7. J. Hilsenrath, C. G. Messina, and W. H. Evans, U.S. Nat. Bur. Stand., Report AD-606163, 1964.
8. L. V. Gurvich, I. V. Veits et al., "Thermodynamic Properties of Individual Substances", 3rd ed., Vol. 1, Nauka, Moscow, 1978.
9. H. M. Rosenstock, K. Draxl et al, J. Phys. Chem. Ref. Data 6, Supp. 1 (1977).
10. D. D. Wagman, W. H. Evans et al., J. Phys. Chem. Ref. Data 11, Supp. 2 (1982).
11. R. D. Levin and S. G. Lias, U. S. Nat. Bur. Stand., NSRDS-NBS-71, 1982.

Neon, Ion (Ne^+) $Ne_1^+(g)$

Enthalpy Reference Temperature = T_r = 298.15 K Standard State Pressure = p° = 0.1 MPa

	J $K^{-1}mol^{-1}$			kJ mol^{-1}			
T/K	$C_p°$	S°	$-[G°-H°(T_r)]/T$	$H°-H°(T_r)$	$\Delta_f H°$	$\Delta_f G°$	Log K_f
0	0.	0.	INFINITE	-6.304	2080.662		
100	20.793	135.147	177.401	-4.225			
200	21.262	149.654	160.304	-2.130			
250	21.715	154.447	158.669	-1.056			
298.15	22.119	158.307	158.307	0.	2086.966	2077.139	-363.906
300	22.133	158.444	158.307	0.041	2087.007	2077.078	-361.651
350	22.448	161.881	158.578	1.156	2088.122	2075.336	-309.727
400	22.650	164.893	159.183	2.284	2089.250	2073.432	-270.762
450	22.755	167.567	159.969	3.419	2090.385	2071.387	-240.440
500	22.788	169.967	160.851	4.558	2091.524	2069.215	-216.169
600	22.718	174.118	162.727	6.835	2093.801	2064.539	-179.734
700	22.562	177.609	164.610	9.099	2096.065	2059.483	-153.681
800	22.382	180.610	166.427	11.346	2098.312	2054.103	-134.119
900	22.207	183.236	168.152	13.576	2100.542	2048.443	-118.889
1000	22.048	185.567	169.779	15.788	2102.754	2042.536	-106.691
1100	21.907	187.662	171.311	17.986	2104.952	2036.408	-96.701
1200	21.784	189.562	172.754	20.170	2107.136	2030.080	-88.367
1300	21.678	191.302	174.115	22.343	2109.309	2023.570	-81.308
1400	21.586	192.905	175.400	24.506	2111.472	2016.894	-75.251
1500	21.507	194.391	176.617	26.661	2113.627	2010.063	-69.997
1600	21.438	195.777	177.772	28.808	2115.774	2003.089	-65.394
1700	21.378	197.075	178.870	30.949	2117.915	1995.980	-61.329
1800	21.325	198.295	179.915	33.084	2120.050	1988.746	-57.712
1900	21.279	199.447	180.913	35.214	2122.180	1981.393	-54.472
2000	21.239	200.537	181.868	37.340	2124.306	1973.928	-51.554
2100	21.203	201.573	182.781	39.462	2126.428	1966.357	-48.910
2200	21.171	202.558	183.658	41.580	2128.546	1958.685	-46.505
2300	21.142	203.499	184.501	43.696	2130.662	1950.917	-44.307
2400	21.117	204.398	185.311	45.809	2132.775	1943.057	-42.290
2500	21.094	205.260	186.092	47.919	2134.885	1935.108	-40.432
2600	21.073	206.086	186.845	50.028	2136.994	1927.076	-38.715
2700	21.054	206.881	187.573	52.134	2139.100	1918.962	-37.125
2800	21.037	207.647	188.276	54.239	2141.205	1910.770	-35.646
2900	21.022	208.385	188.957	56.341	2143.307	1902.503	-34.268
3000	21.008	209.097	189.616	58.443	2145.409	1894.164	-32.980
3100	20.995	209.786	190.256	60.543	2147.509	1885.754	-31.775
3200	20.983	210.452	190.877	62.642	2149.608	1877.277	-30.643
3300	20.972	211.098	191.480	64.740	2151.706	1868.734	-29.580
3400	20.962	211.724	192.066	66.836	2153.802	1860.128	-28.577
3500	20.953	212.331	192.636	68.932	2155.898	1851.460	-27.632
3600	20.945	212.921	193.192	71.027	2157.993	1842.732	-26.737
3700	20.937	213.495	193.733	73.121	2160.087	1833.946	-25.891
3800	20.930	214.053	194.260	75.215	2162.181	1825.103	-25.088
3900	20.923	214.597	194.775	77.307	2164.273	1816.206	-24.325
4000	20.916	215.126	195.277	79.399	2166.365	1807.254	-23.600
4100	20.910	215.643	195.767	81.490	2168.456	1798.250	-22.910
4200	20.905	216.147	196.246	83.581	2170.547	1789.196	-22.252
4300	20.900	216.639	196.715	85.671	2172.637	1780.091	-21.624
4400	20.895	217.119	197.173	87.761	2174.727	1770.938	-21.024
4500	20.890	217.589	197.622	89.850	2176.816	1761.737	-20.450
4600	20.886	218.048	198.061	91.939	2178.905	1752.490	-19.900
4700	20.882	218.497	198.491	94.028	2180.994	1743.198	-19.373
4800	20.878	218.936	198.912	96.116	2183.082	1733.861	-18.868
4900	20.875	219.367	199.325	98.203	2185.169	1724.481	-18.383
5000	20.872	219.789	199.730	100.291	2187.257	1715.058	-17.917
5100	20.868	220.202	200.128	102.378	2189.344	1705.593	-17.469
5200	20.865	220.607	200.518	104.464	2191.430	1696.087	-17.037
5300	20.863	221.004	200.900	106.551	2193.517	1686.542	-16.622
5400	20.860	221.394	201.276	108.637	2195.603	1676.957	-16.221
5500	20.857	221.777	201.646	110.723	2197.689	1667.333	-15.835
5600	20.855	222.153	202.009	112.808	2199.774	1657.671	-15.462
5700	20.853	222.522	202.365	114.894	2201.860	1647.972	-15.102
5800	20.850	222.885	202.716	116.979	2203.945	1638.237	-14.754
5900	20.848	223.241	203.061	119.064	2206.030	1628.465	-14.417
6000	20.846	223.591	203.400	121.149	2208.115	1618.658	-14.092

PREVIOUS: March 1977 (1 atm) CURRENT: March 1982 (1 bar)

Neon, Ion (Ne^+) $Ne_1^+(g)$

NICKEL (Ni)

REFERENCE STATE

0 to 1728	K	crystal
1728 to 3156.584	K	liquid
above 3156.584	K	ideal monatomic gas

Refer to the individual tables for details.

A_r = 58.69 Nickel (Ni) Ni_1(ref)

Enthalpy Reference Temperature = T_r = 298.15 K   Standard State Pressure = p° = 0.1 MPa

T/K	C_p° (J K⁻¹mol⁻¹)	S° (J K⁻¹mol⁻¹)	−[G°−H°(T_r)]/T (J K⁻¹mol⁻¹)	H°−H°(T_r) (kJ mol⁻¹)	$\Delta_f H^\circ$ (kJ mol⁻¹)	$\Delta_f G^\circ$ (kJ mol⁻¹)	Log K_f
0	0.	0.	INFINITE	-4.786	0.	0.	0.
100	13.631	7.454	50.235	-4.278	0.	0.	0.
200	22.468	20.200	32.143	-2.389	0.	0.	0.
298.15	25.987	29.870	29.870	0.	0.	0.	0.
300	26.024	30.031	29.871	0.048	0.	0.	0.
400	28.493	37.863	30.922	2.776	0.	0.	0.
500	31.045	44.473	32.986	5.743	0.	0.	0.
600	34.853	50.440	35.404	9.022	0.	0.	0.
631.000	39.832	52.263	36.187	10.144	Cp LAMBDA MAXIMUM		
631.000	39.832	52.263	36.187	10.144	TRANSITION		
700	30.794	55.575	37.941	12.344	0.	0.	0.
800	31.003	59.697	40.408	15.431	0.	0.	0.
900	31.589	63.382	42.759	18.560	0.	0.	0.
1000	32.217	66.742	44.992	21.750	0.	0.	0.
1100	32.928	69.845	47.112	25.007	0.	0.	0.
1200	33.681	72.742	49.128	28.336	0.	0.	0.
1300	34.518	75.471	51.051	31.746	0.	0.	0.
1400	35.397	78.061	52.888	35.241	0.	0.	0.
1500	36.317	80.534	54.649	38.827	0.	0.	0.
1600	37.279	82.908	56.342	42.506	0.	0.	0.
1700	38.284	85.199	57.972	46.286	0.	0.	0.
1728.000	38.535	85.827	58.419	47.361	CRYSTAL <--> LIQUID		
1728.000	38.911	95.754	58.419	64.516	TRANSITION		
1800	38.911	97.343	59.944	67.317	0.	0.	0.
1900	38.911	99.446	61.968	71.208	0.	0.	0.
2000	38.911	101.442	63.893	75.099	0.	0.	0.
2100	38.911	103.341	65.726	78.991	0.	0.	0.
2200	38.911	105.151	67.477	82.882	0.	0.	0.
2300	38.911	106.881	69.153	86.773	0.	0.	0.
2400	38.911	108.537	70.760	90.664	0.	0.	0.
2500	38.911	110.125	72.303	94.555	0.	0.	0.
2600	38.911	111.651	73.787	98.446	0.	0.	0.
2700	38.911	113.120	75.217	102.337	0.	0.	0.
2800	38.911	114.535	76.596	106.228	0.	0.	0.
2900	38.911	115.900	77.928	110.120	0.	0.	0.
3000	38.911	117.219	79.216	114.011	0.	0.	0.
3100	38.911	118.495	80.462	117.902	0.	0.	0.
3156.584	38.911	119.199	81.151	120.104	LIQUID <--> IDEAL GAS		
3156.584	22.396	238.807	81.151	497.656	FUGACITY = 1 bar		
3200	22.388	239.113	83.292	498.628	0.	0.	0.
3300	22.373	239.802	88.024	500.866	0.	0.	0.
3400	22.366	240.469	92.498	503.103	0.	0.	0.
3500	22.368	241.118	96.735	505.339	0.	0.	0.
3600	22.377	241.748	100.755	507.576	0.	0.	0.
3700	22.395	242.361	104.573	509.815	0.	0.	0.
3800	22.421	242.959	108.207	512.056	0.	0.	0.
3900	22.456	243.542	111.670	514.299	0.	0.	0.
4000	22.499	244.111	114.974	516.547	0.	0.	0.
4100	22.551	244.667	118.130	518.799	0.	0.	0.
4200	22.611	245.211	121.150	521.058	0.	0.	0.
4300	22.681	245.744	124.041	523.322	0.	0.	0.
4400	22.761	246.266	126.813	525.594	0.	0.	0.
4500	22.850	246.779	129.473	527.875	0.	0.	0.
4600	22.949	247.282	132.029	530.164	0.	0.	0.
4700	23.058	247.777	134.486	532.465	0.	0.	0.
4800	23.178	248.263	136.852	534.776	0.	0.	0.
4900	23.309	248.743	139.130	537.101	0.	0.	0.
5000	23.451	249.215	141.327	539.439	0.	0.	0.
5200	23.769	250.141	145.495	544.160	0.	0.	0.
5400	24.136	251.045	149.387	548.950	0.	0.	0.
5600	24.555	251.930	153.034	553.818	0.	0.	0.
5800	25.028	252.799	156.459	558.775	0.	0.	0.
6000	25.560	253.657	159.685	563.833	0.	0.	0.

PREVIOUS: December 1976 (1 atm)   CURRENT: December 1976 (1 bar)

Nickel (Ni) Ni_1(ref)

NICKEL (Ni) CRYSTAL $A_r = 58.69$

$S°(298.15\ K) = 29.870 \pm 0.21\ J\ K^{-1}\ mol^{-1}$
$T_t = 631 \pm 1\ K$
$T_{fus} = 1728 \pm 4\ K$

$\Delta_f H°(0\ K) = 0\ k\ J\ mol^{-1}$
$\Delta_f H°(298.15\ K) = 0\ kJ\ mol^{-1}$
$\Delta_t H° = 0\ kJ\ mol^{-1}$
$\Delta_{fus} H° = 17.15 \pm 0.4\ kJ\ mol^{-1}$

Enthalpy of Formation

Zero by definition.

Heat Capacity and Entropy

The low temperature heat capacity values are based primarily on the studies by Keesom and Clark (1) and Busey and Giaque (2). The adopted values are actually those suggested by Busey and Giauque (2) with slight changes above 280 K so as to smoothly join the high temperature heat capacity data.

There are numerous heat capacity studies in the vicinity of the Curie point. The studies upon which our adopted values are based are as follows:

Source	Year	T, K
(3)	1938	343-1123
(4)	1955	453-1437
(5)	1965	323- 883
(6)	1966	300-1728
(7)	1966	571- 681
(8)	1970	333-1500
(9)	1971	580- 680

The adoped value for the Curie point, $T_t = 631$ K, is from the study of Connelly et al. (9), using an ac calorimetric method to measure relative heat capacity vlaues, and Vollmer et al. (6) using a high temperature adiabatic calorimeter.

Above 700 K, the adopted heat capacity values are those reported by Vollmer et al. (6). Below the Curie point, the above studies and that of Bronson and Wilson (203-383 K, 10) are used to construct a curve which joins smoothly with the low temperature values. The adopted table is identical (purposely) to that suggested by Mah and Pankatz (11).

Fusion Data

See the liquid table for details.

Sublimation Data

The enthalpy of sublimation, $\Delta_{sub}H°(298.15\ K)$, is calculated as the difference between $\Delta_f H°(298.15\ K)$ for Ni(g) and Ni(cr).

References

1. W. H. Keesom and C. W. Clark, Physica 2, 513 (1935).
2. R. H. Busey and W. F. Giauque, J. Am. Chem. Soc. 74, 3157 (1952).
3. C. Sykes and H. Wilkinson, Proc. Phys. Soc. (London) 50, 834 (1938).
4. F. Krauss and H. Warncke, Z. Metallk. 46, 61 (1955).
5. R. E. Pawel and E. E. Stansbury, J. Phys. Chem. Solids 26, 607 (1965).
6. O. Vollmer, R. Kohlhaas, and M. Braun, Z. Naturforsch. 21a, 181 (1965).
7. Ya. A. Kraftmakher, Soviet Physics-Solid State 8, 1048 (1966).
8. T. G. Kollie, ORN-LTH 2649, 1970.
9. D. L. Connelly, J. S. Loomis, and D. E. Mapother, Phys. Rev. 3B, 924 (1971).
10. H. L. Bronson and A. J. C. Wilson, Can. J. Res. 14A, 181 (1936).
11. A. D. Mah and L. B. Pankrantz, U.S. Bur. Mines Bull. 668, 1976.

Nickel (Ni) $Ni_1(cr)$

Enthalpy Reference Temperature = T_r = 298.15 K — Standard State Pressure = p° = 0.1 MPa

T/K	$C_p°$ (J K⁻¹mol⁻¹)	S° (J K⁻¹mol⁻¹)	$-[G°-H°(T_r)]/T$ (J K⁻¹mol⁻¹)	$H°-H°(T_r)$ (kJ mol⁻¹)	$\Delta_f H°$ (kJ mol⁻¹)	$\Delta_f G°$ (kJ mol⁻¹)	Log K_f
0	0.	0.	INFINITE	-4.786	0.	0.	0.
100	13.631	7.454	50.235	-4.278	0.	0.	0.
200	22.468	20.200	32.143	-2.389	0.	0.	0.
250	24.397	25.432	30.290	-1.215	0.	0.	0.
298.15	25.987	29.870	29.870	0.	0.	0.	0.
300	26.024	30.031	29.871	0.048	0.	0.	0.
350	27.294	34.139	30.192	1.381	0.	0.	0.
400	28.493	37.863	30.922	2.776	0.	0.	0.
450	29.623	41.281	31.886	4.228	0.	0.	0.
500	31.045	44.473	32.986	5.743	0.	0.	0.
600	34.853	50.440	35.404	9.022	0.	0.	0.
631.000	39.832	52.263	36.187	10.144	Cp LAMBDA MAXIMUM TRANSITION		
631.000	39.832	52.263	36.187	10.144			
700	30.794	55.575	37.941	12.344	0.	0.	0.
800	31.003	59.697	40.408	15.431	0.	0.	0.
900	31.589	63.382	42.759	18.560	0.	0.	0.
1000	32.217	66.742	44.992	21.750	0.	0.	0.
1100	32.928	69.845	47.112	25.007	0.	0.	0.
1200	33.681	72.742	49.128	28.336	0.	0.	0.
1300	34.518	75.471	51.051	31.746	0.	0.	0.
1400	35.397	78.061	52.888	35.241	0.	0.	0.
1500	36.317	80.534	54.649	38.827	0.	0.	0.
1600	37.279	82.908	56.342	42.506	0.	0.	0.
1700	38.284	85.199	57.972	46.286	0.	0.	0.
1728.000	38.535	85.827	58.419	47.361	------ CRYSTAL <--> LIQUID -----		
1800	39.330	87.411	59.547	50.156	-17.161	0.715	-0.021
1900	42.677	89.637	61.072	54.275	-16.933	1.704	-0.047
2000	43.806	91.855	62.556	58.599	-16.501	2.674	-0.070
2100	44.978	94.021	64.003	63.038	-15.953	3.620	-0.090
2200	46.191	96.141	65.415	67.596	-15.286	4.536	-0.108
2300	47.447	98.222	66.797	72.278	-14.495	5.420	-0.123
2400	48.702	100.268	68.149	77.085	-13.579	6.267	-0.136
2500	49.957	102.281	69.474	82.018	-12.537	7.073	-0.148

PREVIOUS: CURRENT: December 1976

Nickel (Ni) $Ni_1(cr)$

NICKEL (Ni) LIQUID A_r = 58.69

$S°(298.15\ K) = [41.487]\ J\ K^{-1}\ mol^{-1}$

$T_{fus} = 1728 \pm 4\ K$

$\Delta_f H°(298.15\ K) = [17.479]\ kJ\ mol^{-1}$

$\Delta_{fus}H° = 17.15 \pm 0.4\ kJ\ mol^{-1}$

Enthalpy of Formation

The enthalpy of formation of Ni(ℓ) is calculated from that of the crystal by adding $\Delta_{fus}H°$ and the difference in enthalpy, H°(1728 - H°(298.15 K), between the crystal and liquid.

Heat Capacity and Entropy

Vollmer et al. (1), using an adiabatic high temperature calorimeter with an inert gas atmosphere, measured the heat capcity of nickel in the liquid region (1728-1822 K) and reported $C_p°(\ell) = 9.32\ kcal\ mol^{-1}$. Other studies yielding liquid phase heat capacity values are: 7.86 kcal mol^{-1} by Wüst et al. (2), 9.2 kcal mol^{-1} by Umino (3), 10.3 kcal mol^{-1} by Geoffray et al. (4), and 2.58 kcal mol^{-1} by Margrave (5). This latter vlaue (5) is undoubtedly a misprint.

We adopt a constant value, $Cp°(\ell) = 9.3\ kcal\ mol^{-1}$ based on the heat capacity measurements of Vollmer et al. (1). The enthalpy study of Geoffray et al. (1393-2192 K, 4) leads to a reported heat capacity whilch is roughly 10% higher than our adopted value. The adopted value is identical to that chosen by Mah and Pankratz (6).

The entropy at 298.15 K is calculated in a manner analogous to that used for the enthalpy of formation.

Fusion Data

The adopted melting point, T_{fus} = 1728 ± 4 K (based on IPTS-68), is selected from the studies of Vollmer et al. (1727 K, 1), Geoffray et al. (1728 K, 4), Oriani and Jones (1727 K, 7), and Kubachewski (1730 K, 8).

The adopted enthalpy of melting, $\Delta_{fus}H°$ = 4.10 ± 0.1 kcal mol^{-1}, is based on the calorimetric studies by Vollmer et al. (4.039 ± 0.060 kcal mol^{-1}) and Geoffray et al. (4.176 ± 054 kcal mol^{-1}). Other reported results are 3.31 kcal mol^{-1} by Wüst et al. (2), 4.21 kcal mol^{-1} by Umino (3), 3.32 kcal mol^{-1} by Margrave (5), and 4.4 kcal mol^{-1} by Lebedev et al. (9).

References

1. A. Vollmer, R. Kohlhaas, and M. Braun, Z. Naturforsch 21a, 181 (1966).
2. F. Wust, A. Meuthen, and R. Durrer, Forsch. Gebiete. Ingenieurw. VDI 204 (1918).
3. S. Umino, Science Repts. Tohoku Imp. Univ. 15, 597 (1926).
4. H. Geoffray, A. Ferrier, and M. Olette, Compt. Rend. 256, 139 (1963).
5. J. L. Margrave, ORO-2907-92 (1972).
6. A. D. Mah and L. B. Pankratz, U. S. Bur. Mines Bull. 668 (1976).
7. R. A. Oriani and T. S. Jones, Rev. Sci. Inst. 25, 248 (1954).
8. O. Kubaschewski, Z. Elektrochem. 54, 275 (1950).
9. S. V. Lebedev, A. I. Savvatimskii, and Yu. B. Smirnov, High Temp. 9, 578 (1971).

Nickel (Ni) $Ni_1(l)$

Enthalpy Reference Temperature = T_r = 298.15 K   Standard State Pressure = p° = 0.1 MPa

T/K	$C_p°$ (J K⁻¹mol⁻¹)	S°	$-[G°-H°(T_r)]/T$	$H°-H°(T_r)$ (kJ mol⁻¹)	$\Delta_f H°$	$\Delta_f G°$	Log K_f
0							
100							
200							
298.15	26.133	41.487	41.487	0.	17.479	14.016	-2.456
300	26.150	41.648	41.487	0.048	17.480	13.994	-2.437
350	26.568	45.711	41.807	1.366	17.464	13.414	-2.002
400	26.987	49.286	42.523	2.705	17.408	12.839	-1.677
450	27.405	52.489	43.455	4.065	17.316	12.273	-1.425
500	27.824	55.398	44.506	5.446	17.182	11.719	-1.224
600	28.660	60.544	46.761	8.270	16.728	10.665	-0.928
700	29.497	65.025	49.057	11.178	16.313	9.698	-0.724
800	30.334	69.019	51.307	14.169	16.218	8.760	-0.572
900	31.171	72.640	53.479	17.245	16.164	7.831	-0.455
1000	32.008	75.967	55.564	20.404	16.133	6.907	-0.361
1100	32.844	79.057	57.561	23.646	16.119	5.986	-0.284
1200	33.681	81.951	59.474	26.972	16.115	5.065	-0.220
1300	34.518	84.680	61.309	30.382	16.116	4.144	-0.166
1300.000	38.911	84.680	61.309	30.382	GLASS <--> LIQUID		
1300.000	38.911	84.680	61.309	30.382	TRANSITION		
1400	38.911	87.564	63.083	34.273	16.511	3.207	-0.120
1500	38.911	90.248	64.805	38.165	16.817	2.246	-0.078
1600	38.911	92.759	66.475	42.056	17.029	1.267	-0.041
1700	38.911	95.118	68.091	45.947	17.140	0.278	-0.009
1728.000	38.911	95.754	68.534	47.036	------ CRYSTAL <--> LIQUID -----		
1800	38.911	97.343	69.655	49.838	0.	0.	0.
1900	38.911	99.446	71.168	53.729	0.	0.	0.
2000	38.911	101.442	72.632	57.620	0.	0.	0.
2100	38.911	103.341	74.050	61.511	0.	0.	0.
2200	38.911	105.151	75.423	65.402	0.	0.	0.
2300	38.911	106.881	76.753	69.294	0.	0.	0.
2400	38.911	108.537	78.043	73.185	0.	0.	0.
2500	38.911	110.125	79.295	77.076	0.	0.	0.
2600	38.911	111.651	80.510	80.967	0.	0.	0.
2700	38.911	113.120	81.691	84.858	0.	0.	0.
2800	38.911	114.535	82.839	88.749	0.	0.	0.
2900	38.911	115.900	83.955	92.640	0.	0.	0.
3000	38.911	117.219	85.042	96.531	0.	0.	0.
3100	38.911	118.495	86.101	100.423	0.	0.	0.
3156.584	38.911	119.199	86.688	102.624	------- FUGACITY = 1 bar -------		
3200	38.911	119.731	87.133	104.314	-376.835	5.189	-0.085
3300	38.911	120.928	88.139	108.205	-375.182	17.101	-0.271
3400	38.911	122.090	89.120	112.096	-373.527	28.964	-0.445
3500	38.911	123.218	90.078	115.987	-371.873	40.778	-0.609
3600	38.911	124.314	91.014	119.878	-370.219	52.544	-0.762

PREVIOUS: CURRENT: December 1976

Nickel (Ni) $Ni_1(l)$

NICKEL (Ni) CRYSTAL–LIQUID A_r = 58.69

0 to 1728 K crystal
above 1728 K liquid

Refer to the individual tables for details.

Enthalpy Reference Temperature = T_r = 298.15 K   Standard State Pressure = p° = 0.1 MPa

T/K	C_p° (J K⁻¹mol⁻¹)	S° (J K⁻¹mol⁻¹)	−[G°−H°(T_r)]/T (J K⁻¹mol⁻¹)	H°−H°(T_r) (kJ mol⁻¹)	$\Delta_f H^\circ$ (kJ mol⁻¹)	$\Delta_f G^\circ$ (kJ mol⁻¹)	Log K_f
0	0.	0.	INFINITE	-4.786	0.	0.	0.
100	13.631	7.454	50.235	-4.278	0.	0.	0.
200	22.468	20.200	32.143	-2.389	0.	0.	0.
250	24.397	25.432	30.290	-1.215	0.	0.	0.
298.15	25.987	29.870	29.870	0.	0.	0.	0.
300	26.024	30.031	29.871	0.048	0.	0.	0.
350	27.294	34.139	30.192	1.381	0.	0.	0.
400	28.493	37.863	30.922	2.776	0.	0.	0.
450	29.623	41.281	31.886	4.228	0.	0.	0.
500	31.045	44.473	32.986	5.743	0.	0.	0.
600	34.853	50.440	35.404	9.022	0.	0.	0.
631.000	39.832	52.263	36.187	10.144	Cp LAMBDA MAXIMUM TRANSITION		
631.000	39.832	52.263	36.187	10.144			
700	30.794	55.575	37.941	12.344	0.	0.	0.
800	31.003	59.697	40.408	15.431	0.	0.	0.
900	31.589	63.382	42.759	18.560	0.	0.	0.
1000	32.217	66.742	44.992	21.750	0.	0.	0.
1100	32.928	69.845	47.112	25.007	0.	0.	0.
1200	33.681	72.742	49.128	28.336	0.	0.	0.
1300	34.518	75.471	51.051	31.746	0.	0.	0.
1400	35.397	78.061	52.888	35.241	0.	0.	0.
1500	36.317	80.534	54.649	38.827	0.	0.	0.
1600	37.279	82.908	56.342	42.506	0.	0.	0.
1700	38.284	85.199	57.972	46.286	0.	0.	0.
1728.000	38.535	85.827	58.419	47.361	CRYSTAL <--> LIQUID TRANSITION		
1728.000	38.911	95.754	58.419	64.516			
1800	38.911	97.343	59.944	67.317	0.	0.	0.
1900	38.911	99.446	61.968	71.208	0.	0.	0.
2000	38.911	101.442	63.893	75.099	0.	0.	0.
2100	38.911	103.341	65.726	78.991	0.	0.	0.
2200	38.911	105.151	67.477	82.882	0.	0.	0.
2300	38.911	106.881	69.153	86.773	0.	0.	0.
2400	38.911	108.537	70.760	90.664	0.	0.	0.
2500	38.911	110.125	72.303	94.555	0.	0.	0.
2600	38.911	111.651	73.787	98.446	0.	0.	0.
2700	38.911	113.120	75.217	102.337	0.	0.	0.
2800	38.911	114.535	76.596	106.228	0.	0.	0.
2900	38.911	115.900	77.928	110.120	0.	0.	0.
3000	38.911	117.219	79.216	114.011	0.	0.	0.
3100	38.911	118.495	80.462	117.902	0.	0.	0.
3156.584	38.911	119.199	81.151	120.104	------- FUGACITY = 1 bar -------		
3200	38.911	119.731	81.670	121.793	-376.835	5.189	-0.085
3300	38.911	120.928	82.842	125.684	-375.182	17.101	-0.271
3400	38.911	122.090	83.979	129.575	-373.527	28.964	-0.445
3500	38.911	123.218	85.084	133.466	-371.873	40.778	-0.609
3600	38.911	124.314	86.159	137.357	-370.219	52.544	-0.762

PREVIOUS: CURRENT: December 1976

NICKEL (Ni) IDEAL GAS A_r = 58.69

IP(Ni, g) = 61579 ± 10 cm^{-1}

S°(298.15 K) = 182.190 ± 0.08 J K^{-1} mol^{-1}

$\Delta_f H°$(0 K) = 428.1 ± 8.4 kJ mol^{-1}

$\Delta_f H°$(298.15 K) = 430.1 ± 8.4 kJ mol^{-1}

Electronic Levels and Quantum Weights

State	ε_i, cm^{-1}	g_i
$^3F_4$	0.00	9
$^3D_3$	204.786	7
$^3D_2$	879.813	5
$^3F_3$	1332.153	7
$^3D_1$	1713.080	3
$^3F_2$	2216.519	5

Enthalpy of Formation

The enthalpy of formation for Ni(g) is obtained by an analysis of the vapor pressure data of Morris et al. (1). In this study a gas-transport method was used to measure vapor pressures (16 points) over liquid nickel in the range 1816-1895 K. For this vaporization process, we obtain $\Delta_{vap}H°$(298.15 K) = 99.33±1.23 kcal mol^{-1} (2nd law) and 98.54±0.12 cal mol^{-1}(3rd law). The entropy difference (2nd law-3rd law) is only -0.04±0.7 cal K^{-1} mol^{-1} which suggests good consistency between our adopted functions and the vapor pressure data. Using the 3rd law $\Delta_{vap}H°$(298.15 K) and $\Delta_f H°$(Ni, ℓ, 298.15 K) (2), we calculate and adopt $\Delta_f H°$(Ni, g, 298.15 K) = 102.8 kcal mol^{-1}. [This is the same value chosen by Hultgren et al. (4) and Mah and Pankratz (5). Thus we maintain closer similarity with the most recent tabulations.]

Using mass spectrometric intensity data, Farber and Srivastava (3) obtained a 2nd law enthalpy of sublimation of 99.4 kcal mol^{-1} at an average temperature of 1653 K and a 2nd law enthalpy of vaporization of 92.2 kcal mol^{-1} at an average temperature of 1863 K. Using auxiliary data (2), we calculate $\Delta_f H°$(Ni, ℓ, 298.15 K) = 102.1 kcal mol^{-1} and 99.8 kcal mol^{-1} respectively. Rutner and Haury (6) measured vapor pressures over Ni (28 points in the range 1277-1658 K), using the Langmuir technique. Their results yield a 2nd law value of 107.0±5.8 kcal mol^{-1} and a 3rd law value of 103.2±4.0 kcal mol^{-1}. The drift is -2.5±3.9 cal K^{-1} mol^{-1}. In addition Rutner and Haury (6) determined the best values of the heat of sublimation and vaporization and the parameters in a vapor pressure equation by statistically weighting nine sets of data. Although the results of this statistical treatment lead to enthalpies of formation for Ni(g) consistent with the adopted value, the results are not consistent with the melting data. Other vapor pressure studies are discussed by Hultgren et al. (4)

Heat Capacity and Entropy

The electronic energy levels and quantum weights used in the calculation are from Moore (7) and (8). Not all levels are listed above. The tabulated entropy values agree within 0.001 cal K^{-1} mol^{-1} with the recent tabulation by Hultgren et al. (4) and Mah and Pankratz (5). There are predicted levels which have not been observed and/or classified. It is not expected that S°(298.15 K) would be affected by these states, but that in the range 3000-6000K, an error of 0.2-0.3 cal K^{-1} mol^{-1} might result. In our present tabulation, the levels above 20000 cm^{-1} contribute 0.0005 cal K^{-1} mol^{-1} at 3000 K and 0.15 cal K^{-1} mol^{-1} at 6000 K to the entropy.

References

1. J. P. Morris, G. R. Zellars, S. L. Payne, and R. L. Kipp, U. S. Bur. Mines RI5364, 1957.
2. JANAF Thermochemical Tables: Ni(ℓ) and Ni(cr), 12-31-76.
3. M. Farber and R. D. Srivastava, in Analytical Calorimetry, Vol. 3, ed., R. S. Porter and J. F. Johnson, Plenum Press, 1974, pp 731-741.
4. R. Hultgren et al., "Selected Values of the Thermodynamic Properties of the Elements, "American Society for Metals, Metals Park, Ohio, 1973.
5. A. D. Mah and L. B. Pankratz, U. S. Bur. Mines Bull. 668, 1976.
6. E. Rutner and G. L. Haury, AFML-TR-72-217, May, 1973; J. Chem. Eng. Data 19, 19 (1974).
7. C. E. Moore, U.S. Nat. Bur. Stand., NSRDS-NBS 23, Volume II, 1970. [Reprint of NBS Circular 467, Volume II, 1952].
8. C. Roth, J. Research, Natl. Bur. Std. 74A, 715 (1970).

Nickel (Ni) Ni_1(g)

Enthalpy Reference Temperature = T_r = 298.15 K Standard State Pressure = p° = 0.1 MPa

T/K	$C_p°$ (J $K^{-1}mol^{-1}$)	S° (J $K^{-1}mol^{-1}$)	$-[G°-H°(T_r)]/T$ (J $K^{-1}mol^{-1}$)	$H°-H°(T_r)$ (kJ mol^{-1})	$\Delta_f H°$ (kJ mol^{-1})	$\Delta_f G°$ (kJ mol^{-1})	Log K_f
0	0.	0.	INFINITE	-6.825	428.076	428.076	INFINITE
100	23.510	156.498	203.000	-4.650	429.743	414.839	-216.690
200	23.396	172.882	184.321	-2.288	430.216	399.680	-104.386
250	23.275	178.085	182.574	-1.122	430.208	392.044	-81.913
298.15	23.360	182.190	182.190	0.	430.115	384.701	-67.398
300	23.367	182.334	182.190	0.043	430.110	384.419	-66.933
350	23.587	185.952	182.475	1.217	429.951	376.816	-56.237
400	23.859	189.119	183.112	2.403	429.742	369.239	-48.218
450	24.131	191.945	183.939	3.603	429.490	361.691	-41.984
500	24.374	194.500	184.870	4.815	429.187	354.174	-37.000
600	24.734	198.979	186.858	7.272	428.366	339.243	-29.534
700	24.925	202.808	188.870	9.756	427.528	324.465	-24.212
800	24.981	206.141	190.825	12.253	426.937	309.782	-20.227
900	24.942	209.082	192.693	14.749	426.305	295.174	-17.131
1000	24.841	211.705	194.466	17.239	425.605	280.641	-14.659
1100	24.701	214.066	196.142	19.716	424.825	266.182	-12.640
1200	24.538	216.209	197.726	22.178	423.957	251.797	-10.960
1300	24.364	218.166	199.225	24.624	422.993	237.489	-9.542
1400	24.187	219.965	200.643	27.051	421.925	223.259	-8.330
1500	24.012	221.628	201.987	29.461	420.750	209.109	-7.282
1600	23.842	223.172	203.263	31.854	419.463	195.041	-6.367
1700	23.679	224.612	204.477	34.230	418.059	181.057	-5.563
1800	23.524	225.961	205.634	36.590	399.388	167.874	-4.872
1900	23.380	227.229	206.737	38.935	397.842	155.054	-4.263
2000	23.245	228.425	207.792	41.266	396.282	142.316	-3.717
2100	23.121	229.556	208.802	43.584	394.709	129.656	-3.225
2200	23.007	230.629	209.770	45.891	393.124	117.072	-2.780
2300	22.903	231.650	210.699	48.186	391.528	104.560	-2.375
2400	22.810	232.622	211.592	50.472	389.923	92.117	-2.005
2500	22.725	233.552	212.452	52.748	388.308	79.742	-1.666
2600	22.651	234.441	213.281	55.017	386.686	67.431	-1.355
2700	22.585	235.295	214.081	57.279	385.057	55.183	-1.068
2800	22.529	236.115	214.853	59.534	383.421	42.996	-0.802
2900	22.481	236.905	215.600	61.785	381.780	30.866	-0.556
3000	22.441	237.667	216.323	64.031	380.135	18.794	-0.327
3100	22.410	238.402	217.023	66.273	378.487	6.776	-0.114
3156.584	22.396	238.807	217.410	67.541	------- FUGACITY = 1 bar -------		
3200	22.388	239.113	217.703	68.513	0.	0.	0.
3300	22.373	239.802	218.362	70.751	0.	0.	0.
3400	22.366	240.469	219.002	72.988	0.	0.	0.
3500	22.368	241.118	219.625	75.225	0.	0.	0.
3600	22.377	241.748	220.231	77.462	0.	0.	0.
3700	22.395	242.361	220.821	79.700	0.	0.	0.
3800	22.421	242.959	221.395	81.941	0.	0.	0.
3900	22.456	243.542	221.956	84.185	0.	0.	0.
4000	22.499	244.111	222.503	86.433	0.	0.	0.
4100	22.551	244.667	223.036	88.685	0.	0.	0.
4200	22.611	245.211	223.558	90.943	0.	0.	0.
4300	22.681	245.744	224.068	93.208	0.	0.	0.
4400	22.761	246.266	224.566	95.480	0.	0.	0.
4500	22.850	246.779	225.054	97.760	0.	0.	0.
4600	22.949	247.282	225.532	100.050	0.	0.	0.
4700	23.058	247.777	226.000	102.350	0.	0.	0.
4800	23.178	248.263	226.459	104.662	0.	0.	0.
4900	23.309	248.743	226.909	106.986	0.	0.	0.
5000	23.451	249.215	227.350	109.324	0.	0.	0.
5100	23.604	249.681	227.783	111.677	0.	0.	0.
5200	23.769	250.141	228.209	114.045	0.	0.	0.
5300	23.947	250.595	228.627	116.431	0.	0.	0.
5400	24.136	251.045	229.038	118.835	0.	0.	0.
5500	24.339	251.489	229.442	121.258	0.	0.	0.
5600	24.555	251.930	229.840	123.703	0.	0.	0.
5700	24.784	252.366	230.231	126.170	0.	0.	0.
5800	25.028	252.799	230.617	128.660	0.	0.	0.
5900	25.287	253.229	230.996	131.176	0.	0.	0.
6000	25.560	253.657	231.370	133.718	0.	0.	0.

PREVIOUS: December 1976 (1 atm) CURRENT: December 1976 (1 bar)

Nickel (Ni) Ni_1(g)

NICKEL, ION (Ni^+) IDEAL GAS M_r = 58.68945

IP(Ni^+, g) = 146541.56 ± 0.02 cm^{-1}
S°(298.15 K) = 199.72 ± 0.04 J K^{-1} mol^{-1}

$\Delta_f H°$(0 K) = 1164.98 ± 8.4 kJ mol^{-1}
$\Delta_f H°$(298.15 K) = [1179.003] kJ mol^{-1}

Electronic Level and Quantum Weight

State	ε_i, cm^{-1}	g_i
$^2D_{5/2}$	0	6
$^2D_{3/2}$	1606.94	4

Enthalpy of Formation

$\Delta_f H°$(Ni^+, g, 0 K) is calculated from $\Delta_f H°$(Ni, g, 0 K) (1) using the spectroscopic value of IP(Ni) = 61600 ± 10 cm^{-1} (736.90 ± 0.12 kJ mol^{-1}) from Corliss and Sugar (2). The ionization limit is converted from cm^{-1} to kJ mol^{-1} using the factor, 1 cm^{-1} = 0.01196266 kJ cm^{-1}, which is derived from the 1973 CODATA fundamental constants (3). Rosenstock et al. (4) and Levin and Lias (5) have summarized additional ionization and appearance potential data.

$\Delta_f H°$(Ni^+, g, 298.15 K) is calculated from $\Delta_f H°$(Ni, g, 0 K) by using IP(Ni) with JANAF (1) enthalpies, H°(0 K) - H°(298.15 K), for Ni(g), Ni^+(g), and e^-(g). $\Delta_f H°$(Ni → Ni^+ + e^-, 298.15 K) differs from a room temperature threshold energy due to inclusion of these enthalpies and to threshold effects discussed by Rosenstock et al. (4). $\Delta_f H°$(298.15 K) should be changed by -6.197 kJ mol^{-1} if it is to be used in the ion convention that excludes the enthalpy of the electron.

Heat Capacity and Entropy

The information on electronic energy levels and quantum weights, given by Corliss and Sugar (2), is incomplete because many theoretically predicted levels have not been observed. Although we have listed only the ground state for Ni^+(g), all levels listed by Corliss and Sugar (2), as well as estimated levels, are used in the calculation. The observed levels are too numerous to list completely. The calculations indicate that for Ni^+(g), the thermodynamic functions are dependent of the estimated missing levels (for n $\geq$ 3) and the cut-off procedure, the Gibbs energy function shows variations of ~20% at 1000 K. We use the TEMPER-FILL cut-off procedure. The reported uncertainty in S°(298.15 K) is due to uncertainties in the relative ionic mass, and the fundamental constants. Extension of these calculations aboue 6000 K may require consideration of the higher excited states (n > 5), and use of different fill and cut-off procedures (6).

References

1. JANAF Thermochemical Tables: Ni(g), 12-31-76; e^-(g), 3-31-82.
2. C. Corliss and J. Sugar, J. Phys. Chem. Ref. Data 10, 197 (1981).
3. E. R. Cohen and B. N. Taylor, J. Phys. Chem. Ref. Data 2, 663 (1973).
4. H. M. Rosenstock, K. Draxl et al., J. Phys. Chem. Ref. Data 6, Supp. 1 (1977).
5. R. D. Levin and S. G. Lias, U.S. Nat. Bur. Stand., NSRDS-NBS-71, 1982.
6. J. R. Downey, Jr., The Dow Chemical Company, AFOSR-TR-78-0960, Contract No. F44620-75-1-0048, 1978.

Nickel, Ion (Ni^+) $Ni_1^+(g)$

Enthalpy Reference Temperature = T_r = 298.15 K   Standard State Pressure = p° = 0.1 MPa

T/K	$C_p°$ (J $K^{-1}mol^{-1}$)	S°	-[G°-H°(T_r)]/T	H°-H°(T_r) (kJ mol^{-1})	$\Delta_f H°$	$\Delta_f G°$	Log K_f
0	0.	0.	INFINITE	-12.615	1164.976		
100	20.787	151.833	257.198	-10.537			
200	35.551	167.752	208.658	-8.181			
250	86.780	180.587	201.615	-5.257			
298.15	121.588	199.724	199.724	0.	1179.003	1122.106	-196.588
300	121.596	200.476	199.726	0.225	1179.218	1121.752	-195.314
350	96.043	217.764	201.149	5.815	1184.514	1111.737	-165.918
400	62.862	228.293	203.945	9.739	1188.082	1101.076	-143.786
450	43.263	234.441	207.020	12.339	1190.270	1090.058	-126.531
500	33.285	238.423	209.973	14.225	1191.680	1078.842	-112.706
600	25.407	243.650	215.183	17.080	1193.335	1056.100	-91.942
700	22.887	247.345	219.525	19.474	1194.485	1033.143	-77.094
800	21.897	250.327	223.195	21.706	1195.709	1010.009	-65.947
900	21.442	252.877	226.354	23.870	1196.823	986.729	-57.268
1000	21.206	255.122	229.121	26.002	1197.843	963.329	-50.319
1100	21.072	257.137	231.578	28.115	1198.778	939.832	-44.629
1200	20.991	258.967	233.785	30.218	1199.630	916.253	-39.883
1300	20.939	260.645	235.788	32.314	1200.395	892.607	-35.865
1400	20.904	262.195	237.619	34.406	1201.070	868.905	-32.419
1500	20.879	263.636	239.306	36.495	1201.653	845.158	-29.431
1600	20.862	264.983	240.869	38.582	1202.139	821.375	-26.815
1700	20.849	266.248	242.325	40.668	1202.523	797.565	-24.506
1800	20.840	267.439	243.688	42.752	1185.655	774.450	-22.474
1900	20.833	268.566	244.968	44.836	1185.926	751.598	-20.663
2000	20.828	269.634	246.175	46.919	1186.197	728.732	-19.033
2100	20.824	270.650	247.316	49.002	1186.467	705.852	-17.557
2200	20.821	271.619	248.399	51.084	1186.737	682.959	-16.216
2300	20.819	272.544	249.429	53.166	1187.006	660.054	-14.990
2400	20.818	273.430	250.410	55.248	1187.275	637.137	-13.867
2500	20.817	274.280	251.348	57.329	1187.545	614.209	-12.833
2600	20.816	275.097	252.246	59.411	1187.814	591.271	-11.879
2700	20.816	275.882	253.107	61.493	1188.083	568.321	-10.995
2800	20.815	276.639	253.934	63.574	1188.352	545.362	-10.174
2900	20.815	277.370	254.730	65.656	1188.621	522.394	-9.409
3000	20.816	278.075	255.496	67.737	1188.890	499.416	-8.696
3100	20.816	278.758	256.236	69.819	1189.159	476.429	-8.028
3200	20.816	279.419	256.950	71.900	812.593	458.622	-7.486
3300	20.817	280.059	257.640	73.982	814.515	447.531	-7.084
3400	20.817	280.681	258.309	76.064	816.439	436.381	-6.704
3500	20.817	281.284	258.957	78.145	818.363	425.175	-6.345
3600	20.818	281.871	259.585	80.227	820.286	413.914	-6.006
3700	20.819	282.441	260.195	82.309	822.208	402.599	-5.684
3800	20.819	282.996	260.788	84.391	824.127	391.233	-5.378
3900	20.820	283.537	261.364	86.473	826.044	379.816	-5.087
4000	20.820	284.064	261.925	88.555	827.957	368.349	-4.810
4100	20.821	284.578	262.472	90.637	829.865	356.836	-4.546
4200	20.822	285.080	263.004	92.719	831.768	345.275	-4.294
4300	20.822	285.570	263.523	94.801	833.664	333.670	-4.053
4400	20.823	286.049	264.030	96.883	835.553	322.020	-3.823
4500	20.823	286.517	264.524	98.966	837.434	310.328	-3.602
4600	20.824	286.974	265.007	101.048	839.305	298.594	-3.391
4700	20.824	287.422	265.479	103.131	841.165	286.819	-3.188
4800	20.825	287.861	265.941	105.213	843.015	275.005	-2.993
4900	20.826	288.290	266.393	107.296	844.852	263.152	-2.805
5000	20.826	288.711	266.835	109.378	846.675	251.262	-2.625
5100	20.827	289.123	267.268	111.461	848.483	239.336	-2.451
5200	20.827	289.528	267.692	113.543	850.276	227.374	-2.284
5300	20.828	289.924	268.108	115.626	852.052	215.379	-2.123
5400	20.828	290.314	268.516	117.709	853.809	203.349	-1.967
5500	20.829	290.696	268.915	119.792	855.547	191.288	-1.817
5600	20.829	291.071	269.308	121.875	857.264	179.195	-1.671
5700	20.830	291.440	269.693	123.958	858.959	167.071	-1.531
5800	20.830	291.802	270.071	126.041	860.630	154.918	-1.395
5900	20.831	292.158	270.442	128.124	862.276	142.737	-1.264
6000	20.831	292.508	270.807	130.207	863.896	130.527	-1.136

PREVIOUS: CURRENT: June 1984 (1 bar)

Nickel, Ion (Ni^+) $Ni_1^+(g)$

NICKEL, ION (Ni^-) IDEAL GAS M_r = 58.69055

EA(Ni, g) = 1.156 ± 0.010 eV

S°(298.15 K) = 174.577 ± 0.005 J K^{-1} mol^{-1}

$\Delta_f H°$(0 K) = 316.5 ± 1 kJ mol^{-1}

$\Delta_f H°$(298.15 K) = [311.764] kJ mol^{-1}

Electonic Level and Quantum Weight

State	ε_i, cm^{-1}	g_i
$^2D_{5/2}$	0.0	6
$^1D_{3/2}$	1470	4

Enthalpy of Formation

$\Delta_f H°(Ni^-$, g, 0 K) is calculated from $\Delta_f H°$(Ni, g, 0 K) (1) using the adopted electron affinity of EA (Ni) = 1.156 ± 0.010 eV (111.546 ± 0.965 kJ mol^{-1}). This value, recommended by Hotop and Lineberger (2), is based on a laser photodetachment electron spectrometry study (3).

Additional information on Ni^-(g) may be obtained in the critical discussions of Hotop and Lineberger (2, 4), Rosenstock et al. (5) and Massey (6). $\Delta_f H°(Ni^-$, g, 298.15 K) is obtained from $\Delta_f H°$(Ni, g, 0 K) by using EA(Ni) with JANAF (1) enthalpies, H°(0 K) - H°(298.15 K), for Ni^-(g), Ni(g), and e^-(g). $\Delta_r H°(Ni^- \rightarrow Ni + e^-$, 298.15 K) differs from a room-temperature threshold energy due to inclusion of these enthalpies and to threshold effects discussed by Rosenstock et al. (5). $\Delta_f H°$(298.15 K) should be changed by +6.197 kJ mol^{-1} if it is to be used in the ion convention that excludes the enthalpy of the electron.

Heat Capacity and Entropy

The ground state configuration for Ni^-(g) is given by Hotop and Lineberger (2, 4) and Rosenstock et al. (5) and Masey (6). The fine structure separation has been measured experimentally be laser photodetachment electron spectrometry (3) and is that recommended by Hotop and Lineberger (2).

References

1. JANAF Thermochemical Tables: Ni(g), 12-31-76; e^-(g), 3-31-82.
2. H. Hotop and W. C. Lineberger, J. Phys. Chem. Ref. Data, to be published, 1985.
3. R. R. Corderman, P. C. Engleking and W. C. Lineberger, J. Chem. Phys. 70, 4474 (1979).
4. H. Hotop and W. C. Lineberger, J. Phys. Chem. Ref. Data 4, 539 (1975).
5. H. M. Rosenstock, K. Draxl et al., J. Phys. Chem. Ref. Data 6, Supp. 1 (1977).
6. H. S. W. Massey, "Negative Ions", 3rd., Cambridge University Press, Cambridge, 1976.

Nickel, Ion (Ni^-) Ni_1^-(g)

Enthalpy Reference Temperature = T_r = 298.15 K — Standard State Pressure = p° = 0.1 MPa

	J $K^{-1}mol^{-1}$			kJ mol^{-1}			
T/K	$C_p°$	S°	$-[G°-H°(T_r)]/T$	$H°-H°(T_r)$	$\Delta_f H°$	$\Delta_f G°$	Log K_f
0	0.	0.	INFINITE	-6.207	316.540		
100	20.786	151.833	193.117	-4.128			
200	20.802	166.242	176.490	-2.050			
250	20.870	170.890	174.922	-1.008			
298.15	21.017	174.577	174.577	0.	311.764	274.874	-48.157
300	21.025	174.707	174.578	0.039	311.716	274.646	-47.820
350	21.265	177.965	174.834	1.096	310.401	268.570	-40.082
400	21.564	180.824	175.408	2.166	309.037	262.687	-34.303
450	21.886	183.382	176.154	3.253	307.632	256.977	-29.829
500	22.202	185.705	176.995	4.355	306.180	251.427	-26.266
600	22.737	189.802	178.797	6.603	303.071	240.763	-20.960
700	23.099	193.337	180.627	8.896	299.963	230.634	-17.210
800	23.297	196.436	182.414	11.217	297.119	220.923	-14.425
900	23.367	199.185	184.127	13.552	294.245	211.571	-12.279
1000	23.348	201.646	185.758	15.888	291.313	202.542	-10.580
1100	23.274	203.869	187.305	18.219	288.309	193.809	-9.203
1200	23.165	205.889	188.771	20.541	285.223	185.354	-8.068
1300	23.040	207.738	190.160	22.852	282.045	177.160	-7.118
1400	22.908	209.441	191.477	25.149	278.769	169.214	-6.313
1500	22.776	211.017	192.728	27.433	275.389	161.506	-5.624
1600	22.648	212.483	193.918	29.705	271.902	154.027	-5.028
1700	22.526	213.852	195.050	31.963	268.302	146.769	-4.510
1800	22.411	215.136	196.131	34.210	247.439	140.441	-4.076
1900	22.304	216.345	197.163	36.446	243.705	134.599	-3.700
2000	22.206	217.487	198.151	38.671	239.961	128.953	-3.368
2100	22.114	218.568	199.098	40.887	236.207	123.495	-3.072
2200	22.030	219.595	200.006	43.094	232.444	118.216	-2.807
2300	21.952	220.572	200.880	45.293	228.674	113.108	-2.569
2400	21.881	221.505	201.720	47.485	224.896	108.165	-2.354
2500	21.815	222.397	202.529	49.670	221.111	103.379	-2.160
2600	21.755	223.251	203.310	51.848	217.320	98.744	-1.984
2700	21.699	224.071	204.064	54.021	213.522	94.256	-1.823
2800	21.648	224.860	204.792	56.188	209.720	89.908	-1.677
2900	21.600	225.618	205.497	58.350	205.913	85.696	-1.544
3000	21.556	226.350	206.180	60.508	202.101	81.616	-1.421
3100	21.516	227.056	206.843	62.662	198.285	77.662	-1.309
3200	21.478	227.739	207.485	64.811	-182.370	79.021	-1.290
3300	21.443	228.399	208.109	66.958	-184.541	87.223	-1.381
3400	21.411	229.039	208.715	69.100	-186.714	95.491	-1.467
3500	21.380	229.659	209.305	71.240	-188.889	103.823	-1.549
3600	21.352	230.261	209.878	73.376	-191.069	112.217	-1.628
3700	21.326	230.845	210.437	75.510	-193.252	120.671	-1.704
3800	21.301	231.414	210.982	77.642	-195.440	129.185	-1.776
3900	21.278	231.967	211.513	79.770	-197.633	137.756	-1.845
4000	21.257	232.505	212.031	81.897	-199.833	146.384	-1.912
4100	21.237	233.030	212.537	84.022	-202.039	155.067	-1.976
4200	21.218	233.541	213.031	86.145	-204.253	163.803	-2.037
4300	21.200	234.040	213.513	88.265	-206.475	172.593	-2.097
4400	21.183	234.528	213.986	90.385	-208.707	181.434	-2.154
4500	21.167	235.003	214.447	92.502	-210.948	190.326	-2.209
4600	21.152	235.468	214.899	94.618	-213.201	199.268	-2.263
4700	21.138	235.923	215.342	96.733	-215.465	208.259	-2.315
4800	21.125	236.368	215.775	98.846	-217.742	217.299	-2.365
4900	21.113	236.804	216.200	100.958	-220.033	226.386	-2.413
5000	21.101	237.230	216.616	103.068	-222.339	235.520	-2.460
5100	21.090	237.648	217.025	105.178	-224.661	244.700	-2.506
5200	21.079	238.057	217.425	107.286	-227.000	253.926	-2.551
5300	21.069	238.459	217.818	109.394	-229.357	263.197	-2.594
5400	21.059	238.852	218.204	111.500	-231.733	272.513	-2.636
5500	21.050	239.239	218.583	113.606	-234.129	281.873	-2.677
5600	21.041	239.618	218.955	115.710	-236.548	291.276	-2.717
5700	21.033	239.990	219.321	117.814	-238.990	300.723	-2.756
5800	21.025	240.356	219.681	119.917	-241.456	310.213	-2.794
5900	21.018	240.715	220.034	122.019	-243.948	319.746	-2.831
6000	21.010	241.068	220.382	124.120	-246.467	329.322	-2.867

PREVIOUS: CURRENT: June 1984 (1 bar)

Nickel, Ion (Ni^-) Ni_1^-(g)

NICKEL SULFIDE (NiS) CRYSTAL(RHOMBOHEDRAL-HEXAGONAL) M_r = 90.75

$S°(298.15\ K) = 52.992 \pm 0.4\ J\ K^{-1}\ mol^{-1}$
$T_{trs} = 652 \pm 3\ K$
$T_{fus} = 1249 \pm 3\ K$

$\Delta_f H°(0\ K) = -87.13 \pm 6.3\ kJ\ mol^{-1}$
$\Delta_f H°(298.15\ K) = -87.86 \pm 6.3\ kJ\ mol^{-1}$
$\Delta_{trs} H° = 6.443 \pm 0.4\ kJ\ mol^{-1}$
$\Delta_{fus} H° = 30.125 \pm 2.9\ kJ\ mol^{-1}$

Enthalpy of Formation

Rosenqvist (1) studied the sulfur vapor pressure above solid nickel sulfides using the N_2/H_2S equilibrium. Combining his data for several equilibria we calculate 2nd and 3rd law values of $\Delta_f H°(298.15\ K)$ for NiS(cr) to be -20.5±0.7 kcal mol^{-1} and -22.6±3.6 kcal mol^{-1}, respectively. The stoichiometry of NiS varies from at least $NiS_{.98}$ to $NiS_{1.06}$ and Ariya et al. (2) investigated $\Delta_f H°(298.15\ K)$ in the composition range by calorimetric methods. These authors found $\Delta_f H°(298.15\ K)$ = -20.3±0.6 kcal mol^{-1} for stoichiometric NiS with an ill-defined trend toward more exothermic values at higher sulfur content. DeRanter and Breckpot (3) obtained a value of -20.08 kcal mol^{-1} for amorphous $Ni_{.98}S$ from an enthalpometric titration. We calculate $\Delta_f H°(298.15\ K)$ = -21.3 kcal mol^{-1} for the same material based on a 3rd law analysis of DeRanter and Breckpot's emf data (3).

In view of the rather large uncertainties and variations due to stoichiometry we choose to adopt the average of the five values above, $\Delta_f H°(298.15\ K)$ = -21.0±1.5 kcal mol^{-1}. This value is somewhat less negative than values of -22.1 and -22.5 kcal mol^{-1} adopted by Mah and Pankrantz (4) and Mills (5), respectively.

Heat Capacity and Entropy

The low temperature heat capacity data (52-296 K) are taken from Weller and Kelley (6). High temperature enthalpies of $NiS_{1.03}$ were measured by Conard et al. (7) via drop calorimetry. We have joined these values smoothly with the low temperature $C°_p$ measurements of Weller and Kelley to derive the $C°_p$ values from 298-652 K. Values of $C°_p$ above the transition are also taken from our analysis of Conard's et al. enthalpy data, disregarding the points above 1000 K to avoid problems due to incongruent melting. Values above 1000 K are extrapolated.

Phase Data

Both the low (β) and high (α) temperature polymorphs of NiS dissolve excess sulfur to form solid solutions. The β phase shows a limited range of homogeneity but the α phase shows a broad limit extending to approximately $NiS_{1.10}$ (8). The α phase is normally designated α-$Ni_{1-x}S$ to emphasize this feature. However, the sulfur-poor limit of this phase has the stoichiometric composition to at least 873 K; at 1070 K the deviation from stoichiometry is less than 0.05 weight percent nickel (8). Above 1079 K the deviation of the sulfur-poor limit from stoichiometry increases more rapidly and a material of stoichiometry NiS will exist as an equilibrium mixture of liquid Ni_yS_z and solid α-$Ni_{1-x}S$ (8). Thus, NiS melts incongruently.

Transition Data

We adopt 652±3 K as the transition temperature of stoichiometric NiS from rhombohedral (β) to hexagonal (α) form base on the phase diagram of Kullerud and Yund (8). The temperature of this transition is very dependent on the exact stoichiometry of the material (8). The transition enthalpy has been measured via a DTA technique by Conard et al. (7) and we adopt their value of 1.54±0.1 kcal mol^{-1}. This is considerably higher than an older value of 0.63 kcal mol^{-1} measured by Biltz et al. (9, 5) but should be much more accurate due to the calibration technique used. Mah and Pankrantz (4) estimated 0.7 kcal mol^{-1}. This transition enthalpy cannot be measured by drop calorimetry since the α form does not revert to the β form on cooling.

Fusion Data

As stated above, stoichiometric NiS will begin to form a liquid at 1079 K. The conversion to stoichiometric liquid NiS is complete at 1249 K (8). We adopt the latter as the melting point of NiS with an uncertainty estimated to be ±3K. The adopted heat of melting, $\Delta_{fus}H°$ = 7.20±0.7 kcal mol^{-1}, is from our analysis of the high temperature enthalpy data of Conard et al. (7). This value compares favorably with estimated values of 7.4 and 7.0 kcal mol^{-1} due to Nagamori and Ingraham (10) and Mah and Pankrantz (4).

References

1. T. Rosenqvist, J. Iron Steel Inst. (London), 176, 37 (1954).
2. S. M. Ariya, M. P. Morozova, L. A. Pavlinova, and V. L. Ponomareva, Russ. J. Phys. Chem. 45, 1355 (1971).
3. C. DeRanter and R. Breckpot, Bull. Soc. Chim. Belg. 78, 503 (1969).
4. A. D. Mah and L. B. Pankrantz, U. S. Bur. Mines Bull. 668, 1976.
5. K. C. Mills, "Thermodynamic Data for Inorganic Sulphides, Selenides, and Tellurides," Butterworths, London, 1974.
6. W. W. Weller and K. K. Kelley, U. S. Bur. Mines RI 6511, 1964.
7. B. R. Conard, R. Sridhar and J. S. Warner, paper presented at the 106th AIME meeting, March, 1977.
8. G. Kullerud and R. A. Yund, J. Petrol. 3, 126 (1962).
9. W. Biltz, F. Weibke, P. Ehrlich, and K. Meisel, Z. Anorg. Chem. 228, 275 (1936).
10. M. Nagamori and T. R. Ingraham, Met. Trans. 1, 1821 (1970).

Nickel Sulfide (NiS) $Ni_1S_1(cr)$

Enthalpy Reference Temperature = T_r = 298.15 K; Standard State Pressure = p° = 0.1 MPa

T/K	$C°_p$	S°	$-[G°-H°(T_r)]/T$	$H°-H°(T_r)$	$\Delta_f H°$	$\Delta_f G°$	Log K_f
	J K⁻¹mol⁻¹			kJ mol⁻¹			
0	0.	0.	INFINITE	-8.465	-87.131	-87.131	INFINITE
100	23.187	12.857	89.876	-7.702	-87.566	-86.854	45.368
200	40.626	35.370	57.135	-4.353	-87.748	-86.055	22.475
298.15	47.112	52.992	52.992	0.	-87.864	-85.200	14.927
300	47.196	53.284	52.993	0.087	-87.867	-85.184	14.832
400	50.501	67.332	54.886	4.978	-90.301	-84.172	10.992
500	53.053	78.880	58.563	10.158	-92.016	-82.453	8.614
600	55.312	88.757	62.792	15.579	-93.458	-80.402	7.000
652.000	56.372	93.399	65.050	18.483	BETA <--> ALPHA	TRANSITION	
652.000	53.152	103.281	65.050	24.927			
700	54.518	107.105	67.804	27.511	-88.196	-78.598	5.865
800	57.363	114.570	73.189	33.105	-88.900	-77.175	5.039
900	60.208	121.491	78.176	38.983	-142.408	-74.603	4.330
1000	63.053	127.982	82.835	45.146	-141.290	-67.127	3.506
1100	65.898	134.125	87.221	51.594	-139.971	-59.773	2.838
1200	68.743	139.980	91.375	58.326	-138.456	-52.547	2.287
1249.000	70.137	142.759	93.337	61.728	------- ALPHA	<--> LIQUID	------
1300	71.588	145.595	95.332	65.342	-136.751	-45.456	1.826
1400	74.433	151.005	99.116	72.644	-134.864	-38.503	1.437
1500	77.278	156.237	102.751	80.229	-132.797	-31.691	1.104
1600	80.124	161.315	106.253	88.099	-130.554	-25.023	0.817
1700	82.969	166.258	109.638	96.254	-128.142	-18.500	0.568

PREVIOUS: CURRENT: December 1976

Nickel Sulfide (NiS) $Ni_1S_1(cr)$

NICKEL SULFIDE (NiS) LIQUID M_r = 90.75

$S°(298.15\ K)$ = [72.441] J K^{-1} mol^{-1}

T_{fus} = 1249 ± 3 K

$\Delta_f H°(298.15\ K)$ = [-62.474] kJ mol^{-1}

$\Delta_{fus}H°$ = 30.125 ± 2.9 kJ mol^{-1}

Enthalpy of Formation

$\Delta_f H°(\ell, 298.15\ K)$ is calculated from that of NiS(cr) by adding $\Delta_{fus}H°$ and the difference in enthalpy, H°(1249 K)-H°(298.15 K), between the crystal and liquid. Our 2nd and 3rd law analyses of Nagamori and Ingraham's (1) sulfur vapor pressure data yields $\Delta_f H°(298.15\ K)$ = -14.4±0.4 and -16.7±0.4 kcal mol^{-1}, respectively. These values agree with the adopted value within the estimated uncertainty. Meyer et al. (2) and Rosenqvist (3) conducted similar vapor pressure studies at higher and lower temperatures, respectively. Both studies qualitatively show good agreement with Nagamori at overlapping temperatures (2). However, Meyer's data are presented graphically only and Rosenqvist's data cover only compositions near $Ni_3S_2(\ell)$ and is not of high enough quality to extrapolate to the proper composition.

Heat Capacity and Entropy

The heat capacity of liquid NiS above an assumed glass transition at 860 K is taken from our analysis of the enthalpy data of Conard et al. (4). This value has a fairly high uncertainty because of the limited temperature range of the data (35 K). It is in reasonable agreement with the estimated $C_p°$ = 17 cal K^{-1} mol^{-1} of Mah and Pankrantz (5). The heat capacity below the glass transition is that of the crystal. $S°(\ell, 298.15\ K)$ is calculated in a manner similar to that used for the enthalpy of formation.

Fusion Data

See the crystal table for details.

Decomposition Data

The Gibbs energy data indicates that NiS(ℓ) will decompose to the elements, Ni(ℓ) + 1/2 S_2(g), at a temperature of 2350±200 K and a fugacity of 1 atm. In view of the uncertainty in the Gibbs free energy data, computation of the exact zero point does not seem justified. Since the Gibbs energy data for NiS(ℓ) and NiS(g) (6) intersect at 2990±220 K, it seems clear that decomposition will occur rather than boiling.

References

1. M. Nagamori and T. R. Ingraham, Met. Trans. 1, 1821 (1970).
2. G. A. Meyer, J. S. Warner, Y. K. Rao, and H. H. Kellogg. Met. Trans. 6B, 229 (1975).
3. T. Rosengvist, J. Iron Steel Inst. (London) 176, 37 (1954).
4. B. R. Conard, R. Sridhar and J. S. Warner, paper presented at the 106th AIME meeting, March, 1977.
5. A. D. Mah and L. B. Pankrantz, U. S. Bur. Mines Bull. 668, 1976.
6. JANAF Thermochemical Tables: NiS(g), 12-31-76.

Nickel Sulfide (NiS) $Ni_1S_1(l)$

Enthalpy Reference Temperature = T_r = 298.15 K — Standard State Pressure = p° = 0.1 MPa

T/K	$C_p°$ (J K⁻¹mol⁻¹)	S° (J K⁻¹mol⁻¹)	-[G°-H°(T_r)]/T (J K⁻¹mol⁻¹)	H°-H°(T_r) (kJ mol⁻¹)	$\Delta_f H°$ (kJ mol⁻¹)	$\Delta_f G°$ (kJ mol⁻¹)	Log K_f
0							
100							
200							
298.15	47.112	72.441	72.441	0.	-62.474	-65.609	11.494
300	47.196	72.732	72.442	0.087	-62.477	-65.629	11.427
400	50.501	86.780	74.334	4.978	-64.911	-66.561	8.692
500	53.053	98.329	78.012	10.158	-66.627	-66.788	6.977
600	55.312	108.206	82.240	15.579	-68.069	-66.681	5.805
652.000	56.372	112.847	84.498	18.483	BETA <--> ALPHA		
652.000	53.152	122.729	84.498	24.927	TRANSITION		
700	54.518	126.553	87.252	27.511	-62.807	-66.823	4.986
800	57.363	134.019	92.638	33.105	-63.511	-67.344	4.397
860.000	59.070	138.228	95.673	36.598	GLASS <--> LIQUID		
860.000	76.776	138.228	95.673	36.598	TRANSITION		
900	76.776	141.719	97.642	39.669	-116.333	-66.733	3.873
1000	76.776	149.808	102.461	47.346	-113.701	-61.363	3.205
1100	76.776	157.125	107.103	55.024	-111.151	-56.254	2.671
1200	76.776	163.806	111.554	62.702	-108.690	-51.372	2.236
1249.000	76.776	166.878	113.665	66.464	ALPHA <--> LIQUID		
1300	76.776	169.951	115.813	70.379	-106.325	-46.692	1.876
1400	76.776	175.641	119.886	78.057	-104.061	-42.191	1.574
1500	76.776	180.938	123.781	85.735	-101.902	-37.847	1.318
1600	76.776	185.893	127.510	93.412	-99.852	-33.645	1.098
1700	76.776	190.547	131.083	101.090	-97.916	-29.567	0.908
1800	76.776	194.936	134.509	108.768	-113.247	-24.884	0.722
1900	76.776	199.087	137.800	116.445	-111.450	-20.024	0.551
2000	76.776	203.025	140.964	124.123	-109.664	-15.259	0.399
2100	76.776	206.771	144.009	131.800	-107.890	-10.582	0.263
2200	76.776	210.343	146.944	139.478	-106.126	-5.990	0.142
2300	76.776	213.755	149.775	147.156	-104.372	-1.477	0.034
2400	76.776	217.023	152.509	154.833	-102.626	2.959	-0.064
2500	76.776	220.157	155.153	162.511	-100.888	7.322	-0.153
2600	76.776	223.168	157.711	170.189	-99.158	11.617	-0.233
2700	76.776	226.066	160.190	177.866	-97.433	15.845	-0.307
2800	76.776	228.858	162.593	185.544	-95.714	20.009	-0.373
2900	76.776	231.552	164.924	193.222	-94.001	24.111	-0.434
3000	76.776	234.155	167.189	200.899	-92.294	28.155	-0.490
3100	76.776	236.673	169.390	208.577	-90.590	32.142	-0.542
3200	76.776	239.110	171.531	216.254	-465.726	41.262	-0.674
3300	76.776	241.473	173.615	223.932	-462.379	57.054	-0.903
3400	76.776	243.765	175.644	231.610	-459.034	72.744	-1.118

PREVIOUS: CURRENT: December 1976

Nickel Sulfide (NiS) $Ni_1S_1(l)$

NICKEL SULFIDE (NiS) RHOMBOHEDRAL–HEXAGONAL–LIQUID M_r = 90.75

0 to 652 K		crystal, rhombohedral (β)
652 to 1249 K		crystal, hexagonal (α)
above 1249 K		liquid

Refer to the individual tables for details.

Nickel Sulfide (NiS) $Ni_1S_1(cr,l)$

Enthalpy Reference Temperature = T_r = 298.15 K   Standard State Pressure = p° = 0.1 MPa

T/K	C_p° (J K⁻¹mol⁻¹)	S° (J K⁻¹mol⁻¹)	$-[G^\circ - H^\circ(T_r)]/T$ (J K⁻¹mol⁻¹)	$H^\circ - H^\circ(T_r)$ (kJ mol⁻¹)	Δ_fH° (kJ mol⁻¹)	Δ_fG° (kJ mol⁻¹)	Log K_f
0	0.	0.	INFINITE	-8.465	-87.131	-87.131	INFINITE
100	23.187	12.857	89.876	-7.702	-87.566	-86.854	45.368
200	40.626	35.370	57.135	-4.353	-87.748	-86.055	22.475
298.15	47.112	52.992	52.992	0.	-87.864	-85.200	14.927
300	47.196	53.284	52.993	0.087	-87.867	-85.184	14.832
400	50.501	67.332	54.886	4.978	-90.301	-84.172	10.992
500	53.053	78.880	58.563	10.158	-92.016	-82.453	8.614
600	55.312	88.757	62.792	15.579	-93.458	-80.402	7.000
652.000	56.372	93.399	65.050	18.483	BETA <--> ALPHA		
652.000	53.152	103.281	65.050	24.927	TRANSITION		
700	54.518	107.105	67.804	27.511	-88.196	-78.598	5.865
800	57.363	114.570	73.189	33.105	-88.900	-77.175	5.039
900	60.208	121.491	78.176	38.983	-142.408	-74.603	4.330
1000	63.053	127.982	82.835	45.146	-141.290	-67.127	3.506
1100	65.898	134.125	87.221	51.594	-139.971	-59.773	2.838
1200	68.743	139.980	91.375	58.326	-138.456	-52.547	2.287
1249.000	70.137	142.759	93.337	61.728	ALPHA <--> LIQUID		
1249.000	76.776	166.878	93.337	91.853	TRANSITION		
1300	76.776	169.951	96.283	95.769	-106.325	-46.692	1.876
1400	76.776	175.641	101.751	103.447	-104.061	-42.191	1.574
1500	76.776	180.938	106.855	111.124	-101.902	-37.848	1.318
1600	76.776	185.893	111.642	118.802	-99.852	-33.645	1.098
1700	76.776	190.547	116.148	126.479	-97.916	-29.567	0.908
1800	76.776	194.936	120.404	134.157	-113.247	-24.884	0.722
1900	76.776	199.087	124.437	141.835	-111.450	-20.024	0.551
2000	76.776	203.025	128.269	149.512	-109.664	-15.259	0.399
2100	76.776	206.771	131.919	157.190	-107.890	-10.582	0.263
2200	76.776	210.343	135.403	164.868	-106.126	-5.990	0.142
2300	76.776	213.756	138.736	172.545	-104.372	-1.477	0.034
2400	76.776	217.023	141.930	180.223	-102.626	2.959	-0.064
2500	76.776	220.157	144.997	187.901	-100.888	7.322	-0.153
2600	76.776	223.168	147.946	195.578	-99.158	11.617	-0.233
2700	76.776	226.066	150.786	203.256	-97.433	15.845	-0.307
2800	76.776	228.858	153.525	210.933	-95.714	20.008	-0.373
2900	76.776	231.552	156.169	218.611	-94.001	24.111	-0.434
3000	76.776	234.155	158.726	226.289	-92.294	28.155	-0.490
3100	76.776	236.673	161.200	233.966	-90.590	32.142	-0.542
3200	76.776	239.110	163.597	241.644	-465.726	41.262	-0.674
3300	76.776	241.473	165.921	249.322	-462.379	57.054	-0.903
3400	76.776	243.765	168.177	256.999	-459.034	72.744	-1.118

PREVIOUS: CURRENT: December 1976

Nickel Sulfide (NiS) $Ni_1S_1(cr,l)$

NICKEL SULFIDE (NiS) IDEAL GAS $M_r = 90.75$

$S°(298.15\ K) = [252.282 \pm 8.4]\ J\ K^{-1}\ mol^{-1}$

$\Delta_f H°(0\ K) = 357.40 \pm 16.7\ kJ\ mol^{-1}$
$\Delta_f H°(298.15\ K) = 357.42 \pm 16.7\ kJ\ mol^{-1}$

Electronic Levels and Quantum Weights

ε_i, cm^{-1}	g_i
[0]	[9]
[1360.7]	[7]
[2269.6]	[5]
[16661.6]	[5]
[16977.8]	[3]
[17230.7]	[1]

$\omega_e = 544\ cm^{-1}$ $\omega_e x_e = 2.6\ cm^{-1}$ $\sigma = 1$
$B_e = [0.191]\ cm^{-1}$ $\alpha_e = [0.0011]\ cm^{-1}$ $r_e = [2.07]\ Å$

Enthalpy of Formation

Drowart et al. (1) used a modified Knudsen cell and a mass spectrometric technique to study the equilibrium NiS(g) + Mn(g) = MnS(g) + Ni(g). Since they report no data, we adopt their value of $D_0^0 = 81.5 \pm 3.5$ kcal mol^{-1} based on their 3rd law analysis of this equilibrium. With auxiliary JANAF data (2) this yields $\Delta_f H°(298.15\ K) = 85.4 \pm 4.0$ kcal mol^{-1}. Trevedi (3) studied the absorption spectrum of NiS and determined $D_0^0 = 101.6$ kcal mol^{-1} from the onset of continuous absorption due to photodissociation. This value is likely to be too high in view of the experimental difficulties in determining the exact position of the onset of continuous absorption. Complications arise because of background emission from the furnace. A Birge-Sponer extrapolation of the vibrational data to determine D_0^0 is not meaningful since DeVore and Franzen (4) used the reverse procedure to arrive at the reported value of $\omega_e x_e$.

Heat Capacity and Entropy

No experimental information is available on the electronic structure of NiS. The electronic states are estimated to be identical to those of the Ni^{2+} ion (5) following the procedure of Brewer and Rosenblatt (6) for the transition metal oxides. As stated by Brewer and Robenblatt (6), this procedure usually overestimates the contribution of electronic states. Nevertheless, it appears to be the best method available. The uncertainty assigned to S°(298.15 K) is based mainly on the uncertainties in this approximation since the presence of the sulfide ion will undoubtedly alter the electronic structure of the nickel ion. Levels above 50,000 cm^{-1} are not included since they have a negligible effect on the thermodynamic properties. The vibrational constants are taken from a study of the matrix isolation spectrum by DeVore and Franzen (4). The rotational constants are estimated based on an oxide-sulfide correlation due to Barrow and Cousins (7), r(NiS) = 0.237 + 1.116 r(NiO). The value r(NiO) = 1.64 Å is taken from Brewer and Chandraskharaiah (8). The value of α_e is estimated assuming a Morse potential function using the expression given by Herzberg (9).

References

1. J. Drowart, A. Pattoret, and S. Smoes, Proc. Brit. Ceram. Soc. 8, 67 (1967).
2. JANAF Thermochemical Tables: Ni(g, cr), 12-31-76; S(cr), 12-31-65; S(g), 6-30-71.
3. H. Trevedi, Proc. Acad. Sci. United Provinces Agra Oudh, India 5, 34 (1935).
4. T. C. DeVore and H. F. Franzen, High Temp. Sci. 7, 220 (1975).
5. C. E. Moore, U. S. Nat. Bur. Stand. Circ. 467, Vol. 2, 1952.
6. L. Brewer and G. M. Rosenblatt, in "Advances in High Temperature Chemistry," L. Eyring, ed., Vol. 2, Academic Press, New York, 1969.
7. R. F. Barrow and C. Cousins, in "Advances in High Temperature Chemistry," L. Eyring, ed., Vol. 4, Academic Press, New York, 1971.
8. L. Brewer and S. Chandraskharaiah, UCRL-8713 revised, June 1960.
9. G. Herzberg, "Spectra of Diatomic Molecules," Second Edition, Van Nostrand, New York, p. 108, 1950.

Enthalpy Reference Temperature = T_r = 298.15 K   Standard State Pressure = p° = 0.1 MPa

T/K	$C_p°$ (J K^{-1}mol^{-1})	S°	$-[G°-H°(T_r)]/T$	$H°-H°(T_r)$ (kJ mol^{-1})	$\Delta_f H°$	$\Delta_f G°$	Log K_f
0	0.	0.	INFINITE	-9.218	357.398	357.398	INFINITE
100	29.323	218.044	281.104	-6.306	359.113	339.306	-177.235
200	31.871	239.056	255.355	-3.260	358.628	319.584	-83.467
250	33.262	246.320	252.843	-1.631	358.063	309.886	-64.747
298.15	34.452	252.282	252.282	0.	357.419	300.664	-52.675
300	34.495	252.496	252.283	0.064	357.392	300.312	-52.289
350	35.599	257.897	252.707	1.817	356.646	290.857	-43.408
400	36.597	262.717	253.663	3.622	353.625	281.600	-36.773
450	37.497	267.081	254.915	5.475	352.101	272.680	-31.652
500	38.297	271.074	256.334	7.370	350.478	263.944	-27.574
600	39.598	278.177	259.397	11.268	347.513	246.918	-21.496
700	40.520	284.355	262.531	15.277	344.853	230.375	-17.191
800	41.118	289.808	265.607	19.361	342.639	214.174	-13.984
900	41.463	294.673	268.571	23.492	287.384	199.325	-11.569
1000	41.624	299.051	271.404	27.648	286.494	189.588	-9.903
1100	41.658	303.021	274.100	31.813	285.530	179.943	-8.545
1200	41.608	306.644	276.664	35.977	284.477	170.390	-7.417
1300	41.507	309.971	279.099	40.133	283.321	160.929	-6.466
1400	41.377	313.042	281.416	44.277	282.052	151.561	-5.655
1500	41.234	315.892	283.620	48.408	280.664	142.287	-4.955
1600	41.087	318.549	285.721	52.524	279.153	133.111	-4.346
1700	40.944	321.035	287.726	56.625	277.512	124.033	-3.811
1800	40.807	323.371	289.642	60.713	258.592	115.770	-3.360
1900	40.679	325.574	291.476	64.787	256.785	107.885	-2.966
2000	40.562	327.658	293.233	68.849	254.955	100.095	-2.614
2100	40.456	329.634	294.920	72.900	253.102	92.397	-2.298
2200	40.362	331.514	296.541	76.940	251.229	84.789	-2.013
2300	40.278	333.306	298.101	80.972	249.338	77.265	-1.755
2400	40.206	335.019	299.604	84.996	247.430	69.825	-1.520
2500	40.144	336.659	301.053	89.014	245.508	62.464	-1.305
2600	40.092	338.232	302.453	93.026	243.572	55.181	-1.109
2700	40.049	339.745	303.807	97.032	241.626	47.972	-0.928
2800	40.015	341.200	305.116	101.036	239.670	40.835	-0.762
2900	39.990	342.604	306.385	105.036	237.706	33.768	-0.608
3000	39.973	343.960	307.615	109.034	235.734	26.770	-0.466
3100	39.963	345.270	308.809	113.031	233.756	19.837	-0.334
3200	39.961	346.539	309.968	117.027	-145.061	18.156	-0.296
3300	39.964	347.769	311.095	121.023	-145.395	23.262	-0.368
3400	39.974	348.962	312.191	125.020	-145.731	28.378	-0.436
3500	39.990	350.121	313.258	129.018	-146.070	33.504	-0.500
3600	40.011	351.248	314.298	133.018	-146.411	38.639	-0.561
3700	40.036	352.344	315.312	137.020	-146.756	43.784	-0.618
3800	40.066	353.412	316.300	141.025	-147.105	48.938	-0.673
3900	40.100	354.453	317.265	145.034	-147.457	54.102	-0.725
4000	40.138	355.469	318.208	149.046	-147.815	59.275	-0.774
4100	40.178	356.461	319.129	153.061	-148.178	64.457	-0.821
4200	40.222	357.429	320.029	157.081	-148.547	69.647	-0.866
4300	40.269	358.376	320.910	161.106	-148.923	74.847	-0.909
4400	40.318	359.303	321.772	165.135	-149.308	80.055	-0.950
4500	40.369	360.209	322.616	169.169	-149.701	85.272	-0.990
4600	40.422	361.097	323.443	173.209	-150.104	90.498	-1.028
4700	40.476	361.967	324.253	177.254	-150.518	95.733	-1.064
4800	40.532	362.820	325.048	181.304	-150.944	100.977	-1.099
4900	40.589	363.656	325.828	185.360	-151.384	106.230	-1.132
5000	40.647	364.477	326.592	189.422	-151.838	111.492	-1.165
5100	40.706	365.282	327.343	193.490	-152.307	116.763	-1.196
5200	40.765	366.073	328.080	197.563	-152.793	122.044	-1.226
5300	40.825	366.850	328.804	201.643	-153.298	127.334	-1.255
5400	40.886	367.614	329.516	205.728	-153.822	132.634	-1.283
5500	40.946	368.365	330.216	209.820	-154.366	137.943	-1.310
5600	41.007	369.103	330.904	213.918	-154.933	143.263	-1.336
5700	41.067	369.829	331.580	218.021	-155.524	148.593	-1.362
5800	41.128	370.544	332.246	222.131	-156.139	153.934	-1.386
5900	41.188	371.248	332.901	226.247	-156.781	159.286	-1.410
6000	41.249	371.941	333.546	230.369	-157.451	164.648	-1.433

PREVIOUS: December 1976 (1 atm)   CURRENT: December 1976 (1 bar)

NICKEL SULFIDE (NiS_2) CRYSTAL M_r = 122.81

$S°(298.15\ K) = [71.965 \pm 8.4]\ J\ K^{-1}\ mol^{-1}$

$T_{fus} = 1280 \pm 5\ K$

$\Delta_f H°(0\ K)$ = Unknown

$\Delta_f H°(298.15\ K) = -131.38 \pm 16.7\ kJ\ mol^{-1}$

$\Delta_{fus} H° = [65.689 \pm 8.4]\ kJ\ mol^{-1}$

Enthalpy of Formation

Our analysis of equilibrium data involving NiS_2 is summarized below.

Source	Reaction*	T/K	Data Points	$\Delta_r H°(298.15\ K)$, kcal mol^{-1} 2nd law	3rd law	Drift cal K^{-1} mol^{-1}
(1)	A	298.15	1	calorimetric	-35.1	
(2)	B	923-1033	5	-46.3±0.4	-47.1±0.1	-0.8±0.4
(3)	B	673-873	equation	-49.2	-48.7	0.7
(4)	C	500-723	2	13.9	-5.1±9.9	-31.4
(5)	C	673-873	20	-2.9±0.2	-7.5±2.4	-14.1±0.3
(6)	A	298.15	1	emf[ΔG°(298.15 K) = -24.3)]	-25.9	

*A) $Ni(cr) + 2\ S(cr) = NiS_2(cr)$

B) $2\ Ni_{1-x}S(cr) + S_2(g) = 2\ NiS_2(cr)$

C) $Ni_{1-x}S(cr) + H_2S(g) = NiS_2(cr) + H_2(g)$

The studies due to Delafosse and Barret (4) and Rosenqvist (5) show poor agreement between 2nd and 3rd law values of $\Delta_r H°$ as well as a large 3rd law drift and are not considered further. Kullerud and Yund (7) mention that a coating of $NiS_2(cr)$ often forms on $Ni_{1-x}S(cr)$ and it seems possible that this may lead to a slow establishment of equilibrium in this system. DeRanter and Breckpot (6) state that their results may be biased for $NiS_2(cr)$ since it is a semiconductor and this may interfere in emf measurements. Consequently their results are not considered further. 3rd law results of Biltz et al. (2) and Leegaard and Rosenquist (3) yield values of $\Delta_f H°(298.15\ K) = -29.1\pm1.6$ and -29.8 ± 1.7 kcal mol^{-1}, respectively, when combined with auxiliary JANAF data (8). These values are averaged with the calorimetric value due to Ariya (1) of -35.1 kcal mol^{-1} to yield the adopted value of -31.4 kcal mol^{-1}. An uncertainty of ±4 kcal mol^{-1} is assigned due mainly to the variation in stoichiometry of one of the reactants, $Ni_{1-x}S$. The adopted value compares favorably with a value of -32.0 kcal mol^{-1} obtained by Mills (9) in a recent critical evaluation. It also compares favorably with a value of -31.3 kcal mol^{-1} calculated assuming the same contribution per gram-atom as in NiS and Ni_3S_2 (8).

Heat Capacity and Entropy

The only heat capacity measurements on NiS_2 are those of Winterberger and Bonnerot (10) at low temperature (5-10 K). Consequently the heat capacity is estimated via the following procedure. The value at 298.15 K is estimated assuming the same contribution per gram-atom as in NiS and Ni_3S_2 (8) yielding $C_p°(298.15\ K) = 16.88$ cal K^{-1} mol^{-1}. The temperature dependence is estimated using method A of Kubaschewski et al. (11). This yields $C_p° = 15.401 + 4.960 \times 10^{-3}\ T$.

2nd law analyses of equilibrium data of Leegaard and Rosenqvist (3) and Biltz et al. (2) combined with auxiliary JANAF data (8) yield $S°(298.15\ K) = 16.89$ and 17.60 cal K^{-1} mol^{-1}, respectively. We adopt the average of these values, 17.2 cal K^{-1} mol^{-1}, with an uncertainty estimated to be Δ2 cal K^{-1} mol^{-1}. Similar analyses of studies of Delafosse and Barret (4) and Rosenqvist (5) give values of 13.10 and 29.96 cal K^{-1} mol^{-1}, respectively, but as mentioned above, these studies do not appear to be reliable since equilibrium may not have been established. A value of 19.10 cal K^{-1} mol^{-1} is obtained assuming the same contribution per gram-atom as in NiS and Ni_3S_2 (8). This is somewhat higher than the adopted value but this may be reasonable since the crystal symmetry of NiS_2 is higher (cubic) than that of NiS and Ni_3S_2 (hexagonal) (7).

Fusion Data

The melting point of $NiS_2(cr)$ is taken from the phase study by Kullerud and Yund (7). The enthalpy of melting of NiS_2 is unknown and is estimated as follows. Lewis and Randall (12) point out that the sum of the entropy of transition(s) plus the entropy of fusion is approximately constant, on a gram-atom basis, for compounds of similar structure. Using NiS and Ni_3S_2 (8) as base compounds, we calculate the above sums to be 4.06 and 4.13 cal K^{-1} g-atom^{-1}, respectively. Since NiS_2 has no known transition, we use the average value of 4.10 cal K^{-1} g-atom^{-1} to calculate $\Delta_{fus}S°(NiS_2) = 12.3$ cal K^{-1} mol^{-1} which yields $\Delta_{fus}H°$ = 15.7 kcal mol^{-1} with an estimated uncertainty of ±2.0 kcal mol^{-1}.

References

1. S. M. Ariya, M. P. Morozova, L. A. Pavlinova and V. L. Ponomareva, Russ. J. Phys. Chem. 45, 1355 (1971).
2. W. Biltz, A. Voigt, and K. Meisel, Z. Anorg. Chem. 228, 275 (1936).
3. T. Leegaard and T. Rosenqvist, Z. Anorg. Allg. Chem. 328, 294 (1964).
4. D. Delafosse and P. Barret, Compt. Rendu 252, 888 (1961).
5. T. Rosenqvist, J. Iron Steel Inst. (London) 176, 37 (1954).
6. G. DeRanter and R. Breckpot, Bull. Soc. Chim. Belg. 78, 503 (1969).
7. G. Kullerud and R. A. Yund, J. Petrology 3, 126 (1962).
8. JANAF Thermochemical Tables: $S_2(g)$, 12-31-65; $H_2S(g)$, 12-31-65; NiS(cr), 12-31-76; $Ni_3S(cr)$, 12-31-76.
9. K. C. Mills, "Thermodynamic Data for Inorganic Sulphides, Selenides, and Tellurides," Butterworths, London, 1974.
10. M. Winterberger and J. Bonnerot, Colloq. Int. Centre Natl. Rech. Sci. #157, 369 (1967).
11. O. Kubaschewski, E. L. Evans and C. B. Alcock, "Metallurgical Thermochemistry," Pergamon, Oxford, 1967.
12. G. N. Lewis and M. Randall as revised by K. S. Pitzer and L. Brewer, "Thermodynamics," second edition, McGraw-Hill, New York, p. 519, 1961.

Nickel Sulfide (NiS_2) $Ni_1S_2(cr)$

Enthalpy Reference Temperature = T_r = 298.15 K; Standard State Pressure = p° = 0.1 MPa

T/K	$C_p°$ (J K^{-1}mol^{-1})	S°	$-[G°-H°(T_r)]/T$	$H°-H°(T_r)$ (kJ mol^{-1})	$\Delta_f H°$	$\Delta_f G°$	Log K_f
0							
100							
200							
298.15	70.626	71.965	71.965	0.	-131.378	-124.813	21.867
300	70.664	72.402	71.966	0.131	-131.379	-124.772	21.725
400	72.739	93.015	74.763	7.301	-136.131	-122.358	15.978
500	74.814	109.469	80.112	14.678	-139.577	-118.542	12.384
600	76.889	123.292	86.186	22.264	-142.439	-114.057	9.930
700	78.965	135.301	92.363	30.056	-144.663	-109.134	8.144
800	81.040	145.981	98.410	38.057	-146.172	-103.951	6.787
900	83.115	155.646	104.241	46.264	-253.607	-96.438	5.597
1000	85.190	164.510	109.830	54.680	-252.093	-79.055	4.129
1100	87.266	172.727	115.179	63.302	-250.470	-61.829	2.936
1200	89.341	180.409	120.298	72.133	-248.744	-44.755	1.948
1280.000	91.001	186.228	124.238	79.346	------ CRYSTAL <--> LIQUID -----		
1300	91.416	187.642	125.203	81.171	-246.921	-27.829	1.118
1400	93.491	194.493	129.910	90.416	-245.007	-11.047	0.412
1500	95.567	201.014	134.434	99.869	-243.006	5.595	-0.195
1600	97.642	207.248	138.792	109.529	-240.921	22.101	-0.722
1700	99.717	213.229	142.996	119.397	-238.758	38.474	-1.182
1800	101.793	218.988	147.058	129.473	-253.667	55.432	-1.609

PREVIOUS: CURRENT: March 1977

Nickel Sulfide (NiS_2) $Ni_1S_2(cr)$

NICKEL SULFIDE (NiS_2) LIQUID M_r = 122.81

$S°(298.15\ K) = [122.427]\ J\ K^{-1}\ mol^{-1}$

$T_{fus} = 1280 \pm 5\ K$

$\Delta_f H°(298.15\ K) = [-66.568]\ kJ\ mol^{-1}$

$\Delta_{fus} H° = [65.689 \pm 8.4]\ kJ\ mol^{-1}$

Enthalpy of Formation

There are no experimental studies of the enthalpy of formation of $NiS_2(\ell)$. Therefore, $\Delta_f H°(NiS_2, \ell, 298.15\ K)$ is calculated from that of $NiS_2(cr)$ (1) by adding $\Delta_{fus}H°$ and the difference in enthalpy, H°(1280 K)-H°(298.15 K), between the crystal and liquid.

Heat Capacity and Entropy

The heat capacity is assumed to remain constant at 21.75 cal $K^{-1}\ mol^{-1}$ (the value of $NiS_2(cr)$ at the melting point) above an assumed glass transition point of 960 K. Below that temperature the crystal heat capacity (1) is assumed to be valid.

$S°(\ell, 298.15\ K)$ is calculated in a manner similar to that used for the enthalpy of formation.

Fusion Data

See the crystal table for details.

Decomposition Data

Under standard conditions (1 atm. pressure) the use of auxiliary JANAF data (1) indicates that $NiS_2(\ell)$ will decompose spontaneously to $NiS(\ell) + 1/2\ S_2(g)$ at all temperatures above the normal melting point. This appears to be at variance with the phase diagram of Kullerud and Yund (2) which indicates the presence of a liquid of variable composition in the Ni-S system from a stoichiometry less than Ni_3S_2 to just above NiS_2 at temperatures slightly above the melting point. However, the authors point out that this diagram does not refer to a constant pressure system and give a short discussion of the loss of sulfur to the vapor from $NiS_2(cr)$. From the sealed tube experiments of (2) it does appear that $NiS_2(\ell)$ will be stable under its own vapor pressure in a sealed tube.

References

1. JANAF Thermochemical Tables: $NiS_2(cr)$, 3-31-77; $NiS(\ell)$, 12-31-76; $S_2(g)$, 12-31-65.
2. G. Kullerud and R. A. Yund, J. Petrology 3, 126 (1962).

Nickel Sulfide (NiS_2) $Ni_1S_2(l)$

Enthalpy Reference Temperature = T_r = 298.15 K — Standard State Pressure = p° = 0.1 MPa

T/K	$C_p°$ (J K⁻¹mol⁻¹)	S° (J K⁻¹mol⁻¹)	$-[G°-H°(T_r)]/T$ (J K⁻¹mol⁻¹)	$H°-H°(T_r)$ (kJ mol⁻¹)	$\Delta_f H°$ (kJ mol⁻¹)	$\Delta_f G°$ (kJ mol⁻¹)	Log K_f
0							
100							
200							
298.15	70.626	122.427	122.427	0.	-66.568	-75.048	13.148
300	70.664	122.864	122.428	0.131	-66.569	-75.101	13.076
400	72.739	143.477	125.225	7.301	-71.321	-77.733	10.151
500	74.814	159.931	130.574	14.678	-74.767	-78.963	8.249
600	76.889	173.754	136.648	22.264	-77.629	-79.524	6.923
700	78.965	185.763	142.825	30.056	-79.853	-79.647	5.943
800	81.040	196.443	148.872	38.057	-81.362	-79.511	5.192
900	83.115	206.108	154.703	46.264	-188.797	-77.044	4.471
960.000	84.360	211.511	158.086	51.289	GLASS <--> LIQUID TRANSITION		
960.000	91.002	211.511	158.086	51.289			
1000	91.002	215.226	160.298	54.929	-187.034	-64.712	3.380
1100	91.002	223.900	165.692	64.029	-184.934	-52.582	2.497
1200	91.002	231.818	170.877	73.129	-182.937	-40.639	1.769
1280.000	91.002	237.691	174.871	80.409	CRYSTAL <--> LIQUID		
1300	91.002	239.102	175.849	82.229	-181.052	-28.858	1.160
1400	91.002	245.846	180.611	91.329	-179.283	-17.218	0.642
1500	91.002	252.124	185.171	100.430	-177.635	-5.700	0.199
1600	91.002	257.998	189.541	109.530	-176.111	5.712	-0.186
1700	91.002	263.515	193.732	118.630	-174.715	17.032	-0.523
1800	91.002	268.716	197.755	127.730	-190.600	28.989	-0.841

PREVIOUS: CURRENT: March 1977

Nickel Sulfide (NiS_2) $Ni_1S_2(l)$

NICKEL SULFIDE (NiS_2) CRYSTAL-LIQUID M_r = 122.81

0 to 1280 K crystal
above 1280 K liquid

Refer to the individual tables for details.

Nickel Sulfide (NiS_2) Ni_1S_2(cr,l)

Enthalpy Reference Temperature = T_r = 298.15 K   Standard State Pressure = p° = 0.1 MPa

T/K	C_p° (J K⁻¹mol⁻¹)	S° (J K⁻¹mol⁻¹)	$-[G°-H°(T_r)]/T$ (J K⁻¹mol⁻¹)	$H°-H°(T_r)$ (kJ mol⁻¹)	$\Delta_fH°$ (kJ mol⁻¹)	$\Delta_fG°$ (kJ mol⁻¹)	Log K_f
0							
100							
200							
298.15	70.626	71.965	71.965	0.	-131.378	-124.813	21.867
300	70.664	72.402	71.966	0.131	-131.379	-124.772	21.725
400	72.739	93.015	74.763	7.301	-136.131	-122.358	15.978
500	74.814	109.469	80.112	14.678	-139.577	-118.542	12.384
600	76.889	123.292	86.186	22.264	-142.439	-114.057	9.930
700	78.965	135.301	92.363	30.056	-144.663	-109.134	8.144
800	81.040	145.981	98.410	38.057	-146.172	-103.951	6.787
900	83.115	155.646	104.241	46.264	-253.607	-96.438	5.597
1000	85.190	164.510	109.830	54.680	-252.093	-79.055	4.129
1100	87.266	172.727	115.179	63.302	-250.470	-61.829	2.936
1200	89.341	180.409	120.298	72.133	-248.744	-44.755	1.948
1280.000	91.001	186.228	124.238	79.346	CRYSTAL <--> LIQUID	TRANSITION	
1280.000	91.002	237.691	124.238	145.219			
1300	91.002	239.102	125.995	147.039	-181.052	-28.858	1.160
1400	91.002	245.846	134.318	156.140	-179.283	-17.218	0.642
1500	91.002	252.124	141.965	165.240	-177.635	-5.700	0.199
1600	91.002	257.998	149.035	174.340	-176.111	5.712	-0.186
1700	91.002	263.515	155.609	183.440	-174.715	17.032	-0.523
1800	91.002	268.716	161.749	192.540	-190.600	28.989	-0.841

PREVIOUS: CURRENT: March 1977

Nickel Sulfide (NiS_2) Ni_1S_2(cr,l)

NICKEL SULFIDE (Ni_3S_2)　　CRYSTAL　　M_r = 240.19

$S°(298.15\ K) = 133.887 \pm 0.4\ J\ K^{-1}\ mol^{-1}$
$T_{trs} = 829 \pm 3\ K$
$T_{fus} = 1062 \pm 3\ K$

$\Delta_f H°(0\ K) = -214.29 \pm 5.0\ kJ\ mol^{-1}$
$\Delta_f H°(298.15\ K) = -216.31 \pm 5.0\ kJ\ mol^{-1}$
$\Delta_{trs} H° = 56.233 \pm 0.4\ kJ\ mol^{-1}$
$\Delta_{fus} H° = 19.748 \pm 0.4\ kJ\ mol^{-1}$

Enthalpy of Formation

Three sets of authors studied the phase equilibrium between Ni_3S_2(cr) and Ni(cr) by use of the $H_2(g)/H_2S(g)$ method (1, 2, 3). Our 3rd law analyses of the data due to Rosenqvist (1), Sudo (2) and Line and Laffitte (3) yield $\Delta_f H°(298.15\ K) = -51.6\pm1.0$, -50.1 ± 1.6, and -53.4 ± 1.8 kcal mol^{-1}, respectively. We adopt the average of the values, $\Delta_f H°(298.15\ K) = -51.7\pm1.7$ kcal mol^{-1}. Ariya et al. (4) and Vanyukov and Kiseleva (5) reported calorimetric values of -48.3 ± 0.9 and -43.3 kcal mol^{-1}, respectively. Vanuykov's value is undoubtedly incorrect since Mills (6) determined that values for Fes and MnS from the same study are approximately 25% too low. A 3rd law analysis of the emf value reported by DeRanter and Breckpot (7) yields $\Delta_f H°(298.15\ K) = -38.7$ kcal mol^{-1}. This value, however, refers to an amorphous material. At least part of the variation in values of the enthalpy of formation apparently is a result of the very wide range of stoichiometry for the high temperature phase of Ni_3S_2 (8, 9).

Heat Capacity and Entropy

The low temperature heat capacities (52-296 K) are from Weller and Kelley (10). The high temperature heat capacities are from our analysis of the high temperature enthalpy (298.15-1050 K) data of Conard et al. (11). Our analysis differs slightly from that of Conard et al. since we have forced the high and low temperature heat capacities to join smoothly at 298.15 K. Data above 1050 K are extrapolated assuming a constant heat capacity for the high temperature phase of Ni_3S_2. The thermodynamic functions show good agreement with those reported by Mah and Pankrantz (12).

The entropy at 298.15 is calculated based on an extrapolation of $S°(51\ K) = 2.06$ cal $K^{-1}\ mol^{-1}$ by Weller and Kelley (10).

Phase Data

A phase transition is observed at 829 K for Ni_3S_2 (8). Below the transition, Kullerud and Yund (8) reported the material to be stoichiometric within the limits of ±0.3 weight percent while Rau (9) reported the homogeneity range to extend over a smaller range and to the sulfur-rich side only. Above the phase transition, a wide range of homogeneity exists for $Ni_{3\pm x}S_2$ from approximately $Ni_{2.5}S_2$ to $Ni_{3.7}S_2$ (1, 8, 9). The phase diagram indicates that upon heating stoichiometric Ni_3S_2, a two phase region $[Ni_yS_z(\ell) + Ni_{3\pm x}S_2(cr)]$ will be encountered within a few degrees of the formation of stoichiometric liquid Ni_3S_2 (8, 9).

Transition Data

We adopt a value of 829±3 K for the phase transition from Ni_3S_2 to $Ni_{3\pm x}S_2$ based on the detailed study by Kullerud and Yund (8). This value is in fair agreement with values of 838 and 828±5 reported by Rau (9) and Rosenqvist (1), respectively. The adopted heat of transition is from our analysis of the high temperature enthalpy data of Conard et al. (11). Mah and Pankrantz (12) reported the transition at 840 K with $\Delta_f H° = 13.38$ kcal mol^{-1} in excellent agreement with the adopted $\Delta_{trs}H°$.

Fusion Data

Kullerud and Yung (8) reported that stoichiometric Ni_3S_2 begins to melt at 1055 K with melting completed at 1062 K. The cooling curve yielded identical results. In view of the two-phase region in the phase diagram, we adopt the latter temperature as the melting point with an estimated uncertainty of ±3 K. The adopted heat of melting is from our analysis of Conard's et al. high temperature enthalpy data (11) and is in good agreement with a value of 4.70 kcal mol^{-1} reported by Mah and Pankrantz (12).

References

1. T. Rosenqvist, J. Iron Steel Inst. (London), 176, 37 (1954).
2. K. Sudo, Sci. Rep. Res. Inst. Tohoku Univ. Ser. A 4, 182 (1952).
3. G. Line and M. Laffitte, Compt. Rend. 256, 3306 (1963).
4. S. M. Ariya, M. P. Morozova, L. A. Pavlinova, and V. L. Ponomareva, Russ. J. Phys. Chem. 45, 1355 (1971).
5. V. A. Vanyukov and N. A. Kiseleva, Yubeleinyi Sbornik Trudov Kafedry i Lab. Tyazhelykh Metallov Moskov. Inst. Tsvetnykh Metallov i Zolota, 304 (1939); Chem. Abstr. 36, 1234 (1942).
6. K. C. Mills, "Thermodynamic Data for Inorganic Sulphides, Selenides, and Terrurides," Butterworths, London, 1974.
7. C. DeRanter and R. Breckpot, Bull. Soc. Chim. Belg. 78, 503 (1969).
8. G. Kullerud and R. A. Yund, J. Petrol. 3, 126 (1962).
9. H. Rau, J. Phys. Chem. Solids 37, 929 (1976).
10. W. W. Weller and K. K. Kelley, U. S. Bur. Mines RI 6511, 1964.
11. B. R. Conard, R. Sridhar and J. S. Warner, paper presented at the 106th AIME meeting, March 1977.
12. A. D. Mah and L. B. Pankrantz, U. S. Bur. Mines Bull. 668, 1976.

Nickel Sulfide (Ni_3S_2)　　Ni_3S_2(cr)

Enthalpy Reference Temperature = T_r = 298.15 K　　Standard State Pressure = p° = 0.1 MPa

T/K	$C_p°$	S°	$-[G°-H°(T_r)]/T$	$H°-H°(T_r)$	$\Delta_f H°$	$\Delta_f G°$	Log K_f
	J $K^{-1}mol^{-1}$			kJ mol^{-1}			
0	0.	0.	INFINITE	-21.156	-214.288	-214.288	INFINITE
100	58.116	33.666	225.939	-19.227	-215.261	-213.887	111.723
200	101.489	89.917	144.221	-10.861	-215.847	-212.256	55.436
298.15	117.738	133.887	133.887	0.	-216.313	-210.399	36.861
300	117.947	134.616	133.889	0.218	-216.323	-210.362	36.627
400	127.110	169.857	138.632	12.490	-221.430	-208.103	27.175
500	133.972	198.975	147.871	25.552	-225.126	-204.371	21.350
600	139.913	223.932	158.516	39.249	-228.432	-199.906	17.403
700	145.436	245.918	169.462	53.519	-230.824	-194.922	14.545
800	150.708	265.685	180.275	68.328	-231.697	-189.725	12.388
829.000	152.190	271.078	183.357	72.720		I <--> II	
829.000	188.615	338.910	183.357	128.953		TRANSITION	
900	188.615	354.410	196.248	142.345	-279.582	-187.213	10.866
1000	188.615	374.282	213.076	161.206	-274.001	-177.251	9.259
1062.000	188.615	385.628	222.821	172.901	-------- II	<--> LIQUID	--------
1100	188.615	392.259	228.561	180.068	-268.653	-167.837	7.970
1200	188.615	408.671	242.896	198.929	-263.555	-158.899	6.917
1300	188.615	423.768	256.236	217.791	-258.727	-150.376	6.042
1400	188.615	437.746	268.708	236.652	-254.188	-142.213	5.306
1500	188.615	450.759	280.416	255.514	-249.949	-134.365	4.679
1600	188.615	462.932	291.447	274.375	-246.023	-126.789	4.139

PREVIOUS:　　CURRENT: December 1976

Nickel Sulfide (Ni_3S_2)　　Ni_3S_2(cr)

NICKEL SULFIDE (Ni_3S_2) LIQUID M_r = 240.19

$S°(298.15\ K)$ = [215.930] J K^{-1} mol^{-1} $\Delta_f H°(298.15\ K)$ = [-143.926] kJ mol^{-1}

T_{fus} = 1062 ± 3 K $\Delta_{fus}H°$ = 19.748 ± 0.4 kJ mol^{-1}

Enthalpy of Formation

$\Delta_f H°(\ell, 298.15\ K)$ is derived from that of Ni_3S_2(cr) by adding $\Delta_{fus}H°$ and the difference in enthalpy, H°(1062 K)-H°(298.15 K), between crystal and liquid. Our 2nd and 3rd law analyses of Nagamori and Ingraham's equilibrium data (1) yield $\Delta_f H°(298.15\ K)$ = -33.7 ± 6 and -33.9 ± 0.3 kcal mol^{-1}, respectively, in good agreement with the adopted value. Meyer et al. (2) and Rosenqvist (3) conducted similar studies at higher and lower temperatures, respectively. Both studies qualitatively show good agreement with Nagamori and Ingraham at overlapping temperatures (2), but Meyer's data are only presented graphically so a re-analysis is not feasible. Rosenqvist's study included compositions near, but not exactly at, the Ni_3S_2 stoichiometry. Our extrapolation of his data to the proper stoichiometry leads to a higher value of $K°_p$ at 900 K than at 800 or 1000 K. We conclude that Rosenqvist's data are not of sufficient precision for the small number of points at each temperature (3 or 4) to yield meaningful results when extrapolated.

Heat Capacity and Entropy

The enthalpy of $Ni_3S_2(\ell)$ from the melting point to 1250 K was measured by Conard et al. (4). Our analysis of these data yields the adopted constant heat capacity of 45.84 cal K^{-1} mol^{-1}. Above 1250 K the heat capacity is assumed to remain constant. This value is in good agreement with a value of 45.20 cal K^{-1} mol^{-1} reported by Mah and Pankrantz (5).

$S°(\ell, 298.15\ K)$ is calculated in a manner similar to that used for the enthalpy of formation.

Fusion Data

See the crystal table for details.

Decomposition Data

There is no evidence for the existence of Ni_3S_2(g). In the absence of such information we can only point out that $Ni_3S_2(\ell)$ will decompose to the gaseous elements at 3240 ± 15 K according to the Gibbs energy of formation in this table. This seems to yield an excessively long liquid range for Ni_3S_2.

References

1. M. Nagamori and T. R. Ingraham, Met. Trans. 1, 1821 (1970).
2. G. A. Meyer, J. S. Warner, Y. K. Rao, and N. H. Kellogg, Met. Trans. 6B, 229 (1975).
3. T. Rosenqvist, J. Iron Steel Inst. (London) 176, 37 (1954).
4. B. R. Conard, R. Sridhar, and J. S. Warner, paper presented at the 106th AIME meeting, March 1977.
5. A. D. Mah and L. B. Pankrantz, U. S. Bur. Mines Bull. 668, 1976.

Nickel Sulfide (Ni_3S_2) Ni_3S_2(l)

Enthalpy Reference Temperature = T_r = 298.15 K Standard State Pressure = p° = 0.1 MPa

T/K	$C°_p$ (J $K^{-1}mol^{-1}$)	S° (J $K^{-1}mol^{-1}$)	$-[G°-H°(T_r)]/T$ (J $K^{-1}mol^{-1}$)	$H°-H°(T_r)$ (kJ mol^{-1})	$\Delta_f H°$ (kJ mol^{-1})	$\Delta_f G°$ (kJ mol^{-1})	Log K_f
0							
100							
200							
298.15	117.738	215.930	215.930	0.	-143.926	-162.473	28.465
300	117.947	216.659	215.932	0.218	-143.937	-162.588	28.309
400	127.110	251.900	220.675	12.490	-149.043	-168.534	22.008
500	133.972	281.018	229.914	25.552	-152.739	-173.006	18.074
600	139.913	305.975	240.560	39.249	-156.045	-176.745	15.387
700	145.436	327.963	251.505	53.520	-158.436	-179.966	13.429
760.000	148.549	340.051	258.024	62.341	GLASS <--> LIQUID		
760.000	191.795	340.051	258.024	62.341	TRANSITION		
800	191.795	349.889	262.373	70.012	-157.626	-183.018	11.950
900	191.795	372.479	273.377	89.192	-260.348	-184.242	10.693
1000	191.795	392.686	284.315	108.371	-254.450	-176.104	9.199
1062.000	191.795	404.224	290.982	120.263	-------- II <--> LIQUID --------		
1100	191.795	410.966	295.011	127.551	-248.784	-168.546	8.004
1200	191.795	427.655	305.379	146.730	-243.368	-161.493	7.030
1300	191.795	443.006	315.383	165.910	-238.222	-154.880	6.223
1400	191.795	457.220	325.013	185.089	-233.365	-148.654	5.546
1500	191.795	470.452	334.273	204.269	-228.808	-142.764	4.971
1600	191.795	482.830	343.175	223.448	-224.564	-137.168	4.478
1700	191.795	494.458	351.736	242.628	-220.648	-131.827	4.051
1800	191.795	505.421	359.972	261.807	-268.516	-124.563	3.615
1900	191.795	515.790	367.903	280.987	-264.989	-116.662	3.207
2000	191.795	525.628	375.545	300.166	-261.486	-108.946	2.845
2100	191.795	534.986	382.917	319.345	-258.004	-101.404	2.522
2200	191.795	543.908	390.033	338.525	-254.543	-94.028	2.233
2300	191.795	552.434	396.910	357.704	-251.101	-86.809	1.972
2400	191.795	560.596	403.562	376.884	-247.677	-79.740	1.735
2500	191.795	568.426	410.001	396.063	-244.268	-72.812	1.521
2600	191.795	575.948	416.240	415.243	-240.874	-66.021	1.326
2700	191.795	583.187	422.290	434.422	-237.491	-59.360	1.148
2800	191.795	590.162	428.161	453.602	-234.121	-52.825	0.985
2900	191.795	596.892	433.864	472.781	-230.762	-46.409	0.836
3000	191.795	603.394	439.407	491.961	-227.413	-40.109	0.698
3100	191.795	609.683	444.799	511.140	-224.074	-33.921	0.572
3200	191.795	615.772	450.048	530.319	-1351.248	-12.274	0.200
3300	191.795	621.674	455.159	549.499	-1342.966	29.440	-0.466
3400	191.795	627.400	460.141	568.678	-1334.689	70.904	-1.089
3500	191.795	632.959	465.000	587.858	-1326.421	112.125	-1.673
3600	191.795	638.362	469.741	607.037	-1318.162	153.109	-2.222
3700	191.795	643.617	474.370	626.217	-1309.915	193.864	-2.737
3800	191.795	648.732	478.891	645.396	-1301.683	234.397	-3.222

PREVIOUS: CURRENT: December 1976

Nickel Sulfide (Ni_3S_2) Ni_3S_2(l)

NICKEL SULFIDE (Ni_3S_2) CRYSTAL–CRYSTAL–LIQUID M_r = 240.19

0 to 829 K crystal, I
829 to 1062 K crystal, II
above 1062 K liquid

Refer to the individual tables for details.

Nickel Sulfide (Ni_3S_2) Ni_3S_2(cr,l)

Enthalpy Reference Temperature = T_r = 298.15 K — Standard State Pressure = p° = 0.1 MPa

T/K	C_p° (J K⁻¹mol⁻¹)	S° (J K⁻¹mol⁻¹)	−[G°−H°(T_r)]/T (J K⁻¹mol⁻¹)	H°−H°(T_r) (kJ mol⁻¹)	$\Delta_f H^\circ$ (kJ mol⁻¹)	$\Delta_f G^\circ$ (kJ mol⁻¹)	Log K_f
0	0.	0.	INFINITE	-21.156	-214.288	-214.288	INFINITE
100	58.116	33.666	225.939	-19.227	-215.261	-213.887	111.723
200	101.489	89.917	144.221	-10.861	-215.847	-212.256	55.436
298.15	117.738	133.887	133.887	0.	-216.313	-210.399	36.861
300	117.947	134.616	133.889	0.218	-216.323	-210.362	36.627
400	127.110	169.857	138.632	12.490	-221.430	-208.103	27.175
500	133.972	198.975	147.871	25.552	-225.126	-204.371	21.350
600	139.913	223.932	158.516	39.249	-228.432	-199.906	17.403
700	145.436	245.918	169.462	53.519	-230.824	-194.922	14.545
800	150.708	265.685	180.275	68.328	-231.697	-189.725	12.388
829.000	152.190	271.078	183.357	72.720		I <--> II	
829.000	188.615	338.910	183.357	128.953		TRANSITION	
900	188.615	354.410	196.248	142.345	-279.582	-187.213	10.866
1000	188.615	374.282	213.076	161.206	-274.001	-177.251	9.259
1062.000	188.615	385.628	222.821	172.901		II <--> LIQUID	
1062.000	191.795	404.224	222.821	192.649		TRANSITION	
1100	191.795	410.966	229.205	199.937	-248.784	-168.546	8.004
1200	191.795	427.655	245.057	219.117	-243.368	-161.493	7.030
1300	191.795	443.006	259.702	238.296	-238.222	-154.880	6.223
1400	191.795	457.220	273.309	257.476	-233.365	-148.654	5.546
1500	191.795	470.452	286.016	276.655	-228.808	-142.764	4.971
1600	191.795	482.830	297.934	295.834	-224.564	-137.168	4.478
1700	191.795	494.458	309.156	315.014	-220.648	-131.827	4.051
1800	191.795	505.421	319.758	334.193	-268.516	-124.563	3.615
1900	191.795	515.790	329.805	353.373	-264.989	-116.662	3.207
2000	191.795	525.628	339.352	372.552	-261.486	-108.946	2.845
2100	191.795	534.986	348.447	391.732	-258.004	-101.404	2.522
2200	191.795	543.908	357.130	410.911	-254.543	-94.028	2.233
2300	191.795	552.434	365.438	430.091	-251.101	-86.809	1.972
2400	191.795	560.596	373.401	449.270	-247.677	-79.740	1.735
2500	191.795	568.426	381.046	468.450	-244.268	-72.812	1.521
2600	191.795	575.948	388.399	487.629	-240.874	-66.021	1.326
2700	191.795	583.187	395.480	506.809	-237.491	-59.360	1.148
2800	191.795	590.162	402.309	525.988	-234.121	-52.825	0.985
2900	191.795	596.892	408.903	545.167	-230.762	-46.409	0.836
3000	191.795	603.394	415.279	564.347	-227.413	-40.109	0.698
3100	191.795	609.683	421.449	583.526	-224.074	-33.921	0.572
3200	191.795	615.772	427.427	602.706	-1351.248	-12.274	0.200
3300	191.795	621.674	433.224	621.885	-1342.966	29.440	-0.466
3400	191.795	627.400	438.851	641.065	-1334.689	70.904	-1.089
3500	191.795	632.959	444.318	660.244	-1326.421	112.125	-1.673
3600	191.795	638.362	449.634	679.424	-1318.162	153.109	-2.222
3700	191.795	643.617	454.806	698.603	-1309.915	193.864	-2.737
3800	191.795	648.732	459.842	717.783	-1301.683	234.397	-3.222

PREVIOUS: CURRENT: December 1976

Nickel Sulfide (Ni_3S_2) Ni_3S_2(cr,l)

NICKEL SULFIDE (Ni_3S_4) CRYSTAL M_r = 304.31

$\Delta_f H°(0\ K)$ = Unknown

$S°(298.15\ K) = 186.48 \pm 16.7\ J\ K^{-1}\ mol^{-1}$ $\Delta_f H°(298.15\ K) = -301.11 \pm 25.1\ kJ\ mol^{-1}$

Enthalpy of Formation

The only measured value of $\Delta_f H°(298.15\ K)$ for Ni_3S_4 is based on the emf data of DeRanter and Breckpot (1). Our 3rd law analysis of their data yields $\Delta_f H°(298.15\ K) = -71.0\ kcal\ mol^{-1}$. A value of $-72.9\ kcal\ mol^{-1}$ can be estimated assuming the same contribution per gram-atom in NiS and Ni_3S_2 (2). The adopted value, $\Delta_f H°(298.15\ K) = -72.0 \pm 6.0\ kcal\ mol^{-1}$, lies approximately midway between the above values and is chosen so that $\Delta_r G(629\ K) = 0$ for reaction $Ni_3S_4(cr) = 2\ NiS(cr) + NiS_2(cr)$ (2). This temperature represents the decomposition point of $Ni_3S_4(cr)$ (3, 4).

Heat Capacity and Entropy

Since there are no measured values for the heat capacity of Ni_3S_4 it is estimated as follows. The value of $C_p°(298.15\ K)$ is estimated assuming the same contribution per gram-atom as in NiS and Ni_3S_2 (2), yielding a value of $39.39\ cal\ K^{-1}\ mol^{-1}$. The value at the melting point is estimated assuming a contribution of $7.25\ cal\ K^{-1}\ g\text{-}atom^{-1}$ and the heat capacity is assumed to be linear between these two points (5). This leads to $C_p° = 29.153 \pm 0.03434\ T$.

The value of $S°(298.15\ K)$ is estimated assuming the same contribution per gram atom as in NiS and Ni_3S_2 (2). This estimate is subject to a rather large uncertainty since Ni_3S_4 has a spinel structure (6). Navrotsky and Kleppa (7) showed that the entropies of oxide spinels vary by $\pm 3.7\ cal\ K^{-1}\ mol^{-1}$ depending on the distribution of cations among the available octahedral and tetrahedral sites.

Decomposition Data

The decomposition temperature of Ni_3S_4 is $629 \pm 3\ K$ (3, 4). It decomposes according to $Ni_3S_4(cr) = 2\ Ni_{1-x}S(cr) + NiS_2(cr)$. As mentioned above, $\Delta_f H°(298.15\ K)$ for $Ni_3S_4(cr)$ was chosen so that $\Delta_r G°(629\ K)$ for this reaction is zero. The value of $\Delta_r H°(629\ K)$ for this reaction, as calculated using auxiliary JANAF data (2), is $-2.2\ kcal\ mol^{-1}$. Kullerud and Yund (4) calculate $\Delta_r H°(629\ K) = +7.0\ kcal\ mol^{-1}$ for this reaction based on the volume change as calculated from cell dimensions for the three compounds.

Phase Data

The known crystal species in the Ni-S system include NiS, Ni_3S_2, Ni_3S_4, NiS_2, and Ni_7S_6 (3). The first four species occur naturally as minerals and we have prepared JANAF Thermochemical Tables for these species (2). The species Ni_7S_6 (sometimes reported as Ni_9S_8 or Ni_6S_5) does not occur as a mineral and we have not prepared a table since there is a lack of thermochemical data. It is interesting to note that Ni_2S_3 does not occur as a stable species even though stable M_2S_3 species have been observed for Cr, Fe, and Co (8). Estimates of $\Delta_f H°(298.15\ K)$ for Ni_2S_3 range from +225 to +484 $kcal\ mol^{-1}$ (9). Although Moody and Thomas (9) acknowledge large uncertainties in their calculations, they are unlikely to be in error to the extent necessary to predict stability for $Ni_2S_3(cr)$.

References

1. C. DeRanter and R. Breckpot, Bull. Soc. Chim. Belg. 78, 503 (1969).
2. JANAF Thermochemical Tables: NiS(cr), Ni_3S_2(cr), 12-31-76; NiS_2(cr), 3-31-77.
3. G. Kullerud and R. A. Yund, J. Petrology 3, 126 (1962).
4. G. Kullerud and R. A. Yund, Carnegie Inst. Washington, Publ. Geophys. Lab #1363, 176 (1961).
5. O. Kubaschewski, E. L. Evans, and C. B. Alcock, "Metallurgical Thermochemistry," 4th ed., Pergamon, New York, p. 205, 1967.
6. J. D. H. Donnay and M. Ondik, Crystal Data Determinative Tables, 3rd ed., U. S. Nat. Bur. Stand., 1973.
7. A. Navrotsky and O. J. Kleppa, J. Inorg. Nucl. Chem. 29, 2701 (1967).
8. R. P. Elliott, Constitution of Binary Alloys, First Supplement, McGraw-Hill, New York, 1965.
9. G. J. Moody and J. D. R. Thomas, J. Chem. Soc. 1417 (1964).

Nickel Sulfide (Ni_3S_4) Ni_3S_4(cr)

Enthalpy Reference Temperature = T_r = 298.15 K Standard State Pressure = $p°$ = 0.1 MPa

T/K	$C_p°$ (J K⁻¹mol⁻¹)	$S°$ (J K⁻¹mol⁻¹)	$-[G°-H°(T_r)]/T$ (J K⁻¹mol⁻¹)	$H°-H°(T_r)$ (kJ mol⁻¹)	$\Delta_f H°$ (kJ mol⁻¹)	$\Delta_f G°$ (kJ mol⁻¹)	Log K_f
0							
100							
200							
298.15	164.808	186.481	186.481	0.	-301.115	-291.767	51.116
300	165.076	187.501	186.484	0.305	-301.122	-291.709	50.791
400	179.439	236.957	193.130	17.531	-310.469	-288.149	37.628
500	193.807	278.542	206.155	36.193	-316.421	-281.917	29.452
600	208.171	315.147	221.326	56.292	-320.495	-274.604	23.906
700	222.539	348.315	237.133	77.828	-322.315	-266.754	19.905
800	236.906	378.969	252.969	100.800	-321.447	-258.855	16.902
900	251.270	407.702	268.581	125.209	-531.454	-246.842	14.326
1000	265.638	434.919	283.865	151.054	-522.601	-215.682	11.266
1100	280.002	460.911	298.787	178.336	-512.576	-185.466	8.807

PREVIOUS: CURRENT: March 1977

IP(O, g) = 109837.02 ± 0.06 cm^{-1}
S°(298.15 K) = 161.058 ± 0.020 J K^{-1} mol^{-1}

$\Delta_f H°$(0 K) = 246.79 ± 0.10 kJ mol^{-1}
$\Delta_f H°$(298.15 K) = 249.17 ± 0.10 kJ mol^{-1}

Electronic Levels and Quantum Weights

State	ϵ_i, cm^{-1}	g_i
$^3P_2$	0	5
$^3P_1$	158.265	3
$^3P_0$	226.977	1
$^1D_2$	15867.862	5
$^1S_0$	33792.583	1

Enthalpy of Formation

The enthalpy of formation is based on the dissociation energy of O_2(g), 41260 ± 15 cm^{-1}(493.579±0.18 kJ mol^{-1}), as recommended by CODATA (1). This value is derived from the spectroscopic study of Brix and Herzberg (2). The products of dissociation are assumed to be oxygen atoms in their ground states. The justification for this is discussed by Gaydon (3) and Krupinie (4).

Heat Capacity and Entropy

The electronic levles for O(g) are given in the compilation by Moore (5). Our calcualtions indicate that, except for the five states lying below 34000 cm^{-1}, the inclusion of levels up to n=12 has no effect on the thermodynamic functions to 6000 K. This is a result of the high energy of these levels; the fifth excited state lies at 73768.20 cm^{-1} above the ground state. Since the inclusion of these upper levels has no effect on the thermodynamic functions (to 6000 K) we list only the lowest lying status (ϵ_i < 34000 cm^{-1}). The reported uncertainty in S°(298.15 K) is due to uncertainties in the relative atomic mass, the fundamental constants, and the position of the low lying electronic levles. The calculated value of S(298.15 K) is 0.0094 J K^{-1} mol^{-1} greater than that recommended by CODATA (1) due to the use of a 1 bar standard state pressure (CODATA used 1 atm). The difference is due to our use of the current CODATA fundamental constants (6). Extension of these calculations above 6000 K may require consideration of the higher excited states and utilization of proper fill and cut off procedures (7).

References

1. J. D. Cox, Chairman, ICSU-CODATA Task Group on Key Values for Thermodynamics, J. Chem. Thermodynamics 10, 903 (1978).
2. P. Brix and G. Herzberg, Can. J. Phys. 32, 110 (1954).
3. A. G. Gaydon, "Dissociation Energies and Spectra of Diatomic Molecules", 3rd ed., Chapman and Hall Ltd., London, 1968.
4. P. H. Krupenie, J. Phys. Chem. Ref. Data 1, 423 (1972).
5. C. E. Moore, NSRDS-NBS 3, Section 7 (1976).
6. E. R. Cohen and B. N. Taylor, J. Phys. Chem. Ref. Data 2, 663 (1973).
7. J. R. Downey, The Dow Chemcal Company, Report AFOSR-TR-73-0960, Contract No. F44620-75-1-0048, 1978.

Enthalpy Reference Temperature = T_r = 298.15 K   Standard State Pressure = p° = 0.1 MPa

T/K	$C_p°$ (J $K^{-1}mol^{-1}$)	S° (J $K^{-1}mol^{-1}$)	-[G°-H°(T_r)]/T (J $K^{-1}mol^{-1}$)	H°-H°(T_r) (kJ mol^{-1})	$\Delta_f H°$ (kJ mol^{-1})	$\Delta_f G°$ (kJ mol^{-1})	Log K_f
0	0.	0.	INFINITE	-6.725	246.790	246.790	INFINITE
100	23.703	135.947	181.131	-4.518	247.544	242.615	-126.729
200	22.734	152.153	163.085	-2.186	248.421	237.339	-61.986
250	22.246	157.170	161.421	-1.063	248.816	234.522	-49.001
298.15	21.911	161.058	161.058	0.	249.173	231.736	-40.599
300	21.901	161.194	161.059	0.041	249.187	231.628	-40.330
350	21.657	164.551	161.324	1.129	249.537	228.673	-34.128
400	21.482	167.430	161.912	2.207	249.868	225.670	-29.469
450	21.354	169.953	162.668	3.278	250.180	222.626	-25.842
500	21.257	172.197	163.511	4.343	250.474	219.549	-22.936
600	21.124	176.060	165.291	6.462	251.013	213.312	-18.570
700	21.040	179.310	167.067	8.570	251.494	206.990	-15.446
800	20.984	182.116	168.777	10.671	251.926	200.602	-13.098
900	20.944	184.585	170.399	12.767	252.320	194.163	-11.269
1000	20.915	186.790	171.930	14.860	252.682	187.681	-9.803
1100	20.893	188.782	173.373	16.950	253.018	181.165	-8.603
1200	20.877	190.599	174.734	19.039	253.332	174.619	-7.601
1300	20.864	192.270	176.019	21.126	253.627	168.047	-6.752
1400	20.853	193.816	177.236	23.212	253.906	161.453	-6.024
1500	20.845	195.254	178.390	25.296	254.171	154.840	-5.392
1600	20.838	196.599	179.486	27.381	254.421	148.210	-4.839
1700	20.833	197.862	180.530	29.464	254.659	141.564	-4.350
1800	20.830	199.053	181.527	31.547	254.884	134.905	-3.915
1900	20.827	200.179	182.479	33.630	255.097	128.234	-3.525
2000	20.826	201.247	183.391	35.713	255.299	121.552	-3.175
2100	20.827	202.263	184.266	37.796	255.488	114.860	-2.857
2200	20.830	203.232	185.106	39.878	255.667	108.159	-2.568
2300	20.835	204.158	185.914	41.962	255.835	101.450	-2.304
2400	20.841	205.045	186.693	44.045	255.992	94.734	-2.062
2500	20.851	205.896	187.444	46.130	256.139	88.012	-1.839
2600	20.862	206.714	188.170	48.216	256.277	81.284	-1.633
2700	20.877	207.502	188.871	50.303	256.405	74.551	-1.442
2800	20.894	208.261	189.550	52.391	256.525	67.814	-1.265
2900	20.914	208.995	190.208	54.481	256.637	61.072	-1.100
3000	20.937	209.704	190.846	56.574	256.741	54.327	-0.946
3100	20.963	210.391	191.466	58.669	256.838	47.578	-0.802
3200	20.991	211.057	192.068	60.767	256.929	40.826	-0.666
3300	21.022	211.704	192.653	62.867	257.014	34.071	-0.539
3400	21.056	212.332	193.223	64.971	257.094	27.315	-0.420
3500	21.092	212.943	193.777	67.079	257.169	20.555	-0.307
3600	21.130	213.537	194.318	69.190	257.241	13.794	-0.200
3700	21.170	214.117	194.845	71.305	257.309	7.030	-0.099
3800	21.213	214.682	195.360	73.424	257.373	0.265	-0.004
3900	21.257	215.234	195.862	75.547	257.436	-6.501	0.087
4000	21.302	215.772	196.353	77.675	257.496	-13.270	0.173
4100	21.349	216.299	196.834	79.808	257.554	-20.040	0.255
4200	21.397	216.814	197.303	81.945	257.611	-26.811	0.333
4300	21.445	217.318	197.763	84.087	257.666	-33.583	0.408
4400	21.495	217.812	198.213	86.234	257.720	-40.358	0.479
4500	21.545	218.295	198.654	88.386	257.773	-47.133	0.547
4600	21.596	218.769	199.086	90.543	257.825	-53.909	0.612
4700	21.647	219.234	199.510	92.705	257.876	-60.687	0.674
4800	21.697	219.690	199.925	94.872	257.926	-67.465	0.734
4900	21.748	220.138	200.333	97.045	257.974	-74.244	0.791
5000	21.799	220.578	200.734	99.222	258.021	-81.025	0.846
5100	21.849	221.010	201.127	101.405	258.066	-87.806	0.899
5200	21.899	221.435	201.514	103.592	258.110	-94.589	0.950
5300	21.949	221.853	201.893	105.784	258.150	-101.371	0.999
5400	21.997	222.264	202.267	107.982	258.189	-108.155	1.046
5500	22.045	222.668	202.634	110.184	258.224	-114.940	1.092
5600	22.093	223.065	202.995	112.391	258.255	-121.725	1.135
5700	22.139	223.457	203.351	114.602	258.282	-128.510	1.178
5800	22.184	223.842	203.701	116.818	258.304	-135.296	1.218
5900	22.229	224.222	204.046	119.039	258.321	-142.083	1.258
6000	22.273	224.596	204.385	121.264	258.332	-148.869	1.296

PREVIOUS: March 1977 (1 atm)   CURRENT: September 1982 (1 bar)

OXYGEN, ION (O^+) IDEAL GAS Mr = 15.99885

$IP(O^+, g)$ = 283240 ± 10 cm^{-1}

$S°(298.15 K)$ = 154.959 ± 0.02 $JK^{-1} mol^{-1}$

$\Delta_f H°(0 K)$ = 1560.733 ± 0.10 kJ mol^{-1}

$\Delta_f H°(298.15 K)$ = [1568.786] kJ mol^{-1}

Electronic Levels and Quantum Weights

State	ϵ_i, cm^{-1}	g_i
$^4S_{3/2}$	0	4
$^2D_{5/2}$	26810.7	6
$^2D_{3/2}$	26830.5	4
$^2P_{3/2}$	40466.9	4
$^2P_{1/2}$	40468.4	2

Enthalpy of Formation

$\Delta_f H°(O^+, g, 0 K)$ is calculated from $\Delta_f H°(O, g, 0 K)$ (1) using the spectroscopic value of IP(O) = 109837.02 ± 0.06 cm^{-1} (314.040 ± 0.001 kJ mol^{-1}) from Moore (2). The ionization limit is converted from cm^{-1} to kJ mol^{-1} using the factor, 1 cm^{-1} = 0.01196266 kJ cm^{-1}, which is derived from the 1973 CODATA fundamental constants (3). Rosenstock et al. (4) and Levin and Lias (5) have summarized additional ionization and appearance potential data.

$\Delta_f H°(O^+, g, 298.15 K)$ is calculated from $\Delta_f H°(O, g, 0 K)$ by using IP(O) with JANAF (1) enthalpies, H°(0 K) - H°(298.15 K), for O(g), $O^+(g)$, and $e^-(g)$. $\Delta_f H°(O \rightarrow O^+ + e^-, 298.15 K)$ differs from a room temperature threshold energy due to inclusion of these enthalpies and to threshold effects discussed by Rosenstock et al. (4). $\Delta_f H°(298.15 K)$ should be changed by -6.197 kJ mol^{-1} if it is to be used in the ion convention that excludes the enthalpy of the electron.

Heat Capacity and Entropy

The information on electronic energy levels and quantum weights, given by Moore (2, 6) is incomplete because many theoretically predicted levels have not been observed. Our calculations indicate that any reasonable method of filling in these missing levels and cutting off the summation in the partition function (7) has no effect on the thermodynamic functions to 6000 K. This is a result of the high energy of all levels other than the ground state and the first four excited states; the next excited state is 119837.7 cm^{-1} above the ground state. Since inclusion of these excited states has no effect on the thermodynamic functions (to 6000 K), we list only the ground state and the first four excited states. The reported uncertainty in S°(298.15 K) is due to uncertainties in the relative ionic mass and the fundamental constants. Extension of these calculations above 6000 K may require consideration of the excited states and use of different fill and cutoff procedures (7).

References

1. JANAF Thermochemical Tables: O(g), 9-30-82 ; e^-(g), 3-31-82.
2. C. E. Moore, U. S. Nat. Bur. Stand., NSRDS-NBS-34, 1970.
3. E. R. Cohen and B. N. Taylor, J. Phys. Chem. Ref. Data 2, 663 (1973).
4. H. M. Rosenstock, K. Draxl et al., J. Phys. Chem. Ref. Data 6, Supp. 1 (1977).
5. R. D. Levin and S. G. Lias, U. S. Nat. Bur. Stand., NSRDS-NBS-71, 1982.
6. C. E. Moore, U. S. Nat. Bur. Stand., NSRDS-NBS-35, Volume I, 1971 [Repreint of NBS Circular 467, Volume I, 1949].
7. J. R. Downey, Jr., The Dow Chemical Company, AFOSR-TR-0960, Contract No. F44620-75-1-0048, 1978.

Oxygen, Ion (O^+) $O_1^+(g)$

Enthalpy Reference Temperature = T_r = 298.15 K Standard State Pressure = p° = 0.1 MPa

T/K	$C_p°$ (J $K^{-1}mol^{-1}$)	S° (J $K^{-1}mol^{-1}$)	$-[G°-H°(T_r)]/T$ (J $K^{-1}mol^{-1}$)	$H°-H°(T_r)$ (kJ mol^{-1})	$\Delta_f H°$ (kJ mol^{-1})	$\Delta_f G°$ (kJ mol^{-1})	Log K_f
0	0.	0.	INFINITE	-6.197	1560.733		
100	20.786	132.251	173.439	-4.119			
200	20.786	146.659	156.860	-2.040			
250	20.786	151.297	155.301	-1.001			
298.15	20.786	154.959	154.959	0.	1568.786	1546.912	-271.013
300	20.786	155.087	154.959	0.038	1568.835	1546.776	-269.318
350	20.786	158.291	155.212	1.078	1570.176	1542.994	-230.279
400	20.786	161.067	155.774	2.117	1571.507	1539.020	-200.975
450	20.786	163.515	156.501	3.156	1572.827	1534.879	-178.164
500	20.786	165.705	157.314	4.196	1574.135	1530.593	-159.900
600	20.786	169.495	159.038	6.274	1576.712	1521.642	-132.471
700	20.786	172.699	160.766	8.353	1579.242	1512.262	-112.846
800	20.786	175.475	162.435	10.431	1581.731	1502.524	-98.105
900	20.786	177.923	164.023	12.510	1584.185	1492.475	-86.621
1000	20.786	180.113	165.524	14.589	1586.612	1482.155	-77.420
1100	20.786	182.094	166.942	16.667	1589.014	1471.593	-69.880
1200	20.786	183.903	168.281	18.746	1591.397	1460.813	-63.588
1300	20.786	185.566	169.548	20.824	1593.763	1449.834	-58.255
1400	20.786	187.107	170.748	22.903	1596.113	1438.675	-53.678
1500	20.786	188.541	171.887	24.982	1598.450	1427.347	-49.705
1600	20.786	189.882	172.970	27.060	1600.773	1415.865	-46.223
1700	20.786	191.143	174.002	29.139	1603.085	1404.237	-43.147
1800	20.786	192.331	174.988	31.217	1605.384	1392.474	-40.409
1900	20.786	193.455	175.930	33.296	1607.671	1380.583	-37.955
2000	20.786	194.521	176.833	35.375	1609.947	1368.572	-35.743
2100	20.786	195.535	177.700	37.453	1612.212	1356.448	-33.740
2200	20.786	196.502	178.533	39.532	1614.465	1344.216	-31.916
2300	20.786	197.426	179.334	41.611	1616.707	1331.881	-30.248
2400	20.787	198.310	180.107	43.689	1618.938	1319.450	-28.717
2500	20.787	199.159	180.852	45.768	1621.157	1306.925	-27.307
2600	20.788	199.974	181.572	47.847	1623.367	1294.312	-26.003
2700	20.789	200.759	182.268	49.925	1625.565	1281.615	-24.794
2800	20.790	201.515	182.942	52.004	1627.754	1268.836	-23.670
2900	20.792	202.245	183.595	54.083	1629.933	1255.979	-22.623
3000	20.795	202.949	184.229	56.163	1632.102	1243.047	-21.643
3100	20.799	203.631	184.844	58.242	1634.263	1230.043	-20.726
3200	20.804	204.292	185.441	60.323	1636.415	1216.969	-19.865
3300	20.810	204.932	186.022	62.403	1638.558	1203.828	-19.055
3400	20.818	205.553	186.587	64.485	1640.695	1190.622	-18.292
3500	20.827	206.157	187.138	66.567	1642.823	1177.354	-17.571
3600	20.839	206.744	187.674	68.650	1644.946	1164.025	-16.890
3700	20.853	207.315	188.198	70.735	1647.062	1150.636	-16.244
3800	20.870	207.871	188.708	72.821	1649.172	1137.192	-15.632
3900	20.890	208.414	189.206	74.909	1651.278	1123.691	-15.050
4000	20.912	208.943	189.693	76.999	1653.379	1110.136	-14.497
4100	20.938	209.460	190.169	79.091	1655.476	1096.529	-13.970
4200	20.968	209.964	190.634	81.187	1657.569	1082.871	-13.467
4300	21.001	210.458	191.090	83.285	1659.659	1069.163	-12.988
4400	21.038	210.941	191.535	85.387	1661.747	1055.406	-12.529
4500	21.079	211.415	191.972	87.493	1663.832	1041.602	-12.091
4600	21.125	211.879	192.400	89.603	1665.916	1027.752	-11.670
4700	21.174	212.333	192.819	91.718	1667.998	1013.856	-11.268
4800	21.229	212.780	193.230	93.838	1670.079	999.916	-10.881
4900	21.287	213.218	193.634	95.964	1672.160	985.933	-10.510
5000	21.351	213.649	194.030	98.096	1674.240	971.908	-10.153
5100	21.418	214.072	194.418	100.234	1676.319	957.841	-9.810
5200	21.491	214.489	194.800	102.380	1678.399	943.732	-9.480
5300	21.568	214.899	195.176	104.532	1680.479	929.584	-9.162
5400	21.649	215.303	195.545	106.693	1682.560	915.397	-8.855
5500	21.735	215.701	195.908	108.862	1684.640	901.171	-8.559
5600	21.826	216.093	196.265	111.040	1686.721	886.907	-8.273
5700	21.921	216.480	196.616	113.228	1688.803	872.606	-7.997
5800	22.020	216.862	196.962	115.425	1690.885	858.269	-7.730
5900	22.123	217.240	197.302	117.632	1692.967	843.895	-7.471
6000	22.231	217.612	197.638	119.850	1695.049	829.487	-7.221

PREVIOUS: March 1977 (1 atm) CURRENT: September 1982 (1 bar)

Oxygen, Ion (O^+) $O_1^+(g)$

OXYGEN, ION (O^-) IDEAL GAS M_r = 15.99995

Oxygen, Ion (O^-) $O_1^-(g)$

EA(O, g) = 1.4611215 ± 0.0000010 eV

S°(298.15 K) = 157.790 ± 0.001 J K^{-1} mol^{-1}

$\Delta_f H°$(0 K) = 105.814 kJ mol^{-1}

$\Delta_f H°$(298.15 K) = [101.846] kJ mol^{-1}

Electonic Levels and Quantum Weights

State	ε_i, cm^{-1}	g_i
$^2P_{3/2}$	0.0	4
$^2P_{1/2}$	177.08	2

Enthalpy of Formation

$\Delta_f H°(O^-$, g, 0 K) is calculated from $\Delta_f H°$(O, g, 0 K) (1) using the adopted electron affinity of EA(O) = 1.4611215 ± 0.0000010 eV (140.9757 ± 0.0001 kJ mol^{-1}). This value, is assumed to be the same as that recommended by Hotop and Lineberger (2), is based on a tunable laser photodetachment threshold study (3). Additional information on O^-(g) may be obtained in the critical discussions of Hotop and Lineberger (2, 4), Rosenstock et al. (5) and Massey (6).

$\Delta_f H°(O^-$, g, 298.15 K) is obtained from $\Delta_f H°$(O, g, 0 K) by using EA(O) with JANAF (1) enthalpies, H°(0 K) − H°(298.15 K), for O^-(g), O(g), and e^-(g). $\Delta_r H°(O^- \rightarrow O + e^-$, 298.15 K) differs from a room-temperature threshold energy due to inclusion of these enthalpies and to threshold effects discussed by Rosenstock et al. (5). $\Delta_f H°$(298.15 K) should be changed by +6.197 kJ mol^{-1} if it is to be used in the ion convention that excludes the enthalpy of the electron.

Heat Capacity and Entropy

The ground state and the fine-structure separation (177.08 ± 0.05 cm^{-1}) for O^-(g) has been experimentally determined (3); a thorough discussion of the relevant work is given by Hotop and Lineberger (2). Massey (6) listed the hyperfine splitting of the ground state as 226 cm^{-1} based on a quadratic isoelectronic extrapolation. Use of this value would decrease the entropy at 298.15 K by 0.16 J K^{-1} mol^{-1}.

A comparison of the isoelectronic sequence − O^-(g), F(g), Ne^+(g) − would suggest that stable electronic states may exist below EA(O), i.e. below 12000 cm^{-1}. These levels would greatly affect the entropy. However, lacking any experimental evidence as to the stability of these levels, we assume no stable excited states exist.

References

1. JANAF Thermochemical Tables: O(g), 9-30-82; e^-(g), 3-31-82.
2. H. Hotop and W. C. Lineberger, J. Phys. Che. Ref. Data, to be published, 1985.
3. D. N. Neumark, K. R. Lykke, T. Anderson and W. C. Lineberger, Phys. Rev A, to be published, 1985.
4. H. Hotop and W. C. Lineberger, J. Phys. Chem. Ref. Data 4, 539 (1975).
5. H. M. Rosenstock, K. Draxl et al., J. Phys. Chem. Ref. Data 6, Supp. 1 (1977).
6. H. S. W. Massey, "Negative Ions", 3rd ed., Cambridge University Press, Cambridge, 1976.

Enthalpy Reference Temperature = T_r = 298.15 K Standard State Pressure = p° = 0.1 MPa

T/K	$C_p°$ (J $K^{-1}mol^{-1}$)	S° (J $K^{-1}mol^{-1}$)	−[G°−H°(T_r)]/T (J $K^{-1}mol^{-1}$)	H°−H°(T_r) (kJ mol^{-1})	$\Delta_f H°$ (kJ mol^{-1})	$\Delta_f G°$ (kJ mol^{-1})	Log K_f
0	0.	0.	INFINITE	-6.571	105.814		
100	22.725	133.326	177.481	-4.415			
200	22.271	149.016	159.790	-2.155			
250	21.936	153.948	158.148	-1.050			
298.15	21.692	157.790	157.790	0.	101.846	91.638	-16.055
300	21.684	157.924	157.790	0.040	101.821	91.575	-15.945
350	21.500	161.252	158.054	1.119	101.123	89.922	-13.420
400	21.364	164.114	158.636	2.191	100.407	88.371	-11.540
450	21.262	166.624	159.387	3.256	99.675	86.910	-10.088
500	21.184	168.860	160.225	4.318	98.926	85.532	-8.935
600	21.075	172.712	161.995	6.430	97.380	82.997	-7.226
700	21.005	175.955	163.764	8.534	95.778	80.726	-6.024
800	20.958	178.757	165.467	10.632	94.129	78.688	-5.138
900	20.924	181.223	167.083	12.726	92.442	76.859	-4.461
1000	20.899	183.426	168.609	14.817	90.723	75.219	-3.929
1100	20.881	185.417	170.048	16.906	88.979	73.753	-3.502
1200	20.866	187.233	171.406	18.993	87.213	72.446	-3.154
1300	20.855	188.903	172.688	21.079	85.429	71.288	-2.864
1400	20.846	190.448	173.902	23.164	83.629	70.268	-2.622
1500	20.838	191.886	175.054	25.248	81.814	69.377	-2.416
1600	20.832	193.231	176.148	27.332	79.985	68.607	-2.240
1700	20.827	194.494	177.191	29.415	78.144	67.952	-2.088
1800	20.823	195.684	178.185	31.497	76.289	67.406	-1.956
1900	20.819	196.810	179.136	33.580	74.423	66.963	-1.841
2000	20.816	197.878	180.047	35.661	72.545	66.619	-1.740
2100	20.813	198.893	180.920	37.743	70.655	66.369	-1.651
2200	20.811	199.861	181.760	39.824	68.754	66.210	-1.572
2300	20.809	200.786	182.567	41.905	66.841	66.137	-1.502
2400	20.807	201.672	183.345	43.986	64.917	66.148	-1.440
2500	20.805	202.521	184.095	46.066	62.981	66.239	-1.384
2600	20.804	203.337	184.819	48.147	61.035	66.407	-1.334
2700	20.803	204.122	185.520	50.227	59.078	66.651	-1.289
2800	20.801	204.879	186.198	52.307	57.111	66.968	-1.249
2900	20.800	205.609	186.855	54.387	55.133	67.354	-1.213
3000	20.800	206.314	187.492	56.467	53.146	67.810	-1.181
3100	20.799	206.996	188.110	58.547	51.150	68.332	-1.151
3200	20.798	207.656	188.710	60.627	49.144	68.918	-1.125
3300	20.797	208.296	189.294	62.707	47.130	69.567	-1.101
3400	20.797	208.917	189.862	64.787	45.107	70.277	-1.080
3500	20.796	209.520	190.415	66.866	43.076	71.047	-1.060
3600	20.795	210.106	190.954	68.946	41.037	71.875	-1.043
3700	20.795	210.676	191.480	71.025	38.991	72.759	-1.027
3800	20.794	211.230	191.992	73.105	36.938	73.700	-1.013
3900	20.794	211.770	192.492	75.184	34.878	74.694	-1.000
4000	20.794	212.297	192.981	77.264	32.811	75.741	-0.989
4100	20.793	212.810	193.458	79.343	30.737	76.840	-0.979
4200	20.793	213.311	193.925	81.422	28.657	77.990	-0.970
4300	20.793	213.800	194.382	83.502	26.571	79.189	-0.962
4400	20.792	214.278	194.828	85.581	24.479	80.437	-0.955
4500	20.792	214.746	195.266	87.660	22.380	81.732	-0.949
4600	20.792	215.203	195.694	89.739	20.276	83.074	-0.943
4700	20.792	215.650	196.114	91.818	18.165	84.462	-0.939
4800	20.791	216.088	196.526	93.897	16.048	85.895	-0.935
4900	20.791	216.516	196.929	95.977	13.925	87.372	-0.931
5000	20.791	216.936	197.325	98.056	11.795	88.893	-0.929
5100	20.791	217.348	197.714	100.135	9.658	90.456	-0.926
5200	20.791	217.752	198.095	102.214	7.514	92.061	-0.925
5300	20.790	218.148	198.470	104.293	5.363	93.708	-0.924
5400	20.790	218.536	198.838	106.372	3.205	95.395	-0.923
5500	20.790	218.918	199.200	108.451	1.038	97.122	-0.922
5600	20.790	219.292	199.555	110.530	-1.137	98.889	-0.922
5700	20.790	219.660	199.905	112.609	-3.321	100.694	-0.923
5800	20.790	220.022	200.248	114.688	-5.515	102.538	-0.923
5900	20.790	220.377	200.586	116.767	-7.718	104.419	-0.924
6000	20.789	220.727	200.919	118.846	-9.932	106.339	-0.926

PREVIOUS: March 1977 (1 atm) CURRENT: September 1982 (1 bar)

Oxygen, Ion (O^-) $O_1^-(g)$

PHOSPHORUS OXIDE (PO) IDEAL GAS M_r = 46.97316

$S°(298.15\ K) = 222.776 \pm 0.021\ J\ K^{-1}\ mol^{-1}$

$\Delta_fH°(0\ K) = -23.24 \pm 14.6\ kJ\ mol^{-1}$
$\Delta_fH°(298.15\ K) = -23.55 \pm 4.2\ kJ\ mol^{-1}$

Electronic Levels and Quantum Weights

State	ϵ_i, cm^{-1}	g_i
$X^2\Pi_{1/2}$	0	2
$X^2\Pi_{3/2}$	224	2
$B^2\Sigma^+$	30696	2
$B'^2\Pi_i$	32884	4
$D'^2\Pi_r$	38055	4
$A^2\Sigma^+$	40485	2
$C'^2\Sigma^-$	43629	2
$C^2\Sigma^-$	47251	2
$D^2\Pi_r$	48580	4

$\omega_e = 1233.3\ cm^{-1}$ $\omega_ex_e = 6.56\ cm^{-1}$ $\sigma = 1$
$B_e = 0.7337\ cm^{-1}$ $\alpha_e = 0.0056\ cm^{-1}$ $r_e = 1.4757$ Å

Enthalpy of Formation

The dissociation energy of PO has been reported by various investigators to be from 5.15 eV to 7.4 eV. All of these values have been based on the spectroscopic properties of PO. Recently Drowart et al. (1) have measured the dissociation energy by a complex series of exchange reactions in a mass-spectrometer from which they calculate a dissociation energy of 142.2 ± 2.5 kcal mol^{-1} (6.17 cal K^{-1} mol^{-1}). Drowart et al. (1) also point out the agreement with the observed predissociation in the D' state. Couet et al. (2) and Verma and Dixit (3) have observed this predissociation. It was thought to occur between the 0 and 1 levels but recently Verma (4) has shown that the vibrational numbering is in error. This does not affect the predissociation, however, which we take as 49210 ± 350 cm^{-1}. If this is to ground state products, the dissociation energy is 140.7 ± 1.0 kcal mol^{-1}. The linear Birge-Sponer extrapolation (LBX) of the ground state vibrational energy levels yields D_0^0 = 164 kcal mol^{-1}, when corrected by the method of Hildenbrand (5) D_0^0 = 138.1 kcal mol^{-1}, in excellent agreement with the predissociation limit.

The much shorter extrapolation of the $B^2\Sigma^+$ levels yields D_0^0 = 157.5 (3), 155.0 (6) and 152.7 (7) kcal mol^{-1} using the LBX method, a non-linear extrapolation (8) yields 151.0 kcal mol^{-1}. All these values assume ground state atoms as the products, which must be the case since a $^2\Sigma^+$ state exists which correlates with those products and the $B^2\Sigma^+$ state must be this state or be formed by an avoided crossing. It is possible then that these values are high due to a potential maximum in the $B^2\Sigma^+$ state.

The adopted value is D_0^0 = 140.7 ± 1 kcal mol^{-1} and this yields $\Delta_fH°$(PO, g, 298.15 K) = -23.548 kJ mol^{-1}.

Heat Capacity and Entropy

The ground-state constants are from Verma (9) and Verma and Dixit (3). The ground-state splitting is from Rao (10) who also reported the excitation energy of the A state. The energy of the B state is due to Singh (11), while that of the B' state is from Verma (9). The C state energy is due to Dressler (6) while Narisimham (12) reported the C' state energy. Verma and Dixit (3) reported the D state energy and Verma (4) the D' state energy, which is considerably different from earlier reports. The E' state reported by Santaram and Rao (13) has been shown to coincide with the D' state by Verma and Dixit (3). States above 50000 cm^{-1} were not included. Other states in the 40,000 cm^{-1} region have been proposed (14) from perturbation analyses.

References

1. J. Drowart, C. E. Myers, R. Szware and O. M. Uy, 160th Meeting Amer. Chem. Soc., Chicago, Sept. 1970 and personal communication, July, 1971.
2. C. Couet, B. Coquart, N. T. Anh and H. Guenabaut. J. Chim. Phys. 65, 1241 (1968).
3. R. D. Verma and M. N. Dixit, Can. J. Phys. 46, 2079 (1968).
4. R. D. Verma, Can. J. Phys. 49, 279 (1971).
5. D. L. Hildenbrand, J. Chem. Phys. 51, 807 (1969).
6. K. Dressler, Helv. Phys. Acta, 28, 563 (1955).
7. H. Cordes and E. Warkehr, Z. Physik. Chem. 46, 26 (1965).
8. H. Meinel and L. Krauss, Z. Natuforsch. A21, 1520 (1966).
9. R. D. Verma, Can. J. Phys. 48, 2391 (1970).
10. K. S. Rao, Can. J. Phys. 36, 1526 (1958).
11. N. L. Singh, Can. J. Phys. 37, 136 (1959).
12. N. A. Narasimham, M. N. Dixit and V. Sethuraman, Proc. Indian Acad. Sci. A62, 314 (1965).
13. C. V. V. S. N. K. Santaram and P. T. Rao, Indian J. Phys. 37, 14 (1963).
14. C. Coquart, C. Couet, N. T. Anh and H. Guenebaut, J. Chim. Phys. 64, 1197 (1967).

Phosphorus Oxide (PO) $O_1P_1(g)$

Enthalpy Reference Temperature = T_r = 298.15 K Standard State Pressure = p° = 0.1 MPa

T/K	$C_p°$ (J $K^{-1}mol^{-1}$)	S° (J $K^{-1}mol^{-1}$)	$-[G°-H°(T_r)]/T$ (J $K^{-1}mol^{-1}$)	$H°-H°(T_r)$ (kJ mol^{-1})	$\Delta_fH°$ (kJ mol^{-1})	$\Delta_fG°$ (kJ mol^{-1})	Log K_f
0	0.	0.	INFINITE	-9.395	-23.241	-23.241	INFINITE
100	32.287	187.462	251.313	-6.385	-22.571	-30.883	16.131
200	32.204	210.041	225.686	-3.129	-23.033	-39.268	10.256
250	31.821	217.180	223.299	-1.530	-23.255	-43.303	9.048
298.15	31.762	222.776	222.776	0.	-23.548	-47.139	8.258
300	31.765	222.972	222.776	0.059	-23.560	-47.285	8.233
350	31.970	227.882	223.164	1.651	-24.642	-51.140	7.632
400	32.329	232.173	224.027	3.258	-25.098	-54.894	7.168
450	32.754	236.005	225.149	4.885	-25.546	-58.592	6.801
500	33.193	239.479	226.411	6.534	-25.985	-62.240	6.502
600	34.005	245.604	229.113	9.895	-26.836	-69.410	6.043
700	34.678	250.898	231.855	13.330	-27.660	-76.441	5.704
800	35.213	255.565	234.533	16.826	-28.466	-83.354	5.442
900	35.636	259.738	237.106	20.369	-29.258	-90 168	5.233
1000	35.973	263.511	239.561	23.950	-30.040	-96.893	5.061
1100	36.244	266.953	241.897	27.562	-30.816	-103 541	4.917
1200	36.465	270.116	244.118	31.197	-95.159	-109.040	4.746
1300	36.649	273.042	246.232	34.853	-95.153	-110.197	4.428
1400	36.803	275.764	248.245	38.526	-95.150	-111.354	4.155
1500	36.934	278.308	250.166	42.213	-95.150	-112.512	3.918
1600	37.047	280.695	252.000	45.912	-95.155	-113.669	3.711
1700	37.146	282.944	253.755	49.622	-95.164	-114.826	3.528
1800	37.233	285.070	255.436	53.341	-95.179	-115.982	3.366
1900	37.311	287.085	257.049	57.068	-95.201	-117.137	3.220
2000	37.381	289.001	258.599	60.803	-95.229	-118.291	3.089
2100	37.445	290.826	260.091	64.544	-95.263	-119.444	2.971
2200	37.504	292.570	261.528	68.292	-95 305	-120.594	2.863
2300	37.558	294.238	262.914	72.045	-95.355	-121.742	2.765
2400	37.608	295.837	264.253	75.803	-95.412	-122.889	2.675
2500	37.655	297.374	265.547	79.566	-95.478	-124.032	2.592
2600	37.699	298.851	266.800	83.334	-95.551	-125.173	2.515
2700	37.741	300.275	268.013	87.106	-95.631	-126.310	2.444
2800	37.781	301.648	269.190	90.882	-95.720	-127.445	2.378
2900	37.819	302.975	270.332	94.662	-95.816	-128.577	2.316
3000	37.856	304.257	271.442	98.446	-95.919	-129.705	2.258
3100	37.892	305.499	272.521	102.234	-96.030	-130.829	2.204
3200	37.926	306.703	273.570	106.024	-96.147	-131.950	2.154
3300	37.960	307.870	274.592	109.819	-96.272	-133.067	2.106
3400	37.993	309.004	275.587	113.616	-96.402	-134.180	2.061
3500	38.025	310.106	276.558	117.417	-96.539	-135.289	2.019
3600	38.057	311.177	277.505	121.221	-96.681	-136.394	1.979
3700	38.089	312.221	278.429	125.029	-96.829	-137.495	1.941
3800	38.121	313.237	279.332	128.839	-96.982	-138.592	1.905
3900	38.153	314.227	280.214	132.653	-97.141	-139.685	1.871
4000	38.186	315.194	281.076	136.470	-97.303	-140.774	1.838
4100	38.219	316.137	281.920	140.290	-97.471	-141.858	1.807
4200	38.252	317.059	282.746	144.114	-97.642	-142.939	1.778
4300	38.286	317.959	283.554	147.941	-97.818	-144.015	1.749
4400	38.322	318.840	284.346	151.771	-97.998	-145.088	1.722
4500	38.358	319.701	285.122	155.605	-98.181	-146 156	1.697
4600	38.396	320.545	285.883	159.443	-98.368	-147 220	1.672
4700	38.435	321.371	286.630	163.284	-98.558	-148.280	1.648
4800	38.476	322.180	287.362	167.130	-98.752	-149.336	1.625
4900	38.519	322.974	288.080	170.980	-98.949	-150.387	1.603
5000	38.563	323.753	288.786	174.834	-99.150	-151.435	1.582
5100	38.610	324.517	289.479	178.692	-99.354	-152.479	1.562
5200	38.659	325.267	290.160	182.556	-99.561	-153.519	1.542
5300	38.710	326.004	290.830	186.424	-99.772	-154.554	1.523
5400	38.763	326.728	291.488	190.298	-99.987	-155.586	1.505
5500	38.820	327.440	292.135	194.177	-100.205	-156.614	1.487
5600	38.879	328.140	292.772	198.062	-100.427	-157.637	1.470
5700	38.941	328.829	293.398	201.953	-100.653	-158.657	1.454
5800	39.005	329.506	294.015	205.850	-100.883	-159.672	1.438
5900	39.073	330.174	294.622	209.754	-101.118	-160 684	1.423
6000	39.144	330.831	295.220	213.665	-101.357	-161.691	1.408

PREVIOUS: June 1971 (1 atm) CURRENT: June 1971 (1 bar)

LEAD OXIDE, RED (PbO) CRYSTAL(RED) M_r = 223.1994

$S°(298.15\ K) = 66.316 \pm 0.8\ J\ K^{-1}\ mol^{-1}$

$T_{trs} = 762 \pm 1\ K$

$\Delta_f H°(0\ K) = -217.297 \pm 0.8\ kJ\ mol^{-1}$

$\Delta_f H°(298.15\ K) = -219.409 \pm 0.8\ kJ\ mol^{-1}$

$\Delta_{trs} H° = 0.167 \pm 0.29\ kJ\ mol^{-1}$

Enthalpy of Formation

The difference in the enthalpies of formation of red and yellow PbO has been determined directly, using a solution calorimeter, by Knacke and Prescher (1) as 394 ± 70 cal K^{-1} mol^{-1} by dissolving both forms in nitric acid. This difference and $\Delta_f H°$(PbO, yellow, 298.15 K) = -52.12 ± 0.15 kcal mol^{-1} yields $\Delta_f H°$(PbO, red, 298.15 K) = -52.51 ± 0.2 kcal mol^{-1}. Smith and Woods (2) have measured the emf of reversible cells to obtain the Gibbs energy of the reaction PbO(red, cr) + H_2(g) → Pb(cr) + H_2O(ℓ) at two temperatures. From a 2nd law analysis of these data, we obtain $\Delta_r H°$ = -15.91 kcal mol^{-1} which yields $\Delta_f H°$(PbO, red, 298.15 K) = -52.41 ± 0.2 kcal mol^{-1}.

If we assume the entropy to be accurately known, then the enthalpy of formation can be calculated directly from the Gibbs energies of reaction. Smith and Woods (2), for the reaction given above, found $\Delta_r G°$(298.15 K) = -11.513 kcal mol^{-1}, in combination with $\Delta_f G°$(H_2O, ℓ, 298.15 K) = -56.687 kcal mol^{-1} (3) yields $\Delta_f G°$(PbO, red, 298.15 K) = -45.182 kcal mol^{-1}. This Gibbs energy with JANAF entropies (4) is equivalent to $\Delta_f H°$(PbO, red, 298.15 K) = -52.45 kcal mol^{-1}. Spencer and Mote (5) measured the Gibbs energy of the reaction Pb(cr) + HgO(cr) → PbO(red, cr) - Hg(ℓ) by means of a reversible cell and obtained $\Delta_r G°$(298.15 K) = -31.265 kcal mol^{-1}. This value combined with $\Delta_f G°$(HgO, cr, 298.15 K) = -13.983 kcal mol^{-1} (4) yields $\Delta_f G°$(PbO, red, 298.15 K) = -45.248 kcal mol^{-1}, which with JANAF entropies (4) is equivalent to $\Delta_f H°$(PrO, red, 298.15 K) = -52.52 kcal mol^{-1}.

It is apparent that the data for enthalpy, Gibbs energy and entropy are all in excellent agreement, thus we adopt $\Delta_f H°$(PbO, red, 298.15 K) = -52.44 ± 0.2 kcal mol^{-1}. The adopted value was selected within all the uncertainties to assure the correct crossover of the red and yellow forms at 762 K.

Heat Capacity and Entropy

The heat capacity has been reported by King (6) from 53 to 296 K and the entropy was obtained by integration of these data based on S°(51 K) = 2.73 cal K^{-1} mol^{-1}. This extrapolated portion was obtained by use of a Debye-Einstein fit of the measured data. The value was then corrected by analogy with the 0.25 cal K^{-1} mol^{-1} difference between the extrapolation for PbO(cr, yellow) by King (6) and the measured amount. The justification for this procedure is the similarity of the two curves at all other temperatures. The high temperature heat capacities were obtained by fitting an orthogonal polynomial through the low temperature data of King (6) and the enthalpies of Spencer and Spicer (7).

Transition Data

The temperature of transition has been reported by Cohen and Addink (8) and Petersen (9). The enthalpy of transition is calculated from the adopted enthalpy at 298.15 K and the enthalpy difference, H°(762 K)-H°(298.15 K), between PbO yellow and red.

Vaporization Data

The vapor species over PbO(cr, ℓ) consists of $(PbO)_x$ where x = 1-6 (10, 11); thus a value for the heat of sublimation is not given.

References

1. O. Knacke and K. E. Prescher, Z. Erzbergbau und Metallhuttenwesen 17, 28 (1964).
2. D. F. Smith and H. K. Woods, J. Amer. Chem. Soc. 45, 2632 (1923).
3. U. S. Nat. Bur. Stand. Tech. Note 270-3, 1968.
4. JANAF Thermochemical Tables: PbO(cr), 12-31-71; Pb(cr), 12-31-62; O_2(g), 9-30-65; HgO(cr), 6-30-62.
5. H. M. Spencer and J. H. Mote, J. Amer. Chem. Soc. 54, 4618 (1932).
6. E. G. King, J. Amer. Chem. Soc. 80, 2400 (1958).
7. H. M. Spencer and W. M. Spicer, J. Amer. Chem. Soc. 64, 617 (1942).
8. E. Cohen and N. W. H. Addink, Z. Physik. Chem. A168, 202 (1934).
9. M. Petersen, J. Amer. Chem. Soc. 63, 2617 (1941).
10. J. Drowart, R. Colin and G. Exsteen, WADD TR-60-782, AD 612780, 1964.
11. D. M. Chizhikov, E. K. Kazenas and Yu. V. Tsvetkov, Izv. Akad. Nauk, SSSR, Metal. 1969, 57 (1969).

Lead Oxide, Red (PbO) $O_1Pb_1(cr)$

Enthalpy Reference Temperature = T_r = 298.15 K — Standard State Pressure = p° = 0.1 MPa

T/K	$C_p°$ (J K⁻¹mol⁻¹)	S° (J K⁻¹mol⁻¹)	-[G°-H°(T_r)]/T (J K⁻¹mol⁻¹)	H°-H°(T_r) (kJ mol⁻¹)	$\Delta_f H°$ (kJ mol⁻¹)	$\Delta_f G°$ (kJ mol⁻¹)	Log K_f
0	0.000	0.000	INFINITE	-9.109	-217.297	-217.297	INFINITE
100	26.870	24.765	101.583	-7.682	-219.089	-209.218	109.284
200	40.476	48.028	69.262	-4.247	-219.634	-199.037	51.983
298.15	45.773	66.316	66.316	0.000	-219.409	-189.283	33.162
300	45.866	66.600	66.317	0.085	-219.401	-189.096	32.925
400	50.375	80.434	68.173	4.904	-218.796	-179.077	23.385
500	53.430	92.066	71.821	10.123	-217.921	-169.245	17.681
600	55.396	101.992	76.042	15.570	-216.942	-159.601	13.894
700	56.651	110.626	80.380	21.172	-220.789	-149.325	11.143
762.000	57.430	115.466	83.040	24.709	----------	I <--> II	----------
800	57.907	118.273	84.647	26.900	-219.747	-139.186	9.088
900	59.162	125.166	88.773	32.754	-218.581	-129.185	7.498
1000	60.417	131.464	92.731	38.733	-217.288	-119.320	6.233
1100	61.672	137.281	96.520	44.837	-215.863	-109.591	5.204
1200	62.927	142.701	100.145	51.067	-214.301	-99.998	4.353
1300	64.183	147.788	103.616	57.423	-212.608	-90.540	3.638
1400	65.438	152.590	106.944	63.904	-210.796	-81.218	3.030

PREVIOUS: March 1962 CURRENT: December 1971

Lead Oxide, Red (PbO) $O_1Pb_1(cr)$

LEAD OXIDE, YELLOW (PbO) CRYSTAL(YELLOW) M_r = 223.1994

$S°(298.15\ K) = 68.701 \pm 0.21\ J\ K^{-1}\ mol^{-1}$

$T_{fus} = 1159 \pm 5\ K$

$\Delta_f H°(0\ K) = -216.080 \pm 0.63\ kJ\ mol^{-1}$

$\Delta_f H°(298.15\ K) = -218.062 \pm 0.63\ kJ\ mol^{-1}$

$\Delta_{fus} H° = 25.522 \pm 0.4\ kJ\ mol^{-1}$

Enthalpy of Formation

Espada et al. (1) have measured calorimetrically the heat of the reaction PbO(yellow, cr) + H_2(g) → Pb(cr) + H_2O(ℓ) as $\Delta_r H°(298.15\ K) = -16.20 \pm 0.07$ kcal mol^{-1} which yields the adopted enthalpy of formation.

Spencer and Mote (2) have measured the Gibbs energy of the cell reacton Pb(cr) + HgO(cr) → PbO(yellow, cr) + Hg(ℓ) as -31.120 kcal mol^{-1}. With JANAF auxiliary data (3) we obtain $\Delta_f G°$(PbO, yellow, 298.15 K) = -45.103 kcal mol^{-1} which corresponds to $\Delta_f H°$(PbO, yellow, 298.15 K) = -52.13 kcal mol^{-1} in excellent agreement with the adopted value.

Charette and Flengas (13) measured the Gibbs energy of formation directly in a high temperature electrolytic cell. Alcock and Belford (14) measured the Gibbs energy of the reactions (a) PbO(cr) + Ni(cr) → NiO(cr) + Pb(ℓ) and (b) Cu_2O(cr) + Pb(ℓ) = PbO(cr) + 2 Cu(cr) in high temperature electrolytic cells.

Source	T/K	$\Delta_r H°$(298.15 K), kcal mol^{-1} 2nd law	3rd law	Drift cal K^{-1} mol^{-1}	$\Delta_f H°$(298.15 K), kcal mol^{-1}
(13)	770-1160	-52.3	-52.93±0.26	-0.6±0.2	-52.93±0.26
(14a)	780-1070	-3.9	2.27±1.8	6.6±0.2	-54.0 ±0.7
(14b)	710-1050	-12.1	-12.61±0.25	-0.6±0.2	-52.28±0.7

These values indicate that the overall consistency between enthalpies of formation, enthalpies, entropy and Gibbs energies is generally within the experimental error.

Heat Capacity and Entropy

The low temperature heat capacity has been measured from 12.5 K to 302.7 K by Kostryukov and Morozova (4) and from 54.1 K to 296.1 K by King (5). The two sets of data are in reasonable agreement with differences of up to 0.5%. A line fitted through both data sets yields an entropy within the stated uncertainty of the Russian set, whose results we adopt.

The high temperature heat capacity was derived from the enthalpy measurements of Rodigina et al. (6) by curve fitting with orthogonal polynomials. The curve was constrained to join smoothly with the low temperature heat capacities. The adopted values are in good agreement with enthalpies reported by Regnault (7), Magnus (8) and Kopp (9). The enthalpies reported by Spencer and Spicer (10) lie about 3% higher than the adopted values are are admitted to be not too accurate. The values reported by Knacke and Prescher (11) lie 5 to 8% higher than the adopted values.

Fusion Data

The melting point was selected by Schnedier (12) in his review of metal oxide melting points. The enthalpy of melting was calculated from the enthalpy equations derived from the work of Rodigina et al. (6) for crystal and liquid. Knacke and Prescher (11) report a heat of melting of 6.57 kcal mol^{-1} from their enthalpy measurements.

References

1. L. Espada, G. Pilcher and H. A. Skinner, J. Chem. Thermodyn. 2, 647 (1970).
2. H. M. Spencer and J. H. Mote, J. Amer. Chem. Soc. 54, 4618 (1932).
3. JANAF Thermochemical Tables: HgO(cr), 6-30-62; Pb(cr), 3-31-62; O_2(g), 9-30-65.
4. V. N. Kostryukov and G. Kh. Morozova, Russian J. Phys. Chem. 34, 873 (1960).
5. E. G. King, J. Amer. Chem. Soc. 80, 2400 (1958).
6. E. N. Rodigina, K. Z. Gomel'skii and V. F. Luginina, Russ. J. Phys. Chem. 35, 884 (1961).
7. H. V. Regnault, Ann. Chim. Phys. 3, 129 (1841).
8. A. Magnus, Phys. Zeits. 14, 5 (1913).
9. H. Kopp, Annalen, Suppl. 3, 289 (1865).
10. H. M. Spencer and W. M. Spicer, J. Amer. Chem. Soc. 64, 617 (1942).
11. O. Knacke and K. E. Prescher, Z. Erzbergbau und Metallhuttenwesen 17, 28 (1964).
12. S. J. Schneider, U. S. Nat. Bur. Stand. Monograph 68, 1963.
13. G. G. Charette and S. N. Flengas, J. Electrochem. Soc. 115, 746 (1968).
14. C. B. Alcock and T. N. Belford, Trans. Faraday Soc. 60, 822 (1964).

Lead Oxide, Yellow (PbO) $O_1Pb_1(cr)$

Enthalpy Reference Temperature = T_r = 298.15 K   Standard State Pressure = p° = 0.1 MPa

T/K	$C_p°$ (J $K^{-1}mol^{-1}$)	S°	$-[G°-H°(T_r)]/T$	$H°-H°(T_r)$ (kJ mol^{-1})	$\Delta_f H°$	$\Delta_f G°$	Log K_f
0	0.000	0.000	INFINITE	-9.238	-216.080	-216.080	INFINITE
100	27.698	28.284	105.269	-7.699	-217.759	-208.239	108.773
200	40.041	51.547	72.990	-4.289	-218.329	-198.435	51.826
298.15	45.773	68.701	68.701	0.000	-218.062	-188.647	33.050
300	45.834	68.985	68.702	0.085	-218.054	-188.464	32.815
400	48.526	82.565	70.536	4.812	-217.541	-178.675	23.333
500	50.463	93.609	74.080	9.765	-216.931	-169.027	17.658
600	52.103	102.958	78.133	14.895	-216.270	-159.508	13.886
700	53.605	111.104	82.274	20.181	-220.433	-149.304	11.141
762.000	54.495	115.690	84.808	23.532	-----------	I <--> II	----------
800	55.028	118.355	86.339	25.613	-219.687	-139.192	9.088
900	56.409	124.917	90.266	31.185	-218.803	-129.182	7.498
1000	57.764	130.930	94.036	36.894	-217.779	-119.277	6.230
1100	59.099	136.499	97.646	42.738	-216.615	-109.482	5.199
1159.000	59.884	139.607	99.704	46.248	--------	II <--> LIQUID	--------
1200	60.421	141.698	101.103	48.714	-215.307	-99.800	4.344
1300	61.739	146.586	104.415	54.822	-213.861	-90.232	3.626
1400	63.049	151.209	107.594	61.061	-212.291	-80.780	3.014
1500	64.350	155.604	110.650	67.431	-210.606	-71.445	2.488

PREVIOUS: March 1962   CURRENT: December 1971

Lead Oxide, Yellow (PbO) $O_1Pb_1(cr)$

LEAD OXIDE (PbO) LIQUID M_r = 223.1994

$S°(298.15\ K) = [73.376]\ J\ K^{-1}\ mol^{-1}$

$T_{fus} = 1159 \pm 5\ K$

$\Delta_f H°(298.15\ K) = [-202.249]\ kJ\ mol^{-1}$

$\Delta_{fus}H° = 25.522 \pm 0.4\ kJ\ mol^{-1}$

Enthalpy of Formation

$\Delta_f H°(\ell,\ 298.15\ K)$ is calculated from that of PbO (yellow, cr) by adding $\Delta_{fus}H°$ and the difference in enthalpy, H°(1159 K)-H°(298.15 K), between PbO(yellow, cr) and PbO(ℓ). Direct measurements of the Gibbs energy change in the reaction PbO(ℓ) + C(cr) + Pb(ℓ) + CO(g) have been reported by Kvyatkovskii et al. (1). Charette and Flengas (2) and Delimarskii and Andreeva (3) have also determined the Gibbs energy of formation of the liquid directly in a high temperature cell. 2nd- and 3rd-law analysis of these data are summarized below.

Source	T/K	$\Delta_r H°$, kcal mol^{-1} 2nd law	$\Delta_r H°$, kcal mol^{-1} 3rd law	Drift cal K^{-1} mol^{-1}	$\Delta_f H°$(PbO, ℓ, 298.15 K) kcal mol^{-1}
(1)	1173 - 1423	34.5 ± 6.5	32.81 ± 2.49	-8.6 ± 4.9	-48.20 ± 2.5
(2)	1160 - 1371	-49.9 ± 0.2	-49.24 ± 0.13	0.5 ± 0.1	-49.24 ± 0.13
(3)	1173 - 1253	-46.9 ± 5.6	-48.51 ± 0.39	-1.5 ± 4.6	-48.51 ± 0.39

Heat Capacity and Entropy

The heat capacity was obtained from the enthalpy determinations of Rodigina et al. (4) and Knacke and Prescher (5). The absolute enthalpy values differ by 2-3% but the constant heat capacities derived from each set agree very well. The entropy at 298.15 K is calculated from that of the yellow crystal in a manner similar to that used for the enthalpy of formation.

Fusion Data

See the yellow-crystal table for details.

Vaporizaton Data

See the red-crystal table for details.

References

1. A. N. Kvyatkovskii, O. A. Esin and M. A. Abdeev, Russ. J. Phys. Chem. 34, 1162 (1960).
2. C. G. Charette and S. N. Flengas, J. Electrochem Soc. 115, 798 (1968).
3. Yu. K. Delimarskii and V. N. Andreeva, Russ. J. Inorg. Chem. 5, 873 (1960).
4. E. N. Rodigina, K. Z. Gomel'skii and V. F. Luginina, Russ. J. Phys. Chem. 35, 884 (1961).
5. O. Knacke and K. E. Prescher, Erzbergbau und Metallhuttenwesen 17, 28 (1964).

Lead Oxide (PbO) $O_1Pb_1(l)$

Enthalpy Reference Temperature = T_r = 298.15 K   Standard State Pressure = p° = 0.1 MPa

T/K	$C_p°$ (J K^{-1}mol^{-1})	S° (J K^{-1}mol^{-1})	$-[G°-H°(T_r)]/T$ (J K^{-1}mol^{-1})	$H°-H°(T_r)$ (kJ mol^{-1})	$\Delta_f H°$ (kJ mol^{-1})	$\Delta_f G°$ (kJ mol^{-1})	Log K_f
0							
100							
200							
298.15	64.998	73.376	73.376	0.000	-202.249	-174.228	30.524
300	64.998	73.778	73.378	0.120	-202.205	-174.054	30.306
400	64.999	92.478	75.927	6.620	-199.920	-165.018	21.549
500	65.000	106.982	80.742	13.120	-197.763	-156.545	16.354
600	65.000	118.833	86.133	19.620	-195.732	-148.494	12.928
700	65.000	128.853	91.538	26.120	-198.681	-139.976	10.445
800	65.000	137.532	96.757	32.620	-196.867	-131.714	8.600
900	65.000	145.188	101.721	39.120	-195.055	-123.678	7.178
1000	65.000	152.036	106.416	45.620	-193.240	-115.845	6.051
1100	65.000	158.232	110.850	52.120	-191.419	-108.193	5.138
1159.000	65.000	161.628	113.349	55.955	-------- II <--> LIQUID --------		
1200	65.000	163.887	115.037	58.620	-189.588	-100.708	4.384
1300	65.000	169.090	118.998	65.120	-187.750	-93.376	3.752
1400	65.000	173.907	122.750	71.620	-185.919	-86.185	3.216
1500	65.000	178.392	126.312	78.120	-184.104	-79.124	2.755
1600	65.000	182.587	129.699	84.620	-182.310	-72.184	2.357
1700	65.000	186.527	132.927	91.120	-180.542	-65.356	2.008
1800	65.000	190.243	136.009	97.620	-178.801	-58.631	1.701
1900	65.000	193.757	138.957	104.120	-177.091	-52.001	1.430
2000	65.000	197.091	141.781	110.620	-175.412	-45.461	1.187
2100	65.000	200.262	144.491	117.120	-350.987	-31.889	0.793
2200	65.000	203.286	147.095	123.620	-348.984	-16.741	0.397
2300	65.000	206.176	149.602	130.120	-347.080	-1.682	0.038
2400	65.000	208.942	152.017	136.620	-345.278	13.296	-0.289
2500	65.000	211.595	154.347	143.120	-343.578	28.201	-0.589

PREVIOUS: March 1962   CURRENT: December 1971

Lead Oxide (PbO) $O_1Pb_1(l)$

LEAD OXIDE (PbO) RED-YELLOW-LIQUID M_r = 223.1994

0 to 762 K	crystal, red
762 to 1159 K	crystal, yellow
above 1159 K	liquid

Refer to the individual tables for details.

Lead Oxide (PbO) $O_1Pb_1(cr,l)$

Enthalpy Reference Temperature = T_r = 298.15 K   Standard State Pressure = p° = 0.1 MPa

T/K	C_p° (J K⁻¹mol⁻¹)	S° (J K⁻¹mol⁻¹)	$-[G^\circ - H^\circ(T_r)]/T$ (J K⁻¹mol⁻¹)	$H^\circ - H^\circ(T_r)$ (kJ mol⁻¹)	Δ_fH° (kJ mol⁻¹)	Δ_fG° (kJ mol⁻¹)	Log K_f
0	0.000	0.000	INFINITE	-9.109	-217.297	-217.297	INFINITE
100	26.870	24.765	101.583	-7.682	-219.089	-209.218	109.284
200	40.476	48.028	69.262	-4.247	-219.634	-199.037	51.983
298.15	45.773	66.316	66.316	0.000	-219.409	-189.283	33.162
300	45.866	66.600	66.317	0.085	-219.401	-189.096	32.925
400	50.375	80.434	68.173	4.904	-218.796	-179.077	23.385
500	53.430	92.066	71.821	10.123	-217.921	-169.245	17.681
600	55.396	101.992	76.042	15.570	-216.942	-159.601	13.894
700	56.651	110.626	80.380	21.172	-220.789	-149.325	11.143
762.000	57.430	115.466	83.040	24.709		I <--> II	
762.000	54.495	115.690	83.040	24.880		TRANSITION	
800	55.028	118.355	84.654	26.961	-219.687	-139.192	9.088
900	56.409	124.917	88.769	32.533	-218.803	-129.182	7.498
1000	57.764	130.930	92.689	38.242	-217.779	-119.277	6.230
1100	59.099	136.499	96.421	44.085	-216.615	-109.482	5.199
1159.000	59.884	139.607	98.541	47.595		II <--> LIQUID	
1159.000	65.000	161.628	98.541	73.117		TRANSITION	
1200	65.000	163.887	100.735	75.782	-189.586	-100.706	4.384
1300	65.000	169.090	105.796	82.282	-187.748	-93.374	3.752
1400	65.000	173.907	110.491	88.782	-185.918	-86.183	3.216
1500	65.000	178.392	114.870	95.282	-184.103	-79.123	2.755
1600	65.000	182.587	118.973	101.782	-182.309	-72.183	2.357
1700	65.000	186.527	122.832	108.282	-180.540	-65.354	2.008
1800	65.000	190.243	126.475	114.782	-178.800	-58.629	1.701
1900	65.000	193.757	129.924	121.282	-177.089	-52.000	1.430
2000	65.000	197.091	133.200	127.782	-175.410	-45.459	1.187
2100	65.000	200.262	136.318	134.282	-350.985	-31.887	0.793
2200	65.000	203.286	139.294	140.782	-348.982	-16.739	0.397
2300	65.000	206.176	142.140	147.282	-347.078	-1.681	0.038
2400	65.000	208.942	144.866	153.782	-345.276	13.298	-0.289
2500	65.000	211.595	147.482	160.282	-343.577	28.203	-0.589

PREVIOUS: CURRENT: December 1971

Lead Oxide (PbO) $O_1Pb_1(cr,l)$

LEAD OXIDE (PbO) IDEAL GAS M_r = 223.1994

$S°(298.15\ K) = 240.039 \pm 0.021\ J\ K^{-1}\ mol^{-1}$

$\Delta_fH°(0\ K) = 72.55 \pm 7.1\ kJ\ mol^{-1}$
$\Delta_fH°(298.15\ K) = 70.29 \pm 7.1\ kJ\ mol^{-1}$

Electronic States and Molecular Constants (σ = 1)

State	ε_i, cm^{-1}	g_i	r_e, Å	B_e, cm^{-1}	α_e, cm^{-1}	ω_e, cm^{-1}	$\omega_e x_e$, cm^{-1}	$\omega_e x_e/B_e$	$\omega_e/\omega_e x_e$
X 0^+	0	1	1.922	0.3074	0.00212	721.8	3.70	12.036	195.08
A 0^+	20088	1	2.095	0.2587	0.00138	451.7	3.33	12.872	135.65
B 1	22531	2	2.071	0.2646	0.0026	493.5	2.26	8.541	218.36
C 0^+	24108	1	2.11	0.254	0.002	[534]	[3.9]	[15.354]	136.92
C'1	25720	2	2.14	0.248	0.0018	[475]	[2.7]	[10.887]	175.92
D 1	30464	2	2.046	0.2711	0.0031	530.5	2.92	10.771	181.68
E 0^+	34680	1	[2.183]	[0.2382]	[0.0014]	[440]	[2.6]		169.23

Enthalpy of Formation

A value for the dissociation energy can be calculated by applying the Hildenbrand formula (1) to the linear Birge-Sponer extrapolation of the ground state vibrational levels. This procedure yields a D° = 91.4 kcal mol^{-1}, corresponding to $\Delta_fH°$(0 K) = 14.5 kcal mol^{-1}.

Recently, Singh and Nair (3) have calculated D_0^0 = 99.2 kcal mol^{-1} from a special potential energy curve which takes into account the electronegativity of the atoms.

The enthalpy of sublimation of PbO(yellow, cr) to monomeric PbO has been measured mass-spectrometrically by Drowart et al. (2). Our analysis of their vapor pressure data yields $\Delta_{sub}H°$(298.15 K) = 68.9 ± 1.5 kcal mol^{-1}, which yields $\Delta_fH°$(PbO, g, 298.15 K) = 16.8 ± 1.7 kcal mol^{-1}. This value is adopted and corresponds to a dissociation energy at 0 K of 88.6 kcal mol^{-1}.

Heat Capacity and Entropy

The tabulated molecular and electronic constants are taken from Barrow et al. (4). The state designations are those used by Barrow et al. who interpret the spectra in terms of type c coupling, when the state designations Σ, Π, etc. no longer have meaning. The estimated values for $\omega_e x_e$ for the C' and E states were obtained from the B_e values by averaging the $\omega_e x_e/B_e$ ratio for the known states. The functions were calculated from the partition function $Q = Q_t \Sigma Q_r^i Q_v^i g_i \exp(-c_2\varepsilon_i/T)$, the values for Q_r^i and Q_v^i were calculated with first-order anharmonic corrections.

References

1. D. L. Hildenbrand, J. Chem. Phys. 51, 807 (1969).
2. J. Drowart, R. Colin and G. Exsteen, WADD-TR-60-782 part XXVI, Dec. 1964. Also available as AD612780.
3. J. Singh and K. P. R. Nair, Indian J. Pure and Applied Physics 9, 130 (1971).
4. R. F. Barrow, J. L. Deutsch and D. N. Travis, Nature 191, 374 (1961).

Lead Oxide (PbO) $O_1Pb_1(g)$

Enthalpy Reference Temperature = T_r = 298.15 K; Standard State Pressure = p° = 0.1 MPa

T/K	$C_p°$ (J K^{-1}mol^{-1})	S° (J K^{-1}mol^{-1})	$-[G°-H°(T_r)]/T$ (J K^{-1}mol^{-1})	$H°-H°(T_r)$ (kJ mol^{-1})	$\Delta_fH°$ (kJ mol^{-1})	$\Delta_fG°$ (kJ mol^{-1})	Log K_f
0	0.	0.	INFINITE	-8.961	72.550	72.550	INFINITE
100	29.136	207.019	267.535	-6.052	72.241	63.887	-33.371
200	30.424	227.495	242.946	-3.090	71.222	55.926	-14.606
250	31.516	234.401	240.568	-1.542	70.736	52.159	-10.898
298.15	32.508	240.039	240.039	0.	70.291	48.622	-8.518
300	32.544	240.240	240.039	0.060	70.275	48.487	-8.442
350	33.415	245.324	240.439	1.710	69.832	44.891	-6.700
400	34.123	249.834	241.337	3.399	69.399	41.358	-5.401
450	34.692	253.887	242.510	5.120	68.965	37.879	-4.397
500	35.149	257.567	243.834	6.866	68.523	34.448	-3.599
600	35.822	264.039	246.677	10.417	67.605	27.719	-2.413
700	36.281	269.597	249.564	14.024	61.762	21.946	-1.638
800	36.608	274.464	252.378	17.669	60.722	16.329	-1.066
900	36.852	278.791	255.077	21.342	59.707	10.841	-0.629
1000	37.040	282.684	257.646	25.038	58.717	5.465	-0.285
1100	37.191	286.222	260.086	28.749	57.749	0.187	-0.009
1200	37.315	289.463	262.401	32.475	56.807	-5.004	0.218
1300	37.420	292.454	264.599	36.212	55.881	-10.118	0.407
1400	37.510	295.231	266.689	39.958	54.958	-15.160	0.566
1500	37.590	297.821	268.679	43.713	54.029	-20.136	0.701
1600	37.662	300.250	270.577	47.476	53.085	-25.050	0.818
1700	37.727	302.535	272.390	51.245	52.123	-29.904	0.919
1800	37.788	304.693	274.126	55.021	51.139	-34.701	1.007
1900	37.844	306.738	275.789	58.803	50.131	-39.442	1.084
2000	37.898	308.680	277.385	62.590	49.097	-44.130	1.153
2100	37.951	310.530	278.920	66.382	-129.185	-41.650	1.036
2200	38.003	312.297	280.397	70.180	-129.884	-37.465	0.890
2300	38.054	313.988	281.821	73.983	-130.678	-33.247	0.755
2400	38.108	315.608	283.195	77.791	-131.568	-28.993	0.631
2500	38.164	317.165	284.523	81.605	-132.554	-24.699	0.516
2600	38.224	318.663	285.808	85.424	-133.636	-20.363	0.409
2700	38.290	320.107	287.051	89.250	-134.811	-15.985	0.309
2800	38.362	321.501	288.257	93.082	-136.073	-11.561	0.216
2900	38.444	322.848	289.427	96.922	-137.418	-7.091	0.128
3000	38.536	324.153	290.563	100.771	-138.838	-2.573	0.045
3100	38.641	325.418	291.667	104.630	-140.326	1.994	-0.034
3200	38.760	326.647	292.741	108.500	-141.875	6.609	-0.108
3300	38.895	327.842	293.786	112.383	-143.474	11.274	-0.178
3400	39.048	329.005	294.805	116.280	-145.115	15.988	-0.246
3500	39.220	330.139	295.799	120.193	-146.789	20.751	-0.310
3600	39.413	331.247	296.768	124.124	-148.487	25.562	-0.371
3700	39.629	332.330	297.714	128.076	-150.200	30.420	-0.429
3800	39.869	333.390	298.639	132.051	-151.919	35.325	-0.486
3900	40.134	334.429	299.544	136.051	-153.635	40.275	-0.539
4000	40.425	335.448	300.429	140.079	-155.342	45.268	-0.591
4100	40.743	336.450	301.295	144.137	-157.030	50.304	-0.641
4200	41.088	337.436	302.144	148.228	-158.693	55.381	-0.689
4300	41.461	338.407	302.976	152.355	-160.324	60.498	-0.735
4400	41.862	339.365	303.792	156.521	-161.917	65.652	-0.779
4500	42.290	340.311	304.593	160.729	-163.466	70.841	-0.822
4600	42.746	341.245	305.380	164.980	-164.965	76.064	-0.864
4700	43.229	342.169	306.153	169.279	-166.410	81.320	-0.904
4800	43.738	343.085	306.913	173.627	-167.797	86.606	-0.942
4900	44.272	343.992	307.660	178.027	-169.121	91.920	-0.980
5000	44.830	344.892	308.396	182.482	-170.380	97.260	-1.016
5100	45.410	345.786	309.120	186.994	-171.571	102.625	-1.051
5200	46.012	346.673	309.834	191.565	-172.690	108.012	-1.085
5300	46.634	347.555	310.537	196.197	-173.736	113.420	-1.118
5400	47.273	348.433	311.231	200.892	-174.708	118.847	-1.150
5500	47.928	349.306	311.915	205.652	-175.605	124.292	-1.180
5600	48.596	350.176	312.591	210.478	-176.425	129.752	-1.210
5700	49.277	351.042	313.258	215.372	-177.168	135.226	-1.239
5800	49.966	351.905	313.917	220.334	-177.834	140.713	-1.267
5900	50.663	352.765	314.568	225.365	-178.423	146.210	-1.294
6000	51.364	353.623	315.212	230.466	-178.938	151.717	-1.321

PREVIOUS: December 1971 (1 atm) CURRENT: December 1971 (1 bar)

Lead Oxide (PbO) $O_1Pb_1(g)$

$S°(298.15\ K) = 221.944 \pm 0.04\ J\ K^{-1}\ mol^{-1}$

$\Delta_f H°(0\ K) = 5.03 \pm 1.3\ kJ\ mol^{-1}$
$\Delta_f H°(298.15\ K) = 5.01 \pm 1.3\ kJ\ mol^{-1}$

Electronic States and Molecular Constants (σ = 1)

State	ε_i, cm^{-1}	g_i	ω_e, cm^{-1}	$\omega_e x_e$, cm^{-1}	B_e, cm^{-1}	α_e, cm^{-1}	r_e, Å
$X^3\Sigma^-$	0	3	1148.19	6.12	0.72082	0.00574	1.4811
$a^1\Delta$	6350	2	[1148.19]	[6.12]	0.7119	[0.00574]	1.494
$b^1\Sigma^+$	10510	1	1067.66	7.8	0.70261	0.00635	1.5005
$A^3\Pi_0$	38292	2	415.2	1.6	0.6067	0.0194	
$A^3\Pi_1$	38455	2	413.3	1.6	0.6107	0.0194	1.6094
$A^3\Pi_2$	38616	2	412.7	1.7	0.6164	0.0204	
$B^3\Sigma^-$	41629	3	630.4	4.8	0.502	0.0062	1.775
$C^3\Pi$	42200	6	170	[0]	[0.5]	[0]	2.2

Enthalpy of Formation

The selection for $\Delta_f H°(0\ K)$ of SO(g), 1.2 ± 3 kcal mol^{-1}, was derived from one of the two independent means, pointed out by Okabe (1) that employs principally spectroscopic data: $\Delta_f H°(SO, g, 0\ K) = 1/2\ D_0^0(S_2) + 1/2\ D_0^0(O_2) - D_0^0(SO) + 1/2\ \Delta_f H°(S_2, g, 0\ K)$. The data employed are as follows. Ricks and Barrow (2) give $D_0^0(S_2)$ = 100.69 ± 0.01 kcal mol^{-1}. Brix and Herzberg (3) give $D_0^0(O_2)$ = 117.97 ± 0.046 kcal mol^{-1}. On the basis of analysis of spectroscopic data to 1964, JANAF (4) gives $D_0^0(SO)$ = 123.5 ± 0.3 kcal mol^{-1}. NBS (5) gives $\Delta_f H°(S_2, g, 0\ K)$ = 30.65 ± 0.2 kcal mol^{-1}. Assuming the dissociation products of SO are $S(^1D_2)$ and $O(^3P)$, Colin (6) concludes $D_0^0(SO)$ = 123.56 ± 0.23 kcal mol^{-1}. A systematic error of 0.65 kcal mol^{-1} is incurred if the assumption is wrong. Support for this assumption is as follows: 1) Okabe (7) concludes that $D_0^0(OS\text{-}O)$ = 130.06 ± 0.5 kcal mol^{-1} assuming the predissociation of SO_2 is to ground state products. Using $\Delta_f H°(SO_2, g, 0\ K)$ = -70.336 kcal mol^{-1} from (5) and the relation: $\Delta_f H°(SO, g, 0\ K) = D_0^0(OS\text{-}O) - 1/2\ D_0^0(O_2) + \Delta_f H°(SO_2, g, 0\ K)$, one obtains $\Delta_f H°(SO, g, 0\ K)$ = 0.7 ± 0.7 kcal mol^{-1}; 2) Assuming the threshold energy for fluorescence of SO, 217.00 ± 0.7 kcal mol^{-1} (see Okabe (8)), corresponds to the minimum energy for the reaction $OSCl_2 = SO(A^3\Pi_0) + 2\ Cl(^3P_{3/2})$ and using 108.43 kcal mol^{-1} for the difference in energy between the lowest vibrational levels of $SO(A^3\Pi_0)$ and $SO(X^3\Sigma^-)$ (Colin (6)) one obtains $\Delta_f H°(SO, g, 0\ K)$ = 1.3 ± 0.7 kcal mol^{-1}. This is based on $\Delta_f H°(OSCl_2, g, 0\ K)$ = -50.07 kcal mol^{-1} (5) and $D_0^0(Cl_2, 0\ K)$ = 57.177 ± 0.005 kcal mol^{-1} given by Douglas et al. (9); 3) Re-examination of the photoionization-yield curves of SO_2^+ and SO^+ obtained by Dibeler and Liston (10) suggests that their reported threshold for $SO + SO^+ + O + e^-$ should be increased from 15.81 ± 0.02 cal K^{-1} mol^{-1} to 16.00 eV. Using the latter and the ionization energy of SO given by Jonathon et al. (11) of 10.34 ± 0.02 eV and the relation $\Delta_f H°(SO, g, 0\ K) = D_0^0(S\text{-}O) - 1/2\ D_0^0(O_2) + \Delta_f H°(SO_2, g, 0\ K)$, one obtains $\Delta_f H°(SO, g, 0\ K)$ = 1.20 ± 0.5 kcal mol^{-1}.

Heat Capacity and Entropy

The vibrational and rotational constants of the respective electronic levels were taken from Rosen (12). The thermodynamic functions are calculated using first-order anharmonic corrections to Q_r^i and Q_v^i in the partition function $Q = Q_t \Sigma Q_r^i Q_v^i g_i \exp(-c_2\varepsilon_i/T)$. The National Bureau of Standards prepared this table (13) by critical analysis of data existing in 1972. Using molecular constants and $\Delta_f H°$ selected by NBS (13), we recalculate the table in terms of 1973 fundamental constants (14), 1975 atomic weights (15), and current JANAF reference states for the elements.

References

1. H. Okabe, J. Chem. Phys. 56, 3378 (1972).
2. J. M. Ricks and R. F. Barrow, Can. J. Phys. 47, 2423 (1969).
3. P. Brix and G. Herzberg, Can. J. Phys. 32, 110 (1954).
4. JANAF Thermochemical Tables, 2nd ed., NSRDS-NBS 37, 1971.
5. U. S. Nat. Bur. Stand. Tech. Note 270-3, 1968.
6. R. Colin, Can. J. Phys. 47, 979 (1969).
7. H. Okabe, J. Amer. Chem. Soc. 93, 7095 (1971).
8. H. Okabe, J. Chem. Phys. 56, 3378 (1972).
9. A. E. Douglas, C. K. Moeller and B. P. Stoicheff, Can. J. Phys. 41, 1174 (1963).
10. V. H. Dibeler and S. K. Liston, J. Chem. Phys. 49, 482 (1968).
11. N. Jonathan, D. J. Smith and K. J. Ross, Chem. Phys. Letters 9, 217 (1971).
12. B. Rosen, "Spectroscopic Data Relative to Diatomic Molecules," Pergamon Press, Oxford, 1970.
13. S. Abramowitz et al., U. S. Nat. Bur. Stand. Report 10904, 239, 1972.
14. CODATA Task Group on Fundamental Constants, CODATA Bulletin 11, December 1973.
15. IUPAC Commission on Atomic Weights, Pure Appl. Chem. 47, 75 (1976).

Enthalpy Reference Temperature = T_r = 298.15 K — Standard State Pressure = p° = 0.1 MPa

T/K	$C_p°$ (J $K^{-1}mol^{-1}$)	S° (J $K^{-1}mol^{-1}$)	$-[G°-H°(T_r)]/T$ (J $K^{-1}mol^{-1}$)	$H°-H°(T_r)$ (kJ mol^{-1})	$\Delta_f H°$ (kJ mol^{-1})	$\Delta_f G°$ (kJ mol^{-1})	Log K_f
0	0.	0.	INFINITE	-8.733	5.028	5.028	INFINITE
100	29.106	189.916	248.172	-5.826	5.793	-3.281	1.714
200	29.269	210.114	224.668	-2.911	5.610	-12.337	3.222
250	29.634	216.680	222.437	-1.439	5.333	-16.793	3.509
298.15	30.173	221.944	221.944	0.	5.007	-21.026	3.684
300	30 197	222.130	221.944	0.056	4.994	-21.187	3.689
350	30.867	226.835	222.314	1.582	4.616	-25.521	3.809
400	31.560	231.002	223.144	3.143	1.998	-29.711	3.880
450	32.221	234.758	224.230	4.738	0.909	-33.619	3.902
500	32.826	238.184	225.456	6.364	-0.238	-37.391	3.906
600	33.838	244.263	228.097	9.699	-2.067	-44.643	3.887
700	34.612	249.540	230.791	13.124	-3.617	-51.614	3.851
800	35.206	254.202	233.432	16.616	-5.005	-58.374	3.811
900	35.672	258.377	235.976	20.161	-59.419	-63.886	3.708
1000	36.053	262.155	238.408	23.748	-59.420	-64.382	3.363
1100	36.379	265.607	240.726	27.369	-59.424	-64.878	3.081
1200	36.672	268.785	242.933	31.022	-59.432	-65.374	2.846
1300	36.946	271.731	245.037	34.703	-59.446	-65.869	2.647
1400	37.210	274.479	247.043	38.411	-59.462	-66.362	2.476
1500	37.469	277.055	248.959	42.145	-59.482	-66.854	2.328
1600	37.725	279.482	250.791	45.905	-59.505	-67.345	2.199
1700	37.980	281.776	252.547	49.690	-59.528	-67.834	2.084
1800	38.232	283.954	254.232	53.501	-59.552	-68.323	1.983
1900	38.482	286.028	255.851	57.336	-59.575	-68.809	1.892
2000	38.727	288.008	257.410	61.197	-59.597	-69.294	1.810
2100	38.967	289.904	258.912	65.082	-59.618	-69.779	1.736
2200	39.200	291.722	260.363	68.990	-59.636	-70.262	1.668
2300	39.425	293.469	261.764	72.921	-59.652	-70.745	1.607
2400	39.641	295.152	263.121	76.875	-59.666	-71.227	1.550
2500	39.847	296.774	264.434	80.849	-59.678	-71.708	1.498
2600	40.043	298.341	265.709	84.844	-59.687	-72.189	1.450
2700	40.229	299.856	266.945	88.857	-59.694	-72.670	1.406
2800	40.404	301.322	268.147	92.889	-59.699	-73.150	1.365
2900	40.568	302.742	269.316	96.938	-59.702	-73.631	1.326
3000	40.721	304.120	270.453	101.002	-59.705	-74.111	1.290
3100	40.864	305.458	271.561	105 082	-59.707	-74.591	1.257
3200	40.996	306.757	272.640	109.175	-59.708	-75.072	1.225
3300	41.119	308.021	273.693	113.281	-59.710	-75.552	1.196
3400	41.232	309.250	274.721	117.398	-59.712	-76.032	1.168
3500	41.336	310.447	275.725	121.527	-59.716	-76.512	1.142
3600	41.432	311.613	276.706	125.665	-59.722	-76.991	1.117
3700	41.520	312.749	277.665	129.813	-59.730	-77 471	1.094
3800	41.601	313.857	278.603	133.969	-59.741	-77.950	1.071
3900	41.676	314.939	279.520	138.133	-59.755	-78.429	1.050
4000	41.745	315.995	280 419	142.304	-59.774	-78.908	1.030
4100	41.810	317.027	281.299	146.482	-59 796	-79.386	1.011
4200	41.871	318.035	282.162	150.666	-59.824	-79.863	0.993
4300	41.929	319.021	283.008	154.856	-59.857	-80.340	0.976
4400	41.986	319.985	283.837	159.051	-59.896	-80.816	0.959
4500	42.042	320.930	284.651	163.253	-59.941	-81.291	0.944
4600	42.098	321.854	285.450	167.460	-59.992	-81.765	0.928
4700	42.156	322.760	286.234	171.673	-60.049	-82.238	0.914
4800	42.217	323.648	287.004	175.891	-60.113	-82.709	0.900
4900	42.282	324.520	287.761	180.116	-60.183	-83.179	0.887
5000	42.352	325.374	288.505	184.348	-60.259	-83.648	0.874
5100	42.429	326.214	289.236	188.587	-60.342	-84.114	0.862
5200	42.514	327.039	289.955	192.834	-60.430	-84.580	0.850
5300	42.608	327.849	290.663	197.090	-60.524	-85.043	0.838
5400	42.712	328.647	291.359	201.356	-60.623	-85.505	0.827
5500	42.829	329.431	292.044	205.633	-60.726	-85.965	0.816
5600	42.959	330.204	292.718	209.922	-60.832	-86.423	0.806
5700	43.104	330.966	293.383	214.225	-60.941	-86.879	0.796
5800	43.265	331.717	294.037	218.543	-61.051	-87.333	0.787
5900	43.444	332.458	294.682	222.879	-61.162	-87.786	0.777
6000	43.642	333.190	295.318	227.233	-61.271	-88.235	0.768

PREVIOUS: June 1977 (1 atm) CURRENT: June 1977 (1 bar)

SULFUR OXIDE (S_2O) IDEAL GAS M_r = 80.1194

Sulfur Oxide (SSO) $O_1S_2(g)$

$S°(298.15 K) = 267.020 J K^{-1} mol^{-1}$

$\Delta_fH°(0 K) = [-58.86 \pm 33.5] kJ mol^{-1}$

$\Delta_fH°(298.15 K) = [-56.48 \pm 33.5] kJ mol^{-1}$

Vibrational Frequencies and Degeneracies

ν, cm^{-1}
679(1)
388(1)
1165(1)

Ground State Quantum Weight: 1 $\sigma = 1$

Point Group: C_s

Bond Distances: S–S = 1.884 Å S–O = 1.465 Å

Bond Angle: S–S–O = 118°

Product of the Moments of Inertia: $I_AI_BI_C = 6.270091 \times 10^{-115} g^3 cm^6$

Enthalpy of Formation

There is no direct measurement of the standard enthalpy of formation for $S_2O(g)$. Blukis and Myers (1) have discussed the discrepancy of the values of $\Delta_fH°(0 K)$ which vary from -1 to -55 kcal mol^{-1}. Hagman (2) reported $\Delta_fH° = -17$ kcal mol^{-1} from his ionization and appearance potential measurements by mass spectrometry. Steudel and Schenk (3) estimated the enthalpy of formation as -22.7 kcal mol^{-1}. If the bond dissociation energies $D_0^0(S–O) = 123.5$ kcal mol^{-1} and $D_0^0(S–S) = 101$ kcal mol^{-1} were used to estimate the dissociation energy of S_2O (i.e., $D_0^0(S_2O) = 224.5$ kcal mol^{-1}), one will obtain $\Delta_fH = -35.3$ kcal mol^{-1}. Jones (4) reported the predissociation energy, 91 kcal mol^{-1} for sulfur monoxide from the infrared and ultraviolet spectra. [Meschi and Myers (5) have concluded from their experiment that the so-called sulfur monoxide is S_2O.] Therefore, if SO(g) + S(g) were the dissociation products, the $\Delta_fH°$ would be -25.4 kcal mol^{-1} and if $S_2(g)$ + O(g) were the products, the $\Delta_fH°$ would be -1.2 kcal mol^{-1}. (All JANAF auxiliary data used in calculation.)

Dewing and Richardson (6) have investigated the gas phase equilibria in the sulfur-oxygen vapor at 1250° and 1500°C, respectively. Blukis and Myers (1) recalculated their data based on the assumption that the S_2O (instead of SO) was the major product in the equilibrium, and obtained $\Delta_fH° = -13$ kcal mol^{-1} for $S_2O(g)$ which is adopted in this table.

Heat Capacity and Entropy

The bond distance, angle, and vibrational frequencies were obtained from the microwave and infrared spectra measurements by Meschi and Myers (5) and Blukis and Myers (1). The principal moments of inertia are: $I_A = 2.0209 \times 10^{-39}$, $I_B = 16.6326 \times 10^{-39}$ and $I_C = 18.6536 \times 10^{-39}$ g cm^2.

References

1. U. Blukis and R. Myers, J. Phys. Chem. 69, 1154 (1965).
2. R. Hagmann, Compt. Rend. 225, 1102 (1962).
3. R. Steudel and P. W. Schenk, Z. Physik Chem. 43, 33 (1964).
4. A. V. Jones, J. Chem. Phys. 18, 1263 (1950).
5. D. Meschi and R. Myers, J. Mol. Spectrosc. 3, 405 (1959).
6. E. Dewing and F. Richardson, Trans. Faraday Soc. 54, 679 (1958).

Enthalpy Reference Temperature = T_r = 298.15 K Standard State Pressure = p° = 0.1 MPa

	J $K^{-1}mol^{-1}$			kJ mol^{-1}			
T/K	$C_p°$	S°	$-[G°-H°(T_r)]/T$	$H°-H°(T_r)$	$\Delta_fH°$	$\Delta_fG°$	Log K_f
0	0.	0.	INFINITE	-11.128	-58.859	-58.859	INFINITE
100	34.286	225.094	302.941	-7.785	-57.657	-70.249	36.694
200	39.425	250.369	270.911	-4.108	-57.078	-83.076	21.697
250	41.949	259.442	267.734	-2.073	-56.791	-89.608	18.723
298.15	44.114	267.020	267.020	0.	-56.484	-95.956	16.811
300	44.191	267.293	267.020	0.082	-56.472	-96.201	16.750
350	46.136	274.255	267.566	2.341	-56.117	-102.850	15.350
400	47.797	280.527	268.800	4.691	-57.945	-109.465	14.295
450	49.199	286.240	270.425	7.116	-58.203	-115.898	13.453
500	50.379	291.486	272.273	9.607	-58.487	-122.290	12.776
600	52.201	300.842	276.274	14.741	-58.517	-135.040	11.756
700	53.497	308.992	280.378	20.029	-58.203	-147.815	11.030
800	54.436	316.200	284.414	25.429	-57.683	-160.651	10.489
900	55.131	322.653	288.311	30.909	-110.163	-172.478	10.010
1000	55.656	328.491	292.041	36.449	-108.209	-179.507	9.376
1100	56.061	333.815	295.600	42.036	-106.248	-186.732	8.867
1200	56.379	338.707	298.992	47.658	-104.287	-194.135	8.450
1300	56.632	343.230	302.223	53.309	-102.330	-201.702	8.104
1400	56.836	347.435	305.304	58.983	-100.381	-209.419	7.814
1500	57.004	351.362	308.245	64.676	-98.443	-217.275	7.566
1600	57.143	355.045	311.056	70.383	-96.517	-225.260	7.354
1700	57.259	358.513	313.747	76.103	-94.606	-233.365	7.170
1800	57.358	361.789	316.325	81.834	-92.709	-241.582	7.011
1900	57.442	364.893	318.801	87.574	-90.828	-249.904	6.870
2000	57.514	367.841	321.180	93.322	-88.963	-258.325	6.747
2100	57.576	370.648	323.469	99.077	-87.113	-266.839	6.637
2200	57.630	373.328	325.675	104.837	-85.280	-275.440	6.540
2300	57.678	375.891	327.803	110.603	-83.462	-284.124	6.453
2400	57.720	378.347	329.858	116.373	-81.659	-292.888	6.375
2500	57.757	380.704	331.845	122.146	-79.871	-301.726	6.304
2600	57.790	382.970	333.768	127.924	-78.098	-310.635	6.241
2700	57.819	385.151	335.631	133.704	-76.338	-319.612	6.183
2800	57.846	387.254	337.437	139.488	-74.591	-328.654	6.131
2900	57.869	389.285	339.190	145.273	-72.858	-337.759	6.084
3000	57.891	391.247	340.893	151.061	-71.137	-346.922	6.040
3100	57.910	393.145	342.548	156.851	-69.428	-356.143	6.001
3200	57.928	394.984	344.158	162.643	-67.731	-365.420	5.965
3300	57.944	396.767	345.726	168.437	-66.045	-374.749	5.932
3400	57.959	398.497	347.252	174.232	-64.369	-384.128	5.901
3500	57.972	400.177	348.741	180.029	-62.705	-393.557	5.874
3600	57.985	401.811	350.192	185.827	-61.051	-403.034	5.848
3700	57.996	403.400	351.609	191.626	-59.408	-412.556	5.824
3800	58.007	404.946	352.992	197.426	-57.775	-422.122	5.802
3900	58.017	406.453	354.344	203.227	-56.152	-431.731	5.782
4000	58.026	407.922	355.665	209.029	-54.539	-441.382	5.764
4100	58.034	409.355	356.957	214.832	-52.937	-451.073	5.747
4200	58.042	410.754	358.221	220.636	-51.345	-460.803	5.731
4300	58.049	412.120	359.459	226.440	-49.763	-470.570	5.716
4400	58.056	413.454	360.671	232.246	-48.193	-480.375	5.703
4500	58.062	414.759	361.859	238.052	-46.633	-490.215	5.690
4600	58.068	416.035	363.022	243.858	-45.084	-500.090	5.679
4700	58.074	417.284	364.164	249.665	-43.547	-509.998	5.668
4800	58.079	418.507	365.283	255.473	-42.022	-519.938	5.658
4900	58.084	419.704	366.382	261.281	-40.509	-529.910	5.649
5000	58.088	420.878	367.460	267.090	-39.008	-539.913	5.640
5100	58.093	422.028	368.519	272.899	-37.521	-549.946	5.633
5200	58.097	423.156	369.559	278.708	-36.047	-560.008	5.625
5300	58.101	424.263	370.580	284.518	-34.587	-570.098	5.619
5400	58.104	425.349	371.584	290.328	-33.141	-580.216	5.612
5500	58.108	426.415	372.572	296.139	-31.711	-590.360	5.607
5600	58.111	427.462	373.543	301.950	-30.295	-600.530	5.602
5700	58.114	428.491	374.498	307.761	-28.896	-610.725	5.597
5800	58.117	429.502	375.437	313.573	-27.513	-620.945	5.592
5900	58.120	430.495	376.362	319.385	-26.147	-631.189	5.588
6000	58.123	431.472	377.272	325.197	-24.798	-641.455	5.584

PREVIOUS: September 1965 (1 atm) CURRENT: September 1965 (1 bar)

SILICON OXIDE (SiO) IDEAL GAS M_r = 44.0849

$S°(298.15\ K) = 211.579\ J\ K^{-1}\ mol^{-1}$

$\Delta_f H°(0\ K) = -101.57 \pm 8.4\ kJ\ mol^{-1}$

$\Delta_f H°(298.15\ K) = -100.42 \pm 8.4\ kJ\ mol^{-1}$

Electronic Levels and Quantum Weights

State	ε_i, cm^{-1}	g_i
$^1\Sigma^+$	0	1
	[30000]	6
	42640	2

$\omega_e = 1241.44$ $\omega_e x_e = 5.92$ $\sigma = 1$

$B_e = 0.7273$ $\alpha_e = 0.00508$ $r_e = 1.509$ Å

Enthalpy of Formation

The adopted value is based on equilibrium data summarized below. The more reliable results from three different reactions lie in the range from about -23 to -28 kcal mol^{-1}. Extreme values, including -20 and -32 kcal mol^{-1}, can be discounted on experimental grounds. Margrave (1) recently reported $\Delta_r H°(298.15\ K) = 82.8 \pm 1.5$ kcal mol^{-1} for 0.5 SiO_2(s) + 0.5 Si(s) = SiO(g), but details are not yet available. Assuming that this corresponds to reaction A below, we calculate $\Delta_f H(298.15\ K) = -25.4$ kcal mol^{-1}. Reaction A involves the nominal composition SiO(s) which is treated as an intimate mixture of 0.5 Si(cr) + 0.5 SiO_2(high cristobalite). X-ray studies (13, 14) suggest that the condensed "SiO" used in (4, 8) is probably a mixture of Si and amorphous or vitreous SiO_2. The adopted $\Delta_f H°$ corresponds to $D_0^0 = 190 \pm 3$ kcal mol^{-1}. This may be compared with $D_0^0 = 179$ kcal mol^{-1} obtained by linear Birge-Sponer extrapolation of the ground state vibrational constants and with $D_0^0 = 185 \pm 7$ kcal mol^{-1} obtained by Barrow (15) from $D_0'^0(LBX) = 77.5$ kcal mol^{-1} for the excited state at 52580 cm^{-1}. Barrow assumed $D_0^0/D_0'^0(LBX) = 0.456$ based on related molecules.

Source	Reaction	Method	T/K	Data Points	$\Delta_r H°(298.15\ K)$, kcal mol^{-1}		Drift cal K^{-1} mol^{-1}	$\Delta_f H°(298.15\ K)$ kcal mol^{-1}
Ramstad (2)	A	Carrier gas	1510-1758	6	80.2 ± 2.3	85.08	3.0 ± 1.4	-23.1
Bergman (3)	A	$\Delta_{sub}H°$ by explosion	---	-	-	87±2	-	-22
Gunther (4)	A*	Knudsen wt. gain	1202-1520	18	97.0 ± 0.7	86.76	-7.6 ± 0.5	-21.5
Porter (5)	A**	Knudsen mass spec.	1345-1463	2	105.1	87.35	-13	-20.9
Tombs (6)	A	Carrier gas	1573-1920	7	66.4 ± 1.1	84.75	11 ± 1	-23.5
Schafer (7)	A	Knudsen wt. loss	1336-1460	21	78.7 ± 1.5	82.74	2.9 ± 1.0	-25.5
Gel'd (8)	A*	Knudsen wt. loss	1173-1428	11	77.2 ± 1.8	85.87	6.8 ± 1.4	-22.3
Ramstad (2)	B***	Carrier gas	1698-1873	4	142.6 ± 0.7	133.79	-4.9 ± 0.4	-24.8
Bowles (2)	B***	Carrier gas	1703-1898	-	135	132.3	-2	-26.3
Tombs (6)	B***	Carrier gas	1501-1926	23	88 ± 3	127.3	22 ± 2	-31.3
Grube (9)	B***	Carrier gas	1473-1773	4	112 ± 3	134.4	13 ± 2	-24.2
Nesmeyanov (10)	C**	Knudsen collection	1653-1739	24	-	-	-	-20.5
Yang (11)	C	Knudsen wt. loss	1818-1960	15	199.3 ± 3.4	194.5	-2 ± 1.8	-21.9
Porter (5)	C	Knudsen mass spec.	1800-1900	2	130	184.4	29	-32.0
Cochran (12)	D	Knudsen effusion	?	-	-	11.5	-	-25.3

A) 0.5 SiO_2(high cristo.) + 0.5 Si(cr, ℓ) = SiO(g)

B) SiO_2(high cristo.) + H_2(g) = H_2O(g) + SiO(g)

C) SiO_2(high cristo.) = 0.5 O_2(g) + SiO(g)

D) SiO_2(ℓ) + Mg(g) = MgO(cr) + SiO(g)

*Starting material was "SiO" **Starting material was tridymite ***Starting material was quartz

Heat Capacity and Entropy

All constants for the ground state and the $A^1\Pi$ State at 42640 cm^{-1} are taken from the analysis of the A→X system reported by Lagerqvist (16). An intercombination transition has been observed and assigned as $^1\Sigma^- + a^3\Pi_r$ by Verma (17). The value of the a-X excitation energy is unknown but is estimated here as 30000±5000 cm^{-1} by comparison with the a and A states of CO and AℓCl. Several additional states are known (15, 16) to lie near or above the A state, but they are omitted because their contribution to the thermal functions is negligible.

References

1. J. L. Margrave, 5th Meeting ICRPG Thermochemistry Working Group, Linden, New Jersey, March, 1967.
2. H. F. Ramstad, F. D. Richardson and P. J. Bowles, Trans. AIME 221, 1021 (1961). Data given graphically.
3. G. A. Bergman and V. A. Medvedev, Sbornik Trudov Gosudarst. Inst. Priklad. Khim. 1959, 158 (1959).
4. K. G. Gunther, Glastech. Ber. 31, 9 (1958).
5. R. F. Porter, W. A. Chupka and M. G. Inghram, J. Chem. Phys. 23, 216 (1955).
6. N. C. Tombs and A. J. E. Welch, J. Iron Steel Inst. 172, 69 (1952).
7. H. Schafer and R. Hornle, Z. Anorg. Allg. Chem. 263, 261 (1950).
8. P. V. Gel'd and M. I. Kochnev, Zhur. Priklad. Khim. 21, 1249 (1948).
9. G. Grube and H. Speidel, Z. Elektrochem. 53, 341 (1949).
10. An. N. Nesmeyanov and L. P. Firsova, Russ. J. Phys. Chem. 34, 906, 1232, 1279 (1960).
11. L. Yang, C. L. McCabe and R. Miller, "Physical Chemistry of Steelmaking," pp. 63-4, John Wiley and Sons, Inc., New York, 1958. Data given graphically.
12. C. N. Cochran and L. M. Foster, Phys. Chem. 66, 380 (1962).
13. A. N. Gusatinskii, K. M. Kolobova and S. A. Nemnonov, Izv. Akad. Nauk SSSR, Neorg. Materials 1 (6), 877 (1965).
14. G. W. Brady, J. Phys. Chem. 63, 1119 (1959).
15. R. F. Barrow and H. C. Rowlinson, Proc. Soc. (London) A224, 374 (1954).
16. A. Lagerqvist and U. Uhler, Arkiv Fysik 6, 95 (1952).
17. R. D. Verma and R. S. Mulliken, Can. J. Phys. 39, 908 (1961).

Silicon Oxide (SiO) $O_1Si_1(g)$

Enthalpy Reference Temperature = T_r = 298.15 K Standard State Pressure = p° = 0.1 MPa

T/K	$C_p°$ (J $K^{-1}mol^{-1}$)	S° (J $K^{-1}mol^{-1}$)	$-[G°-H°(T_r)]/T$ (J $K^{-1}mol^{-1}$)	$H°-H°(T_r)$ (kJ mol^{-1})	$\Delta_f H°$ (kJ mol^{-1})	$\Delta_f G°$ (kJ mol^{-1})	Log K_f
0	0.	0.	INFINITE	-8.715	-101.571	-101.571	INFINITE
100	29.105	179.626	237.700	-5.807	-100.382	-109.296	57.090
200	29.204	199.813	214.286	-2.894	-100.103	-118.385	30.919
250	29.466	206.354	212.069	-1.429	-100.217	-122.945	25.688
298.15	29.899	211.579	211.579	0.	-100.416	-127.305	22.303
300	29.918	211.764	211.580	0.055	-100.425	-127.472	22.195
350	30.500	216.419	211.946	1.565	-100.688	-131.959	19.694
400	31.135	220.533	212.767	3.106	-100.981	-136.407	17.813
450	31.768	224.236	213.839	4.679	-101.291	-140.817	16.346
500	32.365	227.615	215.050	6.282	-101.612	-145.191	15.168
600	33.399	233.611	217.656	9.572	-102.278	-153 845	13.393
700	34.213	238.823	220.316	12.955	-102.971	-162.385	12.117
800	34.845	243.434	222.923	16.409	-103.694	-170.824	11.154
900	35.336	247.568	225.436	19.919	-104.449	-179.170	10.399
1000	35.722	251.312	227.839	23.473	-105.237	-187.431	9.790
1100	36.030	254.731	230.131	27.061	-106.059	-195.610	9.289
1200	36.279	257.877	232.314	30.677	-106.917	-203.714	8.867
1300	36.484	260.790	234.393	34.315	-107.810	-211.744	8.508
1400	36.655	263.500	236.377	37.972	-108.742	-219.705	8.197
1500	36.800	266 034	238.270	41.645	-109.714	-227.597	7.926
1600	36.923	268.413	240.081	45.332	-110.727	-235.423	7.686
1700	37.031	270.655	241.814	49.029	-161.960	-242.737	7.458
1800	37.125	272.774	243.476	52.737	-162.830	-247.464	7.181
1900	37.209	274.784	245.071	56.454	-163.703	-252.142	6.932
2000	37.283	276.694	246.605	60.179	-164.579	-256.774	6.706
2100	37.351	278.515	248.081	63 910	-165.459	-261.362	6.501
2200	37.412	280.254	249.504	67.649	-166.345	-265.908	6.313
2300	37.469	281.918	250.878	71.393	-167.236	-270.414	6.141
2400	37.521	283.514	252.205	75.142	-168.132	-274.881	5.983
2500	37.569	285.046	253.488	78.897	-169.035	-279.310	5.836
2600	37.615	286.521	254.730	82.656	-169.943	-283.703	5.700
2700	37.658	287.941	255.934	86.420	-170.858	-288.061	5.573
2800	37.700	289.312	257.102	90.188	-171.778	-292.385	5.455
2900	37.739	290.635	258.235	93.960	-172.705	-296.677	5.344
3000	37.778	291.915	259.337	97.735	-173.637	-300.935	5.240
3100	37.816	293.155	260.408	101.515	-174.575	-305.163	5.142
3200	37.853	294.356	261.450	105.299	-175 518	-309.361	5.050
3300	37.891	295.521	262.465	109.086	-176.466	-313.529	4.963
3400	37.929	296.653	263.454	112.877	-177.418	-317.668	4.880
3500	37.968	297.753	264.418	116.672	-178.375	-321.779	4.802
3600	38.009	298.823	265.359	120.470	-563.487	-315.402	4.576
3700	38.052	299.865	266.278	124 273	-564.037	-308.503	4.355
3800	38.097	300.880	267.175	128.081	-564.591	-301.589	4.146
3900	38.145	301.871	268.052	131.893	-565.149	-294.661	3.947
4000	38.196	302.837	268.910	135.710	-565.709	-287.718	3.757
4100	38.251	303.781	269.749	139.532	-566.272	-280.761	3.577
4200	38.311	304.703	270.570	143.360	-566.835	-273.791	3.405
4300	38.375	305.606	271.374	147.195	-567.399	-266.807	3.241
4400	38.445	306.489	272.162	151.036	-567.962	-259.810	3.084
4500	38.521	307.353	272.935	154.884	-568.524	-252.800	2.934
4600	38.602	308.201	273.692	158.740	-569.084	-245.778	2.791
4700	38.690	309.032	274.435	162.604	-569.642	-238.744	2.653
4800	38.785	309.848	275.165	166.478	-570.196	-231.698	2.521
4900	38.887	310.648	275.881	170.362	-570.746	-224.639	2.395
5000	38.997	311.435	276.584	174.256	-571.291	-217.571	2.273
5100	39.114	312.208	277.275	178.161	-571.832	-210.491	2.156
5200	39.239	312.969	277.954	182.079	-572.367	-203.401	2.043
5300	39.372	313.718	278.622	186.009	-572.897	-196.300	1.935
5400	39.513	314.455	279.279	189.953	-573.421	-189.190	1.830
5500	39.662	315.181	279.925	193.912	-573.937	-182.070	1.729
5600	39.820	315.897	280.561	197.886	-574.448	-174.940	1.632
5700	39.986	316.604	281.187	201.876	-574.951	-167.801	1.538
5800	40.161	317.301	281.803	205.884	-575.448	-160.654	1.447
5900	40.343	317.989	282.411	209.909	-575.938	-153.499	1.359
6000	40.534	318.668	283.010	213.953	-576.420	-146.334	1.274

PREVIOUS: September 1967 (1 atm) CURRENT: September 1967 (1 bar)

Silicon Oxide (SiO) $O_1Si_1(g)$

STRONTIUM OXIDE (SrO) CRYSTAL M_r = 103.6194

$S°(298.15\ K) = 55.522 \pm 0.4\ J\ K^{-1}\ mol^{-1}$

$T_{fus} = 2938 \pm 20\ K$

$\Delta_f H°(0\ K) = -589.801 \pm 3.3\ kJ\ mol^{-1}$

$\Delta_f H°(298.15\ K) = -592.036 \pm 3.3\ kJ\ mol^{-1}$

$\Delta_{fus} H° = [75.312]\ kJ\ mol^{-1}$

Enthalpy of Formation

$\Delta_f H°$ is from (1). The adopted value was derived (2) from $\Delta_f H° = -198.1\ kcal\ mol^{-1}$ for $SrCl_2(cr)$, $\Delta_{sol}H°$ of $SrCl_2$ in HCl(aq), and calorimetric data for $\Delta_r H°$ of $SrO(cr) + (x+2)HCl(n\ H_2O) + [SrCl_2(aq) + H_2O(aq) + x\ HCl(n'\ H_2O)]$. Data of Adami and Conway (3) yield $\Delta_f H°(SrO, cr) = -141.3\ kcal\ mol^{-1}$, while those of Mcnaenkova and Vorob'ev (4) yield =141.7 $kcal\ mol^{-1}$. Fldlider et al. (5) derived -140.5 $kcal\ mol^{-1}$ from their (unpublished) data for SrO(cr) and similar measurements for Sr(cr) by Guntz and Benoit (6). Brisi and Abbattistra (7) omit the mole ratios necessary for treatment of their data, but an approximate calculation suggests consistency with the adopted value. We omit from this discussion the early data which Parker (1, 2) reviewed in selecting the adopted value.

Direct measurements of $\Delta_c H°$ of Sr(cr) by Mah (8) gave $-144.44 \pm 0.4\ kcal\ mol^{-1}$. The negative bias of about 3 $kcal\ mol^{-1}$ presumably resulted from inadequate allowance for side reactions, e.g., with combustion products of Mylar used to contain the Sr. Parker (2) noted that the combustion value is incompatible with data for $SrCl_2(cr)$ and related compounds. We find that the combustion value is also less consistent with equilibrium cata for $SrCl_2(\ell$ and g). There is a similar, but much larger, discrepancy for Ba compunds (cf, BaO, crystal). Although impurity effects are of concern in all studies, the evidence predominantly favors the solution calorimetry.

Heat Capacity and Entropy

$C_p°$ is based on data (1.4-310 K) of Gmelin (9) who reported smoothed values only in the range 4 to 300 K. Based on Gmelin's graphs, we have corrected several typographical errors and resmoothed $C_p°$ above 250 K where the scatter increases significantly. The entropy is obtained by integration of our adopted $C_p°$ using $S° = 0.0003\ cal\ K^{-1}\ mol^{-1}$ at 4 K. Gmelin reported $S°(273.15\ K) = 12.40 \pm 0.02\ cal\ K^{-1}\ mol^{-1}$ compared with 12.33 from our integration. $C_p°$ data (58-298 K) of Anderson (10) are lower at all temperatures with deviations ranging from -4.7% (61 K) to -0.3% (223 K) and -0.7% (298 K).

$C_p°$ above 300 K is from constrained fitting of enthalpy data (363-1266 K) of Lander (11). The calorimeter (11) was calibrated with Pt, but details are lacking to assess the bias due to minor changes in the enthalpy of Pt (12). Deviations of Lander's data from the adopted curve range from -0.7 to +1.1%.

Fusion Data

See the liquid table for details.

References

1. U. S. Nat. Bur. Stand. Tech. Note 270-6, 1971.
2. V. B. Parker, U. S. Nat. Bur. Stand. Report 10074, 164, 1968.
3. L. H. Adami and K. C. Conway, U. S. Bur. Mines RI 6822, 1966.
4. A. S. Monaenkova and A. F. Vorob'ev, Izv. Vyssh. Ucheb. Zaved., Khim. Khim. Teknol. 15, 191 (1972).
5. G. V. Fldlider, P. V. Kovtunenko and A. A. Bundel, Russ. J. Phys. Chem. 40, 1168 (1966).
6. A. Guntz and F. Benoit, Ann. Chim. (Paris) [9] 20, 5 (1923).
7. C. Brisi and F. Abbattista, Ann. Chim. (Rome) 50, 165 (1960).
8. A. D. Mah, U. S. Bur. Mines RI 6171, 1963.
9. E. Gmelin, Z. Naturforsch. 24A, 1794 (1969).
10. C. T. Anderson, J. Amer. Chem. Soc. 57, 429 (1935).
11. J. J. Lander, J. Amer. Chem. Soc. 73, 5794 (1951).
12. A. C. Macleod, J. Chem. Thermodyn. 4, 391 (1972).

Strontium Oxide (SrO) $O_1Sr_1(cr)$

Enthalpy Reference Temperature = T_r = 298.15 K; Standard State Pressure = p° = 0.1 MPa

T/K	$C_p°$ (J K⁻¹mol⁻¹)	S°	-[G°-H°(T_r)]/T	H°-H°(T_r) (kJ mol⁻¹)	$\Delta_f H°$	$\Delta_f G°$	Log K_f
0	0.000	0.000	INFINITE	-8.675	-589.801	-589.801	INFINITE
100	24.949	15.242	91.864	-7.662	-591.740	-581.794	303.898
200	40.409	38.334	59.542	-4.242	-592.263	-571.540	149.271
298.15	45.409	55.522	55.522	0.000	-592.036	-561.402	98.355
500	50.534	80.376	60.880	9.748	-590.937	-540.947	56.512
600	52.049	89.728	64.929	14.880	-590.315	-531.007	46.228
700	53.262	97.845	69.064	20.147	-589.709	-521.171	38.890
800	54.300	105.026	73.119	25.526	-589.138	-511.419	33.392
900	55.220	111.476	77.029	31.002	-589.238	-501.664	29.116
1000	56.074	117.338	80.771	36.568	-588.437	-491.977	25.698
1100	56.873	122.720	84.343	42.215	-595.481	-482.006	22.889
1200	57.639	127.702	87.751	47.941	-595.475	-471.690	20.532
1300	58.379	132.345	91.005	53.742	-595.412	-461.376	18.538
1400	59.103	136.698	94.115	59.616	-595.291	-451.070	16.830
1500	59.810	140.800	97.092	65.562	-595.112	-440.775	15.349
1600	60.509	144.682	99.946	71.578	-594.876	-430.493	14.054
1700	61.199	148.371	102.687	77.664	-731.497	-419.048	12.876
1800	61.881	151.889	105.323	83.818	-729.287	-400.732	11.629
1900	62.559	155.253	107.863	90.040	-727.026	-382.541	10.517
2000	63.229	158.479	110.314	96.329	-724.717	-364.469	9.519
2100	63.898	161.580	112.682	102.685	-722.365	-346.515	8.619
2200	64.563	164.568	114.973	109.109	-719.974	-328.672	7.804
2300	65.229	167.452	117.192	115.598	-717.551	-310.940	7.062
2400	65.890	170.242	119.345	122.154	-715.100	-293.314	6.384
2500	66.551	172.945	121.435	128.776	-712.631	-275.791	5.762
2600	67.208	175.568	123.467	135.464	-710.151	-258.366	5.191
2700	67.864	178.117	125.444	142.218	-707.669	-241.036	4.663
2800	68.517	180.597	127.370	149.037	-705.195	-223.799	4.175
2900	69.174	183.013	129.247	155.921	-702.739	-206.650	3.722
2938.000	69.425	183.915	129.948	158.555	------ CRYSTAL <--> LIQUID -----		
3000	69.827	185.369	131.079	162.871	-700.309	-189.585	3.301
3100	70.479	187.669	132.867	169.887	-697.917	-172.600	2.908
3200	71.132	189.917	134.615	176.967	-695.572	-155.692	2.541
3300	71.785	192.116	136.324	184.113	-693.282	-138.857	2.198
3400	72.438	194.269	137.997	191.324	-691.058	-122.089	1.876
3500	73.086	196.378	139.635	198.600	-688.907	-105.387	1.573
3600	73.739	198.446	141.240	205.942	-686.837	-88.744	1.288

PREVIOUS: CURRENT: December 1972

Strontium Oxide (SrO) $O_1Sr_1(cr)$

STRONTIUM OXIDE (SrO) LIQUID M_r = 103.6194

$S°(298.15\ K) = [80.846]\ J\ K^{-1}\ mol^{-1}$ $\Delta_f H°(298.15\ K) = [-517.292]\ kJ\ mol^{-1}$

$T_{fus} = 2938 \pm 20\ K$ $\Delta_{fus}H° = [75.312]\ kJ\ mol^{-1}$

Enthalpy of Formation

$\Delta_f H°(SrO,\ \ell,\ 298.15\ K)$ is calculated from that of the crystal by adding $\Delta_{fus}H°$ and the difference in enthalpy, H°(2938 K)-H°(298.15 K), between the crystal and the liquid.

Heat Capacity and Entropy

$C_p°$ of the liquid is estimated as 8 cal K^{-1} g-atom^{-1} by comparison with other oxides. $C_p°$ is taken from the crystal in the range from 298 K up to the glass transition assumed at 2000 K. S°(298.15 K) is calculated in a manner analogous to that used for $\Delta_f H°$.

Fusion Data

Foex (1, 2) obtained T_{fus} = 2660°C from a solar-furnace study using SrO as its own container. We adopt this result but increase it by 5 for conversion to IPTS-68. The much lower value (2420°C) of Schumacher (3, 4) probably resulted from contamination by WO_3 from tungsten supports of the sample. $\Delta_{fus}H°$ is estimated such that $\Delta_{fus}S°$ is 6 cal K^{-1} g-atom^{-1}. Phase data for the binary system SrO-$SrZrO_3$ (2) yield $\Delta_{fus}H° = 10 \pm 2$ kcal mol^{-1} assuming ideal behavior in the liquid. Comparison with the very non-ideal behavior of other mixed oxides suggests that the SrO data (2) are not inconsistent with $\Delta_{fus}H°$ = 18 kcal mol^{-1}.

References

1. M. Foex, Solar Energy 9, 61 (1965).
2. J. P. Traverse and M. Foex, High Temp.-High Press. 1, 409 (1969).
3. E. E. Schumacher, J. Amer. Chem. Soc. 48, 396 (1926).
4. S. J. Schneider, U.S. Nat. Bur. Stand. Monograph 68, 26, 1963.

Strontium Oxide (SrO) $O_1Sr_1(l)$

Enthalpy Reference Temperature = T_r = 298.15 K Standard State Pressure = p° = 0.1 MPa

T/K	$C_p°$ (J K⁻¹mol⁻¹)	S° (J K⁻¹mol⁻¹)	-[G°-H°(T_r)]/T (J K⁻¹mol⁻¹)	H°-H°(T_r) (kJ mol⁻¹)	$\Delta_f H°$ (kJ mol⁻¹)	$\Delta_f G°$ (kJ mol⁻¹)	Log K_f
0							
100							
200							
298.15	45.409	80.846	80.846	0.000	-517.292	-494.209	86.583
300	45.480	81.128	80.847	0.084	-517.285	-494.066	86.025
400	48.480	94.650	82.671	4.792	-516.792	-486.396	63.517
500	50.534	105.701	86.205	9.748	-516.193	-478.865	50.027
600	52.049	115.053	90.254	14.880	-515.571	-471.458	41.044
700	53.262	123.170	94.389	20.147	-514.965	-464.154	34.636
800	54.300	130.351	98.444	25.526	-514.394	-456.935	29.835
900	55.220	136.800	102.353	31.002	-514.494	-449.713	26.101
1000	56.074	142.663	106.096	36.568	-513.693	-442.557	23.117
1100	56.873	148.045	109.668	42.215	-520.737	-435.119	20.662
1200	57.639	153.027	113.076	47.941	-520.731	-427.335	18.601
1300	58.379	157.670	116.330	53.742	-520.668	-419.555	16.858
1400	59.103	162.023	119.440	59.616	-520.547	-411.781	15.364
1500	59.810	166.125	122.417	65.562	-520.368	-404.018	14.069
1600	60.509	170.007	125.271	71.578	-520.132	-396.269	12.937
1700	61.199	173.696	128.012	77.664	-656.753	-387.356	11.902
1800	61.881	177.213	130.648	83.818	-654.543	-371.573	10.783
1900	62.559	180.577	133.188	90.040	-652.282	-355.914	9.785
2000	63.229	183.803	135.639	96.329	-649.973	-340.375	8.890
2000.010	63.229	183.804	135.639	96.330	GLASS <--> LIQUID	TRANSITION	
2000.010	66.944	183.804	135.639	96.330			
2100	66.944	187.070	138.011	103.024	-647.283	-324.961	8.083
2200	66.944	190.184	140.312	109.718	-644.621	-309.675	7.353
2300	66.944	193.160	142.545	116.412	-641.992	-294.509	6.689
2400	66.944	196.009	144.714	123.107	-639.404	-279.457	6.082
2500	66.944	198.741	146.821	129.801	-636.862	-264.511	5.527
2600	66.944	201.367	148.869	136.496	-634.376	-249.667	5.016
2700	66.944	203.894	150.860	143.190	-631.953	-234.916	4.545
2800	66.944	206.328	152.798	149.884	-629.604	-220.254	4.109
2900	66.944	208.677	154.685	156.579	-627.337	-205.675	3.705
2938.000	66.944	209.549	155.389	159.123	------ CRYSTAL <--> LIQUID -----		
3000	66.944	210.947	156.522	163.273	-625.163	-191.172	3.329
3100	66.944	213.142	158.314	169.968	-623.092	-176.740	2.978
3200	66.944	215.267	160.060	176.662	-621.133	-162.374	2.650
3300	66.944	217.327	161.765	183.356	-619.295	-148.066	2.344
3400	66.944	219.326	163.428	190.051	-617.588	-133.813	2.056
3500	66.944	221.266	165.053	196.745	-616.019	-119.607	1.785
3600	66.944	223.152	166.641	203.440	-614.595	-105.444	1.530
3700	66.944	224.986	168.193	210.134	-613.323	-91.319	1.289
3800	66.944	226.772	169.711	216.828	-612.208	-77.226	1.062
3900	66.944	228.510	171.197	223.523	-611.253	-63.160	0.846
4000	66.944	230.205	172.651	230.217	-610.461	-49.117	0.641
4100	66.944	231.858	174.075	236.912	-609.835	-35.092	0.447
4200	66.944	233.472	175.470	243.606	-609.370	-21.080	0.262
4300	66.944	235.047	176.837	250.300	-609.071	-7.077	0.086
4400	66.944	236.586	178.178	256.995	-608.935	6.921	-0.082
4500	66.944	238.090	179.493	263.689	-608.957	20.918	-0.243
4600	66.944	239.562	180.783	270.384	-609.068	34.911	-0.396
4700	66.944	241.001	182.049	277.078	-609.376	48.914	-0.544
4800	66.944	242.411	183.291	283.772	-609.822	62.924	-0.685
4900	66.944	243.791	184.512	290.467	-610.402	76.946	-0.820
5000	66.944	245.143	185.711	297.161	-611.102	90.980	-0.950

PREVIOUS: CURRENT: December 1972

STRONTIUM OXIDE (SrO) CRYSTAL–LIQUID M_r = 103.6194

0 to 2938 K crystal
above 2938 K liquid

Refer to the individual tables for details.

Strontium Oxide (SrO) $O_1Sr_1(cr,l)$

Enthalpy Reference Temperature = T_r = 298.15 K Standard State Pressure = p° = 0.1 MPa

T/K	C_p° (J K⁻¹mol⁻¹)	S° (J K⁻¹mol⁻¹)	$-[G^\circ-H^\circ(T_r)]/T$ (J K⁻¹mol⁻¹)	$H^\circ-H^\circ(T_r)$ (kJ mol⁻¹)	Δ_fH° (kJ mol⁻¹)	Δ_fG° (kJ mol⁻¹)	Log K_f
0	0.000	0.000	INFINITE	-8.675	-589.801	-589.801	INFINITE
100	24.949	15.242	91.864	-7.662	-591.740	-581.794	303.898
200	40.409	38.334	59.542	-4.242	-592.263	-571.540	149.271
298.15	45.409	55.522	55.522	0.000	-592.036	-561.402	98.355
300	45.480	55.803	55.523	0.084	-592.029	-561.212	97.716
400	48.480	69.325	57.346	4.792	-591.536	-551.010	71.955
500	50.534	80.376	60.880	9.748	-590.937	-540.947	56.512
600	52.049	89.728	64.929	14.880	-590.315	-531.007	46.228
700	53.262	97.845	69.064	20.147	-589.709	-521.171	38.890
800	54.300	105.026	73.119	25.526	-589.138	-511.419	33.392
900	55.220	111.476	77.029	31.002	-589.238	-501.664	29.116
1000	56.074	117.338	80.771	36.568	-588.437	-491.977	25.698
1100	56.873	122.720	84.343	42.215	-595.481	-482.006	22.889
1200	57.639	127.702	87.751	47.941	-595.475	-471.690	20.532
1300	58.379	132.345	91.005	53.742	-595.412	-461.376	18.538
1400	59.103	136.698	94.115	59.616	-595.291	-451.070	16.830
1500	59.810	140.800	97.092	65.562	-595.112	-440.775	15.349
1600	60.509	144.682	99.946	71.578	-594.876	-430.493	14.054
1700	61.199	148.371	102.687	77.664	-731.497	-419.048	12.876
1800	61.881	151.889	105.323	83.818	-729.287	-400.732	11.629
1900	62.559	155.253	107.863	90.040	-727.026	-382.541	10.517
2000	63.229	158.479	110.314	96.329	-724.717	-364.469	9.519
2100	63.898	161.580	112.682	102.685	-722.365	-346.515	8.619
2200	64.563	164.568	114.973	109.109	-719.974	-328.672	7.804
2300	65.229	167.452	117.192	115.598	-717.551	-310.940	7.062
2400	65.890	170.242	119.345	122.154	-715.100	-293.314	6.384
2500	66.551	172.945	121.435	128.776	-712.631	-275.791	5.762
2600	67.208	175.568	123.467	135.464	-710.151	-258.366	5.191
2700	67.864	178.117	125.444	142.218	-707.669	-241.036	4.663
2800	68.517	180.597	127.370	149.037	-705.195	-223.799	4.175
2900	69.174	183.013	129.247	155.921	-702.739	-206.650	3.722
2938.000	69.425	183.915	129.948	158.555	CRYSTAL <--> LIQUID		
2938.000	66.944	209.549	129.948	233.867	TRANSITION		
3000	66.944	210.947	131.608	238.017	-625.163	-191.172	3.329
3100	66.944	213.142	134.203	244.712	-623.092	-176.740	2.978
3200	66.944	215.267	136.703	251.406	-621.133	-162.374	2.650
3300	66.944	217.327	139.115	258.100	-619.295	-148.066	2.344
3400	66.944	219.326	141.445	264.795	-617.588	-133.812	2.056
3500	66.944	221.266	143.698	271.489	-616.019	-119.607	1.785
3600	66.944	223.152	145.879	278.184	-614.595	-105.444	1.530
3700	66.944	224.986	147.992	284.878	-613.323	-91.319	1.289
3800	66.944	226.772	150.042	291.572	-612.208	-77.226	1.062
3900	66.944	228.510	152.032	298.267	-611.253	-63.160	0.846
4000	66.944	230.205	153.965	304.961	-610.461	-49.117	0.641
4100	66.944	231.858	155.845	311.656	-609.835	-35.092	0.447
4200	66.944	233.472	157.674	318.350	-609.370	-21.080	0.262
4300	66.944	235.047	159.455	325.044	-609.071	-7.077	0.086
4400	66.944	236.586	161.191	331.739	-608.935	6.921	-0.082
4500	66.944	238.090	162.883	338.433	-608.957	20.918	-0.243
4600	66.944	239.562	164.534	345.128	-609.068	34.911	-0.396
4700	66.944	241.001	166.146	351.822	-609.376	48.914	-0.544
4800	66.944	242.411	167.720	358.516	-609.822	62.924	-0.685
4900	66.944	243.791	169.258	365.211	-610.402	76.946	-0.820
5000	66.944	245.143	170.762	371.905	-611.102	90.980	-0.950

PREVIOUS: CURRENT: December 1972

Strontium Oxide (SrO) $O_1Sr_1(cr,l)$

STRONTRIUM OXIDE (SrO) IDEAL GAS M_r = 103.6194

$S°(298.15 K) = 230.058 \pm 0.4 J K^{-1} mol^{-1}$

$\Delta_f H°(0 K) = -11.52 \pm 16.7 kJ mol^{-1}$

$\Delta_f H°(298.15 K) = -13.39 \pm 16.7 kJ mol^{-1}$

Electronic States and Molecular Constants (σ-1)

State	ε_i, cm^{-1}	g_i	ω_e, cm^{-1}	$\omega_e x_e$, cm^{-1}	B_e, cm^{-1}	α_e, cm^{-1}	r_e, Å	Source
$X^1\Sigma^+$	0.0	1	653.2	3.92	0.33798	0.00219	1.9199	(1-4)
$a^3\Pi$	9055	6	463.5	1.61	0.2584	0.0020	2.195	(1)
$A'^1\Pi$	9794	2	460.3	1.61	0.2610	0.0023	2.184	(1)
$^3\Sigma^+$	[8000]	3	[619.6]	[0.9]	[0.3047]	[0.0011]	[2.021]	(5)
$A^1\Sigma^+$	10870	1	619.6	0.9	0.30471	0.00112	2.021	(1, 4)
$^3\Sigma^+$	[22000]	3	[480]	[2.6]	[0.272]	[0.002]	[2.14]	(5)
$^3\Delta$	[22000]	6	[480]	[2.6]	[0.272]	[0.002]	[2.14]	(5)
$^1\Delta$	[24000]	2	[480]	[2.6]	[0.272]	[0.002]	[2.14]	(5)
$^3\Sigma^-$	[24000]	3	[480]	[2.6]	[0.272]	[0.002]	[2.14]	(5)
$^1\Sigma^-$	[24000]	1	[480]	[2.6]	[0.272]	[0.002]	[2.14]	(5)
$C^1\Sigma^+$	28546	1	480.2	2.6	0.2742	0.0021	2.132	(4)
$^3\Pi$	[23000]	6	[520]	[3.2]	[0.2936]	[0.0015]	[2.06]	(5)
$B^1\Pi$	24636	2	519.9	3.24	0.2936	0.0015	2.060	(4)

Enthalpy of Formation

We adopt D° = 101.0 ± 4 and $\Delta_f H°(298.15 K) = -3.2 \pm 4 kcal mol^{-1}$ based on the equilibria analyzed below. Greatest weight is given to the conventional mass-spectrometric studies (7, 8). We give "<" or ">" for the resulting D° values which we feel may be biased. The JANAF differences in $\Delta_f H°(298.15 K)$ for MoO_3-MoO_2 and WO_3-WO_2 may be biased by ~4 and ~-3 kcal mol^{-1}, respectively; this would cause bias of the opposite sign in D° values derived from reactions C and D. With these adjustments, our analyses would be comparable with those of Brewer and Rosenblatt (10) based on a $^1\Sigma$ ground state.

Less reliable data were reviewed by Drowart et al. (8) and Schofield (11). Subsequent studies include two new techniques: nonisothermal Langmuir mass-spectrometry (9) and crossed-beam reaction threshold (12). The former yields D° = 98.3 and >96 kcal mol^{-1} (see reactions E and F below). JANAF data for the reaction SrO(cr) = Sr(g) + O(g) suggest that the P_0 assumed (9) for reaction F is too large. JANAF pressures for Sr(g) and O(g) may be combined with the estimated composition (3) of 5% SrO(g) to calculate $K_p°$ for reaction F and D° = 93.5 ± 8 kcal mol^{-1}. D° would increase to 97 kcal mol^{-1} if the beam in the magnetic-deflection experiment (3) contained 10% SrO(g) rather than 5%. The crossed-beam reaction (12) was interpreted to give D° = 112.5(+3.5 to -0.71 kcal mol^{-1}. We assume this value has a positive bias. Finally, Kalff (13) used spectrometry of CO-N_2O flames to derive D° = 93.6 ± 2.3 kcal mol^{-1}. This value becomes ~99 kcal mol^{-1} when adjusted to be consistent with the JANAF free energy functions.

Source	Method	Reaction[a]	T/K	Data Points	δS cal K^{-1} mol^{-1}	$\Delta_r H°$(298.15 K), kcal mol 2nd law	3rd law	$\Delta_f H°$(298.15 K) kcal mol^{-1}	D°
(7)	Knudsen mass spec.	A	1931-2170	5	9.5±7.1	42±15	22.6±2.8	-3.3±3	101.2
(8)	Knudsen mass spec.	B	2156-2321	4	-40.0±10	-73±22	16.3±5.8	-4.1±6	101.9
	Knudsen mass spec.	C	1987-2233	8	-2.5±6.1	40±13	45.4±2.6	1.5±6	>96.4
	Knudsen mass spec.	D	2326	1	-	-	44.9	-4.2±5	<102.0
(9)	Langmuir mass spec.[b]	E	1747-1850	Eqn.	-3.8	134.2	141.0±0.7	-0.5±3	98.3
	Langmuir mass spec.	F[c]	1747-1850	Eqn.	3.3	102.9	97.0	1.8±3	>96.0
(3)	Magnetic deflection	F	~2500	1	-	-	94.5	4.3±8	93.5

[a]Reactions: A) Sr(g) + SO(g) = SrO(g) + S(g) B) Sr(g) + O_2(g) = SrO(g) + O(g) C) Sr(g) + MoO_3(g) = SrO(g) + MoO_2(g)
D) Sr(g) + WO_3(g) = SrO(g) + WO_2(g) E) SrO(cr) = SrO(g) F) SrO(g) = Sr(g) + O(g)

[b]Nonisothermal. [c]$P_0 = 0.4273 P_{Sr}$ assumed.

Heat Capacity and Entropy

Electronic levels (T_0) and vibrational-rotational constants of the observed states are from Field (1) and Rosen (4). The long-sought $a^3\Pi$ and $A^1\Pi$ states of CaO, SrO and BaO were characterized by Field (1) using a new method for assignment of perturbations. This study resolved the long-standing controversy over low-lying electronic levels and confirmed (1) that $X^1\Sigma^+$ is the ground state (3). We estimate the other potentially low-lying state ($^3\Sigma^+$) at 8000 cm^{-1} by assuming that it lies 3000 ± 2500 cm^{-1} (1, 5) below the isoconfigurational A state. Other predicted states and thier vibrational-rotational constants are estimated in isoconfigurational groups by comparison with BeO, MgO, CaO and BaO (5). Comparisons are facilitated by listing the states in the isoconfigurational order of MgO (5, 6). Our thermodynamic functions correspond to an "effective" ground state with 1<g<2 instead of g = 2 (11). Our functions are calculated using first-order anharmonic corrections to Q_r^i and Q_v^i in the partition function $Q = Q_t \sum_i Q_r^i Q_v^i g_i \exp(-c_2\varepsilon_i/T)$.

References

1. R. W. Field, J. Chem. Phys. 60, 2400 (1974).
2. L. Brewer and R. Hauge, J. Mol. Spectrosc. 25, 330 (1968).
3. M. Kaufman, L. Wharton and W. Klemperer, J. Chem. Phys. 43, 943 (1965).
4. B. Rosen, "Spectroscopic Data Relative to Diatomic Molecules," Pergamon Press, New York, p. 386, 1970.
5. JANAF Thermochemical Tables: SrO(cr, ℓ), 12-31-72; SrO(g), BaO(g), 6-30-74; CaO(g), MgO(g), BeO(g), 12-31-74.
6. J. Schamps and H. Lefebvre-Brion, J. Chem. Phys. 56, 573 (1972).
7. R. Colin, P. Goldfinger and M. Jeunehomme, Trans. Faraday Soc. 60, 306 (1964).
8. J. Drowart, G. Exsteen and G. Verhaegen, Trans. Faraday Soc. 60, 1920 (1964).
9. M. Asano, Y. Yamamoto, N. Sasaki and K. Kubo, Bull. Chem. Soc. Japan 45, 82 (1972).
10. L. Brewer and G. M. Rosenblatt, Adv. High Temp. Chem. 2, 1 (1969).
11. K. Schofield, Chem. Rev. 67, 707 (1967).
12. C. Batalli-Cosmovici and K. W. Michel, Chem. Phys. Letters 16, 77 (1972).
13. P. J. Kaliff and C. Th. J. Alkemade, J. Chem. Phys. 59, 2572 (1973); 60, 1698 (1974).

Strontium Oxide (SrO) O_1Sr_1(g)

Enthalpy Reference Temperature = T_r = 298.15 K Standard State Pressure = p° = 0.1 MPa

T/K	$C_p°$ (J K^{-1}mol^{-1})	S°	-[G°-H°(T_r)]/T	H°-H°(T_r) (kJ mol^{-1})	$\Delta_f H°$	$\Delta_f G°$	Log K_f
0	0.	0.	INFINITE	-9.041	-11.519	-11.519	INFINITE
100	29.175	196.665	257.973	-6.131	-11.561	-19.758	10.320
200	30.890	217.295	233.018	-3.145	-12.519	-27.588	7.205
250	32.086	224.318	230.597	-1.570	-12.976	-31.307	6.541
298.15	33.092	230.058	230.058	0.	-13.389	-34.793	6.096
300	33.127	230.263	230.059	0.061	-13.404	-34.926	6.081
350	33.967	235.435	230.465	1.739	-13.817	-38.480	5.743
400	34.627	240.015	231 378	3.455	-14.225	-41.976	5.481
450	35.146	244.125	232.570	5.200	-14.640	-45.420	5.272
500	35.557	247.850	233.915	6.968	-15.070	-48.817	5.100
600	36.154	254.389	236.797	10.555	-15.992	-55.481	4.830
700	36.559	259.995	239 720	14.192	-17.016	-61.983	4.625
800	36.854	264.897	242.567	17.863	-18.153	-68.331	4.462
900	37.089	269.251	245.295	21.561	-20.032	-74.457	4 321
1000	37.309	273.170	247.890	25.281	-21.077	-80.448	4.202
1100	37.561	276.738	250.353	29.024	-30.025	-85.969	4.082
1200	37.894	280.020	252.690	32.796	-31.974	-90.969	3.960
1300	38.360	283.070	254 911	36.607	-33.900	-95.807	3.850
1400	39.004	285.935	257.025	40.474	-35.787	-100.498	3.750
1500	39.862	288.654	259.044	44.415	-37.612	-105.056	3.658
1600	40.957	291.260	260.977	48.454	-39.353	-109.495	3.575
1700	42.293	293.782	262.833	52.614	-177.899	-112.649	3.461
1800	43.859	296.243	264.621	56.920	-177.537	-108.820	3.158
1900	45.625	298.661	266.349	61.393	-177.025	-105.016	2 887
2000	47.549	301.049	268.024	66.051	-176.349	-101.243	2.644
2100	49.576	303.418	269.653	70.906	-175.497	-97.507	2.425
2200	51.648	305.772	271.242	75.967	-174.468	-93 817	2.227
2300	53.703	308.114	272.794	81.235	-173.266	-90.177	2.048
2400	55.681	310.441	274.314	86.705	-171.902	-86.594	1.885
2500	57.532	312.752	275.805	92.367	-170.393	-83.070	1.736
2600	59.211	315.042	277.271	98.206	-168.762	-79.608	1.599
2700	60.688	317.305	278.712	104.203	-167.037	-76.212	1.474
2800	61.941	319.536	280.130	110.336	-165.249	-72.880	1.360
2900	62.963	321.728	281.527	116.583	-163.429	-69.614	1 254
3000	63.752	323.876	282.903	122.921	-161.612	-66.410	1.156
3100	64.320	325.976	284.258	129.326	-159.830	-63.266	1.066
3200	64.681	328.025	285.594	135.778	-158.114	-60.179	0.982
3300	64.855	330.018	286.910	142.256	-156.492	-57.144	0.905
3400	64.865	331.955	288.207	148.743	-154.992	-54.156	0.832
3500	64.736	333.834	289.484	155.225	-153.636	-51 210	0.764
3600	64.489	335.654	290.741	161.687	-152.445	-48.301	0.701
3700	64.148	337.416	291.979	168.119	-151.434	-45.423	0.641
3800	63.734	339.122	293.197	174.514	-150.619	-42.568	0.585
3900	63.265	340.771	294.396	180.864	-150.008	-39.733	0.532
4000	62.758	342.367	295.575	187.166	-149.609	-36.911	0.482
4100	62.225	343.910	296.736	193.415	-149.428	-34.097	0.434
4200	61.680	345.403	297.877	199.610	-149.462	-31.284	0.389
4300	61.131	346.848	298.999	205.751	-149.717	-28.467	0.346
4400	60.586	348.247	300.102	211.837	-150.190	-25.643	0.304
4500	60.051	349.602	301.187	217.868	-150.874	-22.805	0.265
4600	59.531	350.917	302.254	223.847	-151.701	-19.955	0.227
4700	59.029	352.191	303.303	229.775	-152.775	-17.080	0.190
4800	58.547	353.429	304.335	235.654	-154.037	-14.179	0.154
4900	58.088	354.632	305.349	241.485	-155.480	-11.251	0.120
5000	57.651	355.801	306.346	247.272	-157.088	-8.292	0.087
5100	57.237	356.938	307.327	253.016	-158.837	-5.300	0.054
5200	56.846	358.046	308.292	258.720	-160.754	-2.272	0.023
5300	56.477	359.125	309.241	264.386	-162.809	0.796	-0.008
5400	56.130	360.177	310.174	270.016	-164.993	3.902	-0.038
5500	55.804	361.204	311.093	275.613	-167.295	7.051	-0.067
5600	55.498	362.207	311.997	281.178	-169.704	10.243	-0.096
5700	55.210	363.187	312.886	286.713	-172.211	13.478	-0.124
5800	54.940	364.145	313.762	292.220	-174.807	16.759	-0.151
5900	54.686	365.082	314.624	297.702	-177.271	20.060	-0.178
6000	54.448	365.999	315.472	303.158	-179.985	23.427	-0.204

PREVIOUS: June 1974 (1 atm) CURRENT: June 1974 (1 bar)

Strontium Oxide (SrO) O_1Sr_1(g)

TANTALUM OXIDE (TaO) IDEAL GAS M_r = 196.9473

$S°(298.15\ K) = 241.116 \pm 2.1\ J\ K^{-1}\ mol^{-1}$

$\Delta_f H°(0\ K) = 193.72 \pm 62.8\ kJ\ mol^{-1}$

$\Delta_f H°(298.15\ K) = 192.46 \pm 62.8\ kJ\ mol^{-1}$

Electronic Levels and Quantum Weights

State	ε_i, cm^{-1}	g_i	State	ε_i, cm^{-1}	g_i	State	ε_i, cm^{-1}	g_i
$X^2\Delta_{3/2}$	0	2	$^2\Phi_{7/2}$	22918.75	4	$^2\Delta_{5/2}$	27290.63	2
$^2\Delta_{5/2}$	3505.43	2	$^2\Pi_{1/2}$	23341.74	2	$^2\Delta_{3/2}$	32373.60	2
$^2\Delta_{3/2}$	10860.95	4	$^2\Phi_{5/2}$	24058.42	2	$^2\Delta_{5/2}$	35785.83	4
$^2\Pi_{1/2}$	11062.00	4	$^2\Pi_{3/2}$	25593.13	2	$^2\Delta_{5/2}$	35886.20	2
$^2\Phi_{5/2}$	12852.02	4	$^2\Phi_{7/2}$	26121.50	2	$^2\Delta_{5/2}$	36615.00	4
$^2\Delta_{3/2}$	13569.27	4	$^2\Delta_{3/2}$	26673.04	2	$^2\Delta_{5/2}$	36785.00	4
$^2\Phi_{5/2}$	15880.62	4						

$\omega_e = 1028.69\ cm^{-1}$ $\omega_e x_e = 3.51\ cm^{-1}$ $\sigma = 1$

$B_e = 0.402840\ cm^{-1}$ $\alpha_e = 0.00182\ cm^{-1}$ $r_e = 1.6872$ Å

Enthalpy of Formation

Based on the adopted ground state vibrational constants, a linear Birge-Sponer extrapolation yields D_0^0 = 9.28 eV (214.02 kcal mol^{-1}). Applying a correction for the ionicity in the Ta-O bond (1), we recalculate D_0^0 = 8.57 eV (199.94 kcal mol^{-1}). This in turn leads to $\Delta_f H°$(TaO, g, 298.15 K) = 45.52 kcal mol^{-1}, using auxiliary data (2).

Inghram et al. (3) and Krikorian and Carpenter (4) have presented mass spectrometric evidence indicating that the predominant vapors in the vaporization of Ta-Ta_2O_5 mixtures in Ta Knudsen cells are TaO and TaO_2.

Source	Reaction	Data Points	T/K	$\Delta_f H°$(298.15 K), kcal mol^{-1} 2nd law	3rd law	Drift cal K^{-1} mol^{-1}	$\Delta_f H°$(298.15 K), kcal mol^{-1} 2nd law	3rd law
(3)	A*	1	2022	--	147.31	-	-	+49.51
(4)	A	2	1927-2034	162.80	175.47	6.4	+65.00	+77.67
(3)	B*	5	2151-2278	193.95	144.81	-22.3±14.1	+100.30	+51.26
(4)	B	3	2180-2271	166.53	172.90	2.8±7.9	+72.98	+79.35
(16)	C	graph	1900-2900	54.15	93.33	17.5	54.15	93.33

*Recalculated pressures using Otvis and Stevenson method for estimating ionization cross sections (4, 5).

A: 0.6 Ta(cr) + 0.2 Ta_2O_5(cr) = TaO(g) B: 0.6 Ta(cr) + 0.2 Ta_2O_5(ℓ) = TaO(g) C: Ta(cr) + 1/2 O_2(g) = TaO(g)

The discrepancies here are rather severe and indicate considerable uncertainty in the system studied. In addition to the standard difficulties in a Knudsen cell system and the subsequent assumptions used for a 2nd and 3rd law analysis, there is an additional problem of oxygen diffusing through the walls of the Ta Knudsen cell, followed by volatilization of TaO(g) from the outer surfaces (4, 6). Based on JANAF values for Ta_2O_5(cr) and Ta_2O_5(ℓ), $\Delta_r H°$(298.15 K) for reaction (A) should be 4.25 kcal mol^{-1} more positive than reaction (B). The 3rd law results of Krikorian and Carpetner (4) for reactions (A) and (B) differ by 2.57 kcal mol^{-1} but the 2nd law results differ by 3.73 kcal mol^{-1} in the wrong direction. The 3rd law results of Inghram et al. (3) differ by 2.50. It is interesting to note that if the pressures reported by Krikorian and Carpetner (4) are multiplied by a factor of 1000, a 3rd law analysis yields results in excellent agreement with the 3rd law results of Inghram et al. (3). 2nd law values would still be in disagreement, however, and serious 3rd law drifts would still be present. Drowart et al. (16) observed the vapors over a Ta wire at various oxygen pressures in the range 1900-2900 K. Our analysis of their graphically displayed results does not clarify the situation as there is again a large 3rd law drift. Inghram et al. (3) also measured the ion currents corresponding to ZrO(g) and TaO(g) over mixtures of ZrO_2(cr) and Ta(cr) in a Ta Knudsen cell. The reaction was stated as ZrO_2(cr) + Ta(cr) = ZrO(g) + TaO(g). A 2nd law analysis of seven data points yielded $\Delta_r H°$(2200 K) = 317 ± 7 kcal mol^{-1} as reported by Inghram et al. (3). Using auxiliary data (2), this leads to $\Delta_f H°$(TaO, g, 298.15 K) = 51 ± 12 kcal mol^{-1}, where +12 kcal mol^{-1} reflects the additional uncertainty in Δ H°(298.15 K) for ZrO(g). Goldstein et al. (7), in studying the vaporization of Nd_2O_3 from Ta Knudsen cells in the range 2040-2450 K, determined the enthalpy of reaction for Nd_2O_3(cr) + Ta(cr) = 2 NdO(g) + TaO(g). They reported D_0^0= 9.36 ± 0.2 eV for TaO(g). This leads to $\Delta_f H°$(TaO, g, 298.15 K) = 29.6 ± 5.0 kcal mol^{-1}.

Brewer and Rosenblatt (17) have suggested Gibbs energy functions which are 1.38 to 2.14 cal K^{-1} mol^{-1} more positive than those adopted here. The reason for the difference is that Brewer and Rosenblatt (17) based their functions on low lying electronic states as observed in Ta^{++}(g). The $\Delta_f H°$(298.15 K) values derived from the 3rd law would increase by roughly 3.5-4.2 kcal mol^{-1} with the drift becoming more positive by 2-3 cal K^{-1} mol^{-1} if we were to use Brewer and Rosenblatt's valves.

Due to the large scatter in the mass spectrometric results, we prefer not to average the resulting $\Delta_f H°$ values but to rely on correlations with VO(g) and NbO(g). There has been good agreement for VO(g) and NbO(g) between the adopted JANAF $\Delta_f H°$(298.15 K) value and the value derived from the Birge-Sponer result corrected for ionicity (2). Thus we adopt $\Delta_f H°$(TaO, g, 298.15 K) = 46.0 ± 15.0 kcal mol^{-1}, which is a rounded value based on the adjusted Birge-Sponer result. This corresponds to D_0^0 = 8.65 ± 0.65 eV.

Heat Capacity and Entropy

The band spectrum attributable to TaO(g) has been observed by many investigators (8-13). Weltner and McLeod (14) studied TaO spectroscopically in the infrared, visible, and near-ultraviolet regions by means of matrix isolation techniques. Sixteen electronic transitions were observed from the $X^2\Delta_{3/2}$ state, determined to be the ground state. Cheetham and Barrow (15) characterized nineteen electronic states from a rotational analysis of the emission spectrum of TaO in the region 2740-9500 Å. The combination of the results of these latter two studies as given by Cheetham and Barrow (15) provides the adopted vibrational and rotational constants as well as the electronic levels. Cheetham and Barrow (15) indicated the possibility of other low lying states based on the early work by Kiess and Stowell (8).

The electronic states listed by Cheetham and Barrow (15) are actually sublevels as indicated by the Ω value. The quantum weight of 2 is assigned if the two sublevels which are part of the same state are present (i.e. similar ω_e and $\omega_e x_e$ values of rthe same Δ state but different Ω value). If the two sublevels of a particular Δ state are not accounted for, then a quantum weight of 4 is assigned to that sublevel which is tabulated.

References - References are listed on the TaO_2(g) table.

Tantalum Oxide (TaO) $O_1Ta_1(g)$

Enthalpy Reference Temperature = T_r = 298.15 K; Standard State Pressure = p° = 0.1 MPa

T/K	$C_p°$	S°	-[G°-H°(T_r)]/T	H°-H°(T_r)	$\Delta_f H°$	$\Delta_f G°$	Log K_f
	J $K^{-1}$$mol^{-1}$			kJ mol^{-1}			
0	0.	0.	INFINITE	-8.766	193.721	193.721	INFINITE
100	29.105	208.958	267.528	-5.857	194.182	183.566	-95.885
200	29.398	229.177	243.870	-2.939	193.393	173.223	-45.241
250	29.921	235.789	241.615	-1.457	192.921	168.235	-35.151
298.15	30.600	241.116	241.116	0.	192.464	163.522	-28.648
300	30.628	241.305	241.116	0.057	192.447	163.343	-28.441
350	31.394	246.084	241.492	1.607	191.986	158.529	-23.659
400	32.136	250.325	242.336	3.196	191.541	153.780	-20.082
450	32.815	254.150	243.439	4.820	191.108	149.086	-17.305
500	33.424	257.639	244.687	6.476	190.683	144.439	-15.089
600	34.463	263.828	247.375	9.872	189.839	135.270	-11.776
700	35.344	269.208	250.118	13.363	188.999	126.242	-9.420
800	36.137	273.980	252.808	16.938	188.171	117.333	-7.661
900	36.875	278.280	255.403	20.589	187.363	108.527	-6.299
1000	37.564	282.201	257.890	24.311	186.575	99.810	-5.214
1100	38.201	285.812	260.266	28.100	185.799	91.171	-4.329
1200	38.779	289.161	262.536	31.950	185.027	82.603	-3.596
1300	39.297	292.286	264.706	35.854	184.256	74.099	-2.977
1400	39.759	295.215	266.781	39.807	183.492	65.654	-2.450
1500	40.173	297.973	268.770	43.804	182.742	57.263	-1.994
1600	40.549	300.577	270.677	47.840	182.007	48.922	-1.597
1700	40.900	303.046	272.509	51.913	181.279	40.626	-1.248
1800	41.237	305.394	274.272	56.020	180.539	32.373	-0.939
1900	41.572	307.632	275.969	60.160	179.770	24.163	-0.664
2000	41.915	309.773	277.606	64.335	178.971	15.994	-0.418
2100	42.272	311.827	279.187	68.544	178.142	7.865	-0.196
2200	42.650	313.802	280.716	72.790	177.286	-0.223	0.005
2300	43.050	315.707	282.196	77.075	176.402	-8.272	0.188
2400	43.475	317.548	283.631	81.401	175.487	-16.282	0.354
2500	43.923	319.332	285.023	85.771	174.539	-24.253	0.507
2600	44.393	321.063	286.376	90.186	173.551	-32.186	0.647
2700	44.882	322.748	287.692	94.650	172.514	-40.079	0.775
2800	45.385	324.389	288.974	99.163	171.417	-47.933	0.894
2900	45.898	325.991	290.223	103.727	170.249	-55.746	1.004
3000	46.417	327.555	291.441	108.343	168.993	-63.518	1.106
3100	46.937	329.086	292.631	113.011	167.630	-71.246	1.200
3200	47.454	330.584	293.794	117.730	166.142	-78.929	1.288
3300	47.962	332.052	294.931	122.501	128.049	-86.092	1.363
3400	48.459	333.492	296.044	127.322	126.662	-92.560	1.422
3500	48.940	334.903	297.134	132.192	125.316	-98.988	1.477
3600	49.402	336.288	298.202	137.110	124.010	-105.378	1.529
3700	49.843	337.648	299.250	142.072	122.741	-111.732	1.577
3800	50.260	338.983	300.278	147.077	121.509	-118.053	1.623
3900	50.652	340.294	301.288	152.123	120.309	-124.341	1.665
4000	51.018	341.581	302.279	157.207	119.141	-130.599	1.705
4100	51.356	342.845	303.253	162.326	118.002	-136.829	1.743
4200	51.667	344.086	304.210	167.477	116.889	-143.031	1.779
4300	51.950	345.305	305.152	172.658	115.799	-149.206	1.812
4400	52.205	346.502	306.078	177.866	114.730	-155.357	1.844
4500	52.433	347.678	306.990	183.098	113.679	-161.483	1.874
4600	52.635	348.833	307.887	188.352	112.644	-167.587	1.903
4700	52.811	349.967	308.770	193.624	111.621	-173.668	1.930
4800	52.963	351.080	309.640	198.913	110.608	-179.727	1.956
4900	53.091	352.174	310.497	204.216	109.603	-185.765	1.980
5000	53.197	353.247	311.341	209.531	108.603	-191.783	2.004
5100	53.282	354.302	312.173	214.855	107.606	-197.780	2.026
5200	53.348	355.337	312.993	220.187	106.610	-203.759	2.047
5300	53.395	356.354	313.802	225.524	105.612	-209.718	2.067
5400	53.425	357.352	314.599	230.865	104.610	-215.658	2.086
5500	53.440	358.332	315.385	236.208	103.602	-221.580	2.104
5600	53.440	359.295	316.161	241.553	102.586	-227.483	2.122
5700	53.427	360.241	316.926	246.896	101.561	-233.368	2.139
5800	53.401	361.170	317.681	252.238	-638.230	-236.433	2.129
5900	53.365	362.083	318.426	257.576	-639.492	-229.496	2.032
6000	53.318	362.979	319.161	262.910	-640.813	-222.535	1.937

PREVIOUS: December 1973 (1 atm) CURRENT: December 1973 (1 bar)

Tantalum Oxide (TaO) $O_1Ta_1(g)$

TITANIUM OXIDE, ALPHA (α-TiO) CRYSTAL M_r = 63.8794

$\Delta_f H°(0\ K)$ = -539.668 ± 12.6 kJ mol^{-1}

$S°(298.15\ K)$ = 34.769 ± 2.1 J K^{-1} mol^{-1} $\Delta_f H°(298.15\ K)$ = -542.665 ± 12.6 kJ mol^{-1}

T_{trs} = [1265] K $\Delta_{trs}H°$ = [4.2] kJ mol^{-1}

Enthalpy of Formation

Recent data tend to support our speculation (1) that previous calorimetric data refer to ill-defined states of TiO(α, β). New combustion calorimetry (2) at 1100 K gave $\Delta_f H°(298.15\ K)$ = -129.9 ± 1.1 kcal mol^{-1} which is ~5 kcal mol^{-1} more negative than two earlier combustion studies (3, 4) near room temperature. The difference was attributed (2) to use of samples consisting of α-phase (2) or mainly β-phase (3, 4). This interpretation is reasonable but unproved; it is neither confirmed nor refuted by incomplete published descriptions of the samples. Previously, the samples (3, 4) were assumed to be α-phase since similar material gave a discontinuity in enthalpy data above 1200 K.

Lyon (5) proposed that $\Delta_f H°$ may be affected by vacancy concentration, which varies from ~14 to 15% (5) in samples of stoichiometric β-TiO obtained at normal pressure. Samples with vacancy concentrations dow to 0% have been prepared (6) at high pressure. PVT data (6) allowed calculation (5) of values of $\Delta_f H°$ for TiO(β, 0% vacancies) → TiO(β, 14% vacancies). These values, if valid, suggest that $\Delta_f H°$ should be quite different for vacancy-free β-TiO and significantly different even for the normal range of vacancy concentrations. Ideal ordered α-TiO, containing 1/6 or 16.7% vacancies, should involve additional changes in volume (5) and $\Delta_f H°$. In summary, the discrepancy in $\Delta_f H°$ may arise from sample differences - phase, composition and vacancy concentration - or from bias in the reaction calorimetry.

Our choice of $\Delta_f H°$ is influenced by $\Delta_f G°$ values derived from oxygen potentials (7, 8) over TiO_x(0 < x < 1.5). We use solid-state emf data (7) for many compositions in the range 0.6 < x < 1.6; these are converted to $\Delta_f \bar{G}(O_2)$ with the authors' results (7) for the reference couples Ta/Ta_2O_5 and "$TiO_{1.3}$", the integrated over appropriate ranges of x in order to obtain $\Delta_r G°$. For reaction C relating TiO(β) to Ti_2O_3 we obtain $\Delta_r G°$ = -38.20(1300 K), -37.49(1400 K), -36.82(1500 K), -36.180(1600 K) and -35.52(1700 K) kcal mol^{-1} with an estimated uncertainty of ±1 kcal mol^{-1} (~3%). Combination with $\Delta_f G°$ (9) of Ti_2O_3 yields $\Delta_f G°$(TiO, β) = -131.5 + 23.5 (T/1000); hopefully, the uncertainty does not exceed 1.5 kcal mol^{-1}, based on a revised uncertainty of 1 kcal mol^{-1} for Ti_2O_3.

Approximate values of $\Delta_f G°$(TiO) may be derived from $TiO_{0.334}$(cr) and Ti(cr). Calorimetric data (10) yield $\Delta_f G°(TiO_{0.334})$ = -46.1 + 8.40 (T/1000) for 1200 < T < 1700; this equation excludes the added disordering entropy (10) which now is known to be included, at least partially, in order-disorder transitions of the alloy near 800 K (11). $\Delta_f \bar{G}(O_2)$ in the range 0.334 < x < 0.5 is approximated by limiting values, i.e., the mean from x < 0.334 and the univariant value (7) at x = 0.6, in order to get negative and positive limits for $\Delta_f G°$(TiO). These limits differ by 1 to 2.5 kcal mol^{-1} at 1300 to 1700 K; mean $\Delta_f G°$ values are -102.0 (1300 K), -96.8(1400 K), -92.6(1500 K), -89.3(1600 K) and -86.0 kcal mol^{-1}(1700 K). Equilibration with Ca/CaO and Mg/MgO (8) near 1200 K gave $\Delta_f \bar{G}(O_2)$ for x < 0.3; we combine these by interpolation with values (7) at x > 0.6 to give $\Delta_f \bar{G}°$(TiO) = -107.3(1173 K), -106.6(1200 K) and -103.8 kcal mol^{-1}(1273 K). $\Delta_f G°$ values derived from Ti(cr) and $TiO_{0.334}$(cr) are relatively uncertain, perhaps by 5 kcal mol^{-1}, due to gaps in $\Delta_f \bar{G}(O_2)$ and to greater uncertainty in emf and x.

$\Delta_f H°$ values from $\Delta_c H°$ and from 3rd law analyses of $\Delta_r G°$ are compared below. $\Delta_f H°$ from Ti_2O_3 (7, 9) is consistent with the new $\Delta_c H°$ (2) but ~5 kcal mol^{-1} more negative than the earlier value (3) which we assume to refer to β-TiO. $\Delta_f H°$ from Ti_2O_3 (7, 9) and from $\Delta_c H°$ (3) could be made consistent if, as proposed by Charlu et al. (2), $\Delta_t S°$ and S°(β) were increased by 3 cal K^{-1} mol^{-1} [see β-TiO (9)]. If (3) instead of (2) were correct for α-TiO, $\Delta_f \bar{G}(O_2)$ would have to be 14% larger than observed. The more uncertain $\Delta_f H°$ from $TiO_{0.334}$ (7, 9) is very temperature dependent (large δS); it agrees with (2) at 1350 K and with (3) at 1700 K. $\Delta_f H°$ from Ti (7-9) is even more negative than (2). We adopt $\Delta_f H°$ = -129.7 kcal mol^{-1} but increase the uncertainty to +3 kcal mol^{-1} to allow for possible uncertainties in the samples.

Source	Method	Reaction[a]	T/K	δS cal K^{-1} mol^{-1}	$\Delta_r H°$(298.15 K) kcal mol^{-1}	$\Delta_f H°$(298.15 K)[b], kcal mol^{-1} Authors	TiO(α)	TiO(β)
(2)	$\Delta_c H°$	A	1100-298	-	-96.05±1.1	-129.86	-129.7	[-128.7]
(3)	$\Delta_c H°$	B	303-298	-	-101.61±0.12	-123.9±0.3	[-125.2]	-124.2
(4)	$\Delta_c H°$	B	293	-	-99.5	-124.5±0.4	[-127.3]	-126.3
(7,9)From Ti_2O_3	emf, $\Delta_r G°$	C	1300-1700	2.9	-52.77	-	[-130.0]	-129.0
(7,9)From $TiO_{0.334}$	emf, $\Delta_r G°(O_2)$	D	1300-1700	-19.0	-127.5	-	[-128.5]	-127.5
(7,9)From Ti(cr)	emf, $\Delta_f G(O_2)$	D	1273	-	-133.2	-	[-134.2]	-133.2
		E	1173-1200	-2.5	-135.4	-	-135.4	[-134.4]

[a] A) TiO(α?) + 1/2 O_2(g) = TiO_2(rutile) B) TiO(β?) + 1/2 O_2 = TiO_2(rutile) C) TiO(β) + 1/4 O_2(g) = 1/2 Ti_2O_3(cr)
D) Ti(β) + 1/2 O_2(g) = TiO(β) E) Ti(β) + 1/2 O_2(g) = TiO(α?).

[b] $\Delta_f H°$ in brackets calculated from $\Delta_{trs}H°$ = [1.0] kcal mol^{-1}

Heat Capacity and Entropy

$C_p°$ is from data (52.6-296 K) of Shomate (12) and enthalpy data (357-1771 K) of Naylor (13). S°(298.15 K) = 8.31 cal K^{-1} mol^{-1} is calculated from $C_p°$ based on S°(50 K) = 0.23 cal K^{-1} mol^{-1}. Ideal, ordered α-TiO should have no configurational entropy at T/K = 0 (14). We assume that $C_p°(\alpha) = C_p°(\beta)$ and fit the enthalpy data (357-1200 K and 1273-1771 K) by a single curve after subtracting the apparent $\Delta_{trs}H°$ = 0.83 kcal mol^{-1} from points above T_{trs}. The single curve gives larger deviations than separate fits of the two regions, but the increases are of marginal significance.

Recent $C_p°$ data (60-298 K) for $TiO_{1.01}$ (15) are larger at all temperatures with deviations of 2-3% at 110-250 K increasing to 5% at 298 K and 7% at 60 K. The new data yield S°(298.15 K) = 8.55 cal K^{-1} mol^{-1} of TiO. We know nothing about the TiO sample (15), but it also may have been used to obtain enthalpies (700-2300 K) which are not yet published (16). Shomate (12), Naylor (13) and later Humphrey (3, see $\Delta_f H°$) all used the same sample, 99.2% pure with Si and C as impurities. The (O/Ti) ratio was ~0.994 (17). Vacancy concentration is unknown and the phase is uncertain.

Transition Data and Phase Data - See the β-crystal table for details.

Sublimation Data - This is the difference in $\Delta_f H°$ for TiO(g) and TiO(α).

References - References are listed on the liquid table.

Titanium Oxide, Alpha (TiO) O_1Ti_1(cr)

Enthalpy Reference Temperature = T_r = 298.15 K   Standard State Pressure = p° = 0.1 MPa

T/K	$C_p°$ (J $K^{-1}mol^{-1}$)	S° (J $K^{-1}mol^{-1}$)	-[G°-H°(T_r)]/T (J $K^{-1}mol^{-1}$)	H°-H°(T_r) (kJ mol^{-1})	$\Delta_f H°$ (kJ mol^{-1})	$\Delta_f G°$ (kJ mol^{-1})	Log K_f
0	0.000	0.000	INFINITE	-6.174	-539.668	-539.668	INFINITE
100	12.765	5.732	63.375	-5.764	-541.270	-532.352	278.072
200	30.681	20.585	38.164	-3.516	-542.394	-522.918	136.572
298.15	39.957	34.769	34.769	0.000	-542.665	-513.278	89.924
300	40.062	35.017	34.770	0.074	-542.665	-513.096	89.338
400	44.978	47.274	36.408	4.346	-542.491	-503.257	65.719
500	48.242	57.679	39.648	9.015	-542.092	-493.491	51.555
600	50.836	66.709	43.422	13.972	-541.541	-483.821	42.120
700	53.095	74.719	47.331	19.171	-540.857	-474.253	35.389
800	55.187	81.947	51.214	24.586	-540.035	-464.793	30.348
900	57.153	88.561	55.001	30.204	-539.111	-455.442	26.433
1000	59.078	94.682	58.667	36.016	-538.152	-446.197	23.307
1100	60.961	100.402	62.203	42.018	-537.219	-437.047	20.754
1200	62.802	105.785	65.613	48.206	-540.300	-427.861	18.624
1265.000	63.998	109.129	67.764	52.327	--------	ALPHA <--> BETA	-------
1300	64.643	110.884	68.901	54.578	-538.700	-418.555	16.818
1400	66.442	115.741	72.075	61.133	-537.011	-409.376	15.274
1500	68.241	120.386	75.142	67.867	-535.248	-400.320	13.940
1600	70.040	124.848	78.110	74.781	-533.422	-391.384	12.777
1700	71.839	129.148	80.986	81.875	-531.546	-382.564	11.755
1800	73.597	133.304	83.778	89.147	-529.636	-373.856	10.849
1900	75.396	137.331	86.491	96.597	-527.704	-365.254	10.042
2000	77.153	141.243	89.131	104.224	-540.450	-356.300	9.306
2100	78.910	145.050	91.704	112.027	-539.263	-347.121	8.634
2200	80.668	148.762	94.214	120.006	-537.912	-338.002	8.025
2300	82.467	152.387	96.664	128.163	-536.394	-328.949	7.471
2400	84.224	155.934	99.060	136.497	-534.710	-319.965	6.964
2500	85.981	159.408	101.405	145.007	-532.861	-311.055	6.499

PREVIOUS: March 1967   CURRENT: December 1973

Titanium Oxide, Alpha (TiO) O_1Ti_1(cr)

TITANIUM OXIDE, BETA (β-TiO) CRYSTAL M_r = 63.8794

$S°(298.15\ K)$ = [38.077] J K^{-1} mol^{-1}

T_{trs} = [1265] K

T_{fus} = 2023 ± 30 K

$\Delta_f H°(298.15\ K)$ = [-538.481] kJ mol^{-1}

$\Delta_{trs}H°$ = [4.2] kJ mol^{-1}

$\Delta_{fus}H°$ = [41.8] kJ mol^{-1}

Enthalpy of Formation

$\Delta_f H°$ is calculated from that of α-TiO by adding $\Delta_{trs}H°$ to the difference in H°(1265 K)-H°(298.15 K) between α- and β-TiO. $\Delta_f G°$ data for β-TiO are reviewed on the table for α-TiO (9). Values of $\Delta_f H°$ derived from $\Delta_f G°$ depend on the value of S° (see Entropy). There are additional references on the $\Delta_f G°$ which deserve comment. Solid-state emf data of Hoch et al. (18) are insufficient to yield $\Delta_f G°(\beta)$, especially in the direct way used by Drowart et al. (19). Their interpretation is inconsistent with phase diagrams (20, 17) and extensive emf data (7) which show bivariant behavior in which $\Delta_f \bar{G}(O_2)$ is a strong function of (O/Ti). It is not useful to reinterpret the emf data (18); they show a temperature dependence of the wrong sign and we do not know the necessary electrode compositions. The often quoted $\Delta_f G°$ of Kubaschewski and Dench (21) is not an independent value since it assumes the correctness of the calorimetric data of the Bureau of Mines (3, 12, 13). Kubaschewski's reassessment (22, 8) of $\Delta_f \bar{G}°(O_2)$ is superseded (9) due to new data (7).

Heat Capacity and Entropy

We assume $C_p°(\beta) = C_p°(\alpha)$ as discussed on the table for α-TiO. S° is calculated in a manner analogous to that of $\Delta_f H°$. S° might be too low by as much as 3 or 4 cal K^{-1} mol^{-1} (2) due to a possible uncertainty in $\Delta_{trs}H°$ (see Transition Data). The possibility exists because experimental samples (2-4, 12, 13) were not adequately characterized as to phase and vacancy concentration.

Phase Data

Stoichiometric TiO has been identified in two crystalline forms. High-temperature β-TiO has a cubic NaCl-type structure (23), while low-temperature α-TiO has a closely related monoclinic structure (24, 25). Ideal α-TiO is an ordered array in which 1/6 of the lattice sites are vacant; half of the Ti and O atoms are missing alternately in every third (110) plane (24). β-TiO obtained at normal pressures appears to have a lower and somewhat variable (~14-15%) vacancy concentration (5), depending on conditions of preparation. Vacancy concentration can be reduced to arbitrary values (including zero) at high pressure by quenching from high temperature (6). The process is reversed on reheating at normal pressure. Vacancy concentration (6, 26) and vacancy disorder (20, 25-27) in β-TiO appear to increase with increasing temperature, but definite conclusions are hampered since most data are for samples quenched to room temperature. Hilti (25) claimed two transitions: α→cubic (vacancy superlattice)→cubic (random vacancies). Jostsons and McDougall (20) designated the β- and α-phases as γ and γ' in order to avoid confusion with other Ti-O alloys. The relatively wide homogeneity ranges of α- and β-TiO are summarized in recent phase diagrams (20, 17) and emf studies (7). Limits of stability of α-TiO and the identity of its adjacent phases are less well established (20, 17) than for β-phase. TiO is metallic, exhibiting weak paramagnetism (27) and superconductivity with T_c below 1 K (26). Ordering of the vacancies by annealing had little effect on T_c (26).

Transition Data

The nature and kinetics of α⇌β are crucial since we lack definite knowledge of the phases used in thermochemical studies. The transition α→β is more rapid than the reverse one. β-TiO is readily retained by quenching, but α can be obtained only by annealing (23-30) for several hours below T_{trs}. Moreover, there is disagreement as to whether the transition α→β proceeds directly (20) or via an intermediate peritectic decomposition, e.g. $Ti_2O + \beta\text{-}TiO_{1+x}$ (17). Emf data (7) seem more consistent with the latter, but none of the evidence is definitive (17, 20, 24, 27-30). Lyon (5) suggested that peritectic decomposition is an alternative to direct reduction of vacancy concentration on going from α to β (see Phase Data). The difference in vacancy concentration, if real, may preclude direct equilibrium between ideal α and normal β. This could invalidate calculation of $\Delta_{trs}S$ from $\Delta_{trs}H°$ (5).

Approximate ranges for T_{trs} include 1235-1270 K from enthalpy data (13); 1235-1273 K from thermal analysis (29); between 1223 and 1263 K (20, 24), between 1223 and 1288 K (29), and between 1173 and 1223 K (28, 30), each based on the presence of α in samples quenched from various annealing temperatures. Later resistivity data (27) on Pearson's samples (28) suggest that T_{trs} is between 1233 and 1303 K. Hilti (25) reported 1253 K for the first transition and 1523 K for the second. The latter transition is not apparent in the enthalpy data (13).

We adopt T_{trs} = 1265 K and arbitrarily take $\Delta_{trs}H°$ = 1.0 kcal mol^{-1}. The apparent enthalpy difference ($\Delta_t H°$ = 0.83 kcal mol^{-1}) at T_{trs} is equal to $\Delta_{trs}H°$ only if Naylor's sample (13) was originally α-TiO and returned to α-form during drop calorimetry from above T_{trs}. Charlu et al. (2) proposed instead, based on the difference in $\Delta_c H°$ values, that $\Delta_{trs}H°$ = ~5 kcal mol^{-1} and $\Delta_{trs}S° = \Delta_{trs}H°/T_{trs}$ = ~4 cal K^{-1} mol^{-1}. The resulting increase in S°(β) would change our third-law analysis of $\Delta_r G°$ for reactions C and D (see α-TiO). It would minimize the discrepancy between $\Delta_c H°$ (3, 4) and the third-law $\Delta_f H°$ but would more than double the discrepancy (δS) in the temperature dependence of $\Delta_f G°(\beta)$ derived from Ti_2O_3 (7, 9). We prefer to minimize δS even though the larger value is compatible with the estimated uncertainty in $\Delta_f G°(\beta)$.

Entropy calculations are available for processes somewhat related to α⇌β. The configurational entropy of TiO(β) with random vacancies was shown to be 2.0 cal K^{-1} mol^{-1} (14). $\Delta_{fus}S°$ of 1.5 to 2.3 cal K^{-1} mol^{-1} for TiO(β, 0% vacancies) → TiO(β, 14.4% vacancies) was calculated by Lyon (5) from compressibility, thermal expansion and a PV isotherm (6). Taylor and Doyle (6) reported observing β-TiO, presumably with much reduced vacancy concentration, at 56.4 kbar and 1153 K. This yields $\Delta_T \le -80°$ assuming the T_{trs} > 1233 K at 1 atm. We estimate ΔV = -0.16 cm^3 mol^{-1} from the densities (6, 23) observed at 25°C and combine it with ΔP to calculate $\Delta_{trs}S°$ = ~2.6 cal K^{-1} mol^{-1} for α→β. This calculation is highly uncertain; it is invalid if the transition is not an equilibrium one.

Fusion Data - See the liquid table for details.

References - Reference are listed on the liquid table.

Titanium Oxide, Beta (TiO) $O_1Ti_1(cr)$

Enthalpy Reference Temperature = T_r = 298.15 K — Standard State Pressure = p° = 0.1 MPa

T/K	$C_p°$ (J $K^{-1}mol^{-1}$)	S° (J $K^{-1}mol^{-1}$)	$-[G°-H°(T_r)]/T$ (J $K^{-1}mol^{-1}$)	$H°-H°(T_r)$ (kJ mol^{-1})	$\Delta_f H°$ (kJ mol^{-1})	$\Delta_f G°$ (kJ mol^{-1})	Log K_f
0							
100							
200							
298.15	39.957	38.077	38.077	0.000	-538.481	-510.080	89.364
300	40.062	38.324	38.078	0.074	-538.480	-509.904	88.782
400	44.978	50.582	39.715	4.346	-538.307	-500.396	65.345
500	48.242	60.987	42.956	9.015	-537.908	-490.961	51.290
600	50.836	70.016	46.730	13.972	-537.356	-481.621	41.929
700	53.095	78.026	50.639	19.171	-536.673	-472.384	35.250
800	55.187	85.254	54.521	24.586	-535.851	-463.255	30.247
900	57.153	91.868	58.309	30.204	-534.927	-454.235	26.363
1000	59.078	97.990	61.974	36.016	-533.967	-445.321	23.261
1100	60.961	103.709	65.511	42.018	-533.035	-436.501	20.728
1200	62.802	109.093	68.921	48.206	-536.116	-427.646	18.615
1265.000	64.013	112.436	71.071	52.327	-------- ALPHA <--> BETA -------		
1300	64.643	114.192	72.209	54.578	-534.515	-418.671	16.822
1400	66.442	119.048	75.382	61.133	-532.827	-409.822	15.291
1500	68.241	123.694	78.449	67.867	-531.063	-401.098	13.967
1600	70.040	128.155	81.417	74.781	-529.238	-392.492	12.814
1700	71.839	132.456	84.294	81.875	-527.362	-384.003	11.799
1800	73.597	136.612	87.086	89.147	-525.452	-375.625	10.900
1900	75.396	140.639	89.799	96.597	-523.520	-367.354	10.099
2000	77.153	144.551	92.439	104.224	-536.265	-358.732	9.369
2023.000	77.551	145.436	93.037	106.003	------- BETA <--> LIQUID -------		
2100	78.910	148.358	95.012	112.027	-535.079	-349.883	8.703
2200	80.668	152.069	97.521	120.006	-533.728	-341.095	8.099
2300	82.467	155.695	99.972	128.163	-532.210	-332.373	7.548
2400	84.224	159.242	102.368	136.497	-530.526	-323.720	7.046
2500	85.981	162.716	104.713	145.008	-528.677	-315.140	6.584

PREVIOUS: March 1967 CURRENT: December 1973

Titanium Oxide, Beta (TiO) $O_1Ti_1(cr)$

TITANIUM OXIDE (TiO) LIQUID M_r = 63.8794

$S°(298.15\ K) = [60.140]\ J\ K^{-1}\ mol^{-1}$ $\Delta_f H°(298.15\ K) = [-493.939]\ kJ\ mol^{-1}$

$T_{fus} = 2023 \pm 30\ K$ $\Delta_{fus}H° = [41.8]\ kJ\ mol^{-1}$

Enthalpy of Formation

$\Delta_f H°$(TiO, ℓ, 298.15 K) is calculated from that of β-TiO by addition of $\Delta_{fus}H°$ and the enthalpy difference in enthalpy, H°(2023 K)-H°(298.15 K), between β and liquid.

Heat Capacity and Entropy

S°(298.15 K) is calculated in a manner analogous to that used for $\Delta_f H°$. $C_p°$ is estimated as 8 cal K^{-1} g-atom^{-1}. Below the assumed glass transition at 1200 K, $C_p°$ is taken to be the same as that of the crystal.

Fusion Data

T_{fus} was observed to be 1750°C (31), 1760 ± 25°C (32) and >1750°C (33). Incipient melting, partial melting and complete melting were observed (29) at approximately 1710°, 1730° and 1760°C, respectively. We adopt 1750 ± 30°C and estimate $\Delta_{fus}H°$ such that $\Delta_{fus}S°$ is 2.5 cal K^{-1} cal K^{-1}g-atom^{-1}. TiO does not melt congruently (17, 20, 29, 32, 33); there are various interpretations of the liquidus curve.

Vaporization Data

The vapor over TiO(cr, ℓ) is mainly TiO and Ti, so the residue increases in (O/Ti) ratio (34).

References - TiO(α, β, ℓ)

1. JANAF Thermochemical Tables, 2nd ed., NSRDS-NBS 37, 1971.
2. T. V. Charlu, O. J. Kleppa and T. B. Reed, submitted for publication in J. Chem. Thermodyn., Jan., 1974.
3. G. L. Humphrey, J. Amer. Chem. Soc. 73, 1587 (1951).
4. S. M. Ariya, M. P. Morozova and E. Vol'f, Russ. J. Inorg. Chem. 2, 16 (1957).
5. W. G. Lyon and E. F. Westrum, Jr., University of Michigan, personal communication, April, 1974.
6. A. Taylor and N. J. Doyle, High Temp.-High Press. 1, 679 (1969); J. Appl. Cryst. 4, 103, 109 (1971).
7. K. Suzuki and K. Sambongi, Tetsu To Hagane 58, 1579 (1972).
8. K. L. Komarek and M. Silver, "Thermodynamics of Nuclear Materials," International Atomic Energy Agency, Vienna, p. 749, 1962.
9. JANAF Thermochemical Tables: Ti_2O_3(cr), 6-30-73; TiO(α, β, ℓ, g), Ti_3O_5(α, β), Ti_4O_7(cr), 12-31-73.
10. A. D. Mah, K. K. Kelley, N. L. Gellert, E. G. King and C. J. O'Brien, U. S. Bur. of Mines RI 5316, 1957.
11. M. Koiwa and M. Hirabayashi, J. Phys. Soc. Japan 27, 801, 807 (1969); 21, 2096 (1966).
12. C. H. Shomate, J. Amer. Chem. Soc. 68, 310 (1946).
13. B. F. Naylor, J. Amer. Chem. Soc. 68, 1077 (1946).
14. A. Jostsons and A. E. Jenkins, J. Phys. Chem. 73, 749 (1969).
15. V. S. Chernyaev, G. P. Shveikin and P. V. Gel'd, Izv. Akad. Nauk SSSR, Neorg. Mater 8, 459 (1972); Tr. Ural. Politekh. Inst. No. 186, 205, 1970.
16. R. I. Efremova and E. V. Matizen, Bulletin of Thermodynamics and Thermochemistry, No. 14, 40 (1971).
17. P. G. Wahlbeck and P. W. Gilles, J. Amer. Ceram. Soc. 49, 180 (1966).
18. M. Hoch, A. S. Iyer and J. Nelken, J. Phys. Chem. Solids 23, 1463 (1962).
19. J. Drowart, P. Coppens and S. Smoes, J. Chem. Phys. 50, 1046 (1969).
20. A. Jostsons and P. McDougall, Sci., Technol. Appl. Titanium, Proc. Int. Conf., Pergamon (Pub. 1970), pp. 745-763, 1968.
21. O. Kubaschewski and W. A. Dench, J. Inst. Metals 82, 87 (1953).
22. O. Kubaschewski, Rev. Int. Hautes Temp. Refract. 3, 229 (1966).
23. S. Anderson, B. Collen, U. Kuylenstierna and A. Magneli, Acta Chem. Scand. 11, 1641, 1653 (1957); 10, 1195 (1956).
24. D. Watanabe, J. R. Castles, A. Jostsons and A. S. Malin, Acta Cryst. 23, 307 (1967); Nature 210, 934 (1966).
25. E. Hilti and F. Laves, Naturwissenschaften 55, 130, 131 (1968).
26. J. K. Hulm, C. K. Jones, R. A. Hein and J. W. Gibson, J. Low Temp. Phys. 7, 291 (1972).
27. S. P. Denker, J. Appl. Phys. 37, 142 (1966).
28. A. D. Pearson, J. Phys. Chem. Solids, 5, 316 (1958).
29. T. H. Schofield and A. E. Bacon, J. Inst. Metals 84, 47 (1955).
30. C. Wang and N. J. Grant, J. Metals 8, 184 (1956).
31. W. Dawihl and K. Schroter, Z. Anorg. Allg. Chem. 233, 178 (1937).
32. E. S. Bumps, H. D. Kessler and M. Hansen, Trans. Amer. Soc. Metals 45, 1008 (1953).
33. H. Nishimura and H. Kimura, Nippon Kinzoki Gakkaishi 20, 524 (1956); Chem. Abstr. 54, 17211g.
34. P. W. Gilles, K. D. Carlson, H. F. Franzen and P. G. Wahlbeck, J. Chem. Phys. 46, 2461 (1967); 48, 1938 (1968).

References - TiO(g)

1. B. Rosen, "Spectroscopic Data Relative to Diatomic Molecules," Pergamon Press, New York, pp. 395-396, 1970.
2. J. G. Phillips, Astrophy. J. 169, 185 (1971); 157, 449 (1969).
3. N. S. McIntrye, K. R. Thompson and W. Weltner, J. Phys. Chem. 75, 3243 (1971); 69, 3488 (1965).
4. JANAF Thermochemical Tables: TiO(g, ℓ, β, α), 12-31-73.
5. J. G. Phillips and S. P. Davis, Astrophys. J. 167, 209 (1971); 175, 583 (1972).
6. J. G. Phillips, Astrophys. J. 115, 567 (1952).
7. P. W. Merrill, A. J. Deutsch and P. C. Keenan, Astrophys. J. 136, 21 (1962).
8. C. Linton, Can. J. Phys. 50, 312 (1972).
9. B. Lindgren, J. Mol. Spectrosc. 43, 474 (1972).
10. H. B. Palmer and C. J. Hsu, J. Mol. Spectrosc. 43, 320 (1972); 33, 137 (1970).
11. L. Brewer and D. W. Green, High Temp. Sci. 1, 26 (1969).
12. T. Wentink and R. J. Spindler, J. Quant. Spectrosc. Radiat. Transfer 12, 1569 (1972).
13. P. J. Hampson and P. W. Gilles, J. Chem. Phys. 55, 3712 (1971).
14. J. Drowart, P. Coppens and S. Smoes, J. Chem. Phys. 50, 1046 (1969).
15. G. Balducci, G. DeMaria, M. Guido and V. Piacente, J. Chem. Phys. 56, 3422 (1972).
16. H. Y. Wu and P. G. Wahlbeck, J. Chem. Phys. 56, 4634 (1972).
17. W. O. Groves, M. Hoch and H. L. Jonston, J. Phys. Chem. 59, 127 (1955).
18. J. Berkowitz, W. A. Chupka and M. G. Inghram, J. Phys. Chem. 61, 1569 (1957).
19. Q. D. Wheatley, Ph.D. Thesis, Univ. Kansas, 1954; quoted by (13, 21).
20. K. Suzuki and K. Sambongi, Tetsu To Hagane 58, 1579 (1972).
21. P. W. Gilles, J. Chem. Phys. 46, 4987 (1967); 46, 2465 (1967).
22. S. Smoes, J. Drowart and G. Verhaegen, J. Chem. Phys. 43, 732 (1965); Trans. Faraday Soc. 63, 2140 (1967); 61, 1072 (1965).
23. G. Herzberg, "Spectra of Diatomic Molecules," 2nd ed., D. Van Nostrand Co., Inc., New York, 1968.
24. V. S. Kushawaha, J. Phys. Chem. 77, 2885 (1973).

Titanium Oxide (TiO) $O_1Ti_1(l)$

Enthalpy Reference Temperature = T_r = 298.15 K Standard State Pressure = p° = 0.1 MPa

T/K	$C_p°$ (J K⁻¹mol⁻¹)	S° (J K⁻¹mol⁻¹)	−[G°−H°(T_r)]/T (J K⁻¹mol⁻¹)	H°−H°(T_r) (kJ mol⁻¹)	$\Delta_f H°$ (kJ mol⁻¹)	$\Delta_f G°$ (kJ mol⁻¹)	Log K_f
0							
100							
200							
298.15	39.957	60.140	60.140	0.000	-493.939	-472.116	82.713
300	40.062	60.387	60.140	0.074	-493.938	-471.981	82.179
400	44.978	72.644	61.778	4.346	-493.765	-464.679	60.681
500	48.242	83.049	65.018	9.015	-493.366	-457.450	47.790
600	50.836	92.079	68.792	13.972	-492.814	-450.317	39.204
700	53.095	100.089	72.702	19.171	-492.131	-443.286	33.078
800	55.187	107.317	76.584	24.586	-491.309	-436.363	28.492
900	57.153	113.931	80.371	30.204	-490.385	-429.549	24.930
1000	59.078	120.053	84.037	36.016	-489.425	-422.841	22.087
1100	60.961	125.772	87.574	42.018	-488.493	-416.229	19.765
1200	62.802	131.156	90.984	48.206	-491.573	-409.579	17.829
1200.000	62.802	131.156	90.984	48.206	GLASS <--> LIQUID TRANSITION		
1200.000	66.944	131.156	90.984	48.206			
1300	66.944	136.514	94.283	54.901	-489.651	-402.825	16.186
1400	66.944	141.475	97.479	61.595	-487.822	-396.215	14.783
1500	66.944	146.094	100.567	68.290	-486.098	-389.733	13.572
1600	66.944	150.414	103.549	74.984	-484.492	-383.361	12.515
1700	66.944	154.473	106.427	81.678	-483.016	-377.086	11.586
1800	66.944	158.299	109.203	88.373	-481.684	-370.895	10.763
1900	66.944	161.919	111.883	95.067	-480.507	-364.772	10.028
2000	66.944	165.352	114.472	101.762	-494.186	-358.254	9.357
2023.000	66.944	166.118	115.054	103.301	------- BETA <--> LIQUID -------		
2100	66.944	168.619	116.973	108.456	-494.108	-351.460	8.742
2200	66.944	171.733	119.392	115.150	-494.041	-344.668	8.183
2300	66.944	174.709	121.733	121.845	-493.986	-337.880	7.673
2400	66.944	177.558	124.000	128.539	-493.942	-331.094	7.206
2500	66.944	180.291	126.197	135.234	-493.909	-324.309	6.776
2600	66.944	182.916	128.328	141.928	-493.886	-317.526	6.379
2700	66.944	185.443	130.397	148.622	-493.874	-310.743	6.012
2800	66.944	187.877	132.407	155.317	-493.872	-303.960	5.670
2900	66.944	190.226	134.360	162.011	-493.880	-297.178	5.353
3000	66.944	192.496	136.261	168.706	-493.898	-290.395	5.056
3100	66.944	194.691	138.110	175.400	-493.925	-283.611	4.779
3200	66.944	196.816	139.912	182.094	-493.962	-276.826	4.519
3300	66.944	198.876	141.668	188.789	-494.006	-270.040	4.274
3400	66.944	200.875	143.380	195.483	-494.060	-263.252	4.044
3500	66.944	202.815	145.050	202.178	-494.121	-256.463	3.828
3600	66.944	204.701	146.681	208.872	-494.190	-249.672	3.623
3700	66.944	206.535	148.274	215.566	-903.365	-235.090	3.319
3800	66.944	208.321	149.831	222.261	-902.214	-217.044	2.983
3900	66.944	210.060	151.353	228.955	-901.125	-199.027	2.666
4000	66.944	211.754	152.842	235.650	-900.098	-181.038	2.364
4100	66.944	213.407	154.299	242.344	-899.129	-163.074	2.078
4200	66.944	215.021	155.726	249.038	-898.218	-145.133	1.805
4300	66.944	216.596	157.123	255.733	-897.362	-127.212	1.545
4400	66.944	218.135	158.492	262.427	-896.559	-109.311	1.298
4500	66.944	219.639	159.835	269.122	-895.809	-91.428	1.061

PREVIOUS: March 1967 CURRENT: December 1973

Titanium Oxide (TiO) $O_1Ti_1(l)$

TITANIUM OXIDE (TiO) ALPHA-BETA-LIQUID M_r = 63.8794

0 to 1265 K	crystal, alpha
1265 to 2023 K	crystal, beta
above 2023 K	liquid

Refer to the individual tables for details.

Titanium Oxide (TiO) $O_1Ti_1(cr,l)$

Enthalpy Reference Temperature = T_r = 298.15 K — Standard State Pressure = $p°$ = 0.1 MPa

T/K	C_p° (J K⁻¹mol⁻¹)	S° (J K⁻¹mol⁻¹)	$-[G^\circ-H^\circ(T_r)]/T$ (J K⁻¹mol⁻¹)	$H^\circ-H^\circ(T_r)$ (kJ mol⁻¹)	Δ_fH° (kJ mol⁻¹)	Δ_fG° (kJ mol⁻¹)	Log K_f
0	0.000	0.000	INFINITE	-6.174	-539.668	-539.668	INFINITE
100	12.765	5.732	63.375	-5.764	-541.270	-532.352	278.072
200	30.681	20.585	38.164	-3.516	-542.394	-522.918	136.572
298.15	39.957	34.769	34.769	0.000	-542.665	-513.278	89.924
300	40.062	35.017	34.770	0.074	-542.665	-513.096	89.338
400	44.978	47.274	36.408	4.346	-542.491	-503.257	65.719
500	48.242	57.679	39.648	9.015	-542.092	-493.491	51.555
600	50.836	66.709	43.422	13.972	-541.541	-483.821	42.120
700	53.095	74.719	47.331	19.171	-540.857	-474.253	35.389
800	55.187	81.947	51.214	24.586	-540.035	-464.793	30.348
900	57.153	88.561	55.001	30.204	-539.111	-455.442	26.433
1000	59.078	94.682	58.667	36.016	-538.152	-446.197	23.307
1100	60.961	100.402	62.203	42.018	-537.219	-437.047	20.754
1200	62.802	105.785	65.613	48.206	-540.300	-427.861	18.624
1265.000	63.998	109.129	67.764	52.327	ALPHA <--> BETA TRANSITION		
1265.000	64.013	112.436	67.764	56.511			
1300	64.643	114.192	68.990	58.763	-534.515	-418.671	16.822
1400	66.442	119.048	72.394	65.317	-532.827	-409.822	15.291
1500	68.241	123.694	75.660	72.051	-531.063	-401.098	13.967
1600	70.040	128.155	78.802	78.965	-529.238	-392.492	12.814
1700	71.839	132.456	81.833	86.059	-527.362	-384.003	11.799
1800	73.597	136.612	84.761	93.331	-525.452	-375.625	10.900
1900	75.396	140.639	87.597	100.781	-523.520	-367.354	10.099
2000	77.153	144.551	90.347	108.409	-536.265	-358.731	9.369
2023.000	77.551	145.436	90.968	110.188	BETA <--> LIQUID TRANSITION		
2023.000	66.944	166.118	90.968	152.028			
2100	66.944	168.619	93.770	157.182	-494.108	-351.460	8.742
2200	66.944	171.733	97.243	163.877	-494.041	-344.668	8.183
2300	66.944	174.709	100.547	170.571	-493.986	-337.880	7.673
2400	66.944	177.558	103.697	177.266	-493.942	-331.094	7.206
2500	66.944	180.291	106.707	183.960	-493.909	-324.309	6.776
2600	66.944	182.916	109.588	190.654	-493.886	-317.526	6.379
2700	66.944	185.443	112.351	197.349	-493.874	-310.743	6.012
2800	66.944	187.877	115.005	204.043	-493.872	-303.960	5.670
2900	66.944	190.226	117.558	210.738	-493.880	-297.178	5.353
3000	66.944	192.496	120.019	217.432	-493.898	-290.395	5.056
3100	66.944	194.691	122.392	224.126	-493.925	-283.611	4.779
3200	66.944	196.816	124.685	230.821	-493.962	-276.826	4.519
3300	66.944	198.876	126.902	237.515	-494.006	-270.040	4.274
3400	66.944	200.875	129.048	244.210	-494.060	-263.252	4.044
3500	66.944	202.815	131.129	250.904	-494.121	-256.463	3.828
3600	66.944	204.701	133.146	257.598	-494.190	-249.672	3.623
3700	66.944	206.535	135.105	264.293	-903.365	-235.090	3.319
3800	66.944	208.321	137.008	270.987	-902.214	-217.044	2.983
3900	66.944	210.060	138.859	277.682	-901.125	-199.027	2.666
4000	66.944	211.754	140.661	284.376	-900.098	-181.038	2.364
4100	66.944	213.408	142.415	291.070	-899.129	-163.074	2.078
4200	66.944	215.021	144.124	297.765	-898.218	-145.133	1.805
4300	66.944	216.596	145.791	304.459	-897.362	-127.212	1.545
4400	66.944	218.135	147.418	311.154	-896.559	-109.311	1.298
4500	66.944	219.639	149.006	317.848	-895.809	-91.428	1.061

PREVIOUS: CURRENT: December 1973

Titanium Oxide (TiO) $O_1Ti_1(cr,l)$

TITANIUM OXIDE (TiO) IDEAL GAS M_r = 63.8794

$\Delta_f H°(0\ K)$ = 53.93 ± 8.4 kJ mol^{-1}

$S°(298.15\ K)$ = 233.474 ± 0.25 J K^{-1} mol^{-1} $\Delta_f H°(298.15\ K)$ = 54.39 ± 8.4 kJ mol^{-1}

Electronic States and Molecular Constants

State	ε_i, cm^{-1}	g_i	ω_e, cm^{-1}	$\omega_e x_e$, cm^{-1}	α_e, cm^{-1}	B_e, cm^{-1}	r_e, Å	Source
$X^3\Delta$	0	2	1008.26	4.13	0.00303	0.5354	1.620	(1, 2)
	96.4	2						
	197.5	2						
$E^3\Pi$	11800.0	6	919.	[3.8]	[0.0029]	[0.5134]	[1.654]	(3, 4)
$A^3\Phi$	14019.4	2	867.71	3.94	0.00318	0.50734	1.664	(1, 2)
	14192.3	2						
	14360.7	2						
$B^3\Pi$	16226.4	2	873.	4.	[0.0032]	0.5079	1.663	(2, 4)
	16248.0	2						
	16264.2	2						
$C^3\Delta$	19341.7	2	837.86	4.546	0.0029	0.4889	1.695	(1, 2)
	19437.4	2						
	19531.6	2						
$a^1\Delta$	[1000.]	2	1016.3	3.93	[0.003]	0.5377	1.616	(1, 3-7)
$d^1\Sigma^+$	[3219.]	1	1023.8	4.64	0.00337	0.5490	1.600	(1, 4)
$b^1\Pi$	[12273.]	2	918.7	3.75	0.00286	0.5134	1.654	(1, 4)
$c^1\Phi$	[18841.]	2	916.7	4.33	[0.0032]	0.5228	1.639	(1, 4)
$f^1\Delta$	[20069.]	2	[933.]	[4.2]	[0.003]	0.5037	1.670	(8, 5, 4)
$e^1\Sigma$	[27517.]	1	853.9	4.7	0.0025	0.4892	1.694	(9, 5, 4)

Enthalpy of Formation

$\Delta_f H°(298.15\ K)$ = 13.0 ± 2 kcal mol^{-1} and D_0^0 = 158.5 ± 2 kcal mol^{-1} are selected from reactions A, B, C and E (16) which are summarized below. The first three are isomolecular exchange reactions studied by mass spectrometry (13, 14). Reaction E is obtained from univariant Ti_2O_3-Ti_3O_5 by combining Knudsen effusion with the pressure ratio (TiO_2/TiO) = 0.246 obtained from mass spectrometry (16). We use the same analysis (16) for (17, E), although the condensed phase may not be univariant. We presume that the smaller D_0^0 = 145.4 kcal mol^{-1} from reaction D is due to bias in the mass-spectrometric values (15) for the partial pressure of O(g). Larger values of D° from sublimation (17, 19) and vaporization (18) of TiO are subject to bias due to presence of species other than TiO in the vapor and to unknown liquid activities. Activities of TiO in TiO_x(cr) are very close to unity at x=1.0-1.1 and 1900 K according to emf data (20).

The analysis of Drowart et al. (14) was the key to resolution of the discrepancy in D_0^0(21); later studies (13, 16) confirmed their analysis. Our analysis uses D_0^0(GeO, g) = 156.4 ± 1.5, D_0^0(ScO, g) = 160.4 ± 2.0, and D_0^0(YO, g) = 170.4 ± 2.5 kcal mol^{-1} with other auxiliary data and newly calculated Gibbs-energy functions, all of which are essentially consistent with Drowart et al. (14, 22). Our selected D_0^0(TiO) = 158.5 ± 2 kcal mol^{-1} may be compared with 174.5 and 156.3 kcal mol^{-1} calculated by linear Birge-Sponer extrapolation of present (1) and older (23) vibrational constants of $X^3\Delta$. Kushawaha (24) used the older constants in construction of RKRV potential energy curves and thereby estimated D_0^0 = 165.4 kcal mol^{-1}. He also derived D_0^0 = 167.8 ± 20 kcal mol^{-1} from a chemiluminescent reaction (10), but we dismiss this since auxiliary data for KO and TiCl were estimates.

Source	Method	Reaction[a]	T/K	Data Points	δS° cal K^{-1} mol^{-1}	$\Delta_r H°(298.15\ K)$, kcal mol 2nd law	3rd law	$\Delta_f H°(298.15\ K)$ kcal mol^{-1}	D_0^0
(13)	Mass spec.	A	2145-2418	9	-6.4± 5.0	-11.5±11.3	3.2±2.8	11.7±3.4	159.8
	Mass spec.	B	~2023-2417	24[b]	0.2± 1.5	-2.6± 3.3	-3.0±1.5	14.0±2.7	157.4
(14)	Mass spec.	B	1951-2463	Eqn.	0.0± 0.1	-3.0	-3.0	14.0±2.4	157.4
	Mass spec.	C	2313	1	-	-	-12.0	13.0±3.0	158.5
(15)	Mass spec.	D	2015-2240	12, 6	-0.2± 2.6	145.2± 5.4	146.5±1.6	26.1±2	145.4
(16)	Knudsen eff.	E	1937-2044	12	-2.0± 3.6	150.6± 7.1	154.7±0.8	12.3±3.5	159.1
(17)	Knudsen eff.	E	1971-2085	15	0.0±12	152.0± 2.5	152.0±2.8	9.7±3.6	161.8
	Knudsen eff.	F	1847-1968	8	11 ± 3	157.1± 5	135.1±3	6.4±4	165.0
(18)	Mass spec.	G	2027-2072	2	48	225.	125.5±4	7.5±7	164.0

[a] A) TiO(g) + Ge(g) = Ti(g) + GeO(g); B) TiO(g) + Sc(g) = Ti(g) + ScO(g);
C) TiO(g) + Y(g) = Ti(g) + YO(g); D) TiO(g) = Ti(g) + O(g) from difference of two observed reactions $TiO_2(g)$ = O + TiO and TiO_2 + Ti = 2 TiO;
E) 2 Ti_2O_3(cr) = Ti_3O_5(β) + TiO(g); F) TiO(high) = TiO(g)
G) TiO(ℓ) + TiO(g)

[b] Extreme points selected from large data set.

Heat Capacity and Entropy

Electronic levels (T_0) and vibrational-rotational constants are from Rosen (1) as supplemented (4) by extensive new data (2, 3, 5, 8, 9). Additional states near 32000 cm^{-1} (10, 3) are omitted since their effect is negligible. We estimate unknown spectroscopic constants from known values for states arising from like molecular-orbital configurations (11). Although $^3\Delta$ is the ground state (6), several authors (1-3, 12) noted that energy levels of all singlet states are approximate due to uncertainty in the low-lying $a^1\Delta$. We revise $a^1\Delta$ from 581 cm^{-1} (6) to 1000 ± 400 cm^{-1} based on the temperature dependence of band intensities (6, 7), the requirement that $b^1\Pi$ lie above $E^3\Pi$ observed at 11894 cm^{-1} in a Ne-matrix (3), and the assumption that $f^1\Delta$ (8, 4) is the perturber of $C^3\Delta$ (5). The new E and f states confirm empirical predictions (11) which suggest additional unobserved levels at ~13000 ($^3\Sigma$) and 15000 $cm^{-1}\Sigma$). We omit these states since their predicted energies are less well established and their contributions to $C_p°$ and S° would be quite small. The functions are calculated using first-order anharmonic corrections to Q_r^i and Q_v^i in the partition function $Q = Q_t \sum_i Q_r^i Q_v^i g_i \exp(-c_2 \varepsilon_i/T)$.

References - References are listed on the liquid table.

Titanium Oxide (TiO) $O_1Ti_1(g)$

Enthalpy Reference Temperature = T_r = 298.15 K Standard State Pressure = p° = 0.1 MPa

T/K	$C_p°$ (J $K^{-1}mol^{-1}$)	S° (J $K^{-1}mol^{-1}$)	-[G°-H°(T_r)]/T (J $K^{-1}mol^{-1}$)	H°-H°(T_r) (kJ mol^{-1})	$\Delta_f H°$ (kJ mol^{-1})	$\Delta_f G°$ (kJ mol^{-1})	Log K_f
0	0.	0.	INFINITE	-9.630	53.934	53.934	INFINITE
100	33.880	198.046	262.006	-6.396	55.155	44.842	-23.423
200	31.771	220.717	236.408	-3.138	55.040	34.491	-9.008
250	31.908	227.809	234.005	-1.549	54.737	29.387	-6.140
298.15	32.476	233.474	233.474	0.	54.392	24 535	-4.298
300	32.502	233.675	233.474	0.060	54.378	24.350	-4.240
350	33.263	238.741	233.873	1.704	53.997	19.375	-2.892
400	34.021	243.233	234.767	3.386	53.605	14.456	-1.888
450	34.702	247.280	235.937	5.105	53.208	9.586	-1.113
500	35.285	250.968	237.258	6.855	52.804	4.760	-0.497
600	36.176	257.485	240.100	10.431	51.975	-4.771	0.415
700	36.786	263.110	242.995	14.081	51.109	-14.161	1.057
800	37.212	268.051	245.824	17.782	50.217	-23.425	1.529
900	37.518	272.453	248.543	21.519	49.261	-32.573	1.891
1000	37.745	276.418	251.136	25.283	48.172	-41.609	2.173
1100	37.918	280.024	253.600	29.066	46.886	-50.527	2.399
1200	38.052	283.330	255.942	32.865	41.416	-59.198	2.577
1300	38.158	286.380	258.168	36.676	40.455	-67.545	2.714
1400	38.245	289.211	260.285	40.496	39.409	-75.814	2.829
1500	38.319	291.852	262.302	44.324	38.267	-84.005	2.925
1600	38.386	294.327	264.227	48.160	37.014	-92.116	3.007
1700	38.450	296.656	266.067	52.001	35.637	-100.145	3.077
1800	38.516	298.856	267.828	55.850	34.124	-108.089	3.137
1900	38.588	300.940	269.517	59.705	32.461	-115.945	3.188
2000	38.669	302.922	271.138	63.568	15.951	-123.256	3.219
2100	38.762	304.810	272.697	67.439	13.206	-130.149	3.237
2200	38.869	306.616	274.198	71.321	10.459	-136.911	3.251
2300	38.992	308.346	275.645	75.213	7.713	-143.548	3.260
2400	39.132	310.009	277.042	79.120	4.969	-150.066	3.266
2500	39.290	311.609	278.393	83.040	2.229	-156.469	3.269
2600	39.466	313.154	279.701	86.978	-0.505	-162.763	3.270
2700	39.661	314.647	280.967	90.934	-3.232	-168.952	3.269
2800	39.874	316.093	282.196	94.911	-5.948	-175.040	3.265
2900	40.105	317.496	283.389	98.910	-8.651	-181.032	3.261
3000	40.352	318.860	284.549	102.932	-11.341	-186.930	3.255
3100	40.614	320.187	285.677	106.981	-14.014	-192.738	3.248
3200	40.890	321.481	286.776	111.056	-16.670	-198.462	3.240
3300	41.179	322.744	287.847	115.159	-19.306	-204.102	3.231
3400	41.478	323.978	288.892	119.292	-21.921	-209.662	3.221
3500	41.787	325.184	289.911	123.455	-24.513	-215.146	3.211
3600	42.103	326.366	290.908	127.649	-27.082	-220.557	3.200
3700	42.425	327.524	291.882	131.876	-438.725	-218.108	3.079
3800	42.751	328.660	292.835	136.135	-440.010	-212.127	2.916
3900	43.080	329.774	293.768	140.426	-441.324	-206.113	2.761
4000	43.409	330.869	294.681	144.751	-442.666	-200.065	2.613
4100	43.738	331.945	295.577	149.108	-444.034	-193.983	2.471
4200	44.064	333.003	296.456	153.498	-445.427	-187.868	2.336
4300	44.386	334.044	297.318	157.921	-446.843	-181.719	2.207
4400	44.703	335.068	298.164	162.375	-448.281	-175.537	2.084
4500	45.015	336.076	298.996	166.861	-449.739	-169.322	1.965
4600	45.319	337.068	299.812	171.378	-451.216	-163.074	1.852
4700	45.614	338.046	300.616	175.925	-452.711	-156.794	1.743
4800	45.901	339.010	301.405	180.500	-454.222	-150.482	1.638
4900	46.178	339.959	302.183	185.104	-455.750	-144.138	1.537
5000	46.444	340.895	302.947	189.736	-457.293	-137.763	1.439
5100	46.699	341.817	303.701	194.393	-458.850	-131.357	1.345
5200	46.943	342.726	304.442	199.075	-460.421	-124.920	1.255
5300	47.174	343.622	305.173	203.781	-462.005	-118.453	1.167
5400	47.394	344.506	305.893	208.509	-463.602	-111.956	1.083
5500	47.602	345.378	306.603	213.259	-465.211	-105.430	1.001
5600	47.797	346.237	307.303	218.029	-466.833	-98.873	0.922
5700	47.980	347.085	307.994	222.818	-468.467	-92.288	0.846
5800	48.150	347.921	308.675	227.625	-470.113	-85.674	0.772
5900	48.309	348.745	309.347	232.448	-471.772	-79.032	0.700
6000	48.455	349.558	310.011	237.286	-473.442	-72.361	0.630

PREVIOUS: December 1973 (1 atm) CURRENT December 1973 (1 bar)

Titanium Oxide (TiO) $O_1Ti_1(g)$

VANADIUM OXIDE (VO) CRYSTAL M_r = 66.9409

$\Delta_f H°(0\ K) = -429.715 \pm 6.3\ kJ\ mol^{-1}$
$S°(298.15\ K) = 39.011 \pm 0.8\ J\ K^{-1}\ mol^{-1}$ $\Delta_f H°(298.15\ K) = -431.789 \pm 6.3\ kJ\ mol^{-1}$
$T_{fus} = 2063 \pm 10\ K$ $\Delta_{fus}H° = [54.4]\ kJ\ mol^{-1}$

Enthalpy of Formation

The adopted enthalpies of formation for the vanadium oxides are based on the combustion studies by Mah and Kelley (1). The adopted value is $\Delta_f H°$(VO, cr, 298.15 K) = -103.2 kcal mol^{-1}, as reported by Mah and Kelley (1). For more details, refer to the V_2O_5(cr) table (2). We increase their stated uncertainty limits of +0.3 kcal mol^{-1} to ±1.5 kcal mol^{-1} so as to more adequately include all uncertainties in the experimental procedure, rather than just the scatter in the final results.

Other enthalpy of formation results for VO(cr) derived from combustion studies are reported as -102 kcal mol^{-1} by Samsonov (3) and -102 ± 1 kcal mol^{-1} by Vol'f and Ariya (4). However, the latter study (4) also reported a corresponding result for V_2O_5(cr) which is 5.4 kcal mol^{-1} more negative than the adopted JANAF value (2). Older work by Kobayashi (5) involving water vapor equilibrium on vanadium and its oxides and by Mixter (6) involving sodium peroxide fusions are not considered.

Heat Capacity and Entropy

Todd and Bonnickson (7) measured the heat capacity of VO(cr) in the region 55-296 K. The sample, estimated to be 98.2% pure, was prepared by heating a mixture of V(cr) and V_2O_3(cr) in a hydrogen atmosphere. The heat capacity results showed an anomaly similar to that found by Anderson (8) for V_2O_3(cr). Todd and Bonnickson (7) concluded that the ~20 cal K^{-1} mol^{-1} in excess to that expected for a smooth $C_p°$ curve in the region 168-191 K was evidence of either an incomplete reaction in preparing VO(cr) or subsequent disproportionation. Experimental data in this region were not reported. Using the Debye function D(398/T) as suggested by Todd and Bonnickson (7), we calculate S°(50 K) = 0.297 cal K^{-1} mol^{-1} and H°(50 K)-H°(0 K) = 11.051 cal mol^{-1}. The data are graphically curve fitted assuming a smooth sequence of "data points" in the region 168-191 K. Near 298.15 K the heat capacity data of Todd and Bonnickson (7) and the derivative of the smoothed enthalpy data of Orr (9) were constrained to join smoothly. Orr (9) used the same sample as did Todd and Bonnickson (7) and measured the enthalpy (fourteen data points) in the region 394-1698 K. The heat capacities derived from the Orr (9) data were extrapolated to T_{fus} and above.

The heat capacity, entropy and enthalpy values at 298.15 K reported by Chernyaev et al. (10) for VO_x, where x = 0.86, 0.99, 1.24 and 1.30, led to values for VO(cr) which are considerably lower than our adopted values. For example, the S°(298.15 K) value derived from the data of Chernyaev et al. (10, 22) is roughly 1.3 cal K^{-1} mol^{-1} lower than our adopted value.

Phase Data

Vanadium monoxide crystallizes in the cubic NaCl-type structure and is stable over a wide range of composition, roughly between the limit $VO_{0.80}$ and $VO_{1.30}$. Details and many references may be found in Stringer (20) and Alexander and Carlson (21).

Transition Data

There exists a large volume of literature dealing with transitions in the vanadium oxide condensed phases. These various oxides show widely varying electric and magnetic properties, as is discussed by Adler (11). In the case of VO(cr), which has a wide homogeneity range, the reported data in the region of 120 K is conflicting. Sample purity, stoichiometry, and preparation undoubtedly play a major role in the appearance or lack of appearance of a possible transition. Much of the literature does not provide adquate sample characterization for an unequivical analysis of the data.

Conductivity measurements by Morin (12) on single crystals of VO(cr) and numerous measurements, including NMR, by Warren et al. (13) indicated a transition in the region 114-126 K and 88-125 K, respectively. The transition was described by Morin (12) as a semiconductor-metal type. This is further discussed by Zhuze et al. (14) and Hyland (15). However, Kawano et al. (16) and Takei and Koide (17) presented evidence on electric and magnetic properties and thermoelectric power for various vanadium monoxides within its homogeneity range which did not indicate any transition. Recent studies by Warren et al. (18) also indicated no evidence of a transition and, in addition, stated that the sample used in a previous study (13) was mainly V_2O_3(cr) as determined by x-ray analysis. Many more references on this subject may be found in the compilation by Neuberger (19) and in a review of the V-O system by Stringer (20).

We conclude that there is no transition below T_{fus} for VO(cr) based primarily on the work cited above (16-18).

Fusion Data - See the liquid table for details.

Sublimation Data

At temperatures approaching T_{fus}, VO(cr) is reported to yield a vapor containing V(g), VO(g), and VO_2(g) (2).

References

1. A. D. Mah and K. K. Kelley, U. S. Bur. Mines RI 5858, 1961.
2. JANAF Thermochemical Tables: V_2O_5(cr), 6-30-73; VO(g), 12-31-73.
3. G. V. Samsonov, Ukrain. Khim. Zhur. 23, 287 (1957); Chem. Abstr. 52, 1826g (1958).
4. E. Vol'f and S. M. Ariya, J. Gen. Chem. USSR 29, 2433 (1959).
5. M. Kobayashi, Bull. Chem. Soc. Japan 8, 231 (1933); Science Repts. Tohoku Imp. Univ., 1st Ser. 22, 1240 (1933).
6. W. G. Mixter, Amer. J. Sci. 34, 141 (1912).
7. S. S. Todd and K. R. Bonnickson, J. Amer. Chem. Soc. 73, 3894 (1951).
8. C. T. Anderson, J. Amer. Chem. Soc. 58, 564 (1936).
9. R. L. Orr, J. Amer. Chem. Soc. 76, 857 (1954).
10. V. S. Chernyaev, E. N. Shchetnikov and P. V. Gel'd, Tr. Jral. Politekh. Inst., No. 167, 151 (1968); Chem. Abstr. 70, 51339v.
11. D. Adler, Solid State Phys. 21, (1968).
12. F. J. Morin, Phys. Rev. Letters 3, 34 (1959).
13. W. W. Warren, Jr., G. A. Miranda and W. G. Clark, Amer. Phys. Soc. Bull., Ser. 2, 12, 1117 (1967).
14. V. P. Zhuze, A. A. Andreev, and A. I. Shelykh, Soviet Physics-Solid State 10, 2914 (1969).
15. G. J. Hyland, J. Phys. C 1, 189 (1968).
16. S. Kawano, K. Kosuge, and S. Kachi, J. Phys. Soc. Japan 21, 2744 (1966), ibid. 27, 1076 (1969).
17. H. Takei and S. Koide, J. Phys. Soc. Japan 24, 1394 (1968).
18. W. W. Warren, Jr., A. C. Gossard and M. D. Banus, J. Appl. Phys. 41, 881 (1970).
19. M. Neubenger, U. S. Nat. Tech. Inform. Serv., AD 734596, 1971.
20. J. Stringer, J. Less-Common Metals 8, 1 (1965).
21. D. G. Alexander and O. N. Carlson, Met. Trans. 2, 2805 (1971).
22. V. S. Chernyaev, E. N. Shchetnikov, G. P. Shveikin, and P. V. Gel'd, Izv. Akad. Nauk SSR, Neorg. Mater. 4, 2117 (1968).

Vanadium Oxide (VO) O_1V_1(cr)

Enthalpy Reference Temperature = T_r = 298.15 K Standard State Pressure = p° = 0.1 MPa

T/K	$C_p°$ (J K⁻¹mol⁻¹)	S° (J K⁻¹mol⁻¹)	$-[G°-H°(T_r)]/T$ (J K⁻¹mol⁻¹)	$H°-H°(T_r)$ (kJ mol⁻¹)	$\Delta_f H°$ (kJ mol⁻¹)	$\Delta_f G°$ (kJ mol⁻¹)	Log K_f
0	0.000	0.000	INFINITE	-6.908	-429.715	-429.715	INFINITE
100	13.895	6.730	71.088	-6.436	-431.184	-422.473	220.677
200	34.100	23.035	42.859	-3.965	-432.005	-413.352	107.956
298.15	45.501	39.011	39.011	0.000	-431.789	-404.210	70.816
300	45.647	39.293	39.012	0.084	-431.778	-404.039	70.350
400	49.580	53.063	40.863	4.880	-431.032	-394.899	51.569
500	51.651	64.356	44.466	9.945	-430.157	-385.966	40.322
600	53.530	73.940	48.599	15.204	-429.199	-377.216	32.840
700	55.343	82.329	52.831	20.648	-428.158	-368.633	27.508
800	57.099	89.834	56.995	26.271	-427.037	-360.206	23.519
900	58.803	96.658	61.029	32.067	-425.846	-351.922	20.425
1000	60.463	102.940	64.910	38.030	-424.586	-343.776	17.957
1100	62.086	108.779	68.635	44.158	-423.259	-335.758	15.944
1200	63.677	114.250	72.211	50.446	-421.879	-327.864	14.272
1300	65.243	119.408	75.645	56.892	-420.450	-320.087	12.861
1400	66.791	124.300	78.947	63.494	-418.983	-312.422	11.657
1500	68.325	128.961	82.127	70.250	-417.478	-304.863	10.616
1600	69.842	133.419	85.195	77.159	-415.936	-297.405	9.709
1700	71.351	137.698	88.158	84.218	-414.366	-290.045	8.912
1800	72.851	141.819	91.025	91.429	-412.774	-282.778	8.206
1900	74.339	145.798	93.804	98.788	-411.171	-275.599	7.577
2000	75.847	149.649	96.500	106.297	-409.565	-268.506	7.013
2063.000	76.747	152.015	98.160	111.104	------ CRYSTAL <--> LIQUID -----		
2100	77.298	153.385	99.121	113.954	-407.967	-261.492	6.504
2200	78.781	157.015	101.670	121.758	-429.269	-254.449	6.041
2300	80.250	160.549	104.154	129.710	-427.853	-246.534	5.599
2400	81.718	163.996	106.576	137.808	-426.302	-238.684	5.195
2500	83.183	167.361	108.940	146.053	-424.615	-230.900	4.824
2600	84.646	170.652	111.250	154.445	-422.792	-223.187	4.484

PREVIOUS: CURRENT: December 1973

Vanadium Oxide (VO) O_1V_1(cr)

VANADIUM OXIDE (VO) LIQUID M_r = 66.9409

$S°(298.15\ K)$ = [69.137] J K^{-1} mol^{-1}

T_{fus} = 2063 ± 10 K

$\Delta_f H°(298.15\ K)$ = [-370.901] kJ mol^{-1}

$\Delta_{fus}H°$ = [54.4] kJ mol^{-1}

<u>Enthalpy of Formation</u>

$\Delta_f H°$(VO, ℓ, 298.15 K) is calculated from that of the crystal by adding $\Delta_{fus}H°$ and the differences in enthalpy, H°(2063 K)-H°(298.15 K), between the crystal and liquid.

<u>Heat Capacity and Entropy</u>

$C_p°$ is estimated as 7.5 cal K^{-1} g-atom^{-1}. Below the assumed transition temperature at 1200 K, $C_p°$ is assumed to be the same as in the crystal. The entropy at 298.15 K is calculated in a manner analogous to that used for the enthalpy of formation.

<u>Fusion Data</u>

Alexander and Carlson (1) proposed a phase diagram for the V-VO system based on melting point determinations, differential thermal analysis, metallographic observations, and x-ray parametric measurements. These authors (1) reported that VO(cr) melts congruently at 2063 ± 10 K. We adopt this value for T_{fus}. In a phase diagram proposed by Stringer (2) a melting temperature of ~1998 K was suggested but the melting was not congruent.

VO(cr) has a sodium chloride structure and is stable over a wide range of composition, as is discussed, for example by Stringer (2). $\Delta_{fus}H°$ is estimated based on $\Delta_{fus}S°$ = 6.3 cal K^{-1} mol^{-1} as in NaCl (3). Since VO(cr) has a defect structure, a comparison with the $\Delta_{fus}S°$ value of NaCl is not entirely satisfactory. We make this comparison, however, pending further data.

<u>Vaporization Data</u>

There are no data available in the literature dealing with the vaporization of VO(ℓ). We anticipate substantial decomposition to occur during heat treatment of VO(ℓ). See VO(g) table for further information (3).

<u>References</u>

1. D. G. Alexander and O. N. Carlson, Met. Trans. 2, 2805 (1971).
2. J. Stringer, J. Less-Common Metals 8, 1 (1965).
3. JANAF Thermochemical Tables: NaCl(cr), 9-30-64; VO(g), 12-31-73.

Vanadium Oxide (VO) $O_1V_1(l)$

Enthalpy Reference Temperature = T_r = 298.15 K Standard State Pressure = p° = 0.1 MPa

T/K	$C_p°$ (J K⁻¹mol⁻¹)	S° (J K⁻¹mol⁻¹)	$-[G°-H°(T_r)]/T$ (J K⁻¹mol⁻¹)	$H°-H°(T_r)$ (kJ mol⁻¹)	$\Delta_f H°$ (kJ mol⁻¹)	$\Delta_f G°$ (kJ mol⁻¹)	Log K_f
0							
100							
200							
298.15	45.501	69.137	69.137	0.000	-370.901	-352.304	61.722
300	45.647	69.419	69.138	0.084	-370.890	-352.189	61.322
400	49.580	83.188	70.988	4.880	-370.144	-346.062	45.191
500	51.651	94.481	74.592	9.945	-369.269	-340.141	35.534
600	53.530	104.066	78.725	15.204	-368.311	-334.403	29.112
700	55.343	112.454	82.956	20.648	-367.270	-328.833	24.538
800	57.099	119.959	87.121	26.271	-366.149	-323.418	21.117
900	58.803	126.783	91.154	32.067	-364.958	-318.147	18.465
1000	60.463	133.065	95.035	38.030	-363.698	-313.013	16.350
1100	62.086	138.904	98.761	44.158	-362.371	-308.008	14.626
1200	63.676	144.375	102.336	50.446	-360.991	-303.127	13.195
1200.000	63.676	144.375	102.336	50.446	GLASS <--> LIQUID TRANSITION		
1200.000	62.760	144.375	102.336	50.446			
1300	62.760	149.399	105.766	56.722	-359.733	-298.357	11.988
1400	62.760	154.050	109.051	62.998	-358.591	-293.679	10.957
1500	62.760	158.380	112.197	69.274	-357.566	-289.079	10.067
1600	62.760	162.430	115.211	75.550	-356.656	-284.543	9.289
1700	62.760	166.235	118.102	81.826	-355.870	-280.061	8.605
1800	62.760	169.822	120.876	88.102	-355.212	-275.621	7.998
1900	62.760	173.215	123.542	94.378	-354.693	-271.214	7.456
2000	62.760	176.434	126.107	100.654	-354.319	-266.831	6.969
2063.000	62.760	178.381	127.674	104.608	------ CRYSTAL <--> LIQUID -----		
2100	62.760	179.497	128.577	106.930	-354.103	-262.463	6.528
2200	62.760	182.416	130.959	113.206	-376.933	-257.996	6.126
2300	62.760	185.206	133.257	119.482	-377.193	-252.584	5.736
2400	62.760	187.877	135.478	125.758	-377.464	-247.161	5.379
2500	62.760	190.439	137.625	132.034	-377.745	-241.726	5.051
2600	62.760	192.900	139.704	138.310	-378.038	-236.279	4.747
2700	62.760	195.269	141.719	144.586	-378.341	-230.821	4.466
2800	62.760	197.551	143.672	150.862	-378.654	-225.352	4.204
2900	62.760	199.754	145.568	157.138	-378.977	-219.871	3.960
3000	62.760	201.881	147.410	163.414	-379.310	-214.379	3.733
3100	62.760	203.939	149.201	169.690	-379.652	-208.875	3.520
3200	62.760	205.932	150.942	175.966	-380.004	-203.361	3.320
3300	62.760	207.863	152.638	182.242	-380.364	-197.836	3.131
3400	62.760	209.737	154.290	188.518	-380.732	-192.299	2.954
3500	62.760	211.556	155.900	194.794	-381.108	-186.751	2.787

PREVIOUS: CURRENT: December 1973

VANADIUM OXIDE (VO) CRYSTAL–LIQUID M_r = 66.9409

0 to 2063 K crystal
above 2063 K liquid

Refer to the individual tables for details.

Vanadium Oxide (VO) O_1V_1(cr,l)

Enthalpy Reference Temperature = T_r = 298.15 K   Standard State Pressure = p° = 0.1 MPa

T/K	C_p° (J K⁻¹mol⁻¹)	S° (J K⁻¹mol⁻¹)	−[G°−H°(T_r)]/T (J K⁻¹mol⁻¹)	H°−H°(T_r) (kJ mol⁻¹)	$\Delta_f H^\circ$ (kJ mol⁻¹)	$\Delta_f G^\circ$ (kJ mol⁻¹)	Log K_f
0	0.000	0.000	INFINITE	-6.908	-429.715	-429.715	INFINITE
100	13.895	6.730	71.088	-6.436	-431.184	-422.473	220.677
200	34.100	23.035	42.859	-3.965	-432.005	-413.352	107.956
298.15	45.501	39.011	39.011	0.000	-431.789	-404.210	70.816
300	45.647	39.293	39.012	0.084	-431.778	-404.039	70.350
400	49.580	53.063	40.863	4.880	-431.032	-394.899	51.569
500	51.651	64.356	44.466	9.945	-430.157	-385.966	40.322
600	53.530	73.940	48.599	15.204	-429.199	-377.216	32.840
700	55.343	82.329	52.831	20.648	-428.158	-368.633	27.508
800	57.099	89.834	56.995	26.271	-427.037	-360.206	23.519
900	58.803	96.658	61.029	32.067	-425.846	-351.922	20.425
1000	60.463	102.940	64.910	38.030	-424.586	-343.776	17.957
1100	62.086	108.779	68.635	44.158	-423.259	-335.758	15.944
1200	63.677	114.250	72.211	50.446	-421.879	-327.864	14.272
1300	65.243	119.408	75.645	56.892	-420.450	-320.087	12.861
1400	66.791	124.300	78.947	63.494	-418.983	-312.422	11.657
1500	68.325	128.961	82.127	70.250	-417.478	-304.863	10.616
1600	69.842	133.419	85.195	77.159	-415.936	-297.405	9.709
1700	71.351	137.698	88.158	84.218	-414.366	-290.045	8.912
1800	72.851	141.819	91.025	91.429	-412.774	-282.778	8.206
1900	74.339	145.798	93.804	98.788	-411.171	-275.599	7.577
2000	75.847	149.649	96.500	106.297	-409.565	-268.506	7.013
2063.000	76.747	152.015	98.160	111.104	CRYSTAL <--> LIQUID	TRANSITION	
2063.000	62.760	178.381	98.160	165.496			
2100	62.760	179.497	99.583	167.818	-354.103	-262.463	6.528
2200	62.760	182.416	103.282	174.094	-376.933	-257.996	6.126
2300	62.760	185.206	106.784	180.370	-377.193	-252.584	5.736
2400	62.760	187.877	110.108	186.646	-377.464	-247.161	5.379
2500	62.760	190.439	113.270	192.922	-377.745	-241.726	5.051
2600	62.760	192.900	116.286	199.198	-378.038	-236.279	4.747
2700	62.760	195.269	119.167	205.474	-378.341	-230.821	4.466
2800	62.760	197.551	121.926	211.750	-378.654	-225.352	4.204
2900	62.760	199.754	124.572	218.026	-378.977	-219.871	3.960
3000	62.760	201.881	127.114	224.302	-379.310	-214.379	3.733
3100	62.760	203.939	129.559	230.578	-379.652	-208.875	3.520
3200	62.760	205.932	131.915	236.854	-380.004	-203.361	3.320
3300	62.760	207.863	134.187	243.130	-380.364	-197.836	3.131
3400	62.760	209.737	136.382	249.406	-380.732	-192.299	2.954
3500	62.760	211.556	138.504	255.682	-381.108	-186.751	2.787

PREVIOUS: CURRENT: December 1973

Vanadium Oxide (VO) O_1V_1(cr,l)

VANADIUM OXIDE (VO) IDEAL GAS M_r = 66.9409

$\Delta_fH°(0\ K)$ = 127.82 ± 20.9 kJ mol^{-1}

$S°(298.15\ K)$ = 230.901 ± 2.1 J K^{-1} mol^{-1} $\Delta_fH°(298.15\ K)$ = 127.61 ± 20.9 kJ mol^{-1}

Electronic Levels and Quantum Weights

State	ε_i, cm^{-1}	g_i
$X^4\Sigma^-$	0	4
$B^4\Pi$	12706.8	8
$C^4\Sigma^-$	17420.3	4
	23890	[4]

ω_e = 1012.36 cm^{-1} ω_ex_e = 5.26 cm^{-1} σ = 1

B_e = 0.54825 cm^{-1} α_e = 0.00352 cm^{-1} r_e = 1.589 Å

Enthalpy of Formation

Based on the adopted ground state vibrational constants, a linear Birge-Sponer extrapolation yields a dissociation energy of D_0^0 = 6.039 eV (139.3 kcal mol^{-1}) for VO(g). Applying a correction for the ionicity in the V-O bond (1), we recalculate D_0^0 = 6.495 eV (149.8 kcal mol^{-1}). This latter value leads to a $\Delta_fH°(298.15\ K)$ value of 31.6 kcal mol^{-1} using auxiliary data (2).

Berkowitz et al. (3) determined the composition and partial pressures of the vapor over VO(cr) by employing a mass spectrometer to analyze the vapors effusing from a tungsten Knudsen cell containing VO(cr). The vapors detected mass spectrometrically at 1945 K were reported to be V(g), VO(g), and $VO_2(g)$. The VO^+ intensity was determined in two sets of experiments using the temperature variation method. Berkowitz et al. (3) analyzed these results using 2nd law methods and reported $\Delta_rH°(T)$ = 125.2 kcal mol^{-1} (1690-1945 K) and 124.6 kcal mol^{-1} (1680-1950 K) for the reaction VO(cr) = VO(g). Assigning these enthalpies of reaction to an intermediate temperature of 1800 K, we calculate $\Delta_fH°$(VO, g, 298.15 K) = 30.1 and 30.4 kcal mol^{-1}, respectively, using auxiliary data (2). Berkowitz et al. (3) also used the absolute pressure method and determined the partial pressure of VO(g) at two temperatures, 1875 K and 1876 K. The two absolute pressure data points differ in value by ~25%. A 3rd law analysis of these two data points leads to $\Delta_fH°$(VO, g, 298.15 K) of 23.6 and 24.7 kcal mol^{-1}. We prefer the 2nd law results of Berkowtiz et al. (3) since each result is based on multiple data points (~6) which cover a 260 K range.

Coppens et al. (4) used a combination of mass spectrometric and Knudsen cell techniques to determine the partial pressures necessary to define the equilibrium in the system VO(g) + Ge(g) = GeO(g) + V(g). The authors analyzed eight data points (1927-2145 K) by the 3rd law method and reported $\Delta_rH°(0\ K)$ = 7.60 ± 0.38 kcal mol^{-1} which actually applies to the reverse reaction (14). The Gibbs energy functions used by Coppens et al. (4) for VO are very similar to those adopted here. Using auxiliary data (2, 5), we calculate $\Delta_fH°(298.15\ K)$ = 29.1 kcal mol^{-1} for VO(g).

An effusion-mass spectrometric investigation of the vapors in equilibrium with V_2O_3(cr) was conducted by Farber et al. (6). Their study, using an aluminum cell, showed that the vapors at 2270 K contained V(g), VO(g), and $VO_2(g)$ as well as Al(g), AlO(g), and $Al_2O(g)$. A 3rd law analysis of the data at 2270 K yields $\Delta_rH°(298.15\ K)$ = -26.5 kcal mol^{-1} for the reaction AlO(g) + V(g) = VO(g) + Al(g). Using auxiliary data (2) we calculate $\Delta_fH°(298.15\ K)$ = 38.7 kcal mol^{-1} for VO(g). Farber et al. (6), however, reported the use of a more recent determination of $\Delta_fH°(298.15\ K)$ for AlO(g) (7). If this latter value (+16.57 kcal mol^{-1}) is used, then the $\Delta_fH°(298.15\ K)$ value for VO(g) is 35.3 kcal mol^{-1}.

Frantseva and Semenov (8) studied the evaporation of V_2O_3(cr) by a mass spectrometer used in conjunction with a double effusion chamber. They reported a D_0^0 value for VO(g) which is dependent, however, on their D_0^0 value for $VO_2(g)$ (2). The mass spectrometric study by Shchukarev et al. (9) detected VO(g) above V_2O_4(cr) and V_2O_3(cr) but there was insufficient data reported to permit the calculation of a $\Delta_fH°(298.15\ K)$ value for VO(g).

We adopt $\Delta_fH°(298.15\ K)$ = 30.5 kcal mol^{-1} for VO(g). This value is a rounded average of the two 2nd law treatments from Berkowitz et al. (3) and the results from Coppens et al. (4). Each of these three sets of experiments involve multiple data points. Most of the remaining studies each involved only one data point. However, it should be noted that the $\Delta_fH°(298.15\ K)$ derived from the Hildenbrand correction to the linear Birge-Sponer extrapolation is within 1.1 kcal mol^{-1} of the adopted value. The adopted enthalpy of formation leads to a dissociation energy of D_0^0 = 150.9 kcal mol^{-1} or 6.542 eV.

Heat Capacity and Entropy

Much of the spectroscopic work involving VO(g) has been reviewed and referenced by Rosen (10) and Weltner (11). The adopted vibrational and rotational constants as well as the electronic levels are those tabulated by Rosen (10). The EPR work of Kasai (12) and the optical absorption study by Richards and Barrow (13) established that the ground state was $^4\Sigma^-$. As reported in Rosen (10) there is evidence from matrix studies indicating a transition at 23890 cm^{-1}. We assume this level has a quantum weight of 4. the full weight of 8 is assigned to be B $^4\Pi$ state, rather than estimating the splitting. The Gibbs energy functions used here are approximately 3 cal K^{-1} mol^{-1} less than those proposed by Brewer and Rosenblatt (14). The difference arises from the fact that Brewer and Rosenblatt approximated the VO electronic levels by using V^{++} electronic levels (which include many low lying levels) whereas we have used the observed VO electronic levels.

References

1. D. L. Hildenbrand and E. Murad, J. Chem. Phys. 51, 807 (1969).
2. JANAF Thermochemical Tables: V(g), 6-30-73; O(g), 6-30-62; Al(g), 12-31-65; AlO(g), 12-30-70; VO(cr) and $VO_2(g)$, 12-31-73.
3. J. Berkowitz, W. A. Chupka and M. G. Inghram, J. Chem. Phys. 27, 87 (1957).
4. P. Coppens, S. Smoes and J. Drowart, Trans. Faraday Soc. 63, 2140 (1967).
5. U. S. Nat. Bur. Stand. Tech. Note 270-3, 1968.
6. M. Farber, O. M. Uy and R. D. Srivastava, J. Chem. Phys. 56, 5312 (1972).
7. M. Farber, R. D. Srivastava and O. M. Uy, J. Chem. Soc., Faraday I 68, 249 (1972).
8. K. E. Frantseva and G. A. Semenov, High Temperature 7, 52 (1969).
9. S. A. Shchukarev, G. A. Semenov, and K. E. Frantseva, Russ. J. Inorg. Chem. 4, 1217 (1959).
10. B. Rosen, Ed., "Spectroscopic Data Relative to Diatomic Molecules," Pergamon Press, New York, 1970.
11. W. Weltner, Jr. in "Advances in High Temperature Chemistry," Vol. 2, L. Eyring, Ed., Academic Press, New York, pp. 85-105, 1967.
12. P. H. Kasai, J. Chem. Phys. 49, 4979 (1968).
13. D. Richards and R. F. Barrow, Nature 219, 1244 (1968).
14. L. Brewer and G. M. Rosenblatt, in "Advances in High Temperature Chemistry," Vol. 2, L. Eyring, Ed., Academic Press, New York, pp. 1-83, 1967.

Vanadium Oxide (VO) $O_1V_1(g)$

Enthalpy Reference Temperature = T_r = 298.15 K Standard State Pressure = p° = 0.1 MPa

T/K	$C_p°$ (J $K^{-1}mol^{-1}$)	S° (J $K^{-1}mol^{-1}$)	$-[G°-H°(T_r)]/T$ (J $K^{-1}mol^{-1}$)	$H°-H°(T_r)$ (kJ mol^{-1})	$\Delta_fH°$ (kJ mol^{-1})	$\Delta_fG°$ (kJ mol^{-1})	Log K_f
0	0.	0.	INFINITE	-8.773	127.820	127.820	INFINITE
100	29.106	198.710	257.359	-5.865	128.787	118.300	-61.794
200	29.433	218.937	233.663	-2.945	128.415	107.888	-28.178
250	29.991	225.560	231.402	-1.461	128.027	102.800	-21.479
298.15	30.697	230.901	230.901	0.	127.612	97.978	-17.165
300	30.726	231.091	230.902	0.057	127.596	97.795	-17.028
350	31.510	235.886	231.278	1.613	127.147	92.863	-13.859
400	32.258	240.143	232.125	3.207	126.696	87.996	-11.491
450	32.931	243.982	233.233	4.837	126.246	83.186	-9.656
500	33.520	247.483	234.486	6.499	125.798	78.425	-8.193
600	34.459	253.682	237.182	9.900	124.898	69.035	-6.010
700	35.147	259.048	239.931	13.382	123.977	59.797	-4.462
800	35.656	263.777	242.622	16.924	123.016	50.694	-3.310
900	36.040	267.999	245.211	20.509	121.997	41.714	-2.421
1000	36.337	271.813	247.684	24.129	120.914	32.851	-1.716
1100	36.571	275.287	250.038	27.775	119.758	24.100	-1.144
1200	36.761	278.478	252.276	31.442	118.517	15.458	-0.673
1300	36.919	281.427	254.407	35.126	117.184	6.923	-0.278
1400	37.053	284.168	256.436	38.825	115.748	-1.505	0.056
1500	37.171	286.728	258.371	42.536	114.208	-9.827	0.342
1600	37.279	289.131	260.219	46.258	112.565	-18.043	0.589
1700	37.382	291.394	261.987	49.992	110.808	-26.153	0.804
1800	37.484	293.533	263.681	53.735	108.933	-34.156	0.991
1900	37.588	295.563	265.306	57.488	106.930	-42.052	1.156
2000	37.699	297.494	266.867	61.253	104.792	-49.838	1.302
2100	37.819	299.336	268.370	65.029	102.508	-57.514	1.431
2200	37.950	301.098	269.818	68.817	77.190	-64.973	1.543
2300	38.093	302.788	271.215	72.619	74.457	-71.374	1.621
2400	38.251	304.413	272.564	76.436	71.727	-77.656	1.690
2500	38.424	305.978	273.870	80.270	69.003	-83.824	1.751
2600	38.612	307.488	275.134	84.121	66.286	-89.884	1.806
2700	38.815	308.949	276.359	87.992	63.578	-95.839	1.854
2800	39.033	310.365	277.549	91.885	60.881	-101.693	1.897
2900	39.266	311.738	278.704	95.800	58.197	-107.453	1.935
3000	39.512	313.074	279 828	99.738	55.527	-113.119	1.970
3100	39.771	314.374	280.921	103.702	52.872	-118.696	2.000
3200	40.041	315.640	281.986	107.693	50.236	-124.189	2.027
3300	40.321	316.877	283.025	111.711	47.618	-129.599	2.051
3400	40.609	318.085	284.038	115.757	45.020	-134.930	2.073
3500	40.905	319.266	285.028	119.833	42.443	-140.186	2.092
3600	41.206	320.423	285.995	123.939	39.889	-145.367	2.109
3700	41.512	321.556	286.941	128.075	-409.450	-149.278	2.107
3800	41.820	322.667	287.867	132.241	-410.271	-142.235	1.955
3900	42.130	323.757	288.773	136.439	-411.119	-135.170	1.810
4000	42.440	324.828	289.661	140.667	-411.995	-128.084	1.673
4100	42.749	325.880	290.532	144.927	-412.900	-120.975	1.541
4200	43.055	326.913	291.386	149.217	-413.834	-113.843	1.416
4300	43.358	327.930	292.224	153.537	-414.799	-106.689	1.296
4400	43.656	328.930	293.047	157.888	-415.795	-99.513	1.181
4500	43.948	329.915	293.855	162.268	-416.823	-92.313	1.072
4600	44.234	330.884	294.650	166.678	-417.886	-85.090	0.966
4700	44.514	331.838	295.431	171.115	-418.982	-77.844	0.865
4800	44.785	332.778	296.199	175.580	-420.115	-70.573	0.768
4900	45.048	333.704	296.955	180.072	-421.286	-63.279	0.675
5000	45.301	334.617	297.699	184.589	-422.495	-55.960	0.585
5100	45.546	335.516	298.432	189.132	-423.745	-48.617	0.498
5200	45.781	336.403	299.153	193.698	-425.036	-41.250	0.414
5300	46.006	337.277	299.865	198.288	-426.371	-33.856	0.334
5400	46.220	338.139	300.565	202.899	-427.752	-26.438	0.256
5500	46.424	338.989	301.256	207.531	-429.180	-18.993	0.180
5600	46.618	339.828	301.938	212.183	-430.658	-11.521	0.107
5700	46.802	340.654	302.610	216.855	-432.187	-4.024	0.037
5800	46.975	341.470	303.273	221.543	-433.769	3.502	-0.032
5900	47.137	342.274	303.927	226.249	-435.406	11.054	-0.098
6000	47.290	343.068	304.573	230.971	-437.102	18.636	-0.162

PREVIOUS December 1973 (1 atm) CURRENT: December 1973 (1 bar)

Vanadium Oxide (VO) $O_1V_1(g)$

TUNGSTEN OXIDE (WO) IDEAL GAS M_r = 199.8494

$S°(298.15\ K)$ = [248.198] $J\ K^{-1}\ mol^{-1}$

$\Delta_f H°(0\ K)$ = 425.65 ± 41.8 $kJ\ mol^{-1}$

$\Delta_f H°(298.15\ K)$ = 425.09 ± 41.8 $kJ\ mol^{-1}$

Electronic Levels and Quantum Weights

State	ϵ_i, cm^{-1}	g_i
[$^3\Sigma^-$]	[0]	[3]
	[14160]	[1]
	[17211]	[2]
	[19189]	[6]

ω_e = 1065 cm^{-1} $\omega_e x_e$ = [3.85] cm^{-1} σ = 1

B_e = [0.3498] cm^{-1} α_e = [0.0016] cm^{-1} r_e = [1.81] Å

Enthalpy of Formation

The enthalpy of formation, $\Delta_f H°$(WO, g, 298.15 K) = 101.6 kcal mol^{-1}, is derived from the enthalpy of reaction, $\Delta_r H°$(298.15 K) = -42.0 kcal mol^{-1} for WO(g) + W(cr) + O(g) with all JANAF auxiliary data. The value of $\Delta_r H°$(298.15 K) was calculated by the 3rd law method from the partial pressure data which were determined mass-spectrometrically in the temperature range from 2188 to 2475 K by DeMaria et al. (1). The drift in the 3rd law analysis was -20 ± 7 cal $K^{-1}\ mol^{-1}$.

Heat Capacity and Entropy

The fundamental vibrational frequency, the ground state configuration, and the probable electronic levels for tungsten monoxide (g) were obtained from the infrared and ultraviolet spectroscopic investigations by matrix isolation by Weltner and McLeod (2). The rotational constant B_e was calculated from the estimated bond distance of WO(g). The value x_e was calculated from the relation $x_e u^{1/2}$ = 0.014 which was given by Barrow and Caunt (3). The constant (0.014) was estimated based on ZrO(g) and VO(g). The value of α_e was calculated from the Morse potential function.

References

1. G. DeMaria, R. P. Burns, J. Drowart and M. G. Inghram, J. Chem. Phys. 32, 1373 (1960).
2. W. Weltner, Jr. and D. McLeod, Jr., J. Mol. Spectry. 17, 276 (1965).
3. R. F. Barrow and A. D. Caunt, Proc. Roy. Soc. (London) 219A, 120 (1953).

Tungsten Oxide (WO) $O_1W_1(g)$

Enthalpy Reference Temperature = T_r = 298.15 K Standard State Pressure = p° = 0.1 MPa

T/K	$C_p°$ (J K⁻¹mol⁻¹)	S°	$-[G°-H°(T_r)]/T$	$H°-H°(T_r)$ (kJ mol⁻¹)	$\Delta_f H°$	$\Delta_f G°$	Log K_f
0	0.	0.	INFINITE	-8.757	425.652	425.652	INFINITE
100	29.104	216.077	274.560	-5.848	426.456	414.475	-216.500
200	29.360	236.289	250.944	-2.931	425.910	402.656	-105.163
250	29.843	242.888	248.696	-1.452	425.504	396.888	-82.925
298.15	30.490	248.198	248.198	0.	425.094	391.414	-68.574
300	30.517	248.386	248.198	0.056	425.079	391.205	-68.115
350	31.261	253.146	248.573	1.601	424.661	385.593	-57.547
400	31.990	257.369	249.413	3.182	424.255	380.039	-49.628
450	32.659	261.176	250.512	4.799	423.861	374.536	-43.475
500	33.251	264.648	251.754	6.447	423.476	369.076	-38.557
600	34.210	270.800	254.429	9.822	422.714	358.268	-31.190
700	34.919	276.129	257.157	13.280	421.944	347.587	-25.937
800	35.446	280.828	259.828	16.800	421.150	337.019	-22.005
900	35.844	285.027	262.399	20.365	420.324	326.552	-18.953
1000	36.150	288.820	264.854	23.966	419.460	316.178	-16.515
1100	36.391	292.277	267.192	27.593	418.554	305.894	-14.526
1200	36.585	295.452	269.417	31.242	417.604	295.694	-12.871
1300	36.744	298.387	271.534	34.909	416.609	285.574	-11.475
1400	36.878	301.115	273.551	38.590	415.567	275.534	-10.280
1500	36.995	303.663	275.474	42.284	414.478	265.569	-9.248
1600	37.100	306.054	277.311	45.989	413.339	255.678	-8.347
1700	37.199	308.307	279.069	49.704	412.151	245.861	-7.554
1800	37.296	310.436	280.753	53.429	410.914	236.114	-6.852
1900	37.396	312.455	282.369	57.163	409.626	226.438	-6.225
2000	37.502	314.376	283.922	60.908	408.289	216.831	-5.663
2100	37.616	316.208	285.416	64.664	406.902	207.292	-5.156
2200	37.740	317.961	286.856	68.432	405.467	197.821	-4.697
2300	37.878	319.641	288.245	72.212	403.983	188.415	-4.279
2400	38.029	321.257	289.587	76.008	402.454	179.075	-3.897
2500	38.195	322.812	290.885	79.819	400.879	169.800	-3.548
2600	38.376	324.314	292.142	83.647	399.264	160.588	-3.226
2700	38.572	325.766	293.360	87.494	397.583	151.440	-2.930
2800	38.783	327.172	294.543	91.362	395.804	142.356	-2.656
2900	39.009	328.537	295.692	95.252	393.888	133.337	-2.402
3000	39.248	329.864	296.809	99.164	391.810	124.388	-2.166
3100	39.499	331.155	297.896	103.101	389.529	115.511	-1.946
3200	39.761	332.413	298.955	107.064	387.015	106.711	-1.742
3300	40.033	333.640	299.988	111.054	384.244	97.994	-1.551
3400	40.314	334.840	300.995	115.071	381.182	89.365	-1.373
3500	40.602	336.012	301.979	119.117	377.765	80.830	-1.206
3600	40.896	337.160	302.940	123.192	373.899	72.400	-1.050
3700	41.193	338.285	303.880	127.296	334.719	64.275	-0.907
3800	41.494	339.387	304.800	131.431	333.242	56.986	-0.783
3900	41.796	340.469	305.701	135.595	331.789	49.735	-0.666
4000	42.098	341.531	306.584	139.790	330.360	42.521	-0.555
4100	42.398	342.574	307.449	144.015	328.954	35.343	-0.450
4200	42.696	343.600	308.297	148.269	327.572	28.198	-0.351
4300	42.991	344.608	309.130	152.554	326.213	21.086	-0.256
4400	43.281	345.599	309.948	156.868	324.877	14.006	-0.166
4500	43.566	346.575	310.751	161.210	323.564	6.955	-0.081
4600	43.844	347.536	311.540	165.580	322.273	-0.067	0.001
4700	44.115	348.482	312.316	169.978	321.004	-7.060	0.078
4800	44.378	349.413	313.079	174.403	319.754	-14.027	0.153
4900	44.633	350.331	313.830	178.854	318.525	-20.968	0.224
5000	44.878	351.235	314.569	183.329	317.313	-27.884	0.291
5100	45.115	352.126	315.297	187.829	316.120	-34.776	0.356
5200	45.341	353.004	316.014	192.352	314.942	-41.645	0.418
5300	45.558	353.870	316.720	196.897	313.779	-48.491	0.478
5400	45.764	354.724	317.416	201.463	312.630	-55.316	0.535
5500	45.960	355.565	318.102	206.050	311.492	-62.119	0.590
5600	46.146	356.395	318.778	210.655	310.366	-68.901	0.643
5700	46.321	357.213	319.445	215.278	309.248	-75.664	0.693
5800	46.486	358.020	320.103	219.919	308.139	-82.407	0.742
5900	46.640	358.816	320.753	224.575	307.035	-89.132	0.789
6000	46.784	359.601	321.394	229.246	305.936	-95.837	0.834

PREVIOUS: September 1966 (1 atm) CURRENT: September 1966 (1 bar)

Tungsten Oxide (WO) $O_1W_1(g)$

ZIRCONIUM OXIDE (ZrO) IDEAL GAS M_r = 107.2194

$\Delta_f H°(0\ K)$ = 59.63 ± 50.2 kJ mol^{-1}

$S°(298.15\ K)$ = 227.627 ± 8.4 J K^{-1} mol^{-1} $\Delta_f H°(298.15\ K)$ = 58.58 ± 50.2 kJ mol^{-1}

Electronic Levels and Quantum Weights

ε_i, cm^{-1}	g_i	ε_i, cm^{-1}	g_i
0	1	[18000]	2
[4000]	6	[22000]	6
[6000]	2	[25500]	2
15500	2	27300	1
[15700]	6	[28500]	6
[17500]	6		

ω_e = 978 cm^{-1} $\omega_e x_e$ = 4 cm^{-1} σ = 1

B_e = 0.4142 cm^{-1} α_e = 0.0023 cm^{-1} r_e = 1.711 Å

Enthalpy of Formation

$\Delta_f H°(298.15\ K)$ = 14.0 ± 12 kcal mol^{-1} is calculated from $\Delta_r H°(298.15\ K)$ = 290.342 ± 12.8 kcal mol^{-1} for the reaction Zr(cr) + ZrO_2(cr) = 2 ZrO(g). The value of $\Delta_r H°(298.15\ K)$ is obtained by the 3rd law analysis from the mass spectrometric studies of Chupka et al. (1). They observed the variation of the ZrO^+ ion intensity over the system Zr + ZrO_2 and reported equilibrium constants for the reaction Zr + ZrO_2 → 2 ZrO in the range 2124-2322 K. The 3rd law drift is -58 ± 28 cal K^{-1} mol^{-1}.

Heat Capacity and Entropy

The ground state of ZrO(g) was assumed to be $^1\Sigma^+$ as determined from the matrix isolation spectra of Weltner et al. (2). Previously this had been described as the lower state in the A band system reported by Uhler and Akerlind (3).

All molecular constants were obtained from Uhler and Akerlind (3) except ω_e = 978 cm^{-1} and $\omega_e x_e$ = 4 cm^{-1} which were calculated from ω and α_e = 0.0023 cm^{-1} using the equations

$$\alpha_e = \frac{6\ (\omega_e x_e \cdot B_e^3)^{1/2}}{\omega_e} - \frac{6\ B_e^2}{\omega_e} \quad \text{and} \quad \omega = \omega_e - 2\ \omega_e x_e = 969.76\ cm^{-1}$$

The higher electronic levels were obtained or estimated from the correlation diagram given by Weltner and McLeon (2).

References

1. W. A. Chupka, J. Berkowitz and M. G. Inghram, J. Chem. Phys. 26, 1207 (1957).
2. W. Weltner, Jr., and D. McLeod, Jr., J. Phys. Chem. 69, 3488 (1965).
3. V. Uhler and L. Akerlind, Arkiv Fysik 10, 431 (1955).

Zirconium Oxide (ZrO) $O_1Zr_1(g)$

Enthalpy Reference Temperature = T_r = 298.15 K Standard State Pressure = p° = 0.1 MPa

T/K	$C_p°$ (J $K^{-1}mol^{-1}$)	S° (J $K^{-1}mol^{-1}$)	$-[G°-H°(T_r)]/T$ (J $K^{-1}mol^{-1}$)	$H°-H°(T_r)$ (kJ mol^{-1})	$\Delta_f H°$ (kJ mol^{-1})	$\Delta_f G°$ (kJ mol^{-1})	Log K_f
0	0.	0.	INFINITE	-8.786	59.629	59.629	INFINITE
100	29.106	195.385	254.156	-5.877	60.199	50.732	-26.500
200	29.489	215.622	230.400	-2.956	59.472	41.506	-10.840
250	30.099	222.263	228.131	-1.467	59.016	37.067	-7.745
298.15	30.845	227.627	227.627	0.	58.576	32.880	-5.760
300	30.875	227.818	227.628	0.057	58.559	32.721	-5.697
350	31.680	232.639	228.007	1.621	58.114	28.450	-4.246
400	32.439	236.919	228.858	3.224	57.681	24.241	-3.166
450	33.124	240.780	229.972	4.864	57.258	20.087	-2.332
500	33.745	244.303	231.231	6.536	56.838	15.979	-1.669
600	34.906	250.558	233.944	9.968	56.000	7.886	-0.687
700	36.165	256.031	236.716	13.520	55.158	-0.066	0.005
800	37.695	260.956	239.443	17.211	54.330	-7.899	0.516
900	39.539	265.500	242.089	21.070	53.539	-15.629	0.907
1000	41.624	269.772	244.645	25.126	52.809	-23.275	1.216
1100	43.811	273.841	247.116	29.398	52.155	-30.851	1.465
1200	45.942	277.746	249.507	33.887	47.867	-38.149	1.661
1300	47.875	281.501	251.824	38.580	47.900	-45.317	1.821
1400	49.510	285.111	254.074	43.452	48.055	-52.493	1.959
1500	50.791	288.573	256.259	48.470	48.289	-59.683	2.078
1600	51.705	291.882	258.383	53.598	48.556	-66.890	2.184
1700	52.273	295.035	260.447	58.799	48.811	-74.113	2.277
1800	52.538	298.032	262.453	64.042	49.011	-81.350	2.361
1900	52.552	300.874	264.401	69.299	49.116	-88.596	2.436
2000	52.372	303.565	266.292	74.546	49.096	-95.844	2.503
2100	52.049	306.113	268.128	79.768	48.922	-103.087	2.564
2200	51.629	308.525	269.910	84.953	27.249	-109.573	2.602
2300	51.152	310.810	271.639	90.092	26.289	-115.771	2.629
2400	50.648	312.976	273.317	95.182	25.269	-121.926	2.654
2500	50.140	315.034	274.945	100.222	24.187	-128.037	2.675
2600	49.646	316.990	276.525	105.211	23.044	-134.103	2.694
2700	49.179	318.855	278.058	110.152	21.842	-140.125	2.711
2800	48.747	320.636	279.547	115.048	20.585	-146.101	2.726
2900	48.354	322.339	280.994	119.902	19.277	-152.031	2.738
3000	48.004	323.973	282.399	124.720	17.922	-157.915	2.750
3100	47.697	325.542	283.766	129.505	16.525	-163.753	2.759
3200	47.433	327.052	285.095	134.261	15.090	-169.546	2.768
3300	47.209	328.508	286.389	138.993	13.622	-175.294	2.775
3400	47.025	329.914	287.648	143.704	12.126	-180.996	2.781
3500	46.876	331.275	288.875	148.399	10 605	-186.654	2.786
3600	46.762	332.594	290.072	153.080	9.063	-192.268	2.790
3700	46.677	333.874	291.238	157.752	7.504	-197.839	2.793
3800	46.621	335.118	292.377	162.417	5.930	-203.367	2.795
3900	46.589	336.328	293.488	167.077	4.345	-208.855	2.797
4000	46.579	337.508	294.574	171.735	2.752	-214.301	2.798
4100	46.588	338.658	295.635	176.394	1.152	-219.708	2.799
4200	46.614	339.781	296.673	181.054	-0.453	-225.075	2.799
4300	46.654	340.878	297.688	185.717	-2.061	-230.404	2.799
4400	46.707	341.951	298.682	190.385	-3.670	-235.696	2.798
4500	46.770	343.002	299.655	195.059	-5.279	-240.951	2.797
4600	46.841	344.030	300.609	199.739	-6.887	-246.171	2.795
4700	46.918	345.039	301.544	204.427	-8.495	-251.355	2.794
4800	47.001	346.027	302.460	209.123	-571.037	-244.888	2.665
4900	47.088	346.997	303.359	213.827	-572.292	-238.080	2.538
5000	47.176	347.950	304.241	218.540	-573.562	-231.247	2.416
5100	47.267	348.885	305.108	223.263	-574.846	-224.388	2.298
5200	47.357	349.803	305.958	227.994	-576.141	-217.504	2.185
5300	47.447	350.706	306.794	232.734	-577.448	-210.594	2.076
5400	47.536	351.594	307.616	237.483	-578.765	-203.660	1.970
5500	47.622	352.467	308.423	242.241	-580.092	-196.702	1.868
5600	47.707	353.326	309.217	247.008	-581.428	-189.719	1.770
5700	47.788	354.171	309.999	251.782	-582.772	-182.712	1.674
5800	47.865	355.003	310.767	256.565	-584.124	-175.681	1.582
5900	47.939	355.822	311.524	261.355	-585.483	-168.628	1.493
6000	48.008	356.628	312.269	266.153	-586.851	-161.551	1.406

PREVIOUS: December 1965 (1 atm) CURRENT: December 1965 (1 bar)

Zirconium Oxide (ZrO) $O_1Zr_1(g)$

OXYGEN (O_2) (REFERENCE STATE - IDEAL GAS) M_r = 31.9988

0 to 6000 K Ideal Diatomic Gas

S°(298.15 K) = 205.147 ± 0.035

$\Delta_fH°(0\ K) = 0\ kJ\ mol^{-1}$
$\Delta_fH°(298.15\ K) = 0\ kJ\ mol^{-1}$

Electronic and Molecular constants (σ = 2)

State	g_i	T_o, cm^{-1}	ω_e, cm^{-1}	Vibrational Anharmonicity, cm^{-1}	B_e, cm^{-1}	α_i, cm^{-1}	D_e, cm^{-1}	β_i, cm^{-1}
$X\ ^3\Sigma_g^-$	3	0	1580.1932	ω_ex_e 11.980804 ω_ey_e 0.047474736 ω_ez_e -1.272748×10^{-3}	1.445622	0.01593268 6.406456×10^{-5} -2.846158×10^{-6}		
$a\ ^1\Delta_g$	2	7882.39	[1509.3]	ω_ex_e [12.9]	1.4263	0.0171	$[4.97\times10^{-6}]$	
$b\ ^1\Delta_g^+$	1	13120.9085	1432.6661	ω_ex_e 13.9336 ω_ey_e -0.0143	1.4004796	0.018169303 -4.294192×10^{-5}	5.356×10^{-6}	0.077×10^{-6}
$c^1\Sigma_u^-$	1	32664.1	794.29	ω_ex_e 12.736 ω_ey_e -0.2444 ω_ez_e 0.00055	0.9155	0.01391 -0.000740	$[10.5\times10^{-6}]$	
$c^3\Delta_{ui}$	6	34319	[750]	ω_ex_e [14]	[0.91]	[0.014]		
$A^3\Sigma_u^+$	3	35007.15	799.08	ω_ex_e 12.16 ω_ey_e -0.550	0.91053	0.01416 -0.00097	4.79×10^{-6}	-0.30×10^{-6} 0.10×10^{-6}
$B^3\Sigma_u^-$	3	49358.15	709.05770	ω_ex_e 10.614080 ω_ey_e -0.059212435 ω_ez_e -0.023974994	0.818975	0.0119225 -6.30472×10^{-4}		

Enthalpy of Formation

Zero by Definition.

Heat Capacity and Entropy

The thermal functions are calculated using the program of McBride and Gordon (1). The contributions of vibrational anharmonicity, rotation-vibration interaction, and centrifugal stretching are calculated via the procedures given by Pennington and Kobe (3). In this procedure the energy levels are calculated as follows: $\varepsilon = T_o + G(v) - G(o) + B_v J(J+1) - D_v J^2(J+1)^2$ where $G(v) = \omega_e(v+1/2) - \omega_ex_e(v+1/2)^2 + \omega_ey_e(v+1/2)^3 + \omega_ey_e(v+1/2)^4$ and $B_v = B_e - \alpha_1(v+1/2) + \alpha_2(v+1/2)^2 + \alpha_3(v+1/2)^3$ and $D_v = D_e + \beta_i(v+1/2)$. Where a value of D_e is not available it is calculated to be $D_e = 4B_e^3/\omega_e^2$.

Molecular constant data in the table above is from Krupenie (2) with the exception of B_e and α_i for the $C\ ^3\Delta_{ui}$ state which we have estimated by analogy with neighboring states of the same electron configuration. Electronic levels above 50,000 cm^{-1} given by Krupenie are not included since they make negligible contributions to the thermodynamic properties at 6000 K.

The value of S°(298.15 K) agrees exactly with the CODATA recommended value (4). It also agrees with the values of McBride et al. (5) and Gurvich et al. (6) within the stated uncertainty. The uncertainty in S°(298.15 K) is due to uncertainties in fundamental constants (particularly the gas constant), the atomic weight, and the ground state molecular constants. At high temperatures, further uncertainties arise from excited state molecular constants and approximate procedures used to account for higher vibrational-rotational levels of all states. Despite these increased uncertainties, our values of S° are within 0.016 (6000 K) and 0.035 (5000 K) cal K^{-1} mol^{-1} of those calculated by superior methods (5, 6) using older molecular constants.

References

1. B. J. McBride and S. Gordon, NASA TN D-4097, 1967.
2. P. H. Krupenie, J. Phys. Chem. Ref. Data 1, 423 (1972).
3. R. E. Pennington and K. A. Kobe, J. Chem. Phys. 22, 1442 (1954).
4. J. D. Cox, chairman, CODATA Task Group on Key Values for Thermodynamics, J. Chem. Thermodynamics 4, 331 (1972).
5. B. J. McBride, S. Heimel, J. G. Ehlers, and S. Gordon, NASA SP-3001, 1963.
6. L. V. Gurvich, G. A. Khachkuruzov et al., "Thermodynamic Properties of Individual substances. Volume I. Calculation of the Thermodynamic Properties," English Translation, Reports AD-659660, AD-659659, AD-659679 (avail. NTIS), 1967, Volume II, Russian ed., Moscow, 1962.

Oxygen (O_2) O_2(ref)

Enthalpy Reference Temperature = T_r = 298.15 K Standard State Pressure = p° = 0.1 MPa

T/K	$C_p°$ (J K⁻¹mol⁻¹)	S° (J K⁻¹mol⁻¹)	$-[G°-H°(T_r)]/T$ (J K⁻¹mol⁻¹)	$H°-H°(T_r)$ (kJ mol⁻¹)	$\Delta_fH°$ (kJ mol⁻¹)	$\Delta_fG°$ (kJ mol⁻¹)	Log K_f
0	0.	0.	INFINITE	-8.683	0.	0.	0.
100	29.106	173.307	231.094	-5.779	0.	0.	0.
200	29.126	193.485	207.823	-2.868	0.	0.	0.
250	29.201	199.990	205.630	-1.410	0.	0.	0.
298.15	29.376	205.147	205.147	0.	0.	0.	0.
300	29.385	205.329	205.148	0.054	0.	0.	0.
350	29.694	209.880	205.506	1.531	0.	0.	0.
400	30.106	213.871	206.308	3.025	0.	0.	0.
450	30.584	217.445	207.350	4.543	0.	0.	0.
500	31.091	220.693	208.524	6.084	0.	0.	0.
600	32.090	226.451	211.044	9.244	0.	0.	0.
700	32.981	231.466	213.611	12.499	0.	0.	0.
800	33.733	235.921	216.126	15.835	0.	0.	0.
900	34.355	239.931	218.552	19.241	0.	0.	0.
1000	34.870	243.578	220.875	22.703	0.	0.	0.
1100	35.300	246.922	223.093	26.212	0.	0.	0.
1200	35.667	250.010	225.209	29.761	0.	0.	0.
1300	35.988	252.878	227.229	33.344	0.	0.	0.
1400	36.277	255.556	229.158	36.957	0.	0.	0.
1500	36.544	258.068	231.002	40.599	0.	0.	0.
1600	36.796	260.434	232.768	44.266	0.	0.	0.
1700	37.040	262.672	234.462	47.958	0.	0.	0.
1800	37.277	264.796	236.089	51.673	0.	0.	0.
1900	37.510	266.818	237.653	55.413	0.	0.	0.
2000	37.741	268.748	239.160	59.175	0.	0.	0.
2100	37.969	270.595	240.613	62.961	0.	0.	0.
2200	38.195	272.366	242.017	66.769	0.	0.	0.
2300	38.419	274.069	243.374	70.600	0.	0.	0.
2400	38.639	275.709	244.687	74.453	0.	0.	0.
2500	38.856	277.290	245.959	78.328	0.	0.	0.
2600	39.068	278.819	247.194	82.224	0.	0.	0.
2700	39.276	280.297	248.393	86.141	0.	0.	0.
2800	39.478	281.729	249.558	90.079	0.	0.	0.
2900	39.674	283.118	250.691	94.036	0.	0.	0.
3000	39.864	284.466	251.795	98.013	0.	0.	0.
3100	40.048	285.776	252.870	102.009	0.	0.	0.
3200	40.225	287.050	253.918	106.023	0.	0.	0.
3300	40.395	288.291	254.941	110.054	0.	0.	0.
3400	40.559	289.499	255.940	114.102	0.	0.	0.
3500	40.716	290.677	256.916	118.165	0.	0.	0.
3600	40.868	291.826	257.870	122.245	0.	0.	0.
3700	41.013	292.948	258.802	126.339	0.	0.	0.
3800	41.154	294.044	259.716	130.447	0.	0.	0.
3900	41.289	295.115	260.610	134.569	0.	0.	0.
4000	41.421	296.162	261.485	138.705	0.	0.	0.
4100	41.549	297.186	262.344	142.854	0.	0.	0.
4200	41.674	298.189	263.185	147.015	0.	0.	0.
4300	41.798	299.171	264.011	151.188	0.	0.	0.
4400	41.920	300.133	264.821	155.374	0.	0.	0.
4500	42.042	301.076	265.616	159.572	0.	0.	0.
4600	42.164	302.002	266.397	163.783	0.	0.	0.
4700	42.287	302.910	267.164	168.005	0.	0.	0.
4800	42.413	303.801	267.918	172.240	0.	0.	0.
4900	42.542	304.677	268.660	176.488	0.	0.	0.
5000	42.675	305.538	269.389	180.749	0.	0.	0.
5100	42.813	306.385	270.106	185.023	0.	0.	0.
5200	42.956	307.217	270.811	189.311	0.	0.	0.
5300	43.105	308.037	271.506	193.614	0.	0.	0.
5400	43.262	308.844	272.190	197.933	0.	0.	0.
5500	43.426	309.639	272.864	202.267	0.	0.	0.
5600	43.599	310.424	273.527	206.618	0.	0.	0.
5700	43.781	311.197	274.181	210.987	0.	0.	0.
5800	43.973	311.960	274.826	215.375	0.	0.	0.
5900	44.175	312.713	275.462	219.782	0.	0.	0.
6000	44.387	313.457	276.089	224.210	0.	0.	0.

PREVIOUS: March 1977 (1 atm) CURRENT: March 1977 (1 bar)

Oxygen (O_2) O_2(ref)

OXYGEN, ION (O_2^+) IDEAL GAS M_r = 31.99825

$\Delta_f H°(0\ K) = 1164.7 \pm 0.8\ kJ\ mol^{-1}$

$S°(298.15\ K) = 206.2 \pm 0.8\ J\ K\ mol^{-1}$ $\Delta_f H°(298.15\ K) = [1170.888]\ kJ\ mol^{-1}$

Electronic and Molecular Constants (σ = 2)

State	g_i	T_o, cm^{-1}	ω_e, cm^{-1}	$\omega_e x_e$, cm^{-1}	B_e, cm^{-1}	α_e, cm^{-1}	r_e, Å
$X^2\Pi_g$	4	0	1905.13	16.2818	1.68912	0.0195769	1.116384
$a^4\Pi_{ui}$	8	32524	1035.534	10.32194	1.104320	0.01545646	1.31816042
$A^2\Pi_u$	4	40070	898.17	13.568	1.06297	0.02058	1.408220
$b^4\Sigma_g^-$	4	49191	1196.913	17.13456	1.287297	0.02206747	1.2796516

Enthalpy of Formation

The adopted value for the enthalpy of formation of $O_2^+(g)$, $\Delta_f H°(0\ K) = 278.37 \pm 0.2\ kcal\ mol^{-1}$ is that recommended in a critical review on the energetics of gaseous ions by Rosenstock et al. (1). This value is derived from ionization potentials determined in the photoelectron spectrum of O_2 by Edqvist et al. (2), the IP values are 12.071 ± 0.005 ev for the $X^2\Pi_{3/2g}$ state and 12.095 ± 0.005 ev for the $X^2\Pi_{1/2g}$ state.

$\Delta_f H°(O_2^+, g, 298.15\ K)$ is calculated from $\Delta_f H°(O_2, g, 0\ K)$ by using IP(O_2) with JANAF enthalpies, H°(0 K) − H°(298.15 K), for O_2(g), O_2^+(g), and e^-(g). $\Delta_f H°(O_2 \rightarrow O_2^+ + e^-, 298.15\ K)$ differs from a room temperature threshold energy due to inclusion of these enthalpies and to threshold effects discussed by Rosenstock et al. (1). $\Delta_f H°(298.15\ K)$ should be changed by $-6.197\ kJ\ mol^{-1}$ if it is to be used in the ion convention that excludes the enthalpy of the electron.

Heat Capacity and Entropy

The electronic and molecular constants are from the critical review of Krupenie (3). Electronic levels above 50000 cm^{-1} given by Krupenie are not included since they make negligible contributions to the thermodynamic properties at 6000 K. The thermodynamic functions are calculated using first-order anharmonic corrections to Q_r^i and Q_v^i in the partition function $Q = Q_t \Sigma Q_r^i Q_v^i g_i \exp(-c_2 \epsilon_i/T)$. Use of the program of McBride and Gordon (4) yields the same results.

The ground state is taken to be $^2\Pi_{3/2g}$ with a separation of the ground state doublet levels of ~200 cm^{-1}. The hypothetical $^2\Pi$ state without spin-orbit splitting would be 100 cm^{-1} higher (3). Such a shift of the electronic energy levels does not significantly alter the thermal functions. The main source of error in the S°(298.15 K) value lies in the neglect of a more exact treatment of the splitting of the $X^2\Pi$ ground state.

References

1. H. M. Rosenstock, H. Draxl, B. W. Steiner, and J. T. Herron, J. Phys. Chem. Ref. Data 6, Supp. 1 (1977).
2. O. Edqvist, E. Lindholm, L. E. Selin, and L. Asbrink, Physica Scripta 1, 25 (1970).
3. P. H. Krupenie, J. Phys. Chem. Ref. Data 1, 423 (1977).
4. B. J. McBride and A. Gordon, NASA TM D-4097, 1967.

Oxygen, Ion (O_2^+) O_2^+(g)

Enthalpy Reference Temperature = T_r = 298.15 K Standard State Pressure = p° = 0.1 MPa

T/K	$C_p°$ (J K⁻¹mol⁻¹)	S° (J K⁻¹mol⁻¹)	−[G°−H°(T_r)]/T (J K⁻¹mol⁻¹)	H°−H°(T_r) (kJ mol⁻¹)	$\Delta_f H°$ (kJ mol⁻¹)	$\Delta_f G°$ (kJ mol⁻¹)	Log K_f
0	0.	0.	INFINITE	-8.674	1164.700		
100	29.105	174.406	232.113	-5.771			
200	29.111	194.582	208.881	-2.860			
250	29.132	201.079	206.695	-1.404			
298.15	29.195	206.215	206.215	0.	1170.888	1164.315	-203.983
300	29.199	206.396	206.216	0.054	1170.926	1164.274	-202.718
350	29.342	210.906	206.572	1.517	1171.952	1163.084	-173.581
400	29.573	214.839	207.364	2.990	1172.970	1161.748	-151.709
450	29.882	218.339	208.393	4.476	1173.978	1160.285	-134.682
500	30.250	221.506	209.548	5.979	1174.978	1158.710	-121.050
600	31.078	227.093	212.019	9.045	1176.963	1155.269	-100.575
700	31.916	231.947	214.526	12.195	1178.937	1151.497	-85.926
800	32.687	236.260	216.978	15.426	1180.910	1147.442	-74.920
900	33.364	240.150	219.340	18.729	1182.886	1143.140	-66.346
1000	33.945	243.697	221.601	22.095	1184.869	1138.618	-59.475
1100	34.440	246.956	223.760	25.515	1186.859	1133.897	-53.844
1200	34.859	249.971	225.820	28.981	1188.854	1128.994	-49.144
1300	35.216	252.776	227.787	32.485	1190.854	1123.924	-45.160
1400	35.521	255.397	229.667	36.022	1192.856	1118.701	-41.739
1500	35.784	257.857	231.465	39.588	1194.859	1113.334	-38.770
1600	36.011	260.174	233.187	43.178	1196.861	1107.834	-36.167
1700	36.210	262.363	234.840	46.789	1198.859	1102.208	-33.867
1800	36.384	264.438	236.427	50.419	1200.851	1096.465	-31.819
1900	36.538	266.409	237.954	54.065	1202.837	1090.612	-29.983
2000	36.675	268.287	239.424	57.726	1204.813	1084.654	-28.328
2100	36.798	270.079	240.841	61.399	1206.780	1078.598	-26.829
2200	36.908	271.793	242.209	65.085	1208.736	1072.449	-25.463
2300	37.009	273.436	243.532	68.781	1210.680	1066.210	-24.214
2400	37.101	275.013	244.811	72.486	1212.611	1059.887	-23.068
2500	37.185	276.530	246.049	76.201	1214.529	1053.484	-22.011
2600	37.263	277.990	247.250	79.923	1216.434	1047.004	-21.035
2700	37.336	279.397	248.415	83.653	1218.325	1040.452	-20.129
2800	37.403	280.756	249.546	87.390	1220.203	1033.830	-19.286
2900	37.466	282.070	250.645	91.134	1222.068	1027.140	-18.501
3000	37.526	283.341	251.713	94.883	1223.919	1020.387	-17.767
3100	37.583	284.573	252.754	98.639	1225.757	1013.573	-17.079
3200	37.637	285.767	253.767	102.400	1227.583	1006.699	-16.433
3300	37.689	286.926	254.754	106.166	1229.397	999.768	-15.825
3400	37.740	288.051	255.717	109.938	1231.200	992.783	-15.252
3500	37.790	289.146	256.656	113.714	1232.991	985.744	-14.711
3600	37.838	290.211	257.574	117.496	1234.772	978.655	-14.200
3700	37.887	291.249	258.470	121.282	1236.542	971.515	-13.715
3800	37.936	292.260	259.346	125.073	1238.304	964.330	-13.256
3900	37.985	293.246	260.203	128.869	1240.056	957.097	-12.819
4000	38.035	294.208	261.041	132.670	1241.800	949.819	-12.403
4100	38.088	295.148	261.861	136.476	1243.536	942.498	-12.008
4200	38.142	296.067	262.665	140.288	1245.265	935.135	-11.630
4300	38.199	296.965	263.452	144.105	1246.987	927.731	-11.270
4400	38.258	297.844	264.224	147.927	1248.702	920.286	-10.925
4500	38.322	298.704	264.980	151.756	1250.412	912.802	-10.596
4600	38.389	299.547	265.723	155.592	1252.116	905.281	-10.280
4700	38.462	300.373	266.451	159.434	1253.815	897.723	-9.977
4800	38.539	301.184	267.166	163.284	1255.508	890.129	-9.687
4900	38.622	301.979	267.869	167.142	1257.197	882.500	-9.408
5000	38.711	302.761	268.559	171.009	1258.881	874.836	-9.139
5100	38.807	303.528	269.237	174.885	1260.561	867.139	-8.881
5200	38.910	304.283	269.904	178.771	1262.237	859.407	-8.633
5300	39.020	305.025	270.559	182.667	1263.910	851.645	-8.393
5400	39.138	305.755	271.204	186.575	1265.578	843.850	-8.163
5500	39.265	306.475	271.839	190.495	1267.242	836.025	-7.940
5600	39.400	307.183	272.464	194.428	1268.903	828.170	-7.725
5700	39.545	307.882	273.079	198.375	1270.559	820.285	-7.517
5800	39.699	308.571	273.685	202.338	1272.213	812.372	-7.316
5900	39.863	309.251	274.282	206.316	1273.862	804.428	-7.122
6000	40.036	309.922	274.871	210.310	1275.507	796.458	-6.934

PREVIOUS: September 1977 (1 atm) CURRENT: September 1977 (1 bar)

Oxygen, Ion (O_2^+) O_2^+(g)

OXYGEN, ION (O_2^-) IDEAL GAS M_r = 31.99935

$S°(298.15\ K) = 209.591 \pm 0.8\ J\ K^{-1}\ mol^{-1}$

$\Delta_f H°(0\ K) = -42.46 \pm 0.75\ kJ\ mol^{-1}$

$\Delta_f H°(298.15\ K) = [-48.593]\ kJ\ mol^{-1}$

Electronic Levels and Quantum Weights

State	ε_i, cm^{-1}	g_i
$^2\Pi_g$	0	4

$\omega_e = 1089 \pm 40\ cm^{-1}$ $\omega_e x_e = 12.1 \pm 1.6\ cm^{-1}$ $\sigma = 2$

$B_e = 1.17 \pm 0.02\ cm^{-1}$ $\alpha_e = [0.017]\ cm^{-1}$ $r_e = 1.341 \pm 0.010$ Å

Enthalpy of Formation

The enthalpy of formation at 0 K for $O_2^-(g)$ is based n the electron affinity value derived from a thorough analysis of the laser photodetachment measurements of Celotta et al. (1). This study let to $EA(O_2) = 0.440 \pm 0.008$ eV. Using auxiliary data (2), we calculate $\Delta_f H°(O_2^-, g, 0\ K) = 10.15 \pm 0.18\ kcal\ mol^{-1}$ and $D_0^0(O_2^-) = 94.4 \pm 0.3\ kcal\ mol^{-1}$; the latter value assumes dissociation into ground state O(g) and $O^-(g)$. The adopted $EA(O_2)$ value is also recommended by Rosenstock et al. (3) and Massey (4). Both of these latter studies have reviewed and summarized earlier work from which EA values could be derived.

$\Delta_f H°(O_2^-, g, 298.15\ K)$ is obtained from $\Delta_f H°(O_2, g, 0\ K)$ by using $EA(O_2)$ with JANAF (2) enthalpies, H°(0 K) - H°(298.15 K), for $O_2^-(g)$, $O_2(g)$, and $e^-(g)$. $\Delta_r H°(O_2^- \rightarrow O_2 + e^-, 298.15\ K)$ differs from a room-temperature threshold energy due to inclusion of these enthalpies and to threshold energy due to inclusion of these enthalpies and to threshold effects discussed by Rosenstock et al. (3). $\Delta_f H°(298.15\ K)$ should be changed by +6.197 $kJ\ mol^{-1}$ if it is to be used in the ion convention that excludes the enthalpy of the electron.

Heat Capacity and Entropy

The laser photodetachment study of Celotta et al. (1) which yielded a precise value for $EA(O_2)$ also yielded precise values for r_e and B_e as well as values for ω_e and $\omega_e x_e$. We adopt these values and calculate α_e assuming a Morse function for the ground state. Support for these values comes from recent studies based on total scattering cross section (5) and differential eleastic cross section (6, 7). Other (earlier) studies yielding vibrational information are discussed by Krupenie (8). The recent ab initio study by Das et al. (11) suggests slightly different vibrational-rotational constants (ω_e = 1098, $\omega_e x_e$ = 9.04, B_e = 1.135, and $\alpha_e = 0.0151\ cm^{-1}$).

The ground and excited electronic states are discussed by Krupenie (8), Massey (4), and Das et al. (11). The approximate theoretical potential energy diagram given by Krupenie (8) suggests a $^2\Pi_g$ ground state with three rather low lying excited states: $^4\Sigma_u^-$ at ~21000 cm^{-1}, $^2\Pi_u$ at ~28500 cm^{-1}, and $^2\Sigma_u^-$ at ~32300 cm^{-1}. The ab initio study by Das et al. (11) suggests the same ordering of the levels at 19833 cm^{-1}, 27213 cm^{-1}, and 30076 cm^{-1}. The fluorescence observed by Rolfe (9) was assigned as $^2\Pi_{ui} \rightarrow X^2\Pi_{gi}$, but an upper state assignment of $^2\Sigma_u^-$ cannot be ruled out (8). This study indicated that the presumed $^2\Pi_{ui}$ state lies at 29400 cm^{-1} (which we adopt), in good agreement with the calculated potential energy diagram. Hurst and Bortner (10) obtained evidence of excitation to an electronic state of O_2^- (presumable $^4\Sigma_u^-$) which lies approximately 1 eV above the ground state. The potential energy diagrams given by Krupenie (8) and Das et al. (11) suggest this level is significantly higher. The position of this level will not significantly affect the entropy values.

The three excited states are predicted to be quite shallow (binding energies ~ 1 eV) as compared to the ground state. The inclusion in our calcualtion of such states, with the use of appropriate vibrational-rotational constants as given by Das et al. (11) would lead to erroneous heat capacity values above 3000 K. The proper inclusion of these three excited states would require a direct summation technique with rotational cutoff.

References

1. R. J. Celotta, R. A. Bennett, J. L. Hall, M. W. Seigel, and J. Levine, Phys. Rev. A 6, 631 (1972).
2. JANAF Thermochemical Tables: $O_2(g)$, $e^-(g)$, $O^-(g)$ and O(g), 3-31-77.
3. H. M. Rosenstock, K. Draxl, B. W. Steiner, and J. T. Herron, J. Phys. Chem. Ref. Data 6, Supp. 1 (1977).
4. H. S. W. Massey, "Negative Ions, " 3rd ed., Cambridge University Press, Cambridge, 1976.
5. R. L. Gray, H. H. Haselton, D. Krause, Jr., and E. A. Soltysik, Chem. Phys. Letters 13, 51 (1972).
6. R. Linder and H. Schmidt, z. Naturforsch 26a, 1617 (1971).
7. M. G. W. Boness and G. J. Schulz, Phys. Rev. A 2, 2182 (1970).
8. P. H. Krupenie, J. Phys. Chem. Ref. Data 1, 423 (1972).
9. J. Rolfe, J. Chem. Phys. 40, 1664 (1964).
10. G. S. Hurst and T. E. Bortner, Radiation Research, Supp. 1, 547 (1959).
11. G. Das, A. C. Wahl, W. T. Zemke, and W. C. Stwalley, J. Chem. Phys. 68, 4252 (1978).

Oxygen, Ion (O_2^-) $O_2^-(g)$

Enthalpy Reference Temperature = T_r = 298.15 K Standard State Pressure = p° = 0.1 MPa

	J K⁻¹mol⁻¹			kJ mol⁻¹			
T/K	$C_p°$	S°	$-[G°-H°(T_r)]/T$	$H°-H°(T_r)$	$\Delta_f H°$	$\Delta_f G°$	Log K_f
0	0.	0.	INFINITE	-8.752	-42.464		
100	29.110	177.477	235.940	-5.846			
200	29.353	197.692	212.335	-2.929			
250	29.818	204.287	210.089	-1.451			
298.15	30.453	209.591	209.591	0.	-48.593	-43.663	7.650
300	30.480	209.780	209.592	0.056	-48.629	-43.633	7.597
350	31.224	214.534	209.966	1.599	-49.603	-42.723	6.376
400	31.962	218.752	210.805	3.179	-50.557	-41.674	5.442
450	32.648	222.557	211.903	4.794	-51.498	-40.507	4.702
500	33.262	226.029	213.144	6.442	-52.431	-39.236	4.099
600	34.271	232.187	215.818	9.821	-54.290	-36.422	3.171
700	35.032	237.530	218.546	13.288	-56.156	-33.297	2.485
800	35.609	242.247	221.220	16.821	-58.038	-29.904	1.953
900	36.055	246.468	223.795	20.406	-59.938	-26.273	1.525
1000	36.407	250.285	226.256	24.029	-61.855	-22.430	1.172
1100	36.691	253.769	228.601	27.685	-63.787	-18.393	0.873
1200	36.926	256.972	230.834	31.366	-65.734	-14.181	0.617
1300	37.123	259.936	232.960	35.069	-67.693	-9.805	0.394
1400	37.293	262.693	234.986	38.790	-69.664	-5.279	0.197
1500	37.441	265.271	236.920	42.527	-71.647	-0.611	0.021
1600	37.572	267.692	238.769	46.277	-73.642	4.190	-0.137
1700	37.690	269.973	240.538	50.041	-75.649	9.116	-0.280
1800	37.798	272.131	242.234	53.815	-77.669	14.160	-0.411
1900	37.897	274.177	243.861	57.600	-79.702	19.317	-0.531
2000	37.989	276.123	245.426	61.394	-81.749	24.582	-0.642
2100	38.075	277.979	246.933	65.197	-83.810	29.949	-0.745
2200	38.157	279.752	248.384	69.009	-85.885	35.415	-0.841
2300	38.234	281.450	249.785	72.829	-87.975	40.975	-0.931
2400	38.308	283.079	251.139	76.656	-90.079	46.626	-1.015
2500	38.380	284.644	252.448	80.490	-92.198	52.366	-1.094
2600	38.448	286.151	253.715	84.332	-94.332	58.190	-1.169
2700	38.515	287.603	254.944	88.180	-96.479	64.098	-1.240
2800	38.580	289.005	256.135	92.035	-98.641	70.085	-1.307
2900	38.644	290.360	257.292	95.896	-100.816	76.148	-1.372
3000	38.706	291.671	258.416	99.763	-103.004	82.288	-1.433
3100	38.767	292.941	259.510	103.637	-105.204	88.501	-1.491
3200	38.826	294.173	260.574	107.517	-107.417	94.784	-1.547
3300	38.885	295.368	261.610	111.402	-109.641	101.137	-1.601
3400	38.943	296.530	262.620	115.294	-111.876	107.558	-1.652
3500	39.001	297.660	263.605	119.191	-114.121	114.045	-1.702
3600	39.057	298.759	264.567	123.094	-116.376	120.596	-1.750
3700	39.113	299.830	265.505	127.002	-118.641	127.209	-1.796
3800	39.169	300.874	266.422	130.916	-120.914	133.885	-1.840
3900	39.224	301.892	267.319	134.836	-123.195	140.620	-1.883
4000	39.279	302.886	268.196	138.761	-125.484	147.413	-1.925
4100	39.333	303.857	269.054	142.692	-127.780	154.264	-1.965
4200	39.387	304.805	269.894	146.628	-130.084	161.171	-2.004
4300	39.440	305.732	270.716	150.569	-132.395	168.133	-2.042
4400	39.494	306.640	271.523	154.516	-134.713	175.149	-2.079
4500	39.547	307.528	272.313	158.468	-137.037	182.217	-2.115
4600	39.599	308.398	273.088	162.425	-139.369	189.337	-2.150
4700	39.652	309.250	273.848	166.387	-141.708	196.508	-2.184
4800	39.704	310.085	274.594	170.355	-144.053	203.729	-2.217
4900	39.756	310.904	275.327	174.328	-146.407	210.999	-2.249
5000	39.808	311.708	276.047	178.307	-148.768	218.317	-2.281
5100	39.860	312.497	276.754	182.290	-151.138	225.683	-2.311
5200	39.912	313.271	277.449	186.279	-153.516	233.094	-2.341
5300	39.963	314.032	278.132	190.272	-155.904	240.552	-2.371
5400	40.015	314.780	278.803	194.271	-158.302	248.054	-2.399
5500	40.066	315.514	279.464	198.275	-160.711	255.601	-2.427
5600	40.117	316.237	280.114	202.285	-163.131	263.193	-2.455
5700	40.168	316.947	280.754	206.299	-165.565	270.827	-2.482
5800	40.219	317.646	281.385	210.318	-168.011	278.505	-2.508
5900	40.270	318.334	282.005	214.343	-170.473	286.223	-2.534
6000	40.321	319.011	282.616	218.372	-172.950	293.986	-2.559

PREVIOUS: September 1977 (1 atm) CURRENT: September 1977 (1 bar)

Oxygen, Ion (O_2^-) $O_2^-(g)$

PHOSPHORUS OXIDE (PO_2) IDEAL GAS M_r = 62.97256

Phosphorus Oxide (PO_2) $O_2P_1(g)$

$\Delta_fH°$(0 K) =[-311.142] kJ mol^{-1}

$S°$(298.15 K) = [253.685] J K^{-1} mol^{-1} $\Delta_fH°$(298.15 K) =[-314.524] kJ mol^{-1}

Vibrational Frequencies and Degeneracies

ν, cm^{-1}
[980](1)
[515](1)
[1044](1)

Ground State Quantum Weight: 2 σ = 2

Point Group: C_{2v}

Bond Distance: P-O = [1.485] Å

Bond Angle: O-P-O = [134°4']

Product of the Moments of Inertia: $I_AI_BI_C$ = [9.42392 x 10^{-117}] g^3 cm^6

Enthalpy of Formation

$\Delta_fH°$(298.15 K) is calculated from an estimated enthalpy of dissociation (270 ± 25 kcal mol^{-1}) reported by Brewer and Rosenblatt (1)

Heat Capacity and Entropy

The molecular constants selected for this table were estimated by Gordon (2). The molecular constants for PO_2(g) have been estimated by Gordon (2), Potter et al. (3) and Henderson and Scheffee (4). The entropies at 3000 K calculated from these three estimates differ by less than 0.8 cal K^{-1} mol^{-1}. The principal moments of inertia are: I_A = [0.8775 x 10^{-39}], I_B = [9.9337 x 10^{-39}], and I_C = [10.8112 x 10^{-39}] g^3 cm^6

References

1. L. Brewer and G. M. Rosenblatt, Chem. Reviews 61, 257 (1961).
2. J. S. Gordon, "Thermodynamic Data for Combustion Products," January, 1960, Thiokol Chemical Corp., Reaction Motor Division, Denville, N. J.
3. R. L. Potter, V. N. DiStefano and S. N. Fox, "The Thermodynamic Functions for Some Combustion Products Containing Phosphorus-I", American Cyanamid Company.
4. C. B. Henderson and R. S. Scheffee, "Survey of Thermochemical Data", Atlantic Research Corp., Alexandria, Virginia, January, 1960.

Enthalpy Reference Temperature = T_r = 298.15 K Standard State Pressure = p° = 0.1 MPa

T/K	$C_p°$ (J $K^{-1}mol^{-1}$)	S°	-[G°-H°(T_r)]/T	H°-H°(T_r) (kJ mol^{-1})	$\Delta_fH°$	$\Delta_fG°$	Log K_f
0	0.	0.	INFINITE	-10.662	-311.142	-311.142	INFINITE
100	33.536	214.043	287.365	-7.332	-311.604	-313.909	163.969
200	36.825	238.134	257.323	-3.838	-313.285	-315.789	82.476
250	39.137	246.597	254.354	-1.939	-313.936	-316.339	66.095
298.15	41.407	253.685	253.685	0.	-314.524	-316.748	55.493
300	41.493	253.942	253.686	0.077	-314.546	-316.762	55.153
350	43.699	260.506	254.199	2.207	-315.827	-317.015	47.312
400	45.659	266.472	255.366	4.442	-316.403	-317.145	41.415
450	47.347	271.950	256.909	6.769	-316.910	-317.206	36.820
500	48.779	277.015	258.669	9.173	-317.364	-317.214	33.139
600	50.998	286.116	262.503	14.168	-318.161	-317.107	27.607
700	52.574	294.103	266.458	19.351	-318.865	-316.875	23.646
800	53.710	301.201	270.366	24.668	-319.517	-316.546	20.668
900	54.548	307.578	274.152	30.083	-320.140	-316.137	18.348
1000	55.178	313.359	277.788	35.571	-320.747	-315.660	16.488
1100	55.663	318.642	281.266	41.114	-321.346	-315.122	14.964
1200	56.042	323.502	284.586	46.700	-385.513	-313.451	13.644
1300	56.344	328.001	287.754	52.320	-385.334	-307.453	12.354
1400	56.587	332.185	290.780	57.967	-385.164	-301.469	11.248
1500	56.786	336.096	293.672	63.636	-385.003	-295.497	10.290
1600	56.951	339.767	296.440	69.323	-384.853	-289.535	9.452
1700	57.089	343.224	299.091	75.025	-384.716	-283.581	8.713
1800	57.206	346.490	301.635	80.740	-384.593	-277.636	8.057
1900	57.305	349.586	304.078	86.466	-384.486	-271.697	7.469
2000	57.390	352.527	306.427	92.201	-384.395	-265.763	6.941
2100	57.464	355.329	308.690	97.944	-384.321	-259.833	6.463
2200	57.528	358.004	310.871	103.693	-384.265	-253.906	6.028
2300	57.584	360.563	312.976	109.449	-384.227	-247.982	5.632
2400	57.634	363.014	315.010	115.210	-384.208	-242.059	5.268
2500	57.678	365.368	316.978	120.976	-384.209	-236.136	4.934
2600	57.717	367.631	318.883	126.745	-384.228	-230.213	4.625
2700	57.751	369.810	320.729	132.519	-384.266	-224.289	4.339
2800	57.782	371.911	322.520	138.295	-384.322	-218.362	4.074
2900	57.810	373.939	324.258	144.075	-384.398	-212.435	3.826
3000	57.836	375.899	325.947	149.857	-384.491	-206.503	3.596
3100	57.859	377.796	327.589	155.642	-384.602	-200.568	3.380
3200	57.880	379.633	329.187	161.429	-384.730	-194.630	3.177
3300	57.899	381.415	330.743	167.218	-384.875	-188.687	2.987
3400	57.916	383.143	332.258	173.009	-385.037	-182.739	2.807
3500	57.932	384.822	333.736	178.801	-385.214	-176.787	2.638
3600	57.947	386.455	335.178	184.595	-385.406	-170.829	2.479
3700	57.960	388.043	336.586	190.390	-385.613	-164.866	2.327
3800	57.972	389.588	337.960	196.187	-385.834	-158.896	2.184
3900	57.984	391.094	339.303	201.985	-386.070	-152.921	2.048
4000	57.995	392.563	340.617	207.784	-386.318	-146.940	1.919
4100	58.004	393.995	341.901	213.584	-386.580	-140.953	1.796
4200	58.014	395.393	343.158	219.385	-386.855	-134.958	1.678
4300	58.022	396.758	344.389	225.187	-387.143	-128.957	1.567
4400	58.030	398.092	345.594	230.989	-387.443	-122.950	1.460
4500	58.038	399.396	346.775	236.793	-387.756	-116.936	1.357
4600	58.045	400.672	347.933	242.597	-388.082	-110.914	1.259
4700	58.051	401.920	349.069	248.401	-388.420	-104.885	1.166
4800	58.057	403.142	350.183	254.207	-388.771	-98.848	1.076
4900	58.063	404.339	351.276	260.013	-389.136	-92.804	0.989
5000	58.069	405.513	352.349	265.819	-389.515	-86.752	0.906
5100	58.074	406.662	353.402	271.627	-389.908	-80.693	0.826
5200	58.078	407.790	354.437	277.434	-390.315	-74.627	0.750
5300	58.083	408.897	355.455	283.242	-390.737	-68.551	0.676
5400	58.087	409.982	356.454	289.051	-391.176	-62.469	0.604
5500	58.091	411.048	357.437	294.860	-391.632	-56.378	0.535
5600	58.095	412.095	358.404	300.669	-392.105	-50.277	0.469
5700	58.099	413.123	359.355	306.479	-392.597	-44.169	0.405
5800	58.102	414.134	360.291	312.289	-393.108	-38.051	0.343
5900	58.106	415.127	361.212	318.099	-393.640	-31.927	0.283
6000	58.109	416.104	362.119	323.910	-394.193	-25.790	0.225

PREVIOUS: September 1962 (1 atm) CURRENT: September 1962 (1 bar)

LEAD OXIDE (PbO_2) CRYSTAL M_r = 239.1988

$\Delta_f H°(0\ K) = -269.874 \pm 2.9\ kJ\ mol^{-1}$
$\Delta_f H°(298.15\ K) = -274.470 \pm 2.9\ kJ\ mol^{-1}$

$S°(298.15\ K) = 71.797 \pm 0.42\ J\ K^{-1}\ mol^{-1}$

Enthalpy of Formation

Espada et al. (1) have calorimetrically determined the heat of the reaction $PbO_2(cr) + 2\ H_2(g) \rightarrow Pb(cr) + 2\ H_2O(\ell)$ as -71.02 ± 0.16 kcal mol^{-1} from which they derive $\Delta_f H°(PbO_2$, cr, 298.15 K) = -65.61 ± 0.32 kcal mol^{-1}. This value assumes that the PbO_2 was 100% pure, in fact it is virtually impossible (2) to prepare a sample free of water and fully oxidized. Thus, Espada et al. (1) have increased their uncertainty to ±0.7 kcal mol^{-1} to cover the likely spread of values assuming typical compositions.

Millar (3) has recalculated the data of Glasstone for the Gibbs energy of the cell reaction $PbO_2(cr) + H_2(g) \rightarrow PbO(red) + H_2O(\ell)$ as $\Delta_f G°(298.15\ K)$ = -49.600 kcal mol^{-1}. With auxiliary Gibbs energy data (4, 5) we calculate $\Delta_f G°(PbO_2$, cr, 298.15 K) = -52.33 kcal mol^{-1} which corresponds to $\Delta_f H°(PbO_2$, cr, 298.15 K) = -66.45 kcal mol^{-1} in reasonable agreement with the value reported by Espada et al. (1).

We adopt the value $\Delta_f H°(PbO_2$, cr, 298.15 K) = -65.6 ± 0.7 kcal mol^{-1}.

Heat Capacity and Entropy

Duisman and Giauque (2) have reported the heat capacity of PbO_2 from 15 to 318 K and have calculated the entropy at 298.15 K based on S°(15 K) = 0.079 cal K^{-1} mol^{-1}. They had considerable difficulty in preparing a suitable sample of PbO and finally used a sample containing PbO and H_2O, as impurities, for which significant corrections were made to the measured data. Millar (3) also has measured the heat capacity from 70 to 298 K but is 7% higher than the adopted values.

High temperature enthalpy data have been reported by Palmaer (6) and Bousquet et al. (7) but these data suffer from unknown amounts of impurity, for which the correction is probably substantial, and decomposition to intermediate oxides. Thus, the heat capacity above room temperature was estimated by a graphical extrapolation of the low temperature heat capacity.

Decomposition Data

Numerous investigators have reported the decomposition of PbO_2 at elevated temperatures, however, the exact processes are still not well defined. Otto (8) indicates a three step process to Pb_3O_4, but the intermediate compositions are not established. The temperature at which the decomposition pressure of oxygen reaches 0.2 atm was reported as 260°C by Otto while White and Roy (9) indicate 293°C. The enthalpies and entropies reported are in serious disagreement with adopted values and indicate failure to attain equilibrium.

References

1. L. Espada, G. Pilcher and H. A. Skinner, J. Chem. Thermodyn. 2, 647 (1970).
2. J. A. Duisman and W. F. Giauque, J. Phys. Chem. 72, 562 (1968).
3. R. W. Millar, J. Amer. Chem. Soc. 51, 207 (1929).
4. JANAF Thermochemical Tables: PbO(red, cr), 12-31-71.
5. U. S. Nat. Bur. Stand. Tech. Note 270-3, 1968.
6. W. Palmaer, Z. Elektrochem. 29, 45 (1923).
7. J. Bousquet, J. M. Blanchard and G. Perachon, Bull. Soc. Chim. Fr. 1969, 733 (1969).
8. E. M. Otto, J. Electrochem. Soc. 113, 525 (1966).
9. W. B. White and R. Roy, J. Amer. Ceram. Soc. 47, 242 (1964).

Lead Oxide (PbO_2) $O_2Pb_1(cr)$

Enthalpy Reference Temperature = T_r = 298.15 K Standard State Pressure = p° = 0.1 MPa

T/K	$C_p°$ (J K⁻¹mol⁻¹)	S° (J K⁻¹mol⁻¹)	-[G°-H°(T_r)]/T (J K⁻¹mol⁻¹)	H°-H°(T_r) (kJ mol⁻¹)	$\Delta_f H°$ (kJ mol⁻¹)	$\Delta_f G°$ (kJ mol⁻¹)	Log K_f
0	0.000	0.000	INFINITE	-10.965	-269.874	-269.874	INFINITE
100	29.878	21.983	117.951	-9.597	-273.176	-254.361	132.865
200	49.957	49.530	77.063	-5.507	-274.522	-234.876	61.343
298.15	61.170	71.797	71.797	0.000	-274.470	-215.397	37.737
300	61.318	72.176	71.799	0.113	-274.461	-215.030	37.440
400	67.613	90.754	74.289	6.586	-273.688	-195.323	25.507
500	71.698	106.314	79.180	13.567	-272.580	-175.855	18.371
600	74.517	119.649	84.841	20.885	-271.311	-156.628	13.636
700	76.210	131.267	90.661	28.424	-274.848	-136.820	10.210
800	77.571	141.536	96.391	36.116	-273.510	-117.192	7.652
900	78.602	150.736	101.927	43.927	-272.090	-97.737	5.673
1000	79.301	159.056	107.231	51.825	-270.609	-78.443	4.097
1100	79.669	166.634	112.292	59.777	-269.091	-59.300	2.816
1200	79.705	173.570	117.114	67.748	-267.562	-40.296	1.754

PREVIOUS: March 1962 CURRENT: December 1971

Lead Oxide (PbO_2) $O_2Pb_1(cr)$

SULFUR DIOXIDE (SO_2) IDEAL GAS M_r = 64.0588

Sulfur Dioxide (SO_2) $O_2S_1(g)$

$S°(298.15\ K) = 248.212 \pm 0.08\ J\ K^{-1}\ mol^{-1}$

$\Delta_f H°(0\ K) = -294.299 \pm 0.21\ kJ\ mol^{-1}$
$\Delta_f H°(298.15\ K) = -296.842 \pm 0.21\ kJ\ mol^{-1}$

Vibrational Levels and Degeneracies

ν, cm^{-1}
1151.38(1)
517.69(1)
1361.76(1)

Point Group: C_{2v} $\sigma = 2$
Bond Distance: S–O = 1.432 ± 0.003 Å
Bond Angle: O–S–O = 119.53 ± 0.40
Product of the Moments of Inertia: $I_A I_B I_C = 1.070 \times 10^{-115}\ g^3\ cm^6$

Enthalpy of Formation

Eckman and Rossini (1) measured the enthalpy of combustion of rhombic sulfur to SO_2. Changes due to different atomic weights and electrical constants, though negligible, have been made.

Heat Capacity and Entropy

The functions were calculated by Gordon (2) using the method of Pennington and Kobe (3). Structural parameters are given by Sirvetz (4), obtained from microwave measurements. Frequencies were measured by Shelton et al. (5). A 3rd law value of S°(298.15 K) = 59.24 cal K^{-1} mol^{-1} was obtained by Giauque and Stevenson (6). The principal moments of inertia are (all in g cm^2): $I_A = 1.380 \times 10^{-39}$, $I_B = 8.131 \times 10^{-39}$, and $I_C = 9.534 \times 10^{-39}$.

References

1. J. R. Eckman and F. D. Rossini, J. Res. Nat. Bur. Stand. 3, 597 (1929).
2. J. Gordon, personal communication, Feb. 2, 1961.
3. R. E. Pennington and K. A. Kobe, J. Chem. Phys. 22, 1442 (1954).
4. M. H. Sirvetz, J. Chem. Phys. 19, 938 (1951).
5. R. D. Shelton, A. H. Nielsen and W. H. Fletcher, J. Chem. Phys. 21, 2178 (1953); ibid., 22, 1791 (1954).
6. W. F. Giauque and C. C. Stevenson, J. Amer. Chem. Soc. 60, 1389 (1938).

Enthalpy Reference Temperature = T_r = 298.15 K Standard State Pressure = p° = 0.1 MPa

T/K	$C_p°$ (J $K^{-1}mol^{-1}$)	S° (J $K^{-1}mol^{-1}$)	$-[G°-H°(T_r)]/T$ (J $K^{-1}mol^{-1}$)	$H°-H°(T_r)$ (kJ mol^{-1})	$\Delta_f H°$ (kJ mol^{-1})	$\Delta_f G°$ (kJ mol^{-1})	Log K_f
0	0.000	0.000	INFINITE	-10.552	-294.299	-294.299	INFINITE
100	33.526	209.025	281.199	-7.217	-294.559	-296.878	155.073
200	36.372	233.033	251.714	-3.736	-295.631	-298.813	78.042
298.15	39.878	248.212	248.212	0.000	-296.842	-300.125	52.581
300	39.945	248.459	248.213	0.074	-296.865	-300.145	52.260
400	43.493	260.448	249.824	4.250	-300.257	-300.971	39.303
500	46.576	270.495	252.979	8.758	-302.736	-300.871	31.432
600	49.049	279.214	256.641	13.544	-304.694	-300.305	26.144
700	50.961	286.924	260.427	18.548	-306.291	-299.444	22.345
800	52.434	293.829	264.178	23.721	-307.667	-298.370	19.482
900	53.580	300.073	267.825	29.023	-362.026	-296.051	17.182
1000	54.484	305.767	271.339	34.428	-361.940	-288.725	15.081
1100	55.204	310.995	274.710	39.914	-361.835	-281.409	13.363
1200	55.794	315.824	277.937	45.464	-361.720	-274.102	11.931
1300	56.279	320.310	281.026	51.069	-361.601	-266.806	10.720
1400	56.689	324.496	283.983	56.718	-361.484	-259.518	9.683
1500	57.036	328.419	286.816	62.404	-361.372	-252.239	8.784
1600	57.338	332.110	289.533	68.123	-361.268	-244.967	7.997
1700	57.601	335.594	292.141	73.870	-361.176	-237.701	7.304
1800	57.831	338.893	294.647	79.642	-361.096	-230.440	6.687
1900	58.040	342.026	297.059	85.436	-361.031	-223.183	6.136
2000	58.229	345.007	299.383	91.250	-360.981	-215.929	5.639
2100	58.400	347.853	301.624	97.081	-360.948	-208.678	5.191
2200	58.555	350.573	303.787	102.929	-360.931	-201.427	4.782
2300	58.702	353.179	305.878	108.792	-360.930	-194.177	4.410
2400	58.840	355.680	307.902	114.669	-360.947	-186.927	4.068
2500	58.965	358.085	309.861	120.559	-360.980	-179.675	3.754
2600	59.086	360.400	311.761	126.462	-361.030	-172.422	3.464
2700	59.199	362.632	313.604	132.376	-361.095	-165.166	3.195
2800	59.308	364.787	315.394	138.302	-361.175	-157.908	2.946
2900	59.413	366.870	317.133	144.238	-361.270	-150.648	2.713
3000	59.513	368.886	318.825	150.184	-361.379	-143.383	2.497
3100	59.609	370.839	320.471	156.140	-361.502	-136.114	2.294
3200	59.706	372.733	322.075	162.106	-361.638	-128.842	2.103
3300	59.794	374.572	323.638	168.081	-361.786	-121.565	1.924
3400	59.881	376.358	325.162	174.065	-361.946	-114.283	1.756
3500	59.969	378.095	326.650	180.057	-362.118	-106.996	1.597
3600	60.053	379.786	328.103	186.058	-362.300	-99.704	1.447
3700	60.137	381.432	329.522	192.068	-362.493	-92.408	1.305
3800	60.216	383.037	330.909	198.086	-362.697	-85.105	1.170
3900	60.296	384.602	332.266	204.111	-362.911	-77.796	1.042
4000	60.375	386.130	333.593	210.145	-363.134	-70.484	0.920
4100	60.450	387.621	334.893	216.186	-363.368	-63.164	0.805
4200	60.530	389.079	336.166	222.235	-363.611	-55.840	0.694
4300	60.605	390.504	337.413	228.292	-363.864	-48.508	0.589
4400	60.676	391.898	338.636	234.356	-364.128	-41.172	0.489
4500	60.752	393.263	339.834	240.427	-364.401	-33.829	0.393
4600	60.823	394.599	341.010	246.506	-364.686	-26.480	0.301
4700	60.894	395.908	342.165	252.592	-364.981	-19.125	0.213
4800	60.969	397.190	343.298	258.685	-365.288	-11.763	0.128
4900	61.036	398.448	344.410	264.785	-365.607	-4.394	0.047
5000	61.107	399.682	345.504	270.893	-365.938	2.981	-0.031
5100	61.178	400.893	346.578	277.007	-366.283	10.364	-0.106
5200	61.250	402.082	347.634	283.128	-366.641	17.751	-0.178
5300	61.317	403.249	348.672	289.257	-367.014	25.148	-0.248
5400	61.388	404.396	349.693	295.392	-367.403	32.550	-0.315
5500	61.455	405.523	350.698	301.534	-367.807	39.960	-0.380
5600	61.522	406.631	351.687	307.683	-368.230	47.379	-0.442
5700	61.588	407.720	352.661	313.838	-368.671	54.804	-0.502
5800	61.655	408.792	353.619	320.000	-369.131	62.238	-0.561
5900	61.727	409.846	354.563	326.169	-369.611	69.677	-0.617
6000	61.793	410.884	355.493	332.346	-370.113	77.128	-0.671

PREVIOUS: June 1961 (1 atm) CURRENT: June 1961 (1 bar)

Sulfur Dioxide (SO_2) $O_2S_1(g)$

$S°(298.15\ K) = 50.053\ J\ K^{-1}\ mol^{-1}$

T_{trs1} = 543 ± 3 K(low cristobalite → high cristobalite)

T_{trs2} = 1079 ± 250 K(high quartz → high cristobalite)

T_{fus} = 1996 ± 5 K(high cristobalite → liquid)

$\Delta_f H°(298.15\ K) = -905.489\ kJ\ mol^{-1}$

$\Delta_{trs1}H° = 1.343 \pm 0.25\ kJ\ mol^{-1}$

$\Delta_{trs2}H° = 2.008 \pm 0.63\ kJ\ mol^{-1}$

$\Delta_{fus}H° = 9.581 \pm 2.1\ kJ\ mol^{-1}$

Enthalpy of Formation

The enthalpy of formation is calculated from that of low cristobalite by addition of $\Delta_{trs}H° = 0.321\ kcal\ mol^{-1}$ and the difference between H°(543 K)-H°(298.15 K) for low and high forms.

Heat Capacity and Entropy

$C_p°$ is derived from enthalpies of Mosesman (1) and White (2) whose data deviate from the adopted functions by -0.6 to +0.5% and +0.9 to 0.0%, respectively. White's deviations for cristobalite have a magnitude and trend similar to his deviations for quartz and vitreous silica. Data of Wietzel (3) for his sample prepared at 1600°C deviate by +3% at 673 K and +1.6% at 1973 K. $C_p°$ below 543 K is smoothly extrapolated to give the proper trend in the Gibbs energy, but the heat capacity at 298 K is not realistic. The entropy is calculated in a manner analogous to that of the enthalpy of formation.

Transition Data

Literature on the polymorphism of SiO_2 is voluminous but is summarized in detail in recent books by Sosman (4) and Eitel (5). The known phases of SiO_2 include quartz, tridymite(?), cristobalite, vitreous silica and the more recently discovered high pressures phases. Muan (6) concisely reviewed the still controversial question of whether tridymite is stabilized only by the presence of foreign ions or whether it is a stable phase of SiO_2 intermediate between quartz and cristobalite. Transitions among the different phases are generally sluggish because of large energy barriers to the structural change and small differences in Gibbs free energies. As a result, the phases often persist far into metastable regions. In contrast, transitions within a particular phase are rapid and reversible (i.e., the low → high inversions of quartz or cristobalite).

High cristobalite is the stable phase near T_{fus}. It becomes metastable with respect to quartz at lower temperatures but persists down to the high → low inversion at 543 K (see table for low cristobalite). The high form is cubic and the low form tetragonal (12, 13).

The adopted tables give 1079 ± 250 K as the temperature at which $\Delta_r G° = 0$ for high quartz → high cristobalite; $\Delta_{trs}H° = 0.48 \pm 0.15\ kcal\ mol^{-1}$ is the corresponding enthalpy difference. The uncertainty in T_{trs} is derived from that in $\Delta_{trs}H°$ which, in turn, is taken from $\Delta_r H°(970\ K) = 0.45 \pm 0.15\ kcal\ mol^{-1}$ obtained by Holm (7) from $\Delta_{sol}H°$ measurements in an oxide melt. It is apparent that T_{trs} is extremely sensitive to small changes in G or H. Holmquist (8) tentatively placed this inversion at 1298 ± 25 K but did observe cristobalite formation at temperatures as low as 1171 K.

Fusion Data

Melting point data were reviewed by Schneider (9) and adjusted to the 1948 International Temperature Scale. The adopted melting point for cristobalite is based on Greig (10). $\Delta_{fus}H°$ is calculated as the difference in $\Delta_f H°$ for liquid and crystal at T_{fus}. Mackenzie et al. (11) showed that cristobalite may be superheated by as much as 40° above T_{fus} due to its slow rate of fusion.

References

1. M. A. Mosesman and K. S. Pitzer, J. Amer. Chem. Soc. 63, 2348 (1941). enthalpy (542-622 K).
2. W. P. White, Amer. J. Sci. 47, 1 (1919). enthalpy (573-1673 K).
3. R. Wietzel, Z. Anorg. Allg. Chem. 116, 71 (1921). enthalpy (673-1973 K).
4. R. B. Sosman, "Phases of Silica," 388 pp., Rutgers Univ. Press, New Brunswick, New Jersey, 1965.
5. W. Eitel, "Silicate Science," Vol. III, pp. 44-157, Academic Press, New York, 1965.
6. A. Muan and E. F. Osborn, "Phase Equilibria Among Oxides in Steelmaking," pp. 17-20, Addison-Wesley Publishing Co., Reading, Mass., 1965.
7. J. L. Holm, O. J. Kleppa and E. F. Westrum, Jr., Geochim. Cosmochim. Acta, to be published. $\Delta_{sol}H°$.
8. S. B. Holmquist, J. Amer. Ceram. Soc. 44, 82 (1961). T_{trs}.
9. S. J. Schneider, U. S. Nat. Bur. Stand. Monograph 68, 1963.
10. J. W. Greig, Amer. J. Sci. 13, 1 (1927). T_{fus}.
11. N. G. Ainslee, J. D. Mackenzie and D. Turnbull, J. Phys. Chem. 65, 1718 (1961); J. Amer. Ceram. Soc. 43, 615 (1960).
12. C. Berger, L. Eyrand, M. Richard and R. Riviere, Bull. Soc. Chim. France 1966, 628 (1966). Crystal structure.
13. W. A. Dollase, Z. Krist. 121, 369 (1965). Crystal structure.

Enthalpy Reference Temperature = T_r = 298.15 K   Standard State Pressure = p° = 0.1 MPa

T/K	$C_p°$ (J K⁻¹mol⁻¹)	S° (J K⁻¹mol⁻¹)	$-[G°-H°(T_r)]/T$ (J K⁻¹mol⁻¹)	$H°-H°(T_r)$ (kJ mol⁻¹)	$\Delta_f H°$ (kJ mol⁻¹)	$\Delta_f G°$ (kJ mol⁻¹)	Log K_f
0							
100							
200							
298.15	26.581	50.053	50.053	0.000	-905.489	-853.636	14[illegible].554
300	27.154	50.219	50.054	0.050	-905.530	-853.315	148.575
400	47.413	61.218	51.416	3.921	-906.752	-835.678	109.128
500	56.856	72.923	54.555	9.184	-906.826	-817.885	85.444
600	62.049	83.788	58.536	15.151	-906.394	-800.132	69.658
700	65.229	93.609	62.857	21.526	-905.721	-782.473	58.389
800	67.337	102.466	67.264	28.161	-904.932	-764.919	49.944
900	68.823	110.487	71.628	34.973	-904.088	-747.468	43.382
1000	69.923	117.798	75.885	41.913	-903.221	-730.112	38.137
1100	70.772	124.504	80.004	48.949	-902.350	-712.843	33.850
1200	71.446	130.691	83.974	56.061	-901.485	-695.653	30.281
1300	71.998	136.432	87.791	63.234	-900.636	-678.535	27.264
1400	72.467	141.786	91.459	70.458	-899.808	-661.482	24.680
1500	72.869	146.799	94.983	77.725	-899.006	-644.487	22.443
1600	73.220	151.514	98.370	85.030	-898.235	-627.544	20.487
1700	73.534	155.962	101.628	92.368	-947.674	-610.202	18.749
1800	73.814	160.173	104.765	99.735	-946.742	-590.378	17.132
1900	74.074	164.171	107.787	107.130	-945.806	-570.605	15.687
2000	74.316	167.977	110.702	114.549	-944.869	-550.882	14.388
2100	74.542	171.608	113.517	121.992	-943.931	-531.206	13.213
2200	74.751	175.081	116.237	129.457	-942.994	-511.573	12.146
2300	74.952	178.408	118.868	136.942	-942.059	-491.985	11.173
2400	75.145	181.602	121.416	144.447	-941.126	-472.437	10.282
2500	75.333	184.674	123.885	151.971	-940.197	-452.927	9.463
2600	75.509	187.632	126.280	159.513	-939.271	-433.454	8.708
2700	75.684	190.485	128.606	167.073	-938.348	-414.017	8.010
2800	75.852	193.240	130.865	174.650	-937.428	-394.614	7.362
2900	76.015	195.905	133.062	182.243	-936.512	-375.245	6.759
3000	76.178	198.485	135.200	189.853	-935.599	-355.906	6.197

PREVIOUS: CURRENT: June 1967

CRISTOBALITE, LOW (SiO_2) CRYSTAL M_r = 60.0843

Silicon Oxide, Cristobalite, Low (SiO_2) $O_2Si_1(cr)$

$S°(298.15\ K) = 43.396 \pm 0.13\ J\ K^{-1}\ mol^{-1}$

$T_{trs} = 543 \pm 3\ K$ (low cristobalite → high cristobalite)

$\Delta_fH°(0\ K) = -903.482 \pm 2.1\ kJ\ mol^{-1}$

$\Delta_fH°(298.15\ K) = -908.346 \pm 2.1\ kJ\ mol^{-1}$

$\Delta_{trs}H° = 1.343 \pm 0.25\ kJ\ mol^{-1}$

Enthalpy of Formation

The enthalpy of formation is calculated from that of quartz using $\Delta_rH°(970\ K) = 0.45 \pm 0.15\ kcal\ mol^{-1}$ for high quartz → high cristobalite, as determined by Holm (1) from $\Delta_{sol}H°$ is an oxide melt. This value reduces to $\Delta_rH°(298.15\ K) = 0.60\ kcal\ mol^{-1}$ for low quartz → low cristobalite. Kracek (2) obtained essentially the same value, $\Delta_rH°(298.15\ K) = 0.63\ kcal\ mol^{-1}$, from $\Delta_{sol}H°$ in HF(aq, 74.7°C), while Hummel (3) found $\Delta_rH°(298.15\ K) = 1.80\ kcal\ mol^{-1}$ from $\Delta_{sol}H°$ in HF(aq, 26.5°C). The larger $\Delta_rH°$ corresponds to a less negative $\Delta_fH°$ (i.e., less stable) and suggests that the sample of Hummel has disorder approaching that of vitreous silica. Presumably, this disagreement arises from different methods of preparation of cristobalite from quartz. Holm heated for 24 hours at 1743 K, while Hummel heated for 4 hours at 1973 K. Hummel's sample was then ground and treated with dilute HF to remove amorphous surface layers created by grinding; particles smaller than 2.5 microns were used in order to obtain adequate rates of dissolution. Neither author gave X-ray or DTA evidence to confirm that the samples were the more stable, ordered form of cristobalite (15).

The calorimetric data are summarized below along with equilibrium data which confirm the adopted values of $\Delta_fH°$ for cristobalite and quartz. Data from Matoba (5) and Rein (6) were reduced to the standard state of Si(ℓ) using activity coefficients from the curves of (10).

Source	Method	Reaction	T/K	Data Points	$\Delta_rH(T)$ kcal mol^{-1}	$\Delta_rH°(298.15\ K)$ kcal mol^{-1}	Drift cal K^{-1} mol^{-1}	$\Delta_fH°(298.15\ K)$ kcal mol^{-1}
(1)	$\Delta_{sol}H°$ in oxide melt	A	970	3	0.45±0.15	0.6	-	-217.10
(2)	$\Delta_{sol}H°$ in HF(aq, 26.5°C)	A	299.65	8	1.80±0.2	1.80	-	-215.90
(3)	$\Delta_{sol}H°$ in HF(aq, 74.7°C)	A	298-348	-	-	0.63±0.05	-	-217.07
(4)	$\Delta_{sol}H°$ in HF(aq, 73.7°C)	A	298-347	6	-	0.93±0.1	-	-216.77
(5)	Keq + γ(Si)	B	1843-1953	3	104.5 ±0.6*	112.48*	0.8	-217.17
(6)	Keq + γ(Si)	C	1773-1798	5	-	175.42*	31±37	-217.38
	Keq + γ(Si)	D	1823-1873	6	-	228.5*	10±6	-212.3
(7)	$K_p°$	E	1703-1829	14	145.9 ±1.3*	146.92*	-2.3±0.7	-217.94
(8)	$K_p°$	E	1673-1853	5	145.5 ±3.5*	147.43*	-1.9±1.9	-218.45

*Values at T are from 2nd law analysis while those at 298.15 K are from 3rd law.

Reactions:
A) SiO_2(quartz, low or high) = SiO_2(cristobalite, low or high)
B) SiO_2(high cristobalite) + 2 H_2(g) = Si(ℓ) + 2 H_2O(g)
C) SiO_2(high cristobalite) + 2 C(graph) = Si(ℓ) + 2 CO(g)
D) SiO_2(high cristobalite) + 2 SiC(β) = 3 Si(ℓ) + 2 CO(g)
E) SiO_2(high cristobalite) + 3 C(graph) = SiC(β) + 2 CO(g)

Heat Capacity and Entropy

$C_p°$ below 300 K is based on data of Westrum (11). Earlier data of Anderson (12) are in satisfactory agreement. The entropy is obtained from $C_p°$ using $S°(5\ K) = 0.0007\ cal\ K^{-1}\ mol^{-1}$. $C_p°$ above 300 K is derived from enthalpy data of Mosesman (13) and White (14). The former deviate from the adopted functions by -0.7 to 0.5% and the latter by +0.4%. Although the functions above 300 K appear to be reliable, it would be desirable to have new data for a well-characterized sample (see Transition Data).

Transition Data

Low cristobalite is metastable with respect to quartz but persists up to T_{trs}. T_{trs} is taken as the temperature at the peak in the heating curve for well-ordered cristobalite; however, see (15, 16) for detaled discussions of temperatures about 25 lower which were found from cooling curves. $\Delta_{trs}H°$ is derived from enthalpy data (13, 14) for high cristobalite using the adopted enthalpy for the low form. The enthalpy data of Mosesman (13) gave a lower T_{trs} which suggests that the sample was less well-ordered than desirable. $\Delta_{trs}H°$ is confirmed, however, by other data which are summarized below. Additional polymorphism is discussed on the table for high cristobalite.

Source	Method	$\Delta_{trs}H°$, kcal mol^{-1}
(13)	Enthalpy data	0.321
(17)	Microcalorimetric thermal analysis	0.318±0.018
(18)	ΔV and dT/DP	0.27
(19)	Differential thermal analysis	0.26±0.01

References

1. J. L. Holm, O. J. Kleppa and E. F. Westrum, Jr., Geochim. Cosmochim. Acta, to be published. $\Delta_{sol}H°$.
2. C. Hummel and H. E. Schwiete, Glastech. Ber. 32, 327 (1959). $\Delta_{sol}H°$.
3. F. C. Kracek, Ann. Rept. Director of the Geophysical Laboratory, No. 1215, 69 (1953). $\Delta_{sol}H°$.
4. G. L. Humphrey and E. G. King, J. Amer. Chem. Soc. 74, 2041 (1952); 73, 656 (1951). $\Delta_{sol}H°$.
5. S. Matoba, K. Gunji and T. Kuwana, Tetsu to Hagane 45, 229 (1959); Stahl Eisen 80, 299 (1960). Keq.
6. R. H. Rein and J. Chipman, J. Phys. Chem. 67, 839 (1963). Keq.
7. D. A. R. Kay and J. Taylor, Trans. Faraday Soc. 56, 1372 (1960). $K_p°$.
8. J. D. Baird and J. Taylor, Trans. Faraday Soc. 54, 526 (1958). $K_p°$.
9. W. Eitel, "Silicate Science," Vol. III, pp. 135-8, Academic Press, New York, 1965.
10. J. Chipman and R. Baschwitz, Trans. AIME 227, 473 (1963); 233, 415 (1965). Si act. coeff.
11. E. F. Westrum, Jr. et al., Univ. Michigan, personal comm., May 19, 1960; also Phys. Chem. Glasses 4, 188 (1963); $C_p°$(5-300 K).
12. C. T. Anderson, J. Amer. Chem. Soc. 58, 568 (1936). $C_p°$(55-297 K).
13. M. A. Mosesman and K. S. Pitzer, J. Amer. Chem. Soc. 63, 2348 (1941); enthalpy(367-498 K).
14. W. P. White, Amer. J. Sci. 47, 1 (1919). enthalpy(373 K).
15. V. G. Hill and R. Roy, J. Amer. Ceram. Soc. 41, 532-7 (1958). Transition data.
16. R. F. Walker, S. Zerfoss, S. F. Holley and L. J. Gross, J. Res. Nat. Bur. Stand. 61, 251 (1958). T_{trs}.
17. C. Berger, M. Richard and L. Eyraud. Bull. Soc. Chim. France 1965, 1491 (1965). $\Delta_{trs}H°$.
18. A. J. Majumdar, H. A. McKinstry and R. Roy, J. Phys. Chem. Solids 25, 1487 (1964). $\Delta_{trs}H°$.
19. G. Sabatier, Bull. Soc. Franc., Mineral. Crist. 80, 444 (1957). $\Delta_{trs}H°$.

Enthalpy Reference Temperature = T_r = 298.15 K; Standard State Pressure = p° = 0.1 MPa

T/K	$C_p°$ (J K^{-1}mol^{-1})	S° (J K^{-1}mol^{-1})	$-[G°-H°(T_r)]/T$ (J K^{-1}mol^{-1})	$H°-H°(T_r)$ (kJ mol^{-1})	$\Delta_fH°$ (kJ mol^{-1})	$\Delta_fG°$ (kJ mol^{-1})	Log K_f
0	0.000	0.000	INFINITE	-7.037	-903.482	-903.482	INFINITE
100	16.000	11.247	74.760	-6.351	-905.967	-889.378	464.563
200	32.995	27.878	47.145	-3.853	-907.559	-872.105	227.770
298.15	44.953	43.396	43.396	0.000	-908.346	-854.509	149.706
300	45.145	43.675	43.397	0.083	-908.355	-854.175	148.725
400	53.145	57.848	45.275	5.029	-908.501	-836.079	109.181
500	58.672	70.326	49.064	10.631	-908.236	-817.997	85.456
600	62.969	81.426	53.549	16.726	-907.677	-799.998	69.646
700	65.689	91.348	58.254	23.166	-906.940	-782.108	58.362
800	67.572	100.247	62.956	29.832	-906.119	-764.330	49.906
900	69.036	108.292	67.554	36.664	-905.254	-746.658	43.335
1000	70.291	115.633	72.000	43.632	-904.359	-729.084	38.083

PREVIOUS: CURRENT: June 1967

Silicon Oxide, Cristobalite, Low (SiO_2) $O_2Si_1(cr)$

$S°(298.15\ K) = 41.463 \pm 0.13\ J\ K^{-1}\ mol^{-1}$
$T_{trs1} = 847 \pm 1.5$ K (low quartz → high quartz)
$T_{trs2} = 1079 \pm 250$ K (high quartz → high cristobalite)
$T_{fus} = 1696 \pm 50$ K (high quartz → liquid)

$\Delta_fH°(0\ K) = -905.872 \pm 1.7\ kJ\ mol^{-1}$
$\Delta_fH°(298.15\ K) = -910.857 \pm 1.7\ kJ\ mol^{-1}$
$\Delta_{trs1}H° = 0.728 \pm 0.17\ kJ\ mol^{-1}$
$\Delta_{trs2}H° = 2.008 \pm 0.63\ kJ\ mol^{-1}$
$\Delta_{fus}H° = 7.699 \pm 0.8\ kJ\ mol^{-1}$

Enthalpy of Formation

The enthalpy of formation is based on data of Wise (1) for combustion of SiO_2(low quartz) and Si(cr) in F_2 to form SiF_4(g). Good (2) burned a mixture of Si(cr) and vinylidene fluoride polymer in oxygen in the presence of HF(aq) to form H_2SiF_6(aq, HF). Combining this result with a previous heat of solution of quartz (3), the authors derived $\Delta_fH°(298.15\ K) = -217.5 \pm 0.5$ kcal mol^{-1}. Agreement with the adopted value is excellent, considering the complexity of the latter scheme. Both schemes are independent of $\Delta_fH°$(HF), although the polymer-Si scheme depends on the heat of dilution of HF(aq). Wise (1) has reviewed other data which confirm the adopted value rather than the oxygen combustion value (4) of -209.9 kcal mol^{-1}. Pertinent equilibrium data are analyzed on the table of SiO_2(low cristobalite), while the calorimetric results are summarized below.

Source	Reaction	$\Delta_rH°(298.15\ K)$, kcal mol^{-1}	$\Delta_fH°(298.15\ K)$, kcal mol^{-1}
(1)	SiO_2(low quartz) + 2 F_2(g) → SiF_4(g) + O_2(g)	-168.26 ± 0.28	-217.72 ± 0.34
(2)	See text.	--	-217.5 ± 0.5
(3)	Si(cr) + 2 O_2(g) → SiO_2(low quartz)	-209.9	-209.9

Heat Capacity and Entropy

Heat capacities below 300 K are based on data of Jones (5) and Westrum (6). Earlier data of Anderson (7) are in good agreement. The entropy is obtained from $C_p°$ using $S°(2\ K) = 0.00001$ cal K^{-1} mol^{-1} from 300 to 847 K is derived from adiabatic calorimetric data of Moser (8) and Sinel'nikov (9). The adopted $C_p°$ is discontinuous at 847 K and $\Delta C_p° = -2.17$ cal K mol . Values above the transition temperature are based on data of Moser and on data up to 1473 K obtained with a thermal analysis method by Leonidov (10). $C_p°$ is extrapolated linearly to higher temperatures.

From a review of data available before 1953, Kelley (11) selected almost identical enthalpies below 847 K but a value higher by 96 cal K^{-1} mol^{-1} at 900 K. The higher value, due to use of a larger $\Delta_{trs1}H°$, was presumably based on Roth (12), Wietzel (13) and White (14) whose enthalpy data were given the most weight by Kelley. Near 900 K these three sets of data deviate from the adopted function by -150±30, +130±40 and +15±35 kcal mol^{-1}, respectively; however, a recent enthalpy datum (15) obtained by transposed temperature drop calorimetry deviates +75 cal K^{-1} mol^{-1} at 968 K.

Transition Data

$T_{trs1} = 847 \pm 1.5$ K is based on thermal analysis of many specimens by Keith and Tuttle (16). Wide variations in T_{trs1} for some specimens were attributed by the authors to small amounts of impurities. The discontinuity in the volume curve (17) during inversion from trigonal low quartz to hexagonal high quartz suggests that the transition should be treated as first order. Thus, the enthalpy of transition is derived as (1011-854+17) = 174 cal K^{-1} mol^{-1}, where 1011 ± 6 is H°(873.15 K) - H°(823.15 K) measured by Moser (8), 854 is the corresponding increment calculated from the JANAF $C_p°$ and 17 is an endpoint correction for the difference between the observed and adopted curves below 823.15 K. Some recent data which confirm the adopted $\Delta_{trs1}H°$ are summarized below.

Source	Method	$\Delta_{trs1}H°$, kcal mol^{-1}
(8)	H and $C_p°$ from adiabatic calorimetry	0.174
(9)	Integration of $C_p°$ from adiabatic calorimetry	0.143
(10)	Thermal analysis	0.198
(17)	ΔV with dT/dP = 0.026 K/atm	0.12
(18)	Microcalorimetric thermal analysis, Single crystal	0.156, 0.165, 0.178 ± -.010
	Various crushed samples	0.124 to 0.193

Additional polymorphism of SiO_2 is discussed on the table for high cristobalite. Quartz has been superheated through the cristobalite region to some 300 above its metastable melting point. This melting point is calculated from the adopted tables as 1696 ± 50 K, which agrees well with the range 1673 - 1723 K indicated by the data of Mackenzie (19). The adopted tables predict the inversion of high quartz to high cristobalite at 1079 ± 250 K. Holmquist (20) tentatively placed this inversion at 1298 ± 25 K. $\Delta_{trs2}H°$ and $\Delta_{fus}H°$ are calculated from the differences in the $\Delta_fH°$ values of the appropriate phases at T_{trs2} and T_{fus}.

References

1. S. S. Wise, J. L. Margrave, H. M. Feder and W. N. Hubbard, J. Phys. Chem. 67, 815 (1963). $\Delta_rH°$.
2. W. D. Good, J. L. Lacina, B. L. DePrater and J. P. McCullough, J. Phys. Chem. 68, 579 (1964); 66, 380 (1962), $\Delta_rH°$.
3. E. G. King, J. Amer. Chem. Soc. 73, 656 (1951). $\Delta_rH°$.
4. G. L. Humphrey and E. G. King, J. Amer. Chem. Soc. 74, 2041 (1952). $\Delta_rH°$.
5. G. H. S. Jones and A. C. H. Hallett, Can. J. Phys. 38, 696 (1960); $C_p°$ (2-4 K).
6. E. F. Westrum, Jr. et al., Univ. Michigan, personal communication, May 19, 1960; see also A. J. Leadbetter and J. A. Morrison, Phys. Chem. Glasses 4, 188 (1963). $C_p°$ (5-300 K).
7. C. T. Anderson, J. Amer. Chem. Soc. 58, 568 (1936). $C_p°$ (53-296 K).
8. H. Moser, Physik. Z. 37, 737 (1936). $C_p°$ (317-949 K).
9. N. N. Sinel'nikov, Dokl. Akad. Nauk SSSR 92, 369 (1953). $C_p°$ (300-900 K), given graphically.
10. V. Ya. Leonidov, Yu. P. Barskii and N. I. Khitarov, Geochemistry 1964, 409 (1964). $C_p°$ (373-1473 K), graphical.
11. K. K. Kelley, U. S. Bur. Mines Bull. 584, 1960.
12. W. A. Roth and W. W. Bertram, Z. Elektrochem. 35, 297 (1929). enthalpy (369-1171 K).
13. R. Wietzel, Z. Anorg. Allg. Chem. 116, 71 (1921). enthalpy (373-1673 K).
14. W. P. White, Amer. J. Sci. 47, 1 (1919). enthalpy (373-1373 K).
15. J. L. Holm, O. J. Kleppa and E. F. Westrum, Jr., Geochim. Cosmochim. Acta, to be published. enthalpy (968 K).
16. M. L. Keith and O. F. Tuttle, Amer. J. Sci., Bowen Vol. 1952, 203 (1952). T_{trs}.
17. C. Berger, L. Eyraud, M. Richard and R. Riviere, Bull. Soc. Chim. France 1966, 628 (1966). Transition data.
18. C. Berger, M. Richard and L. Eyraud, Bull. Soc. Chim. France 1965, 1491 (1965). $\Delta_{trs}H°$.
19. J. D. Mackenzie, J. Amer. Ceram. Soc. 43, 615 (1960). T_{fus}.
20. S. B. Holmquist, J. Amer. Ceram. Soc. 44, 82 (1961). T_{trs2}.

Enthalpy Reference Temperature = T_r = 298.15 K — Standard State Pressure = p° = 0.1 MPa

T/K	$C_p°$ (J $K^{-1}mol^{-1}$)	S° (J $K^{-1}mol^{-1}$)	-[G°-H°(T_r)]/T (J $K^{-1}mol^{-1}$)	H°-H°(T_r) (kJ mol^{-1})	$\Delta_fH°$ (kJ mol^{-1})	$\Delta_fG°$ (kJ mol^{-1})	Log K_f
0	0.000	0.000	INFINITE	-6.916	-905.872	-905.872	INFINITE
100	15.686	9.694	72.496	-6.280	-908.406	-891.662	465.756
200	32.639	26.087	45.187	-3.820	-910.036	-874.223	228.324
298.15	44.589	41.463	41.463	0.000	-910.857	-856.443	150.045
300	44.769	41.740	41.464	0.083	-910.866	-856.106	149.061
400	53.430	55.874	43.333	5.017	-911.024	-837.812	109.407
500	59.643	68.499	47.127	10.686	-910.692	-819.539	85.617
600	64.417	79.808	51.648	16.896	-910.017	-801.368	69.765
700	68.768	90.063	56.414	23.554	-909.062	-783.331	58.453
800	73.701	99.555	61.220	30.668	-907.794	-765.452	49.979
847.000	76.509	103.840	63.467	34.196	I <--> II		
847.000	67.417	104.700	63.467	34.924	TRANSITION		
900	67.948	108.807	66.017	38.512	-905.918	-747.785	43.400
1000	68.952	116.018	70.662	45.357	-905.145	-730.256	38.145
1100	69.956	122.637	75.090	52.302	-904.365	-712.805	33.848
1200	70.961	128.767	79.311	59.348	-903.567	-695.426	30.271
1300	71.965	134.487	83.338	66.494	-902.744	-678.114	27.247
1400	72.969	139.857	87.185	73.741	-901.893	-660.866	24.657
1500	73.973	144.925	90.867	81.088	-901.011	-643.681	22.415
1600	74.977	149.731	94.397	88.535	-900.097	-626.555	20.455
1696.000	75.940	154.128	97.654	95.779	-------- II <--> LIQUID	--------	
1700	75.981	154.307	97.787	96.083	-949.326	-609.041	18.714
1800	76.986	158.678	101.050	103.732	-948.113	-589.059	17.094
1900	77.990	162.868	104.194	111.480	-946.823	-569.146	15.647
2000	78.994	166.894	107.229	119.330	-945.456	-549.303	14.346

PREVIOUS: December 1962 CURRENT: June 1967

SILICON OXIDE (SiO_2) LIQUID M_r = 64.0843

$\Delta_f H°(0\ K) = 897.756\ kJ\ mol^{-1}$
$S°(298.15\ K) = 47.928 \pm 1.3\ J\ K^{-1}\ mol^{-1}$ $\Delta_f H°(298.15\ K) = -902.661\ kJ\ mol^{-1}$
T_{fus} = 1996 ± 5 K(high cristobalite → liquid) $\Delta_{fus}H° = 9.581 \pm 2.1\ kJ\ mol^{-1}$

Enthalpy of Formation

$\Delta_f H°$(ℓ, supercooled, 298.15 K) is calculated from that of quartz using $\Delta_r H°$(970 K) = 1.45 Δ 0.15 kcal mol^{-1} for high quartz A SiO_2(ℓ), as determined by Holm (1) from $\Delta_{sol}H°$ in an oxide melt. This value reduces to $\Delta_r H°$(298.15 K) = 1.96 kcal mol^{-1} for low quartz → SiO_2(ℓ) and is consistent within ±0.3 kcal mol^{-1} with other calorimetric data summarized below. Earlier results were reviewed by Hummel (4).

Source	Method	Reaction	T/K	$\Delta_r H°(T)$ kcal mol^{-1}	$\Delta_r H°$(298.15 K) kcal mol^{-1}	$\Delta_f H°$(298.15 K) kcal mol^{-1}
(1)	$\Delta_{sol}H°$ in oxide melt	A	970	1.45±0.15	1.96	-215.74
(2)	$\Delta_c H°$ in flourine	B	298	1.78±0.46	1.78	-215.92
(3)	$\Delta_{sol}H°$ in HF(aq, 74.7°C)	A	298-348	-	2.18±0.05	-215.52
(4)	$\Delta_{sol}H°$ in HF(aq, 26.5°C)	A	300	2.27±0.2	2.27	-215.43

A. SiO_2(quartz, low or high) → SiO_2(ℓ)
B. SiO_2(ℓ) + 2 F_2(g) → SiF_4(g) + O_2(g)

Heat Capacity and Entropy

$C_p°$ below 300 K is based on data of Flubacher (5) and Westrum (6) for samples annealed at 1100 and 1070°C, respectively. Data of Turdakin (7) are higher by about 1% in the range 60-160 K and at 300 K but are in agreement elsewhere. The entropy is calculated from $C_p°$ using S°(2.34 K) -S°(0 K) = 0.0001 cal $K^{-1} mol^{-1}$ and S°(0 K) = 1.098 cal K^{-1} mol^{-1}. $C_p°$ data (6) for a different sample annealed at 1300°C yield an entropy larger by <0.04 cal K^{-1} mol^{-1}. Earlier data were reviewed by Kelley (8). The residual entropy is calculated from S°(0 K) = S°(1996 K) - S°(1996;0 K) = 41.258 - 40.160 = 1.098 cal K^{-1} mol^{-1}, where 40.160 is obtained from the adopted functions, while 41.258 is the sum of $\Delta_{fus}S°$ = 1.14 cal K^{-1} mol^{-1} and S°(1996 K) = 40.111 cal K^{-1} mol^{-1} for high cristobalite. We estimate the uncertainty as ±0.3 cal K^{-1} mol^{-1}, primarily from the uncertainty in $\Delta_{fus}H°$.

$C_p°$ above 300 K is based on adiabatic calorimetric data of Moser (9) up to 900 K. Enthalpy data of White (10) and Fischer (11) are in excellent agreement. $C_p°$ from 900 to 1500 K is derived from enthalpy data of Southard (12) and White (10). Maximum deviation of the data from the adopted values in this range is 0.5%. Enthalpy data of Egan (13) and Kelley (14) deviate from the adopted values by less than 1% in the range 1000-1500 K, although large positive deviations appear at lower temperatures.

In the region 1500-1700 K, the adopted $C_p°$ shows a rapid, sigmoidal rise which is suggested by the enthalpy data of Wietzel (15). This is presumed to be the glass transition region in which $C_p°$ rises to that of the stable liquid. $C_p°$ above 1700 K is estimated as 20.5 cal K^{-1} mol^{-1} or 6.83 cal K^{-1} g-$atom^{-1}$. This value is consistent with the enthalpy data (15) and with liquid heat capacities of Al_2O_3, $Na_2Si_2O_5$, and Na_2SiO_3.

Fusion Data

The melting point of high cristobalite is 1996 ± 5 K, while the metastable melting point of high quartz is 1696 ± 50 K. $\Delta_{fus}H°$ is the difference between $\Delta_f H°$ for liquid and high cristobalite at T_{fus}. Values ranging from 1.8 to 3.6 kcal mol^{-1} have been derived from various interpretations of phase diagrams (1).

References

1. J. L. Holm, O. J. Kleppa and E. F. Westrum, Jr., Geochim. Cosmochim. Acta, to be published. $\Delta_{sol}H°$.
2. S. S. Wise, J. L. Margrave, H. M. Feder and W. N. Hubbard, J. Phys. Chem. 67, 815 (1963). $\Delta_r H°$.
3. F. C. Kracek, Ann. Rept. Director of the Geophysical Laboratory, No. 1215, 69 (1953). $\Delta_{sol}H°$.
4. C. Hummel and H. E. Schwiete, Glastech. Ber. 32, 327 (1959). $\Delta_{sol}H°$.
5. P. Flubacher, A. J. Leadbetter, J. A. Morrison and B. P. Stoicheff, Phys. Chem. Solids 12, 53 (1959). $C_p°$ (2.3-19 K).
6. E. F. Westrum, Jr., et al., Univ. Michigan, personal communication, May 19, 1960. $C_p°$(5-300 K)
7. V. V. Tarassov, Phys. Status Solidi 20, 37 (1967). $C_p°$(50-300 K) measured by Turdakin.
8. K. K. Kelley and E. G. King, U. S. Bur. Mines Bull. 592, 1961.
9. H. Moser, Physik. Z. 37, 737 (1936). $C_p°$(324-936 K).
10. W. P. White, J. Amer. Chem. Soc. 55, 1047 (1933). enthalpy (573-1173 K).
11. W. Fischer, Z. Anorg. Allg. Chem. 200, 335 (1931). enthalpy (321-577 K).
12. J. C. Southard, J. Amer. Chem. Soc. 63, 3142 (1941). enthalpy (506-1522 K).
13. E. P. Egan, Jr., Z. T. Wakefield and K. L. Elmore, J. Amer. Chem. Soc. 72, 2418 (1950). enthalpy (763-1470 K).
14. K. K. Kelley, B. F. Naylor and C. H. Shomate, U. S. Bur. Mines Tech. Paper 686, 1946. enthalpy (400-1500 K).
15. R. Wietzel, Z. Anorg. Allg. Chem. 116, 71 (1921). H°(373-1973).

Silicon Oxide (SiO_2) $O_2Si_1(l)$

Enthalpy Reference Temperature = T_r = 298.15 K; Standard State Pressure = p° = 0.1 MPa

T/K	$C_p°$ (J $K^{-1}mol^{-1}$)	S° (J $K^{-1}mol^{-1}$)	$-[G°-H°(T_r)]/T$ (J $K^{-1}mol^{-1}$)	$H°-H°(T_r)$ (kJ mol^{-1})	$\Delta_f H°$ (kJ mol^{-1})	$\Delta_f G°$ (kJ mol^{-1})	Log K_f
0	0.000	4.594	INFINITE	-6.996	-897.756	-897.756	INFINITE
100	16.154	15.983	78.952	-6.297	-900.228	-884.112	461.813
200	32.736	32.602	51.618	-3.803	-901.824	-867.314	226.519
298.15	44.183	47.928	47.928	0.000	-902.661	-850.175	148.947
300	44.367	48.202	47.929	0.082	-902.671	-849.850	147.972
400	52.518	62.156	49.776	4.952	-902.893	-832.194	108.673
500	58.082	74.510	53.513	10.499	-902.684	-814.537	85.094
600	62.045	85.468	57.943	16.515	-902.203	-796.949	69.381
700	64.973	95.263	62.587	22.873	-901.548	-779.457	58.164
800	67.208	104.091	67.233	29.486	-900.780	-762.067	49.758
900	68.956	112.112	71.780	36.298	-899.936	-744.777	43.226
1000	70.350	119.452	76.186	43.266	-899.041	-727.585	38.005
1100	71.475	126.211	80.430	50.359	-898.113	-710.484	33.738
1200	72.396	132.471	84.509	57.554	-897.165	-693.469	30.186
1300	73.157	138.296	88.425	64.832	-896.211	-676.533	27.183
1400	73.848	143.743	92.184	72.183	-895.256	-659.670	24.613
1500	74.475	148.852	95.794	79.587	-894.317	-642.876	22.387
1600	80.040	153.767	99.263	87.206	-893.232	-626.146	20.442
1696.000	85.772	158.655	102.487	95.261	-------- II	<--> LIQUID	--------
1700	85.772	158.857	102.619	95.604	-941.610	-609.059	18.714
1800	85.772	163.759	105.881	104.181	-939.468	-589.559	17.109
1900	85.772	168.397	109.050	112.759	-937.350	-570.178	15.675
2000	85.772	172.796	112.128	121.336	-935.255	-550.907	14.388
2100	85.772	176.981	115.118	129.913	-933.183	-531.741	13.226
2200	85.772	180.971	118.021	138.490	-931.133	-512.672	12.172
2300	85.772	184.784	120.842	147.067	-929.106	-493.697	11.212
2400	85.772	188.434	123.582	155.645	-927.102	-474.809	10.334
2500	85.772	191.936	126.247	164.222	-925.119	-456.004	9.528
2600	85.772	195.300	128.839	172.799	-923.158	-437.279	8.785
2700	85.772	198.537	131.360	181.376	-921.217	-418.628	8.099
2800	85.772	201.656	133.816	189.953	-919.297	-400.048	7.463
2900	85.772	204.666	136.207	198.531	-917.397	-381.538	6.872
3000	85.772	207.574	138.538	207.108	-915.517	-363.092	6.322
3100	85.772	210.386	140.810	215.685	-913.655	-344.708	5.808
3200	85.772	213.109	143.027	224.262	-911.811	-326.385	5.328
3300	85.772	215.749	145.191	232.839	-909.985	-308.119	4.877
3400	85.772	218.309	147.304	241.417	-908.175	-289.907	4.454
3500	85.772	220.796	149.369	249.994	-906.381	-271.749	4.056
3600	85.772	223.212	151.387	258.571	-1288.754	-243.181	3.528
3700	85.772	225.562	153.360	267.148	-1286.577	-214.168	3.024
3800	85.772	227.849	155.290	275.725	-1284.416	-185.212	2.546
3900	85.772	230.077	157.179	284.303	-1282.269	-156.313	2.094
4000	85.772	232.249	159.029	292.880	-1280.137	-127.470	1.665
4100	85.772	234.367	160.841	301.457	-1278.019	-98.680	1.257
4200	85.772	236.434	162.616	310.034	-1275.914	-69.941	0.870
4300	85.772	238.452	164.356	318.611	-1273.822	-41.252	0.501
4400	85.772	240.424	166.063	327.189	-1271.742	-12.612	0.150
4500	85.772	242.351	167.737	335.766	-1269.674	15.981	-0.186

PREVIOUS: December 1962 CURRENT: June 1967

Silicon Oxide (SiO_2) $O_2Si_1(l)$

SILICON OXIDE (SiO_2) CRYSTAL–CRYSTAL–CRYSTAL–LIQUID M_r = 60.0843 Silicon Oxide (SiO_2) O_2Si_1(cr,l)

0 to 847 K		crystal, low quartz (trigonal)
847 to 1079 K		crystal, high quartz (hexagonal)
1079 to 1996 K		crystal, high crystabolite
above 1996 K		liquid

Refer to the individual tables for details.

Enthalpy Reference Temperature = T_r = 298.15 K   Standard State Pressure = p° = 0.1 MPa

T/K	C_p° (J K⁻¹mol⁻¹)	S° (J K⁻¹mol⁻¹)	$-[G°-H°(T_r)]/T$ (J K⁻¹mol⁻¹)	$H°-H°(T_r)$ (kJ mol⁻¹)	$\Delta_fH°$ (kJ mol⁻¹)	$\Delta_fG°$ (kJ mol⁻¹)	Log K_f
0	0.000	0.000	INFINITE	-6.916	-905.872	-905.872	INFINITE
100	15.686	9.694	72.496	-6.280	-908.406	-891.662	465.756
200	32.639	26.087	45.187	-3.820	-910.036	-874.223	228.324
298.15	44.589	41.463	41.463	0.000	-910.857	-856.443	150.045
300	44.769	41.740	41.464	0.083	-910.866	-856.106	149.061
400	53.430	55.874	43.333	5.017	-911.024	-837.812	109.407
500	59.643	68.499	47.127	10.686	-910.692	-819.539	85.617
600	64.417	79.808	51.648	16.896	-910.017	-801.368	69.765
700	68.768	90.063	56.414	23.554	-909.062	-783.331	58.453
800	73.701	99.555	61.220	30.668	-907.794	-765.452	49.979
847.000	76.509	103.840	63.467	34.196	I <--> II		
847.000	67.417	104.700	63.467	34.924	TRANSITION		
900	67.948	108.807	66.017	38.512	-905.918	-747.785	43.400
1000	68.952	116.018	70.662	45.357	-905.145	-730.256	38.145
1100	69.956	122.637	75.090	52.302	-904.365	-712.805	33.848
1200	70.961	128.767	79.311	59.348	-903.567	-695.426	30.271
1300	71.965	134.487	83.338	66.494	-902.744	-678.114	27.247
1400	72.969	139.857	87.185	73.741	-901.893	-660.866	24.657
1500	73.973	144.925	90.867	81.088	-901.011	-643.681	22.415
1600	74.977	149.731	94.397	88.535	-900.097	-626.555	20.455
1696.000	75.940	154.128	97.654	95.779	II <--> LIQUID		
1696.000	85.772	158.655	97.654	103.456	TRANSITION		
1700	85.772	158.857	97.798	103.799	-941.610	-609.059	18.714
1800	85.772	163.759	101.328	112.377	-939.468	-589.559	17.109
1900	85.772	168.397	104.737	120.954	-937.350	-570.178	15.675
2000	85.772	172.796	108.031	129.531	-935.255	-550.907	14.388
2100	85.772	176.981	111.215	138.108	-933.183	-531.741	13.226
2200	85.772	180.971	114.296	146.685	-931.133	-512.672	12.172
2300	85.772	184.784	117.278	155.263	-929.106	-493.697	11.212
2400	85.772	188.434	120.168	163.840	-927.102	-474.809	10.334
2500	85.772	191.936	122.969	172.417	-925.119	-456.004	9.528
2600	85.772	195.300	125.687	180.994	-923.158	-437.279	8.785
2700	85.772	198.537	128.325	189.571	-921.217	-418.628	8.099
2800	85.772	201.656	130.889	198.149	-919.297	-400.048	7.463
2900	85.772	204.666	133.381	206.726	-917.397	-381.538	6.872
3000	85.772	207.574	135.806	215.303	-915.517	-363.092	6.322
3100	85.772	210.386	138.167	223.880	-913.655	-344.708	5.808
3200	85.772	213.109	140.466	232.457	-911.811	-326.385	5.328
3300	85.772	215.749	142.708	241.035	-909.985	-308.119	4.877
3400	85.772	218.309	144.894	249.612	-908.175	-289.907	4.454
3500	85.772	220.796	147.027	258.189	-906.381	-271.749	4.056
3600	85.772	223.212	149.110	266.766	-1288.754	-243.181	3.528
3700	85.772	225.562	151.145	275.343	-1286.577	-214.168	3.024
3800	85.772	227.849	153.133	283.921	-1284.416	-185.212	2.546
3900	85.772	230.077	155.078	292.498	-1282.269	-156.313	2.094
4000	85.772	232.249	156.980	301.075	-1280.137	-127.470	1.665
4100	85.772	234.367	158.842	309.652	-1278.019	-98.680	1.257
4200	85.772	236.434	160.665	318.229	-1275.914	-69.941	0.870
4300	85.772	238.452	162.450	326.807	-1273.822	-41.252	0.501
4400	85.772	240.424	164.200	335.384	-1271.742	-12.612	0.150
4500	85.772	242.351	165.916	343.961	-1269.674	15.981	-0.186

PREVIOUS: CURRENT: June 1967

Silicon Oxide (SiO_2) O_2Si_1(cr,l)

SILICON OXIDE (SiO_2) IDEAL GAS M_r = 60.0843

$S°(298.15\ K) = [228.976]\ J\ K^{-1}\ mol^{-1}$

$\Delta_f H°(0\ K) = -304.11 \pm 33.5\ kJ\ mol^{-1}$

$\Delta_f H°(298.15\ K) = -305.43 \pm 33.5\ kJ\ mol^{-1}$

Vibrational Frequencies and Degeneracies

ν, cm^{-1}
[980] (1)
[370] (2)
[1430](1)

Ground State Quantum Weight: [1] σ = [2]
Point Group: $[D_{\infty h}]$
Bond Angle: O-Si-O = [180]°
Bond Distance: Si-O = [1.55] Å
Rotational Constant: B_0 = [0.219280] cm^{-1}

Enthalpy of Formation

The adopted value is based on the data summarized below. Margrave (1) has reported a value of $\Delta_{sub}H°(298.15\ K) = 140.9\pm1\ kcal\ mol^{-1}$ but details are not yet available. Assuming that this corresponds to reaction A below, we calculate $\Delta_f H°(298.15\ K) = -75.5\ kcal\ mol^{-1}$. Firsova (2) reported two values of the sublimation pressure which were derived from Knudsen effusion data using auxiliary data for SiO(g). Porter (4) derived two pressures from a mass spectrometric study and also obtained a 2nd law value for $\Delta_{sub}H°$ from ion intensities. The resulting range for $\Delta_{sub}H°(298.15\ K)$, 136 to 141 kcal mol^{-1}, is consistent with the lower limit of 130 kcal mol^{-1}, obtained by Bergman (3) from a method involving explosion in a spherical bomb. An alternative analysis of the mass spectrometric data (4) using the gas phase reaction B suggests an even higher $\Delta_{sub}H°$ of 148 kcal mol^{-1}. The adopted $\Delta_f H°$ corresponds to $\Delta_{sub}H°(298.15\ K) = 143.4\pm8\ kcal\ mol^{-1}$ and to $\Delta_{at}H(0\ K) = 297.3 \pm 8\ kJ\ mol^{-1}$.

Source	Reaction	Method	T/K	$\Delta_f H°(298.15\ K)$ kcal mol^{-1} 2nd law	3rd law	Drift cal K^{-1} mol^{-1}	Δ H°(298.15 K), kcal mol^{-1}
Margrave (1967)	A	-	-	-	140.9±1	-	-75.5
Firsova (1960)	A	Knudsen, calculation	1600-1800	-	137.0±2	-	-79.4
Bergman (1959)	A	Explosion	-	-	>130	-	>-86
Porter (1955)	A	Knudsen mass spec.	1800-1900	128	139.6±0.3	6	-76.8
	A	Ion intensities	1750-1980	136±8	-	-	-80
	B	Knudsen mass spec.	1800-1900	-2	-44.8±1.2	-23	-68.8

A. SiO_2(high cristobalite)= SiO_2(g) B. $SiO(g)+1/2O_2(g) = SiO_2(g)$

Heat Capacity and Entropy

Point group, ground state quantum weight and the linear symmetrical configuration are estimated by analogy with the properties of CO_2. From the bond lengths in SiO, CO_2 and CO, the length in SiO_2 is estimated as 1.509 x 1.160/1.128. Bond lengths and atomization energies for these molecules are reasonably consistent with the existence of six-electron bonds in the monoxides and four-electron bonds in the dioxides (5). Vibrational frequencies are estimated from a valence bond calculation using force constants, $k = 9 \times 10^5$ and $k\delta/\ell^2 = 0.3 \times 10^5$ dyne/cm, which are obtained from those of SiO, CO_2 and CO.

References

1. J. L. Margave, 5th Meeting ICRPG Thermochemistry Working Group, Linden, New Jersey, March 1967.
2. L. P. Firsova and An. N. Nesmeyanov, Russ. J. Phys. Chem. 34, 1279, 1279, 1232, 906 (1960).
3. G. A. Bergman and V. A. Medvedev, Sbornik Trudox Gosudarst. Inst. Priklad. Khim. 1959, 158 (1959).
4. R. F. Porter, W. A. Chupka and M. G. Inghram, J. Chem. Phys. 23, 216 (1955).
5. J. W. Linnett, "The Electronic Structure of Molecules," pp. 43-55, Methuen and Co. Ltd., London, 1964.

Silicon Oxide (SiO_2) O_2Si_1(g)

Enthalpy Reference Temperature = T_r = 298.15 K; Standard State Pressure = p° = 0.1 MPa

T/K	$C_p°$ (J $K^{-1}mol^{-1}$)	S°	$-[G°-H°(T_r)]/T$	$H°-H°(T_r)$ (kJ mol^{-1})	$\Delta_f H°$	$\Delta_f G°$	Log K_f
0	0.	0.	INFINITE	-10.581	-304.112	-304.112	INFINITE
100	31.422	188.175	264.459	-7.628	-304.330	-305.433	159.542
200	38.997	212.419	232.851	-4.086	-304.878	-306.331	80.006
250	41.746	221.426	229.688	-2.066	-305.165	-306.662	64.074
298.15	44.007	228.976	228.976	0.	-305.432	-306.926	53.772
300	44.089	229.249	228.977	0.081	-305.442	-306.935	53.442
350	46.161	236.204	229.522	2.339	-305.696	-307.163	45.842
400	48.007	242.491	230.756	4.694	-305.922	-307.357	40.137
450	49.641	248.241	232.384	7.136	-306.121	-307.524	35.696
500	51.076	253.547	234.238	9.655	-306.298	-307.670	32.142
600	53.416	263.076	238.269	14.885	-306.604	-307.915	26.806
700	55.180	271.449	242.423	20.319	-306.872	-308.112	22.992
800	56.513	278.909	246.526	25.906	-307.130	-308.272	20.128
900	57.531	285.627	250.503	31.611	-307.394	-308.399	17.899
1000	58.317	291.730	254.326	37.405	-307.672	-308.496	16.114
1100	58.934	297.319	257.984	43.269	-307.974	-308.563	14.652
1200	59.425	302.468	261.479	49.187	-308.302	-308.603	13.433
1300	59.820	307.241	264.818	55.150	-308.663	-308.613	12.400
1400	60.143	311.687	268.009	61.149	-309.060	-308.595	11.514
1500	60.409	315.845	271.061	67.177	-309.497	-308.547	10.745
1600	60.631	319.751	273.983	73.229	-309.978	-308.468	10.070
1700	60.817	323.433	276.785	79.302	-360.682	-307.911	9.461
1800	60.976	326.914	279.474	85.392	-361.028	-304.797	8.845
1900	61.111	330.214	282.058	91.497	-361.383	-301.663	8.293
2000	61.228	333.352	284.545	97.614	-361.747	-298.511	7.796
2100	61.330	336.342	286.941	103.742	-362.125	-295.340	7.346
2200	61.418	339.197	289.252	109.879	-362.515	-292.150	6.937
2300	61.496	341.929	291.483	116.025	-362.919	-288.943	6.562
2400	61.564	344.547	293.640	122.178	-363.339	-285.718	6.218
2500	61.625	347.062	295.727	128.337	-363.774	-282.475	5.902
2600	61.679	349.480	297.748	134.503	-364.225	-279.214	5.609
2700	61.727	351.809	299.707	140.673	-364.691	-275.935	5.338
2800	61.770	354.054	301.609	146.848	-365.173	-272.639	5.086
2900	61.809	356.223	303.455	153.027	-365.672	-269.326	4.851
3000	61.845	358.319	305.249	159.210	-366.186	-265.995	4.631
3100	61.877	360.347	306.993	165.396	-366.715	-262.646	4.426
3200	61.906	362.312	308.692	171.585	-367.259	-259.281	4.232
3300	61.933	364.217	310.346	177.777	-367.818	-255.898	4.051
3400	61.957	366.067	311.957	183.971	-368.390	-252.497	3.879
3500	61.979	367.863	313.529	190.168	-368.977	-249.081	3.717
3600	62.000	369.609	315.063	196.367	-753.729	-235.186	3.412
3700	62.019	371.308	316.560	202.568	-753.928	-220.779	3.117
3800	62.036	372.962	318.023	208.771	-754.141	-206.366	2.837
3900	62.052	374.574	319.452	214.975	-754.367	-191.948	2.571
4000	62.067	376.145	320.850	221.181	-754.606	-177.524	2.318
4100	62.081	377.678	322.217	227.389	-754.858	-163.094	2.078
4200	62.094	379.174	323.556	233.597	-755.122	-148.658	1.849
4300	62.106	380.635	324.866	239.807	-755.396	-134.215	1.630
4400	62.117	382.063	326.150	246.018	-755.683	-119.766	1.422
4500	62.128	383.459	327.408	252.231	-755.979	-105.310	1.222
4600	62.137	384.825	328.641	258.444	-756.287	-90.848	1.032
4700	62.147	386.161	329.851	264.658	-756.606	-76.378	0.849
4800	62.155	387.470	331.038	270.873	-756.936	-61.902	0.674
4900	62.163	388.752	332.203	277.089	-757.278	-47.418	0.505
5000	62.171	390.007	333.346	283.306	-757.632	-32.927	0.344
5100	62.178	391.239	334.469	289.523	-757.997	-18.429	0.189
5200	62.185	392.446	335.573	295.742	-758.376	-3.925	0.039
5300	62.192	393.631	336.657	301.960	-758.769	10.588	-0.104
5400	62.198	394.793	337.723	308.180	-759.177	25.108	-0.243
5500	62.203	395.935	338.771	314.400	-759.599	39.635	-0.376
5600	62.209	397.055	339.802	320.621	-760.038	54.171	-0.505
5700	62.214	398.157	340.816	326.842	-760.495	68.714	-0.630
5800	62.219	399.239	341.814	333.063	-760.972	83.267	-0.750
5900	62.224	400.302	342.796	339.286	-761.468	97.825	-0.866
6000	62.228	401.348	343.763	345.508	-761.985	112.395	-0.978

PREVIOUS: September 1967 (1 atm) CURRENT: September 1967 (1 bar)

Silicon Oxide (SiO_2) O_2Si_1(g)

$\Delta_f H°(0\ K) = -197.73 \pm 62.8\ kJ\ mol^{-1}$

$S°(298.15\ K) = [280.306 \pm 11.72]\ J\ K^{-1}\ mol^{-1}$ $\Delta_f H°(298.15\ K) = -200.83 \pm 62.8\ kJ\ mol^{-1}$

Electronic Levels and Quantum Weights

State	ε_i, cm^{-1}	g_i
X^2B_1	0	2
$^2A_1$	11618	2
$^2A_2$	16236	2

Vibrational Frequencies and Degeneracies

ν, cm^{-1}
971 (1)
[300](1)
912 (1)

Point Group: C_{2v} σ = 2

Bond Distance: Ta-O = [1.69] Å Bond Angle: O-Ta-O = [110]°

Product of the Moments of Inertia: $I_AI_BI_C = [6.232542 \times 10^{-115}]\ g^3\ cm^6$

Enthalpy of Formation

Inghram, et al. (1) and Krikorian and Carpenter (2) have presented mass spectrometric evidence indicating that the predominant vapors in the vaporization of Ta-Ta_2O_5 mixtures in Ta Knudsen cells are TaO and TaO_2. As in the case with TaO(g) (5), the absolute pressures of TaO_2(g) calculated from both studies differ by a factor of roughly 1000 in the range 2022-2278 K (IPTS-68). Results of our 2nd and 3rd law analysis of these and other data are presented below.

Source	Reaction	Data Points	T/K	$\Delta_rH°$(298.15 K), kcal mol^{-1} 2nd law	3rd law	Drift cal K^{-1} mol^{-1}	$\Delta_fH°$(298.15 K), kcal mol^{-1} 2nd law	3rd law
(1)	A*	2	2022-2023	-	156.13		-	-39.47
(1)	B*	4	2151-2209	51.67	150.86	45 ± 41	-	-36.25
(2)	B	1	2271	-	184.67		-	-2.44
(1)	C*	4	2022-2179	70.72	70.05	-0.3 ± 4.5	-49.44	-48.10
(2)	C	1	2271	-	81.03		-	-70.06
(3)	D	graph	1900-2860	-19.83	26.38	20.6 ± 2.8	-19.83	26.38

*Recalculated pressures using Otvis and Stevenson method for estimating ionization cross sections (2, 4); procedure increases $\Delta_rH°$ by ~ 3 kcal mol^{-1}.

A: 0.2 Ta(cr) + 0.4 Ta_2O_5(cr) = TaO_2(g) C: 0.5 Ta(cr) + 0.5 TaO_2(g) = TaO(g)

B: 0.2 Ta(cr) + 0.4 Ta_2O_5(ℓ) = TaO_2(g) D: Ta(cr) + O_2(g) = TaO_2(g)

As was observed in the TaO(g) table (5), the results of the 2nd and 3rd law analysis show considerable scatter. We adopt $\Delta_fH°(TaO_2$, g, 298.15 K) = -48.0 kcal mol^{-1} based primarily on our analysis of the data from Inghram et al. (1) for the reaction 0.5 Ta(cr) + 0.5 TaO_2(g) = TaO(g). For this reaction we adopt a rounded result of $\Delta_rH°$(298.15 K) = 70.0 kcal mol^{-1}. The adopted $\Delta_fH°$(298.15 K) value leads to a dissocation energy D_0^0 = 15.3 eV (3520 kcal mol^{-1}) for the process TaO_2(g) = Ta(g) + 2 O(g). This value is 1.8 times the dissociation energy of TaO(g) (5).

Heat Capacity and Entropy

The electronic levels and vibrational frequencies are those reported by Weltner and McLeod (7). The results were based on a matrix isolation study of TaO_2 trapped in argon and neon matrices at 20 K and 4 K, respectively. The infrared spectrum was interpreted to favor a bent molecule in the ground state. Kaufman et al. (8) reported that TaO_2(g) is bent based on the deflection of molecular beams of TaO_2 by an electronic quadrupole. This latter result is dependent, however, on there being no low lying electronic states. The bending frequency for the ground state has not been observed but is estiamted to be 300 cm^{-1} based on the observed ν_2 = 285 cm^{-1} value for the $^2A_1$ state (7). The Ta-O bond distance is estimated to be the same as in TaO(g) (5) whereas the bond angle is estimated to be 110°.

It should be noted that the Gibbs energy functions used here are roughly 2-4 cal K^{-1} g-$atom^{-1}$ more positive those of Brewer and Rosenblatt (6) for linear TaO_2; the 3rd law $\Delta_rH°$(298.15 K) values in the above table would change by ~4.5 kcal mol^{-1}.

References - TaO_2(g)

1. M. G. Inghram, W. A. Chupka, and J. Berkowitz, J. Chem. Phys. 27, 569 (1957).
2. O. H. Krikorian and J. H. Carpenter, J. Phys. Chem. 69, 4399 (1965).
3. J. Drowart, A. Pattoret, S. Smoes, F. Degreve, and D. Detry in "Advances in Mass Spectrometry," Vol. 3, W. L. Mead, Ed., The Institute of Petroleum, London, pp. 932-943, 1966.
4. J. W. Otvis and D. R. Stevenson, J. Amer. Chem. Soc. 78, 546 (1956).
5. JANAF Thermochemical Tables: TaO(g), 12-31-73.
6. L. Brewer and G. M. Rosenblatt, Chem. Rev. 61, 257 (1961).
7. W. Weltner, Jr. and D. McLeod, Jr., J. Chem. Phys. 42, 882 (1965).
8. M. Kaufman, J. Muenter, and W. Klemperer, J. Chem. Phys. 47, 3365 (1967).

References - TaO(g)

1. D. L. Hildenbrand and E. Murad, J. Chem. Phys. 51, 807 (1969).
2. JANAF Thermochemical Tables: O(g), 6-30-62; ZrO_2 and ZrO(g), 12-31-65; Ta(cr, g), Ta_2O_5(cr, ℓ), VO(g) and NbO(g), 12-31-73.
3. M. G. Inghram, W. A. Chupka, and J. Berkowitz, J. Chem. Phys. 27, 569 (1957).
4. O. H. Krikorian and J. H. Carpenter, J. Phys. Chem. 69, 4399 (1965).
5. J. W. Otvis and D. P. Stevenson, J. Amer. Chem. Soc. 78, 546 (1956).
6. E. D. Cater, E. R. Plante, and P. W. Gilles, J. Chem. Phys. 32, 1269 (1960).
7. H. W. Goldstein, P. N. Walsh and D. White, U. S. At. Energy Comm. AFOSR-TN-58-291, 1958 [Chem. Abstr. 55, 18227h (1961)].
8. C. C. Kiess and E. Z. Stowell, J. Res. Nat. Bur. Stand. 12, 459 (1934).
9. S. G. Krishnamurty and I. G. Fernando, Curr. Sci. 18, 371 (1949).
10. D. Premaswarup, Nature 175, 1003 (1955).
11. D. Premaswarup, Indian J. Phys. 29, 109 (1955).
12. D. Premaswarup, Curr. Sci. 24, 45 (1955).
13. D. Premaswarup and R. F. Barrow, Nature 180, 602 (1957).
14. W. Weltner Jr. and D. McLeod Jr., J. Chem. Phys. 42, 882 (1965).
15. C. J. Cheetham and R. F. Barrow, Trans. Faraday Soc. 63, 1835 (1967).
16. J. Drowart, A. Pattoret, S. Smoes, F. Degreve, and D. Detry in "Advances in Mass Spectrometry," Vol. 3, W. L. Mead, Ed., The Institute of Petroleum, London, pp. 931-943, 1966.
17. L. Brewer and G. M. Rosenblatt in "Advances in High Temperature Chemistry," Vol. 2, L. Eyring, Ed., Academic Press, New York, pp. 1-83, 1969.

Enthalpy Reference Temperature = T_r = 298.15 K Standard State Pressure = p° = 0.1 MPa

T/K	$C_p°$ (J $K^{-1}mol^{-1}$)	S°	-[G°-H°(T_r)]/T	H°-H°(T_r) (kJ mol^{-1})	$\Delta_fH°$	$\Delta_fG°$	Log K_f
0	0.	0.	INFINITE	-11.262	-197.730	-197.730	INFINITE
100	35.386	237.647	316.528	-7.888	-198.255	-203.075	106.076
200	39.861	263.617	284.196	-4.116	-199.646	-207.356	54.156
250	41.977	272.738	281.019	-2.070	-200.283	-209.208	43.712
298.15	44.000	280.306	280.306	0.	-200.832	-210.876	36.945
300	44.076	280.578	280.306	0.081	-200.852	-210.938	36.728
350	46.019	287.521	280.850	2.335	-201.348	-212.579	31.726
400	47.726	293.780	282.082	4.679	-201.784	-214.153	27.966
450	49.181	299.488	283.703	7.103	-202.176	-215.675	25.035
500	50.404	304.735	285.548	9.594	-202.538	-217.155	22.686
600	52.279	314.100	289.545	14.733	-203.218	-220.015	19.154
700	53.596	322.264	293.648	20.031	-203.879	-222.762	16.623
800	54.538	329.486	297.685	25.440	-204.540	-225.414	14.718
900	55.229	335.951	301.584	30.930	-205.212	-227.983	13.232
1000	55.746	341.798	305.318	36.480	-205.903	-230.477	12.039
1100	56.143	347.131	308.880	42.076	-206.627	-232.899	11.059
1200	56.453	352.030	312.275	47.706	-207.393	-235.254	10.240
1300	56.701	356.559	315.509	53.364	-208.202	-237.543	9.545
1400	56.903	360.768	318.593	59.045	-209.045	-239.769	8.946
1500	57.073	364.700	321.538	64.744	-209.914	-241.933	8.425
1600	57.218	368.388	324.352	70.458	-210.804	-244.039	7.967
1700	57.348	371.861	327.045	76.187	-211.722	-246.088	7.561
1800	57.467	375.142	329.627	81.928	-212.686	-248.083	7.199
1900	57.581	378.253	332.105	87.680	-213.712	-250.021	6.874
2000	57.692	381.209	334.487	93.444	-214.804	-251.904	6.579
2100	57.805	384.026	336.779	99.219	-215.960	-253.731	6.311
2200	57.922	386.718	338.989	105.005	-217.180	-255.501	6.066
2300	58.043	389.296	341.120	110.803	-218.466	-257.215	5.842
2400	58.170	391.768	343.179	116.614	-219.822	-258.871	5.634
2500	58.305	394.146	345.171	122.437	-221.254	-260.469	5.442
2600	58.447	396.435	347.099	128.275	-222.768	-262.008	5.264
2700	58.596	398.644	348.967	134.127	-224.376	-263.486	5.097
2800	58.753	400.778	350.780	139.994	-226.087	-264.904	4.942
2900	58.917	402.842	352.540	145.878	-227.914	-266.259	4.796
3000	59.088	404.843	354.250	151.778	-229.875	-267.548	4.658
3100	59.264	406.783	355.913	157.696	-231.985	-268.769	4.529
3200	59.445	408.667	357.533	163.631	-234.265	-269.920	4.406
3300	59.631	410.499	359.110	169.585	-273.190	-270.526	4.282
3400	59.820	412.282	360.648	175.557	-275.449	-270.411	4.154
3500	60.011	414.019	362.148	181.549	-277.706	-270.230	4.033
3600	60.205	415.712	363.612	187.560	-279.958	-269.985	3.917
3700	60.398	417.365	365.043	193.590	-282.206	-269.677	3.807
3800	60.592	418.978	366.441	199.639	-284.449	-269.308	3.702
3900	60.785	420.554	367.809	205.708	-286.686	-268.880	3.601
4000	60.976	422.096	369.147	211.796	-288.918	-268.395	3.505
4100	61.165	423.604	370.456	217.903	-291.143	-267.854	3.413
4200	61.350	425.080	371.739	224.029	-293.363	-267.260	3.324
4300	61.532	426.525	372.997	230.173	-295.576	-266.612	3.239
4400	61.710	427.942	374.230	236.335	-297.784	-265.913	3.157
4500	61.882	429.331	375.439	242.515	-299.986	-265.164	3.078
4600	62.050	430.693	376.625	248.712	-302.184	-264.367	3.002
4700	62.212	432.029	377.790	254.925	-304.377	-263.520	2.929
4800	62.368	433.340	378.933	261.154	-306.567	-262.628	2.858
4900	62.518	434.628	380.057	267.398	-308.755	-261.689	2.790
5000	62.662	435.892	381.161	273.657	-310.940	-260.706	2.724
5100	62.799	437.135	382.246	279.931	-313.126	-259.680	2.660
5200	62.930	438.355	383.314	286.217	-315.312	-258.612	2.598
5300	63.053	439.555	384.364	292.516	-317.499	-257.499	2.538
5400	63.170	440.735	385.397	298.827	-319.691	-256.347	2.480
5500	63.280	441.895	386.413	305.150	-321.886	-255.155	2.423
5600	63.384	443.036	387.414	311.483	-324.088	-253.920	2.368
5700	63.480	444.159	388.400	317.826	-326.298	-252.648	2.315
5800	63.570	445.264	389.371	324.179	-1067.272	-248.535	2.238
5900	63.653	446.351	390.327	330.540	-1069.715	-234.400	2.075
6000	63.730	447.422	391.270	336.910	-1072.215	-220.220	1.917

PREVIOUS: December 1973 (1 atm) CURRENT: December 1973 (1 bar)

ANATASE (TiO_2) CRYSTAL M_r = 79.8788

$S°(298.15 K) = 49.907 \pm 0.33\ J\ K^{-1}\ mol^{-1}$
T_{fus}(anatase → liquid) = [1949 ± 60 K]

$\Delta_f H°(0\ K) = -933.837 \pm 2.1\ kJ\ mol^{-1}$
$\Delta_f H°(298.15\ K) = -938.722 \pm 2.1\ kJ\ mol^{-1}$
$\Delta_{fus} H° = [57.990]\ kJ\ mol^{-1}$

Enthalpy of Formation

$\Delta_f H°$ is calculated from that of TiO_2(rutile) using $\Delta_r H°(968\ K) = -1.57 \pm 0.19\ kcal\ mol^{-1}$ for anatase → rutile. Navrotsky and Kleppa (1), using oxide-melt calorimetry, derived $\Delta_r H°$ from the difference in $\Delta_{sol} H°$ between anatase and rutile. Reduction of their value using JANAF enthalpies yields $\Delta_r H°(298.15\ K) = -1.44 \pm 0.3$ and $\Delta_f H°$(anatase, 298.15 K) = $-224.36 \pm 0.5\ kcal\ mol^{-1}$. Uncertainties in the enthalpies and in $\Delta_f H°$(rutile) contribute significantly to the overall uncertainty. Kybett and Margrave (2) derived $\Delta_f H°(298.15\ K) = -223 \pm 1\ kcal\ mol^{-1}$ from fluorine calorimetry. Their reaction was TiO_2(anatase) + 2 F_2(g) → TiF_4(cr) + O_2(g), but they gave no experimental data. Kybett's results for anatase and rutile may be combined to yield $\Delta_r H°(298.15\ K) = -2 \pm 1.4\ kcal\ mol^{-1}$ for anatase → rutile; this is consistent with the adopted value. Values for $\approx_r G° = \approx_r H°$ of -0.2 to -0.3 kcal mol^{-1} for anatase → rutile were derived by Schuiling and Vink (12) from relative solubilities in aqueous NaCl at 200° and 300°C. We suspect that this less negative result arises from saturating solid phases which differ in physical form and chemical composition from standard-state macrocrystalline anatase (see Phase Data).

Heat Capacity and Entropy

$C_p°$ below 300 K is based on data of Shomate (3, 52-296 K) and his Debye-Einstein extrapolation to absolute zero. The entropy is calculated by polynomial integration of $C_p°$ based on S°(50 K) = 0.45 cal $K^{-1}\ mol^{-1}$.

$C_p°$ above 300 K is from constrained fitting of enthalpy data (739-1305 K) of Naylor (4). Naylor's points at 416 and 546 K are omitted due to positive bias in this temperature range. Enthalpy data (580-1001 K) of Lietz (5) deviate by -3.7 → 1.2% at all temperatures. Similar discrepancies exist in Lietz' data for TiO_2(rutile). These discrepancies and the bias in Naylor's data near room temperature are discussed on the table for rutile (6). The choice between Naylor (4) and Lietz (5) is not clearcut, but we feel that Naylor's data lead to more reasonable values of $C_p°$. New data on a well-characterized sample are needed.

The two discrepant studies (4, 5) do agree that the enthalpy of anatase is slightly larger than that of rutile at all temperatures in the observed region. Our adopted $C_p°$ reproduces the approximate enthalpy difference between anatase and rutile as observed by each author (4, 5).

Phase Data

Both anatase and rutile have tetragonal crystal structures, but in anatase the elongated unit cell contains four molecules compared with two in the unit cell of rutile. Anatase transforms exothermally and irreversibly to rutile at temperatures above ≈1200 K (7-9). The rate of transformation is relatively slow and is governed by the nature and amount of impurities (7). Certain impurities markedly decrease the transformation temperature which can be as low as 700 K in the presence of an alkali flux (10). These observations led to the conclusion that anatase is a metastable phase which persists up to a temperature sufficient to activate the crystal rearrangement. This was confirmed by the calorimetric data of Navrotsky and Kleppa (1), who concluded that anatase is metastable with respect to rutile at all temperatures.

Lietz (5) reported a transition in anatase at ≈918 K based on his enthalpy data and on changes in prism angles measured by Schroder (11). We reject Lietz' T_{trs}, since his $\approx_{trs} H°$ is not distinguishable from his experimental scatter. Lietz' data for rutile and anatase exhibit remarkably similar deviations in this temperature region. Schroder's data have the same deficiency. Lattice parameter data (13) show scatter but no obvious change in this temperature range.

Characterization of anatase samples is particularly important for interpreting physical and chemical properties. Anatase precipitated from acidic sulfate solutions consists of agglomerates of very small crystals with large amounts of water and SO_3 (14). X-ray and electron diffraction show the major lines of anatase but fail to show the impurities, suggesting that the latter are present in noncrystalline form. These precipitates (14) lose water near 150°C and SO_3 near 650°C; crystal growth starts near 600°C. Precipitates from other media give anatase at lower temperatures (15), but calcination temperatures of at least 650°C are needed fror complete conversion of these precipitates into pure, macrocrystalline anatase. This suggests an alternative explanation for reports (5, 11) of T_{trs} ~645°C.

Fusion Data

T_{fus} is hypothetical in the sense that metastable anatase probably will convert to rutile (see Phase Data) before it can be superheated to the melting point. T_{fus} is calculated as the temperature at which $\Delta_r G° = 0$ for TiO_2(anatase) → TiO_2(ℓ). $\Delta_{fus} H°$ is the calculated difference in $\Delta_f H°$ at T_{fus}.

References

1. A. Navrotsky and O. J. Kleppa, J. Amer. Ceram. Soc. 50, 626 (1967).
2. B. D. Kybett and J. L. Margrave, Rice University, AFML-TR-65-123, Aug. 1965; Rev. Sci. Instr. 37, 675 (1966).
3. C. H. Shomate, J. Amer. Chem. Soc. 69, 218 (1947).
4. B. F. Naylor, J. Amer. Chem. Soc. 68, 1077 (1946).
5. J. Lietz, Hamburger Beitr. Angew. Mineral, U. Kirsallphys. 1, 229 (1956).
6. JANAF Thermochemical Tables: TiO_2(rutile), 12-31-73.
7. R. D. Shannon and J. A. Pask, J. Amer. Ceram. Soc. 48, 391 (1965).
8. F. W. Vahldiek, J. Less-Common Metals 11, 99 (1966).
9. L. D. Afanasenko and V. V. Klimov, Russ. J. Phys. Chem. 46, 878 (1972).
10. E. N. Bunting, J. Res. Nat. Bur. Stand. 11, 719 (1933).
11. A. Schroder, Z. Kryst. 66, 493 (1928); 67, 485 (1928).
12. R. D. Schuiling and B. W. Vink, Geochim. Cosmochim. Acta 31, 2399 (1967).
13. K. V. Krishna Rao, S. V. Nagender Naidu and L. Iyengar, J. Amer. Ceram. Soc. 53, 124 (1970).
14. W. F. Sullivan and S. S. Cole, J. Amer. Ceram. Soc. 42, 127 (1959).
15. M. N. Danchevskaya et al., Russ. J. Phys. Chem. 43, 444 (1969); I. A. Bondar et al., Inorganic Materials (USSR) 7, 1183 (1971).

Titanium Oxide, Anatase (TiO_2) O_2Ti_1(cr)

Enthalpy Reference Temperature = T_r = 298.15 K; Standard State Pressure = p° = 0.1 MPa

T/K	$C_p°$ (J K⁻¹mol⁻¹)	S° (J K⁻¹mol⁻¹)	$-[G°-H°(T_r)]/T$ (J K⁻¹mol⁻¹)	$H°-H°(T_r)$ (kJ mol⁻¹)	$\Delta_f H°$ (kJ mol⁻¹)	$\Delta_f G°$ (kJ mol⁻¹)	Log K_f
0	0.000	0.000	INFINITE	-8.627	-933.837	-933.837	INFINITE
100	19.159	9.623	89.203	-7.958	-936.632	-919.438	480.2[illegible]
200	41.865	30.501	54.580	-4.816	-938.318	-901.476	235.441
298.15	55.271	49.907	49.907	0.000	-938.722	-883.266	154.745
300	55.471	50.249	49.908	0.102	-938.721	-882.922	153.730
400	63.593	67.433	52.195	6.095	-938.313	-864.368	112.875
500	68.145	82.156	56.753	12.702	-937.506	-845.970	88.378
600	70.890	94.842	62.069	19.664	-936.529	-827.753	72.062
700	72.655	105.911	67.559	26.846	-935.488	-809.706	60.421
800	73.860	115.696	72.976	34.176	-934.421	-791.810	51.700
900	74.726	124.448	78.218	41.607	-933.386	-774.046	44.924
1000	75.354	132.355	83.242	49.112	-932.464	-756.393	39.510
1100	75.814	139.559	88.039	56.672	-931.729	-738.823	35.084
1200	76.191	146.172	92.612	64.272	-935.171	-721.191	31.393
1300	76.484	152.283	96.970	71.907	-934.101	-703.403	28.263
1400	76.735	157.960	101.126	79.568	-933.112	-685.695	25.584
1500	76.944	163.262	105.094	87.252	-932.219	-668.054	23.264
1600	77.111	168.233	108.886	94.955	-931.438	-650.470	21.236
1700	77.237	172.912	112.516	102.673	-930.785	-632.930	19.448
1800	77.362	177.330	115.995	110.403	-930.274	-615.425	17.859
1900	77.446	181.515	119.335	118.143	-929.921	-597.943	16.439
2000	77.530	185.490	122.544	125.892	-944.427	-580.022	15.1[illegible]

PREVIOUS: September 1966 CURRENT: December 1973

Titanium Oxide, Anatase (TiO_2) O_2Ti_1(cr)

$S°(298.15 K) = 50.292 \pm 0.17\ J\ K^{-1}\ mol^{-1}$

$T_{fus} = 2130 \pm 20\ K$ (rutile → liquid)

$\Delta_fH°(0 K) = -939.870 \pm 1.3\ kJ\ mol^{-1}$

$\Delta_fH°(298.15 K) = -944.747 \pm 1.3\ kJ\ mol^{-1}$

$\Delta_{fus}H° = [66.944 \pm 16.7]\ kJ\ mol^{-1}$

Enthalpy of Formation

$\Delta_fH°$ is the value obtained by Mah et al. (1) a study by oxygen calorimetry. Eighteen combustions (1) of three well-characterized samples of titanium metal led to a selected value of -225.8 ± 0.1 kcal mol^{-1}. This is comparable with other oxygen-combustion values of -225.5 ± 0.2 (2), -224.9 ± 0.4 (3) and -225.3 ± 0.3 (4) and a fluorine-combustion value of -225 ± 1 (5) derived from the reaction TiO_2(rutile) + 2 F_2(g) → TiF_4(cr) + O_2(g). No experimental data were given in the latter study (5).

The 1967 JANAF Table (6) assigned an uncertainty of ±1 kcal mol^{-1} to $\Delta_fH°$ due to possible uncertainty in corrections for oxygen deficiency in the rutile produced by combusion (1, 3). The atomic ratio O/Ti in the combustion product was determined by ignition to constant weight in air (1, 2) or in wet oxygen (3) at 1000°C. Ratios obtained were 1.985 to 1.994 (1), 2.00 (2) and about 1.95 (3). Mah's corrections to the stoichiometric composition correspond to about -0.8 to -1.7 kcal mol^{-1} in $\Delta_fH°$. Our opinion of the reliability of these corrections has changed due to new information on the crystal structure, $\Delta_fG°$ and $\Delta_fS°$ of oxygen-deficient compositions (7, 8). Mah's O/Ti ratios are at compositions where there appears to be hysteresis in measurements of the oxygen potential near 1300 K (7). The full range of the hysteresis corresponds to less than 20% of the oxygen potential. We assume that this percentage is an upper limit for error which might be introduced in the corrections of $\Delta_fG°$ and $\Delta_fH°$ from the observed O/Ti ratios to the stoichiometric composition. Thus, we assign an uncertainty of ±0.3 kcal mol^{-1} to $\Delta_fH°$ = -225.8, but note that the less negative value of -224.9 ± 0.4 kcal mol^{-1} (3) is one of several alternatives for minimizing the deviation from integral values of $\Delta_rG°$ (7) relating rutile to Ti_3O_5 at 1304 K.

Heat Capacity and Entropy

$C_p°$ below 300 K is based on data of Sandin and Keesom (9, 0.3-20 K), Dugdale et al. (10, 12-270 K) and Shomate (11, 52-298 K). Smoothed values rather than data points were reported in the first two studies (9, 10) and only in the range 20-50 K by Dugdale (10). Dugdale's deviation plot indicates that the smooth curves (10, 11) agree within about ±0.3% above 77 K, but at lower temperatures the data of Shomate show a positive bias increasing to over 2% at 52 K. We adopt Dugdale's curve below 80 K and calculate S° by polynomial integrations of $C_p°$. The extrapolation below 0.3 K is insignificant. $C_p°$ data (69-295 K) of McDonald and Seltz (12) are omitted since they are roughly 3% higher than the adopted curve.

$C_p°$ above 300 K is from constrained fitting of enthalpy data (763-1746 K) of Naylor (13). The point at 394 K is omitted due to positive bias (~6%) in the drop calorimeter at the lowest temperature; a similar bias is observed in Naylor's data for the four other titanium oxides and in materials such as MgF_2 studied at nearly the same time. This bias and the low purity (97.9%) of Naylor's sample are unfortunate; they favor the selection of two more recent enthalpy studies (14, 15) which conflict with that of Naylor. The adopted curve is favored, however, by four enthalpy points observed at 965 K by Navrotsky et al. (16) and by comparisons with $C_p°$ for GeO_2, MnO_2, and V_2O_4, which have the same tetragonal-SnO_2 crystal structure as rutile.

Enthalpy data of Lietz (15, 578-1283 K) deviate from the adopted curve by ~-3% at all temperatures; similar deviations occur in his data for TiO_2(anatase). Data of Slyusar et al. (14, 645-1990 K) deviate by ~-4% below 1136 K and by -1.7 to +3.4% at higher temperatures. The discontinuity of almost 3% in the data near 1100 K seems to correspond to the temperature at which the authors switched container materials and temperature sensors (14). Data for Ti_2O_3, Ti_3O_5 and Ti_4O_7 (14) also have discontinuities in this region; the data below the discontinuity for Ti_2O_3 show a negative bias like rutile. Moreoever, the authors (14) report T_{fus} = 2014 K, which is ~100 too low for rutile. Other data are too scattered (17) and too uncertain (18, 19) to help resolve the difference with Naylor (13).

These discrepancies cause us to reject the data of Slyusar et al. (14) and to consider as coincidence the agreement with Lietz (15). These two data sets yield $C_p°$ values deviating from the adopted curve by -0.8 cal K^{-1} mol^{-1} at 700 K, 0.0 at 1100 K, and +3 at 2000 K. New data on a pure sample of rutile are needed to resolve the disagreement. We note that use of the rejected $C_p°$ curve is one of several alternatives for minimizing the deviation from equilibria (7) relating rutile to Ti_3O_5 at 1304 K.

Fusion Data - See the liquid table for details.

Sublimation Data

$\Delta_{sub}H°(298.15 K) = 152.8 \pm 3$ kcal mol^{-1} is calculated from $\Delta_fH°(TiO_2, g)$. Sublimation makes TiO_2(cr) deficient in oxygen.

References

1. A. D. Mah, K. K. Kelley, N. L. Gellert, E. G. King and C. J. O'Brien, U. S. Bur. Mines RI-5316, 1957.
2. G. L. Humphrey, J. Amer. Chem. Soc. 73, 1587 (1951).
3. S. M. Ariya, M. P. Morozova, and E. Vol'f, Russ. J. Inorg. Chem. 2, 16 (1957).
4. B. Neumann, C. Korger and H. Kunz, Z. Anorg. Allg. Chem. 218, 379 (1934).
5. B. D. Kybett and J. L. Margrave, Rice University, AFML-TR-65-123, Aug., 1965; Rev. Sci. Instr. 37, 675 (1966).
6. JANAF Thermochemical Tables, 2nd ed., NSRDS-NBS 37, 1971.
7. R. R. Merritt, B. G. Hyde, L. A. Bursill and D. K. Philp, Phil. Trans. Roy. Soc. London, Ser. A 274, 627 (1973).
8. K. Suzuki and K. Sambongi, Tetsu To Hagane 58, 1579 (1972).
9. T. R. Sandin and P. H. Keesom, Phys. Rev. 177, 1370 (1969).
10. J. S. Dugdale, J. A. Morrison and D. Patterson, Proc. Roy. Soc. (London) A224, 228 (1954).
11. C. H. Shomate, J. Amer. Chem. Soc. 69, 218 (1947).
12. H. J. McDonald and H. Seltz, J. Amer. Chem. Soc. 61, 2405 (1939).
13. B. F. Naylor, J. Amer. Chem. Soc. 68, 1077 (1946).
14. N. P. Slyusar, A. D. Krivorontenko, E. N. Fomichev, A. A. Kalashnik and V. P. Bondarenko, Teplofiz. Vys. Temp. 11, 213 (1973).
15. J. Lietz, Hamburger Beitr. Angew. Mineral. u. Kristallphys. 1, 229 (1956).
16. A. Navrotsky, J. C Jamieson and O. J. Kleppa, Science 158, 388 (1967).
17. J. S. Arthur, J. Appl. Phys. 21, 732 (1950).
18. M. S. Erofeeva, N. L. Lukinykh and S. M. Ariya, Russ. J. Phys. Chem. 35, 375 (1961).
19. L. F. Nilson and O. Pettersson, Z. Phys. Chem. 1, 27 (1887).

Enthalpy Reference Temperature = T_r = 298.15 K; Standard State Pressure = p° = 0.1 MPa

T/K	$C_p°$ (J K⁻¹mol⁻¹)	S° (J K⁻¹mol⁻¹)	-[G°-H°(T_r)]/T (J K⁻¹mol⁻¹)	H°-H°(T_r) (kJ mol⁻¹)	$\Delta_fH°$ (kJ mol⁻¹)	$\Delta_fG°$ (kJ mol⁻¹)	Log K_f
0	0.000	0.000	INFINITE	-8.636	-939.870	-939.870	INFINITE
100	18.502	10.142	89.638	-7.950	-942.649	-925.506	483.435
200	42.012	30.807	54.969	-4.833	-944.360	-907.579	237.035
298.15	55.103	50.292	50.292	0.000	-944.747	-889.406	155.820
300	55.287	50.633	50.293	0.102	-944.746	-889.063	154.800
400	62.835	67.671	52.564	6.043	-944.390	-870.541	113.681
500	67.203	82.201	57.077	12.562	-943.670	-852.157	89.024
600	69.931	94.712	62.331	19.429	-942.789	-833.936	72.601
700	71.764	105.638	67.754	26.519	-941.841	-815.868	60.881
800	73.078	115.311	73.106	33.764	-940.857	-797.939	52.100
900	74.057	123.977	78.285	41.122	-939.896	-780.132	45.278
1000	74.852	131.822	83.253	48.569	-939.032	-762.428	39.825
1100	75.479	138.986	87.999	56.086	-938.339	-744.803	35.368
1200	76.023	145.577	92.526	63.662	-941.807	-727.113	31.650
1300	76.525	151.683	96.844	71.290	-940.742	-709.265	28.499
1400	76.944	157.370	100.967	78.964	-939.741	-691.497	25.800
1500	77.320	162.691	104.906	86.677	-938.819	-673.798	23.464
1600	77.655	167.692	108.676	94.426	-937.992	-656.158	21.421
1700	77.990	172.410	112.287	102.209	-937.274	-638.566	19.621
1800	78.283	176.876	115.753	110.022	-936.679	-621.013	18.021
1900	78.576	181.117	119.082	117.865	-936.224	-603.489	16.591
2000	78.868	185.155	122.286	125.738	-950.606	-585.531	15.293
2100	79.161	189.010	125.372	133.640	-951.213	-567.263	14.110
2130.000	79.228	190.133	126.276	136.016	------ CRYSTAL <--> LIQUID -----		
2200	79.412	192.698	128.349	141.568	-951.817	-548.964	13.034
2300	79.663	196.234	131.224	149.522	-952.417	-530.640	12.051
2400	79.914	199.630	134.004	157.501	-953.015	-512.289	11.150
2500	80.165	202.897	136.695	165.505	-953.610	-493.913	10.320

PREVIOUS: September 1966 CURRENT: December 1973

TITANIUM OXIDE (TiO_2) LIQUID M_r = 79.8788

$S°$(298.15 K) = [72.345] J K^{-1} mol^{-1} $\Delta_f H°$(298.15 K) = [-894.055] kJ mol^{-1}

T_{fus} = 2130 ± 20 K (rutile + liquid) $\Delta_{fus}H°$ = [66.944 ± 16.7] kJ mol^{-1}

Enthalpy of Formation

$\Delta_f H°$(TiO_2, ℓ, 298.15 K) is calculated from that of TiO_2(rutile) by adding $\Delta_{fus}H°$ to the difference in enthalpy, H°(2130 K)-H°(298.15 K), between rutile and liquid.

Heat Capacity and Entropy

$C_p°$ is estimated as 8 cal K^{-1} g-$atom^{-1}$ by comparison with MgF_2, $CaCl_2$, V_2O_4 and WO_3. Below the assumed glass transition at 1400 K, $C_p°$ is taken to be the same as that of TiO_2(rutile). S°(298.15 K) is calculated in a manner analogous to that used for $\Delta_f H°$.

Fusion Data

Schneider (1) reviewed eleven melting point studies and converted the values to IPTS-48. Eight of the results fall in the range 2113 ± 10 K for measurements in air. Brauer and Littke (2), using an atmosphere of oxygen-argon mixtures, found 2103 ± 15 K in air, 2113 ± 15 K with oxygen at 300 torr, 2133 ± 15 K at 500 torr and 2143 ± 15 K at higher oxygen pressures. The authors (2) interpreted the sharp increase in T_{fus} with increased oxygen pressure as being due to suppression of the oxygen deficiency in TiO_2. The residue of the sample melted in air gave an oxygen stoichiometry of 1.996 compared with 2.000 at higher oxygen pressures; the color showed a corresponding change from dark blue to bright yellow. The oxygen deficiency may have been even greater if there was significant reoxidation during cooling to room temperature. Schneider (1) questioned the attainment of black-body conditions in the temperature measurements; in fact, Brauer and Littke (2) tried to compensate for the change in sample color by use of similarly colored compounds in calibrations of the pyrometer. We wonder if this eliminated all relative bias in the melting temperatures.

T_{fus} = 2130 ± 20 K is adopted. This is intermediate between the data in air and those at highest oxygen pressures. $\Delta_{fus}H°$ is estimated such that $\Delta_{fus}S°$ = 7.5 cal K^{-1} mol^{-1}, or 2.5 cal K^{-1} g-$atom^{-1}$, based on observed data for isostructural V_2O_4. The resulting $\Delta_{fus}H°$ of 16 ± 4 kcal mol^{-1} may be compared with approximate values of $\Delta_{fus}H°$ derived from binary phase data. We obtain 12 kcal mol^{-1} from the binary with Al_2O_3 (3), 14 kcal mol^{-1} from BaO (4) and ZrO_2 (5), 16 kcal mol^{-1} from Ta_2O_5 (6) and 21 kcal mol^{-1} from WO_3 (7). These values are derived by calculating apparent $\Delta_{fus}H°$ values assuming activity coefficients of unity and then extrapolating the apparent values to (T_{fus}-T) = 0 in order to minimize activity effects. We omit other binary systems which deviate grossly from ideality. Uncertainties in the binary results are estimated to be about ±5 kcal mol^{-1}.

References - TiO_2(ℓ)

1. S. J. Schneider, U. S. Nat. Bur. Stand. Monograph 68, 1963.
2. G. Brauer and W. Littke, J. Inorg. Nucl. Chem. 16, 67 (1960).
3. E. N. Bunting, J. Res. Nat. Bur. Stand. 11, 719 (1933).
4. D. E. Rase and R. Roy, J. Amer. Ceram. Soc. 38, 111 (1955).
5. L. W. Coughanour, R. S. Roth and V. A. DeProsse, J. Res. Nat. Bur. Stand. 52, 37 (1954).
6. J. L. Waring and R. S. Roth, J. Res. Nat. Bur. Stand. 72A, 175 (1968).
7. L. L. Y. Chang, M. G. Scroger and B. Phillips, J. Less-Common Metals 12, 51 (1967).

References - TiO_2(g)

1. P. J. Hampson and P. W. Gilles, J. Chem. Phys. 55, 3712 (1971).
2. G. Balducci, G. DeMaria, M. Guido and V. Piacente, J. Chem. Phys. 56, 3422 (1972).
3. J. Berkowitz, W. A. Chupka and M. G. Inghram, J. Phys. Chem. 61, 1569 (1957).
4. P. G. Wahlbeck and P. W. Gilles, J. Chem. Phys. 46, 2465 (1967).
5. W. O. Groves, M. Hoch and H. L. Johnston, J. Phys. Chem. 59, 127 (1955).
6. G. A. Semenov, Inorganic Materials (USSR) 5, 54 (1969).
7. JANAF Thermochemical Tables: TiO(g), TiO_2(rutile and g), Ti_3O_5(α, β), Ti_4O_7(cr), 12-31-73; O(g), 6-30-62.
8. JANAF Thermochemical Tables, 2nd ed., NSRDS-NBS 37, 1971.
9. K. Suzuki and K. Sambongi, Tetsu To Hagane 58, 1579 (1972).
10. R. R. Merritt, B. G. Hyde, L. A. Bursill and D. K. Philp, Phil. Trans. Roy. Soc. London, Ser. A 274, 627 (1973).
11. J. Drowart, P. Coppens and S. Smoes, J. Chem. Phys. 50, 1046 (1969).
12. B. R. Conard and P. W. Gilles, presented at 162nd Natl. Meeting, Amer. Chem. Soc., Washington, DC, 1971.
13. E. G. Rauh and R. J. Ackermann, J. Chem. Phys. 60, 1396 (1974).
14. N. S. McIntyre, K. R. Thompson and W. Weltner, J. Phys. Chem. 75, 3243 (1971); 69, 3488 (1965).
15. R. L. DeKock and W. Weltner, J. Phys. Chem. 75, 514 (1971).
16. M. Kaufman, J. Muenter and W. Klemperer, J. Chem. Phys. 47, 3365 (1967).
17. G. Herzberg, "Infrared and Raman Spectra of Polyatomic Molecules," D. Van Nostrand Co., Inc., New York, pp. 186-187, 1962.
18. V. Calder, D. E. Mann, K. S. Seshadri, M. Allavena and D. White, J. Chem. Phys. 51, 2093 (1969).
19. W. Weltner and D. McLeod, J. Chem. Phys. 42, 882 (1965); J. Mol. Spectrosc. 17, 276 (1965).

Titanium Oxide (TiO_2) O_2Ti_1(l)

Enthalpy Reference Temperature = T_r = 298.15 K Standard State Pressure = p° = 0.1 MPa

T/K	$C_p°$ (J K⁻¹mol⁻¹)	S° (J K⁻¹mol⁻¹)	-[G°-H°(T_r)]/T (J K⁻¹mol⁻¹)	H°-H°(T_r) (kJ mol⁻¹)	$\Delta_f H°$ (kJ mol⁻¹)	$\Delta_f G°$ (kJ mol⁻¹)	Log K_f
0							
100							
200							
298.15	55.103	72.345	72.345	0.000	-894.055	-845.289	148.091
300	55.287	72.686	72.346	0.102	-894.054	-844.987	147.125
400	62.835	89.724	74.617	6.043	-893.698	-828.670	108.213
500	67.203	104.254	79.130	12.562	-892.978	-812.492	84.880
600	69.931	116.765	84.384	19.429	-892.097	-796.475	69.339
700	71.764	127.691	89.807	26.519	-891.149	-780.613	58.250
800	73.078	137.364	95.159	33.764	-890.166	-764.889	49.942
900	74.057	146.030	100.338	41.122	-889.204	-749.288	43.488
1000	74.852	153.875	105.306	48.569	-888.340	-733.789	38.329
1100	75.479	161.039	110.052	56.086	-887.647	-718.369	34.113
1200	76.023	167.630	114.579	63.662	-891.115	-702.884	30.596
1300	76.525	173.736	118.897	71.290	-890.050	-687.242	27.614
1400	76.944	179.423	123.020	78.964	-889.049	-671.679	25.061
1400.000	76.944	179.423	123.020	78.964	GLASS <--> LIQUID TRANSITION		
1400.000	100.416	179.423	123.020	78.964			
1500	100.416	186.351	127.014	89.006	-885.798	-656.268	22.853
1600	100.416	192.832	130.927	99.047	-882.679	-641.068	20.929
1700	100.416	198.919	134.749	109.089	-879.701	-626.059	19.236
1800	100.416	204.659	138.475	119.131	-876.879	-611.221	17.737
1900	100.416	210.088	142.103	129.172	-874.225	-596.536	16.400
2000	100.416	215.239	145.632	139.214	-886.438	-581.532	15.188
2100	100.416	220.138	149.064	149.255	-884.906	-566.324	14.087
2130.000	100.416	221.563	150.075	152.268	------ CRYSTAL <--> LIQUID -----		
2200	100.416	224.809	152.402	159.297	-883.396	-551.189	13.087
2300	100.416	229.273	155.648	169.339	-881.909	-536.122	12.176
2400	100.416	233.547	158.805	179.380	-880.444	-521.120	11.342
2500	100.416	237.646	161.877	189.422	-879.001	-506.177	10.576
2600	100.416	241.584	164.868	199.463	-877.579	-491.293	9.870
2700	100.416	245.374	167.780	209.505	-876.179	-476.462	9.218
2800	100.416	249.026	170.617	219.547	-874.799	-461.682	8.613
2900	100.416	252.550	173.381	229.588	-873.438	-446.953	8.050
3000	100.416	255.954	176.077	239.630	-872.098	-432.269	7.526
3100	100.416	259.247	178.707	249.671	-870.775	-417.630	7.037
3200	100.416	262.435	181.274	259.713	-869.471	-403.034	6.579
3300	100.416	265.525	183.781	269.755	-868.184	-388.478	6.149
3400	100.416	268.522	186.229	279.796	-866.914	-373.960	5.745
3500	100.416	271.433	188.622	289.838	-865.660	-359.480	5.365
3600	100.416	274.262	190.962	299.879	-864.422	-345.035	5.006
3700	100.416	277.013	193.251	309.921	-1272.296	-322.836	4.558
3800	100.416	279.691	195.491	319.963	-1269.853	-297.207	4.085
3900	100.416	282.300	197.683	330.004	-1267.478	-271.642	3.638
4000	100.416	284.842	199.830	340.046	-1265.171	-246.138	3.214

PREVIOUS: September 1966 CURRENT: December 1973

Titanium Oxide (TiO_2) O_2Ti_1(l)

TITANIUM OXIDE (TiO_2) RUTILE–LIQUID M_r = 79.8788

0 to 2130 K crystal, rutile
above 2130 K liquid

Refer to the individual tables for details.

Titanium Oxide (TiO_2) $O_2Ti_1(cr,l)$

Enthalpy Reference Temperature = T_r = 298.15 K Standard State Pressure = $p°$ = 0.1 MPa

T/K	$C_p°$ (J K⁻¹mol⁻¹)	$S°$ (J K⁻¹mol⁻¹)	$-[G°-H°(T_r)]/T$ (J K⁻¹mol⁻¹)	$H°-H°(T_r)$ (kJ mol⁻¹)	$\Delta_fH°$ (kJ mol⁻¹)	$\Delta_fG°$ (kJ mol⁻¹)	Log K_f
0	0.000	0.000	INFINITE	-8.636	-939.870	-939.870	INFINITE
100	18.502	10.142	89.638	-7.950	-942.649	-925.506	483.435
200	42.012	30.807	54.969	-4.833	-944.360	-907.579	237.035
298.15	55.103	50.292	50.292	0.000	-944.747	-889.406	155.820
500	67.203	82.201	57.077	12.562	-943.670	-852.157	89.024
600	69.931	94.712	62.331	19.429	-942.789	-833.936	72.601
700	71.764	105.638	67.754	26.519	-941.841	-815.868	60.881
800	73.078	115.311	73.106	33.764	-940.857	-797.939	52.100
900	74.057	123.977	78.285	41.122	-939.896	-780.132	45.278
1000	74.852	131.822	83.253	48.569	-939.032	-762.428	39.825
1100	75.479	138.986	87.999	56.086	-938.339	-744.803	35.368
1200	76.023	145.577	92.526	63.662	-941.807	-727.113	31.650
1300	76.525	151.683	96.844	71.290	-940.742	-709.265	28.499
1400	76.944	157.370	100.967	78.964	-939.741	-691.497	25.800
1500	77.320	162.691	104.906	86.677	-938.819	-673.798	23.464
1600	77.655	167.692	108.676	94.426	-937.992	-656.158	21.421
1700	77.990	172.410	112.287	102.209	-937.274	-638.566	19.621
1800	78.283	176.876	115.753	110.022	-936.679	-621.013	18.021
1900	78.576	181.117	119.082	117.865	-936.224	-603.489	16.591
2000	78.868	185.155	122.286	125.738	-950.606	-585.531	15.293
2100	79.161	189.010	125.372	133.640	-951.213	-567.263	14.110
2130.000	79.228	190.133	126.276	136.016	CRYSTAL <--> LIQUID		
2130.000	100.416	221.563	126.276	202.960	TRANSITION		
2200	100.416	224.809	129.360	209.989	-883.396	-551.188	13.087
2300	100.416	229.273	133.608	220.030	-881.909	-536.122	12.176
2400	100.416	233.547	137.683	230.072	-880.444	-521.120	11.342
2500	100.416	237.646	141.601	240.114	-879.001	-506.177	10.576
2600	100.416	241.584	145.371	250.155	-877.579	-491.293	9.870
2700	100.416	245.374	149.005	260.197	-876.179	-476.462	9.218
2800	100.416	249.026	152.512	270.238	-874.799	-461.682	8.613
2900	100.416	252.550	155.901	280.280	-873.438	-446.953	8.050
3000	100.416	255.954	159.180	290.322	-872.098	-432.269	7.526
3100	100.416	259.247	162.355	300.363	-870.775	-417.630	7.037
3200	100.416	262.435	165.433	310.405	-869.471	-403.034	6.579
3300	100.416	265.525	168.420	320.446	-868.184	-388.478	6.149
3400	100.416	268.522	171.320	330.488	-866.914	-373.960	5.745
3500	100.416	271.433	174.139	340.530	-865.660	-359.480	5.365
3600	100.416	274.262	176.881	350.571	-864.422	-345.035	5.006
3700	100.416	277.013	179.550	360.613	-1272.296	-322.836	4.558
3800	100.416	279.691	182.151	370.654	-1269.853	-297.207	4.085
3900	100.416	282.300	184.685	380.696	-1267.478	-271.642	3.638
4000	100.416	284.842	187.157	390.738	-1265.171	-246.138	3.214

PREVIOUS: CURRENT: December 1973

Titanium Oxide (TiO_2) $O_2Ti_1(cr,l)$

TITANIUM OXIDE (TiO_2) IDEAL GAS M_r = 79.8788

$\Delta_fH°$(0 K) = -303.28 ± 12.6 kJ mol^{-1}

$S°$(298.15 K) = 260.148 ± 4.2 J K^{-1} mol^{-1} $\Delta_fH°$(298.15 K) = -305.43 ± 12.6 kJ mol^{-1}

Electronic Levels and Quantum Weights

ε_i, cm^{-1}	g_i	ε_i, cm^{-1}	g_i
0	1	18880	[3]

Vibrational Frequencies and Degeneracies

ν, cm^{-1}	ν, cm^{-1}	ν, cm^{-1}
962 (1)	[270] (1)	935 (1)

Point Group: C_{2v} σ = 2

Bond Distance: TiO = [1.62 ± 0.08] Å Bond Angle: O-Ti-O = 110° ± 15°

Product of the Moments of Inertia: $I_AI_BI_C$ = [3.115429 x 10^{-115}] g^3 cm^6

Enthalpy of Formation

$\Delta_fH°$ = -73.0 ± 3 kcal mol^{-1} is selected from the ten data sets analyzed below. These include four different reactions studied by Knudsen effusion or mass spectrometry. Our adopted value is consistent with all but two of the data sets, namely reactions C (2) and D (5). We presume that the less negative $\Delta_fH°$ = -59.1 from reaction C is due to bias in the mass-spectrometric values (2) for the partial pressure of O(g). The more negative $\Delta_fH°$ = -86.0 from the sublimation reaction D corresponds to effusion pressures too large by a factor of ~25; allowance for the activity of TiO_2 in the oxygen-deficient condensed phase (5) increases the discrepancy. Gross mass transport from the cell by species other than TiO_2 may have occurred (3), since mass-spectrometric pressures (3, 6) are fairly consistent with our $\Delta_fH°$. We give little weight to reactions D and E due to problems in interpretation of the vapor composition (5, 6) and the activity of oxygen-deficient TiO_2(cr, ℓ).

Activity effects appear to be minor for TiO_2 in TiO_x(cr) in the range 1.833 < x < 1.889. Univariant, two-phase systems involving the crystals Ti_6O_{11}, Ti_7O_{13}, Ti_8O_{15} and Ti_9O_{17} occur in this region. Our calculations based on emf data (9) yield a(TiO_2) ≥ ~0.7 at 1900 K; these values change only slowly with x and T. Sublimation experiments in this region may involve hysteresis, but the degree should be less than that found in equilibrium data (10) near 1300 K.

Drowart et al. (11) provided the key improvement in $\Delta_fH°$ by reinterpretation of the congruent vaporization (4) of Ti_3O_5(cr). Gilles and co-workers (1, 12) confirmed the congruency of vaporization and gave supportive mass-spectrometric data. We use JANAF auxiliary data (7) with D°(TiO) = 158.5 ± 2 kcal mol^{-1} to recalculate the vaporization (4) as described in (1, 11). Resulting partial pressures of TiO_2, TiO and O(g) are analyzed in terms of reaction C. Third-law analysis yields our selected value of $\Delta_fH°$ and D°(TiO_2) = 302.8 ± 3 kcal mol^{-1}. D° would decrease by ~1 kcal mol^{-1} if D°(TiO) were increased by 2 kcal mol^{-1} in the recalculation. The conclusion (1) that the ionization cross-section ratio (TiO/TiO_2) = 2.2 is unchanged in our recalculation; even larger ratios may be expected at low ionizing potentials (13).

Source	Method	Reaction[a]	T/K	Data Points	δS° cal K^{-1} mol^{-1}	$\Delta_rH°$(298.15 K), kcal mol^{-1} 2nd law	3rd law	$\Delta_fH°$(298.15) kcal mol^{-1}	D_0^0
(1) Hampson	Mass spec.	A	2145-2318	5	-3.2±4.0	-19.7±8.9	-12.6±1.0	-74.4±3	304.2
(2) Balducci	Mass spec.	A	2165-2500	6	4.6±0.4	-4.2±0.9	-14.8±1.2	-72.2±3	302.0
(3) Berkowitz	Mass spec.	A	2194	1	-	-	-10.8	-76.2±4	306.0
(1) Hampson	Mass spec.	B	1893	7	-	-	2.5±0.5	-72.8±7	302.6
(2) Balducci	Mass spec.	C	2015-2240	12	4.4±2.5	141.0±5.3	131.6±1.0	-59.1±3	288.9
(4) Wahlbeck	Effusion	C	1837-2014	Eqn.	3.5	152.3	145.6±0.7	-73.0±3	302.8
(5) Groves	Effusion	D	1849-2010	14	-5.5±4.0	129.2±7.7	139.8±1.3	-86.0	315.8
(3) Berkowitz	Mass spec.	D	1881	3	-	-	152.3±3.2	-73.5	303.3
(6) Semenov	Mass spec.	D	1850-2050	Eqn.	-3.6	142.7	149.8±1.0	-76.0	305.8
(6) Semenov	Mass spec.	E	2263-2538	6	-7.2±2.9	122.8±6.9	139.9±1.8	-73.8	303.6

[a]Reactions are A) TiO_2(g) + Ti(g) = 2 TiO(g) B) TiO(g) + WO_3(g) = TiO_2(g) + WO_2(g) C) TiO_2(g) = TiO(g) + O(g)
D) TiO_2(rutile) = TiO_2(g) E) TiO_2(ℓ) = TiO_2(g)

Heat Capacity and Entropy

The electronic levels, bond angle and stretching frequencies are from infrared and emission spectra of TiO_2 isolated in rare-gas matrices (14). We assume the emission at 5295 Å to be a forbidden triplet-singlet transition due to its similarity with CeO_2 (15). Frequency ν_1 was observed both in infrared and emission spectra. The bond angle was derived (14) from ν_3 using splittings observed for the five Ti isotopes. Earlier electric-deflection experiments (16) at 2500 K had shown that the populated electronic states were polar (i.e., nonlinear). We estimate the bond length to be the same as that of ground-state TiO (7). Principal moments of inertia are: I_A = 2.7500 x 10^{-39}, I_B = 9.3573 x 10^{-39} and I_C = 12.1072 x 10^{-39} g cm^2.

The bending frequency is estimated as 270 ± 80 cm^{-1} from the following information. A value of 272 cm^{-1} derived from the three-constant valence-force field (17) using ν_1, ν_3 and $(k\delta/\ell^2)/k_1$ = 0.040 transferred from isoelectronic CaF_2 (18). The resulting value of k_{12}/k_1 is similar to that of CaF_2. The inferior two-constant central-force field gives ν_2 = 325 cm^{-1}. Bending modes have been observed (19) for electronic states of TaO_2 (285 and 281 cm^{-1}) and WO_2 (300 cm^{-1}). These molecules have stretching frequencies similar to TiO_2, but the excited states may be linear rather than bent as are the ground states. We estimate that S°(298.15 K) is uncertain by about 0.7 and 0.3 cal K^{-1} mol^{-1} due to uncertainties in ν_2 and the geometry, respectively.

References - References are listed on the liquid table.

Titanium Oxide (TiO_2) O_2Ti_1(g)

Enthalpy Reference Temperature = T_r = 298.15 K Standard State Pressure = p° = 0.1 MPa

T/K	$C_p°$ (J $K^{-1}mol^{-1}$)	S°	-[G°-H°(T_r)]/T	H°-H°(T_r) (kJ mol^{-1})	$\Delta_fH°$	$\Delta_fG°$	Log K_f
0	0.	0.	INFINITE	-11.357	-303.276	-303.276	INFINITE
100	35.950	217.025	296.659	-7.963	-303.347	-306.893	160.304
200	40.232	243.358	264.059	-4.140	-304.352	-310.081	80.985
250	42.214	252.547	260.864	-2.079	-304.913	-311.449	65.074
298.15	44.149	260.148	260.148	0.	-305.432	-312.660	54.777
300	44.222	260.422	260.149	0.082	-305.451	-312.705	54.447
350	46.106	267.382	260.695	2.341	-305.956	-313.873	46.843
400	47.777	273.651	261.929	4.689	-306.429	-314.971	41.131
450	49.210	279.363	263.553	7.114	-306.878	-316.012	36.682
500	50.419	284.612	265.400	9.606	-307.311	-317.004	33.117
600	52.280	293.979	269.402	14.746	-308.156	-318.863	27.759
700	53.591	302.143	273.509	20.044	-309.001	-320.581	23.922
800	54.532	309.363	277.548	25.452	-309.854	-322.177	21.036
900	55.222	315.828	281.448	30.942	-310.761	-323.664	18.785
1000	55.740	321.674	285.183	36.491	-311.795	-325.043	16.979
1100	56.136	327.006	288.747	42.086	-313.024	-326.310	15.495
1200	56.446	331.905	292.142	47.716	-318.438	-327.337	14.249
1300	56.693	336.433	295.377	53.373	-319.344	-328.042	13.181
1400	56.891	340.642	298.462	59.052	-320.337	-328.674	12.263
1500	57.053	344.573	301.406	64.750	-321.431	-329.233	11.465
1600	57.188	348.259	304.220	70.462	-322.641	-329.714	10.764
1700	57.300	351.730	306.914	76.187	-323.980	-330.116	10.143
1800	57.396	355.008	309.496	81.922	-325.465	-330.435	9.589
1900	57.478	358.113	311.973	87.666	-327.108	-330.667	9.091
2000	57.550	361.063	314.355	93.417	-343.612	-330.354	8.628
2100	57.614	363.873	316.646	99.175	-346.363	-329.624	8.199
2200	57.673	366.554	318.854	104.940	-349.130	-328.761	7.806
2300	57.727	369.119	320.985	110.710	-351.915	-327.773	7.444
2400	57.781	371.577	323.042	116.485	-354.716	-326.664	7.110
2500	57.833	373.937	325.031	122.266	-357.534	-325.437	6.800
2600	57.888	376.206	326.956	128.052	-360.368	-324.098	6.511
2700	57.945	378.392	328.820	133.843	-363.217	-322.648	6.242
2800	58.006	380.500	330.629	139.641	-366.081	-321.093	5.990
2900	58.072	382.537	332.384	145.445	-368.959	-319.436	5.754
3000	58.144	384.507	334.088	151.256	-371.849	-317.679	5.531
3100	58.223	386.415	335.746	157.074	-374.750	-315.825	5.322
3200	58.310	388.265	337.358	162.900	-377.661	-313.879	5.124
3300	58.405	390.060	338.928	168.736	-380.580	-311.841	4.936
3400	58.509	391.806	340.458	174.582	-383.505	-309.713	4.758
3500	58.622	393.503	341.949	180.438	-386.437	-307.501	4.589
3600	58.743	395.156	343.404	186.306	-389.371	-305.204	4.428
3700	58.874	396.768	344.825	192.187	-801.407	-295.037	4.165
3800	59.013	398.339	346.213	198.081	-803.110	-281.328	3.867
3900	59.161	399.874	347.569	203.990	-804.869	-267.574	3.584
4000	59.318	401.374	348.895	209.914	-806.679	-253.775	3.314
4100	59.483	402.841	350.193	215.854	-808.539	-239.929	3.057
4200	59.655	404.276	351.464	221.811	-810.446	-226.038	2.811
4300	59.834	405.682	352.709	227.785	-812.397	-212.100	2.577
4400	60.019	407.060	353.928	233.778	-814.390	-198.117	2.352
4500	60.210	408.411	355.124	239.789	-816.421	-184.088	2.137
4600	60.406	409.736	356.297	245.820	-818.489	-170.014	1.931
4700	60.607	411.037	357.448	251.871	-820.591	-155.894	1.733
4800	60.812	412.315	358.578	257.941	-822.725	-141.729	1.542
4900	61.019	413.571	359.687	264.033	-824.889	-127.518	1.359
5000	61.229	414.806	360.777	270.145	-827.081	-113.264	1.183
5100	61.440	416.021	361.849	276.279	-829.299	-98.966	1.014
5200	61.653	417.216	362.902	282.433	-831.542	-84.625	0.850
5300	61.865	418.392	363.938	288.609	-833.807	-70.238	0.692
5400	62.077	419.551	364.957	294.806	-836.095	-55.810	0.540
5500	62.289	420.692	365.960	301.025	-838.403	-41.340	0.393
5600	62.498	421.816	366.947	307.264	-840.731	-26.826	0.250
5700	62.706	422.924	367.920	313.524	-843.079	-12.272	0.112
5800	62.911	424.016	368.877	319.805	-845.444	2.325	-0.021
5900	63.113	425.093	369.821	326.106	-847.828	16.961	-0.150
6000	63.311	426.156	370.751	332.428	-850.230	31.640	-0.275

PREVIOUS: December 1973 (1 atm) CURRENT: December 1973 (1 bar)

Titanium Oxide (TiO_2) O_2Ti_1(g)

VANADIUM OXIDE (VO_2) IDEAL GAS M_r = 82.9403

$\Delta_f H°(0\ K) = -230.56 \pm 41.8\ kJ\ mol^{-1}$

$S°(298.15\ K) = [265.269 \pm 5.9]\ J\ K^{-1}\ mol^{-1}$ $\Delta_f H°(298.15\ K) = -232.63 \pm 41.8\ kJ\ mol^{-1}$

Vibrational Frequencies and Degeneracies

ν, cm^{-1}
[970](1)
[300](1)
[930](1

Ground State Quantum Weight: [2]

Point Group: [C_{2v}] σ = [2]

Bond Distance: V-O = [1.589] Å

Bond Angle: O-V-O = [110]°

Product of the Moments of Inertia: $I_A I_B I_C = [2.858786 \times 10^{-115}]\ g^3\ cm^6$

Enthalpy of Formation

Four independent mass spectrometric studies lead to values for $\Delta_f H°(VO_2,\ g,\ 298.15\ K)$ (1, 2, 3, 4). Shchukarev et al. (2) studied the composition of the vapor above V_2O_4 during evaporation from the surface of a platinum strip. A 2nd law sublimation heat, $\Delta_{sub}H° = 104 \pm 2\ kcal\ mol^{-1}$, was reported for the range 1500-1800 K. Assuming the reported value corresponds to 1700 K and using auxiliary data (5), we calculate $\Delta_f H°(VO_2,\ g,\ 298.15\ K) = -62.3 \pm 2.0\ kcal\ mol^{-1}$. The sublimation was described by 0.5 $V_2O_4(cr) = VO_2(g)$. More recent related work (3) yields a $\Delta_f H°(298.15\ K)$ value approximately 6 kcal mol^{-1} more positive. This latter work and other investigations are summarized in the following table.

Reference	Reaction	Data Points	T/K	$\Delta_r H°(298.15\ K)$, kcal mol^{-1} 2nd law	3rd law	Drift cal $K^{-1}\ mol^{-1}$	$\Delta_f H°(298.15\ K)$, kcal mol^{-1} 2nd law	3rd law
(3)	A	Equation	1673-1818	114.29	114.32	0.02	-56.26	-56.23
(3)	B	Equation	1818-2273	104.90	103.77	-0.58	-54.30	-55.43
(1)	C	2	1948	-	-16.23	-	-	-45.97
(4)	C	1	2270	-	-12.86	-	-	-49.34
(4)	D	1	2270	-	-40.14	-	-	-32.94
								(-39.79)**

**Value is based on recent determination by Farber et al. (8) for $\Delta_f H°(AlO,\ g,\ 298.15\ K)$.

Reactions: (A) 0.5 $V_2O_4(cr) = VO_2(g)$ (C) $V(g) + VO_2(g) = 2\ VO(g)$

(B) 0.5 $V_2O_4(\ell) = VO_2(g)$ (D) $2\ AlO(g) + V(g) = 2\ Al(g) + VO_2(g)$

We adopt $\Delta_f H°(298.15\ K) = -55.6\ kcal\ mol^{-1}$ for $VO_2(g)$. This value is an average of the 2nd and 3rd law results of Frantseva and Semenov (3). These results are independent of the VO(g) data and show small third law drifts. The results of our analysis for reaction (C) for the data of Berkowitz et al. (1) and Farber et al. (4) show good agreement in $\Delta_r H°(298.15\ K)$ but lead to values for $\Delta_f H°(298.15\ K)$ which are 6-10 kcal mol^{-1} less negative than our adopted value. Killingbeck (6) studied the evaporation of V_2O_3 from a molybdenum effusion chamber and reported $\Delta_r H°(298.15\ K) = 279.5 \pm 5\ kcal\ mol^{-1}$ for $V_2O_3(cr) = VO(g) + VO_2(g)$. Using auxiliary data (5) we calculate $\Delta_f H°(298.15\ K) = -42.3 \pm 5.0\ kcal\ mol^{-1}$ for $VO_2(g)$.

The adopted $\Delta_f H°(298.15\ K)$ value corresponds to a dissociation energy D° = 295.5 kcal mol^{-1} (12.8 eV) for the process $VO_2(g) = V(g) + 2\ O(g)$. This D° value is 1.96 times the dissociation energy of VO(g) (5).

Heat Capacity and Entropy

The vibrational frequencies and geometry of $VO_2(g)$ are estimated based on existing and estimated data for related metal dioxides such as $TiO_2(g)$, $TaO_2(g)$ and $WO_2(g)$ (5). The V-O bond distance is assumed to be the same as in VO(g) (5).

The ground state quantum weight of 2 is adopted so as to be consistent with $TaO_2(g)$ (5). This introduces an uncertainty of ±1.38 cal $K^{-1}\ mol^{-1}$ in S°(298.15 K). The estimation of ν_2 at 300 cm^{-1} results in an uncertainty of 0.72 cal $K^{-1}\ mol^{-1}$ in S°(298.15 K) for ±100 cm^{-1}. The estimates for $VO_2(g)$ are such that the 3rd law drifts as indicated in the above table are small. Nevertheless, until more data are available for $VO_2(g)$, the uncertainty in the entropy must be considered as ±1.4 cal $K^{-1}\ mol^{-1}$ for a bent molecule. The free energy functions suggested by Brewer and Rosenblatt (7) are roughly 3-4 cal $K^{-1}\ mol^{-1}$ lower than those adopted here. The main reason for the difference is that Brewer and Rosenblatt (7) treated $VO_2(g)$ as a linear molecule.

The electronic levels of the V^{+4} ion as reported by Moore (9) consist of a ground state with quantum weight four and an excited state at 620 cm^{-1} with a quantum weight six. The next excited state is listed at ~148000 cm^{-1}. Inclusion of an estimated state for $VO_2(g)$ at 620 cm^{-1} with a quantum weight of six would raise S°(298.15 K) by 1.06 cal $K^{-1}\ mol^{-1}$. This is within the estimated uncertainty of S°(298.15 K).

References

1. J. Berkowitz, W. A. Chupka, and M. G. Inghram, J. Chem. Phys. 27, 87 (1957).
2. S. A. Shchukarev, G. A. Semenov and K. E. Frantseva, Russ. J. Inorg. Chem. 4, 1217 (1959).
3. K. E. Frantseva and G. A. Semenov, High Temp. 7, 52 (1969).
4. M. Farber, O. M. Uly, and R. D. Srivastava, J. Chem. Phys. 56, 5312 (1972).
5. JANAF Thermochemical Tables: V_2O_4(cr), V_2O_3(cr), VO(g), TaO(g), and TiO_2(g), 12-31-73.
6. S. Killingbeck, Ph.D. Thesis, Univ. Kansas, 1964; Diss Abstr. 65, 1985 (1964).
7. L. Brewer and G. M. Rosenblatt, Chem. Rev. 61, 257 (1961).
8. M. Farber, R. D. Srivastava and O. M. Uy, J. Chem. Soc. Faraday I 68, 249 (1972).
9. C. E. Moore, U. S. Nat. Bur. Stand. Circ. 467, Vol. I, 1949.

Vanadium Oxide (VO_2) $O_2V_1(g)$

Enthalpy Reference Temperature = T_r = 298.15 K Standard State Pressure = p° = 0.1 MPa

T/K	$C_p°$ (J $K^{-1}mol^{-1}$)	S° (J $K^{-1}mol^{-1}$)	$-[G°-H°(T_r)]/T$ (J $K^{-1}mol^{-1}$)	$H°-H°(T_r)$ (kJ mol^{-1})	$\Delta_f H°$ (kJ mol^{-1})	$\Delta_f G°$ (kJ mol^{-1})	Log K_f
0	0.	0.	INFINITE	-11.254	-230.561	-230.561	INFINITE
100	35.385	222.647	301.443	-7.880	-230.580	-234.796	122.645
200	39.819	248.608	269.153	-4.109	-231.557	-238.670	62.334
250	41.903	257.716	265.981	-2.066	-232.116	-240.383	50.225
298.15	43.907	265.269	265.269	0.	-232.630	-241.928	42.385
300	43.982	265.541	265.270	0.081	-232.650	-241.986	42.134
350	45.916	272.468	265.812	2.330	-233.144	-243.503	36.341
400	47.622	278.714	267.041	4.669	-233.597	-244.951	31.987
450	49.082	284.409	268.659	7.088	-234.017	-246.345	28.595
500	50.311	289.646	270.499	9.573	-234.412	-247.693	25.876
600	52.201	298.996	274.489	14.705	-235.162	-250.278	21.789
700	53.531	307.149	278.585	19.995	-235.902	-252.739	18.860
800	54.485	314.363	282.615	25.398	-236.669	-255.092	16.656
900	55.184	320.822	286.507	30.884	-237.491	-257.347	14.936
1000	55.708	326.665	290.236	36.429	-238.380	-259.506	13.555
1100	56.110	331.994	293.793	42.021	-239.344	-261.572	12.421
1200	56.424	336.891	297.183	47.649	-240.399	-263.547	11.472
1300	56.674	341.417	300.414	53.304	-241.552	-265.430	10.665
1400	56.875	345.625	303.495	58.982	-242.816	-267.220	9.970
1500	57.039	349.554	306.436	64.678	-244.192	-268.916	9.364
1600	57.175	353.240	309.247	70.388	-245.681	-270.516	8.831
1700	57.288	356.710	311.938	76.112	-247.293	-272.020	8.358
1800	57.384	359.987	314.517	81.846	-249.035	-273.425	7.935
1900	57.466	363.092	316.993	87.588	-250.919	-274.729	7.553
2000	57.536	366.041	319.372	93.338	-252.953	-275.931	7.207
2100	57.597	368.850	321.662	99.095	-255.148	-277.026	6.891
2200	57.649	371.531	323.868	104.857	-280.396	-277.908	6.598
2300	57.695	374.094	325.997	110.625	-283.080	-277.735	6.308
2400	57.736	376.551	328.052	116.396	-285.781	-277.445	6.038
2500	57.772	378.908	330.040	122.172	-288.501	-277.042	5.788
2600	57.804	381.175	331.963	127.951	-291.239	-276.530	5.556
2700	57.832	383.357	333.826	133.732	-293.995	-275.912	5.338
2800	57.858	385.461	335.633	139.517	-296.768	-275.191	5.134
2900	57.881	387.491	337.387	145.304	-299.559	-274.372	4.942
3000	57.902	389.454	339.090	151.093	-302.368	-273.455	4.761
3100	57.921	391.353	340.745	156.884	-305.193	-272.444	4.591
3200	57.938	393.192	342.356	162.677	-308.034	-271.343	4.429
3300	57.953	394.975	343.923	168.472	-310.891	-270.153	4.276
3400	57.967	396.705	345.450	174.268	-313.763	-268.875	4.131
3500	57.981	398.386	346.939	180.065	-316.650	-267.513	3.992
3600	57.992	400.020	348.391	185.864	-319.551	-266.068	3.861
3700	58.004	401.609	349.808	191.664	-769.272	-263.342	3.718
3800	58.014	403.156	351.191	197.465	-770.514	-249.651	3.432
3900	58.023	404.663	352.543	203.266	-771.819	-235.927	3.160
4000	58.032	406.132	353.865	209.069	-773.188	-222.169	2.901
4100	58.040	407.565	355.157	214.873	-774.623	-208.376	2.655
4200	58.047	408.964	356.421	220.677	-776.124	-194.547	2.420
4300	58.055	410.330	357.659	226.482	-777.691	-180.681	2.195
4400	58.061	411.664	358.872	232.288	-779.325	-166.779	1.980
4500	58.067	412.969	360.059	238.094	-781.026	-152.839	1.774
4600	58.073	414.245	361.223	243.901	-782.795	-138.860	1.577
4700	58.078	415.494	362.365	249.709	-784.634	-124.841	1.387
4800	58.083	416.717	363.485	255.517	-786.541	-110.783	1.206
4900	58.088	417.915	364.583	261.326	-788.518	-96.684	1.031
5000	58.092	419.089	365.662	267.135	-790.566	-82.544	0.862
5100	58.097	420.239	366.720	272.944	-792.686	-68.362	0.700
5200	58.101	421.367	367.761	278.754	-794.878	-54.140	0.544
5300	58.104	422.474	368.782	284.564	-797.144	-39.872	0.393
5400	58.108	423.560	369.787	290.375	-799.485	-25.563	0.247
5500	58.111	424.626	370.774	296.186	-801.902	-11.210	0.106
5600	58.114	425.673	371.745	301.997	-804.396	3.190	-0.030
5700	58.117	426.702	372.700	307.809	-806.969	17.633	-0.162
5800	58.120	427.713	373.640	313.621	-809.622	32.124	-0.289
5900	58.123	428.706	374.565	319.433	-812.356	46.658	-0.413
6000	58.126	429.683	375.476	325.245	-815.175	61.243	-0.533

PREVIOUS: December 1973 (1 atm) CURRENT: December 1973 (1 bar)

Vanadium Oxide (VO_2) $O_2V_1(g)$

$\Delta_f H°(0\ K) = -584.745 \pm 0.88\ kJ\ mol^{-1}$

$S°(298.15\ K) = 50.528 \pm 0.29\ J\ K^{-1}\ mol^{-1}$ $\Delta_f H°(298.15\ K) = -589.693 \pm 0.88\ kJ\ mol^{-1}$

$\Delta_{sub} H°(298.15\ K) = 666.260\ kJ\ mol^{-1}$

Enthalpy of Formation

The adopted enthalpy of formation, $-140.94 \pm 0.21\ kcal\ mol^{-1}$, was determined by combustion calorimetry by Mah (1). Griffis (2) has measured the enthalpy of combustion to WO_3(cr) which yields $\Delta_f H°(WO_2$, cr, 298.15 K) = $-139.8 \pm 1.5\ kcal\ mol^{-1}$. When recalculated with $\Delta_f H°(WO_3$, cr, 298.15 K) = $-201.46\ kcal\ mol^{-1}$. This result is less certain because of the incomplete characterization of the final products.

Recent equilibrium data yield essentially the same average of the enthalpy of formation by the 3rd law method. The results are presented as follows.

Source	Reaction	T/K	Data Points	$\Delta_r H°(298.15\ K)$, kcal mol^{-1} 3rd law	2nd law	Drift cal K^{-1} mol^{-1}	$\Delta_f H°(298.15\ K)$ kcal mol^{-1}
(3)	A	1275-1497	10	2.672	0.036	1.9 ± 0.6	-140.62
(3)	B	973-1467	6	11.994	12.076	0.2 ± 0.76	-139.58
(4)	C	923-1223	7	-70.281	-67.943	-2.2 ± 0.2	-140.56
(5)	B	1075-1210	5	12.949	13.051	-0.1 ± 0.4	-141.49
(6)	B	873-1273	30	12.550	10.295	2.1 ± 0.1	-140.70

Reactions: A) $1/2\ WO_2(cr) + CO(g) = 1/2\ W(cr) + CO_2(g)$
B) $1/2\ WO_2(cr) + H_2(g) = 1/2\ W(cr) + H_2O(g)$
C) $1/2\ W(cr) + 1/2\ O_2(g) = 1/2\ WO_2(cr)$

Heat Capacity and Entropy

King et al. (7) have measured the low temperature heat capacities (53-297 K) and high temperature enthalpy changes (396-1800 K) by drop calorimetry. The low temperature and high temperature heat capacities were joined smoothly at 298.15 K. The entropy was obtained from the heat capacities based on $S°(51\ K) = 0.52\ cal\ K^{-1}\ mol^{-1}$. It is possible that at temperatures below 50 K the entropy due to the uncoupling of the d^2 electrons will appear contributing an additional 1.3 cal K^{-1} mol^{-1}.

Decomposition Data

Tungsten dioxide (cr) has no melting point since disproportionation occurs to yield W(cr) and WO_3(ℓ) before melting starts. T_{dcm} = 1997 K is calculated as the temperature at which $\Delta_f G° = 0$ for $3\ WO_2(cr) = 2\ WO_3(\ell) + W(cr)$.

Sublimation Data

The enthalpy of sublimation is calculated from the difference between the enthalpies of formation of WO_2(g) and WO_2(cr).

References

1. A. D. Mah, J. Amer. Chem. Soc. 81, 1mg† k/xmxu5
2. R. C. Griffis, J. Electrochem. Soc. 105, 398 (1958).
3. G. R. St. Pierre, W. T. Ebihara, M. J. Pool and R. Speiser, Trans. AIME 224, 259 (1962).
4. Ya. I. Gerasimov, I. A. Vasil'eva, T. P. Chusova, V. W. Geiderikh, and M. A. Timofeeva, Zh. Fiz. Khim. 36, 358 (1962).
5. I. A. Vasil'eva, Ya. I. Gerasimov and Yu. P. Simanov, Zh. Fiz. Khim. 34, 1811 (1960).
6. R. C. Griffis, J. Electrochem. Soc. 106, 418 (1959).
7. G. G. King, W. W. Weller and A. V. Christensen, U. S. Bur. Mines RI 5664, 1960.

Enthalpy Reference Temperature = T_r = 298.15 K; Standard State Pressure = p° = 0.1 MPa

T/K	$C_p°$ (J $K^{-1}mol^{-1}$)	S°	$-[G°-H°(T_r)]/T$	$H°-H°(T_r)$ (kJ mol^{-1})	$\Delta_f H°$	$\Delta_f G°$	Log K_f
0	0.000	0.000	INFINITE	-8.708	-584.745	-584.745	INFINITE
100	19.106	9.991	90.199	-8.021	-587.614	-570.321	297.905
200	42.211	30.893	55.247	-4.871	-589.383	-552.210	144.222
298.15	55.739	50.528	50.528	0.000	-589.693	-533.856	93.529
300	55.898	50.873	50.529	0.103	-589.689	-533.509	92.892
400	63.429	68.063	52.822	6.096	-589.131	-514.850	67.233
500	68.157	82.763	57.377	12.693	-588.107	-496.391	51.858
600	71.302	95.484	62.692	19.675	-586.843	-478.164	41.628
700	73.623	106.657	68.191	26.926	-585.447	-460.160	34.338
800	75.474	116.613	73.633	34.384	-583.971	-442.362	28.883
900	76.978	125.592	78.915	42.009	-582.440	-424.752	24.652
1000	78.174	133.767	83.998	49.769	-580.876	-407.315	21.276
1100	79.122	141.264	88.868	57.635	-579.298	-390.035	18.521
1200	79.968	148.184	93.526	65.590	-577.716	-372.899	16.232
1300	80.983	154.622	97.981	73.634	-576.126	-355.895	14.300
1400	82.969	160.690	102.245	81.823	-574.466	-339.016	12.649
1500	85.103	166.491	106.336	90.232	-572.661	-322.261	11.222
1600	88.031	172.080	110.272	98.893	-570.676	-305.631	9.978
1700	91.044	177.508	114.068	107.848	-568.471	-289.132	8.884
1800	93.847	182.791	117.740	117.092	-566.047	-272.769	7.916
1900	96.169	187.929	121.299	126.597	-563.434	-256.546	7.053
2000	98.198	192.914	124.756	136.316	-560.678	-240.465	6.280
2100	99.998	197.750	128.117	146.227	-557.802	-224.525	5.585
2200	101.755	202.442	131.390	156.315	-554.822	-208.723	4.956
2300	103.512	207.004	134.579	166.579	-551.738	-193.061	4.385
2400	105.228	211.445	137.689	177.014	-548.553	-177.534	3.864
2500	106.943	215.776	140.726	187.623	-545.268	-162.142	3.388
2600	108.617	220.003	143.695	198.401	-541.881	-146.884	2.951
2700	110.290	224.133	146.598	209.346	-538.423	-131.757	2.549
2800	111.964	228.175	149.439	220.459	-534.926	-116.759	2.178
2900	113.470	232.130	152.223	231.733	-531.436	-101.886	1.835
3000	115.102	236.004	154.951	243.160	-527.988	-87.133	1.517

PREVIOUS: June 1962 CURRENT: September 1966

TUNGSTEN OXIDE (WO_2) IDEAL GAS M_r = 215.8488

$S°(298.15 K) = [285.499] J K^{-1} mol^{-1}$

$\Delta_f H°(0 K) = 78.98 \pm 29.3 kJ mol^{-1}$
$\Delta_f H°(298.15 K) = 76.57 \pm 29.3 kJ mol^{-1}$

Electronic Levels and Quantum Weights

ε_i, cm^{-1}	g_i
0	[3]
[25,000]	[10]

Vibrational Frequencies and Degeneracies

ν, cm^{-1}
992 (1)
[300](1)
928 (1)

Ground State Quantum Weight: [3] σ = [2]
Point Group: [C_{2v}]
Bond Distance: W-O = [1.81] Å
Bond Angle: O-W-O = [110]°
Product of the Moments of Inertia: $I_A I_B I_C = [9.435202 \times 10^{-115}] g^3 cm^6$

Enthalpy of Formation

DeMaria et al. (1) have reported the partial pressures of WO(g), WO_2(g), WO_3(g), and O(g) in a mass spectrometric study of aluminum oxide vaporized from a tungsten cell from 2188 to 2475 K. The adopted value, $\Delta_f H°(WO_2, g, 298.15 K) = 18.3 \pm 7$ kcal mol^{-1}, was calculated from $\Delta_r H°(298.15 K) = -100.9$ kcal mol^{-1} for W(cr) + 2 O(g) → WO_2(g), which was obtained from 3rd law treatment of their partial pressure data, using all JANAF functions. The 3rd law drift is 15.0 ± 5.2 cal K^{-1} mol^{-1}. Partial pressures of WO_2(g), WO_3(g) and O(g) at 2242 K have also been reported by Chupka et al. (2) in a mass spectrometric study of beryllium oxide in a tungsten cell. Similar calculation gives $\Delta_r H°(298.15 K) = -105.9$ kcal mol^{-1} which yields $\Delta_f H°(WO_2, g, 298.15 K) = 13.2$ kcal mol^{-1}.

Heat Capacity and Entropy

The WO_2(g) was assumed to have a C_{2v} symmetry with a bond angle O-W-O = 110°. The bond distance W-O was estimated to be the same as that in WO(g). The symmetric and asymmetric stretching frequencies were obtained from the infrared spectrum of WO_2 in a neon matrix by Weltner and McLeod (3). The bending vibrational frequency (300 cm^{-1}) was estimated by comparison with other dioxides. The electronic levels and quantum weights were estimated from W^{+4} ion by analogy with Mo^{+4} ion. The principal moments of inertia are: $I_A = 4.8780 \times 10^{-39}$, $I_B = 11.6809 \times 10^{-39}$ and $I_C = 16.5589 \times 10^{-39}$ g cm^2.

References

1. G. DeMaria, R. P. Burns, J. Drowart and M. G. Inghram, J. Chem. Phys. 32, 1373 (1960).
2. W. A. Chupka, J. Berkowitz and C. F. Giese, J. Chem. Phys. 30, 827 (1959).
3. W. Weltner, Jr., and D. McLeod, Jr., J. Mol. Spectry. 17, 276 (1965).

Tungsten Oxide (WO_2) O_2W_1(g)

Enthalpy Reference Temperature = T_r = 298.15 K; Standard State Pressure = p° = 0.1 MPa

T/K	$C_p°$ (J K⁻¹mol⁻¹)	S° (J K⁻¹mol⁻¹)	-[G°-H°(T_r)]/T (J K⁻¹mol⁻¹)	H°-H°(T_r) (kJ mol⁻¹)	$\Delta_f H°$ (kJ mol⁻¹)	$\Delta_f G°$ (kJ mol⁻¹)	Log K_f
0	0.	0.	INFINITE	-11.246	78.977	78.977	INFINITE
100	35.385	242.910	321.628	-7.872	78.795	72.796	-38.025
200	39.783	268.864	289.376	-4.102	77.646	67.224	-17.557
250	41.834	277.961	286.210	-2.062	77.071	64.686	-13.515
298.15	43.812	285.499	285.499	0.	76.567	62.348	-10.923
300	43.887	285.771	285.500	0.081	76.549	62.259	-10.840
350	45.806	292.682	286.042	2.324	76.091	59.915	-8.942
400	47.508	298.913	287.267	4.658	75.691	57.632	-7.526
450	48.970	304.595	288.882	7.071	75.335	55.396	-6.430
500	50.204	309.820	290.718	9.551	75.011	53.199	-5.558
600	52.109	319.152	294.698	14.673	74.415	48.893	-4.256
700	53.454	327.292	298.785	19.954	73.842	44.684	-3.334
800	54.420	334.496	302.807	25.351	73.256	40.559	-2.648
900	55.129	340.949	306.693	30.830	72.641	36.508	-2.119
1000	55.662	346.786	310.415	36.371	71.986	32.528	-1.699
1100	56.071	352.111	313.967	41.958	71.285	28.616	-1.359
1200	56.390	357.004	317.353	47.582	70.536	24.769	-1.078
1300	56.644	361.528	320.579	53.234	69.735	20.988	-0.843
1400	56.849	365.734	323.656	58.909	68.880	17.270	-0.644
1500	57.016	369.662	326.593	64.602	67.970	13.614	-0.474
1600	57.154	373.346	329.401	70.311	67.001	10.022	-0.327
1700	57.270	376.814	332.089	76.033	65.974	6.492	-0.199
1800	57.368	380.091	334.666	81.765	64.886	3.024	-0.088
1900	57.451	383.195	337.139	87.506	63.735	-0.382	0.010
2000	57.523	386.143	339.516	93.254	62.520	-3.725	0.097
2100	57.585	388.952	341.804	99.010	61.240	-7.006	0.174
2200	57.639	391.632	344.008	104.771	59.894	-10.224	0.243
2300	57.686	394.195	346.135	110.537	58.481	-13.381	0.304
2400	57.728	396.651	348.189	116.308	57.001	-16.474	0.359
2500	57.766	399.008	350.175	122.083	55.452	-19.504	0.408
2600	57.801	401.275	352.097	127.861	53.839	-22.470	0.451
2700	57.833	403.457	353.959	133.643	52.134	-25.373	0.491
2800	57.863	405.560	355.765	139.428	50.303	-28.210	0.526
2900	57.892	407.591	357.517	145.216	48.306	-30.980	0.558
3000	57.920	409.555	359.219	151.006	46.118	-33.677	0.586
3100	57.949	411.454	360.874	156.800	43.696	-36.297	0.612
3200	57.978	413.294	362.483	162.596	41.008	-38.836	0.634
3300	58.009	415.079	364.050	168.395	38.031	-41.287	0.654
3400	58.042	416.811	365.577	174.198	34.731	-43.641	0.670
3500	58.077	418.494	367.065	180.004	31.042	-45.894	0.685
3600	58.115	420.131	368.516	185.813	26.871	-48.034	0.697
3700	58.156	421.724	369.933	191.627	-12.647	-49.861	0.704
3800	58.202	423.275	371.316	197.445	-14.494	-50.841	0.699
3900	58.252	424.788	372.668	203.267	-16.350	-51.773	0.693
4000	58.307	426.263	373.989	209.095	-18.214	-52.658	0.688
4100	58.367	427.704	375.282	214.929	-20.086	-53.496	0.682
4200	58.432	429.111	376.547	220.769	-21.963	-54.288	0.675
4300	58.502	430.487	377.785	226.615	-23.847	-55.035	0.669
4400	58.579	431.832	378.998	232.470	-25.735	-55.739	0.662
4500	58.661	433.150	380.187	238.331	-27.628	-56.400	0.655
4600	58.749	434.440	381.353	244.202	-29.524	-57.019	0.647
4700	58.844	435.704	382.496	250.082	-31.423	-57.596	0.640
4800	58.944	436.944	383.617	255.971	-33.325	-58.132	0.633
4900	59.050	438.161	384.718	261.871	-35.230	-58.629	0.625
5000	59.162	439.355	385.799	267.781	-37.136	-59.087	0.617
5100	59.280	440.528	386.860	273.703	-39.045	-59.507	0.609
5200	59.404	441.680	387.904	279.637	-40.956	-59.890	0.602
5300	59.532	442.813	388.929	285.584	-42.868	-60.235	0.594
5400	59.666	443.927	389.937	291.544	-44.783	-60.546	0.586
5500	59.805	445.023	390.929	297.517	-46.700	-60.820	0.578
5600	59.949	446.102	391.904	303.505	-48.620	-61.059	0.570
5700	60.097	447.164	392.865	309.507	-50.543	-61.264	0.561
5800	60.249	448.211	393.810	315.525	-52.470	-61.435	0.553
5900	60.405	449.242	394.741	321.557	-54.401	-61.575	0.545
6000	60.565	450.258	395.657	327.606	-56.337	-61.679	0.537

PREVIOUS: September 1966 (1 atm) CURRENT: September 1966 (1 bar)

Tungsten Oxide (WO_2) O_2W_1(g)

TUNGSTEN OXIDE ($WO_{2.72}$) CRYSTAL M_r = 227.3684

$S°(298.15 K) = [68.425] J K^{-1} mol^{-1}$

$\Delta_f H°(0 K)$ = Unknown
$\Delta_f H°(298.15 K) = -781.153 \pm 2.1 kJ mol^{-1}$

Enthalpy of Formation

The adopted enthalpy of formation, -186.7 ± 0.5 kcal mol^{-1}, is derived from recent equilibrium data which are in excellent agreement, the results are summarized below. In addition, Griffis (1) measured the enthalpy of combustion to WO_3(cr) which yields $\Delta_f H°(WO_{2.72}$, cr, 298.15 K) = -185.4 ± 1 kcal mol^{-1} when recalculated with $\Delta_f H°(WO_3$, cr, 298.15 K) = -201.46 kcal mol^{-1}. This result is less certain because of the incomplete characterization of the final products. The emf data reported by Gerasimov et al. (2) give the value of -186.89 kcal mol^{-1} for the enthalpy of formation at 298.15 K. Ackermann and Rauh (3) have investigated the tungsten-oxygen system by mass effusion, mass spectroscopy and X-ray diffraction. In the temperature range from 1300 to 1600 K, they have derived indirectly from their measurements the enthalpy of formation -182.2 kcal mol^{-1} (corresponding to -187.2 kcal mol^{-1} at 298.15 K, using the JANAF functions).

Source	Reaction*	T/K	Data Points	$\Delta_r H°$(298.15 K), kcal mol^{-1} 2nd law	3rd law	Drift cal K^{-1} mol^{-1}	$\Delta_f H°$(298.15 K) kcal mol^{-1}
(4)	A	830-1048	5	2.951	4.095	1.2 ± 0.2	-186.65
(5)	B	1173-1423	-	-45.739	-	-	-186.68
(6)	A	904-1066	4	2.676	4.279	1.6 ± 0.3	-186.83
(7)	A	873-1173	24	4.314	4.314	-0.01± 0.06	-186.87

A) $100/72\ WO_{2.72}(cr) + H_2(g) = 100/72\ WO_2(cr) + H_2O(g)$

B) $WO_2(cr) + 0.72\ O_2(g) = WO_{2.72}(cr)$

Heat Capacity and Entropy

The monoclinic unit cell of $WO_{2.72}$ corresponds to $W_{18}O_{49}$ as shown by the X-ray diffraction studies of Magneli et al. (8). The tabulated heat capacities of $WO_{2.72}$(cr) were estimated from those of WO_2(cr) and WO_3(cr) because all three have similar monoclinic structures as identified by Magneli et al.

The entropy, S°(298.15 K) = 16.35 cal K^{-1} mol^{-1}, was calculated from $\Delta_r S°$(298.15 K) = 7.96 ± 0.09 cal K^{-1} mol^{-1} for reaction A, using all JANAF values. The value of $\Delta_r S°$(298.15 K) was derived from the 2nd law analysis of Griffis' data (7).

References

1. R. C. Griffis, J. Electrochem. Soc. 105, 398 (1958).
2. Ya. I. Gerasimov, I. A. Vasil'eva, T. P. Chusova, V. A. Geiderikh, and M. A. Timofeeva, Zh. Fiz. Khim. 36, 358 (1962).
3. R. J. Ackermann and E. G. Rauh, J. Phys. Chem. 67, 2596 (1963).
4. J. Bousquet and G. Perachon, Compt. rend. 258, 3869 (1964).
5. G. R. St. Pierre, W. T. Ebihara, M. J. Pool, and R. Speiser, Trans. AIME 224, 259 (1962).
6. I. A. Vasil'eva, Ya. I. Gerasimov and Yu. P. Simanov, Zh. Fiz. Khim. 34, 1811 (1960).
7. R. C. Griffis, J. Electrochem. Soc. 106, 418 (1959).
8. A. Magneli, G. Anderson, B. Blomberg and L. Kihlborg, Anal. Chem. 24, 1998 (1952).

Tungsten Oxide ($WO_{2.72}$) $O_{2.72}W_1$(cr)

Enthalpy Reference Temperature = T_r = 298.15 K Standard State Pressure = p° = 0.1 MPa

T/K	$C°_p$ (J $K^{-1}mol^{-1}$)	S°	$-[G°-H°(T_r)]/T$	$H°-H°(T_r)$ (kJ mol^{-1})	$\Delta_f H°$	$\Delta_f G°$	Log K_f
0							
100							
200							
298.15	68.283	68.425	68.425	0.000	-781.153	-708.632	124.149
300	68.408	68.848	68.426	0.126	-781.145	-708.183	123.306
400	76.902	89.731	71.216	7.406	-780.370	-683.959	89.316
500	82.927	107.581	76.748	15.417	-779.034	-660.002	68.950
600	86.985	123.079	83.207	23.923	-777.382	-636.347	55.399
700	89.810	136.719	89.897	32.775	-775.557	-612.984	45.741
800	91.839	148.850	96.522	41.862	-773.653	-589.889	38.516
900	93.471	159.765	102.953	51.131	-771.705	-567.035	32.910
1000	95.019	169.696	109.138	60.558	-769.720	-544.400	28.437
1100	96.441	178.819	115.064	70.131	-767.698	-521.965	24.786
1200	97.822	187.269	120.733	79.843	-765.637	-499.717	21.752
1300	99.286	195.155	126.158	89.697	-763.526	-477.642	19.192
1400	100.751	202.568	131.353	99.701	-761.353	-455.732	17.004
1500	102.006	209.562	136.336	109.838	-759.130	-433.979	15.113
1600	103.261	216.185	141.122	120.102	-756.864	-412.376	13.463
1700	104.516	222.483	145.724	130.491	-754.552	-390.916	12.011
1800	105.772	228.493	150.157	141.005	-752.196	-369.594	10.725
1900	107.027	234.245	154.432	151.645	-749.794	-348.404	9.578
2000	108.282	239.767	158.562	162.410	-747.347	-327.341	8.549
2100	109.537	245.080	162.556	173.301	-744.854	-306.402	7.621
2200	110.792	250.205	166.424	184.318	-742.316	-285.581	6.781
2300	112.048	255.158	170.175	195.460	-739.732	-264.879	6.016
2400	113.303	259.953	173.816	206.727	-737.103	-244.290	5.317
2500	114.558	264.603	177.355	218.120	-734.428	-223.811	4.676
2600	115.813	269.121	180.798	229.639	-731.704	-203.440	4.087
2700	117.068	273.515	184.151	241.283	-728.957	-183.173	3.544
2800	118.324	277.795	187.420	253.053	-726.221	-163.009	3.041
2900	119.579	281.970	190.608	264.948	-723.535	-142.943	2.575
3000	120.834	286.045	193.722	276.968	-720.925	-122.967	2.141

PREVIOUS: CURRENT: September 1966

Tungsten Oxide ($WO_{2.72}$) $O_{2.72}W_1$(cr)

TUNGSTEN OXIDE ($WO_{2.90}$) CRYSTAL M_r = 230.2483

$S°(298.15\ K)$ = [73.371] J K^{-1} mol^{-1}

$\Delta_f H°(0\ K)$ = Unknown
$\Delta_f H°(298.15\ K)$ = -820.064 ± 0.8 kJ mol^{-1}

Enthalpy of Formation

The adopted enthalpy of formation, -196.0 ± 3 kcal mol^{-1}, is a weighted average of those derived from equilibrium data, in addition the following results have been reported.

Griffis (1) has measured the enthalpy of combustion to WO_3(cr) which yields -195.6 kcal mol^{-1} when recalculated with $\Delta_f H°(WO_3$, cr, 298.15 K) = -201.46 kcal mol^{-1}. This result is less certain because of the incomplete characterization of the final products. Ackermann and Rauh (2) have investigated the tungsten-oxygen system by mass effusion, mass spectroscopy, and x-ray diffraction. In the temperature range from 1300 to 1600 K, they have derived indirectly two values of $\Delta_f H°(WO_{2.90}$, cr) = -190.9 and -188.0 kcal mol^{-1} (corresponding to -196.1 and -193.3 kcal mol^{-1}, respectively, at 298.15 K, using all JANAF functions).

Source	Reaction	T/K	Data Points	$\Delta_r H°$(298.15 K), kcal mol^{-1} 2nd law	3rd law	Drift cal K^{-1} mol^{-1}	$\Delta_f H°$(298.15 K) kcal mol^{-1}
(3)	A	773-898	5	2.168	0.270	-2.3 ± 0.07	-195.4
(4)	A	897-993	4	2.178	0.240	-2.1 ± 0.01	-195.4
(5)	A	873-1073	18	1.504	0.284	-1.3 ± 0.02	-195.4
(3)	B	844-934	5	2.657	0.754	-2.1 ± 0.09	-197.9
(4)	B	915-1021	5	1.585	0.834	-0.76± 0.1	-197.9
(5)	B	873-1073	18	3.388	0.589	-2.88± 0.02	-197.7
(3)	C	737-798	5	6.762	5.013	-2.3 ± 0.9	-198.0
(6)	D	1023-1273	-	-7.97	-	-	-193.5
(6)	E	1173-1322	-	-13.67	-	-	-200.4

Reactions:
A) $WO_3(cr) + 0.1\ H_2(g) = WO_{2.90}(cr) + 0.1\ H_2O(g)$
B) $WO_{2.90}(cr) + 0.18\ H_2(g) = WO_{2.72}(cr) + 0.18\ H_2O(g)$
C) $WO_{2.90}(cr) + 0.9\ H_2(g) = WO_2(cr) + 0.9\ H_2O(g)$
D) $WO_{2.90}(cr) + 0.05\ O_2(g) = WO_3(cr)$
E) $WO_{2.72}(cr) + 0.09\ O_2(g) = WO_{2.90}(cr)$

Heat Capacity and Entropy

The monoclinic unit cell of $WO_{2.90}$ corresponds to $W_{20}O_{58}$ as shown by the X-ray diffraction studies of Magneli et al. (7). The tabulated heat capacities and the entropy, $S°(298.15\ K)$ = 17.536 cal K^{-1} mol^{-1}, were estimated from those of WO_2(cr) and WO_3(cr) because all three have the similar monoclinic structure as identified by Magneli et al. (7).

References

1. R. C. Griffis, J. Electrochem. Soc. 105, 398 (1958).
2. R. J. Ackermann and E. G. Rauh, J. Phys. Chem. 67, 2596 (1963).
3. J. Bousquet and G. Perachon, Compt. rend. 258, 3869 (1964); 256, 694 (1963).
4. I. A. Vasil'eva, Ya. I. Perasimov and Yu. P. Simanov, Zh. Fiz. Khim. 34, 1811 (1960).
5. R. C. Griffis, J. Electrochem. Soc. 106, 418 (1959).
6. G. R. St. Pierre, W. T. Ebihara, M. J. Pool and R. Speiser, Trans. AIME 224, 259 (1962).
7. A. Magneli, G. Anderson, B. Blomberg and L. Kihlborg, Anal. Chem. 24, 1998 (1952).

Tungsten Oxide ($WO_{2.90}$) $O_{2.90}W_1$(cr)

Enthalpy Reference Temperature = T_r = 298.15 K   Standard State Pressure = p° = 0.1 MPa

T/K	$C_p°$ (J $K^{-1}mol^{-1}$)	S°	$-[G°-H°(T_r)]/T$	$H°-H°(T_r)$ (kJ mol^{-1})	$\Delta_f H°$	$\Delta_f G°$	Log K_f
0							
100							
200							
298.15	71.379	73.371	73.371	0.000	-820.064	-743.513	130.260
300	71.546	73.813	73.372	0.132	-820.056	-743.038	129.374
400	80.333	95.634	76.287	7.739	-819.221	-717.472	93.692
500	86.609	114.277	82.067	16.105	-817.804	-692.189	72.313
600	90.918	130.473	88.815	24.995	-816.053	-667.227	58.087
700	93.931	144.726	95.805	34.245	-814.124	-642.573	47.949
800	95.981	157.409	102.727	43.746	-812.106	-618.203	40.365
900	97.655	168.813	109.447	53.430	-810.049	-594.088	34.480
1000	99.244	179.189	115.910	63.279	-807.953	-570.205	29.784
1100	100.562	188.692	122.101	73.251	-805.849	-546.531	25.953
1200	101.880	197.496	128.021	83.370	-803.700	-523.051	22.768
1300	103.303	205.707	133.685	93.629	-801.506	-499.753	20.080
1400	104.558	213.411	139.107	104.025	-799.266	-476.625	17.783
1500	105.688	220.663	144.305	114.537	-796.996	-453.658	15.798
1600	106.818	227.520	149.293	125.162	-794.699	-430.844	14.066
1700	108.073	234.033	154.088	135.906	-792.364	-408.174	12.542
1800	109.328	240.246	158.703	146.777	-789.987	-385.643	11.191
1900	110.583	246.191	163.153	157.772	-787.566	-363.245	9.986
2000	111.838	251.895	167.448	168.893	-785.101	-340.976	8.905
2100	113.094	257.382	171.601	180.140	-782.593	-318.832	7.931
2200	114.349	262.672	175.621	191.512	-780.042	-296.806	7.047
2300	115.604	267.782	179.517	203.009	-777.448	-274.900	6.243
2400	116.859	272.729	183.299	214.633	-774.810	-253.107	5.509
2500	118.114	277.525	186.972	226.381	-772.128	-231.424	4.835
2600	119.370	282.182	190.545	238.255	-769.399	-209.849	4.216
2700	120.625	286.710	194.023	250.255	-766.648	-188.380	3.644
2800	121.880	291.120	197.413	262.380	-763.911	-167.013	3.116
2900	123.135	295.419	200.718	274.631	-761.226	-145.743	2.625
3000	124.390	299.614	203.945	287.007	-758.618	-124.564	2.169

PREVIOUS: CURRENT: September 1966

Tungsten Oxide ($WO_{2.90}$) $O_{2.90}W_1$(cr)

TUNGSTEN OXIDE ($WO_{2.96}$) CRYSTAL M_r = 231.2082

$\Delta_fH°$(0 K) = Unknown

S°(298.15 K) = [74.894] J K^{-1} mol^{-1} $\Delta_fH°$(298.15 K) = -834.959 ± 12.6 kJ mol^{-1}

Enthalpy of Formation

The adopted value, $\Delta_fH°(WO_{2.96}$, cr, 298.15 K) = -199.56 kcal mol^{-1}, was obtained by 3rd law treatment of $\Delta_fG°$ = 191,600 + 54.0 T (cal K^{-1} mol^{-1}) in the temperature range from 1300 - 1550 K, using all JANAF functions. The equation for $\Delta_fG°$ was reported by Ackermann and Rauh (1), based on studies by mass spectroscopy and X-ray diffraction. In this study, $\Delta_fG°$ was obtained by comparison of the Gibbs free energies of sublimation (to gaseous W_2O_6, W_3O_8, W_3O_9 and W_4O_{12}) over the two systems W-WO_2(cr) and $WO_{2.96}$(cr) and $WO_{2.96}$(cr); thus $WO_{2.96}$(cr) was related to WO_2(cr) through the gaseous species.

Battles (2) has reported partial pressures of W_3O_9(g) and W_4O_{12}(g) in equilibrium with WO_3 - $WO_{2.90}$(cr). In light of Ackermann and Rauh's observation, the condensed phase was presumed to have the composition of $WO_{2.96}$(cr). The pressure data are in reasonable agreement with the studies of Ackermann and Rauh as summarized below. Values of $\Delta_fH°(WO_{2.96}$, cr, 298.15 K) derived from the data of Battles are based on the same author's data for W_3O_9(g), and W_4O_{12}(g) in order to relate $WO_{2.96}$ to WO_2 directly.

Source	Reaction	T/K	$\Delta_rH°$(298.15 K), kcal mol^{-1} 2nd law	3rd law	Drift cal K^{-1} mol^{-1}	$\Delta_fH°$(298.15 K) kcal mol^{-1}
(1)	A	1300-1550	56.32	60.82	3.2 ± 0.6	-199.57
(1)	B	1300-1550	38.35	41.37	2.09± 0.09	-199.58
(2)	B	1274-1418	43.28	40.82	-1.85± 0.05	-199.72
(1)	C	1300-1550	33.58	35.81	1.55± 0.09	-199.60
(2)	C	1333-1418	38.74	35.37	-2.46± 0.1	-199.95

Reactions: A) $WO_{2.96}$(cr) = 0.44 W_2O_6(g) + 0.04 W_3O_8(g)
B) $WO_{2.96}$(cr) = 0.29333 W_3O_9(g) + 0.04 W_3O_8(g)
C) $WO_{2.96}$(cr) = 0.22 W_4O_{12}(g) + 0.04 W_3O_8(g)

Heat Capacity and Entropy

The tabulated heat capacities and the entropy at 298.15 K were estiamted from those of WO_2(cr) and WO_3(cr).

Ackermann and Rauh (1) have found that the composition $WO_{2.96}$ probably represents the azeotropic composition of the WO_{3-x} solid solution and is the only single phase which evaporates congruently below 1550 K.

References

1. R. J. Ackermann and E. G. Rauh, J. Phys. Chem. 67, 2596 (1963).
2. J. E. Battles, Ph.D. Thesis, "A Mass Spectrometric Investigation of Tungsten Dioxide and Tungsten Trioxide," Ohio State University (1964).

Tungsten Oxide ($WO_{2.96}$) $O_{2.96}W_1$(cr)

Enthalpy Reference Temperature = T_r = 298.15 K; Standard State Pressure = p° = 0.1 MPa

T/K	$C_p°$ (J $K^{-1}mol^{-1}$)	S°	-[G°-H°(T_r)]/T	H°-H°(T_r) (kJ mol^{-1})	$\Delta_fH°$	$\Delta_fG°$	Log K_f
0							
100							
200							
298.15	72.383	74.894	74.894	0.000	-834.959	-757.027	132.628
300	72.634	75.342	74.895	0.134	-834.950	-756.544	131.726
400	81.504	97.478	77.852	7.850	-834.095	-730.517	95.396
500	87.864	116.394	83.716	16.339	-832.648	-704.781	73.628
600	92.257	132.825	90.562	25.358	-830.863	-679.371	59.145
700	95.312	147.287	97.653	34.744	-828.895	-654.276	48.823
800	97.404	160.156	104.677	44.384	-826.838	-629.471	41.100
900	99.370	171.735	111.495	54.216	-824.734	-604.925	35.109
1000	100.709	182.264	118.053	64.211	-822.598	-580.616	30.328
1100	102.215	191.936	124.336	74.359	-820.421	-556.523	26.427
1200	103.512	200.886	130.347	84.647	-818.211	-532.630	23.185
1300	104.851	209.224	136.097	95.065	-815.965	-508.922	20.449
1400	106.106	217.041	141.603	105.613	-813.681	-485.389	18.110
1500	107.194	224.399	146.880	116.278	-811.368	-462.020	16.089
1600	108.282	231.351	151.944	127.052	-809.032	-438.806	14.326
1700	109.537	237.954	156.811	137.943	-806.662	-415.740	12.774
1800	110.792	244.250	161.495	148.959	-804.249	-392.814	11.399
1900	112.048	250.274	166.010	160.101	-801.794	-370.023	10.173
2000	113.303	256.053	170.369	171.369	-799.296	-347.363	9.072
2100	114.558	261.612	174.582	182.762	-796.755	-324.828	8.080
2200	115.813	266.970	178.660	194.280	-794.172	-302.415	7.180
2300	117.068	272.145	182.613	205.924	-791.546	-280.122	6.362
2400	118.324	277.154	186.449	217.694	-788.877	-257.944	5.614
2500	119.579	282.010	190.174	229.589	-786.165	-235.877	4.928
2600	120.834	286.724	193.798	241.610	-783.406	-213.920	4.298
2700	122.089	291.308	197.325	253.756	-780.627	-192.069	3.716
2800	123.344	295.771	200.761	266.028	-777.862	-170.321	3.177
2900	124.600	300.121	204.113	278.425	-775.148	-148.672	2.678
3000	125.855	304.367	207.384	290.948	-772.513	-127.114	2.213

PREVIOUS: CURRENT: September 1966

Tungsten Oxide ($WO_{2.96}$) $O_{2.96}W_1$(cr)

ZIRCONIUM OXIDE (ZrO_2) CRYSTAL M_r = 123.2188

$S°(298.15\ K) = [50.359 \pm 0.33]\ J\ K^{-1}\ mol^{-1}$
$T_{trs} = 1478 \pm 5\ K$
$T_{fus} = 2950\ K$

$\Delta_f H°(0\ K) = -1092.032 \pm 1.7\ kJ\ mol^{-1}$
$\Delta_f H°(298.15\ K) = -1097.463 \pm 1.7\ kJ\ mol^{-1}$
$\Delta_{trs} H° = 5.941 \pm 0.4\ kJ\ mol^{-1}$
$\Delta_{fus} H° = [87.027]\ kJ\ mol^{-1}$

Enthalpy of Formation

$\Delta_f H°(ZrO_2$, cr, 298.15 K) = -262.3 ± 0.4 kcal mol^{-1} is an average of the values 261.5 ± 0.2 kcal mol^{-1} and -263.1 ± 0.5 kcal mol^{-1} measured (by combustion calorimetry) by Humphrey (1) and Huber et al. (2), respectively. Also the enthalpy of formation was measured calorimetrically by Neumann et al. (3), Roth et al. (4) and by Sieverts et al. (5) and was found to be -258.2, -258.8 and -256.1 kcal mol^{-1}, respectively. However, more weight was given to the Humphrey and Huber values.

Heat Capacity and Entropy

The low temperature heat capacities, 54.3-295 K, were measured by Kelley (6). The heat capacities in the temperature range 298-1478 K were calculated by using the Coughlin and King (7) equation, $C°_p = 16.64 + 1.80 \times 10^{-3} T - 3.36 \times 10^5\ T^{-2}$, obtained from their measured enthalpy data (296.8-1841 K). The values from the two sources join smoothly at 298.15 K. Above the transition 1478 K, the heat capacity was taken as constant at 17.80 cal $K^{-1}\ mol^{-1}$. The entropy was calculated at 54.30 K using the Debye and Einstein function D(345/T) + E(513/T) + E(861/T) given by Kelley (6). The value, S°(54.3 K) = 0.5507 cal $K^{-1}\ mol^{-1}$, is thus calculated.

Transition Data

T_{trs} and $\Delta_{trs}H°$ were taken from Coughlin and King (7).

Fusion Data

T_{fus} and $\Delta_{fus}H°$ were taken from U. S. Nat. Bur. Stand. (8).

References

1. G. L. Humphrey, J. Amer. Chem. Soc. 76, 978 (1954).
2. E. J. Huber, Jr., E. L. Head and C. E. Holley, Jr., J. Phys. Chem. 68, 3040 (1964).
3. B. Neumann, C. Kroger and H. Kunz, Z. Anorg. Chem. 218, 379 (1934).
4. W. A. Roth, E. Borger and H. Siemonsen, Z. Anorg. Chem. 239, 321 (1938).
5. A. Sieverts, A. Gotta and S. Halberstadt, Z. Anorg. Chem. 187, 155 (1930).
6. K. K. Kelley, Ind. Eng. Chem. 36, 377 (1944).
7. J. P. Coughlin and E. G. King, J. Amer. Chem. Soc. 72, 2262 (1950).
8. U. S. Nat. Bur. Stand. Circ. 500, 1952.

Zirconium Oxide (ZrO_2) O_2Zr_1(cr)

Enthalpy Reference Temperature = T_r = 298.15 K Standard State Pressure = p° = 0.1 MPa

T/K	$C°_p$ (J K⁻¹mol⁻¹)	S° (J K⁻¹mol⁻¹)	-[G°-H°(T_r)]/T (J K⁻¹mol⁻¹)	H°-H°(T_r) (kJ mol⁻¹)	$\Delta_f H°$ (kJ mol⁻¹)	$\Delta_f G°$ (kJ mol⁻¹)	Log K_f
0	0.000	0.000	INFINITE	-8.749	-1092.032	-1092.032	INFINITE
100	18.927	9.488	90.390	-8.090	-1095.164	-1077.376	562.764
200	42.778	30.546	55.114	-4.914	-1097.091	-1058.694	276.502
298.15	56.191	50.359	50.359	0.000	-1097.463	-1039.724	182.155
300	56.262	50.707	50.360	0.104	-1097.460	-1039.366	180.970
400	63.848	68.064	52.675	6.155	-1096.940	-1020.063	133.207
500	67.764	82.769	57.264	12.753	-1096.026	-1000.945	104.568
600	70.237	95.356	62.589	19.660	-1094.969	-982.027	85.493
700	72.023	106.323	68.070	26.777	-1093.873	-963.289	71.882
800	73.450	116.036	73.471	34.053	-1092.785	-944.709	61.683
900	74.663	124.759	78.693	41.459	-1091.731	-926.264	53.759
1000	75.747	132.682	83.702	48.981	-1090.727	-907.933	47.426
1100	76.743	139.949	88.489	56.606	-1089.783	-889.699	42.248
1200	77.684	146.667	93.061	64.327	-1092.612	-871.328	37.928
1300	78.580	152.921	97.428	72.141	-1091.250	-852.942	34.272
1400	79.450	158.776	101.603	80.042	-1089.872	-834.662	31.142
1478.000	80.108	163.101	104.735	86.265		I <--> II	
1478.000	74.475	167.121	104.735	92.207		TRANSITION	
1500	74.475	168.222	105.658	93.845	-1082.674	-816.569	28.435
1600	74.475	173.028	109.720	101.293	-1081.921	-798.854	26.080
1700	74.475	177.543	113.578	108.740	-1081.266	-781.183	24.003
1800	74.475	181.800	117.251	116.188	-1080.720	-763.547	22.158
1900	74.475	185.827	120.756	123.635	-1080.293	-745.939	20.507
2000	74.475	189.647	124.105	131.083	-1079.995	-728.349	19.023
2100	74.475	193.280	127.314	138.530	-1079.836	-710.772	17.679
2200	74.475	196.745	130.392	145.978	-1101.150	-692.452	16.441
2300	74.475	200.056	133.349	153.425	-1101.717	-673.863	15.304
2400	74.475	203.225	136.195	160.873	-1102.306	-655.248	14.261
2500	74.475	206.265	138.937	168.320	-1102.918	-636.608	13.301
2600	74.475	209.186	141.583	175.768	-1103.550	-617.943	12.415
2700	74.475	211.997	144.140	183.215	-1104.204	-599.253	11.593
2800	74.475	214.706	146.612	190.663	-1104.878	-580.539	10.830
2900	74.475	217.319	149.005	198.110	-1105.572	-561.801	10.119
2950.000	74.475	218.592	150.174	201.834	--------	II <--> LIQUID	--------
3000	74.475	219.844	151.325	205.558	-1106.286	-543.038	9.455
3100	74.475	222.286	153.575	213.005	-1107.018	-524.250	8.834
3200	74.475	224.650	155.759	220.453	-1107.768	-505.440	8.250
3300	74.475	226.942	157.881	227.900	-1108.536	-486.606	7.702
3400	74.475	229.165	159.945	235.348	-1109.320	-467.747	7.186
3500	74.475	231.324	161.954	242.795	-1110.120	-448.866	6.699

PREVIOUS: June 1961 CURRENT: December 1965

Zirconium Oxide (ZrO_2) O_2Zr_1(cr)

ZIRCONIUM OXIDE (ZrO_2) LIQUID M_r = 123.2188

$S°(298.15\ K) = [74.657]\ J\ K^{-1}\ mol^{-1}$ $\Delta_f H°(298.15\ K) = [-1023.155]\ kJ\ mol^{-1}$

$T_{fus} = 2950\ K$ $\Delta_{fus}H° = [87.027]\ kJ\ mol^{-1}$

Enthalpy of Formation

$\Delta_f H°(ZrO_2, \ell, 298.15\ K)$ is calculated from $\Delta_f H°(ZrO_2, cr, 298.15\ K)$ by adding $\Delta_{fus}H°$ and the difference in enthalpy, H°(2950 K)-H°(298.15 K), between the crystal and liquid.

Heat Capacity and Entropy

A glass transition is assumed at 2000 K. The heat capacity below 2000 K was obtained from the heat capacity of the crystal. Above 2000 K the heat capacity was assumed constant and estimated to be 21.0 cal $K^{-1}\ mol^{-1}$ or 7 cal $K^{-1}\ mol^{-1}$ per g-atom as suggested by Kubaschewski and Evans (1). The entropy at 298.15 K is calculated in a manner analogous to that used for the enthalpy of formation.

Fusion Data

See the crystal table for details.

Vaporization Data

$T_{vap}(1\ atm)$ = 4548 K is calculated as the temperature at which the fugacity is 1 atm for the reaction $ZrO_2(\ell) \rightarrow ZrO_2(g)$. The difference between $\Delta_f H°(4548\ K)$ for $ZrO_2(g)$ and $ZrO_2(\ell)$ gives $\Delta_{vap}H°(4548\ K) = 149.2\ kcal\ mol^{-1}$.

Reference

1. O. Kubaschewski and E. LL. Evans, "Metallurgical Thermochemistry", Pergamon Press, New York (1958).

Zirconium Oxide (ZrO_2) $O_2Zr_1(l)$

Enthalpy Reference Temperature = T_r = 298.15 K Standard State Pressure = p° = 0.1 MPa

T/K	$C_p°$ (J K⁻¹mol⁻¹)	S° (J K⁻¹mol⁻¹)	$-[G°-H°(T_r)]/T$ (J K⁻¹mol⁻¹)	$H°-H°(T_r)$ (kJ mol⁻¹)	$\Delta_f H°$ (kJ mol⁻¹)	$\Delta_f G°$ (kJ mol⁻¹)	Log K_f
0							
100							
200							
298.15	56.191	74.657	74.657	0.000	-1023.155	-972.661	170.406
300	56.262	75.004	74.658	0.104	-1023.152	-972.347	169.301
400	63.848	92.361	76.972	6.155	-1022.632	-955.474	124.772
500	67.764	107.066	81.561	12.753	-1021.718	-938.785	98.074
600	70.237	119.653	86.886	19.660	-1020.661	-922.297	80.293
700	72.023	130.621	92.367	26.777	-1019.565	-905.989	67.606
800	73.450	140.333	97.768	34.053	-1018.477	-889.839	58.101
900	74.663	149.056	102.990	41.459	-1017.423	-873.823	50.715
1000	75.747	156.979	107.999	48.981	-1016.420	-857.922	44.813
1100	76.743	164.246	112.786	56.606	-1015.475	-842.118	39.989
1200	77.684	170.964	117.358	64.327	-1018.304	-826.176	35.963
1300	78.580	177.218	121.725	72.141	-1016.942	-810.221	32.555
1400	79.450	183.073	125.900	80.042	-1015.565	-794.371	29.638
1478.000	80.107	187.398	129.032	86.265	I <--> II TRANSITION		
1478.000	74.475	191.418	129.032	92.207			
1500	74.475	192.519	129.955	93.845	-1008.367	-778.707	27.117
1600	74.475	197.325	134.017	101.293	-1007.613	-763.421	24.923
1700	74.475	201.840	137.875	108.740	-1006.958	-748.180	22.989
1800	74.475	206.097	141.548	116.188	-1006.412	-732.974	21.270
1900	74.475	210.124	145.053	123.635	-1005.985	-717.795	19.734
2000	74.475	213.944	148.403	131.083	-1005.687	-702.636	18.351
2000.000	74.475	213.944	148.403	131.083	GLASS <--> LIQUID TRANSITION		
2000.000	87.864	213.944	148.403	131.083			
2100	87.864	218.231	151.626	139.869	-1004.189	-687.521	17.101
2200	87.864	222.318	154.748	148.655	-1024.164	-671.727	15.949
2300	87.864	226.224	157.771	157.442	-1023.393	-655.725	14.892
2400	87.864	229.963	160.702	166.228	-1022.643	-639.756	13.924
2500	87.864	233.550	163.544	175.015	-1021.916	-623.817	13.034
2600	87.864	236.996	166.304	183.801	-1021.209	-607.908	12.213
2700	87.864	240.312	168.984	192.587	-1020.524	-592.024	11.453
2800	87.864	243.508	171.588	201.374	-1019.860	-576.166	10.749
2900	87.864	246.591	174.122	210.160	-1019.215	-560.332	10.093
2950.000	87.864	248.093	175.363	214.553	-------- II <--> LIQUID --------		
3000	87.864	249.570	176.587	218.947	-1018.589	-544.518	9.481
3100	87.864	252.451	178.988	227.733	-1017.983	-528.726	8.909
3200	87.864	255.240	181.328	236.519	-1017.394	-512.953	8.373
3300	87.864	257.944	183.609	245.306	-1016.823	-497.199	7.870
3400	87.864	260.567	185.834	254.092	-1016.268	-481.460	7.397
3500	87.864	263.114	188.006	262.879	-1015.729	-465.739	6.951
3600	87.864	265.589	190.127	271.665	-1015.206	-450.032	6.530
3700	87.864	267.997	192.199	280.451	-1014.698	-434.340	6.132
3800	87.864	270.340	194.224	289.238	-1014.204	-418.661	5.755
3900	87.864	272.622	196.206	298.024	-1013.724	-402.995	5.398
4000	87.864	274.847	198.144	306.811	-1013.257	-387.341	5.058
4100	87.864	277.016	200.041	315.597	-1012.803	-371.699	4.736
4200	87.864	279.133	201.899	324.383	-1012.362	-356.068	4.428
4300	87.864	281.201	203.720	333.170	-1011.933	-340.446	4.136
4400	87.864	283.221	205.504	341.956	-1011.517	-324.836	3.856
4500	87.864	285.195	207.253	350.743	-1011.112	-309.235	3.590
4600	87.864	287.127	208.968	359.529	-1010.720	-293.642	3.334
4700	87.864	289.016	210.651	368.315	-1010.340	-278.057	3.090
4800	87.864	290.866	212.303	377.102	-1570.909	-250.863	2.730
4900	87.864	292.678	213.925	385.888	-1570.207	-223.369	2.381
5000	87.864	294.453	215.518	394.675	-1569.534	-195.889	2.046

PREVIOUS: June 1961 CURRENT: December 1965

Zirconium Oxide (ZrO_2) $O_2Zr_1(l)$

ZIRCONIUM OXIDE (ZrO_2) CRYSTAL-CRYSTAL-LIQUID M_r = 123.2188 Zirconium Oxide (ZrO_2) O_2Zr_1(cr,l)

0 to 1478 K		crystal, I
1478 to 2950 K		crystal, II
above 2950 K		liquid

Refer to the individual tables for details.

Enthalpy Reference Temperature = T_r = 298.15 K   Standard State Pressure = p° = 0.1 MPa

T/K	C_p° (J K⁻¹mol⁻¹)	S° (J K⁻¹mol⁻¹)	$-[G°-H°(T_r)]/T$ (J K⁻¹mol⁻¹)	$H°-H°(T_r)$ (kJ mol⁻¹)	$\Delta_fH°$ (kJ mol⁻¹)	$\Delta_fG°$ (kJ mol⁻¹)	Log K_f
0	0.000	0.000	INFINITE	-8.749	-1092.032	-1092.032	INFINITE
100	18.927	9.488	90.390	-8.090	-1095.164	-1077.376	562.764
200	42.778	30.546	55.114	-4.914	-1097.091	-1058.694	276.502
298.15	56.191	50.359	50.359	0.000	-1097.463	-1039.724	182.155
300	56.262	50.707	50.360	0.104	-1097.460	-1039.366	180.970
400	63.848	68.064	52.675	6.155	-1096.940	-1020.063	133.207
500	67.764	82.769	57.264	12.753	-1096.026	-1000.945	104.568
600	70.237	95.356	62.589	19.660	-1094.969	-982.027	85.493
700	72.023	106.323	68.070	26.777	-1093.873	-963.289	71.882
800	73.450	116.036	73.471	34.053	-1092.785	-944.709	61.683
900	74.663	124.759	78.693	41.459	-1091.731	-926.264	53.759
1000	75.747	132.682	83.702	48.981	-1090.727	-907.933	47.426
1100	76.743	139.949	88.489	56.606	-1089.783	-889.699	42.248
1200	77.684	146.667	93.061	64.327	-1092.612	-871.328	37.928
1300	78.580	152.921	97.428	72.141	-1091.250	-852.942	34.272
1400	79.450	158.776	101.603	80.042	-1089.872	-834.662	31.142
1478.000	80.107	163.101	104.735	86.265		I <--> II	
1478.000	74.475	167.121	104.735	92.207		TRANSITION	
1500	74.475	168.222	105.658	93.845	-1082.675	-816.569	28.435
1600	74.475	173.028	109.720	101.293	-1081.921	-798.854	26.080
1700	74.475	177.543	113.578	108.740	-1081.266	-781.183	24.003
1800	74.475	181.800	117.251	116.188	-1080.720	-763.547	22.158
1900	74.475	185.827	120.756	123.635	-1080.293	-745.939	20.507
2000	74.475	189.647	124.105	131.083	-1079.995	-728.349	19.023
2100	74.475	193.280	127.314	138.530	-1079.836	-710.772	17.679
2200	74.475	196.745	130.392	145.978	-1101.150	-692.452	16.441
2300	74.475	200.056	133.349	153.425	-1101.717	-673.862	15.304
2400	74.475	203.225	136.195	160.873	-1102.306	-655.248	14.261
2500	74.475	206.265	138.937	168.320	-1102.918	-636.608	13.301
2600	74.475	209.186	141.583	175.768	-1103.550	-617.943	12.415
2700	74.475	211.997	144.140	183.215	-1104.204	-599.253	11.593
2800	74.475	214.706	146.612	190.663	-1104.878	-580.539	10.830
2900	74.475	217.319	149.005	198.110	-1105.572	-561.801	10.119
2950.000	74.475	218.592	150.174	201.834		II <--> LIQUID	
2950.000	87.864	248.093	150.174	288.861		TRANSITION	
3000	87.864	249.570	151.818	293.254	-1018.589	-544.518	9.481
3100	87.864	252.451	155.018	302.041	-1017.983	-528.726	8.909
3200	87.864	255.240	158.107	310.827	-1017.394	-512.953	8.373
3300	87.864	257.944	161.091	319.614	-1016.823	-497.199	7.870
3400	87.864	260.567	163.979	328.400	-1016.268	-481.460	7.397
3500	87.864	263.114	166.775	337.186	-1015.729	-465.739	6.951
3600	87.864	265.589	169.486	345.973	-1015.206	-450.032	6.530
3700	87.864	267.997	172.116	354.759	-1014.698	-434.340	6.132
3800	87.864	270.340	174.670	363.546	-1014.204	-418.661	5.755
3900	87.864	272.622	177.152	372.332	-1013.724	-402.995	5.398
4000	87.864	274.847	179.567	381.118	-1013.257	-387.341	5.058
4100	87.864	277.016	181.917	389.905	-1012.803	-371.699	4.736
4200	87.864	279.133	184.207	398.691	-1012.362	-356.068	4.428
4300	87.864	281.201	186.439	407.478	-1011.933	-340.446	4.136
4400	87.864	283.221	188.615	416.264	-1011.517	-324.836	3.856
4500	87.864	285.195	190.740	425.050	-1011.112	-309.234	3.590
4600	87.864	287.127	192.814	433.837	-1010.720	-293.642	3.334
4700	87.864	289.016	194.841	442.623	-1010.340	-278.057	3.090
4800	87.864	290.866	196.822	451.410	-1570.909	-250.863	2.730
4900	87.864	292.678	198.760	460.196	-1570.207	-223.369	2.381
5000	87.864	294.453	200.656	468.982	-1569.534	-195.889	2.046

PREVIOUS: CURRENT: December 1965

Zirconium Oxide (ZrO_2) O_2Zr_1(cr,l)

ZIRCONIUM OXIDE (ZrO_2) IDEAL GAS M_r = 123.2188

$S°(298.15\ K) = 273.746\ J\ K^{-1}\ mol^{-1}$

$\Delta_f H°(0\ K) = -284.01 \pm 46.0\ kJ\ mol^{-1}$
$\Delta_f H°(298.15\ K) = -286.19 \pm 46.0\ kJ\ mol^{-1}$

Vibrational Frequencies and Degeneracies

ν, cm^{-1}
884 (1)
[137](1)
818 (1)

Ground State Quantum Weight: 1 $\sigma = 2$
Point Group: C_{2v}
Bond Distance: O-Zr = 1.711 Å
Bond Angle: O-Zr-O = 109°
Product of the Moments of Inertia: $I_A I_B I_C = 5.682779 \times 10^{-115}\ g^3\ cm^6$

Enthalpy of Formation

$\Delta_f H°(ZrO_2$, g, 298.15 K) = $-68.4 \pm 11\ kcal\ mol^{-1}$ was calculated from the $\Delta_r H°$(298.15 K) = $193.867 \pm 11.25\ kcal\ mol^{-1}$ for the reaction $ZrO_2(cr) \rightarrow ZrO_2(g)$. The value of $\Delta_r H°$(298.15 K) was obtained by the 3rd law method using the determined vapor pressures (at the range 2331-2480 K) by Chupka et al. (1). They observed the variation of the ZrO_2^+ ion intensity over ZrO_2 and reported equilibrium constants for the reaction $ZrO_2(cr) \rightarrow ZrO_2(g)$ in the range 2331-2480 K. The 3rd law drift is $80 \pm 23\ cal\ K^{-1}\ mol^{-1}$.

Heat Capacity and Entropy

The vibrational frequencies $\nu_1 = 884$, $\nu_2 = [137]$, $\nu_3 = 818$ and angle O-Zr-O = 109° were obtained from Linevsky (2). The bond distance r_e(O-Zr) = 1.711 Å was obtained from Uhler and Akerline (3). The principal moments of inertia are: $I_A = 3.8834 \times 10^{-39}$, $I_B = 10.3101 \times 10^{-39}$, $I_C = 14.1935 \times 10^{-39}\ g\ cm^2$.

References

1. W. A. Chupka, J. Berkowitz and M. G. Inghram, J. Chem. Phys. 26, 1207 (1957).
2. M. J. Linevsky, Spectroscopic Studies of the Vaporization of Refractory Materials, AFML TR-64-420, Air Force Materials Lab. Research and Technology Division, Wright Patterson Air Force Base, Ohio (1965).
3. U. Uhler and L. Akerlind, Arkiv Fysik 10, 431 (1955).

Zirconium Oxide (ZrO_2) $O_2Zr_1(g)$

Enthalpy Reference Temperature = T_r = 298.15 K Standard State Pressure = p° = 0.1 MPa

T/K	$C_p°$ (J K⁻¹mol⁻¹)	S° (J K⁻¹mol⁻¹)	$-[G°-H°(T_r)]/T$ (J K⁻¹mol⁻¹)	$H°-H°(T_r)$ (kJ mol⁻¹)	$\Delta_f H°$ (kJ mol⁻¹)	$\Delta_f G°$ (kJ mol⁻¹)	Log K_f
0	0.	0.	INFINITE	-12.008	-284.013	-284.013	INFINITE
100	39.345	227.980	312.150	-8.417	-284.213	-288.275	150.579
200	42.320	256.161	277.836	-4.335	-285.235	-291.960	76.252
250	44.193	265.801	274.494	-2.173	-285.747	-293.581	61.340
298.15	46.061	273.746	273.746	0.	-286.186	-295.049	51.691
300	46.131	274.031	273.747	0.085	-286.201	-295.105	51.382
350	47.910	281.279	274.315	2.437	-286.597	-296.556	44.259
400	49.444	287.779	275.599	4.872	-286.945	-297.955	38.909
450	50.727	293.679	277.285	7.377	-287.262	-299.312	34.743
500	51.787	299.081	279.198	9.941	-287.560	-300.635	31.407
600	53.383	308.673	283.332	15.205	-288.147	-303.195	26.395
700	54.482	316.990	287.559	20.601	-288.772	-305.654	22.808
800	55.258	324.318	291.705	26.090	-289.469	-308.019	20.112
900	55.822	330.861	295.699	31.646	-290.267	-310.291	18.009
1000	56.242	336.765	299.515	37.250	-291.180	-312.468	16.322
1100	56.561	342.141	303.149	42.891	-292.220	-314.548	14.937
1200	56.810	347.074	306.607	48.560	-297.101	-316.305	13.768
1300	57.007	351.629	309.897	54.251	-297.862	-317.875	12.772
1400	57.165	355.860	313.031	59.960	-298.677	-319.384	11.916
1500	57.294	359.808	316.019	65.683	-299.559	-320.833	11.172
1600	57.401	363.509	318.873	71.418	-300.518	-322.221	10.519
1700	57.490	366.992	321.602	77.163	-301.566	-323.545	9.941
1800	57.565	370.280	324.216	82.916	-302.714	-324.806	9.426
1900	57.629	373.394	326.723	88.676	-303.975	-325.999	8.962
2000	57.684	376.352	329.131	94.441	-305.358	-327.123	8.544
2100	57.731	379.167	331.447	100.212	-306.876	-328.175	8.163
2200	57.772	381.854	333.678	105.987	-329.863	-328.404	7.797
2300	57.808	384.423	335.829	111.766	-332.098	-328.289	7.456
2400	57.840	386.884	337.905	117.549	-334.353	-328.075	7.140
2500	57.868	389.246	339.912	123.334	-336.626	-327.767	6.848
2600	57.893	391.516	341.853	129.122	-338.918	-327.367	6.577
2700	57.915	393.701	343.733	134.912	-341.229	-326.879	6.324
2800	57.935	395.808	345.556	140.705	-343.559	-326.305	6.087
2900	57.952	397.841	347.324	146.499	-345.906	-325.648	5.866
3000	57.969	399.806	349.041	152.295	-348.271	-324.908	5.657
3100	57.983	401.707	350.709	158.093	-350.653	-324.090	5.461
3200	57.997	403.548	352.332	163.892	-353.052	-323.196	5.276
3300	58.009	405.333	353.911	169.692	-355.466	-322.226	5.100
3400	58.020	407.065	355.449	175.494	-357.897	-321.181	4.934
3500	58.030	408.747	356.948	181.296	-360.342	-320.066	4.777
3600	58.039	410.382	358.409	187.100	-362.802	-318.881	4.627
3700	58.048	411.972	359.836	192.904	-365.276	-317.627	4.484
3800	58.056	413.520	361.228	198.709	-367.763	-316.305	4.348
3900	58.063	415.028	362.588	204.515	-370.263	-314.918	4.218
4000	58.070	416.498	363.918	210.322	-372.776	-313.468	4.093
4100	58.076	417.932	365.218	216.129	-375.301	-311.954	3.974
4200	58.082	419.332	366.490	221.937	-377.839	-310.378	3.860
4300	58.087	420.699	367.735	227.745	-380.388	-308.741	3.750
4400	58.092	422.034	368.954	233.554	-382.949	-307.046	3.645
4500	58.097	423.340	370.148	239.364	-385.521	-305.292	3.544
4600	58.102	424.617	371.318	245.174	-388.106	-303.481	3.446
4700	58.106	425.866	372.465	250.984	-390.702	-301.613	3.352
4800	58.110	427.090	373.591	256.795	-954.246	-288.073	3.135
4900	58.113	428.288	374.695	262.606	-956.519	-274.170	2.923
5000	58.117	429.462	375.778	268.418	-958.821	-260.221	2.719
5100	58.120	430.613	376.842	274.230	-961.152	-246.226	2.522
5200	58.123	431.741	377.887	280.042	-963.511	-232.186	2.332
5300	58.126	432.849	378.914	285.854	-965.897	-218.098	2.149
5400	58.129	433.935	379.923	291.667	-968.310	-203.967	1.973
5500	58.131	435.002	380.914	297.480	-970.749	-189.790	1.802
5600	58.134	436.049	381.890	303.293	-973.213	-175.568	1.638
5700	58.136	437.078	382.849	309.107	-975.703	-161.303	1.478
5800	58.138	438.089	383.793	314.920	-978.218	-146.992	1.324
5900	58.140	439.083	384.721	320.734	-980.757	-132.641	1.174
6000	58.142	440.060	385.636	326.548	-983.321	-118.243	1.029

PREVIOUS: December 1965 (1 atm) CURRENT: December 1965 (1 bar)

Zirconium Oxide (ZrO_2) $O_2Zr_1(g)$

OZONE (O_3) IDEAL GAS M_r = 47.9982

$\Delta_f H°(0\ K) = 145.348 \pm 1.7\ kJ\ mol^{-1}$

$S°(298.15\ K) = 238.932\ J\ K^{-1}\ mol^{-1}$ $\Delta_f H°(298.15\ K) = 142.674 \pm 1.7\ kJ\ mol^{-1}$

Vibrational Levels and Degeneracies

ν, cm^{-1}
1110 (1)
705 (1)
1043 (1)

Ground State Quantum Weight: 1 $\sigma = 2$

Point Group: C_{2v}

Rotational Constants: $A_{000} = 3.55381\ cm^{-1}$ $B_{000} = 0.44530\ cm^{-1}$ $C_{000} = 0.39477\ cm^{-1}$

Anharmonicity Constants: $X_{11} = -3.8\ cm^{-1}$ $X_{33} = -4.0\ cm^{-1}$ $X_{23} = -3.5\ cm^{-1}$

$X_{22} = -3.5\ cm^{-1}$ $X_{12} = -2.0\ cm^{-1}$ $X_{31} = -9.0\ cm^{-1}$

Enthalpy of Formation

Gunther et al. (1) measured ΔE for the reaction $O_3(g) + 3/2\ O_2(g)$ in a calorimeter calibrated by means of the reaction CO(g) + 1/2 $O_2(g) \rightarrow CO_2(g)$ and found the ratio of the ΔE values for these reactions to be 0.5077 ± 0.0024. The temperature of the experiments is not reported and is assumed to have been 18°C. From the above ratio, $\Delta_f H°(298.15\ K)$ is calculated to be 33.89 ± 0.36 cal mol^{-1}. Kailan and Jahn (2) measured $\Delta_r H°$ at an unspecified temperature for the reaction $O_3(g) \rightarrow 3/2\ O_2(g)$ in an electrically calibrated calorimeter. If the temperature of the experiments is assumed to have been 18°C and the current joule-calorie conversion factor is used, $\Delta_f H°(298.15\ K)$ is found to be 34.3 ± 0.6 kcal mol^{-1}. Variability in the O_3 content of the feed gas is allowed for in the uncertainty. The value adopted here is 34.1 ± 0.4 kcal mol^{-1}.

Heat Capacity and Entropy

Gordon (3) has calculated the thermodynamic functions from 298.15 to 6000 K by the method of Pennington and Kobe (4) which takes vibration-rotation interaction and anharmonicity into account. Gordon's data are from Pierce (5). The functions below 298.15 K have been calculated for a rigidly rotating harmonic oscillator.

References

1. P. Gunther, E. Wassmuth and L. A. Schryver, Z. Phys. Chem. 158, 297 (1932).
2. A. Kailan and S. Jahn, Z. Anorg. Chem. 68, 243 (1910).
3. J. S. Gordon, Thiokol Chemical Corp., personal communication, April, 1961.
4. R. E. Pennington and K. A. Kobe, J. Chem. Phys. 22, 1442 (1954).
5. L. Pierce, J. Chem. Phys. 24, 139 (1956).

Ozone (O_3) $O_3(g)$

Enthalpy Reference Temperature = T_r = 298.15 K Standard State Pressure = p° = 0.1 MPa

T/K	$C_p°$	S°	$-[G°-H°(T_r)]/T$	$H°-H°(T_r)$	$\Delta_f H°$	$\Delta_f G°$	Log K_f
	J $K^{-1}mol^{-1}$			kJ mol^{-1}			
0	0.000	0.000	INFINITE	-10.351	145.348	145.348	INFINITE
100	33.292	200.791	271.040	-7.025	144.318	150.235	-78.474
200	35.058	224.221	242.401	-3.636	143.340	156.541	-40.884
298.15	39.238	238.932	238.932	0.000	142.674	163.184	-28.589
300	39.330	239.175	238.933	0.073	142.665	163.311	-28.435
400	43.744	251.116	240.531	4.234	142.370	170.247	-22.232
500	47.262	261.272	243.688	8.792	142.340	177.224	-18.514
600	49.857	270.129	247.373	13.654	142.462	184.191	-16.035
700	51.752	277.963	251.194	18.738	142.665	191.130	-14.262
800	53.154	284.969	254.986	23.986	142.907	198.037	-12.931
900	54.208	291.292	258.674	29.356	143.169	204.913	-11.893
1000	55.024	297.048	262.228	34.819	143.439	211.759	-11.061
1100	55.660	302.323	265.637	40.355	143.711	218.578	-10.379
1200	56.174	307.189	268.899	45.947	143.980	225.372	-9.810
1300	56.593	311.702	272.020	51.586	144.245	232.144	-9.328
1400	56.948	315.909	275.007	57.264	144.502	238.896	-8.913
1500	57.245	319.849	277.866	62.974	144.750	245.629	-8.554
1600	57.501	323.551	280.607	68.711	144.987	252.347	-8.238
1700	57.722	327.044	283.237	74.473	145.211	259.050	-7.960
1800	57.919	330.349	285.763	80.255	145.419	265.740	-7.712
1900	58.095	333.485	288.193	86.056	145.611	272.419	-7.489
2000	58.250	336.469	290.533	91.873	145.784	279.089	-7.289
2100	58.396	339.315	292.789	97.705	145.938	285.750	-7.108
2200	58.526	342.035	294.966	103.552	146.072	292.406	-6.943
2300	58.647	344.639	297.069	109.410	146.185	299.054	-6.792
2400	58.764	347.137	299.104	115.281	146.276	305.698	-6.653
2500	58.869	349.538	301.073	121.163	146.346	312.339	-6.526
2600	58.969	351.849	302.982	127.055	146.393	318.978	-6.408
2700	59.066	354.077	304.833	132.956	146.419	325.616	-6.299
2800	59.158	356.226	306.631	138.868	146.424	332.253	-6.198
2900	59.245	358.304	308.377	144.788	146.408	338.889	-6.104
3000	59.329	360.314	310.075	150.716	146.371	345.527	-6.016
3100	59.409	362.260	311.727	156.653	146.314	352.167	-5.934
3200	59.488	364.148	313.336	162.598	146.238	358.808	-5.857
3300	59.563	365.980	314.904	168.551	146.144	365.451	-5.785
3400	59.639	367.759	316.432	174.511	146.033	372.100	-5.717
3500	59.714	369.489	317.923	180.479	145.905	378.750	-5.653
3600	59.781	371.172	319.379	186.453	145.761	385.405	-5.592
3700	59.852	372.811	320.801	192.435	145.601	392.063	-5.535
3800	59.919	374.408	322.191	198.424	145.427	398.728	-5.481
3900	59.986	375.965	323.550	204.419	145.239	405.396	-5.430
4000	60.053	377.485	324.879	210.421	145.038	412.069	-5.381
4100	60.120	378.968	326.181	216.429	144.823	418.747	-5.335
4200	60.183	380.418	327.455	222.445	144.597	425.431	-5.291
4300	60.245	381.835	328.703	228.466	144.358	432.121	-5.249
4400	60.308	383.220	329.926	234.494	144.106	438.815	-5.209
4500	60.371	384.576	331.126	240.528	143.844	445.516	-5.171
4600	60.434	385.904	332.302	246.568	143.568	452.222	-5.135
4700	60.492	387.204	333.457	252.614	143.281	458.935	-5.100
4800	60.555	388.479	334.590	258.667	142.981	465.655	-5.067
4900	60.614	389.728	335.702	264.725	142.667	472.381	-5.036
5000	60.672	390.953	336.795	270.789	142.340	479.113	-5.005
5100	60.731	392.155	337.869	276.859	141.999	485.853	-4.976
5200	60.789	393.335	338.924	282.935	141.643	492.597	-4.948
5300	60.848	394.493	339.962	289.017	141.270	499.351	-4.921
5400	60.906	395.631	340.982	295.105	140.880	506.110	-4.896
5500	60.965	396.749	341.986	301.199	140.472	512.876	-4.871
5600	61.024	397.848	342.974	307.298	140.045	519.653	-4.847
5700	61.078	398.929	343.946	313.403	139.596	526.434	-4.824
5800	61.137	399.992	344.903	319.514	139.126	533.226	-4.802
5900	61.191	401.037	345.846	325.630	138.631	540.023	-4.781
6000	61.250	402.066	346.774	331.752	138.111	546.832	-4.761

PREVIOUS: June 1961 (1 atm) CURRENT: June 1961 (1 bar)

Ozone (O_3) $O_3(g)$

LEAD SILICATE ($PbSiO_3$) CRYSTAL M_r = 283.2837

$S°(298.15\ K) = 109.927 \pm 1.3\ J\ K^{-1}\ mol^{-1}$

$T_{fus} = 1037\ K$

$\Delta_f H°(0\ K) = -1138.091 \pm 4.2\ kJ\ mol^{-1}$

$\Delta_f H°(298.15\ K) = -1144.993 \pm 4.2\ kJ\ mol^{-1}$

Enthalpy of Formation

The enthalpy of the reaction PbO(red) + SiO_2(cr) → $PbSiO_3$(cr) was reported by Kelley (1). This value for the enthalpy of reaction of the oxides has been converted to the enthalpy of formation using JANAF auxiliary data (4). The above $\Delta_r H°(298.15\ K) = -3.75 \pm 0.12\ kcal\ mol^{-1}$.

Heat Capacity and Entropy

In the low temperature region (53,25-298.15 K) the $C_p°$ values are those determined by King (2). Above 298.15 $C_p°$ values were estimated by summing the values for the constituent oxides and graphically smoothing these into the low temperature measurements. The entropy was calculated at 53.25 K using the Debye and Einstein functions D(140/T) + E(213/T) + 2E(596/T) + E(1440/T) given by King (2). $S°(53.25\ K) = 3.662\ cal\ K^{-1}\ mol^{-1}$.

Fusion Data

T_{fus} was taken from Geller et al. (3).

References

1. K. K. Kelley, U. S. Bur. Mines RI 5901, 1962.
2. E. G. King, J. Amer. Chem. Soc. 81, 799-800 (1959).
3. Geller, Creamer and Bunting, J. Res. Nat. Bur. Stand. 13, 237 (1934).
4. JANAF Thermochemical Tables: PbO(cr, red), 3-31-62; SiO_2(cr), 12-31-62.

Lead Silicate ($PbSiO_3$) $O_3Pb_1Si_1$(cr)

Enthalpy Reference Temperature = T_r = 298.15 K Standard State Pressure = p° = 0.1 MPa

T/K	$C_p°$ (J K⁻¹mol⁻¹)	S° (J K⁻¹mol⁻¹)	$-[G°-H°(T_r)]/T$ (J K⁻¹mol⁻¹)	$H°-H°(T_r)$ (kJ mol⁻¹)	$\Delta_f H°$ (kJ mol⁻¹)	$\Delta_f G°$ (kJ mol⁻¹)	Log K_f
0	0.000	0.000	INFINITE	-16.219	-1138.091	-1138.091	INFINITE
100	44.683	37.168	177.245	-14.008	-1142.269	-1115.924	582.899
200	72.562	77.500	117.638	-8.028	-1144.359	-1088.626	284.320
298.15	90.040	109.927	109.927	0.000	-1144.993	-1061.094	185.899
300	90.333	110.485	109.929	0.167	-1144.995	-1060.574	184.662
400	102.843	138.266	113.631	9.854	-1144.615	-1032.467	134.826
500	111.964	162.238	121.011	20.614	-1143.535	-1004.544	104.944
600	118.742	183.282	129.672	32.166	-1141.987	-976.885	85.045
700	123.428	201.958	138.690	44.288	-1145.017	-948.711	70.794
800	126.566	218.655	147.661	56.795	-1143.041	-920.801	60.122
900	128.784	233.697	156.399	69.568	-1140.924	-893.146	51.837
1000	130.332	247.349	164.822	82.527	-1138.724	-865.733	45.221
1100	131.545	259.829	172.900	95.622	-1136.473	-838.542	39.819
1200	132.591	271.321	180.629	108.831	-1134.180	-811.559	35.326
1300	133.428	281.967	188.020	122.132	-1131.864	-784.767	31.532
1400	134.223	291.884	195.089	135.514	-1129.547	-758.154	28.287
1500	135.018	301.173	201.854	148.978	-1127.234	-731.708	25.480
1600	135.645	309.907	208.338	162.512	-1124.940	-705.414	23.029
1700	136.189	318.147	214.557	176.103	-1172.856	-678.816	20.857
1800	136.775	325.949	220.530	189.753	-1170.401	-649.826	18.857
1900	137.152	333.354	226.275	203.450	-1167.953	-620.972	17.072
2000	137.444	340.397	231.807	217.180	-1165.526	-592.246	15.468

PREVIOUS: CURRENT: June 1965

Lead Silicate ($PbSiO_3$) $O_3Pb_1Si_1$(cr)

SULFUR TRIOXIDE (SO_3) IDEAL GAS M_r = 80.0582

$S°(298.15 K) = 256.769 J K^{-1} mol^{-1}$

$\Delta_f H°(0 K) = -390.025 \pm 0.71 kJ mol^{-1}$
$\Delta_f H°(298.15 K) = -395.765 \pm 0.71 kJ mol^{-1}$

Vibrational Frequencies and Degeneracies

ν, cm^{-1}
1068 (1)
495 (1)
1391 (2)
529 (2)

Ground State Quantum Weight: 1 $\sigma = 6$
Point Group: D_{3h}
Bond Distance: S–O = 1.43 Å
Bond Angle: O–S–O = 120°
Product of the Moments of Inertia: $I_A I_B I_C = 1.0824273 \times 10^{-114} g^3 cm^6$

Enthalpy of Formation

The enthalpy of formation is calculated from the reaction $SO_2 + 1/2\, O_2 \rightarrow SO_3$ for which three sets of equilibrium constants are available. A summary of the 2nd and 3rd law treatment of the data is given below.

Source	T/K	Data Points	$\Delta_f H°$(298.15 K), kcal mol^{-1} 2nd law	3rd law	Drift cal K^{-1} mol^{-1}
(1)	933- 945 K	5	-27.67±2.7	-23.608	3.7±2.9
(2)	801-1170 K	8	-23.45±0.1	-23.661	-0.8±0.1
(3)	850-1001 K	11	-22.74±0.6	-23.614	-1.5±0.6
*	801-1170 K	23*	-23.40±0.2	-23.638	-0.9±0.2

*Combination of above references with one point omitted due to failure of a statistical test.

The combined set was adopted though it appears that the three sets are not in excellent agreement, and the combination heavily favors reference 2. Although there is a definite trend in all the data, it cannot be considered definitive enough for use in the modification of the SO_3 functions.

Heat Capacity and Entropy

The bond length and angles are taken from the electron diffraction data of Palmer (4). The vibrational frequencies are taken from the infrared studies of Lovejoy et al. (5) which are in excellent accord with those of Bent and Ladner (6) using different techniques. These values differ from the assignment used by Stockmayer et al. (7) in their analysis of the thermodynamic properties of SO_3.

The principal moments of inertia are: $I_A = I_B = 8.1493 \times 10^{-39} g\ cm^2$ and $I_C = 16.2987 \times 10^{-39} g\ cm^2$.

References

1. G. B. Taylor and S. Lehner, Z. Physik. Chem. B30 (1931).
2. M. Bodenstein and W. Pohl, Z. Electrochem. 11, 373 (1905).
3. A. F. Kapustinsky and L. M. Shamovsky, Acta Physicochem. USSR 4, 791 (1936).
4. K. J. Palmer, J. Amer. Chem. Soc. 60, 2360 (1938).
5. R. W. Lovejoy, J. H. Colwell, D. F. Eggers and G. D. Halsey, J. Chem. Phys. 36, 612 (1962).
6. R. Bent and W. R. Ladner, Spectrochim. Acta 19, 931 (1963).
7. W. H. Stockmayer, G. M. Kavanagh and H. S. Mickley, J. Chem. Phys. 12, 408 (1944).

Sulfur Trioxide (SO_3) $O_3S_1(g)$

Enthalpy Reference Temperature = T_r = 298.15 K; Standard State Pressure = p° = 0.1 MPa

T/K	$C_p°$ (J $K^{-1}mol^{-1}$)	S°	$-[G°-H°(T_r)]/T$	$H°-H°(T_r)$ (kJ mol^{-1})	$\Delta_f H°$	$\Delta_f G°$	Log K_f
0	0.	0.	INFINITE	-11.697	-390.025	-390.025	INFINITE
100	34.076	212.371	295.976	-8.361	-391.735	-385.724	201.481
200	42.336	238.259	261.145	-4.577	-393.960	-378.839	98.943
250	46.784	248.192	257.582	-2.348	-394.937	-374.943	78.340
298.15	50.661	256.769	256.769	0.	-395.765	-371.016	65.000
300	50.802	257.083	256.770	0.094	-395.794	-370.862	64.573
350	54.423	265.191	257.402	2.726	-396.543	-366.646	54.719
400	57.672	272.674	258.849	5.530	-399.412	-362.242	47.304
450	60.559	279.637	260.777	8.487	-400.656	-357.529	41.501
500	63.100	286.152	262.992	11.580	-401.878	-352.668	36.843
600	67.255	298.041	267.862	18.107	-403.675	-342.647	29.830
700	70.390	308.655	272.945	24.997	-405.014	-332.365	24.801
800	72.761	318.217	278.017	32.160	-406.068	-321.912	21.019
900	74.570	326.896	282.973	39.531	-460.062	-310.258	18.007
1000	75.968	334.828	287.768	47.060	-459.581	-293.639	15.338
1100	77.065	342.122	292.382	54.714	-459.063	-277.069	13.157
1200	77.937	348.866	296.811	62.466	-458.521	-260.548	11.341
1300	78.639	355.133	301.060	70.296	-457.968	-244.073	9.807
1400	79.212	360.983	305.133	78.189	-457.413	-227.640	8.493
1500	79.685	366.465	309.041	86.135	-456.863	-211.247	7.356
1600	80.079	371.620	312.793	94.124	-456.323	-194.890	6.363
1700	80.410	376.485	316.398	102.149	-455.798	-178.567	5.487
1800	80.692	381.090	319.865	110.204	-455.293	-162.274	4.709
1900	80.932	385.459	323.203	118.286	-454.810	-146.009	4.014
2000	81.140	389.616	326.421	126.390	-454.351	-129.768	3.389
2100	81.319	393.579	329.525	134.513	-453.919	-113.549	2.824
2200	81.476	397.366	332.523	142.653	-453.514	-97.350	2.311
2300	81.614	400.990	335.422	150.807	-453.137	-81.170	1.843
2400	81.735	404.466	338.227	158.975	-452.790	-65.006	1.415
2500	81.843	407.805	340.944	167.154	-452.472	-48.855	1.021
2600	81.939	411.017	343.578	175.343	-452.183	-32.716	0.657
2700	82.025	414.111	346.133	183.541	-451.922	-16.587	0.321
2800	82.102	417.096	348.614	191.748	-451.690	-0.467	0.009
2900	82.171	419.978	351.026	199.961	-451.487	15.643	-0.282
3000	82.234	422.765	353.371	208.182	-451.311	31.748	-0.553
3100	82.290	425.462	355.653	216.408	-451.161	47.849	-0.806
3200	82.342	428.076	357.876	224.640	-451.038	63.943	-1.044
3300	82.389	430.610	360.042	232.876	-450.940	80.034	-1.267
3400	82.432	433.070	362.153	241.117	-450.866	96.124	-1.477
3500	82.472	435.460	364.214	249.363	-450.817	112.210	-1.675
3600	82.508	437.784	366.225	257.612	-450.792	128.297	-1.862
3700	82.542	440.045	368.190	265.864	-450.789	144.382	-2.038
3800	82.573	442.247	370.110	274.120	-450.809	160.469	-2.206
3900	82.601	444.392	371.987	282.379	-450.850	176.556	-2.365
4000	82.628	446.484	373.824	290.640	-450.914	192.643	-2.516
4100	82.653	448.524	375.621	298.904	-450.999	208.733	-2.659
4200	82.676	450.516	377.381	307.171	-451.105	224.825	-2.796
4300	82.697	452.462	379.104	315.439	-451.234	240.921	-2.927
4400	82.717	454.364	380.793	323.710	-451.383	257.019	-3.051
4500	82.735	456.223	382.449	331.983	-451.555	273.120	-3.170
4600	82.753	458.041	384.072	340.257	-451.749	289.225	-3.284
4700	82.769	459.821	385.665	348.533	-451.965	305.336	-3.393
4800	82.785	461.564	387.228	356.811	-452.205	321.452	-3.498
4900	82.799	463.271	388.763	365.090	-452.469	337.573	-3.599
5000	82.813	464.944	390.270	373.371	-452.757	353.700	-3.695
5100	82.825	466.584	391.750	381.653	-453.071	369.832	-3.788
5200	82.837	468.192	393.205	389.936	-453.412	385.969	-3.877
5300	82.849	469.770	394.634	398.220	-453.780	402.116	-3.963
5400	82.860	471.319	396.040	406.505	-454.178	418.268	-4.046
5500	82.870	472.840	397.423	414.792	-454.605	434.427	-4.126
5600	82.879	474.333	398.783	423.079	-455.065	450.597	-4.203
5700	82.889	475.800	400.121	431.368	-455.557	466.773	-4.277
5800	82.897	477.242	401.439	439.657	-456.084	482.960	-4.350
5900	82.906	478.659	402.735	447.947	-456.647	499.153	-4.419
6000	82.913	480.052	404.012	456.238	-457.248	515.359	-4.487

PREVIOUS: September 1965 (1 atm) CURRENT: September 1965 (1 bar)

Sulfur Trioxide (SO_3) $O_3S_1(g)$

TITANIUM OXIDE (Ti_2O_3) CRYSTAL M_r = 143.7582

$S°(298.15\ K) = 77.253 \pm 0.21\ J\ K^{-1}\ mol^{-1}$
$T_{trs} = 470 \pm 20\ K$
$T_{fus} = 2115 \pm 10\ K$

$\Delta_f H°(0\ K) = -1512.287 \pm 8.4\ kJ\ mol^{-1}$
$\Delta_f H°(298.15\ K) = -1520.884 \pm 8.4\ kJ\ mol^{-1}$
$\Delta_{trs} H° = 1.138\ kJ\ mol^{-1}$
$\Delta_{fus} H° = [105]\ kJ\ mol^{-1}$

Enthalpy of Formation

Humphrey (1) measured $\Delta_r H°(298.15\ K) = -88.11 \pm 0.1\ kcal\ mol^{-1}$ for the reaction $Ti_2O_3(cr) + 0.5\ O_2(g) = 2\ TiO_2$(rutile) using bomb calorimetry. When combined with $\Delta_f H°(298.15\ K) = -225.8\ kcal\ mol^{-1}$ for rutile, this yields $-363.5 \pm 2\ kcal\ mol^{-1}$ as the adopted $\Delta_f H°(298.15\ K)$ for Ti_2O_3. The less negative value of $-362.8 \pm 0.8\ kcal\ mol^{-1}$ was obtained by Ariya et al. (2) from their heats of combustion of various compositions in the Ti-O system. Ariya et al. (2) used an estimated correction for the formation of TiO_{2-x}, while Humphrey (1) obtained values of 0.1 to 1.3 kcal mol^{-1} by grinding the product and reburning it with while oil. Most of the assigned uncertainty arises from this problem [see TiO_2(rutile) for further details].

Heat Capacity and Entropy

$C_p°$ is based on data of Sjostrand and Keesom (3, 0.4-20 K) and Paukov and Berezovskii (4, 12-305 K). The two sets of data disagree in the region of overlap. Paukov (4) appears to have a positive bias at the lowest temperatures (12-18 K), so we adopt a curve which shifts gradually from Sjostrand at 10 K to Paukov at 30 K. Deviations of Paukov (4) from the adopted $C_p°$ are +36% (12 K) and +3% (20 K). Earlier data of Shomate (5, 53-296 K) deviate by +0.6% to +4.1% with extreme deviations of 3.0% at 53 K and 2.9 to 4.1% in a "hump" observed by Shomate near 240 K. No hump was found by Paukov. S° is derived from $C_p°$ based on an extrapolation which is negligible.

$C_p°$ between 298 K and 470 K is a smooth extrapolation of the low-temperature curve using the approximate shape found by Barros et al. (6, 150-700 K) via differential scanning calorimetry (Perkin-Elmer DSC-1B). The DSC data (6) have a positive bias of ~10%, over twice as large as we might expect for this apparatus so the results do not help establish $C_p°$ between 298 K and T_{trs}. Data measured by a conduction-type calorimeter (7, 323-523 K) have a large negative bias near 400 K. Drop-calorimetric data (8, 375-472 K) deviate from the adopted enthalpies by +5 to +2%.

$C_p°$ above the λ-transition is derived from enthalpy data (8, 487-1750 K) by fitting only the region above 549 K. Maximum deviations from the adopted cruve are ±0.4% in the fitted region and +0.9% at 526 K. Enthalpy data (493-1082 K) reported graphically by Erofeeva et al. (9) deviate by about -6 to -2%.

Slyusar et al. (15) recently reported new enthalpy data for Ti_2O_3, Ti_3O_5, Ti_4O_7 and TiO_2. The new data for Ti_3O_5 and TiO_2 are in reasonable agreement with existing JANAF Tables (16), but those for Ti_2O_3 cross the adopted values near 1100 K and deviate by -4.5 ± 1% (672-878 K) and +3.4 ± 1.1% (1158-2027 K). The transition vanishes for Ti_2O_3 samples doped with 4% V_2O_3 (6) and may be affected by other impurities or non-stoichiometry. We feel that Slyusar et al. (15) have not characterized their sample adequately. We tentatively reject their data pending new measurements on a well-characterized sample.

Transition Data

We arbitrarily choose T_{trs} = 470 ± 20 K and derive $\Delta_{trs}H°$ as the difference between the adopted enthalpy curves at that temperature. Reported values of $\Delta_{trs}H°$ include 0.380 ± 0.025 (6), 0.036 (7) and 0.215 kcal mol^{-1} (8); each derives from different data and difference base lines.

Convenience dictates that we treat the electrical transition as first order although recent $C_p°$ data (6) suggest a gradual λ-type anomaly peaking near 470 K and extending from about 380 to 560 K. There is no change in the corundum-type symmetry, but the lattice parameters shift rapidly in this temperature range (10, 11). Honig (11) summarized evidence on the nature of the change from semiconductor to semimetal and noted that the transition can be explained without invoking antiferromagnetic ordering. Recent studies (6, 12-14) are consistent with this interpretation.

Fusion Data

See the liquid table for details.

References

1. G. L. Humphrey, J. Amer. Chem. Soc. 73, 1587 (1951).
2. S. M. Ariya, M. P. Morozova and E. Vol'f, Russ. J. Inorg. Chem. 2, 16 (1957).
3. M. E. Sjostrand and P. H. Keesom, Phys. Rev. Lett. 27, 1434 (1971); Phys. Rev. B7, 3558 (1973).
4. I. E. Paukov and G. A. Berezovskii, Russ. J. Phys. Chem. 46, 1538 (1972).
5. C. H. Shomate, J. Amer. Chem. Soc. 68, 310 (1946).
6. H. L. Barros, G. V. Chandrashekhar, T. C. Chi, J. M. Honig and R. J. Sladek, Phys. Rev. B7, 5147 (1973).
7. S. Nomura, T. Kawakubo and T. Yanagi, J. Phys. Soc. Jap. 16, 706 (1961).
8. B. F. Naylor, J. Amer. Chem. Soc. 68, 1077 (1946).
9. M. S. Erofeeva, N. L. Lukinykh and S. M. Ariya, Russ. J. Phys. Chem. 35, 375 (1961).
10. A. D. Pearson, J. Phys. Chem. Solids 5, 316 (1958).
11. J. M. Honig, Rev. Mod. Phys. 40, 748 (1968).
12. S. H. Shin, G. V. Chandrashekhar, R. E. Loehman and J. M. Honig, Phys. Rev. B8, 15 August (1973), in press.
13. T. C. Chi and R. J. Sladek, Phys. Rev. B7, 15 June (1973), in press.
14. A. Mooradian and P. M. Raccah, Phys. Rev. B3, 4253 (1971).
15. N. P. Slyusar, A. D. Krivorotenko, E. N. Fomichev, A. A. Kalashnik and V. P. Bondarenko, Teplofiz. Vys. Temp. 11, 213 (1973).
16. JANAF Thermochemical Tables, 2nd ed., NSRDS-NBS 37, 1971: Ti_3O_5(cr, α, β), 3-31-67; TiO_2(cr, rutile), 3-31-67.

Titanium Oxide (Ti_2O_3) $O_3Ti_2(cr)$

Enthalpy Reference Temperature = T_r = 298.15 K — Standard State Pressure = p° = 0.1 MPa

T/K	$C_p°$	S°	$-[G°-H°(T_r)]/T$	$H°-H°(T_r)$	$\Delta_f H°$	$\Delta_f G°$	Log K_f
	J K⁻¹mol⁻¹			kJ mol⁻¹			
0	0.000	0.000	INFINITE	-14.088	-1512.287	-1512.287	INFINITE
100	26.175	11.058	143.900	-13.284	-1516.961	-1490.419	778.515
200	70.793	43.832	85.316	-8.297	-1520.175	-1462.405	381.941
298.15	95.814	77.253	77.253	0.000	-1520.884	-1433.829	251.201
300	96.190	77.847	77.255	0.178	-1520.881	-1433.288	249.558
400	117.529	108.277	81.264	10.805	-1519.938	-1404.187	183.368
470.010	141.716	128.963	86.822	19.807	ALPHA <--> BETA		
470.010	127.653	131.384	86.822	20.945	TRANSITION		
500	130.248	139.362	89.735	24.813	-1515.999	-1375.626	143.711
600	136.440	163.704	100.081	38.174	-1513.028	-1347.825	117.339
700	140.331	185.051	110.727	52.027	-1509.833	-1320.542	98.540
800	142.967	203.971	121.223	66.199	-1506.516	-1293.726	84.472
900	144.892	220.928	131.375	80.597	-1503.208	-1267.327	73.554
1000	146.356	236.272	141.110	95.162	-1500.078	-1241.288	64.838
1100	147.528	250.277	150.407	109.858	-1497.276	-1215.547	57.722
1200	148.532	263.158	159.273	124.662	-1502.785	-1189.806	51.791
1300	149.369	275.081	167.729	139.558	-1499.224	-1163.870	46.765
1400	150.080	286.177	175.798	154.531	-1495.789	-1138.203	42.467
1500	150.750	296.554	183.506	169.573	-1492.509	-1112.777	38.750
1600	151.377	306.304	190.879	184.680	-1489.413	-1087.564	35.505
1700	151.921	315.497	197.942	199.845	-1486.531	-1062.538	32.648
1800	152.465	324.197	204.716	215.065	-1483.892	-1037.675	30.113
1900	152.925	332.452	211.224	230.334	-1481.528	-1012.950	27.848
2000	153.427	340.309	217.483	245.652	-1508.838	-987.436	25.789
2100	153.888	347.806	223.512	261.018	-1508.597	-961.372	23.913
2115.000	153.953	348.902	224.398	263.327	BETA <--> LIQUID		
2200	154.306	354.975	229.326	276.427	-1508.348	-935.318	22.207
2300	154.766	361.844	234.939	291.881	-1508.087	-909.278	20.650
2400	155.185	368.440	240.365	307.379	-1507.817	-883.249	19.223
2500	155.603	374.783	245.616	322.919	-1507.537	-857.231	17.911
2600	155.980	380.894	250.702	338.497	-1507.250	-831.224	16.700
2700	156.398	386.788	255.634	354.117	-1506.954	-805.229	15.578
2800	156.774	392.483	260.420	369.775	-1506.649	-779.244	14.537
2900	157.151	397.991	265.070	385.471	-1506.337	-753.272	13.568
3000	157.569	403.325	269.590	401.207	-1506.015	-727.309	12.664

PREVIOUS: June 1967 — CURRENT: June 1973

Titanium Oxide (Ti_2O_3) $O_3Ti_2(cr)$

TITANIUM OXIDE (Ti_2O_3) LIQUID M_r = 143.7582

S°(298.15 K) = [127.126] J K^{-1} mol^{-1} $\Delta_f H°$(298.15 K) = [-1418.459] kJ mol^{-1}

T_{fus} = 2115 ± 10 K $\Delta_{fus}H°$ = [105] kJ mol^{-1}

Enthalpy of Formation

$\Delta_f H°$(Ti_2O_3, ℓ, 298.15 K) is calculated from that of the crystal by adding $\Delta_{fus}H°$ and the difference in enthalpy, H°(2115 K)-H°(298.15 K), between the crystal and liquid.

Heat Capacity and Entropy

$C_p°$ is estimated as 7.5 cal K^{-1} g-atom^{-1}. Below the assumed glass transition at 1400 K, $C_p°$ is taken from the crystal; the high-temperature curve is extrapolated down to 298.15 K. S°(298.15 K) is calculated in a manner analogous to that used for $\Delta_f H°$(298.15 K).

Fusion Data

T_{fus} is the value quoted by Wahlbeck and Gilles (1) based on work of K. D. Carlson. The adopted value is converted to IPTS-68. $\Delta_{fus}H°$ is estimated such that $\Delta_{fus}S°$ is 12 cal K^{-1} mol^{-1} as observed (2) for Al_2O_3(α, corundum). Other reported melting points were summarized earlier (3). Slyusar et al. (4) recently obtained T_{fus} = 2043 K in an enthalpy study, but we tentatively reject this work (see Ti_2O_3, crystal).

References

1. P. G. Wahlbeck and P. W. Gilles, J. Amer. Ceram. Soc. 49, 180 (1966).
2. JANAF Thermochemical Tables: Al_2O_3(cr, α), Al_2O_3(ℓ), 6-30-72.
3. JANAF Thermochemical Tables, 2nd ed., NSRDS-NBS 37, 1971; Ti_2O_3(ℓ), 6-30-67.
4. N. P. Slyusar, A. D. Krivorotenko, E. N. Fomichev, A. A. Kalashnik and V. P. Bondarenko, Teplofiz, Vys. Temp. 11, 213 (1973).

Titanium Oxide (Ti_2O_3) O_3Ti_2(l)

Enthalpy Reference Temperature = T_r = 298.15 K Standard State Pressure = p° = 0.1 MPa

T/K	$C_p°$ (J K⁻¹mol⁻¹)	S° (J K⁻¹mol⁻¹)	$-[G°-H°(T_r)]/T$ (J K⁻¹mol⁻¹)	$H°-H°(T_r)$ (kJ mol⁻¹)	$\Delta_f H°$ (kJ mol⁻¹)	$\Delta_f G°$ (kJ mol⁻¹)	Log K_f
0							
100							
200							
298.15	94.851	127.126	127.126	0.000	-1418.459	-1346.273	235.861
300	95.521	127.714	127.127	0.176	-1418.458	-1345.825	234.329
400	119.118	158.902	131.219	11.073	-1417.245	-1321.744	172.602
500	130.248	186.808	139.609	23.600	-1414.787	-1298.137	135.615
600	136.440	211.151	149.550	36.961	-1411.816	-1275.081	111.006
700	140.331	232.498	159.907	50.814	-1408.621	-1252.543	93.466
800	142.967	251.418	170.186	64.986	-1405.305	-1230.471	80.342
900	144.892	268.374	180.170	79.384	-1401.996	-1208.817	70.158
1000	146.356	283.719	189.770	93.949	-1398.867	-1187.523	62.030
1100	147.528	297.724	198.956	108.644	-1396.065	-1166.527	55.394
1200	148.532	310.605	207.731	123.449	-1401.573	-1145.531	49.864
1300	149.369	322.528	216.109	138.345	-1398.013	-1124.339	45.177
1400	150.080	333.624	224.111	153.318	-1394.577	-1103.417	41.169
1400.000	150.080	333.624	224.111	153.318	GLASS <--> LIQUID		
1400.000	156.900	333.624	224.111	153.318	TRANSITION		
1500	156.900	344.449	231.777	169.008	-1390.649	-1082.759	37.705
1600	156.900	354.575	239.139	184.698	-1386.969	-1062.354	34.682
1700	156.900	364.087	246.212	200.388	-1383.562	-1042.171	32.022
1800	156.900	373.055	253.012	216.078	-1380.453	-1022.181	29.663
1900	156.900	381.538	259.555	231.768	-1377.668	-1002.355	27.557
2000	156.900	389.586	265.857	247.458	-1404.606	-981.758	25.641
2100	156.900	397.241	271.933	263.148	-1404.042	-960.630	23.894
2115.000	156.900	398.358	272.825	265.502	BETA <--> LIQUID		
2200	156.900	404.540	277.796	278.838	-1403.512	-939.527	22.307
2300	156.900	411.515	283.459	294.528	-1403.015	-918.449	20.859
2400	156.900	418.192	288.935	310.218	-1402.552	-897.391	19.531
2500	156.900	424.597	294.234	325.908	-1402.122	-876.351	18.310
2600	156.900	430.751	299.367	341.598	-1401.724	-855.329	17.184
2700	156.900	436.673	304.344	357.288	-1401.357	-834.320	16.141
2800	156.900	442.379	309.172	372.978	-1401.021	-813.324	15.173
2900	156.900	447.885	313.861	388.668	-1400.715	-792.342	14.272
3000	156.900	453.204	318.418	404.358	-1400.438	-771.368	13.431
3100	156.900	458.348	322.849	420.048	-1400.189	-750.402	12.644
3200	156.900	463.330	327.162	435.738	-1399.967	-729.446	11.907
3300	156.900	468.158	331.361	451.428	-1399.771	-708.496	11.215
3400	156.900	472.842	335.454	467.118	-1399.600	-687.550	10.563
3500	156.900	477.390	339.445	482.808	-1399.454	-666.610	9.949

PREVIOUS: June 1967 CURRENT: June 1973

Titanium Oxide (Ti_2O_3) O_3Ti_2(l)

TITANIUM OXIDE (Ti_2O_3) CRYSTAL–CRYSTAL–LIQUID M_r = 143.7582

0 to 470 K crystal, I
470 to 2115 K crystal, II
above 2115 K liquid

Refer to the individual tables for details.

Titanium Oxide (Ti_2O_3) O_3Ti_2(cr,l)

Enthalpy Reference Temperature = T_r = 298.15 K Standard State Pressure = $p°$ = 0.1 MPa

	J K⁻¹mol⁻¹			kJ mol⁻¹			
T/K	$C_p°$	S°	$-[G°-H°(T_r)]/T$	$H°-H°(T_r)$	$\Delta_f H°$	$\Delta_f G°$	Log K_f
0	0.000	0.000	INFINITE	-14.088	-1512.287	-1512.287	INFINITE
100	26.175	11.058	143.900	-13.284	-1516.961	-1490.419	778.515
200	70.793	43.832	85.316	-8.297	-1520.175	-1462.405	381.941
298.15	95.814	77.253	77.253	0.000	-1520.884	-1433.829	251.201
300	96.190	77.847	77.255	0.178	-1520.881	-1433.288	249.558
400	117.529	108.277	81.264	10.805	-1519.938	-1404.187	183.368
470.010	141.716	128.963	86.822	19.807	ALPHA <--> BETA		
470.010	127.653	131.384	86.822	20.945	TRANSITION		
500	130.248	139.362	89.735	24.813	-1515.999	-1375.626	143.711
600	136.440	163.704	100.081	38.174	-1513.028	-1347.825	117.339
700	140.331	185.051	110.727	52.027	-1509.833	-1320.542	98.540
800	142.967	203.971	121.223	66.199	-1506.516	-1293.726	84.472
900	144.892	220.928	131.375	80.597	-1503.208	-1267.327	73.554
1000	146.356	236.272	141.110	95.162	-1500.078	-1241.288	64.838
1100	147.528	250.277	150.407	109.858	-1497.276	-1215.547	57.722
1200	148.532	263.158	159.273	124.662	-1502.785	-1189.806	51.791
1300	149.369	275.081	167.729	139.558	-1499.224	-1163.870	46.765
1400	150.080	286.177	175.798	154.531	-1495.789	-1138.203	42.467
1500	150.750	296.554	183.506	169.573	-1492.509	-1112.777	38.750
1600	151.377	306.304	190.879	184.680	-1489.413	-1087.564	35.505
1700	151.921	315.497	197.942	199.845	-1486.531	-1062.538	32.648
1800	152.465	324.197	204.716	215.065	-1483.892	-1037.675	30.113
1900	152.925	332.452	211.224	230.334	-1481.528	-1012.950	27.848
2000	153.427	340.309	217.483	245.652	-1508.838	-987.436	25.789
2100	153.888	347.806	223.512	261.018	-1508.597	-961.372	23.913
2115.000	153.953	348.902	224.398	263.327	BETA <--> LIQUID		
2115.000	156.900	398.358	224.398	367.927	TRANSITION		
2200	156.900	404.540	231.239	381.263	-1403.512	-939.527	22.307
2300	156.900	411.515	238.927	396.953	-1403.015	-918.449	20.859
2400	156.900	418.192	246.258	412.643	-1402.552	-897.391	19.531
2500	156.900	424.597	253.264	428.333	-1402.122	-876.351	18.310
2600	156.900	430.751	259.973	444.023	-1401.724	-855.329	17.184
2700	156.900	436.673	266.408	459.713	-1401.357	-834.320	16.141
2800	156.900	442.379	272.592	475.403	-1401.021	-813.324	15.173
2900	156.900	447.885	278.542	491.093	-1400.715	-792.342	14.272
3000	156.900	453.204	284.276	506.783	-1400.438	-771.368	13.431
3100	156.900	458.348	289.809	522.473	-1400.189	-750.402	12.644
3200	156.900	463.330	295.154	538.163	-1399.967	-729.446	11.907
3300	156.900	468.158	300.324	553.853	-1399.771	-708.496	11.215
3400	156.900	472.842	305.329	569.543	-1399.600	-687.550	10.563
3500	156.900	477.390	310.180	585.233	-1399.454	-666.610	9.949

PREVIOUS: CURRENT: June 1973

Titanium Oxide (Ti_2O_3) O_3Ti_2(cr,l)

VANADIUM OXIDE (V_2O_3) CRYSTAL M_r = 149.8812

$S°(298.15\ K) = 98.067 \pm 1.3\ J\ K^{-1}\ mol^{-1}$

$T_{trs} = 168.8 \pm 0.2\ K$

$T_{fus} = 2340 \pm 20\ K$

$\Delta_f H°(0\ K) = -1213.971 \pm 6.3\ kJ\ mol^{-1}$

$\Delta_f H°(298.15\ K) = -1218.799 \pm 6.3\ kJ\ mol^{-1}$

$\Delta_{trs} H° = 1.623\ kJ\ mol^{-1}$

$\Delta_{fus} H° = [117.152]\ kJ\ mol^{-1}$

Enthalpy of Formation

The adopted enthalpies of formation for the vanadium oxides, and in particular V_2O_3(cr), are based on the combustion studies by Mah and Kelley (1). The adopted value is $\Delta_f H°(V_2O_3$, cr, 298.15 K) = -291.3 kcal mol^{-1} as reported by Mah and Kelley (1). For more details refer to the V_2O_5(cr) table (2). We assign an uncertainty limit of ±1.5 kcal mol^{-1}.

Charlu and Kleppa (3) reported a enthalpy of formation value of -291.0 ± 0.9 kcal mol^{-1} for V_2O_3 based on oxidation studies to V_2O_5(cr) in a high temperature microcalorimeter. An advantage of this technique is that complete oxidation to V_2O_5(cr) was achieved whereas in the study by Mah and Kelley (1) a mixture of the two oxides V_2O_4 and V_2O_5 was obtained. Other combustion studies have been reported by Siemonsen and Ulich (4) and Vol'f and Ariya (5). Additional enthalpy of formation values may be obtained from the sodium peroxide fusion studies by Mixter (6) and Ruff and Friedrich (7) and the following equilibrium studies involving vanadium and its oxides: H_2O/H_2 by Kobayashi (8), Muller (9), and Karasev, Polyakov and Samarin (10); and CO/CO_2 by Spencer and Justice (11). As an example, the study by Spencer and Justice (11) leads to $\Delta_f H°$(298.15 K) = -287.9 kcal mol^{-1} for V_2O_3(cr). The analysis, based on data in the range 1024-1169 K, indicates a third law drift of -0.96 ± 0.43 cal K^{-1} mol^{-1}.

Heat Capacity and Entropy

Anderson (12) measured the heat capacity of V_2O_3(cr) in the range 57-287 K. The data indicated an anomaly in the region 165-182 K. These heat capacity data are joined smoothly at 298 K with the high temperature heat capacity values as derived from the enthalpy measurements of Cook (13). The adopted $C_p°$ values are based on these two studies (12, 13). Using the combination of Debye and Einstein functions as suggested by Anderson (12), we calculate S°(50 K) = 0.783 cal K^{-1} mol^{-1} and H°(50;0) = 0.0284 kcal mol^{-1}. There is considerable scatter in the data of Cook (13); the deviations from the adopted values range from -0.8 to 0.6% except for the data point at 369.1 K which is -1.4% low (~25 cal K^{-1} mol^{-1}).

Khlyustov et al. (14) measured the heat capacity of V_2O_3 in the region 120-190 K. For $T < T_{trs}$, this latter study lies above the Anderson data by ~7% while for $T > T_{trs}$, it is 10% higher. Three other studies examined the heat capacity in the range 343-573 K (15, 16, 17). These latter three studies indicated irregularities in the $C_p°$ values in the region of investigation. We do not adopt these works, pending further independent verification.

Phase Data

A review of the literature of the V-O system by Stringer (18) stated that V_2O_3 has a homogeneity range of roughly $VO_{1.44}$ to $VO_{1.65}$. On cooling V_2O_3(cr) through T_{trs}, the crystal undergoes a structural distortion from hexagonal corundum to monoclinic. Numerous references as to the homogeneity range and crystal structure are contained, for example, in the reviews by Stringer (18) and Adler (19) as well as in the x-ray study by Minomura and Nagasaki (20).

Transition Data

The heat capacity data of Anderson (12) indicated an anomaly at 168.8 K. The anomaly appeared similar to a λ-type transition and a smooth curve is easily drawn joining the heat capacity data on either side of the transition. Anderson (12) made five measurements to determine the energy necessary to heat V_2O_3(cr) through the transition. The mean enthalpy of the five measurements from 161.3 to 180.0 K was reported as 692.2 ± 2.4 cal K^{-1} mol^{-1}. Subtracting the calculated base heat in this same region, we calculate $\Delta_{trs}H°$ = 0.388 kcal mol^{-1}. We adopt T_{trs} = 168.8 K and $\Delta_{trs}H°$ = 0.388 kcal mol^{-1}, and consider it to be first order.

Khlyustov et al. (14) reported heat capacity data in the region 123-184 K and reported a transition which resembled a λ-type with a maximum at 165.66 K with $\Delta_{trs}H°$ = 0.356 kcal mol^{-1}. Although both heat capacity studies (12, 14) appeared to indicate a second order transition, Minomura and Nagasaki (20) determined through x-ray measurements that V_2O_3(cr) undergoes a discontinuous volume contraction (~3%) at T_{trs}, indicative of a first order transition. Austin (21) and Feinlieb and Paul (22) studied the pressure dependence of T_{trs} and reported $(dT_{trs}/dP)_{trs}$ values. Using the Clausius-Clapeyron equation for a first order transition with Δ_{vap} from Minomura and Nagasaki (20), we calculate $\Delta_{trs}H°$ = 0.863 and 1.020 kcal mol^{-1}, respectively. The calorimetric value for $\Delta_{trs}H°$ is more reliable than these latter two determinations. For further information on this transition, the review by Adler (19) is suggested. There are indications of possible transitional behavior above room temperature (15, 16, 17, 19). We adopt only regular behavior above 298.15 K.

Fusion Data

See the liquid table for details.

Sublimation Data

The experimental evidence indicated that, although V_2O_3(cr) sublimes congruently, the vapors do not contain V_2O_3(g) (23).

References

References are listed on the liquid table.

Vanadium Oxide (V_2O_3) O_3V_2(cr)

Enthalpy Reference Temperature = T_r = 298.15 K Standard State Pressure = p° = 0.1 MPa

T/K	$C_p°$ (J $K^{-1}mol^{-1}$)	S° (J $K^{-1}mol^{-1}$)	$-[G°-H°(T_r)]/T$ (J $K^{-1}mol^{-1}$)	$H°-H°(T_r)$ (kJ mol^{-1})	$\Delta_f H°$ (kJ mol^{-1})	$\Delta_f G°$ (kJ mol^{-1})	Log K_f
0	0.000	0.000	INFINITE	-17.476	-1213.971	-1213.971	INFINITE
100	28.627	14.163	179.288	-16.512	-1218.342	-1192.325	622.807
168.800	63.053	38.652	116.725	-13.179		I <--> II	
168.800	63.136	48.269	116.725	-11.555		TRANSITION	
200	80.107	60.767	107.019	-9.250	-1219.118	-1165.403	304.372
298.15	104.963	98.067	98.067	0.000	-1218.799	-1139.036	199.554
300	105.260	98.717	98.069	0.194	-1218.779	-1138.542	198.238
400	117.540	130.898	102.373	11.410	-1217.149	-1112.018	145.215
500	123.750	157.887	110.854	23.516	-1214.951	-1085.983	113.452
600	127.326	180.791	120.651	36.084	-1212.566	-1060.412	92.317
700	130.071	200.631	130.691	48.958	-1210.125	-1035.245	77.251
800	132.639	218.168	140.550	62.094	-1207.662	-1010.430	65.974
900	135.246	233.940	150.065	75.488	-1205.180	-985.925	57.222
1000	137.957	248.329	159.182	89.147	-1202.658	-961.698	50.234
1100	140.786	261.609	167.897	103.083	-1200.079	-937.726	44.529
1200	143.718	273.984	176.228	117.307	-1197.445	-913.992	39.785
1300	146.741	285.607	184.199	131.830	-1194.750	-890.480	35.780
1400	149.841	296.594	191.839	146.658	-1191.997	-867.177	32.355
1500	153.005	307.040	199.173	161.800	-1189.178	-844.074	29.393
1600	156.221	317.017	206.229	177.261	-1186.283	-821.161	26.808
1700	159.481	326.585	213.029	193.045	-1183.324	-798.431	24.533
1800	162.782	335.794	219.595	209.158	-1180.305	-775.877	22.515
1900	166.112	344.684	225.946	225.603	-1177.243	-753.492	20.715
2000	169.468	353.290	232.099	242.382	-1174.151	-731.269	19.099
2100	172.847	361.640	238.070	259.497	-1171.047	-709.202	17.640
2200	176.246	369.759	243.872	276.952	-1213.709	-687.072	16.313
2300	179.655	377.669	249.518	294.747	-1210.901	-663.196	15.062
2340.000	181.021	380.778	251.736	301.960	-------- II	<--> LIQUID	--------
2400	183.086	385.388	255.019	312.884	-1207.784	-639.449	13.917
2500	186.526	392.931	260.386	331.364	-1204.357	-615.837	12.867
2600	189.978	400.314	265.626	350.189	-1200.617	-592.369	11.901
2700	193.437	407.549	270.749	369.360	-1196.563	-569.051	11.009
2800	196.907	414.646	275.762	388.877	-1192.193	-545.888	10.184

PREVIOUS: CURRENT: December 1973

Vanadium Oxide (V_2O_3) O_3V_2(cr)

VANADIUM OXIDE (V_2O_3) LIQUID M_r = 149.8812

$S°(298.15\ K)$ = [152.250] J K^{-1} mol^{-1} $\Delta_fH°(298.15\ K)$ = [-1093.054] kJ mol^{-1}

T_{fus} = 2340 ± 20 K $\Delta_{fus}H°$ = [117.152] kJ mol^{-1}

Enthalpy of Formation

$\Delta_fH°(V_2O_3,\ \ell,\ 298.15\ K)$ is calculated from that of the crystal by adding $\Delta_{fus}H°$ and the difference in enthalpy, H°(2340 K)-H°(298.15 K), between the crystal and the liquid.

Heat Capacity and Entropy

$C_p°$ for the liquid phase is estimated to be a constant 7.5 cal K^{-1} g-atom^{-1}. Below the assumed glass transition temperature at 1600 K, $C_p°$ is taken from the crystal. The entropy at 298.15 K is calculated in a manner analogous to that used for the enthalpy of formation.

Fusion Data

T_{fus} = 2340 ± 20 K is the value extracted from the V-O phase diagram proposed by Stringer (1). This value is adopted and corrected to IPTS-68. $\Delta_{fus}H°$ is estimated such that $\Delta_{fus}S°$ is 12 cal K^{-1} mol^{-1} as observed (2) for Al_2O_3(α, corundum).

Vaporization Data

There is no evidence as to the existence of V_2O_3(g) and thus we do not report any heat of vaporization. See VO(g) and VO_2(g) tables for more information (2).

References - $V_2O_3(\ell)$

1. J. Stringer, J. Less-Common Metals 8, 1 (1965).
2. JANAF Thermochemical Tables: Al_2O_3(cr, α), 6-30-72; VO(g) and VO_2(g), 12-31-73.

References - V_2O_3(cr)

1. A. D. Mah and K. K. Kelley, U. S. Bur. Mines RI 5858, 1961.
2. JANAF Thermochemical Tables: V_2O_5(cr), 6-30-73.
3. F. V. Charlu and O. J. Kleppa, High Temp. Sci. 5, 260 (1973).
4. H. Siemonsen and H. Ulich, Z. Electrochem. 46, 141 (1940).
5. E. Vol'f and S. M. Ariya, Zhur. Obshchei. Khim. 29, 2470 (1959).
6. W. G. Mixter, Amer. J. Sci. 34, 141 (1912).
7. O. Ruff and L. Friedrich, Z. Anorg. Chem. 89, 279 (1914).
8. M. Kobayashi, Bull. Chem. Soc. Japan 8, 231 (1933).
9. H. Muller, W. Biltz's Festschrift, Hannover, 1927; quoted by J. W. Mellor, "Comprehensive Treatise on Inorganic and Theoretical Chemistry," Vol. IX, Longman, Green and Co., London, 1929.
10. R. A. Karasev, A. Yu. Polyakov and A. M. Samarin, Dokl. Akad. Nauk (SSSR) 85, 1313 (1952).
11. H. M. Spencer and J. L. Justice, J. Amer. Chem. Soc. 56, 2306 (1934).
12. C. T. Anderson, J. Amer. Chem. Soc. 58, 564 (1936).
13. O. A. Cook, J. Amer. Chem. Soc. 69, 331 (1947).
14. V. G. Khlyustov, A. S. Borukhovich, and V. A. Perelyaev, Soviet Physics-Solid State 14, 1848 (1973).
15. J. Jaffray and R. Lyand, Comp. rend 233, 133 (1951).
16. M. Foex, S. Goldztaub, R. Wey, J. Jeffray, and J. Wucher, J. recherches centre natl. recherche sci., Lab. Bellevue (Paris) 21, 237 (1952); Chem. Abstr. 47, 9080i (1953).
17. J. Jaffray and A. Dumas, J. recherches centre natl. recherche sci., Lab. Bellevue (Paris) 5, 360 (1954); Chem. Abstr. 49, 6759d (1955).
18. J. Stringer, J. Less-Common Metals 8, 1 (1965).
19. D. Adler, Solid State Phys. 21, 1 (1968).
20. S. Minomura and H. Nagasaki, J. Phys. Soc. Japan 19, 131 (1964).
21. I. G. Austin, Phil. Mag. 7, 961 (1962).
22. J. Feinleib and W. Paul, Phys. Rev. 155, 841 (1967).
23. M. Farber, O. M. Uy, and R. D. Srivastava, J. Chem. Phys. 56, 5312 (1972).

Vanadium Oxide (V_2O_3) $O_3V_2(1)$

Enthalpy Reference Temperature = T_r = 298.15 K Standard State Pressure = p° = 0.1 MPa

T/K	$C_p°$ (J $K^{-1}mol^{-1}$)	S° (J $K^{-1}mol^{-1}$)	$-[G°-H°(T_r)]/T$ (J $K^{-1}mol^{-1}$)	$H°-H°(T_r)$ (kJ mol^{-1})	$\Delta_fH°$ (kJ mol^{-1})	$\Delta_fG°$ (kJ mol^{-1})	Log K_f
0							
100							
200							
298.15	104.964	152.250	152.250	0.000	-1093.054	-1029.446	180.355
300	105.260	152.900	152.252	0.194	-1093.033	-1029.051	179.174
400	117.540	185.081	156.556	11.410	-1091.403	-1007.946	131.624
500	123.750	212.069	165.037	23.516	-1089.206	-987.329	103.146
600	127.326	234.974	174.833	36.084	-1086.821	-967.176	84.200
700	130.071	254.813	184.873	48.958	-1084.380	-947.427	70.698
800	132.639	272.350	194.733	62.094	-1081.917	-928.031	60.594
900	135.246	288.123	204.247	75.488	-1079.435	-908.944	52.754
1000	137.957	302.511	213.365	89.147	-1076.912	-890.135	46.496
1100	140.786	315.792	222.080	103.083	-1074.334	-871.582	41.388
1200	143.718	328.167	230.411	117.307	-1071.700	-853.265	37.142
1300	146.741	339.789	238.382	131.830	-1069.004	-835.172	33.558
1400	149.841	350.777	246.021	146.658	-1066.252	-817.287	30.493
1500	152.464	361.210	253.356	161.782	-1063.451	-799.602	27.845
1600	157.846	371.198	260.410	177.261	-1060.537	-782.105	25.533
1600.000	157.846	371.198	260.410	177.261	——— GLASS <--> LIQUID ———		
1600.000	156.900	371.198	260.410	177.261	TRANSITION		
1700	156.900	380.710	267.210	192.951	-1057.674	-764.793	23.499
1800	156.900	389.678	273.767	208.641	-1055.077	-747.641	21.696
1900	156.900	398.162	280.093	224.331	-1052.770	-730.625	20.086
2000	156.900	406.209	286.199	240.021	-1050.767	-713.723	18.641
2100	156.900	413.865	292.098	255.711	-1049.088	-696.914	17.335
2200	156.900	421.164	297.800	271.401	-1093.514	-679.966	16.144
2300	156.900	428.138	303.316	287.091	-1092.811	-661.185	15.016
2340.000	156.900	430.843	305.473	293.367	-------- II <--> LIQUID --------		
2400	156.900	434.816	308.657	302.781	-1092.141	-642.434	13.982
2500	156.900	441.221	313.832	318.471	-1091.505	-623.709	13.032
2600	156.900	447.374	318.851	334.161	-1090.900	-605.009	12.155
2700	156.900	453.296	323.722	349.851	-1090.326	-586.332	11.343
2800	156.900	459.002	328.452	365.541	-1089.784	-567.675	10.590
2900	156.900	464.508	333.049	381.231	-1089.271	-549.038	9.889
3000	156.900	469.827	337.520	396.921	-1088.787	-530.417	9.235
3100	156.900	474.972	341.871	412.611	-1088.332	-511.812	8.624
3200	156.900	479.953	346.109	428.301	-1087.903	-493.223	8.051
3300	156.900	484.781	350.238	443.991	-1087.500	-474.646	7.513
3400	156.900	489.465	354.265	459.681	-1087.123	-456.079	7.007
3500	156.900	494.013	358.193	475.371	-1086.769	-437.525	6.530
3600	156.900	498.433	362.027	491.061	-1086.439	-418.979	6.079
3700	156.900	502.732	365.772	506.751	-1979.745	-398.042	5.619
3800	156.900	506.916	369.432	522.441	-1976.085	-355.342	4.885
3900	156.900	510.992	373.010	538.131	-1972.548	-312.736	4.189
4000	156.900	514.964	376.509	553.821	-1969.135	-270.222	3.529

PREVIOUS: CURRENT: December 1973

Vanadium Oxide (V_2O_3) $O_3V_2(1)$

VANADIUM OXIDE (V_2O_3) CRYSTAL–CRYSTAL–LIQUID M_r = 149.8812

Vanadium Oxide (V_2O_3) O_3V_2(cr,l)

0 to 168.8 K		crystal, I
168.8 to 2340	K	crystal, II
above 2340	K	liquid

Refer to the individual tables for details.

Enthalpy Reference Temperature = T_r = 298.15 K   Standard State Pressure = p° = 0.1 MPa

T/K	C_p° (J K⁻¹mol⁻¹)	S° (J K⁻¹mol⁻¹)	$-[G°-H°(T_r)]/T$ (J K⁻¹mol⁻¹)	$H°-H°(T_r)$ (kJ mol⁻¹)	$\Delta_fH°$ (kJ mol⁻¹)	$\Delta_fG°$ (kJ mol⁻¹)	Log K_f
0	0.000	0.000	INFINITE	-17.476	-1213.971	-1213.971	INFINITE
100	28.627	14.163	179.288	-16.512	-1218.342	-1192.325	622.807
168.800	63.053	38.652	116.725	-13.179		I <--> II	
168.800	63.136	48.269	116.725	-11.555		TRANSITION	
200	80.107	60.767	107.019	-9.250	-1219.118	-1165.403	304.372
298.15	104.963	98.067	98.067	0.000	-1218.799	-1139.036	199.554
300	105.260	98.717	98.069	0.194	-1218.779	-1138.542	198.238
400	117.540	130.898	102.373	11.410	-1217.149	-1112.018	145.215
500	123.750	157.887	110.854	23.516	-1214.951	-1085.983	113.452
600	127.326	180.791	120.651	36.084	-1212.566	-1060.412	92.317
700	130.071	200.631	130.691	48.958	-1210.125	-1035.245	77.251
800	132.639	218.168	140.550	62.094	-1207.662	-1010.430	65.974
900	135.246	233.940	150.065	75.488	-1205.180	-985.925	57.222
1000	137.957	248.329	159.182	89.147	-1202.658	-961.698	50.234
1100	140.786	261.609	167.897	103.083	-1200.079	-937.726	44.529
1200	143.718	273.984	176.228	117.307	-1197.445	-913.992	39.785
1300	146.741	285.607	184.199	131.830	-1194.750	-890.480	35.780
1400	149.841	296.594	191.839	146.658	-1191.997	-867.177	32.355
1500	153.005	307.040	199.173	161.800	-1189.178	-844.074	29.393
1600	156.221	317.017	206.229	177.261	-1186.283	-821.161	26.808
1700	159.481	326.585	213.029	193.045	-1183.324	-798.431	24.533
1800	162.782	335.794	219.595	209.158	-1180.305	-775.877	22.515
1900	166.112	344.684	225.946	225.603	-1177.243	-753.492	20.715
2000	169.468	353.290	232.099	242.382	-1174.151	-731.269	19.099
2100	172.847	361.640	238.070	259.497	-1171.047	-709.202	17.640
2200	176.246	369.759	243.872	276.952	-1213.709	-687.072	16.313
2300	179.655	377.669	249.518	294.747	-1210.901	-663.196	15.062
2340.000	181.021	380.778	251.736	301.960		II <--> LIQUID	
2340.000	156.900	430.843	251.736	419.112		TRANSITION	
2400	156.900	434.816	256.263	428.526	-1092.141	-642.434	13.982
2500	156.900	441.221	263.534	444.216	-1091.505	-623.709	13.032
2600	156.900	447.374	270.487	459.906	-1090.900	-605.009	12.155
2700	156.900	453.296	277.149	475.596	-1090.326	-586.332	11.343
2800	156.900	459.002	283.543	491.286	-1089.784	-567.675	10.590
2900	156.900	464.508	289.688	506.976	-1089.271	-549.038	9.889
3000	156.900	469.827	295.605	522.666	-1088.787	-530.418	9.235
3100	156.900	474.972	301.308	538.356	-1088.332	-511.812	8.624
3200	156.900	479.953	306.814	554.046	-1087.903	-493.223	8.051
3300	156.900	484.781	312.134	569.736	-1087.500	-474.646	7.513
3400	156.900	489.465	317.281	585.426	-1087.123	-456.079	7.007
3500	156.900	494.013	322.266	601.116	-1086.769	-437.525	6.530
3600	156.900	498.433	327.098	616.806	-1086.439	-418.979	6.079
3700	156.900	502.732	331.787	632.496	-1979.745	-398.042	5.619
3800	156.900	506.916	336.341	648.186	-1976.085	-355.342	4.885
3900	156.900	510.992	340.767	663.876	-1972.548	-312.736	4.189
4000	156.900	514.964	345.073	679.566	-1969.135	-270.222	3.529

PREVIOUS: CURRENT: December 1973

Vanadium Oxide (V_2O_3) O_3V_2(cr,l)

TUNGSTEN OXIDE (WO_3) CRYSTAL M_r = 231.8482

$S°(298.15 K) = 75.912 \pm 1.3 J K^{-1} mol^{-1}$

T_{trs} = 1050 K

T_{fus} = 1745 K

$\Delta_f H°(0 K) = -837.259 \pm 0.8 kJ mol^{-1}$

$\Delta_f H°(298.15 K) = -842.909 \pm 0.8 kJ mol^{-1}$

$\Delta_{trs} H° = 1.485 kJ mol^{-1}$

$\Delta_{fus} H° = 73.429 kJ mol^{-1}$

Enthalpy of Formation

The selected enthalpy of formation, $\Delta_f H°(WO_3$, cr, 298.15 K) = -201.46±0.2 kcal mol^{-1}, was determined by Mah (1) using bomb calorimetry. Huff et al. (2) have reviewed the literature enthalpy of formation data and also measured calorimetrically the enthalpy of formation, -201.84 ± 0.1 kcal mol^{-1}, which is in good agreement with the value selected. Other combustion values were reported as $\Delta_f H°$(298.15 K) = -199 ± 1 kcal mol^{-1} by Griffis (3); and $\Delta_f H°$(298.15 K) = -205.3 kcal mol^{-1} by Vasil'eva et al. (4). Griffis (5) also derived the enthalpy of formation, -202.8 ± 1 kcal mol^{-1}, from equilibrium data.

Heat Capacity and Entropy

The low temperature (55-296.7 K) heat capacities and high temperature (399.6 - 1835.8 K) enthalpy data have been measured by King et al. (6), and these data have been used to derive the tabulated heat capacities. Low temperature and high temperature values were joined smoothly at 298.15 K. The entropy was calculated from the heat capacities based on S°(51 K) = 1.31 cal $K^{-1} mol^{-1}$.

King et al. (6) were apparently unaware of the transition near 593 K which was reported by Perri et al. (7), and they did not take sufficient points in that region for the precise enthalpy curve. We have adopted only one curve through the region because of the insufficiency of the data and also because the heat effects appear to be small or gradual.

Transition and Fusion Data

The enthalpy of transition, the transition point, the enthalpy of melting and the melting point were derived from the high temperature enthalpy data of King et al. (6).

The melting point, 1746 ± 1 K, has also been quoted by Schneider (8).

In the high temperature x-ray diffractometric studies, Perri et al. (7) indicate that WO_3 undergoes a phase transition from monoclinic to orthorhombic at approximately 593 K and from orthorhombic to tetragonal at 993 K.

Sublimation Data

The values of $\Delta_{sub} H°$(298.15 K) are calculated as the enthalpy of sublimation of one mole of crystal to one mole of monomer, 1/2 mole of dimer, 1/3 mole of trimer and 1/4 mole of tetramer, respectively.

References

1. A. D. Mah, J. Amer. Chem. Soc. 81, 1582 (1959).
2. G. Huff, E. Squitieri and P. E. Snyder, J. Amer. Chem. Soc. 70, 3380 (1948).
3. R. C. Griffis, J. Electrochem. Soc. 105, 398 (1958).
4. I. A. Vasil'eva, V. I. Gerassimov, and Y. P. Simanov, Zh. Fiz. Khim. 31, 682 (1957).
5. R. C. Griffis, J. Electrochem. Soc. 106, 418 (1959).
6. E. G. King, W. W. Weller and A. U. Christensen, U. S. Bur. Mines RI 5664, 1960.
7. J. A. Perri, E. Banks, and B. Post, J. Appl. Phys. 28, 1272 (1957).
8. S. J. Schneider, U.S. Nat. Bur. Stand. Monograph 68, 1963.

Tungsten Oxide (WO_3) O_3W_1(cr)

Enthalpy Reference Temperature = T_r = 298.15 K   Standard State Pressure = p° = 0.1 MPa

	J K⁻¹mol⁻¹			kJ mol⁻¹			
T/K	$C_p°$	S°	-[G°-H°(T_r)]/T	H°-H°(T_r)	$\Delta_f H°$	$\Delta_f G°$	Log K_f
0	0.000	0.000	INFINITE	-12.349	-837.259	-837.259	INFINITE
100	30.449	18.862	130.106	-11.124	-841.044	-815.973	426.221
200	58.196	49.480	82.212	-6.546	-842.841	-790.036	206.336
298.15	73.139	75.912	75.912	0.000	-842.909	-764.057	133.860
300	73.304	76.365	75.914	0.135	-842.900	-763.568	132.949
400	82.209	98.738	78.904	7.934	-842.022	-737.237	96.273
500	88.641	117.815	84.825	16.495	-840.563	-711.200	74.299
600	93.113	134.397	91.736	25.597	-838.758	-685.493	59.677
700	96.141	148.993	98.894	35.069	-836.769	-660.104	49.258
800	98.239	161.975	105.983	44.794	-834.694	-635.007	41.462
900	99.923	173.645	112.864	54.703	-832.582	-610.173	35.414
1000	101.671	184.261	119.480	64.781	-830.431	-585.575	30.587
1050.000	102.508	189.242	122.684	69.886		I <--> II	
1050.000	98.144	190.655	122.684	71.369		TRANSITION	
1100	98.963	195.239	125.879	76.296	-826.958	-561.262	26.652
1200	100.601	203.920	132.025	86.275	-825.128	-537.188	23.383
1300	102.238	212.037	137.871	96.417	-823.230	-513.269	20.623
1400	103.875	219.674	143.444	106.722	-821.261	-489.499	18.263
1500	105.512	226.896	148.769	117.192	-819.216	-465.873	16.223
1600	107.149	233.758	153.868	127.825	-817.094	-442.386	14.442
1700	108.787	240.303	158.761	138.621	-814.892	-419.034	12.875
1745.000	109.537	243.155	160.901	143.534	--------	II <--> LIQUID	--------
1800	110.458	246.569	163.467	149.583	-812.608	-395.814	11.486
1900	112.131	252.586	168.000	160.713	-810.240	-372.723	10.247
2000	113.805	258.380	172.375	172.010	-807.788	-349.758	9.135
2100	115.478	263.973	176.604	183.474	-805.252	-326.919	8.132
2200	117.152	269.383	180.699	195.105	-802.632	-304.201	7.223
2300	118.826	274.628	184.670	206.904	-799.928	-281.606	6.395
2400	120.499	279.720	188.524	218.870	-797.139	-259.130	5.640
2500	122.173	284.673	192.272	231.004	-794.266	-236.772	4.947
2600	123.846	289.498	195.919	243.305	-791.305	-214.530	4.310
2700	125.520	294.203	199.472	255.773	-788.282	-192.404	3.722
2800	127.194	298.798	202.938	268.409	-785.231	-170.389	3.179
2900	128.867	303.291	206.321	281.212	-782.191	-148.486	2.675
3000	130.541	307.688	209.627	294.182	-779.188	-126.684	2.206

PREVIOUS: March 1963   CURRENT: September 1966

Tungsten Oxide (WO_3) O_3W_1(cr)

TUNGSTEN OXIDE (WO_3) LIQUID M_r = 231.8482

$S°(298.15\ K) = [103.508]\ J\ K^{-1}\ mol^{-1}$

$T_{fus} = 1745\ K$

$\Delta_f H°(298.15\ K) = [-788.915]\ kJ\ mol^{-1}$

$\Delta_{fus}H° = 73.429\ kJ\ mol^{-1}$

Enthalpy of Formation

$\Delta_f H°(\ell,\ 298.15\ K)$ is calculated from that of the crystal by adding $\Delta_{fus}H°$ and the difference in enthalpy, H°(1745 K)-H°(298.15 K), between the crystal and liquid.

Heat Capacity and Entropy

The heat capacity between the melting point and 1836 K was determined from the enthalpy measurement by King et al. (1). The heat capacity was assumed constant above and below the measured range. At 1000 K, a glass transition was assumed below which the heat capacity was assumed to be that of the crystal. The entropy at 298.15 K is calculated in a manner analogous to that used for the enthalpy of formation.

Fusion and Vaporization Data

The heat and temperature of melting were obtained from King et al. (1).

T_{vap}(to equilibrium vapor) = 2110 K is the temperature at which the total pressure reaches one atm based on the calculated partial pressures of $WO_3(g)$, $W_2O_6(g)$, $W_3O_8(g)$, $W_3O_9(g)$, and $W_4O_{12}(g)$. $\Delta_{vap}H°$(to equilbrium vapor) = 18.3 kcal mol^{-1} at the boiling point is calculated as the enthalpy of vaporization of one mole of the liquid to vapor containing 23.7 mole percent of $W_2O_6(g)$, 0.57 mole percent of $W_3O_8(g)$, 58.2 mole percent of $W_3O_9(g)$ and 17.53 mole percent of $W_4O_{12}(g)$.

Reference

1. E. G. King, W. W. Weller and A. U. Christensen, U. S. Bur. Mines RI 5664, 1960.

Tungsten Oxide (WO_3) $O_3W_1(l)$

Enthalpy Reference Temperature = T_r = 298.15 K Standard State Pressure = p° = 0.1 MPa

T/K	$C_p°$	S°	$-[G°-H°(T_r)]/T$	$H°-H°(T_r)$	$\Delta_f H°$	$\Delta_f G°$	Log K_f
	J K⁻¹mol⁻¹			kJ mol⁻¹			
0							
100							
200							
298.15	73.140	103.508	103.508	0.000	-788.915	-718.291	125.842
300	73.304	103.961	103.509	0.135	-788.906	-717.853	124.989
400	82.209	126.334	106.499	7.934	-788.028	-694.281	90.664
500	88.641	145.410	112.421	16.495	-786.570	-671.004	70.099
600	93.113	161.993	119.332	25.597	-784.764	-648.056	56.418
700	96.141	176.589	126.490	35.069	-782.775	-625.427	46.670
800	98.239	189.571	133.579	44.794	-780.701	-603.090	39.378
900	99.923	201.241	140.459	54.703	-778.588	-581.015	33.721
1000	101.671	211.857	147.076	64.781	-776.437	-559.177	29.208
1000.000	101.671	211.857	147.076	64.781	GLASS <--> LIQUID		
1000.000	131.796	211.857	147.076	64.781	TRANSITION		
1100	131.796	224.418	153.545	77.961	-771.300	-537.701	25.533
1200	131.796	235.886	159.936	91.140	-766.268	-516.687	22.491
1300	131.796	246.435	166.190	104.320	-761.333	-496.090	19.933
1400	131.796	256.203	172.275	117.499	-756.490	-475.869	17.755
1500	131.796	265.296	178.176	130.679	-751.735	-455.991	15.879
1600	131.796	273.802	183.890	143.859	-747.066	-436.427	14.248
1700	131.796	281.792	189.416	157.038	-742.481	-417.153	12.818
1745.000	131.796	285.235	191.843	162.969	-------- II <--> LIQUID --------		
1800	131.796	289.325	194.759	170.218	-737.980	-398.147	11.554
1900	131.796	296.451	199.926	183.397	-733.562	-379.388	10.430
2000	131.796	303.211	204.922	196.577	-729.227	-360.860	9.425
2100	131.796	309.641	209.757	209.757	-724.975	-342.546	8.520
2200	131.796	315.772	214.438	222.936	-720.807	-324.433	7.703
2300	131.796	321.631	218.972	236.116	-716.722	-306.508	6.961
2400	131.796	327.240	223.367	249.295	-712.720	-288.759	6.285
2500	131.796	332.620	227.630	262.475	-708.802	-271.175	5.666
2600	131.796	337.790	231.769	275.655	-704.962	-253.746	5.098
2700	131.796	342.764	235.788	288.834	-701.227	-236.462	4.575
2800	131.796	347.557	239.695	302.014	-697.633	-219.314	4.091
2900	131.796	352.182	243.494	315.193	-694.216	-202.294	3.644
3000	131.796	356.650	247.192	328.373	-691.004	-185.386	3.228

PREVIOUS: March 1963 CURRENT: September 1966

Tungsten Oxide (WO_3) $O_3W_1(l)$

TUNGSTEN OXIDE (WO_3) CRYSTAL–CRYSTAL–LIQUID M_r = 231.8482

0 to 1050 K crystal, I
1050 to 1745 K crystal, II
above 1745 K liquid

Refer to the individual tables for details.

Tungsten Oxide (WO_3) O_3W_1(cr,l)

Enthalpy Reference Temperature = T_r = 298.15 K Standard State Pressure = p° = 0.1 MPa

T/K	C_p° (J K⁻¹mol⁻¹)	S° (J K⁻¹mol⁻¹)	$-[G^\circ-H^\circ(T_r)]/T$ (J K⁻¹mol⁻¹)	$H^\circ-H^\circ(T_r)$ (kJ mol⁻¹)	Δ_fH° (kJ mol⁻¹)	Δ_fG° (kJ mol⁻¹)	Log K_f
0	0.000	0.000	INFINITE	-12.349	-837.259	-837.259	INFINITE
100	30.449	18.862	130.106	-11.124	-841.044	-815.973	426.221
200	58.196	49.480	82.212	-6.546	-842.841	-790.036	206.336
298.15	73.139	75.912	75.912	0.000	-842.909	-764.057	133.860
300	73.304	76.365	75.914	0.135	-842.900	-763.568	132.949
400	82.209	98.738	78.904	7.934	-842.022	-737.237	96.273
500	88.641	117.815	84.825	16.495	-840.563	-711.200	74.299
600	93.113	134.397	91.736	25.597	-838.758	-685.493	59.677
700	96.141	148.993	98.894	35.069	-836.769	-660.104	49.258
800	98.239	161.975	105.983	44.794	-834.694	-635.007	41.462
900	99.923	173.645	112.864	54.703	-832.582	-610.173	35.414
1000	101.671	184.261	119.480	64.781	-830.431	-585.575	30.587
1050.000	102.508	189.242	122.684	69.886		I <--> II	
1050.000	98.144	190.655	122.684	71.369		TRANSITION	
1100	98.963	195.239	125.879	76.296	-826.958	-561.262	26.652
1200	100.601	203.920	132.025	86.275	-825.128	-537.188	23.383
1300	102.238	212.037	137.871	96.417	-823.230	-513.269	20.623
1400	103.875	219.674	143.444	106.722	-821.261	-489.499	18.263
1500	105.512	226.896	148.769	117.192	-819.216	-465.873	16.223
1600	107.149	233.758	153.868	127.825	-817.094	-442.386	14.442
1700	108.787	240.303	158.761	138.621	-814.892	-419.034	12.875
1745.000	109.537	243.155	160.901	143.534		II <--> LIQUID	
1745.000	131.796	285.235	160.901	216.963		TRANSITION	
1800	131.796	289.325	164.763	224.212	-737.980	-398.147	11.554
1900	131.796	296.451	171.508	237.391	-733.562	-379.388	10.430
2000	131.796	303.211	177.926	250.571	-729.227	-360.860	9.425
2100	131.796	309.641	184.046	263.750	-724.975	-342.546	8.520
2200	131.796	315.772	189.895	276.930	-720.807	-324.433	7.703
2300	131.796	321.631	195.496	290.110	-716.722	-306.508	6.961
2400	131.796	327.240	200.870	303.289	-712.720	-288.759	6.285
2500	131.796	332.620	206.033	316.469	-708.802	-271.175	5.666
2600	131.796	337.790	211.002	329.648	-704.962	-253.746	5.098
2700	131.796	342.764	215.790	342.828	-701.227	-236.462	4.575
2800	131.796	347.557	220.411	356.008	-697.633	-219.314	4.091
2900	131.796	352.182	224.876	369.187	-694.216	-202.294	3.644
3000	131.796	356.650	229.194	382.367	-691.004	-185.386	3.228

PREVIOUS: CURRENT: September 1966

Tungsten Oxide (WO_3) O_3W_1(cr,l)

TUNGSTEN OXIDE (WO_3) IDEAL GAS M_r = 231.8482

$S°(298.15\ K) = [286.440]\ J\ K^{-1}\ mol^{-1}$

$\Delta_fH°(0\ K) = -288.47 \pm 29.3\ kJ\ mol^{-1}$
$\Delta_fH°(298.15\ K) = -292.88 \pm 29.3\ kJ\ mol^{-1}$

Vibrational Frequencies and Degeneracies

ν, cm^{-1}	ν, cm^{-1}
[564](1)	1040 (2)
[347](1)	[320](2)

Ground State Quantum Weight: 1 $\sigma = [6]$
Point Group: $[D_{3h}]$
Bond Distance: W-O = [1.81] Å
Bond Angle: O-W-O = [120]°
Product of the Moments of Inertia: $I_AI_BI_C = [4.4509666 \times 10^{-114}]\ g^3\ cm^6$

Enthalpy of Formation

The Gibbs energy of formation of gaseous tungsten trioxide, $\Delta_fG°(WO_3, g) = -75,700 + 15.54\ T\ cal\ K^{-1}\ mol^{-1}$, over the temperature range 1920 - 2020 K has been reported by Ackermann and Thorn (1), using the reaction 3 MgO(cr) + W(cr) + 3 Mg(g) + WO_3(g). 3rd law calculation gives $\Delta_fH°(WO_3, g, 298.15\ K) = -71.3\ kcal\ mol^{-1}$ using JANAF functions. The 2nd law value of $\Delta_fH°(298.15\ K)$ is $-74.8\ kcal\ mol^{-1}$.

DeMaria et al. (2) have reported the partial pressures of WO(g), WO_2(g), WO_3(g) and O(g) in a mass spectrometric study of aluminum oxide vaporized from a tungsten cell from 2188 to 2475 K. 3rd law treatment of their partial pressure data gives $\Delta_rH°(298.15\ K) = -245.2\ kcal\ mol^{-1}$ for W(cr) + 3 O(g) + WO(g) which yields $\Delta_fH°(WO_3, g, 298.15\ K) = -66.5\ kcal\ mol^{-1}$. The 3rd law drift is $21 \pm 7\ cal\ K^{-1}\ mol^{-1}$. Partial pressures of WO_2(g), WO_3(g), and O(g) at 2242 K have also been reported by Chupka et al. (3) in a mass spectrometric study of beryllium oxide in a tungsten cell. Similar calculation gives $\Delta_rH°(WO_3, g, 298.15\ K) = -73.0\ kcal\ mol^{-1}$.

The value, $-70 \pm 7\ kcal\ mol^{-1}$, is selected for the enthalpy of formation at 298.15 K.

Heat Capacity and Entropy

The WO_3(g) was assumed to be a planar D_{3h} molecule with a bond angle O-W-O = 120°. The bond distance W-O was estimated to be the same as that in WO(g). Weltner and McLeod (4) have observed the vibrational frequency of 1040 cm^{-1} in the infrared spectrum by matrix isolation, and also estimated the frequencies 347 and 564 cm^{-1} based on the stretching force constant of WO(g). The estimation of 320 cm^{-1} was obtained from DeMaria et al (2). These frequencies have been adopted in the tabulation. The principal moments of inertia are $I_A = I_B = 13.0559 \times 10^{-39}$ and $I_C = 26.1119 \times 10^{-39}\ g\ cm^2$.

References

1. R. J. Ackermann and R. J. Thorn, J. Phys. Chem. 64, 350 (1960).
2. G. DeMaria, R. P. Burns, J. Drowart and M. G. Inghram, J. Chem. Phys. 32, 1373 (1960).
3. W. A. Chupka, J. Berkowitz and C. F. Giese, J. Chem. Phys. 30, 827 (1959).
4. W. Weltner, Jr., and D. McLeod, Jr., J. Mol. Spectrosc. 17, 276 (1965).

Tungsten Oxide (WO_3) O_3W_1(g)

Enthalpy Reference Temperature = T_r = 298.15 K Standard State Pressure = p° = 0.1 MPa

T/K	$C_p°$ (J K⁻¹mol⁻¹)	S° (J K⁻¹mol⁻¹)	$-[G°-H°(T_r)]/T$ (J K⁻¹mol⁻¹)	$H°-H°(T_r)$ (kJ mol⁻¹)	$\Delta_fH°$ (kJ mol⁻¹)	$\Delta_fG°$ (kJ mol⁻¹)	Log K_f
0	0.	0.	INFINITE	-13.592	-288.474	-288.474	INFINITE
100	38.449	232.689	334.270	-10.158	-290.049	-286.361	149.580
200	52.190	263.843	291.760	-5.583	-291.849	-281.917	73.629
250	57.112	276.037	287.423	-2.847	-292.456	-279.361	58.369
298.15	61.005	286.440	286.440	0.	-292.880	-276.798	48.494
300	61.141	286.817	286.441	0.113	-292.894	-276.698	48.177
350	64.479	296.501	287.198	3.256	-293.189	-273.973	40.888
400	67.241	305.297	288.919	6.551	-293.376	-271.214	35.417
450	69.521	313.353	291.193	9.972	-293.482	-268.437	31.159
500	71.402	320.778	293.785	13.497	-293.533	-265.651	27.752
600	74.248	334.064	299.417	20.788	-293.538	-260.073	22.641
700	76.231	345.667	305.213	28.318	-293.491	-254.499	18.991
800	77.646	355.944	310.925	36.015	-293.444	-248.932	16.254
900	78.681	365.152	316.447	43.834	-293.422	-243.370	14.125
1000	79.457	373.484	321.741	51.743	-293.440	-237.808	12.422
1100	80.052	381.086	326.796	59.720	-293.506	-232.241	11.028
1200	80.516	388.072	331.615	67.749	-293.625	-226.667	9.867
1300	80.885	394.532	336.209	75.820	-293.799	-221.081	8.883
1400	81.182	400.538	340.592	83.924	-294.031	-215.478	8.040
1500	81.425	406.147	344.778	92.054	-294.325	-209.858	7.308
1600	81.626	411.409	348.779	100.207	-294.683	-204.216	6.667
1700	81.794	416.363	352.611	108.379	-295.106	-198.549	6.101
1800	81.936	421.042	356.284	116.565	-295.598	-192.855	5.597
1900	82.057	425.475	359.810	124.765	-296.159	-187.132	5.145
2000	82.160	429.687	363.199	132.976	-296.793	-181.378	4.737
2100	82.250	433.698	366.461	141.197	-297.501	-175.590	4.368
2200	82.328	437.526	369.605	149.426	-298.283	-169.766	4.031
2300	82.396	441.187	372.638	157.662	-299.141	-163.906	3.722
2400	82.456	444.695	375.568	165.904	-300.077	-158.007	3.439
2500	82.509	448.062	378.401	174.153	-301.089	-152.067	3.177
2600	82.557	451.299	381.143	182.406	-302.175	-146.085	2.935
2700	82.599	454.416	383.799	190.664	-303.363	-140.058	2.710
2800	82.637	457.420	386.375	198.926	-304.686	-133.986	2.500
2900	82.671	460.321	388.876	207.191	-306.184	-127.865	2.303
3000	82.701	463.124	391.304	215.460	-307.882	-121.687	2.119
3100	82.729	465.836	393.665	223.731	-309.824	-115.449	1.945
3200	82.754	468.463	395.961	232.005	-312.041	-109.145	1.782
3300	82.778	471.010	398.197	240.282	-314.556	-102.767	1.627
3400	82.799	473.481	400.375	248.561	-317.404	-96.306	1.480
3500	82.818	475.882	402.499	256.842	-320.650	-89.758	1.340
3600	82.836	478.215	404.570	265.124	-324.387	-83.109	1.206
3700	82.852	480.485	406.591	273.409	-363.482	-76.159	1.075
3800	82.867	482.695	408.565	281.695	-364.915	-68.373	0.940
3900	82.881	484.847	410.493	289.982	-366.367	-60.549	0.811
4000	82.894	486.946	412.378	298.271	-367.838	-52.690	0.688
4100	82.906	488.993	414.222	306.561	-369.328	-44.793	0.571
4200	82.917	490.991	416.026	314.852	-370.835	-36.859	0.458
4300	82.927	492.942	417.792	323.144	-372.359	-28.889	0.351
4400	82.937	494.849	419.522	331.438	-373.901	-20.884	0.248
4500	82.946	496.713	421.217	339.732	-375.461	-12.844	0.149
4600	82.955	498.536	422.878	348.027	-377.037	-4.769	0.054
4700	82.963	500.320	424.507	356.323	-378.632	3.341	-0.037
4800	82.970	502.067	426.104	364.619	-380.244	11.485	-0.125
4900	82.977	503.778	427.672	372.917	-381.875	19.664	-0.210
5000	82.984	505.454	429.211	381.215	-383.524	27.876	-0.291
5100	82.990	507.097	430.722	389.513	-385.194	36.121	-0.370
5200	82.996	508.709	432.206	397.813	-386.883	44.396	-0.446
5300	83.001	510.290	433.665	406.113	-388.594	52.708	-0.519
5400	83.006	511.841	435.098	414.413	-390.328	61.050	-0.591
5500	83.011	513.365	436.507	422.714	-392.085	69.424	-0.659
5600	83.016	514.860	437.893	431.015	-393.867	77.833	-0.726
5700	83.021	516.330	439.257	439.317	-395.675	86.271	-0.791
5800	83.025	517.774	440.598	447.619	-397.510	94.743	-0.853
5900	83.029	519.193	441.918	455.922	-399.375	103.244	-0.914
6000	83.033	520.588	443.218	464.225	-401.270	111.780	-0.973

PREVIOUS: September 1966 (1 atm) CURRENT: September 1966 (1 bar)

Tungsten Oxide (WO_3) O_3W_1(g)

LEAD SILICATE (Pb_2SiO_4) CRYSTAL M_r = 506.4831

$\Delta_fH°(0\ K) = -1369.466 \pm 14.6\ kJ\ mol^{-1}$

$S°(298.15\ K) = 186.832 \pm 2.1\ J\ K^{-1}\ mol^{-1}$ $\Delta_fH°(298.15\ K) = -1377.875 \pm 14.6\ kJ\ mol^{-1}$

T_{fus} = 1016 K

Enthalpy of Formation

The enthalpy of the reaction 2 PbO(cr) + SiO_2(cr) + Pb_2SiO_4(cr) was reported by Kubaschewski and Evans (1). This value for the enthalpy of reaction of the oxides has been converted to the enthalpy of formation using JANAF auxiliary data (4). The above $\Delta_rH°(298.15\ K) = -7.00 \pm 3.5\ kcal\ mol^{-1}$

Heat Capacity and Entropy

In the low temperature region (53.07-298.15 K) the $C_p°$ values are those determined by King (2). Above 298.15, $C_p°$ values were estimated by summing the values for the constituent oxides and graphically smoothing these into the low temperature measurements. The entropy was calculated at 53.07 K using the Debye and Einstein functions D(78.5/T) + 2E(164/T) + 3E(547/T) + E(1379/T) given by King (2). $S°(53.07\ K) = 8.248\ cal\ K^{-1}\ mol^{-1}$.

Fusion Data

T_{fus} was taken from Geller et al. (3).

References

1. O. Kubaschewski and E. L. Evans, Metallurgical Thermochemistry, Pergamon Press, New York (1958).
2. E. G. King, J. Amer. Chem. Soc. 81, 799-800 (1959).
3. Geller, Creamer and Bunting, J. Res. Nat. Bur. Stand. 13, 237 (1934).
4. JANAF Thermochemical Tables: PbO(cr), 3-31-62; SiO_2(cr), 12-31-62.

Lead Silicate (Pb_2SiO_4) $O_4Pb_2Si_1$(cr)

Enthalpy Reference Temperature = T_r = 298.15 K Standard State Pressure = p° = 0.1 MPa

	J K⁻¹mol⁻¹			kJ mol⁻¹			
T/K	$C_p°$	S°	$-[G°-H°(T_r)]/T$	$H°-H°(T_r)$	$\Delta_fH°$	$\Delta_fG°$	Log K_f
0	0.000	0.000	INFINITE	-25.932	-1369.466	-1369.466	INFINITE
100	73.789	72.007	291.667	-21.966	-1375.108	-1339.898	699.891
200	113.968	136.700	198.686	-12.397	-1377.589	-1303.493	340.437
298.15	136.984	186.832	186.832	0.000	-1377.875	-1267.007	221.974
300	137.235	187.680	186.835	0.254	-1377.866	-1266.320	220.486
400	152.005	229.255	192.403	14.741	-1376.901	-1229.256	160.524
500	163.762	264.478	203.383	30.547	-1375.118	-1192.537	124.583
600	173.343	295.224	216.181	47.426	-1372.712	-1156.238	100.659
700	180.121	322.489	229.458	65.122	-1379.617	-1118.781	83.484
800	184.180	346.822	242.636	83.349	-1376.607	-1081.721	70.629
900	186.774	368.668	255.447	101.899	-1373.401	-1045.052	60.653
1000	189.117	388.473	267.775	120.699	-1370.045	-1008.747	52.692
1100	190.790	406.580	279.582	139.698	-1366.569	-972.784	46.194
1200	192.046	423.232	290.868	158.836	-1363.015	-937.142	40.793
1300	193.719	438.670	301.651	178.125	-1359.374	-901.800	36.235
1400	195.309	453.086	311.959	197.578	-1355.656	-866.741	32.338
1500	196.648	466.607	321.823	217.177	-1351.892	-831.950	28.971
1600	197.903	479.339	331.273	236.906	-1348.109	-797.410	26.033
1700	199.033	491.371	340.340	256.753	-1394.502	-762.660	23.434
1800	200.121	502.778	349.050	276.710	-1390.499	-725.609	21.057
1900	201.292	513.629	357.429	296.780	-1386.467	-688.780	18.936
2000	202.506	523.985	365.500	316.970	-1382.402	-652.165	17.033

PREVIOUS: CURRENT: June 1965

Lead Silicate (Pb_2SiO_4) $O_4Pb_2Si_1$(cr)

LEAD OXIDE (Pb_3O_4) CRYSTAL M_r = 685.5976

$S°(298.15\ K) = 211.961 \pm 6.7\ J\ K^{-1}\ mol^{-1}$

$\Delta_fH°(0\ K) = -710.871 \pm 6.2\ kJ\ mol^{-1}$
$\Delta_fH°(298.15\ K) = -718.686 \pm 6.2\ kJ\ mol^{-1}$

Enthalpy of Formation

Espada et al. (1) have measured the enthalpy of the reduction reaction $PbO_{1.3}(cr) + 1.3\ H_2(g) \rightarrow Pb(cr) + 1.3\ H_2O(\ell)$ as -32.07 kcal mol^{-1} as -32.07 kcal mol^{-1}. They assumed that the sample consisted of 90.1% mol percent $PbO_{1.333}$ and 9.9 mol PbO and corrected the observed enthalpy of reaction to -33.82 ± 0.25 kcal mol^{-1} for the reaction $PbO_{1.333}(cr) + 1.333\ H_2(g) \rightarrow Pb(cr) + 1.333\ H_2O(\ell)$. This leads to $\Delta_fH°(Pb_3O_4, cr) = -171.77 \pm 1.5$ kcal mol^{-1} which is adopted.

Andrews and Brown (2) used reversible cells to obtain $\Delta_rG°(298.15\ K) = -6940$ cal K^{-1} mol^{-1} for the reaction $Hg(\ell) + Pb_3O_4(cr) \rightarrow HgO(cr) + 3\ PbO(red, cr)$, with JANAF auxiliary data (3) this yields $\Delta_fG°(Pb_3O_4, cr, 298.15\ K) = -142.76 \pm 0.6$ kcal mol^{-1} which is equivalent to $\Delta_fH°(Pb_3O_4, cr, 298.15\ K) = -170.73 \pm 1.1$ kcal mol^{-1}. This value is in agreement with the adopted value within the combined uncertainties.

Reinders and Hamburger (4) and Otto (5) studied the decomposition pressure of the reaction $Pb_3O_4(cr) \rightarrow 3\ PbO(cr) + 0.5\ O_2(g)$ a 2nd and 3rd law analysis of their data is given below.

Source	Data Points	T/K	$\Delta_rH°(298.15\ K)$ kcal mol^{-1} 2nd law	3rd law	Drift cal K^{-1} mol^{-1}	$\Delta_fH°(298.15\ K)$ kcal mol^{-1}
(4)	16	718-880	20.19±0.14	19.08±0.17	-1.4±0.2	-175.44
(5)	19	757-911	20.84±0.10	18.98±0.22	-2.2±0.1	-175.34

Although these two sets of data are in good agreement, they are outside the adopted enthalpy of formation uncertainty limits, and realistic adjustments of the entropy or heat capacity fail to eliminate this discrepancy.

In addition there have been cell measurements linking PbO_2 and Pb_3O_4, which serve to indicate the overall consistency of the lead oxygen system. The absolute uncertainty in the enthalpy of formation and the difficulty of preparing pure PbO_2 make it a poor choice on which to base the heat of formation of Pb_3O_4. Andrews and Brown (2) obtained $\Delta_rG°(298.15\ K) = -18.12$ kcal mol^{-1} for the reaction $2\ Hg(\ell) + 3\ PbO_2(cr) \rightarrow Pb_3O_4(cr) + 2\ HgO(cr)$. This yields $\Delta_fG°(Pb_3O_4, cr, 298.15\ K) = -144.62 \pm 2.1$ kcal mol^{-1} with JANAF auxiliary data (3), which corresponds to $\Delta_fH°(Pb_3O_4, cr, 298.15\ K) = -172.59 \pm 2.6$ kcal mol^{-1}. Millar (6) has recalculated the data of Glasstone for the reaction $3\ PbO_2(cr) + 2\ H_2(g) \rightarrow Pb_3O_4 + 2\ H_2O(\ell)$ as $\Delta_rG°(298.15\ K) = -104.36$ kcal mol^{-1} which with auxiliary data (3, 7) yields $\Delta_fG°(Pb_3O_4, cr, 298.15\ K) = -145.45 \pm 2.1$ kcal mol^{-1} which corresponds to $\Delta_fH°(Pb_3O_4, cr, 298.15\ K) = -173.42 \pm 2.6$ kcal mol^{-1}.

Heat Capacity and Entropy

The low temperature enthalpy capacity from 71.5 K to 298.6 K has been reported by Millar (6). These values were used to calculate $S°(298.15\ K)$ based on $S°(70.8\ K) = 12.67$ cal K^{-1} mol^{-1} (7). Enthalpies in the range 365 to 781 K have been reported by Bousquet et al. (8). These data encompass the range where the dissociation pressure is significant (76 mm) and we have not used the data above 600 K. The mean heat capacities were plotted at the mean temperatures for the five points used and a smooth curve was drawn graphically through them and the low temperature data set. The four highest points of Millar (6) were not used since at least two were reported as bad points and they did not fit onto the smooth curve.

Decomposition Data

The decomposition pressure reaches 1 atm at 869.7 according to Reinders and Hamburger (4) and at 867.2 K according to Otto (5). This table indicates a temperature of 690 K for the decomposition pressure to reach 1 atm. The significant difference is due to the fact that heat and entropy derived from the dissociation cannot be simultaneously satisfied by any realistic variation of the entropy and heat capacity of Pb_3O_4.

References

1. L. Espada, G. Pilcher and H. A. Skinner, J. Chem. Thermodyn. 2, 647 (1970).
2. L. V. Andrews and D. J. Brown, J. Amer. Chem. Soc. 56, 388 (1934).
3. JANAF Thermochemical Tables: HgO(cr), 6-30-62; PbO(red, cr), 12-31-71; PbO_2(cr), 12-31-71.
4. W. Reinders and L. Hamburger, Z. Anorg. Chem. 89, 71 (1914).
5. E. M. Otto, J. Electrochem. Soc. 113, 525 (1966).
6. R. W. Millar, J. Amer. Chem. Soc. 51, 207 (1929).
7. K. K. Kelley and E. G. King, U. S. Bur. Mines Bull. 592, 1961.
8. J. Bousquet, J. M. Blanchard and G. Perachon, Bull. Soc. Chim. Fr. 1969, 733 (1969).

Lead Oxide (Pb_3O_4) $O_4Pb_3(cr)$

Enthalpy Reference Temperature = T_r = 298.15 K; Standard State Pressure = p° = 0.1 MPa

T/K	$C_p°$ (J $K^{-1}mol^{-1}$)	S°	$-[G°-H°(T_r)]/T$	$H°-H°(T_r)$ (kJ mol^{-1})	$\Delta_fH°$	$\Delta_fG°$	Log K_f
0	0.000	0.000	INFINITE	-30.188	-710.871	-710.871	INFINITE
100	87.195	79.860	331.486	-25.163	-716.955	-679.231	354.794
200	130.122	155.101	225.392	-14.058	-719.246	-640.309	167.232
298.15	154.934	211.961	211.961	0.000	-718.686	-601.606	105.399
300	155.344	212.921	211.964	0.287	-718.656	-600.879	104.622
400	173.050	260.195	218.289	16.762	-716.310	-561.936	73.381
500	184.096	300.077	230.766	34.656	-712.976	-523.715	54.712
600	190.790	334.272	245.237	53.421	-709.197	-486.213	42.329
700	194.974	363.996	260.126	72.709	-719.884	-446.962	33.353
800	199.158	390.305	274.785	92.416	-715.902	-408.241	26.655
900	203.342	414.004	288.958	112.541	-711.544	-370.041	21.477
1000	207.526	435.645	302.560	133.084	-706.789	-332.348	17.360
1100	211.710	455.620	315.578	154.046	-701.619	-295.150	14.016
1200	215.894	474.221	328.032	175.426	-696.018	-258.442	11.250
1300	220.078	491.667	339.955	197.225	-689.997	-222.219	8.929
1400	224.262	508.130	351.385	219.442	-683.595	-186.474	6.957
1500	228.446	523.745	362.360	242.078	-676.835	-151.199	5.265

PREVIOUS: March 1962 CURRENT: December 1971

Lead Oxide (Pb_3O_4) $O_4Pb_3(cr)$

ZINC SULFATE ($ZnSO_4$) CRYSTAL M_r = 161.4376

$S°(298.15 K) = 110.541 \pm 0.8 J K^{-1} mol^{-1}$
$T_{trs1}(\alpha \to \alpha'$, proposed) = [540] K
$T_{trs2}(\alpha' \to \beta) = 1013 \pm 5 K$

$\Delta_f H°(0 K) = -969.934 \pm 0.10 kJ mol^{-1}$
$\Delta_f H°(298.15 K) = -980.144 \pm 0.10 kJ mol^{-1}$
$\Delta_{trs1}H°(\alpha \to \alpha') = [5.021] kJ mol^{-1}$
$\Delta_{trs2}H°(\alpha' \to \beta) = 20.334 \pm 0.8 kJ mol^{-1}$

Enthalpy of Formation

Adami and King (1) determined calorimetrically the enthalpies of solution of crystalline ZnO and $ZnSO_4$ in hydrochloric acid solution. From these measurements they derived the enthalpy for the reaction ZnC(cr) + $H_2SO_4 \cdot 7.068 H_2O(\ell) \to ZnSO_4(cr) + 8.068 H_2O(\ell)$ as $\Delta_r H°(298.15 K) = -9.250 \pm 0.080 kcal mol^{-1}$. Combining this result with the enthalpies of formation of ZnO (2), $H_2SO_4 \cdot 7.068 H_2O$ (3, 4), and H_2O (4), we derive $\Delta_f H°(ZnSO_4$, cr, 298.15 K) $-234.26 \pm 0.25 kcal mol^{-1}$ which is the value adopted for this tabulation.

The enthalpy of solution at infinite dilution, $\Delta_{sol}H°(ZnSO_4$, cr, 298.15 K) = $-19.9 \pm 0.2 kcal mol^{-1}$ was taken from the work of Larson et al. (5). When this result is combined with enthalpies of formation of the infinitely dilute ions from (4), we obtain $\Delta_f H°(ZnSO_4$, aq, ∞, 298.15 K) = $-254.06 \pm 0.14 kcal mol^{-1}$ and $\Delta_f H°(ZnSO_4$, cr, 298.15 K) = $-234.16 \pm 0.34 kcal mol^{-1}$.

Ishikawa and Murooka (6) determined $\Delta_r H°(298.15 K) = -56.330 kcal mol^{-1}$ from cell measurements for the reaction Zn(cr) + $Hg_2SO_4(cr) = ZnSO_4(cr) + 2 Hg(\ell)$ as reported by Kelley (7). When this result is combined with JANAF auxiliary data for Zn(cr) (8) and with data from CODATA (9) for $Hg(\ell)$ and $Hg_2SO_4(cr)$, we obtain $\Delta_f H°(ZnSO_4$, cr, 298.15 K) = $-234.01 \pm 1.0 kcal mol^{-1}$.

Heat Capacity and Entropy

Low temperature heat capacities of $ZnSO_4(cr, \alpha)$ have been measured by Weller (10) from 51.7 - 296.5 K. A small heat capacity maximum was observed at 124.37 K. Our adopted value of S°(298.15 K) = $26.42 \pm 0.3 cal K^{-1} mol^{-1}$ obtained from $C_p°$ is based on S°(51 K) = $2.27 cal K^{-1} mol^{-1}$ obtained by Weller (10) by extrapolation of the measured heat capacity with a combination of Debye and Einstein functions. We have smoothed the data of Weller (10) by fitting the data with orthogonal polynomials over selected overlapping temperature intervals.

High temperature enthalpy data have been measured by Hosmer and Krikorian (11, 510 - 1168 K) as well as by Voskresenskaya and Patsukova (12, 523-806 K), Krestovnikov and Feigina (13, 573-1073 K), and by Chizhikov and Khirik (14, 373-973 K). The enthalpy data of Hosmer and Krikorian (11) is higher than the data of Voskresenskaya and Patsukova (12) or Krestovnikov and Feigina (13). Wet-chemical and x-ray analysis of their sample (11) indicate that the sample was pure zinc sulfate. The lower results of (12) and (13) may be due to decomposition to the oxy-sulfate which can occur in vacuum dehydration (11). We adopt the enthalpy results from Hosmer and Krikorian (11). Problems remain in matching the low temperature heat capacity data to the high temperature enthalpy data (300 - 600 K region). There are at least three possibilities for the mismatch: 1) an unobserved transition occurs between 300-600 K, 2) the data of (10) and (11) are essentially correct but were measured on different materials, or 3) a bias exists in the data from one of the calorimeters. In the absence of other information, we arbitrarily adopt a transition at 540 K with $\Delta_{trs}H° = 1.20 kcal mol^{-1}$. Below 540 K, we adopt heat capacities based on an extrapolation of the work of Weller (10). Above 540 K, we adopt heat capacities based on the enthalpy data of Hosmer and Krikorian (11).

Transition Data

$ZnSO_4(cr, \alpha)$ is the low-temperature form of zinc sulfate, also known as the mineral zinkosite and $ZnSO_4$(cr, II). The crystals are orthorhombic, space group D_{2h}^{16} - Pnma (15). The transition to the high-temperature (β-phase) cubic form (space group F43m, 16) occurs at 1013±5 K (15). The heat of transition is based on the drop calorimetry measurement of Hosmer and Krikorian (11) which is in good agreement with the DTA measurement of Ingraham and Marier (17, 4.74 kcal mol^{-1}). Decomposition occurs through an oxy-sulfate intermediate which has the probable composition ZnO · 2 $ZnSO_4$ (11).

Hosmer and Krikorian (11) have also measured enthalpies of the high-temperature $ZnSO_4(cr, \beta)$ phase of zinc sulfate. We adopt their results (1038 - 1168 K), which imply a constant heat capacity, and extrapolate this value to 2000 K.

References

1. L. H. Adami and E. G. King, U. S. Bur. Mines RI 6617, 965.
2. R. A. Robie, B. S. Hemingway and J. R. Fisher, U. S. Geol. Sur. Bull. 1452, 1978.
3. U. S. Nat. Bur. Stand. Tech. Note 270-3, 1968.
4. U. S. Nat. Bur. Stand. Interim Report, 75-968, 1976.
5. J. W. Larson, P. Cerutti, H. K. Garber, and L. G. Hepler, J. Phys. Chem. 72, 2902 (1968).
6. F. Ishikawa and T. Murooka, Sci. Rept. Tohoku Imp. Univ. 22, 138 (1933).
7. K. K. Kelley, U. S. Bur. Mines Bull. 406, 1937.
8. JANAF Thermochemical Tables: Zn(cr), 12-31-78.
9. ICSU-CODATA Task Group on Key Values for Thermodynamics, J. Chem. Thermodyn. 10, 903 (1978).
10. W. W. Weller, U. S. Bur. Mines RI 6669, 1965.
11. P. K. Hosmer and O. H. Krikorian, UCRL Preprint 84562, paper submitted to "Proceedings of the 7th European Thermophysical Properties Conference, June 30, 1980, Antwerp, Belgium.
12. N. K. Voskresenskaya and N. N. Patsukova, Izvest. Sektora Fiz.-Khim. Anal., Inst. Obshchei i Neorg. Khim., Akad. Nauk. S.S.S.R. 25, 159 (1954); see Chem. Abstr. 49, 14460 (1955).
13. A. N. Krestovnikov and E. I. Feigina, J. Gen. Chem. USSR 6, 1481 (1936); Chem. Abstr. 31, 2079 (1937).
14. D. M. Chizhikov and A. S. Khirik, Tr. Inst. Met. Im. A. A. Baikova, Akad. Nauk SSSR 1963, 79 (1963); Chem. Abstr. 59, 5800 (1963).
15. C. N. R. Rao and B. Prakash, NSRDS-NBS 56, 1975.
16. M. Spiess and R. Gruehn, Naturwissenschaffen 65, 594 (1978).
17. T. R. Ingraham and P. Marier, Can. Met. Quart. 4, 169 (1965).

Zinc Sulfate ($ZnSO_4$) $O_4S_1Zn_1(cr)$

Enthalpy Reference Temperature = T_r = 298.15 K; Standard State Pressure = p° = 0.1 MPa

T/K	$C_p°$ (J K⁻¹mol⁻¹)	S° (J K⁻¹mol⁻¹)	$-[G°-H°(T_r)]/T$ (J K⁻¹mol⁻¹)	$H°-H°(T_r)$ (kJ mol⁻¹)	$\Delta_f H°$ (kJ mol⁻¹)	$\Delta_f G°$ (kJ mol⁻¹)	Log K_f
0	0.000	0.000	INFINITE	-17.238	-969.934	-969.934	INFINITE
100	47.585	31.581	183.544	-15.196	-975.390	-940.983	491.519
200	78.053	75.224	118.947	-8.745	-978.636	-905.195	236.413
298.15	99.035	110.541	110.541	0.000	-980.144	-868.777	152.206
300	99.370	111.155	110.543	0.184	-980.158	-868.086	151.147
400	116.022	142.074	114.647	10.971	-982.499	-830.588	108.464
500	131.210	169.620	122.933	23.344	-982.857	-792.557	82.798
540.000	136.817	179.933	126.775	28.705	I <--> II TRANSITION		
540.000	136.817	189.231	126.775	33.726			
600	137.444	203.678	133.755	41.954	-976.948	-755.128	65.740
700	138.574	224.950	145.301	55.755	-983.260	-718.162	53.590
800	139.704	243.528	156.442	69.668	-982.369	-680.351	44.422
900	140.875	260.050	167.053	83.698	-1034.546	-641.575	37.236
1000	142.005	274.954	177.110	97.844	-1032.317	-598.031	31.238
1013.000	142.130	276.789	178.377	99.691	ALPHA <--> BETA TRANSITION		
1013.000	145.185	296.862	178.377	120.025			
1100	145.185	308.825	188.228	132.656	-1009.533	-556.465	26.424
1200	145.185	321.457	198.812	147.175	-1122.238	-513.448	22.350
1300	145.185	333.078	208.699	161.693	-1118.867	-462.853	18.598
1400	145.185	343.838	217.972	176.212	-1115.572	-412.515	15.391
1500	145.185	353.854	226.701	190.730	-1112.348	-362.410	12.620
1600	145.185	363.224	234.944	205.249	-1109.190	-312.517	10.203
1700	145.185	372.026	242.751	219.767	-1106.097	-262.820	8.075
1800	145.185	380.325	250.166	234.286	-1103.065	-213.304	6.190
1900	145.185	388.174	257.225	248.804	-1100.093	-163.954	4.507
2000	145.185	395.622	263.960	263.323	-1097.180	-114.759	2.997

PREVIOUS: CURRENT: March 1979

Zinc Sulfate ($ZnSO_4$) $O_4S_1Zn_1(cr)$

ZIRCONIUM SILICATE ($ZrSiO_4$) CRYSTAL M_r = 183.3031

$\Delta_f H°(0\ K) = -2012.774 \pm 2.9\ kJ\ mol^{-1}$

$S°(298.15\ K) = 84.026 \pm 1.3\ J\ K^{-1}\ mol^{-1}$ $\Delta_f H°(298.15\ K) = -2023.951 \pm 2.9\ kJ\ mol^{-1}$

Enthalpy of Formation

$\Delta_f H°(ZrSiO_4$, cr, 298.15 K) was calculated from the $\Delta_r H°$(298.15 K) of its constituent oxides $ZrO_2(cr) + SiO(cr) = ZrSiO_4(cr)$ and the $\Delta_f H°$(298.15 K) values for ZrO_2(cr) and SiO_2(cr) in JANAF tables (4). The $\Delta_r H°$(298.15 K) = -4.736 kcal mol^{-1} was based on the Gibbs energy of formation of its constituent oxides at the decomposition temperature 1811 K.

Heat Capacity and Entropy

At low temperatures, $C_p°$ were used from Kelley (1). From 298-940 K, $C_p°$ was calculated by using the Coughlin and King (2) equation $C_p° = 31.48 + 3.92 \times 10^{-3}T - 8.08 \times 10^5 T^{-2}$ which joined smoothly with the low temperature data. Above 940 K the data were estimated by extending the curve smoothly. The entropy was calculated at 52.7 K using the Debye and Einstein (3) functions D(321/T) + 2E(407/T) + 2E(796/T) + E(1624/T) given by Kelley (2). S°(52.7 K) = 0.65404 cal $K^{-1}\ mol^{-1}$.

Decomposition Data

T_{dcm} is the temperature at which $\Delta_r G° = 0$ for the reaction $ZrSiO_4(cr) \rightarrow ZrO_2(cr) + SiO_2(cr)$. T_{dcm} = 1811 K was taken from Curtis and Sowman (3).

References

1. K. K. Kelley, J. Amer. Chem. Soc. 63, 2750 (1941).
2. J. P. Coughlin and E. G. King, J. Amer. Chem. Soc. 72, 2262 (1950).
3. C. E. Curtis and H. G. Sowman, J. Amer. Chem. Soc. 36, 190 (1953).
4. JANAF Thermochemical Tables: ZrO_2(cr), 6-30-61; SiO_2(cr), 12-31-62.

Zirconium Silicate ($ZrSiO_4$) $O_4Si_1Zr_1(cr)$

Enthalpy Reference Temperature = T_r = 298.15 K Standard State Pressure = p° = 0.1 MPa

T/K	$C_p°$ (J K⁻¹mol⁻¹)	S° (J K⁻¹mol⁻¹)	-[G°-H°(T_r)]/T (J K⁻¹mol⁻¹)	H°-H°(T_r) (kJ mol⁻¹)	$\Delta_f H°$ (kJ mol⁻¹)	$\Delta_f G°$ (kJ mol⁻¹)	Log K_f
0	0.000	0.000	INFINITE	-14.905	-2012.774	-2012.774	INFINITE
100	31.034	14.070	153.089	-13.902	-2018.733	-1983.690	1036.173
200	73.298	49.742	92.291	-8.510	-2022.535	-1946.946	508.490
298.15	98.659	84.026	84.026	0.000	-2023.951	-1909.475	334.532
300	99.119	84.638	84.028	0.183	-2023.961	-1908.764	332.345
400	114.642	115.379	88.114	10.906	-2023.861	-1870.349	244.243
500	125.938	142.241	96.311	22.965	-2022.823	-1832.076	191.396
600	133.679	165.925	105.980	35.967	-2021.207	-1794.071	156.188
700	138.825	186.935	116.073	49.603	-2019.294	-1756.364	131.061
800	142.674	205.738	126.127	63.689	-2017.241	-1718.942	112.235
900	145.268	222.698	135.931	78.091	-2015.160	-1681.779	97.608
1000	147.277	238.112	145.390	92.723	-2013.119	-1644.847	85.918
1100	148.741	252.222	154.470	107.528	-2011.159	-1608.115	76.363
1200	149.704	265.209	163.164	122.454	-2013.031	-1571.335	68.398
1300	150.164	277.211	171.481	137.449	-2010.811	-1534.618	61.662
1400	150.415	288.349	179.436	152.478	-2008.702	-1498.068	55.894
1500	150.624	298.735	187.047	167.532	-2006.718	-1461.664	50.900
1600	150.624	308.456	194.335	182.595	-2004.883	-1425.388	46.534
1700	150.624	317.588	201.319	197.657	-2053.390	-1388.775	42.672
1800	150.624	326.197	208.020	212.719	-2051.665	-1349.731	39.168
1900	150.624	334.341	214.456	227.782	-2050.081	-1310.778	36.036
2000	150.624	342.067	220.645	242.844	-2048.650	-1271.905	33.219
2100	150.624	349.416	226.603	257.907	-2047.382	-1233.100	30.672
2200	150.624	356.423	232.346	272.969	-2067.609	-1193.604	28.340
2300	150.624	363.118	237.887	288.031	-2067.111	-1153.889	26.206
2400	150.624	369.529	243.240	303.094	-2066.658	-1114.194	24.250
2500	150.624	375.678	248.415	318.156	-2066.249	-1074.516	22.451
2600	150.624	381.585	253.424	333.219	-2065.883	-1034.854	20.790
2700	150.624	387.270	258.277	348.281	-2065.559	-995.205	19.253
2800	150.624	392.748	262.982	363.343	-2065.275	-955.567	17.826
2900	150.624	398.033	267.549	378.406	-2065.032	-915.940	16.498
3000	150.624	403.140	271.984	393.468	-2064.827	-876.320	15.258

PREVIOUS: September 1961 CURRENT: June 1965

Zirconium Silicate ($ZrSiO_4$) $O_4Si_1Zr_1(cr)$

VANADIUM OXIDE (V_2O_4) CRYSTAL M_r = 165.8806

$S°(298.15\ K) = 103.525 \pm 2.1\ J\ K^{-1}\ mol^{-1}$

$T_{trs} = 340 \pm 5\ K$

$T_{fus} = 1818 \pm 15\ K$

$\Delta_f H°(0\ K) = -1418.617 \pm 6.3\ kJ\ mol^{-1}$

$\Delta_f H°(298.15\ K) = -1427.162 \pm 6.3\ kJ\ mol^{-1}$

$\Delta_{trs} H° = 8.996 \pm 0.8\ kJ\ mol^{-1}$

$\Delta_{fus} H° = 112.068 \pm 2.5\ kJ\ mol^{-1}$

Enthalpy of Formation

The adopted $\Delta_f H°$ values for the vanadium oxides, and in particular V_2O_4(cr) are based on the combustion studies by Mah and Kelley (1). The adopted value is -341.1 kcal mol^{-1} which was obtained by rounding the reported value of Mah and Kelley (1). For more details refer to the V_2O_5(cr) table (2). Charlu and Kleppa (3) reported a enthalpy of formation value at $\Delta_f H°$(298.15 K) = -342.4 ± 0.78 kcal mol^{-1} based on oxidation studies to V_2O_5(cr) in a high temperature microcalorimeter. The combustion by this technique yielded complete oxidation to V_2O_5(cr) as opposed to the method of Mah and Kelley (1), whereby $\Delta_f H°$ for V_2O_4(cr) was determined simultaneously with that for V_2O_5(cr) due to incomplete combustion of V(cr). A combustion study by Siemonsen and Ulich (4) led to the reported value of -342 ± 2 kcal mol^{-1} for $\Delta_f H°(V_2O_4$, cr, 293 K). Additional thermodynamic data which relate V_2O_3(cr) with V_2O_4(cr) or V_2O_5(cr) is contained in the sodium oxide fusion studies by Mixter (5), the H_2O-H_2 equilibrium study by Muller (6), the CO-CO_2 equilibrium study by Spencer and Justice (7), and the reduction of V_2O_5(cr) with SO_2 by Flood and Kleppa (8). See V_2O_3(cr) table for some additional information (2).

Heat Capacity and Entropy

Anderson (9) measured the heat capacity of V_2O_4 (reported purity of 99.5% or better) in the range 61.4-279.4 K. Using the combination of Debye and Einstein functions suggested by Anderson (9), we calculate S°(60 K) = 1.48 cal $K^{-1} mol^{-1}$ and H°(60;0) = 63.66 cal K^{-1} mol^{-1}. The data of Anderson (9) show considerable scatter from our adopted values, with a ±2.0% deviation being typical for this data. Cook (10) measured the enthalpy of V_2O_4 in the range 315-1856 K. The enthalpy data for T < T_{trs} appears too high to be fully consistent with $C_p°$ data of Anderson (9). The adopted $C_p°$ values in the region 280-340 K join smoothly with a positive power polynomial fit of the data of Anderson (4) and lead to enthalpy values at 315.5, 322.8, and 323.0 K which are roughly 48, 34, and 39 kcal mol^{-1}, respectively, less than the data of Cook (10). This is a difference of 5-10%. The enthalpy data in the region T_{trs}-T_{fus} are fit to a Kelley type equation and then differentiated to yield our adopted $C_p°$ values. The data of Cook (10) also shows considerable scatter, with deviations ranging from -0.69% to +1.12%.

Three $C_p°$ studies concentrated on the region near T_{trs} but presented only graphical data (11, 12, 13). Kawakubo and Nakagawa (11) reported $C_p°$ values between 30°C and 90°C which are roughly 18% higher than our adopted values below the transition and 25% higher above the transition. Ryder et al. (12) and Chandrashekhar et al. (13) measured $C_p°$ values in the range 40-375 K and 150-730 K. These latter two works agree well graphically in their region of overlap. The $C_p°$ values above T_{trs} are very similar to our adopted values; at 500 K the adopted values are less than 0.1 cal K^{-1} mol^{-1} higher while at 700 K, roughly 0.75 cal K^{-1} mol^{-1} lower. Below T_{trs}, the differences are more substantial; the adopted values lying approximately 1.5 cal K^{-1} mol^{-1} higher.

Phase Data

Minomura and Nagasaki (14) referenced earlier x-ray data and provided additional data to show that V_2O_4 undergoes a structural distortion to lower symmetry on cooling through a transition temperature of 340 K; tetragonal rutile to a monoclinic structure with the unit cell doubling in size. Recent crystallographic investigations by Kawada et al. (15) verified the doubling of the size of the unit cell and also detected considerable distortion of the shape of the unit cell. The literature review on the V-O system by Stringer (16) indicated that there are a series of oxides between V_3O_5 and V_2O_4. These oxides may be considered as members of a homologous series with $V_2O_4(VO_2)$ being the series limit (16). Numerous references on phase data are contained in the reviews by Stringer (16), Adler (17) and Neuberger (18).

Transition Data

Minomura and Nagasaki (14) repored that the structural distortion occurs at 340 K. This value is consistent with the enthalpy data of Cook (10). Kawakubo and Nakagawa (11) measured the temperature dependence of five physical properties through the transition point and concluded that the transition occurs at 338 K. We adopt T_{trs} = 340 + 5 K.

Using our adopted enthalpy values for T < T_{trs} and T > T_{trs}, we calculate and adopt $\Delta_{fus}H°$ = 2.15 ± 0.20 kcal mol^{-1}. Ryder et al. (12) reported a value of 2.040 ± 0.010 kcal mol^{-1} and Chandrashekhar et al. (13), a value of 2.24 kcal mol^{-1}. The values of $\Delta_{fus}H°$ = 1.60 and 1.50 kcal mol^{-1} reported by Klemm and Grimm (19) and Kawakubo and Nakagawa (11), respectively, are considered too low. In our tabulation, the transition is treated as being first order. Controversy exists concerning the order of the transition, as discussed by Adler (17). Additional references may be found in Stringer (16), Adler (17) and Neuberger (18).

Fusion Data

See the liquid table for details.

References

References are listed on the liquid table.

Vanadium Oxide (V_2O_4) O_4V_2(cr)

Enthalpy Reference Temperature = T_r = 298.15 K — Standard State Pressure = p° = 0.1 MPa

T/K	$C_p°$ (J K⁻¹mol⁻¹)	S° (J K⁻¹mol⁻¹)	$-[G°-H°(T_r)]/T$ (J K⁻¹mol⁻¹)	$H°-H°(T_r)$ (kJ mol⁻¹)	$\Delta_f H°$ (kJ mol⁻¹)	$\Delta_f G°$ (kJ mol⁻¹)	Log K_f
0	0.000	0.000	INFINITE	-18.100	-1418.617	-1418.617	INFINITE
100	36.907	18.652	186.891	-16.824	-1424.127	-1389.894	726.006
200	90.274	62.333	113.357	-10.205	-1427.002	-1354.252	353.694
298.15	115.399	103.525	103.525	0.000	-1427.162	-1318.445	230.986
300	115.787	104.240	103.527	0.214	-1427.150	-1317.770	229.444
340.000	123.336	119.218	104.501	5.004	I <--> II TRANSITION		
340.000	127.621	145.675	104.501	13.999			
400	135.281	167.065	112.309	21.902	-1416.532	-1283.094	167.555
500	143.269	198.162	126.456	35.853	-1414.020	-1250.016	130.588
600	148.448	224.758	140.677	50.448	-1411.187	-1217.477	105.991
700	152.175	247.927	154.379	65.483	-1408.212	-1185.426	88.458
800	155.457	268.466	167.380	80.868	-1405.169	-1153.807	75.336
900	158.157	286.934	179.655	96.551	-1402.101	-1122.571	65.152
1000	160.682	303.729	191.236	112.494	-1399.025	-1091.677	57.023
1100	162.987	319.153	202.173	128.678	-1395.953	-1061.091	50.387
1200	165.209	333.430	212.523	145.089	-1392.908	-1030.784	44.869
1300	167.330	346.738	222.341	161.716	-1389.898	-1000.729	40.210
1400	169.406	359.215	231.677	178.553	-1386.944	-970.904	36.225
1500	171.429	370.972	240.575	195.595	-1384.045	-941.289	32.779
1600	173.427	382.100	249.076	212.838	-1381.201	-911.865	29.769
1700	175.302	392.669	257.214	230.273	-1378.439	-882.616	27.119
1800	177.343	402.746	265.022	247.904	-1375.760	-853.528	24.769
1818.000	177.728	404.513	266.394	251.100	-------- II <--> LIQUID --------		
1900	179.276	412.387	272.526	265.737	-1373.179	-824.586	22.669
2000	181.192	421.632	279.752	283.760	-1370.723	-795.777	20.784
2100	183.092	430.518	286.721	301.975	-1368.413	-767.087	19.080
2200	184.975	439.079	293.453	320.378	-1412.030	-738.293	17.529

PREVIOUS: June 1973 CURRENT: December 1973

Vanadium Oxide (V_2O_4) O_4V_2(cr)

VANADIUM OXIDE (V_2O_4) LIQUID M_r = 165.8806

S°(298.15 K) = [173.473] J K^{-1} mol^{-1}
T_{fus} = 1818 ± 15 K

$\Delta_f H°$(298.15 K) = [-1332.151] kJ mol^{-1}
$\Delta_{fus}H°$ = 112.068 ± 2.5 kJ mol^{-1}

Enthalpy of Formation

$\Delta_f H°(V_2O_4$, ℓ, 298.15 K) is calculated from that of the crystal by adding $\Delta_{fus}H°$ and the difference in enthalpy, H°(1818 K)-H°(298.15 K), between V_2O_4(cr) and V_2O_4(ℓ).

Heat Capacity and Entropy

Cook (1) measured the enthalpy of V_2O_4(ℓ) in the region 1829.7-1856.8 K and represented the data by the equation H°(T)-H°(298.15 K) = 51.00 T-5910 cal K^{-1} mol^{-1}. Thus $C_p°$ for the liquid is adopted as 51.00 cal K^{-1} mol^{-1}. Since the enthalpy measurements only cover a range of 39 K in the liquid region, the value of $C_p°$ should not be viewed as a precise value. A glass transition is assumed at 1200 K. Below 1200 K, the $C_p°$ values are those of the crystal with the exception that the values of $C_p°$ between 298 and 340 K are a smooth extrapolation of the crystal values above T_{trs}. The entropy at 298.15 K is calculated in a manner analogous to that used for the enthalpy of formation.

Fusion Data

The melting point of V_2O_4(cr) is adopted as 1818 K, based on the enthalpy data of Cook (1). The enthalpy of melting is calculated from the difference in H°(1818 K)-H°(298.15 K) for V_2O_4(ℓ) and V_2O_4(cr).

References - V_2O_4(ℓ)
1. O. A. Cook, J. Amer. Chem. Soc. 69, 331 (1947).

References - V_2O_4(cr)
1. A. D. Mah and K. K. Kelley, U. S. Bur. Mines RI 5858, 1961.
2. JANAF Thermochemical Tables: V_2O_5(cr), 6-30-73; V_2O_3(cr), 12-31-73.
3. F. V. Charlu and O. J. Kleppa, High Temp. Soc. 5, 260 (1973).
4. H. Siemonsen and H. Ulich, Z. Electrochem. 46, 141 (1940).
5. W. G. Mixter, Amer. J. Sci. 34, 141 (1912).
6. H. Muller, W. Biltz's Festschrift, Hannover, 1927; quoted by J. W. Mellor, "Comprehensive Treatise on Inorganic and Theoretical Chemistry," Vol. IX, Longman, Green and Co., London, 1929.
7. H. M. Spencer and J. L. Justice, J. Amer. Chem. Soc. 56, 2306 (1934).
8. H. Flood and O. J. Kleppa, J. Amer. Chem. Soc. 69, 998 (1947).
9. C. T. Anderson, J. Amer. Chem. Soc. 58, 564 (1936).
10. O. A. Cook, J. Amer. Chem. Soc. 69, 331 (1947).

Vanadium Oxide (V_2O_4) O_4V_2(l)

Enthalpy Reference Temperature = T_r = 298.15 K Standard State Pressure = p° = 0.1 MPa

T/K	$C_p°$ (J K⁻¹mol⁻¹)	S°	-[G°-H°(T_r)]/T	H°-H°(T_r) (kJ mol⁻¹)	$\Delta_f H°$	$\Delta_f G°$	Log K_f
0							
100							
200							
298.15	120.081	173.473	173.473	0.000	-1332.151	-1244.289	217.994
300	120.499	174.217	173.475	0.223	-1332.130	-1243.744	216.555
400	135.281	211.142	178.409	13.093	-1330.330	-1214.523	158.600
500	143.269	242.239	188.151	27.044	-1327.818	-1185.852	123.885
600	148.448	268.835	199.436	41.639	-1324.985	-1157.722	100.789
700	152.175	292.004	211.041	56.674	-1322.010	-1130.079	84.327
800	155.457	312.543	222.469	72.060	-1318.967	-1102.867	72.010
900	158.157	331.011	233.520	87.742	-1315.899	-1076.039	62.452
1000	160.682	347.807	244.122	103.685	-1312.823	-1049.552	54.823
1100	163.033	363.233	254.258	119.872	-1309.748	-1023.374	48.596
1200	165.209	377.513	263.942	136.286	-1306.700	-997.475	43.419
1200.000	165.209	377.513	263.942	136.286	GLASS <--> LIQUID TRANSITION		
1200.000	213.384	377.513	263.942	136.286			
1300	213.384	394.593	273.344	157.624	-1298.979	-972.021	39.056
1400	213.384	410.407	282.576	178.963	-1291.523	-947.152	35.339
1500	213.384	425.129	291.595	200.301	-1284.329	-922.807	32.135
1600	213.384	438.900	300.375	221.639	-1277.389	-898.933	29.347
1700	213.384	451.836	308.908	242.978	-1270.723	-875.485	26.900
1800	213.384	464.033	317.191	264.316	-1264.336	-852.422	24.737
1818.000	213.384	466.156	318.655	268.157	-------- II <--> LIQUID --------		
1900	213.384	475.570	325.226	285.655	-1258.250	-829.705	22.810
2000	213.384	486.515	333.019	306.993	-1252.480	-807.301	21.085
2100	213.384	496.926	340.578	328.331	-1247.045	-785.177	19.530
2200	213.384	506.853	347.912	349.670	-1287.728	-763.093	18.118
2300	213.384	516.338	355.030	371.008	-1283.291	-739.346	16.791
2400	213.384	525.420	361.942	392.347	-1278.900	-715.792	15.579
2500	213.384	534.131	368.657	413.685	-1274.552	-692.418	14.467
2600	213.384	542.500	375.183	435.023	-1270.247	-669.218	13.445
2700	213.384	550.553	381.530	456.362	-1265.984	-646.182	12.501
2800	213.384	558.313	387.706	477.700	-1261.761	-623.303	11.628
2900	213.384	565.801	393.719	499.039	-1257.579	-600.576	10.818
3000	213.384	573.035	399.576	520.377	-1253.435	-577.991	10.064

PREVIOUS: CURRENT: June 1973

Vanadium Oxide (V_2O_4) O_4V_2(l)

VANADIUM OXIDE (V_2O_4) CRYSTAL–CRYSTAL–LIQUID M_r = 165.8806

0 to 340 K	crystal, I
340 to 1818 K	crystal, II
above 1818 K	liquid

Refer to the individual tables for details.

Vanadium Oxide (V_2O_4) O_4V_2(cr,l)

Enthalpy Reference Temperature = T_r = 298.15 K; Standard State Pressure = p° = 0.1 MPa

T/K	C_p° (J K⁻¹mol⁻¹)	S° (J K⁻¹mol⁻¹)	$-[G^\circ-H^\circ(T_r)]/T$ (J K⁻¹mol⁻¹)	$H^\circ-H^\circ(T_r)$ (kJ mol⁻¹)	Δ_fH° (kJ mol⁻¹)	Δ_fG° (kJ mol⁻¹)	Log K_f
0	0.000	0.000	INFINITE	-18.100	-1418.617	-1418.617	INFINITE
100	36.907	18.652	186.891	-16.824	-1424.127	-1389.894	726.006
200	90.274	62.333	113.357	-10.205	-1427.002	-1354.252	353.694
298.15	115.399	103.525	103.525	0.000	-1427.162	-1318.445	230.986
300	115.787	104.240	103.527	0.214	-1427.150	-1317.770	229.444
340.000	123.336	119.218	104.501	5.004		I <--> II TRANSITION	
340.000	127.621	145.675	104.501	13.999			
400	135.281	167.065	112.309	21.902	-1416.532	-1283.094	167.555
500	143.269	198.162	126.456	35.853	-1414.020	-1250.016	130.588
600	148.448	224.758	140.677	50.448	-1411.187	-1217.477	105.991
700	152.175	247.927	154.379	65.483	-1408.212	-1185.426	88.458
800	155.457	268.466	167.380	80.868	-1405.169	-1153.807	75.336
900	158.157	286.934	179.655	96.551	-1402.101	-1122.571	65.152
1000	160.682	303.729	191.236	112.494	-1399.025	-1091.677	57.023
1100	162.987	319.153	202.173	128.678	-1395.953	-1061.091	50.387
1200	165.209	333.430	212.523	145.089	-1392.908	-1030.784	44.869
1300	167.330	346.738	222.341	161.716	-1389.898	-1000.729	40.210
1400	169.406	359.215	231.677	178.553	-1386.944	-970.904	36.225
1500	171.429	370.972	240.575	195.595	-1384.045	-941.289	32.779
1600	173.427	382.100	249.076	212.838	-1381.201	-911.865	29.769
1700	175.302	392.669	257.214	230.273	-1378.439	-882.616	27.119
1800	177.343	402.746	265.022	247.904	-1375.760	-853.528	24.769
1818.000	177.728	404.513	266.394	251.100		II <--> LIQUID TRANSITION	
1818.000	213.384	466.156	266.394	363.168			
1900	213.384	475.570	275.220	380.666	-1258.250	-829.705	22.810
2000	213.384	486.515	285.513	402.004	-1252.480	-807.301	21.085
2100	213.384	496.926	295.335	423.342	-1247.045	-785.177	19.530
2200	213.384	506.853	304.725	444.681	-1287.728	-763.093	18.118
2300	213.384	516.338	313.721	466.019	-1283.291	-739.346	16.791
2400	213.384	525.420	322.354	487.358	-1278.900	-715.792	15.579
2500	213.384	534.131	330.652	508.696	-1274.552	-692.418	14.467
2600	213.384	542.500	338.640	530.034	-1270.247	-669.218	13.445
2700	213.384	550.553	346.341	551.373	-1265.984	-646.182	12.501
2800	213.384	558.313	353.773	572.711	-1261.761	-623.303	11.628
2900	213.384	565.801	360.956	594.050	-1257.579	-600.576	10.818
3000	213.384	573.035	367.906	615.388	-1253.435	-577.991	10.064

PREVIOUS: CURRENT: December 1973

Vanadium Oxide (V_2O_4) O_4V_2(cr,l)

$S°(298.15\ K) = 143.114 \pm 1.3\ J\ K^{-1}\ mol^{-1}$
$T_{fus} = 2058 \pm 30\ K$

$\Delta_f H°(0\ K) = -2035.947 \pm 4.2\ kJ\ mol^{-1}$
$\Delta_f H°(298.15\ K) = -2045.976 \pm 4.2\ kJ\ mol^{-1}$
$\Delta_{fus} H° = [120.081 \pm 16.7]\ kJ\ mol^{-1}$

Enthalpy of Formation

$\Delta_f H°(Ta_2O_5$, cr, 298.15 K) values show scatter due to uncertainties concerning the polymorphic state of the sample and/or the incomplete specification of sample purity (1). The reported values for $\Delta_f H°(Ta_2O_5$, cr, 298.15 K) based on bomb calorimetry studies are as follows:

Source	Ta(cr) purity, %	% run*	$\Delta_f H°(Ta_2O_5$, cr, 298.15 K) kcal mol^{-1}
(2)			309.5
(3)	99.96	>99.0	-498.3
(4)			-480.5
(5)	99.9	72.92-93.83	-486.0±0.5
(6)			-499.9±1.0
(7)	99.89	97.23-99.99	-489.0±0.5**
(8)	99.80		-487.7±0.9
(9)	99.16, 99.7	99.5 -99.9	-489.3±0.4
(10)			-480.0±0.8

*Refers to extent of reaction 2 Ta(cr) + 5/2 O_2(g) = Ta_2O_5(cr)
**The reported value of -488.8±0.5 was corrected by Huber et al. (8) in order to have the value based on the current atomic weight of Ta.

The value chosen for $\Delta_f H°(Ta_2O_5$, cr, 298.15 K) is -489.0 ± 0.5 based primarily on the work by Humphrey (7) and Kornilov et al. (9). Worrell (17), using a high temperature galvanic cell, obtained $\Delta_f G°$ values of Ta_2O_5 in the range 1050-1300 K which are approximately 0.6-1.1 kcal mol^{-1} more negative than the values tabulated here. Ignatowicz and Davies (12) using a galvanic cell approach reported $\Delta_f G°$ values for Ta_2O_5 in the range 1073-1373 K. Barbi (11, 13) determined $\Delta_f G°$ based on emf measurements in the range 1079-1776 K while Rezukhina and Kravchenko (16) worked in the range 1200-1350 K. A third law analysis of the reported smoothed results yields the following.

Source	$\Delta_f H°$(298.15 K), kcal mol^{-1}	Drift cal K^{-1} mol^{-1}
(11)	-434.8	0.85
(12)	-437.9	-0.65
(13)*	-434.7	+0.95
(16)	-437.9	+0.83

*Actual experimental data was analyzed.

These latter three works (12, 13, 16) did not specify purity of sample or the state of polymorphism.

Based on the recent critical review by Reisman and Holtzberg (1), this discussion has been concerned with the low temperature β-phase of Ta_2O_5. Even though two $\Delta_f H°$(298.15 K) studies (9, 10) apparently deal with the high temperature α-phase, there does not appear to be any difference in the $\Delta_f H°$(298.15 K) values for the α-phase or β-phase, within experimental error (1).

Heat Capacity and Entropy

Kelley (14) measured the heat capacity of Ta_2O_5(cr) from 53.4 K to 294.2 K and fitted the data (29 data points) with a combination of Debye and Einstein functions. These functions were used to calculate S°(50 K) = 2.47 cal K^{-1} mol^{-1}. Orr (15) measured the enthalpy of Ta_2O_5(cr) from 395.6 K to 1802.6 K (15 data points). Nine data points at the high temperature end of the Kelley data were fitted by a four term polynomial, and integrated to yield enthalpy values in the vicinity of 298.15 K. These enthalpy values were combined with the Orr data, fitted to a six term polynomial, and extrapolated to 2800 K. The complete Kelley data was also fitted by a six term polynomial. The Nb_2O_5 heat capacity and enthalpy data presented in these tables are treated in the same manner so as to maintain consistency between the results for Nb_2O_5 and Ta_2O_5.

The Ta_2O_5(cr) sample used by Kelley (14) and Orr (15) was the same. Presumably both were using the low temperature β-phase of Ta_2O_5, as discussed by Reisman and Holtzberg (1) and Orr (15). The enthalpy data of Orr (15) did not conclusively indicate any phase transitions, i.e., the β → α transition, in the vicinity of 1633 K, as reported by Reisman and Holtzberg (1).

Fusion Data

See the liquid table for details.

References

1. A. Reisman and F. Holtzberg, High Temperature Oxides 2, 217 (1970).
2. W. Muthmann, L. Weiss, and R. Riedelbanch, Ann. Chem. 55, 67, 82 (1907).
3. J. E. Moose and S. W. Parr, J. Amer. Chem. Soc. 46, 2656 (1924).
4. A. Sieverts, A. Gotta and S. Halberstadt, Z. Anorg. Chem. 187, 159 (1930).
5. G. Becker and W. A. Roth, Z. Phys. Chem. 167A, 16 (1933).
6. B. Neumann, C. Kroger and H. Kunz, Z. Anorg. Allg. Chem. 218, 379 (1934).
7. G. H. Humphrey, J. Amer. Chem. Soc. 76, 978 (1954).
8. E. J. Huber, Jr., E. L. Head, C. E. Holly, Jr., and A. L. Bowman, J. Phys. Chem. 67, 793 (1963).
9. A. N. Kornilov, V. Ya. Leonidov and S. M. Skuratov, Zh. Fiz. Khim. 38, 2008 (1964).
10. V. I. Smirnova and B. F. Ormont, Russian J. Phys. Chem. 30, 1327 (1956).
11. G. B. Barbi, J. Less-Common Metals 22, 487 (1970).
12. S. Ignatowicz and M. W. Davies, J. Less-Common Metals 15, 100 (1968).
13. G. B. Barbi, Z. Naturforsch. 25A, 1515 (1970).
14. K. K. Kelley, J. Amer. Chem. Soc. 62, 818 (1940).
15. R. L. Orr, J. Amer. Chem. Soc. 75, 2808 (1953).
16. T. N. Rezukhina and L. J. Kravchenko, J. Chem. Thermodyn. 4, 655 (1972).
17. W. L. Worrell, in "Thermodynamics," Vol. 1, International Atomic Energy Agency, Vienna, 1966.

Enthalpy Reference Temperature = T_r = 298.15 K; Standard State Pressure = p° = 0.1 MPa

T/K	$C_p°$ (J $K^{-1}mol^{-1}$)	S° (J $K^{-1}mol^{-1}$)	-[G°-H°(T_r)]/T (J $K^{-1}mol^{-1}$)	H°-H°(T_r) (kJ mol^{-1})	$\Delta_f H°$ (kJ mol^{-1})	$\Delta_f G°$ (kJ mol^{-1})	Log K_f
0	0.000	0.000	INFINITE	-23.041	-2035.947	-2035.947	INFINITE
100	58.082	37.108	243.530	-20.642	-2042.799	-1999.955	1044.669
200	108.311	94.161	154.766	-12.121	-2046.060	-1955.515	510.728
298.15	135.034	143.114	143.114	0.000	-2045.976	-1911.004	334.800
300	135.342	143.950	143.116	0.250	-2045.956	-1910.167	332.590
400	147.515	184.681	148.588	14.437	-2044.314	-1865.126	243.561
500	156.862	218.636	159.293	29.671	-2041.946	-1820.591	190.196
600	164.431	247.930	171.680	45.750	-2039.087	-1776.582	154.665
700	170.431	273.746	184.453	62.505	-2035.876	-1733.081	129.324
800	175.197	296.825	197.083	79.794	-2032.397	-1690.059	110.350
900	179.038	317.683	209.342	97.507	-2028.710	-1647.487	95.618
1000	182.753	336.753	221.144	115.610	-2024.820	-1605.335	83.854
1100	184.824	354.267	232.461	133.987	-2020.837	-1563.579	74.248
1200	187.071	370.448	243.294	152.584	-2016.807	-1522.187	66.259
1300	189.021	385.500	253.661	171.391	-2012.725	-1481.134	59.513
1400	190.740	399.572	263.586	190.380	-2008.590	-1440.398	53.742
1500	192.280	412.785	273.096	209.533	-2004.394	-1399.959	48.751
1600	193.673	425.239	282.220	228.831	-2000.138	-1359.802	44.393
1700	194.953	437.020	290.982	248.263	-1995.845	-1319.912	40.556
1800	196.138	448.197	299.409	267.819	-1991.558	-1280.276	37.153
1900	197.246	458.831	307.522	287.488	-1987.315	-1240.876	34.114
2000	198.288	468.975	315.343	307.265	-1983.130	-1201.698	31.385
2058.000	198.870	474.652	319.753	318.783	------ CRYSTAL <--> LIQUID -----		
2100	199.278	478.674	322.891	327.144	-1979.005	-1162.728	28.921
2200	200.225	487.967	330.185	347.120	-1974.946	-1123.952	26.686
2300	201.133	496.887	337.240	367.188	-1970.962	-1085.361	24.649
2400	202.012	505.466	344.072	387.345	-1967.065	-1046.941	22.786
2500	202.859	513.730	350.694	407.589	-1963.269	-1008.681	21.075
2600	203.685	521.702	357.119	427.916	-1959.594	-970.570	19.499
2700	204.491	529.404	363.358	448.325	-1956.062	-932.597	18.042
2800	205.275	536.855	369.422	468.814	-1952.699	-894.753	16.692

PREVIOUS: CURRENT: December 1972

TANTALUM OXIDE (Ta_2O_5) LIQUID M_r = 441.89128

S°(298.15 K) = [183.050] J K^{-1} mol^{-1} $\Delta_f H$°(298.15 K) = [-1957.171] kJ mol^{-1}

T_{fus} = 2058 ± 30 K $\Delta_{fus}H$° = [120.081 ± 16.7] kJ mol^{-1}

Enthalpy of Formation

$\Delta_f H$°(Ta_2O_5, ℓ, 298.15 K) is calculated from that of the crystal by adding $\Delta_{fus}H$° and the difference in enthalpy, H°(2058 K)-H°(298.15 K), between Ta_2O_5(cr) and Ta_2O_5(ℓ).

Heat Capacity and Entropy

The heat capacity for Ta_2O_5(ℓ) is estimated to be 58.00 cal K^{-1} mol^{-1} by analogy with Nb_2O_5(ℓ). A glass transition temperature is assumed at 1400 K. The heat capacity values used below 1400 K are those of Ta_2O_5(cr). The entropy at 298.15 K is calculated in a manner analogous to that used in calculating the enthalpy of formation.

Fusion Data

The melting point of β-Ta_2O_5 is chosen as 2058 ± 30 K based on work by Reisman et al. (1). The melting point data and the Ta_2O_5 polymorphism are discussed by Reisman and Holtzberg (2). The heat of fusion, $\Delta_{fus}H$°, is estimated from the entropy of fusion for Nb_2O_5 as reported in these tables. Using the Nb_2O_5 entropy of fusion value of 24.92 cal K^{-1} mol^{-1} and a melting temperature of 2058 K, the enthalpy of fusion for Ta_2O_5 is calculated to be 28.7 ± 4 kcal mol^{-1}.

References

1. A. Reisman, F. Holtzberg, M. Berkenblit, and M. Berry, J. Amer. Chem. Soc. 78, 4514 (1956).
2. A. Reisman and F. Holtzberg, High Temperature Oxides 2, 217 (1970).

Tantalum Oxide (Ta_2O_5) O_5Ta_2(l)

Enthalpy Reference Temperature = T_r = 298.15 K; Standard State Pressure = p° = 0.1 MPa

T/K	C_p° (J K⁻¹mol⁻¹)	S° (J K⁻¹mol⁻¹)	-[G°-H°(T_r)]/T (J K⁻¹mol⁻¹)	H°-H°(T_r) (kJ mol⁻¹)	$\Delta_f H$° (kJ mol⁻¹)	$\Delta_f G$° (kJ mol⁻¹)	Log K_f
0							
100							
200							
298.15	135.034	183.050	183.050	0.000	-1957.171	-1834.107	321.328
300	135.342	183.886	183.052	0.250	-1957.151	-1833.343	319.213
400	147.515	224.617	188.524	14.437	-1955.509	-1792.296	234.050
500	156.862	258.572	199.230	29.671	-1953.141	-1751.754	183.005
600	164.431	287.866	211.616	45.750	-1950.282	-1711.739	149.020
700	170.431	313.682	224.389	62.505	-1947.071	-1672.231	124.783
800	175.197	336.761	237.019	79.794	-1943.592	-1633.204	106.637
900	179.038	357.619	249.279	97.507	-1939.905	-1594.624	92.550
1000	182.753	376.690	261.080	115.610	-1936.016	-1556.467	81.301
1100	184.824	394.203	272.397	133.987	-1932.032	-1518.704	72.117
1200	187.071	410.384	283.230	152.584	-1928.002	-1481.306	64.480
1300	189.021	425.436	293.597	171.391	-1923.920	-1444.247	58.031
1400	190.740	439.508	303.522	190.381	-1919.785	-1407.503	52.515
1400.000	190.740	439.508	303.522	190.381	GLASS <--> LIQUID		
1400.000	242.672	439.508	303.522	190.381	TRANSITION		
1500	242.672	456.251	313.152	214.648	-1910.474	-1371.238	47.751
1600	242.672	471.913	322.591	238.915	-1901.249	-1335.590	43.603
1700	242.672	486.624	331.811	263.182	-1892.121	-1300.516	39.960
1800	242.672	500.495	340.801	287.450	-1883.122	-1265.977	36.738
1900	242.672	513.616	349.554	311.717	-1874.282	-1231.933	33.868
2000	242.672	526.063	358.071	335.984	-1865.606	-1198.350	31.298
2058.000	242.672	533.001	362.904	350.059	------ CRYSTAL <--> LIQUID -----		
2100	242.672	537.903	366.355	360.251	-1857.093	-1165.197	28.983
2200	242.672	549.192	374.411	384.518	-1848.743	-1132.445	26.888
2300	242.672	559.980	382.247	408.786	-1840.560	-1100.071	24.983
2400	242.672	570.308	389.869	433.053	-1832.552	-1068.049	23.245
2500	242.672	580.214	397.286	457.320	-1824.733	-1036.356	21.653
2600	242.672	589.732	404.506	481.587	-1817.118	-1004.971	20.190
2700	242.672	598.890	411.537	505.854	-1809.728	-973.875	18.841
2800	242.672	607.716	418.386	530.122	-1802.587	-943.049	17.593
2900	242.672	616.231	425.063	554.389	-1795.721	-912.475	16.435
3000	242.672	624.458	431.573	578.656	-1789.164	-882.130	15.359
3100	242.672	632.415	437.924	602.923	-1782.950	-851.997	14.356
3200	242.672	640.120	444.123	627.190	-1777.119	-822.063	13.419
3300	242.672	647.587	450.176	651.458	-1844.626	-791.361	12.526
3400	242.672	654.832	456.089	675.725	-1838.846	-759.529	11.669
3500	242.672	661.866	461.869	699.992	-1833.107	-727.869	10.863
3600	242.672	668.703	467.519	724.259	-1827.406	-696.371	10.104
3700	242.672	675.352	473.047	748.526	-1821.742	-665.033	9.389
3800	242.672	681.823	478.456	772.794	-1816.114	-633.844	8.713
3900	242.672	688.127	483.752	797.061	-1810.520	-602.804	8.074
4000	242.672	694.271	488.939	821.328	-1804.960	-571.909	7.468
4100	242.672	700.263	494.020	845.595	-1799.432	-541.151	6.894
4200	242.672	706.111	499.001	869.862	-1793.936	-510.528	6.349
4300	242.672	711.821	503.884	894.130	-1788.471	-480.034	5.831
4400	242.672	717.400	508.673	918.397	-1783.036	-449.670	5.338
4500	242.672	722.853	513.372	942.664	-1777.632	-419.428	4.869
4600	242.672	728.187	517.984	966.931	-1772.258	-389.306	4.421
4700	242.672	733.406	522.513	991.198	-1766.916	-359.298	3.993
4800	242.672	738.515	526.960	1015.466	-1761.604	-329.405	3.585
4900	242.672	743.519	531.328	1039.733	-1756.324	-299.621	3.194
5000	242.672	748.421	535.621	1064.000	-1751.077	-269.945	2.820

PREVIOUS: CURRENT: December 1972

Tantalum Oxide (Ta_2O_5) O_5Ta_2(l)

TANTALUM OXIDE (Ta_2O_5) CFYSTAL-LIQUID M_r = 441.8928 Tantalum Oxide (Ta_2O_5) O_5Ta_2(cr,l)

0 tc 2058 K crystal
above 2058 K liquid

Refer to the individual tables for details.

Enthalpy Reference Temperature = T_r = 298.15 K; Standard State Pressure = p° = 0.1 MPa

T/K	C_p° (J K⁻¹mol⁻¹)	S° (J K⁻¹mol⁻¹)	$-[G^\circ-H^\circ(T_r)]/T$ (J K⁻¹mol⁻¹)	$H^\circ-H^\circ(T_r)$ (kJ mol⁻¹)	Δ_fH° (kJ mol⁻¹)	Δ_fG° (kJ mol⁻¹)	Log K_f
0	0.000	0.000	INFINITE	-23.041	-2035.947	-2035.947	INFINITE
100	58.082	37.108	243.530	-20.642	-2042.799	-1999.955	1044.669
200	108.311	94.161	154.766	-12.121	-2046.060	-1955.515	510.728
298.15	135.034	143.114	143.114	0.000	-2045.976	-1911.004	334.800
300	135.342	143.950	143.116	0.250	-2045.956	-1910.167	332.590
400	147.515	184.681	148.588	14.437	-2044.314	-1865.126	243.561
500	156.862	218.636	159.293	29.671	-2041.946	-1820.591	190.196
600	164.431	247.930	171.680	45.750	-2039.087	-1776.582	154.665
700	170.431	273.746	184.453	62.505	-2035.876	-1733.081	129.324
800	175.197	296.825	197.083	79.794	-2032.397	-1690.059	110.350
900	179.038	317.683	209.342	97.507	-2028.710	-1647.487	95.618
1000	182.753	336.753	221.144	115.610	-2024.820	-1605.335	83.854
1100	184.824	354.267	232.461	133.987	-2020.837	-1563.579	74.248
1200	187.071	370.448	243.294	152.584	-2016.807	-1522.187	66.259
1300	189.021	385.500	253.661	171.391	-2012.725	-1481.134	59.513
1400	190.740	399.572	263.586	190.380	-2008.590	-1440.398	53.742
1500	192.280	412.785	273.096	209.533	-2004.394	-1399.959	48.751
1600	193.673	425.239	282.220	228.831	-2000.138	-1359.802	44.393
1700	194.953	437.020	290.982	248.263	-1995.845	-1319.912	40.556
1800	196.138	448.197	299.409	267.819	-1991.558	-1280.276	37.153
1900	197.246	458.831	307.522	287.488	-1987.315	-1240.876	34.114
2000	198.288	468.975	315.343	307.265	-1983.130	-1201.698	31.385
2058.000	198.870	474.652	319.753	318.783	CRYSTAL <--> LIQUID	TRANSITION	
2058.000	242.672	533.001	319.753	438.864			
2100	242.672	537.903	324.067	449.056	-1857.093	-1165.197	28.983
2200	242.672	549.192	334.045	473.323	-1848.743	-1132.445	26.888
2300	242.672	559.980	343.636	497.591	-1840.560	-1100.071	24.983
2400	242.672	570.308	352.867	521.858	-1832.552	-1068.049	23.245
2500	242.672	580.214	361.764	546.125	-1824.733	-1036.355	21.653
2600	242.672	589.732	370.350	570.392	-1817.118	-1004.971	20.190
2700	242.672	598.890	378.646	594.659	-1809.728	-973.875	18.841
2800	242.672	607.716	386.670	618.927	-1802.587	-943.049	17.593
2900	242.672	616.231	394.440	643.194	-1795.721	-912.475	16.435
3000	242.672	624.458	401.971	667.461	-1789.164	-882.130	15.359
3100	242.672	632.415	409.277	691.728	-1782.950	-851.997	14.356
3200	242.672	640.120	416.371	715.995	-1777.119	-822.063	13.419
3300	242.672	647.587	423.265	740.263	-1844.626	-791.361	12.526
3400	242.672	654.832	429.970	764.530	-1838.846	-759.529	11.669
3500	242.672	661.866	436.496	788.797	-1833.107	-727.869	10.863
3600	242.672	668.703	442.851	813.064	-1827.406	-696.371	10.104
3700	242.672	675.352	449.046	837.331	-1821.742	-665.033	9.389
3800	242.672	681.823	455.087	861.599	-1816.114	-633.844	8.713
3900	242.672	688.127	460.982	885.866	-1810.520	-602.804	8.074
4000	242.672	694.271	466.737	910.133	-1804.960	-571.909	7.468
4100	242.672	700.263	472.360	934.400	-1799.432	-541.151	6.894
4200	242.672	706.111	477.857	958.667	-1793.936	-510.528	6.349
4300	242.672	711.821	483.231	982.935	-1788.471	-480.034	5.831
4400	242.672	717.400	488.490	1007.202	-1783.036	-449.670	5.338
4500	242.672	722.853	493.638	1031.469	-1777.632	-419.428	4.869
4600	242.672	728.187	498.679	1055.736	-1772.258	-389.306	4.421
4700	242.672	733.406	503.618	1080.003	-1766.916	-359.298	3.993
4800	242.672	738.515	508.459	1104.271	-1761.604	-329.405	3.585
4900	242.672	743.519	513.205	1128.538	-1756.324	-299.621	3.194
5000	242.672	748.421	517.860	1152.805	-1751.077	-269.945	2.820

PREVIOUS: CURRENT: December 1972

Tantalum Oxide (Ta_2O_5) O_5Ta_2(cr,l)

TITANIUM OXIDE, ALPHA (α-Ti_3O_5) CRYSTAL M_r = 223.6370

$\Delta_fH°(0\ K) = -2446.057\ kJ\ mol^{-1}$

$S°(298.15\ K) = 129.369 \pm 1.7\ J\ K^{-1}\ mol^{-1}$ $\Delta_fH°(298.15\ K) = -2459.146 \pm 4.2\ kJ\ mol^{-1}$

$T_{trs}(\alpha \to \beta) = 450 \pm 20\ K$ $\Delta_{trs}H° = [13.263 \pm 5.0]\ kJ\ mol^{-1}$

Enthalpy of Formation

$\Delta_fH°(298.15\ K) = -587.75\ kcal\ mol^{-1}$ is calculated from that of TiO_2(rutile) using $\Delta_cH° = -89.65 \pm 0.05\ kcal\ mol^{-1}$ determined by Humphrey (1) from combustion calorimetry. Ariya et al. (2) reported $\Delta_fH° = -587.0 \pm 1.0\ kcal\ mol^{-1}$ derived from their own values of $\Delta_cH°$ for $Ti_3O_{5.03}$ and Ti(cr). Tests for incomplete combustion were made by grinding the product and igniting it to constant weight in air (1) or in moist oxygen (2). Humphrey (1) reported that his rutile was bright yellow in color and did not gain weight. This implies an atomic ratio (O/Ti) $\geq$ 2.00, compared with the value of ~1.95 reported by Ariya et al. (2). New high-temperature calorimetry by Charlu et al. (9) gave $\Delta_cH°(1100) = -93.57 \pm 1.3\ kcal\ mol^{-1}$ after linear correction to (O/Ti) = 2.00 from 1.97 produced by combustion. Our reduction to 298 K and $Ti_3O_5(\alpha)$ yields $\Delta_cH° = -90.36 \pm 2.5\ kcal\ mol^{-1}$, including an assumed uncertainty allowance of 1% for the weight gains.

The calorimetric data are compared below with new high-temperature data for $\Delta_rG°$ (3, 4). Emf data (4) for reaction B are consistent with the adopted $\Delta_fH°$, but integrated values of $\Delta_rG°$ for reaction A yield $\Delta_fH°$ more negative by 1.8 kcal mol^{-1}. Perhaps coincidentally, the $\Delta_cH°$ of Ariya et al. (2) yields $\Delta_fH°$ more negative by 1.9 kcal mol^{-1} when combined with the JANAF $\Delta_fH°$ (5) for TiO_2(rutile). Merritt et al. (3) obtained $\Delta_rG°$ for the combustion reaction A by integrating extensive data for equilibrium oxygen potentials of the many stable phases and solid solutions between TiO_2 and Ti_3O_5. They determined the oxide composition by mass during equilibration with (H_2 + H_2O) gas mixtures of known oxygen potential. Gross hysteresis was observed in three regions of composition. Oxygen potentials in these regions were deduced with a reasonable assumption, combined with Zador's emf data for oxygen-deficient rutile, and integrated to yield $\Delta_rG°(1304\ K) = -56.14 \pm 0.43\ kcal\ mol^{-1}$ for reaction A. We obtain $\Delta_rG° = -57.30\ kcal\ mol^{-1}$ at 1273 K and -51.15 at 1473 K by similar integrations of oxygen potentials derived from emf data relative to Fe/FeO and Cr/Cr_2O_3 reported by Suzuki and Sambongi (4).

Suzuki and Sambongi (4) also reported solid-state emf data which relate Ti_3O_5 to Ti_2O_3. We convert their (graphical) emf data (4) to oxygen potentials using the authors' self-consistent data for the reference couples. The results for reaction B yield $\Delta_rH°$ and $\Delta_rS°$ values consistent with our adopted tables. 3rd law values of $\Delta_rH°(298.15\ K)$ of -42.29, -42.15 and -42.13 kcal mol^{-1} agree with -42.52 (1), -42.8 ± 1.3 (2) and -42.0 (9) which can be calculated from combustion calorimetry.

The discrepancy in reaction A may be due to undetected bias (2% too small in magnitude) in $\Delta_rG°$, but alternative possibilities include bias in S° (see Ti_3O_5, β), $\Delta_cH°$ or enthalpy (see TiO_2, rutile).

Source	Method	Reaction[a]	T/K	δS cal K^{-1} mol^{-1}	$\Delta_rH°(298.15\ K)$ kcal mol^{-1}	$\Delta_fH°(298.15\ K)$, kcal mol^{-1} Authors	JANAF
(1)	$\Delta_cH°$	A	303→298	-	-89.65±0.05	-586.9±0.7	-587.75±09
(2)	$\Delta_cH°$	A	293	-	-87.7	-587.0±1.0	-589.7
(9)	$\Delta_cH°$	A	1100→298	-	-90.36±2.5	-586.3	-587.0 ±2.7
(3)	$\Delta_rG(O_2)$	A	1304	-	-87.81±0.6	-	-589.6 ±1.1
(4)	$\Delta_rG(O_2)$, emf	A	1273-1473	-3.5	-87.81±1	-	-589.6 ±1.4
(4)	emf, Cr/Cr_2O_3	B[b]	~940-1740	0.6	-42.29	-	-587.5[d]
	emf, Ta/Ta_2O_5	B[b]	~1100-1750	0.0	-42.15	-	-587.4[d]
	emf, "$TiO_{1.3}$"	B[b]	~1300-1700	0.0	-42.13	-	-587.4[d]

[a]Reactions: A) $Ti_3O_5(\alpha, \beta) + 1/2\ O_2 \to 3\ TiO_2$(rutile); B) $3/2\ Ti_2O_3(\alpha, \beta) + 1/4\ O_2 \to Ti_3O_5(\alpha, \beta)$.

[b]$\Delta_rG°$ from integration of $\Delta_r\bar{G}(O_2)$ assuming negligible homogeneity range for Ti_2O_3. Alternative $\Delta_rG°$, assuming a linear variation of $\Delta_f\bar{G}°(O_2)$ for 1.42 < x < 1.56, changes δS by ~0.2 and $\Delta_rH°$ and $\Delta_fH°$ by ~-0.3.

Heat Capacity and Entropy

$C_p°$ below 300 K is based on data of Shomate (6, 53-297 K) and his Debye-Einstein extrapolation to absolute zero. The entropy is calculated by polynomial integration of $C_p°$ based on $S°(50\ K) = 0.87\ cal\ K^{-1}\ mol^{-1}$. Our assigned entropy uncertainty of ±0.4 cal K^{-1} mol^{-1} includes an additional contribution because the sample gave weak, fuzzy x-ray lines and was only 99.1% pure. Apparently there are new $C_p°$ data (12-300 K) which are unpublished and not yet available (10). $C_p°$ is extrapolated smoothly from 300 K to 450 K and to higher temperatures where $Ti_3O_5(\beta)$ is the stable form. Our extrapolation probably represents the lower limit for the true $C_p°$ below T_{trs}, while the curve of Furukawa and Reilly (7) is an upper limit. The equations reported by Naylor (8) yield even higher enthalpies and fail to join properly with the low temperature data at 298 K. This is due to complications in the enthalpy data (see below) and to positive bias in the drop calorimeter in this temperature range (5).

Naylor (8) measured enthalpy data (326-1340 K) which revealed a transition at ~450 K. When the samples were heated above this temperature, they did not return to the original state on cooling. Enthalpies were then systematically higher by up to 10%. The change in the enthalpy curve seemed to depend on both the temperature and time of heating above T_{trs}. Regrinding was the only method found for returning the sample to its original state. X-ray diffraction at room temperature revealed no difference between these two states, but failure to detect a difference may be due to the weak, fuzzy x-ray lines (6, 8). Possible explanations for the anomalous enthalpy data are noted in Transition Data.

Transition Data

See the β-crystal table for details

References

1. G. L. Humphrey, J. Amer. Chem. Soc. 73, 1587 (1951).
2. S. M. Ariya, M. P. Morozova, and E. Vol'f, Russ. J. Inorg. Chem. 2, 16 (1957).
3. R. R. Merritt, B. G. Hyde, L. A. Bursill and D. K. Philp, Phil. Trans. Roy, Soc. London, Ser. A 274, (1245), 627 (1973).
4. K. Suzuki and K. Sambongi, Tetsu To Hagane 58, 1579 (1972).
5. JANAF Thermochemical Tables: TiO_2(rutile), 12-31-73; Ti_2O_3, 6-30-73.
6. C. H. Shomate, J. Amer. Chem. Soc. 68, 310 (1946).
7. G. T. Furukawa and M. L. Rilly, U. S. Nat. Bur. Stand. Report 7093, 177, 1961.
8. B. F. Naylor, J. Amer. Chem. Soc. 68, 1077 (1946).
9. T. V. Chgarlu, O. J. Kleppa, and T. B. Reed, submitted for publication in J. Chem. Thermodyn.
10. G. A. Berezovskii, V. A. Ganenko and I. E. Paukov, Bulletin of Thermodynamics and Thermochemistry, No. 14, 39 (1971).

Titanium Oxide, Alpha (Ti_3O_5) O_5Ti_3(cr)

Enthalpy Reference Temperature = T_r = 298.15 K Standard State Pressure = p° = 0.1 MPa

T/K	$C_p°$ (J $K^{-1}mol^{-1}$)	S° (J $K^{-1}mol^{-1}$)	$-[G°-H°(T_r)]/T$ (J $K^{-1}mol^{-1}$)	$H°-H°(T_r)$ (kJ mol^{-1})	$\Delta_fH°$ (kJ mol^{-1})	$\Delta_fG°$ (kJ mol^{-1})	Log K_f
0	0.000	0.000	INFINITE	-23.108	-2446.057	-2446.057	INFINITE
100	46.116	21.062	237.166	-21.610	-2453.501	-2409.803	1258.752
200	114.445	75.659	142.352	-13.339	-2458.258	-2363.911	617.390
298.15	154.808	129.369	129.369	0.000	-2459.146	-2317.293	405.980
300	155.477	130.329	129.372	0.287	-2459.135	-2316.413	403.323
400	182.841	179.228	135.852	17.350	-2457.340	-2269.052	296.308
450	189.954	201.217	141.909	26.688	-2455.868	-2245.602	260.663
500	194.138	221.452	148.866	36.293	-2454.266	-2222.324	232.165
600	200.832	257.461	164.038	56.054	-2450.880	-2176.251	189.459
700	206.062	288.828	179.673	76.409	-2447.326	-2130.758	158.999
800	210.037	316.611	195.087	97.219	-2443.632	-2085.786	136.188
900	213.384	341.551	209.999	118.397	-2439.941	-2041.278	118.473
1000	215.894	364.163	224.302	139.861	-2436.495	-1997.170	104.321
1100	218.405	384.861	237.971	161.579	-2433.495	-1953.386	92.759
1200	220.497	403.954	251.017	183.524	-2442.906	-1909.496	83.118
1300	222.589	421.689	263.472	205.682	-2438.647	-1865.218	74.945
1400	224.262	438.246	275.371	228.025	-2434.515	-1821.264	67.952
1500	225.936	453.775	286.752	250.535	-2430.558	-1777.600	61.902

PREVIOUS: March 1967 CURRENT: December 1973

Titanium Oxide, Alpha (Ti_3O_5) O_5Ti_3(cr)

TITANIUM OXIDE, BETA (β-Ti_3O_5) CRYSTAL M_r = 223.6370

$S°$(298.15 K) = [157.617] J K^{-1} mol^{-1}

T_{trs}($\alpha \rightarrow \beta$) = 450 ± 20 K

T_{fus} = 2050 ± 20 K

$\Delta_f H°$(298.15 K) = [-2446.189] kJ mol^{-1}

$\Delta_{trs}H°$[13.263 ± 5.0] kJ mol^{-1}

$\Delta_{fus}H°$ = [171.544] kJ mol^{-1}

Enthalpy of Formation

$\Delta_f H°$ is calculated from that of $Ti_3O_5(\alpha)$ by adding $\Delta_{trs}H°$ to the difference in H°(450 K)-H°(298.15 K) between the α- and β-phases. Experimental data for $\Delta_r G°$ of the β-phase are analyzed on the table for the α-phase.

Heat Capacity and Entropy

$C°_p$ of the β-phase is derived from fitting of enthalpy data of Naylor (1, 504-1340 K) and Slyusar et al. (2, 580-1994 K). Naylor's data below 482 K are omitted due to thermal hysteresis in the transition; even the points near 500 K show unusual scatter. The new data of Slyusar et al. (2) are suspect due to a discontinuity of almost 4% at ~1250 K where the authors switched container materials and temperature sensors. Their data (2) for Ti_2O_3 and TiO_2(rutile) show similar discontinuities and appear to have a bias of about -4% below 1100 K. Their data for Ti_3O_5 do not differ significantly from those of Naylor (1), so we combine the two sets.

S° is calculated in a manner analogous to that of $\Delta_f H°$. The uncertainty in S° may be as large as ±2.6 cal K^{-1} mol^{-1} due to uncertainty in $\Delta_{trs}H°$, if significant fractions of the β-phase failed to revert to the α-phase during drop calorimetry (1, 2). Naylor's data (1) suggest that this may be the case, but Slyusar et al. (2) make no mention of the problem (see Transition Data). Excluding this possible uncertainty, we estimate that S° of $Ti_3O_5(\beta)$ is uncertain by ±0.6 cal K^{-1} mol^{-1}.

Transition Data and Phase Data

Ti_3O_5 has been identified in three crystalline structures. Low-temperature α-Ti_3O_5 is monoclinic (3), high-temperature β-Ti_3O_5 is a pseudobrookite type with slight monoclinic distortion (3) and metastable γ-Ti_3O_5 (4) is analogous to V_3O_5 (3, 4). Based on its similarity to V_3O_5, γ is the lowest member of the homologous series of phases Ti_nO_{2n-1} with structures related to rutile (9, 10). The magnetic anomaly, observed near 250 K in γ-Ti_3O_5 (4), corresponds to those observed at lower temperatures in Ti_4O_7 and higher homologues. We give the subsequent phase observations because they may have bearing on the anomalous enthalpy data and the discrepancy in $\Delta_f G°$ at high temperatures.

Our designations for the stable phases α and β are reversed from those of Asbrink et al. (4). These phases were called M-type (our α), D-type and D'-type (our β) by Iwasaki et al. (11). D- and D'-types differed only in their low-temperature oxidation products; D-type gave rutile, while D'-type gave mixtures of anatase and rutile (11). D-type was obtained by H_2-reduction of anatase and D'-type by action of H_2 on rutile and Ti or CO on either anatase or rutile. D-type presumably is metastable; Iwasaki et al. (11) postulated free energy diagrams to explain conversion of D-type to other forms. The observations also can be explained be assuming that conversion of D-type is hindered kinetically, like TiO_2(anatase) → TiO_2(rutile), until gradually activated at elevated temperature. X-ray patterns of D-type, though not well defined, were similar to those of D'-type (11). Thus, D-type is not the Ti_nO_{2n-1} homologue (γ-Ti_3O_5), as postulated (11) on the basis of oxidation behavior. D- and D'-types may have the same crystal structure but differ in degree of ordering or content of impurities.

Enthalpy data of Naylor (1) gave T_{trs} ~450 K for α → β. X-ray diffraction (4-6), DTA (5), magnetic susceptibility (4, 6), and electrical conductivity (5) confirm the transition temperature. The transition is reversible (3, 5-7) but exhibits thermal hysteresis of about ±20° (5, 6). More important to interpretation of enthalpy measurements is the observation that the β-phase is stabilized (5, 3) at room temperature by small amounts of impurities such as Fe, W and Mg. Samples doped with 5% Fe did not transform; those with 2% Fe gave a much-reduced endotherm (DTA) at ~75° below the normal T_{trs} (5). Accidental contamination by W stabilized the β-form (5).

$\Delta_{trs}H°$ = 3.17 kcal mol^{-1} is calculated at T_{trs} from the difference in enthalpy between our β-phase curve fit and our adopted α-phase. Alternative enthalpy curves (8) gave $\Delta_{trs}H°$ = 2.48 kcal mol^{-1}. These calculations assume that the β-phase reverted completely to α-phase at room temperature during drop calorimetry (1, 2). Naylor's anomalous enthalpies (1) for the α-phase suggest that part of his sample may have been stabilized in the β-form due to impurities. Other metastable phases may have been present. If the drop-calorimetric samples did not revert completely to α-phase, the true value of $\Delta_{trs}H°$ should be larger than the value calculated on the assumption of complete reversion. $\Delta_r G°$ data (see Ti_3O_5, α) tend to favor an entropy greater than or equal to the adopted value, but DTA data (5) gave $\Delta_{trs}H°$ = 1.6 ± 0.4 kcal mol^{-1}, opposite to the expected direction. We emphasize this unsatisfactory situation by placing $\Delta_{trs}H°$ in brackets and assigning an uncertainty of ±1.2 kcal mol^{-1}. The seeming contradictions should be resolved easily by study of a well-characterized sample in an adiabatic calorimeter or an accurate differential-scanning calorimeter.

Fusion Data and Sublimation Data

See the liquid table for details.

References

1. B. F. Naylor, J. Amer. Chem. Soc. 68, 1077 (1946).
2. N. P. Slyusar, A. D. Krivorotenko, E. N. Fomichev, A. A. Kalashnik and V. P. Bondarenko, Teplofiz. Vys. Temp. 11, 213 (1973).
3. S. Asbrink and A. Magneli, Acta Cryst. 12, 575, (1959); Acta Chem. Scand. 13, 603 (1959).
4. G. Asbrink, S. Asbrink, A. Magneli, H. Okinaka, K. Kosuge and S. Kachi, Acta Chem. Scand. 25, 3889 (1971).
5. C. N. R. Rao, S. Ramdas, R. E. Loehman and J. M. Honig, J. Solid State Chem. 3, 83 (1971).
6. W. J. Danley and L. N. Mulay, Mat. Res. Bull. 7, 739 (1972); J. Appl. Phys. 41, 877 (1970).
7. P. G. Wahlbeck and P. W. Gilles, J. Amer. Chem. Soc. 49, 180 (1966).
8. G. T. Furukawa and M. L. Reilly, U. S. Nat. Bur. Stand. Report 7093, 177, 1961.
9. L. A. Bursill and B. G. Hyde, Progr. Solid State Chem. 7 177 (1972).
10. R. R. Merritt, B. G. Hyde, L. A. Bursill and D. K. Philp Phil. Trans. Roy. Soc. London, Ser. A 274, (1245), 627 (1973).
11. H. Iwasaki, N. F. H. Bright, and J. F. Rowland, J. Less-Common Metals 17, 99 (1969).

Titanium Oxide, Beta (Ti_3O_5) O_5Ti_3(cr)

Enthalpy Reference Temperature = T_r = 298.15 K — Standard State Pressure = p° = 0.1 MPa

T/K	$C°_p$ (J $K^{-1}mol^{-1}$)	S° (J $K^{-1}mol^{-1}$)	-[G°-H°(T_r)]/T (J $K^{-1}mol^{-1}$)	H°-H°(T_r) (kJ mol^{-1})	$\Delta_f H°$ (kJ mol^{-1})	$\Delta_f G°$ (kJ mol^{-1})	Log K_f
0							
100							
200							
298.15	173.962	157.617	157.617	0.000	-2446.189	-2312.759	405.186
300	174.054	158.694	157.621	0.322	-2446.143	-2311.931	402.543
400	179.075	209.454	164.508	17.978	-2443.756	-2267.558	296.113
450	181.586	230.691	170.702	26.995	-2442.604	-2245.602	260.663
500	184.096	249.952	177.679	36.137	-2441.465	-2223.774	232.316
600	189.117	283.961	192.632	54.798	-2439.180	-2180.450	189.825
700	194.138	313.490	207.833	73.960	-2436.818	-2137.514	159.503
800	199.158	339.742	222.710	93.625	-2434.270	-2094.928	136.785
900	204.179	363.489	237.053	113.792	-2431.589	-2052.670	119.134
1000	209.200	385.261	250.800	134.461	-2428.938	-2010.711	105.029
1100	214.221	405.436	263.952	155.632	-2426.485	-1969.009	93.500
1200	219.242	424.291	276.536	177.305	-2436.168	-1927.162	83.887
1300	224.262	442.038	288.591	199.480	-2431.892	-1884.917	75.737
1400	229.283	458.841	300.157	222.158	-2427.425	-1843.008	68.763
1500	234.304	474.831	311.273	245.337	-2422.799	-1801.425	62.731
1600	239.325	490.113	321.976	269.018	-2418.050	-1760.154	57.463
1700	244.346	504.773	332.301	293.202	-2413.214	-1719.183	52.824
1800	249.366	518.881	342.277	317.888	-2408.329	-1678.499	48.709
1900	254.387	532.498	351.932	343.075	-2403.433	-1638.087	45.034
2000	259.408	545.674	361.292	368.765	-2442.627	-1596.570	41.698
2100	264.429	558.452	370.378	394.957	-2440.070	-1554.329	38.662
2200	269.450	570.869	379.210	421.651	-2437.067	-1512.218	35.905
2300	274.470	582.958	387.807	448.847	-2433.619	-1470.256	33.391
2400	279.491	594.745	396.185	476.545	-2429.725	-1428.453	31.089
2500	284.512	606.256	404.358	504.745	-2425.383	-1386.821	28.976

PREVIOUS: March 1967 CURRENT: December 1973

Titanium Oxide, Beta (Ti_3O_5) O_5Ti_3(cr)

TITANIUM OXIDE (Ti_3O_5) LIQUID M_r = 223.6370

$S°(298.15\ K)$ = [232.446] $J\ K^{-1}\ mol^{-1}$ $\Delta_f H°(298.15\ K)$ = [-2289.059] $kJ\ mol^{-1}$

T_{fus} = 2050 ± 20 K $\Delta_{fus}H°$ = [171.544] $kJ\ mol^{-1}$

Enthalpy of Formation

$\Delta_f H°(Ti_3O_5$, ℓ, 298.15 K) is calculated from that of $Ti_3O_5(\beta)$ by adding $\Delta_{fus}H°$ and the difference in enthalpy, H°(2050 K)-H°(298.15 K), between the β-phase and liquid.

Heat Capacity and Entropy

$C_p°$ is estimated as 8 cal K^{-1} g-atom^{-1}. Below the assumed glass transition at 1400 K, $C_p°$ is taken to be the same as that of $Ti_3O_5(\beta)$. S°(298.15 K) is calculated in a manner analogous to that used for $\Delta_f H°$.

Fusion Data

$Ti_3O_5(\beta)$ melts incongruently at 2047 ± 10 K according to Wahlbeck and Gilles (1). Their phase diagram (1) suggests that melting starts at 2047 K and completes near 2077 K. The latter value is close to the melting point of 2083 K reported by Slyusar et al. (2), but is almost 90° above the value of ~1990 which may be interpolated from the liquidus curve of Brauer and Littke (3) for the Ti-O system. Conversion of the value of Wahlbeck and Gilles (1) to IPTS-68 yields 2050 K, which is adopted.

Sublimation and Vaporization Data

$Ti_3O_5(\beta)$ is the congruently vaporizing crystalline phase in the Ti-O system according to the studies of Gilles and coworkers (4-6). Studies of liquid samples (7) showed that the congruently vaporizing composition is between Ti_3O_5 and Ti_4O_7. The principal reaction (4) in congruent vaporization is $Ti_3O_5 \rightarrow TiO(g) + 2\ TiO_2(g)$.

References

1. P. G. Wahlbeck and P. W. Gilles, J. Amer. Ceram. Soc. 49, 180 (1966).
2. N. P. Slyusar, A. D. Krivorotenko, E. N. Fomichev, A. A. Kalashnik and V. P. Bondarenko, Teplofiz. Vys. Temp. 11, 213 (1973).
3. G. Brauer and W. Littke, J. Inorg. Nucl. Chem. 16, 67 (1960).
4. P. J. Hampson and P. W. Gilles, J. Chem. Phys. 55, 3712 (1971).
5. B. R. Conard and P. W. Gilles, "High Temperature Vaporization and Thermodynamics of Trititanium Pentoxide," presented at 162nd National Meeting, Amer. Chem. Soc., Washington, D.C., Sept., 1971.
6. P. W. Gilles, K. D. Carlson, H. F. Franzen and P. G. Wahlbeck, J. Chem. Phys. 46, 2461 (1967).
7. P. W. Gilles, H. F. Franzen, G. D. Stone and P. G. Wahlbeck, J. Chem. Phys. 48, 1938 (1968).

Titanium Oxide (Ti_3O_5) $O_5Ti_3(l)$

Enthalpy Reference Temperature = T_r = 298.15 K Standard State Pressure = p° = 0.1 MPa

T/K	$C_p°$ (J K⁻¹mol⁻¹)	S° (J K⁻¹mol⁻¹)	$-[G°-H°(T_r)]/T$ (J K⁻¹mol⁻¹)	$H°-H°(T_r)$ (kJ mol⁻¹)	$\Delta_f H°$ (kJ mol⁻¹)	$\Delta_f G°$ (kJ mol⁻¹)	Log K_f
0							
100							
200							
298.15	173.962	232.446	232.446	0.000	-2289.059	-2177.939	381.566
300	174.054	233.522	232.449	0.322	-2289.013	-2177.250	379.093
400	179.075	284.282	239.336	17.978	-2286.625	-2140.359	279.502
500	184.096	324.781	252.507	36.137	-2284.335	-2104.058	219.809
600	189.117	358.789	267.460	54.798	-2282.050	-2068.217	180.054
700	194.138	388.319	282.661	73.960	-2279.687	-2032.764	151.687
800	199.158	414.570	297.539	93.625	-2277.139	-1997.660	130.434
900	204.179	438.318	311.882	113.792	-2274.459	-1962.886	113.923
1000	209.200	460.090	325.629	134.461	-2271.808	-1928.410	100.730
1100	214.221	480.264	338.781	155.632	-2269.355	-1894.191	89.948
1200	219.242	499.119	351.365	177.305	-2279.038	-1859.826	80.956
1300	224.262	516.866	363.420	199.480	-2274.762	-1825.064	73.332
1400	229.283	533.669	374.985	222.158	-2270.295	-1790.638	66.810
1400.000	229.283	533.669	374.985	222.158	GLASS <--> LIQUID TRANSITION		
1400.000	267.776	533.669	374.985	222.158			
1500	267.776	552.144	386.187	248.935	-2262.071	-1756.666	61.173
1600	267.776	569.426	397.105	275.713	-2254.226	-1723.250	56.258
1700	267.776	585.660	407.724	302.490	-2246.795	-1690.273	51.936
1800	267.776	600.965	418.039	329.268	-2239.818	-1657.741	48.106
1900	267.776	615.443	428.051	356.046	-2233.333	-1625.582	44.690
2000	267.776	629.178	437.767	382.823	-2271.438	-1592.390	41.589
2050.000	267.776	635.790	442.516	396.212	------- BETA <--> LIQUID -------		
2100	267.776	642.243	447.195	409.601	-2268.296	-1558.516	38.766
2200	267.776	654.700	456.346	436.378	-2265.210	-1524.788	36.203
2300	267.776	666.603	465.231	463.156	-2262.180	-1491.201	33.866
2400	267.776	678.000	473.861	489.934	-2259.206	-1457.745	31.727
2500	267.776	688.931	482.246	516.711	-2256.287	-1424.411	29.761
2600	267.776	699.433	490.399	543.489	-2253.421	-1391.193	27.949
2700	267.776	709.539	498.329	570.266	-2250.608	-1358.083	26.274
2800	267.776	719.278	506.048	597.044	-2247.846	-1325.077	24.720
2900	267.776	728.674	513.563	623.822	-2245.133	-1292.170	23.275
3000	267.776	737.752	520.886	650.599	-2242.470	-1259.354	21.927
3100	267.776	746.533	528.024	677.377	-2239.852	-1226.626	20.669
3200	267.776	755.034	534.986	704.154	-2237.280	-1193.984	19.490
3300	267.776	763.274	541.779	730.932	-2234.751	-1161.421	18.384
3400	267.776	771.268	548.412	757.710	-2232.264	-1128.932	17.344
3500	267.776	779.030	554.891	784.487	-2229.818	-1096.518	16.365
3600	267.776	786.574	561.222	811.265	-2227.409	-1064.171	15.441
3700	267.776	793.910	567.412	838.042	-3452.333	-1008.528	14.238
3800	267.776	801.051	573.467	864.820	-3446.295	-942.558	12.956
3900	267.776	808.007	579.392	891.598	-3440.457	-876.746	11.743
4000	267.776	814.787	585.193	918.375	-3434.815	-811.083	10.592

PREVIOUS: March 1967 CURRENT: December 1973

Titanium Oxide (Ti_3O_5) $O_5Ti_3(l)$

TITANIUM OXIDE (Ti_3O_5) ALPHA-BETA-LIQUID M_r = 223.6370 Titanium Oxide (Ti_3O_5) O_5Ti_3(cr,l)

0 to 450 K	crystal, alpha
450 to 2050 K	crystal, beta
above 2050 K	liquid

Refer to the individual tables for details.

Enthalpy Reference Temperature = T_r = 298.15 K — Standard State Pressure = $p°$ = 0.1 MPa

T/K	$C_p°$ (J K⁻¹mol⁻¹)	$S°$ (J K⁻¹mol⁻¹)	$-[G°-H°(T_r)]/T$ (J K⁻¹mol⁻¹)	$H°-H°(T_r)$ (kJ mol⁻¹)	$\Delta_fH°$ (kJ mol⁻¹)	$\Delta_fG°$ (kJ mol⁻¹)	Log K_f
0	0.000	0.000	INFINITE	-23.108	-2446.057	-2446.057	INFINITE
100	46.116	21.062	237.166	-21.610	-2453.501	-2409.803	1258.752
200	114.445	75.659	142.352	-13.339	-2458.258	-2363.911	617.390
298.15	154.808	129.369	129.369	0.000	-2459.146	-2317.293	405.980
300	155.477	130.329	129.372	0.287	-2459.135	-2316.413	403.323
400	182.841	179.228	135.852	17.350	-2457.340	-2269.052	296.308
450	189.954	201.217	141.909	26.688	-2455.868	-2245.602	260.663
450.000	181.586	230.691	141.909	39.952	-------- ALPHA <--> BETA -------		
500	184.096	249.952	151.765	49.094	-2441.465	-2223.774	232.316
600	189.117	283.961	171.037	67.754	-2439.180	-2180.450	189.825
700	194.138	313.490	189.323	86.917	-2436.818	-2137.514	159.503
800	199.158	339.742	206.514	106.582	-2434.270	-2094.928	136.785
900	204.179	363.489	222.657	126.749	-2431.589	-2052.670	119.134
1000	209.200	385.261	237.844	147.418	-2428.938	-2010.711	105.029
1100	214.221	405.436	252.173	168.589	-2426.485	-1969.009	93.500
1200	219.242	424.291	265.739	190.262	-2436.168	-1927.162	83.887
1300	224.262	442.038	278.624	212.437	-2431.892	-1884.917	75.737
1400	229.283	458.841	290.902	235.114	-2427.425	-1843.008	68.763
1500	234.304	474.831	302.635	258.294	-2422.799	-1801.425	62.731
1600	239.325	490.113	313.878	281.975	-2418.050	-1760.154	57.463
1700	244.346	504.773	324.679	306.159	-2413.214	-1719.183	52.824
1800	249.366	518.881	335.079	330.844	-2408.329	-1678.499	48.709
1900	254.387	532.498	345.113	356.032	-2403.433	-1638.087	45.034
2000	259.408	545.674	354.813	381.722	-2442.627	-1596.570	41.698
2050.000	261.918	552.111	359.547	394.755	BETA <--> LIQUID TRANSITION		
2050.000	267.776	635.791	359.547	566.299			
2100	267.776	642.243	366.201	579.688	-2268.296	-1558.516	38.766
2200	267.776	654.700	379.034	606.465	-2265.210	-1524.788	36.203
2300	267.776	666.603	391.280	633.243	-2262.180	-1491.201	33.866
2400	267.776	678.000	402.991	660.020	-2259.206	-1457.745	31.727
2500	267.776	688.931	414.212	686.798	-2256.287	-1424.411	29.761
2600	267.776	699.433	424.981	713.576	-2253.421	-1391.193	27.949
2700	267.776	709.539	435.334	740.353	-2250.608	-1358.083	26.274
2800	267.776	719.278	445.302	767.131	-2247.846	-1325.077	24.720
2900	267.776	728.674	454.913	793.908	-2245.133	-1292.170	23.275
3000	267.776	737.752	464.190	820.686	-2242.470	-1259.354	21.927
3100	267.776	746.533	473.157	847.464	-2239.852	-1226.626	20.669
3200	267.776	755.034	481.834	874.241	-2237.280	-1193.984	19.490
3300	267.776	763.274	490.238	901.019	-2234.751	-1161.421	18.384
3400	267.776	771.268	498.387	927.796	-2232.264	-1128.932	17.344
3500	267.776	779.030	506.295	954.574	-2229.818	-1096.518	16.365
3600	267.776	786.574	513.976	981.352	-2227.409	-1064.171	15.441
3700	267.776	793.910	521.443	1008.129	-3452.333	-1008.528	14.238
3800	267.776	801.051	528.708	1034.907	-3446.295	-942.558	12.956
3900	267.776	808.007	535.780	1061.684	-3440.457	-876.746	11.743
4000	267.776	814.787	542.671	1088.462	-3434.815	-811.083	10.592

PREVIOUS: CURRENT: December 1973

Titanium Oxide (Ti_3O_5) O_5Ti_3(cr,l)

VANADIUM OXIDE (V_2O_5) CRYSTAL M_r = 181.8800

$S°(298.15 K) = 130.549 \pm 2.1 J K^{-1} mol^{-1}$

$T_{fus} = 943 \pm 10 K$

$\Delta_f H°(0 K) = -1541.109 \pm 6.3 kJ mol^{-1}$

$\Delta_f H°(298.15 K) = -1550.590 \pm 6.3 kJ mol^{-1}$

$\Delta_{fus} H° = 64.517 \pm 3.14 kJ mol^{-1}$

Enthalpy of Formation

A review (1) of early experimental studies led to a value of -373 kcal mol^{-1} for $\Delta_f H°(V_2O_5$, cr, 298.15 K). More recent combustion studies on V(cr) have led to the following $\Delta_f H°$(298.15 K) values for V_2O_5(cr).

Source	Year	$\Delta_f H°(V_2O_5$, cr, 298.15 K), kcal mol^{-1}
(2)	1969	-376 ± 2
(3)	1960	-370 ± 1
(4)	1961	-370.6 ± 0.5
(5)	1963	-371.9 ± 1.0

We adopt -370.6 ± 1 kcal mol^{-1} as the enthalpy of formation of V_2O_5(cr) at 298.15 K. This value is the Mah and Kelley result (4) but it is also in good agreement with the results of Golutvin and Kozlovskaya (3) and Volkova and Gel'd (5). The same value was recently adopted by NBS (6).

The combustion of vanadium and its oxides, as reported by Mah and Kelley (4), led to two problems: (1) the oxide products liquefy before the pentoxide composition is obtained and (2) the oxide products are corrosive and cause a container problem. The combustion of vanadium led to varying mixtures of V_2O_4(rutile) and V_2O_5. This was verified by x-ray diffraction. The enthalpy of formation values for V_2O_4 and V_2O_5 were derived from the same set of experiments (4).

Heat Capacity and Entropy

Using a V_2O_5 sample with a reported purity of 99.5% or better, Anderson (7) measured the heat capacity in the range 56.8-289.5 K. Using the combination of Debye and Einstein functions suggested by Anderson (7), we calculate S°(60 K) = 2.81 cal K^{-1} mol^{-1} and H°(60 K)-H°(0 K) = 130.59 cal mol^{-1}. Cook (8) measured the enthalpy of V_2O_5 in the range 372-1513 K. The enthalpy data are fit to a Kelley type equation with the constraint that the data join smoothly with a portion of the integrated data of Anderson (7). The enthalpy data appear to lie too high in comparison with the integrated data of Anderson (7). A better fit is obtained by neglecting the 289.5 data point of Anderson. The Anderson data show considerable scatter and deviate from our adopted $C_p°$ values by ±4% below 65 K and ±1.8% between 65 and 280 K. The 289.5 K data point is 2.7% lower than the adopted value. The Cook data deviates from the adopted enthalpy values by roughly 8% (200 cal mol^{-1}) at 371.6 K with the difference decreasing to 2 cal K^{-1} mol^{-1} at 754.4 K. In the range 771.1-921.1 K the enthalpy difference ranges from -121 to +139 cal K^{-1} mol^{-1}.

Fusion Data

The melting point of V_2O_5 is adopted as T_{fus} = 943 ± 10 K as derived from the enthalpy studies of Cook (8). The heat of melting is calculated to be $\Delta_{fus}H°$ = 15.42 ± 0.75 kcal mol^{-1} based on our analyses of Cook's enthalpy data. Refer to the heat capacity discussion for V_2O_5(cr) and V_2O_5(ℓ) for details. By way of comparison, Kohlmuller and Leitner (9) reported T_{fus} = 937.9 K and $\Delta_{fus}H°$ = 14.5 kcal mol^{-1} while Pantony and Vasu (10) reported T_{fus} = 951.12 K and $\Delta_{fus}H°$ = 13.8 kcal mol^{-1}. The latter two studies (9, 10) involved cryoscopic techniques. Murphy and West (11) reported T_{fus} = 943 K based on DTA measurements and Holtzberg et al. (12) determined T_{fus} = 947 ± 5 K in their study of the phases in the K_2O-V_2O_5 system.

The variance in the reported T_{fus} values is probably attributable to the presence of V_2O_4(cr). Refer to the discussion in the section on Sublimation Data in the V_2O_5(cr) table and Decomposition Data in the V_2O_5(ℓ) table.

Sublimation Data

Using the combination of Knudsen effusion and mass spectrometric techniques, Berkowitz et al. (13) concluded that V_2O_5(cr) did not "vaporize" invariantly. The temperature range investigated was not reported. Their conclusion was based on two observations: (1) some reduced vanadium oxide species were present in the gas phase, and (2) oxygen was also found to be effusing from the Knudsen cell. Chizhikov et al. (14) studied the dissociation of V_2O_5 at temperatures below T_{fus} using mass spectrometry. They assumed reduced vanadium oxides, in addition to V_4O_{10}, were present in the vapor upon dissociation of V_2O_5(cr) in vacuum. We conclude that there is no invariant sublimation process for V_2O_5(cr) and thus we do not calculate a $\Delta_{sub}H°$(298.15 K) value. For related discussion refer to V_2O_5(ℓ) table.

References

1. U. S. Nat. Bur. Stand. Circ. 500, 1952.
2. E. Vol'f and S. M. Ariya, Zhur. Obshchei Khim. 29, 2470 (1959).
3. Yu. M. Golutvin and T. M. Kozlovskaya, Russ. J. Phys. Chem. 34, 1116 (1960).
4. A. D. Mah and K. K. Kelley, U. S. Bur. Mines RI 5858, 1961.
5. N. M. Volkova and P. V. Gel'd, Izv. Vysshikh. Uchebn. Zavedenii, Tsvetn. Met. 6, 89 (1963); Chem. Abstr. 60, 3554f (1964).
6. U. S. Nat. Bur. Stand. Tech. Note 270-5, 1971.
7. C. T. Anderson, J. Amer. Chem. Soc. 58, 564 (1936).
8. O. A. Cook, J. Amer. Chem. Soc. 69, 331 (1947).
9. R. Kohlmuller and N. Leitner, Rev. Chim. Minerale (Paris) 1, 275 (1964).
10. D. A. Pantony and K. I. Vasu, J. Inorg. Nucl. Chem. 30, 433 (1968).
11. C. B. Murphy and R. R. West, Ind. Eng. Chem. 51, 952 (1959).
12. F. Holtzberg, A. Reisman, M. Berry and M. Birkenblit, J. Amer. Chem. Soc. 78, 1536 (1956).
13. J. Berkowitz, W. A. Chupka and M. G. Inghram, J. Chem. Phys. 27, 87 (1957).
14. D. M. Chizhikov, Yu. A. Pavlov, Yu. V. Tsvetkov, E. K. Kazenas, and P. A. Nesterenko, Izv. Vyssh. Ucheb. Zaved., Chem. Met. 13, 5 (1970); Chem. Abstr. 74, 6712h (1971).

Vanadium Oxide (V_2O_5) O_5V_2(cr)

Enthalpy Reference Temperature = T_r = 298.15 K   Standard State Pressure = p° = 0.1 MPa

T/K	$C_p°$ (J $K^{-1}mol^{-1}$)	S° (J $K^{-1}mol^{-1}$)	$-[G°-H°(T_r)]/T$ (J $K^{-1}mol^{-1}$)	$H°-H°(T_r)$ (kJ mol^{-1})	$\Delta_f H°$ (kJ mol^{-1})	$\Delta_f G°$ (kJ mol^{-1})	Log K_f
0	0.000	0.000	INFINITE	-21.506	-1541.109	-1541.109	INFINITE
100	52.647	31.564	238.170	-20.661	-1548.502	-1506.895	787.121
200	101.123	84.203	141.649	-11.489	-1550.281	-1462.556	381.980
298.15	130.599	130.549	130.549	0.000	-1550.590	-1419.348	248.664
300	131.027	131.358	130.552	0.242	-1550.577	-1418.533	246.989
400	150.950	171.988	135.954	14.414	-1548.962	-1374.719	179.520
500	161.753	206.948	146.747	30.100	-1546.243	-1331.458	139.096
600	168.314	237.044	159.348	46.617	-1543.068	-1288.795	112.200
700	173.234	263.370	172.367	63.702	-1539.671	-1246.682	93.029
800	177.251	286.771	185.232	81.231	-1536.152	-1205.066	78.683
900	180.749	307.858	197.705	99.138	-1532.562	-1163.895	67.551
943.000	182.004	316.323	202.922	106.937	------ CRYSTAL <--> LIQUID -----		
1000	183.678	327.054	209.694	117.359	-1528.939	-1123.126	58.666
1100	186.606	344.697	221.176	135.874	-1525.292	-1082.721	51.414
1200	189.535	361.060	232.159	154.681	-1521.624	-1042.649	45.385
1300	192.464	376.346	242.669	173.781	-1517.934	-1002.884	40.296
1400	195.393	390.717	252.736	193.173	-1514.230	-963.403	35.945
1500	198.322	404.297	262.391	212.859	-1510.509	-924.189	32.183

PREVIOUS: CURRENT: June 1973

Vanadium Oxide (V_2O_5) O_5V_2(cr)

VANADIUM OXIDE (V_2O_5) LIQUID M_r = 181.8880

$S°(298.15\ K)$ = [191.956] $J\ K^{-1}\ mol^{-1}$
T_{fus} = 943 ± 10 K
T_{dcm} = 1963 K

$\Delta_f H°(298.15\ K)$ = [-1491.209] $kJ\ mol^{-1}$
$\Delta_{fus} H°$ = 64.517 ± 3.14 $kJ\ mol^{-1}$

Enthalpy of Formation

$\Delta_f H°(V_2O_5, \ell, 298.15\ K)$ is calculated from that of the crystal by adding $\Delta_{fus}H°$ and the difference in enthalpy, H°(943 K)-H°(298.15 K), between $V_2O_5(cr)$ and $V_2O_5(\ell)$.

Heat Capacity and Entropy

Cook (1) measured the enthalpy of V_2O_5 in the range 373-1513 K. In the liquid region (T > 943 K), Cook (1) represented his data by a straight line with a slope of $C_p°$ = 45.60 cal $K^{-1}\ mol^{-1}$. We adopt this value of $V_2O_5(\ell)$. A glass transition is assumed at 600 K. The heat capacity values used below 600 K are those of $V_2O_5(cr)$. The entropy at 298.15 K is calculated in a manner analogous to that used in calculating the enthalpy of formation.

Decomposition Data

Polyakov (2) measured the weight loss of molten V_2O_5 in the range 700-1200°C for various oxygen flows. The oxygen pressure was greater than the dissociation pressure of V_2O_5. The vapor pressure was calculated by extrapolating to zero oxygen flow and assuming the vapor was monomeric $V_2O_5(g)$. Polyakov (2) assumed no loss of oxygen. Pak (3) proposed a method of calculating vapor pressures from experimental results obtained by the entrainment method with allowance for diffusion. Using Polystov's data and assuming that the vapor phase is the dimeric molecule $V_4O_{10}(g)$, Pak (3) calculated a heat of vaporization $\Delta_{vap}H°(298.15\ K)$ which is 1.4 kcal mol^{-1} more positive than Polyakov's reported value of 32.3 kcal mol^{-1} V_2O_5. Semenov et al. (4) determined the vapor pressure of $V_4O_{10}(g)$ over $V_2O_5(\ell)$ at 1215-1530 K by an improved transfer technique in an oxygen carrier stream. The reported heat of vaporization was calculated to be 3.5 kcal mol^{-1} more positive than Polyakov's value.

Pantony and Vasu (5) stated that molten V_2O_5 attains a constant weight under fixed oxygen pressure and temperature in a reasonably short length of time while in pure nitrogen it is converted into $V_2O_4(cr)$. Pantony and Vasu (5) studied gravimetrically the dissociation equilibrium of $V_2O_5(\ell)$ and confirmed its existence by cryoscopic methods. They concluded that the vapor pressure ascribed to $V_2O_5(g)$ or $V_4O_{10}(g)$ and measured by a transpiration-weight loss method is due to the loss of oxygen arising from the following dissociation $V_2O_5(soln) = V_2O_4(soln) + 1/2\ O_2(g)$.

An effusion-mass spectrometric investigation of the vapor in equilibrium with V_2O_5 in the range 1000-1200 K led Farber et al. (6) to the conclusion that the vapor contained not only $V_4O_{10}(g)$, significant amounts of $V_4O_8(g)$, but also other vanadium oxides. We conclude that there is decomposition upon heating $V_2O_5(\ell)$ in an inert atmosphere or in vacuo. The use of an oxygen atmosphere suppresses decomposition but in the transpiration-weight loss experiment it has not been verified that the weight loss is indeed $V_4O_{10}(g)$. We calculate a decomposition temperature, T_{dcm}, as that temperature at which the free energy approaches zero for the reaction $V_2O_5(\ell) = V_2O_4(\ell) + 1/2\ O_2(g)$.

References

1. O. A. Cook, J. Amer. Chem. Soc. 69, 331 (1947).
2. A. Yu. Polyakov, J. Phys. Chem. (USSR) 20, 1021 (1946).
3. T. A. Pak, J. Phys. Chem. 46, 1210 (1972).
4. G. A. Semenov, K. E. Frantseva, and E. K. Shalkova, Vestn. Leningrad. Univ. Fiz. Khim. 1970, 82 (1970); Chem. Abstr. 74, 34934a (1971).
5. D. A. Pantomy and I. I. Vasu, J. Inorg. Nucl. Chem. 30, 433 (1968).
6. M. Farber, O. M. Uy and R. D. Srivastava, J. Chem. Phys. 56, 5312 (1972).

Vanadium Oxide (V_2O_5) $O_5V_2(l)$

Enthalpy Reference Temperature = T_r = 298.15 K Standard State Pressure = p° = 0.1 MPa

T/K	$C_p°$ (J K⁻¹mol⁻¹)	S° (J K⁻¹mol⁻¹)	-[G°-H°(T_r)]/T (J K⁻¹mol⁻¹)	H°-H°(T_r) (kJ mol⁻¹)	$\Delta_f H°$ (kJ mol⁻¹)	$\Delta_f G°$ (kJ mol⁻¹)	Log K_f
0							
100							
200							
298.15	130.599	191.956	191.956	0.000	-1491.209	-1378.274	241.468
300	131.027	192.765	191.959	0.242	-1491.195	-1377.574	239.857
400	150.950	233.396	197.362	14.414	-1489.580	-1339.900	174.973
500	161.753	268.355	208.154	30.100	-1486.861	-1302.780	136.100
600	168.314	298.477	220.757	46.632	-1483.672	-1266.258	110.238
600.000	168.314	298.477	220.757	46.632	GLASS <--> LIQUID TRANSITION		
600.000	190.790	298.477	220.757	46.632			
700	190.790	327.887	234.015	65.711	-1478.280	-1230.454	91.818
800	190.790	353.364	247.377	84.790	-1473.211	-1195.400	78.052
900	190.790	375.836	260.426	103.869	-1468.449	-1160.962	67.381
943.000	190.790	384.740	265.893	112.073	CRYSTAL <--> LIQUID		
1000	190.790	395.938	272.990	122.948	-1463.969	-1127.040	58.871
1100	190.790	414.122	285.006	142.027	-1459.757	-1093.553	51.928
1200	190.790	430.723	296.468	161.106	-1455.817	-1060.437	46.160
1300	190.790	445.994	307.390	180.185	-1452.148	-1027.640	41.291
1400	190.790	460.133	317.802	199.264	-1448.758	-995.114	37.128
1500	190.790	473.296	327.734	218.343	-1445.643	-962.822	33.528
1600	190.790	485.610	337.221	237.422	-1442.797	-930.728	30.385
1700	190.790	497.176	346.293	256.501	-1440.236	-898.804	27.617
1800	190.790	508.082	354.981	275.580	-1437.966	-867.023	25.160
1900	190.790	518.397	363.313	294.659	-1436.009	-835.358	22.966
2000	190.790	528.183	371.314	313.738	-1434.379	-803.788	20.993
2100	190.790	537.492	379.008	332.817	-1433.097	-772.292	19.210
2200	190.790	546.368	386.415	351.896	-1477.943	-740.637	17.585
2300	190.790	554.849	393.555	370.975	-1477.681	-707.130	16.059
2400	190.790	562.969	400.446	390.054	-1477.475	-673.634	14.661
2500	190.790	570.757	407.104	409.133	-1477.324	-640.143	13.375
2600	190.790	578.240	413.543	428.212	-1477.227	-606.658	12.188
2700	190.790	585.440	419.777	447.292	-1477.182	-573.175	11.089
2800	190.790	592.379	425.818	466.371	-1477.188	-539.693	10.068
2900	190.790	599.074	431.678	485.450	-1477.243	-506.212	9.118
3000	190.790	605.542	437.366	504.529	-1477.348	-472.725	8.231

PREVIOUS: CURRENT: June 1973

Vanadium Oxide (V_2O_5) $O_5V_2(l)$

VANADIUM OXIDE (V_2O_5) CRYSTAL–LIQUID M_r = 181.8800

0 to 943 K crystal
above 943 K liquid

Refer to the individual tables for details.

Vanadium Oxide (V_2O_5) O_5V_2(cr,l)

Enthalpy Reference Temperature = T_r = 298.15 K Standard State Pressure = p° = 0.1 MPa

T/K	C_p° (J K⁻¹mol⁻¹)	S° (J K⁻¹mol⁻¹)	$-[G^\circ-H^\circ(T_r)]/T$ (J K⁻¹mol⁻¹)	$H^\circ-H^\circ(T_r)$ (kJ mol⁻¹)	Δ_fH° (kJ mol⁻¹)	Δ_fG° (kJ mol⁻¹)	Log K_f
0	0.000	0.000	INFINITE	-21.506	-1541.109	-1541.109	INFINITE
100	52.647	31.564	238.170	-20.661	-1548.502	-1506.895	787.121
200	101.123	84.203	141.649	-11.489	-1550.281	-1462.556	381.980
298.15	130.599	130.549	130.549	0.000	-1550.590	-1419.348	248.664
300	131.027	131.358	130.552	0.242	-1550.577	-1418.533	246.989
400	150.950	171.988	135.954	14.414	-1548.962	-1374.719	179.520
500	161.753	206.948	146.747	30.100	-1546.243	-1331.458	139.096
600	168.314	237.044	159.348	46.617	-1543.068	-1288.795	112.200
700	173.234	263.370	172.367	63.702	-1539.671	-1246.682	93.029
800	177.251	286.771	185.232	81.231	-1536.152	-1205.066	78.683
900	180.749	307.858	197.705	99.138	-1532.562	-1163.895	67.551
943.000	182.004	316.323	202.922	106.937	CRYSTAL <--> LIQUID	TRANSITION	
943.000	190.790	384.740	202.922	171.455			
1000	190.790	395.938	213.608	182.330	-1463.969	-1127.040	58.871
1100	190.790	414.122	231.023	201.409	-1459.757	-1093.553	51.928
1200	190.790	430.723	246.983	220.488	-1455.817	-1060.437	46.160
1300	190.790	445.994	261.712	239.567	-1452.148	-1027.640	41.291
1400	190.790	460.133	275.386	258.646	-1448.758	-995.114	37.128
1500	190.790	473.296	288.146	277.725	-1445.643	-962.822	33.528
1600	190.790	485.610	300.107	296.804	-1442.797	-930.728	30.385
1700	190.790	497.176	311.363	315.883	-1440.236	-898.804	27.617
1800	190.790	508.082	321.992	334.962	-1437.966	-867.023	25.160
1900	190.790	518.397	332.060	354.041	-1436.009	-835.358	22.966
2000	190.790	528.183	341.623	373.120	-1434.379	-803.788	20.993
2100	190.790	537.492	350.731	392.199	-1433.097	-772.292	19.210
2200	190.790	546.368	359.423	411.278	-1477.943	-740.637	17.585
2300	190.790	554.849	367.737	430.357	-1477.681	-707.130	16.059
2400	190.790	562.969	375.703	449.436	-1477.475	-673.634	14.661
2500	190.790	570.757	383.351	468.515	-1477.324	-640.143	13.375
2600	190.790	578.240	390.704	487.594	-1477.227	-606.658	12.188
2700	190.790	585.440	397.784	506.673	-1477.182	-573.175	11.089
2800	190.790	592.379	404.610	525.752	-1477.188	-539.693	10.068
2900	190.790	599.074	411.201	544.831	-1477.243	-506.212	9.118
3000	190.790	605.542	417.572	563.910	-1477.348	-472.725	8.231

PREVIOUS: CURRENT: June 1973

Vanadium Oxide (V_2O_5) O_5V_2(cr,l)

PHOSPHORUS OXIDE $((P_2O_3)_2)$ IDEAL GAS $M_r = 219.89144$

$S°(298.15\ K) = 345.641 \pm 33.5\ J\ K^{-1}\ mol^{-1}$

$\Delta_f H°(0\ K) = -2189.97 \pm 3.35\ kJ\ mol^{-1}$

$\Delta_f H°(298.15\ K) = -2214.31 \pm 33.5\ kJ\ mol^{-1}$

Vibrational Frequencies and Degeneracies

ν, cm^{-1}	ν, cm^{-1}
613 (1)	465 (2)
613 (1)	643 (3)
302 (3)	919 (3)
370 (3)	1029 (2)
407 (3)	636 (3)

Ground State Quantum Weight: 1 $\sigma = 12$

Point Group: T_d

Bond Distances: P–O = 1.65 ± 0.02 Å

Bond Angles: P–O–P = 127.5 ± 1.0° O–P–O = 99° ± 1°

Product of the Moments of Inertia: $I_A I_B I_C = 4.838017 \times 10^{-112}\ g^3\ cm^6$

Enthalpy of Formation

$\Delta_f H°(P_4O_6$, ℓ, 298.15 K) was measured by Koerner and Daniels (1). The measurement was made by burning samples of red phosphorus in nitric oxide in a flow calorimeter. Calculation of the enthalpy of formation of the gas was made using $\Delta_{vap}H°(298.15\ K) = 11.46\ kcal\ mol^{-1}$ as reported by Farr (2).

Heat Capacity and Entropy

The vibrational levels are from the Raman data of Gerding et al. (3) and infrared spectra data of Sidorov and Sobolev (4). Bond distances and angles are from electron diffraction data by Hampson and Stosick (5).

References

1. W. E. Koerner and F. Daniels, J. Chem. Phys. 20, 113 (1952).
2. T. D. Farr, Tenn. Valley Authority, Chem. Engr. Report No. 8 (1950).
3. H. Gerding, H. van Brederode and H. C. J. de Decker, Rec. Trav. Chim. 61, 549 (1942).
4. T. A. Sidorov and N. N. Sobolev, Optics and Spect. 2, 710 (1957).
5. G. C. Hampson and A. J. Stosick, J. Amer. Chem. Soc. 60, 1814 (1938).

Phosphorus Oxide $((P_2O_3)_2)$ $O_6P_4(g)$

Enthalpy Reference Temperature = T_r = 298.15 K; Standard State Pressure = p° = 0.1 MPa

T/K	$C°_p$ (J K⁻¹mol⁻¹)	S° (J K⁻¹mol⁻¹)	$-[G°-H°(T_r)]/T$ (J K⁻¹mol⁻¹)	$H°-H°(T_r)$ (kJ mol⁻¹)	$\Delta_f H°$ (kJ mol⁻¹)	$\Delta_f G°$ (kJ mol⁻¹)	Log K_f
0	0.	0.	INFINITE	-23.153	-2189.970	-2189.970	INFINITE
100	47.025	247.701	443.275	-19.557	-2198.637	-2164.340	1130.535
200	101.296	296.685	357.555	-12.174	-2209.040	-2126.586	555.407
250	125.151	321.933	347.911	-6.494	-2212.101	-2105.599	439.941
298.15	143.937	345.641	345.641	0.	-2214.307	-2084.877	365.262
300	144.581	346.534	345.644	0.267	-2214.379	-2084.074	362.870
350	160.035	370.028	347.464	7.897	-2218.922	-2061.933	307.727
400	172.246	392.227	351.686	16.216	-2220.352	-2039.401	266.319
450	181.911	413.094	357.362	25.080	-2221.305	-2016.721	234.095
500	189.609	432.674	363.925	34.375	-2221.900	-1993.955	208.307
600	200.809	468.305	378.414	53.935	-2222.349	-1948.310	169.615
700	208.312	499.860	393.554	74.414	-2222.164	-1902.645	141.977
800	213.527	528.037	408.635	95.521	-2221.598	-1857.035	121.252
900	217.274	553.414	423.335	117.071	-2220.794	-1811.511	105.137
1000	220.045	576.457	437.513	138.944	-2219.838	-1766.085	92.251
1100	222.147	597.533	451.116	161.058	-2218.781	-1720.759	81.712
1200	223.776	616.935	464.137	183.358	-2471.947	-1671.220	72.746
1300	225.062	634.899	476.590	205.802	-2467.683	-1604.666	64.476
1400	226.095	651.617	488.502	228.362	-2463.415	-1538.440	57.400
1500	226.935	667.246	499.903	251.014	-2459.153	-1472.520	51.278
1600	227.628	681.914	510.825	273.744	-2454.906	-1406.883	45.930
1700	228.206	695.732	521.299	296.536	-2450.683	-1341.511	41.220
1800	228.693	708.790	531.356	319.382	-2446.490	-1276.387	37.040
1900	229.107	721.166	541.023	342.272	-2442.334	-1211.495	33.306
2000	229.462	732.927	550.327	365.201	-2438.217	-1146.820	29.952
2100	229.768	744.130	559.291	388.163	-2434.145	-1082.351	26.922
2200	230.034	754.825	567.937	411.154	-2430.121	-1018.072	24.172
2300	230.267	765.056	576.287	434.169	-2426.148	-953.979	21.666
2400	230.471	774.861	584.358	457.206	-2422.226	-890.057	19.372
2500	230.652	784.273	592.168	480.262	-2418.358	-826.296	17.265
2600	230.813	793.322	599.732	503.336	-2414.543	-762.689	15.323
2700	230.957	802.036	607.064	526.424	-2410.783	-699.227	13.527
2800	231.085	810.438	614.178	549.527	-2407.077	-635.903	11.863
2900	231.201	818.549	621.086	572.641	-2403.425	-572.714	10.316
3000	231.305	826.389	627.800	595.766	-2399.825	-509.646	8.874
3100	231.399	833.975	634.329	618.902	-2396.278	-446.697	7.527
3200	231.485	841.323	640.683	642.046	-2392.781	-383.865	6.266
3300	231.564	848.447	646.872	665.199	-2389.333	-321.141	5.083
3400	231.635	855.361	652.903	688.358	-2385.933	-258.518	3.972
3500	231.700	862.076	658.784	711.525	-2382.581	-195.997	2.925
3600	231.761	868.605	664.522	734.698	-2379.273	-133.570	1.938
3700	231.816	874.955	670.124	757.877	-2376.009	-71.235	1.006
3800	231.867	881.138	675.596	781.061	-2372.789	-8.985	0.124
3900	231.914	887.162	680.943	804.251	-2369.609	53.180	-0.712
4000	231.958	893.034	686.173	827.444	-2366.471	115.261	-1.505
4100	231.999	898.762	691.288	850.642	-2363.372	177.266	-2.258
4200	232.036	904.353	696.295	873.844	-2360.312	239.195	-2.975
4300	232.071	909.813	701.197	897.049	-2357.291	301.054	-3.657
4400	232.104	915.149	705.999	920.258	-2354.308	362.840	-4.307
4500	232.135	920.365	710.705	943.470	-2351.363	424.559	-4.928
4600	232.163	925.468	715.319	966.685	-2348.457	486.213	-5.521
4700	232.190	930.461	719.843	989.903	-2345.590	547.806	-6.088
4800	232.216	935.350	724.282	1013.123	-2342.762	609.338	-6.631
4900	232.239	940.138	728.639	1036.346	-2339.975	670.813	-7.151
5000	232.262	944.830	732.916	1059.571	-2337.229	732.230	-7.650
5100	232.283	949.430	737.116	1082.798	-2334.527	793.593	-8.128
5200	232.302	953.940	741.243	1106.027	-2331.869	854.900	-8.588
5300	232.321	958.365	745.298	1129.258	-2329.258	916.161	-9.029
5400	232.339	962.708	749.284	1152.491	-2326.696	977.370	-9.454
5500	232.355	966.971	753.203	1175.726	-2324.185	1038.532	-9.863
5600	232.371	971.158	757.058	1198.962	-2321.727	1099.653	-10.257
5700	232.386	975.271	760.850	1222.200	-2319.326	1160.726	-10.637
5800	232.400	979.313	764.582	1245.440	-2316.984	1221.761	-11.003
5900	232.414	983.286	768.255	1268.680	-2314.704	1282.749	-11.357
6000	232.427	987.192	771.872	1291.922	-2312.491	1343.707	-11.698

PREVIOUS: December 1962 (1 atm) CURRENT: December 1962 (1 bar)

Phosphorus Oxide $((P_2O_3)_2)$ $O_6P_4(g)$

TUNGSTEN OXIDE $((WO_3)_2)$ IDEAL GAS $M_r = 463.6964$

$S°(298.15\ K) = [415.601]\ J\ K^{-1}\ mol^{-1}$

$\Delta_f H°(0\ K) = -1157.96\ kJ\ mol^{-1}$
$\Delta_f H°(298.15\ K) = -1163.99 \pm 41.8\ kJ\ mol^{-1}$

Vibrational Frequencies and Degeneracies

ν, cm^{-1}
[600] (4)
[300] (8)
[172] (6)

Ground State Quantum Weight: [1] $\sigma = [4]$
Point Group: $[D_{2h}]$
Bond Distance: W–O = [1.81] Å
Bond Angles: O–W–O = [109.47]° $W–O_{bridge}–W$ = [70.53]°
Product of the Moments of Inertia: $I_A I_B I_C = [5.918184 \times 10^{-112}]\ g^3\ cm^6$

Enthalpy of Formation

The adopted value, $\Delta_f H°(W_2O_6,\ g,\ 298.15\ K) = -278.2\ kcal\ mol^{-1}$, was reduced from $\Delta_f H°(1450\ K) = -272.5\ kcal\ mol^{-1}$, reported by Ackermann and Rauh (1). They have studied the sublimation behavior of the tungsten-oxygen system over the temperature range from 1300 to 1600 K by mass spectrometry, mass effusion, and X-ray diffraction, and also have determined $\Delta_f G°(W_2O_6,\ g) = 272,500 + 54.98T\ cal\ K^{-1}\ mol^{-1}$ from partial pressure measurements over the $W\text{-}WO_2(cr)$ system.

Battles (2) has also studied the same system in the temperature range from 1418 to 1527 K. Third law calculation of the partial pressure data gives $\Delta_r H°(298.15\ K) = 144.0\ kcal\ mol^{-1}$ for $3\ WO_2(cr) = W_2O_6(g) + W(cr)$ which yields $\Delta_f H°(W_2O_6,\ g,\ 298.15\ K) = -278.8\ kcal\ mol^{-1}$ using all JANAF functions. This value is in good agreement with the value adopted.

Norman and Staley (3) have obtained $\Delta_r H°(1800\ K) = -127\ kcal\ mol^{-1}$ for $2\ WO_3(g) = W_2O_6(g)$ by mass spectrometric study of the vapor over WO_3 dissolved in $CaO\text{-}Al_2O_3\text{-}SiO_2$. The value of $\Delta_r H°$ was determined by the 2nd law method from ion intensities and leads to $\Delta_f H°(W_2O_6,\ g,\ 298.15\ K) = -275\ kcal\ mol^{-1}$, using all JANAF functions (dated Sept. 30, 1966). The absolute pressure values reported by Norman and Staley appear to be quite inconsistent with the JANAF functions since they lead to $\Delta_f H°(W_2O_6,\ g,\ 298.15\ K) = -246\ kcal\ mol^{-1}$ by the 3rd law method.

Source	Reaction	T/K	$\Delta_r H°(298.15\ K)$, kcal mol^{-1} 2nd law	3rd law	Drift cal K^{-1} mol^{-1}	$\Delta_f H°(298.15\ K)$ kcal mol^{-1}
(1)	$3\ WO_2(cr) = W_2O_6(g) + W(cr)$	1300-1600	144.54	144.63	0.01	-278.2*
(2)	$3\ WO_2(cr) = W_2O_6(g) + W(cr)$	1418-1527	142.66	143.97	0.85 ± 0.40	-278.8*
(3)	$2\ WO_3(g) = W_2O_6(g)$	1700-1900	-135	-106		-275**

*3rd law value adopted in calculation **2nd law value adopted in calculation

Heat Capacity and Entropy

The heat capacity was approximated by making a tentative choice of the vibrational frequencies of 600 cm^{-1} (4), 300 cm^{-1} (8), and 172 cm^{-1} (6) in order to give the entropy at 1450 K, $S°(1450\ K) = 164.66\ cal\ K^{-1}\ mol^{-1}$, corresponding to $\Delta_f S°(1450\ K) = 54.98\ cal\ K^{-1}\ mol^{-1}$ which was reported by Ackermann and Rauh (1). The adopted value, $S°(W_2O_6,\ g,\ 298.15\ K) = 99.3\ cal\ K^{-1}\ mol^{-1}$, was reduced from the entropy at 1450 K, using JANAF functions. Second law calculation of the partial pressure data of Battles gives $\Delta_r S°(1470\ K) = 63.5\ cal\ K^{-1}\ mol^{-1}$ for $3\ WO_2(cr) = W_2O_6(g) + W(cr)$. This yields $S°(W_2O_6,\ g,\ 298.15\ K) = 97.9\ cal\ K^{-1}\ mol^{-1}$.

The molecular configuration for $W_2O_6(g)$ was estimated by assuming that the oxygens are tetrahedrally located around the tungsten atoms, and two of these tetrahedra share a common oxygen edge to form a planar four-member ring. The bond distance W-O was estimated as 1.81 Å and the bond angle O-W-O = 109.47° and $W\text{-}O_{bridge}\text{-}W = 70.53°$. The principal moments of inertia are: $I_A = 34.8158 \times 10^{-39}$, $I_B = 124.7049 \times 10^{-39}$, and $I_C = 136.3101 \times 10^{-39}\ g\ cm^2$.

References

1. R. J. Ackermann and E. G. Rauh, J. Phys. Chem. 67, 2596 (1963).
2. J. E. Battles, Ph.D. dissertation, "A Mass Spectrometric Investigation of Tungsten Dioxide and Tungsten Trioxide," the Ohio State University, 1964.
3. J. H. Norman and H. G. Staley, J. Chem. Phys. 43, 3804 (1965).

Tungsten Oxide $((WO_3)_2)$ $O_6W_2(g)$

Enthalpy Reference Temperature = T_r = 298.15 K Standard State Pressure = p° = 0.1 MPa

T/K	$C_p°$ (J $K^{-1}mol^{-1}$)	S°	$-[G°-H°(T_r)]/T$	$H°-H°(T_r)$ (kJ mol^{-1})	$\Delta_f H°$	$\Delta_f G°$	Log K_f
0	0.	0.	INFINITE	-29.966	-1157.958	-1157.958	INFINITE
100	81.358	284.301	535.415	-25.111	-1163.122	-1137.638	594.241
200	131.491	358.417	429.062	-14.129	-1164.889	-1111.172	290.208
250	144.838	389.289	418.085	-7.199	-1164.646	-1097.760	229.364
298.15	153.665	415.601	415.601	0.	-1163.989	-1084.931	190.075
300	153.946	416.552	415.604	0.285	-1163.957	-1084.441	188.818
350	160.314	440.789	417.503	8.150	-1162.969	-1071.263	159.877
400	164.881	462.511	421.795	16.286	-1161.796	-1058.240	138.192
450	168.240	482.134	427.427	24.618	-1160.519	-1045.372	121.344
500	170.768	499.997	433.805	33.096	-1159.192	-1032.649	107.880
600	174.231	531.464	447.530	50.360	-1156.521	-1007.592	87.719
700	176.420	558.498	461.497	67.901	-1153.947	-982.975	73.351
800	177.883	582.157	475.132	85.620	-1151.527	-958.718	62.598
900	178.907	603.171	488.213	103.463	-1149.279	-934.754	54.252
1000	179.649	622.061	500.669	121.392	-1147.203	-911.031	47.587
1100	180.205	639.211	512.496	139.386	-1145.294	-887.507	42.144
1200	180.630	654.910	523.719	157.429	-1143.547	-864.150	37.615
1300	180.963	669.382	534.374	175.509	-1141.956	-840.932	33.789
1400	181.229	682.802	544.503	193.619	-1140.519	-817.832	30.514
1500	181.444	695.314	554.145	211.753	-1139.234	-794.829	27.678
1600	181.621	707.029	563.338	229.907	-1138.101	-771.906	25.200
1700	181.767	718.045	572.117	248.077	-1137.121	-749.050	23.015
1800	181.890	728.438	580.516	266.260	-1136.295	-726.248	21.075
1900	181.995	738.275	588.562	284.454	-1135.623	-703.486	19.340
2000	182.084	747.612	596.283	302.658	-1135.109	-680.756	17.780
2100	182.161	756.498	603.703	320.870	-1134.753	-658.047	16.368
2200	182.228	764.974	610.842	339.090	-1134.556	-635.351	15.085
2300	182.286	773.076	617.721	357.316	-1134.519	-612.662	13.914
2400	182.338	780.835	624.357	375.547	-1134.644	-589.972	12.840
2500	182.383	788.279	630.766	393.783	-1134.929	-567.271	11.852
2600	182.423	795.433	636.963	412.023	-1135.368	-544.557	10.940
2700	182.459	802.319	642.960	430.268	-1136.015	-521.821	10.095
2800	182.491	808.955	648.771	448.515	-1136.937	-499.056	9.310
2900	182.520	815.359	654.405	466.766	-1138.213	-476.256	8.578
3000	182.546	821.547	659.874	485.019	-1139.894	-453.401	7.894
3100	182.569	827.533	665.187	503.275	-1142.066	-430.482	7.254
3200	182.590	833.330	670.351	521.533	-1144.789	-407.488	6.652
3300	182.610	838.949	675.375	539.793	-1148.112	-384.399	6.085
3400	182.628	844.401	680.267	558.055	-1152.105	-361.196	5.549
3500	182.644	849.695	685.032	576.318	-1156.894	-337.868	5.042
3600	182.659	854.840	689.678	594.583	-1162.669	-314.387	4.562
3700	182.673	859.845	694.210	612.850	-1239.161	-290.351	4.099
3800	182.685	864.717	698.633	631.118	-1240.331	-264.689	3.638
3900	182.697	869.462	702.953	649.387	-1241.541	-238.998	3.201
4000	182.708	874.088	707.174	667.657	-1242.790	-213.277	2.785
4100	182.718	878.599	711.300	685.928	-1244.078	-187.524	2.389
4200	182.727	883.003	715.336	704.201	-1245.402	-161.738	2.012
4300	182.736	887.302	719.285	722.474	-1246.762	-135.919	1.651
4400	182.744	891.504	723.152	740.748	-1248.159	-110.072	1.307
4500	182.752	895.610	726.939	759.023	-1249.591	-84.191	0.977
4600	182.759	899.627	730.649	777.298	-1251.059	-58.278	0.662
4700	182.765	903.558	734.287	795.574	-1252.563	-32.331	0.359
4800	182.772	907.406	737.853	813.851	-1254.104	-6.352	0.069
4900	182.778	911.174	741.352	832.129	-1255.683	19.661	-0.210
5000	182.783	914.867	744.786	850.407	-1257.300	45.705	-0.477
5100	182.788	918.487	748.156	868.685	-1258.958	71.782	-0.735
5200	182.793	922.036	751.466	886.964	-1260.656	97.888	-0.983
5300	182.798	925.518	754.717	905.244	-1262.398	124.034	-1.222
5400	182.802	928.935	757.912	923.524	-1264.187	150.208	-1.453
5500	182.806	932.289	761.052	941.804	-1266.022	176.416	-1.675
5600	182.810	935.583	764.139	960.085	-1267.907	202.662	-1.890
5700	182.814	938.819	767.176	978.366	-1269.846	228.937	-2.098
5800	182.817	941.998	770.162	996.648	-1271.840	255.252	-2.299
5900	182.821	945.123	773.101	1014.930	-1273.892	281.593	-2.493
6000	182.824	948.196	775.994	1033.212	-1276.007	307.978	-2.681

PREVIOUS: September 1966 (1 atm) CURRENT: September 1966 (1 bar)

Tungsten Oxide $((WO_3)_2)$ $O_6W_2(g)$

TITANIUM OXIDE (Ti_4O_7) CRYSTAL M_r = 303.5158

$S°(298.15\ K) = 198.740 \pm 12.6\ J\ K^{-1}\ mol^{-1}$

$T_{trs1} = 130 \pm 8\ K$

$T_{trs2} = 150 \pm 3\ K$

$T_{fus} = 1950 \pm 20\ K$

$\Delta_f H°(0\ K)$ = Unknown

$\Delta_f H°(298.15\ K) = -3404.521 \pm 6.3\ kJ\ mol^{-1}$

$\Delta_{trs1}H° = 95 \pm 5\ kJ\ mol^{-1}$

$\Delta_{trs2}H° = 468 \pm 5\ kJ\ mol^{-1}$

$\Delta_{fus}H° = [225.936]\ kJ\ mol^{-1}$

Enthalpy of Formation

$\Delta_f H°$ is calculated from that of $Ti_3O_5(\beta)$ using $\Delta_r H°(298.15\ K) = -34.16\ kcal\ mol^{-1}$ for 4/3 $Ti_3O_5(\beta)$ + 1/6 $O_2(g)$ + $Ti_4O_7(cr)$. $\Delta_r H°$ is derived by the 3rd law method from two concordant studies (1, 2) of high-temperature equilibria. Merritt et al. (1) obtained isothermal gravimetric equilibrium over the two-phase system Ti_3O_5-Ti_4O_7 using (H_2 + H_2O) gas mixtures of known oxygen potential. Integration yields $\Delta_r G°(1304\ K) = -20.29 \pm 0.16\ kcal\ mol^{-1}$. Suzuki and Sambongi (2) reported solid-state emf data relative to the couple Cr/Cr_2O_3 over a range of ~500°. We convert their (graphical) emf data (2) to oxygen potentials, using the authors' self-consistent data for Cr/Cr_2O_3, and integrate to get $\Delta_r G°$. Vasil'eva and Shaulova (3) reported an equation for the oxygen potential based on their emf data relative to $Fe/Fe_{0.95}O$ over a range of ~90°. Equilibration data of Porter (4) are not considered because of the likelihood of oxidation during quenching of the composition (1).

3rd law analyses of the equilibria for 4/3 $Ti_3O_5(\beta)$ + 1/6 $O_2(g)$ + $Ti_4O_7(cr)$ are summarized below. Merritt et al. (1) agree closely with Suzuki and Sambongi (2) but the earlier emf study (3) yields a $\Delta_f H°$ which is 1.9 $kcal\ mol^{-1}$ less negative. All three studies agree with the earlier conclusion of Wahlbeck and Gilles (5) that Ti_4O_7 is stable with respect to disproportionation into $Ti_3O_5(\beta)$ and TiO_2(rutile). We dismiss the $\Delta_f H°$ derived from Vasil'eva (3) since it implies the contrary when combined with our adopted S° and $C_p°$. We adopt $\Delta_f H° = -813.7 \pm 1.5\ kcal\ mol^{-1}$, where the uncertainty includes the possibility that $\Delta_{trs}H°$ and S° of $Ti_3O_5(\beta)$ are biased (6).

Source	Method	T/K	δS cal K^{-1} mol^{-1}	$\Delta_r H°$(298.15 K), kcal mol^{-1} 2nd law	3rd law	$\Delta_f H°$(298.15 K) kcal mol^{-1}
(1)	Gravimetric	1304	-	-	-34.10	-813.63
(2)	emf	~1273-1773	0.1	-34.0	-34.20	-813.74
(3)	emf	1168-1257	2.0	-29.8	-32.31	-811.84

Heat Capacity and Entropy

$C_p°$ is based on constrained fitting of enthalpy data (679-1937 K) of Slyusar et al. (6). The constraints required that the enthalpy be zero at the reference temperature (298.15 K) and that $C_p° = 50\ cal\ K^{-1}\ mol^{-1}$ and $dC_p°/dT$ ~0.09 cal K^{-1} mol^{-1} at 300 K. The latter values are estimated by comparison with TiO_2, Ti_3O_5, Ti_2O_3 and TiO (6). Deviations of the data from the resulting fit have a pattern (6) similar to that found with the authors' data (7) for $Ti_3O_5(\beta)$. The enthalpies of Slyusar et al. (7) for TiO_2, Ti_2O_3, Ti_3O_5 and Ti_4O_7 are less than satisfactory due to inconsistencies in H near 1200 K and in the reported values of T_{fus} (6). Our constraints for Ti_4O_7 are an attempt to minimize the effects of the inconsistencies. They lead to a reasonable variation of $C_p°$ with temperature. We assume that there are no phase transitions above 298 K.

$S°(298.15\ K) = 47.5 \pm 3\ cal\ K^{-1}\ mol^{-1}$ is derived from the emf data of Suzuki and Sambongi (2). The uncertainty includes the possibility that $\Delta_{trs}H°$ and S° of $Ti_3O_5(\beta)$ are biased (6); excluding this possibility, we estimate an uncertainty of ±1.5 cal K^{-1} mol^{-1}. $S°(298.15\ K) = 49.5\ cal\ K^{-1}\ mol^{-1}$ can be derived from the less reliable data of Vasil'eva and Shaulova (3). These authors (3) derived $S°(298.15\ K) = 42\ cal\ K^{-1}\ mol^{-1}$ which is clearly too low. Unfortunately, the entropy derived form $C_p°$ data (12-300 K) of Berezovskii et al. (11) is unpublished and not yet available. These workers also measured $C_p°$ of the higher monologues Ti_5O_9, Ti_6O_{11} and Ti_9O_{17}.

Transition Data and Phase Data

Transition temperatures near 130 K and 150 K were found by use of DTA (8), elecrical conductivity (8), magnetic susceptibility (9) and x-ray diffraction (8, 10). The three crystalline phases are triclinic; the two transitions involve no change in symmetry but only rearrangements of atoms within the unit cell (10). There is obvious thermal hysteresis in the transition near 130 K. DTA endotherms (8) suggest that $\Delta_{trs}H°$ (150 K) is considerably larger than $\Delta_{trs}H°$(130 K). $\Delta_{trs}H°$ values presumably were obtained with the $C_p°$ data (11), but these are not yet available.

Ti_4O_7 is the lowest stable phase in the homologous series of structures Ti_nO_{2n-1} related to rutile. Phases exist with n=4,5,6,7,8,9 and certain higher integer values. These were reviewed in detail by Bursill and Hyde (12) in terms of the persistent structural phenomenon - crystallographic shear. Equilibrium data (1, 2) emphasize that $\Delta_r G°$ is negative, but were small in magnitude, for formation of Ti_nO_{2n-1} from its two nearest neighbors. Kinetic limitations can cause hysteresis in measured properties (1).

Fusion Data

See the liquid table for details.

Sublimation Data

Ti_4O_7(cr) vaporizes incongruently such that the residue approaches the congruently vaporizing composition of $Ti_3O_5(\beta)$.

References

1. R. R. Merritt, B. G. Hyde, L. A. Bursill and D. K. Philp, Phil. Trans. Roy. Soc. London, Ser. A 274, (1245), 627 (1973).
2. K. Suzuki and K, Sambongi, Tetsu To Hagane 58, 1579 (1972).
3. I. A. Vasil'eva and E. Yu. Shaulova, Russ. J. Phys. Chem. 43, 1713 (1969).
4. V. R. Porter, Thesis, Pennsylvania State Univ., 1965; Amer. Ceram. Soc. Bull. 43 (4), 263 (1964).
5. P. G. Wahlbeck and P. W. Gilles, J. Chem. Phys. 46, 2465 (1967).
6. JANAF Thermochemical Tables: TiO_2(rutile and anatase), $Ti_3O_5(\alpha,\beta)$, TiO(α), 12-31-73; Ti_2O_3, 6-30-73.
7. N. P. Slyusar, A. D. Krivorotenko, E. N. Fomichev, A. A. Kalashnik and V. P. Bondarenko, Teplofiz. Vys. Temp. 11, 213 (1978).
8. R. F. Bartholomew and D. R. Frankl, Phys. Rev. 187, 828 (1969).
9. W. J. Danley and L. N. Mulay, Mat. Res. Bull. 7, 739 (1972); J. Appl. Phys. 41, 877 (1970).
10. M. Marezio, D. B. McWhan, P. D. Dernier and J. P. Remeika, J. Solid State Chem. 6, 213 (1973).
11. G. A. Berezovskii, V. E. Ganenko and I. E. Paukov, Bulletin of Thermodynamics and Thermochemistry, No. 14, 19 (1971); No. 16, 75 (1973).
12. L. A. Bursill and B. G. Hyde, Progr. Solid State Chem. 7, 177 (1972).

Titanium Oxide (Ti_4O_7) O_7Ti_4(cr)

Enthalpy Reference Temperature = T_r = 298.15 K — Standard State Pressure = p° = 0.1 MPa

T/K	$C_p°$ (J $K^{-1}mol^{-1}$)	S°	$-[G°-H°(T_r)]/T$	$H°-H°(T_r)$ (kJ mol^{-1})	$\Delta_f H°$	$\Delta_f G°$	Log K_f
0							
100							
200							
298.15	208.489	198.740	198.740	0.000	-3404.521	-3213.016	562.907
300	209.200	200.032	198.744	0.386	-3404.512	-3211.828	559.229
400	239.911	264.765	207.361	22.961	-3402.790	-3147.798	411.060
500	258.864	320.495	224.551	47.972	-3399.447	-3084.414	322.226
600	271.165	368.848	244.663	74.511	-3395.268	-3021.790	263.070
700	279.659	411.321	265.500	102.075	-3390.647	-2959.904	220.871
800	285.976	449.097	286.132	130.372	-3385.729	-2898.703	189.266
900	290.830	483.070	306.158	159.221	-3380.763	-2838.124	164.720
1000	294.805	513.923	325.416	188.508	-3376.077	-2778.086	145.112
1100	298.152	542.182	343.855	218.159	-3371.968	-2718.492	129.090
1200	301.081	568.252	361.483	248.123	-3384.403	-2658.763	115.733
1300	303.675	592.456	378.330	278.363	-3378.626	-2598.529	104.410
1400	306.018	615.047	394.441	308.849	-3373.023	-2538.733	94.721
1500	308.235	636.236	409.861	339.563	-3367.654	-2479.330	86.338
1600	310.285	656.196	424.639	370.490	-3362.581	-2420.276	79.014
1700	312.210	675.065	438.820	401.615	-3357.867	-2361.528	72.561
1800	314.051	692.962	452.446	432.929	-3353.574	-2303.047	66.833
1900	315.850	709.991	465.557	464.424	-3349.758	-2244.790	61.714
1950.000	316.729	718.207	471.930	480.239	------ CRYSTAL <--> LIQUID -----		
2000	317.607	726.237	478.188	496.097	-3405.222	-2184.907	57.064
2100	319.281	741.773	490.372	527.942	-3405.522	-2123.883	52.829
2200	320.913	756.664	502.140	559.952	-3405.736	-2062.845	48.978
2300	322.545	770.965	513.520	592.125	-3405.865	-2001.803	45.462
2400	324.134	784.726	524.535	624.459	-3405.911	-1940.757	42.239

PREVIOUS: CURRENT: December 1973

Titanium Oxide (Ti_4O_7) O_7Ti_4(cr)

TITANIUM OXIDE (Ti_4O_7) LIQUID M_r = 303.5158

S°(298.15 K) = [291.066] J K^{-1} mol^{-1} Δ_fH°(298.15 K) = [-3216.035] kJ mol^{-1}

T_{fus} = 1950 ± 20 K $\Delta_{fus}H$° = [225.936] kJ mol^{-1}

Enthalpy of Formation

Δ_fH°(Ti_4O_7, ℓ, 298.15 K) is calculated from that of Ti_4O_7(cr) by adding $\Delta_{fus}H$° and the difference in enthalpy, H°(1950 K)-H°(298.15 K), between the crystal and liquid.

Heat Capacity and Entropy

C_p° is estimated as 8 cal K^{-1} g-$atoms^{-1}$. Below the assumed glass transition at 1300 K, C_p° is taken to be the same as that of the crystal. S°(298.15 K) is calculated in a manner analogous to that used for Δ_fH°.

Fusion Data

Hampson and Gilles (1) reported that T_{fus} < 1943 K, which is in reasonable agreement with T_{fus} ~1955 K obtained from the liquidus curve of Brauer and Littke (2). The value of 1993 K reported by Slyusar et al. (3) is over 40° higher than our adopted value of 1950 ± 20 K. Ti_4O_7 probably melts incongruently, but the data (2) are not completely conclusive on this point.

Vaporization Data

Studies of liquid samples (4) showed that the congruently vaporizing composition is between Ti_3O_5 and Ti_4O_7. The principal products (1) of vaporization are TiO_2(g) and TiO(g).

References

1. P. J. Hampson and P. W. Gilles, J. Chem. Phys. 55, 3712 (1971).
2. G. Brauer and W. Littke, J. Inorg. Nucl. Chem. 16, 67 (1960).
3. N. P. Slyusar, A. D. Krivorotenko, E. N. Fomichev, A. A. Kalashnik and V. P. Bondarenko, Teplofiz. Vys. Temp. 11, 213 (1973).
4. P. W. Gilles, H. F. Franzen, G. D. Stone and P. G. Wahlbeck, J. Chem. Phys. 48, 1938 (1968).

Titanium Oxide (Ti_4O_7) O_7Ti_4(l)

Enthalpy Reference Temperature = T_r = 298.15 K Standard State Pressure = p° = 0.1 MPa

T/K	C_p° (J K⁻¹mol⁻¹)	S° (J K⁻¹mol⁻¹)	−[G°−H°(T_r)]/T (J K⁻¹mol⁻¹)	H°−H°(T_r) (kJ mol⁻¹)	Δ_fH° (kJ mol⁻¹)	Δ_fG° (kJ mol⁻¹)	Log K_f
0							
100							
200							
298.15	208.489	291.066	291.066	0.000	-3216.035	-3052.056	534.708
300	209.200	292.358	291.070	0.386	-3216.026	-3051.039	531.233
400	239.911	357.091	299.687	22.961	-3214.304	-2996.243	391.269
500	258.864	412.821	316.877	47.972	-3210.960	-2942.090	307.358
600	271.165	461.174	336.989	74.511	-3206.781	-2888.699	251.484
700	279.659	503.647	357.826	102.075	-3202.160	-2836.046	211.628
800	285.976	541.423	378.458	130.372	-3197.242	-2784.078	181.782
900	290.830	575.396	398.484	159.221	-3192.277	-2732.732	158.604
1000	294.805	606.249	417.742	188.508	-3187.591	-2681.926	140.089
1100	298.152	634.508	436.181	218.159	-3183.482	-2631.564	124.963
1200	301.081	660.578	453.809	248.123	-3195.917	-2581.068	112.351
1300	303.675	684.782	470.656	278.364	-3190.139	-2530.066	101.659
1300.000	303.675	684.782	470.656	278.364	GLASS <--> LIQUID TRANSITION		
1300.000	368.192	684.782	470.656	278.364			
1400	368.192	712.068	486.937	315.183	-3178.202	-2479.742	92.520
1500	368.192	737.471	502.803	352.002	-3166.728	-2430.256	84.629
1600	368.192	761.233	518.220	388.821	-3155.763	-2381.518	77.749
1700	368.192	783.555	533.178	425.641	-3145.356	-2333.450	71.698
1800	368.192	804.600	547.678	462.460	-3135.557	-2285.977	66.337
1900	368.192	824.507	561.729	499.279	-3126.417	-2239.030	61.555
1950.000	368.192	834.071	568.590	517.689	------ CRYSTAL <--> LIQUID -----		
2000	368.192	843.393	575.344	536.098	-3176.735	-2190.733	57.216
2100	368.192	861.357	588.539	572.917	-3172.060	-2141.548	53.268
2200	368.192	878.485	601.332	609.737	-3167.465	-2092.582	49.684
2300	368.192	894.852	613.741	646.556	-3162.948	-2043.826	46.417
2400	368.192	910.522	625.783	683.375	-3158.509	-1995.265	43.426
2500	368.192	925.553	637.475	720.194	-3154.147	-1946.887	40.678
2600	368.192	939.993	648.834	757.013	-3149.859	-1898.681	38.145
2700	368.192	953.889	659.877	793.833	-3145.645	-1850.638	35.803
2800	368.192	967.279	670.618	830.652	-3141.503	-1802.750	33.631
2900	368.192	980.200	681.072	867.471	-3137.430	-1755.012	31.611
3000	368.192	992.682	691.252	904.290	-3133.426	-1707.410	29.729
3100	368.192	1004.755	701.171	941.109	-3129.486	-1659.939	27.970
3200	368.192	1016.445	710.842	977.929	-3125.610	-1612.599	26.323
3300	368.192	1027.774	720.275	1014.748	-3121.795	-1565.377	24.778
3400	368.192	1038.766	729.482	1051.567	-3118.037	-1518.268	23.325
3500	368.192	1049.439	738.472	1088.386	-3114.337	-1471.271	21.958
3600	368.192	1059.811	747.254	1125.205	-3110.690	-1424.377	20.667
3700	368.192	1069.899	755.839	1162.025	-4743.489	-1346.433	19.008
3800	368.192	1079.718	764.233	1198.844	-4735.007	-1254.732	17.247
3900	368.192	1089.282	772.446	1235.663	-4726.794	-1163.252	15.580
4000	368.192	1098.604	780.484	1272.482	-4718.845	-1071.983	13.999

PREVIOUS: CURRENT: December 1973

Titanium Oxide (Ti_4O_7) O_7Ti_4(l)

TITANIUM OXIDE (Ti_4O_7) CRYSTAL–CRYSTAL–CRYSTAL–LIQUID M_r = 303.5158

0 to 130 K	crystal, I
130 to 150 K	crystal, II
150 to 1950 K	crystal, III
above 1950 K	liquid

Refer to the individual tables for details.

Titanium Oxide (Ti_4O_7) O_7Ti_4(cr,l)

Enthalpy Reference Temperature = T_r = 298.15 K Standard State Pressure = p° = 0.1 MPa

T/K	C_p° (J K⁻¹mol⁻¹)	S° (J K⁻¹mol⁻¹)	$-[G^\circ-H^\circ(T_r)]/T$ (J K⁻¹mol⁻¹)	$H^\circ-H^\circ(T_r)$ (kJ mol⁻¹)	Δ_fH° (kJ mol⁻¹)	Δ_fG° (kJ mol⁻¹)	Log K_f
0							
100							
200							
298.15	208.489	198.740	198.740	0.000	-3404.521	-3213.016	562.907
300	209.200	200.032	198.744	0.386	-3404.512	-3211.828	559.229
400	239.911	264.765	207.361	22.961	-3402.790	-3147.798	411.060
500	258.864	320.495	224.551	47.972	-3399.447	-3084.414	322.226
600	271.165	368.848	244.663	74.511	-3395.268	-3021.790	263.070
700	279.659	411.321	265.500	102.075	-3390.647	-2959.904	220.871
800	285.976	449.097	286.132	130.372	-3385.729	-2898.703	189.266
900	290.830	483.070	306.158	159.221	-3380.763	-2838.124	164.720
1000	294.805	513.923	325.416	188.508	-3376.077	-2778.086	145.112
1100	298.152	542.182	343.855	218.159	-3371.968	-2718.492	129.090
1200	301.081	568.252	361.483	248.123	-3384.403	-2658.763	115.733
1300	303.675	592.456	378.330	278.363	-3378.626	-2598.529	104.410
1400	306.018	615.047	394.441	308.849	-3373.023	-2538.733	94.721
1500	308.235	636.236	409.861	339.563	-3367.654	-2479.330	86.338
1600	310.285	656.196	424.639	370.490	-3362.581	-2420.276	79.014
1700	312.210	675.065	438.820	401.615	-3357.867	-2361.528	72.561
1800	314.051	692.962	452.446	432.929	-3353.574	-2303.047	66.833
1900	315.850	709.991	465.557	464.424	-3349.758	-2244.790	61.714
1950.000	316.729	718.207	471.930	480.239	CRYSTAL <--> LIQUID		
1950.000	368.192	834.071	471.930	706.175	TRANSITION		
2000	368.192	843.393	481.101	724.584	-3176.735	-2190.732	57.216
2100	368.192	861.357	498.784	761.404	-3172.060	-2141.548	53.268
2200	368.192	878.485	515.657	798.223	-3167.465	-2092.582	49.684
2300	368.192	894.852	531.790	835.042	-3162.948	-2043.826	46.417
2400	368.192	910.522	547.247	871.861	-3158.509	-1995.265	43.426
2500	368.192	925.553	562.080	908.680	-3154.147	-1946.887	40.678
2600	368.192	939.993	576.340	945.500	-3149.859	-1898.681	38.145
2700	368.192	953.889	590.067	982.319	-3145.645	-1850.638	35.803
2800	368.192	967.279	603.301	1019.138	-3141.503	-1802.750	33.631
2900	368.192	980.200	616.076	1055.957	-3137.430	-1755.012	31.611
3000	368.192	992.682	628.423	1092.776	-3133.426	-1707.410	29.729
3100	368.192	1004.755	640.369	1129.596	-3129.486	-1659.939	27.970
3200	368.192	1016.445	651.940	1166.415	-3125.610	-1612.599	26.323
3300	368.192	1027.774	663.158	1203.234	-3121.795	-1565.377	24.778
3400	368.192	1038.766	674.044	1240.053	-3118.037	-1518.268	23.325
3500	368.192	1049.439	684.618	1276.872	-3114.337	-1471.271	21.958
3600	368.192	1059.811	694.897	1313.692	-3110.690	-1424.377	20.667
3700	368.192	1069.899	704.896	1350.511	-4743.489	-1346.433	19.008
3800	368.192	1079.718	714.632	1387.330	-4735.007	-1254.732	17.247
3900	368.192	1089.282	724.116	1424.149	-4726.794	-1163.252	15.580
4000	368.192	1098.604	733.362	1460.968	-4718.845	-1071.983	13.999

PREVIOUS: CURRENT: December 1973

Titanium Oxide (Ti_4O_7) O_7Ti_4(cr,l)

TUNGSTEN OXIDE (W_3O_8) IDEAL GAS M_r = 679.5452

$S°(298.15\ K)$ = [494.013] J K^{-1} mol^{-1}

$\Delta_fH°(0\ K)$ = -1700.31 kJ mol^{-1}

$\Delta_fH°(298.15\ K)$ = -1710.00 ± 41.8 kJ mol^{-1}

Vibrational Frequencies and Degeneracies

ν, cm^{-1}	ν, cm^{-1}
[800] (6)	[199] (3)
[350] (4)	[150] (6)
[242] (8)	

Ground State Quantum Weight: [1] σ = [4]

Point Group: [D_{2d}]

Bond Distance: W-O = [1.81] Å

Bond Angle: O-W-O = [109.47]°

Product of the Moments of Inertia: $I_AI_BI_C$ = [7.651013 x 10^{-111}] g^3 cm^6

Enthalpy of Formation

Ackermann and Rauh (1) have studied the sublimation behavior of the tungsten-oxygen system over the temperature range from 1300 to 1600 K by mass spectrometry, mass effusion and x-ray diffracton, and also have determined the $\Delta_fG°(W_3O_8, g)$ = -400,900 + 9054 T cal K^{-1} mol^{-1} from partial pressure measurements over W-WO_2(cr). The adopted value, $\Delta_fH°(W_3O_8, g, 298.15\ K)$ = -408.7 kcal mol^{-1} was reduced from $\Delta_fH°(W_3O_8, g, 1450\ K)$ = -400.9 kcal mol^{-1}, using JANAF functions.

Source	Reaction	T/K	$\Delta_rH°(298.15\ K)$, kcal mol^{-1} 2nd law	3rd law	Drift cal K^{-1} mol^{-1}	$\Delta_fH°(298.15\ K)$ kcal mol^{-1}
(1)	4 WO_2(cr) = W_3O_8(g) + W(cr)	1300-1600	155.05	155.09	-0.04	-408.7

Heat Capacity and Entropy

The heat capacity was approximated by making a tentative choice of the vibrational frequencies of 800 cm^{-1}(6), 350 cm^{-1}(4), 242 cm^{-1}(8), 199 cm^{-1}(3) and 150 cm^{-1}(6) in order to give the entropy at 1450 K, S°(1450 K) = 208.18 cal K^{-1} mol^{-1}, corresponding to $\Delta_fS°(1450\ K)$ = -90.54 cal K^{-1} mol^{-1}, which was reported by Ackermann and Rauh (1).

The molecular configuration for W_3O_8(g) was estimated by assuming that the oxygens are tetrahedrally located around the tungsten atoms, the three tetrahedra are connected by two common edges, and also the three tungsten atoms are in the same axis. The bond distance W-O was estimated as 1.81 Å and the bond angle ∠ OWO = 109.47°. The principal moments of inertia are: I_A = 46.4211 x 10^{-39} and $I_B = I_C$ = 405.9772 x 10^{-39} g cm^2.

Reference

1. R. J. Ackermann and E. G. Rauh, J. Phys. Chem. 67, 2596 (1963).

Tungsten Oxide (W_3O_8) O_8W_3(g)

Enthalpy Reference Temperature = T_r = 298.15 K   Standard State Pressure = p° = 0.1 MPa

T/K	$C_p°$ (J $K^{-1}mol^{-1}$)	S°	-[G°-H°(T_r)]/T	H°-H°(T_r) (kJ mol^{-1})	$\Delta_fH°$	$\Delta_fG°$	Log K_f
0	0.	0.	INFINITE	-39.964	-1700.312	-1700.312	INFINITE
100	112.720	315.859	655.406	-33.955	-1707.878	-1667.258	870.886
200	177.022	417.309	512.062	-18.951	-1710.541	-1625.251	424.472
250	194.050	458.740	497.345	-9.651	-1710.544	-1603.915	335.120
298.15	206.277	494.013	494.013	0.	-1710.001	-1583.420	277.409
300	206.684	495.291	494.017	0.382	-1709.971	-1582.635	275.561
350	216.267	527.904	496.571	10.967	-1708.965	-1561.487	233.039
400	223.618	557.284	502.354	21.972	-1707.657	-1540.504	201.169
450	229.316	583.965	509.963	33.301	-1706.151	-1519.699	176.402
500	233.784	608.366	518.601	44.883	-1704.524	-1499.068	156.607
600	240.175	651.597	537.257	68.604	-1701.114	-1458.296	126.956
700	244.386	688.958	556.321	92.846	-1697.693	-1418.098	105.820
800	247.281	721.792	574.995	117.438	-1694.384	-1378.384	89.999
900	249.344	751.043	592.960	142.275	-1691.235	-1339.074	77.718
1000	250.861	777.396	610.108	167.288	-1688.271	-1300.106	67.911
1100	252.006	801.362	626.422	192.434	-1685.499	-1261.424	59.900
1200	252.891	823.329	641.928	217.681	-1682.920	-1222.988	53.235
1300	253.588	843.600	656.672	243.006	-1680.538	-1184.758	47.604
1400	254.146	862.414	670.704	268.394	-1678.352	-1146.704	42.784
1500	254.599	879.964	684.076	293.832	-1676.367	-1108.801	38.612
1600	254.973	896.408	696.838	319.311	-1674.586	-1071.022	34.965
1700	255.284	911.875	709.037	344.824	-1673.011	-1033.348	31.751
1800	255.546	926.474	720.715	370.366	-1671.646	-995.763	28.896
1900	255.768	940.297	731.911	395.932	-1670.495	-958.246	26.344
2000	255.958	953.421	742.662	421.519	-1669.561	-920.784	24.048
2100	256.123	965.913	752.998	447.123	-1668.849	-883.363	21.972
2200	256.265	977.831	762.948	472.743	-1668.359	-845.969	20.086
2300	256.390	989.226	772.541	498.376	-1668.095	-808.596	18.364
2400	256.500	1000.140	781.798	524.020	-1668.057	-771.230	16.785
2500	256.597	1010.613	790.743	549.675	-1668.247	-733.859	15.333
2600	256.683	1020.678	799.394	575.339	-1668.654	-696.476	13.992
2700	256.760	1030.367	807.770	601.011	-1669.359	-659.071	12.750
2800	256.829	1039.706	815.888	626.691	-1670.466	-621.633	11.597
2900	256.890	1048.720	823.762	652.377	-1672.090	-584.151	10.522
3000	256.946	1057.430	831.407	678.069	-1674.311	-546.598	9.517
3100	256.997	1065.856	838.834	703.766	-1677.257	-508.959	8.576
3200	257.043	1074.016	846.057	729.468	-1681.020	-471.217	7.692
3300	257.084	1081.926	853.085	755.174	-1685.674	-433.343	6.859
3400	257.123	1089.601	859.929	780.885	-1691.321	-395.309	6.073
3500	257.158	1097.055	866.598	806.599	-1698.154	-357.096	5.329
3600	257.190	1104.300	873.101	832.316	-1706.457	-318.666	4.624
3700	257.219	1111.347	879.445	858.037	-1820.827	-279.412	3.945
3800	257.247	1118.207	885.639	883.760	-1822.207	-237.730	3.268
3900	257.272	1124.890	891.688	909.486	-1823.639	-196.013	2.625
4000	257.295	1131.403	897.600	935.214	-1825.122	-154.263	2.014
4100	257.317	1137.757	903.380	960.945	-1826.655	-112.472	1.433
4200	257.337	1143.958	909.035	986.678	-1828.236	-70.644	0.879
4300	257.356	1150.013	914.569	1012.412	-1829.865	-28.776	0.350
4400	257.373	1155.930	919.987	1038.149	-1831.542	13.127	-0.156
4500	257.390	1161.714	925.295	1063.887	-1833.265	55.069	-0.639
4600	257.405	1167.372	930.496	1089.627	-1835.035	97.050	-1.102
4700	257.419	1172.908	935.595	1115.368	-1836.854	139.073	-1.546
4800	257.433	1178.327	940.596	1141.111	-1838.720	181.134	-1.971
4900	257.446	1183.635	945.502	1166.854	-1840.637	223.236	-2.380
5000	257.457	1188.837	950.317	1192.600	-1842.604	265.376	-2.772
5100	257.469	1193.935	955.044	1218.346	-1844.625	307.557	-3.150
5200	257.479	1198.935	959.686	1244.093	-1846.700	349.773	-3.514
5300	257.489	1203.839	964.247	1269.842	-1848.832	392.037	-3.864
5400	257.499	1208.652	968.728	1295.591	-1851.026	434.335	-4.201
5500	257.508	1213.377	973.134	1321.341	-1853.281	476.676	-4.527
5600	257.516	1218.017	977.465	1347.093	-1855.605	519.063	-4.842
5700	257.524	1222.575	981.725	1372.845	-1857.998	561.487	-5.145
5800	257.532	1227.054	985.917	1398.597	-1860.464	603.959	-5.439
5900	257.539	1231.457	990.041	1424.351	-1863.009	646.464	-5.723
6000	257.546	1235.785	994.101	1450.105	-1865.636	689.023	-5.998

PREVIOUS: September 1966 (1 atm)   CURRENT: September 1966 (1 bar)

Tungsten Oxide (W_3O_8) O_8W_3(g)

TUNGSTEN OXIDE $((WO_3)_3)$ IDEAL GAS M_r = 695.5446

S°(298.15 K) = [504.674] J K^{-1} mol^{-1}

$\Delta_fH°$(0 K) = -2011.05 kJ mol^{-1}

$\Delta_fH°$(298.15 K) = -2023.38 ± 41.8 kJ mol^{-1}

Vibrational Frequencies and Degeneracies

ν, cm^{-1}	ν, cm^{-1}
[800](6)	[200](6)
[396](3)	[150](6)
[330](9)	

Ground State Quantum Weight: [1] σ = [6]

Point Group: $[D_{3h}]$

Bond Distance: W-O = [1.81] Å

Bond Angles: O-W-O = [109.47]° W-O_{bridge}-W = [130.53]°

Product of the Moments of Inertia: $I_AI_BI_C$ = [3.865176 x 10^{-110}] g^3 cm^6

Enthalpy of Formation

The adopted value, $\Delta_fH°(W_3O_9$, g, 298.15 K) = -483.6 kcal mol^{-1}, was reduced from $\Delta_fH°(W_3O_9$, g, 1450 K) = -474.1 kcal mol^{-1} of Ackermann and Rauh (1), who studied the sublimation behavior of the tungsten-oxygen system over the temperature range from 1300 to 16000 K by mass spectrometry, mass effusion and x-ray diffraction, and determined $\Delta_fG°(W_3O_9$, g) = -474,100 + 110.26T cal K^{-1} mol^{-1} from partial pressure measurements over W-WO_2(cr). They reviewed the previous determinations (1, 2, 3, 4) of the thermodynamic properties of sublimation of WO_3 and also pointed out that it was not possible to make quantitative observations on stoichiometric WO_3 in vacuo since it is a "bivariant" system and the ion current is not constant until the composition of the sample reaches $WO_{2.96}$. This is probably the cause of previous discrepancies in sublimation studies.

Battles (2) has also studied that the partial pressure of W_3O_9(g) over W-WO_2(cr) in the temperature range from 1389 to 1527 K. 3rd law calculation of the partial pressure data gives $\Delta_rH°$(298.15 K) = 148.2 kcal mol^{-1} for 4.5 WO_2(cr) = W_3O_9(g) + 1.5 W(cr) which yields $\Delta_fH°(W_3O_9$, g, 298.15 K) = -486.0 kcal mol^{-1}, using all JANAF functions.

Normand and Staley (3) have obtained $\Delta_rH°$(1800 K) = -240 kcal mol^{-1} for 3 WO_3(g) = W_3O_9(g) by mass spectrometric study of the vapor over WO_3 dissolved in CaO-Al_2O_3-SiO_2. The value of $\Delta_rH°$ was determined by the 2nd law method from ion intensities and leads to $\Delta_fH°(W_3O_9$, g) = -463.4 kcal mol^{-1}. Using all JANAF functions (dated September 30, 1966). The absolute pressure values reported by Norman and Staley also appear to be quite inconsistent with the JANAF functions since they lead to -443 kcal mol^{-1} by the 3rd law method.

Source	Reaction	T/K	$\Delta_rH°$(298.15 K), kcal mol^{-1} 2nd law	3rd law	Drift cal K^{-1} mol^{-1}	$\Delta_fH°$(298.15 K) kcal mol^{-1}
(1)	4.5 WO_2(cr) = W_3O_9(g) + 1.5 W(cr)	1300-1600	150.61	150.64	-0.05	-483.6*
(2)	4.5 WO_2(cr) = W_3O_9(g) + 1.5 W(cr)	1389-1527	148.20	148.20	-1.91 ± 0.24	-486.0*
(3)	3 WO_3(g) = W_3O_9(g)	1700-1900	-253.4	-233		-463.4**

*3rd law value adopted in calculation ** 2nd law value adopted in calculation

Heat Capacity and Entropy

The heat capacity was approximated by making a tenative choice of the vibrational frequencies of 800 cm^{-1}(6), 396 cm^{-1}(3), 330 cm^{-1}(9), 200 cm^{-1}(6), and 150 cm^{-1}(6) in order to give the entropy at 1450 K. S°(1450 = 219.14 cal K^{-1} mol^{-1}, corresponding to $\Delta_fS°$(1450 K) = 110.26 cal K^{-1} mol^{-1}, reported by Ackermann and Rauh (1). The adopted value, S°(W_3O_9, g, 298.15 K) = 120.6 cal K^{-1} mol^{-1}, was reduced from the entropy at 1450 K, using JANAF functions. 2nd law calculation of the partial pressure data of Battles (2) gives $\Delta_fS°$(1455 K) = 71.7 cal K^{-1} mol^{-1} for 4.5 WO_2(cr) = W_3O_9(g) + 1.5 W(cr). This yields S°(W_3O_9, g, 298.15 K) = 122.4 cal K^{-1} mol^{-1}.

The molecular configuration for W_3O_9(g) is estimated by assuming that the oxygens are tetrahedrally located around the tungsten atoms, and that each tetrahedra shares two oxygen atoms to form a planar six-member ring. The bond distance W-O is estimated as 1.81 Å and the bond angles O-W-O = 109.47° and W-O_{bridge}-W = 130.53°. The principal moments of inertia are: $I_A = I_B$ = 280.4787 x 10^{-39} and I_C = 491.3257 x 10^{-39} g cm^2.

References

1. R. J. Ackermann and E. G. Rauh, J. Phys. Chem. 67, 2596 (1963).
2. J. E. Battles, Ph.D. dissertation, "A Mass Spectrometric Investigation of Tungsten Dioxide and Tungsten Trioxide," the Ohio State University, 1964.
3. J. H. Norman and H. G. Staley, J. Chem. Phys. 43, 3804 (1965).
4. K. Ueno, J. Chem. Soc. Japan 62, 990 (1941).
5. J. Berkowitz, W. A. Chupka and M. G. Inghram, J. Chem. Phys. 27, 85 (1957).
6. P. E. Blackburn, J. Hoch and H. L. Johnston, J. Phys. Chem. 62, 769 (1958); P. E. Blackburn, WADC TR 59 - 575, part 1, Mar. 1960; and part 2, Dec. 1960.
7. G. Meyer, J. F. Oosterom, and J. L. DeRoo, Rec. Trav. Chim. 78, 412 (1959).

Tungsten Oxide $((WO_3)_3)$ O_9W_3(g)

Enthalpy Reference Temperature = T_r = 298.15 K Standard State Pressure = p° = 0.1 MPa

T/K	$C_p°$ (J $K^{-1}mol^{-1}$)	S°	-[G°-H°(T_r)]/T	H°-H°(T_r) (kJ mol^{-1})	$\Delta_fH°$	$\Delta_fG°$	Log K_f
0	0.	0.	INFINITE	-41.659	-2011.047	-2011.047	INFINITE
100	111.355	318.753	675.866	-35.711	-2020.127	-1971.131	1029.613
200	186.893	422.590	524.066	-20.295	-2023.834	-1920.252	501.518
250	208.026	466.694	508.266	-10.393	-2023.962	-1894.323	395.797
298.15	222.942	504.674	504.674	0.	-2023.382	-1869.398	327.511
300	223.433	506.055	504.678	0.413	-2023.349	-1868.443	325.325
350	234.911	541.401	507.443	11.885	-2022.194	-1842.710	275.009
400	243.591	573.362	513.718	23.858	-2020.665	-1817.170	237.298
450	250.250	602.454	521.986	36.211	-2018.894	-1791.837	207.991
500	255.430	629.100	531.384	48.858	-2016.973	-1766.710	184.567
600	262.784	676.370	551.710	74.796	-2012.925	-1717.036	149.481
700	267.594	717.264	572.506	101.331	-2008.839	-1668.045	124.471
800	270.884	753.225	592.894	128.265	-2004.856	-1619.634	105.751
900	273.223	785.273	612.521	155.476	-2001.036	-1571.712	91.220
1000	274.939	814.153	631.264	182.889	-1997.404	-1524.206	79.616
1100	276.232	840.421	649.102	210.450	-1993.970	-1477.053	70.139
1200	277.230	864.501	666.063	238.125	-1990.738	-1430.205	62.255
1300	278.015	886.723	682.193	265.889	-1987.708	-1383.619	55.594
1400	278.644	907.350	697.548	293.723	-1984.883	-1337.257	49.894
1500	279.154	926.592	712.183	321.614	-1982.266	-1291.092	44.960
1600	279.575	944.622	726.153	349.551	-1979.861	-1245.093	40.648
1700	279.924	961.582	739.508	377.526	-1977.669	-1199.238	36.848
1800	280.219	977.591	752.294	405.534	-1975.697	-1153.508	33.474
1900	280.469	992.749	764.554	433.569	-1973.946	-1107.879	30.458
2000	280.683	1007.140	776.327	461.627	-1972.423	-1062.336	27.745
2100	280.867	1020.839	787.647	489.704	-1971.130	-1016.865	25.293
2200	281.028	1033.909	798.546	517.799	-1970.069	-971.447	23.065
2300	281.168	1046.404	809.053	545.909	-1969.243	-926.076	21.032
2400	281.291	1058.374	819.193	574.032	-1968.653	-880.736	19.169
2500	281.400	1069.859	828.992	602.167	-1968.301	-835.414	17.455
2600	281.496	1080.897	838.470	630.312	-1968.175	-790.102	15.873
2700	281.583	1091.523	847.646	658.466	-1968.357	-744.787	14.409
2800	281.660	1101.765	856.540	686.628	-1968.949	-699.459	13.049
2900	281.729	1111.650	865.168	714.798	-1970.069	-654.106	11.782
3000	281.792	1121.202	873.544	742.974	-1971.795	-608.699	10.598
3100	281.849	1130.443	881.683	771.156	-1974.254	-563.221	9.490
3200	281.900	1139.392	889.597	799.343	-1977.538	-517.657	8.450
3300	281.947	1148.067	897.299	827.536	-1981.721	-471.975	7.471
3400	281.990	1156.485	904.799	855.732	-1986.906	-426.148	6.547
3500	282.029	1164.659	912.107	883.933	-1993.284	-380.155	5.674
3600	282.065	1172.605	919.233	912.138	-2001.139	-333.958	4.846
3700	282.098	1180.334	926.186	940.346	-2115.069	-286.949	4.051
3800	282.129	1187.857	932.974	968.558	-2116.015	-237.524	3.265
3900	282.157	1195.186	939.603	996.772	-2117.019	-188.075	2.519
4000	282.184	1202.330	946.083	1024.989	-2118.081	-138.604	1.810
4100	282.208	1209.298	952.418	1053.209	-2119.200	-89.104	1.135
4200	282.231	1216.099	958.615	1081.431	-2120.372	-39.575	0.492
4300	282.252	1222.740	964.681	1109.655	-2121.599	9.984	-0.121
4400	282.271	1229.229	970.620	1137.881	-2122.879	59.568	-0.707
4500	282.290	1235.573	976.437	1166.109	-2124.210	109.182	-1.267
4600	282.307	1241.777	982.139	1194.339	-2125.596	158.827	-1.804
4700	282.323	1247.849	987.728	1222.570	-2127.035	208.505	-2.317
4800	282.338	1253.793	993.209	1250.803	-2128.529	258.213	-2.810
4900	282.352	1259.615	998.587	1279.038	-2130.079	307.956	-3.283
5000	282.366	1265.319	1003.864	1307.274	-2131.686	357.728	-3.737
5100	282.378	1270.911	1009.046	1335.511	-2133.353	407.534	-4.174
5200	282.390	1276.394	1014.135	1363.749	-2135.081	457.367	-4.594
5300	282.401	1281.773	1019.134	1391.989	-2136.873	507.244	-4.999
5400	282.412	1287.052	1024.047	1420.230	-2138.735	557.147	-5.389
5500	282.422	1292.234	1028.876	1448.471	-2140.667	607.087	-5.766
5600	282.431	1297.323	1033.624	1476.714	-2142.674	657.068	-6.129
5700	282.440	1302.322	1038.294	1504.958	-2144.760	707.079	-6.480
5800	282.449	1307.234	1042.889	1533.202	-2146.929	757.134	-6.819
5900	282.457	1312.063	1047.411	1561.447	-2149.185	807.216	-7.147
6000	282.465	1316.810	1051.861	1589.693	-2151.535	857.349	-7.464

PREVIOUS: September 1966 (1 atm) CURRENT: September 1966 (1 bar)

Tungsten Oxide $((WO_3)_3)$ O_9W_3(g)

PHOSPHORUS OXIDE, DIMER ($(P_2O_5)_2$) CRYSTAL M_r = 283.88904 Phosphorus Oxide ($(P_2O_5)_2$) $O_{10}P_4(cr)$

$S°(298.15\ K) = 228.781 \pm 0.42\ J\ K^{-1}\ mol^{-1}$

$T_{sub} = 631\ K$

$\Delta_fH°(0\ K) = -2979.041 \pm 8.9\ kJ\ mol^{-1}$

$\Delta_fH°(298.15\ K) = -3009.936 \pm 8.9\ kJ\ mol^{-1}$

Enthalpy of Formation

The enthalpy of formation, $\Delta_fH°(298.15\ K)$ of $P_4O_{10}(cr)$ has been reported by many investigators. The most recent value, $\Delta_rH°(298.15\ K) = -719.4 \pm 2.0\ kcal\ mol^{-1}$ for the reaction $P_4(\alpha,\ white,\ cr) + 5\ O_2(g) = P_4O_{10}(hex,\ cr)$ was determined by Egan and Luff (1). This $\Delta_rH°(298.15\ K)$ value was derived from energy of combustion, $\Delta_cE = -716.397\ kcal\ mol^{-1}$, which is an average value of many measurements using three different sample forms, i.e., uncoated phosphorus, lucite-coated phosphorus and cellulose acetate-coated phosphorus. The adopted $\Delta_fH°(P_4O_{10},\ cr,\ 298.15\ K)$ value was recalculated based on the P(red, V, cr) as reference state.

The enthalpy of combustion of white phosphorus was determined as $-713.2 \pm 1.0\ kcal\ mol^{-1}$ by Holmes (2) yielding $\Delta_fH°(P_4O_{10},\ cr,\ 298.15\ K) = -696.5 \pm 1.3\ kcal\ mol^{-1}$, based on P(red, V, cr) as reference state. The enthalpy of formation of P_2O_5 from red phosphorus was reported to be $-355 \pm 2\ kcal\ mol^{-1}$ by Koerner and Daniels (3). The product, P_2O_5, was considered to be a mixture of two forms, i.e., crystalline and amorphous. The other $\Delta_fH°(P_4O_{10},\ cr,\ 298.15\ K)$ values reported by early investigators were reviewed by Egan and Luff (1).

Heat Capacity and Entropy

The heat capacities (12-324 K) and S°(298.15 K) adopted were taken from Andon et al. (4). The $C_p°$ values above 324 K were obtained by graphical extrapolation. The S°(298.15 K) value was calculated based on $S°(10\ K) = 0.36\ cal\ K^{-1}\ mol^{-1}$.

Low temperature (15-310 K) and high temperature (298.15-623.15 K) heat capacities were also measured by Thermal Laboratory (5) and Frandsen (6), respectively. The low temperature data were not adopted because they are less satisfactory than the data of Andon et al. (4). The high temperature data were not used because the sample employed for enthalpy measurements was not explicitly reported as pure hexagonal form.

Sublimation Data

The value of $\Delta_{sub}H°(298.15\ K)$ was derived from the vapor pressure data by both the 2nd and 3rd law analyses. See the $P_4O_{10}(g)$ table for details. T_{sub} is the temperature at which the Gibbs energy change of the reaction $P_4O_{10}(hex,\ cr) = P_4O_{10}(g)$ approaches zero. Sublimation temperature for $P_4O_{10}(cr)$ was derived to be 632 K, 636 K (average of three determinations) and 623.7 K (average of two determinations) by Frandsen (6) from the vapor pressure data reported by Hoeflake and Scheffer (7), Smits (8) and Smits and Rutgers (9), respectively.

References

1. E. P. Egan, Jr. and B. B. Luff, "Heats of Formation of Phosphorus Oxides," 1963, Tennessee Valley Authority, Office of Agriculture and Chemical Development, Fundamental Research Branch, Wilson Dam, Alabama, June 1, 1963 to Nov. 30, 1963.
2. W. S. Holmes, Trans. Faraday Soc. 58, 1916 (1962).
3. W. E. Koerner and F. Daniels, J. Chem. Phys. 20, 113 (1952).
4. R. J. L. Andon, J. F. Counsell, H. McKerrell and J. P. Martin, Trans. Faraday Soc. 59, 2702 (1963).
5. Thermal Laboratory, The Dow Chemical Company, personal communication, June 18, 1962.
6. M. Frandsen, J. Res. Nat. Bur. Stand. 10, 35 (1933).
7. J. M. Hoeflake and M. F. Scheffer, Rec. Trav. Chim. 45, 191 (1926).
8. A. Smits, Z. Physik. Chem. A149, 337 (1930).
9. A. Smits and A. J. Rutgers, J. Chem. Soc. 125, 2573 (1924).

Enthalpy Reference Temperature = T_r = 298.15 K Standard State Pressure = p° = 0.1 MPa

T/K	$C_p°$ (J K⁻¹mol⁻¹)	S°	$-[G°-H°(T_r)]/T$	$H°-H°(T_r)$ (kJ mol⁻¹)	$\Delta_fH°$	$\Delta_fG°$	Log K_f
0	0.000	0.000	INFINITE	-33.962	-2979.041	-2979.041	INFINITE
100	77.613	77.530	376.309	-29.878	-2993.030	-2907.054	1518.489
200	154.557	155.896	246.521	-18.125	-3004.886	-2816.879	735.693
298.15	211.710	228.781	228.781	0.000	-3009.936	-2723.335	477.117
300	212.673	230.094	228.785	0.393	-3009.992	-2721.557	473.865
400	260.245	297.978	237.719	24.104	-3014.145	-2624.398	342.711
500	299.993	360.390	256.096	52.147	-3011.926	-2527.145	264.009
600	335.975	418.323	278.362	83.977	-3006.425	-2430.655	211.607
700	368.192	472.581	302.272	119.216	-2997.988	-2335.322	174.264
800	396.643	523.641	326.783	157.486	-2986.933	-2241.381	146.347
900	421.747	571.849	351.360	198.440	-2973.536	-2148.968	124.723
1000	442.667	617.389	375.707	241.682	-2958.135	-2058.158	107.507
1100	461.077	660.477	399.655	286.904	-2940.989	-1968.976	93.499
1200	475.302	701.230	423.104	333.751	-3176.705	-1877.108	81.708
1300	486.181	739.718	445.993	381.843	-3153.960	-1769.725	71.108
1400	494.967	776.080	468.284	430.914	-3130.407	-1664.124	62.089
1500	502.080	810.481	489.961	480.780	-3106.213	-1560.231	54.332

PREVIOUS: December 1962 CURRENT: December 1965

Phosphorus Oxide ($(P_2O_5)_2$) $O_{10}P_4(cr)$

PHOSPHORUS OXIDE $((P_2O_5)_2)$ IDEAL GAS M_r = 283.88904

$S°(298.15\ K) = [403.964]\ J\ K^{-1}\ mol^{-1}$

$\Delta_f H°(0\ K) = -2870.18 \pm 8.9\ kJ\ mol^{-1}$
$\Delta_f H°(298.15\ K) = -2904.08 \pm 8.9\ kJ\ mol^{-1}$

Vibrational Frequencies and Degeneracies

ν, cm^{-1}	ν, cm^{-1}	ν, cm^{-1}	ν, cm^{-1}
424(1)	650 (2)	750(3)	764(3)
721(1)	952 (2)	257(3)	1015(3)
1417(1)	[170](3)	329(3)	1390(3)
278(2)	[470](3)	573(3)	

Ground State Quantum Weight: [1] σ = 12

Point Group: T_d

Bond Distances: $P\text{-}O_{apical}$ = 1.40 ± 0.03 Å   P-O = 1.60 ± 0.01 Å   P-P = 2.83 ± 0.03 Å

Bond Angles: $O_{apical}\text{-}P\text{-}O$ = 117 ± 1°   O-P-O = 101 ± 1°   P-O-P = 124°30' ± 1°

Product of the Moments of Inertia: $I_A I_B I_C = 2.939188 \times 10^{-111}\ g^3\ cm^6$

Enthalpy of Formation

$\Delta_f H°(P_4O_{10}$, g, 298.15 K) was calculated from the enthalpies of formation and sublimation for P_4O_{10}(hex, cr). The enthalpy of sublimation, $\Delta_{sub}H°$(298.15 K), was derived from vapor pressure data by the second and third law methods. The results obtained are presented as follows.

Source	T/K	$\Delta_{sub}H°$(298.15 K), kcal mol^{-1} 2nd law	3rd law	Drift cal K^{-1} mol^{-1}
(1)	260.0-401.0	25.42±0.07	25.33	-0.39±0.13
(2)	216.2-300.4	25.59±0.06	25.35	-0.18±0.14
	283.6-389.3	25.45±0.07	25.35	-0.10±0.07

The value of $\Delta_{sub}H°$(298.15 K) adopted was 25.34 + 0.08 kcal mol^{-1}. There are many vapor pressure measurements reported but only the above two over P_4O_{10}(hex, cr).

It has been shown that P_4O_{10} has three solid phases and two liquid phases. The three solid phases are hexagonal, orthorhombic and tetragonal. As temperature increases, the thermal stability of P_4O_{10} follows the same order. The metastable liquid, according to data of Hoeflake and Scheffer (1) is from hexagonal form. The stable liquid is from tetragonal form. Hill et al. (3) reported that molecules of P_4O_{10} having the same structure as in the vapor are present in the hexagonal form of the solid. The vapor pressures of P_4O_{10} over different condensed phases have been measured by many investigators and reviewed by Farr (4). The complexity of the solid state of P_4O_{10} has also been discussed by Smits and Rutgers (5), Smits (6), Smits et al. (7) and Smits (8).

Heat Capacity and Entropy

The molecular structure and constants were obtained from Akishin et al. (9). The molecular structure of P_4O_{10}(g) has also been studied by the electron diffraction method by Maxwell et al. (10) and Hampson and Stosick (11). The model of the P_4O_{10} molecule was later confirmed by the x-ray method reported by Decker and MacGillavry (12) and spectroscopic studies reported by Gerding and DeDecker (13) and Sidorov and Sobolev (14). All vibrational frequencies were obtained from Egan (15) except nine frequencies, i.e. 170(3), 470(3), and 750(3) cm^{-1} which were estimated in order to make the second and third law values of $\Delta_{sub}H°$(298.15 K) agree. The principal moments of inertia are: $I_A = I_B = I_C = 143.2430 \times 10^{-39}\ g\ cm^2$. The heat capacities (549-1098°C or 822-1371 K) of P_4O_{10}(g) have been determined by Frandsen (16). His results are in fair agreement with the calculated ones.

References

1. J. M. A. Hoeflake and F. E. C. Scheffer, Rec. Trav. Chim. 45, 191 (1926).
2. J. C. Southard and R. A. Nelson, J. Amer. Chem. Soc. 59 911 (1937).
3. W. L. Hill, G. T. Faust and S. B. Hendricks, J. Amer. Chem. Soc. 65, 794 (1943).
4. T. D. Farr, "Phosphorus, Properties of the Elements and Some of Its Compounds," Tennessee Valley Authority, Wilson Dam, Alabama, 1950.
5. A. Smits and A. J. Rutgers, J. Chem. Soc. 125, 2573 (1924).
6. A. Smits, Z. Physik. Chem. A149, 337 (1930).
7. A. Smits, J. A. A. Ketelaar and J. L. Meyering, Z. physik. Chem., B41, 87 (1938).
8. A. Smits, Z. physik. Chem., B46, 43 (1940).
9. P. A. Akishin, N. G. Rambidi and E. Z. Zasorin, Kristallografiya 4, 360 (1959).
10. L. R. Maxwell, S. B. Hendricks and L. S. Deming, J. Chem. Phys. 5, 626 (1937).
11. G. C. Hampson and A. J. Stosick, J. Amer. Chem. Soc. 60 1814 (1938).
12. H. C. J. Decker and C. H. MacGillavry, Rec. Trav. Chim. 60, 153 (1941).
13. H. Gerding and H. C. J. DeDecker, Rec. Trav. Chim. 64, 191 (1945).
14. T. A. Sidorov and N. N. Sobolev, Opt. Spectr. (USSR) 2, 717 (1957).
15. E. P. Egan, Jr., Tennesse Valley Authority personal communication, Feb. 23, 1960.
16. M. Frandsen, J. Res. Nat. Bur. Stand. 10, 35 (1933).

Phosphorus Oxide $((P_2O_5)_2)$ $O_{10}P_4$(g)

Enthalpy Reference Temperature = T_r = 298.15 K   Standard State Pressure = p° = 0.1 MPa

T/K	$C_p°$ (J $K^{-1}mol^{-1}$)	S°	$-[G°-H°(T_r)]/T$	$H°-H°(T_r)$ (kJ mol^{-1})	$\Delta_f H°$	$\Delta_f G°$	Log K_f
0	0.	0.	INFINITE	-30.951	-2870.176	-2870.176	INFINITE
100	70.439	269.420	534.866	-26.545	-2883.841	-2817.054	1471.478
200	136.770	339.152	419.662	-16.102	-2897.007	-2745.652	717.091
250	165.183	372.796	406.944	-8.537	-2901.099	-2707.317	565.663
298.15	188.804	403.964	403.964	0.	-2904.081	-2669.711	467.722
300	189.639	405.134	403.967	0.350	-2904.179	-2668.257	464.585
350	210.273	435.963	406.355	10.363	-2909.292	-2628.464	392.277
400	227.511	465.202	411.900	21.321	-2911.072	-2588.215	337.986
450	241.860	492.853	419.373	33.066	-2912.178	-2547.785	295.739
500	253.807	518.973	428.039	45.467	-2912.751	-2507.262	261.932
600	272.134	566.965	447.269	71.818	-2912.729	-2426.144	211.215
700	285.128	609.946	467.494	99.717	-2911.633	-2345.122	174.995
800	294.543	648.667	487.761	128.724	-2909.840	-2264.308	147.844
900	301.519	683.781	507.621	158.544	-2907.577	-2183.748	126.741
1000	306.801	715.835	526.863	188.972	-2904.990	-2103.459	109.873
1100	310.877	745.276	545.399	219.864	-2902.173	-2023.439	96.085
1200	314.081	772.469	563.203	251.119	-3153.482	-1939.372	84.419
1300	316.638	797.713	580.283	282.659	-3147.288	-1838.447	73.870
1400	318.708	821.257	596.664	314.430	-3141.035	-1738.001	64.846
1500	320.406	843.305	612.380	346.388	-3134.750	-1638.004	57.040
1600	321.814	864.030	627.467	378.501	-3128.454	-1538.426	50.224
1700	322.995	883.576	641.962	410.744	-3122.165	-1439.242	44.223
1800	323.993	902.067	655.904	443.094	-3115.899	-1340.429	38.898
1900	324.845	919.608	669.325	475.537	-3109.669	-1241.961	34.144
2000	325.577	936.290	682.260	508.059	-3103.484	-1143.821	29.874
2100	326.211	952.190	694.738	540.650	-3097.355	-1045.988	26.018
2200	326.763	967.379	706.788	573.299	-3091.289	-948.444	22.519
2300	327.246	981.915	718.436	606.000	-3085.291	-851.178	19.331
2400	327.672	995.851	729.707	638.746	-3079.366	-754.172	16.414
2500	328.050	1009.235	740.622	671.533	-3073.517	-657.409	13.736
2600	328.385	1022.108	751.203	704.355	-3067.746	-560.879	11.268
2700	328.685	1034.507	761.467	737.208	-3062.056	-464.568	8.988
2800	328.954	1046.466	771.434	770.091	-3056.445	-368.466	6.874
2900	329.196	1058.014	781.118	802.998	-3050.914	-272.569	4.909
3000	329.415	1069.178	790.535	835.929	-3045.464	-176.855	3.079
3100	329.613	1079.982	799.698	868.881	-3040.091	-81.321	1.370
3200	329.793	1090.450	808.622	901.851	-3034.796	14.035	-0.229
3300	329.957	1100.601	817.316	934.839	-3029.575	109.229	-1.729
3400	330.108	1110.453	825.794	967.842	-3024.427	204.269	-3.138
3500	330.245	1120.024	834.064	1000.860	-3019.352	299.155	-4.465
3600	330.372	1129.329	842.138	1033.891	-3014.344	393.899	-5.715
3700	330.488	1138.383	850.022	1066.934	-3009.405	488.502	-6.896
3800	330.596	1147.198	857.727	1099.988	-3004.531	582.980	-8.014
3900	330.695	1155.787	865.260	1133.053	-2999.720	677.326	-9.072
4000	330.788	1164.160	872.629	1166.127	-2994.972	771.546	-10.075
4100	330.873	1172.329	879.839	1199.210	-2990.286	865.651	-11.029
4200	330.953	1180.304	886.898	1232.301	-2985.658	959.642	-11.935
4300	331.027	1188.092	893.813	1265.400	-2981.091	1053.526	-12.798
4400	331.096	1195.703	900.588	1298.507	-2976.583	1147.299	-13.620
4500	331.161	1203.144	907.229	1331.620	-2972.132	1240.972	-14.405
4600	331.221	1210.423	913.741	1364.739	-2967.742	1334.546	-15.154
4700	331.278	1217.547	920.130	1397.864	-2963.413	1428.029	-15.871
4800	331.331	1224.523	926.399	1430.994	-2959.145	1521.419	-16.556
4900	331.381	1231.355	932.553	1464.130	-2954.941	1614.724	-17.213
5000	331.429	1238.050	938.596	1497.270	-2950.801	1707.940	-17.843
5100	331.473	1244.614	944.532	1530.416	-2946.730	1801.076	-18.447
5200	331.515	1251.051	950.365	1563.565	-2942.728	1894.126	-19.027
5300	331.554	1257.366	956.098	1596.718	-2938.801	1987.109	-19.584
5400	331.592	1263.564	961.735	1629.876	-2934.952	2080.012	-20.120
5500	331.627	1269.648	967.278	1663.037	-2931.182	2172.844	-20.636
5600	331.660	1275.624	972.731	1696.201	-2927.500	2265.616	-21.133
5700	331.692	1281.495	978.097	1729.369	-2923.907	2358.316	-21.612
5800	331.722	1287.264	983.378	1762.539	-2920.408	2450.960	-22.073
5900	331.751	1292.934	988.576	1795.713	-2917.010	2543.532	-22.519
6000	331.778	1298.510	993.696	1828.890	-2913.718	2636.061	-22.949

PREVIOUS: December 1965 (1 atm)   CURRENT: December 1965 (1 bar)

Phosphorus Oxide $((P_2O_5)_2)$ $O_{10}P_4$(g)

TUNGSTEN OXIDE $((WO_3)_4)$ IDEAL GAS $M_r = 927.3928$

$\Delta_f H°(0\ K) = -2789.53 \pm 41.8\ kJ\ mol^{-1}$

$S°(298.15\ K) = [605.272]\ J\ K^{-1}\ mol^{-1}$ $\Delta_f H°(298.15\ K) = -2804.12 \pm 41.8\ kJ\ mol^{-1}$

Vibrational Frequencies and Degeneracies

ν, cm^{-1}	ν, cm^{-1}
[800](8)	[240](8)
[300](8)	[200](18)

Ground State Quantum Weight: [1] $\sigma = [8]$

Point Group: $[D_{4h}]$

Bond Distance: W-O = [1.81] Å

Bond Angles: O-W-O = [109.47]° $W\text{-}O_{bridge}\text{-}W$ = [160.53]°

Product of the Moments of Inertia: $I_A I_B I_C = [3.854 \times 10^{-109}]\ g^3\ cm^6$

Enthalpy of Formation

The adopted value, $\Delta_f H°(W_4O_{12}$, g, 298.15 K) = -670.2 kcal mol^{-1}, was reduced from $\Delta_f H°(W_4O_{12}$, g, 1450 K) = -655.6 kcal mol^{-1} of Ackermann and Rauh (1) who studied the sublimation behavior of the tungsten-oxygen system over the temperature range from 1300 to 1600 K by mass spectrometry, mass effusion and x-ray diffraction, and determined $\Delta_f G°(W_4O_{12}$, g) = -655,600 + 160.29 T cal K^{-1} mol^{-1} from partial pressure measurements over the $W\text{-}WO_2$(cr). They reviewed the previous determinations (1, 2, 3, 4) of the thermodynamic properties of sublimation of WO_3 and also pointed out that it was not possible to make quantitative observations of stoichiometric WO_3 in vacuo since it is a "bivariant" system and the ion current is not constant until the composition of the sample reaches $WO_{2.96}$. This is probably the cause of previous discrepancies in sublimation studies.

Battles (2) has also studied the partial pressure of W_4O_{12}(g) over $W\text{-}WO_2$(cr) in the temperature range from 1449 to 1527 K. 3rd law calculation of the partial pressure data gives $\Delta_r H°$(298.15 K) = 172.0 kcal mol^{-1} for 6 WO_2(cr) = W_4O_{12}(g) + 2 W(cr) which yields $\Delta_f H°(W_4O_{12}$, g, 298.15 K) = -673.6 kcal mol^{-1}, using all JANAF functions.

Source	Reaction	T/K	$\Delta_r H°$(298.15 K), kcal mol^{-1} 2nd law	3rd law	Drift cal K^{-1} mol^{-1}	$\Delta_f H°$, 298.15 K) kcal mol^{-1}
(1)	6 WO_2(cr) = W_4O_{12}(g) + 2 W(cr)	1300-1600	175.41	175.55	0.01	-670.2
(2)	6 WO_2(cr) = W_4O_{12}(g) + 2 W(cr)	1449-1257	176.16	172.01	-2.90 ± 0.39	-673.6

Heat Capacity and Entropy

The heat capacity was approximated by making a tentative choice of the vibrational frequencies of 800 cm^{-1}(8), 300 cm^{-1}(8), 240 cm^{-1}(8), and 200 cm^{-1}(18) in order to give the entropy at 1450 K, S°(1450 K) = 278.9 cal K^{-1} mol^{-1}, corresponding to $\Delta_f S°$(1450 K) = -160.29 cal K^{-1} mol^{-1}, reported by Ackermann and Rauh (1). The adopted value, $S°(W_4O_{12}$, g, 298.15 K) = 144.6 cal K^{-1} mol^{-1}, was reduced from the entropy at 1450 K, using JANAF functions. 2nd law calculation of the partial pressure data of Battles gives $\Delta_f S°$(1487 K) = 82.3 cal K^{-1} mol^{-1} for 6 WO_2(cr) = W_4O_{12}(g) + 2 W(cr). This yields $S°(W_4O_{12}$, g, 298.15 K) = 147.4 cal K^{-1} mol^{-1}.

The molecular configuration for W_4O_{12}(g) is estimated by assuming that the oxygens are tetrahedrally located around the tungsten atoms, and that each tetrahedra shares two oxygen atoms to form a planar eight-member ring. The bond distance W-O is estimated as 1.81 Å and the bond angles O-W-O = 109.47° and $W\text{-}O_{bridge}\text{-}W$ = 160.53°. The principal moments of inertia are: $I_A = I_B = 5.935 \times 10^{-37} g\ cm^2$, and $I_C = 1.094 \times 10^{-36}\ g\ cm^2$.

References

1. R. J. Ackermann and E. G. Rauh, J. Phys. Chem. 67, 2596 (1963).
2. J. E. Battles, Ph.D. dissertation, "A Mass Spectroscopic Investigation of Tungsten Dioxide and Tungsten Trioxide," the Ohio State University, 1964.
3. K. Ueno, J. Chem. Soc. Japan 62, 990 (1941).
4. J. Berkowitz, W. A. Chupka and M. G. Inghram, J. Chem. Phys. 27, 85 (1957).
5. P. E. Blackburn, M. Hoch, and H. L. Johnston, J. Phys. Chem. 62, 769 (1958); P. E. Blackburn, WADC TR 59-575, part 1, Mar. 1960; and part 2, Dec. 1960.
6. G. Meyer, J. F. Oosterom and J. L. DeRoo, Rec. Trav. Chim. 78, 412 (1959).

Tungsten Oxide $((WO_3)_4)$ $O_{12}W_4$(g)

Enthalpy Reference Temperature = T_r = 298.15 K Standard State Pressure = p° = 0.1 MPa

T/K	$C_p°$ (J $K^{-1}mol^{-1}$)	S° (J $K^{-1}mol^{-1}$)	$-[G°-H°(T_r)]/T$ (J $K^{-1}mol^{-1}$)	$H°-H°(T_r)$ (kJ mol^{-1})	$\Delta_f H°$ (kJ mol^{-1})	$\Delta_f G°$ (kJ mol^{-1})	Log K_f
0	0.	0.	INFINITE	-57.403	-2789.527	-2789.527	INFINITE
100	155.404	342.363	846.055	-50.369	-2802.530	-2728.938	1425.451
200	264.535	490.351	632.324	-28.395	-2806.053	-2653.323	692.977
250	290.904	552.391	610.266	-14.469	-2805.502	-2615.182	546.412
298.15	309.177	605.272	605.272	0.	-2804.117	-2578.641	451.767
300	309.778	607.186	605.277	0.573	-2804.051	-2577.242	448.738
350	323.782	656.042	609.104	16.428	-2801.950	-2539.600	379.015
400	334.379	700.000	617.765	32.894	-2799.410	-2502.290	326.766
450	342.521	739.875	629.152	49.825	-2796.588	-2465.317	286.167
500	348.867	776.305	642.072	67.117	-2793.598	-2428.668	253.721
600	357.895	840.770	669.957	102.488	-2787.414	-2356.261	205.131
700	363.813	896.414	698.424	138.593	-2781.241	-2284.891	170.501
800	367.868	945.276	726.289	175.189	-2775.246	-2214.397	144.585
900	370.754	988.780	753.082	212.128	-2769.495	-2144.639	124.472
1000	372.873	1027.958	778.644	249.314	-2764.016	-2075.507	108.413
1100	374.471	1063.575	802.952	286.685	-2758.816	-2006.909	95.300
1200	375.705	1096.213	826.049	324.197	-2753.895	-1938.773	84.393
1300	376.676	1126.326	848.004	361.817	-2749.253	-1871.037	75.179
1400	377.453	1154.270	868.895	399.525	-2744.890	-1803.647	67.295
1500	378.085	1180.334	888.798	437.303	-2740.810	-1736.560	60.472
1600	378.605	1204.752	907.790	475.138	-2737.017	-1669.735	54.511
1700	379.038	1227.718	925.941	513.021	-2733.513	-1603.139	49.258
1800	379.402	1249.394	943.314	550.944	-2730.304	-1536.742	44.595
1900	379.712	1269.915	959.968	588.900	-2727.394	-1470.513	40.427
2000	379.976	1289.399	975.957	626.885	-2724.788	-1404.430	36.680
2100	380.205	1307.944	991.328	664.894	-2722.491	-1338.470	33.293
2200	380.404	1325.636	1006.124	702.925	-2720.506	-1272.608	30.216
2300	380.577	1342.549	1020.387	740.974	-2718.836	-1206.836	27.408
2400	380.730	1358.750	1034.150	779.039	-2717.481	-1141.129	24.836
2500	380.864	1374.294	1047.447	817.119	-2716.444	-1075.470	22.471
2600	380.984	1389.235	1060.307	855.212	-2715.711	-1009.846	20.288
2700	381.091	1403.615	1072.758	893.316	-2715.388	-944.241	18.267
2800	381.187	1417.476	1084.823	931.430	-2715.614	-878.640	16.391
2900	381.273	1430.854	1096.526	969.553	-2716.543	-813.024	14.644
3000	381.350	1443.781	1107.887	1007.684	-2718.281	-747.356	13.013
3100	381.420	1456.287	1118.925	1045.822	-2720.997	-681.614	11.485
3200	381.484	1468.397	1129.658	1083.968	-2724.814	-615.773	10.051
3300	381.542	1480.137	1140.101	1122.119	-2729.830	-549.794	8.703
3400	381.596	1491.528	1150.271	1160.276	-2736.182	-483.637	7.430
3500	381.644	1502.590	1160.180	1198.438	-2744.126	-417.277	6.228
3600	381.689	1513.342	1169.841	1236.605	-2754.039	-350.659	5.088
3700	381.730	1523.801	1179.267	1274.776	-2905.385	-282.975	3.995
3800	381.768	1533.981	1188.468	1312.951	-2906.086	-212.085	2.915
3900	381.803	1543.898	1197.455	1351.129	-2906.866	-141.178	1.891
4000	381.835	1553.565	1206.237	1389.311	-2907.723	-70.256	0.917
4100	381.866	1562.994	1214.824	1427.496	-2908.655	0.693	-0.009
4200	381.894	1572.196	1223.224	1465.684	-2909.660	71.664	-0.891
4300	381.920	1581.183	1231.445	1503.875	-2910.737	142.663	-1.733
4400	381.944	1589.963	1239.493	1542.068	-2911.885	213.682	-2.537
4500	381.967	1598.547	1247.377	1580.264	-2913.103	284.729	-3.305
4600	381.988	1606.942	1255.103	1618.461	-2914.392	355.804	-4.040
4700	382.008	1615.158	1262.677	1656.661	-2915.754	426.912	-4.745
4800	382.027	1623.201	1270.104	1694.863	-2917.187	498.048	-5.420
4900	382.044	1631.078	1277.391	1733.066	-2918.696	569.218	-6.068
5000	382.061	1638.796	1284.542	1771.272	-2920.281	640.416	-6.690
5100	382.076	1646.362	1291.563	1809.479	-2921.947	711.648	-7.289
5200	382.091	1653.782	1298.457	1847.687	-2923.694	782.906	-7.864
5300	382.105	1661.060	1305.230	1885.897	-2925.527	854.210	-8.419
5400	382.118	1668.202	1311.886	1924.108	-2927.452	925.540	-8.953
5500	382.131	1675.214	1318.429	1962.320	-2929.471	996.907	-9.468
5600	382.142	1682.100	1324.861	2000.534	-2931.590	1068.320	-9.965
5700	382.153	1688.863	1331.188	2038.749	-2933.815	1139.764	-10.445
5800	382.164	1695.510	1337.413	2076.965	-2936.150	1211.255	-10.909
5900	382.174	1702.043	1343.537	2115.182	-2938.602	1282.773	-11.357
6000	382.184	1708.466	1349.566	2153.399	-2941.178	1354.349	-11.791

PREVIOUS: September 1966 (1 atm) CURRENT: September 1966 (1 bar)

Tungsten Oxide $((WO_3)_4)$ $O_{12}W_4$(g)

PHOSPHORUS (P) REFERENCE STATE A_r = 30.97376 Phosphorus (P) P_1(ref)

0 to 195.4 K	crystal, white, beta
195.4 to 317.3 K	crystal, white, alpha
317.3 to 1180.008 K	liquid
above 1180.008 K	ideal diatomic gas

Refer to the individual tables for details.

Enthalpy Reference Temperature = T_r = 298.15 K — Standard State Pressure = p° = 0.1 MPa

	J K⁻¹mol⁻¹			kJ mol⁻¹			
T/K	C_p°	S°	$-[G^\circ-H^\circ(T_r)]/T$	$H^\circ-H^\circ(T_r)$	Δ_fH°	Δ_fG°	Log K_f
0	0.	0.	INFINITE	-5.360	0.	0.	INFINITE
100	13.728	17.688	62.415	-4.473	0.	0.	0.
195.400	21.117	28.969	43.439	-2.827	ALPHA <--> BETA		
195.400	20.941	31.637	43.439	-2.306	TRANSITION		
200	21.092	32.126	43.173	-2.209	0.	0.	0.
298.15	23.825	41.077	41.077	0.	0.	0.	0.
300	23.870	41.225	41.078	0.044	0.	0.	0.
317.300	24.351	42.574	41.123	0.460	BETA <--> LIQUID		
317.300	26.326	44.649	41.123	1.119	TRANSITION		
400	26.326	50.747	42.507	3.296	0.	0.	0.
500	26.326	56.621	44.764	5.929	0.	0.	0.
600	26.326	61.421	47.152	8.561	0.	0.	0.
700	26.326	65.479	49.488	11.194	0.	0.	0.
800	26.326	68.995	51.711	13.826	0.	0.	0.
900	26.326	72.095	53.807	16.459	0.	0.	0.
1000	26.326	74.869	55.777	19.092	0.	0.	0.
1100	26.326	77.378	57.629	21.724	0.	0.	0.
1180.008	26.326	79.226	59.031	23.830	LIQUID <--> IDEAL GAS		
1180.008	18.540	133.233	59.031	87.558	FUGACITY = 1 bar		
1200	18.551	133.544	60.270	87.929	0.	0.	0.
1300	18.604	135.031	65.964	89.787	0.	0.	0.
1400	18.648	136.412	70.948	91.649	0.	0.	0.
1500	18.686	137.699	75.355	93.516	0.	0.	0.
1600	18.720	138.907	79.290	95.386	0.	0.	0.
1700	18.750	140.042	82.831	97.260	0.	0.	0.
1800	18.776	141.115	86.039	99.136	0.	0.	0.
1900	18.801	142.131	88.965	101.015	0.	0.	0.
2000	18.823	143.096	91.647	102.896	0.	0.	0.
2100	18.844	144.015	94.119	104.780	0.	0.	0.
2200	18.864	144.892	96.407	106.665	0.	0.	0.
2300	18.883	145.731	98.534	108.553	0.	0.	0.
2400	18.901	146.535	100.517	110.442	0.	0.	0.
2500	18.918	147.306	102.373	112.333	0.	0.	0.
2600	18.934	148.049	104.116	114.225	0.	0.	0.
2700	18.950	148.764	105.756	116.119	0.	0.	0.
2800	18.965	149.453	107.305	118.015	0.	0.	0.
2900	18.980	150.119	108.770	119.912	0.	0.	0.
3000	18.994	150.763	110.159	121.811	0.	0.	0.
3100	19.009	151.386	111.479	123.711	0.	0.	0.
3200	19.022	151.989	112.735	125.613	0.	0.	0.
3300	19.036	152.575	113.934	127.516	0.	0.	0.
3400	19.050	153.143	115.079	129.420	0.	0.	0.
3500	19.063	153.696	116.174	131.326	0.	0.	0.
3600	19.076	154.233	117.224	133.233	0.	0.	0.
3700	19.089	154.756	118.231	135.141	0.	0.	0.
3800	19.102	155.265	119.199	137.050	0.	0.	0.
3900	19.114	155.761	120.130	138.961	0.	0.	0.
4000	19.127	156.245	121.027	140.873	0.	0.	0.
4100	19.140	156.718	121.892	142.787	0.	0.	0.
4200	19.152	157.179	122.727	144.701	0.	0.	0.
4300	19.164	157.630	123.533	146.617	0.	0.	0.
4400	19.176	158.071	124.313	148.534	0.	0.	0.
4500	19.189	158.502	125.068	150.452	0.	0.	0.
4600	19.201	158.924	125.799	152.372	0.	0.	0.
4700	19.213	159.337	126.509	154.292	0.	0.	0.
4800	19.225	159.741	127.197	156.214	0.	0.	0.
4900	19.237	160.138	127.865	158.137	0.	0.	0.
5000	19.248	160.527	128.514	160.062	0.	0.	0.
5200	19.272	161.282	129.760	163.914	0.	0.	0.
5400	19.296	162.010	130.941	167.771	0.	0.	0.
5600	19.319	162.712	132.063	171.632	0.	0.	0.
5800	19.342	163.390	133.132	175.498	0.	0.	0.
6000	19.366	164.047	134.152	179.369	0.	0.	0.

PREVIOUS: June 1961 (1 atm) CURRENT: June 1961 (1 bar)

Phosphorus (P) P_1(ref)

PHOSPHORUS, BLACK (P) CRYSTAL A_r = 30.97376

$S°(298.15\ K) = 22.59 \pm 0.04\ J\ K^{-1}\ mol^{-1}$

$\Delta_f H°(0\ K) = [-11.15]\ kJ\ mol^{-1}$
$\Delta_f H°(298.15\ K) = [-12.85]\ kJ\ mol^{-1}$

Enthalpy of Formation

The enthalpy of formation is derived from $\Delta_r G°(820\ K) = 0$ for the process P(black) → P(red, IV). This calculation is based on the assumption that at 820 K the conversion of black to red phosphorus is known to occur (1).

Heat Capacity and Entropy

The thermal functions are derived from the heat capacity measurments of Stephenson et al (1) who reported measured values from 17.70 to 295.89 K.

Reference

1. C. C. Stephenson, R. L. Potter, T. G. Maple and J. C. Morrow, J. Chem. Thermodynamics 1, 59 (1969).

Phosphorus, Black (P) $P_1(cr)$

Enthalpy Reference Temperature = T_r = 298.15 K Standard State Pressure = p° = 0.1 MPa

	J K⁻¹mol⁻¹			kJ mol⁻¹			
T/K	$C_p°$	S°	$-[G°-H°(T_r)]/T$	$H°-H°(T_r)$	$\Delta_f H°$	$\Delta_f G°$	Log K_f
0	0.	0.	INFINITE	-3.660	-11.151	-11.151	INFINITE
100	9.063	5.673	38.599	-3.293	-11.671	-10.469	5.469
200	17.343	14.800	24.438	-1.928	-12.569	-9.104	2.378
250	19.778	18.944	22.931	-0.997	-12.730	-8.218	1.717
298.15	21.547	22.586	22.586	0.	-12.851	-7.338	1.286
300	21.602	22.719	22.586	0.040	-12.855	-7.304	1.272
350	22.644	26.129	22.853	1.147	-13.684	-6.298	0.940
400	23.531	29.213	23.458	2.302	-13.845	-5.232	0.683
450	24.183	32.022	24.256	3.495	-13.969	-4.147	0.481
500	24.836	34.604	25.164	4.720	-14.060	-3.051	0.319
600	26.146	39.246	27.132	7.268	-14.144	-0.839	0.073
700	27.547	43.382	29.163	9.953	-14.092	1.376	-0.103
800	28.970	47.153	31.179	12.779	-13.899	3.575	-0.233
900	30.393	50.647	33.151	15.747	-13.563	5.740	-0.333
1000	31.815	53.923	35.066	18.857	-13.085	7.861	-0.411

PREVIOUS: CURRENT: June 1969

Phosphorus, Black (P) $P_1(cr)$

PHOSPHORUS, RED IV (P) CRYSTAL $A_r = 30.97376$

$S°(298.15\ K) = 23.197 \pm 0.08\ J\ K^{-1}\ mol^{-1}$

$\Delta_f H°(0\ K) = [-10.71]\ kJ\ mol^{-1}$
$\Delta_f H°(298.15\ K) = [-12.44]\ kJ\ mol^{-1}$

Enthalpy of Formation

Stevenson et al. (1) have referenced the fluorine bomb calorimetry study of O'Hare and Hubbard (2) which yielded $\Delta_r H° = -1.2 \pm 1.0\ kcal\ mol^{-1}$ for P(red, IV) → P(red, V).

Heat Capacity and Entropy

The thermal functions are derived from the heat capacity measurments of Stephenson et al (1) who reported measured values from 16.15 to 304.93 K.

Reference

1. C. C. Stephenson, R. L. Potter, T. G. Maple and J. C. Morrow, J. Chem. Thermodynamics 1, 59 (1969).
2. P. A. G. O'Hare and W. N. Hubbard, Trans. Faraday Soc. 62, 2709 (1966).

Phosphorus, Red, IV (P) $P_1(cr)$

Enthalpy Reference Temperature = T_r = 298.15 K — Standard State Pressure = p° = 0.1 MPa

T/K	$C_p°$ (J K⁻¹mol⁻¹)	S°	$-[G°-H°(T_r)]/T$	$H°-H°(T_r)$ (kJ mol⁻¹)	$\Delta_f H°$	$\Delta_f G°$	Log K_f
0	0.	0.	INFINITE	-3.635	-10.714	-10.714	INFINITE
100	8.807	6.537	39.017	-3.248	-11.214	-10.099	5.275
200	17.175	15.494	25.028	-1.907	-12.136	-8.810	2.301
250	19.581	19.599	23.538	-0.985	-12.306	-7.958	1.663
298.15	21.263	23.197	23.197	0.	-12.439	-7.108	1.245
300	21.313	23.329	23.197	0.039	-12.444	-7.075	1.232
350	22.356	26.694	23.461	1.132	-13.287	-6.099	0.910
400	23.242	29.739	24.058	2.272	-13.463	-5.060	0.661
450	23.894	32.514	24.846	3.451	-13.601	-4.001	0.464
500	24.548	35.066	25.742	4.662	-13.706	-2.928	0.306
600	25.857	39.655	27.687	7.181	-13.819	-0.760	0.066
700	27.259	43.746	29.694	9.837	-13.796	1.417	-0.106
800	28.681	47.479	31.687	12.634	-13.632	3.581	-0.234
900	30.104	50.939	33.636	15.573	-13.325	5.715	-0.332
1000	31.526	54.185	35.530	18.655	-12.876	7.808	-0.408

PREVIOUS: CURRENT: June 1969

Phosphorus, Red, IV (P) $P_1(cr)$

PHOSPHORUS, RED V, (P) CRYSTAL $A_r = 30.97376$

$S°(298.15\ K) = 22.85 \pm 0.08\ J\ K^{-1}\ mol^{-1}$

$T_{fus} = 870\ K$

$\Delta_f H°(0\ K) = -15.707\ kJ\ mol^{-1}$

$\Delta_f H°(298.15\ K) = -17.460\ kJ\ mol^{-1}$

$\Delta_{fus} H° = 18.8 \pm 0.8\ kJ\ mol^{-1}$

Enthalpy of Formation

The enthalpy of formation is obtained from the enthalpies of sublimation represented by reaction I and II at 298.15 K.

(I) $4P(cr, white) + P_4(g)$ 14.08 kcal

(II) $4P(cr, red\ V) + P_4(g)$ 30.771 Kcal

Of the calorimetric determinations only those of Giran (12) result in a reliable value for the enthalpy of formation, 4.4 ± 0.4 kcal mol^{-1}. A comprehensive review is given by Rodewald (13).

Heat Capacity and Entropy

The only reliable low temperature measurements, 15 to 305 K, are reported by Potter (1). A value at 10 K was obtained by graphical extrapolation and a T^3 function was assumed from 0 to 10 K. High temperature heat capacities were obtained from a smooth curve joining the measurements of Kubashewski et al (2) in the range 573 to 773 K and that of Regnault (3) at 331 K with those of Potter (1). Measurements of Wigand (4) are too high. The uncertainty in the high temperature heat capacity is about 2%.

Fusion Data

The values of $\Delta_{fus}H$ and T_{fus} were calculated from the functions of the red (V) crystal and liquid. This may be compared with a reported triple point of 862.7 K by Smits and Bokhorst (5).

Sublimation Data

Vapor pressure measurements on the triclinic allotrope, T. V. A. designation V, are reported by Farr (6). These data lead to a sublimation point of 704 K and $\Delta_{sub}H(298.15\ K)$ of 30.77 ± 0.4 kcal mol^{-1} of P_4. Smits and Bokhorst (7) measured the vapor pressure of a sample whose preparation indicates that is was probably P(cr, red, V). These results lead to $\Delta_{sub}H(298.15\ K) = 30.6 \pm 1$ kcal mol^{-1} of $P_4(g)$. An unpublished value of 30.84 kcal mol^{-1} of P_4 by Stephenson (8) is quoted by Kane.

Stevenson and Yost (9) show that the vapor density measurements of Stock et al. (10) are best explained by a vapor essentially composed of only $P_4(g)$ below 800 K. This conclusion is supported by recent mass spectrometer measurements of Kane and Reynolds (11).

References

1. R. L. Potter, Dissertation, Massachusetts Institute of Technology (1946).
2. O. Kubashewski and G. Schrag, Z. Elektrochem. 46, 675 (1940).
3. V. Regnault, Ann. chim. phys. 9, 322 (1843).
4. A. Wigan, Ann. Physik, 22, 64 (1907).
5. A. Smits and S. C. Bokhorst, Verslag. Akad. Wetenschap. 23, 930 (1914).
6. T. D. Farr, Tennessee Valley Authority Chemical Engineering Report No. 8 (1950).
7. A. Smits and S. C. Bokhorst, Z. phys. Chem. 91, 248 (1916).
8. C. C. Stephenson, quoted by J. S. Kane, Thesis Univ. of California (1955).
9. D. P. Stevenson and D. M. Yost, J. Chem. Phys. 9, 403 (1941).
10. A. Stock, G. E. Gibson and E. Stamm, Ber. 45, 3527 (1912).
11. J. S. Kane and J. H. Reynolds, J. Chem. Phys. 25, 342 (1956).
12. H. Giran, Ann. Chem. Phys. 30, 203 (1903).
13. H. J. Rodewald, Helv. Chem. Acta 43, 878 (1960).

Phosphorus, Red, V (P) $P_1(cr)$

Enthalpy Reference Temperature = T_r = 298.15 K Standard State Pressure = p° = 0.1 MPa

T/K	$C_p°$	S°	$-[G°-H°(T_r)]/T$	$H°-H°(T_r)$	$\Delta_f H°$	$\Delta_f G°$	Log K_f
	J K⁻¹mol⁻¹			kJ mol⁻¹			
0	0.	0.	INFINITE	-3.607	-15.707	-15.707	INFINITE
100	8.724	6.303	38.578	-3.227	-16.215	-15.076	7.875
200	17.046	15.193	24.675	-1.896	-17.147	-13.760	3.594
250	19.464	19.270	23.192	-0.980	-17.323	-12.892	2.694
298.15	21.187	22.853	22.853	0.	-17.460	-12.026	2.107
300	21.251	22.984	22.853	0.039	-17.465	-11.992	2.088
350	22.293	26.340	23.116	1.128	-18.311	-10.999	1.642
400	23.179	29.376	23.711	2.266	-18.490	-9.942	1.298
450	23.832	32.144	24.497	3.441	-18.631	-8.864	1.029
500	24.485	34.689	25.390	4.649	-18.739	-7.773	0.812
600	25.794	39.267	27.330	7.162	-18.859	-5.566	0.485
700	27.196	43.241	29.324	9.742	-18.912	-3.345	0.250
800	36.987	47.698	31.335	13.091	-18.196	-1.158	0.076
900	30.041	51.571	33.380	16.372	-17.547	0.925	-0.054
1000	31.464	54.736	35.359	19.378	-17.174	2.959	-0.155

PREVIOUS: CURRENT: June 1961

Phosphorus, Red, V (P) $P_1(cr)$

PHOSPHORUS, WHITE (P) CRYSTAL $A_r = 30.97376$

$S°(298.15\ K) = 41.077 \pm 0.08\ J\ K^{-1}\ mol^{-1}$
$T_{trs} = 195.4 \pm 0.1\ K$ (white β → white α)
$T_{fus} = 317.30 \pm 0.05\ K$

$\Delta_f H°(0\ K) = 0\ kJ\ mol^{-1}$
$\Delta_f H°(298.15\ K) = 0\ kJ\ mol^{-1}$
$\Delta_{trs} H° = 0.521 \pm 0.001\ kJ\ mol^{-1}$
$\Delta_{fus} H° = 0.659 \pm 0.002\ kJ\ mol^{-1}$

Enthalpy of Formation

Zero by definition. The enthalpies of white and red V phosphorus are tied together by the enthalpies of sublimation as determined by Giran (1) . A comprehensive review is given by Rodewald (2).

Heat Capacity and Entropy

Low temperature measurements, 15-315 K, were made by Maples (3). A Debye temperature, θ_D 124 K, was used to extrapolate to T/K = 0. Young and Hildenbrand (4) obtained heat capacities from enthalpy measurements that are 4% higher than those from the low temperature calorimeter which are preferred because of the proximity to the melting point.

Transition Data

Observed by Maples (3).

Fusion Data

The selected enthalpy of fusion is that observed by Maples (3). Young and Hildenbrand (4) obtain 125 cal mol^{-1} for the same quantity. Maples' (3) value is consistent with vapor pressures and other thermochemical data; see the liquid and P_4(g) tables for details.

Sublimation Data

The vapor pressure measurements of Dainton and Kimberly (5) are the most precise and lead to $\Delta_{sub}H°(298.15\ K) = 14.08 \pm 0.2$ kcal mol^{-1} of P_4(g). Measurements of Jolibois (6), Smits and Bokhorst (7), and Centzwer (8) lead to values that agree within the assigned uncertainty.

References

1. H. Giran, Ann. Chim. Phys. 30, 203 (1903).
2. H. J. Rodewald, Helv. Chim. Acta 43, 878 (1960).
3. R. T. Maples, Dissertation, Massachusette Institute of Technology (1949).
4. F. E. Young and J. H. Hildenbrand, J. Am. Chem. Soc. 64, 839 (1942).
5. F. S. Dainton and H. M. Kimberly, Trans. Faraday Soc. 46, 912 (1950).
6. A. Jolibois, Compt. rend. 149, 287 (1908).
7. A. Smits and S. C. Bokhorst, Z. Phys. Chem. 91, 248 (1916).
8. M. Centzwer, Z. phys. Chem. 85, 99 (1913).

Phosphorus, White (P) P_1(cr)

Enthalpy Reference Temperature = T_r = 298.15 K   Standard State Pressure = p° = 0.1 MPa

	J K⁻¹mol⁻¹			kJ mol⁻¹			
T/K	$C_p°$	S°	$-[G°-H°(T_r)]/T$	$H°-H°(T_r)$	$\Delta_f H°$	$\Delta_f G°$	Log K_f
0	0.	0.	INFINITE	-5.360	0.	0.	INFINITE
100	13.728	17.688	62.415	-4.473	0.	0.	0.
195.400	21.117	28.969	43.439	-2.827	ALPHA <--> BETA		
195.400	20.941	31.637	43.439	-2.306	TRANSITION		
200	21.092	32.126	43.173	-2.209	0.	0.	0.
250	22.564	36.992	41.463	-1.118	0.	0.	0.
298.15	23.825	41.077	41.077	0.	0.	0.	0.
300	23.870	41.225	41.078	0.044	0.	0.	0.
317.300	24.351	42.574	41.123	0.460	BETA <--> LIQUID		
350	24.731	44.981	41.372	1.263	-0.717	0.071	-0.011
400	25.313	48.321	42.036	2.514	-0.782	0.188	-0.025
450	25.997	51.344	42.905	3.798	-0.815	0.312	-0.036
500	26.485	54.110	43.889	5.110	-0.818	0.437	-0.046
600	26.485	58.939	46.007	7.759	-0.802	0.687	-0.060
700	26.485	63.021	48.154	10.407	-0.786	0.934	-0.070
800	26.485	66.558	50.238	13.056	-0.771	1.179	-0.077
900	26.485	69.677	52.228	15.704	-0.755	1.422	-0.083
1000	26.485	72.468	54.115	18.353	-0.739	1.662	-0.087

PREVIOUS: CURRENT: June 1961

Phosphorus, White (P) P_1(cr)

PHOSPHORUS (P) LIQUID A_r = 30.97376

S°(298.15 K) = [43.01] J K^{-1} mol^{-1}

T_{fus}(white → ℓ) = 317.30 ± 0.05 K

T_{fus}(red, V → ℓ) = 870 K

$\Delta_f H°$(298.15 K) = [0.615] kJ mol^{-1}

$\Delta_{fus}H$(white → ℓ) = 0.659 kJ mol^{-1}

$\Delta_{fus}H$(red, V → ℓ) = 18.8 ± 0.002 kJ mol^{-1}

Enthalpy of Formation

The enthalpy of formation for P(ℓ) is calculated from that of P(white) by adding the enthalpy of fusion (white to liquid) and the difference in enthalpy, H°(317.30 K) - H°(298.15 K), between the white crystal and liquid.

Heat Capacity and Entropy

Young and Hildenbrand (1) obtained an equation from enthalpy measurements which decreased with temperature. However, a better fit to the thermochemical data results if a constant heat capacity is assumed.

The entropy is derived from low temperature measurements on white phosphorus. See the white phosphorus table for details.

Vaporization Data

MacRae and Van Voorhis' (2), determinations of the vapor pressure, 44 to 150°C, are the most precise and lead to a $\Delta_{vap}H$(298.15 K) of 13.504 ± 0.1 kcal mol^{-1} of P_4. A vapor pressure equation derived by Farr (3), to fit the measurements reported in the literature leads to a value of 13.6 ± 0.1 kcal mol^{-1} of P_4. The boiling point is calculated from the functions of P(liq) and P_4(g), and may be compared with the values of 552.7 and 553.7 resulting from Farr's (3) equation and observations of Smits and Bokhorst (4), respectively.

References

1. F. E. Young and J. H. Hildenbrand, J. Am. Chem. Sec. 64, 839 (1942).
2. D. MacRae and C. C. Van Voorhis, J. Am. Chem. Soc. 43, 547 (1921).
3. T. D. Farr, Tennessee Valley Authority Chemical Engineering Report No. 8(1950).
4. A. Smits and S. C. Bokhorst, Proc. Akad. Wentenschop. 18, 106 (1915).

Phosphorus (P) P_1(l)

Enthalpy Reference Temperature = T_r = 298.15 K Standard State Pressure = p° = 0.1 MPa

T/K	$C_p°$ (J K⁻¹mol⁻¹)	S° (J K⁻¹mol⁻¹)	-[G°-H°(T_r)]/T (J K⁻¹mol⁻¹)	H°-H°(T_r) (kJ mol⁻¹)	$\Delta_f H°$ (kJ mol⁻¹)	$\Delta_f G°$ (kJ mol⁻¹)	Log K_f
0							
100							
200							
250							
298.15	26.326	43.011	43.011	0.	0.615	0.038	-0.007
300	26.326	43.173	43.011	0.049	0.619	0.035	-0.006
317.300	26.326	44.649	43.061	0.504	------- BETA <--> LIQUID -------		
350	26.326	47.232	43.332	1.365	0.	0.	0.
400	26.326	50.747	44.044	2.681	0.	0.	0.
450	26.326	53.848	44.964	3.998	0.	0.	0.
500	26.326	56.621	45.994	5.314	0.	0.	0.
600	26.326	61.421	48.177	7.946	0.	0.	0.
700	26.326	65.479	50.366	10.579	0.	0.	0.
800	26.326	68.995	52.480	13.212	0.	0.	0.
900	26.326	72.095	54.491	15.844	0.	0.	0.
1000	26.326	74.869	56.392	18.477	0.	0.	0.
1100	26.326	77.378	58.188	21.109	0.	0.	0.
1180.008	26.326	79.226	59.552	23.216	------- FUGACITY = 1 bar -------		
1200	26.326	79.669	59.884	23.742	-63.572	1.078	-0.047
1300	26.326	81.776	61.488	26.374	-62.797	6.434	-0.259
1400	26.326	83.727	63.008	29.007	-62.027	11.731	-0.438
1500	26.326	85.543	64.450	31.640	-61.262	16.973	-0.591
1600	26.326	87.242	65.822	34.272	-60.499	22.164	-0.724
1700	26.326	88.838	67.129	36.905	-59.740	27.307	-0.839
1800	26.326	90.343	68.378	39.537	-58.984	32.405	-0.940
1900	26.326	91.766	69.572	42.170	-58.230	37.462	-1.030
2000	26.326	93.117	70.715	44.802	-57.479	42.479	-1.109

PREVIOUS: CURRENT: June 1961

Phosphorus (P) P_1(l)

PHOSPHORUS (P) BETA–ALPHA–LIQUID A_r = 30.97376

0 to 195.4 K crystal, white, beta
195.4 to 317.3 K crystal, white, alpha
above 317.3 K liquid

Refer to the individual tables for details.

Phosphorus (P) P_1(cr,l)

Enthalpy Reference Temperature = T_r = 298.15 K; Standard State Pressure = p° = 0.1 MPa

T/K	C_p° (J K⁻¹mol⁻¹)	S° (J K⁻¹mol⁻¹)	$-[G^\circ - H^\circ(T_r)]/T$ (J K⁻¹mol⁻¹)	$H^\circ - H^\circ(T_r)$ (kJ mol⁻¹)	$\Delta_f H^\circ$ (kJ mol⁻¹)	$\Delta_f G^\circ$ (kJ mol⁻¹)	Log K_f
0	0.	0.	INFINITE	-5.360	0.	0.	INFINITE
100	13.728	17.688	62.415	-4.473	0.	0.	0.
195.400	21.117	28.969	43.439	-2.827	ALPHA <--> BETA		
195.400	20.941	31.637	43.439	-2.306	TRANSITION		
200	21.092	32.126	43.173	-2.209	0.	0.	0.
250	22.564	36.992	41.463	-1.118	0.	0.	0.
298.15	23.825	41.077	41.077	0.	0.	0.	0.
300	23.870	41.225	41.078	0.044	0.	0.	0.
317.300	24.351	42.574	41.123	0.460	BETA <--> LIQUID		
317.300	26.326	44.649	41.123	1.119	TRANSITION		
350	26.326	47.232	41.575	1.980	0.	0.	0.
400	26.326	50.747	42.507	3.296	0.	0.	0.
450	26.326	53.848	43.598	4.612	0.	0.	0.
500	26.326	56.621	44.764	5.929	0.	0.	0.
600	26.326	61.421	47.152	8.561	0.	0.	0.
700	26.326	65.479	49.488	11.194	0.	0.	0.
800	26.326	68.995	51.711	13.826	0.	0.	0.
900	26.326	72.095	53.807	16.459	0.	0.	0.
1000	26.326	74.869	55.777	19.092	0.	0.	0.
1100	26.326	77.378	57.629	21.724	0.	0.	0.
1180.008	26.326	79.226	59.031	23.830	------- FUGACITY = 1 bar -------		
1200	26.326	79.669	59.371	24.357	-63.572	1.078	-0.047
1300	26.326	81.776	61.015	26.989	-62.797	6.434	-0.259
1400	26.326	83.727	62.568	29.622	-62.027	11.731	-0.438
1500	26.326	85.543	64.040	32.254	-61.262	16.973	-0.591
1600	26.326	87.242	65.438	34.887	-60.499	22.164	-0.724
1700	26.326	88.838	66.768	37.520	-59.740	27.307	-0.839
1800	26.326	90.343	68.036	40.152	-58.984	32.405	-0.940
1900	26.326	91.766	69.248	42.785	-58.230	37.462	-1.030
2000	26.326	93.117	70.408	45.417	-57.479	42.479	-1.109

PREVIOUS: CURRENT: June 1961

Phosphorus (P) P_1(cr,l)

PHOSPHORUS (P) IDEAL GAS M_r = 30.97376

IP(P, g) = 84580 ± 10 cm^{-1}

S°(298.15 K) = 163.198 ± 0.02 J K^{-1} mol^{-1}

$\Delta_f H°$(0 K) = 315.55 ± 1.0 kJ mol^{-1}

$\Delta_f H°$(298.15 K) = 316.39 ± 1.0 kJ mol^{-1}

Electronic Levels and Quantum Weights

State	ε_i, cm^{-1}	g_i
$^4S_{3/2}$	0.0	4
$^2D_{3/2}$	11360.80	4
$^2D_{5/2}$	11376.40	6
$^2P_{1/2}$	18722.65	2
$^2P_{3/2}$	18747.95	4

Enthalpy of Formation

The value adopted for the enthalpy of formation, $\Delta_f H°$(298.15 K) = 316.39 ± 1.0 kJ mol^{-1}, of the monatomic gas is that recommended by CODATA (1). It is derived from $\Delta_f H°(P_2$, g, 298.15 K), (1, 2) and the dissociation energy $D_0^0(P_2)$ = 40596 ± 20 cm^{-1} obtained by Herzberg (3).

Heat Capacity and Entropy

The electronic energy levels are given in the compilations by Moore (4, 5). There are 2.45 levels observed below the ionization potential. Only the ground state and the four lowest lying levels are used in the calculation. These are the contributions of the levels arising from three terms - $^4S°$, $^2D°$, and $^2P°$ - of the $3s^2 3p^3$ configuration. The remaining levels lie above 55000 cm^{-1} and do not contribute significantly even at 6000 K. The reported uncertainty in S°(298.15 K) is due to uncertainties in the relative atomic mass and the fundamental constants. Extension of these calculations beyond 6000 K may require consideration of higher excited states, estimation of missing levels, and utilization of proper cut-off procedures (6).

CODATA (1) used a slightly different values for the relative atomic mass and the fundamental constants. This causes a difference in the S°(298.15 K) value of 0.004 J K^{-1} mol^{-1}.

References

1. J. D. Cox, chairman, CODATA Task Group on Key Values for Thermodynamics, J. Chem. Thermodynamics 10, 903 (1978).
2. JANAF Thermochemical Tables: P_2(g), 12-30-82.
3. G. Herzberg, Ann. Phys. 15, 677 (1932).
4. C. E. Moore, U. S. Nat. Bur. Stand., NSRDS - NBS 35, 1971.
5. C. E. Moore, U. S. Nat. Bur. Stand., NSRDS-NBS 34, Vol. II, 1970. [Reprint of NBS Circ. 497, Vol II, 1952].
6. J. R. Downey, Jr., AFOSR Rept. TR-78-0960, 1978.

Phosphorus (P) P_1(g)

Enthalpy Reference Temperature = T_r = 298.15 K Standard State Pressure = p° = 0.1 MPa

T/K	$C_p°$ (J $K^{-1}mol^{-1}$)	S° (J $K^{-1}mol^{-1}$)	-[G°-H°(T_r)]/T (J $K^{-1}mol^{-1}$)	H°-H°(T_r) (kJ mol^{-1})	$\Delta_f H°$ (kJ mol^{-1})	$\Delta_f G°$ (kJ mol^{-1})	Log K_f
0	0.	0.	INFINITE	-6.197	315.553	315.553	INFINITE
100	20.786	140.490	181.678	-4.119	316.744	304.464	-159.036
200	20.786	154.898	165.099	-2.040	316.559	292.005	-76.264
250	20.786	159.536	163.540	-1.001	316.507	285.871	-59.729
298.15	20.786	163.198	163.198	0.	316.390	279.980	-49.051
300	20.786	163.326	163.198	0.038	316.384	279.754	-48.709
350	20.786	166.530	163.451	1.078	315.488	273.733	-40.852
400	20.786	169.306	164.013	2.117	315.211	267.787	-34.969
450	20.786	171.754	164.740	3.156	314.934	261.876	-30.398
500	20.786	173.944	165.553	4.196	314.657	255.996	-26.744
600	20.786	177.734	167.277	6.274	314.103	244.315	-21.270
700	20.786	180.938	169.005	8.353	313.549	232.728	-17.366
800	20.786	183.714	170.674	10.431	312.995	221.220	-14.444
900	20.786	186.162	172.262	12.510	312.441	209.781	-12.175
1000	20.786	188.352	173.763	14.589	311.887	198.404	-10.364
1100	20.788	190.333	175.181	16.667	311.333	187.083	-8.884
1200	20.791	192.142	176.520	18.746	247.207	176.890	-7.700
1300	20.797	193.806	177.787	20.826	247.429	171.021	-6.872
1400	20.810	195.348	178.987	22.906	247.647	165.135	-6.161
1500	20.831	196.785	180.126	24.988	247.862	159.234	-5.545
1600	20.865	198.130	181.210	27.073	248.076	153.319	-5.005
1700	20.914	199.396	182.242	29.161	248.291	147.390	-4.529
1800	20.981	200.593	183.229	31.256	248.510	141.448	-4.105
1900	21.068	201.730	184.173	33.358	248.733	135.494	-3.725
2000	21.178	202.813	185.078	35.470	248.964	129.528	-3.383
2100	21.312	203.850	185.948	37.595	249.205	123.551	-3.073
2200	21.471	204.845	186.784	39.734	249.458	117.561	-2.791
2300	21.653	205.803	187.590	41.890	249.727	111.560	-2.534
2400	21.859	206.729	188.369	44.065	250.013	105.547	-2.297
2500	22.088	207.626	189.121	46.262	250.320	99.521	-2.079
2600	22.337	208.497	189.850	48.483	250.648	93.483	-1.878
2700	22.605	209.345	190.556	50.730	251.001	87.431	-1.691
2800	22.890	210.172	191.242	53.005	251.380	81.366	-1.518
2900	23.189	210.981	191.909	55.309	251.786	75.287	-1.356
3000	23.500	211.772	192.558	57.643	252.222	69.194	-1.205
3100	23.821	212.548	193.190	60.009	252.688	63.085	-1.063
3200	24.148	213.309	193.807	62.407	253.185	56.961	-0.930
3300	24.480	214.057	194.409	64.839	253.713	50.821	-0.804
3400	24.814	214.793	194.998	67.303	254.273	44.664	-0.686
3500	25.148	215.517	195.574	69.802	254.866	38.491	-0.574
3600	25.480	216.230	196.138	72.333	255.490	32.300	-0.469
3700	25.808	216.933	196.690	74.897	256.146	26.091	-0.368
3800	26.129	217.625	197.232	77.494	256.834	19.864	-0.273
3900	26.444	218.308	197.764	80.123	257.552	13.619	-0.182
4000	26.749	218.982	198.286	82.783	258.299	7.355	-0.096
4100	27.045	219.646	198.799	85.473	259.076	1.072	-0.014
4200	27.329	220.301	199.303	88.191	259.880	-5.231	0.065
4300	27.601	220.947	199.799	90.938	260.711	-11.553	0.140
4400	27.860	221.585	200.287	93.711	261.567	-17.894	0.212
4500	28.106	222.214	200.767	96.510	262.447	-24.255	0.282
4600	28.338	222.834	201.240	99.332	263.350	-30.637	0.348
4700	28.555	223.446	201.706	102.177	264.274	-37.037	0.412
4800	28.758	224.049	202.165	105.042	265.218	-43.458	0.473
4900	28.947	224.644	202.618	107.928	266.180	-49.899	0.532
5000	29.121	225.231	203.064	110.831	267.160	-56.360	0.589
5100	29.281	225.809	203.505	113.752	268.154	-62.840	0.644
5200	29.426	226.379	203.939	116.687	269.163	-69.340	0.697
5300	29.557	226.941	204.368	119.636	270.185	-75.859	0.748
5400	29.675	227.494	204.791	122.598	271.218	-82.398	0.797
5500	29.779	228.040	205.209	125.571	272.260	-88.956	0.845
5600	29.871	228.577	205.621	128.553	273.311	-95.533	0.891
5700	29.950	229.107	206.029	131.545	274.370	-102.129	0.936
5800	30.016	229.628	206.431	134.543	275.435	-108.743	0.979
5900	30.072	230.142	206.828	137.547	276.505	-115.376	1.021
6000	30.116	230.647	207.221	140.557	277.578	-122.027	1.062

PREVIOUS: June 1961 (1 atm) CURRENT: December 1982 (1 bar)

Phosphorus (P) P_1(g)

PHOSPHORUS, ION (P^+) IDEAL GAS M_r = 30.97321 Phosphorus, Ion (P^+) $P_1^+(g)$

IP(P^+, g) = 159100 ± 100 cm^{-1}
S°(298.15 K) = 166.970 ± 0.02 J K^{-1} mol^{-1}

$\Delta_fH°$(0 K) = 1327.35 ± 0.42 kJ mol^{-1}
$\Delta_fH°$(298.15 K) = [1336.333] kJ mol^{-1}

Electronic Levels and Quantum Weights

State	ε_i cm^{-1}	g_i
$^3P_0$	0	1
$^3P_1$	164.8	3
$^3P_2$	469.0	5
$^1D_2$	8882.6	5
$^1S_0$	21576.2	1

Enthalpy of Formation

$\Delta_fH°$(P^+, g, 0 K) is calculated from $\Delta_fH°$(P, g, 0 K) (1) using the spectroscopic value of IP(P) = 84580 ± 10 cm^{-1} (1011.80 ± 0.12 kJ mol^{-1}) from Moore (2). The ionization limit is converted from cm^{-1} to kJ mol^{-1} using the factor, 1 cm^{-1} = 0.01196266 kJ cm^{-1}, which is derived from the 1973 CODATA fundamental constants (3). Rosenstock et al. (4) and Levin and Lias (5) have summarized additional ionization and appearance potential data.

$\Delta_fH°$(P^+, g, 298.15 K) is calculated from $\Delta_fH°$(P, g, 0 K) by using IP(P) with JANAF (1) enthalpies, H°(0 K) - H°(298.15 K), for P(g), P^+(g), and e^-(g). $\Delta_fH°$(P → P^+ + e^-, 298.15 K) differs from a room temperature threshold energy due to inclusion of these enthalpies and to threshold effects discussed by Rosenstock et al. (4). $\Delta_fH°$(298.15 K) should be changed by -6.197 kJ mol^{-1} if it is to be used in the ion convention that excludes the enthalpy of the electron.

Heat Capacity and Entropy

The information on electronic energy levels and quantum weights, given by Moore (2, 6) is incomplete because many theoretically predicted levels have not been observed. Our calculations indicate that any reasonable method of filling in these missing levels and cutting off the summation in the partition function (7) has no effect on the thermodynamic functions to 6000 K. This is a result of the high energy of all levels other than the ground state and the first four excited states; the next excited state is approximately 52000 cm^{-1} above the ground state. Since inclusion of these excited states has no effect on the thermodynamic functions (to 6000 K), we list only the ground state and the first four excited states. The reported uncertainty in S°(298.15 K) is due to uncertainties in the relative ionic mass and the fundamental constants. Extension of these calculations above 6000 K may require consideration of the excited states and use of different fill and cutoff procedures (7).

References

1. JANAF Thermochemical Tables: P (g), 6-30-62; e^-(g), 3-31-82.
2. C. E. Moore, U. S. Nat. Bur. Stand., NSRDS-NBS-34, 1970.
3. E. R. Cohen and B. N. Taylor, J. Phys. Chem. Ref. Data 2, 663 (1973).
4. H. M. Rosenstock, K. Draxl et al., J. Phys. Chem. Ref. Data 6, Supp. 1 (1977).
5. R. D. Levin and S. G. Lias, U. S. Nat. Bur. Stand., NSRDS-NBS-71, 1982.
6. C. E. Moore, U. S. Nat. Bur. Stand., NSRDS-NBS-35, Volume I, 1971 [Reprint of NBS Circular 467, Volume I, 1949].
7. J. R. Downey, Jr., The Dow Chemical Company, AFOSR-TR-0960, Contract No. F44620-75-1-0048, 1978.

Enthalpy Reference Temperature = T_r = 298.15 K Standard State Pressure = p° = 0.1 MPa

	J $K^{-1}$$mol^{-1}$			kJ mol^{-1}			
T/K	$C_p°$	S°	-[G°-H°(T_r)]/T	H°-H°(T_r)	$\Delta_fH°$	$\Delta_fG°$	Log K_f
0	0.	0.	INFINITE	-8.141	1327.354		
100	30.205	135.606	191.683	-5.608			
200	28.395	156.122	169.413	-2.658			
250	27.038	162.312	167.402	-1.273			
298.15	25.856	166.970	166.970	0.	1336.333	1292.543	-226.448
300	25.815	167.130	166.971	0.048	1336.375	1292.272	-225.004
350	24.826	171.032	167.281	1.313	1336.744	1284.905	-191.762
400	24.058	174.295	167.959	2.534	1337.688	1277.434	-166.816
450	23.466	177.093	168.822	3.722	1338.599	1269.848	-147.400
500	23.010	179.540	169.774	4.883	1339.483	1262.161	-131.857
600	22.372	183.675	171.759	7.150	1341.196	1246.535	-108.520
700	21.966	187.091	173.712	9.365	1342.857	1230.626	-91.830
800	21.694	190.005	175.571	11.547	1344.485	1214.481	-79.297
900	21.505	192.548	177.319	13.707	1346.091	1198.134	-69.538
1000	21.371	194.807	178.957	15.850	1347.680	1181.609	-61.721
1100	21.274	196.839	180.492	17.982	1349.258	1164.926	-55.318
1200	21.206	198.687	181.932	20.106	1287.256	1149.177	-50.022
1300	21.160	200.382	183.287	22.224	1289.595	1137.576	-45.708
1400	21.134	201.949	184.565	24.338	1291.925	1125.795	-42.004
1500	21.126	203.407	185.773	26.451	1294.250	1113.847	-38.788
1600	21.134	204.771	186.918	28.564	1296.571	1101.744	-35.968
1700	21.156	206.053	188.006	30.679	1298.891	1089.497	-33.476
1800	21.192	207.263	189.043	32.796	1301.210	1077.112	-31.257
1900	21.239	208.410	190.032	34.917	1303.531	1064.599	-29.268
2000	21.296	209.501	190.979	37.044	1305.855	1051.964	-27.474
2100	21.362	210.541	191.886	39.177	1308.183	1039.212	-25.849
2200	21.435	211.537	192.756	41.317	1310.516	1026.349	-24.369
2300	21.514	212.491	193.594	43.464	1312.855	1013.380	-23.015
2400	21.596	213.409	194.400	45.619	1315.200	1000.309	-21.771
2500	21.681	214.292	195.179	47.783	1317.551	987.140	-20.625
2600	21.768	215.144	195.930	49.956	1319.910	973.877	-19.565
2700	21.855	215.967	196.657	52.137	1322.275	960.524	-18.582
2800	21.941	216.763	197.361	54.327	1324.648	947.082	-17.668
2900	22.026	217.535	198.043	56.525	1327.028	933.556	-16.815
3000	22.110	218.283	198.706	58.732	1329.414	919.947	-16.018
3100	22.190	219.009	199.349	60.947	1331.808	906.259	-15.270
3200	22.268	219.715	199.974	63.170	1334.208	892.493	-14.568
3300	22.343	220.401	200.583	65.400	1336.614	878.652	-13.908
3400	22.414	221.069	201.176	67.638	1339.026	864.738	-13.285
3500	22.481	221.720	201.754	69.883	1341.444	850.753	-12.697
3600	22.545	222.354	202.317	72.134	1343.867	836.699	-12.140
3700	22.605	222.973	202.867	74.392	1346.295	822.578	-11.613
3800	22.660	223.576	203.404	76.655	1348.727	808.391	-11.112
3900	22.712	224.166	203.929	78.924	1351.164	794.140	-10.636
4000	22.760	224.741	204.442	81.197	1353.604	779.826	-10.183
4100	22.804	225.304	204.944	83.476	1356.048	765.451	-9.752
4200	22.845	225.854	205.435	85.758	1358.494	751.017	-9.340
4300	22.882	226.392	205.916	88.045	1360.943	736.524	-8.947
4400	22.916	226.918	206.388	90.334	1363.395	721.975	-8.571
4500	22.946	227.434	206.850	92.628	1365.848	707.369	-8.211
4600	22.973	227.938	207.303	94.924	1368.303	692.709	-7.866
4700	22.998	228.433	207.747	97.222	1370.760	677.996	-7.535
4800	23.019	228.917	208.183	99.523	1373.217	663.230	-7.217
4900	23.038	229.392	208.611	101.826	1375.676	648.414	-6.912
5000	23.054	229.858	209.031	104.131	1378.135	633.546	-6.619
5100	23.068	230.314	209.444	106.437	1380.594	618.630	-6.336
5200	23.080	230.762	209.850	108.744	1383.054	603.666	-6.064
5300	23.090	231.202	210.249	111.053	1385.513	588.654	-5.802
5400	23.097	231.634	210.641	113.362	1387.972	573.596	-5.548
5500	23.103	232.058	211.026	115.672	1390.430	558.492	-5.304
5600	23.107	232.474	211.406	117.983	1392.888	543.344	-5.068
5700	23.110	232.883	211.779	120.294	1395.345	528.152	-4.840
5800	23.111	233.285	212.146	122.605	1397.801	512.916	-4.619
5900	23.110	233.680	212.508	124.916	1400.256	497.639	-4.406
6000	23.108	234.068	212.864	127.227	1402.710	482.319	-4.199

PREVIOUS: CURRENT: December 1982 (1 bar)

Phosphorus, Ion (P^+) $P_1^+(g)$

PHOSPHORUS, ION (P^-) IDEAL GAS M_r = 30.97431

EA(P, g) = 0.7465 ± 0.0003 eV
S°(298.15 K) = 169.124 ± 0.001 J K^{-1} mol^{-1}

$\Delta_f H°$(0 K) = 243.527 kJ mol^{-1}
$\Delta_f H°$(298.15 K) = [238.717] kJ mol^{-1}

Electonic Levels and Quantum Weights

State	ϵ_i, cm^{-1}	g_i
$^3P_2$	0.0	5
$^3P_1$	181.0	3
$^3P_0$	263.0	1
$^1D_2$	6675	5
$^1S_0$	16131	1

Enthalpy of Formation

$\Delta_f H°(P^-$, g, 0 K) is calculated from $\Delta_f H°$(P, g, 0 K) (1) using the adopted electron affinity of EA(P) = 0.7465 ± 0.0003 eV (72.026 ± 0.029 kJ mol^{-1}). This value, recommended by Hotop and Lineberger (2), is based on tunable laser photodetachment threshold studies (3, 4). Additional information on P^-(g) may be obtained in the critical discussions of Hotop and Lineberger (2, 5), Rosenstock et al. (6) and Massey (7).

$\Delta_f H°(P^-$, g, 298.15 K) is obtained from $\Delta_f H°$(P, g, 0 K) by using EA(P) with JANAF (1) enthalpies, H°(0 K) - H°(298.15 K), for P^-(g), P(g), and e^-(g). $\Delta_r H°(P^- \rightarrow P + e^-$, 298.15 K) differs from a room-temperature threshold energy due to inclusion of these enthalpies and to threshold effects discussed by Rosenstock et al. (6). $\Delta_f H°$(298.15 K) should be changed by +6.197 kJ mol^{-1} if it is to be used in the ion convention that excludes the enthalpy of the electron.

Heat Capacity and Entropy

The ground state electronic configuration for P^-(g) is given by Hotop and Lineberger (2, 5), Rosenstock (6), and Massey (7). The fine sturcture seperation has been measured experimentally via a tunetable laser photodetachment threshold technique (3, 4) and is that recommended by Hotop and Lineberger (2).

References

1. JANAF Thermochemical Tables: P(g), 12-31-82; e^-(g), 3-31-82.
2. H. Hotop and W. C. Lineberger, J. Phys. Chem. Ref. Data, to be published, 1985.
3. D. Feldmann, Z. Physik A 277, 19 (1976).
4. J. Slater and W. C. Lineberger, Phys. Rev. A 15, 2277 (1977).
5. H. Hotop and W. C. Lineberger, J. Phys. Chem. Ref. Data 4, 539 (1975).
6. H. M. Rosenstock, K. Draxl et al., J. Phys. Chem. Ref. Data 6, Supp. 1 (1977).
7. H. S. W. Massey, "Negative Ions", 3rd ed., Cambridge University Press, Cambridge, 1976.

Phosphorus, Ion (P^-) P_1^-(g)

Enthalpy Reference Temperature = T_r = 298.15 K Standard State Pressure = p° = 0.1 MPa

	J $K^{-1}mol^{-1}$			kJ mol^{-1}			
T/K	$C_p°$	S°	-[G°-H°(T_r)]/T	H°-H°(T_r)	$\Delta_f H°$	$\Delta_f G°$	Log K_f
0	0.	0.	INFINITE	-6.748	243.527		
100	23.554	143.795	189.435	-4.564			
200	23.042	160.101	171.176	-2.215			
250	22.540	165.187	169.490	-1.076			
298.15	22.168	169.124	169.124	0.	238.717	206.795	-36.230
300	22.156	169.261	169.124	0.041	238.676	206.597	-35.972
350	21.873	172.654	169.393	1.141	236.801	201.412	-30.059
400	21.664	175.560	169.986	2.230	235.534	196.443	-25.653
450	21.508	178.102	170.750	3.309	234.257	191.634	-22.244
500	21.388	180.362	171.600	4.381	232.974	186.966	-19.532
600	21.222	184.245	173.394	6.511	230.393	178.007	-15.497
700	21.116	187.508	175.184	8.627	227.798	169.481	-12.647
800	21.047	190.323	176.904	10.735	225.195	161.328	-10.534
900	21.004	192.799	178.535	12.838	222.586	153.501	-8.909
1000	20.983	195.011	180.074	14.937	219.974	145.965	-7.624
1100	20.982	197.011	181.525	17.035	217.361	138.690	-6.586
1200	21.002	198.837	182.892	19.134	151.176	132.732	-5.778
1300	21.042	200.520	184.184	21.236	149.342	131.269	-5.274
1400	21.100	202.081	185.408	23.343	147.508	129.948	-4.848
1500	21.174	203.539	186.568	25.456	145.676	128.758	-4.484
1600	21.261	204.909	187.672	27.578	143.849	127.689	-4.169
1700	21.360	206.200	188.725	29.709	142.028	126.735	-3.894
1800	21.467	207.424	189.730	31.850	140.214	125.888	-3.653
1900	21.579	208.588	190.692	34.003	138.409	125.142	-3.440
2000	21.694	209.698	191.615	36.166	136.612	124.490	-3.251
2100	21.810	210.759	192.501	38.341	134.826	123.928	-3.083
2200	21.924	211.776	193.354	40.528	133.048	123.450	-2.931
2300	22.036	212.753	194.177	42.726	131.280	123.054	-2.795
2400	22.144	213.693	194.970	44.935	129.522	122.734	-2.671
2500	22.246	214.599	195.738	47.155	127.772	122.487	-2.559
2600	22.344	215.474	196.480	49.384	126.030	122.310	-2.457
2700	22.435	216.319	197.199	51.623	124.296	122.200	-2.364
2800	22.520	217.136	197.897	53.871	122.570	122.154	-2.279
2900	22.598	217.928	198.574	56.127	120.850	122.169	-2.201
3000	22.670	218.695	199.232	58.390	119.136	122.244	-2.128
3100	22.735	219.440	199.872	60.661	117.427	122.376	-2.062
3200	22.794	220.162	200.495	62.937	115.724	122.563	-2.001
3300	22.847	220.865	201.101	65.219	114.024	122.803	-1.944
3400	22.894	221.547	201.693	67.506	112.328	123.094	-1.891
3500	22.936	222.212	202.269	69.798	110.636	123.435	-1.842
3600	22.972	222.858	202.832	72.093	108.946	123.825	-1.797
3700	23.004	223.488	203.382	74.392	107.258	124.262	-1.754
3800	23.030	224.102	203.919	76.694	105.571	124.744	-1.715
3900	23.053	224.700	204.445	78.998	103.886	125.270	-1.678
4000	23.072	225.284	204.958	81.304	102.201	125.840	-1.643
4100	23.087	225.854	205.461	83.612	100.517	126.452	-1.611
4200	23.098	226.411	205.953	85.921	98.834	127.105	-1.581
4300	23.106	226.954	206.435	88.232	97.149	127.798	-1.552
4400	23.112	227.486	206.908	90.543	95.465	128.530	-1.526
4500	23.115	228.005	207.371	92.854	93.779	129.301	-1.501
4600	23.115	228.513	207.825	95.165	92.093	130.108	-1.477
4700	23.113	229.010	208.270	97.477	90.405	130.953	-1.455
4800	23.109	229.497	208.708	99.788	88.716	131.834	-1.435
4900	23.104	229.973	209.137	102.099	87.025	132.749	-1.415
5000	23.096	230.440	209.558	104.409	85.332	133.700	-1.397
5100	23.088	230.897	209.972	106.718	83.637	134.684	-1.379
5200	23.077	231.345	210.379	109.026	81.940	135.701	-1.363
5300	23.066	231.785	210.779	111.333	80.241	136.752	-1.348
5400	23.053	232.216	211.172	113.639	78.539	137.834	-1.333
5500	23.040	232.639	211.558	115.944	76.835	138.947	-1.320
5600	23.025	233.054	211.938	118.247	75.128	140.092	-1.307
5700	23.010	233.461	212.312	120.549	73.419	141.267	-1.295
5800	22.994	233.861	212.680	122.849	71.707	142.473	-1.283
5900	22.977	234.254	213.043	125.148	69.992	143.707	-1.272
6000	22.960	234.640	213.399	127.445	68.274	144.971	-1.262

PREVIOUS: CURRENT: December 1982 (1 bar)

Phosphorus, Ion (P^-) P_1^-(g)

PHOSPHORUS SULFIDE (PS) IDEAL GAS M_r = 63.03376

$S°(298.15\ K) = 234.241\ J\ K^{-1}\ mol^{-1}$

$\Delta_f H°(0\ K) = 138.8 \pm 41.8\ kJ\ mol^{-1}$
$\Delta_f H°(298.15\ K) = 138.6 \pm 41.8\ kJ\ mol^{-1}$

Electronic Levels and Quantum Weights

State	ε_i, cm^{-1}	g_i
$^2\Pi$	0	2
$A^2\Sigma$	324	2
$B^2\Pi$	[20000]	2
$C^2\Sigma$	22000	4
	34594	2

$\omega_e = 743.5 \pm 3.8\ cm^{-1}$ $\omega_e x_e = 3.8 \pm 0.9\ cm^{-1}$ $\sigma = 1$
$B_e = [0.29]\ cm^{-1}$ $\alpha_e = [0.0016]\ cm^{-1}$ $r_e = [1.92]$ Å

Enthalpy of Formation

The dissociaton energy, D_0^0, of PS(g) is estimated to be 140 ± 25 kcal mol^{-1} by comparison with the corresponding quantity for PO(g) and from a linear Birge-Sponer extrapolation of the third excited ($C^2\Sigma$) electronic state potential function. It is assumed that the $C^2\Sigma$ state dissociates to S(g) in its ground (^3P) state and P(g) in its first excited (^2D) state. The fundamental vibrational frequency in the $C^2\Sigma$ state (ω_e') is 531.8 ± 2.5 cm^{-1} and the corresponding anharmonic term ($\omega_e x_e$) is 2.6 ± 0.5 cm^{-1}. These constants are calculated from a reanalysis of the spectral data of Dressler (1). The corresponding enthalpy of formation of PS(g) is 37.3 ± 25 kcal mol^{-1}. The excited state ($C^2\Sigma$) data are used to calculate D_0^0 instead of the ground state ($X^2\Pi$) data because the vibrational constants of the upper level are known more accurately and because the extrapolation is 50 kcal mol^{-1} smaller for this level.

Heat Capacity and Entropy

The spectra of PS(g) have been studied by Dressler (1) in the near UV and visible regions. He reported the above electronic levels which correspond to the doublet pi ground state and two of the excited levels ($B^2\Pi$, $C^2\Sigma$). The level of the first excited electronic state ($A^2\Sigma$) is estimated by analogy with NO(g).

Dressler (1) estimated the value of the rotational constant B_e. The value of α_e is calculated from the Morse potential function. The bond distance is calculated from B_e. The fundamental vibrational frequency ω_e and the anharmonic vibrational term $\omega_e x_e$ are calculated from a reanalysis of the spectral data reported by Dressler (1).

Reference

1. K. Dressler, Helv. Phys. Acta. 28, 563 (1955).

Phosphorus Sulfide (PS) $P_1S_1(g)$

Enthalpy Reference Temperature = T_r = 298.15 K Standard State Pressure = p° = 0.1 MPa

T/K	$C_p°$ (J K⁻¹mol⁻¹)	S° (J K⁻¹mol⁻¹)	$-[G°-H°(T_r)]/T$ (J K⁻¹mol⁻¹)	$H°-H°(T_r)$ (kJ mol⁻¹)	$\Delta_f H°$ (kJ mol⁻¹)	$\Delta_f G°$ (kJ mol⁻¹)	Log K_f
0	0.	0.	INFINITE	-9.610	138.765	138.765	INFINITE
100	30.805	197.931	264.579	-6.665	140.134	123.361	-64.437
200	33.947	220.417	237.429	-3.402	139.491	106.560	-27.831
250	34.711	228.080	234.819	-1.685	139.097	98.370	-20.553
298.15	35.243	234.241	234.241	0.	138.603	90.569	-15.867
300	35.260	234.459	234.242	0.065	138.582	90.271	-15.718
350	35.679	239.928	234.673	1.839	137.254	82.337	-12.288
400	36.008	244.714	235.635	3.632	134.300	74.630	-9.746
450	36.268	248.971	236.885	5.439	132.865	67.248	-7.806
500	36.478	252.803	238.288	7.258	131.365	60.040	-6.272
600	36.790	259.484	241.280	10.922	128.812	46.022	-4.007
700	37.008	265.172	244.297	14.613	126.523	32.407	-2.418
800	37.169	270.125	247.223	18.322	124.389	19.108	-1.248
900	37.292	274.510	250.016	22.045	69.223	7.153	-0.415
1000	37.392	278.445	252.665	25.779	68.468	0.296	-0.015
1100	37.474	282.012	255.173	29.523	67.708	-6.484	0.308
1200	37.544	285.276	257.548	33.274	3.367	-12.117	0.527
1300	37.606	288.284	259.798	37.031	3.364	-13.407	0.539
1400	37.662	291.073	261.934	40.795	3.347	-14.697	0.548
1500	37.712	293.673	263.964	44.564	3.316	-15.984	0.557
1600	37.759	296.108	265.898	48.337	3.271	-17.270	0.564
1700	37.803	298.399	267.743	52.115	3.213	-18.552	0.570
1800	37.845	300.561	269.506	55.898	3.142	-19.830	0.575
1900	37.885	302.608	271.195	59.684	3.061	-21.104	0.580
2000	37.924	304.552	272.815	63.475	2.968	-22.373	0.584
2100	37.962	306.403	274.371	67.269	2.867	-23.638	0.588
2200	38.000	308.170	275.867	71.067	2.757	-24.898	0.591
2300	38.037	309.860	277.309	74.869	2.639	-26.152	0.594
2400	38.074	311.480	278.699	78.675	2.515	-27.401	0.596
2500	38.112	313.035	280.041	82.484	2.385	-28.645	0.599
2600	38.150	314.530	281.339	86.297	2.250	-29.884	0.600
2700	38.190	315.971	282.595	90.114	2.110	-31.117	0.602
2800	38.231	317.361	283.812	93.935	1.968	-32.345	0.603
2900	38.274	318.703	284.993	97.760	1.822	-33.568	0.605
3000	38.319	320.001	286.138	101.590	1.675	-34.786	0.606
3100	38.367	321.259	287.251	105.424	1.525	-35.999	0.607
3200	38.418	322.477	288.333	109.263	1.375	-37.207	0.607
3300	38.471	323.660	289.385	113.108	1.225	-38.410	0.608
3400	38.528	324.810	290.410	116.958	1.074	-39.609	0.609
3500	38.588	325.927	291.409	120.814	0.924	-40.803	0.609
3600	38.652	327.015	292.383	124.676	0.775	-41.993	0.609
3700	38.720	328.075	293.334	128.544	0.626	-43.179	0.610
3800	38.792	329.109	294.262	132.420	0.480	-44.361	0.610
3900	38.868	330.117	295.168	136.303	0.335	-45.539	0.610
4000	38.948	331.103	296.054	140.193	0.192	-46.714	0.610
4100	39.032	332.065	296.921	144.092	0.051	-47.885	0.610
4200	39.120	333.007	297.769	148.000	-0.087	-49.052	0.610
4300	39.212	333.928	298.599	151.916	-0.223	-50.216	0.610
4400	39.307	334.831	299.412	155.842	-0.356	-51.377	0.610
4500	39.407	335.715	300.209	159.778	-0.485	-52.536	0.610
4600	39.510	336.583	300.991	163.724	-0.612	-53.691	0.610
4700	39.617	337.434	301.757	167.680	-0.735	-54.843	0.610
4800	39.727	338.269	302.509	171.647	-0.854	-55.993	0.609
4900	39.840	339.089	303.247	175.626	-0.970	-57.141	0.609
5000	39.957	339.895	303.972	179.615	-1.082	-58.286	0.609
5100	40.076	340.688	304.684	183.617	-1.191	-59.429	0.609
5200	40.197	341.467	305.384	187.631	-1.295	-60.570	0.608
5300	40.321	342.234	306.072	191.657	-1.395	-61.709	0.608
5400	40.447	342.989	306.749	195.695	-1.491	-62.846	0.608
5500	40.575	343.732	307.414	199.746	-1.583	-63.981	0.608
5600	40.705	344.464	308.070	203.810	-1.670	-65.115	0.607
5700	40.836	345.186	308.714	207.887	-1.753	-66.247	0.607
5800	40.968	345.897	309.349	211.977	-1.831	-67.378	0.607
5900	41.102	346.599	309.975	216.081	-1.905	-68.508	0.607
6000	41.236	347.291	310.591	220.198	-1.974	-69.636	0.606

PREVIOUS: June 1967 (1 atm) CURRENT: June 1967 (1 bar)

Phosphorus Sulfide (PS) $P_1S_1(g)$

PHOSPHORUS (P_2) IDEAL GAS M_r = 61.94752

$S°(298.15\ K) = 218.13 \pm 0.4\ J\ K^{-1}\ mol^{-1}$

$\Delta_f H°(0\ K) = 145.47 \pm 2.1\ kJ\ mol^{-1}$
$\Delta_f H°(298.15\ K) = 143.65 \pm 2.1\ kJ\ mol^{-1}$

Electronic Level and Quantum Weight

State	ϵ_i, cm^{-1}	g_i
$^1\Sigma$	0	1

$\omega_e = 780.43\ cm^{-1}$ $\omega_e x_e = 2.804\ cm^{-1}$ $\sigma = 2$
$B_e = 0.30327\ cm^{-1}$ $\alpha_e = 0.00142\ cm^{-1}$ $r_e = 1.8943\ Å$

Enthalpy of Formation

Stevenson and Yost (1) have shown that the vapor density measurements of Stock et al. (2) best fit a vapor composed of P_2(g) and P_4(g). From these data, Farr (3) calculated the equilibrium constants which lead to $\Delta_r H°(298.15\ K) = 54.59 \pm 0.01\ kcal\ mol^{-1}$ for the process $P_4(g) \rightarrow P_2(g)$; from which the value for $\Delta_f H°(P_2, g, 298.15\ K)$ is calculated using $\Delta_f H°(P_4, g, 298.15\ K) = 30.77\ kcal\ mol^{-1}$.

Heat Capacity and Entropy

Molecular and spectroscopic constants are those listed by Herzberg (4). More recent measurements of Dressler (5) are in agreement with the values selected by Herzberg (4).

The dissociation energy, $D_0^0 = 116 \pm 1\ kcal\ mol^{-1}$, is given by Gaydon (6).

References

1. D. P. Stevenson and D. M. Yost, J. Chem. Phys. 9, 403 (1941).
2. A. Stock, G. E. Gibson, and E. Stamm, Ber. 45, 3527 (1912).
3. T. D. Farr, Tennessee Valley Authority, Chemical Engineering Report No. 8,
4. G. Herzberg, "Spectra of Diatomic Molecules," D. Van Nostrand Co., Inc., New York, 1945.
5. K. Dressler, Helv. Phys. Acta 28, 563 (1955).
6. A. G. Gaydon, "Dissociation Energies," Chapman and Hall, Ltd., London, 1953.

Enthalpy Reference Temperature = T_r = 298.15 K Standard State Pressure = p° = 0.1 MPa

T/K	$C_p°$ (J K^{-1}mol^{-1})	S°	$-[G°-H°(T_r)]/T$	$H°-H°(T_r)$ (kJ mol^{-1})	$\Delta_f H°$	$\Delta_f G°$	Log K_f
0	0.	0.	INFINITE	-8.904	145.469	145.469	INFINITE
100	29.120	185.370	245.321	-5.995	146.604	131.604	-68.743
200	30.103	205.754	220.998	-3.049	145.023	116.723	-30.485
250	31.082	212.574	218.653	-1.520	144.369	109.721	-22.925
298.15	32.034	218.131	218.131	0.	143.653	103.112	-18.065
300	32.069	218.329	218.132	0.059	143.625	102.861	-17.910
350	32.945	223.340	218.525	1.685	141.379	96.272	-14.368
400	33.679	227.789	219.410	3.351	140.413	89.894	-11.739
450	34.280	231.792	220.568	5.051	139.479	83.636	-9.708
500	34.770	235.430	221.875	6.778	138.574	77.480	-8.094
600	35.499	241.838	224.682	10.293	136.824	65.427	-5.696
700	36.000	247.350	227.536	13.870	135.136	53.661	-4.004
800	36.356	252.182	230.321	17.489	133.489	42.135	-2.751
900	36.618	256.480	232.993	21.138	131.873	30.813	-1.788
1000	36.819	260.348	235.538	24.810	130.280	19.670	-1.027
1100	36.976	263.865	237.956	28.500	128.705	8.685	-0.412
1180.008	37.079	266.465	239.802	31.463	------- FUGACITY = 1 bar -------		
1200	37.103	267.088	240.251	32.204	0.	0.	0.
1300	37.208	270.062	242.432	35.920	0.	0.	0.
1400	37.296	272.823	244.505	39.645	0.	0.	0.
1500	37.373	275.399	246.480	43.379	0.	0.	0.
1600	37.440	277.813	248.363	47.120	0.	0.	0.
1700	37.499	280.085	250.163	50.867	0.	0.	0.
1800	37.553	282.230	251.886	54.619	0.	0.	0.
1900	37.602	284.261	253.537	58.377	0.	0.	0.
2000	37.647	286.191	255.121	62.139	0.	0.	0.
2100	37.689	288.029	256.645	65.906	0.	0.	0.
2200	37.728	289.783	258.112	69.677	0.	0.	0.
2300	37.766	291.461	259.526	73.452	0.	0.	0.
2400	37.801	293.069	260.890	77.230	0.	0.	0.
2500	37.835	294.613	262.208	81.012	0.	0.	0.
2600	37.868	296.098	263.483	84.797	0.	0.	0.
2700	37.899	297.527	264.718	88.586	0.	0.	0.
2800	37.930	298.906	265.914	92.377	0.	0.	0.
2900	37.960	300.238	267.075	96.171	0.	0.	0.
3000	37.989	301.525	268.202	99.969	0.	0.	0.
3100	38.017	302.771	269.297	103.769	0.	0.	0.
3200	38.045	303.979	270.362	107.572	0.	0.	0.
3300	38.072	305.150	271.399	111.378	0.	0.	0.
3400	38.099	306.287	272.408	115.187	0.	0.	0.
3500	38.126	307.392	273.392	118.998	0.	0.	0.
3600	38.152	308.466	274.351	122.812	0.	0.	0.
3700	38.178	309.512	275.288	126.628	0.	0.	0.
3800	38.204	310.530	276.202	130.448	0.	0.	0.
3900	38.229	311.523	277.095	134.269	0.	0.	0.
4000	38.254	312.491	277.968	138.093	0.	0.	0.
4100	38.279	313.436	278.821	141.920	0.	0.	0.
4200	38.304	314.359	279.656	145.749	0.	0.	0.
4300	38.328	315.260	280.474	149.581	0.	0.	0.
4400	38.353	316.142	281.275	153.415	0.	0.	0.
4500	38.377	317.004	282.059	157.251	0.	0.	0.
4600	38.401	317.848	282.828	161.090	0.	0.	0.
4700	38.425	318.674	283.582	164.932	0.	0.	0.
4800	38.449	319.483	284.321	168.775	0.	0.	0.
4900	38.473	320.276	285.047	172.621	0.	0.	0.
5000	38.497	321.053	285.759	176.470	0.	0.	0.
5100	38.521	321.816	286.459	180.321	0.	0.	0.
5200	38.544	322.564	287.146	184.174	0.	0.	0.
5300	38.568	323.299	287.821	188.030	0.	0.	0.
5400	38.591	324.020	288.485	191.888	0.	0.	0.
5500	38.615	324.728	289.138	195.748	0.	0.	0.
5600	38.638	325.424	289.779	199.611	0.	0.	0.
5700	38.661	326.108	290.411	203.476	0.	0.	0.
5800	38.685	326.781	291.032	207.343	0.	0.	0.
5900	38.708	327.442	291.644	211.212	0.	0.	0.
6000	38.731	328.093	292.246	215.084	0.	0.	0.

PREVIOUS: June 1961 (1 atm) CURRENT: June 1961 (1 bar)

PHOSPHORUS (P_4) IDEAL GAS M_r = 123.89504

Phosphorus (P_4) P_4(g)

$S°(298.15 K) = 280.0 \pm 0.4 J K^{-1} mol^{-1}$

$\Delta_f H°(0 K) = 66.2 \pm 2.1 kJ mol^{-1}$
$\Delta_f H°(298.15 K) = 58.9 \pm 2.1 kJ mol^{-1}$

Vibrational Levels and Degeneracies

ν, cm^{-1}
606 (1)
363 (2)
464.5(3)

Point Group: T_d σ = 12
Bond Length: P-P = 2.21 ± 0.2 Å
Bond Angle: P-P-P = 109.4712°
Product of the Principal Moments of Inertia: $I_A I_B I_C = 15.8528 \times 10^{-114} g^3 cm^6$

Enthalpy of Formation

Taken as the enthalpy of formation of white phosphorus.

Heat Capacity and Entropy

The frequencies and degeneracies are those assigned by Gutowsky and Hoffman (1). Molecular parameters of the gas were determined from electron diffraction measurements by Maxwell et al. (2). Thomas and Gingrich (3) obtained a P-P distance of 2.25 Å from X-ray studies on the liquid. On the basis of the selected values for P(white), a 3rd law entropy of 66.89 cal $K^{-1} mol^{-1}$ (for p=1 atm) is derived for P_4(g). The principal moments of inertia are: $I_A = I_B = I_C = 25.1209 \times 10^{-39} g cm^2$.

References

1. H. S. Gutowsky and G. S. Hoffman, J. Am. Chem. Soc. 72, 5751(1950).
2. L. R. Maxwell, S. B. Hendricks, and V. M. Moseley, J. Chem. Phys. 3, 699(1935).
3. C. D. Thomas and N. S. Gingrich, J. Chem. Phys. 6, 659(1938).

Enthalpy Reference Temperature = T_r = 298.15 K Standard State Pressure = p° = 0.1 MPa

	J K⁻¹mol⁻¹			kJ mol⁻¹			
T/K	$C_p°$	S°	$-[G°-H°(T_r)]/T$	$H°-H°(T_r)$	$\Delta_f H°$	$\Delta_f G°$	Log K_f
0	0.	0.	INFINITE	-14.133	66.214	66.214	INFINITE
100	37.232	223.901	331.278	-10.738	66.060	50.745	-26.506
200	55.620	255.381	285.819	-6.087	61.657	36.281	-9.476
250	62.447	268.567	281.073	-3.127	60.250	30.101	-6.289
298.15	67.158	279.992	279.992	0.	58.907	24.416	-4.278
300	67.309	280.408	279.993	0.124	58.854	24.202	-4.214
350	70.765	291.058	280.827	3.581	54.568	18.822	-2.809
400	73.260	300.679	282.717	7.185	52.907	13.830	-1.806
450	75.100	309.419	285.206	10.896	51.353	9.040	-1.049
500	76.486	317.407	288.032	14.687	49.879	4.418	-0.462
600	78.386	331.534	294.136	22.439	47.100	-4.410	0.384
700	79.586	343.714	300.369	30.341	44.472	-12.785	0.954
800	80.388	354.397	306.468	38.343	41.943	-20.791	1.358
900	80.949	363.899	312.331	46.411	39.482	-28.485	1.653
1000	81.356	372.450	317.923	54.527	37.068	-35.907	1.876
1100	81.660	380.219	323.238	62.679	34.689	-43.089	2.046
1200	81.893	387.335	328.287	70.857	-221.952	-45.742	1.991
1300	82.075	393.897	333.085	79.056	-221.184	-31.089	1.249
1400	82.221	399.985	337.649	87.271	-220.420	-16.495	0.615
1500	82.338	405.662	341.996	95.499	-219.659	-1.956	0.068
1600	82.435	410.979	346.143	103.738	-218.902	12.533	-0.409
1700	82.515	415.979	350.106	111.985	-218.148	26.975	-0.829
1800	82.583	420.698	353.898	120.240	-217.398	41.372	-1.201
1900	82.640	425.164	357.532	128.502	-216.653	55.728	-1.532
2000	82.689	429.405	361.020	136.768	-215.911	70.045	-1.829
2100	82.731	433.440	364.374	145.039	-215.174	84.324	-2.097
2200	82.767	437.289	367.601	153.314	-214.440	98.569	-2.340
2300	82.799	440.969	370.712	161.592	-213.711	112.780	-2.561
2400	82.827	444.494	373.713	169.874	-212.987	126.960	-2.763
2500	82.852	447.876	376.612	178.158	-212.266	141.110	-2.948
2600	82.874	451.126	379.416	186.444	-211.550	155.230	-3.119
2700	82.893	454.254	382.130	194.733	-210.839	169.324	-3.276
2800	82.911	457.269	384.760	203.023	-210.131	183.391	-3.421
2900	82.927	460.178	387.311	211.315	-209.429	197.433	-3.556
3000	82.941	462.990	389.787	219.608	-208.730	211.451	-3.682
3100	82.954	465.710	392.193	227.903	-208.036	225.445	-3.799
3200	82.966	468.344	394.531	236.199	-207.346	239.418	-3.908
3300	82.976	470.897	396.807	244.496	-206.661	253.368	-4.010
3400	82.986	473.374	399.023	252.794	-205.980	267.298	-4.107
3500	82.995	475.780	401.182	261.093	-205.303	281.208	-4.197
3600	83.003	478.118	403.286	269.393	-204.631	295.099	-4.282
3700	83.010	480.392	405.340	277.694	-203.964	308.972	-4.362
3800	83.017	482.606	407.344	285.995	-203.300	322.826	-4.438
3900	83.024	484.762	409.302	294.297	-202.641	336.663	-4.509
4000	83.030	486.864	411.215	302.600	-201.987	350.483	-4.577
4100	83.035	488.915	413.085	310.903	-201.337	364.286	-4.641
4200	83.040	490.916	414.914	319.207	-200.692	378.074	-4.702
4300	83.045	492.870	416.704	327.511	-200.051	391.847	-4.760
4400	83.050	494.779	418.457	335.816	-199.414	405.604	-4.815
4500	83.054	496.645	420.174	344.121	-198.782	419.348	-4.868
4600	83.058	498.471	421.856	352.426	-198.154	433.077	-4.918
4700	83.061	500.257	423.506	360.732	-197.531	446.793	-4.966
4800	83.065	502.006	425.123	369.039	-196.912	460.495	-5.011
4900	83.068	503.719	426.709	377.345	-196.298	474.185	-5.055
5000	83.071	505.397	428.266	385.652	-195.688	487.862	-5.097
5100	83.074	507.042	429.795	393.960	-195.082	501.527	-5.137
5200	83.076	508.655	431.296	402.267	-194.481	515.180	-5.175
5300	83.079	510.238	432.771	410.575	-193.885	528.822	-5.212
5400	83.081	511.791	434.220	418.883	-193.293	542.452	-5.247
5500	83.084	513.315	435.644	427.191	-192.705	556.071	-5.281
5600	83.086	514.812	437.044	435.500	-192.122	569.680	-5.314
5700	83.088	516.283	438.422	443.808	-191.543	583.279	-5.345
5800	83.090	517.728	439.777	452.117	-190.969	596.867	-5.375
5900	83.091	519.148	441.110	460.426	-190.399	610.446	-5.404
6000	83.093	520.545	442.422	468.735	-189.834	624.014	-5.433

PREVIOUS: June 1961 (1 atm) CURRENT: June 1961 (1 bar)

Phosphorus (P_4) P_4(g)

PHOSPHORUS SULFIDE (P_4S_3) CRYSTAL M_r = 220.07504

$S°(298.15\ K) = [200.832]\ J\ K^{-1}\ mol^{-1}$

$T_{fus} = 440\ K$

$\Delta_f H°(298.15\ K) = [-224.647]\ kJ\ mol^{-1}$

$\Delta_{fus} H° = [9.204]\ kJ\ mol^{-1}$

Enthalpy of Formation

$\Delta_f H°(298.15\ K)$ is estimated from vapor density reported by Yost and Russell (1). T_{fus} is from Yost and Russell (1). Other data estimated.

Reference

1. D. M. Yost and H. Russell, "Systematic Inorganic Chemistry of the Fifth- and Sixth-Group Nonmetallic Elements," Prentice-Hall, New York (1944).

Phosphorus Sulfide (P_4S_3) P_4S_3(cr)

Enthalpy Reference Temperature = T_r = 298.15 K Standard State Pressure = p° = 0.1 MPa

	J K⁻¹mol⁻¹			kJ mol⁻¹			
T/K	$C_p°$	S°	$-[G°-H°(T_r)]/T$	$H°-H°(T_r)$	$\Delta_f H°$	$\Delta_f G°$	Log K_f
0							
100							
200							
250							
298.15	146.440	200.832	200.832	0.	-224.647	-206.864	36.242
300	146.440	201.738	200.835	0.271	-224.679	-206.754	35.999
400	146.440	243.866	206.579	14.915	-236.834	-199.434	26.043
440.000	146.440	257.823	210.613	20.773	------ CRYSTAL <--> LIQUID -----		
500	146.440	276.543	217.425	29.559	-244.505	-189.235	19.769
600	146.440	303.242	229.571	44.203	-251.145	-177.539	15.456
700	146.440	325.816	241.749	58.847	-257.072	-164.797	12.297
800	146.440	345.370	253.507	73.491	-262.592	-151.234	9.875
900	146.440	362.619	264.691	88.135	-427.249	-133.753	7.763
1000	146.440	378.048	275.269	102.779	-428.703	-101.065	5.279
1100	146.440	392.005	285.257	117.423	-430.204	-68.228	3.240
1200	146.440	404.747	294.691	132.067	-686.040	-30.940	1.347
1300	146.440	416.468	303.614	146.711	-684.535	23.590	-0.948
1400	146.440	427.321	312.067	161.355	-683.095	78.007	-2.910
1500	146.440	437.424	320.091	175.999	-681.718	132.323	-4.608
1600	146.440	446.875	327.723	190.643	-680.401	186.549	-6.090
1700	146.440	455.753	334.996	205.287	-679.139	240.694	-7.396
1800	146.440	464.123	341.939	219.931	-677.929	294.767	-8.554
1900	146.440	472.041	348.580	234.575	-676.769	348.774	-9.588
2000	146.440	479.552	354.943	249.219	-675.654	402.721	-10.518

PREVIOUS: CURRENT: December 1960

Phosphorus Sulfide (P_4S_3) P_4S_3(cr)

PHOSPHORUS SULFIDE (P_4S_3) LIQUID M_r = 220.07504

$S°(298.15\ K) = [207.097]\ J\ K^{-1}\ mol^{-1}$ $\Delta_fH°(298.15\ K) = [-220.784]\ kJ\ mol^{-1}$

$T_{fus} = 440\ K$ $\Delta_{fus}H° = [9.204]\ kJ\ mol^{-1}$

Enthalpy of Formation

$\Delta_fH°(P_4S_3, \ell, 298.15\ K)$ is calculated from that of the crystal by adding $\Delta_{fus}H°$ and the difference in enthalpy, H°(440 K)-H°(298.15 K), between the crystal and liquid.

Fusion Data

T_{fus} is from Yost and Russell (1).

Vaporization Data

T_{vap}(1 atm) = 680 K is given by Yost and Russell (1).

Reference

1. D. M. Yost and H. Russell, "Systematic Inorganic Chemistry of the Fifth- and Sixth-Group Nonmetallic Elements," Prentice-Hall, New York (1944).

Phosphorus Sulfide (P_4S_3) $P_4S_3(l)$

Enthalpy Reference Temperature = T_r = 298.15 K Standard State Pressure = p° = 0.1 MPa

T/K	$C_p°$	$S°$	$-[G°-H°(T_r)]/T$	$H°-H°(T_r)$	$\Delta_fH°$	$\Delta_fG°$	Log K_f
	J K⁻¹mol⁻¹			kJ mol⁻¹			
0							
100							
200							
250							
298.15	184.096	207.097	207.097	0.	-220.784	-204.869	35.892
300	184.096	208.236	207.101	0.341	-220.746	-204.770	35.654
400	184.096	261.197	214.322	18.750	-229.136	-198.668	25.943
440.000	184.096	278.743	219.393	26.114	------ CRYSTAL <--> LIQUID -----		
500	184.096	302.277	227.957	37.160	-233.041	-190.638	19.916
600	184.096	335.842	243.226	55.569	-235.916	-181.869	15.833
700	184.096	364.220	258.536	73.979	-238.077	-172.684	12.886
800	184.096	388.803	273.317	92.389	-239.831	-163.219	10.657
900	184.096	410.486	287.377	110.798	-400.722	-150.307	8.724
1000	184.096	429.882	300.675	129.208	-398.411	-122.607	6.404
1100	184.096	447.429	313.231	147.617	-396.146	-95.137	4.518
1200	184.096	463.447	325.091	166.027	-648.216	-63.557	2.767
1300	184.096	478.183	336.308	184.437	-642.946	-15.050	0.605
1400	184.096	491.826	346.936	202.846	-637.741	33.054	-1.233
1500	184.096	504.527	357.023	221.256	-632.598	80.788	-2.813
1600	184.096	516.408	366.617	239.665	-627.515	128.181	-4.185
1700	184.096	527.569	375.760	258.075	-622.487	175.258	-5.385
1800	184.096	538.092	384.489	276.485	-617.512	222.041	-6.443
1900	184.096	548.045	392.838	294.894	-612.586	268.548	-7.383
2000	184.096	557.488	400.836	313.304	-607.706	314.797	-8.222
2100	184.096	566.470	408.511	331.713	-602.867	360.803	-8.974
2200	184.096	575.034	415.887	350.123	-598.067	406.579	-9.653
2300	184.096	583.218	422.986	368.533	-593.303	452.138	-10.268
2400	184.096	591.053	429.827	386.942	-588.572	497.491	-10.828
2500	184.096	598.568	436.427	405.352	-583.872	542.647	-11.338
2600	184.096	605.788	442.803	423.761	-579.201	587.615	-11.805
2700	184.096	612.736	448.969	442.171	-574.554	632.404	-12.235
2800	184.096	619.431	454.938	460.581	-569.931	677.022	-12.630
2900	184.096	625.891	460.722	478.990	-565.330	721.474	-12.995
3000	184.096	632.133	466.333	497.400	-560.752	765.769	-13.333

PREVIOUS: CURRENT: December 1960

Phosphorus Sulfide (P_4S_3) $P_4S_3(l)$

PHOSPHORUS SULFIDE (P_4S_3) CRYSTAL-LIQUID M_r = 220.07504 Phosphorus Sulfide (P_4S_3) P_4S_3(cr,l)

0 to 440 K crystal
above 440 K liquid

Refer to the individual tables for details.

Enthalpy Reference Temperature = T_r = 298.15 K Standard State Pressure = p° = 0.1 MPa

T/K	C_p° (J K⁻¹mol⁻¹)	S° (J K⁻¹mol⁻¹)	−[G°−H°(T_r)]/T (J K⁻¹mol⁻¹)	H°−H°(T_r) (kJ mol⁻¹)	Δ_fH° (kJ mol⁻¹)	Δ_fG° (kJ mol⁻¹)	Log K_f
0							
100							
200							
250							
298.15	146.440	200.832	200.832	0.	-224.647	-206.864	36.242
300	146.440	201.738	200.835	0.271	-224.679	-206.754	35.999
400	146.440	243.866	206.579	14.915	-236.834	-199.434	26.043
440.000	146.440	257.823	210.613	20.773	CRYSTAL <--> LIQUID	TRANSITION	
440.000	184.096	278.743	210.613	29.977			
500	184.096	302.277	220.231	41.023	-233.041	-190.638	19.916
600	184.096	335.842	236.787	59.433	-235.916	-181.869	15.833
700	184.096	364.220	253.017	77.842	-238.077	-172.684	12.886
800	184.096	388.803	268.488	96.252	-239.831	-163.219	10.657
900	184.096	410.486	283.084	114.661	-400.722	-150.307	8.724
1000	184.096	429.882	296.811	133.071	-398.411	-122.607	6.404
1100	184.096	447.429	309.719	151.481	-396.146	-95.137	4.518
1200	184.096	463.447	321.872	169.890	-648.216	-63.557	2.767
1300	184.096	478.183	333.337	188.300	-642.946	-15.050	0.605
1400	184.096	491.826	344.176	206.709	-637.741	33.054	-1.233
1500	184.096	504.527	354.448	225.119	-632.598	80.788	-2.813
1600	184.096	516.408	364.203	243.529	-627.515	128.181	-4.185
1700	184.096	527.569	373.488	261.938	-622.487	175.258	-5.385
1800	184.096	538.092	382.343	280.348	-617.512	222.040	-6.443
1900	184.096	548.045	390.804	298.757	-612.586	268.548	-7.383
2000	184.096	557.488	398.905	317.167	-607.706	314.797	-8.222
2100	184.096	566.470	406.672	335.577	-602.867	360.803	-8.974
2200	184.096	575.034	414.131	353.986	-598.067	406.579	-9.653
2300	184.096	583.218	421.306	372.396	-593.303	452.138	-10.268
2400	184.096	591.053	428.217	390.805	-588.572	497.491	-10.828
2500	184.096	598.568	434.882	409.215	-583.872	542.647	-11.338
2600	184.096	605.788	441.317	427.625	-579.201	587.615	-11.805
2700	184.096	612.736	447.538	446.034	-574.554	632.404	-12.235
2800	184.096	619.431	453.558	464.444	-569.931	677.022	-12.630
2900	184.096	625.891	459.390	482.853	-565.330	721.474	-12.995
3000	184.096	632.133	465.045	501.263	-560.752	765.769	-13.333

PREVIOUS: CURRENT: December 1960

Phosphorus Sulfide (P_4S_3) P_4S_3(cr,l)

PHOSPHORUS SULFIDE (P_4S_3) IDEAL GAS M_r = 220.07504

S°(298.15 K) = [319.265] J K^{-1} mol^{-1}

$\Delta_fH°$(298.15 K) = [-151.042] kJ mol^{-1}

Enthalpy of Formation

All data estimated.

Heat Capacity and Entropy

All data estimated.

Phosphorus Sulfide (P_4S_3) P_4S_3(g)

Enthalpy Reference Temperature = T_r = 298.15 K Standard State Pressure = p° = 0.1 MPa

T/K	$C_p°$ (J $K^{-1}mol^{-1}$)	S° (J $K^{-1}mol^{-1}$)	$-[G°-H°(T_r)]/T$ (J $K^{-1}mol^{-1}$)	$H°-H°(T_r)$ (kJ mol^{-1})	$\Delta_fH°$ (kJ mol^{-1})	$\Delta_fG°$ (kJ mol^{-1})	Log K_f
0							
100							
200							
250							
298.15	154.808	319.265	319.265	0.	-151.042	-168.570	29.533
300	154.808	320.223	319.268	0.286	-151.059	-168.679	29.370
400	154.808	364.758	325.340	15.767	-162.377	-173.334	22.635
500	154.808	399.302	336.806	31.248	-169.211	-175.321	18.316
600	154.808	427.527	349.646	46.729	-175.015	-175.979	15.320
700	154.808	451.391	362.520	62.210	-180.105	-175.731	13.113
800	154.808	472.063	374.950	77.690	-184.788	-174.783	11.412
900	154.808	490.297	386.773	93.171	-348.607	-170.022	9.868
1000	154.808	506.607	397.955	108.652	-349.225	-150.146	7.843
1100	154.808	521.362	408.514	124.133	-349.889	-130.206	6.183
1200	154.808	534.832	418.487	139.614	-604.888	-105.891	4.609
1300	154.808	547.223	427.920	155.094	-602.546	-64.403	2.588
1400	154.808	558.696	436.856	170.575	-600.270	-23.093	0.862
1500	154.808	569.376	445.339	186.056	-598.056	18.056	-0.629
1600	154.808	579.368	453.407	201.537	-595.902	59.060	-1.928
1700	154.808	588.753	461.095	217.018	-593.803	99.930	-3.070
1800	154.808	597.601	468.435	232.498	-591.757	140.679	-4.082
1900	154.808	605.971	475.456	247.979	-589.760	181.315	-4.985
2000	154.808	613.912	482.182	263.460	-587.808	221.847	-5.794
2100	154.808	621.465	488.636	278.941	-585.898	262.283	-6.524
2200	154.808	628.667	494.839	294.422	-584.027	302.628	-7.185
2300	154.808	635.548	500.808	309.902	-582.191	342.889	-7.787
2400	154.808	642.137	506.560	325.383	-580.389	383.072	-8.337
2500	154.808	648.456	512.111	340.864	-578.619	423.179	-8.842
2600	154.808	654.528	517.472	356.345	-576.875	463.217	-9.306
2700	154.808	660.371	522.657	371.826	-575.157	503.188	-9.735
2800	154.808	666.001	527.677	387.306	-573.463	543.095	-10.132
2900	154.808	671.433	532.541	402.787	-571.792	582.943	-10.500
3000	154.808	676.681	537.258	418.268	-570.142	622.733	-10.843
3100	154.808	681.757	541.838	433.749	-568.512	662.469	-11.163
3200	154.808	686.672	546.288	449.230	-566.900	702.152	-11.461
3300	154.808	691.436	550.615	464.710	-565.307	741.785	-11.741
3400	154.808	696.057	554.825	480.191	-563.732	781.370	-12.004
3500	154.808	700.545	558.924	495.672	-562.175	820.909	-12.251
3600	154.808	704.906	562.919	511.153	-560.635	860.404	-12.484
3700	154.808	709.148	566.814	526.634	-559.113	899.857	-12.704
3800	154.808	713.276	570.614	542.114	-557.609	939.268	-12.911
3900	154.808	717.297	574.324	557.595	-556.123	978.640	-13.107
4000	154.808	721.217	577.948	573.076	-554.655	1017.974	-13.293
4100	154.808	725.039	581.489	588.557	-553.207	1057.272	-13.470
4200	154.808	728.770	584.951	604.038	-551.778	1096.535	-13.637
4300	154.808	732.412	588.338	619.518	-550.369	1135.764	-13.797
4400	154.808	735.971	591.653	634.999	-548.981	1174.960	-13.949
4500	154.808	739.450	594.899	650.480	-547.615	1214.125	-14.093
4600	154.808	742.853	598.079	665.961	-546.270	1253.260	-14.231
4700	154.808	746.182	601.195	681.442	-544.948	1292.366	-14.363
4800	154.808	749.441	604.249	696.922	-543.650	1331.444	-14.489
4900	154.808	752.634	607.245	712.403	-542.375	1370.495	-14.610
5000	154.808	755.761	610.184	727.884	-541.124	1409.520	-14.725
5100	154.808	758.827	613.069	743.365	-539.898	1448.521	-14.836
5200	154.808	761.833	615.901	758.846	-538.698	1487.498	-14.942
5300	154.808	764.782	618.682	774.326	-537.523	1526.452	-15.044
5400	154.808	767.675	621.415	789.807	-536.375	1565.384	-15.142
5500	154.808	770.516	624.100	805.288	-535.253	1604.295	-15.236
5600	154.808	773.305	626.739	820.769	-534.157	1643.186	-15.327
5700	154.808	776.045	629.335	836.250	-533.088	1682.058	-15.414
5800	154.808	778.738	631.888	851.730	-532.046	1720.911	-15.498
5900	154.808	781.384	634.399	867.211	-531.031	1759.746	-15.580
6000	154.808	783.986	636.871	882.692	-530.044	1798.564	-15.658

PREVIOUS: CURRENT: December 1960

Phosphorus Sulfide (P_4S_3) P_4S_3(g)

LEAD (Pb)

REFERENCE STATE

A_r = 207.2

0 to 600.6 K	crystal	
600.6 to 2019.022 K	liquid	
above 2019.022 K	ideal monatomic gas	

Refer to the individual tables for details.

Lead (Pb) Pb_l(ref)

Enthalpy Reference Temperature = T_r = 298.15 K Standard State Pressure = p° = 0.1 MPa

T/K	C_p° (J K⁻¹mol⁻¹)	S° (J K⁻¹mol⁻¹)	$-[G^\circ-H^\circ(T_r)]/T$ (J K⁻¹mol⁻¹)	$H^\circ-H^\circ(T_r)$ (kJ mol⁻¹)	Δ_fH° (kJ mol⁻¹)	Δ_fG° (kJ mol⁻¹)	Log K_f
0	0.	0.	INFINITE	-6.878	0.	0.	0.
100	24.426	36.829	87.950	-5.112	0.	0.	0.
200	25.870	54.272	67.210	-2.587	0.	0.	0.
298.15	26.836	64.785	64.785	0.	0.	0.	0.
300	26.853	64.951	64.786	0.050	0.	0.	0.
400	27.723	72.796	65.849	2.779	0.	0.	0.
500	28.549	79.071	67.886	5.592	0.	0.	0.
600	29.404	84.336	70.201	8.481	0.	0.	0.
600.600	29.414	84.366	70.215	8.499	CRYSTAL <--> LIQUID		
600.600	30.627	92.314	70.215	13.273	TRANSITION		
700	30.332	96.984	73.693	16.303	0.	0.	0.
800	30.016	101.013	76.863	19.321	0.	0.	0.
900	29.698	104.530	79.746	22.306	0.	0.	0.
1000	29.397	107.643	82.383	25.261	0.	0.	0.
1100	29.091	110.431	84.808	28.185	0.	0.	0.
1200	28.786	112.949	87.050	31.079	0.	0.	0.
1300	28.652	115.247	89.132	33.950	0.	0.	0.
1400	28.623	117.368	91.074	36.812	0.	0.	0.
1500	28.673	119.344	92.893	39.677	0.	0.	0.
1600	28.786	121.198	94.605	42.549	0.	0.	0.
1700	28.928	122.948	96.221	45.435	0.	0.	0.
1800	29.109	124.606	97.753	48.336	0.	0.	0.
1900	29.296	126.185	99.208	51.256	0.	0.	0.
2000	29.497	127.693	100.595	54.196	0.	0.	0.
2019.022	29.541	127.972	100.851	54.758	LIQUID <--> IDEAL GAS		
2019.022	24.831	215.927	100.851	232.340	FUGACITY = 1 bar		
2100	25.499	216.916	105.308	234.378	0.	0.	0.
2200	26.365	218.122	110.408	236.971	0.	0.	0.
2300	27.261	219.314	115.118	239.652	0.	0.	0.
2400	28.173	220.493	119.484	242.423	0.	0.	0.
2500	29.084	221.662	123.547	245.286	0.	0.	0.
2600	29.979	222.820	127.343	248.240	0.	0.	0.
2700	30.844	223.968	130.901	251.281	0.	0.	0.
2800	31.669	225.105	134.245	254.407	0.	0.	0.
2900	32.444	226.230	137.398	257.613	0.	0.	0.
3000	33.160	227.342	140.377	260.894	0.	0.	0.
3100	33.813	228.440	143.200	264.243	0.	0.	0.
3200	34.398	229.523	145.881	267.654	0.	0.	0.
3300	34.913	230.589	148.432	271.120	0.	0.	0.
3400	35.360	231.639	150.864	274.635	0.	0.	0.
3500	35.737	232.669	153.186	278.190	0.	0.	0.
3600	36.048	233.680	155.408	281.780	0.	0.	0.
3700	36.296	234.672	157.537	285.398	0.	0.	0.
3800	36.485	235.642	159.580	289.037	0.	0.	0.
3900	36.618	236.592	161.542	292.693	0.	0.	0.
4000	36.700	237.520	163.430	296.359	0.	0.	0.
4100	36.737	238.427	165.248	300.031	0.	0.	0.
4200	36.733	239.312	167.001	303.705	0.	0.	0.
4300	36.692	240.176	168.693	307.377	0.	0.	0.
4400	36.619	241.019	170.327	311.042	0.	0.	0.
4500	36.518	241.841	171.907	314.699	0.	0.	0.
4600	36.395	242.642	173.436	318.345	0.	0.	0.
4700	36.252	243.423	174.917	321.978	0.	0.	0.
4800	36.093	244.185	176.352	325.595	0.	0.	0.
4900	35.923	244.927	177.744	329.196	0.	0.	0.
5000	35.743	245.651	179.095	332.779	0.	0.	0.
5100	35.557	246.357	180.407	336.344	0.	0.	0.
5200	35.367	247.046	181.682	339.890	0.	0.	0.
5300	35.175	247.717	182.922	343.417	0.	0.	0.
5400	34.985	248.373	184.128	346.925	0.	0.	0.
5500	34.796	249.013	185.302	350.414	0.	0.	0.
5600	34.612	249.639	186.445	353.885	0.	0.	0.
5700	34.434	250.250	187.559	357.337	0.	0.	0.
5800	34.263	250.847	188.645	360.772	0.	0.	0.
5900	34.097	251.431	189.704	364.189	0.	0.	0.
6000	33.942	252.003	190.738	367.590	0.	0.	0.

PREVIOUS: March 1962 (1 atm) CURRENT: March 1962 (1 bar)

Lead (Pb) Pb_l(ref)

LEAD (Pb) CRYSTAL $A_r = 207.26$

$S°(298.15 K) = 64.785 \pm 0.54 J K^{-1} mol^{-1}$
$T_{fus} = 600.58 \pm 0.00 K$

$\Delta_f H°(0 K) = 0 kJ mol^{-1}$
$\Delta_f H°(298.15 K) = 0 kJ mol^{-1}$
$\Delta_{fus} H° = 4.77 \pm 0.01 kJ mol^{-1}$

Enthalpy of Formation

Zero by definition.

Heat Capacity and Entropy

Heat Capacity measurements have been reported by Horowitz, et al. (1-4 K, 14-21 K, 64-77 K, 1), Meads, et al. (15-300 K, 2), Douglas and Devir (298.15 K), 3), and many others whose work is listed by Kelley (4).

Heat capacities from 15 to 300 K and 1 to 4 K (in superconducting state) were taken from Meads et al. (2) and Horowitz et al. (1), respectively. The missing data below the melting point were obtained by graphical extrapolation and joined smoothly to the data of Douglas and Dever (3) on approaching the melting point.

Fusion Data

The melting point of pure Pb(cr) was found by McLaren and Mardock (5) to be 327.426 ± 0.002°C. $\Delta_{fus}H°$ was given by Douglas and Dever (3).

Sublimation Data

The enthalpy of sublimation, $\Delta_{sub}H°(298.15 K)$, is the enthalpy of formation of the ideal gas, $\Delta_f H°(298.15 K)$. See the ideal gas table for details.

References

1. M. Horowitz, A. A. Silvidi, S. F. Malaker and J. G. Daunt, Phys. Rev. 88, 1182(1952).
2. P. V. Meads, W. R. Forsythe and W. F. Giaugue, J. Am. chem. Soc. 63, 1902(1941).
3. T. B. Douglas and J. L. Dever, J. Am. Chem. Soc. 76, 4824(1954).
4. K. K. Kelley, U. S. Bur. Mines Bull. 584, 1960.
5. E. H. McLaren and E. G. Mardock, Can. J. Phys. 38, 577(1930).

Lead (Pb) $Pb_1(cr)$

Enthalpy Reference Temperature = T_r = 298.15 K — Standard State Pressure = p° = 0.1 MPa

T/K	$C_p°$ (J K⁻¹mol⁻¹)	S° (J K⁻¹mol⁻¹)	$-[G°-H°(T_r)]/T$ (J K⁻¹mol⁻¹)	$H°-H°(T_r)$ (kJ mol⁻¹)	$\Delta_f H°$ (kJ mol⁻¹)	$\Delta_f G°$ (kJ mol⁻¹)	Log K_f
0	0.	0.	INFINITE	-6.878	0.	0.	0.
100	24.426	36.829	87.950	-5.112	0.	0.	0.
200	25.870	54.272	67.210	-2.587	0.	0.	0.
250	26.363	60.096	65.225	-1.282	0.	0.	0.
298.15	26.836	64.785	64.785	0.	0.	0.	0.
300	26.853	64.951	64.786	0.050	0.	0.	0.
350	27.288	69.124	65.114	1.403	0.	0.	0.
400	27.723	72.796	65.849	2.779	0.	0.	0.
450	28.139	76.085	66.807	4.175	0.	0.	0.
500	28.549	79.071	67.886	5.592	0.	0.	0.
600	29.404	84.336	70.201	8.481	0.	0.	0.
600.600	29.414	84.366	70.215	8.499	------ CRYSTAL <--> LIQUID -----		
700	30.367	88.941	72.556	11.470	-4.833	0.796	-0.059
800	31.342	93.060	74.866	14.556	-4.765	1.598	-0.104
900	32.313	96.808	77.098	17.738	-4.568	2.382	-0.138
1000	33.288	100.263	79.244	21.018	-4.242	3.138	-0.164
1100	34.263	103.481	81.303	24.396	-3.789	3.856	-0.183
1200	35.238	106.504	83.278	27.871	-3.208	4.526	-0.197
1300	36.208	109.363	85.176	31.443	-2.506	5.143	-0.207
1400	37.183	112.082	87.001	35.113	-1.700	5.701	-0.213
1500	38.158	114.681	88.761	38.880	-0.797	6.199	-0.216

PREVIOUS: CURRENT: March 1962

Lead (Pb) $Pb_1(cr)$

LEAD (Pb) LIQUID $A_r = 207.2$

$S°(298.15\ K) = [71.710]\ J\ K^{-1}\ mol^{-1}$ $\Delta_f H°(298.15\ K) = [4.283]\ kJ\ mol^{-1}$

$T_{fus} = 600.58 \pm 0.00\ K$ $\Delta_{fus}H° = 4.77 \pm 0.01\ kJ\ mol^{-1}$

Enthalpy of Formation

The enthalpy of formation of the liquid, $\Delta_f H°(298.15\ K)$, is calculated from that of the crystal by adding $\Delta_{fus}H°$ and the difference in enthalpy, H°(600.58 K) - H°(298.15 K), between the crystal and liquid.

Heat Capacity and Entropy

Heat capacity values from T_{fus} to 1200 K were taken from Douglas and Dever (1). Heat capacity values below T_{fus} and above 1200 K were extrapolated, a glass transition being assumed at 400 K.

The entropy is calculated in a manner anologous to that used for the enthalpy of formation.

Vaporization Data

T_{vap} is calculated as the temperature at which $\Delta_r G° = 0$ for the process Pb(ℓ) = Pb(g). $\Delta_{vap}H°$ is calculated as the corresponding enthalpy of formation difference at T_{vap}. At temperatures below T_{vap}, the amount of $Pb_2(g)$ in the vapor is negligible.

References

1. T. B. Douglas and J. L. Dever, J. Am. Chem. Soc. 76, 4824(1954).

Lead (Pb) $Pb_1(l)$

Enthalpy Reference Temperature = T_r = 298.15 K — Standard State Pressure = p° = 0.1 MPa

T/K	$C_p°$ (J K⁻¹mol⁻¹)	S°	$-[G°-H°(T_r)]/T$	$H°-H°(T_r)$ (kJ mol⁻¹)	$\Delta_f H°$	$\Delta_f G°$	Log K_f
0							INFINITE
100							
200							
250							
298.15	26.836	71.710	71.710	0.	4.283	2.218	-0.389
300	26.853	71.876	71.710	0.050	4.283	2.205	-0.384
400	27.723	79.720	72.774	2.779	4.283	1.513	-0.198
400.000	27.723	79.720	72.774	2.779	GLASS <--> LIQUID TRANSITION		
400.000	31.254	79.720	72.774	2.779			
500	30.945	86.661	74.884	5.889	4.579	0.784	-0.082
600	30.635	92.284	77.330	8.972	4.773	0.005	0.
600.600	30.627	92.314	77.345	8.990	CRYSTAL <--> LIQUID		
700	30.332	96.984	79.811	12.021	0.	0.	0.
800	30.016	101.013	82.216	15.038	0.	0.	0.
900	29.698	104.530	84.504	18.023	0.	0.	0.
1000	29.397	107.643	86.665	20.978	0.	0.	0.
1100	29.091	110.431	88.701	23.902	0.	0.	0.
1200	28.786	112.949	90.619	26.796	0.	0.	0.
1300	28.652	115.247	92.426	29.667	0.	0.	0.
1400	28.623	117.368	94.133	32.530	0.	0.	0.
1500	28.673	119.344	95.748	35.394	0.	0.	0.
1600	28.786	121.198	97.282	38.267	0.	0.	0.
1700	28.928	122.948	98.741	41.152	0.	0.	0.
1800	29.109	124.606	100.132	44.054	0.	0.	0.
1900	29.296	126.185	101.462	46.974	0.	0.	0.
2000	29.497	127.693	102.736	49.914	0.	0.	0.
2019.022	29.541	127.972	102.972	50.475	FUGACITY = 1 bar		
2100	29.706	129.137	103.959	52.874	-177.221	7.115	-0.177
2200	29.920	130.524	105.135	55.855	-176.833	15.884	-0.377
2300	30.133	131.859	106.268	58.858	-176.511	24.636	-0.560
2400	30.347	133.146	107.361	61.882	-176.259	33.376	-0.726
2500	30.556	134.389	108.418	64.927	-176.077	42.107	-0.880
2600	30.761	135.591	109.440	67.993	-175.964	50.832	-1.021
2700	30.962	136.756	110.430	71.079	-175.920	59.553	-1.152
2800	31.171	137.885	111.391	74.186	-175.939	68.275	-1.274
2900	31.376	138.983	112.323	77.313	-176.018	76.998	-1.387
3000	31.581	140.050	113.230	80.461	-176.151	85.725	-1.493

PREVIOUS: CURRENT: March 1962

Lead (Pb) $Pb_1(l)$

LEAD (Pb) CRYSTAL–LIQUID A_r = 207.2

0 to 600.6 K crystal
above 600.6 K liquid

Refer to the individual tables for details.

Lead (Pb) Pb_1(cr,l)

Enthalpy Reference Temperature = T_r = 298.15 K Standard State Pressure = p° = 0.1 MPa

T/K	C_p° (J K⁻¹mol⁻¹)	S° (J K⁻¹mol⁻¹)	$-[G^\circ-H^\circ(T_r)]/T$ (J K⁻¹mol⁻¹)	$H^\circ-H^\circ(T_r)$ (kJ mol⁻¹)	Δ_fH° (kJ mol⁻¹)	Δ_fG° (kJ mol⁻¹)	Log K_f
0	0.	0.	INFINITE	-6.878	0.	0.	0.
100	24.426	36.829	87.950	-5.112	0.	0.	0.
200	25.870	54.272	67.210	-2.587	0.	0.	0.
298.15	26.836	64.785	64.785	0.	0.	0.	0.
300	26.853	64.951	64.786	0.050	0.	0.	0.
400	27.723	72.796	65.849	2.779	0.	0.	0.
500	28.549	79.071	67.886	5.592	0.	0.	0.
600	29.404	84.336	70.201	8.481	0.	0.	0.
600.600	29.414	84.366	70.215	8.499	CRYSTAL <--> LIQUID		
600.600	30.627	92.314	70.215	13.273	TRANSITION		
700	30.332	96.984	73.693	16.303	0.	0.	0.
800	30.016	101.013	76.863	19.321	0.	0.	0.
900	29.698	104.530	79.746	22.306	0.	0.	0.
1000	29.397	107.643	82.383	25.261	0.	0.	0.
1100	29.091	110.431	84.808	28.185	0.	0.	0.
1200	28.786	112.949	87.050	31.079	0.	0.	0.
1300	28.652	115.247	89.132	33.950	0.	0.	0.
1400	28.623	117.368	91.074	36.812	0.	0.	0.
1500	28.673	119.344	92.893	39.677	0.	0.	0.
1600	28.786	121.198	94.605	42.549	0.	0.	0.
1700	28.928	122.948	96.221	45.435	0.	0.	0.
1800	29.109	124.606	97.753	48.336	0.	0.	0.
1900	29.296	126.185	99.208	51.256	0.	0.	0.
2000	29.497	127.693	100.595	54.196	0.	0.	0.
2019.022	29.541	127.972	100.851	54.758	------- FUGACITY = 1 bar -------		
2100	29.706	129.137	101.920	57.156	-177.221	7.115	-0.177
2200	29.920	130.524	103.189	60.138	-176.833	15.884	-0.377
2300	30.133	131.859	104.406	63.140	-176.511	24.636	-0.560
2400	30.347	133.146	105.577	66.164	-176.259	33.376	-0.726
2500	30.556	134.389	106.705	69.210	-176.077	42.107	-0.880
2600	30.761	135.591	107.793	72.275	-175.964	50.832	-1.021
2700	30.962	136.756	108.844	75.362	-175.920	59.553	-1.152
2800	31.171	137.885	109.861	78.468	-175.939	68.275	-1.274
2900	31.376	138.983	110.847	81.595	-176.018	76.998	-1.387
3000	31.581	140.050	111.802	84.743	-176.151	85.725	-1.493

PREVIOUS: CURRENT: March 1962

Lead (Pb) Pb_1(cr,l)

LEAD (Pb) IDEAL GAS A_r = 207.2

IP(Pb, g) = 59819.4 ± 0.3 cm^{-1}
S°(298.15 K) = 175.374 ± 0.020 J K^{-1} mol^{-1}

$\Delta_f H°$(0 K) = 195.88 ± 0.80 kJ mol^{-1}
$\Delta_f H°$(298.15 K) = 195.20 ± 0.80 kJ mol^{-1}

Electronic Levels and Quantum Weights

State	ε_i, cm^{-1}	g_i
$^3P_0$	0.00	1
$^3P_1$	7819.2626	3
$^3P_2$	10650.3271	5
$^1D_2$	21457.90	5
$^1S_0$	29466.81	1

Enthalpy of Formation

The adopted enthalpy of formation value for Pb(g), $\Delta_f H°$(298.15 K) = 195.20 ± 0.80 kJ mol^{-1}, is that recommended by CODATA (1). This value was calculated, according to CODATA, from the vapor pressure measurements cited by Hultgren et al (2) and the more recent measurements by Andon et al. (3), Chang, et al. (4), Kim and Cosgarea (5), and Shiu and Miner (6).

Heat Capacity and Entropy

The information on electronic energy levels and quantum weights, given by Moore (7, 8), is incomplete because many theoretically predicted levels have not been observed. Our calcualtions indicate that any reasonable method of filling in these missing levels (for n<21) and cutting off the summation in the partition function (9) has little effect on the thermodynamic functions to 6000 K. This is a result of the high energy of all levels other than those listed above; the next excited state is approximately 34950 cm^{-1} above the ground state. Although we list only a few levels above, all levels listed by Moore (7, 8) are considered in the calculation. The reported uncertainty in S°(298.15 K) is due to uncertainties in the relative atomic mass and the fundamental constants. Extension of these calcualtions above 6000 K may require consideration of the excited states (n>21) and use of different fill and cutoff procedures (9).

The thermal functions at 298.15 K differ from the CODATA recommendations (1) for two reasons. First, the entropy differs by 0.1094 J K^{-1} mol^{-1} because this table uses a reference pressure of 1 bar, whereas CODATA recommendations are based on 1 atm. Second, small changes, 0.005 J K^{-1} mol^{-1}, arise due to the use of slightly different values for the fundamental constants.

References

1. J. D. Cox, chairman, CODATA Task Group on Key Values for Thermodynamics, J. Chem. Thermodynamics 10, 903 (1978); CODATA Bulletin No 28, 1978.
2. R. Hultgren, P. D. Desai, et al., "Selected Values of the Thermodynamic Properties of the Elements, 11 American Society for Metals, Metals Park, Ohio, 1973.
3. R. J. L. Andon, J. F. Martin and K. C. Mills, J. Chem. Soc. A1788 (1971).
4. Y. A. Chang, G. C. Wilhelm, M. Lathrop and I. Gyuk, Acta Met. 19, 795 (1971).
5. J. H. Kim and A. Cosgarea, Jr., J. Chem. Phys. 44, 806 (1966).
6. D. H. Shiu and Z. A. Munir, Metallurg. Trans. 2, 2953 (1973).
7. C. E. Moore, U. S. Nat. Bur. Stnad., NSRDS-NBS-35, Volume III, 1970, [Reprint of NBS Circular 467, Volume III, 1958].
8. C. E. Moore, U. S. Nat. Bur. Stand., NSRDS-34, 1970.
9. J. R. Downey, Jr. The Dow Chemical Company Co., Rept. AFOSR-TR-78-0960, Contract No. F44670-75-1-0048, 1978.

Lead (Pb) $Pb_1(g)$

Enthalpy Reference Temperature = T_r = 298.15 K Standard State Pressure = p° = 0.1 MPa

T/K	$C_p°$ (J $K^{-1}mol^{-1}$)	S° (J $K^{-1}mol^{-1}$)	$-[G°-H°(T_r)]/T$ (J $K^{-1}mol^{-1}$)	$H°-H°(T_r)$ (kJ mol^{-1})	$\Delta_f H°$ (kJ mol^{-1})	$\Delta_f G°$ (kJ mol^{-1})	Log K_f
0	0.	0.	INFINITE	-6.197	195.881	195.881	INFINITE
100	20.786	152.667	193.855	-4.119	196.193	184.610	-96.430
200	20.786	167.075	177.276	-2.040	195.747	173.187	-45.232
250	20.786	171.713	175.716	-1.001	195.481	167.577	-35.013
298.15	20.786	175.374	175.374	0.	195.200	162.228	-28.422
300	20.786	175.503	175.375	0.038	195.189	162.023	-28.211
350	20.786	178.707	175.628	1.078	194.874	156.520	-23.359
400	20.786	181.483	176.190	2.117	194.538	151.064	-19.727
450	20.786	183.931	176.917	3.156	194.181	145.651	-16.907
500	20.786	186.121	177.730	4.196	193.803	140.278	-14.655
600	20.786	189.911	179.453	6.274	192.993	129.648	-11.287
700	20.787	193.115	181.182	8.353	187.250	119.958	-8.951
800	20.790	195.891	182.851	10.432	186.311	110.409	-7.209
900	20.801	198.340	184.439	12.511	185.405	100.976	-5.861
1000	20.829	200.533	185.940	14.592	184.532	91.642	-4.787
1100	20.888	202.520	187.359	16.678	183.693	82.394	-3.913
1200	20.991	204.342	188.699	18.771	182.892	73.221	-3.187
1300	21.155	206.028	189.968	20.878	182.129	64.113	-2.576
1400	21.394	207.604	191.172	23.005	181.393	55.063	-2.054
1500	21.717	209.091	192.317	25.160	180.683	46.064	-1.604
1600	22.132	210.505	193.410	27.352	180.002	37.112	-1.212
1700	22.640	211.862	194.456	29.589	179.355	28.201	-0.867
1800	23.239	213.172	195.460	31.883	178.746	19.327	-0.561
1900	23.922	214.447	196.426	34.240	178.184	10.486	-0.288
2000	24.680	215.693	197.358	36.669	177.673	1.674	-0.044
2019.022	24.831	215.927	197.532	37.140	------- FUGACITY = 1 bar -------		
2100	25.499	216.916	198.260	39.178	0.	0.	0.
2200	26.365	218.122	199.136	41.771	0.	0.	0.
2300	27.261	219.314	199.987	44.452	0.	0.	0.
2400	28.173	220.493	200.817	47.224	0.	0.	0.
2500	29.084	221.662	201.627	50.086	0.	0.	0.
2600	29.979	222.820	202.420	53.040	0.	0.	0.
2700	30.844	223.968	203.197	56.081	0.	0.	0.
2800	31.669	225.105	203.959	59.207	0.	0.	0.
2900	32.444	226.230	204.708	62.413	0.	0.	0.
3000	33.160	227.342	205.444	65.694	0.	0.	0.
3100	33.813	228.440	206.168	69.043	0.	0.	0.
3200	34.398	229.523	206.881	72.454	0.	0.	0.
3300	34.913	230.589	207.583	75.921	0.	0.	0.
3400	35.360	231.639	208.275	79.435	0.	0.	0.
3500	35.737	232.669	208.958	82.990	0.	0.	0.
3600	36.048	233.680	209.630	86.580	0.	0.	0.
3700	36.296	234.672	210.294	90.198	0.	0.	0.
3800	36.485	235.642	210.948	93.837	0.	0.	0.
3900	36.618	236.592	211.594	97.493	0.	0.	0.
4000	36.700	237.520	212.230	101.159	0.	0.	0.
4100	36.737	238.427	212.858	104.831	0.	0.	0.
4200	36.733	239.312	213.477	108.505	0.	0.	0.
4300	36.692	240.176	214.088	112.177	0.	0.	0.
4400	36.619	241.019	214.691	115.842	0.	0.	0.
4500	36.518	241.841	215.285	119.499	0.	0.	0.
4600	36.395	242.642	215.871	123.145	0.	0.	0.
4700	36.252	243.423	216.449	126.778	0.	0.	0.
4800	36.093	244.185	217.019	130.395	0.	0.	0.
4900	35.923	244.927	217.581	133.996	0.	0.	0.
5000	35.743	245.651	218.135	137.579	0.	0.	0.
5100	35.557	246.357	218.682	141.144	0.	0.	0.
5200	35.367	247.046	219.221	144.690	0.	0.	0.
5300	35.175	247.717	219.752	148.217	0.	0.	0.
5400	34.985	248.373	220.276	151.725	0.	0.	0.
5500	34.796	249.013	220.793	155.214	0.	0.	0.
5600	34.612	249.639	221.302	158.685	0.	0.	0.
5700	34.434	250.250	221.805	162.137	0.	0.	0.
5800	34.263	250.847	222.300	165.572	0.	0.	0.
5900	34.097	251.431	222.789	168.989	0.	0.	0.
6000	33.942	252.003	223.271	172.390	0.	0.	0.

PREVIOUS: March 1962 (1 atm) CURRENT: March 1983 (1 bar)

Lead (Pb) $Pb_1(g)$

LEAD, ION (Pb^+) IDEAL GAS M_r = 207.19945

Lead, Ion (Pb^+) $Pb_1^+(g)$

IP(Pb^+, g) = 121243 ± 3 cm^{-1}

S°(298.15 K) = 181.137 ± 0.02 JK^{-1} mol^{-1}

$\Delta_f H°$(0 K) = 911.480 ± 0.5 kJ mol^{-1}

$\Delta_f H°$(298.15 K) = [916.996] kJ mol^{-1}

Electronic Levels and Quantum Weights

State	ϵ_i, cm^{-1}	g_i
$^2P_{1/2}$	0	2
$^3P_{3/2}$	14081.074	4

Enthalpy of Formation

$\Delta_f H°(Pb^+, g, 0 K)$ is calculated from $\Delta_f H°(Pb, g, 0 K)$ (1) using the spectroscopic value of IP(Pb) = 59819.4 ± 0.3 cm^{-1} (715.599 ± 0.004 kJ mol^{-1}) from Moore (2). The ionization limit is converted from cm^{-1} to kJ mol^{-1} using the factor, 1 cm^{-1} = 0.01196266 kJ cm^{-1}, which is derived from the 1973 CODATA fundamental constants (3). Rosenstock et al. (4) and Levin and Lias (5) have summarized additional ionization and appearance potential data.

$\Delta_f H°(Pb^+, g, 298.15 K)$ is calculated from $\Delta_f H°(Pb, g, 0 K)$ by using IP(Pb) with JANAF (1) enthalpies, H°(0 K) - H°(298.15 K), for Pb(g), Pb^+(g), and e^-(g). $\Delta_f H°(Pb \rightarrow Pb^+ + e^-$, 298.15 K) differs from a room temperature threshold energy due to inclusion of these enthalpies and to threshold effects discussed by Rosenstock et al. (4). $\Delta_f H°$(298.15 K) should be changed by -6.197 kJ mol^{-1} if it is to be used in the ion convention that excludes the enthalpy of the electron.

Heat Capacity and Entropy

The information on electronic energy levels and quantum weights, given by Moore (2, 6) is incomplete because many theoretically predicted levels have not been observed. Our calculations indicate that any reasonable method of filling in these missing levels and cutting off the summation in the partition function (7) has no effect on the thermodynamic functions to 6000 K. This is a result of the high energy of all levels other than the ground state and the first excited state; the next excited state is 57911 cm^{-1}, above the ground state. Since inclusion of these excited states has no effect on the thermodynamic functions (to 6000 K), we list only the ground state and the first excited states. The reported uncertainty in S°(298.15 K) is due to uncertainties in the relative ionic mass and the fundamental constants. Extension of these calculations above 6000 K may require consideration of the excited states and use of different fill and cutoff procedures (7).

References

1. JANAF Thermochemical Tables: Pb(g), 3-31-62 ; e^-(g), 3-31-82.
2. C. E. Moore, U. S. Nat. Bur. Stand., NSRDS-NBS-34 (1970).
3. E. R. Cohen and B. N. Taylor, J. Phys. Chem. Ref. Data 2, 663 (1973).
4. H. M. Rosenstock, K. Draxl et al., J. Phys. Chem. Ref. Data 6, Supp. 1 (1977).
5. R. D. Levin and S. G. Lias, U. S. Nat. Bur. Stand., NSRDS-NBS-71, 1982.
6. C. E. Moore, U. S. Nat. Bur. Stand., NSRDS-NBS-35, Volume 3, 1971 [Reprint of NBS Circular 467, Volume 3, 1958].
7. J. R. Downey, Jr., The Dow Chemical Company, AFOSR-TR-0960, Contract No. F44620-75-1-0048, 1978.

Enthalpy Reference Temperature = T_r = 298.15 K Standard State Pressure = p° = 0.1 MPa

T/K	$C_p°$ (J K⁻¹mol⁻¹)	S° (J K⁻¹mol⁻¹)	-[G°-H°(T_r)]/T (J K⁻¹mol⁻¹)	H°-H°(T_r) (kJ mol⁻¹)	$\Delta_f H°$ (kJ mol⁻¹)	$\Delta_f G°$ (kJ mol⁻¹)	Log K_f
0	0.	0.	INFINITE	-6.197	911.480		
100	20.786	158.430	199.618	-4.119			
200	20.786	172.838	183.039	-2.040			
250	20.786	177.476	181.480	-1.001			
298.15	20.786	181.137	181.137	0.	916.996	876.051	-153.481
300	20.786	181.266	181.138	0.038	917.024	875.797	-152.490
350	20.786	184.470	181.391	1.078	917.749	868.868	-129.672
400	20.786	187.246	181.953	2.117	918.452	861.837	-112.544
450	20.786	189.694	182.680	3.156	919.134	854.719	-99.213
500	20.786	191.884	183.493	4.196	919.795	847.526	-88.540
600	20.786	195.674	185.217	6.274	921.064	832.952	-72.515
700	20.786	198.878	186.945	8.353	917.399	818.970	-61.112
800	20.786	201.653	188.614	10.431	918.539	804.831	-52.550
900	20.786	204.102	190.202	12.510	919.711	790.547	-45.882
1000	20.786	206.292	191.703	14.589	920.913	776.132	-40.541
1100	20.786	208.273	193.121	16.667	922.146	761.594	-36.165
1200	20.786	210.081	194.460	18.746	923.409	746.943	-32.514
1300	20.787	211.745	195.726	20.825	924.696	732.185	-29.420
1400	20.788	213.286	196.926	22.903	925.991	717.328	-26.764
1500	20.790	214.720	198.065	24.982	927.284	702.379	-24.459
1600	20.794	216.062	199.149	27.061	928.569	687.343	-22.439
1700	20.802	217.323	200.181	29.141	929.842	672.228	-20.655
1800	20.813	218.512	201.167	31.222	931.100	657.037	-19.067
1900	20.830	219.638	202.109	33.304	932.340	641.778	-17.644
2000	20.854	220.707	203.013	35.388	933.563	626.453	-16.361
2100	20.886	221.725	203.880	37.475	757.547	618.183	-15.376
2200	20.927	222.698	204.713	39.566	759.123	611.510	-14.519
2300	20.979	223.629	205.516	41.661	760.616	604.766	-13.735
2400	21.041	224.523	206.289	43.762	762.024	597.960	-13.014
2500	21.116	225.383	207.036	45.870	763.347	591.096	-12.350
2600	21.202	226.213	207.757	47.985	764.588	584.181	-11.736
2700	21.301	227.015	208.456	50.110	765.751	577.220	-11.167
2800	21.411	227.792	209.133	52.246	766.839	570.217	-10.638
2900	21.534	228.545	209.789	54.393	767.858	563.177	-10.144
3000	21.667	229.278	210.427	56.553	768.816	556.103	-9.683
3100	21.811	229.990	211.046	58.727	769.719	548.997	-9.251
3200	21.964	230.685	211.649	60.915	770.576	541.863	-8.845
3300	22.126	231.364	212.236	63.120	771.392	534.703	-8.464
3400	22.296	232.027	212.809	65.341	772.178	527.519	-8.104
3500	22.472	232.675	213.367	67.579	772.939	520.312	-7.765
3600	22.653	233.311	213.912	69.835	773.684	513.084	-7.445
3700	22.839	233.934	214.445	72.110	774.420	505.834	-7.141
3800	23.029	234.546	214.966	74.403	775.152	498.566	-6.853
3900	23.220	235.146	215.476	76.716	775.888	491.277	-6.580
4000	23.413	235.737	215.975	79.048	776.632	483.970	-6.320
4100	23.606	236.317	216.464	81.398	777.389	476.644	-6.073
4200	23.798	236.888	216.944	83.769	778.164	469.300	-5.837
4300	23.989	237.451	217.414	86.158	778.960	461.937	-5.611
4400	24.177	238.004	217.876	88.566	779.782	454.554	-5.396
4500	24.362	238.550	218.329	90.993	780.630	447.153	-5.190
4600	24.543	239.087	218.774	93.439	781.508	439.733	-4.993
4700	24.720	239.617	219.212	95.902	782.418	432.294	-4.804
4800	24.892	240.139	219.643	98.383	783.360	424.834	-4.623
4900	25.059	240.654	220.066	100.880	784.335	417.355	-4.449
5000	25.220	241.162	220.483	103.394	785.344	409.855	-4.282
5100	25.375	241.663	220.894	105.924	786.388	402.335	-4.121
5200	25.523	242.157	221.298	108.469	787.465	394.794	-3.966
5300	25.665	242.645	221.696	111.028	788.576	387.232	-3.816
5400	25.800	243.126	222.088	113.602	789.720	379.649	-3.672
5500	25.928	243.600	222.475	116.188	790.896	372.045	-3.533
5600	26.049	244.069	222.857	118.787	792.103	364.418	-3.399
5700	26.164	244.531	223.233	121.398	793.340	356.770	-3.269
5800	26.271	244.987	223.604	124.020	794.606	349.100	-3.144
5900	26.371	245.437	223.970	126.652	795.900	341.408	-3.023
6000	26.465	245.881	224.332	129.294	797.219	333.693	-2.905

PREVIOUS: CURRENT: March 1983 (1 bar)

Lead, Ion (Pb^+) $Pb_1^+(g)$

LEAD, ION (Pb^-) IDEAL GAS M_r = 207.20055

Lead, Ion (Pb^-) $Pb_1^-(g)$

EA(Pb, g) = 0.364 ± 0.008 eV

S°(298.15 K) = 186.900 ± 0.002 J K^{-1} mol^{-1}

$\Delta_f H°$(0 K) = 160.76 ± 3 kJ mol^{-1}

$\Delta_f H°$(298.15 K) = [153.882] kJ mol^{-1}

Electonic Level and Quantum Weight

State	ε_i, cm^{-1}	g_i
$^4S_{3/2}$	0.0	4

Enthalpy of Formation

$\Delta_f H°(Pb^-$, g, 0 K) is calculated from $\Delta_f H°$(Pb, g, 0 K) (1) using the adopted electron affinity of EA(Pb) = 0.364 ± 0.008 eV (35.120 ± 0.772 kJ mol^{-1}). This value, recommended by Hotop and Lineberger (2), is based on a laser photodetachment electron spectrometry study (3). Additional information on Pb^-(g) may be obtained in the critical discussions of Hotop and Lineberger (2, 4), Rosenstock et al. (5) and Massey (6).

$\Delta_f H°(Pb^-$, g, 298.15 K) is obtained from $\Delta_f H°$(Pb, g, 0 K) by using EA(Pb) with JANAF (1) enthalpies, H°(0 K) - H°(298.15 K), for Pb^-(g), Pb(g), and e^-(g). $\Delta_r H°(Pb^- \rightarrow Pb + e^-$, 298.15 K) differs from a room-temperature threshold energy due to inclusion of those enthalpies and to threshold effects discussed by Rosenstock et al. (5). $\Delta_f H°$(298.15 K) should be changed by +6.197 kJ mol^{-1} if it is to be used in the ion convention that excludes the enthalpy of the electron.

Heat Capacity and Entropy

The ground state electronic configuration for Pb^-(g) is given by Hotop and Lineberger (2, 4) and Rosenstock et al. (5). Lacking any experimental evidence as to the stability of any excited states, we assume that no stable excited states exist.

References

1. JANAF Thermochemical Tables: Pb(g), 3-31-83; e^-(g), 3-31-82.
2. H. Hotop and W. C. Lineberger, J. Phys. Chem. Ref. Data, to be published, 1985.
3. C. S. Feigerle, R. R. Cordermann and W. C. Lineberger, J. Chem. Phys. 74, 1513 (1981).
4. H. Hotop and W. C. Lineberger, J. Phys. Chem. kRef. Data 4, 539 (1975).
5. H. M. Rosenstock, K. Draxl et al., J. Phys. kChem. Ref. Data 6, Supp. 1 (1977).
6. H. S. W. Massey, "Negative Ions", 3rd ed., Cambridge University Press, Cambridge, 1976.

Enthalpy Reference Temperature = T_r = 298.15 K Standard State Pressure = p° = 0.1 MPa

T/K	$C_p°$ (J $K^{-1}mol^{-1}$)	S° (J $K^{-1}mol^{-1}$)	-[G°-H°(T_r)]/T (J $K^{-1}mol^{-1}$)	H°-H°(T_r) (kJ mol^{-1})	$\Delta_f H°$ (kJ mol^{-1})	$\Delta_f G°$ (kJ mol^{-1})	Log K_f
0	0.	0.	INFINITE	-6.197	160.761		
100	20.786	164.193	205.381	-4.119			
200	20.786	178.601	188.802	-2.040			
250	20.786	183.239	187.243	-1.001			
298.15	20.786	186.900	186.900	0.	153.882	123.728	-21.677
300	20.786	187.029	186.901	0.038	153.833	123.541	-21.510
350	20.786	190.233	187.154	1.078	152.479	118.600	-17.700
400	20.786	193.009	187.716	2.117	151.104	113.853	-14.868
450	20.786	195.457	188.443	3.156	149.707	109.281	-12.685
500	20.786	197.647	189.256	4.196	148.290	104.864	-10.955
600	20.786	201.437	190.980	6.274	145.401	96.449	-8.397
700	20.786	204.641	192.708	8.353	137.579	89.322	-6.665
800	20.786	207.417	194.377	10.431	134.562	82.635	-5.396
900	20.786	209.865	195.965	12.510	131.576	76.324	-4.430
1000	20.786	212.055	197.466	14.589	128.622	70.343	-3.674
1100	20.786	214.036	198.884	16.667	125.697	64.657	-3.070
1200	20.786	215.845	200.223	18.746	122.803	59.236	-2.578
1300	20.786	217.508	201.490	20.824	119.933	54.055	-2.172
1400	20.786	219.049	202.689	22.903	117.070	49.095	-1.832
1500	20.786	220.483	203.828	24.982	114.206	44.340	-1.544
1600	20.786	221.824	204.912	27.060	111.333	39.775	-1.299
1700	20.786	223.085	205.944	29.139	108.448	35.391	-1.087
1800	20.786	224.273	206.930	31.217	105.546	31.178	-0.905
1900	20.786	225.396	207.872	33.296	102.626	27.126	-0.746
2000	20.786	226.463	208.775	35.375	99.686	23.228	-0.607
2100	20.786	227.477	209.642	37.453	-80.496	26.593	-0.661
2200	20.786	228.444	210.475	39.532	-83.088	31.753	-0.754
2300	20.786	229.368	211.276	41.610	-85.770	37.032	-0.841
2400	20.786	230.252	212.049	43.689	-88.541	42.431	-0.923
2500	20.786	231.101	212.794	45.768	-91.404	47.946	-1.002
2600	20.786	231.916	213.514	47.846	-94.357	53.578	-1.076
2700	20.786	232.701	214.210	49.925	-97.399	59.326	-1.148
2800	20.786	233.457	214.884	52.004	-100.525	65.187	-1.216
2900	20.786	234.186	215.537	54.082	-103.731	71.162	-1.282
3000	20.786	234.891	216.170	56.161	-107.012	77.248	-1.345
3100	20.786	235.572	216.785	58.239	-110.361	83.445	-1.406
3200	20.786	236.232	217.383	60.318	-113.772	89.751	-1.465
3300	20.786	236.872	217.964	62.397	-117.238	96.165	-1.522
3400	20.786	237.492	218.529	64.475	-120.752	102.684	-1.578
3500	20.786	238.095	219.079	66.554	-124.308	109.308	-1.631
3600	20.786	238.680	219.616	68.632	-127.898	116.033	-1.684
3700	20.786	239.250	220.139	70.711	-131.515	122.859	-1.734
3800	20.786	239.804	220.649	72.790	-135.155	129.782	-1.784
3900	20.786	240.344	221.147	74.868	-138.810	136.802	-1.832
4000	20.786	240.870	221.634	76.947	-142.477	143.916	-1.879
4100	20.786	241.384	222.109	79.025	-146.149	151.121	-1.925
4200	20.786	241.885	222.574	81.104	-149.823	158.416	-1.970
4300	20.786	242.374	223.029	83.183	-153.494	165.798	-2.014
4400	20.786	242.852	223.474	85.261	-157.160	173.266	-2.057
4500	20.786	243.319	223.910	87.340	-160.817	180.817	-2.099
4600	20.786	243.776	224.337	89.418	-164.463	188.449	-2.140
4700	20.786	244.223	224.755	91.497	-168.095	196.160	-2.180
4800	20.786	244.660	225.165	93.576	-171.713	203.949	-2.219
4900	20.786	245.089	225.567	95.654	-175.313	211.812	-2.258
5000	20.786	245.509	225.962	97.733	-178.897	219.749	-2.296
5100	20.786	245.920	226.349	99.811	-182.462	227.757	-2.333
5200	20.786	246.324	226.730	101.890	-186.008	235.836	-2.369
5300	20.786	246.720	227.103	103.969	-189.535	243.982	-2.405
5400	20.786	247.108	227.470	106.047	-193.043	252.194	-2.439
5500	20.786	247.490	227.831	108.126	-196.532	260.471	-2.474
5600	20.786	247.864	228.185	110.204	-200.002	268.812	-2.507
5700	20.786	248.232	228.533	112.283	-203.455	277.214	-2.540
5800	20.786	248.594	228.876	114.362	-206.890	285.677	-2.573
5900	20.786	248.949	229.213	116.440	-210.306	294.199	-2.605
6000	20.786	249.298	229.545	118.519	-213.708	302.778	-2.636

PREVIOUS: CURRENT: March 1983 (1 bar)

Lead, Ion (Pb^-) $Pb_1^-(g)$

LEAD SULFIDE (PbS) CRYSTAL M_r = 239.26

$S°(298.15\ K) = 91.34 \pm 1.7\ J\ K^{-1}\ mol^{-1}$

$T_{fus} = 1386.5 \pm 1.5\ K$

$\Delta_f H°(0\ K) = -98.55 \pm 2.1\ kJ\ mol^{-1}$

$\Delta_f H°(298.15\ K) = -98.32 \pm 2.1\ kJ\ mol^{-1}$

$\Delta_{fus} H° = 18.83 \pm 6.3\ kJ\ mol^{-1}$

Enthalpy of Formation

2nd and 3rd law analyses are made of the equilibrium data of Jellinek et al. (1), Sudo (2), and Stubbles and Birchenall (3), the Gibbs free energies (via emf measurements) of Kapustinskii and Makolkin (4) and of Kiukkola and Wagner (5) and the emf data reported by Thompson and Flengas (6). The results of our analyses using auxiliary data (7) are summarized in the following table.

Source	Reaction*	T/K	Data Points	$\Delta_r H°$(298.15 K), kcal mol^{-1} 2nd law	3rd law	Drift cal K^{-1} mol^{-1}	$\Delta_f H°$(298.15 K) kcal mol^{-1}
(1)	A	788-1273	7	-18.4±0.8	-18.60	-0.16±0.80	-22.5
(2)	A	855-1048	Equation	-18.0	-20.12	-2.2	-24.0
(3)	A	859-1193	30	-19.3±0.1	-19.76	-0.42±0.12	-23.6
(4)	B	288-308	3	-17.7±0.5	-16.63	3.7 ±1.6	-21.5
(5)	C	523-573	2	-24.7	-23.84	1.5	-23.5
(5)	D	600-700	4	-24.5±0.2	-24.79	-0.45±0.32	-23.4
(6)	E	773-1071	12	-40.7±0.5	-39.63	1.2 ±0.6	-23.2

*A) $Pb(\ell) + H_2S(g) = PbS(cr) + H_2(g)$ D) $Pb(\ell) + S(\ell) = PbS(cr)$

B) $Pb(cr) + H_2S(g) = PbS(cr) + H_2(g)$ E) $Pb(\ell) + 1/2\ S_2(g) = PbS(cr)$

C) $Pb(cr) + S(\ell) = PbS(cr)$

The experiences of Jellinek et al. (1) are unsatisfactory (8) and are superseded by two later studies (2, 3) of the same equilibrium. The Gibbs energies of Kapustinskii and Makolkin (4) may be dismissed as biased on the basis of inconsistencies in both $\Delta_f H°$ and entropy (drift). The other studies (2, 3, 5, 6) are consistent with the adopted $\Delta_f H°(298.15\ K)$ of -23.5 ± 0.5 kcal mol^{-1}.

We also calculate $\Delta_f H° = -23.5$ kcal mol^{-1} from $\Delta_r H°(293\ K) = -13.38 \pm 0.2$ kcal mol^{-1} for the reaction $Pb(NO_3)_2(aq, 3000\ H_2O) + H_2S(aq) + PbS(cr) + 2\ HNO_3(aq, 2000\ H_2O)$ studied calorimetrically by Zeumer and Roth (9). The uncertainty of this calorimetric $\Delta_f H°$ may approach ±1 kcal mol^{-1} due to the auxiliary data (10) and approximations used in our calculation. Recent DTA data (11) for $Pb(cr) + S(\ell) + PbS(cr)$ gave $\Delta_r H°(480\ K) = -23.4 \pm 0.3$ kcal mol^{-1} which we reduce to $\Delta_f H°(298.15\ K) = -22.5 \pm 1$ kcal mol^{-1}. The uncertainty assigned (11) to $\Delta_r H°$ appears to be too optimistic. Korshunov (12) obtained $\Delta_f H° = -22.38$ kcal mol^{-1} by combustion of S-Mg-Pb mixtures in a bomb, but the chemical reactions probably were not characterized adequately.

Heat Capacity and Entropy

$C_p°$ is from the work of Parkinson and Quarrington (13, $C_p°$ given as smoothed values, 20-260 K) and of Anderson (14, 54-289 K) which were graphically smoothed and extrapolated to 300 K. The data of Parkinson and Quarrington and Anderson are in good agreement; Anderson's values show more scatter. The largest deviation of the Parkinson and Quarrington values from the adopted values is +0.4% at 210 K. Anderson's points are mostly within less than ±1% of the selected values.

Above 300 K, $C_p°$ values are derived from the drop-calorimetric enthalpies of Thompson and Flengas (15, 353-1281 K) with the constraint to join smoothly with the low temperature data. The observed enthalpies differ from the adopted values by +4% at 400 K; above 500 K the deviation is generally within ±0.4%.

Also considered were the low temperature heat capacities measured by Eastman and Rodebush (16, 64-283 K) and Kelley's equations (17) which are based on the work of Bornemann and Hengstenberg (18, 273-873 K). The Eastman and Rodebush data appears to be high above 200 K. At 900 K the enthalpy calculated from Kelley's equation is 4% higher than the adopted value.

$S°(20\ K)$ of 0.83 cal K^{-1} mol^{-1} given by Kelley and King (19) is adopted here.

Fusion Data - See the liquid table for details.

Sublimation Data - See the ideal gas table for details.

References

1. K. Jellinek and A. Deubel, Z. Electrochem. 35, 451 (1929); K. Jellinek and J. Zakowski, Z. Anorg. Chem. 142, 26 (1925).
2. K. Sudo, Sci. Repts. Research Insts., Tohoku Univ. A2, 325 (1950); Chem. Abstr. 45, 10011h (1951).
3. J. R. Stubbles and C. E. Birchenall, Trans. AIME 215, 535 (1959).
4. A. F. Kapustinskii and I. A. Makolkin, J. Phys. Chem. USSR 12, 371 (1938).
5. K. Kiukkola and C. Wagner, J. Electrochem. Soc. 104, 379 (1957).
6. W. T. Thompson and S. N. Flengas, J. Electrochem. Soc. 118, 419 (1971).
7. JANAF Thermochemical Tables: Pb(ℓ), 3-31-62; $H_2S(g)$, S(ℓ), $S_2(g)$, 12-31-65.
8. F. D. Richardson and J. H. E. Jeffes, J. Iron Steel Inst. (London) 171, 165 (1952).
9. H. Zeumer and W. A. Roth, Z. Phys. Chem. A173, 365 (1935).
10. U. S. Nat. Bur. Stand. Tech. Note 270-3, 1968.
11. Shamsuddin and S. Misra, Curr. Sci. 42, 119 (1973).
12. I. A. Korshunov, J. Gen. Chem. (USSR) 10, 2087 (1940).
13. D. H. Parkinson and J. E. Quarrington, Proc. Phys. Soc. (London) 67A, 569 (1954).
14. C. T. Anderson, J. Amer. Chem. Soc. 54, 107 (1932).
15. W. T. Thompson and S. N. Flengas, Can. J. Chem. 49, 1550 (1971).
16. E. D. Eastman and W. H. Rodebush, J. Amer. Chem. Soc. 40, 489 (1918).
17. K. K. Kelley, U. S. Bur. Mines Bull. 584, 1960.
18. K. Bornemann and O. Hengstenberg, Metall u. Erz. 17, 313, 339 (1920).
19. K. K. Kelley and E. G. King, U. S. Bur. Mines Bull. 592, 1961.

Lead Sulfide (PbS) $Pb_1S_1(cr)$

Enthalpy Reference Temperature = T_r = 298.15 K   Standard State Pressure = p° = 0.1 MPa

T/K	$C_p°$ (J $K^{-1}mol^{-1}$)	S°	$-[G°-H°(T_r)]/T$	$H°-H°(T_r)$ (kJ mol^{-1})	$\Delta_f H°$	$\Delta_f G°$	Log K_f
0	0.	0.	INFINITE	-11.511	-98.545	-98.545	INFINITE
100	39.748	41.296	133.734	-9.244	-98.734	-97.928	51.152
200	47.614	71.920	95.823	-4.781	-98.437	-97.239	25.396
298.15	49.438	91.343	91.343	0.	-98.324	-96.685	16.939
300	49.455	91.649	91.344	0.091	-98.324	-96.675	16.833
400	50.459	106.008	93.295	5.085	-100.657	-96.024	12.539
500	51.463	117.375	97.013	10.181	-102.302	-94.688	9.892
600	52.384	126.839	101.217	15.373	-103.584	-93.039	8.100
700	53.346	134.988	105.473	20.661	-109.465	-90.400	6.746
800	54.308	142.176	109.620	26.045	-110.310	-87.616	5.721
900	55.271	148.627	113.602	31.523	-164.074	-83.658	4.855
1000	56.191	154.498	117.402	37.096	-163.312	-74.763	3.905
1100	57.153	159.900	121.023	42.764	-162.439	-65.949	3.132
1200	58.074	164.912	124.474	48.525	-161.459	-57.220	2.491
1300	59.036	169.599	127.767	54.382	-160.375	-48.577	1.952
1400	59.957	174.008	130.914	60.332	-159.206	-40.020	1.493
1500	60.877	178.176	133.927	66.374	-157.962	-31.550	1.099
1600	61.714	182.132	136.817	72.503	-156.653	-23.164	0.756
1700	62.551	185.898	139.595	78.716	-155.288	-14.863	0.457
1800	63.388	189.497	142.268	85.013	-153.869	-6.643	0.193

PREVIOUS: CURRENT: June 1973

Lead Sulfide (PbS) $Pb_1S_1(cr)$

LEAD SULFIDE (PbS) — LIQUID — $M_r = 239.26$

$S°(298.15\ K) = [100.795]\ J\ K^{-1}\ mol^{-1}$ — $\Delta_f H°(298.15\ K) = [-84.064]\ kJ\ mol^{-1}$

$T_{fus} = 1386.5 \pm 1.5\ K$ — $\Delta_{fus}H° = 18.83 \pm 6.3\ kJ\ mol^{-1}$

Enthalpy of Formation

$\Delta_f H°$(PbS, ℓ, 298.15 K) is calculated from that of the crystal by adding $\Delta_{fus}H°$ and the difference in enthalpy, H°(1386.5 K)-H°(298.15 K), between the crystal and liquid.

Heat Capacity and Entropy

The heat capacity of PbS(ℓ) is estimated to be 16.0 cal $K^{-1}\ mol^{-1}$. This appears reasonable as the $C_p°$ value for PbO(ℓ) is 15.535 cal $K^{-1}\ mol^{-1}$ (12). A glass transition is assumed at 900 K. Below this temperature, the heat capacity values used are those of PbS(cr). S°(298.15 K) is calculated in a manner similar to that used for $\Delta_f H°$(298.15 K).

Fusion Data

Employing cryoscopic measurements, Pelton and Flengas (1) and Bell and Flengas (2) determined the melting point of PbS to be 1113.4°C and 1111.9°C, respectively. Pelton and Flengas (1) estimated an accuracy of ±1°C. Miller and Komarek (3), in studying the retrograde solubility in the Pb-S system, reported T_{fus} = 1113.3 ± 0.5°C. The latter authors have summarized many early determinations of T_{fus} which were in the range 1103-1135°C. We adopt 1386.5 ± 1.5 K for T_{fus} of PbS.

Kelley (4) adopted $\Delta_{fus}H°$ = 4.150 kcal mol^{-1} based on melting point data obtained by Friedrich (5) for the PbS-Cu_2S system and by Truthe (6) for the PbS-$PbCl_2$ system. Maier (7) recalculated $\Delta_{fus}H°$ from Friedrich's data and obtained 4.155 kcal mol^{-1}. Kelley (4) also calculated $\Delta_{fus}H°$ = 5.03 kcal mol^{-1} from data on the PbS-Ag_2S system obtained by Friedrich. Other inconclusive works are discussed by Kelley (4). In the preceding calculations, $\Delta_{fus}H°$ was derived from a graph of log x versus 1/T where x is the mole fraction of PbS in the melt and T is the liquidus temperature.

We calculate a $\Delta_{fus}H°$ value for each melting point-composition data pair and plot this value against ΔT, the melting point depression of PbS. The resulting plot is extrapolated to ΔT = 0 to obtain $\Delta_{fus}H°$. Friedrich's data then yields $\Delta_{fus}H°$ = 4.33 ± 0.15 kcal mol^{-1}. Truthe's data has considerable scatter but indicates a value of 4-6 kcal mol^{-1} for $\Delta_{fus}H°$. An analysis of the data reported by Kohlmeyer and Monzer (8) for the PbS-PbO system yields $\Delta_{fus}H°$ = 5 ± 1 kcal mol^{-1}. Unfortunately, these three studies necessitate long extrapolations as the ΔT values are in the range 60-550 K. However, these three binary systems form simple, single eutectic phase diagrams and all are consistent with a $\Delta_{fus}H°$ value of 4.5 kcal mol^{-1}.

Recent work by Pelton and Flengas (1) on PbS-alkali halide systems involved ΔT values of 2-17 K so that a much shorter extrapolation is possible. An analysis of their data is tabulated below.

System	Data Points	$\Delta_{fus}H°$, kcal mol^{-1}
PbS-NaCl	3	8.5
PbS-KCl	10	4.5
PbS-RbCl	2	4.3
PbS-CsCl	2	4.6

Even in the PbS-KCl system, the scatter is such that it is impossible to precisely define $\Delta_{fus}H°$. We adopt a value of $\Delta_{fus}H°$ = 4.5 ± 1.5 kcal mol^{-1}. This value is consistent with all systems discussed above except for PbS-NaCl.

It should be noted that our analyses treated all systems as though no association, dissociation, or compound formation occurs in the melt. It is possible that the occurrence of any or all of these factors would cause the data to be consistent with a $\Delta_{fus}H°$ which might be larger than the adopted value and even outside our uncertainty limits. For example, Pelton and Flengas (1) discussed their results in the PbS-alkali halide systems in terms of an estimated $\Delta_{fus}H°$ = 8.7 kcal mol^{-1}. Their conclusions need modifications with use of our adopted $\Delta_{fus}H°$ = 4.5 kcal mol^{-1}. In addition Blachnik and Kluge (9) have compared SnS and PbS; a difficulty arises in that PbS does not have the same crystal structure as GeS and SnS (9). Thus, $\Delta_{fus}S°$ is not necessarily expected to the same for all three species.

Decomposition Data

Kohlmeyer (10, 11) reported that, although PbS(ℓ) had an apparent boiling point of 1617 K, the vaporization did not yield solely PbS(g). Decomposition occurred to yield sulfide molecules in the melt which were rich in Pb. The temperature at which the fugacity of PbS(g) is 1 atm for PbS(ℓ) = PbS(g) is calculated to be 1593 K. A similar calculation for the reaction PbS(ℓ) = Pb(ℓ) + 1/2 S S_2(g) yields 1999 K.

References

1. A. D. Pelton and S. N. Flengas, Can. J. Chem. 48, 2016 (1970).
2. M. C. Bell and S. N. Flengas, J. Electrochem. Soc. 113, 27 (1966).
3. E. Miller and K. L. Komarek, Trans. Met. Soc. AIME 236, 832 (1966).
4. K. K. Kelley, U. S. Bur. Mines Bull. 393, 1936.
5. K. Friedrich, Metallurgie 4, 671 (1907); data given in ref (7).
6. W. Truthe, Z. Anorg. Chem. 76, 161 (1912).
7. C. G. Maier, U. S. Bur. Mines RI 3263, 1934.
8. E. J. Kohlmeyer and W. Monzer, Z. Anorg. Chem. 252, 74 (1944).
9. R. Blachnik and W. Kluge, Thermochim. Acta 3, 317 (1972).
10. E. J. Kohlmeyer, Metall u. Erg. 29, 105 (1932).
11. E. J. Kohlmeyer and H. Brinkman, Z. Anorg. Chem. 292, 275 (1957).
12. JANAF Thermochemical Tables: PbO(ℓ), 12-31-71; PbS(g), 6-30-73; Pb(ℓ), 3-31-62; S_2(g), 12-31-65.

Lead Sulfide (PbS) — $Pb_1S_1(l)$

Enthalpy Reference Temperature = T_r = 298.15 K — Standard State Pressure = p° = 0.1 MPa

T/K	$C_p°$ (J K⁻¹mol⁻¹)	S° (J K⁻¹mol⁻¹)	$-[G°-H°(T_r)]/T$ (J K⁻¹mol⁻¹)	$H°-H°(T_r)$ (kJ mol⁻¹)	$\Delta_f H°$ (kJ mol⁻¹)	$\Delta_f G°$ (kJ mol⁻¹)	Log K_f
0							
100							
200							
298.15	49.438	100.795	100.795	0.	-84.064	-85.242	14.934
300	49.455	101.101	100.796	0.091	-84.064	-85.250	14.843
400	50.459	115.459	102.746	5.085	-86.396	-85.544	11.171
500	51.463	126.827	106.464	10.181	-88.042	-85.153	8.896
600	52.384	136.291	110.668	15.373	-89.323	-84.449	7.352
700	53.346	144.440	114.924	20.661	-95.205	-82.755	6.175
800	54.308	151.627	119.071	26.045	-96.049	-80.916	5.283
900	55.271	158.078	123.053	31.522	-149.814	-77.904	4.521
900.000	55.271	158.078	123.053	31.522	GLASS <--> LIQUID		
900.000	66.944	158.078	123.053	31.522	TRANSITION		
1000	66.944	165.131	126.914	38.217	-147.930	-70.014	3.657
1100	66.944	171.512	130.683	44.911	-146.032	-62.315	2.959
1200	66.944	177.336	134.332	51.606	-144.118	-54.788	2.385
1300	66.944	182.695	137.849	58.300	-142.197	-47.422	1.905
1386.500	66.944	187.007	140.782	64.091	------ CRYSTAL <--> LIQUID -----		
1400	66.944	187.656	141.231	64.994	-140.283	-40.204	1.500
1500	66.944	192.275	144.482	71.689	-138.386	-33.122	1.153
1600	66.944	196.595	147.606	78.383	-136.513	-26.165	0.854
1700	66.944	200.654	150.608	85.078	-134.667	-19.325	0.594
1800	66.944	204.480	153.495	91.772	-132.850	-12.593	0.365
1900	66.944	208.099	156.275	98.466	-131.065	-5.961	0.164
2000	66.944	211.533	158.953	105.161	-129.312	0.578	-0.015

PREVIOUS: — CURRENT: June 1973

Lead Sulfide (PbS) — $Pb_1S_1(l)$

LEAD SULFIDE (PbS) CRYSTAL-LIQUID M_r = 239.26

0 to 1386.5 K crystal
above 1386.5 K liquid

Refer to the individual tables for details.

Lead Sulfide (PbS) Pb_1S_1(cr,l)

Enthalpy Reference Temperature = T_r = 298.15 K Standard State Pressure = p° = 0.1 MPa

T/K	C_p° (J K⁻¹mol⁻¹)	S° (J K⁻¹mol⁻¹)	$-[G^\circ - H^\circ(T_r)]/T$ (J K⁻¹mol⁻¹)	$H^\circ - H^\circ(T_r)$ (kJ mol⁻¹)	$\Delta_f H^\circ$ (kJ mol⁻¹)	$\Delta_f G^\circ$ (kJ mol⁻¹)	Log K_f
0	0.	0.	INFINITE	-11.511	-98.545	-98.545	INFINITE
100	39.748	41.296	133.734	-9.244	-98.734	-97.928	51.152
200	47.614	71.920	95.823	-4.781	-98.437	-97.239	25.396
298.15	49.438	91.343	91.343	0.	-98.324	-96.685	16.939
300	49.455	91.649	91.344	0.091	-98.324	-96.675	16.833
400	50.459	106.008	93.295	5.085	-100.657	-96.024	12.539
500	51.463	117.375	97.013	10.181	-102.302	-94.688	9.892
600	52.384	126.839	101.217	15.373	-103.584	-93.039	8.100
700	53.346	134.988	105.473	20.661	-109.465	-90.400	6.746
800	54.308	142.176	109.620	26.045	-110.310	-87.616	5.721
900	55.271	148.627	113.602	31.523	-164.074	-83.658	4.855
1000	56.191	154.498	117.402	37.096	-163.312	-74.763	3.905
1100	57.153	159.900	121.023	42.764	-162.439	-65.949	3.132
1200	58.074	164.912	124.474	48.525	-161.459	-57.220	2.491
1300	59.036	169.599	127.767	54.382	-160.375	-48.577	1.952
1386.500	59.832	173.428	130.497	59.523	CRYSTAL <--> LIQUID		
1386.500	66.944	187.007	130.497	78.351	TRANSITION		
1400	66.944	187.656	131.045	79.255	-140.283	-40.204	1.500
1500	66.944	192.275	134.975	85.949	-138.386	-33.122	1.153
1600	66.944	196.595	138.693	92.644	-136.513	-26.165	0.854
1700	66.944	200.654	142.219	99.338	-134.667	-19.325	0.594
1800	66.944	204.480	145.573	106.033	-132.850	-12.593	0.365
1900	66.944	208.099	148.769	112.727	-131.065	-5.961	0.164
2000	66.944	211.533	151.823	119.421	-129.312	0.578	-0.015

PREVIOUS: CURRENT: June 1973

Lead Sulfide (PbS) Pb_1S_1(cr,l)

LEAD SULFIDE (PbS) IDEAL GAS $M_r = 239.26$

$\Delta_f H°(0\ K) = 133.66 \pm 6.3\ kJ\ mol^{-1}$

$S°(298.15\ K) = 251.41 \pm 0.21\ J\ K^{-1}\ mol^{-1}$ $\Delta_f H°(298.15\ K) = 131.80 \pm 6.3\ kJ\ mol^{-1}$

Electronic State and Molecular Constants(σ=1)

State	ε_i, cm^{-1}	g_i	ω_e, cm^{-1}	$\omega_e x_e$, cm^{-1}	B_e, cm^{-1}	α_e, cm^{-1}	r_e, Å
X 0^+	0	1	429.40	1.30	0.11633	0.000426	2.287
a 1	14821.9	2	285.9	0.88	0.09267	0.000374	2.562
A 0^+	18768.9	1	260.83	0.363	0.09634	0.000262	2.513
B 1	21774.5	2	282.17	0.856	0.09992	0.000602	2.467
C 0^+	23150.7	1	303.93	1.436	[0.09561]	[0.000347]	[2.522]
C'1	24952.3	2	283.95	1.171	[0.09358]	[0.000297]	[2.549]
D 1	29587.4	2	297.83	1.365	0.10160	0.00064	2.447
E 0^+	34000	1	[262.32]	[1.114]	[0.09095]	[0.000429]	[2.585]

Enthalpy of Formation

Vago and Barrow (1) observed the convergence in the E-X system at 2715 Å. The dissociation energy for PbS(g) is calculated to be 81.85 ± 0.40 kcal mol^{-1}. Barrow et al. (2) analyzed the PbS absorption spectrum in terms of type C coupling and assigned the electronically excited E state to be 0^+. If the non-crossing rule is applicable (2), then the most likely dissociation products are $Pb(^3P_1)$ and $S(^3P_1)$. Using auxiliary data (3), $\Delta_f H°$(PbS, g, 298.15 K) = 32.00 ± 0.40 kcal mol^{-1}.

Colin and Drowart (4) have studied the sublimation and dissociation of PbS by means of a Knudsen effusion cell coupled with a mass spectrometer. A 2nd and 3rd law analysis by Colin and Drowart (4) of four sets of experiments (865-1100 K) yielded an average value of 55.7 ± 1.4 kcal mol^{-1} for the enthalpy of sublimation. It is anticipated that this result is roughly 0.5 kcal mol^{-1} too large since the Gibbs energy functions for the sublimation used by Colin and Drowart (4) are approximately 0.5 cal K^{-1} mol^{-1} greater than those presently adopted at 1000 K (3).

Colin and Drowart (4) presented data for 13 experiments in which either Pb, S_2 or both Pb and S_2 partial pressures were measured in addition to the PbS partial pressure. The missing pressure is calculated using the formula $[P(Pb)/P(S_2)][M_r(S_2)/M_r(Pb)]^{1/2} = 2$ as in Colin and Drowart (4).

Source	Reaction*	T/K	Data Points	Method	$\Delta_r H°$(298.15 K), kcal mol^{-1} 2nd law	3rd law	Drift cal K^{-1} mol^{-1}	$\Delta_f H°$(298.15 K) kcal mol^{-1}
(4)	A	990-1043	5	calc. $P(S_2)$	30.20±14.63	29.04	-1.1±14.3	33.12
	A	979-1118	4	calc. P(Pb)	36.48±10.22	28.31	-7.9±9.9	33.86
	A	1012-1182	4	all P meas.	21.97±4.67	29.32	6.8±4.3	32.85
	B	979-1182	13*	Knudsen-mass spec.	55.25±0.69	55.55	0.3±0.7	32.05
(5)	B	1123-1268	12*	Entrainment	59.48±0.66	55.00	-3.6±0.5	31.50
(6)	B	878-1069	6	Knudsen	55.38±0.44	54.88	-0.5±0.4	31.38
(7)	B	776-927	**	Langmuir	53.16	57.24	4.8	33.74
(8)	B	825-1025	**		54.56	54.95	0.4	31.45
(9)	B	1048-1193	10	Knudsen	57.66±1.16	56.12	-1.3±1.0	32.62
(10)	B	1113-1378	11	Entrainment	57.27±1.35	54.34	-2.3±1.1	30.84
(11)	B	877-1020	12	torsion-effusion	55.99±0.87	54.55	-1.5±0.9	31.05
			17*	torsion-effusion	57.71±0.53	54.55	-3.2±0.5	31.05
	B	860-975	20	torsion-Langmuir	53.58±1.94	54.68	1.2±2.2	31.18
			7	torsion-Langmuir	53.69±0.70	54.55	0.9±0.7	31.05

* A) PbS(g) = Pb(g) + 1/2 S_2(g) B) PbS(cr) = PbS(g)

* One point rejected due to statistical test. ** Actual data not presented, smoothed data used.

The value adopted for $\Delta_f H°$(PbS, g, 298.15 K) is 31.5 ±1.5 kcal mol^{-1}. Most of the uncertainty is attributable to the possible decomposition and/or reaction during sublimation. Colin and Drowart (4) and Isakova et al. (10) considered the effect of decomposition whereas many of the other investigators analyzed their data in terms of PbS(g) formation only. In addition, Hansen and Munir (11) attest to the congruency of sublimation using x-ray observations on samples which had lost at least 30% of their initial weight as a result of heating. The adopted $\Delta_f H°$(298.15 K) is somewhat of a compromise between the spectroscopic work (1), the mass spectrometric work (4) and the bulk of the sublimation studies. Correspondingly, D_0^0(PbS, g) = 82.4 ± 1.5 kcal mol^{-1}.

D_0^0(PbS, g) = 86.5 kcal mol^{-1} is calculated by applying the Hildenbrand formula (12) to the linear Birge-Sponer extrapolation of the ground state vibrational levels. Using auxiliary data (3), $\Delta_f H°$(298.15 K) = 27.4 kcal mol^{-1} is obtained.

Heat Capacity and Entropy

The tabulated electronic and molecular constants are taken from Rosen (13) as reviewed and compiled by Barrow. The constants came from the work of Bell and Harvey (14), Rochester and Howell (15), Vago and Barrow (1) and Barrow et al. (2). The state designations are those which are consistent with type C coupling. In this case, the state designations Σ, Π, etc., no longer have meaning. The estimated values in the tabulation are derived from comparisons of known values in the PbS(g) and PbO(g) systems (3). The functions are calculated from the partition function $Q = Q_t \Sigma Q_r^i Q_v^i g_i e^{-c_2 \varepsilon_i/T}$; the values for Q_r^i and Q_v^i were calculated with first order anharmonic corrections.

References

1. E. E. Vago and R. F. Barrow, Proc. Phys. Soc. (London) 59, 449 (1947).
2. R. F. Barrow, P. W. Fry and R. C. LeBargy, Proc. Phys. Soc. (London) 81, 697 (1963).
3. JANAF Thermochemical Tables: Pb(g), 3-31-62; S(g), 6-30-71; PbS(cr), 6-30-73; S_2(g), 12-31-65; PbO(g), 12-31-71.
4. R. Colin and J. Drowart, J. Chem. Phys. 37, 1120 (1962).
5. R. Schenck and A. Albers, Z. Anorg. Chem. 105, 145 (1919).
6. B. K. Veselowskii, J. Appl. Chem. (USSR) 15, 398 (1942).
7. C. M. Hsiao and A. W. Schlechten, J. Metals 4, 65 (1952).
8. O. G. Miller and M. A. Abdeev, Tr. Altaisk Gorno Met. Nauch. - Issledo vatd' Inst. akad. Nauk Kazakh SSR 7, 182 (1958); Chem. Abstr. 54, 4297f (1960).
9. K. Sudo, Sci. Repts. Research Insts., Tohoku Univ. A12, 54 (1960).
10. R. A. Isakova, V. N. Nesterov, and A. S. Shendyapin, Russ. J. Inorg. Chem. 8, 8 (1963).
11. E. E. Hansen and Z. A. Munir, High Temp. Sci. 2, 169 (1970).
12. D. L. Hildenbrand and E. Murad, J. Chem. Phys. 51, 807 (1969).
13. B. Rosen, Ed., "Spectroscopic Data Relative to Diatomic Molecules," Pergamon Press, New York, 1970.
14. H. Bell and A. Harvey, Proc. Phys. Soc. (London) 50, 427 (1938).
15. G. D. Rochester and H. G. Howell, Proc. Roy. Soc. (London) 148, 157 (1935).

Lead Sulfide (PbS) $Pb_1S_1(g)$

Enthalpy Reference Temperature = T_r = 298.15 K Standard State Pressure = p° = 0.1 MPa

T/K	$C_p°$ (J $K^{-1}mol^{-1}$)	S° (J $K^{-1}mol^{-1}$)	$-[G°-H°(T_r)]/T$ (J $K^{-1}mol^{-1}$)	$H°-H°(T_r)$ (kJ mol^{-1})	$\Delta_f H°$ (kJ mol^{-1})	$\Delta_f G°$ (kJ mol^{-1})	Log K_f
0	0.	0.	INFINITE	-9.430	133.656	133.656	INFINITE
100	29.787	216.077	281.169	-6.509	134.121	117.448	-61.349
200	33.139	237.771	254.569	-3.360	133.104	101.132	-26.413
250	34.305	245.299	251.986	-1.672	132.467	93.211	-19.475
298.15	35.085	251.412	251.412	0.	131.796	85.711	-15.016
300	35.110	251.629	251.413	0.065	131.769	85.425	-14.874
350	35.671	257.086	251.842	1.835	131.020	77.759	-11.605
400	36.073	261.877	252.803	3.629	128.008	70.292	-9.179
450	36.370	266.144	254.053	5.441	126.497	63.159	-7.331
500	36.596	269.988	255.457	7.265	124.902	56.210	-5.872
600	36.912	276.690	258.454	10.942	122.104	42.739	-3.721
700	37.121	282.397	261.477	14.644	114.638	30.517	-2.277
800	37.272	287.364	264.409	18.364	112.130	18.673	-1.219
900	37.386	291.761	267.208	22.097	56.621	8.216	-0.477
1000	37.477	295.705	269.864	25.841	55.553	2.895	-0.151
1100	37.554	299.280	272.378	29.592	54.509	-2.320	0.110
1200	37.620	302.551	274.758	33.351	53.487	-7.441	0.324
1300	37.679	305.564	277.014	37.116	52.479	-12.477	0.501
1400	37.734	308.359	279.154	40.887	51.469	-17.436	0.651
1500	37.785	310.964	281.189	44.663	50.447	-22.322	0.777
1600	37.836	313.404	283.127	48.444	49.407	-27.140	0.886
1700	37.889	315.699	284.976	52.230	48.345	-31.891	0.980
1800	37.945	317.867	286.744	56.022	47.259	-36.580	1.062
1900	38.008	319.920	288.436	59.819	46.147	-41.207	1.133
2000	38.081	321.871	290.060	63.624	45.010	-45.776	1.196
2100	38.166	323.731	291.619	67.436	-133.372	-43.171	1.074
2200	38.266	325.509	293.119	71.257	-134.166	-38.858	0.923
2300	38.386	327.213	294.565	75.090	-135.047	-34.506	0.784
2400	38.527	328.849	295.960	78.935	-136.013	-30.115	0.655
2500	38.693	330.425	297.307	82.796	-137.064	-25.681	0.537
2600	38.887	331.946	298.610	86.675	-138.195	-21.203	0.426
2700	39.112	333.418	299.872	90.574	-139.398	-16.681	0.323
2800	39.368	334.845	301.096	94.498	-140.669	-12.113	0.226
2900	39.659	336.232	302.284	98.449	-141.997	-7.499	0.135
3000	39.986	337.581	303.438	102.431	-143.375	-2.837	0.049
3100	40.349	338.898	304.560	106.447	-144.791	1.871	-0.032
3200	40.750	340.186	305.654	110.502	-146.235	6.625	-0.108
3300	41.189	341.446	306.719	114.599	-147.696	11.424	-0.181
3400	41.666	342.683	307.759	118.741	-149.164	16.268	-0.250
3500	42.179	343.898	308.774	122.933	-150.628	21.155	-0.316
3600	42.729	345.094	309.766	127.178	-152.077	26.083	-0.378
3700	43.314	346.272	310.737	131.480	-153.502	31.052	-0.438
3800	43.932	347.436	311.688	135.842	-154.892	36.059	-0.496
3900	44.581	348.585	312.619	140.268	-156.239	41.101	-0.550
4000	45.259	349.722	313.532	144.759	-157.536	46.178	-0.603
4100	45.964	350.848	314.429	149.320	-158.773	51.286	-0.653
4200	46.692	351.965	315.309	153.953	-159.946	56.424	-0.702
4300	47.440	353.072	316.175	158.659	-161.047	61.589	-0.748
4400	48.206	354.172	317.026	163.442	-162.072	66.778	-0.793
4500	48.985	355.264	317.863	168.301	-163.017	71.990	-0.836
4600	49.774	356.349	318.688	173.239	-163.877	77.222	-0.877
4700	50.570	357.428	319.501	178.256	-164.651	82.472	-0.917
4800	51.368	358.501	320.302	183.353	-165.337	87.738	-0.955
4900	52.166	359.568	321.093	188.530	-165.932	93.016	-0.992
5000	52.959	360.630	321.873	193.786	-166.437	98.306	-1.027
5100	53.745	361.687	322.643	199.121	-166.851	103.605	-1.061
5200	54.519	362.738	323.404	204.535	-167.175	108.912	-1.094
5300	55.279	363.784	324.156	210.025	-167.410	114.223	-1.126
5400	56.022	364.824	324.900	215.590	-167.559	119.539	-1.156
5500	56.744	365.858	325.635	221.228	-167.622	124.856	-1.186
5600	57.443	366.887	326.363	226.938	-167.603	130.174	-1.214
5700	58.117	367.910	327.082	232.716	-167.504	135.490	-1.242
5800	58.764	368.926	327.795	238.561	-167.329	140.805	-1.268
5900	59.382	369.936	328.501	244.468	-167.081	146.115	-1.294
6000	59.969	370.939	329.200	250.436	-166.764	151.421	-1.318

PREVIOUS: June 1973 (1 atm) CURRENT: June 1973 (1 bar)

Lead Sulfide (PbS) $Pb_1S_1(g)$

LEAD (Pb_2) IDEAL GAS M_r = 414.4

$S°(298.15\ K) = [281.334]\ J\ K^{-1}\ mol^{-1}$

$\Delta_f H°(0\ K) = 336.4 \pm 19.2\ kJ\ mol^{-1}$
$\Delta_f H°(298.15\ K) = 332.6 \pm 19.2\ kJ\ mol^{-1}$

Electronic Levels and Quantum Weights

State	ϵ_i, cm^{-1}	g_i
$[^3\Sigma_g^-]$	0	[3]

$\omega_e = 256.5\ cm^{-1}$ $\omega_e x_e = 2.96\ cm^{-1}$ $\sigma = 2$
$B_e = [0.01727]\ cm^{-1}$ $\alpha_e = [0.00020]\ cm^{-1}$ $r_e = [3.08]$ Å

Enthalpy of Formation

The enthalpy of formation is calculated from the dissociation energy of $Pb_2(g)$ as reported by Gaydon (1).

Heat Capacity and Entropy

The values of ω_e and $\omega_e x_e$ are obtained from Herzberg (2). Those for B_e and α_e are estimated according to the method suggested by Herzberg. The bond distance (r_e) is calculated from the moment of inertia, $I = 1630 \times 10^{-40}\ g\ cm^2$, estimated by Kelley and King (3). The ground state configuration is assumed to be $^3\Sigma$ by comparison to that for $Si_2(g)$ reported by Douglas (4).

References

1. A. G. Gaydon, "Dissociation Energies", Chapman and Hall, Ltd., London, 1953.
2. G. Herzberg, "Spectra of Diatomic Molecules", D. Van Nostrand Company, Inc., New York, 1950.
3. K. K. Kelley and E. G. King, U. S. Bur. Mines Bull. 592, 1961.
4. A. E. Douglas, Can. J. Phys. 33, 801 (1955).

Lead (Pb_2) $Pb_2(g)$

Enthalpy Reference Temperature = T_r = 298.15 K Standard State Pressure = p° = 0.1 MPa

	J K⁻¹mol⁻¹			kJ mol⁻¹			
T/K	$C_p°$	S°	$-[G°-H°(T_r)]/T$	$H°-H°(T_r)$	$\Delta_f H°$	$\Delta_f G°$	Log K_f
0	0.	0.	INFINITE	-10.004	336.381	336.381	INFINITE
100	32.267	243.142	313.231	-7.009	335.843	318.895	-166.574
200	35.750	266.807	284.681	-3.575	334.228	302.576	-79.025
250	36.473	274.869	281.939	-1.768	333.425	294.755	-61.586
298.15	36.921	281.334	281.334	0.	332.628	287.380	-50.348
300	36.935	281.562	281.335	0.068	332.597	287.099	-49.988
350	37.258	287.281	281.785	1.924	331.745	279.583	-41.725
400	37.502	292.273	282.791	3.793	330.864	272.191	-35.544
450	37.697	296.702	284.095	5.673	329.951	264.911	-30.750
500	37.861	300.682	285.558	7.562	329.005	257.735	-26.925
600	38.133	307.610	288.673	11.362	327.027	243.665	-21.213
700	38.362	313.506	291.810	15.187	315.209	231.532	-17.277
800	38.568	318.642	294.850	19.034	313.021	219.728	-14.347
900	38.761	323.196	297.751	22.900	310.917	208.194	-12.083
1000	38.945	327.290	300.504	26.786	308.893	196.889	-10.284
1100	39.123	331.010	303.111	30.689	306.947	185.784	-8.822
1200	39.298	334.422	305.580	34.610	305.080	174.852	-7.611
1300	39.470	337.574	307.921	38.549	303.277	164.073	-6.593
1400	39.640	340.505	310.145	42.504	301.508	153.432	-5.725
1500	39.808	343.246	312.261	46.477	299.751	142.916	-4.977
1600	39.976	345.820	314.279	50.466	297.996	132.518	-4.326
1700	40.142	348.249	316.207	54.472	296.231	122.229	-3.756
1800	40.308	350.548	318.051	58.494	294.450	112.045	-3.251
1900	40.473	352.732	319.819	62.533	292.649	101.961	-2.803
2000	40.638	354.812	321.517	66.589	290.825	91.972	-2.402
2100	40.803	356.799	323.151	70.661	-65.467	96.305	-2.395
2200	40.967	358.701	324.724	74.750	-66.564	104.033	-2.470
2300	41.131	360.525	326.241	78.855	-67.821	111.815	-2.539
2400	41.295	362.279	327.706	82.976	-69.243	119.655	-2.604
2500	41.459	363.968	329.123	87.114	-70.831	127.558	-2.665
2600	41.622	365.598	330.495	91.268	-72.584	135.528	-2.723
2700	41.785	367.172	331.824	95.438	-74.496	143.568	-2.777
2800	41.949	368.694	333.114	99.625	-76.562	151.681	-2.830
2900	42.112	370.169	334.366	103.828	-78.771	159.872	-2.880
3000	42.275	371.599	335.584	108.047	-81.113	168.140	-2.928
3100	42.438	372.988	336.768	112.283	-83.576	176.489	-2.974
3200	42.601	374.338	337.921	116.535	-86.146	184.919	-3.018
3300	42.764	375.652	339.045	120.803	-88.810	193.430	-3.062
3400	42.927	376.931	340.140	125.087	-91.554	202.024	-3.104
3500	43.089	378.177	341.209	129.388	-94.364	210.700	-3.145
3600	43.252	379.393	342.253	133.705	-97.227	219.456	-3.184
3700	43.415	380.581	343.273	138.039	-100.129	228.293	-3.223
3800	43.578	381.741	344.270	142.388	-103.058	237.208	-3.261
3900	43.740	382.875	345.245	146.754	-106.003	246.201	-3.297
4000	43.903	383.984	346.200	151.136	-108.954	255.269	-3.333
4100	44.066	385.070	347.135	155.535	-111.900	264.411	-3.369
4200	44.228	386.134	348.051	159.949	-114.833	273.625	-3.403
4300	44.391	387.177	348.949	164.380	-117.745	282.909	-3.437
4400	44.553	388.199	349.829	168.828	-120.629	292.260	-3.470
4500	44.716	389.202	350.693	173.291	-123.480	301.676	-3.502
4600	44.878	390.187	351.541	177.771	-126.292	311.155	-3.533
4700	45.041	391.154	352.373	182.267	-129.060	320.694	-3.564
4800	45.204	392.104	353.191	186.779	-131.783	330.292	-3.594
4900	45.366	393.037	353.995	191.307	-134.456	339.947	-3.624
5000	45.529	393.955	354.785	195.852	-137.078	349.655	-3.653
5100	45.691	394.859	355.562	200.413	-139.647	359.415	-3.681
5200	45.854	395.747	356.326	204.990	-142.162	369.225	-3.709
5300	46.016	396.622	357.078	209.584	-144.623	379.084	-3.736
5400	46.179	397.484	357.819	214.194	-147.029	388.987	-3.763
5500	46.341	398.333	358.548	218.820	-149.381	398.935	-3.789
5600	46.503	399.169	359.266	223.462	-151.680	408.926	-3.814
5700	46.666	399.994	359.973	228.120	-153.926	418.956	-3.839
5800	46.828	400.807	360.670	232.795	-156.121	429.026	-3.864
5900	46.991	401.609	361.357	237.486	-158.263	439.133	-3.888
6000	47.153	402.400	362.034	242.193	-160.359	449.276	-3.911

PREVIOUS: September 1963 (1 atm) CURRENT: September 1963 (1 bar)

Lead (Pb_2) $Pb_2(g)$

RUBIDUIM (Rb) REFERENCE STATE A_r = 85.4678

0 to 312.65 K crystal
312.5 to 970.385 K liquid
above 970.385 K ideal monatomic gas

Refer to the individual tables for details.

Rubidium (Rb) Rb_1(ref)

Enthalpy Reference Temperature = T_r = 298.15 K; Standard State Pressure = p° = 0.1 MPa

T/K	C_p° (J K⁻¹mol⁻¹)	S° (J K⁻¹mol⁻¹)	$-[G^\circ-H^\circ(T_r)]/T$ (J K⁻¹mol⁻¹)	$H^\circ-H^\circ(T_r)$ (kJ mol⁻¹)	$\Delta_f H^\circ$ (kJ mol⁻¹)	$\Delta_f G^\circ$ (kJ mol⁻¹)	Log K_f
0	0.	0.	INFINITE	-7.490	0.	0.	0.
100	25.510	47.022	101.811	-5.479	0.	0.	0.
200	27.447	65.316	79.458	-2.828	0.	0.	0.
298.15	31.062	76.778	76.778	0.	0.	0.	0.
300	31.188	76.971	76.779	0.058	0.	0.	0.
312.650	32.401	78.284	76.813	0.460	CRYSTAL <--> LIQUID		
312.650	32.307	85.300	76.813	2.653	TRANSITION		
400	31.715	93.188	79.566	5.449	0.	0.	0.
500	31.197	100.207	83.020	8.593	0.	0.	0.
600	30.850	105.861	86.371	11.694	0.	0.	0.
700	30.674	110.601	89.503	14.769	0.	0.	0.
800	30.669	114.695	92.402	17.835	0.	0.	0.
900	30.835	118.315	95.084	20.908	0.	0.	0.
970.385	31.053	120.645	96.854	23.086	LIQUID <--> IDEAL GAS		
970.385	20.786	194.623	96.854	94.873	FUGACITY = 1 bar		
1000	20.786	195.248	99.759	95.489	0.	0.	0.
1100	20.786	197.229	108.531	97.567	0.	0.	0.
1200	20.787	199.037	115.999	99.646	0.	0.	0.
1300	20.790	200.701	122.452	101.725	0.	0.	0.
1400	20.795	202.242	128.097	103.804	0.	0.	0.
1500	20.805	203.677	133.088	105.884	0.	0.	0.
1600	20.821	205.020	137.542	107.965	0.	0.	0.
1700	20.847	206.283	141.549	110.048	0.	0.	0.
1800	20.886	207.476	145.179	112.135	0.	0.	0.
1900	20.941	208.607	148.488	114.226	0.	0.	0.
2000	21.015	209.683	151.521	116.324	0.	0.	0.
2100	21.110	210.710	154.315	118.430	0.	0.	0.
2200	21.231	211.695	156.901	120.547	0.	0.	0.
2300	21.379	212.642	159.304	122.677	0.	0.	0.
2400	21.558	213.555	161.546	124.824	0.	0.	0.
2500	21.769	214.439	163.644	126.989	0.	0.	0.
2600	22.016	215.298	165.614	129.178	0.	0.	0.
2700	22.294	216.134	167.470	131.393	0.	0.	0.
2800	22.616	216.950	169.222	133.638	0.	0.	0.
2900	22.980	217.750	170.882	135.917	0.	0.	0.
3000	23.389	218.536	172.457	138.235	0.	0.	0.
3100	23.846	219.310	173.956	140.596	0.	0.	0.
3200	24.355	220.075	175.385	143.006	0.	0.	0.
3300	24.850	220.827	176.751	145.453	0.	0.	0.
3400	25.442	221.578	178.058	147.966	0.	0.	0.
3500	26.088	222.324	179.312	150.542	0.	0.	0.
3600	26.766	223.067	180.517	153.179	0.	0.	0.
3700	27.516	223.810	181.677	155.893	0.	0.	0.
3800	28.322	224.555	182.796	158.684	0.	0.	0.
3900	29.184	225.302	183.876	161.559	0.	0.	0.
4000	30.060	226.048	184.921	164.507	0.	0.	0.
4100	31.023	226.802	185.933	167.561	0.	0.	0.
4200	32.038	227.561	186.915	170.714	0.	0.	0.
4300	33.104	228.328	187.869	173.970	0.	0.	0.
4400	33.569	229.027	188.791	177.038	0.	0.	0.
4500	34.345	229.761	189.690	180.318	0.	0.	0.
4600	35.326	230.521	190.569	183.780	0.	0.	0.
4700	36.365	231.292	191.427	187.364	0.	0.	0.
4800	37.419	232.069	192.266	191.053	0.	0.	0.
4900	38.481	232.851	193.086	194.848	0.	0.	0.
5000	39.145	233.583	193.883	198.498	0.	0.	0.
5100	40.162	234.368	194.669	202.463	0.	0.	0.
5200	41.169	235.158	195.440	206.530	0.	0.	0.
5300	42.161	235.951	196.197	210.697	0.	0.	0.
5400	43.133	236.749	196.941	214.961	0.	0.	0.
5500	44.078	237.549	197.672	219.322	0.	0.	0.
5600	44.545	238.273	198.381	223.394	0.	0.	0.
5700	45.394	239.069	199.088	227.891	0.	0.	0.
5800	46.204	239.866	199.784	232.472	0.	0.	0.
5900	46.972	240.662	200.470	237.131	0.	0.	0.
6000	47.693	241.458	201.147	241.864	0.	0.	0.

PREVIOUS: CURRENT: December 1983 (1 bar)

Rubidium (Rb) Rb_1(ref)

RUBIDIUM (Rb) CRYSTAL A_r = 85.4678

$S°(298.15\ K) = 76.78 \pm 0.30\ J\ K^{-1}\ mol^{-1}$

$T_{fus} = 312.65 \pm 0.1\ K$

$\Delta_f H°(0\ K) = 0\ kJ\ mol^{-1}$

$\Delta_f H°(298.15\ K) = 0\ kJ\ mol^{-1}$

$\Delta_{fus} H° = 2.19 \pm 0.09\ kJ\ mol^{-1}$

Enthalpy of Formation

Zero by definition.

Heat Capacity and Entropy

The thermal functions for Rb(cr) are derived following CODATA recommendations (1). That is, these functions are based on the previous evaluation of Hultgren et al. (2), Martin (3, 4), and Shpil'rain et al. (5). The prime sources of data are: Lien and Phillips (0.19-4.1 K, 6), Filby and Martin (0.4-1.5 K, 3-320 K, 7), McCallum and Silsbee (1.3-12 K, 8), Dauphinee, Martin, and Preston-Thomas (25-300 K, 9), and Martin (0.4-3 K, 4).

Fusion Data

The adopted fusion temperature, T_{fus} = 312.65±0.1 K, is that recommended by Ohse et al. (10) based on their analysis of seven experimental studies: Rengade (312.15±0.05 K, 11), Dauphine et al. (311.90±0.5 K, 9), Weatherford et al. (312.54 K, 12), Filby and Martin (312.65±0.01 K, 7), Basin et al. (312.31±0.05 K, 13), Ott et al. (312.45 K, 14), and Martin (312.47 K, 4).

The adopted enthalpy of fusion, $\Delta_{fus}H°$, 2.19±0.09 kJ mol^{-1}, is derived from the studies of Rengade (11), Filby and Martin (7), and Martin (4); all three results being within 1 cal mol^{-1}.

Sublimation Data

Refer to the ideal gas tables for details.

References

1. J. D. Cox, Chairman, ICSU-CODATA Task Group on Key Values for Thermodynamics, J. Chem. Thermodynamics 10, 903 (1978).
2. R. Hultgren, P. D. Desai, et al., "Selected Values of the Thermodynamic Properties of the Elements", American Society for Metals, Metals Park, Ohio, 1973.
3. D. L. Martin, Phys. Rev. 139A, 150 (1965).
4. D. L. Martin, Can. J. Phys. 48, 1327 (1970).
5. E. E. Shpil'rain, K. A. Yakimovich, et al., "Thermophysical Properties of the Alkali Metals", STANDARDS, Moscow, 1970.
6. W. H. Lien and N. E. Phillips, Phys. Rev. 133, A1370 (1964).
7. J. D. Filby and D. L. Martin, Proc. Roy. Soc. A284, 83 (1965).
8. D. C. McCollum, Jr., and H. B. Silsbee, Phys. Rev. 127, 119 (1962).
9. T. M. Dauphinee, D. L. Martin and H. Preston-Thomas, Proc. Roy. Soc. A233, 214 (1955).
10. R. W. Ohse, J.-F. Babelot, et al., "An Assessment of the Melting, Boiling, and Critical Point Data of the Alkali Metals", Chapter 6.1 in "Handbook of Thermodynamic and Transport Properties of Alkali Metals", R. W. Ohse, ed., Blackwell Scientific Publishers, London, 1985.
11. E. Rengade, Compt. Rend. 156, 1897 (1913).
12. W. A. Weatherford, Jr., R. K. Johnston, et al., Aeronautical Systems Division, Wright-Patterson Air Force Base, Report ASD-TR-63, 413, 1963.
13. A. S. Basin, S. P. Volchkova, and A. N. Soloviev, J. Appl. Mech. and Tech. Phys. 10, 961 (1969).
14. J. B. Ott and J. R. Goates, Tech. Prog. Report COO-1707-9 (1970); J. R. Goates, J. B. Ott and C. C. Hsu, Trans. Faraday Soc. 66, 25 (1970).

Rubidium (Rb) $Rb_1(cr)$

Enthalpy Reference Temperature = T_r = 298.15 K Standard State Pressure = p° = 0.1 MPa

T/K	$C_p°$ (J K⁻¹mol⁻¹)	S° (J K⁻¹mol⁻¹)	−[G°−H°(T_r)]/T (J K⁻¹mol⁻¹)	H°−H°(T_r) (kJ mol⁻¹)	$\Delta_fH°$ (kJ mol⁻¹)	$\Delta_fG°$ (kJ mol⁻¹)	Log K_f
0	0.	0.	INFINITE	-7.490	0.	0.	0.
100	25.510	47.022	101.811	-5.479	0.	0.	0.
200	27.447	65.316	79.458	-2.828	0.	0.	0.
250	28.690	71.560	77.272	-1.428	0.	0.	0.
298.15	31.062	76.778	76.778	0.	0.	0.	0.
300	31.188	76.971	76.779	0.058	0.	0.	0.
312.650	32.401	78.284	76.813	0.460	------ CRYSTAL <--> LIQUID -----		
350	35.848	82.132	77.176	1.735	-2.121	0.259	-0.039
400	40.465	87.220	78.114	3.642	-1.806	0.581	-0.076
450	45.082	92.253	79.406	5.781	-1.246	0.848	-0.098
500	49.699	97.242	80.940	8.151	-0.443	1.040	-0.109
600	58.932	107.119	84.482	13.582	1.888	1.134	-0.099

PREVIOUS: CURRENT: December 1983

Rubidium (Rb) $Rb_1(cr)$

RUBIDIUM (Rb) LIQUID A_r = 85.4678

$S°(298.15\ K) = [83.764]\ J\ K^{-1}\ mol^{-1}$

$\Delta_f H°(298.15\ K) = [2.184]\ kJ\ mol^{-1}$

$T_{fus} = 312.65 \pm 0.1\ K$

$\Delta_{fus}H° = 2.19 \pm 0.09\ kJ\ mol^{-1}$

Enthalpy of Formation

The enthalpy of formation is calculated from that of the crystal by adding the enthalpy of fusion and the difference in enthalpy, H°(298.15 K) - H°(298.15 K), between the crystal and liquid.

Heat Capacity and Entropy

The adopted heat capacity values for Rb(ℓ) are those recommended by Fink and Leibowitz (1). These authors reviewed the liquid phase heat capacity and enthalpy studies for all alkali metals and recommended a Cp° equation for Rb(ℓ) based on their evaluation of four enthalpy (2, 3, 4, 5) and three heat capacity (6, 7, 8) experimental studies. These studies provided information in the liquid region up to 1334 K.

The equation recommened by Fink and Liebowitz (1) is $Cp° = 35.494 - 0.012864\ T + 0.85409 \times 10^{-5}\ T^2\ (J\ K^{-1}\ mol^{-1})$.

Vaporization Data

Refer to the ideal gas table for details.

References

1. J. K. Fink and L. Leibowitz, "Entalpy, Entropy, and Specific Heat-Data Assessment," Chapter 6.3.2 in "Handbook of Thermodynamic and Transport Properties of Alkalj Metals," R. W. Ohse, ed., Blackwell Scientific rublications, London, 1985
2. F. Tepper, A. Murchison, et al., MSA Research Corporation Report ML-TDR-64-42, 1964.
3. P. Y. Achener, Aerojet-General Nucleonics Report AGN-8090, 1964.
4. I. I. Novikov, V. V. Roshchupkin, and L. K. Fordeeva, Teplofiz. Vys. Temp. 14, 75(1976).
5. E. E. Shpilrain and D. N. Kagam, Teplofiz. Vys. Temp. 7, 362(1969).
6. I. T. Aladyev and I. M. Pchelkin, in "Thermophysical Properties of Alkali Metals, E. E. Shpilrain et al., eds., Izd.-Vo Standartov, pp 130, 140-1, 147, 155-57, Moscow, 1970.
7. J. D. Filby and D. L. Martin, Proc. Roy. Soc. (London) A284, 83(1965).
8. P. F. Young, Aerojet-General Nucleonics Report AGN-8034, 1962).

Rubidium (Rb) $Rb_1(l)$

Enthalpy Reference Temperature = T_r = 298.15 K Standard State Pressure = p° = 0.1 MPa

T/K	$C_p°$ (J K⁻¹mol⁻¹)	S° (J K⁻¹mol⁻¹)	$-[G°-H°(T_r)]/T$ (J K⁻¹mol⁻¹)	$H°-H°(T_r)$ (kJ mol⁻¹)	$\Delta_f H°$ (kJ mol⁻¹)	$\Delta_f G°$ (kJ mol⁻¹)	Log K_f
0							
100							
200							
250							
298.15	32.418	83.764	83.764	0.	2.184	0.102	-0.018
300	32.403	83.965	83.765	0.060	2.187	0.089	-0.015
312.650	32.307	85.300	83.800	0.469	------ CRYSTAL <--> LIQUID -----		
350	32.038	88.932	84.158	1.671	0.	0.	0.
400	31.715	93.188	85.027	3.264	0.	0.	0.
450	31.435	96.907	86.145	4.843	0.	0.	0.
500	31.197	100.207	87.389	6.409	0.	0.	0.
600	30.850	105.861	90.012	9.510	0.	0.	0.
700	30.674	110.601	92.624	12.584	0.	0.	0.
800	30.669	114.695	95.133	15.650	0.	0.	0.
900	30.835	118.315	97.511	18.724	0.	0.	0.
970.385	31.053	120.645	99.105	20.901	------- FUGACITY = 1 bar -------		
1000	31.171	121.580	99.757	21.823	-71.481	2.186	-0.114
1100	31.678	124.573	101.879	24.964	-70.419	9.502	-0.451
1200	32.356	127.358	103.888	28.164	-69.297	16.718	-0.728
1300	33.205	129.980	105.795	31.441	-68.100	23.838	-0.958
1400	34.225	132.477	107.612	34.811	-66.809	30.863	-1.152
1500	35.415	134.878	109.350	38.291	-65.408	37.791	-1.316
1600	36.776	137.206	111.019	41.899	-63.881	44.622	-1.457

PREVIOUS: CURRENT: December 1983

Rubidium (Rb) $Rb_1(l)$

RUBIDIUM (Rb) CRYSTAL-LIQUID A_r = 85.4678

0 to 312.65 K crystal
above 312.65 K liquid

Refer to the individual tables for details.

Rubidium (Rb) $Rb_1(cr,l)$

Enthalpy Reference Temperature = T_r = 298.15 K — Standard State Pressure = p° = 0.1 MPa

T/K	C_p° (J K⁻¹mol⁻¹)	S° (J K⁻¹mol⁻¹)	$-[G^\circ - H^\circ(T_r)]/T$ (J K⁻¹mol⁻¹)	$H^\circ - H^\circ(T_r)$ (kJ mol⁻¹)	$\Delta_f H^\circ$ (kJ mol⁻¹)	$\Delta_f G^\circ$ (kJ mol⁻¹)	Log K_f
0	0.	0.	INFINITE	-7.490	0.	0.	0.
100	25.510	47.022	101.811	-5.479	0.	0.	0.
200	27.447	65.316	79.458	-2.828	0.	0.	0.
250	28.690	71.560	77.272	-1.428	0.	0.	0.
298.15	31.062	76.778	76.778	0.	0.	0.	0.
300	31.188	76.971	76.779	0.058	0.	0.	0.
312.650	32.401	78.284	76.813	0.460	CRYSTAL <--> LIQUID	TRANSITION	
312.650	32.307	85.300	76.813	2.653			
350	32.038	88.932	77.917	3.855	0.	0.	0.
400	31.715	93.188	79.566	5.449	0.	0.	0.
450	31.435	96.907	81.291	7.028	0.	0.	0.
500	31.197	100.207	83.020	8.593	0.	0.	0.
600	30.850	105.861	86.371	11.694	0.	0.	0.
700	30.674	110.601	89.503	14.769	0.	0.	0.
800	30.669	114.695	92.402	17.835	0.	0.	0.
900	30.835	118.315	95.084	20.908	0.	0.	0.
970.385	31.053	120.645	96.854	23.086	------- FUGACITY	= 1 bar -------	
1000	31.171	121.580	97.573	24.007	-71.481	2.186	-0.114
1100	31.678	124.573	99.893	27.148	-70.419	9.502	-0.451
1200	32.356	127.358	102.067	30.349	-69.297	16.719	-0.728
1300	33.205	129.980	104.114	33.625	-68.100	23.839	-0.958
1400	34.225	132.477	106.052	36.995	-66.809	30.863	-1.152
1500	35.415	134.878	107.894	40.476	-65.408	37.791	-1.316
1600	36.776	137.206	109.653	44.084	-63.881	44.622	-1.457

PREVIOUS: CURRENT: December 1983

Rubidium (Rb) $Rb_1(cr,l)$

RUBIDIUM (Rb) IDEAL GAS A_r = 85-4678

IP(Rb, g) = 33690.81 ± 0.01 cm^{-1}

S°(298.15 K) = 170.093 ± 0.025 J K^{-1} mol^{-1}

$\Delta_fH°$(298.15 K) = 82.2 ± 0.4 kJ mol^{-1}

$\Delta_fH°$(298.15 K) = 80.9 ± 0.4 kJ mol^{-1}

Electronic Levels and Quantum Weights

State	ϵ_i, cm^{-1}	g_i
$^2S_{1/2}$	0.00	2
$^2P_{1/2}$	12578.96	2
$^2P_{3/2}$	12816.56	4
$^2D_{5/2}$	19355.01	6
$^2D_{3/2}$	19355.45	4
.	.	.
.	.	.
.	.	.
$^2P_{3/2}$	33671.07	4
IP	33690.81	

Enthalpy of Formation

The enthalpy of formation of rubidium gas is chosen to be the value recommended by CODATA (1). This value was obtained from 2nd and 3rd law analysis of the vapor pressure studies reported by Hultgren et al. (2). As stated by CODATA (1), the correction for the presence of the dimer in the vapor phase was made using a virial treatment, since the dissociation energy of Rb_2(g) was not well established.

The six studies used by Hultgren et al. (1) are: Volyak et al. (594-970 K, 3), Achener (745-1260 K, 4), Tepper et al. (716-1353 K, 5), Buck and Pauly (307-363 K, 6), Killian (312-377 K, 7), and Scott (365-400 K, 8).

Heat Capacity and Entropy

The thermal functions for the five alkali metal monatomic gases are calculated by the same procedure. Observed and estimated atomic energy levels are included in the partition function calculation, using an ionization potential lowering (IP-kT) technique as the cutoff procedure in the energy level summation (11).

The lowest lying levels for these metals [14904 cm^{-1} (Li), 16956 cm^{-1} (Na), 12985 cm^{-1} (K), 12578 cm^{-1} (Rb), and 11178 cm^{-1} (Cs)] do not contribute to the thermal functions below ~1000 K; there is only a translational contribution below this temperature. Above this approximate temperature, the thermal functions become increasingly sensitive to the partition function cutoff procedure used, due to the combined effect of the observation of atomic energy levels of high principal quantum number and a low ionization potential. Energy levels have been observed up to n = 42 (Li), 59(Na), 79(K), 77(Rb), and 73(Cs). However, not all predicted levels have been observed for each of these principal quantum numbers. The ionization potentials vary from 43487.29 cm^{-1} for Li to 31406.1 cm^{-1} for Cs. In calculating the thermal functions with the inclusion of missing levels up to the high principal quantum numbers just mentioned, the Gibbs energy functions show significant differences (depending on the cutoff procedure) above 3000 K (11).

The atomic energy levels have been taken from the compilation of Moore (9, 10). Our calculated values for the thermal functions are similar to those recommended by CODATA (1). They do differ for two reasons, however. First, the entropy differs by 0.1094 J K^{-1} mol^{-1} because this table uses a reference pressure of 1 bar, whereas the CODATA recommendations are based on 1 atm. Second, the entropies at 298.15 K for all alkali metal gases differs by ~0.004 J K^{-1} mol^{-1}, presumably due to the use of slightly different values for auxiliary data.

References

1. J. D. Cox, chairman, CODATA Task Group on Key values for Thermodynamics, J. Chem. Thermodynamics 10, 903 (2978); CODATA Special Report No. 8, 1980.
2. R. Hultgren, P. D. Desai, et al., "Selected Values of the Thermodynamic Properties of the Elements," American Society for Metals, Metals Park, Ohio, 1973.
3. L. D. Volyak, Yu. K. Vinogradov, and V. M. Anisimov, High Temp. 6, 719 (1968).
4. P. Y. Achener, U. S. AEC Report AGN-8090, 1964.
5. F. Tepper, A. Murchison, J. Zelinek, and F. Roehlick, AFML Report RTD-TDR-63-4018, Part I, 1963.
6. U. Buck and H. Pauley, Z. Physik. Chem. 44, 345 (1965).
7. T. J. Killian, Phys. Rev. 27, 578 (1926).
8. D. H. Scott, Phil. Mag. 47, 32 (1924).
9. C. E. Moore, U. S. Nat. Bur. Stand., NSRDS-NBS 34, 1970.
10. C. E. Moore, U. S. Nat. Bur. Stand., NSRDS-NBS 35 Volume II, 1970 [Reprint of NBS circular 467, Volume II, 1952].
11. J. R. Downey, Jr., The Dow Chemical Company, Rept. AFOSR-TR-78-0960, Contract No. F44620-75-1-0048, 1978.

Rubidium (Rb) Rb_1(g)

Enthalpy Reference Temperature = T_r = 298.15 K; Standard State Pressure = p° = 0.1 MPa

T/K	$C_p°$ (J $K^{-1}mol^{-1}$)	S° (J $K^{-1}mol^{-1}$)	$-[G°-H°(T_r)]/T$ (J $K^{-1}mol^{-1}$)	$H°-H°(T_r)$ (kJ mol^{-1})	$\Delta_fH°$ (kJ mol^{-1})	$\Delta_fG°$ (kJ mol^{-1})	Log K_f
0	0.	0.	INFINITE	-6.197	82.192	82.192	INFINITE
100	20.786	147.386	188.573	-4.119	82.260	72.224	-37.726
200	20.786	161.794	171.994	-2.040	81.688	62.393	-16.295
250	20.786	166.432	170.435	-1.001	81.327	57.609	-12.037
298.15	20.786	170.093	170.093	0.	80.900	53.078	-9.299
300	20.786	170.222	170.094	0.038	80.881	52.906	-9.212
350	20.786	173.426	170.347	1.078	78.122	48.550	-7.246
400	20.786	176.202	170.909	2.117	77.568	44.363	-5.793
450	20.786	178.650	171.636	3.156	77.029	40.245	-4.671
500	20.786	180.840	172.448	4.196	76.503	36.186	-3.780
600	20.786	184.630	174.172	6.274	75.480	28.219	-2.457
700	20.786	187.834	175.901	8.353	74.484	20.421	-1.524
800	20.786	190.609	177.570	10.431	73.497	12.766	-0.834
900	20.786	193.058	179.157	12.510	72.502	5.234	-0.304
970.385	20.786	194.623	180.223	13.973	------- FUGACITY = 1 bar -------		
1000	20.786	195.248	180.659	14.589	0.	0.	0.
1100	20.786	197.229	182.077	16.667	0.	0.	0.
1200	20.787	199.037	183.416	18.746	0.	0.	0.
1300	20.790	200.701	184.682	20.825	0.	0.	0.
1400	20.795	202.242	185.882	22.904	0.	0.	0.
1500	20.805	203.677	187.021	24.984	0.	0.	0.
1600	20.821	205.020	188.105	27.065	0.	0.	0.
1700	20.847	206.283	189.137	29.149	0.	0.	0.
1800	20.886	207.476	190.123	31.235	0.	0.	0.
1900	20.941	208.607	191.067	33.326	0.	0.	0.
2000	21.015	209.683	191.971	35.424	0.	0.	0.
2100	21.110	210.710	192.839	37.530	0.	0.	0.
2200	21.231	211.695	193.674	39.647	0.	0.	0.
2300	21.379	212.642	194.478	41.777	0.	0.	0.
2400	21.558	213.555	195.254	43.924	0.	0.	0.
2500	21.769	214.439	196.004	46.090	0.	0.	0.
2600	22.016	215.298	196.729	48.278	0.	0.	0.
2700	22.294	216.134	197.433	50.493	0.	0.	0.
2800	22.616	216.950	198.115	52.738	0.	0.	0.
2900	22.980	217.750	198.778	55.017	0.	0.	0.
3000	23.389	218.536	199.424	57.335	0.	0.	0.
3100	23.846	219.310	200.053	59.696	0.	0.	0.
3200	24.355	220.075	200.667	62.106	0.	0.	0.
3300	24.850	220.827	201.266	64.553	0.	0.	0.
3400	25.442	221.578	201.852	67.066	0.	0.	0.
3500	26.088	222.324	202.427	69.642	0.	0.	0.
3600	26.766	223.067	202.989	72.279	0.	0.	0.
3700	27.516	223.810	203.542	74.993	0.	0.	0.
3800	28.322	224.555	204.085	77.784	0.	0.	0.
3900	29.184	225.302	204.620	80.659	0.	0.	0.
4000	30.060	226.048	205.146	83.607	0.	0.	0.
4100	31.023	226.802	205.665	86.661	0.	0.	0.
4200	32.038	227.561	206.177	89.814	0.	0.	0.
4300	33.104	228.328	206.683	93.070	0.	0.	0.
4400	33.569	229.027	207.177	96.138	0.	0.	0.
4500	34.345	229.761	207.668	99.418	0.	0.	0.
4600	35.326	230.521	208.156	102.880	0.	0.	0.
4700	36.365	231.292	208.640	106.464	0.	0.	0.
4800	37.419	232.069	209.120	110.154	0.	0.	0.
4900	38.481	232.851	209.596	113.948	0.	0.	0.
5000	39.145	233.583	210.063	117.598	0.	0.	0.
5100	40.162	234.368	210.532	121.563	0.	0.	0.
5200	41.169	235.158	210.998	125.630	0.	0.	0.
5300	42.161	235.951	211.461	129.797	0.	0.	0.
5400	43.133	236.749	211.922	134.062	0.	0.	0.
5500	44.078	237.549	212.381	138.422	0.	0.	0.
5600	44.545	238.273	212.828	142.494	0.	0.	0.
5700	45.394	239.069	213.281	146.991	0.	0.	0.
5800	46.204	239.866	213.733	151.572	0.	0.	0.
5900	46.972	240.662	214.182	156.231	0.	0.	0.
6000	47.693	241.458	214.630	160.964	0.	0.	0.

PREVIOUS: CURRENT: December 1983 (1 bar)

Rubidium (Rb) Rb_1(g)

RUBIDIUM, ION (Rb^+) IDEAL GAS M_r = 85.46725

$IP(Rb^+, g)$ = 220048 ± 30 cm^{-1}

$S°(298.15\ K)$ = 164.330 ± 0.02 $JK^{-1}\ mol^{-1}$

$\Delta_f H°(0\ K)$ = 485.224 kJ mol^{-1}

$\Delta_f H°(298.15\ K)$ = [490.129] kJ mol^{-1}

Electronic Level and Quantum Weight

State	ε_i cm^{-1}	g_i
$^1S_0$	0	1

Enthalpy of Formation

$\Delta_f H°(Rb^+, g, 0\ K)$ is calculated from $\Delta_f H°(Rb, g, 0\ K)$ (1) using the spectroscopic value of IP(Rb) = 33690.81 ± 0.01 cm^{-1} (403.032 ± .0001 kJ mol^{-1}) from Moore (2). The ionization limit is converted from cm^{-1} to kJ mol^{-1} using the factor, 1 cm^{-1} = 0.01196266 kJ cm^{-1}, which is derived from the 1973 CODATA fundamental constants (3). Rosenstock et al. (4) and Levin and Lias (5) have summarized additional ionization and appearance potential data.

$\Delta_f H°(Rb^+, g, 298.15\ K)$ is calculated from $\Delta_f H°(Rb, g, 0\ K)$ by using IP(Rb) with JANAF (1) enthalpies, $H°(0\ K) - H°(298.15\ K)$, for Rb(g), $Rb^+(g)$, and $e^-(g)$. $\Delta_r H°(Rb \rightarrow Rb^+ + e^-, 298.15\ K)$ differs from a room temperature threshold energy due to inclusion of these enthalpies and to threshold effects discussed by Rosenstock et al. (4). $\Delta_f H°(298.15\ K)$ should be changed by -6.197 kJ mol^{-1} if it is to be used in the ion convention that excludes the enthalpy of the electron.

Heat Capacity and Entropy

The information on electronic energy levels and quantum weights, given by Moore (6), is incomplete because many theoretically predicted levels have not been observed. Our calculations indicate that any reasonable method of filling in these missing levels and cutting off the summation in the partition function (7) has no effect on the thermodynamic functions to 6000 K. This is a result of the high energy of all levels other than the ground state; the first excited state is approximately 126454 cm^{-1} above the ground state. Since inclusion of these excited states has no effect on the thermodynamic functions (to 6000 K), we list only the ground state. The reported uncertainty in S°(298.15 K) is due to uncertainties in the relative ionic mass and the fundamental constants. Extension of these calculations above 6000 K may require consideration of the excited states and use of different fill and cutoff procedures (7).

References

1. JANAF Thermochemical Tables: Rb(g), 12-31-83; $e^-(g)$, 3-31-82.
2. C. E. Moore, U. S. Nat. Bur. Stand., NSRDS-NBS-34, 1970.
3. E. R. Cohen and B. N. Taylor, J. Phys. Chem. Ref. Data 2, 663 (1973).
4. H. M. Rosenstock, K. Draxl et al., J. Phys. Chem. Ref. Data 6, Supp. 1 (1977).
5. R. D. Levin and S. G. Lias, U. S. Nat. Bur. Stand., NSRDS-NBS-71, 1982.
6. C. E. Moore, U. S. Nat. Bur. Stand., NSRDS-NBS-35, Volume II, 1970 (Reprint of NBS Circular 467, Volume II, 1952).
7. J. R. Downey, Jr., The Dow Chemical Company, AFOSR-TR-78-0960, Contract No. F44620-75-1-0048, 1978.

Rubidium, Ion (Rb^+) $Rb_1^+(g)$

Enthalpy Reference Temperature = T_r = 298.15 K Standard State Pressure = p° = 0.1 MPa

T/K	$C_p°$ (J $K^{-1}mol^{-1}$)	S°	$-[G°-H°(T_r)]/T$	$H°-H°(T_r)$ (kJ mol^{-1})	$\Delta_f H°$	$\Delta_f G°$	Log K_f
0	0.	0.	INFINITE	-6.197	485.224		
100	20.786	141.623	182.810	-4.119			
200	20.786	156.031	166.231	-2.040			
250	20.786	160.669	164.672	-1.001			
298.15	20.786	164.330	164.330	0.	490.129	457.771	-80.200
300	20.786	164.459	164.330	0.038	490.148	457.570	-79.670
350	20.786	167.663	164.583	1.078	488.429	452.364	-67.512
400	20.786	170.438	165.146	2.117	488.914	447.179	-58.396
450	20.786	172.887	165.872	3.156	489.414	441.933	-51.298
500	20.786	175.077	166.685	4.196	489.927	436.630	-45.614
600	20.786	178.866	168.409	6.274	490.983	425.871	-37.075
700	20.786	182.071	170.138	8.353	492.066	414.934	-30.963
800	20.786	184.846	171.807	10.431	493.157	403.841	-26.368
900	20.786	187.294	173.394	12.510	494.241	392.611	-22.787
1000	20.786	189.484	174.896	14.589	423.818	383.448	-20.029
1100	20.786	191.465	176.313	16.667	425.896	379.310	-18.012
1200	20.786	193.274	177.653	18.746	427.975	374.984	-16.323
1300	20.786	194.938	178.919	20.824	430.053	370.483	-14.886
1400	20.786	196.478	180.119	22.903	432.131	365.823	-13.649
1500	20.786	197.912	181.258	24.982	434.209	361.014	-12.572
1600	20.786	199.254	182.341	27.060	436.285	356.067	-11.624
1700	20.786	200.514	183.374	29.139	438.358	350.990	-10.785
1800	20.786	201.702	184.359	31.217	440.429	345.791	-10.035
1900	20.786	202.826	185.302	33.296	442.495	340.477	-9.360
2000	20.786	203.892	186.205	35.375	444.555	335.054	-8.751
2100	20.786	204.906	187.071	37.453	446.606	329.528	-8.197
2200	20.786	205.873	187.904	39.532	448.646	323.906	-7.690
2300	20.786	206.797	188.706	41.610	450.673	318.190	-7.226
2400	20.786	207.682	189.478	43.689	452.684	312.387	-6.799
2500	20.786	208.530	190.223	45.768	454.675	306.500	-6.404
2600	20.786	209.346	190.943	47.846	456.643	300.534	-6.038
2700	20.786	210.130	191.639	49.925	458.586	294.493	-5.697
2800	20.786	210.886	192.313	52.004	460.498	288.380	-5.380
2900	20.786	211.615	192.966	54.082	462.376	282.200	-5.083
3000	20.786	212.320	193.600	56.161	464.215	275.955	-4.805
3100	20.786	213.002	194.215	58.239	466.011	269.650	-4.544
3200	20.786	213.662	194.812	60.318	467.759	263.288	-4.298
3300	20.786	214.301	195.393	62.397	469.469	256.871	-4.066
3400	20.786	214.922	195.959	64.475	471.113	250.404	-3.847
3500	20.786	215.524	196.509	66.554	472.694	243.889	-3.640
3600	20.786	216.110	197.045	68.632	474.215	237.330	-3.444
3700	20.786	216.679	197.568	70.711	475.658	230.729	-3.257
3800	20.786	217.234	198.079	72.790	477.024	224.091	-3.080
3900	20.786	217.774	198.577	74.868	478.307	217.418	-2.912
4000	20.786	218.300	199.063	76.947	479.515	210.712	-2.752
4100	20.786	218.813	199.539	79.025	480.619	203.978	-2.599
4200	20.786	219.314	200.004	81.104	481.623	197.218	-2.453
4300	20.786	219.803	200.458	83.183	482.524	190.436	-2.313
4400	20.786	220.281	200.904	85.261	483.613	183.603	-2.180
4500	20.786	220.748	201.339	87.340	484.491	176.763	-2.052
4600	20.786	221.205	201.766	89.418	485.186	169.914	-1.929
4700	20.786	221.652	202.185	91.497	485.759	163.054	-1.812
4800	20.786	222.090	202.595	93.576	486.227	156.183	-1.700
4900	20.786	222.518	202.997	95.654	486.589	149.303	-1.592
5000	20.786	222.938	203.392	97.733	487.097	142.386	-1.487
5100	20.786	223.350	203.779	99.811	487.289	135.490	-1.388
5200	20.786	223.753	204.159	101.890	487.379	128.591	-1.292
5300	20.786	224.149	204.533	103.969	487.370	121.691	-1.199
5400	20.786	224.538	204.900	106.047	487.262	114.792	-1.110
5500	20.786	224.919	205.260	108.126	487.058	107.896	-1.025
5600	20.786	225.294	205.614	110.204	487.144	100.950	-0.942
5700	20.786	225.662	205.963	112.283	486.804	94.057	-0.862
5800	20.786	226.023	206.306	114.362	486.381	87.170	-0.785
5900	20.786	226.379	206.643	116.440	485.879	80.291	-0.711
6000	20.786	226.728	206.975	118.519	485.302	73.422	-0.639

PREVIOUS: CURRENT: December 1983 (1 bar)

Rubidium, Ion (Rb^+) $Rb_1^+(g)$

RUBIDIUM, ION (Rb^-) IDEAL GAS M_r = 85.46835

EA(Rb, g) = 0.48592 ± 0.00002 eV

S°(298.15 K) = 164.330 ± 0.005 J K^{-1} mol^{-1}

$\Delta_f H°$(0 K) = 35.380 kJ mol^{-1}

$\Delta_f H°$(298.15 K) = [27.819] kJ mol^{-1}

Electronic Level and Quantum Weight

State	ε_i, cm^{-1}	g_i
$^1S_0$	0.0	1

Enthalpy of Formation

The enthalpy of formation, at 0 K, for Rb^-(g) is calculated from the adopted electron affinity, EA(Rb, g) = 0.48592 ± 0.00002 eV (46.884±0.002 kJ mol^{-1}). This value, recommended by Hotop and Lineberger (6), was measured by a tunable laser photodetachment threshold technique (2). Additional discussion on Rb^-(g) may be obtained in the critical discussions of Hotop and Lineberger (1, 6), Rosenstock et al. (3), and Massey (4).

$\Delta_f H°$(Rb^-, g, 298.15 K) is obtained from $\Delta_f H°$(Rb, g, 0 K) by using EA(Rb, g) with JANAF (5) enthalpies, H°(0 K)-H°(298.15 K), for Rb^-(g), Rb(g), and e^-(g). $\Delta_r H°$(Rb^- + Rb + e^-, 298.15 K) differs from a room-temperature threshold energy due to inclusion of these enthalpies and to threshold effects discussed by Rosenstock et al. (3). $\Delta_f H°$(298.15 K) should be changed by +6.197 kJ mol^{-1} if it is to be used in the ion convention which excludes the enthalpy of the electron.

Heat Capacity and Entropy

The ground state configuration for Rb^-(g) is given by Hotop and Lineberger (1, 6) and Rosenstock et al. (3). Lacking any experimental evidence as to the stability of any excited states, we assume that no stable excited states exist.

References

1. H. Hotop and W. C. Lineberger, J. Phys. Chem. Ref. Data 4, 539 (1975).
2. P. Frey, F. Breyer, and H. Hotop, J. Phys. B: Atom. Molec. Phys. 11, 1-589 (1978).
3. H. M. Rosenstock, K. Draxl et al., J. Phys. Chem. Ref. Data 6, Supp. 1 (1977).
4. H. S. W. Massey, "Negative Ions," 3rd ed., Cambridge University Press, Cambridge, 1976.
5. JANAF Thermochemical Tables: Rb(g), 3-31-82; e^-(g), 3-31-82.
6. H. Hotop and W. C. Lineberger, J. Phys. Chem. Ref. Data, to be published, 1985.

Rubidium, Ion (Rb^-) Rb_1^-(g)

Enthalpy Reference Temperature = T_r = 298.15 K Standard State Pressure = p° = 0.1 MPa

T/K	$C_p°$ (J $K^{-1}mol^{-1}$)	S°	$-[G°-H°(T_r)]/T$	$H°-H°(T_r)$ (kJ mol^{-1})	$\Delta_f H°$	$\Delta_f G°$	Log K_f
0	0.	0.	INFINITE	-6.197	35.308		
100	20.786	141.623	182.810	-4.119			
200	20.786	156.031	166.231	-2.040			
250	20.786	160.669	164.672	-1.001			
298.15	20.786	164.330	164.330	0.	27.819	7.970	-1.396
300	20.786	164.459	164.331	0.038	27.761	7.847	-1.366
350	20.786	167.663	164.584	1.078	23.964	4.917	-0.734
400	20.786	170.438	165.146	2.117	22.370	2.305	-0.301
450	20.786	172.887	165.873	3.156	20.791	-0.109	0.013
500	20.786	175.077	166.685	4.196	19.226	-2.347	0.245
600	20.786	178.866	168.409	6.274	16.125	-6.369	0.555
700	20.786	182.071	170.138	8.353	13.050	-9.875	0.737
800	20.786	184.846	171.807	10.431	9.984	-12.941	0.845
900	20.786	187.295	173.394	12.510	6.910	-15.622	0.907
1000	20.786	189.485	174.896	14.589	-67.670	-15.774	0.824
1100	20.786	191.466	176.314	16.667	-69.748	-10.483	0.498
1200	20.786	193.274	177.653	18.746	-71.827	-5.004	0.218
1300	20.786	194.938	178.919	20.824	-73.906	0.649	-0.026
1400	20.786	196.478	180.119	22.903	-75.985	6.462	-0.241
1500	20.786	197.913	181.258	24.982	-78.065	12.424	-0.433
1600	20.786	199.254	182.341	27.060	-80.146	18.524	-0.605
1700	20.786	200.514	183.374	29.139	-82.230	24.755	-0.761
1800	20.786	201.702	184.359	31.217	-84.316	31.108	-0.903
1900	20.786	202.826	185.302	33.296	-86.407	37.577	-1.033
2000	20.786	203.892	186.205	35.375	-88.505	44.157	-1.153
2100	20.786	204.906	187.072	37.453	-90.611	50.842	-1.265
2200	20.786	205.873	187.904	39.532	-92.728	57.628	-1.368
2300	20.786	206.797	188.706	41.610	-94.858	64.510	-1.465
2400	20.786	207.682	189.478	43.689	-97.005	71.484	-1.556
2500	20.786	208.531	190.223	45.768	-99.171	78.549	-1.641
2600	20.786	209.346	190.943	47.846	-101.360	85.701	-1.722
2700	20.786	210.130	191.640	49.925	-103.574	92.938	-1.798
2800	20.786	210.886	192.314	52.004	-105.819	100.257	-1.870
2900	20.786	211.616	192.967	54.082	-108.098	107.657	-1.939
3000	20.786	212.320	193.600	56.161	-110.416	115.136	-2.005
3100	20.786	213.002	194.215	58.239	-112.777	122.693	-2.067
3200	20.786	213.662	194.812	60.318	-115.187	130.327	-2.127
3300	20.786	214.301	195.393	62.397	-117.634	138.037	-2.185
3400	20.786	214.922	195.959	64.475	-120.147	145.822	-2.240
3500	20.786	215.524	196.509	66.554	-122.724	153.682	-2.294
3600	20.786	216.110	197.046	68.632	-125.360	161.616	-2.345
3700	20.786	216.680	197.569	70.711	-128.074	169.624	-2.395
3800	20.786	217.234	198.079	72.790	-130.865	177.707	-2.443
3900	20.786	217.774	198.577	74.868	-133.740	185.865	-2.489
4000	20.786	218.300	199.063	76.947	-136.688	194.096	-2.535
4100	20.786	218.813	199.539	79.025	-139.742	202.403	-2.579
4200	20.786	219.314	200.004	81.104	-142.895	210.786	-2.622
4300	20.786	219.803	200.459	83.183	-146.151	219.246	-2.663
4400	20.786	220.281	200.904	85.261	-149.219	227.751	-2.704
4500	20.786	220.748	201.339	87.340	-152.499	236.343	-2.743
4600	20.786	221.205	201.766	89.418	-155.961	245.020	-2.782
4700	20.786	221.652	202.185	91.497	-159.545	253.775	-2.820
4800	20.786	222.090	202.595	93.576	-163.235	262.608	-2.858
4900	20.786	222.518	202.997	95.654	-167.029	271.519	-2.894
5000	20.786	222.938	203.392	97.733	-170.679	280.478	-2.930
5100	20.786	223.350	203.779	99.811	-174.644	289.540	-2.965
5200	20.786	223.754	204.159	101.890	-178.711	298.681	-3.000
5300	20.786	224.150	204.533	103.969	-182.878	307.901	-3.035
5400	20.786	224.538	204.900	106.047	-187.143	317.201	-3.068
5500	20.786	224.919	205.260	108.126	-191.503	326.581	-3.102
5600	20.786	225.294	205.615	110.204	-195.575	335.986	-3.134
5700	20.786	225.662	205.963	112.283	-200.072	345.518	-3.166
5800	20.786	226.023	206.306	114.362	-204.653	355.130	-3.198
5900	20.786	226.379	206.643	116.440	-209.312	364.821	-3.230
6000	20.786	226.728	206.975	118.519	-214.045	374.592	-3.261

PREVIOUS: CURRENT: December 1983 (1 bar)

Rubidium, Ion (Rb^-) Rb_1^-(g)

RUBIDIUM (Rb_2) IDEAL GAS M_r = 169.8234

D_0^0 = 46.91 ± 1.9 kJ mol^{-1} (natural abundance) $\Delta_f H°(0\ K)$ = 117.34 ± 2.5 kJ mol^{-1}

S°(298.15 K) = 271.07 ± 0.08 J K^{-1} mol^{-1} $\Delta_f H°(298.15\ K)$ = 113.29 ± 2.5 kJ mol^{-1}

Spectroscopic Data for $^{85}Rb_2$ (Ground State: $^1\Sigma_g^+$) in cm^{-1}

ω_e = 57.747 $\omega_e x_e$ = 0.1582

B_e = 0.02278 α_e = 4.7 x 10^{-5}

D_e = 1.5 x 10^{-8} r_e = 4.173 Å

Heat of Formation

Using the adopted value of the dissociation energy for Rb_2(g), D_0^0 = 46.91 kJ mol^{-1} (natural abundance Rb_2) as deduced from the ^{85}Rb dissociation energy from reference (4) (D° = 47.277 ± 1.93 kJ mol^{-1}) and the recommended $\Delta_f H°$(Rb, g, 0 K) = 82.17 kJ mol^{-1} from NBS (5), we calculate $\Delta_f H°(Rb_2$, g, 0 K) = 117.34 kJ mol^{-1}. Using the NBS enthalpy value for Rb(cr) (H°(298.15 K)-H°(0 K) = 7.489 kJ mol^{-1}) and the current JANAF value for H°(298.15 K)-H°(0 K) of Rb_2(g), we obtain $\Delta_f H°(Rb_2$, g, 298.15 K) = 113.29 kJ mol^{-1}. Note that in a mass spectrometric study by Piacente et al. (6), the dissociation energy of Rb_2(g) was determined as 41.84 ± 2.1 kJ mol^{-1}. This value appears to be too low when compared to the alkali metal group trends apparent in plots of D_0^0 versus r_e or D_0^0 versus group member.

Heat Capacity and Entropy

The heat capacity and entropy were calculated using a direct summation technique similar to the Li_2(g) species (7) using the data for the $^{85}Rb_2$ species given above (1,2). Since insufficient spectroscopic data exist for the lower lying excited electronic states, we performed the calculation on the ground state only as discussed in the K_2(g) table (7). Separate calculations were performed on the isotopic species $^{85}Rb_2$, $^{87}Rb_2$, and $^{85}Rb^{87}Rb$ and the results combined according to the natural atomic isotopic abundances (3): ^{85}Rb, 72.165%; ^{87}Rb, 27.835%. Spectroscopic input data for the calculation of the isotopic species were calculated from the $^{85}Rb_2$(g) data using reduced mass scaling. The rotational levels were weighted according to the nuclear spins of the atoms (^{85}Rb, I=5/2; ^{87}Rb, I=3/2) as follows: $^{85}Rb_2$, even J weight = 0.583, odd J weight = 0.471; $^{87}Rb_2$, even J weight = 0.625, odd J weight = 0.375; $^{85}Rb^{87}Rb$, equal weights of 0.5.

Based on the results for the Cs_2(g) calculation (7), neglect of the excited states probably results in negligible uncertainties in the thermodynamic functions (except for Cp) at temperatures below ca. 2000 K. We estimate that the entropy and Gibbs energy function at 3000 K are too low by ca. 0.4 J K^{-1} mol^{-1} due to the neglect of the excited states. The adopted value of S°(298.15 K) is 0.07 J K^{-1} mol^{-1} larger than that adopted by Gurvich, et al. (8).

References

1. W. C. Stwalley, Director: Iowa Laser Facility and Professor of Chemistry and Physics, University of Iowa, personal communcation (1982).
2. C. D. Caldwell, F. Engelke and H. Hage, Chem. Phys. 54, 21 (1980).
3. Handbook of Chemistry and Physics, R. C. Weast, Editor, CRC Press, 57th edition, p. B-288 (1977).
4. Ny Tsi-ze and T. S. Tsiang, Phys. Rev. 52, 91 (1937).
5. D. D. Wagman, et al., J. Phys. Chem. Ref. Data 11, Supp. 2 (1982).
6. V. Piacente, G. Bardi and L. Malaspina, J. Chem. Thermodynamics 5, 219 (1973).
7. JANAF Thermochemical Tables: Li_2(g), 12-31-82; K_2, 12-31-82; Cs_2, 12-31-82.
8. L. V. Gurvich, et al., "Thermodynamic Properties of Individual Substances," Volume IV, 3rd ed., Nauka, Moscow, 1982.

Rubidium (Rb_2) Rb_2(g)

Enthalpy Reference Temperature = T_r = 298.15 K Standard State Pressure = p° = 0.1 MPa

T/K	$C_p°$	S°	-[G°-H°(T_r)]/T	H°-H°(T_r)	$\Delta_f H°$	$\Delta_f G°$	Log K_f
	J K^{-1}mol^{-1}			kJ mol^{-1}			
0	0.	0.	INFINITE	-10.934	117.335	117.335	INFINITE
100	37.185	229.924	304.707	-7.478	116.769	103.181	-53.896
200	37.777	255.918	274.542	-3.725	115.222	90.165	-23.549
298.15	38.116	271.067	271.067	0.	113.290	78.254	-13.710
300	38.123	271.302	271.067	0.071	113.245	78.037	-13.587
400	38.440	282.314	272.567	3.899	106.291	67.916	-8.869
500	38.631	290.917	275.407	7.754	103.858	58.607	-6.123
600	38.475	297.953	278.597	11.614	101.515	49.777	-4.333
700	37.845	303.842	281.795	15.433	99.186	41.338	-3.085
800	36.797	308.831	284.871	19.168	96.789	33.236	-2.170
900	35.490	313.092	287.776	22.784	94.257	25.443	-1.477
1000	34.084	316.758	290.495	26.263	-51.424	22.313	-1.166
1100	32.696	319.941	293.031	29.602	-52.243	29.725	-1.412
1200	31.398	322.730	295.392	32.805	-53.196	37.218	-1.620
1300	30.221	325.196	297.592	35.885	-54.274	44.795	-1.800
1400	29.176	327.396	299.643	38.854	-55.464	52.460	-1.957
1500	28.258	329.377	301.561	41.725	-56.753	60.213	-2.097
1600	27.457	331.175	303.356	44.510	-58.131	68.055	-2.222
1700	26.760	332.818	305.042	47.220	-59.587	75.986	-2.335
1800	26.154	334.330	306.627	49.865	-61.115	84.005	-2.438
1900	25.626	335.729	308.123	52.453	-62.710	92.110	-2.532
2000	25.165	337.032	309.536	54.992	-64.366	100.301	-2.620
2100	24.762	338.250	310.874	57.488	-66.082	108.576	-2.701
2200	24.408	339.393	312.145	59.946	-67.858	116.935	-2.776
2300	24.096	340.471	313.354	62.371	-69.693	125.375	-2.847
2400	23.821	341.491	314.505	64.766	-71.591	133.897	-2.914
2500	23.577	342.458	315.604	67.136	-73.553	142.499	-2.977
2600	23.360	343.379	316.654	69.483	-75.584	151.181	-3.037
2700	23.166	344.257	317.661	71.809	-77.686	159.942	-3.094
2800	22.992	345.096	318.625	74.117	-79.869	168.783	-3.149
2900	22.836	345.900	319.552	76.408	-82.137	177.703	-3.201
3000	22.695	346.672	320.443	78.684	-84.496	186.703	-3.251
3100	22.568	347.414	321.302	80.947	-86.955	195.783	-3.299
3200	22.453	348.128	322.129	83.198	-89.524	204.945	-3.345
3300	22.348	348.818	322.927	85.438	-92.177	214.186	-3.390
3400	22.253	349.483	323.699	87.668	-94.974	223.511	-3.434
3500	22.166	350.127	324.445	89.889	-97.906	232.920	-3.476
3600	22.086	350.750	325.167	92.102	-100.966	242.415	-3.517
3700	22.013	351.354	325.866	94.307	-104.188	251.997	-3.558
3800	21.945	351.941	326.545	96.505	-107.573	261.669	-3.597
3900	21.884	352.510	327.203	98.696	-111.132	271.431	-3.635
4000	21.826	353.063	327.843	100.881	-114.843	281.285	-3.673
4100	21.773	353.601	328.465	103.061	-118.771	291.237	-3.710
4200	21.724	354.126	329.069	105.236	-122.901	301.287	-3.747
4300	21.679	354.636	329.658	107.406	-127.244	311.437	-3.783
4400	21.636	355.134	330.231	109.572	-131.214	321.630	-3.818
4500	21.597	355.620	330.790	111.734	-135.612	331.947	-3.853
4600	21.560	356.094	331.335	113.892	-140.378	342.385	-3.888
4700	21.525	356.558	331.867	116.046	-145.393	352.933	-3.922
4800	21.493	357.010	332.386	118.197	-150.620	363.590	-3.957
4900	21.463	357.453	332.893	120.344	-156.062	374.359	-3.991
5000	21.435	357.887	333.389	122.489	-161.216	385.179	-4.024
5100	21.408	358.311	333.873	124.631	-167.005	396.164	-4.058
5200	21.383	358.726	334.347	126.771	-172.999	407.266	-4.091
5300	21.360	359.133	334.811	128.908	-179.195	418.483	-4.124
5400	21.338	359.532	335.265	131.043	-185.590	429.820	-4.158
5500	21.317	359.924	335.710	133.176	-192.179	441.277	-4.191
5600	21.297	360.308	336.146	135.306	-198.192	452.746	-4.223
5700	21.278	360.684	336.573	137.435	-205.058	464.431	-4.256
5800	21.261	361.054	336.992	139.562	-212.091	476.236	-4.289
5900	21.244	361.418	337.403	141.687	-219.284	488.166	-4.322
6000	21.228	361.775	337.806	143.811	-226.628	500.217	-4.355

PREVIOUS: CURRENT: December 1983 (1 bar)

Rubidium (Rb_2) Rb_2(g)

RADON (Rn) REFERENCE STATE A_r = 222

0 to 6000 K Ideal Monatomic Gas

IP(Rn, g) = 86692.5 ± 0.2 cm^{-1}

S°(298.15 K) = 176.235 ± 0.003 J K^{-1} mol^{-1}

$\Delta_f H°(0\ K)$ = 0 kJ mol^{-1}

$\Delta_f H°(298.15\ K)$ = 0 kJ mol^{-1}

Electronic Levels and Quantum Weights

State	ϵ_i, cm^{-1}	g_i
$^1S_0$	0	1

Enthalpy of Formation

Zero by definition.

Heat Capacity and Entropy

Information on the electronic energy levels and quantum weights is taken from Moore (1, 9). Many of the theoretically predicted levels have not been observed. Our calculations indicate that any reasonable method of filling in these missing levels and cutting off the summation in the partition function (2) has little effect on the thermodynamic properties to 6000 K. This is undoubtedly a result of the high energy of these levels; the first excited level is over 54620 cm^{-1} above the ground state. Therefore, we list the ground state only. Extension to higher temperatures may require consideration of excited states and utilization of different fill and cutoff procedures (2).

Thermodynamic functions at 298.15 K are recommended by CODATA (3) for all rare gases except radon. This table agrees with other tabulations except for two minor changes. First, the entropy differs by 0.1094 J K^{-1} mol^{-1} because this table uses a reference pressure of 1 bar, whereas the other tabulations are based on 1 atm. Second, entropy differences of the order of 0.001-0.004 J K^{-1} mol^{-1} for the rare gases arise due to the use of slightly different values for R and the relative atomic mass; this table uses R = 8.31441 J K^{-1} mol^{-1}. With some minor changes, our tables agree within the estimated uncertainty with those by Hilsenrath et al. (5), Gurvich et al. (8), and Wagman et al. (10). The estimated uncertainty is due to uncertainties in the relative atomic mass and fundamental constants which in this tabulation are based on the 1981 scale (6) and the 1973 values (7), respectively.

Phase Data

Hultgren et al. (4) had recommended a melting point of 202 K and a boiling point of 211 K (1 atm). These values are provided for the convenience of the reader and have not been evaluated by the present authors. As a result of the low values, the reference state for radon is chosen to be the ideal gas at all temperatures. This may differ from the choice of other authors.

References

1. C. E. Moore, U. S. Nat. Bur. Stand., NSRDS-NBS-35, Volume III, 1970 [Reprint of NBS Circular 467, Volume III, 1958].
2. J. R. Downey, Jr., The Dow Chemical Co., AFOSR-TR-78-0960, Contract No. F44620-75-1-0048, 1978.
3. J. D. Cox, chairman, ICSU-CODATA Task Group on Key Values for Thermodynamics, J. Chem. Thermodynamics 10, 903 (1978).
4. R. Hultgren, P. D. Desai et al. "Selected Values of the Thermodynamic Properties of the Elements", American Society for Metals, Metals Park, Ohio, 1973.
5. J. Hilsenrath, C. G Messina and W. H. Evans, U.S. Nat. Bur. Stand., Report AD-606163, 1964.
6. N. E. Holden and R. L. Martin, Pure Appl. Chem. 51, 405 (1979).
7. E. R. Cohen and B. N. Taylor, J. Phys. Chem. Ref. Data 2, 663 (1973).
8. L. V. Gurvich, I. V. Veits et al., "Thermodynamic Properties of Individual Substances," 3rd ed., Vol. I, Nauka, Moscow, 1978.
9. C. E. Moore, U. S. Nat. Bur. Stand., NSRDS-NBS-34, 1970.
10. D. D. Wagman, W. H. Evans et al., J. Phys. Chem. Ref. Data 11, Supp. 2 (1982).

Radon (Rn) Rn_1(ref)

Enthalpy Reference Temperature = T_r = 298.15 K Standard State Pressure = p° = 0.1 MPa

T/K	$C_p°$ (J $K^{-1}mol^{-1}$)	S° (J $K^{-1}mol^{-1}$)	$-[G°-H°(T_r)]/T$ (J $K^{-1}mol^{-1}$)	$H°-H°(T_r)$ (kJ mol^{-1})	$\Delta_f H°$ (kJ mol^{-1})	$\Delta_f G°$ (kJ mol^{-1})	Log K_f
0	0.	0.	INFINITE	-6.197	0.	0.	0.
100	20.786	153.527	194.715	-4.119	0.	0.	0.
200	20.786	167.935	178.136	-2.040	0.	0.	0.
250	20.786	172.574	176.577	-1.001	0.	0.	0.
298.15	20.786	176.235	176.235	0.	0.	0.	0.
300	20.786	176.363	176.235	0.038	0.	0.	0.
350	20.786	179.567	176.488	1.078	0.	0.	0.
400	20.786	182.343	177.050	2.117	0.	0.	0.
450	20.786	184.791	177.777	3.156	0.	0.	0.
500	20.786	186.981	178.590	4.196	0.	0.	0.
600	20.786	190.771	180.314	6.274	0.	0.	0.
700	20.786	193.975	182.043	8.353	0.	0.	0.
800	20.786	196.751	183.711	10.431	0.	0.	0.
900	20.786	199.199	185.299	12.510	0.	0.	0.
1000	20.786	201.389	186.800	14.589	0.	0.	0.
1100	20.786	203.370	188.218	16.667	0.	0.	0.
1200	20.786	205.179	189.557	18.746	0.	0.	0.
1300	20.786	206.843	190.824	20.824	0.	0.	0.
1400	20.786	208.383	192.024	22.903	0.	0.	0.
1500	20.786	209.817	193.163	24.982	0.	0.	0.
1600	20.786	211.159	194.246	27.060	0.	0.	0.
1700	20.786	212.419	195.278	29.139	0.	0.	0.
1800	20.786	213.607	196.264	31.217	0.	0.	0.
1900	20.786	214.731	197.206	33.296	0.	0.	0.
2000	20.786	215.797	198.110	35.375	0.	0.	0.
2100	20.786	216.811	198.976	37.453	0.	0.	0.
2200	20.786	217.778	199.809	39.532	0.	0.	0.
2300	20.786	218.702	200.610	41.610	0.	0.	0.
2400	20.786	219.587	201.383	43.689	0.	0.	0.
2500	20.786	220.435	202.128	45.768	0.	0.	0.
2600	20.786	221.250	202.848	47.846	0.	0.	0.
2700	20.786	222.035	203.544	49.925	0.	0.	0.
2800	20.786	222.791	204.218	52.004	0.	0.	0.
2900	20.786	223.520	204.871	54.082	0.	0.	0.
3000	20.786	224.225	205.505	56.161	0.	0.	0.
3100	20.786	224.906	206.120	58.239	0.	0.	0.
3200	20.786	225.566	206.717	60.318	0.	0.	0.
3300	20.786	226.206	207.298	62.397	0.	0.	0.
3400	20.786	226.826	207.863	64.475	0.	0.	0.
3500	20.786	227.429	208.414	66.554	0.	0.	0.
3600	20.786	228.015	208.950	68.632	0.	0.	0.
3700	20.786	228.584	209.473	70.711	0.	0.	0.
3800	20.786	229.138	209.983	72.790	0.	0.	0.
3900	20.786	229.678	210.481	74.868	0.	0.	0.
4000	20.786	230.205	210.968	76.947	0.	0.	0.
4100	20.786	230.718	211.443	79.025	0.	0.	0.
4200	20.786	231.219	211.908	81.104	0.	0.	0.
4300	20.786	231.708	212.363	83.183	0.	0.	0.
4400	20.786	232.186	212.808	85.261	0.	0.	0.
4500	20.786	232.653	213.244	87.340	0.	0.	0.
4600	20.786	233.110	213.671	89.418	0.	0.	0.
4700	20.786	233.557	214.089	91.497	0.	0.	0.
4800	20.786	233.994	214.499	93.576	0.	0.	0.
4900	20.786	234.423	214.902	95.654	0.	0.	0.
5000	20.786	234.843	215.296	97.733	0.	0.	0.
5100	20.786	235.255	215.684	99.811	0.	0.	0.
5200	20.786	235.658	216.064	101.890	0.	0.	0.
5300	20.786	236.054	216.437	103.969	0.	0.	0.
5400	20.786	236.443	216.804	106.047	0.	0.	0.
5500	20.786	236.824	217.165	108.126	0.	0.	0.
5600	20.786	237.199	217.519	110.204	0.	0.	0.
5700	20.786	237.566	217.868	112.283	0.	0.	0.
5800	20.786	237.928	218.210	114.362	0.	0.	0.
5900	20.786	238.283	218.548	116.440	0.	0.	0.
6000	20.786	238.633	218.879	118.519	0.	0.	0.

PREVIOUS: CURRENT: March 1982 (1 bar)

Radon (Rn) Rn_1(ref)

RADON, ION (Rn^+) IDEAL GAS M_r = 221.99945

$S°(298.15\ K) = 187.761 \pm 0.003\ J\ K^{-1}\ mol^{-1}$

$\Delta_f H°(0\ K) = 1037.073 \pm 0.002\ kJ\ mol^{-1}$
$\Delta_f H°(298.15\ K) = [1043.270]\ kJ\ mol^{-1}$

Electronic Levels and Quantum Weights

State	ϵ_i, cm^{-1}	g_i
$^2P_{3/2}$	0	4
$^2P_{1/2}$	30895.1	2

Enthalpy of Formation

The ionization limit of neutral radon (86692.5±0.2 cm^{-1}) reported by Moore (1) is adopted as $\Delta_f H°(0\ K)$ for $Rn^+(g)$. The ionization limit is converted from cm^{-1} to kJ mol^{-1} using the factor, 1 cm^{-1} = 0.01196266 kJ mol^{-1}, which is derived from the 1973 CODATA fundamental constants (2). The uncertainty in the ionization limit is estimated to be ±0.2 cm^{-1} which corresponds to an uncertainty of ±0.002 kJ mol^{-1} in the enthalpy of formation. Gurvich et al. (8) adopted the same ionization potential, but the use of slightly different value by Wagman et al. (10) results in a enthalpy of formation difference of 0.010 kJ mol^{-1}.

$\Delta_f H°(Rn^+, g, 298.15\ K)$ is obtained from $\Delta_f H°(Rn, g, 0\ K)$ by using IP(Rn) with JANAF (3) enthalpies H°(0 K)-H°(298.15 K) for $Rn^+(g)$, Rn(r), and $e^-(r)$. $\Delta_r H°(Rn \rightarrow Rn^+ + e^-, 298.15\ K)$ differs from a room temperature threshold energy due to inclusion of these enthalpies and to threshold effects discussed by Rosenstock et al. (9). $\Delta_f H°(298.15\ K)$ should be changed by -6.197 kJ mol^{-1} if it is to be used in the ion convention that excludes the enthalpy of the electron.

Heat Capacity and Entropy

The information on electronic energy levels and quantum weights of Moore (4) is incomplete because no levels have been observed. Our calculations indicate that any reasonable method of filling in missing levels and cutting off the summation in the partition function (5) has no effect on the thermodynamic functions to 6000 K. This is a result of the high energy of all levels other than the ground state and the $^2P_{1/2}$ level; the next lowest level is estimated to be over 80000 cm^{-1} above the ground state. Since inclusion of these upper levels has no effect on the thermodynamic functions (to 6000 K) we list only the ground state and the $^2P_{1/2}$ state, with the energy of the latter state taken from a more recent study by Moore (1). The reported uncertainty in S°(298.15 K) is due to uncertainties in the relative ionic mass and fundamental constants. Extension of these calculations above 6000 K may require consideration of the higher excited states and use of different fill and cutoff procedures (5).

The thermodynamic functions reported here agree with those of Green et al. (6), Hilsenrath et al. (7), and Gurvich et al. (8) except for two minor changes. First, the entropy differs by 0.1094 J K^{-1} mol^{-1} because this table uses a standard state pressure of 1 bar, whereas the cited references used a pressure of 1 atm. Second, smaller differences arise from the use of different values for the fundamental constants, the relative ionic mass, and the position of the $^2P_{1/2}$ electronic level.

References

1. C. E. Moore, U. S. Nat. Bur. Stand., NSRDS-NBS-34, 1970.
2. E. R. Cohen and B. N. Taylor, J. Phys. Chem. Ref. Data 2, 663 (1973).
3. JANAF Thermochemical Tables: Rn(r), 3-31-82; e^-(r), 3-31-82.
4. C. E. Moore, U. S. Nat. Bur. Stand., NSRDS-NBS-35, Volume III, 1970 [Reprint of NBS Circular 467, Volume III, 1958].
5. J. R. Downey, Jr., The Dow Chemical Co., AFOSR-TR-78-0960, Contract No. F44620-75-1-0048, 1978.
6. J. W. Green, D. E. Poland and J. L. Margrave, University of Wisconsin, Report AD-275542, 1961.
7. J. Hilsenrath, C. G. Messina, and W. H. Evans, U.S. Nat. Bur. Stand., Report AD-606163, 1964.
8. L. V. Gurvich, I. V. Veits et al., "Thermodynamic Properties of Individual Substances", 3rd ed., Vol. 1, Nauka, Moscow, 1978.
9. H. M. Rosenstock, K. Draxl et al., J. Phys. Chem. Ref. Data 6, Supp. 1 (1977).
10. D. D. Wagman, W. H. Evans et al., J. Phys. Chem. Ref. Data 11, Supp. 2 (1982).

Radon, Ion (Rn^+) $Rn_1^+(g)$

Enthalpy Reference Temperature = T_r = 298.15 K; Standard State Pressure = p° = 0.1 MPa

T/K	$C_p°$ (J $K^{-1}mol^{-1}$)	S°	-[G°-H°(T_r)]/T	H°-H°(T_r) (kJ mol^{-1})	$\Delta_f H°$	$\Delta_f G°$	Log K_f
0	0.	0.	INFINITE	-6.197	1037.073		
100	20.786	165.054	206.241	-4.119			
200	20.786	179.461	189.662	-2.040			
250	20.786	184.100	188.103	-1.001			
298.15	20.786	187.761	187.761	0.	1043.270	1033.579	-181.079
300	20.786	187.889	187.761	0.038	1043.309	1033.519	-179.952
350	20.786	191.094	188.014	1.078	1044.348	1031.805	-153.988
400	20.786	193.869	188.577	2.117	1045.387	1029.942	-134.497
450	20.786	196.317	189.303	3.156	1046.427	1027.949	-119.321
500	20.786	198.507	190.116	4.196	1047.466	1025.840	-107.169
600	20.786	202.297	191.840	6.274	1049.545	1021.320	-88.914
700	20.786	205.501	193.569	8.353	1051.623	1016.451	-75.849
800	20.786	208.277	195.238	10.431	1053.702	1011.285	-66.030
900	20.786	210.725	196.825	12.510	1055.780	1005.858	-58.378
1000	20.786	212.915	198.327	14.589	1057.859	1000.200	-52.245
1100	20.786	214.896	199.744	16.667	1059.938	994.333	-47.217
1200	20.786	216.705	201.083	18.746	1062.016	988.277	-43.019
1300	20.786	218.369	202.350	20.824	1064.095	982.048	-39.459
1400	20.786	219.909	203.550	22.903	1066.173	975.659	-36.402
1500	20.786	221.343	204.689	24.982	1068.252	969.121	-33.748
1600	20.786	222.685	205.772	27.060	1070.331	962.445	-31.421
1700	20.786	223.945	206.804	29.139	1072.409	955.638	-29.363
1800	20.786	225.133	207.790	31.217	1074.488	948.709	-27.531
1900	20.786	226.257	208.733	33.296	1076.566	941.665	-25.888
2000	20.786	227.323	209.636	35.375	1078.645	934.511	-24.407
2100	20.786	228.337	210.502	37.453	1080.724	927.253	-23.064
2200	20.786	229.304	211.335	39.532	1082.802	919.896	-21.841
2300	20.786	230.228	212.137	41.610	1084.881	912.445	-20.722
2400	20.786	231.113	212.909	43.689	1086.959	904.903	-19.695
2500	20.786	231.961	213.654	45.768	1089.038	897.275	-18.748
2600	20.786	232.777	214.374	47.846	1091.117	889.563	-17.872
2700	20.786	233.561	215.070	49.925	1093.195	881.772	-17.059
2800	20.786	234.317	215.744	52.004	1095.274	873.903	-16.303
2900	20.786	235.046	216.397	54.082	1097.352	865.961	-15.598
3000	20.786	235.751	217.031	56.161	1099.431	857.946	-14.938
3100	20.787	236.433	217.646	58.239	1101.510	849.862	-14.320
3200	20.787	237.093	218.243	60.318	1103.588	841.711	-13.740
3300	20.787	237.732	218.824	62.397	1105.667	833.496	-13.193
3400	20.788	238.353	219.389	64.476	1107.746	825.217	-12.678
3500	20.788	238.955	219.940	66.554	1109.825	816.877	-12.191
3600	20.789	239.541	220.476	68.633	1111.903	808.478	-11.731
3700	20.790	240.111	220.999	70.712	1113.982	800.020	-11.294
3800	20.791	240.665	221.510	72.791	1116.061	791.507	-10.880
3900	20.792	241.205	222.008	74.870	1118.140	782.939	-10.486
4000	20.794	241.732	222.494	76.949	1120.220	774.318	-10.112
4100	20.796	242.245	222.970	79.029	1122.299	765.644	-9.754
4200	20.798	242.746	223.435	81.109	1124.379	756.920	-9.414
4300	20.800	243.236	223.889	83.189	1126.459	748.147	-9.088
4400	20.803	243.714	224.335	85.269	1128.539	739.325	-8.777
4500	20.807	244.181	224.770	87.349	1130.619	730.456	-8.479
4600	20.811	244.639	225.197	89.430	1132.700	721.540	-8.193
4700	20.815	245.086	225.616	91.511	1134.782	712.579	-7.919
4800	20.820	245.525	226.026	93.593	1136.863	703.574	-7.656
4900	20.825	245.954	226.428	95.675	1138.946	694.526	-7.404
5000	20.831	246.375	226.823	97.758	1141.028	685.435	-7.161
5100	20.838	246.787	227.211	99.842	1143.112	676.303	-6.927
5200	20.845	247.192	227.591	101.926	1145.196	667.129	-6.701
5300	20.853	247.589	227.964	104.011	1147.281	657.916	-6.484
5400	20.861	247.979	228.332	106.096	1149.367	648.663	-6.275
5500	20.870	248.362	228.692	108.183	1151.453	639.371	-6.072
5600	20.879	248.738	229.047	110.270	1153.541	630.042	-5.877
5700	20.890	249.108	229.396	112.359	1155.629	620.675	-5.688
5800	20.901	249.471	229.739	114.448	1157.719	611.272	-5.505
5900	20.912	249.828	230.076	116.539	1159.809	601.832	-5.328
6000	20.924	250.180	230.408	118.631	1161.901	592.357	-5.157

PREVIOUS: CURRENT: March 1982 (1 bar)

Radon, Ion (Rn^+) $Rn_1^+(g)$

SULFUR (S) REFERENCE STATE A_r = 32.06

0 to 368.3	K	crystal, alpha
368.3 to 388.36	K	crystal, beta
388.36 to 882.117	K	liquid
above 882.117	K	ideal diatomic gas (see below)

The normal boiling point of sulfur at 1 atm, 717.824 K, is a secondary standard on the International Practical Temperature Scale of 1968. The vapor composition at this temperature is a mixture of several sulfur species, the predominant species being S_8(g), S_7(g), and S_6(g). In our reference state for sulfur, we have arbitrarily chosen 0.5 S_2(g) to be the gas phase species.

Sulfur (S) S_1(ref)

Enthalpy Reference Temperature = T_r = 298.15 K Standard State Pressure = p° = 0.1 MPa

T/K	C_p° (J K⁻¹mol⁻¹)	S° (J K⁻¹mol⁻¹)	$-[G^\circ - H^\circ(T_r)]/T$ (J K⁻¹mol⁻¹)	$H^\circ - H^\circ(T_r)$ (kJ mol⁻¹)	$\Delta_f H^\circ$ (kJ mol⁻¹)	$\Delta_f G^\circ$ (kJ mol⁻¹)	Log K_f
0	0.	0.	INFINITE	-4.412	0.	0.	0.
100	12.770	12.522	49.744	-3.722	0.	0.	0.
200	19.368	23.637	34.038	-2.080	0.	0.	0.
298.15	22.698	32.056	32.056	0.	0.	0.	0.
300	22.744	32.196	32.056	0.042	0.	0.	0.
368.300	24.246	37.015	32.540	1.648	ALPHA <--> BETA		
368.300	24.773	38.103	32.540	2.049	TRANSITION		
388.360	25.167	39.427	32.861	2.550	BETA <--> LIQUID		
388.360	31.058	43.859	32.861	4.271	TRANSITION		
400	32.162	44.793	33.195	4.639	0.	0.	0.
432.020	53.808	47.431	34.151	5.737	Cp LAMBDA MAXIMUM		
432.020	53.806	47.431	34.151	5.737	TRANSITION		
500	37.986	53.532	36.398	8.567	0.	0.	0.
600	34.308	60.078	39.825	12.152	0.	0.	0.
700	32.681	65.241	43.099	15.499	0.	0.	0.
800	31.699	69.530	46.143	18.710	0.	0.	0.
882.117	31.665	72.624	48.467	21.310	LIQUID <--> IDEAL GAS		
882.117	18.454	133.077	48.467	74.636	FUGACITY = 1 bar		
900	18.483	133.448	50.152	74.967	0.	0.	0.
1000	18.638	135.403	58.581	76.823	0.	0.	0.
1100	18.792	137.187	65.647	78.694	0.	0.	0.
1200	18.947	138.829	71.678	80.581	0.	0.	0.
1300	19.103	140.352	76.903	82.484	0.	0.	0.
1400	19.257	141.773	81.486	84.402	0.	0.	0.
1500	19.409	143.107	85.550	86.335	0.	0.	0.
1600	19.556	144.364	89.187	88.283	0.	0.	0.
1700	19.697	145.554	92.468	90.246	0.	0.	0.
1800	19.830	146.684	95.449	92.223	0.	0.	0.
1900	19.956	147.759	98.174	94.212	0.	0.	0.
2000	20.072	148.786	100.679	96.213	0.	0.	0.
2100	20.176	149.768	102.993	98.226	0.	0.	0.
2200	20.274	150.708	105.141	100.248	0.	0.	0.
2300	20.364	151.612	107.142	102.280	0.	0.	0.
2400	20.448	152.480	109.013	104.321	0.	0.	0.
2500	20.526	153.316	110.769	106.370	0.	0.	0.
2600	20.589	154.123	112.421	108.426	0.	0.	0.
2700	20.650	154.901	113.980	110.488	0.	0.	0.
2800	20.707	155.653	115.455	112.555	0.	0.	0.
2900	20.762	156.381	116.853	114.629	0.	0.	0.
3000	20.813	157.085	118.183	116.708	0.	0.	0.
3100	20.855	157.768	119.449	118.791	0.	0.	0.
3200	20.898	158.431	120.657	120.879	0.	0.	0.
3300	20.940	159.075	121.811	122.971	0.	0.	0.
3400	20.983	159.701	122.916	125.067	0.	0.	0.
3500	21.025	160.310	123.976	127.167	0.	0.	0.
3600	21.065	160.902	124.994	129.272	0.	0.	0.
3700	21.106	161.480	125.972	131.380	0.	0.	0.
3800	21.149	162.044	126.914	133.493	0.	0.	0.
3900	21.194	162.594	127.822	135.610	0.	0.	0.
4000	21.240	163.131	128.698	137.732	0.	0.	0.
4200	21.337	164.169	130.362	141.989	0.	0.	0.
4400	21.443	165.164	131.922	146.267	0.	0.	0.
4600	21.558	166.120	133.388	150.567	0.	0.	0.
4800	21.680	167.040	134.771	154.891	0.	0.	0.
5000	21.810	167.928	136.080	159.240	0.	0.	0.
5200	21.948	168.786	137.321	163.615	0.	0.	0.
5400	22.090	169.617	138.502	168.019	0.	0.	0.
5600	22.236	170.423	139.628	172.452	0.	0.	0.
5800	22.385	171.206	140.703	176.914	0.	0.	0.
6000	22.538	171.967	141.733	181.406	0.	0.	0.

PREVIOUS: September 1977 (1 atm) CURRENT: September 1977 (1 bar)

Sulfur (S) S_1(ref)

SULFUR, MONOCLINIC (S) CRYSTAL A_r = 32.06

$S°(298.15\ K) = 33.028 \pm 0.050\ J\ K^{-1}\ mol^{-1}$
$T_{trs} = 368.3 \pm 0.3\ K\ (\alpha-\beta)$
$T_{fus} = 388.36 \pm 0.1\ K\ (\beta-\ell)$

$\Delta_f H°(0\ K) = 0.247\ kJ\ mol^{-1}$
$\Delta_f H°(298.15\ K) = 0.360\ kJ\ mol^{-1}$
$\Delta_{trs} H° = 0.4004 \pm 0.0029\ kJ\ mol^{-1}$
$\Delta_{fus} H° = 1.721 \pm 0.008\ kJ\ mol^{-1}$

Enthalpy of Formation

The enthalpy of formation of monoclinic sulfur at 298.15 K is calculated from that of orthorhombic sulfur by adding $\Delta_{trs}H°$ and the difference in enthalpy, H°(368.3) - H°(298.15 K), between orthorhobic and monoclinic sulfur.

Heat Capacity and Entropy

Experimentally determined heat capacity values for monoclinic (β) sulfur have been reported by Eastman and McGavock (1, 64.83-376.16 K), West (2, 368-388 K), and Montgomery (3, 12.41-386.83 K). The thermochemical functions tabulated by Hultgren et al. (4) are based on the work of Eastman and McGavock (1) and West (2).

We adopt the heat capacity values tabulated (in 10 K increments) by Montgomery (3). The heat capacity values above T_{fus} = 388.36 K are obtained by linear extrapolation of the tabulated heat capacity values (3) above 360 K. Using S°(10 K) = 0.063 cal K^{-1} mol^{-1} and integrating the heat capacity values, we calculate S°(298.15 K) = 7.893 cal K^{-1} mol^{-1}. The S°(10) value contains a contribution of 0.007 cal K^{-1} mol^{-1} which is needed to make S°(S, cr, β, T) = S°(S, cr, α, T) + $\Delta_{trs}H°/T_{trs}$. Since both forms of sulfur are perfectly ordered crystals at 0 K (3), the S°(0 K) value should be zero, by definition. The contribution of 0.007 cal K^{-1} mol^{-1} is arbitrarily added at 10 K although it could appear in any temperature region at or below 368.3 K. Part of this entropy may be in the region of the lambda anomaly at 198.3 K. This contribution is comparable to the uncertainty expected from this adiabatic calorimeter.

Phase Data

The α-β transition at 368.3 K is discussed on the orthorhombic sulfur table (5). The heat capacity data of Montgomery (3) did not support the transition at 374 K reported by West (2).

Montgomery (3) reported a lambda anomaly in the heat capacity of monoclinic sulfur with a peak at 198.3 K. The entropy change in the lambda anomaly was 0.052 ± 0.005 cal K^{-1} mol^{-1}, this was interpreted as due to the disordering of the structure of monoclinic sulfur on heating (7).

Fusion Data

The melting of sulfur has been studied by many workers. An excellent review by Meyer (6) summarizes these studies. In addition, Hultgren et al. (4) tabulated numerous enthalpy of melting determinations. We adopt the results obtained by Montgomery (3) via adiabatic calorimetry: T_{fus} = 388.36 ± 0.02 K and $\Delta_{fus}H°$ = 0.4128 ± 0.0006 kcal mol^{-1}. The only other precise study reported in the literature is that of West (2) who obtained $\Delta_{fus}H°$ = 0.41051 ± 0.0005 kcal mol^{-1} by adiabatic calorimetry.

References

1. E. D. Eastman and W. C. McGavock, J. Am. Chem. Soc. 59, 145 (1937).
2. E. D. West, J. Am. Chem. Soc. 81, 29 (1959).
3. R. L. Montgomery, Ph.D. Dissertation, Oklahoma State University, 1976. [Diss. Abst. 76-09730].
4. R. L. Hultgren et al., "Selected Values of the Thermodynamic Properties of the Elements," American Society for Metals, Metals Park, Ohio, 1973.
5. JANAF Thermochemical Tables: S(cr, α), 9-30-77.
6. B. Meyer, Chem. Rev. 76, 367 (1976).
7. R. L. Montgonery, Science 184, 562 (1974).

Sulfur, Monoclinic (S) $S_1(cr)$

Enthalpy Reference Temperature = T_r = 298.15 K — Standard State Pressure = p° = 0.1 MPa

T/K	$C_p°$ (J K⁻¹mol⁻¹)	S° (J K⁻¹mol⁻¹)	-[G°-H°(T_r)]/T (J K⁻¹mol⁻¹)	H°-H°(T_r) (kJ mol⁻¹)	$\Delta_f H°$ (kJ mol⁻¹)	$\Delta_f G°$ (kJ mol⁻¹)	Log K_f
0	0.	0.	INFINITE	-4.525	0.247	0.247	INFINITE
100	12.874	12.964	51.217	-3.825	0.257	0.213	-0.111
198.300	21.966	24.242	35.146	-2.162	Cp LAMBDA MAXIMUM TRANSITION		
198.300	21.966	24.242	35.146	-2.162			
200	20.154	24.421	35.054	-2.127	0.313	0.157	-0.041
250	21.736	29.066	33.402	-1.084	0.336	0.115	-0.024
298.15	23.225	33.028	33.028	0.	0.360	0.070	-0.012
300	23.276	33.172	33.028	0.043	0.361	0.068	-0.012
350	24.414	36.849	33.316	1.237	0.388	0.017	-0.003
368.300	24.773	38.103	33.523	1.687	ALPHA <--> BETA		
388.360	25.167	39.427	33.794	2.188	BETA <--> LIQUID		
400	25.397	40.173	33.969	2.482	-1.797	0.050	-0.007
450	26.380	43.222	34.830	3.776	-2.428	0.310	-0.036
500	27.363	46.052	35.812	5.120	-3.087	0.653	-0.068
600	29.330	51.215	37.957	7.955	-3.838	1.481	-0.129
700	31.296	55.884	40.189	10.986	-4.153	2.397	-0.179
800	33.263	60.191	42.424	14.214	-4.136	3.335	-0.218

PREVIOUS: CURRENT: September 1977

SULFUR, ORTHORHOMBIC (S) CRYSTAL $A_r = 32.06$

$S°(298.15\ K) = 32.056 \pm 0.050\ J\ K^{-1}\ mol^{-1}$
$T_{trs} = 368.3 \pm 0.3\ K\ (\alpha-\beta)$

$\Delta_f H°(0\ K) = 0\ kJ\ mol^{-1}$
$\Delta_f H°(298.15\ K) = 0\ kJ\ mol^{-1}$
$\Delta_{trs} H° = 0.4004 \pm 0.0029\ kJ\ mol^{-1}$

Enthalpy of Formation

Zero by definition.

Heat Capacity and Entropy

Experimentally determined heat capacity vlaues for orthorhombic (α) sulfur have been reported by Nernst (1, 23-202 K), Eastman and McGavock (2, 12.68-365.60 K), West (3, 2981-368 K), Mal'tsev and Demidenko (4, 53-305 K), Finke (5, 12-432 K), Montgomery (6, 12.34-360.99 K), and Paukov (7). The thermochemical functions tabulated by Hultgren et al. (8) are based primarily on the work of Eastman and McGavock (2) and unpublished work of Finke (5), the latter presumably being the preliminary results of Montgomery (6). The tentative values reported by CODATA (9) for S°(298.15 K) and H°(298.15 K) - H°(0 K) are based on Eastman and McGavock (2), Montgomery (6), and Paukov (7).

We adopt the heat capacity values tabulated (in 10 K increments) by Montgomery (6). Using $S°(10) = 0.036\ cal\ K^{-1}\ mol^{-1}$ and integrating the heat capacity values, we calculate $S°(298.15\ K) = 7.661\ cal\ K^{-1}\ mol^{-1}$, in agreement with CODATA (9). The heat capacity values above $T_{trs} = 368.3$ K are obtained by linear extrapolation of the tabulated values (6) above 340 K.

The various heat capacity studies are in reasonably good agreement. The values recommended by Eastman and McGavock (2), West (3), and Mal'tsev and Demidenko (4) agree with Montgomery (6) to within 0.5% except for the ranges 15-50 K (2) and 53-80 K (3). The smoothed value at 190 K reported by Mal'tsev and Demidenko (4) is undoubtedly a typographical error. The selected values of Hultgren et al. (8) agree within 0.5% with the data of Montgomery (6) except below 35 K.

Fusion Data

The orthorhomic-monoclinic (α-β) phase change has been studied by many workers. An excellent review by Meyer (10) summarizes these studies. We adopt the results obtained by Montgomery (6) via adiabatic calorimetry: $T_{trs} = 368.3 \pm 0.3$ K and $\Delta_{trs}H° = 0.0957 \pm 0.0007\ kcal\ mol^{-1}$. The only other precise study reported in the literature is that of West (3) who obtained $\Delta_{trs}H° = 0.0960 \pm 0.0005\ kcal\ mol^{-1}$ by adiabatic calorimetry. Other drop calorimetric (11, 12) and vapor pressure (13, 14) studies are much less precise with an uncertainty of $\pm 5\ cal\ mol^{-1}$ or greater.

Sublimation Data

Since the reference state of sulfur involves 0.5 $S_2(g)$ rather than S(g), we define the enthalpy of sublimation (at 298.15 K) of orthorhombic sulfur as the process S(cr, α) = 0.5 $S_2(g)$. Thus the enthalpy of sublimation of orthorhombic sulfur is one-half the enthalpy of formation of $S_2(g)$. Since the dominant gaseous sulfur species at room temperature is actually $S_8(g)$, the enthalpy of sublimation to the equilibrium gas is significantly smaller.

References

1. W. Nernst, Ann. Physik. 36, 395 (1911).
2. E. D. Eastman and W. C. McGavock, J. Am. Chem. Soc. 59, 145 (1937).
3. E. D. West, J. Am. Chem. Soc. 81, 29 (1959).
4. A. K. Mal'tsev and A. F. Demidenko, Tr. Mosk. Khim. - Tekhnol. Inst., No. 51, 136 (1967).
5. H. L. Finke, U. S. Bureau of Mines, Unpublished Work, Quoted in Reference (8).
6. R. L. Montgomery, Ph.D. Dissertation, Oklahoma State University, 1976.
7. I. E. Paukov, to be published, 1977; mentioned in reference (9).
8. R. Hultgen et al., "Selected Values of the Thermodynamic Properties of the Elements," American Society for Metals, Metals Park, Ohio, 1973.
9. ICUS-CODATA Task Group Key Values for Thermodynamics, CODATA Special Report 4, March, 1977.
10. B. Meyer, Chem. Rev. 76, 367 (1976).
11. P. Mondain-Monval, Bull. Soc. Chim. France 39, 1349 (1926).
12. J. N. Bronsted, Z. Physik. Chem. 55, 371 (1906).
13. M. Thackray, Nature 203, 1278 (1964); J. Chem. Eng. Data 15, 495 (1970).
14. K. Neumann, Z. Physik. Chem. A171, 416 (1934).

Sulfur, Orthorhombic (S) $S_1(cr)$

Enthalpy Reference Temperature = T_r = 298.15 K Standard State Pressure = p° = 0.1 MPa

T/K	$C_p°$	S°	$-[G°-H°(T_r)]/T$	$H°-H°(T_r)$	$\Delta_f H°$	$\Delta_f G°$	Log K_f
	J K⁻¹mol⁻¹			kJ mol⁻¹			
0	0.	0.	INFINITE	-4.412	0.	0.	0.
100	12.770	12.522	49.744	-3.722	0.	0.	0.
200	19.368	23.637	34.038	-2.080	0.	0.	0.
250	21.297	28.179	32.422	-1.061	0.	0.	0.
298.15	22.698	32.056	32.056	0.	0.	0.	0.
300	22.744	32.196	32.056	0.042	0.	0.	0.
350	23.870	35.789	32.337	1.208	0.	0.	0.
368.300	24.246	37.015	32.540	1.648	-------- ALPHA <--> BETA -------		
400	24.895	39.044	32.975	2.427	-2.212	0.088	-0.011
450	25.920	42.035	33.818	3.698	-2.867	0.406	-0.047
500	26.945	44.819	34.780	5.019	-3.548	0.809	-0.084
600	28.995	49.913	36.886	7.816	-4.336	1.764	-0.154
700	31.045	54.537	39.082	10.818	-4.680	2.812	-0.210
800	33.095	58.816	41.284	14.025	-4.685	3.887	-0.254

PREVIOUS: December 1965 CURRENT: September 1977

Sulfur, Orthorhombic (S) $S_1(cr)$

SULFUR (S) LIQUID A_r = 32.06

$S°(298.15\ K) = [36.825]\ J\ K^{-1}\ mol^{-1}$
$T_{trs} = 432.02 \pm 0.2\ K$
$T_{fus} = 388.36\ K$

$\Delta_fH°(298.15\ K) = [1.854]\ kJ\ mol^{-1}$
$\Delta_{trs}H° = 0.0\ kJ\ mol^{-1}$
$\Delta_{fus}H° = 1.7272 \pm 0.0025\ kJ\ mol^{-1}$

Enthalpy of Formation

The enthalpy of formation of liquid sulfur at 298.15 K is calculated from that of monoclinic sulfur by adding $\Delta_{fus}H°$ and the difference, in enthalpy, H°(388.36 K) - H°(298.15 K), between monoclinic and liquid sulfur.

Heat Capacity and Entropy

The adopted heat capacity values are based on the studies by Montgomery (1, 405.79-433.31 K) and West (2, 373-678 K). Liquid sulfur undergoes a second order transition with a maximum reported at 432.02 ± 0.20 K (1) and 432.25 ± 0.30 K (2); this has been attributed to the polymerization of S_8 molecules (3). We adopt the tabulated heat capacity values of Montgomery (1) up to 434 K and those of West (2) above 434 K. The heat capacity is assumed to be constant at 7.568 cal K^{-1} mol^{-1} above 810 K. Below T_{fus} = 388.36 K, the heat capacity values are obtained by linear extrapolation using the slope of the values in the region T_{fus} to 420 K. The entropy is calculated in a manner similar to that used for the enthalpy of formation.

Vaporization Data

The normal boiling point of sulfur, T_{vap} = 717.824 K (1 atm) is a secondary reference on the International Practical Temperature Scale of 1968. At this temperature, equilibrium sulfur vapor contains monomeric and several polymeric sulfur species: $S_2(g)$ and S(g) predominate above 1000 K while $S_6(g)$, $S_7(g)$, and $S_8(g)$ dominate at T_{vap}.

Since our reference state for sulfur is arbitrarily defined to involve 0.5 $S_2(g)$ as the vapor species we calculate $\Delta_{vap}H°(T_{vap}) = [10.839]$ k cal mol^{-1} for the process $S(\ell) = 0.5\ S_2(g)$. The brackets are used to indicate the arbitrary nature of this value.

References

1. R. L. Montgomery, Ph.D. Dissertation, Oklahoma State University, 1976.
2. E. D. West, J. Am. Chem. Soc. 81, 29 (1959).
3. B. Meyer, Chem. Rev. 76, 367 (1976).

Sulfur (S) $S_1(l)$

Enthalpy Reference Temperature = T_r = 298.15 K Standard State Pressure = p° = 0.1 MPa

T/K	$C_p°$ (J K⁻¹mol⁻¹)	S° (J K⁻¹mol⁻¹)	$-[G°-H°(T_r)]/T$ (J K⁻¹mol⁻¹)	$H°-H°(T_r)$ (kJ mol⁻¹)	$\Delta_fH°$ (kJ mol⁻¹)	$\Delta_fG°$ (kJ mol⁻¹)	Log K_f
0							
100							
200							
250							
298.15	22.531	36.825	36.825	0.	1.854	0.432	-0.076
300	22.707	36.965	36.826	0.042	1.854	0.423	-0.074
350	27.434	40.821	37.120	1.295	1.941	0.180	-0.027
388.360	31.058	43.859	37.635	2.417	------- BETA <--> LIQUID -------		
400	32.162	44.793	37.829	2.785	0.	0.	0.
432.020	53.829	47.431	38.442	3.884	Cp LAMBDA MAXIMUM TRANSITION		
432.020	53.808	47.431	38.442	3.884			
450	43.046	49.308	38.840	4.711	0.	0.	0.
500	37.986	53.532	40.106	6.713	0.	0.	0.
600	34.308	60.078	42.915	10.298	0.	0.	0.
700	32.681	65.241	45.748	13.645	0.	0.	0.
800	31.699	69.530	48.460	16.856	0.	0.	0.
882.117	31.665	72.624	50.568	19.456	------- FUGACITY = 1 bar -------		
900	31.665	73.260	51.012	20.023	-53.090	1.079	-0.063
1000	31.665	76.596	53.407	23.189	-51.780	7.028	-0.367
1100	31.665	79.614	55.654	26.356	-50.485	12.846	-0.610
1200	31.665	82.369	57.767	29.522	-49.205	18.546	-0.807
1300	31.665	84.904	59.758	32.689	-47.941	24.141	-0.970
1400	31.665	87.250	61.639	35.855	-46.693	29.639	-1.106
1500	31.665	89.435	63.420	39.021	-45.460	35.048	-1.220

PREVIOUS: December 1965 CURRENT: September 1977

Sulfur (S) $S_1(l)$

SULFUR (S) ALPHA-BETA-LIQUID A_r = 32.06

0 to 368.3 K crystal, alpha
368.3 to 388.36 K crystal, beta
above 388.36 K liquid

Refer to the individual tables for details.

Sulfur (S) S_1(cr,l)

Enthalpy Reference Temperature = T_r = 298.15 K Standard State Pressure = $p°$ = 0.1 MPa

T/K	$C_p°$ (J K⁻¹mol⁻¹)	$S°$ (J K⁻¹mol⁻¹)	$-[G°-H°(T_r)]/T$ (J K⁻¹mol⁻¹)	$H°-H°(T_r)$ (kJ mol⁻¹)	$\Delta_f H°$ (kJ mol⁻¹)	$\Delta_f G°$ (kJ mol⁻¹)	Log K_f
0	0.	0.	INFINITE	-4.412	0.	0.	0.
100	12.770	12.522	49.744	-3.722	0.	0.	0.
200	19.368	23.637	34.038	-2.080	0.	0.	0.
250	21.297	28.179	32.422	-1.061	0.	0.	0.
298.15	22.698	32.056	32.056	0.	0.	0.	0.
300	22.744	32.196	32.056	0.042	0.	0.	0.
350	23.870	35.789	32.337	1.208	0.	0.	0.
368.300	24.246	37.015	32.540	1.648	ALPHA <--> BETA TRANSITION		
368.300	24.773	38.103	32.540	2.049			
388.360	25.167	39.427	32.861	2.550	BETA <--> LIQUID TRANSITION		
388.360	31.062	43.859	32.861	4.271			
400	32.162	44.793	33.195	4.639	0.	0.	0.
432.020	53.808	47.431	34.151	5.737	Cp LAMBDA MAXIMUM TRANSITION		
432.020	53.806	47.431	34.151	5.737			
450	43.046	49.308	34.720	6.564	0.	0.	0.
500	37.986	53.532	36.398	8.567	0.	0.	0.
600	34.308	60.078	39.825	12.152	0.	0.	0.
700	32.681	65.241	43.099	15.499	0.	0.	0.
800	31.699	69.530	46.143	18.710	0.	0.	0.
882.117	31.665	72.624	48.467	21.310	------- FUGACITY = 1 bar -------		
900	31.665	73.260	48.952	21.877	-53.090	1.079	-0.063
1000	31.665	76.596	51.553	25.043	-51.780	7.028	-0.367
1100	31.665	79.614	53.969	28.209	-50.485	12.846	-0.610
1200	31.665	82.369	56.222	31.376	-49.205	18.546	-0.807
1300	31.665	84.904	58.332	34.542	-47.941	24.141	-0.970
1400	31.665	87.250	60.315	37.709	-46.693	29.639	-1.106
1500	31.665	89.435	62.185	40.875	-45.460	35.048	-1.220

PREVIOUS: CURRENT: September 1977

Sulfur (S) S_1(cr,l)

SULFUR (S) IDEAL GAS A_r = 32.06

IP(S, g) = 83558.0 ± 1 cm^{-1}

S°(298.15 K) = 167.828 ± 0.035 J K^{-1} mol^{-1}

$\Delta_f H°$(0 K) = 274.73 ± 0.25 kJ mol^{-1}

$\Delta_f H°$(298.15 K) = 276.98 ± 0.25 kJ mol^{-1}

Electronic Levels and Quantum Weights

State	ε_i, cm^{-1}	g_i
$^3P_2$	0.00	5
$^3P_1$	396.09	3
$^3P_0$	573.65	1
$^1D_2$	9238.58	5
$^1S_0$	22179.99	1

Enthalpy of Formation

The enthalpy of formation is based on the dissociation energy for S_2(g) as recommended by CODATA (1). The adopted value for $\Delta_f H°$(S, g, 298.15 K) is calculated using $D_0°(S_2)$ = 35216.4 ± 2.5 cm^{-1} (421.282 ± 0.03 kJ mol^{-1}) from the study of Ricks and Barrow (2).

The selected dissociation energy, obtained from precise spectroscopic measurements (2), is supported by additional studies on thermochemical cycles (3), photoionization (4, 5) and torsion-effusion measurements (6). Earlier measurements have been reviewed by Herzberg (7), Gaydon (8), Brewer (9), and Drowart and Goldfinger (3).

Heat Capacity and Entropy

The electronic levels for S(g) are given in the compilation by Moore (10, 11). Our calculations indicate that, except for the five levels lying below 23000 cm^{-1}, the inclusion of levels up to n = 10 has no effect on the thermodynamic functions to 6000 K. This is a result of the high energy of these levels; the fifth excited state lies at 52623.88 cm^{-1} above the ground state. Since the inclusion of these upper levels has no effect on the thermodynamic functions (to 6000 K) we list only the lowest lying states (ε_i < 23000 cm^{-1}). The reported uncertainty in S°(298.15 K) is due to uncertainties in the relative atomic mass and the fundeamental constants. Extension of these calculations above 6000 K may require consideration of the higher excited states and utilization of proper fill and cutoff procedures (12). In fact, the inclusion of these higher excited states and consideration of various cutoff procedures leads to calculational differences in the Gibbs energy function of ~0.4 J K^{-1} mol^{-1} or greater at temperatures in excess of 11500 K.

The thermal functions at 298.15 K agree with CODATA recommendations (1) except for two minor differences. First, the entropy differs by 0.1094 J K^{-1} mol^{-1} because this table uses a standard state pressure of 1 bar, whereas the CODATA recommendations are based on 1 atm. Second, an entropy difference of 0.004 J K^{-1} mol^{-1} arises due to the use of slightly different values for the fundamental constants.

References

1. J. D. Cox, Chairman, ICSU-CODATA Task Group on Key Values for Thermodynamics, J. Chem. Thermodynamics 10, 903 (1978).
2. J. M. Ricks and R. F. Barrow, Can. J. Phys. 47, 2423 (1969).
3. J. Drowart and P. Goldfinger, Quart. Rev. 20, 545 (1966).
4. J. Berkowitz and W. A. Chupka, J. Chem. Phys. 50, 4245 (1969).
5. V. H. Dibeler and S. K. Liston, J. Chem. Phys. 49, 482 (1968).
6. P. Budininkas, R. K. Edwards, and P. G. Wahlbeck, J. Chem. Phys. 48, 2859 (1968).
7. G. Herzberg, "Spectra of Diatomic Molecules", 2nd ed., D. Van Nostrand Co., Inc., New York, 1963.
8. A. H. Gaydon, "Dissociation Energies and Spectra of Diatomic Molecules", 3rd ed., Chapman and Hall Ltd., London, 1968.
9. L. Brewer, J. Chem. Phys. 31, 1143 (1959).
10. C. E. Moore, U.S. Nat. Bur. Stand., NSRDS-NBS 35, Volume I, 1970 [Reprint of NBS Circular 467, Volume I, 1949
11. C. E. Moore, U.S. Nat. Bur. Stand., NSRDS-NBS-34, 1970.
12. J. R. Downey, Jr., The Dow Chemical Company, AFOSR-TR-78-0960, Contract No. F44620-75-1-0048, 1978.

Sulfur (S) S_1(g)

Enthalpy Reference Temperature = T_r = 298.15 K Standard State Pressure = p° = 0.1 MPa

T/K	$C_p°$ (J $K^{-1}mol^{-1}$)	S° (J $K^{-1}mol^{-1}$)	$-[G°-H°(T_r)]/T$ (J $K^{-1}mol^{-1}$)	$H°-H°(T_r)$ (kJ mol^{-1})	$\Delta_f H°$ (kJ mol^{-1})	$\Delta_f G°$ (kJ mol^{-1})	Log K_f
0	0.	0.	INFINITE	-6.657	274.735	274.735	INFINITE
100	21.356	142.891	188.580	-4.569	276.133	263.096	-137.427
200	23.388	158.392	169.994	-2.320	276.740	249.789	-65.238
250	23.696	163.653	168.218	-1.141	276.899	243.031	-50.779
298.15	23.673	167.828	167.828	0.	276.980	236.500	-41.434
300	23.669	167.974	167.828	0.044	276.982	236.248	-41.135
350	23.480	171.610	168.116	1.223	276.995	229.458	-34.245
400	23.233	174.729	168.752	2.391	274.732	222.757	-29.089
450	22.979	177.451	169.571	3.546	273.962	216.297	-25.107
500	22.741	179.859	170.481	4.689	273.102	209.938	-21.932
600	22.338	183.968	172.398	6.942	271.770	197.436	-17.188
700	22.031	187.388	174.302	9.160	270.641	185.138	-13.815
800	21.800	190.314	176.125	11.351	269.621	172.994	-11.295
900	21.624	192.871	177.847	13.522	215.535	162.055	-9.405
1000	21.489	195.142	179.465	15.677	215.834	156.096	-8.154
1100	21.386	197.185	180.984	17.821	216.106	150.109	-7.128
1200	21.307	199.042	182.413	19.955	216.354	144.098	-6.272
1300	21.249	200.745	183.758	22.083	216.579	138.067	-5.548
1400	21.209	202.318	185.029	24.205	216.784	132.020	-4.926
1500	21.186	203.781	186.231	26.325	216.970	125.959	-4.386
1600	21.178	205.148	187.371	28.443	217.140	119.886	-3.914
1700	21.184	206.432	188.455	30.561	217.295	113.803	-3.497
1800	21.203	207.643	189.487	32.680	217.438	107.711	-3.126
1900	21.234	208.790	190.473	34.802	217.570	101.611	-2.793
2000	21.276	209.880	191.417	36.927	217.694	95.505	-2.494
2100	21.327	210.920	192.321	39.058	217.812	89.392	-2.224
2200	21.386	211.913	193.189	41.193	217.925	83.275	-1.977
2300	21.452	212.865	194.024	43.335	218.035	77.152	-1.752
2400	21.523	213.780	194.828	45.484	218.143	71.024	-1.546
2500	21.598	214.660	195.604	47.640	218.250	64.892	-1.356
2600	21.676	215.508	196.353	49.803	218.358	58.755	-1.180
2700	21.756	216.328	197.078	51.975	218.467	52.615	-1.018
2800	21.837	217.121	197.780	54.155	218.579	46.470	-0.867
2900	21.919	217.888	198.460	56.343	218.694	40.321	-0.726
3000	21.999	218.633	199.120	58.538	218.811	34.169	-0.595
3100	22.078	219.355	199.761	60.742	218.931	28.012	-0.472
3200	22.155	220.058	200.384	62.954	219.055	21.851	-0.357
3300	22.230	220.740	200.991	65.173	219.183	15.687	-0.248
3400	22.303	221.405	201.582	67.400	219.313	9.518	-0.146
3500	22.372	222.053	202.157	69.634	219.447	3.346	-0.050
3600	22.439	222.684	202.719	71.874	219.583	-2.830	0.041
3700	22.502	223.300	203.267	74.121	219.721	-9.011	0.127
3800	22.561	223.900	203.802	76.375	219.862	-15.194	0.209
3900	22.618	224.487	204.325	78.634	220.003	-21.382	0.286
4000	22.670	225.061	204.836	80.898	220.146	-27.573	0.360
4100	22.720	225.621	205.336	83.167	220.289	-33.768	0.430
4200	22.766	226.169	205.826	85.442	220.432	-39.966	0.497
4300	22.808	226.705	206.305	87.720	220.575	-46.168	0.561
4400	22.847	227.230	206.775	90.003	220.716	-52.373	0.622
4500	22.884	227.744	207.235	92.290	220.856	-58.581	0.680
4600	22.917	228.247	207.686	94.580	220.993	-64.792	0.736
4700	22.947	228.740	208.129	96.873	221.127	-71.006	0.789
4800	22.974	229.224	208.563	99.169	221.258	-77.223	0.840
4900	22.999	229.698	208.990	101.468	221.386	-83.443	0.890
5000	23.021	230.163	209.409	103.769	221.509	-89.665	0.937
5100	23.040	230.619	209.820	106.072	221.628	-95.890	0.982
5200	23.057	231.066	210.224	108.377	221.741	-102.117	1.026
5300	23.072	231.505	210.622	110.683	221.849	-108.346	1.068
5400	23.085	231.937	211.013	112.991	221.952	-114.577	1.108
5500	23.096	232.361	211.397	115.300	222.048	-120.810	1.147
5600	23.105	232.777	211.775	117.610	222.138	-127.044	1.185
5700	23.112	233.186	212.147	119.921	222.222	-133.281	1.221
5800	23.117	233.588	212.513	122.232	222.299	-139.518	1.256
5900	23.121	233.983	212.874	124.544	222.368	-145.757	1.290
6000	23.124	234.372	213.229	126.857	222.431	-151.997	1.323

PREVIOUS: September 1977 (1 atm) CURRENT: September 1982 (1 bar)

Sulfur (S) S_1(g)

SULFUR, ION (S^+) IDEAL GAS Mr = 32.05945

$IP(S^+, g) = 188200 \pm 100\ cm^{-1}$

$S°(298.15\ K) = 163.627 \pm 0.03\ J\ K^{-1}\ mol^{-1}$

$\Delta_f H°(0\ K) = 1274.311 \pm 0.4\ kJ\ mol^{-1}$

$\Delta_f H°(298.15\ K) = [1282.293]\ kJ\ mol^{-1}$

Electronic Levels and Quantum Weights

State	ε_i, cm^{-1}	g_i
$^4S_{3/2}$	0	4
$^2D_{3/2}$	14853.0	4
$^2D_{5/2}$	14884.8	6
$^2P_{1/2}$	24524.9	2
$^2P_{3/2}$	24571.6	4

Enthalpy of Formation

$\Delta_f H°(S^+, g, 0\ K)$ is calculated from $\Delta_f H°(S, g, 0\ K)$ (1) using the spectroscopic value of IP(S) = 83558.0 ± 1 cm^{-1} (238.904 ± 0.01 kJ mol^{-1}) from Moore (2). The ionization limit is converted from cm^{-1} to kJ mol^{-1} using the factor, 1 cm^{-1} = 0.01196266 kJ cm^{-1}, which is derived from the 1973 CODATA fundamental constants (3). Rosenstock et al. (4) and Levin and Lias (5) have summarized additional ionization and appearance potential data.

$\Delta_f H°(S^+, g, 298.15\ K)$ is calculated from $\Delta_f H°(S, g, 0\ K)$ by using IP(S) with JANAF (1) enthalpies, H°(0 K) − H°(298.15 K), for S(g), S^+(g), and e^-(g). $\Delta_r H°(S \rightarrow S^+ + e^-, 298.15\ K)$ differs from a room temperature threshold energy due to inclusion of these enthalpies and to threshold effects discussed by Rosenstock et al. (4). $\Delta_f H°(298.15\ K)$ should be changed by −6.197 kJ mol^{-1} if it is to be used in the ion convention that excludes the enthalpy of the electron.

Heat Capacity and Entropy

The information on electronic energy levels and quantum weights, given by Moore (2, 6), is incomplete because many theoretically predicted levels have not been observed. Our calculations indicate that any reasonable method of filling in these missing levels and cutting off the summation in the partition function (7) has no effect on the thermodynamic functions to 6000 K. This is a result of the high energy of all levels other than the ground state and the first four excited states; the next excited state is approximately 79000 cm^{-1} above the ground state. Since inclusion of these excited states has no effect on the thermodynamic functions (to 6000 K), we list only the ground state and the first four excited states. The reported uncertainty in S°(298.15 K) is due to uncertainties in the relative ionic mass and the fundamental constants. Extension of these calculations above 6000 K may require consideration of the excited states and use of different fill and cutoff procedures (7).

References

1. JANAF Thermochemical Tables: S(g), 9-30-82; e^-(g), 3-31-82.
2. C. E. Moore, U. S. Nat. Bur. Stand., NSRDS-NBS-34, 1970.
3. E. R. Cohen and B. N. Taylor, J. Phys. Chem. Ref. Data 2, 663 (1973).
4. H. M. Rosenstock, K. Draxl et al., J. Phys. Chem. Ref. Data 6, Supp. 1 (1977).
5. R. D. Levin and S. G. Lias, U. S. Nat. Bur. Stand., NSRDS-NBS-71, 1982.
6. C. E. Moore, U. S. Nat. Bur. Stand., NSRDS-NBS-35, Volume I, 1971 [Reprinted from NBS Circular 467, Volume I, 1949].
7. J. R. Downey, Jr., The Dow Chemical Company, AFOSR-TR-78-0960, Contract No. F44620-75-1-0048, 1978.

Sulfur, Ion (S^+) $S_1^+(g)$

Enthalpy Reference Temperature = T_r = 298.15 K Standard State Pressure = p° = 0.1 MPa

T/K	$C_p°$ (J K^{-1}mol^{-1})	S°	−[G°−H°(T_r)]/T	H°−H°(T_r) (kJ mol^{-1})	$\Delta_f H°$	$\Delta_f G°$	Log K_f
0	0.	0.	INFINITE	-6.197	1274.311		
100	20.786	140.920	182.108	-4.119			
200	20.786	155.328	165.529	-2.040			
250	20.786	159.966	163.970	-1.001			
298.15	20.786	163.627	163.627	0.	1282.293	1236.811	-216.684
300	20.786	163.756	163.628	0.038	1282.328	1236.528	-215.299
350	20.786	166.960	163.881	1.078	1283.241	1228.822	-183.392
400	20.786	169.736	164.443	2.117	1281.888	1221.076	-159.456
450	20.786	172.184	165.170	3.156	1282.042	1213.457	-140.854
500	20.786	174.374	165.983	4.196	1282.118	1205.834	-125.973
600	20.786	178.164	167.707	6.274	1282.690	1190.530	-103.645
700	20.786	181.368	169.435	8.353	1283.500	1175.108	-87.688
800	20.786	184.143	171.104	10.431	1284.446	1159.560	-75.711
900	20.786	186.592	172.692	12.510	1232.347	1144.969	-66.452
1000	20.786	188.782	174.193	14.589	1234.648	1135.137	-59.293
1100	20.786	190.763	175.611	16.667	1236.934	1125.075	-53.425
1200	20.786	192.571	176.950	18.746	1239.204	1114.805	-48.526
1300	20.786	194.235	178.216	20.825	1241.459	1104.347	-44.373
1400	20.787	195.776	179.416	22.903	1243.698	1093.716	-40.807
1500	20.789	197.210	180.555	24.982	1245.922	1082.926	-37.711
1600	20.792	198.552	181.639	27.061	1248.131	1071.987	-34.997
1700	20.797	199.812	182.671	29.140	1250.326	1060.911	-32.598
1800	20.806	201.001	183.656	31.221	1252.509	1049.706	-30.462
1900	20.820	202.127	184.599	33.302	1254.679	1038.380	-28.547
2000	20.840	203.195	185.503	35.385	1256.839	1026.939	-26.821
2100	20.867	204.212	186.369	37.470	1258.991	1015.391	-25.256
2200	20.904	205.184	187.203	39.558	1261.135	1003.741	-23.832
2300	20.951	206.114	188.005	41.651	1263.275	991.994	-22.529
2400	21.009	207.007	188.778	43.749	1265.410	980.153	-21.332
2500	21.080	207.866	189.525	45.853	1267.545	968.223	-20.230
2600	21.164	208.694	190.246	47.965	1269.679	956.208	-19.210
2700	21.262	209.495	190.944	50.086	1271.817	944.111	-18.265
2800	21.374	210.270	191.621	52.218	1273.960	931.935	-17.385
2900	21.500	211.022	192.277	54.362	1276.108	919.682	-16.565
3000	21.640	211.753	192.914	56.519	1278.265	907.354	-15.798
3100	21.795	212.465	193.533	58.690	1280.432	894.955	-15.080
3200	21.962	213.160	194.136	60.878	1282.611	882.486	-14.405
3300	22.142	213.839	194.722	63.083	1284.802	869.948	-13.770
3400	22.335	214.502	195.295	65.307	1287.009	857.343	-13.171
3500	22.538	215.153	195.853	67.550	1289.230	844.674	-12.606
3600	22.751	215.791	196.398	69.815	1291.469	831.941	-12.071
3700	22.973	216.417	196.930	72.101	1293.725	819.145	-11.564
3800	23.204	217.033	197.451	74.410	1296.000	806.288	-11.083
3900	23.441	217.638	197.961	76.742	1298.293	793.371	-10.626
4000	23.684	218.235	198.460	79.098	1300.606	780.395	-10.191
4100	23.931	218.823	198.950	81.479	1302.939	767.361	-9.776
4200	24.182	219.403	199.430	83.884	1305.292	754.269	-9.381
4300	24.436	219.975	199.901	86.315	1307.665	741.122	-9.003
4400	24.691	220.539	200.364	88.772	1310.059	727.919	-8.641
4500	24.946	221.097	200.818	91.253	1312.472	714.661	-8.296
4600	25.201	221.648	201.265	93.761	1314.905	701.350	-7.964
4700	25.454	222.193	201.705	96.294	1317.358	687.985	-7.646
4800	25.705	222.731	202.137	98.852	1319.830	674.568	-7.341
4900	25.953	223.264	202.563	101.434	1322.320	661.099	-7.047
5000	26.197	223.791	202.982	104.042	1324.828	647.580	-6.765
5100	26.436	224.312	203.395	106.674	1327.354	634.010	-6.494
5200	26.671	224.827	203.803	109.329	1329.897	620.390	-6.232
5300	26.899	225.338	204.204	112.008	1332.456	606.721	-5.980
5400	27.121	225.842	204.600	114.709	1335.030	593.004	-5.736
5500	27.337	226.342	204.991	117.432	1337.619	579.239	-5.501
5600	27.545	226.837	205.377	120.176	1340.222	565.427	-5.274
5700	27.746	227.326	205.757	122.940	1342.838	551.568	-5.055
5800	27.940	227.810	206.133	125.725	1345.466	537.663	-4.842
5900	28.125	228.289	206.505	128.528	1348.106	523.713	-4.637
6000	28.302	228.763	206.872	131.350	1350.756	509.718	-4.437

PREVIOUS: September 1977 (1 atm) CURRENT: September 1982 (1 bar)

Sulfur, Ion (S^+) $S_1^+(g)$

SULFUR, ION (S^-) IDEAL GAS M_r = 32.06055

EA(S, g) = 2.077120 ± 0.000001 eV
S°(298.15 K) = 164.918 ± 0.001 J K^{-1} mol^{-1}

$\Delta_f H°$(0 K) = 74.325 kJ mol^{-1}
$\Delta_f H°$(298.15 K) = [70.180] kJ mol^{-1}

Electonic Levels and Quantum Weights

State	ε_i, cm^{-1}	g_i
$^2P_{3/2}$	0.0	4
$^2P_{1/2}$	483.54	2

Enthalpy of Formation

$\Delta_f H°(S^-$, g, 0 K) is calculated from $\Delta_f H°$(S, g, 0 K) (1) using the adopted electron affinity of EA(S) = 2.077120 ± 0.000001 eV (200.4100 ± 0.0001 kJ mol^{-1}). This value, recommended by Hotop and Lineberger (2), is based on a tunable laser photodetachment threshold study (3). Additional information on S^-(g) may be obtained in the critical discussions of Hotop and Lineberger (2, 4), Rosenstock et al. (5) and Massey (6).

$\Delta_f H°(S^-$, g, 298.15 K) is obtained from $\Delta_f H°$(S, g, 0 K) by using EA(S) with JANAF (1) enthalpies, H°(0 K) - H°(298.15 K), for S^-(g), S(g), and e^-(g). $\Delta_r H°(S^- \rightarrow S + e^-$, 298.15 K) differs from a room-temperature threshold energy due to inclusion of these enthalpies and to threshold effects discussed by Rosenstock et al. (5). $\Delta_f H°$(298.15 K) should be changed by +6.197 kJ mol^{-1} if it is to be used in the ion convention that excludes the enthalpy of the electron.

Heat Capacity and Entropy

The ground state and the fine-structure separation (483.54 ± 0.01 cm^{-1}) for S^-(g) has been experimentally determined (3); a discussion of the relevant work is given by Hotop and Lineberger (2) and Massey (6). Lacking any experimental evidence as to the stability of any excited states, we assume no stable excited states exist.

References

1. JANAF Thermochemical Tables: S(g), 9-30-82; e^-(g), 3-31-82.
2. H. Hotop and W. C. Lineberger, J. Phys. Chem. Ref. Data, to be published, 1985.
3. R. D. Mead, K. R. Lykke and W. C. Lineberger, unpublished, referenced in (2).
4. H. Hotop and W. C. Lineberger, J. Phys. Chem. Ref. Data 4, 539 (1975).
5. H. M. Rosenstock, K. Draxl et al., J. Phys. Chem. Ref. Data 6, Supp. 1 (1977).
6. H. S. W. Massey, "Negative Ions", 3rd ed., Cambridge University Press, Cambridge, 1976.

Sulfur, Ion (S^-) S_1^-(g)

Enthalpy Reference Temperature = T_r = 298.15 K   Standard State Pressure = p° = 0.1 MPa

T/K	$C_p°$ (J $K^{-1}mol^{-1}$)	S° (J $K^{-1}mol^{-1}$)	$-[G°-H°(T_r)]/T$ (J $K^{-1}mol^{-1}$)	$H°-H°(T_r)$ (kJ mol^{-1})	$\Delta_f H°$ (kJ mol^{-1})	$\Delta_f G°$ (kJ mol^{-1})	Log K_f
0	0.	0.	INFINITE	-6.465	74.325		
100	20.977	140.952	184.787	-4.383			
200	22.291	155.895	166.994	-2.220			
250	22.660	160.914	165.293	-1.095			
298.15	22.783	164.918	164.918	0.	70.180	36.822	-6.451
300	22.784	165.059	164.919	0.042	70.142	36.615	-6.375
350	22.757	168.571	165.196	1.181	69.075	31.111	-4.643
400	22.653	171.603	165.812	2.317	65.741	25.851	-3.376
450	22.515	174.264	166.606	3.446	63.905	20.966	-2.434
500	22.370	176.628	167.493	4.568	61.985	16.300	-1.703
600	22.096	180.682	169.364	6.791	58.545	7.492	-0.652
700	21.868	184.071	171.229	8.989	55.317	-0.760	0.057
800	21.687	186.978	173.021	11.166	52.205	-8.558	0.559
900	21.543	189.524	174.716	13.327	-3.969	-14.889	0.864
1000	21.429	191.788	176.312	15.476	-5.756	-16.007	0.836
1100	21.338	193.826	177.813	17.614	-7.568	-16.944	0.805
1200	21.264	195.679	179.226	19.744	-9.403	-17.716	0.771
1300	21.203	197.378	180.558	21.867	-11.261	-18.334	0.737
1400	21.153	198.948	181.816	23.985	-13.140	-18.807	0.702
1500	21.112	200.406	183.007	26.098	-15.039	-19.146	0.667
1600	21.076	201.767	184.138	28.207	-16.956	-19.357	0.632
1700	21.047	203.044	185.213	30.313	-18.891	-19.448	0.598
1800	21.021	204.246	186.237	32.417	-20.843	-19.425	0.564
1900	20.999	205.382	187.215	34.518	-22.810	-19.292	0.530
2000	20.980	206.459	188.151	36.617	-24.791	-19.056	0.498
2100	20.963	207.482	189.047	38.714	-26.785	-18.720	0.466
2200	20.949	208.457	189.907	40.809	-28.791	-18.289	0.434
2300	20.936	209.388	190.734	42.904	-30.807	-17.767	0.403
2400	20.924	210.279	191.530	44.997	-32.833	-17.156	0.373
2500	20.914	211.133	192.297	47.089	-34.869	-16.461	0.344
2600	20.905	211.953	193.037	49.180	-36.912	-15.685	0.315
2700	20.897	212.742	193.753	51.270	-38.963	-14.829	0.287
2800	20.890	213.501	194.445	53.359	-41.020	-13.898	0.259
2900	20.883	214.234	195.114	55.448	-43.083	-12.893	0.232
3000	20.877	214.942	195.764	57.536	-45.153	-11.817	0.206
3100	20.871	215.627	196.393	59.623	-47.227	-10.671	0.180
3200	20.866	216.289	197.005	61.710	-49.307	-9.459	0.154
3300	20.862	216.931	197.599	63.796	-51.391	-8.181	0.130
3400	20.858	217.554	198.177	65.882	-53.479	-6.841	0.105
3500	20.854	218.158	198.739	67.968	-55.573	-5.439	0.081
3600	20.850	218.746	199.287	70.053	-57.671	-3.976	0.058
3700	20.847	219.317	199.820	72.138	-59.773	-2.456	0.035
3800	20.844	219.873	200.341	74.222	-61.880	-0.879	0.012
3900	20.841	220.414	200.849	76.307	-63.991	0.754	-0.010
4000	20.838	220.942	201.344	78.391	-66.108	2.441	-0.032
4100	20.836	221.457	201.829	80.474	-68.229	4.181	-0.053
4200	20.834	221.959	202.302	82.558	-70.355	5.973	-0.074
4300	20.832	222.449	202.765	84.641	-72.487	7.816	-0.095
4400	20.830	222.928	203.218	86.724	-74.624	9.708	-0.115
4500	20.828	223.396	203.661	88.807	-76.767	11.648	-0.135
4600	20.826	223.854	204.095	90.890	-78.916	13.637	-0.155
4700	20.824	224.301	204.520	92.972	-81.070	15.672	-0.174
4800	20.823	224.740	204.937	95.055	-83.232	17.753	-0.193
4900	20.821	225.169	205.345	97.137	-85.399	19.880	-0.212
5000	20.820	225.590	205.746	99.219	-87.573	22.050	-0.230
5100	20.819	226.002	206.139	101.301	-89.754	24.264	-0.249
5200	20.818	226.406	206.525	103.383	-91.943	26.521	-0.266
5300	20.816	226.803	206.904	105.464	-94.138	28.820	-0.284
5400	20.815	227.192	207.276	107.546	-96.340	31.161	-0.301
5500	20.814	227.574	207.642	109.627	-98.550	33.542	-0.319
5600	20.813	227.949	208.001	111.709	-100.767	35.964	-0.335
5700	20.812	228.317	208.354	113.790	-102.992	38.425	-0.352
5800	20.811	228.679	208.701	115.871	-105.224	40.926	-0.369
5900	20.811	229.035	209.043	117.952	-107.464	43.465	-0.385
6000	20.810	229.385	209.379	120.033	-109.711	46.042	-0.401

PREVIOUS: September 1977 (1 atm)   CURRENT: September 1982 (1 bar)

Sulfur, Ion (S^-) S_1^-(g)

SILICON SULFIDE (SiS) IDEAL GAS M_r = 60.1455

$S°(298.15\ K) = 223.80 \pm 2.0\ J\ K^{-1}\ mol^{-1}$

$\Delta_f H°(0\ K) = 104.6 \pm 12.6\ kJ\ mol^{-1}$
$\Delta_f H°(298.15\ K) = 106.0 \pm 12.6\ kJ\ mol^{-1}$

Electronic Levels and Quantum Weights

State	ε_i, cm^{-1}	g_i
$X^1\Sigma^+$	0	1
$[a^3\Pi]$	[24700]	6
$D^1\Pi$	35029	2
$[e^3\Sigma^-]$	[35140]	3
F	37114	2
$I^1\Sigma^-$	[37290]	1
$E^1\Sigma^+$	41924	1

$\omega_e = 740.34\ cm^{-1}$ $\omega_e x_e = 2.544\ cm^{-1}$ $\sigma = 1$
$B_e = 0.29930\ cm^{-1}$ $\alpha_e = 0.00145\ cm^{-1}$ $r_e = 1.92932$ Å

Enthalpy of Formation

Herzberg (2) suggests a dissociation energy for SiS(g) of 6.6 eV. This result was derived from a spectroscopic analysis of the D-X and E-X band systems by Barrow and Jevons (4), Vago and Barrow (5), and Barrow (6). Robinson and Barrow (8) extended the work of Vago and Barrow (5) and arrived at a value of 147.4 ± 3.0 kcal mol^{-1} (6.39 eV). In this latter work, a graphical Birge-Sponer extrapolation was made using 28 vibrational levels of the E state. The highest level is quite near the convergence limit and thus only a short extrapolation is necessary. The extrapolation graphically indicates the absolute minimum and maximum to be 6.28 eV and 6.57 eV, respectively. This amounts to a range of maximum uncertainty of approximately 7 kcal mol^{-1}. The products of the dissociation must be Si(^{3}p) and S(^{3}p) but it is not certain as to which of the sublevels of the ^{3}p states are involved. This leads to an uncertainty of ± 1.2 kcal mol^{-1} (8). The result is in agreement with a rotational analysis of the $D^1\Pi$-$X^1\Sigma^+$ system by Lagerqvist, Nilheden, and Barrow (7) which led to a value of 6.47 eV (149.2 kcal mol^{-1}).

The value of the dissociation energy chosen is 147.4 ± 3.0 kcal mol^{-1}, based on the work by Robinson and Barrow (8). Gaydon (1) and Vedeneyev et al. (3) suggest a similar value, based on the work by Robinson and Barrow (8) and others. This value is also consistent with the tabulated dissociation energies of Gaydon (1) as far as trends which might be expected in the oxides, sulfides, and selenides of C, Si, Ge, Sn, and Pb. The resulting enthalpy of formation is $\Delta_f H°(0\ K)$ = 25.01 ± 3.00 kcal mol^{-1}.

Rosenqvist and Tungesvik (13) studied two reaction systems by a transportation method. By combining these two systems, the standard free energy of the reaction 2 Si(s) + S_2(g) = 2 SiS(g) was calculated. Using JANAF values (14) for Si(s) and S_2(g), $\Delta_f G°$ (1 atm) values for SiS(g) at 0 K and 300 K are 27.652 kcal mol^{-1} and 15.933 kcal mol^{-1}, respectively. These values are within the error range of the values in this tabulation.

Heat Capacity and Entropy

The spectroscopic constants, corrected for the natural abundances of Si and S, are from Hoeft (9), and Lagerqvist, Nilheden, and Barrow (7).

Herzberg (2) tabulates the $X^1\Sigma^+$, $D^1\Pi$, and E states and comments on unresolved bands at 16000-29000 cm^{-1}. The $a^3\Pi$ state is predicted by an interpolation of similar $a^3\Pi$-$X^1\Sigma^+$ splittings in the sulfides and oxides of C, Ge, Sn, (10, 11). The remaining states are deduced by a rotational analysis of the E-X system of SiS (7, 12). It is not known conclusively whether the F state is $^1\Pi$ or $^1\Delta$. The energy of the e and I states are upper bound estimates.

References

1. A. G. Gaydon, "Dissociation Energies and Spectra of Diatomic Molecules," Chapman and Hall, Ltd., London, 1968.
2. G. Herzberg, "Spectra of Diatomic Molecules," 2nd Ed., D. Van Nostrand Company, New York, 1950.
3. V. I. Vedeneyev, L. V. Gurvich, V. N. Kondrat'yev, V. A. Medvedev, and Ye. L. Frankevich, "Bond Energies, Ionization Potentials, and Electron Affinities," St. Martin's Press, New York, 1966.
4. R. F. Barrow and W. Jevons, Proc. Roy. Soc. (London) 169, 45 (1938).
5. E. E. Vago and R. F. Barrow, Proc. Phys. Soc. (London) 58, 538 (1946).
6. R. F. Barrow, Proc. Phys. Soc. (London) 58, 606 (1946).
7. A. Lagerqvist, G. Nilheden, and R. F. Barrow, Proc. Phys. Soc. (London) A65, 419 (1952).
8. S. J. Q. Robinson and R. F. Barrow, Proc. Phys. Soc. (London) A67, 95 (1954).
9. J. Hoeft, Z. Naturforsch. 20A, 1327 (1965).
10. H. B. Palmer, J. Chem. Phys. 54, 3244 (1971).
11. B. Meyer, J. J. Smith and K. Spitzer, J. Chem. Phys. 53, 3616 (1970).
12. R. F. Barrow, J. L. Deutsch, A. Lagerqvist, and B. Westerlund, Proc. Phys. Soc. (London) 78, 1307 (1961).
13. R. Rosenqvist and K. Tungesvik, Trans. Faraday Soc. 67, 2945 (1971); Chem. Abstr. 75, 144493j.
14. JANAF Thermochemical Tables: Si(s), 12-31-66; S_2(g), 12-31-65.

Silicon Sulfide (SiS) S_1Si_1(g)

Enthalpy Reference Temperature = T_r = 298.15 K Standard State Pressure = p° = 0.1 MPa

T/K	$C_p°$ (J K^{-1}mol^{-1})	S° (J K^{-1}mol^{-1})	$-[G°-H°(T_r)]/T$ (J K^{-1}mol^{-1})	$H°-H°(T_r)$ (kJ mol^{-1})	$\Delta_f H°$ (kJ mol^{-1})	$\Delta_f G°$ (kJ mol^{-1})	Log K_f
0	0.	0.	INFINITE	-8.940	104.650	104.650	INFINITE
100	29.129	190.877	251.180	-6.030	106.603	89.151	-46.568
200	30.303	211.317	226.691	-3.075	106.738	71.535	-18.683
250	31.355	218.191	224.325	-1.534	106.410	62.769	-13.115
298.15	32.332	223.799	223.799	0.	105.960	54.403	-9.531
300	32.368	223.999	223.799	0.060	105.941	54.083	-9.417
350	33.239	229.056	224.197	1.701	105.380	45.483	-6.788
400	33.953	233.542	225.090	3.381	102.543	37.056	-4.839
450	34.531	237.576	226.257	5.094	101.206	28.942	-3.359
500	34.996	241.239	227.575	6.832	99.789	20.990	-2.193
600	35.681	247.684	230.404	10.368	97.364	5.464	-0.476
700	36.148	253.222	233.277	13.961	95.162	-9.676	0.722
800	36.478	258.072	236.080	17.594	93.074	-24.510	1.600
900	36.722	262.383	238.767	21.254	37.916	-38.005	2.206
1000	36.908	266.262	241.326	24.936	37.131	-46.399	2.424
1100	37.054	269.787	243.755	28.634	36.302	-54.712	2.598
1200	37.173	273.016	246.061	32.346	35.428	-62.947	2.740
1300	37.271	275.995	248.251	36.068	34.507	-71.108	2.857
1400	37.355	278.761	250.332	39.800	33.538	-79.197	2.955
1500	37.427	281.340	252.315	43.539	32.519	-87.214	3.037
1600	37.491	283.758	254.205	47.285	31.451	-95.161	3.107
1700	37.548	286.033	256.011	51.037	-19.845	-102.594	3.152
1800	37.600	288.180	257.739	54.794	-20.783	-107.435	3.118
1900	37.647	290.214	259.395	58.556	-21.730	-112.223	3.085
2000	37.692	292.147	260.985	62.323	-22.684	-116.961	3.055
2100	37.733	293.987	262.513	66.095	-23.645	-121.651	3.026
2200	37.773	295.743	263.984	69.870	-24.612	-126.295	2.999
2300	37.811	297.423	265.401	73.649	-25.584	-130.896	2.973
2400	37.848	299.033	266.769	77.432	-26.561	-135.453	2.948
2500	37.885	300.579	268.091	81.219	-27.543	-139.971	2.925
2600	37.922	302.065	269.369	85.009	-28.528	-144.448	2.902
2700	37.959	303.497	270.607	88.803	-29.516	-148.888	2.880
2800	37.998	304.878	271.806	92.601	-30.505	-153.291	2.860
2900	38.040	306.212	272.970	96.403	-31.496	-157.659	2.840
3000	38.084	307.503	274.100	100.209	-32.489	-161.992	2.821
3100	38.132	308.752	275.197	104.020	-33.481	-166.293	2.802
3200	38.185	309.964	276.265	107.836	-34.472	-170.561	2.784
3300	38.243	311.140	277.304	111.657	-35.462	-174.799	2.767
3400	38.308	312.282	278.316	115.484	-36.451	-179.006	2.750
3500	38.380	313.394	279.303	119.319	-37.436	-183.185	2.734
3600	38.460	314.476	280.265	123.161	-422.570	-176.874	2.566
3700	38.549	315.531	281.204	127.011	-423.134	-170.041	2.401
3800	38.648	316.560	282.121	130.871	-423.694	-163.194	2.243
3900	38.758	317.566	283.017	134.741	-424.250	-156.331	2.094
4000	38.880	318.548	283.893	138.623	-424.800	-149.454	1.952
4100	39.013	319.510	284.750	142.517	-425.342	-142.564	1.816
4200	39.160	320.452	285.589	146.426	-425.876	-135.660	1.687
4300	39.320	321.375	286.410	150.350	-426.399	-128.744	1.564
4400	39.493	322.281	287.215	154.290	-426.912	-121.816	1.446
4500	39.681	323.171	288.004	158.249	-427.411	-114.876	1.333
4600	39.884	324.045	288.778	162.227	-427.897	-107.926	1.226
4700	40.101	324.905	289.538	166.226	-428.367	-100.965	1.122
4800	40.334	325.752	290.283	170.248	-428.821	-93.994	1.023
4900	40.582	326.586	291.016	174.294	-429.256	-87.014	0.928
5000	40.844	327.408	291.735	178.365	-429.672	-80.025	0.836
5100	41.122	328.220	292.443	182.463	-430.067	-73.028	0.748
5200	41.415	329.021	293.139	186.590	-430.440	-66.024	0.663
5300	41.723	329.813	293.823	190.746	-430.791	-59.012	0.582
5400	42.045	330.596	294.497	194.935	-431.116	-51.995	0.503
5500	42.381	331.370	295.160	199.156	-431.416	-44.971	0.427
5600	42.730	332.137	295.814	203.411	-431.689	-37.942	0.354
5700	43.092	332.897	296.458	207.702	-431.935	-30.909	0.283
5800	43.467	333.649	297.092	212.030	-432.152	-23.871	0.215
5900	43.853	334.396	297.718	216.396	-432.340	-16.830	0.149
6000	44.251	335.136	298.336	220.801	-432.496	-9.787	0.085

PREVIOUS: December 1971 (1 atm) CURRENT: December 1971 (1 bar)

Silicon Sulfide (SiS) S_1Si_1(g)

STRONTIUM SULFIDE (SrS) CRYSTAL M_r = 119.68

$S°(298.15 K) = 68.41 \pm 0.8 J K^{-1} mol^{-1}$

$\Delta_f H°(0 K) = -467.4 \pm 17 kJ mol^{-1}$
$\Delta_f H°(298.15 K) = -468.6 \pm 17 kJ mol^{-1}$

Enthalpy of Formation

Sabatier (1) measured the enthalpy of solution of SrS in hydrochloric acid solution as $\Delta_r H(283.6 K) = -26.9 \pm 1.0 kcal mol^{-1}$ for $SrS(cr) + 2 HCl(aq, 100 H_2O) \rightarrow SrCl_2(aq, 100 H_2O) + H_2S(aq)$. We derive $\Delta_f H°(SrS, cr, 298.15 K) = -113.1 \pm 2.0 kcal mol^{-1}$ based on the following auxiliary data: $\Delta_f H°(HCl \cdot aq, 100 H_2O, 298.15 K) = -39.657 kcal mol^{-1}$ (3), $\Delta_f H°(H_2S, aq, 298.15 K) = -9.5 kcal mol^{-1}$ (2), and $\Delta_f H°(SrCl_2 \cdot aq, 200 H_2O, 298.15 K) = -209.78 kcal mol^{-1}$ (3). The uncertainty reflects the correction to 298.15 K and the ambiguity of the exact solution composition. Mourlot (4) also measured the enthalpy of solution of SrS in hydrochloric acid solution as $\Delta_r H° = -27.1 \pm 1.0 kcal mol^{-1}$ fromwhich we calculate $\Delta_f H°(SrS, cr, 298.15 K) = -112.8 \pm 2.0 kcal mol^{-1}$, using the same auxiliary data.

Cater and Johnson (5) determined that SrS sublimes congruently and derived partial pressures of the various gaseous species Sr(g), S(g), and SrS(g) in an effusion and mass-spectrometric study. We analyze their data and the equilibrium study of Schenck and Hammerschmidt (6) below. Other mass-spectrometric studies of Berkowitz and Marquart (7) and Colin et al. (8) are also listed. The calculated 3rd law $\Delta_r H°(298.15 K)$ may have an uncertainty of 0.5 kcal mol^{-1} since the JANAF Gibbs energy functions are partially based on the estimated $C_p°$ data (above 300 K).

The enthalpy of solution studies (1, 4) show wide disagreement from the mass spectrometric studies (5, 7, 8). Our experience tends to indicate that mass spectrometric experiments provide very good data on gas phase species for homogeneous reactions, but are not as reliable for condensed phases in heterogeneous reactions. A weighted average of the equilibrium and enthalpy of solution studies, $\Delta_f H°(SrS, cr, 298.15 K) = -112.0 \pm 4 kcal mol^{-1}$ is adopted in the tabulation.

Source	Method	Reaction*	T/K	Data Points	δS cal $K^{-1} mol^{-1}$	$\Delta_r H°(298.15 K)$, kcal mol^{-1} 2nd law	3rd law	$\Delta_f H°(298.15 K)$ kcal mol^{-1}
(5)	Knudsen mass spec.	A	1825-2181	Equation	1.26	212.70	210.18±0.38	-104.70±2
(5)	Knudsen mass spec.	B	1825-2181	Equation	0.23	131.75	131.29±0.08	-105.43±2
(6)	Equilibrium	C	1237-1383	3	8.8±1.8	-53.0±2.4	-64.55±1.3	-110.02±4
(7)	Mass spec.	B	1809	1			210.18±1.0	-104.7 ±2
(8)	Mass spec.	D	1857-2170	9	-12.6±2.8	-3.3±5.5	21.8 ±2.4	-119.8 ±16
(9)	Vaporization	B	1400-1700	Equation	-3.3	102.6	107.7 ±0.9	-82 ±30

A) $SrS(cr) = Sr(g) + S(g)$
B) $SrS(cr) = SrS(g)$
C) $SrS(cr) + 2 SO_2(g) = SrSO_4(cr) + S_2(g)$
D) $SrS(cr) + 3 S(g) = Sr(g) + 2 S_2(g)$

Heat Capacity and Entropy

King and Weller (10) measured the low temperature heat capacities from 53-296 K. Their smooth values are adopted in the tabulation. The entropy, $S°(298.15 K) = 16.3 \pm 0.2 cal K^{-1} mol^{-1}$, is based on $S°(51 K) = 1.59 cal K^{-1} mol^{-1}$ as extrapolated by King and Weller using Debye and Einstein functions ($\theta_D = 208$, $\theta_E = 311$). The $C_p°$ values above 300 K are estimated by graphical extrapolation combined with method B of Kubaschewski et al. (11).

Fusion Data

Literature melting data for SrS(cr) are not available. The value quoted by Mills (12) for the melting point, 2275 K, seems too low when compared to the alkaline earth oxide melting points (13).

References

1. M. Sabatier, Ann. Chim. Phys. 22, 5 (1881)
2. U. S. Nat. Bur. Stand. Tech. Note 270-3, 1968.
3. U. S. Nat. Bur. Stand. Tech. Note 270-6, 1971.
4. M. A. Mourlot, Ann. Chim. Phys. 17, 510 (1899).
5. E. D. Cater and E. W. Johnson, J. Chem. Phys. 47, 5353 (1967).
6. R. Schenck and F. Hammerschmidt, Z. Anorg. Allgem. Chem. 210, 313 (1933).
7. J. Berkowitz and J. R. Marquart, J. Chem. Phys. 39, 275 (1963).
8. R. Colin, P. Goldfinger, and M. Jeunehomme, Trans. Faraday Soc. 60, 306 (1964).
9. B. P. Nikonov and N. G. Otmakhova, Zhur. Fiz. Khim. 35, 1494 (1961).
10. E. G. King and W. W. Weller, U. S. Bur. Mines RI 5590, 1960.
11. O. Kubaschewski, E. LL. Evans, and C. B. Alcock, "Metallurgical Thermochemistry," Pergamon, Oxford, 1967.
12. K. C. Mills, "Thermodynamic Data for Inorganic Sulphides, Selenides, and Tellurides," Butterworths, London, 1974.
13. JANAF Thermochemical Tables: BaO(cr), 6-30-74; CaO(cr), 6-30-73; SrO(cr), 12-31-72.

Strontium Sulfide (SrS) $S_1Sr_1(cr)$

Enthalpy Reference Temperature = T_r = 298.15 K   Standard State Pressure = p° = 0.1 MPa

T/K	$C_p°$ (J $K^{-1}mol^{-1}$)	S°	$-[G°-H°(T_r)]/T$	$H°-H°(T_r)$ (kJ mol^{-1})	$\Delta_f H°$	$\Delta_f G°$	Log K_f
0	0.	0.	INFINITE	-9.811	-467.438	-467.438	INFINITE
100	32.008	22.471	108.640	-8.617	-468.434	-466.624	243.739
200	44.936	49.669	72.763	-4.619	-468.566	-464.731	121.375
298.15	48.702	68.410	68.410	0.	-468.608	-462.842	81.088
300	48.727	68.711	68.411	0.090	-468.609	-462.806	80.582
400	50.208	82.927	70.339	5.035	-470.990	-460.763	60.169
500	51.882	94.318	74.032	10.143	-472.639	-458.026	47.850
600	53.179	103.897	78.232	15.399	-473.897	-454.979	39.609
700	54.099	112.165	82.503	20.764	-474.913	-451.744	33.710
800	54.894	119.443	86.674	26.215	-475.813	-448.372	29.276
900	55.522	125.944	90.683	31.735	-530.424	-443.737	25.754
1000	56.233	131.832	94.508	37.324	-529.724	-434.143	22.677
1100	56.777	137.217	98.149	42.974	-536.882	-424.255	20.146
1200	57.321	142.180	101.614	48.679	-537.010	-414.010	18.021
1300	57.870	146.790	104.914	54.439	-537.099	-403.756	16.223
1400	58.409	151.099	108.061	60.253	-537.150	-393.496	14.682
1500	58.937	155.146	111.066	66.120	-537.162	-383.235	13.345
1600	59.455	158.967	113.942	72.040	-537.137	-372.974	12.176
1700	59.951	162.586	116.698	78.010	-673.990	-361.536	11.109
1800	60.459	166.027	119.343	84.030	-672.032	-343.212	9.960
1900	60.982	169.310	121.887	90.103	-670.040	-324.999	8.935
2000	61.505	172.451	124.338	96.227	-668.017	-306.891	8.015
2100	62.028	175.465	126.701	102.403	-665.965	-288.885	7.186
2200	62.551	178.362	128.984	108.632	-663.886	-270.977	6.434
2300	63.074	181.154	131.192	114.914	-661.788	-253.165	5.750
2400	63.597	183.850	133.330	121.247	-659.674	-235.444	5.124
2500	64.120	186.456	135.403	127.633	-657.552	-217.811	4.551
2600	64.643	188.981	137.416	134.071	-655.430	-200.263	4.023
2700	65.166	191.431	139.371	140.562	-653.314	-182.797	3.536
2800	65.689	193.810	141.273	147.104	-651.216	-165.409	3.086
2900	66.212	196.124	143.125	153.699	-649.143	-148.095	2.667
3000	66.735	198.378	144.929	160.347	-647.107	-130.853	2.278

PREVIOUS: CURRENT: September 1977

Strontium Sulfide (SrS) $S_1Sr_1(cr)$

STRONTIUM SULFIDE (SrS) IDEAL GAS M_r = 119.68

$S°$(298.15 K) = 243.107 ± 0.42 J K^{-1} mol^{-1}

$\Delta_f H°$(0 K) = 109.6 ± 8.4 kJ mol^{-1}
$\Delta_f H°$(298.15 K) = 108.2 ± 8.4 kJ mol^{-1}

Electronic States and Molecular Constants

State	ε_i, cm^{-1}	g_i	ω_e, cm^{-1}	$\omega_e x_e$, cm^{-1}	B_e, cm^{-1}	α_e, cm^{-1}	r_e, Å	Source
$X^1\Sigma^+$	0.0	1	388.38	1.310	0.12072	0.00044	2.4405	(7)
$^3\pi$	[7800.]	6	[270]	[1.1]	[0.0943]	[0.0005]	[2.76]	(6)
$^3\Sigma^+$	[8000.]	3	[330]	[1.2]	[0.114]	[0.0005]	[2.51]	(6)
$^1\pi$	[8100.]	2	[270]	[1.1]	[0.0943]	[0.0005]	[2.76]	(6)
$^1\Sigma^+$	[11000.]	1	[330]	[1.2]	[0.114]	[0.0005]	[2.51]	(6)
$B^1\Sigma^+$	26531.2	1	286.8	0.84	0.10566	0.00032	2.609	(7, 8)

Enthalpy of Formation

The adopted value for the enthalpy of formation, $\Delta_f H°$(SrS, g, 298.15 K) = 25.86±2 kcal mol^{-1}, is based on the Knudsen mass spectrometric studies analyzed below. The three independent studies are in very good agreement and we have chosen the average of the measurements. Our value is in very good agreement with $\Delta_f H°$(298.15 K) = 25.0 kcal mol^{-1} adopted by NBS (4) and is slightly greater than $\Delta_f H°$(298.15 K) = 24.1±4 kcal mol^{-1} obtained by Mills (5) (using different Gibbs energy functions) in a recent critical compilation. Using auxiliary JANAF data (6) and a recommended value for $D_0^0(S_2, g,)$ = 100.69±0.01 kcal mol^{-1} (6), we calculate D_0^0(SrS, g) = 78.8±2.0 kcal mol^{-1}.

Source	Method	Reaction*	T/K	Data Points	δS cal K^{-1} mol^{-1}	$\Delta_r H°$(298.15 K), kcal mol^{-1} 2nd law	3rd law	$\Delta_f H°$(298.15 K) kcal mol^{-1}	D_0^0 kcal mol^{-1}
Colin (1)	Mass spec	A	2010-2170	4	-1.86±8.5	17.42±17.9	21.310±1.60	25.06±1.66	79.63±2.0
Marquart (2)	Mass spec	A	1809-1934	2	6.85	34.19	21.375±1.21	25.12±1.3	79.56±1.5
Cater (3)	Mass spec	A	1825-2181	Equation	-1.16	20.85	23.07 ±0.34	26.82±0.6	77.87±1.0
Cater (3)	Mass spec	B	1825-2181	Equation	-1.03	-80.95	-78.89 ±0.31	26.60±0.6	78.09±0.3

*A) Sr(g) + S_2(g) = SrS(g) + S(g) B) Sr(g) + S(g) = SrS(g)

Heat Capacity and Entropy

Electronic levels (T_∞) and vibrational-rotational constants for the observed states are from the optical study of Barrow (7, 8). Other low-lying electronic states and their vibrational-rotational constants are estimated in isoconfigurational groups by analogy with SrO (6) and from trends observed in the known states of the other alkaline-earth oxides and sulfides (6). Uncertainty in the energy and constants for the estimated states may contribute as much as 2-3 cal K^{-1} mol^{-1} to S° at 3000 K. The molecular constants have been corrected to the natural isotopic abundances. The thermodynamic functions are calculated using first-order anharmonic corrections to Q_r^i and Q_v^i in the partition function $Q = Q_t \Sigma_i Q_r^i Q_v^i g_i \exp(-c_2\varepsilon_i/T)$.

References

1. R. Colin, P. Goldfinger and M. Jeunehomme, Trans. Faraday Soc. 60, 306 (1964).
2. J. R. Marquart and J. Berkowitz, J. Chem. Phys. 39, 283 (1963).
3. E. D. Cater and E. W. Johnson, J. Chem. Phys. 47, 5353 (1967).
4. U. S. Nat. Bur. Stand. Tech. Note 270-6, 1971.
5. K. C. Mills, "Thermodynamic Data for Inorganic Sulphies, Selenides, and Tellurides," Butterworths, London, 1974.
6. JANAF Thermochemical Tables: Sr(g), 12-31-70; S(g), 6-30-71; S_2(g), 12-31-65; BaO(g), SrO(g), 6-30-74; BeO(g), CaO(g), MgO(g), 12-31-74; BeS(g), MgS(g), CaS(g), BaS(g), 9-30-77.
7. "Diatomic Molecules, A Critical Bibliography of Spectroscopic Data," I-1973, R. F. Barrow, ed., Centre National de la Recherche Scientifique, Paris, p. 428, 1973.

Strontium Sulfide (SrS) S_1Sr_1(g)

Enthalpy Reference Temperature = T_r = 298.15 K   Standard State Pressure = p° = 0.1 MPa

T/K	$C_p°$ (J $K^{-1}mol^{-1}$)	S°	$-[G°-H°(T_r)]/T$	$H°-H°(T_r)$ (kJ mol^{-1})	$\Delta_f H°$	$\Delta_f G°$	Log K_f
0	0.	0.	INFINITE	-9.536	109.642	109.642	INFINITE
100	30.117	207.226	273.308	-6.608	110.381	93.716	-48.952
200	33.694	229.274	246.305	-3.406	109.453	77.367	-20.206
250	34.778	236.918	243.688	-1.692	108.845	69.410	-14.502
298.15	35.477	243.107	243.107	0.	108.198	61.878	-10.841
300	35.499	243.327	243.108	0.066	108.172	61.591	-10.724
350	35.993	248.838	243.542	1.854	107.441	53.884	-8.042
400	36.343	253.669	244.512	3.663	104.443	46.374	-6.056
450	36.601	257.965	245.773	5.486	102.940	39.197	-4.550
500	36.796	261.832	247.188	7.322	101.346	32.202	-3.364
600	37.072	268.567	250.207	11.016	98.526	18.642	-1.623
700	37.262	274.296	253.249	14.733	95.863	5.540	-0.413
800	37.414	279.282	256.198	18.467	93.246	-7.185	0.469
900	37.578	283.698	259.013	22.216	36.864	-18.428	1.070
1000	37.811	287.668	261.683	25.985	35.743	-24.512	1.280
1100	38.185	291.288	264.212	29.783	26.734	-30.118	1.430
1200	38.773	294.634	266.610	33.629	24.746	-35.198	1.532
1300	39.639	297.770	268.887	37.547	22.816	-40.115	1.612
1400	40.828	300.749	271.057	41.568	20.972	-44.886	1.675
1500	42.357	303.616	273.133	45.724	19.249	-49.529	1.725
1600	44.211	306.407	275.126	50.050	17.680	-54.062	1.765
1700	46.345	309.150	277.047	54.576	-120.618	-57.323	1.761
1800	48.689	311.865	278.906	59.326	-119.930	-53.618	1.556
1900	51.155	314.563	280.711	64.318	-119.019	-49.958	1.373
2000	53.642	317.250	282.471	69.558	-117.880	-46.352	1.211
2100	56.053	319.926	284.191	75.044	-116.518	-42.808	1.065
2200	58.297	322.586	285.876	80.763	-114.950	-39.334	0.934
2300	60.298	325.223	287.530	86.695	-113.200	-35.935	0.816
2400	62.001	327.826	289.154	92.813	-111.302	-32.616	0.710
2500	63.372	330.386	290.753	99.084	-109.295	-29.378	0.614
2600	64.398	332.893	292.325	105.475	-107.219	-26.223	0.527
2700	65.086	335.337	293.873	111.952	-105.117	-23.147	0.448
2800	65.456	337.712	295.397	118.482	-103.032	-20.150	0.376
2900	65.538	340.011	296.896	125.034	-101.002	-17.226	0.310
3000	65.370	342.231	298.370	131.581	-99.066	-14.370	0.250
3100	64.991	344.369	299.820	138.101	-97.255	-11.577	0.195
3200	64.443	346.424	301.244	144.574	-95.598	-8.840	0.144
3300	63.762	348.397	302.643	150.985	-94.120	-6.152	0.097
3400	62.983	350.289	304.017	157.323	-92.841	-3.506	0.054
3500	62.136	352.102	305.365	163.580	-91.778	-0.895	0.013
3600	61.248	353.840	306.688	169.749	-90.945	1.689	-0.025
3700	60.339	355.506	307.985	175.828	-90.349	4.254	-0.060
3800	59.427	357.103	309.257	181.817	-89.998	6.805	-0.094
3900	58.526	358.635	310.503	187.714	-89.896	9.351	-0.125
4000	57.645	360.106	311.725	193.523	-90.045	11.897	-0.155
4100	56.793	361.519	312.922	199.244	-90.443	14.450	-0.184
4200	55.975	362.877	314.096	204.882	-91.085	17.016	-0.212
4300	55.195	364.185	315.246	210.440	-91.972	19.599	-0.238
4400	54.455	365.446	316.372	215.923	-93.097	22.207	-0.264
4500	53.756	366.661	317.476	221.333	-94.451	24.842	-0.288
4600	53.098	367.836	318.558	226.675	-95.962	27.505	-0.312
4700	52.481	368.971	319.619	231.954	-97.733	30.208	-0.336
4800	51.903	370.070	320.659	237.172	-99.702	32.950	-0.359
4900	51.364	371.134	321.678	242.336	-101.861	35.736	-0.381
5000	50.861	372.167	322.677	247.446	-104.191	38.567	-0.403
5100	50.394	373.169	323.658	252.509	-106.669	41.444	-0.424
5200	49.960	374.144	324.619	257.526	-109.320	44.374	-0.446
5300	49.557	375.091	325.563	262.502	-112.113	47.356	-0.467
5400	49.183	376.014	326.488	267.439	-115.037	50.392	-0.487
5500	48.837	376.913	327.397	272.339	-118.080	53.484	-0.508
5600	48.517	377.790	328.289	277.207	-121.231	56.631	-0.528
5700	48.221	378.647	329.165	282.044	-124.479	59.836	-0.548
5800	47.948	379.483	330.026	286.852	-127.814	63.099	-0.568
5900	47.696	380.300	330.871	291.634	-131.016	66.395	-0.588
6000	47.464	381.100	331.701	296.392	-134.465	69.770	-0.607

PREVIOUS: September 1977 (1 atm)   CURRENT: September 1977 (1 bar)

Strontium Sulfide (SrS) S_1Sr_1(g)

$S°(298.15\ K) = 228.165 \pm 0.05\ J\ K^{-1}\ mol^{-1}$

$\Delta_f H°(0\ K) = 128.30 \pm 0.30\ kJ\ mol^{-1}$

$\Delta_f H°(298.15\ K) = 128.60 \pm 0.30\ kJ\ mol^{-1}$

Electronic and Molecular Constants

State	g_i	ε_i, cm^{-1}	ω_e, cm^{-1}	$\omega_e x_e$, cm^{-1}	B_e, cm^{-1}	α_e, cm^{-1}	D_e, cm^{-1}	r_e, Å
$X^3\Sigma_g^-$	3	0	724.67	2.836	0.2946	0.00157	2.134×10^{-7}	1.889
$a^1\Delta_g$	2	4700	702.35	3.09	0.2923	0.0017	2.040×10^{-7}	1.889
$b^1\Delta_g$	1	8500	700.87	3.47	[0.29]	[0.0016]	$[2.0 \times 10^{-7}]$	--
$A^3\Delta$	6	21855	488.6	2.63	0.2284	0.0014	1.996×10^{-7}	2.148

Enthalpy of Formation

The adopted enthalpy of formation is that selected by CODATA (1). The value was calculated from measurments of the equilibrium $H_2S(g) = H_2(g) \pm 0.05\ S_2(g)$ reported by Randall and Bichowsky (2), Preuner (3), and Preuner and Schupp (4), and from measurements of the partial pressure of $S_2(g)$ over sulfur reported by Drowart, Goldfinger, Detry, Rickert, and Keller (5).

The dissociation energy, $D_0^0 = 35216.4\pm2.5\ cm^{-1}$ ($100.689\pm0.021\ kcal\ mol^{-1}$) is from the spectroscopic study by Ricks and Barrow (12).

Heat Capacity and Entropy

The thermal functions above 600 K are calculated using the program of McBride and Gordon (6). The contribution of vibrational anharmonicity, rotation-vibration interaction, and centrifugal stretching are calculated via the procedures given by Pennington and Kobe (7). The electronic and molecular constant data in the above table are that summarized by Rosen (8) and Suchard (9) with the exception that the position of the $a^1\Delta_g$ state at $4700\pm800\ cm^{-1}$ is from the more recent absorption study by Carleer and Colin (10).

The thermal functions below 600 K are calculated by direct summation over vibrational-rotational energy levels of the ground state (including the proper treatment of the splitting of this level, i.e. $\alpha = -0.0066\ cm^{-1}$ and $\lambda = 11.84\ cm^{-1}$, and three excited states. We performed the direct summation with a program written by W. H. Evans of the U. S. Bureau of Standards. The direct summation results are not extended to temperatures above 600 K since the enrgy levels derived from the reported vibrational-rotational constants do not converge to the dissociation energy. For these higher temperatures, the program of McBride and Gordon is used, as in the case of gaseous O_2 (11).

The value of S°(298.15 K) agrees exactly with the CODATA recommended value (1). The uncertainty in S°(298.15 K) is due to uncertainties in fundamental constants, the atomic weight, and the ground state molecular constants. The contributions of the excited states above 22000 cm^{-1} do not affect the thermal functions below 2000 K but do become significant at higher temperatures. However, these states are not included for two reasons. First, the electronic and molecular constants are not all well-defined. Second, many of these levels have a sufficiently shallow potential energy well, which would lead to unreasonably large heat capacity values at high temperatures.

References

1. J. D. Cox, chairman, CODATA Task Group on Key Values in Thermodynamics, J. Chem. Thermodyn. 4, 331 (1972).
2. M. Randall and F. R. Bichowsky, J. Am. Chem. Soc. 40, 368 (1918).
3. G. Preuner, Z. Anorg. Chem. 55, 279 (1907).
4. G. Preuner and W. Schupp, Z. Physik Chem. 68, 157 (1910).
5. J. Drowart, P. Goldfinger, D. Detrey, H. Rickert, H. Keller, Advan. Mass Spectroscopy 4, 499 (1968).
6. B. J. McBride and S. Gordon, NASA TN D-4097, 1967.
7. R. E. Pennington and K. A. Kobe, J. Chem. Phys. 22, 1442 (1954).
8. T. B. Rosen, Ed., "Spectroscopic Data Relative to Diatomic Molecules," Pergamon Press, New York, 1970.
9. S. N. Suchard and J. E. Melzer, "Spectroscopic Data," Vol. 2 (Honomuclear Diatomic Molecules), IFI Plenum, New York, 1976.
10. M. Carleer and R. Colin, J. Phys. B 3, 1715 (1970).
11. JANAF Thermochemical Tables: $O_2(g)$, 3-31-77.
12. J. M. Ricks and R. F. Barrow, Can. J. Phys. 47, 2423 (1969).

Enthalpy Reference Temperature = T_r = 298.15 K; Standard State Pressure = p° = 0.1 MPa

T/K	$C_p°$ (J K^{-1}mol^{-1})	S° (J K^{-1}mol^{-1})	$-[G°-H°(T_r)]/T$ (J K^{-1}mol^{-1})	$H°-H°(T_r)$ (kJ mol^{-1})	$\Delta_f H°$ (kJ mol^{-1})	$\Delta_f G°$ (kJ mol^{-1})	Log K_f
0	0.	0.	INFINITE	-9.124	128.300	128.300	INFINITE
100	29.367	195.067	255.684	-6.062	129.983	112.980	-59.015
200	30.452	215.621	231.072	-3.090	129.670	96.001	-25.073
250	31.513	222.529	228.695	-1.541	129.180	87.637	-18.311
298.15	32.490	228.165	228.165	0.	128.600	79.687	-13.961
300	32.525	228.366	228.166	0.060	128.576	79.384	-13.822
350	33.387	233.447	228.565	1.709	127.892	71.238	-10.632
400	34.090	237.953	229.462	3.396	122.718	63.371	-8.275
450	34.656	242.002	230.634	5.115	120.586	56.063	-6.508
500	35.111	245.678	231.958	6.860	118.326	49.019	-5.121
600	35.781	252.142	234.798	10.407	114.703	35.511	-3.092
700	36.260	257.695	237.681	14.009	111.612	22.562	-1.684
800	36.637	262.562	240.493	17.655	108.835	10.033	-0.655
882.117	36.908	266.155	242.717	20.674	------- FUGACITY = 1 bar -------		
900	36.966	266.896	243.191	21.335	0.	0.	0.
1000	37.277	270.807	245.760	25.047	0.	0.	0.
1100	37.584	274.374	248.201	28.790	0.	0.	0.
1200	37.894	277.658	250.521	32.564	0.	0.	0.
1300	38.205	280.703	252.727	36.369	0.	0.	0.
1400	38.514	283.546	254.828	40.205	0.	0.	0.
1500	38.818	286.213	256.832	44.072	0.	0.	0.
1600	39.112	288.728	258.748	47.968	0.	0.	0.
1700	39.394	291.108	260.582	51.894	0.	0.	0.
1800	39.661	293.367	262.341	55.847	0.	0.	0.
1900	39.911	295.518	264.031	59.825	0.	0.	0.
2000	40.144	297.571	265.657	63.828	0.	0.	0.
2100	40.352	299.535	267.224	67.853	0.	0.	0.
2200	40.547	301.417	268.736	71.898	0.	0.	0.
2300	40.729	303.223	270.196	75.962	0.	0.	0.
2400	40.897	304.960	271.609	80.044	0.	0.	0.
2500	41.052	306.633	272.977	84.141	0.	0.	0.
2600	41.179	308.246	274.302	88.253	0.	0.	0.
2700	41.299	309.802	275.588	92.377	0.	0.	0.
2800	41.414	311.306	276.837	96.512	0.	0.	0.
2900	41.523	312.761	278.051	100.659	0.	0.	0.
3000	41.626	314.171	279.232	104.817	0.	0.	0.
3100	41.711	315.537	280.381	108.984	0.	0.	0.
3200	41.795	316.863	281.500	113.159	0.	0.	0.
3300	41.880	318.150	282.592	117.343	0.	0.	0.
3400	41.966	319.402	283.656	121.535	0.	0.	0.
3500	42.051	320.619	284.695	125.736	0.	0.	0.
3600	42.130	321.805	285.709	129.945	0.	0.	0.
3700	42.212	322.960	286.700	134.162	0.	0.	0.
3800	42.298	324.087	287.669	138.387	0.	0.	0.
3900	42.388	325.187	288.617	142.622	0.	0.	0.
4000	42.481	326.261	289.545	146.865	0.	0.	0.
4100	42.575	327.312	290.454	151.118	0.	0.	0.
4200	42.675	328.339	291.343	155.380	0.	0.	0.
4300	42.778	329.344	292.215	159.653	0.	0.	0.
4400	42.886	330.329	293.070	163.936	0.	0.	0.
4500	42.997	331.294	293.909	168.230	0.	0.	0.
4600	43.115	332.240	294.732	172.536	0.	0.	0.
4700	43.236	333.169	295.540	176.853	0.	0.	0.
4800	43.361	334.080	296.334	181.183	0.	0.	0.
4900	43.488	334.975	297.113	185.525	0.	0.	0.
5000	43.619	335.855	297.879	189.881	0.	0.	0.
5100	43.757	336.721	298.632	194.250	0.	0.	0.
5200	43.897	337.572	299.373	198.632	0.	0.	0.
5300	44.038	338.409	300.102	203.029	0.	0.	0.
5400	44.181	339.234	300.819	207.440	0.	0.	0.
5500	44.325	340.046	301.525	211.865	0.	0.	0.
5600	44.472	340.845	302.220	216.305	0.	0.	0.
5700	44.620	341.634	302.904	220.760	0.	0.	0.
5800	44.770	342.411	303.579	225.229	0.	0.	0.
5900	44.922	343.178	304.243	229.714	0.	0.	0.
6000	45.075	343.934	304.899	234.213	0.	0.	0.

PREVIOUS: September 1977 (1 atm) CURRENT: September 1977 (1 bar)

SILICON SULFIDE (SiS_2) CRYSTAL M_r = 92.2055

$S°(298.15\ K) = [80.333 \pm 4.2]\ J\ K^{-1}\ mol^{-1}$

$T_{fus} = 1363\ K$

$\Delta_f H°(0\ K)$ = Unknown

$\Delta_f H°(298.15\ K) = -213.384 \pm 20.9\ kJ\ mol^{-1}$

$\Delta_{fus} H° = [8.4 \pm 4.2]\ kJ\ mol^{-1}$

Enthalpy of Formation

Rocquet and Ancey-Haret (1) determined the enthalpy of reaction for the following:

$SiS_2(cr) + 6\ HF(20\%) + H_2SiF_6(<20\%HF) + 2\ H_2S(g)$, $\Delta_r H°(298.15\ K) = -72.8 \pm 2.2\ kcal\ mol^{-1}$.

A related system was examined by Kilday and Prosen (2):

$SiO_2\ (\alpha\text{-quartz}) + 6\ HF(20.0\%) + H_2SiF_6(<20.0\%\ HF) + 2\ H_2O(\ell)$, $\Delta_r H°(298.15\ K) = -32.65 \pm 0.02\ kcal\ mol^{-1}$.

Combining these two systems mathematically yields approximately the following:

$SiO_2\ (\alpha\text{-quartz}) + 2\ H_2S(g) + SiS_2(cr) + 2\ H_2O(\ell)$, $\Delta_r H°(298.15\ K) = 40.2 \pm 3.0\ kcal\ mol^{-1}$.

Using JANAF (4) and NBS (6) auxiliary data, $\Delta_f H°(SiS_2,\ cr,\ 298.15\ K)$ is equal to $-51.1 \pm 3.0\ kcal\ mol^{-1}$. In addition to the normal experimental error in each study, another source of error arises from the saturated solution of H_2S in the former work but not the latter.

Emmons and Theisen (3) measured the equilibrium vapor pressures for the system $1/2\ SiS_2(cr) + 1/2\ Si(cr) + SiS(g)$ and presented graphically log P(SiS) vs 1/T data. Using these data, a 3rd law analysis yields $\Delta_r H°(298.15\ K) = 55.27\ kcal\ mol^{-1}$ with a drift of $5.1 \pm 1.2\ cal\ K^{-1}\ mol^{-1}$. A 2nd law analysis yields $\Delta_r H°(298.15\ K) = 50.09\ kcal\ mol^{-1}$. These results lead, respectively, to $\Delta_f H°(SiS_2,\ cr,\ 298.15\ K) = -59.84\ kcal\ mol^{-1}$ and $\Delta_f H°(SiS_2,\ cr,\ 298.15\ K) = -49.53\ kcal\ mol^{-1}$, with JANAF (4) auxiliary data.

Fruehan and Turkdogan (5) studied the same reaction by means of a silica Knudsen cell-mass specrometer combination. A 2nd law analysis of their graphical data for two different orifice sizes yields values of $\Delta_r H°(298.15\ K) = 53.74\ kcal\ mol^{-1}$ and 46.9 $kcal\ mol^{-1}$. Again, using JANAF auxiliary data (4), $\Delta_f H°(SiS_2,\ cr,\ 298.15\ K) = -56.83\ kcal\ mol^{-1}$ and $-42.33\ kcal\ mol^{-1}$ respectively, or an average value of $-49.58\ kcal\ mol^{-1}$.

The value chosen for the enthalpy of formation is $\Delta_f H°(SiS_2,\ cr,\ 298.15\ K) = -51.0 \pm 5.0$ kcal mol. This value is a rounded value of the work of Rocquet and Ancey-Maret (1). Berezknoi (7) references five reported values for the enthalpy of formation (through 1954) and recommended the work of Rocquet and Ancey-Maret (1) as the most accurate. Kubaschewski et al. (8) in their compilation of enthlpies of formation (through 1965) also rely on the data from Rocquet and Ancey-Maret (1).

Heat Capacity and Entropy

The heat capacity is assumed to be given by the relation $C_p°/cal\ K^{-1}\ mol^{-1} = 17.72 \pm 2.96 \times 10^{-3}T$ for $298 \le T \le 1363$ K. This relation was suggested by Rasch (9) following procedures described by Kubaschewski et al. (8). This equation was linearly extrapolated to yield $C_p°$ values up to 2000 K. Kubaschewski et al. (8) suggested $S°(298.15\ K) = 19.2\ cal\ K^{-1}\ mol^{-1}$.

Fusion Data

The melting point of $SiS_2(cr)$ was determined to be T_{fus} = 1363 K by Tiede and Thimann (10). $\Delta_{fus}H°$ was estimated, based on the entropy of melting for SiO_2 (quartz and cristobolite).

References

1. P. Rocquet and H. F. Ancey-Maret, Bull. Soc. Chim. France 1954, 1038 (1954).
2. M. V. Kilday and E. J. Prosen, U. S. Nat. Bur. Stand., personal communication, July 26, 1972.
3. H. H. Emmons and L. Theisen, Monatsh. Chem. 103, 62 (1972).
4. JANAF Thermochemical Tables: SiS(g), 12-31-71; SiO_2(cr), 6-30-67; H_2S(g), 12-31-65.
5. R. J. Fruehan and E. T. Turkdogan, Met. Trans. 2, 895 (1971).
6. U. S. Nat. Bur. Stand. Tech. Note 270-3, 1968.
7. A. S. Berezknoi, "Silicon and its Binary Systems," Consultants Bureau, New York, 1960 (Russ. trans.).
8. O. Kubaschewski, E. L. Evans, and C. B. Alcock, "Metallurgical Thermochemistry," Pergamon Press, New York, 1967.
9. R. Rasch, Glas-Email-Keramo-Tech. 20, 297 (1969).
10. E. Tiede and H. Thimann, Ber. 59, II B, 1703 (1926).

Silicon Sulfide (SiS_2) $S_2Si_1(cr)$

Enthalpy Reference Temperature = T_r = 298.15 K; Standard State Pressure = p° = 0.1 MPa

T/K	$C_p°$ (J K⁻¹mol⁻¹)	S° (J K⁻¹mol⁻¹)	$-[G°-H°(T_r)]/T$ (J K⁻¹mol⁻¹)	$H°-H°(T_r)$ (kJ mol⁻¹)	$\Delta_f H°$ (kJ mol⁻¹)	$\Delta_f G°$ (kJ mol⁻¹)	Log K_f
0							
100							
200							
298.15	77.488	80.333	80.333	0.	-213.384	-212.609	37.248
300	77.508	80.812	80.334	0.143	-213.362	-212.605	37.018
400	78.617	103.260	83.387	7.949	-216.872	-212.329	27.727
500	79.789	120.915	89.189	15.863	-219.092	-210.962	22.039
600	81.714	135.657	95.737	23.952	-220.548	-209.185	18.211
700	82.006	148.265	102.363	32.131	-221.511	-207.211	15.462
800	83.136	159.289	108.804	40.388	-222.185	-205.119	13.393
900	84.266	169.146	114.970	48.758	-328.891	-200.795	11.654
1000	85.395	178.083	120.842	57.241	-326.731	-186.677	9.751
1100	86.525	186.275	126.423	65.837	-324.534	-172.778	8.205
1200	87.655	193.852	131.730	74.546	-322.297	-159.081	6.925
1300	88.784	200.913	136.783	83.368	-320.021	-145.571	5.849
1363.000	89.496	205.131	139.846	88.984	------ CRYSTAL <--> LIQUID -----		
1400	89.915	207.534	141.603	92.303	-317.704	-132.239	4.934
1500	91.044	213.776	146.208	101.351	-315.347	-119.073	4.146
1600	92.154	219.687	150.618	110.511	-312.950	-106.066	3.463
1700	93.270	225.307	154.847	119.782	-360.689	-92.764	2.850
1800	94.391	230.670	158.912	129.165	-357.979	-77.081	2.237
1900	95.518	235.804	162.824	138.660	-355.182	-61.552	1.692
2000	96.650	240.732	166.597	148.269	-352.296	-46.172	1.206

PREVIOUS: December 1960 CURRENT: June 1972

Silicon Sulfide (SiS_2) $S_2Si_1(cr)$

SILICON SULFIDE (SiS_2) LIQUID M_r = 92.2055

Silicon Sulfide (SiS_2) S_2Si_1(l)

S°(298.15 K) = [84.687] J K^{-1} mol^{-1}

T_{fus} = 1363 K

$\Delta_f H°$(298.15 K) = [-206.924] kJ mol^{-1}

$\Delta_{fus} H°$ = [8.4 ± 4.2] kJ mol^{-1}

Enthalpy of Formation

$\Delta_f H°$(ℓ, 298.15 K) is calculated from $\Delta_f H°$(cr, 298.15 K) by adding $\Delta_{fus} H°$ and the difference in enthalpy, H°(1363 K)-H°(298.15 K), between the crystal and the liquid.

Heat Capacity and Entropy

Using methods suggested by Kubaschewski et al. (1), the heat capacity of the liquid is assumed to be a constant 21.75 cal K^{-1} mol^{-1} from 1363 K to 1403 K. S°(ℓ, 298.15 K) is calculated in a manner similar to that used for the enthalpy of formation. A glass transition is assumed at 900 K.

Fusion Data

See the crystal table for details.

References

1. O. Kubaschewski, E. L. Evans, and C. B. Alcock, "Metallurgical Thermochemistry," Pergamon Press, New York, 1967.

Enthalpy Reference Temperature = T_r = 298.15 K Standard State Pressure = p° = 0.1 MPa

T/K	$C_p°$ (J $K^{-1}mol^{-1}$)	S° (J $K^{-1}mol^{-1}$)	$-[G°-H°(T_r)]/T$ (J $K^{-1}mol^{-1}$)	$H°-H°(T_r)$ (kJ mol^{-1})	$\Delta_f H°$ (kJ mol^{-1})	$\Delta_f G°$ (kJ mol^{-1})	Log K_f
0							
100							
200							
298.15	77.488	84.687	84.687	0.	-206.924	-207.448	36.344
300	77.508	85.166	84.688	0.143	-206.902	-207.451	36.120
400	78.617	107.614	87.742	7.949	-210.412	-207.610	27.111
500	79.789	125.269	93.543	15.863	-212.632	-206.679	21.592
600	81.714	140.011	100.091	23.952	-214.089	-205.337	17.876
700	82.006	152.619	106.717	32.131	-215.051	-203.799	15.208
800	83.136	163.643	113.158	40.388	-215.725	-202.142	13.199
900	84.266	173.500	119.324	48.758	-322.431	-198.254	11.506
900.000	84.266	173.500	119.324	48.758	GLASS <--> LIQUID TRANSITION		
900.000	91.002	173.500	119.324	48.758			
1000	91.002	183.088	125.230	57.858	-319.654	-184.606	9.643
1100	91.002	191.762	130.890	66.958	-316.952	-171.232	8.131
1200	91.002	199.680	136.297	76.059	-314.325	-158.102	6.882
1300	91.002	206.964	141.457	85.159	-311.770	-145.187	5.834
1363.000	91.002	211.270	144.585	90.892	------ CRYSTAL <--> LIQUID -----		
1400	91.002	213.708	146.380	94.259	-309.288	-132.467	4.942
1500	91.002	219.986	151.080	103.359	-306.879	-119.921	4.176
1600	91.002	225.859	155.572	112.459	-304.541	-107.534	3.511
1700	91.002	231.376	159.871	121.560	-352.452	-94.844	2.914
1800	91.002	236.578	163.989	130.660	-350.024	-79.760	2.315
1900	91.002	241.498	167.940	139.760	-347.622	-64.811	1.782
2000	91.002	246.166	171.736	148.860	-345.244	-49.988	1.306

PREVIOUS: December 1960 CURRENT: June 1972

Silicon Sulfide (SiS_2) S_2Si_1(l)

SILICON SULFIDE (SiS_2) CRYSTAL–LIQUID M_r = 92.2055

0 to 1363 K crystal
above 1363 K liquid

Refer to the individual tables for details.

Silicon Sulfide (SiS_2) S_2Si_1(cr,l)

Enthalpy Reference Temperature = T_r = 298.15 K — Standard State Pressure = p° = 0.1 MPa

T/K	C_p° (J K⁻¹mol⁻¹)	S° (J K⁻¹mol⁻¹)	$-[G^\circ-H^\circ(T_r)]/T$ (J K⁻¹mol⁻¹)	$H^\circ-H^\circ(T_r)$ (kJ mol⁻¹)	Δ_fH° (kJ mol⁻¹)	Δ_fG° (kJ mol⁻¹)	Log K_f
0							
100							
200							
298.15	77.488	80.333	80.333	0.	-213.384	-212.609	37.248
300	77.508	80.812	80.334	0.143	-213.362	-212.605	37.018
400	78.617	103.260	83.387	7.949	-216.872	-212.329	27.727
500	79.789	120.915	89.189	15.863	-219.092	-210.962	22.039
600	81.714	135.657	95.737	23.952	-220.548	-209.185	18.211
700	82.006	148.265	102.363	32.131	-221.511	-207.211	15.462
800	83.136	159.289	108.804	40.388	-222.185	-205.119	13.393
900	84.266	169.146	114.970	48.758	-328.891	-200.795	11.654
1000	85.395	178.083	120.842	57.241	-326.731	-186.677	9.751
1100	86.525	186.275	126.423	65.837	-324.534	-172.778	8.205
1200	87.655	193.852	131.730	74.546	-322.297	-159.081	6.925
1300	88.784	200.913	136.783	83.368	-320.021	-145.571	5.849
1363.000	89.496	205.131	139.846	88.984	CRYSTAL <--> LIQUID		
1363.000	91.002	211.270	139.846	97.352	TRANSITION		
1400	91.002	213.708	141.766	100.719	-309.288	-132.467	4.942
1500	91.002	219.986	146.773	109.819	-306.879	-119.921	4.176
1600	91.002	225.859	151.535	118.919	-304.541	-107.534	3.511
1700	91.002	231.376	156.071	128.020	-352.452	-94.844	2.914
1800	91.002	236.578	160.400	137.120	-350.024	-79.761	2.315
1900	91.002	241.498	164.540	146.220	-347.622	-64.811	1.782
2000	91.002	246.166	168.506	155.320	-345.244	-49.988	1.306

PREVIOUS: CURRENT: June 1972

Silicon Sulfide (SiS_2) S_2Si_1(cr,l)

$S°(298.15 K) = [269.5 \pm 4]\ J\ K^{-1}\ mol^{-1}$ $\Delta_f H°(298.15 K) = [141.5 \pm 8]\ kJ\ mol^{-1}$

Enthalpy of Formation

The enthalpy of formation for each vapor phase sulfur species S_3 to S_7 is calculated by a third law calculation using the thermal functions described below and assuming that the vapor pressure data given by Hultgren et al. (1) and Rau et al. (2) are correct. These values for the enthalpy of formation are dependent on the somewhat arbitrary thermal functions adopted for the polyatomic sulfur gases.

Heat Capacity and Entropy

The thermal functions for the polyatomic species, S_3 to S_7, are estimated following the suggestion of Rau et al. (2). Using thermal functions calculated for S(g), S_2(g), and S_8(g), the heat capacity values for the remaining species were calculated by Rau et al. assuming that the vibrational heat capacity per degree of freedom is the same for all sulfur polyatomic species. The resulting heat capacity equations are given by Rau et al. (2).

The use of the procedure of Rau et al. (2) leads to a set of thermal functions and associated enthalpies of formation which reproduce the observed sulfur vapor pressure data. That is, the sum of the calculated partial pressures of all eight sulfur vapor species, S_1(g) to S_8(g), does closely reproduce the observed vapor pressure. [A difference between the calculated boiling point (at 1 atm) and the secondary reference temperature boiling point is due to the difference between the ideal gas calculation and the real observed value.]

The thermal fuctions for S_8(g) were calculated (3) by adjusting the low valued vibrational frequencies such that S(298.15 K) and $\Delta_f H°$(298.15 K) values could be found which reproduced the partial pressure of the octamer. The entropy of S_2(g) is well-known (3), being based on reliable data and sound statistical mechanical calculations.

The thermal functions which would result from the use of the available structural and vibrational inbformation for the remaining polyatomic species, are biased in the same way as for S_8(g). Some of the necessary information is derived from condensed phase spectra (rather than gas phase spectra). as with the octamer, the low-valued vibrational frequencies (and possibly the structure) may need to be adjusted. In any event, the use of the existing spectroscopic data (unaltered) leads to calculated entropies which are unreasonable when used in a 3rd law analysis to mesh in the vapor pressure with these thermal functions.

References

1. R. Hultgren, P. D. Desai, et al., "Selected Values of the Thermodynamic Properties of the Elements", American Society for Metals, Metals Park, Ohio, 1973.
2. H. Rau, R. N. Kutty and J. R. F. Guedes de Carvalho, J. Chem. Thermodynamics 5, 833 (1973).
3. JANAF Thermochemical Tables: S_2(g), 9-30-77.

Enthalpy Reference Temperature = T_r = 298.15 K Standard State Pressure = p° = 0.1 MPa

T/K	$C_p°$ (J K⁻¹mol⁻¹)	S° (J K⁻¹mol⁻¹)	$-[G°-H°(T_r)]/T$ (J K⁻¹mol⁻¹)	$H°-H°(T_r)$ (kJ mol⁻¹)	$\Delta_f H°$ (kJ mol⁻¹)	$\Delta_f G°$ (kJ mol⁻¹)	Log K_f
0							
100							
200							
298.15	47.764	269.517	269.517	0.	141.461	89.777	-15.729
300	47.862	269.813	269.518	0.088	141.423	89.456	-15.576
350	49.996	277.363	270.109	2.539	140.375	80.877	-12.070
400	51.458	284.140	271.447	5.077	132.621	72.716	-9.496
450	52.528	290.265	273.203	7.678	129.446	65.392	-7.590
500	53.356	295.844	275.192	10.326	126.086	58.462	-6.107
600	54.586	305.687	279.477	15.727	120.732	45.460	-3.958
700	55.500	314.173	283.841	21.233	116.198	33.281	-2.483
800	56.246	321.635	288.108	26.821	112.152	21.717	-1.418
900	56.895	328.298	292.210	32.479	-50.960	13.882	-0.806
1000	57.482	334.323	296.125	38.198	-50.809	21.078	-1.101
1100	58.030	339.827	299.851	43.974	-50.648	28.259	-1.342
1200	58.551	344.899	303.397	49.803	-50.480	35.425	-1.542
1300	59.053	349.606	306.772	55.684	-50.307	42.577	-1.711
1400	59.541	354.000	309.990	61.613	-50.131	49.715	-1.855
1500	60.019	358.124	313.063	67.591	-49.953	56.841	-1.979
1600	60.489	362.013	316.002	73.617	-49.772	63.955	-2.088
1700	60.953	365.694	318.818	79.689	-49.588	71.057	-2.183
1800	61.413	369.191	321.520	85.807	-49.399	78.148	-2.268
1900	61.869	372.524	324.118	91.972	-49.203	85.229	-2.343
2000	62.321	375.709	326.618	98.181	-48.998	92.299	-2.411
2100	62.772	378.760	329.029	104.436	-48.781	99.359	-2.471
2200	63.220	381.691	331.356	110.735	-48.549	106.407	-2.526
2300	63.666	384.511	333.607	117.080	-48.300	113.445	-2.576
2400	64.112	387.230	335.785	123.469	-48.033	120.472	-2.622
2500	64.556	389.856	337.895	129.902	-47.746	127.487	-2.664
2600	64.998	392.397	339.943	136.380	-47.436	134.491	-2.702
2700	65.441	394.858	341.931	142.902	-47.100	141.482	-2.737
2800	65.882	397.246	343.864	149.468	-46.738	148.460	-2.770
2900	66.323	399.565	345.745	156.078	-46.348	155.424	-2.799
3000	66.763	401.821	347.577	162.732	-45.930	162.375	-2.827
3100	67.203	404.018	349.363	169.431	-45.482	169.311	-2.853
3200	67.642	406.158	351.104	176.173	-45.002	176.232	-2.877
3300	68.081	408.246	352.804	182.959	-44.492	183.138	-2.899
3400	68.520	410.285	354.465	189.789	-43.950	190.028	-2.919
3500	68.958	412.278	356.088	196.663	-43.378	196.901	-2.939
3600	69.396	414.226	357.676	203.581	-42.773	203.758	-2.956
3700	69.834	416.134	359.231	210.542	-42.138	210.597	-2.973
3800	70.271	418.002	360.753	217.547	-41.471	217.419	-2.989
3900	70.709	419.833	362.244	224.596	-40.773	224.223	-3.003
4000	71.146	421.629	363.706	231.689	-40.045	231.008	-3.017
4100	71.583	423.391	365.141	238.825	-39.288	237.776	-3.029
4200	72.020	425.121	366.548	246.006	-38.502	244.524	-3.041
4300	72.457	426.821	367.930	253.229	-37.687	251.253	-3.052
4400	72.894	428.492	369.288	260.497	-36.844	257.963	-3.062
4500	73.330	430.135	370.622	267.808	-35.974	264.653	-3.072
4600	73.767	431.751	371.933	275.163	-35.077	271.324	-3.081
4700	74.203	433.342	373.223	282.561	-34.155	277.975	-3.089
4800	74.639	434.909	374.492	290.004	-33.208	284.606	-3.097
4900	75.076	436.452	375.740	297.489	-32.236	291.217	-3.104
5000	75.512	437.974	376.970	305.019	-31.239	297.808	-3.111
5100	75.948	439.473	378.181	312.592	-30.219	304.379	-3.117
5200	76.384	440.952	379.374	320.208	-29.177	310.930	-3.123
5300	76.820	442.411	380.549	327.869	-28.112	317.460	-3.129
5400	77.256	443.851	381.708	335.572	-27.024	323.970	-3.134
5500	77.692	445.273	382.851	343.320	-25.915	330.460	-3.138
5600	78.128	446.677	383.978	351.111	-24.784	336.929	-3.143
5700	78.564	448.063	385.091	358.945	-23.631	343.378	-3.147
5800	79.000	449.434	386.188	366.824	-22.457	349.806	-3.150
5900	79.435	450.788	387.272	374.745	-21.262	356.215	-3.154
6000	79.871	452.126	388.341	382.711	-20.046	362.602	-3.157

PREVIOUS: December 1977 (1 atm) CURRENT: September 1977 (1 bar)

SULFUR (S_4) IDEAL GAS M_r = 128.24

$S°(298.15\ K) = [310.6 \pm 4]\ J\ K^{-1}\ mol^{-1}$ $\Delta_f H°(298.15\ K) = [145.8 \pm 8]\ kJ\ mol^{-1}$

Enthalpy of Formation

The enthalpy of formation for each vapor phase sulfur species S_3 to S_7 is calculated by a third law calculation using the thermal functions described below and assuming that the vapor pressure data given by Hultgren et al. (1) and Rau et al. (2) are correct. These values for the enthalpy of formation are dependent on the somewhat arbitrary thermal functions adopted for the polyatomic sulfur gases.

Heat Capacity and Entropy

The thermal functions for the polyatomic species, S_3 to S_7, are estimated following the suggestion of Rau et al. (2). Using thermal functions calculated for S(g), S_2(g), and S_8(g), the heat capacity values for the remaining species were calculated by Rau et al. assuming that the vibrational heat capacity per degree of freedom is the same for all sulfur polyatomic species. The resulting heat capacity equations are given by Rau et al. (2).

The use of the procedure of Rau et al. (2) leads to a set of thermal functions and associated enthalpies of formation which reproduce the observed sulfur vapor pressure data. That is, the sum of the calculated partial pressures of all eight sulfur vapor species, S_1(g) to S_8(g), does closely reproduce the observed vapor pressure. [A difference between the calculated boiling point (at 1 atm) and the secondary reference temperature boiling point is due to the difference between the ideal gas calculation and the real observed value.]

The thermal fuctions for S_8(g) were calculated (3) by adjusting the low valued vibrational frequencies such that S(298.15 K) and $\Delta_f H°$(298.15 K) values could be found which reproduced the partial pressure of the octamer. The entropy of S_2(g) is well-known (3), being based on reliable data and sound statistical mechanical calculations.

The thermal functions which would result from the use of the available structural and vibrational inbformation for the remaining polyatomic species, are biased in the same way as for S_8(g). Some of the necessary information is derived from condensed phase spectra (rather than gas phase spectra). As with the octamer, the low-valued vibrational frequencies (and possibly the structure) may need to be adjusted. In any event, the use of the existing spectroscopic data (unaltered) leads to calculated entropies which are unreasonable when used in a 3rd law analysis to mesh in the vapor pressure with these thermal functions.

References

1. R. Hultgren, P. D. Desai, et al., "Selected Values of the Thermodynamic Properties of the Elements", American Society for Metals, Metals Park, Ohio, 1973.
2. H. Rau, R. N. Kutty and J. R. F. Guedes de Carvalho, J. Chem. Thermodynamics 5, 833 (1973).
3. JANAF Thermochemical Tables: S_2(g), 9-30-77.

Sulfur (S_4) S_4(g)

Enthalpy Reference Temperature = T_r = 298.15 K Standard State Pressure = p° = 0.1 MPa

T/K	$C_p°$ (J K⁻¹mol⁻¹)	S° (J K⁻¹mol⁻¹)	$-[G°-H°(T_r)]/T$ (J K⁻¹mol⁻¹)	$H°-H°(T_r)$ (kJ mol⁻¹)	$\Delta_f H°$ (kJ mol⁻¹)	$\Delta_f G°$ (kJ mol⁻¹)	Log K_f
0							
100							
200							
298.15	67.585	310.646	310.646	0.	145.771	91.381	-16.010
300	67.754	311.065	310.647	0.125	145.728	91.044	-15.852
350	71.396	321.803	311.487	3.611	144.548	82.023	-12.241
400	73.817	331.505	313.393	7.245	134.459	73.525	-9.601
450	75.529	340.303	315.902	10.981	130.493	66.111	-7.674
500	76.799	348.330	318.749	14.790	126.292	59.192	-6.184
600	78.569	362.500	324.892	22.565	119.728	46.416	-4.041
700	79.766	374.707	331.156	30.485	114.261	34.640	-2.585
800	80.658	385.419	337.283	38.508	109.439	23.600	-1.541
900	81.373	394.962	343.172	46.611	-107.485	17.462	-1.013
1000	81.977	403.567	348.788	54.779	-106.741	31.305	-1.635
1100	82.509	411.406	354.129	63.004	-106.002	45.074	-2.140
1200	82.993	418.606	359.206	71.280	-105.275	58.776	-2.558
1300	83.442	425.267	364.035	79.602	-104.563	72.418	-2.910
1400	83.865	431.466	368.633	87.967	-103.869	86.006	-3.209
1500	84.271	437.266	373.017	96.374	-103.196	99.545	-3.466
1600	84.662	442.718	377.205	104.821	-102.542	113.039	-3.690
1700	85.042	447.862	381.211	113.306	-101.908	126.494	-3.887
1800	85.414	452.733	385.050	121.829	-101.291	139.911	-4.060
1900	85.779	457.361	388.735	130.389	-100.688	153.295	-4.214
2000	86.138	461.770	392.278	138.984	-100.098	166.647	-4.352
2100	86.493	465.981	395.688	147.616	-99.517	179.970	-4.477
2200	86.845	470.013	398.975	156.283	-98.940	193.265	-4.589
2300	87.193	473.881	402.149	164.985	-98.366	206.535	-4.691
2400	87.539	477.599	405.216	173.721	-97.792	219.779	-4.783
2500	87.882	481.180	408.183	182.493	-97.216	232.999	-4.868
2600	88.224	484.633	411.057	191.298	-96.634	246.196	-4.946
2700	88.564	487.969	413.844	200.137	-96.042	259.371	-5.018
2800	88.903	491.196	416.550	209.011	-95.440	272.523	-5.084
2900	89.241	494.322	419.178	217.918	-94.827	285.654	-5.145
3000	89.578	497.353	421.734	226.859	-94.201	298.764	-5.202
3100	89.914	500.296	424.221	235.834	-93.560	311.852	-5.255
3200	90.249	503.156	426.643	244.842	-92.903	324.919	-5.304
3300	90.584	505.938	429.004	253.883	-92.229	337.966	-5.350
3400	90.917	508.647	431.306	262.958	-91.538	350.992	-5.392
3500	91.251	511.287	433.554	272.067	-90.831	363.997	-5.432
3600	91.584	513.863	435.749	281.209	-90.108	376.982	-5.470
3700	91.916	516.377	437.895	290.384	-89.367	389.946	-5.505
3800	92.248	518.832	439.992	299.592	-88.610	402.891	-5.538
3900	92.580	521.233	442.045	308.833	-87.837	415.815	-5.569
4000	92.911	523.581	444.054	318.108	-87.049	428.719	-5.598
4100	93.243	525.879	446.022	327.415	-86.247	441.603	-5.626
4200	93.574	528.130	447.950	336.756	-85.431	454.468	-5.652
4300	93.904	530.336	449.840	346.130	-84.602	467.313	-5.677
4400	94.235	532.498	451.694	355.537	-83.761	480.138	-5.700
4500	94.565	534.620	453.514	364.977	-82.910	492.945	-5.722
4600	94.895	536.702	455.300	374.450	-82.048	505.732	-5.743
4700	95.225	538.746	457.053	383.956	-81.177	518.500	-5.762
4800	95.555	540.754	458.776	393.495	-80.297	531.250	-5.781
4900	95.885	542.728	460.470	403.067	-79.410	543.982	-5.799
5000	96.214	544.669	462.134	412.672	-78.516	556.695	-5.816
5100	96.544	546.577	463.771	422.310	-77.616	569.390	-5.832
5200	96.873	548.455	465.382	431.981	-76.710	582.068	-5.847
5300	97.202	550.303	466.967	441.684	-75.800	594.728	-5.861
5400	97.531	552.123	468.527	451.421	-74.885	607.371	-5.875
5500	97.860	553.916	470.063	461.191	-73.966	619.997	-5.888
5600	98.189	555.682	471.576	470.993	-73.043	632.606	-5.901
5700	98.518	557.423	473.067	480.829	-72.117	645.199	-5.913
5800	98.847	559.139	474.536	490.697	-71.188	657.775	-5.924
5900	99.176	560.832	475.985	500.598	-70.256	670.335	-5.935
6000	99.505	562.501	477.413	510.532	-69.321	682.880	-5.945

PREVIOUS: September 1977 (1 atm) CURRENT: September 1977 (1 bar)

Sulfur (S_4) S_4(g)

SULFUR (S_5) IDEAL GAS M_r = 160.30

$S°(298.15 K) = [308.6 \pm 4] J K^{-1} mol^{-1}$ $\Delta_f H°(298.15 K) = [109.4 \pm 8] kJ mol^{-1}$

Enthalpy of Formation

The enthalpy of formation for each vapor phase sulfur species S_3 to S_7 is calculated by a third law calculation using the thermal functions described below and assuming that the vapor pressure data given by Hultgren et al. (1) and Rau et al. (2) are correct. These values for the enthalpy of formation are dependent on the somewhat arbitrary thermal functions adopted for the polyatomic sulfur gases.

Heat Capacity and Entropy

The thermal functions for the polyatomic species, S_3 to S_7, are estimated following the suggestion of Rau et al. (2). Using thermal functions calculated for S(g), S_2(g), and S_8(g), the heat capacity values for the remaining species were calculated by Rau et al. assuming that the vibrational heat capacity per degree of freedom is the same for all sulfur polyatomic species. The resulting heat capacity equations are given by Rau et al. (2).

The use of the procedure of Rau et al. (2) leads to a set of thermal functions and associated enthalpies of formation which reproduce the observed sulfur vapor pressure data. That is, the sum of the calculated partial pressures of all eight sulfur vapor species, S_1(g) to S_8(g), does closely reproduce the observed vapor pressure. [A difference between the calculated boiling point (at 1 atm) and the secondary reference temperature boiling pcint is due to the difference between the ideal gas calculation and the real observed value.]

The thermal fuctions for S_8(g) were calculated (3) by adjusting the low valued vibrational frequencies such that S(298.15 K) and $\Delta_f H°$(298.15 K) values could be found which reproduced the partial pressure of the octamer. The entropy of S_2(g) is well-known (3), being based on reliable data and sound statistical mechanical calculations.

The thermal functions which would result from the use of the available structural and vibrational inbformation for the remaining polyatomic species, are biased in the same way as for S_8(g). Some of the necessary information is derived from condensed phase spectra (rather than gas phase spectra). As with the octamer, the low-valued vibrational frequencies (and possibly the structure) may need to be adjusted. In any event, the use of the existing spectroscopic data (unaltered) leads to calculated entropies which are unreasonable when used in a 3rd law analysis to mesh in the vapor pressure with these thermal functions.

References

1. R. Hultgren, P. D. Desai, et al., "Selected Values of the Thermodynamic Properties of the Elements", American Society for Metals, Metals Park, Ohio, 1973.
2. H. Rau, R. N. Kutty and J. R. F. Guedes de Carvalho, J. Chem. Thermodynamics 5, 833 (1973).
3. JANAF Thermochemical Tables: S_2(g), 9-30-77.

Sulfur (S_5) S_5(g)

Enthalpy Reference Temperature = T_r = 298.15 K   Standard State Pressure = p° = 0.1 MPa

T/K	$C_p°$	S°	$-[G°-H°(T_r)]/T$	$H°-H°(T_r)$	$\Delta_f H°$	$\Delta_f G°$	Log K_f
	J K⁻¹mol⁻¹			kJ mol⁻¹			
0							
100							
200							
298.15	89.501	308.638	308.638	0.	109.370	65.137	-11.412
300	89.721	309.192	308.639	0.166	109.325	64.862	-11.294
350	94.425	323.403	309.751	4.778	108.107	57.548	-8.588
400	97.497	336.226	312.272	9.581	95.756	50.851	-6.640
450	99.619	347.839	315.589	14.512	91.060	45.475	-5.279
500	101.153	358.419	319.351	19.534	86.068	40.689	-4.251
600	103.187	377.057	327.458	29.759	78.370	32.370	-2.818
700	104.456	393.065	335.714	40.146	72.022	25.218	-1.882
800	105.316	407.073	343.777	50.637	66.457	18.919	-1.235
900	105.939	419.515	351.514	61.201	-204.262	18.690	-1.085
1000	106.415	430.703	358.883	71.820	-202.924	43.391	-2.266
1100	106.795	440.864	365.881	82.481	-201.620	67.959	-3.227
1200	107.109	450.170	372.522	93.177	-200.359	92.410	-4.022
1300	107.377	458.754	378.830	103.902	-199.147	116.758	-4.691
1400	107.612	466.720	384.826	114.651	-197.988	141.014	-5.261
1500	107.821	474.152	390.537	125.423	-196.883	165.190	-5.752
1600	108.012	481.117	395.982	136.215	-195.832	189.293	-6.180
1700	108.188	487.670	401.185	147.025	-194.836	213.333	-6.555
1800	108.353	493.859	406.163	157.852	-193.891	237.315	-6.887
1900	108.509	499.722	410.934	168.695	-192.994	261.247	-7.182
2000	108.657	505.291	415.514	179.554	-192.143	285.132	-7.447
2100	108.800	510.596	419.917	190.427	-191.333	308.975	-7.685
2200	108.937	515.661	424.154	201.314	-190.558	332.781	-7.901
2300	109.071	520.506	428.239	212.214	-189.818	356.553	-8.098
2400	109.201	525.151	432.181	223.128	-189.108	380.293	-8.277
2500	109.329	529.611	435.989	234.054	-188.425	404.003	-8.441
2600	109.453	533.902	439.673	244.993	-187.765	427.688	-8.592
2700	109.576	538.035	443.240	255.945	-187.123	451.347	-8.732
2800	109.697	542.022	446.697	266.909	-186.499	474.982	-8.861
2900	109.817	545.873	450.051	277.884	-185.890	498.596	-8.981
3000	109.935	549.598	453.308	288.872	-185.297	522.188	-9.092
3100	110.052	553.205	456.472	299.871	-184.714	545.762	-9.196
3200	110.168	556.701	459.550	310.882	-184.142	569.316	-9.293
3300	110.283	560.093	462.546	321.905	-183.579	592.853	-9.384
3400	110.397	563.387	465.463	332.939	-183.025	616.373	-9.469
3500	110.511	566.588	468.307	343.984	-182.482	639.877	-9.550
3600	110.624	569.703	471.081	355.041	-181.948	663.365	-9.625
3700	110.736	572.736	473.787	366.109	-181.422	686.839	-9.696
3800	110.848	575.690	476.430	377.188	-180.907	710.298	-9.764
3900	110.959	578.571	479.013	388.279	-180.402	733.745	-9.827
4000	111.070	581.382	481.537	399.380	-179.909	757.178	-9.888
4100	111.181	584.126	484.006	410.493	-179.429	780.599	-9.945
4200	111.291	586.806	486.421	421.616	-178.961	804.009	-9.999
4300	111.401	589.426	488.787	432.751	-178.508	827.407	-10.051
4400	111.511	591.989	491.103	443.896	-178.070	850.796	-10.100
4500	111.620	594.496	493.373	455.053	-177.649	874.174	-10.147
4600	111.729	596.950	495.598	466.220	-177.245	897.543	-10.192
4700	111.838	599.354	497.780	477.399	-176.861	920.904	-10.235
4800	111.947	601.710	499.921	488.588	-176.496	944.257	-10.276
4900	112.056	604.019	502.022	499.788	-176.152	967.603	-10.315
5000	112.164	606.284	504.084	510.999	-175.829	990.941	-10.352
5100	112.273	608.507	506.110	522.221	-175.530	1014.274	-10.388
5200	112.381	610.688	508.100	533.454	-175.254	1037.600	-10.423
5300	112.489	612.829	510.056	544.697	-175.002	1060.922	-10.456
5400	112.597	614.933	511.979	555.952	-174.775	1084.239	-10.488
5500	112.705	617.000	513.870	567.217	-174.573	1107.552	-10.519
5600	112.812	619.032	515.730	578.492	-174.397	1130.862	-10.548
5700	112.920	621.029	517.559	589.779	-174.247	1154.169	-10.577
5800	113.027	622.994	519.360	601.076	-174.123	1177.473	-10.604
5900	113.135	624.927	521.133	612.384	-174.026	1200.776	-10.631
6000	113.242	626.830	522.879	623.703	-173.957	1224.077	-10.657

PREVIOUS: September 1977 (1 atm)   CURRENT: September 1977 (1 bar)

Sulfur (S_5) S_5(g)

SULFUR (S_6) IDEAL GAS M_r = 160.30

S°(298.15 K) = [354.1 ± 4] J K^{-1} mol^{-1} $\Delta_fH°$(298.15 K) = [101.9 ± 8] kJ mol^{-1}

Enthalpy of Formation

The enthalpy of formation for each vapor phase sulfur species S_3 to S_7 is calculated by a third law calculation using the thermal functions described below and assuming that the vapor pressure data given by Hultgren et al. (1) and Rau et al. (2) are correct. These values for the enthalpy of formation are dependent on the somewhat arbitrary thermal functions adopted for the polyatomic sulfur gases.

Heat Capacity and Entropy

The thermal functions for the polyatomic species, S_3 to S_7, are estimated following the suggestion of Rau et al. (2). Using thermal functions calculated for S(g), S_2(g), and S_8(g), the heat capacity values for the remaining species were calculated by Rau et al. assuming that the vibrational heat capacity per degree of freedom is the same for all sulfur polyatomic species. The resulting heat capacity equations are given by Rau et al. (2).

The use of the procedure of Rau et al. (2) leads to a set of thermal functions and associated enthalpies of formation which reproduce the observed sulfur vapor pressure data. That is, the sum of the calculated partial pressures of all eight sulfur vapor species, S_1(g) to S_8(g), does closely reproduce the observed vapor pressure. [A difference between the calculated boiling point (at 1 atm) and the secondary reference temperature boiling point is due to the difference between the ideal gas calculation and the real observed value.]

The thermal fuctions for S_8(g) were calculated (3) by adjusting the low valued vibrational frequencies such that S(298.15 K) and $\Delta_fH°$(298.15 K) values could be found which reproduced the partial pressure of the octamer. The entropy of S_2(g) is well-known (3), being based on reliable data and sound statistical mechanical calculations.

The thermal functions which would result from the use of the available structural and vibrational inbformation for the remaining polyatomic species, are biased in the same way as for S_8(g). Some of the necessary information is derived from condensed phase spectra (rather than gas phase spectra). As with the octamer, the low-valued vibrational frequencies (and possibly the structure) may need to be adjusted. In any event, the use of the existing spectroscopic data (unaltered) leads to calculated entropies which are unreasonable when used in a 3rd law analysis to mesh in the vapor pressure with these thermal functions.

References

1. R. Hultgren, P. D. Desai, et al., "Selected Values of the Thermodynamic Properties of the Elements", American Society for Metals, Metals Park, Ohio, 1973.
2. H. Rau, R. N. Kutty and J. R. F. Guedes de Carvalho, J. Chem. Thermodynamics 5, 833 (1973).
3. JANAF Thermochemical Tables: S_2(g), 9-30-77.

Sulfur (S_6) S_6(g)

Enthalpy Reference Temperature = T_r = 298.15 K Standard State Pressure = p° = 0.1 MPa

T/K	$C_p°$	S°	$-[G°-H°(T_r)]/T$	$H°-H°(T_r)$	$\Delta_fH°$	$\Delta_fG°$	Log K_f
	J $K^{-1}mol^{-1}$			kJ mol^{-1}			
0							
100							
200							
298.15	111.571	354.076	354.076	0.	101.922	53.699	-9.408
300	111.826	354.767	354.078	0.207	101.877	53.400	-9.298
350	117.278	372.446	355.462	5.945	100.618	45.419	-6.778
400	120.826	388.354	358.596	11.903	85.991	38.152	-4.982
450	123.265	402.735	362.715	18.009	80.544	32.445	-3.766
500	125.018	415.817	367.381	24.218	74.738	27.426	-2.865
600	127.318	438.833	377.424	36.845	65.856	18.838	-1.640
700	128.725	458.573	387.640	49.653	58.583	11.592	-0.865
800	129.656	475.826	397.608	62.575	52.237	5.320	-0.347
900	130.310	491.137	407.165	75.575	-272.303	6.293	-0.365
1000	130.792	504.893	416.262	88.631	-270.383	37.144	-1.940
1100	131.162	517.377	424.895	101.729	-268.514	67.806	-3.220
1200	131.455	528.802	433.085	114.861	-266.705	98.300	-4.279
1300	131.694	539.334	440.858	128.019	-264.962	128.646	-5.169
1400	131.894	549.101	448.245	141.198	-263.290	158.861	-5.927
1500	132.066	558.207	455.276	154.397	-261.692	188.958	-6.580
1600	132.215	566.735	461.979	167.611	-260.167	218.951	-7.148
1700	132.347	574.755	468.379	180.839	-258.715	248.851	-7.646
1800	132.466	582.323	474.501	194.080	-257.333	278.668	-8.087
1900	132.575	589.488	480.366	207.332	-256.017	308.410	-8.479
2000	132.675	596.291	485.994	220.594	-254.763	338.084	-8.830
2100	132.768	602.766	491.401	233.867	-253.566	367.696	-9.146
2200	132.855	608.945	496.605	247.148	-252.420	397.253	-9.432
2300	132.937	614.852	501.618	260.437	-251.322	426.759	-9.692
2400	133.016	620.512	506.455	273.735	-250.269	456.218	-9.929
2500	133.091	625.943	511.127	287.040	-249.256	485.634	-10.147
2600	133.164	631.164	515.644	300.353	-248.278	515.010	-10.347
2700	133.234	636.191	520.016	313.673	-247.330	544.349	-10.531
2800	133.302	641.038	524.252	327.000	-246.410	573.654	-10.702
2900	133.368	645.717	528.361	340.333	-245.518	602.926	-10.860
3000	133.432	650.239	532.348	353.673	-244.650	632.168	-11.007
3100	133.496	654.616	536.222	367.020	-243.804	661.381	-11.144
3200	133.558	658.855	539.988	380.373	-242.978	690.567	-11.272
3300	133.619	662.966	543.653	393.731	-242.170	719.728	-11.392
3400	133.679	666.955	547.221	407.096	-241.382	748.865	-11.505
3500	133.738	670.831	550.698	420.467	-240.614	777.978	-11.611
3600	133.796	674.600	554.087	433.844	-239.864	807.070	-11.710
3700	133.854	678.266	557.394	447.226	-239.133	836.142	-11.804
3800	133.911	681.837	560.622	460.615	-238.421	865.194	-11.893
3900	133.968	685.316	563.775	474.008	-237.730	894.227	-11.977
4000	134.024	688.708	566.856	487.408	-237.060	923.243	-12.056
4100	134.080	692.018	569.869	500.813	-236.414	952.243	-12.132
4200	134.135	695.250	572.816	514.224	-235.790	981.227	-12.203
4300	134.190	698.407	575.700	527.640	-235.192	1010.196	-12.271
4400	134.245	701.493	578.524	541.062	-234.619	1039.152	-12.336
4500	134.299	704.510	581.290	554.489	-234.075	1068.095	-12.398
4600	134.353	707.462	584.001	567.922	-233.559	1097.026	-12.457
4700	134.407	710.352	586.659	581.360	-233.073	1125.947	-12.513
4800	134.461	713.183	589.265	594.803	-232.619	1154.857	-12.567
4900	134.514	715.956	591.823	608.252	-232.198	1183.759	-12.619
5000	134.567	718.674	594.333	621.706	-231.810	1212.652	-12.668
5100	134.621	721.339	596.797	635.165	-231.457	1241.538	-12.716
5200	134.673	723.954	599.217	648.630	-231.140	1270.417	-12.761
5300	134.726	726.519	601.595	662.100	-230.860	1299.290	-12.805
5400	134.779	729.038	603.932	675.575	-230.618	1328.159	-12.847
5500	134.831	731.512	606.229	689.056	-230.413	1357.023	-12.888
5600	134.884	733.942	608.488	702.542	-230.247	1385.884	-12.927
5700	134.936	736.330	610.710	716.033	-230.119	1414.742	-12.965
5800	134.988	738.677	612.896	729.529	-230.032	1443.598	-13.001
5900	135.040	740.985	615.048	743.030	-229.984	1472.454	-13.036
6000	135.092	743.255	617.165	756.537	-229.977	1501.308	-13.070

PREVIOUS: September 1977 (1 atm) CURRENT: September 1977 (1 bar)

Sulfur (S_6) S_6(g)

SULFUR (S_7) IDEAL GAS M_r = 224.42

S°(298.15 K) = [407.7 ± 4] J K^{-1} mol^{-1} Δ_fH°(298.15 K) = [113.7 ± 8] kJ mol^{-1}

Enthalpy of Formation

The enthalpy of formation for each vapor phase sulfur species S_3 to S_7 is calculated by a third law calculation using the thermal functions described below and assuming that the vapor pressure data given by Hultgren et al. (1) and Rau et al. (2) are correct. These values for the enthalpy of formation are dependent on the somewhat arbitrary thermal functions adopted for the polyatomic sulfur gases.

Heat Capacity and Entropy

The thermal functions for the polyatomic species, S_3 to S_7, are estimated following the suggestion of Rau et al. (2). Using thermal functions calculated for S(g), S_2(g), and S_8(g), the heat capacity values for the remaining species were calculated by Rau et al. assuming that the vibrational heat capacity per degree of freedom is the same for all sulfur polyatomic species. The resulting heat capacity equations are given by Rau et al. (2).

The use of the procedure of Rau et al. (2) leads to a set of thermal functions and associated enthalpies of formation which reproduce the observed sulfur vapor pressure data. That is, the sum of the calculated partial pressures of all eight sulfur vapor species, S_1(g) to S_8(g), does closely reproduce the observed vapor pressure. [A difference between the calculated boiling point (at 1 atm) and the secondary reference temperature boiling point is due to the difference between the ideal gas calculation and the real observed value.]

The thermal fuctions for S_8(g) were calculated (3) by adjusting the low valued vibrational frequencies such that S(298.15 K) and Δ_fH°(298.15 K) values could be found which reproduced the partial pressure of the octamer. The entropy of S_2(g) is well-known (3), being based on reliable data and sound statistical mechanical calculations.

The thermal functions which would result from the use of the available structural and vibrational inbformation for the remaining polyatomic species, are biased in the same way as for S_8(g). Some of the necessary information is derived from condensed phase spectra (rather than gas phase spectra). As with the octamer, the low-valued vibrational frequencies (and possibly the structure) may need to be adjusted. In any event, the use of the existing spectroscopic data (unaltered) leads to calculated entropies which are unreasonable when used in a 3rd law analysis to mesh in the vapor pressure with these thermal functions.

References

1. R. Hultgren, P. D. Desai, et al., "Selected Values of the Thermodynamic Properties of the Elements", American Society for Metals, Metals Park, Ohio, 1973.
2. H. Rau, R. N. Kutty and J. R. F. Guedes de Carvalho, J. Chem. Thermodynamics 5, 833 (1973).
3. JANAF Thermochemical Tables: S_2(g), 9-30-77.

Sulfur (S_7) S_7(g)

Enthalpy Reference Temperature = T_r = 298.15 K Standard State Pressure = p° = 0.1 MPa

T/K	C_p°	S°	$-[G°-H°(T_r)]/T$	$H°-H°(T_r)$	$\Delta_f H°$	$\Delta_f G°$	Log K_f
	J K⁻¹mol⁻¹			kJ mol⁻¹			
0							
100							
200							
298.15	133.502	407.673	407.673	0.	113.679	59.034	-10.342
300	133.780	408.500	407.675	0.247	113.632	58.695	-10.220
350	139.733	429.603	409.328	7.096	112.318	49.641	-7.409
400	143.642	448.534	413.067	14.187	95.393	41.399	-5.406
450	146.363	465.619	417.972	21.441	89.169	34.960	-4.058
500	148.345	481.147	423.525	28.811	82.520	29.310	-3.062
600	151.017	508.450	435.467	43.790	72.406	19.665	-1.712
700	152.729	531.867	447.605	58.983	64.171	11.543	-0.861
800	153.931	552.344	459.445	74.319	57.029	4.522	-0.295
900	154.836	570.529	470.796	89.759	-321.328	5.919	-0.344
1000	155.556	586.881	481.600	105.280	-318.800	42.144	-2.201
1100	156.155	601.736	491.857	120.867	-316.314	78.117	-3.709
1200	156.672	615.346	501.588	136.508	-313.881	113.867	-4.956
1300	157.132	627.904	510.828	152.199	-311.508	149.416	-6.004
1400	157.549	639.565	519.612	167.933	-309.200	184.784	-6.894
1500	157.936	650.448	527.976	183.708	-306.958	219.990	-7.661
1600	158.299	660.652	535.953	199.520	-304.785	255.049	-8.326
1700	158.643	670.260	543.573	215.367	-302.676	289.974	-8.910
1800	158.974	679.337	550.866	231.248	-300.630	324.776	-9.425
1900	159.293	687.941	557.856	247.161	-298.642	359.466	-9.882
2000	159.603	696.119	564.566	263.106	-296.708	394.053	-10.292
2100	159.905	703.914	571.018	279.082	-294.820	428.545	-10.659
2200	160.201	711.359	577.229	295.087	-292.972	462.947	-10.992
2300	160.492	718.487	583.217	311.122	-291.161	497.267	-11.293
2400	160.779	725.324	588.996	327.185	-289.382	531.508	-11.568
2500	161.063	731.893	594.582	343.278	-287.632	565.675	-11.819
2600	161.343	738.215	599.985	359.398	-285.902	599.774	-12.050
2700	161.621	744.310	605.218	375.546	-284.187	633.805	-12.262
2800	161.896	750.192	610.292	391.722	-282.487	667.774	-12.457
2900	162.170	755.878	615.214	407.925	-280.798	701.682	-12.639
3000	162.442	761.381	619.995	424.156	-279.118	735.532	-12.807
3100	162.712	766.711	624.643	440.414	-277.445	769.326	-12.963
3200	162.981	771.882	629.163	456.698	-275.773	803.066	-13.109
3300	163.249	776.901	633.565	473.010	-274.105	836.754	-13.245
3400	163.516	781.778	637.852	489.348	-272.440	870.392	-13.372
3500	163.782	786.522	642.033	505.713	-270.778	903.980	-13.491
3600	164.048	791.140	646.111	522.105	-269.118	937.521	-13.603
3700	164.312	795.638	650.091	538.523	-267.460	971.016	-13.708
3800	164.576	800.024	653.980	554.967	-265.804	1004.466	-13.807
3900	164.840	804.302	657.779	571.438	-264.154	1037.873	-13.901
4000	165.103	808.479	661.495	587.935	-262.508	1071.237	-13.989
4100	165.365	812.559	665.130	604.458	-260.869	1104.560	-14.072
4200	165.627	816.547	668.688	621.008	-259.239	1137.844	-14.151
4300	165.889	820.447	672.172	637.584	-257.617	1171.088	-14.226
4400	166.150	824.264	675.585	654.186	-256.006	1204.296	-14.297
4500	166.411	828.000	678.931	670.814	-254.407	1237.466	-14.364
4600	166.672	831.661	682.211	687.468	-252.823	1270.601	-14.428
4700	166.932	835.248	685.429	704.148	-251.254	1303.702	-14.489
4800	167.192	838.765	688.587	720.854	-249.702	1336.770	-14.547
4900	167.452	842.215	691.688	737.587	-248.168	1369.806	-14.602
5000	167.712	845.601	694.732	754.345	-246.654	1402.810	-14.655
5100	167.971	848.925	697.723	771.129	-245.160	1435.785	-14.705
5200	168.231	852.189	700.662	787.939	-243.690	1468.730	-14.754
5300	168.490	855.396	703.552	804.775	-242.242	1501.647	-14.800
5400	168.749	858.548	706.393	821.637	-240.818	1534.538	-14.844
5500	169.008	861.646	709.187	838.525	-239.419	1567.401	-14.886
5600	169.267	864.694	711.937	855.439	-238.045	1600.240	-14.926
5700	169.525	867.692	714.644	872.378	-236.696	1633.055	-14.965
5800	169.784	870.643	717.308	889.344	-235.374	1665.846	-15.003
5900	170.042	873.547	719.931	906.335	-234.078	1698.615	-15.038
6000	170.301	876.408	722.516	923.352	-232.810	1731.361	-15.073

PREVIOUS: September 1977 (1 atm) CURRENT: September 1977 (1 bar)

Sulfur (S_7) S_7(g)

SULFUR (S_8) IDEAL GAS M_r = 256.48

$S°(298.15\ K) = 430.31 \pm 0.05\ J\ K^{-1}\ mol^{-1}$

$\Delta_f H°(0\ K) = 104.39 \pm 0.63\ kJ\ mol^{-1}$

$\Delta_f H°(298.15\ K) = 100.42 \pm 0.63\ kJ\ mol^{-1}$

Vibrational Frequencies and Degeneracies

ν, cm^{-1}	ν, cm^{-1}	ν, cm^{-1}
475(1)	475(2)	243(1)
218(1)	152(2)	437(2)
471(2)	56(2)	248(2)
191(2)	411(1)	

Ground State Quantum Weight: [1] $\sigma = 8$

Point Group: D_{4d}

Bond Distance: S–S = 2.059 ± 0.002 Å

Bond Angle: S–S–S = 107.9 ± 0.6°

Product of the Moments of Inertia: $I_A I_B I_C = 3.90326 \times 10^{-111}\ g^3\ cm^6$

Enthalpy of Formation

The vapor pressure of S(cr) has been measured by many investigators. Using the seven more recent vapor pressure data, the respective enthalpy of sublimation was calculated by both the 2nd and 3rd law methods. The values obtained are tabulated as follows:

Source	T/K	$\Delta_{sub}H°(298.15)$, kcal mol^{-1} 3rd Law	2nd Law
West-Menzies (1)	376.6 - 448.8	24.29	21.29
Neumann (2)	332.6 - 362.1	24.35	26.47
Fouretier (3)	304.1 - 351.6	24.36	24.10
Taillade (4)	303.6 - 352.5	24.29	24.87
Bradley (5)	288.3 - 305.7	24.29	23.80
Magee (6)	331.9 - 368.0	24.11	24.44
	332.7 - 367.4	24.12	23.88
Briske (7)	273.2 - 313.2	24.28	24.00

Due to the presence of $S_6(g)$ in $S_8(g)$, the measured total pressures were corrected to give the parttial pressures of $S_8(g)$ for the first five sets of vapor pressure data, according to Scott (8). However, the magnitude of the correction is only 0.1 - 0.2 kcal mol^{-1}. The adopted value is the weighted average of both the 2nd and 3rd law values.

Heat Capacity and Entropy

The vibrational frequencies were taken from Scott, McCullough and Kruse (9). Four of the eighteen frequencies, i.e. 86(2) and 218(2) cm^{-1}, were changed to 56(2) and 248 (2) cm^{-1}, respectively, as suggested by Scott (10). The molecular structure and bond distance and angle were reported by Donohur, Caron and Goldish (11). The principle moments of inertia are: $I_A = I_B = 1.28594 \times 10^{-37}$ and $I_C = 2.36042 \times 10^{-37}\ g\ cm^2$.

References

1. W. A. West and A. W. Menzies, J. Phys. Chem. 33, 1880 (1929).
2. K. Neumann, Z. Physik. Chem. A171, 416 (1934).
3. G. Fouretier, Compt. Rend. 218, 194 (1944).
4. M. Taillade, Compt. Rend 218, 836 (1944).
5. R. S. Bradley, Proc. Roy. Soc. (London) A 205, 553 (1951).
6. D. W. Magee, Ph. D. Thessis, The Ohio State University, 1955.
7. C. Briske, N. H. Hartshorne and D. R. Stranks, J. Chem. Soc. (London), 1200 (1960).
8. D. W. Scott, U. S. Bur. Mines, Bartlesville, Oklahoma, personal communication, May 22, 1964.
9. D. W. Scott, J. P. McCullough and F. P. Kruse, "Vibrational Assignment and Force Constants of S_8(condensed states) from a Normal-Coordinate Treatment," Prepublication Report 64, U. S. Bur. Mines, May 20, 1963.
10. D. W. Scott, U. S. Bur. Mines, Bartlesville, Oklahoma, April 23, 1964.
11. J. Donohur, A. Caron and E. Goldish, J. Am. Chem. Soc. 83, 3748 (1961).

Sulfur (S_8) $S_8(g)$

Enthalpy Reference Temperature = T_r = 298.15 K Standard State Pressure = p° = 0.1 MPa

T/K	$C_p°$ (J $K^{-1}mol^{-1}$)	S°	$-[G°-H°(T_r)]/T$	$H°-H°(T_r)$ (kJ mol^{-1})	$\Delta_f H°$	$\Delta_f G°$	Log K_f
0	0.	0.	INFINITE	-31.324	104.388	104.388	INFINITE
100	87.155	295.550	552.531	-25.698	104.495	84.958	-44.377
200	134.162	372.097	444.003	-14.381	102.677	66.077	-17.258
250	147.427	403.559	432.835	-7.319	101.582	57.051	-11.920
298.15	156.043	430.311	430.311	0.	100.416	48.578	-8.511
300	156.314	431.277	430.314	0.289	100.369	48.257	-8.402
350	162.400	455.858	432.241	8.266	99.016	39.677	-5.921
400	166.689	477.839	436.591	16.499	79.803	32.003	-4.179
450	169.799	497.662	442.294	24.915	72.816	26.376	-3.062
500	172.113	515.677	448.745	33.466	65.345	21.636	-2.260
600	175.248	547.358	462.613	50.847	54.048	14.009	-1.220
700	177.206	574.531	476.707	68.477	44.903	8.078	-0.603
800	178.506	598.284	490.450	86.267	37.003	3.368	-0.220
900	179.411	619.364	503.625	104.165	-395.153	8.246	-0.479
1000	180.064	638.302	516.162	122.140	-392.026	52.900	-2.763
1100	180.552	655.488	528.059	140.172	-388.966	97.243	-4.618
1200	180.925	671.215	539.342	158.247	-385.987	141.312	-6.151
1300	181.216	685.709	550.051	176.355	-383.099	185.136	-7.439
1400	181.448	699.147	560.227	194.488	-380.310	228.741	-8.534
1500	181.636	711.672	569.910	212.643	-377.622	272.150	-9.477
1600	181.790	723.400	579.141	230.814	-375.037	315.383	-10.296
1700	181.918	734.425	587.954	249.000	-372.553	358.458	-11.014
1800	182.025	744.826	596.383	267.197	-370.167	401.389	-11.648
1900	182.116	754.670	604.457	285.404	-367.875	444.191	-12.212
2000	182.194	764.013	612.203	303.620	-365.671	486.874	-12.716
2100	182.261	772.904	619.646	321.843	-363.548	529.448	-13.169
2200	182.319	781.384	626.806	340.072	-361.499	571.924	-13.579
2300	182.370	789.490	633.705	358.306	-359.520	614.307	-13.951
2400	182.414	797.253	640.359	376.546	-357.607	656.607	-14.291
2500	182.453	804.700	646.784	394.789	-355.753	698.827	-14.601
2600	182.488	811.857	652.996	413.036	-353.952	740.975	-14.886
2700	182.519	818.744	659.009	431.286	-352.198	783.054	-15.149
2800	182.547	825.383	664.833	449.540	-350.488	825.069	-15.392
2900	182.572	831.789	670.480	467.796	-348.819	867.023	-15.617
3000	182.595	837.979	675.961	486.054	-347.191	908.921	-15.826
3100	182.615	843.966	681.284	504.315	-345.598	950.765	-16.020
3200	182.634	849.764	686.459	522.577	-344.037	992.558	-16.202
3300	182.651	855.385	691.493	540.841	-342.508	1034.303	-16.372
3400	182.666	860.837	696.394	559.107	-341.011	1076.002	-16.531
3500	182.680	866.133	701.169	577.375	-339.547	1117.657	-16.680
3600	182.693	871.279	705.823	595.643	-338.114	1159.272	-16.821
3700	182.705	876.285	710.362	613.913	-336.713	1200.846	-16.953
3800	182.716	881.158	714.793	632.184	-335.344	1242.384	-17.078
3900	182.726	885.904	719.120	650.456	-334.008	1283.885	-17.196
4000	182.736	890.530	723.348	668.729	-332.709	1325.353	-17.307
4100	182.744	895.042	727.481	687.003	-331.446	1366.789	-17.413
4200	182.752	899.446	731.523	705.278	-330.221	1408.194	-17.513
4300	182.760	903.747	735.478	723.554	-329.036	1449.571	-17.609
4400	182.767	907.948	739.350	741.830	-327.892	1490.921	-17.699
4500	182.774	912.056	743.143	760.107	-326.791	1532.245	-17.786
4600	182.780	916.073	746.859	778.385	-325.736	1573.545	-17.868
4700	182.786	920.004	750.501	796.663	-324.728	1614.823	-17.947
4800	182.791	923.852	754.073	814.942	-323.768	1656.079	-18.022
4900	182.796	927.621	757.576	833.221	-322.859	1697.317	-18.094
5000	182.801	931.314	761.014	851.501	-322.000	1738.536	-18.162
5100	182.805	934.934	764.389	869.782	-321.195	1779.739	-18.228
5200	182.810	938.484	767.703	888.062	-320.445	1820.926	-18.291
5300	182.814	941.966	770.958	906.343	-319.751	1862.100	-18.352
5400	182.817	945.383	774.157	924.625	-319.113	1903.261	-18.410
5500	182.821	948.738	777.300	942.907	-318.532	1944.410	-18.466
5600	182.824	952.032	780.391	961.189	-318.009	1985.550	-18.520
5700	182.828	955.268	783.431	979.472	-317.545	2026.681	-18.572
5800	182.831	958.448	786.421	997.755	-317.140	2067.804	-18.623
5900	182.834	961.573	789.363	1016.038	-316.795	2108.921	-18.671
6000	182.836	964.646	792.259	1034.321	-316.510	2150.032	-18.718

PREVIOUS: September 1977 (1 atm) CURRENT: September 1977 (1 bar)

Sulfur (S_8) $S_8(g)$

SILICON (Si) REFERENCE STATE A_r = 28.0855

0 to 1685 K	crystal
1685 to 3504.616 K	liquid
above 3504.616 K	ideal monatomic gas

Refer to the individual tables for details.

Silicon (Si) Si_1(ref)

Enthalpy Reference Temperature = T_r = 298.15 K — Standard State Pressure = $p°$ = 0.1 MPa

T/K	$C_p°$ (J K⁻¹mol⁻¹)	$S°$ (J K⁻¹mol⁻¹)	$-[G°-H°(T_r)]/T$ (J K⁻¹mol⁻¹)	$H°-H°(T_r)$ (kJ mol⁻¹)	$\Delta_f H°$ (kJ mol⁻¹)	$\Delta_f G°$ (kJ mol⁻¹)	Log K_f
0	0.	0.	INFINITE	-3.218	0.	0.	0.
100	7.268	3.833	33.351	-2.952	0.	0.	0.
200	15.636	11.665	20.531	-1.773	0.	0.	0.
298.15	20.000	18.820	18.820	0.	0.	0.	0.
300	20.050	18.943	18.820	0.037	0.	0.	0.
400	22.142	25.032	19.636	2.159	0.	0.	0.
500	23.330	30.110	21.237	4.436	0.	0.	0.
600	24.154	34.440	23.086	6.812	0.	0.	0.
700	24.803	38.212	24.983	9.260	0.	0.	0.
800	25.359	41.562	26.850	11.769	0.	0.	0.
900	25.874	44.579	28.655	14.331	0.	0.	0.
1000	26.338	47.329	30.387	16.942	0.	0.	0.
1100	26.778	49.860	32.044	19.598	0.	0.	0.
1200	27.196	52.208	33.627	22.297	0.	0.	0.
1300	27.614	54.401	35.142	25.037	0.	0.	0.
1400	28.033	56.463	36.592	27.820	0.	0.	0.
1500	28.451	58.411	37.982	30.644	0.	0.	0.
1600	28.870	60.261	39.317	33.510	0.	0.	0.
1685.000	29.225	61.765	40.412	35.979	CRYSTAL <--> LIQUID		
1685.000	27.196	91.562	40.412	86.187	TRANSITION		
1700	27.196	91.803	40.864	86.595	0.	0.	0.
1800	27.196	93.357	43.738	89.315	0.	0.	0.
1900	27.196	94.827	46.388	92.034	0.	0.	0.
2000	27.196	96.222	48.846	94.754	0.	0.	0.
2100	27.196	97.549	51.133	97.473	0.	0.	0.
2200	27.196	98.815	53.272	100.193	0.	0.	0.
2300	27.196	100.023	55.279	102.913	0.	0.	0.
2400	27.196	101.181	57.167	105.632	0.	0.	0.
2500	27.196	102.291	58.950	108.352	0.	0.	0.
2600	27.196	103.358	60.638	111.071	0.	0.	0.
2700	27.196	104.384	62.239	113.791	0.	0.	0.
2800	27.196	105.373	63.762	116.511	0.	0.	0.
2900	27.196	106.327	65.214	119.230	0.	0.	0.
3000	27.196	107.249	66.600	121.950	0.	0.	0.
3100	27.196	108.141	67.925	124.669	0.	0.	0.
3200	27.196	109.005	69.196	127.389	0.	0.	0.
3300	27.196	109.842	70.415	130.109	0.	0.	0.
3400	27.196	110.653	71.586	132.828	0.	0.	0.
3500	27.196	111.442	72.714	135.548	0.	0.	0.
3504.616	27.196	111.478	72.765	135.673	LIQUID <--> IDEAL GAS		
3504.616	23.023	221.204	72.765	520.221	FUGACITY = 1 bar		
3600	23.046	221.822	76.706	522.419	0.	0.	0.
3700	23.066	222.454	80.637	524.725	0.	0.	0.
3800	23.081	223.070	84.377	527.032	0.	0.	0.
3900	23.093	223.669	87.941	529.341	0.	0.	0.
4000	23.102	224.254	91.341	531.651	0.	0.	0.
4100	23.108	224.825	94.590	533.961	0.	0.	0.
4200	23.111	225.381	97.698	536.272	0.	0.	0.
4300	23.113	225.925	100.673	538.583	0.	0.	0.
4400	23.112	226.457	103.526	540.895	0.	0.	0.
4500	23.111	226.976	106.264	543.206	0.	0.	0.
4600	23.108	227.484	108.893	545.517	0.	0.	0.
4700	23.104	227.981	111.422	547.827	0.	0.	0.
4800	23.100	228.467	113.855	550.138	0.	0.	0.
4900	23.096	228.943	116.199	552.447	0.	0.	0.
5000	23.092	229.410	118.459	554.757	0.	0.	0.
5100	23.089	229.867	120.639	557.066	0.	0.	0.
5200	23.087	230.316	122.744	559.374	0.	0.	0.
5300	23.086	230.755	124.777	561.683	0.	0.	0.
5400	23.086	231.187	126.744	563.992	0.	0.	0.
5500	23.087	231.610	128.647	566.300	0.	0.	0.
5600	23.091	232.026	130.489	568.609	0.	0.	0.
5700	23.098	232.435	132.274	570.918	0.	0.	0.
5800	23.108	232.837	134.005	573.228	0.	0.	0.
5900	23.121	233.232	135.683	575.539	0.	0.	0.
6000	23.135	233.621	137.312	577.851	0.	0.	0.

PREVIOUS: March 1967 (1 atm) CURRENT: March 1967 (1 bar)

Silicon (Si) Si_1(ref)

SILICON (Si) CRYSTAL A_r = 28.0855

$S°(298.15\ K) = 18.820 \pm 0.04\ J\ K^{-1}\ mol^{-1}$

$T_{fus} = 1685 \pm 3\ K$

$\Delta_f H°(0\ K) = 0\ kJ\ mol^{-1}$

$\Delta_f H°(298.15\ K) = 0\ kJ\ mol^{-1}$

$\Delta_{fus} H° = 50.2 \pm 0.4\ kJ\ mol^{-1}$

Enthalpy of Formation

Zero by definition.

Heat Capacity and Entropy

Low temperature heat capacities are based on the precise data of Flubacher et al. (8-300 K, 1). The entropy is obtained from the heat capacity using $S°(8\ K) = 0.0003\ cal\ K^{-1}\ mol^{-1}$. The selected values are consistent with recent data of Keesom and Seidel (1.2-4.2 K, 2) and Kalishevich et al. (60-300 K, 3). Earlier data have been reviewed by Hultgren et al. (10).

High temperature studies are summarized below along with the pertinent low temperatures studies. The selected heat capacities above 300 K are obtained from a Shomate plot of the adopted low temperature heat capacities and the enthalpies reported or derived from the work of Dennison (5), Kantor et al. (6), Olette (7), Serebrennikov and Gel'd (8) and Magnus (9). Discrepancies are apparent in the enthalpy data, particularly in the range of 370-1100 K where only the recent study of Dennison (5) made use of a high purity sample. Enthalpies from Dennison (5) join most smoothly with the low temperature heat capacity but deviate increasingly at higher temperatures, being 1-2% lower than the other measurements above 1000 K. The selected functions agree with Dennison (5) below 600 K, with Magnus (1) from 600 to 1200 K, and with Kantor et al. (6) above 1200 K. The resulting heat capacities are slightly different from those selected by Hultgren, et al. (10); the latter are $0.07\ cal\ K^{-1}\ mol^{-1}$ higher at 800 K and $0.21\ cal\ K^{-1}\ mol^{-1}$ lower at the melting point.

Source	Method	Quantity Reported	T/K	Sample
(1)	Calorimetry	Cp°	7.7 - 300	Single crystal, 3mm pieces, Resistance (300 K) = 100 ohm cm, Carrier Concentration $\leq 10^{15}\ cm^{-3}$
(2)	Calorimetry	Specific Heat	1.2 - 4.2	
(3)	Calorimetry	Cp°	60 - 300	99.999%
(4)	Temperature Modulation	Cp°	300 - 900	Resistance (300 K) = 0.027 ohm cm, Carrier Concentration = $10^{19}\ cm^{-3}$
(5)	Drop Calorimetry	Specific Heat	273 - 1373	Resistance (300 K) = 107 ohm cm
(6)	Drop Calorimetry	Enthalpy Equation	1148 - 1685	"Highly purified"
(7)	Drop Calorimetry	Specific Enthalpy	1467 - 1685	99.99%
(8)	Drop Calorimetry	Specific Enthalpy	302 - 1556	99.3%
(9)	Drop Calorimetry	Specific Enthalpy	372 - 1175	99.2%

Fusion Data

See the liquid table for details.

Sublimation Data

See the ideal gas table for details.

References

1. P. Flubacher, A. J. Leadbetter and J. A. Morrison, Phil. Mag. 4, 275 (1959).
2. P. H. Keesom and G. Seidel, Phys. Rev. 113, 33 (1959).
3. G. I. Kalishevich, P. V. Gel'd. and R. P. Krentsis, Russ. J. Phys. Chem. 39, 1602 (1965).
4. D. Gerlich, B. Abeles and R. E. Miller, J. Appl. Phys. 36, 76 (1965).
5. D. H. Dennison, quoted by H. R. Shanks et al., Phys. Rev. 130, 1743 (1963).
6. P. B. Kantor, A. M. Kisil and E. M. Fomichev, Ukrain. Fiz. Zhur. 5, 358 (1960).
7. M. Olette, Phys. Chem. Steelmaking, Proc. Dedham, Mass. 1956, 18-26 (Pub. 1958).
8. N. N. Serebrennikov and P. v. Gel'd, Dokl. Akad. Nauk SSSR 87, 1021 (1952).
9. A. Magnus, Ann. Physik 70, 303 (1923).
10. R. Hultgren, R. L. Orr, and K. K. Kelley, "Supplement to Selected Values of Thermodynamic Properties of Metals and Alloys, University of California, Berkeley, 1965.

Silicon (Si) $Si_1(cr)$

Enthalpy Reference Temperature = T_r = 298.15 K Standard State Pressure = p° = 0.1 MPa

T/K	$C_p°$ (J K⁻¹mol⁻¹)	S°	$-[G°-H°(T_r)]/T$	$H°-H°(T_r)$ (kJ mol⁻¹)	$\Delta_f H°$	$\Delta_f G°$	Log K_f
0	0.	0.	INFINITE	-3.218	0.	0.	0.
100	7.268	3.833	33.351	-2.952	0.	0.	0.
200	15.636	11.665	20.531	-1.773	0.	0.	0.
250	18.221	15.446	19.138	-0.923	0.	0.	0.
298.15	20.000	18.820	18.820	0.	0.	0.	0.
300	20.050	18.943	18.820	0.037	0.	0.	0.
350	21.276	22.132	19.069	1.072	0.	0.	0.
400	22.142	25.032	19.636	2.159	0.	0.	0.
450	22.803	27.680	20.385	3.283	0.	0.	0.
500	23.330	30.110	21.237	4.436	0.	0.	0.
600	24.154	34.440	23.086	6.812	0.	0.	0.
700	24.803	38.212	24.983	9.260	0.	0.	0.
800	25.359	41.562	26.850	11.769	0.	0.	0.
900	25.874	44.579	28.655	14.331	0.	0.	0.
1000	26.338	47.329	30.387	16.942	0.	0.	0.
1100	26.778	49.860	32.044	19.598	0.	0.	0.
1200	27.196	52.208	33.627	22.297	0.	0.	0.
1300	27.614	54.401	35.142	25.037	0.	0.	0.
1400	28.033	56.463	36.592	27.820	0.	0.	0.
1500	28.451	58.411	37.982	30.644	0.	0.	0.
1600	28.870	60.261	39.317	33.510	0.	0.	0.
1685.000	29.225	61.765	40.412	35.979	------ CRYSTAL <--> LIQUID -----		
1700	29.288	62.024	40.602	36.418	-50.177	0.447	-0.014
1800	29.706	63.710	41.839	39.368	-49.947	3.418	-0.099
1900	30.125	65.327	43.033	42.359	-49.675	6.376	-0.175
2000	30.543	66.883	44.187	45.393	-49.361	9.318	-0.243
2100	30.962	68.383	45.303	48.468	-49.006	12.243	-0.305
2200	31.380	69.833	46.386	51.585	-48.608	15.151	-0.360
2300	31.798	71.237	47.436	54.744	-48.169	18.039	-0.410
2400	32.217	72.600	48.456	57.945	-47.688	20.908	-0.455
2500	32.635	73.923	49.448	61.187	-47.165	23.755	-0.496

PREVIOUS: December 1962 CURRENT: December 1966

Silicon (Si) $Si_1(cr)$

SILICON (Si) LIQUID A_r = 28.0855

$S°(298.15\ K) = [44.460]\ J\ K^{-1}\ mol^{-1}$

$T_{fus} = 1685 \pm 3\ K$

$\Delta_f H°(298.15\ K) = [48.470]\ kJ\ mol^{-1}$

$\Delta_{fus} H° = 50.2 \pm 0.4\ kJ\ mol^{-1}$

Enthalpy of Formation

The enthalpy of formation, $\Delta_f H°(Si, \ell, 298.15\ K)$ is calculated from that of the crystal by adding $\Delta_{fus}H°$ and the difference in enthalpy, H°(1685 K) − H°(298.15 K), between the crystal and liquid.

Heat Capacity and Entropy

Enthalpy data for high purity samples in quartz or vitreous silica capsules have been reported for the range 1698-1915 K by Kantor et al. (1) and for the range 1686-1825 K by Olette (2). Due to the limited temperature range and the experimental uncertainty, the data do not appear to justify more than a constant heat capacity. A value of 6.5 cal $K^{-1}\ mol^{-1}$ is selected, intermediate between the values of 6.75 and 6.15 obtained from separate experiments.

The entropy is calculated in a manner analogous to that used for the enthalpy of formation.

Fusion Data

Modern determinations of the melting point range from 1683 ± 1 to 1690 ± 4 K, the former from the specific volume study of Lucas and Urbain (3) and the later from the enthalpy study of Kantor et al. (1). The selected value, T_{fus} = 1685 K, is taken from Hultgren et al. (4), who have reviewed the melting data.

The enthalpy of melting is calculated from the selected enthalpy of the crystal and the liquid enthalpy measurements of Kantor et al. (1) and Olette (2), who reported comparable values of $\Delta_{fus}H° = 11.95 \pm 0.18$ and $12.095 \pm 0.1\ kcal\ mol^{-1}$, respectively.

Vaporization Data

T_{vap} is calculated as the temperature for which $\Delta_r G° = 0$ for Si(ℓ) = Si(g). $\Delta_{vap}H°$ is calculated as the difference between $\Delta_f H°(g)$ and $\Delta_f H°(\ell)$ at T_{vap}. The normal boiling for the equilibrium vapor is calculated as the temperature at which Si, Si_2, and Si_3 attain a total pressure of 1 atm; the vapor containing 92.1%, 7.0% and 0.9% (mole) of monomer, dimer, and trimer, repectively. tetramer and high polymers, which are ignored in this calculation, are probably negligible at this temperature (~ 3490 K).

References

1. P. B. Kantor, A. M. Kisil, and E. M. Fomichev, Ukrain. Fiz. Zhur. 5, 358 (1960).
2. M. Olette, Phys. Chem. Steelmaking, Proc. Dedham, Mass. 1956, 18-26 (Pub. 1958).
3. L. D. Lucas and G. Urbain, Compt. Rend. 255, 2414 (1962).
4. R. Hultgren, R. L. Orr and K. K. Kelley, "Supplement to Selected Values of Thermodynamic Properties of Metals and Alloys," Univ. of California, Berkeley, 1965.

Silicon (Si) $Si_1(l)$

Enthalpy Reference Temperature = T_r = 298.15 K — Standard State Pressure = p° = 0.1 MPa

T/K	$C_p°$ (J K⁻¹mol⁻¹)	S° (J K⁻¹mol⁻¹)	−[G°−H°(T_r)]/T (J K⁻¹mol⁻¹)	H°−H°(T_r) (kJ mol⁻¹)	$\Delta_f H°$ (kJ mol⁻¹)	$\Delta_f G°$ (kJ mol⁻¹)	Log K_f
0							
100							
200							
250							
298.15	27.196	44.460	44.460	0.	48.470	40.826	-7.152
300	27.196	44.628	44.461	0.050	48.483	40.778	-7.100
350	27.196	48.821	44.792	1.410	48.808	39.467	-5.890
400	27.196	52.452	45.527	2.770	49.082	38.114	-4.977
450	27.196	55.655	46.478	4.130	49.317	36.728	-4.263
500	27.196	58.521	47.542	5.490	49.523	35.318	-3.690
600	27.196	63.479	49.797	8.209	49.867	32.443	-2.824
700	27.196	67.671	52.059	10.929	50.139	29.517	-2.203
800	27.196	71.303	54.243	13.648	50.349	26.556	-1.734
900	27.196	74.506	56.320	16.368	50.507	23.572	-1.368
1000	27.196	77.372	58.284	19.088	50.616	20.573	-1.075
1100	27.196	79.964	60.139	21.807	50.679	17.565	-0.834
1200	27.196	82.330	61.891	24.527	50.700	14.554	-0.634
1300	27.196	84.507	63.548	27.246	50.679	11.542	-0.464
1400	27.196	86.522	65.118	29.966	50.616	8.534	-0.318
1500	27.196	88.399	66.608	32.686	50.512	5.531	-0.193
1600	27.196	90.154	68.026	35.405	50.365	2.537	-0.083
1685.000	27.196	91.562	69.178	37.717	------ CRYSTAL <--> LIQUID -----		
1700	27.196	91.803	69.376	38.125	0.	0.	0.
1800	27.196	93.357	70.666	40.844	0.	0.	0.
1900	27.196	94.827	71.899	43.564	0.	0.	0.
2000	27.196	96.222	73.081	46.284	0.	0.	0.
2100	27.196	97.549	74.215	49.003	0.	0.	0.
2200	27.196	98.814	75.304	51.723	0.	0.	0.
2300	27.196	100.023	76.353	54.442	0.	0.	0.
2400	27.196	101.181	77.363	57.162	0.	0.	0.
2500	27.196	102.291	78.338	59.882	0.	0.	0.
2600	27.196	103.358	79.280	62.601	0.	0.	0.
2700	27.196	104.384	80.191	65.321	0.	0.	0.
2800	27.196	105.373	81.073	68.040	0.	0.	0.
2900	27.196	106.327	81.928	70.760	0.	0.	0.
3000	27.196	107.249	82.756	73.480	0.	0.	0.
3100	27.196	108.141	83.561	76.199	0.	0.	0.
3200	27.196	109.005	84.343	78.919	0.	0.	0.
3300	27.196	109.842	85.103	81.638	0.	0.	0.
3400	27.196	110.653	85.842	84.358	0.	0.	0.
3500	27.196	111.442	86.562	87.078	0.	0.	0.
3504.616	27.196	111.478	86.595	87.203	------- FUGACITY = 1 bar -------		
3600	27.196	112.208	87.264	89.797	-384.152	10.461	-0.152
3700	27.196	112.953	87.949	92.517	-383.738	21.417	-0.302
3800	27.196	113.678	88.616	95.236	-383.326	32.361	-0.445
3900	27.196	114.385	89.268	97.956	-382.915	43.295	-0.580
4000	27.196	115.073	89.904	100.676	-382.505	54.218	-0.708
4100	27.196	115.745	90.526	103.395	-382.096	65.131	-0.830
4200	27.196	116.400	91.135	106.115	-381.687	76.034	-0.946
4300	27.196	117.040	91.730	108.834	-381.279	86.927	-1.056
4400	27.196	117.665	92.312	111.554	-380.871	97.811	-1.161
4500	27.196	118.277	92.882	114.274	-380.462	108.686	-1.262

PREVIOUS: December 1966 CURRENT: March 1967

Silicon (Si) $Si_1(l)$

SILICON (Si) CRYSTAL–LIQUID $A_r = 28.0855$

0 to 1685 K crystal
above 1685 K liquid

Refer to the individual tables for details.

Enthalpy Reference Temperature = T_r = 298.15 K Standard State Pressure = $p°$ = 0.1 MPa

T/K	$C_p°$ (J K⁻¹mol⁻¹)	$S°$ (J K⁻¹mol⁻¹)	$-[G°-H°(T_r)]/T$ (J K⁻¹mol⁻¹)	$H°-H°(T_r)$ (kJ mol⁻¹)	$\Delta_f H°$ (kJ mol⁻¹)	$\Delta_f G°$ (kJ mol⁻¹)	Log K_f
0	0.	0.	INFINITE	-3.218	0.	0.	0.
100	7.268	3.833	33.351	-2.952	0.	0.	0.
200	15.636	11.665	20.531	-1.773	0.	0.	0.
250	18.221	15.446	19.138	-0.923	0.	0.	0.
298.15	20.000	18.820	18.820	0.	0.	0.	0.
300	20.050	18.943	18.820	0.037	0.	0.	0.
350	21.276	22.132	19.069	1.072	0.	0.	0.
400	22.142	25.032	19.636	2.159	0.	0.	0.
450	22.803	27.680	20.385	3.283	0.	0.	0.
500	23.330	30.110	21.237	4.436	0.	0.	0.
600	24.154	34.440	23.086	6.812	0.	0.	0.
700	24.803	38.212	24.983	9.260	0.	0.	0.
800	25.359	41.562	26.850	11.769	0.	0.	0.
900	25.874	44.579	28.655	14.331	0.	0.	0.
1000	26.338	47.329	30.387	16.942	0.	0.	0.
1100	26.778	49.860	32.044	19.598	0.	0.	0.
1200	27.196	52.208	33.627	22.297	0.	0.	0.
1300	27.614	54.401	35.142	25.037	0.	0.	0.
1400	28.033	56.463	36.592	27.820	0.	0.	0.
1500	28.451	58.411	37.982	30.644	0.	0.	0.
1600	28.870	60.261	39.317	33.510	0.	0.	0.
1685.000	29.225	61.765	40.412	35.979	CRYSTAL <--> LIQUID		
1685.000	27.196	91.562	40.412	86.187	TRANSITION		
1700	27.196	91.803	40.864	86.595	0.	0.	0.
1800	27.196	93.357	43.738	89.315	0.	0.	0.
1900	27.196	94.827	46.388	92.034	0.	0.	0.
2000	27.196	96.222	48.846	94.754	0.	0.	0.
2100	27.196	97.549	51.133	97.473	0.	0.	0.
2200	27.196	98.815	53.272	100.193	0.	0.	0.
2300	27.196	100.023	55.279	102.913	0.	0.	0.
2400	27.196	101.181	57.167	105.632	0.	0.	0.
2500	27.196	102.291	58.950	108.352	0.	0.	0.
2600	27.196	103.358	60.638	111.071	0.	0.	0.
2700	27.196	104.384	62.239	113.791	0.	0.	0.
2800	27.196	105.373	63.762	116.511	0.	0.	0.
2900	27.196	106.327	65.214	119.230	0.	0.	0.
3000	27.196	107.249	66.600	121.950	0.	0.	0.
3100	27.196	108.141	67.925	124.669	0.	0.	0.
3200	27.196	109.005	69.196	127.389	0.	0.	0.
3300	27.196	109.842	70.415	130.109	0.	0.	0.
3400	27.196	110.653	71.586	132.828	0.	0.	0.
3500	27.196	111.442	72.714	135.548	0.	0.	0.
3504.616	27.196	111.478	72.765	135.673	------- FUGACITY = 1 bar -------		
3600	27.196	112.208	73.800	138.267	-384.152	10.461	-0.152
3700	27.196	112.953	74.848	140.987	-383.738	21.417	-0.302
3800	27.196	113.678	75.861	143.707	-383.326	32.361	-0.445
3900	27.196	114.385	76.840	146.426	-382.915	43.295	-0.580
4000	27.196	115.073	77.787	149.146	-382.505	54.218	-0.708
4100	27.196	115.745	78.704	151.865	-382.096	65.131	-0.830
4200	27.196	116.400	79.594	154.585	-381.687	76.034	-0.946
4300	27.196	117.040	80.458	157.305	-381.279	86.927	-1.056
4400	27.196	117.665	81.296	160.024	-380.871	97.811	-1.161
4500	27.196	118.277	82.111	162.744	-380.462	108.686	-1.262

PREVIOUS: CURRENT: March 1967

SILICON (Si) IDEAL GAS A_r = 28.0855

IP(Si, g) = 65747.5 ± 0.6 cm^{-1}

S(298.15 K) = 167.980 ± 0.035 J K^{-1} mol^{-1}

$\Delta_f H°$(0 K) = 446 ± 8 kJ mol^{-1}

$\Delta_f H°$(298.15 K) = 450 ± 8 kJ mol^{-1}

Electronic Levels and Quantum Weights

State	ε_i, cm^{-1}	g_i
$^3P_2$	0.00	5
$^3P_1$	77.115	3
$^3P_0$	223.157	1
$^1D_2$	6298.850	5
$^1S_0$	15394.370	1

Enthalpy of Formation

The enthalpy of formation of silicon gas is chosen to be the value recommended by CODATA (1). This value was calculated from the sublimation and decomposition measurements of Davis et al. (1848-2003 K, 2), Grievison and Alcock(1940-2054 K, 3), Drowart et al. (4), Gulbranson et al. (1373-1623 K, 5), Batdorf and Smits (1493-1601 K, 6), and Zenbov et al. (7).

Heat Capacity and Entropy

The information on electronic energy levels and quantum weights, given by Moore (8, 9), is incomplete because many theoretically predicted levels have not been observed. Our calculations indicate that any reasonable method of filling in these missing levels (for n<11) and cutting off the summation in the partition function (10) has no effect on the thermodynamic functions to 6000 K. This is a result of the high energy of all levels other than those listed above; the next excited state is approximately 33326 cm^{-1} above the ground state. Although we list only a few levels, all levels given by Moore (8, 9) and estimated levels (for n<11) are considered in the calculation. The reported uncertainty in S°(298.15 K) is due to uncertainties in the relative atomic mass and the fundamental constants. Extension of these calculations above 6000 K may require consideration of the excited states (for n>10) and use of different fill and cutoff procedures (10).

The thermal functions at 298.15 K closely agree with the recent CODATA recommendations (1) except for two minor changes. First, the entropy differs by 0.1094 J K^{-1} mol^{-1} because this table used a reference pressure of 1 bar, wheras CODATA recommendations are based on 1 atm. Second, the entropy differs by ~0.001 J K^{-1} mol^{-1} due to the use of slightly different values for the fundamental constants.

References

1. J. D. Cox, chairman, CODATA Task Group on Key Values for Thermodynamics, J. Chem. Thermodynamics 10, 903 (1978).
2. S. G. Davis, D. F. Anthrop, and A. W. Searcy, J. Chem. Phys. 34, 659 (1961).
3. P. Grieveson and L. B. Alcock, in "Special Ceramics," P. Popper, ed., Academic Press, London, 1960.
4. J. Drowart, G. DeMaria and M. Inghram, J. Chem. Phys. 29, 1015 (1958).
5. E. A. Gulbranson, K. F. Andrew and F. A. Brassart, J. Electrochem. Soc. 113, 834 (1966).
6. R. L. Batdorf and F. M. Smits, J. Appl. Phys. 10, 259 (1959).
7. K. F. Zonbov, L. L. Ames and J. L. Margrave, High Temp. Sci. 5, 235 (1973).
8. C. E. Moore, U. S. Nat. Bur. Stand., NSRDS-NBS 34, 1970.
9. C. E. Moore, U. S. Nat. Bur. Stand., NSRDS-NBS 3, Section 2, 1967.
10. J. R. Downey, Jr., The Dow Chemical Company, Rept. AFOSR-TR-78-0960, Contract No. F44620-75-1-0048, 1978.

Silicon (Si) Si_1(g)

Enthalpy Reference Temperature = T_r = 298.15 K Standard State Pressure = p° = 0.1 MPa

T/K	$C_p°$ (J $K^{-1}mol^{-1}$)	S° (J $K^{-1}mol^{-1}$)	$-[G°-H°(T_r)]/T$ (J $K^{-1}mol^{-1}$)	$H°-H°(T_r)$ (kJ mol^{-1})	$\Delta_f H°$ (kJ mol^{-1})	$\Delta_f G°$ (kJ mol^{-1})	Log K_f
0	0.	0.	INFINITE	-7.550	445.668	445.668	INFINITE
100	28.022	140.886	188.980	-4.809	448.142	434.437	-226.927
200	23.796	158.817	170.054	-2.247	449.526	420.095	-109.718
250	22.821	164.013	168.349	-1.084	449.839	412.697	-86.228
298.15	22.251	167.980	167.980	0.	450.000	405.528	-71.047
300	22.234	168.117	167.980	0.041	450.004	405.252	-70.561
350	21.861	171.515	168.249	1.143	450.071	397.787	-59.366
400	21.613	174.417	168.843	2.229	450.071	390.317	-50.970
450	21.440	176.952	169.606	3.305	450.023	382.850	-44.440
500	21.316	179.204	170.456	4.374	449.938	375.391	-39.217
600	21.153	183.075	172.246	6.497	449.684	360.504	-31.385
700	21.057	186.328	174.032	8.607	449.347	345.666	-25.794
800	21.000	189.135	175.748	10.710	448.940	330.881	-21.604
900	20.971	191.607	177.376	12.808	448.476	316.151	-18.349
1000	20.968	193.816	178.911	14.905	447.962	301.476	-15.747
1100	20.989	195.815	180.358	17.002	447.404	286.854	-13.622
1200	21.033	197.643	181.724	19.103	446.806	272.284	-11.852
1300	21.099	199.329	183.014	21.210	446.172	257.766	-10.357
1400	21.183	200.896	184.236	23.323	445.504	243.298	-9.078
1500	21.282	202.360	185.396	25.447	444.803	228.879	-7.970
1600	21.394	203.737	186.500	27.580	444.070	214.508	-7.003
1700	21.513	205.038	187.552	29.726	393.131	200.631	-6.165
1800	21.638	206.271	188.558	31.883	392.569	189.323	-5.494
1900	21.764	207.444	189.522	34.053	392.019	178.047	-4.895
2000	21.889	208.564	190.446	36.236	391.482	166.799	-4.356
2100	22.011	209.635	191.334	38.431	390.958	155.578	-3.870
2200	22.129	210.662	192.190	40.638	390.445	144.382	-3.428
2300	22.241	211.648	193.014	42.857	389.944	133.208	-3.025
2400	22.346	212.597	193.811	45.086	389.454	122.056	-2.656
2500	22.444	213.511	194.581	47.326	388.974	110.924	-2.318
2600	22.535	214.393	195.326	49.575	388.503	99.812	-2.005
2700	22.617	215.245	196.048	51.832	388.041	88.717	-1.716
2800	22.692	216.069	196.748	54.098	387.587	77.639	-1.448
2900	22.759	216.866	197.428	56.370	387.140	66.578	-1.199
3000	22.819	217.639	198.089	58.649	386.700	55.531	-0.967
3100	22.872	218.388	198.732	60.934	386.265	44.500	-0.750
3200	22.918	219.115	199.358	63.223	385.835	33.482	-0.547
3300	22.958	219.821	199.967	65.517	385.409	22.477	-0.356
3400	22.993	220.507	200.561	67.815	384.987	11.486	-0.176
3500	23.022	221.174	201.141	70.116	384.568	0.507	-0.008
3504.616	23.023	221.204	201.167	70.222	------- FUGACITY = 1 bar -------		
3600	23.046	221.822	201.706	72.419	0.	0.	0.
3700	23.066	222.454	202.258	74.725	0.	0.	0.
3800	23.081	223.070	202.798	77.032	0.	0.	0.
3900	23.093	223.669	203.325	79.341	0.	0.	0.
4000	23.102	224.254	203.841	81.651	0.	0.	0.
4100	23.108	224.825	204.346	83.961	0.	0.	0.
4200	23.111	225.381	204.840	86.272	0.	0.	0.
4300	23.113	225.925	205.324	88.583	0.	0.	0.
4400	23.112	226.457	205.799	90.895	0.	0.	0.
4500	23.111	226.976	206.264	93.206	0.	0.	0.
4600	23.108	227.484	206.719	95.517	0.	0.	0.
4700	23.104	227.981	207.167	97.827	0.	0.	0.
4800	23.100	228.467	207.605	100.138	0.	0.	0.
4900	23.096	228.943	208.036	102.447	0.	0.	0.
5000	23.092	229.410	208.459	104.757	0.	0.	0.
5100	23.089	229.867	208.874	107.066	0.	0.	0.
5200	23.087	230.316	209.282	109.375	0.	0.	0.
5300	23.086	230.755	209.683	111.683	0.	0.	0.
5400	23.086	231.187	210.077	113.992	0.	0.	0.
5500	23.087	231.610	210.465	116.300	0.	0.	0.
5600	23.091	232.026	210.846	118.609	0.	0.	0.
5700	23.098	232.435	211.221	120.918	0.	0.	0.
5800	23.108	232.837	211.591	123.228	0.	0.	0.
5900	23.121	233.232	211.954	125.539	0.	0.	0.
6000	23.135	233.621	212.312	127.851	0.	0.	0.

PREVIOUS: March 1967 (1 atm) CURRENT: March 1983 (1 bar)

Silicon (Si) Si_1(g)

SILICON, ION (Si^+) IDEAL GAS M_r = 28.08495

Silicon, Ion (Si^+) $Si_1^+(g)$

IP(Si^+, g) = 131838.4 ± 0.1 cm^{-1}

S°(298.15 K) = 163.426 ± 0.02 JK^{-1} mol^{-1}

$\Delta_fH°$(0 K) = 1232.2 ± 4 kJ mol^{-1}

$\Delta_fH°$(298.15 K) = [1242.505] kJ mol^{-1}

Electronic Levels and Quantum Weights

State	η_i cm^{-1}	g_i
$^2P_{1/2}$	0	2
$^2P_{3/2}$	287.32	4
$^4P_{1/2}$	42824.35	4
$^4P_{3/2}$	42932.68	6
$^4P_{5/2}$	43107.97	4

Enthalpy of Formation

$\Delta_fH°(Si^+$, g, 0 K) is calculated from $\Delta_fH°$(Si, g, 0 K) (1) using the spectroscopic value of IP(Si) = 65747.5 ± 0.6 cm^{-1} (786.515 ± 0.001 kJ mol^{-1}) from Moore (2, 6). The ionization limit is converted from cm^{-1} to kJ mol^{-1} using the factor, 1 cm^{-1} = 0.01196266 kJ cm^{-1}, which is derived from the 1973 CODATA fundamental constants (3). Rosenstock et al. (4) and Levin and Lias (5) have summarized additional ionization and appearance potential data.

$\Delta_fH°(Si^+$, g, 298.15 K) is calculated from $\Delta_fH°$(Si, g, 0 K) by using IP(Si) with JANAF (1) enthalpies, H°(0 K) - H°(298.15 K), for Si(g), Si^+(g), and e^-(g). $\Delta_fH°(Si \rightarrow Si^+ + e^-$, 298.15 K) differs from a room temperature threshold energy due to inclusion of these enthalpies and to threshold effects discussed by Rosenstock et al. (4). $\Delta_fH°$(298.15 K) should be changed by -6.197 kJ mol^{-1} if it is to be used in the ion convention that excludes the enthalpy of the electron.

Heat Capacity and Entropy

The information on electronic energy levels and quantum weights, given by Moore (2, 6) is incomplete because many theoretically predicted levels have not been observed. Our calculations indicate that any reasonable method of filling in these missing levels and cutting off the summation in the partition function (7) has no effect on the thermodynamic functions to 6000 K. This is a result of the high energy of all levels other than the ground state and the first four excited states; the next excited state is 55309.61 cm^{-1} above the ground state. Since inclusion of these excited states has no effect on the thermodynamic functions (to 6000 K), we list only the ground state and the first four excited states. The reported uncertainty in S°(298.15 K) is due to uncertainties in the relative ionic mass and the fundamental constants. Extension of these calculations above 6000 K may require consideration of the excited states and use of different fill and cutoff procedures (7).

References

1. JANAF Thermochemical Tables: Si (g), 3-31-83 ; e^-(g), 3-31-82.
2. C. E. Moore, U. S. Nat. Bur. Stand., NSRDSS-NBS-34 (1970).
3. E. R. Cohen and B. N. Taylor, J. Phys. Chem. Ref. Data 2, 663 (1973).
4. H. M. Rosenstock, K. Draxl et al., J. Phys. Chem. Ref. Data 6, Supp. 1 (1977).
5. R. D. Levin and S. G. Lias, U. S. Nat. Bur. Stand., NSRDS-NBS-71, 1982.
6. C. E. Moore, U. S. Nat. Bur. Stand., NSRDS-NBS-3, Section 1 (1965) and Section II (1967).
7. J. R. Downey, Jr., The Dow Chemical Company, AFOSR-TR-0960, Contract No. F44620-75-1-0048, 1978.

Enthalpy Reference Temperature = T_r = 298.15 K Standard State Pressure = p° = 0.1 MPa

	J $K^{-1}mol^{-1}$			kJ mol^{-1}			
T/K	$C_p°$	S°	$-[G°-H°(T_r)]/T$	$H°-H°(T_r)$	$\Delta_fH°$	$\Delta_fG°$	Log K_f
0	0.	0.	INFINITE	-7.343	1232.183		
100	25.060	134.836	186.411	-5.158			
200	26.512	153.262	165.719	-2.491			
250	25.337	159.052	163.832	-1.195			
298.15	24.338	163.426	163.426	0.	1242.505	1193.136	-209.032
300	24.304	163.577	163.427	0.045	1242.551	1192.829	-207.690
350	23.520	167.261	163.719	1.240	1243.750	1184.446	-176.769
400	22.943	170.362	164.361	2.400	1244.864	1175.897	-153.556
450	22.518	173.039	165.180	3.536	1245.915	1167.213	-135.487
500	22.201	175.394	166.086	4.654	1246.918	1158.414	-121.019
600	21.773	179.400	167.982	6.851	1248.818	1140.532	-99.292
700	21.509	182.735	169.858	9.014	1250.611	1122.342	-83.750
800	21.337	185.596	171.651	11.156	1252.323	1103.900	-72.077
900	21.219	188.101	173.342	13.283	1253.967	1085.248	-62.986
1000	21.134	190.332	174.932	15.401	1255.552	1066.416	-55.704
1100	21.072	192.344	176.425	17.511	1257.085	1047.427	-49.738
1200	21.025	194.175	177.829	19.615	1258.570	1028.302	-44.761
1300	20.989	195.857	179.152	21.716	1260.008	1009.054	-40.544
1400	20.960	197.411	180.401	23.814	1261.402	989.697	-36.926
1500	20.937	198.856	181.584	25.908	1262.751	970.243	-33.787
1600	20.918	200.207	182.706	28.001	1264.056	950.699	-31.037
1700	20.903	201.475	183.773	30.092	1215.141	931.522	-28.622
1800	20.890	202.669	184.790	32.182	1216.590	914.797	-26.547
1900	20.879	203.798	185.761	34.270	1218.037	897.991	-24.688
2000	20.870	204.869	186.690	36.358	1219.484	881.109	-23.012
2100	20.862	205.887	187.580	38.444	1220.929	864.155	-21.495
2200	20.855	206.857	188.434	40.530	1222.374	847.132	-20.113
2300	20.849	207.784	189.256	42.615	1223.818	830.043	-18.851
2400	20.844	208.671	190.046	44.700	1225.262	812.892	-17.692
2500	20.839	209.522	190.808	46.784	1226.705	795.680	-16.625
2600	20.835	210.339	191.544	48.868	1228.148	778.410	-15.638
2700	20.831	211.126	192.255	50.951	1229.590	761.085	-14.724
2800	20.828	211.883	192.942	53.034	1231.032	743.707	-13.874
2900	20.825	212.614	193.608	55.117	1232.474	726.277	-13.082
3000	20.823	213.320	194.254	57.199	1233.915	708.797	-12.341
3100	20.820	214.003	194.880	59.281	1235.356	691.270	-11.648
3200	20.818	214.664	195.488	61.363	1236.797	673.696	-10.997
3300	20.816	215.304	196.078	63.445	1238.238	656.076	-10.385
3400	20.814	215.926	196.653	65.526	1239.678	638.414	-9.808
3500	20.813	216.529	197.212	67.608	1241.119	620.708	-9.264
3600	20.811	217.115	197.757	69.689	858.407	613.423	-8.901
3700	20.810	217.685	198.288	71.770	860.261	606.592	-8.564
3800	20.809	218.240	198.806	73.851	862.113	599.711	-8.244
3900	20.808	218.781	199.311	75.932	863.964	592.782	-7.939
4000	20.807	219.308	199.805	78.013	865.814	585.805	-7.650
4100	20.807	219.821	200.286	80.093	867.662	578.782	-7.374
4200	20.806	220.323	200.758	82.174	869.511	571.714	-7.110
4300	20.806	220.812	201.218	84.254	871.359	564.601	-6.859
4400	20.806	221.291	201.669	86.335	873.207	557.446	-6.618
4500	20.806	221.758	202.110	88.416	875.055	550.249	-6.387
4600	20.806	222.216	202.542	90.496	876.903	543.011	-6.166
4700	20.807	222.663	202.966	92.577	878.751	535.732	-5.954
4800	20.808	223.101	203.381	94.658	880.601	528.414	-5.750
4900	20.809	223.530	203.788	96.738	882.450	521.058	-5.555
5000	20.811	223.951	204.187	98.819	884.300	513.664	-5.366
5100	20.813	224.363	204.578	100.901	886.151	506.233	-5.185
5200	20.816	224.767	204.963	102.982	888.002	498.765	-5.010
5300	20.819	225.163	205.340	105.064	889.854	491.262	-4.842
5400	20.822	225.553	205.711	107.146	891.706	483.724	-4.679
5500	20.827	225.935	206.075	109.228	893.559	476.152	-4.522
5600	20.832	226.310	206.433	111.311	895.412	468.546	-4.370
5700	20.838	226.679	206.785	113.395	897.265	460.907	-4.224
5800	20.845	227.041	207.131	115.479	899.117	453.236	-4.082
5900	20.852	227.398	207.472	117.564	900.969	445.532	-3.944
6000	20.861	227.748	207.807	119.649	902.822	437.797	-3.811

PREVIOUS: December 1971 (1 atm) CURRENT: March 1983 (1 bar)

Silicon, Ion (Si^+) $Si_1^+(g)$

SILICON, ION (Si^-) IDEAL GAS M_r = 28.08605

EA(Si, g) = 1.385 ± 0.005 eV

S°(298.15 K) = 161.977 ± 0.001 J K^{-1} mol^{-1}

$\Delta_f H°$(0 K) = 312.04 ± 3 kJ mol^{-1}

$\Delta_f H°$(298.15 K) = [308.819] kJ mol^{-1}

Electonic Levels and Quantum Weights

State	ε_i, cm^{-1}	g_i
$^4S_{3/2}$	0.0	4
^2D	6952	10
^2P	10937	6

Enthalpy of Formation

$\Delta_f H°$(Si^-, g, 0 K) is calculated from $\Delta_f H°$(Si, g, 0 K) (1) using the adopted electron affinity of EA(Si) = 1.385 ± 0.005 eV (133.631 ± 0.482 kJ mol^{-1}). This value, recommended by Hotop and Lineberger (2), is based on a laser photodetachment electron spectrometry study (3). Additional information on Si^-(g) may be obtained in the critical discussions of Hotop and Lineberger (2, 4), Rosenstock et al. (5) and Massey (6).

$\Delta_f H°$(Si^-, g, 298.15 K) is obtained from $\Delta_f H°$(Si, g, 0 K) by using EA(Si) with JANAF (1) enthalpies, H°(0 K) - H°(298.15 K), for Si^-(g), Si(g), and e^-(g). $\Delta_r H°$(Si^- → Si + e^-, 298.15 K) differs from a room-temperature threshold energy due to inclusion of these enthalpies and to threshold effects discussed by Rosenstock et al. (5). $\Delta_f H°$(298.15 K) should be changed by +6.197 kJ mol^{-1} if it is to be used in the ion convention that excludes the enthalpy of the electron.

Heat Capacity and Entropy

Electronic energy levels are from the laser photodetachment electron spectrometry study of Kasdan et al. (3). These workers found that the $Si^-(^2D)$ and $Si^-(^2P)$ states have binding energies of 0.523 ± 0.005 eV and 0.029 ± 0.005 eV, respectively, relative to ground state $Si(^3P_0)$. The electronic energies of 6952 cm^{-1} and 10937 cm^{-1} are the difference in binding energy between ground and excited states. We neglect the spin-orbit splitting of the ^2D and ^2P states; its effect is negligible.

References

1. JANAF Thermochemical Tables: Si(g), 3-31-83; e^-(g), 3-31-82.
2. H. Hotop and W. C. Lineberger, J. Phys. Chem. Ref. Data to be published, 1985.
3. A. Kasdan, E. Herbst and W. C. Lineberger, J. Chem. Phys. 62, 541 (1975).
4. H. Hotop and W. C. Lineberger, J. Phys. Chem. Ref. Data 4, 539 (1975).
5. H. M. Rosenstock, K. Draxl et al., J. Phys. Chem. Ref. Data 6, Supp. 1 (1977).
6. H. S. W. Massey, "Negative Ions", 3rd ed., Cambridge University Press, Cambridge, 1976.

Silicon, Ion (Si^-) Si_1^-(g)

Enthalpy Reference Temperature = T_r = 298.15 K Standard State Pressure = p° = 0.1 MPa

T/K	$C_p°$ (J $K^{-1}mol^{-1}$)	S° (J $K^{-1}mol^{-1}$)	-[G°-H°(T_r)]/T (J $K^{-1}mol^{-1}$)	H°-H°(T_r) (kJ mol^{-1})	$\Delta_f H°$ (kJ mol^{-1})	$\Delta_f G°$ (kJ mol^{-1})	Log K_f
0	0.	0.	INFINITE	-6.197	312.037		
100	20.786	139.270	180.457	-4.119			
200	20.786	153.678	163.878	-2.040			
250	20.786	158.316	162.319	-1.001			
298.15	20.786	161.977	161.977	0.	308.819	272.391	-47.722
300	20.786	162.106	161.977	0.038	308.782	272.165	-47.388
350	20.786	165.310	162.231	1.078	307.747	266.143	-39.720
400	20.786	168.085	162.793	2.117	306.660	260.274	-33.988
450	20.786	170.534	163.520	3.156	305.536	254.543	-29.547
500	20.786	172.724	164.332	4.196	304.382	248.938	-26.006
600	20.786	176.513	166.056	6.274	302.006	238.071	-20.726
700	20.787	179.718	167.785	8.353	299.558	227.608	-16.984
800	20.791	182.494	169.454	10.432	297.050	217.500	-14.201
900	20.803	184.943	171.041	12.511	294.489	207.709	-12.055
1000	20.831	187.136	172.543	14.593	291.881	198.207	-10.353
1100	20.887	189.124	173.961	16.678	289.232	188.967	-8.973
1200	20.981	190.945	175.302	18.771	286.547	179.971	-7.834
1300	21.122	192.629	176.571	20.876	283.833	171.199	-6.879
1400	21.318	194.202	177.775	22.998	281.094	162.638	-6.068
1500	21.572	195.681	178.920	25.142	278.335	154.273	-5.372
1600	21.884	197.082	180.011	27.314	275.562	146.092	-4.769
1700	22.253	198.420	181.055	29.520	222.605	138.532	-4.257
1800	22.674	199.704	182.056	31.766	220.053	133.661	-3.879
1900	23.140	200.942	183.017	34.057	217.545	128.930	-3.545
2000	23.645	202.141	183.944	36.396	215.086	124.329	-3.247
2100	24.179	203.308	184.838	38.787	212.679	119.851	-2.981
2200	24.735	204.445	185.704	41.232	210.326	115.486	-2.742
2300	25.303	205.558	186.543	43.734	208.030	111.227	-2.526
2400	25.876	206.646	187.358	46.293	205.790	107.066	-2.330
2500	26.444	207.714	188.151	48.909	203.608	102.998	-2.152
2600	27.002	208.762	188.923	51.581	201.483	99.015	-1.989
2700	27.542	209.792	189.677	54.309	199.412	95.114	-1.840
2800	28.058	210.803	190.414	57.089	197.394	91.288	-1.703
2900	28.548	211.796	191.134	59.920	195.426	87.533	-1.577
3000	29.005	212.772	191.839	62.797	193.506	83.846	-1.460
3100	29.428	213.730	192.530	65.719	191.629	80.222	-1.352
3200	29.815	214.670	193.207	68.682	189.794	76.657	-1.251
3300	30.164	215.593	193.871	71.681	187.995	73.150	-1.158
3400	30.475	216.498	194.524	74.714	186.229	69.696	-1.071
3500	30.748	217.386	195.164	77.775	184.492	66.294	-0.989
3600	30.982	218.255	195.794	80.862	-201.371	73.402	-1.065
3700	31.180	219.107	196.412	83.970	-202.647	81.052	-1.144
3800	31.343	219.941	197.020	87.097	-203.906	88.736	-1.220
3900	31.472	220.757	197.619	90.238	-205.153	96.454	-1.292
4000	31.568	221.555	198.207	93.390	-206.389	104.203	-1.361
4100	31.635	222.335	198.786	96.550	-207.617	111.983	-1.427
4200	31.674	223.098	199.356	99.716	-208.841	119.793	-1.490
4300	31.687	223.843	199.917	102.884	-210.063	127.632	-1.550
4400	31.676	224.572	200.469	106.053	-211.284	135.500	-1.609
4500	31.643	225.283	201.012	109.219	-212.508	143.395	-1.664
4600	31.591	225.978	201.548	112.381	-213.736	151.317	-1.718
4700	31.521	226.657	202.075	115.536	-214.969	159.267	-1.770
4800	31.435	227.320	202.594	118.684	-216.210	167.242	-1.820
4900	31.336	227.967	203.105	121.823	-217.460	175.244	-1.868
5000	31.224	228.599	203.609	124.951	-218.720	183.271	-1.915
5100	31.101	229.216	204.105	128.067	-219.991	191.323	-1.960
5200	30.969	229.819	204.593	131.171	-221.275	199.401	-2.003
5300	30.829	230.407	205.075	134.261	-222.572	207.503	-2.045
5400	30.683	230.982	205.549	137.337	-223.884	215.630	-2.086
5500	30.531	231.544	206.017	140.397	-225.209	223.781	-2.125
5600	30.374	232.092	206.478	143.443	-226.552	231.957	-2.164
5700	30.214	232.629	206.932	146.472	-227.910	240.156	-2.201
5800	30.051	233.153	207.379	149.485	-229.286	248.380	-2.237
5900	29.886	233.665	207.821	152.482	-230.679	256.627	-2.272
6000	29.720	234.166	208.255	155.463	-232.089	264.899	-2.306

PREVIOUS: CURRENT: March 1983 (1 bar)

Silicon, Ion (Si^-) Si_1^-(g)

SILICON (Si_2) IDEAL GAS M_r = 56.1710

$S°(298.15\ K) = 229.79\ J\ K^{-1}\ mol^{-1}$

$\Delta_f H°(0\ K) = 587.1 \pm 13\ kJ\ mol^{-1}$
$\Delta_f H°(298.15\ K) = 589.9 \pm 13\ kJ\ mol^{-1}$

Electronic Levels and Quantum Weights

State	ε_i, cm^{-1}	g_i	State	ε_i, cm^{-1}	g_i	State	ε_i, cm^{-1}	g_i
$^3\Sigma_g^-$	0	3	$[^1\Delta_g]$	[10000]	[2]	$^3\Pi_g$	[30000]	6
$^3\Pi_u$	[2000]	6	$[^1\Sigma_g^+]$	[13000]	[1]	$^3\Sigma_u^-$	30769	3
$[^1\Sigma_g^+]$	[6000]	[1]	$[^3\Sigma_u^+]$	[20000]	[3]	$[^1\Pi_g]$	[40000]	[2]
$[^1\Pi_u]$	[8000]	[2]	$^3\Sigma_u^-$	24583	3	$[^1\Sigma_u^+]$	[45000]	[1]
						$^3\Sigma_u^-$	46762	3

$\omega_e = 510.98\ cm^{-1}$ $\omega_e x_e = 2.02\ cm^{-1}$ $\sigma = 2$
$B_e = 0.2390\ cm^{-1}$ $\alpha_e = 0.0013\ cm^{-1}$ $r_e = 2.246$ Å

Enthalpy of Formation

The selected value is based on the spectroscopic and equilibrium data summarized below. Verma and Warsop (1) concluded from analysis of the absorption spectra that $D_0^0 = 70 \pm 4\ kcal\ mol^{-1}$. Predissociation of the H state suggested the upper limit $D_0^0 \leq 74.0$, while linear Birge-Sponer extrapolations of the H and K states gave 78.4 and 68.9 $kcal\ mol^{-1}$ for the ground state. By assuming that the true extrapolation would not reduce the linear extrapolation for the H state by more than half, i.e. from 26 to 13 $kcal\ mol^{-1}$, the authors arrived at the lower limit of $D_0^0 \geq 65.5$ for the ground state.

Drowart and co-workers have used the Knudsen effusion-mass spectrometric technique to determine the vapor equilibria over the systems SiC-graphite (2), SiC-silicon (3) and boron-carbon-silicon (4). Third law analysis of the partial pressures of Si_2 and Si over the three systems yields D_0^0 values of 73.3, 74.3 and 70.4 $kcal\ mol^{-1}$, which are in good agreement with the spectroscopic values. The selected value, $\Delta_f H°(298.15\ K) = 141 \pm 3\ kcal\ mol^{-1}$, corresponds to $D_0^0 = 73.0\ kcal\ mol^{-1}$.

Source	Method	T/K	Data Points	$\Delta_r H°(298.15\ K)$, $kcal\ mol^{-1}$ 2nd law	3rd law	Drift $cal\ K^{-1}\ mol^{-1}$	$\Delta_f H°(298.15\ K)$ $kcal\ mol^{-1}$
Verma (1963)	Prediss. of H state		-		≤75.4	-	≥140.0
	LEX of K state		-		70.3		145.1
	Extrap. of H state		-		≥66.9		≤148.5
Drowart (1958)	Mass spec.	2149-2316	7	95.3±7.1	74.68	-9±3	140.7
Drowart (1960)	Mass spec.	1703-2160	9	81.0±1.1	75.71	-2.7±0.6	139.7
Verhaegen (1964)	Mass spec.	2166-2344	4	68.0±16	71.76	2±7	143.6

*For the reaction $Si_2(g) = 2\ Si(g)$

Heat Capacity and Entropy

Vibrational and rotational constants are those obtained by Verma and Warsop (1) through combination of their data for the H-X system with those of Douglas (5). Observation of the same H-X system in matrix isolation by Weltner and McLeod (6) confirms that the lower state is the ground state. Comparison with the isoelectronic molecules C_2, BN, BeO and MgO (7, 8, 9, 10) suggest that there are several possible low-lying excited states in Si_2. Tentative estimates for these levels are given above, based on this comparison and on the observed states (1, 5). These estimates are relatively uncertain and probably yield an upper limit for the entropy at temperatures where Si_2 is significant. A probable lower limit may be obtained by increasing the two $^3\Pi$ states by 8000 cm^{-1} and omitting the other estimated levels. This would reduce the entropy by 1.7 $cal\ K^{-1}\ mol^{-1}$ at 2000 K, corresponding to a change of 3.4 $kcal\ mol^{-1}$ in $\Delta_r H°$ values based on equilibrium data.

References

1. R. D. Verma and P. A. Warsop, Can. J. Phys. 41, 152 (1963).
2. J. Drowart, G. DeMaria and M. G. Inghram, J. Chem. Phys. 29, 1015 (1958).
3. J. Drowart and G. DeMaria, pp. 16-23 in "Silicon Carbide", Edited by J. R. O'Connor and J. Smiltens, Pergamon Press, New York, 1960.
4. G. Verhaegen, F. E. Stafford and J. Drowart, J. Chem. Phys. 40, 1622 (1964).
5. A. E. Douglas, Can. J. Phys. 33, 801 (1955).
6. W. Weltner, Jr., and D. McLeod, Jrs., J. Chem. Phys. 41, 235 (1964).
7. P. F. Fougere and R. K. Nesbet, J. Chem. Phys. 44, 285 (1966).
8. R. S. Mulliken, Phys. Rev. 56, 778 (1939).
9. G. Verhaegen, W. G. Richards and C. M. Moser, J. Chem. Phys. 46, 160 (1967); 45, 3226 (1966); 45, 1828 (1966).
10. L. Brewer and S. Trajmar, J. Chem. Phys. 36, 1585 (1962).

Silicon (Si_2) $Si_2(g)$

Enthalpy Reference Temperature = T_r = 298.15 K Standard State Pressure = p° = 0.1 MPa

T/K	$C_p°$ (J K⁻¹mol⁻¹)	S°	$-[G°-H°(T_r)]/T$	$H°-H°(T_r)$ (kJ mol⁻¹)	$\Delta_f H°$	$\Delta_f G°$	Log K_f
0	0.	0.	INFINITE	-9.263	587.118	587.118	INFINITE
100	29.412	195.317	258.806	-6.349	589.499	570.734	-298.121
200	32.184	216.488	232.879	-3.278	590.212	551.581	-144.058
250	33.463	223.812	230.355	-1.636	590.154	541.924	-113.229
298.15	34.452	229.793	229.793	0.	589.944	532.653	-93.319
300	34.487	230.007	229.794	0.064	589.934	532.298	-92.681
350	35.374	235.390	230.217	1.811	589.611	522.716	-78.011
400	36.227	240.170	231.168	3.601	589.228	513.185	-67.015
450	37.096	244.487	232.412	5.434	588.812	503.705	-58.469
500	37.986	248.441	233.820	7.311	588.382	494.272	-51.636
600	39.743	255.524	236.860	11.198	587.517	475.531	-41.399
700	41.286	261.770	239.981	15.252	586.675	456.933	-34.097
800	42.472	267.365	243.061	19.443	585.848	438.455	-28.628
900	43.274	272.417	246.046	23.734	585.015	420.081	-24.381
1000	43.738	277.003	248.916	28.087	584.146	401.801	-20.988
1100	43.941	281.183	251.663	32.472	583.220	383.611	-18.216
1200	43.963	285.008	254.284	36.869	582.219	365.508	-15.910
1300	43.869	288.524	256.785	41.261	581.130	347.493	-13.962
1400	43.709	291.769	259.169	45.640	579.945	329.564	-12.296
1500	43.520	294.779	261.444	50.002	578.658	311.724	-10.855
1600	43.324	297.581	263.616	54.344	577.268	293.974	-9.597
1700	43.135	300.202	265.692	58.667	475.421	277.207	-8.518
1800	42.962	302.662	267.678	62.972	474.287	265.579	-7.707
1900	42.810	304.981	269.581	67.260	473.136	254.016	-6.983
2000	42.679	307.174	271.406	71.534	471.971	242.513	-6.334
2100	42.569	309.253	273.160	75.797	470.794	231.069	-5.748
2200	42.479	311.231	274.846	80.049	469.607	219.682	-5.216
2300	42.407	313.118	276.469	84.293	468.412	208.348	-4.732
2400	42.352	314.922	278.034	88.531	467.210	197.067	-4.289
2500	42.311	316.650	279.544	92.764	466.004	185.836	-3.883
2600	42.282	318.308	281.003	96.993	464.795	174.653	-3.509
2700	42.264	319.904	282.415	101.221	463.583	163.517	-3.163
2800	42.255	321.441	283.781	105.446	462.369	152.425	-2.844
2900	42.254	322.923	285.106	109.672	461.155	141.377	-2.546
3000	42.259	324.356	286.390	113.897	459.942	130.371	-2.270
3100	42.271	325.742	287.637	118.124	458.729	119.405	-2.012
3200	42.287	327.084	288.849	122.352	457.518	108.479	-1.771
3300	42.307	328.386	290.028	126.581	456.308	97.590	-1.545
3400	42.331	329.649	291.174	130.813	455.101	86.738	-1.333
3500	42.358	330.876	292.291	135.048	453.896	75.921	-1.133
3600	42.388	332.070	293.380	139.285	-315.609	86.060	-1.249
3700	42.420	333.232	294.441	143.525	-315.980	97.223	-1.373
3800	42.455	334.364	295.477	147.769	-316.351	108.396	-1.490
3900	42.492	335.467	296.488	152.016	-316.721	119.578	-1.602
4000	42.531	336.543	297.476	156.267	-317.090	130.770	-1.708
4100	42.571	337.594	298.442	160.523	-317.456	141.971	-1.809
4200	42.614	338.620	299.386	164.782	-317.819	153.181	-1.905
4300	42.658	339.623	300.311	169.045	-318.177	164.399	-1.997
4400	42.704	340.605	301.215	173.313	-318.532	175.626	-2.085
4500	42.751	341.565	302.101	177.586	-318.881	186.861	-2.169
4600	42.800	342.505	302.969	181.864	-319.226	198.103	-2.250
4700	42.850	343.426	303.820	186.146	-319.565	209.353	-2.327
4800	42.902	344.329	304.655	190.434	-319.897	220.610	-2.401
4900	42.955	345.214	305.474	194.726	-320.224	231.874	-2.472
5000	43.010	346.082	306.277	199.025	-320.545	243.145	-2.540
5100	43.066	346.934	307.066	203.328	-320.859	254.422	-2.606
5200	43.123	347.771	307.841	207.638	-321.167	265.705	-2.669
5300	43.181	348.593	308.602	211.953	-321.469	276.994	-2.730
5400	43.241	349.401	309.350	216.274	-321.765	288.288	-2.789
5500	43.302	350.195	310.086	220.601	-322.054	299.588	-2.845
5600	43.364	350.976	310.809	224.935	-322.338	310.893	-2.900
5700	43.427	351.744	311.520	229.274	-322.618	322.204	-2.953
5800	43.492	352.500	312.220	233.620	-322.892	333.519	-3.004
5900	43.557	353.244	312.909	237.973	-323.162	344.838	-3.053
6000	43.623	353.976	313.588	242.332	-323.427	356.163	-3.101

PREVIOUS: March 1967 (1 atm) CURRENT: March 1967 (1 bar)

Silicon (Si_2) $Si_2(g)$

SILICON (Si_3) IDEAL GAS M_r = 84.2565

$S°(298.15\ K) = 267.90\ J\ K^{-1}\ mol^{-1}$

$\Delta_fH°(0\ K) = 632.7 \pm 42\ kJ\ mol^{-1}$
$\Delta_fH°(298.15\ K) = 636.0 \pm 42\ kJ\ mol^{-1}$

Electronic Levels and Quantum Weights

State	ϵ_i, cm^{-1}	g_i
$^3\Sigma^-$	0	3
$^3\Pi_u$	[10000]	[6]
$^1\Pi_u$	[18000]	[2]
$^3\Sigma^-_u$	21460	3

Vibrational Frequencies and Degeneracies

ν, cm^{-1}
[360](1)
[200](2)
[630](1)

Point Group: [$D_{\infty h}$] σ = [2]
Bond Distance: [2.25] Å
Bond Angle: [180]°
Rotational Constant: B_0 = [0.059282] cm^{-1}

Enthalpy of Formation

The selected value is an average based on the equilibrim data summarized below. Drowart and co-workers have used the Knudsen effusion-mass spectrometric technique to determine the vapor species over the systems SiC-graphite (1) and SiC-silicon (2). Third law analysis of the partial pressures of Si_3 and Si yields the values 154.0 and 149.9 kcal mol^{-1}. Both drifts suggest that the entropy may be lower than the tabulated values. It is unlikely that the entropy is in error by more than 5 cal K^{-1} mol^{-1} so that most of the drift is inherent in the data. The adopted value of $\Delta_fH°(298.15\ K)$ = 152 ± 10 kcal mol^{-1} includes allowance for an error of up to 5 cal K^{-1} mol^{-1}.

Source	Method	T/K	Data Points	$\Delta_rH°(298.15\ K)$, kcal mol^{-1}* 2nd law	3rd law	cal K^{-1} mol^{-1} Drift	$\Delta_fH°(298.15\ K)$ kcal, mol^{-1}
Drowart (1958)	Mass Spec.	2230-2316	2	209	169.1	-18	154.0
Drowart (1960)	Mass Spec.	1703-1890	4	204±3	173.2	-17±2	149.9

*For reaction $Si_3(g) = 3\ Si(g)$

Heat Capacity and Entropy

Weltner and McLeod (3) observed an absorption band near 4660 Å in matrix isolation studies. Their tenative assignment of this band as the $^3\Sigma_u^- \leftarrow {}^3\Sigma_g^-$ transition of Si_3 is adopted here. A $^1\Pi_u$ state is assumed at 18000 cm^{-1}, which is 7000 and 2000 cm^{-1} below the analogous levels[3] for C_3 and C_2Si. Also a $^3\Pi_u$ level is assumed at 10000 cm^{-1}, presumably arising from the same molecular orbital configuration as the $^1\Pi_u$ state. The molecule is assumed to be linear with a bond distance equal to that in Si_2. Vibrational frequencies are estimated from a valence bond calculation using $k_1 = 2.16 \times 10^5$ and $k\sigma/l^2 = 0.11 \times 10^5$ dyn cm^{-1}. The stretching force constant is obtained from Si_2, while the bending force constant is based on C_2Si and the $^1\Pi_u$ excited state[3] of C_3.

References

1. J. Drowart, G. DeMaria and M. G. Inghram, J. Chem. Phys. 29, 1015 (1958).
2. J. Drowart and G. DeMaria, pp. 16-23 in "Silicon Carbide", Edited by J. R. O'Connor and J. Smiltens, Pergamon Press, New York, 1960.
3. W. Weltner, Jr. and D. McLeod, Jr., J. Chem. Phys. 41, 235 (1964); 40, 1305 (1964); 45, 3096 (1966).

Silicon (Si_3) $Si_3(g)$

Enthalpy Reference Temperature = T_r = 298.15 K; Standard State Pressure = p° = 0.1 MPa

T/K	$C_p°$ (J $K^{-1}mol^{-1}$)	S°	$-[G°-H°(T_r)]/T$	$H°-H°(T_r)$ (kJ mol^{-1})	$\Delta_fH°$	$\Delta_fG°$	Log K_f
0	0.	0.	INFINITE	-12.919	632.703	632.703	INFINITE
100	39.149	216.004	312.988	-9.698	635.125	614.674	-321.073
200	49.903	246.896	272.794	-5.180	636.108	593.728	-155.066
250	52.965	258.381	268.794	-2.603	636.134	583.123	-121.837
298.15	55.055	267.898	267.898	0.	635.968	572.927	-100.374
300	55.122	268.239	267.899	0.102	635.959	572.536	-99.687
350	56.664	276.859	268.577	2.899	635.651	561.989	-83.872
400	57.787	284.502	270.099	5.761	635.254	551.492	-72.017
450	58.621	291.359	272.087	8.673	634.792	541.048	-62.803
500	59.255	297.570	274.329	11.620	634.279	530.660	-55.438
600	60.129	308.457	279.135	17.593	633.123	510.042	-44.403
700	60.686	317.770	284.005	23.636	631.823	489.629	-36.537
800	61.060	325.900	288.745	29.724	630.384	469.412	-30.649
900	61.323	333.108	293.281	35.844	628.818	449.383	-26.082
1000	61.516	339.579	297.593	41.986	627.128	429.536	-22.437
1100	61.663	345.450	301.681	48.146	625.319	409.863	-19.463
1200	61.781	350.820	305.555	54.318	623.395	390.360	-16.992
1300	61.884	355.769	309.230	60.501	621.357	371.022	-14.908
1400	61.981	360.359	312.720	66.695	619.203	351.846	-13.128
1500	62.081	364.639	316.040	72.898	616.934	332.827	-11.590
1600	62.190	368.649	319.204	79.111	614.549	313.964	-10.250
1700	62.312	372.423	322.225	85.336	461.519	296.594	-9.113
1800	62.451	375.988	325.114	91.574	459.598	286.948	-8.327
1900	62.610	379.369	327.881	97.827	457.692	277.408	-7.626
2000	62.789	382.585	330.536	104.097	455.803	267.969	-6.999
2100	62.989	385.653	333.089	110.385	453.933	258.623	-6.433
2200	63.208	388.588	335.545	116.695	452.084	249.366	-5.921
2300	63.445	391.403	337.913	123.028	450.258	240.192	-5.455
2400	63.699	394.109	340.198	129.385	448.456	231.098	-5.030
2500	63.966	396.714	342.407	135.768	446.681	222.078	-4.640
2600	64.244	399.228	344.545	142.178	444.932	213.128	-4.282
2700	64.531	401.658	346.615	148.617	443.212	204.246	-3.951
2800	64.824	404.011	348.623	155.085	441.521	195.426	-3.646
2900	65.121	406.291	350.573	161.582	439.860	186.666	-3.362
3000	65.420	408.503	352.467	168.109	438.228	177.963	-3.099
3100	65.718	410.653	354.309	174.666	436.626	169.314	-2.853
3200	66.013	412.744	356.103	181.252	435.054	160.717	-2.623
3300	66.304	414.780	357.850	187.868	433.511	152.168	-2.409
3400	66.589	416.764	359.554	194.513	431.997	143.665	-2.207
3500	66.867	418.698	361.216	201.186	430.511	135.206	-2.018
3600	67.136	420.586	362.839	207.886	-723.403	158.172	-2.295
3700	67.396	422.429	364.425	214.613	-723.593	182.663	-2.579
3800	67.646	424.229	365.975	221.365	-723.763	207.158	-2.848
3900	67.886	425.990	367.492	228.142	-723.913	231.658	-3.103
4000	68.114	427.711	368.976	234.942	-724.042	256.162	-3.345
4100	68.330	429.396	370.429	241.764	-724.151	280.668	-3.576
4200	68.535	431.045	371.853	248.607	-724.241	305.177	-3.795
4300	68.727	432.660	373.248	255.471	-724.312	329.688	-4.005
4400	68.908	434.242	374.616	262.353	-724.363	354.200	-4.205
4500	69.077	435.792	375.959	269.252	-724.398	378.713	-4.396
4600	69.233	437.312	377.276	276.167	-724.415	403.227	-4.579
4700	69.379	438.803	378.569	283.098	-724.416	427.741	-4.754
4800	69.512	440.265	379.839	290.043	-724.402	452.255	-4.922
4900	69.635	441.699	381.087	297.000	-724.374	476.769	-5.082
5000	69.747	443.107	382.314	303.969	-724.333	501.281	-5.237
5100	69.848	444.490	383.519	310.949	-724.280	525.793	-5.385
5200	69.939	445.847	384.705	317.939	-724.217	550.304	-5.528
5300	70.020	447.180	385.871	324.937	-724.145	574.813	-5.665
5400	70.091	448.489	387.018	331.942	-724.065	599.321	-5.797
5500	70.154	449.776	388.148	338.955	-723.976	623.827	-5.925
5600	70.208	451.041	389.260	345.973	-723.885	648.332	-6.047
5700	70.254	452.284	390.354	352.996	-723.790	672.835	-6.166
5800	70.291	453.506	391.433	360.023	-723.693	697.336	-6.280
5900	70.322	454.708	392.495	367.054	-723.596	721.836	-6.391
6000	70.345	455.890	393.542	374.087	-723.498	746.334	-6.497

PREVIOUS: March 1967 (1 atm) CURRENT: March 1967 (1 bar)

Silicon (Si_3) $Si_3(g)$

STRONTIUM (Sr) REFERENCE STATE A_r = 87.62

0 to 820	K	crystal, alpha
820 to 1050	K	crystal, beta
1050 to 1685.492	K	liquid
above 1685.492	K	ideal monatomic gas

Refer to the individual tables for details.

Strontium (Sr) Sr_1(ref)

Enthalpy Reference Temperature = T_r = 298.15 K   Standard State Pressure = $p°$ = 0.1 MPa

	J K⁻¹mol⁻¹			kJ mol⁻¹			
T/K	$C_p°$	$S°$	$-[G°-H°(T_r)]/T$	$H°-H°(T_r)$	$\Delta_fH°$	$\Delta_fG°$	Log K_f
0	0.	0.	INFINITE	-6.568	0.	0.	0.
100	23.481	28.050	78.738	-5.069	0.	0.	0.
200	25.757	45.209	58.112	-2.581	0.	0.	0.
298.15	26.791	55.694	55.694	0.	0.	0.	0.
300	26.809	55.859	55.694	0.050	0.	0.	0.
400	27.777	63.704	56.757	2.779	0.	0.	0.
500	28.785	70.010	58.797	5.607	0.	0.	0.
600	29.814	75.349	61.122	8.536	0.	0.	0.
700	30.864	80.023	63.495	11.570	0.	0.	0.
800	31.935	84.214	65.827	14.710	0.	0.	0.
820.000	32.151	85.005	66.285	15.351	ALPHA <--> BETA		
820.000	29.790	86.026	66.285	16.188	TRANSITION		
900	30.125	88.814	68.165	18.584	0.	0.	0.
1000	30.543	92.010	70.393	21.618	0.	0.	0.
1050.000	30.752	93.505	71.458	23.150	BETA <--> LIQUID		
1050.000	39.463	100.582	71.458	30.581	TRANSITION		
1100	39.463	102.418	72.824	32.554	0.	0.	0.
1200	39.463	105.852	75.435	36.500	0.	0.	0.
1300	39.463	109.010	77.898	40.446	0.	0.	0.
1400	39.463	111.935	80.226	44.393	0.	0.	0.
1500	39.463	114.658	82.432	48.339	0.	0.	0.
1600	39.463	117.204	84.526	52.285	0.	0.	0.
1685.492	39.463	119.259	86.236	55.659	LIQUID <--> IDEAL GAS		
1685.492	20.834	200.650	86.236	192.844	FUGACITY = 1 bar		
1700	20.839	200.829	87.214	193.146	0.	0.	0.
1800	20.882	202.021	93.559	195.232	0.	0.	0.
1900	20.949	203.152	99.297	197.323	0.	0.	0.
2000	21.049	204.229	104.517	199.423	0.	0.	0.
2100	21.191	205.259	109.290	201.534	0.	0.	0.
2200	21.383	206.249	113.675	203.662	0.	0.	0.
2300	21.636	207.205	117.721	205.813	0.	0.	0.
2400	21.959	208.132	121.469	207.992	0.	0.	0.
2500	22.361	209.036	124.953	210.207	0.	0.	0.
2600	22.849	209.923	128.204	212.467	0.	0.	0.
2700	23.430	210.796	131.247	214.780	0.	0.	0.
2800	24.109	211.660	134.104	217.157	0.	0.	0.
2900	24.889	212.519	136.793	219.606	0.	0.	0.
3000	25.770	213.377	139.331	222.138	0.	0.	0.
3100	26.753	214.238	141.734	224.763	0.	0.	0.
3200	27.834	215.104	144.013	227.492	0.	0.	0.
3300	29.008	215.978	146.181	230.333	0.	0.	0.
3400	30.270	216.863	148.246	233.296	0.	0.	0.
3500	31.611	217.759	150.220	236.389	0.	0.	0.
3600	33.022	218.670	152.108	239.620	0.	0.	0.
3700	34.493	219.594	153.920	242.995	0.	0.	0.
3800	36.011	220.534	155.660	246.520	0.	0.	0.
3900	37.565	221.490	157.336	250.199	0.	0.	0.
4000	39.142	222.461	158.952	254.034	0.	0.	0.
4100	40.730	223.447	160.513	258.027	0.	0.	0.
4200	42.304	224.446	162.023	262.176	0.	0.	0.
4300	43.874	225.460	163.487	266.485	0.	0.	0.
4400	45.418	226.486	164.907	270.950	0.	0.	0.
4500	46.927	227.524	166.287	275.568	0.	0.	0.
4600	48.214	228.556	167.628	280.269	0.	0.	0.
4700	49.584	229.608	168.936	285.159	0.	0.	0.
4800	50.884	230.665	170.211	290.182	0.	0.	0.
4900	52.114	231.727	171.455	295.332	0.	0.	0.
5000	53.254	232.791	172.671	300.596	0.	0.	0.
5200	55.250	234.914	175.024	311.429	0.	0.	0.
5400	56.945	237.032	177.282	322.655	0.	0.	0.
5600	58.296	239.129	179.453	334.184	0.	0.	0.
5800	59.317	241.193	181.547	345.951	0.	0.	0.
6000	59.709	243.172	183.564	357.650	0.	0.	0.

PREVIOUS: December 1970 (1 atm)   CURRENT: September 1983 (1 bar)

Strontium (Sr) Sr_1(ref)

STRONTIUM (α-Sr) CRYSTAL A_r = 87.62

Pearson Notation: CF4

$S°(298.15\ K) = 55.69 \pm 0.1\ J\ K^{-1}\ mol^{-1}$

$T_{trs}(\alpha \rightarrow \beta) = 820 \pm 8\ K$

$\Delta_f H°(0\ K) = 0\ kJ\ mol^{-1}$

$\Delta_f H°(298.15\ K) = 0\ kJ\ mol^{-1}$

$\Delta_{trs} H°(820\ K) = 0.8368 \pm 0.17\ kJ\ mol^{-1}$

Enthalpy of Formation

Zero by definition.

Heat Capacity and Entropy

The adopted heat capacities are derived from the heat capacity (1, 2) and enthalpy (2) data by fitting the experimental data with orthogonal polynomials over selected overlapping temperature intervals. The temperature region of 1-9 K is described by the equation: $C_p° = 3.281 \times 10^{-3}\ T + 1.89 \times 10^{-4}\ T^3\ cal\ K^{-1}\ mol^{-1}$. The S°(298.15 K) value calculated from the polynomials agrees to within 0.002 $cal\ K^{-1}\ mol^{-1}$ with that given by Boerio and Westrum (1) and adopted by Gurvich et al. (3).

Boerio and Westrum (1) used adiabatic-shield calorimetry to measure the low temperature heat capacity, 83 data points in the range 6.28 to 346.42 K. Two other studies have also been made: Kreplovich and Paukov (5-300 K (4, 5)) and Roberts (1-20 K (6)). In the temperature region below 20 K, both these studies deviate from the adopted values similarly, from ~44% low at 5 K to ~15% low at 20 K. Since the sample used by Kreplovich and Paukov (4, 5) was shown to contain a substantial oxygen impurity, of the order of 6 ± 3 mole percent (1), the sample used by Roberts (6) probably had similar impurities. Roberts reported a purity of 98.25% but only metallic impurities were analyzed.

Stephens and Roth (2) measured enthalpy (350-1191 K) by drop calorimetry and heat capacity (298-785 K) by differential scanning calorimetry. We fit the 14 enthalpy points at 400-812 K with a polynomial, using the adiabatic (1) and DSC (2) data as a guide to the shape of the heat capacity curve. We adopt this fit only above 350 K, although the derived $C_p°$ values agree closely (±0.3%) with the adiabatic data (1) at 250 to 320 K. Between 320 and 400 K the adopted $C_p°$ values are a compromise between those of Boerio and Westrum and those of Stephens and Roth. This compromise yields an upward inflection point in $C_p°$ versus T near 330 K. The largest deviation of Boerio and Westrum (1) above 320 K is -0.7% at 343.42 K. The DSC data (2) deviate by -0.8 to +2.0% except for +3.8% at 785 K. The enthalpy data (2) deviate by -1.6 to +1.8% (400-812 K) with larger deviations at 350 and 550 K.

Phase Data

According to Pearson (7), α-strontium has the face-centered-cubic structure, isotypic with Cu, designated as CF4. In the alkaline earth metal series, only α-calcium has the same structure.

Transition Data

See the β-crystal table for details.

Sublimation Data

The enthalpy of sublimation of α-strontium is the enthalpy of formation of the gas. The experimental studies are discussed in the ideal gas table.

References

1. J. Boerio and E. F. Westrum, J. Chem. Thermodynaics 10, 1 (1978).
2. H. J. Stephens and E. P. Roth, Sandia National Laboratories, Albuquerque, New Mexico, personal communication, May 3, 1982.
3. L. V. Gurvich, I. V. Veits et al., "Thermodynamic Properties of Individual Substances," 3rd ed., Vol. III, Nauka, Moscow, 1981.
4. L. M. Khriplovich and I. E. Paukov, Zh. Fiz. Khim. 50, 537 (1976).
5. L. M. Khriplovich and I. E. Paukov, VINITI Documentary Deposit No. 3190-75, November, 11, 1975.
6. L. M. Roberts, Proc. Phys. Soc. (London) B70, 733 (1957).
7. W. B. Pearson, "A Handbook of Lattice Spacing and Structures of Metals and Alloys," Pergamon Press, London, 1958, 1967.

Strontium, Alpha (Sr) $Sr_1(cr)$

Enthalpy Reference Temperature = T_r = 298.15 K — Standard State Pressure = p° = 0.1 MPa

T/K	$C_p°$ (J K⁻¹mol⁻¹)	S° (J K⁻¹mol⁻¹)	−[G°−H°(T_r)]/T (J K⁻¹mol⁻¹)	H°−H°(T_r) (kJ mol⁻¹)	$\Delta_f H°$ (kJ mol⁻¹)	$\Delta_f G°$ (kJ mol⁻¹)	Log K_f
0	0.	0.	INFINITE	-6.568	0.	0.	0.
100	23.481	28.050	78.738	-5.069	0.	0.	0.
200	25.757	45.209	58.112	-2.581	0.	0.	0.
298.15	26.791	55.694	55.694	0.	0.	0.	0.
300	26.809	55.859	55.694	0.050	0.	0.	0.
350	27.298	60.029	56.022	1.402	0.	0.	0.
400	27.777	63.704	56.757	2.779	0.	0.	0.
450	28.278	67.005	57.716	4.180	0.	0.	0.
500	28.785	70.010	58.797	5.607	0.	0.	0.
600	29.814	75.349	61.122	8.536	0.	0.	0.
700	30.864	80.023	63.495	11.570	0.	0.	0.
800	31.935	84.214	65.827	14.710	0.	0.	0.
820.000	32.151	85.005	66.285	15.351	-------- ALPHA	<-->	BETA -------
900	33.027	88.038	68.085	17.958	-0.626	0.072	-0.004
1000	34.139	91.576	70.260	21.316	-0.302	0.133	-0.007
1100	35.273	94.882	72.349	24.786	-7.768	0.522	-0.025
1200	36.428	98.001	74.358	28.371	-8.129	1.292	-0.056
1300	37.604	100.963	76.292	32.073	-8.374	2.088	-0.084
1400	38.801	103.794	78.156	35.893	-8.500	2.898	-0.108
1500	40.019	106.512	79.956	39.834	-8.505	3.713	-0.129

PREVIOUS: December 1970 CURRENT: September 1983

Strontium, Alpha (Sr) $Sr_1(cr)$

STRONTIUM (β-Sr) CRYSTAL A_r = 87.62

Pearson Notation: cI2

S°(298.15 K) = [57.1748] J K^{-1} mol^{-1}

T_{trs}(α→β) = 820 ± 8 K

T_{fus}(β→ℓ) = 1050 ± 10 K

$\Delta_f H°$(298.15 K) = [1.21126] kJ mol^{-1}

$\Delta_{trs} H°$(820 K) = 0.8368 ± 0.17 kJ mol^{-1}

$\Delta_{fus} H°$(1050 K) = 7.4308 ± 0.8 kJ mol^{-1}

Enthalpy of Formation

$\Delta_f H°$(Sr, cr, β, 298.15 K) is calculated from $\Delta_f H°$(Sr, α, cr, 298.15 K) by adding the enthalpy of transition, $\Delta_{trs} H°$, and the difference in enthalpy, H°(820 K)-H°(298.15 K), between α-Sr and β-Sr.

Heat Capacity and Entropy

The drop-calorimetry data of Stephens and Roth (1) contain 9 points (819.6-1048.8 K) that appear to be in the region of stability of β-strontium. Their values of [H°(T)-H°(298.15 K)]/(T-298.15) show systematic curvature suggestive of the effects of incomplete transition or premelting near the extreme temperatures. Constant $C_p°$ values ranging from ~6.3 to 9.3 cal K^{-1} mol^{-1} can be derived by various selections and weightings of the points. We arbitrarily select a $C_p°$ near 7.3 cal K^{-1} mol^{-1} with a linear temperature dependence that maintains the proper stability relationships when extrapolated below T_{trs} and above T_{fus}. This curve reproduces the enthalpy data within ±1.2% (819.6-990.5 K) and corresponds to premelting effects of ~140 to 450 cal/mol in the region from 1029 to 1049 K. S°(298.15 K) is calculated from that of α-strontium using entropy increments analogous to the enthalpy increments used to calculate $\Delta_f H°$(298.15 K).

Phase Data

According to Pearson (2), β-strontium has the body-centered-cubic structure, isotypic with W, designated as CI_2. In the alkaline earth metal series, β-calcium, β-beryllium, and barium have the same structure. Note that we now use the name beta (3) for the phase of Sr stable near T_{fus}, unlike Pearson who retains the historical name gamma. The change arose when Peterson and Colburn (3) found that an intermediate hexagonal-close-packed phase does not occur in Sr containing a smaller amount of H_2 impurity.

Transition Data

We select T_{trs} = 820 ± 8 K based on enthalpy data of Stephens and Roth (1) and DTA data of Peterson and Colburn (3). DTA gave ~813 K on cooling and 828 K (after correction for H impurity) on heating. We adopt an enthalpy of transition of 200 ± 40 cal/mol derived from our fits of the enthalpy data (1) for β and α phases; Stephens and Roth selected the same value. Almost 170 out of the 200 cal/mol appeared between 812 and 819.6 K. Another 30 cal/mol appeared between 819.6 and 824.9 K, while possibly 40 cal/mol appeared between 824.9 and 835.6 K. The latter enthalpy increments are uncertain because of almost equal-sized scatter in the data and because of uncertainty in $C_p°$(β). Our adopted value of T_{trs} = 820 K is in the middle of the DTA hystersis range. The temperature range up to T_{trs} includes most of the transition enthalpy seen in the sample of Stephens and Roth. Their sample was identical with that used by Boerio and Westrum (4) to measure the low temperature $C_p°$ and S° of α phase.

Peterson and Colburn (3) found that T_{trs} increased rapidly with increasing hydrogen content. They also observed the intermediate hexagonal-close-packed phase at hydrogen contents of >3 mole % but not at 0.4 mole %, suggesting that this phase does not occur in pure strontium. The enthalpy and heat capacity data (1) are consistent with this interpretation.

Sublimation Data

The enthaply of sublimation of β-strontium is calculated as the difference in the enthlapies of formation of the gas and β-strontium. The experimental studies are discussed in the Sr(g) table.

References

1. H. J. Stephens and E. P. Roth, Sandia National Laboratories, Albuquerque, New Mexico, personal communication, May 3, 1982.
2. W. B. Pearson, "A Handbook of Lattice Spacing and Structures of Metals and Alloys," Pergamon Press, London, 1958, 1967.
3. D. T. Peterson and R. P. Colburn, J. Phys. Chem. 70, 463 (1966).
4. J. Boerio and E. F. Westrum, J. Chem. Thermodynamics 10, 1 (1978).

Strontium, Beta (Sr) Sr_1(cr)

Enthalpy Reference Temperature = T_r = 298.15 K Standard State Pressure = p° = 0.1 MPa

T/K	$C_p°$ (J $K^{-1}mol^{-1}$)	S° (J $K^{-1}mol^{-1}$)	$-[G°-H°(T_r)]/T$ (J $K^{-1}mol^{-1}$)	$H°-H°(T_r)$ (kJ mol^{-1})	$\Delta_f H°$ (kJ mol^{-1})	$\Delta_f G°$ (kJ mol^{-1})	Log K_f
0							
100							
200							
250							
298.15	27.607	57.175	57.175	0.	1.211	0.770	-0.135
300	27.614	57.346	57.175	0.051	1.213	0.767	-0.134
400	28.033	65.347	58.263	2.833	1.266	0.609	-0.080
500	28.451	71.647	60.332	5.658	1.262	0.444	-0.046
600	28.870	76.872	62.665	8.524	1.199	0.285	-0.025
700	29.288	81.353	65.022	11.432	1.073	0.142	-0.011
800	29.706	85.291	67.315	14.381	0.883	0.021	-0.001
820.000	29.790	86.026	67.762	14.976	-------- ALPHA <--> BETA -------		
900	30.125	88.814	69.511	17.373	0.	0.	0.
1000	30.543	92.010	71.604	20.406	0.	0.	0.
1050.000	30.752	93.505	72.611	21.939	------- BETA <--> LIQUID -------		
1100	30.962	94.941	73.594	23.481	-7.861	0.364	-0.017
1200	31.380	97.653	75.487	26.599	-8.690	1.149	-0.050
1300	31.798	100.181	77.291	29.757	-9.478	2.001	-0.080
1400	32.217	102.553	79.011	32.958	-10.223	2.912	-0.109
1500	32.635	104.790	80.656	36.201	-10.927	3.875	-0.135

PREVIOUS: CURRENT: September 1983

Strontium, Beta (Sr) Sr_1(cr)

$S°(298.15\ K) = [50.901]\ J\ K^{-1}\ mol^{-1}$

$T_{fus}(\beta \to l) = 1050 \pm 10\ K$

$\Delta_f H°(298.15\ K) = [0.911]\ kJ\ mol^{-1}$

$\Delta_{fus} H°(1050\ K) = 7.4308 \pm 0.8\ kJ\ mol^{-1}$

Enthalpy of Formation

$\Delta_f H°(Sr, l, 298.15\ K)$ is calculated from $\Delta_f H°(Sr, cr, \beta, 298.15\ K)$ by adding the enthalpy of fusion, $\Delta_{fus}H°$, and the difference in enthalpy, H°(1050 K)-H°(298.15 K), between β-Sr and the liquid.

Heat Capacity and Entropy

$C_p°$ is the constant value derived from three enthalpy points (1059-1191 K) measured by Stephens and Roth (1). Their points deviate from the adopted enthalpies by -0.8 to +0.5%. We assume a glass transition with zero enthalpy at 650 K; below 650 K we assume that $C_p°$ (glass) is equal to the $C_p°$ (alpha) curve derived from the enthalpy data for alpha strontium (see that table). These or similar assumptions are necessary to ensure that S°(298.15 K) of the supercooled liquid is larger than that of alpha α-strontium. S°(298.15 K) is calculated from that of β-strontium using entropy increments analogous to the enthalpy increments used to calculate $\Delta_f H°(298.15\ K)$.

Fusion Data

T_{fus} = 1050±10 K is the value selected by Stephens and Roth (1) based on their enthalpy study. Previous studies by thermal analysis (2-3) gave 1042±2 K (IPTS-68) for high-purity strontium. The reason for the difference is not clear; although it probably arises from differences in sample purity or measurement method. The observed enthalpies (1) showed that melting of the sample was complete at 1059.1 K but only ~20 to 30% complete at 1048.8 K. This suggests either that the sample needed a much longer equilibration time at 1048.8 K or that it actually melted at least 7 degrees higher than 1042 K. The enthalpy at 1035.1 K showed a small premelting effect. A melting point of 1048 K (IPTS-68) was found by Kanda and coworkers (4-6) for three samples that, based on their transition behavior, probably contained over 3 mole % of hydrogen. Although hydrogen increases both T_{fus} and T_{trs}, the changes are greatest in the transition region (2). The sample in the enthalpy study had normal transition behavior without the changes typical of hydrogen impurity. This sample was identical to that used in measurement of $C_p°$ and S° at low temperature; analytical data (7) showed it to be of high purity. We choose to base all properties, including melting point, on the data for this same sample

The enthalply of fusion is derived from our fits of the enthalpy data (1) in the liquid and beta regions. Stephens and Roth (1) selected the value 1871±55 cal/mol based on a somewhat different fit in the beta region.

Vaporization Data

T_{vap} is calculated as the temperature for which $\Delta_{vap}G° = 0$ for the liquid vaporizing to the ideal monatomic gas. T_{vap} corresponds to a fugacity of one bar. The enthalpy of vaporization is calculated from the enthalpy of formation of the liquid and the gas at T_{vap}. Vaporization studies are discussed in the ideal gas table (8).

References

1. H. J. Stephens and E. P. Roth, Sandia National Laboratories, Albuquerque, New Mexico, personal communication, May 3, 1982.
2. D. T. Peterson and R. P. Colburn, J. Phys. Chem. 70, 468 (1966).
3. A. S. Dworkin, H. R. Bronstein and M. A. Bredig, J. Phys. Chem. 72, 1892 (1968).
4. W. O. Roberts, Ph.D. Dissertation, Syracuse Univ., 1964; Diss. Abstr. 25, 6218 (1965).
5. F. E. Wang, F. A. Kanda and A. J. King, J. Phys. Chem. 66, 2138 (1962).
6. J. C. Schottmiller, A. J. King and F. A. Kanda, J. Phys. Chem. 62, 1446 (1958).
7. J. Boerio and E. F. Westrum, Jr., J. Chem. Thermodynamics 10, 1 (1978).
8. JANAF Thermochemical Tables: Sr(g), 9-30-83.

Enthalpy Reference Temperature = T_r = 298.15 K; Standard State Pressure = p° = 0.1 MPa

T/K	$C_p°$ (J K⁻¹mol⁻¹)	S° (J K⁻¹mol⁻¹)	$-[G°-H°(T_r)]/T$ (J K⁻¹mol⁻¹)	$H°-H°(T_r)$ (kJ mol⁻¹)	$\Delta_f H°$ (kJ mol⁻¹)	$\Delta_f G°$ (kJ mol⁻¹)	Log K_f
0							
100							
200							
250							
298.15	39.463	50.901	50.901	0.	0.911	2.340	-0.410
300	39.463	51.145	50.902	0.073	0.934	2.348	-0.409
400	39.463	62.498	52.450	4.019	2.151	2.634	-0.344
500	39.463	71.304	55.373	7.966	3.270	2.623	-0.274
600	39.463	78.498	58.645	11.912	4.286	2.396	-0.209
700	39.463	84.582	61.927	15.858	5.199	2.008	-0.150
800	39.463	89.851	65.096	19.804	6.005	1.496	-0.098
900	39.463	94.499	68.110	23.751	6.077	0.961	-0.056
1000	39.463	98.657	70.960	27.697	6.990	0.343	-0.018
1050.000	39.463	100.582	72.325	29.670	------- BETA <--> LIQUID -------		
1100	39.463	102.418	73.652	31.643	0.	0.	0.
1200	39.463	105.852	76.194	35.589	0.	0.	0.
1300	39.463	109.010	78.598	39.536	0.	0.	0.
1400	39.463	111.935	80.877	43.482	0.	0.	0.
1500	39.463	114.658	83.039	47.428	0.	0.	0.
1600	39.463	117.204	85.095	51.374	0.	0.	0.
1685.492	39.463	119.259	86.777	54.748	------- FUGACITY = 1 bar -------		
1700	39.463	119.597	87.055	55.321	-136.914	1.180	-0.036
1800	39.463	121.853	88.926	59.267	-135.054	9.249	-0.268
1900	39.463	123.986	90.716	63.213	-133.199	17.215	-0.473
2000	39.463	126.010	92.431	67.159	-131.352	25.084	-0.655
2100	39.463	127.936	94.076	71.106	-129.518	32.861	-0.817
2200	39.463	129.771	95.657	75.052	-127.700	40.550	-0.963
2300	39.463	131.526	97.179	78.998	-125.904	48.158	-1.094
2400	39.463	133.205	98.645	82.944	-124.137	55.688	-1.212
2500	39.463	134.816	100.060	86.891	-122.406	63.145	-1.319
2600	39.463	136.364	101.427	90.837	-120.720	70.533	-1.417
2700	39.463	137.853	102.748	94.783	-119.086	77.858	-1.506
2800	39.463	139.288	104.028	98.729	-117.516	85.123	-1.588
2900	39.463	140.673	105.268	102.676	-116.019	92.334	-1.663
3000	39.463	142.011	106.470	106.622	-114.605	99.494	-1.732

PREVIOUS: December 1970 CURRENT: September 1983

STRONTIUM (Sr) ALPHA-BETA-LIQUID A_r = 87.62

0 to 820 K crystal, alpha
820 to 1050 K crystal, beta
above 1050 K liquid

Refer to the individual tables for details.

Strontium (Sr) Sr_1(cr,l)

Enthalpy Reference Temperature = T_r = 298.15 K — Standard State Pressure = p° = 0.1 MPa

T/K	C_p° (J K⁻¹mol⁻¹)	S° (J K⁻¹mol⁻¹)	−[G°−H°(T_r)]/T (J K⁻¹mol⁻¹)	H°−H°(T_r) (kJ mol⁻¹)	$\Delta_f H^\circ$ (kJ mol⁻¹)	$\Delta_f G^\circ$ (kJ mol⁻¹)	Log K_f
0	0.	0.	INFINITE	-6.568	0.	0.	0.
100	23.481	28.050	78.738	-5.069	0.	0.	0.
200	25.757	45.209	58.112	-2.581	0.	0.	0.
298.15	26.791	55.694	55.694	0.	0.	0.	0.
300	26.809	55.859	55.694	0.050	0.	0.	0.
350	27.298	60.029	56.022	1.402	0.	0.	0.
400	27.777	63.704	56.757	2.779	0.	0.	0.
450	28.278	67.005	57.716	4.180	0.	0.	0.
500	28.785	70.010	58.797	5.607	0.	0.	0.
600	29.814	75.349	61.122	8.536	0.	0.	0.
700	30.864	80.023	63.495	11.570	0.	0.	0.
800	31.935	84.214	65.827	14.710	0.	0.	0.
820.000	32.151	85.005	66.285	15.351	ALPHA <--> BETA		
820.000	29.790	86.026	66.285	16.188	TRANSITION		
900	30.125	88.814	68.165	18.584	0.	0.	0.
1000	30.543	92.010	70.393	21.618	0.	0.	0.
1050.000	30.752	93.505	71.458	23.150	BETA <--> LIQUID		
1050.000	39.463	100.582	71.458	30.581	TRANSITION		
1100	39.463	102.418	72.824	32.554	0.	0.	0.
1200	39.463	105.852	75.435	36.500	0.	0.	0.
1300	39.463	109.010	77.898	40.446	0.	0.	0.
1400	39.463	111.935	80.226	44.393	0.	0.	0.
1500	39.463	114.658	82.432	48.339	0.	0.	0.
1600	39.463	117.204	84.526	52.285	0.	0.	0.
1685.492	39.463	119.259	86.236	55.659	FUGACITY = 1 bar		
1700	39.463	119.597	86.520	56.231	-136.914	1.180	-0.036
1800	39.463	121.853	88.421	60.178	-135.054	9.249	-0.268
1900	39.463	123.986	90.237	64.124	-133.199	17.215	-0.473
2000	39.463	126.010	91.975	68.070	-131.352	25.084	-0.655
2100	39.463	127.936	93.642	72.016	-129.518	32.861	-0.817
2200	39.463	129.771	95.243	75.963	-127.700	40.550	-0.963
2300	39.463	131.526	96.783	79.909	-125.904	48.158	-1.094
2400	39.463	133.205	98.266	83.855	-124.137	55.688	-1.212
2500	39.463	134.816	99.696	87.801	-122.406	63.145	-1.319
2600	39.463	136.364	101.076	91.748	-120.720	70.533	-1.417
2700	39.463	137.853	102.411	95.694	-119.086	77.858	-1.506
2800	39.463	139.288	103.703	99.640	-117.516	85.123	-1.588
2900	39.463	140.673	104.954	103.586	-116.019	92.334	-1.663
3000	39.463	142.011	106.167	107.533	-114.605	99.494	-1.732

PREVIOUS: CURRENT: September 1983

Strontium (Sr) Sr_1(cr,l)

STRONTIUM (Sr) IDEAL GAS A_r = 87.62

IP(Sr, g) = 45932.0 ± 0.2 cm^{-1}
S°(298.15 K) = 164.64 ± 0.02 J K^{-1} mol^{-1}

$\Delta_f H°$(0 K) = 164.4 ± 1.7 kJ mol^{-1}
$\Delta_f H°$(298.15 K) = 164.0 ± 1.7 kJ mol^{-1}

Electronic Levels and Quantum Weights

State	ε_i, cm^{-1}	g_i
$^5S_0$	0.00	1
$^3P_0$	14317.520	1
$^3P_1$	14504.351	3
$^3P_2$	14898.563	5
$^3D_1$	18159.056	3
$^3D_2$	18218.795	5
$^3D_3$	18319.267	7

Enthalpy of Formation

The enthalpy of formation is the enthalpy of sublimation, 164.0±1.7 kJ mol^{-1}. The adopted value is derived mainly from the boiling point study of Bohdansky and Schins (1). It is confirmed by the early data of Hartmann and Schneider (2) and by the Knudsen effusion - mass - spectrometric study of Boerboom et al. (4). Data of Ruff and Hartmann (3) are readily dismissed due to the large entropy discrepancy.

A complete reanalysis of all the alkaline earth metals is in progress.

Source	Method	T/K	$\Delta_f H°$(298.15 K), kcal mol^{-1} 2nd Law	3rd Law	δS cal K mol
(1) Bohdansky (1967)	Boiling Point	1218-1982	40.1±0.4	39.16±0.46	0.6±0.3
(2) Hartmann (1929)	Boiling Point	1200-1380	40.0±0.6	39.46±0.14	0.4±0.4
(3) Ruff (1924)	Boiling Point	1218-1412	77 ±4	38.0 ±5	29 ±3
(4) Boerboom (1964)	Effusion-Mass Spec.	772-923	36.4	39.04±0.39	-3.1
(5) Priselkov (1954)	Effusion	673-873	35.7±0.5	38.65±0.57	-3.8±0.6

[a]Temperatures are adjusted to IPTS-68 by assuming published values to be IPTS-48.

Heat Capacity and Entropy

Observed energy levels and quantum weights are from Moore (6) as modified by Garton et al. (7, 8). Energies of unobserved but predicted terms (6) are estimated by comparison of Ca I, Sr I, Ba I and their isolectronic ions. The most important energies are for terms of the 4d5s' configuration which are erstimated at 25000 ± 10000 cm^{-1}. We adopt an energy-level cutoff which is about kTmax (Tmax - 6000 K) below each series limit. For Sr I this corresponds to omitting levels above 8s, 7p, 6d and 5f. Levels above 30000 cm^{-1} are averaged. The adopted functions are essentially identical with earlier tables up to 3000 K; however, the entropies at 6000 K differ as follows: Hilsenrath et al. (9), 57.49; Gurvich et al. (10), 58.80 cal K^{-1} mol^{-1}. These differences arise from different methods of cutoff and different degrees of accounting for the unobserved terms.

References

1. J. Bohdansky and H. E. J. Schins, J. Phys. Chem. 71, 215 (1967).
2. H. Hartmann and R. Schneider, Z. Anorg. Chem. 180, 275 (1929).
3. O. Ruff and H. Hartmann, Z. Anorg. Chem. 133, 29 (19240.
4. A. J. H. Boerboom. H. W. Reyn and J. Kistemaker, Physica 30, 254 (1964).
5. Yu. A. Priselkov and An. N. Nesmeyanov, dokl. Akad. Nauk s. S. S. R. 95, 1207 (1954).
6. C. E. Moore, "Atomic Energy Levels," Vol. 2, U. S. Natl. Bur. Std. Circ. 467, 1952.
7. W. R. S. Garton and K. Codling, J. Phys. B 1, 106 (1968).
8. W. R. S. Garton, G. L. Grasdalen, W. H. Parkinson and E. M. Reeves, J. Phys. B 1, 114 (1968).
9. J. Hilsenrath, C. G. Messina and W. H. Evans, AD606163, 1964.
10. L. V. Gurvich et al., "Thermodynamic Properties of Individual Substances," Vol. I and II, Moscow, 1962; English transl., AD659659, March, 1967.

Strontium (Sr) $Sr_1(g)$

Enthalpy Reference Temperature = T_r = 298.15 K — Standard State Pressure = p° = 0.1 MPa

T/K	$C_p°$ (J $K^{-1}mol^{-1}$)	S°	$-[G°-H°(T_r)]/T$	$H°-H°(T_r)$ (kJ mol^{-1})	$\Delta_f H°$	$\Delta_f G°$	Log K_f
0	0.	0.	INFINITE	-6.197	164.371	164.371	INFINITE
100	20.786	141.933	183.121	-4.119	164.950	153.562	-80.212
200	20.786	156.341	166.542	-2.040	164.540	142.314	-37.169
250	20.786	160.979	164.982	-1.001	164.277	136.782	-28.579
298.15	20.786	164.640	164.640	0.	164.000	131.518	-23.041
300	20.786	164.769	164.641	0.038	163.989	131.316	-22.864
350	20.786	167.973	164.894	1.078	163.675	125.895	-18.789
400	20.786	170.749	165.456	2.117	163.338	120.521	-15.738
450	20.786	173.197	166.183	3.156	162.976	115.190	-13.371
500	20.786	175.387	166.996	4.196	162.589	109.901	-11.481
600	20.786	179.177	168.719	6.274	161.738	99.441	-8.657
700	20.786	182.381	170.448	8.353	160.783	89.133	-6.651
800	20.786	185.156	172.117	10.431	159.722	78.968	-5.156
900	20.786	187.605	173.705	12.510	157.926	69.015	-4.006
1000	20.786	189.795	175.206	14.589	156.971	59.187	-3.092
1100	20.786	191.776	176.624	16.667	148.113	49.820	-2.366
1200	20.787	193.584	177.963	18.746	146.246	40.967	-1.783
1300	20.788	195.248	179.229	20.825	144.378	32.269	-1.297
1400	20.791	196.789	180.429	22.904	142.511	23.715	-0.885
1500	20.798	198.224	181.568	24.983	140.644	15.295	-0.533
1600	20.813	199.566	182.652	27.063	138.778	6.999	-0.229
1685.492	20.834	200.650	183.537	28.844	------- FUGACITY = 1 bar -------		
1700	20.839	200.829	183.684	29.146	0.	0.	0.
1800	20.882	202.021	184.670	31.232	0.	0.	0.
1900	20.949	203.152	185.613	33.323	0.	0.	0.
2000	21.049	204.229	186.517	35.423	0.	0.	0.
2100	21.191	205.259	187.385	37.534	0.	0.	0.
2200	21.383	206.249	188.220	39.663	0.	0.	0.
2300	21.636	207.205	189.025	41.813	0.	0.	0.
2400	21.959	208.132	189.802	43.992	0.	0.	0.
2500	22.361	209.036	190.553	46.207	0.	0.	0.
2600	22.849	209.923	191.281	48.467	0.	0.	0.
2700	23.430	210.796	191.988	50.780	0.	0.	0.
2800	24.109	211.660	192.675	53.157	0.	0.	0.
2900	24.889	212.519	193.345	55.606	0.	0.	0.
3000	25.770	213.377	193.998	58.138	0.	0.	0.
3100	26.753	214.238	194.637	60.763	0.	0.	0.
3200	27.834	215.104	195.263	63.492	0.	0.	0.
3300	29.008	215.978	195.878	66.333	0.	0.	0.
3400	30.270	216.863	196.482	69.296	0.	0.	0.
3500	31.611	217.759	197.077	72.389	0.	0.	0.
3600	33.022	218.670	197.664	75.620	0.	0.	0.
3700	34.493	219.594	198.244	78.996	0.	0.	0.
3800	36.011	220.534	198.818	82.520	0.	0.	0.
3900	37.565	221.490	199.387	86.199	0.	0.	0.
4000	39.142	222.461	199.952	90.034	0.	0.	0.
4100	40.730	223.447	200.513	94.028	0.	0.	0.
4200	42.304	224.446	201.071	98.176	0.	0.	0.
4300	43.874	225.460	201.626	102.485	0.	0.	0.
4400	45.418	226.486	202.180	106.950	0.	0.	0.
4500	46.927	227.524	202.731	111.568	0.	0.	0.
4600	48.214	228.556	203.280	116.269	0.	0.	0.
4700	49.584	229.608	203.829	121.159	0.	0.	0.
4800	50.884	230.665	204.377	126.182	0.	0.	0.
4900	52.114	231.727	204.925	131.333	0.	0.	0.
5000	53.254	232.791	205.471	136.597	0.	0.	0.
5100	54.274	233.851	206.017	141.953	0.	0.	0.
5200	55.250	234.914	206.563	147.429	0.	0.	0.
5300	56.141	235.975	207.108	153.000	0.	0.	0.
5400	56.945	237.032	207.652	158.655	0.	0.	0.
5500	57.662	238.084	208.196	164.386	0.	0.	0.
5600	58.296	239.129	208.739	170.184	0.	0.	0.
5700	58.846	240.166	209.281	176.042	0.	0.	0.
5800	59.317	241.193	209.822	181.951	0.	0.	0.
5900	59.417	242.171	210.359	187.693	0.	0.	0.
6000	59.709	243.172	210.897	193.650	0.	0.	0.

PREVIOUS: December 1970 (1 atm) CURRENT: September 1983 (1 bar)

Strontium (Sr) $Sr_1(g)$

STRONTIUM, ION (Sr^+) IDEAL GAS M_r = 87.61945

$IP(Sr^+, g) = 88964.0 \pm 0.5\ cm^{-1}$

$S°(298.15\ K) = 170.403 \pm 0.02\ J\ K^{-1}\ mol^{-1}$

$\Delta_f H°(0\ K) = 713.840\ kJ\ mol^{-1}$

$\Delta_f H°(298.15\ K) = [719.666]\ kJ\ mol^{-1}$

Electronic Levels and Quantum Weights

State	ϵ_i, cm^{-1}	g_i
$^2S_{1/2}$	0	2
$^2D_{3/2}$	14555.90	4
$^2D_{5/2}$	14836.24	6
$^2P_{1/2}$	23715.19	2
$^2P_{3/2}$	24516.65	4
$^2S_{1/2}$	47736.53	2

Enthalpy of Formation

$\Delta_f H°(Sr^+, g, 0\ K)$ is calculated from $\Delta_f H°(Sr, g, 0\ K)$ (1) using the spectroscopic value of $IP(Sr) = 45932.0 \pm 0.2\ cm^{-1}$ ($549.469 \pm 0.002\ kJ\ mol^{-1}$) from Moore (2). The ionization limit is converted from cm^{-1} to $kJ\ mol^{-1}$ using the factor, $1\ cm^{-1} = 0.01196266\ kJ\ mol^{-1}$, which is derived from the 1973 CODATA fundamental constants (3). Rosenstock et al. (4) and Levin and Lias (5) have summarized additional ionization and appearance potential data.

$\Delta_f H°(Sr^+, g, 298.15\ K)$ is calculated from $\Delta_f H°(Sr, g, 0\ K)$ by using IP(Sr) with JANAF (1) enthalpies, $H°(0\ K) - H°(298.15\ K)$, for Sr(g), Sr^+(g), and e^-(g). $\Delta_r H°(Sr \rightarrow Sr^+ + e^-, 298.15\ K)$ differs from a room temperature threshold energy due to inclusion of these enthalpies and to threshold effects discussed by Rosenstock et al. (4). $\Delta_f H°(298.15\ K)$ should be changed by $-6.197\ kJ\ mol^{-1}$ if it is to be used in the ion convention that excludes the enthalpy of the electron.

Heat Capacity and Entropy

The information on electronic energy levels and quantum weights, given by Moore (2), is incomplete because many theoretically predicted levels have not been observed. Our calculations indicate that any reasonable method of filling in these missing levels and cutting off the summation in the partition function (7) has no effect on the thermodynamic functions to 6000 K. This is a result of the high energy of all levels other than the ground state and the levels listed above; the next excited state is approximately 53286 cm^{-1} above the ground state. Since inclusion of these higher excited states have little effect on the thermodynamic functions (to 6000 K), we list only the ground state and the first four excited states. The reported uncertainty in S°(298.15 K) is due to uncertainties in the relative ionic mass and the fundamental constants. Extension of these calculations above 6000 K may require consideration of the excited states and use of different fill and cutoff procedures (7).

References

1. JANAF Thermochemical Tables: Sr(g), 9-30-83; e^-(g), 3-31-82.
2. C. E. Moore, U. S. Nat. Bur. Stand., NSRDS-NBS-34, 1970.
3. E. R. Cohen and B. N. Taylor, J. Phys. Chem. Ref. Data 2, 663 (1973).
4. H. M. Rosenstock, K. Draxl et al., J. Phys. Chem. Ref. Data 6, Supp. 1 (1977).
5. R. D. Levin and S. G. Lias, U. S. Nat. Bur. Stand., NSRDS-NBS-71, 1982.
6. C. E. Moore, U.S. Nat. Bur. Stand., NSRDS-NBS 35, volume II, 1970 [Reprint of NBS Circular 467, Volume II, 1952].
7. J. R. Downey, Jr., The Dow Chemical Company, AFOSR-TRS-78-0960, Contract No. F44620-75-1-0048, 1978.

Strontium, Ion (Sr^+) $Sr_1^+(g)$

Enthalpy Reference Temperature = T_r = 298.15 K; Standard State Pressure = p° = 0.1 MPa

T/K	$C_p°$ (J K⁻¹mol⁻¹)	S° (J K⁻¹mol⁻¹)	$-[G°-H°(T_r)]/T$ (J K⁻¹mol⁻¹)	$H°-H°(T_r)$ (kJ mol⁻¹)	$\Delta_f H°$ (kJ mol⁻¹)	$\Delta_f G°$ (kJ mol⁻¹)	Log K_f
0	0.	0.	INFINITE	-6.197	713.840		
100	20.786	147.696	188.884	-4.119			
200	20.786	162.104	172.305	-2.040			
250	20.786	166.742	170.745	-1.001			
298.15	20.786	170.403	170.403	0.	719.666	679.211	-118.995
300	20.786	170.532	170.404	0.038	719.694	678.960	-118.217
350	20.786	173.736	170.657	1.078	720.419	672.113	-100.307
400	20.786	176.512	171.219	2.117	721.122	665.164	-86.861
450	20.786	178.960	171.946	3.156	721.799	658.128	-76.394
500	20.786	181.150	172.759	4.196	722.451	651.018	-68.011
600	20.786	184.940	174.483	6.274	723.678	636.615	-55.422
700	20.786	188.144	176.211	8.353	724.802	622.014	-46.415
800	20.786	190.919	177.880	10.431	725.819	607.259	-39.650
900	20.786	193.368	179.468	12.510	726.102	592.456	-34.385
1000	20.786	195.558	180.969	14.589	727.226	577.546	-30.168
1100	20.786	197.539	182.387	16.667	720.447	562.889	-26.729
1200	20.786	199.347	183.726	18.746	720.658	548.556	-23.878
1300	20.787	201.011	184.992	20.825	720.869	534.206	-21.465
1400	20.789	202.552	186.192	22.903	721.080	519.839	-19.395
1500	20.792	203.986	187.331	24.982	721.291	505.457	-17.602
1600	20.799	205.328	188.415	27.062	721.503	491.061	-16.032
1700	20.811	206.589	189.447	29.142	584.802	477.832	-14.682
1800	20.831	207.780	190.433	31.224	586.876	471.480	-13.682
1900	20.861	208.907	191.376	33.309	588.948	465.012	-12.784
2000	20.904	209.978	192.279	35.397	591.015	458.436	-11.973
2100	20.963	210.999	193.146	37.490	593.075	451.756	-11.237
2200	21.041	211.976	193.980	39.590	595.126	444.978	-10.565
2300	21.140	212.913	194.783	41.699	597.163	438.108	-9.950
2400	21.264	213.815	195.557	43.819	599.182	431.149	-9.384
2500	21.414	214.686	196.305	45.953	601.179	424.107	-8.861
2600	21.590	215.530	197.029	48.103	603.148	416.985	-8.377
2700	21.795	216.348	197.729	50.272	605.082	409.788	-7.928
2800	22.028	217.145	198.408	52.463	606.976	402.520	-7.509
2900	22.290	217.922	199.068	54.678	608.821	395.185	-7.118
3000	22.580	218.683	199.709	56.922	610.611	387.788	-6.752
3100	22.897	219.428	200.333	59.195	612.338	380.332	-6.409
3200	23.239	220.161	200.941	61.502	613.994	372.821	-6.086
3300	23.605	220.881	201.535	63.844	615.574	365.260	-5.782
3400	23.993	221.592	202.114	66.223	617.069	357.652	-5.495
3500	24.400	222.293	202.681	68.643	618.474	350.001	-5.223
3600	24.824	222.986	203.235	71.104	619.782	342.312	-4.967
3700	25.263	223.672	203.778	73.608	620.990	334.587	-4.724
3800	25.714	224.352	204.311	76.157	622.093	326.832	-4.493
3900	26.174	225.026	204.833	78.751	623.087	319.048	-4.273
4000	26.640	225.694	205.346	81.392	623.971	311.241	-4.064
4100	27.110	226.358	205.851	84.079	624.744	303.413	-3.866
4200	27.581	227.017	206.347	86.814	625.408	295.567	-3.676
4300	28.051	227.672	206.835	89.596	625.959	287.707	-3.495
4400	28.517	228.322	207.316	92.424	626.401	279.835	-3.322
4500	28.976	228.968	207.790	95.299	626.737	271.955	-3.157
4600	29.428	229.610	208.258	98.219	627.035	264.063	-2.999
4700	29.869	230.247	208.719	101.184	627.188	256.171	-2.847
4800	30.299	230.881	209.174	104.193	627.252	248.276	-2.702
4900	30.715	231.510	209.623	107.243	627.231	240.380	-2.562
5000	31.115	232.134	210.067	110.335	627.137	232.486	-2.429
5100	31.500	232.754	210.506	113.466	626.991	224.592	-2.300
5200	31.868	233.369	210.940	116.634	626.761	216.704	-2.177
5300	32.217	233.980	211.369	119.839	626.474	208.821	-2.058
5400	32.548	234.585	211.793	123.077	626.136	200.944	-1.944
5500	32.859	235.185	212.213	126.348	625.754	193.074	-1.834
5600	33.151	235.780	212.628	129.648	625.335	185.210	-1.728
5700	33.423	236.369	213.040	132.977	624.884	177.355	-1.625
5800	33.675	236.953	213.447	136.332	624.409	169.508	-1.527
5900	33.908	237.530	213.850	139.712	624.125	161.645	-1.431
6000	34.121	238.102	214.250	143.113	623.649	153.810	-1.339

PREVIOUS: CURRENT: September 1983 (1 bar)

Strontium, Ion (Sr^+) $Sr_1^+(g)$

TANTALUM (Ta) REFERENCE STATE A_r = 180.9479

0 to 3258 K		crystal
3258 to 5778.070 K		liquid
above 5778.070 K		ideal monatomic gas

Refer to the individual tables for details.

Tantalum (Ta) Ta_1(ref)

Enthalpy Reference Temperature = T_r = 298.15 K Standard State Pressure = $p°$ = 0.1 MPa

T/K	$C_p°$ (J K⁻¹mol⁻¹)	$S°$ (J K⁻¹mol⁻¹)	$-[G°-H°(T_r)]/T$ (J K⁻¹mol⁻¹)	$H°-H°(T_r)$ (kJ mol⁻¹)	$\Delta_fH°$ (kJ mol⁻¹)	$\Delta_fG°$ (kJ mol⁻¹)	Log K_f
0	0.	0.	INFINITE	-5.681	0.	0.	0.
100	19.744	16.143	63.003	-4.686	0.	0.	0.
200	24.085	31.586	43.756	-2.434	0.	0.	0.
250	24.861	37.049	41.886	-1.209	0.	0.	0.
298.15	25.295	41.471	41.471	0.	0.	0.	0.
300	25.307	41.628	41.472	0.047	0.	0.	0.
350	25.598	45.552	41.781	1.320	0.	0.	0.
400	25.840	48.986	42.471	2.606	0.	0.	0.
450	26.088	52.044	43.368	3.904	0.	0.	0.
500	26.349	54.806	44.376	5.215	0.	0.	0.
600	26.843	59.655	46.530	7.875	0.	0.	0.
700	27.214	63.822	48.709	10.579	0.	0.	0.
800	27.459	67.473	50.831	13.313	0.	0.	0.
900	27.668	70.718	52.864	16.069	0.	0.	0.
1000	27.933	73.647	54.798	18.848	0.	0.	0.
1100	28.281	76.325	56.635	21.659	0.	0.	0.
1200	28.662	78.802	58.380	24.507	0.	0.	0.
1300	28.989	81.110	60.041	27.390	0.	0.	0.
1400	29.202	83.267	61.624	30.301	0.	0.	0.
1500	29.319	85.286	63.135	33.227	0.	0.	0.
1600	29.439	87.182	64.579	36.164	0.	0.	0.
1700	29.688	88.973	65.962	39.119	0.	0.	0.
1800	30.124	90.682	67.288	42.108	0.	0.	0.
1900	30.665	92.325	68.563	45.148	0.	0.	0.
2000	31.191	93.911	69.791	48.240	0.	0.	0.
2100	31.713	95.445	70.976	51.386	0.	0.	0.
2200	32.252	96.933	72.122	54.584	0.	0.	0.
2300	32.828	98.379	73.233	57.837	0.	0.	0.
2400	33.459	99.789	74.310	61.151	0.	0.	0.
2500	34.167	101.169	75.357	64.532	0.	0.	0.
2600	34.970	102.525	76.376	67.988	0.	0.	0.
2700	35.890	103.861	77.369	71.530	0.	0.	0.
2800	36.946	105.185	78.339	75.170	0.	0.	0.
2900	38.158	106.502	79.287	78.924	0.	0.	0.
3000	39.546	107.819	80.216	82.808	0.	0.	0.
3100	41.130	109.141	81.128	86.840	0.	0.	0.
3200	42.930	110.474	82.024	91.041	0.	0.	0.
3258.000	44.080	111.256	82.538	93.564	CRYSTAL <--> LIQUID		
3258.000	41.840	122.480	82.538	130.132	TRANSITION		
3300	41.840	123.016	83.049	131.889	0.	0.	0.
3400	41.840	124.265	84.243	136.073	0.	0.	0.
3500	41.840	125.478	85.404	140.257	0.	0.	0.
3600	41.840	126.656	86.534	144.441	0.	0.	0.
3700	41.840	127.803	87.634	148.625	0.	0.	0.
3800	41.840	128.919	88.706	152.809	0.	0.	0.
3900	41.840	130.005	89.751	156.993	0.	0.	0.
4000	41.840	131.065	90.770	161.177	0.	0.	0.
4100	41.840	132.098	91.766	165.361	0.	0.	0.
4200	41.840	133.106	92.738	169.545	0.	0.	0.
4300	41.840	134.091	93.688	173.729	0.	0.	0.
4400	41.840	135.052	94.618	177.913	0.	0.	0.
4500	41.840	135.993	95.527	182.097	0.	0.	0.
4600	41.840	136.912	96.416	186.281	0.	0.	0.
4700	41.840	137.812	97.288	190.465	0.	0.	0.
4800	41.840	138.693	98.141	194.649	0.	0.	0.
4900	41.840	139.556	98.978	198.833	0.	0.	0.
5000	41.840	140.401	99.798	203.017	0.	0.	0.
5200	41.840	142.042	101.391	211.385	0.	0.	0.
5400	41.840	143.621	102.926	219.753	0.	0.	0.
5600	41.840	145.143	104.407	228.121	0.	0.	0.
5778.070	41.840	146.452	105.682	235.572	LIQUID <--> IDEAL GAS		
5778.070	43.671	274.300	105.682	974.285	FUGACITY = 1 bar		
5800	43.762	274.465	106.320	975.244	0.	0.	0.
6000	44.636	275.963	111.950	984.082	0.	0.	0.

PREVIOUS: December 1972 (1 atm) CURRENT: December 1972 (1 bar)

Tantalum (Ta) Ta_1(ref)

TANTALUM (Ta) CRYSTAL A_r = 180.9479

$S°(298.15\ K) = 41.47 \pm 0.21\ J\ K^{-1}\ mol^{-1}$

$T_{fus} = 3258 \pm 10\ K$

$\Delta_f H°(0\ K) = 0\ kJ\ mol^{-1}$

$\Delta_f H°(298.15\ K) = 0\ kJ\ mol^{-1}$

$\Delta_{fus} H° = 36.57 \pm 4.2\ kJ\ mol^{-1}$

Enthalpy of Formation

Zero by definition.

Heat Capacity and Entropy

The heat capacity values for $T \leq 10$ K are chosen to be the same as those adopted by Hultgren et al. (1). A graphical integration of these Cp° data yields $S°(10\ K) = 0.025\ cal\ K^{-1} mol^{-1}$. The adopted Cp° values for the range $10 \leq T \leq T_{fus}$ are based on the following experimental data.

Source	T/K	Method
(2)	53-295	Cp°
(3)	10-273	Cp°
(4)	12-550	Cp°
(5)	300-1554	drop, enthalpy
(6)	533-1383	drop, enthalpy
(7)	1900-3200	Cp°, pulse heating

The data of Cezairliyan (7), reported as smoothed Cp° values and represented by a third-order polynomial, yields Cp° values from 3200 K to T_{fus} by extrapolation. These smooth Cp° values, as reported by Cezirliyan (7), are adopted for the range 1900-3200 K. Using a polynomial fit for each enthalpy data set (5, 6), Cp° values are obtained which define the region 298-1500 K. A polynomial, constrained at 10 K, is used to produce smoothed Cp° values in the range 10-298 K based on the three low temperature studies (3, 4, 5).

For T < 100 K, the tabulated values do not agree well with the experimental data. In this region the deviations from the tabulated values are ±5% for Clusius and Losa (3), -0.3 to +1.9% for Kelley (2), and -1.7 to 2.8% for Sterrett and Wallace (4). The deviations are due in part to experimental scatter (3) and a poor match of experimental data at 10 K.

Many other experimental studies have been referenced by Gmelin (8) and Hultgren et al. (1). In addition, Cezairliyan (7) has compared graphically many of the investigations. The Cp° data and enthalpy data are all in generally good agreement. The importance of the recent Cezairliyan study (7) is that it gives added evidence to the deviation from a linear Cp-T relationship above 1000 K.

Fusion Data

The melting point of Ta is chosen as 3258 ± 10 K based on the subsecond pulse heating technique of Cezairliyan (7). This choice is made so as to have a T_{fus} consistent with Cp° data at temperatures near T_{fus} (see heat capacity discussion). Other T_{fus} values covering the range 3053-3273 K are referenced by Charlesworth (9); while additional references are found in Gmelin (8). Hultgren et al. (1) recommends a T_{fus} value of 3287 K.

The enthalpy of fusion, $\Delta_{fus}H°$, for Ta has been measured by Lebedev, Savvatimskii, and Smirnov (10). They reported a $\Delta_{fus}H°$ value of $8.74 \pm 0.5\ kcal\ mol^{-1}$ based on measurements made during electrical explosion in thin wires. We adopt this value for $\Delta_{fus}H°$. Based on $\Delta_{fus}H° = 8.74\ kcal\ mol^{-1}$ and $T_{fus} = 3258$ K, the entropy of melting, $\Delta_{fus}S°$, is calculated to be 2.68 cal $K^{-1}mol^{-1}$. This compares with $\Delta_{fus}S° = 2.33\ cal\ K^{-1}\ mol^{-1}$ for Nb, as reported in these tables.

References

1. R. Hultgren, R. L. Orr, and K. K. Kelley, Supplement to Selected Values of Thermodynamic Properties of Metals and Alloys, Ta table, August, 1971.
2. K. K. Kelley, J. Chem. Phys. 8, 316 (1940).
3. K Clusius and C. G. Losa, Z. Naturforsch. 10A, 939 (1955).
4. K. F. Sterrett and W. E. Wallace, J. Amer. Chem. Soc. 80, 3176 (1958).
5. F. M. Jaeger and W. A. Veenstra, Rec. Trav. Chim. 53, 677 (1934).
6. F. L. Oetting and J. D. Navratil, J. Chem. Eng. Data 17, 230 (1972).
7. A. Cezairliyan, U. S. Natl. Bur. Stands. Rept. 10326, July, 1970, Ch. 7, Ch. 8.
8. "Gmelins Handbuch der Anorganishen Chemie," 8th Edition, Tantal, Vol. 50, Pt. A, No. 2, Verlag Chemie, Weinheim (1969).
9. J. H. Charlesworth, Melting Points of Metallic Elements and Selected Compounds, AFML-TR-70-137, October (1970).
10. S. V. Lebedev, A. I. Savvatimskii, and Yu. B. Smirnov, High Temperature USSR 9, 635 (1971).

Tantalum (Ta) $Ta_1(cr)$

Enthalpy Reference Temperature = T_r = 298.15 K — Standard State Pressure = p° = 0.1 MPa

T/K	$C_p°$ (J K⁻¹mol⁻¹)	S° (J K⁻¹mol⁻¹)	$-[G°-H°(T_r)]/T$ (J K⁻¹mol⁻¹)	$H°-H°(T_r)$ (kJ mol⁻¹)	$\Delta_f H°$ (kJ mol⁻¹)	$\Delta_f G°$ (kJ mol⁻¹)	Log K_f
0	0.	0.	INFINITE	-5.681	0.	0.	
100	19.744	16.143	63.003	-4.686	0.	0.	0.
200	24.085	31.586	43.756	-2.434	0.	0.	0.
250	24.861	37.049	41.886	-1.209	0.	0.	0.
298.15	25.295	41.471	41.471	0.	0.	0.	0.
300	25.307	41.628	41.472	0.047	0.	0.	0.
350	25.598	45.552	41.781	1.320	0.	0.	0.
400	25.840	48.986	42.471	2.606	0.	0.	0.
450	26.088	52.044	43.368	3.904	0.	0.	0.
500	26.349	54.806	44.376	5.215	0.	0.	0.
600	26.843	59.655	46.530	7.875	0.	0.	0.
700	27.214	63.822	48.709	10.579	0.	0.	0.
800	27.459	67.473	50.831	13.313	0.	0.	0.
900	27.668	70.718	52.864	16.069	0.	0.	0.
1000	27.933	73.647	54.798	18.848	0.	0.	0.
1100	28.281	76.325	56.635	21.659	0.	0.	0.
1200	28.662	78.802	58.380	24.507	0.	0.	0.
1300	28.989	81.110	60.041	27.390	0.	0.	0.
1400	29.202	83.267	61.624	30.301	0.	0.	0.
1500	29.319	85.286	63.135	33.227	0.	0.	0.
1600	29.439	87.182	64.579	36.164	0.	0.	0.
1700	29.688	88.973	65.962	39.119	0.	0.	0.
1800	30.124	90.682	67.288	42.108	0.	0.	0.
1900	30.665	92.325	68.563	45.148	0.	0.	0.
2000	31.191	93.911	69.791	48.240	0.	0.	0.
2100	31.713	95.445	70.976	51.386	0.	0.	0.
2200	32.252	96.933	72.122	54.584	0.	0.	0.
2300	32.828	98.379	73.233	57.837	0.	0.	0.
2400	33.459	99.789	74.310	61.151	0.	0.	0.
2500	34.167	101.169	75.357	64.532	0.	0.	0.
2600	34.970	102.525	76.376	67.988	0.	0.	0.
2700	35.890	103.861	77.369	71.530	0.	0.	0.
2800	36.946	105.185	78.339	75.170	0.	0.	0.
2900	38.158	106.502	79.287	78.924	0.	0.	0.
3000	39.546	107.819	80.216	82.808	0.	0.	0.
3100	41.130	109.141	81.128	86.840	0.	0.	0.
3200	42.930	110.474	82.024	91.041	0.	0.	0.
3258.000	44.080	111.256	82.538	93.564	------ CRYSTAL <--> LIQUID -----		
3300	44.965	111.826	82.907	95.433	-36.456	0.471	-0.007
3400	47.256	113.202	83.778	100.042	-36.031	1.584	-0.024
3500	49.823	114.608	84.638	104.894	-35.363	2.681	-0.040
3600	52.685	116.051	85.491	110.017	-34.424	3.756	-0.054
3700	55.864	117.537	86.337	115.441	-33.184	4.800	-0.068
3800	59.377	119.073	87.178	121.201	-31.609	5.806	-0.080

PREVIOUS: CURRENT: December 1972

Tantalum (Ta) $Ta_1(cr)$

TANTALUM (Ta) LIQUID A_r = 180.9479

$S°(298.15\ K) = [50.366]\ J\ K^{-1}\ mol^{-1}$

$T_{fus} = 3258 \pm 10\ K$

$\Delta_f H°(298.15\ K) = [30.795]\ kJ\ mol^{-1}$

$\Delta_{fus}H° = 36.57 \pm 4.2\ kJ\ mol^{-1}$

Enthalpy of Formation

The enthalpy of formation of Ta(ℓ) at 298.15 K is calculated from that of the crystal by adding $\Delta_{fus}H°$ and the difference in enthalpy, H°(3258 K) − H°(298.15 K), between the crystal and liquid.

Heat Capacity and Entropy

The heat capacity for Ta(ℓ) is estimated as 10.0 cal K^{-1} mol^{-1} by analogy with other monatomic metals. The same value was adopted by Hultgren et al. (1) and Stull and Sinke (2). A glass transition is assumed at 2150 K. Below this temperature, the heat capacity values used are those of Ta(cr). The entropy at 298.15 K is calculated in a manner similar to that used for the enthalpy of formation.

Fusion Data

See the crystal table for details.

Vaporization Data

T_{vap} is the temperature at which the Gibbs energy change for the reaction Ta(ℓ) → Ta(g) is zero. The difference between $\Delta_f H°$(Ta, g) and $\Delta_f H°$(Ta, ℓ) at T_{vap} is $\Delta_{vap}H°$. The uncertainty in the boiling point is probably of the order of ±200 K.

References

1. R. Hultgren, R. L. Orr, and K. K. Kelley, Supplement to Selected Values of Thermodynamic Properties of Metals and Alloys, Ta table, August, 1971.
2. D. R. Stull and G. C. Sinke, "Thermodynamic Properties of the Elements," American Chemical Society, Washington, D. C., 1956.

Tantalum (Ta) $Ta_1(l)$

Enthalpy Reference Temperature = T_r = 298.15 K   Standard State Pressure = p° = 0.1 MPa

T/K	$C_p°$	S° (J K⁻¹mol⁻¹)	−[G°−H°(T_r)]/T	H°−H°(T_r) (kJ mol⁻¹)	$\Delta_f H°$	$\Delta_f G°$	Log K_f
0							
100							
200							
250							
298.15	25.295	50.366	50.366	0.	30.795	28.143	-4.931
300	25.307	50.522	50.366	0.047	30.795	28.127	-4.897
350	25.598	54.447	50.676	1.320	30.795	27.682	-4.131
400	25.840	57.881	51.366	2.606	30.795	27.237	-3.557
450	26.088	60.938	52.263	3.904	30.795	26.793	-3.110
500	26.349	63.701	53.271	5.215	30.795	26.348	-2.753
600	26.843	68.550	55.424	7.875	30.795	25.458	-2.216
700	27.214	72.717	57.604	10.579	30.795	24.569	-1.833
800	27.459	76.367	59.726	13.313	30.795	23.680	-1.546
900	27.668	79.613	61.758	16.069	30.795	22.790	-1.323
1000	27.933	82.541	63.693	18.848	30.795	21.901	-1.144
1100	28.281	85.220	65.530	21.659	30.795	21.011	-0.998
1200	28.662	87.697	67.275	24.507	30.795	20.122	-0.876
1300	28.989	90.005	68.936	27.390	30.795	19.232	-0.773
1400	29.202	92.162	70.518	30.301	30.795	18.343	-0.684
1500	29.319	94.181	72.029	33.227	30.795	17.453	-0.608
1600	29.439	96.076	73.474	36.164	30.795	16.564	-0.541
1700	29.688	97.868	74.856	39.119	30.795	15.674	-0.482
1800	30.124	99.576	76.183	42.108	30.795	14.785	-0.429
1900	30.665	101.219	77.457	45.148	30.795	13.896	-0.382
2000	31.191	102.806	78.685	48.240	30.795	13.006	-0.340
2100	31.713	104.340	79.871	51.386	30.795	12.117	-0.301
2150.000	31.980	105.089	80.448	52.978	GLASS <--> LIQUID		
2150.000	41.840	105.089	80.448	52.978	TRANSITION		
2200	41.840	106.051	81.019	55.070	31.281	11.222	-0.266
2300	41.840	107.911	82.148	59.254	32.212	10.289	-0.234
2400	41.840	109.692	83.259	63.438	33.082	9.317	-0.203
2500	41.840	111.400	84.351	67.622	33.885	8.310	-0.174
2600	41.840	113.041	85.423	71.806	34.613	7.272	-0.146
2700	41.840	114.620	86.475	75.990	35.256	6.208	-0.120
2800	41.840	116.141	87.508	80.174	35.799	5.122	-0.096
2900	41.840	117.610	88.521	84.358	36.229	4.018	-0.072
3000	41.840	119.028	89.514	88.542	36.529	2.902	-0.051
3100	41.840	120.400	90.488	92.726	36.681	1.778	-0.030
3200	41.840	121.728	91.444	96.910	36.664	0.652	-0.011
3258.000	41.840	122.480	91.990	99.337	CRYSTAL <--> LIQUID		
3300	41.840	123.016	92.381	101.094	0.	0.	0.
3400	41.840	124.265	93.301	105.278	0.	0.	0.
3500	41.840	125.478	94.203	109.462	0.	0.	0.
3600	41.840	126.656	95.088	113.646	0.	0.	0.
3700	41.840	127.803	95.957	117.830	0.	0.	0.
3800	41.840	128.919	96.810	122.014	0.	0.	0.
3900	41.840	130.005	97.647	126.198	0.	0.	0.
4000	41.840	131.065	98.469	130.382	0.	0.	0.
4100	41.840	132.098	99.277	134.566	0.	0.	0.
4200	41.840	133.106	100.070	138.750	0.	0.	0.
4300	41.840	134.091	100.850	142.934	0.	0.	0.
4400	41.840	135.052	101.617	147.118	0.	0.	0.
4500	41.840	135.993	102.370	151.302	0.	0.	0.
4600	41.840	136.912	103.111	155.486	0.	0.	0.
4700	41.840	137.812	103.840	159.670	0.	0.	0.
4800	41.840	138.693	104.557	163.854	0.	0.	0.
4900	41.840	139.556	105.262	168.038	0.	0.	0.
5000	41.840	140.401	105.957	172.222	0.	0.	0.
5200	41.840	142.042	107.313	180.590	0.	0.	0.
5400	41.840	143.621	108.629	188.958	0.	0.	0.
5600	41.840	145.143	109.906	197.326	0.	0.	0.
5778.070	41.840	146.452	111.012	204.776	FUGACITY = 1 bar		
5800	41.840	146.611	111.146	205.694	-738.755	2.801	-0.025
6000	41.840	148.029	112.352	214.062	-739.225	28.380	-0.247

PREVIOUS: CURRENT: December 1972

Tantalum (Ta) $Ta_1(l)$

TANTALUM (Ta) CRYSTAL–LIQUID A_r = 180.9479

0 to 3258 K crystal
above 3258 K liquid

Refer to the individual tables for details.

Tantalum (Ta) $Ta_1(cr,l)$

Enthalpy Reference Temperature = T_r = 298.15 K Standard State Pressure = p° = 0.1 MPa

T/K	C_p° (J K⁻¹mol⁻¹)	S° (J K⁻¹mol⁻¹)	$-[G°-H°(T_r)]/T$ (J K⁻¹mol⁻¹)	$H°-H°(T_r)$ (kJ mol⁻¹)	$\Delta_fH°$ (kJ mol⁻¹)	$\Delta_fG°$ (kJ mol⁻¹)	Log K_f
0	0.	0.	INFINITE	-5.681	0.	0.	0.
100	19.744	16.143	63.003	-4.686	0.	0.	0.
200	24.085	31.586	43.756	-2.434	0.	0.	0.
250	24.861	37.049	41.886	-1.209	0.	0.	0.
298.15	25.295	41.471	41.471	0.	0.	0.	0.
300	25.307	41.628	41.472	0.047	0.	0.	0.
350	25.598	45.552	41.781	1.320	0.	0.	0.
400	25.840	48.986	42.471	2.606	0.	0.	0.
450	26.088	52.044	43.368	3.904	0.	0.	0.
500	26.349	54.806	44.376	5.215	0.	0.	0.
600	26.843	59.655	46.530	7.875	0.	0.	0.
700	27.214	63.822	48.709	10.579	0.	0.	0.
800	27.459	67.473	50.831	13.313	0.	0.	0.
900	27.668	70.718	52.864	16.069	0.	0.	0.
1000	27.933	73.647	54.798	18.848	0.	0.	0.
1100	28.281	76.325	56.635	21.659	0.	0.	0.
1200	28.662	78.802	58.380	24.507	0.	0.	0.
1300	28.989	81.110	60.041	27.390	0.	0.	0.
1400	29.202	83.267	61.624	30.301	0.	0.	0.
1500	29.319	85.286	63.135	33.227	0.	0.	0.
1600	29.439	87.182	64.579	36.164	0.	0.	0.
1700	29.688	88.973	65.962	39.119	0.	0.	0.
1800	30.124	90.682	67.288	42.108	0.	0.	0.
1900	30.665	92.325	68.563	45.148	0.	0.	0.
2000	31.191	93.911	69.791	48.240	0.	0.	0.
2100	31.713	95.445	70.976	51.386	0.	0.	0.
2200	32.252	96.933	72.122	54.584	0.	0.	0.
2300	32.828	98.379	73.233	57.837	0.	0.	0.
2400	33.459	99.789	74.310	61.151	0.	0.	0.
2500	34.167	101.169	75.357	64.532	0.	0.	0.
2600	34.970	102.525	76.376	67.988	0.	0.	0.
2700	35.890	103.861	77.369	71.530	0.	0.	0.
2800	36.946	105.185	78.339	75.170	0.	0.	0.
2900	38.158	106.502	79.287	78.924	0.	0.	0.
3000	39.546	107.819	80.216	82.808	0.	0.	0.
3100	41.130	109.141	81.128	86.840	0.	0.	0.
3200	42.930	110.474	82.024	91.041	0.	0.	0.
3258.000	44.080	111.256	82.538	93.564	CRYSTAL <--> LIQUID		
3258.000	41.840	122.480	82.538	130.132	TRANSITION		
3300	41.840	123.016	83.049	131.889	0.	0.	0.
3400	41.840	124.265	84.243	136.073	0.	0.	0.
3500	41.840	125.478	85.404	140.257	0.	0.	0.
3600	41.840	126.656	86.534	144.441	0.	0.	0.
3700	41.840	127.803	87.634	148.625	0.	0.	0.
3800	41.840	128.919	88.706	152.809	0.	0.	0.
3900	41.840	130.005	89.751	156.993	0.	0.	0.
4000	41.840	131.065	90.770	161.177	0.	0.	0.
4100	41.840	132.098	91.766	165.361	0.	0.	0.
4200	41.840	133.106	92.738	169.545	0.	0.	0.
4300	41.840	134.091	93.688	173.729	0.	0.	0.
4400	41.840	135.052	94.618	177.913	0.	0.	0.
4500	41.840	135.993	95.527	182.097	0.	0.	0.
4600	41.840	136.912	96.416	186.281	0.	0.	0.
4700	41.840	137.812	97.288	190.465	0.	0.	0.
4800	41.840	138.693	98.141	194.649	0.	0.	0.
4900	41.840	139.556	98.978	198.833	0.	0.	0.
5000	41.840	140.401	99.798	203.017	0.	0.	0.
5200	41.840	142.042	101.391	211.385	0.	0.	0.
5400	41.840	143.621	102.926	219.753	0.	0.	0.
5600	41.840	145.143	104.407	228.121	0.	0.	0.
5778.070	41.840	146.452	105.682	235.572	------- FUGACITY = 1 bar -------		
5800	41.840	146.611	105.837	236.489	-738.755	2.801	-0.025
6000	41.840	148.029	107.220	244.857	-739.225	28.380	-0.247

PREVIOUS: CURRENT: December 1972

Tantalum (Ta) $Ta_1(cr,l)$

TANTALUM (Ta) IDEAL GAS A_r = 180.9479

IP(Br, g) = 63600 ± 100 cm^{-1}

S°(298.15 K) = 185.22 J K^{-1} mol^{-1}

$\Delta_f H°$(0 K) = 781.47 kJ mol^{-1}

$\Delta_f H°$(298.15 K) = 781.99 kJ mol^{-1}

Electronic Levels and Quantum Weights

State	ε_i, cm^{-1}	g_i
$^4F_{3/2}$	0.00	4
$^4F_{5/2}$	2010.10	6
$^4F_{7/2}$	3363.92	8
$^4F_{9/2}$	5321.04	10

Enthalpy of Formation

The enthalpy of sublimation of tantalum has been derived from 2nd and 3rd law analysis of the vapor pressure data of several investigators. These four studies are based on the Langmuir method.

Source	T/K	Data Points	$\Delta_{sub}H°$(298.15 K), kcal mol^{-1} 2nd Law	3rd Law	Drift cal K^{-1} mol^{-1}
(1)	2628-2952	7	190.0±3.0	186.9	-1.1±1.1
(2)	2004-3269	16*	183.0	186.9	1.5
(3)	2637-2854	7	151.9±10.1	182.6	11.2±3.7
(4)	2904-3155	8**	233.1	190.3	-14.2

*Only smooth data was reported; however, 16 refers to the number of experimental determinations.

**Only smooth data reported.

Babeliowsky (5) reported an enthalpy of sublimation value of 188.4 ± 4.0 kcal mol^{-1}. This value is based on mass spectrometric vaporization studies and is calculated by the 2nd law method. Sasaki et al. (6), also using a mass spectrometer, reported an enthalpy of sublimation value of 185.4 ± 0.3 kcal mol^{-1} as an average value in the range 2550-2770 K. The value chosen for the enthalpy of sublimation which, in this case, is the enthalpy of formation of Ta(g) is 186.9 ± 0.5 kcal mol^{-1}, based on the work of Edwards et al. (1) and Langmuir and Malter (2).

Heat Capacity and Entropy

The electronic levels and quantum weights are obtained from Moore (7). Although not all levels are listed, all levels given by Moore are used in the calculation. There are predicted electronic levels which have not been observed (7). These levels are assumed to lie above 20000 cm^{-1} and thus will not significantly affect the entropy at temperatures below 3000 K. The heat capacity and entropy values are very similar to those adopted by Hultgren et al. (8), being identical at 298.15 K and differing by 0.154 cal K^{-1} mol^{-1} in Cp° at 6000 K.

References

1. J. W. Edwards, H. L. Johnston, and P. E. Blackburn, J. Amer. Chem. Soc. 73, 4727 (1951).
2. D. B. Langmuir and L. Malter, Phys. Rev. 55, 748 (1939).
3. R. Fiske, Phys. Rev. 61, 513 (1942).
4. E. Gebhardt, H. Seghezzi, and H. Keil, Z. Metallk. 53, 524 (1962).
5. T. P. Babeliowsky, Physica 28, 1160 (1962).
6. C. Sasaki, K. Kubo, and M. Asano, Shitsuryo Bunsiki 18, 1189 (1970).
7. C. E. Moore, U. S. Natl. Bur. Std. Circ. 467, Vol II, 1952.
8. R. Hultgren, R. L. Orr, and K. K. Kelley, Supplement to Selected Values of Thermodynamic Properties of Metals and Alloys, Ta table, August, 1971.

Tantalum (Ta) Ta_1(g)

Enthalpy Reference Temperature = T_r = 298.15 K   Standard State Pressure = p° = 0.1 MPa

T/K	$C_p°$ (J $K^{-1}mol^{-1}$)	S°	$-[G°-H°(T_r)]/T$	$H°-H°(T_r)$ (kJ mol^{-1})	$\Delta_f H°$	$\Delta_f G°$	Log K_f
0	0.	0.	INFINITE	-6.200	781.471	781.471	INFINITE
100	20.786	162.504	203.713	-4.121	782.555	767.919	-401.120
200	20.787	176.912	187.123	-2.042	782.381	753.316	-196.746
250	20.802	181.551	185.562	-1.003	782.196	746.071	-155.883
298.15	20.858	185.219	185.219	0.	781.990	739.131	-129.493
300	20.861	185.348	185.219	0.039	781.981	738.865	-128.648
350	21.006	188.573	185.474	1.085	781.754	731.697	-109.200
400	21.259	191.394	186.041	2.141	781.525	724.562	-94.618
450	21.624	193.918	186.778	3.213	781.298	717.455	-83.280
500	22.082	196.219	187.609	4.305	781.080	710.373	-74.212
600	23.183	200.340	189.395	6.567	780.681	696.270	-60.616
700	24.381	204.003	191.225	8.945	780.355	682.229	-50.909
800	25.562	207.336	193.033	11.442	780.119	668.228	-43.631
900	26.676	210.412	194.796	14.055	779.975	654.251	-37.972
1000	27.701	213.276	196.502	16.774	779.916	640.286	-33.445
1100	28.633	215.961	198.150	19.592	779.923	626.323	-29.742
1200	29.470	218.489	199.741	22.498	779.981	612.357	-26.655
1300	30.212	220.878	201.276	25.483	780.082	598.385	-24.043
1400	30.865	223.141	202.757	28.537	780.226	584.403	-21.804
1500	31.437	225.291	204.189	31.653	780.416	570.409	-19.863
1600	31.937	227.336	205.572	34.822	780.648	556.401	-18.165
1700	32.377	229.286	206.910	38.039	780.909	542.378	-16.665
1800	32.768	231.147	208.205	41.296	781.177	528.339	-15.332
1900	33.119	232.929	209.460	44.591	781.433	514.285	-14.139
2000	33.441	234.636	210.676	47.919	781.668	500.219	-13.064
2100	33.739	236.275	211.856	51.278	781.882	486.141	-12.092
2200	34.021	237.851	213.002	54.666	782.072	472.053	-11.208
2300	34.289	239.369	214.116	58.082	782.234	457.958	-10.401
2400	34.548	240.834	215.199	61.524	782.362	443.856	-9.660
2500	34.799	242.249	216.253	64.991	782.449	429.750	-8.979
2600	35.045	243.619	217.279	68.484	782.486	415.641	-8.350
2700	35.287	244.946	218.279	72.000	782.460	401.532	-7.768
2800	35.525	246.234	219.255	75.541	782.360	387.425	-7.227
2900	35.761	247.484	220.207	79.105	782.171	373.323	-6.724
3000	35.994	248.701	221.136	82.693	781.875	359.229	-6.255
3100	36.226	249.885	222.045	86.304	781.454	345.148	-5.816
3200	36.457	251.039	222.933	89.938	780.887	331.082	-5.404
3300	36.688	252.164	223.802	93.595	743.696	317.507	-5.026
3400	36.918	253.263	224.652	97.276	743.192	304.600	-4.680
3500	37.149	254.336	225.485	100.979	742.712	291.707	-4.353
3600	37.381	255.386	226.301	104.706	742.254	278.828	-4.046
3700	37.614	256.413	227.101	108.455	741.820	265.961	-3.755
3800	37.849	257.419	227.886	112.228	741.409	253.106	-3.479
3900	38.085	258.406	228.656	116.025	741.022	240.261	-3.218
4000	38.324	259.373	229.412	119.846	740.658	227.425	-2.970
4100	38.565	260.322	230.154	123.690	740.319	214.599	-2.734
4200	38.810	261.254	230.883	127.559	740.003	201.780	-2.510
4300	39.057	262.171	231.600	131.452	739.713	188.969	-2.296
4400	39.309	263.071	232.305	135.370	739.447	176.164	-2.091
4500	39.566	263.958	232.999	139.314	739.207	163.365	-1.896
4600	39.827	264.830	233.681	143.284	738.992	150.571	-1.710
4700	40.096	265.689	234.353	147.280	738.804	137.781	-1.531
4800	40.371	266.536	235.015	151.303	738.644	124.995	-1.360
4900	40.655	267.372	235.667	155.354	738.511	112.212	-1.196
5000	40.948	268.196	236.309	159.434	738.407	99.432	-1.039
5100	41.251	269.010	236.942	163.544	738.333	86.653	-0.888
5200	41.566	269.814	237.567	167.685	738.289	73.875	-0.742
5300	41.893	270.609	238.183	171.858	738.278	61.098	-0.602
5400	42.235	271.395	238.791	176.064	738.300	48.321	-0.467
5500	42.592	272.173	239.390	180.305	738.358	35.543	-0.338
5600	42.964	272.944	239.983	184.583	738.451	22.764	-0.212
5700	43.354	273.708	240.568	188.899	738.583	9.983	-0.091
5778.070	43.671	274.300	241.020	192.296	------- FUGACITY = 1 bar -------		
5800	43.762	274.465	241.146	193.254	0.	0.	0.
5900	44.189	275.217	241.717	197.652	0.	0.	0.
6000	44.636	275.963	242.281	202.093	0.	0.	0.

PREVIOUS: December 1972 (1 atm)   CURRENT: December 1972 (1 bar)

Tantalum (Ta) Ta_1(g)

TANTALUM, ION (Ta^+) IDEAL GAS M_r = 180.94735 Tantalum, Ion (Ta^+) $Ta_1^+(g)$

IP(Ta^+, g) = [130660 ± 4000] cm^{-1} $\Delta_fH°$(0 K) = 1542.3 ± 2 kJ mol^{-1}

S°(298.15 K) = 183.38 ± 0.2 J K^{-1} mol^{-1} $\Delta_fH°$(298.15 K) = [1549.150] kJ mol^{-1}

Electronic Levels and Quantum Weights

State	ε_i, cm^{-1}	g_i
$^5F_1$	0	3
$^5F_2$	1031.33	5
$^5F_3$	2642.19	7
$^5F_4$	4415.70	9
$^5F_5$	6186.72	11

Enthalpy of Formation

$\Delta_fH°$(Ta^+, g, 0 K) is calculated from $\Delta_fH°$(Ta, g, 0 K) (1) using the spectroscopic value of IP(Ta) = 63600 ± 100 cm^{-1} (760.825 ± 1.20 kJ mol^{-1}) from Moore (2). The ionization limit is converted from cm^{-1} to kJ mol^{-1} using the factor, 1 cm^{-1} = 0.01196266 kJ cm^{-1}, which is derived from the 1973 CODATA fundamental constants (3). Rosenstock et al. (4) and Levin and Lias (5) have summarized additional ionization and appearance potential data.

$\Delta_fH°$(Ta^+, g, 298.15 K) is calculated from $\Delta_fH°$(Ta, g, 0 K) by using IP(Ta) with JANAF (1) enthalpies, H°(0 K) - H°(298.15 K), for Ta(g), Ta^+(g), and e^-(g). $\Delta_rH°$(Ta → Ta^+ + e^-, 298.15 K) differs from a room temperature threshold energy due to inclusion of these enthalpies and to threshold effects discussed by Rosenstock et al. (4). $\Delta_fH°$(298.15 K) should be changed by -6.197 kJ mol^{-1} if it is to be used in the ion convention that excludes the enthalpy of the electron.

Heat Capacity and Entropy

The information on electronic energy levels and quantum weights, given by Moore (2), is incomplete because many theoretically predicted levels have not been observed. Although we have only a few of the atomic energy levels for Ta^+(g), all levels listed by Moore (2), as well as estimated levels, are used in the calculation. The observed levels are too numerous to list completely. The calculations indicate that for Ta^+(g), the thermodynamic functions are independent of the estimated missing levels (for n=6), and the cut-off procedure up to 6000 K; the Gibbs energy function showing no significant variations at this temperature. The reported uncertainty is S°(298.15 K) is due to uncertainties in the relative ionic mass, and the fundamental constants. Extension of these calculations above 6000 K may require consideration of the higher excited states (n>6), and use of different fill and cut-off procedures (6)u

References

1. JANAF Thermochemical Tables: Ta(g), 12-31-72; e^-(g), 3-31-82.
2. C. E. Moore, U. S. Nat. Bur. Stand., NSRDS-NBS-35, Volume III, 1970 [Reprint of NBS Circular 467, Volume III, 1958].
3. E. R. Cohen and B. N. Taylor, J. Phys. Chem. Ref. Data 2, 663 (1973).
4. H. M. Rosenstock, K. Draxl et al., J. Phys. Chem. Ref. Data 6, Supp. 1 (1977).
5. R. D. Levin and S. G. Lias, U. S. Natl. Bur. Stand. NSRDS-NBS-71, 1982.
6. J. R. Downey, Jr., The Dow Chemical Company, AFOSR-TR-78-0960, Contract No. F44620-75-1-0048, 1978.

Enthalpy Reference Temperature = T_r = 298.15 K Standard State Pressure = p° = 0.1 MPa

	J $K^{-1}mol^{-1}$			kJ mol^{-1}			
T/K	$C_p°$	S°	-[G°-H°(T_r)]/T	H°-H°(T_r)	$\Delta_fH°$	$\Delta_fG°$	Log K_f
0	0.	0.	INFINITE	-6.338	1542.296		
100	20.787	160.112	202.703	-4.259			
200	21.242	174.589	185.431	-2.168			
250	22.067	179.411	183.759	-1.087			
298.15	23.109	183.385	183.385	0.	1549.150	1500.584	-262.896
300	23.151	183.528	183.385	0.043	1549.185	1500.283	-261.222
350	24.256	187.181	183.671	1.228	1550.136	1492.057	-222.677
400	25.237	190.485	184.320	2.466	1551.128	1483.693	-193.750
450	26.048	193.506	185.175	3.749	1552.152	1475.203	-171.237
500	26.708	196.286	186.149	5.069	1553.200	1466.597	-153.214
600	27.734	201.251	188.262	7.793	1555.342	1449.076	-126.153
700	28.581	205.591	190.434	10.610	1557.534	1431.192	-106.797
800	29.369	209.459	192.575	13.507	1559.776	1412.991	-92.259
900	30.120	212.962	194.648	16.482	1562.074	1394.506	-80.935
1000	30.818	216.172	196.642	19.530	1564.420	1375.762	-71.862
1100	31.441	219.139	198.554	22.643	1566.802	1356.781	-64.428
1200	31.971	221.899	200.386	25.815	1569.205	1337.582	-58.223
1300	32.399	224.475	202.141	29.034	1571.619	1318.182	-52.965
1400	32.725	226.889	203.824	32.291	1574.044	1298.596	-48.451
1500	32.956	229.155	205.438	35.576	1576.481	1278.836	-44.533
1600	33.103	231.287	206.987	38.880	1578.926	1258.914	-41.099
1700	33.181	233.297	208.476	42.194	1581.365	1238.838	-38.065
1800	33.204	235.194	209.908	45.514	1583.774	1218.620	-35.363
1900	33.188	236.989	211.287	48.834	1586.133	1198.269	-32.943
2000	33.145	238.690	212.615	52.151	1588.435	1177.795	-30.761
2100	33.085	240.306	213.895	55.462	1590.680	1157.208	-28.784
2200	33.017	241.844	215.131	58.767	1592.866	1136.515	-26.984
2300	32.947	243.310	216.325	62.066	1594.989	1115.724	-25.339
2400	32.881	244.710	217.478	65.357	1597.045	1094.842	-23.829
2500	32.821	246.051	218.595	68.642	1599.028	1073.876	-22.437
2600	32.768	247.338	219.676	71.921	1600.930	1052.832	-21.152
2700	32.724	248.574	220.723	75.196	1602.742	1031.716	-19.960
2800	32.689	249.763	221.739	78.467	1604.450	1010.535	-18.852
2900	32.660	250.910	222.725	81.734	1606.042	989.296	-17.819
3000	32.639	252.016	223.683	84.999	1607.502	968.003	-16.854
3100	32.623	253.086	224.615	88.262	1608.812	946.665	-15.951
3200	32.610	254.122	225.521	91.524	1609.951	925.286	-15.104
3300	32.601	255.125	226.403	94.784	1574.442	904.346	-14.315
3400	32.594	256.098	227.262	98.044	1575.596	884.023	-13.581
3500	32.586	257.043	228.099	101.303	1576.750	863.666	-12.890
3600	32.579	257.961	228.916	104.561	1577.903	843.276	-12.236
3700	32.569	258.853	229.713	107.818	1579.055	822.853	-11.617
3800	32.558	259.722	230.492	111.075	1580.206	802.400	-11.030
3900	32.544	260.567	231.252	114.330	1581.355	781.917	-10.473
4000	32.527	261.391	231.995	117.584	1582.504	761.404	-9.943
4100	32.506	262.194	232.722	120.835	1583.650	740.862	-9.439
4200	32.482	262.977	233.433	124.085	1584.794	720.292	-8.958
4300	32.453	263.741	234.129	127.331	1585.935	699.695	-8.500
4400	32.421	264.487	234.811	130.575	1587.074	679.072	-8.062
4500	32.386	265.215	235.478	133.816	1588.209	658.423	-7.643
4600	32.347	265.926	236.132	137.052	1589.340	637.748	-7.242
4700	32.305	266.622	236.774	140.285	1590.467	617.049	-6.858
4800	32.260	267.301	237.403	143.513	1591.590	596.327	-6.489
4900	32.212	267.966	238.020	146.737	1592.708	575.580	-6.136
5000	32.163	268.616	238.625	149.955	1593.821	554.811	-5.796
5100	32.112	269.253	239.219	153.169	1594.930	534.020	-5.469
5200	32.060	269.876	239.803	156.378	1596.033	513.207	-5.155
5300	32.008	270.486	240.376	159.581	1597.131	492.373	-4.853
5400	31.955	271.084	240.939	162.779	1598.224	471.518	-4.561
5500	31.904	271.670	241.493	165.972	1599.311	450.644	-4.280
5600	31.853	272.244	242.037	169.160	1600.394	429.749	-4.009
5700	31.805	272.807	242.572	172.343	1601.471	408.835	-3.747
5800	31.758	273.360	243.098	175.521	863.789	390.703	-3.519
5900	31.715	273.903	243.615	178.695	864.644	382.539	-3.387
6000	31.675	274.435	244.125	181.864	865.451	374.361	-3.259

PREVIOUS: CURRENT: March 1984 (1 bar)

Tantalum, Ion (Ta^+) $Ta_1^+(g)$

TANTALUM, ION (Ta^-) IDEAL GAS M_r = 180.94855

EA(Ta, g) = 0.322 ± 0.012 eV

S°(298.15 K) = 174.560 ± 0.002 J K^{-1} mol^{-1}

$\Delta_f H°$(0 K) = 750.403 ± 1.5 kJ mol^{-1}

$\Delta_f H°$(298.15 K) = [744.941] kJ mol^{-1}

Electonic Levels and Quantum Weights

State	ε_i, cm^{-1}	g_i
$^5D_0$	0.0	1
$^5D_1$	1070	3
$^5D_2$	2240	5

Enthalpy of Formation

$\Delta_f H°(Ta^-$, g, 0 K) is calculated from $\Delta_f H°$(Ta, g, 0 K) (1) using the adopted electron affinity of EA(Ta) = 0.322 ± 0.012 eV (31.068 ± 1.158 kJ mol^{-1}). This value, recommended by Hotop and Lineberger (2), is based on a laser photodetachment electron spectrometry study (3). Additional information on Ta^-(g) may be obtained in the critical discussions of Hotop and Lineberger (2, 4), Rosenstock et al. (5) and Massey (6).

$\Delta_f H°(Ta^-$, g, 298.15 K) is obtained from $\Delta_f H°$(Ta, g, 0 K) by using EA(Ta) with JANAF (1) enthalpies, H°(0 K) - H°(298.15 K), for Ta^-(g), Ta(g), and e^-(g). $\Delta_r H°(Ta^- \rightarrow Ta + e^-$, 298.15 K) differs from a room-temperature threshold energy due to inclusion of these enthalpies and to threshold effects discussed by Rosenstock et al. (5). $\Delta_f H°$(298.15 K) should be changed by +6.197 kJ mol^{-1} if it is to be used in the ion convention that excludes the enthalpy of the electron.

Heat Capacity and Entropy

The ground state electronic configuration for Ta^-(g) is given by Hotop and Lineberger (2, 4), Rosenstock et al (5), and Massey (6). The fine structure separation has been calculated by an isoelectronic extrapolation of ratios of fine structure separations (3) and is that recommended by Hotop and Lineberger (2).

References

1. JANAF Thermochemical Tables: Ta(g), 12-31-72; e^-(g), 3-31-82.
2. H. Hotop and W. C. Lineberger, J. Phys. Chem. Ref. Data, to be published, 1985.
3. C. S. Feigerle, R. R. Cordermann, S. V. Bobashev, and W. C. Lineberger, J. Chem. Phys. 74, 1580 (1981).
4. H. Hotop and W. C. Lineberger, J. Phys. Chem. Ref. Data 4, 539 (1975).
5. H. M. Rosenstock, K. Draxl et al., J. Phys. Chem. Ref. Data 6, Supp. 1 (1977).
6. H. S. W. Massey, "Negative Ions", 3rd ed., Cambridge University Press, Cambridge, 1976.

Tantalum, Ion (Ta^-) Ta_1^-(g)

Enthalpy Reference Temperature = T_r = 298.15 K Standard State Pressure = p° = 0.1 MPa

T/K	$C_p°$ (J $K^{-1}mol^{-1}$)	S° (J $K^{-1}mol^{-1}$)	$-[G°-H°(T_r)]/T$ (J $K^{-1}mol^{-1}$)	$H°-H°(T_r)$ (kJ mol^{-1})	$\Delta_f H°$ (kJ mol^{-1})	$\Delta_f G°$ (kJ mol^{-1})	Log K_f
0	0.	0.	INFINITE	-6.416	750.403		
100	20.787	150.978	194.351	-4.337			
200	21.456	165.484	176.691	-2.241			
250	22.780	170.401	174.954	-1.138			
298.15	24.558	174.560	174.560	0.	744.941	711.515	-124.654
300	24.632	174.712	174.561	0.045	744.901	711.308	-123.850
350	26.631	178.660	174.868	1.327	743.870	705.791	-105.334
400	28.486	182.340	175.575	2.706	742.924	700.417	-91.465
450	30.041	185.788	176.520	4.171	742.051	695.157	-80.692
500	31.236	189.018	177.610	5.704	741.234	689.991	-72.083
600	32.583	194.852	180.009	8.906	739.697	679.888	-59.189
700	32.816	199.904	182.499	12.184	738.192	670.038	-49.999
800	32.336	204.261	184.954	15.446	736.642	660.407	-43.120
900	31.479	208.022	187.313	18.638	735.000	650.975	-37.782
1000	30.476	211.288	189.551	21.737	733.240	641.732	-33.521
1100	29.464	214.145	191.660	24.733	731.348	632.672	-30.043
1200	28.511	216.667	193.641	27.631	729.320	623.790	-27.153
1300	27.649	218.914	195.500	30.439	727.165	615.082	-24.714
1400	26.885	220.935	197.246	33.164	724.902	606.545	-22.630
1500	26.216	222.766	198.887	35.819	722.551	598.172	-20.830
1600	25.633	224.439	200.432	38.411	720.127	589.959	-19.260
1700	25.127	225.977	201.890	40.948	717.631	581.900	-17.880
1800	24.688	227.401	203.269	43.438	715.053	573.989	-16.657
1900	24.306	228.725	204.574	45.887	712.384	566.225	-15.567
2000	23.972	229.963	205.813	48.301	709.626	558.603	-14.589
2100	23.680	231.126	206.991	50.683	706.785	551.122	-13.708
2200	23.424	232.221	208.113	53.038	703.863	543.777	-12.911
2300	23.199	233.258	209.184	55.369	700.862	536.568	-12.186
2400	23.000	234.241	210.208	57.679	697.779	529.490	-11.524
2500	22.823	235.176	211.188	59.970	694.611	522.543	-10.918
2600	22.666	236.068	212.128	62.244	691.351	515.724	-10.361
2700	22.525	236.921	213.030	64.504	687.990	509.033	-9.848
2800	22.399	237.737	213.898	66.750	684.517	502.468	-9.374
2900	22.286	238.521	214.734	68.984	680.918	496.029	-8.934
3000	22.184	239.275	215.539	71.207	677.180	489.717	-8.527
3100	22.091	240.001	216.317	73.421	673.283	483.532	-8.147
3200	22.008	240.701	217.068	75.626	669.208	477.476	-7.794
3300	21.931	241.377	217.794	77.823	628.478	472.020	-7.471
3400	21.862	242.031	218.498	80.012	624.405	467.340	-7.180
3500	21.799	242.664	219.179	82.195	620.325	462.780	-6.907
3600	21.740	243.277	219.840	84.372	616.239	458.337	-6.650
3700	21.687	243.872	220.482	86.543	612.148	454.007	-6.409
3800	21.638	244.449	221.105	88.710	608.052	449.787	-6.183
3900	21.593	245.011	221.711	90.871	603.951	445.676	-5.969
4000	21.551	245.557	222.300	93.028	599.845	441.670	-5.768
4100	21.512	246.089	222.874	95.182	595.736	437.766	-5.577
4200	21.476	246.607	223.433	97.331	591.623	433.963	-5.397
4300	21.443	247.112	223.977	99.477	587.506	430.258	-5.227
4400	21.412	247.604	224.509	101.620	583.386	426.649	-5.065
4500	21.383	248.085	225.027	103.759	579.263	423.133	-4.912
4600	21.356	248.555	225.534	105.896	575.138	419.709	-4.766
4700	21.331	249.014	226.029	108.031	571.009	416.375	-4.627
4800	21.307	249.463	226.512	110.163	566.879	413.128	-4.496
4900	21.285	249.902	226.985	112.292	562.746	409.968	-4.370
5000	21.264	250.332	227.448	114.420	558.610	406.892	-4.251
5100	21.245	250.752	227.900	116.545	554.473	403.898	-4.137
5200	21.227	251.165	228.344	118.669	550.334	400.986	-4.028
5300	21.209	251.569	228.778	120.790	546.193	398.154	-3.924
5400	21.193	251.965	229.204	122.911	542.051	395.399	-3.825
5500	21.178	252.354	229.621	125.029	537.907	392.722	-3.730
5600	21.163	252.735	230.031	127.146	533.761	390.120	-3.639
5700	21.149	253.110	230.432	129.262	529.614	387.591	-3.552
5800	21.136	253.478	230.827	131.376	-213.289	387.937	-3.494
5900	21.124	253.839	231.214	133.489	-217.652	398.340	-3.527
6000	21.112	254.194	231.594	135.601	-222.060	408.818	-3.559

PREVIOUS: CURRENT: March 1984 (1 bar)

Tantalum, Ion (Ta^-) Ta_1^-(g)

TITANIUM (Ti) (REFERENCE STATE) A_r = 47.88

0 -1166 K Crystal, alpha
1166 - 1939 K Crystal, beta
1939 - 3630.956 K Liquid
3630.956 - 6000 K Ideal Monatomic Gas

Refer to the individual tables for details

Titanium (Ti) Ti_1(ref)

Enthalpy Reference Temperature = T_r = 298.15 K Standard State Pressure = p° = 0.1 MPa

T/K	C_p°	S°	$-[G^\circ-H^\circ(T_r)]/T$	$H^\circ-H^\circ(T_r)$	$\Delta_f H^\circ$	$\Delta_f G^\circ$	Log K_f
	J K⁻¹mol⁻¹			kJ mol⁻¹			
0	0.	0.	INFINITE	-4.830	0.	0.	0.
100	14.334	8.261	50.955	-4.269	0.	0.	0.
200	22.367	21.227	32.989	-2.352	0.	0.	0.
298.15	25.238	30.759	30.759	0.	0.	0.	0.
300	25.276	30.915	30.760	0.047	0.	0.	0.
400	26.862	38.423	31.772	2.660	0.	0.	0.
500	27.877	44.534	33.733	5.401	0.	0.	0.
600	28.596	49.683	35.973	8.226	0.	0.	0.
700	29.135	54.134	38.257	11.114	0.	0.	0.
800	29.472	58.039	40.490	14.039	0.	0.	0.
900	30.454	61.561	42.639	17.030	0.	0.	0.
1000	32.074	64.848	44.697	20.151	0.	0.	0.
1100	34.334	68.006	46.673	23.466	0.	0.	0.
1166.000	36.175	70.058	47.938	25.791	ALPHA <--> BETA		
1166.000	29.245	73.636	47.938	29.963	TRANSITION		
1200	29.459	74.479	48.679	30.961	0.	0.	0.
1300	30.175	76.864	50.756	33.941	0.	0.	0.
1400	31.023	79.131	52.702	37.000	0.	0.	0.
1500	32.003	81.304	54.537	40.150	0.	0.	0.
1600	33.115	83.404	56.276	43.405	0.	0.	0.
1700	34.359	85.448	57.932	46.778	0.	0.	0.
1800	35.736	87.451	59.517	50.281	0.	0.	0.
1900	37.244	89.422	61.039	53.929	0.	0.	0.
1939.000	37.868	90.186	61.617	55.394	BETA <--> LIQUID		
1939.000	47.237	97.481	61.617	69.540	TRANSITION		
2000	47.237	98.944	62.734	72.421	0.	0.	0.
2100	47.237	101.249	64.513	77.145	0.	0.	0.
2200	47.237	103.446	66.233	81.869	0.	0.	0.
2300	47.237	105.546	67.897	86.592	0.	0.	0.
2400	47.237	107.557	69.508	91.316	0.	0.	0.
2500	47.237	109.485	71.069	96.040	0.	0.	0.
2600	47.237	111.338	72.582	100.764	0.	0.	0.
2700	47.237	113.120	74.051	105.487	0.	0.	0.
2800	47.237	114.838	75.477	110.211	0.	0.	0.
2900	47.237	116.496	76.863	114.935	0.	0.	0.
3000	47.237	118.097	78.211	119.659	0.	0.	0.
3100	47.237	119.646	79.523	124.382	0.	0.	0.
3200	47.237	121.146	80.800	129.106	0.	0.	0.
3300	47.237	122.600	82.045	133.830	0.	0.	0.
3400	47.237	124.010	83.259	138.554	0.	0.	0.
3500	47.237	125.379	84.443	143.277	0.	0.	0.
3600	47.237	126.710	85.598	148.001	0.	0.	0.
3630.956	47.237	127.114	85.951	149.463	LIQUID <--> IDEAL GAS		
3630.956	34.219	240.028	85.951	559.447	FUGACITY = 1 bar		
3700	34.613	240.676	88.832	561.823	0.	0.	0.
3800	35.173	241.607	92.840	565.313	0.	0.	0.
3900	35.721	242.527	96.667	568.857	0.	0.	0.
4000	36.254	243.439	100.324	572.456	0.	0.	0.
4100	36.772	244.340	103.826	576.108	0.	0.	0.
4200	37.273	245.232	107.182	579.810	0.	0.	0.
4300	37.757	246.115	110.403	583.562	0.	0.	0.
4400	38.223	246.988	113.497	587.361	0.	0.	0.
4500	38.670	247.852	116.473	591.206	0.	0.	0.
4600	39.097	248.707	119.339	595.094	0.	0.	0.
4700	39.505	249.552	122.100	599.025	0.	0.	0.
4800	39.892	250.388	124.764	602.995	0.	0.	0.
4900	40.259	251.214	127.336	607.002	0.	0.	0.
5000	40.606	252.031	129.822	611.046	0.	0.	0.
5200	41.240	253.637	134.554	619.232	0.	0.	0.
5400	41.796	255.204	138.993	627.537	0.	0.	0.
5600	42.277	256.733	143.171	635.945	0.	0.	0.
5800	42.686	258.224	147.113	644.443	0.	0.	0.
6000	43.029	259.677	150.841	653.015	0.	0.	0.

PREVIOUS: June 1979 (1 atm) CURRENT: June 1979 (1 bar)

Titanium (Ti) Ti_1(ref)

TITANIUM, ALPHA (α-Ti) CRYSTAL A_r = 47.88

$S°(298.15\ K) = 30.759 \pm 0.2\ J\ K^{-1}\ mol^{-1}$

$T_{trs} = 1166 \pm 10\ K\ (\alpha-\beta)$

$\Delta_f H°(0\ K) = 0\ kJ\ mol^{-1}$

$\Delta_f H°(298.15\ K) = 0\ kJ\ mol^{-1}$

$\Delta_{trs} H° = 0.997 \pm 0.03\ kJ\ mol^{-1}$

Enthalpy of Formation

Zero by definition

Heat Capacity and Entropy

The adopted thermal functions for α-Ti(cr) are derived from the studies of Collings and Ho (1), Clusius and Franzosini (2), Stalinski and Bieganski (3), Scott (4), Backhurst (5) and Cezairliyan (6). The mathematical and graphical treatment of these four studies yields a continuous and smooth heat capacity curve.

There are ten studies (1, 7 - 15) on the heat capacity of titanium in the range 1-5 K. Wolcott (8) and Aven et al. (7) reported the experimental data. In the other studies, an equation (with two constants) was given to describe the entire set of experimental data. The constants given by Collings and Ho (1) agree with the experimental data of Wolcott (8) and of Aven et al. (7). The maximum difference in the calculated values for S°(5 K) is 0.00071 cal $K^{-1}\ mol^{-1}$, with the adopted value being S°(5 K) = 0.0043 cal $K^{-1}\ mol^{-1}$. Similarly for Cp°(5 K) the maximum difference is 0.00211 cal $K^{-1}\ mol^{-1}$, with the adopted value being Cp° (5 K) = 0.0049 cal $K^{-1}\ mol^{-1}$.

In the region above 5 K and below 360 K, there are seven heat capacity studies (2, 3, 8, 9 - 18. The adopted values were based on the work of Clusius and Franzosini (2) and Stalinski and Bieganski (3). The maximum deviation of all studies from the adopted is 2.5% over this temperature range.

Above 300 K, there are eleven heat capacity studies (4, 5, 6, 18 - 25) and seven enthalpy studies (26 - 32). The heat capacity studies all have a similar temperature dependence in comparison with the adopted values. However, the studies of Holland (19), Affortit (23), Arutyunov (24), and Rumyantsev (25) lie roughly 9-33% higher than the adopted values. There is much scatter in all studies within 50 K of the α-β transition temperature. The adopted values were based on Scott (4), Backhurst (5), and Cezairliyan (6). The enthalpy studies show a large scatter within each study as well as a general lack of agreement between the studies.

Transition Data

Refer to the β-crystal table for details.

Sublimation Data

There are no sublimation studies involving α-Ti(cr). The enthalpy of sublimation is calculated as the difference between the enthalpy of formation of the monatomic gas and the enthalpy of formation of α-Ti(cr).

References

1. E. W. Collins and J. C. Ho, Phys. Rev. B4, 235 (1970).
2. K. Clusius and P. Franzosini, Z. Physik. Chem. (Frankfurt) 16, 194 (1958).
3. B. Stalinski and Z. Bieganski, Rocznike Chem. 35, 273 (1961).
4. J. L. Scott, AEC Rept. ORNL-2328, 1957.
5. I. Backhurst, J. Iron Steel Inst. (London) 189, 124 (1958).
6. A. Cezairliyan and A. P. Miiller, High-Temp. -High Pressures 9, 319 (1977).
7. I. Estermann, S. A. Friedberg and J. E. Goldman, Phys. Rev. 87, 582 (1952).
8. M. H. Aven, R. S. Craig, T. R. Waite and W. E. Wallace, Phys. Rev. 102, 1263 (1956).
9. N. M. Wolcott, Phil. Mag. 2, 1246 (1957).
10. G. D. Kneip, J. O. Betterton, Jr. and J. O. Scarbrough, Phys. Rev. 139, 1687 (1963).
11. R. R. Hake and J. A. Cape, Phys. Rev. 135A, 1151 (1964).
12. F. Heiniger and J. Muller, Phys. Rev. 134, A1407 (1964).
13. S. Danner and G. Dummer, Z. Phys. 222, 243 (1969).
14. E. W. Collins and J. C. Ho, Phys. Rev. B4, 349 (1971).
15. K. L. Agarwall and J. O. Betterton, Jr., Proc. Int. Conf. Low Temp. Phys. 13, 597 (1972).
16. K. K. Kelley, Ind. Eng. Chem. 36, 865 (1944).
17. C. W. Kothen and H. L. Johnston, J. Am. Chem. Soc. 75, 3101 (1953).
18. D. L. Burk, I. Estermann, and S. A. Friedberg, Z. Physik. Chem. (Frankfurt) 16, 183 (1958).
19. L. R. Holland, J. Applied Phys. 34, 2350 (1963).
20. R. Kohlhass, M. Braun and G. Vollmer, Z. Naturforsch. 20a, 1077 (1965).
21. V. O. Shestopal, Fiz. Tverd. Tela 7, 3461 (1965).
22. G. G. Zaitseva and Ya. A. Kraftmakher, Zh. Prikl. Mekhan. I Tekh. Fiz. 3, 117 (1965).
23. C. Affortit, Commis. Energ. At. Rapp. CEA-R 3287, 1967.
24. A. V. Arutyunov, S. N. Banchilaand L. P. Filippov, Teplofiz. Vys. Temp. 9, 535 (1979).
25. A. V. Rumyantsev, I. N. Makarenko and S. N. Banchila, J. Eng. Phys. 36, 377 (1979).
26. F. M. Jaeger, E. Rosenbohm and R. Fonteyne, Rec. Trav. Chim. 55, 615 (1936).
27. C. W. Kothen, Ohio State University, Thesis, 1952.
28. Yu. M. Golutvin, Russ. J. Phys. Chem. 33, 164 (1959).
29. N. N. Serebryannikov and P. V. Gel'd, Izvest. Vysshikh. Ucheb. Zavedenii, Tsvetnaya Met. 4, 80 (1961).
30 P. V. Gel'd and Yu. V. Putintsev, Tr. Ural. Politekhn. 186, 199 91970).
31. J. A. Treverton and J. L. Margave, H. Chem. Thermodyn. 3, 473 (1971).
32. B. Ya. Berezin, V. Ya. Chekhovskoy, S. A. Katz and M. M. Kenisarin, Proc. Symp. Thermophys. Prop. 6th, 263 91973.

Titanium, Alpha (Ti) $Ti_1(cr)$

Enthalpy Reference Temperature = T_r = 298.15 K; Standard State Pressure = p° = 0.1 MPa

T/K	$C_p°$ (J K⁻¹mol⁻¹)	S° (J K⁻¹mol⁻¹)	$-[G°-H°(T_r)]/T$ (J K⁻¹mol⁻¹)	$H°-H°(T_r)$ (kJ mol⁻¹)	$\Delta_f H°$ (kJ mol⁻¹)	$\Delta_f G°$ (kJ mol⁻¹)	Log K_f
0	0.	0.	INFINITE	-4.830	0.	0.	0.
100	14.334	8.261	50.955	-4.269	0.	0.	0.
200	22.367	21.227	32.989	-2.352	0.	0.	0.
250	24.074	26.414	31.169	-1.189	0.	0.	0.
298.15	25.238	30.759	30.759	0.	0.	0.	0.
300	25.276	30.915	30.760	0.047	0.	0.	0.
350	26.169	34.882	31.071	1.334	0.	0.	0.
400	26.862	38.423	31.772	2.660	0.	0.	0.
450	27.418	41.620	32.692	4.018	0.	0.	0.
500	27.877	44.534	33.733	5.401	0.	0.	0.
600	28.596	49.683	35.973	8.226	0.	0.	0.
700	29.135	54.134	38.257	11.114	0.	0.	0.
800	29.472	58.039	40.490	14.039	0.	0.	0.
900	30.454	61.561	42.639	17.030	0.	0.	0.
1000	32.074	64.848	44.697	20.151	0.	0.	0.
1100	34.334	68.006	46.673	23.466	0.	0.	0.
1166.000	36.175	70.058	47.938	25.791	-------- ALPHA <--> BETA -------		
1200	37.233	71.113	48.580	27.039	-3.922	0.118	-0.005
1300	40.771	74.228	50.433	30.934	-3.007	0.420	-0.017
1400	44.948	77.399	52.245	35.214	-1.786	0.640	-0.024
1500	49.764	80.660	54.030	39.945	-0.205	0.760	-0.026
1600	55.220	84.043	55.800	45.189	1.784	0.762	-0.025
1700	61.314	87.570	57.564	51.010	4.232	0.625	-0.019

PREVIOUS: September 1966 CURRENT: June 1979

Titanium, Alpha (Ti) $Ti_1(cr)$

TITANIUM, BETA (β-Ti) CRYSTAL A_r = 47.88

$S°(298.15\ K) = [38.122]\ J\ K^{-1}\ mol^{-1}$
$T_{trs} = 1166 \pm 10\ K\ (\alpha - \beta)$
$T_{fus} = 1939 \pm 10\ K\ (\beta - \ell)$

$\Delta_f H°(298.15\ K) = [6.860]\ kJ\ mol^{-1}$
$\Delta_{trs} H° = 4.172 \pm 0.13\ kJ\ mol^{-1}$
$\Delta_{fus} H° = 14.146 \pm 5.0\ kJ\ mol^{-1}$

Enthalpy of Formation

The enthalpy of formation is calculated from that of the α-crystal by addition of $\Delta_{trs}H°$ and the difference in enthalpy, H°(1166 K) - H°(298.15 K), between the α-crystal and the β-crystal.

Heat Capacity and Entropy

For the β-crystal there are ten heat capacity studies and six enthalpy studies. The discussion on heat capacity and enthalpy on the α-Ti(cr) table also applies to the β-Ti(cr).

The entropy at 298.15 K is calculated in a manner similar to that for the enthalpy of formation.

Phase Data

Titanium, at ambient pressures, exists in two crystal modifications. The low temperature form, α-Ti, is hexagonal close packed, an hcp (A3) structure isotypic with Mg. The high temperature form, β-Ti, is body centered cubic, a bcc (A2) structure isotypic with W.

Transition Data

The following table indicates the values of transition temperatures and enthalpies of transition reported by various investigators. The value of T_{trs} adoped is 1166±10 K, and the value of $\Delta_{trs}H°$ is selected as 0.997±0.03 kcal mol^{-1}, based on the work of Cezairliyan and Miller (1).

Source	T_{trs}/K	$\Delta_{trs}H°$, kcal mol^{-1}	Purity, %	Method
Fast (2)	1158±10		Iodide	electrical resistance
Greiner and Ellis (3)	1158±2		99.9	electrical resitance
McQuillan (4)	1155.7±1	0.67	99.93	hydrogen solubility
Worner (5)	1158±2		99.93	thermoelectric power
Dewez (6)	1155±4		Iodide	cooling curve
Kothen (7)	1154	0.943	99.96	drop calorimetry
Edwards et al. (8)	1157±3.5		99.88	cooling curve
Schofield (9)	1158	0.814	--	rate of heating
Scott (10)	1156±2	0.978±0.025	Iodide	adiabatic calorimetry
Backhurst (11)	1165-1179	0.88	Commercial	adiabatic calorimetry
Golutvin (12)	1155	0.82±.020	Iodide	drop calorimetry
Kohlhaas et al. (13)	1167	0.992	99.8	adiabatic calorimetry
Gel'd and Putintsev (14)		0.998		
Martynyuk and Tsapkov (15)		1.028	99.74	pulsed heating
Harmelin and Lehr (16)		1.007		DTA
Etchessahar and Debuiqne (17)	1154			dilatometric
Cezairliyan and Miller (1)	1166	0.997	99.9+	electrical resistance

Fusion Data

Refer to the liquid table for details.

Sublimation Data

There are three sublimation studies involving β-Ti(cr). They are summarized on the ideal gas table. The enthalpy of sublimation for β-Ti(cr) is calculated as the difference in the enthalpies of formation at 298.15 K of Ti(g) and β-Ti(cr).

References

1. A. Cezairliyan and A. P. Miller, J. Res. Nat. Bur. Stand. 83, 127 (1978).
2. J. D. Fast, Rec. Trav. Chim. 58, 973 (1938).
3. E. S. Greiner and W. C. Ellis, Am. Inst. Mining Met. Engrs. Inst. Metals Div., Metals Technol. 15, Tech. Pub. No. 2466(9), 1948.
4. A. D. McQuillan, J. Inst. Metals 78, 249 (1950).
5. H. W. Worner, Proc. Australian Inst. Mining and Met. 158/159/, 89 (1950).
6. P. Duwez, J. Metals 3, Trans. AIME, 765 (1951).
7. C. W. Kothan, Ohio State University, Thesis, 1952.
8. J. W. Edwards, H. L. Johnston and W. E. Ditmars, J. Am. Chem. Soc. 75, 2467 (1953).
9. T. H. Schofield, J. Inst. Metals 85, 68 (1956).
10. J. L. Scott, AEC Rept. ORNL-2328, 1957.
11. I. Backhurst, J. Iron Steel Inst. (London) 189, 124 (1958).
12. Y. M. Golutvin, Russ. J. Phys. Chem. 33, 164 (1959).
13. R. Kohlhass, M. Braun and O. Vollmer, Z. Naturforsch. 20a, 1077 (1965).
14. R. V. Gel'd and Yu. V. Putintsev, Tr. Ural. Politekhn. Inst. 186, 199 (1970).
15. M. M. Martynyuk and V. I. Tsapkov, Izv. Akad. Nauk SSR, Metal 2, 181 (1974).
16. M. Harmelin and P. Lehr, C. R. Hebd. Seances Acad. Sci. C 280, 1011 (1975).
17. E. Etchessar and J. Debuigne, C. R. Hebd. Seances Acad. Sci. C 283, 63 (1976).

Titanium, Beta (Ti) $Ti_1(cr)$

Enthalpy Reference Temperature = T_r = 298.15 K   Standard State Pressure = p° = 0.1 MPa

T/K	$C_p°$ (J K⁻¹mol⁻¹)	S°	$-[G°-H°(T_r)]/T$	$H°-H°(T_r)$ (kJ mol⁻¹)	$\Delta_f H°$	$\Delta_f G°$	Log K_f
0							
100							
200							
298.15	23.995	38.122	38.122	0.	6.860	4.665	-0.817
300	24.006	38.271	38.123	0.044	6.858	4.652	-0.810
350	24.309	41.994	38.416	1.252	6.779	4.290	-0.640
400	24.611	45.260	39.071	2.475	6.675	3.941	-0.515
450	24.914	48.176	39.924	3.713	6.556	3.606	-0.419
500	25.216	50.816	40.883	4.967	6.427	3.285	-0.343
600	25.821	55.467	42.937	7.519	6.153	2.682	-0.234
700	26.426	59.493	45.020	10.131	5.877	2.126	-0.159
800	27.031	63.061	47.057	12.804	5.625	1.607	-0.105
900	27.636	66.280	49.017	15.537	5.368	1.120	-0.065
1000	28.241	69.223	50.892	18.331	5.040	0.665	-0.035
1100	28.846	71.943	52.684	21.185	4.580	0.248	-0.012
1166.000	29.245	73.636	53.822	23.102	-------- ALPHA <--> BETA -------		
1200	29.459	74.479	54.396	24.100	0.	0.	0.
1300	30.175	76.864	56.033	27.081	0.	0.	0.
1400	31.023	79.131	57.603	30.140	0.	0.	0.
1500	32.003	81.304	59.111	33.290	0.	0.	0.
1600	33.115	83.404	60.564	36.544	0.	0.	0.
1700	34.359	85.448	61.968	39.917	0.	0.	0.
1800	35.736	87.451	63.328	43.421	0.	0.	0.
1900	37.244	89.422	64.650	47.069	0.	0.	0.
1939.000	37.868	90.186	65.155	48.533	------- BETA <--> LIQUID -------		
2000	38.885	91.374	65.937	50.874	-14.687	0.454	-0.012
2100	40.657	93.314	67.195	54.850	-15.435	1.230	-0.031
2200	42.562	95.248	68.426	59.010	-15.998	2.037	-0.048
2300	44.599	97.185	69.634	63.367	-16.365	2.866	-0.065
2400	46.768	99.128	70.822	67.934	-16.522	3.706	-0.081
2500	49.069	101.084	71.994	72.725	-16.455	4.549	-0.095

PREVIOUS: September 1966   CURRENT: June 1979

Titanium, Beta (Ti) $Ti_1(cr)$

TITANIUM (Ti) LIQUID A_r = 47.88

$S°(298.15\ K)$ = [39.181] $J\ K^{-1}\ mol^{-1}$ $\Delta_f H°(298.15\ K)$ = [13.652] $kJ\ mol^{-1}$

T_{fus} = 1936 ± 10 K $\Delta_{fus}H°$ = 14.146 ± 5.0 $kJ\ mol^{-1}$

Enthalpy of Formation

The enthalpy of formation is calculated from that of the β-crystal by addition of $\Delta_{fus}H°$ and the difference in enthalpy, H°(1936 K) - H°(198.15 K), between the β-crystal and the liquid.

Heat Capacity and Entropy

There are two enthalpy studies (1, 2) covering the liquid region of titanium. The work of Treverton and Margrave (1) was adopted. A glass transition is assumed at 1300 K below which extrapolated β-Ti heat capacity values are used. The entropy is calculated in a manner similar to that used for the enthalpy of formation.

Fusion Data

The investigators reporting melting temperatures and heats of melting are listed below. The value of T_{fus} adopted is 1939±10 K, and the value of $\Delta_{fus}H°$ is selected as 3.381±0.03 kcal mol^{-1}, based on the work of Berezin et al. [164].

Source	T_{fus}/K	$\Delta_{fus}H°$ kcal mol^{-1}
Burgess and Waltenburg (3)	2068	
Fast (4)	1993	
Adenstedt et al. (5)	1973	
Hansen et al. (6)	1993	
Maykuth et al. (7)	1953	
Schofield and Bacon (8)	1933	
Oriani and Jones (9)	1945	
Deardorff and Hayes (10)	1941	
Westrum and Feick (11)	1933	
Elyutin et al. (12)	1941	4.1±0.1
Treverton and Margrave (1)	1943	3.161±0.12
Berezin et al. (13)	1939	3.381±0.115

Vaporization Data

The two vaporization studies are summarized on the ideal gas table. The boiling point is calculated as that temperature for which ΔG° = 0 for Ti(ℓ) = Ti(g). $\Delta_{vap}H°$ is the corresponding enthalpy change. T_{vap} is the temperature at which fugacity is one bar. The normal boiling point (p=1 bar) would be a slightly lower temperature.

References

1. J. A. Treverton and J. L. Margrave, J. Chem. Thermodyn. 3, 473 (1971).
2. B. Ya. Berezin, V. Ya. Chekhovskoy, S. A. Kats and M. M. Kenisarin, Proc. Symp. Thermophys. Prop. 6th, 263 (1973).
3. G. K. Burgess and R. G. Waltenberg, Z. Anorg. Allgem. Chem. 82, 361 (1913).
4. J. D. Fast, Z. Anorg. Allgem. Chem. 241, 41 (1939).
5. H. K. Adenstedt, J. R. Pequingnot and J. M. Raymer, Trans. Am. Soc. Metals 44, 990 (1952).
6. M. Hansen, H. D. Kessler and D. H. McPherson, Trans. Am. Soc. Metals, Eng. 44, 518(1952).
7. D. J. Maykuth, H. R. Ogden and R. I. Jaffee, Trans. Am. Inst. Min. Met. Eng. 197, 231 (1953).
8. T. H. Schofield and A. E. Bacon, J. Inst. Metals 82, 167 (1953).
9. R. A. Oriani and T. S. Jones, Rev. Sci. Instr. 25, 248 (1954).
10. D. K. Deardorff and E. T. Hayes, J. Metals, AIME Trans. 206, 509 (1956).
11. E. F. Westrum, Jr. and G. Feick, ASD-TDR-62-204, Part II, 1963.
12. V. P. Elyutin, M. A. Maurakh, VC. Ya. Mishuk and G. M. Sverdlov, Izv. Vyssh. Ucheb. Zaved., Tsvet. Met. 10, 60 (1967).
13. B. Ya. Berezin, V. Ya. Chekhovskoi and S. A. Kats, Conf. Int. Thermodyn. Chim. (C. R.) 4, 77 (1975).

Titanium (Ti) $Ti_1(l)$

Enthalpy Reference Temperature = T_r = 298.15 K   Standard State Pressure = p° = 0.1 MPa

	J $K^{-1}mol^{-1}$			kJ mol^{-1}			
T/K	$C_p°$	S°	$-[G°-H°(T_r)]/T$	$H°-H°(T_r)$	$\Delta_f H°$	$\Delta_f G°$	Log K_f
0							
100							
200							
298.15	27.638	39.181	39.181	0.	13.652	11.141	-1.952
300	27.639	39.352	39.181	0.051	13.657	11.126	-1.937
350	27.667	43.614	39.518	1.434	13.752	10.696	-1.596
400	27.694	47.310	40.266	2.818	13.810	10.255	-1.339
450	27.722	50.574	41.234	4.203	13.838	9.809	-1.139
500	27.750	53.496	42.316	5.590	13.842	9.360	-0.978
600	27.805	58.560	44.614	8.368	13.794	8.468	-0.737
700	27.860	62.851	46.921	11.151	13.689	7.587	-0.566
800	27.916	66.573	49.150	13.939	13.552	6.725	-0.439
900	28.104	69.871	51.272	16.739	13.361	5.882	-0.341
1000	28.423	72.847	53.283	19.564	13.065	5.066	-0.265
1100	28.875	75.576	55.188	22.428	12.614	4.286	-0.204
1200	29.459	78.113	56.993	25.343	8.035	3.674	-0.160
1300	30.175	80.498	58.711	28.324	8.035	3.311	-0.133
1400	31.023	82.765	60.348	31.383	8.035	2.948	-0.110
1460.000	31.595	84.078	61.297	33.261	GLASS <--> LIQUID TRANSITION		
1460.000	47.237	84.078	61.297	33.261			
1500	47.237	85.355	61.921	35.150	8.652	2.576	-0.090
1600	47.237	88.404	63.482	39.874	10.121	2.122	-0.069
1700	47.237	91.267	65.033	44.598	11.473	1.580	-0.049
1800	47.237	93.967	66.566	49.321	12.693	0.962	-0.028
1900	47.237	96.521	68.076	54.045	13.768	0.281	-0.008
1939.000	47.237	97.481	68.658	55.887	BETA <--> LIQUID		
2000	47.237	98.944	69.560	58.769	0.	0.	0.
2100	47.237	101.249	71.014	63.493	0.	0.	0.
2200	47.237	103.446	72.439	68.216	0.	0.	0.
2300	47.237	105.546	73.833	72.940	0.	0.	0.
2400	47.237	107.557	75.197	77.664	0.	0.	0.
2500	47.237	109.485	76.530	82.388	0.	0.	0.
2600	47.237	111.338	77.833	87.111	0.	0.	0.
2700	47.237	113.120	79.107	91.835	0.	0.	0.
2800	47.237	114.838	80.353	96.559	0.	0.	0.
2900	47.237	116.496	81.571	101.283	0.	0.	0.
3000	47.237	118.097	82.762	106.006	0.	0.	0.
3100	47.237	119.646	83.927	110.730	0.	0.	0.
3200	47.237	121.146	85.067	115.454	0.	0.	0.
3300	47.237	122.600	86.182	120.178	0.	0.	0.
3400	47.237	124.010	87.274	124.901	0.	0.	0.
3500	47.237	125.379	88.343	129.625	0.	0.	0.
3600	47.237	126.710	89.391	134.349	0.	0.	0.
3630.956	47.237	127.114	89.711	135.811	FUGACITY = 1 bar		
3700	47.237	128.004	90.417	139.072	-409.098	7.788	-0.110
3800	47.237	129.264	91.423	143.796	-407.864	19.039	-0.262
3900	47.237	130.491	92.409	148.520	-406.685	30.258	-0.405
4000	47.237	131.687	93.376	153.244	-405.560	41.447	-0.541
4100	47.237	132.853	94.324	157.967	-404.488	52.609	-0.670
4200	47.237	133.991	95.255	162.691	-403.467	63.745	-0.793
4300	47.237	135.103	96.169	167.415	-402.495	74.858	-0.909
4400	47.237	136.189	97.066	172.139	-401.570	85.948	-1.020
4500	47.237	137.250	97.948	176.862	-400.691	97.018	-1.126

PREVIOUS: September 1966   CURRENT: June 1979

Titanium (Ti) $Ti_1(l)$

TITANIUM (Ti) ALPHA-BETA-LIQUID A_r = 47.88

0 to 1166 K Crystal, alpha
1166 to 1939 K Crystal, beta
above 1939 K Liquid

Refer to the individual tables for details

Titanium (Ti) $Ti_1(cr,l)$

Enthalpy Reference Temperature = T_r = 298.15 K Standard State Pressure = $p°$ = 0.1 MPa

T/K	$C_p°$ (J K⁻¹mol⁻¹)	$S°$ (J K⁻¹mol⁻¹)	$-[G°-H°(T_r)]/T$ (J K⁻¹mol⁻¹)	$H°-H°(T_r)$ (kJ mol⁻¹)	$\Delta_f H°$ (kJ mol⁻¹)	$\Delta_f G°$ (kJ mol⁻¹)	Log K_f
0	0.	0.	INFINITE	-4.830	0.	0.	0.
100	14.334	8.261	50.955	-4.269	0.	0.	0.
200	22.367	21.227	32.989	-2.352	0.	0.	0.
250	24.074	26.414	31.169	-1.189	0.	0.	0.
298.15	25.238	30.759	30.759	0.	0.	0.	0.
300	25.276	30.915	30.760	0.047	0.	0.	0.
350	26.169	34.882	31.071	1.334	0.	0.	0.
400	26.862	38.423	31.772	2.660	0.	0.	0.
450	27.418	41.620	32.692	4.018	0.	0.	0.
500	27.877	44.534	33.733	5.401	0.	0.	0.
600	28.596	49.683	35.973	8.226	0.	0.	0.
700	29.135	54.134	38.257	11.114	0.	0.	0.
800	29.472	58.039	40.490	14.039	0.	0.	0.
900	30.454	61.561	42.639	17.030	0.	0.	0.
1000	32.074	64.848	44.697	20.151	0.	0.	0.
1100	34.334	68.006	46.673	23.466	0.	0.	0.
1166.000	36.175	70.058	47.938	25.791	ALPHA <--> BETA		
1166.000	29.245	73.636	47.938	29.963	TRANSITION		
1200	29.459	74.479	48.679	30.961	0.	0.	0.
1300	30.175	76.864	50.756	33.941	0.	0.	0.
1400	31.023	79.131	52.702	37.000	0.	0.	0.
1500	32.003	81.304	54.537	40.150	0.	0.	0.
1600	33.115	83.404	56.276	43.405	0.	0.	0.
1700	34.359	85.448	57.932	46.778	0.	0.	0.
1800	35.736	87.451	59.517	50.281	0.	0.	0.
1900	37.244	89.422	61.039	53.929	0.	0.	0.
1939.000	37.868	90.186	61.617	55.394	BETA <--> LIQUID		
1939.000	47.237	97.481	61.617	69.540	TRANSITION		
2000	47.237	98.944	62.734	72.421	0.	0.	0.
2100	47.237	101.249	64.513	77.145	0.	0.	0.
2200	47.237	103.446	66.233	81.869	0.	0.	0.
2300	47.237	105.546	67.897	86.592	0.	0.	0.
2400	47.237	107.557	69.508	91.316	0.	0.	0.
2500	47.237	109.485	71.069	96.040	0.	0.	0.
2600	47.237	111.338	72.582	100.764	0.	0.	0.
2700	47.237	113.120	74.051	105.487	0.	0.	0.
2800	47.237	114.838	75.477	110.211	0.	0.	0.
2900	47.237	116.496	76.863	114.935	0.	0.	0.
3000	47.237	118.097	78.211	119.659	0.	0.	0.
3100	47.237	119.646	79.523	124.382	0.	0.	0.
3200	47.237	121.146	80.800	129.106	0.	0.	0.
3300	47.237	122.600	82.045	133.830	0.	0.	0.
3400	47.237	124.010	83.259	138.554	0.	0.	0.
3500	47.237	125.379	84.443	143.277	0.	0.	0.
3600	47.237	126.710	85.598	148.001	0.	0.	0.
3630.956	47.237	127.114	85.951	0.149	------- FUGACITY = 1 bar -------		
3700	47.237	128.004	86.727	152.725	-409.098	7.788	-0.110
3800	47.237	129.264	87.830	157.449	-407.864	19.039	-0.262
3900	47.237	130.491	88.908	162.172	-406.685	30.258	-0.405
4000	47.237	131.687	89.963	166.896	-405.560	41.447	-0.541
4100	47.237	132.853	90.995	171.620	-404.488	52.609	-0.670
4200	47.237	133.991	92.005	176.343	-403.467	63.745	-0.793
4300	47.237	135.103	92.994	181.067	-402.495	74.857	-0.909
4400	47.237	136.189	93.964	185.791	-401.570	85.948	-1.020
4500	47.237	137.250	94.914	190.515	-400.691	97.018	-1.126

PREVIOUS: CURRENT: June 1979

Titanium (Ti) $Ti_1(cr,l)$

TITANIUM (Ti) IDEAL GAS A_r = 47.88

IP(Ti, g) = 55010 ± 100 cm^{-1}
S°(298.15K) = 180.297 ± 0.03 J K^{-1} mol^{-1}

$\Delta_f H°$(0 K) = 470.92 ± 16.7 kJ mol^{-1}
$\Delta_f H°$(298.15K) = 473.63 ± 16.7 kJ mol^{-1}

Electronic Levels and Quantum Weights

State	ε_i, cm^{-1}	g_i
$^3F_2$	0.00	5
$^3F_3$	170.132	7
$^3F_4$	386.874	9
.	.	.
.	.	.
.	.	.
.	.	.
.	.	.
$^1P_1$	53663.32	3
IP	55010	

Enthalpy of Formation

The value adopted for the enthalpy of formation of Ti(g) is that recommended by CODATA (1). It is obtained from a 2nd and 3rd law evaluation of the vapor pressure data of Wu and Wahlbeck (2), Berstein and Kaufman (3), Koch et al. (4), Strassmair and Stark (5), and Edwards, Johnston, and Ditmars (6).

Source	Reaction	T/K	Data Points	$\Delta_r H°$(298.15K), kcal mol^{-1} 2nd Law	3rd Law	Drift cal K^{-1} mol^{-1}	$\Delta_f H°$(298.15K) kcal mol^{-1}
Edwards et al. (6)	a	1587 - 1764	8	112.0±1.1	111.5±0.2	-0.3±0.6	113.1
Strassmair and Stark (5)	a	1554 - 1772	8	111.4±3.5	109.8±0.8	-1.0±2.1	111.4
Wu and Wahlbeck (2)	a	1830 - 1923	16	112.2±9.3	111.6±1.0	-0.3±4.9	113.2
Koch et al (4)	b	1955 - 2379	21*	110.9±2.4	109.2±1.3	-0.8±1.1	112.46
Bernstein and Kaufman (3)	b	2000 - 3000	Smooth	110.6	109.0	-0.6	112.3

*Two points deleted due to a statistical test   Reaction: a) Ti(β) = Ti(g)   b) Ti(ℓ) = Ti(g)

Heat Capacity and Entropy

The electronic energy levels are given in the recent compilation by Corliss and Sugar (7). Although we have listed only the ground state, first two excited states, the highest observed excited state, and the ionization potential for Ti(g), all levels listed by Corliss and Sugar (7), as well as estimated levels, are used in the calculation. The observed levels are too numerous to list completely. In our calculations, the missing levels for n=4, n=5, and n=6 have been arbitrarily added near the ionization limit. Our calculations indicate that for Ti(g) the thermochemical functions are independent of the estimated missing levels (for n=4,5,6) and the cut-off procedure up to 3000 K. The Gibbs energy function is essentially unaffected (<0.002 cal K^{-1} mol^{-1}) at 6000 K. The reported uncertainty in S°(298.15K) is due to uncertainties in the relative atomic mass and the fundamental constants. Extension of these calculations above 6000 K may require consideration of the higher excited states (n>6), more precise estimation of missing states (all n), and utilization of proper fill and cut-off procedures (8). In fact, the inclusion of some higher states (n=4,5,6) and consideration of various cut-off procedures leads to calculational differences in the Gibbs energy function of rougly 0.1 cal K^{-1} mol^{-1}) or greater at temperatures near 10000 K.

References

1. ICSU-CODATA Task Group, CODATA Special Report No. 7 (April, 1978).
2. H. Y. Wu and P. G. Wahlbeck, High Temp. Sci. 3, 469 (1971).
3. H. Bernstein and L. Kaufman, Rept. AFML-TR66-193 (1966).
4. R. K. Koch, E. D. Clavert, C. R. Thomas, R. A. Beall, U.S. Bur. Mines RI 7271, 1969.
5. H. Strassmair and D. Stark, Z. Angew. Phys. 23, 40 (1967).
6. J. W. Edwards, H. L. Johnston, and W. E. Ditmars, J. Am. Chem. Soc. 75, 2467 (1953).
7. C. Corliss and J. Sugar, J. Phys. Chem. Ref. Data 8, 1 (1979).
8. J. R. Downey, Jr., The Dow Chemical Company, Thermal Research, AFOSR-TR-78-0960, March, 1978.

Titanium (Ti) Ti_1(g)

Enthalpy Reference Temperature = T_r = 298.15 K   Standard State Pressure = p° = 0.1 MPa

T/K	$C_p°$ (J $K^{-1}mol^{-1}$)	S°	$-[G°-H°(T_r)]/T$	$H°-H°(T_r)$ (kJ mol^{-1})	$\Delta_f H°$	$\Delta_f G°$	Log K_f
0	0.	0.	INFINITE	-7.539	470.920	470.920	INFINITE
100	26.974	151.246	203.383	-5.214	472.684	458.386	-239.436
200	26.487	170.126	182.593	-2.493	473.488	443.708	-115.885
250	25.355	175.913	180.704	-1.198	473.620	436.245	-91.148
298.15	24.430	180.297	180.297	0.	473.629	429.044	-75.167
300	24.399	180.448	180.297	0.045	473.627	428.768	-74.655
350	23.661	184.151	180.591	1.246	473.541	421.297	-62.875
400	23.104	187.272	181.236	2.414	473.383	413.843	-54.042
450	22.683	189.968	182.060	3.558	473.169	406.413	-47.175
500	22.360	192.340	182.972	4.684	472.912	399.009	-41.684
600	21.913	196.374	184.881	6.896	472.299	384.284	-33.455
700	21.632	199.729	186.769	9.072	471.587	369.670	-27.585
800	21.454	202.605	188.573	11.226	470.816	355.163	-23.190
900	21.353	205.126	190.275	13.366	469.965	340.756	-19.777
1000	21.323	207.373	191.875	15.499	468.977	326.451	-17.052
1100	21.362	209.407	193.377	17.633	467.795	312.254	-14.828
1200	21.474	211.270	194.792	19.774	462.442	298.293	-12.984
1300	21.657	212.996	196.127	21.930	461.617	284.647	-11.437
1400	21.910	214.609	197.390	24.107	460.736	271.066	-10.114
1500	22.228	216.132	198.589	26.314	459.793	257.551	-8.969
1600	22.604	217.578	199.731	28.555	458.779	244.101	-7.969
1700	23.029	218.961	200.822	30.836	457.688	230.717	-7.089
1800	23.497	220.290	201.867	33.162	456.510	217.399	-6.309
1900	23.999	221.574	202.870	35.537	455.237	204.149	-5.612
2000	24.529	222.818	203.837	37.963	439.171	191.423	-4.999
2100	25.080	224.028	204.769	40.443	436.927	179.091	-4.455
2200	25.648	225.208	205.672	42.979	434.740	166.864	-3.962
2300	26.228	226.361	206.546	45.573	432.610	154.736	-3.514
2400	26.819	227.489	207.396	48.225	430.538	142.699	-3.106
2500	27.416	228.596	208.221	50.937	428.526	130.748	-2.732
2600	28.019	229.683	209.026	53.709	426.574	118.875	-2.388
2700	28.626	230.752	209.811	56.541	424.683	107.077	-2.072
2800	29.235	231.804	210.578	59.434	422.852	95.347	-1.779
2900	29.846	232.841	211.328	62.388	421.082	83.682	-1.507
3000	30.456	233.863	212.062	65.403	419.374	72.077	-1.255
3100	31.065	234.871	212.781	68.479	417.726	60.528	-1.020
3200	31.671	235.867	213.487	71.616	416.139	49.031	-0.800
3300	32.273	236.851	214.180	74.813	414.612	37.582	-0.595
3400	32.870	237.823	214.861	78.071	413.146	26.179	-0.402
3500	33.460	238.785	215.531	81.387	411.739	14.818	-0.221
3600	34.042	239.736	216.190	84.762	410.390	3.497	-0.051
3630.956	34.219	240.028	216.392	85.819	------- FUGACITY = 1 bar -------		
3700	34.613	240.676	216.840	88.195	0.	0.	0.
3800	35.173	241.607	217.479	91.685	0.	0.	0.
3900	35.721	242.527	218.110	95.229	0.	0.	0.
4000	36.254	243.439	218.731	98.828	0.	0.	0.
4100	36.772	244.340	219.345	102.480	0.	0.	0.
4200	37.273	245.232	219.951	106.182	0.	0.	0.
4300	37.757	246.115	220.549	109.934	0.	0.	0.
4400	38.223	246.988	221.140	113.733	0.	0.	0.
4500	38.670	247.852	221.724	117.578	0.	0.	0.
4600	39.097	248.707	222.301	121.466	0.	0.	0.
4700	39.505	249.552	222.872	125.397	0.	0.	0.
4800	39.892	250.388	223.437	129.367	0.	0.	0.
4900	40.259	251.214	223.995	133.374	0.	0.	0.
5000	40.606	252.031	224.548	137.418	0.	0.	0.
5100	40.933	252.839	225.095	141.495	0.	0.	0.
5200	41.240	253.637	225.636	145.604	0.	0.	0.
5300	41.528	254.425	226.172	149.742	0.	0.	0.
5400	41.796	255.204	226.702	153.909	0.	0.	0.
5500	42.046	255.973	227.227	158.101	0.	0.	0.
5600	42.277	256.733	227.747	162.317	0.	0.	0.
5700	42.490	257.483	228.263	166.556	0.	0.	0.
5800	42.686	258.224	228.773	170.815	0.	0.	0.
5900	42.866	258.955	229.278	175.092	0.	0.	0.
6000	43.029	259.677	229.779	179.387	0.	0.	0.

PREVIOUS: June 1979 (1 atm)   CURRENT: June 1979 (1 bar)

Titanium (Ti) Ti_1(g)

TITANIUM ION (Ti^+) (IDEAL GAS) M_r = 47.87945

IP(Ti^+, g) = 109500 ± 1000 cm^{-1}

S°(298.15K) = 183.594 ± 0.04 J K^{-1} mol^{-1}

$\Delta_f H°$(0 K)= 1128.985 ± 6.3 kJ mol^{-1}

$\Delta_f H°$(298.15K) = [1138.252] kJ mol^{-1}

Electronic Levels and Quantum Weights

State	ε_i, cm^{-1}	g_i
$^4F_{3/2}$	0.00	4
$^4F_{5/2}$	93.94	6
$^4F_{7/2}$	225.47	8
$^4F_{9/2}$	393.22	10
.	.	.
.	.	.
$^2F_{7/2}$	70893.00	8
IP	109500	

Enthalpy of Formation

$\Delta_f H°$(Ti^+, g, 0 K) is calaculated from $\Delta_f H°$(Ti, g, 0 K) (1) using the spectroscopic value of IP(Ti) = 55010 ± 100 cm^{-1} (658.07 ± 0.4 kJ mol^{-1}) from Corliss & Sugar (2). The ionization limit is converted from cm^{-1} to kJ mol^{-1} using the factor, 1 cm^{-1} = 0.01196266 kJ cm^{-1}, which is derived from the 1973 CODATA fundamental constants (3). Rosenstock et al. (4) and Levon amd Lias (5) have summarized additional ionization and appearance potential data.

$\Delta_f H°$(Ti^+, g, 298.15K) is calculated from $\Delta_f H°$(Ti, g, 0 K) by using IP(Ti) with JANAF (1) enthalpies, H°(0 K) - H°(298.15K), for Ti(g), Ti^+(g), and e^-(g). $\Delta_f H°$($Ti \rightarrow Ti^+ + e^-$, 298.15K) differs from a room temperature threshold energy due to inclusion of these enthalpies and to threshold effects discussed by Rosenstock et al. (4). $\Delta_f H°$(298.15 K) should be changed by -6.197 kJ mol^{-1} if it is to be used in this ion convention that excludes the enthalpy of the electron.

Heat Capacity and Entropy

The electronic energy levels are given in the recent compilation by Corliss and Sugar (2). Although we have listed only the ground state, the first three excited states, the highest observed excited state, and the ionization potential for Ti^+(g), all levels listed by Corliss and Sugar (2), as well as estimated levels, are used in the calculation. The observed levels are too numerous to list completely. In our calculations the missing levels for n=4,5 have been arbitrarily added near the ionization limit. The calculations indicate that for Ti^+(g), the thermochemical functions are independent of the estimated missing levels for n=4,5,the cut-off procedure, and the inclusion of n=6 levels up to 6000 K. The Gibbs energy function is essentially unaffected (by less than 0.2%) up to 20000 K. The reported uncertainty in S°(298.15K) is due to uncertainties in the relative atomic mass and the fundamental constants. Extension of these calculations above 6000 K may require consideration of the higher excited states (n>6), and use of proper fill and cut-off procedures (6).

References

1. JANAF Thermochemical Tables: Ti(g), 6-30-79; e^-(g), 3-31-82.
2. C. Corliss and J. Sugar, J. Phys. Chem. Ref. Data 8, 1 (1979).
3. E. R. Cohen and B. N. Taylor, J. Phys. Chem. Ref. Data 2, 663 (1973).
4. H. M. Rosenstock, K. Draxl, B. W. Steiner, and J. T. Herron, J. Phys. Chem. Ref. Data 6, Supp. 1 (1977).
5. R. D. Levine and S. G. Lias, U. S. Nat. Bur. Stand., NSRDS-NBS-71, 1982.
6. J. R. Downey, Jr., The Dow Chemical Company, Thermal Research, AFOSR-TR-78-0960, March 1, 1978.

Titanium, Ion (Ti^+) Ti_1^+(g)

Enthalpy Reference Temperature = T_r = 298.15 K Standard State Pressure = p° = 0.1 MPa

T/K	$C_p°$ (J $K^{-1}mol^{-1}$)	S° (J $K^{-1}mol^{-1}$)	$-[G°-H°(T_r)]/T$ (J $K^{-1}mol^{-1}$)	$H°-H°(T_r)$ (kJ mol^{-1})	$\Delta_f H°$ (kJ mol^{-1})	$\Delta_f G°$ (kJ mol^{-1})	Log K_f
0	0.	0.	INFINITE	-7.899	1128.985		
100	28.375	153.817	207.355	-5.354			
200	26.772	173.059	185.998	-2.588			
250	26.305	178.974	184.025	-1.263			
298.15	26.181	183.594	183.594	0.	1138.252	1086.429	-190.338
300	26.180	183.756	183.595	0.048	1138.292	1086.108	-189.108
350	26.166	187.790	183.913	1.357	1139.353	1077.326	-160.782
400	26.128	191.282	184.621	2.665	1140.373	1068.395	-139.518
450	26.021	194.354	185.535	3.969	1141.359	1059.338	-122.965
500	25.843	197.087	186.556	5.265	1142.312	1050.173	-109.711
600	25.356	201.758	188.714	7.826	1144.126	1031.573	-89.806
700	24.817	205.626	190.862	10.335	1145.826	1012.678	-75.567
800	24.320	208.907	192.917	12.791	1147.436	993.546	-64.872
900	23.903	211.746	194.855	15.202	1148.934	974.218	-56.542
1000	23.572	214.247	196.672	17.575	1150.264	954.732	-49.870
1100	23.322	216.481	198.373	19.919	1151.372	935.123	-44.405
1200	23.145	218.502	199.968	22.242	1148.279	915.543	-39.853
1300	23.030	220.350	201.465	24.550	1149.685	896.091	-36.005
1400	22.970	222.054	202.876	26.850	1151.004	876.534	-32.704
1500	22.959	223.638	204.208	29.146	1152.229	856.886	-29.839
1600	22.992	225.121	205.469	31.443	1153.350	837.159	-27.330
1700	23.065	226.517	206.667	33.745	1154.358	817.365	-25.115
1800	23.173	227.838	207.806	36.057	1155.245	797.516	-23.143
1900	23.313	229.095	208.894	38.381	1156.000	777.621	-21.378
2000	23.480	230.295	209.934	40.721	1141.926	758.144	-19.801
2100	23.670	231.445	210.931	43.078	1141.638	738.962	-18.381
2200	23.879	232.551	211.889	45.455	1141.370	719.793	-17.090
2300	24.104	233.617	212.811	47.854	1141.124	700.636	-15.912
2400	24.340	234.648	213.699	50.276	1140.901	681.489	-14.832
2500	24.583	235.646	214.557	52.722	1140.702	662.351	-13.839
2600	24.832	236.615	215.387	55.193	1140.527	643.221	-12.922
2700	25.082	237.557	216.191	57.689	1140.378	624.096	-12.074
2800	25.331	238.474	216.970	60.209	1140.254	604.977	-11.286
2900	25.576	239.367	217.727	62.755	1140.154	585.862	-10.553
3000	25.816	240.238	218.463	65.324	1140.078	566.750	-9.868
3100	26.048	241.088	219.179	67.918	1140.026	547.640	-9.228
3200	26.272	241.919	219.877	70.534	1139.997	528.531	-8.627
3300	26.486	242.731	220.557	73.172	1139.990	509.423	-8.063
3400	26.689	243.524	221.221	75.831	1140.004	490.315	-7.533
3500	26.881	244.301	221.870	78.509	1140.037	471.206	-7.032
3600	27.060	245.061	222.503	81.206	1140.089	452.096	-6.560
3700	27.228	245.804	223.123	83.921	731.060	440.772	-6.223
3800	27.384	246.533	223.730	86.652	732.380	432.909	-5.951
3900	27.528	247.246	224.323	89.397	733.660	425.011	-5.692
4000	27.660	247.944	224.905	92.157	734.899	417.081	-5.447
4100	27.781	248.629	225.476	94.929	736.098	409.121	-5.212
4200	27.890	249.300	226.035	97.712	737.258	401.132	-4.989
4300	27.989	249.957	226.584	100.507	738.379	393.115	-4.775
4400	28.078	250.602	227.122	103.310	739.462	385.073	-4.571
4500	28.157	251.234	227.651	106.122	740.507	377.007	-4.376
4600	28.227	251.853	228.170	108.941	741.517	368.918	-4.189
4700	28.289	252.461	228.681	111.767	742.491	360.808	-4.010
4800	28.343	253.057	229.182	114.599	743.431	352.677	-3.838
4900	28.390	253.642	229.676	117.435	744.339	344.527	-3.673
5000	28.430	254.216	230.161	120.276	745.215	336.358	-3.514
5100	28.464	254.779	230.638	123.121	746.061	328.173	-3.361
5200	28.493	255.332	231.107	125.969	746.879	319.971	-3.214
5300	28.516	255.875	231.570	128.819	747.669	311.753	-3.073
5400	28.535	256.408	232.025	131.672	748.434	303.521	-2.936
5500	28.550	256.932	232.473	134.526	749.175	295.275	-2.804
5600	28.561	257.447	232.914	137.382	749.893	287.016	-2.677
5700	28.569	257.952	233.349	140.238	750.589	278.744	-2.554
5800	28.575	258.449	233.778	143.096	751.266	270.460	-2.436
5900	28.578	258.938	234.200	145.953	751.925	262.165	-2.321
6000	28.579	259.418	234.616	148.811	752.566	253.858	-2.210

PREVIOUS: June 1979 (1 atm) CURRENT: March 1984 (1 bar)

Titanium, Ion (Ti^+) Ti_1^+(g)

TITANIUM, ION (Ti^-) IDEAL GAS M_r = 47.88055

EA(Ti, g) = 0.079 ± 0.014 eV

S°(298.15 K) = 183.715 ± 0.001 J K^{-1} mol^{-1}

$\Delta_f H°$(0 K) = 463.3 ± 4 kJ mol^{-1}

$\Delta_f H°$(298.15 K) = [459.833] kJ mol^{-1}

Electonic Levels and Quantum Weights

State	ε_i, cm^{-1}	g_i
$^4F_{3/2}$	0.0	4
$^4F_{5/2}$	72	6
$^4F_{7/2}$	171	8
$^4F_{9/2}$	295	10

Enthalpy of Formation

$\Delta_f H°(Ti^-$, g, 0 K) is calculated from $\Delta_f H°$(Ti, g, 0 K) (1) using the adopted electron affinity of EA(Ti) = 0.079 ± 0.014 eV (7.622 ± 1.351 kJ mol^{-1}). This value, recommended by Hotop and Lineberger (2), is based on a laser photodetachment electron spectrometry study (3). Additional information on Ti^-(g) may be obtained in the critical discussions of Hotop and Lineberger (2, 4), Rosenstock et al. (5) and Massey (6).

$\Delta_f H°(Ti^-$, g, 298.15 K) is obtained from $\Delta_f H°$(Ti, g, 0 K) by using EA(Ti) with JANAF (1) enthalpies, H°(0 K) − H°(298.15 K), for Ti^-(g), Ti(g), and e^-(g). $\Delta_r H°(Ti^- \rightarrow Ti + e^-$, 298.15 K) differs from a room-temperature threshold energy due to inclusion of these enthalpies and to threshold effects discussed by Rosenstock et al. (5). $\Delta_f H°$(298.15 K) should be changed by +6.197 kJ mol^{-1} if it is to be used in the ion convention that excludes the enthalpy of the electron.

Heat Capacity and Entropy

The ground state electronic configuration for Ti^-(g) is given by Hotop and Lineberger (2, 4), Rosenstock et al. (5) and Massey(6). The fine structure has been calculated via an isoelectronic extrapolation from a logarithmic plot (3) and is that recommended by Hotop and Lineberger(2).

References

1. JANAF Thermochemical Tables: Ti(g), 6-30-79; e^-(g), 3-31-82.
2. H. Hotop and W. C. Lineberger, J. Phys. Chem. Ref. Data, to be published, 1985.
3. C. S. Feigerle, R. R. Cordermann, S. V. Bobashev and W. C. Lineberger J. Chem. Phys. 74, 1580 (1981).
4. H. Hotop and W. C. Lineberger, J. Phys. Chem. Ref. Data 4, 539 (1975).
5. H. M. Rosenstock, K. Draxl et al., J. Phys. Chem. Ref. Data 6, Supp. 1 (1977).
6. H. S. W. Massey, "Negative Ions", 3rd ed., Cambridge University Press, Cambridge, 1976.

Titanium, Ion (Ti^-) Ti_1^-(g)

Enthalpy Reference Temperature = T_r = 298.15 K Standard State Pressure = p° = 0.1 MPa

	J $K^{-1}mol^{-1}$			kJ mol^{-1}			
T/K	$C_p°$	S°	−[G°−H°(T_r)]/T	H°−H°(T_r)	$\Delta_f H°$	$\Delta_f G°$	Log K_f
0	0.	0.	INFINITE	-7.563	463.297		
100	28.077	155.897	205.362	-4.947			
200	24.629	174.264	185.851	-2.317			
250	23.533	179.634	184.094	-1.115			
298.15	22.835	183.715	183.715	0.	459.833	420.484	-73.667
300	22.813	183.856	183.715	0.042	459.790	420.240	-73.170
350	22.330	187.334	183.991	1.170	458.591	413.742	-61.748
400	21.995	190.293	184.598	2.278	457.333	407.420	-53.204
450	21.756	192.869	185.377	3.371	456.030	401.259	-46.577
500	21.580	195.151	186.243	4.454	454.691	395.245	-41.291
600	21.344	199.063	188.064	6.600	451.932	383.613	-33.396
700	21.199	202.342	189.876	8.726	449.092	372.450	-27.793
800	21.103	205.166	191.615	10.841	446.203	361.698	-23.616
900	21.037	207.647	193.261	12.948	443.241	351.311	-20.390
1000	20.990	209.861	194.812	15.049	440.142	341.261	-17.826
1100	20.955	211.860	196.273	17.146	436.846	331.531	-15.743
1200	20.928	213.682	197.649	19.240	429.366	322.230	-14.026
1300	20.907	215.356	198.947	21.332	426.399	313.422	-12.593
1400	20.890	216.905	200.175	23.422	423.351	304.845	-11.374
1500	20.877	218.346	201.339	25.510	420.211	296.489	-10.325
1600	20.866	219.693	202.445	27.597	416.965	288.346	-9.414
1700	20.857	220.958	203.497	29.683	413.600	280.410	-8.616
1800	20.849	222.150	204.500	31.769	410.103	272.676	-7.913
1900	20.843	223.277	205.459	33.853	406.461	265.139	-7.289
2000	20.837	224.346	206.377	35.937	387.974	258.253	-6.745
2100	20.832	225.362	207.257	38.021	383.255	251.883	-6.265
2200	20.828	226.331	208.102	40.104	378.536	245.737	-5.835
2300	20.825	227.257	208.915	42.186	373.816	239.807	-5.446
2400	20.821	228.143	209.698	44.269	369.096	234.082	-5.095
2500	20.819	228.993	210.453	46.351	364.376	228.553	-4.775
2600	20.816	229.810	211.182	48.432	359.655	223.213	-4.484
2700	20.814	230.595	211.886	50.514	354.934	218.055	-4.219
2800	20.812	231.352	212.568	52.595	350.213	213.072	-3.975
2900	20.810	232.082	213.228	54.676	345.492	208.257	-3.751
3000	20.809	232.788	213.869	56.757	340.770	203.605	-3.545
3100	20.807	233.470	214.490	58.838	336.049	199.111	-3.355
3200	20.806	234.131	215.094	60.919	331.327	194.769	-3.179
3300	20.805	234.771	215.680	62.999	326.605	190.575	-3.017
3400	20.804	235.392	216.251	65.080	321.884	186.523	-2.866
3500	20.803	235.995	216.806	67.160	317.162	182.611	-2.725
3600	20.802	236.581	217.348	69.240	312.439	178.834	-2.595
3700	20.801	237.151	217.875	71.320	-101.381	182.976	-2.583
3800	20.800	237.706	218.390	73.400	-104.869	190.708	-2.621
3900	20.799	238.246	218.892	75.480	-108.413	198.532	-2.659
4000	20.799	238.773	219.382	77.560	-112.010	206.448	-2.696
4100	20.798	239.286	219.862	79.640	-115.660	214.455	-2.732
4200	20.798	239.787	220.330	81.720	-119.362	222.551	-2.768
4300	20.797	240.277	220.788	83.800	-123.112	230.736	-2.803
4400	20.797	240.755	221.237	85.879	-126.910	239.009	-2.837
4500	20.796	241.222	221.676	87.959	-130.754	247.368	-2.871
4600	20.796	241.679	222.106	90.038	-134.642	255.814	-2.905
4700	20.795	242.126	222.527	92.118	-138.571	264.344	-2.938
4800	20.795	242.564	222.940	94.197	-142.540	272.959	-2.970
4900	20.794	242.993	223.345	96.277	-146.547	281.656	-3.002
5000	20.794	243.413	223.742	98.356	-150.590	290.436	-3.034
5100	20.794	243.825	224.132	100.436	-154.666	299.297	-3.065
5200	20.794	244.229	224.514	102.515	-158.774	308.238	-3.096
5300	20.793	244.625	224.890	104.594	-162.912	317.259	-3.127
5400	20.793	245.013	225.259	106.674	-167.077	326.357	-3.157
5500	20.793	245.395	225.622	108.753	-171.269	335.534	-3.187
5600	20.792	245.770	225.978	110.832	-175.485	344.786	-3.216
5700	20.792	246.138	226.329	112.912	-179.722	354.114	-3.245
5800	20.792	246.499	226.673	114.991	-183.981	363.517	-3.274
5900	20.792	246.855	227.012	117.070	-188.258	372.993	-3.302
6000	20.792	247.204	227.346	119.149	-192.552	382.542	-3.330

PREVIOUS: June 1979 (1 atm) CURRENT: March 1984 (1 bar)

Titanium, Ion (Ti^-) Ti_1^-(g)

VANADIUM (V) REFERENCE STATE A_r = 50.9415

0 to 2190 K		crystal
2190 to 3690.080 K		liquid
above 3690.080 K		ideal monatomic gas

Refer to the individual tables for details.

Vanadium (V) V_1(ref)

Enthalpy Reference Temperature = T_r = 298.15 K Standard State Pressure = $p°$ = 0.1 MPa

T/K	$C_p°$ (J K⁻¹mol⁻¹)	$S°$ (J K⁻¹mol⁻¹)	$-[G°-H°(T_r)]/T$ (J K⁻¹mol⁻¹)	$H°-H°(T_r)$ (kJ mol⁻¹)	$\Delta_f H°$ (kJ mol⁻¹)	$\Delta_f G°$ (kJ mol⁻¹)	Log K_f
0	0.	0.	INFINITE	-4.640	0.	0.	0.
100	13.119	7.185	48.695	-4.151	0.	0.	0.
200	21.876	19.558	31.133	-2.315	0.	0.	0.
298.15	24.896	28.936	28.936	0.	0.	0.	0.
300	24.928	29.090	28.936	0.046	0.	0.	0.
400	26.234	36.459	29.932	2.611	0.	0.	0.
500	26.945	42.392	31.850	5.271	0.	0.	0.
600	27.489	47.353	34.032	7.993	0.	0.	0.
700	28.033	51.630	36.247	10.768	0.	0.	0.
800	28.660	55.413	38.411	13.602	0.	0.	0.
900	29.372	58.830	40.493	16.504	0.	0.	0.
1000	30.083	61.961	42.485	19.476	0.	0.	0.
1100	30.878	64.864	44.389	22.523	0.	0.	0.
1200	31.798	67.590	46.210	25.656	0.	0.	0.
1300	32.740	70.172	47.955	28.882	0.	0.	0.
1400	33.807	72.638	49.630	32.210	0.	0.	0.
1500	34.811	75.004	51.244	35.640	0.	0.	0.
1600	35.857	77.283	52.800	39.173	0.	0.	0.
1700	37.028	79.492	54.306	42.817	0.	0.	0.
1800	38.200	81.641	55.765	46.577	0.	0.	0.
1900	39.539	83.742	57.182	50.464	0.	0.	0.
2000	40.920	85.805	58.562	54.485	0.	0.	0.
2100	42.468	87.837	59.908	58.652	0.	0.	0.
2190.000	44.141	89.653	61.093	62.547	CRYSTAL <--> LIQUID		
2190.000	46.204	100.085	61.093	85.392	TRANSITION		
2200	46.204	100.295	61.271	85.854	0.	0.	0.
2300	46.204	102.349	63.012	90.474	0.	0.	0.
2400	46.204	104.315	64.693	95.095	0.	0.	0.
2500	46.204	106.202	66.316	99.715	0.	0.	0.
2600	46.204	108.014	67.885	104.335	0.	0.	0.
2700	46.204	109.757	69.403	108.956	0.	0.	0.
2800	46.204	111.438	70.875	113.576	0.	0.	0.
2900	46.204	113.059	72.302	118.197	0.	0.	0.
3000	46.204	114.626	73.687	122.817	0.	0.	0.
3100	46.204	116.141	75.032	127.437	0.	0.	0.
3200	46.204	117.607	76.339	132.058	0.	0.	0.
3300	46.204	119.029	77.612	136.678	0.	0.	0.
3400	46.204	120.409	78.850	141.299	0.	0.	0.
3500	46.204	121.748	80.057	145.919	0.	0.	0.
3600	46.204	123.049	81.233	150.539	0.	0.	0.
3690.080	46.204	124.191	82.268	154.701	LIQUID <--> IDEAL GAS		
3690.080	29.042	245.321	82.268	601.678	FUGACITY = 1 bar		
3700	29.090	245.399	82.705	601.967	0.	0.	0.
3800	29.591	246.181	86.997	604.901	0.	0.	0.
3900	30.105	246.956	91.088	607.885	0.	0.	0.
4000	30.631	247.725	94.995	610.922	0.	0.	0.
4100	31.166	248.488	98.729	614.012	0.	0.	0.
4200	31.710	249.246	102.304	617.156	0.	0.	0.
4300	32.260	249.998	105.730	620.354	0.	0.	0.
4400	32.817	250.746	109.017	623.608	0.	0.	0.
4500	33.379	251.490	112.175	626.918	0.	0.	0.
4600	33.946	252.230	115.211	630.284	0.	0.	0.
4700	34.517	252.966	118.135	633.707	0.	0.	0.
4800	35.093	253.699	120.951	637.187	0.	0.	0.
4900	35.672	254.428	123.668	640.726	0.	0.	0.
5000	36.255	255.155	126.290	644.322	0.	0.	0.
5100	36.843	255.878	128.824	647.977	0.	0.	0.
5200	37.435	256.600	131.274	651.691	0.	0.	0.
5300	38.031	257.318	133.646	655.464	0.	0.	0.
5400	38.632	258.035	135.943	659.297	0.	0.	0.
5500	39.238	258.749	138.169	663.190	0.	0.	0.
5600	39.849	259.462	140.329	667.144	0.	0.	0.
5700	40.465	260.172	142.425	671.160	0.	0.	0.
5800	41.087	260.881	144.461	675.237	0.	0.	0.
5900	41.712	261.589	146.440	679.376	0.	0.	0.
6000	42.343	262.295	148.365	683.579	0.	0.	0.

PREVIOUS: June 1973 (1 atm) CURRENT: June 1973 (1 bar)

Vanadium (V) V_1(ref)

VANADIUM (V) CRYSTAL A_r = 50.9415

$S°(298.15\ K) = 28.936 \pm 0.42\ J\ K^{-1}\ mol^{-1}$

$T_{fus} = 2190 \pm 20\ K$

$\Delta_f H° = 0\ kJ\ mol^{-1}$

$\Delta_f H°(298.15\ K) = 0\ kJ\ mol^{-1}$

$\Delta_{fus} H° = 22.84 \pm 6.28\ kJ\ mol^{-1}$

Enthalpy of Formation

Zero by definition.

Heat Capacity and Entropy

The heat capacity values for temperatures below 10 K are those chosen by Hultgren (1). A graphical integration yeilds S°(10 K) = 0.0253 cal K^{-1} mol^{-1} and H°(10 K) - H°(0 K) = 0.147 cal mol^{-1}. The adopted heat capacity values for 10 ≤ T ≤ 320 K are obtained graphically, based on the following three works.

Source	Year	T/K	Purity of Vanadium
(2)	1936	54-297	reported ≥ 99.5%
(3)	1960	10-274	99.5%; remainder oxygen
(4)	1961	25-340	99.8%; metallic impurities <0.08%

The low temperature (T < 50 K) results of Clusius et al. (3) do not join smoothly with the values adopted for T < 10 K. This mismatch is in part due to the oxygen impurity in the vanadium sample used by Clusius et al. (3). In order to effect a smooth joining of the Cpd values, a graphical representation of Cp(lattice)/T^3 versus T for the Hultgren data (1) and the Clusius et al. data (3) was used to derive adjusted Cp° values. The three data sets (2, 3, 4) are in reasonably good agreement; two sets (2, 4) lie predominantly above the adopted values while the remaining set (3) lies below. Above 100 K, the three data sets differ from the adopted Cp° values by roughly 1%, with the difference decreasing as T increases.

For the region 320≤ T ≤ 2190, there are four works upon which the adopted Cp° values are based.

Source	Year	Method	T/K	Purity of Vandadium
(5)	1965	Cp°	320-1800	99.74%
(6)	1934	drop	565-1928	reported as "purest"
(7)	1962	drop	488-1486	0.1%C, 0.07, 0.03%N
(8)	1971	levitation	1989-2284	0.06% total impurities

The heat capacity values reported by Kohlhaas et al. (5) are 1-2% lower than the adopted values below 950 K; 0.3% higher in the region 950-1300 K; and 7% lower above 1600 K. The results of Golutvin and Kozlowskaya (7) show considerable scatter and lie roughly 2-10% above the tabulated enthalpy values. The enthalpy data of Jaeger and Veenstra (6) do not join smoothly with that of Berezin et al. (8). In the region 1500-2000 K the adopted enthalpy values are a compromise between the latter two sets of data. Below roughly 1100 K the adopted enthalpy and that reported by Jaeger and Veenstra (6) differ at most by ± 15 cal mol^{-1}. From 1300-1800 K the difference increases from 70 cal mol^{-1} to 308 cal mol^{-1} or from 1.0% to 2.6%. The enthalpy at 2190 K (T_{fus}) is very close to that which is obtained from a linear fit of the Berezin et al. data (8). Whereas the two sets of data (6, 8) separately indicate a Cp° value near T_{fus} to be at least 11 cal K^{-1} mol^{-1}, the compromise yields 10.5 cal K^{-1} mol^{-1}.

Fusion Data

See the liquid for details.

Transition Data

The literature indicates differences in opinion as to the possibility of a transition in the region 175-325 K and near 1825 K. Although there are no detailed enthalpy measurements in the 1825 K region, Barnes (9) used field-electron emission techniques to show that an allotropic trnsformation does not occur in vanadium in this region if the samples are of high purity. Earlier work concerning a possible transition in this regilon was referenced by Barnes (9).

Documentation exists in the literature as to the observation of anomalies in the temperature dependence of some physical properties of vanadium in the range 175-325 K. Although the anomaly was attributed by different workers to an antiferromagnetic transition, a small distortion of the body-centered cubic crystal structure, and impurities, Finkel' et al. (10) recently ascribed the anomaly to a second order phase transition at 230 K. Using low temperature x-ray diffraction techniques in the study of a single crystal of vanadium, Finkel' et al. (10) observed a decrease in crystal lattice symmetry form body-centered cubic (T > 230 K) to tetragonal (T < 230 K). Nevertheless, Crangle and Smith (11) measured the specific heat of polycrystalline vanadium and confirmed the results of Clusius et al. (3) that no appreciable anomaly in Cp° exists in vanadium between 175 and 265 K. We do not adopt a ΔCp or $\Delta_t H$ value in the two previously mentioned regions.

Sublimation Data

See the ideal gas table for details.

References

1. R. Hultgren, R. L. Orr, and K. K. Kelley, Supplement to Selected Values of Thermodynamic Properties of Metals and Alloys, V table, May, 1966.
2. C. T. Anderson, J. Amer. Chem. Soc. 58, 564 (1936).
3. K. Clusius, P. Franzosini, and U. Piesbergen, Z. Naturforsch. 15A, 728 (1960).
4. Z. Bieganski and B. Stalinski, Bull. Acad. Polon. Sci., Ser. Sci. Chim. 9, 367 (1961).
5. R. Kohlhaas, M. Braun, and O. Vollmer, Z. Naturforsch. 20A, 1077 (1965).
6. F. M. Jaeger and W. A. Veenstra, Rec. Trav. Chim. 53, 677 (1934); Proc. Acad. Sci. Amsterdam 37, 61 (1934).
7. Yu. M. Golutvin and T. M. Koslovskaya, Russ. J. Phys. Chem. 36, 183 (1962).
8. B. Ya. Berezin, V. Ya. Chekshovskoii, and A. E. Sheindlin, Dokl. Akad. Nauk SSSR 201, 583 (1971).
9. G. Barnes, Met. Trans. 4, 549 (1973).
10. V. A. Kinkel', V. I. Glamazda, and G. P. Kovtun, Soviet Physics JETP 30, 581 (1970).
11. J. Crangle and T. F. Smith, Phys. Rev. Letters 9, 86 (1962).

Vanadium (V) $V_1(cr)$

Enthalpy Reference Temperature = T_r = 298.15 K Standard State Pressure = p° = 0.1 MPa

T/K	$C_p°$ (J K⁻¹mol⁻¹)	S° (J K⁻¹mol⁻¹)	$-[G°-H°(T_r)]/T$ (J K⁻¹mol⁻¹)	$H°-H°(T_r)$ (kJ mol⁻¹)	$\Delta_f H°$ (kJ mol⁻¹)	$\Delta_f G°$ (kJ mol⁻¹)	Log K_f
0	0.	0.	INFINITE	-4.640	0.	0.	0.
100	13.119	7.185	48.695	-4.151	0.	0.	0.
200	21.876	19.558	31.133	-2.315	0.	0.	0.
250	23.703	24.655	29.340	-1.171	0.	0.	0.
298.15	24.896	28.936	28.936	0.	0.	0.	0.
300	24.928	29.090	28.936	0.046	0.	0.	0.
350	25.681	32.992	29.243	1.312	0.	0.	0.
400	26.234	36.459	29.932	2.611	0.	0.	0.
450	26.610	39.571	30.833	3.932	0.	0.	0.
500	26.945	42.392	31.850	5.271	0.	0.	0.
600	27.489	47.353	34.032	7.993	0.	0.	0.
700	28.033	51.630	36.247	10.768	0.	0.	0.
800	28.660	55.413	38.411	13.602	0.	0.	0.
900	29.372	58.830	40.493	16.504	0.	0.	0.
1000	30.083	61.961	42.485	19.476	0.	0.	0.
1100	30.878	64.864	44.389	22.523	0.	0.	0.
1200	31.798	67.590	46.210	25.656	0.	0.	0.
1300	32.740	70.172	47.955	28.882	0.	0.	0.
1400	33.807	72.638	49.630	32.210	0.	0.	0.
1500	34.811	75.004	51.244	35.640	0.	0.	0.
1600	35.857	77.283	52.800	39.173	0.	0.	0.
1700	37.028	79.492	54.306	42.817	0.	0.	0.
1800	38.200	81.641	55.765	46.577	0.	0.	0.
1900	39.539	83.742	57.182	50.464	0.	0.	0.
2000	40.920	85.805	58.562	54.485	0.	0.	0.
2100	42.468	87.837	59.908	58.652	0.	0.	0.
2190.000	44.141	89.653	61.093	62.547	------ CRYSTAL	<--> LIQUID	-----
2200	44.329	89.856	61.223	62.991	-22.863	0.104	-0.002
2300	46.254	91.869	62.512	67.520	-22.954	1.151	-0.026
2400	48.179	93.878	63.777	72.242	-22.853	2.198	-0.048
2500	50.103	95.884	65.021	77.156	-22.559	3.236	-0.068
2600	52.028	97.886	66.247	82.263	-22.073	4.259	-0.086
2700	53.932	99.886	67.456	87.561	-21.395	5.259	-0.102

PREVIOUS: CURRENT: June 1973

Vanadium (V) $V_1(cr)$

VANADIUM (V) LIQUID $A_r = 50.9415$

$S°(298.15\ K) = [36.136]\ J\ K^{-1}\ mol^{-1}$ $\Delta_f H°(298.15\ K) = [17.288]\ kJ\ mol^{-1}$

$T_{fus} = 2190 \pm 20\ K$ $\Delta_{fus}H° = 22.84 \pm 6.28\ kJ\ mol^{-1}$

Enthalpy of Formation

The enthalpy of formation of V(ℓ) at 298.15 K is calculated from that of the crystal by adding $\Delta_{fus}H°$ and the difference in enthalpy H°(2190) - H°(298.15 K), between the crystal and liquid. Refer to V(g) table for an additional $\Delta_f H°(298.15\ K)$ value as derived from effusion-mass spectrometric work (10).

Heat Capacity and Entropy

Enthalpy data in the liquid phase have been reported by Treverton and Margrave (1) and Berezin et al. (2). Using levitation calorimetry, Treverton and Margrave (1) determined the enthalpy of V(ℓ) in the range 2205-2638 K. They reported a least squares analysis of the enthalpy data in the form H°(T) - H°(298.15 K) = 11.6527T - 5837.7 where T is in Kelvins and the enthalpy difference is in cal mol^{-1}. The standard deviation of the experimental points from the calculated values was ±167 cal mol^{-1}. Berezin et al. (2) also used a form of levitation calorimetry and measured enthalpies in the region 1990-2325 K. For the liquid region their data were represented by H°(T) - H°(298.15 K) = 11.043T - 3776, with a standard deviation of ±90 cal mol^{-1}. These two works agree remarkably well for the Cp° value of V(ℓ). In the measured enthalpy region, the Treverton and Margrave data (1) lie lower than the Berezin et al. data (2) by roughly 600-700 cal mol^{-1}. The two linear representations of the data intersect at 3380 K. We adopt the enthalpy values of Berezin et al. (2) since this work contains enthalpy data surrounding the melting point and a T_{fus} value which is our adopted value.

Fusion Data

The adopted value of the enthalpy of fusion, $\Delta_{fus}H° = 5.46 \pm 1.50$ kcal mol^{-1} is based on the work by Berezin et al. (2). The actual value is calculated by taking the difference in the adopted value of H°(2190) - H°(298.15 K) for V(ℓ) and V(cr). Refer to the heat capacity discussion in the tables for V(cr) and V(ℓ).

Preliminary results of experiments by Berezin et al. (2) indicated that T_{fus} for 99.96% pure vanadium was close to 2190 K. Representative values of T_{fus} reported in the literature are tabulated below.

Source	Purity of V	T_{fus}, K
(3)	0.33% O, 0.078% N 0.15% of 7 other impurities	2161 ± 10
(4)	99.8-99.9%	2192 ± 2
(5)	purity not given	2223
(6)	purified by electron zone fusion 99.8%	2173 ± 25
(7)	purity not given	2193

We adopt $T_{fus} = 2190 \pm 20$ K principally since it is consistent with the work upon which $\Delta_{fus}H°$ is based. The adopted T_{fus} value is also suggested by Charlesworth (8) in his compilation of elemental melting points.

Vaporization Data

The vaporization of V(ℓ) was studied by Farber and Srivastava (10). Refer to V(g) table for details. Kant and Lin (9), using a combination of effusion and mass-spectrometric techniques, reported the ratio of observed ion intensities relating to V_2(g) and V(g) and corresponding to the vaporization of vanadium in the range 2060-2316 K. This ratio is of the order of 10^{-6} which suggests that at least in this region the contribution of V_2(g) in the vapor phase may be neglected. Linearly extrapolating these data to T_{vap}, the V_2 partial pressure is of the order of 10^5 smaller than the V partial pressure.

T_{vap} is calculated as the temperature at which the Gibbs energy change for the reaction V(ℓ) → V(g) is zero. The difference between $\Delta_f H°$(V, g) and $\Delta_f H°$(V, ℓ) at T_{vap} is $\Delta_{vap}H°$. The uncertainty in T_{vap} is probably of the order of ±50 K.

References

1. J. A. Treverton and J. L. Margrave, J. Chem. Thermodynamics 3, 473 (1971).
2. B. Ya. Berezin, V. Ya. Chekhooskoy, and A. E. Sheindlin, High Temp. Sci., 4, 478 (1972); Dokl. Acad. Nauk USSR 201, 583 (1971).
3. E. K. Storms and R. J. McNeal, J. Phys. Chem. 66, 1401 (1962).
4. R. A. Oriani and T. S. Jones, Rev. Sci. Instr. 25, 248 (1954).
5. Yu. A. Kocherzhinskii, G. F. Kolzenko, V. M. Pan, V. K. Sviridenko, and L. M. Yupko, Sb. Nauchn. Tr. Inst. Metallofiz., Akad. Nauk. Ukr. SSR 17, 209 (1963); Chem. Abstr., 60, 10227b (1964).
6. H. K. Adenstedt, J. R. Pequignot, and J. M. Raymer, Trans. Amer. Soc. Metals 44, 990 (1952).
7. D. J. McPherson, Trans. Amer. Soc. Metals 44, 1029 (1952).
8. J. H. Charlesworth, AFML-TR-70-137, October, 1970.
9. A. Kant and S. S. Lin, J. Chem. Phys. 51, 1644 (1969).
10. M. Farber and R. D. Srivastava, J. Chem. Soc., Faraday Trans. I 69, 390 (1973).

Vanadium (V) $V_1(l)$

Enthalpy Reference Temperature = T_r = 298.15 K Standard State Pressure = p° = 0.1 MPa

T/K	$C_p°$ (J K⁻¹mol⁻¹)	S° (J K⁻¹mol⁻¹)	$-[G°-H°(T_r)]/T$ (J K⁻¹mol⁻¹)	$H°-H°(T_r)$ (kJ mol⁻¹)	$\Delta_f H°$ (kJ mol⁻¹)	$\Delta_f G°$ (kJ mol⁻¹)	Log K_f
0							
100							
200							
250							
298.15	24.896	36.136	36.136	0.	17.288	15.141	-2.653
300	24.928	36.290	36.137	0.046	17.288	15.128	-2.634
350	25.681	40.192	36.443	1.312	17.288	14.768	-2.204
400	26.234	43.659	37.132	2.611	17.288	14.408	-1.881
450	26.610	46.771	38.033	3.932	17.288	14.048	-1.631
500	26.945	49.592	39.050	5.271	17.288	13.688	-1.430
600	27.489	54.553	41.232	7.993	17.288	12.968	-1.129
700	28.033	58.830	43.447	10.768	17.288	12.248	-0.914
800	28.660	62.613	45.611	13.602	17.288	11.528	-0.753
900	29.372	66.030	47.693	16.504	17.288	10.808	-0.627
1000	30.083	69.161	49.685	19.476	17.288	10.088	-0.527
1100	30.878	72.064	51.589	22.523	17.288	9.368	-0.445
1200	31.798	74.790	53.410	25.656	17.288	8.648	-0.376
1300	32.740	77.372	55.155	28.882	17.288	7.928	-0.319
1400	33.807	79.838	56.831	32.210	17.288	7.208	-0.269
1450.000	34.304	81.033	57.645	33.913	GLASS <--> LIQUID TRANSITION		
1450.000	46.204	81.033	57.645	33.913			
1500	46.204	82.599	58.451	36.223	17.871	6.478	-0.226
1600	46.204	85.581	60.054	40.844	18.959	5.682	-0.185
1700	46.204	88.382	61.639	45.464	19.935	4.822	-0.148
1800	46.204	91.023	63.199	50.084	20.795	3.907	-0.113
1900	46.204	93.521	64.729	54.705	21.529	2.948	-0.081
2000	46.204	95.891	66.229	59.325	22.128	1.954	-0.051
2100	46.204	98.146	67.695	63.946	22.582	0.934	-0.023
2190.000	46.204	100.085	68.987	68.104	CRYSTAL <--> LIQUID		
2200	46.204	100.295	69.129	68.566	0.	0.	0.
2300	46.204	102.349	70.529	73.186	0.	0.	0.
2400	46.204	104.315	71.896	77.807	0.	0.	0.
2500	46.204	106.202	73.231	82.427	0.	0.	0.
2600	46.204	108.014	74.534	87.048	0.	0.	0.
2700	46.204	109.757	75.806	91.668	0.	0.	0.
2800	46.204	111.438	77.049	96.288	0.	0.	0.
2900	46.204	113.059	78.263	100.909	0.	0.	0.
3000	46.204	114.626	79.449	105.529	0.	0.	0.
3100	46.204	116.141	80.608	110.150	0.	0.	0.
3200	46.204	117.607	81.742	114.770	0.	0.	0.
3300	46.204	119.029	82.850	119.390	0.	0.	0.
3400	46.204	120.409	83.935	124.011	0.	0.	0.
3500	46.204	121.748	84.996	128.631	0.	0.	0.
3600	46.204	123.049	86.035	133.251	0.	0.	0.
3690.080	46.204	124.191	86.953	137.414	FUGACITY = 1 bar		
3700	46.204	124.315	87.053	137.872	-446.807	1.201	-0.017
3800	46.204	125.548	88.050	142.492	-445.120	13.286	-0.183
3900	46.204	126.748	89.027	147.113	-443.485	25.328	-0.339
4000	46.204	127.918	89.984	151.733	-441.901	37.329	-0.487
4100	46.204	129.058	90.923	156.353	-440.370	49.290	-0.628
4200	46.204	130.172	91.845	160.974	-438.894	61.216	-0.761
4300	46.204	131.259	92.749	165.594	-437.472	73.106	-0.888
4400	46.204	132.321	93.636	170.215	-436.105	84.964	-1.009
4500	46.204	133.360	94.507	174.835	-434.795	96.792	-1.124

PREVIOUS: CURRENT: June 1973

Vanadium (V) $V_1(l)$

VANADIUM (V) CRYSTAL–LIQUID A_r = 50.9415 Vanadium (V) V_1(cr,l)

0 to 2190 K crystal
above 2190 K liquid

Refer to the individual tables for details.

Enthalpy Reference Temperature = T_r = 298.15 K Standard State Pressure = p° = 0.1 MPa

T/K	C_p° (J K⁻¹mol⁻¹)	S° (J K⁻¹mol⁻¹)	$-[G^\circ-H^\circ(T_r)]/T$ (J K⁻¹mol⁻¹)	$H^\circ-H^\circ(T_r)$ (kJ mol⁻¹)	Δ_fH° (kJ mol⁻¹)	Δ_fG° (kJ mol⁻¹)	Log K_f
0	0.	0.	INFINITE	-4.640	0.	0.	0.
100	13.119	7.185	48.695	-4.151	0.	0.	0.
200	21.876	19.558	31.133	-2.315	0.	0.	0.
250	23.703	24.655	29.340	-1.171	0.	0.	0.
298.15	24.896	28.936	28.936	0.	0.	0.	0.
300	24.928	29.090	28.936	0.046	0.	0.	0.
350	25.681	32.992	29.243	1.312	0.	0.	0.
400	26.234	36.459	29.932	2.611	0.	0.	0.
450	26.610	39.571	30.833	3.932	0.	0.	0.
500	26.945	42.392	31.850	5.271	0.	0.	0.
600	27.489	47.353	34.032	7.993	0.	0.	0.
700	28.033	51.630	36.247	10.768	0.	0.	0.
800	28.660	55.413	38.411	13.602	0.	0.	0.
900	29.372	58.830	40.493	16.504	0.	0.	0.
1000	30.083	61.961	42.485	19.476	0.	0.	0.
1100	30.878	64.864	44.389	22.523	0.	0.	0.
1200	31.798	67.590	46.210	25.656	0.	0.	0.
1300	32.740	70.172	47.955	28.882	0.	0.	0.
1400	33.807	72.638	49.630	32.210	0.	0.	0.
1500	34.811	75.004	51.244	35.640	0.	0.	0.
1600	35.857	77.283	52.800	39.173	0.	0.	0.
1700	37.028	79.492	54.306	42.817	0.	0.	0.
1800	38.200	81.641	55.765	46.577	0.	0.	0.
1900	39.539	83.742	57.182	50.464	0.	0.	0.
2000	40.920	85.805	58.562	54.485	0.	0.	0.
2100	42.468	87.837	59.908	58.652	0.	0.	0.
2190.000	44.141	89.653	61.093	62.547	CRYSTAL <--> LIQUID		
2190.000	46.204	100.085	61.093	85.392	TRANSITION		
2200	46.204	100.295	61.271	85.854	0.	0.	0.
2300	46.204	102.349	63.012	90.474	0.	0.	0.
2400	46.204	104.315	64.693	95.095	0.	0.	0.
2500	46.204	106.202	66.316	99.715	0.	0.	0.
2600	46.204	108.014	67.885	104.335	0.	0.	0.
2700	46.204	109.757	69.403	108.956	0.	0.	0.
2800	46.204	111.438	70.875	113.576	0.	0.	0.
2900	46.204	113.059	72.302	118.197	0.	0.	0.
3000	46.204	114.626	73.687	122.817	0.	0.	0.
3100	46.204	116.141	75.032	127.437	0.	0.	0.
3200	46.204	117.607	76.339	132.058	0.	0.	0.
3300	46.204	119.029	77.612	136.678	0.	0.	0.
3400	46.204	120.409	78.850	141.299	0.	0.	0.
3500	46.204	121.748	80.057	145.919	0.	0.	0.
3600	46.204	123.049	81.233	150.539	0.	0.	0.
3690.080	46.204	124.191	82.268	154.701	------- FUGACITY = 1 bar -------		
3700	46.204	124.315	82.380	155.160	-446.807	1.201	-0.017
3800	46.204	125.548	83.500	159.780	-445.120	13.286	-0.183
3900	46.204	126.748	84.594	164.401	-443.485	25.328	-0.339
4000	46.204	127.918	85.662	169.021	-441.901	37.329	-0.487
4100	46.204	129.058	86.707	173.641	-440.370	49.291	-0.628
4200	46.204	130.172	87.729	178.262	-438.894	61.216	-0.761
4300	46.204	131.259	88.728	182.882	-437.472	73.106	-0.888
4400	46.204	132.321	89.707	187.503	-436.105	84.964	-1.009
4500	46.204	133.360	90.666	192.123	-434.795	96.792	-1.124

PREVIOUS: CURRENT: June 1973

Vanadium (V) V_1(cr,l)

VANADIUM (V) IDEAL GAS $A_r = 50.9414$

$IP(V, g) = 54360 \pm 16\ cm^{-1}$

$S°(298.15\ K) = 182.30 \pm 0.8\ J\ K^{-1}\ mol^{-1}$

$\Delta_f H°(0\ K) = 512.2 \pm 8\ kJ\ mol^{-1}$

$\Delta_f H°(298.15\ K) = 515.5 \pm 8\ kJ\ mol^{-1}$

Electronic Levels and Quantum Weights

State	ϵ_i, cm^{-1}	g_i
$^4F_{3/2}$	0	4
$^4F_{5/2}$	137.38	6
$^4F_{7/2}$	323.42	8
$^4F_{9/2}$	553.02	10
.	.	.
.	.	.
.	.	.
$^2H_{11/2}$	54251.26	12
IP	54360	

Enthalpy of Formation

The vapor pressure over vanadium was measured by Edwards et al. (1) using the Knudsen method and by Farber and Srivastava (6) using effusion-mass spectrometric techniques. Our 2nd and 3rd law analysis is tabulated below, where reaction (A) refers to the sublimation V(cr) = V(g) and reaction (B), the vaporization V(ℓ) = V(g).

Source	Reaction	Data Points	T/K	$\Delta_r H°(298.15\ K)$, kcal mol^{-1} 2nd Law	3rd Law	drift cal K^{-1} mol^{-1}
1	A	12*	1666-1882	121.67±0.99	122.95	0.72±0.55
6	A	7	1900-2155	121.60±1.48	123.45	0.91±0.73
6	B	6	2198-2412	113.04±5.34	119.18	2.65±2.30

*A statistical test discounted the 1766 K data point.

We adopt $\Delta_f H°(298.15\ K) = 123.2$ kcal mol^{-1} for V(g). This value, which is identical to $\Delta_{sub} H°(298.15\ K)$, is a median of the two sublimation studies. Hultgren et al. (2) reported a 3rd law value of 123.089 ± 0.250 kcal mol^{-1} based on a Knudsen vapor pressure study in the range 1771-1880 K by Saxer (3). This result is consistent with our adopted value.

The results of Kant and Lin (5) indicated that the partial pressure of $V_2(g)$ is negligible in the temperature ranges studied above.

Heat Capacity and Entropy

The electronic energy levels and quantum weights are obtained from Moore (4). Only a few of the levels are listed above. All levels reported by Moore (4) are considered in the calculation. The heat capacity and entropy values are very similar to those adopted by Hultgren et al. (2), being identical at 298.15 K and differing by 0.045 cal K^{-1} mol^{-1} in Cp° at 3800 K. There are predicted levels which have not been observed and/or classified. It is not expected that S°(298.15 K) would be affected by the missing states, but that in the range 3000-6000 K, an error of 0.2-0.3 cal K^{-1} mol^{-1} might result.

References

1. J. W. Edwards, H. L. Johnston, and P. E. Blackburn, J. Amer. Chem. Soc. 73, 4727 (1951).
2. R. Hutgren, R. L. Orr, and K. K. Kelley, Supplement to Selected Values of Thermodynamic Properties of Metals and Alloys, V table, May, 1966.
3. R. K. Saxer, Ph. D. Thesis, Ohio State University, 1962.
4. C. E. Moore, U. S. Nat. Bur. Stand., NSRDS-NBS 35, Volume I, 1970 [Reprint of NBS Circular 467, Volume I, 1949].
5. A. Kant and S. Lin, J. Chem. Phys. 51, 1644 (1969).
6. M. Farber and R. D. Srivastava, J. Chem. Soc., Faraday Trans. I 69, 390 (1973).

Vanadium (V) $V_1(g)$

Enthalpy Reference Temperature = T_r = 298.15 K; Standard State Pressure = p° = 0.1 MPa

T/K	$C_p°$ (J $K^{-1}mol^{-1}$)	S°	$-[G°-H°(T_r)]/T$	$H°-H°(T_r)$ (kJ mol^{-1})	$\Delta_f H°$	$\Delta_f G°$	Log K_f
0	0.	0.	INFINITE	-7.907	512.201	512.201	INFINITE
100	28.040	151.831	206.684	-5.485	514.134	499.670	-261.001
200	27.960	171.498	184.740	-2.649	515.135	484.747	-126.603
250	26.963	177.633	182.732	-1.275	515.365	477.121	-99.689
298.15	26.012	182.298	182.298	0.	515.469	469.744	-82.297
300	25.978	182.459	182.299	0.048	515.471	469.460	-81.740
350	25.191	186.402	182.612	1.326	515.483	461.790	-68.918
400	24.647	189.727	183.299	2.571	515.430	454.122	-59.302
450	24.330	192.610	184.177	3.795	515.332	446.464	-51.824
500	24.196	195.165	185.150	5.007	515.205	438.819	-45.843
600	24.283	199.579	187.198	7.428	514.905	423.569	-36.875
700	24.582	203.343	189.242	9.871	514.572	408.372	-30.473
800	24.890	206.647	191.215	12.345	514.212	393.225	-25.675
900	25.117	209.592	193.097	14.846	513.811	378.125	-21.946
1000	25.237	212.246	194.881	17.365	513.358	363.073	-18.965
1100	25.262	214.653	196.571	19.890	512.837	348.069	-16.528
1200	25.214	216.849	198.171	22.415	512.227	333.116	-14.500
1300	25.117	218.864	199.686	24.931	511.518	318.218	-12.786
1400	24.994	220.721	201.123	27.437	510.696	303.379	-11.319
1500	24.864	222.441	202.488	29.930	509.758	288.603	-10.050
1600	24.739	224.042	203.785	32.410	508.706	273.893	-8.942
1700	24.630	225.538	205.021	34.878	507.530	259.252	-7.966
1800	24.545	226.943	206.201	37.337	506.229	244.685	-7.101
1900	24.487	228.269	207.327	39.788	504.793	230.193	-6.328
2000	24.460	229.524	208.406	42.235	503.219	215.781	-5.636
2100	24.467	230.717	209.440	44.682	501.498	201.450	-5.011
2200	24.510	231.856	210.434	47.130	476.745	187.310	-4.447
2300	24.588	232.948	211.389	49.585	474.579	174.203	-3.956
2400	24.702	233.996	212.309	52.049	472.423	161.189	-3.508
2500	24.853	235.008	213.197	54.526	470.280	148.265	-3.098
2600	25.040	235.986	214.055	57.021	468.154	135.426	-2.721
2700	25.261	236.935	214.885	59.535	466.048	122.669	-2.373
2800	25.517	237.858	215.689	62.074	463.967	109.990	-2.052
2900	25.806	238.758	216.469	64.640	461.912	97.384	-1.754
3000	26.127	239.639	217.227	67.236	459.888	84.849	-1.477
3100	26.477	240.501	217.963	69.866	457.898	72.380	-1.220
3200	26.855	241.348	218.681	72.533	455.944	59.975	-0.979
3300	27.259	242.180	219.381	75.238	454.029	47.631	-0.754
3400	27.687	243.000	220.063	77.985	452.156	35.344	-0.543
3500	28.136	243.809	220.730	80.776	450.326	23.112	-0.345
3600	28.605	244.608	221.382	83.613	448.543	10.931	-0.159
3690.080	29.042	245.321	221.958	86.210	------- FUGACITY = 1 bar -------		
3700	29.090	245.399	222.021	86.498	0.	0.	0.
3800	29.591	246.181	222.646	89.432	0.	0.	0.
3900	30.105	246.956	223.260	92.417	0.	0.	0.
4000	30.631	247.725	223.862	95.453	0.	0.	0.
4100	31.166	248.488	224.453	98.543	0.	0.	0.
4200	31.710	249.246	225.034	101.687	0.	0.	0.
4300	32.260	249.998	225.606	104.885	0.	0.	0.
4400	32.817	250.746	226.169	108.139	0.	0.	0.
4500	33.379	251.490	226.723	111.449	0.	0.	0.
4600	33.946	252.230	227.270	114.815	0.	0.	0.
4700	34.517	252.966	227.809	118.238	0.	0.	0.
4800	35.093	253.699	228.341	121.719	0.	0.	0.
4900	35.672	254.428	228.866	125.257	0.	0.	0.
5000	36.255	255.155	229.384	128.853	0.	0.	0.
5100	36.843	255.878	229.896	132.508	0.	0.	0.
5200	37.435	256.600	230.403	136.222	0.	0.	0.
5300	38.031	257.318	230.904	139.995	0.	0.	0.
5400	38.632	258.035	231.400	143.828	0.	0.	0.
5500	39.238	258.749	231.891	147.721	0.	0.	0.
5600	39.849	259.462	232.377	151.675	0.	0.	0.
5700	40.465	260.172	232.858	155.691	0.	0.	0.
5800	41.087	260.881	233.335	159.768	0.	0.	0.
5900	41.712	261.589	233.808	163.908	0.	0.	0.
6000	42.343	262.295	234.277	168.110	0.	0.	0.

PREVIOUS: June 1973 (1 atm) CURRENT: June 1973 (1 bar)

Vanadium (V) $V_1(g)$

VANADIUM, ION (V^+) IDEAL GAS M_r = 50.94095

$IP(V^+, g) = 118200 \pm 200\ cm^{-1}$

$S°(298.15\ K) = 183.38 \pm 0.8\ J\ K^{-1}\ mol^{-1}$

$\Delta_f H°(0\ K) = 1162.5 \pm 8\ kJ\ mol^{-1}$

$\Delta_f H°(298.15\ K) = [1171.947]\ kJ\ mol^{-1}$

Electronic Levels and Quantum Weights

State	ϵ, cm^{-1}	g_i
$^5D_0$	0.00	1
$^5D_1$	36.05	3
$^5D_2$	103.83	5
$^5D_3$	203.89	7
$^5D_4$	339.21	9
$^5F_1$	2604.82	3

Enthalpy of Formation

$\Delta_f H°(V^+, g, 0\ K)$ is calculated from $\Delta_f H°(V, g, 0\ K)$ (1) using the spectroscopic value of IP(V) = 54360 ± 16 cm^{-1} (650.290 ± 0.19 kJ mol^{-1}) from Sugar and Corliss (2). The ionization limit is converted from cm^{-1} to kJ mol^{-1} using the factor, 1 cm^{-1} = 0.01196266 kJ cm^{-1}, which is derived from the 1973 CODATA fundamental constants (3). Rosenstock et al. (4) and Levin and Lias (5) have summarized additional ionization and appearance potential data.

$\Delta_f H°(V^+, g, 298.15\ K)$ is calculated from $\Delta_f H°(V, g, 0\ K)$ by using IP(V) with JANAF (1) enthalpies, H°(0 K) - H°(298.15 K), for V(g), $V^+(g)$, and $e^-(g)$. $\Delta_f H°(V \rightarrow V^+ + e^-, 298.15\ K)$ differs from a room temperature threshold energy due to inclusion of these enthalpies and to threshold effects discussed by Rosenstock et al. (4). $\Delta_f H°(298.15\ K)$ should be changed by -6.197 kJ mol^{-1} if it is to be used in the ion convention that excludes the enthalpy of the electron.

Heat Capacity and Entropy

The information on electronic energy levels and quantum weights, given by Sugar and Corliss (2), is incomplete because many theoretically predicted levels have not been observed. Although we have listed only the ground, the first excited state, the highest observed excited state, and the ionization potential for $V^+(g)$, all levels listed by Sugar and Corliss (2), as well as estimated levels, are used in the calculation. The observed levels are too numerous to list completely. The calculations indicate that for $V^+(g)$, the thermodynamic functions are independent of the estimated missing levels (for n=4, 5), the cut-off procedure, and the inclusion of n=5 levels up to 6000 K; the Gibbs energy function showing variations of 0.001 J $K^{-1}mol^{-1}$ at this temperature. The reported uncertainty in S°(298.15 K) is due to uncertainties in the relative ionic mass, and the fundamental constants. Extension of these calculations above 6000 K may require consideration of the higher excited states (n>5), and use of different fill and cut-off procedures (6).

References

1. JANAF Thermochemical Tables: V(g), 3-30-73; e^-(g), 3-31-32.
2. J. Sugar and C. Corliss, J. Phys. Chem. Ref. Data 7, 1194 (1978).
3. E. R. Cohen and B. N. Taylor, J. Phys. Chem. Ref. Data 2, 663 (1973).
4. H. M. Rosenstock, K. Draxl et al., J. Phys. Chem. Ref. Data 6, Supp. 1 (1977).
5. R. D. Levin and S. G. Lias, U. S. Natl. Bur. Stand. NSRDS-NBS-71, 1982.
6. J. R. Downey, Jr., The Dow Chemical Company, AFOSR-TR-78-0960, Contract No. F44620-75-1-0048, 1978.

Vanadium, Ion (V^+) $V_1^+(g)$

Enthalpy Reference Temperature = T_r = 298.15 K Standard State Pressure = p° = 0.1 MPa

T/K	$C_p°$ (J $K^{-1}mol^{-1}$)	S° (J $K^{-1}mol^{-1}$)	$-[G°-H°(T_r)]/T$ (J $K^{-1}mol^{-1}$)	$H°-H°(T_r)$ (kJ mol^{-1})	$\Delta_f H°$ (kJ mol^{-1})	$\Delta_f G°$ (kJ mol^{-1})	Log K_f
0	0.	0.	INFINITE	-7.898	1162.492		
100	28.792	154.942	205.447	-5.051			
200	25.153	173.759	185.546	-2.357			
250	23.935	179.232	183.761	-1.132			
298.15	23.149	183.376	183.376	0.	1171.947	1119.646	-196.157
300	23.125	183.519	183.376	0.043	1171.982	1119.321	-194.891
350	22.591	187.041	183.656	1.185	1172.898	1110.471	-165.729
400	22.253	190.033	184.271	2.305	1173.759	1101.494	-143.840
450	22.066	192.642	185.059	3.413	1174.584	1092.411	-126.804
500	22.004	194.963	185.935	4.514	1175.386	1083.238	-113.165
600	22.179	198.985	187.785	6.720	1176.949	1064.661	-92.687
700	22.631	202.435	189.637	8.959	1178.491	1045.824	-78.040
800	23.221	205.495	191.431	11.251	1180.028	1026.767	-67.041
900	23.836	208.266	193.150	13.604	1181.558	1007.517	-58.475
1000	24.400	210.807	194.790	16.016	1183.077	988.098	-51.613
1100	24.872	213.155	196.355	18.481	1184.573	968.527	-45.992
1200	25.238	215.336	197.847	20.987	1186.024	948.822	-41.301
1300	25.502	217.367	199.271	23.525	1187.415	928.998	-37.328
1400	25.677	219.264	200.632	26.085	1188.725	909.070	-33.918
1500	25.782	221.039	201.934	28.658	1189.946	889.052	-30.960
1600	25.834	222.705	203.181	31.239	1191.074	868.955	-28.368
1700	25.850	224.272	204.376	33.824	1192.093	848.791	-26.080
1800	25.846	225.749	205.523	36.409	1192.996	828.570	-24.044
1900	25.834	227.147	206.624	38.993	1193.772	808.302	-22.222
2000	25.823	228.471	207.684	41.575	1194.412	787.997	-20.580
2100	25.821	229.731	208.704	44.157	1194.906	767.663	-19.095
2200	25.833	230.933	209.687	46.740	1172.365	747.415	-17.746
2300	25.864	232.082	210.636	49.325	1172.408	728.098	-16.536
2400	25.915	233.183	211.553	51.914	1172.455	708.779	-15.426
2500	25.988	234.243	212.439	54.509	1172.508	689.458	-14.405
2600	26.084	235.264	213.298	57.112	1172.570	670.135	-13.463
2700	26.201	236.250	214.130	59.726	1172.642	650.809	-12.591
2800	26.339	237.206	214.937	62.353	1172.727	631.480	-11.780
2900	26.497	238.133	215.721	64.994	1172.827	612.148	-11.026
3000	26.672	239.034	216.483	67.653	1172.944	592.813	-10.322
3100	26.864	239.911	217.224	70.329	1173.078	573.473	-9.663
3200	27.070	240.768	217.947	73.026	1173.233	554.128	-9.045
3300	27.288	241.604	218.651	75.744	1173.409	534.779	-8.465
3400	27.516	242.422	219.338	78.484	1173.608	515.423	-7.919
3500	27.752	243.223	220.009	81.247	1173.829	496.062	-7.403
3600	27.993	244.008	220.665	84.034	1174.075	476.693	-6.917
3700	28.239	244.778	221.306	86.846	727.537	458.519	-6.473
3800	28.486	245.535	221.934	89.682	729.518	451.221	-6.202
3900	28.735	246.278	222.549	92.543	731.473	443.872	-5.945
4000	28.981	247.008	223.151	95.429	733.401	436.473	-5.700
4100	29.226	247.727	223.742	98.340	735.300	429.027	-5.466
4200	29.467	248.434	224.321	101.274	737.170	421.534	-5.243
4300	29.702	249.130	224.890	104.233	739.008	413.997	-5.029
4400	29.932	249.816	225.449	107.214	740.815	406.418	-4.825
4500	30.155	250.491	225.998	110.219	742.588	398.798	-4.629
4600	30.372	251.156	226.538	113.245	744.327	391.139	-4.442
4700	30.580	251.812	227.068	116.293	746.030	383.442	-4.261
4800	30.780	252.458	227.591	119.361	747.696	375.710	-4.089
4900	30.972	253.094	228.105	122.449	749.324	367.943	-3.922
5000	31.156	253.722	228.611	125.555	750.913	360.144	-3.762
5100	31.331	254.340	229.109	128.680	752.461	352.313	-3.608
5200	31.498	254.950	229.600	131.821	753.968	344.452	-3.460
5300	31.657	255.552	230.084	134.979	755.431	336.563	-3.317
5400	31.808	256.145	230.561	138.152	756.850	328.646	-3.179
5500	31.952	256.730	231.032	141.340	758.223	320.704	-3.046
5600	32.090	257.307	231.496	144.542	759.550	312.737	-2.917
5700	32.221	257.876	231.954	147.758	760.828	304.746	-2.793
5800	32.347	258.438	232.406	150.986	762.058	296.734	-2.672
5900	32.468	258.992	232.851	154.227	763.238	288.701	-2.556
6000	32.585	259.538	233.292	157.480	764.367	280.649	-2.443

PREVIOUS: CURRENT: March 1984 (1 bar)

Vanadium, Ion (V^+) $V_1^+(g)$

VANADIUM, ION (V^-) IDEAL GAS M_r = 50.94205

EA(V, g) = 0.525 ± 0.012 eV
S°(298.15 K) = 183.436 ± 0.002 $J\ K^{-1}\ mol^{-1}$

$\Delta_f H°$(0 K) = 461.547 ± 2.5 $kJ\ mol^{-1}$
$\Delta_f H°$(298.15 K) = [458.588] $kJ\ mol^{-1}$

Electonic Levels and Quantum Weights

State	ε_i, cm^{-1}	g_i
$^5D_0$	0.0	1
$^5D_1$	35	3
$^5D_2$	105	5
$^5D_3$	205	7
$^5D_4$	330	9

Enthalpy of Formation

$\Delta_f H°(V^-$, g, 0 K) is calculated from $\Delta_f H°$(V, g 0 K) (1) using the adopted electron affinity of EA(V) = 0.525 ± 0.012 eV (50.654 ± 1.158 $kJ\ mol^{-1}$). This value, recommended by Hotop and Lineberger (2), is based on a laser photodetachment electron spectrometry study (3). Additional information on V^-(g) may be obtained in the critical discussions of Hotop and Lineberger (2, 4), Rosenstock et al. (5) and Massey (6).

$\Delta_f H°(V^-$, g, 298.15 K) is obtained from $\Delta_f H°$(V, g, 0 K) by using EA(V) with JANAF (1) enthalpies, H°(0 K) - H°(298.15 K), for V^-(g), V(g), and e^-(g). $\Delta_r H°(V^- \rightarrow V + e^-$, 298.15 K) differs from a room-temperature threshold energy due to inclusion of these enthalpies and to threshold effects discussed by Rosenstock et al. (5). $\Delta_f H°$(298.15 K) should be changed by +6.197 $kJ\ mol^{-1}$ if it is to be used in the ion convention that excludes the enthalpy of the electron.

Heat Capacity and Entropy

The ground state electronic configuration for V^-(g) is given by Hotop and Lineberger (2, 4), Rosenstock et al (5), and Massey(6). The fine structure separation has been calculated by an isoelectronic extrapolation of ratios of fine structure separations (3) and is that recommended by Hotop and Lineberger (2).

References

1. JANAF Thermochemical Tables: V(g), 6-30-73 ; e^-(g), 3-31-82.
2. H. Hotop and W. C. Lineberger, J. Phys. Chem. Ref. Data, to be published, 1985.
3. C. S. Feigerle, R. R. Corderman, S. V. Bobaskev and W. C. Lineberger, J. Chem. Phys. 74, 1580 (1981).
4. H. Hotop and W. C. Lineberger, J. Phys. Chem. Ref. Data 4, 539 (1975).
5. H. M. Rosenstock, K. Draxl et al., J. Phys. Chem. Ref. Data 6, Supp. 1 (1977).
6. H. S. W. Massey, "Negative Ions", 3rd ed., Cambridge University Press, Cambridge, 1976.

Vanadium, Ion (V^-) V_1^-(g)

Enthalpy Reference Temperature = T_r = 298.15 K Standard State Pressure = p° = 0.1 MPa

T/K	$C_p°$ (J K⁻¹mol⁻¹)	S° (J K⁻¹mol⁻¹)	$-[G°-H°(T_r)]/T$ (J K⁻¹mol⁻¹)	$H°-H°(T_r)$ (kJ mol⁻¹)	$\Delta_f H°$ (kJ mol⁻¹)	$\Delta_f G°$ (kJ mol⁻¹)	Log K_f
0	0.	0.	INFINITE	-7.878	461.547		
100	28.791	155.101	205.411	-5.031			
200	25.030	173.865	185.596	-2.346			
250	23.821	179.311	183.819	-1.127			
298.15	23.049	183.436	183.436	0.	458.588	418.779	-73.368
300	23.025	183.579	183.436	0.043	458.546	418.532	-72.873
350	22.490	187.085	183.715	1.180	457.378	411.954	-61.481
400	22.119	190.062	184.327	2.294	456.155	405.548	-52.959
450	21.854	192.652	185.111	3.393	454.893	399.298	-46.349
500	21.659	194.943	185.982	4.481	453.603	393.190	-41.076
600	21.399	198.867	187.813	6.633	450.954	381.354	-33.200
700	21.239	202.153	189.633	8.764	448.231	369.969	-27.607
800	21.134	204.982	191.379	10.882	445.437	358.978	-23.439
900	21.061	207.467	193.031	12.992	442.566	348.342	-20.217
1000	21.009	209.683	194.588	15.095	439.619	338.030	-17.657
1100	20.970	211.683	196.052	17.194	436.593	328.017	-15.576
1200	20.941	213.507	197.432	19.290	433.476	318.283	-13.855
1300	20.918	215.182	198.734	21.382	430.264	308.814	-12.408
1400	20.900	216.731	199.965	23.473	426.949	299.595	-11.178
1500	20.885	218.173	201.131	25.562	423.529	290.617	-10.120
1600	20.873	219.520	202.239	27.650	420.006	281.871	-9.202
1700	20.863	220.785	203.293	29.737	416.370	273.348	-8.399
1800	20.855	221.978	204.298	31.823	412.617	265.043	-7.691
1900	20.848	223.105	205.259	33.908	408.737	256.949	-7.064
2000	20.842	224.174	206.178	35.993	404.721	249.064	-6.505
2100	20.836	225.191	207.059	38.076	400.560	241.382	-6.004
2200	20.832	226.160	207.906	40.160	373.362	234.008	-5.556
2300	20.828	227.086	208.720	42.243	368.746	227.777	-5.173
2400	20.825	227.972	209.503	44.325	364.130	221.747	-4.826
2500	20.822	228.822	210.259	46.408	359.513	215.909	-4.511
2600	20.819	229.639	210.989	48.490	354.896	210.256	-4.224
2700	20.816	230.425	211.694	50.572	350.279	204.780	-3.962
2800	20.814	231.182	212.377	52.653	345.662	199.476	-3.721
2900	20.812	231.912	213.038	54.734	341.044	194.337	-3.500
3000	20.811	232.618	213.679	56.816	336.426	189.356	-3.297
3100	20.809	233.300	214.301	58.897	331.808	184.530	-3.109
3200	20.808	233.961	214.905	60.977	327.190	179.853	-2.936
3300	20.806	234.601	215.492	63.058	322.572	175.320	-2.775
3400	20.805	235.222	216.063	65.139	317.953	170.927	-2.626
3500	20.804	235.825	216.620	67.219	313.335	166.670	-2.487
3600	20.803	236.411	217.161	69.300	308.716	162.545	-2.358
3700	20.802	236.981	217.689	71.380	-142.709	159.749	-2.255
3800	20.801	237.536	218.204	73.460	-145.642	167.963	-2.309
3900	20.801	238.076	218.707	75.540	-148.625	176.254	-2.361
4000	20.800	238.603	219.198	77.620	-151.660	184.623	-2.411
4100	20.799	239.116	219.677	79.700	-154.749	193.068	-2.460
4200	20.799	239.617	220.146	81.780	-157.891	201.590	-2.507
4300	20.798	240.107	220.605	83.860	-161.088	210.186	-2.553
4400	20.797	240.585	221.053	85.940	-164.341	218.858	-2.598
4500	20.797	241.052	221.493	88.019	-167.650	227.604	-2.642
4600	20.796	241.509	221.923	90.099	-171.015	236.425	-2.685
4700	20.796	241.957	222.344	92.179	-174.437	245.319	-2.726
4800	20.796	242.395	222.757	94.258	-177.917	254.287	-2.767
4900	20.795	242.823	223.163	96.338	-181.454	263.327	-2.807
5000	20.795	243.243	223.560	98.417	-185.049	272.441	-2.846
5100	20.795	243.655	223.950	100.497	-188.703	281.627	-2.884
5200	20.794	244.059	224.333	102.576	-192.416	290.885	-2.922
5300	20.794	244.455	224.709	104.655	-196.188	300.215	-2.959
5400	20.794	244.844	225.078	106.735	-200.021	309.617	-2.995
5500	20.793	245.225	225.441	108.814	-203.913	319.091	-3.030
5600	20.793	245.600	225.798	110.893	-207.867	328.636	-3.065
5700	20.793	245.968	226.148	112.973	-211.882	338.252	-3.100
5800	20.793	246.330	226.493	115.052	-215.958	347.938	-3.134
5900	20.792	246.685	226.832	117.131	-220.097	357.696	-3.167
6000	20.792	247.035	227.166	119.211	-224.299	367.525	-3.200

PREVIOUS: CURRENT: March 1984 (1 bar)

Vanadium, Ion (V^-) V_1^-(g)

TUNGSTEN (W) REFERENCE STATE A_r = 183.85

0 to 3680 K crystal
3680 to 6000 K liquid

Refer to the individual tables for details.

T_{vap}(1 bar) > 6000 K

Tungsten (W) W_1(ref)

Enthalpy Reference Temperature = T_r = 298.15 K | Standard State Pressure = p° = 0.1 MPa

T/K	C_p° (J K⁻¹mol⁻¹)	S° (J K⁻¹mol⁻¹)	$-[G°-H°(T_r)]/T$ (J K⁻¹mol⁻¹)	$H°-H°(T_r)$ (kJ mol⁻¹)	$\Delta_f H°$ (kJ mol⁻¹)	$\Delta_f G°$ (kJ mol⁻¹)	Log K_f
0	0.	0.	INFINITE	-4.973	0.	0.	0.
100	16.033	9.612	52.820	-4.321	0.	0.	0.
200	22.489	23.273	34.839	-2.313	0.	0.	0.
250	23.686	28.431	33.056	-1.156	0.	0.	0.
298.15	24.295	32.660	32.660	0.	0.	0.	0.
300	24.313	32.810	32.660	0.045	0.	0.	0.
350	24.644	36.583	32.957	1.269	0.	0.	0.
400	24.928	39.893	33.622	2.509	0.	0.	0.
450	25.144	42.842	34.485	3.760	0.	0.	0.
500	25.359	45.502	35.456	5.023	0.	0.	0.
600	25.790	50.163	37.529	7.580	0.	0.	0.
700	26.229	54.172	39.627	10.181	0.	0.	0.
800	26.669	57.703	41.670	12.826	0.	0.	0.
900	27.112	60.870	43.631	15.515	0.	0.	0.
1000	27.564	63.750	45.501	18.249	0.	0.	0.
1100	28.017	66.398	47.282	21.028	0.	0.	0.
1200	28.472	68.855	48.978	23.852	0.	0.	0.
1300	28.930	71.152	50.596	26.722	0.	0.	0.
1400	29.393	73.313	52.143	29.639	0.	0.	0.
1500	29.862	75.357	53.623	32.601	0.	0.	0.
1600	30.334	77.299	55.042	35.611	0.	0.	0.
1700	30.807	79.152	56.407	38.668	0.	0.	0.
1800	31.284	80.927	57.720	41.773	0.	0.	0.
1900	31.765	82.631	58.986	44.925	0.	0.	0.
2000	32.254	84.273	60.210	48.126	0.	0.	0.
2100	32.744	85.858	61.394	51.376	0.	0.	0.
2200	33.238	87.393	62.541	54.675	0.	0.	0.
2300	33.736	88.881	63.654	58.024	0.	0.	0.
2400	34.233	90.328	64.735	61.422	0.	0.	0.
2500	34.736	91.735	65.787	64.870	0.	0.	0.
2600	35.246	93.106	66.812	68.366	0.	0.	0.
2700	36.192	94.453	67.811	71.935	0.	0.	0.
2800	37.447	95.791	68.786	75.613	0.	0.	0.
2900	39.120	97.133	69.740	79.440	0.	0.	0.
3000	41.003	98.490	70.676	83.442	0.	0.	0.
3100	43.430	99.873	71.595	87.662	0.	0.	0.
3200	46.024	101.292	72.501	92.132	0.	0.	0.
3300	48.953	102.752	73.396	96.877	0.	0.	0.
3400	52.300	104.261	74.281	101.933	0.	0.	0.
3500	56.484	105.835	75.160	107.363	0.	0.	0.
3600	61.714	107.499	76.035	113.270	0.	0.	0.
3680.000	66.149	108.904	76.734	118.383	CRYSTAL <--> LIQUID		
3680.000	35.564	118.522	76.734	153.780	TRANSITION		
3700	35.564	118.715	76.961	154.491	0.	0.	0.
3800	35.564	119.663	78.072	158.048	0.	0.	0.
3900	35.564	120.587	79.150	161.604	0.	0.	0.
4000	35.564	121.488	80.197	165.161	0.	0.	0.
4100	35.564	122.366	81.215	168.717	0.	0.	0.
4200	35.564	123.223	82.205	172.273	0.	0.	0.
4300	35.564	124.060	83.169	175.830	0.	0.	0.
4400	35.564	124.877	84.108	179.386	0.	0.	0.
4500	35.564	125.676	85.023	182.943	0.	0.	0.
4600	35.564	126.458	85.915	186.499	0.	0.	0.
4700	35.564	127.223	86.786	190.055	0.	0.	0.
4800	35.564	127.972	87.636	193.612	0.	0.	0.
4900	35.564	128.705	88.467	197.168	0.	0.	0.
5000	35.564	129.423	89.279	200.725	0.	0.	0.
5100	35.564	130.128	90.073	204.281	0.	0.	0.
5200	35.564	130.818	90.850	207.837	0.	0.	0.
5300	35.564	131.496	91.610	211.394	0.	0.	0.
5400	35.564	132.161	92.355	214.950	0.	0.	0.
5500	35.564	132.813	93.085	218.507	0.	0.	0.
5600	35.564	133.454	93.800	222.063	0.	0.	0.
5700	35.564	134.083	94.501	225.619	0.	0.	0.
5800	35.564	134.702	95.189	229.176	0.	0.	0.
5900	35.564	135.310	95.864	232.732	0.	0.	0.
6000	35.564	135.908	96.526	236.289	0.	0.	0.

PREVIOUS: June 1966 (1 atm) CURRENT: June 1966 (1 bar)

Tungsten (W) W_1(ref)

TUNGSTEN (W) CRYSTAL A_r = 183.85

$S°(298.15\ K) = 32.660 \pm 0.10\ J\ K^{-1}\ mol^{-1}$

$T_{fus} = 3680 \pm 20\ K$

$\Delta_f H°(0\ K) = 0\ kJ\ mol^{-1}$

$\Delta_f H°(298.15\ K) = 0\ kJ\ mol^{-1}$

$\Delta_{fus} H° = 35.40 \pm 10.5\ kJ\ mol^{-1}$

Enthalpy of Formation

Zero by definition.

Heat Capacity and Entropy

There have been several investigations of the low temperature heat capacity of tungsten; Lange (1) covered the range 26-91 K. Horowitz and Daunt (2) reported values in the range 1-77 K, while Waite et al. (3) worked from 4-15° and DeSorbo (4) from 15-90 K. Zwikker and Schmidt (5) covered the range from 90 to 2521 K. Clusius et al. (6) made a thorough study of the heat capacity in the range 12-274 K, and their measurements were adopted leading to S°(298.15 K) = 7.806 cal $K^{-1}\ mol^{-1}$ based on S°(12.5) = 0.0088 cal $K^{-1}\ mol^{-1}$. This value is in disagreement with the quoted value of 7.83 cal $K^{-1}\ mol^{-1}$ but agrees exactly with a separate integration by Kirillin et al. (7). In the intermediate temperature range the adiabatic heat capacity measurements of Bronson et al. (8) from 253 to 773 K joined well with the low temperature measurements and were adopted. The high temperature enthalpies have been measured by several investigators from which were selected the values of Jaeger and Rosenbohm (9) from 273-1800 K; Magnus and Holzmann (10) who covered the range 373 to 1173 K; Hoch and Johnston, (11) who made measurements between 1382 and 2900 K; Kirillin et al. (7) who have made several determinations over the range 600 to 3100 K, and have also analysed the above data and have presented smooth functions from 0 to 3500 K. The present table agreess with that of Kirillin et al. (7) up to 2700 K, above this temperature the values of Cp adopted follow those reported by Novikov et al. (12) measured by an electric modulation method, up to the melting point. These values rise rapidly above 2700 K and are not inconsistent with the individual measurements of Kirillin et al (7) at their highest temperatures.

Fusion Data

Langmuir (13) determined the melting point as 3540 K from intrinsic brilliance measurements; this was later corrected to 3655 ± 30 K by Jones et al. (14), and Pirani and Alterthum (15) from pyrometer measurements on a black body hole found 3660 ± 60 K. Using the same technique Zalabak (16) reports 3680 K on a low carbon specimen. He reports a decrease of the melting point with increasing carbon content. Rudy and Windisch (17) report a melting point of 3696 ± 20 K. The value adopted is 3680 K ± 20 K. The enthalpy of melting is obtained by assuming an entropy of melting of 2.3 cal $K^{-1}\ mol^{-1}$ obtained from a comparison of several high melting metals (Fe, Cu, Co, Mg, Al).

Sublimation Data

See ideal gas table for details.

References

1. F. Lange, Z. physik. Chem. 110, 343 (1924).
2. M. Horowitz and J. G. Daunt, Phys. Rev. 91, 1099 (1953).
3. T. R. Waite, R. S. Craig, and W. E. Wallace, Phys. Rev. 104, 1240 (1956).
4. W. DeSorbo, J. Phys. Chem. 62, 965 (1958).
5. C. Zwikker and G. Schmidt, Z. Physik. 52, 668 (1928).
6. K. Clusius and P. Franzozini, Z. Naturforsch. 14A, 99 (1959).
7. V. A. Kirillin, A. E. Sheindlin, V. Ya. Chekhovskoi and V. A. Petrov, Zhur. Fiz. Khim. 37, 2249 (1963).
8. H. L. Bronson, H. M. Chisholm and S. M. Dockerty, Can. J. Research 8, 282 (1933).
9. F. M. Jaeger and E. Rosenbohm, Rec. trav. chim. 51, 1 (1932).
10. A. Magnus and H. Holzmann, Ann. Physik. Ser 5, 3, 585 (1929).
11. M. Hoch and H. L. Johnston, J. Phys. Chem. 65, 855 (1961).
12. I. I. Novikov and P. G. Strelkov, Vestnik. Akad. Nauk SSSR 34, 26 (1964).
13. I. Langmuir, Phys. Rev. 6, 138 (1915).
14. H. A. Jones, I. Langmuir and G. M. J. Mackay, Phys. Rev. 30, 201 (1927).
15. M. Pirani and H. Alterthum, Z. Elektrochem. 29, 5 (1923).
16. C. F. zalabak, NASA Tech. Note D-761 (1961).
17. E. Rudy and S. Windisch, Aerojet-General Tech. Rept. No. AFML- TR-65-2 Part 1, Vol. III July 1965.

Tungsten (W) W_1(cr)

Enthalpy Reference Temperature = T_r = 298.15 K Standard State Pressure = p° = 0.1 MPa

	J K⁻¹mol⁻¹			kJ mol⁻¹			
T/K	$C_p°$	S°	$-[G°-H°(T_r)]/T$	$H°-H°(T_r)$	$\Delta_f H°$	$\Delta_f G°$	Log K_f
0	0.	0.	INFINITE	-4.973	0.	0.	0.
100	16.033	9.612	52.820	-4.321	0.	0.	0.
200	22.489	23.273	34.839	-2.313	0.	0.	0.
250	23.686	28.431	33.056	-1.156	0.	0.	0.
298.15	24.295	32.660	32.660	0.	0.	0.	0.
300	24.313	32.810	32.660	0.045	0.	0.	0.
350	24.644	36.583	32.957	1.269	0.	0.	0.
400	24.928	39.893	33.622	2.509	0.	0.	0.
450	25.144	42.842	34.485	3.760	0.	0.	0.
500	25.359	45.502	35.456	5.023	0.	0.	0.
600	25.790	50.163	37.529	7.580	0.	0.	0.
700	26.229	54.172	39.627	10.181	0.	0.	0.
800	26.669	57.703	41.670	12.826	0.	0.	0.
900	27.112	60.870	43.631	15.515	0.	0.	0.
1000	27.564	63.750	45.501	18.249	0.	0.	0.
1100	28.017	66.398	47.282	21.028	0.	0.	0.
1200	28.472	68.855	48.978	23.852	0.	0.	0.
1300	28.930	71.152	50.596	26.722	0.	0.	0.
1400	29.393	73.313	52.143	29.639	0.	0.	0.
1500	29.862	75.357	53.623	32.601	0.	0.	0.
1600	30.334	77.299	55.042	35.611	0.	0.	0.
1700	30.807	79.152	56.407	38.668	0.	0.	0.
1800	31.284	80.927	57.720	41.773	0.	0.	0.
1900	31.765	82.631	58.986	44.925	0.	0.	0.
2000	32.254	84.273	60.210	48.126	0.	0.	0.
2100	32.744	85.858	61.394	51.376	0.	0.	0.
2200	33.238	87.393	62.541	54.675	0.	0.	0.
2300	33.736	88.881	63.654	58.024	0.	0.	0.
2400	34.233	90.328	64.735	61.422	0.	0.	0.
2500	34.736	91.735	65.787	64.870	0.	0.	0.
2600	35.246	93.106	66.812	68.366	0.	0.	0.
2700	36.192	94.453	67.811	71.935	0.	0.	0.
2800	37.447	95.791	68.786	75.613	0.	0.	0.
2900	39.120	97.133	69.740	79.440	0.	0.	0.
3000	41.003	98.490	70.676	83.442	0.	0.	0.
3100	43.430	99.873	71.595	87.662	0.	0.	0.
3200	46.024	101.292	72.501	92.132	0.	0.	0.
3300	48.953	102.752	73.396	96.877	0.	0.	0.
3400	52.300	104.261	74.281	101.933	0.	0.	0.
3500	56.484	105.835	75.160	107.363	0.	0.	0.
3600	61.714	107.499	76.035	113.270	0.	0.	0.
3680.000	66.149	108.904	76.734	118.383	------ CRYSTAL <--> LIQUID -----		
3700	67.205	109.265	76.909	119.716	-34.775	0.191	-0.003
3800	72.697	111.130	77.785	126.712	-31.336	1.091	-0.015
3900	78.188	113.089	78.665	134.256	-27.348	1.893	-0.025
4000	83.680	115.138	79.551	142.349	-22.811	2.587	-0.034
4100	89.171	117.272	80.445	150.992	-17.725	3.160	-0.040
4200	94.663	119.487	81.348	160.184	-12.090	3.602	-0.045
4300	100.154	121.778	82.261	169.924	-5.905	3.904	-0.047
4400	105.646	124.144	83.186	180.214	0.828	4.055	-0.048
4500	111.137	126.579	84.123	191.054	8.111	4.047	-0.047
4600	116.629	129.082	85.073	202.442	15.943	3.872	-0.044
4700	122.120	131.649	86.037	214.379	24.324	3.520	-0.039
4800	127.612	134.278	87.014	226.866	33.254	2.984	-0.032

PREVIOUS: December 1961 CURRENT: June 1966

Tungsten (W) W_1(cr)

TUNGSTEN (W) LIQUID A_r = 183.85

$S°$(298.15 K) = [45.678] J K^{-1} mol^{-1} $\Delta_f H°$(298.15 K) = [46.896] kJ mol^{-1}

T_{fus} = 3680 ± 20 K $\Delta_{fus}H°$ = 35.40 ± 10.5 kJ mol^{-1}

Enthalpy of Formation

The enthalpy of formation at 298.15 K was calculated from that of the crystal by adding $\Delta_{fus}H°$ and the difference in enthalpy, H°(3680 K) - H°(298.15 K), between the crystal and liquid.

Heat Capacity and Entropy

The heat capacity was estimated as 8.5 cal mol^{-1} K^{-1} by analogy with other monatomic metals. The entropy at 298.15 K was calculated in manner analogous to that used for the enthalpy of formation. At 2450 K a glass transition is assumed below which the heat capacity was that of the crystal.

Fusion Data

See the crystal table for details.

Vaporization Data

The boiling point and enthalpy of vaporization are calculated from the adopted functions and enthalpy of sublimation in order to maintain proper thermodynamic consistency.

Tungsten (W) W_1(l)

Enthalpy Reference Temperature = T_r = 298.15 K Standard State Pressure = p° = 0.1 MPa

T/K	$C_p°$ (J K⁻¹mol⁻¹)	S° (J K⁻¹mol⁻¹)	−[G°−H°(T_r)]/T (J K⁻¹mol⁻¹)	H°−H°(T_r) (kJ mol⁻¹)	$\Delta_f H°$ (kJ mol⁻¹)	$\Delta_f G°$ (kJ mol⁻¹)	Log K_f
0							
100							
200							
250							
298.15	24.296	45.678	45.678	0.	46.896	43.015	-7.536
300	24.310	45.828	45.678	0.045	46.896	42.991	-7.485
350	24.642	49.601	45.975	1.269	46.896	42.340	-6.319
400	24.928	52.911	46.640	2.508	46.896	41.689	-5.444
450	25.144	55.859	47.503	3.760	46.896	41.038	-4.764
500	25.359	58.519	48.474	5.023	46.896	40.388	-4.219
600	25.790	63.181	50.547	7.580	46.896	39.086	-3.403
700	26.229	67.190	52.645	10.181	46.896	37.784	-2.819
800	26.669	70.721	54.688	12.826	46.896	36.482	-2.382
900	27.112	73.887	56.648	15.515	46.896	35.180	-2.042
1000	27.564	76.767	58.518	18.249	46.896	33.879	-1.770
1100	28.017	79.416	60.299	21.028	46.896	32.577	-1.547
1200	28.472	81.873	61.996	23.852	46.896	31.275	-1.361
1300	28.930	84.170	63.614	26.722	46.896	29.973	-1.204
1400	29.393	86.331	65.160	29.638	46.896	28.672	-1.070
1500	29.862	88.375	66.640	32.601	46.896	27.370	-0.953
1600	30.334	90.317	68.060	35.611	46.896	26.068	-0.851
1700	30.806	92.170	69.424	38.668	46.896	24.766	-0.761
1800	31.284	93.944	70.737	41.772	46.896	23.465	-0.681
1900	31.768	95.649	72.004	44.925	46.896	22.163	-0.609
2000	32.254	97.291	73.228	48.126	46.897	20.861	-0.545
2100	32.745	98.876	74.411	51.376	46.897	19.559	-0.487
2200	33.238	100.411	75.558	54.675	46.897	18.258	-0.433
2300	33.735	101.899	76.672	58.024	46.897	16.956	-0.385
2400	34.233	103.345	77.753	61.422	46.897	15.654	-0.341
2450.000	34.483	104.054	78.283	63.140	GLASS <--> LIQ TRANSITION		
2450.000	35.564	104.054	78.283	63.140			
2500	35.564	104.772	78.805	64.918	46.944	14.352	-0.300
2600	35.564	106.167	79.831	68.475	47.005	13.047	-0.262
2700	35.564	107.509	80.831	72.031	46.992	11.740	-0.227
2800	35.564	108.803	81.807	75.587	46.870	10.437	-0.195
2900	35.564	110.051	82.760	79.144	46.600	9.140	-0.165
3000	35.564	111.256	83.690	82.700	46.155	7.855	-0.137
3100	35.564	112.423	84.598	86.257	45.491	6.589	-0.111
3200	35.564	113.552	85.485	89.813	44.577	5.348	-0.087
3300	35.564	114.646	86.352	93.369	43.388	4.139	-0.066
3400	35.564	115.708	87.200	96.926	41.889	2.972	-0.046
3500	35.564	116.739	88.030	100.482	40.015	1.853	-0.028
3600	35.564	117.741	88.841	104.039	37.665	0.795	-0.012
3680.000	35.564	118.522	89.478	106.884	------ CRYSTAL <--> LIQUID -----		
3700	35.564	118.715	89.635	107.595	0.	0.	0.
3800	35.564	119.663	90.413	111.151	0.	0.	0.
3900	35.564	120.587	91.175	114.708	0.	0.	0.
4000	35.564	121.488	91.922	118.264	0.	0.	0.
4100	35.564	122.366	92.653	121.821	0.	0.	0.
4200	35.564	123.223	93.371	125.377	0.	0.	0.
4300	35.564	124.060	94.075	128.933	0.	0.	0.
4400	35.564	124.877	94.766	132.490	0.	0.	0.
4500	35.564	125.676	95.444	136.046	0.	0.	0.
4600	35.564	126.458	96.110	139.603	0.	0.	0.
4700	35.564	127.223	96.764	143.159	0.	0.	0.
4800	35.564	127.972	97.406	146.715	0.	0.	0.
4900	35.564	128.705	98.037	150.272	0.	0.	0.
5000	35.564	129.423	98.658	153.828	0.	0.	0.
5200	35.564	130.818	99.868	160.941	0.	0.	0.
5400	35.564	132.161	101.039	168.054	0.	0.	0.
5600	35.564	133.454	102.174	175.167	0.	0.	0.
5800	35.564	134.702	103.274	182.279	0.	0.	0.
6000	35.564	135.908	104.342	189.392	0.	0.	0.

PREVIOUS: December 1961 CURRENT: June 1966

Tungsten (W) W_1(l)

TUNGSTEN (W) CRYSTAL–LIQUID A_r = 183.85

0 to 3680 K crystal
above 3680 K liquid

Refer to the individual tables for details.

Enthalpy Reference Temperature = T_r = 298.15 K — Standard State Pressure = $p°$ = 0.1 MPa

T/K	$C_p°$ (J K⁻¹mol⁻¹)	$S°$ (J K⁻¹mol⁻¹)	$-[G°-H°(T_r)]/T$ (J K⁻¹mol⁻¹)	$H°-H°(T_r)$ (kJ mol⁻¹)	$\Delta_f H°$ (kJ mol⁻¹)	$\Delta_f G°$ (kJ mol⁻¹)	Log K_f
0	0.	0.	INFINITE	-4.973	0.	0.	0.
100	16.033	9.612	52.820	-4.321	0.	0.	0.
200	22.489	23.273	34.839	-2.313	0.	0.	0.
250	23.686	28.431	33.056	-1.156	0.	0.	0.
298.15	24.295	32.660	32.660	0.	0.	0.	0.
300	24.313	32.810	32.660	0.045	0.	0.	0.
350	24.644	36.583	32.957	1.269	0.	0.	0.
400	24.928	39.893	33.622	2.509	0.	0.	0.
450	25.144	42.842	34.485	3.760	0.	0.	0.
500	25.359	45.502	35.456	5.023	0.	0.	0.
600	25.790	50.163	37.529	7.580	0.	0.	0.
700	26.229	54.172	39.627	10.181	0.	0.	0.
800	26.669	57.703	41.670	12.826	0.	0.	0.
900	27.112	60.870	43.631	15.515	0.	0.	0.
1000	27.564	63.750	45.501	18.249	0.	0.	0.
1100	28.017	66.398	47.282	21.028	0.	0.	0.
1200	28.472	68.855	48.978	23.852	0.	0.	0.
1300	28.930	71.152	50.596	26.722	0.	0.	0.
1400	29.393	73.313	52.143	29.639	0.	0.	0.
1500	29.862	75.357	53.623	32.601	0.	0.	0.
1600	30.334	77.299	55.042	35.611	0.	0.	0.
1700	30.807	79.152	56.407	38.668	0.	0.	0.
1800	31.284	80.927	57.720	41.773	0.	0.	0.
1900	31.765	82.631	58.986	44.925	0.	0.	0.
2000	32.254	84.273	60.210	48.126	0.	0.	0.
2100	32.744	85.858	61.394	51.376	0.	0.	0.
2200	33.238	87.393	62.541	54.675	0.	0.	0.
2300	33.736	88.881	63.654	58.024	0.	0.	0.
2400	34.233	90.328	64.735	61.422	0.	0.	0.
2500	34.736	91.735	65.787	64.870	0.	0.	0.
2600	35.246	93.106	66.812	68.366	0.	0.	0.
2700	36.192	94.453	67.811	71.935	0.	0.	0.
2800	37.447	95.791	68.786	75.613	0.	0.	0.
2900	39.120	97.133	69.740	79.440	0.	0.	0.
3000	41.003	98.490	70.676	83.442	0.	0.	0.
3100	43.430	99.873	71.595	87.662	0.	0.	0.
3200	46.024	101.292	72.501	92.132	0.	0.	0.
3300	48.953	102.752	73.396	96.877	0.	0.	0.
3400	52.300	104.261	74.281	101.933	0.	0.	0.
3500	56.484	105.835	75.160	107.363	0.	0.	0.
3600	61.714	107.499	76.035	113.270	0.	0.	0.
3680.000	66.149	108.904	76.734	118.383	CRYSTAL <--> LIQUID		
3680.000	35.564	118.522	76.734	153.780	TRANSITION		
3700	35.564	118.715	76.961	154.491	0.	0.	0.
3800	35.564	119.663	78.072	158.048	0.	0.	0.
3900	35.564	120.587	79.150	161.604	0.	0.	0.
4000	35.564	121.488	80.197	165.161	0.	0.	0.
4100	35.564	122.366	81.215	168.717	0.	0.	0.
4200	35.564	123.223	82.205	172.273	0.	0.	0.
4300	35.564	124.060	83.169	175.830	0.	0.	0.
4400	35.564	124.877	84.108	179.386	0.	0.	0.
4500	35.564	125.676	85.023	182.943	0.	0.	0.
4600	35.564	126.458	85.915	186.499	0.	0.	0.
4700	35.564	127.223	86.786	190.055	0.	0.	0.
4800	35.564	127.972	87.636	193.612	0.	0.	0.
4900	35.564	128.705	88.467	197.168	0.	0.	0.
5000	35.564	129.423	89.279	200.725	0.	0.	0.
5100	35.564	130.128	90.073	204.281	0.	0.	0.
5200	35.564	130.818	90.850	207.837	0.	0.	0.
5300	35.564	131.496	91.610	211.394	0.	0.	0.
5400	35.564	132.161	92.355	214.950	0.	0.	0.
5500	35.564	132.813	93.085	218.507	0.	0.	0.
5600	35.564	133.454	93.800	222.063	0.	0.	0.
5700	35.564	134.083	94.501	225.619	0.	0.	0.
5800	35.564	134.702	95.189	229.176	0.	0.	0.
5900	35.564	135.310	95.864	232.732	0.	0.	0.
6000	35.564	135.908	96.526	236.289	0.	0.	0.

PREVIOUS: CURRENT: June 1966

TUNGSTEN (W) IDEAL GAS A_r = 183.85

IP(W,g) = 64400 ± 100 cm^{-1}
$S°(298.15\ K)$ = 173.95 ± 0.08 J K^{-1} mol^{-1}

$\Delta_f H°(0\ K)$ = 849.8 ± 6.3 kJ mol^{-1}
$\Delta_f H°(298.15\ K)$ = 851.0 ± 6.3 kJ mol^{1}

Electronic Levels and Quantum Weights

State	ε_i, cm^{-1}	g_i
$^5D_0$	0.00	1
$^5D_1$	1670.29	3
$^5D_2$	3325.53	5
$^5D_3$	4830.00	7
$^5D_4$	6219.33	9
.	.	
.	.	
.	.	
-	62154.50	11
IP	66400.00	

Enthalpy of Formation

The enthalpy of formation is the enthalpy of sublimation at 298.15 K; this has been obtained by a 2nd and 3rd law analysis of the vapor pressure data of three investigators.

Source	T/K**	Data Points	$\Delta_{sub}H°(298.15\ K)$, kcal mol^{-1} 2nd law	3rd law	Drift cal K^{-1} mol^{-1}
1	2511-3053	12*	200.9 ± 3.6	201.57 ± 1.6	0.1 ± 1.4
2	2383-3123	14	214.8 ± 1.4	202.17 ± 2.5	-4.7 ± 0.5
3	2574-3183	10	202.4 ± 2.7	203.35 ± 1.2	0.4 ± 1.0

* 1 point rejected due to failure of a statistical test.
** All temperatures are taken from Szwarc (3) who has corrected the earlier measurements.

The values are remarkably good since even at the highest temperature the pressures are less than 10^{-6} atm; all workers used the Langmuir technique. Since the data of Refs. 1 and 3 do not drift, it may be assumed that the accommodation coefficient is unity or close to unity. The drift in the second set cannot be eliminated by assuming a non-unity coefficient, nor does there appear to be a constant pressure error. Most probably the drift is due to slight errors in temperature, for example if the readings were 24 K high at the low end and 10 K low at the high end the drift would be eliminated; errors of this magnitude are quite possible. The value adopted for $\Delta_{sub}H°(298.15\ K)$ is 203.4 ± 1.5 kcal mol^{-1}.

Heat Capacity and Entropy

The electronic energy levels are taken from those listed by Moore (4). Levels above 25000 cm^{-1} were averaged.

References

1. H. A. Jones, I. Langmuir and G. M. J. Mackay, Phys. Rev. 30, 201 (1927).
2. C. Zwikker, Physica 5, 249 (1925).
3. R. Szwarc, E. R. Plante, and J. J. Diamond, J. Research Nat. Bur. Stand. 69A, 417 (1965).

Tungsten (W) $W_1(g)$

Enthalpy Reference Temperature = T_r = 298.15 K Standard State Pressure = p° = 0.1 MPa

T/K	$C_p°$ (J $K^{-1}mol^{-1}$)	S° (J $K^{-1}mol^{-1}$)	$-[G°-H°(T_r)]/T$ (J $K^{-1}mol^{-1}$)	$H°-H°(T_r)$ (kJ mol^{-1})	$\Delta_f H°$ (kJ mol^{-1})	$\Delta_f G°$ (kJ mol^{-1})	Log K_f
0	0.	0.	INFINITE	-6.216	849.782	849.782	INFINITE
100	20.786	151.176	192.553	-4.138	851.209	837.052	-437.231
200	20.808	165.586	175.879	-2.059	851.280	822.818	-214.898
250	20.940	170.240	174.303	-1.016	851.166	815.714	-170.434
298.15	21.297	173.954	173.954	0.	851.026	808.899	-141.716
300	21.316	174.086	173.955	0.039	851.020	808.637	-140.796
350	22.004	177.420	174.217	1.121	850.878	801.585	-119.630
400	22.968	180.418	174.807	2.244	850.762	794.551	-103.758
450	24.103	183.188	175.586	3.421	850.686	787.530	-91.414
500	25.285	185.789	176.478	4.656	850.658	780.515	-81.540
600	27.406	190.594	178.437	7.294	850.740	766.481	-66.728
700	28.913	194.940	180.489	10.116	850.960	752.422	-56.146
800	29.823	198.866	182.545	13.057	851.256	738.326	-48.208
900	30.328	202.411	184.559	16.067	851.577	724.190	-42.031
1000	30.606	205.622	186.507	19.115	851.892	710.019	-37.088
1100	30.771	208.548	188.380	22.184	852.182	695.817	-33.042
1200	30.877	211.230	190.174	25.267	852.440	681.591	-29.669
1300	30.944	213.704	191.890	28.359	852.662	667.344	-26.814
1400	30.980	215.999	193.531	31.455	852.842	653.082	-24.367
1500	30.989	218.137	195.101	34.554	852.978	638.808	-22.245
1600	30.978	220.136	196.604	37.652	853.067	624.527	-20.389
1700	30.958	222.014	198.044	40.749	853.107	610.242	-18.750
1800	30.944	223.783	199.425	43.844	853.097	595.956	-17.294
1900	30.951	225.456	200.752	46.939	853.039	581.672	-15.991
2000	30.996	227.045	202.027	50.036	852.935	567.392	-14.819
2100	31.094	228.559	203.255	53.140	852.789	553.118	-13.758
2200	31.257	230.009	204.438	56.256	852.607	538.852	-12.794
2300	31.496	231.403	205.580	59.393	852.395	524.595	-11.914
2400	31.818	232.750	206.684	62.558	852.162	510.348	-11.107
2500	32.229	234.057	207.753	65.760	851.915	496.110	-10.366
2600	32.729	235.331	208.790	69.007	851.667	481.883	-9.681
2700	33.319	236.577	209.796	72.309	851.399	467.666	-9.048
2800	33.993	237.800	210.774	75.674	851.086	453.459	-8.459
2900	34.747	239.006	211.727	79.110	850.696	439.264	-7.912
3000	35.572	240.198	212.656	82.625	850.209	425.085	-7.401
3100	36.459	241.379	213.564	86.227	849.590	410.924	-6.924
3200	37.397	242.551	214.451	89.919	848.813	396.786	-6.477
3300	38.375	243.717	215.320	93.707	847.856	382.674	-6.057
3400	39.381	244.877	216.173	97.595	846.688	368.595	-5.663
3500	40.402	246.033	217.009	101.584	845.246	354.553	-5.291
3600	41.426	247.186	217.832	105.675	843.431	340.558	-4.941
3700	42.442	248.335	218.640	109.869	806.403	326.810	-4.614
3800	43.440	249.480	219.437	114.163	807.141	313.838	-4.314
3900	44.408	250.621	220.222	118.556	807.977	300.846	-4.029
4000	45.338	251.757	220.996	123.043	808.909	287.831	-3.759
4100	46.222	252.887	221.760	127.622	809.931	274.792	-3.501
4200	47.054	254.011	222.515	132.286	811.038	261.726	-3.255
4300	47.829	255.128	223.260	137.031	812.227	248.634	-3.020
4400	48.543	256.236	223.997	141.850	813.489	235.512	-2.796
4500	49.193	257.334	224.726	146.737	814.820	222.362	-2.581
4600	49.778	258.422	225.446	151.686	816.213	209.181	-2.375
4700	50.297	259.498	226.159	156.691	817.661	195.969	-2.178
4800	50.752	260.562	226.865	161.744	819.158	182.726	-1.988
4900	51.142	261.612	227.564	166.839	820.696	169.451	-1.806
5000	51.471	262.649	228.255	171.970	822.271	156.144	-1.631
5100	51.741	263.671	228.939	177.131	823.876	142.806	-1.463
5200	51.955	264.678	229.617	182.317	825.505	129.435	-1.300
5300	52.117	265.669	230.288	187.520	827.152	116.034	-1.144
5400	52.230	266.644	230.952	192.738	828.814	102.601	-0.992
5500	52.298	267.603	231.610	197.965	830.484	89.137	-0.847
5600	52.325	268.546	232.261	203.196	832.159	75.643	-0.706
5700	52.315	269.472	232.906	208.429	833.835	62.119	-0.569
5800	52.271	270.382	233.544	213.658	835.508	48.565	-0.437
5900	52.197	271.275	234.176	218.882	837.175	34.983	-0.310
6000	52.097	272.151	234.802	224.097	838.834	21.373	-0.186

PREVIOUS: June 1966 (1 atm) CURRENT: June 1966 (1 bar)

Tungsten (W) $W_1(g)$

TUNGSTEN, ION (W^+) IDEAL GAS M_r = 183.84945

$IP(W^+, g) = 142800 \pm 4000\ cm^{-1}$
$S°(298.15\ K) = 180.00 \pm 0.08\ J\ K^{-1}\ mol^{-1}$

$\Delta_f H°(0\ K) = 1620.2 \pm 7\ kJ\ mol^{-1}$
$\Delta_f H°(298.15\ K) = [1627.691]\ kJ\ mol^{-1}$

Electronic Levels and Quantum Weights

State	ε_i, cm^{-1}	g_i
$^6D_{1/2}$	0.0	2
$^6D_{3/2}$	1518.78	4
$^6D_{5/2}$	3172.52	6
$^6D_{7/2}$	4716.32	8
$^6D_{9/2}$	6147.16	10
.	.	.
.	.	.
.	.	.
$^4H_{9/2}$	64516.37	10
IP	[142800]	

Enthalpy of Formation

$\Delta_f H°(W^+, g, 0\ K)$ is calculated from $\Delta_f H°(W, g, 0\ K)$ (1) using the spectroscopic value of IP(W) = 64400 ± 100 cm^{-1} (770.40 ± 1.20 kJ mol^{-1}) from Moore (2). The ionization limit is converted from cm^{-1} to kJ mol^{-1} using the factor, 1 cm^{-1} = 0.01196266 kJ mol^{-1}, which is derived from the 1973 CODATA fundamental constants (3). Rosenstock et al. (4) and Levin and Lias (5) have summarized additional ionization and appearance potential data.

$\Delta_f H°(W^+, g, 298.15\ K)$ is calculated from $\Delta_f H°(W, g, 0\ K)$ by using IP(W) with JANAF (1) enthalpies, $H°(0\ K) - H°(298.15\ K)$, for W(g), $W^+(g)$, and $e^-(g)$. $\Delta_f H°(W + W^+ + e^-, 298.15\ K)$ differs from a room temperature threshold energy due to inclusion of these enthalpies and to threshold effects discussed by Rosenstock et al. (4). $\Delta_f H°(298.15\ K)$ should be changed by -6.197 kJ mol^{-1} if it is to be used in the ion convention that excludes the enthalpy of the electron.

Heat Capacity and Entropy

The information on electronic energy levels and quantum weights, given by Moore (2), is incomplete because many theoretically predicted levels have not been observed. Although we have listed only the ground, the first four excited states, the highest observed excited state, and the ionization potential for $W^+(g)$, all levels listed by Moore (6), as well as estimated levels, are used in the calculation. The observed levels are too numerous to list completely. The calculations indicate that for $W^+(g)$, the thermodynamic functions are independent of the estimated missing levels (for n=6) and the cut-off procedure; the Gibbs energy function showing variations of 0.8% at this temperature. The reported uncertainty is S°(298.15 K) is due to uncertainties in the relative ionic mass, the fundamental constants, and the low lying atomic energy levels. Extension of these calculations above 6000 K may require consideration of the higher excited states (n>6), and use of different fill and cut-off procedures (7).

References

1. JANAF Thermochemical Tables: W(g), 6-30-66; $e^-(g)$, 3-31-82.
2. C. E. Moore, U. S. Nat. Bur. Stand., NSRDS-NBS-34, 1970.
3. E. R. Cohen and B. N. Taylor, J. Phys. Chem. Ref. Data 2, 663 (1973).
4. H. M. Rosenstock, K. Draxl et al., J. Phys. Chem. Ref. Data 6, Supp. 1 (1977).
5. R. D. Levin and S. G. Lias, U. S. Natl. Bur. Stand. NSRDS-NBS-71, 1982.
6. C. E. Moore, U. S. Nat. Bur. Stand., NSRDS-NBS-35, Volume III, 1970 [Reprint of NBS Circular 467, Volume III, 1958].
7. J. R. Downey, Jr., The Dow Chemical Company, AFOSR-TR-78-0960, Contract No. F44620-75-1-0048, 1978.

Tungsten, Ion (W^+) $W_1^+(g)$

Enthalpy Reference Temperature = T_r = 298.15 K Standard State Pressure = p° = 0.1 MPa

	J K⁻¹mol⁻¹			kJ mol⁻¹			
T/K	$C_p°$	S°	$-[G°-H°(T_r)]/T$	$H°-H°(T_r)$	$\Delta_f H°$	$\Delta_f G°$	Log K_f
0	0.	0.	INFINITE	-6.289	1620.178		
100	20.786	156.939	199.044	-4.211			
200	20.930	171.361	182.008	-2.129			
250	21.575	176.088	180.366	-1.070			
298.15	22.995	179.997	179.997	0.	1627.691	1577.508	-276.373
300	23.066	180.139	179.997	0.043	1627.727	1577.196	-274.614
350	25.368	183.859	180.286	1.251	1628.750	1568.695	-234.115
400	28.116	187.424	180.956	2.587	1629.887	1560.039	-203.720
450	30.836	190.895	181.869	4.062	1631.149	1551.234	-180.063
500	33.146	194.268	182.941	5.664	1632.527	1542.282	-161.121
600	35.866	200.592	185.365	9.136	1635.521	1523.955	-132.672
700	36.298	206.176	187.948	12.760	1638.622	1505.116	-112.313
800	35.433	210.976	190.535	16.353	1641.649	1485.835	-97.015
900	34.117	215.075	193.040	19.832	1644.518	1466.184	-85.095
1000	32.806	218.600	195.424	23.177	1647.207	1446.224	-75.543
1100	31.682	221.673	197.673	26.399	1649.730	1426.002	-67.715
1200	30.784	224.389	199.789	29.521	1652.105	1405.557	-61.182
1300	30.093	226.825	201.776	32.563	1654.356	1384.919	-55.647
1400	29.573	229.035	203.646	35.545	1656.501	1364.112	-50.896
1500	29.187	231.061	205.407	38.482	1658.554	1343.155	-46.773
1600	28.902	232.936	207.069	41.386	1660.526	1322.064	-43.161
1700	28.694	234.681	208.643	44.265	1662.427	1300.852	-39.970
1800	28.545	236.317	210.135	47.127	1664.263	1279.529	-37.131
1900	28.442	237.857	211.554	49.976	1666.038	1258.107	-34.588
2000	28.375	239.314	212.906	52.816	1667.756	1236.592	-32.296
2100	28.337	240.698	214.197	55.651	1669.420	1214.992	-30.221
2200	28.326	242.015	215.432	58.484	1671.033	1193.315	-28.333
2300	28.337	243.275	216.615	61.317	1672.596	1171.565	-26.607
2400	28.370	244.481	217.751	64.153	1674.111	1149.749	-25.024
2500	28.423	245.641	218.844	66.992	1675.581	1127.870	-23.566
2600	28.494	246.757	219.896	69.838	1677.009	1105.933	-22.218
2700	28.585	247.834	220.911	72.692	1678.373	1083.943	-20.970
2800	28.693	248.875	221.891	75.555	1679.636	1061.903	-19.810
2900	28.818	249.884	222.839	78.431	1680.764	1039.821	-18.729
3000	28.958	250.863	223.757	81.319	1681.729	1017.703	-17.720
3100	29.114	251.815	224.647	84.223	1682.491	995.555	-16.775
3200	29.283	252.742	225.510	87.143	1683.020	973.387	-15.889
3300	29.465	253.646	226.349	90.080	1683.290	951.206	-15.056
3400	29.658	254.529	227.165	93.036	1683.269	929.021	-14.273
3500	29.860	255.391	227.959	96.012	1682.893	906.842	-13.534
3600	30.069	256.235	228.733	99.008	1682.061	884.679	-12.836
3700	30.285	257.062	229.488	102.026	1645.937	862.738	-12.180
3800	30.506	257.873	230.224	105.065	1647.498	841.550	-11.568
3900	30.729	258.668	230.943	108.127	1649.082	820.320	-10.987
4000	30.953	259.449	231.646	111.211	1650.688	799.049	-10.435
4100	31.177	260.216	232.334	114.318	1652.317	777.738	-9.909
4200	31.400	260.970	233.006	117.447	1653.968	756.387	-9.407
4300	31.619	261.711	233.665	120.598	1655.641	734.996	-8.928
4400	31.834	262.441	234.311	123.770	1657.336	713.566	-8.471
4500	32.043	263.158	234.944	126.964	1659.052	692.098	-8.034
4600	32.246	263.865	235.565	130.179	1660.789	670.591	-7.615
4700	32.441	264.561	236.175	133.413	1662.546	649.046	-7.213
4800	32.628	265.246	236.773	136.667	1664.321	627.463	-6.828
4900	32.806	265.920	237.361	139.938	1666.115	605.843	-6.458
5000	32.975	266.585	237.939	143.227	1667.927	584.187	-6.103
5100	33.134	267.239	238.507	146.533	1669.754	562.494	-5.761
5200	33.283	267.884	239.066	149.854	1671.597	540.765	-5.432
5300	33.423	268.519	239.616	153.189	1673.455	519.000	-5.115
5400	33.552	269.145	240.157	156.538	1675.326	497.201	-4.809
5500	33.672	269.762	240.689	159.899	1677.210	475.366	-4.515
5600	33.782	270.370	241.214	163.272	1679.105	453.498	-4.230
5700	33.883	270.969	241.731	166.655	1681.010	431.595	-3.955
5800	33.976	271.559	242.240	170.048	1682.925	409.659	-3.689
5900	34.061	272.140	242.742	173.450	1684.849	387.689	-3.432
6000	34.137	272.713	243.237	176.860	1686.782	365.687	-3.184

PREVIOUS: CURRENT: March 1984 (1 bar)

Tungsten, Ion (W^+) $W_1^+(g)$

TUNGSTEN, ION (W^-) IDEAL GAS M_r = 183.85055

EA(W, g) = 0.815 ± 0.008 eV
S°(298.15 K) = 188.781 J K^{-1} mol^{-1}

$\Delta_f H°$(0 K) = 771.147 ± 2.0 kJ mol^{-1}
$\Delta_f H°$(298.15 K) = [766.174] kJ mol^{-1}

Electonic Level and Quantum Weight

State	ε_i, cm^{-1}	g_i
$^6S_{5/2}$	0.0	6

Enthalpy of Formation

$\Delta_f H°(W^-$, g, 0 K) is calculated from $\Delta_f H°$(W, g, 0 K) (1) using the adopted electron affinity of EA(W) = 0.815 ± 0.008 eV (78.635 ± 0.772 kJ mol^{-1}). This value, recommended by Hotop and Lineberger (2), is based on a laser photodetachment electron spectrometry study (3). Additional information on W^-(g) may be obtained in the critical discussions of Hotop and Lineberger (2, 4), Rosenstock et al. (5) and Massey (6).

$\Delta_f H°(W^-$, g, 298.15 K) is obtained from $\Delta_f H°$(W, g, 0 K) by using EA(W) with JANAF (1) enthalpies, H°(0 K) - H°(298.15 K), for W^-(g), W(g), and e^-(g). $\Delta_r H°(W^- \rightarrow W + e^-$, 298.15 K) differs from a room-temperature threshold energy due to inclusion of these enthalpies and to threshold effects discussed by Rosenstock et al. (5). $\Delta_f H°$(298.15 K) should be changed by +6.197 kJ mol^{-1} if it is to be used in the ion convention that excludes the enthalpy of the electron.

Heat Capacity and Entropy

The ground state electronic configuration for W^-(g) is given by Hotop and Lineberger (2, 4) and Rosenstock et al. (5). Lacking any experimental evidence as to the stability of any excited states, we assume that no stable excited states exist.

References

1. JANAF Thermochemical Tables: W(g), 6-30-66; e^-(g), 3-31-82.
2. H. Hotop and W. C. Lineberger, J. Phys. Chem. Ref. Data, to be published, 1985.
3. C. S. Feigerle, R. R. Cordermann, S. V. Bobashev and W. C. Lineberger, J. Chem. Phys. 74, 1580 (1981).
4. H. Hotop and W. C. Lineberger, J. Phys. Chem. Ref. Data 4, 539 (1975).
5. H. M. Rosenstock, K. Draxl et al. J. Phys. Chem. Ref. Data 6, Supp. 1 (1977).
6. H. S. W. Massey, "Negative Ions", 3rd ed., Cambridge University Press, Cambridge, 1976.

Tungsten, Ion (W^-) W_1^-(g)

Enthalpy Reference Temperature = T_r = 298.15 K Standard State Pressure = p° = 0.1 MPa

T/K	$C_p°$ (J $K^{-1}mol^{-1}$)	S° (J $K^{-1}mol^{-1}$)	$-[G°-H°(T_r)]/T$ (J $K^{-1}mol^{-1}$)	$H°-H°(T_r)$ (kJ mol^{-1})	$\Delta_f H°$ (kJ mol^{-1})	$\Delta_f G°$ (kJ mol^{-1})	Log K_f
0	0.	0.	INFINITE	-6.197	771.147		
100	20.786	166.073	207.261	-4.119			
200	20.786	180.481	190.682	-2.040			
250	20.786	185.119	189.123	-1.001			
298.15	20.786	188.781	188.781	0.	766.174	725.882	-127.171
300	20.786	188.909	188.781	0.038	766.129	725.632	-126.344
350	20.786	192.113	189.034	1.078	764.905	718.979	-107.302
400	20.786	194.889	189.596	2.117	763.666	712.502	-93.043
450	20.786	197.337	190.323	3.156	762.414	706.182	-81.971
500	20.786	199.527	191.136	4.196	761.151	700.001	-73.129
600	20.786	203.317	192.860	6.274	758.594	688.011	-59.897
700	20.786	206.521	194.588	8.353	755.993	676.452	-50.477
800	20.786	209.297	196.257	10.431	753.348	665.269	-43.438
900	20.786	211.745	197.845	12.510	750.659	654.420	-37.982
1000	20.786	213.935	199.346	14.589	747.925	643.873	-33.632
1100	20.786	215.916	200.764	16.667	745.146	633.602	-30.087
1200	20.786	217.725	202.103	18.746	742.322	623.586	-27.144
1300	20.786	219.388	203.370	20.824	739.452	613.807	-24.663
1400	20.786	220.929	204.569	22.903	736.536	604.251	-22.545
1500	20.786	222.363	205.708	24.982	733.573	594.906	-20.716
1600	20.786	223.704	206.792	27.060	730.563	585.759	-19.123
1700	20.786	224.965	207.824	29.139	727.506	576.802	-17.723
1800	20.786	226.153	208.810	31.217	724.402	568.027	-16.484
1900	20.786	227.277	209.752	33.296	721.249	559.425	-15.380
2000	20.786	228.343	210.655	35.375	718.049	550.990	-14.390
2100	20.786	229.357	211.522	37.453	714.799	542.717	-13.499
2200	20.786	230.324	212.355	39.532	711.500	534.600	-12.693
2300	20.786	231.248	213.156	41.610	708.151	526.634	-11.960
2400	20.786	232.132	213.929	43.689	704.752	518.814	-11.292
2500	20.786	232.981	214.674	45.768	701.304	511.138	-10.680
2600	20.786	233.796	215.394	47.846	697.809	503.600	-10.117
2700	20.786	234.581	216.090	49.925	694.239	496.198	-9.600
2800	20.786	235.337	216.764	52.004	690.561	488.930	-9.121
2900	20.786	236.066	217.417	54.082	686.734	481.796	-8.678
3000	20.786	236.771	218.050	56.161	682.733	474.797	-8.267
3100	20.786	237.452	218.665	58.239	678.512	467.934	-7.885
3200	20.786	238.112	219.263	60.318	674.042	461.212	-7.529
3300	20.786	238.752	219.844	62.397	669.297	454.634	-7.196
3400	20.786	239.372	220.409	64.475	664.241	448.204	-6.886
3500	20.786	239.975	220.960	66.554	658.811	441.928	-6.595
3600	20.786	240.560	221.496	68.632	652.904	435.813	-6.323
3700	20.786	241.130	222.019	70.711	611.683	430.061	-6.071
3800	20.786	241.684	222.529	72.790	608.127	425.200	-5.845
3900	20.786	242.224	223.027	74.868	604.570	420.433	-5.631
4000	20.786	242.750	223.514	76.947	601.014	415.757	-5.429
4100	20.786	243.264	223.989	79.025	597.457	411.169	-5.238
4200	20.786	243.765	224.454	81.104	593.901	406.669	-5.058
4300	20.786	244.254	224.909	83.183	590.345	402.253	-4.886
4400	20.786	244.732	225.354	85.261	586.788	397.920	-4.724
4500	20.786	245.199	225.790	87.340	583.232	393.667	-4.570
4600	20.786	245.656	226.217	89.418	579.675	389.494	-4.423
4700	20.786	246.103	226.635	91.497	576.119	385.398	-4.283
4800	20.786	246.540	227.045	93.576	572.563	381.377	-4.150
4900	20.786	246.969	227.448	95.654	569.006	377.431	-4.023
5000	20.786	247.389	227.842	97.733	565.450	373.558	-3.903
5100	20.786	247.800	228.229	99.811	561.893	369.755	-3.787
5200	20.786	248.204	228.610	101.890	558.337	366.022	-3.677
5300	20.786	248.600	228.983	103.969	554.781	362.358	-3.571
5400	20.786	248.988	229.350	106.047	551.224	358.761	-3.470
5500	20.786	249.370	229.711	108.126	547.668	355.229	-3.374
5600	20.786	249.744	230.065	110.204	544.111	351.762	-3.281
5700	20.786	250.112	230.414	112.283	540.555	348.359	-3.192
5800	20.786	250.474	230.756	114.362	536.999	345.018	-3.107
5900	20.786	250.829	231.093	116.440	533.442	341.739	-3.026
6000	20.786	251.178	231.425	118.519	529.886	338.520	-2.947

PREVIOUS: CURRENT: March 1984 (1 bar)

Tungsten, Ion (W^-) W_1^-(g)

XENON (Xe) REFERENCE STATE A_r = 131.29

0 to 6000 K Ideal Monatomic Gas

IP(Xe, g) = 97834.0 ± 0.1 cm^{-1}
S°(298.15 K) = 169.684 ± 0.003 J K^{-1} mol^{-1}

$\Delta_f H°$(0 K) = 0 kJ mol^{-1}
$\Delta_f H°$(298.15 K) = 0 kJ mol^{-1}

Electronic Levels and Quantum Weights

State	ε_i, cm^{-1}	g_i
$^1S_0$	0	1

Enthalpy of Formation

Zero by definition.

Heat Capacity and Entropy

Information on the electronic energy levels and quantum weights is taken from Moore (1, 9). Many of the theoretically predicted levels have not been observed. Our calculations indicate that any reasonable method of filling in these missing levels and cutting off the summation in the partition function (2) has no effect on the thermodynamic properties to 6000 K. This is undoubtedly a result of the high energy of these levels; the first excited level is over 67000 cm^{-1} above the ground state. Therefore, we list the ground state only. Extension to higher temperatures may require consideration of excited states and utilization of different fill and cutoff procedures (2).

The thermodynamic functions at 298.15 K agree exactly with recent CODATA recommendations (3) except for two minor changes. First, the entropy differs by 0.1094 J K^{-1} mol^{-1} because this table uses a reference pressure of 1 bar, whereas the CODATA recommendations are based on 1 atm. Second, entropy differences of the order of 0.001-0.004 J K^{-1} mol^{-1} for the rare gases arise due to the use of slightly different values for R and the relative atomic mass; this table uses R = 8.31441 J K^{-1} mol^{-1}. Within these same minor changes, the tables agree within the estimated uncertainty with those by Hultgren et al. (4), Hilsenrath et al. (5), Gurvich et al. (8), and Wagman et al. (12). The estimated uncertainty is due to uncertainties in the relative atomic mass and fundamental constants which are based on the 1981 scale (6) and the 1973 values (7), respectively.

Phase Data

The triple point, 161.388 K, and boiling point, 165.060 K, are secondary fixed points of IPTS-68 (10, 11). Hultgren et al. (4) had recommended a triple point of 161.36 K (0.8059 atm) and a boiling point of 165.060 K (1 atm). These values are provided for the convenience of the reader and have not been evaluated by the present authors. As a result of the low values, the reference state for xenon is chosen to be the ideal gas at all temperatures. This may differ from the choice of other authors.

References

1. C. E. Moore, U. S. Nat. Bur. Stand., NSRDS-NBS-35, Volume III, 1970 [Reprint of NBS Circular 467, Volume III, 1958].
2. J. R. Downey, Jr., The Dow Chemical Co., AFOSR-TR-78-0960, Contract No. F44620-75-1-0048, 1978.
3. J. D. Cox, ICSU-CODATA Task Group, J. Chem. Thermodynamics 10, 903 (1978).
4. R. Hultgren, P. D. Desai et al, "Selected Values of the Thermodynamic Properties of the Elements", American Society for Metals, Metals Park, Ohio, 1973.
5. J. Hilsenrath, C. G Messina, and W. H. Evans, U.S. Nat. Bur. Stand., Report AD-606163 (avail. NTIS), 1964.
6. N. E. Holden and R. L. Martin, Pure Appl. Chem. 51, 405 (1979).
7. E. R. Cohen and B. N. Taylor, J. Phys. Chem. Ref. Data 2, 663 (1973).
8. L. V. Gurvich, I. V. Veits et al., "Thermodynamic Properties of Individual Substances," 3rd ed., Volume I, Nauka, Moscow, 1978.
9. C. E. Moore, U. S. Nat. Bur. Stand., NSRDS-NBS-34, 1970.
10. H. Preston-Thomas, Metrologia 12, 7 (1976).
11. L. Crovini, R. E. Bedford and A. Moser, Metrologia 13, 197 (1977).
12. D. D. Wagman, W. H. Evans et al., J. Phys. Chem. Ref. Data 11, Supp. 2 (1982).

Xenon (Xe) Xe_1(ref)

Enthalpy Reference Temperature = T_r = 298.15 K Standard State Pressure = p° = 0.1 MPa

	J $K^{-1}mol^{-1}$			kJ mol^{-1}			
T/K	$C_p°$	S°	$-[G°-H°(T_r)]/T$	$H°-H°(T_r)$	$\Delta_f H°$	$\Delta_f G°$	Log K_f
0	0.	0.	INFINITE	-6.197	0.	0.	0.
100	20.786	146.977	188.164	-4.119	0.	0.	0.
200	20.786	161.384	171.585	-2.040	0.	0.	0.
250	20.786	166.023	170.026	-1.001	0.	0.	0.
298.15	20.786	169.684	169.684	0.	0.	0.	0.
300	20.786	169.812	169.684	0.038	0.	0.	0.
350	20.786	173.017	169.937	1.078	0.	0.	0.
400	20.786	175.792	170.499	2.117	0.	0.	0.
450	20.786	178.240	171.226	3.156	0.	0.	0.
500	20.786	180.430	172.039	4.196	0.	0.	0.
600	20.786	184.220	173.763	6.274	0.	0.	0.
700	20.786	187.424	175.492	8.353	0.	0.	0.
800	20.786	190.200	177.161	10.431	0.	0.	0.
900	20.786	192.648	178.748	12.510	0.	0.	0.
1000	20.786	194.838	180.249	14.589	0.	0.	0.
1100	20.786	196.819	181.667	16.667	0.	0.	0.
1200	20.786	198.628	183.006	18.746	0.	0.	0.
1300	20.786	200.292	184.273	20.824	0.	0.	0.
1400	20.786	201.832	185.473	22.903	0.	0.	0.
1500	20.786	203.266	186.612	24.982	0.	0.	0.
1600	20.786	204.608	187.695	27.060	0.	0.	0.
1700	20.786	205.868	188.727	29.139	0.	0.	0.
1800	20.786	207.056	189.713	31.217	0.	0.	0.
1900	20.786	208.180	190.655	33.296	0.	0.	0.
2000	20.786	209.246	191.559	35.375	0.	0.	0.
2100	20.786	210.260	192.425	37.453	0.	0.	0.
2200	20.786	211.227	193.258	39.532	0.	0.	0.
2300	20.786	212.151	194.059	41.610	0.	0.	0.
2400	20.786	213.036	194.832	43.689	0.	0.	0.
2500	20.786	213.884	195.577	45.768	0.	0.	0.
2600	20.786	214.699	196.297	47.846	0.	0.	0.
2700	20.786	215.484	196.993	49.925	0.	0.	0.
2800	20.786	216.240	197.667	52.004	0.	0.	0.
2900	20.786	216.969	198.320	54.082	0.	0.	0.
3000	20.786	217.674	198.954	56.161	0.	0.	0.
3100	20.786	218.355	199.569	58.239	0.	0.	0.
3200	20.786	219.015	200.166	60.318	0.	0.	0.
3300	20.786	219.655	200.747	62.397	0.	0.	0.
3400	20.786	220.276	201.312	64.475	0.	0.	0.
3500	20.786	220.878	201.863	66.554	0.	0.	0.
3600	20.786	221.464	202.399	68.632	0.	0.	0.
3700	20.786	222.033	202.922	70.711	0.	0.	0.
3800	20.786	222.587	203.432	72.790	0.	0.	0.
3900	20.786	223.127	203.930	74.868	0.	0.	0.
4000	20.786	223.654	204.417	76.947	0.	0.	0.
4100	20.786	224.167	204.892	79.025	0.	0.	0.
4200	20.786	224.668	205.357	81.104	0.	0.	0.
4300	20.786	225.157	205.812	83.183	0.	0.	0.
4400	20.786	225.635	206.257	85.261	0.	0.	0.
4500	20.786	226.102	206.693	87.340	0.	0.	0.
4600	20.786	226.559	207.120	89.418	0.	0.	0.
4700	20.786	227.006	207.538	91.497	0.	0.	0.
4800	20.786	227.443	207.949	93.576	0.	0.	0.
4900	20.786	227.872	208.351	95.654	0.	0.	0.
5000	20.786	228.292	208.745	97.733	0.	0.	0.
5100	20.786	228.704	209.133	99.811	0.	0.	0.
5200	20.786	229.107	209.513	101.890	0.	0.	0.
5300	20.786	229.503	209.886	103.969	0.	0.	0.
5400	20.786	229.892	210.253	106.047	0.	0.	0.
5500	20.786	230.273	210.614	108.126	0.	0.	0.
5600	20.786	230.648	210.968	110.204	0.	0.	0.
5700	20.786	231.015	211.317	112.283	0.	0.	0.
5800	20.786	231.377	211.659	114.362	0.	0.	0.
5900	20.786	231.732	211.997	116.440	0.	0.	0.
6000	20.786	232.082	212.329	118.519	0.	0.	0.

PREVIOUS: March 1977 (1 atm) CURRENT: March 1982 (1 bar)

Xenon (Xe) Xe_1(ref)

XENON, ION (Xe^+) IDEAL GAS M_r = 131.28945

IP(Xe^+, g) = 171068.4 ± 0.2 cm^{-1}
S°(298.15 K) = 181.210 ± 0.003 J K^{-1} mol^{-1}

$\Delta_f H°$(0 K) = 1170.355 ± 0.001 kJ mol^{-1}
$\Delta_f H°$(298.15 K) = [1176.552] kJ mol^{-1}

Electronic Levels and Quantum Weights

State	ε_i, cm^{-1}	g_i
$^2P_{3/2}$	0	4
$^2P_{1/2}$	10537.01	2

Enthalpy of Formation

The ionization limit of neutral xenon (97834.0±0.1 cm^{-1}) reported by Moore (1) is adopted as $\Delta_f H°$(0 K) for Xe^+(g). The ionization limit is converted from cm^{-1} to kJ mol^{-1} using the factor, 1 cm^{-1} = 0.01196266 kJ mol^{-1}, which is derived from the latest CODATA fundamental constants (2). The uncertainty in the ionization limit is estimated to be ±0.1 cm^{-1} which corresponds to an uncertainty of ±0.001 kJ mol^{-1} in the heat of formation. Rosenstock et al. (9) and Levine and Lias (11) have summarized additional ionization potential and appearance potential data. Gurvich et al. (8) and Wagman et al. (10) adopted the same ionization potential, but the use of slightly different fundamental constants by Wagman et al. (10) results in a heat of formation difference of 0.011 kJ mol^{-1}.

$\Delta_f H°$(Xe^+, g, 298.15 K) is obtained from $\Delta_f H°$(Xe, g, 0 K) by using IP(Xe) with JANAF (3) enthalpies H°(0 K)−H°(298.15 K) for Xe^+(g), Xe(r), and e^-(r). $\Delta_r H°$(Xe → Xe^+ + e^-, 298.15 K) differs from a room temperature threshold energy due to inclusion of these enthalpies and to threshold effects discussed by Rosenstock et al. (9). $\Delta_f H°$(298.15 K) should be changed by −6.197 kJ mol^{-1} if it is to be used in the ion convention that excludes the enthalpy of the electron.

Heat Capacity and Entropy

The information on electronic energy levels and quantum weights of Moore (4) is incomplete because many theoretically predicted levels have not been observed. Our calculations indicate that any reasonable method of filling in these missing levels and cutting off the summation in the partition function (5) has no effect on the thermodynamic functions to 6000 K. This is a result of the high energy of all levels other than the ground state and the $^2P_{1/2}$ level; the next lowest level is over 90000 cm^{-1} above the ground state. Since inclusion of these upper levels has no effect on the thermodynamic functions (to 6000 K) we list only the ground state and the $^2P_{1/2}$ state, with the energy of the latter state taken from a more recent study by Moore (1). The reported uncertainty in S°(298.15 K) is due to uncertainties in the relative ionic mass and fundamental constants. Extension of these calculations above 6000 K may require consideration of the higher excited states and use of different fill and cutoff procedures (5).

The thermodynamic functions reported here agree with those of Green et al. (6), Hilsenrath et al. (7), and Gurvich et al. (8) except for two minor changes. First, the entropy differs by 0.1094 J K^{-1} mol^{-1} because this table uses a standard state pressure of 1 bar, whereas the cited references used a pressure of 1 atm. Second, smaller differences arise from the use of different values for the fundamental constants, the relative ionic mass, and the position of the $^2P_{1/2}$ electronic level.

References

1. C. E. Moore, U. S. Nat. Bur. Stand., NSRDS-NBS-34, 1970.
2. E. R. Cohen and B. N. Taylor, J. Phys. Chem. Ref. Data 2, 663 (1973).
3. JANAF Thermochemical Tables: Xe(r), 3-31-82; e^-(r), 3-31-82.
4. C. E. Moore, U. S. Nat. Bur. Stand., NSRDS-NBS-35, Volume III, 1970 [Reprint of NBS Circular 467, Volume III, 1958].
5. J. R. Downey, Jr., The Dow Chemical Co., AFOSR-TR-78-0960, Contract No. F44620-75-1-0048, 1978.
6. J. W. Green, D. E. Poland and J. L. Margrave, University of Wisconsin, ARL-191 (AD-275542, avail. NTIS), Contract No. AF 33(616)-6384, 1961.
7. J. Hilsenrath, C. G. Messina and W. H. Evans, U.S. Nat. Bur. Stand., AD-606163 (avail. NTIS), 1964.
8. L. V. Gurvich, I. V. Veits et al., "Thermodynamic Properties of Individual Substances", 3rd ed., Volume I, Nauka, Moscow, 1978.
9. H. M. Rosenstock, K. Draxl et al, J. Phys. Chem. Ref. Data 6, Supp. 1 (1977).
10. D. D. Wagman, W. H. Evans et al., J. Phys. Chem. Ref. Data 11, Supp. 2 (1982).
11. R. D. Levine and S. G. Lias, NSRDS-NBS-71, 1982.

Xenon, Ion (Xe^+) Xe_1^+(g)

Enthalpy Reference Temperature = T_r = 298.15 K Standard State Pressure = p° = 0.1 MPa

T/K	$C_p°$ (J $K^{-1}mol^{-1}$)	S° (J $K^{-1}mol^{-1}$)	−[G°−H°(T_r)]/T (J $K^{-1}mol^{-1}$)	H°−H°(T_r) (kJ mol^{-1})	$\Delta_f H°$ (kJ mol^{-1})	$\Delta_f G°$ (kJ mol^{-1})	Log K_f
0	0.	0.	INFINITE	-6.197	1170.355		
100	20.786	158.503	199.690	-4.119			
200	20.786	172.910	183.111	-2.040			
250	20.786	177.549	181.552	-1.001			
298.15	20.786	181.210	181.210	0.	1176.552	1166.861	-204.429
300	20.786	181.338	181.210	0.038	1176.591	1166.801	-203.158
350	20.786	184.543	181.463	1.078	1177.630	1165.087	-173.880
400	20.786	187.318	182.026	2.117	1178.669	1163.224	-151.902
450	20.786	189.766	182.752	3.156	1179.709	1161.231	-134.792
500	20.786	191.957	183.565	4.196	1180.748	1159.123	-121.093
600	20.786	195.746	185.289	6.274	1182.826	1154.602	-100.517
700	20.786	198.950	187.018	8.353	1184.905	1149.734	-85.794
800	20.786	201.726	188.687	10.431	1186.984	1144.568	-74.733
900	20.786	204.174	190.274	12.510	1189.062	1139.141	-66.114
1000	20.786	206.364	191.776	14.589	1191.141	1133.483	-59.207
1100	20.787	208.345	193.193	16.667	1193.220	1127.616	-53.546
1200	20.788	210.154	194.532	18.746	1195.298	1121.561	-48.820
1300	20.791	211.818	195.799	20.825	1197.377	1115.331	-44.815
1400	20.796	213.359	196.999	22.904	1199.457	1108.942	-41.375
1500	20.803	214.794	198.138	24.984	1201.536	1102.404	-38.389
1600	20.815	216.137	199.221	27.065	1203.617	1095.728	-35.772
1700	20.830	217.399	200.254	29.147	1205.700	1088.921	-33.458
1800	20.851	218.591	201.240	31.231	1207.784	1081.991	-31.399
1900	20.877	219.719	202.183	33.318	1209.870	1074.946	-29.552
2000	20.908	220.790	203.087	35.407	1211.959	1067.791	-27.888
2100	20.945	221.811	203.954	37.499	1214.052	1060.531	-26.379
2200	20.987	222.787	204.788	39.596	1216.148	1053.171	-25.005
2300	21.033	223.720	205.591	41.697	1218.249	1045.716	-23.749
2400	21.085	224.617	206.366	43.803	1220.355	1038.170	-22.595
2500	21.141	225.479	207.113	45.914	1222.466	1030.535	-21.532
2600	21.200	226.309	207.835	48.031	1224.583	1022.816	-20.549
2700	21.262	227.110	208.534	50.154	1226.706	1015.016	-19.637
2800	21.326	227.884	209.212	52.283	1228.836	1007.136	-18.788
2900	21.392	228.634	209.869	54.419	1230.972	999.181	-17.997
3000	21.460	229.360	210.506	56.562	1233.114	991.152	-17.257
3100	21.528	230.065	211.126	58.711	1235.264	983.051	-16.564
3200	21.596	230.750	211.729	60.868	1237.420	974.880	-15.913
3300	21.664	231.415	212.315	63.031	1239.583	966.643	-15.301
3400	21.732	232.063	212.886	65.200	1241.753	958.339	-14.723
3500	21.798	232.694	213.443	67.377	1243.929	949.972	-14.178
3600	21.863	233.309	213.987	69.560	1246.112	941.542	-13.661
3700	21.927	233.909	214.517	71.749	1248.302	933.052	-13.172
3800	21.988	234.494	215.035	73.945	1250.497	924.502	-12.708
3900	22.048	235.066	215.541	76.147	1252.699	915.895	-12.267
4000	22.105	235.625	216.037	78.355	1254.907	907.230	-11.847
4100	22.160	236.172	216.521	80.568	1257.120	898.511	-11.447
4200	22.213	236.706	216.995	82.787	1259.339	889.738	-11.066
4300	22.263	237.230	217.460	85.011	1261.563	880.911	-10.701
4400	22.311	237.742	217.915	87.239	1263.791	872.033	-10.352
4500	22.356	238.244	218.361	89.473	1266.025	863.105	-10.019
4600	22.398	238.736	218.799	91.710	1268.263	854.126	-9.699
4700	22.438	239.218	219.228	93.952	1270.504	845.099	-9.392
4800	22.476	239.691	219.649	96.198	1272.750	836.024	-9.098
4900	22.511	240.154	220.063	98.447	1274.999	826.902	-8.815
5000	22.543	240.610	220.470	100.700	1277.252	817.735	-8.543
5100	22.573	241.056	220.869	102.956	1279.508	808.522	-8.281
5200	22.601	241.495	221.261	105.214	1281.767	799.265	-8.029
5300	22.626	241.926	221.647	107.476	1284.028	789.964	-7.786
5400	22.650	242.349	222.027	109.740	1286.292	780.621	-7.551
5500	22.671	242.765	222.400	112.006	1288.558	771.236	-7.325
5600	22.690	243.173	222.767	114.274	1290.826	761.810	-7.106
5700	22.707	243.575	223.129	116.544	1293.096	752.343	-6.894
5800	22.722	243.970	223.485	118.815	1295.367	742.836	-6.690
5900	22.736	244.359	223.835	121.088	1297.640	733.290	-6.492
6000	22.747	244.741	224.180	123.362	1299.914	723.706	-6.300

PREVIOUS: March 1977 (1 atm) CURRENT: March 1982 (1 bar)

Xenon, Ion (Xe^+) Xe_1^+(g)

ZINC (Zn) REFERENCE STATE A_r = 65.38

0 to 692.73 K crystal
692.73 to 1180.173 K liquid
above 1180.173 K ideal monatomic gas

Refer to the individual tables for details.

Zinc (Zn) Zn_1(ref)

Enthalpy Reference Temperature = T_r = 298.15 K Standard State Pressure = p° = 0.1 MPa

	J K⁻¹mol⁻¹			kJ mol⁻¹			
T/K	C_p°	S°	$-[G^\circ-H^\circ(T_r)]/T$	$H^\circ-H^\circ(T_r)$	$\Delta_f H^\circ$	$\Delta_f G^\circ$	Log K_f
0	0.	0.	INFINITE	-5.669	0.	0.	0.
100	19.455	16.523	63.224	-4.670	0.	0.	0.
200	24.050	31.820	44.005	-2.437	0.	0.	0.
298.15	25.387	41.717	41.717	0.	0.	0.	0.
300	25.406	41.874	41.717	0.047	0.	0.	0.
400	26.346	49.314	42.725	2.636	0.	0.	0.
500	27.386	55.301	44.660	5.321	0.	0.	0.
600	28.588	60.399	46.868	8.118	0.	0.	0.
692.730	29.802	64.591	48.965	10.825	CRYSTAL <--> LIQUID		
692.730	31.380	75.161	48.965	18.147	TRANSITION		
700	31.380	75.489	49.239	18.375	0.	0.	0.
800	31.380	79.679	52.788	21.513	0.	0.	0.
900	31.380	83.375	55.985	24.651	0.	0.	0.
1000	31.380	86.681	58.892	27.789	0.	0.	0.
1100	31.380	89.672	61.557	30.927	0.	0.	0.
1180.173	31.380	91.880	63.542	33.443	LIQUID <--> IDEAL GAS		
1180.173	20.786	189.586	63.542	148.754	FUGACITY = 1 bar		
1200	20.786	189.933	65.628	149.166	0.	0.	0.
1300	20.786	191.597	75.255	151.244	0.	0.	0.
1400	20.786	193.137	83.620	153.323	0.	0.	0.
1500	20.786	194.571	90.970	155.402	0.	0.	0.
1600	20.786	195.913	97.487	157.480	0.	0.	0.
1700	20.786	197.173	103.315	159.559	0.	0.	0.
1800	20.786	198.361	108.562	161.637	0.	0.	0.
1900	20.786	199.485	113.318	163.716	0.	0.	0.
2000	20.786	200.551	117.653	165.795	0.	0.	0.
2100	20.786	201.565	121.625	167.873	0.	0.	0.
2200	20.786	202.532	125.281	169.952	0.	0.	0.
2300	20.786	203.456	128.660	172.030	0.	0.	0.
2400	20.786	204.341	131.795	174.109	0.	0.	0.
2500	20.786	205.189	134.714	176.188	0.	0.	0.
2600	20.786	206.004	137.440	178.266	0.	0.	0.
2700	20.787	206.789	139.994	180.345	0.	0.	0.
2800	20.787	207.545	142.393	182.424	0.	0.	0.
2900	20.788	208.274	144.653	184.502	0.	0.	0.
3000	20.789	208.979	146.785	186.581	0.	0.	0.
3100	20.791	209.661	148.803	188.660	0.	0.	0.
3200	20.793	210.321	150.715	190.739	0.	0.	0.
3300	20.796	210.961	152.531	192.819	0.	0.	0.
3400	20.800	211.582	154.258	194.899	0.	0.	0.
3500	20.806	212.185	155.905	196.979	0.	0.	0.
3600	20.814	212.771	157.476	199.060	0.	0.	0.
3700	20.824	213.341	158.979	201.142	0.	0.	0.
3800	20.836	213.897	160.417	203.225	0.	0.	0.
3900	20.852	214.438	161.795	205.309	0.	0.	0.
4000	20.871	214.966	163.117	207.395	0.	0.	0.
4100	20.894	215.482	164.388	209.483	0.	0.	0.
4200	20.922	215.986	165.611	211.574	0.	0.	0.
4300	20.955	216.478	166.788	213.668	0.	0.	0.
4400	20.994	216.961	167.923	215.765	0.	0.	0.
4500	21.039	217.433	169.018	217.867	0.	0.	0.
4600	21.091	217.896	170.076	219.973	0.	0.	0.
4700	21.150	218.350	171.098	222.085	0.	0.	0.
4800	21.218	218.796	172.087	224.204	0.	0.	0.
4900	21.295	219.234	173.045	226.329	0.	0.	0.
5000	21.381	219.665	173.973	228.463	0.	0.	0.
5100	21.478	220.090	174.873	230.606	0.	0.	0.
5200	21.585	220.508	175.746	232.759	0.	0.	0.
5300	21.703	220.920	176.595	234.923	0.	0.	0.
5400	21.833	221.327	177.420	237.100	0.	0.	0.
5500	21.976	221.729	178.222	239.290	0.	0.	0.
5600	22.131	222.126	179.002	241.496	0.	0.	0.
5700	22.300	222.519	179.762	243.717	0.	0.	0.
5800	22.482	222.909	180.503	245.956	0.	0.	0.
5900	22.680	223.295	181.225	248.214	0.	0.	0.
6000	22.892	223.678	181.929	250.492	0.	0.	0.

PREVIOUS: December 1978 (1 atm) CURRENT: December 1978 (1 bar)

Zinc (Zn) Zn_1(ref)

ZINC (Zn) CRYSTAL A_r = 65.38

Zinc (Zn) $Zn_1(cr)$

Pearson Notation: hP2
S°(298.15 K) = 41.717 ± 0.17 J K^{-1} mol^{-1}
T_{fus} = 692.73 K

$\Delta_fH°$(0 K) = 0 kJ mol^{-1}
$\Delta_fH°$(298.15 K) = 0 kJ mol^{-1}
$\Delta_{fus}H°$ = 7.322 ± 0.105 kJ mol^{-1}

Enthalpy Reference Temperature = T_r = 298.15 K; Standard State Pressure = p° = 0.1 MPa

T/K	$C_p°$ (J K⁻¹mol⁻¹)	S° (J K⁻¹mol⁻¹)	-[G°-H°(T_r)]/T (J K⁻¹mol⁻¹)	H°-H°(T_r) (kJ mol⁻¹)	$\Delta_fH°$ (kJ mol⁻¹)	$\Delta_fG°$ (kJ mol⁻¹)	Log K_f
0	0.	0.	INFINITE	-5.669	0.	0.	0.
100	19.455	16.523	63.224	-4.670	0.	0.	0.
200	24.050	31.820	44.005	-2.437	0.	0.	0.
250	25.039	37.290	42.132	-1.211	0.	0.	0.
298.15	25.387	41.717	41.717	0.	0.	0.	0.
300	25.406	41.874	41.717	0.047	0.	0.	0.
350	25.889	45.828	42.028	1.330	0.	0.	0.
400	26.346	49.314	42.725	2.636	0.	0.	0.
450	26.843	52.446	43.634	3.965	0.	0.	0.
500	27.386	55.301	44.660	5.321	0.	0.	0.
600	28.588	60.399	46.868	8.118	0.	0.	0.
692.730	29.802	64.591	48.965	10.825	------ CRYSTAL <--> LIQUID -----		
700	29.900	64.903	49.129	11.042	-7.333	0.077	-0.006
800	31.284	68.985	51.360	14.101	-7.412	1.143	-0.075
900	32.716	72.752	53.530	17.300	-7.351	2.210	-0.128
1000	34.178	76.275	55.630	20.645	-7.144	3.262	-0.170

PREVIOUS: CURRENT: December 1978

Enthalpy of Formation

Zero by definition.

Heat Capacity and Entropy

The adopted heat capacity values are derived from the low temperature heat capacity studies of Martin (1) and Zabetakis (2) and four high temperature enthalpy studies (3, 4, 5, 6). The heat capacity values at the temperature where these two different types of study join are forced to be smooth and continuous.

The results of Martin (1) differ from the adopted values by less than 1% in the range 14-30 K. The results of Zabetakis (2) also agree within 0.4% in the region 23-195 K. In the region 200-260 K, the data of Zabetakis (2) indicated a slight depression of ~0.2 cal K^{-1} mol^{-1} at the mid-point of this region. Other low temperature studies did not support this result, so that the adopted heat capacity values are smoothed in this region to be in better agreement with the studies of Behrens and Drucker (7), Bronson and Wilson (8), and Eichenauer and Schulze (9). The enthalpy data of Jaeger and Poppema (6), Eastman et al. (4), Ruer and Kremers (5), and Schubel (3) all agree within 1% of the adoptec enthalpy vlaues.

The thermodynamic functions are based on S°(12) = 0.0230 cal K^{-1} mol^{-1} and H°(12 K) - H°(0 K) = 0.20 cal mol^{-1}. These values are obtained by integration of smoothed low temperature data and are in agreement with the values used by Martin (1) and Zabetakis (2). Below 0.850 ± 0.01 K, zinc is a superconductive element; this behavior is summarized in a recent review by Roberts (10). The literature survey for Zn was done in part by Hultgren (11), his help is greatly appreciated.

Fusion Data

We adopt $\Delta_{fus}H°$ = 1.750 ± 0.025 kcal mol^{-1}. This value is derived primarily from the adiabatic calorimetric study of Chiotti et al. (12). Numerous other studies (over 20) reported values in the range 1.504-1.850 kcal mol^{-1} with a mean of 1.735 kcal mol^{-1}.

The adopted value for T_{fus} = 692.73 K is a primary reference point on the IPTS-68 (13).

Sublimation Data

The sublimation studies are summarized on the Zn(g) table.

References

1. D. L. Martin, Phys. Rev. 167, 640 (1968).
2. M. G. Zabetakis, Ph. D. Thesis, Univ. Pittsburg, 1956.
3. P. Schubel, Z. Anorg. Chem. 87, 81 (1914).
4. E. D. Eastman, A. M. Williams, and T. F. Young, J. Am. Chem. Soc. 46, 1178 (1924).
5. R. Ruer and K. Kremers, Z. Anorg. Chem. 184, 193 (1929).
6. F. M. Jager and T. J. Poppema, Rec. Trav. Chim. 55, 492 (1936).
7. W. U. Behrens and C. Drucker, Z. Phsik. Chem. 113, 79 (1924).
8. H. L. Bronson and A. J. Wilson, Can. J. Res. 14A, 181 (1936).
9. W. Eichenauer and M. Schulze, Z. Naturforsch. 14a, 28 (1959).
10. B. W. Roberts, J. Phys. Chem. Ref. Data 5, 581 (1976).
11. R. Hultgren, Univ. California (Berkeley), personal communication, 1978.
12. P. Chiotti, G. Gartner, E. R. Stevens, and Y. Saito, J. Chem. Eng. Data 11, 571 (1966).
13. The International Practical Temperature Scale, Metrologia 5, 35 (1969).

ZINC (Zn) LIQUID A_r = 65.38

$S°(298.15\ K) = [50.792]\ J\ K^{-1}\ mol^{-1}$
$T_{fus} = 692.73\ K$

$\Delta_f H°(298.15\ K) = [6.519]\ kJ\ mol^{-1}$
$\Delta_{fus} H° = 7.322 \pm 0.105\ kJ\ mol^{-1}$

Enthalpy of Formation

The enthalpy of formation of Zn(ℓ) at 298.15 K is calculated from that of the crystal by adding $\Delta_{fus}H°$ and the difference in enthalpy, H°(692.73 K) - H°(298.15 K), between the crystal and liquid.

Heat Capacity and Entropy

The adopted heat capacity for Zn(ℓ) is 7.5 cal K^{-1} mol^{-1}. It is taken as a constant for the real liquid region. There are five enthalpy and two heat capacity studies on liquid zinc. The enthalpy studies are summarized in the following table. The Cp°(ℓ) value given is that derived from the enthalpy data assuming a linear temperature dependence.

Source	Year	Reference Temperature	No. of Observations Liquid	Total	T/K	Cp° (ℓ) cal K^{-1} mol^{-1}
(1)	1919	293	28	47	469-992	8.14
(2)	1920	293	15	18	675-1140	7.24
(3)	1926	273	6	14	323-1123	8.20
(4)	1926	291	4	9	483-914	6.69
(5)	1960	471	2	13	481-723	7.0

Forster and Tschentke (6) reported heat capacity data graphically in the region 723-973 K. Their study showed a heat capacity of approximately 8.2 cal K^{-1} mol^{-1} at the melting point with a steadily decreasing trend to 7.0 cal K^{-1} mol^{-1} at 973 K. Cordoba (7) measured heat capacities in the crystal and liquid region, including 22 experimental points in the liquid region up to 798.2 K. This study resulted in a parabolic temperature dependence, with a minimum at 730 K of Cp° = 6.8 cal K^{-1} mol^{-1} and rapidly increasing values on either side of this minimum.

All these heat capacity and enthalpy studies do not conclusively define the temperature dependence of the heat capacity of liquid zinc. In addition, some of these heat capacity and enthalpy studies (3, 4, 7) do not show good agreement with the adopted values in the crystal region. However, the adopted constant value is consistent with the experimental data (within the uncertainty of the data) and with the vaporization data. The literature survey for Zn was done in part by Hultgren (8).

Fusion Data

Refer to the crystal table for details.

Vaporization Data

There are numerous studies on the vaporization of liquid zinc (2, 9-22). These studies, when depicted graphically in a log p vs 1/T plot, show remarkably good agreement in the region T_{fus} to 1666 K (approximately 0.2 to 23000 mmHg) with the exception of the studies by Ruff and Bergdahl (10) and Piacente et al. (22). A 2nd and 3rd law analysis of all zinc vaporization studies up to 400 mmHg is summarized in the following table.

Vaporization Studies

Source	Year	Number Observations	Method	T/K	Drift cal mol^{-1} K^{-1}	$\Delta_{vap}H°$(298.15 K), kcal mol^{-1} 2nd law	3rd law
(9)	1919	7	boiling point	926-1198	-1.3±1.0	31.91±1.02	30.58±0.40
(10)	1920	5	transport	761-995	-0.1±0.1	29.67±0.10	29.62±0.04
(11)	1925	30	boiling point	857-960	-0.4±0.1	29.94±0.07	29.61±0.04
(12)	1926	21	static	903-1259	-0.6±0.1	30.22±0.14	29.58±0.10
(14)	1934	1	transport	954			29.68
(15)	1934	6	static	955-1089	0.6±0.7	28.90±0.66	29.52±0.16
(16)	1955	eqn	torsion effusion	700-900	0.0	29.60	29.60
(17)	1957	3	transport	862-1020	-0.2±0.1	29.76±0.05	29.62±0.02
(18)	1959	3	transport	823-923	0.6±0.1	29.08±0.04	29.62±0.06
(22)	1973	10	transpiration	826-1041	-0.1±0.3	29.94±0.25	29.86±0.11

Not including the results of Ruff and Bergdahl (10) and Piacente et al. (22), the mean of the tabulated 3rd law values is $\Delta_{vap}H°(298.15\ K) = 29.59 \pm 0.1\ kcal\ mol^{-1}$. This may be compared with our adopted value of 29.531 derived from $\Delta_{fus}H°$ and $\Delta_f H°$(Zn, g, 298.15 K) = 31.1±0.2 kcal mol^{-1}. This difference of 0.059 kcal mol^{-1} is within the uncertainty of the vapor pressure studies. Refer to the ideal gas table for more details.

The boiling point is calculated as that temperature for which ΔG° = 0 for Zn(ℓ) = Zn(g). T_{vap} is the temperature for which the fugacity is one bar.

References

References are listed on the Zn(cr,l) table.

Zinc (Zn) $Zn_1(l)$

Enthalpy Reference Temperature = T_r = 298.15 K — Standard State Pressure = p° = 0.1 MPa

T/K	$C_p°$ (J K^{-1} mol^{-1})	S°	$-[G°-H°(T_r)]/T$	$H°-H°(T_r)$ (kJ mol^{-1})	$\Delta_f H°$	$\Delta_f G°$	Log K_f
0							
100							
200							
250							
298.15	25.387	50.792	50.792	0.	6.519	3.813	-0.668
300	25.406	50.949	50.793	0.047	6.519	3.796	-0.661
350	25.889	54.903	51.104	1.330	6.519	3.343	-0.499
400	26.346	58.390	51.801	2.636	6.519	2.889	-0.377
440.000	26.740	60.919	52.516	3.697	GLASS <--> LIQUID		
440.000	31.380	60.919	52.516	3.697	TRANSITION		
450	31.380	61.624	52.711	4.011	6.565	2.435	-0.283
500	31.380	64.930	53.770	5.580	6.778	1.964	-0.205
600	31.380	70.652	56.122	8.718	7.119	0.967	-0.084
692.730	31.380	75.161	58.376	11.628	------ CRYSTAL <--> LIQUID -----		
700	31.380	75.489	58.552	11.856	0.	0.	0.
800	31.380	79.679	60.937	14.994	0.	0.	0.
900	31.380	83.375	63.228	18.132	0.	0.	0.
1000	31.380	86.681	65.411	21.270	0.	0.	0.
1100	31.380	89.672	67.483	24.408	0.	0.	0.
1180.173	31.380	91.880	69.066	26.924	------- FUGACITY = 1 bar -------		
1200	31.380	92.402	69.448	27.546	-115.101	1.935	-0.084
1300	31.380	94.914	71.311	30.684	-114.041	11.646	-0.468
1400	31.380	97.240	73.081	33.822	-112.982	21.274	-0.794
1500	31.380	99.405	74.765	36.960	-111.923	30.827	-1.073
1600	31.380	101.430	76.369	40.098	-110.863	40.309	-1.316
1700	31.380	103.332	77.899	43.236	-109.804	49.725	-1.528
1800	31.380	105.126	79.363	46.374	-108.744	59.078	-1.714
1900	31.380	106.823	80.764	49.512	-107.685	68.373	-1.880
2000	31.380	108.432	82.107	52.650	-106.626	77.612	-2.027

PREVIOUS: CURRENT: December 1978

Zinc (Zn) $Zn_1(l)$

ZINC (Zn) CRYSTAL–LIQUID $A_r = 65.38$

0 to 692.73 K crystal
above 692.73 K liquid

Refer to the individual tables for details.

References for Zn(ℓ) table

1. I. Iitake, Sic. Repts., Tohoku Univ. 8, 99 (1913).
2. H. Braune, Z. Anorg. Chem. 111, 109 (1920).
3. S. Umino, Sci. Repts., Tohoku Imp. Univ. 15, 597 (1926).
4. J. H. Awbery and E. Griffiths, Proc. Phys. Soc. (London) 38, 378 (1926).
5. M. Genot and R. Hagege, Compt. rend. 251, 2901 (1960).
6. F. Forster and G. Tschentke, Z. Metallk. 32, 191 (1940).
7. G. Cordoba, Ph.D. Dissertation, Univ. Tennessee, 1970.
8. R. Hultgren, Univ. California (Berkeley), 1978.
9. C. T. Heycock and F. E. E. Lamplough, Proc. Chem. Soc. (London) 28, 3 (1912).
10. O. Ruff and B. Bergdahl, Z. Anorg. Chem. 106, 76 (1919).
11. W. H. Rodebush and A. L. Dixon, J. Am. Chem. Soc. 47, 1033 (1925).
12. C. H. M. Jenkins, Proc. Roy. Soc. (London) 110A, 456 (1923).
13. W. Leitgebel, Z. Anorg. Chem. 202, 305 (1931).
14. E. Brumeister and K. Jellinek, Z. Physik. Chem. A165, 121 (1934).
15. E. Baur and R. Bruner, Helv. Chim. Acta 17, 958 (1934).
16. R. F. Barrow, P. G. Dodsworth, et al., Trans. Farday Soc. 51, 1354 (1955).
17. L. Everett, P. W. M. Jacobs, and J. Kitchener, Acta Met. 5, 281 (1957).
18. G. Gattow and A. Schneider, Angew. Chem. 71, 181 (1959).
19. V. S. Chernyaev and S. Ershova, Izv. Vysshikh Uchebn. Zavedenii, Tsvetnaya Met. (1), 76 (1964).
20. E. H. Baker, J. Appl. Chem. 16, 321 (1966).
21. L. Malaspina, R. Gigli, and G. Bardi, Rev. Int. Hautes temp. Refract. 9, 131 (1972).
22. V. Piacente, A. Desideri, L. Malaspina, and A. Hallgass, Rev. Int. Hautes Temp. Refract. 10, 85 (1973).

Zinc (Zn) $Zn_1(cr,l)$

Enthalpy Reference Temperature = T_r = 298.15 K — Standard State Pressure = p° = 0.1 MPa

T/K	C_p° (J K⁻¹mol⁻¹)	S° (J K⁻¹mol⁻¹)	$-[G^\circ-H^\circ(T_r)]/T$ (J K⁻¹mol⁻¹)	$H^\circ-H^\circ(T_r)$ (kJ mol⁻¹)	Δ_fH° (kJ mol⁻¹)	Δ_fG° (kJ mol⁻¹)	Log K_f
0	0.	0.	INFINITE	-5.669	0.	0.	0.
100	19.455	16.523	63.224	-4.670	0.	0.	0.
200	24.050	31.820	44.005	-2.437	0.	0.	0.
250	25.039	37.290	42.132	-1.211	0.	0.	0.
298.15	25.387	41.717	41.717	0.	0.	0.	0.
300	25.406	41.874	41.717	0.047	0.	0.	0.
350	25.889	45.828	42.028	1.330	0.	0.	0.
400	26.346	49.314	42.725	2.636	0.	0.	0.
450	26.843	52.446	43.634	3.965	0.	0.	0.
500	27.386	55.301	44.660	5.321	0.	0.	0.
600	28.588	60.399	46.868	8.118	0.	0.	0.
692.730	29.802	64.591	48.965	10.825	CRYSTAL <--> LIQUID		
692.730	31.380	75.161	48.965	18.147	TRANSITION		
700	31.380	75.489	49.239	18.375	0.	0.	0.
800	31.380	79.679	52.788	21.513	0.	0.	0.
900	31.380	83.375	55.985	24.651	0.	0.	0.
1000	31.380	86.681	58.892	27.789	0.	0.	0.
1100	31.380	89.672	61.557	30.927	0.	0.	0.
1180.173	31.380	91.880	63.542	33.443	------- FUGACITY = 1 bar -------		
1200	31.380	92.402	64.015	34.065	-115.101	1.936	-0.084
1300	31.380	94.914	66.297	37.203	-114.041	11.646	-0.468
1400	31.380	97.240	68.425	40.341	-112.982	21.274	-0.794
1500	31.380	99.405	70.419	43.479	-111.923	30.827	-1.073
1600	31.380	101.430	72.294	46.617	-110.863	40.309	-1.316
1700	31.380	103.332	74.065	49.755	-109.804	49.725	-1.528
1800	31.380	105.126	75.741	52.893	-108.744	59.078	-1.714
1900	31.380	106.823	77.333	56.031	-107.685	68.373	-1.880
2000	31.380	108.432	78.848	59.169	-106.626	77.612	-2.027

PREVIOUS: CURRENT: December 1978

Zinc (Zn) $Zn_1(cr,l)$

IP(Zn, g) = 75768.10 ± 0.5 cm^{-1}

S(298.15 K) = 160.989 ± 0.025 K^{-1} mol^{-1}

$\Delta_f H°$(0 K) = 129.89 ± 0.2 kJ mol^{-1}

$\Delta_f H°$(298.15 K) = 130.42 ± 0.2 kJ mol^{-1}

Electronic Levels and Quantum Weights

State	ε_i, cm^{-1}	g_i
^1S_o	0	1
^3P_o	32311.308	1
$^3P_1$	32501.390	3
$^3P_2$	32890.317	5

Enthalpy of Formation

The adopted value for the enthalpy of formation is that recommended by CODATA (1). This value is based on the reanalysis of the vapor pressure data cited by Hultgren et al. (2), and Gurvich et al. (3). The studies by Mar and Searcy (512-666 K, 4) and McCreary and Thorn (610 - 690 K, 5) are also considered. There are twelve sublimation and eleven vaporizastion studies which lead to enthalpy of formation values showing excellent agreement, which the exception of two studies.

Heat Capacity and Entropy

The information on electronic energy levels and quantum weights, given by Moore (6), is incomplete because many theoretically predicted levels have not been observed. Our calculations indicate that any reasonable method of filling in these missing levels (for n<10) and cutting off the summation in the partition function (7) has no effect on the thermodynamic functions to 6000 K. This is a result of the high energy of all levels other than the ground state and the levels listed above; the next excited state is at approximately 46745 cm^{-1} above the ground state. Although we list only a few electronic levels, all levels reported by Moore (6) are considered in the calculation. The reported uncertainty in S°(298.15 K) is due to uncertainties in he relative atomic mass and the fundamental constants. Extension of these calculations above 6000 K may require consideration of the excited states (n>10) and use of different fill and cutoff procedures (7).

The thermal functions at 298.15 K differ from the CODATA recommendations (1) for two reasons. First, the entropy differs by 0.1094 J K^{-1} mol^{-1} because this table uses a reference pressure of 1 bar, wheras CODATA recommendations are based on 1 atm. Second, small changes, ~0.004 J K^{-1} mol^{-1}, arise presumably due to the use of slightly different values for the fundamental constants.

References

1. J. D. Cox, chairman, CODATA Task Group on Key Values for Thermodynamics, J. Chem. Thermodynamics 10, 903 (1978).
2. R. Hultgren, P. D. Desai, et al., "Selected Values of the Thermodynamic Properties of the Elements," American Society for Metals, Metals Park, Ohio, 1973.
3. V. P. Gluskko, V. A. Medvedevc, et al., "Thermal Constants of Substances," Part IV, VINITI, Moscow, 1972.
4. R. W. Man and A. W. searcy, J. Chem. Phys. 53, 3076 (1970).
5. J. R. McCreary and R. J. Thorn, J. Chem. Phys. 50, 3725 (1969).
6. C. E. Moore, U. S. Nat. Bur. Stand., NSRDS-NBS 35, Volume II, 1971 [Reprint of NBS Circular 467, Volume II, 1952].
7. J. R. Downey, Jr., The Dow Chemical Company, Rept. AFOSR-TR-78-0960, Contract No. F44620-75-1-0048, 1978.

Enthalpy Reference Temperature = T_r = 298.15 K   Standard State Pressure = p° = 0.1 MPa

	J $K^{-1}mol^{-1}$			kJ mol^{-1}			
T/K	$C_p°$	S°	$-[G°-H°(T_r)]/T$	$H°-H°(T_r)$	$\Delta_f H°$	$\Delta_f G°$	Log K_f
0	0.	0.	INFINITE	-6.197	129.892	129.892	INFINITE
100	20.786	138.281	179.469	-4.119	130.971	118.795	-62.052
200	20.786	152.689	162.890	-2.040	130.817	106.643	-27.852
250	20.786	157.327	161.331	-1.001	130.630	100.620	-21.023
298.15	20.786	160.989	160.989	0.	130.420	94.859	-16.619
300	20.786	161.117	160.989	0.038	130.411	94.638	-16.478
350	20.786	164.321	161.242	1.078	130.168	88.695	-13.237
400	20.786	167.097	161.804	2.117	129.902	82.788	-10.811
450	20.786	169.545	162.531	3.156	129.611	76.916	-8.928
500	20.786	171.735	163.344	4.196	129.295	71.078	-7.425
600	20.786	175.525	165.068	6.274	128.576	59.500	-5.180
700	20.786	178.729	166.797	8.353	120.398	48.130	-3.591
800	20.786	181.505	168.465	10.431	119.339	37.878	-2.473
900	20.786	183.953	170.053	12.510	118.279	27.759	-1.611
1000	20.786	186.143	171.554	14.589	117.220	17.758	-0.928
1100	20.786	188.124	172.972	16.667	116.160	7.863	-0.373
1180.173	20.786	189.586	174.052	18.334	------- FUGACITY = 1 bar -------		
1200	20.786	189.933	174.311	18.746	0.	0.	0.
1300	20.786	191.597	175.578	20.824	0.	0.	0.
1400	20.786	193.137	176.778	22.903	0.	0.	0.
1500	20.786	194.571	177.917	24.982	0.	0.	0.
1600	20.786	195.913	179.000	27.060	0.	0.	0.
1700	20.786	197.173	180.032	29.139	0.	0.	0.
1800	20.786	198.361	181.018	31.217	0.	0.	0.
1900	20.786	199.485	181.960	33.296	0.	0.	0.
2000	20.786	200.551	182.863	35.375	0.	0.	0.
2100	20.786	201.565	183.730	37.453	0.	0.	0.
2200	20.786	202.532	184.563	39.532	0.	0.	0.
2300	20.786	203.456	185.364	41.610	0.	0.	0.
2400	20.786	204.341	186.137	43.689	0.	0.	0.
2500	20.786	205.189	186.882	45.768	0.	0.	0.
2600	20.786	206.004	187.602	47.846	0.	0.	0.
2700	20.787	206.789	188.298	49.925	0.	0.	0.
2800	20.787	207.545	188.972	52.004	0.	0.	0.
2900	20.788	208.274	189.625	54.082	0.	0.	0.
3000	20.789	208.979	190.259	56.161	0.	0.	0.
3100	20.791	209.661	190.874	58.240	0.	0.	0.
3200	20.793	210.321	191.471	60.319	0.	0.	0.
3300	20.796	210.961	192.052	62.399	0.	0.	0.
3400	20.800	211.582	192.617	64.479	0.	0.	0.
3500	20.806	212.185	193.168	66.559	0.	0.	0.
3600	20.814	212.771	193.704	68.640	0.	0.	0.
3700	20.824	213.341	194.227	70.722	0.	0.	0.
3800	20.836	213.897	194.738	72.805	0.	0.	0.
3900	20.852	214.438	195.236	74.889	0.	0.	0.
4000	20.871	214.966	195.722	76.975	0.	0.	0.
4100	20.894	215.482	196.198	79.064	0.	0.	0.
4200	20.922	215.986	196.663	81.154	0.	0.	0.
4300	20.955	216.478	197.118	83.248	0.	0.	0.
4400	20.994	216.961	197.564	85.345	0.	0.	0.
4500	21.039	217.433	198.000	87.447	0.	0.	0.
4600	21.091	217.896	198.428	89.553	0.	0.	0.
4700	21.150	218.350	198.847	91.665	0.	0.	0.
4800	21.218	218.796	199.258	93.784	0.	0.	0.
4900	21.295	219.234	199.661	95.909	0.	0.	0.
5000	21.381	219.665	200.057	98.043	0.	0.	0.
5100	21.478	220.090	200.445	100.186	0.	0.	0.
5200	21.585	220.508	200.827	102.339	0.	0.	0.
5300	21.703	220.920	201.202	104.503	0.	0.	0.
5400	21.833	221.327	201.571	106.680	0.	0.	0.
5500	21.976	221.729	201.934	108.870	0.	0.	0.
5600	22.131	222.126	202.291	111.076	0.	0.	0.
5700	22.300	222.519	202.643	113.297	0.	0.	0.
5800	22.482	222.909	202.989	115.536	0.	0.	0.
5900	22.680	223.295	203.330	117.794	0.	0.	0.
6000	22.892	223.678	203.666	120.072	0.	0.	0.

PREVIOUS: December 1978 (1 atm)   CURRENT: September 1984 (1 bar)

ZINC, ION (Zn^+) — IDEAL GAS — M_r = 65.37945

IP(Zn^+, g) = 144892.6 ± 2 cm^{-1}
S°(298.15 K) = 166.751 ± 0.02 $J\ K^{-1}\ mol^{-1}$

$\Delta_f H°$(0 K) = 1030.6 ± 0.84 $kJ\ mol^{-1}$
$\Delta_f H°$(298.15 K) = [1037.336] $kJ\ mol^{-1}$

Electronic Levels and Quantum Weights

State	ε_i, cm^{-1}	g_i
$^2S_{1/2}$	0.0	2
$^2P_{1/2}$	48480.6	2
$^2P_{1/2}$	49354.4	4

Enthalpy of Formation

$\Delta_f H°$(Zn^+, g, 0 K) is calculated from $\Delta_f H°$(Zn, g, 0 K) (1) using the spectroscopic value of IP(Zn) = 75768.10 ± 0.5 cm^{-1} (906.388 ± 0.006 $kJ\ mol^{-1}$) from Moore (2). The ionization limit is converted from cm^{-1} to $kJ\ mol^{-1}$ using the factor, 1 cm^{-1} = 0.01196266 $kJ\ cm^{-1}$, which is derived from the 1973 CODATA fundamental constants (3). Rosenstock et al. (4) and Levin and Lias (5) have summarized additional ionization and appearance potential data.

$\Delta_f H°$(Zn^+, g, 298.15 K) is calculated from $\Delta_f H°$(Zn, g, 0 K) by using IP(Zn) with JANAF (1) enthalpies, H°(0 K) - H°(298.15 K), for Zn(g), Zn^+(g), and e^-(g). $\Delta_f H°$(Zn → Zn^+ + e^-, 298.15 K) differs from a room temperature threshold energy due to inclusion of these enthalpies and to threshold effects discussed by Rosenstock et al. (4). $\Delta_f H°$(298.15 K) should be changed by -6.197 $kJ\ mol^{-1}$ if it is to be used in the ion convention that excludes the enthalpy of the electron.

Heat Capacity and Entropy

The information on electronic energy levels and quantum weights, given by Moore (2, 6) is incomplete because many theoretically predicted levels have not been observed. Our calculations indicate that any reasonable method of filling in these missing levels and cutting off the summation in the partition function (7) has no effect on the thermodynamic functions to 6000 K. This is a result of the high energy of all levels other than the ground state and the first two excited states; the next excited state is 62721.9 cm^{-1} above the ground state. Since inclusion of these excited states has no effect on the thermodynamic functions (to 6000 K), we list only the ground state and the first two excited states. The reported uncertainty in S°(298.15 K) is due to uncertainties in the relative ionic mass and the fundamental constants. Extension of these calculations above 6000 K may require consideration of the excited states and use of different fill and cutoff procedures (7).

References

1. JANAF Thermochemical Tables: Zn(g), 9-30-84; e^-(g), 3-31-82.
2. C. E. Moore, U. S. Nat. Bur. Stand., NSRDS-NBS-34, 1970.
3. E. R. Cohen and B. N. Taylor, J. Phys. Chem. Ref. Data 2, 663 (1973).
4. H. M. Rosenstock, K. Draxl et al., J. Phys. Chem. Ref. Data 6, Supp. 1 (1977).
5. R. D. Levin and S. G. Lias, U. S. Nat. Bur. Stand., NSRDS-NBS-71, 1982.
6. C. E. Moore, U. S. Nat. Bur. Stand., NSRDS-NBS-35, Volume I, 1971 [Reprint of NBS Circular 467, Volume I, 1949].
7. J. R. Downey, Jr., The Dow Chemical Company, AFOSR-TR-0960, Contract No. F44620-75-1-0048, 1978.

Zinc, Ion (Zn^+) — Zn_1^+(g)

Enthalpy Reference Temperature = T_r = 298.15 K — Standard State Pressure = p° = 0.1 MPa

	J K⁻¹mol⁻¹			kJ mol⁻¹			
T/K	$C_p°$	S°	-[G°-H°(T_r)]/T	H°-H°(T_r)	$\Delta_f H°$	$\Delta_f G°$	Log K_f
0	0.	0.	INFINITE	-6.197	1030.611		
100	20.786	144.043	185.231	-4.119			
200	20.786	158.451	168.652	-2.040			
250	20.786	163.090	167.093	-1.001			
298.15	20.786	166.751	166.751	0.	1037.336	993.803	-174.110
300	20.786	166.879	166.751	0.038	1037.366	993.532	-172.989
350	20.786	170.083	167.004	1.078	1038.162	986.164	-147.177
400	20.786	172.859	167.566	2.117	1038.935	978.682	-127.803
450	20.786	175.307	168.293	3.156	1039.684	971.105	-112.723
500	20.786	177.497	169.106	4.196	1040.407	963.446	-100.651
600	20.786	181.287	170.830	6.274	1041.767	947.925	-82.524
700	20.786	184.491	172.559	8.353	1035.667	932.262	-69.566
800	20.786	187.267	174.227	10.431	1036.686	917.420	-59.901
900	20.786	189.715	175.815	12.510	1037.705	902.451	-52.377
1000	20.786	191.905	177.316	14.589	1038.725	887.368	-46.351
1100	20.786	193.886	178.734	16.667	1039.744	872.183	-41.417
1200	20.786	195.695	180.073	18.746	925.662	858.840	-37.384
1300	20.786	197.359	181.340	20.824	927.741	853.188	-34.282
1400	20.786	198.899	182.540	22.903	929.819	847.375	-31.616
1500	20.786	200.333	183.679	24.982	931.898	841.413	-29.301
1600	20.786	201.675	184.762	27.060	933.977	835.313	-27.270
1700	20.786	202.935	185.794	29.139	936.055	829.083	-25.475
1800	20.786	204.123	186.780	31.217	938.134	822.731	-23.875
1900	20.786	205.247	187.722	33.296	940.212	816.263	-22.441
2000	20.786	206.313	188.626	35.375	942.291	809.685	-21.147
2100	20.786	207.327	189.492	37.453	944.370	803.004	-19.974
2200	20.786	208.294	190.325	39.532	946.448	796.224	-18.905
2300	20.786	209.218	191.126	41.610	948.527	789.349	-17.927
2400	20.786	210.103	191.899	43.689	950.605	782.383	-17.028
2500	20.786	210.951	192.644	45.768	952.684	775.331	-16.200
2600	20.786	211.766	193.364	47.846	954.763	768.196	-15.433
2700	20.786	212.551	194.060	49.925	956.841	760.981	-14.722
2800	20.786	213.307	194.734	52.004	958.920	753.689	-14.060
2900	20.786	214.036	195.387	54.082	960.998	746.323	-13.443
3000	20.786	214.741	196.021	56.161	963.077	738.885	-12.865
3100	20.786	215.422	196.636	58.239	965.155	731.377	-12.324
3200	20.786	216.082	197.233	60.318	967.233	723.803	-11.815
3300	20.786	216.722	197.814	62.397	969.311	716.164	-11.336
3400	20.786	217.343	198.379	64.475	971.388	708.461	-10.884
3500	20.786	217.945	198.930	66.554	973.465	700.698	-10.457
3600	20.786	218.531	199.466	68.632	975.541	692.875	-10.053
3700	20.786	219.100	199.989	70.711	977.616	684.995	-9.670
3800	20.786	219.654	200.499	72.790	979.691	677.058	-9.307
3900	20.786	220.194	200.997	74.868	981.763	669.067	-8.961
4000	20.786	220.721	201.484	76.947	983.835	661.023	-8.632
4100	20.786	221.234	201.959	79.025	985.904	652.927	-8.318
4200	20.786	221.735	202.424	81.104	987.970	644.781	-8.019
4300	20.787	222.224	202.879	83.183	990.033	636.585	-7.733
4400	20.787	222.702	203.324	85.261	992.093	628.342	-7.459
4500	20.787	223.169	203.760	87.340	994.149	620.051	-7.197
4600	20.787	223.626	204.187	89.419	996.200	611.716	-6.946
4700	20.788	224.073	204.605	91.497	998.245	603.335	-6.705
4800	20.788	224.511	205.015	93.576	1000.284	594.911	-6.474
4900	20.789	224.939	205.418	95.655	1002.316	586.445	-6.252
5000	20.790	225.359	205.812	97.734	1004.340	577.937	-6.038
5100	20.791	225.771	206.200	99.813	1006.355	569.389	-5.832
5200	20.792	226.175	206.580	101.892	1008.360	560.802	-5.633
5300	20.794	226.571	206.953	103.972	1010.353	552.176	-5.442
5400	20.796	226.959	207.320	106.051	1012.334	543.512	-5.257
5500	20.798	227.341	207.681	108.131	1014.302	534.812	-5.079
5600	20.800	227.716	208.035	110.211	1016.256	526.076	-4.907
5700	20.803	228.084	208.384	112.291	1018.193	517.306	-4.741
5800	20.807	228.446	208.727	114.371	1020.113	508.502	-4.580
5900	20.811	228.801	209.064	116.452	1022.015	499.665	-4.424
6000	20.815	229.151	209.396	118.533	1023.896	490.795	-4.273

PREVIOUS: December 1978 (1 atm) — CURRENT: September 1984 (1 bar)

Zinc, Ion (Zn^+) — Zn_1^+(g)

ZINC, ION (Zn^-) IDEAL GAS M_r = 65.38055

EA(Zn, g) = 0.9 ± 0.20 eV

S°(298.15 K) = 160.99 ± 0.03 J K^{-1} mol^{-1}

$\Delta_f H°$(0 K) = 115.5 ± 20 kJ mol^{-1}

$\Delta_f H°$(298.15 K) = [109.870] kJ mol^{-1}

Electonic Level and Quantum Weight

State	ϵ_i, cm^{-1}	g_i
$^2S_{1/2}$	0	2

Enthalpy of Formation

$\Delta_f H°$(Zn^-, g, 0 K) is calculated from $\Delta_f H°$(Zn, g, 0 K) (1) using the adopted electron affinity of EA(Zn) = 0.09 ± 0.20 eV -0.67 ± 19.3 kJ mol^{-1}). This value is calculated by Zollwig (2) using a horizontal analysis technique for the ground state $3d^{10}4s^25s(^2S_{1/2})$. There is a possibility that the ground state may be $3d^{10}4s^24p$ (^2P_o) in which case Zollwig (2) calculated an electron affinity EA = -0.67 eV. Hotop and Lineberger (3), in a recent review, recommends that Zn^-(g) has a metastable state, $4s4p^24p$, with EA>0 (2, 4). Additional information on Zn^- may be obtained in the critical discussions of Hotop and Lineberger (3, 4), Rosenstock et al. (6), and Massey (7); and in the articles by Gohel et al. (8) and Kaiser et al. (9).

$\Delta_f H°$(Zn^-, g, 298.15 K) is obtained from $\Delta_f H°$(Zn, g, 0 K) by using EA(Zn) with JANAF (1) enthalpies, H°(0 K) - H°(298.15 K), for Zn^-(g), Zn(g), and e^-(g). $\Delta_r H°$(Zn^- → Zn + e^-, 298.15 K) differs from a room-temperature threshold energy due to inclusion of these enthalpies and to threshold effects discussed by Rosenstock et al (6). $\Delta_f H°$(298.15 K) should be changed by +6.197 kJ mol^{-1} if it is to be used in the ion convention that excludes the enthalpy of the electron.

Heat Capacity and Entropy

The ground state electronic configuration for Zn^-(g) is that given by Zollwig (2). Lacking any experimental evidence as to the stability of any excited states, we assume that no stable states exist.

References

1. JANAF Thermochemical Tables: Zn(g), 9-30-84; e^-(g), 3-31-82.
2. R. J. Zollwig, J. Chem. Phys. 50, 4251 (1969).
3. H. Hotop and W. C. Lineberger, J. Phys. Chem. Ref. Data, to be published, 1985.
4. P. D. Burrow, J. A. Mickejda, and J. Comer, J. Phys. B: Atom, Molec. Phys. 9, 3225 (1976).
5. H. Hotop and W. C. Lineberger, J. Phys. Chem. Ref. Data 4, 539 (1975).
6. H. M. Rosenstock, K. Draxel et al., J. Phys. Chem. Ref. Data 6, Supp. 1 (1977).
7. H. S. W. Massey, "Negative Ions", 3rd ed., Cambridge University Press, Cambridge, 1976.
8. V. B. Gohel and V. M. Trivedi, Indian J. Pure Appl. Phys. 10, 474 (1972).
9. H. J. Kaiser, E. Heinicke, H. Baumann, and K. Bethge, Z. Physik, 243, 46 (1971).

Zinc, Ion (Zn^-) Zn_1^-(g)

Enthalpy Reference Temperature = T_r = 298.15 K Standard State Pressure = p° = 0.1 MPa

T/K	$C_p°$	S°	-[G°-H°(T_r)]/T	H°-H°(T_r)	$\Delta_f H°$	$\Delta_f G°$	Log K_f
	J $K^{-1}mol^{-1}$			kJ mol^{-1}			
0	0.	0.	INFINITE	-6.197	115.539		
100	20.786	138.282	179.469	-4.119			
200	20.786	152.689	162.890	-2.040			
250	20.786	157.328	161.331	-1.001			
298.15	20.786	160.989	160.989	0.	109.870	80.564	-14.114
300	20.786	161.117	160.989	0.038	109.823	80.382	-13.996
350	20.786	164.321	161.242	1.078	108.540	75.576	-11.279
400	20.786	167.097	161.804	2.117	107.234	70.956	-9.266
450	20.786	169.545	162.531	3.156	105.905	66.501	-7.719
500	20.786	171.735	163.344	4.196	104.549	62.195	-6.497
600	20.786	175.525	165.068	6.274	101.752	53.985	-4.700
700	20.786	178.729	166.797	8.353	91.495	46.330	-3.457
800	20.786	181.505	168.466	10.431	88.357	40.092	-2.618
900	20.786	183.953	170.053	12.510	85.219	34.247	-1.988
1000	20.786	186.143	171.554	14.589	82.081	28.752	-1.502
1100	20.786	188.124	172.972	16.667	78.943	23.571	-1.119
1200	20.786	189.933	174.311	18.746	-39.296	20.611	-0.897
1300	20.786	191.597	175.578	20.824	-41.374	25.688	-1.032
1400	20.786	193.137	176.778	22.903	-43.453	30.924	-1.154
1500	20.786	194.571	177.917	24.982	-45.532	36.310	-1.264
1600	20.786	195.913	179.000	27.060	-47.610	41.834	-1.366
1700	20.786	197.173	180.032	29.139	-49.689	47.487	-1.459
1800	20.786	198.361	181.018	31.217	-51.767	53.264	-1.546
1900	20.786	199.485	181.960	33.296	-53.846	59.155	-1.626
2000	20.786	200.551	182.864	35.375	-55.925	65.157	-1.702
2100	20.786	201.565	183.730	37.453	-58.003	71.262	-1.773
2200	20.786	202.532	184.563	39.532	-60.082	77.466	-1.839
2300	20.786	203.456	185.364	41.610	-62.160	83.765	-1.902
2400	20.786	204.341	186.137	43.689	-64.239	90.154	-1.962
2500	20.786	205.189	186.882	45.768	-66.318	96.630	-2.019
2600	20.786	206.004	187.602	47.846	-68.396	103.189	-2.073
2700	20.786	206.789	188.298	49.925	-70.475	109.827	-2.125
2800	20.786	207.545	188.972	52.004	-72.554	116.543	-2.174
2900	20.786	208.274	189.625	54.082	-74.632	123.334	-2.221
3000	20.786	208.979	190.259	56.161	-76.711	130.195	-2.267
3100	20.786	209.660	190.874	58.239	-78.790	137.127	-2.311
3200	20.786	210.320	191.471	60.318	-80.869	144.125	-2.353
3300	20.786	210.960	192.052	62.397	-82.949	151.188	-2.393
3400	20.786	211.581	192.617	64.475	-85.029	158.314	-2.432
3500	20.786	212.183	193.168	66.554	-87.109	165.502	-2.470
3600	20.786	212.769	193.704	68.632	-89.190	172.749	-2.507
3700	20.786	213.338	194.227	70.711	-91.272	180.053	-2.542
3800	20.786	213.892	194.737	72.790	-93.355	187.414	-2.576
3900	20.786	214.432	195.235	74.868	-95.439	194.830	-2.609
4000	20.786	214.959	195.722	76.947	-97.525	202.299	-2.642
4100	20.786	215.472	196.197	79.025	-99.614	209.821	-2.673
4200	20.786	215.973	196.662	81.104	-101.704	217.393	-2.704
4300	20.786	216.462	197.117	83.183	-103.798	225.016	-2.733
4400	20.786	216.940	197.562	85.261	-105.895	232.687	-2.762
4500	20.786	217.407	197.998	87.340	-107.997	240.406	-2.791
4600	20.786	217.864	198.425	89.418	-110.103	248.171	-2.818
4700	20.786	218.311	198.843	91.497	-112.215	255.982	-2.845
4800	20.786	218.748	199.253	93.576	-114.334	263.839	-2.871
4900	20.786	219.177	199.656	95.654	-116.459	271.739	-2.897
5000	20.786	219.597	200.050	97.733	-118.593	279.683	-2.922
5100	20.786	220.009	200.438	99.811	-120.736	287.670	-2.946
5200	20.786	220.412	200.818	101.890	-122.889	295.699	-2.970
5300	20.786	220.808	201.191	103.969	-125.053	303.770	-2.994
5400	20.786	221.197	201.558	106.047	-127.230	311.881	-3.017
5500	20.786	221.578	201.919	108.126	-129.420	320.033	-3.039
5600	20.786	221.953	202.273	110.204	-131.626	328.224	-3.062
5700	20.786	222.320	202.622	112.283	-133.847	336.456	-3.083
5800	20.786	222.682	202.964	114.362	-136.086	344.726	-3.105
5900	20.786	223.037	203.302	116.440	-138.344	353.035	-3.126
6000	20.786	223.387	203.634	118.519	-140.622	361.383	-3.146

PREVIOUS: December 1978 (1 atm) CURRENT: September 1984 (1 bar)

Zinc, Ion (Zn^-) Zn_1^-(g)

ZIRCONIUM (Zr) REFERENCE STATE A_r = 91.22 Zirconium (Zr) Zr_1(ref)

0 - 1135 K Crystal, alpha
1135 - 2125 K Crystal, beta
2125 - 4702.633 K Liquid
4702.633 - 6000 K Ideal Monatomic Gas

Refer to the individual tables for details

Enthalpy Reference Temperature = T_r = 298.15 K — Standard State Pressure = p° = 0.1 MPa

T/K	C_p° (J K⁻¹mol⁻¹)	S° (J K⁻¹mol⁻¹)	$-[G^\circ-H^\circ(T_r)]/T$ (J K⁻¹mol⁻¹)	$H^\circ-H^\circ(T_r)$ (kJ mol⁻¹)	Δ_fH° (kJ mol⁻¹)	Δ_fG° (kJ mol⁻¹)	Log K_f
0	0.	0.	INFINITE	-5.497	0.	0.	0.
100	18.617	14.055	60.166	-4.611	0.	0.	0.
200	23.873	29.050	41.140	-2.418	0.	0.	0.
298.15	25.202	38.869	38.869	0.	0.	0.	0.
300	25.218	39.025	38.870	0.047	0.	0.	0.
400	25.935	46.384	39.868	2.606	0.	0.	0.
500	26.564	52.238	41.776	5.231	0.	0.	0.
600	27.281	57.143	43.939	7.922	0.	0.	0.
700	28.053	61.406	46.136	10.689	0.	0.	0.
800	28.966	65.210	48.287	13.538	0.	0.	0.
900	30.003	68.680	50.363	16.486	0.	0.	0.
1000	31.128	71.899	52.357	19.542	0.	0.	0.
1100	32.306	74.921	54.272	22.713	0.	0.	0.
1135.000	32.724	75.939	54.925	23.851		ALPHA <--> BETA	
1135.000	28.329	79.478	54.925	27.868		TRANSITION	
1200	28.511	81.060	56.298	29.715	0.	0.	0.
1300	28.879	83.356	58.292	32.584	0.	0.	0.
1400	29.353	85.513	60.160	35.494	0.	0.	0.
1500	29.934	87.558	61.919	38.458	0.	0.	0.
1600	30.621	89.511	63.583	41.485	0.	0.	0.
1700	31.414	91.390	65.164	44.585	0.	0.	0.
1800	32.314	93.211	66.672	47.771	0.	0.	0.
1900	33.320	94.985	68.115	51.052	0.	0.	0.
2000	34.433	96.722	69.502	54.439	0.	0.	0.
2100	35.652	98.431	70.839	57.942	0.	0.	0.
2125.000	35.973	98.854	71.166	58.837		BETA <--> LIQUID	
2125.000	41.840	108.699	71.166	79.757		TRANSITION	
2200	41.840	110.150	72.471	82.895	0.	0.	0.
2300	41.840	112.010	74.150	87.079	0.	0.	0.
2400	41.840	113.791	75.765	91.263	0.	0.	0.
2500	41.840	115.499	77.320	95.447	0.	0.	0.
2600	41.840	117.140	78.820	99.631	0.	0.	0.
2700	41.840	118.719	80.269	103.815	0.	0.	0.
2800	41.840	120.241	81.669	107.999	0.	0.	0.
2900	41.840	121.709	83.025	112.183	0.	0.	0.
3000	41.840	123.127	84.338	116.367	0.	0.	0.
3100	41.840	124.499	85.612	120.551	0.	0.	0.
3200	41.840	125.827	86.848	124.735	0.	0.	0.
3300	41.840	127.115	88.049	128.919	0.	0.	0.
3400	41.840	128.364	89.216	133.103	0.	0.	0.
3500	41.840	129.577	90.352	137.287	0.	0.	0.
3600	41.840	130.756	91.458	141.471	0.	0.	0.
3700	41.840	131.902	92.536	145.655	0.	0.	0.
3800	41.840	133.018	93.586	149.839	0.	0.	0.
3900	41.840	134.105	94.611	154.023	0.	0.	0.
4000	41.840	135.164	95.612	158.207	0.	0.	0.
4100	41.840	136.197	96.589	162.391	0.	0.	0.
4200	41.840	137.205	97.544	166.575	0.	0.	0.
4300	41.840	138.190	98.478	170.759	0.	0.	0.
4400	41.840	139.152	99.392	174.943	0.	0.	0.
4500	41.840	140.092	100.286	179.127	0.	0.	0.
4600	41.840	141.011	101.161	183.311	0.	0.	0.
4700	41.840	141.911	102.019	187.495	0.	0.	0.
4702.633	41.840	141.935	102.041	187.605		LIQUID <--> IDEAL GAS	
4702.633	38.083	261.292	102.041	748.899		FUGACITY = 1 bar	
4800	38.268	262.074	105.279	752.616	0.	0.	0.
4900	38.445	262.865	108.487	756.451	0.	0.	0.
5000	38.608	263.643	111.583	760.304	0.	0.	0.
5200	38.894	265.163	117.460	768.055	0.	0.	0.
5400	39.126	266.636	122.958	775.858	0.	0.	0.
5600	39.306	268.062	128.115	783.702	0.	0.	0.
5800	39.437	269.444	132.965	791.577	0.	0.	0.
6000	39.522	270.782	137.537	799.474	0.	0.	0.

PREVIOUS: June 1979 (1 atm) CURRENT: June 1979 (1 bar)

Zirconium (Zr) Zr_1(ref)

ZIRCONIUM, ALPHA (α-Zr) CRYSTAL A_r = 91.22

$S°(298.15\ K) = 38.87 \pm 0.2\ J\ K^{-1}\ mol^{-1}$

$T_{trs} = 1135 \pm 10\ K\ (\alpha-\beta)$

$\Delta_f H°(0\ K) = 0\ kJ\ mol^{-1}$

$\Delta_f H°(298.15\ K) = 0\ kJ\ mol^{-1}$

$\Delta_{trs} H° = 4.017 \pm 0.3\ kJ\ mol^{-1}$

Enthalpy of Formation

Zero by definition.

Heat Capacity and Entropy

The adopted thermal functions for α-Zr(cr) are derived from the studies of Collings and Ho (1), Wolcott (2), Burk et al. (3), Skinner and Johnston (4), Todd (5), and Scott (6). The mathematical and graphical treatment of these four studies yields a continuous and smooth heat capacity curve.

There are six studies (1, 2, 7 - 10) on the heat capacity of zirconium in the region 1-5 K. Only Wolcott (2) reported the experimental data (60 data points). In the other studies, an equation (with two constants) was given to describe the entire set of experimental data. The constants given by Collings and Ho (1) and the data of Wolcott (2) were used in this region. The maximum difference in the calculated values for S°(5 K) is 0.00029 cal $mol^{-1}\ K^{-1}$, with the adopted value being S°(5 K) = 0.0043 cal $mol^{-1}\ K^{-1}$. Similarly for Cp°(5 K) the maximum difference is 0.00073 cal $mol^{-1}\ K^{-1}$, with the adopted value being Cp°(5 K) = 0.00610 cal $mol^{-1}\ K^{-1}$.

In the region above 5 K and below 300 K, there are four heat capacity studies (2 - 5). There is much scatter among the different studies, between 150 and 300 K. From 150-200 K, we adopt graphically smoothed Cp values intermediate between Burk et al. (2), Skinner and Johnston (4), and Todd (5). From 200-300 K, we adopt the data of Skinner and Johnston (4), because it meshes better with the high temperature data. Burk's data is 1.4% higher than the adopted at 170 K, Todd's data is 0.7% higher and Skinner's data ia 1% lower. Todd's data is 3% higher than the adopted at 300 K.

Between 300 K and 1135 K there are three heat capacity studies (6, 11, 12) and seven enthalpy studies (13 - 19). The three heat capacity studies show much scatter and disagreement. The data of Scott (6) was adopted, although only the data from runs 17 and 64, out of a total of nine runs, was actually used in the fit. None of Scott's data below 540 K were used in the fit, becuase the data had a hump in this region and appeared to be high. The region between the low temperature data (300 K) and 540 K was graphically smoothed to assure continuity in the Cp curve. The data of Scott (6) is about 5% higher than the adopted between 300 and 540 K. Between 540 and 960 K it is within ±3% of the adopted. From 960 K to the transition temperature, there is much scatter; some of the data are as much as 20% higher than the adopted. The data of Vollmer et al. (12) are about 3% lower than the adopted at 300 K. Between 540 and 580 it agrees with the adopted. From 580-1000 K it is about 1.5% higher than the adopted. The data of Klein and Danielson (11) was given only as a graph. It is considerably lower than the adopted values.

The enthalpy studies show an unusually large scatter within each study, as well as a lack of agreement between the studies. When plotted as mean heat capacity, the seven studies are all higher than the adopted values, by 2 to 40%.

Transition Data

Refer to the β-Zr(cr) table for details.

Sublimation Data

There are no experimental sublimation studies involving α-Zr(cr). The enthalpy of sublimation is calculated as the difference between the enthalpy of formation of the monatomic gas and the enthalpy of formation of α-Zr(cr).

References

1. E. W. Collins and J. C. Ho, Phys. Rev. B 4, 349 (1971).
2. N. W. Wolcott, Phil. Mag. 2, 1246 (1957).
3. D. L. Burk, I. Estermann and S. A. Friedberg, Z. Physik. Chem. 16, 183 (1958).
4. G. B. Skinner and H. L. Johnston, J. Am. Chem. Soc. 73, 4549 (1951).
5. S. S. Todd, J. Am. Chem. Soc. 72, 2914 (1950).
6. J. L. Scott, AEC Report ORNL-2328, 1957.
7. I. Esterman, S. A. Friedberg and J. E. Goldman, Phys. Rev. 87, 582 (1952).
8. T. S. Smith and J. G. Daunt, Phys. Rev. 88, 1172 (1952).
9. G. D. Kneip, Jr., J. O. Betterton and J. O. Scarbrough, Phys. Rev. 130, 1687 (1963).
10. G. Dummer, Z. Phys. 186, 249 (1965).
11. A. H. Klein and G. C. Danielson, AEC Acession No. 42016, Rept. NP-14281 (1962).
12. O. Vollmer, M. Braun and R. Kohlhass, Z. Naturforsch. 22A, 833 (1967).
13. F. M. Jaeger and W. A. Veenstrra, Rec. Trav. Chim. 53, 917 (1934).
14. J. P. Coughlin and E. G. King, J. Am. Chem. Soc. 72, 2262 (1950).
15. G. B. Skinner, Ohio State University, Thesis, 1951.
16. H. K. Adenstedt, Trans. Am. Soc. Metals 44, 949 (1952).
17. R. F. Redmond and J. Lones, AEC Report ORNL-1342, 1952.
18. T. B. Douglas and A. C. Victor, Nat. Bur. Stand. Res. J. 61, 13 (1958).
19. I. B. Fieldhouse and J. I. Lang, WADD Tech. Rept. 60-904, 1961.

Zirconium, Alpha (Zr) $Zr_1(cr)$

Enthalpy Reference Temperature = T_r = 298.15 K Standard State Pressure = p° = 0.1 MPa

T/K	$C_p°$ (J K⁻¹mol⁻¹)	S° (J K⁻¹mol⁻¹)	$-[G°-H°(T_r)]/T$ (J K⁻¹mol⁻¹)	$H°-H°(T_r)$ (kJ mol⁻¹)	$\Delta_f H°$ (kJ mol⁻¹)	$\Delta_f G°$ (kJ mol⁻¹)	Log K_f
0	0.	0.	INFINITE	-5.497	0.	0.	0.
100	18.617	14.055	60.166	-4.611	0.	0.	0.
200	23.873	29.050	41.140	-2.418	0.	0.	0.
250	24.693	34.474	39.281	-1.202	0.	0.	0.
298.15	25.202	38.869	38.869	0.	0.	0.	0.
300	25.218	39.025	38.870	0.047	0.	0.	0.
350	25.606	42.943	39.178	1.318	0.	0.	0.
400	25.935	46.384	39.868	2.606	0.	0.	0.
450	26.246	49.457	40.766	3.911	0.	0.	0.
500	26.564	52.238	41.776	5.231	0.	0.	0.
600	27.281	57.143	43.939	7.922	0.	0.	0.
700	28.053	61.406	46.136	10.689	0.	0.	0.
800	28.966	65.210	48.287	13.538	0.	0.	0.
900	30.003	68.680	50.363	16.486	0.	0.	0.
1000	31.128	71.899	52.357	19.542	0.	0.	0.
1100	32.306	74.921	54.272	22.713	0.	0.	0.
1135.000	32.724	75.939	54.925	23.851	-------- ALPHA <--> BETA -------		
1200	33.502	77.783	56.113	26.004	-3.711	0.222	-0.010
1300	34.681	80.511	57.886	29.413	-3.170	0.528	-0.021
1400	35.807	83.123	59.596	32.938	-2.556	0.790	-0.029
1500	36.846	85.630	61.249	36.572	-1.886	1.006	-0.035
1600	37.762	88.038	62.848	40.304	-1.181	1.176	-0.038
1700	38.520	90.351	64.398	44.119	-0.466	1.301	-0.040

PREVIOUS: December 1967 CURRENT: June 1979

Zirconium, Alpha (Zr) $Zr_1(cr)$

ZIRCONIUM, BETA (β-Zr) CRYSTAL A_r = 91.22

$S°(298.15\ K) = [43.192]\ J\ K^{-1}\ mol^{-1}$

$T_{trs} = 1135 \pm 10\ K\ (\alpha-\beta)$

$T_{fus} = 2125 \pm 15\ K\ (\beta-\ell)$

$\Delta_f H°(298.15\ K) = [4.810]\ kJ\ mol^{-1}$

$\Delta_{trs} H° = 4.017 \pm 0.3\ kJ\ mol^{-1}$

$\Delta_{fus} H° = 20.92 \pm 6.3\ kJ\ mol^{-1}$

Enthalpy of Formation

The enthalpy of formation is calculated from that of the α-crystal by addition of $\Delta_{trs}H°$ and the difference in enthalpy, H°(1135 K) - H°(298.15 K), between the α-crystal and the β-crystal.

Heat Capacity and Entropy

There are five heat capacity studies (1 - 5) and five enthalpy studies (6 - 10) over the temperature range 1135-2125 K. The adopted values were based on the heat capcity study of Cezairliyan and Righini (4). The data of Scott (1) showed much scatter with a maximum deviation of 7% from the adopted. The data of Rumyantsu et al. (5) were largely scattered and about 15% high. The data of Vollmer et al. (3) and of Klein and Danielson (2) were 5 - 10% lower than the adopted. The enthalpy studies show an unusually large scatter within each study, as well as lack of agreement between the studies. Like the α-Zr(cr), the mean heat capacities are all higher than the adopted values, by 3 to 10%. The entropy at 298.15 K is calculated in a manner similar to that for the enthalpy of formation.

Phase Data

Zirconium, at ambient pressures, exists in two crystal modifications. The low temperature form, α-Zr, is hexagonal close packed, an hcp (A3) structure isotypic with Mg. The high temperature form, β-Zr, is body centered cubic, a bcc (A2) structure.

Transition Data

The transition temperature and enthalpies of transition are summarized in the table below. The adopted value of T_{trs} is 1135±10 K. The adopted $\Delta_{trs}H°$ is 4.017 kJ mol^{-1}.

Source	T_{trs}/K	$\Delta_{trs}H°$, kcal mol^{-1}	Purity	Method
Vogel and Tonn (11)	1135±5	-	-	cooling and dilatometric curves
Jaeger and Veenstra (12)	<1150	-	-	drop calorimetry
Coughlin and King (6)	1135	0.920	97.85	drop calorimetry
Dewey (13)	1138±10	-	-	cooling curve
Skinner (7)	1143±5	1.042	99.05	drop calorimetry
Komar and Shrednik (14)	1135	-	-	electron microscope
Scott (1)	1143±2	0.993±0.025	-	adiabatic calorimetry
Douglas and Victor (9)	1136	0.894	99.91	drop calorimetry
Gertzricken and Slyusar (15)	1135	0.712±0.060	-	differential calorimetry
Klein and Danielson (2)	1138	-	-	resistivity
Rudy et al. (16)	1145±15	-	99.81	differential thermal analysis
Vollmer et al. (3)	1155	0.950	99.8	adiabatic calorimetry
Peletskii et al. (17)	1148±5	-	-	resistivity
Martynyuk and Tsapkov (18)	-	1.076	99.76	resistivity

Fusion Data

Refer to the liquid table for details.

Sublimation Data

The sublimation studies involving β-Zr(cr) are summarized on the Zr(g) table. The enthalpy of sublimation for β-Zr(cr) is calculated as the difference in the enthalpies of formation at 298.15 K of Zr(g) and β-Zr(cr).

References

1. J. L. Scott, AEC Rept. ORNL-2328, 1957.
2. A. H. Klein and G. C. Danielson, AEC Acession No. 42016, Rept. NP-14281, 1962.
3. O. Vollmer, M. Braun and R. Kohlhass, Z. Naturforsch. 22A, 833 (1967).
4. A. Cezairliyan and F. Righini, J. Res. Nat. Bur. Stand. 78A, 509 (1974).
5. A. V. Rumyantsev, I. N. Makarenko, and S. N. Banchila, Izh. Fiz. Zh. 36, 581 (1979).
6. J. P. Coughlin and E. G. King, J. Am. Chem. Soc. 72, 2262 (1950).
7. G. B. Skinner, Ohio State University, Thesis, 1951.
8. R. F. Redmond and J. Lones, AEC Rept. ORNL-1342, 1952.
9. T. B. Douglas and A. C. Victor, Nat. Bur. Stand. Res. J. 61, 13 (1958).
10. I. B. Fieldhouse and J. I. Lang, WADD Tech. Rept. 60-904, 1961.
11. R. Vogel and W. Tonn, Z. Anorg. Allgem. Chem. 202, 292 (1931).
12. F. M. Jaeger and W. A. Veenstra, Rec. Trav. Chim. 53, 917 (1934).
13. P. Duwez, Trans. AIME, J. Met. 3, 765 (1951).
14. A. P. Komar and V. N. Shrednik, Soviet Phys. JETP 5, 127 (1957).
15. S. D. Gertsricken and B. P. Slyusar, Ukrain. Fiz. Zhur. 7, 439 (1962).
16. E. Rudy, D. P. Harmon, and C. E. Brukl, AFML-TR-65-2, Part I, Vol. II, 1965.
17. V. E. Peletskii, V. P. Druzhinin, and Ya. G. Sobol, Teplofiz. Vys. Temp. 8, 774 (1970).
18. M. M. Martynyuk and V. I. Tsapkov, Izv. Akad. Nauk SSSR, Metal (2), 181 (1974).

Zirconium, Beta (Zr) $Zr_1(cr)$

Enthalpy Reference Temperature = T_r = 298.15 K Standard State Pressure = p° = 0.1 MPa

T/K	$C_p°$ (J $K^{-1}mol^{-1}$)	S° (J $K^{-1}mol^{-1}$)	$-[G°-H°(T_r)]/T$ (J $K^{-1}mol^{-1}$)	$H°-H°(T_r)$ (kJ mol^{-1})	$\Delta_fH°$ (kJ mol^{-1})	$\Delta_fG°$ (kJ mol^{-1})	Log K_f
0							
100							
200							
250							
298.15	25.202	43.192	43.192	0.	4.810	3.521	-0.617
300	25.218	43.348	43.193	0.047	4.810	3.513	-0.612
350	25.587	47.264	43.501	1.317	4.809	3.297	-0.492
400	25.935	50.703	44.191	2.605	4.809	3.081	-0.402
450	26.238	53.775	45.088	3.909	4.808	2.865	-0.333
500	26.564	56.557	46.098	5.229	4.808	2.649	-0.277
600	27.281	61.462	48.261	7.921	4.809	2.217	-0.193
700	28.053	65.724	50.457	10.687	4.808	1.785	-0.133
800	28.966	69.535	52.608	13.542	4.813	1.353	-0.088
900	28.473	72.916	54.681	16.412	4.736	0.923	-0.054
1000	28.238	75.902	56.656	19.245	4.513	0.511	-0.027
1100	28.260	78.592	58.530	22.068	4.164	0.126	-0.006
1135.000	28.329	79.478	59.163	23.058	-------- ALPHA <--> BETA -------		
1200	28.511	81.060	60.306	24.905	0.	0.	0.
1300	28.879	83.356	61.992	27.774	0.	0.	0.
1400	29.353	85.513	63.596	30.684	0.	0.	0.
1500	29.934	87.558	65.126	33.648	0.	0.	0.
1600	30.621	89.511	66.589	36.675	0.	0.	0.
1700	31.414	91.390	67.993	39.776	0.	0.	0.
1800	32.314	93.211	69.344	42.961	0.	0.	0.
1900	33.320	94.985	70.647	46.242	0.	0.	0.
2000	34.433	96.722	71.907	49.629	0.	0.	0.
2100	35.652	98.431	73.130	53.132	0.	0.	0.
2125.000	35.973	98.854	73.430	54.027	------- BETA <--> LIQUID -------		
2200	36.977	100.119	74.318	56.763	-21.323	0.746	-0.018
2300	38.409	101.794	75.476	60.531	-21.738	1.759	-0.040
2400	39.947	103.461	76.608	64.448	-22.005	2.786	-0.061
2500	41.591	105.125	77.715	68.524	-22.113	3.822	-0.080
2600	43.342	106.790	78.801	72.770	-22.052	4.859	-0.098

PREVIOUS: December 1967 CURRENT: June 1979

Zirconium, Beta (Zr) $Zr_1(cr)$

ZIRCONIUM (Zr) LIQUID A_r = 91.22

$S°(298.15\ K) = [47.605]\ J\ K^{-1}\ mol^{-1}$ $\Delta_f H°(298.15\ K) = [17.404]\ kJ\ mol^{-1}$

$T_{fus} = 2125 \pm 15\ K$ $\Delta_{fus}H° = 20.92 \pm 6.3\ kJ\ mol^{-1}$

Enthalpy of Formation

The enthalpy of formation is calculated from that of the β-crystal by addition of $\Delta_{fus}H°$ and the difference in enthalpy, H°(2125 K) - H°(298.15 K), between the β-crystal and the liquid.

Heat Capacity and Entropy

There are no heat capacity or enthalpy studies covering the liquid region of zirconium. A constant value of Cp° = 41.84 J $K^{-1}\ mol^{-1}$ is assumed for the region 2125-5500 K. A glass transition is assumed at 1580 K below which extrapolated β-Zr heat capacity values are used. The entropy is calculated in a manner similar to that used for the enthalpy of formation.

Fusion Data

The available fusion temperature and enthalpy of fusion data are summarized in the table below. We adopt T_{fus} = 2125 ± 15 K and $\Delta_{fus}H° = 20.92 \pm 6.3\ kJ\ mol^{-1}$.

Source	T_{fus}/k	$\Delta_{fus}H°$, kcal mol^{-1}
deBoer and Fast (1)	2130	
Adenstedt (2)	2113	
Oriani and Jones (3)	2125 ± 2	
Deardorff and Hayes (4)	2128	
Scott (5)	2118	
Klein and Danielson (6)	2118	
Rudy et al (7)	2150 ± 20	
Sara (8)	2133 ± 15	
Rudy and Windisch (9)	2149 ± 4	
Elyutin et al (10)		5.0 ± 0.3
Ackerman and Rauk (11)	2134	
Martynyuk and Tsapkov (12)		5.139

Vaporization Data

The vaporization studies are summarized on the Zr(g) table. The boiling point is calculated as that temperature for which $\Delta_r G° = 0$ for Zr(ℓ) = Zr(g). $\Delta_{vap}H°$ is the corresponding enthalpy change. T_{vap} is the temperature at which the fugacity is one bar.

References

1. J. H. de Boer and J. D. Fast, Z. Anorg. Allgem. Chem. 187, 193 (1930).
2. H. K. Adenstedt, Trans. Am. Soc. Metals 44, 949 (1952).
3. R. A. Oriani and T. S. Jones, Rev. Sci. Inst. 25, 245 (1954).
4. D. K. Deardorff and E. T. Hayes, J. Metals AIME Trans. 206, 509 (1956).
5. J. L. Scott, AEC rept. ORNL-2328, 1957.
6. a. H. Klein and g. C. Danielson, AEC Acession No. NP-14281, 1962.
7. E. Rudy, D. P. Harmon and c. E. Brukl, AFML-TR-65-2, Part I, Vol. II, 1965.
8. R. V. Sara, J. A,. Ceram. Soc. 48, 243 (1965).
9. E. Rudy and S. Windisch, AFML-TR-65-2, Part I, Vol. III, 1966.
10. V. P. Elyutin, M. A. Maurakh and G. M. Sverdlov, Izv. Vyssh. Ucheb. Zaved., Tsvet. Met. 10, 87 (1967).
11. R. J. Ackermann and E. G. Rauh, J. Chem. Thermodyn. 4, 521 (1972).
12. M. M. Martynyuk and V. I. Tsapkov, Izv. Akad. Nauk SSSR, Metal (2), 181 (1974).

Zirconium (Zr) $Zr_1(l)$

Enthalpy Reference Temperature = T_r = 298.15 K Standard State Pressure = p° = 0.1 MPa

T/K	$C_p°$ (J $K^{-1}mol^{-1}$)	S° (J $K^{-1}mol^{-1}$)	$-[G°-H°(T_r)]/T$ (J $K^{-1}mol^{-1}$)	$H°-H°(T_r)$ (kJ mol^{-1})	$\Delta_f H°$ (kJ mol^{-1})	$\Delta_f G°$ (kJ mol^{-1})	Log K_f
0							
100							
200							
250							
298.15	25.202	47.605	47.605	0.	17.404	14.799	-2.593
300	25.218	47.761	47.605	0.047	17.404	14.783	-2.574
350	25.606	51.678	47.914	1.318	17.404	14.346	-2.141
400	25.935	55.120	48.604	2.606	17.404	13.910	-1.816
450	26.246	58.192	49.502	3.911	17.404	13.473	-1.564
500	26.564	60.974	50.512	5.231	17.404	13.036	-1.362
600	27.281	65.878	52.675	7.922	17.404	12.162	-1.059
700	28.053	70.141	54.872	10.689	17.404	11.289	-0.842
800	28.966	73.945	57.022	13.538	17.404	10.415	-0.680
900	30.003	77.416	59.098	16.486	17.404	9.542	-0.554
1000	31.128	80.635	61.093	19.542	17.404	8.668	-0.453
1100	32.306	83.656	63.008	22.713	17.404	7.795	-0.370
1200	33.502	86.519	64.849	26.004	13.693	7.143	-0.311
1300	34.681	89.247	66.622	29.413	14.234	6.575	-0.264
1400	35.807	91.859	68.332	32.938	14.848	5.964	-0.223
1500	36.846	94.365	69.984	36.572	15.518	5.306	-0.185
1580.000	37.614	96.300	71.268	39.551	GLASS <--> LIQUID TRANSITION		
1580.000	41.840	96.300	71.268	39.551			
1600	41.840	96.826	71.584	40.387	16.307	4.602	-0.150
1700	41.840	99.363	73.144	44.571	17.390	3.837	-0.118
1800	41.840	101.754	74.668	48.755	18.388	3.010	-0.087
1900	41.840	104.016	76.154	52.939	19.291	2.131	-0.059
2000	41.840	106.163	77.601	57.123	20.089	1.207	-0.032
2100	41.840	108.204	79.010	61.307	20.769	0.245	-0.006
2125.000	41.840	108.699	79.356	62.353	------- BETA <--> LIQUID -------		
2200	41.840	110.150	80.382	65.491	0.	0.	0.
2300	41.840	112.010	81.717	69.675	0.	0.	0.
2400	41.840	113.791	83.016	73.859	0.	0.	0.
2500	41.840	115.499	84.282	78.043	0.	0.	0.
2600	41.840	117.140	85.514	82.227	0.	0.	0.
2700	41.840	118.719	86.715	86.411	0.	0.	0.
2800	41.840	120.241	87.885	90.595	0.	0.	0.
2900	41.840	121.709	89.026	94.779	0.	0.	0.
3000	41.840	123.127	90.139	98.963	0.	0.	0.
3100	41.840	124.499	91.226	103.147	0.	0.	0.
3200	41.840	125.828	92.286	107.331	0.	0.	0.
3300	41.840	127.115	93.322	111.515	0.	0.	0.
3400	41.840	128.364	94.335	115.699	0.	0.	0.
3500	41.840	129.577	95.324	119.883	0.	0.	0.
3600	41.840	130.756	96.292	124.067	0.	0.	0.
3700	41.840	131.902	97.239	128.251	0.	0.	0.
3800	41.840	133.018	98.166	132.435	0.	0.	0.
3900	41.840	134.105	99.074	136.619	0.	0.	0.
4000	41.840	135.164	99.963	140.803	0.	0.	0.
4100	41.840	136.197	100.834	144.987	0.	0.	0.
4200	41.840	137.205	101.688	149.171	0.	0.	0.
4300	41.840	138.190	102.526	153.355	0.	0.	0.
4400	41.840	139.152	103.347	157.539	0.	0.	0.
4500	41.840	140.092	104.153	161.723	0.	0.	0.
4600	41.840	141.011	104.945	165.907	0.	0.	0.
4700	41.840	141.911	105.722	170.091	0.	0.	0.
4702.633	41.840	141.935	105.742	170.201	------- FUGACITY = 1 bar -------		
4800	41.840	142.792	106.485	174.275	-560.936	11.617	-0.126
4900	41.840	143.655	107.235	178.459	-560.588	23.542	-0.251
5000	41.840	144.500	107.971	182.643	-560.257	35.459	-0.370
5100	41.840	145.329	108.696	186.827	-559.941	47.371	-0.485
5200	41.840	146.141	109.408	191.011	-559.640	59.276	-0.595
5300	41.840	146.938	110.109	195.195	-559.352	71.175	-0.701
5400	41.840	147.720	110.798	199.379	-559.075	83.069	-0.804
5500	41.840	148.488	111.476	203.563	-558.809	94.958	-0.902

PREVIOUS: December 1967 CURRENT: June 1979

Zirconium (Zr) $Zr_1(l)$

ZIRCONIUM (Zr) ALPLHA-BETA-LIQUID A_r = 91.22

0 to 1135 K crystal, alpha
1135 to 2125 K crystal, beta
above 2125 K liquid

Refer to the individual tables for details.

Zirconium (Zr) Zr_1(cr,l)

Enthalpy Reference Temperature = T_r = 298.15 K; Standard State Pressure = p° = 0.1 MPa

T/K	C_p° (J K⁻¹mol⁻¹)	S° (J K⁻¹mol⁻¹)	-[G°-H°(T_r)]/T (J K⁻¹mol⁻¹)	H°-H°(T_r) (kJ mol⁻¹)	$\Delta_f H^\circ$ (kJ mol⁻¹)	$\Delta_f G^\circ$ (kJ mol⁻¹)	Log K_f
0	0.	0.	INFINITE	-5.497	0.	0.	0.
100	18.617	14.055	60.166	-4.611	0.	0.	0.
200	23.873	29.050	41.140	-2.418	0.	0.	0.
250	24.693	34.474	39.281	-1.202	0.	0.	0.
298.15	25.202	38.869	38.869	0.	0.	0.	0.
300	25.218	39.025	38.870	0.047	0.	0.	0.
350	25.606	42.943	39.178	1.318	0.	0.	0.
400	25.935	46.384	39.868	2.606	0.	0.	0.
450	26.246	49.457	40.766	3.911	0.	0.	0.
500	26.564	52.238	41.776	5.231	0.	0.	0.
600	27.281	57.143	43.939	7.922	0.	0.	0.
700	28.053	61.406	46.136	10.689	0.	0.	0.
800	28.966	65.210	48.287	13.538	0.	0.	0.
900	30.003	68.680	50.363	16.486	0.	0.	0.
1000	31.128	71.899	52.357	19.542	0.	0.	0.
1100	32.306	74.921	54.272	22.713	0.	0.	0.
1135.000	32.724	75.939	54.925	23.851	ALPHA <--> BETA		
1135.000	28.329	79.478	54.925	27.868	TRANSITION		
1200	28.511	81.060	56.298	29.715	0.	0.	0.
1300	28.879	83.356	58.292	32.584	0.	0.	0.
1400	29.353	85.513	60.160	35.494	0.	0.	0.
1500	29.934	87.558	61.919	38.458	0.	0.	0.
1600	30.621	89.511	63.583	41.485	0.	0.	0.
1700	31.414	91.390	65.164	44.585	0.	0.	0.
1800	32.314	93.211	66.672	47.771	0.	0.	0.
1900	33.320	94.985	68.115	51.052	0.	0.	0.
2000	34.433	96.722	69.502	54.439	0.	0.	0.
2100	35.652	98.431	70.839	57.942	0.	0.	0.
2125.000	35.973	98.854	71.166	58.837	BETA <--> LIQUID		
2125.000	41.840	108.699	71.166	79.757	TRANSITION		
2200	41.840	110.150	72.471	82.895	0.	0.	0.
2300	41.840	112.010	74.150	87.079	0.	0.	0.
2400	41.840	113.791	75.765	91.263	0.	0.	0.
2500	41.840	115.499	77.320	95.447	0.	0.	0.
2600	41.840	117.140	78.820	99.631	0.	0.	0.
2700	41.840	118.719	80.269	103.815	0.	0.	0.
2800	41.840	120.241	81.669	107.999	0.	0.	0.
2900	41.840	121.709	83.025	112.183	0.	0.	0.
3000	41.840	123.127	84.338	116.367	0.	0.	0.
3100	41.840	124.499	85.612	120.551	0.	0.	0.
3200	41.840	125.827	86.848	124.735	0.	0.	0.
3300	41.840	127.115	88.049	128.919	0.	0.	0.
3400	41.840	128.364	89.216	133.103	0.	0.	0.
3500	41.840	129.577	90.352	137.287	0.	0.	0.
3600	41.840	130.756	91.458	141.471	0.	0.	0.
3700	41.840	131.902	92.536	145.655	0.	0.	0.
3800	41.840	133.018	93.586	149.839	0.	0.	0.
3900	41.840	134.105	94.611	154.023	0.	0.	0.
4000	41.840	135.164	95.612	158.207	0.	0.	0.
4100	41.840	136.197	96.589	162.391	0.	0.	0.
4200	41.840	137.205	97.544	166.575	0.	0.	0.
4300	41.840	138.190	98.478	170.759	0.	0.	0.
4400	41.840	139.152	99.392	174.943	0.	0.	0.
4500	41.840	140.092	100.286	179.127	0.	0.	0.
4600	41.840	141.011	101.161	183.311	0.	0.	0.
4700	41.840	141.911	102.019	187.495	0.	0.	0.
4702.633	41.840	141.935	102.041	187.605	------- FUGACITY = 1 bar -------		
4800	41.840	142.792	102.859	191.679	-560.936	11.617	-0.126
4900	41.840	143.655	103.683	195.863	-560.588	23.542	-0.251
5000	41.840	144.500	104.491	200.047	-560.257	35.459	-0.370
5100	41.840	145.329	105.283	204.231	-559.941	47.371	-0.485
5200	41.840	146.141	106.061	208.415	-559.640	59.276	-0.595
5300	41.840	146.938	106.825	212.599	-559.352	71.175	-0.701
5400	41.840	147.720	107.575	216.783	-559.075	83.069	-0.804
5500	41.840	148.488	108.312	220.967	-558.809	94.958	-0.902

PREVIOUS: CURRENT: June 1979

Zirconium (Zr) Zr_1(cr,l)

ZIRCONIUM (Zr) IDEAL GAS A_r = 91.22

IP(Zr, g) = 55145 ± 500 cm^{-1}
S°(298.15 K) = 183.027 ± 0.04 J K^{-1} mol^{-1}

$\Delta_f H°$(0 K)= 602.1 ± 8.4 kJ mol^{-1}
$\Delta_f H°$(298.15 K) = 610.0 ± 8.4 kJ mol^{-1}

Electronic Levels and Quantum Weights

State	ϵ_i, cm^{-1}	g_i
$^3F_2$	0.00	5
$^3F_3$	570.41	7
$^3F_4$	1240.84	9
.	.	.
$^1P^0_1$	51899.40	3
IP	55145	

Enthalpy of Formation

Vapor pressures over Zr(β) and Zr(ℓ) were measured, using the Langmuir free-vaporization method, by Zwikker (1), Skinner et al. (2), Fidorov and Smirnov (3, 4) and Koch and Anable (5). Trulson and Goldstein (6) used a mass spectrometric method with a high temperature Knudson effusion source to obtain the vapor pressure for Zr(β) and Zr(ℓ). Similarly, Ackermann and Rauh (7) determined the vapor pressure for Zr(ℓ) by a combination of mass effusion and mass spectrometric techniques. In this latter study the reported values (equation only) were obtained from vapor pressure measurements over Zr(ℓ) saturated with tungsten. These values were then corrected for the dissolved tungsten by the corresponding mole fraction.

A plot of log p vs 1/T for the above mentioned studies reveals considerable discrepancy. The results of Fredorov and Smirnov (3, 4) are an order of magnitude higher than those of Zwikker (1); however, both studies appear inconsistent with the other studies involving Zr(β) (2, 6, 7). The sublimation studies of Golubtsov et al. (9) and Sumin and Peizularz (10) reported $\Delta_f H°$(0 K) values of 143.1 ± 0.22 and 145.6 ± 0.4 kcal mol^{-1}, respectively. The experimental data is not given in either study.

A second and third law evaluation of seven studies yields an adopted enthalpy of formation value, $\Delta_f H°$(298.15 K, Zr, g) = 145.8 ± 2.0 kcal mol^{-1}. This adopted value is a weighted average of four vapor pressure studies (2, 5, 6, 7). Hultgren (8), using different thermodynamic functions and data available through 1967, recommended a enthalpy of foramtion (at 298.15 K) of 145.5 ± 1.0 kcal mol^{-1}, based primarily on the data of Skinner et al. (2).

Investigator	Reaction	T/K	Data Points	$\Delta_r H°$(298.15 K), kcal mol^{-1} 2nd law	3rd law	Drift cal K^{-1} mol^{-1}	$\Delta_f H°$(298.15 K) kcal mol^{-1}
Zwikker (1)	a	1600-2100	Eqn	62.9	139.6	41.5	140.8
Skinner et al. (2)	a	1949-2054	12	144.2±3.0	146.5±0.3	1.2±1.0	147.7
Trulson and Goldstein (6)	a	1968-2112	14	143.0±3.6	142.8±0.5	-0.1±1.8	144.0
Federov et al. (3, 4)	a	1498-1723	6	59.9±0.1	122.4±6.8	39.0±0.1	123.6
Ackermann and Rauh (7)	a	1975-2100	Eqn	148.5	144.4	-2.0	145.6
Trulson and Goldstein (6)	b	2148-2274	8	129.7±4.9	137.7±0.6	3.6±2.2	142.7
Koch and Anable (5)	b	2229-2795	16	144.5±4.2	144.1±2.4	-0.2±1.7	149.1
Ackermann and Rauh (7)	b	2150-2500	Eqn	139.6	139.3	-0.1	144.3

Reactions: a. Zr(β) = Zr(g) b. Zr(ℓ) = Zr(g)

Heat Capacity and Entropy

The electronic energy levels are given in the compilation by Moore (11, 12). Although we have only listed the ground state, first two excited states, the highest observed excited state, and the ionization limit for Zr(g), all levels listed by Moore (11, 12), as well as estimated missing levels for n = 5 and 6, are used in our calculations. The observed levels are too numerous to list completely. Our calculations indicate that for Zr(g) the thermochemical functions are independent of the estimated missing levels for n = 5 and 6 (levels estimated and included above 52000 cm^{-1}) and the cut-off procedure up to 4000 K. The Gibbs energy function is essentially unaffected (<0.001 kcal mol^{-1}) up to 6000 K. The reported uncertainty in S°(298.15 K) is due to uncertainties in the relative atomic mass and the fundamental constants. Extension of these calculations above 6000 K may require consideration of the more exact estimation of excited states for n = 5 and 6, the consideration of states for n > 6, and the utilization of proper fill and cut-off procedures (13).

References

1. C. Zwikker, Versl. Kon. Ak. van Wetenschappen, Amsterdam 35, 336 (1926); Proc. Roy. Ac. Amsterdam 29, 792 (1926); Physica 8, 240 (1928); and J. H. deBoer and J. D. Fack, Z. Anorg. Allgem. Chem. 187, 193 (1930).
2. G. B. Skinner, J. W. Edwards, and H. L. Johnston, J. Am. Chem. Soc. 73, 174 (1934); G. B. Skinner, Ph.D. Thesis, The Ohio State University, 1951.
3. G. B. Federov and E. A. Smirnov, Thermody. Nucl. Mater, Proc. Symp., Vienna 1962, 285 (1962).
4. G. B. Federov, P. I. Kalinin, E. A. Smirnov, and V. N. Bespalov, Met. Metalloved. Chist. Metal. No. 9, 24 (1971).
5. R. K. Koch and W. E. Anable, U.S. Bur. Mines RI 7063, 1968.
6. O. C. Trulson and H. W. Goldstein, J. Phys. Chem. 69, 2531 (1965).
7. R. J. Ackermann and E. G. Rauk, J. Chem. Thermodynamics 4, 521 (1972).
8. R. Hultgren, P. D. Desai, D. T. Hawkins, M. Gleiser, K. K. Kelley, and D. D. Wagman, "Selected Values of the Thermodynamic Properties of the Elements," American Society for Metals, Metals Park, Ohio (1973).
9. I. V. Golubtsov and G. Yu. Mikul'skaya, V.Sb., Termodinam. Svoistva Metal. Splavov 75, 62 (1975).
10. V. V. Sumin and Sh. I. Reizulaw, Russ. J. Phys. Chem. 47, 910 (1973).
11. C. E. Moore, NSRDS-NBS 34 (1970).
12. C. E. Moore, NSRDS-NBS 35, Vol II, Reissued 1971.
13. J. R. Downey, Jr., The Dow Chemical Company, Thermal Research, AFOSR-TR-78-0960, March, 1978.

Zirconium (Zr) Zr_1(g)

Enthalpy Reference Temperature = T_r = 298.15 K   Standard State Pressure = p° = 0.1 MPa

T/K	$C_p°$ (J $K^{-1}mol^{-1}$)	S°	$-[G°-H°(T_r)]/T$	$H°-H°(T_r)$ (kJ mol^{-1})	$\Delta_f H°$	$\Delta_f G°$	Log K_f
0	0.	0.	INFINITE	-13.407	602.117	602.117	INFINITE
100	20.850	158.622	203.665	-4.504	610.134	595.677	-311.150
200	22.745	173.518	185.239	-2.344	610.101	581.207	-151.796
250	23.969	178.730	183.431	-1.175	610.054	573.990	-119.929
298.15	24.775	183.027	183.027	0.	610.027	567.046	-99.344
300	24.798	183.180	183.028	0.046	610.026	566.780	-98.685
350	25.196	187.038	183.331	1.297	610.007	559.574	-83.512
400	25.271	190.410	184.010	2.560	609.981	552.370	-72.132
450	25.147	193.381	184.890	3.821	609.938	545.172	-63.282
500	24.923	196.020	185.873	5.073	609.869	537.979	-56.202
600	24.431	200.519	187.952	7.540	609.645	523.620	-45.585
700	24.105	204.257	190.022	9.965	609.304	509.307	-38.005
800	24.054	207.470	192.006	12.371	608.859	495.051	-32.324
900	24.285	210.313	193.885	14.785	608.326	480.857	-27.908
1000	24.752	212.895	195.659	17.235	607.721	466.725	-24.379
1100	25.392	215.283	197.336	19.741	607.055	452.657	-21.495
1200	26.132	217.523	198.926	22.317	602.629	438.874	-19.104
1300	26.911	219.645	200.439	24.969	602.413	425.237	-17.086
1400	27.678	221.668	201.883	27.699	602.232	411.615	-15.358
1500	28.399	223.603	203.267	30.503	602.073	398.005	-13.860
1600	29.053	225.457	204.596	33.376	601.919	384.405	-12.550
1700	29.632	227.236	205.876	36.311	601.753	370.816	-11.394
1800	30.134	228.944	207.111	39.300	601.556	357.237	-10.367
1900	30.567	230.585	208.303	42.336	601.311	343.670	-9.448
2000	30.940	232.163	209.457	45.411	601.000	330.118	-8.622
2100	31.265	233.680	210.575	48.522	600.607	316.583	-7.875
2200	31.553	235.142	211.658	51.663	578.795	303.814	-7.213
2300	31.817	236.550	212.710	54.832	577.780	291.338	-6.616
2400	32.064	237.909	213.732	58.026	576.790	278.905	-6.070
2500	32.304	239.223	214.725	61.244	575.824	266.514	-5.568
2600	32.543	240.495	215.692	64.487	574.883	254.160	-5.106
2700	32.784	241.728	216.634	67.753	573.965	241.842	-4.679
2800	33.031	242.924	217.551	71.044	573.072	229.557	-4.282
2900	33.286	244.088	218.447	74.360	572.204	217.304	-3.914
3000	33.549	245.221	219.320	77.701	571.361	205.081	-3.571
3100	33.821	246.325	220.174	81.070	570.546	192.885	-3.250
3200	34.100	247.403	221.008	84.466	569.758	180.715	-2.950
3300	34.385	248.457	221.824	87.890	568.998	168.569	-2.668
3400	34.674	249.488	222.622	91.343	568.267	156.446	-2.404
3500	34.967	250.497	223.404	94.825	567.565	144.344	-2.154
3600	35.260	251.486	224.171	98.336	566.892	132.262	-1.919
3700	35.552	252.456	224.922	101.877	566.249	120.198	-1.697
3800	35.842	253.408	225.659	105.446	565.634	108.150	-1.487
3900	36.127	254.343	226.383	109.045	565.049	96.119	-1.287
4000	36.406	255.261	227.093	112.672	564.492	84.102	-1.098
4100	36.677	256.163	227.791	116.326	563.962	72.099	-0.919
4200	36.940	257.050	228.477	120.007	563.459	60.109	-0.748
4300	37.194	257.923	229.152	123.714	562.982	48.130	-0.585
4400	37.437	258.781	229.816	127.445	562.529	36.162	-0.429
4500	37.668	259.624	230.469	131.200	562.100	24.204	-0.281
4600	37.888	260.455	231.112	134.978	561.694	12.255	-0.139
4700	38.078	261.271	231.745	138.772	561.304	0.315	-0.004
4702.633	38.083	261.292	231.761	138.872	------- FUGACITY = 1 bar -------		
4800	38.268	262.074	232.368	142.589	0.	0.	0.
4900	38.445	262.865	232.982	146.425	0.	0.	0.
5000	38.608	263.643	233.588	150.278	0.	0.	0.
5100	38.758	264.409	234.185	154.146	0.	0.	0.
5200	38.894	265.163	234.773	158.029	0.	0.	0.
5300	39.017	265.905	235.354	161.924	0.	0.	0.
5400	39.126	266.636	235.926	165.832	0.	0.	0.
5500	39.223	267.355	236.491	169.749	0.	0.	0.
5600	39.306	268.062	237.049	173.676	0.	0.	0.
5700	39.378	268.759	237.599	177.610	0.	0.	0.
5800	39.437	269.444	238.142	181.551	0.	0.	0.
5900	39.485	270.118	238.678	185.497	0.	0.	0.
6000	39.522	270.782	239.208	189.448	0.	0.	0.

PREVIOUS: June 1979 (1 atm)   CURRENT: June 1979 (1 bar)

Zirconium (Zr) Zr_1(g)

ZIRCONIUM, ION (Zr^+) IDEAL GAS M_r = 91.21945

$IP(Zr^+, g) = 105900 \pm 100\ cm^{-1}$
$S°(298.15\ K) = 183.639 \pm 0.04\ J\ K^{-1}\ mol^{-1}$

$\Delta_f H°(0\ K) = 1269.7 \pm 25.1\ kJ\ mol^{-1}$
$\Delta_f H°(298.15\ K) = [1277.880]\ kJ\ mol^{-1}$

Electronic Levels and Quantum Weights

State	ϵ_i, cm^{-1}	g_i
$^4F_{3/2}$	0.00	4
$^4F_{5/2}$	314.67	6
$^4F_{7/2}$	763.44	8
$^4F_{9/2}$	1322.91	10
.	.	.
.	.	.
.	.	.
$^2H_{11/2}$	91737.40	12
IP	105900	

Enthalpy of Formation

$\Delta_f H°(Zr^+, g, 0\ K)$ is calculated from $\Delta_f H°(Zr, g, 0\ K)$ (1) using the spectroscopic value of $IP(Zr) = 55145 \pm 5\ cm^{-1}$ ($659.68 \pm 0.06\ kJ\ mol^{-1}$) from Moore (2). The ionization limit is converted from cm^{-1} to $kJ\ mol^{-1}$ using the factor, $1\ cm^{-1} = 0.01196266\ kJ\ mol^{-1}$, which is derived from the 1973 CODATA fundamental constants (3). Rosenstock et al. (4) and Levin and Lias (5) have summarized additional ionization and appearance potential data.

$\Delta_f H°(Zr^+, g, 298.15\ K)$ is calculated from $\Delta_f H°(Zr, g, 0\ K)$ by using IP(Zr) with JANAF (1) enthalpies, $H°(0\ K) - H°(298.15\ K)$, for Zr(g), Zr^+(g), and e^-(g). $\Delta_f H°(Zr + Zr^+ + e^-, 298.15\ K)$ differs from a room temperature threshold energy due to inclusion of these enthalpies and to threshold effects discussed by Rosenstock et al. (4). $\Delta_f H°(298.15\ K)$ should be changed by $-6.197\ kJ\ mol^{-1}$ if it is to be used in the ion convention that excludes the enthalpy of the electron.

Heat Capacity and Entropy

The electronic energy levels are given in the compilations by Moore (2, 6). Although we have only listed the ground state, first three excited states, the highest observed excited state, and the ionization limit for Zr^+(g), all levels listed by Moore (2, 6), as well as estimated missing levels for n = 5 and 6, are used in our calculations. The observed levels are too numerous to list completely. Our calculations indicate that for Zr^+(g) the thermochemical functions are independent of the estimated missing levels for n = 5 and 6 (levels estimated and included above 90000 cm^{-1}) and the cut-off procedure up to 6000 K. The Gibbs energy function is essentially unaffected (<0.001 cal $mol^{-1}\ K^{-1}$)) up to 12000 K. The reported uncertainty in S°(298.15 K) is due to uncertainties in the relative atomic mass and the fundamental constants. Extension of these calculations above 6000 K may require consideration of the more exact estimation of excited states for n = 5 and 6, the consideration of states for n>6, and the utilization of proper fill and cut-off procedures (7).

References

1. JANAF Thermochemical Tables: Zr(g), 6-30-79; e^-(g), 3-31-82.
2. C. E. Moore, U.S. Nat. Bur. Stand., NSRDS-NBS-34, 1970.
3. E. R. Cohen and B. N. Taylor, J. Phys. Chem. Ref. Data 2, 663 (1973).
4. H. M. Rosenstock, K. Draxl, et al., J. Phys. Chem. Ref. Data 6, Supp. 1 (1977).
5. R. D. Levin and Lias,. U.S. Nat. Bur. Stand., NSRDS-NBS-71, 1982.
6. C. E. Moore, U.S. Nat. Bur. Stand., NSRS-NBS-35, Volume II, 1970 [Reprint of NBS Circular 467, Volume II, 1952].
7. J. R. Downey, Jr., The Dow Chemical Company, AFOSR-TR-78-0960, Contract No. F44620-75-1-0048, 1978.

Zirconium, Ion (Zr^+) $Zr_1^+(g)$

Enthalpy Reference Temperature = T_r = 298.15 K Standard State Pressure = p° = 0.1 MPa

	J K⁻¹mol⁻¹			kJ mol⁻¹			
T/K	$C_p°$	S°	$-[G°-H°(T_r)]/T$	$H°-H°(T_r)$	$\Delta_f H°$	$\Delta_f G°$	Log K_f
0	0.	0.	INFINITE	-7.472	1269.708		
100	23.495	154.699	208.027	-5.333			
200	27.404	172.491	186.207	-2.743			
250	28.010	178.679	184.104	-1.356			
298.15	28.282	183.639	183.639	0.	1277.880	1228.462	-215.221
300	28.288	183.814	183.640	0.052	1277.924	1228.155	-213.841
350	28.358	188.182	183.984	1.469	1279.109	1219.766	-182.040
400	28.280	191.965	184.751	2.886	1280.276	1211.209	-158.168
450	28.112	195.287	185.741	4.296	1281.421	1202.507	-139.583
500	27.905	198.238	186.846	5.696	1282.541	1193.678	-124.703
600	27.513	203.290	189.179	8.466	1284.698	1175.701	-102.354
700	27.268	207.510	191.505	11.204	1286.748	1157.371	-86.364
800	27.188	211.145	193.738	13.925	1288.698	1138.755	-74.353
900	27.232	214.348	195.853	16.646	1290.549	1119.899	-64.997
1000	27.351	217.223	197.849	19.374	1292.301	1100.844	-57.502
1100	27.502	219.837	199.731	22.117	1293.951	1081.617	-51.362
1200	27.658	222.237	201.508	24.875	1291.785	1062.467	-46.248
1300	27.803	224.456	203.189	27.648	1293.769	1043.276	-41.919
1400	27.928	226.521	204.782	30.435	1295.723	1023.934	-38.203
1500	28.031	228.452	206.297	33.233	1297.637	1004.453	-34.978
1600	28.110	230.264	207.739	36.040	1299.496	984.847	-32.152
1700	28.169	231.970	209.114	38.854	1301.287	965.126	-29.655
1800	28.208	233.581	210.429	41.673	1302.999	945.302	-27.432
1900	28.233	235.107	211.688	44.495	1304.619	925.385	-25.441
2000	28.244	236.555	212.896	47.319	1306.135	905.386	-23.646
2100	28.245	237.933	214.055	50.144	1307.535	885.314	-22.021
2200	28.238	239.247	215.171	52.968	1287.485	865.923	-20.560
2300	28.225	240.502	216.245	55.791	1288.202	846.745	-19.230
2400	28.208	241.703	217.281	58.613	1288.919	827.536	-18.011
2500	28.187	242.854	218.281	61.433	1289.633	808.297	-16.888
2600	28.164	243.959	219.248	64.250	1290.345	789.030	-15.852
2700	28.139	245.022	220.183	67.065	1291.055	769.735	-14.891
2800	28.114	246.045	221.088	69.878	1291.762	750.414	-13.999
2900	28.087	247.031	221.966	72.688	1292.467	731.067	-13.168
3000	28.060	247.983	222.817	75.496	1293.169	711.697	-12.392
3100	28.033	248.902	223.644	78.300	1293.868	692.303	-11.665
3200	28.006	249.792	224.447	81.102	1294.565	672.886	-10.984
3300	27.979	250.653	225.228	83.901	1295.259	653.448	-10.343
3400	27.951	251.488	225.989	86.698	1295.950	633.989	-9.740
3500	27.924	252.298	226.729	89.492	1296.638	614.509	-9.171
3600	27.897	253.084	227.450	92.283	1297.324	595.010	-8.633
3700	27.869	253.848	228.153	95.071	1298.007	575.492	-8.124
3800	27.842	254.591	228.839	97.857	1298.687	555.956	-7.642
3900	27.816	255.314	229.509	100.640	1299.364	536.401	-7.184
4000	27.789	256.018	230.163	103.420	1300.039	516.830	-6.749
4100	27.763	256.704	230.802	106.197	1300.711	497.241	-6.335
4200	27.737	257.372	231.426	108.972	1301.381	477.636	-5.940
4300	27.712	258.025	232.037	111.745	1302.048	458.015	-5.564
4400	27.687	258.661	232.635	114.515	1302.712	438.379	-5.204
4500	27.664	259.283	233.221	117.282	1303.375	418.727	-4.860
4600	27.641	259.891	233.794	120.047	1304.034	399.061	-4.531
4700	27.619	260.485	234.355	122.810	1304.692	379.381	-4.216
4800	27.599	261.067	234.906	125.571	744.411	371.304	-4.041
4900	27.579	261.635	235.446	128.330	745.413	363.520	-3.875
5000	27.561	262.192	235.975	131.087	746.396	355.716	-3.716
5100	27.545	262.738	236.494	133.843	747.361	347.893	-3.563
5200	27.530	263.273	237.004	136.596	748.311	340.051	-3.416
5300	27.517	263.797	237.505	139.349	749.246	332.191	-3.274
5400	27.505	264.311	237.997	142.100	750.168	324.314	-3.137
5500	27.495	264.816	238.480	144.850	751.079	316.419	-3.005
5600	27.487	265.311	238.954	147.599	751.981	308.508	-2.878
5700	27.481	265.798	239.421	150.347	752.873	300.581	-2.755
5800	27.477	266.276	239.880	153.095	753.759	292.638	-2.635
5900	27.475	266.745	240.331	155.843	754.639	284.680	-2.520
6000	27.475	267.207	240.775	158.590	755.515	276.707	-2.409

PREVIOUS: June 1979 (1 atm) CURRENT: March 1984 (1 bar)

Zirconium, Ion (Zr^+) $Zr_1^+(g)$

ZIRCONIUM, ION (Zr^-) IDEAL GAS M_r = 91.22055

EA(Zr, g) = 0.426 ± 0.014 eV

S°(298.15 K) = 185.762 ± 0.001 $J\ K^{-1}\ mol^{-1}$

$\Delta_f H°$(0 K) = 568.92 ± 1.5 $kJ\ mol^{-1}$

$\Delta_f H°$(298.15 K) = [564.979] $kJ\ mol^{-1}$

Electonic Levels and Quantum Weights

State	ε_i, cm^{-1}	g_i
$^4F_{3/2}$	0.0	4
$^4F_{5/2}$	250	6
$^4F_{7/2}$	580	8
$^4F_{9/2}$	950	10

Enthalpy of Formation

$\Delta_f H°(Zr^-$, g, 0 K) is calculated from $\Delta_f H°$(Zr, g, 0 K) (1) using the adopted electron affinity of EA(Zr) = 0.426 ± 0.014 eV (41.102 ± 1.351 $kJ\ mol^{-1}$). This value, recommended by Hotop and Lineberger (2), is based on a laser photodetachment electron spectrometry study (3). Additional information on Zr^-(g) may be obtained in the critical discussions of Hotop and Lineberger (2, 4), Rosenstock et al. (5) and Massey (6).

$\Delta_f H°(Zr^-$, g, 298.15 K) is obtained from $\Delta_f H°$(Zr, g, 0 K) by using EA(Zr) with JANAF (1) enthalpies, H°(0 K) - H°(298.15 K), for Zr^-(g), Zr(g), and e^-(g). $\Delta_r H°(Zr^- \rightarrow Zr + e^-$, 298.15 K) differs from a room-temperature threshold energy due to inclusion of these enthalpies and to threshold effects discussed by Rosenstock et al. (5). $\Delta_f H°$(298.15 K) should be changed by +6.197 kJ mol^{-1} if it is to be used in the ion convention that excludes the enthalpy of the electron.

Heat Capacity and Entropy

The ground state electronic configuration for Zr^-(g) is given by Hotop and Lineberger (2, 4), Rosenstock et al. (5) and Massey (6). The fine structure separation has been calculated by an isoelectronic extrapolation of ratios of fine struture separations and is that recommended by Hotop and Lineberger (2).

References

1. JANAF Thermochemical Tables: Zr(g), 6-30-79; e^-(g), 3-31-82.
2. H. Hotop and W. C. Lineberger, J. Phys. Chem. Ref. Data, to be published, 1985.
3. C. S. Feigerle, R. R. Cordermann, S. V. Bobashev and W. C. Lineberger, J. Chem. Phys. 74, 1580 (1981).
4. H. Hotop and W. C. Lineberger, J. Phys. Chem. Ref. Data 4, 539 (1975).
5. H. M. Rosenstock, K. Draxl et al., J. Phys. Chem. Ref. Data 6, Supp. 1 (1977).
6. H. S. W. Massey, "Negative Ions", 3rd ed., Cambridge University Press, Cambridge, 1976.

Zirconium, Ion (Zr^-) Zr_1^-(g)

Enthalpy Reference Temperature = T_r = 298.15 K Standard State Pressure = p° = 0.1 MPa

T/K	$C_p°$ ($J\ K^{-1}mol^{-1}$)	S° ($J\ K^{-1}mol^{-1}$)	-[G°-H°(T_r)]/T ($J\ K^{-1}mol^{-1}$)	H°-H°(T_r) ($kJ\ mol^{-1}$)	$\Delta_f H°$ ($kJ\ mol^{-1}$)	$\Delta_f G°$ ($kJ\ mol^{-1}$)	Log K_f
0	0.	0.	INFINITE	-7.749	568.925		
100	25.123	155.513	211.008	-5.550			
200	28.509	174.284	188.395	-2.822			
250	28.830	180.691	186.236	-1.386			
298.15	28.693	185.762	185.762	0.	564.979	527.438	-92.405
300	28.681	185.940	185.763	0.053	564.947	527.205	-91.795
350	28.231	190.330	186.110	1.477	564.061	520.984	-77.753
400	27.627	194.061	186.877	2.874	563.130	514.894	-67.238
450	26.976	197.278	187.858	4.239	562.151	508.922	-59.074
500	26.339	200.087	188.944	5.571	561.124	503.063	-52.555
600	25.216	204.786	191.207	8.147	558.930	491.653	-42.802
700	24.332	208.604	193.428	10.623	556.561	480.626	-35.865
800	23.656	211.807	195.531	13.021	554.030	469.948	-30.684
900	23.139	214.562	197.496	15.359	551.343	459.598	-26.674
1000	22.742	216.978	199.326	17.652	548.501	449.555	-23.482
1100	22.433	219.131	201.030	19.911	545.509	439.804	-20.885
1200	22.188	221.072	202.621	22.141	538.660	430.554	-18.742
1300	21.993	222.840	204.109	24.350	535.921	421.655	-16.942
1400	21.834	224.464	205.506	26.541	533.123	412.970	-15.408
1500	21.704	225.965	206.820	28.718	530.258	404.487	-14.086
1600	21.597	227.363	208.061	30.883	527.317	396.198	-12.935
1700	21.507	228.669	209.235	33.038	524.293	388.095	-11.925
1800	21.430	229.896	210.349	35.184	521.175	380.173	-11.032
1900	21.365	231.053	211.409	37.324	517.956	372.427	-10.239
2000	21.310	232.148	212.419	39.458	514.624	364.853	-9.529
2100	21.262	233.186	213.383	41.586	511.171	357.449	-8.891
2200	21.220	234.174	214.306	43.710	486.263	350.958	-8.333
2300	21.183	235.117	215.190	45.830	482.120	344.901	-7.833
2400	21.151	236.017	216.039	47.947	477.974	339.024	-7.379
2500	21.122	236.880	216.856	50.061	473.825	333.319	-6.964
2600	21.097	237.708	217.642	52.172	469.674	327.781	-6.585
2700	21.075	238.504	218.400	54.280	465.520	322.403	-6.237
2800	21.054	239.270	219.132	56.387	461.363	317.178	-5.917
2900	21.036	240.009	219.839	58.491	457.205	312.102	-5.622
3000	21.020	240.721	220.523	60.594	453.046	307.169	-5.348
3100	21.005	241.410	221.186	62.695	448.884	302.375	-5.095
3200	20.991	242.077	221.829	64.795	444.721	297.716	-4.860
3300	20.979	242.723	222.452	66.894	440.557	293.186	-4.641
3400	20.968	243.349	223.058	68.991	436.392	288.783	-4.437
3500	20.958	243.957	223.646	71.087	432.226	284.502	-4.246
3600	20.948	244.547	224.218	73.183	428.058	280.340	-4.068
3700	20.940	245.121	224.776	75.277	423.890	276.294	-3.901
3800	20.932	245.679	225.318	77.370	419.721	272.361	-3.744
3900	20.924	246.223	225.847	79.463	415.551	268.537	-3.597
4000	20.917	246.752	226.364	81.555	411.381	264.821	-3.458
4100	20.911	247.269	226.867	83.647	407.210	261.209	-3.328
4200	20.905	247.773	227.359	85.738	403.038	257.698	-3.205
4300	20.900	248.264	227.839	87.828	398.865	254.287	-3.089
4400	20.894	248.745	228.309	89.917	394.693	250.973	-2.979
4500	20.890	249.214	228.768	92.007	390.519	247.753	-2.876
4600	20.885	249.673	229.218	94.095	386.345	244.627	-2.778
4700	20.881	250.123	229.658	96.184	382.171	241.591	-2.685
4800	20.877	250.562	230.089	98.272	-182.940	250.261	-2.723
4900	20.873	250.993	230.511	100.359	-186.767	259.326	-2.764
5000	20.870	251.414	230.925	102.446	-190.611	268.469	-2.805
5100	20.867	251.827	231.331	104.533	-194.471	277.689	-2.844
5200	20.864	252.233	231.729	106.620	-198.346	286.985	-2.883
5300	20.861	252.630	232.119	108.706	-202.234	296.355	-2.921
5400	20.858	253.020	232.503	110.792	-206.134	305.799	-2.958
5500	20.855	253.403	232.879	112.877	-210.045	315.315	-2.995
5600	20.853	253.778	233.249	114.963	-213.964	324.903	-3.031
5700	20.850	254.147	233.613	117.048	-217.892	334.560	-3.066
5800	20.848	254.510	233.970	119.133	-221.827	344.286	-3.101
5900	20.846	254.866	234.321	121.218	-225.767	354.081	-3.135
6000	20.844	255.217	234.666	123.302	-229.711	363.942	-3.168

PREVIOUS: June 1979 (1 atm) CURRENT: March 1984 (1 bar)

Zirconium, Ion (Zr^-) Zr_1^-(g)

Journal of Physical and Chemical Reference Data
Cumulative Listing of Reprints and Supplements

Reprints from Volume 1

1. Gaseous Diffusion Coefficients, *T.R. Marrero and E.A. Mason,* Vol. 1, No. 1, pp. 1–118 (1972) — $7.00
2. Selected Values of Critical Supersaturation for Nucleation of Liquids from the Vapor, *G.M. Pound,* Vol. 1, No. 1, pp. 119–134 (1972) — $3.00
3. Selected Values of Evaporation and Condensation Coefficients for Simple Substances, *G.M. Pound,* Vol. 1, No. 1, pp. 135–146 (1972) — $3.00
4. Atlas of the Observed Absorption Spectrum of Carbon Monoxide between 1060 and 1900 Å, *S.G. Tilford and J.D. Simmons,* Vol. 1, No. 1, pp. 147–188 (1972) — $4.50
5. Tables of Molecular Vibrational Frequencies, Part 5, *T. Shimanouchi,* Vol. 1, No. 1, pp. 189–216 (1972) (superseded by No.103) — $4.00
6. Selected Values of Heats of Combustion and Heats of Formation of Organic Compounds Containing the Elements C, H, N, O, P, and S, *Eugene S. Domalski,* Vol. 1, No. 2, pp. 221–278 (1972) — $5.00
7. Thermal Conductivity of the Elements, *C.Y. Ho, R.W. Powell, and P.E. Liley,* Vol. 1, No. 2, pp. 279–422 (1972) — $7.50
8. The Spectrum of Molecular Oxygen, *Paul H. Krupenie,* Vol. 1, No. 2, pp. 423–534 (1972) — $6.50
9. A Critical Review of the Gas-Phase Reaction Kinetics of the Hydroxyl Radical, *Wm. E. Wilson, Jr.,* Vol. 1, No. 2, pp. 535–574 (1972) — $4.50
10. Molten Salts: Volume 3, Nitrates, Nitrites, and Mixtures, Electrical Conductance, Density, Viscosity, and Surface Tension Data, *G.J. Janz, Ursula Krebs, H.F. Siegenthaler, and R.P.T. Tomkins,* Vol. 1, No. 3, pp. 581–746 (1972) — $8.50
11. High Temperature Properties and Decomposition of Inorganic Salts—Part 3. Nitrates and Nitrites, *Kurt H. Stern,* Vol. 1, No. 3, pp. 747–772 (1972) — $4.00
12. High-Pressure Calibration: A Critical Review, *D.L. Decker, W.A. Bassett, L. Merrill, H.T. Hall, and J.D. Barnett,* Vol. 1, No. 3, pp. 773–836 (1972) — $5.00
13. The Surface Tension of Pure Liquid Compounds, *Joseph J. Jasper,* Vol. 1, No. 4, pp. 841–1009 (1972) — $8.50
14. Microwave Spectra of Molecules of Astrophysical Interest, I. Formaldehyde, Formamide, and Thioformaldehyde, *Donald R. Johnson, Frank J. Lovas, and William H. Kirchhoff,* Vol. 1, No. 4, pp. 1011–1046 (1972) — $4.50
15. Osmotic Coefficients and Mean Activity Coefficients of Uni-univalent Electrolytes in Water at 25° C, *Walter J. Hamer and Yung-Chi Wu,* Vol. 1, No. 4, pp. 1047-1099 (1972) — $5.00
16. The Viscosity and Thermal Conductivity Coefficients of Gaseous and Liquid Fluorine, *H.J.M. Hanley and R. Prydz,* Vol. 1, No. 4, pp. 1101-1113 (1972) — $3.00

Reprints from Volume 2

17. Microwave Spectra of Molecules of Astrophysical Interest, II. Methylenimine, *William H. Kirchhoff, Donald R. Johnson, and Frank J. Lovas,* Vol. 2, No. 1, pp. 1-10 (1973) — $3.00
18. Analysis of Specific Heat Data in the Critical Region of Magnetic Solids, *F.J. Cook,* Vol. 2, No. 1, pp. 11-24 (1973) — $3.00
19. Evaluated Chemical Kinetic Rate Constants for Various Gas Phase Reactions, *Keith Schofield,* Vol. 2, No. 1, pp. 25–84 (1973) — $5.00
20. Atomic Transition Probabilities for Forbidden Lines of the Iron Group Elements. (A Critical Data Compilation for Selected Lines), *M.W. Smith and W.L. Wiese,* Vol. 2, No. 1, pp. 85–120 (1973) — $4.50
21. Tables of Molecular Vibrational Frequencies, Part 6, *T. Shimanouchi,* Vol. 2, No. 1, pp. 121–162 (1973) (superseded by No. 103) — $4.50
22. Compilation of Energy Band Gaps in Elemental and-Binary Compound Semiconductors and Insulators, *W.H. Strehlow and E.L. Cook,* Vol. 2, No. 1, pp. 163–200 (1973) — $4.50
23. Microwave Spectra of Molecules of Astrophysical Interest, III. Methanol, *R.M. Lees, F.J. Lovas, W.H. Kirchhoff, and D.R. Johnson,* Vol. 2, No. 2, pp. 205–214 (1973) — $3.00
24. Microwave Spectra of Molecules of Astrophysical Interest, IV. Hydrogen Sulfide, *Paul Helminger, Frank C. De Lucia, and William H. Kirchhoff,* Vol. 2, No. 2, pp. 215–224 (1973) — $3.00

Journal of Physical and Chemical Reference Data
Reprint and Supplement Orders

To: American Chemical Society
Distribution Office
1155 Sixteenth Street, N.W.
Washington, D.C. 20036

Name: ____________________

Title: ____________________

Organization: ____________________

Address: ____________________

City: __________ State: __________

Country: __________ Zip: __________

I am a member of ____________________
(ACS or AIP Member or Affiliated Society)

ORDERS FOR REPRINTS AND SUPPLEMENTS MUST BE PREPAID. *Foreign orders for Reprints, add $2.50 for each reprint for postage and handling. Foreign orders for Reprint Packages, add $4.00 for each Reprint Package for postage and handling. California residents add 6% state use tax. Make checks payable to the American Chemical Society.

BULK RATES: Subtract 20% from the listed price for orders of 50 or more of any one item.

Please ship the following reprints and supplements:

Reprint No./Package ______, ______ copies $ ______

Reprint No./Package ______, ______ copies $ ______

Reprint No./Package ______, ______ copies $ ______

Reprint No./Package ______, ______ copies $ ______

Vol. 2, Suppl. 1	☐ Hardcover ☐ Softcover	______ copies	$ ______
Vol. 3, Suppl. 1	☐ Hardcover ☐ Softcover	______ copies	$ ______
Vol. 6, Suppl. 1	☐ Hardcover ☐ Softcover	______ copies	$ ______
Vol. 10, Suppl. 1	☐ Hardcover	______ copies	$ ______
Vol. 11, Suppl. 1	☐ Hardcover	______ copies	$ ______
Vol. 11, Suppl. 2	☐ Hardcover	______ copies	$ ______
Vol. 13, Suppl. 1	☐ Hardcover	______ copies	$ ______
Other Suppl.:	______	______ copies	$ ______

Total Enclosed $ ______

(Continuation of Cumulative Listing of Reprints)

25. Tables of Molecular Vibrational Frequencies, Part 7, *T. Shimanouchi,* Vol. 2, No. 2, pp. 225–256 (1973) (superseded by No. 103) — $4.00
26. Energy Levels of Neutral Helium (^{4}He I), *W.C. Martin,* Vol. 2, No. 2, pp. 257–266 (1973) — $3.00
27. Survey of Photochemical and Rate Data for Twenty-eight Reactions of Interest in Atmospheric Chemistry, *R.F. Hampson, Editor, W. Braun, R.L. Brown, D. Garvin, J.T. Herron, R.E. Huie, M.J. Kurylo, A.H. Laufer, J.D. McKinley, H. Okabe, M.D. Scheer, W. Tsang, and D.H. Stedman,* Vol. 2, No. 2, pp. 267–312 (1973) — $4.50
28. Compilation of the Static Dielectric Constant of Inorganic Solids, *K.F. Young and H.P.R. Frederikse,* Vol. 2, No. 2, pp. 313–410 (1973) — $6.50
29. Soft X-Ray Emission Spectra of Metallic Solids: Critical Review of Selected Systems, *A.J. McAlister, R.C. Dobbyn, J.R. Cuthill, and M.L. Williams,* Vol. 2, No. 2, pp. 411–426 (1973) — $3.00
30. Ideal Gas Thermodynamic Properties of Ethane and Propane, *J. Chao, R.C. Wilhoit, and B.J. Zwolinski,* Vol. 2, No. 2, pp. 427–438 (1973) — $3.00
31. An Analysis of Coexistence Curve Data for Several Binary Liquid Mixtures Near Their Critical Points, *A. Stein and G.F. Allen,* Vol. 2, No. 3, pp. 443–466 (1973) — $4.00
32. Rate Constants for the Reactions of Atomic Oxygen (O ^{3}P) with Organic Compounds in the Gas Phase, *John T. Herron and Robert E. Huie,* Vol. 2, No. 3, pp. 467–518 (1973) — $5.00
33. First Spectra of Neon, Argon, and Xenon 136 in the 1.2-4.0 μm Region, *Curtis J. Humphreys,* Vol. 2, No. 3, pp. 519–530 (1973) — $3.00
34. Elastic Properties of Metals and Alloys, I. Iron, Nickel, and Iron-Nickel Alloys, *H.M. Ledbetter and R.P. Reed,* Vol. 2, No. 3, pp. 531–618 (1973) — $6.00
35. The Viscosity and Thermal Conductivity Coefficients of Dilute Argon, Krypton, and Xenon, *H.J.M. Hanley,* Vol. 2, No. 3, pp. 619–642 (1973) — $4.00
36. Diffusion in Copper and Copper Alloys, Part I. Volume and Surface Self-Diffusion in Copper, *Daniel B. Butrymowicz, John R. Manning, and Michael E. Read,* Vol. 2, No. 3, pp. 643–656 (1973) — $3.00
37. The 1973 Least-Squares Adjustment of the Fundamental Constants, *E. Richard Cohen and B.N. Taylor,* Vol. 2, No. 4, pp. 663–734 (1973) — $5.50
38. The Viscosity and Thermal Conductivity Coefficients of Dilute Nitrogen and Oxygen, *H.J.M. Hanley and James F. Ely,* Vol. 2, No. 4, pp. 735–756 (1973) — $4.00
39. Thermodynamic Properties of Nitrogen Including Liquid and Vapor Phases from 63 K to 2000 K with Pressures to 10,000 Bar, *Richard T. Jacobsen and Richard B. Stewart,* Vol. 2, No. 4, pp. 757–922 (1973) — $8.50
40. Thermodynamic Properties of Helium 4 from 2 to 1500 K at Pressures to 10^8 Pa, *Robert T. McCarty,* Vol. 2, No. 4, pp. 923–1042 (1973) — $7.00

Reprints from Volume 3

41. Molten Salts: Volume 4, Part 1, Fluorides and Mixtures, Electrical Conductance, Density, Viscosity, and Surface Tension Data, *G.J. Janz, G.L. Gardner, Ursula Krebs, and R.P.T. Tomkins,* Vol. 3, No. 1, pp. 1–115 (1974) — $7.00
42. Ideal Gas Thermodynamic Properties of Eight Chloro- and Fluoromethanes, *A.S. Rodgers, J. Chao, R. C. Wilhoit, and B.J. Zwolinski,* Vol. 3, No. 1, pp. 117–140 (1974) — $4.00
43. Ideal Gas Thermodynamic Properties of Six Chloroethanes, *J. Chao, A.S. Rodgers, R.C. Wilhoit, and B.J. Zwolinski,* Vol. 3, No. 1, pp. 141–162 (1974) — $4.00
44. Critical Analysis of Heat-Capacity Data and Evaluation of Thermodynamic Properties of Ruthenium, Rhodium, Palladium, Iridium, and Platinum from 0 to 300 K. A Survey of the Literature Data on Osmium, *George T. Furukawa, Martin L. Reilly, and John S. Gallagher,* Vol. 3, No. 1, pp. 163–209 (1974) — $4.50
45. Microwave Spectra of Molecules of Astrophysical Interest, V. Water Vapor, *Frank C. De Lucia, Paul Helminger, and William H. Kirchhoff,* Vol. 3, No. 1, pp. 211–219 (1974) — $3.00
46. Microwave Spectra of Molecules of Astrophysical Interest, VI. Carbonyl Sulfide and Hydrogen Cyanide, *Arthur G. Maki,* Vol. 3, No. 1, pp. 221–244 (1974) — $4.00
47. Microwave Spectra of Molecules of Astrophysical Interest, VII. Carbon Monoxide, Carbon Monosulfide, and Silicon Monoxide, *Frank J. Lovas and Paul H. Krupenie,* Vol. 3, No. 1, pp. 245–257 (1974) — $3.00
48. Microwave Spectra of Molecules of Astrophysical Interest, VIII. Sulfur Monoxide, *Eberhard Tiemann,* Vol. 3, No. 1, pp. 259–268 (1974) — $3.00
49. Tables of Molecular Vibrational Frequencies, Part 8, *T. Shimanouchi,* Vol. 3, No. 1, pp. 269–308 (1974) (superseded by No. 103) — $4.50
50. JANAF Thermochemical Tables, 1974 Supplement, *M.W. Chase, J.L. Curnutt, A.T. Hu, H. Prophet, A.N. Syverud, and L.C. Walker,* Vol. 3, No. 2, pp. 311–480 (1974) — $8.50
51. High Temperature Properties and Decomposition of Inorganic Salts, Part 4. Oxy-Salts of the Halogens, *Kurt H. Stern,* Vol. 3, No. 2, pp. 481–526 (1974) — $4.50
52. Diffusion in Copper and Copper Alloys, Part II. Copper-Silver and Copper-Gold Systems, *Daniel B. Butrymowicz, John R. Manning, and Michael E. Read,* Vol. 3, No. 2, pp. 527–602 (1974) — $5.50
53. Microwave Spectral Tables I. Diatomic Molecules, *Frank J. Lovas and Eberhard Tiemann,* Vol. 3, No. 3, pp. 609–770 (1974) — $8.50
54. Ground Levels and Ionization Potentials for Lanthanide and Actinide Atoms and Ions, *W.C. Martin, Lucy Hagan, Joseph Reader, and Jack Sugar,* Vol. 3, No. 3, pp. 771–780 (1974) — $3.00
55. Behavior of the Elements at High Pressures, *John Francis Cannon,* Vol. 3, No. 3, pp. 781–824 (1974) — $4.50
56. Reference Wavelengths from Atomic Spectra in the Range 15 Å to 25000 Å, *Victor Kaufman and Bengt Edlén,* Vol. 3, No. 4, pp. 825–895 (1974) — $5.50
57. Elastic Properties of Metals and Alloys. II. Copper, *H.M. Ledbetter and E.R. Naimon,* Vol. 3, No. 4, pp.897–935 (1974) — $4.50
58. A Critical Review of H-Atom Transfer in the Liquid Phase: Chlorine Atom, Alkyl, Trichloromethyl, Alkoxy, and Alkylperoxy Radicals, *D.G. Hendry, T. Mill, L. Piszkiewicz, J.A. Howard, and H.K. Eigenmann,* Vol. 3, No. 4, pp. 937–978 (1974) — $4.50
59. The Viscosity and Thermal Conductivity Coefficients for Dense Gaseous and Liquid Argon, Krypton, Xenon, Nitrogen, and Oxygen, *H.J.M. Hanley, R.D. McCarty, and W.M. Haynes,* Vol. 3, No. 4, pp. 979–1017 (1974) — $4.50

Reprints from Volume 4

60. JANAF Thermochemical Tables, 1975 Supplement, *M.W. Chase, J.L. Curnutt, H. Prophet, R.A. McDonald, and A.N. Syverud,* Vol. 4, No. 1, pp. 1–175 (1975) — $8.50
61. Diffusion in Copper and Copper Alloys, Part III. Diffusion in Systems Involving Elements of the Groups IA, IIA, IIIB, IVB, VB, VIB, and VIIB, *Daniel B. Butrymowicz, John R. Manning, and Michael E. Read,* Vol. 4, No. 1, pp. 177–249 (1975) — $6.00

(Continuation of Cumulative Listing of Reprints)

62. Ideal Gas Thermodynamic Properties of Ethylene and Propylene, *Jing Chao and Bruno J. Zwolinski*, Vol. 4, No. 1, pp. 251–261 (1975) $3.00

63. Atomic Transition Probabilities for Scandium and Titanium (A Critical Data Compilation of Allowed Lines), *W.L. Wiese and J.R. Fuhr*, Vol. 4, No. 2, pp. 263–352 (1975) $6.00

64. Energy Levels of Iron, Fe I through Fe XXVI, *Joseph Reader and Jack Sugar*, Vol. 4, No. 2, pp. 353–440 (1975) $6.00

65. Ideal Gas Thermodynamic Properties of Six Fluoroethanes, *S.S. Chen, A.S. Rodgers, J. Chao, R.C. Wilhoit, and B.J. Zwolinski*, Vol. 4, No. 2, pp. 441–456 (1975) $3.00

66. Ideal Gas Thermodynamic Properties of the Eight Bromo- and Iodomethanes, *S.A. Kudchadker and A.P. Kudchadker*, Vol. 4, No. 2, pp. 457–470 (1975) $3.00

67. Atomic Form Factors, Incoherent Scattering Functions, and Photon Scattering Cross Sections, *J.H. Hubbell, Wm.J. Veigele, E.A. Briggs, R.T. Brown, D.T. Cromer, and R.J. Howerton*, Vol. 4, No. 3, pp. 471–538 (1975) $5.50

68. Binding Energies in Atomic Negative Ions, *H. Hotop and W.C. Lineberger*, Vol. 4, No. 3, pp. 539–576 (1975) $4.50

69. A Survey of Electron Swarm Data, *J. Dutton*, Vol. 4, No. 3, pp. 577–856 (1975) $12.00

70. Ideal Gas Thermodynamic Properties and Isomerization of *n*-Butane and Isobutane, *S.S. Chen, R.C. Wilhoit, and B.J. Zwolinski*, Vol. 4, No. 4, pp. 859–869 (1975) $3.00

71. Molten Salts: Volume 4, Part 2, Chlorides and Mixtures, Electrical Conductance, Density, Viscosity, and Surface Tension Data, *G.J. Janz, R.P.T. Tomkins, C.B. Allen, J.R. Downey, Jr., G.L. Gardner, U. Krebs, and S.K. Singer*, Vol. 4, No. 4, pp. 871–1178 (1975) $13.00

72. Property Index to Volumes 1–4 (1972–1975), Vol. 4, No. 4, pp. 1179–1192 (1975) $3.00

Reprints from Volume 5

73. Scaled Equation of State Parameters for Gases in the Critical Region, *J.M.H. Levelt Sengers, W.L. Greer, and J.V. Sengers*, Vol. 5, No. 1, pp. 1–51 (1976) $5.00

74. Microwave Spectra of Molecules of Astrophysical Interest, IX. Acetaldehyde, *A. Bauder, F.J. Lovas, and D.R. Johnson*, Vol. 5, No. 1, pp. 53–77 (1976) $4.00

75. Microwave Spectra of Molecules of Astrophysical Interest, X. Isocyanic Acid, *G. Winnewisser, W.H. Hocking, and M.C.L. Gerry*, Vol. 5, No. 1, pp. 79–101 (1976) $4.00

76. Diffusion in Copper and Copper Alloys, Part IV. Diffusion in Systems Involving Elements of Group VIII, *Daniel B. Butrymowicz, John R. Manning, and Michael E. Read*, Vol. 5, No. 1, pp. 103–200 (1976) $6.50

77. A Critical Review of the Stark Widths and Shifts of Spectral Lines from Non-Hydrogenic Atoms, *N. Konjevic and J.R. Roberts*, Vol. 5, No. 2, pp. 209–257 (1976) $5.00

78. Experimental Stark Widths and Shifts for Non-Hydrogenic Spectral Lines of Ionized Atoms (A Critical Review and Tabulation of Selected Data), *N. Konjevic and W.L. Wiese*, Vol. 5, No. 2, pp. 259–308 (1976) $5.00

79. Atlas of the Absorption Spectrum of Nitric Oxide (NO) between 1420 and 1250 Å, *E. Miescher and F. Alberti*, Vol. 5, No. 2, pp. 309–317 (1976) $3.00

80. Ideal Gas Thermodynamic Properties of Propanone and 2-Butanone, *Jing Chao and Bruno J. Zwolinski*, Vol. 5, No. 2, pp. 319–328 (1976) $3.00

81. Refractive Index of Alkali Halides and Its Wavelength and Temperature Derivatives, *H.H. Li*, Vol. 5, No. 2, pp. 329–528 (1976) $9.50

82. Tables of Critically Evaluated Oscillator Strengths for the Lithium Isoelectronic Sequence, *G.A. Martin and W.L. Wiese*, Vol. 5, No. 3, pp. 537–570 (1976) $4.50

83. Ideal Gas Thermodynamic Properties of Six Chlorofluoromethanes, *S.S. Chen, R. C. Wilhoit, and B.J. Zwolinski*, Vol. 5, No. 3, pp. 571–580 (1976) $3.00

84. Survey of Superconductive Materials and Critical Evaluation of Selected Properties, *B.W. Roberts*, Vol. 5, No. 3, pp. 581–821 (1976) $12.50

85. Nuclear Spins and Moments, *Gladys H. Fuller*, Vol. 5, No. 4, pp. 835–1092 (1976) $11.50

86. Nuclear Moments and Moment Ratios as Determined by Mössbauer Spectroscopy, *J.G. Stevens and B.D. Dunlap*, Vol. 5, No. 4, pp. 1093–1121 (1976) $4.00

87. Rate Coefficients for Ion-Molecule Reactions, I. Ions Containing C and H, *L. Wayne Sieck and Sharon G. Lias*, Vol. 5, No. 4, pp. 1123–1146 (1976) $4.00

88. Microwave Spectra of Molecules of Astrophysical Interest, XI. Silicon Sulfide, *Eberhard Tiemann*, Vol. 5, No. 4, pp. 1147–1156 (1976) $3.00

89. Property Index and Author Index to Volumes 1–5 (1972–1976), Vol. 5, No. 4, pp. 1161–1183 $4.00

Reprints from Volume 6

90. Diffusion in Copper and Copper Alloys, Part V. Diffusion in Systems Involving Elements of Group VA, *Daniel B. Butrymowicz, John R. Manning, and Michael E. Read*, Vol. 6, No. 1, pp. 1–50 (1977) $5.00

91. The Calculated Thermodynamic Properties of Superfluid Helium-4, *James S. Brooks and Russell J. Donnelly*, Vol. 6, No. 1, pp. 51–104 (1977) $5.00

92. Thermodynamic Properties of Normal and Deuterated Methanols, *S.S. Chen, R.C. Wilhoit, and B.J. Zwolinski*, Vol. 6, No. 1, pp. 105–112 (1977) $3.00

93. The Spectrum of Molecular Nitrogen, *Alf Lofthus and Paul H. Krupenie*, Vol. 6, No. 1, pp. 113–307 (1977) $9.50

94. Energy Levels of Chromium, Cr I through Cr XXIV, *Jack Sugar and Charles Corliss*, Vol. 6, No. 2, pp. 317–383 (1977) $5.50

95. The Activity and Osmotic Coefficients of Aqueous Calcium Chloride at 298.15 K, *Bert R. Staples and Ralph L. Nuttall*, Vol. 6, No. 2, pp. 385–407 (1977) $4.00

96. Molten Salts: Volume 4, Part 3, Bromides and Mixtures; Iodides and Mixtures–Electrical Conductance, Density, Viscosity, and Surface Tension Data, *G.J. Janz, R.P.T. Tomkins, C.B. Allen, J.R. Downey, Jr., and S.K. Singer*, Vol. 6, No. 2, pp. 409–596 (1977) $9.00

97. The Viscosity and Thermal Conductivity Coefficients for Dense Gaseous and Liquid Methane, *H.J.M. Hanley, W.M. Haynes, and R.D. McCarty*, Vol. 6, No. 2, pp. 597–609 (1977) $3.00

98. Phase Diagrams and Thermodynamic Properties of Ternary Copper-Silver Systems, *Y. Austin Chang, Daniel Goldberg, and Joachim P. Neumann*, Vol. 6, No. 3, pp. 621–673 (1977) $5.00

99. Crystal Data Space-Group Tables, *Alan D. Mighell, Helen M. Ondik, and Bettijoyce Breen Molino*, Vol. 6, No. 3, pp. 675–829 (1977) $8.00

100. Energy Levels of One-Electron Atoms, *Glen W. Erickson*, Vol. 6, No. 3, pp. 831–869 (1977) $4.50

101. Rate Constants for Reactions of ClO_x of Atmospheric Interest, *R.T. Watson*, Vol. 6, No. 3, pp. 871–917 (1977) $4.50

102. NMR Spectral Data: A Compilation of Aromatic Proton Chemical Shifts in Mono- and Di-Substituted Benzenes, *B.L. Shapiro and L.E. Mohrmann*, Vol. 6, No. 3, pp. 919–991 (1977) $5.50

(Continuation of Cumulative Listing of Reprints)

103. Tables of Molecular Vibrational Frequencies. Consolidated Volume II. *T. Shimanouchi,* Vol. 6, No. 3, pp. 993–1102 (1977) (supersedes Nos. 5, 21, 25, 49) $6.50

104. Effects of Isotopic Composition, Temperature, Pressure, and Dissolved Gases on the Density of Liquid Water, *George S. Kell,* Vol. 6, No. 4, pp. 1109–1131 (1977) $4.00

105. Viscosity of Water Substance–New International Formulation and Its Background, *A. Nagashima,* Vol. 6, No. 4, pp. 1133–1166 (1977) $4.50

106. A Correlation of the Existing Viscosity and Thermal Conductivity Data of Gaseous and Liquid Ethane, *H.J.M. Hanley, K.E. Gubbins, and S. Murad,* Vol. 6, No. 4, pp. 1167–1180 (1977) $3.00

107. Elastic Properties of Zinc: A Compilation and a Review, *H.M. Ledbetter,* Vol. 6, No. 4, pp. 1181–1203 (1977) $4.00

108. Behavior of the AB-Type Compounds at High Pressures and High Temperatures, *Leo Merrill,* Vol. 6, No. 4, pp. 1205–1252 (1977) $4.50

109. Energy Levels of Manganese, Mn I through Mn XXV, *Charles Corliss and Jack Sugar,* Vol. 6, No. 4, pp. 1253–1329 (1977) $5.50

Reprints from Volume 7

110. Tables of Atomic Spectral Lines for the 10 000 Å to 40 000 Å Region, *Michael Outred,* Vol. 7, No. 1, pp. 1–262 (1978) $11.50

111. Evaluated Activity and Osmotic Coefficients for Aqueous Solutions: The Alkaline Earth Metal Halides, *R.N. Goldberg and R.L. Nuttall,* Vol. 7, No. 1, pp. 263–310 (1978) $4.50

112. Microwave Spectra of Molecules of Astrophysical Interest XII. Hydroxyl Radical, *Robert A. Beaudet and Robert L. Poynter,* Vol. 7, No. 1, pp. 311–362 (1978) $5.00

113. Ideal Gas Thermodynamic Properties of Methanoic and Ethanoic Acids, *Jing Chao and Bruno J. Zwolinski,* Vol. 7, No. 1, pp. 363–377 (1978) $3.00

114. Critical Review of Hydrolysis of Organic Compounds in Water Under Environmental Conditions, *W. Mabey and T. Mill,* Vol. 7, No. 2, pp. 383–415 (1978) $4.50

115. Ideal Gas Thermodynamic Properties of Phenol and Creosols, *S.A. Kudchadker, A.P. Kudchadker, R.C. Wilhoit, and B.J. Zwolinski,* Vol. 7, No. 2, pp. 417–423 (1978) $3.00

116. Densities of Liquid $CH_{4-a}X_a$ (X = Br,I) and $CH_{4-(a+b+c+d)}F_aCl_bBr_cI_d$ Halomethanes, *A.P. Kudchadker, S.A. Kudchadker, P.R. Patnaik, and P.P. Mishra,* Vol. 7, No. 2, pp. 425–439 (1978) $3.00

117. Microwave Spectra of Molecules of Astrophysical Interest XIII. Cyanoacetylene, *W.J. Lafferty and F.J. Lovas,* Vol. 7, No. 2, pp. 441–493 (1978) $5.00

118. Atomic Transition Probabilities for Vanadium, Chromium, and Manganese (A Critical Data Compilation of Allowed Lines), *S.M. Younger, J.R. Fuhr, G.A. Martin, and W.L. Wiese,* Vol. 7, No. 2, pp. 495–629 (1978) $7.50

119. Thermodynamic Properties of Ammonia, *Lester Haar and John S. Gallagher,* Vol. 7, No. 3, pp. 635–792 (1978) $8.00

120. JANAF Thermochemical Tables, 1978 Supplement, *M.W. Chase, Jr., J.L. Curnutt, R.A. McDonald, and A.N. Syverud,* Vol. 7, No. 3, pp. 793–940 (1978) $8.00

121. Viscosity of Liquid Water in the Range −8 °C to 150 °C, *Joseph Kestin, Mordechai Sokolov, and William A. Wakeham,* Vol. 7, No. 3, pp. 941–948 (1978) $3.00

122. The Molar Volume (Density) of Solid Oxygen in Equilibrium with Vapor, *H.M. Roder,* Vol. 7, No. 3, pp. 949–957 (1978) $3.00

123. Thermal Conductivity of Ten Selected Binary Alloy Systems, *C.Y. Ho, M.W. Ackerman, K.Y. Wu, S.G. Oh, and T.N. Havill,* Vol. 7, No. 3, pp. 959–1177 (1978) $10.00

124. Semi-Empirical Extrapolation and Estimation of Rate Constants for Abstraction of H from Methane by H, O, HO, and O_2, *Robert Shaw,* Vol. 7, No. 3, pp. 1179–1190 (1978) $3.00

125. Energy Levels of Vanadium, V I through V XXIII, *Jack Sugar and Charles Corliss,* Vol. 7, No. 3, pp. 1191–1262 (1978) $5.50

126. Recommended Atomic Electron Binding Energies, $1s$ to $6p_{3/2}$, for the Heavy Elements, $Z = 84$ to 103, *F.T. Porter and M.S. Freedman,* Vol. 7, No. 4, pp. 1267–1284 (1978) $4.00

127. Ideal Gas Thermodynamic Properties of $CH_{4-(a+b+c+d)}F_aCl_bBr_cI_d$ Halomethanes, *Shanti A. Kudchadker and Arvind P. Kudchadker,* Vol. 7, No. 4, pp. 1285–1307 (1978) $4.00

128. Critical Review of Vibrational Data and Force Field Constants for Polyethylene, *John Barnes and Bruno Fanconi,* Vol. 7, No. 4, pp. 1309–1321 (1978) $3.00

129. Tables of Molecular Vibrational Frequencies, Part 9, *Takehiko Shimanouchi, Hiroatsu Matsuura, Yoshiki Ogawa, and Issei Harada,* Vol. 7, No. 4, pp. 1323–1443 (1978) $7.00

130. Microwave Spectral Tables. II. Triatomic Molecules, *Frank J. Lovas,* Vol. 7, No. 4, pp. 1445–1750 (1978) $13.00

Reprints from Volume 8

131. Energy Levels of Titanium, Ti I through Ti XXII, *Charles Corliss and Jack Sugar,* Vol. 8, No. 1, pp. 1–62 (1979) $5.00

132. The Spectrum and Energy Levels of the Neutral Atom of Boron (B I), *G.A. Odintzova and A.R. Striganov,* Vol. 8, No. 1, pp. 63–67 (1979) $3.00

133. Relativistic Atomic Form Factors and Photon Coherent Scattering Cross Sections, *J.H. Hubbell and I. Øverbø,* Vol. 8, No. 1, pp. 69–105 (1979) $4.50

134. Microwave Spectra of Molecules of Astrophysical Interest. XIV. Vinyl Cyanide (Acrylonitrile), *M.C.L. Gerry, K. Yamada, and G. Winnewisser,* Vol. 8, No. 1, pp. 107–123 (1979) $4.00

135. Molten Salts: Volume 4, Part 4, Mixed Halide Melts. Electrical Conductance, Density, Viscosity, and Surface Tension Data, *G.J. Janz, R.P.T. Tomkins, and C.B. Allen,* Vol. 8, No. 1, pp. 125–302 (1979) $9.00

136. Atomic Radiative and Radiationless Yields for K and L Shells, *M.O. Krause,* Vol. 8, No. 2, pp. 307–327 (1979) $4.00

137. Natural Widths of Atomic K and L Levels, $K\alpha$ X-ray Lines and Several KLL Auger Lines, *M.O. Krause and J.H. Oliver,* Vol. 8, No. 2, pp. 329–338 (1979) $3.00

138. Electrical Resistivity of Alkali Elements, *T.C. Chi,* Vol. 8, No. 2, pp. 339–438 (1979) $6.50

139. Electrical Resistivity of Alkaline Earth Elements, *T.C. Chi,* Vol. 8, No. 2, pp. 439–497 (1979) $5.00

140. Vapor Pressures and Boiling Points of Selected Halomethanes, *A.P. Kudchadker, S.A. Kudchadker, R.P. Shukla, and P.R. Patnaik,* Vol. 8, No. 2, pp. 499–517 (1979) $4.00

141. Ideal Gas Thermodynamic Properties of Selected Bromoethanes and Iodoethane, *S.A. Kudchadker and A.P. Kudchadker,* Vol. 8, No. 2, pp. 519–526 (1979) $3.00

142. Thermodynamic Properties of Normal and Deuterated Naphthalenes, *S.S. Chen, S.A. Kudchadker, and R.C. Wilhoit,* Vol. 8, No. 2, pp. 527–535 (1979) $3.00

(Continuation of Cumulative Listing of Reprints)

143. Microwave Spectra of Molecules of Astrophysical Interest. XV. Propyne, *A. Bauer, D. Boucher, J. Burie, J. Demaison, and A. Dubrulle,* Vol. 8, No. 2, pp. 537–558 (1979) $4.00

144. A Correlation of the Viscosity and Thermal Conductivity Data of Gaseous and Liquid Propane, *P.M. Holland, H.J.M. Hanley, K.E. Gubbins, and J.M. Haile,* Vol. 8, No. 2, pp. 559–575 (1979) $4.00

145. Microwave Spectra of Molecules of Astrophysical Interest. XVI. Methyl Formate, *A. Bauder,* Vol. 8, No. 3, pp. 583–618 (1979) $4.50

146. Molecular Structures of Gas-Phase Polyatomic Molecules Determined by Spectroscopic Methods, *Marlin D. Harmony, Victor W. Laurie, Robert L. Kuczkowski, R.H. Schwendeman, D.A. Ramsay, Frank J. Lovas, Walter J. Lafferty, and Arthur G. Maki,* Vol. 8, No. 3, pp. 619–721 (1979) $6.50

147. Critically Evaluated Rate Constants for Gaseous Reactions of Several Electronically Excited Species, *Keith Schofield,* Vol. 8, No. 3, pp. 723–798 (1979) $5.50

148. A Review, Evaluation, and Correlation of the Phase Equilibria, Heat of Mixing, and Change in Volume on Mixing for Liquid Mixtures of Methane + Ethane, *M.J. Hiza, R.C. Miller, and A.J. Kidnay,* Vol. 8, No. 3, pp. 799–816 (1979) $4.00

149. Energy Levels of Aluminum, Al I through Al XIII, *W.C. Martin and Romuald Zalubas,* Vol. 8, No. 3, pp. 817–864 (1979) $4.50

150. Energy Levels of Calcium, Ca I through Ca XX, *Jack Sugar and Charles Corliss,* Vol. 8, No. 3, pp. 865–916 (1979) $5.00

151. Evaluated Activity and Osmotic Coefficients for Aqueous Solutions: Iron Chloride and the Bi-univalent Compounds of Nickel and Cobalt, *R.N. Goldberg, R.L. Nutall, and B.R. Staples,* Vol. 8, No. 4, pp. 923–1003 (1979) $6.00

152. Evaluated Activity and Osmotic Coefficients for Aqueous Solutions: Bi-univalent Compounds of Lead, Copper, Manganese, and Uranium, *R.N. Goldberg,* Vol. 8, No. 4, pp. 1005–1050 (1979) $4.50

153. Microwave Spectra of Molecules of Astrophysical Interest. XVII. Dimethyl Ether, *F.J. Lovas, H. Lutz, and H. Dreizler,* Vol. 8, No. 4, pp. 1051–1107 (1979) $5.00

154. Energy Levels of Potassium, K I through K XIX, *Charles Corliss and Jack Sugar,* Vol. 8, No. 4, pp. 1109–1145 (1979) $4.50

155. Electrical Resistivity of Copper, Gold, Palladium, and Silver, *R.A. Matula,* Vol. 8, No. 4, pp. 1147–1298 (1979) $8.00

Reprints from Volume 9

156. Energy Levels of Magnesium, Mg I through Mg XII, *W.C. Martin and Romuald Zalubas,* Vol 9, No. 1, pp. 1–58(1980) $6.00

157. Microwave Spectra of Molecules of Astrophysical Interest. XVIII. Formic Acid, *Edmond Willemot, Didier Dangoisse, Nicole Monnanteuil, and Jean Bellet,* Vol. 9, No. 1, pp. 59–160 (1980) $7.50

158. Refractive Index of Alkaline Earth Halides and Its Wavelength and Temperature Derivatives, *H.H. Li,* Vol. 9, No. 1, pp. 161–289 (1980). $8.50

159. Evaluated Kinetic and Photochemical Data for Atmospheric Chemistry, *D.L. Baulch, R.A. Cox, R.F. Hampson, Jr., J.A. Kerr, J. Troe, and R.L. Watson,* Vol. 9, No. 2, pp. 295–471 (1980) $10.00

160. Energy Levels of Scandium, Sc I through Sc XXII, *Jack Sugar and Charles Corliss,* Vol. 9, No. 2, pp. 473–511 (1980) $5.50

161. A Compilation of Kinetic Parameters for the Thermal Degradation of *n*-Alkane Molecules, *D.L. Allara and Robert Shaw,* Vol. 9, No. 3, pp. 523–559 (1980) $5.50

162. Refractive Index of Silicon and Germanium and Its Wavelength and Temperature Derivatives, *H.H. Li,* Vol. 9, No. 3, pp. 561–658 (1980) $7.50

163. Microwave Spectra of Molecules of Astrophysical Interest XIX. *Methyl Cyanide, D. Boucher, J. Burie, A. Bauer, A. Dubrulle, and J. Demaison,* Vol. 9, No. 3, pp. 659–719 (1980). $6.00

164. A Review, Evaluation, and Correlation of the Phase Equilibria, Heat of Mixing, and Change in Volume on Mixing for Liquid Mixtures of Methane + Propane, *R.C. Miller, A.J. Kidnay, and M.J. Hiza,* Vol. 9, No. 3, pp. 721–734 (1980) $4.00

165. Saturation States of Heavy Water, *P.G. Hill and R.D. Chris MacMillan,* Vol. 9, No. 3, pp. 735–749 (1980) $4.00

166. The Solubility of Some Sparingly Soluble Lead Salts: An Evaluation of the Solubility in Water and Aqueous Electrolyte Solution, *H. Lawrence Clever and Francis J. Johnston,* Vol. 9, No. 3, pp. 751–784 (1980) $5.50

167. Molten Salts Data as Reference Standards for Density, Surface Tension, Viscosity, and Electrical Conductance: KNO_3 and NaCl, *George J. Janz,* Vol. 9, No. 4, pp. 791–829 (1980) $5.50

168. Molten Salts: Volume 5, Part 1, Additional Single and Multi-Component Salt Systems. Electrical Conductance, Density, Viscosity, and Surface Tension Data, *G.J. Janz and R.P. Tomkins,* Vol. 9, No. 4, pp. 831–1021 (1980) $10.50

169. Pair, Triplet, and Total Atomic Cross Sections (and Mass Attenuation Coefficients) for 1 MeV–100 GeV Photons in Elements $Z=1$ to 100, *J.H. Hubbell, H.A. Gimm, and I. Øverbø,* Vol. 9, No. 4, pp. 1023–1147 (1980) $8.00

170. Tables of Molecular Vibrational Frequencies, Part 10, *Takehiko Shimanouchi, Hiroatsu Matsuura, Yoshiki Ogawa, and Issei Harada,* Vol. 9, No. 4, pp. 1149–1254 (1980) $7.50

171. An Improved Representative Equation for the Dynamic Viscosity of Water Substance, *J.T.R. Watson, R.S. Basu, and J.V. Sengers,* Vol. 9, No. 4, pp. 1255–1290 (1980) $5.50

172. Static Dielectric Constant of Water and Steam, *M. Uematsu and E. U. Franck,* Vol. 9, No. 4, pp. 1291–1306 (1980) $4.00

173. Compilation and Evaluation of Solubility Data in the Mercury (I) Chloride-Water System, *Y. Marcus,* Vol. 9, No. 4, pp. 1307–1329 (1980) $5.00

Reprints from Volume 10

174. Evaluated Activity and Osmotic Coefficients for Aqueous Solutions: Bi-Univalent Compounds of Zinc, Cadmium, and Ethylene Bis(Trimethylammonium) Chloride and Iodide, *R. N. Goldberg,* Vol. 10, No. 1, pp. 1–55 (1981) $6.00

175. Tables of the Dynamic and Kinematic Viscosity of Aqueous KCl Solutions in the Temperature Range 25–150 °C and the Pressure Range 0.1–35 MPa, *Joseph Kestin, H. Ezzat Khalifa, and Robert J. Correia,* Vol. 10, No. 1, pp. 57–70 (1981) $4.00

176. Tables of the Dynamic and Kinematic Viscosity of Aqueous NaCl Solutions in the Temperature Range 20–150 °C and the Pressure Range 0.1–35 MPa, *Joseph Kestin, H. Ezzat Khalifa, and Robert J. Correia,* Vol. 10, No. 1, pp. 71–87 (1981) $5.00

177. Heat Capacity and Other Thermodynamic Properties of Linear Macromolecules. I. Selenium, *Umesh Gaur, Hua-Cheng Shu, Aspy Mehta, and Bernhard Wunderlich,* Vol. 10, No. 1, pp. 89–117 (1981) $5.00

(Continuation of Cumulative Listing of Reprints)

178. Heat Capacity and Other Thermodynamic Properties of Linear Macromolecules. II. Polyethylene, *Umesh Gaur and Bernhard Wunderlich*, Vol. 10, No. 1, pp. 119–152 (1981) $5.50

179. Energy Levels of Sodium, Na I through Na XI, *W. C. Martin and Romuald Zalubas*, Vol. 10, No. 1, pp. 153–195 (1981) $5.50

180. Energy Levels of Nickel, Ni I through Ni XXVIII, *Charles Corliss and Jack Sugar*, Vol. 10, No. 1, pp. 197–289 (1981) $7.00

181. Ion Product of Water Substance, 0–1000 °C, 1–10,000 bars New International Formulation and Its Background, *William L. Marshall and E. U. Franck*, Vol. 10, No. 2, pp. 295–304 (1981) $4.00

182. Atomic Transition Probabilities for Iron, Cobalt, and Nickel (A Critical Data Compilation of Allowed Lines), *J. R. Fuhr, G. A. Martin, W. L. Wiese, and S. M. Younger*, Vol. 10, No. 2, pp. 305–565 (1981) $12.50

183. Thermodynamic Tabulations for Selected Phases in the System CaO-$Al_2O_3SiO_2$-H_2O at 101.325 kPa (1 atm) between 273.15 and 1800 K, *John L. Haas, Jr., Gilpin R. Robinson, Jr., and Bruce S. Hemingway*, Vol. 10, No. 3, pp. 575–669 (1981) $7.00

184. Evaluated Activity and Osmotic Coefficients for Aqueous Solutions: Thirty-Six Uni-Bivalent Electrolytes, R. N. Goldberg, Vol. 10, No. 3, pp. 671–764 (1981) $7.00

185. Activity and Osmotic Coefficients of Aqueous Alkali Metal Nitrites, *Bert R. Staples*, Vol. 10, No. 3, pp. 765–778 (1981) $4.00

186. Activity and Osmotic Coefficients of Aqueous Sulfuric Acid at 298.15 K, *Bert R. Staples*, Vol. 10, No. 3, pp. 779–798 (1981) $5.00

187. Rate Constants for the Decay and Reactions of the Lowest Electronically Excited Singlet State of Molecular Oxygen in Solution, *Francis Wilkinson and James G. Brummer*, Vol. 10, No. 4, pp. 809–999 (1981) $10.00

188. Heat Capacity and Other Thermodynamic Properties of Linear Macromolecules. III. Polyoxides, *Umesh Gaur and Bernhard Wunderlich*, Vol. 10, No. 4, pp. 1001–1049 (1981) $5.50

189. Heat Capacity and Other Thermodynamic Properties of Linear Macromolecules. IV. Polypropylene, *Umesh Gaur and Bernhard Wunderlich*, Vol. 10, No. 4, pp. 1051–1064 (1981) $4.00

190. Tables of N_2O Absorption Lines for the Calibration of Tunable Infrared Lasers from 522 cm^{-1} to 657 cm^{-1} and from 1115 cm^{-1} to 1340 cm^{-1}, *W. B. Olson, A. G. Maki, and W. J. Lafferty*, Vol. 10, No. 4, pp. 1065–1084 (1981) $5.00

191. Microwave Spectra of Molecules of Astrophysical Interest. XX. Methane, *I. Ozier, M. C. L. Gerry, and A. G. Robiette*, Vol. 10, No. 4, pp. 1085–1095 (1981) $4.00

192. Energy Levels of Cobalt, Co I through Co XXVII,, *Jack Sugar and Charles Corliss*, Vol. 10, No. 4, pp. 1097–1174 (1981) $6.50

193. A Critical Review of Henry's Law Constants for Chemicals of Environmental Interest, *Donald Mackay and Wan Ying Shiu*, Vol. 10, No. 4, pp. 1175–1199 (1981) $5.00

194. Property, Materials, and Author Indexes to the Journal of Physical and Chemical Reference Data, Vol. 1–10, pp. 1205–1225 (1972–1981) $5.00

Reprints from Volume 11

195. A Fundamental Equation of State for Heavy Water, *P. G. Hill, R. D. Chris MacMillan, and V. Lee*, Vol. 11, No. 1, pp. 1–14 (1982) $5.00

196. Volumetric Properties of Aqueous Sodium Chloride Solutions, *P. S. Z. Rogers and Kenneth S. Pitzer*, Vol. 11, No. 1, pp. 15–81 (1982) $9.00

197. Ideal Gas Thermodynamic Properties of CH_3, CD_3, CD_4, C_2D_2, C_2D_4, C_2D_6, C_2H_6, $CH_3N_2CH_3$, and $CD_3N_2CD_3$, *Krishna M. Pamidimukkala, David Rogers, and Gordon B. Skinner*, Vol. 11, No. 1, pp. 83–99 (1982) $6.00

198. Peak Absorption Coefficients of Microwave Absorption Lines of Carbonyl Sulphide, *Z. Kisiel and D. J. Millen*, Vol. 11, No. 1, pp. 99–116 (1982) $6.00

199. Vibrational Contributions to Molecular Dipole Polarizabilities, *David M. Bishop and Lap M. Cheung*, Vol. 11, No. 1, pp. 119–133 (1982) $5.00

200. Energy Levels of Iron, Fe I through Fe XXVI, *Charles Corliss and Jack Sugar*, Vol. 11, No. 1, pp. 135–241 (1982) $11.00

201. Microwave Spectra of Molecules of Astrophysical Interest. XXI. Ethanol(C_2H_5OH) and Propionitrile (C_2H_5CN), *Frank J. Lovas*, Vol. 11, No. 2, pp. 251–312 (1982) $8.00

202. Heat Capacity and Other Thermodynamic Properties of Linear Macromolecules, V. Polystyrene, *Umesh Gaur and Bernhard Wunderlich*, Vol. 11, No. 2, pp. 313–325 (1982) $5.00

203. Evaluated Kinetic and Photochemical Data for Atmospheric Chemistry: Supplement 1, CODATA Task Group on Chemical Kinetics, *D. L. Baulch, R. A. Cox, P. J. Crutzen, R. F. Hampson, Jr., J. A. Kerr (Chairman), J. Troe, and R. T. Watson*, Vol. 11, No. 2, pp. 327–496 (1982) $15.00

204. Molten Salts Data: Diffusion Coefficients in Single and Multi-Component Salt Systems, *G. J. Janz and N. P. Bansal*, Vol. 11, No. 3, pp. 505–693 (1982) $16.00

205. JANAF Thermochemical Tables, 1982 Supplement, *M. W. Chase, Jr., J. L. Curnutt, J. R. Downey, Jr., R. A. McDonald, A. N. Syverud, and E. A. Valenzuela*, Vol. 11, No. 3, pp. 695–940 (1982) $20.00

206. Critical Evaluation of Vapor-Liquid Equilibrium, Heat of Mixing, and Volume Change of Mixing Data. General Procedures, *Buford D. Smith, Ol Muthu, Ashok Dewan, and Matthew Gierlach*, Vol. 11, No. 3, pp. 941–951 (1982) $5.00

207. Rate Coefficients for Vibrational Energy Transfer Involving the Hydrogen Halides, *Stephen R. Leone*, Vol. 11, No. 3, pp. 953–996 (1982) $7.00

208. Behavior of the AB_2-Type Compounds at High Pressures and High Temperatures, *Leo Merrill*, Vol. 11, No. 4, pp. 1005–1064 (1982) $8.00

209. Heat Capacity and Other Thermodynamic Properties of Linear Macromolecules. VI. Acrylic Polymers, *Umesh Gaur, Suk-fai Lau, Brent B. Wunderlich, and Bernhard Wunderlich*, Vol. 11, No. 4, pp. 1065–1089 (1982) $6.00

210. Molecular Form Factors and Photon Coherent Scattering Cross Sections of Water, *L. R. M. Morin*, Vol. 11, No. 4, pp. 1091–1098 (1982) $5.00

211. Evaluation of Binary *PTxy* Vapor–Liquid Equilibrium Data for C_6 Hydrocarbons. Benzene + Cyclohexane, *Buford D. Smith, Ol Muthu, Ashok Dewan, and Matthew Gierlach*, Vol. 11, No. 4, pp. 1099–1126 (1982) $6.00

212. Evaluation of Binary Excess Enthalpy Data for C_6 Hydrocarbons. Benzene + Cyclohexane, *Buford D. Smith, Ol Muthu, Ashok Dewan, and Matthew Gierlach*, Vol. 11, No. 4, pp. 1127–1149 (1982) $6.00

213. Evaluation of Binary Excess Volume Data for C_6 Hydrocarbons. Benzene + Cyclohexane, *Buford D. Smith, Ol Muthu, Ashok Dewan, and Matthew Gierlach*, Vol. 11, No. 4, pp. 1151–1169 (1982) $6.00

Reprints from Volume 12

214. Thermodynamic Properties of Steam in the Critical Region, *J. M. H. Levelt Sengers, B. Kamgar-Parsi, F. W. Balfour, and J. V. Sengers*, Vol. 12, No. 1, pp. 1–28 (1983) $6.00

(Continuation of Cumulative Listing of Reprints)

215. Heat Capacity and Other Thermodynamic Properties of Linear Macromolecules. VII. Other Carbon Backbone Polymers, *Umesh Gaur, Brent B. Wunderlich, and Bernhard Wunderlich,* Vol. 12, No. 1, pp. 29–63 (1983) $7.00

216. Heat Capacity and Other Thermodynamic Properties of Linear Macromolecules. VIII. Polyesters and Polyamides, *Umesh Gaur, Suk-fai Lau, Brent B. Wunderlich, and Bernhard Wunderlich,* Vol. 12, No. 1, pp. 65–89 (1983) $6.00

217. Heat Capacity and Other Thermodynamic Properties of Linear Macromolecules. IX. Final Group of Aromatic and Inorganic Polymers, *Umesh Gaur, Suk-fai Lau, and Bernhard Wunderlich,* Vol. 12, No. 1, pp. 91–108 (1983) $6.00

218. An Annotated Compilation and Appraisal of Electron Swarm Data in Electronegative Gases, *J. W. Gallagher, E. C. Beaty, J. Dutton, and L. C. Pitchford,* Vol. 12, No. 1, pp. 109–152 (1983) $7.00

219. The Solubility of Oxygen and Ozone in Liquids, *Rubin Battino, Timothy R. Rettich, and Toshihiro Tominaga,* Vol. 12, No. 2, pp. 163–178 (1983) $5.00

220. Recommended Values for the Thermal Expansivity of Silicon from 0 to 1000 K, *C. A. Swenson,* Vol. 12, No. 2, pp. 179–182 (1983) $5.00

221. Electrical Resistivity of Ten Selected Binary Alloy Systems, *C. Y. Ho, M. W. Ackerman, K. Y. Wu, T. N. Havill, R. H. Bogaard, R. A. Matula, S. G. Oh, and H. M. James,* Vol. 12, No. 2, pp. 183–322 (1983) $13.00

222. Energy Levels of Silicon, Si I through Si XIV, *W. C. Martin and Romuald Zalubas,* Vol. 12, No. 2, pp. 323–380 (1983) $8.00

223. Evaluation of Binary *PTxy* Vapor–Liquid Equilibrium Data for C_6 Hydrocarbons. Benzene + Hexane, *Buford D. Smith, Ol Muthu, and Ashok Dewan,* Vol. 12, No. 2, pp. 381–387 (1983) $5.00

224. Evaluation of Binary Excess Enthalpy Data for C_6 Hydrocarbons. Benzene + Hexane, *Buford D. Smith, Ol Muthu, and Ashok Dewan,* Vol. 12, No. 2, pp, 389–393 (1983) $5.00

225. Evaluation of Binary Excess Volume Data for C_6 Hydrocarbons. Benzene + Hexane, *Buford D. Smith, Ol Muthu, and Ashok Dewan,* Vol. 12, No. 2, pp. 395–401 (1983) $5.00

226. Atlas of the High-Temperature Water Vapor Spectrum in the 3000 to 4000 cm^{-1} Region, *A. S. Pine, M. J. Coulombe, C. Camy-Peyret, and J-M. Flaud,* Vol. 12, No. 3, pp. 413–465 (1983) $8.00

227. Small-Angle Rayleigh Scattering of Photons at High Energies: Tabulations of Relativistic HFS Modified Atomic Form Factors, *D. Schaupp, M. Schumacher, F. Smend, P. Rullhusen, and J. H. Hubbell,* Vol. 12, No. 3, pp. 467–512 (1983) $7.00

228. Thermodynamic Properties of D_2O in the Critical Region, *B. Kamgar-Parsi, J. M. H. Levelt Sengers, and J. V. Sengers,* Vol. 12, No. 3, pp. 513–529 (1983) $6.00

229. Chemical Kinetic Data Sheets for High-Temperature Chemical Reactions, *N. Cohen and K. R. Westberg,* Vol. 12, No. 3, pp. 531–590 (1983) $8.00

230. Molten Salts: Volume 5, Part 2. Additional Single and Multi-Component Salt Systems. Electrical Conductance, Density, Viscosity and Surface Tension Data, *G. J. Janz and R. P. T. Tomkins,* Vol. 12, No. 3, pp. 591–815 (1983) $19.00

231. International Tables of the Surface Tension of Water, *N. B. Vargaftik, B. N. Volkov, and L. D. Voljak,* Vol. 12, No. 3, pp. 817–820 (1983) $5.00

232. Evaluated Theoretical Cross Section Data for Charge Exchange of Multiply Charged Ions with Atoms. I. Hydrogen Atom-Fully Stripped Ion Systems, *R. K. Janev, B. H. Bransden, and J. W. Gallagher,* Vol. 12, No. 4, pp. 829–872 (1983) $7.00

233. Evaluated Theoretical Cross Section Data for Charge Exchange of Multiply Charged Ions with Atoms. II. Hydrogen Atom-Partially Stripped Ion Systems, *J. W. Gallagher, B. H. Bransden, and R. K. Janev,* Vol. 12, No. 4, pp. 873–890 (1983) $6.00

234. Recommended Data on the Electron Impact Ionization of Light Atoms and Ions, *K. L. Bell, H. B. Gilbody, J. G. Hughes, A. E. Kingston, and F. J. Smith,* Vol. 12, No. 4, pp. 891–916 (1983) $6.00

235. A Correlation of the Viscosity and Thermal Conductivity Data of Gaseous and Liquid Ethylene, *P. M. Holland, B. E. Eaton, and H. J. M. Hanley,* Vol. 12, No. 4, pp. 917–932 (1983) $5.00

236. Transport Properties of Liquid and Gaseous D_2O over a Wide Range of Temperature and Pressure, *N. Matsunaga and A. Nagashima,* Vol. 12, No. 4, pp. 933–966 (1983) $7.00

237. Thermochemical Data for Gaseous Monoxides, *J. B. Pedley and E. M. Marshall,* Vol. 12, No. 4, pp. 967–1031 (1983) $9.00

238. Vapor Pressure of Coal Chemicals, *J. Chao, C. T. Lin, and T. H. Chung,* Vol. 12, No. 4, pp. 1033–1063 (1983) $6.00

Reprints from Volume 13

239. Thermodynamic Properties of Aqueous Sodium Chloride Solutions, *Kenneth S. Pitzer, J. Christopher Peiper, and R. H. Busey,* Vol. 13, No. 1, pp. 1–102 (1984) $11.00

240. Refractive Index of ZnS, ZnSe, and ZnTe and Its Wavelength and Temperature Derivatives, *H. H. Li,* Vol. 13, No. 1, pp. 103–150 (1984) $7.00

241. High Temperature Vaporization Behavior of Oxides. I. Alkali Metal Binary Oxides, *R. H. Lamoreaux and D. L. Hildenbrand,* Vol. 13, No. 1, pp. 151–173 (1984) $6.00

242. Thermophysical Properties of Fluid H_2O, *J. Kestin, J. V. Sengers, B. Kamgar-Parsi, and J. M. H. Levelt Sengers,* Vol. 13, No. 1, pp. 175–183 (1984) $5.00

243. Representative Equations for the Viscosity of Water Substance, *J. V. Sengers and B. Kamgar-Parsi,* Vol. 13, No. 1, pp. 185–205 (1984) $6.00

244. Atlas of the Schumann–Runge Absorption Bands of O_2 in the Wavelength Region 175–205 nm, *K. Yoshino, D. E. Freeman, and W. H. Parkinson,* Vol. 13, No. 1, pp. 207–227 (1984) $6.00

245. Equilibrium and Transport Properties of the Noble Gases and Their Mixtures at Low Density, *J. Kestin, K. Knierim, E. A. Mason, B. Najafi, S. T. Ro, and M. Waldman,* Vol. 13, No. 1, pp. 229–303 (1984) $9.00

246. Evaluation of Kinetic and Mechanistic Data For Modeling of Photochemical Smog, *Roger Atkinson and Alan C. Lloyd,* Vol. 13, No. 2, pp. 315–444 (1984) $13.00

247. Rate Data for Inelastic Collision Processes in the Diatomic Halogen Molecules, *J. I. Steinfeld,* Vol. 13, No. 2, pp. 445–553 (1984) $11.00

248. Water Solubilities of Polynuclear Aromatic and Heteroaromatic Compounds, *Robert S. Pearlman, Samuel H. Yalkowsky, and Sujit Banerjee,* Vol. 13, No. 2, pp. 555–562 (1984) $5.00

249. The Solubility of Nitrogen and Air in Liquids, *Rubin Battino, Timothy R. Rettich, and Toshihiro Tominaga,* Vol. 13, No. 2, pp. 563–600 (1984) $7.00

250. Thermophysical Properties of Fluid D_2O, *J. Kestin, J. V. Sengers, B. Kamgar-Parsi, and J. M. H. Levelt Sengers,* Vol. 13, No. 2, pp. 601–609 (1984) $5.00

251. Experimental Stark Widths and Shifts for Spectral Lines of Neutral Atoms (A Critical Review of Selected Data for the Period 1976 to 1982), *N. Konjević, M. S. Dimitrijević, and W. L. Wiese,* Vol. 13, No. 3, pp. 619–647 (1984) $6.00

(Continuation of Cumulative Listing of Reprints)

252. Experimental Stark Widths and Shifts for Spectral Lines of Positive Ions (A Critical Review and Tabulation of Selected Data for the Period 1976 to 1982), *N. Konjević, M. S. Dimitrijević, and W. L. Wiese*, Vol. 13, No. 3, pp. 649–686 (1984) $7.00

253. A Review of Deuterium Triple-Point Temperatures, *L. A. Schwalbe and E. R. Grilly*, Vol. 13, No. 3, pp. 687–693 (1984) $5.00

254. Evaluated Gas Phase Basicities and Proton Affinities of Molecules; Heats of Formation of Protonated Molecules, *Sharon G. Lias, Joel F. Liebman, and Rhoda D. Levin*, Vol. 13, No. 3, pp. 695–808 (1984) $12.00

255. Isotopic Abundances and Atomic Weights of the Elements, *Paul De Bièvre, Marc Gallet, Norman E. Holden, and I. Lynus Barnes*, Vol. 13, No. 3, pp. 809–891 (1984) $10.00

256. Representative Equations for the Thermal Conductivity of Water Substance, *J. V. Sengers, J. T. R. Watson, R. S. Basu, B. Kamgar-Parsi, and R. C. Hendricks*, Vol. 13, No. 3, pp. 893–933 (1984) $7.00

257. Ground-State Vibrational Energy Levels of Polyatomic Transient Molecules, *Marilyn E. Jacox*, Vol. 13, No. 4, pp. 945–1068 (1984) $12.00

258. Electrical Resistivity of Selected Elements, *P. D. Desai, T. K. Chu, H. M. James, and C. Y. Ho*, Vol. 13, No. 4, pp. 1069–1096 (1984) $6.00

259. Electrical Resistivity of Vanadium and Zirconium, *P. D. Desai, H. M. James, and C. Y. Ho*, Vol. 13, No. 4, pp. 1097–1130 (1984) $7.00

260. Electrical Resistivity of Aluminum and Manganese, *P. D. Desai, H. M. James, and C. Y. Ho*, Vol. 13, No. 4, pp. 1131–1172 (1984) $7.00

261. Standard Chemical Thermodynamic Properties of Alkane Isomer Groups, *Robert A. Alberty and Catherine A. Gehrig*, Vol. 13, No. 4, pp. 1173–1197 (1984) $6.00

262. Evaluated Theoretical Cross-Section Data for Charge Exchange of Multiply Charged Ions with Atoms. III. Nonhydrogenic Target Atoms, *R. K. Janev and J. W. Gallagher*, Vol. 13, No. 4, pp. 1199–1249 (1984) $8.00

263. Heat Capacity of Reference Materials: Cu and W, *G. K. White and S. J. Collocott*, Vol. 13, No. 4, pp. 1251–1257 (1984) $5.00

264. Evaluated Kinetic and Photochemical Data for Atmospheric Chemistry: Supplement II. CODATA Task Group on Gas Phase Chemical Kinetics, *D. L. Baulch, R. A. Cox, R. F. Hampson, Jr., J. A. Kerr (Chairman), J. Troe, and R. T. Watson*, Vol. 13, No. 4, pp. 1259–1380 (1984) $12.00

Reprints from Volume 14

265. Thermodynamic Properties of Key Organic Oxygen Compounds in the Carbon Range C_1 to C_4. Part 1. Properties of Condensed Phases, *Randolph C. Wilhoit, Jing Chao, and Kenneth R. Hall*, Vol. 14, No. 1, pp. 1–175 (1985) $15.00

266. Standard Chemical Thermodynamic Properties of Alkylbenzene Isomer Groups, *Robert A. Alberty*, Vol. 14, No. 1, pp. 177–192 (1985) $5.00

267. Assessment of Critical Parameter Values for H_2O and D_2O, *J. M. H. Levelt Sengers, J. Straub, K. Watanabe, and P. G. Hill*, Vol. 14, No. 1, pp. 193–207 (1985) $5.00

268. The Viscosity of Nitrogen, Oxygen, and Their Binary Mixtures in the Limit of Zero Density, *Wendy A. Cole and William A. Wakeham*, Vol. 14, No. 1, pp. 209–226 (1985) $6.00

269. The Thermal Conductivity of Fluid Air, *K. Stephan and A. Laesecke*, Vol. 14, No. 1, pp. 227–234 (1985) $5.00

270. The Electronic Spectrum and Energy Levels of the Deuterium Molecule, *Robert S. Freund, James A. Schiavone, and H. M. Crosswhite*, Vol. 14, No. 1, pp. 235–383 (1985) $14.00

271. Microwave Spectra of Molecules of Astrophysical Interest. XXII. Sulfur Dioxide (SO_2), *F. J. Lovas*, Vol. 14, No. 2, pp. 395–488 (1985) $10.00

272. Evaluation of the Thermodynamic Functions for Aqueous Sodium Chloride from Equilibrium and Calorimetric Measurements below 154 °C, *E. Colin W. Clarke and David N. Glew*, Vol. 14, No. 2, pp. 489–610 (1985) $12.00

273. The Mark–Houwink–Sakurada Equation for the Viscosity of Linear Polyethylene, *Herman L. Wagner*, Vol. 14, No. 2, pp. 611–617 (1985) $5.00

274. The Solubility of Mercury and Some Sparingly Soluble Mercury Salts in Water and Aqueous Electrolyte Solutions, *H. Lawrence Clever, Susan A. Johnson, and M. Elizabeth Derrick*, Vol. 14, No. 3, pp. 631–680 (1985) $8.00

275. A Review and Evaluation of the Phase Equilibria, Liquid-Phase Heats of Mixing and Excess Volumes, and Gas-Phase *PVT* Measurements for Nitrogen + Methane, *A. J. Kidnay, R. C. Miller, E. D. Sloan, and M. J. Hiza*, Vol. 14, No. 3, pp. 681–694 (1985) $5.00

276. The Homogeneous Nucleation Limits of Liquids, *C. T. Avedisian*, Vol. 14, No. 3, pp. 695–729 (1985) $7.00

277. Binding Energies in Atomic Negative Ions: II, *H. Hotop and W. C. Lineberger*, Vol. 14, No. 3, pp. 731–750 (1985) $6.00

278. Energy Levels of Phosphorus, P I through P XV, *W. C. Martin, Romuald Zalubas, and Arlene Musgrove*, Vol. 14, No. 3, pp. 751–802 (1985) $8.00

279. Standard Chemical Thermodynamic Properties of Alkene Isomer Groups, *Robert A. Alberty and Catherine A. Gehrig*, Vol. 14, No. 3, pp. 803–820 (1985) $6.00

280. Standard Chemical Thermodynamic Properties of Alkylnaphthalene Isomer Groups, *Robert A. Alberty and Theodore M. Bloomstein*, Vol. 14, No. 3, pp. 821–837 (1985) $6.00

281. Carbon Monoxide Thermophysical Properties from 68 to 1000 K at Pressures to 100 MPa, *Robert D. Goodwin*, Vol. 14, No. 4, pp. 849–932 (1985) $10.00

282. Refractive Index of Water and Its Dependence on Wavelength, Temperature, and Density, *I. Thormählen, J. Straub, and U. Grigull*, Vol. 14, No. 4, pp. 933–945 (1985) $5.00

283. Viscosity and Thermal Conductivity of Dry Air in the Gaseous Phase, *K. Kadoya, N. Matsunaga, and A. Nagashima*, Vol. 14, No. 4, pp. 947–970 (1985) $6.00

284. Charge Transfer of Hydrogen Ions and Atoms in Metal Vapors, *T. J. Morgan, R. E. Olson, A. S. Schlachter, and J. W. Gallagher*, Vol. 14, No. 4, pp. 971–1040 (1985) $9.00

285. Reactivity of HO_2/O_2^- Radicals in Aqueous Solution, *Benon H. J. Bielski, Diane E. Cabelli, Ravindra L. Arudi, and Alberta B. Ross*, Vol. 14, No. 4, pp. 1041–1100 (1985) $8.00

286. The Mark–Houwink–Sakurada Equation for the Viscosity of Atactic Polystyrene, *Herman L. Wagner*, Vol. 14, No. 4, pp. 1101–1106 (1985) $5.00

287. Standard Chemical Thermodynamic Properties of Alkylcyclopentane Isomer Groups, Alkylcyclohexane Isomer Groups, and Combined Isomer Groups, *Robert A. Alberty and Young S. Ha*, Vol. 14, No. 4, pp. 1107–1132 (1985) $6.00

Reprints from Volume 15

288. Triplet–Triplet Absorption Spectra of Organic Molecules in Condensed Phases, *Ian Carmichael and Gordon L. Hug*, Vol. 15, No. 1, pp. 1–250 (1986) $20.00

(Continuation of Cumulative Listing of Reprints)

289. Recommended Rest Frequencies for Observed Interstellar Molecular Microwave Transitions—1985 Revision, *F. J. Lovas,* Vol. 15, No. 1, pp. 251–303 (1986) $6.00

290. New International Formulations for the Thermodynamic Properties of Light and Heavy Water, *J. Kestin and J. V. Sengers,* Vol. 15, No. 1, pp. 305–320 (1986) $5.00

291. Forbidden Lines in ns^2np^k Ground Configurations and *nsnp* Excited Configurations of Beryllium through Molybdenum Atoms and Ions, *Victor Kaufman and Jack Sugar,* Vol. 15, No. 1, pp. 321–426 (1986) $11.00

292. Thermodynamic Properties of Twenty-One Monocyclic Hydrocarbons, *O. V. Dorofeeva, L. V. Gurvich, and V. S. Jorish,* Vol. 15, No. 2, pp. 437–464 (1986) $6.00

293. Evaluated Kinetic Data for High-Temperature Reactions. Volume 5. Part 1. Homogeneous Gas Phase Reactions of the Hydroxyl Radical with Alkanes, *D. L. Baulch, M. Bowers, D. G. Malcolm, and R. T. Tuckerman,* Vol. 15, No. 2, pp. 465–592 (1986) $12.00

294. Thermodynamic Properties of Ethylene from the Freezing Line to 450 K at Pressures to 260 MPa, *Majid Jahangiri, Richard T Jacobsen, Richard B. Stewart, and Robert D. McCarty,* Vol. 15, No. 2, pp. 593–734 (1986) $13.00

295. Thermodynamic Properties of Nitrogen from the Freezing Line to 2000 K at Pressures to 1000 MPa, *Richard T Jacobsen, Richard B. Stewart, and Majid Jahangiri,* Vol. 15, No. 2, pp. 735–909 (1986) $15.00

296. A Critical Review of Aqueous Solubilities, Vapor Pressures, Henry's Law Constants, and Octanol–Water Partition Coefficients of the Polychlorinated Biphenyls, *Wan Ying Shiu and Donald Mackay,* Vol. 15, No. 2, pp. 911–929 (1986) $6.00

297. Computer Methods Applied to the Assessment of Thermochemical Data. Part I. The Establishment of a Computerized Thermochemical Data Base Illustrated by Data for $TiCl_4$(g), $TiCl_4$(l), $TiCl_3$(cr), and $TiCl_2$(cr), *S. P. Kirby, E. M. Marshall, and J. B. Pedley,* Vol. 15, No. 3, pp. 943–965 (1986) $6.00

298. Thermodynamic Properties of Iron and Silicon, *P. D. Desai,* Vol. 15, No. 3, pp. 967–983 (1986) $6.00

299. Cross Sections for Collisions of Electrons and Photons with Nitrogen Molecules, *Y. Itikawa, M. Hayashi, A. Ichimura, K. Onda, K. Sakimoto, K. Takayanagi, M. Nakamura, H. Nishimura, and T. Takayanagi,* Vol. 15, No. 3, pp. 985–1010 (1986) $6.00

300. Thermochemical Data on Gas-Phase Ion-Molecule Association and Clustering Reactions, *R. G. Keesee and A. W. Castleman, Jr.,* Vol. 15, No. 3, pp. 1011–1071 (1986) $8.00

301. Standard Reference Data for the Thermal Conductivity of Liquids, *C. A. Nieto de Castro, S. F. Y. Li, A. Nagashima, R. D. Trengove, and W. A. Wakeham,* Vol. 15, No. 3, pp. 1073–1086 (1986) $5.00

302. Chemical Kinetic Data Base for Combustion Chemistry. Part I. Methane and Related Compounds, *W. Tsang and R. F. Hampson,* Vol. 15, No. 3, pp. 1087–1279 (1986) $17.00

Special Reprints Packages

These special reprints packages offer selected articles in specific subject areas from the JOURNAL OF PHYSICAL AND CHEMICAL REFERENCE DATA, and they are offered at a better rate than when purchased individually. You will have available a complete library of literature for your specific requirements at a fraction of the cost of purchasing back issues of the journal.

Look over the reprints packages available—they are listed by subject area. In the Cumulative Listing of Reprints you will find the titles corresponding to the reprint numbers. You are sure to find building your information bank in this manner to be thorough and economical.

Package C1 (5 Parts) MOLECULAR VIBRATIONAL FREQUENCIES. Consisting of Reprint Nos. 103, 129, 170, 257, NSRD 39.
If purchased individually: $ 33.00
Special package price: **$ 26.00**

Package C2 (22 Parts) ATOMIC ENERGY LEVELS. Consisting of Reprint Nos. 26, 54, 64, 68, 94, 100, 109, 125, 126, 131, 132, 149, 150, 154, 156, 160, 179, 180, 192, 200, 222, 278.
If purchased individually: $121.00
Special package price: **$ 96.00**

Package C3 (6 Parts) ATOMIC SPECTRA. Consisting of Reprint Nos. 33, 56, 77, 78, 110, 132.
If purchased individually: $ 33.00
Special package price: **$ 27.00**

Package C4 (5 Parts) ATOMIC TRANSITION PROBABILITIES. Consisting of Reprint Nos. 20, 63, 82, 118, 182.
If purchased individually: $ 35.00
Special package price: **$ 28.00**

Package C5 (7 Parts) MOLECULAR SPECTRA. Consisting of Reprint Nos. 4, 8, 53, 79, 93, 130, 146.
If purchased individually: $ 51.50
Special package price: **$ 41.00**

Package C6 (9 Parts) THERMODYNAMIC PROPERTIES OF ELECTROLYTE SOLUTIONS. Consisting of Reprint Nos. 15, 95, 111, 151, 152, 174, 184, 185, 186.
If purchased individually: $ 46.00
Special package price: **$ 37.00**

Package C7 (12 Parts) IDEAL GAS THERMODYNAMIC PROPERTIES. Consisting of Reprint Nos. 30, 42, 43, 62, 65, 66, 70, 80, 83, 113, 115, 141.
If purchased individually: $ 38.00
Special package price: **$ 31.00**

Package C8 (7 Parts) RESISTIVITY. Consisting of Reprint Nos. 138, 139, 155, 221, 258, 259, 260.
If purchased individually: $ 47.50
Special package price: **$ 39.00**

Package C9 (7 Parts) MOLTEN SALTS. Consisting of Reprint Nos. 10, 41, 71, 96, 135, 167, 168.
If purchased individually: $ 62.50
Special package price: **$ 44.00**

Package C10 (4 Parts) REFRACTIVE INDEX. Consisting of Reprint Nos. 81, 158, 162, 240.
If purchased individually: $ 32.50
Special package price: **$ 26.00**

Supplements to JPCRD

When the topic demands it, and the quality of the data justifies it, the JOURNAL OF PHYSICAL AND CHEMICAL REFERENCE DATA issues a special Supplement. Each Supplement is a monograph—collected tables of highly significant physical or chemical property data in one complete volume. Listed below are the special Supplements to JPCRD that have been published. Each is a valuable resource for the physical chemist and chemical physicist.

ATOMIC ENERGY LEVELS OF THE IRON-PERIOD ELEMENTS: POTASSIUM THROUGH NICKEL by J. Sugar and C. Corliss. (Supplement No. 2 to Volume 14) 1985, 664 pages. Hardcover.
U.S. & Canada: $50.00
Abroad: $58.00

HEAT CAPACITIES AND ENTROPIES OF ORGANIC COMPOUNDS IN THE CONDENSED PHASE by E.S. Domalski, W.H. Evans, and E.D. Hearing. (Supplement No. 1 to Volume 13) 1984, 288 pages. Hardcover.
U.S. & Canada: $40.00
Abroad: $48.00

THE NBS TABLES OF CHEMICAL THERMODYNAMIC PROPERTIES. SELECTED VALUES FOR INORGANIC AND C_1 AND C_2 ORGANIC SUBSTANCES IN SI UNITS by D.D. Wagman, W.H. Evans, V.B. Parker, R.H. Schumm, I. Halow, S.M. Bailey, K.L. Churney, and R.L. Nuttall. (Supplement No. 2 to Volume 11) 1982, 394 pages. Hardcover.
U.S. & Canada: $40.00
Abroad: $48.00

THERMOPHYSICAL PROPERTIES OF FLUIDS. 1. ARGON, ETHYLENE, PARAHYDROGEN, NITROGEN, NITROGEN TRIFLUORIDE, AND OXYGEN by B.A. Younglove. (Supplement No. 1 to Volume 11) 1982, 368 pages. Hardcover.
U.S. & Canada: $40.00
Abroad: $48.00

EVALUATED KINETIC DATA FOR HIGH TEMPERATURE REACTIONS: VOLUME 4, HOMOGENEOUS GAS PHASE REACTIONS OF HALOGEN- AND CYANIDE-CONTAINING SPECIES by D.L. Baulch, J. Duxbury, S.J. Grant, and D.C. Montague. (Supplement No. 1 to Volume 10) 1981, 721 pages. Hardcover.
U.S. & Canada: $80.00
Abroad: $96.00

ENERGETICS OF GASEOUS IONS by H.M. Rosenstock, K. Draxl, B.W. Steiner, and J.T. Herron. (Supplement No. 1 to Volume 6) 1977, 783 pages.*
U.S. & Canada: $70/$65
Abroad: $84/$78

THERMAL CONDUCTIVITY OF THE ELEMENTS: A COMPREHENSIVE REVIEW by C.Y. Ho, R.W. Powell, and P.E. Liley. (Supplement No. 1 to Volume 3) 1974, 796 pages.*
U.S. & Canada: $60/$55
Abroad: $72/$66

PHYSICAL AND THERMODYNAMIC PROPERTIES OF ALIPHATIC ALCOHOLS by R.C. Wilhoit and B.J. Zwolinski. (Supplement No. 1 to Volume 2) 1973, 420 pages.*
U.S. & Canada: $33/$30
Abroad: $40/$36

*Prices are for hardcover/softcover.